CHILTON'S

ANTI-LOCK BRAKES AND TRACTION-CONTROL MANUAL

Publisher and Editor-in-Chief Kerry A. Freeman, S.A.E
Executive Editors Dean F. Morgantini, Jr., S.A.E. □ W. Calvin Settle, Jr., S.A.E
Managing Editor Nicholas L. D'Andrea
Senior Editors Jacques Gordon □ Ben Greisler, S.A.E.
Michael L. Grady □ Kevin M. G. Maher □ Debra McCall □ Richard J. Rivele, S.A.E.
Richard T. Smith □ Jim Taylor □ Ronald T. Webb
Project Managers Lawrence C. Braun, S.A.E. □ Thomas P. Browne III
Robert E. Doughten □ Joseph L. DeFrancesco, A.S.E., S.A.E. □ Martin J. Gunther
Will Kessler □ Craig P. Nangle, A.S.E. □ Ernest H. Ralph, A.S.E., S.A.E.
Richard Schwartz □ Paul Shanahan
Service Editors Jaffer A. Ahmad □ Bradley D. Bower □ Robert Chabot □ William C. Cottman Jr., A.S.E.
Leonard Davis, A.S.E. □ Michael C. DiFurio Jr., S.A.E. □ Robert F. Dougherty, Jr. □ Sam Fiorani
Matthew E. Frederick □ Bill C. Friedauer, A.S.E. □ Edward J. Giacomucci, A.S.E., S.A.E. □ Al Gibbs
Herb Guie □ George B. Heinrich, III □ Dawn M. Hoch □ Daniel Howells □ David E. Jester, A.S.E.
Lori L. Johnson □ Kenneth F. Konzelman □ Neil J. Leonard, A.S.E. □ James R. Marotta □ Robert McAnally
Thomas A. Mellon □ Raymond K. Moore □ Norman D. Norville, A.S.E. □ Christine L. Nuckowski
Eric S. Peterson, A.S.E. □ Charles T. Ramsey □ William Renn □ Roy Ripple, A.S.E.
George Ritter □ Richard Saxton, A.S.E. □ Derek A. Scott □ Larry E. Stiles
Gordon L. Tobias, S.A.E. □ Albert A. Wood, A.S.E.

Director of Manufacturing Mike D'Imperio
Manager of Manufacturing Robin S. Norman
Manager, Product Systems Development Robert E. Maxey
Production Manager Andrea M. Steiger
Assistant Production Manager Marsha Park Herman
Production Specialists Christina Davis □ Kimberly T. Hayes □ Joseph C. McGinty □ Liz Thompson

National Sales Manager Lawrence R. Rufo
National Administrator, Sales Mark Amentt
Eastern Regional Manager Howie Freeman
Wes
Mi
South

Manufactured in USA , Pa. 19089
ISBN 0-8019-8900-0 -0682

SAFETY NOTICE

Proper service and repair procedures are vital to the safe, reliable operation of all motor vehicles as well as the personal safety of those performing repairs. This manual outlines procedures for servicing and repairing vehicles using safe, effective methods. The procedures contain many NOTES, CAUTIONS and WARNINGS which should be followed along with standard safety procedures to eliminate the possibility of personal injury or improper service which could damage the vehicle or compromise its safety.

It is important to note that the repair procedures and techniques, tools and parts for servicing motor vehicles, as well as the skill and experience of the individual performing the work vary widely. It is not possible to anticipate all of the conceivable ways or conditions under which vehicles may be serviced, or to provide cautions as to all of the possible hazards that may result. Standard and accepted safety precautions and equipment should be used when handling toxic or flammable fluids, and safety glasses or other protection should be used during cutting, grinding, chiseling, or any other process that can cause material removal or projectiles.

Some procedures require the use of tools specially designed for a specific purpose. Before substituting another tool or procedure, you must be completely satisfied that neither your personal safety, nor the performance of the vehicle will be endangered.

PART NUMBERS

Part numbers listed in this reference are not recommendations by Chilton for any product by brand name. They are references that can be used with interchange manuals and aftermarket supplier catalogs to locate each brand supplier's discrete part number.

BOSCH 2U ANTI-LOCK BRAKE SYSTEM

General Description

The Bosch ABS systems are designed to prevent wheel lock-up under heavy braking on almost any road surface. Under normal conditions, the system functions in the same fashion as any conventional vacuum-boost brake system.

The Bosch 2U ABS system is used on Caprice, Custom Cruiser, Eldorado, Fleetwood Brougham, Impala SS, Riviera, Seville and Roadmaster. Rear wheel drive vehicles provide 3-wheel speed input signals to the controller; each front wheel is monitored separately and the signal for the rear wheels is taken at the rear axle. Front wheel drive vehicles provide 4 separate wheel speed signals. All systems use a 3-channel output, controlling the front wheels individually and the rear wheels together. This combination of inputs and outputs results in stable braking while allowing the operator to maintain steering control.

SYSTEM OPERATION

If a wheel begins to lock under braking, the system enters the anti-lock mode. Hydraulic pressure in the individual wheel circuits is modulated to prevent any wheel from locking. A separate hydraulic line and 2 solenoid valves run from the hydraulic modulator to each front wheel. The rear wheels share the remaining solenoid valve. Some models use one hydraulic line to the rear, the remainder use two individual lines from the brake modulator. If a single line is used, it splits to each rear wheel at the rear axle tee fitting. The double line systems are still controlled by a single solenoid.

The ABS system can apply, reduce or hold hydraulic line pressure to the wheels; it cannot increase pressure above that developed within the master cylinder. During anti-lock braking, a series of rapid pedal pulsations will be felt by the operator. This is caused by the rapid cycling of the solenoids as they control line pressure according to the direction of the electronic brake control module (EBCM). The pedal vibration will stop when normal braking resumes or when the vehicle comes to a stop. Under some conditions, a ticking or popping noise may be heard within the vehicle when ABS is engaged; this is a normal noise and is caused by the solenoids and other components operating.

SYSTEM COMPONENTS

Electronic Brake Control Module (EBCM)

The EBCM is a small computer used to control the ABS system. On the 1993 models, it is located in the luggage compartment, above the left wheel well. On 1994 models, it is mounted under the left dash, above the brake pedal on the EBCM bracket. On the 1995 models, the EBCM is integral with the Brake Pressure Modulator (PBM) valve located in the left front corner of the engine compartment, next to the steering gear.

The EBCM performs 3 primary functions: detecting wheel lock based on signals from the wheel speed sensors, controlling the brake system pressures by operating the solenoid valves when anti-lock is engaged and monitoring the function of the related electrical components within the system.

The EBCM can disable the ABS system if a fault is detected during monitoring. If this occurs, the amber ANTI-LOCK warning lamp will illuminate on the dashboard. When the ABS is disabled, the vehicle's normal braking system will still function but without anti-lock capability.

Once the EBCM has detected a fault, it will generate a fault code and store it for later retrieval. The codes may be read by putting the EBCM into the diagnostic mode. While some or all of the codes may be read through the flashing of a dashboard lamp, the use of a hand-scanner such as the Tech 1 or its equivalent is strongly recommended. The Bosch system adapter is required when linking the scan tool to the ABS system at the ALDL connector.

Wheel Speed Sensors

Information on wheel speed is transmitted from sensors at each front wheel. Rear wheel drive vehicles employ a sensor on the rear axle differential; front drive vehicles use a sensor at each rear wheel. A small AC voltage is generated as a toothed wheel turns past a fixed sensor. The voltage varies with the speed; the control module uses this voltage to sense wheel speed. The signal is passed to the EBCM through a pair of wires which are shielded to avoid interference.

Brake Pressure Modulator (BPM) Valve Assembly

Mounted in the engine compartment, the brake pressure modulator valve assembly provides pressure control for each of the wheel circuits as required during anti-lock braking. The front and rear circuits are hydraulically isolated so that a leak or malfunction in 1 circuit will allow continued braking ability on the other.

The modulator solenoid valves open or close to control passage of pressurized brake fluid into the wheel circuits. Although operation of the solenoids can reduce fluid pressure to the wheel(s) to control locking, the modulator cannot develop or apply more pressure than that developed by the master cylinder.

With the exception of the ABS relay and the pump motor relay (under the cover) the brake pressure modulator is an integral, non-serviceable unit. It must be replaced as a unit and must never be disassembled.

Additional components contained within the hydraulic modulator are: solenoid valves, pump and motor and relays.

Warning System

The anti-lock system continuously monitors all of its components and will notify the operator of a malfunction. When the vehicle is started, a functional check of the ABS electrical circuitry is performed. After the vehicle speed reaches 4 mph, each valve in the hydraulic modulator is cycled and the pump is momentarily activated. The clicking and buzzing of these components may be heard by the operator as the vehicle begins to move after initial start-up. This self test is performed only once per ignition cycle; the noises of the components are considered normal operation.

Brake Warning Light

The red BRAKE warning light in the dashboard will illuminate to warn the operator of conditions within the system which may affect braking ability. The lamp will illuminate if the parking brake is not fully released or it there is unequal pressure within the 2 circuits of the hydraulic system.

The BRAKE lamp will also illuminate for a bulb test when the ignition switch is turned to the **START** position; the lamp should go out when the ignition switch is returned to the **RUN** position.

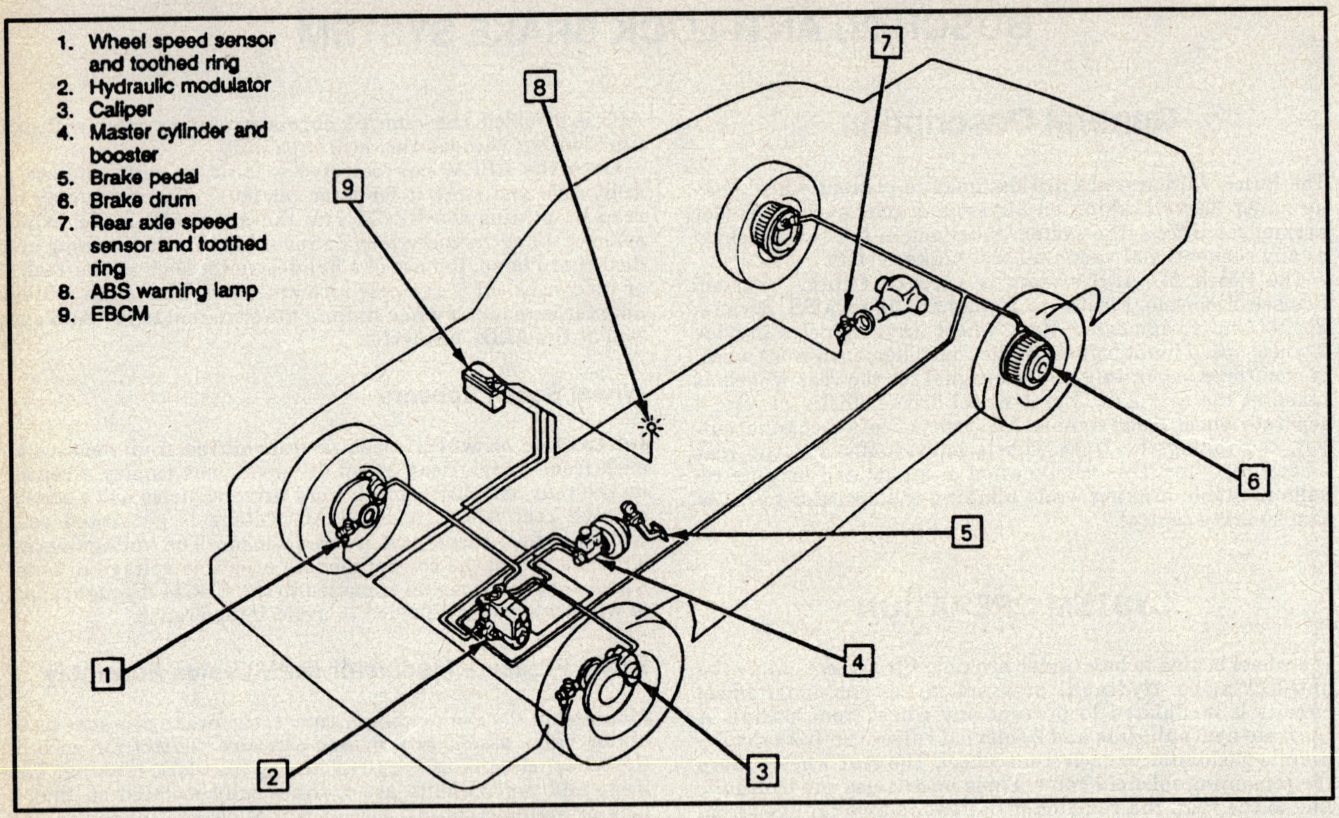

ABS System Components — 1993 Rear Wheel Drive

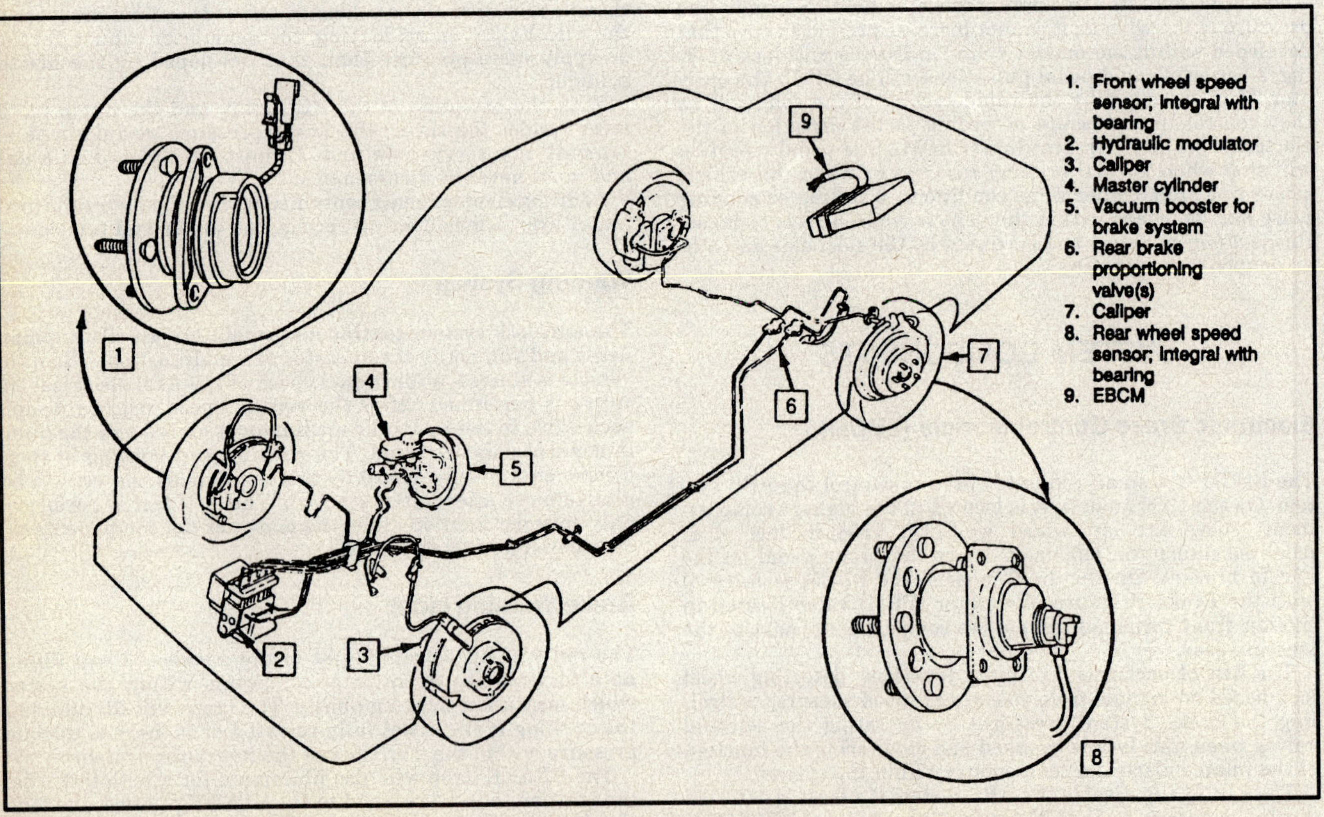

ABS System Components — 1993 Front Wheel Drive

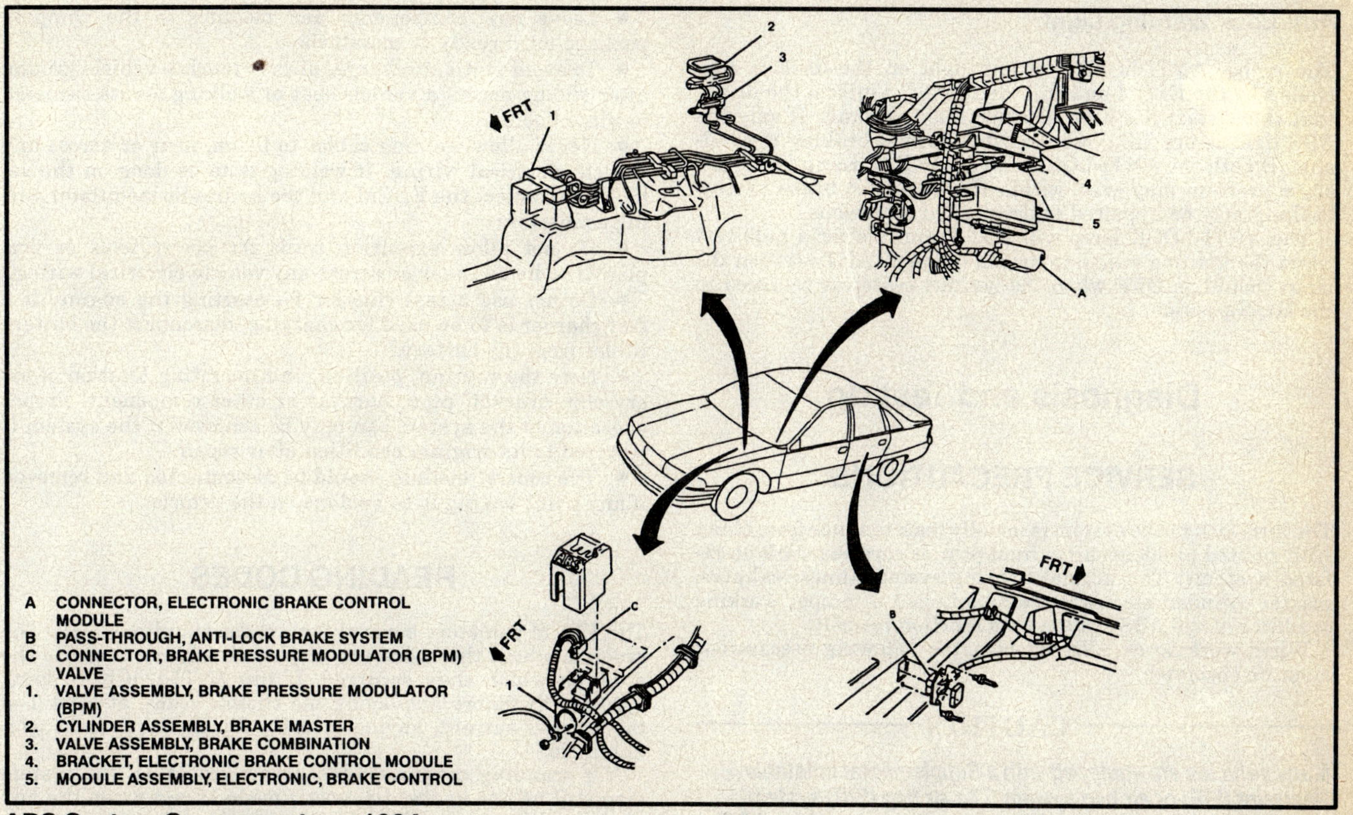

ABS System Components — 1994

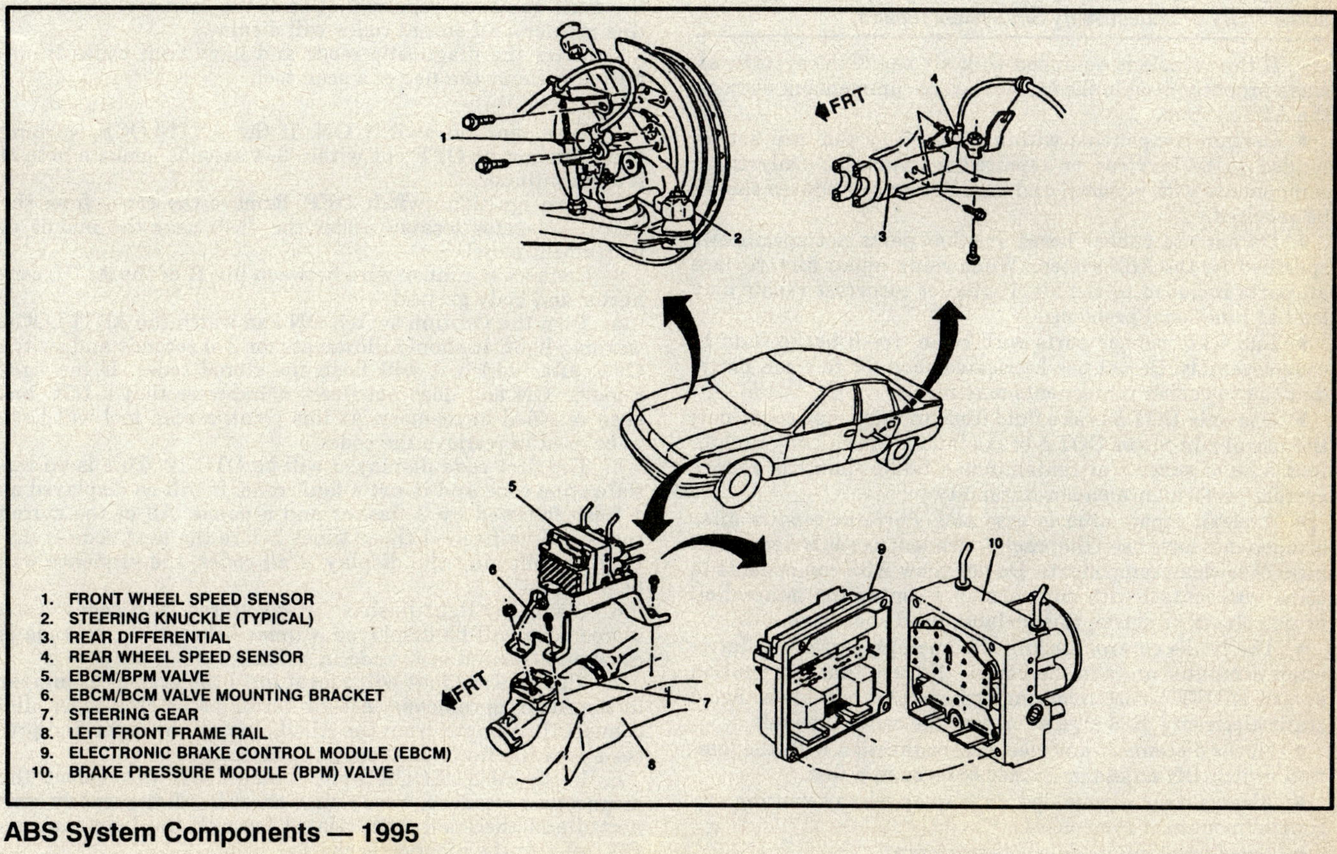

ABS System Components — 1995

Anti-Lock Warning Light

The amber ANTI-LOCK warning light on the dash is controlled by the EBCM and will illuminate to inform the driver that the system has been disabled due to a fault. If only the ABS light is on, full power assisted normal braking is available. If both the ANTI-LOCK and BRAKE warning lights are on, a problem may exist within the hydraulic brake system braking may be impaired under normal conditions.

The ANTI-LOCK lamp will also illuminate for a bulb test when the ignition switch is turned to the **START** position the lamp should go **OFF** when the ignition switch is returned to the **RUN** position.

Diagnosis and Testing

SERVICE PRECAUTIONS

The anti-lock brake system is usually maintenance free. Many ABS related problems arise from poor or careless work on related systems. The normal braking system (lines, calipers, master cylinder, etc.) must be maintained in proper working condition if the ABS system is to function reliably.

When working on ABS vehicles, the following precautions must be observed.

CAUTION

Many vehicles are equipped with a Supplemental Inflatable Restraint (SIR) or air bag system. The air bag (SIR) system must be disabled before performing service on or around SIR components or wiring. Failure to follow safety and disabling procedures could result in possible air bag deployment, personal injury or unnecessary SIR system repairs.

- If the vehicle is equipped with air bag (SIR) system, always properly disable the system before commencing work on the ABS system.
- Certain components within the ABS system are not intended to be serviced or repaired individually. Only those components with removal and installation procedures should be serviced.
- Do not use rubber hoses or other parts not specifically specified for the ABS system. When using repair kits, replace all parts included in the kit. Partial or incorrect repair may lead to functional problems.
- Lubricate rubber parts with clean, fresh brake fluid to ease assembly. Do not use lubricated shop air to clean parts; damage to rubber components may result.
- Use only DOT 3 brake fluid from an unopened container; the use of silicone or DOT 5 brake fluid is specifically prohibited. Use of suspect or contaminated brake fluid can reduce system performance and/or durability.
- A clean repair area is essential. Perform repairs after components have been thoroughly cleaned; use only denatured alcohol to clean components. Do not allow ABS components to come into contact with any substance containing mineral oil or petroleum products; this includes used shop rage.
- The brake control module is a microprocessor similar to other computer units in the vehicle. Ensure that the ignition switch is **OFF** before removing or installing harnesses. Avoid static electricity discharge at or near the control module.
- Never disconnect any electrical connection with the ignition switch **ON** unless instructed to do so in a test.
- Always wear a grounded wrist strap when servicing any control module or component.
- Avoid touching module connector pins.
- Leave new components and modules in the shipping package until ready to install them.
- To avoid static discharge, always touch a vehicle ground after sliding across a vehicle seat or walking across carpeted or vinyl floors.
- Never allow welding cables to lie on, near or across any vehicle electrical wiring. If welding is to be done on the vehicle, disconnect the EBCM and the hydraulic modulator wiring connectors.
- Do not allow extension cords for power tools or droplights to lie on, near or across any vehicle electrical wiring.
- Do not use a fast charger for starting the engine. If a fast charger is to be used for charging, disconnect the battery cables from the battery.
- Note the routing, position, and mounting location of every clip, bracket, pipe, harness or other component. Proper operation of the system can only be achieved if the system is restored to its original condition after repair.
- The control module should be disconnected and removed if any paint baking is to be done on the vehicle.

READING CODES

The EBCM can store several fault codes simultaneously. The order in which the codes are displayed may not indicate the order in which they occurred. If the ANTI-LOCK warning light is **ON** before displaying the trouble codes, at least 1 of the codes is current, having occurred during the present ignition cycle.

If a scanning diagnostic tool is not available, the system may still be put in the diagnostic mode except when the any of the current or hard faults involve the solenoid valves or pump motor. If 1 or more of the codes is/are 41, 45, 55, 61 or 63, diagnostic display of the trouble codes cannot occur until the fault has been repaired. (Use of the scan tool eliminates this problem; all stored codes will display.)

To enter the diagnostic mode and have fault codes transmitted without the use of a scan tool:

1. Turn ignition switch **ON**. If the ANTI-LOCK warning light does not go **OFF** out within 3–4 seconds, make a note of it and continue.
2. Turn ignition switch **OFF**. Remove the cover from the ALDL connector located under the dash near the middle of instrument panel.
3. Connect a jumper wire between pin **H** of the ALDL connector and body ground.
4. Turn the ignition switch **ON** and watch the ANTI-LOCK warning light. It should illuminate for 3–4 seconds and switch **OFF**, after which it will flash the stored codes. If the light remains **ON** and does not flash, it indicates that a DTC has been retained in memory. At this point, a scan tool will have to be used to retrieve the codes.
5. The first code displayed will be DTC 12. This is an initialization code and is not a fault code. It will be displayed as 1 flash followed by 2 flashes and a pause. All of the stored codes will be flashed three times before the next code is displayed. Following the display of all codes, the sequence will begin again.
6. Count the light flashes for any stored fault codes. Each stored code will be displayed 3 times before the next code is displayed. Record each code in writing.
7. The code readout will repeat until the ground or jumper at the ALDL is removed. After recording all stored codes, disconnect the jumper from the ALDL, install the cover and proceed with the functional check as instructed.
8. If the ANTI-LOCK light remains **ON** throughout the diagnostic procedure, verify that the EBCM is properly connected and check all ABS related fuses. If the light remains **ON**, refer to the diagnostic charts.

NOTE: Code 72 may be set if power to the EBCM is interrupted while the scan tool is connected. This may occur if the ignition is cycled while the scan tool is powered. Communication between the EBCM and the scan tool is impossible when the code is present. To re-enable the system, cycle the power to the scan tool and allow it to reset. Although not a serious condition, the code should be cleared before completing diagnosis.

CLEARING CODES

The control module will not permit codes to be cleared until all have been read. Codes cannot be cleared by unplugging the module, disconnecting the battery or turning the ignition **OFF**.

If using the Tech I diagnostic tool, codes may be cleared using the Clear Codes function of the tool.

To clear codes via the ALDL connector:

1. Turn the ignition switch **OFF**.
2. Ground pin **H** on the ALDL connector.
3. Turn the ignition switch **ON**.
4. Wait for codes to begin flashing.
5. Disconnect pin **H** for at least 1 second, then reconnect the ground. This must be done 3 times within 10 seconds.
6. Wait at least 15 seconds. Repeat Steps 1–4 of this procedure: confirm that code 12 is the only code being flashed. If not, either codes were not cleared properly or an ABS fault still exists.

Additionally, if the vehicle ignition is cycled a pre-determined number of times without a particular fault reappearing, the related fault code will be erased from the control module memory and the ignition cycle counter will reset to zero. The reset threshold is usually either 50 or 100 cycles, depending on the model and the system fault.

INTERMITTENTS

Many intermittent problems are caused by faulty electrical connections or wiring. Always check for poor mating of connector halves or terminals not fully seated in connector bodies, deformed or damaged terminals and poor terminal to wire connections.

Most failures within the ABS will disable the anti-lock function for the entire ignition cycle, even if the fault clears before the next key-off occurrence. One situation will allow the ABS to re-engage if the condition corrects during the ignition cycle; if the EBCM detects low system voltage, the ANTI-LOCK warning lamp is illuminated. If correct minimum voltage is restored to the EBCM, normal ABS function resumes.

Although the ABS trouble codes stored by the EBCM cannot be identified as current or history codes without the use of the scanner, these codes may still be useful in diagnosing intermittent conditions.

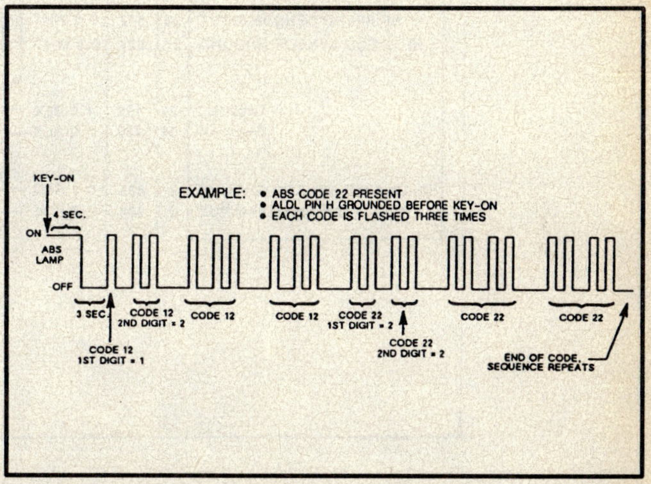

Interpreting ABS Codes Displayed Through Dashboard Warning Lamp

DIAGNOSTIC CHART DTC 21 — 1993 FLEETWOOD BROUGHAM

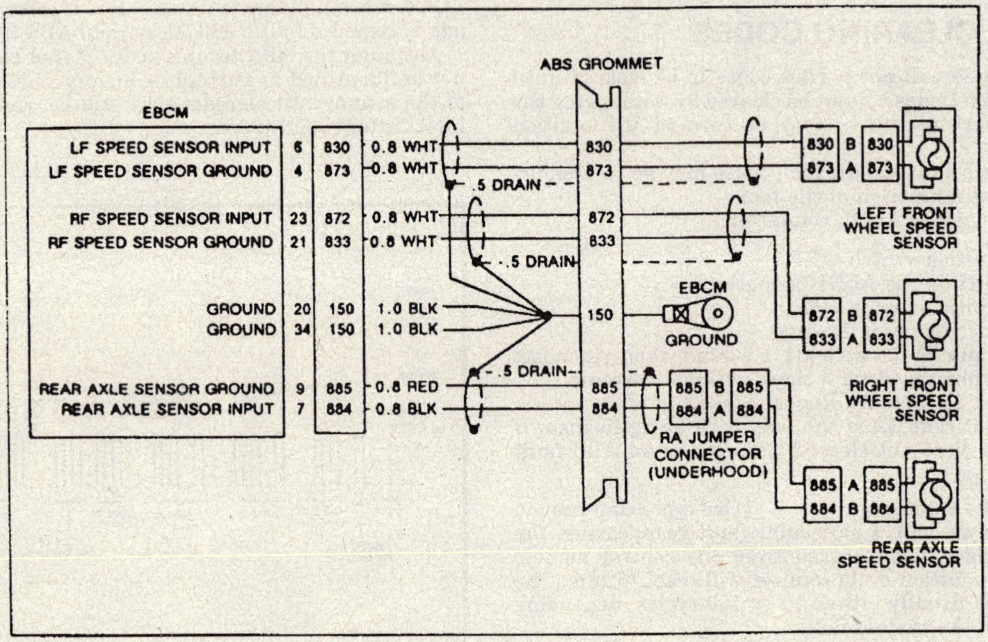

CIRCUIT DESCRIPTION

As each wheel turns, the wheel speed sensor for that wheel creates a small AC voltage as a toothed sensor ring, mounted on the brake rotor (in the front) or differential pinion shaft (in the rear), is passed by the stationary sensor. The sensor consists of a permanent magnet and coil and generates voltage by magnetic induction. The frequency of the wheel speed sensor signal is proportional to wheel speed.

The EBCM uses the wheel speed sensor signal to calculate vehicle reference speeds and individual speed, acceleration and slip values for each wheel. These values are used to determine when anti-lock control is required.

The EBCM performs two basic types of checks on the wheel speed sensors, these are sensor continuity and sensor output.

WHEEL SPEED SENSOR CABLES

In order to prevent electromagnetic interference from disturbing the wheel speed sensor signal, the sensor cables are protected with grounded shielding (indicated by dashed lines in the above schematic). This shield surrounds the two individual sensor wires. A black conduit surrounds the wires and shield. When servicing the sensor cables, the sensor wires are accessible by separating the shielding. If the shielding is disturbed, repair as required.

NOTES ON CODE 21 CHART

1. This section of the fault tree determines if the code stored is due to a current fault condition. 'Current' wheel speed sensor faults will be detected when the vehicle is driven and reaches a speed of 4 mph (6 kph). If the ABS light turns on, the fault is current. The TECH 1 will indicate ignition cycle since code was set. If the ignition cycle counter is at zero or one the code is current.
2. This section of the fault tree checks for an open or short in the wheel speed sensor wiring or sensor coil.
3. This section of the fault tree checks for conditions which may cause intermittent code setting. Some causes of intermittent wheel speed sensor codes to set are: improper routing or sensor cables next to spark plug wires, cables not retained in brackets, loose fitting or improperly mounted sensors, damaged sensors or toothed rings, poor terminal engagement of connectors, or intermittent shorts or opens in wiring.

DIAGNOSTIC CHART DTC 21 — 1993 FLEETWOOD BROUGHAM, CONTINUED

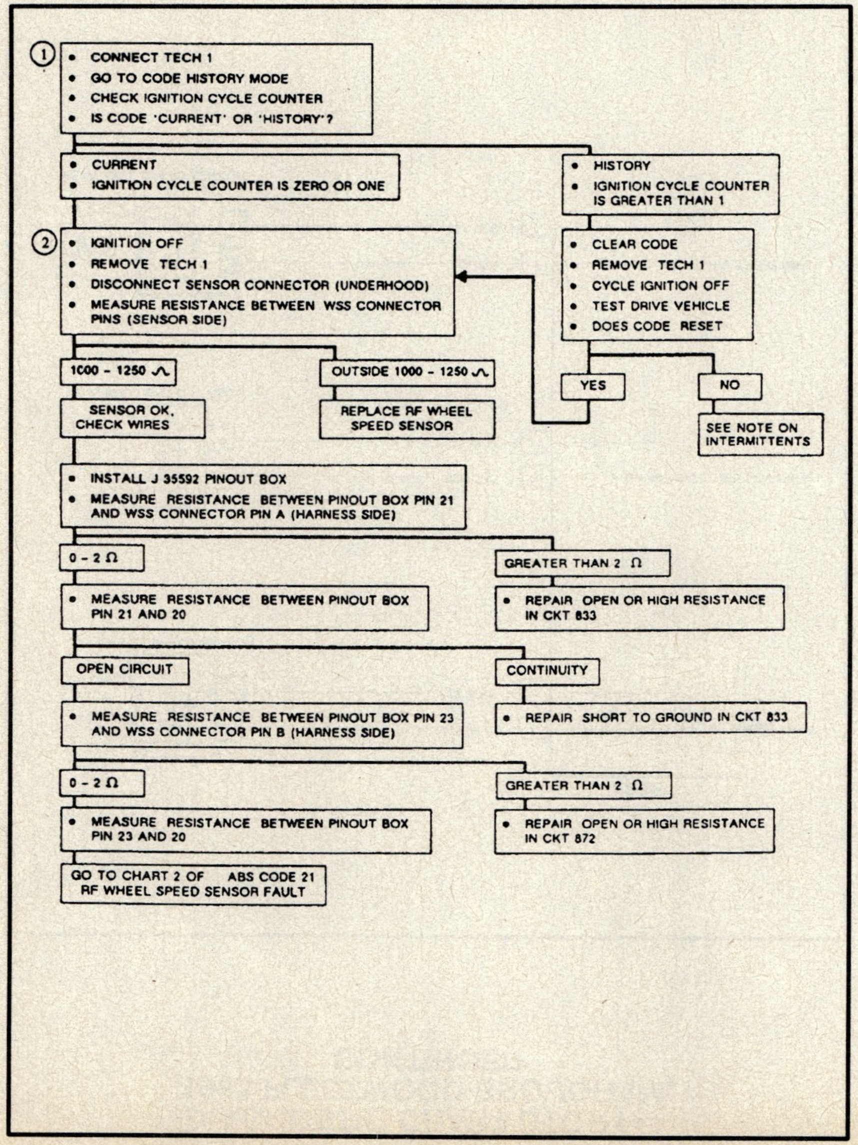

DIAGNOSTIC CHART DTC 21 — 1993 FLEETWOOD BROUGHAM, CONTINUED

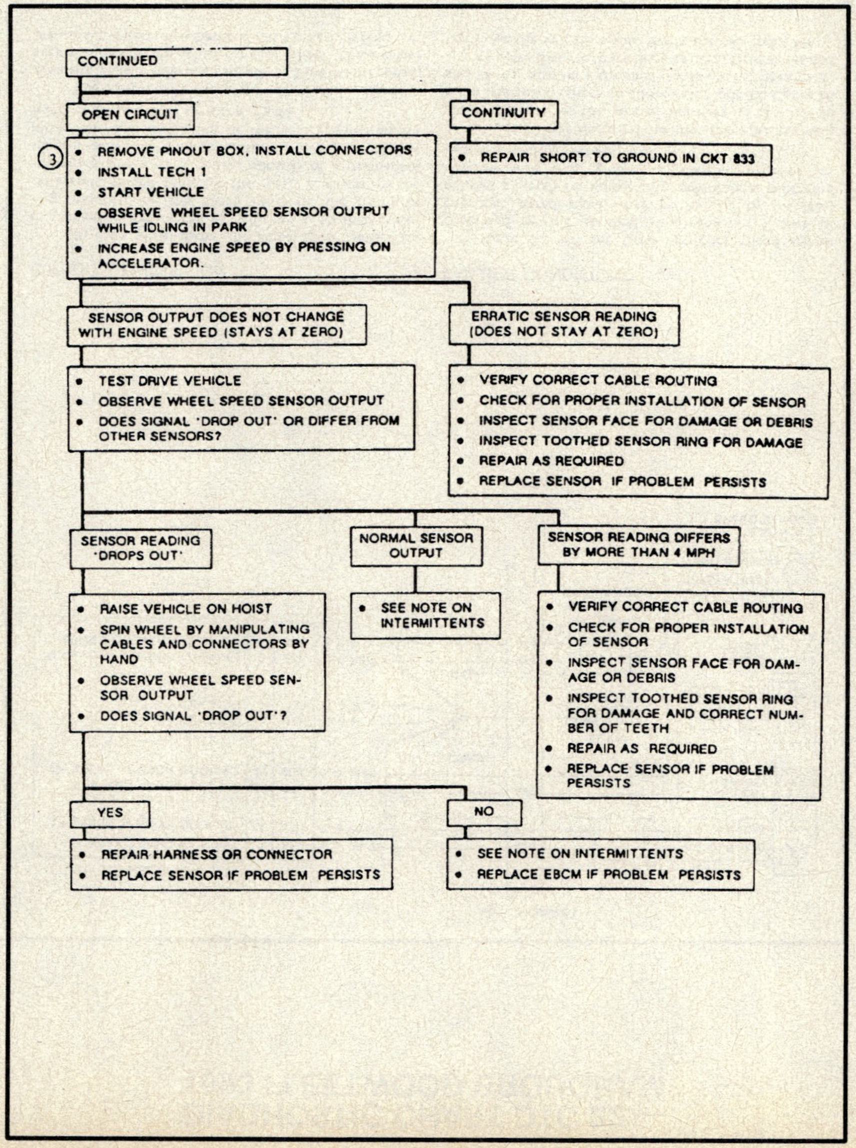

DIAGNOSTIC CHART DTC 21 — 1993 FLEETWOOD BROUGHAM, CONTINUED

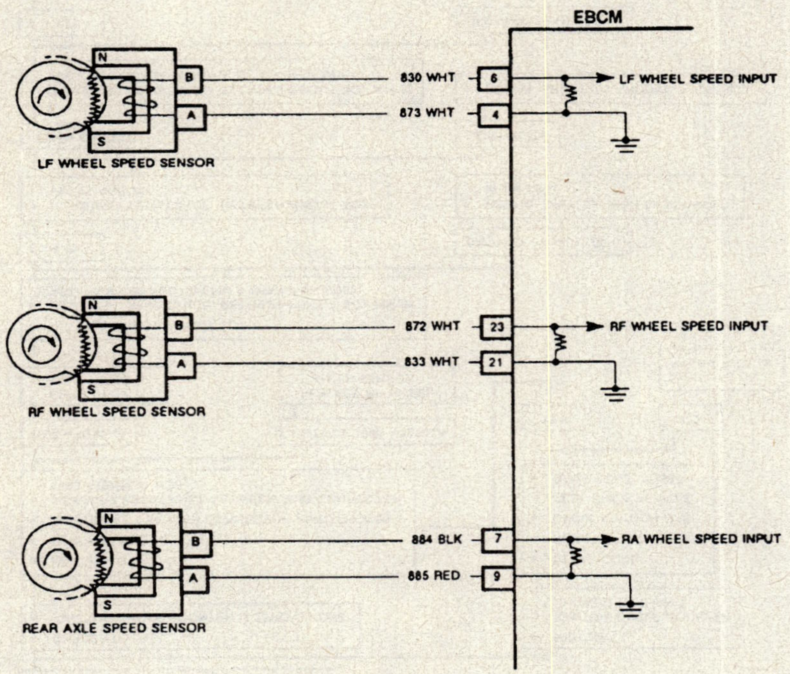

DIAGNOSTIC CHART DTC 22 — 1993 FLEETWOOD BROUGHAM

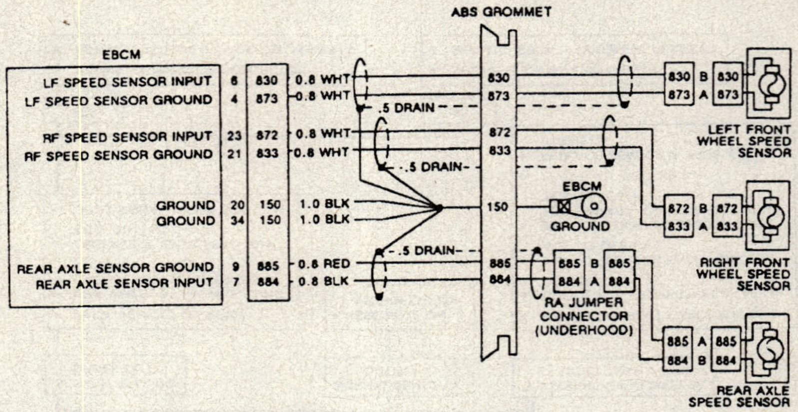

CIRCUIT DESCRIPTION

As each wheel turns, the wheel speed sensor for that wheel creates a small AC voltage as a toothed sensor ring, mounted on the brake rotor (in the front) or differential pinion shaft (in the rear), is passed by the stationary sensor. The sensor consists of a permanent magnet and coil and generates voltage by magnetic induction. The frequency of the wheel speed sensor signal is proportional to wheel speed.

The EBCM uses the wheel speed sensor signal to calculate vehicle reference speeds and individual speed, acceleration and slip values for each wheel. These values are used to determine when anti-lock control is required.

FAILURE CONDITIONS

Code 22 will set from improper speed signals generated by the toothed ring sensor. Causes of improper speed signal may be caused by incorrect number of teeth on sensor ring, sensor rings which are covered with dirt, grease or metallic particles, or damaged toothed sensor ring.

Code 22 may also set if the mini-spare tire has been previously used or the size of the tires on the vehicle differ from each other. In addition to Code 22, Codes 26 and 36 may also set if the mini-spare tire has been used. If it is known that the spare tire was used, clear codes and test drive the vehicle again, verify that no codes set.

DIAGNOSTIC CHART DTC 22 — 1993 FLEETWOOD BROUGHAM, CONTINUED

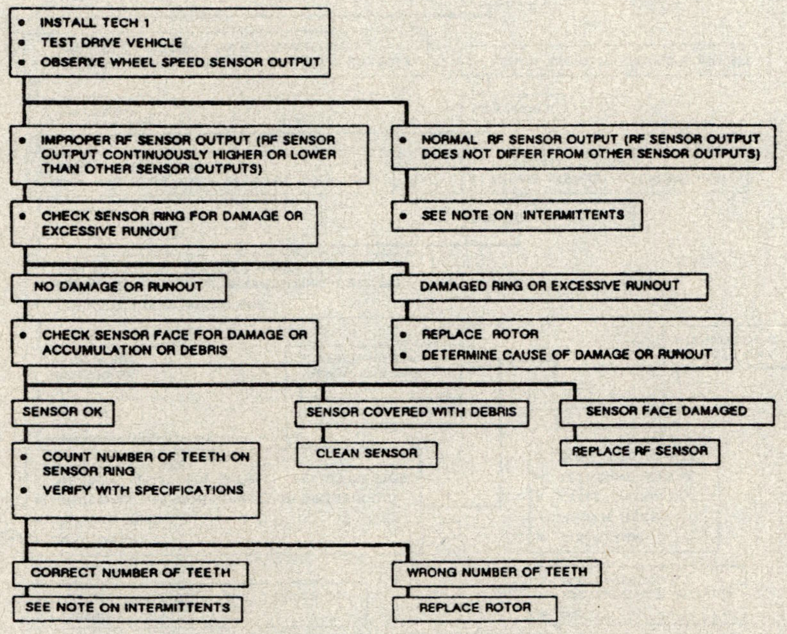

DIAGNOSTIC CHART DTC 25 — 1993 FLEETWOOD BROUGHAM

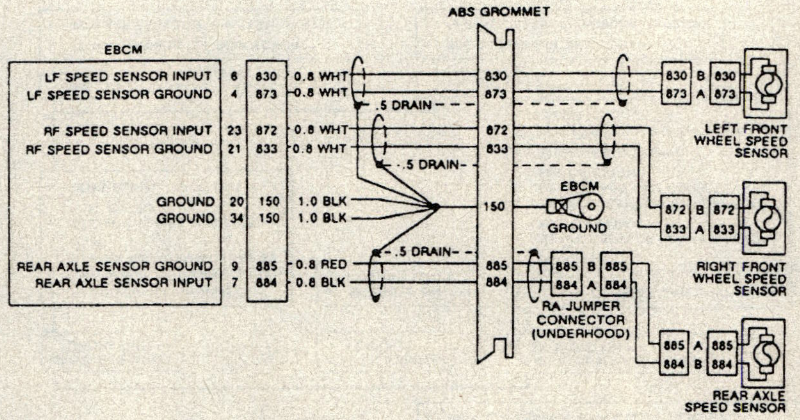

CIRCUIT DESCRIPTION

As each wheel turns, the wheel speed sensor for that wheel creates a small AC voltage as a toothed sensor ring, mounted on the brake rotor (in the front) or differential pinion shaft (in the rear), is passed by the stationary sensor. The sensor consists of a permanent magnet and coil and generates voltage by magnetic induction. The frequency of the wheel speed sensor signal is proportional to wheel speed.

The EBCM uses the wheel speed sensor signal to calculate vehicle reference speeds and individual speed, acceleration and slip values for each wheel. These values are used to determine when anti-lock control is required.

The EBCM performs two basic types of checks on the wheel speed sensors, these are sensor continuity and sensor output.

WHEEL SPEED SENSOR CABLES

In order to prevent electromagnetic interference from disturbing the wheel speed sensor signal, the sensor cables are protected with grounded shielding (indicated by dashed lines in the above schematic). This shield surrounds the two individual sensor wires. A black conduit surrounds the wires and shield. When servicing the sensor cables, the sensor wires are accessible by separating the shielding. If the shielding is disturbed, repair as required.

NOTES ON CODE 25 CHART

1. This section of the fault tree determines if the code stored is due to a current fault condition. 'Current' wheel speed sensor faults will be detected when the vehicle is driven and reaches a speed of 4 mph (6 kph). If the ABS light turns on, the fault is current. The TECH 1 will indicate ignition cycle since code was set. If the ignition cycle counter is at zero or one the code is current.
2. This section of the fault tree checks for an open or short in the wheel speed sensor wiring or sensor coil.
3. This section of the fault tree checks for conditions which may cause intermittent code setting. Some causes of intermittent wheel speed sensor codes to set are: improper routing or sensor cables next to spark plug wires, cables not retained in brackets, loose fitting or improperly mounted sensors, damaged sensors or toothed rings, poor terminal engagement of connectors, or intermittent shorts or opens in wiring.

DIAGNOSTIC CHART DTC 25 — 1993 FLEETWOOD BROUGHAM, CONTINUED

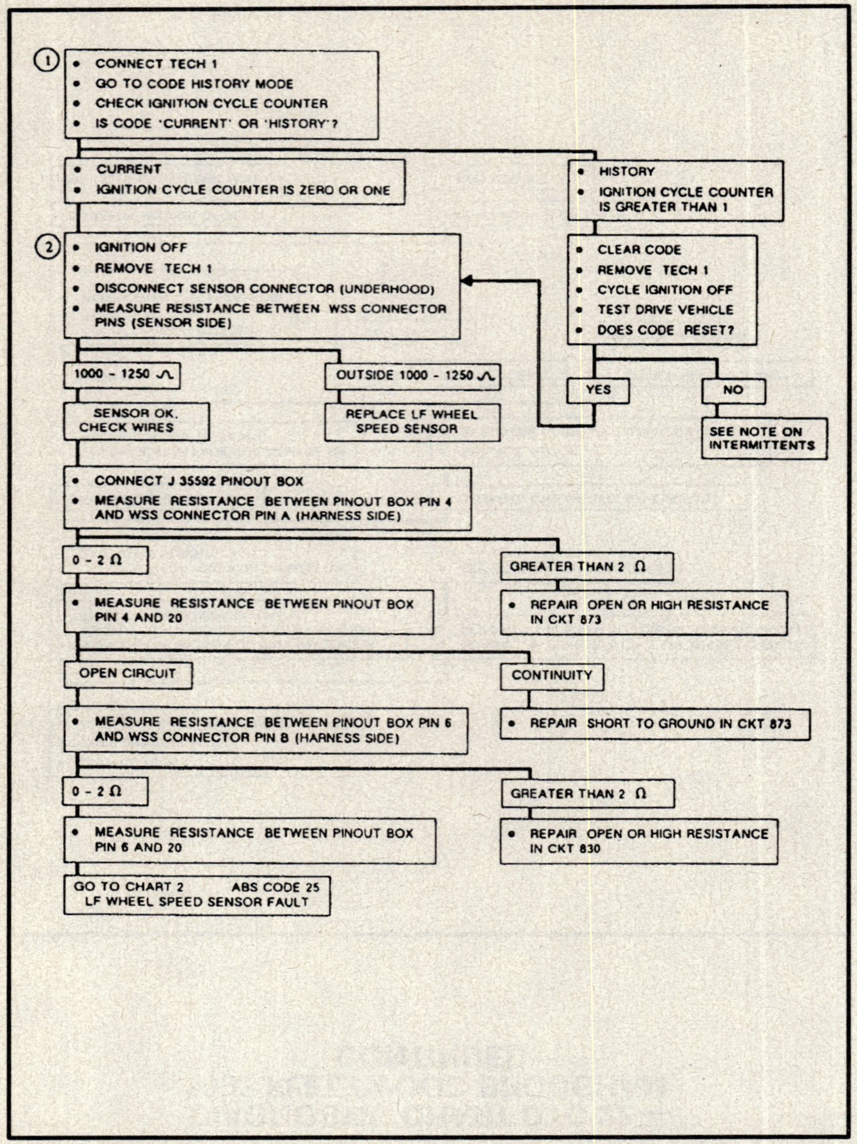

DIAGNOSTIC CHART DTC 25 — 1993 FLEETWOOD BROUGHAM, CONTINUED

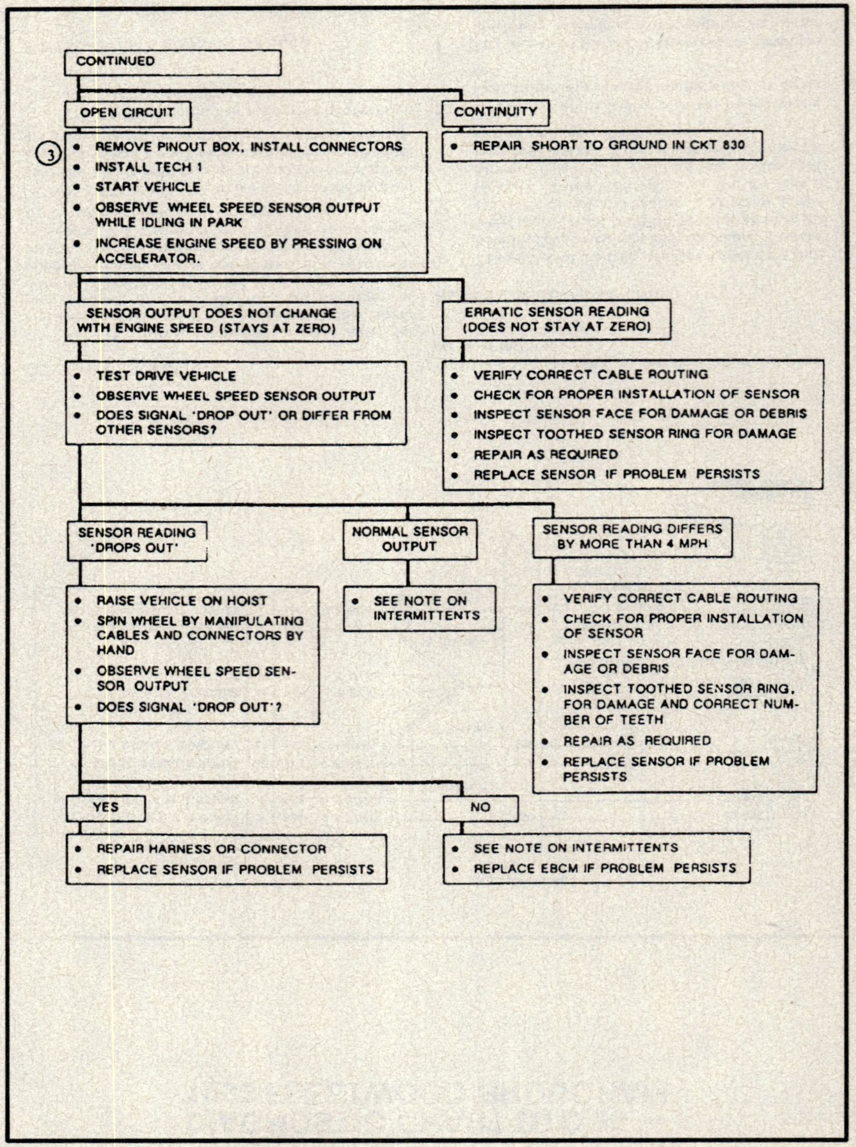

DIAGNOSTIC CHART DTC 25 — 1993 FLEETWOOD BROUGHAM, CONTINUED

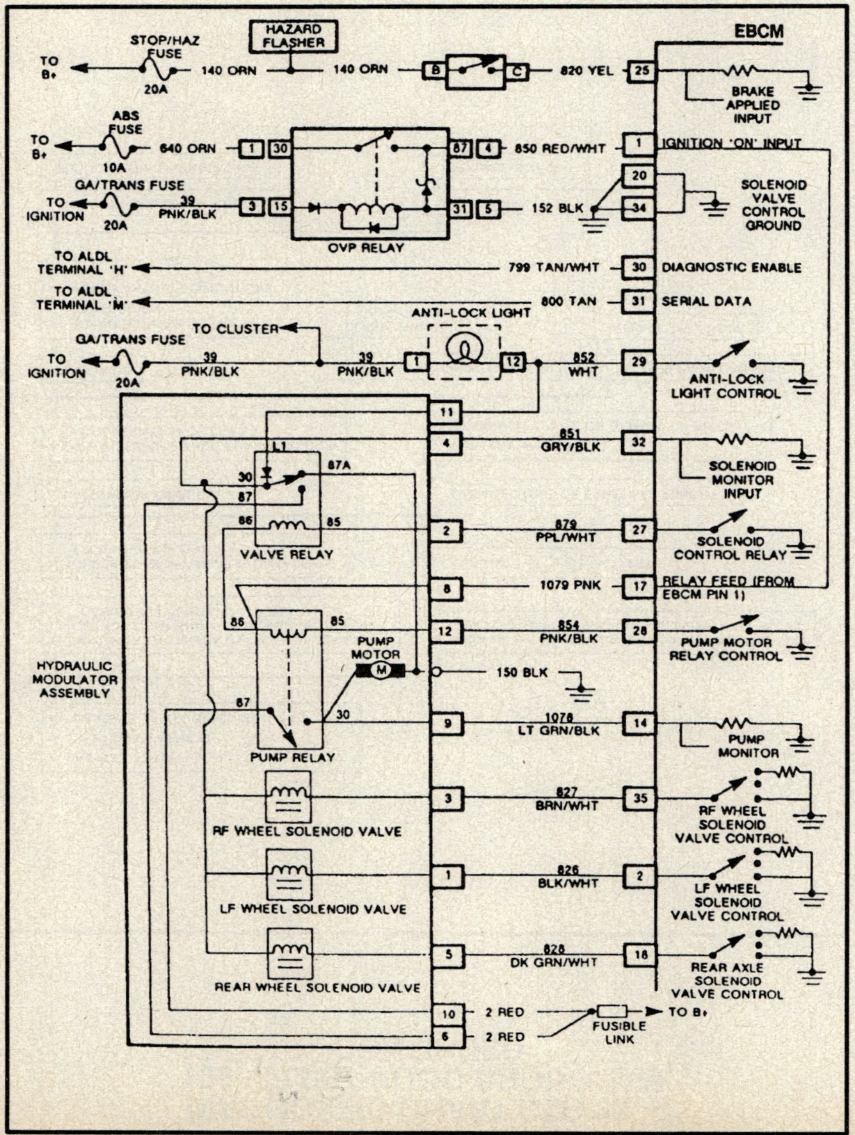

DIAGNOSTIC CHART DTC 26 — 1993 FLEETWOOD BROUGHAM

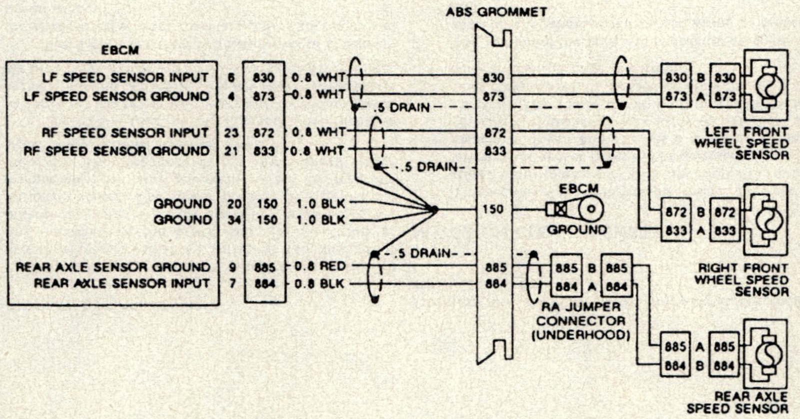

CIRCUIT DESCRIPTION

As each wheel turns, the wheel speed sensor for that wheel creates a small AC voltage as a toothed sensor ring, mounted on the brake rotor (in the front) or differential pinion shaft (in the rear), is passed by the stationary sensor. The sensor consists of a permanent magnet and coil and generates voltage by magnetic induction. The frequency of the wheel speed sensor signal is proportional to wheel speed.

The EBCM uses the wheel speed sensor signal to calculate vehicle reference speeds and individual speed, acceleration and slip values for each wheel. These values are used to determine when anti-lock control is required.

FAILURE CONDITIONS

Code 26 will set from improper speed signals generated by the toothed ring sensor. Causes of improper speed signal may be caused by incorrect number of teeth on sensor ring, sensor rings which are covered with dirt, grease or metallic particles, or damaged toothed sensor ring.

Code 26 may also set if the mini-spare tire has been previously used or the size of the tires on the vehicle differ from each other. In addition to Code 26, Codes 22 and 36 may also set if the mini-spare tire has been used. If it is known that the spare tire was used, clear codes and test drive the vehicle again, verify that no codes set.

DIAGNOSTIC CHART DTC 26 — 1993 FLEETWOOD BROUGHAM, CONTINUED

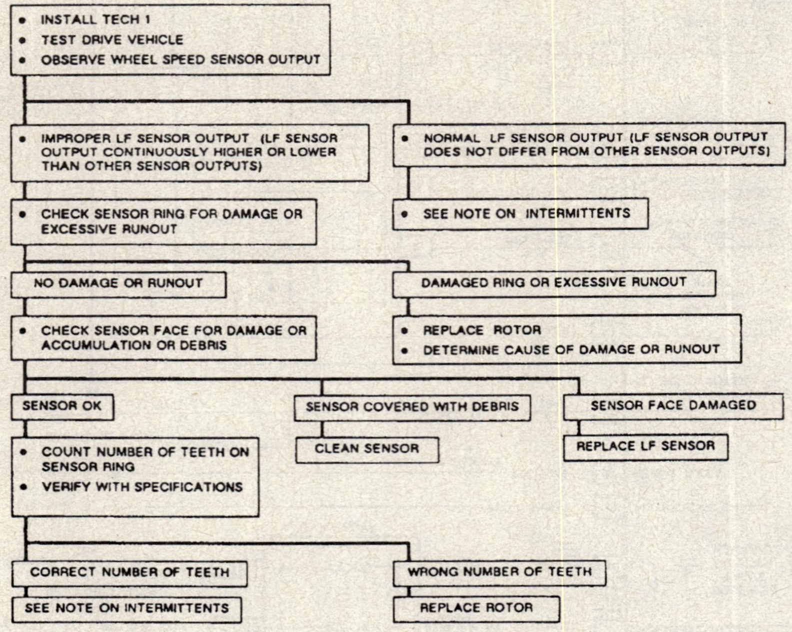

DIAGNOSTIC CHART DTC 35 — 1993 FLEETWOOD BROUGHAM

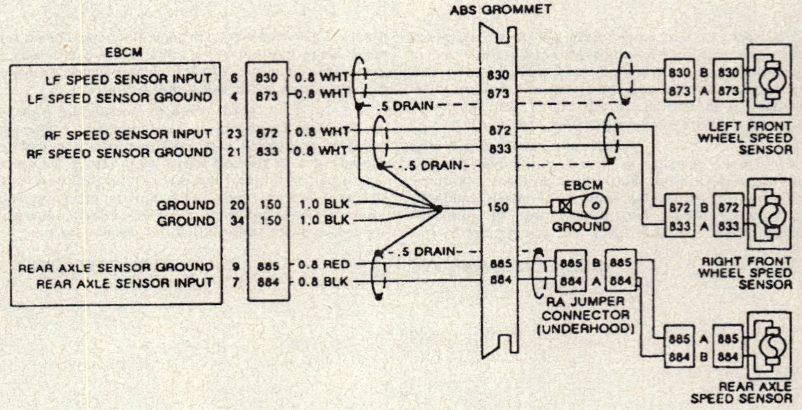

CIRCUIT DESCRIPTION

As each wheel turns, the wheel speed sensor for that wheel creates a small AC voltage as a toothed sensor ring, mounted on the brake rotor (in the front) or differential pinion shaft (in the rear), is passed by the stationary sensor. The sensor consists of a permanent magnet and coil and generates voltage by magnetic induction. The frequency of the wheel speed sensor signal is proportional to wheel speed.

The EBCM uses the wheel speed sensor signal to calculate vehicle reference speeds and individual speed, acceleration and slip values for each wheel. These values are used to determine when anti-lock control is required.

The EBCM performs two basic types of checks on the wheel speed sensors, these are sensor continuity and sensor output.

WHEEL SPEED SENSOR CABLES

In order to prevent electromagnetic interference from disturbing the wheel speed sensor signal, the sensor cables are protected with grounded shielding (indicated by dashed lines in the above schematic). This shield surrounds the two individual sensor wires. A black conduit surrounds the wires and shield. When servicing the sensor cables, the sensor wires are accessible by separating the shielding. If the shielding is disturbed, repair as required.

NOTES ON CODE 35 CHART

1. This section of the fault tree determines if the code stored is due to a current fault condition. 'Current' wheel speed sensor faults will be detected when the vehicle is driven and reaches a speed of 4 mph (6 kph). If the ABS light turns on, the fault is current. The TECH 1 will indicate ignition cycle since code was set. If the ignition cycle counter is at zero or one the code is current.
2. This section of the fault tree checks for an open or short in the wheel speed sensor wiring or sensor coil.
3. This section of the fault tree checks for conditions which may cause intermittent code setting. Some causes of intermittent wheel speed sensor codes to set are: improper routing or sensor cables next to spark plug wires, cables not retained in brackets, loose fitting or improperly mounted sensors, damaged sensors or toothed rings, poor terminal engagement of connectors, or intermittent shorts or opens in wiring.

DIAGNOSTIC CHART DTC 35 — 1993 FLEETWOOD BROUGHAM, CONTINUED

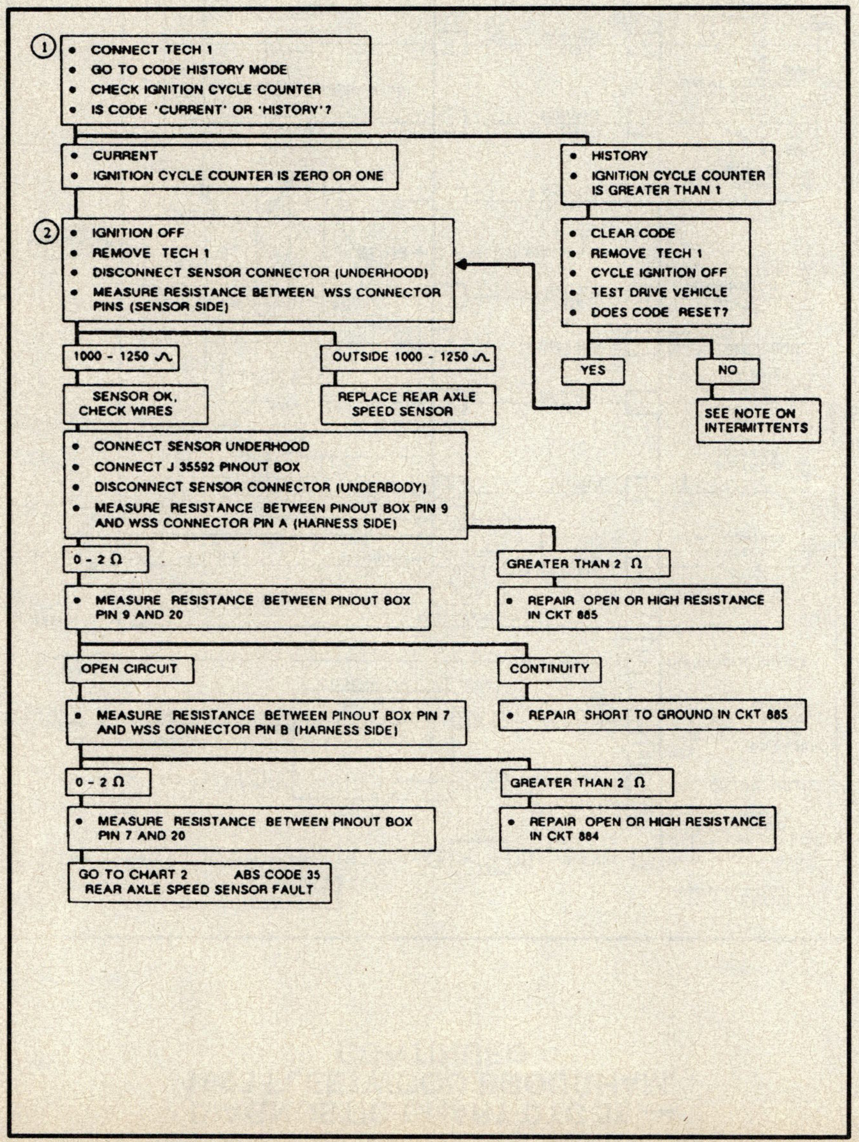

DIAGNOSTIC CHART DTC 35 — 1993 FLEETWOOD BROUGHAM, CONTINUED

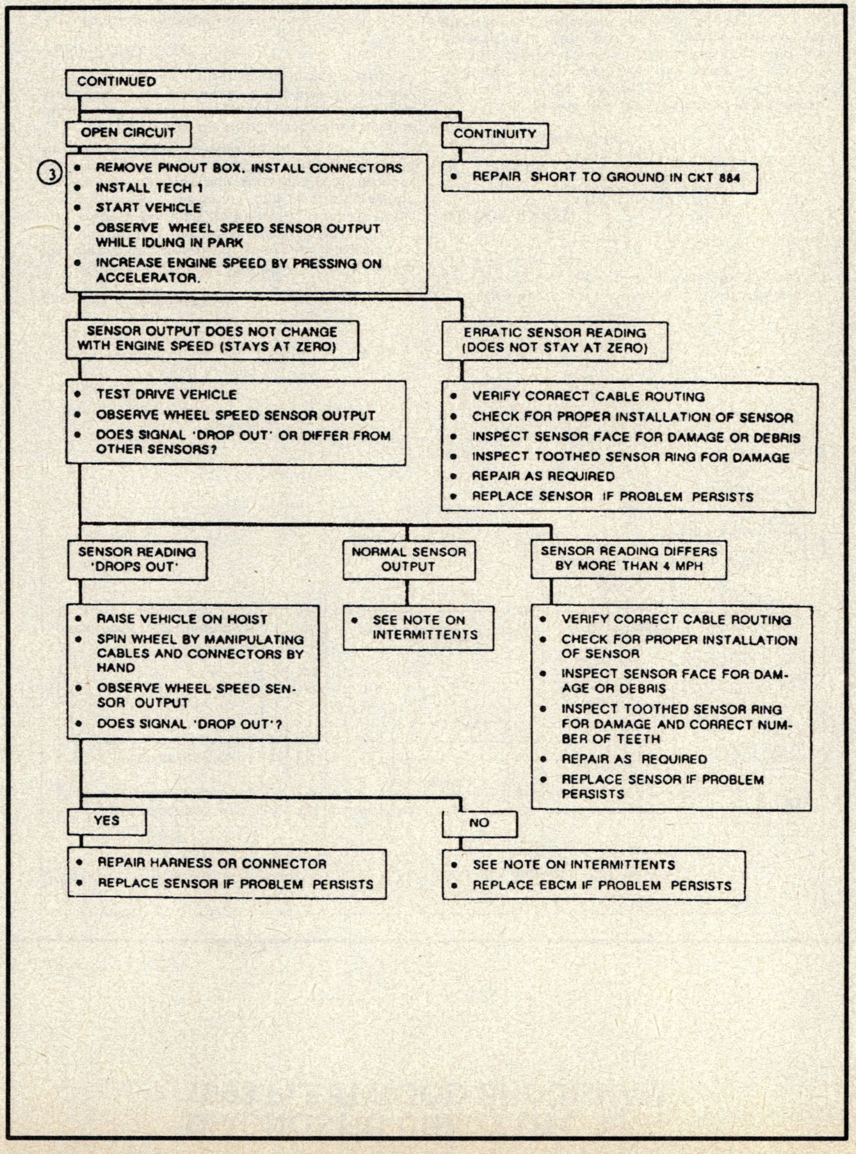

DIAGNOSTIC CHART DTC 35 — 1993 FLEETWOOD BROUGHAM, CONTINUED

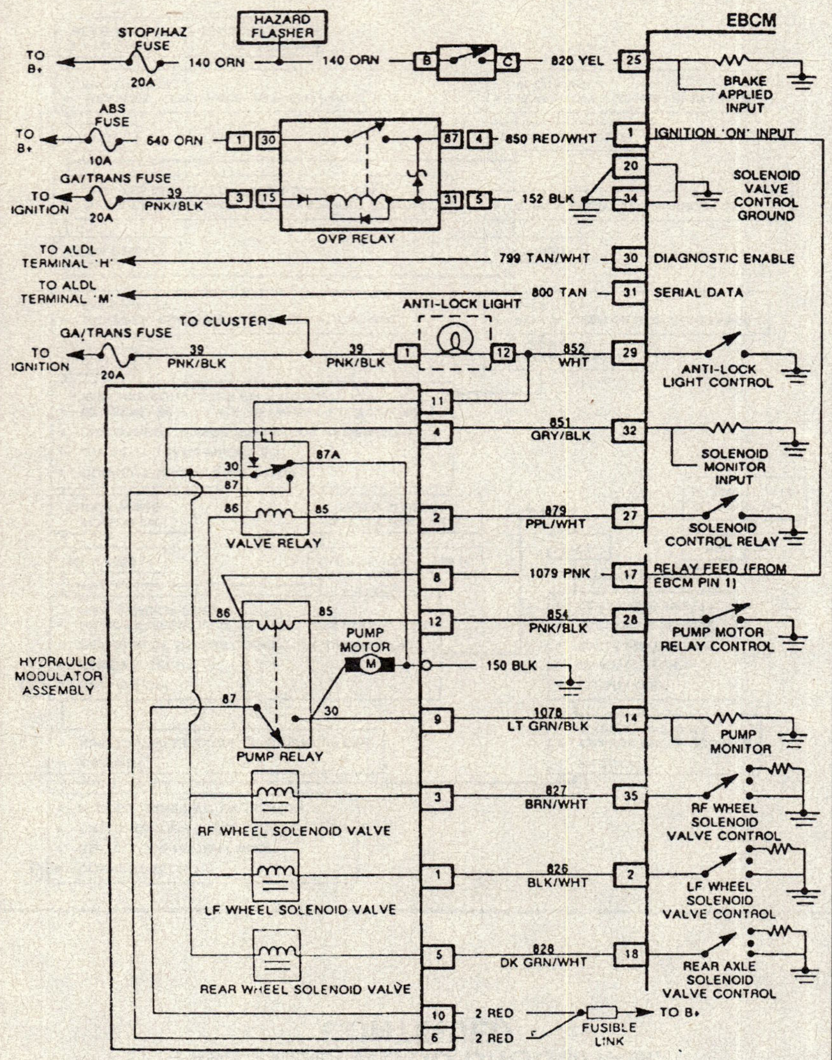

DIAGNOSTIC CHART DTC 36 — 1993 FLEETWOOD BROUGHAM

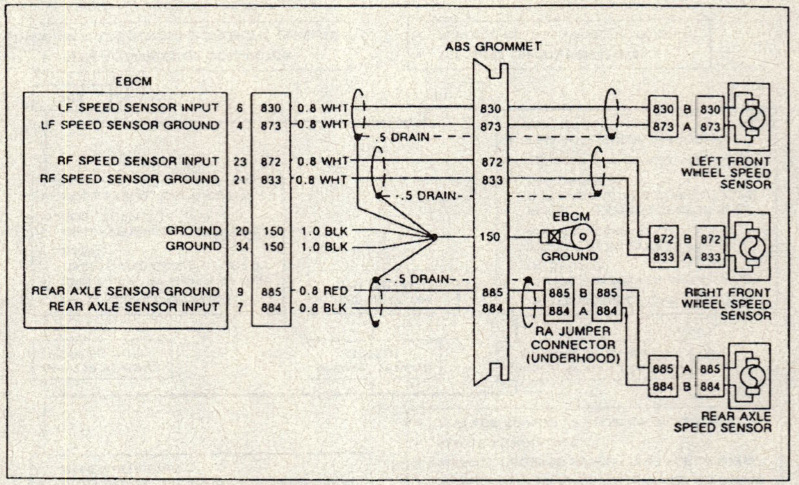

CIRCUIT DESCRIPTION

As each wheel turns, the wheel speed sensor for that wheel creates a small AC voltage as a toothed sensor ring, mounted on the brake rotor (in the front) or differential pinion shaft (in the rear), is passed by the stationary sensor. The sensor consists of a permanent magnet and coil and generates voltage by magnetic induction. The frequency of the wheel speed sensor signal is proportional to wheel speed.

The EBCM uses the wheel speed sensor signal to calculate vehicle reference speeds and individual speed, acceleration and slip values for each wheel. These values are used to determine when antilock control is required.

TEST CONDITIONS

1. Tested continuously.

FAILURE CONDITIONS

1. While Driving: One wheel is detected at least 30 kph or 19 mph faster than the remaining wheels for an extended period of time.

– or –

One wheel is at least 6 kph or 4 mph less than or greater than the remaining wheels for an extended period of time.

ACTION TAKEN

1. Code set.
2. ABS switched 'OFF'.

Code 36 will set from improper speed signals generated by the toothed ring sensor. Causes of improper speed signal may be caused by incorrect number of teeth on sensor ring, sensor rings which are covered with dirt, grease or metallic particles, or damaged toothed sensor ring.

Code 36 may also set if the mini-spare tire has been previously used or the size of the tires on the vehicle differ from each other. In addition to Code 36, Codes 22 and 26 may also set if the mini-spare tire has been used. If it is known that the spare tire was used, clear codes and test drive the vehicle again, verify that no codes set.

DIAGNOSTIC CHART DTC 36 — 1993 FLEETWOOD BROUGHAM, CONTINUED

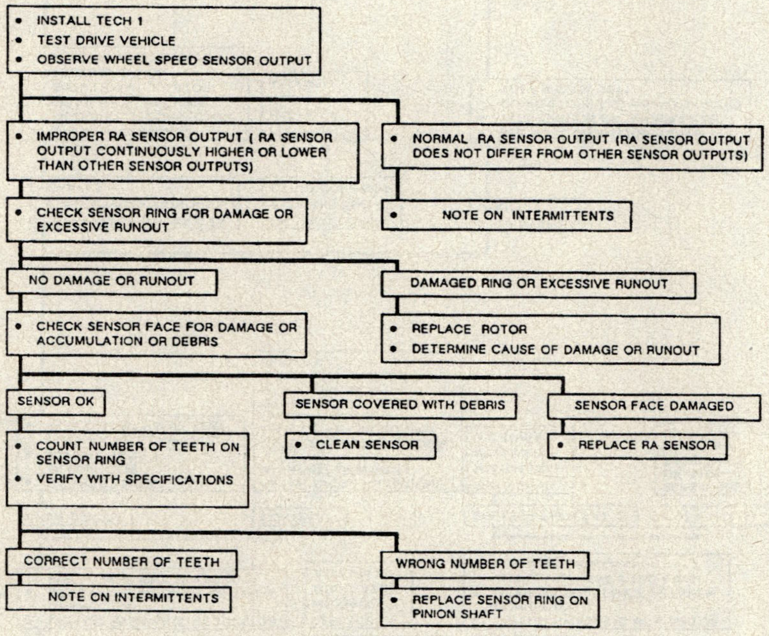

DIAGNOSTIC CHART DTC 41 — 1993 FLEETWOOD BROUGHAM

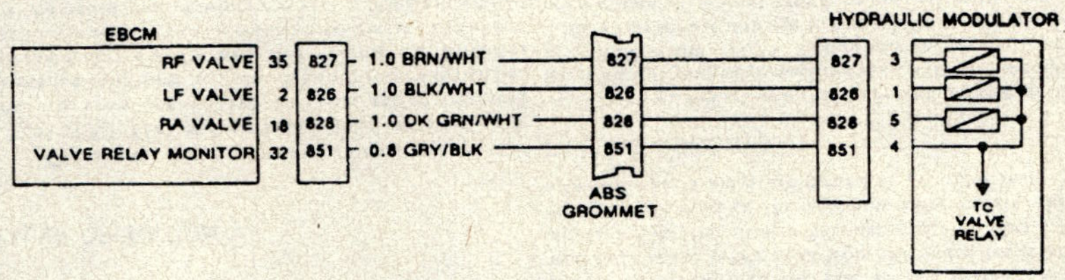

CIRCUIT DESCRIPTION

The right front solenoid valve in the hydraulic modulator may be commanded to three different positions by the EBCM. Valve position is determined by the amount of current which is allowed to flow through the solenoid valve coil.

The solenoid valve circuits receive power through the valve relay on the hydraulic modulator. The valve relay is engaged at key-on and remains engaged throughout the ignition cycle. Solenoid valve lines should have battery voltage available at all times.

When the key is turned to RUN, power is supplied to the EBCM. The EBCM will complete an internal self-check before providing ground to the valve relay. When the valve relay is energized, battery power is supplied to the solenoid valves.

VALVE OPERATION

The valves are continuously monitored by the EBCM (CKT 851) and each individual solenoid valve line. If valve line voltage does not match expected values, an ABS code will set. When the vehicle reaches approximately 4 MPH after key-on, the EBCM cycles each valve and checks for a proper response. This check may be heard and felt while it is taking place and should be considered a normal mode of operation.

Pressure Increase

The pressure increase position is the normal operating position of the valves for braking. Brake fluid flows thru the modulator from the master cylinder to the wheel brake. No current is supplied to the valve coil during pressure increase.

Pressure Hold

The pressure hold position is commanded when the EBCM desires to hold the pressure in the wheel circuit constant. The modulator will not allow pressure flow from the master cylinder to the wheel brake. The pressure hold position is commanded by passing approximately 2.0 amps to the valve coil.

Pressure Reduction

The pressure reduction position is used to decrease hydraulic pressure in a specific wheel circuit without decreasing pressure in the rest of the system. The modulator takes brake fluid from the wheel brake and injects it into the master cylinder circuit. This fully energized position of the solenoid valve occurs when approximately 5 amps are passed to the valve coil.

FAILURE CONDITIONS

Code 41 will set when the expected position of the right front solenoid valve does not match the commanded position from the EBCM. Conditions which could cause Code 41 to set are damage to the right front solenoid, an open circuit or a short circuit to ground or battery on CKT 827.

DIAGNOSTIC CHART DTC 41 — 1993 FLEETWOOD BROUGHAM, CONTINUED

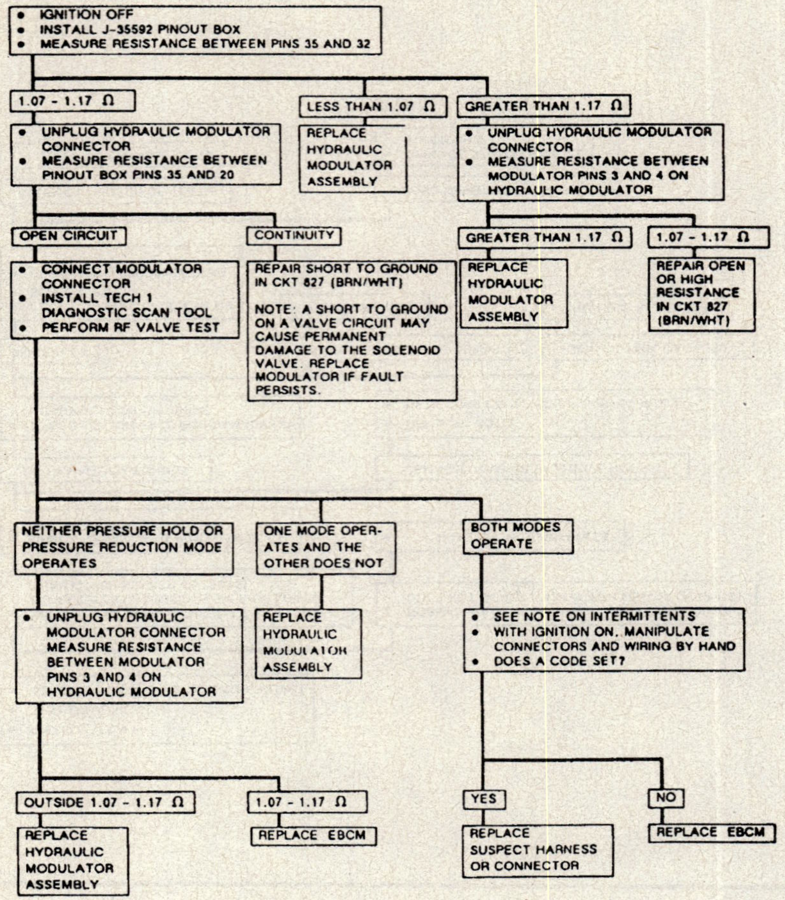

DIAGNOSTIC CHART DTC 45 — 1993 FLEETWOOD BROUGHAM

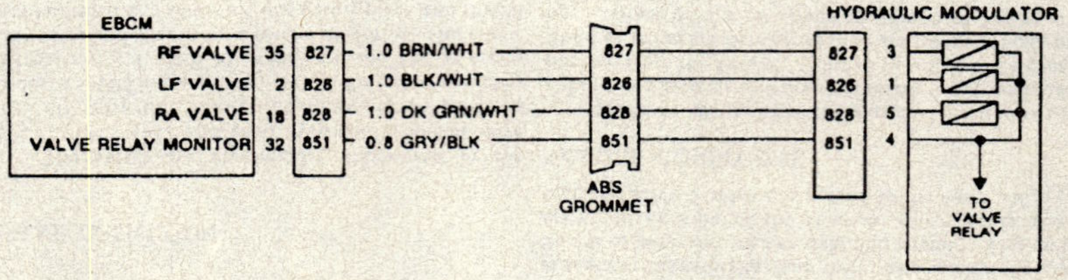

CIRCUIT DESCRIPTION

The left front solenoid valve in the hydraulic modulator may be commanded to three different positions by the EBCM. Valve position is determined by the amount of current which is allowed to flow through the solenoid valve coil.

The solenoid valve circuits receive power through the valve relay on the hydraulic modulator. The valve relay is engaged at key-on and remains engaged throughout the ignition cycle. Solenoid valve lines should have battery voltage available at all times.

When the key is turned to RUN, power is supplied to the EBCM. The EBCM will complete an internal self-check before providing ground to the valve relay. When the valve relay is energized, battery power is supplied to the solenoid valves.

VALVE OPERATION

The valves are continuously monitored by the EBCM (CKT 851) and each individual solenoid valve line. If valve line voltage does not match expected values, an ABS code will set. When the vehicle reaches approximately 4 MPH after key-on, the EBCM cycles each valve and checks for a proper response. This check may be heard and felt while it is taking place and should be considered a normal mode of operation.

Pressure Increase

The pressure increase position is the normal operating position of the valves for braking. Brake fluid flows thru the modulator from the master cylinder to the wheel brake. No current is supplied to the valve coil during pressure increase.

Pressure Hold

The pressure hold position is commanded when the EBCM desires to hold the pressure in the wheel circuit constant. The modulator will not allow pressure flow from the master cylinder to the wheel brake. The pressure hold position is commanded by passing approximately 2.0 amps to the valve coil.

Pressure Reduction

The pressure reduction position is used to decrease hydraulic pressure in a specific wheel circuit without decreasing pressure in the rest of the system. The modulator takes brake fluid from the wheel brake and injects it into the master cylinder circuit. This fully energized position of the solenoid valve occurs when approximately 5 amps are passed to the valve coil.

FAILURE CONDITIONS

Code 45 will set when the expected position of the left front solenoid valve does not match the commanded position from the EBCM. Conditions which could cause Code 45 to set are damage to the left front solenoid, an open circuit or a short circuit to ground or battery on CKT 826.

DIAGNOSTIC CHART DTC 45 — 1993 FLEETWOOD BROUGHAM, CONTINUED

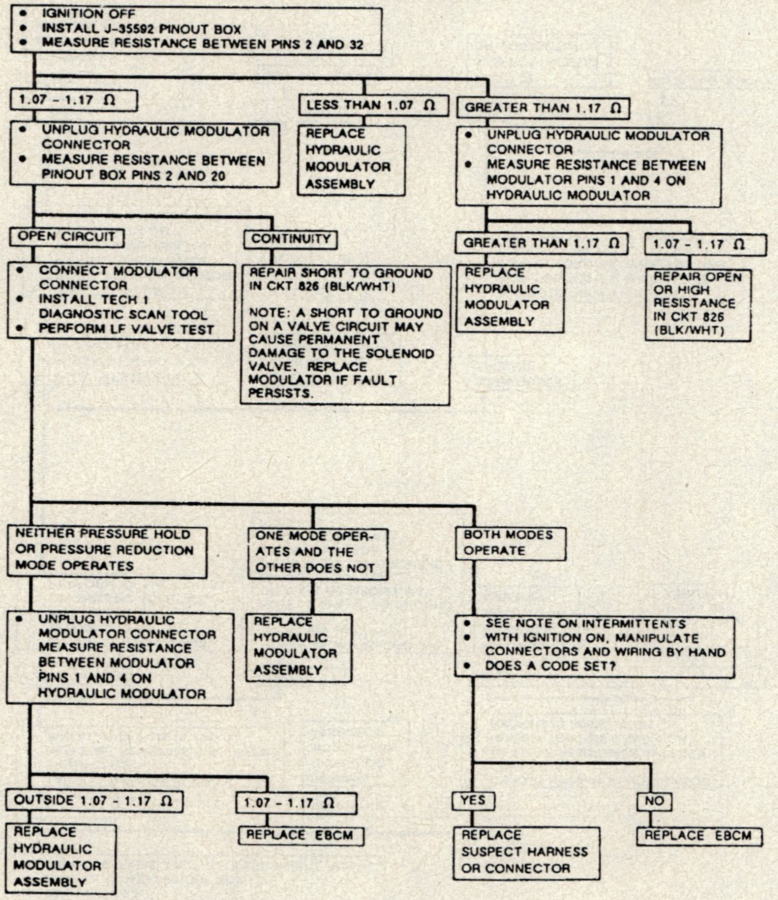

DIAGNOSTIC CHART DTC 55 — 1993 FLEETWOOD BROUGHAM

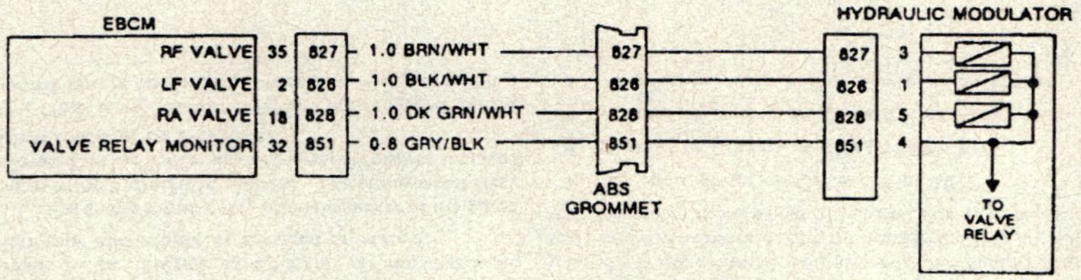

CIRCUIT DESCRIPTION

The rear axle solenoid valve in the hydraulic modulator may be commanded to three different positions by the EBCM. The rear axle solenoid valve modulates hydraulic brake pressure to both rear wheel circuits. Valve position is determined by the amount of current which is allowed to flow through the solenoid valve coil.

The solenoid valve circuits receive power through the valve relay on the hydraulic modulator. The valve relay is engaged at key-on and remains engaged throughout the ignition cycle. Solenoid valve lines should have battery voltage available at all times.

When the key is turned to RUN, power is supplied to the EBCM. The EBCM will complete an internal self-check before providing ground to the valve relay. When the valve relay is energized, battery power is supplied to the solenoid valves.

VALVE OPERATION

The valves are continuously monitored by the EBCM (CKT 851) and each individual solenoid valve line. If valve line voltage does not match expected values, an ABS code will set. When the vehicle reaches approximately 4 mph after key-on, the EBCM cycles each valve and checks for a proper response. This check may be heard and felt while it is taking place and should be considered a normal mode of operation.

Pressure Increase

The pressure increase position is the normal operating position of the valves for braking. Brake fluid flows thru the modulator from the master cylinder to the wheel brake. No current is supplied to the valve coil during pressure increase.

Pressure Hold

The pressure hold position is commanded when the EBCM desires to hold the pressure in the wheel circuit constant. The modulator will not allow pressure flow from the master cylinder to the wheel brake. The pressure hold position is commanded by passing approximately 2.0 amps to the valve coil.

Pressure Reduction

The pressure reduction position is used to decrease hydraulic pressure in a specific wheel circuit without decreasing pressure in the rest of the system. The modulator takes brake fluid from the wheel brake and injects it into the master cylinder circuit. This fully energized position of the solenoid valve occurs when approximately 5 amps are passed to the valve coil.

FAILURE CONDITIONS

Code 55 will set when the expected position of the rear axle solenoid valve does not match the commanded position from the EBCM. Conditions which could cause Code 55 to set are damage to the rear axle solenoid, an open circuit or a short circuit to ground or battery on CKT 828.

DIAGNOSTIC CHART DTC 55 — 1993 FLEETWOOD BROUGHAM, CONTINUED

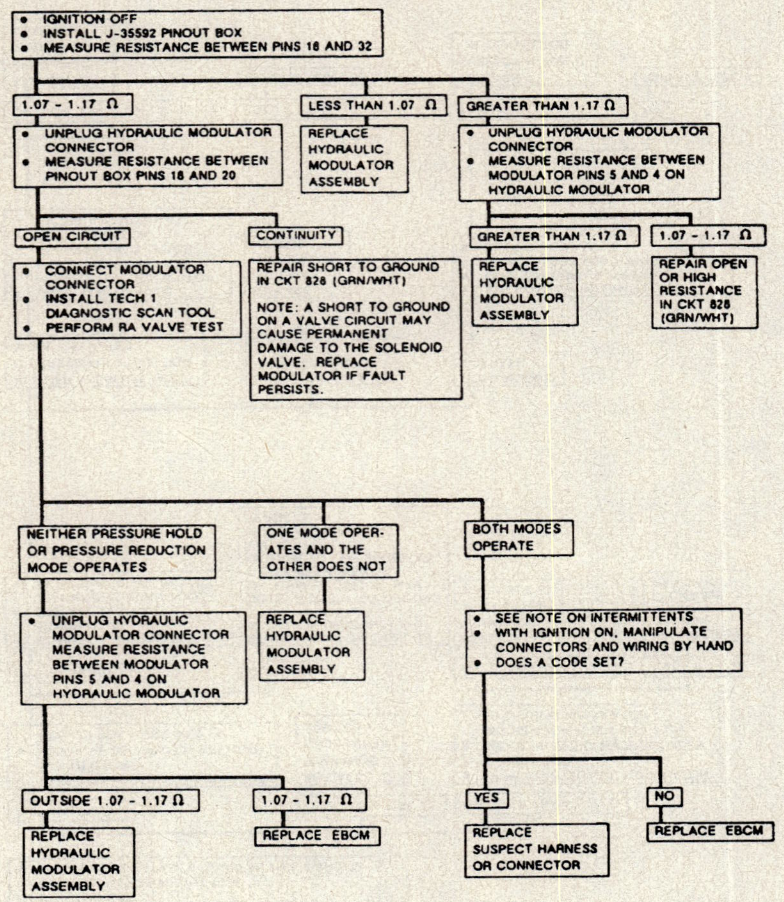

DIAGNOSTIC CHART DTC 61 — 1993 FLEETWOOD BROUGHAM

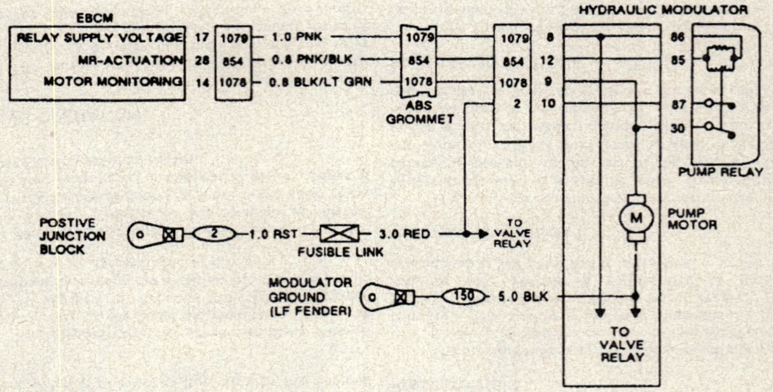

CIRCUIT DESCRIPTION

The pump motor returns brake fluid to the master cylinder brake circuit at the hydraulic modulator during anti-lock braking. During normal braking, the pump does not operate. When the vehicle begins to move after start-up, the EBCM will turn on the pump motor and perform a self-check of the pump motor and pump motor circuit. This check may be felt and heard by the driver when the vehicle begins to move. The self-check should be considered normal operation. The pump motor is an integral component of the hydraulic modulator and cannot be serviced separately.

The pump motor relay provides power to the pump motor in the hydraulic modulator. The pump motor relay is located on the hydraulic modulator an may be replaced if it is found to be defective.

The pump motor relay is not engaged during normal system operation. When anti-lock operation is required, the motor relay actuation line on CKT 854 (EBCM pin 28) is pulled to ground. The pump motor relay is engaged by the relay supply voltage line on CKT 1079 (EBCM pin 17). When the relay switches, battery power is provided to the pump motor.

FAILURE CONDITIONS

When the pump motor relay is engaged, the motor monitor line should be at battery voltage. With the pump motor relay disengaged, the monitoring line should be at ground. If the commanded position of the pump motor relay and motor monitor line do not agree, Code 61 will set. Some conditions which will cause Code 61 to set are:

- An open circuit on CKTs 854 or 1078
- A short to voltage on CKTs 854 or 1078
- A short to ground on CKTs 854 or 1078
- A defective pump motor relay or pump motor

DIAGNOSTIC CHART DTC 61 — 1993 FLEETWOOD BROUGHAM, CONTINUED

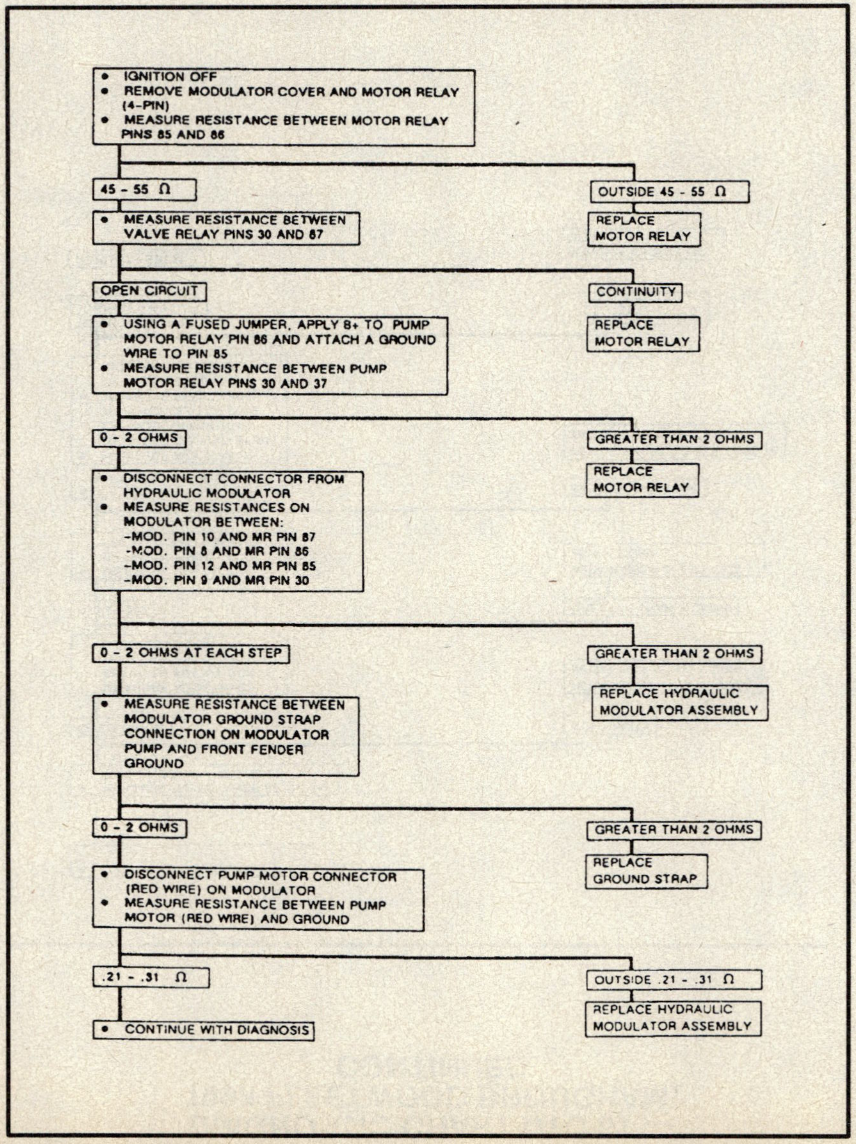

DIAGNOSTIC CHART DTC 61 — 1993 FLEETWOOD BROUGHAM, CONTINUED

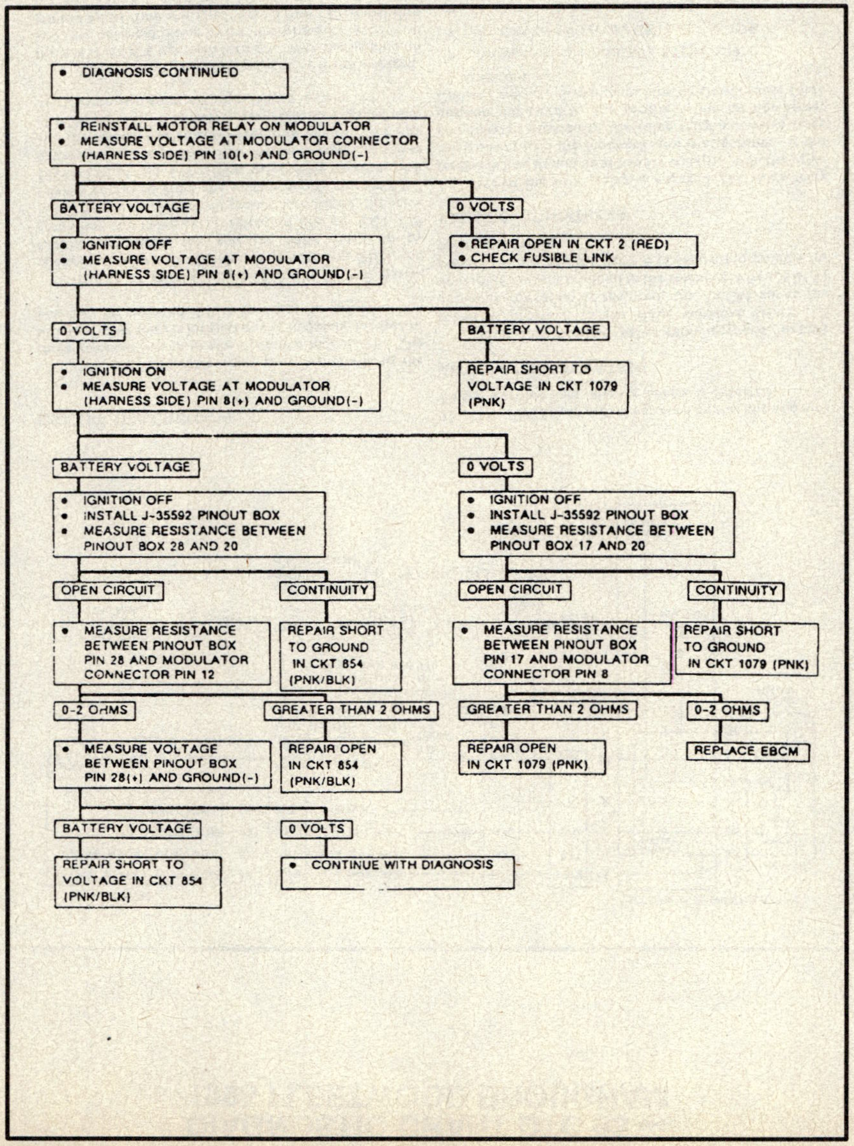

DIAGNOSTIC CHART DTC 61 — 1993 FLEETWOOD BROUGHAM, CONTINUED

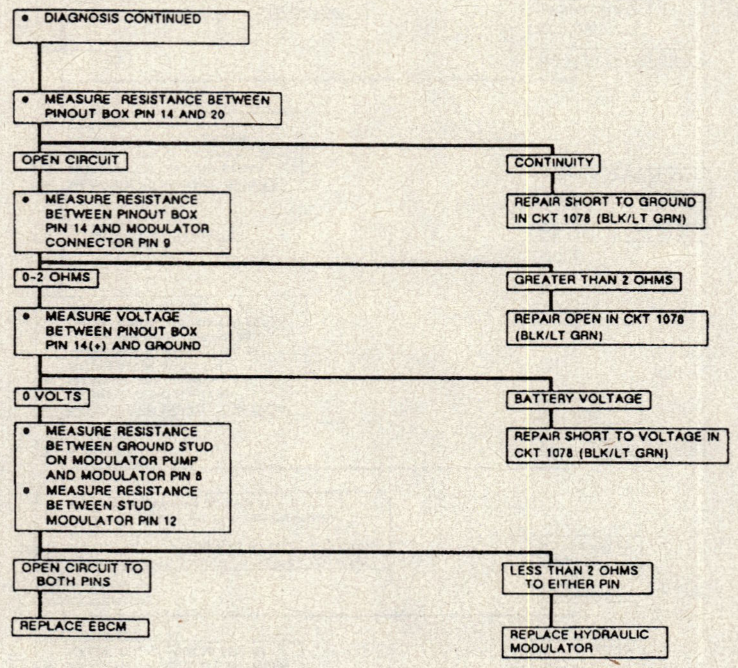

DIAGNOSTIC CHART DTC 63 — 1993 FLEETWOOD BROUGHAM

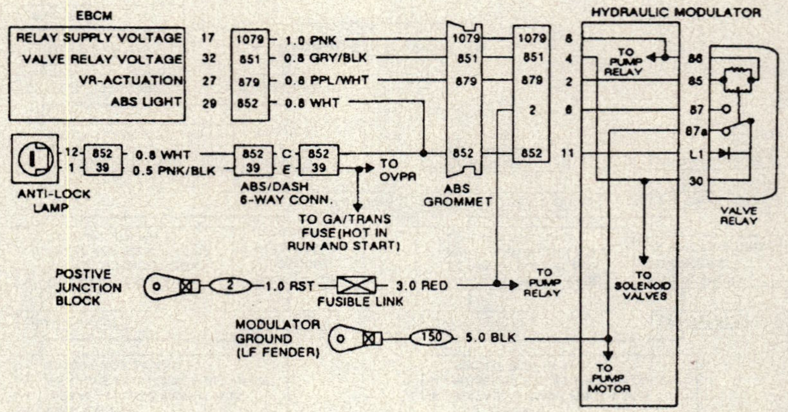

CIRCUIT DESCRIPTION

The solenoid valve relay provides power to the three solenoid valves in the hydraulic modulator. The solenoid valve relay is located on the hydraulic modulator and may be replaced if it is found to be defective.

The valve relay is engaged during normal system operation. When the ignition is switched to RUN, the EBCM commands the solenoid valve relay on by grounding the valve relay actuation line on CKT 879 (EBCM pin 27). When this ground is provided, the valve relay is energized from the voltage supply line on CKT 1079 (EBCM pin 17). The valve relay switches and battery voltage is provided to the three solenoid valves and the solenoid valve relay feedback line on CKT 851 (EBCM pin 32). The valve relay remains engaged until the ignition is turned off or a failure is detected.

Whenever the solenoid valve relay is not engaged, the ANTI-LOCK light will turn on. With the ignition on and the solenoid valve relay not engaged, a path to ground exists from the GA/TRANS fuse (20 amps) thru the ANTI-LOCK light on CKT 39 which turns on the warning light. The ANTI-LOCK light can also be commanded on by the EBCM on CKT 852 (EBCM pin 29). The EBCM will provide ground on this line and turn on the warning light when a failure is detected.

ANTI-LOCK OPERATION

With the solenoid valve relay engaged, battery power is provided to the three solenoid valves. If anti-lock operation is required, the EBCM varies the amount of current supplied to the solenoid valves (EBCM pins 2, 18, 35) and the valve is positioned to provide for optimum braking.

FAILURE CONDITIONS

When the valve relay is engaged, the valve relay feedback line should be at battery voltage. With the valve relay disengaged the feedback line is at ground. If the commanded position of the valve relay and valve relay position indicated by the feedback line do not agree, Code 63 will set. Some conditions which will cause Code 63 to set are:

- An open circuit on CKTs 851 or 879
- A short to voltage on CKTs 851 or 879
- A short to ground on CKTs 851 or 879
- A defective solenoid valve relay
- Open circuit on CKT 1079.

DIAGNOSTIC CHART DTC 63 — 1993 FLEETWOOD BROUGHAM, CONTINUED

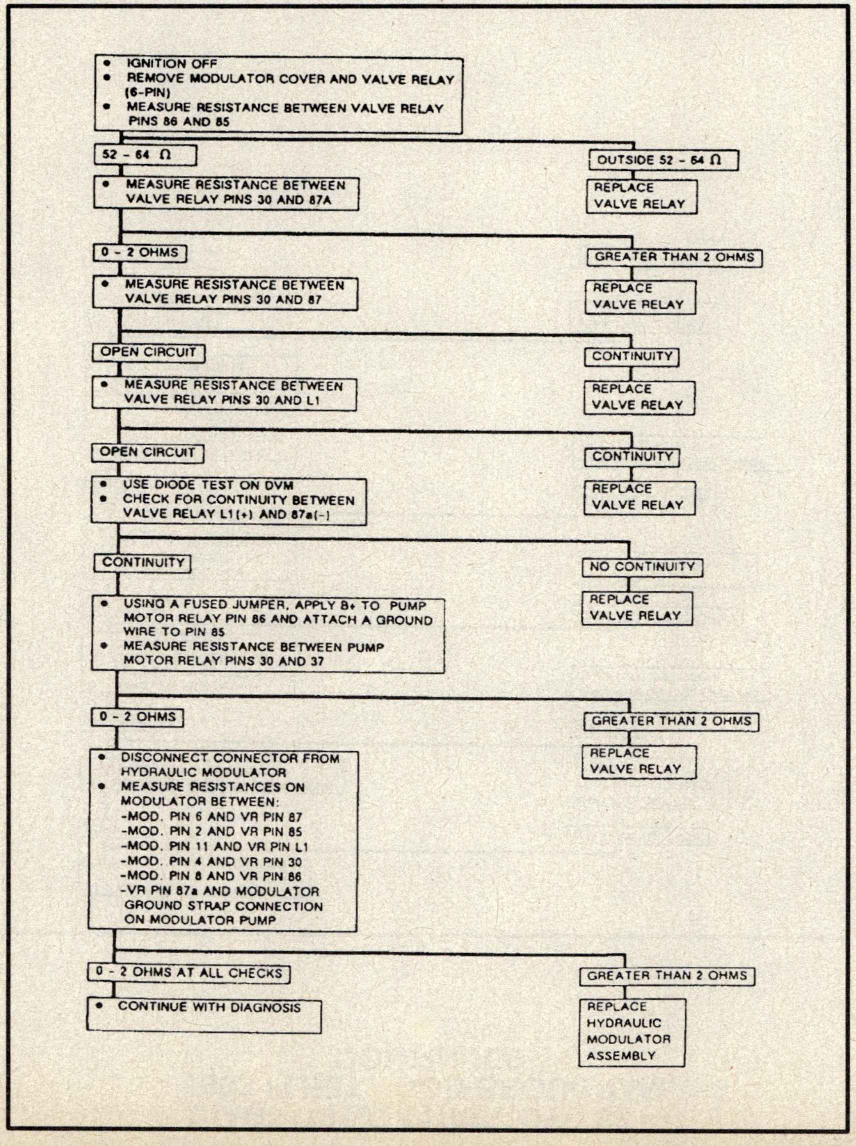

DIAGNOSTIC CHART DTC 63 — 1993 FLEETWOOD BROUGHAM, CONTINUED

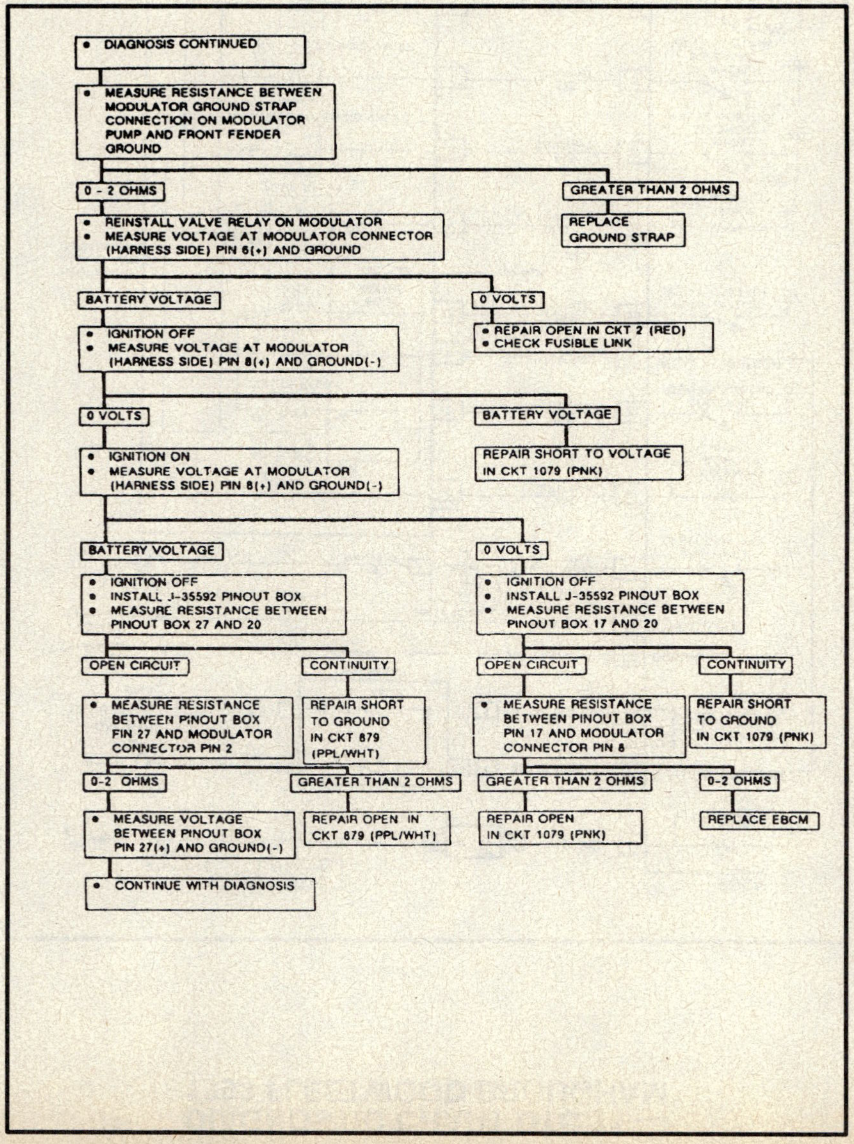

DIAGNOSTIC CHART DTC 63 — 1993 FLEETWOOD BROUGHAM, CONTINUED

- DIAGNOSIS CONTINUED
  - 0 VOLTS
    - MEASURE RESISTANCE BETWEEN PINOUT BOX PIN 32 AND 20
      - OPEN CIRCUIT
        - MEASURE RESISTANCE BETWEEN PINOUT BOX PIN 32 AND MODULATOR CONNECTOR PIN 4
          - 0–2 OHMS
            - MEASURE VOLTAGE BETWEEN PINOUT BOX PIN 32(+) AND GROUND
              - 0 VOLTS
                - MEASURE RESISTANCE BETWEEN GROUND STUD ON MODULATOR PUMP AND MODULATOR PIN 8
                - MEASURE RESISTANCE BETWEEN STUD MODULATOR PIN 12
                  - OPEN CIRCUIT TO BOTH PINS
                    - REPLACE EBCM
                  - LESS THAN 2 OHMS TO EITHER PIN
                    - REPLACE HYDRAULIC MODULATOR
              - BATTERY VOLTAGE
                - REPAIR SHORT TO VOLTAGE IN CKT 851 (GRY/BLK)
          - GREATER THAN 2 OHMS
            - REPAIR OPEN IN CKT 851 (GRY/BLK)
      - CONTINUITY
        - REPAIR SHORT TO GROUND IN CKT 851 (GRY/BLK)
  - BATTERY VOLTAGE
    - REPAIR SHORT TO VOLTAGE IN CKT 879 (PPL/WHT)

DIAGNOSTIC CHART DTC 71 — 1993 FLEETWOOD BROUGHAM

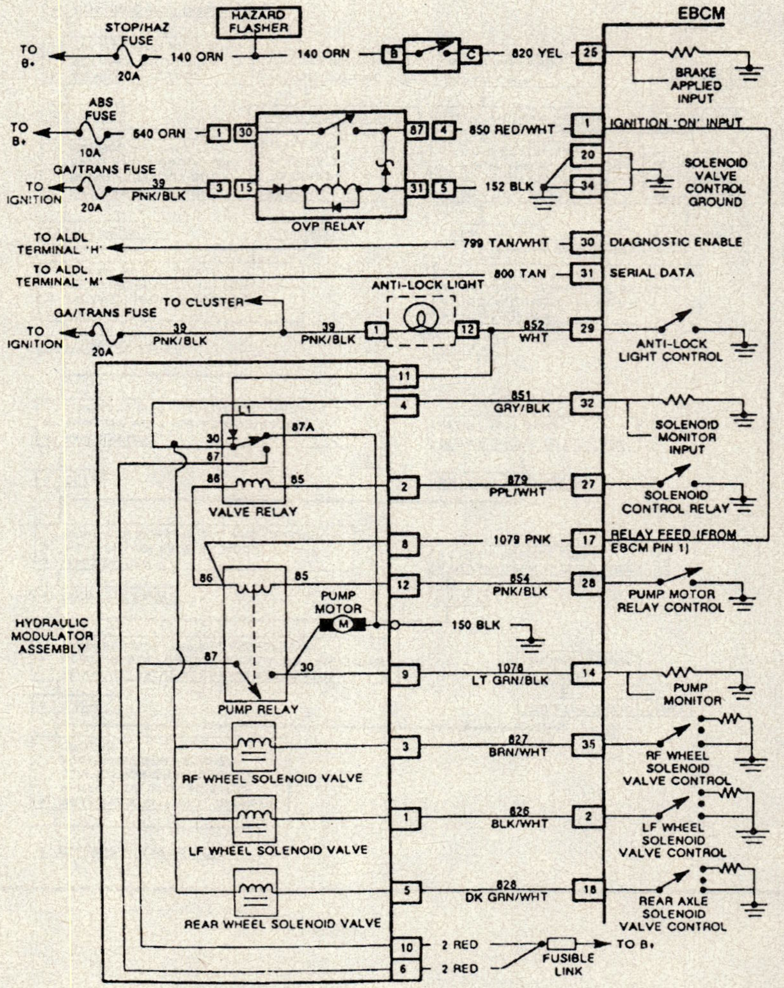

DIAGNOSTIC CHART DTC 71 — 1993 FLEETWOOD BROUGHAM, CONTINUED

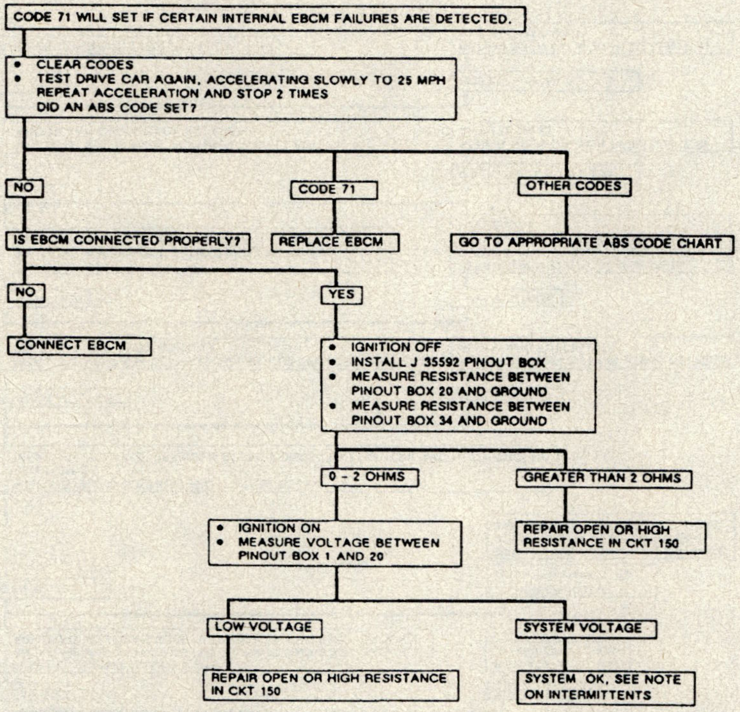

DIAGNOSTIC CHART A — 1993 FLEETWOOD BROUGHAM

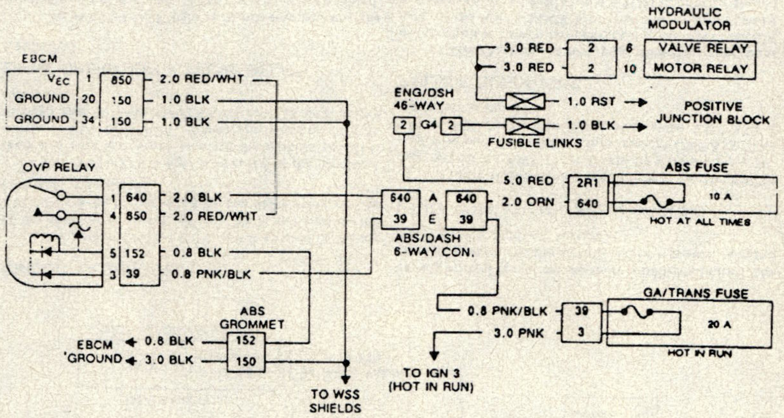

CIRCUIT DESCRIPTION

The EBCM receives power from the overvoltage protection relay on CKT 850 (EBCM Pin 1). When the ignition is turned to RUN, the OVP relay is energized by the GA/TRANS fuse; the normally open relay contacts of the OVP relay close. Battery voltage is supplied to the EBCM thru the ABS fuse (10 amp) and CKT 640. If high system voltages occur, the ABS fuse will blow in order to prevent damage to the Anti-Lock Brake System. The OVP relay is located under the dash, mounted near the EBCM bracket behind the glovebox.

FAILURE CONDITIONS

Intermittent operation of the ANTI-LOCK light may be caused by intermittent voltage levels which do not correspond to the proper operating range of the Anti-Lock Brake System. Some conditions which may cause a lack of system power are:

- Improper electrical contact of the EBCM connector.
- Low system voltage.
- Improper electrical contact of the vehicle power and ground connections.

DIAGNOSTIC CHART A — 1993 FLEETWOOD BROUGHAM, CONTINUED

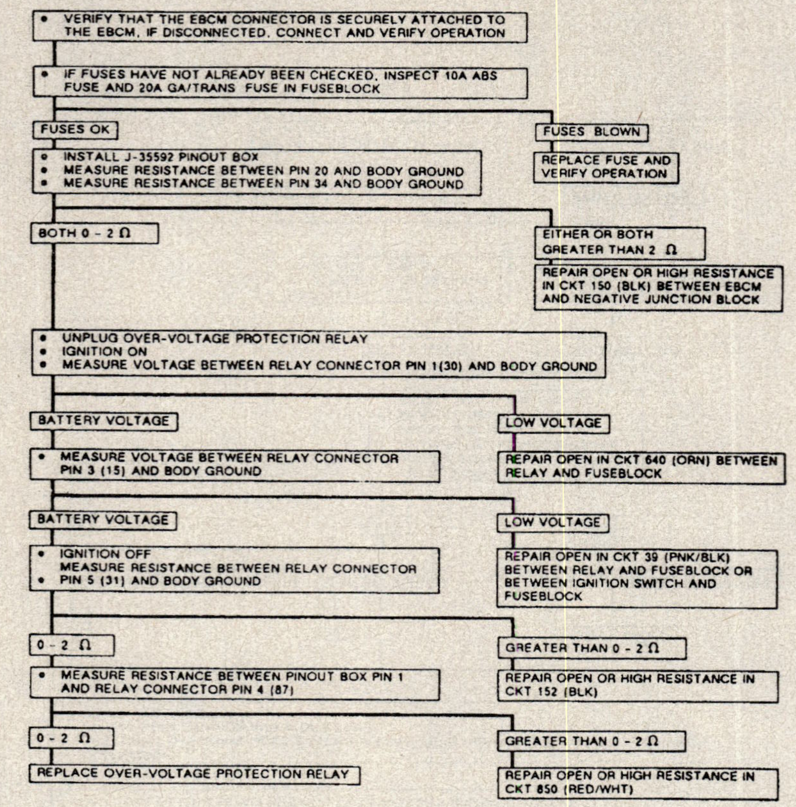

DIAGNOSTIC CHART B — 1993 FLEETWOOD BROUGHAM

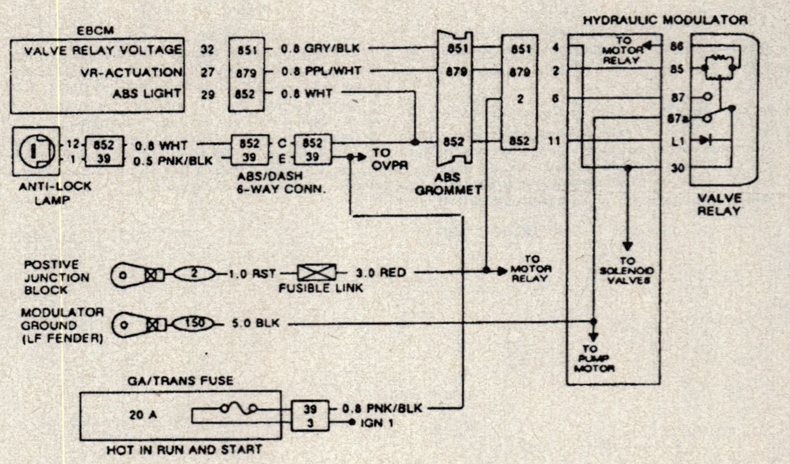

CIRCUIT DESCRIPTION

The amber ANTI-LOCK light is located in the top of the instrument panel. It may be illuminated by the EBCM or a path to ground in the valve relay on the hydraulic modulator.

The ANTI-LOCK light is powered by the Ignition 1 feed and receives power anytime the ignition switch is turned to RUN or START. Power is provided on CKT 39 through the 20 amp GA/TRANS fuse located in the fuseblock.

LIGHT OPERATION - VALVE RELAY GROUND

The ANTI-LOCK light is illuminated anytime the valve relay is not enabled. When the relay is not enabled, the light is grounded through CKT 852 and the valve relay to the modulator ground on the left front fender.

During normal operation (no fault conditions found), the valve will switch and the ground will be removed, and the ANTI-LOCK light will turn off.

LIGHT OPERATION - EBCM GROUND

If the EBCM detects a fault in the ABS, the EBCM can turn on the ANTI-LOCK light by grounding pin 29. A ground path exists for the light through CKT 852. The EBCM will disable the valve relay when CKT 852 is grounded at pin 29.

FAILURE CONDITIONS

The chart on the next page addresses conditions which result in continuous operation of the ANTI-LOCK light. If the ANTI-LOCK light is on and no codes can be extracted from the EBCM, the EBCM connector should be checked for proper connection and system power should be checked.

DIAGNOSTIC CHART B — 1993 FLEETWOOD BROUGHAM, CONTINUED

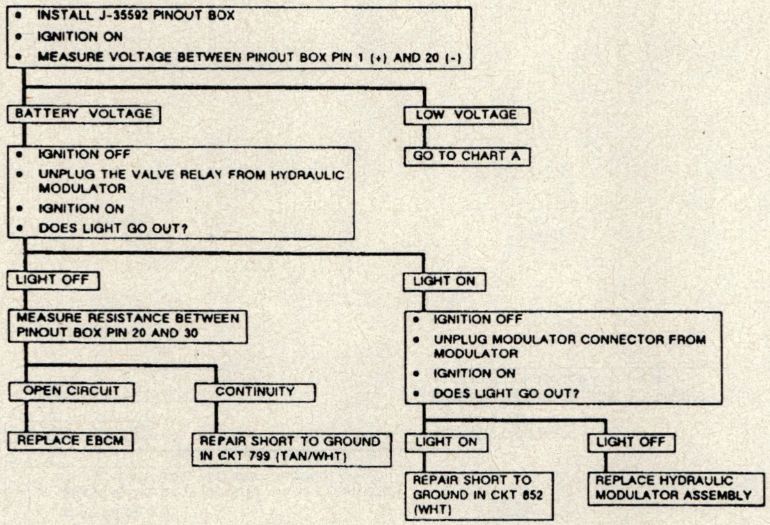

DIAGNOSTIC CHART C — 1993 FLEETWOOD BROUGHAM

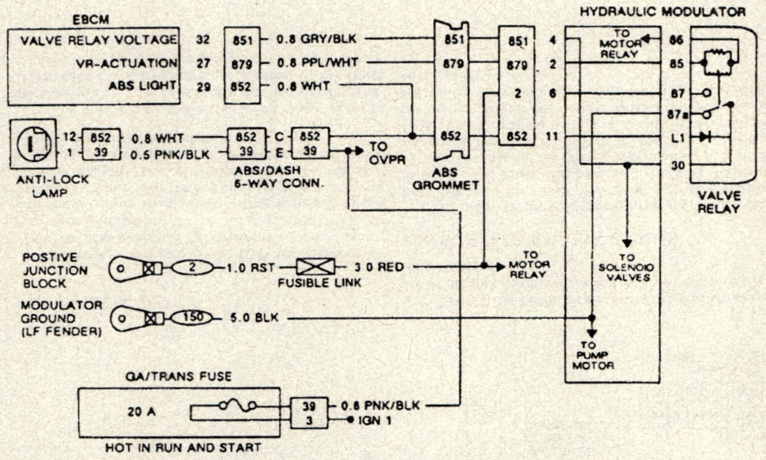

CIRCUIT DESCRIPTION

The amber ANTI-LOCK light is located in the top of the instrument panel. It may be illuminated by the EBCM or a path to ground in the valve relay on the hydraulic modulator.

The ANTI-LOCK light is powered by the Ignition 1 feed and receives power anytime the ignition switch is turned to RUN or START. Power is provided on CKT 39 through the 20 amp GA/TRANS fuse located in the fuseblock.

LIGHT OPERATION – VALVE RELAY GROUND

The ANTI-LOCK light is illuminated anytime the valve relay is not enabled. When the relay is not enabled, the light is grounded through CKT 852 and the valve relay to the modulator ground on the left front fender. During normal operation (no fault conditions found), the valve will switch and the ground will be removed, and the ANTI-LOCK light will turn off.

LIGHT OPERATION – EBCM GROUND

Whether the valve relay is engaged or not, the EBCM can turn on the ANTI-LOCK light by grounding pin 29. A ground path exists for the light through CKT 852.

FAILURE CONDITIONS

The chart on the next page addresses conditions which result in no ANTI-LOCK light operation. As described at the beginning of this section, the ANTI-LOCK light should illuminate for a minimum of four seconds when the key is moved from OFF to RUN. This chart addresses conditions such as lack of power to the ANTI-LOCK light which would prevent the light from turning on. This condition could prevent the EBCM from turning on the light in the event of an anti-lock failure.

DIAGNOSTIC CHART C — 1993 FLEETWOOD BROUGHAM, CONTINUED

CHECK GA/TRANS FUSE IN FUSE BLOCK

- FUSE OK
  - IGNITION OFF
  - INSTALL J 35592 PINOUT BOX
  - UNPLUG SOLENOID VALVE RELAY FROM HYDRAULIC MODULATOR
  - IGNITION ON
  - MEASURE VOLTAGE BETWEEN PINOUT BOX PINS 1 (+) AND 29 (-)
    - LOW VOLTAGE
      - IGNITION OFF
      - DISCONNECT ABS-DASH 6-WAY CONNECTOR
      - IGNITION ON
      - MEASURE VOLTAGE BETWEEN 6-WAY CONNECTOR PIN 'C' (DASH SIDE) AND BODY GROUND
        - LOW VOLTAGE → DIAGNOSE TELLTALES
        - BATTERY VOLTAGE → REPAIR OPEN CKT 852 (WHT) BETWEEN ABS/DASH CONNECTOR AND EBCM.
    - BATTERY VOLTAGE → REPLACE EBCM
- FUSE BLOWN → REPLACE FUSE

DIAGNOSTIC CHART D — 1993 FLEETWOOD BROUGHAM

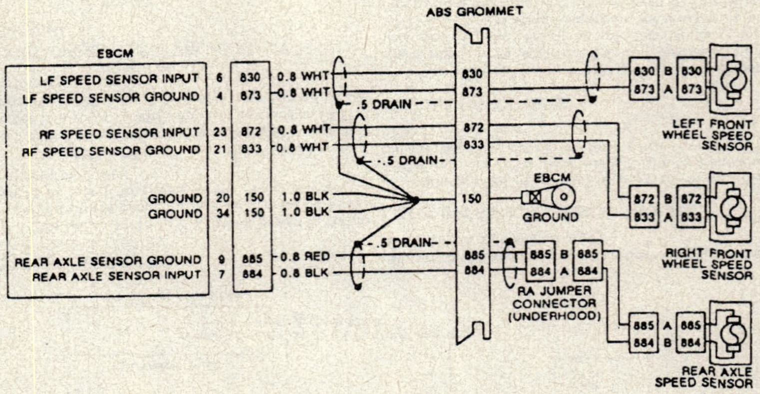

CIRCUIT DESCRIPTION

As each wheel turns, the wheel speed sensor for that wheel creates a small AC voltage as a toothed sensor ring, mounted on the brake rotor (in the front) or differential pinion shaft (in the rear), is passed by the stationary sensor. The sensor consists of a permanent magnet and coil and generates voltage by magnetic induction. The frequency of the wheel speed sensor signal is proportional to wheel speed.

The EBCM uses the wheel speed sensor signal to calculate vehicle reference speeds and individual speed, acceleration and slip values for each wheel. These values are used to determine when anti-lock control is required.

The EBCM performs two basic types of checks on the wheel speed sensors, these are sensor continuity and sensor output.

WHEEL SPEED SENSOR CABLES

In order to prevent electromagnetic interference from disturbing the wheel speed sensor signal, the sensor cables are protected with grounded shielding (indicated by dashed lines in the above schematic). This shield surrounds the two individual sensor wires. A black conduit surrounds the wires and shield. When servicing the sensor cables, the sensor wires are accessible by separating the shielding. If the shielding is disturbed, repair as required.

DIAGNOSTIC CHART D — 1993 FLEETWOOD BROUGHAM, CONTINUED

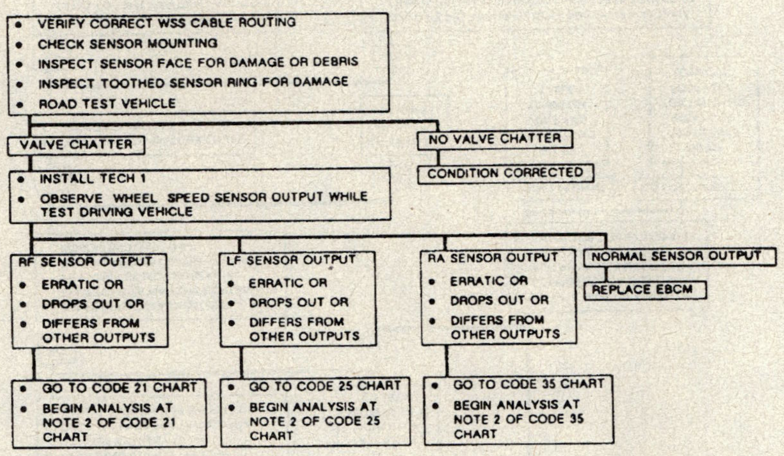

DIAGNOSTIC CHART ABS FUNCTIONAL CHECK — 1993 CAPRICE, CUSTOM CRUISER AND ROADMASTER

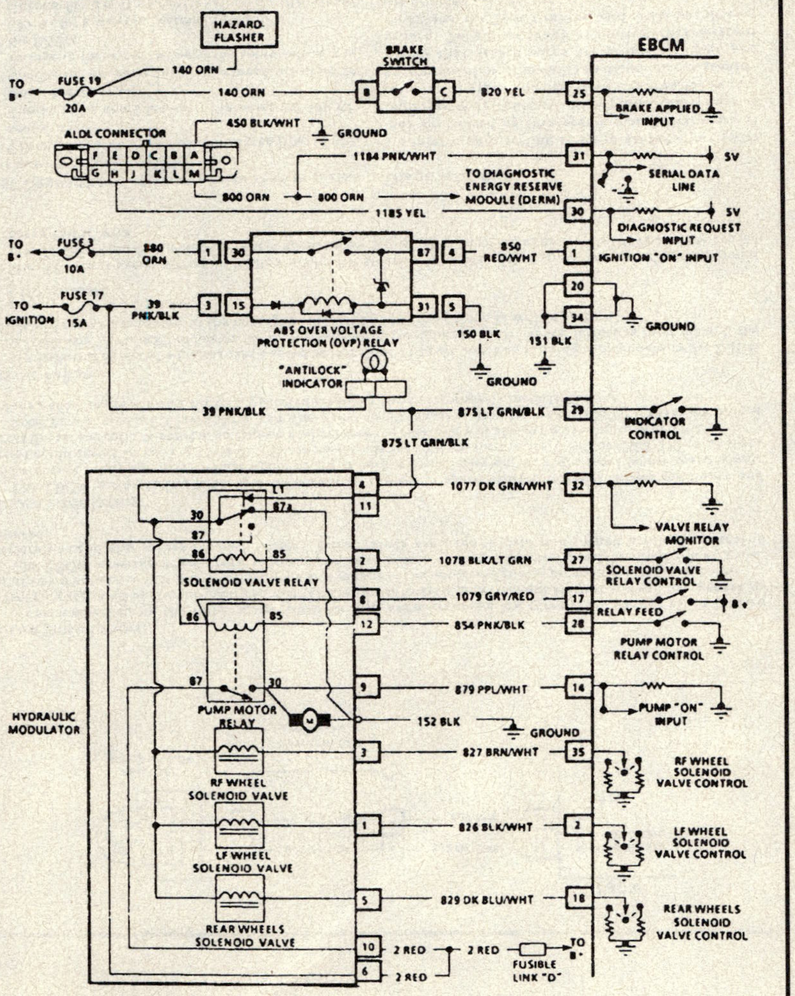

DIAGNOSTIC CHART ABS FUNCTIONAL CHECK — 1993 CAPRICE, CUSTOM CRUISER AND ROADMASTER, CONTINUED

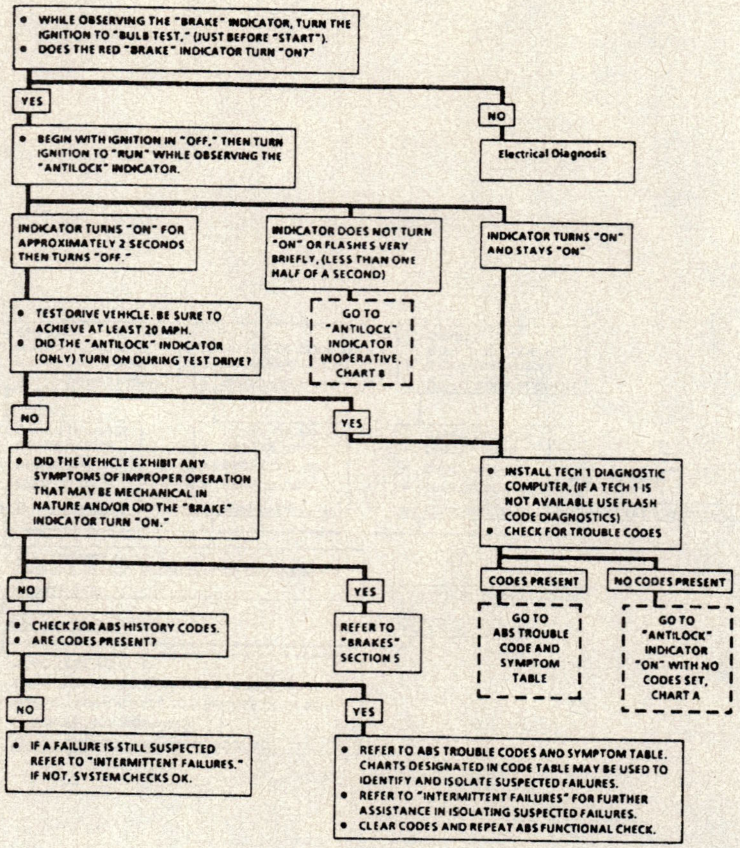

DIAGNOSTIC CHART DTC 21 — 1993 CAPRICE, CUSTOM CRUISER AND ROADMASTER

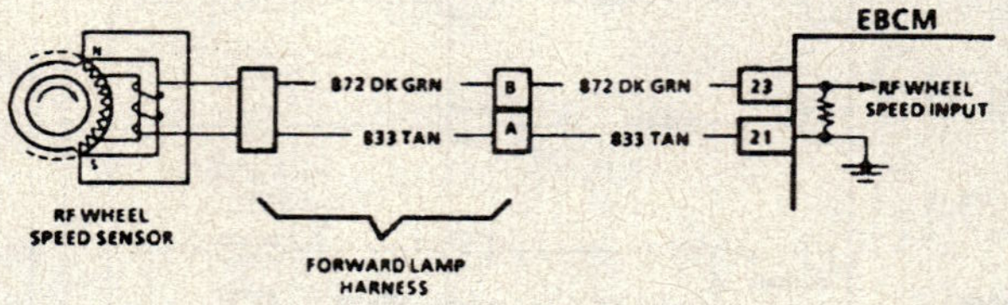

Circuit Description:

As each wheel turns, the wheel speed sensor for that wheel creates a low current AC voltage as the toothed sensor ring is passed by the stationary sensor. The sensor consists of a permanent magnet and coil that generates voltage by magnetic induction. The frequency and voltage of the induced signal is proportional to wheel speed.

The EBCM uses the wheel speed sensor signal to calculate vehicle reference speeds and individual speed, acceleration and slip values for each wheel. These values are used to determine when antilock control is required.

Failure Conditions:

The EBCM performs two basic types of checks on the wheel speed sensors to check sensor continuity and sensor output. Any condition which would result in lack of continuity on the right front wheel speed sensor circuit could result in setting Code 21. These conditions include an open in CKT 872 or CKT 833 or an open across the sensor coil. A short to voltage on a wheel speed sensor line will also generate this type of failure.

Code 21 will also set if low output from the wheel speed sensor is detected by the EBCM. These conditions include a short to ground on CKT 833, a shorted sensor coil or an improperly installed sensor.

Action Taken:

If a failure is detected which causes a speed sensor Code 21 to set, antilock braking is disabled and the EBCM turns "ON" the "Antilock" indicator for the remainder of the ignition cycle. If the failure is intermittent, the EBCM will enable the system at the next ignition cycle and a history Code 21 will be present.

Important

- The RF wheel speed sensor is located in the RF steering knuckle. In order to prevent electromagnetic interference from disturbing the wheel speed sensor signal, the sensor wires are twisted together 9 times for every foot of wire.

Test Description: Numbers below refer to circled numbers on the diagnostic chart.

1. Checks for a correct resistance reading of the sensor itself.
2. Checks for a short between the wires for the RF wheel speed sensor.
3. Checks for an open high resistance or short to voltage in the wires between the speed sensor and the EBCM.
4. Checks for a short to ground in CKT 872.
5. Checks for an "Intermittent" in the RF wheel speed sensor circuitry. If no "Intermittent" is found in the wiring, it may be in the sensor itself. Replace the sensor and road test the vehicle. If the code returns, replace the EBCM.

Diagnostic Aids:

Intermittent setting of speed sensor error codes may be caused by improperly mounted sensors or improper wire routing. Verify that the sensor is mounted correctly and the sensor face does not contain an accumulation of metallic particles, dirt or grease. Verify that sensor wires are properly routed and secure. To aid in troubleshooting intermittent conditions, the Tech 1 can be used while test driving the vehicle.

DIAGNOSTIC CHART DTC 21 — 1993 CAPRICE, CUSTOM CRUISER AND ROADMASTER, CONTINUED

BE SURE SENSOR IS NOT VISIBLY DAMAGED AND THAT IT IS PROPERLY MOUNTED. IF IT IS DAMAGED OR NOT CORRECTLY MOUNTED AND RETAINED, REPLACE THE SENSOR AND/OR PROPERLY ATTACH IT, CLEAR CODES, AND ROAD TEST TO SEE IF CODE RESETS.

(1)
- TURN IGNITION "OFF."
- DISCONNECT SENSOR PIGTAIL CONNECTOR ON WHEEL HOUSING.
- USING A DVOM, CHECK SENSOR RESISTANCE BY PROBING SENSOR PIGTAIL CONNECTOR TERMINALS ON SENSOR SIDE.
- IS IT APPROX 980 TO 1130 OHMS?

NO: REPLACE SENSOR.

YES: CHECK BOTH TERMINALS OF SENSOR PIGTAIL CONNECTOR (SENSOR SIDE) FOR CONTINUITY TO GROUND.

CONTINUITY: REPLACE SENSOR.

NO CONTINUITY:

(2)
- DISCONNECT EBCM CONNECTOR.
- INSTALL EBCM PINOUT BOX J 35592.
- CHECK FOR CONTINUITY BETWEEN CONNECTOR TERMINALS "21" AND "23."

CONTINUITY: REPAIR SHORT BETWEEN CKTS 872 AND 833.

NO CONTINUITY:

(3)
- CHECK CKT 872 AND 833 FOR OPEN, HIGH RESISTANCE OR SHORT TO VOLTAGE BETWEEN SENSOR CONNECTOR AND EBCM CONNECTOR.
- ARE THEY OK?

NO: REPAIR OPEN, HIGH RESISTANCE OR SHORT TO VOLTAGE IN WIRING OR CONNECTORS.

YES:

(4)
- CHECK CKT 872 FOR SHORT TO GROUND.
- IS IT OK?

NO: REPAIR SHORT TO GROUND IN CKT 872.

YES:

(5)
- RECONNECT SENSOR PIGTAIL CONNECTOR.
- MEASURE RESISTANCE FROM PINOUT BOX TERMINALS "21" TO "23."
- OBSERVE RESISTANCE VALUE WHILE WIGGLING CKT 833, CKT 872 AND THEIR CONNECTORS. THESE CIRCUITS ARE PART OF THE ABS AND FORWARD LAMP HARNESSES.

RESISTANCE FLUCTUATING: INTERMITTENT CONNECTION, REPAIR HARNESS OR CONNECTOR AS NECESSARY.

RESISTANCE STAYS CONSTANT AT APPROX 980 TO 1130 OHMS:
- INSPECT TOOTHED SENSOR RING FOR DAMAGE, PROPER MOUNTING OR EXCESSIVE RUNOUT.
- INSPECT SENSOR FACE FOR AN ACCUMULATION OF METALLIC PARTICLES, GREASE OR DIRT.
- REPAIR/REPLACE AS NECESSARY.
- IF OK, REPLACE EBCM.

DIAGNOSTIC CHART DTC 22 — 1993 CAPRICE, CUSTOM CRUISER AND ROADMASTER

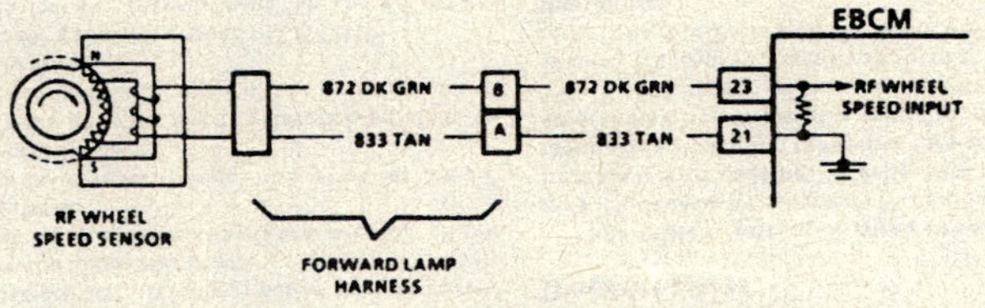

Circuit Description:

As each wheel turns, the wheel speed sensor for that wheel creates a low current AC voltage as the toothed sensor ring is passed by the stationary sensor. The sensor consists of a permanent magnet and coil that generates voltage by magnetic induction. The frequency and voltage of the induced signal is proportional to wheel speed.

The EBCM uses the wheel speed sensor signal to calculate vehicle reference speeds and individual speed, acceleration and slip values for each wheel. These values are used to determine when antilock control is required.

Failure Conditions:

Code 22 will set if improper speed signals are being generated by the toothed wheel. Causes of improper speed signals include; an incorrect number of teeth or damaged teeth present on the toothed wheel, toothed wheel or sensor face covered with dirt, grease or metallic particles that may impair magnetic induction.

Code 22 may also set if the mini-spare tire has previously been used or the size of the tires on the vehicle differ. In addition to Code 22, Codes 26 and 36 may also set if the mini-spare tire has been used. If it is known that the spare tire was used, clear codes and test drive the vehicle again. Verify that codes do not set.

Action Taken:

If a failure is detected which causes a speed sensor Code 22 to set, antilock braking is disabled and the EBCM turns "ON" the "Antilock" indicator for the remainder of the ignition cycle. If the failure is intermittent, the EBCM will enable the system at the next ignition cycle and a history Code 22 will be present.

Test Description: Numbers below refer to circled numbers on the diagnostic chart.

1. Checks for a physical problem with the toothed wheel.
2. Checks for properly mounted sensor and mounting torque.

Diagnostic Aids:

A worn hub/bearing assembly may cause this fault in extreme cases; the bearing play allows the sensor-to-toothed ring gap to change excessively.

Check for a buildup of foreign material in the gaps between teeth on the toothed wheel, this material may cause this error.

Also, check the toothed wheel for any large grooves, gouges, marks, etc., that might influence the tooth's signal at the wheel speed sensor.

DIAGNOSTIC CHART DTC 22 — 1993 CAPRICE, CUSTOM CRUISER AND ROADMASTER, CONTINUED

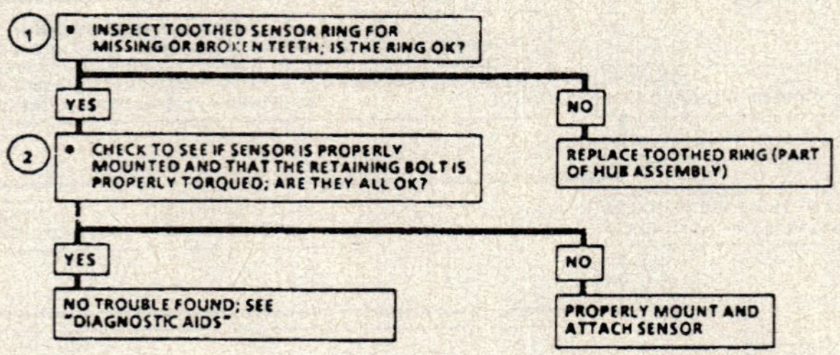

DIAGNOSTIC CHART DTC 25 — 1993 CAPRICE, CUSTOM CRUISER AND ROADMASTER

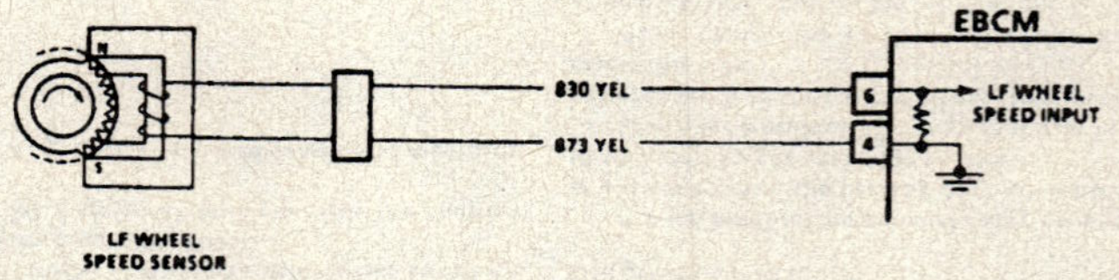

Circuit Description:

As each wheel turns, the wheel speed sensor for that wheel creates a low current AC voltage as the toothed sensor ring is passed by the stationary sensor. The sensor consists of a permanent magnet and coil that generates voltage by magnetic induction. The frequency and voltage of the induced signal is proportional to wheel speed.

The EBCM uses the wheel speed sensor signal to calculate vehicle reference speeds and individual speed, acceleration and slip values for each wheel. These values are used to determine when antilock control is required.

Failure Conditions:

The EBCM performs two basic types of checks on the wheel speed sensors to check sensor continuity and sensor output. Any condition which would result in lack of continuity on the left front wheel speed sensor circuit could result in setting Code 25. These conditions include an open in CKT 873 or CKT 830 or an open across the sensor coil. A short to voltage on a wheel speed sensor line will also generate this type of failure.

Code 25 will also set if low output from the wheel speed sensor is detected by the EBCM. These conditions include a short to ground on CKT 873, a shorted sensor coil or an improperly installed sensor.

Action Taken:

If a failure is detected which causes a speed sensor Code 25 to set, antilock braking is disabled and the EBCM turns "ON" the "Antilock" indicator for the remainder of the ignition cycle. If the failure is intermittent, the EBCM will enable the system at the next ignition cycle and a history Code 25 will be present.

Important

- The LF wheel speed sensor is located in the LF steering knuckle. In order to prevent electromagnetic interference from disturbing the wheel speed sensor signal, the sensor wires are twisted together 9 times for every foot of wire.

Test Description: Numbers below refer to circled numbers on the diagnostic chart.

1. Checks for a correct resistance reading of the sensor itself.
2. Checks for a short between the wires for the LF wheel speed sensor.
3. Checks for an open high resistance or short to voltage in the wires between the speed sensor and the EBCM.
4. Checks for a short to ground in CKT 830.
5. Checks for an "Intermittent" in the LF wheel speed sensor circuitry. If no "Intermittent" is found in the wiring, it may be in the sensor itself. Replace the sensor and road test the vehicle. If the code returns, replace the EBCM.

Diagnostic Aids:

Intermittent setting of speed sensor error codes may be caused by improperly mounted sensors or improper wire routing. Verify that the sensor is mounted correctly and the sensor face does not contain an accumulation of metallic particles, dirt or grease. Verify that sensor wires are properly routed and secure. To aid in troubleshooting intermittent conditions, the Tech 1 can be used while test driving the vehicle.

DIAGNOSTIC CHART DTC 25 — 1993 CAPRICE, CUSTOM CRUISER AND ROADMASTER, CONTINUED

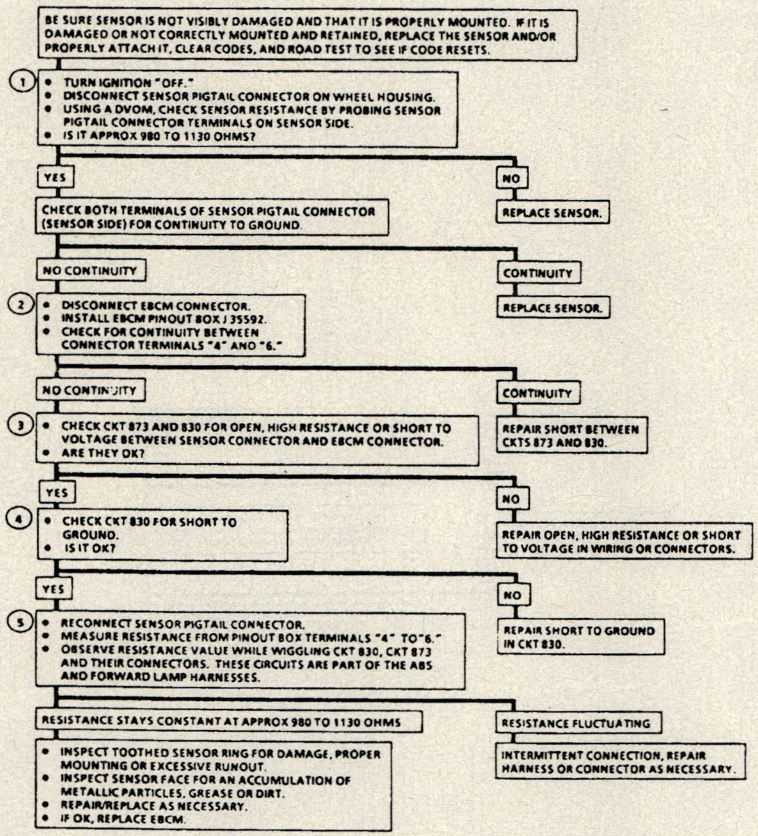

DIAGNOSTIC CHART DTC 26 — 1993 CAPRICE, CUSTOM CRUISER AND ROADMASTER

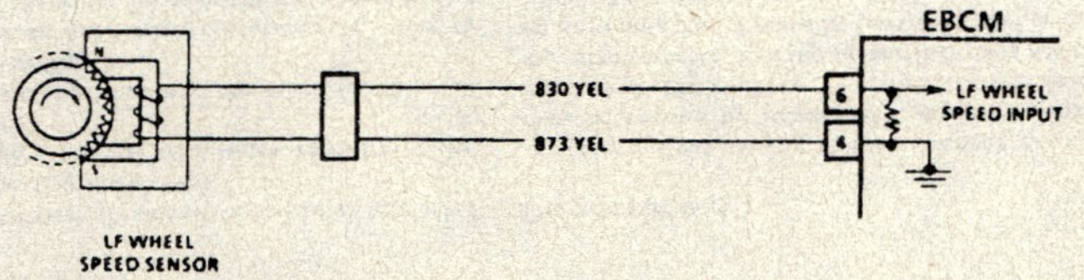

Circuit Description:

As each wheel turns, the wheel speed sensor for that wheel creates a low current AC voltage as the toothed sensor ring is passed by the stationary sensor. The sensor consists of a permanent magnet and coil that generates voltage by magnetic induction. The frequency and voltage of the induced signal is proportional to wheel speed.

The EBCM uses the wheel speed sensor signal to calculate vehicle reference speeds and individual speed, acceleration and slip values for each wheel. These values are used to determine when antilock control is required.

Failure Conditions:

Code 26 will set if improper speed signals are being generated by the toothed wheel. Causes of improper speed signals include: an incorrect number of teeth or damaged teeth present on the toothed wheel, toothed wheel or sensor face covered with dirt, grease or metallic particles that may impair magnetic induction.

Code 26 may also set if the mini-spare tire has previously been used or the size of the tires on the vehicle differ. In addition to Code 26, Codes 22 and 36 may also set if the mini-spare tire has been used. If it is known that the spare tire was used, clear codes and test drive the vehicle again. Verify that codes do not set.

Action Taken:

If a failure is detected which causes a speed sensor Code 26 to set, antilock braking is disabled and the EBCM turns "ON" the "Antilock" indicator for the remainder of the ignition cycle. If the failure is intermittent, the EBCM will enable the system at the next ignition cycle and a history Code 26 will be present.

Test Description: Numbers below refer to circled numbers on the diagnostic chart.

1. Checks for a physical problem with the toothed wheel.
2. Checks for properly mounted sensor and mounting torque.

Diagnostic Aids:

A worn hub/bearing assembly may cause this fault in extreme cases; the bearing play allows the sensor-to-toothed ring gap to change excessively.

Check for a buildup of foreign material in the gaps between teeth on the toothed wheel; this material may cause this error.

Also, check the toothed wheel for any large grooves, gouges, marks, etc., that might influence the tooth's signal at the wheel speed sensor.

DIAGNOSTIC CHART DTC 26 — 1993 CAPRICE, CUSTOM CRUISER AND ROADMASTER, CONTINUED

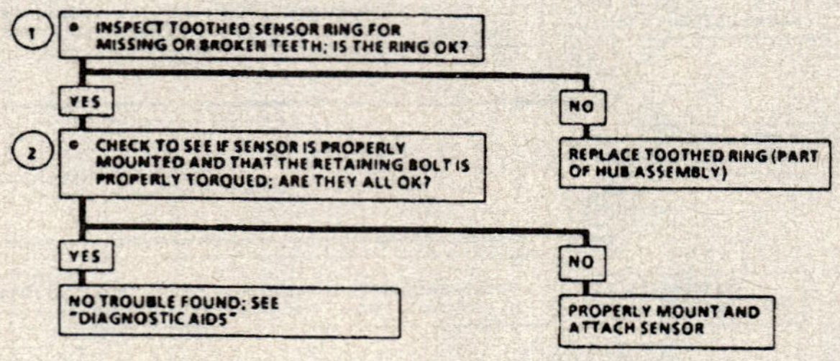

DIAGNOSTIC CHART DTC 35 — 1993 CAPRICE, CUSTOM CRUISER AND ROADMASTER

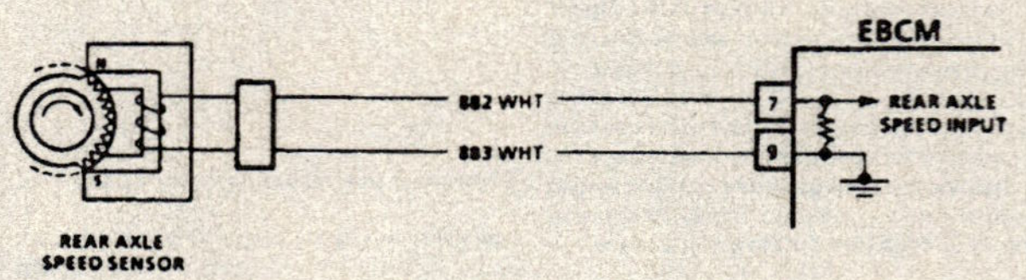

Circuit Description:

As the differential pinion gear turns, the rear axle speed sensor creates a low current AC voltage as the toothed sensor ring (mounted to the differential pinion gear) is passed by the stationary sensor. The sensor consists of a permanent magnet and coil that generates voltage by magnetic induction. The frequency and voltage of the induced signal is proportional to rear axle speed.

The EBCM uses the rear axle speed sensor signal to calculate vehicle reference speeds and individual speed, acceleration and slip values for each wheel. These values are used to determine when antilock control is required.

Failure Conditions:

The EBCM performs two basic types of checks on the rear axle speed sensor to check sensor continuity and sensor output. Any condition which would result in lack of continuity on the rear axle speed sensor circuit could result in setting Code 35. These conditions include an open in CKT 882 or CKT 883 or an open across the sensor coil. A short to voltage on a rear axle speed sensor line will also generate this type of failure.

Code 35 will also set if low output from the rear axle speed sensor is detected by the EBCM. These conditions include a short to ground on CKT 883, a shorted sensor coil or an improperly installed sensor.

Action Taken:

If a failure is detected which causes a speed sensor Code 35 to set, antilock braking is disabled and the EBCM turns "ON" the "Antilock" indicator for the remainder of the ignition cycle. If the failure is intermittent, the EBCM will enable the system at the next ignition cycle and a history Code 35 will be present.

Important

- The rear axle speed sensor is located in the rear axle housing. In order to prevent electromagnetic interference from disturbing the rear axle speed sensor signal, the sensor wires are twisted together 9 times for every foot of wire.

Test Description: Numbers below refer to circled numbers on the diagnostic chart.

1. Checks for a correct resistance reading of the sensor itself.
2. Checks for a short between the wires for the rear axle speed sensor.
3. Checks for an open high resistance or short to voltage in the wires between the speed sensor and the EBCM.
4. Checks for a short to ground in CKT 882.
5. Checks for an "Intermittent" in the rear axle speed sensor circuitry. If no intermittent is found in the wiring, it may be in the sensor itself. Replace the sensor and road test the vehicle. If the code returns, replace the EBCM.

Diagnostic Aids:

Intermittent setting of speed sensor error codes may be caused by improperly mounted sensors or improper wire routing. Verify that the sensor is mounted correctly and the sensor face does not contain an accumulation of metallic particles, dirt or grease. Verify that sensor wires are properly routed and secure. To aid in troubleshooting intermittent conditions, the Tech 1 can be used while test driving the vehicle.

DIAGNOSTIC CHART DTC 35 — 1993 CAPRICE, CUSTOM CRUISER AND ROADMASTER, CONTINUED

BE SURE SENSOR IS NOT VISIBLY DAMAGED AND THAT IT IS PROPERLY MOUNTED. IF IT IS DAMAGED OR NOT CORRECTLY MOUNTED AND RETAINED, REPLACE THE SENSOR AND/OR PROPERLY ATTACH IT, CLEAR CODES, AND ROAD TEST TO SEE IF CODE RESETS.

1.
   - TURN IGNITION "OFF."
   - DISCONNECT SENSOR PIGTAIL CONNECTOR ON LEFT FRAME RAIL.
   - USING A DVOM, CHECK SENSOR RESISTANCE BY PROBING SENSOR PIGTAIL CONNECTOR TERMINALS ON SENSOR SIDE.
   - IS IT APPROX 980 TO 1130 OHMS?

   - YES → CHECK BOTH TERMINALS OF SENSOR PIGTAIL CONNECTOR (SENSOR SIDE) FOR CONTINUITY TO GROUND.
     - NO CONTINUITY → go to 2.
     - CONTINUITY → REPLACE SENSOR.
   - NO → REPLACE SENSOR.

2.
   - DISCONNECT EBCM CONNECTOR.
   - INSTALL EBCM PINOUT BOX J 35592.
   - CHECK FOR CONTINUITY BETWEEN CONNECTOR TERMINALS "7" AND "9."

   - NO CONTINUITY → go to 3.
   - CONTINUITY → REPAIR SHORT BETWEEN CKTS 882 AND 883.

3.
   - CHECK CKT 882 AND 883 FOR OPEN, HIGH RESISTANCE OR SHORT TO VOLTAGE BETWEEN SENSOR CONNECTOR AND EBCM CONNECTOR.
   - ARE THEY OK?

   - YES → go to 4.
   - NO → REPAIR OPEN, HIGH RESISTANCE OR SHORT TO VOLTAGE IN WIRING OR CONNECTORS.

4.
   - CHECK CKT 882 FOR SHORT TO GROUND.
   - IS IT OK?

   - YES → go to 5.
   - NO → REPAIR SHORT TO GROUND IN CKT 882.

5.
   - RECONNECT SENSOR PIGTAIL CONNECTOR.
   - MEASURE RESISTANCE FROM PINOUT BOX TERMINALS "7" TO "9."
   - OBSERVE RESISTANCE VALUE WHILE WIGGLING CKT 882, CKT 883 AND THEIR CONNECTORS.

   - RESISTANCE STAYS CONSTANT AT APPROX 980 TO 1130 OHMS →
     - INSPECT TOOTHED SENSOR RING FOR DAMAGE AND PROPER MOUNTING.
     - INSPECT SENSOR FACE FOR AN ACCUMULATION OF METALLIC PARTICLES, GREASE OR DIRT.
     - REPAIR/REPLACE AS NECESSARY.
     - IF OK, REPLACE EBCM.
   - RESISTANCE FLUCTUATING → INTERMITTENT CONNECTION, REPAIR HARNESS OR CONNECTOR AS NECESSARY.

DIAGNOSTIC CHART DTC 36 — 1993 CAPRICE, CUSTOM CRUISER AND ROADMASTER

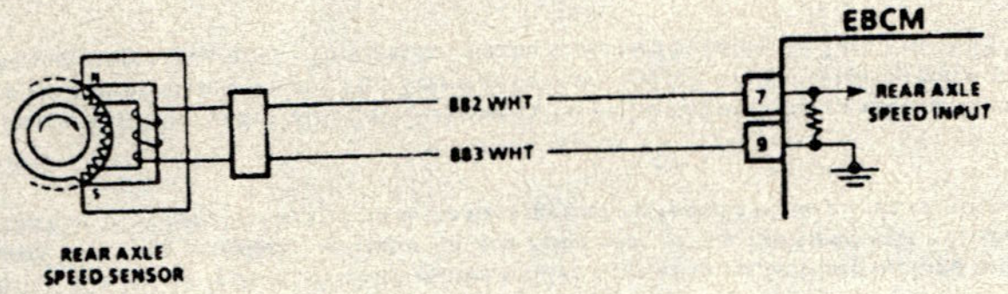

Circuit Description:

As the differential pinion gear turns, the rear axle speed sensor creates a low current AC voltage as the toothed sensor ring (mounted to the differential pinion gear) is passed by the stationary sensor. The sensor consists of a permanent magnet and coil that generates voltage by magnetic induction. The frequency and voltage of the induced signal is proportional to rear axle speed.

The EBCM uses the rear axle speed sensor signal to calculate vehicle reference speeds and individual speed, acceleration and slip values for each wheel. These values are used to determine when antilock control is required.

Failure Conditions:

Code 36 will set if improper speed signals are being generated by the toothed wheel. Causes of improper speed signals include; an incorrect number of teeth or damaged teeth present on the toothed wheel, toothed wheel or sensor face covered with dirt, grease or metallic particles that may impair magnetic induction.

Code 36 may also set if the mini spare tire has previously been used or the size of the tires on the vehicle differ. In addition to Code 36, Codes 22 and 26 may also set if the mini-spare tire has been used. If it is known that the spare tire was used, clear codes and test drive the vehicle again. Verify that codes do not set.

Action Taken:

If a failure is detected which causes a speed sensor Code 36 to set, antilock braking is disabled and the EBCM turns "ON" the "Antilock" indicator for the remainder of the ignition cycle. If the failure is intermittent, the EBCM will enable the system at the next ignition cycle and a history Code 36 will be present.

Test Description: Numbers below refer to circled numbers on the diagnostic chart.

1. Checks for a physical problem with the toothed wheel.
2. Checks for properly mounted sensor and mounting torque.

Diagnostic Aids:

Check for a buildup of foreign material in the gaps between teeth on the toothed wheel, this material may cause this error.

Also, check the toothed wheel for any large grooves, gouges, marks, etc., that might influence the tooth's signal at the speed sensor.

DIAGNOSTIC CHART DTC 36 — 1993 CAPRICE, CUSTOM CRUISER AND ROADMASTER, CONTINUED

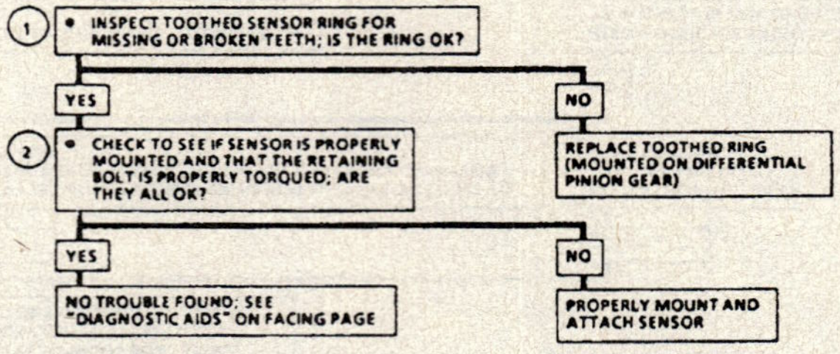

DIAGNOSTIC CHART DTC 41 — 1993 CAPRICE, CUSTOM CRUISER AND ROADMASTER

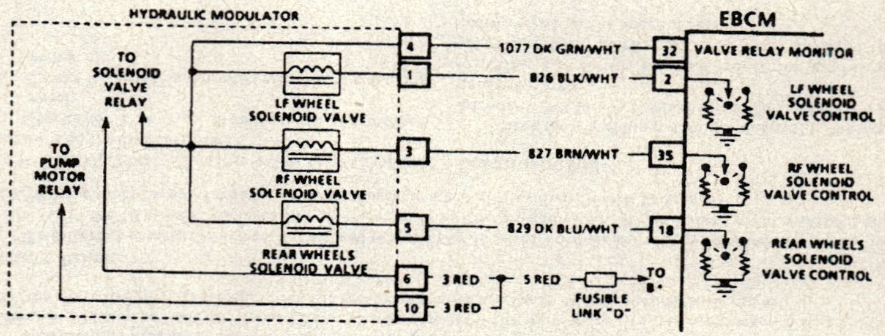

Circuit Description:

With the ignition in Run, Bulb Test or start the EBCM applies voltage to the Coil of the solenoid valve relay. The EBCM will complete an internal self-check before providing ground to the coil. Once energized the relay will provide B+ to the solenoid valves. The solenoid valve position can be commanded to three different positions depending on the amount of current the EBCM allows to flow through the solenoid valve coil.

Pressure Increase position is the normal operating position of the valves for braking. Brake fluid flows through the modulator from the master cylinder to the wheel brake. No current flows through the valve coil during pressure increase.

Pressure Hold position is commanded when the EBCM desires to hold the pressure in the wheel hydraulic circuit constant. The modulator will not allow pressure flow from the master cylinder to the wheel brake. The pressure hold position is commanded by passing approximately 2.0 amps through the valve coil.

Pressure Reduction position is used to decrease hydraulic pressure in a specific wheel hydraulic circuit without decreasing pressure in the rest of the system. The modulator takes brake fluid from the wheel brake and injects it into the master cylinder hydraulic circuit. This fully energized position of the solenoid valve occurs when approximately 5 amps are passed through the valve coil.

Failure Conditions: The solenoid valves are continuously monitored by the EBCM through CKT 1077 and through each individual valve line. Code 41 will set when the expected position of the right front solenoid valve does not match the commanded position from the EBCM (if valve line voltage does not match expected value). Conditions which could cause Code 41 to set are damage to the right front solenoid, an open or short to ground or battery voltage on CKT 827.

Action Taken: If a failure is detected which causes a valve Code 41 to set, antilock braking is disabled and the EBCM turns "ON" the "Antilock" indicator for the remainder of the ignition cycle. If the failure is intermittent, the EBCM will enable the system at the next ignition cycle and a history Code 41 will be present.

[!] Important:

The Auto test (as described under "ABS Self Diagnostics") is performed once during each ignition cycle when the vehicle reaches approximately 4 mph. This Test may be heard and felt while it is taking place and should be considered a normal mode of operation. This test can also be performed by using the Tech 1 in "ABS Tests" mode F4.

Test Description: Numbers below refer to circled numbers on the diagnostic chart.

1. Checks the resistance in the RF Solenoid Valve circuitry.
2. Checks for a short to ground in CKT 827.
3. This test uses the Tech 1 Diagnostic Computer to check for proper RF Solenoid Valve operation.
4. This test determines of a problem found in Step 3 is due to the EBCM or the Hydraulic Modulator.
5. This test determines if a problem found in Step 1 is due to the Hydraulic Modulator or a problem in CKT 827.
6. This test determines if a problem found in Step 3 is due to an intermittent in the wiring and/or connectors, or a fault with the EBCM.

DIAGNOSTIC CHART DTC 41 — 1993 CAPRICE, CUSTOM CRUISER AND ROADMASTER, CONTINUED

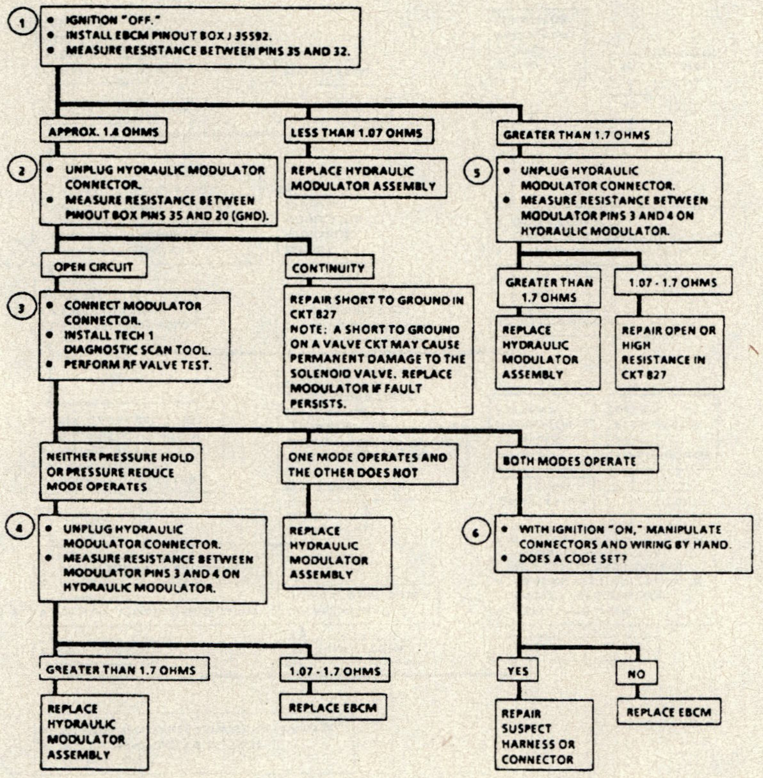

DIAGNOSTIC CHART DTC 45 — 1993 CAPRICE, CUSTOM CRUISER AND ROADMASTER

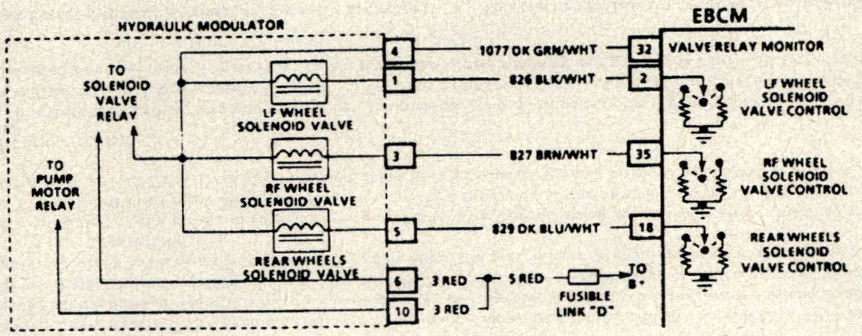

Circuit Description:

With the ignition in Run, Bulb Test or start the EBCM applies voltage to the Coil of the solenoid valve relay. The EBCM will complete an internal self-check before providing ground to the coil. Once energized the relay will provide B+ to the solenoid valves. The solenoid valve positionn can be commanded to three different positions depending on the amount of current the EBCM allows to flow through the solenoid valve coil.

Pressure Increase position is the normal operating position of the valves for braking. Brake fluid flows through the modulator from the master cylinder to the wheel brake. No current flows through the valve coil during pressure increase.

Pressure Hold position is commanded when the EBCM desires to hold the pressure in the wheel hydraulic circuit constant. The modulator will not allow pressure flow from the master cylinder to the wheel brake. The pressure hold position is commanded by passing approximately 2.0 amps through the valve coil.

Pressure Reduction position is used to decrease hydraulic pressure in a specific wheel hydraulic circuit without decreasing pressure in the rest of the system. The modulator takes brake fluid from the wheel brake and injects it into the master cylinder hydraulic circuit. This fully energized position of the solenoid valve occurs when approximately 5 amps are passed through the valve coil.

Failure Conditions: The solenoid valves are continuously monitored by the EBCM through CKT 1077 and through each individual valve line. Code 45 will set when the expected position of the left front solenoid valve does not match the commanded position from the EBCM (if valve line voltage does not match expected value). Conditions which could cause Code 45 to set are damage to the left front solenoid, an open or short to ground or battery voltage on CKT 826.

Action Taken: If a failure is detected which causes a valve Code 45 to set, antilock braking is disabled and the EBCM turns "ON" the "Antilock" indicator for the remainder of the ignition cycle. If the failure is intermittent, the EBCM will enable the system at the next ignition cycle and a history Code 45 will be present.

[?] Important:

The Auto test (as described under "ABS Self Diagnostics") is performed once during each ignition cycle when the vehicle reaches approximately 4 mph. This Test may be heard and felt while it is taking place and should be considered a normal mode of operation. This test can also be performed by using the Tech 1 in "ABS Tests" mode F4.

Test Description: Numbers below refer to circled numbers on the diagnostic chart.

1. Checks the resistance in the LF Solenoid Valve circuitry.
2. Checks for a short to ground in CKT 826.
3. This test uses the Tech 1 Diagnostic Computer to check for proper LF Solenoid Valve operation.
4. This test determines of a problem found in Step 3 is due to the EBCM or the Hydraulic Modulator.
5. This test determines if a problem found in Step 1 is due to the Hydraulic Modulator or a problem in CKT 826.
6. This test determines if a problem found in Step 3 is due to an intermittent in the wiring and/or connectors, or a fault with the EBCM.

DIAGNOSTIC CHART DTC 45 — 1993 CAPRICE, CUSTOM CRUISER AND ROADMASTER, CONTINUED

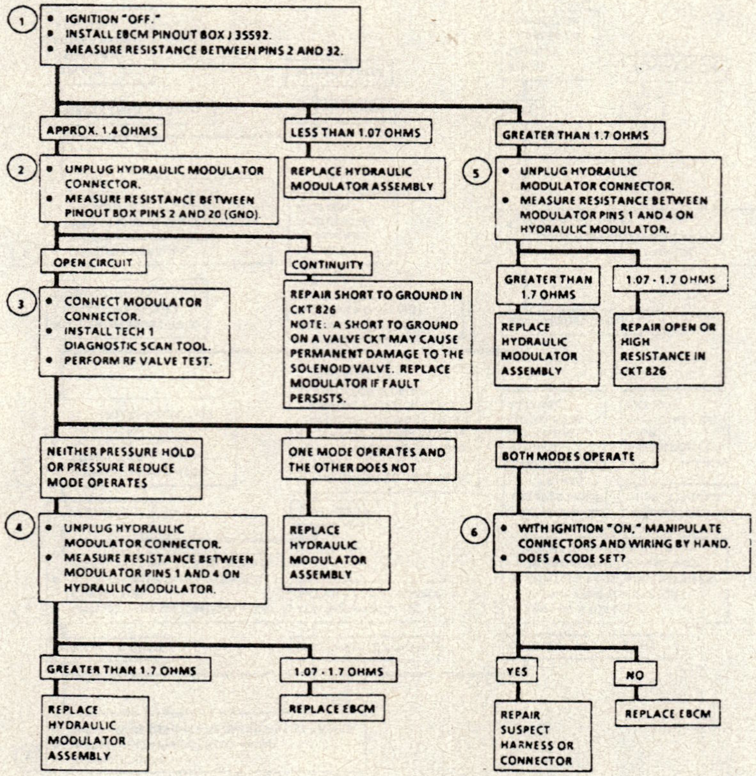

DIAGNOSTIC CHART DTC 55 — 1993 CAPRICE, CUSTOM CRUISER AND ROADMASTER

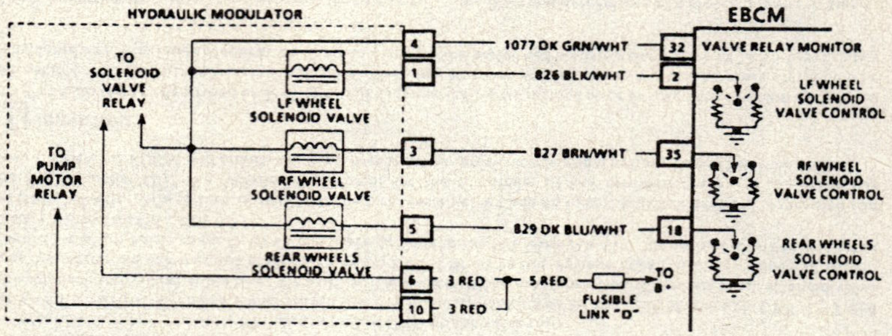

Circuit Description:

With the ignition in Run, Bulb Test or start the EBCM applies voltage to the Coil of the solenoid valve relay. The EBCM will complete an internal self-check before providing ground to the coil. Once energized the relay will provide B+ to the solenoid valves. The solenoid valve position can be commanded to three different positions depending on the amount of current the EBCM allows to flow through the solenoid valve coil.

Pressure Increase position is the normal operating position of the valves for braking. Brake fluid flows through the modulator from the master cylinder to the wheel brake. No current flows through the valve coil during pressure increase.

Pressure Hold position is commanded when the EBCM desires to hold the pressure in the wheel hydraulic circuit constant. The modulator will not allow pressure flow from the master cylinder to the wheel brake. The pressure hold position is commanded by passing approximately 2.0 amps through the valve coil.

Pressure Reduction position is used to decrease hydraulic pressure in a specific wheel hydraulic circuit without decreasing pressure in the rest of the system. The modulator takes brake fluid from the wheel brake and injects it into the master cylinder hydraulic circuit. This fully energized position of the solenoid valve occurs when approximately 5 amps are passed through the valve coil.

Failure Conditions: The solenoid valves are continuously monitored by the EBCM through CKT 1077 and through each individual valve line. Code 55 will set when the expected position of the rear wheels solenoid valve does not match the commanded position from the EBCM (if valve line voltage does not match expected value). Conditions which could cause Code 55 to set are damage to the rear wheels solenoid, an open or short to ground or battery voltage on CKT 829.

Action Taken: If a failure is detected which causes a valve Code 55 to set, antilock braking is disabled and the EBCM turns "ON" the "Antilock" indicator for the remainder of the ignition cycle. If the failure is intermittent, the EBCM will enable the system at the next ignition cycle and a history Code 55 will be present.

Important:

The Auto test (as described under "ABS Self Diagnostics") is performed once during each ignition cycle when the vehicle reaches approximately 4 mph. This Test may be heard and felt while it is taking place and should be considered a normal mode of operation. This test can also be performed by using the Tech 1 in "ABS Tests" mode F4.

Test Description: Numbers below refer to circled numbers on the diagnostic chart.

1. Checks the resistance in the rear wheels Solenoid Valve circuitry.
2. Checks for a short to ground in CKT 829.
3. This test uses the Tech 1 Diagnostic Computer to check for proper rear wheels Solenoid Valve operation.
4. This test determines of a problem found in Step 3 is due to the EBCM or the Hydraulic Modulator.
5. This test determines if a problem found in Step 1 is due to the Hydraulic Modulator or a problem in CKT 829.
6. This test determines if a problem found in Step 3 is due to an intermittent in the wiring and/or connectors, or a fault with the EBCM.

DIAGNOSTIC CHART DTC 55 — 1993 CAPRICE, CUSTOM CRUISER AND ROADMASTER, CONTINUED

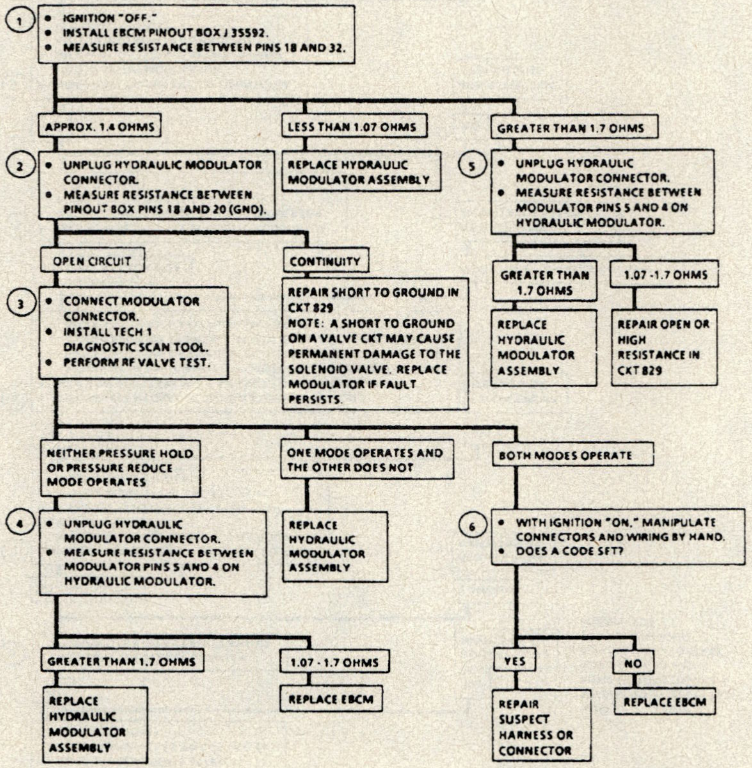

DIAGNOSTIC CHART DTC 61 — 1993 CAPRICE, CUSTOM CRUISER AND ROADMASTER

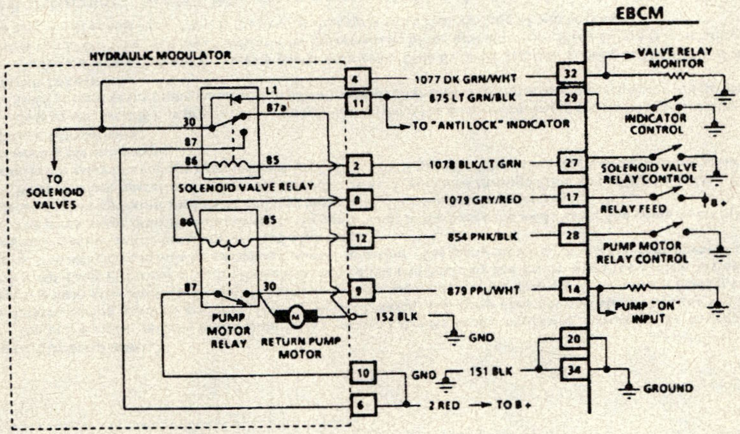

Circuit Description:

The pump motor returns brake fluid to the master cylinder brake hydraulic circuit at the hydraulic modulator, during antilock braking. During normal braking, the pump does not operate. When the vehicle begins to move after start-up, the EBCM will turn on the pump motor and perform a self-check of the pump motor and pump motor electrical circuit. This check may be felt and heard by the driver when the vehicle begins to move. The self-check should be considered normal operation. The pump motor is an integral component of the hydraulic modulator and cannot be serviced separately.

The pump motor relay provides power to the pump motor in the hydraulic modulator. The pump motor relay is located on the hydraulic modulator and may be replaced if it is found to be defective.

When antilock operation is required, the motor relay is grounded through CKT 854 (EBCM pin 28). The pump motor relay has voltage supplied from CKT 1079 (EBCM pin 17). When the relay toggles, battery power is provided to the pump motor.

Failure Conditions:

Code 61 will set if B+ is present at the pump motor without the pump motor being requested to activate, or if B+ is not present at the pump motor within 60 milliseconds after the pump motor has been requested to activate.

Action Taken:

If a failure is detected which causes a pump motor/relay Code 61 to set antilock braking is disabled and the EBCM turns "ON" the "Antilock" indicator for the remainder of the ignition cycle. If the failure is intermittent, the EBCM will enable the system at the next ignition cycle and a History Code 61 will be present.

Test Description: Numbers below refer to circled numbers on the diagnostic chart.

1. Checks the relay for correct resistance.
2. Checks if relay is stuck closed.
3. Checks for proper switching action in pump motor relay.
4. Checks for an open in the Hydraulic Modulator Circuitry.
5. Checks for a good ground.
6. Checks the pump motor for correct resistance.

DIAGNOSTIC CHART DTC 61 — 1993 CAPRICE, CUSTOM CRUISER AND ROADMASTER, CONTINUED

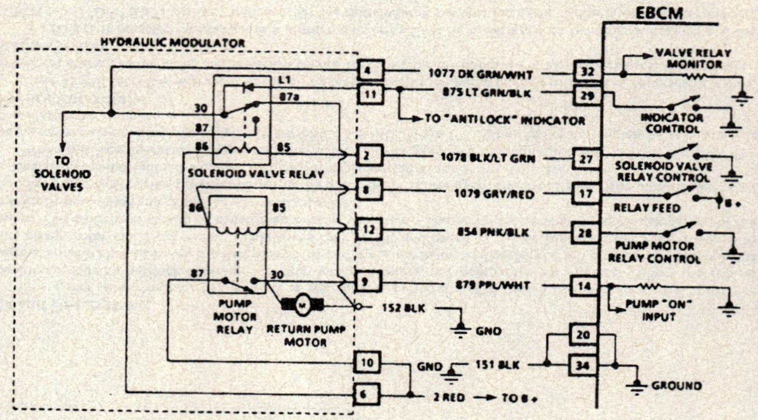

Circuit Description:

The pump motor returns brake fluid to the master cylinder brake hydraulic circuit at the hydraulic modulator, during antilock braking. During normal braking, the pump does not operate. When the vehicle begins to move after start-up, the EBCM will turn on the pump motor and perform a self-check of the pump motor and pump motor electrical circuit. This check may be felt and heard by the driver when the vehicle begins to move. The self-check should be considered normal operation. The pump motor is an integral component of the hydraulic modulator and cannot be serviced separately.

The pump motor relay provides power to the pump motor in the hydraulic modulator. The pump motor relay is located on the hydraulic modulator and may be replaced if it is found to be defective.

When antilock operation is required, the motor relay is grounded through CKT 854 (EBCM pin 28). The pump motor relay has voltage supplied from CKT 1079 (EBCM pin 17). When the relay toggles, battery power is provided to the pump motor.

Failure Conditions:

Code 61 will set if B+ is present at the pump motor without the pump motor being requested to activate, or if B+ is not present at the pump motor within 60 milliseconds after the pump motor has been requested to activate.

Action Taken:

If a failure is detected which causes a pump motor/relay Code 61 to set antilock braking is disabled and the EBCM turns "ON" the "Antilock" indicator for the remainder of the ignition cycle. If the failure is intermittent, the EBCM will enable the system at the next ignition cycle and a History Code 61 will be present.

Test Description: Numbers below refer to circled numbers on the diagnostic chart.

7. Determines if battery voltage is available to the pump motor relay.
8. Checks for a short to B+ in the relay feed circuit.
9. Checks for a potential open or short to ground in CKT 1079.
10. Checks for a short to ground in CKT 854.
11. Checks for an open in CKT 854.
12. Checks for a short to B+ in CKT 854.
13. Determines if a problem found in Step 9 is a short to ground.
14. Determines if a problem found in Step 13 is due to a problem in the circuit itself or a faulty EBCM.

DIAGNOSTIC CHART DTC 61 — 1993 CAPRICE, CUSTOM CRUISER AND ROADMASTER, CONTINUED

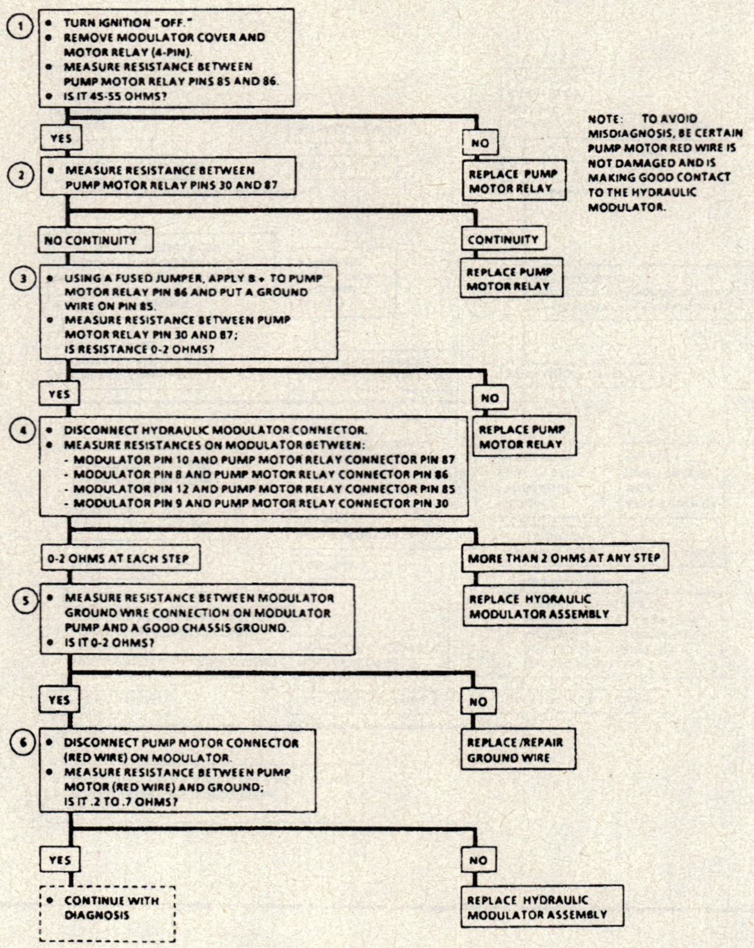

DIAGNOSTIC CHART DTC 61 — 1993 CAPRICE, CUSTOM CRUISER AND ROADMASTER, CONTINUED

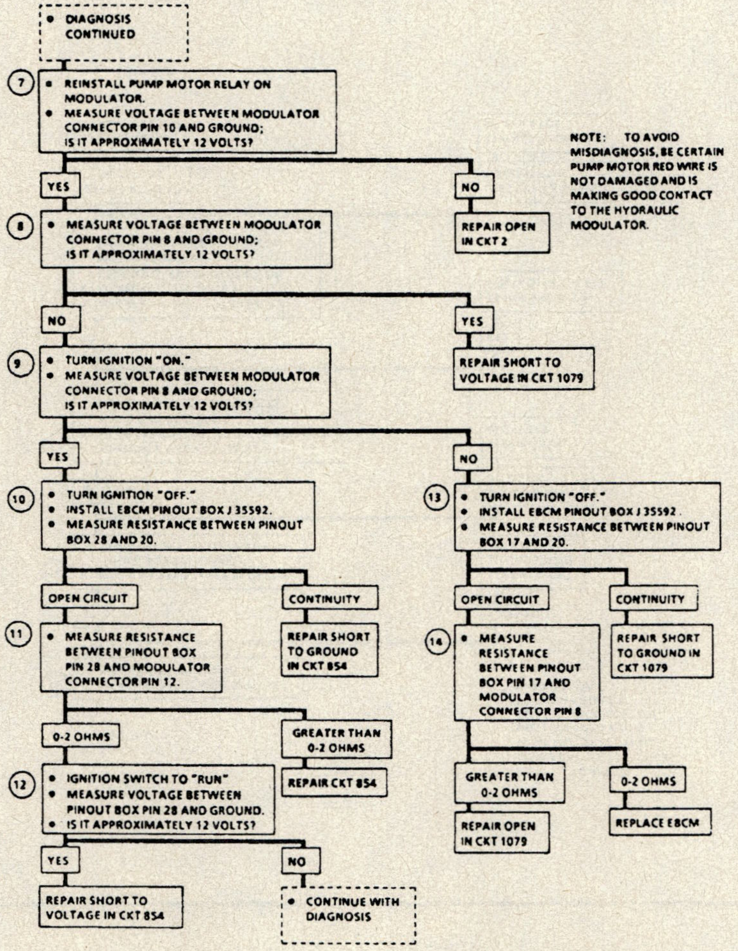

DIAGNOSTIC CHART DTC 61 — 1993 CAPRICE, CUSTOM CRUISER AND ROADMASTER, CONTINUED

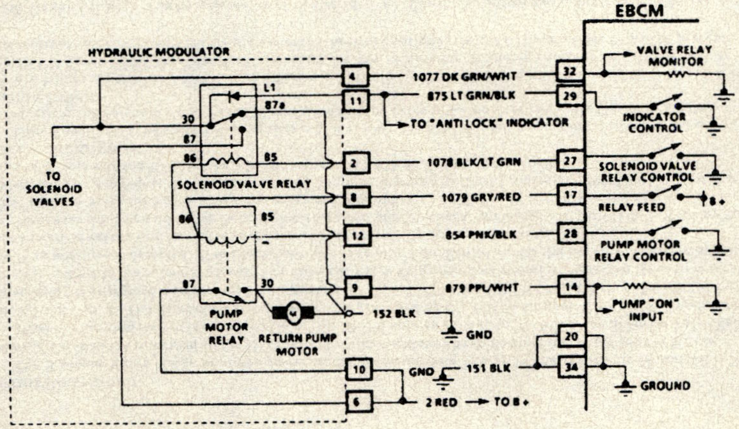

Circuit Description:

The pump motor returns brake fluid to the master cylinder brake hydraulic circuit at the hydraulic modulator, during antilock braking. During normal braking, the pump does not operate. When the vehicle begins to move after start-up, the EBCM will turn on the pump motor and perform a self-check of the pump motor and pump motor electrical circuit. This check may be felt and heard by the driver when the vehicle begins to move. The self-check should be considered normal operation. The pump motor is an integral component of the hydraulic modulator and cannot be serviced separately.

The pump motor relay provides power to the pump motor in the hydraulic modulator. The pump motor relay is located on the hydraulic modulator and may be replaced if it is found to be defective.

When antilock operation is required, the motor relay is grounded through CKT 854 (EBCM pin 28). The pump motor relay has voltage supplied from CKT 1079 (EBCM pin 17). When the relay toggles, battery power is provided to the pump motor.

Failure Conditions:

Code 61 will set if B+ is present at the pump motor without the pump motor being requested to activate, or if B+ is not present at the pump motor within 60 milliseconds after the pump motor has been requested to activate.

Action Taken:

If a failure is detected which causes a pump motor/relay Code 61 to set antilock braking is disabled and the EBCM turns "ON" the "Antilock" indicator for the remainder of the ignition cycle. If the failure is intermittent, the EBCM will enable the system at the next ignition cycle and a History Code 61 will be present.

Test Description: Numbers below refer to circled numbers on the diagnostic chart.

15. Checks for a short to ground in CKT 879.
16. Checks for an open in CKT 879.
17. Determines if CKT 879 is shorted to voltage or EBCM is faulty.
18. Determines if fault is in Hydraulic Modulator or EBCM.

DIAGNOSTIC CHART DTC 61 — 1993 CAPRICE, CUSTOM CRUISER AND ROADMASTER, CONTINUED

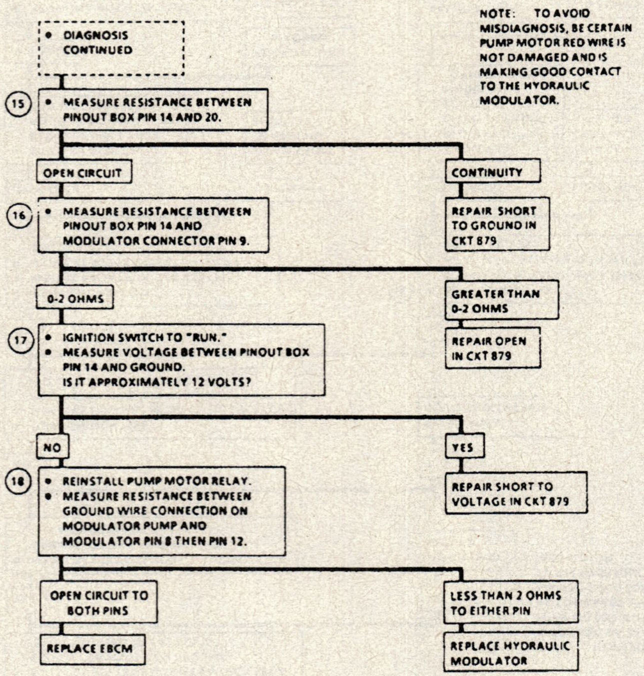

DIAGNOSTIC CHART DTC 63 — 1993 CAPRICE, CUSTOM CRUISER AND ROADMASTER

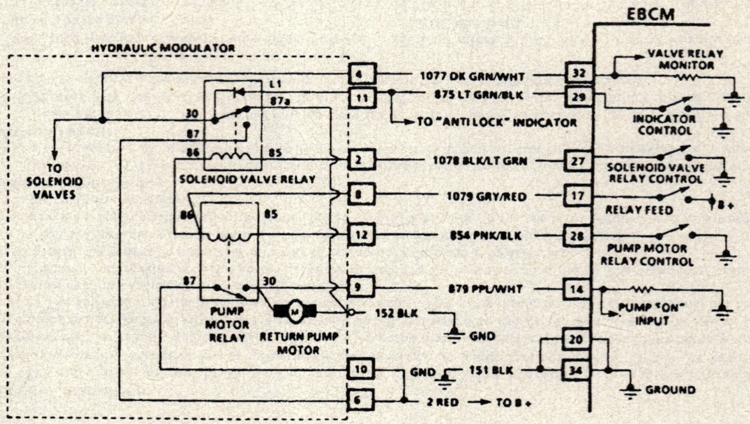

Circuit Description:

The Solenoid Valve Relay provides power to the three solenoid valves in the Hydraulic Modulator. The Solenoid Valve Relay is located on the Hydraulic Modulator and may be replaced if it is found to be defective.

When the ignition switch is turned to "Run," the EBCM applies voltage to the coil of the Solenoid Valve Relay through CKT 1079. The Solenoid Valve Relay is then grounded by the EBCM through CKT 1078. The Solenoid Valve Relay is energized and provides voltage through CKT 1077 to the EBCM (Valve Relay Monitor) and also provides voltage to the solenoid valves. The Solenoid Valve Relay remains energized until the ignition is turned "OFF" or a failure is detected. If antilock operation is required the EBCM varies the amount of current supplied to the solenoid valves and the valves are positioned to provide for optimum braking.

Whenever the Solenoid Valve Relay is not energized, the "Antilock" indicator will turn "ON." With ignition "ON" and the Solenoid Valve Relay not energized, the indicator CKT 875 is pulled low through the Solenoid Valve Relay to the hydraulic modulator ground, illuminating the "Antilock" indicator. The "Antilock" indicator can also be commanded "ON" by the EBCM through CKT 875 if a failure is detected.

Failure Conditions:

Code 63 will set if the valve relay monitor is at B + when the EBCM is not requesting it to be, or if the EBCM is requesting the Solenoid Valve Relay be energized and the valve relay monitor voltage is less than 5 volts.

Action Taken:

If a failure is detected which causes a Solenoid Valve Relay Code 63 to set, antilock braking is disabled and the EBCM turns "ON" the "Antilock" indicator for the remainder of the ignition cycle. If the failure is intermittent, the EBCM will enable the system at the next ignition cycle and a History Code 63 will be present.

Test Description: Numbers below refer to circled numbers on the diagnostic chart.

1. Checks the relay coil for correct resistance.
2. Checks the relay for a poor internal contact.
3. Checks the relay for a stuck internal contact
4. Checks for proper switching action in Relay
5. Checks for faulty diode in Solenoid Valve Relay.
6. Checks circuitry in Hydraulic Modulator.
7. Checks for a good ground.

DIAGNOSTIC CHART DTC 63 — 1993 CAPRICE, CUSTOM CRUISER AND ROADMASTER, CONTINUED

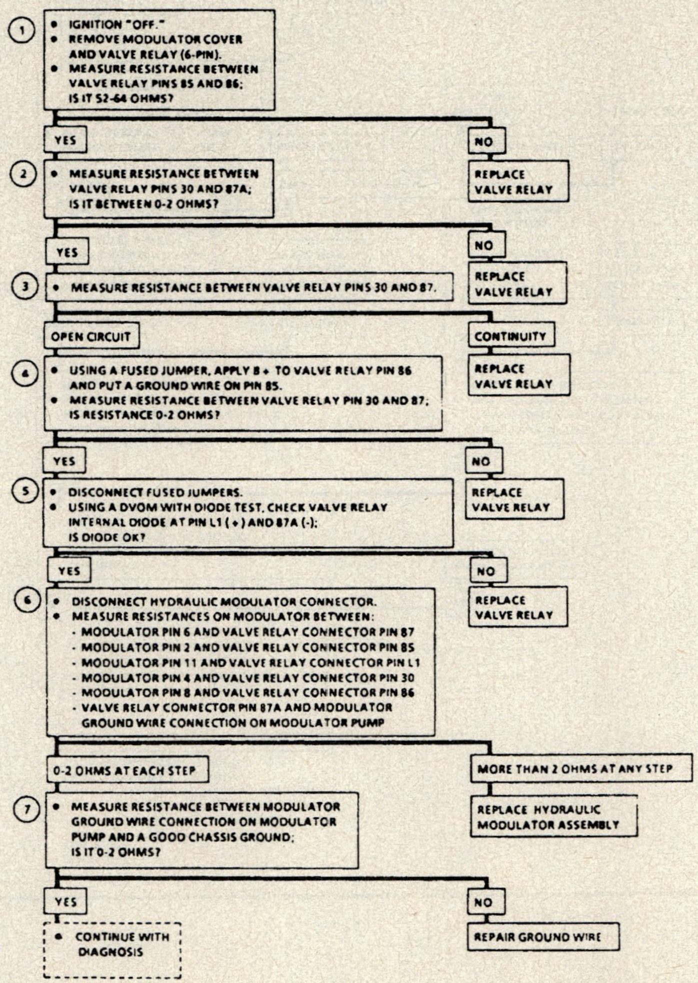

DIAGNOSTIC CHART DTC 63 — 1993 CAPRICE, CUSTOM CRUISER AND ROADMASTER, CONTINUED

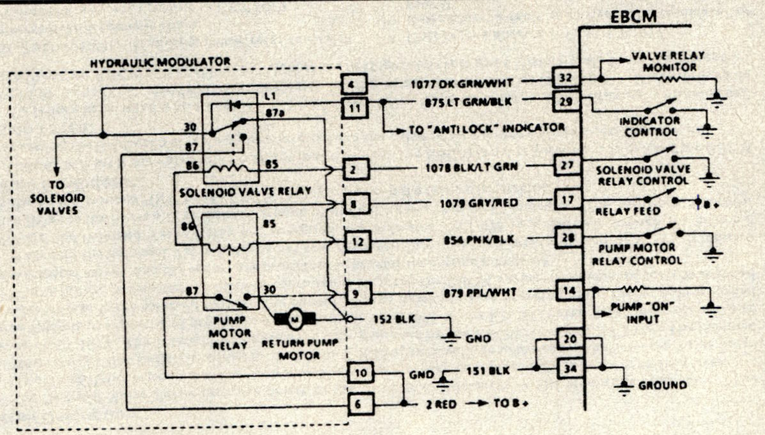

Circuit Description:

The Solenoid Valve Relay provides power to the three solenoid valves in the Hydraulic Modulator. The Solenoid Valve Relay is located on the Hydraulic Modulator and may be replaced if it is found to be defective.

When the ignition switch is turned to "Run," the EBCM applies voltage to the coil of the Solenoid Valve Relay through CKT 1079. The Solenoid Valve Relay is then grounded by the EBCM through CKT 1078. The Solenoid Valve Relay is energized and provides voltage through CKT 1077 to the EBCM (Valve Relay Monitor) and also provides voltage to the solenoid valves. The Solenoid Valve Relay remains energized until the ignition is turned "OFF" or a failure is detected. If antilock operation is required the EBCM varies the amount of current supplied to the solenoid valves and the valves are positioned to provide for optimum braking.

Whenever the Solenoid Valve Relay is not energized, the "Antilock" indicator will turn "ON." With ignition "ON" and the Solenoid Valve Relay not energized, the indicator CKT 875 is pulled low through the Solenoid Valve Relay to the hydraulic modulator ground, illuminating the "Antilock" indicator. The "Antilock" indicator can also be commanded "ON" by the EBCM through CKT 875 if a failure is detected.

Failure Conditions:

Code 63 will set if the valve relay monitor is at B+ when the EBCM is not requesting it to be, or if the EBCM is requesting the Solenoid Valve Relay be energized and the valve relay monitor voltage is less than 5 volts.

Action Taken:

If a failure is detected which causes a Solenoid Valve Relay Code 63 to set, antilock braking is disabled and the EBCM turns "ON" the "Antilock" indicator for the remainder of the ignition cycle. If the failure is intermittent, the EBCM will enable the system at the next ignition cycle and a History Code 63 will be present.

Test Description: Numbers below refer to circled numbers on the diagnostic chart.

8. Checks power feed to solenoid valve relay.
9. Checks for a short to B+ in the relay feed circuit.
10. Checks for a potential open or short to ground in CKT 1079
11. Checks for a short to ground in CKT 1078.
12. Checks for an open in CKT 1078.
13. Checks for a short to B+ in CKT 1078.
14. Checks CKT 1079 for short to ground.
15. Determines if fault is in CKT 1079 or in EBCM.

DIAGNOSTIC CHART DTC 63 — 1993 CAPRICE, CUSTOM CRUISER AND ROADMASTER, CONTINUED

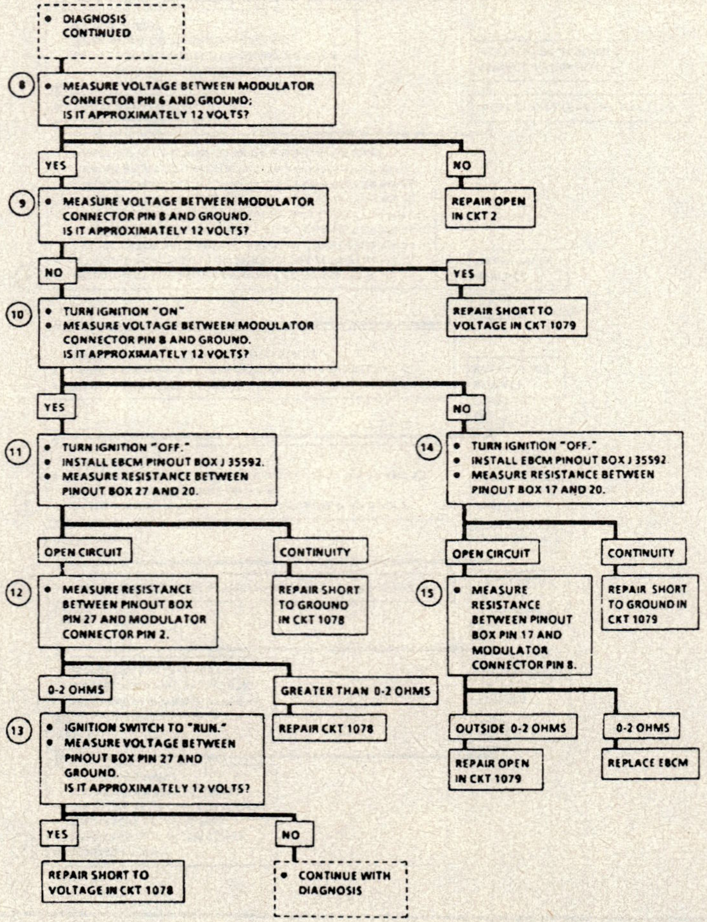

DIAGNOSTIC CHART DTC 63 — 1993 CAPRICE, CUSTOM CRUISER AND ROADMASTER, CONTINUED

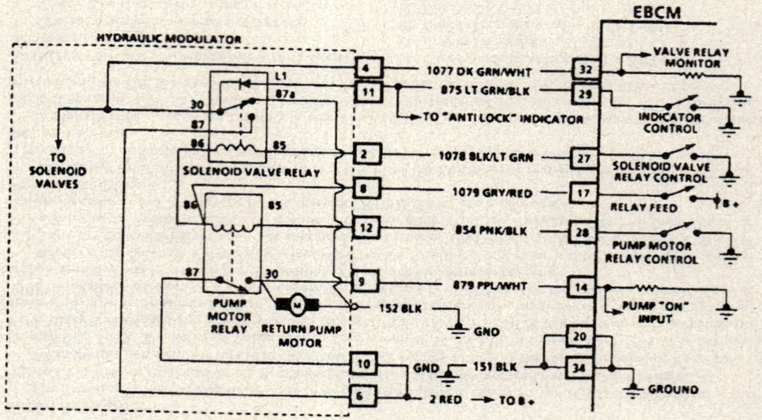

Circuit Description:

The Solenoid Valve Relay provides power to the three solenoid valves in the Hydraulic Modulator. The Solenoid Valve Relay is located on the Hydraulic Modulator and may be replaced if it is found to be defective.

When the ignition switch is turned to "Run," the EBCM applies voltage to the coil of the Solenoid Valve Relay through CKT 1079. The Solenoid Valve Relay is then grounded by the EBCM through CKT 1078. The Solenoid Valve Relay is energized and provides voltage through CKT 1077 to the EBCM (Valve Relay Monitor) and also provides voltage to the solenoid valves. The Solenoid Valve Relay remains energized until the ignition is turned "OFF" or a failure is detected. If antilock operation is required the EBCM varies the amount of current supplied to the solenoid valves and the valves are positioned to provide for optimum braking.

Whenever the Solenoid Valve Relay is not energized, the "Antilock" indicator will turn "ON." With ignition "ON" and the Solenoid Valve Relay not energized, the indicator CKT 875 is pulled low through the Solenoid Valve Relay to the hydraulic modulator ground, illuminating the "Antilock" indicator. The "Antilock" indicator can also be commanded "ON" by the EBCM through CKT 875 if a failure is detected.

Failure Conditions:

Code 63 will set if the valve relay monitor is at B+ when the EBCM is not requesting it to be, or if the EBCM is requesting the Solenoid Valve Relay be energized and the valve relay monitor voltage is less than 5 volts.

Action Taken:

If a failure is detected which causes a Solenoid Valve Relay Code 63 to set, antilock braking is disabled and the EBCM turns "ON" the "Antilock" indicator for the remainder of the ignition cycle. If the failure is intermittent, the EBCM will enable the system at the next ignition cycle and a History Code 63 will be present.

Test Description: Numbers below refer to circled numbers on the diagnostic chart.

16. Checks for a short to ground in CKT 1077.
17. Checks for an open in CKT 1077.
18. Checks for a short to B+ in CKT 1077.
19. Determines if fault is in hydraulic modulator or EBCM.

DIAGNOSTIC CHART DTC 63 — 1993 CAPRICE, CUSTOM CRUISER AND ROADMASTER, CONTINUED

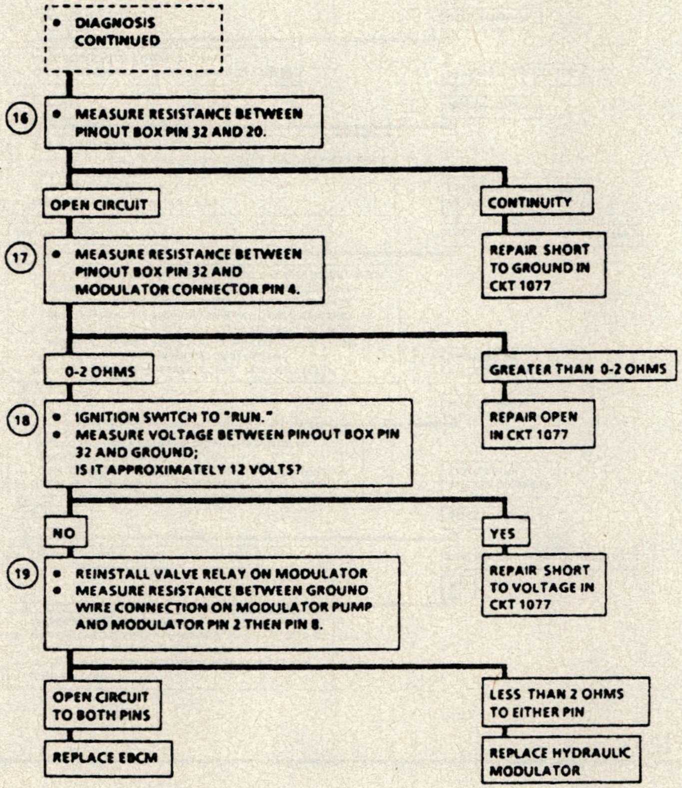

DIAGNOSTIC CHART DTC 63 — 1993 CAPRICE, CUSTOM CRUISER AND ROADMASTER, CONTINUED

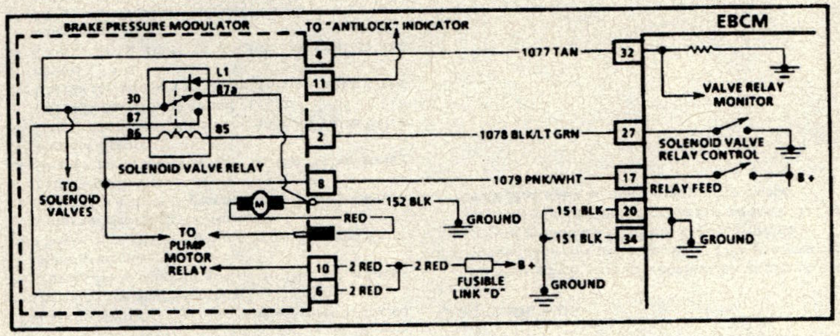

SOLENOID VALVE RELAY
(RELAY FAULT)

Circuit Description:

The Solenoid Valve Relay has a dual function. With ignition "ON" and the relay coil not energized, the relay provides ground to the "Antilock" indicator causing it to illuminate. This allows illumination of the "Antilock" indicator when the EBCM is disconnected or disabled. With ignition "ON" and the relay coil energized, the relay supplies battery voltage to the solenoid valves. This allows the EBCM to control actuation and position of the solenoid valves by controlling the ground path. CKT 1077 allows the EBCM to monitor the state of the Solenoid Valve Relay to compare with the requested state.

Failure Conditions:

Code 63 will set if the valve relay monitor is at B+ when the EBCM is not requesting it to be, or if the EBCM is requesting the Solenoid Valve Relay be energized and the valve relay monitor voltage is less than 5 volts.

Action Taken:

If a failure is detected which causes a Solenoid Valve Relay Code 63 to set, antilock braking is disabled and the EBCM turns "ON" the "Antilock" indicator for the remainder of the ignition cycle. If the failure is intermittent, the EBCM will enable the system at the next ignition cycle and a History Code 63 will be present.

Test Description: Number(s) below refer to circled number(s) on the diagnostic chart.

1. Checks for proper resistance of the relay coil.
2. Checks for the relay contacts stuck in the normally open position.
3. Checks for the relay contacts stuck in the normally closed position.
4. Checks for integrity of the circuitry internal to the modulator.
5. Checks for a good pump motor ground circuit.
6. Checks for an open in the Valve Relay Monitor circuit.
7. Checks for an open in the Solenoid Valve Relay coil switched ground circuit.
8. Checks for an open in the Relay Feed (coil battery feed) circuit.

DIAGNOSTIC CHART DTC 63 — 1993 CAPRICE, CUSTOM CRUISER AND ROADMASTER, CONTINUED

1. • IGNITION "OFF."
   • REMOVE COVER AND SOLENOID VALVE RELAY (6-PIN) FROM BRAKE PRESSURE MODULATOR.
   • MEASURE RESISTANCE BETWEEN RELAY PINS 85 AND 86. IS RESISTANCE 52 TO 64 OHMS?

   - YES → step 2
   - NO → REPLACE SOLENOID VALVE RELAY.

2. • CHECK CONTINUITY BETWEEN RELAY PINS 30 AND 87A.

   - CONTINUITY → step 3
   - NO CONTINUITY → REPLACE SOLENOID VALVE RELAY.

3. • USING A FUSED JUMPER SUCH AS J 36169 WITH A 3 AMP FUSE, CONNECT GROUND TO RELAY PIN 85 AND B+ THROUGH THE FUSED JUMPER TO PIN 86.
   • CHECK CONTINUITY BETWEEN RELAY PINS 30 AND 87.

   - CONTINUITY → step 4
   - NO CONTINUITY → REPLACE SOLENOID VALVE RELAY.

4. • DISCONNECT BRAKE PRESSURE MODULATOR CONNECTOR.
   • CHECK CONTINUITY ON MODULATOR BETWEEN:
     - MODULATOR PIN 2 AND VALVE RELAY CONNECTOR PIN 85.
     - MODULATOR PIN 4 AND VALVE RELAY CONNECTOR PIN 30.
     - MODULATOR PIN 6 AND VALVE RELAY CONNECTOR PIN 87.
     - MODULATOR PIN 8 AND VALVE RELAY CONNECTOR PIN 86.
     - MODULATOR PIN 11 AND VALVE RELAY CONNECTOR PIN L1.
     - PUMP MOTOR GROUND AND VALVE RELAY CONNECTOR PIN 87A.

   - CONTINUITY AT EACH STEP → step 5
   - LACK OF CONTINUITY AT ANY STEP. → REPLACE BRAKE PRESSURE MODULATOR.

5. • REINSTALL SOLENOID VALVE RELAY.
   • DISCONNECT EBCM CONNECTOR.
   • INSTALL PINOUT BOX J 35592.
   • CHECK CONTINUITY BETWEEN PUMP MOTOR GROUND AND PINOUT BOX PIN 20 (GROUND).

   - CONTINUITY → step 6
   - NO CONTINUITY → REPAIR OPEN IN CKT 152.

6. • CHECK CONTINUITY BETWEEN PINOUT BOX PIN 32 AND MODULATOR CONNECTOR PIN 4.

   - CONTINUITY → step 7
   - NO CONTINUITY → REPAIR OPEN IN CKT 1077.

7. • CHECK CONTINUITY BETWEEN PINOUT BOX PIN 27 AND MODULATOR CONNECTOR PIN 2.

   - CONTINUITY → step 8
   - NO CONTINUITY → REPAIR OPEN IN CKT 1078.

8. • CHECK CONTINUITY BETWEEN PINOUT BOX PIN 17 AND MODULATOR CONNECTOR PIN 8.

   - CONTINUITY → • CONTINUE WITH DIAGNOSIS ON NEXT PAGE.
   - NO CONTINUITY → REPAIR OPEN IN CKT 1079.

WHEN ALL DIAGNOSIS AND REPAIRS ARE COMPLETED, CLEAR CODES AND VERIFY OPERATION.

DIAGNOSTIC CHART DTC 63 — 1993 CAPRICE, CUSTOM CRUISER AND ROADMASTER, CONTINUED

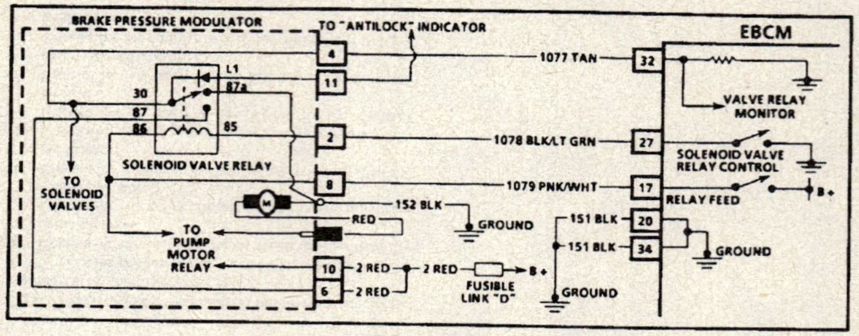

CODE 63

SOLENOID VALVE RELAY (RELAY FAULT)

Circuit Description:

The Solenoid Valve Relay has a dual function. With ignition "ON" and the relay coil not energized, the relay provides ground to the "Antilock" indicator causing it to illuminate. This allows illumination of the "Antilock" indicator when the EBCM is disconnected or disabled. With ignition "ON" and the relay coil energized, the relay supplies battery voltage to the solenoid valves. This allows the EBCM to control actuation and position of the solenoid valves by controlling the ground path. CKT 1077 allows the EBCM to monitor the state of the Solenoid Valve Relay to compare with the requested state.

Failure Conditions:

Code 63 will set if the valve relay monitor is at B+ when the EBCM is not requesting it to be, or if the EBCM is requesting the Solenoid Valve Relay be energized and the valve relay monitor voltage is less than 5 volts.

Action Taken:

If a failure is detected which causes a Solenoid Valve Relay Code 63 to set, antilock braking is disabled and the EBCM turns "ON" the "Antilock" indicator for the remainder of the ignition cycle. If the failure is intermittent, the EBCM will enable the system at the next ignition cycle and a History Code 63 will be present.

Test Description: Number(s) below refer to circled number(s) on the diagnostic chart.

9. Checks for a short to ground in the Valve Relay Monitor circuit.
10. Checks for a short to ground in the Solenoid Valve Relay coil switch ground circuit.
11. Checks for a short to ground in the Relay Feed (coil battery feed) circuit.
12. Checks for a short to B+ in the Valve Relay Monitor circuit.
13. Checks for a short B+ in the Solenoid Valve Relay coil switched ground circuit.
14. Checks for an open in the modulator battery feed circuit.
15. Checks for a short to ground internal to the modulator.
16. Checks for improper connections which may have set the code.
17. Checks for a code which may have set falsely

Diagnostic Aids:

All tests which require a connection to pinout box Pin 20 are using this pin as ground. This assumes the integrity of ground CKT 151 has been maintained.

A disconnected or improperly seated Brake Pressure Modulator connector may set this code.

DIAGNOSTIC CHART DTC 63 — 1993 CAPRICE, CUSTOM CRUISER AND ROADMASTER, CONTINUED

- DIAGNOSIS CONTINUED FROM PREVIOUS PAGE.

(9) • CHECK CONTINUITY BETWEEN PINOUT BOX PINS 32 AND 20 (GROUND).
- CONTINUITY → REPAIR SHORT FROM CKT 1077 TO GROUND.
- NO CONTINUITY →

(10) • CHECK CONTINUITY BETWEEN PINOUT BOX PINS 27 AND 20 (GROUND).
- CONTINUITY → REPAIR SHORT FROM CKT 1078 TO GROUND.
- NO CONTINUITY →

(11) • CHECK CONTINUITY BETWEEN PINOUT BOX PINS 17 AND 20 (GROUND).
- CONTINUITY → REPAIR SHORT FROM CKT 1079 TO GROUND.
- NO CONTINUITY →

(12) • IGNITION "ON."
• MEASURE VOLTAGE FROM PINOUT BOX PIN 32 TO PIN 20 (GROUND). IS VOLTAGE ABOUT 12 VOLTS?
- YES → REPAIR SHORT FROM CKT 1077 TO B+.
- NO →

(13) • MEASURE VOLTAGE FROM PINOUT BOX PIN 27 TO PIN 20 (GROUND). IS VOLTAGE ABOUT 12 VOLTS?
- YES → REPAIR SHORT FROM CKT 1078 TO B+.
- NO →

(14) • MEASURE VOLTAGE FROM MODULATOR CONNECTOR PIN 6 TO PINOUT BOX PIN 20 (GROUND). IS VOLTAGE ABOUT 12 VOLTS?
- NO → REPAIR OPEN IN CKT 2.
- YES →

(15) • IGNITION "OFF."
• CHECK CONTINUITY BETWEEN PUMP MOTOR GROUND AND MODULATOR PIN 2 AND THEN PIN 8.
- CONTINUITY AT EITHER OR BOTH STEPS → REPLACE BRAKE PRESSURE MODULATOR.
- NO CONTINUITY AT EITHER STEP →

(16) • IGNITION "OFF."
• REMOVE J 35592 PINOUT BOX.
• RECONNECT EBCM CONNECTOR.
• RECONNECT BRAKE PRESSURE MODULATOR CONNECTOR AND INSTALL COVER.
• INSPECT ALL WIRING AND CONNECTORS FOR CONDITIONS WHICH COULD CAUSE INTERMITTENTS. ARE ALL WIRES IN GOOD CONDITION, TERMINALS CLEAN AND MAKING GOOD CONTACT?
- NO → REPAIR DEFECTIVE WIRING AND/OR CONNECTOR.
- YES →

(17) • CLEAR CODE AND ROAD TEST VEHICLE. DOES CODE 63 RESET?
- YES → REPLACE EBCM.
- NO → NO TROUBLE FOUND, SEE "DIAGNOSTIC AIDS" ON FACING PAGE.

WHEN ALL DIAGNOSIS AND REPAIRS ARE COMPLETED, CLEAR CODES AND VERIFY OPERATION.

DIAGNOSTIC CHART DTC 71 — 1993 CAPRICE, CUSTOM CRUISER AND ROADMASTER

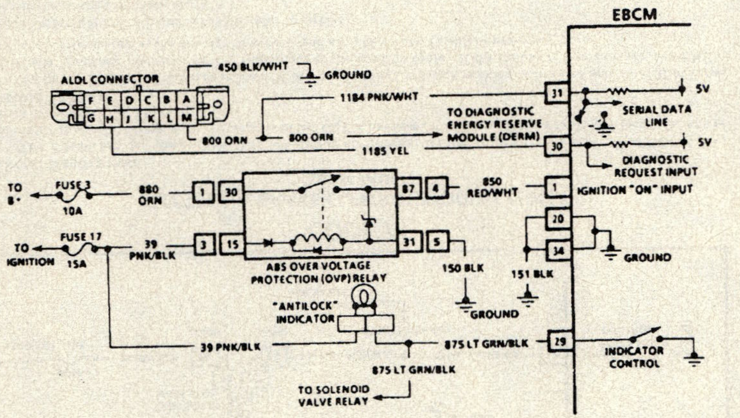

Circuit Description:

The EBCM has self diagnostics that can detect an internal failure within the module.

If a failure is detected which causes a EBCM Code 71 to set, antilock braking is disabled and the EBCM turns "ON" the "Antilock" indicator for the remainder of the ignition cycle. If the failure is "Intermittent" the EBCM will enable the system at the next ignition cycle and a History Code 71 will be present.

DIAGNOSTIC CHART DTC 71 — 1993 CAPRICE, CUSTOM CRUISER AND ROADMASTER, CONTINUED

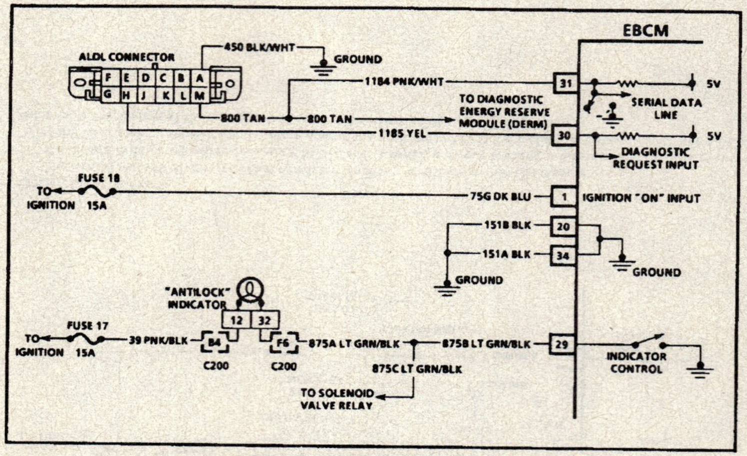

CODE 71

ELECTRONIC BRAKE CONTROL MODULE (EBCM)
(EBCM FAULT)

Circuit Description:

The EBCM has self diagnostics that can detect an internal failure within the module. The EBCM also monitors its supply voltage and the integrity of the EBCM harness connection. When a fault is detected for any of these conditions Code 71 is set.

Action Taken:

If a failure is detected which causes a EBCM Code 71 to set, antilock braking is disabled and the EBCM turns "ON" the "Antilock" indicator for the remainder of the ignition cycle. If the failure is intermittent, the EBCM will enable the system at the next ignition cycle and a History Code 71 will be present.

Test Description: Number(s) below refer to circled number(s) on the diagnostic chart.

1. Checks for additional faults which may be causing Code 71 to set.
2. Checks the integrity of the EBCM ground circuit.
3. Checks for the minimum supply voltage necessary for proper ABS operation.
4. Checks for intermittent connections of the EBCM harness connector and terminals to the EBCM.

DIAGNOSTIC CHART DTC 71 — 1993 CAPRICE, CUSTOM CRUISER AND ROADMASTER, CONTINUED

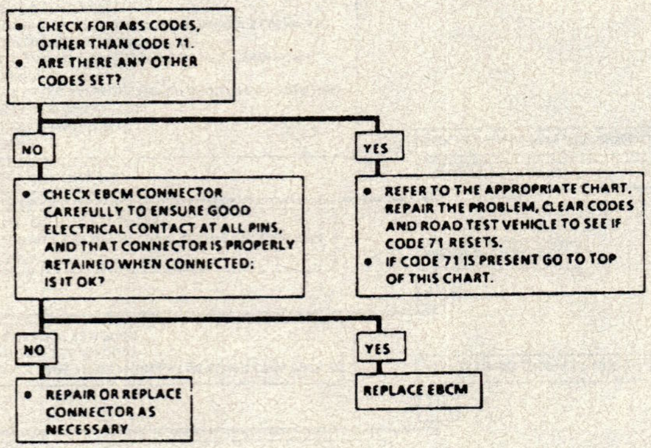

DIAGNOSTIC CHART DTC 71 — 1993 CAPRICE, CUSTOM CRUISER AND ROADMASTER, CONTINUED

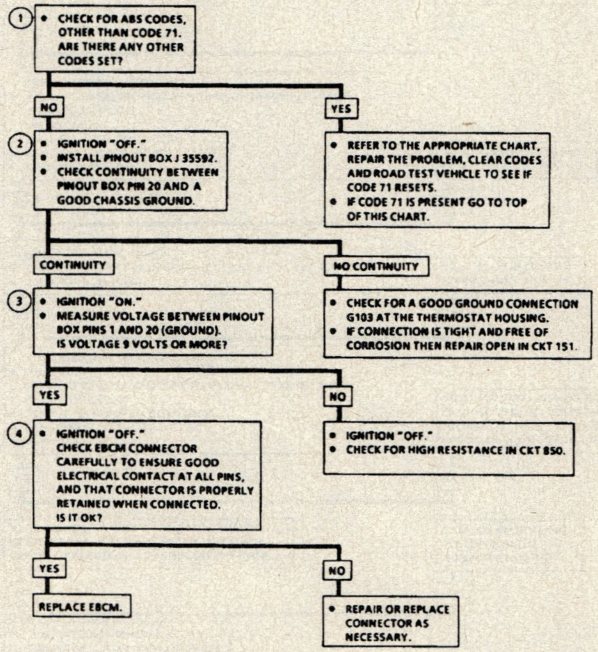

WHEN ALL DIAGNOSIS AND REPAIRS ARE COMPLETED, CLEAR CODES AND VERIFY OPERATION

DIAGNOSTIC CHART DTC 72 — 1993 CAPRICE, CUSTOM CRUISER AND ROADMASTER, CONTINUED

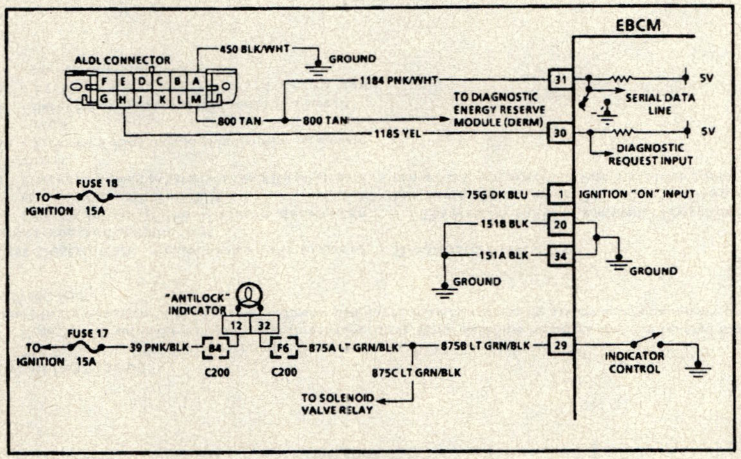

CODE 72

ELECTRONIC BRAKE CONTROL MODULE (EBCM)
(SERIAL DATA LINE FAULT)

Circuit Description:

The Serial Data Line is an asynchronous line operating at 8192 bits per second.

Code 72 will set if the EBCM detects three consecutive serial data line messages that are ignored due to errors in transmission.

Important

- The "Antilock" indicator will not be illuminated and the system will remain operational with this fault.

Test Description: Number(s) below refer to circled number(s) on the diagnostic chart.

1. Checks to see if the Tech 1 can communicate with the EBCM.
2. Determines if the serial data circuit is open.
3. Checks for a short to battery in the ABS serial data circuit.
4. Checks for a short to ground in the ABS serial data circuit.
5. Checks to see if the Tech 1 can communicate with the EBCM.

Diagnostic Aids:

The problem may be intermittent. Try performing the tests shown while "wiggling" wiring and connectors; this can often cause the fault to appear.

DIAGNOSTIC CHART DTC 72 — 1993 CAPRICE, CUSTOM CRUISER AND ROADMASTER, CONTINUED

IMPORTANT: THIS CODE WILL SET ONLY WHEN AN OFF-BOARD SERIAL DATA DEVICE, SUCH AS THE TECH 1 (T-1) IS ATTEMPTING TO COMMUNICATE WITH THE EBCM. SINCE, BY DEFINITION, THIS CODE MEANS YOU CANNOT COMMUNICATE WITH THE EBCM USING AN OFF-BOARD SERIAL DATA DEVICE, YOU MUST READ AND CLEAR THE CODE USING ALDL FLASH CODE AND CODE CLEARING TECHNIQUES.

(1)
- TURN IGNITION "OFF."
- INSTALL TECH 1.
- TURN IGNITION "ON."
- ATTEMPT TO ESTABLISH COMMUNICATIONS WITH THE EBCM. CAN YOU COMMUNICATE WITH IT?

YES: NO TROUBLE FOUND. SEE "DIAGNOSTIC AIDS" ON FACING PAGE.

NO: (2) DID THE EBCM ENTER FLASH CODE DIAGNOSTICS (FLASHING THE "ANTILOCK" INDICATOR) WHEN ATTEMPTING TO COMMUNICATE WITH THE TECH 1?

YES: CHECK FOR POOR CONNECTION AT EBCM PIN 31. IF OK, REPAIR OPEN IN CKT 1184 OR IN CKT 800 BETWEEN ALDL TERMINAL "M" AND SPLICE.

NO: (3)
- TURN IGNITION "OFF."
- DISCONNECT EBCM CONNECTOR.
- INSTALL EBCM PINOUT BOX J 35592.
- TURN IGNITION "ON."
- CONNECT DVM BETWEEN PINOUT BOX PIN 31 AND GROUND.
- NOTE VOLTAGE. IS IT APPROXIMATELY 12 VOLTS?

YES: REPAIR SHORT TO B + IN CKT 1184 OR IN CKT 800.

NO: (4)
- TURN IGNITION "OFF."
- CHECK RESISTANCE BETWEEN PINOUT BOX PIN 31 AND GROUND. IS IT LESS THAN 5 OHMS?

YES: REPAIR SHORT TO GROUND IN CKT 1184 OR IN CKT 800.

NO: (5)
- TURN IGNITION "OFF."
- REMOVE EBCM PINOUT BOX J 35592.
- CONNECT EBCM CONNECTOR.
- INSTALL TECH 1.
- TURN IGNITION "ON."
- ATTEMPT TO ESTABLISH COMMUNICATIONS WITH THE EBCM. CAN YOU COMMUNICATE WITH IT?

YES: NO TROUBLE FOUND. SEE "DIAGNOSTIC AIDS" ON FACING PAGE.

NO:
- CHECK FOR POOR CONNECTIONS AT EBCM AND ALDL CONNECTORS. IF OK, REPLACE EBCM.

WHEN ALL DIAGNOSIS AND REPAIRS ARE COMPLETED, CLEAR CODES AND VERIFY OPERATION

DIAGNOSTIC CHART DTC 72 — 1993 CAPRICE, CUSTOM CRUISER AND ROADMASTER, CONTINUED

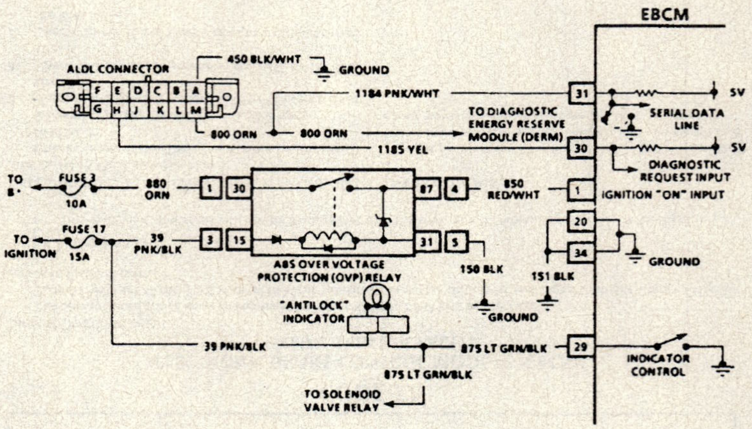

Circuit Description:

The Serial Data Line is an asynchronous line operating at 8192 bits per second

Code 72 will set if the EBCM detects three consecutive serial data line messages that are ignored due to errors in transmission. *The "Antilock" indicator will not be illuminated and the system will remain operational with this fault.*

Test Description:

Numbers below refer to circled numbers on the diagnostic chart.

1. Checks to see if the off-board device can communicate with the EBCM.
2. Checks for a short to battery in the ABS serial data circuit.
3. Checks for a short to ground in the ABS serial data circuit.
4. Checks for an open in the ABS serial data circuit.
5. Checks to see if the off board device can communicate with the EBCM.

Diagnostic Aids:

The problem may be intermittent. Try performing the tests shown while "wiggling" wiring and connectors; this can often cause the fault to appear.

DIAGNOSTIC CHART DTC 72 — 1993 CAPRICE, CUSTOM CRUISER AND ROADMASTER, CONTINUED

NOTE: THIS CODE WILL SET ONLY WHEN AN OFF-BOARD SERIAL DATA DEVICE, SUCH AS THE TECH 1 (T-1) OR CAMS (T-100), IS ATTEMPTING TO COMMUNICATE WITH THE EBCM. SINCE, BY DEFINITION, THIS CODE MEANS YOU CANNOT COMMUNICATE WITH THE EBCM USING AN OFF-BOARD SERIAL DATA DEVICE, YOU MUST READ AND CLEAR THE CODE USING ALDL FLASH CODE AND CODE CLEARING TECHNIQUES.

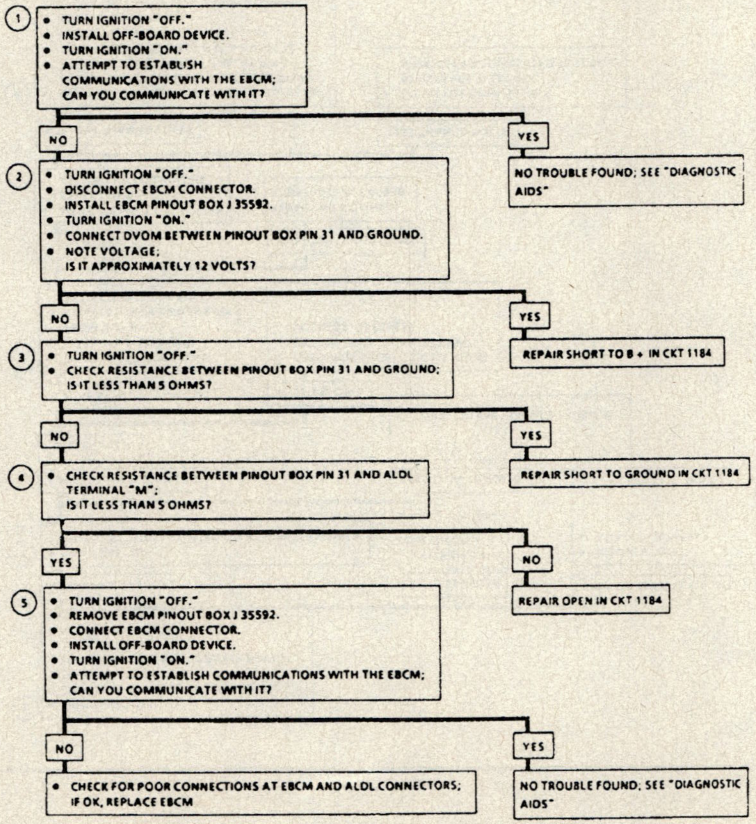

DIAGNOSTIC CHART — 1993 CAPRICE, CUSTOM CRUISER AND ROADMASTER

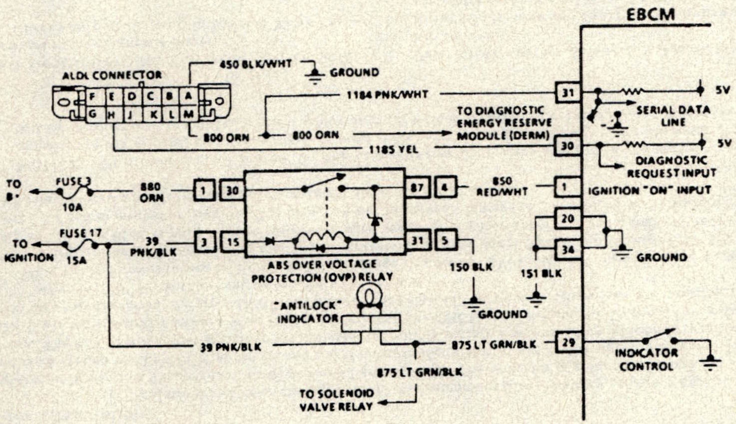

Circuit Description:

The "Antilock" Indicator is located on the right side of the instrument panel. Battery voltage is applied with ignition in "Run, Bulb Test or Start" through fuse 17 and CKT 39. The indicator is grounded through CKT 875 and is controlled by the EBCM at pin 29 and the Hydraulic Modulator at pin 11 through the Solenoid Valve Relay. The EBCM will supply ground to the indicator for approximately 2 seconds when ignition is first turned "ON." During the first 2 seconds, if no faults are detected, the EBCM will energize the Solenoid Valve Relay. With the relay energized, the ground CKT 875 through the relay will be opened. This allows the indicator to go "OFF" when the EBCM indicator control line at pin 29 is opened.

When the ignition is turned to "Run," the Over Voltage Protection (OVP) relay coil is energized by Fuse 17. When energized the OVP closes its normally open contacts, supplying power from fuse 3 through CKT 880 and CKT 850 to the EBCM at pin 1. If high system voltages occur (at or above approx. 16 volts) the OVP relay will short CKT 880 to ground causing fuse 3 to blow in order to prevent damage to the antilock brake system. The OVP relay is located inside the luggage compartment above the LH wheelhouse.

NOTICE: Do not use a quick charger for engine starting or for battery charging if the battery is still connected to the vehicle's electrical system. Always disconnect the negative battery cable when charging a battery. Failure to do so may cause fuse 3 to blow or cause serious damage to other electronic components.

Test Description: Numbers below refer to circled numbers on the diagnostic chart.

1. Determines if EBCM has proper supply voltage.
2. Determines if fault is in charging system or power feed to OVP relay (Fuse 3).
3. Checks for open or high resistance in ckt 850.
4. Determines if OVP relay is faulty.
5. Determines weather open is in CKT 150 or CKT 39.

DIAGNOSTIC CHART — 1993 CAPRICE, CUSTOM CRUISER AND ROADMASTER, CONTINUED

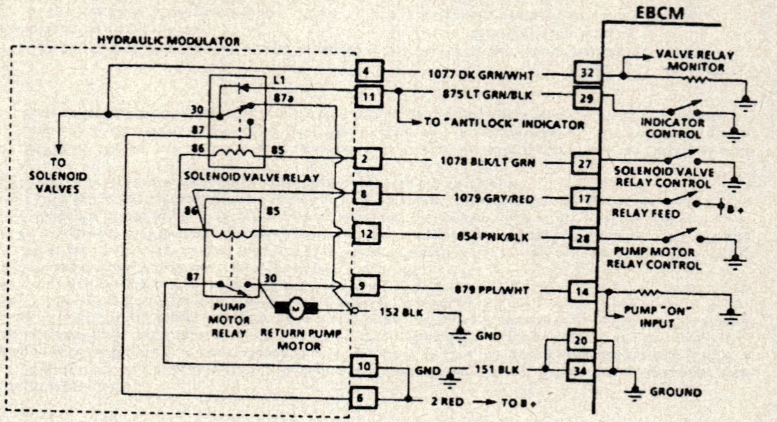

Circuit Description:

The "Antilock" Indicator is located on the right side of the instrument panel. Battery voltage is applied with ignition in "Run, Bulb Test or Start" through fuse 17 and CKT 39. The indicator is grounded through CKT 875 and is controlled by the EBCM at pin 29 and the Hydraulic Modulator at pin 11 through the Solenoid Valve Relay. The EBCM will supply ground to the indicator for approximately 2 seconds when ignition is first turned "ON." During the first 2 seconds, if no faults are detected, the EBCM will energize the Solenoid Valve Relay. With the relay energized, the ground CKT 875 through the relay will be opened. This allows the indicator to go "OFF" when the EBCM indicator control line at pin 29 is opened.

When the ignition is turned to "Run," the Over Voltage Protection (OVP) relay coil is energized by Fuse 17. When energized the OVP closes its normally open contacts, supplying power from fuse 3 through CKT 880 and CKT 850 to the EBCM at pin 1. If high system voltages occur (at or above approx. 16 volts) the OVP relay will short CKT 880 to ground causing fuse 3 to blow in order to prevent damage to the antilock brake system. The OVP relay is located inside the luggage compartment above the LH wheelhouse.

NOTICE: Do not use a quick charger for engine starting or for battery charging if the battery is still connected to the vehicle's electrical system. Always disconnect the negative battery cable when charging a battery. Failure to do so may cause fuse 3 to blow or cause serious damage to other electronic components.

Test Description: Numbers below refer to circled numbers on the diagnostic chart.

6. Checks for possible fault in Solenoid Valve Relay or EBCM
7. Determines weather Solenoid Valve Relay or EBCM is faulty.
8. Determines weather fault is in hydraulic modulator or short to ground in CKT 875.

DIAGNOSTIC CHART — 1993 CAPRICE, CUSTOM CRUISER AND ROADMASTER, CONTINUED

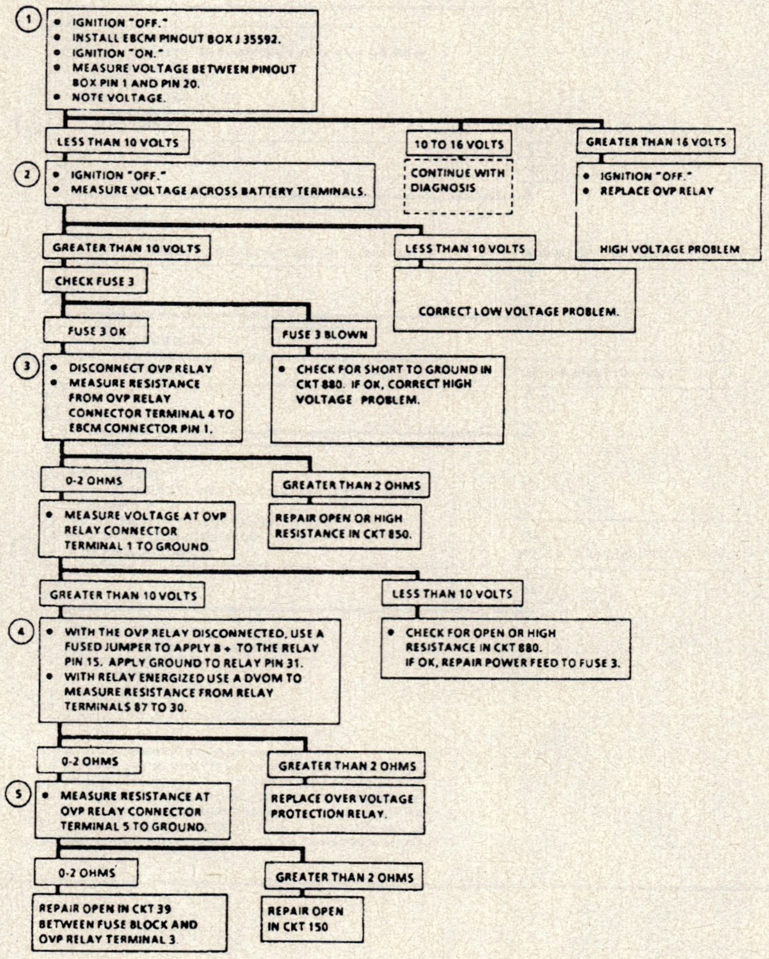

DIAGNOSTIC CHART — 1993 CAPRICE, CUSTOM CRUISER AND ROADMASTER, CONTINUED

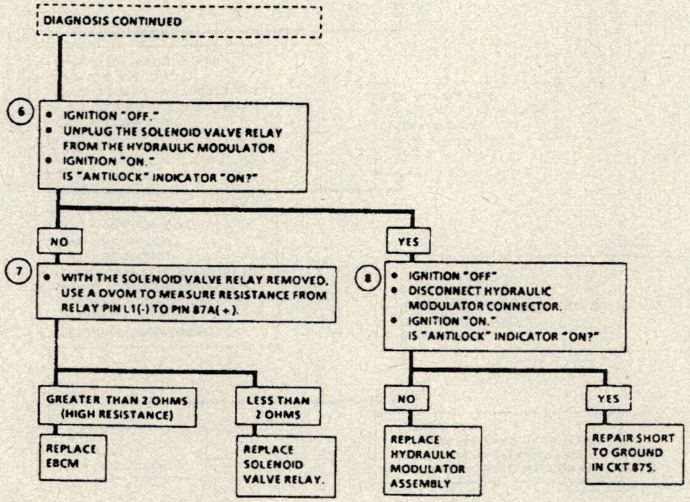

DIAGNOSTIC CHART — 1993 CAPRICE, CUSTOM CRUISER AND ROADMASTER, CONTINUED

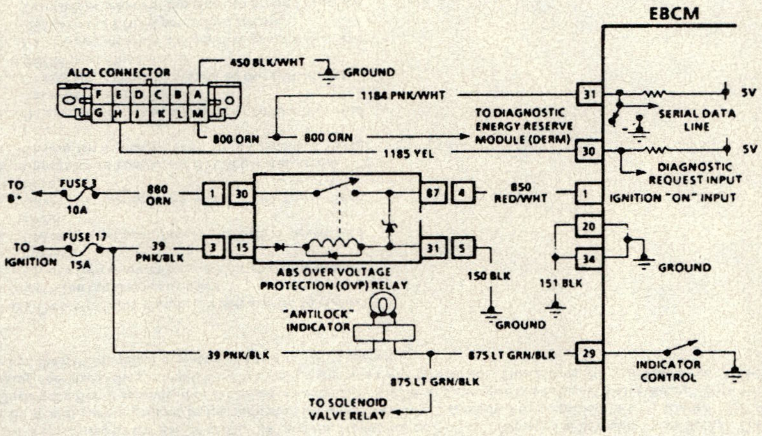

Circuit Description:

The "Antilock" Indicator is located on the right side of the instrument panel. Battery voltage is applied with ignition in "Run, Bulb Test or Start" through fuse 17 and CKT 39. The indicator is grounded through CKT 875 and is controlled by the EBCM at pin 29 and the Hydraulic Modulator at pin 11 through the Solenoid Valve Relay. The EBCM will supply ground to the indicator for approximately 2 seconds when ignition is first turned "ON." During the first 2 seconds, if no faults are detected, the EBCM will energize the Solenoid Valve Relay. With the relay energized, the ground CKT 875 through the relay will be opened. This allows the indicator to go "OFF" when the EBCM indicator control line at pin 29 is opened.

Test Description: Numbers below refer to circled numbers on the diagnostic chart.

1. Verifies an inoperative indicator.
2. Checks for a faulty EBCM indicator control.
3. Checks Fuse 17.
4. Checks for a faulty EBCM or connection at EBCM pin 29. (This test will detect a bad EBCM even if the Solenoid Valve Relay energized before the indicator had a chance to flash in Step 2)
5. Checks for open in CKT 875 between EBCM and splice.
6. Checks for open in CKT 39 or CKT 875. At this point if replacing the lamp does not correct the problem, either CKT 39 or CKT 875 is open

Diagnostic Aids:

If indicator failure is intermittent, install the Tech 1 diagnostic tool and select any ABS test mode so the "Antilock" indicator will be turned "ON." While observing the indicator, go through the indicators power and ground circuits wiggling the wiring and connector's. Watch for flickering or failure of the indicator to pin point problem area.

DIAGNOSTIC CHART A — 1993 CAPRICE, CUSTOM CRUISER AND ROADMASTER

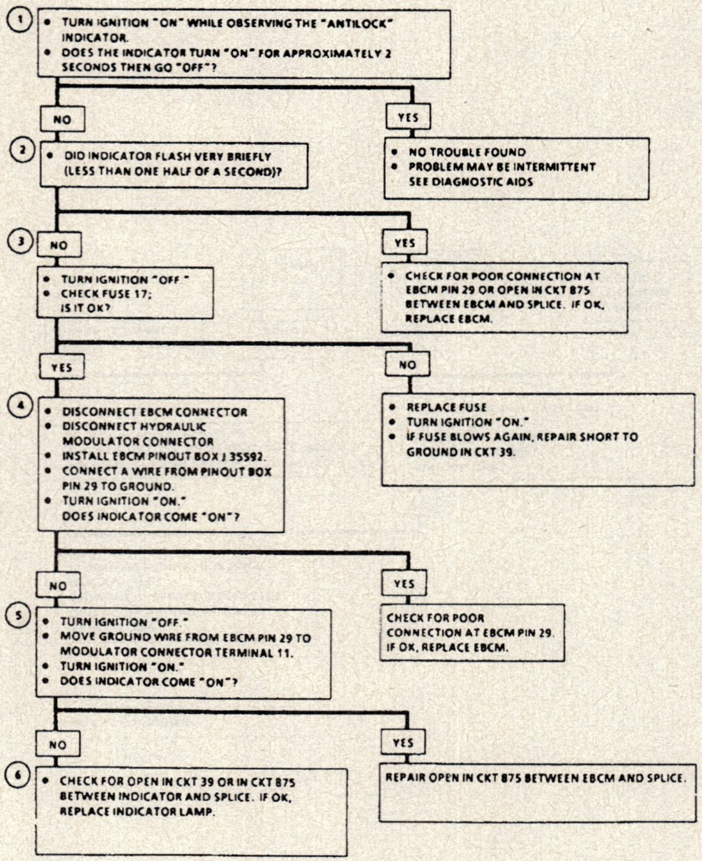

DIAGNOSTIC CHART A — 1993 CAPRICE, CUSTOM CRUISER AND ROADMASTER, CONTINUED

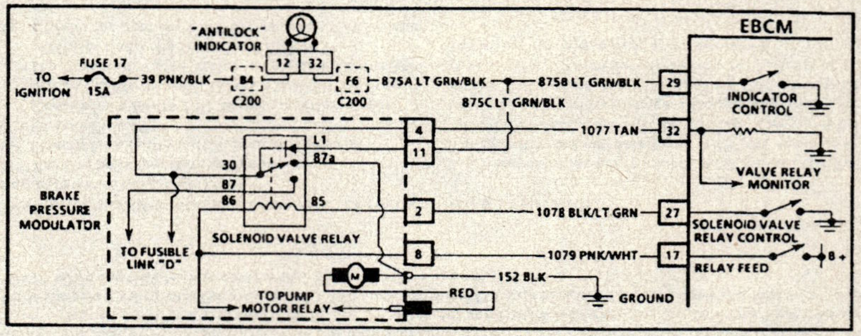

CHART A

"ANTILOCK" INDICATOR "ON" WITH NO CODES SET OR WITH TECH 1 UNABLE TO RECEIVE DATA

Circuit Description:

The "Antilock" Indicator is located on the right side of the instrument panel. Battery voltage is applied with ignition in "Run, Bulb Test or Start" through Fuse 17 and CKT 39. The indicator is grounded through CKT 875 and is controlled by the EBCM at Pin "29" and the Modulator at Pin "11" through the Solenoid Valve Relay. The EBCM will supply ground to the indicator for approximately 2 seconds when ignition is first turned "ON." During the first 2 seconds, if no faults are detected, the EBCM will energize the Solenoid Valve Relay. With the relay energized, the ground CKT 875 through the relay will be opened. This allows the indicator to go "OFF" when the EBCM indicator control line at Pin "29" is opened.

Test Description: Number(s) below refer to circled number(s) on the diagnostic chart.

1. Checks for proper battery and/or charging system operation.
2. Checks for an open fuse in the EBCM power feed circuit
3. Checks for an open in the EBCM power feed circuit
4. Checks for an open in the EBCM ground circuit.
5. Checks for a possible fault in the Solenoid Valve Relay.
6. Determines whether the fault is in the Solenoid Valve Relay or the EBCM.
7. Checks for a short to ground in the EBCM or the EBCM power feed circuit.
8. Determines whether the short to ground is in the EBCM or the EBCM power feed circuit.
9. Determines whether the short to ground is in the Brake Pressure Modulator or the indicator control circuit.

DIAGNOSTIC CHART A — 1993 CAPRICE, CUSTOM CRUISER AND ROADMASTER, CONTINUED

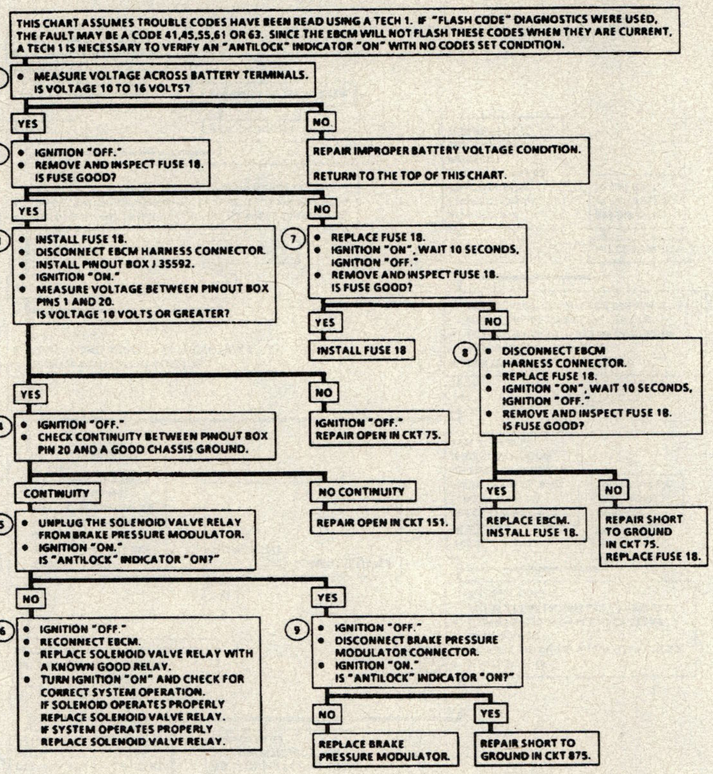

WHEN ALL DIAGNOSIS AND REPAIRS ARE COMPLETED, VERIFY PROPER OPERATION

DIAGNOSTIC CHART B — 1993 CAPRICE, CUSTOM CRUISER AND ROADMASTER

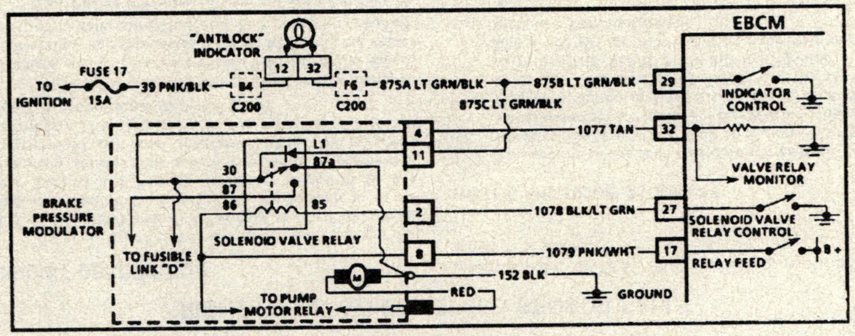

CHART B

"ANTILOCK" INDICATOR INOPERATIVE OR INDICATOR FLASHES VERY BRIEFLY AT IGNITION "ON" (LESS THAN ONE HALF OF A SECOND)

Circuit Description:

The "Antilock" Indicator is located on the right side of the instrument panel. Battery voltage is applied with ignition in "Run, Bulb Test or Start" through Fuse 17 and CKT 39. The indicator is grounded through CKT 875 and is controlled by the EBCM at Pin "29" and the Hydraulic Modulator at Pin "11" through the Solenoid Valve Relay. The EBCM will supply ground to the indicator for approximately 2 seconds when ignition is first turned "ON." During the first 2 seconds, if no faults are detected, the EBCM will energize the Solenoid Valve Relay. With the relay energized, the ground CKT 875 through the relay will be opened. This allows the indicator to go "OFF" when the EBCM indicator control line at Pin "29" is opened.

Test Description: Number(s) below refer to circled number(s) on the diagnostic chart.

1. Verifies an improperly operating indicator.
2. Determines whether the fault is an inoperative indicator or an improperly operating indicator.
3. Checks for an inoperative Instrument Cluster.
4. Determines whether the fault is in the Instrument Cluster portion of the circuitry or in the ABS portion.
5. Checks for a bad bulb.
6. Determines whether the fault is in the wiring, connections, or in the Instrument Cluster printed circuits. For LTZ only the Instrument Cluster will have different terminals for the "Antilock" indicator.
7. Determines whether the fault is in the wiring, connections, or Solenoid Valve Relay.
8. Determines whether the fault is in the wiring or connections.

Diagnostic Aids:

If indicator failure in "Intermittent," install the Tech 1 diagnostic tool and select any "ABS test mode so the "Antilock" indicator will be turned "ON." While observing the indicator, go through the indicators power and ground circuits wiggling the wiring and connectors. Watch for flickering or failure of the indicator to pin point roblem area.

DIAGNOSTIC CHART B — 1993 CAPRICE, CUSTOM CRUISER AND ROADMASTER, CONTINUED

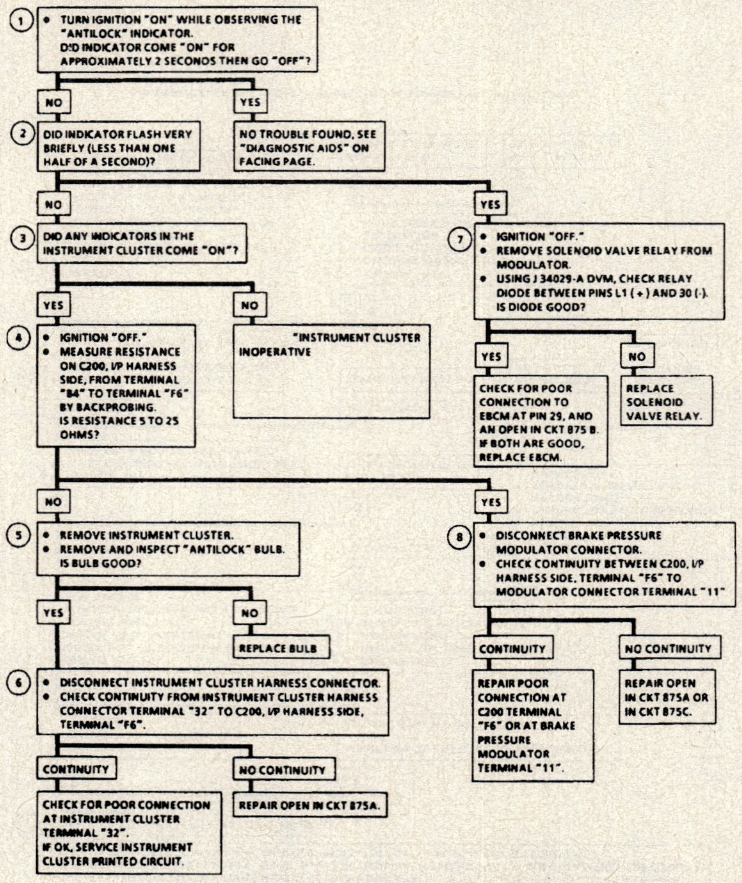

WHEN ALL DIAGNOSIS AND REPAIRS ARE COMPLETED, VERIFY PROPER OPERATION.

DIAGNOSTIC CHART DTC 21 — 1993 ELDORADO, RIVIERA AND SEVILLE

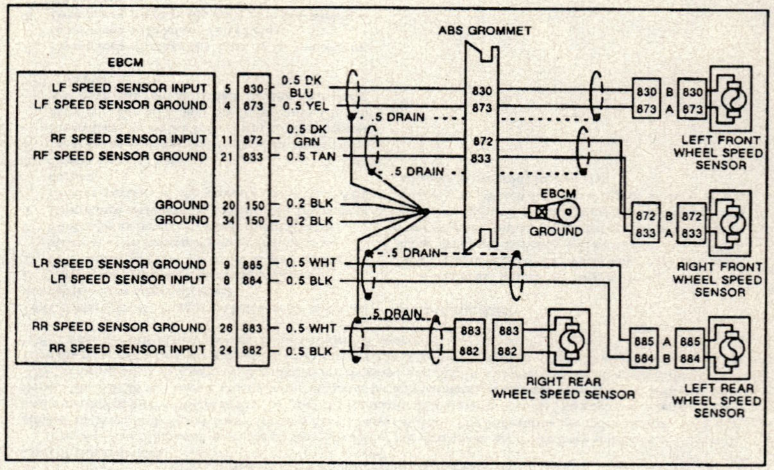

ABS CODE 21

RIGHT FRONT WHEEL SPEED SENSOR FAULT

CIRCUIT DESCRIPTION

As each wheel turns, the wheel speed sensor for that wheel creates a small AC voltage as a toothed sensor ring, mounted on the front suspension knuckle or in the rear integral hub-bearing assembly, is passed by the stationary sensor. The sensor consists of a permanent magnet and coil and generates voltage by magnetic induction. The frequency of the wheel speed sensor signal is proportional to wheel speed.

The EBCM uses the wheel speed sensor signal to calculate vehicle reference speeds and individual speed, acceleration and slip values for each wheel. These values are used to determine when anti-lock control is required.

The EBCM performs two basic types of checks on the wheel speed sensors, these are sensor continuity and sensor output.

WHEEL SPEED SENSOR CABLES

In order to prevent electromagnetic interference from disturbing the wheel speed sensor signal, the sensor cables are protected with grounded shielding (indicated by dashed lines in the above schematic). This shield surrounds the two individual sensor wires. A black conduit surrounds the wires and shield. When servicing the sensor cables, the sensor wires are accessible by separating the shielding. If the shielding is disturbed, repair as required.

NOTES ON CODE 21 CHART

1. This section of the fault tree determines if the code stored is due to a current fault condition. 'Current' wheel speed sensor faults will be detected when the vehicle is driven and reaches a speed of 4 mph (6 kph). If the ABS light turns on, the fault is current. The TECH 1 will indicate ignition cycle since code was set. If the ignition cycle counter is at zero, the code is current.
2. This section of the fault tree checks for an open or short in the wheel speed sensor wiring or sensor coil.
3. This section of the fault tree checks for conditions which may cause intermittent code setting. Some causes of intermittent wheel speed sensor codes to set are: improper routing or sensor cables next to spark plug wires, cables not retained in brackets, loose fitting or improperly mounted sensors, damaged sensors or toothed rings, poor terminal engagement of connectors, or intermittent shorts or opens in wiring.

DIAGNOSTIC CHART DTC 21 — 1993 ELDORADO, RIVIERA AND SEVILLE, CONTINUED

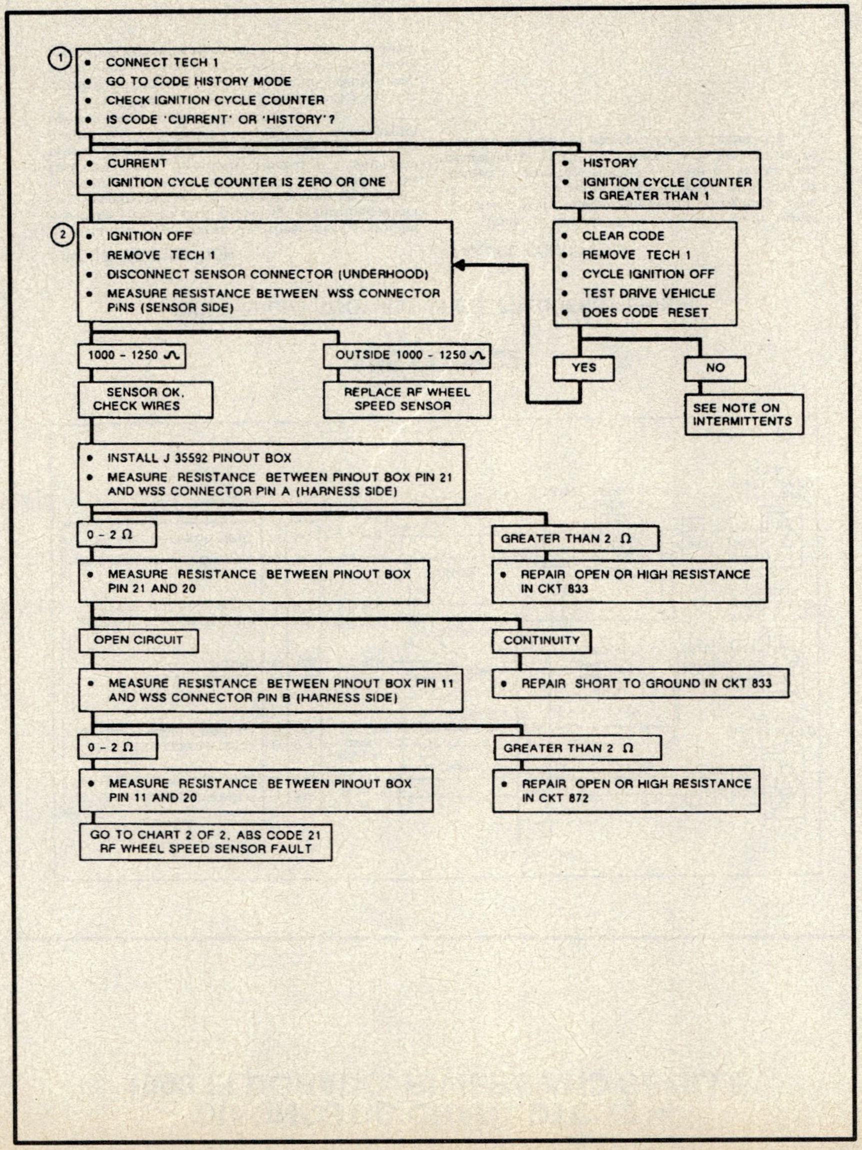

DIAGNOSTIC CHART DTC 21 — 1993 ELDORADO, RIVIERA AND SEVILLE, CONTINUED

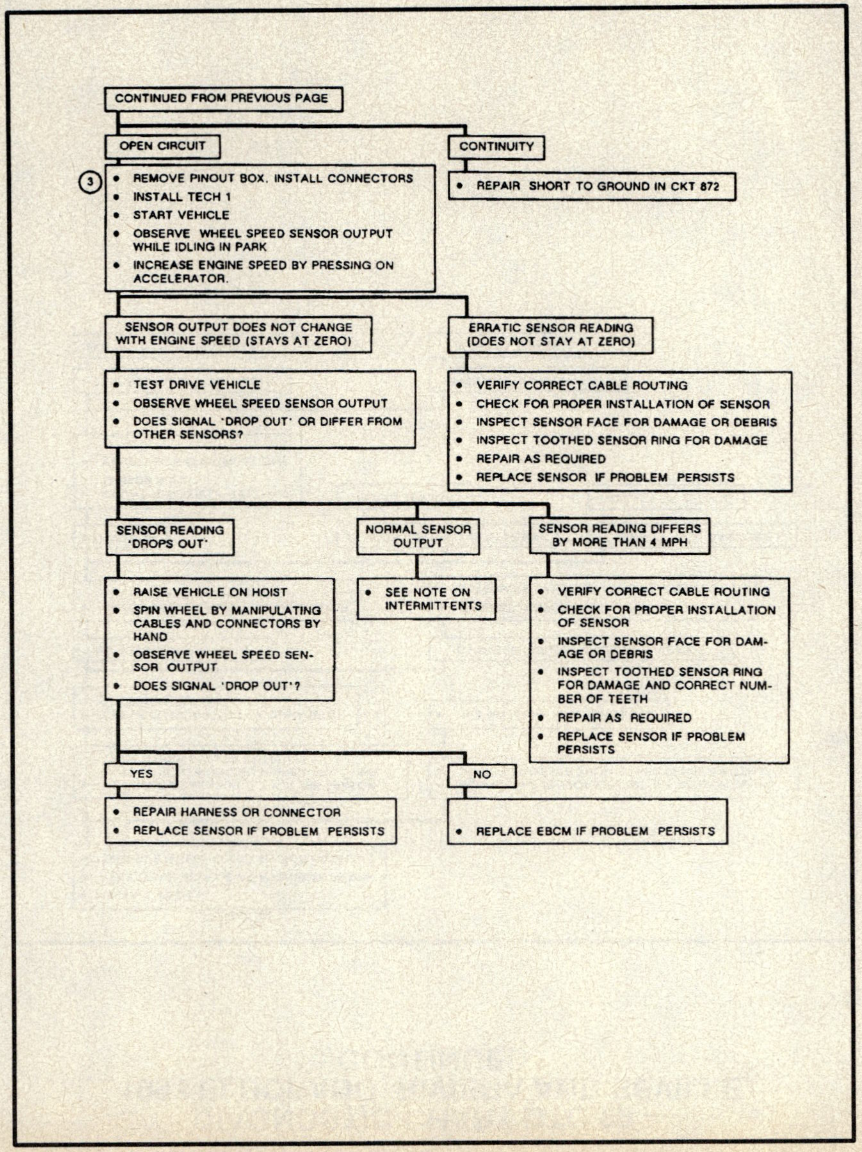

DIAGNOSTIC CHART DTC 22 — 1993 ELDORADO, RIVIERA AND SEVILLE

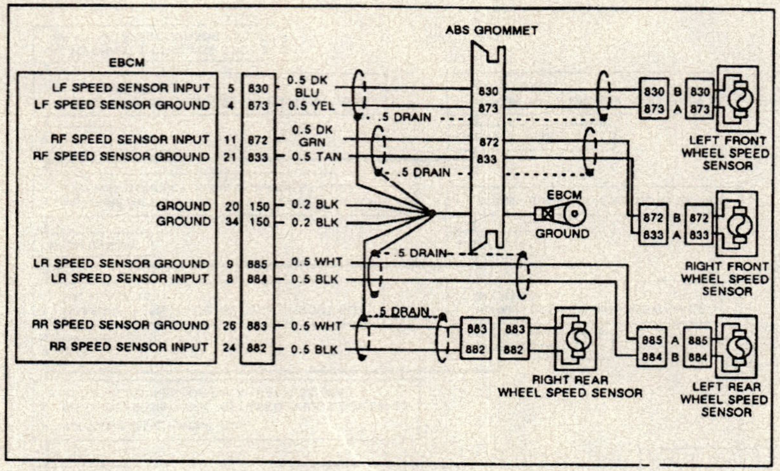

ABS CODE 22

RIGHT FRONT TOOTHED WHEEL FREQUENCY ERROR

CIRCUIT DESCRIPTION

As each wheel turns, the wheel speed sensor for that wheel creates a small AC voltage as a toothed sensor ring, mounted on the front suspension knuckle or in the rear integral hub-bearing assembly, is passed by the stationary sensor. The sensor consists of a permanent magnet and coil and generates voltage by magnetic induction. The frequency of the wheel speed sensor signal is proportional to wheel speed.

The EBCM uses the wheel speed sensor signal to calculate vehicle reference speeds and individual speed, acceleration and slip values for each wheel. These values are used to determine when anti-lock control is required.

FAILURE CONDITIONS

Code 22 will set from improper speed signals generated by the toothed ring sensor lasting longer than 20 seconds. Causes of improper speed signal may be caused by incorrect number of teeth on sensor ring, sensor rings which are covered with dirt, grease or metallic particles, or damaged toothed sensor ring.

DIAGNOSTIC CHART DTC 22 — 1993 ELDORADO, RIVIERA AND SEVILLE, CONTINUED

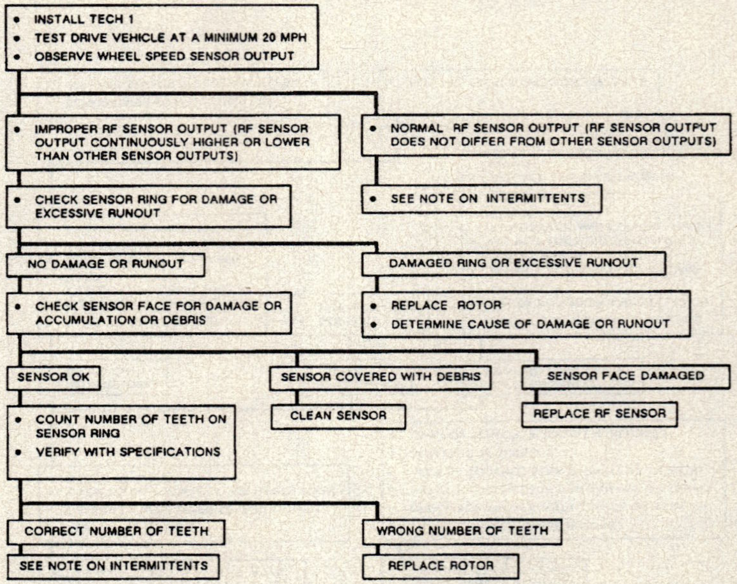

DIAGNOSTIC CHART DTC 25 — 1993 ELDORADO, RIVIERA AND SEVILLE

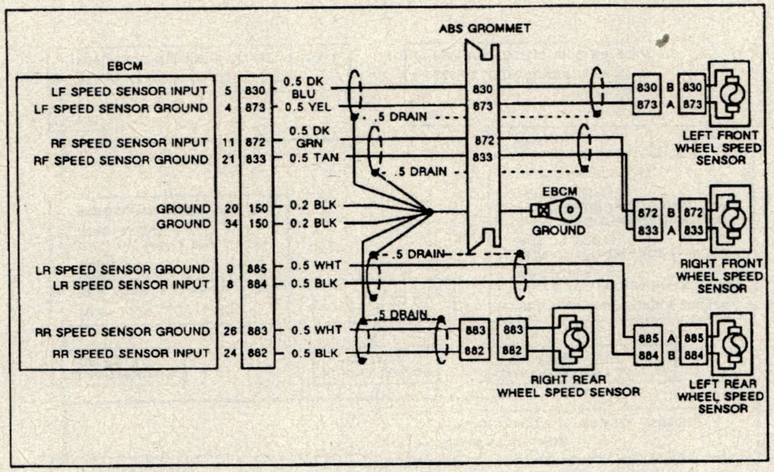

ABS CODE 25

LEFT FRONT WHEEL SPEED FAULT

CIRCUIT DESCRIPTION

As each wheel turns, the wheel speed sensor for that wheel creates a small AC voltage as a toothed sensor ring, mounted on the front suspension knuckle or in the rear integral hub-bearing assembly, is passed by the stationary sensor. The sensor consists of a permanent magnet and coil and generates voltage by magnetic induction. The frequency of the wheel speed sensor signal is proportional to wheel speed.

The EBCM uses the wheel speed sensor signal to calculate vehicle reference speeds and individual speed, acceleration and slip values for each wheel. These values are used to determine when anti-lock control is required.

The EBCM performs two basic types of checks on the wheel speed sensors, these are sensor continuity and sensor output.

WHEEL SPEED SENSOR CABLES

In order to prevent electromagnetic interference from disturbing the wheel speed sensor signal, the sensor cables are protected with grounded shielding (indicated by dashed lines in the above schematic). This shield surrounds the two individual sensor wires. A black conduit surrounds the wires and shield. When servicing the sensor cables, the sensor wires are accessible by separating the shielding. If the shielding is disturbed, repair as required.

NOTES ON CODE 25 CHART

1. This section of the fault tree determines if the code stored is due to a current fault condition. 'Current' wheel speed sensor faults will be detected when the vehicle is driven and reaches a speed of 4 mph (6 kph). If the ABS light turns on, the fault is current. The TECH 1 will indicate ignition cycle since code was set. If the ignition cycle counter is at zero, the code is current.
2. This section of the fault tree checks for an open or short in the wheel speed sensor wiring or sensor coil.
3. This section of the fault tree checks for conditions which may cause intermittent code setting. Some causes of intermittent wheel speed sensor codes to set are: improper routing or sensor cables next to spark plug wires, cables not retained in brackets, loose fitting or improperly mounted sensors, damaged sensors or toothed rings, poor terminal engagement of connectors, or intermittent shorts or opens in wiring.

DIAGNOSTIC CHART DTC 25 — 1993 ELDORADO, RIVIERA AND SEVILLE, CONTINUED

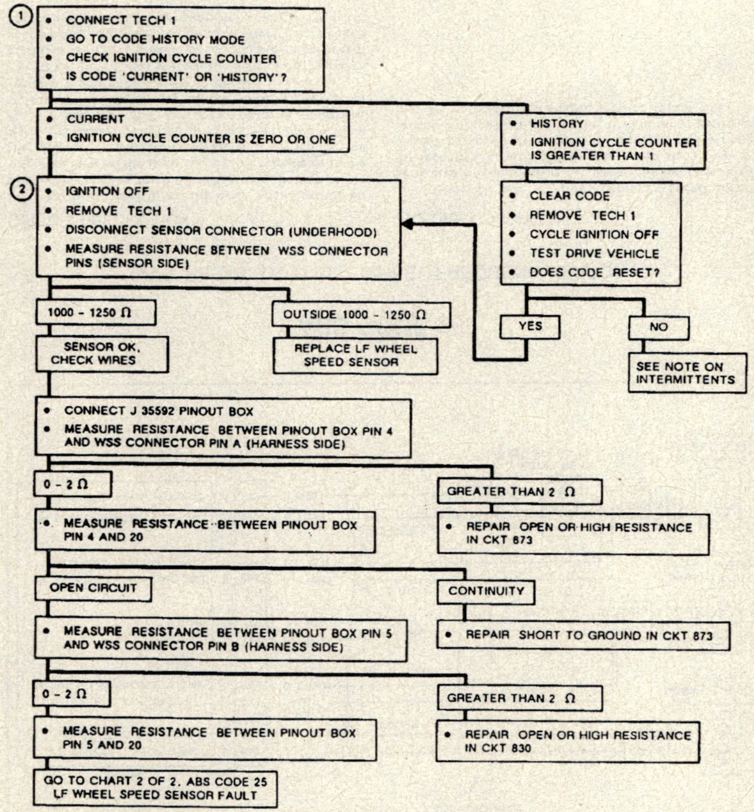

DIAGNOSTIC CHART DTC 25 — 1993 ELDORADO, RIVIERA AND SEVILLE, CONTINUED

CONTINUED FROM PREVIEW PAGE

OPEN CIRCUIT

(3)
- REMOVE PINOUT BOX, INSTALL CONNECTORS
- INSTALL TECH 1
- START VEHICLE
- OBSERVE WHEEL SPEED SENSOR OUTPUT WHILE IDLING IN PARK
- INCREASE ENGINE SPEED BY PRESSING ON ACCELERATOR.

CONTINUITY

- REPAIR SHORT TO GROUND IN CKT 830

SENSOR OUTPUT DOES NOT CHANGE WITH ENGINE SPEED (STAYS AT ZERO)

- TEST DRIVE VEHICLE
- OBSERVE WHEEL SPEED SENSOR OUTPUT
- DOES SIGNAL 'DROP OUT' OR DIFFER FROM OTHER SENSORS?

ERRATIC SENSOR READING (DOES NOT STAY AT ZERO)

- VERIFY CORRECT CABLE ROUTING
- CHECK FOR PROPER INSTALLATION OF SENSOR
- INSPECT SENSOR FACE FOR DAMAGE OR DEBRIS
- INSPECT TOOTHED SENSOR RING FOR DAMAGE
- REPAIR AS REQUIRED
- REPLACE SENSOR IF PROBLEM PERSISTS

SENSOR READING 'DROPS OUT'

- RAISE VEHICLE ON HOIST
- SPIN WHEEL BY MANIPULATING CABLES AND CONNECTORS BY HAND
- OBSERVE WHEEL SPEED SENSOR OUTPUT
- DOES SIGNAL 'DROP OUT'?

YES

- REPAIR HARNESS OR CONNECTOR
- REPLACE SENSOR IF PROBLEM PERSISTS

NO

- SEE NOTE ON INTERMITTENTS
- REPLACE EBCM IF PROBLEM PERSISTS

NORMAL SENSOR OUTPUT

- SEE NOTE ON INTERMITTENTS

SENSOR READING DIFFERS BY MORE THAN 4 MPH

- VERIFY CORRECT CABLE ROUTING
- CHECK FOR PROPER INSTALLATION OF SENSOR
- INSPECT SENSOR FACE FOR DAMAGE OR DEBRIS
- INSPECT TOOTHED SENSOR RING FOR DAMAGE AND CORRECT NUMBER OF TEETH
- REPAIR AS REQUIRED
- REPLACE SENSOR IF PROBLEM PERSISTS

DIAGNOSTIC CHART DTC 26 — 1993 ELDORADO, RIVIERA AND SEVILLE

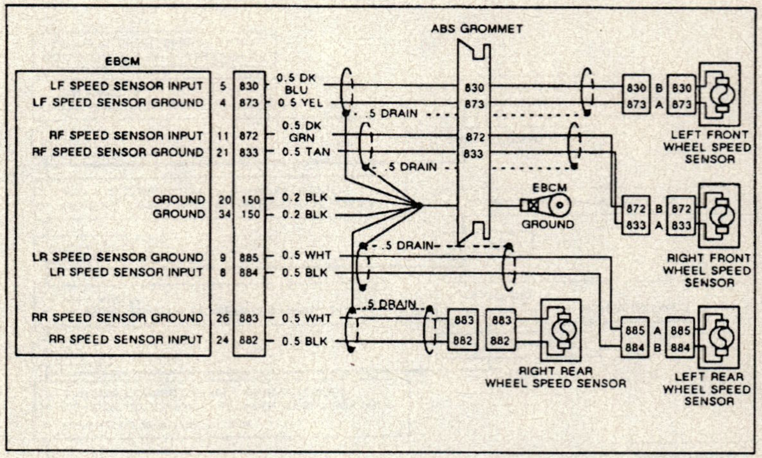

ABS CODE 26

LEFT FRONT TOOTHED WHEEL FREQUENCY ERROR

CIRCUIT DESCRIPTION

As each wheel turns, the wheel speed sensor for that wheel creates a small AC voltage as a toothed sensor ring, mounted on the front suspension knuckle or in the rear integral hub-bearing assembly, is passed by the stationary sensor. The sensor consists of a permanent magnet and coil and generates voltage by magnetic induction. The frequency of the wheel speed sensor signal is proportional to wheel speed.

The EBCM uses the wheel speed sensor signal to calculate vehicle reference speeds and individual speed, acceleration and slip values for each wheel. These values are used to determine when anti-lock control is required.

FAILURE CONDITIONS

Code 26 will set from improper speed signals generated by the toothed ring sensor lasting longer than 20 seconds. Causes of improper speed signal may be caused by incorrect number of teeth on sensor ring, sensor rings which are covered with dirt, grease or metallic particles, or damaged toothed sensor ring.

DIAGNOSTIC CHART DTC 26 — 1993 ELDORADO, RIVIERA AND SEVILLE, CONTINUED

- INSTALL TECH 1
- TEST DRIVE VEHICLE
- OBSERVE WHEEL SPEED SENSOR OUTPUT

Left branch:

- IMPROPER LF SENSOR OUTPUT (LF SENSOR OUTPUT CONTINUOUSLY HIGHER OR LOWER THAN OTHER SENSOR OUTPUTS)
- CHECK SENSOR RING FOR DAMAGE OR EXCESSIVE RUNOUT

Right branch:

- NORMAL LF SENSOR OUTPUT (LF SENSOR OUTPUT DOES NOT DIFFER FROM OTHER SENSOR OUTPUTS)
- SEE NOTE ON INTERMITTENTS

From the sensor ring check:

NO DAMAGE OR RUNOUT

- CHECK SENSOR FACE FOR DAMAGE OR ACCUMULATION OR DEBRIS

DAMAGED RING OR EXCESSIVE RUNOUT

- REPLACE ROTOR
- DETERMINE CAUSE OF DAMAGE OR RUNOUT

From the sensor face check:

SENSOR OK

- COUNT NUMBER OF TEETH ON SENSOR RING
- VERIFY WITH SPECIFICATIONS

SENSOR COVERED WITH DEBRIS

CLEAN SENSOR

SENSOR FACE DAMAGED

REPLACE LF SENSOR

From the teeth count:

CORRECT NUMBER OF TEETH

SEE NOTE ON INTERMITTENTS

WRONG NUMBER OF TEETH

REPLACE ROTOR

DIAGNOSTIC CHART DTC 31 — 1993 ELDORADO, RIVIERA AND SEVILLE

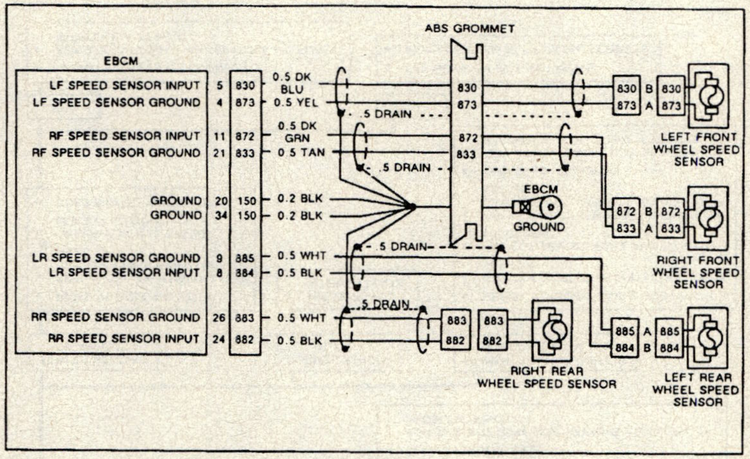

ABS CODE 31

RIGHT REAR WHEEL SPEED FAULT

CIRCUIT DESCRIPTION

As each wheel turns, the wheel speed sensor for that wheel creates a small AC voltage as a toothed sensor ring, mounted on the front suspension knuckle or in the rear integral hub-bearing assembly, is passed by the stationary sensor. The sensor consists of a permanent magnet and coil and generates voltage by magnetic induction. The frequency of the wheel speed sensor signal is proportional to wheel speed.

The EBCM uses the wheel speed sensor signal to calculate vehicle reference speeds and individual speed, acceleration and slip values for each wheel. These values are used to determine when anti-lock control is required.

The EBCM performs two basic types of checks on the wheel speed sensors, these are sensor continuity and sensor output.

WHEEL SPEED SENSOR CABLES

In order to prevent electromagnetic interference from disturbing the wheel speed sensor signal, the sensor cables are protected with grounded shielding (indicated by dashed lines in the above schematic). This shield surrounds the two individual sensor wires. A black conduit surrounds the wires and shield. When servicing the sensor cables, the sensor wires are accessible by separating the shielding. If the shielding is disturbed, repair as required.

NOTES ON CODE 31 CHART

1. This section of the fault tree determines if the code stored is due to a current fault condition. 'Current' wheel speed sensor faults will be detected when the vehicle is driven and reaches a speed of 4 mph (6 kph). If the ABS light turns on, the fault is current. The TECH 1 will indicate ignition cycle since code was set. If the ignition cycle counter is at zero, the code is current.
2. This section of the fault tree checks for an open or short in the wheel speed sensor wiring or sensor coil.
3. This section of the fault tree checks for conditions which may cause intermittent code setting. Some causes of intermittent wheel speed sensor codes to set are: improper routing or sensor cables next to spark plug wires, cables not retained in brackets, loose fitting or improperly mounted sensors, damaged sensors or toothed rings, poor terminal engagement of connectors, or intermittent shorts or opens in wiring.

DIAGNOSTIC CHART DTC 31 — 1993 ELDORADO, RIVIERA AND SEVILLE, CONTINUED

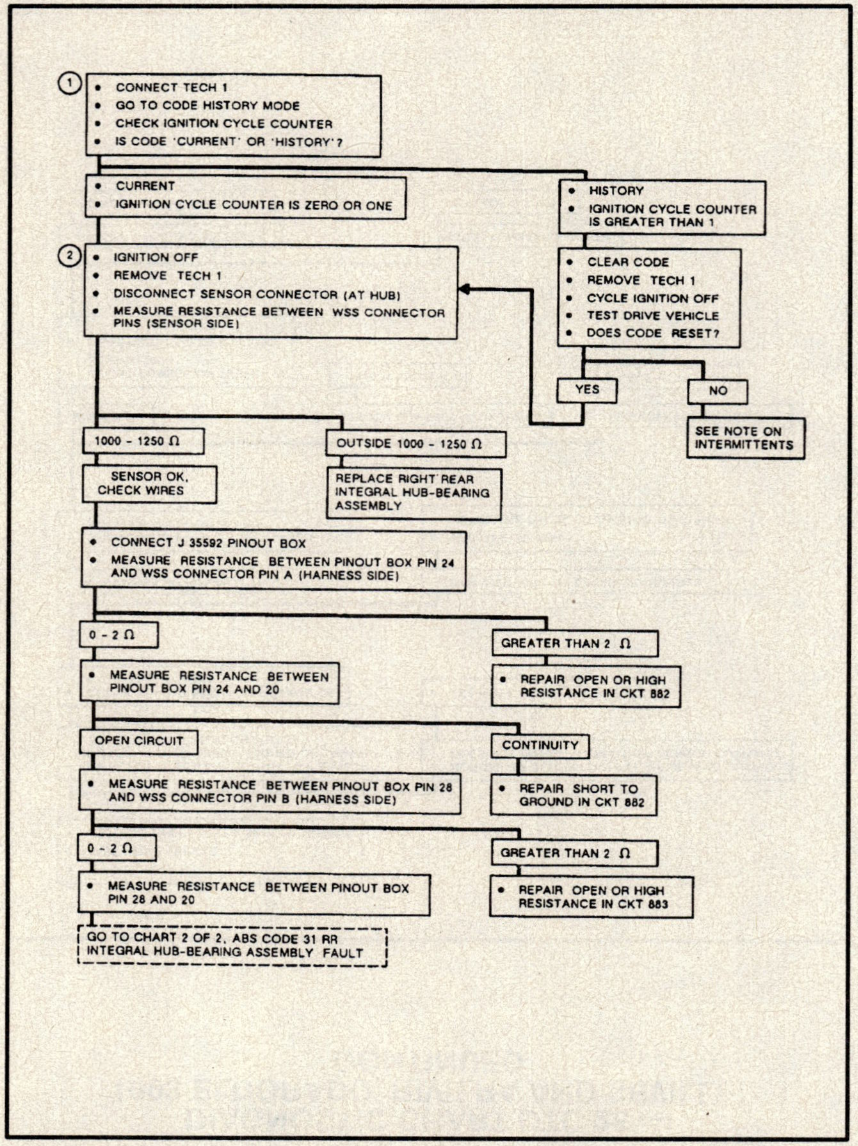

DIAGNOSTIC CHART DTC 31 — 1993 ELDORADO, RIVIERA AND SEVILLE, CONTINUED

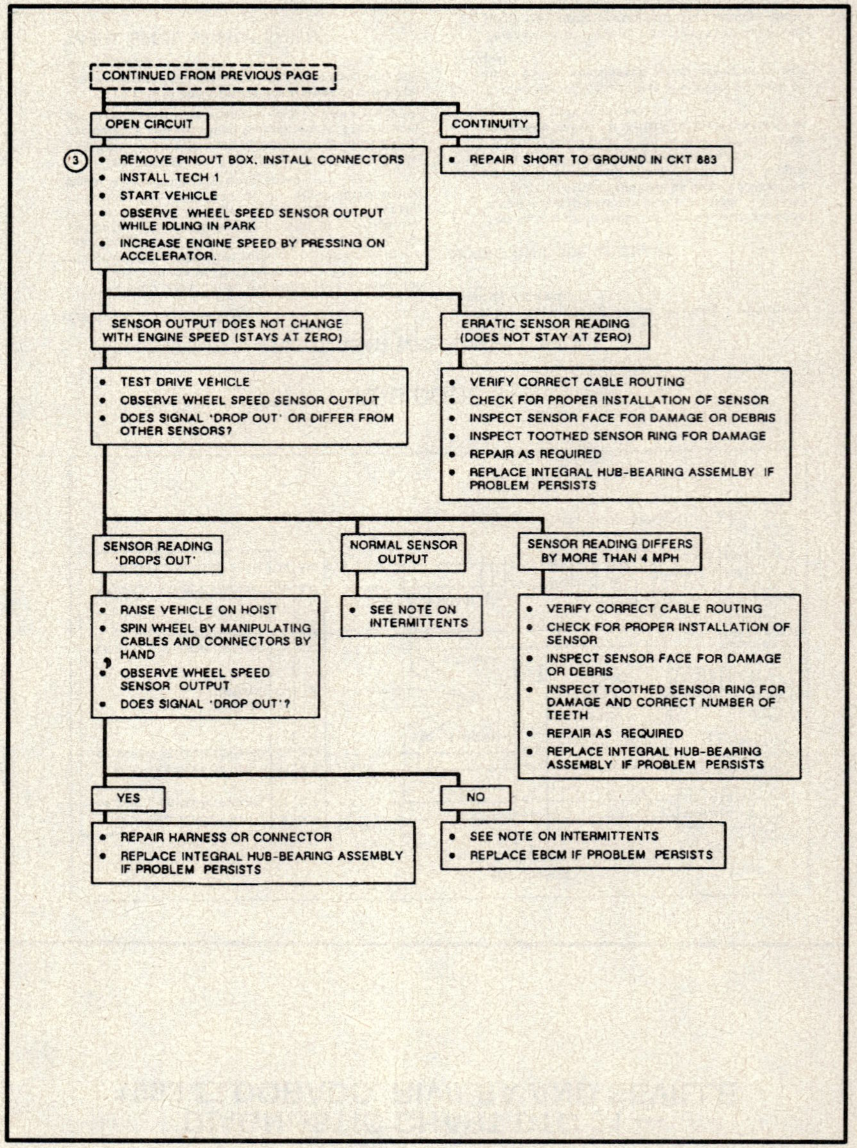

DIAGNOSTIC CHART DTC 32 — 1993 ELDORADO, RIVIERA AND SEVILLE

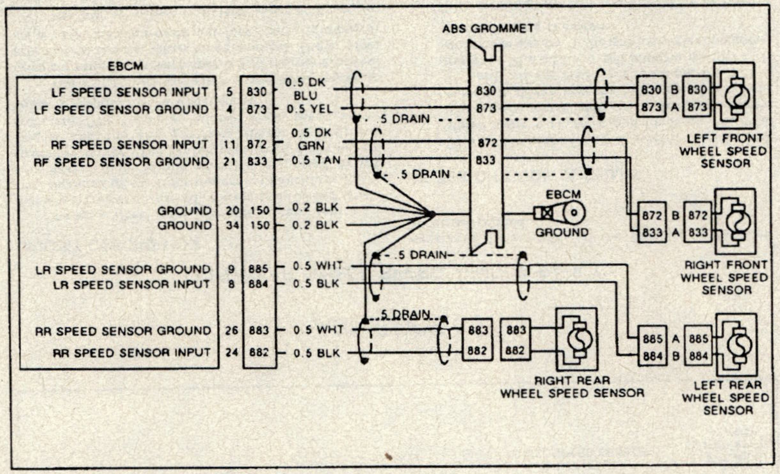

ABS CODE 32

RIGHT REAR TOOTHED WHEEL FREQUENCY ERROR

CIRCUIT DESCRIPTION

As each wheel turns, the wheel speed sensor for that wheel creates a small AC voltage as a toothed sensor ring, mounted on the front suspension knuckle or in the rear integral hub-bearing assembly, is passed by the stationary sensor. The sensor consists of a permanent magnet and coil and generates voltage by magnetic induction. The frequency of the wheel speed sensor signal is proportional to wheel speed.

The EBCM uses the wheel speed sensor signal to calculate vehicle reference speeds and individual speed, acceleration and slip values for each wheel. These values are used to determine when anti-lock control is required.

FAILURE CONDITIONS

Code 32 will set from improper speed signals generated by the toothed ring sensor lasting longer than 20 seconds. Causes of improper speed signal may be caused by incorrect number of teeth on sensor ring, sensor rings which are covered with dirt, grease or metallic particles, or damaged toothed sensor ring.

DIAGNOSTIC CHART DTC 32 — 1993 ELDORADO, RIVIERA AND SEVILLE, CONTINUED

- INSTALL TECH 1
- TEST DRIVE VEHICLE
- OBSERVE WHEEL SPEED SENSOR OUTPUT

- IMPROPER RR SENSOR OUTPUT (RR SENSOR OUTPUT CONTINUOUSLY HIGHER OR LOWER THAN OTHER SENSOR OUTPUTS)
  - CHECK SENSOR RING FOR DAMAGE OR EXCESSIVE RUNOUT
    - NO DAMAGE OR RUNOUT
      - CHECK SENSOR FACE FOR DAMAGE OR ACCUMULATION OR DEBRIS
        - SENSOR OK
          - COUNT NUMBER OF TEETH ON SENSOR RING
          - VERIFY WITH SPECIFICATIONS
            - CORRECT NUMBER OF TEETH
              - SEE NOTE ON INTERMITTENTS
            - WRONG NUMBER OF TEETH
              - REPLACE RIGHT REAR INTEGRAL HUB-BEARING ASSEMBLY
        - SENSOR COVERED WITH DEBRIS
          - CLEAN SENSOR
        - SENSOR FACE DAMAGED
          - REPLACE RIGHT REAR INTEGRAL HUB-BEARING ASSEMBLY
    - DAMAGED RING OR EXCESSIVE RUNOUT
      - REPLACE RIGHT REAR INTEGRAL HUB-BEARING ASSEMBLY
      - DETERMINE CAUSE OF DAMAGE OR RUNOUT
- NORMAL RR SENSOR OUTPUT (RR SENSOR OUTPUT DOES NOT DIFFER FROM OTHER SENSOR OUTPUTS)
  - SEE NOTE ON INTERMITTENTS

DIAGNOSTIC CHART DTC 35 — 1993 ELDORADO, RIVIERA AND SEVILLE

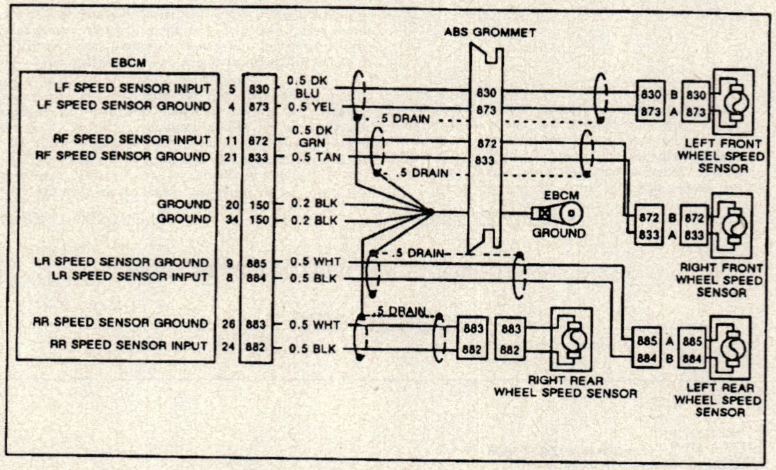

ABS CODE 35

LEFT REAR WHEEL SPEED SENSOR FAULT

CIRCUIT DESCRIPTION

As each wheel turns, the wheel speed sensor for that wheel creates a small AC voltage as a toothed sensor ring, mounted on the front suspension knuckle or in the rear integral hub-bearing assembly, is passed by the stationary sensor. The sensor consists of a permanent magnet and coil and generates voltage by magnetic induction. The frequency of the wheel speed sensor signal is proportional to wheel speed.

The EBCM uses the wheel speed sensor signal to calculate vehicle reference speeds and individual speed, acceleration and slip values for each wheel. These values are used to determine when anti-lock control is required.

The EBCM performs two basic types of checks on the wheel speed sensors, these are sensor continuity and sensor output.

WHEEL SPEED SENSOR CABLES

In order to prevent electromagnetic interference from disturbing the wheel speed sensor signal, the sensor cables are protected with grounded shielding (indicated by dashed lines in the above schematic). This shield surrounds the two individual sensor wires. A black conduit surrounds the wires and shield. When servicing the sensor cables, the sensor wires are accessible by separating the shielding. If the shielding is disturbed, repair as required.

NOTES ON CODE 35 CHART

1. This section of the fault tree determines if the code stored is due to a current fault condition. 'Current' wheel speed sensor faults will be detected when the vehicle is driven and reaches a speed of 4 mph (6 kph). If the ABS light turns on, the fault is current. The TECH 1 will indicate ignition cycle since code was set. If the ignition cycle counter is at zero, the code is current.
2. This section of the fault tree checks for an open or short in the wheel speed sensor wiring or sensor coil.
3. This section of the fault tree checks for conditions which may cause intermittent code setting. Some causes of intermittent wheel speed sensor codes to set are: improper routing or sensor cables next to spark plug wires, cables not retained in brackets, loose fitting or improperly mounted sensors, damaged sensors or toothed rings, poor terminal engagement of connectors, or intermittent shorts or opens in wiring.

DIAGNOSTIC CHART DTC 35 — 1993 ELDORADO, RIVIERA AND SEVILLE, CONTINUED

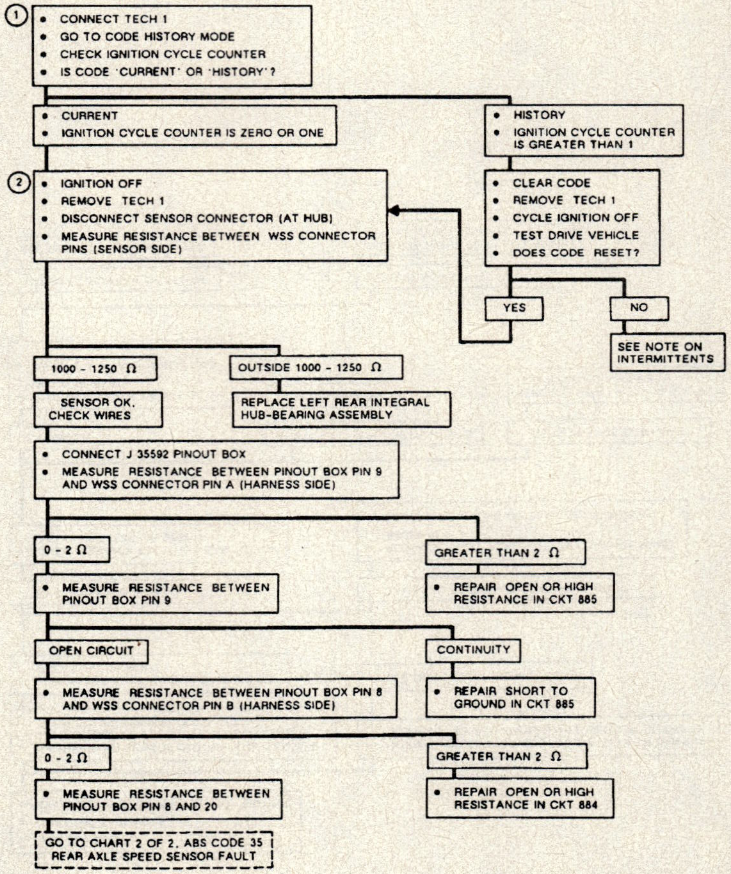

DIAGNOSTIC CHART DTC 35 — 1993 ELDORADO, RIVIERA AND SEVILLE, CONTINUED

CONTINUED FROM PREVIOUS PAGE

OPEN CIRCUIT

(3)
- REMOVE PINOUT BOX, INSTALL CONNECTORS
- INSTALL TECH 1
- START VEHICLE
- OBSERVE WHEEL SPEED SENSOR OUTPUT WHILE IDLING IN PARK
- INCREASE ENGINE SPEED BY PRESSING ON ACCELERATOR.

CONTINUITY

- REPAIR SHORT TO GROUND IN CKT 884

SENSOR OUTPUT DOES NOT CHANGE WITH ENGINE SPEED (STAYS AT ZERO)

- TEST DRIVE VEHICLE
- OBSERVE WHEEL SPEED SENSOR OUTPUT
- DOES SIGNAL 'DROP OUT' OR DIFFER FROM OTHER SENSORS?

ERRATIC SENSOR READING (DOES NOT STAY AT ZERO)

- VERIFY CORRECT CABLE ROUTING
- CHECK FOR PROPER INSTALLATION OF SENSOR
- INSPECT SENSOR FACE FOR DAMAGE OR DEBRIS
- INSPECT TOOTHED SENSOR RING FOR DAMAGE
- REPAIR AS REQUIRED
- REPLACE INTEGRAL HUB-BEARING ASSEMBLY IF PROBLEM PERSISTS

SENSOR READING 'DROPS OUT'

- RAISE VEHICLE ON HOIST
- SPIN WHEEL BY MANIPULATING CABLES AND CONNECTORS BY HAND
- OBSERVE WHEEL SPEED SENSOR OUTPUT
- DOES SIGNAL 'DROP OUT'?

YES

- REPAIR HARNESS OR CONNECTOR
- REPLACE INTEGRAL HUB-BEARING ASSEMBLY IF PROBLEM PERSISTS

NO

- SEE NOTE ON INTERMITTENTS
- REPLACE EBCM IF PROBLEM PERSISTS

NORMAL SENSOR OUTPUT

- SEE NOTE ON INTERMITTENTS

SENSOR READING DIFFERS BY MORE THAN 4 MPH

- VERIFY CORRECT CABLE ROUTING
- CHECK FOR PROPER INSTALLATION OF SENSOR
- INSPECT SENSOR FACE FOR DAMAGE OR DEBRIS
- INSPECT TOOTHED SENSOR RING FOR DAMAGE AND CORRECT NUMBER OF TEETH
- REPAIR AS REQUIRED
- REPLACE INTEGRAL HUB-BEARING ASSEMBLY IF PROBLEM PERSISTS

DIAGNOSTIC CHART DTC 36 — 1993 ELDORADO, RIVIERA AND SEVILLE

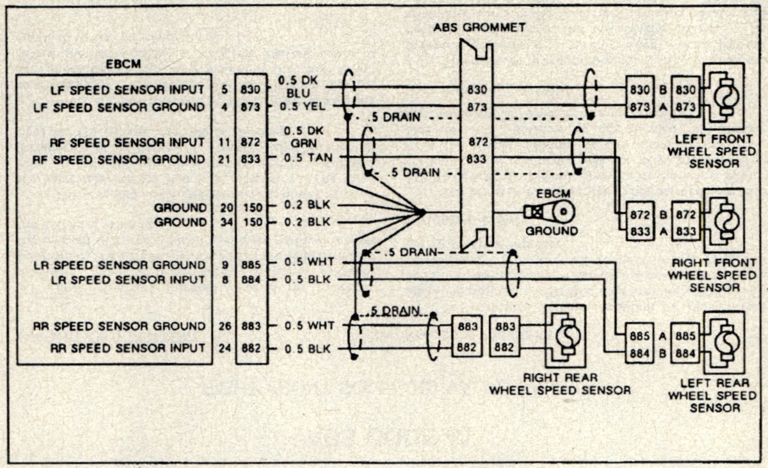

ABS CODE 36

LEFT REAR TOOTHED WHEEL FREQUENCY ERROR

CIRCUIT DESCRIPTION

As each wheel turns, the wheel speed sensor for that wheel creates a small AC voltage as a toothed sensor ring, mounted on the front suspension knuckle or in the rear integral hub-bearing assembly, is passed by the stationary sensor. The sensor consists of a permanent magnet and coil and generates voltage by magnetic induction. The frequency of the wheel speed sensor signal is proportional to wheel speed.

The EBCM uses the wheel speed sensor signal to calculate vehicle reference speeds and individual speed, acceleration and slip values for each wheel. These values are used to determine when anti-lock control is required.

FAILURE CONDITIONS

Code 36 will set from improper speed signals generated by the toothed ring sensor. Causes of improper speed signal may be caused by incorrect number of teeth on sensor ring, sensor rings which are covered with dirt, grease or metallic particles, or damaged toothed sensor ring.

DIAGNOSTIC CHART DTC 36 — 1993 ELDORADO, RIVIERA AND SEVILLE, CONTINUED

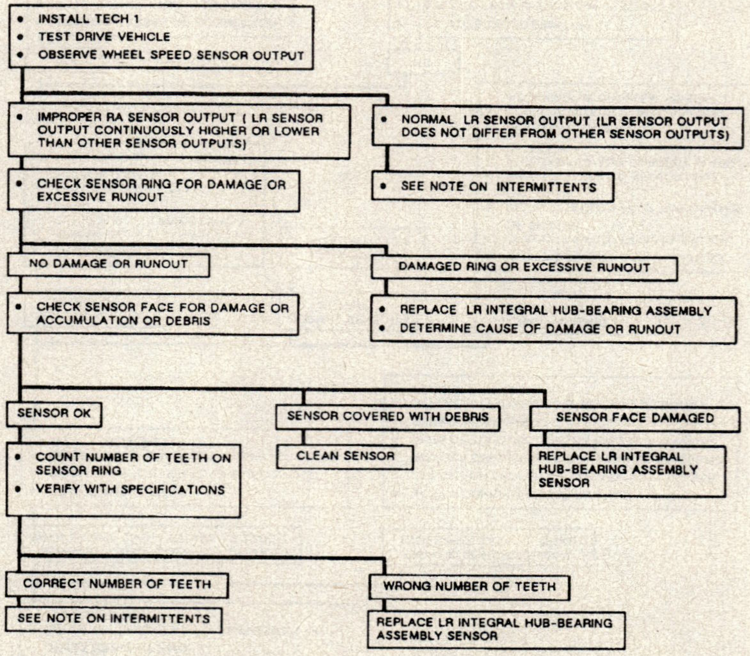

DIAGNOSTIC CHART DTC 41 — 1993 ELDORADO, RIVIERA AND SEVILLE

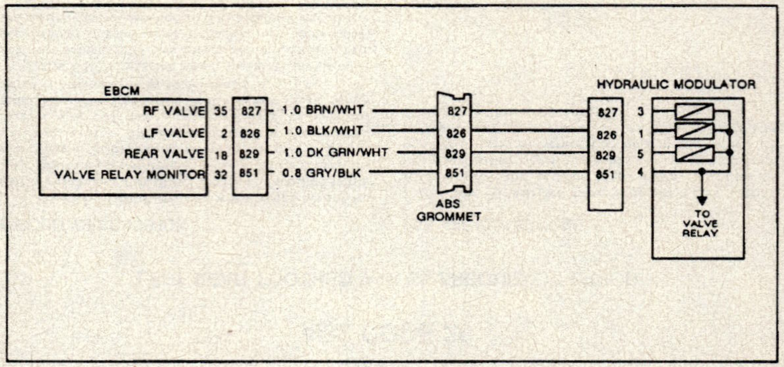

ABS CODE 41

RIGHT FRONT SOLENOID VALVE FAULT

CIRCUIT DESCRIPTION

The right front solenoid valve in the hydraulic modulator may be commanded to three different positions by the EBCM. Valve position is determined by the amount of current which is allowed to flow through the solenoid valve coil.

The solenoid valve circuits receive power through the valve relay on the hydraulic modulator. The valve relay is engaged at key-on and remains engaged throughout the ignition cycle. Solenoid valve lines should have battery voltage available at all times.

When the key is turned to RUN, power is supplied to the EBCM. The EBCM will complete an internal self-check before providing ground to the valve relay. When the valve relay is energized, battery power is supplied to the solenoid valves.

VALVE OPERATION

The valves are continuously monitored by the EBCM (CKT 851) and each individual solenoid valve line. If valve line voltage does not match expected values, an ABS code will set. When the vehicle reaches approximately 4 mph after key-on, the EBCM cycles each valve and checks for a proper response. This check may be heard and felt while it is taking place and should be considered a normal mode of operation.

Pressure Increase

The pressure increase position is the normal operating position of the valves for braking. Brake fluid flows thru the modulator from the master cylinder to the wheel brake. No current is supplied to the valve coil during pressure increase.

Pressure Hold

The pressure hold position is commanded when the EBCM desires to hold the pressure in the wheel circuit constant. The modulator will not allow pressure flow from the master cylinder to the wheel brake. The pressure hold position is commanded by passing approximately 2.0 amps to the valve coil.

Pressure Reduction

The pressure reduction position is used to decrease hydraulic pressure in a specific wheel circuit without decreasing pressure in the rest of the system. The modulator takes brake fluid from the wheel brake and injects it into the master cylinder circuit. This fully energized position of the solenoid valve occurs when approximately 5 amps are passed to the valve coil.

FAILURE CONDITIONS

Code 41 will set when the expected position of the right front solenoid valve does not match the commanded position from the EBCM. Conditions which could cause Code 41 to set are damage to the right front solenoid, an open circuit or a short circuit to ground or battery on CKT 827.

DIAGNOSTIC CHART DTC 41 — 1993 ELDORADO, RIVIERA AND SEVILLE, CONTINUED

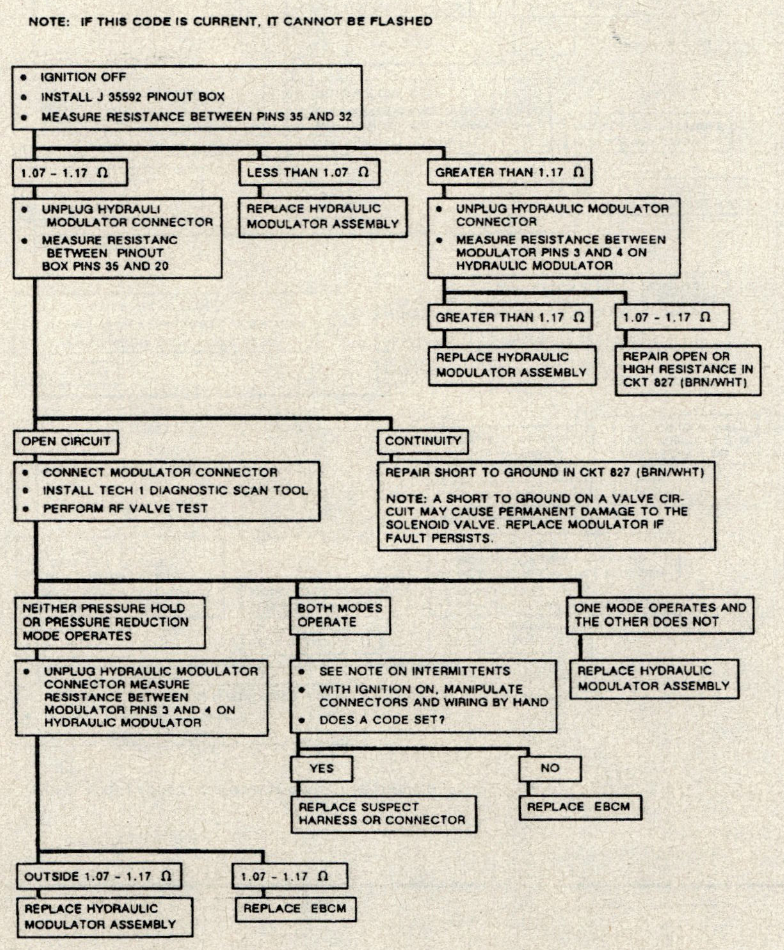

DIAGNOSTIC CHART DTC 45 — 1993 ELDORADO, RIVIERA AND SEVILLE

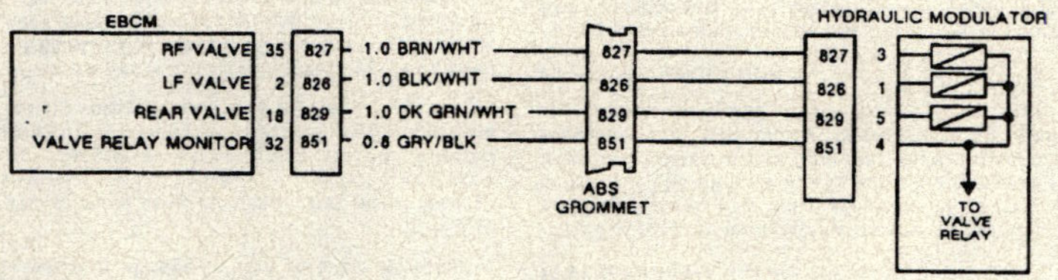

ABS CODE 45

LEFT FRONT SOLENOID VALVE FAULT

CIRCUIT DESCRIPTION

The left front solenoid valve in the hydraulic modulator may be commanded to three different positions by the EBCM. Valve position is determined by the amount of current which is allowed to flow through the solenoid valve coil.

The solenoid valve circuits receive power through the valve relay on the hydraulic modulator. The valve relay is engaged at key-on and remains engaged throughout the ignition cycle. Solenoid valve lines should have battery voltage available at all times.

When the key is turned to RUN, power is supplied to the EBCM. The EBCM will complete an internal self-check before providing ground to the valve relay. When the valve relay is energized, battery power is supplied to the solenoid valves.

VALVE OPERATION

The valves are continuously monitored by the EBCM (CKT 851) and each individual solenoid valve line. If valve line voltage does not match expected values, an ABS code will set. When the vehicle reaches approximately 4 mph after key-on, the EBCM cycles each valve and checks for a proper response. This check may be heard and felt while it is taking place and should be considered a normal mode of operation.

Pressure Increase

The pressure increase position is the normal operating position of the valves for braking. Brake fluid flows thru the modulator from the master cylinder to the wheel brake. No current is supplied to the valve coil during pressure increase.

Pressure Hold

The pressure hold position is commanded when the EBCM desires to hold the pressure in the wheel circuit constant. The modulator will not allow pressure flow from the master cylinder to the wheel brake. The pressure hold position is commanded by passing approximately 2.0 amps to the valve coil.

Pressure Reduction

The pressure reduction position is used to decrease hydraulic pressure in a specific wheel circuit without decreasing pressure in the rest of the system. The modulator takes brake fluid from the wheel brake and injects it into the master cylinder circuit. This fully energized position of the solenoid valve occurs when approximately 5 amps are passed to the valve coil.

FAILURE CONDITIONS

Code 45 will set when the expected position of the left front solenoid valve does not match the commanded position from the EBCM. Conditions which could cause Code 45 to set are damage to the left front solenoid, an open circuit or a short circuit to ground or battery on CKT 826.

DIAGNOSTIC CHART DTC 45 — 1993 ELDORADO, RIVIERA AND SEVILLE, CONTINUED

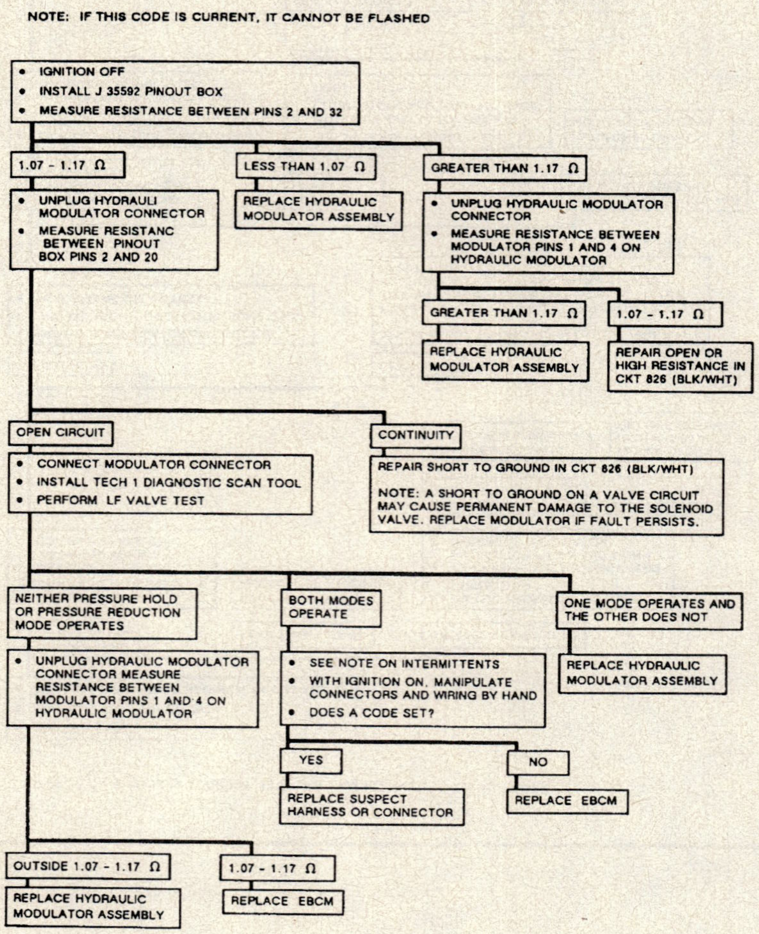

DIAGNOSTIC CHART DTC 55 — 1993 ELDORADO, RIVIERA AND SEVILLE

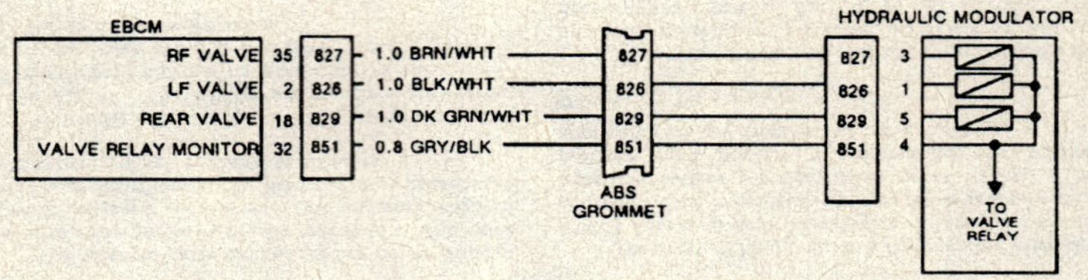

ABS CODE 55

REAR WHEEL SOLENOID VALVE FAULT

CIRCUIT DESCRIPTION

The rear axle solenoid valve in the hydraulic modulator may be commanded to three different positions by the EBCM. The rear axle solenoid valve modulates hydraulic brake pressure to both rear wheel circuits. Valve position is determined by the amount of current which is allowed to flow through the solenoid valve coil.

The solenoid valve circuits receive power through the valve relay on the hydraulic modulator. The valve relay is engaged at key-on and remains engaged throughout the ignition cycle. Solenoid valve lines should have battery voltage available at all times.

When the key is turned to RUN, power is supplied to the EBCM. The EBCM will complete an internal self-check before providing ground to the valve relay. When the valve relay is energized, battery power is supplied to the solenoid valves.

VALVE OPERATION

The valves are continuously monitored by the EBCM (CKT 851) and each individual solenoid valve line. If valve line voltage does not match expected values, an ABS code will set. When the vehicle reaches approximately 4 mph after key-on, the EBCM cycles each valve and checks for a proper response. This check may be heard and felt while it is taking place and should be considered a normal mode of operation.

Pressure Increase

The pressure increase position is the normal operating position of the valves for braking. Brake fluid flows thru the modulator from the master cylinder to the wheel brake. No current is supplied to the valve coil during pressure increase.

Pressure Hold

The pressure hold position is commanded when the EBCM desires to hold the pressure in the wheel circuit constant. The modulator will not allow pressure flow from the master cylinder to the wheel brake. The pressure hold position is commanded by passing approximately 2.0 amps to the valve coil.

Pressure Reduction

The pressure reduction position is used to decrease hydraulic pressure in a specific wheel circuit without decreasing pressure in the rest of the system. The modulator takes brake fluid from the wheel brake and injects it into the master cylinder circuit. This fully energized position of the solenoid valve occurs when approximately 5 amps are passed to the valve coil.

FAILURE CONDITIONS

Code 55 will set when the expected position of the rear axle solenoid valve does not match the commanded position from the EBCM. Conditions which could cause Code 55 to set are damage to the rear axle solenoid, an open circuit or a short circuit to ground or battery on CKT 829.

DIAGNOSTIC CHART DTC 55 — 1993 ELDORADO, RIVIERA AND SEVILLE, CONTINUED

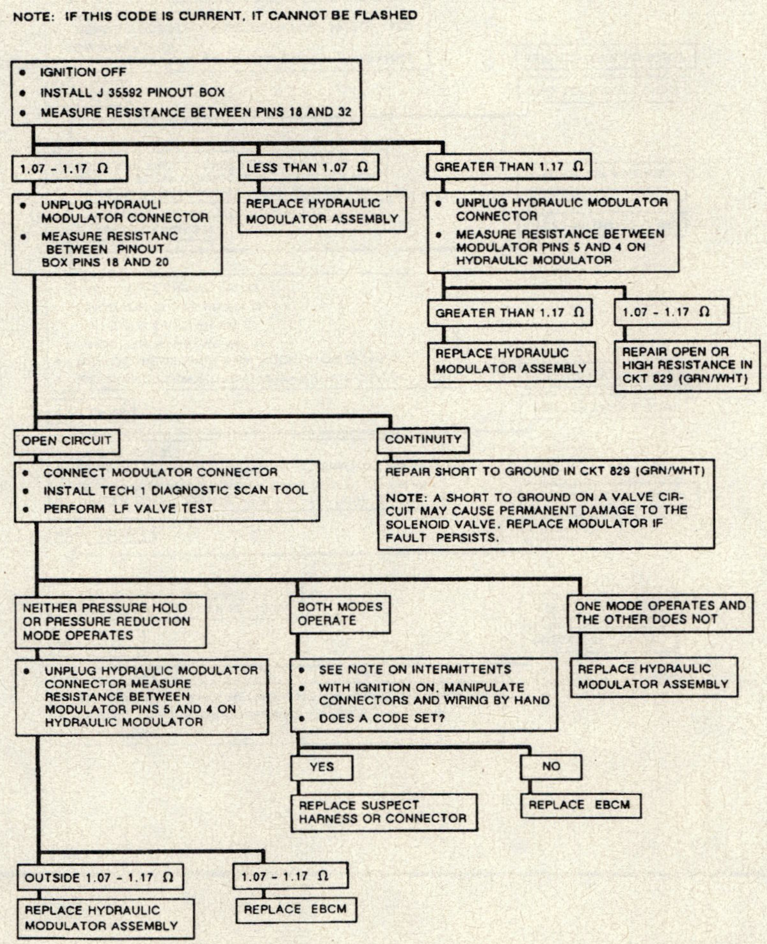

DIAGNOSTIC CHART DTC 61 — 1993 ELDORADO, RIVIERA AND SEVILLE

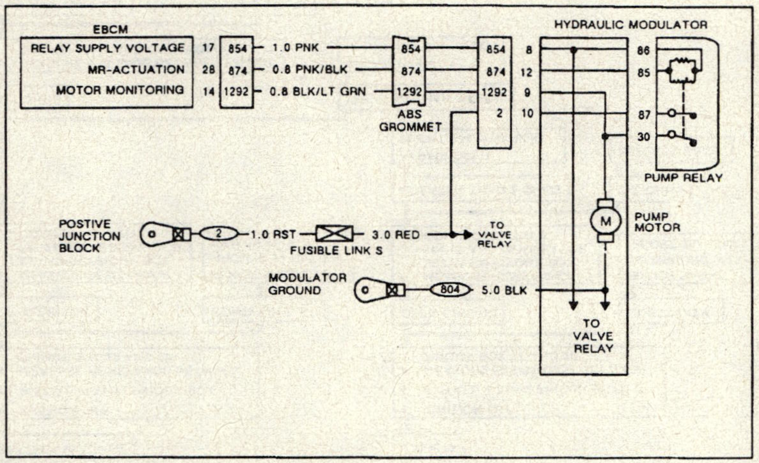

ABS CODE 61

PUMP MOTOR CIRCUIT FAULT

CIRCUIT DESCRIPTION

The pump motor returns brake fluid to the master cylinder brake circuit at the hydraulic modulator during anti-lock braking. During normal braking, the pump does not operate. When the vehicle begins to move after start-up, the EBCM will turn on the pump motor and perform a self-check of the pump motor and pump motor circuit. This check may be felt and heard by the driver when the vehicle begins to move. The self-check should be considered normal operation. The pump motor is an integral component of the hydraulic modulator and cannot be serviced separately.

The pump motor relay provides power to the pump motor in the hydraulic modulator. The pump motor relay is located on the hydraulic modulator an may be replaced if it is found to be defective.

The pump motor relay is not engaged during normal system operation. When anti-lock operation is required, the motor relay actuation line on CKT 874 (EBCM pin 28) is pulled to ground. The pump motor relay is engaged by the relay supply voltage line on CKT 854 (EBCM pin 17). When the relay switches, battery power is provided to the pump motor.

FAILURE CONDITIONS

When the pump motor relay is engaged, the motor monitor line should be at battery voltage. With the pump motor relay disengaged, the monitoring line should be at ground. If the commanded position of the pump motor relay and motor monitor line do not agree, Code 61 will set. Some conditions which will cause Code 61 to set are:

- An open circuit on CKTs 874 or 1292
- A short to voltage on CKTs 874 or 1292
- A short to ground on CKTs 874 or 1292
- A defective pump motor relay or pump motor

DIAGNOSTIC CHART DTC 61 — 1993 ELDORADO, RIVIERA AND SEVILLE, CONTINUED

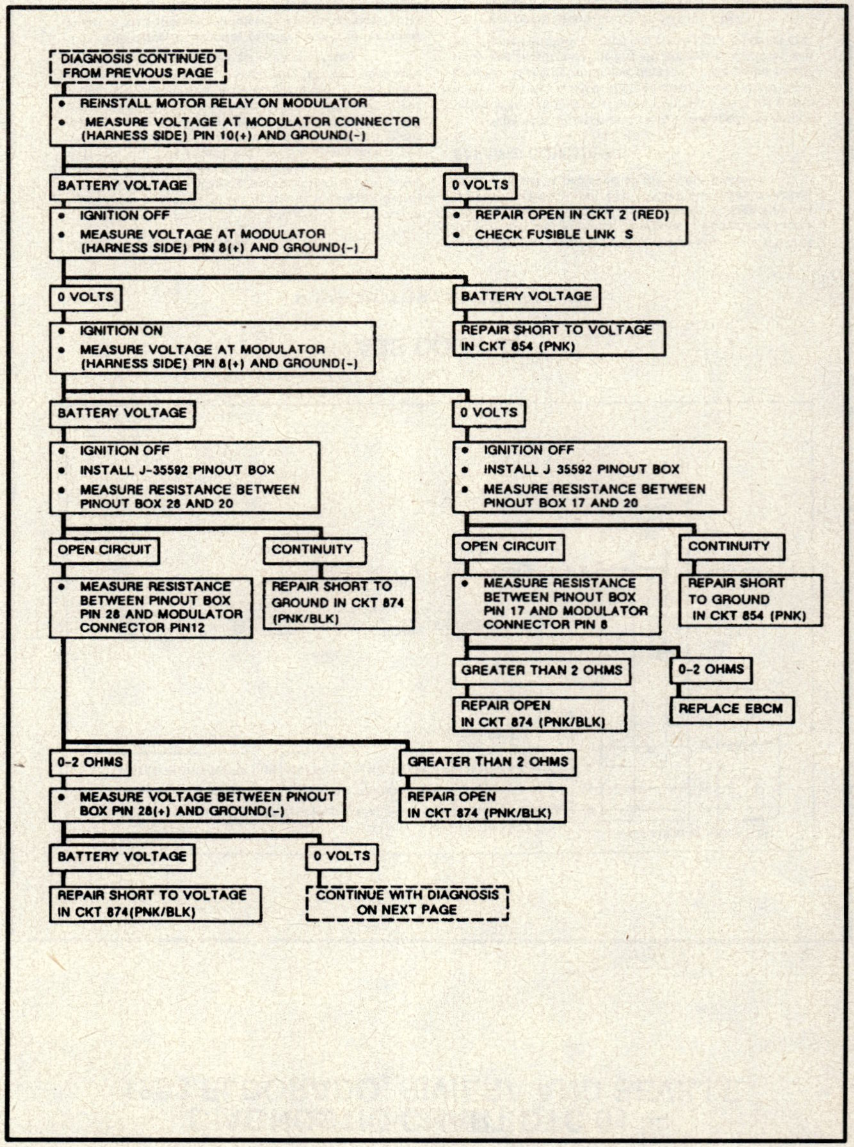

DIAGNOSTIC CHART DTC 61 — 1993 ELDORADO, RIVIERA AND SEVILLE, CONTINUED

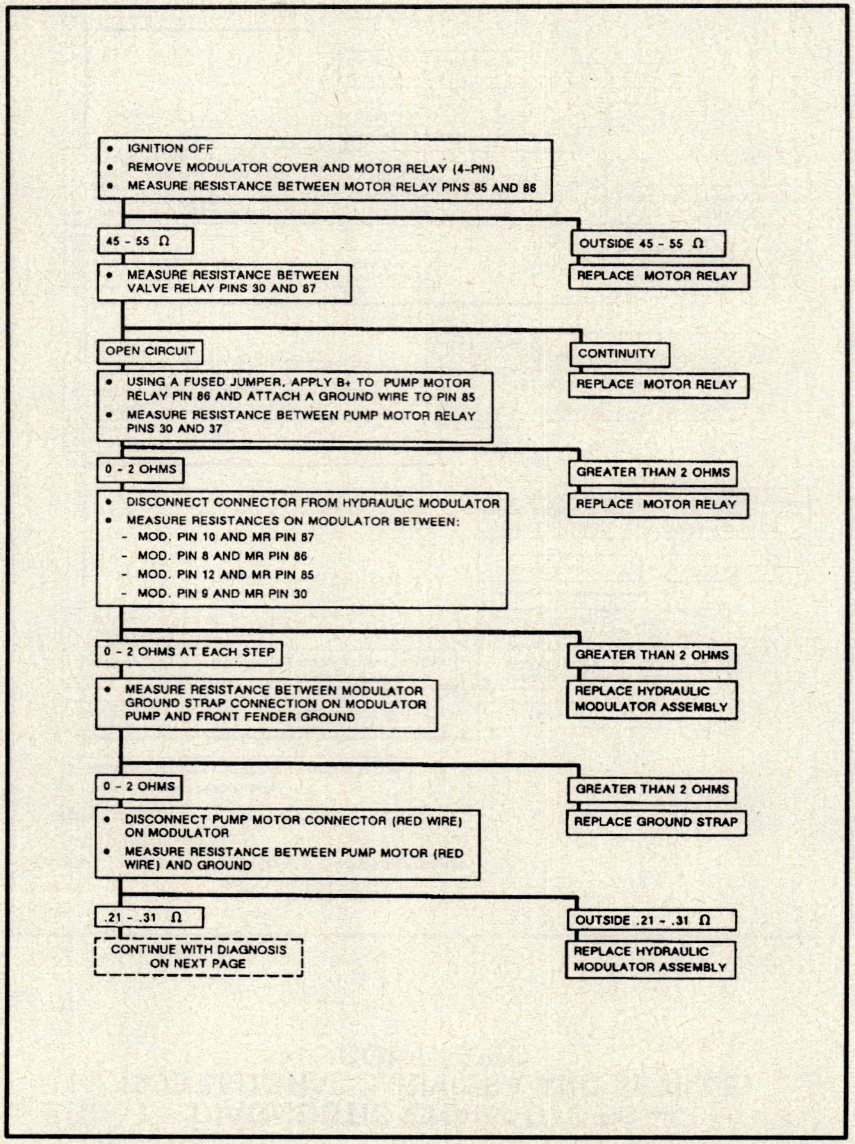

DIAGNOSTIC CHART DTC 61 — 1993 ELDORADO, RIVIERA AND SEVILLE, CONTINUED

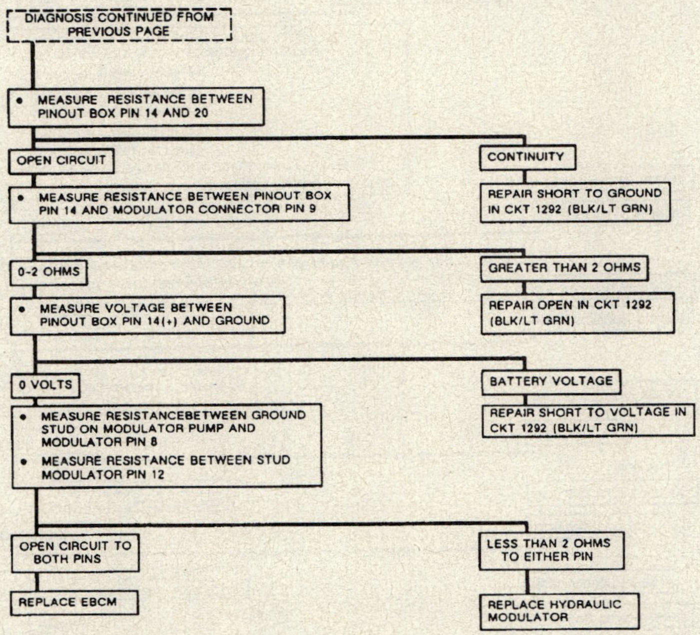

DIAGNOSTIC CHART DTC 63 — 1993 ELDORADO, RIVIERA AND SEVILLE

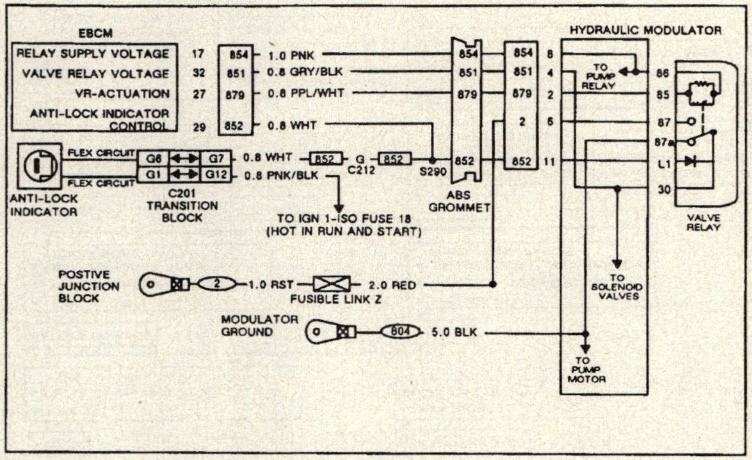

ABS CODE 63

SOLENOID VALVE RELAY FAULT

CIRCUIT DESCRIPTION

The solenoid valve relay provides power to the three solenoid valves in the hydraulic modulator. The solenoid valve relay is located on the hydraulic modulator and may be replaced if it is found to be defective.

The valve relay is engaged during normal system operation. When the ignition is switched to RUN, the EBCM commands the solenoid valve relay on by grounding the valve relay actuation line on CKT 879 (EBCM pin 27). When this ground is provided, the valve relay is energized from the voltage supply line on CKT 854 (EBCM pin 17). The valve relay switches and battery voltage is provided to the three solenoid valves and the solenoid valve relay feedback line on CKT 851 (EBCM pin 32). The valve relay remains engaged until the ignition is turned off or a failure is detected.

Whenever the solenoid valve relay is not engaged, the ANTI-LOCK light will turn on. With the ignition on and the solenoid valve relay not engaged, a path to ground exists from the IGN1-ISO fuse (10 amps) thru the ANTI-LOCK light on CKT 139 which turns on the warning light. The ANTI-LOCK light can also be commanded on by the EBCM on CKT 852 (EBCM pin 29). The EBCM will provide ground on this line and turn on the warning light when a failure is detected.

ANTI-LOCK OPERATION

With the solenoid valve relay engaged, battery power is provided to the three solenoid valves. If anti-lock operation is required, the EBCM varies the amount of current supplied to the solenoid valves (EBCM pins 2, 18, 35) and the valve is positioned to provide for optimum braking.

FAILURE CONDITIONS

When the valve relay is engaged, the valve relay feedback line should be at battery voltage. With the valve relay disengaged the feedback line is at ground. If the commanded position of the valve relay and the position indicated by the feedback line do not agree, Code 63 will set. Some conditions which will cause Code 63 to set are:

- An open circuit on CKTs 851 or 879
- A short to voltage on CKTs 851 or 879
- A short to ground on CKTs 851 or 879
- A defective solenoid valve relay
- Open circuit on CKT 854.

DIAGNOSTIC CHART DTC 63 — 1993 ELDORADO, RIVIERA AND SEVILLE, CONTINUED

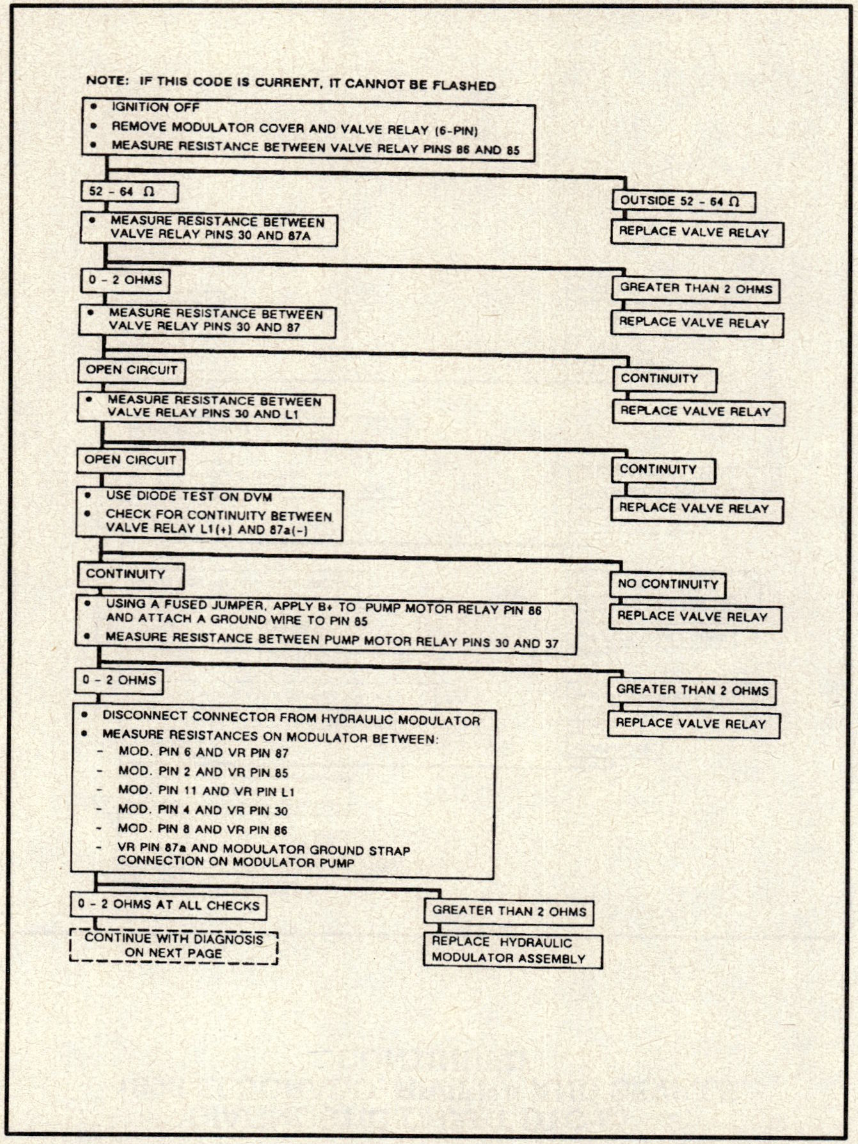

DIAGNOSTIC CHART DTC 63 — 1993 ELDORADO, RIVIERA AND SEVILLE, CONTINUED

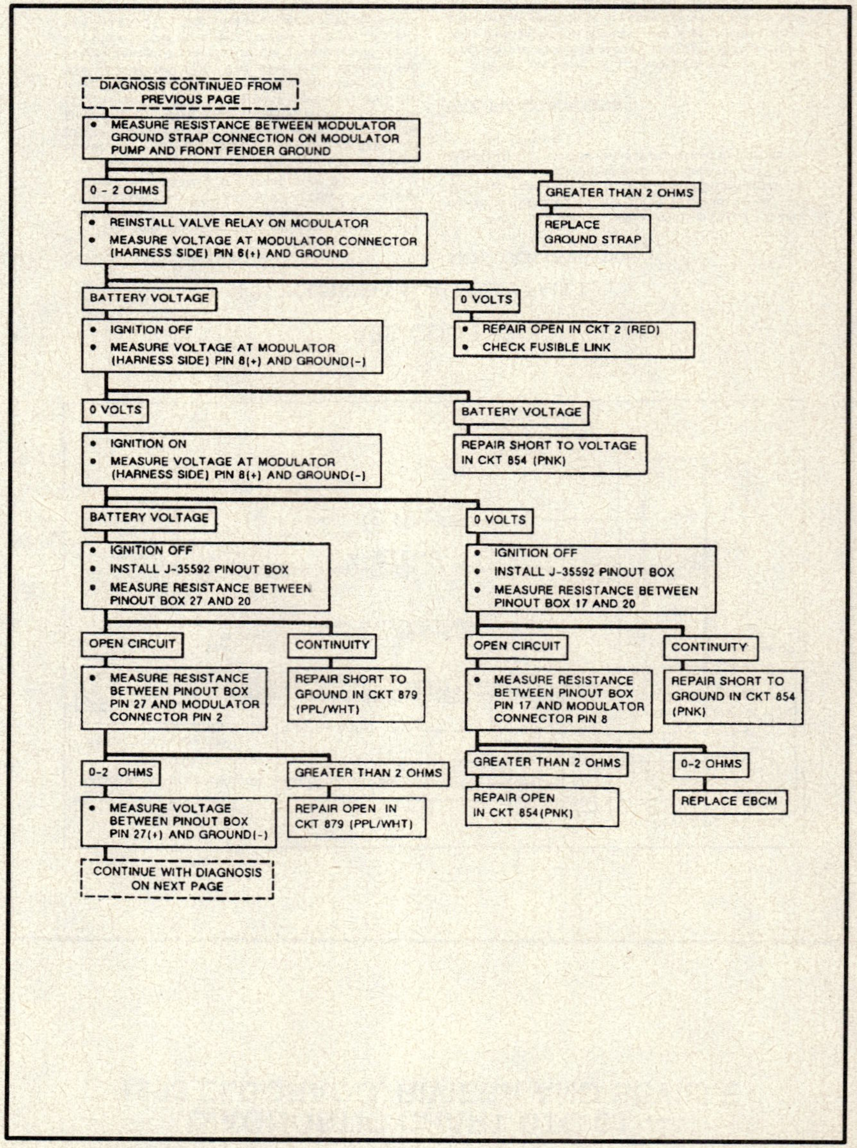

DIAGNOSTIC CHART DTC 63 — 1993 ELDORADO, RIVIERA AND SEVILLE, CONTINUED

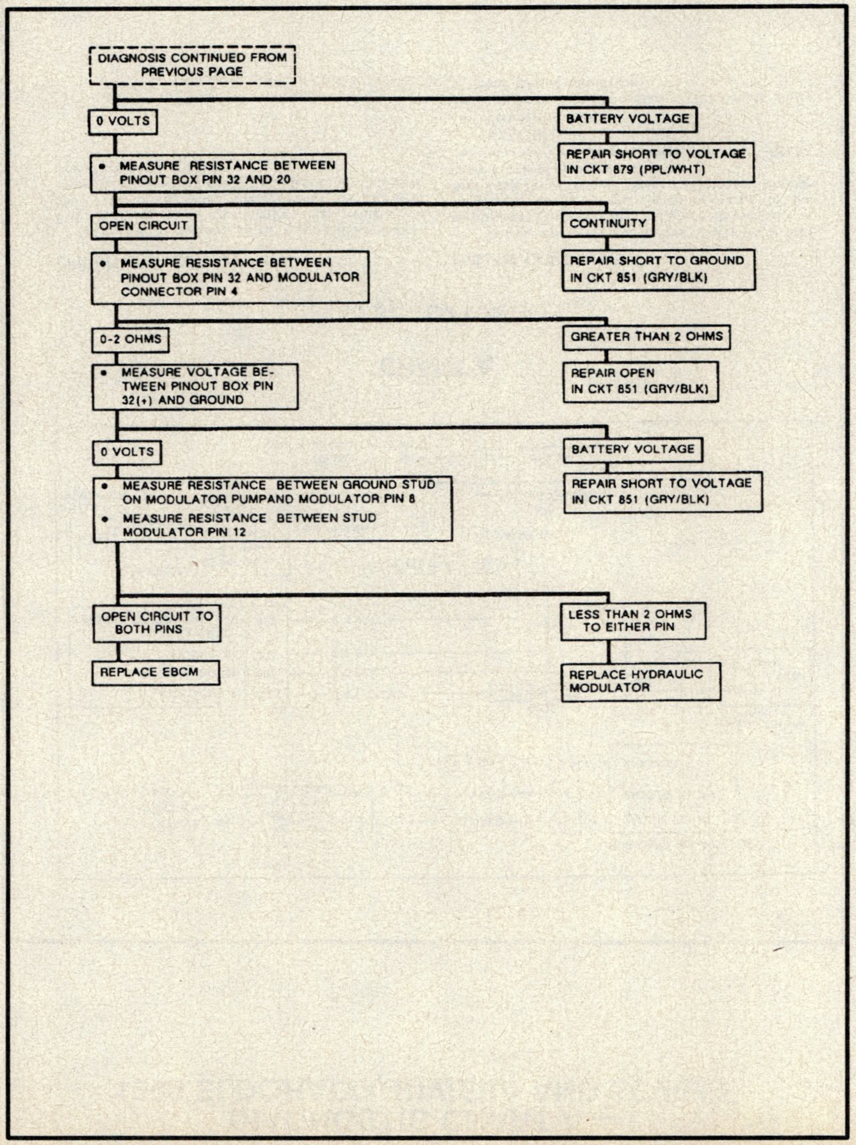

DIAGNOSTIC CHART DTC 71 — 1993 ELDORADO, RIVIERA AND SEVILLE

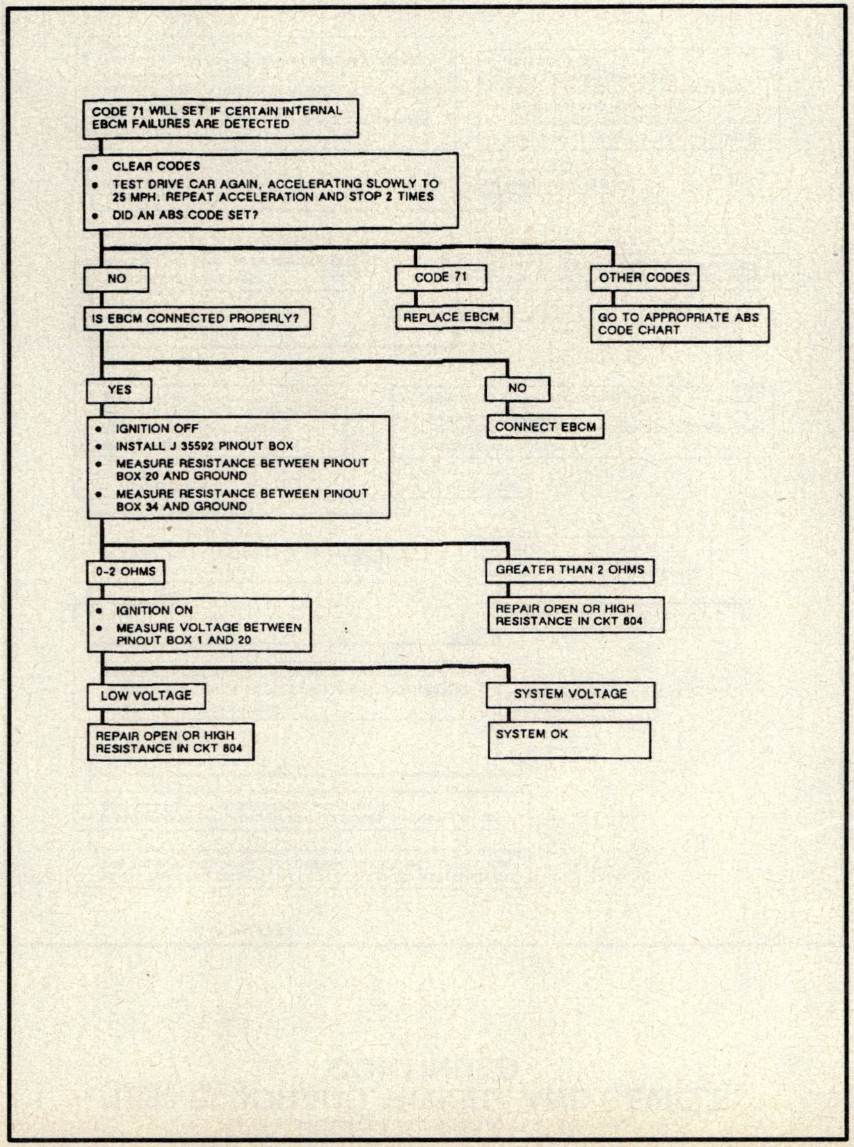

DIAGNOSTIC CHART A — 1993 ELDORADO, RIVIERA AND SEVILLE

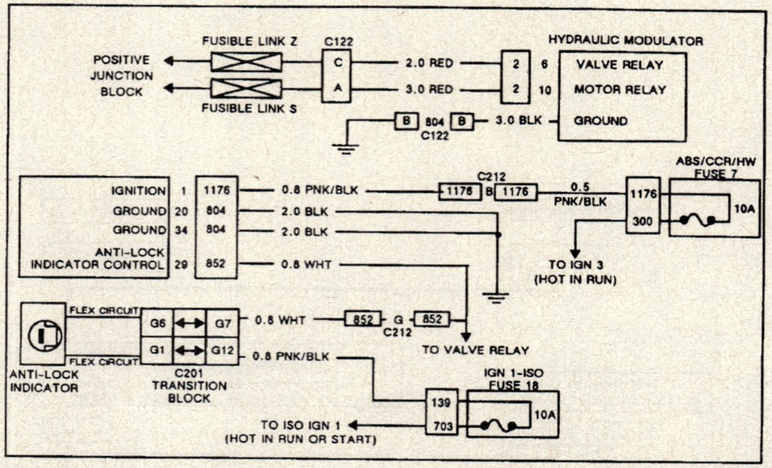

CHART A

NO SYSTEM POWER

CIRCUIT DESCRIPTION

The EBCM receives power from ABS/CCR/HW Fuse 7, (located in the interior relay center) on CKT 1176. If high system voltages occur, the EBCM is internally protected by a network of diodes which directs excess voltage to ground.

FAILURE CONDITIONS

Intermittent operation of the ANTI-LOCK light may be caused by intermittent voltage levels which do not correspond to the proper operating range of the Anti-Lock Brake System. Some conditions which may cause a lack of system power are:

- Improper electrical contact of the EBCM connector.
- Low system voltage.
- Improper electrical contact of the vehicle power and ground connections.

DIAGNOSTIC CHART A — 1993 ELDORADO, RIVIERA AND SEVILLE, CONTINUED

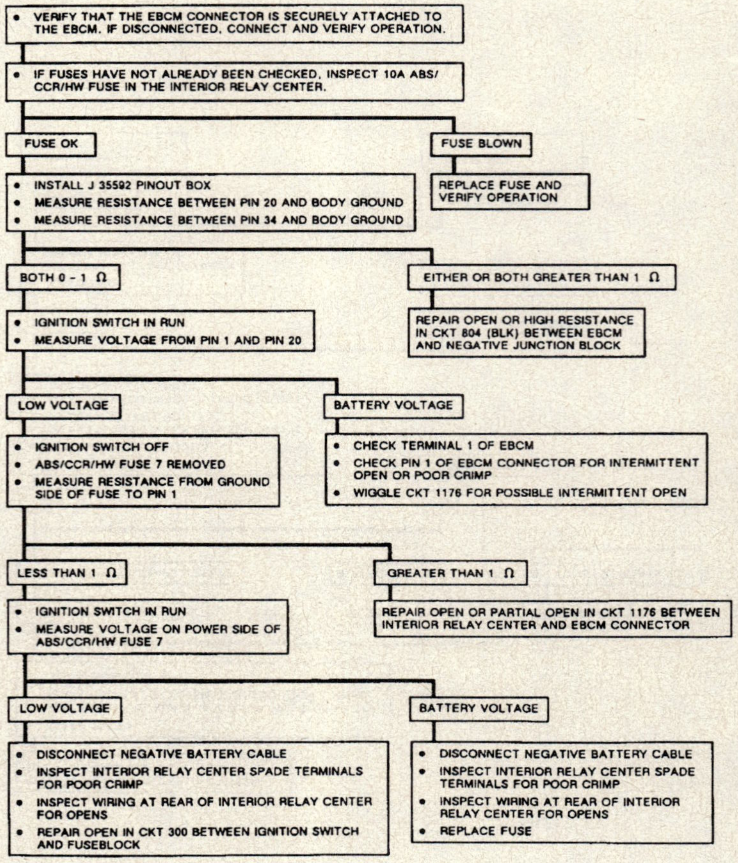

DIAGNOSTIC CHART B — 1993 ELDORADO, RIVIERA AND SEVILLE

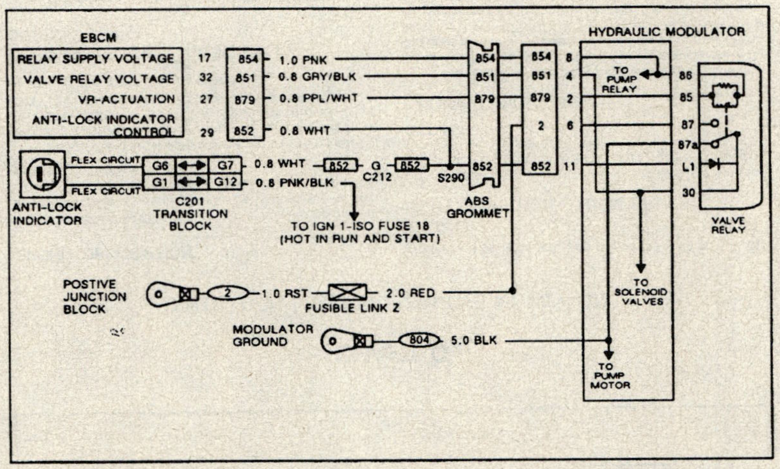

CHART B

ABS LIGHT ON, NO CODES SET

CIRCUIT DESCRIPTION

The amber ANTI-LOCK light is located below the instrument panel cluster. It may be illuminated by the EBCM or a path to ground in the valve relay on the hydraulic modulator.

The ANTI-LOCK light is powered by the Ignition 1 feed and receives power anytime the ignition switch is turned to RUN or START. Power is provided on CKT 139 through the 10 amp IGN 1-ISO fuse 18 located in the interior relay center.

LIGHT OPERATION – VALVE RELAY GROUND

The ANTI-LOCK light is illuminated anytime the valve relay is not enabled. When the relay is not enabled, the light is grounded through CKT 852 and the valve relay to the modulator ground. During normal operation (no fault conditions found), the valve will switch and the ground will be removed, and the ANTI-LOCK light will turn off. The ABS light is on whenever the EBCM is disconnected.

LIGHT OPERATION – EBCM GROUND

If the EBCM detects a fault in the ABS, the EBCM can turn on the ANTI-LOCK light by grounding pin 29. A ground path exists for the light through CKT 852. The EBCM will disable the valve relay when CKT 852 is grounded at pin 29.

NOTE: It is best to check for codes using a TECH-1. It is possible that a valve (Code 41, 45, 55 and 63) may have set. It is NOT possible to flash codes when a valve code is present.

FAILURE CONDITIONS

The chart on the next page addresses conditions which result in continuous operation of the ANTI-LOCK light. If the ANTI-LOCK light is on and no codes can be extracted from the EBCM, the EBCM connector should be checked for proper connection and system power should be checked.

DIAGNOSTIC CHART B — 1993 ELDORADO, RIVIERA AND SEVILLE, CONTINUED

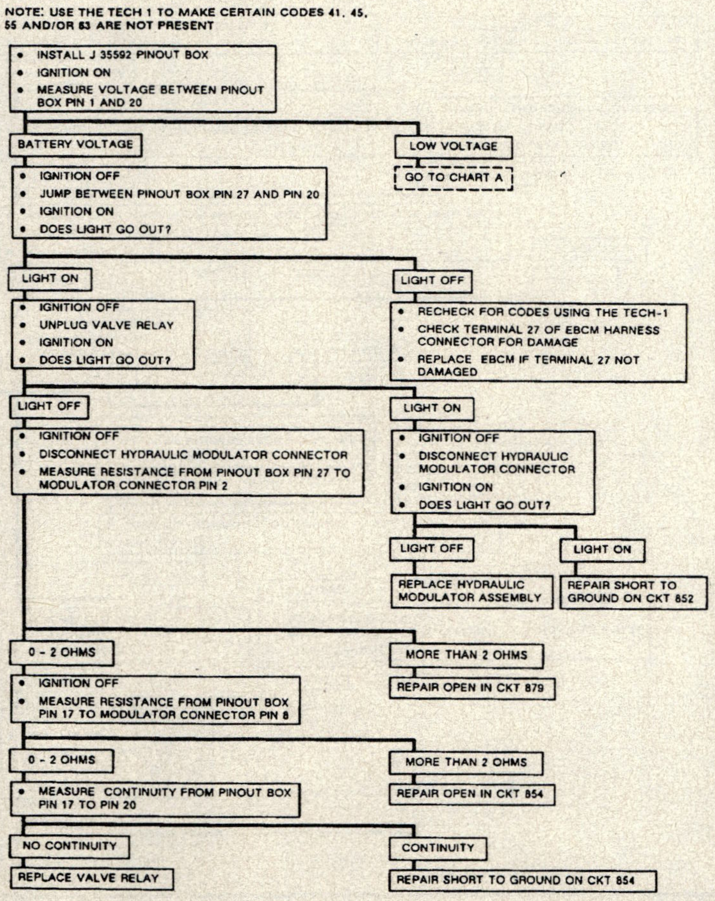

DIAGNOSTIC CHART C — 1993 ELDORADO, RIVIERA AND SEVILLE

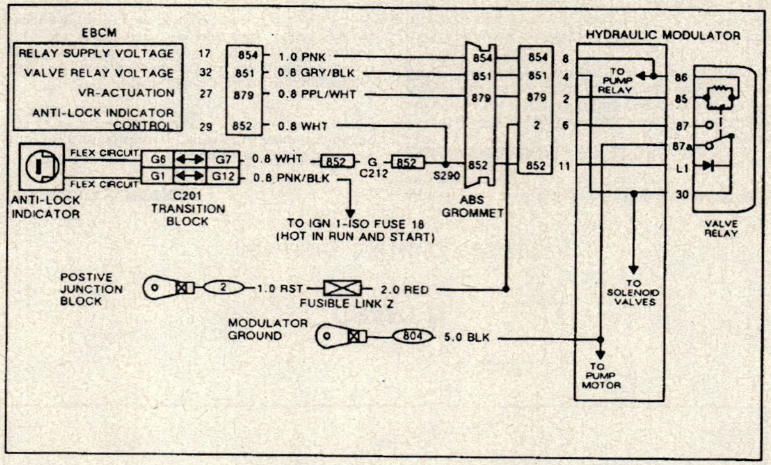

CHART C

ANTI-LOCK LIGHT INOPERATIVE AT KEY-ON

CIRCUIT DESCRIPTION

The amber ANTI-LOCK light is located below the instrument panel cluster. It may be illuminated by the EBCM or a path to ground in the valve relay on the hydraulic modulator.

The ANTI-LOCK light is powered by the Ignition 1 feed and receives power anytime the ignition switch is turned to RUN or START. Power is provided on CKT 139 through the 10 amp IGN-1 ISO fuse 18 located in the interior relay center.

LIGHT OPERATION - VALVE RELAY GROUND

The ANTI-LOCK light is illuminated anytime the valve relay is not enabled. When the relay is not enabled, the light is grounded through CKT 852 and the valve relay to the modulator ground. During normal operation (no fault conditions found), the valve will switch and the ground will be removed, and the ANTI-LOCK light will turn off.

LIGHT OPERATION - EBCM GROUND

Whether the valve relay is engaged or not, the EBCM can turn on the ANTI-LOCK light by grounding pin 29. A ground path exists for the light through CKT 852.

FAILURE CONDITIONS

The chart on the next page addresses conditions which result in no ANTI-LOCK light operation; the ABS operates properly. As described at the beginning of this section, the ANTI-LOCK light should illuminate for a minimum of four seconds when the key is moved from 'OFF' to RUN. This chart addresses conditions such as lack of power to the ANTI-LOCK light which would prevent the light from turning on. This condition could prevent the EBCM from turning on the light in the event of an anti-lock failure.

DIAGNOSTIC CHART C — 1993 ELDORADO, RIVIERA AND SEVILLE, CONTINUED

- CHECK IGN 1-ISO FUSE 18
  - FUSE OK
    - IGNITION OFF
    - INSTALL J 35592 PINOUT BOX
    - CONNECT TECH-1
    - IGNITION ON
    - DOES LIGHT COME ON?
      - LIGHT OFF
        - IGNITION ON
        - MEASURE VOLTAGE FROM PINOUT BOX PIN 1 TO PIN 20
          - LOW VOLTAGE
            - IGNITION OFF
            - MEASURE RESISTANCE TO GROUND AT PINOUT BOX PIN 20
              - 0 - 2 OHMS
                - IGNITION ON
                - MEASURE VOLTAGE FROM PINOUT BOX PIN 1 TO BATTERY NEGATIVE
                  - LOW VOLTAGE
                    - GO TO CHART A
              - MORE THAN 2 OHMS
                - REPAIR OPEN IN CKT 804
          - BATTERY VOLTAGE
            - JUMP FROM PINOUT BOX PIN 29 TO BATTERY NEGATIVE
            - DOES LIGHT COME ON?
              - LIGHT OFF
                - IGNITION OFF
                - REMOVE IGN 1-ISO FUSE 18
                - MEASURE RESISTANCE FROM 'GROUND' SIDE OF IGN 1-ISO FUSE 18 TO PIN 29
                  - 0 - 7 OHMS
                    - IGNITION ON
                    - MEASURE VOLTAGE ON 'HOT' SIDE OF IGN 1-ISO FUSE 18
                      - BATTERY VOLTAGE
                        - CHECK FUSE BLOCK TERMINAL CONNECTIONS
                        - REPLACE 'ANTI-LOCK' INDICATOR BULB
                      - LOW VOLTAGE
                        - CHECK FOR AN OPEN OR HIGH RESISTANCE IN CKT 703 BETWEEN FUSE AND IGNITION SWITCH
                  - MORE THAN 7 OHMS
                    - REPAIR OPEN IN CKT 139/852
                    - CHECK SPLICE S290
                    - REPLACE 'ANTI-LOCK' INDICATOR BULB
              - LIGHT ON
                - REPLACE EBCM
      - LIGHT ON
        - CONDITION NOT PRESENT
        - CHECK FOR INTERMITTENT LOSS OF EBCM POWER (PIN 1) OR GROUND (PIN 20)
        - CHECK TERMINAL 29 OF EBCM HARNESS CONNECTOR FOR DAMAGE
  - FUSE BLOWN
    - REPLACE FUSE

DIAGNOSTIC CHART D — 1993 ELDORADO, RIVIERA AND SEVILLE

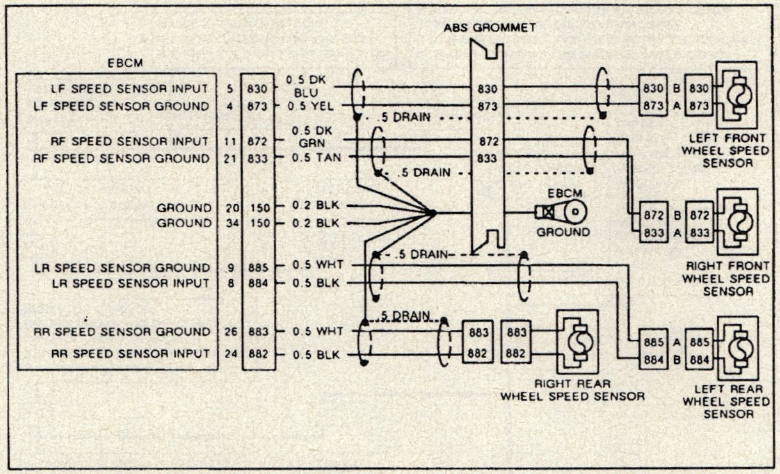

CHART D

VALVE CYCLING (CHATTER) DURING NORMAL STOPS

CIRCUIT DESCRIPTION

As each wheel turns, the wheel speed sensor for that wheel creates a small AC voltage as a toothed sensor ring, mounted on the front suspension knuckle or in the rear integral hub-bearing assembly, is passed by the stationary sensor. The sensor consists of a permanent magnet and coil and generates voltage by magnetic induction. The frequency of the wheel speed sensor signal is proportional to wheel speed.

The EBCM uses the wheel speed sensor signal to calculate vehicle reference speeds and individual speed, acceleration and slip values for each wheel. These values are used to determine when anti-lock control is required.

The EBCM performs two basic types of checks on the wheel speed sensors, these are sensor continuity and sensor output.

WHEEL SPEED SENSOR CABLES

In order to prevent electromagnetic interference from disturbing the wheel speed sensor signal, the sensor cables are protected with grounded shielding (indicated by dashed lines in the above schematic). This shield surrounds the two individual sensor wires. A black conduit surrounds the wires and shield. When servicing the sensor cables, the sensor wires are accessible by separating the shielding. If the shielding is disturbed, repair as required.

Notes On Chart D

1. The wheel speed sensor output should gradually increase with vehicle speed. There should be no skips or jumps. All four wheel speed sensor outputs should be identical. On a slow down, one of the signals may drop off faster than the others, indicating a lock-up condition or a problem with the speed sensor air gap, wiring, or sensor itself.

DIAGNOSTIC CHART D — 1993 ELDORADO, RIVIERA AND SEVILLE, CONTINUED

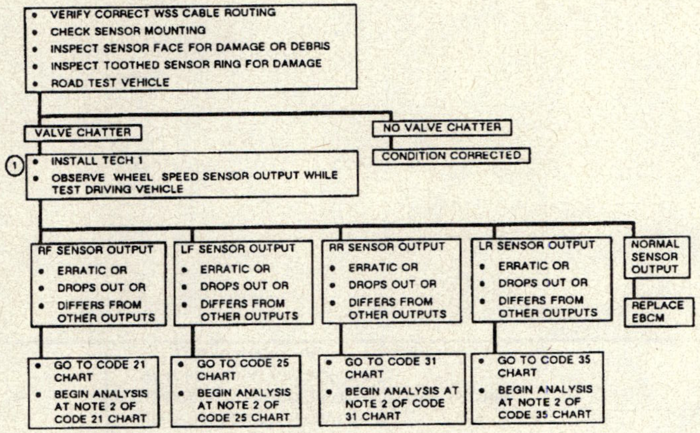

DIAGNOSTIC SYSTEM CHECK — 1994

DIAGNOSTIC SYSTEM CHECK

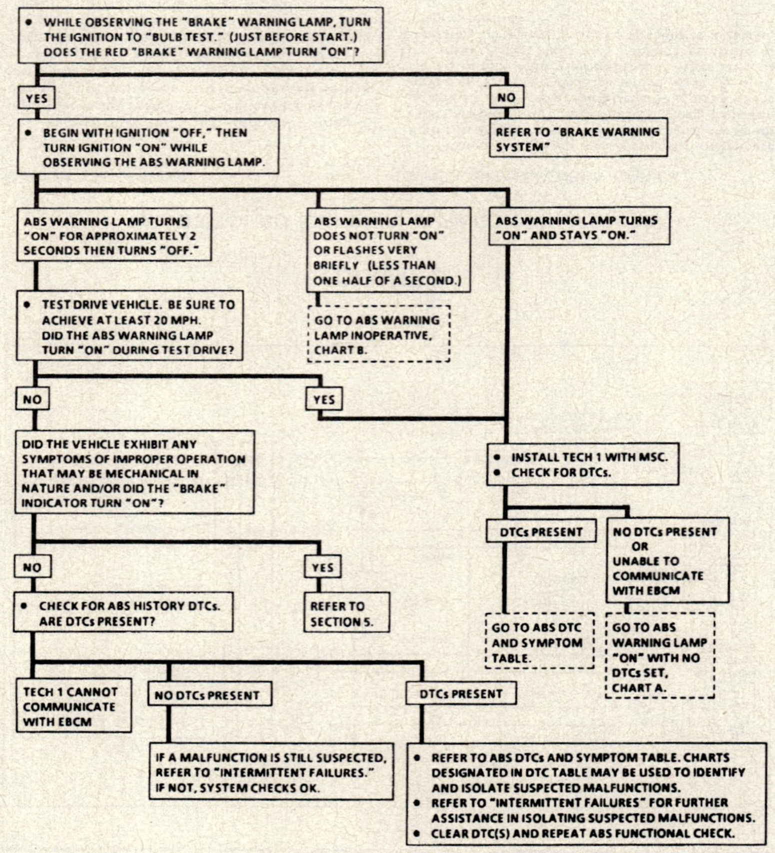

WHEN ALL DIAGNOSIS AND REPAIRS ARE COMPLETED, CLEAR DTCs AND VERIFY OPERATION.

DIAGNOSTIC TROUBLE CODE CHART — 1994

ABS DIAGNOSTIC TROUBLE CODE AND SYMPTOM TABLE

DTC	DESCRIPTION
21	RF WHEEL SPEED SENSOR CIRCUIT MALFUNCTION
23	RF WHEEL SPEED SENSOR CIRCUIT CONTINUITY MALFUNCTION
25	LF WHEEL SPEED SENSOR CIRCUIT MALFUNCTION
27	LF WHEEL SPEED SENSOR CIRCUIT CONTINUITY MALFUNCTION
28	WHEEL SPEED SENSOR CIRCUIT FREQUENCY MALFUNCTION
35	REAR WHEEL SPEED SENSOR CIRCUIT MALFUNCTION
37	REAR WHEEL SPEED SENSOR CIRCUIT CONTINUITY MALFUNCTION
41	RF VALVE SOLENOID MALFUNCTION
45	LF VALVE SOLENOID MALFUNCTION
55	REAR VALVE SOLENOID MALFUNCTION
61	BPM VALVE PUMP MOTOR OR PUMP MOTOR RELAY MALFUNCTION (1 of 2)
63	BPM VALVE RELAY MALFUNCTION (1 of 2)
71	EBCM INTERNAL MALFUNCTION (EBCM FAULT)
CHART	**SYMPTOM**
A	ABS WARNING LAMP "ON" WITH NO DTCs SET OR WITH TECH 1 UNABLE TO RECEIVE DATA
B	ABS WARNING LAMP INOPERATIVE OR LAMP FLASHES VERY BRIEFLY AT IGNITION "ON" (LESS THAN ONE HALF OF A SECOND)

DIAGNOSTIC CHART DTC 21 — 1994

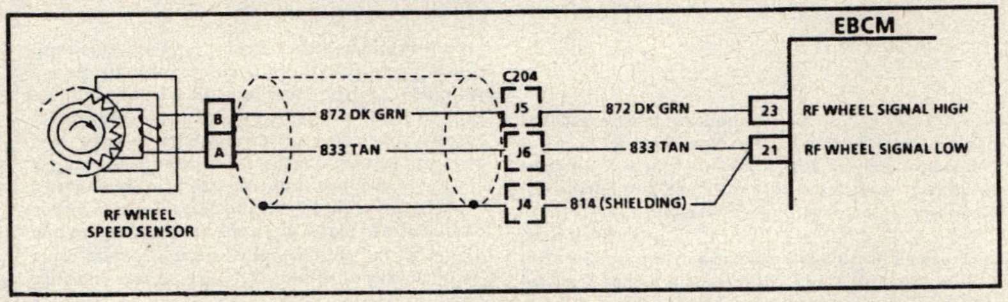

DTC 21

RF WHEEL SPEED SENSOR CIRCUIT MALFUNCTION

Circuit Description:

As each wheel turns, the speed sensor for that wheel creates an AC voltage as the toothed ring rotates by the stationary sensor. The sensor consists of a permanent magnet and coil generating voltage by magnetic induction. The frequency and voltage of the induced signal is proportional to wheel speed.

The EBCM uses the speed sensor signal to calculate vehicle reference speed, individual speed, acceleration and slip values for each wheel which determine when antilock control is required.

DTC 21 Will Set When: There is a short to voltage or ground in CKT 872 or 833, or a malfunctioning speed sensor. The testing for this malfunction occurs when the vehicle is in motion; it will not set with the ignition "ON" and the vehicle at rest.

Action Taken: If a malfunction is detected which causes a speed sensor DTC 21 to set, antilock braking is disabled and the EBCM turns "ON" the ABS warning lamp for the remainder of the ignition cycle. If the malfunction is intermittent, the EBCM will enable the system at the next ignition cycle and a history DTC 21 will be present.

Important

- The RF wheel speed sensor is located in the RF steering knuckle. In order to prevent electromagnetic interference from disturbing the speed sensor signal, the sensor wires are twisted together 9 times for every foot of wire.

DTC Chart Test Description: Number(s) below refer to circled number(s) on the diagnostic chart.

1. Checks for a short to battery in either of the speed sensor circuit wires.
2. Checks for a short between the speed sensor circuit wires.
3. Checks for a short to ground in the speed sensor input circuit wire. The ground may be either a "hard" short to ground, or a resistive (partial) short. A short with a resistance less than 2 megohms, though not a hard short, can still cause DTC 21 to set.
4. Checks if the DTC resets during a road test. If so, since tests 1-4 have validated that the circuitry and components are good, intermittent malfunctions are suspected.
5. Checks wiring and connectors for intermittents.
6. Replace the wheel bearing/speed sensor assembly, as it is likely the cause of a DTC reset experienced during the road test, since other portions of the circuit have checked out OK. If the DTC resets after sensor replacement, the EBCM must be concluding there is a malfunction present in the speed sensor circuit when there is not.
7. Ensures DTC was not set due to electromagnetic interference.

Diagnostic Aids: Setting of speed sensor DTCs may be caused by improperly mounted sensors or improper wire routing. Verify that the sensor is properly mounted and free of foreign material such as metallic particles, dirt or grease. Verify that sensor wires are properly routed and secure. Improper RF wheel speed sensor wire routing may cause DTC 25 and/or DTC 35 to set. To aid in troubleshooting intermittent conditions the Tech 1 can be used while test driving the vehicle however, antilock braking will be disabled. Refer to the Tech 1 Brake Cartridge Operators Manual for additional information. If this does not identify the intermittent, wet the speed sensor harness on the underside of the vehicle and road test monitoring wheel speeds with the Tech 1.

DIAGNOSTIC CHART DTC 21 — 1994, CONTINUED

DTC 21
RF WHEEL SPEED SENSOR CIRCUIT MALFUNCTION

ENSURE SENSOR IS NOT VISIBLY DAMAGED; IF IT IS DAMAGED, REPLACE THE WHEEL BEARING/SENSOR ASSEMBLY, CLEAR DTCs, AND ROAD TEST TO SEE IF DTC RESETS.

(1)
- DISCONNECT SENSOR CONNECTOR.
- CONNECT J 39200 BETWEEN SENSOR HARNESS CONNECTOR TERMINAL "A" AND GROUND (USE DC VOLTAGE SETTING)
- IGNITION "ON."
- NOTE VOLTAGE; REPEAT BETWEEN TERMINAL "B" AND GROUND.

IS IT ABOUT B + ON EITHER TERMINAL "A" OR "B"?

- YES: REPAIR SHORT TO B + IN CKT 872 OR 833.
- NO: go to (2)

(2)
- IGNITION "OFF."
- DISCONNECT EBCM.
- INSTALL J 39700 UNIVERSAL BREAKOUT BOX TO EBCM HARNESS CONNECTOR ONLY.
- USING J 39200, MEASURE RESISTANCE BETWEEN J 39700 TERMINALS "21" AND "23".

IS IT LESS THAN 5 OHMS?

- YES: REPAIR SHORT BETWEEN CKTS 872 & 833.
- NO: go to (3)

(3)
- USING J 39200, MEASURE RESISTANCE BETWEEN J 39700 TERMINAL "23" AND GROUND.

IS IT GREATER THAN 2 MEGOHMS?

- NO:
  - IF 5 OHMS OR LESS TO GROUND, REPAIR HARD SHORT IN HARNESS CKT 872.
  - IF BETWEEN 5 OHMS AND 2 MEGOHMS, REPAIR RESISTIVE (PARTIAL) SHORT IN HARNESS CKT 872.
- YES: go to (4)

(4)
- IGNITION "OFF."
- REMOVE J 39700.
- RECONNECT EBCM CONNECTOR.
- CLEAR DTCs, IGNITION "OFF."
- DISCONNECT TECH I, IF CONNECTED.
- ROAD TEST VEHICLE, REACHING AT LEAST 25 km/h (15 MPH).

DOES DTC 21 RESET?

- NO: NO TROUBLE FOUND; SEE "DIAGNOSTIC AIDS"
- YES: go to (5)

(5)
- CHECK ALL WIRING, CONNECTORS AND TERMINALS THOROUGHLY FOR INTERMITTENTS OR CORROSION.

IS ALL OK WITH CONNECTIONS AND WIRING?

- NO: REPAIR WIRING AND/OR CONNECTIONS AS NEEDED;
- YES: go to (6)

(6)
- CLEAR DTCs, IGNITION "OFF."
- REPLACE WHEEL BEARING/SPEED SENSOR ASSEMBLY
- DISCONNECT TECH I, IF CONNECTED.
- ROAD TEST VEHICLE, REACHING AT LEAST 25 km/h (15 MPH).

DOES DTC 21 RESET?

- NO: SYSTEM OK; SEE "DIAGNOSTIC AIDS"
- YES: go to (7)

(7) PERFORM EMI TEST. REFER TO EMI TEST FOUND IN "INTERMITTENTS AND POOR CONNECTIONS" IN THIS SECTION. WERE EMI TESTS COMPLETED WITH NO TROUBLE FOUND?

- YES: REPLACE EBCM.
- NO: PERFORM NECESSARY REPAIRS.

WHEN ALL DIAGNOSIS AND REPAIRS ARE COMPLETED, CLEAR DTCs AND VERIFY OPERATION.

DIAGNOSTIC CHART DTC 23 — 1994

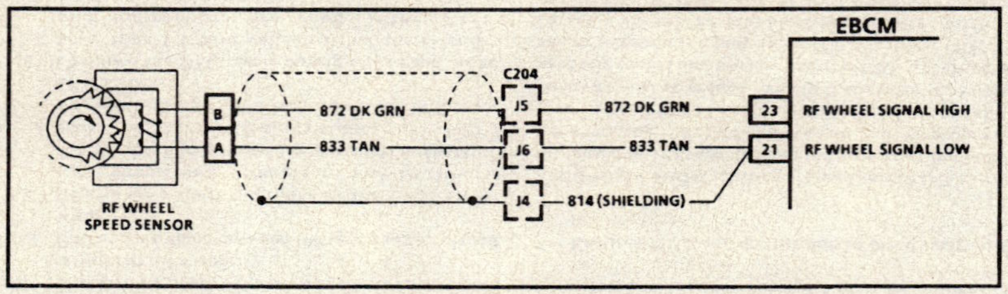

DTC 23

RF WHEEL SPEED SENSOR CIRCUIT CONTINUITY MALFUNCTION

Circuit Description:

The toothed wheel generates a voltage pulse as it moves past the sensor; each tooth-gap-tooth series on the wheel generates these pulses. The frequency of these pulses is used by the EBCM to determine wheel speed. The amount of voltage generated in each pulse depends on the air gap between the sensor and the toothed wheel, and on wheel speed.

DTC 23 Will Set When: There is a short to voltage or an open in CKT 872 or 833, or a malfunctioning speed sensor. The testing for this malfunction occurs with the ignition "ON" and the vehicle at rest.

Action Taken: Antilock braking is disabled. The ABS warning lamp will be "ON" for the remainder of the ignition cycle. If the malfunction is intermittent, the EBCM will enable the system at the next ignition cycle and a history DTC 23 will be present.

DTC Chart Test Description: Number(s) below refer to circled number(s) on the diagnostic chart.

1. Checks for a short to battery in either of the speed sensor circuit wires.
2. Checks for an open in both the speed sensor circuit wires.
3. Manipulates wiring and connectors, trying to induce a intermittent malfunction not currently present.
4. This test checks if the DTC resets on key-up. If so, since tests 1-4 have validated that the circuitry and components are good, the wheel bearing/speed sensor assembly may be causing the malfunction.
5. Checks if DTC 21 sets during a road test.
6. DTC 23 sets when the vehicle is at rest; DTC 21 sets with a malfunction in the speed sensor circuitry with the vehicle in motion. If malfunctions are still present, DTC 21 would set during the road test, not a 23. If the DTC 21 sets at this point, the DTC 21 diagnostics should be used.
7. The wheel speed sensor may be causing an intermittent malfunction. If DTC 23 resets after sensor replacement, the EBCM must be concluding there is a malfunction present in the speed sensor circuit when there is not.
8. Ensures DTC was not set due to electromagnetic interference.

Diagnostic Aids: Be sure the speed sensor wiring is properly routed and retained. This will help prevent false signals due to electrical noise being picked up by the wiring.

It is very important that a thorough inspection of the wiring and connectors be performed. Failure to carefully and fully inspect wiring and connectors may result in misdiagnosis, causing part replacement with reappearance of the malfunction.

DIAGNOSTIC CHART DTC 23 — 1994, CONTINUED

DTC 23

RF WHEEL SPEED SENSOR CIRCUIT CONTINUITY MALFUNCTION

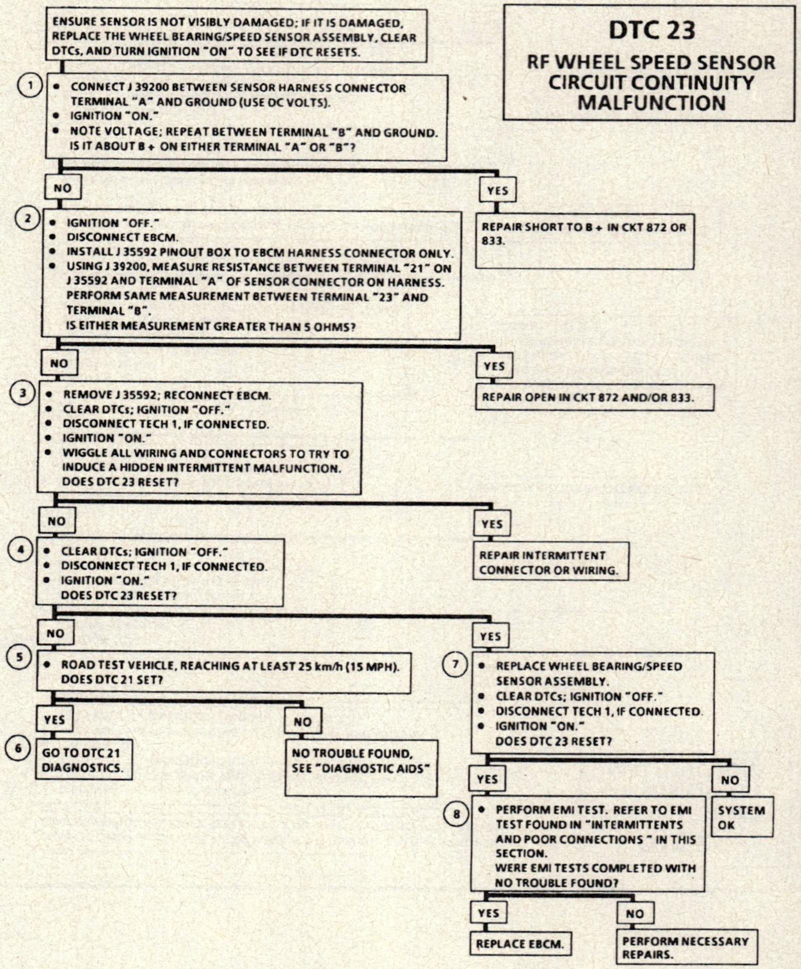

WHEN ALL DIAGNOSIS AND REPAIRS ARE COMPLETED, CLEAR DTCs AND VERIFY PROPER OPERATION.

DIAGNOSTIC CHART DTC 25 — 1994

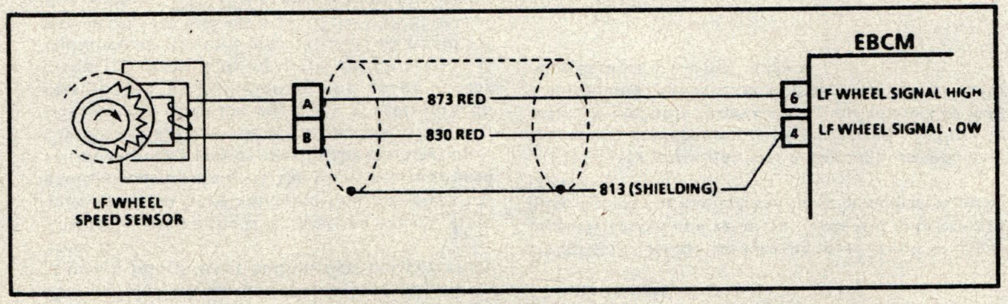

DTC 25

LF WHEEL SPEED SENSOR CIRCUIT MALFUNCTION

Circuit Description:

As each wheel turns, the speed sensor for that wheel creates an AC voltage as the toothed ring rotates by the stationary sensor. The sensor consists of a permanent magnet and coil generating voltage by magnetic induction. The frequency and voltage of the induced signal is proportional to wheel speed.

The EBCM uses the speed sensor signal to calculate vehicle reference speed, individual speed, acceleration and slip values for each wheel which determine when antilock control is required.

DTC 25 Will Set When: There is a short to voltage or ground in CKT 830 or 873, or a malfunctioning speed sensor. The testing for this malfunction occurs when the vehicle is in motion; it will not set with the ignition "ON" and the vehicle at rest.

Action Taken: If a malfunction is detected which causes a speed sensor DTC 25 to set, antilock braking is disabled and the EBCM turns "ON" the ABS warning lamp for the remainder of the ignition cycle. If the malfunction is intermittent, the EBCM will enable the system at the next ignition cycle and a history DTC 25 will be present.

Important

- The LF wheel speed sensor is located in the LF steering knuckle. In order to prevent electromagnetic interference from disturbing the wheel speed sensor signal, the sensor wires are twisted together 9 times for every foot of wire.

DTC Chart Test Description: Number(s) below refer to circled number(s) on the diagnostic chart.

1. Checks for a short to battery in either of the speed sensor circuit wires.
2. Checks for a short between the speed sensor circuit wires.
3. Checks for a short to ground in the speed sensor input circuit wire. The ground may be either a "hard" short to ground, or a resistive (partial) short. A short with a resistance less than 2 megohms, though not a hard short, can still cause DTC 25 to set.
4. Checks if the DTC resets during a road test. If so, since tests 1-4 have validated that the circuitry and components are good, intermittent malfunctions are suspected.
5. Checks wiring and connectors for intermittents.
6. Replace the wheel bearing/speed sensor assembly, as it is likely the cause of a DTC reset experienced during the road test, since other portions of the circuit have checked out OK. If the DTC resets after sensor replacement, the EBCM must be concluding there is a malfunction present in the speed sensor circuit when there is not.
7. Ensures DTC was not set due to electromagnetic interference.

Diagnostic Aids: Setting of speed sensor DTCs may be caused by improperly mounted sensors or improper wire routing. Verify that the sensor is properly mounted and free of foreign material such as metallic particles, dirt or grease. Verify that sensor wires are properly routed and secure. Improper LF wheel speed sensor wire routing may cause DTC 21 and/or DTC 35 to set. To aid in troubleshooting intermittent conditions the Tech 1 can be used while test driving the vehicle however, antilock braking will be disabled. Refer to the Tech 1 Brake Cartridge Operators Manual for additional information. If this does not identify the intermittent, wet the speed sensor harness on the underside of the vehicle and road test monitoring wheel speeds with the Tech 1.

DIAGNOSTIC CHART DTC 25 — 1994, CONTINUED

DTC 25
LF WHEEL SPEED SENSOR CIRCUIT MALFUNCTION

ENSURE SENSOR IS NOT VISIBLY DAMAGED; IF IT IS DAMAGED, REPLACE THE WHEEL BEARING/SENSOR ASSEMBLY, CLEAR DTCs, AND ROAD TEST TO SEE IF DTC RESETS.

(1)
- CONNECT J 39200 BETWEEN SENSOR HARNESS CONNECTOR TERMINAL "A" AND GROUND (USE DC VOLTS).
- IGNITION "ON."
- NOTE VOLTAGE; REPEAT BETWEEN TERMINAL "B" AND GROUND. IS IT ABOUT B + ON EITHER TERMINAL "A" OR "B"?

YES: REPAIR SHORT TO B + IN CKT 830 OR 873.

NO:

(2)
- IGNITION "OFF."
- DISCONNECT EBCM.
- INSTALL J 35592 PINOUT BOX TO EBCM HARNESS CONNECTOR ONLY.
- USING J 39200, MEASURE RESISTANCE BETWEEN J 35592 TERMINALS "4" AND "6". IS IT LESS THAN 5 OHMS?

YES: REPAIR SHORT BETWEEN CKTs 830 AND 873.

NO:

(3)
- USING J 39200, MEASURE RESISTANCE BETWEEN J 35592 TERMINAL "6" AND GROUND. IS IT GREATER THAN 2 MEGOHMS?

NO:
- IF 5 OHMS OR LESS TO GROUND, REPAIR HARD SHORT IN HARNESS CKT 830.
- IF BETWEEN 5 OHMS AND 2 MEGOHMS, REPAIR RESISTIVE (PARTIAL) SHORT IN HARNESS CKT 830.

YES:

(4)
- IGNITION "OFF."
- REMOVE J 35592.
- RECONNECT EBCM CONNECTOR.
- CLEAR DTCs, IGNITION "OFF."
- DISCONNECT TECH 1, IF CONNECTED.
- ROAD TEST VEHICLE, REACHING AT LEAST 25 km/h (15 MPH). DOES DTC 25 RESET?

NO: NO TROUBLE FOUND; SEE "DIAGNOSTIC AIDS".

YES:

(5)
- CHECK ALL WIRING, CONNECTORS AND TERMINALS THOROUGHLY FOR INTERMITTENTS OR CORROSION. REFER TO "INTERMITTENTS AND POOR CONNECTIONS" IN SECTION 8A-4. IS ALL OK WITH CONNECTIONS AND WIRING?

NO: REPAIR WIRING AND/OR CONNECTIONS AS NEEDED;

YES:

(6)
- CLEAR DTCs, IGNITION "OFF."
- REPLACE WHEEL BEARING/SPEED SENSOR ASSEMBLY
- DISCONNECT TECH I, IF CONNECTED.
- ROAD TEST VEHICLE, REACHING AT LEAST 25 km/h (15 MPH). DOES DTC 25 RESET?

NO: SYSTEM OK; SEE "DIAGNOSTIC AIDS".

YES:

(7) PERFORM EMI TEST. REFER TO EMI TEST FOUND IN "INTERMITTENT FAILURES" IN SECTION. WERE EMI TESTS COMPLETED WITH NO TROUBLE FOUND?

YES: REPLACE EBCM.

NO: PERFORM NECESSARY REPAIRS.

WHEN ALL DIAGNOSIS AND REPAIRS ARE COMPLETED, CLEAR DTCs AND VERIFY OPERATION.

DIAGNOSTIC CHART DTC 27 — 1994

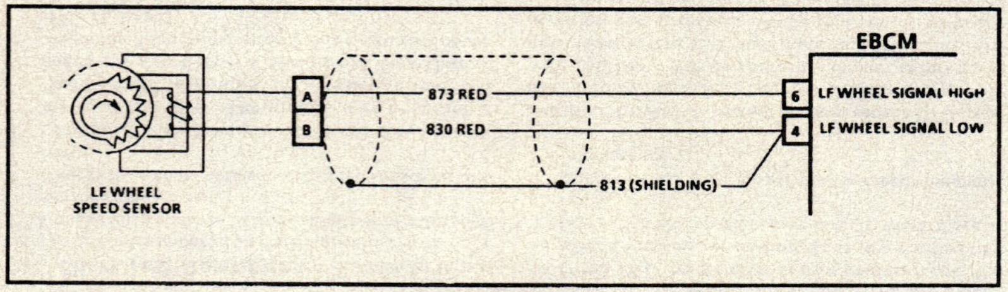

DTC 27

LF WHEEL SPEED SENSOR CIRCUIT CONTINUITY MALFUNCTION

Circuit Description:

The toothed wheel generates a voltage pulse as it moves past the sensor; each tooth-gap-tooth series on the wheel generates these pulses. The frequency of these pulses is used by the EBCM to determine wheel speed. The amount of voltage generated in each pulse depends on the air gap between the sensor and the toothed wheel, and on wheel speed.

DTC 27 Will Set When: There is a short to voltage or an open in CKT 830 or 873, or a malfunctioning speed sensor. The testing for this malfunction occurs with the ignition "ON" and the vehicle at rest.

Action Taken: Antilock braking is disabled. The ABS warning lamp will be "ON" for the remainder of the ignition cycle. If the malfunction is intermittent, the EBCM will enable the system at the next ignition cycle and a history DTC 27 will be present.

DTC Chart Test Description: Number(s) below refer to circled number(s) on the diagnostic chart.

1. Checks for a short to battery in either of the speed sensor circuit wires.
2. Checks for an open in both the speed sensor circuit wires.
3. Manipulates wiring and connectors, trying to induce a intermittent malfunction not currently present.
4. This test checks if the DTC resets on key-up. If so, since tests 1-4 have validated that the circuitry and components are good, the wheel bearing/speed sensor assembly may be causing the malfunction.
5. Checks if DTC 25 sets during a road test.
6. DTC 27 sets when the vehicle is at rest; DTC 25 sets with a malfunction in the speed sensor circuitry with the vehicle in motion. If malfunctions are still present, DTC 25 would set during the road test, not a 27. If the DTC 25 sets at this point, the DTC 25 diagnostics should be used.
7. The wheel speed sensor may be causing an intermittent malfunction. If DTC 27 resets after sensor replacement, the EBCM must be concluding there is a malfunction present in the speed sensor circuit when there is not.
8. Ensures DTC was not set due to electromagnetic interference.

Diagnostic Aids: Be sure the speed sensor wiring is properly routed and retained. This will help prevent false signals due to electrical noise being picked up by the wiring.

It is very important that a thorough inspection of the wiring and connectors be performed. Failure to carefully and fully inspect wiring and connectors may result in misdiagnosis, causing part replacement with reappearance of the malfunction.

DIAGNOSTIC CHART DTC 27 — 1994, CONTINUED

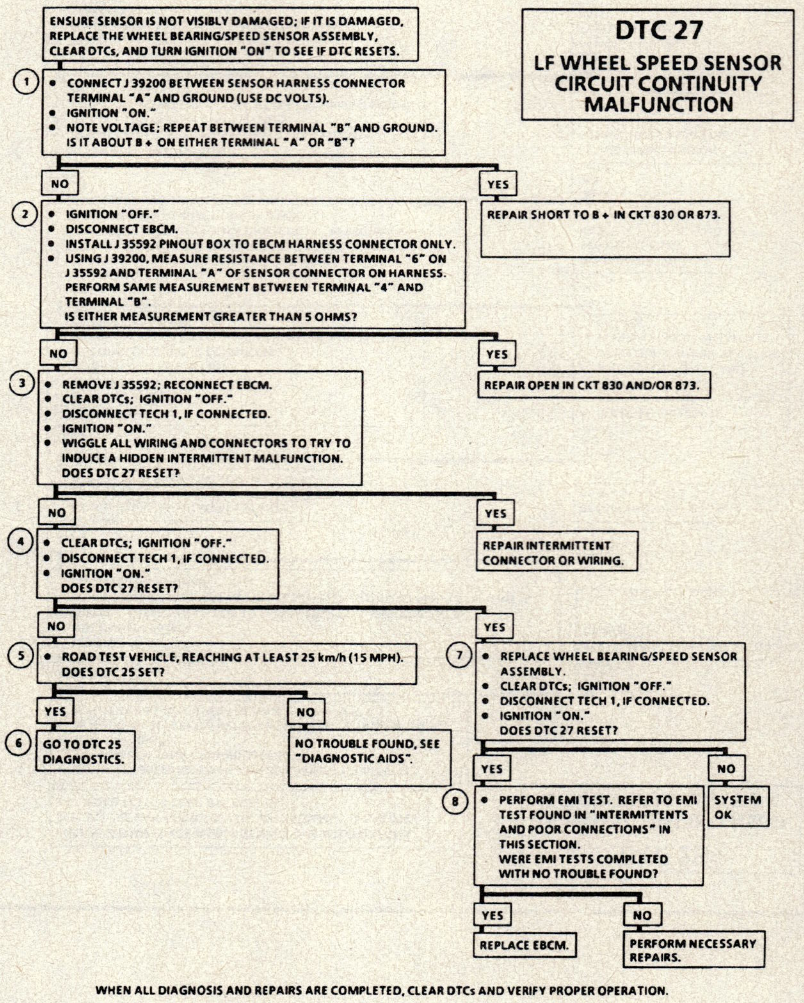

DIAGNOSTIC CHART DTC 28 — 1994

DTC 28

WHEEL SPEED SENSOR CIRCUIT FREQUENCY MALFUNCTION

Circuit Description:

The toothed wheel generates a voltage pulse as it moves past the sensor; each tooth-gap-tooth series on the wheel generates these pulses. The frequency of these pulses is used by the EBCM to determine wheel speed. The amount of voltage generated in each pulse depends on the air gap between the sensor and the toothed wheel, and on wheel speed.

DTC 28 Will Set When: The EBCM cannot specifically identify which wheel speed sensor is causing the frequency error malfunction. If it can define the specific speed sensor causing the malfunction, the sensor DTC associated with the sensor (21, 25, 31, 35) will be set instead of DTC 28.

Action Taken: Antilock braking is disabled. The ABS warning lamp will be "ON" for the remainder of the ignition cycle. If the malfunction is intermittent, the EBCM will enable the system at the next ignition cycle and a history DTC 28 will be present.

DTC Chart Test Description: Number(s) below refer to circled number(s) on the diagnostic chart.

The ABS warning lamp will be turned "ON" and ABS will be disabled.

1. Checks wiring and connections for malfunctions. It is critical that this step is performed thoroughly, as wiring/connector malfunctions are the most likely cause for this DTC.
2. Uses the Tech 1 to monitor for vehicle electrical/electronic system "noise" being picked up by the speed sensor circuits.
3. Uses the Tech 1 to monitor the wheel speed sensors while the vehicle is operating. If the Tech 1 auto-trigger snapshot mode triggers, the speed sensor or the wiring and/or connectors are going intermittent during the road test.
4. Checks for proper resistance in the sensor itself.
5a. Replace rear wheel speed sensor as the likely cause of the triggering intermittent.
5b. Replaces front wheel speed sensor as the likely cause of the triggering intermittent.
6. Ensures malfunction is not due to suspected wheel speed sensor.
7. Checks for a short to ground in the speed sensor input circuit wires. The circuit to be tested depends on which speed sensor triggered the Tech 1. The ground may be either a "hard" short to ground, or a resistive (partial) short. A short with a resistance less than 2 megohms, though not a hard short, can still cause DTC 28 to set.
8. Ensures DTC was not set due to electromagnetic interference.

Diagnostic Aids: Check the toothed wheel for any large grooves, gouges, marks, etc. that might influence the tooth's signal at the wheel speed sensor. Also check for a buildup of foreign material in the gaps between teeth in the toothed wheel; this material may cause this malfunction.

A worn hub/bearing assembly may cause this malfunction in extreme cases; the bearing play allows the sensor-to-toothed-ring gap to change excessively.

DIAGNOSTIC CHART DTC 28 — 1994, CONTINUED

DTC 28
WHEEL SPEED SENSOR CIRCUIT FREQUENCY MALFUNCTION

(1) VERY THOROUGHLY INSPECT ALL WIRING AND CONNECTIONS FOR INTERMITTENTS, CHAFING, POOR TERMINAL CONTACT TENSION, CORROSION, AND MISROUTING AT EBCM, ALL SENSORS, AND ALL HARNESS INTERCONNECTS. REPAIR ANY MALFUNCTIONS, CLEAR DTCs, THEN ROAD TEST VEHICLE, REACHING AT LEAST 25 km/h (15 MPH). DOES DTC RESET?

- NO: NO TROUBLE FOUND. SEE "DIAGNOSTIC AIDS".
- YES:

(2)
- INSTALL TECH 1 WITH MSC; CLEAR DTCs.
- SET TECH 1 AUTO TRIGGER SNAPSHOT MODE.
- START VEHICLE AND ALLOW TO IDLE FOR ONE MINUTE. DOES TECH 1 TRIGGER?

- YES:

(7)
- NOTE TECH 1, WHICH SENSOR TRIGGERED?
- IGNITION "OFF."
- DISCONNECT EBCM.
- INSTALL J 35592 PINOUT BOX TO EBCM HARNESS CONNECTOR.
- USING TABLE BELOW, MEASURE RESISTANCE TO GROUND AT THE TERMINAL ON J 35592 THAT CORRESPONDS TO THE SPEED SENSOR THE TECH 1 TRIGGERED ON:

LEFT FRONT TERMINAL "6"
RIGHT FRONT TERMINAL "23"
REAR TERMINAL "7"

- IF RESISTANCE IS 5 OHMS OR LESS TO GROUND, REPAIR HARD SHORT; IF BETWEEN 5 OHMS AND 2 MEGOHMS, REPAIR RESISTIVE (PARTIAL) SHORT IN CKT:

LEFT FRONT CKT 873
RIGHT FRONT CKT 872
REAR CKT 1811

- NO:

(3)
- ROAD TEST VEHICLE REACHING AT LEAST 25 km/h (15 MPH). DOES TECH 1 TRIGGER?

- NO: NO TROUBLE FOUND; SEE "DIAGNOSTIC AIDS" ON FACING PAGE.
- YES: NOTE TECH 1, WHICH SENSOR TRIGGERED?

(4)
- IGNITION "OFF."
- DISCONNECT SENSOR CONNECTOR.
- USING J 39200, MEASURE RESISTANCE BETWEEN SENSOR TERMINALS, USING ADAPTERS FROM J 35616-A IF MEASURING A FRONT SENSOR. IS RESISTANCE WITHIN RANGE SHOWN?

FRONT	REAR
990-1210	980-1130

- YES: REPLACE EBCM
- NO:
  - REAR: (5A) REPLACE SPEED SENSOR.
  - FRONT: (5B) REPLACE WHEEL BEARING/SPEED SENSOR ASSEMBLY.

(6)
- CLEAR DTCs, IGNITION "OFF."
- DISCONNECT TECH 1, IF CONNECTED.
- ROAD TEST VEHICLE, REACHING AT LEAST 25 km/h (15 MPH). DOES DTC 28 RESET?

- NO: SYSTEM OK
- YES:

(8) PERFORM EMI TEST. REFER TO EMI TEST FOUND IN "INTERMITTENTS AND POOR CONNECTIONS" IN THIS SECTION. WERE EMI TESTS COMPLETED WITH NO TROUBLE FOUND?

- YES: REPLACE EBCM.
- NO: PERFORM NECESSARY REPAIRS.

WHEN ALL DIAGNOSIS AND REPAIRS ARE COMPLETED, CLEAR DTCs AND VERIFY PROPER OPERATION.

DIAGNOSTIC CHART DTC 35 — 1994

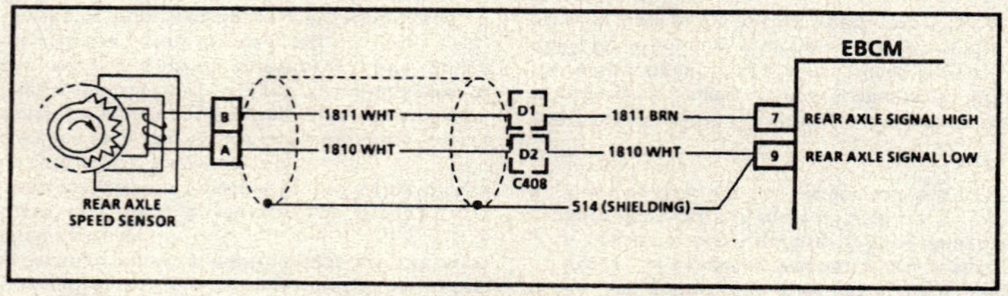

DTC 35

REAR WHEEL SPEED SENSOR CIRCUIT MALFUNCTION

Circuit Description:
As the differential pinion gear turns, the speed sensor creates an AC voltage as the toothed ring (mounted to the differential pinion gear) rotates by the stationary sensor. The sensor consists of a permanent magnet and coil generating voltage by magnetic induction. The frequency and voltage of the induced signal is proportional to rear axle speed.

The EBCM uses the speed sensor signal to calculate vehicle reference speed, individual speed, acceleration and slip values for the rear wheels which determine when antilock control is required.

DTC 35 Will Set When: There is a short to voltage or ground in CKT 1811 or 1810, or a malfunctioning speed sensor. The testing for this malfunction occurs when the vehicle is in motion; it will not set with the ignition "ON" and the vehicle at rest.

Action Taken: If a malfunction is detected which causes a speed sensor DTC 35 to set, antilock braking is disabled and the EBCM turns "ON" the ABS warning lamp for the remainder of the ignition cycle. If the malfunction is intermittent, the EBCM will enable the system at the next ignition cycle and a history DTC 35 will be present.

Important

- The rear axle speed sensor is located in the rear axle housing. In order to prevent electromagnetic interference from disturbing the speed sensor signal, the sensor wires are twisted together 9 times for every foot of wire.

DTC Chart Test Description: Number(s) below refer to circled number(s) on the diagnostic chart.

1. Checks for a short to battery in either of the speed sensor circuit wires.
2. Checks for a short between the speed sensor circuit wires.
3. Checks for a short to ground in the speed sensor input circuit wire. The ground may be either a "hard" short to ground, or a resistive (partial) short. A short with a resistance less than 2 megohms, though not a hard short, can still cause DTC 35 to set.
4. Checks if the DTC resets during a road test. If so, since tests 1-4 have validated that all of the circuitry and components are good, intermittent malfunctions are suspected.
5. Checks wiring and connectors for intermittents.
6. Checks for sensor mounting, toothed wheel, and rear wheel bearing malfunctions which may be causing the malfunction. Check the toothed wheel for any large grooves, gouges, marks, etc. that might influence the tooth's signal at the wheel speed sensor. Also check for a buildup of foreign material in the gaps between teeth in the toothed wheel; this material may cause this malfunction.
7. Replace the speed sensor assembly, as it is likely the cause of a DTC reset experienced during the road test, since other portions of the circuit have checked out OK.
8. Ensures DTC was not set due to electromagnetic interference.

Diagnostic Aids: Setting of speed sensor DTCs may be caused by improperly mounted sensors or improper wire routing. Verify that the sensor is properly mounted and free of foreign material such as metallic particles, dirt or grease. If rear axle service was performed and the rear axle toothed wheel was removed it must be replaced. Verify that sensor wires are properly routed and secure. Improper rear axle wheel speed sensor wire routing may cause DTC 21 and/or DTC 25 to set. To aid in troubleshooting intermittent conditions the Tech 1 can be used while test driving the vehicle however, antilock braking will be disabled. Refer to the Tech 1 Brake Cartridge Operators Manual for additional information. If this does not identify the intermittent, wet the speed sensor harness on the underside of the vehicle and road test monitoring wheel speeds with the Tech 1.

DIAGNOSTIC CHART DTC 35 — 1994, CONTINUED

DTC 35
REAR WHEEL SPEED SENSOR CIRCUIT MALFUNCTION

ENSURE SENSOR IS NOT VISIBLY DAMAGED, AND THAT IT IS PROPERLY MOUNTED. IF IT IS NOT PROPERLY MOUNTED OR RETAINED, PROPERLY ATTACH IT; IF IT IS DAMAGED, REPLACE IT. THEN, CLEAR DTCs AND ROAD TEST TO SEE IF DTC RESETS.

(1)
- CONNECT J 39200 BETWEEN SENSOR HARNESS CONNECTOR TERMINAL "A" AND GROUND (USE DC VOLTS).
- IGNITION "ON."
- NOTE VOLTAGE; REPEAT BETWEEN TERMINAL "B" AND GROUND. IS IT ABOUT B+ ON EITHER TERMINAL "A" OR "B"?

YES: REPAIR SHORT TO B+ IN CKT 1811 OR 833.

NO:

(2)
- IGNITION "OFF."
- DISCONNECT EBCM.
- INSTALL J 35592 PINOUT BOX TO EBCM HARNESS CONNECTOR ONLY.
- USING J 39200, MEASURE RESISTANCE BETWEEN J 35592 TERMINALS "9" AND "7." IS IT LESS THAN 5 OHMS?

YES: REPAIR SHORT BETWEEN CKTS 1811 AND 1810.

NO:

(3)
- USING J 39200, MEASURE RESISTANCE BETWEEN J 35592 TERMINAL "7" AND GROUND. IS IT GREATER THAN 2 MEGOHMS?

NO:
- IF 5 OHMS OR LESS TO GROUND, REPAIR HARD SHORT IN HARNESS CKT 1811.
- IF BETWEEN 5 OHMS AND 2 MEGOHMS, REPAIR RESISTIVE (PARTIAL) SHORT IN HARNESS CKT 1811.

YES:

(4)
- IGNITION "OFF."
- REMOVE J 35592.
- RECONNECT EBCM CONNECTOR.
- CLEAR DTCs, IGNITION "OFF."
- DISCONNECT TECH 1, IF CONNECTED.
- ROAD TEST VEHICLE, REACHING AT LEAST 25 km/h (15 MPH). DOES DTC 31 RESET?

NO: NO TROUBLE FOUND; SEE "DIAGNOSTIC AIDS".

YES:

(5)
- CHECK ALL WIRING, CONNECTORS AND TERMINALS THOROUGHLY FOR INTERMITTENTS OR CORROSION.

IS ALL OK WITH CONNECTIONS AND WIRING?

NO: REPAIR WIRING AND/OR CONNECTIONS AS NEEDED

YES:

(6)
- CHECK SENSOR MOUNTING, TOOTHED WHEEL FOR DAMAGE OR FOREIGN MATERIAL, AND FOR EXCESSIVELY WORN REAR WHEEL BEARING; SEE "DIAGNOSTIC AIDS". IS ALL OK?

NO: REPAIR AS NECESSARY.

YES:

(7)
- REPLACE SPEED SENSOR
- CLEAR DTCs, IGNITION "OFF."
- DISCONNECT TECH 1, IF CONNECTED.
- ROAD TEST VEHICLE, REACHING AT LEAST 25 km/h (15 MPH). DOES DTC 35 RESET?

NO: SYSTEM OK; SEE "DIAGNOSTIC AIDS".

YES:

(8) PERFORM EMI TEST.

NO: PERFORM NECESSARY REPAIRS.

YES: REPLACE EBCM.

WHEN ALL DIAGNOSIS AND REPAIRS ARE COMPLETED, CLEAR DTCs AND VERIFY OPERATION.

DIAGNOSTIC CHART DTC 37 — 1994

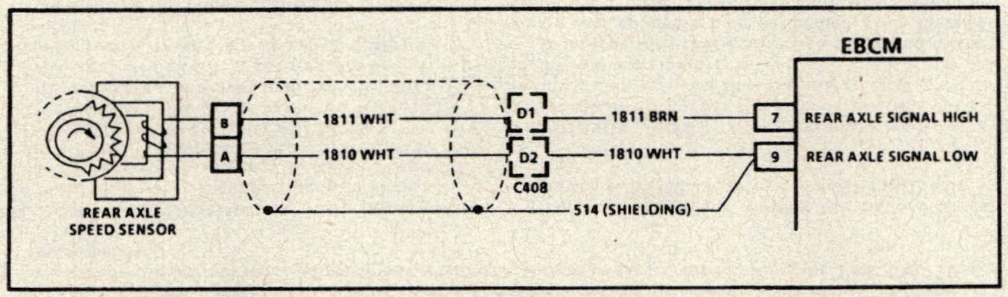

DTC 37

REAR WHEEL SPEED SENSOR CIRCUIT CONTINUITY MALFUNCTION

Circuit Description:

The toothed wheel generates a voltage pulse as it moves past the sensor; each tooth-gap-tooth series on the wheel generates these pulses. The frequency of these pulses is used by the EBCM to determine wheel speed. The amount of voltage generated in each pulse depends on the air gap between the sensor and the toothed wheel, and on wheel speed.

DTC 37 Will Set When: There is a short to voltage or an open in CKT 1811 or 1810, or a malfunctioning speed sensor. The testing for this malfunction occurs with the ignition "ON" and the vehicle at rest.

Action Taken: Antilock braking is disabled. The ABS warning lamp will be "ON" for the remainder of the ignition cycle. If the malfunction is intermittent, the EBCM will enable the system at the next ignition cycle and a history DTC 37 will be present.

DTC Chart Test Description: Number(s) below refer to circled number(s) on the diagnostic chart.

1. Checks for a short to battery in either of the speed sensor circuit wires.
2. Checks for an open in both the speed sensor circuit wires.
3. Manipulates wiring and connectors, trying to induce a intermittent malfunction not currently present.
4. This test checks if the DTC resets on key-up. If so, since tests 1-4 have validated that the circuitry and components are good, the speed sensor may be causing the malfunction.
5. Checks if DTC 35 sets during a road test.
6. DTC 37 sets when the vehicle is at rest; DTC 35 sets with a malfunction in the speed sensor circuitry with the vehicle in motion. If malfunctions are still present, DTC 35 would set during the road test, not a 37. If the DTC 37 sets at this point, the DTC 35 diagnostics should be used.
7. The wheel speed sensor may be causing an intermittent malfunction. If DTC 37 resets after sensor replacement, the EBCM must be concluding there is a malfunction present in the speed sensor circuit when there is not.
8. Ensures DTC was not set due to electromagnetic interference.

Diagnostic Aids: Be sure the speed sensor wiring is properly routed and retained. This will help prevent false signals due to electrical noise being picked up by the wiring.

It is very important that a thorough inspection of the wiring and connectors be performed. Failure to carefully and fully inspect wiring and connectors may result in misdiagnosis, causing part replacement with reappearance of the malfunction.

DIAGNOSTIC CHART DTC 37 — 1994, CONTINUED

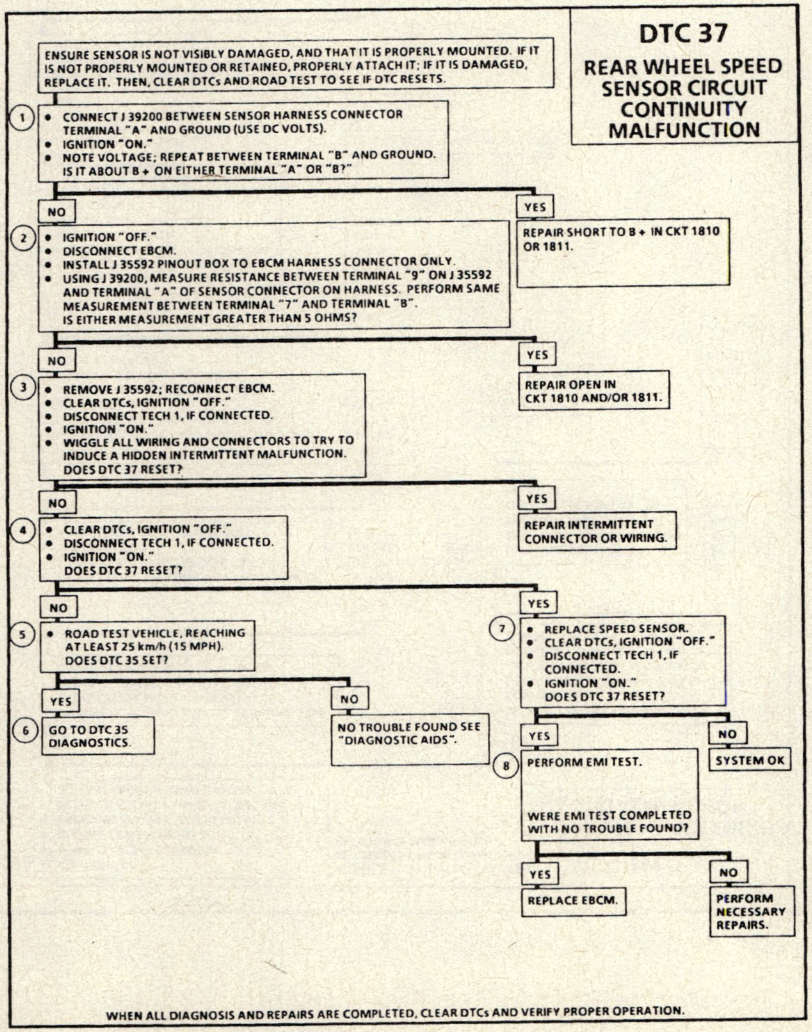

DIAGNOSTIC CHART DTC 41 — 1994

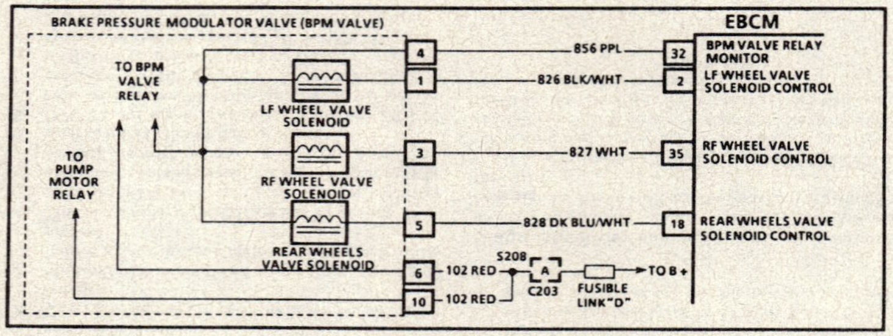

DTC 41
RF VALVE SOLENOID MALFUNCTION

Circuit Description:
With the ignition in "Run," "Bulb Test" or "Start," the EBCM applies voltage to the coil of the BPM valve relay. The EBCM will complete an internal self-check before providing ground to the coil. Once energized, the relay will provide B + to the valve solenoids.

DTC 41 Will Set When: The expected position of the right front valve solenoid does not match the commanded position from the EBCM (if valve line voltage does not match expected value). Conditions which could cause DTC 41 to set are damage to the right front solenoid, an open or short to ground or battery voltage on CKT 827.

Action Taken: If a malfunction is detected which causes a valve DTC 41 to set, antilock braking is disabled and the EBCM turns "ON" the ABS warning lamp for the remainder of the ignition cycle. If the malfunction is intermittent, the EBCM will enable the system at the next ignition cycle and a history DTC 41 will be present.

Important
- The auto test (as described under "ABS Self Diagnostics") is performed once during each ignition cycle when the vehicle reaches approximately 4 mph. This test may be heard and felt while it is taking place and should be considered a normal mode of operation. This test can also be performed by using the Tech 1 in "Misc Tests" mode F4.

DTC Chart Test Description: Number(s) below refer to circled number(s) on the diagnostic chart.
1. This test checks the integrity of CKTs 827 and 856, the RF valve solenoid circuitry internal to the BPM valve and the RF valve solenoid coil.
2. This test checks for a short to ground in CKT 827.
3. This test checks for a short to B + in CKT 827.
4. This test checks for connector or wiring malfunctions which may cause an intermittent condition.
5. This test uses the Tech 1 to check for proper valve solenoid operation.
6. This test determines whether the DTC was set by an intermittent condition or an EBCM malfunction.
7. This test determines whether the malfunction revealed in Step 1 is due to a malfunctioning BPM valve or a wiring malfunction.
8. This test determines whether the malfunction is an open CKT 827 or a short from CKT 827 to CKT 856.

Diagnostic Aids: All tests which require a connection to Pinout Box Pin 20 are using this pin as ground. This assumes that the integrity of ground CKT 51 has been maintained.

It is important that a thorough inspection of the wiring and connectors be performed prior to road testing the vehicle. This DTC can only be set when the vehicle is in motion. Failure to properly inspect wiring and connectors may result in a misdiagnosis of a malfunctioning BPM valve or EBCM.

If DTCs 41 and 45 are both set the malfunction is likely to be a short to B + on CKT 827.

If DTCs 41 and 55 are both set the malfunction is likely to be a short to B + on CKT 828.

DIAGNOSTIC CHART DTC 41 — 1994, CONTINUED

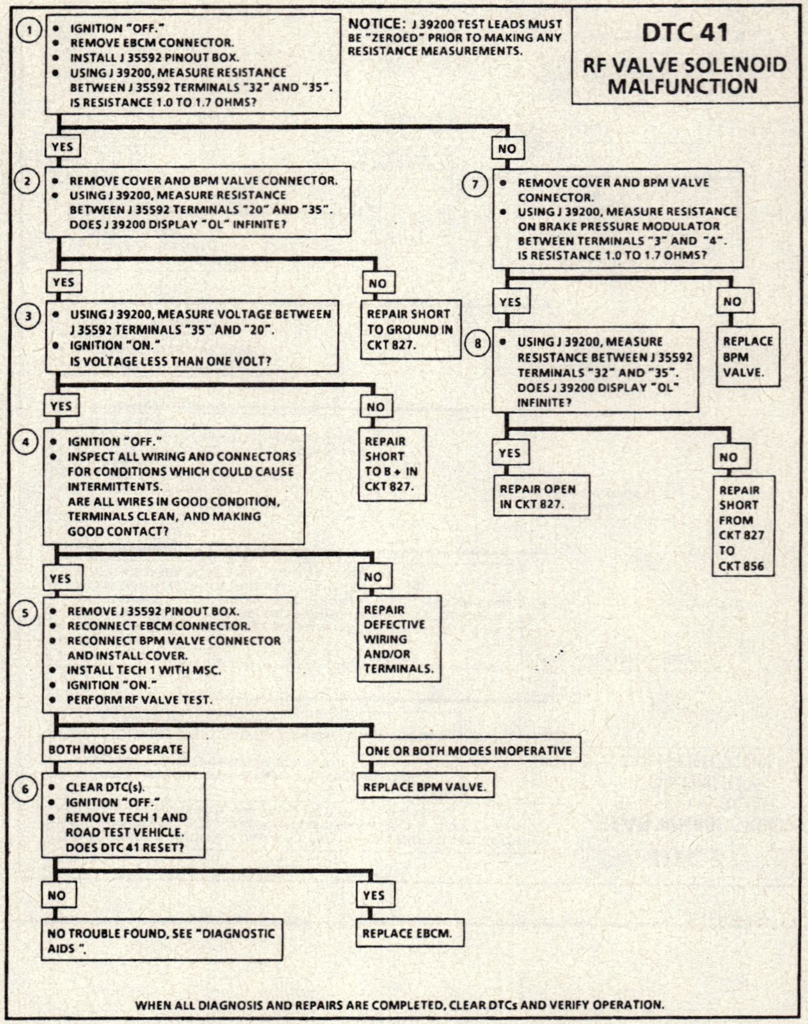

DIAGNOSTIC CHART DTC 45 — 1994

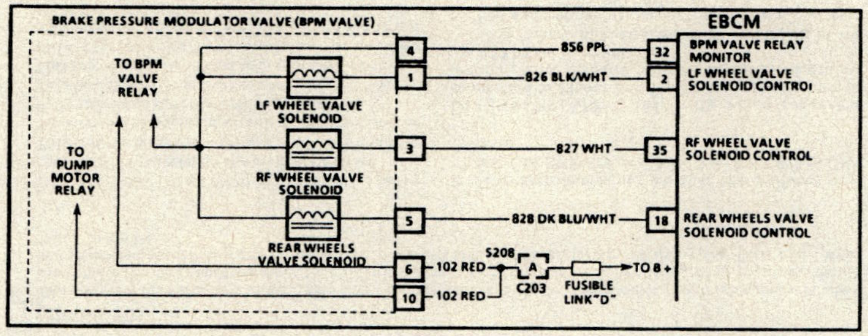

DTC 45

LF VALVE SOLENOID MALFUNCTION

Circuit Description:
With the ignition in "Run," "Bulb Test" or "Start," the EBCM applies voltage to the coil of the BPM valve relay. The EBCM will complete an internal self-check before providing ground to the coil. Once energized, the relay will provide B+ to the valve solenoids.

DTC 45 Will Set When: The expected position of the left front valve solenoid does not match the commanded position from the EBCM (if valve line voltage does not match expected value). Conditions which could cause DTC 45 to set are damage to the left front solenoid, an open or short to ground or battery voltage on CKT 826.

Action Taken: If a malfunction is detected which causes a valve DTC 45 to set, antilock braking is disabled and the EBCM turns "ON" the ABS warning lamp for the remainder of the ignition cycle. If the malfunction is intermittent, the EBCM will enable the system at the next ignition cycle and a history DTC 45 will be present.

Important
- The auto test (as described under "ABS Self Diagnostics") is performed once during each ignition cycle when the vehicle reaches approximately 4 mph. This test may be heard and felt while it is taking place and should be considered a normal mode of operation. This test can also be performed by using the Tech 1 in "Misc. Tests" mode F4.

DTC Chart Test Description: Number(s) below refer to circled number(s) on the diagnostic chart.
1. This test checks the integrity of CKTs 826 and 856, the LF valve solenoid circuitry internal to the BPM valve and the LF valve solenoid coil.
2. This test checks for a short to ground in CKT 826.
3. This test checks for a short to B+ in CKT 826.
4. This test checks for connector or wiring malfunctions which may cause an intermittent condition.
5. This test uses the Tech 1 to check for proper valve solenoid operation.
6. This test determines whether the DTC was set by an intermittent condition or an EBCM malfunction.
7. This test determines whether the malfunction revealed in Step 1 is due to a malfunctioning BPM valve or a wiring malfunction.
8. This test determines whether the malfunction is an open CKT 826 or a short from CKT 826 to CKT 856.

Diagnostic Aids: All tests which require a connection to Pinout Box Pin 20 are using this pin as ground. This assumes that the integrity of ground CKT 51 has been maintained.

It is important that a thorough inspection of the wiring and connectors be performed prior to road testing the vehicle. This DTC can only be set when the vehicle is in motion. Failure to properly inspect wiring and connectors may result in a misdiagnosis of a faulty BPM valve or EBCM.

If DTCs 41 and 45 are both set the malfunction is likely to be a short to B+ on CKT 827.

If DTCs 45 and 55 are both set the malfunction is likely to be a short to B+ on CKT 826.

DIAGNOSTIC CHART DTC 45 — 1994, CONTINUED

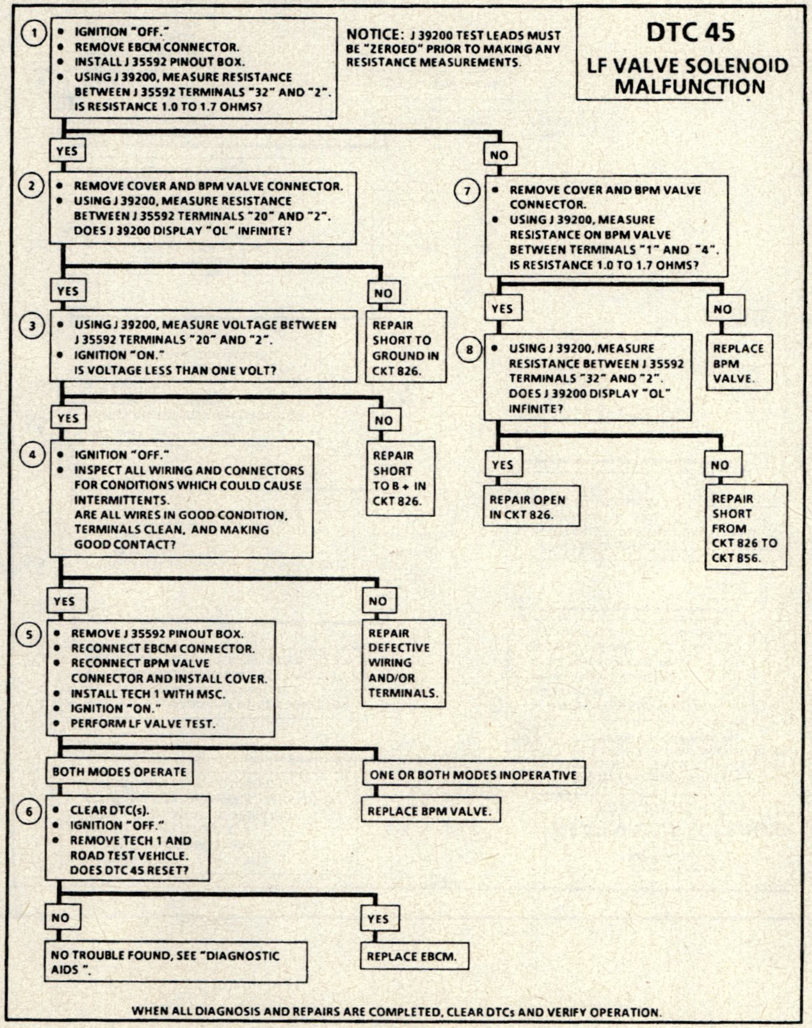

DIAGNOSTIC CHART DTC 55 — 1994

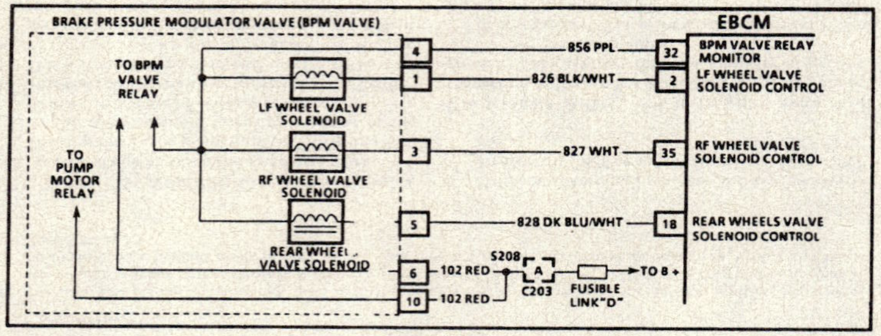

DTC 55

REAR VALVE SOLENOID MALFUNCTION

Circuit Description: With the ignition in "Run," "Bulb Test" or "Start," the EBCM applies voltage to the coil of the BPM valve relay. The EBCM will complete an internal self-check before providing ground to the coil. Once energized, the relay will provide B+ to the valve solenoids.

DTC 55 Will Set When: The expected position of the rear valve solenoid does not match the commanded position from the EBCM (if valve line voltage does not match expected value). Conditions which could cause DTC 55 to set are damage to the rear valve solenoid, an open or short to ground or battery voltage on CKT 828.

Action Taken: If a malfunction is detected which causes a valve DTC 55 to set, antilock braking is disabled and the EBCM turns "ON" the ABS warning lamp for the remainder of the ignition cycle. If the malfunction is intermittent, the EBCM will enable the system at the next ignition cycle and a history DTC 55 will be present

Important

- The auto test (as described under "ABS Self Diagnostics") is performed once during each ignition cycle when the vehicle reaches approximately 4 mph. This test may be heard and felt while it is taking place and should be considered a normal mode of operation. This test can also be performed by using the Tech 1 in "Misc. Tests" mode F4.

DTC Chart Test Description: Number(s) below refer to circled number(s) on the diagnostic chart.

1. This test checks the integrity of CKTs 828 and 856, the rear valve solenoid circuitry internal to the BPM valve and the rear valve solenoid coil.
2. This test checks for a short to ground in CKT 828.
3. This test checks for a short to B+ in CKT 828.
4. This test checks for connector or wiring malfunctions which may cause an intermittent condition.
5. This test uses the Tech 1 to check for proper valve solenoid operation.
6. This test determines whether the DTC was set by an intermittent condition or an EBCM malfunction.
7. This test determines whether the malfunction revealed in Step 1 is due to a malfunctioning BPM valve or a wiring malfunction.
8. This test determines whether the malfunction is an open CKT 828 or a short from CKT 828 to CKT 856.

Diagnostic Aids: All tests which require a connection to Pinout Box Pin 20 are using this pin as ground. This assumes that the integrity of ground CKT 51 has been maintained.

It is important that a thorough inspection of the wiring and connectors be performed prior to road testing the vehicle. This DTC can only be set when the vehicle is in motion. Failure to properly inspect wiring and connectors may result in a misdiagnosis of a faulty BPM valve or EBCM.

If DTCs 41 and 55 are both set the malfunction is likely to be a short to B+ on CKT 828.

If DTCs 45 and 55 are both set the malfunction is likely to be a short to B+ on CKT 826.

DIAGNOSTIC CHART DTC 55 — 1994, CONTINUED

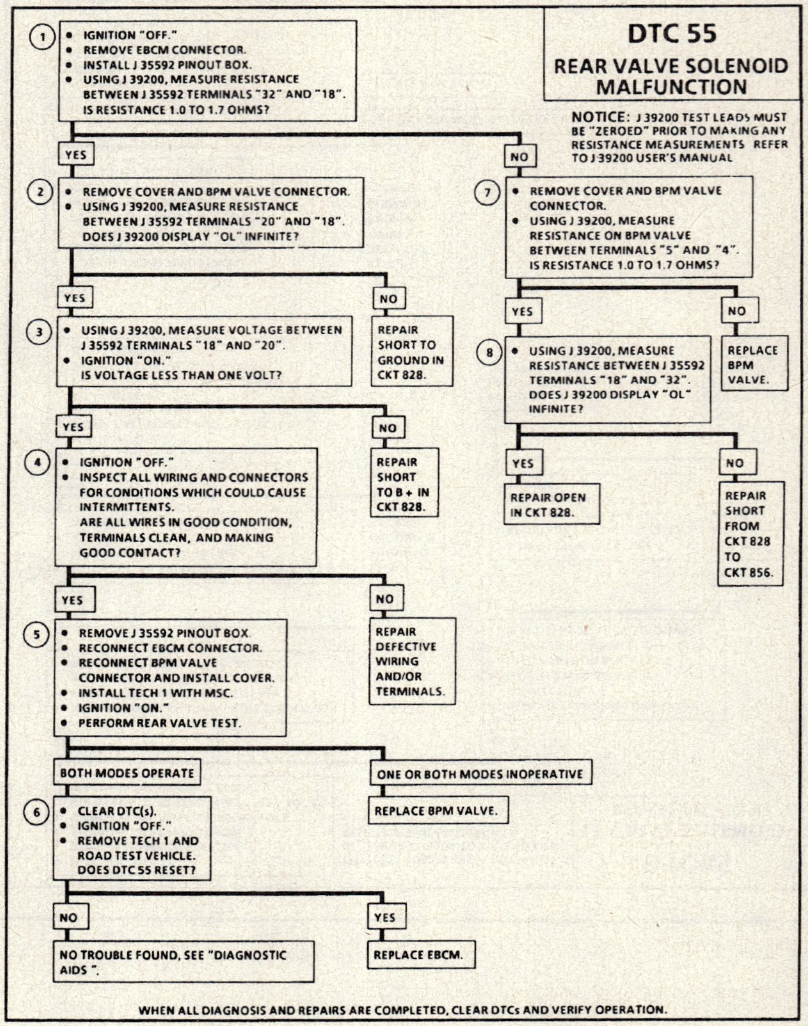

DIAGNOSTIC CHART DTC 61 — 1994

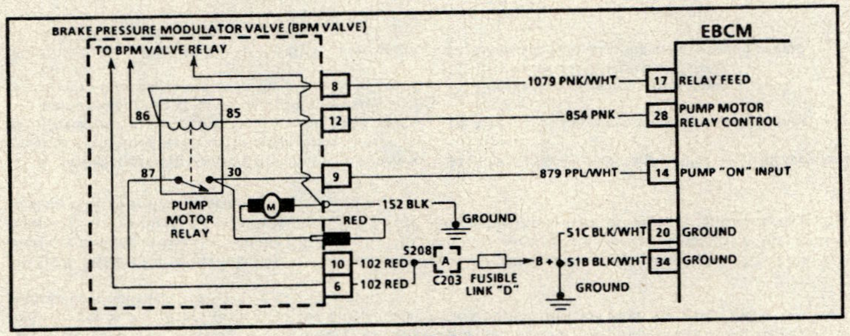

DTC 61

(Page 1 of 2)

BPM VALVE PUMP MOTOR OR PUMP MOTOR RELAY MALFUNCTION

Circuit Description:

The pump motor returns brake fluid to the master cylinder brake hydraulic circuit at the BPM valve during antilock braking. During normal braking, the pump does not operate. When the vehicle begins to move after start-up, the EBCM will turn "ON" the pump motor and perform a self-check of the pump motor and pump motor electrical circuit. This check may be felt and heard by the driver when the vehicle begins to move. The self-check should be considered normal operation. The pump motor is an integral component of the BPM valve and cannot be serviced separately.

The BPM valve pump motor relay provides power to the pump motor in the BPM valve. The BPM valve pump motor relay is located on the BPM valve and may be replaced if it is found to be defective.

DTC 61 Will Set When: B+ is present at the pump motor without the pump motor being requested to activate, or if B+ is not present at the pump motor within 60 milliseconds after the pump motor has been requested to activate.

Action Taken: If a malfunction is detected which causes a pump motor/relay DTC 61 to set, antilock braking is disabled and the EBCM turns "ON" the ABS warning lamp for the remainder of the ignition cycle. If the malfunction is intermittent, the EBCM will enable the system at the next ignition cycle and a history DTC 61 will be present.

DTC Chart Test Description: Number(s) below refer to circled number(s) on the diagnostic chart.

1. Checks for proper resistance of the relay coil.
2. Checks for relay contacts stuck closed.
3. Checks for relay contacts stuck open.
4. Checks the integrity of the circuits internal to the BPM valve.
5. Checks for an open in the pump "ON" input circuit.
6. Checks for an open in the pump motor relay coil switched ground circuit.
7. Checks for a short to ground in the pump "ON" input circuit.

DIAGNOSTIC CHART DTC 61 — 1994, CONTINUED

DTC 61

(Page 1 of 2)

BPM VALVE PUMP MOTOR OR PUMP MOTOR RELAY MALFUNCTION

Important

- TO AVOID MISDIAGNOSIS, ENSURE PUMP MOTOR RED WIRE IS NOT DAMAGED AND IS MAKING GOOD CONTACT WITH THE BRAKE PRESSURE MODULATOR.

(1)
- IGNITION "OFF."
- REMOVE COVER AND BPM VALVE PUMP MOTOR RELAY (4 PIN) FROM BPM VALVE.
- USING J 39200, MEASURE RESISTANCE BETWEEN BPM VALVE PUMP MOTOR RELAY TERMINALS "85" AND "86".

IS RESISTANCE 45 TO 55 OHMS?

NO: REPLACE BPM VALVE PUMP MOTOR RELAY.

YES:

(2)
- USING J 39200, MEASURE RESISTANCE BETWEEN BPM VALVE PUMP MOTOR RELAY TERMINALS "30" AND "87".

DOES J 39200 DISPLAY "OL" (INFINITE)?

NO: REPLACE BPM VALVE PUMP MOTOR RELAY.

YES:

(3)
- USING AN ALLIGATOR CLIP JUMPER WIRE, CONNECT BPM VALVE PUMP MOTOR RELAY TERMINAL "85" TO GROUND THEN, USING A FUSED JUMPER WIRE, SUCH AS J 36169 WITH A 3 AMP FUSE, CONNECT BPM VALVE PUMP MOTOR RELAY TERMINAL "86" TO B+.
- USING J 39200, MEASURE RESISTANCE BETWEEN BPM VALVE PUMP MOTOR RELAY TERMINALS "30" AND "87"

IS RESISTANCE 5 OHMS OR LESS?

NO: REPLACE BPM VALVE PUMP MOTOR RELAY.

YES:

(4)
- DISCONNECT BPM VALVE CONNECTOR.
- USING J 39200, MEASURE RESISTANCE ON BPM VALVE BETWEEN:
  - BPM VALVE TERMINAL "8" AND BPM VALVE PUMP MOTOR RELAY CONNECTOR TERMINAL "86".
  - BPM VALVE TERMINAL "9" AND BPM VALVE PUMP MOTOR RELAY CONNECTOR TERMINAL "30".
  - BPM VALVE TERMINAL "10" AND BPM VALVE PUMP MOTOR RELAY CONNECTOR TERMINAL "87".
  - BPM VALVE TERMINAL "12" AND BPM VALVE PUMP MOTOR RELAY CONNECTOR TERMINAL "85".

IS RESISTANCE 5 OHMS OR LESS AT EACH STEP?

NO: REPLACE BPM VALVE.

YES:

(5)
- REINSTALL BPM VALVE PUMP MOTOR RELAY ON BPM VALVE.
- DISCONNECT EBCM CONNECTOR.
- INSTALL PINOUT BOX J 35592.
- USING J 39200, MEASURE RESISTANCE BETWEEN J 35592 TERMINAL "14" AND BPM VALVE HARNESS CONNECTOR TERMINAL "9".

IS RESISTANCE 5 OHMS OR LESS?

NO: REPAIR OPEN IN CKT 879.

YES:

(6)
- USING J 39200, MEASURE RESISTANCE BETWEEN J 35592 TERMINAL "28" AND BPM VALVE HARNESS CONNECTOR TERMINAL "12".

IS RESISTANCE 5 OHMS OR LESS?

NO: REPAIR OPEN IN CKT 854.

YES:

(7)
- IGNITION "ON."
- USING J 39200, MEASURE RESISTANCE BETWEEN J 35592 TERMINALS "20" AND "14".

DOES J 39200 DISPLAY "OL" (INFINITE)?

NO: REPAIR SHORT TO GROUND IN CKT 879.

YES: CONTINUE WITH DIAGNOSIS ON NEXT PAGE

WHEN ALL DIAGNOSIS AND REPAIRS ARE COMPLETED, CLEAR DTCs AND VERIFY OPERATION.

DIAGNOSTIC CHART DTC 61 — 1994, CONTINUED

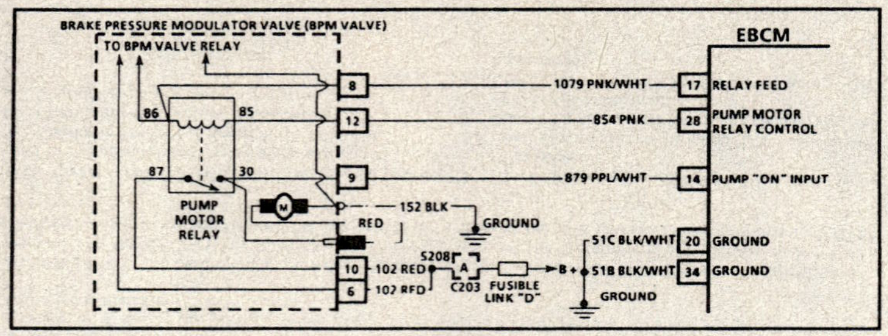

DTC 61
(Page 2 of 2)

BPM VALVE PUMP MOTOR OR PUMP MOTOR RELAY MALFUNCTION

Circuit Description:

The pump motor returns brake fluid to the master cylinder brake hydraulic circuit at the BPM valve during antilock braking. During normal braking, the pump does not operate. When the vehicle begins to move after start-up, the EBCM will turn "ON" the pump motor and perform a self-check of the pump motor and pump motor electrical circuit. This check may be felt and heard by the driver when the vehicle begins to move. The self-check should be considered normal operation. The pump motor is an integral component of the BPM valve and cannot be serviced separately.

The BPM valve pump motor relay provides power to the pump motor in the BPM valve. The BPM valve pump motor relay is located on the BPM valve and may be replaced if it is found to be defective.

DTC 61 Will Set When: B+ is present at the pump motor without the pump motor being requested to activate, or if B+ is not present at the pump motor within 60 milliseconds after the pump motor has been requested to activate.

Action Taken: If a malfunction is detected which causes a BPM valve pump motor/relay DTC 61 to set, antilock braking is disabled and the EBCM turns "ON" the ABS warning lamp for the remainder of the ignition cycle. If the malfunction is intermittent, the EBCM will enable the system at the next ignition cycle and a history DTC 61 will be present.

DTC Chart Test Description: Number(s) below refer to circled number(s) on the diagnostic chart.

8. Checks for a short to ground in the BPM valve pump motor relay coil switched ground circuit.
9. Checks for a short to B+ in the pump "ON" input circuit.
10. Checks for a short to B+ in the BPM valve pump motor relay switched ground circuit.
11. Checks for an open in the BPM valve battery feed circuit.
12. Checks for improper connections which may have set the DTC.
13. Checks that the pump motor is operational.
14. Checks for a DTC which set falsely.
15. Checks for a good pump motor ground circuit.

DIAGNOSTIC CHART DTC 61 — 1994, CONTINUED

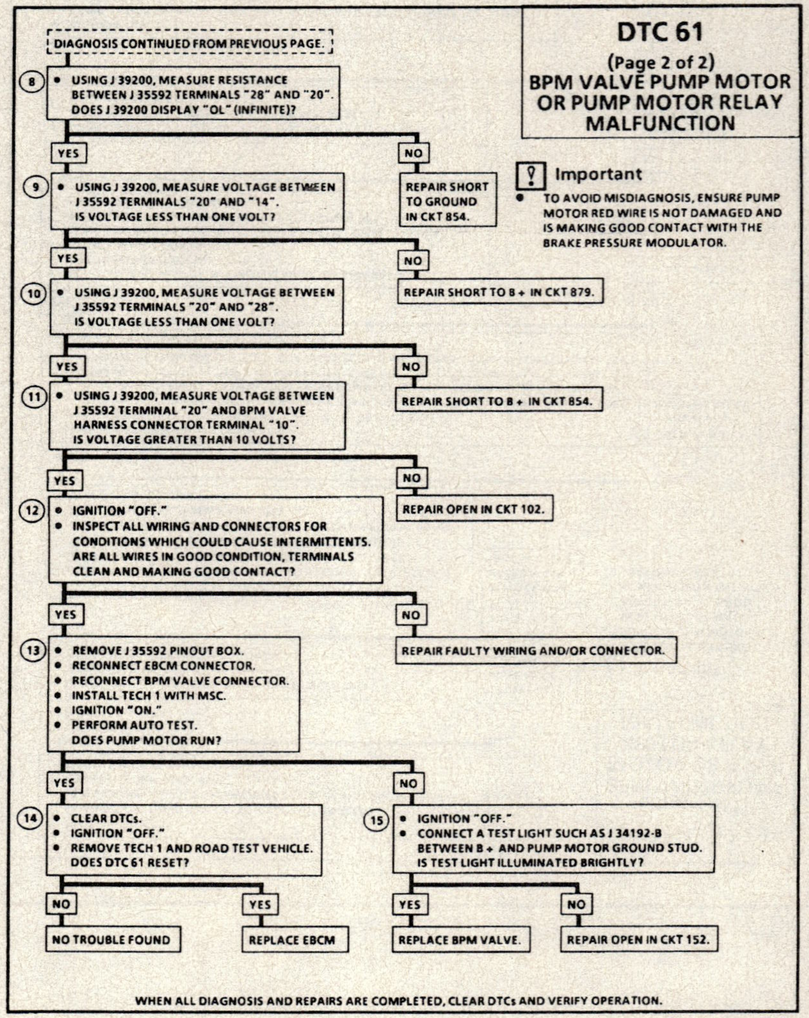

DIAGNOSTIC CHART DTC 63 — 1994

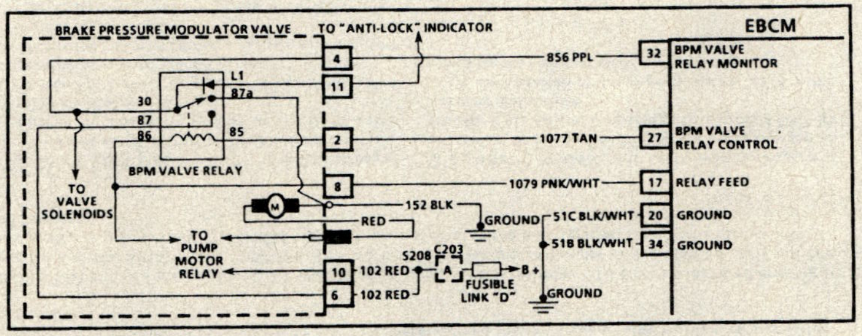

DTC 63
(Page 1 of 2)
BPM VALVE RELAY MALFUNCTION

Circuit Description:
The BPM valve relay has a dual function. With ignition "ON" and the relay coil not energized, the relay provides ground to the ABS warning lamp causing it to illuminate. This allows illumination of the ABS warning lamp when the EBCM is disconnected or disabled. With ignition "ON" and the relay coil energized, the relay supplies battery voltage to the valve solenoids. This allows the EBCM to control actuation and position of the valve solenoids by controlling the ground path. CKT 856 allows the EBCM to monitor the state of the BPM valve relay to compare with the requested state.

DTC 63 Will Set When: The BPM valve relay monitor is at B+ when the EBCM is not requesting it to be, or if the EBCM is requesting the BPM valve relay be energized and the BPM valve relay monitor voltage is less than 5 volts.

Action Taken: If a malfunction is detected which causes a BPM valve relay DTC 63 to set, antilock braking is disabled and the EBCM turns "ON" the ABS warning lamp for the remainder of the ignition cycle. If the malfunction is intermittent, the EBCM will enable the system at the next ignition cycle and a history DTC 63 will be present.

DTC Chart Test Description: Number(s) below refer to circled number(s) on the diagnostic chart.
1. Checks for proper resistance of the relay coil.
2. Checks for the relay contacts stuck in the normally open position.
3. Checks for the relay contacts stuck in the normally closed position.
4. Checks for integrity of the circuitry internal to the BPM valve.
5. Checks for a good pump motor ground circuit.
6. Checks for an open in the BPM valve relay monitor circuit.
7. Checks for an open in the BPM valve relay coil switched ground circuit.
8. Checks for an open in the relay feed (coil battery feed) circuit.

DIAGNOSTIC CHART DTC 63 — 1994, CONTINUED

DTC 63
(Page 1 of 2)
BPM VALVE RELAY MALFUNCTION

(1)
- IGNITION "OFF."
- REMOVE COVER AND BPM VALVE RELAY (6-PIN) FROM BPM VALVE.
- USING J 39200, MEASURE RESISTANCE BETWEEN RELAY TERMINALS "85" AND "86". IS RESISTANCE 52 TO 64 OHMS?

NO → REPLACE BPM VALVE RELAY.

YES ↓

(2)
- USING J 39200, MEASURE RESISTANCE BETWEEN BPM VALVE RELAY TERMINALS "30" AND "87A". IS RESISTANCE 5 OHMS OR LESS?

NO → REPLACE BPM VALVE RELAY.

YES ↓

(3)
- USING AN ALLIGATOR CLIP JUMPER WIRE, CONNECT RELAY TERMINAL "85" TO GROUND. THEN, USING A FUSED JUMPER WIRE, SUCH AS J 36169 WITH A 3 AMP FUSE, CONNECT RELAY TERMINAL "86" TO B+.
- USING J 39200, MEASURE RESISTANCE BETWEEN RELAY TERMINALS "30" AND "87". IS RESISTANCE 5 OHMS OR LESS?

NO → REPLACE BPM VALVE RELAY.

YES ↓

(4)
- DISCONNECT BPM VALVE CONNECTOR.
- USING J 39200, MEASURE RESISTANCE ON BPM VALVE BETWEEN:
  - BPM VALVE TERMINAL "2" AND BPM VALVE RELAY CONNECTOR TERMINAL "85".
  - BPM VALVE TERMINAL "4" AND BPM VALVE RELAY CONNECTOR TERMINAL "30".
  - BPM VALVE TERMINAL "6" AND BPM VALVE RELAY CONNECTOR TERMINAL "87".
  - BPM VALVE TERMINAL "8" AND BPM VALVE RELAY CONNECTOR TERMINAL "86".
  - BPM VALVE TERMINAL "11" AND BPM VALVE RELAY CONNECTOR TERMINAL "L1".
  - PUMP MOTOR GROUND AND BPM VALVE RELAY CONNECTOR TERMINAL "87A".

  IS RESISTANCE 5 OHMS OR LESS AT EACH STEP?

NO → REPLACE BPM VALVE.

YES ↓

(5)
- REINSTALL BPM VALVE RELAY.
- DISCONNECT EBCM CONNECTOR.
- INSTALL PINOUT BOX J 35592.
- USING J 39200, MEASURE RESISTANCE BETWEEN PUMP MOTOR GROUND AND J 35592 TERMINAL "20" (GROUND). IS RESISTANCE 5 OHMS OR LESS?

NO → REPAIR OPEN IN CKT 152.

YES ↓

(6)
- USING J 39200, MEASURE RESISTANCE BETWEEN J 35592 TERMINAL "32" AND BPM VALVE HARNESS CONNECTOR TERMINAL "4". IS RESISTANCE 5 OHMS OR LESS?

NO → REPAIR OPEN IN CKT 856.

YES ↓

(7)
- USING J 39200, MEASURE RESISTANCE BETWEEN J 35592 TERMINAL "27" AND BPM VALVE HARNESS CONNECTOR TERMINAL "2". IS RESISTANCE 5 OHMS OR LESS?

NO → REPAIR OPEN IN CKT 1077.

YES ↓

(8)
- USING J 39200, MEASURE RESISTANCE BETWEEN J 35592 TERMINAL "17" AND BPM VALVE HARNESS CONNECTOR TERMINAL "8". IS RESISTANCE 5 OHMS OR LESS?

NO → REPAIR OPEN IN CKT 1079.

YES → DIAGNOSTICS CONTINUED ON NEXT PAGE.

WHEN ALL DIAGNOSIS AND REPAIRS ARE COMPLETED, CLEAR DTCs AND VERIFY OPERATION.

DIAGNOSTIC CHART DTC 63 — 1994, CONTINUED

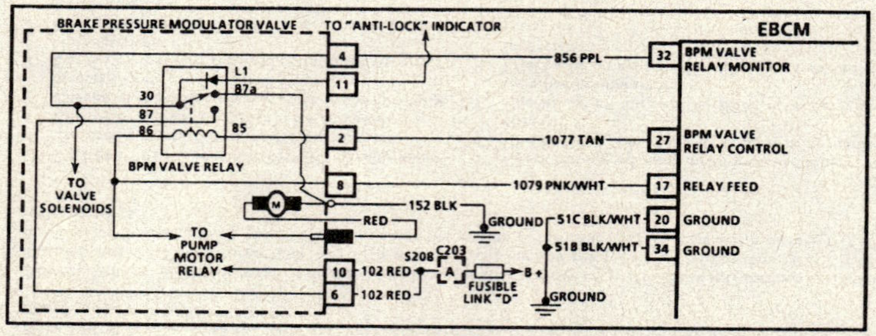

DTC 63
(Page 2 of 2)
BPM VALVE RELAY MALFUNCTION

Circuit Description:
The BPM valve relay has a dual function With ignition "ON" and the relay coil not energized, the relay provides ground to the ABS warning lamp causing it to illuminate. This allows illumination of the ABS warning lamp when the EBCM is disconnected or disabled. With ignition "ON" and the relay coil energized, the relay supplies battery voltage to the valve solenoids. This allows the EBCM to control actuation and position of the valve solenoids by controlling the ground path. CKT 856 allows the EBCM to monitor the state of the BPM valve relay to compare with the requested state.

DTC 63 Will Set When: The BPM valve relay monitor is at B+ when the EBCM is not requesting it to be, or if the EBCM is requesting the BPM valve relay be energized and the BPM valve relay monitor voltage is less than 5 volts.

Action Taken: If a malfunction is detected which causes a BPM valve relay DTC 63 to set, antilock braking is disabled and the EBCM turns "ON" the ABS warning lamp for the remainder of the ignition cycle. If the malfunction is intermittent, the EBCM will enable the system at the next ignition cycle and a history DTC 63 will be present.

DTC Chart Test Description: Number(s) below refer to circled number(s) on the diagnostic chart.

9. Checks for a short to ground in the BPM valve relay monitor circuit.
10. Checks for a short to ground in the BPM valve relay coil switched ground circuit.
11. Checks for a short to ground in the relay feed (coil battery feed) circuit.
12. Checks for a short to B+ in the BPM valve relay monitor circuit.
13. Checks for a short B+ in the BPM valve relay coil switched ground circuit.
14. Checks for an open in the BPM valve battery feed circuit.
15. Checks for a short to ground internal to the BPM valve.
16. Checks for improper connections which may have set the DTC.
17. Checks for a DTC which may have set falsely.

Diagnostic Aids: All tests which require a connection to pinout box Pin 20 are using this pin as ground. This assumes the integrity of ground CKT 51 has been maintained.

A disconnected or improperly seated BPM valve connector may set this DTC.

DIAGNOSTIC CHART DTC 63 — 1994, CONTINUED

DTC 63
(Page 2 of 2)
BPM VALVE RELAY MALFUNCTION

DIAGNOSIS CONTINUED FROM PREVIOUS PAGE.

(9)
- USING J 39200, MEASURE RESISTANCE BETWEEN J 35592 TERMINALS "32" AND "20". DOES J 39200 DISPLAY "OL" INFINITE?

NO: REPAIR SHORT TO GROUND IN CKT 856.
YES:

(10)
- USING J 39200, MEASURE RESISTANCE BETWEEN J 35592 TERMINALS "27" AND "20". DOES J 39200 DISPLAY "OL" INFINITE?

NO: REPAIR SHORT TO GROUND IN CKT 1077.
YES:

(11)
- USING J 39200, MEASURE RESISTANCE BETWEEN J 35592 TERMINALS "17" AND "20". DOES J 39200 DISPLAY "OL" INFINITE?

NO: REPAIR SHORT TO GROUND IN CKT 1079.
YES:

(12)
- IGNITION "ON."
- USING J 39200, MEASURE VOLTAGE BETWEEN J 35592 TERMINALS "32" AND "20". IS VOLTAGE LESS THAN ONE VOLT?

NO: REPAIR SHORT TO B+ IN CKT 856.
YES:

(13)
- USING J 39200, MEASURE VOLTAGE BETWEEN J 35592 TERMINALS "27" AND "20". IS VOLTAGE LESS THAN ONE VOLT?

NO: REPAIR SHORT TO B+ IN CKT 1077.
YES:

(14)
- CONNECT TEST LIGHT BETWEEN PINOUT BOX TERMINAL "20" (GROUND) AND BPM VALVE CONNECTOR TERMINAL "6". IS TEST LIGHT "ON"?

NO: REPAIR OPEN IN CKT 102.
YES:

(15)
- IGNITION "OFF."
- USING J 39200, MEASURE THE RESISTANCE BETWEEN PUMP MOTOR GROUND AND BPM VALVE TERMINALS "2" AND "8". IS RESISTANCE INFINITE (OPEN) AT BOTH STEPS?

NO: REPLACE BPM VALVE.
YES:

(16)
- REMOVE J 35592 PINOUT BOX.
- RECONNECT EBCM CONNECTOR.
- RECONNECT BPM VALVE CONNECTOR AND INSTALL COVER.
- INSPECT ALL WIRING AND CONNECTORS FOR CONDITIONS WHICH COULD CAUSE INTERMITTENTS. ARE ALL WIRES IN GOOD CONDITION, TERMINALS CLEAN AND MAKING GOOD CONTACT?

NO: REPAIR FAULTY WIRING AND/OR CONNECTOR.
YES:

(17)
- CLEAR DTCs AND ROAD TEST VEHICLE. DOES DTC 63 RESET?

YES: REPLACE EBCM.
NO:
- NO TROUBLE FOUND. SEE "DIAGNOSTIC AIDS".

WHEN ALL DIAGNOSIS AND REPAIRS ARE COMPLETED, CLEAR DTCs AND VERIFY OPERATION.

DIAGNOSTIC CHART DTC 71 — 1994

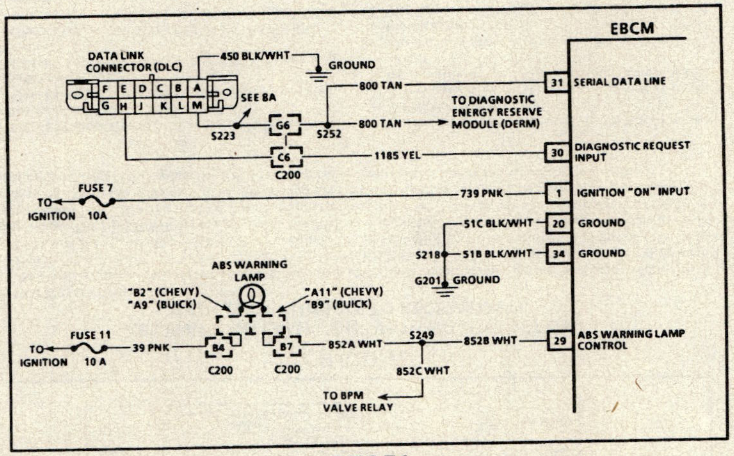

DTC 71
EBCM INTERNAL MALFUNCTION (EBCM FAULT)

Circuit Description:
The EBCM has self diagnostics that can detect an internal malfunction within the module. The EBCM also monitors its supply voltage and the integrity of the EBCM harness connection. When a malfunction is detected for any of these conditions DTC 71 is set.

Action Taken: If a malfunction is detected which causes a EBCM DTC 71 to set, antilock braking is disabled and the EBCM turns "ON" the ABS warning lamp for the remainder of the ignition cycle. If the malfunction is intermittent, the EBCM will enable the system at the next ignition cycle and a history DTC 71 will be present.

DTC Chart Test Description: Number(s) below refer to circled number(s) on the diagnostic chart.

1. Checks for additional malfunctions which may be causing DTC 71 to set.
2. Checks the integrity of the EBCM ground circuit.
3. Checks for the minimum supply voltage necessary for proper ABS operation.
4. Checks for intermittent connections of the EBCM harness connector and terminals to the EBCM.

DIAGNOSTIC CHART DTC 71 — 1994, CONTINUED

DTC 71
EBCM INTERNAL MALFUNCTION (EBCM FAULT)

(1)
- CHECK FOR ABS DTC(s), OTHER THAN DTC 71. ARE THERE ANY OTHER DTC(s) SET?

YES:
- REFER TO THE APPROPRIATE CHART, REPAIR THE PROBLEM, CLEAR DTC(s) AND ROAD TEST VEHICLE TO SEE IF DTC 71 RESETS.
- IF DTC 71 IS PRESENT, GO TO TOP OF THIS CHART.

NO:

(2)
- IGNITION "OFF."
- INSTALL PINOUT BOX J 35592.
- CONNECT A TEST LIGHT SUCH AS J 34192-B BETWEEN B+ AND J 35592 TERMINAL "20" AND GROUND. IS TEST LIGHT ILLUMINATED BRIGHTLY?

NO:
- CHECK FOR A GOOD GROUND CONNECTION G201 AT THE THERMOSTAT HOUSING.
- IF CONNECTION IS TIGHT AND FREE OF CORROSION THEN REPAIR OPEN IN CKT 51.

YES:

(3)
- IGNITION "ON."
- USING J 39200, MEASURE VOLTAGE BETWEEN J 35592 TERMINALS "20" AND "1". IS VOLTAGE 10 VOLTS OR ABOVE?

NO:
- IGNITION "OFF."
- CHECK FOR HIGH RESISTANCE IN CKT 739.

YES:

(4)
- IGNITION "OFF."
- CHECK EBCM CONNECTOR CAREFULLY TO ENSURE GOOD ELECTRICAL CONTACT AT ALL PINS AND THAT CONNECTOR IS PROPERLY RETAINED WHEN CONNECTED. IS IT OK?

YES: REPLACE EBCM.

NO: REPAIR OR REPLACE CONNECTOR AS NECESSARY.

WHEN ALL DIAGNOSIS AND REPAIRS ARE COMPLETED, CLEAR DTCs AND VERIFY OPERATION.

DIAGNOSTIC CHART A — 1994

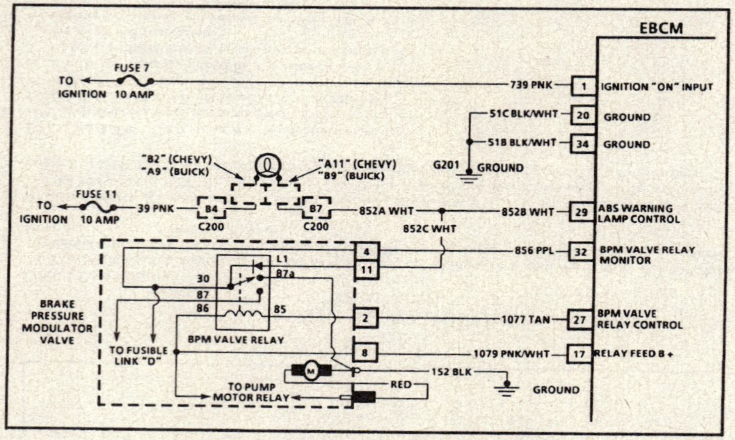

CHART A

ABS WARNING LAMP "ON" WITH NO DTCs SET OR WITH TECH 1 UNABLE TO RECEIVE DATA

Circuit Description:

The ABS warning lamp is located on the right side of the instrument panel. Battery voltage is applied with ignition in "Run," "Bulb," or "Start" through Fuse 11 and CKT 39. The ABS warning lamp is grounded through CKT 852 and is controlled by the EBCM at terminal "29" and the BPM valve at terminal "11" through the BPM valve relay. The EBCM will supply ground to the ABS warning lamp for approximately 2 seconds when ignition is first turned "ON." During the first 2 seconds, if no malfunctions are detected, the EBCM will energize the BPM valve relay. With the relay energized, the ground CKT 852 through the relay will be open. This allows the ABS warning lamp to go "OFF" when the EBCM ABS warning lamp control line at terminal "29" is opened.

Chart Test Description: Number(s) below refer to circled number(s) on the diagnostic chart.

1. Checks for proper battery and/or charging system operation.
2. Checks for an open fuse in the EBCM power feed circuit.
3. Checks for an open in the EBCM ground circuit.
4. Checks for an open in the EBCM power feed circuit.
5. Checks for a possible malfunction in the BPM valve relay.
6. Determines whether the malfunction is in the BPM valve relay or the EBCM.
7. Checks for a short to ground in the EBCM or the EBCM power feed circuit.
8. Determines whether the short to ground is in the BPM valve or the ABS warning lamp control circuit.
9. Checks for a short to ground in the ignition "ON" input circuit.
10. Checks for a short to ground in the BPM valve and pump motor relay feed circuitry.
11. Determines whether the short to ground is in the BPM valve or its relay feed circuitry.

DIAGNOSTIC CHART A — 1994, CONTINUED

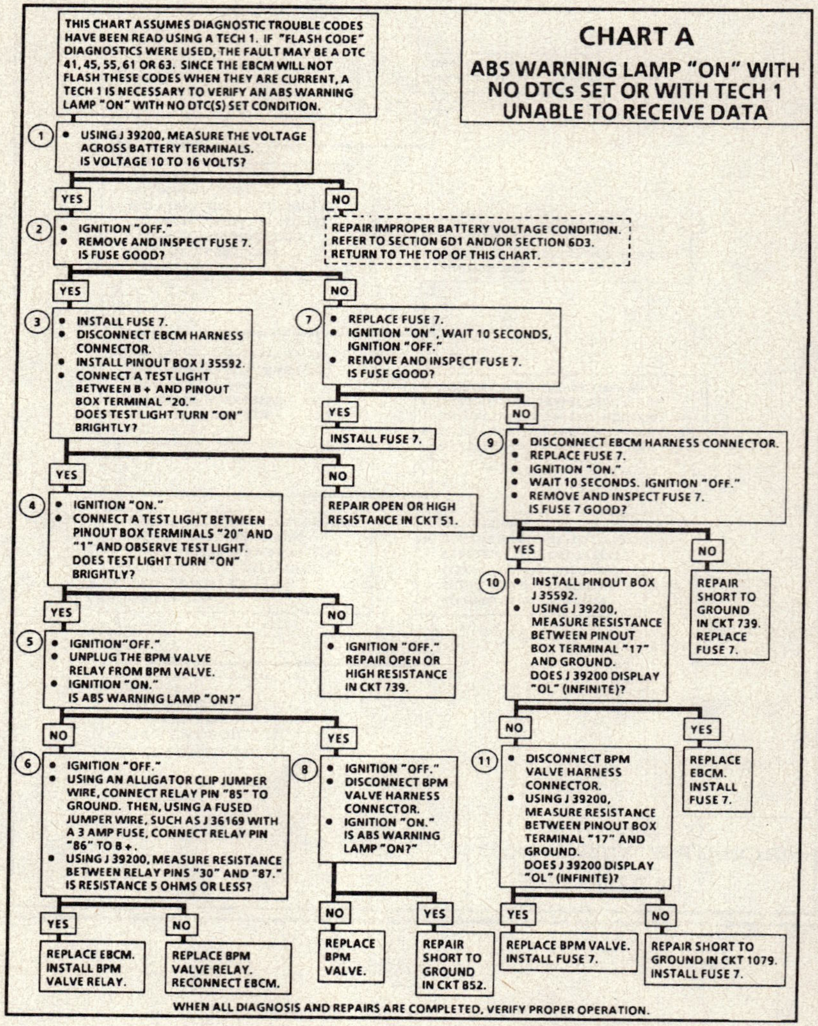

DIAGNOSTIC CHART B — 1994

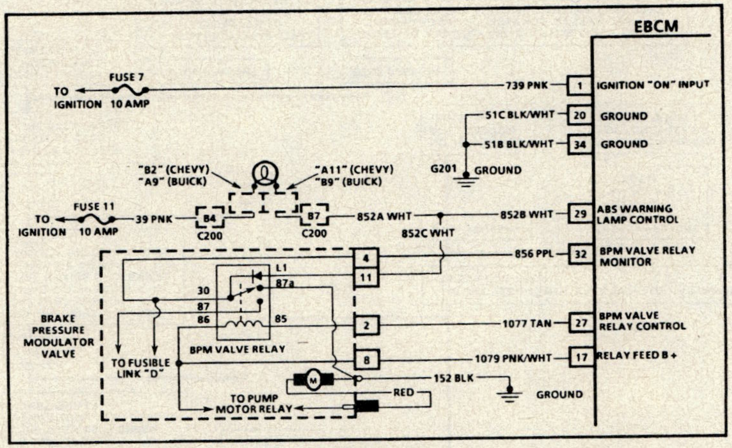

CHART B

ABS WARNING LAMP INOPERATIVE OR LAMP FLASHES VERY BRIEFLY AT IGNITION "ON" (LESS THAN ONE HALF OF A SECOND)

Circuit Description:

The ABS warning lamp is located on the right side of the instrument panel. Battery voltage is applied with ignition in "Run, Bulb Test or Start" through Fuse 11 and CKT 39. The ABS warning lamp is grounded through CKT 852 and is controlled by the EBCM at terminal "29" and the BPM valve at terminal "11" through the BPM valve relay. The EBCM will supply ground to the ABS warning lamp for approximately 2 seconds when ignition is first turned "ON." During the first 2 seconds, if no malfunctions are detected, the EBCM will energize the BPM valve relay. With the relay energized, the ground CKT 852 through the relay will be opened. This allows the ABS warning lamp to go "OFF" when the EBCM warning lamp control line at terminal "29" is opened.

Chart Test Description: Number(s) below refer to circled number(s) on the diagnostic chart.

1. Verifies an improperly operating ABS warning lamp.
2. Determines whether the malfunction is an inoperative ABS warning lamp or an improperly operating ABS warning lamp.
3. Checks for an inoperative instrument cluster.
4. Determines whether the malfunction is in the instrument cluster portion of the circuitry or in the ABS portion.
5. Checks for a bad bulb.
6. Determines whether the malfunction is in the wiring, connections, or in the instrument cluster printed circuits.
7. Determines whether the malfunction is in the wiring, connections, or BPM valve relay.
8. Determines whether the malfunction is in the wiring or connections.

Diagnostic Aids: If a lamp malfunction is intermittent, install the Tech 1 diagnostic tool and select any ABS test mode so the ABS warning lamp will be turned "ON." While observing the lamp, go through the lamp's power and ground circuits wiggling the wiring and connectors. Watch for flickering or failure of the lamp to pin point problem area.

DIAGNOSTIC CHART B — 1994, CONTINUED

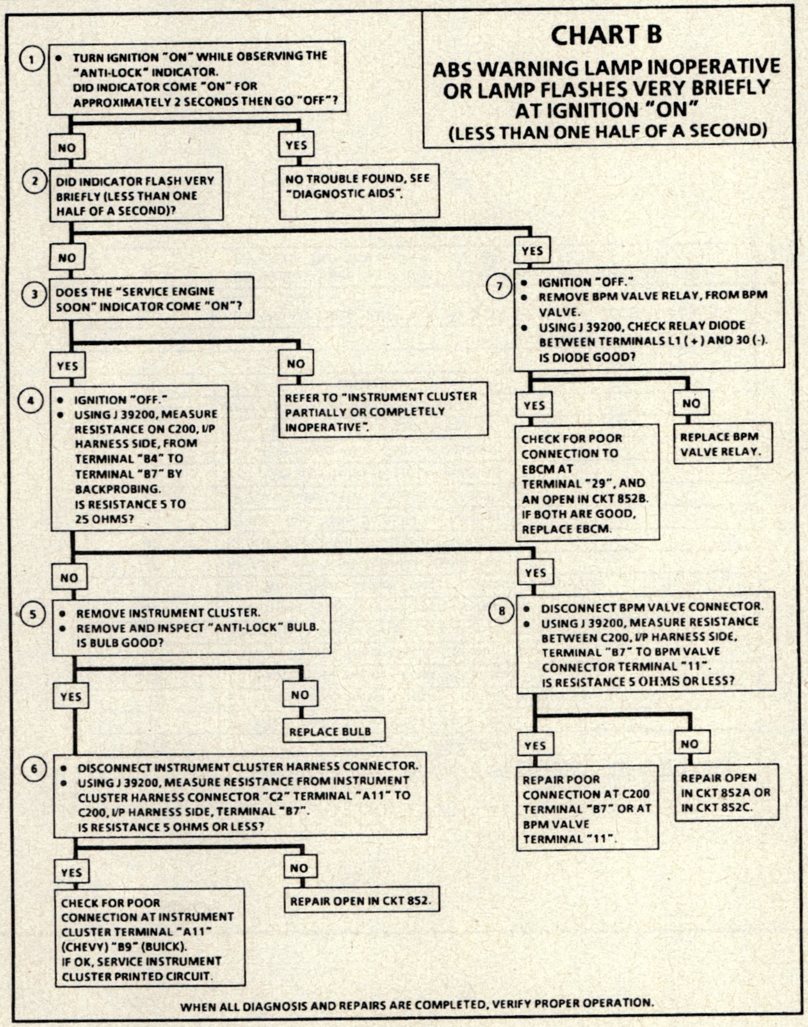

DIAGNOSTIC SYSTEM CHECK — 1995

DIAGNOSTIC SYSTEM CHECK

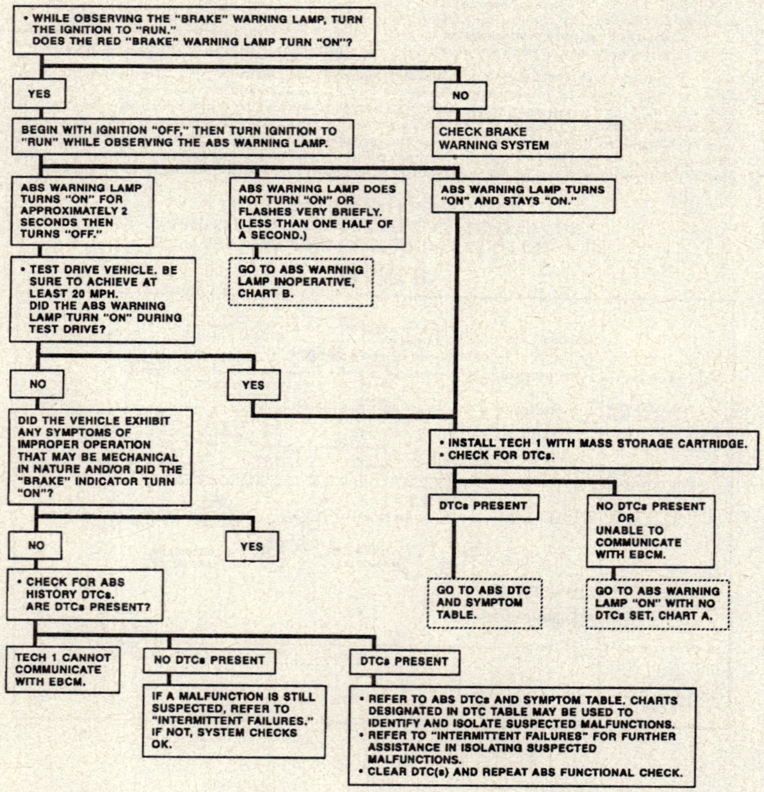

WHEN ALL DIAGNOSIS AND REPAIRS ARE COMPLETED, CLEAR DTCs AND VERIFY OPERATION.

DIAGNOSTIC TROUBLE CODE CHART — 1995

ABS DIAGNOSTIC TROUBLE CODE AND SYMPTOM TABLE

DTC	DESCRIPTION
21	RF Wheel Speed Sensor Circuit Malfunction
23	RF Wheel Speed Sensor Circuit Continuity Malfunction
25	LF Wheel Speed Sensor Circuit Malfunction
27	LF Wheel Speed Sensor Circuit Continuity Malfunction
28	Wheel Speed Sensor Circuit Frequency Malfunction
35	Rear Wheel Speed Sensor Circuit Malfunction
37	Rear Wheel Speed Sensor Circuit Continuity Malfunction
41	RF ABS Inlet Valve Solenoid Malfunction
42	RF ABS Outlet Valve Solenoid Malfunction
45	LF ABS Inlet Valve Solenoid Malfunction
46	LF ABS Outlet Valve Solenoid Malfunction
55	Rear ABS Inlet Valve Solenoid Malfunction
56	Rear ABS Outlet Valve Solenoid Malfunction
61	BPM Valve Pump Motor Malfunction
63	BPM Valve Power Supply Malfunction
71	EBCM Internal Malfunction
85	Low Voltage
CHART	**SYMPTOM**
A	ABS Warning Lamp "ON" With No DTCs Set Or With TECH1 Unable To Receive Data
B	ABS Warning Lamp Inoperative Or Lamp Flashes Very Briefly At Ingition "ON" (Less Than One Half Of A Second)

ELECTRICAL CONNECTOR TERMINAL IDENTIFICATION — 1995

C200

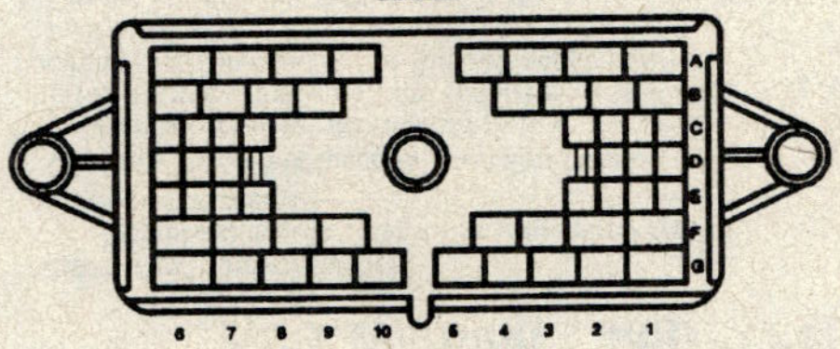

12034257
56 – WAY F METRI – PACK MIXED SERIES
BLK

C402B

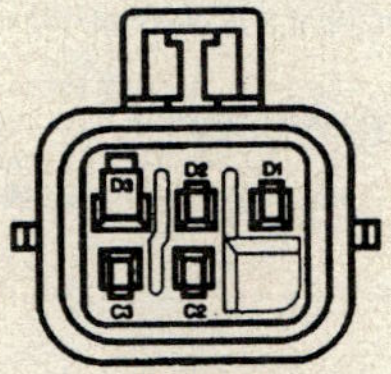

12110481
5 – WAY F METRI – PACK 150, 280 SERIES
RED
ANTI LOCK BRAKE TO BODY HARNESS SYSTEM HARNESS

C204

BODY HARNESS TO ENGINE HARNESS

12020916
37 – WAY F METRI – PACK MIXED SERIES
BLK

C402

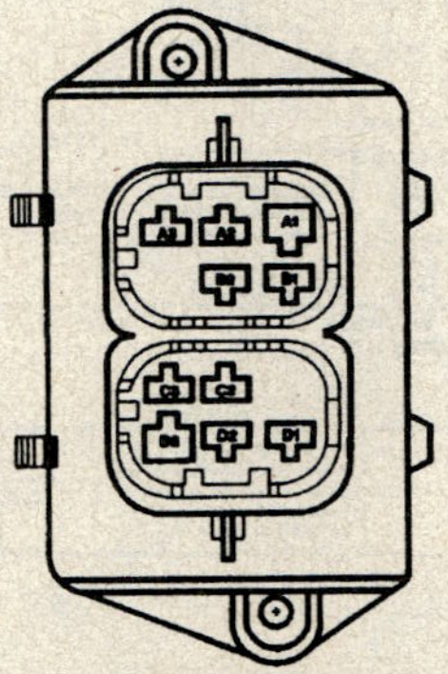

12092789
10 – WAY M METRI – PACK 150, 280 SERIES
BLK
ANTILOCK BRAKE SYSTEM PASS THRU BODY HARNESS TO ELECTRONIC BRAKE CONTROL HARNESS

ELECTRICAL CONNECTOR TERMINAL IDENTIFICATION — 1995, CONTINUED

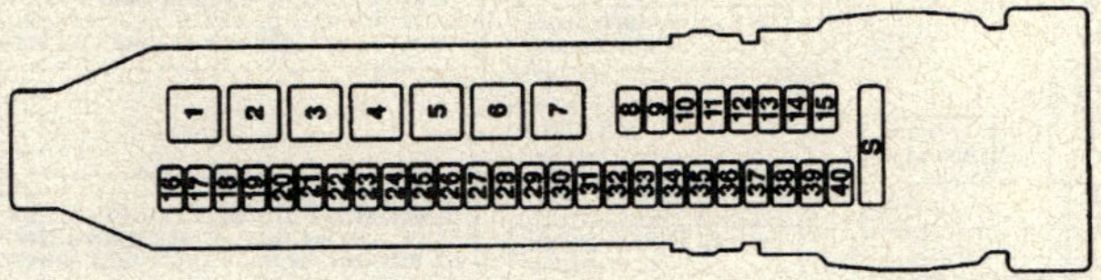

12124389
40 – WAY F
NAT
BRAKE PRESSURE MODULATOR (BPM) VALVE
(WITH ELECTRONIC BRAKE CONTROL MODULE (EBCM))

CAVITIES NOT LISTED ARE NOT USED

CAVITY	WIRE COLOR	CKT	DESCRIPTION
1	BLK/WHT	51	Ground
2	RED	102	Voltage Feed to BPMV Pump Motor Relay
3	RED	102	Voltage Feed to BPMV Solenoid Valve Relay
4	PNK	539	Ignition Feed
15	BLK/WHT	451	Ground
16	LT BLU	830	LH Front Wheel Speed Sensor Input
17	YEL	873	LH Front Wheel Speed Sensor Input Ground
18	WHT	1810	Rear Wheel Speed Sensor Input
19	BRN	1811	Rear Wheel Speed Sensor Input Ground
23	DK GRN	872	RH Front Wheel Speed Sensor Input
24	TAN	833	RH Front Wheel Speed Sensor Input Ground
28	WHT	17	Brake Applied Signal
29	BRN	800	Serial Data Line
40	WHT	852	"ABS" Indicator Control
S	–	–	Shorting Bar

DIAGNOSTIC CHART A — 1995

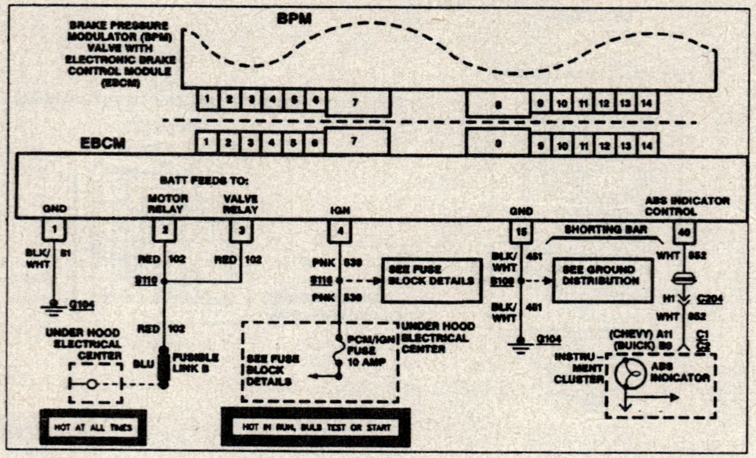

CHART A
(Page 1 of 2)
ABS Warning Lamp "ON" With No DTC(s) Set

Chart Test Description:

Number(s) below refer to circled number(s) on the diagnostic chart.

1. Confirms that a malfunction condition exists.
2. Determines whether the malfunction is due to a DTC setting condition, an indicator low side circuit malfunction, or an EBCM supply power malfunction.
3. Checks whether the malfunction is due to an open fuse.
4. Checks for an open power feed circuit.
5. Checks for an open in the ground feed to the EBCM.
6. Checks for proper connection of EBCM ground CKT 51.
7. Checks whether the fuse malfunction is due to a short to ground.
8. Identifies if the short to ground found in Step 7 is in CKT 539 or the EBCM.
9. Checks for proper retention of EBCM harness connector.
10. Checks for damaged EBCM harness connector.
11. Checks for a damaged connector (broken lift finger) on the EBCM itself, or for a short to ground in circuit 852.

DIAGNOSTIC CHART A — 1995, CONTINUED

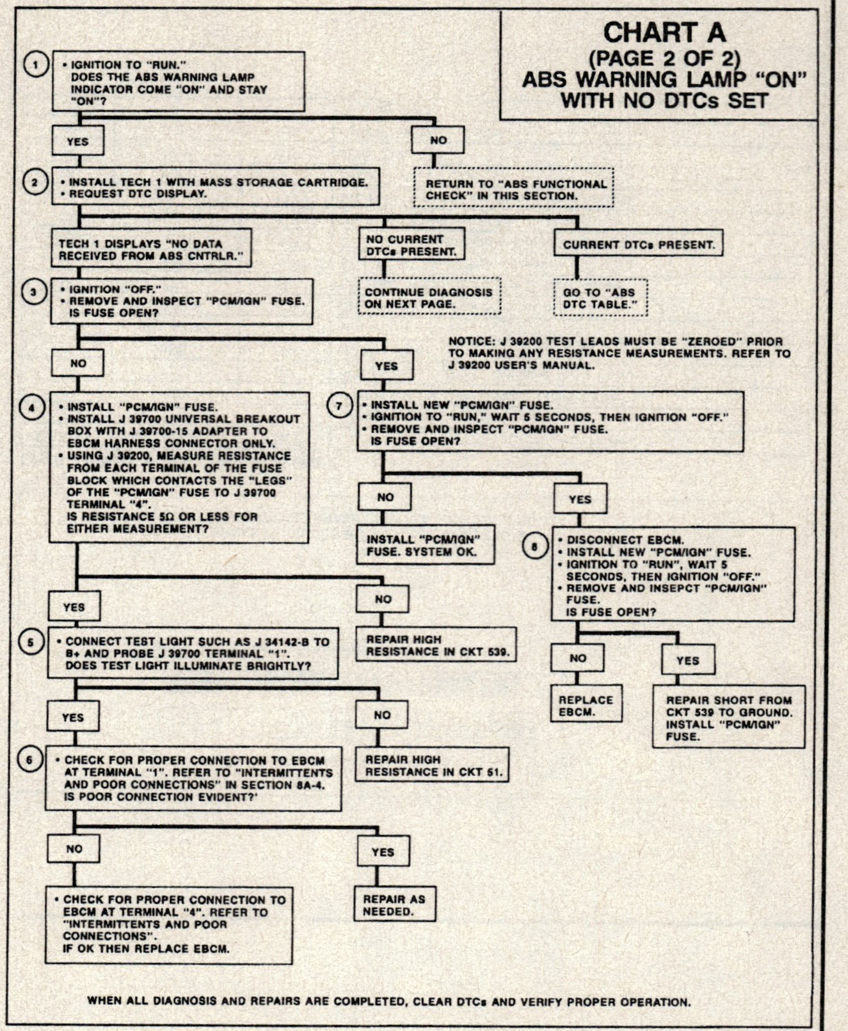

DIAGNOSTIC CHART A — 1995, CONTINUED

CHART A
(PAGE 2 OF 2)
ABS WARNING LAMP "ON" WITH NO DTCs SET

DIAGNOSIS CONTINUED FROM PREVIOUS PAGE.

(9)
- IGNITION "OFF."
- CHECK EBCM HARNESS CONNECTOR CAREFULLY TO ENSURE IT IS PROPERLY RETAINED TO THE EBCM.

IS CONNECTOR PROPERLY RETAINED?

- YES → go to step 10
- NO → RE-ENGAGE CONNECTOR TO EBCM TO OBTAIN PROPER CONNECTOR RETENTION.

(10)
- IGNITION "OFF."
- DISCONNECT J 39700.
- INSPECT HARNESS CONNECTOR FOR DAMAGE AT LOCATION "S" (SHORTING BAR) SUCH AS: BENT CONTACTS, OR BROKEN/BENT LIFT TAB.

IS DAMAGE FOUND?

- NO → go to step 11
- YES → REPAIR OR REPLACE CONNECTOR.

(11)
- IGNITION "OFF."
- DISCONNECT EBCM CONNECTOR.
- CONNECT J 39700 UNIVERSAL BREAKOUT BOX WITH J 39700-15 ADAPTER TO HARNESS ONLY.
- TURN IGNITION TO "RUN".

IS ABS WARNING LAMP "ON"?

- YES → REPAIR SHORT TO GROUND IN CKT 852.
- NO → REPLACE EBCM.

WHEN ALL DIAGNOSIS AND REPAIRS ARE COMPLETED, CLEAR DTCs AND VERIFY PROPER OPERATION.

DIAGNOSTIC CHART B — 1995

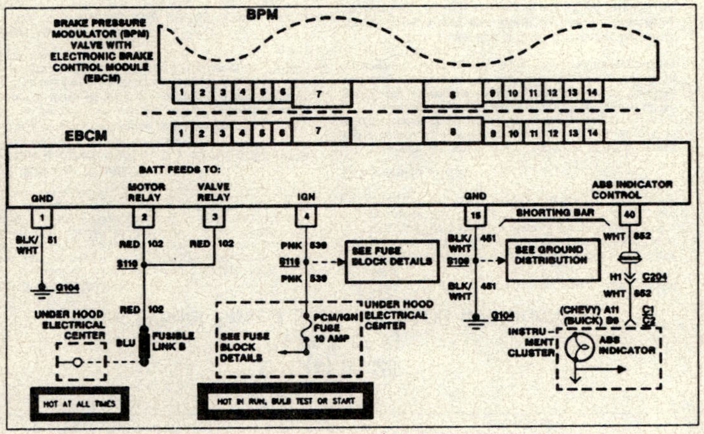

CHART B
ABS Warning Lamp Inoperative Or Flickers Briefly At Ignition "RUN"

Chart Test Description:
Number(s) below refer to circled number(s) on the diagnostic chart.

1. Confirms that a malfunction condition exists.
2. Check to see if the indicator can be turned "ON" manually.
3. Checks the integrity of the indicator bulb.
4. Checks for an indicator low side circuit open.
5. Checks for an indicator low side circuit short to voltage.

DIAGNOSTIC CHART B — 1995, CONTINUED

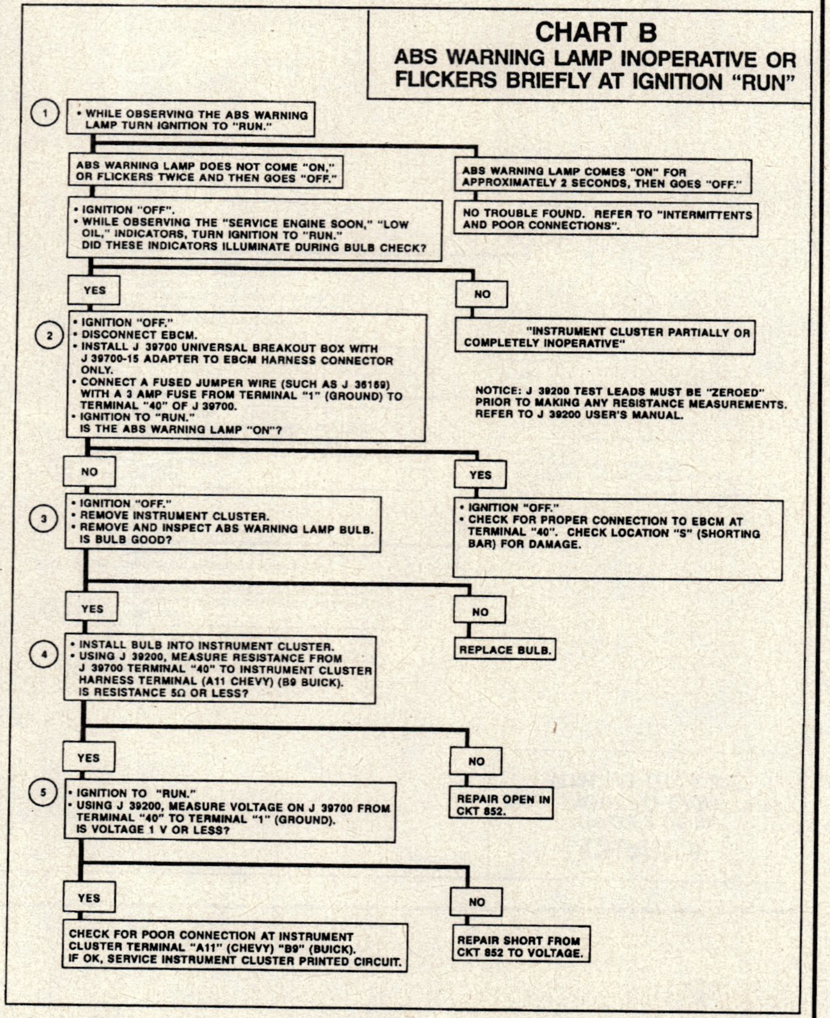

DIAGNOSTIC CHART DTC 21 AND 25 — 1995

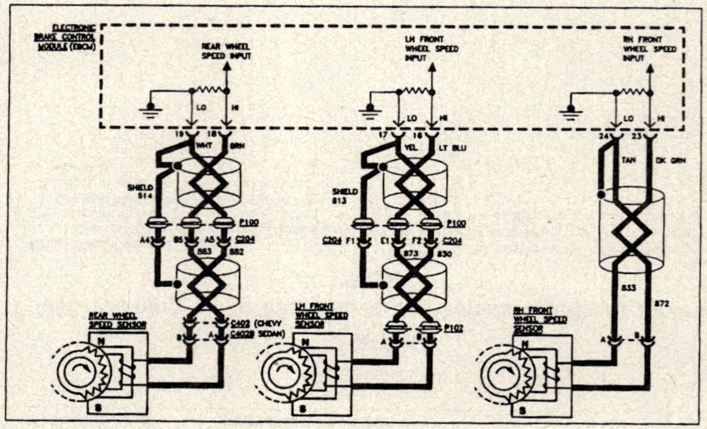

DTC(s) 21,25
(Page 1 of 2)
Front Wheel Speed Sensor Circuit Malfunction

Circuit Description:

The toothed wheel generates a voltage pulse as it moves past the sensor; each tooth-gap-tooth series on the wheel generates a pulse. The frequency of these pulses is used by the EBCM to determine wheel speed. The amplitude of the voltage generated in each pulse depends on the air gap between the sensor and the toothed wheel, and on wheel speed.

DTC Will Set When:

- There is a short to voltage or ground in (CKT 872 or 833 RF)(CKT 873 or 830 LF).
- There is a malfunctioning speed sensor.
- It is possible that driving with a compact spare at speeds above 40 km/h (24 mph) may cause this fault.

The test for this malfunction occurs when the vehicle is in motion; it will not set with the ignition in the "RUN" position and the vehicle at rest.

Action Taken:

The ABS warning lamp will be "ON". ABS is disabled.

DTC Chart Test Description:

Number(s) below refer to circled number(s) on the diagnostic chart.

1. Checks for proper resistance in the sensor itself.
2. Checks for a short to voltage in either of the speed sensor circuit wires.
3. Checks for a short between the speed sensor circuit wires.
4. Checks for a short to ground in the speed sensor input circuit wire. The ground may be either a "hard" short to ground, or a resistive (partial) short. A short with a resistance less than 2 megohms, though not a hard short, can still cause the DTC to set.

Diagnostic Aids:

Be sure the speed sensor wiring is properly routed and retained. This will help prevent false signals due to electrical noise being picked up by the wiring.

It is very important that a thorough inspection of the wiring and connectors be performed. Failure to carefully and fully inspect wiring and connectors may result in misdiagnosis, causing part replacement with reappearance of the malfunction.

The Tech 1 can also be used to monitor the wheel speeds during a road test. Watch the wheel speeds being displayed on the Tech 1 to see if any of the readings are unusual, such as one sensor varying in speed from the other three, a signal going intermittently high or low, etc. If this does not identify the intermittent, wet the speed sensor harness on the underside of the vehicle and road test monitoring wheel speeds with the Tech 1.

DIAGNOSTIC CHART DTC 21 AND 25 — 1995, CONTINUED

DTC(s) 21, 25 (PAGE 1 OF 2) FRONT WHEEL SPEED SENSOR CIRCUIT MALFUNCTION

ENSURE SENSOR IS NOT VISIBLY DAMAGED; IF IT IS DAMAGED, REPLACE THE WHEEL BEARING/SENSOR ASSEMBLY, CLEAR DTCs, AND ROAD TEST TO SEE IF DTC RESETS.

(1)
- IGNITION "OFF".
- DISCONNECT SENSOR CONNECTOR.
- USING J 39200 AND ADAPTER FROM J 35616-A, CHECK SENSOR RESISTANCE BY PROBING SENSOR CONNECTOR TERMINALS.

IS RESISTANCE 1015-1245 Ω?

- NO → REPLACE SENSOR.
- YES → (2)

(2)
- CONNECT J 39200 BETWEEN SENSOR HARNESS CONNECTOR TERMINAL "A" AND GROUND (USE DC VOLTAGE SETTING).
- IGNITION TO "RUN".
- NOTE VOLTAGE; REPEAT BETWEEN TERMINAL "B" AND GROUND.

IS VOLTAGE MORE THAN 1 V ON EITHER TERMINAL "A" OR "B"?

- YES → REPAIR SHORT TO VOLTAGE IN CKT: RF - 872 OR 833; LF 873 OR 830.
- NO → (3)

(3)
- IGNITION "OFF."
- DISCONNECT EBCM.
- INSTALL J 39700 UNIVERSAL BREAKOUT BOX WITH ADAPTER J 39700-15 TO EBCM HARNESS CONNECTOR ONLY.
- USING J 39200, MEASURE RESISTANCE BETWEEN J 39700 TERMINALS: RF - "23" AND "24", LF - "16" AND "17".

IS IT LESS THAN 5Ω?

- YES → REPAIR SHORT BETWEEN CKTS: 872 AND 833 FOR RF; 873 AND 830 FOR LF.
- NO → (4)

(4)
- USING J 39200, MEASURE RESISTANCE BETWEEN J 39700 TERMINAL "23" FOR RF OR "16" FOR LF, AND GROUND.

IS IT GREATER THAN 2 MEGOHMS?

- YES → CONTINUE DIAGNOSTICS ON NEXT PAGE.
- NO →
  - IF 5Ω OR LESS TO GROUND, REPAIR HARD SHORT IN HARNESS CKT 872 OR 830.
  - IF BETWEEN 5Ω AND 2 MEGOHMS, REPAIR RESISTIVE (PARTIAL) SHORT IN HARNESS CKT 872 OR 830.

WHEN ALL DIAGNOSIS AND REPAIRS ARE COMPLETED, CLEAR DTCs AND VERIFY PROPER OPERATION.

DIAGNOSTIC CHART DTC 21 AND 25 — 1995, CONTINUED

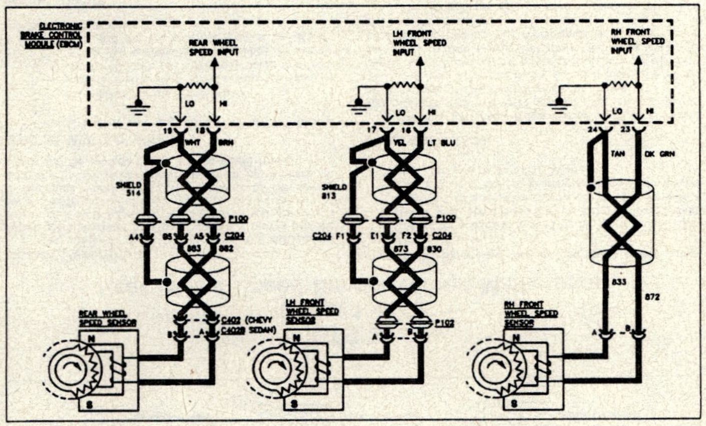

DTC(s) 21,25 (Page 2 of 2) Front Wheel Speed Sensor Circuit Malfunction

Circuit Description:

The toothed wheel generates a voltage pulse as it moves past the sensor; each tooth-gap-tooth series on the wheel generates a pulse. The frequency of these pulses is used by the EBCM to determine wheel speed. The amplitude of the voltage generated in each pulse depends on the air gap between the sensor and the toothed wheel, and on wheel speed.

DTC Will Set When:

- There is a short to voltage or ground in (CKT 872 or 833 RF)(CKT 873 or 830 LF).
- There is a malfunctioning speed sensor.
- It is possible that driving with a compact spare at speeds above 40 km/h (24 mph) may cause this fault.

The test for this malfunction occurs when the vehicle is in motion; it will not set with the ignition in the "RUN" position and the vehicle at rest.

Action Taken:

The ABS warning lamp will be "ON". ABS is disabled.

DTC Chart Test Description:

Number(s) below refer to circled number(s) on the diagnostic chart.

5. Checks if the DTC resets during a road test. If so, since tests 1-4 have validated that the circuitry and components are good, intermittent malfunctions are suspected.
6. Checks wiring and connectors for intermittents.
7. Replace the wheel bearing/speed sensor assembly, as it is likely the cause of a DTC reset experienced during the road test, since other portions of the circuit have checked out OK. If the DTC resets after sensor replacement, the EBCM must be concluding there is a malfunction present in the speed sensor circuit when there is not.
8. Ensures DTC was not set due to electromagnetic interference.

Diagnostic Aids:

Be sure the speed sensor wiring is properly routed and retained. This will help prevent false signals due to electrical noise being picked up by the wiring.

It is very important that a thorough inspection of the wiring and connectors be performed. Failure to carefully and fully inspect wiring and connectors may result in misdiagnosis, causing part replacement with reappearance of the malfunction.

The Tech 1 can also be used to monitor the wheel speeds during a road test. Watch the wheel speeds being displayed on the Tech 1 to see if any of the readings are unusual, such as one sensor varying in speed from the other three, a signal going intermittently high or low, etc. If this does not identify the intermittent, wet the speed sensor harness on the underside of the vehicle and road test monitoring wheel speeds with the Tech 1.

DIAGNOSTIC CHART DTC 21 AND 25 — 1995, CONTINUED

DTC(s) 21, 25 (PAGE 2 OF 2) FRONT WHEEL SPEED SENSOR CIRCUIT MALFUNCTION

DIAGNOSTIC CONTINUED FROM PREVIOUS PAGE.

(5)
- IGNITION "OFF."
- REMOVE J 39700.
- RECONNECT EBCM CONNECTOR.
- CLEAR DTCs, IGNITION "OFF."
- DISCONNECT TECH 1, IF CONNECTED.
- ROAD TEST VEHICLE, REACHING AT LEAST 25 km/h (15 MPH).

DOES DTC 21 OR 25 RESET?

- YES → (6)
- NO → NO TROUBLE FOUND; SEE "DIAGNOSTIC AIDS" ON FACING PAGE.

(6)
- CHECK ALL WIRING, CONNECTORS AND TERMINALS THOROUGHLY FOR INTERMITTENTS OR CORROSION. REFER TO "INTERMITTENTS AND POOR CONNECTIONS".

ARE ALL CONNECTIONS AND WIRING GOOD?

- YES → (7)
- NO → REPAIR WIRING AND/OR CONNECTIONS AS NEEDED.

(7)
- CLEAR DTCs, IGNITION "OFF."
- REPLACE SENSOR.
- DISCONNECT TECH 1, IF CONNECTED.
- ROAD TEST VEHICLE, REACHING AT LEAST 25 km/h (15 MPH).

DOES DTC 21 OR 25 RESET?

- YES → (8)
- NO → SYSTEM OK; SEE "DIAGNOSTIC AIDS".

(8) PERFORM EMI TEST. REFER TO EMI TEST FOUND IN "INTERMITTENTS AND POOR CONNECTIONS" IN THIS SECTION. WERE EMI TESTS COMPLETED WITH NO TROUBLE FOUND?

- YES → REPLACE EBCM.
- NO → PERFORM NECESSARY REPAIRS.

WHEN ALL DIAGNOSIS AND REPAIRS ARE COMPLETED, CLEAR DTCs AND VERIFY PROPER OPERATION.

DIAGNOSTIC CHART DTC 35 — 1995

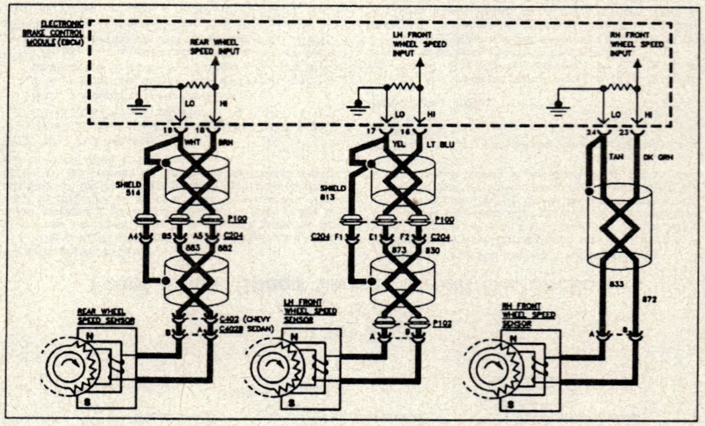

DTC 35 (Page 1 of 2) Rear Wheel Speed Sensor Circuit Malfunction

Circuit Description:

The toothed wheel (mounted to the differential pinion gear) generates a voltage pulse as it moves past the sensor; each tooth-gap-tooth series on the wheel generates a pulse. The frequency of these pulses is used by the EBCM to determine wheel speed. The amplitude of the voltage generated in each pulse depends on the air gap between the sensor and the toothed wheel, and on wheel speed.

DTC 35 Will Set When:

- There is a short to voltage or ground in CKT 883 or 882.
- there is a malfunctioning speed sensor.
- It is possible that driving with a compact spare at speeds above 40 km/h (24 mph) may cause this fault

The test for this malfunction occurs when the vehicle is in motion; it will not set with the ignition in "RUN" and the vehicle at rest.

Action Taken:

The ABS warning lamp will be "ON". ABS is disabled.

DTC Chart Test Description:

Number(s) below refer to circled number(s) on the diagnostic chart.

1. Checks for proper resistance in the sensor itself.
2. Checks for a short to voltage in either of the speed sensor circuit wires.
3. Checks for a short between the speed sensor circuit wires.
4. Checks for a short to ground in the speed sensor input circuit wire. The ground may be either a "hard" short to ground, or a resistive (partial) short. A short with a resistance less than 2 megohms, though not a hard short, can still cause DTC 35 to set.

Diagnostic Aids:

Be sure the speed sensor wiring is properly routed and retained. This will help prevent false signals due to electrical noise being picked up by the wiring.

It is very important that a thorough inspection of the wiring and connectors be performed. Failure to carefully and fully inspect wiring and connectors may result in misdiagnosis, causing part replacement with reappearance of the malfunction.

The Tech 1 can also be used to monitor the wheel speeds during a road test. Watch the wheel speeds being displayed on the Tech 1 to see if any of the readings are unusual, such as one sensor varying in speed from the other three, a signal going intermittently high or low, etc. If this does not identify the intermittent, wet the speed sensor harness on the underside of the vehicle and road test monitoring wheel speeds with the Tech 1.

DIAGNOSTIC CHART DTC 35 — 1995, CONTINUED

DTC 35
(PAGE 1 OF 2)
REAR WHEEL SPEED SENSOR CIRCUIT MALFUNCTION

ENSURE SENSOR IS NOT VISIBLY DAMAGED, AND THAT IT IS PROPERLY MOUNTED. IF IT IS NOT PROPERLY MOUNTED OR RETAINED, PROPERLY ATTACH IT; IF IT IS DAMAGED, REPLACE IT. THEN, CLEAR DTCs AND ROAD TEST TO SEE IF DTC RESETS.

(1)
- IGNITION "OFF."
- DISCONNECT SENSOR CONNECTOR.
- USING J 39200, MEASURE SENSOR RESISTANCE BY PROBING SENSOR CONNECTOR PIGTAIL TERMINALS.

IS RESISTANCE 1470-1730 Ω?

- YES → go to (2)
- NO → REPLACE SPEED SENSOR.

(2)
- CONNECT J 39200 BETWEEN SENSOR HARNESS CONNECTOR TERMINAL "A" AND GROUND (USE DC VOLTS).
- IGNITION TO "RUN."
- NOTE VOLTAGE; REPEAT BETWEEN TERMINAL "B" AND GROUND.

IS VOLTAGE MORE THAN ONE VOLT ON EITHER TERMINAL "A" OR "B"?

- NO → go to (3)
- YES → REPAIR SHORT TO VOLTAGE IN CKT 883 OR 882.

(3)
- IGNITION "OFF."
- DISCONNECT EBCM.
- INSTALL J 39700 UNIVERSAL BREAKOUT BOX WITH J 39700-15 ADAPTER TO EBCM HARNESS CONNECTOR ONLY.
- USING J 39200, MEASURE RESISTANCE BETWEEN J 39700 TERMINALS "18" AND "19."

IS IT LESS THAN 5 Ω?

- NO → go to (4)
- YES → REPAIR SHORT BETWEEN CKTS 883 AND 882.

(4)
- USING J 39200 MEASURE RESISTANCE BETWEEN J 39700 TERMINAL "18" AND GROUND.

IS IT GREATER THAN 2 MEGOHMS?

- YES → CONTINUE DIAGNOSTICS ON NEXT PAGE.
- NO →
  - IF 5 Ω OR LESS TO GROUND, REPAIR HARD SHORT IN HARNESS CKT 882.
  - IF BETWEEN 5 Ω AND 2 MEGOHMS, REPAIR RESISTIVE (PARTIAL) SHORT IN HARNESS CKT 882.

WHEN ALL DIAGNOSIS AND REPAIRS ARE COMPLETED, CLEAR DTCs AND VERIFY PROPER OPERATION.

DIAGNOSTIC CHART DTC 35 — 1995, CONTINUED

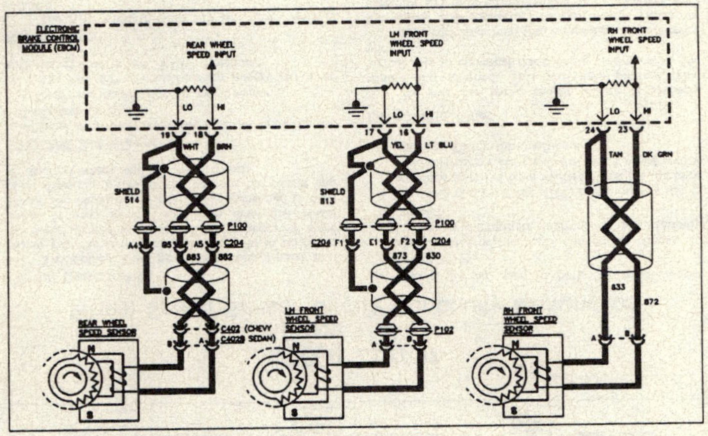

DTC 35
(Page 2 of 2)
Rear Wheel Speed Sensor Circuit Malfunction

Circuit Description:

The toothed wheel (mounted to the differential pinion gear) generates a voltage pulse as it moves past the sensor; each tooth-gap-tooth series on the wheel generates a pulse. The frequency of these pulses is used by the EBCM to determine wheel speed. The amplitude of the voltage generated in each pulse depends on the air gap between the sensor and the toothed wheel, and on wheel speed.

DTC 35 Will Set When:

- There is a short to voltage or ground in CKT 883 or 882.
- There is a malfunctioning speed sensor.
- It is possible that driving with a compact spare at speeds above 40 km/h (24 mph) may cause this fault

The test for this malfunction occurs when the vehicle is in motion; it will not set with the ignition in "RUN" and the vehicle at rest.

Action Taken:

The ABS warning lamp will be "ON". ABS is disabled.

DTC Chart Test Description:

Number(s) below refer to circled number(s) on the diagnostic chart.

5. Checks if the DTC resets during a road test. If so, since tests 1-4 have validated that all of the circuitry and components are good, intermittent malfunctions are suspected.
6. Checks wiring and connectors for intermittents.
7. Checks for sensor mounting and for damage to the toothed wheel which may be causing the malfunction. Check the toothed wheel for any large grooves, gouges, marks, etc. that might influence the tooth's signal at the wheel speed sensor.
8. Replace the speed sensor as it is likely the cause of a DTC reset experienced during the road test, since other portions of the circuit have checked out OK.

Diagnostic Aids:

Be sure the speed sensor wiring is properly routed and retained. This will help prevent false signals due to electrical noise being picked up by the wiring.

It is very important that a thorough inspection of the wiring and connectors be performed. Failure to carefully and fully inspect wiring and connectors may result in misdiagnosis, causing part replacement with reappearance of the malfunction.

The Tech 1 can also be used to monitor the wheel speeds during a road test. Watch the wheel speeds being displayed on the Tech 1 to see if any of the readings are unusual, such as one sensor varying in speed from the other three, a signal going intermittently high or low, etc. If this does not identify the intermittent, wet the speed sensor harness on the underside of the vehicle and road test monitoring wheel speeds with the Tech 1.

DIAGNOSTIC CHART DTC 35 — 1995, CONTINUED

DTC 35 (PAGE 2 OF 2) REAR WHEEL SPEED SENSOR CIRCUIT MALFUNCTION

DIAGNOSTICS CONTINUED FROM PREVIOUS PAGE.

(5)
- IGNITION "OFF."
- REMOVE J 39700.
- RECONNECT EBCM CONNECTOR.
- CLEAR DTCs, IGNITION "OFF."
- DISCONNECT TECH 1, IF CONNECTED.
- ROAD TEST VEHICLE, REACHING AT LEAST 25 km/h (15 MPH).

DOES DTC 31 RESET?

- YES → go to (6)
- NO → NO TROUBLE FOUND; SEE "DIAGNOSTIC AIDS"

(6)
- CHECK ALL WIRING, CONNECTORS AND TERMINALS THOROUGHLY FOR INTERMITTENTS OR CORROSION.

ARE CONNECTIONS AND WIRING GOOD?

- YES → go to (7)
- NO → REPAIR WIRING AND/OR CONNECTIONS AS NEEDED

(7)
- CHECK SENSOR MOUNTING, TOOTHED WHEEL FOR DAMAGE OR FOREIGN MATERIAL. SEE "DIAGNOSTIC AIDS"

IS DAMAGE OR FOREIGN MATERIAL EVIDENT?

- NO → go to (8)
- YES → REPAIR AS NECESSARY.

(8)
- REPLACE SPEED SENSOR.
- CLEAR DTCs, IGNITION "OFF."
- DISCONNECT TECH 1, IF CONNECTED.
- ROAD TEST VEHICLE, REACHING AT LEAST 25 km/h (15 MPH).

DOES DTC 35 RESET?

- YES → REPLACE EBCM.
- NO → SYSTEM OK; SEE "DIAGNOSTIC AIDS".

WHEN ALL DIAGNOSIS AND REPAIRS ARE COMPLETED, CLEAR DTCs AND VERIFY PROPER OPERATION.

DIAGNOSTIC CHART DTC 23, 27 AND 37 — 1995

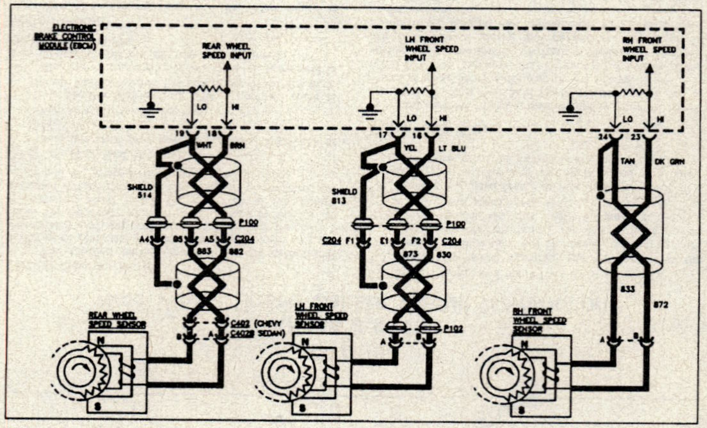

DTC(s) 23,27,37 (Page 1 of 2) Wheel Speed Sensor Circuit Continuity Malfunction

Circuit Description:

The toothed wheel generates a voltage pulse as it moves past the sensor; each tooth-gap-tooth series on the wheel generates a pulse. The frequency of these pulses is used by the EBCM to determine wheel speed. The amplitude of the voltage generated in each pulse depends on the air gap between the sensor and the toothed wheel, and on wheel speed.

DTC Will Set When:

- There is a short to voltage or an open in the affected circuit.
- There is a malfunctioning speed sensor.

The test for this malfunction occurs any time with the ignition in the "RUN" position.

Action Taken:

The ABS warning lamp will be "ON". ABS is disabled.

DTC Chart Test Description:

Number(s) below refer to circled number(s) on the diagnostic chart.

1. Check for proper resistance in the sensor itself.
2. Checks for a short to voltage in either of the speed sensor circuit wires.
3. Checks for an open in both the speed sensor circuit wires.
4. Manipulates wiring and connectors, trying to induce an intermittent malfunction not currently present.
5. This test checks if the DTC resets on key-up. If so, since tests 1-4 have validated that the circuitry and components are good, the wheel speed sensor may be causing the malfunction.

Diagnostic Aids:

Be sure the speed sensor wiring is properly routed and retained. This will help prevent false signals due to electrical noise being picked up by the wiring.

It is very important that a thorough inspection of the wiring and connectors be performed. Failure to carefully and fully inspect wiring and connectors may result in misdiagnosis, causing part replacement with reappearance of the malfunction.

DIAGNOSTIC CHART DTC 23, 27 AND 37 — 1995, CONTINUED

DTC(s) 23, 27, 37 (PAGE 1 OF 2) WHEEL SPEED SENSOR CIRCUIT CONTINUITY MALFUNCTION

ENSURE SENSOR IS NOT VISIBLY DAMAGED; IF IT IS DAMAGED, REPLACE THE SENSOR, CLEAR DTCs, AND TURN IGNITION TO "RUN" TO SEE IF DTC RESETS.

(1)
- IGNITION "OFF."
- DISCONNECT SENSOR CONNECTOR.
- USING J 39200 AND ADAPTERS FROM J 35616-A, CHECK SENSOR RESISTANCE BY PROBING SENSOR CONNECTOR TERMINALS.

IS RESISTANCE 1015-1245 Ω FOR FRONT, OR 1470-1730 FOR REAR?

- NO → REPLACE SENSOR.
- YES ↓

(2)
- CONNECT J 39200 BETWEEN SENSOR HARNESS CONNECTOR TERMINAL "A" AND GROUND (USE DC VOLTS).
- IGNITION TO "RUN."
- NOTE VOLTAGE; REPEAT BETWEEN TERMINAL "B" AND GROUND.

IS VOLTAGE MORE THAN 1 V ON EITHER TERMINAL "A" OR "B"?

- YES → REPAIR SHORT TO VOLTAGE IN THE AFFECTED CIRCUIT.
- NO ↓

(3)
- IGNITION "OFF."
- DISCONNECT EBCM.
- INSTALL J 39700 UNIVERSAL BREAKOUT BOX WITH ADAPTER J 39700-15 TO EBCM HARNESS CONNECTOR ONLY.
- USING J 39200, MEASURE RESISTANCE BETWEEN J39700 TERMINAL ANS SENSOR HARNESS AS SHOWN BELOW:

J 39700 TERMINAL	SENSOR HARNESS	CKT	FOR DTC
24	A	833	23
23	B	872	23
17	A	873	27
16	B	830	27
18	A	882	37
19	B	883	37

IS MEASUREMENT GREATER THAN 5Ω?

- YES → REPAIR OPEN IN THE AFFECTED CIRCUIT.
- NO ↓

(4)
- REMOVE J 39700; RECONNECT EBCM.
- CLEAR DTCs; IGNITION "OFF."
- DISCONNECT TECH 1, IF CONNECTED.
- IGNITION TO "RUN."
- WIGGLE ALL WIRING AND CONNECTORS TO TRY TO INDUCE A HIDDEN INTERMITTENT MALFUNCTION.

DOES DTC RESET?

- YES → REPAIR INTERMITTENT IN CONNECTOR OR WIRING.
- NO ↓

(5)
- CLEAR DTCs; IGNITION "OFF."
- DISCONNECT TECH 1, IF CONNECTED.
- IGNITION TO "RUN."

DOES DTC RESET?

CONTINUE DIAGNOSTICS ON NEXT PAGE

WHEN ALL DIAGNOSIS AND REPAIRS ARE COMPLETED, CLEAR DTCs AND VERIFY PROPER OPERATION.

DIAGNOSTIC CHART DTC 23, 27 AND 37 — 1995, CONTINUED

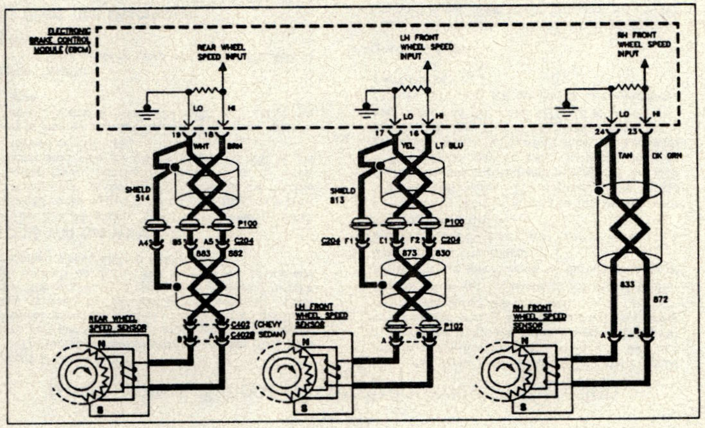

DTC(s) 23,27,37 (Page 2 of 2)
Wheel Speed Sensor Circuit Continuity Malfunction

Circuit Description:

The toothed wheel generates a voltage pulse as it moves past the sensor; each tooth-gap-tooth series on the wheel generates a pulse. The frequency of these pulses is used by the EBCM to determine wheel speed. The amplitude of the voltage generated in each pulse depends on the air gap between the sensor and the toothed wheel, and on wheel speed.

DTC Will Set When:

- There is a short to voltage or an open in the affected circuit.
- There is a malfunctioning speed sensor.

The test for this malfunction occurs any time with the ignition in the "RUN" position.

Action Taken:

The ABS warning lamp will be "ON". ABS is disabled.

DTC Chart Test Description:

Number(s) below refer to circled number(s) on the diagnostic chart.

6. Checks if DTC (21 RF),(25 LF) or (35 R) sets during a road test.
7. If the circuit malfunction DTC sets at this point, diagnostics for circuit malfunctions should be used.
8. The wheel speed sensor may be causing an intermittent malfunction. If DTC resets after sensor replacement, the EBCM must be concluding there is a malfunction present in the speed sensor circuit when there is not.

Diagnostic Aids:

Be sure the speed sensor wiring is properly routed and retained. This will help prevent false signals due to electrical noise being picked up by the wiring.

It is very important that a thorough inspection of the wiring and connectors be performed. Failure to carefully and fully inspect wiring and connectors may result in misdiagnosis, causing part replacement with reappearance of the malfunction.

DIAGNOSTIC CHART DTC 23, 27 AND 37 — 1995, CONTINUED

DTC(s) 23, 27, 37
(PAGE 2 OF 2)
WHEEL SPEED SENSOR CIRCUIT CONTINUITY MALFUNCTION

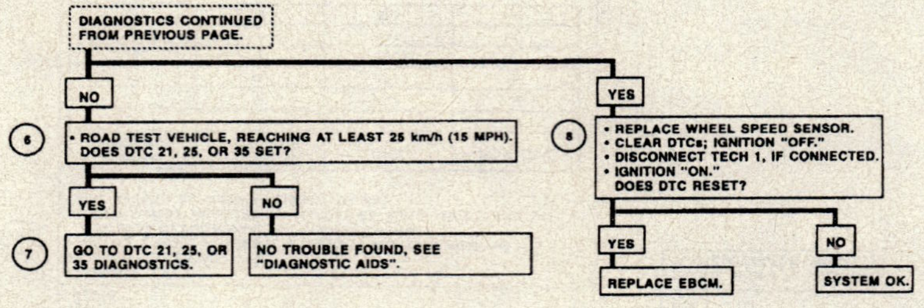

WHEN ALL DIAGNOSIS AND REPAIRS ARE COMPLETED, CLEAR DTCs AND VERIFY PROPER OPERATION.

DIAGNOSTIC CHART DTC 28 — 1995

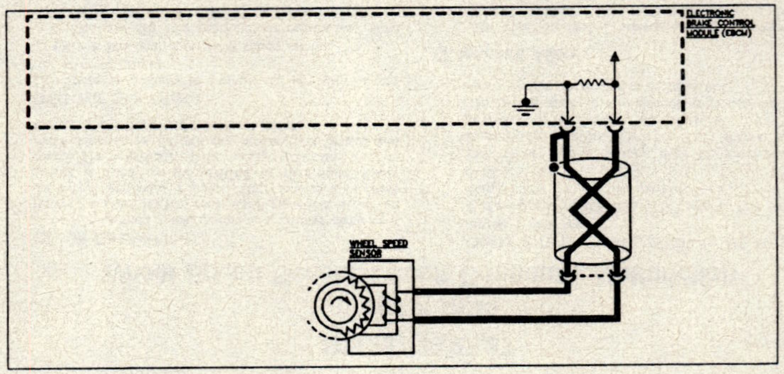

DTC 28
Wheel Speed Sensor Circuit Frequency Malfunction

Circuit Description:

The toothed wheel generates a voltage pulse as it moves past the sensor; each tooth-gap-tooth series on the wheel generates a pulse. The frequency of these pulses is used by the EBCM to determine wheel speed. The amplitude of the voltage generated in each pulse depends on the air gap between the sensor and the toothed wheel, and on wheel speed.

DTC 28 Will Set When:

- The EBCM cannot specifically identify which wheel speed sensor is causing the frequency error malfunction. If the EBCM can define the specific speed sensor causing the malfunction, the DTC associated with the sensor (21,25,35) will be set instead of DTC 28.
- It is possible that driving with a compact spare at speeds above 40 km/h (24 mph) may cause this fault.

Action Taken:

The ABS warning lamp will be "ON". ABS is disabled.

DTC Chart Test Description:

Number(s) below refer to circled number(s) on the diagnostic chart.

1. Checks wiring and connections for malfunctions. It is critical that this step is performed thoroughly, as wiring/connector malfunctions are the most likely case for this DTC.
2. Uses the Tech 1 to monitor for vehicle electrical/electronic system "noise" being picked up by the speed sensor circuits.
3. Uses the Tech 1 to monitor the wheel speed sensors while the vehicle is operating. If the Tech 1 autotrigger snapshot mode triggers, the speed sensor or the wiring and/or connectors are going intermittent during the road test.
4. Checks for proper resistance in the sensor itself.
5. Replace wheel speed sensor as the likely cause of the triggering intermittent.
6. Ensures DTC does not reset due to a malfunctioning EBCM.
7. Checks for a short to ground in the speed sensor input circuit wires. The circuit to be tested depends on which speed sensor triggered the Tech 1. The ground may be either a "hard" short to ground, or a resistive (partial) short. A short with a resistance less than 2 megohms, though not a hard short, can still cause DTC 28 to set.
8. Ensures DTC was not set due to electromagnetic interference.

Diagnostic Aids:

Check the toothed wheel for any large grooves, gouges, marks, etc. That might influence the tooth's signal at the wheel speed sensor. Also check for a buildup of foreign material in the gaps between teeth in the toothed wheel; this material may cause this malfunction.

A worn hub/bearing assembly may cause this malfunction in extreme cases; the bearing play allows the sensor-toothed-ring gap to change excessively.

DIAGNOSTIC CHART DTC 28 — 1995, CONTINUED

DTC 28
WHEEL SPEED SENSOR CIRCUIT FREQUENCY MALFUNCTION

(1) VERY THOROUGHLY INSPECT ALL WIRING AND CONNECTIONS FOR INTERMITTENTS, CHAFING, POOR TERMINAL CONTACT TENSION, CORROSION, AND MISROUTING AT EBCM, ALL SENSORS, AND ALL HARNESS INTERCONNECTS. REPAIR ANY MALFUNCTIONS, CLEAR DTCs, THEN ROAD TEST VEHICLE, REACHING AT LEAST 25km/h (15 MPH). DOES DTC RESET?

- NO → NO TROUBLE FOUND. SEE "DIAGNOSTIC AIDS"
- YES →

(2)
- INSTALL TECH 1 WITH MSC; CLEAR DTCs.
- SET TECH 1 AUTO TRIGGER SNAPSHOT MODE.
- START VEHICLE AND ALLOW TO IDLE FOR ONE MINUTE.

DOES TECH 1 TRIGGER?

- YES → (7)
- NO →

(3)
- ROAD TEST VEHICLE REACHING AT LEAST 25 km/h (15 MPH).

DOES TECH 1 TRIGGER?

- NO → NO TROUBLE FOUND; SEE "DIAGNOSTIC AIDS"
- YES → NOTE TECH 1, WHICH SENSOR TRIGGERED?

(4)
- IGNITION "OFF."
- DISCONNECT SENSOR CONNECTOR.
- USING J 39200, MEASURE RESISTANCE BETWEEN SENSOR TERMINALS, USING ADAPTERS FROM J 35616-A IF MEASURING A FRONT SENSOR.

IS RESISTANCE WITHIN RANGE SHOWN?

FRONT	REAR
1015-1245Ω	1470-1730Ω

- NO → (5) REPLACE SPEED SENSOR.
- YES → (8)

(7)
- NOTE TECH 1, WHICH SENSOR TRIGGERED?
- IGNITION "OFF."
- DISCONNECT EBCM.
- INSTALL J 39700 PINOUT BOX TO EBCM HARNESS CONNECTOR USING ADAPTER 39700-15.
- USING TABLE BELOW, MEASURE RESISTANCE TO GROUND AT THE TERMINAL ON J 39700 THAT CORRESPONDS TO THE SPEED SENSOR THE TECH 1 TRIGGERED ON:

LEFT FRONT TERMINAL "15"
RIGHT FRONT TERMINAL "23"
REAR TERMINAL "18"

- IF RESISTANCE IS 5Ω OR LESS TO GROUND, REPAIR HARD SHORT, IF BETWEEN 5Ω AND 2 MEGOHMS, REPAIR RESISTIVE (PARTIAL) SHORT IN CKT:

LEFT FRONT CKT 830
RIGHT FRONT CKT 872
REAR CKT 882

(8) PERFORM EMI TEST.

WERE EMI TESTS COMPLETED WITH NO TROUBLE FOUND?

- YES → REPLACE EBCM.
- NO → PERFORM NECESSARY REPAIRS.

(6)
- CLEAR DTCs, IGNITION "OFF."
- DISCONNECT TECH 1, IF CONNECTED.
- ROAD TEST VEHICLE, REACHING AT LEAST 25 km/h (15 MPH).

DOES DTC 28 RESET?

- YES → REPLACE EBCM.
- NO → SYSTEM OK.

WHEN ALL DIAGNOSIS AND REPAIRS ARE COMPLETED, CLEAR DTCs AND VERIFY PROPER OPERATION.

DIAGNOSTIC CHART DTC 41, 45 AND 55 — 1995

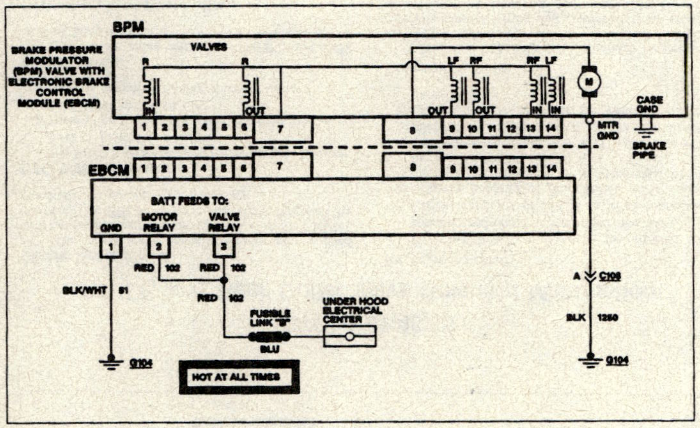

DTC(s) 41,45,55
RF, LF, and Rear Inlet Valve Solenoid Malfunction

Circuit Description:
The wheel valve solenoid circuits are supplied with battery power when the ignition is in the "RUN" position. The EBCM controls the valve functions by grounding the circuit when necessary.

DTC Will Set When:
- The EBCM senses a discrepancy such as an open, short to ground or voltage.
- There is poor connection in the circuit internal to the EBCM and BPM.
- There is a malfunctioning solenoid valve.

Action Taken:
The valve relay will turn "OFF", the ABS warning lamp will come "ON". ABS is disabled.

DTC Chart Test Description:
Number(s) below refer to circled number(s) on the diagnostic chart.

1. Uses the Tech 1 to exercise the affected valve solenoid and check it for proper operation.
2. Checks condition of EBCM to BPM connector.
3. Checks integrity of the affected valve solenoid circuitry internal to the BPM valve assembly, and the valve solenoid coil.
4. Determines whether the DTC was set by an intermittent condition or an EBCM malfunction.
5. Checks if DTC 71 was set concurrently with valve DTC.

Diagnostic Aids:
It is very important that a thorough inspection of the wiring and connectors be performed. Failure to carefully inspect wiring and connectors may result in misdiagnosis, causing part replacement with reappearance of the malfunction.

DIAGNOSTIC CHART DTC 41, 45 AND 55 — 1995, CONTINUED

DTC(s) 41, 45, 55
RF, LF, AND REAR INLET SOLENOID VALVE MALFUNCTION

(1)
- INSTALL TECH 1 WITH MASS STORAGE CARTRIDGE.
- IGNITION TO "RUN."
- PERFORM TECH 1 VALVE SOLENOID TEST TO AFFECTED VALVE. DOES "HOLD PRESSURE" MODE OPERATE PROPERLY?

YES → (2)

NO → (5) IS DTC 71 SET CONCURRENTLY WITH VALVE DTC?
- YES → REPLACE EBCM.
- NO →
  - IGNITION "OFF".
  - REMOVE EBCM.
  - INSPECT EBCM TO BPM CONNECTOR FOR DAMAGE, CORROSION OR POOR TERMINAL CONTACT, REPAIR OR REPLACE COMPONENTS AS NEEDED.
  - IF CONNECTOR IS OK, REPLACE BPM.

(2)
- IGNITION "OFF."
- DISCONNECT EBCM CONNECTOR.
- REMOVE EBCM FROM BPM VALVE ASSEMBLY.
- INSPECT EBCM TO BPM CONNECTOR FOR CONDITIONS WHICH COULD CAUSE INTERMITTENTS, SUCH AS DAMAGE, CORROSION, OR POOR TERMINAL CONTACT.

IS CONNECTOR OK?

YES → (3)

NO → REPAIR OR REPLACE COMPONENTS NEEDED.

(3)
- INSTALL J 41247 BPM PINOUT BOX TO BPM CONNECTOR.
- USING J 39200, MEASURE RESISTANCE BETWEEN J 41247 TERMINAL "7" AND TERMINAL OF AFFECTED SOLENOID. IS IT 8.5 - 9.5 Ω?

YES → (4)

NO → REPLACE BPM VALVE

(4)
- REMOVE J 41247.
- REPLACE EBCM TO BPM.
- PERFORM "TECH 1 AUTO" TEST. DOES DTC RESET?

YES → REPLACE EBCM.

NO → NO TROUBLE FOUND SEE "DIAGNOSTIC AIDS"

NOTICE: J 39200 TEST LEADS MUST BE "ZEROED" PRIOR TO MAKING ANY RESISTANCE MEASUREMENTS. REFER FO J 39200 USER'S MANUAL.

DIAGNOSTIC CHART DTC 42, 26 AND 56 — 1995

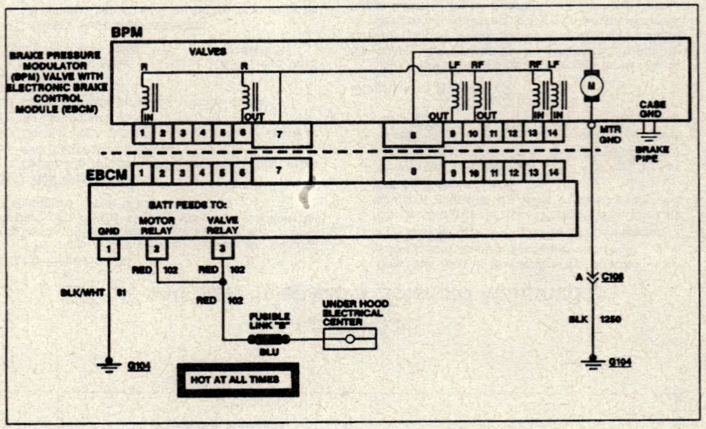

DTC(s) 42,46,56
RF, LF, and Rear Outlet Valve Solenoid Malfunction

Circuit Description:
The wheel valve solenoid circuits are supplied with battery power when the ignition is in the "RUN" position. The EBCM controls the valve functions by grounding the circuit when necessary.

DTC Will Set When:
- The EBCM senses a discrepancy such as an open, short to ground or voltage.
- There is poor connection in the circuit internal to the EBCM and BPM.
- There is a malfunctioning solenoid valve.

Action Taken:
The valve relay will turn "OFF", the ABS warning lamp will come "ON". ABS is disabled.

DTC Chart Test Description:
Number(s) below refer to circled number(s) on the diagnostic chart.

1. Uses the Tech 1 to exercise the affected valve solenoid and check it for proper operation.
2. Checks condition of EBCM to BPM connector.
3. Checks integrity of the affected valve solenoid circuitry internal to the BPM valve assembly, and the valve solenoid coil.
4. Determines whether the DTC was set by an intermittent condition or an EBCM malfunction.
5. Checks if DTC 71 was set concurrently with valve DTC.

Diagnostic Aids:
It is very important that a thorough inspection of the wiring and connectors be performed. Failure to carefully inspect wiring and connectors may result in misdiagnosis, causing part replacement with reappearance of the malfunction.

DIAGNOSTIC CHART DTC 42, 26 AND 56 — 1995, CONTINUED

DTC(s) 42, 46, 56
RF, LF, AND REAR OUTLET SOLENOID VALVE MALFUNCTION

1. • INSTALL TECH 1 WITH MASS STORAGE CARTRIDGE.
   • IGNITION TO "RUN".
   • PERFORM TECH 1 VALVE SOLENOID TEST TO AFFECTED VALVE.
   DOES "RELEASE PRESSURE" MODE OPERATE PROPERLY?
   - YES → 2
   - NO → 5

2. • IGNITION "OFF."
   • DISCONNECT EBCM CONNECTOR.
   • REMOVE EBCM FROM BPM VALVE ASSEMBLY.
   • INSPECT EBCM TO BPM CONNECTOR FOR CONDITIONS WHICH COULD CAUSE INTERMITTENTS, SUCH AS DAMAGE, CORROSION, OR POOR TERMINAL CONTACT.
   IS CONNECTOR OK?
   - YES → 3
   - NO → REPAIR OR REPLACE COMPONENTS AS NEEDED.

3. • INSTALL J 41247 BPM PINOUT BOX TO BPM CONNECTOR.
   • USING J 39200, MEASURE RESISTANCE BETWEEN J 41247 TERMINAL "7" AND TERMINAL OF AFFECTED SOLENOID.
   IS IT 4.2 - 5.2 Ω?
   - YES → 4
   - NO → REPLACE BPM VALVE

4. • REMOVE J 41247.
   • REPLACE EBCM TO BPM.
   • PERFORM "TECH 1 AUTO" TEST.
   DOES DTC RESET?
   - YES → REPLACE EBCM.
   - NO → NO TROUBLE FOUND SEE "DIAGNOSTIC AIDS"

5. IS DTC 71 SET CONCURRENTLY WITH VALVE DTC?
   - YES → REPLACE EBCM.
   - NO → • IGNITION "OFF".
     • REMOVE EBCM.
     • INSPECT EBCM TO BPM CONNECTOR FOR DAMAGE, CORROSION OR POOR TERMINAL CONTACT, REPAIR OR REPLACE COMPONENTS AS NEEDED.
     • IF CONNECTOR IS OK, REPLACE BPM.

NOTICE: J 39200 TEST LEADS MUST BE "ZEROED" PRIOR TO MAKING ANY RESISTANCE MEASUREMENTS. REFER FO J 39200 USER'S MANUAL.

DIAGNOSTIC CHART DTC 61 — 1995

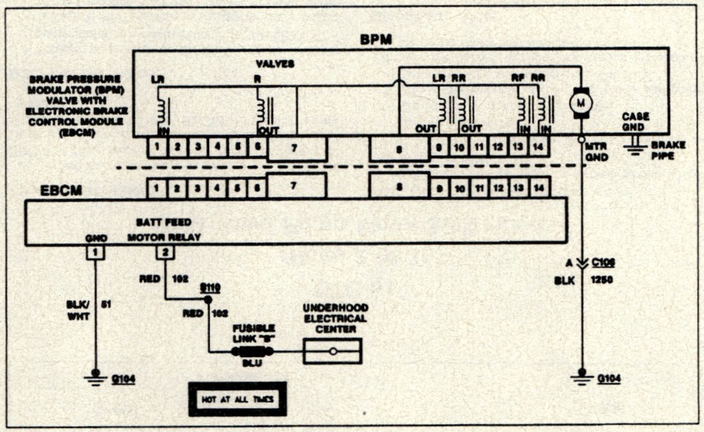

DTC 61
(page 1 of 2)
BPM Valve Pump Motor Malfunction

Circuit Description:

When the pump motor relay is grounded by the EBCM, it closes and provides voltage to operate the pump. Once the motor circuit is energized, a "Pump ON" signal is sensed by the EBCM to verify pump operation.

DTC Will Set When:

- Voltage is present at the pump motor without pump motor relay activation from the EBTCM.
- Voltage is NOT present at the pump motor within 40 milliseconds after the EBCM requests pump motor relay activation.
- The motor GND circuit is open.
- There is a pump motor malfunction.

Action Taken:

The ABS warning lamp will be "ON". ABS is disabled.

DTC Chart Test Description:

Number(s) below refer to circled number(s) on the diagnostic chart.

1. Checks that Fusible Link "B" feeding the EBCM Motor Relay circuit is good.
2. Checks for an open in Pump Motor Ground CKT 1250.
3. Measures for appropriate voltage from the EBCM Motor Relay Circuit, using the EBCM ground line to complete the Circuit.
4. Checks EBCM connector terminals for good contact, i.e.not spread open, backed out of connector, corroded, etc.

Diagnostic Aids:

All tests using J 39700 Universal Breakout Box terminal "1" are using the terminal as ground These tests assume that the ground CKT at terminal "1" is good.

It is very important that a thorough inspection of the wiring and connectors be performed. Failure to carefully and fully inspect wiring and connectors may result in misdiagnosis, causing part replacement with reappearance of the malfunction.

DIAGNOSTIC CHART DTC 61 — 1995, CONTINUED

DTC 61 (PAGE 1 OF 2) BPM VALVE PUMP MOTOR MALFUNCTION

BEFORE BEGINNING TESTING, ENSURE THAT BATTERY IS FULLY CHARGED, AND THAT CHARGING SYSTEM IS OPERATING PROPERLY IF DISCHARGED BATTERY IS NOTED. REFER TO SECTION 6D1 FOR BATTERY INFORMATION AND/OR SECTION 6D3 FOR CHARGING SYSTEM INFORMATION.

1. • ENSURE FUSIBLE LINK "B" IS GOOD. IS FUSIBLE LINK "B" OPEN?
   - YES: REPAIR SHORT TO GROUND IN CKT 102.
   - NO: go to 2.
2. • USING J 39200, MEASURE RESISTANCE BETWEEN PUMP MOTOR GROUND STUD AND A GOOD CHASSIS GROUND. IS IT 5Ω OR LESS?
   - NO: REPAIR OPEN OR LOOSE CONNECTION IN CKT 1250.
   - YES: go to 3.
3. • IGNITION "OFF."
   • DISCONNECT EBCM CONNECTOR.
   • INSTALL J 39700 UNIVERSAL BREAKOUT BOX WITH J 39700-15 ADAPTER TO EBCM HARNESS CONNECTOR ONLY.
   • CONNECT J 39200 BETWEEN J 39700 TERMINALS "2" AND "1". NOTE VOLTAGE. WAS VOLTAGE MEASUREMENT LESS THAN 9 V?
   - YES: REPAIR OPEN OR POOR CONNECTION IN CKT 102.
   - NO: go to 4.
4. • DISCONNECT EBCM HARNESS CONNECTOR FROM J 39700.
   • CHECK EBCM HARNESS CONNECTOR CAREFULLY TO ENSURE GOOD ELECTRICAL CONTACT AT ALL TERMINALS, ESPECIALLY TERMINALS "2" AND "1".
   • ARE TERMINALS CORRODED OR DAMAGED?
   - YES: REPAIR/REPLACE TERMINALS AND/OR CONNECTORS AS NECESSARY.
   - NO: CONTINUE DIAGNOSTICS ON NEXT PAGE.

WHEN ALL DIAGNOSIS AND REPAIRS ARE COMPLETED, CLEAR DTCs AND VERIFY PROPER OPERATION.

DIAGNOSTIC CHART DTC 61 — 1995, CONTINUED

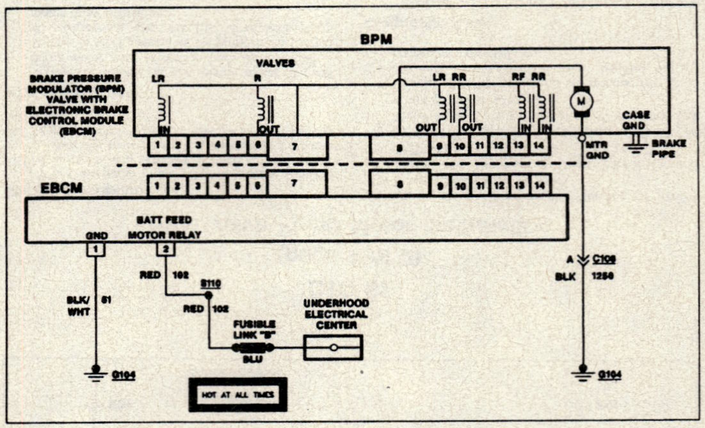

DTC 61 (Page 2 of 2) BPM Valve Pump Motor Malfunction

Circuit Description:

When the pump motor relay is grounded by the EBCM, it closes and provides B+ to operate the pump. Once the motor circuit is energized, a "Pump ON" signal is sensed by the EBCM to verify pump operation.

DTC Will Set When:

- Voltage is present at the pump motor without pump motor relay activation from the EBTCM.
- Voltage is NOT present at the pump motor within 40 milliseconds after the EBCM requests pump motor relay activation.
- The motor GND circuit is open.
- There is a pump motor malfunction.

Action Taken:

The ABS warning lamp will be "ON". ABS is disabled.

DTC Chart Test Description:

Number(s) below refer to circled number(s) on the diagnostic chart.

5. Checks EBCM connector for good retention to EBCM when installed.
6. Checks for condition of Connector.
7. Checks the integrity of the pump motor and pump motor circuitry internal to the BPM valve
8. Uses the Tech 1 to check for proper operation of the pump motor and associated circuitry.

Diagnostic Aids:

All tests using J 39700 Universal Breakout Box terminal "1" are using the terminal as ground These tests assume that the ground CKT at terminal "1" is good.

It is very important that a thorough inspection of the wiring and connectors be performed. Failure to carefully and fully inspect wiring and connectors may result in misdiagnosis, causing part replacement with reappearance of the malfunction.

DIAGNOSTIC CHART DTC 61 — 1995, CONTINUED

DTC 61 (PAGE 2 OF 2) BPM VALVE PUMP MOTOR MALFUNCTION

DIAGNOSIS CONTINUED FROM PREVIOUS PAGE.

(5)
- REMOVE J 39700.
- RECONNECT EBCM.

DOES CONNECTOR SEEM TO BE PROPERLY RETAINED TO EBCM WHEN CONNECTED?

NO → REPLACE OR REPAIR WIRING HARNESS CONNECTOR OR EBCM AS NEEDED TO OBTAIN PROPER CONNECTOR RETENTION.

YES →

(6)
- IGNITION "OFF."
- DISCONNECT EBCM CONNECTOR.
- REMOVE EBCM FROM BPM VALVE ASSEMBLY.
- INSPECT EBCM TO BPM CONNECTOR FOR CONDITIONS WHICH COULD CAUSE INTERMITTENTS SUCH AS DAMAGE, CORROSION, OR POOR TERMINAL CONTACT.

IS CONNECTOR OK?

NO → REPAIR OR REPLACE COMPONENTS AS NEEDED.

YES →

(7)
- INSTALL J 41247 BPM PINOUT BOX TO BPM CONNECTOR.
- USING J 39200, MEASURE RESISTANCE BETWEEN J 41247 TERMINAL "8" AND PUMP MOTOR GROUND STUD.

IS RESISTANCE .05-10 Ω?

NO → REPLACE BPM VALVE ASSEMBLY.

YES →

(8)
- REMOVE J 41247.
- RE-INSTALL EBCM TO BPM.
- RE-CONNECT EBCM CONNECTOR.
- IGNITION TO "RUN".
- PERFORM TECH 1 AUTO TEST.

DOES PUMP MOTOR RUN?

YES → NO TROUBLE FOUND SEE "DIAGNOSTIC AIDS"

NO → REPLACE EBCM.

NOTICE: J39200 TEST LEADS MUST BE "ZEROED" PRIOR TO MAKING ANY RESISTANCE MEASUREMENTS. REFER TO J 39200 USER'S MANUAL.

WHEN ALL DIAGNOSIS AND REPAIRS ARE COMPLETED, CLEAR DTCs AND VERIFY PROPER OPERATION.

DIAGNOSTIC CHART DTC 63 — 1995

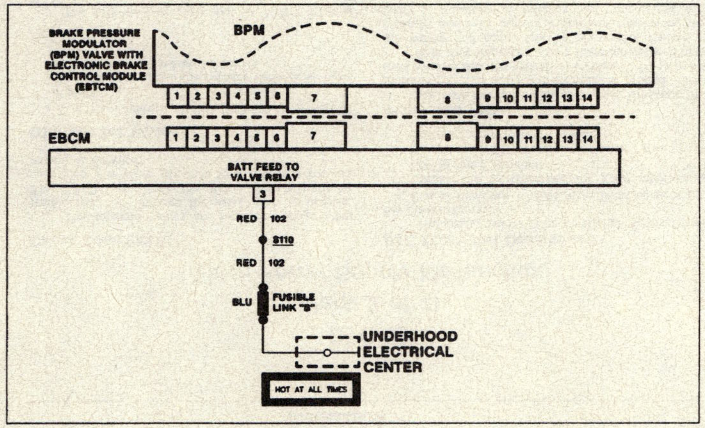

DTC 63 (Page 1 of 2) Valve Power Supply Malfunction

Circuit Description:

The solenoid valves are supplied voltage from fusible link "B" through the valve relay to operate. The EBCM monitors the valve feed circuit to determine if a valve response to a EBCM activation request is detected.

DTC Will Set When:

- 2 or more valves do not respond to a EBCM request.
- The EBCM senses a discrepancy such as an open or a short to ground in the circuit.

Action Taken:

The ABS warning lamp will be "ON". ABS is disabled.

DTC Chart Test Description:

Number(s) below refer to circled number(s) on the diagnostic chart.

1. Checks that Fusible Link "B" fuse feeding the Valve Relay circuit is good.
2. Measures for appropriate voltage from the valve circuit supply voltage.
3. Checks EBCM connector terminals for good contact, i.e.not spread open, backed out of connector, corroded, etc.

Diagnostic Aids:

All tests using J 39700 Universal Breakout Box terminal "1" are using that terminal as ground. These tests assume that the ground at terminal "1" is good.

It is very important that a thorough inspection of the wiring and connectors be performed. Failure to carefully and fully inspect wiring and connectors may result in misdiagnosis, causing part replacement with reappearance of the malfunction.

DIAGNOSTIC CHART DTC 63 — 1995, CONTINUED

DTC 63 (PAGE 1 OF 2) VALVE POWER SUPPLY MALFUNCTION

BEFORE BEGINNING TESTING, ENSURE THAT BATTERY IS FULLY CHARGED, AND THAT CHARGING SYSTEM IS OPERATING PROPERLY IF DISCHARGED BATTERY IS NOTED. REFER TO SECTION 6D1 FOR BATTERY INFORMATION AND/OR SECTION 6D3 FOR CHARGING SYSTEM INFORMATION.

(1)
- ENSURE FUSIBLE LINK "B" IS GOOD. IS FUSIBLE LINK "B" OPEN?

YES: REPAIR SHORT TO GROUND IN CKT 102.

NO:

(2)
- IGNITION "OFF."
- DISCONNECT EBCM CONNECTOR.
- INSTALL J 39700 UNIVERSAL BREAKOUT BOX WITH J 39700-15 ADAPTER TO EBCM HARNESS CONNECTOR ONLY.
- CONNECT J 39200 BETWEEN J 39700 TERMINALS "3" AND "1". NOTE VOLTAGE. WAS VOLTAGE MEASUREMENT LESS THAN 9 V?

YES: REPAIR OPEN OR POOR CONNECTION IN CKT 102.

NO:

(3)
- IGNITION "OFF."
- DISCONNECT EBCM HARNESS CONNECTOR FROM J 39700.
- CHECK EBCM HARNESS CONNECTOR CAREFULLY TO ENSURE GOOD ELECTRICAL CONTACT AT ALL TERMINALS, ESPECIALLY TERMINALS "3" AND "1". ARE TERMINALS CORRODED OR DAMAGED?

NO: CONTINUE DIAGNOSTICS ON NEXT PAGE.

YES: REPAIR/REPLACE TERMINALS AND/OR CONNECTOR AS NECESSARY.

WHEN ALL DIAGNOSIS AND REPAIRS ARE COMPLETED, CLEAR DTCs AND VERIFY PROPER OPERATION.

DIAGNOSTIC CHART DTC 63 — 1995, CONTINUED

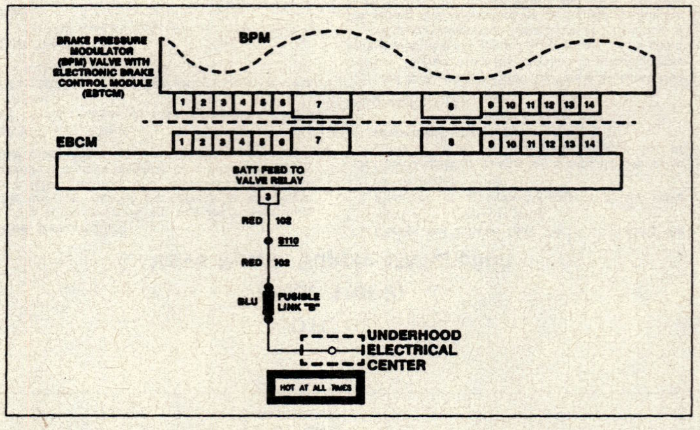

DTC 63 (Page 2 of 2) Valve Power Supply Malfunction

Circuit Description:

The solenoid valves are supplied voltage from fusible link "B" through the valve relay to operate. The EBCM monitors the valve feed circuit to determine if a valve response to a EBCM activation request is detected.

DTC Will Set When:

- 2 or more valves do not respond to a EBCM request.
- The EBCM senses a discrepancy such as an open or a short to ground in the circuit.

Action Taken:

The ABS warning lamp will be "ON". ABS is disabled.

DTC Chart Test Description:

Number(s) below refer to circled number(s) on the diagnostic chart.

4. Checks condition of EBCM to BPM connector.
5. Checks EBCM connector for good retention to EBCM when installed.
6. Ensures DTC does not reset due to a malfunctioning EBCM.

Diagnostic Aids:

All tests using J 39700 Universal Breakout Box terminal "1" are using that terminal as ground. These tests assume that the ground at terminal "1" is good.

It is very important that a thorough inspection of the wiring and connectors be performed. Failure to carefully and fully inspect wiring and connectors may result in misdiagnosis, causing part replacement with reappearance of the malfunction.

DIAGNOSTIC CHART DTC 63 — 1995, CONTINUED

DTC 63
(PAGE 2 OF 2)
VALVE POWER SUPPLY MALFUNCTION

DIAGNOSTICS CONTINUED FROM PREVIOUS PAGE

(4)
- IGNITION "OFF."
- DISCONNECT EBCM CONNECTOR.
- REMOVE EBCM FROM BPM VALVE ASSEMBLY.
- INSPECT EBCM TO BPM CONNECTOR FOR CONDITIONS WHICH COULD CAUSE INTERMITTENTS, SUCH AS DAMAGE, CORROSION, OR POOR TERMINAL CONTACT.

IS CONNECTOR OK?

NO → REPAIR OR REPLACE COMPONENTS AS NEEDED.

YES →

(5)
- REMOVE J 39700.
- RE-INSTALL EBCM TO BPM.
- RECONNECT EBCM.

IS CONNECTOR PROPERLY RETAINED TO EBCM WHEN CONNECTED?

NO → REPLACE OR REPAIR WIRING HARNESS CONNECTOR OR EBCM AS NEEDED TO OBTAIN PROPER CONNECTOR RETENTION.

YES →

(6)
- CLEAR DTCS, IGNITION "OFF."
- DISCONNECT TECH 1, IF CONNECTED.
- IGNITION TO "RUN."

DOES DTC 63 RESET?

YES → REPLACE EBCM.

NO → NO TROUBLE FOUND; SEE "DIAGNOSTIC AIDS"

WHEN ALL DIAGNOSIS AND REPAIRS ARE COMPLETED, CLEAR DTCs AND VERIFY PROPER OPERATION.

DIAGNOSTIC CHART DTC 71 — 1995

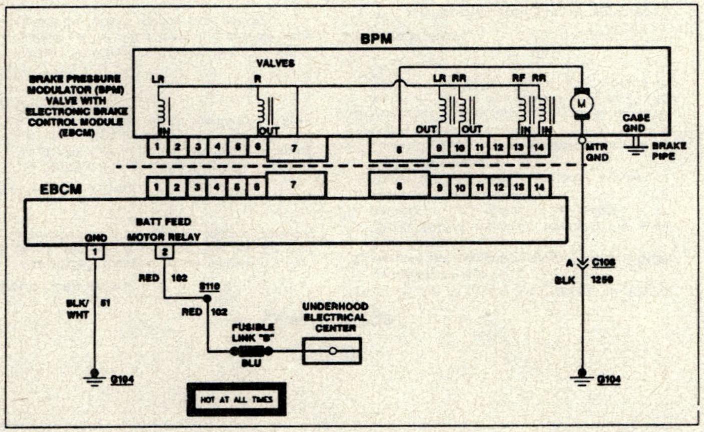

DTC 71
EBCM Internal Malfunction

Circuit Description:

The EBCM performs various self diagnostics that can detect an internal malfunction within the module. If it detects a malfunction, DTC 71 is set.

DTC Will Set When:

- The EBTCM during self diagnostics finds an internal malfunction.
- The pump motor supply line circuit is open or shorted to ground (open fuse).

Action Taken:

The ABS warning lamp will be "ON". ABS is disabled.

DTC Chart Test Description:

Number(s) below refer to circled number(s) on the diagnostic chart.

1. Checks that fusible link "B" feeding the pump motor relay is good.
2. Measures for appropriate voltage from the pump motor supply circuit using the EBCM ground to complete the circuit.
3. Checks for good connections from the harness to the EBCM itself.
4. Checks to see if the malfunction was false. If DTC 71 resets, replace the EBCM

Diagnostic Aids:

All tests using J 39700 Universal Breakout Box terminal "1" are using that terminal as ground. these tests assume that the ground at terminal "1" is good.

It is very important that a thorough inspection of the wiring and connectors be performed. Failure to carefully and fully inspect wiring and connectors may result in misdiagnosis causing part replacement with reappearance of the malfunction.

DIAGNOSTIC CHART DTC 71 — 1995, CONTINUED

DTC 71 EBCM INTERNAL MALFUNCTION

BEFORE BEGINNING TESTING, ENSURE THAT BATTERY IS FULLY CHARGED, AND THAT CHARGING SYSTEM IS OPERATING PROPERLY IF DISCHARGED BATTERY IS NOTED.

(1)
- ENSURE FUSIBLE LINK "B" IS GOOD. IS FUSIBLE LINK "B" OPEN?

YES: REPAIR SHORT TO GROUND IN CKT 102.

NO:

(2)
- IGNITION "OFF."
- DISCONNECT EBCM.
- INSTALL J 39700 UNIVERSAL BREAKOUT BOX WITH J 39700-15 ADAPTER TO EBCM HARNESS CONNECTOR ONLY.
- CONNECT J 39200 BETWEEN J 39700 TERMINALS "2" AND "1". NOTE VOLTAGE. WAS VOLTAGE MEASUREMENT LESS THAN 9 V?

YES: REPAIR OPEN OR POOR CONNECTION IN CKT 102.

NO:

(3)
- IGNITION "OFF."
- CHECK EBCM CONNECTOR CAREFULLY TO ENSURE GOOD ELECTRICAL CONTACT AT ALL TERMINALS, AND THAT CONNECTOR IS PROPERLY RETAINED WHEN CONNECTED. IS POOR CONNECTION EVIDENT?

YES: REPAIR TERMINAL(S), CONNECTOR OR CONNECTOR RETENTION AS NECESSARY.

NO:

(4)
- CLEAR DTCs, IGNITION "OFF."
- DISCONNECT TECH 1, IF CONNECTED.
- IGNITION TO "RUN". DOES DTC 71 RESET?

YES: REPLACE EBCM.

NO: NO TROUBLE FOUND; SEE "DIAGNOSTIC AIDS"

WHEN ALL DIAGNOSIS AND REPAIRS ARE COMPLETED, CLEAR DTCs AND VERIFY PROPER OPERATION.

DIAGNOSTIC CHART DTC 85 — 1995

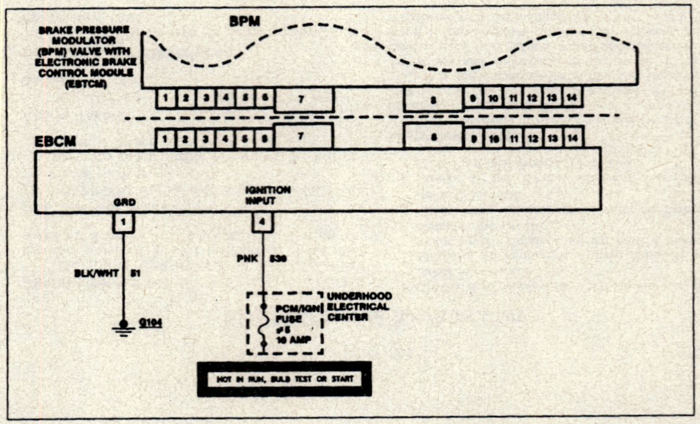

DTC 85
Low Voltage

Circuit Description:

The EBCM requires a certain minimum voltage to operate properly. It monitors the ignition feed circuit to determine if voltage falls below a minimum level.

DTC Will Set When:

- The EBCM operating voltage falls below 9.4 volts out of ABS control, 8.8 volts during ABS control and low voltage is detected for more than 500 milliseconds with the vehicle speed above 6 km/h (3.6 mph).

Action Taken:

When low voltage is detected, the ABS valves are switched passive, which prevents a possible pump start-up and the "ABS" warning lamp will be "ON".

DTC Chart Test Description:

Number(s) below refer to circled number(s) on the diagnostic chart.

1. Checks that PCM/IGN fuse feeding the EBCM ignition circuit is good.
2. Measures for appropriate voltage from the EBCM ignition circuit.
3. Checks EBCM connector terminals for good contact, i.e., not spread open, backed out of connector, corroded, etc.
4. Checks EBCM connector for good retention to EBCM when installed.
5. Checks for DTC reset after other portions of the system have tested as OK.
6. Determines if a malfunction found in step 2 is due to an open or poor connection in the EBCM ignition feed CKT 539, or poor ground in CKT 51.

Diagnostic Aids:

It is very important that a thorough inspection of the wiring and connectors be performed. Failure to carefully and fully inspect wiring and connectors may result in misdiagnosis, causing part replacement with reappearance of the malfunction.

DIAGNOSTIC CHART DTC 85 — 1995, CONTINUED

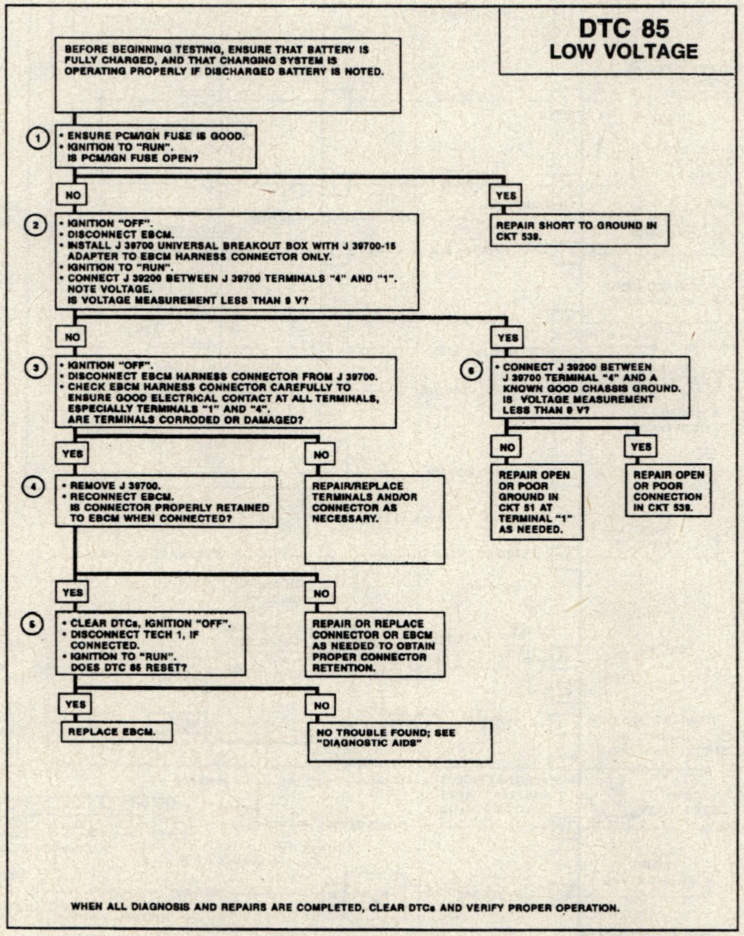

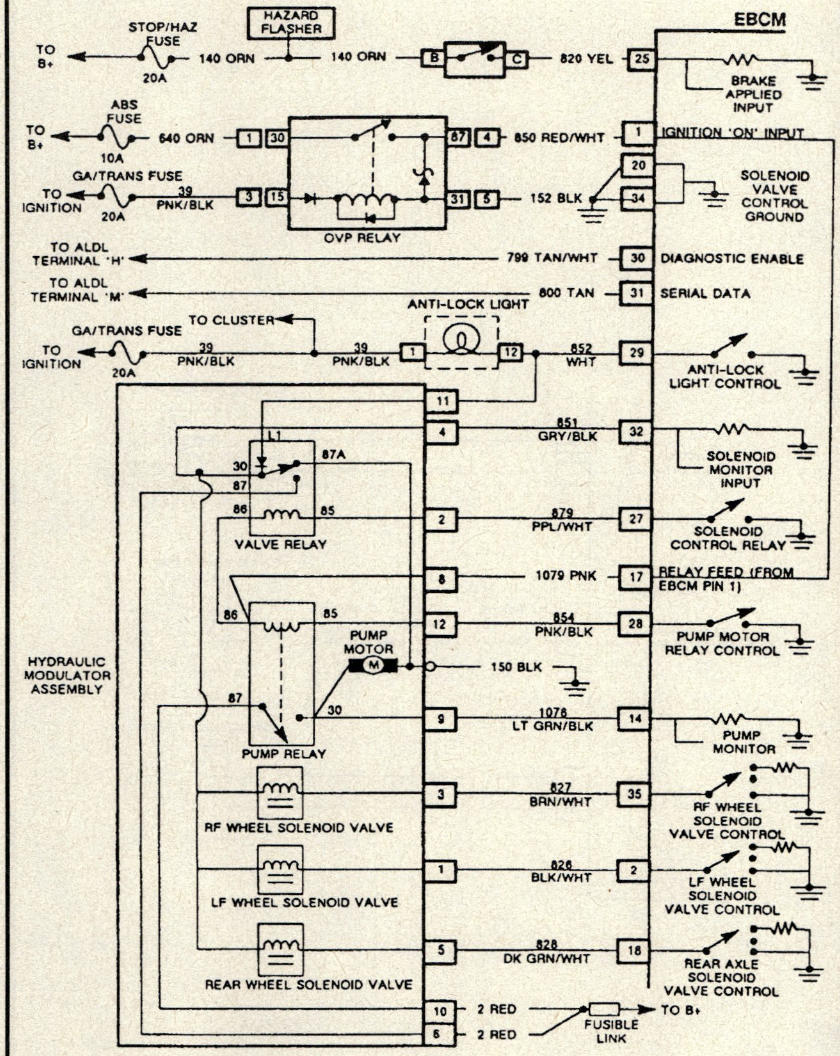

ABS electrical schematic — 1993 Fleetwood Brougham

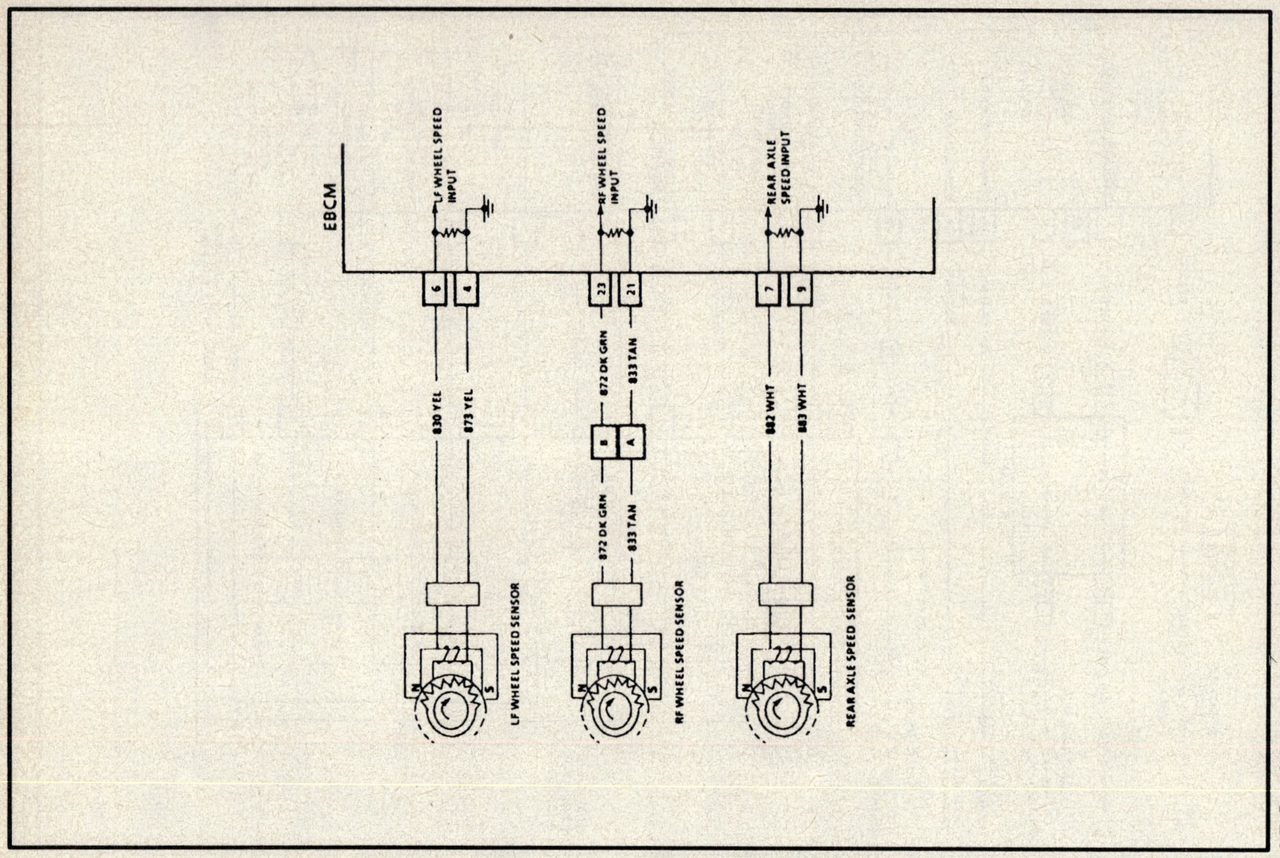

ABS wiring schematic — 1993 Caprice, Custom Cruiser and Roadmaster

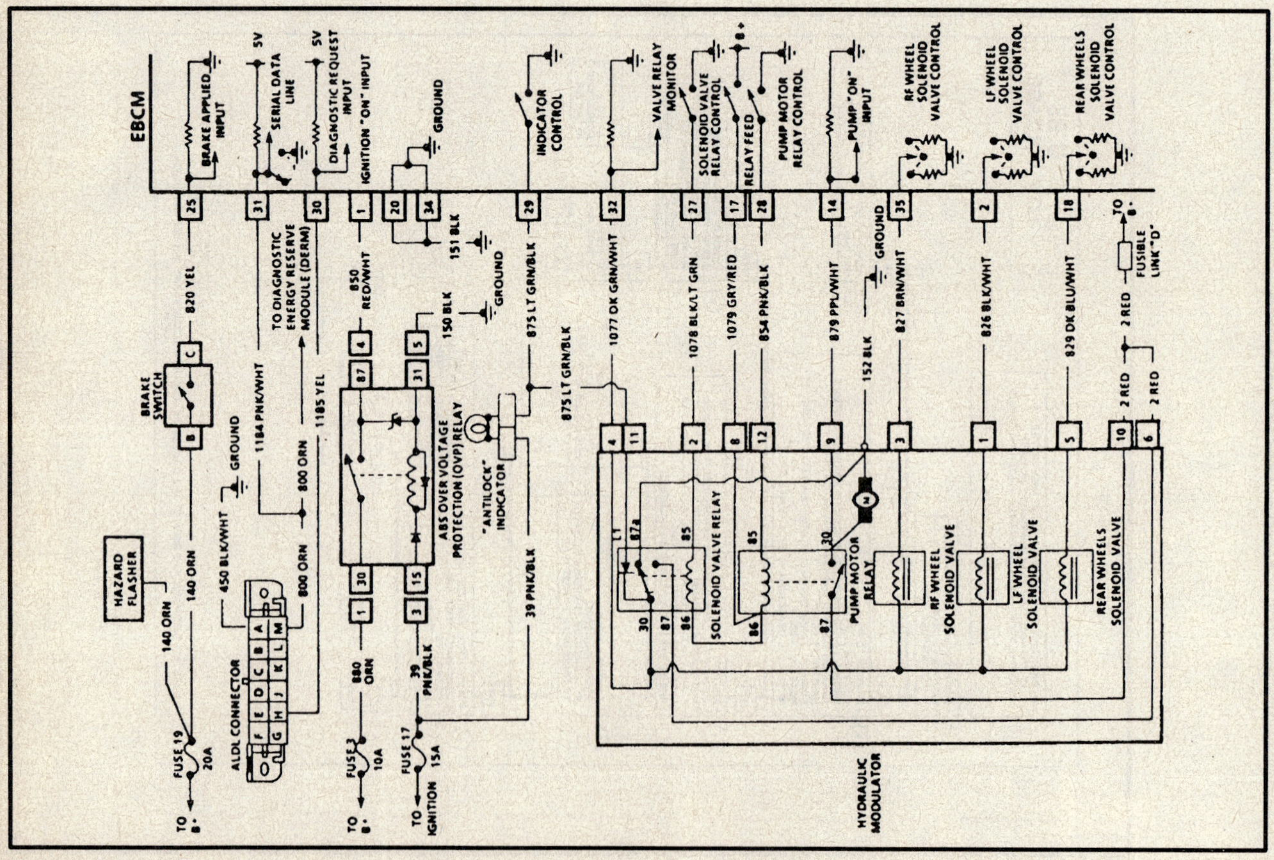

ABS wiring schematic — 1993 Caprice, Custom Cruiser and Roadmaster

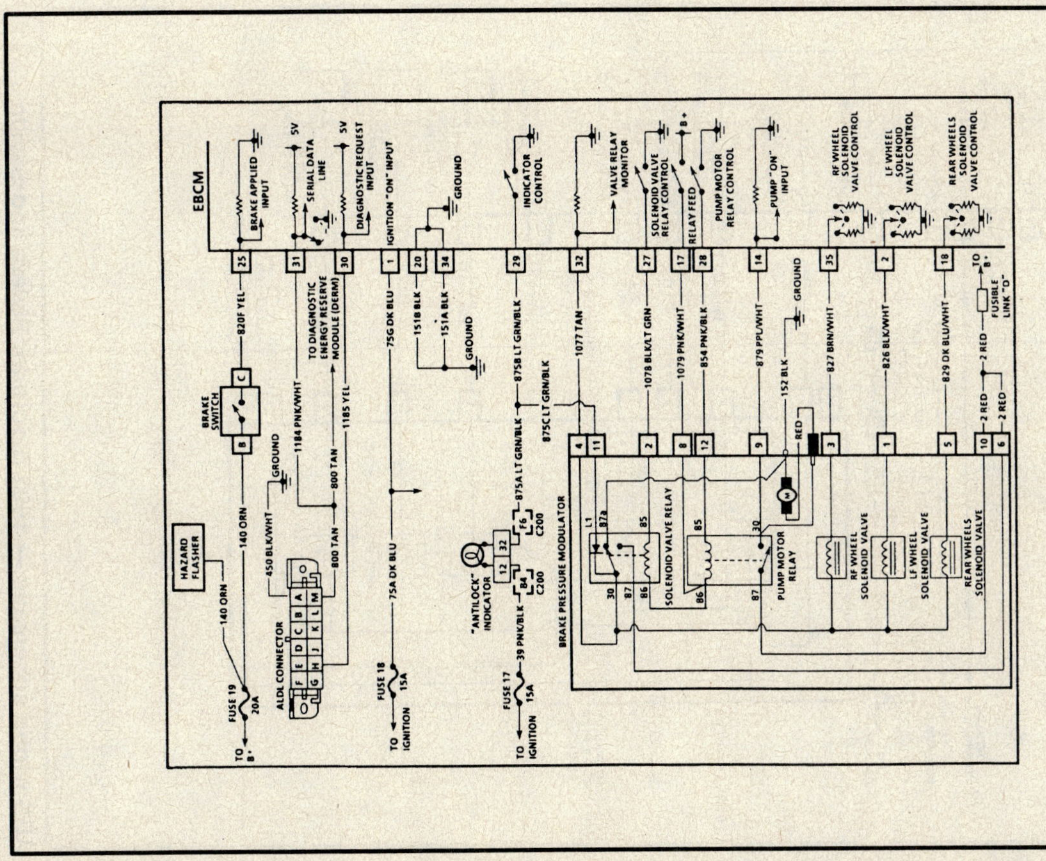

ABS wiring schematic — 1993 Caprice, Custom Cruiser and Roadmaster

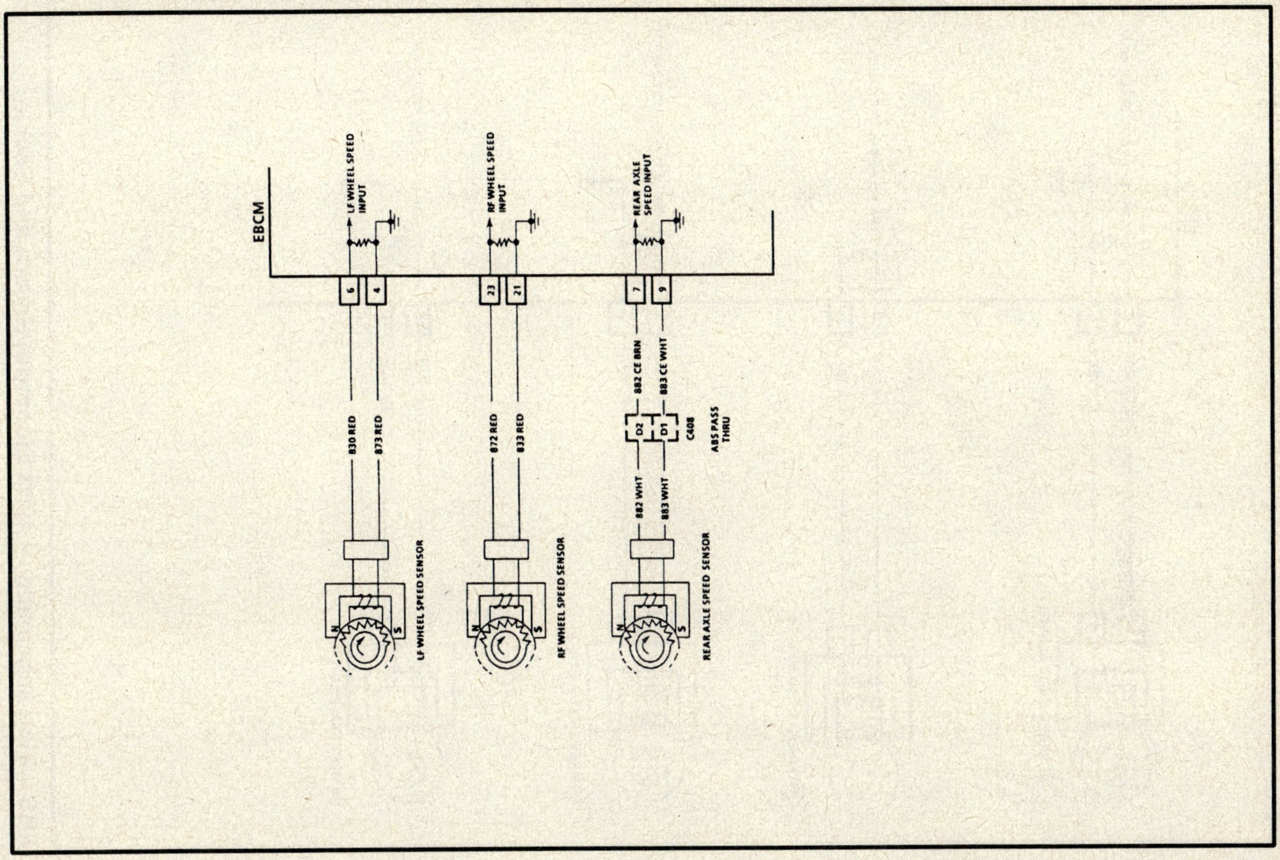

ABS wiring schematic — 1993 Caprice, Custom Cruiser and Roadmaster

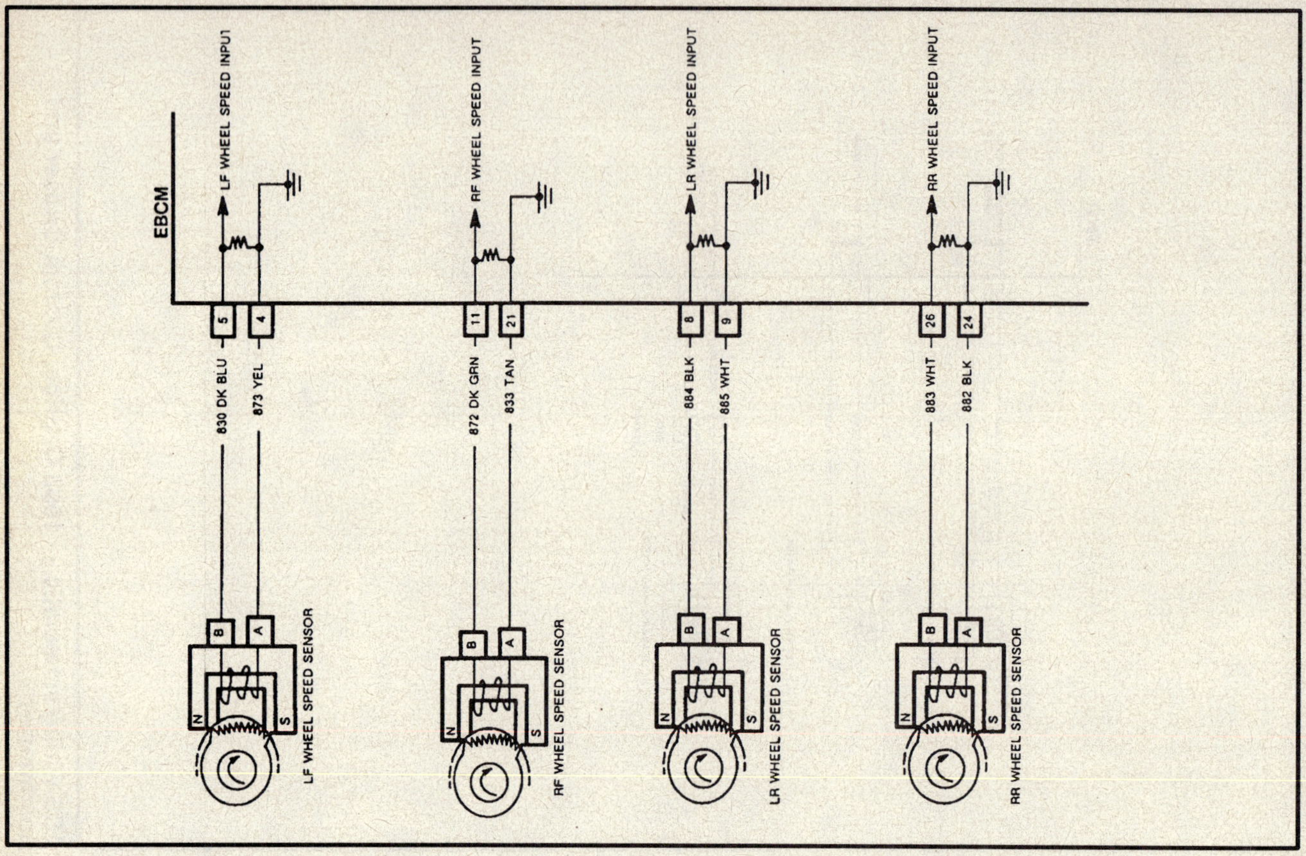

ABS wiring schematic — 1993 Eldorado, Riviera and Seville

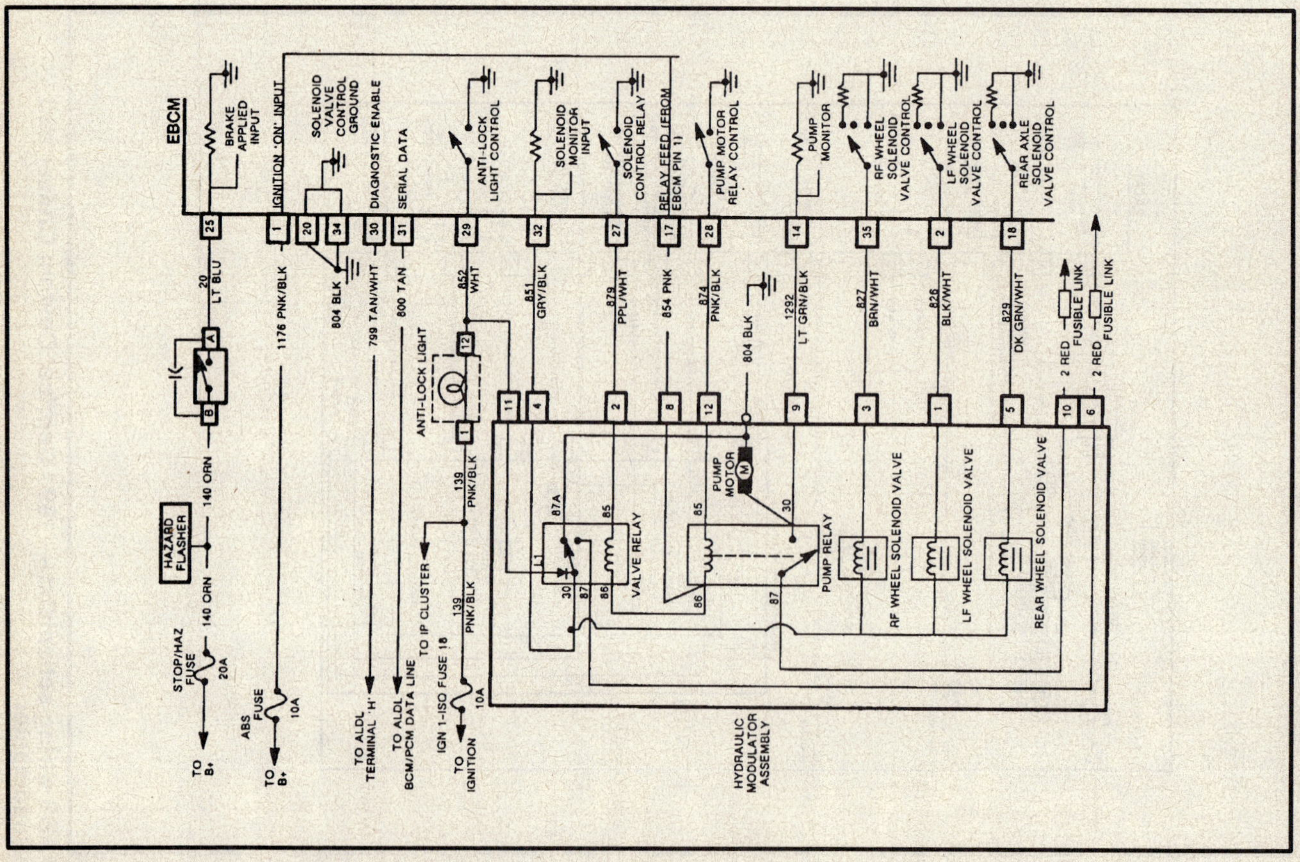

ABS wiring schematic — 1993 Eldorado, Riviera and Seville

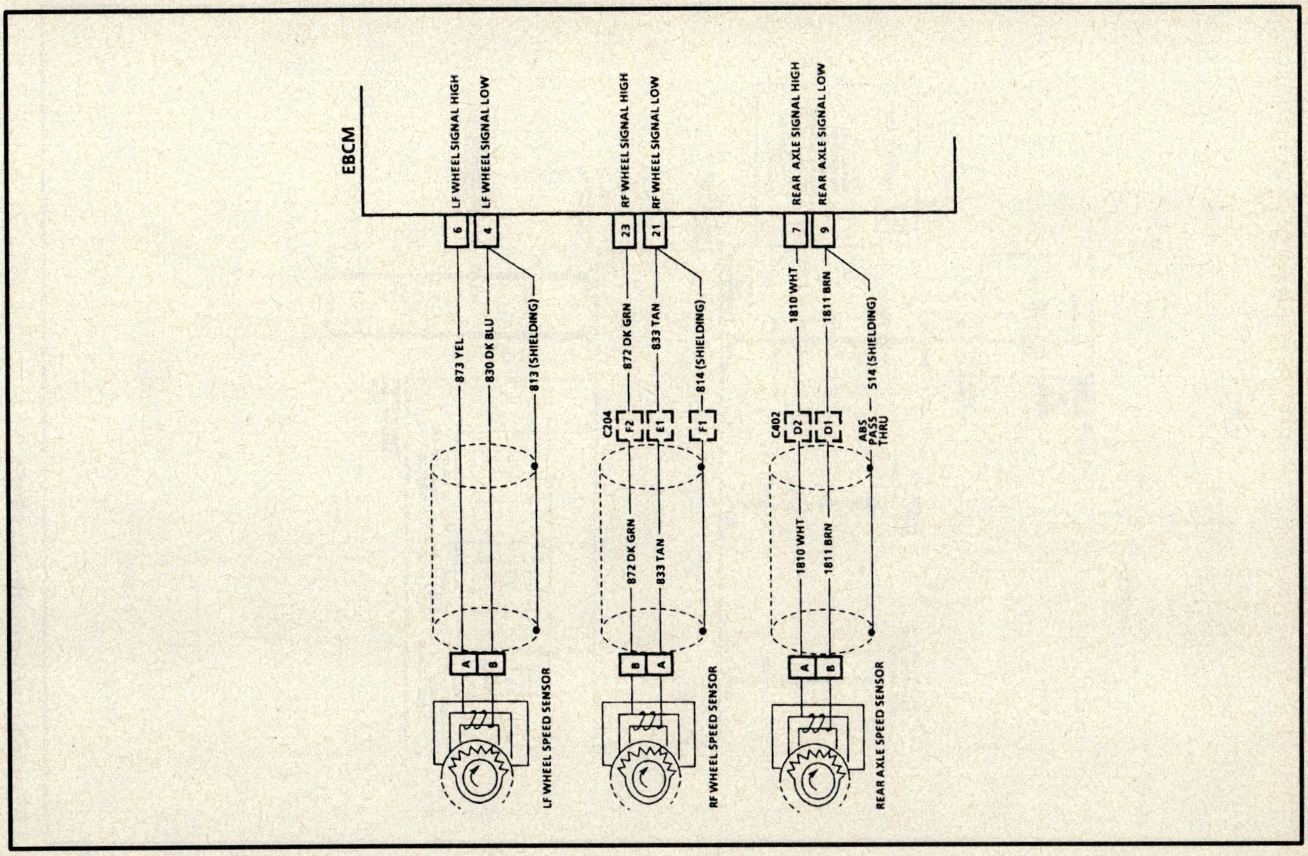

ABS wiring schematic — 1994

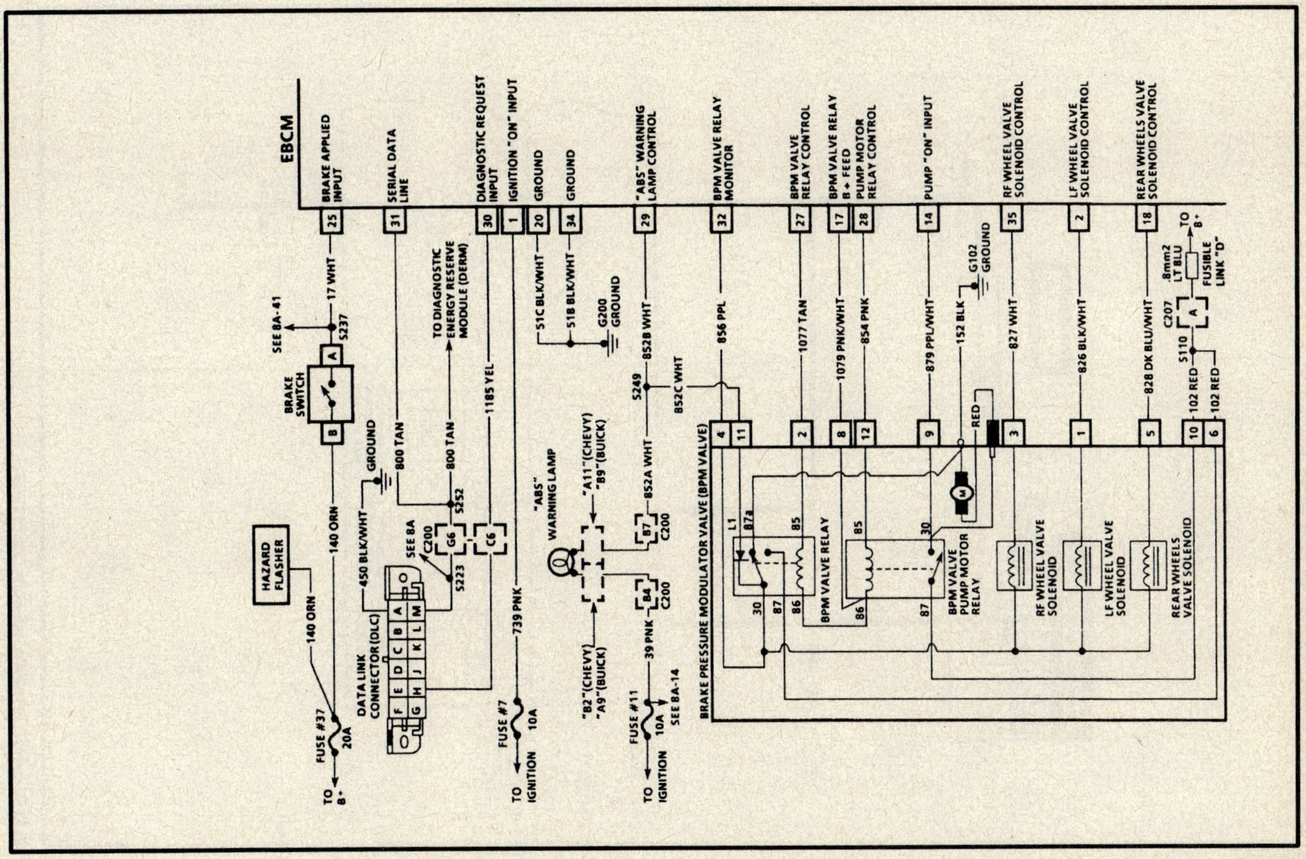

ABS wiring schematic — 1994

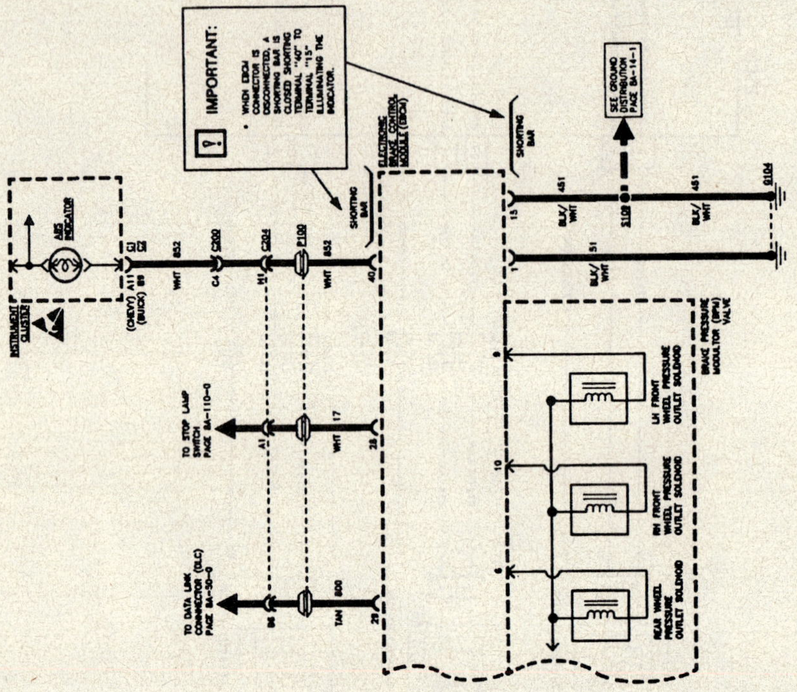

ABS wiring schematic — 1995

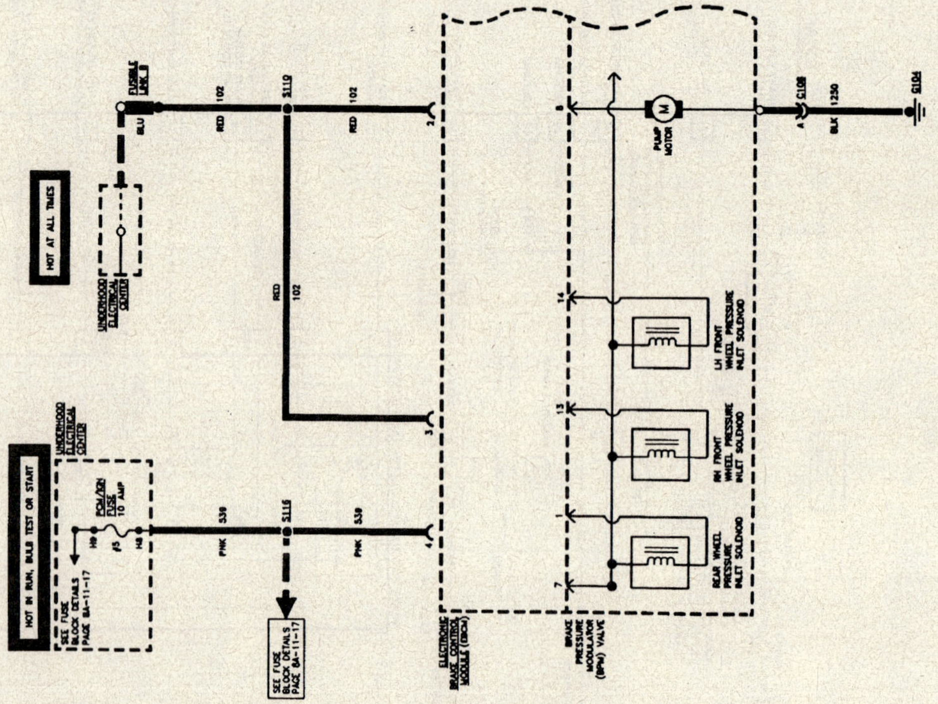

ABS wiring schematic — 1995

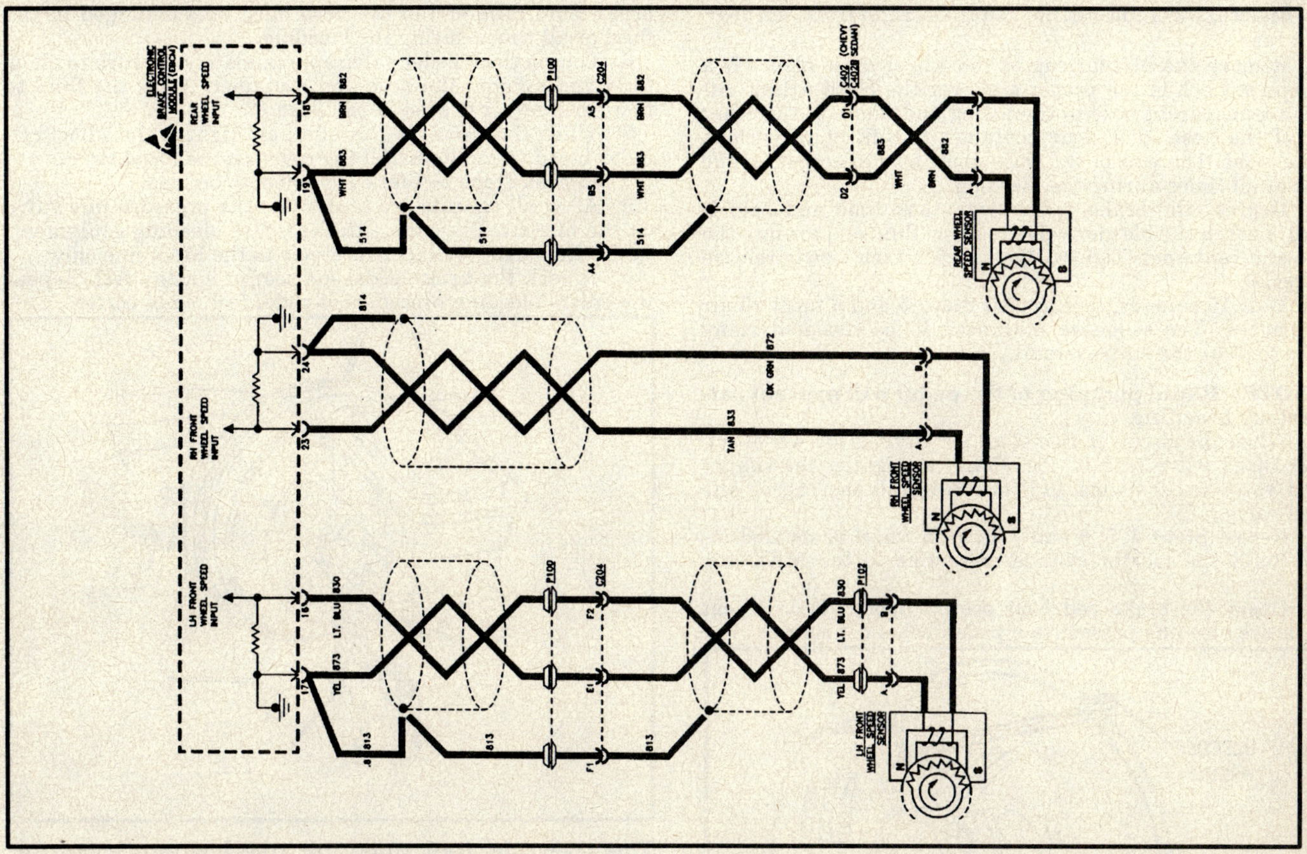

ABS wiring schematic — 1995

Component Replacement

CAUTION

Some vehicles are equipped with a Supplemental Inflatable Restraint (SIR) or air hag system. The air bag (SIR) system must be disabled before performing service on or around SIR components or wiring. Failure to follow safety and disabling procedures could result in possible air bag deployment, personal injury or unnecessary SIR system repairs.

SYSTEM REFILL

The fluid level must be maintained at the **MAX** mark on the master cylinder reservoir. A system operating properly and free of leaks should not need fluid added to it. The fluid level will appear to drop in the reservoir as the brakes wear, but will return to normal after brake linings are replaced. Adding fluid when not needed may result in fluid overflow.

Fluid may be added to the reservoir in normal fashion. The ignition must be **OFF**. Carefully wipe the reservoir cap and surrounding area free of dirt and debris before removing the cap. Use only DOT3 brake fluid from an unopened container. The use of DOT 5 or silicone brake fluid is specifically prohibited in this system. Fill the reservoir only to the **MAX** mark; do not overfill. Make certain the reservoir cap is securely installed.

SYSTEM BLEEDING

Brake system bleeding is performed to remove air present in the lines or components. When a line is disconnected at a wheel, only that caliper and line must be bled. If a line is disconnected at a fitting between the master cylinder and a wheel, the brake system served by that line must be bled. Four wheel bleeding is required only if the primary or secondary lines are disconnected at the master cylinder or hydraulic modulator or if air has entered due to a low fluid condition.

CAUTION

Do not move the vehicle until a firm brake pedal is obtained. An improperly bled system may cause reduced braking, collision or personal injury.

Manual Bleeding Procedure

1. Remove the booster vacuum reserve by applying the brake pedal several times with the ignition **OFF**. The master cylinder must be free of air before bleeding the system. If the master cylinder contains air, it must be bled.
   a. Loosen the forward brake line connection at the master cylinder. Allow fluid to flow from the connection; maintain the fluid level in the reservoir.
   b. Tighten the fitting at the master cylinder, stopping the flow of fluid.
   c. Have an assistant depress the brake pedal slowly one time and hold the pedal down. Loosen the forward fitting again, purging air from the cylinder bore. Tighten the fitting, then release the pedal slowly.
   d. Wait 15 seconds. Repeat Steps **c** and **d**, including the 15-second pause, until all air is removed from the bore.
   e. Tighten the fitting and check for leakage.
   f. Repeat the entire procedure, Steps **a** through **f**, at the rearward brake line fitting.

2. Maintain the correct fluid level in the reservoir during the procedure.
3. If more than one circuit is to be bled, do so in this order:
   - Caprice, Roadmaster, Custom Cruiser, Fleetwood Brougham: right rear, left rear, right front and left front

- Riviera, Eldorado, Seville: left front, right front, left rear, right rear

4. Remove the bleeder cap at the wheel to be bled. Fit a box-end wrench of the proper size over the bleed fitting. Attach a transparent hose to the fitting and immerse the other end of the hose in a clear container half-filled with clean brake fluid. The end of the hose must be submerged in the fluid at all times during the bleeding.
5. Depress the brake pedal slowly one time and hold it down. Loosen the bleeder valve to allow fluid and air into the hose and container. Tighten the bleeder, then slowly release the pedal.
6. Wait 15 seconds, then repeat Steps **5** and **6** until all air is removed. The sequence may need to be repeated many times to clear the entire circuit.

NOTE: Rapid pumping of the pedal will prevent satisfactory bleeding.

7. When the circuit is free of all air, tighten the wheel cylinder bleed valve to 5 ft. lbs. (7Nm) or tighten the caliper bleed valve to 10 ft. lbs. (13Nm). Reinstall the cap on the bleed valve.
8. Repeat Steps **4**, **5**, **6** and **7** for each wheel being bled.
9. Refill the master cylinder reservoir to the **MAX** line only.
10. Check the brake pedal for poor or spongy feel. Repeat the entire bleeding procedure if pedal feel is not correct.

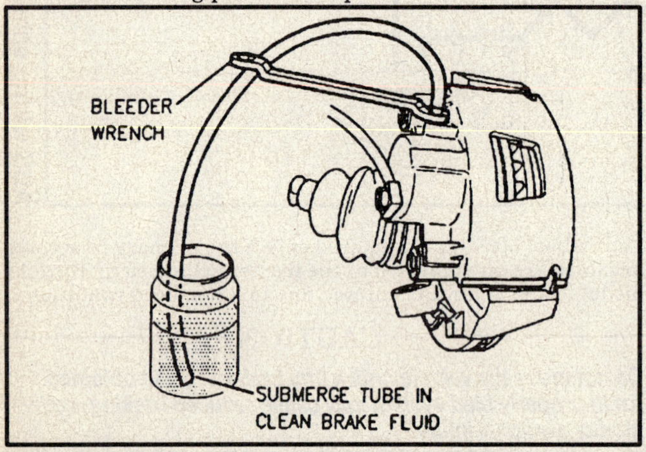

Bleeding the brakes

Pressure Bleeding Procedure

NOTE: Use of a diaphragm pressure bleeder is required. The air chamber must be separated from the brake fluid to prevent entry of dirt and moisture. The correct bleeder adapter must be used to prevent damage to the reservoir.

1. Fill the master cylinder reservoir to the proper level.
2. Install the bleeder adapter or cap to the reservoir.
3. Pressurize the bleeder at 20–25 psi (140–172 kPa). Connect the bleeder pressure hose to the adapter on the reservoir; open the line valve on the bleeder.

NOTE: Vehicles with rear drum brakes require a manual override to permit fluid flow to the front during pressure bleeding. Use tool J-35856 or its equivalent to hold the valve stem of the combination valve open during pressure bleeding.

4. If more than one circuit is to be bled, do so in the order given.
5. Remove the bleeder cap at the wheel to be bled. Fit a box-end wrench of the proper size over the bleed fitting. Attach a transparent hose to the fitting and immerse the other end of the hose in a clear container half-filled with clean brake fluid. The end of the hose must be submerged in the fluid at all times during the bleeding.
6. Loosen the bleeder valve approximately 3/4 turn to allow fluid and air into the hose and container. Allow the fluid to flow until no air is seen in the fluid.
7. When the circuit is free of all air, tighten the wheel cylinder bleed valve. Reinstall the cover on the bleed valve.
8. Repeat Steps **5–7** for each wheel to be bled.
9. After all circuits are bled, close the pressure line valve on the bleeder. Disconnect the pressure bleeding equipment. Refill the master cylinder reservoir to the **MAX** line only.
10. Check the brake pedal for poor or spongy feel. Repeat the entire bleeding procedure if pedal feel is not correct.

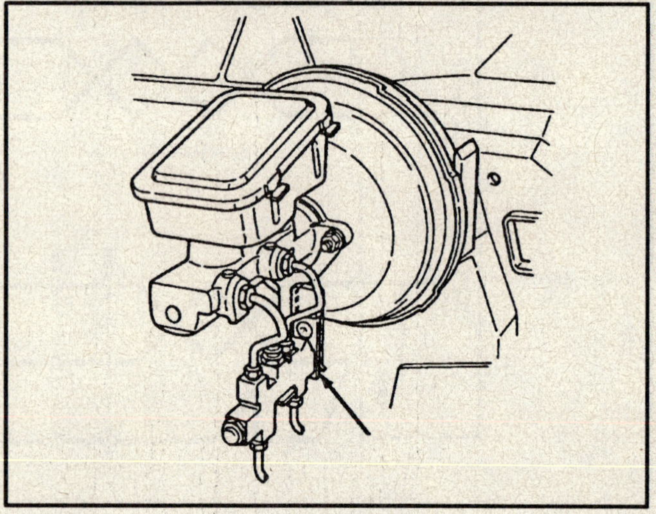

The override tool must be used when pressure bleeding vehicles with drum rear brakes

BRAKE PRESSURE MODULATOR (BPM) VALVE ASSEMBLY

NOTE: The brake pressure modulator (BPM) valve assembly is not serviceable. Do not attempt disassembly or repair on the modulator unit.

Removal and Installation

1993–94

1. Disconnect the negative battery cable.
2. Remove the air intake duct and the resonator. If necessary, remove the upper coolant hose and canister purge hose and move them out of the way. It may also be necessary to remove the left front radiator brace.
3. Remove the cover from the modulator.
4. Release the lock tab on the modulator harness connector and carefully remove the connector from the modulator.
5. Disconnect the modulator ground wire or strap.
6. Label and carefully disconnect each brake line at the modulator. Plug each line immediately to prevent entry of dirt and debris. Contain any fluid spillage.
7. Remove the modulator mounting nuts. Remove the modulator unit from its bracket. Remove the mounting insulators or bushings from the modulator; inspect them for wear or deformation.

To install:

8. Reinstall the bushings on the modulator. Install the modulator on the bracket and tighten the nuts to 89 inch lbs. (10Nm).
9. Connect the brake lines to the correct ports on the modulator. Tighten each line to 11 ft. lbs. (15Nm).

10. Attach the ground strap and harness connector to the modulator. Make certain the multi-pin connector is squarely set and firmly seated; the connector lock must be firmly engaged.
11. Install the relay cover.
12. Install the air cleaner intake hose and radiator brace if it was removed.
13. Install the canister purge line and air intake duct if they were removed.
14. Fill and bleed the entire brake system.
15. Inspect the system for leakage; test drive the vehicle to confirm proper operation.
16. Remove the mounting insulators from the BPM.
17. Remove the brake lines from the BPM. Plug the pipes and fittings to make sure that no dirt enters the brake system.

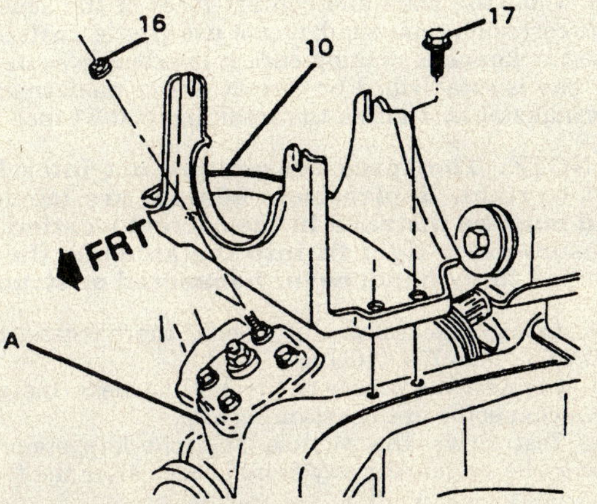

A GEAR ASSEMBLY, STEERING
10 BRACKET, BRAKE PRESSURE MODULATOR (BPM) VALVE
16 NUT, BRAKE PRESSURE MODULATOR (BPM) VALVE BRACKET, 24 N•m (18 LB. FT.)
17 BOLT/SCREW, BRAKE PRESSURE MODULATOR (BPM) VALVE BRACKET, 24 N•m (18 LB. FT.)

Brake pressure modulator valve bracket

1995

1. Disconnect the negative battery cable.
2. Remove the EBCM from the BPM assembly.
3. Remove the ground cable.
4. Carefully lift the BPM from the vehicle.

To install:

5. Place the BPM into the vehicle and install the ground cable. Tighten the nut to 25 inch lbs. (2.8Nm).
6. Attach the EBCM to the PBM and tighten the mounting bolts to 18 inch lbs. (2Nm).
7. Install the insulator bolt to 11 ft. lbs. (15Nm).
8. Tighten the mounting nuts to 89 inch lbs. (10Nm).
9. Install the brake lines into the BPM and tighten them to 12 ft. lbs. (16Nm).
10. Connect the electrical connector.
11. Install the negative battery cable.
12. Fill and bleed the hydraulic system. Perform a visual inspection of the ABS system and road test to verify correct system operation.

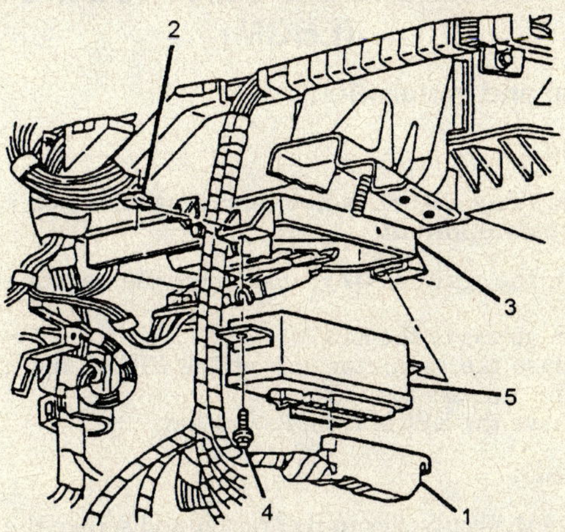

1 CONNECTOR, ELECTRONIC BRAKE CONTROL HARNESS
2 NUT, ELECTRONIC BRAKE CONTROL MODULE
3 BRACKET, ELECTRONIC BRAKE CONTROL MODULE
4 BOLT/SCREW, ELECTRONIC BRAKE CONTROL MODULE
5 MODULE ASSEMBLY, ELECTRONIC BRAKE CONTROL

Electronic brake control module assembly

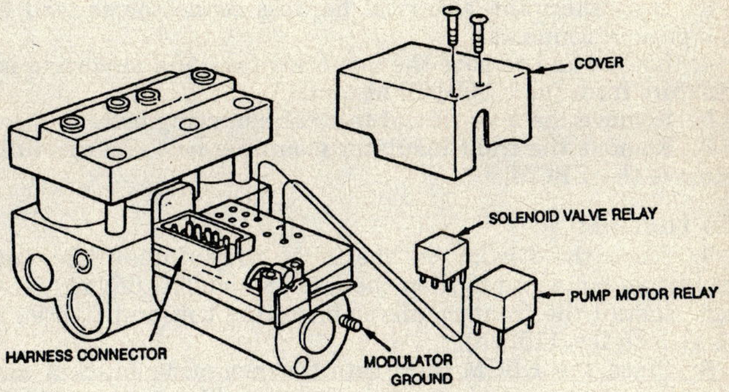

ABS Brake pressure modulator valve and components

MODULATOR RELAYS

The solenoid valve relay and pump motor relay are located on the top of the modulator unit. After removing the relay cover, the relays may be changed with the modulator in place. The solenoid valve relay has a silver case and a 6 pin base. The 4 pin relay in the black case controls the pump motor. On 1995 vehicles, the relays are an integral part of the EBCM and cannot be replaced individually.

ELECTRONIC BRAKE CONTROL MODULE (EBCM)

Removal and Installation

1993–94

NOTE: The EBCM is not serviceable; it is a replaceable only component.

1. With the ignition **OFF**, disconnect the negative battery cable.
2. Gain access to the EBCM.
3. Release the connector lock at the EBCM and carefully remove the harness connector.
4. Remove the EBCM from its bracket.

To install:

5. Place the controller in its bracket and secure the mounting screws. The unit must be correctly placed so that the wiring harness connector is on the proper side.
6. Connect the wiring harness connector. Make certain the connector is squarely placed and securely mated to the EBCM. The connector lock must be engaged.
7. Reinstall or reposition any panels, carpeting or components moved to gain access.
8. Test drive the vehicle and check system operation. Check the system for any stored codes after the test drive.

1995

NOTE: The EBCM is not serviceable; it is a replaceable only component.

1. Disconnect the negative battery cable.
2. Remove the air duct resonator assembly. Carefully clean the area around the EBCM/BPM unit so that dirt and other contaminants will not enter any open brake lines.
3. Disconnect the electrical harness connector as well as the ground connector.
4. Loosen and remove the mounting nuts and mounting insulators from the EBCM/BPM unit.
5. Remove the four EBCM to BPM mounting bolts.
6. Remove the front insulator mounting bolt and carefully remove the EBCM.

To install:

7. Place the EBCM on the BPM body. Install the four mounting bolts and tighten them to 18 inch lbs. (2Nm).
8. Install the front insulator mounting bolt and tighten it to 11 inch lbs. (15Nm).
9. Place the EBCM/BPM into the mounting bracket and install the mounting insulators and nuts. Tighten them to 89 inch lbs. (10Nm).

FRONT WHEEL SPEED SENSORS

Removal and Installation

CAPRICE, CUSTOM CRUISER AND ROADMASTER

1. With the ignition **OFF**, elevate and safely support the vehicle.
2. Remove the sensor harness connector from its retaining clip or bracket. Disconnect the sensor from the harness.
3. Remove the bolt and bracket holding the harness to the frame rail.
4. Free the sensor cable from all clips and retainers. Take careful note of the cable routing; the harness must be reinstalled in its exact original position.
5. Remove the sensor retaining bolt and remove the sensor.
6. Inspect the face of the sensor for dirt, debris or damage. There should be no sign of contact between the sensor and the toothed wheel. Clean or replace the sensor as necessary.

To install:

7. Route the sensor cable correctly and secure the clips and retainers. Double check that the wiring will be clear of moving parts with enough slack to allow for full suspension travel.
8. Coat the body and contact faces of the sensor with an anti-corrosion compound; do not use grease. Install the sensor into the knuckle, making sure it is properly seated. The sensor gap is established by correct sensor mounting; the gap is not adjustable. Tighten the retaining bolt 71 inch lbs. (8Nm).

NOTE: The speed sensors are not interchangeable left to right. Replacement sensors are labeled L or R and must be installed in the correct location. All speed sensors are a tight fit into the knuckle; they must be pushed in by hand, never hammered or struck.

9. Install the frame clip to hold the harness; tighten the bolt to 89 inch lbs. (10Nm).
10. Connect the sensor to the ABS harness. Install the complete connector in the retaining clip.
11. Test drive the vehicle and check system operation. Check the system for any stored codes after the test drive.

ELDORADO, REATTA, RIVIERA AND SEVILLE

The front wheel speed sensors on these vehicles are integrated with the hub and bearing assembly. The sensor may be replaced and serviced separately from the hub and bearing assembly.

NOTE: The sensor must be replaced whenever it is separated from the hub or bearing. Do not attempt to reuse the sensor once it is removed.

1. With the ignition **OFF**, disconnect the wheel speed sensor connector.
2. Remove the hub and bearing assembly. Clean all dirt from the sensor housing area.
3. Gently pry the slinger from the speed sensor assembly. Discard the used slinger.
4. Carefully pry the speed sensor from the bearing assembly. Do not allow dirt to enter the bearing when the sensor is removed. Do not add lubricant through the sensor opening; the bearing is permanently lubricated.
5. Inspect the bearing and sensor for any signs of rust or water intrusion; if such damage is found, replace the hub and bearing assembly. Do not clean the grease from the toothed sensor ring; grease does not affect operation of the sensor.
6. Use a clean, lint-free cloth to remove sealant from the outer diameter of the bearing hub.

1. EBCM/BPM VALVE
2. EBCM/BPM VALVE MOUNTING BRACKET
3. BRACKET MOUNTING BOLT
4. EBCM/BPM VALVE BRACKET TO STEERING GEAR UNIT
5. EBCM/BPM MOUNTING INSULATOR
6. EBCM/BPM VALVE TO BRACKET NUT
7. ELECTRONIC BRAKE CONTROL MODULE (EBCM)
8. BRAKE PRESSURE MODULATOR (BPM) VALVE
9. PUMP MOTOR (NOT SERVICEABLE, INTEGRAL TO BPM)
10. VALVE RELAY (NOT SERVICEABLE, INTEGRAL TO EBCM)
11. PUMP MOTOR RELAY (NOT SERVICEABLE, INTEGRAL TO EBCM)
12. EBCM SEAL
13. EBCM TO BPM VALVE BOLT
14. INSULATOR MOUNTING BOLT

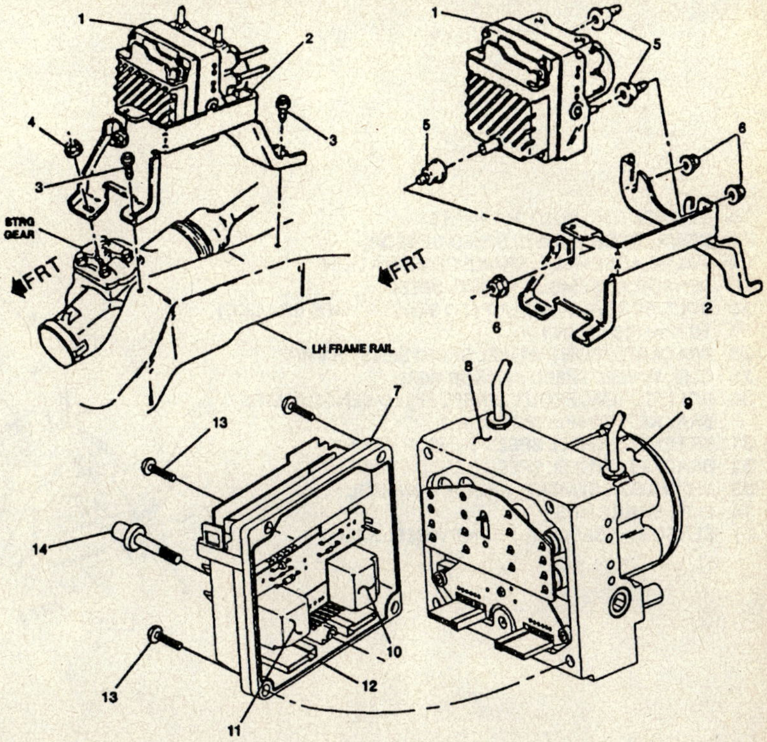

Electronic brake control module and brake pressure modulator valve

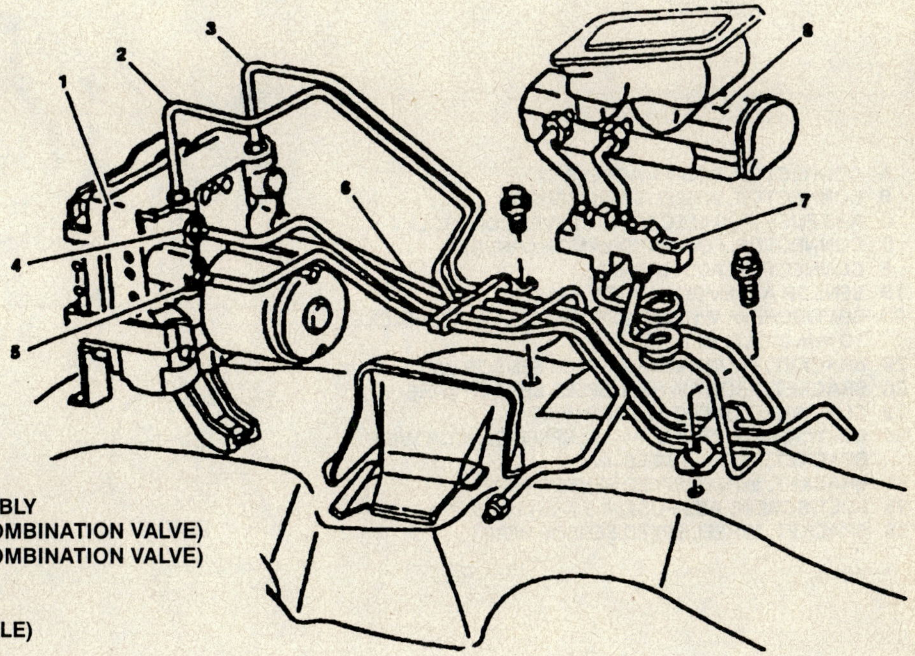

1. EBCM/BPM VALVE ASSEMBLY
2. FRONT BRAKE PIP (TO COMBINATION VALVE)
3. REAR BRAKE PIPE (TO COMBINATION VALVE)
4. FRONT BRAKE PIPE (LH)
5. FRONT BRAKE PIPE (RH)
6. REAR BRAKE PIPE (TO AXLE)
7. COMBINATION VALVE
8. MASTER CYLINDER

Brake pipe locations — 1995 models

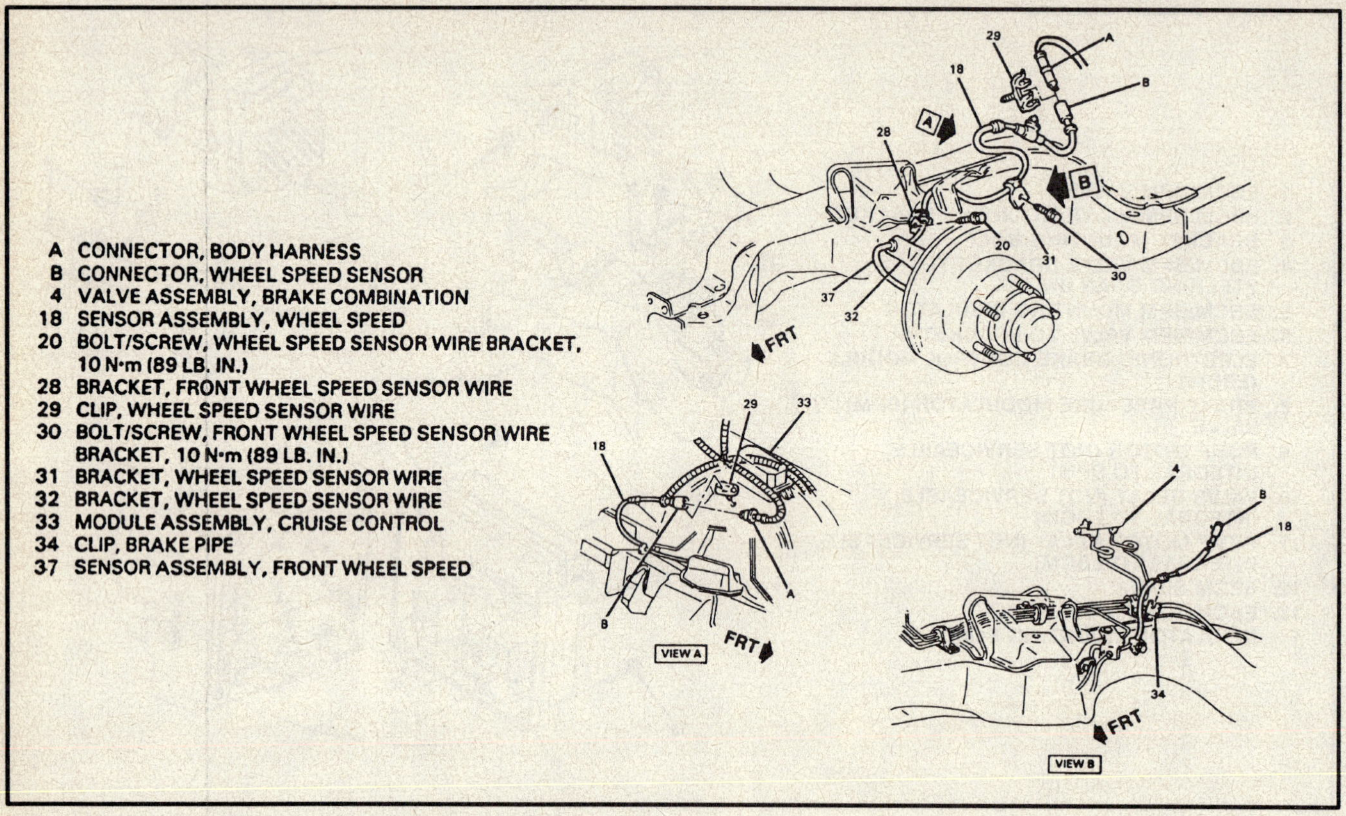

Left front wheel speed sensor assembly — 1994

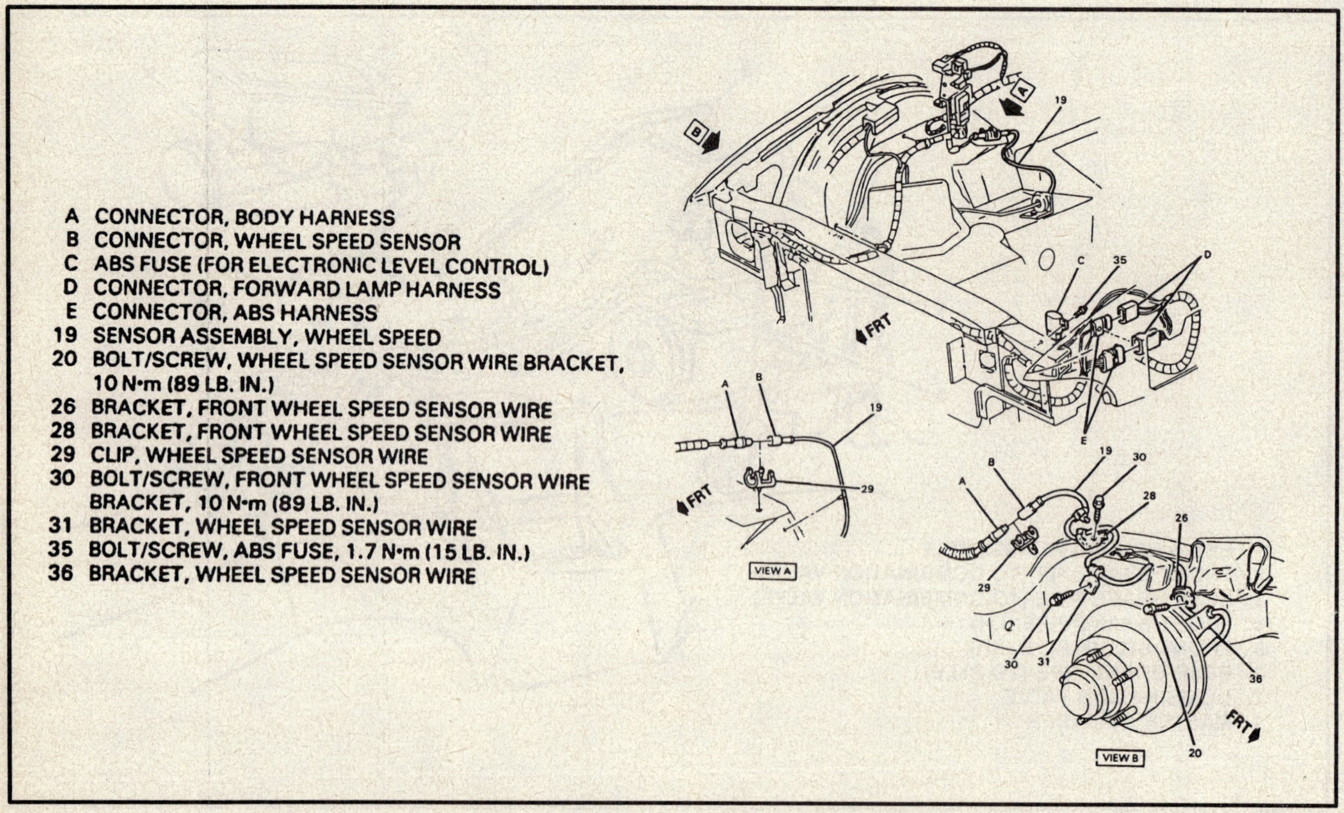

Right front wheel speed sensor assembly — 1994

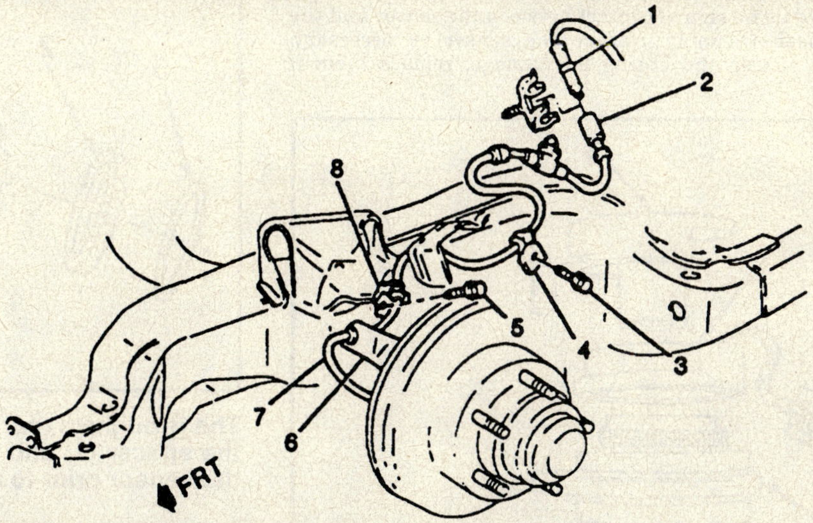

1. CONNECTOR, BODY HARNESS
2. CONNECTOR, WHEEL SPEED SENSOR
3. BOLT/SCREW, FRONT WHEEL SPEED SENSOR WIRE BRACKET
4. BOLT/SCREW, WHEEL SPEED WIRE
5. BOLT/SCREW, WHEEL SPEED SENSOR WIRE BRACKET
6. BRACKET, WHEEL SPEED SENSOR WIRE
7. SENSOR, FRONT WHEEL SPEED
8. BRACKET, FRONT WHEEL SPEED SENSOR WIRE

Front wheel speed sensor assembly — 1995

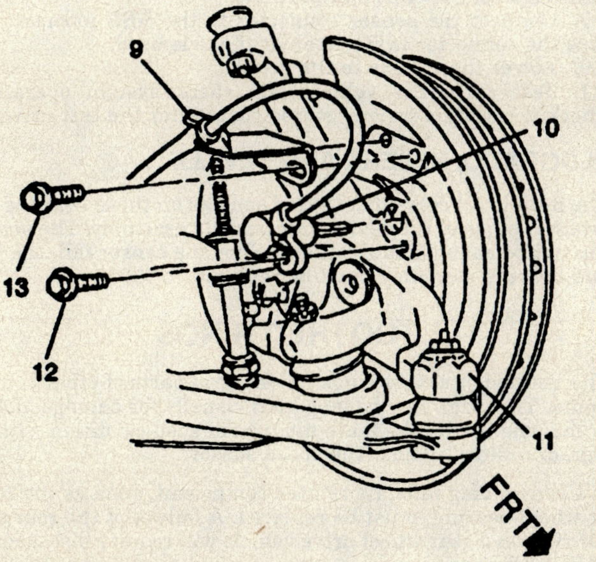

9. BRACKET, WHEEL SPEED SENSOR WIRE
10. SENSOR, FRONT WHEEL SPEED
11. KNUCKLE, STEERING
12. BOLT/SCREW, WHEEL SPEED SENSOR
13. BOLT/SCREW, WIRE SPEED SENSOR WIRE BRACKET

Front wheel speed sensor assembly (typical) — 1994–95

To install:

14. Apply a sealant, 620 Loctite® or its equivalent, to the groove in the outer diameter of the bearing hub.
15. Place the hub and bearing on tool J–38764–4 or its equivalent to prevent damage to the studs.
16. Install the sensor using a press and tool J–38764–1 or its equivalent.
17. Install the hub and bearing assembly.
18. Connect the sensor wiring to the ABS harness.
19. Test drive the vehicle and check system operation. Check the system for any stored codes after the test drive.

REAR WHEEL SPEED SENSORS

Removal and Installation

Excessive runout of the speed sensor ring can cause erratic wheel speed sensor signals. If runout of the ring exceeds 0.010 inch (0.25mm), the ring and component will need to be replaced.

CAPRICE, CUSTOM CRUISER, FLEETWOOD BROUGHAM AND ROADMASTER

1. With the ignition **OFF**, elevate and safely support the vehicle.
2. The rear speed sensor is located in the differential housing, not at the wheels. Disconnect the sensor connector from the ABS harness.
3. Free the sensor wiring from the clips and retainers holding it.
4. Remove the sensor retaining bolt; remove the sensor from the differential housing. The Fleetwood Brougham sensor includes a plastic spacer and O-ring which must be removed with the sensor.

5. Inspect the face of the sensor for dirt, debris or damage. There should be no sign of contact between the sensor and the toothed wheel. Clean or replace the sensor as necessary. Inspect the spacer and O-ring for damage; replace them if necessary.

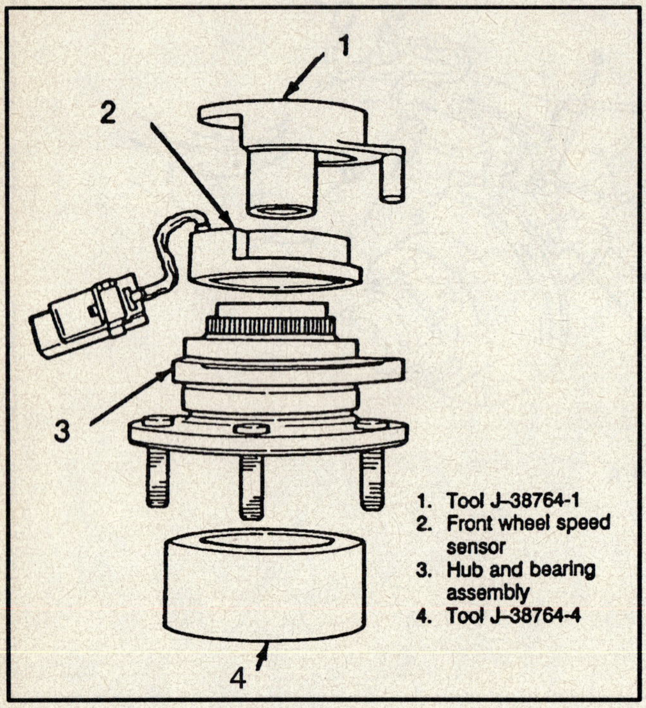

Installation of wheel speed sensor using press

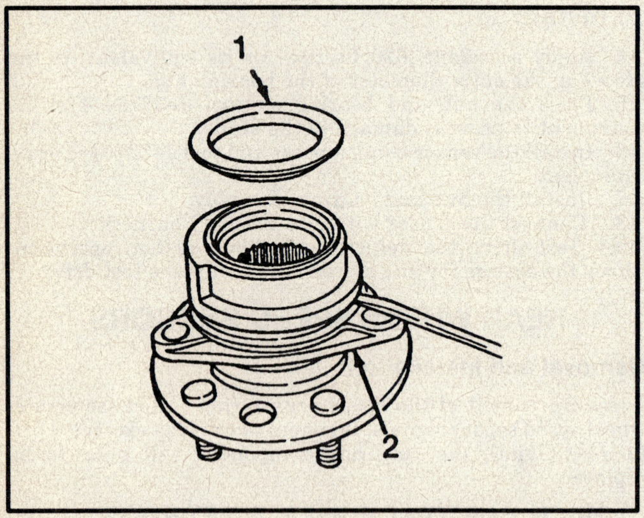

After the slinger (1) is removed, carefully pry the sensor from the hub and bearing assembly (2)

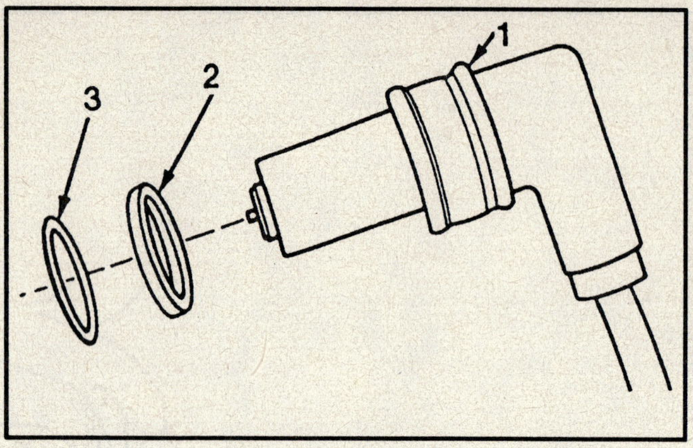

The Brougham rear wheel speed sensor (1) requires the spacer (2) and the O-ring (3) to be installed on the sensor prior to the installation in the spindle

To install:

6. Position the sensor in the differential. On Fleetwood Brougham, make certain the O-ring and spacer are in place.

NOTE: The speed sensor is designed to fit tightly in its bore. It must be pressed in by hand, never hammered or struck. The sensor gap is determined by correct installation; the gap is not adjustable.

7. Install the sensor retaining bolt and tighten it to 71 inch lbs. (8Nm).
8. Route the sensor wiring harness correctly; secure the harness with the clips and brackets.
9. Connect the sensor connector to the ABS harness. Secure the connector in its retainer if one is used.
10. Lower the vehicle to the ground.
11. Test drive the vehicle and check system operation. Check the system for any stored codes after the test drive.

ELDORADO, RIVIERA AND SEVILLE

The individual rear wheel speed sensors on these vehicles are integral parts of the rear wheel bearing assembly. The sensor can not be changed individually. Should a sensor fail, the hub and bearing assembly must be replaced.

TOOTHED RINGS

The toothed or sensor rings are integral parts of other components. The rings may be inspected visually for damage, debris or missing teeth. The teeth must be in good condition with no sign of contact against the speed sensor.

If a sensor ring fails, its related component, such as the rotor or wheel bearing, must be replaced. A failure of the rear sensor ring on a rear wheel drive vehicle will require disassembly of the differential.

A PAN, FLOOR
B PASS-THROUGH, ANTILOCK BRAKE SYSTEM
C CONNECTOR, ELECTRONIC BRAKE CONTROL WIRING HARNESS
D CONNECTOR, WHEEL SPEED SENSOR
E COVER, ANTILOCK BRAKE SYSTEM PASS-THROUGH
F CONNECTOR, HEIGHT SENSOR RELAY
40 SENSOR ASSEMBLY, REAR WHEEL SPEED
45 CLIP, WHEEL SPEED SENSOR WIRE
65 HARNESS ASSEMBLY, ELECTRONIC BRAKE CONTROL WIRING
67 BOLT/SCREW, ELECTRONIC BRAKE CONTROL WIRING HARNESS CLIP, 4 N•m (35 LB. IN.)
68 CLIP, ELECTRONIC BRAKE CONTROL WIRING HARNESS
71 HARNESS ASSEMBLY, ELECTRONIC BRAKE CONTROL WIRING

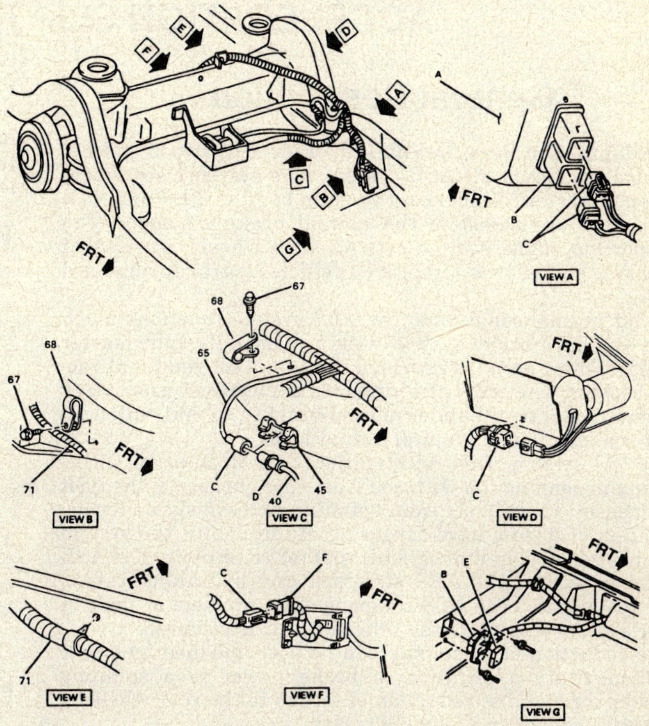

Rear wheel speed sensor wiring with electronic level control system — typical

A PAN, FLOOR
B PASS-THROUGH, ANTILOCK BRAKE SYSTEM
C RIVET
D COVER, ANTILOCK BRAKE SYSTEM PASS-THROUGH
E CONNECTOR, ELECTRONIC BRAKE CONTROL WIRING HARNESS
F CONNECTOR, WHEEL SPEED SENSOR
40 SENSOR ASSEMBLY, REAR WHEEL SPEED
45 CLIP, WHEEL SPEED SENSOR WIRE
65 HARNESS ASSEMBLY, ELECTRONIC BRAKE CONTROL WIRING
67 BOLT/SCREW, ELECTRONIC BRAKE CONTROL WIRING HARNESS CLIP, 4 N•m (35 LB. IN.)
68 CLIP, ELECTRONIC BRAKE CONTROL WIRING HARNESS
71 HARNESS ASSEMBLY, ELECTRONIC BRAKE CONTROL WIRING

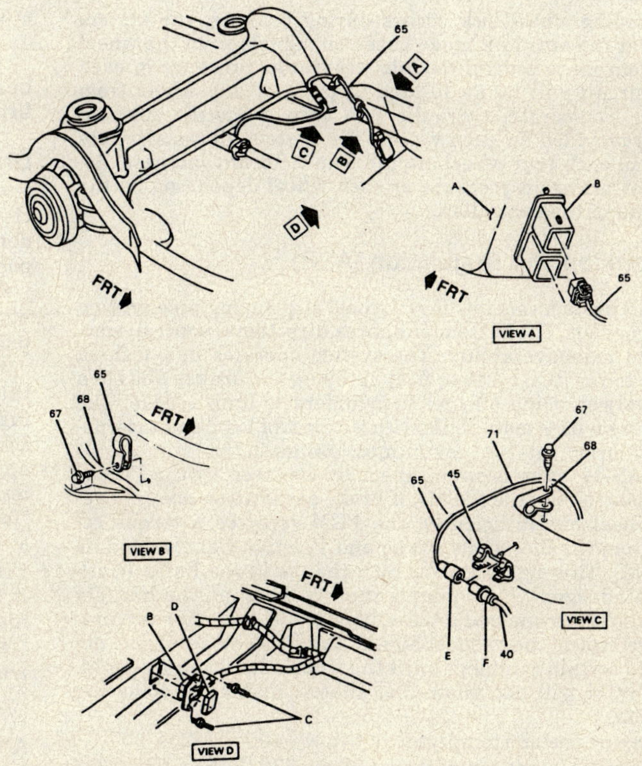

Rear wheel sensor without electronic level control system (typical) — 1994–95

BOSCH 2U ANTI-LOCK SYSTEM WITH ACCELERATION SLIP REGULATION (ASR)

General Description

The Allante, Concours, DeVille, Eldorado, Fleetwood and Seville use the Bosch 2U/ASR system. The system uses a conventional master cylinder and vacuum booster. ABS and ASR components are added to the normal braking system. The ABS portion of the system acts on all 4 wheels, preventing wheel lock and allowing improved vehicle control during hard braking.

Under normal conditions, the ABS system functions much like a standard brake system with a diagonally split master cylinder. Power assist is provided through a conventional vacuum booster. The ABS also provides a means of preventing excessive pedal travel under normal braking so that sufficient pedal remains in the event of a fluid leak.

The 2U system uses Acceleration Slip Regulation (ASR) working in conjunction with the ABS. The purpose of the ASR is to prevent the wheels from spinning excessively under acceleration or during hard cornering. Additionally, the system prevents wheel spin during foul weather conditions. The ASR system is part of the ABS electrical and hydraulic system. Each system uses the ABS components; a problem or fault in either system will affect the other until it is repaired.

The reduction of wheel slip can be accomplished by either the momentary application of brake power to a spinning wheel or through the reduction of torque to the rear wheel by a selective reduction of engine power.

SYSTEM OPERATION

Anti-lock Brake Systems

If impending wheel lock occurs during braking, the system will enter the anti-lock mode based on signals from the wheel speed sensors to control module. Hydraulic pressure in each wheel circuit will be modulated to prevent any wheel from locking. A separate hydraulic line with a specific solenoid valve is provided for each wheel. The system allows separate control of each rear wheel circuit. The ABS can increase, decrease or maintain pressure at each wheel depending on the command of the controller.

Acceleration Slip Regulation (ASR)

The ASR system controls drive wheel slip during acceleration and cornering. The system enhances directional control, traction and maneuverability. The system operates in a fashion similar to the TCS system, first applying the brakes and then selectively shutting off fuel to cylinders to limit torque. The ASR system uses more sophisticated control logic and incorporates different internal communication methods.

The ASR system is simultaneously overseen by the Powertrain Control Module (PCM); if brake application does not reduce wheelspin sufficiently, the PCM receives a torque request from the Electronic Brake and Traction Control Module (EBCTM). This signal is sent by a change in the Pulse Width Modulated signal. By changing the PWM signal, the EBCTM can request torque reductions between 0 and 100 percent. Upon receipt of the request signal, the PCM disables fuel delivery to certain cylinders until the wheel slippage is reduced.

The PCM will not allow fuel shutoff under the following conditions:

- Engine coolant temperature below –400°F (–204°C)
- Engine coolant temperature above 278°F (137°C)
- Low coolant level is detected
- Engine speed below 600 rpm

Whenever the ASR system is functioning, the TRAC CNTRL lamp on the instrument panel will illuminate. The system is not disabled when the mini-spare is in use but it may record a trouble code for a defective wheel speed sensor due to the difference in rotational speeds of the wheels.

Timed Cylinder Cut-Out

The traction control system will operate only when the system senses that one or both of the front wheel is slipping or beginning to loose traction. If the system senses that excessive wheel slip is present, the TCS pilot valve in the BPM valve assembly will activate and will increase, hold or decrease hydraulic pressure to the front wheel brakes, therefore keeping an optimum slip rate on the front wheels.

WARNING SYSTEM

The BRAKE warning lamp in the instrument cluster will light to warn the operator of conditions within the system which may reduce braking ability. Conditions causing the BRAKE lamp to illuminate are:

- Parking brake not fully released.
- Low brake fluid. The fluid level sensor is located in the hydraulic unit reservoir cap. In this case, the BRAKE warning lamp is triggered by the BCM via the serial data link.
- Traction Control fault codes. When any diagnostic code is current, except Codes 67 and 73, the BRAKE warning lamp will be lit.
- During system repairs when put into the diagnostic mode or connected to a hand scanner.
- Bulb check during the starting cycle. This is normal operation and confirms that the lamp is operating.

NOTE: Any time the BRAKE warning lamp is lit, the braking system may be impaired or provide reduced braking function.

Driver Information Center Messages

- LOW BRAKE FLUID — The message is commanded by the BCM when low fluid level is detected in the master cylinder reservoir. The message is displayed during the key **ON** portion of starting. It is usually followed by other messages.
- SERVICE BRAKE FLUID LEVEL — message turns **ON** when an open circuit in the brake fluid level system is detected.
- ANTI-LOCK DISABLED — A fault has been detected in the system by either the EBCTM or the BCM. The anti-lock function is disabled; normal braking is available if the red BRAKE lamp is not lit. Additionally, this message appears any time a fault code (except 67 or 73) is current in the system.
- TRACTION DISABLED — message is displayed any time a traction control fault is stored as a current fault. The EBCTM or the BCM has disabled the traction control system. If this is the only message displayed, it is possible that ABS function remains available.
- TRACTION ACTIVE — message is displayed anytime the traction control system is activated while driving.

NOTE: For 1993, all situations which cause the ANTI-LOCK DISABLED and/or TRACTION DISABLED message will also turn ON the BRAKE lamp. Conversely, any time the BRAKE warning lamp is lit, the anti-lock and traction control systems are disabled.

1 FRONT WHEEL SPEED SENSOR
2 STEERING KNUCKLE (TYPICAL)
3 REAR BRAKE AND DRUM
4 REAR WHEEL SPEED SENSOR
5 EBTCM/BPM VALVE
6 EBTCM/BPM VALVE MOUNTING BRACKET
7 STEERING GEAR
8 LEFT FRONT FRAME RAIL
9 ELECTRONIC BRAKE TRACTION CONTROL MODULE (EBTCM)
10 BRAKE PRESSURE MODULATOR (BPM) VALVE

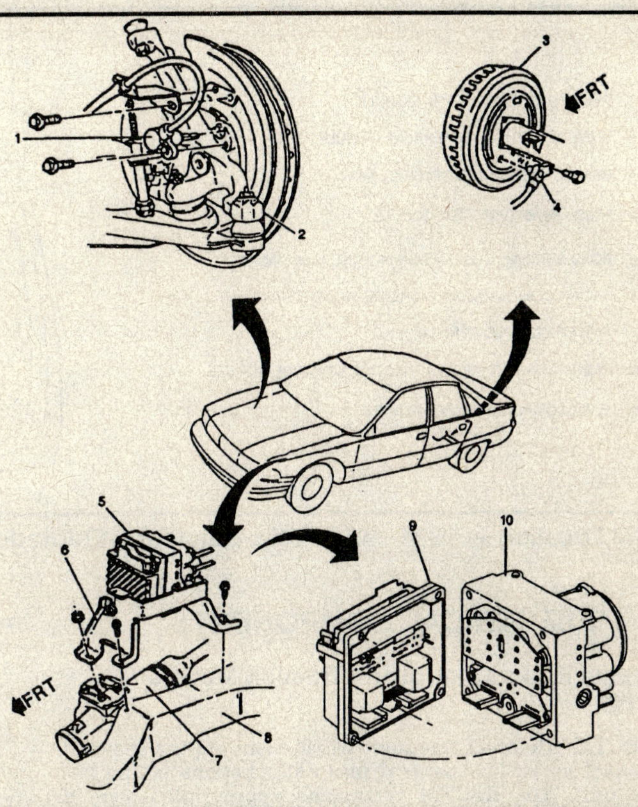

ABS and ASR Component locations — 1994–95 Fleetwood

1 BRAKE PRESSURE MODULATOR (BPM) VALVE
2 FRAME
3 BPM VALVE MOUNTING BRACKET
4 BRAKE LINES
5 LF HUB AND ROTOR
6 BPM VALVE GROUNDSTRAP
7 BPM VALVE HARNESS CONNECTOR
8 VALVE RELAY
9 VALVE RELAY
10 TCS PRIME PIPE BLEED SCREW
11 TCS PRIME PIPE
12 MOUNTING BRACKET NUT

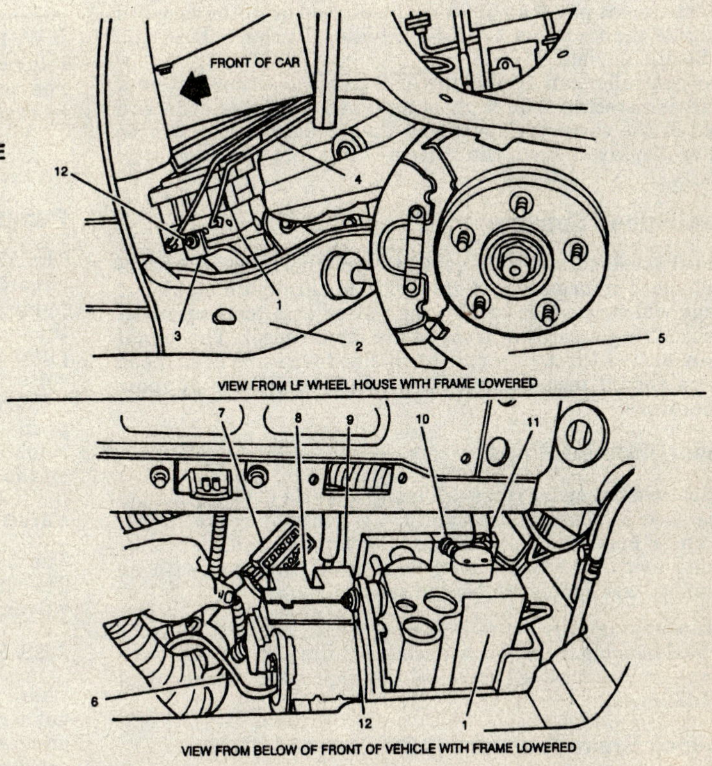

ABS and ASR Component locations — 1994–95 DeVille, Concours, Eldorado and Seville

1 REMOTE KEYLESS ENTRY MODULE
2 UNIVERSAL THEFT DETERRENT MODULE
3 HEATED WINDSHIELD CONTROL MODULE
4 PASS–Key® DECODER MODULE
5 ROAD SENSING SUSPENSION MODULE (4.6L ONLY)
6 TRUNK COMPARTMENT MICRO RELAY CENTER
7 POWER DOOR LOCK RELAY
8 ELECTRONIC BRAKE AND TRACTION CONTROL MODULE
9 RAP/ILLUMINATED ENTRY MODULE

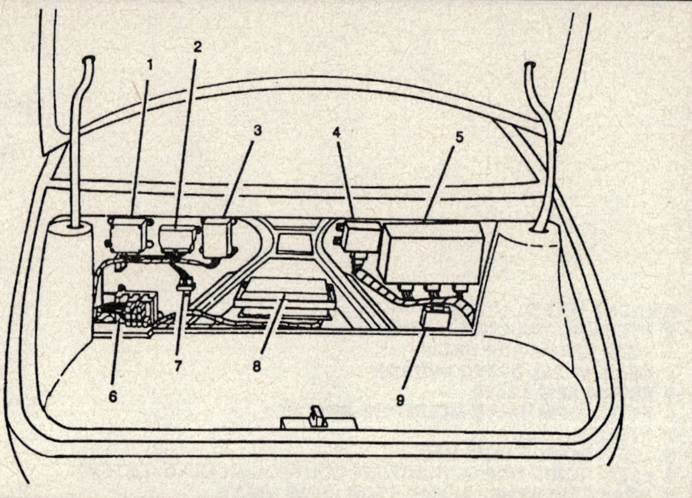

EBTCM Location — 1994–95 DeVille, Concours, Eldorado and Seville

ELECTRICAL COMPONENTS

Electronic Brake Control and Traction Module (EBCTM)

The EBCTM is a small computer which monitors the speed of each wheel as well as several internal functions of the hydraulic unit. The EBCTM commands appropriate solenoid valve positions to prevent wheel lock-up based on signals from each wheel sensor. With the ASR System, the EBCTM also reacts to slippage at the wheels and commands suitable valve response to slow the wheel(s). The EBTCM is located behind the instrument panel on 1993 models or under the left side of the instrument panel on 1994–95 Fleetwood or in the forward portion of the trunk on 1994–95 DeVille, Concours, Eldorado and Seville models.

The controller can disable the ABS and/or traction system if a fault is detected. The fault is assigned a trouble code and stored in the computer's memory. The trouble code(s) can be read or displayed when the system is put into the diagnostic mode.

Wheel Speed Sensors

A wheel speed sensor is present at each wheel and transmits wheel speed information to the EBTCM through a small AC voltage which is generated by the sensor as a toothed wheel or sensor ring rotates in front of the fixed sensor. The signal is transmitted through a pair of shielded wires. Wheel speed sensors should measure 900–1300 ohms resistance at room temperature.

Sensor/Toothed Rings

The toothed or sensor rings are integral parts of other components such as the wheel bearings, CV joints or brake rotors. The rings may be inspected visually for damage, debris or missing teeth. The teeth must be in good condition with no sign of contact against the speed sensor.

If a sensor ring fails, its related component, such as the rotor or wheel bearing, must be replaced. A failure of the rear sensor ring on a rear wheel drive vehicle requires disassembly of the differential.

Anti-lock Brake Pressure (ABP) Valve — 1993

Formerly called the modulator or pressure modulator valve, this unit is mounted on the left side of the engine cradle. This unit contains the solenoids and pilot valves which operate the ABS and ASR. With the exception of the 2 external relays, the ABP valve is non-serviceable; it must never be disassembled. Several components are mounted within the ABP valve.

Brake Pressure Modulator (BPM) Valve — 1994–95

Formerly referred to as the hydraulic modulator or pressure modulator valve, the BPM is located on the left front side of the engine cradle in the engine compartment. It provides brake fluid modulation for each of the individual wheel circuits as required during anti-lock braking and traction control. During the anti-lock mode, the BPM can maintain or reduce brake fluid pressure independent of the pressure generated by the brake master cylinder.

With the exception of the valve relay and the pump motor relay on 1993 and 1994 models, the BPM valve is an integral, non-serviceable component. It cannot be serviced and can only be replaced if a malfunction should occur.

HYDRAULIC COMPONENTS

Pump and Motor

The ABP or BPM contains a recirculation pump with separate circuits for each wheel circuit. During ABS operation, the pump transfers released brake fluid from the brake lines to the master cylinder reservoir. When ASR is engaged, the pump transfers fluid from the reservoir to the front brake circuits.

During ASR operation, the pump draws fluid through a prime pipe connected to a separate reservoir in the master cylinder. This prime pipe may be bled through a bleeder screw on the ABP valve assembly.

Valve and Pump Relays

The 2 relays are located under the cover on the ABP valve. The valve relay provides power to the 4 relays and the 2 pilot valves; the pump relay energizes the pump motor circuit.

ABS Solenoids

These valves are contained within the ABP or BPM and are not serviceable. During normal operation, the solenoid valves, one per wheel, are open, allowing pressurized brake fluid to act directly on the brake system. When ABS is engaged, the EBCTM commands the solenoid(s) into position to hold or reduce the brake fluid pressure in the lines.

Pilot Valves

There are two TCS pilot valves within the BPM. These are normally closed and are open during an ASR event, allowing the pump to draw fluid from the master cylinder reservoir, through the compensating ports in the master cylinder bore. The pilot valves also do not modulate.

Isolation Valves

The 2 traction control pilot valves within the BPM which operate at the command of the EBCTM. When traction control is necessary, these valves isolate the master cylinder so the pump motor can build brake fluid pressure to the front brakes. Unlike the ABS solenoids, the isolation valves do not modulate; they are normally open and then close during traction control operation.

Prime Pipe and Nozzle

The 6mm prime pipe draws fluid from the master cylinder reservoir. This low pressure rubber hose provides fluid to the ABP during traction control operation.

Prime Pipe Bleed Screw

Located on the side of the ABP or BPM, the bleed screw allows removal of air from the prime pipe. The prime pipe must be bled anytime the master cylinder, reservoir or ABP valve is removed or the lines opened. It must also be bled if the reservoir fluid level has become too low and admitted air into the system.

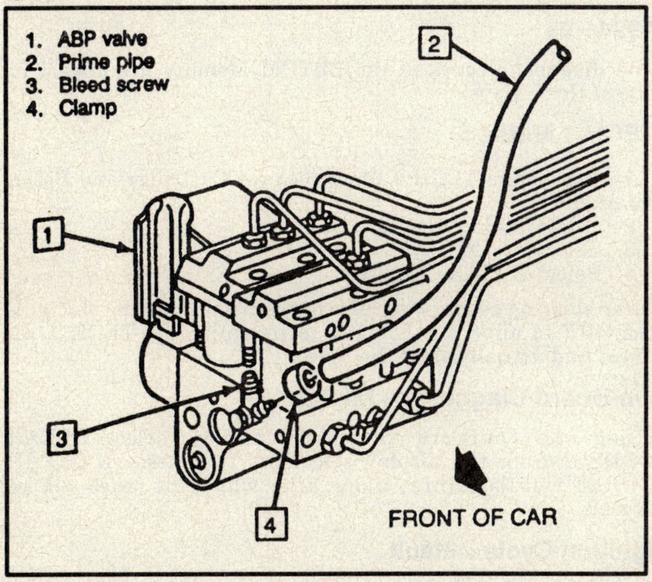

The prime pipe bleed screw is located on the ABP valve

Diagnosis and Testing

SERVICE PRECAUTIONS

• If the vehicle is equipped with air bag (SIR) system, always properly disable the system before commencing work on the ABS system.

• Certain components within the ABS system are not intended to be serviced or repaired individually. Only those components with removal and installation procedures should be serviced.

• Do not use rubber hoses or other parts not specifically specified for the Bosch ABS or ASR system. When using repair kits, replace all parts included in the kit. Partial or incorrect repair may lead to functional problems and require the replacement of the hydraulic unit.

• Lubricate rubber parts with clean, fresh brake fluid to ease assembly. Do not use lubricated shop air to clean parts; damage to rubber components may result.

• Use only brake fluid from an unopened container. Use of suspect or contaminated brake fluid can reduce system performance and/or durability.

• When any hydraulic component or line is removed or replaced, it may be necessary to bleed the entire system.

• A clean repair area is essential. Perform repairs after components have been thoroughly cleaned; use only denatured alcohol to clean components. Do not allow ABS components to come into contact with any substance containing mineral oil; this includes used shop rags.

• The control module is a microprocessor similar to other computer units in the vehicle. Ensure that the ignition switch is **OFF** before removing or installing controller harnesses. Avoid static electricity discharge at or near the controller.

• Never disconnect any electrical connection with the ignition switch **ON** unless instructed to do so in a test.

• Always wear a grounded wrist strap when servicing any control module or component labeled with a Electrostatic Discharge (ESD) symbol.

• Avoid touching module connector pins.

• Leave new components and modules in the shipping package until ready to install them.

• To avoid static discharge, always touch a vehicle ground after sliding across a vehicle seat or walking across carpeted or vinyl floors.

• Never allow welding cables to lie on, near or across any vehicle electrical wiring.

• Do not allow extension cords for power tools or droplights to lie on, near or across any vehicle electrical wiring.

VISUAL INSPECTION

Before any system diagnosis is begun, the brake system should be inspected visually for common faults which could disable the ABS and/or Traction Control or cause a code to set.

Check the vehicle carefully for any sign of: binding parking brake or faulty parking brake switch, low brake fluid, system fluid leaks including the pump/motor area, hydraulic unit or brake fluid reservoir. Additionally, check for failed fuses or fusible links, failed relays, loose or damaged wiring including connectors, harnesses, and insulation wear.

FUNCTIONAL CHECK

Once the visual check has been performed, perform the functional check to determine if the problem is truly ABS related or arising from other common faults.

Performance of the functional check will prompt the reading of stored codes at the appropriate time.

INTERMITTENTS

Most intermittent problems are caused by faulty electrical connections or wiring. Always check for poor mating of connector halves or terminals not fully seated in connector bodies, deformed or damaged terminals and poor terminal to wire connections.

Most failures within the ABS or ASR will disable the anti-lock function for the entire ignition cycle, even if the fault clears before the next key-off occurrence. There are situations which will allow the ABS to re-engage if the condition corrects during the ignition cycle. These conditions may cause the BRAKE light and/or the Driver Information Center to activate.

- Low system voltage: If the EBCTM detects low voltage, the ANTI-LOCK DISABLED and TRACTION DISABLED messages appear. If correct minimum voltage is restored to the EBCTM, normal ABS function resumes.
- Low brake fluid level: Once detected by the fluid level sensor, this condition illuminates the BRAKE warning lamp and triggers the ANTI-LOCK DISABLED and TRACTION DISABLED messages. When the sensor indicates acceptable fluid level, the normal function resumes.

READING CODES

NOTE: The use of a hand scanner, such as TECH-1 or its equivalent, is strongly recommended when diagnosing this system.

The EBCTM can display fault codes through the use of its On-Board diagnostic program or through the use of a hand scanner such as General Motors TECH-1. The system can store as many as 3 codes at one time. The order of the codes does not indicate the order of occurrence. If either the ANTI-LOCK DISABLED or TRACTION DISABLED warning lamps are lit before the codes are retrieved, at least one of the stored codes is current. All codes for either ABS or ASR are prefixed by the letter **T**.

The On-Board diagnostic system can be used to display or erase stored codes; it has no other diagnostic capabilities or test functions. The TECH-1 or equivalent tool, when equipped with the correct system cartridge, cannot only read and clear codes but will also operate in several helpful diagnostic modes: data, history, codes (DTC), snapshot and system test. The testing and code charts in this section refer frequently to the use of a hand scanner.

If codes must be read without the use of a hand scanner:

1. Switch the ignition **ON**. If the vehicle is to remain in diagnostic mode for any length of time, connect an auxiliary battery charger to prevent battery discharge.
2. Press the **OFF** and **WARM** buttons on the Climate Control Panel simultaneously and hold them until the segment check function begins on the instrument and combo panel. This places the system in diagnostic mode.
3. After the segment check, any codes stored in the memory will be displayed. All stored codes will be displayed, including those for lighting, PCM, BCM, SIR and other systems as well as ABS/ASR.
4. Each code consists of a system identifier (T for ABS/ASR), the 3-digit code number and a **C** or **H** suffix. The **C** indicates the code is current; it was present the last time the EBCTM checked the system. An **H** suffix indicates a history code; the fault was not present during the last self-check.
5. If no codes are stored for a particular system, the display will state that in a message such as NO PCM CODE or NO T CODE.
6. Once all relevant codes are recorded, press the **RESET** button on the Driver Information Center to return the system to normal operation. Although the onboard diagnostic system can perform tests and snapshots for other systems, no testing functions are available for the ABS/ASR system.

CLEARING CODES

1993

On 1993 vehicles, disconnecting the battery or EBCTM will not clear stored codes. The preferred method is to clear codes through the use of the hand scanner, selecting the CLEAR CODES program from the menu. Codes may also be cleared through the On-Board diagnostic system. Any code which is set but does not recur through 50 ignition cycles will be automatically cleared by the EBCTM.

To clear codes through the on-board system:

1. After stored codes have been displayed, the first system available for testing will be displayed (PCM?). Press the **LO** button to page through the systems until TCS? appears.
2. Press the **HI** button to select this system. Since no tests are available, the message TCS CLEAR CODES? will appear as the only choice.
3. Press the **HI** button to select the clear codes screen. After approximately 3 seconds, the message TCS CODES CLEAR should appear. Note that this selection clears only the codes related to this system; other stored codes, such as SIR or BCM remain stored and available.
4. After clearing the codes, the system returns to the next available test.
5. Press the **RESET** button on the Driver Information Center to exit diagnostics.

1994–95

The diagnostic codes in the EBTCM Memory are erased in one of three ways:

TECH 1 Method

1. Install the TECH 1 Mass Storage Cartridge and Bosch Adapter
2. Select TCS system features
3. Select F2: DTC
4. Select CLEAR CODES

After clearing codes, wait 15 seconds before turning the ignition**OFF** to allow the EBTCM to recognize the TECH 1 request and actually clear the codes.

On-Board Diagnostics Method

Using the On-Board Diagnostic display, select CLEAR CODES? from the display selection. The message CLEAR CODES will flash three times, after which the codes will be erased.

Ignition Cycle Default

If the ignition cycles 50 times without a particular fault reappearing, the Diagnostic Trouble Code will be erased from the EBTCM memory and the ignition cycle counter in the EBTCM will be reset to zero.

ABS/TCS FUNCTIONAL CHECK CHART — 1993 ALLANTE

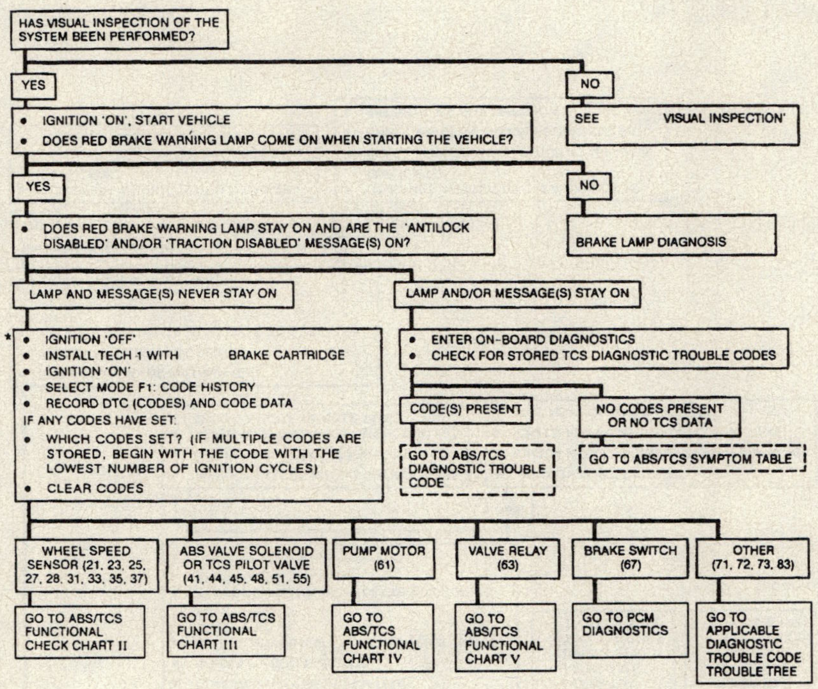

* FROM THIS POINT ON, IT IS ASSUMED THAT THE CONDITION(S) RESPONSIBLE FOR SETTING THE DIAGNOSTIC TROUBLE CODE(S) IS INTERMITTENT AND THE CODE IS NEVER SET CURRENT IN THE SHOP. MOST OF THE FOLLOWING TESTS ARE PERFORMED IN ORDER TO EXERCISE THE ABS/TCS CIRCUITS AND GET A CODE TO SET WITHOUT HAVING TO LEAVE THE BAY IF AT ANYTIME A DIAGNOSTIC TROUBLE CODE IS SET 'HARD', REFER IMMEDIATELY TO THE 'ABS/TCS DIAGNOSTIC TROUBLE CODE TABLE'

ABS/TCS FUNCTIONAL CHECK CHART — 1993 ALLANTE, CONTINUED

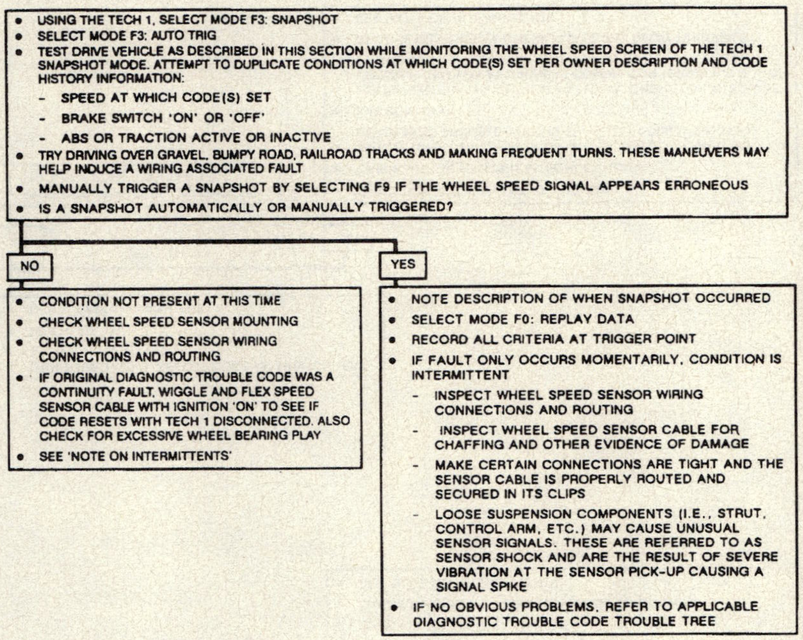

ABS/TCS FUNCTIONAL CHECK CHART — 1993 ALLANTE, CONTINUED

- USING THE TECH 1, SELECT MODE F4: ABS/TCS TEST
- SELECT THE APPROPRIATE TEST:
  - AUTO TEST (CODES 41, 45, 51, 55)
  - TCS TEST (CODES 44, 48)
- PERFORM TEST
- DOES SYSTEM PASS TEST?

TCS PASSES TEST

- CONDITION MAY BE INTERMITTENT
- SELECT F3: PILOT VALVE TEST
- SELECT APPROPRIATE PILOT VALVE TO BE TESTED:
  F0: LF
  F1: RF
- DOES SYSTEM PASS TEST?

NO

- CONDITION INTERMITTENT BUT CAN BE DUPLICATED
- REFER TO APPROPRIATE DIAGNOSTIC TROUBLE CODE TROUBLE TREE

YES

- CONDITION INTERMITTENT BUT CANNOT BE DUPLICATED
- CHECK HARNESS CONNECTION AT ABP VALVE
- CHECK ABP VALVE GROUND CONNECTION
- SEE 'NOTE ON INTERMITTENTS'

AUTO TEST PASSES

- CONDITION MAY BE INTERMITTENT
- SELECT F0: SOLENOID TEST
- SELECT APPROPRIATE SOLENOID TO BE TESTED
  F0: LF F4: LR
  F1: RF F5: RR
- PERFORM PRESSURE RELEASE TEST OF THAT VALVE SOLENOID
- DOES SYSTEM PASS TEST?

NO

- INSPECT BASE BRAKE SYSTEM FOR LEAKS
- INSPECT FOR LEAKS AT ABP VALVE
- REFER TO APPROPRIATE DIAGNOSTIC TROUBLE CODE TROUBLE TREE

YES

- CONDITION INTERMITTENT
- CHECK HARNESS CONNECTION AT ABP VALVE
- CHECK ABP VALVE GROUND CONNECTION
- SEE 'NOTE ON INTERMITTENTS'

TCS TEST FAILS

- REFER TO APPLICABLE DIAGNOSTIC TROUBLE CODE TROUBLE TREE

AUTO TEST FAILS

- REFER TO APPLICABLE DIAGNOSTIC TROUBLE CODE TROUBLE TREE

ABS/TCS FUNCTIONAL CHECK CHART — 1993 ALLANTE, CONTINUED

- USING THE TECH 1, SELECT MODE F4: ABS/TCS TEST
- SELECT F1: AUTO TEST
- PERFORM TEST
- DOES SYSTEM PASS TEST?

NO

- REFER TO TROUBLE TREE FOR DIAGNOSTIC TROUBLE CODE 61

YES

- CONDITION NOT PRESENT AT THIS TIME
- CHECK ABP VALVE HARNESS CONNECTION
- CHECK FOR CLEAN AND TIGHT ABP VALVE GROUND CONNECTIONS AT GROUND STUD AND ON FRAME
- CHECK PUMP RELAY CONNECTION
- CHECK PUMP RELAY FOR POOR/LOOSE TERMINALS
- SEE 'NOTE ON INTERMITTENTS'

- USING THE TECH 1, SELECT MODE F4: ABS/TCS TEST
- SELECT F1: AUTO TEST
- PERFORM TEST
- DOES SYSTEM PASS TEST?

NO

- REFER TO APPLICABLE DIAGNOSTIC TROUBLE CODE TROUBLE TREE

YES

- CONDITION NOT PRESENT AT THIS TIME
- CHECK ABP VALVE HARNESS CONNECTION
- WIGGLE HARNESS CONNECTOR WHILE CONNECTED AND SEE IF DIAGNOSTIC TROUBLE CODE IS SET (TECH 1 DISCONNECTED)
- IGNITION 'OFF'
- DISCONNECT HARNESS FROM ABP VALVE. CHECK FOR ADEQUATE TERMINAL TENSION AT HARNESS CONNECTOR TERMINALS '2, 4, 6, 10, AND 12'
- REMOVE VALVE RELAY. CHECK FOR LOOSE RELAY TERMINALS
- SEE 'NOTE ON INTERMITTENTS'

DIAGNOSTIC CHART DTC 21 — 1993 ALLANTE

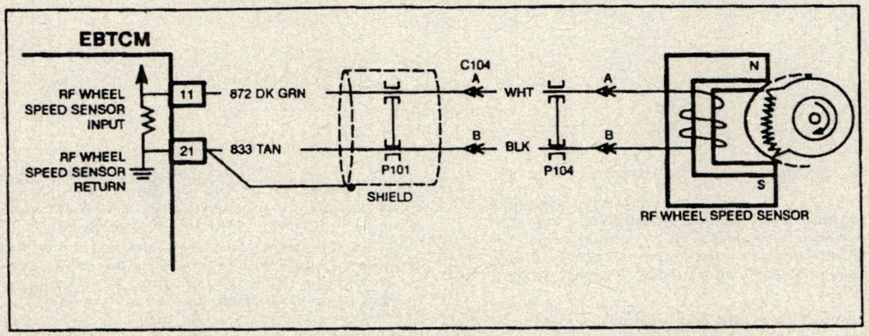

CODE 21

RIGHT FRONT WHEEL SPEED SENSOR FAULT

CIRCUIT DESCRIPTION:

As each wheel turns, the wheel speed sensor for that wheel creates a small AC voltage as a toothed sensor ring, mounted in the front integral hub/bearing assembly, is passed by the stationary sensor. The sensor consists of a permanent magnet and coil and generates voltage by magnetic induction. The frequency of the wheel speed sensor signal is proportional to wheel speed.

The EBTCM uses the wheel speed sensor signal to calculate vehicle reference speeds and individual speed, acceleration and slip values for each wheel. These values are used to determine when antilock control is required.

The EBTCM will only set one wheel speed sensor fault code at a time, even if multiple faults exist. Detection is prioritized accordingly: LF, RR, LR, RF. Always check for additional stored codes after making repairs.

TEST CONDITION(S):

1. Tested continuously.

FAILURE CONDITION(S):

Sensor Output

1. Drive-away: Speed signal is not detected when vehicle reaches 12 kph or 7.5 mph.
2. While Driving: Speed signal is lost while vehicle speed is greater than 40 kph or 25 mph.

ACTION TAKEN:

1. Code set.
2. ABS/TCS switched 'OFF'.
3. Red BRAKE warning lamp 'ON'.
4. ABS and TCS 'DISABLED' messages displayed.

Code 31 is set if there is a short to voltage or ground in CKT 882 or 883, or faulty speed sensor. The testing for this fault occurs when the vehicle is in motion, it will not set with the ignition 'ON' and the vehicle at rest.

NOTICE: A Code 21 may set if CKT 872 or 833 lose continuity and the WSS signal is lost while driving. A Code 23 should accompany this code on the subsequent ignition cycle. If both Code 21 and Code 23 are set concurrently, diagnose Code 23 first.

TEST DESCRIPTION: Test numbers refer to circled numbers on the diagnostic chart.

1. Checks if the code resets during a road test.
2. Use the TECH 1 to check for sources of induced electrical noise. Some sources of induced electrical noise which may cause intermittent wheel speed sensor codes to set are: improper routing or sensor cables next to spark plug wires; cables not retained in brackets; loose fitting or improperly mounted sensors; damaged sensors or toothed rings; poor terminal engagement of connectors; or intermittent shorts or opens in wiring; or a loss of sensor circuit shield.
3. Checks for damaged wheel speed sensor circuit shield. The shield is a foil insulator that is wrapped around the wheel speed sensor circuits. Inside the EBTCM harness connecter there is an uninsulated drain wire that is attached to the shield and jumped to the wheel speed sensor return line and taken to ground through the EBTCM. A shield which has lost its ground through the EBTCM leaves the sensor signal circuit susceptible to noise. It may be necessary to disassemble the EBTCM connector to check if the drain wire has a poor connection with the wheel speed sensor return connector terminal. The shield can be repaired by reterminating the sensor return line circuit, including the drain wire.
4. Checks for intermittent opens/shorts in wheel speed sensor circuits and cables.

DIAGNOSTIC CHART DTC 21 — 1993 ALLANTE, CONTINUED

NOTE: DIAGNOSE CODE 23 FIRST IF SET CONCURRENTLY.

ENSURE SENSOR IS NOT VISIBLY DAMAGED; IF IT IS DAMAGED, REPLACE THE HUB/BEARING/SENSOR ASSEMBLY. CLEAR CODES, AND ROAD TEST TO SEE IF CODE RESETS.

(1)
- CLEAR CODES
- ROAD TEST VEHICLE, REACHING AT LEAST 25 KPH (15 MPH)
- DOES CODE 21 RESET?

YES

(2)
- INSTALL TECH 1
- SELECT MODE F0: DATA LIST
- OBSERVE WHEEL SPEED SENSOR OUTPUT WHILE IDLING IN PARK
- INCREASE ENGINE SPEED BY PRESSING ON ACCELERATOR

NO
- NO TROUBLE FOUND.

SENSOR OUTPUT DOES NOT CHANGE WITH ENGINE SPEED (STAYS AT ZERO)

(3)
- TEST DRIVE VEHICLE
- OBSERVE WHEEL SPEED SENSOR OUTPUT
- DOES SIGNAL 'DROP OUT' OR DIFFER FROM OTHER SENSORS?

ERRATIC SENSOR READING (DOES NOT STAY AT ZERO)
- VERIFY CORRECT CABLE ROUTING
- CHECK FOR PROPER INSTALLATION OF SENSOR
- REPAIR AS REQUIRED
- INSPECT EBTCM HARNESS CONNECTOR FOR AN OPEN FROM WHEEL SPEED SENSOR WIRING SHIELD TO WHEEL SPEED SENSOR RETURN LINE

SENSOR READING 'DROP OUT'

(4)
- RAISE VEHICLE ON HOIST
- SPIN WHEEL WHILE MANIPULATING CABLES AND CONNECTORS BY HAND
- OBSERVE WHEEL SPEED SENSOR OUTPUT
- DOES SIGNAL 'DROP OUT'?

CONTINUED ON PAGE 4 OF 4

NORMAL SENSOR OUTPUT

SENSOR READING DIFFERS BY MORE THAN 4 MPH
- VERIFY CORRECT CABLE ROUTING
- CHECK FOR PROPER INSTALLATION OF SENSOR
- CHECK FOR OPEN IN WHEEL SPEED SENSOR WIRING SHIELD
- REPAIR AS REQUIRED
- REPLACE SENSOR IF PROBLEM PERSISTS

DIAGNOSTIC CHART DTC 21 — 1993 ALLANTE, CONTINUED

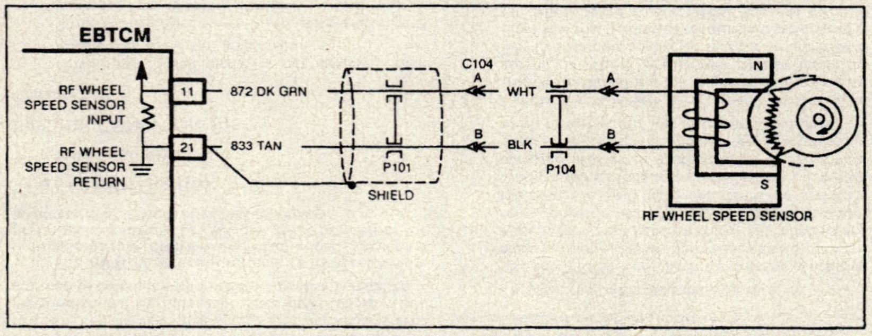

CODE 21

RIGHT FRONT WHEEL SPEED SENSOR FAULT

CIRCUIT DESCRIPTION:

As each wheel turns, the wheel speed sensor for that wheel creates a small AC voltage as a toothed sensor ring, mounted in the front integral hub/bearing assembly, is passed by the stationary sensor. The sensor consists of a permanent magnet and coil and generates voltage by magnetic induction. The frequency of the wheel speed sensor signal is proportional to wheel speed.

The EBTCM uses the wheel speed sensor signal to calculate vehicle reference speeds and individual speed, acceleration and slip values for each wheel. These values are used to determine when antilock control is required.

The EBTCM will only set one wheel speed sensor fault code at a time, even if multiple faults exist. Detection is prioritized accordingly: LF, RR, LR, RF. Always check for additional stored codes after making repairs.

TEST DESCRIPTION: Test numbers refer to circled numbers on the diagnostic chart.

5. Checks for proper resistance in the sensor itself.
6. Checks for a short to battery in either of the speed sensor circuit wires.
7. Checks for a short between the speed sensor circuit wires.
8. Checks for a short to ground in the speed sensor input circuit wire. The ground may be either a 'hard' short to ground, or a resistive short. A short with a resistance less than 2 megaohms, though not a hard short, can still cause Code 21 to set.
9. Checks wiring and connectors for intermittents.

DIAGNOSTIC CHART DTC 21 — 1993 ALLANTE, CONTINUED

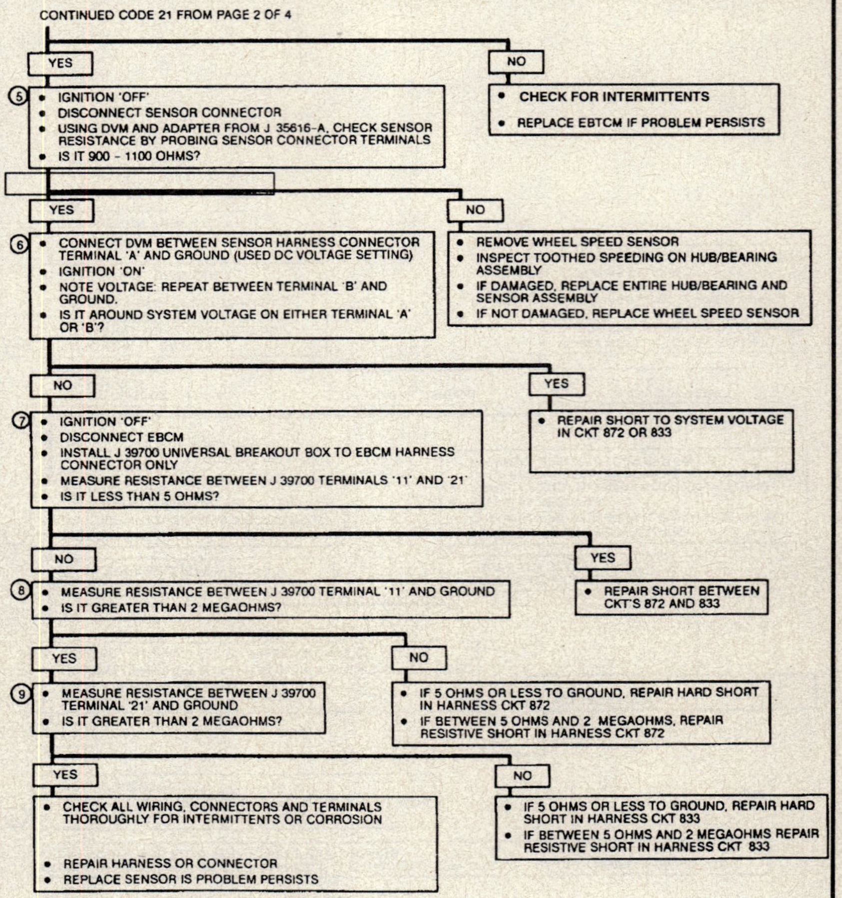

DIAGNOSTIC CHART DTC 23 — 1993 ALLANTE

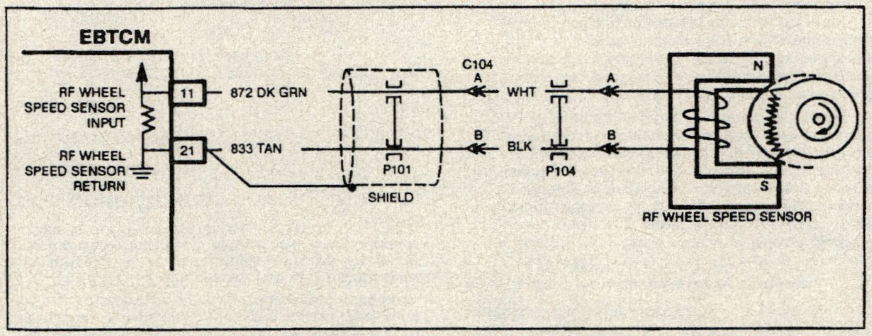

CODE 23

RIGHT FRONT WHEEL SPEED SENSOR CONTINUITY FAULT

CIRCUIT DESCRIPTION:

As each wheel turns, the wheel speed sensor for that wheel creates a small AC voltage as a toothed sensor ring, mounted in the front integral hub/bearing assembly, is passed by the stationary sensor. The sensor consists of a permanent magnet and coil and generates voltage by magnetic induction. The frequency of the wheel speed sensor signal is proportional to wheel speed.

The EBTCM uses the wheel speed sensor signal to calculate vehicle reference speeds and individual speed, acceleration and slip values for each wheel. These values are used to determine when antilock control is required.

The EBTCM will only set one wheel speed sensor fault code at a time, even if multiple faults exist. Detection is prioritized accordingly: LF, RR, LR, RF. Always check for additional stored codes after making repairs.

TEST CONDITION(S):

1. Tested continuously.

FAILURE CONDITION(S):

Continuity

1. While standing: Open circuit is seen while the vehicle is standing still with the ignition 'ON'

ACTION TAKEN:

1. Code set.
2. ABS/TCS switched 'OFF'.
3. Red BRAKE warning lamp 'ON'.
4. ABS and TCS 'DISABLED' messages displayed.

Code 23 is set if there is a short to voltage or ground in CKT 872 or 883, or faulty speed sensor. The testing for this fault occurs with the ignition 'ON' and the vehicle at rest.

NOTICE: A Code 21 may set if CKT 872 or 833 lose continuity and the WSS signal is lost while driving. A Code 23 should accompany this code on the subsequent ignition cycle. If both Code 21 and Code 23 are set concurrently, diagnose Code 23 first.

TEST DESCRIPTION: Test numbers refer to circled numbers on the diagnostic chart.

1. Checks for proper resistance in the sensor itself.
2. Checks for a short to battery in either of the speed sensor circuit wires. There should be 5V on CKT 872. If there is, CKT 872 has no hard opens or shorts. The condition may be intermittent.
3. Checks for an open in both the speed sensor circuit wires.
4. Manipulates wiring and connectors, trying to induce an intermittent fault not currently present.
5. This test checks if the code resets on key-up. If so, since Tests 1-4 have validated that the circuitry and components are good, the Wheel Bearing/Speed Sensor assembly may be causing the fault.
6. Checks if the Code 21 sets during a road test.
7. Code 23 sets when the vehicle is at rest: Code 21 sets with a problem in the speed sensor circuitry with the vehicle in motion. If problems are still present, Code 21 would set during the road test, not a 23. If Code 21 sets at this point, Code 21 diagnostics should be used.
8. The wheel speed sensor may be causing an intermittent fault. If Code 23 resets after sensor replacement, the EBTCM must be concluding there is a problem present in the speed sensor circuit when there is not.

DIAGNOSTIC AIDS:

Be sure the speed sensor wiring is properly routed and retained. This will help prevent false signals due to electrical noise being picked up by the wiring.

It is very important that a thorough inspection of the wiring and connectors be performed. Failure to carefully and fully inspect wiring and connectors may result in misdiagnosis, causing part replacement with reappearance of the fault.

DIAGNOSTIC CHART DTC 23 — 1993 ALLANTE, CONTINUED

ENSURE SENSOR IS NOT VISIBLY DAMAGED; IF IT IS DAMAGED, REPLACE THE WHEEL BEARING/SPEED/SENSOR ASSEMBLY. CLEAR CODES, AN TURN IGNITION 'ON' TO SEE IF CODE RESETS.

(1)
- IGNITION 'OFF'
- DISCONNECT SENSOR CONNECTOR
- USING DVM AND ADAPTER FROM J 35616-A, CHECK SENSOR RESISTANCE BY PROBING SENSOR CONNECTOR TERMINALS.
- IS IT 900 - 1100 OHMS?

YES → (2) | NO → • REPLACE WHEEL SPEED SENSOR

(2)
- CONNECT DVM BETWEEN SENSOR HARNESS CONNECTOR TERMINAL 'A' AND GROUND (USED DC VOLTAGE SETTING)
- IGNITION 'ON'
- NOTE VOLTAGE: REPEAT BETWEEN TERMINAL 'B' AND GROUND.
- IS IT AROUND SYSTEM VOLTAGE ON EITHER TERMINAL 'A' OR 'B'?

NO → (3) | YES → • REPAIR SHORT TO SYSTEM VOLTAGE IN CKT 872 OR 883

(3)
- IGNITION 'OFF'
- DISCONNECT EBTCM
- INSTALL J 39700 UNIVERSAL BREAKOUT BOX TO EBTCM HARNESS CONNECTOR ONLY
- MEASURE RESISTANCE BETWEEN PIN 11 ON J 39700 TERMINAL 'A' OF SENSOR CONNECTOR ON HARNESS. PERFORM SAME MEASUREMENT BETWEEN PIN 21 AND TERMINAL 'B'.
- IS EITHER MEASUREMENT GREATER THAN 5 OHMS?

NO → (4) | YES → • REPAIR OPEN IN CKT 872 AND/OR 833

(4)
- REMOVE J 39700; RECONNECT EBCM
- CLEAR CODES; IGNITION 'OFF'
- DISCONNECT TECH 1, IF CONNECTED
- IGNITION 'ON'
- WIGGLE ALL WIRING AND CONNECTORS TO TRY TO DUPLICATE AN INTERMITTENT FAULT
- DOES CODE 23 RESET?

NO → (5) | YES → • REPAIR WIRING AND/OR CONNECTIONS

(5)
- CLEAR CODES; IGNITION 'OFF'
- DISCONNECT TECH 1, IF CONNECTED
- IGNITION 'ON'
- DOES CODE 23 RESET?

YES → (8) | NO → (6)

(8)
- REPLACE WHEEL SPEED SENSOR
- CLEAR CODES; IGNITION 'OFF'
- DISCONNECT TECH 1, IF CONNECTED
- IGNITION 'ON'
- DOES CODE 23 RESET?

YES → • REPLACE EBTCM | NO → • SYSTEM OK

(6)
- ROAD TEST VEHICLE, REACHING AT LEAST 25 KPH (15 MPH).
- DOES CODE 21 SET?

YES → (7) GO TO CODE 21 DIAGNOSTICS | NO → • NO TROUBLE FOUND. SEE 'DIAGNOSTIC AIDS' ON FACING PAGE.

WHEN ALL DIAGNOSIS AND REPAIRS ARE COMPLETED, CLEAR CODES AND VERIFY OPERATION

DIAGNOSTIC CHART DTC 25 — 1993 ALLANTE

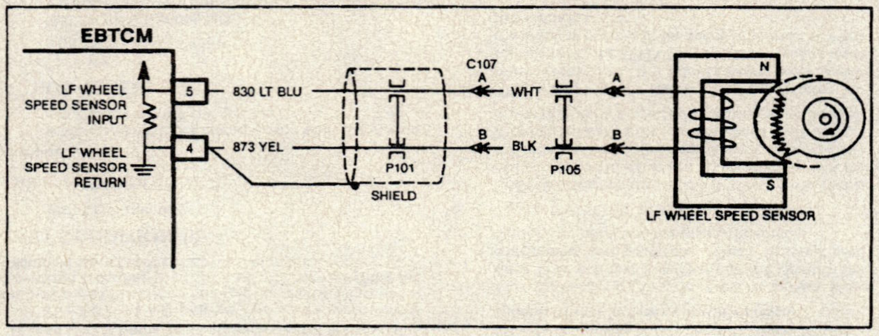

CODE 25

LEFT FRONT WHEEL SPEED SENSOR FAULT

CIRCUIT DESCRIPTION:

As each wheel turns, the wheel speed sensor for that wheel creates a small AC voltage as a toothed sensor ring, mounted in the front integral hub/bearing assembly, is passed by the stationary sensor. The sensor consists of a permanent magnet and coil and generates voltage by magnetic induction. The frequency of the wheel speed sensor signal is proportional to wheel speed.

The EBTCM uses the wheel speed sensor signal to calculate vehicle reference speeds and individual speed, acceleration and slip values for each wheel. These values are used to determine when antilock control is required.

The EBTCM will only set one wheel speed sensor fault code at a time, even if multiple faults exist. Detection is prioritized accordingly: LF, RR, LR, RF. Always check for additional stored codes after making repairs.

TEST CONDITION(S):

1. Tested continuously.

FAILURE CONDITION(S):

Sensor Output

1. Drive-away: Speed signal is not detected when vehicle reaches 12 kph or 7.5 mph.

OR

2. While Driving: Speed signal is lost while vehicle speed is greater than 40 kph or 25 mph

ACTION TAKEN:

1. Code set.
2. ABS/TCS switched 'OFF'.
3. Red BRAKE warning lamp 'ON'.
4. ABS and TCS 'DISABLED' messages displayed.

Code 25 is set if there is a short to voltage or ground in CKT 830 or 873, or faulty speed sensor. The testing for this fault occurs when vehicle is in motion; it will not set with the ignition 'ON' and the vehicle at rest.

NOTICE: A Code 25 may set if CKT 830 or 873 lose continuity and the WSS signal is lost while driving. A Code 27 should accompany this code on the subsequent ignition cycle. If both Code 25 and Code 27 are set concurrently, diagnose Code 27 first.

TEST DESCRIPTION: Test numbers refer to circled numbers on the diagnostic chart.

1. Checks if the code resets during a road test.
2. Use the TECH 1 to check for sources of induced electrical noise. Some sources of induced electrical noise which may cause intermittent wheel speed sensor codes to set are: improper routing or sensor cables next to spark plug wires; cables not retained in brackets; loose fitting or improperly mounted sensors; damaged sensors or toothed rings; poor terminal engagement of connectors; or intermittent shorts or opens in wiring; or a loss of sensor circuit shield.
3. Checks for damaged wheel speed sensor circuit shield. The shield is a foil insulator that is wrapped around the wheel speed sensor circuits. Inside the EBTCM harness connecter there is an uninsulated drain wire that is attached to the shield and jumped to the wheel speed sensor return line and taken to ground through the EBTCM. A shield which has lost its ground through the EBTCM leaves the sensor signal circuit susceptible to noise. It may be necessary to disassemble the EBTCM connector to check if the drain wire has a poor connection with the wheel speed sensor return connector terminal. The shield can be repaired by reterminating the sensor return line circuit, including the drain wire.
4. Checks for intermittent opens/shorts in wheel speed sensor circuits and cables.

DIAGNOSTIC CHART DTC 25 — 1993 ALLANTE, CONTINUED

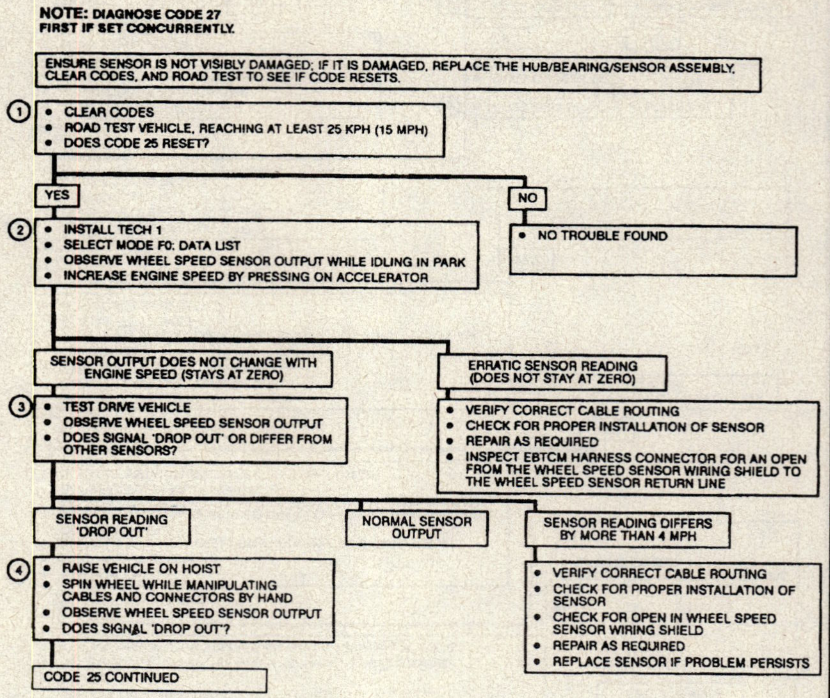

DIAGNOSTIC CHART DTC 25 — 1993 ALLANTE, CONTINUED

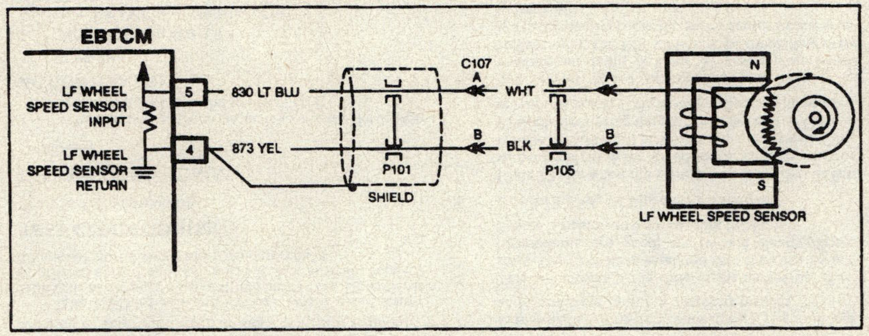

CODE 25

LEFT FRONT WHEEL SPEED SENSOR FAULT

CIRCUIT DESCRIPTION:

As each wheel turns, the wheel speed sensor for that wheel creates a small AC voltage as a toothed sensor ring, mounted in the front integral hub/bearing assembly, is passed by the stationary sensor. The sensor consists of a permanent magnet and coil and generates voltage by magnetic induction. The frequency of the wheel speed sensor signal is proportional to wheel speed.

The EBTCM uses the wheel speed sensor signal to calculate vehicle reference speeds and individual speed, acceleration and slip values for each wheel. These values are used to determine when antilock control is required.

The EBTCM will only set one wheel speed sensor fault code at a time, even if multiple faults exist. Detection is prioritized accordingly: LF, RR, LR, RF. Always check for additional stored codes after making repairs.

TEST DESCRIPTION: Test numbers refer to circled numbers on the diagnostic chart.

5. Checks for proper resistance in the sensor itself.
6. Checks for a short to battery in either of the speed sensor circuit wires.
7. Checks for a short between the speed sensor circuit wires.
8. Checks for a short to ground in the speed sensor input circuit wire. The ground may be either a 'hard' short to ground, or a resistive short. A short with a resistance less than 2 megaohms, though not a hard short, can still cause Code 21 to set.
9. Checks wiring and connectors for intermittents.

DIAGNOSTIC CHART DTC 25 — 1993 ALLANTE, CONTINUED

CONTINUED FROM CODE 25

YES

(5)
- IGNITION 'OFF'
- DISCONNECT SENSOR CONNECTOR
- USING DVM AND ADAPTER FROM J 36616-A, CHECK SENSOR RESISTANCE BY PROBING SENSOR CONNECTOR TERMINALS

NO
- CHECK FOR INTERMITTENTS
- REPLACE EBTCM IF PROBLEM PERSISTS

YES

(6)
- CONNECT DVM BETWEEN SENSOR HARNESS CONNECTOR TERMINAL 'A' AND GROUND (USE DC VOLTAGE SETTING)
- IGNITION 'ON'
- NOTE VOLTAGE: REPEAT BETWEEN TERMINAL 'B' AND GROUND
- IS IT AROUND SYSTEM VOLTAGE ON EITHER TERMINAL 'A' OR 'B'?

NO
- REMOVE WHEEL SPEED SENSOR
- INSPECT TOOTHED SPEEDING ON HUB/BEARING ASSEMBLY
- IF DAMAGED, REPLACE ENTIRE HUB/BEARING AND SENSOR ASSEMBLY
- IF NOT DAMAGED, REPLACE WHEEL SPEED SENSOR

NO

(7)
- IGNITION 'OFF'
- DISCONNECT EBTCM
- INSTALL J 39700 UNIVERSAL BREAKOUT BOX TO EBTCM HARNESS CONNECTOR ONLY
- MEASURE RESISTANCE BETWEEN J 39700 TERMINALS '4' AND '5'
- IS IT LESS THAN 5 OHMS?

YES
- REPAIR SHORT TO SYSTEM VOLTAGE IN CKTS 830 AND 873

NO

(8)
- MEASURE RESISTANCE BETWEEN J 39700 TERMINAL '4' AND GROUND
- IS IT GREATER THAN 2 MEGAOHMS?

YES
- REPAIR SHORT BETWEEN CKTS 830 AND 873

YES

(9)
- MEASURE RESISTANCE BETWEEN J 39700 TERMINAL '5' AND GROUND
- IS IT GREATER THAN 2 MEGAOHMS?

NO
- IF 5 OHMS OR LESS TO GROUND, REPAIR HARD SHORT IN HARNESS CKT 873
- IF BETWEEN 5 OHMS AND 2 MEGAOHMS, REPAIR RESISTIVE SHORT IN HARNESS CKT 873

YES
- CHECK ALL WIRING, CONNECTORS AND TERMINALS THOROUGHLY FOR INTERMITTENTS OR CORROSION
- CHECK FOR INTERMITTENTS AND POOR CONNECTIONS
- REPAIR HARNESS OR CONNECTOR
- REPLACE SENSOR IF PROBLEM PERSISTS

NO
- IF 5 OHMS OR LESS TO GROUND, REPAIR HARD SHORT IN HARNESS CKT 830
- IF BETWEEN 5 OHMS AND 2 MEGAOHMS, REPAIR RESISTIVE SHORT IN HARNESS CKT 830

DIAGNOSTIC CHART DTC 27 — 1993 ALLANTE

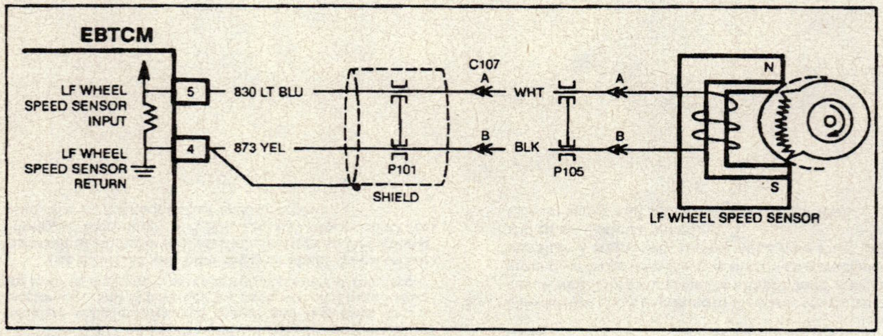

CODE 27

LEFT FRONT WHEEL SPEED SENSOR CONTINUITY FAULT

CIRCUIT DESCRIPTION:

As each wheel turns, the wheel speed sensor for that wheel creates a small AC voltage as a toothed sensor ring, mounted in the front integral hub/bearing assembly, is passed by the stationary sensor. The sensor consists of a permanent magnet and coil and generates voltage by magnetic induction. The frequency of the wheel speed sensor signal is proportional to wheel speed.

The EBTCM uses the wheel speed sensor signal to calculate vehicle reference speeds and individual speed, acceleration and slip values for each wheel. These values are used to determine when antilock control is required.

The EBTCM will only set one wheel speed sensor fault code at a time, even if multiple faults exist. Detection is prioritized accordingly: LF, RR, LR, RF. Always check for additional stored codes after making repairs.

TEST CONDITION(S):

1. Tested continuously.

FAILURE CONDITION(S):

Continuity

1. While Standing: Open circuit is seen while the vehicle is standing still with the ignition 'ON'.

ACTION TAKEN:

1. Code set.
2. ABS/TCS switched 'OFF'.
3. Red BRAKE warning lamp 'ON'.
4. ABS and TCS 'DISABLED' messages displayed.

Code 27 is set if there is a short to voltage or an open in CKT 830 or 873, or faulty speed sensor. The testing for this fault occurs with the ignition 'ON' and the vehicle at rest.

NOTICE: A Code 25 may set CKT 830 or 873 lose continuity and the WSS signal is lost while driving. A Code 27 should accompany this code on the subsequent ignition cycle. If both Code 25 and Code 27 are set concurrently, diagnose Code 27 first.

TEST DESCRIPTION: Test numbers refer to circled numbers on the diagnostic chart.

1. Checks for proper resistance in the sensor itself.
2. Checks for a short to battery in either of the speed sensor circuit wires. There should be 5V on CKT 830. If there is, CKT 830 has no hard opens or shorts. The condition may be intermittent.
3. Checks for an open in both the speed sensor circuit wires.
4. Manipulates wiring and connectors, trying to induce an intermittent fault not currently present.
5. This test checks if the code resets on key-up. If so, since Tests 1–4 have validated that the circuitry and components are good, the Wheel Bearing/Speed Sensor assembly may be causing the fault.
6. Checks if Code 25 sets during a road test.
7. Code 27 sets when the vehicle is at rest: Code 25 sets with a problem in the speed sensor circuitry with the vehicle in motion. If problems are still present, Code 25 would set during the road test, not a 27. If Code 25 sets at this point, Code 25 diagnostics should be used.
8. The wheel speed sensor may be causing an intermittent fault. If Code 27 resets after sensor replacement, the EBTCM must be concluding there is a problem present in the speed sensor circuit when there is not.

DIAGNOSTIC AIDS:

Be sure the speed sensor wiring is properly routed and retained. This will help prevent false signals due to electrical noise being picked up by the wiring.

It is very important that a thorough inspection of the wiring and connectors be performed. Failure to carefully and fully inspect wiring and connectors may result in misdiagnosis, causing part replacement with reappearance of the fault.

DIAGNOSTIC CHART DTC 27 — 1993 ALLANTE, CONTINUED

ENSURE SENSOR IS NOT VISIBLY DAMAGED; IF IT IS DAMAGED, REPLACE THE WHEEL BEARING/SPEED/SENSOR ASSEMBLY. CLEAR CODES, AN TURN IGNITION 'ON' TO SEE IF CODE RESETS.

(1)
- IGNITION 'OFF'
- DISCONNECT SENSOR CONNECTOR
- USING DVM AND ADAPTER FROM J 35616-A, CHECK SENSOR RESISTANCE BY PROBING SENSOR CONNECTOR TERMINALS.
- IS IT 900–1100 OHMS?

NO:
- REPLACE WHEEL SPEED SENSOR

YES:

(2)
- CONNECT DVM BETWEEN SENSOR HARNESS CONNECTOR TERMINAL 'A' AND GROUND (USED DC VOLTAGE SETTING)
- IGNITION 'ON'
- NOTE VOLTAGE: REPEAT BETWEEN TERMINAL 'B' AND GROUND.
- IS IT AROUND SYSTEM VOLTAGE ON EITHER TERMINAL 'A' OR 'B'?

YES:
- REPAIR SHORT TO SYSTEM VOLTAGE IN CKT 830 OR 873

NO:

(3)
- IGNITION 'OFF'
- DISCONNECT EBTCM
- INSTALL J 39700 UNIVERSAL BREAKOUT BOX TO EBTCM HARNESS CONNECTOR ONLY
- MEASURE RESISTANCE BETWEEN PIN 5 ON J 39700 TERMINAL 'A' OF SENSOR CONNECTOR ON HARNESS. PERFORM SAME MEASUREMENT BETWEEN PIN 4 AND TERMINAL 'B'.
- IS EITHER MEASUREMENT GREATER THAN 5 OHMS?

YES:
- REPAIR OPEN IN CKT 830 AND/OR 873

NO:

(4)
- REMOVE J 39700; RECONNECT EBCM
- CLEAR CODES; IGNITION 'OFF'
- DISCONNECT TECH 1, IF CONNECTED
- IGNITION 'ON'
- WIGGLE ALL WIRING AND CONNECTORS TO TRY TO DUPLICATE AN INTERMITTENT FAULT
- DOES CODE 27 RESET?

YES:
- REPAIR WIRING AND/OR CONNECTIONS

NO:

(5)
- CLEAR CODES; IGNITION 'OFF'
- DISCONNECT TECH 1, IF CONNECTED
- IGNITION 'ON'
- DOES CODE 27 RESET?

YES:

(8)
- REPLACE WHEEL SPEED SENSOR
- CLEAR CODES; IGNITION 'OFF'
- DISCONNECT TECH 1, IF CONNECTED
- IGNITION 'ON'
- DOES CODE 27 RESET?

YES:
- REPLACE EBTCM

NO:
- SYSTEM OK

NO:

(6)
- ROAD TEST VEHICLE, REACHING AT LEAST 25 KPH (15 MPH).
- DOES CODE 25 SET?

YES:

(7) GO TO CODE 25 DIAGNOSTICS

NO:
- NO TROUBLE FOUND. SEE 'DIAGNOSTIC AIDS' ON FACING PAGE.

WHEN ALL DIAGNOSIS AND REPAIRS ARE COMPLETED. CLEAR CODES AND VERIFY OPERATION

DIAGNOSTIC CHART DTC 28 — 1993 ALLANTE

CODE 28
WHEEL SPEED SENSOR FREQUENCY ERROR

CIRCUIT DESCRIPTION:

TEST CONDITION(S):

1. Tested continuously.

FAILURE CONDITION(S):

Continuity

1. While Driving: One wheel is detected at least 30 kph or 19 mph faster than the remaining wheels for an extended period of time.

   OR

   One wheel is at least 6 kph or 4 mph less than or greater than the remaining wheels for an extended period of time.

ACTION TAKEN:

1. Code set.
2. ABS/TCS switched 'OFF'.
3. Red BRAKE warning lamp 'ON'.
4. ABS and TCS 'DISABLED' messages displayed.

Code 28 is set if the EBTCM cannot specifically identify which wheel speed sensor is causing the frequency error problem. If it can define the specific speed sensor causing the problem, the Sensor Fault code associated with the sensor (21, 25, 31, 35) will be set instead of Code 28.

TEST DESCRIPTION: Test numbers refer to circled numbers on the diagnostic chart.

1. Checks wiring and connections for problems. It is critical that this step is performed thoroughly, as wiring/connector problems are the most likely cause for this code.
2. Use the TECH 1 to monitor for vehicle electrical/electronic system 'noise' being picked up by the speed sensor circuits.
3. Use the TECH 1 to monitor the wheel speed sensors while the vehicle is operating. If the TECH 1 Auto-Trigger Snapshot mode triggers, the speed sensor or the wiring and/or connectors are going intermittent during the road test.
4. Checks for proper resistance in the sensor itself.

5a. Replace rear wheel speed sensor as the likely cause of the triggering intermittent.

5b. Replaces front wheel speed sensor as the likely cause of the triggering intermittent.

6. Use the TECH 1 to monitor the wheel speed sensors while the vehicle is operating. If the TECH 1 Auto-Snapshot mode triggers, the EBTCM must be concluding there is a problem with wheel speed sensor frequencies when there is not.
7. Checks for a short to ground in the speed sensor input circuit wires. The circuit to be tested depends on which speed sensor triggered the TECH 1. The ground may be either a 'hard' short to ground, or a resistive short. A short with a resistance less that 2 megaohms, though not a hard short, can still cause Code 28 to set.

DIAGNOSTIC AIDS:

Be sure the speed sensor wiring is properly routed and retained. This will help prevent false signals due to electrical noise being picked up by the wiring.

It is very important that a thorough inspection of the wiring and connectors be performed. Failure to carefully and fully inspect wiring and connectors may result in misdiagnosis, causing part replacement with reappearance of the fault.

Front Speed Sensors Only:

Check the toothed wheel for any large grooves, gouges, marks, etc. that might influence the tooth's signal at the wheel speed sensor. Also check for a buildup of foreign material in the gaps between teeth in the toothed wheel; this material may cause this fault.

A worn hub/bearing assembly may cause this fault in extreme cases; the bearing play allows the sensor-to-toothed-ring gap to change excessively. Never reuse a wheel speed sensor once it has been separated from the front hub/bearing assembly.

DIAGNOSTIC CHART DTC 28 — 1993 ALLANTE, CONTINUED

(1)
- VERY THOROUGHLY INSPECT ALL WIRING AND CONNECTIONS FOR INTERMITTENTS, CHAFING, POOR TERMINAL CONTACT TENSION, CORROSION, AND MISROUTING, AT EBTCM, ALL SENSORS, AND ALL HARNESS INTERCONNECTS. REPAIR ANY PROBLEMS CLEAR CODES, THEN ROAD TEST VEHICLE, REACHING AT LEAST 25 KPH (15 MPH).
- DOES CODE RESET?

YES → (2)
NO → • NO TROUBLE FOUND; SEE 'DIAGNOSTIC AIDS' ON FACING PAGE.

(2)
- INSTALL TECH 1 AND '88–'93 BRAKE CARTRIDGE; CLEAR CODES
- SET TECH 1 AUTO TRIGGER SNAPSHOT MODE
- START VEHICLE AND ALLOW TO IDLE FOR ONE MINUTE
- DOES TECH 1 TRIGGER?

NO → (3)
NO →
- NOTE TECH1, WHICH SENSOR TRIGGERED?
- IGNITION 'OFF'
- DISCONNECT EBTCM
- INSTALL J 39700 UNIVERSAL BREAKOUT BOX TO EBTCM AND EBTCM HARNESS CONNECTOR.
- USING TABLE BELOW, MEASURE RESISTANCE TO GROUND AT THE TERMINAL ON J 39700 THAT CORRESPONDS TO THE SPEED SENSOR THE TECH 1 TRIGGERED ON:

  LF-TERMINAL 5
  RF-TERMINAL 11
  LR-TERMINAL 8
  RR-TERMINAL 24
- IF RESISTANCE IS 5 OHMS OR LESS TO GROUND, REPAIR HARD SHORT; IF BETWEEN 5 OHMS AND 2 MEGAOHMS, REPAIR RESISTIVE SHORT IN CKT.

  LF-CKT 830
  RF-CKT 872
  LR-CKT 885
  RR-CKT 882

(3)
- ROAD TEST VEHICLE REACHING AT LEAST 25 KPH (15 MPH).
- DOES TECH 1 TRIGGER?

YES → • NOTE TECH 1, WHICH SENSOR TRIGGERED? → (4)
NO → • NO TROUBLE FOUND; SEE DIAGNOSTIC AIDS ON FACING PAGE.

(4)
- IGNITION 'OFF'
- DISCONNECT SENSOR CONNECTOR
- MEASURE RESISTANCE BETWEEN SENSOR TERMINALS, USING ADAPTERS FROM J 35616-A IF FRONT SENSOR.
- IS RESISTANCE WITH IN RANGE SHOWN?

REAR	FRONT
990–1210	900–1100

NO → FRONT / REAR
YES → • REPLACE EBTCM

FRONT → (5a) • REPLACE WHEEL SPEED SENSOR
REAR → (5b) • REPLACE HUB/BEARING/SENSOR ASSEMBLY

(6)
- CLEAR CODES, IGNITION 'OFF'
- DISCONNECT TECH 1, IF CONNECTED
- ROAD TEST VEHICLE, REACHING AT LEAST 25 KPH (15 MPH)
- DOES CODE 28 RESET?

YES → • REPLACE EBTCM
NO → • SYSTEM OK

WHEN ALL DIAGNOSIS AND REPAIRS ARE COMPLETED, CLEAR CODES AND VERIFY OPERATION

DIAGNOSTIC CHART DTC 31 — 1993 ALLANTE

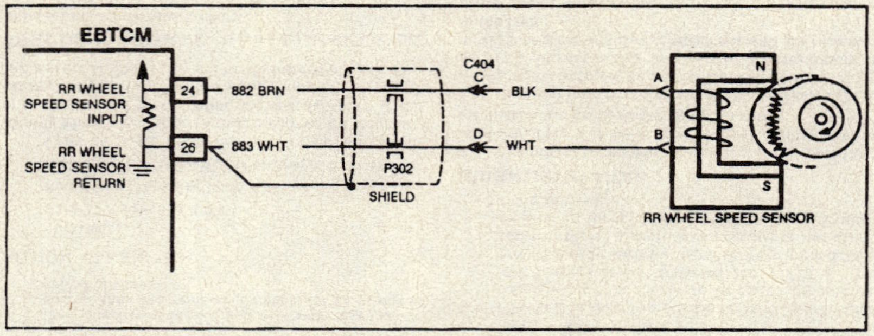

CODE 31
RIGHT REAR WHEEL SPEED SENSOR FAULT

CIRCUIT DESCRIPTION:

As each wheel turns, the wheel speed sensor for that wheel creates a small AC voltage as a toothed sensor ring, mounted in the front integral hub/bearing assembly, is passed by the stationary sensor. The sensor consists of a permanent magnet and coil and generates voltage by magnetic induction. The frequency of the wheel speed sensor signal is proportional to wheel speed.

The EBTCM uses the wheel speed sensor signal to calculate vehicle reference speeds and individual speed, acceleration and slip values for each wheel. These values are used to determine when antilock control is required.

The EBTCM will only set one wheel speed sensor fault code at a time, even if multiple faults exist. Detection is prioritized accordingly: LF, RR, LR, RF. Always check for additional stored codes after making repairs.

TEST CONDITION(S):

1. Tested continuously.

FAILURE CONDITION(S):

Sensor Output

1. Drive-away: Speed signal is not detected when vehicle reaches 12 kph or 7.5 mph.
2. While Driving: Speed signal is lost while vehicle speed is greater than 40 kph or 25 mph

ACTION TAKEN:

1. Code set.
2. ABS/TCS switched 'OFF'.
3. Red BRAKE warning lamp 'ON'.
4. ABS and TCS 'DISABLED' messages displayed.

Code 21 is set if there is a short to voltage or ground in CKT 882 or 883, or faulty speed sensor. The testing for this fault occurs when the vehicle is in motion, it will not set with the ignition 'ON' and the vehicle at rest.

NOTICE: A Code 31 may set if CKT 882 or 883 lose continuity and the WSS signal is lost while driving. A Code 33 should accompany this code on the subsequent ignition cycle. If both Code 31 and Code 33 are set concurrently, diagnose Code 33 first.

TEST DESCRIPTION: Test numbers refer to circled numbers on the diagnostic chart.

1. Checks if the code resets during a road test.
2. Use the TECH 1 to check for sources of induced electrical noise. Some sources of induced electrical noise which may cause intermittent wheel speed sensor codes to set are: improper routing or sensor cables next to spark plug wires; cables not retained in brackets; loose fitting or improperly mounted sensors; damaged sensors or toothed rings; poor terminal engagement of connectors; or intermittent shorts or opens in wiring; or a loss of sensor circuit shield.
3. Checks for damaged wheel speed sensor circuit shield. The shield is a foil insulator that is wrapped around the wheel speed sensor circuits. Inside the EBTCM harness connecter there is an uninsulated drain wire that is attached to the shield and jumped to the wheel speed sensor return line and taken to ground through the EBTCM. A shield which has lost its ground through the EBTCM leaves the sensor signal circuit susceptible to noise. It may be necessary to disassemble the EBTCM connector to check if the drain wire has a poor connection with the wheel speed sensor return connector terminal. The shield can be repaired by reterminating the sensor return line circuit, including the drain wire.
4. Checks for intermittent opens/shorts in wheel speed sensor circuits and cables.

DIAGNOSTIC CHART DTC 31 — 1993 ALLANTE, CONTINUED

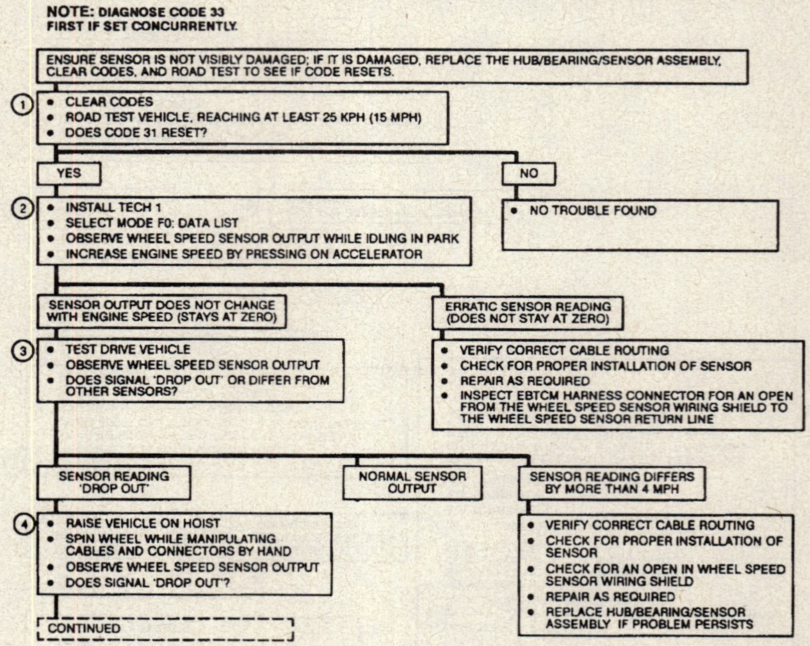

DIAGNOSTIC CHART DTC 31 — 1993 ALLANTE, CONTINUED

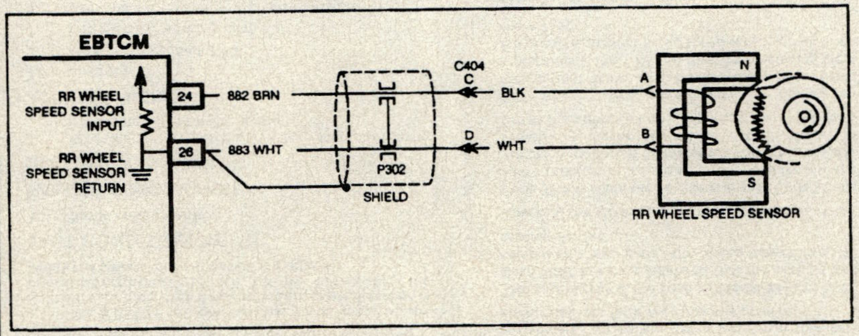

CODE 31

RIGHT REAR WHEEL SPEED SENSOR FAULT

CIRCUIT DESCRIPTION:

As each wheel turns, the wheel speed sensor for that wheel creates a small AC voltage as a toothed sensor ring, mounted in the front integral hub/bearing assembly, is passed by the stationary sensor. The sensor consists of a permanent magnet and coil and generates voltage by magnetic induction. The frequency of the wheel speed sensor signal is proportional to wheel speed.

The EBTCM uses the wheel speed sensor signal to calculate vehicle reference speeds and individual speed, acceleration and slip values for each wheel. These values are used to determine when antilock control is required.

The EBTCM will only set one wheel speed sensor fault code at a time, even if multiple faults exist. Detection is prioritized accordingly: LF, RR, LR, RF. Always check for additional stored codes after making repairs.

TEST DESCRIPTION: Test numbers refer to circled numbers on the diagnostic chart.

5. Checks for proper resistance in the sensor itself.
6. Checks for a short to battery in either of the speed sensor circuit wires.
7. Checks for a short between the speed sensor circuit wires.
8. Checks for a short to ground in the speed sensor input circuit wire. The ground may be either a 'hard' short to ground, or a resistive short. A short with a resistance less than 2 megaohms, through not a hard short, can still cause Code 21 to set.
9. Checks wiring and connectors for intermittents.

DIAGNOSTIC CHART DTC 31 — 1993 ALLANTE, CONTINUED

CONTINUED FROM CODE 31

(5)
- IGNITION 'OFF'
- DISCONNECT SENSOR CONNECTOR
- USING DVM AND ADAPTER FROM J 35616-A, CHECK SENSOR RESISTANCE BY PROBING SENSOR CONNECTOR TERMINALS.
- IS IT 990 - 1210 OHMS?

NO:
- REPLACE EBTCM IF PROBLEM PERSISTS

YES:

(6)
- CONNECT DVM BETWEEN SENSOR HARNESS CONNECTOR TERMINAL 'A' AND GROUND (USE DC VOLTAGE SETTING)
- IGNITION 'ON'
- NOTE VOLTAGE: REPEAT BETWEEN TERMINAL 'B' AND GROUND.
- IS IT AROUND SYSTEM VOLTAGE ON EITHER TERMINAL 'A' OR 'D'?

NO:
- REPLACE HUB/BEARING/SENSOR ASSEMBLY

YES:
- REPAIR SHORT TO SYSTEM VOLTAGE IN CKT 882 OR 883

NO:

(7)
- IGNITION 'OFF'
- DISCONNECT EBTCM
- INSTALL J 39700 UNIVERSAL BREAKOUT BOX TO EBTCM HARNESS CONNECTOR ONLY
- MEASURE RESISTANCE BETWEEN J 39700 TERMINALS '26' AND '24'
- IS IT LESS THAN 5 OHMS?

YES:
- REPAIR SHORT BETWEEN CKT'S 882 AND 883

NO:

(8)
- MEASURE RESISTANCE BETWEEN J 39700 TERMINAL '24' AND GROUND
- IS IT GREATER THAN 2 MEGAOHMS?

NO:
- IF 5 OHMS OR LESS TO GROUND, REPAIR HARD SHORT IN HARNESS CKT 882
- IF BETWEEN 5 OHMS AND 2 MEGAOHMS, REPAIR RESISTIVE SHORT IN HARNESS CKT 882

YES:

(9)
- MEASURE RESISTANCE BETWEEN J 39700 TERMINAL '26' AND GROUND
- IS IT GREATER THAN 2 MEGAOHMS?

NO:
- IF 5 OHMS OR LESS TO GROUND, REPAIR HARD SHORT IN HARNESS CKT 883
- IF BETWEEN 5 OHMS AND 2 MEGAOHMS REPAIR RESISTIVE SHORT IN HARNESS CKT 883

YES:
- CHECK ALL WIRING, CONNECTORS AND TERMINALS THOROUGHLY FOR INTERMITTENTS OR CORROSION
- REPAIR HARNESS OR CONNECTOR
- REPLACE SENSOR IF PROBLEM PERSISTS

DIAGNOSTIC CHART DTC 33 — 1993 ALLANTE

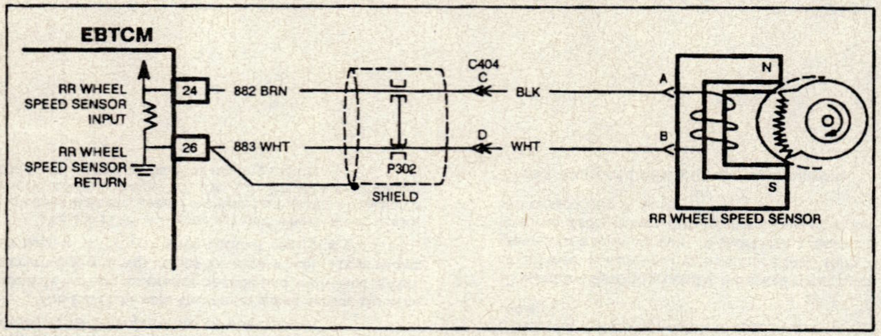

CODE 33

RIGHT REAR WHEEL SPEED SENSOR CONTINUITY FAULT

CIRCUIT DESCRIPTION:

As each wheel turns, the wheel speed sensor for that wheel creates a small AC voltage as a toothed sensor ring, mounted in the front integral hub/bearing assembly, is passed by the stationary sensor. The sensor consists of a permanent magnet and coil and generates voltage by magnetic induction. The frequency of the wheel speed sensor signal is proportional to wheel speed.

The EBTCM uses the wheel speed sensor signal to calculate vehicle reference speeds and individual speed, acceleration and slip values for each wheel. These values are used to determine when antilock control is required.

The EBTCM will only set one wheel speed sensor fault code at a time, even if multiple faults exist. Detection is prioritized accordingly: LF, RR, LR, RF. Always check for additional stored codes after making repairs.

TEST CONDITION(S):

1. Tested continuously.

FAILURE CONDITION(S):

Continuity

1. While standing: Open circuit is seen while the vehicle is standing still with the ignition 'ON'.

ACTION TAKEN:

1. Code set.
2. ABS/TCS switched 'OFF'.
3. Red BRAKE warning lamp 'ON'.
4. ABS and TCS 'DISABLED' messages displayed.

Code 33 is set if there is a short to voltage or an open in CKT 882 or 883, or faulty speed sensor. The testing for this fault occurs with the ignition 'ON' and the vehicle at rest.

NOTICE: A Code 31 may set if CKT 882 or 883 lose continuity and the WSS signal is lost while driving. A Code 33 should accompany this code on the subsequent ignition cycle. If both Code 31 and Code 33 are set concurrently, diagnose Code 33 first.

TEST DESCRIPTION: Test numbers refer to circled numbers on the diagnostic chart.

1. Checks for proper resistance in the sensor itself.
2. Checks for a short to battery in either of the speed sensor circuit wires. There should be 5V on CKT 882. If there is, CKT 882 has no hard opens or shorts. The condition may be intermittent.
3. Checks for an open in both the speed sensor circuit wires.
4. Manipulates wiring and connectors, trying to induce an intermittent fault not currently present.
5. This test checks if the code resets on key-up. If so, since Tests 1–4 have validated that the circuitry and components are good, the speed sensor may be causing the fault.
6. Checks if the Code 31 sets during a road test.
7. Code 33 sets when the vehicle is at rest: Code 31 sets with a problem in the speed sensor circuitry with the vehicle in motion. If problems are still present, Code 31 would set during the road test, not a 33. If Code 33 sets at this point, Code 31 diagnostics should be used.
8. The wheel speed sensor may be causing an intermittent fault. If Code 33 resets after sensor replacement, the EBTCM must be concluding there is a problem present in the speed sensor circuit when there is not.

DIAGNOSTIC AIDS:

Be sure the speed sensor wiring is properly routed and retained. This will help prevent false signals due to electrical noise being picked up by the wiring.

It is very important that a thorough inspection of the wiring and connectors be performed. Failure to carefully and fully inspect wiring and connectors may result in misdiagnosis, causing part replacement with reappearance of the fault.

DIAGNOSTIC CHART DTC 33 — 1993 ALLANTE, CONTINUED

ENSURE SENSOR IS NOT VISIBLY DAMAGED; IF IT IS DAMAGED, REPLACE THE HUB/BEARING/SPEED SENSOR ASSEMBLY. CLEAR CODES, AN TURN IGNITION 'ON' TO SEE IF CODE RESETS.

(1)
- IGNITION 'OFF'
- DISCONNECT SENSOR CONNECTOR
- USING DVM AND ADAPTER FROM J 35616-A, CHECK SENSOR RESISTANCE BY PROBING SENSOR CONNECTOR TERMINALS.
- IS IT 990 - 1210 OHMS?

NO:
- REPLACE HUB/BEARING/SENSOR ASSEMBLY

YES:

(2)
- CONNECT DVM BETWEEN SENSOR HARNESS CONNECTOR TERMINAL 'C' AND GROUND (USED DC VOLTAGE SETTING)
- IGNITION 'ON'
- NOTE VOLTAGE: REPEAT BETWEEN TERMINAL 'D' AND GROUND.
- IS IT AROUND SYSTEM VOLTAGE ON EITHER TERMINAL 'C' OR 'D'?

YES:
- REPAIR SHORT TO SYSTEM VOLTAGE IN CKT 882 OR 883

NO:

(3)
- IGNITION 'OFF'
- DISCONNECT EBTCM
- INSTALL J 39700 UNIVERSAL BREAKOUT BOX TO EBTCM HARNESS CONNECTOR ONLY
- MEASURE RESISTANCE BETWEEN PIN 24 ON J 39700 TERMINAL 'A' OF SENSOR CONNECTOR ON HARNESS. PERFORM SAME MEASUREMENT BETWEEN TERMINAL 26 AND TERMINAL 'B'.
- IS EITHER MEASUREMENT GREATER THAN 5 OHMS?

YES:
- REPAIR OPEN IN CKT 882 AND/OR 883

NO:

(4)
- REMOVE J 39700; RECONNECT EBTCM
- CLEAR CODES; IGNITION 'OFF'
- DISCONNECT TECH 1, IF CONNECTED
- IGNITION 'ON'
- WIGGLE ALL WIRING AND CONNECTORS TO TRY TO DUPLICATE AN INTERMITTENT FAULT
- DOES CODE 33 RESET?

YES:
- REPAIR WIRING AND/OR CONNECTIONS

NO:

(5)
- CLEAR CODES; IGNITION 'OFF'
- DISCONNECT TECH 1, IF CONNECTED
- IGNITION 'ON'
- DOES CODE 33 RESET?

YES:

(8)
- REPLACE HUB/BEARING/SENSOR ASSEMBLY
- CLEAR CODES; IGNITION 'OFF'
- DISCONNECT TECH 1, IF CONNECTED
- IGNITION 'ON'
- DOES CODE 33 RESET?

YES:
- REPLACE EBTCM

NO:
- SYSTEM OK

NO:

(6)
- ROAD TEST VEHICLE, REACHING AT LEAST 25 KPH (15 MPH).
- DOES CODE 31 SET?

YES:

(7) GO TO CODE 31 DIAGNOSTICS

NO:
- NO TROUBLE FOUND. SEE 'DIAGNOSTIC AIDS' ON FACING PAGE.

WHEN ALL DIAGNOSIS AND REPAIRS ARE COMPLETED, CLEAR CODES AND VERIFY OPERATION

DIAGNOSTIC CHART DTC 35 — 1993 ALLANTE

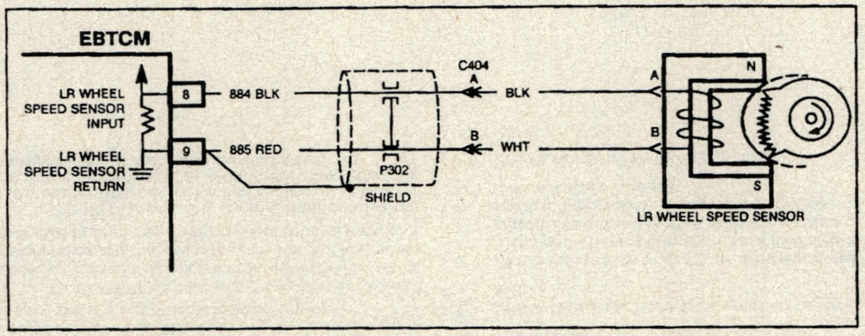

CODE 35

LEFT REAR WHEEL SPEED SENSOR FAULT

CIRCUIT DESCRIPTION:

As each wheel turns, the wheel speed sensor for that wheel creates a small AC voltage as a toothed sensor ring, mounted in the front integral hub/bearing assembly, is passed by the stationary sensor. The sensor consists of a permanent magnet and coil and generates voltage by magnetic induction. The frequency of the wheel speed sensor signal is proportional to wheel speed.

The EBTCM uses the wheel speed sensor signal to calculate vehicle reference speeds and individual speed, acceleration and slip values for each wheel. These values are used to determine when antilock control is required.

The EBTCM will only set one wheel speed sensor fault code at a time, even if multiple faults exist. Detection is prioritized accordingly: LF, RR, LR, RF. Always check for additional stored codes after making repairs.

TEST CONDITION(S):

1. Tested continuously.

FAILURE CONDITION(S):

Sensor Output

1. Drive-away: Speed signal is not detected when vehicle reaches 12 kph or 7.5 mph.
2. While Driving: Speed signal is lost while vehicle speed is greater than 40 kph or 25 mph.

ACTION TAKEN:

1. Code set.
2. ABS/TCS switched 'OFF'.
3. Red BRAKE warning lamp 'ON'.
4. ABS and TCS 'DISABLED' messages displayed.

Code 35 is set if there is a short to voltage or ground in CKT 885 or 884, or faulty speed sensor. The testing for this fault occurs when the vehicle is in motion, it will not set with the ignition 'ON' and the vehicle at rest.

NOTICE: A Code 35 may set if CKT 884 or 885 lose continuity and the WSS signal is lost while driving. A Code 37 should accompany this code on the subsequent ignition cycle. If both Code 35 and Code 37 are set concurrently, diagnose Code 37 first.

TEST DESCRIPTION: Test numbers refer to circled numbers on the diagnostic chart.

1. Checks if the code resets during a road test.
2. Use the TECH 1 to check for sources of induced electrical noise. Some sources of induced electrical noise which may cause intermittent wheel speed sensor codes to set are: improper routing or sensor cables next to spark plug wires; cables not retained in brackets; loose fitting or improperly mounted sensors; damaged sensors or toothed rings; poor terminal engagement of connectors; or intermittent shorts or opens in wiring; or a loss of sensor circuit shield.
3. Checks for damaged wheel speed sensor circuit shield. The shield is a foil insulator that is wrapped around the wheel speed sensor circuits. Inside the EBTCM harness connecter there is an uninsulated drain wire that is attached to the shield and jumped to the wheel speed sensor return line and taken to ground through the EBTCM. A shield which has lost its ground through the EBTCM leaves the sensor signal circuit susceptible to noise. It may be necessary to disassemble the EBTCM connector to check if the drain wire has a poor connection with the wheel speed sensor return connector terminal. The shield can be repaired by reterminating the sensor return line circuit, including the drain wire.
4. Checks for intermittent opens/shorts in wheel speed sensor circuits and cables.

DIAGNOSTIC CHART DTC 35 — 1993 ALLANTE, CONTINUED

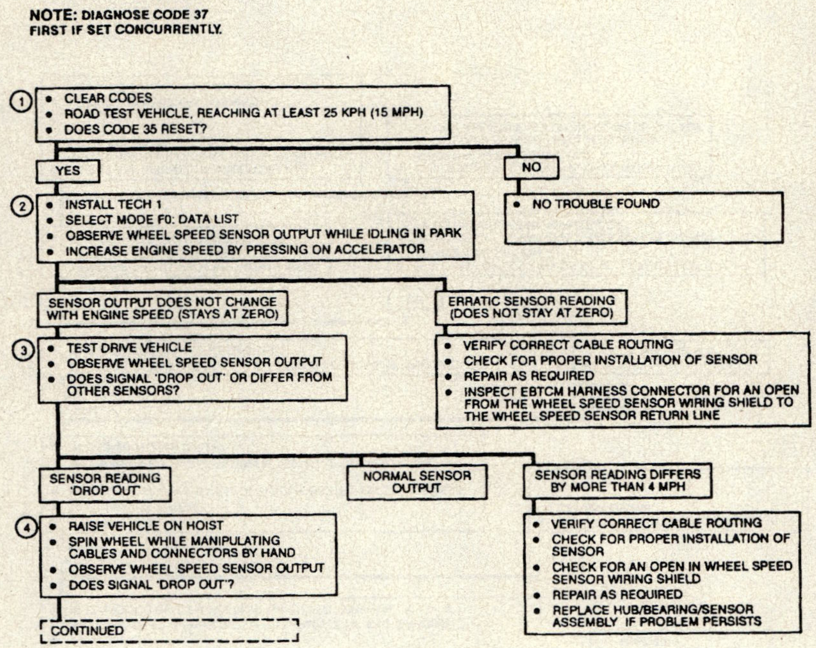

DIAGNOSTIC CHART DTC 35 — 1993 ALLANTE, CONTINUED

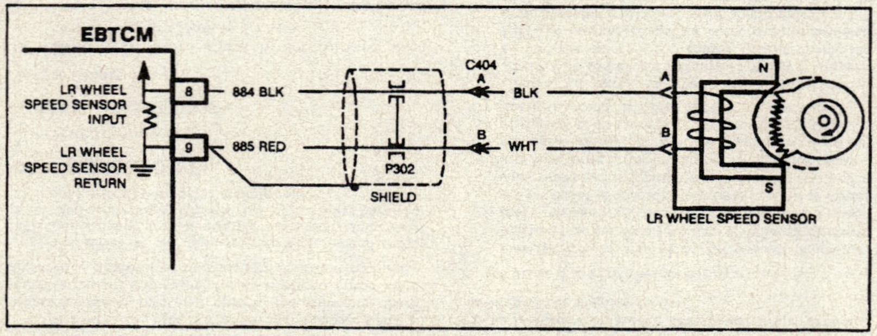

CODE 35

LEFT REAR WHEEL SPEED SENSOR FAULT

CIRCUIT DESCRIPTION:

As each wheel turns, the wheel speed sensor for that wheel creates a small AC voltage as a toothed sensor ring, mounted in the front integral hub/bearing assembly, is passed by the stationary sensor. The sensor consists of a permanent magnet and coil and generates voltage by magnetic induction. The frequency of the wheel speed sensor signal is proportional to wheel speed.

The EBTCM uses the wheel speed sensor signal to calculate vehicle reference speeds and individual speed, acceleration and slip values for each wheel. These values are used to determine when antilock control is required.

The EBTCM will only set one wheel speed sensor fault code at a time, even if multiple faults exist. Detection is prioritized accordingly: LF, RR, LR, RF. Always check for additional stored codes after making repairs.

TEST DESCRIPTION: Test numbers refer to circled numbers on the diagnostic chart.

5. Checks for proper resistance in the sensor itself.
6. Checks for a short to battery in either of the speed sensor circuit wires.
7. Checks for a short between the speed sensor circuit wires.
8. Checks for a short to ground in the speed sensor input circuit wire. The ground may be either a 'hard' short to ground, or a resistive short. A short with a resistance less than 2 megaohms, through not a hard short, can still cause Code 21 to set.
9. Checks wiring and connectors for intermittents.

DIAGNOSTIC CHART DTC 35 — 1993 ALLANTE, CONTINUED

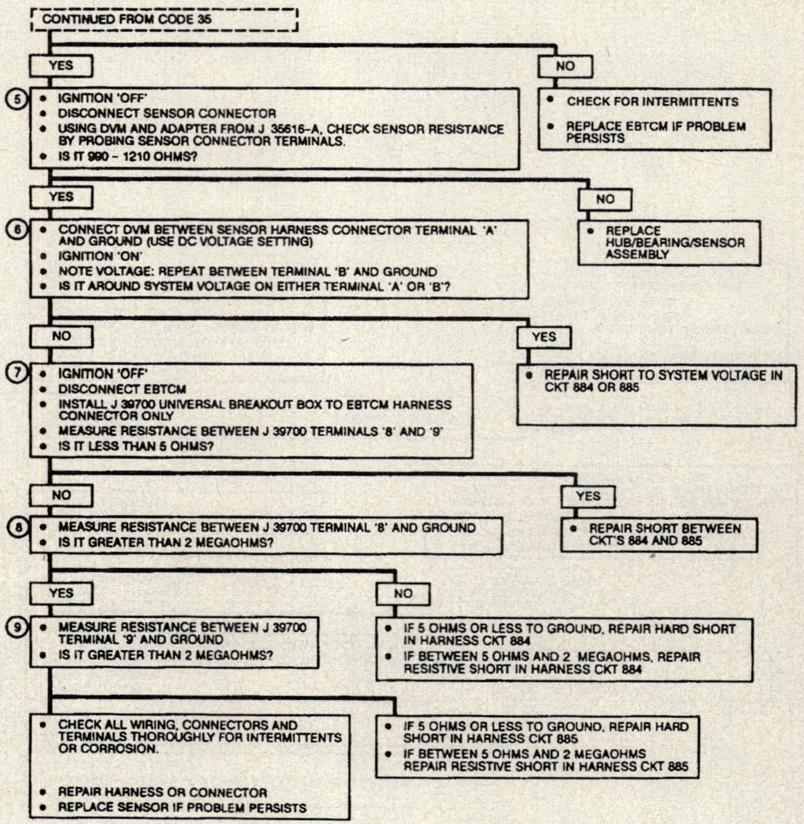

DIAGNOSTIC CHART DTC 37 — 1993 ALLANTE

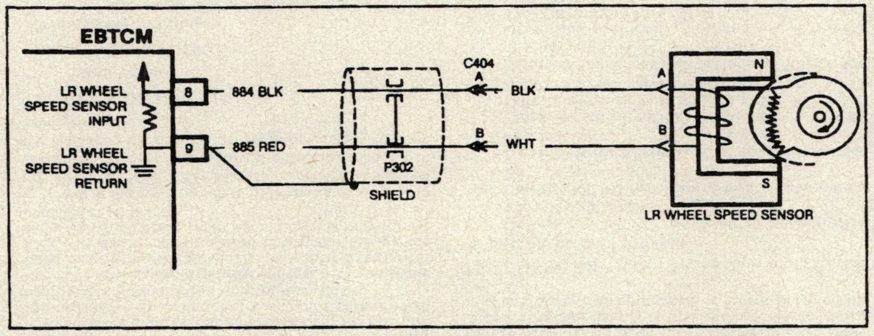

CODE 37

LEFT REAR WHEEL SPEED SENSOR CONTINUITY FAULT

CIRCUIT DESCRIPTION:

As each wheel turns, the wheel speed sensor for that wheel creates a small AC voltage as a toothed sensor ring, mounted in the front integral hub/bearing assembly, is passed by the stationary sensor. The sensor consists of a permanent magnet and coil and generates voltage by magnetic induction. The frequency of the wheel speed sensor signal is proportional to wheel speed.

The EBTCM uses the wheel speed sensor signal to calculate vehicle reference speeds and individual speed, acceleration and slip values for each wheel. These values are used to determine when antilock control is required.

The EBTCM will only set one wheel speed sensor fault code at a time, even if multiple faults exist. Detection is prioritized accordingly: LF, RR, LR, RF. Always check for additional stored codes after making repairs.

TEST CONDITION(S):

1. Tested continuously.

FAILURE CONDITION(S):

Continuity

1. While standing: Open circuit is seen while the vehicle is standing still with the ignition 'ON'

ACTION TAKEN:

1. Code set.
2. ABS/TCS switched 'OFF'.
3. Red BRAKE warning lamp 'ON'.
4. ABS and TCS 'DISABLED' messages displayed.

Code 37 is set if there is a short to voltage or an open in CKT 885 or 884, or faulty speed sensor. The testing for this fault occurs with the ignition 'ON' and the vehicle at rest.

NOTICE: A Code 35 may set if CKT 884 or 885 lose continuity and the WSS signal is lost while driving. A Code 37 should accompany this code on the subsequent ignition cycle. If both Code 35 and Code 37 are set concurrently, diagnose Code 37 first.

TEST DESCRIPTION: Test numbers refer to circled numbers on the diagnostic chart.

1. Checks for proper resistance in the sensor itself.
2. Checks for a short to battery in either of the speed sensor circuit wires. There should be 5V on CKT 884. If there is, CKT 884 has no hard opens or shorts. The condition may be intermittent.
3. Checks for an open in both the speed sensor circuit wires.
4. Manipulates wiring and connectors, trying to induce an intermittent fault not currently present.
5. This test checks if the code resets on key-up. If so, since Tests 1–4 have validated that the circuitry and components are good, the Speed Sensor may be causing the fault.
6. Checks if the Code 35 sets during a road test.
7. Code 37 sets when the vehicle is at rest: Code 35 sets with a problem in the speed sensor circuitry with the vehicle in motion. If problems are still present, Code 35 would set during the road test, not a 37. If Code 35 sets at this point, Code 35 diagnostics should be used.
8. The wheel speed sensor may be causing an intermittent fault. If Code 37 resets after sensor replacement, the EBTCM must be concluding there is a problem present in the speed sensor circuit when there is not.

DIAGNOSTIC AIDS:

Be sure the speed sensor wiring is properly routed and retained. This will help prevent false signals due to electrical noise being picked up by the wiring.

It is very important that a thorough inspection of the wiring and connectors be performed. Failure to carefully and fully inspect wiring and connectors may result in misdiagnosis, causing part replacement with reappearance of the fault.

DIAGNOSTIC CHART DTC 37 — 1993 ALLANTE, CONTINUED

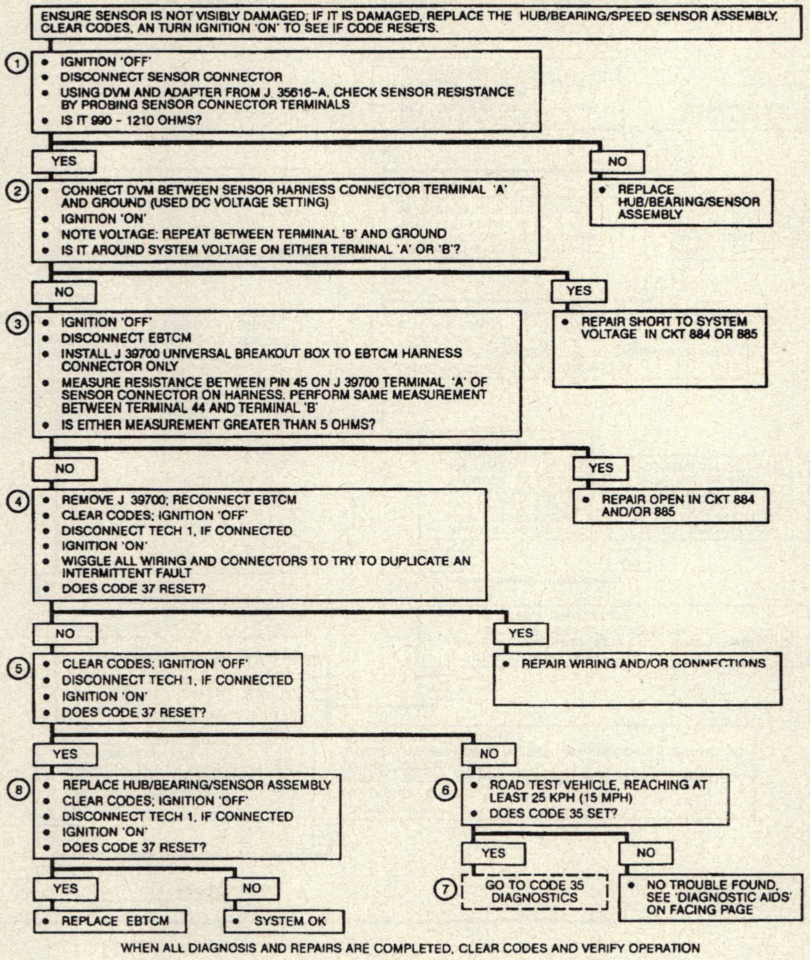

DIAGNOSTIC CHART DTC 41 — 1993 ALLANTE

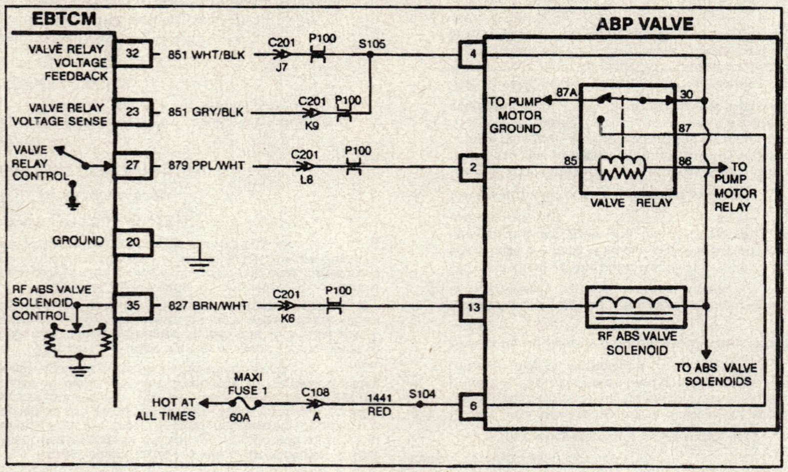

CODE 41

RIGHT FRONT VALVE SOLENOID FAULT

CIRCUIT DESCRIPTION:

The right front valve solenoid in the ABP valve may be commanded to three different positions by the EBTCM. Valve position is determined by the amount of current which is allowed to flow through the valve solenoid coil.

The valve solenoid circuits receive power through the valve relay on the ABP valve. The valve relay is engaged at KEY–ON and remains engaged throughout the ignition cycle. Valve solenoid circuits should have battery voltage available at all times.

When the key is turned to RUN, power is supplied to the EBTCM. The EBTCM will complete an internal self-check before providing ground to the valve relay. When the valve relay is energized, battery power is supplied to the valve solenoids.

TEST CONDITION(S):

1. Tested continuously.

FAILURE CONDITION(S):

While standing or driving:

1. Valve solenoid cannot be activated.

OR

2. Valve solenoid is activated in a defective way.

ACTION TAKEN:

1. Code set.
2. ABS/TCS switched 'OFF'.
3. Red BRAKE warning lamp 'ON'.
4. ABS and TCS 'DISABLED' messages displayed.

Code 41 will set when the expected position of the right front valve solenoid does not match the commanded position from the EBTCM. Conditions which could cause Code 41 to set are: damage to the right front valve solenoid; an open circuit; or an open circuit to ground or battery on CKT 827.

TEST DESCRIPTION: Test numbers refer to circled numbers on the diagnostic chart.

1. Damage to CKT 851 may result in multiple valve solenoid codes setting.
2. Checks CKT 851 for opens or high resistance from EBTCM terminal '23' to '32'.
3. Determines which branch of CKT 851 has an open or high resistance.
4. Checks the integrity of CKTs 851 and 827, the RF ABS Valve Solenoid circuitry internal to the ABP valve, and the RF ABS Valve Solenoid coil.
5. Determines whether a problem found in Step 4 is due to an open in CKT 827, 851, or ABP valve.
6. Determines whether the problem found in Step 4 is due to an open in CKT 827 or 851.

DIAGNOSTIC AIDS:

All tests using J 39700 Universal Breakout Box terminal '20' are using the terminal as ground. These tests, of course, assume that the ground CKT 804 terminal '20' is good.

If Codes 41 and 45 are both set, the fault is likely to be an open or high resistance on CKT 851.

If Codes 41 and 55 are both set, the fault is likely to be a short to B + on CKT 828.

DIAGNOSTIC CHART DTC 41 — 1993 ALLANTE, CONTINUED

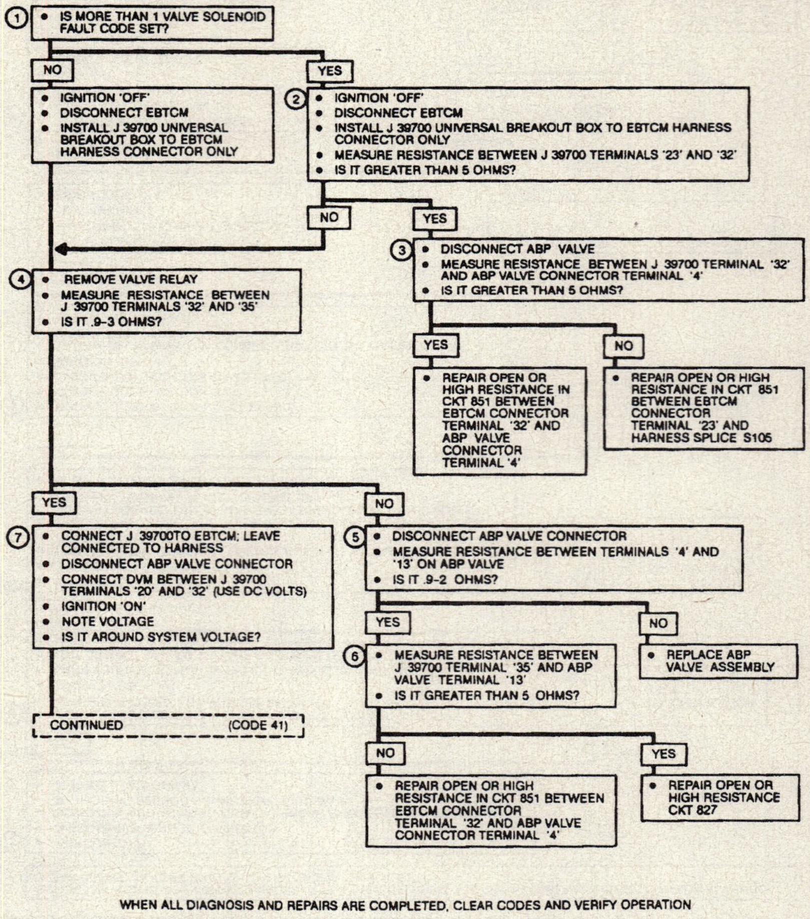

DIAGNOSTIC CHART DTC 41 — 1993 ALLANTE, CONTINUED

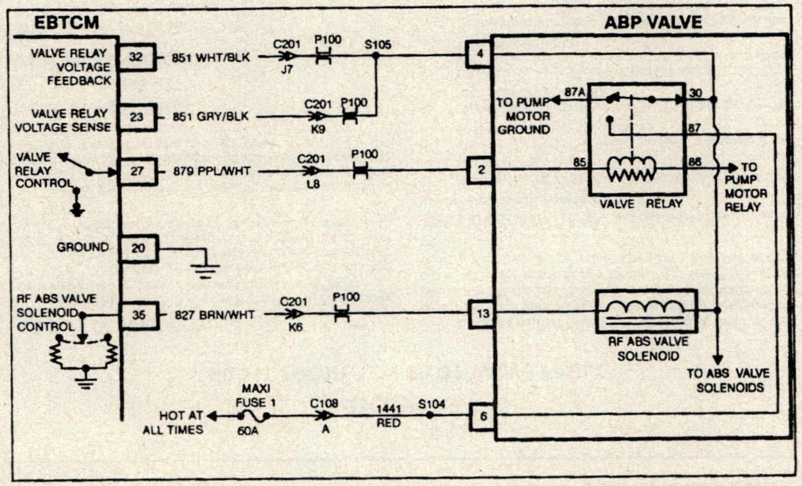

CODE 41

RIGHT FRONT VALVE SOLENOID FAULT

CIRCUIT DESCRIPTION:

The right front valve solenoid in the ABP valve may be commanded to three different positions by the EBTCM. Valve position is determined by the amount of current which is allowed to flow through the valve solenoid coil.

The valve solenoid circuits receive power through the valve relay on the ABP valve. The valve relay is engaged at KEY–ON and remains engaged throughout the ignition cycle. Valve solenoid circuits should have battery voltage available at all times.

When the key is turned to RUN, power is supplied to the EBTCM. The EBTCM will complete an internal self-check before providing ground to the valve relay. When the valve relay is energized, battery power is supplied to the valve solenoids.

TEST DESCRIPTION: Test numbers refer to circled numbers on the diagnostic chart.

8. Checks for a short to battery in CKT 827 using EBTCM terminal '20' as ground.
9. Checks for a short to ground in CKT 851 using EBTCM terminal '20' as ground.
10. Checks for a short to ground in CKT 827 using EBTCM terminal '20' as ground.
11. Checking wiring and connectors for intermittents.
12. Use the TECH 1 to exercise the RF ABS Valve Solenoid and check it for proper operation.
13. Determines whether the code was set by an intermittent condition or an EBTCM fault.

DIAGNOSTIC AIDS:

All tests using J 39700 Universal Breakout Box terminal '20' are using the terminal as ground. These tests, of course, assume that the ground CKT 804 terminal '20' is good.

If Codes 41 and 45 are both set, the fault is likely to be an open or high resistance on CKT 851.

If Codes 41 and 55 are both set, the fault is likely to be a short to B+ on CKT 828.

DIAGNOSTIC CHART DTC 41 — 1993 ALLANTE, CONTINUED

CONTINUED FROM (CODE 41)

(8)
- CONNECT DVM BETWEEN J 39700 TERMINALS '20' AND '35'
- NOTE VOLTAGE
- IS IT AROUND SYSTEM VOLTAGE?

YES: • REPAIR SHORT TO SYSTEM VOLTAGE CKT 851

NO: (9)
- IGNITION 'OFF'
- MEASURE RESISTANCE BETWEEN J 39700 TERMINALS '20' AND '32'
- IS IT 5 OHMS OR LESS?

YES:
- REPAIR SHORT TO SYSTEM VOLTAGE ON CKT 827
- REFER TO EBTCM SWITCH FAULTS IN THIS SECTION

NO: (10)
- MEASURE RESISTANCE BETWEEN J 39700 TERMINALS '20' AND '35'
- IS IT 5 OHMS OR LESS?

YES: • REPAIR SHORT TO GROUND ON CKT 851

NO: (11)
- INSPECT ALL WIRING AND CONNECTORS FOR CONDITIONS WHICH COULD CAUSE INTERMITTENTS. SUCH AS CHAFING, CORROSION, OR POOR TERMINAL CONTACT
- IS ALL OK?

YES: • REPAIR SHORT TO GROUND ON CKT 827

NO: • REPAIR WIRING, CONNECTORS OR TERMINALS AS NEEDED

YES: (12)
- REMOVE J 39700
- RECONNECT EBTCM AND ABP VALVE CONNECTORS
- INSTALL TECH 1 AND BRAKE CARTRIDGE
- IGNITION 'ON'
- PERFORM TECH 1 RF ABS VALVE TEST
- DO BOTH TEST MODES OPERATE?

NO: • REPLACE ABP VALVE ASSEMBLY

YES: (13)
- CLEAR CODES; IGNITION 'OFF'
- DISCONNECT TECH 1, IF CONNECTED
- IGNITION 'ON'
- DRIVE VEHICLE AT LEAST 7 KPH (4 MPH) TO ACTIVATE AUTO TEST SEQUENCE.
- DOES CODE 41 RESET?

NO: • NO TROUBLE FOUND; SEE DIAGNOSTIC AIDS ON FACING PAGE

YES: • REPLACE EBTCM

WHEN ALL DIAGNOSIS AND REPAIRS ARE COMPLETED, CLEAR CODES AND VERIFY OPERATION

DIAGNOSTIC CHART DTC 44 — 1993 ALLANTE

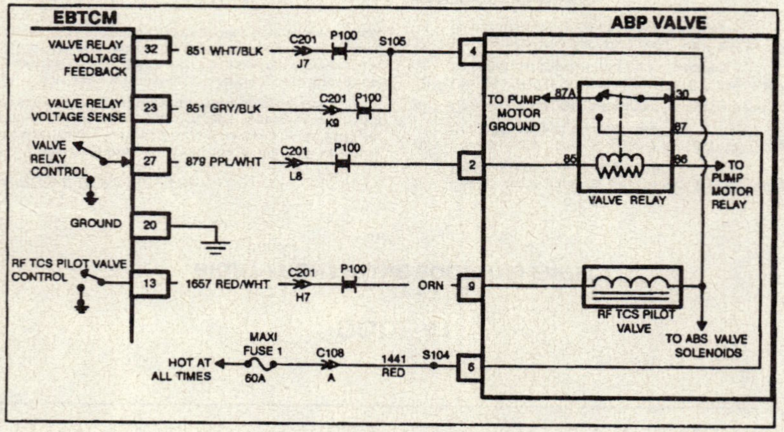

CODE 44

RIGHT FRONT TCS PILOT VALVE FAULT

CIRCUIT DESCRIPTION:

The right front TCS pilot valve in the ABP valve may be commanded to close, isolating the master cylinder.

The pilot valve circuits receive power through the valve relay on the ABP valve. The valve relay is engaged at KEY-ON and remains engaged throughout the ignition cycle. Valve circuits should have battery voltage available at all times.

When the key is turned to RUN, power is supplied to the EBTCM. The EBTCM will complete an internal self-check before providing ground to the valve relay. When the valve relay is energized, battery power is supplied to the pilot valves.

TEST CONDITION(S):

1. Tested continuously.

FAILURE CONDITION(S):

While standing or driving:

1. Valve cannot be activated.

OR

2. Valve is activated in a defective way.

ACTION TAKEN:

1. Code set.
2. ABS/TCS switched 'OFF'.
3. Red BRAKE warning lamp 'ON'.
4. ABS and TCS 'DISABLED' messages displayed.

Code 44 will set when the expected position of the right front TCS pilot valve does not match the commanded position from the EBTCM. Conditions which could cause Code 44 to set are; damage to the right front TCS pilot valve, an open circuit or a short circuit to ground or battery on CKT 1657.

TEST DESCRIPTION: Test numbers refer to circled numbers on the diagnostic chart.

1. Damage to CKT 851 may result in multiple pilot valve codes setting.
2. Checks CKT 851 for opens or high resistance from EBTCM terminal '23' to '32'.
3. Determines which branch of CKT 851 has an open or high resistance.
4. Checks the integrity of CKTs 1657 and 851, the TCS Pilot Valve circuitry internal to the ABP valve, and the TCS Pilot Valve coil.
5. Determines whether a problem found in Step 4 is due to an open in CKTs 851, 1657 or a faulty ABP valve.
6. Determines whether the problem found in Step 4 is due to an open in CKT 851 or 1657.

DIAGNOSTIC AIDS:

All tests using J 39700 Universal Breakout Box terminal '20' are using the terminal as ground. These tests, of course, assume that the ground CKT 804 terminal '20' is good.

If Codes 44 and 48 are both set, the fault is likely to be an open or short in CKT 851.

DIAGNOSTIC CHART DTC 44 — 1993 ALLANTE, CONTINUED

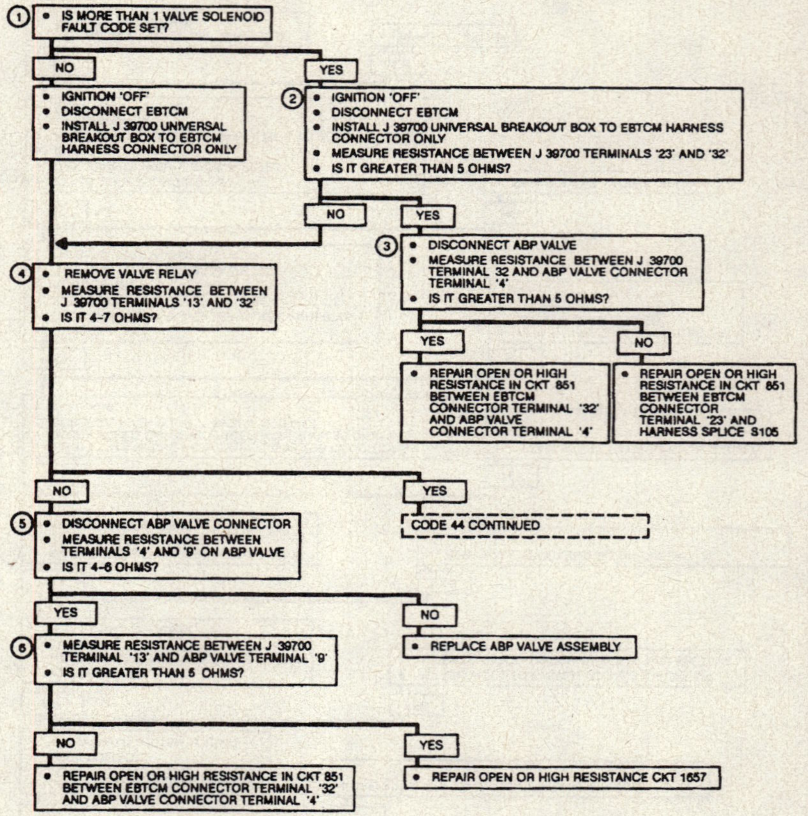

WHEN ALL DIAGNOSIS AND REPAIRS ARE COMPLETED, CLEAR CODES AND VERIFY OPERATION

DIAGNOSTIC CHART DTC 44 — 1993 ALLANTE, CONTINUED

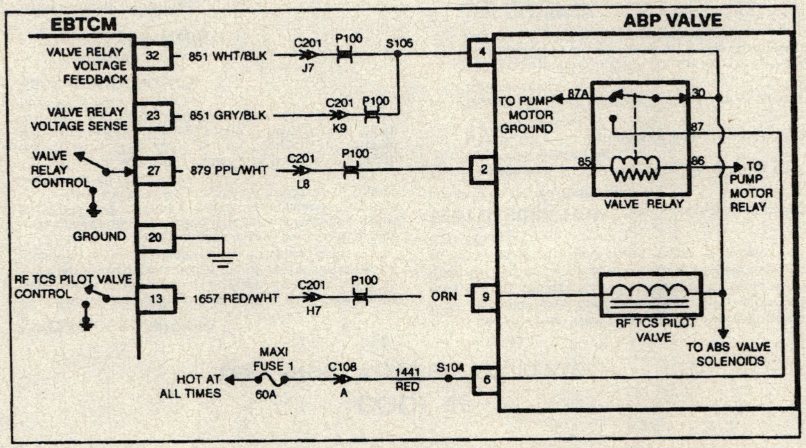

CODE 44
RIGHT FRONT TCS PILOT VALVE FAULT

CIRCUIT DESCRIPTION:

The right front TCS pilot valve in the ABP valve may be commanded to close, isolating the master cylinder.

The pilot valve circuits receive power through the valve relay on the ABP valve. The valve relay is engaged at KEY-ON and remains engaged throughout the ignition cycle. Valve circuits should have battery voltage available at all times.

When the key is turned to RUN, power is supplied to the EBTCM. The EBTCM will complete an internal self-check before providing ground to the valve relay. When the valve relay is energized, battery power is supplied to the pilot valves.

TEST DESCRIPTION: Test numbers refer to circled numbers on the diagnostic chart.

7. Checks for a short to battery in CKT 851 using EBTCM terminal '20' as ground.
8. Checks for a short to battery in CKT 1657, using EBTCM terminal '20' as ground.
9. Checks for a short to ground in CKT 851 using EBTCM terminal '20' as ground.
10. Checks for a short to ground in CKT 1657, using EBTCM terminal '20' as ground.
11. Checks for connector or wiring problems which could cause intermittents.
12. Uses the TECH 1 to exercise the TCS pilot valve and check it for proper operation.
13. Determines whether the code was set by an intermittent conditions or an EBTCM fault.

DIAGNOSTIC AIDS:

All tests using J 39700 Universal Breakout Box terminal '20' are using the terminal as ground. These tests, of course, assume that the ground CKT 804 terminal '20' is good.

If Codes 44 and 48 are both set, the fault is likely to be an open or short in CKT 851.

DIAGNOSTIC CHART DTC 44 — 1993 ALLANTE, CONTINUED

CODE 44 CONTINUED

(7)
- CONNECT J 39700TO EBTCM; LEAVE CONNECTED TO HARNESS
- DISCONNECT ABP VALVE CONNECTOR
- CONNECT DVM BETWEEN J 39700 TERMINALS '20' AND '32' (USE DC VOLTS)
- IGNITION 'ON'
- NOTE VOLTAGE
- IS IT AROUND SYSTEM VOLTAGE?

YES:
- REPAIR SHORT TO SYSTEM VOLTAGE CKT 851

NO:

(8)
- CONNECT DVM BETWEEN J 39700 TERMINALS '13' AND '20'
- NOTE VOLTAGE
- IS IT AROUND SYSTEM VOLTAGE?

YES:
- REPAIR SHORT TO SYSTEM VOLTAGE ON CKT 1657

NO:

(9)
- IGNITION 'OFF'
- MEASURE RESISTANCE BETWEEN J 39700 TERMINALS '20' AND '32'
- IS IT 5 OHMS OR LESS?

YES:
- REPAIR SHORT TO GROUND ON CKT 851

NO:

(10)
- MEASURE RESISTANCE BETWEEN J 39700 TERMINALS '13' AND '20'
- IS IT 5 OHMS OR LESS?

YES:
- REPAIR SHORT TO GROUND ON CKT 1657

NO:

(11)
- INSPECT ALL WIRING AND CONNECTORS FOR CONDITIONS WHICH COULD CAUSE INTERMITTENTS. SUCH AS CHAFING, CORROSION, OR POOR TERMINAL CONTACT
- IS ALL OK?

NO:
- REPAIR WIRING, CONNECTORS OR TERMINALS AS NEEDED

YES:

(12)
- REMOVE J 39700
- RECONNECT EBTCM AND ABP VALVE CONNECTORS
- INSTALL TECH 1 AND BRAKE CARTRIDGE
- IGNITION 'ON'
- PERFORM TECH 1 RF TCS VALVE TEST
- DOES TEST MODE OPERATE?

NO:
- REPLACE ABP VALVE ASSEMBLY

YES:

(13)
- CLEAR CODES, IGNITION 'OFF'
- DISCONNECT TECH 1, IF CONNECTED
- IGNITION 'ON'
- DRIVE VEHICLE AT LEAST 7 KPH (4 MPH) TO ACTIVATE AUTO TEST SEQUENCE
- DOES CODE 44 RESET?

NO:
- NO TROUBLE FOUND; SEE DIAGNOSTIC AIDS ON FACING PAGE

YES:
- REPLACE EBTCM

WHEN ALL DIAGNOSIS AND REPAIRS ARE COMPLETED, CLEAR CODES AND VERIFY OPERATION

DIAGNOSTIC CHART DTC 45 — 1993 ALLANTE

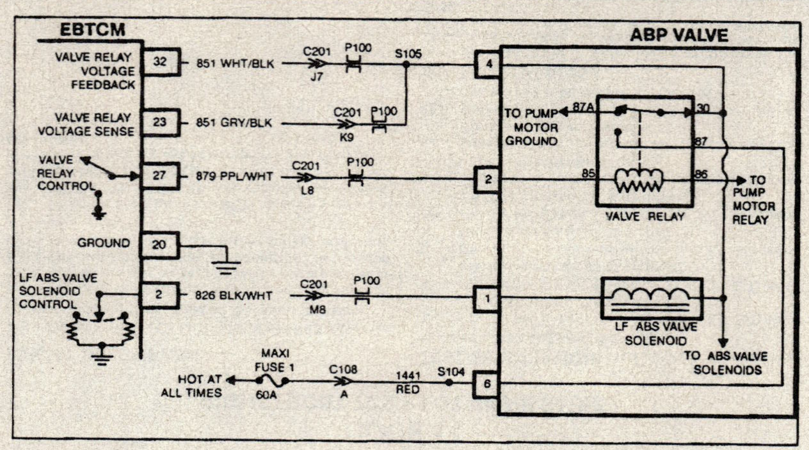

CODE 45
LEFT FRONT VALVE SOLENOID FAULT

CIRCUIT DESCRIPTION:

The left front valve solenoid in the ABP valve may be commanded to three different positions by the EBTCM. Valve position is determined by the amount of current which is allowed to flow through the valve solenoid coil.

The valve solenoid circuits receive power through the valve relay on the ABP valve. The valve relay is engaged at KEY-ON and remains engaged throughout the ignition cycle. Valve solenoid circuits should have battery voltage available at all times.

When the key is turned to RUN, power is supplied to the EBTCM. The EBTCM will complete an internal self-check before providing ground to the valve relay. When the valve relay is energized, battery power is supplied to the valve solenoids.

TEST CONDITION(S):

1. Tested continuously.

FAILURE CONDITION(S):

While standing or driving:

1. Valve solenoid cannot be activated.

OR

2. Valve solenoid is activated in a defective way.

ACTION TAKEN:

1. Code set.
2. ABS/TCS switched 'OFF'.
3. Red BRAKE warning lamp 'ON'.
4. ABS and TCS 'DISABLED' messages displayed.

Code 45 will set when the expected position of the left front solenoid valve does not match the commanded position from the EBTCM. Conditions which could cause Code 45 to set are: damage to the left front valve solenoid; an open circuit; or a short circuit to ground or battery on CKT 826.

TEST DESCRIPTION: Test numbers refer to circled numbers on the diagnostic chart.

1. Damage to CKT 851 may result in multiple valve solenoid codes setting.
2. Checks CKT 851 for opens or high resistance from EBTCM terminal '23' to '32'.
3. Determines which branch of CKT 851 has an open or high resistance.
4. Checks the integrity of CKTs 826 and 851, the LF ABS Valve Solenoid circuitry internal to the ABP valve, and the LF ABS Valve Solenoid coil.
5. Determines whether a problem found in Step 4 is due to an open in CKT 826, 851 or a faulty ABP valve.
6. Determines whether the problem found in Step 4 is due to an open in CKT 826 or 851.

DIAGNOSTIC AIDS:

All tests using J 39700 Universal Breakout Box terminal '20' are using the terminal as ground. These tests, of course, assume that the ground CKT 804 terminal '20' is good.

If Codes 41 and 45 are both set, the fault is likely to be an open or high resistance on CKT 851.

DIAGNOSTIC CHART DTC 45 — 1993 ALLANTE, CONTINUED

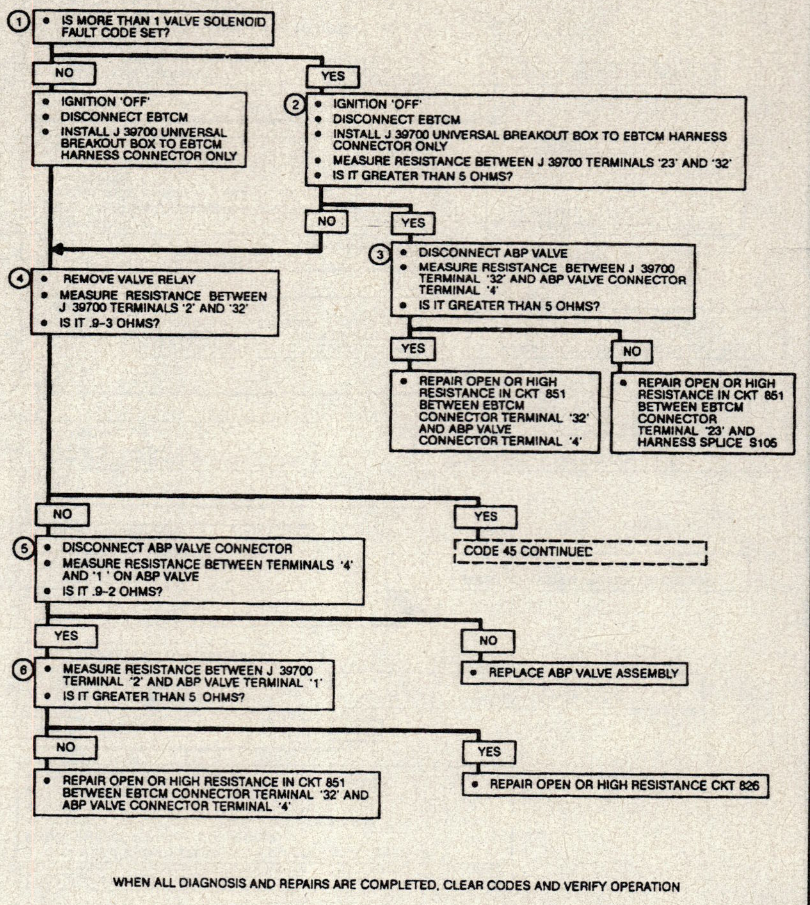

DIAGNOSTIC CHART DTC 45 — 1993 ALLANTE, CONTINUED

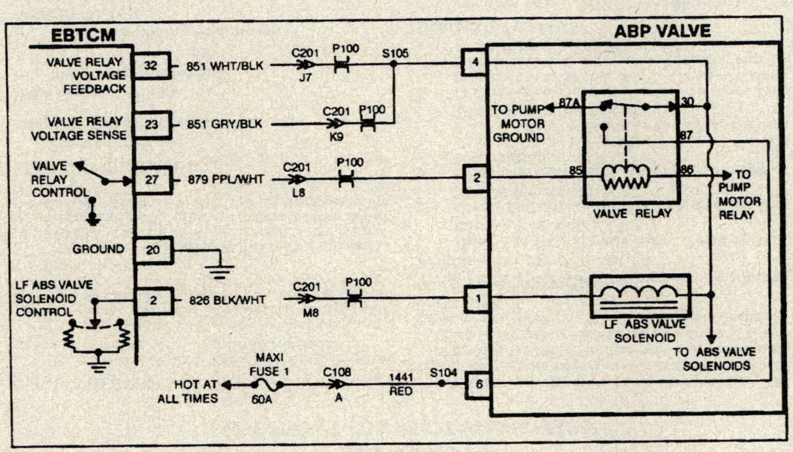

CODE 45

LEFT FRONT VALVE SOLENOID FAULT

CIRCUIT DESCRIPTION:

The left front valve solenoid in the ABP valve may be commanded to three different positions by the EBTCM. Valve position is determined by the amount of current which is allowed to flow through the solenoid valve coil.

The valve solenoid circuits receive power through the valve relay on the ABP valve. The valve relay is engaged at KEY-ON and remains engaged throughout the ignition cycle. Valve solenoid circuits should have battery voltage available at all times.

When the key is turned to RUN, power is supplied to the EBTCM. The EBTCM will complete an internal self-check before providing ground to the valve relay. When the valve relay is energized, battery power is supplied to the valve solenoids.

TEST DESCRIPTION: Test numbers refer to circled numbers on the diagnostic chart.

7. Checks for a short to battery in CKT 851 using EBTCM terminal '20' as ground.
8. Checks for a short to battery in CKT 826, using EBTCM terminal '20' as ground.
9. Checks for a short to ground in CKT 851 using EBTCM terminal '20' as ground.
10. Checks for a short to ground in CKT 826, using EBTCM terminal '20' as ground.
11. Checks wiring and connectors for intermittents.
12. Uses the TECH 1 to exercise the LF ABS valve solenoid and check it for proper operation.
13. Determines whether the code was set by an intermittent condition or an EBTCM fault.

DIAGNOSTIC AIDS:

All tests using J 39700 Universal Breakout Box terminal '20' are using the terminal as ground. These tests, of course, assume that the ground CKT 804 terminal '20' is good.

If Codes 41 and 45 are both set, the fault is likely to be an open or high resistance on CKT 851.

DIAGNOSTIC CHART DTC 45 — 1993 ALLANTE, CONTINUED

CODE 45 CONTINUED

(7)
- CONNECT J 39700TO EBTCM; LEAVE CONNECTED TO HARNESS
- DISCONNECT ABP VALVE CONNECTOR
- CONNECT DVM BETWEEN J 39700 TERMINALS '20' AND '32' (USE DC VOLTS)
- IGNITION 'ON'
- NOTE VOLTAGE
- IS IT AROUND SYSTEM VOLTAGE?

YES → • REPAIR SHORT TO SYSTEM VOLTAGE CKT 851

NO ↓

(8)
- CONNECT DVM BETWEEN J 39700 TERMINALS '2' AND '20'
- NOTE VOLTAGE
- IS IT AROUND SYSTEM VOLTAGE?

YES → • REPAIR SHORT TO SYSTEM VOLTAGE ON CKT 826

NO ↓

(9)
- IGNITION 'OFF'
- MEASURE RESISTANCE BETWEEN J 39700 TERMINALS '20' AND '32'
- IS IT 5 OHMS OR LESS?

YES → • REPAIR SHORT TO GROUND ON CKT 851

NO ↓

(10)
- MEASURE RESISTANCE BETWEEN J 39700 TERMINALS '2' AND '20'
- IS IT 5 OHMS OR LESS?

YES → • REPAIR SHORT TO GROUND ON CKT 826

NO ↓

(11)
- INSPECT ALL WIRING AND CONNECTORS FOR CONDITIONS WHICH COULD CAUSE INTERMITTENTS. SUCH AS CHAFING, CORROSION, OR POOR TERMINAL CONTACT
- IS ALL OK?

NO → • REPAIR WIRING, CONNECTORS OR TERMINALS AS NEEDED

YES ↓

(12)
- REMOVE J 39700
- RECONNECT EBTCM AND ABP VALVE CONNECTORS
- INSTALL TECH 1 AND BRAKE CARTRIDGE
- IGNITION 'ON'
- PERFORM TECH 1 LF ABS VALVE TEST
- DOES TEST MODE OPERATE?

NO → • REPLACE ABP VALVE ASSEMBLY

YES ↓

(13)
- CLEAR CODES, IGNITION 'OFF'
- DISCONNECT TECH 1, IF CONNECTED
- IGNITION 'ON'
- DRIVE VEHICLE AT LEAST 7 KPH (4 MPH) TO ACTIVATE AUTO TEST SEQUENCE.
- DOES CODE 45 RESET?

NO → • NO TROUBLE FOUND; SEE 'DIAGNOSTIC AIDS' ON FACING PAGE

YES → • REPLACE EBTCM

WHEN ALL DIAGNOSIS AND REPAIRS ARE COMPLETED, CLEAR CODES AND VERIFY OPERATION

DIAGNOSTIC CHART DTC 48 — 1993 ALLANTE

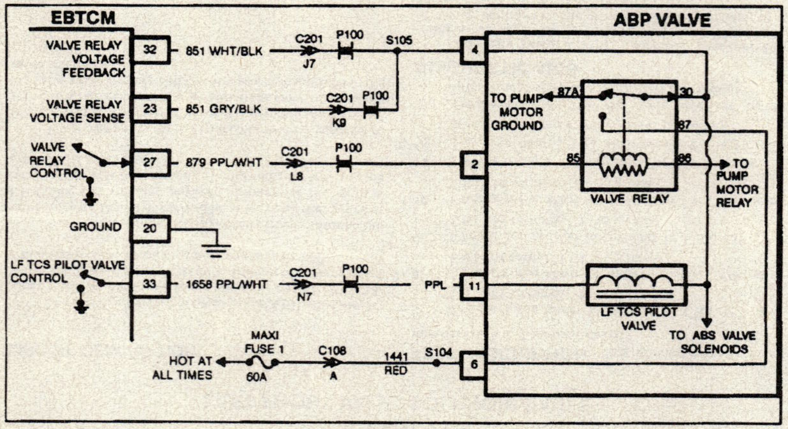

CODE 48

LEFT FRONT TCS PILOT VALVE FAULT

CIRCUIT DESCRIPTION:

The left front TCS pilot valve in the ABP valve maybe commanded to close, isolating the master cylinder.

The pilot valve circuits receive power through the valve relay on the ABP valve. The valve relay is engaged at KEY-ON and remains engaged throughout the ignition cycle. Pilot valve circuits should have battery voltage available at all times.

When the key is turned to RUN, power is supplied to the EBTCM. The EBTCM will complete an internal self-check before providing ground to the valve relay. When the valve relay is energized, battery power is supplied to the pilot valves.

TEST CONDITION(S):

1. Tested continuously.

FAILURE CONDITION(S):

While standing or driving:

1. Valve cannot be activated.

OR

2. Valve is activated in a defective way.

ACTION TAKEN:

1. Code set.
2. ABS/TCS switched 'OFF'.
3. Red BRAKE warning lamp 'ON'.
4. ABS and TCS 'DISABLED' messages displayed.

Code 48 will set when the expected position of the right front TCS pilot valve does not match the commanded position from the EBTCM. Conditions which could cause Code 48 to set are: damage to the right front TCS pilot valve; an open circuit; or a short circuit to ground or battery on CKT 1658.

TEST DESCRIPTION: Test numbers refer to circled numbers on the diagnostic chart.

1. Damage to CKT 851 may result in multiple valve solenoid codes setting.
2. Checks CKT 851 for opens or high resistance from EBTCM terminal '23' to '32'.
3. Determines which branch of CKT 851 has an open or high resistance.
4. Checks the integrity of CKTs 1658 and 851, the TCS Pilot Valve circuitry internal to the ABP valve, and the TCS Pilot Valve Solenoid coil.
5. Determines whether a problem found in Step 4 is due to an open in CKTs 851, 1658 or a faulty ABP valve.
6. Determines whether the problem found in Step 4 is due to an open in CKT 851 or 1658.

DIAGNOSTIC AIDS:

All tests using J 39700 Universal Breakout Box terminal '20' are using the terminal as ground. These tests, of course, assume that the ground CKT 804 terminal '20' is good.

If Codes 44 and 48 are both set, the fault is likely to be an open or short in CKT 851.

DIAGNOSTIC CHART DTC 48 — 1993 ALLANTE, CONTINUED

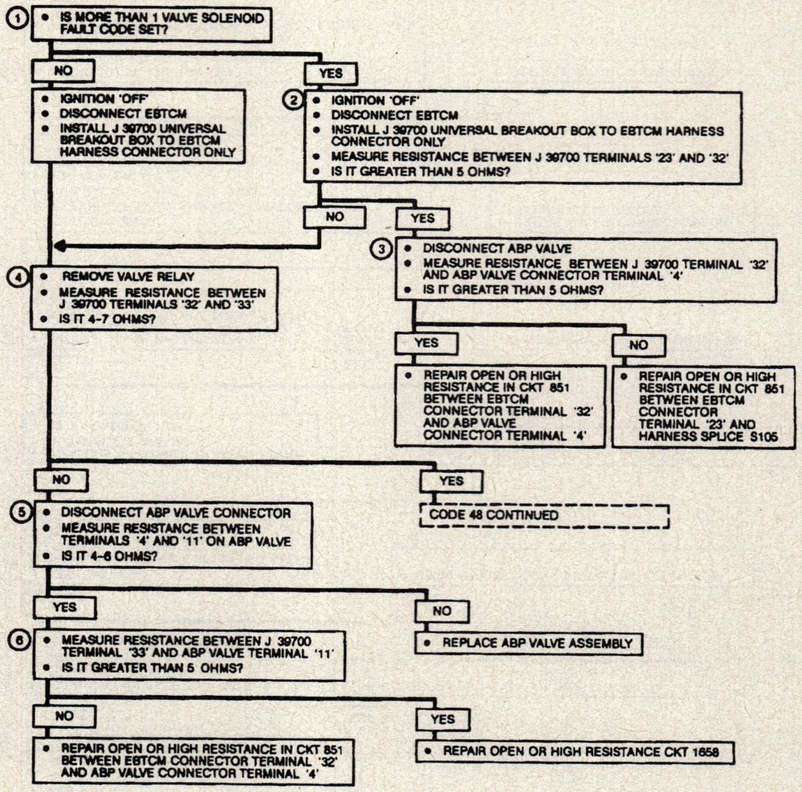

WHEN ALL DIAGNOSIS AND REPAIRS ARE COMPLETED, CLEAR CODES AND VERIFY OPERATION

DIAGNOSTIC CHART DTC 48 — 1993 ALLANTE, CONTINUED

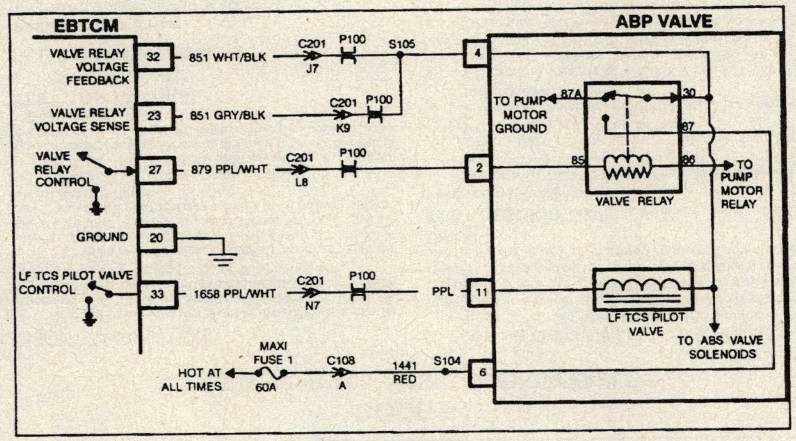

CODE 48
LEFT FRONT TCS PILOT VALVE FAULT

CIRCUIT DESCRIPTION:

The left front TCS pilot valve in the ABP valve may be commanded to close, isolating the master cylinder.

The pilot valve circuits receive power through the valve relay on the ABP valve. The valve relay is engaged at KEY-ON and remains engaged throughout the ignition cycle. Pilot valve circuits should have battery voltage available at all times.

When the key is turned to RUN, power is supplied to the EBTCM. The EBTCM will complete an internal self-check before providing ground to the valve relay. When the valve relay is energized, battery power is supplied to the pilot valves.

TEST DESCRIPTION: Test numbers refer to circled numbers on the diagnostic chart.

7. Checks for a short to battery in CKT 851 using EBTCM terminal '20' as ground.
8. Checks for a short to battery in CKT 1658, using EBTCM terminal '20' as ground.
9. Checks for a short to ground in CKT 851 using EBTCM terminal '20' as ground.
10. Checks for a short to ground in CKT 1658, using EBTCM terminal '20' as ground.
11. Checks for connector or wiring problems which could cause intermittents.
12. Uses the TECH 1 to exercise the TCS pilot valve solenoid and check it for proper operation.
13. Determines whether the code was set by an intermittent conditions or an EBTCM fault.

DIAGNOSTIC AIDS:

All tests using J 39700 Universal Breakout Box terminal '20' are using the terminal as ground. These tests, of course, assume that the ground CKT 804 terminal '20' is good.

If Codes 44 and 48 are both set, the fault is likely to be an open or short in CKT 851.

DIAGNOSTIC CHART DTC 48 — 1993 ALLANTE, CONTINUED

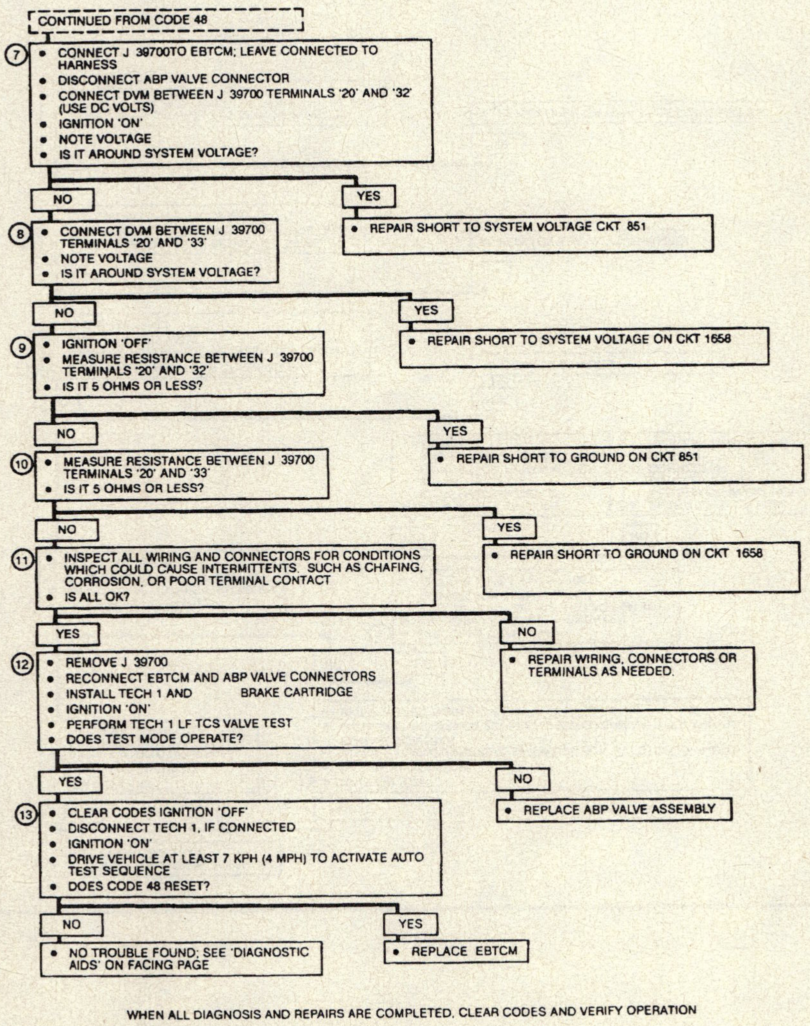

DIAGNOSTIC CHART DTC 51 — 1993 ALLANTE

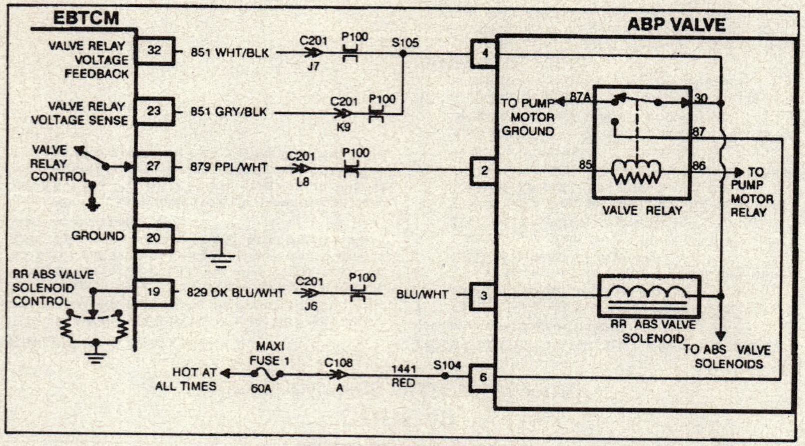

CODE 51
RIGHT REAR ABS VALVE SOLENOID FAULT

CIRCUIT DESCRIPTION:

The right rear valve solenoid in the ABP valve may be commanded to three different positions by the EBTCM. Valve position is determined by the amount of current which is allowed to flow through the valve solenoid coil.

The valve solenoid circuits receive power through the valve relay on the ABP valve. The valve relay is engaged at KEY-ON and remains engaged throughout the ignition cycle. Valve solenoid circuits should have battery voltage available at all times.

When the key is turned to RUN, power is supplied to the EBTCM. The EBTCM will complete an internal self-check before providing ground to the valve relay. When the valve relay is energized, battery power is supplied to the valve solenoids.

TEST CONDITION(S):

1. Tested continuously.

FAILURE CONDITION(S):

While standing or driving:

1. Valve solenoid cannot be activated.

OR

2. Valve solenoid is activated in a defective way.

ACTION TAKEN:

1. Code set.
2. ABS/TCS switched 'OFF'.
3. Red BRAKE warning lamp 'ON'.
4. ABS and TCS 'DISABLED' messages displayed.

Code 51 will set when the expected position of the right rear valve solenoid does not match the commanded position from the EBTCM. Conditions which could cause Code 51 to set are: damage to the right rear valve solenoid; and open circuit; or a short circuit to ground or battery on CKT 829.

TEST DESCRIPTION: Test numbers refer to circled numbers on the diagnostic chart.

1. Damage to CKT 851 may result in multiple valve solenoid codes setting.
2. Checks CKT 851 for opens or high resistance from EBTCM terminal '23' to '32'.
3. Determines which branch of CKT 851 has an open or high resistance.
4. Checks the integrity of CKTs 829 and 851, the RR ABS Valve Solenoid circuitry internal to the ABP valve, and the RR ABS Valve Solenoid coil.
5. Determines whether a problem found in Step 4 is due to an open in CKT 829, 851, or a faulty ABP valve.
6. Determines whether the problem found in Step 4 is due to an open in CKT 829 or 851.

DIAGNOSTIC AIDS:

All tests using J 39700 Universal Breakout Box terminal '20' are using the terminal as ground. These tests, of course, assume that the ground CKT 804 terminal '20' is good.

DIAGNOSTIC CHART DTC 51 — 1993 ALLANTE, CONTINUED

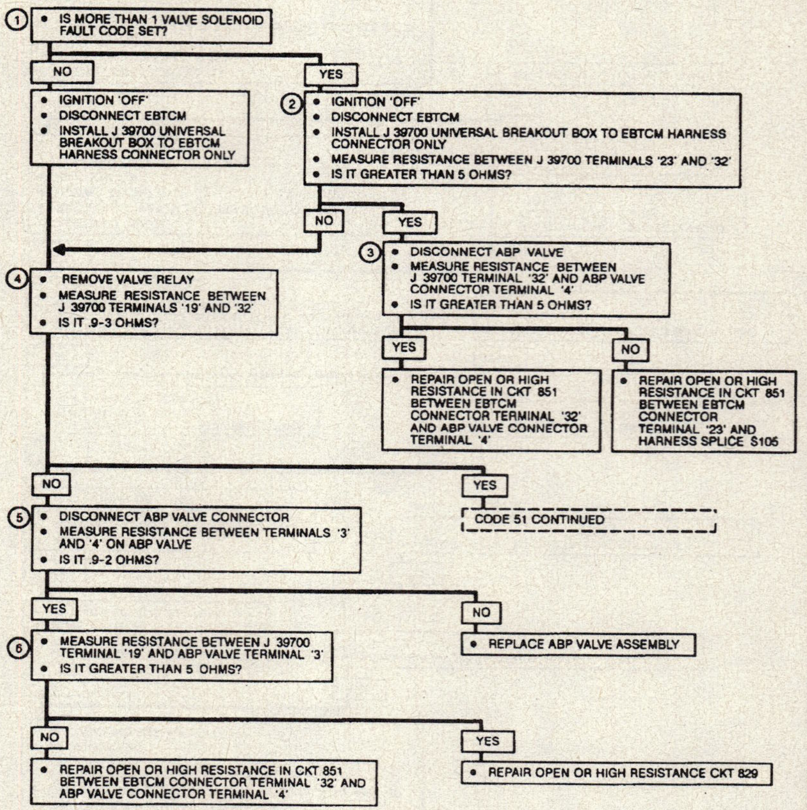

WHEN ALL DIAGNOSIS AND REPAIRS ARE COMPLETED, CLEAR CODES AND VERIFY OPERATION

DIAGNOSTIC CHART DTC 51 — 1993 ALLANTE, CONTINUED

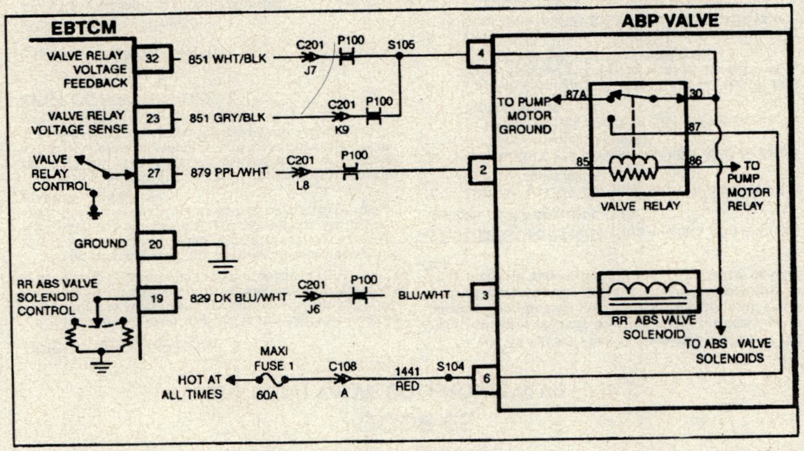

CODE 51

RIGHT REAR ABS VALVE SOLENOID FAULT

CIRCUIT DESCRIPTION:

The right rear valve solenoid in the ABP valve may be commanded to three different positions by the EBTCM. Valve position is determined by the amount of current which is allowed to flow through the valve solenoid coil.

The valve solenoid circuits receive power through the valve relay on the ABP valve. The valve relay is engaged at KEY-ON and remains engaged throughout the ignition cycle. Valve solenoid circuits should have battery voltage available at all times.

When the key is turned to RUN, power is supplied to the EBTCM. The EBTCM will complete an internal self-check before providing ground to the valve relay. When the valve relay is energized, battery power is supplied to the valve solenoids.

TEST DESCRIPTION: Test numbers refer to circled numbers on the diagnostic chart.

7. Checks for a short to battery in CKT 851 using EBTCM terminal '20' as ground.
8. Checks for a short to battery in CKT 829, using EBTCM terminal '20' as ground.
9. Checks for a short to ground in CKT 851 using EBTCM terminal '20' as ground.
10. Checks for a short to ground in CKT 829, using EBTCM terminal '20' as ground.
11. Checks wiring and connectors for intermittents.
12. Uses the TECH 1 to exercise the RR ABS valve solenoid and check it for proper operation.
13. Determines whether the code was set by an intermittent conditions or an EBTCM fault.

DIAGNOSTIC AIDS:

All tests using J 39700 Universal Breakout Box terminal '20' are using the terminal as ground. These tests, of course, assume that the ground CKT 804 terminal '20' is good.

DIAGNOSTIC CHART DTC 51 — 1993 ALLANTE, CONTINUED

CODE 51 CONTINUED

(7)
- CONNECT J 39700TO EBTCM; LEAVE CONNECTED TO HARNESS
- DISCONNECT ABP VALVE CONNECTOR
- CONNECT DVM BETWEEN J 39700 TERMINALS '20' AND '32' (USE DC VOLTS)
- IGNITION 'ON'
- NOTE VOLTAGE
- IS IT AROUND SYSTEM VOLTAGE?

YES → REPAIR SHORT TO SYSTEM VOLTAGE CKT 851

NO ↓

(8)
- CONNECT DVM BETWEEN J 39700 TERMINALS '19' AND '20'
- NOTE VOLTAGE
- IS IT AROUND SYSTEM VOLTAGE?

YES → REPAIR SHORT TO SYSTEM VOLTAGE ON CKT 829

NO ↓

(9)
- IGNITION 'OFF'
- MEASURE RESISTANCE BETWEEN J 39700 TERMINALS '20' AND '32'
- IS IT 5 OHMS OR LESS?

YES → REPAIR SHORT TO GROUND ON CKT 851

NO ↓

(10)
- MEASURE RESISTANCE BETWEEN J 39700 TERMINALS '19' AND '20'
- IS IT 5 OHMS OR LESS?

YES → REPAIR SHORT TO GROUND ON CKT 829

NO ↓

(11)
- INSPECT ALL WIRING AND CONNECTORS FOR CONDITIONS WHICH COULD CAUSE INTERMITTENTS. SUCH AS CHAFING, CORROSION, OR POOR TERMINAL CONTACT
- IS ALL OK?

NO → REPAIR WIRING, CONNECTORS OR TERMINALS AS NEEDED

YES ↓

(12)
- REMOVE J 39700
- RECONNECT EBTCM AND ABP VALVE CONNECTORS
- INSTALL TECH 1 AND BRAKE CARTRIDGE
- IGNITION 'ON'
- PERFORM TECH 1 RR ABS VALVE TEST
- DO BOTH TEST MODES OPERATE?

NO → REPLACE ABP VALVE ASSEMBLY

YES ↓

(13)
- CLEAR CODES; IGNITION 'OFF'
- DISCONNECT TECH 1, IF CONNECTED
- IGNITION 'ON'
- DRIVE VEHICLE AT LEAST 7 KPH (4 MPH) TO ACTIVATE AUTO TEST SEQUENCE
- DOES CODE 51 RESET?

NO → NO TROUBLE FOUND; SEE DIAGNOSTIC AIDS ON FACING PAGE

YES → REPLACE EBTCM

WHEN ALL DIAGNOSIS AND REPAIRS ARE COMPLETED, CLEAR CODES AND VERIFY OPERATION

DIAGNOSTIC CHART DTC 55 — 1993 ALLANTE

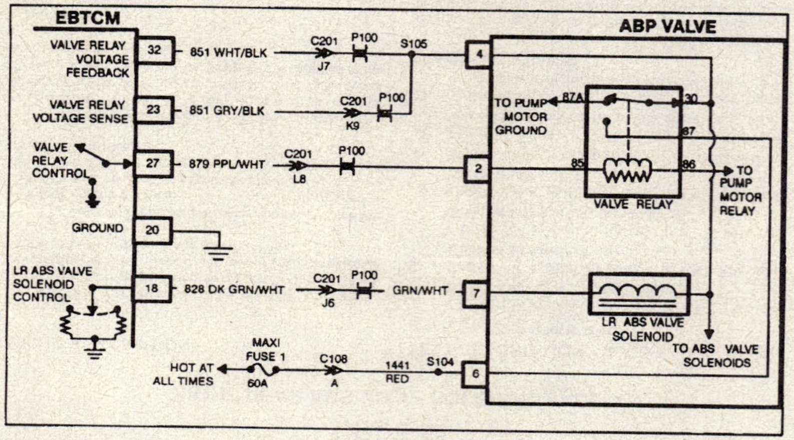

CODE 55

LEFT REAR VALVE SOLENOID VALVE FAULT

CIRCUIT DESCRIPTION:

The left rear valve solenoid in the ABP valve may be commanded to three different positions by the EBTCM. Valve position is determined by the amount of current which is allowed to flow through the valve solenoid coil.

The valve solenoid circuits receive power through the valve relay on the ABP valve. The valve relay is engaged at KEY-ON and remains engaged throughout the ignition cycle. Valve solenoid circuits should have battery voltage available at all times.

When the key is turned to RUN, power is supplied to the EBTCM. The EBTCM will complete an internal self-check before providing ground to the valve relay. When the valve relay is energized, battery power is supplied to the valve solenoids.

TEST CONDITION(S):

1. Tested continuously.

FAILURE CONDITION(S):

While standing or driving:

1. Valve solenoid cannot be activated.

OR

2. Valve solenoid is activated in a defective way.

ACTION TAKEN:

1. Code set.
2. ABS/TCS switched 'OFF'.
3. Red BRAKE warning lamp 'ON'.
4. ABS and TCS 'DISABLED' messages displayed.

Code 55 will set when the expected position of the right rear valve solenoid does not match the commanded position from the EBTCM. Conditions which could cause Code 55 to set are: damage to the left rear valve solenoid; an open circuit; or a short circuit to ground or battery on CKT 828.

TEST DESCRIPTION: Test numbers refer to circled numbers on the diagnostic chart.

1. Damage to CKT 851 may result in multiple valve solenoid codes setting.
2. Checks CKT 851 for opens or high resistance from EBTCM terminal '23' to '32'.
3. Determines which branch of CKT 851 has an open or high resistance.
4. Checks the integrity of CKTs 828 and 851, the LR ABS Valve Solenoid circuitry internal to the ABP valve, and the LR ABS Valve Solenoid coil.
5. Determines whether a problem found in Step 4 is due to an open in CKT 828, 851, or a faulty ABP valve.
6. Determines whether the problem found in Step 4 is due to an open in CKT 828 or 851.

DIAGNOSTIC AIDS:

All tests using J 39700 Universal Breakout Box terminal '20' are using the terminal as ground. These tests, of course, assume that the ground CKT 804 terminal '20' is good.

If Codes 41 and 55 are both set, the fault is likely to be a short to B+ on CKT 828.

DIAGNOSTIC CHART DTC 55 — 1993 ALLANTE, CONTINUED

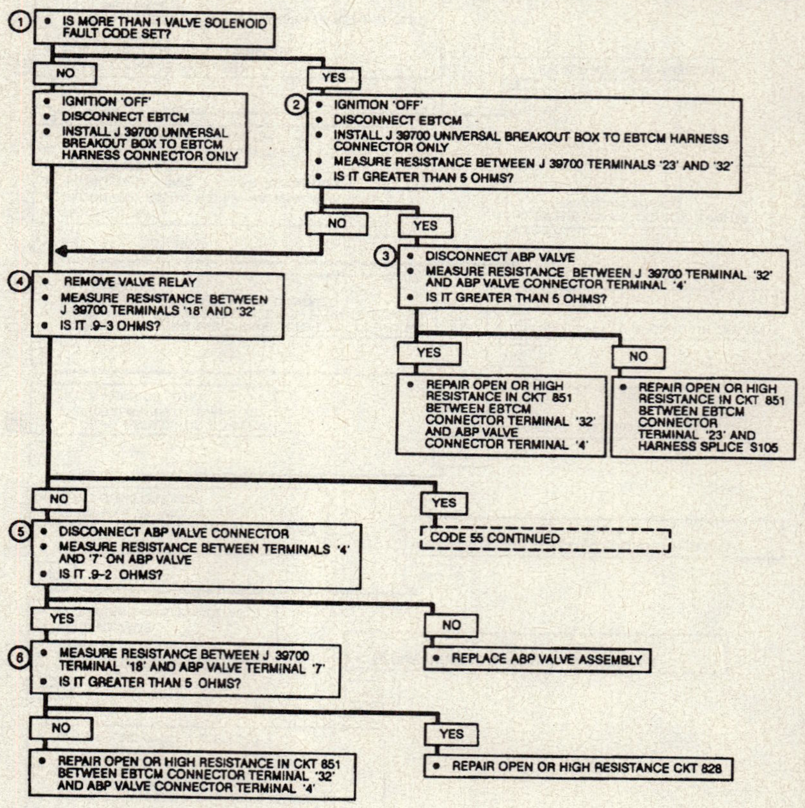

WHEN ALL DIAGNOSIS AND REPAIRS ARE COMPLETED, CLEAR CODES AND VERIFY OPERATION

DIAGNOSTIC CHART DTC 55 — 1993 ALLANTE, CONTINUED

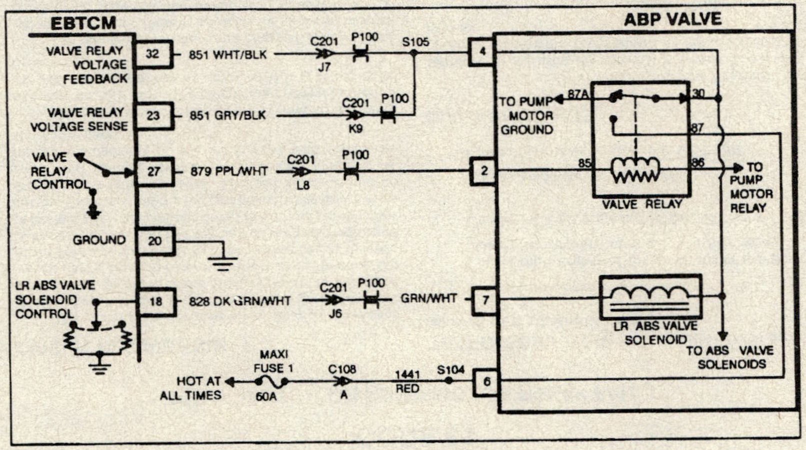

CODE 55

LEFT REAR VALVE SOLENOID VALVE FAULT

CIRCUIT DESCRIPTION:

The left rear valve solenoid in the ABP valve may be commanded to three different positions by the EBTCM. Valve position is determined by the amount of current which is allowed to flow through the valve solenoid coil.

The valve solenoid circuits receive power through the valve relay on the ABP valve. The valve relay is engaged at KEY–ON and remains engaged throughout the ignition cycle. Valve solenoid circuits should have battery voltage available at all times.

When the key is turned to RUN, power is supplied to the EBTCM. The EBTCM will complete an internal self-check before providing ground to the valve relay. When the valve relay is energized, battery power is supplied to the valve solenoids.

TEST DESCRIPTION: Test numbers refer to circled numbers on the diagnostic chart.

7. Checks for a short to battery in CKT 851 using EBTCM terminal '20' as ground.
8. Checks for a short to battery in CKT 828, using EBTCM terminal '20' as ground.
9. Checks for a short to ground in CKT 851 using EBTCM terminal '20' as ground.
10. Checks for a short to ground in CKT 828, using EBTCM terminal '20' as ground.
11. Checks wiring and connectors for intermittents.
12. Uses the TECH 1 to exercise the LR ABS valve solenoid and check it for proper operation.
13. Determines whether the code was set by an intermittent conditions or an EBTCM fault.

DIAGNOSTIC AIDS:

All tests using J 39700 Universal Breakout Box terminal '20' are using the terminal as ground. These tests, of course, assume that the ground CKT 804 terminal '20' is good.

If Codes 41 and 55 are both set, the fault is likely to be a short to B+ on CKT 828.

DIAGNOSTIC CHART DTC 55 — 1993 ALLANTE, CONTINUED

CODE 55 CONTINUED

(7)
- CONNECT J 39700TO EBTCM; LEAVE CONNECTED TO HARNESS
- DISCONNECT ABP VALVE CONNECTOR
- CONNECT DVM BETWEEN J 39700 TERMINALS '20' AND '32' (USE DC VOLTS)
- IGNITION 'ON'
- NOTE VOLTAGE
- IS IT AROUND SYSTEM VOLTAGE?

YES:
- REPAIR SHORT TO SYSTEM VOLTAGE CKT 851

NO:

(8)
- CONNECT DVM BETWEEN J 39700 TERMINALS '18' AND '20'
- NOTE VOLTAGE
- IS IT AROUND SYSTEM VOLTAGE?

YES:
- REPAIR SHORT TO SYSTEM VOLTAGE ON CKT 828

NO:

(9)
- IGNITION 'OFF'
- MEASURE RESISTANCE BETWEEN J 39700 TERMINALS '20' AND '32'
- IS IT 5 OHMS OR LESS?

YES:
- REPAIR SHORT TO GROUND ON CKT 851

NO:

(10)
- MEASURE RESISTANCE BETWEEN J 39700 TERMINALS '18' AND '20'
- IS IT 5 OHMS OR LESS?

YES:
- REPAIR SHORT TO GROUND ON CKT 828

NO:

(11)
- INSPECT ALL WIRING AND CONNECTORS FOR CONDITIONS WHICH COULD CAUSE INTERMITTENTS, SUCH AS CHAFING, CORROSION, OR POOR TERMINAL CONTACT
- IS ALL OK?

NO:
- REPAIR WIRING, CONNECTORS OR TERMINALS AS NEEDED

YES:

(12)
- REMOVE J 39700
- RECONNECT EBTCM AND ABP VALVE CONNECTORS
- INSTALL TECH 1 AND BRAKE CARTRIDGE
- IGNITION 'ON'
- PERFORM TECH 1 LR ABS VALVE TEST
- DO BOTH TEST MODES OPERATE?

NO:
- REPLACE ABP VALVE ASSEMBLY

YES:

(13)
- CLEAR CODES, IGNITION 'OFF'
- DISCONNECT TECH 1, IF CONNECTED
- IGNITION 'ON'
- DRIVE VEHICLE AT LEAST 7 KPH (4 MPH) TO ACTIVATE AUTO TEST SEQUENCE
- DOES CODE 55 RESET?

NO:
- NO TROUBLE FOUND; SEE DIAGNOSTIC AIDS ON FACING PAGE

YES:
- REPLACE EBTCM

WHEN ALL DIAGNOSIS AND REPAIRS ARE COMPLETED, CLEAR CODES AND VERIFY OPERATION

DIAGNOSTIC CHART DTC 61 — 1993 ALLANTE

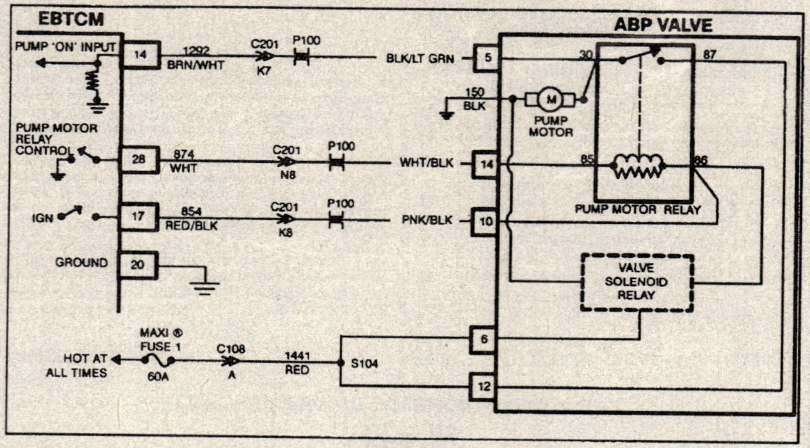

CODE 61

PUMP MOTOR OR PUMP MOTOR RELAY FAULT

CIRCUIT DESCRIPTION:

The pump motor returns brake fluid to the master cylinder brake circuit at the ABP valve during antilock braking and provides brake fluid pressure to the front wheels during traction control. During normal braking, the pump does not operate. When the vehicle begins to move after start-up, the EBTCM will turn on the pump motor and perform a self-check of the pump motor and pump motor circuit. This check may be felt and heard by the driver when the vehicle begins to move. The self-check should be considered normal operation. The pump motor is an integral component of the ABP valve and cannot be serviced separately.

The pump motor relay provides power to the pump motor in the ABP valve. The pump motor relay is located on the ABP valve and may be replaced if it is found to be defective.

The pump motor relay is not engaged during normal system operation. When antilock or traction control operation is required, the motor relay actuation line on CKT 874 (EBTCM pin '28') is pulled to ground. The pump motor relay is engaged by the relay supply voltage line on CKT 854 (EBTCM pin '17'). When the relay switches, battery power is provided to the pump motor.

TEST DESCRIPTION: Test numbers refer to circled numbers on the diagnostic chart.

4. Checks for an open in power feed CKT 1441.
5. Checks the integrity of the pump motor and pump motor relay circuitry internal to the ABP valve.
6. Checks for a good ground to the ABP valve (CKT 150).
7. Checks for an open in CKT 1292.
8. Checks for a short to ground in CKT 1292.

DIAGNOSTIC AIDS:

All tests using J 39700 Universal Breakout Box terminal '20' are using the terminal as ground. These tests, of course, assume that the ground CKT 804 at terminal '20' is good.

It is very important that a thorough inspection of the wiring and connectors be performed. Failure to carefully and fully inspect wiring and connectors may result in misdiagnosis, causing part replacement with reappearance of the fault.

DIAGNOSTIC CHART DTC 61 — 1993 ALLANTE, CONTINUED

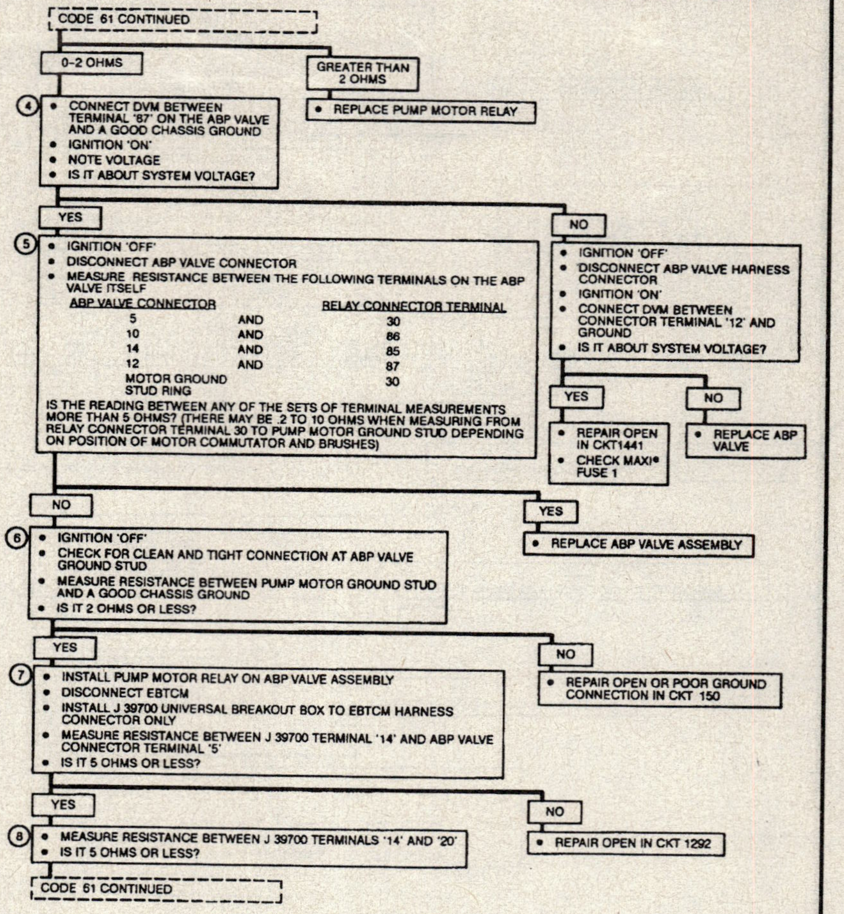

WHEN ALL DIAGNOSIS AND REPAIRS ARE COMPLETED, CLEAR CODES AND VERIFY OPERATION

DIAGNOSTIC CHART DTC 61 — 1993 ALLANTE, CONTINUED

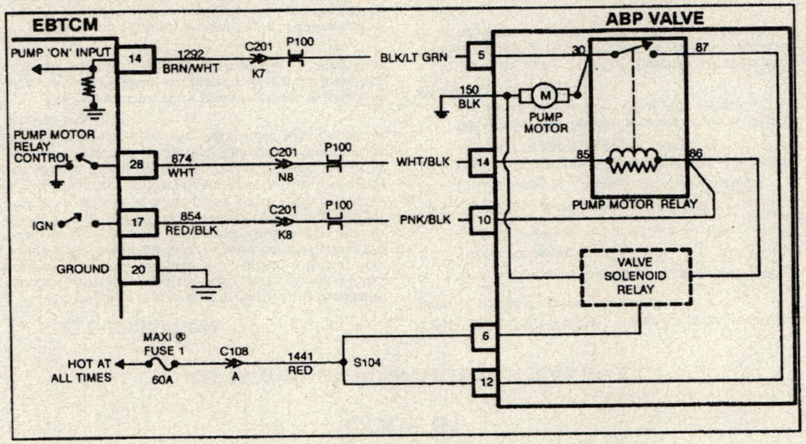

CODE 61
PUMP MOTOR OR PUMP MOTOR RELAY FAULT

CIRCUIT DESCRIPTION:

The pump motor returns brake fluid to the master cylinder brake circuit at the ABP valve during antilock braking and provides brake fluid pressure to the front wheels during traction control. During normal braking, the pump does not operate. When the vehicle begins to move after start-up, the EBTCM will turn on the pump motor and perform a self-check of the pump motor and pump motor circuit. This check may be felt and heard by the driver when the vehicle begins to move. The self-check should be considered normal operation. The pump motor is an integral component of the ABP valve and cannot be serviced separately.

The pump motor relay provides power to the pump motor in the ABP valve. The pump motor relay is located on the ABP valve and may be replaced if it is found to be defective.

The pump motor relay is not engaged during normal system operation. When antilock or traction control operation is required, the motor relay actuation line on CKT 874 (EBTCM pin '28') is pulled to ground. The pump motor relay is engaged by the relay supply voltage line on CKT 854 (EBTCM pin '17'). When the relay switches, battery power is provided to the pump motor.

TEST CONDITION(S):

1. Tested continuously.

FAILURE CONDITION(S):

1. Drive Away: Motor voltage is not detected during first drive away after engine start.
2. While Driving: Motor relay is energized and motor voltage is not detected.
3. Voltage at EBTCM pin '14' below 9 volts.
4. Pump runs continuously.

ACTION TAKEN:

1. Code set.
2. ABS/TCS switched 'OFF'.
3. Red BRAKE warning lamp 'ON'.
4. ABS and TCS 'DISABLED' messages displayed.

When the pump motor relay is engaged, the motor monitor line should be at battery voltage. With the pump motor relay disengaged, the monitoring line should be at ground. If the commanded position of the pump motor relay and motor monitor line do not agree, Code 61 will set. Some conditions which will cause Code 61 to set are:

- An open circuit on CKTs 874 or 1292
- A short to voltage on CKTs 874 or 1292
- A short to ground on CKTs 874 or 1292
- A defective pump motor relay or pump motor

TEST DESCRIPTION: Test numbers refer to circled numbers on the diagnostic chart.

1. Checks the pump motor relay coil for proper resistance.
2. Determines if the pump motor relay contacts are stuck closed.
3. Determines if the pump motor relay contacts are stuck open.

DIAGNOSTIC AIDS:

All tests using J39700 Universal Breakout Box terminal '20' are using the terminal as ground. These tests, of course, assume that the ground CKT 804 at terminal '20' is good.

It is very important that a through inspection of the wiring and connectors be performed. Failure to carefully and fully inspect wiring and connectors may result in misdiagnosis, causing part replacement with reappearance of the fault.

DIAGNOSTIC CHART DTC 61 — 1993 ALLANTE, CONTINUED

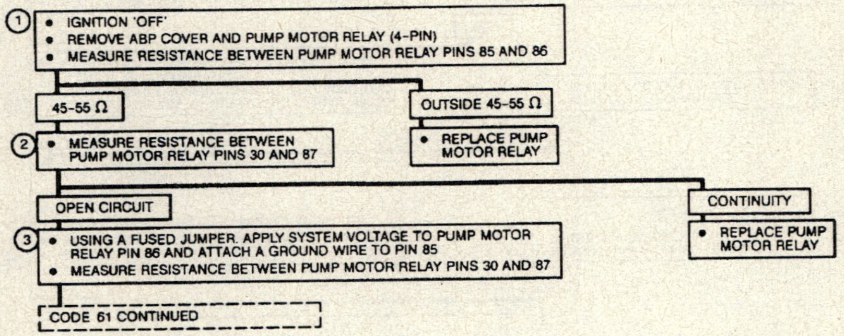

WHEN ALL DIAGNOSIS AND REPAIRS ARE COMPLETED, CLEAR CODES AND VERIFY OPERATION

DIAGNOSTIC CHART DTC 61 — 1993 ALLANTE, CONTINUED

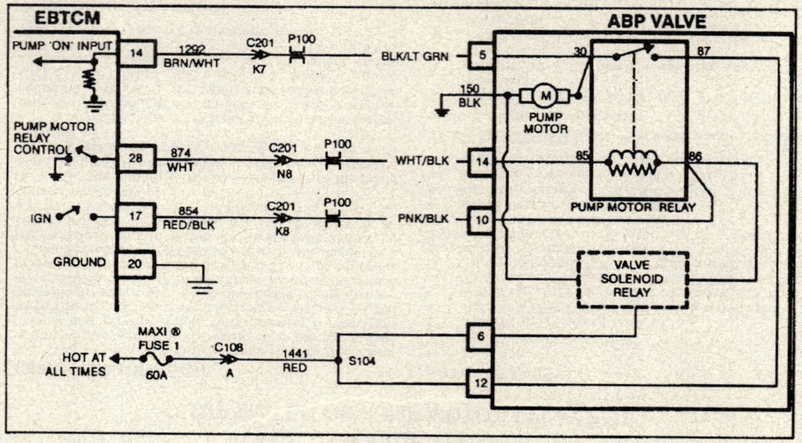

CODE 61

PUMP MOTOR OR PUMP MOTOR RELAY FAULT

CIRCUIT DESCRIPTION:

The pump motor returns brake fluid to the master cylinder brake circuit at the ABP valve during antilock braking and provides brake fluid pressure to the front wheels during traction control. During normal braking, the pump does not operate. When the vehicle begins to move after start-up, the EBTCM will turn on the pump motor and perform a self-check of the pump motor and pump motor circuit. This check may be felt and heard by the driver when the vehicle begins to move. The self-check should be considered normal operation. The pump motor is an integral component of the ABP valve and cannot be serviced separately.

The pump motor relay provides power to the pump motor in the ABP valve. The pump motor relay is located on the ABP valve and may be replaced if it is found to be defective.

The pump motor relay is not engaged during normal system operation. When antilock or traction control operation is required, the motor relay actuation line on CKT 874 (EBTCM pin '28') is pulled to ground. The pump motor relay is engaged by the relay supply voltage line on CKT 854 (EBTCM pin '17'). When the relay switches, battery power is provided to the pump motor.

TEST DESCRIPTION: Test numbers refer to circled numbers on the diagnostic chart.

9. Checks for a short to system voltage in CKT 1292.
10. Checks for an open in CKT 874.
11. Checks for a short to ground on CKT 874.
12. Checks for a short to B+ in CKT 874.
13. Checks wiring and connectors for intermittents.
14. Uses the TECH 1 to check for proper operation of the pump motor and associated circuitry.
15. Determines whether the problem found in Step 13 was due to a faulty EBTCM or ABP valve.

DIAGNOSTIC AIDS:

All tests using J 39700 Universal Breakout Box terminal '20' are using the terminal as ground. These tests, of course, assume that the ground CKT 804 at terminal '20' is good.

It is very important that a thorough inspection of the wiring and connectors be performed. Failure to carefully and fully inspect wiring and connectors may result in misdiagnosis, causing part replacement with reappearance of the fault.

DIAGNOSTIC CHART DTC 61 — 1993 ALLANTE, CONTINUED

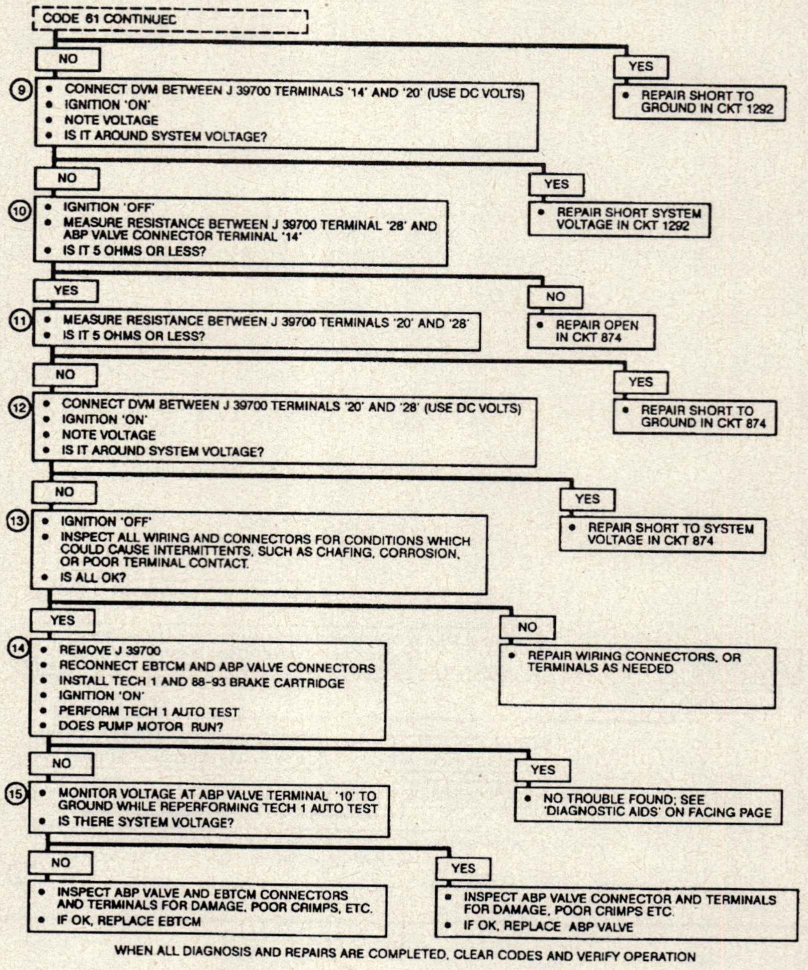

DIAGNOSTIC CHART DTC 63 — 1993 ALLANTE

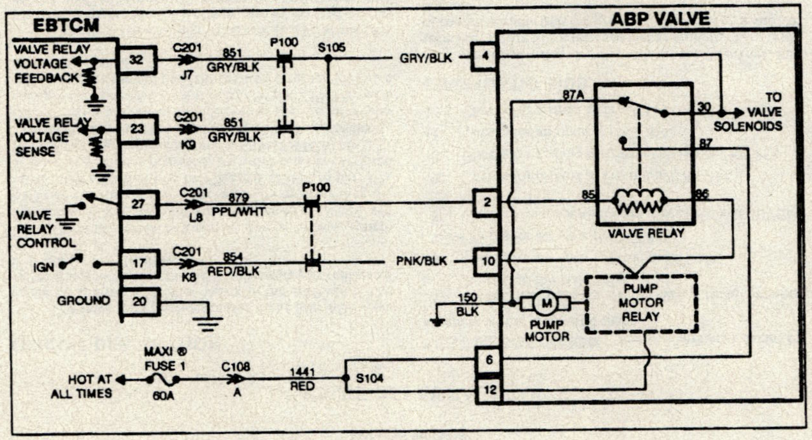

CODE 63
VALVE RELAY FAULT

CIRCUIT DESCRIPTION:

The valve relay provides power to the four ABS valve solenoids and two TCS Pilot Valves in the ABP valve. The valve relay is located on the ABP valve an may be replaced if it is found to be defective.

The valve relay is engaged during normal system operation. When the ignition is switched to RUN, the EBTCM commands the valve relay ON by grounding the valve relay actuation line on CKT 879 (EBTCM pin '27'). When this ground is provided, the valve relay is energized from the voltage supply line on CKT 854 (EBTCM pin '17'). The valve relay switches and battery voltage is provided to the four valve solenoids, the pilot valves and the valve relay voltage sense line on CKT 851 (EBTCM pin '23'). The valve relay remains engaged until the ignition is turned OFF or a failure is detected.

The valve relay voltage feedback (EBTCM pin '32') provides suppression of voltage spikes created during operation of both front ABS valve solenoids during antilock and traction control.

TEST CONDITION(S):

1. Tested continuously.

FAILURE CONDITION(S):

1. Less than 5 volts at EBTCM pin '23', valve relay sense;

OR

2. Voltage at EBTCM pin '23', valve relay sense remains high, although voltage at EBTCM pin '27', valve relay control, is low (valve relay sticks).

ACTION TAKEN:

1. Code set.
2. ABS/TCS switched 'OFF'.
3. Red BRAKE warning lamp 'ON'.
4. ABS and TCS 'DISABLED' messages displayed.

When the valve relay is engaged, the valve relay sense line should be at battery voltage. With the valve relay disengaged, the sense line should be at ground. If the commanded position of the valve relay and the position indicated by the sense line do not agree, Code 63 will set. Some conditions which will cause Code 63 to set are:

- An open circuit on CKTs 851 or 879.
- A short to voltage on CKTs 851 or 879.
- A short to ground on CKTs 851 or 879.
- A defective solenoid valve relay.
- Open circuit on CKT 854.

TEST DESCRIPTION: Test numbers refer to circled numbers on the diagnostic chart.

1. Checks for availability of battery voltage to the Valve Relay through Maxi® fuse 1 and CKT 1441.
2. Checks the Valve Relay coil for proper resistance.
3. Checks for the Valve Relay contacts being internally open when in the relay unenergized position.
4. Checks for the Valve Relay contacts being stuck open when in the relay-energized position.

DIAGNOSTIC AIDS:

All tests using J 39700 Universal Breakout Box terminal '20' are using the terminal as ground. These tests, of course, assume that the ground CKT 804 at terminal '20' is good.

It is very important that a thorough inspection of the wiring and connectors be performed. Failure to carefully and fully inspect wiring and connectors may result in misdiagnosis, causing part replacement with reappearance of the fault.

DIAGNOSTIC CHART DTC 63 — 1993 ALLANTE, CONTINUED

(1)
- IGNITION 'OFF'
- DISCONNECT ABP VALVE CONNECTOR
- CONNECT DVM BETWEEN ABP VALVE CONNECTOR TERMINAL 6 AND A GOOD CHASSIS GROUND
- NOTE VOLTAGE
- IS IT AROUND SYSTEM VOLTAGE?

NO:
- REPAIR OPEN CKT 1441
- CHECK MAX1® FUSE 1

YES:

(2)
- REMOVE VALVE RELAY (6-PIN)
- MEASURE RESISTANCE BETWEEN RELAY PINS 85 AND 86
- IS IT 52-64 OHMS?

NO:
- REPLACE VALVE SOLENOID RELAY

YES:

(3)
- MEASURE RESISTANCE BETWEEN RELAY PINS 30 AND 87A
- IS IT 2 OHMS OR LESS?

NO:
- REPLACE VALVE SOLENOID RELAY

YES:

(4)
- CONNECT RELAY PIN 85 TO GROUND
- USING A FUSED JUMPER WIRE (SUCH AS J 36169) WITH A3 AMP FUSE CONNECT RELAY PIN 86 TO SYSTEM VOLTAGE
- MEASURE RESISTANCE BETWEEN RELAY PINS 30 AND 87
- IS IT 2 OHMS OR LESS?

CODE 63 CONTINUED

WHEN ALL DIAGNOSIS AND REPAIRS ARE COMPLETED, CLEAR CODES AND VERIFY OPERATION

DIAGNOSTIC CHART DTC 63 — 1993 ALLANTE, CONTINUED

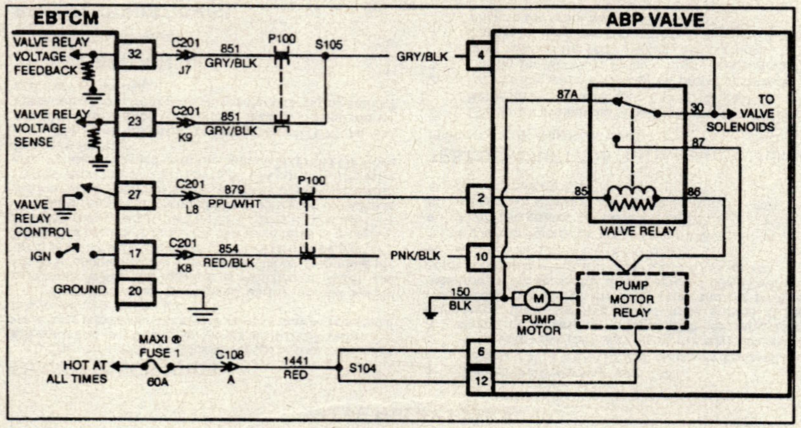

CODE 63

VALVE RELAY FAULT

CIRCUIT DESCRIPTION:

The valve relay provides power to the four ABS valve solenoids and two TCS Pilot Valves in the ABP valve. The valve relay is located on the ABP valve an may be replaced if it is found to be defective.

The valve relay is engaged during normal system operation. When the ignition is switched to RUN, the EBTCM commands the valve relay ON by grounding the valve relay actuation line on CKT 879 (EBTCM pin '27'). When this ground is provided, the valve relay is energized from the voltage supply line on CKT 854 (EBTCM pin '17'). The valve relay switches and battery voltage is provided to the four valve solenoids, the pilot valves and the valve relay voltage sense line on CKT 851 (EBTCM pin '23'). The valve relay remains engaged until the ignition is turned OFF or a failure is detected.

The valve relay voltage feedback (EBTCM pin '32') provides suppression of voltage spikes created during operation of both front ABS valve solenoids during antilock and traction control.

TEST DESCRIPTION: Test numbers refer to circled numbers on the diagnostic chart.

5. Checks the integrity of the Valve Relay circuitry internal to the ABP valve.
6. Checks for an open in ground CKT 150.
7. Checks for an open in CKT 851.
8. Checks for an open between splice S105 and EBTCM terminal '23'.
9. Checks for a short to ground in CKT 851.
10. Checks for a short to system voltage in CKT 851.
11. Checks for an open in CKT 854.
12. Checks for a short to ground in CKT 854.

DIAGNOSTIC AIDS:

All tests using J 39700 Universal Breakout Box terminal '20' are using the terminal as ground. These tests, of course, assume that the ground CKT 804 at terminal '20' is good.

It is very important that a thorough inspection of the wiring and connectors be performed. Failure to carefully and fully inspect wiring and connectors may result in misdiagnosis, causing part replacement with reappearance of the fault.

DIAGNOSTIC CHART DTC 63 — 1993 ALLANTE, CONTINUED

CODE 63 CONTINUED

YES

(5)
- MEASURE RESISTANCE BETWEEN THE FOLLOWING TERMINALS ON THE ABP VALVE ASSEMBLY ITSELF.

HYDRAULIC ABP VALVE CONNECTOR		RELAY CONNECTOR TERMINAL
2	AND	85
4	AND	30
6	AND	87
MOTOR GROUND STUD RING	AND	87A
10	AND	86

IS THE READING BETWEEN ANY OF THE SETS OF TERMINAL MEASUREMENTS MORE THAN 2 OHMS? (THERE MAY BE .2 TO 10 OHMS WHEN MEASURING FROM RELAY CONNECTOR TERMINAL '30' TO PUMP MOTOR GROUND STUD DEPENDING ON POSITION OF MOTOR COMMUTATOR AND BRUSHES)

NO → • REPLACE VALVE RELAY

NO

(6)
- DISCONNECT EBTCM
- INSTALL J 39700 UNIVERSAL BREAKOUT BOX TO EBTCM HARNESS CONNECTOR ONLY
- MEASURE RESISTANCE BETWEEN PUMP MOTOR GROUND STUD AND J 39700 TERMINAL '20'
- IS IT 2 OHMS OR LESS?

YES → • REPLACE ABP VALVE

YES

(7)
- INSTALL VALVE SOLENOID RELAY
- MEASURE RESISTANCE BETWEEN ABP VALVE CONNECTOR TERMINAL '4' AND J 39700 TERMINAL '32'
- IS IT 2 OHMS OR LESS?

NO → • REPAIR OPEN IN CKT 150 TO PUMP MOTOR GROUND STUD

YES

(8)
- MEASURE RESISTANCE BETWEEN ABP VALVE CONNECTOR TERMINAL '4' AND J 39700 TERMINAL '23'
- IS IT 2 OHMS OR LESS?

NO → • REPAIR OPEN IN CKT 851

YES

(9)
- MEASURE RESISTANCE BETWEEN J 39700 TERMINALS '20' AND '32'
- IS IT 5 OHMS OR LESS?

NO → • REPAIR OPEN IN CKT 851 BETWEEN TERMINAL '23' SPLICE S105 AND THE EBTCM

NO

(10)
- CONNECT DVM BETWEEN J 39700 TERMINALS '20' AND '32' (USE DC VOLTS)
- IGNITION 'ON'
- NOTE VOLTAGE
- IS IT AROUND SYSTEM VOLTAGE?

YES → • REPAIR SHORT GROUND IN CKT 851

NO

(11)
- IGNITION 'OFF'
- MEASURE RESISTANCE BETWEEN ABP VALVE CONNECTOR TERMINAL '10' AND J 39700 TERMINAL '17'
- IS IT 5 OHMS OR LESS?

YES → • REPAIR SHORT TO SYSTEM VOLTAGE ON CKT 851

YES

(12)
- MEASURE RESISTANCE BETWEEN J 39700 TERMINALS '17' AND '20'
- IS IT 5 OHMS OR LESS?

NO → • REPAIR OPEN IN CKT 854

YES → CODE 63 CONTINUED

WHEN ALL DIAGNOSIS AND REPAIRS ARE COMPLETED, CLEAR CODES AND VERIFY OPERATION

DIAGNOSTIC CHART DTC 63 — 1993 ALLANTE, CONTINUED

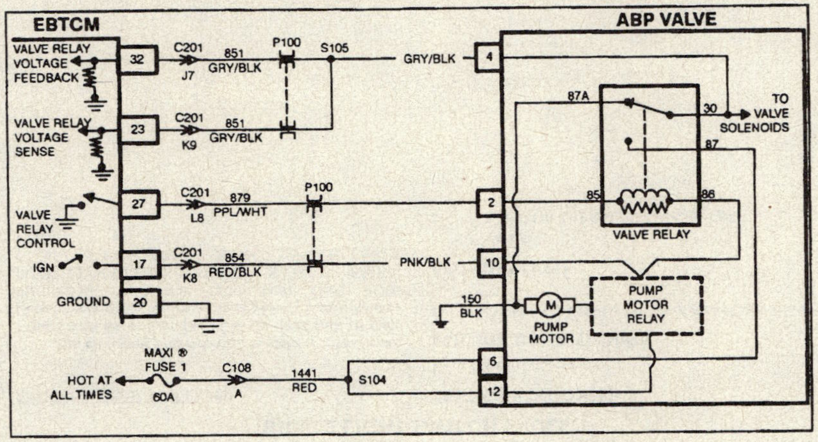

CODE 63

VALVE RELAY FAULT

CIRCUIT DESCRIPTION:

The valve relay provides power to the four ABS valve solenoids and two TCS Pilot Valves in the ABP valve. The valve relay is located on the ABP valve an may be replaced if it is found to be defective.

The valve relay is engaged during normal system operation. When the ignition is switched to RUN, the EBTCM commands the valve relay ON by grounding the valve relay actuation line on CKT 879 (EBTCM pin '27'). When this ground is provided, the valve relay is energized from the voltage supply line on CKT 854 (EBTCM pin '17'). The valve relay switches and battery voltage is provided to the four valve solenoids, the pilot valves and the valve relay voltage sense line on CKT 851 (EBTCM pin '23'). The valve relay remains engaged until the ignition is turned OFF or a failure is detected.

The valve relay voltage feedback (EBTCM pin '32') provides suppression of voltage spikes created during operation of both front ABS valve solenoids during antilock and traction control.

TEST DESCRIPTION: Test numbers refer to circled numbers on the diagnostic chart.

13. Checks for a short to system voltage in CKT 854.
14. Checks for an open in CKT 879.
15. Checks for a short to ground in CKT 879.
16. Checks for a short to system voltage in CKT 879.
17. Checks wiring and connectors for intermittents.
18. Determines whether the code was set due to an intermittent condition or an EBTCM fault.
19. Checks if problem found in step 13 is due to a short to system voltage in CKT 874, 879, or 854.

DIAGNOSTIC AIDS:

All tests using J 39700 Universal Breakout Box terminal '20' are using the terminal as ground. These tests, of course, assume that the ground CKT 804 at terminal '20' is good.

It is very important that a thorough inspection of the wiring and connectors be performed. Failure to carefully and fully inspect wiring and connectors may result in misdiagnosis, causing part replacement with reappearance of the fault.

DIAGNOSTIC CHART DTC 63 — 1993 ALLANTE, CONTINUED

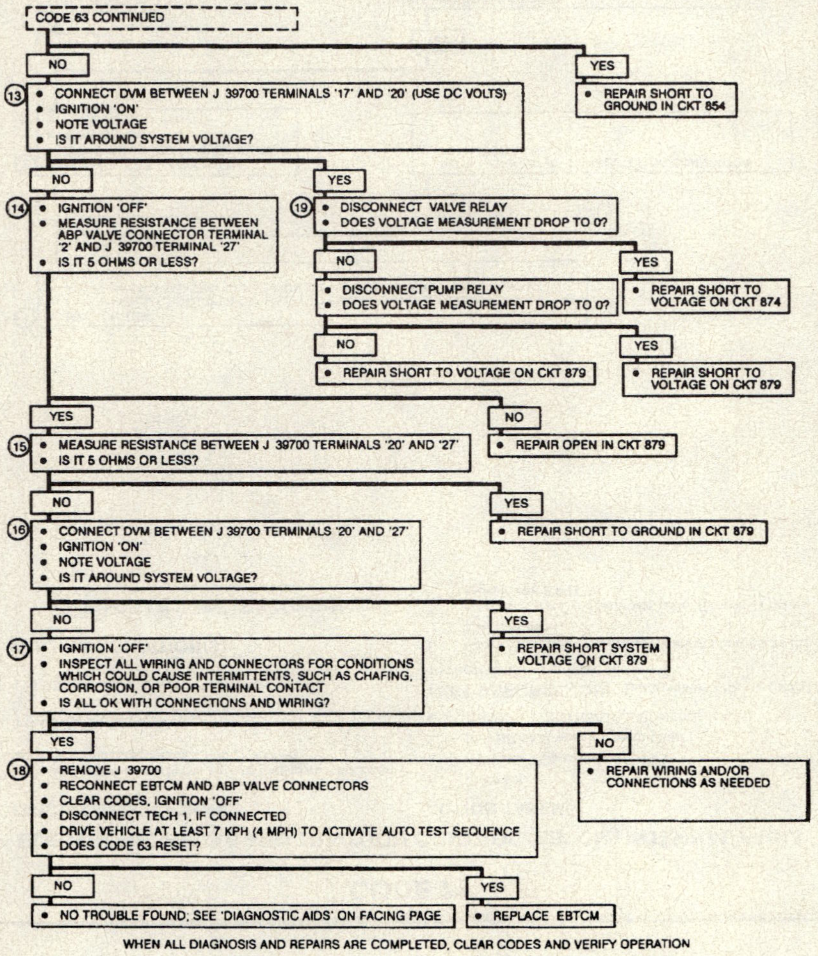

DIAGNOSTIC CHART DTC 67 — 1993 ALLANTE

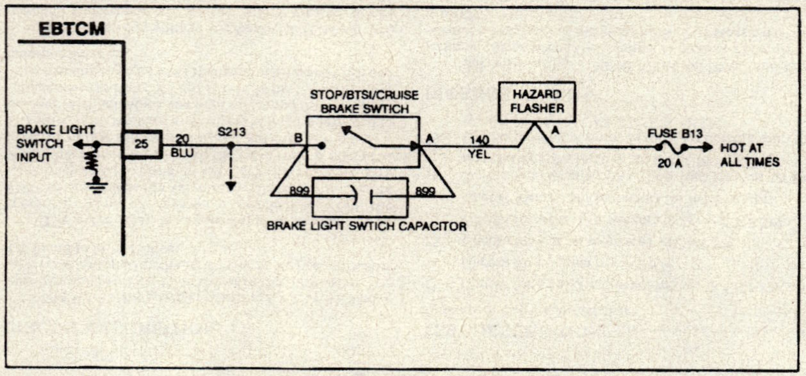

CODE 67

BRAKE LIGHT SWITCH FAULT

CIRCUIT DESCRIPTION:

The electronic brake control module receives Brake Light Switch status from the PCM via serial data. If fault Code 67 sets, it may be indicative of a loss of switch input to the PCM, check for PCM fault Code PO90 (STOP/BTSI/CRUISE BRAKE SWITCH FAULT CODE) or P106 (BRAKE LIGHT SWITCH FAULT CODE)

TEST CONDITION(S):

1. Tested continuously.

FAILURE CONDITION(S):

1. EBTCM receives failure status of Brake Light Switch from PCM.

ACTION TAKEN:

1. Code set.
2. 'TRACTION DISABLED' message ON.

DIAGNOSTIC CHART DTC 71 — 1993 ALLANTE

CODE 71

ELECTRONIC BRAKE AND TRACTION CONTROL (EBTCM) INTERNAL FAULT

CIRCUIT DESCRIPTION:

The EBTCM performs various diagnostic checks on itself. If it finds a problem, Code 71 will set.

TEST CONDITION(S):

1. Tested continuously.

FAILURE CONDITION(S):

1. While Standing or Driving: EBTCM internal error is detected.

ACTION TAKEN:

1. Code set.
2. ABS and TCS switched 'OFF'.
3. Red BRAKE warning lamp ON.
4. ABS and TCS 'DISABLED' messages ON.

TEST DESCRIPTION: Test numbers refer to circled numbers on the diagnostic chart.

1. Checks for good connections from the harness to the EBTCM itself.
2. Checks to see if the fault was false. If Code 71 resets, replace the EBTCM.

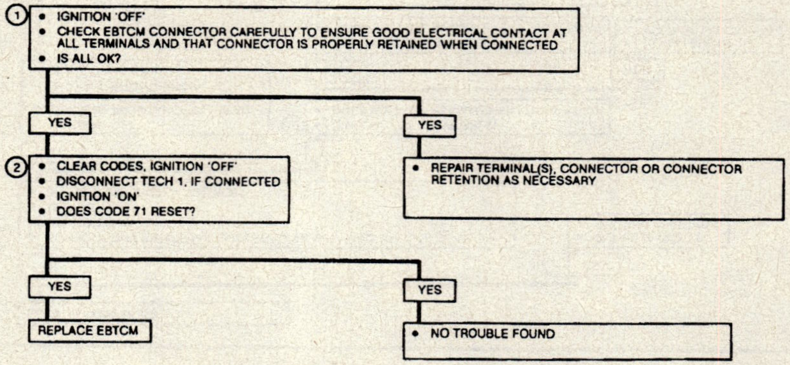

WHEN ALL DIAGNOSIS AND REPAIRS ARE COMPLETED, CLEAR CODES AND VERIFY OPERATION

DIAGNOSTIC CHART DTC 72 — 1993 ALLANTE

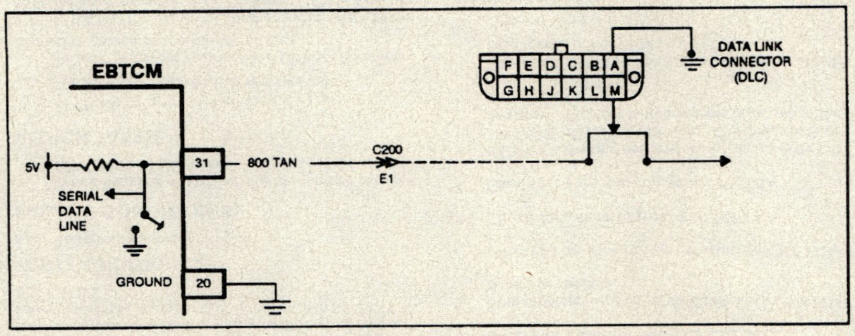

CODE 72

SERIAL DATA LINK FAULT

CIRCUIT DESCRIPTION:

The Serial Data Link is an asynchronous link operating at 8192 bits per second.

Code 72 will set and the red brake warning lamp will illuminate if the EBTCM does not get polled by the BCM for information, or if it does not see PCM information on the data link.

TEST CONDITION(S):

1. Tested continuously.

FAILURE CONDITION(S):

1. While Standing or Driving: Serial Data communications are lost.

ACTION TAKEN:

1. Code set.
2. ABS and TCS switched 'OFF'.
3. Red brake warning lamp ON.
4. ABS and TCS 'DISABLED' messages ON.

TEST DESCRIPTION: Test numbers refer to circled numbers on the diagnostic chart.

1. This test checks to see if the off-board device can communicate with the EBTCM.
2. This test checks for a short to battery in the serial data circuit.
3. This test checks for a short to ground in the serial data circuit.
4. This test checks for an open in the serial data circuit.
5. This test checks to see if the off-board device can communicate with the EBTCM.
6. If the connections at the EBTCM and DLC connector are OK, replace the EBTCM.

DIAGNOSTIC AIDS:

The problem may be intermittent. Try performing the tests shown while 'wiggling' wiring and connectors; this can often cause the fault to appear.

NOTICE: Diagnostic Trouble Code TO72 will never be displayed while serial data fault is current. A current Code 72 will result in a 'NO TCS DATA', 'ANTILOCK DISABLED' and 'TCS DISABLED' messaged as well as BCM Diagnostic Trouble Code B332 and PCM Code P133.

DIAGNOSTIC CHART DTC 72 — 1993 ALLANTE, CONTINUED

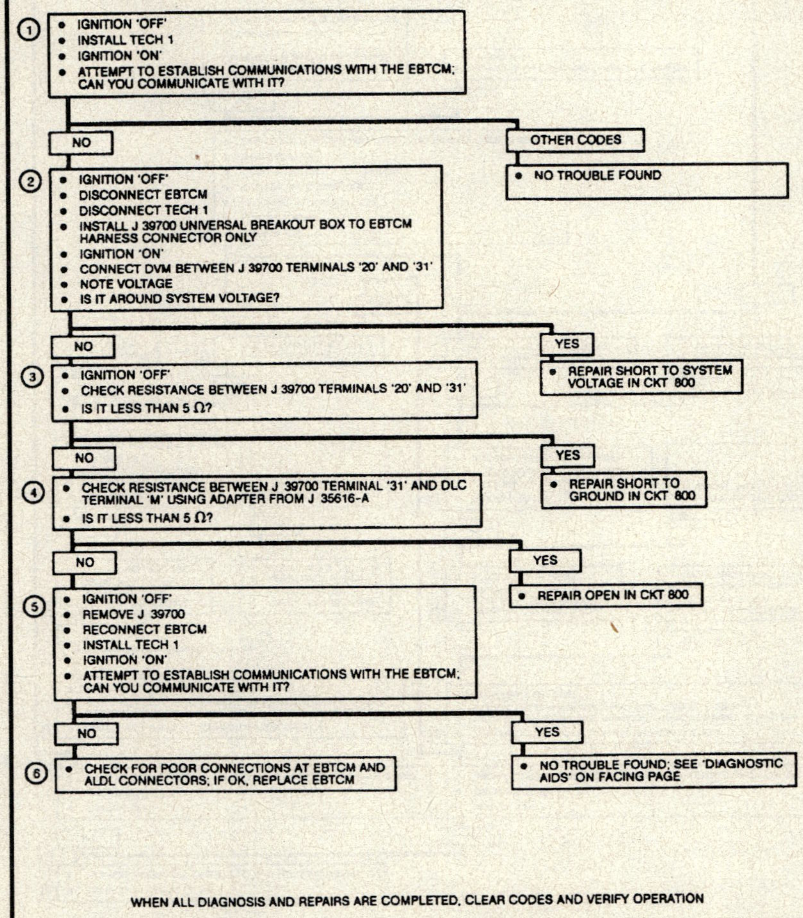

DIAGNOSTIC CHART DTC 73 — 1993 ALLANTE

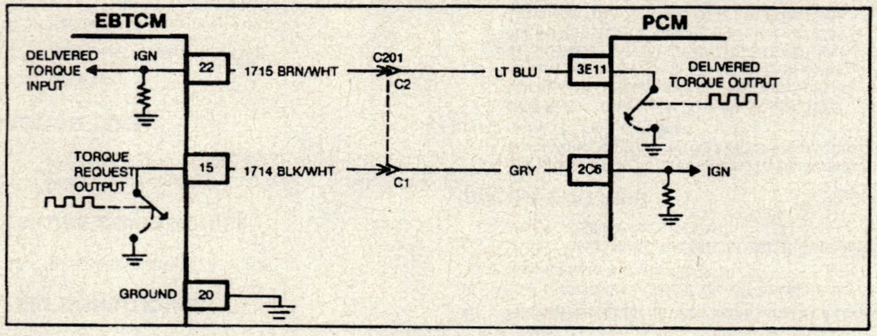

CODE 73

PCM–EBTCM PWM SIGNAL FAULT

CIRCUIT DESCRIPTION:

Traction control is simultaneously controlled by the EBTCM and the PCM. The PCM receives a TORQUE request via a pulse width modulated (PWM) signal from the EBTCM (CKT 1714) requesting the desired torque level for proper traction control system operation. Torque reduction will be indicated if the duty cycle of the PWM signal is between 10% (100 % torque reduction) and 90% (no torque reduction) and at least one (up to five) cylinder is requested to be disabled. The percent duty cycle will indicate the percent torque required. The PCM will selectively disable fuel to certain cylinders to achieve this reduction and communicate the actual torque output back to the EBTCM (CKT 1715).

TEST CONDITION(S):

1. Tested continuously.

FAILURE CONDITION(S):

1. While Standing or Driving: PWM interface between EBTCM and PCM is lost.

ACTION TAKEN:

1. Code set.
2. 'TRACTION DISABLED' message ON.

A Code 73 may set if either CKT 1714 or 1715 is open, shorted to ground or shorted to voltage.

If CKT 1715, Delivered Torque, is damaged and the EBTCM loses this input, a TCS Diagnostic Code 73 will be set. The PCM monitors the serial data line to see if the traction control has been disabled. If it has, the PCM will not set Code 93.

If CKT 1714, Torque Request, is damaged and the PCM loses this input, the PCM monitors the serial data line to see if the traction control has been disabled. If not, the PCM sets a Code 93 and then tells the EBTCM, via serial data, that there is a PWM circuit problem and the EBTCM sets a Code 73.

TEST DESCRIPTION: Test numbers refer to circled numbers on the diagnostic chart.

1. Checks if the PCM Diagnostic Trouble Code 93 is set. If it is, there is likely a problem with CKT 1714.
2. Checks for an open or high resistance in CKT 1715.
3. Checks for a short to ground in CKT 1715.
4. Checks for a short to voltage in CKT 1715.
5. If no problem is found up until this point, CKT 1714 should be checked.
6. Checks for an open or high resistance in CKT 1714.
7. Checks for a short to ground in CKT 1714.
8. Checks for a short to voltage in CKT 1714.
9. If CKT 1715 has not been checked, proceed to the left hand side of the trouble tree. If it has, continue downward. Problem is intermittent or in one of the controllers.
10. Checks if Code 73 and/or 93 are intermittent. If not, trouble is in either PCM or EBTCM.

DIAGNOSTIC AIDS:

- It is theoretically impossible for only a PCM Code 93 to set. If only a PCM Code 93 is present, either the TCS Code 73 has been cleared or there is a problem with the PCM
- A Code 73 may set by itself if the PCM PROM has been improperly installed.

DIAGNOSTIC CHART DTC 73 — 1993 ALLANTE, CONTINUED

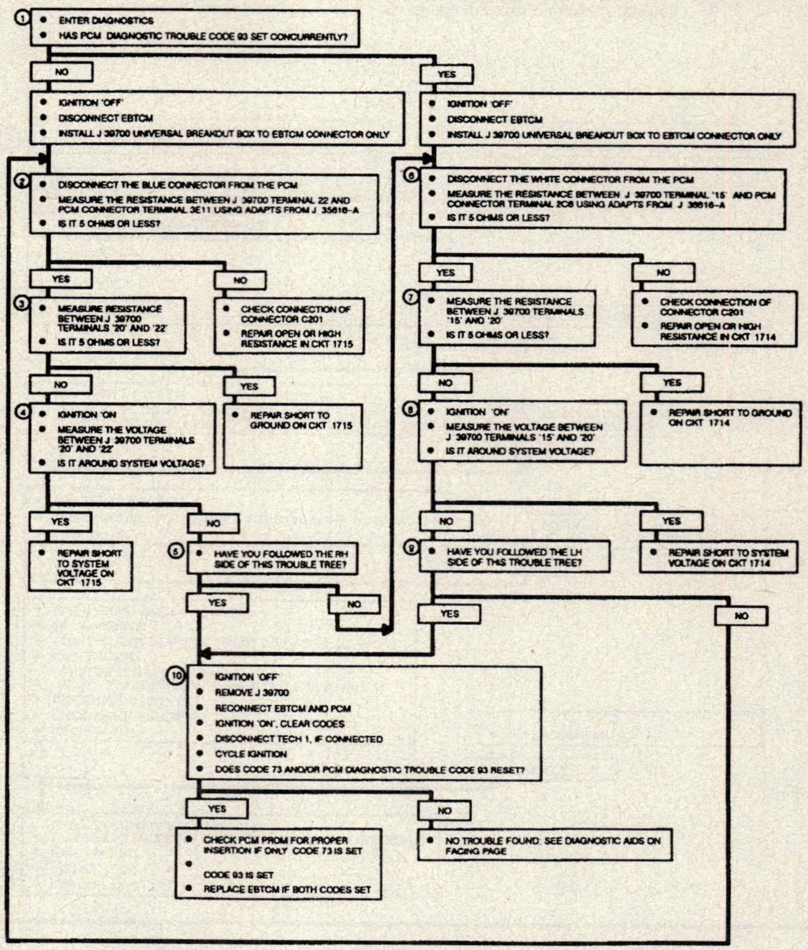

WHEN ALL DIAGNOSIS AND REPAIRS ARE COMPLETED, CLEAR CODES AND VERIFY OPERATION

DIAGNOSTIC CHART DTC 83 — 1993 ALLANTE

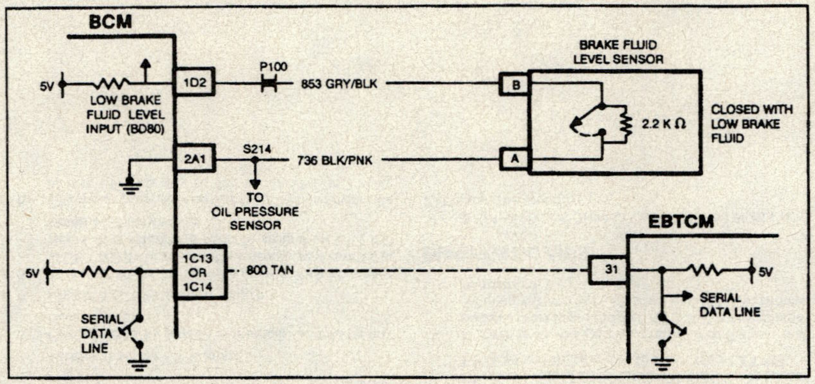

CODE 83

LOW BRAKE FLUID LEVEL

CIRCUIT DESCRIPTION:

The BCM outputs 5 volts on CKT 853 to the brake fluid level sensor. The BCM determines the level of brake fluid by the voltage drop it detects across the sensor circuit. This information is delivered to the EBTCM via the serial data line.

TEST CONDITION(S):

1. Tested continuously.

FAILURE CONDITION(S):

1. While Standing or Driving: Brake fluid falls below acceptable level.
2. Brake fluid level sensor disconnected.
3. Open, short to ground or short to battery on CKT 853.

ACTION TAKEN:

1. Code set.
2. ABS and TCS switched 'OFF'.
3. Red brake warning lamp ON.
4. ABS and TCS 'DISABLED' messages ON.

TEST DESCRIPTION: Test numbers refer to circled numbers on the diagnostic chart.

1. Check the brake fluid level in master cylinder reservoir.
2. Check brake fluid level as detected by BCM.
3. Brake fluid level not acceptable. Fill and recheck.
4. Brake fluid level as detected by BCM. See 'DIAGNOSTIC AIDS'.
5. Checks BCM input parameter BD80 with sensor disconnected for a short to ground in CKT 853, or bad sensor.
6. Checks BCM input parameter BD80 with sensor disconnected and connector shorted.
7. BCM detects acceptable brake fluid level. Problem not present.
8. Checks sensor resistor and checks for a short between CKTs 853 and 736.
9. Checks for open in CKT 853 and 736.

DIAGNOSTIC AIDS:

BCM data parameter BD80 will display the following:

- 3.6–5V – open in sensor circuit.
- 1.5–3.5V – circuit OK. Problem not present.
- 0–1.4V – short between CKT 853 and 736.

DIAGNOSTIC CHART DTC 83 — 1993 ALLANTE, CONTINUED

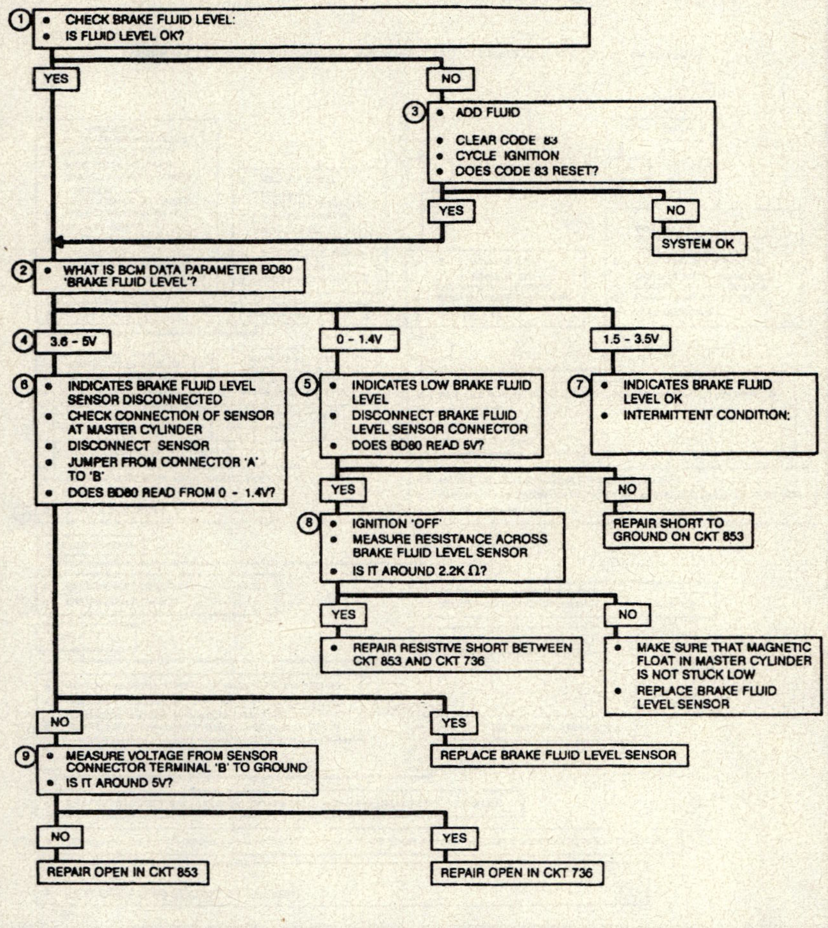

WHEN ALL DIAGNOSIS AND REPAIRS ARE COMPLETED, CLEAR CODES AND VERIFY OPERATION

DIAGNOSTIC CHART A — 1993 ALLANTE

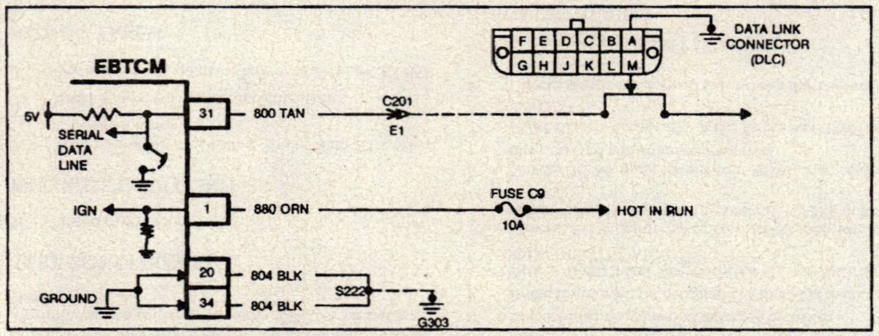

CHART A

'ANTILOCK DISABLED' AND 'TRACTION DISABLED' MESSAGES 'ON', NO CODES SET / NO TCS DATA

TEST DESCRIPTION: Test numbers refer to circled numbers on the diagnostic chart.

1. Checks if condition is due to a poor connection at the EBTCM. Make sure the connecter is not 'cocked' on module.
2. Verifies that fuse C9 is intact.
3. Checks EBTCM ground resistances. Should be less than 5 ohms.
4. Checks EBTCM ignition voltage.
5. This step checks for possible open or high resistance in CKT 880 if there is insufficient voltage measured in step 4. It is assumed a short to ground on CKT 880 would have blown fuse C9.
6. Checks for possible insufficient voltage on power side of fuse C9. This may be due to Maxi® fuse 2 being damaged or an intermittent connection in the ignition switch.
7. Checks for EBTCM terminal '1' having a poor solder joint or for being partially backed out.
8. Checks if a loss of serial data communication between EBTCM and BCM is due to a damaged CKT 800.
9. If there is no voltage measured in step 8, there is a possible short to ground on CKT 800. If not, regulated 5V supply from EBTCM is lost and EBTCM should be replaced.

DIAGNOSTIC AIDS:

BCM Diagnostic Trouble Code B332 and PCM Code P137 should be present.

DIAGNOSTIC CHART A — 1993 ALLANTE, CONTINUED

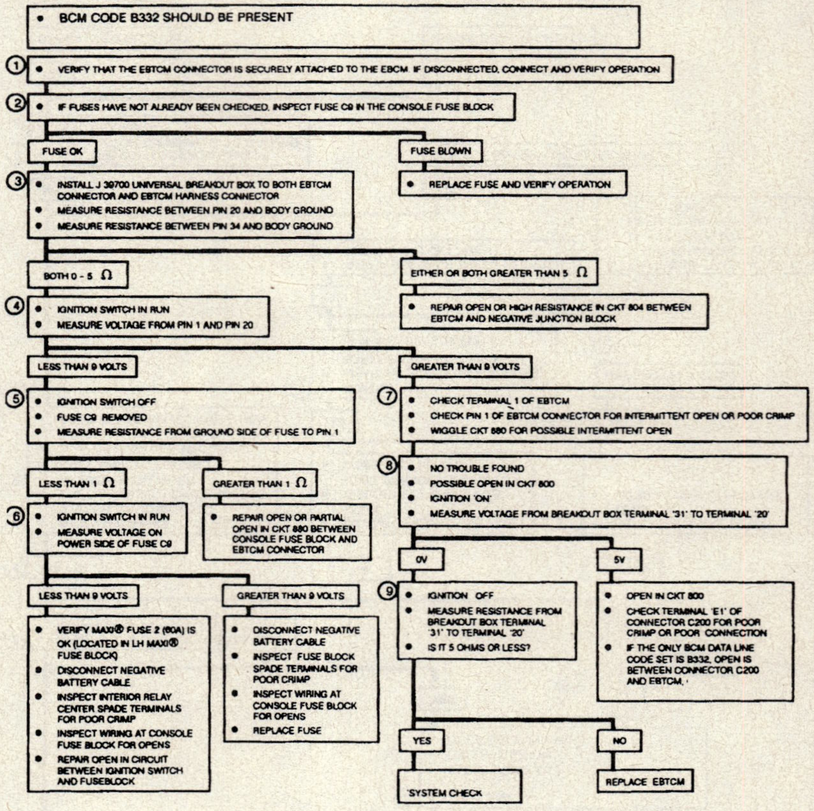

WHEN ALL DIAGNOSIS AND REPAIRS ARE COMPLETED, CLEAR CODES AND VERIFY OPERATION

DIAGNOSTIC CHART B — 1993 ALLANTE

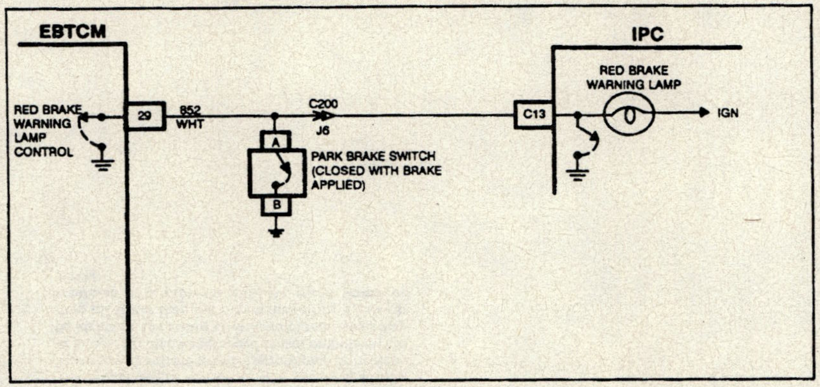

CHART B

RED BRAKE WARNING LAMP 'ON', NO TCS CODES SET

TEST DESCRIPTION: Test numbers refer to circled numbers on the diagnostic chart.

1. Checks to be certain light is not on because parking brake in not fully released.
2. Checks if red BRAKE warning lamp is ON due to a loss of communication between the BCM and EBTCM.
3. Checks if light is ON due to a shorted parking brake switch.
4. Checks if light is ON for a legitimate base brake condition.
5. Checks for a short to ground in CKT 852.
6. Final check for diagnostic trouble codes.

DIAGNOSTIC AIDS:

The EBTCM controls the red BRAKE warning lamp only in instances where serial data communication with the BCM is lost. In all other instances the IPC controls this lamp.

DIAGNOSTIC CHART B — 1993 ALLANTE, CONTINUED

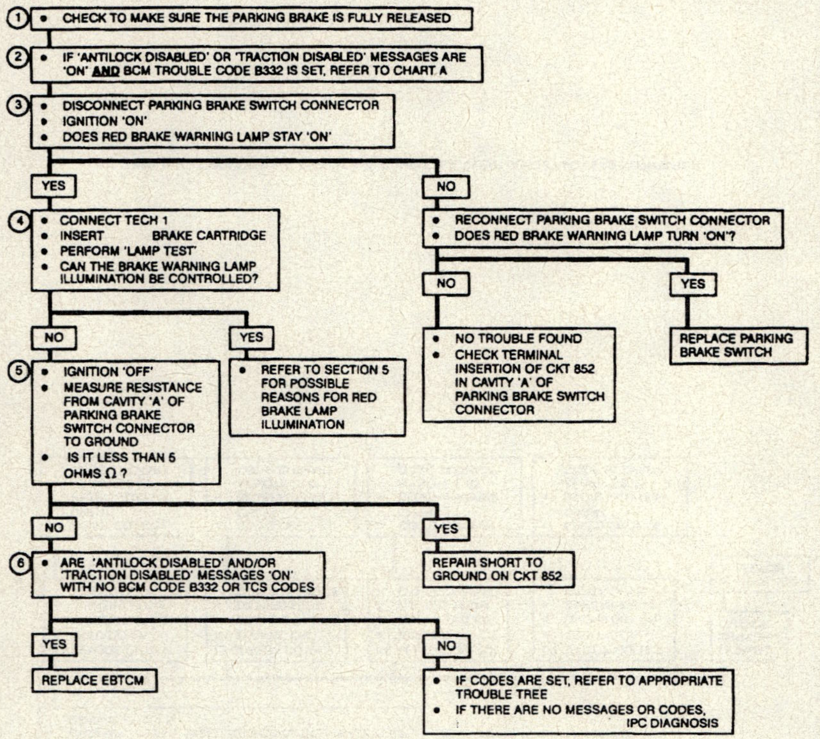

WHEN ALL DIAGNOSIS AND REPAIRS ARE COMPLETED, CLEAR CODES AND VERIFY OPERATION

DIAGNOSTIC CHART C — 1993 ALLANTE

CHART C

VALVE CYCLING (CHATTER) DURING NORMAL STOPS

TEST DESCRIPTION: Test numbers refer to circled numbers on the diagnostic chart.

1. Checks if valve activation due to erroneous 'noise' induced onto wheel speed sensor circuits.
2. The wheel speed sensor output should gradually increase with vehicle speed. There should be no skips or jumps. All four wheel speed sensor outputs should be identical. On a slow down, one of the signals may drop off faster than the others, indicating a lock-up condition or a problem with the speed sensor or wiring.

DIAGNOSTIC CHART C — 1993 ALLANTE, CONTINUED

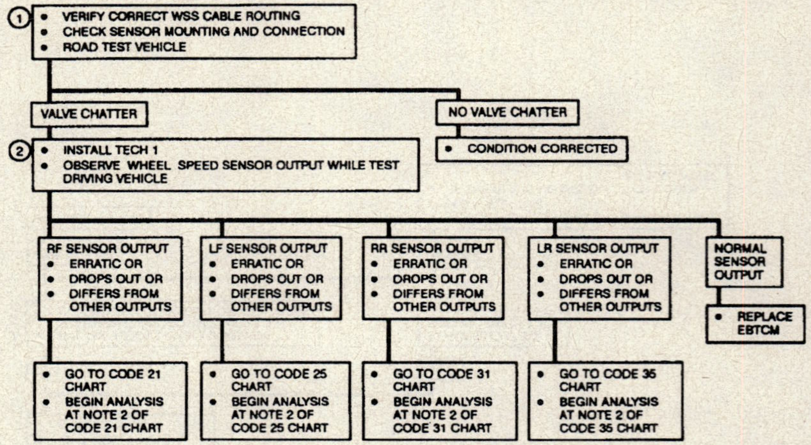

WHEN ALL DIAGNOSIS AND REPAIRS ARE COMPLETED, CLEAR CODES AND VERIFY OPERATION

DIAGNOSTIC CHART D — 1993 ALLANTE

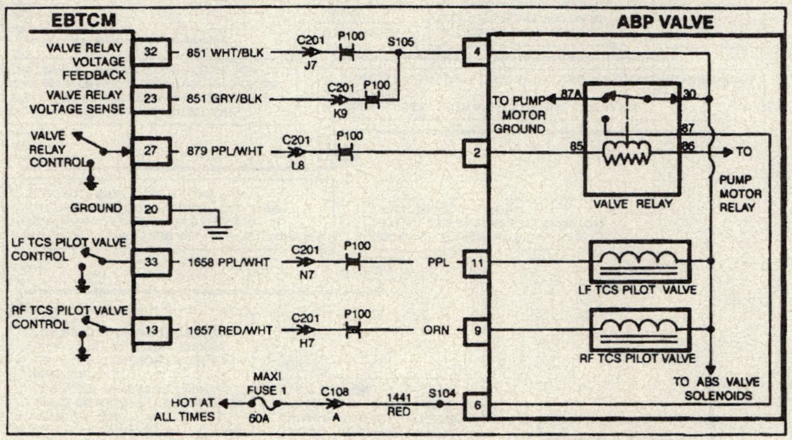

CHART D

TRACTION CONTROL OPERATION CHECK (NO DIAGNOSTIC TROUBLE CODES SET)

TEST DESCRIPTION: Test numbers refer to circled numbers on the diagnostic chart.

1. Traction control system should interpret the elevated front wheels as an icy or free spin condition. At $^1/_2$ throttle a properly operating system will allow both front wheels to spin up quickly to about 20–25 MPH and then slow down (but not stop) to about 5–10 MPH and have a jerky appearance. The traction control 'ACTIVE' lamp should illuminate on the IPC. The pump motor should also be heard.
2. System is building up pressure to the front brake and operating correctly.
3. System is not building up pressure to the front brakes. Possible air or dirt in the system.
4. No dirt in reservoir and the fluid level is OK. Air may still be trapped in system. The TCS hydraulic circuit should be bled using TECH 1 TCS test and bleed procedure One of the front caliper bleed screws must be open to allow air to escape.
5. Fluid level OK. Air may still be trapped in the system. The TCS hydraulic circuit should be bled using TECH 1 TCS test and bleed procedure One of the front caliper bleed screws must be open to allow air to escape.
6. Reservoir TCS prime pipe may be contaminated. Flush pipe and reservoir to remove any debris. Air may still be trapped in system. The TCS hydraulic circuit should be bled using TECH 1 TCS test and bleed procedure One of the front caliper bleed screws must be open to allow air to escape.

DIAGNOSTIC AIDS:

- The Brake Light Switch must be properly adjusted to prevent an erroneous brake application input to the EBTCM. Misadjustment may prevent brake intervention during traction control operation.
- The PCM shall not allow traction control fuel shutoff if:
  - Engine Coolant Temperature < -40°C
  - Engine Coolant Temperature ≥ 131°C
  - Low Coolant Level Detected
- Cylinders shall be disabled if the engine speed ≥ 600 RPM.
- The PCM can disable traction control for up to 11 ignition cycles if override PS24, first gear enable, is activated. Several different powertrain failures disable first gear starts to reduce transaxle stress (e.g. PCM Code P039). If a powertrain failure has resulted in loss of first gear start, PS24 may be used to re-enable first gear starts and disable traction control. When PS24 is selected, the display will alternate between '—' for one second and the current state of first gear enable override ('99' for override active and '00' for override not active). Pressing the WARMER button will enable first gear starts and disable traction control for the next eleven ignition cycles and display '99'. The 'TRACTION DISABLED' message will be displayed and no TCS diagnostic trouble codes will be set. Pressing the COOLER button will allow normal program control of first gear starts and traction control and display '00'.

DIAGNOSTIC CHART D — 1993 ALLANTE, CONTINUED

NOTE: SOME OF THESE CHECKS WILL CAUSE DIAGNOSTIC TROUBLE CODES TO SET. BE SURE TO VERIFY AUTHENTICITY OF ANY CODES AND CORRECT IF NECESSARY. BE SURE TO CLEAR ANY ERRONEOUS OR CORRECTED CODES.

(1)
- CLEAR CODES
- IGNITION 'OFF'
- DISCONNECT TECH 1
- RAISE FRONT WHEELS 4" TO 5" ON HOIST
- IGNITION 'ON', ENGINE 'ON'
- WITH TRANSMISSION IN DRIVE, ACCELERATE TO $^1/_2$ THROTTLE WHILE AN OBSERVER MONITORS THE DRIVE WHEELS
- ARE THE FRONT BRAKES MODULATED?

YES → (2)
- SYSTEM IS OPERATING CORRECTLY

NO → (3)
- CHECK FLUID LEVEL IN MASTER CYLINDER RESERVOIR
- CHECK FOR DIRT OR DEBRIS IN RESERVOIR FLUID

LEVEL OK, FLUID NOT DIRTY → (4)
- POSSIBLE AIR IN SYSTEM
- BLEED TCS PRIME PIPE AT ABP VALVE
- PERFORM THE TCS TEST USING THE TECH 1

  ONE OF THE FRONT CALIPER BLEED SCREWS MUST BE 'OPENED'
- RETEST TRACTION CONTROL OPERATION ON HOIST

LOW → (7)
- ADD FLUID
- POSSIBLE AIR IN SYSTEM
- BLEED TCS PRIME PIPE AT ABP VALVE
- PERFORM THE TCS TEST USING THE TECH 1

  ONE OF THE FRONT CALIPER BLEED SCREWS MUST BE 'OPENED'
- RETEST TRACTION CONTROL OPERATION ON HOIST

DIRTY → (6)
- DISCONNECT TCS PRIME PIPE AT ABP VALVE
- ALLOW FLUID TO DRAIN FROM RESERVOIR
- FLUSH RESERVOIR AND TCS PRIME PIPE WITH FRESH BRAKE FLUID
- RECONNECT TCS PRIME PIPE
- REFILL
- POSSIBLE AIR IN SYSTEM
- BLEED TCS PRIME PIPE AT ABP VALVE
- PERFORM THE TCS TEST USING THE TECH 1

  ONE OF THE FRONT CALIPER BLEED SCREWS MUST BE 'OPENED'
- RETEST TRACTION CONTROL OPERATION ON HOIST

WHEN ALL DIAGNOSIS AND REPAIRS ARE COMPLETED, CLEAR CODES AND VERIFY OPERATION

DIAGNOSTIC CHART ABS/ASR DIAGNOSTIC SYSTEM CHECK — 1994 FLEETWOOD

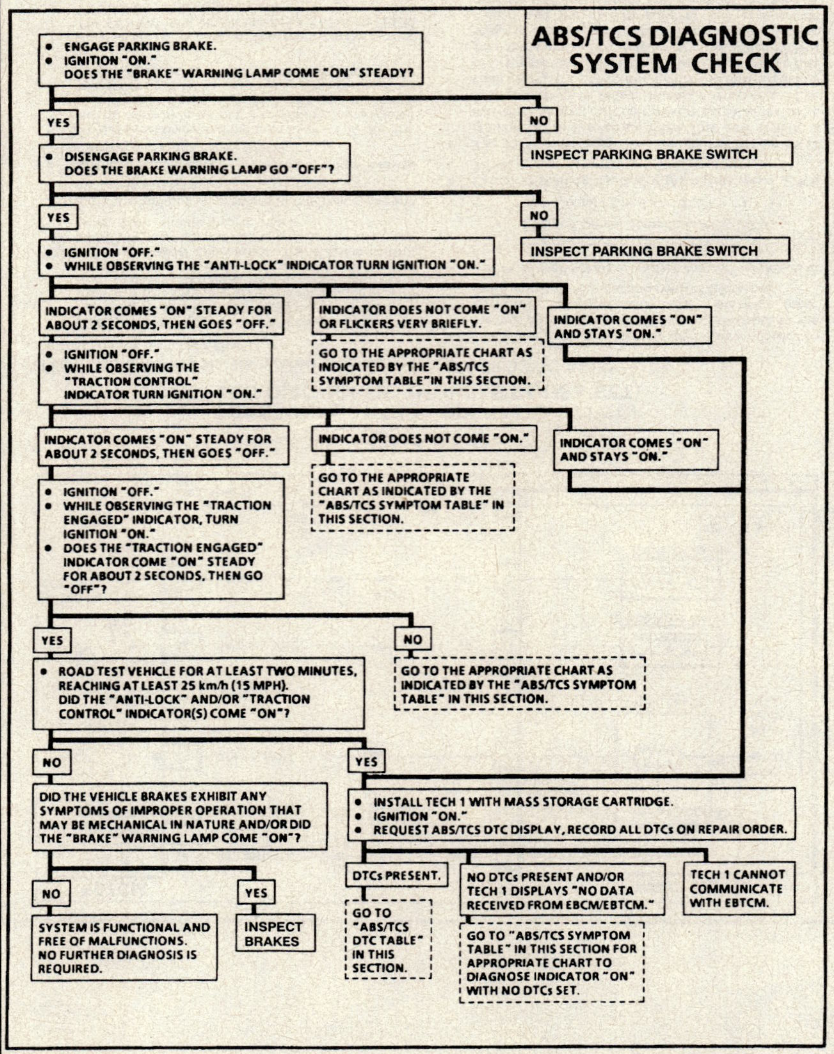

DIAGNOSTIC CHART ABS/ASR SYMPTOM TABLE — 1994 FLEETWOOD

ABS/TCS SYMPTOM TABLE

CHART	SYMPTOM
A	"ANTI-LOCK" INDICATOR "ON" WITH NO DTCs SET
B	"ANTI-LOCK" INDICATOR INOPERATIVE OR FLICKERS BRIEFLY AT IGNITION "ON"
C	"TRACTION CONTROL" INDICATOR "ON" WITH NO DTCs SET
D	"TRACTION CONTROL" INDICATOR INOPERATIVE
E	"TRACTION ENGAGED" INDICATOR "ON" WITH NO DTCs SET
F	"TRACTION ENGAGED" INDICATOR INOPERATIVE WITH NO DTCs SET
	"BRAKE" WARNING LAMP "ON"
	"BRAKE" WARNING LAMP INOPERATIVE

When the "ANTI-LOCK," "TRACTION CONTROL," and "TRACTION ENGAGED" indicators are all inoperative, inspect for open EBTCM/TPS fuse #7.

When the "ANTI-LOCK," "TRACTION CONTROL," and "TRACTION ENGAGED" indicators all flash "ON" and "OFF" repeatedly, repair an intermittent open in CKT 739.

DIAGNOSTIC TROUBLE CODE CHART — 1994 FLEETWOOD

ABSTCS DTC TABLE	
DIAGNOSTIC TROUBLE CODE	DESCRIPTION
21	RF WHEEL SPEED SENSOR CIRCUIT MALFUNCTION
23	RF WHEEL SPEED SENSOR CIRCUIT CONTINUITY MALFUNCTION
25	LF WHEEL SPEED SENSOR CIRCUIT MALFUNCTION
27	LF WHEEL SPEED SENSOR CIRCUIT CONTINUITY MALFUNCTION
28	WHEEL SPEED SENSOR CIRCUIT FREQUENCY MALFUNCTION
31	RR WHEEL SPEED SENSOR CIRCUIT MALFUNCTION
33	RR WHEEL SPEED SENSOR CIRCUIT CONTINUITY MALFUNCTION
35	LR WHEEL SPEED SENSOR CIRCUIT MALFUNCTION
37	LR WHEEL SPEED SENSOR CIRCUIT CONTINUITY MALFUNCTION
41	RF VALVE SOLENOID MALFUNCTION
44	PILOT VALVE SOLENOID MALFUNCTION
45	LF VALVE SOLENOID MALFUNCTION
51	RR VALVE SOLENOID MALFUNCTION
55	LR VALVE SOLENOID MALFUNCTION
58	EBTCM INTERNAL ADJUSTER ASSEMBLY MALFUNCTION
61	PUMP MOTOR OR PUMP MOTOR RELAY MALFUNCTION
62	TRACTION CONTROL SWITCH USED OR RPM SIGNAL MALFUNCTION
63	BPMV RELAY MALFUNCTION
64	THROTTLE POSITION SIGNAL MALFUNCTION
65	ADJUSTER ASSEMBLY CIRCUIT MALFUNCTION
66	ADJUSTER ASSEMBLY CONTROL MALFUNCTION
71	EBTCM INTERNAL MALFUNCTION
72	SERIAL DATA LINE MALFUNCTION
83	BRAKE FLUID PRESSURE DIFFERENTIAL MALFUNCTION

ELECTRICAL CONNECTOR TERMINAL IDENTIFICATION — 1994 FLEETWOOD

C200

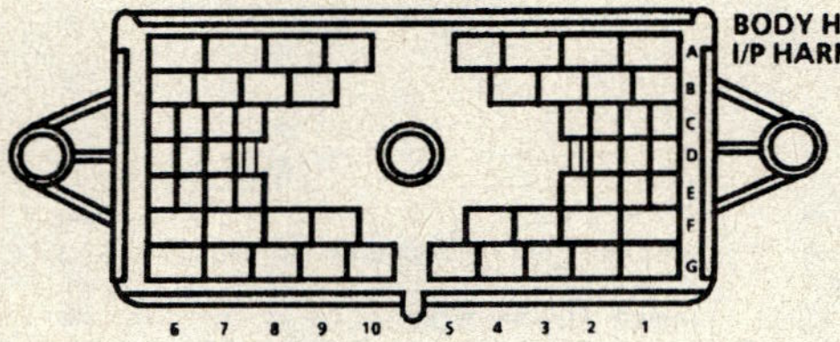

BODY HARNESS TO I/P HARNESS

12034257
57-WAY F METRI-PACK MIXED SERIES
BLK

C202

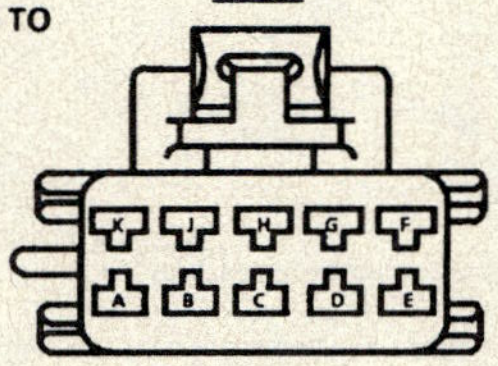

12064769
10-WAY F METRI-PACK 150 SERIES
NAT
I/P HARNESS TO BODY HARNESS

C204

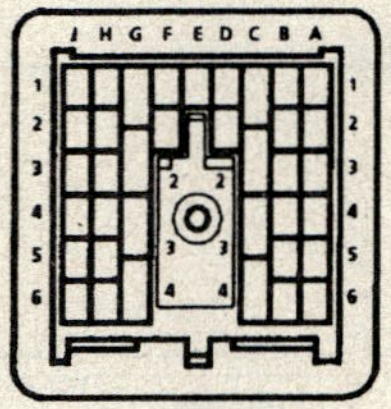

BODY HARNESS TO ENGINE HARNESS

12020916
37-WAY F METRI-PACK MIXED SERIES
BLK

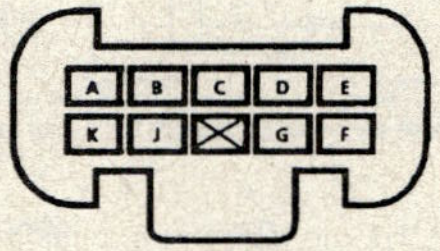

12065425
10-WAY F METRI-PACK 150 SERIES
BLK
THROTTLE POSITION (TP) SENSOR INTERFACE MODULE

C402

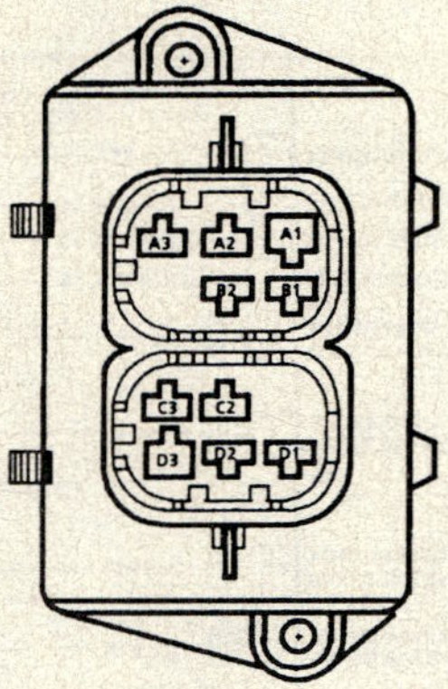

12129525
10-WAY M METRI-PACK 150, 280 SERIES
BLK
ANTILOCK BRAKE SYSTEM PASS THRU BODY HARNESS TO ELECTRONIC BRAKE CONTROL HARNESS

ELECTRICAL CONNECTOR TERMINAL IDENTIFICATION — 1994 FLEETWOOD, CONTINUED

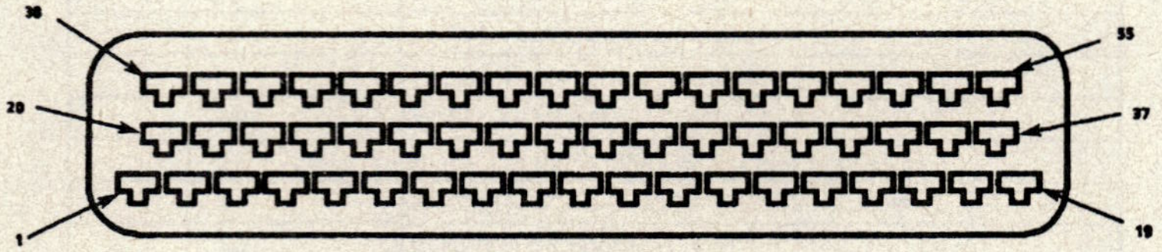

12103563
55-WAY F CONNECTOR BODY
BLK
ELECTRONIC BRAKE AND TRACTION CONTROL MODULE (EBTCM)

12103561
14-WAY F CONNECTOR BODY
BLK
BRAKE PRESSURE MODULATOR VALVE (BPMV)

DIAGNOSTIC CHART A — 1994 FLEETWOOD

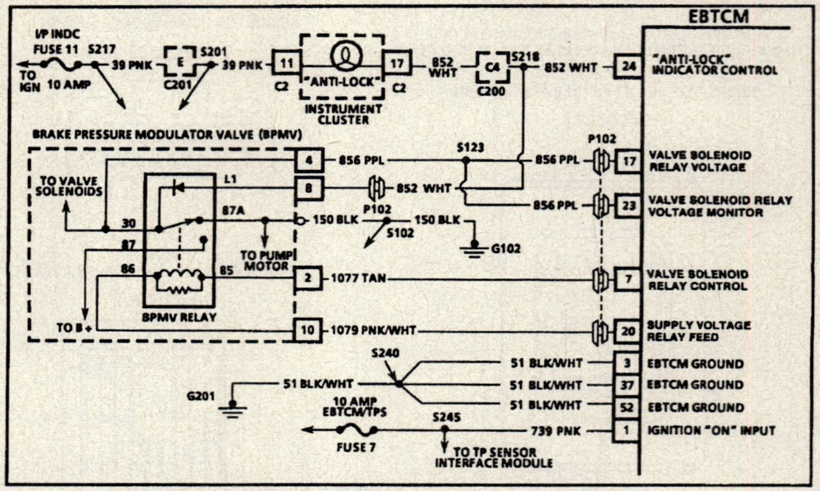

CHART A
(Page 1 of 4)
"ANTI-LOCK" INDICATOR "ON" WITH NO DTCs SET

Chart Test Description: Number(s) below refer to circled number(s) on the diagnostic chart.

1. Confirms that a malfunction condition exists.
2. Determines whether the malfunction is due to a DTC setting condition, an indicator low side circuit malfunction, or an EBTCM supply power malfunction.
3. Checks whether the malfunction is due to an open fuse.
4. Checks for an open power feed circuit.
5. Checks for an open in the ground feed to the EBTCM.
6. Verifies proper connection of the EBTCM ground CKT 51.
7. Checks whether the fuse failure is due to a short to ground.
8. Identifies if the short to ground found in Step 7 is in CKT 739 or the EBTCM.

DIAGNOSTIC CHART A — 1994 FLEETWOOD, CONTINUED

CHART A
(Page 2 of 4)
"ANTI-LOCK" INDICATOR "ON" WITH NO DTCs SET

(1)
- IGNITION "ON."
  DOES THE "ANTI-LOCK" INDICATOR COME "ON" AND STAY "ON"?

YES → (2) / NO → RETURN TO "ABS/ASR FUNCTIONAL CHECK" IN THIS SECTION.

(2)
- INSTALL TECH 1 WITH MSC.
- REQUEST DTC DISPLAY.

Results:
- TECH 1 DISPLAYS "NO DATA RECEIVED FROM ABS CNTRLR." → (3)
- NO CURRENT DTCs PRESENT. → CONTINUE DIAGNOSIS ON NEXT PAGE.
- CURRENT DTCs PRESENT. → GO TO "ABS/ASR DTC TABLE."

(3)
- IGNITION "OFF."
- REMOVE AND INSPECT "EBTCM/TPS" FUSE.
  IS FUSE GOOD?

YES → (4) / NO → (7)

(4)
- INSTALL "EBTCM/TPS" FUSE.
- INSTALL J 39700 UNIVERSAL BREAKOUT BOX TO EBTCM HARNESS CONNECTOR ONLY.
- USING J 39200, MEASURE RESISTANCE FROM EACH TERMINAL OF THE FUSE BLOCK WHICH CONTACTS THE "LEGS" OF THE "EBTCM/TPS" FUSE TO J 39700 TERMINAL "1".
  IS RESISTANCE 5 OHMS OR LESS FOR EITHER MEASUREMENT?

YES → (5) / NO → REPAIR HIGH RESISTANCE IN CKT 739.

(7)
- INSTALL NEW "EBTCM/TPS" FUSE.
- IGNITION "ON," WAIT 5 SECONDS, THEN IGNITION "OFF."
- REMOVE AND INSPECT "EBTCM/TPS" FUSE.
  IS FUSE GOOD?

YES → INSTALL "EBTCM/TPS" FUSE. SYSTEM OK. / NO → (8)

(8)
- DISCONNECT EBTCM.
- INSTALL NEW "EBTCM/TPS" FUSE.
- IGNITION "ON", WAIT 5 SECONDS, THEN IGNITION "OFF."
- REMOVE AND INSPECT "EBTCM/TPS" FUSE.
  IS FUSE GOOD?

YES → INSPECT EBTCM HARNESS CONNECTOR FOR PHYSICAL DAMAGE WHICH MAY RESULT IN A SHORT TO GROUND WITH THE EBTCM CONNECTED. IF OK, THEN REPLACE EBTCM.
NO → REPAIR SHORT FROM CKT 739 TO GROUND. INSTALL "EBTCM/TPS" FUSE.

(5)
- CONNECT TEST LIGHT SUCH AS J 34142-B TO B+ AND PROBE J 39700 TERMINAL "3".
  DOES TEST LIGHT ILLUMINATE BRIGHTLY?

YES → (6) / NO → REPAIR HIGH RESISTANCE IN CKT 739.

(6)
- CHECK FOR PROPER CONNECTION TO EBTCM AT TERMINAL "3", "37" AND "52".
  OK?

YES → next step / NO → REPAIR HIGH RESISTANCE IN CKT 51.

- CHECK FOR PROPER CONNECTION TO EBTCM AT TERMINAL "1".
  IF OK THEN REPLACE EBTCM.

NO → REPAIR AS NEEDED.

WHEN ALL DIAGNOSIS AND REPAIRS ARE COMPLETED, CLEAR DTC(s) AND VERIFY PROPER OPERATION.

DIAGNOSTIC CHART A — 1994 FLEETWOOD, CONTINUED

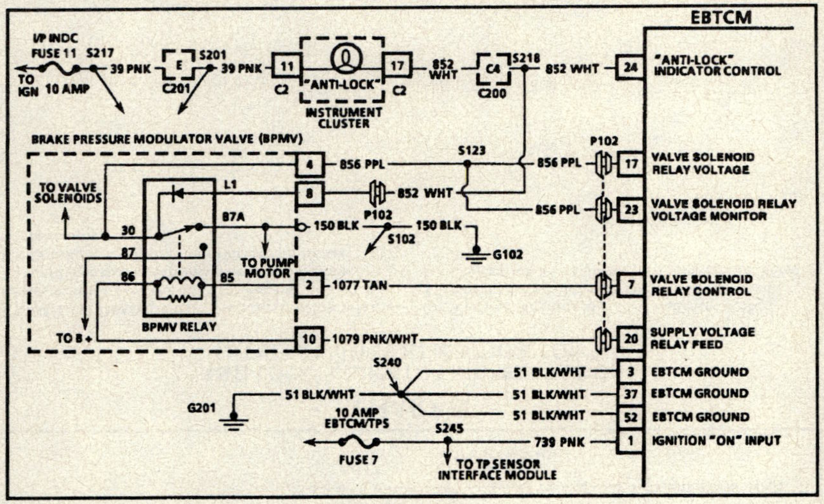

CHART A
(Page 3 of 4)
"ANTI-LOCK" INDICATOR "ON" WITH NO DTCs SET

Chart Test Description: Number(s) below refer to circled number(s) on the diagnostic chart.

9. Determines whether the malfunction is due to a short to ground in CKT 852.
10. Ensures the malfunction is not due to an internal short to ground in the BPMV.
11. Checks for an internally shorted BPMV relay.
12. Determines whether the malfunction is due to physical damage of the EBTCM, BPMV connectors, or the EBTCM.

DIAGNOSTIC CHART A — 1994 FLEETWOOD, CONTINUED

CHART A
(Page 4 of 4)
"ANTI-LOCK" INDICATOR "ON" WITH NO DTCs SET

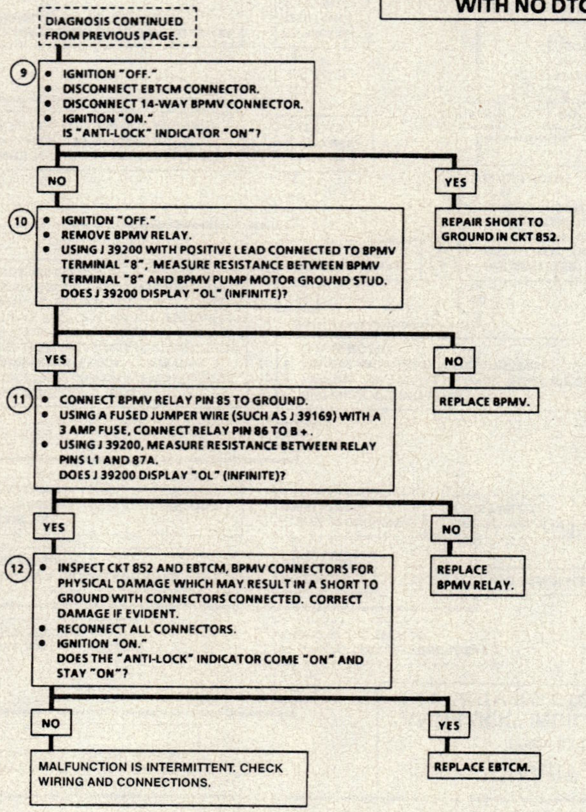

DIAGNOSTIC CHART B — 1994 FLEETWOOD

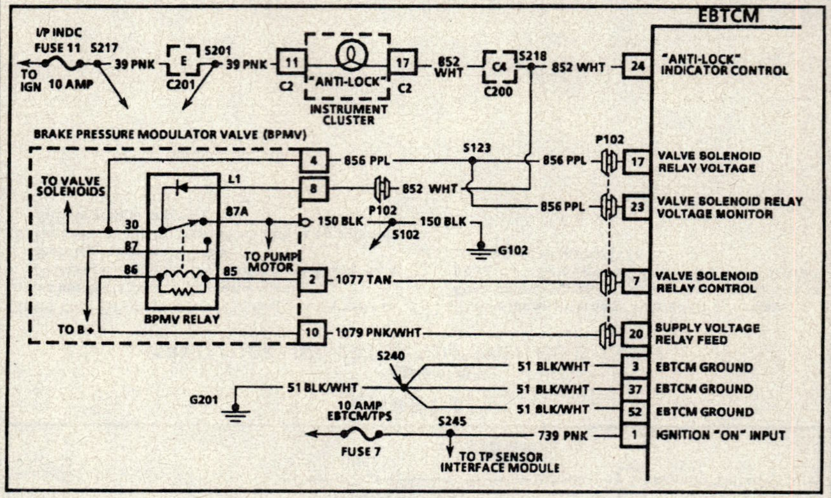

CHART B
"ANTI-LOCK" INDICATOR INOPERATIVE OR FLICKERS BRIEFLY AT IGNITION "ON"

Chart Test Description: Number(s) below refer to circled number(s) on the diagnostic chart.

1. Confirms that a malfunction condition exists.
2. Checks to see if the indicator can be turned "ON" manually.
3. Verifies the integrity of the indicator bulb.
4. Checks for an indicator low side circuit open.
5. Checks for an indicator low side circuit short to B+.

DIAGNOSTIC CHART B — 1994 FLEETWOOD, CONTINUED

CHART B
"ANTI-LOCK" INDICATOR INOPERATIVE OR FLICKERS BRIEFLY AT IGNITION "ON"

(1)
- WHILE OBSERVING THE "ANTI-LOCK" INDICATOR, TURN IGNITION "ON."

"ANTI-LOCK" INDICATOR DOES NOT COME "ON," OR FLICKERS TWICE THEN GOES "OFF."

- IGNITION "OFF."
- WHILE OBSERVING THE "SERVICE ENGINE SOON," "LOW OIL LEVEL," "AIR BAG," AND "LOW COOLANT" INDICATORS, TURN IGNITION "ON." DO INDICATORS ILLUMINATE DURING BULB CHECK?

NO → INSPECT INSTRUMENT CLUSTER

YES →

(2)
- IGNITION "OFF."
- DISCONNECT EBTCM.
- INSTALL J 39700 UNIVERSAL BREAKOUT BOX TO EBTCM HARNESS CONNECTOR ONLY.
- CONNECT A FUSED JUMPER WIRE (SUCH AS J 36169) WITH A 3 AMP FUSE ON J 39700 FROM TERMINAL "3" (GROUND) TO TERMINAL "24".
- IGNITION "ON."

IS THE "ANTI-LOCK" INDICATOR "ON"?

YES →
- IGNITION "OFF."
- CHECK FOR PROPER CONNECTION TO EBTCM AT TERMINAL "24".

IF OK, THEN REPLACE EBTCM.

NO →

(3)
- IGNITION "OFF."
- REMOVE INSTRUMENT CLUSTER.
- REMOVE AND INSPECT "ANTI-LOCK" INDICATOR BULB.

IS BULB GOOD?

NO → REPLACE BULB.

YES →

(4)
- REINSTALL BULB INTO INSTRUMENT CLUSTER.
- USING J 39200, MEASURE RESISTANCE FROM J 39700 TERMINAL "24" TO INSTRUMENT CLUSTER HARNESS CONNECTOR TERMINAL "17".

NO → REPAIR OPEN IN CKT 852.

YES →

(5)
- IGNITION "ON."
- USING J 39200, MEASURE VOLTAGE ON J 39700 FROM TERMINAL "24" TO TERMINAL "3" (GROUND).

IS VOLTAGE 1 VOLT OR LESS?

NO → REPAIR SHORT FROM CKT 852 TO B+. REPLACE BPMV RELAY.

YES →
- IGNITION "OFF."
- CHECK FOR PROPER CONNECTION TO INSTRUMENT CLUSTER AT TERMINAL "17" OF THE INSTRUMENT CLUSTER HARNESS CONNECTOR.

IF OK, THEN SERVICE INSTRUMENT CLUSTER.
- RECONNECT EBTCM.

WHEN ALL DIAGNOSIS AND REPAIRS ARE COMPLETED, CLEAR DTC(s) AND VERIFY PROPER OPERATION.

DIAGNOSTIC CHART C — 1994 FLEETWOOD

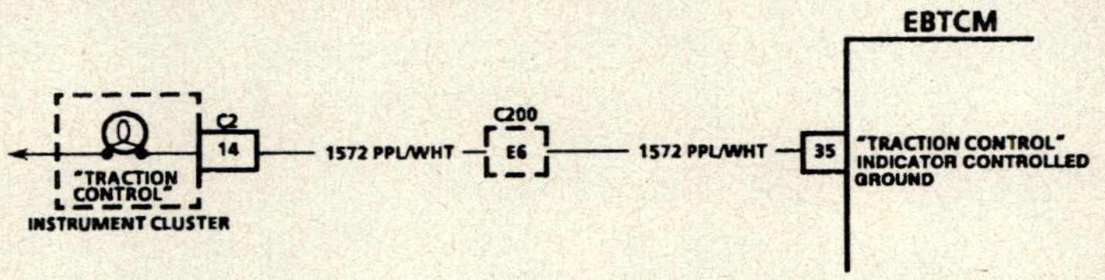

CHART C
"TRACTION CONTROL" INDICATOR "ON" WITH NO DTCs SET

Chart Test Description: Number(s) below refer to circled number(s) on the diagnostic chart.

1. Confirms that a malfunction condition exists.
2. Checks if the malfunction is due to a DTC-setting condition.
3. Determines whether the malfunction is due to an indicator low side circuit short to ground or an EBTCM circuit malfunction.

DIAGNOSTIC CHART C — 1994 FLEETWOOD, CONTINUED

CHART C

"TRACTION CONTROL" INDICATOR "ON" WITH NO DTCs SET

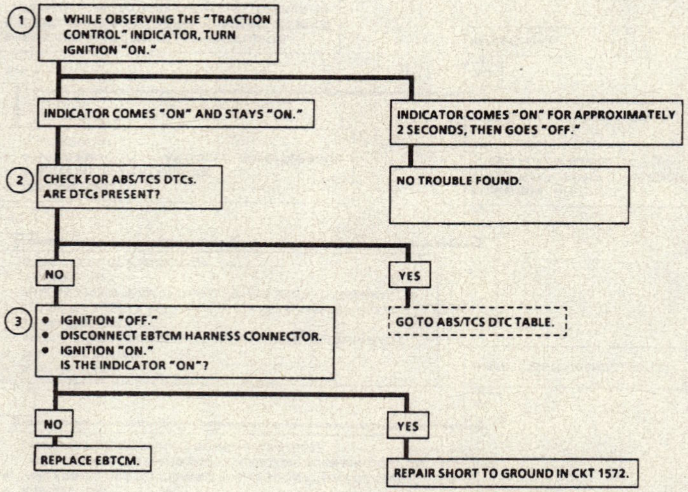

WHEN ALL DIAGNOSIS AND REPAIRS ARE COMPLETED, CLEAR DTC(s) AND VERIFY PROPER OPERATION.

DIAGNOSTIC CHART D — 1994 FLEETWOOD

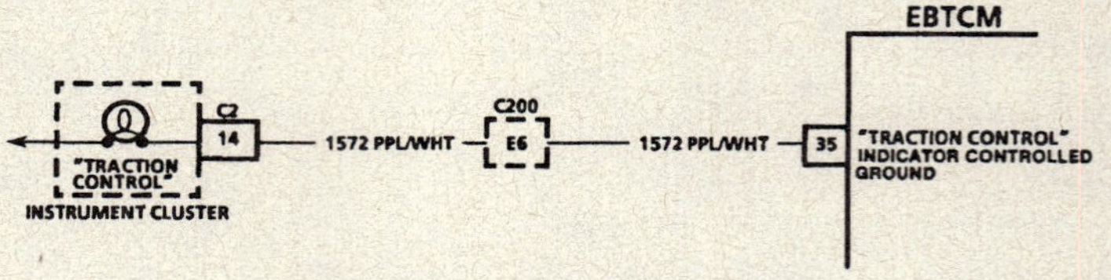

CHART D

"TRACTION CONTROL" INDICATOR INOPERATIVE

Chart Test Description: Number(s) below refer to circled number(s) on the diagnostic chart.

1. Confirms that a malfunction condition exists.
2. Checks to see if the indicator can be turned "ON" manually.
3. Verifies the integrity of the indicator bulb.
4. Checks for an indicator low side circuit open.
5. Checks for an indicator low side circuit short to B+.

DIAGNOSTIC CHART D — 1994 FLEETWOOD, CONTINUED

CHART D
"TRACTION CONTROL" INDICATOR INOPERATIVE

(1) • WHILE OBSERVING THE "TRACTION CONTROL" INDICATOR, TURN THE IGNITION "ON."

- INDICATOR DOES NOT COME "ON." → go to (2)
- INDICATOR COMES "ON" FOR APPROXIMATELY 2 SECONDS, THEN GOES "OFF." → NO TROUBLE FOUND. INSPECT FOR POOR CONNECTIONS.

(2)
- IGNITION "OFF."
- DISCONNECT EBTCM HARNESS CONNECTOR.
- INSTALL J 39700 UNIVERSAL BREAKOUT BOX TO EBTCM HARNESS CONNECTOR ONLY.
- CONNECT A FUSED JUMPER WIRE (SUCH AS J 36169) WITH A 3 AMP FUSE FROM J 39700 TERMINAL "3" (GROUND) TO TERMINAL "35".
- IGNITION "ON."

IS "TRACTION CONTROL" INDICATOR "ON"?

- NO → go to (3)
- YES →
  - IGNITION "OFF."
  - CHECK FOR PROPER CONNECTION TO EBTCM AT TERMINAL "35".

  IF OK THEN REPLACE EBTCM.

(3)
- IGNITION "OFF."
- REMOVE INSTRUMENT CLUSTER, REFER TO SECTION 8C.
- REMOVE AND INSPECT THE "TRACTION CONTROL" INDICATOR BULB.

IS BULB GOOD?

- YES → go to (4)
- NO → REPLACE BULB.

(4)
- REINSTALL BULB INTO INSTRUMENT CLUSTER.
- USING J 39200, MEASURE RESISTANCE FROM J 39700 TERMINAL "35" TO INSTRUMENT CLUSTER HARNESS CONNECTOR TERMINAL "14".

IS IT 5 OHMS OR LESS?

- YES → go to (5)
- NO → REPAIR OPEN IN CKT 1572.

(5)
- IGNITION "ON."
- USING J 39200, MEASURE VOLTAGE ON J 39700 FROM TERMINAL "35" TO TERMINAL "3" (GROUND).

IS IT 1 VOLT OR LESS?

- YES → INSPECT INSTRUMENT CLUSTER.
- NO → REPAIR SHORT TO B+ IN CKT 1572.

WHEN ALL DIAGNOSIS AND REPAIRS ARE COMPLETED, CLEAR DTC(s) AND VERIFY PROPER OPERATION.

DIAGNOSTIC CHART E — 1994 FLEETWOOD

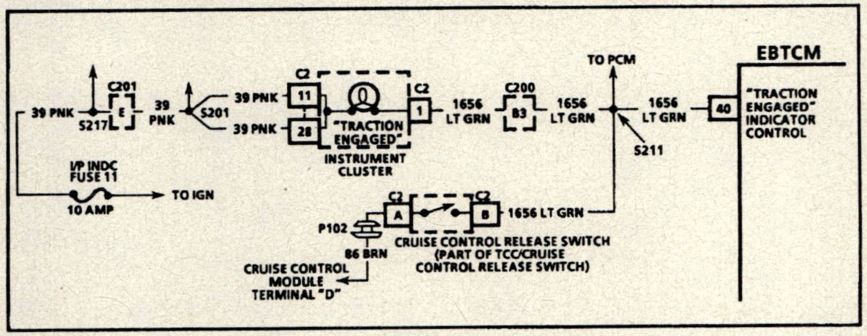

CHART E
(Page 1 of 4)

"TRACTION ENGAGED" INDICATOR "ON" WITH NO DTCs SET

Chart Test Description: Number(s) below refer to circled number(s) on the diagnostic chart.

1. Confirms that a malfunction condition exists.
2. Checks whether the malfunction is due to a DTC-setting condition.
3. Checks to see if the EBTCM is incorrectly commanding the indicator "ON" when it should not.
4. Checks to see if the PCM is internally shorted to ground, causing the indicator to remain "ON."
5. Determines whether the malfunction is due to the cruise control portion of the circuitry.
6. This test determines if the malfunction is due to an indicator low side circuit short to ground, or an internal short to ground in the instrument cluster.

DIAGNOSTIC CHART E — 1994 FLEETWOOD, CONTINUED

CHART E
(Page 2 of 4)
"TRACTION ENGAGED" INDICATOR "ON" WITH NO DTCs SET

1. • IGNITION "ON."
   DOES THE "TRACTION ENGAGED" INDICATOR COME "ON" AND STAY "ON"?
   - NO: RETURN TO "ABS/TCS FUNCTIONAL CHECK" IN THIS SECTION.
   - YES:
2. • IGNITION "OFF."
   • INSTALL TECH 1 WITH MSC.
   • IGNITION "ON."
   • REQUEST ABS/TCS DTC DISPLAY.
   - DTCs PRESENT: GO TO ABS/TCS DTC TABLE.
   - NO DTCs PRESENT:
3. • IGNITION "OFF."
   • DISCONNECT EBTCM HARNESS CONNECTOR.
   • IGNITION "ON."
   DOES INDICATOR COME "ON" AND STAY "ON"?
   - NO: INSPECT EBTCM HARNESS CONNECTOR FOR PHYSICAL DAMAGE WHICH MAY RESULT IN A SHORT TO GROUND WITH THE EBTCM CONNECTED. IF OK, THEN REPLACE EBTCM.
   - YES:
4. • IGNITION "OFF."
   • DISCONNECT PCM HARNESS CONNECTOR.
   • IGNITION "ON."
   DOES INDICATOR COME "ON" AND STAY "ON"?
   - NO: INSPECT PCM HARNESS CONNECTOR FOR PHYSICAL DAMAGE WHICH MAY RESULT IN A SHORT TO GROUND WITH THE PCM CONNECTED. IF OK, THEN REPLACE PCM.
   - YES:
5. • IGNITION "OFF."
   • DISCONNECT CRUISE CONTROL RELEASE SWITCH HARNESS CONNECTOR.
   • IGNITION "ON."
   DOES INDICATOR COME "ON" AND STAY "ON"?
   - NO: CONTINUE DIAGNOSIS ON NEXT PAGE.
   - YES:
6. • IGNITION "OFF."
   • REMOVE INSTRUMENT CLUSTER.
   • USING J 39200, MEASURE RESISTANCE BETWEEN INSTRUMENT HARNESS CLUSTER CONNECTOR TERMINAL "1" AND GROUND.
   DOES J 39200 DISPLAY "OL" (INFINITE)?
   - NO: REPAIR SHORT TO GROUND IN CKT 1656.
   - YES: SERVICE INSTRUMENT CLUSTER.

WHEN ALL DIAGNOSIS AND REPAIRS ARE COMPLETED, CLEAR DTC(s) AND VERIFY PROPER OPERATION.

DIAGNOSTIC CHART E — 1994 FLEETWOOD, CONTINUED

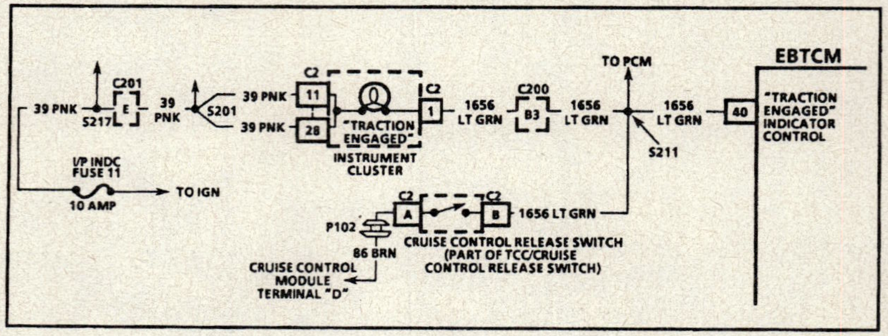

CHART E
(Page 3 of 4)
"TRACTION ENGAGED" INDICATOR "ON" WITH NO DTCs SET

Chart Test Description: Number(s) below refer to circled number(s) on the diagnostic chart.

7. Ensures the cruise control release switch is not shorted to ground.
8. Checks CKT 86 for short to ground.
9. Checks to see if the cruise control module is internally shorted to ground, causing the indicator to remain "ON."

DIAGNOSTIC CHART E — 1994 FLEETWOOD, CONTINUED

CHART E
(Page 4 of 4)
"TRACTION ENGAGED" INDICATOR "ON" WITH NO DTCs SET

CONTINUED FROM PREVIOUS PAGE.

(7)
- IGNITION "OFF."
- USING J 39200, MEASURE RESISTANCE BETWEEN CRUISE CONTROL RELEASE SWITCH TERMINAL "A" AND GROUND. REPEAT BETWEEN TERMINAL "B" AND GROUND. DOES J 39200 DISPLAY "OL" (INFINITE) FOR BOTH MEASUREMENTS?

NO → REPLACE CRUISE CONTROL RELEASE SWITCH.

YES →

(8)
- DISCONNECT CRUISE CONTROL MODULE.
- USING J 39200, MEASURE RESISTANCE BETWEEN CRUISE CONTROL MODULE HARNESS CONNECTOR TERMINAL "D" AND GROUND. DOES J 39200 DISPLAY "OL" (INFINITE)?

NO → REPAIR SHORT TO GROUND IN CKT 86.

YES →

(9)
- RECONNECT CRUISE CONTROL MODULE.
- RECONNECT CRUISE CONTROL RELEASE SWITCH.
- IGNITION "ON." DOES "TRACTION ENGAGED" INDICATOR COME "ON" AND STAY "ON"?

YES → REPLACE CRUISE CONTROL MODULE.

NO → THOROUGHLY INSPECT ALL WIRING AND CONNECTIONS FOR INTERMITTENTS, CHAFING, AND MISROUTING AT EBTCM, INSTRUMENT CLUSTER, CRUISE CONTROL MODULE AND ALL INTERCONNECTS. REPAIR ANY PROBLEMS FOUND.

DIAGNOSTIC CHART F — 1994 FLEETWOOD

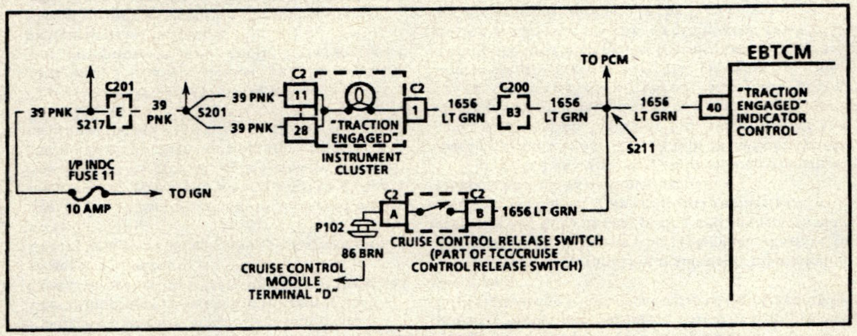

CHART F
"TRACTION ENGAGED" INDICATOR INOPERATIVE WITH NO DTCs SET

Chart Test Description: Number(s) below refer to circled number(s) on the diagnostic chart.

1. Checks for open "I/P INDC #11" fuse causing inoperative indicator.
2. Checks that other indicators fed from CKT 39 are working properly. If not, CKT 39 the instrument cluster is open.
3. Ensures indicator can be turned "ON" manually. If the indicator illuminates, malfunction is due to poor terminal contact at EBTCM harness connector or EBTCM.
4. Checks "TRACTION ENGAGED" bulb.
5. Checks for open in CKT 1656 between EBTCM and instrument cluster.
6. Checks for short to B+ on CKT 1656 causing inoperative bulb.
7. Checks for short to ground in CKT 39.
8. Checks for proper "TRACTION ENGAGED" lamp operation after fuse replacement.

DIAGNOSTIC CHART F — 1994 FLEETWOOD, CONTINUED

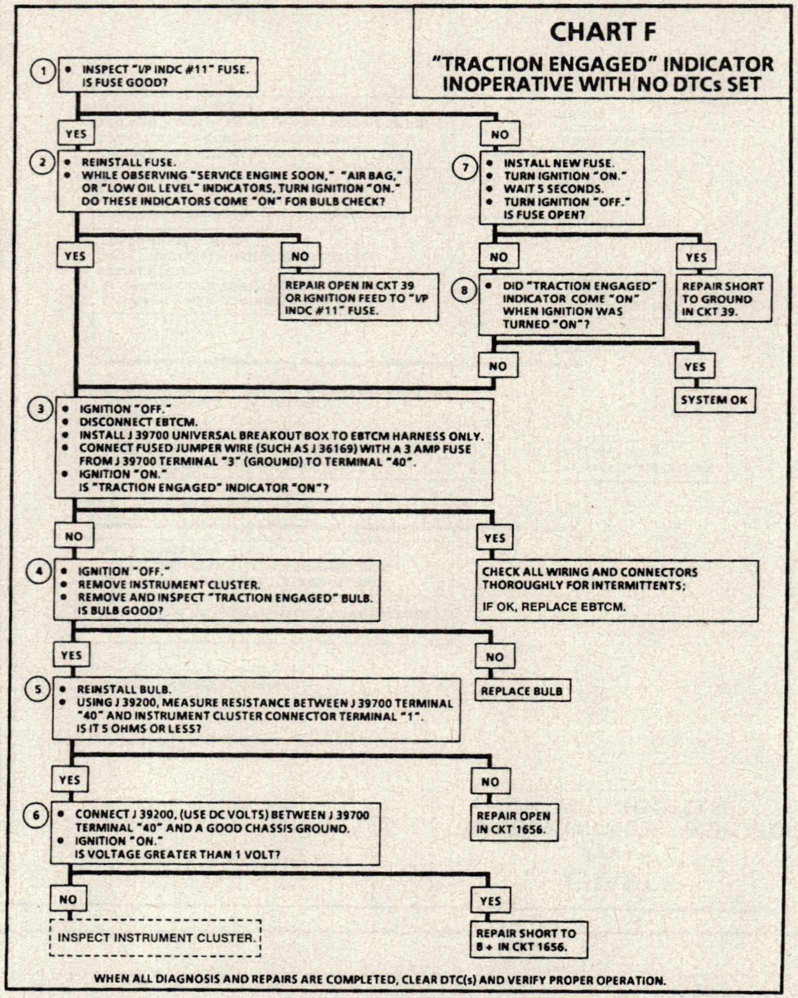

DIAGNOSTIC CHART DTC 21 — 1994 FLEETWOOD

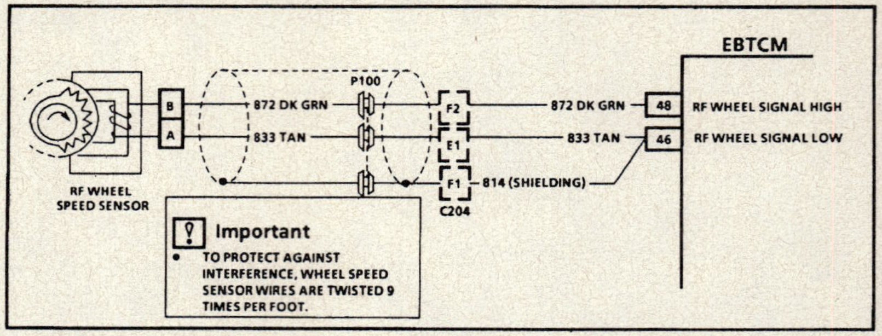

DTC 21

RF WHEEL SPEED SENSOR CIRCUIT MALFUNCTION

Circuit Description:

The toothed wheel generates a voltage pulse as it moves past the sensor; each tooth-gap-tooth series on the wheel generates these pulses. The frequency of these pulses is used by the EBTCM to determine wheel speed. The amplitude of the voltage generated in each pulse depends on the air gap between the sensor and the toothed wheel, and on wheel speed.

DTC Will Set When: DTC 21 is set if there is a short to voltage or ground in CKT 872 or 833, or a malfunctioning speed sensor. The test for this malfunction occurs when the vehicle is in motion; it will not set with the ignition "ON" and the vehicle at rest.

Action Taken: The "ANTI-LOCK" and "TRACTION CONTROL" lamps will be "ON."

DTC Chart Test Description: Number(s) below refer to circled number(s) on the diagnostic chart.

1. Checks for proper resistance in the sensor itself.
2. Checks for a short to battery in either of the speed sensor circuit wires.
3. Checks for a short between the speed sensor circuit wires.
4. Checks for a short to ground in the speed sensor input circuit wire. The ground may be either a "hard" short to ground, or a resistive (partial) short. A short with a resistance less than 2 megohms, though not a hard short, can still cause DTC 21 to set.
5. Checks if the DTC resets during a road test. If so, since tests 1-4 have validated that the circuitry and components are good, intermittent malfunctions are suspected.
6. Checks wiring and connectors for intermittents.
7. Replace the wheel speed sensor, as it is likely the cause of a DTC reset experienced during the road test, since other portions of the circuit have checked out OK. If the DTC resets after sensor replacement, the EBTCM must be concluding there is a malfunction present in the speed sensor circuit when there is not.

Diagnostic Aids: Be sure the speed sensor wiring is properly routed and retained. This will help prevent false signals due to electrical noise being picked up by the wiring.

It is very important that a thorough inspection of the wiring and connectors be performed. Failure to carefully and fully inspect wiring and connectors may result in misdiagnosis, causing part replacement with reappearance of the malfunction.

The Tech 1 can also be used to monitor the wheel speeds during a road test. Watch the wheel speeds being displayed on the Tech 1 to see if any of the readings are unusual, such as one sensor varying in speed from the other three, a signal going intermittently high or low, etc. If this does not identify the intermittent, wet the speed sensor harness on the underside of the vehicle and road test monitoring wheel speeds with the Tech 1.

DIAGNOSTIC CHART DTC 21 — 1994 FLEETWOOD, CONTINUED

DTC 21
RF WHEEL SPEED SENSOR CIRCUIT MALFUNCTION

ENSURE SENSOR IS NOT VISIBLY DAMAGED; IF IT IS DAMAGED, REPLACE THE SENSOR, CLEAR DTCs, AND ROAD TEST TO SEE IF DTC RESETS.

(1)
- IGNITION "OFF."
- DISCONNECT SENSOR CONNECTOR.
- USING J 39200 AND ADAPTER FROM J 35616-A, CHECK SENSOR RESISTANCE BY PROBING SENSOR CONNECTOR TERMINALS. IS RESISTANCE 1015-1245 OHMS?

YES → go to 2. NO → REPLACE SENSOR.

(2)
- CONNECT J 39200 BETWEEN SENSOR HARNESS CONNECTOR TERMINAL "A" AND GROUND (USE DC VOLTAGE SETTING).
- IGNITION "ON."
- NOTE VOLTAGE; REPEAT BETWEEN TERMINAL "B" AND GROUND. IS VOLTAGE GREATER THAN 2 VOLTS BETWEEN EITHER TERMINAL "A" OR "B" AND GROUND?

NO → go to 3. YES → REPAIR SHORT TO B+ IN CKT 872 OR 833; REFER TO "EBTCM-SWITCHED FAULTS" IN THIS SECTION.

(3)
- IGNITION "OFF."
- DISCONNECT EBTCM.
- INSTALL J 39700 UNIVERSAL BREAKOUT BOX TO EBTCM HARNESS CONNECTOR ONLY.
- USING J 39200, MEASURE RESISTANCE BETWEEN J 39700 TERMINALS "46" AND "48". DOES J 39200 DISPLAY RESISTANCE AS "OL" (INFINITE)?

YES → go to 4. NO → REPAIR SHORT BETWEEN CKTs 872 & 833.

(4)
- USING J 39200, MEASURE RESISTANCE BETWEEN J 39700 TERMINAL "48" AND GROUND AND TERMINAL "46" AND GROUND. DOES J 39200 DISPLAY RESISTANCE AS "OL" (INFINITE)?

YES → go to 5. NO → REPAIR SHORT TO GROUND IN CKT 872 OR CKT 833.

(5)
- REMOVE J 39700.
- RECONNECT ALL CONNECTORS.
- CLEAR DTCs, IGNITION "OFF."
- DISCONNECT TECH 1, IF CONNECTED.
- ROAD TEST VEHICLE, REACHING AT LEAST 25 km/h (15 MPH). DOES DTC 21 RESET?

YES → go to 6. NO → NO TROUBLE FOUND; SEE "DIAGNOSTIC AIDS" ON FACING PAGE.

(6)
- CHECK ALL WIRING, CONNECTORS AND TERMINALS THOROUGHLY FOR INTERMITTENTS OR CORROSION.

ARE ALL CONNECTIONS AND WIRING GOOD?

YES → go to 7. NO → REPAIR WIRING AND/OR CONNECTIONS AS NEEDED;

(7)
- CLEAR DTCs, IGNITION "OFF."
- REPLACE SENSOR.
- DISCONNECT TECH 1, IF CONNECTED.
- ROAD TEST VEHICLE, REACHING AT LEAST 25 km/h (15 MPH). DOES DTC 21 RESET?

YES → REPLACE EBTCM. NO → SYSTEM OK; SEE "DIAGNOSTIC AIDS" ON FACING PAGE.

WHEN ALL DIAGNOSIS AND REPAIRS ARE COMPLETED, CLEAR DTC(s) AND VERIFY PROPER OPERATION.

DIAGNOSTIC CHART DTC 23 — 1994 FLEETWOOD

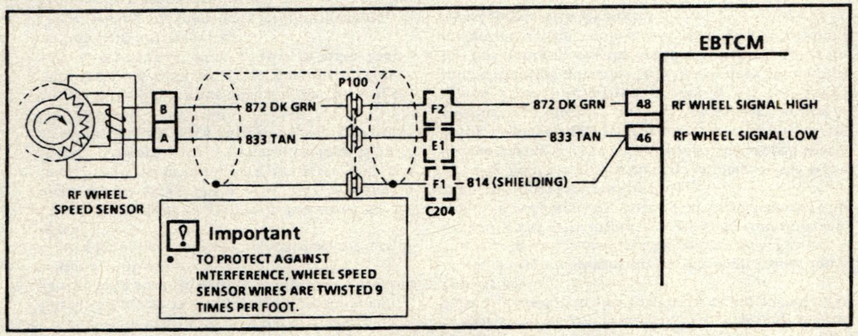

DTC 23
RF WHEEL SPEED SENSOR CIRCUIT CONTINUITY MALFUNCTION

Circuit Description:
The toothed wheel generates a voltage pulse as it moves past the sensor; each tooth-gap-tooth series on the wheel generates these pulses. The frequency of these pulses is used by the EBTCM to determine wheel speed. The amplitude of the voltage generated in each pulse depends on the air gap between the sensor and the toothed wheel, and on wheel speed.

DTC Will Set When: DTC 23 is set if there is a short to voltage or an open in CKT 872 or 833, or a malfunctioning speed sensor. The test for this malfunction occurs with the ignition "ON" and the vehicle at rest.

Action Taken: The "ANTI-LOCK" and "TRACTION CONTROL" lamps will be "ON."

DTC Chart Test Description: Number(s) below refer to circled number(s) on the diagnostic chart.

1. Checks for proper resistance in the sensor itself.
2. Checks for a short to battery in either of the speed sensor circuit wires.
3. Checks for an open in both the speed sensor circuit wires.
4. Manipulates wiring and connectors, trying to induce an intermittent malfunction not currently present.
5. This test checks if the DTC resets on key-up. If so, since tests 1-4 have validated that the circuitry and components are good, the wheel speed sensor may be causing the malfunction.
6. Checks if DTC 21 sets during a road test.
7. DTC 23 sets when the vehicle is at rest; DTC 21 sets with a malfunction in the speed sensor circuitry when the vehicle is in motion. If a malfunction is still present, DTC 21 would set during the road test, not DTC 23. If the DTC 21 sets at this point, the DTC 21 diagnostics should be used.
8. The wheel speed sensor may be causing an intermittent malfunction. If DTC 23 resets after sensor replacement, the EBTCM must be concluding there is a malfunction present in the speed sensor circuit when there is not.

Diagnostic Aids: Be sure the speed sensor wiring is properly routed and retained. This will help prevent false signals due to electrical noise being picked up by the wiring.

It is very important that a thorough inspection of the wiring and connectors be performed. Failure to carefully and fully inspect wiring and connectors may result in misdiagnosis, causing part replacement with reappearance of the malfunction.

DIAGNOSTIC CHART DTC 23 — 1994 FLEETWOOD, CONTINUED

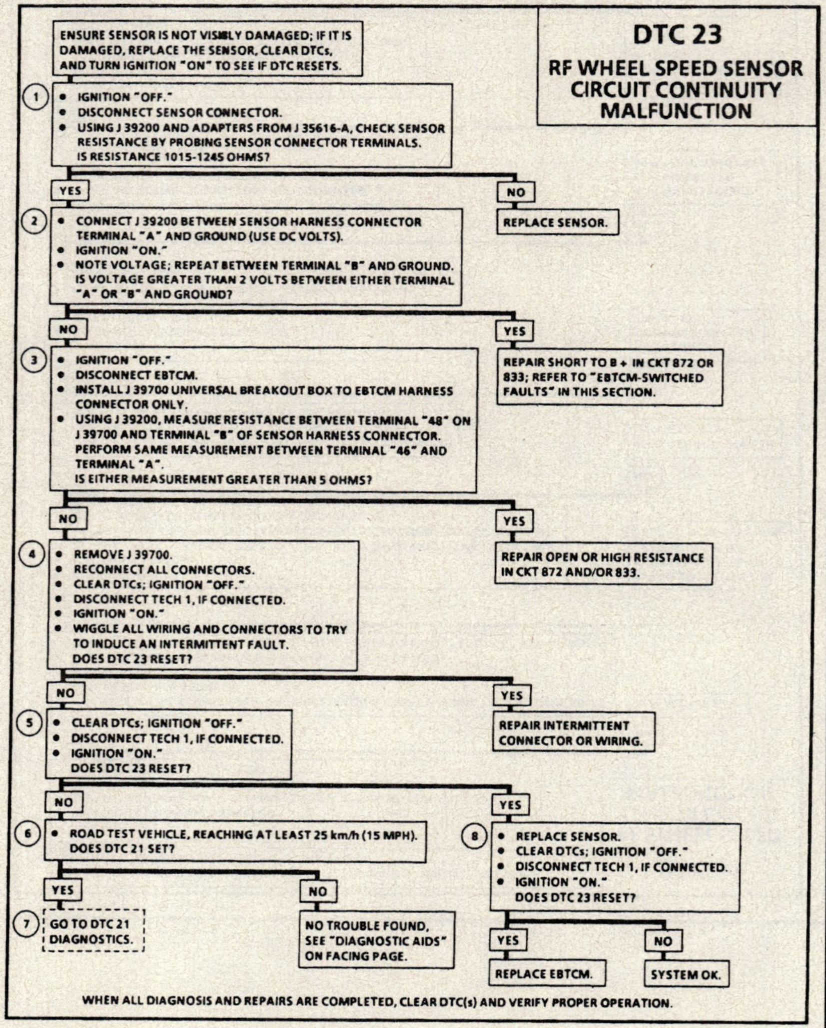

DIAGNOSTIC CHART DTC 25 — 1994 FLEETWOOD

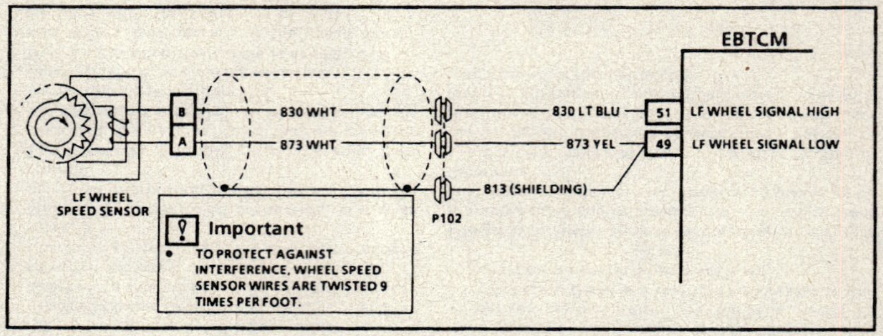

DTC 25

LF WHEEL SPEED SENSOR CIRCUIT MALFUNCTION

Circuit Description:

The toothed wheel generates a voltage pulse as it moves past the sensor; each tooth-gap-tooth series on the wheel generates these pulses. The frequency of these pulses is used by the EBTCM to determine wheel speed. The amplitude of the voltage generated in each pulse depends on the air gap between the sensor and the toothed wheel, and on wheel speed.

DTC Will Set When: DTC 25 is set if there is a short to voltage or ground in CKT 830 or 873, or a malfunctioning speed sensor. The test for this malfunction occurs when the vehicle is in motion; it will not set with the ignition "ON" and the vehicle at rest.

Action Taken: The "ANTI-LOCK" and "TRACTION CONTROL" lamps will be "ON."

DTC Chart Test Description: Number(s) below refer to circled number(s) on the diagnostic chart.

1. Checks for proper resistance in the sensor itself.
2. Checks for a short to battery in either of the speed sensor circuit wires.
3. Checks for a short between the speed sensor circuit wires.
4. Checks for a short to ground in the speed sensor input circuit wire. The ground may be either a "hard" short to ground, or a resistive (partial) short. A short with a resistance less than 2 megohms, though not a hard short, can still cause DTC 25 to set.
5. Checks if the DTC resets during a road test. If so, since tests 1-4 have validated that the circuitry and components are good, intermittent malfunctions are suspected.
6. Checks wiring and connectors for intermittents.
7. Replace the wheel speed sensor, as it is likely the cause of a DTC reset experienced during the road test, since other portions of the circuit have checked out OK. If the DTC resets after sensor replacement, the EBTCM must be concluding there is a malfunction present in the speed sensor circuit when there is not.

Diagnostic Aids: Be sure the speed sensor wiring is properly routed and retained. This will help prevent false signals due to electrical noise being picked up by the wiring.

It is very important that a thorough inspection of the wiring and connectors be performed. Failure to carefully and fully inspect wiring and connectors may result in misdiagnosis, causing part replacement with reappearance of the malfunction.

The Tech 1 can also be used to monitor the wheel speeds during a road test. Watch the wheel speeds being displayed on the Tech 1 to see if any of the readings are unusual, such as one sensor varying in speed from the other three, a signal going intermittently high or low, etc. If this does not identify the intermittent, wet the speed sensor harness on the underside of the vehicle and road test monitoring wheel speeds with the Tech 1.

DIAGNOSTIC CHART DTC 25 — 1994 FLEETWOOD, CONTINUED

DTC 25
LF WHEEL SPEED SENSOR CIRCUIT MALFUNCTION

ENSURE SENSOR IS NOT VISIBLY DAMAGED; IF IT IS DAMAGED, REPLACE THE SENSOR, CLEAR DTCs, AND ROAD TEST TO SEE IF DTC RESETS.

(1)
- IGNITION "OFF."
- DISCONNECT SENSOR CONNECTOR.
- USING J 39200 AND ADAPTERS FROM J 35616-A, CHECK SENSOR RESISTANCE BY PROBING SENSOR CONNECTOR TERMINALS. IS RESISTANCE 1015-1245 OHMS?

YES → go to (2). NO → REPLACE SENSOR.

(2)
- CONNECT J 39200 BETWEEN SENSOR HARNESS CONNECTOR TERMINAL "A" AND GROUND (USE DC VOLTS).
- IGNITION "ON."
- NOTE VOLTAGE; REPEAT BETWEEN TERMINAL "B" AND GROUND. IS VOLTAGE GREATER THAN 2 VOLTS BETWEEN EITHER TERMINAL "A" OR "B" AND GROUND?

NO → go to (3). YES → REPAIR SHORT TO B+ IN CKT 830 OR 873.

(3)
- IGNITION "OFF."
- DISCONNECT EBTCM.
- INSTALL J 39700 UNIVERSAL BREAKOUT BOX TO EBTCM HARNESS CONNECTOR ONLY.
- USING J 39200, MEASURE RESISTANCE BETWEEN J 39700 TERMINALS "49" AND "51". DOES J 39200 DISPLAY RESISTANCE AS "OL" (INFINITE)?

YES → go to (4). NO → REPAIR SHORT BETWEEN CKTs 830 AND 873.

(4)
- USING J 39200, MEASURE RESISTANCE BETWEEN J 39700 TERMINAL "51" AND GROUND AND TERMINAL "49" AND GROUND. DOES J 39200 DISPLAY RESISTANCE AS "OL" (INFINITE)?

YES → go to (5). NO → REPAIR SHORT TO GROUND IN CKT 830 OR CKT 873.

(5)
- REMOVE J 39700.
- RECONNECT ALL CONNECTORS.
- CLEAR DTCs, IGNITION "OFF."
- DISCONNECT TECH 1, IF CONNECTED.
- ROAD TEST VEHICLE, REACHING AT LEAST 25 km/h (15 MPH). DOES DTC 25 RESET?

YES → go to (6). NO → NO TROUBLE FOUND; SEE "DIAGNOSTIC AIDS" ON FACING PAGE.

(6)
- CHECK ALL WIRING, CONNECTORS AND TERMINALS THOROUGHLY FOR INTERMITTENTS OR CORROSION.

ARE ALL CONNECTORS AND WIRING GOOD?

YES → go to (7). NO → REPAIR WIRING AND/OR CONNECTIONS AS NEEDED.

(7)
- CLEAR DTCs, IGNITION "OFF."
- REPLACE SENSOR.
- DISCONNECT TECH 1, IF CONNECTED.
- ROAD TEST VEHICLE, REACHING AT LEAST 25 km/h (15 MPH). DOES DTC 25 RESET?

YES → REPLACE EBTCM. NO → SYSTEM OK; SEE "DIAGNOSTIC AIDS" ON FACING PAGE.

WHEN ALL DIAGNOSIS AND REPAIRS ARE COMPLETED, CLEAR DTC(s) AND VERIFY PROPER OPERATION.

DIAGNOSTIC CHART DTC 27 — 1994 FLEETWOOD

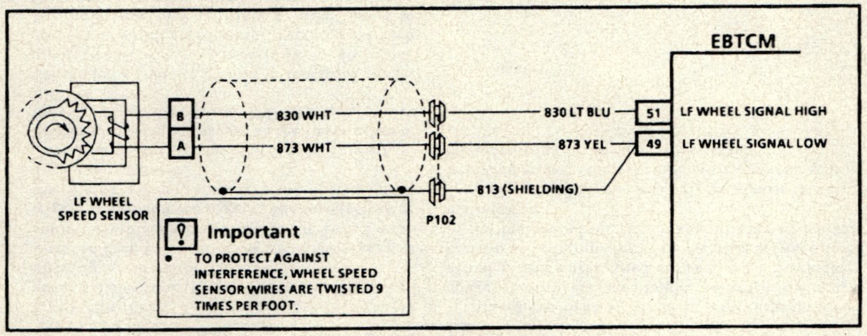

DTC 27
LF WHEEL SPEED SENSOR CIRCUIT CONTINUITY MALFUNCTION

Circuit Description:

The toothed wheel generates a voltage pulse as it moves past the sensor; each tooth-gap-tooth series on the wheel generates these pulses. The frequency of these pulses is used by the EBTCM to determine wheel speed. The amplitude of the voltage generated in each pulse depends on the air gap between the sensor and the toothed wheel, and on wheel speed.

DTC Will Set When: DTC 27 is set if there is a short to voltage or an open in CKT 830 or 873, or a malfunctioning speed sensor. The test for this malfunction occurs with the ignition "ON" and the vehicle at rest.

Action Taken: The "ANTI-LOCK" and "TRACTION CONTROL" lamps will be "ON."

DTC Chart Test Description: Number(s) below refer to circled number(s) on the diagnostic chart.

1. Checks for proper resistance in the sensor itself.
2. Checks for a short to battery in either of the speed sensor circuit wires.
3. Checks for an open in both the speed sensor circuit wires.
4. Manipulates wiring and connectors, trying to induce an intermittent malfunction not currently present.
5. This test checks if the DTC resets on key-up. If so, since Tests 1-4 have validated that the circuitry and components are good, the wheel speed sensor may be causing the malfunction.
6. Checks if DTC 25 sets during a road test.
7. DTC 27 sets when the vehicle is at rest; DTC 25 sets with a malfunction in the speed sensor circuitry when the vehicle is in motion. If a malfunction is still present, DTC 25 would set during the road test, not DTC 27. If the DTC 25 sets at this point, the DTC 25 diagnostics should be used.
8. The wheel speed sensor may be causing an intermittent malfunction. If DTC 27 resets after sensor replacement, the EBTCM must be concluding there is a malfunction present in the speed sensor circuit when there is not.

Diagnostic Aids: Be sure the speed sensor wiring is properly routed and retained. This will help prevent false signals due to electrical noise being picked up by the wiring.

It is very important that a thorough inspection of the wiring and connectors be performed. Failure to carefully and fully inspect wiring and connectors may result in misdiagnosis, causing part replacement with reappearance of the malfunction.

DIAGNOSTIC CHART DTC 27 — 1994 FLEETWOOD, CONTINUED

DTC 27
LF WHEEL SPEED SENSOR CIRCUIT CONTINUITY MALFUNCTION

ENSURE SENSOR IS NOT VISIBLY DAMAGED; IF IT IS DAMAGED, REPLACE THE SENSOR, CLEAR DTCs, AND TURN IGNITION "ON" TO SEE IF DTC RESETS.

(1)
- IGNITION "OFF."
- DISCONNECT SENSOR CONNECTOR.
- USING J 39200 AND ADAPTERS FROM J 35616-A, CHECK SENSOR RESISTANCE BY PROBING SENSOR CONNECTOR TERMINALS.

IS RESISTANCE 1015-1245 OHMS?

- NO: REPLACE SENSOR.
- YES: go to (2)

(2)
- CONNECT J 39200 BETWEEN SENSOR HARNESS CONNECTOR TERMINAL "A" AND GROUND (USE DC VOLTS).
- IGNITION "ON."
- NOTE VOLTAGE; REPEAT BETWEEN TERMINAL "B" AND GROUND.

IS VOLTAGE GREATER THAN 2 VOLTS BETWEEN EITHER TERMINAL "A" OR "B" AND GROUND?

- YES: REPAIR SHORT TO B+ IN CKT 830 OR 873.
- NO: go to (3)

(3)
- IGNITION "OFF."
- DISCONNECT EBTCM.
- INSTALL J 39700 UNIVERSAL BREAKOUT BOX TO EBTCM HARNESS CONNECTOR ONLY.
- USING J 39200, MEASURE RESISTANCE BETWEEN TERMINAL "51" ON J 39700 AND TERMINAL "B" OF SENSOR HARNESS CONNECTOR.
- PERFORM SAME MEASUREMENT BETWEEN TERMINAL "49" AND TERMINAL "A".

IS EITHER MEASUREMENT GREATER THAN 5 OHMS?

- YES: REPAIR OPEN OR HIGH RESISTANCE IN CKT 830 AND/OR 873.
- NO: go to (4)

(4)
- REMOVE J 39700.
- RECONNECT ALL CONNECTORS.
- CLEAR DTCs; IGNITION "OFF."
- DISCONNECT TECH 1, IF CONNECTED.
- IGNITION "ON."
- WIGGLE ALL WIRING AND CONNECTORS TO TRY TO INDUCE AN INTERMITTENT FAULT.

DOES DTC 27 RESET?

- YES: REPAIR INTERMITTENT CONNECTOR OR WIRING.
- NO: go to (5)

(5)
- CLEAR DTCs; IGNITION "OFF."
- DISCONNECT TECH 1, IF CONNECTED.
- IGNITION "ON."

DOES DTC 27 RESET?

- YES: go to (8)
- NO: go to (6)

(6)
- ROAD TEST VEHICLE, REACHING AT LEAST 25 km/h (15 MPH).

DOES DTC 25 SET?

- YES: (7) GO TO DTC 25 DIAGNOSTICS.
- NO: NO TROUBLE FOUND, SEE "DIAGNOSTIC AIDS" ON FACING PAGE.

(8)
- REPLACE SENSOR.
- CLEAR DTCs; IGNITION "OFF."
- DISCONNECT TECH 1, IF CONNECTED.
- IGNITION "ON."

DOES DTC 27 RESET?

- YES: REPLACE EBTCM.
- NO: SYSTEM OK.

WHEN ALL DIAGNOSIS AND REPAIRS ARE COMPLETED, CLEAR DTC(s) AND VERIFY PROPER OPERATION.

DIAGNOSTIC CHART DTC 28 — 1994 FLEETWOOD

DTC 28
WHEEL SPEED SENSOR CIRCUIT FREQUENCY MALFUNCTION

Circuit Description:

The toothed wheel generates a voltage pulse as it moves past the sensor; each tooth-gap-tooth series on the wheel generates these pulses. The frequency of these pulses is used by the EBTCM to determine wheel speed. The amplitude of the voltage generated in each pulse depends on the air gap between the sensor and the toothed wheel, and on wheel speed.

DTC Will Set When: DTC 28 is set if the EBTCM cannot specifically identify which wheel speed sensor is causing the frequency malfunction. If the EBTCM can define the specific speed sensor causing the malfunction, the sensor DTC associated with the sensor (21, 25, 31, 35) will be set instead of DTC 28.

Action Taken: The "ANTI-LOCK" and "TRACTION CONTROL" lamps will be "ON."

DTC Chart Test Description: Number(s) below refer to circled number(s) on the diagnostic chart.

1. Checks wiring and connections for malfunctions. It is critical that this step is performed thoroughly, as wiring/connector malfunctions are the most likely cause for this DTC being set.
2. Uses the Tech 1 to monitor for vehicle electrical/electronic system "noise" being picked up by the speed sensor circuits.
3. Uses the Tech 1 to monitor the wheel speed sensors while the vehicle is operating. If the Tech 1 auto-trigger snapshot mode triggers, the speed sensor or the wiring and/or connectors are intermittent during the road test.
4. Checks for proper resistance in the sensor itself.
5. Replaces wheel speed sensor as the likely cause of the triggering intermittent. If DTC resets, replace EBTCM.
6. Checks for a short to ground in the speed sensor input circuit wires. The circuit to be tested depends on which speed sensor triggered the Tech 1. The ground may be either a "hard" short to ground, or a resistive (partial) short. A short with a resistance less than 2 megohms, though not a hard short, can still cause DTC 28 to set.

Diagnostic Aids: DTC 28 may be set by running the Tech 1 "Auto Test" if the throttle angle readings are not updating while in the data list mode. If this is the case, clear DTCs, disconnect Tech 1, and road test to at least 25 km/h (15 mph) to see if DTC resets.

Rear Speed Sensors Only:

Check the toothed wheel for any large grooves, gouges, marks, etc. that might influence the tooth's signal at the wheel speed sensor. Also check for a buildup of foreign material in the gaps between teeth in the toothed wheel; this material may cause this malfunction.

A worn hub/bearing assembly may cause this fault in extreme cases; the bearing play allows the sensor-to-toothed-ring gap to change excessively.

DIAGNOSTIC CHART DTC 28 — 1994 FLEETWOOD, CONTINUED

DTC 28
WHEEL SPEED SENSOR CIRCUIT FREQUENCY MALFUNCTION

(1) VERY THOROUGHLY INSPECT ALL WIRING AND CONNECTIONS FOR INTERMITTENTS, CHAFING, POOR TERMINAL CONTACT TENSION, CORROSION, AND MISROUTING AT EBTCM, ALL SENSORS, AND ALL HARNESS INTERCONNECTS. REPAIR ANY PROBLEMS, CLEAR DTCs, THEN ROAD TEST VEHICLE, REACHING AT LEAST 25 km/h (15 MPH). DOES DTC RESET?

- YES → (2)
- NO → NO TROUBLE FOUND. SEE "DIAGNOSTIC AIDS" ON FACING PAGE.

(2)
- INSTALL TECH 1 AND MSC. CLEAR DTCs.
- SET TECH 1 AUTO TRIGGER SNAPSHOT MODE.
- START VEHICLE AND ALLOW TO IDLE FOR ONE MINUTE. DOES TECH 1 TRIGGER?

- NO → (3)
- YES → (6)

(3)
- ROAD TEST VEHICLE REACHING AT LEAST 25 km/h (15 MPH). DOES TECH 1 TRIGGER?

- YES → NOTE TECH 1, WHICH SENSOR TRIGGERED? → (4)
- NO → NO TROUBLE FOUND; SEE "DIAGNOSTIC AIDS" ON FACING PAGE.

(4)
- IGNITION "OFF."
- DISCONNECT SENSOR CONNECTOR.
- USING J 39200, MEASURE RESISTANCE BETWEEN SENSOR TERMINALS, USING ADAPTERS FROM J 35616-A IF MEASURING A FRONT SENSOR. IS RESISTANCE 1015-1245 OHMS?

- NO → (5)
- YES → REPLACE EBTCM.

(5)
- REPLACE SPEED SENSOR.
- CLEAR DTCs, IGNITION "OFF."
- DISCONNECT TECH 1, IF CONNECTED.
- ROAD TEST VEHICLE, REACHING AT LEAST 25 km/h (15 MPH). DOES DTC 28 RESET?

- YES → REPLACE EBTCM.
- NO → SYSTEM OK.

(6)
- NOTE WITH TECH 1 WHICH SENSOR TRIGGERED.
- IGNITION "OFF."
- DISCONNECT EBTCM.
- INSTALL J 39700 UNIVERSAL BREAKOUT BOX TO EBTCM AND EBTCM HARNESS CONNECTOR.
- USING TABLE BELOW, WITH J 39200, MEASURE RESISTANCE TO GROUND AT THE TERMINAL ON J 39700 THAT CORRESPONDS TO THE SPEED SENSOR THE TECH 1 TRIGGERED ON:

LF-TERMINAL "51"
RF-TERMINAL "48"
LR-TERMINAL "45"
RR-TERMINAL "43"

- IF RESISTANCE IS 5 OHMS OR LESS TO GROUND, REPAIR HARD SHORT; IF BETWEEN 5 OHMS AND 2 MEGOHMS, REPAIR RESISTIVE (PARTIAL) SHORT IN CKT:

LF-CKT 830
RF-CKT 872
LR-CKT 884
RR-CKT 882

WHEN ALL DIAGNOSIS AND REPAIRS ARE COMPLETED, CLEAR DTC(s) AND VERIFY PROPER OPERATION.

DIAGNOSTIC CHART DTC 31 — 1994 FLEETWOOD

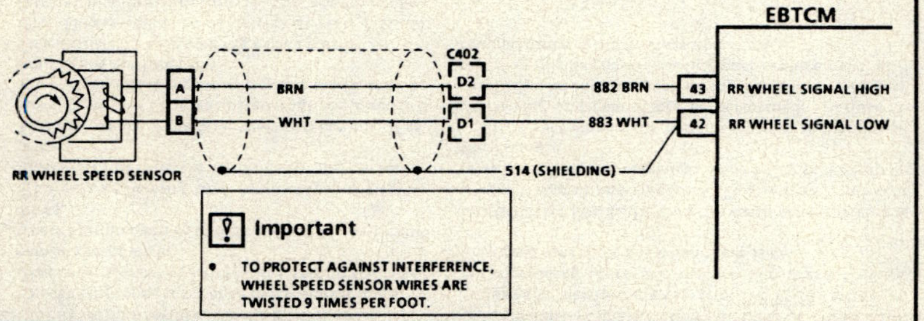

DTC 31
RR WHEEL SPEED SENSOR CIRCUIT MALFUNCTION

Circuit Description:

The toothed wheel generates a voltage pulse as it moves past the sensor; each tooth-gap-tooth series on the wheel generates these pulses. The frequency of these pulses is used by the EBTCM to determine wheel speed. The amplitude of the voltage generated in each pulse depends on the air gap between the sensor and the toothed wheel, and on wheel speed.

DTC Will Set When: DTC 31 is set if there is a short to voltage or ground in CKT 882 or 883, or a malfunctioning speed sensor. The test for this malfunction occurs when the vehicle is in motion; it will not set with the ignition "ON" and the vehicle at rest.

Action Taken: The "ANTI-LOCK" and "TRACTION CONTROL" lamps will be "ON."

DTC Chart Test Description: Number(s) below refer to circled number(s) on the diagnostic chart.

1. Checks for proper resistance in the sensor itself.
2. Checks for a short to battery in either of the speed sensor circuit wires.
3. Checks for a short between the speed sensor circuit wires.
4. Checks for a short to ground in the speed sensor input circuit wire. The ground may be either a "hard" short to ground, or a resistive (partial) short. A short with a resistance less than 2 megohms, though not a hard short, can still cause DTC 31 to set.
5. Checks if the DTC resets during a road test. If so, since tests 1-4 have validated that all of the circuitry and components are good, intermittent malfunctions are suspected.
6. Checks wiring and connectors for intermittents.
7. Checks for sensor mounting, toothed wheel, and rear wheel bearing problems which may be causing the malfunction. Check the toothed wheel for any large grooves, gouges, marks, etc. that might influence the tooth's signal at the wheel speed sensor. Also check for a buildup of foreign material in the gaps between teeth in the toothed wheel; this material may cause this malfunction.
8. Replace the speed sensor assembly, as it is likely the cause of a DTC reset experienced during the road test, since other portions of the circuit have checked out OK.

Diagnostic Aids: A worn hub-bearing assembly may cause this malfunction in extreme cases; the bearing play allows the sensor-to-toothed-ring gap to change excessively.

Be sure the speed sensor wiring is properly routed and retained. This will help prevent false signals due to electrical noise being picked up by the wiring.

It is very important that a thorough inspection of the wiring and connectors be performed. Failure to carefully and fully inspect wiring and connectors may result in misdiagnosis, causing part replacement with reappearance of the malfunction.

The Tech 1 can also be used to monitor the wheel speeds during a road test. Watch the wheel speeds being displayed on the Tech 1 to see if any of the readings are unusual, such as one sensor varying in speed from the other three, a signal going intermittently high or low, etc. If this does not identify the intermittent, wet the speed sensor harness on the underside of the vehicle and road test monitoring wheel speeds with the Tech 1.

DIAGNOSTIC CHART DTC 31 — 1994 FLEETWOOD, CONTINUED

DTC 31
RR WHEEL SPEED SENSOR CIRCUIT MALFUNCTION

ENSURE SENSOR IS NOT VISIBLY DAMAGED, AND THAT IT IS PROPERLY MOUNTED. IF IT IS NOT PROPERLY MOUNTED OR RETAINED, PROPERLY ATTACH IT; IF IT IS DAMAGED, REPLACE IT. THEN, CLEAR DTCs AND ROAD TEST TO SEE IF DTC RESETS.

(1)
- IGNITION "OFF."
- DISCONNECT SENSOR CONNECTOR.
- USING J 39200, MEASURE SENSOR RESISTANCE BY PROBING SENSOR CONNECTOR PIGTAIL TERMINALS. IS RESISTANCE 1015-1245 OHMS?

NO: REPLACE SPEED SENSOR.

YES:

(2)
- CONNECT J 39200 BETWEEN SENSOR HARNESS CONNECTOR TERMINAL "A" AND GROUND (USE DC VOLTS).
- IGNITION "ON."
- NOTE VOLTAGE; REPEAT BETWEEN TERMINAL "B" AND GROUND. IS VOLTAGE GREATER THAN 2 VOLTS BETWEEN EITHER TERMINAL "A" OR "B" AND GROUND?

YES: REPAIR SHORT TO B+ IN CKT 882 OR 883.

NO:

(3)
- IGNITION "OFF."
- DISCONNECT EBTCM.
- INSTALL J 39700 UNIVERSAL BREAKOUT BOX TO EBTCM HARNESS CONNECTOR ONLY.
- USING J 39200, MEASURE RESISTANCE BETWEEN J 39700 TERMINALS "42" AND "43". DOES J 39200 DISPLAY RESISTANCE AS "OL" (INFINITE)?

NO: REPAIR SHORT BETWEEN CKTs 882 AND 883.

YES:

(4)
- USING J 39200, MEASURE RESISTANCE BETWEEN J 39700 TERMINAL "43" AND GROUND, AND TERMINAL "42" AND GROUND. DOES J 39200 DISPLAY RESISTANCE AS "OL" (INFINITE)?

NO: REPAIR SHORT TO GROUND IN CKT 882 OR CKT 883.

YES:

(5)
- REMOVE J 39700.
- RECONNECT ALL CONNECTORS.
- CLEAR DTCs, IGNITION "OFF."
- DISCONNECT TECH 1, IF CONNECTED.
- ROAD TEST VEHICLE, REACHING AT LEAST 25 km/h (15 MPH). DOES DTC 31 RESET?

NO: NO TROUBLE FOUND; SEE "DIAGNOSTIC AIDS" ON FACING PAGE.

YES:

(6)
- CHECK ALL WIRING, CONNECTORS AND TERMINALS THOROUGHLY FOR INTERMITTENTS OR CORROSION.

ARE ALL CONNECTIONS AND WIRING GOOD?

NO: REPAIR WIRING AND/OR CONNECTIONS AS NEEDED.

YES:

(7)
- CHECK SENSOR MOUNTING, TOOTHED WHEEL FOR DAMAGE OR FOREIGN MATERIAL, AND FOR EXCESSIVELY WORN REAR WHEEL BEARING; SEE "DIAGNOSTIC AIDS" ON FACING PAGE. IS THERE PHYSICAL DAMAGE/WEAR INDICATED?

YES: REPAIR AS NECESSARY.

NO:

(8)
- REPLACE SPEED SENSOR
- CLEAR DTCs, IGNITION "OFF."
- DISCONNECT TECH 1, IF CONNECTED.
- ROAD TEST VEHICLE, REACHING AT LEAST 25 km/h (15 MPH). DOES DTC 31 RESET?

YES: REPLACE EBTCM.

NO: SYSTEM OK; SEE "DIAGNOSTIC AIDS" ON FACING PAGE.

WHEN ALL DIAGNOSIS AND REPAIRS ARE COMPLETED, CLEAR DTC(s) AND VERIFY PROPER OPERATION.

DIAGNOSTIC CHART DTC 33 — 1994 FLEETWOOD

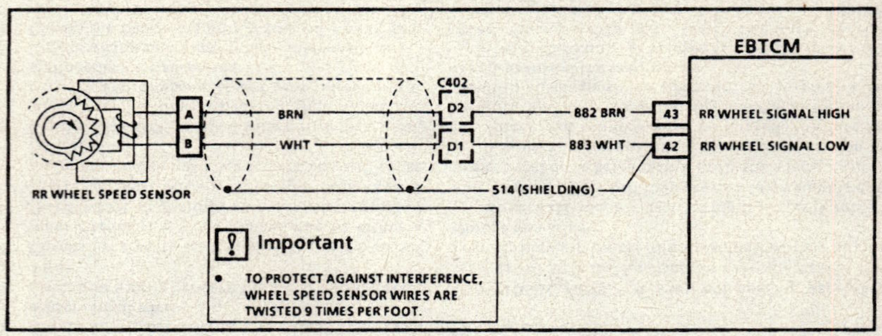

DTC 33
RR WHEEL SPEED SENSOR CIRCUIT CONTINUITY MALFUNCTION

Circuit Description:

The toothed wheel generates a voltage pulse as it moves past the sensor; each tooth-gap-tooth series on the wheel generates these pulses. The frequency of these pulses is used by the EBTCM to determine wheel speed. The amplitude of the voltage generated in each pulse depends on the air gap between the sensor and the toothed wheel, and on wheel speed.

DTC Will Set When: DTC 33 is set if there is a short to voltage or an open in CKT 882 or 883, or a malfunctioning speed sensor. The test for this malfunction occurs with the ignition "ON" and the vehicle at rest.

Action Taken: The "ANTI-LOCK" and "TRACTION CONTROL" lamps will be "ON."

DTC Chart Test Description: Number(s) below refer to circled number(s) on the diagnostic chart.

1. Checks for proper resistance in the sensor itself.
2. Checks for a short to battery in either of the speed sensor circuit wires.
3. Checks for an open in both the speed sensor circuit wires.
4. Manipulates wiring and connectors, trying to induce an intermittent malfunction not currently present.
5. This test checks if the DTC resets on key-up. If so, since Tests 1-4 have validated that the circuitry and components are good, the speed sensor may be causing the malfunction.
6. Checks if DTC 31 sets during a road test.
7. DTC 33 sets when the vehicle is at rest; DTC 31 sets with a malfunction in the speed sensor circuitry when the vehicle is in motion. If a malfunction is still present, DTC 31 would set during the road test, not DTC 33. If the DTC 33 sets at this point, the DTC 31 diagnostics should be used.
8. The wheel speed sensor may be causing an intermittent malfunction. If DTC 33 resets after sensor replacement, the EBTCM must be concluding there is a malfunction present in the speed sensor circuit when there is not.

Diagnostic Aids: Be sure the speed sensor wiring is properly routed and retained. This will help prevent false signals due to electrical noise being picked up by the wiring.

It is very important that a thorough inspection of the wiring and connectors be performed. Failure to carefully and fully inspect wiring and connectors may result in misdiagnosis, causing part replacement with reappearance of the malfunction.

DIAGNOSTIC CHART DTC 33 — 1994 FLEETWOOD, CONTINUED

DTC 33
RR WHEEL SPEED SENSOR CIRCUIT CONTINUITY MALFUNCTION

ENSURE SENSOR IS NOT VISIBLY DAMAGED, AND THAT IT IS PROPERLY MOUNTED. IF IT IS NOT PROPERLY MOUNTED OR RETAINED, PROPERLY ATTACH IT; IF IT IS DAMAGED, REPLACE IT. THEN, CLEAR DTCs AND ROAD TEST TO SEE IF DTC RESETS.

(1)
- IGNITION "OFF."
- DISCONNECT SENSOR CONNECTOR.
- USING J 39200, MEASURE SENSOR RESISTANCE BY PROBING SENSOR CONNECTOR PIGTAIL TERMINALS. IS RESISTANCE 1015-1245 OHMS?

NO: REPLACE SPEED SENSOR.

YES:

(2)
- CONNECT J 39200 BETWEEN SENSOR HARNESS CONNECTOR TERMINAL "A" AND GROUND (USE DC VOLTS).
- IGNITION "ON."
- NOTE VOLTAGE; REPEAT BETWEEN TERMINAL "B" AND GROUND. IS VOLTAGE GREATER THAN 2 VOLTS BETWEEN EITHER TERMINAL "A" OR "B" AND GROUND?

YES: REPAIR SHORT TO B+ IN CKT 882 OR 883.

NO:

(3)
- IGNITION "OFF."
- DISCONNECT EBTCM.
- INSTALL J 39700 UNIVERSAL BREAKOUT BOX TO EBTCM HARNESS CONNECTOR ONLY.
- USING J 39200, MEASURE RESISTANCE BETWEEN TERMINAL "43" ON J 39700 AND TERMINAL "A" OF SENSOR HARNESS CONNECTOR. PERFORM SAME MEASUREMENT BETWEEN TERMINAL "42" AND TERMINAL "B". IS EITHER MEASUREMENT GREATER THAN 5 OHMS?

YES: REPAIR OPEN OR HIGH RESISTANCE IN CKT 882 AND/OR 883.

NO:

(4)
- REMOVE J 39700; RECONNECT ALL CONNECTORS.
- CLEAR DTCs, IGNITION "OFF."
- DISCONNECT TECH 1, IF CONNECTED.
- IGNITION "ON."
- WIGGLE ALL WIRING AND CONNECTORS TO TRY TO INDUCE AN INTERMITTENT MALFUNCTION. DOES DTC 33 RESET?

YES: REPAIR INTERMITTENT CONNECTOR OR WIRING.

NO:

(5)
- CLEAR DTCs, IGNITION "OFF."
- DISCONNECT TECH 1, IF CONNECTED.
- IGNITION "ON." DOES DTC 33 RESET?

YES:

(8)
- REPLACE SPEED SENSOR.
- CLEAR DTCs, IGNITION "OFF."
- DISCONNECT TECH 1, IF CONNECTED.
- IGNITION "ON." DOES DTC 33 RESET?

YES: REPLACE EBTCM

NO: SYSTEM OK

NO:

(6)
- ROAD TEST VEHICLE, REACHING AT LEAST 25 km/h (15 MPH). DOES DTC 31 SET?

YES: (7) GO TO DTC 31 DIAGNOSTICS.

NO: NO TROUBLE FOUND SEE "DIAGNOSTIC AIDS" ON FACING PAGE.

WHEN ALL DIAGNOSIS AND REPAIRS ARE COMPLETED, CLEAR DTC(s) AND VERIFY PROPER OPERATION.

DIAGNOSTIC CHART DTC 35 — 1994 FLEETWOOD

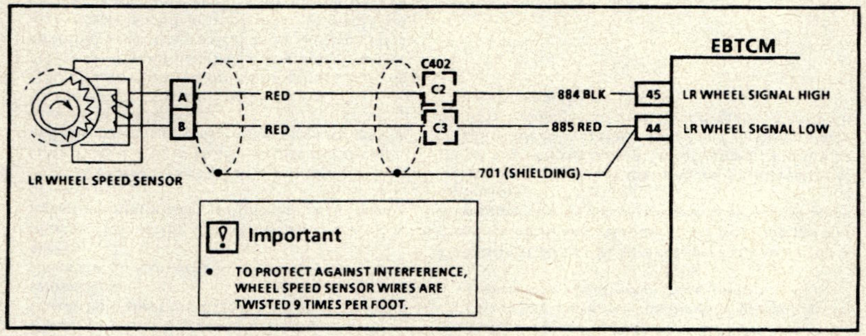

DTC 35
LR WHEEL SPEED SENSOR CIRCUIT MALFUNCTION

Circuit Description:
The toothed wheel generates a voltage pulse as it moves past the sensor; each tooth-gap-tooth series on the wheel generates these pulses. The frequency of these pulses is used by the EBTCM to determine wheel speed. The amplitude of the voltage generated in each pulse depends on the air gap between the sensor and the toothed wheel, and on wheel speed.

DTC Will Set When: DTC 35 is set if there is a short to voltage or ground in CKT 885 or 884, or a malfunctioning speed sensor. The test for this malfunction occurs when the vehicle is in motion; it will not set with the ignition "ON" and the vehicle at rest.

Action Taken: The "ANTI-LOCK" and "TRACTION CONTROL" lamps will be "ON."

DTC Chart Test Description: Number(s) below refer to circled number(s) on the diagnostic chart.

1. Checks for proper resistance in the sensor itself.
2. Checks for a short to battery in either of the speed sensor circuit wires.
3. Checks for a short between the speed sensor circuit wires.
4. Checks for a short to ground in the speed sensor input circuit wire. The ground may be either a "hard" short to ground, or a resistive (partial) short. A short with a resistance less than 2 megohms, though not a hard short, can still cause DTC 35 to set.
5. Checks if the DTC resets during a road test. If so, since tests 1-4 have validated that all of the circuitry and components are good, intermittent malfunctions are suspected.
6. Checks wiring and connectors for intermittents.
7. Checks for sensor mounting, toothed wheel, and rear wheel bearing problems which may be causing the malfunction. Check the toothed wheel for any large grooves, gouges, marks, etc. that might influence the tooth's signal at the wheel speed sensor. Also check for a buildup of foreign material in the gaps between teeth in the toothed wheel; this material may cause this malfunction.
8. Replace the speed sensor assembly, as it is likely the cause of a DTC reset experienced during the road test, since other portions of the circuit have checked out OK.

Diagnostic Aids: A worn hub-bearing assembly may cause this malfunction in extreme cases; the bearing play allows the sensor-to-toothed-ring gap to change excessively.

Be sure the speed sensor wiring is properly routed and retained. This will help prevent false signals due to electrical noise being picked up by the wiring.

It is very important that a thorough inspection of the wiring and connectors be performed. Failure to carefully and fully inspect wiring and connectors may result in misdiagnosis, causing part replacement with reappearance of the malfunction.

The Tech 1 can also be used to monitor the wheel speeds during a road test. Watch the wheel speeds being displayed on the Tech 1 to see if any of the readings are unusual, such as one sensor varying in speed from the other three, a signal going intermittently high or low, etc. If this does not identify the intermittent, wet the speed sensor harness on the underside of the vehicle and road test monitoring wheel speeds with the Tech 1.

DIAGNOSTIC CHART DTC 35 — 1994 FLEETWOOD, CONTINUED

DTC 35
LR WHEEL SPEED SENSOR CIRCUIT MALFUNCTION

ENSURE SENSOR IS NOT VISIBLY DAMAGED, AND THAT IT IS PROPERLY MOUNTED. IF IT IS NOT PROPERLY MOUNTED OR RETAINED, PROPERLY ATTACH IT; IF IT IS DAMAGED, REPLACE IT. THEN, CLEAR DTCs AND ROAD TEST TO SEE IF DTC RESETS.

1
- IGNITION "OFF."
- DISCONNECT SENSOR CONNECTOR.
- USING J 39200, MEASURE SENSOR RESISTANCE BY PROBING SENSOR CONNECTOR PIGTAIL TERMINALS.

IS RESISTANCE 1015-1245 OHMS?

NO: REPLACE SPEED SENSOR.

YES:

2
- CONNECT J 39200 BETWEEN SENSOR HARNESS CONNECTOR TERMINAL "A" AND GROUND (USE DC VOLTS).
- IGNITION "ON."
- NOTE VOLTAGE; REPEAT BETWEEN TERMINAL "B" AND GROUND.

IS VOLTAGE GREATER THAN 2 VOLTS BETWEEN EITHER TERMINAL "A" OR "B"?

YES: REPAIR SHORT TO B + IN CKT 885 OR 884.

NO:

3
- IGNITION "OFF."
- DISCONNECT EBTCM.
- INSTALL J 39700 UNIVERSAL BREAKOUT BOX TO EBTCM HARNESS CONNECTOR ONLY.
- USING J 39200, MEASURE RESISTANCE BETWEEN J 39700 TERMINALS "44" AND "45".

DOES J 39200 DISPLAY RESISTANCE AS "OL" (INFINITE)?

NO: REPAIR SHORT BETWEEN CKTs 885 AND 884.

YES:

4
- USING J 39200, MEASURE RESISTANCE BETWEEN J 39700 TERMINAL "45" AND GROUND AND TERMINAL "44" AND GROUND.

DOES J 39200 DISPLAY RESISTANCE AS "OL" (INFINITE)?

NO: REPAIR SHORT TO GROUND IN CKT 884 OR CKT 885.

YES:

5
- REMOVE J 39700.
- RECONNECT ALL CONNECTORS.
- CLEAR DTCs, IGNITION "OFF."
- DISCONNECT TECH 1, IF CONNECTED.
- ROAD TEST VEHICLE, REACHING AT LEAST 25 km/h (15 MPH).

DOES DTC 35 RESET?

NO: NO TROUBLE FOUND; SEE "DIAGNOSTIC AIDS" ON FACING PAGE.

YES:

6
- CHECK ALL WIRING, CONNECTORS AND TERMINALS THOROUGHLY FOR INTERMITTENTS OR CORROSION.

ARE ALL CONNECTIONS AND WIRING GOOD?

NO: REPAIR WIRING AND/OR CONNECTIONS AS NEEDED.

YES:

7
- CHECK SENSOR MOUNTING, TOOTHED WHEEL FOR DAMAGE OR FOREIGN MATERIAL, AND FOR EXCESSIVELY WORN REAR WHEEL BEARING; SEE "DIAGNOSTIC AIDS" ON FACING PAGE.

IS THERE PHYSICAL DAMAGE/WEAR INDICATED?

YES: REPAIR AS NECESSARY.

NO:

8
- REPLACE SPEED SENSOR
- CLEAR DTCs, IGNITION "OFF."
- DISCONNECT TECH 1, IF CONNECTED.
- ROAD TEST VEHICLE, REACHING AT LEAST 25 km/h (15 MPH).

DOES DTC 35 RESET?

YES: REPLACE EBTCM.

NO: SYSTEM OK; SEE "DIAGNOSTIC AIDS" ON FACING PAGE.

WHEN ALL DIAGNOSIS AND REPAIRS ARE COMPLETED, CLEAR DTC(s) AND VERIFY PROPER OPERATION.

DIAGNOSTIC CHART DTC 37 — 1994 FLEETWOOD

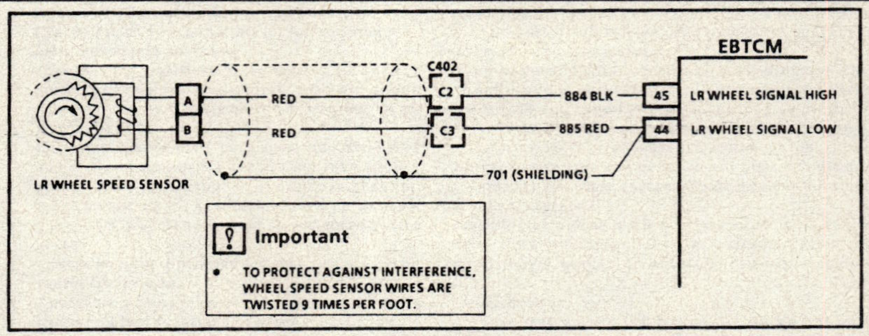

DTC 37
LR WHEEL SPEED SENSOR CIRCUIT CONTINUITY MALFUNCTION

Circuit Description:

The toothed wheel generates a voltage pulse as it moves past the sensor; each tooth-gap-tooth series on the wheel generates these pulses. The frequency of these pulses is used by the EBTCM to determine wheel speed. The amplitude of the voltage generated in each pulse depends on the air gap between the sensor and the toothed wheel, and on wheel speed.

DTC Will Set When: DTC 37 is set if there is a short to voltage or an open in CKT 885 or 884, or a malfunctioning speed sensor. The test for this malfunction occurs with the ignition "ON" and the vehicle at rest.

Action Taken: The "ANTI-LOCK" and "TRACTION CONTROL" lamps will be "ON."

DTC Chart Test Description: Number(s) below refer to circled number(s) on the diagnostic chart.

1. Checks for proper resistance in the sensor itself.
2. Checks for a short to battery in either of the speed sensor circuit wires.
3. Checks for an open in both the speed sensor circuit wires.
4. Manipulates wiring and connectors, trying to induce an intermittent malfunction not currently present.
5. This test checks if the DTC resets on key-up. If so, since Tests 1-4 have validated that the circuitry and components are good, the speed sensor may be causing the malfunction.
6. Checks if DTC 35 sets during a road test.
7. DTC 37 sets when the vehicle is at rest; DTC 35 sets with a malfunction in the speed sensor circuitry when the vehicle is in motion. If a malfunction is still present, DTC 35 would set during the road test, not DTC 37. If the DTC 35 sets at this point, the DTC 35 diagnostics should be used.
8. The wheel speed sensor may be causing an intermittent malfunction. If DTC 37 resets after sensor replacement, the EBTCM must be concluding there is a malfunction present in the speed sensor circuit when there is not.

Diagnostic Aids: Be sure the speed sensor wiring is properly routed and retained. This will help prevent false signals due to electrical noise being picked up by the wiring.

It is very important that a thorough inspection of the wiring and connectors be performed. Failure to carefully and fully inspect wiring and connectors may result in misdiagnosis, causing part replacement with reappearance of the malfunction.

DIAGNOSTIC CHART DTC 37 — 1994 FLEETWOOD, CONTINUED

DTC 37
LR WHEEL SPEED SENSOR CIRCUIT CONTINUITY MALFUNCTION

ENSURE SENSOR IS NOT VISIBLY DAMAGED, AND THAT IT IS PROPERLY MOUNTED. IF IT IS NOT PROPERLY MOUNTED OR RETAINED, PROPERLY ATTACH IT; IF IT IS DAMAGED, REPLACE IT. THEN, CLEAR DTCs AND ROAD TEST TO SEE IF DTC RESETS.

(1)
- IGNITION "OFF."
- DISCONNECT SENSOR CONNECTOR.
- USING J 39200, MEASURE SENSOR RESISTANCE BY PROBING SENSOR CONNECTOR PIGTAIL TERMINALS. IS RESISTANCE 1015-1245 OHMS?

NO → REPLACE SPEED SENSOR.

YES ↓

(2)
- CONNECT J 39200 BETWEEN SENSOR HARNESS CONNECTOR TERMINAL "A" AND GROUND (USE DC VOLTS).
- IGNITION "ON."
- NOTE VOLTAGE; REPEAT BETWEEN TERMINAL "B" AND GROUND. IS VOLTAGE GREATER THAN 2 VOLTS BETWEEN EITHER TERMINAL "A" OR "B" AND GROUND?

YES → REPAIR SHORT TO B + IN CKT 885 OR 884.

NO ↓

(3)
- IGNITION "OFF."
- DISCONNECT EBTCM.
- INSTALL J 39700 UNIVERSAL BREAKOUT BOX TO EBTCM HARNESS CONNECTOR ONLY.
- USING J 39200, MEASURE RESISTANCE BETWEEN TERMINAL "45" ON J 39700 AND TERMINAL "A" OF SENSOR HARNESS CONNECTOR. PERFORM SAME MEASUREMENT BETWEEN TERMINAL "44" AND TERMINAL "B". IS EITHER MEASUREMENT GREATER THAN 5 OHMS?

YES → REPAIR OPEN OR HIGH RESISTANCE IN CKT 885 AND/OR 884.

NO ↓

(4)
- REMOVE J 39700; RECONNECT ALL CONNECTORS.
- CLEAR DTCs, IGNITION "OFF."
- DISCONNECT TECH 1, IF CONNECTED.
- IGNITION "ON."
- WIGGLE ALL WIRING AND CONNECTORS TO TRY TO INDUCE AN INTERMITTENT MALFUNCTION. DOES DTC 37 RESET?

YES → REPAIR INTERMITTENT CONNECTOR OR WIRING.

NO ↓

(5)
- CLEAR DTCs, IGNITION "OFF."
- DISCONNECT TECH 1, IF CONNECTED.
- IGNITION "ON." DOES DTC 37 RESET?

YES → (8)
- REPLACE SPEED SENSOR.
- CLEAR DTCs, IGNITION "OFF."
- DISCONNECT TECH 1, IF CONNECTED.
- IGNITION "ON." DOES DTC 37 RESET?

YES → REPLACE EBTCM

NO → SYSTEM OK

NO ↓

(6)
- ROAD TEST VEHICLE, REACHING AT LEAST 25 km/h (15 MPH).
- DOES DTC 35 SET?

YES → (7) GO TO DTC 35 DIAGNOSTICS.

NO → NO TROUBLE FOUND SEE "DIAGNOSTIC AIDS" ON FACING PAGE.

WHEN ALL DIAGNOSIS AND REPAIRS ARE COMPLETED, CLEAR DTC(s) AND VERIFY PROPER OPERATION.

DIAGNOSTIC CHART DTC 41 — 1994 FLEETWOOD, CONTINUED

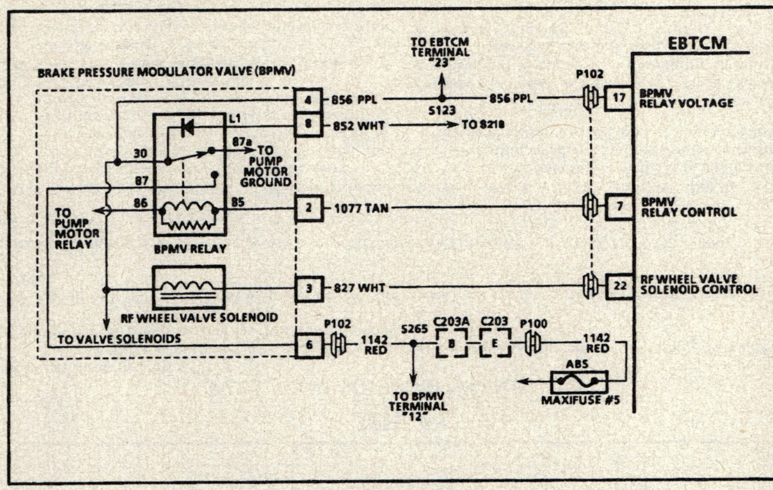

DTC 41
RF VALVE SOLENOID MALFUNCTION

Circuit Description:
The wheel valve solenoid circuits are supplied with battery power when the ignition is "ON." The EBTCM controls the valve functions by permitting one of three levels of current flow (0, 2.5, 5 amps) to the solenoid.

DTC Will Set When: DTC 41 will set if the EBTCM senses a discrepancy such as an open or ground in CKT 827.

Action Taken: The BPMV relay will turn "OFF," the "ANTI-LOCK" and "TRACTION CONTROL" lamps will come "ON."

DTC Chart Test Description: Number(s) below refer to circled number(s) on the diagnostic chart.

1. Checks the integrity of CKTs 827 and 856, the RF valve solenoid circuitry internal to the modulator valve assembly, and the RF valve solenoid coil.
2. Checks for a short to battery in CKT 827, using EBTCM terminal "3" as ground.
3. Checks for a short to ground in CKT 827, using EBTCM terminal "3" as ground.
4. Checks wiring and connectors for intermittents.
5. Uses the Tech 1 to exercise the RF valve solenoid and check it for proper operation.
6. Determines whether the DTC was set by an intermittent condition or an EBTCM malfunction.
7. Determines whether a problem found in Step 1 is due to an open in CKT 827 or a malfunctioning BPMV.

Diagnostic Aids: All tests using J 39700 Universal Breakout Box terminal "3" are using the terminal as ground. These tests, of course, assume that the ground at terminal "3" is good.

If DTCs 41 and 45 are both set the malfunction is likely to be a short to B + on CKT 827.

If DTCs 41 and 55 are both set the malfunction is likely to be a short to B + on CKT 828.

DIAGNOSTIC CHART DTC 41 — 1994 FLEETWOOD, CONTINUED

DTC 41
RF VALVE SOLENOID MALFUNCTION

NOTICE: J 39200 TEST LEADS MUST BE "ZEROED" PRIOR TO MAKING ANY RESISTANCE MEASUREMENTS. REFER TO J 39200 USER'S MANUAL.

(1)
- IGNITION "OFF."
- DISCONNECT EBTCM.
- INSTALL J 39700, UNIVERSAL BREAKOUT BOX TO EBTCM HARNESS CONNECTOR ONLY.
- USING J 39200, MEASURE RESISTANCE BETWEEN J 39700 TERMINALS "17" AND "22".

IS RESISTANCE 1.7-2.4 OHMS?

YES → (2); NO → (7)

(2)
- CONNECT J 39700 TO EBTCM; LEAVE CONNECTED TO HARNESS.
- DISCONNECT BPMV CONNECTOR.
- CONNECT J 39200 BETWEEN J 39700 TERMINALS "22" AND "3" (USE DC VOLTS).
- IGNITION "ON."

IS VOLTAGE GREATER THAN ONE VOLT?

NO → (3); YES → REPAIR SHORT TO B + IN CKT 827.

(7)
- DISCONNECT BPMV CONNECTOR.
- USING J 39200, MEASURE RESISTANCE BETWEEN TERMINALS "3" AND "4" ON BPMV.

IS RESISTANCE 0.9-1.6 OHMS?

YES → REPAIR OPEN IN CKT 827.
NO → REPLACE BPMV.

(3)
- IGNITION "OFF."
- DISCONNECT EBTCM.
- USING J 39200, MEASURE RESISTANCE BETWEEN J 39700 TERMINALS "22" AND "3".

DOES J 39200 DISPLAY "OL" (INFINITE)?

YES → (4); NO → REPAIR SHORT TO GROUND IN CKT 827.

(4)
- INSPECT ALL WIRING AND CONNECTORS FOR CONDITIONS WHICH COULD CAUSE INTERMITTENTS, SUCH AS CHAFING, CORROSION, OR POOR TERMINAL CONTACT.

ARE ALL CONNECTIONS AND WIRING GOOD?

YES → (5); NO → REPAIR WIRING, CONNECTORS OR TERMINALS AS NEEDED.

(5)
- REMOVE J 39700.
- RECONNECT EBTCM AND BPMV CONNECTORS.
- INSTALL TECH 1 WITH MSC.
- IGNITION "ON."
- PERFORM TECH 1 RF SOLENOID TEST.

DO BOTH MODES OPERATE?

YES → (6); NO → REPLACE BPMV.

(6)
- CLEAR DTCs, IGNITION "OFF."
- DISCONNECT TECH 1, IF CONNECTED.
- DRIVE VEHICLE AT LEAST 7 km/h (4 MPH) TO ACTIVATE AUTO TEST SEQUENCE.

DOES DTC 41 RESET?

NO → NO TROUBLE FOUND; SEE "DIAGNOSTIC AIDS" ON FACING PAGE.
YES → REPLACE EBTCM.

WHEN ALL DIAGNOSIS AND REPAIRS ARE COMPLETED, CLEAR DTC(s) AND VERIFY PROPER OPERATION.

DIAGNOSTIC CHART DTC 44 — 1994 FLEETWOOD

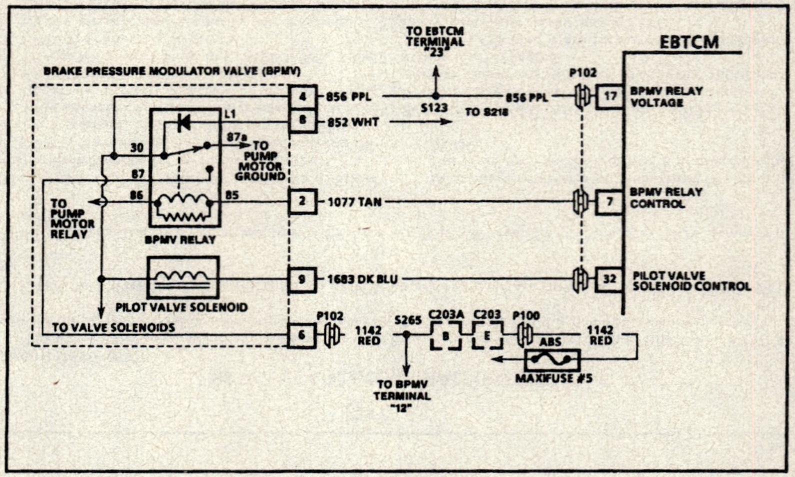

DTC 44
PILOT VALVE SOLENOID MALFUNCTION

Circuit Description:
The pilot valve solenoid circuit is supplied with battery power when the ignition is "ON." The EBTCM controls the valve by grounding the circuit when necessary.

DTC Will Set When: DTC 44 will set if the EBTCM senses a discrepancy such as an open or ground in CKT 1683.

Action Taken: The BPMV relay will turn "OFF," the "ANTI-LOCK" and "TRACTION CONTROL" lamps will come "ON."

DTC Chart Test Description: Number(s) below refer to circled number(s) on the diagnostic chart.

1. Checks the integrity of CKTs 1683 and 856, the pilot valve solenoid circuitry internal to the BPM valve assembly, and the pilot valve solenoid coil.
2. Checks for a short to battery in CKT 1683, using EBTCM terminal "3" as ground.
3. Checks for a short to ground in CKT 1683, using EBTCM terminal "3" as ground.
4. Checks for connector or wiring malfunctions which could cause intermittents.
5. Uses the Tech 1 to exercise the pilot valve solenoid and check it for proper operation.
6. Determines whether the DTC was set by an intermittent condition or an EBTCM malfunction.
7. Determines whether a malfunction found in Step 1 is due to an open in CKT 1683 or a malfunctioning BPMV.

Diagnostic Aids: All tests using J 39700 Universal Breakout Box terminal "3" are using the terminal as ground. These tests, of course, assume that the ground at terminal "3" is good.

DIAGNOSTIC CHART DTC 44 — 1994 FLEETWOOD, CONTINUED

DTC 44
PILOT VALVE SOLENOID MALFUNCTION

NOTICE: J 39200 TEST LEADS MUST BE "ZEROED" PRIOR TO MAKING ANY RESISTANCE MEASUREMENTS. REFER TO J 39200 USER'S MANUAL.

(1)
- IGNITION "OFF."
- DISCONNECT EBTCM.
- INSTALL J 39700 UNIVERSAL BREAKOUT BOX TO EBTCM HARNESS CONNECTOR ONLY.
- USING J 39200, MEASURE RESISTANCE BETWEEN J 39700 TERMINALS "17" AND "32". IS RESISTANCE 4.85-5.75 OHMS?

YES → (2); NO → (7)

(2)
- CONNECT J 39700 TO EBTCM; LEAVE CONNECTED TO HARNESS.
- DISCONNECT BPMV CONNECTOR.
- CONNECT J 39200 BETWEEN J 39700 TERMINALS "32" AND "3" (USE DC VOLTS).
- IGNITION "ON." IS VOLTAGE GREATER THAN 1 VOLT?

NO → (3); YES → REPAIR SHORT TO B+ IN CKT 1683.

(3)
- IGNITION "OFF."
- DISCONNECT EBTCM.
- USING J 39200, MEASURE RESISTANCE BETWEEN J 39700 TERMINALS "32" AND "3". DOES J 39200 DISPLAY "O.L" (INFINITE)?

YES → (4); NO → REPAIR SHORT TO GROUND IN CKT 1683;

(4)
- IGNITION "OFF."
- INSPECT ALL WIRING AND CONNECTORS FOR CONDITIONS WHICH COULD CAUSE INTERMITTENTS, SUCH AS CHAFING, CORROSION, OR POOR TERMINAL CONTACT.

ARE ALL CONNECTIONS AND WIRING GOOD?

YES → (5); NO → REPAIR WIRING, CONNECTORS OR TERMINALS AS NEEDED.

(5)
- REMOVE J 39700.
- RECONNECT EBTCM AND BPMV CONNECTORS.
- INSTALL TECH 1 WITH MSC.
- IGNITION "ON."
- PERFORM TECH 1 PILOT VALVE TEST. DOES TEST MODE OPERATE?

YES → (6); NO → REPLACE BPMV.

(6)
- CLEAR DTCs, IGNITION "OFF."
- DISCONNECT TECH 1, IF CONNECTED.
- DRIVE VEHICLE AT LEAST 7 km/h (4 MPH) TO ACTIVATE AUTO TEST SEQUENCE. DOES DTC 44 RESET?

NO → NO TROUBLE FOUND; SEE "DIAGNOSTIC AIDS" ON FACING PAGE.
YES → REPLACE EBTCM.

(7)
- DISCONNECT BPM VALVE CONNECTOR.
- USING J 39200, MEASURE RESISTANCE BETWEEN TERMINALS "4" AND "9" ON BPMV. IS RESISTANCE 4.05-4.95 OHMS?

YES → REPAIR OPEN IN CKT 1683.; NO → REPLACE BPMV.

WHEN ALL DIAGNOSIS AND REPAIRS ARE COMPLETED, CLEAR DTC(s) AND VERIFY PROPER OPERATION.

DIAGNOSTIC CHART DTC 45 — 1994 FLEETWOOD

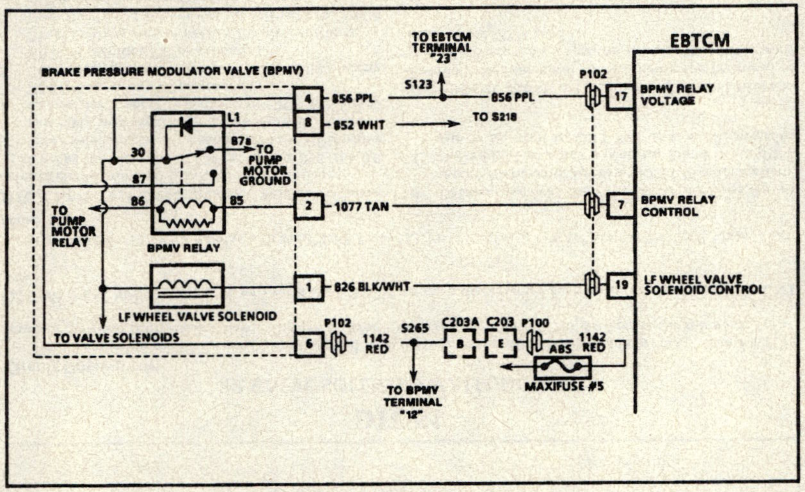

DTC 45
LF VALVE SOLENOID MALFUNCTION

Circuit Description:
The wheel valve solenoid circuits are supplied with battery power when the ignition is "ON." The EBTCM controls the valve functions by permitting one of three levels of current flow (0, 2.5, 5 amps) to the solenoid.

DTC Will Set When: DTC 45 will set if the EBTCM senses a discrepancy such as an open or ground in CKT 826.

Action Taken: The BPMV relay will turn "OFF," the "ANTI-LOCK" and "TRACTION CONTROL" lamps will come "ON."

DTC Chart Test Description: Number(s) below refer to circled number(s) on the diagnostic chart.

1. Checks the integrity of CKTs 826 and 856, the LF valve solenoid circuitry internal to the BPM valve assembly, and the LF valve solenoid coil.
2. Checks for a short to battery in CKT 826, using EBTCM terminal "3" as ground.
3. Checks for a short to ground in CKT 826, using EBTCM terminal "3" as ground.
4. Checks wiring and connectors for intermittents.
5. Uses the Tech 1 to exercise the LF valve solenoid and check it for proper operation.
6. Determines whether the DTC was set by an intermittent condition or an EBTCM malfunction.
7. Determines whether a problem found in Step 1 is due to an open in CKT 826 or a malfunctioning BPMV.

Diagnostic Aids: All tests using J 39700 Universal Breakout Box terminal "3" are using the terminal as ground. These tests, of course, assume that the ground at terminal "3" is good.

If DTCs 41 and 45 are both set, the malfunction is likely to be a short to B+ on CKT 827 or 826.

DIAGNOSTIC CHART DTC 45 — 1994 FLEETWOOD, CONTINUED

DTC 45
LF VALVE SOLENOID MALFUNCTION

NOTICE: J 39200 TEST LEADS MUST BE "ZEROED" PRIOR TO MAKING ANY RESISTANCE MEASUREMENTS. REFER TO J 39200 USER'S MANUAL.

(1)
- IGNITION "OFF."
- DISCONNECT EBTCM.
- INSTALL J 39700 UNIVERSAL BREAKOUT BOX TO EBTCM HARNESS CONNECTOR ONLY.
- USING J 39200 MEASURE RESISTANCE BETWEEN J 39700 TERMINALS "17" AND "19". IS RESISTANCE 1.7-2.4 OHMS?

YES → (2) / NO → (7)

(2)
- CONNECT J 39700 TO EBTCM; LEAVE CONNECTED TO HARNESS.
- DISCONNECT BPMV CONNECTOR.
- CONNECT J 39200 BETWEEN J 39700 TERMINALS "19" AND "3" (USE DC VOLTS).
- IGNITION "ON." IS VOLTAGE GREATER THAN ONE VOLT?

NO → (3) / YES → REPAIR SHORT TO B+ IN CKT 826.

(7)
- DISCONNECT BPMV CONNECTOR.
- USING J 39200, MEASURE RESISTANCE BETWEEN TERMINALS "1" AND "4" ON BPMV. IS RESISTANCE 0.9-1.6 OHMS?

YES → REPAIR OPEN IN CKT 826. / NO → REPLACE BPM VALVE.

(3)
- IGNITION "OFF."
- DISCONNECT EBTCM.
- USING J 39200, MEASURE RESISTANCE BETWEEN J 39700 TERMINALS "19" AND "3". DOES J 39200 DISPLAY "O.L" (INFINITE)?

YES → (4) / NO → REPAIR SHORT TO GROUND IN CKT 826.

(4)
- IGNITION "OFF."
- INSPECT ALL WIRING AND CONNECTORS FOR CONDITIONS WHICH COULD CAUSE INTERMITTENTS, SUCH AS CHAFING, CORROSION, OR POOR TERMINAL CONTACT.

ARE ALL CONNECTIONS AND WIRING GOOD?

YES → (5) / NO → REPAIR WIRING, CONNECTORS OR TERMINALS AS NEEDED.

(5)
- REMOVE J 39700.
- RECONNECT EBTCM AND BPMV CONNECTORS.
- INSTALL TECH 1 WITH MSC.
- IGNITION "ON."
- PERFORM TECH 1 LF SOLENOID TEST. DO BOTH MODES OPERATE?

YES → (6) / NO → REPLACE BPMV.

(6)
- CLEAR DTCs, IGNITION "OFF."
- DISCONNECT TECH 1, IF CONNECTED.
- DRIVE VEHICLE AT LEAST 7 km/h (4 MPH) TO ACTIVATE AUTO TEST SEQUENCE. DOES DTC 45 RESET?

NO → NO TROUBLE FOUND; SEE "DIAGNOSTIC AIDS" ON FACING PAGE. / YES → REPLACE EBTCM.

WHEN ALL DIAGNOSIS AND REPAIRS ARE COMPLETED, CLEAR DTC(s) AND VERIFY PROPER OPERATION.

DIAGNOSTIC CHART DTC 51 — 1994 FLEETWOOD

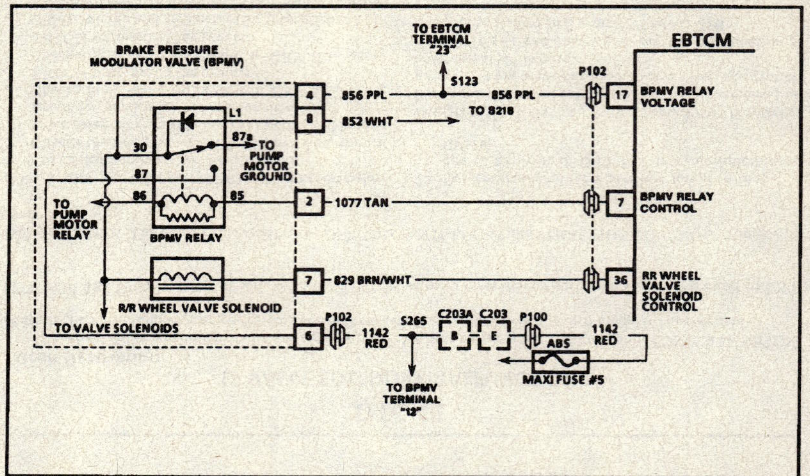

DTC 51
RR VALVE SOLENOID MALFUNCTION

Circuit Description:
The wheel valve solenoid circuits are supplied with battery power when the ignition is "ON." The EBTCM controls the valve functions by permitting one of three levels of current flow (0, 2.5, 5 amps) to the solenoid.

DTC Will Set When: DTC 51 will set if the EBTCM senses a discrepancy such as an open or ground in CKT 829.

Action Taken: The BPMV relay will turn "OFF," the "ANTI-LOCK" and "TRACTION CONTROL" lamps will come "ON."

DTC Chart Test Description: Number(s) below refer to circled number(s) on the diagnostic chart.

1. Checks the integrity of CKTs 829 and 856, the RR valve solenoid circuitry internal to the BPM valve assembly, and the RR valve solenoid coil.
2. Checks for a short to battery in CKT 829, using EBTCM terminal "3" as ground.
3. Checks for a short to ground in CKT 829, using EBTCM terminal "3" as ground.
4. Checks wiring and connectors for intermittents.
5. Uses the Tech 1 to exercise the RR valve solenoid and check it for proper operation.
6. Determines whether the DTC was set by an intermittent condition or an EBTCM malfunction.
7. Determines whether a problem found in Step 1 is due to an open in CKT 829 or a malfunctioning BPMV.

Diagnostic Aids: All tests using J 39700 Universal Breakout Box terminal "3" are using the terminal as ground. These tests, of course, assume that the ground at terminal "3" is good.

DIAGNOSTIC CHART DTC 51 — 1994 FLEETWOOD, CONTINUED

DTC 51
RR VALVE SOLENOID MALFUNCTION

NOTICE: J 39200 TEST LEADS MUST BE "ZEROED" PRIOR TO MAKING ANY RESISTANCE MEASUREMENTS. REFER TO J 39200 USER'S MANUAL.

(1)
- IGNITION "OFF."
- DISCONNECT EBTCM.
- INSTALL J 39700 UNIVERSAL BREAKOUT BOX TO EBTCM HARNESS CONNECTOR ONLY.
- MEASURE RESISTANCE BETWEEN J 39700 TERMINALS "17" AND "36".

IS RESISTANCE 1.7-2.4 OHMS?

YES → (2); NO → (7)

(2)
- CONNECT J 39700 TO EBTCM; LEAVE CONNECTED TO HARNESS.
- DISCONNECT BPM VALVE CONNECTOR.
- CONNECT J 39200 BETWEEN J 39700 TERMINALS "36" AND "3" (USE DC VOLTS).
- IGNITION "ON."

IS VOLTAGE GREATER THAN ONE VOLT?

YES: REPAIR SHORT TO B+ IN CKT 829.

NO → (3)

(7)
- DISCONNECT BPMV CONNECTOR.
- USING J 39200, MEASURE RESISTANCE BETWEEN TERMINALS "4" AND "7" ON BPMV.

IS RESISTANCE 0.9-1.6 OHMS?

YES: REPAIR OPEN IN CKT 829.

NO: REPLACE BPMV.

(3)
- IGNITION "OFF."
- DISCONNECT EBTCM.
- USING J 39200, MEASURE RESISTANCE BETWEEN J 39700 TERMINALS "36" AND "3".

DOES J 39200 DISPLAY "OL" (INFINITE)?

YES → (4)

NO: REPAIR SHORT TO GROUND IN CKT 829.

(4)
- INSPECT ALL WIRING AND CONNECTORS FOR CONDITIONS WHICH COULD CAUSE INTERMITTENTS, SUCH AS CHAFING, CORROSION, OR POOR TERMINAL CONTACT.

ARE ALL CONNECTIONS AND WIRING GOOD?

YES → (5)

NO: REPAIR WIRING, CONNECTORS OR TERMINALS AS NEEDED.

(5)
- REMOVE J 39700.
- RECONNECT EBTCM AND BPMV CONNECTORS.
- INSTALL TECH 1 WITH MSC.
- IGNITION "ON."
- PERFORM TECH 1 RR SOLENOID TEST.

DO BOTH MODES OPERATE?

YES → (6)

NO: REPLACE BPMV.

(6)
- CLEAR DTCs, IGNITION "OFF."
- DISCONNECT TECH 1, IF CONNECTED.
- DRIVE VEHICLE AT LEAST 7 km/h (4 MPH) TO ACTIVATE AUTO TEST SEQUENCE.

DOES DTC 51 RESET?

NO: NO TROUBLE FOUND; SEE "DIAGNOSTIC AIDS" ON FACING PAGE.

YES: REPLACE EBTCM.

WHEN ALL DIAGNOSIS AND REPAIRS ARE COMPLETED, CLEAR DTC(s) AND VERIFY PROPER OPERATION.

DIAGNOSTIC CHART DTC 55 — 1994 FLEETWOOD

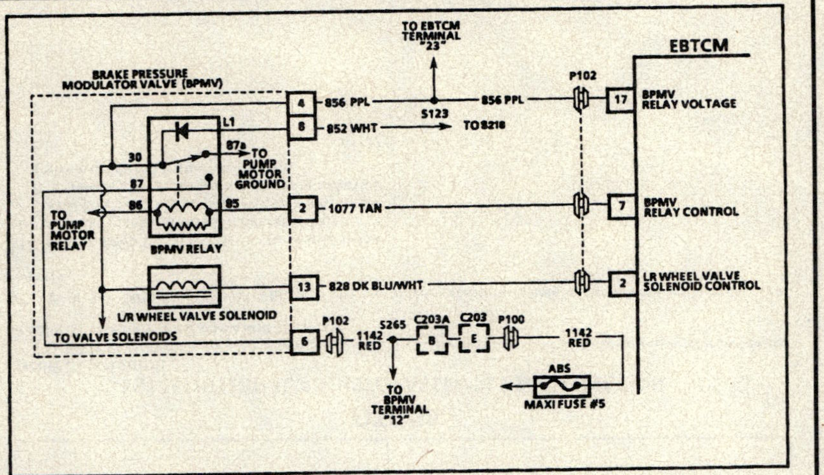

DTC 55
LR VALVE SOLENOID MALFUNCTION

Circuit Description:
The wheel valve solenoid circuits are supplied with battery power when the ignition is "ON." The EBTCM controls the valve functions by permitting one of three levels of current flow (0, 2.5, 5 amps) to the solenoid.

DTC Will Set When: DTC 55 will set if the EBTCM senses a discrepancy such as an open or ground in CKT 828.

Action Taken: The BPMV relay will turn "OFF," the "ANTI-LOCK" and "TRACTION CONTROL" lamps will come "ON."

DTC Chart Test Description: Number(s) below refer to circled number(s) on the diagnostic chart.

1. Checks the integrity of CKTs 828 and 856, the LR valve solenoid circuitry internal to the BPM valve assembly, and the LR valve solenoid coil.
2. Checks for a short to battery in CKT 828, using EBTCM terminal "3" as ground.
3. Checks for a short to ground in CKT 828, using EBTCM terminal "3" as ground.
4. Checks wiring and connectors for intermittents.
5. Uses the Tech 1 to exercise the LR valve solenoid and check it for proper operation.
6. Determines whether the DTC was set by an intermittent condition or an EBTCM malfunction.
7. Determines whether a problem found in Step 1 is due to an open in CKT 828 or a malfunctioning BPMV.

Diagnostic Aids: All tests using J 39700 Universal Breakout Box terminal "3" are using the terminal as ground. These tests, of course, assume that the ground at terminal "3" is good.

If DTCs 41 and 55 are both set the malfunction is likely to be a short to B+ on CKT 828.

DIAGNOSTIC CHART DTC 55 — 1994 FLEETWOOD, CONTINUED

DTC 55
LR VALVE SOLENOID MALFUNCTION

NOTICE: J 39200 TEST LEADS MUST BE "ZEROED" PRIOR TO MAKING ANY RESISTANCE MEASUREMENTS. REFER TO J 39200 USER'S MANUAL.

(1)
- IGNITION "OFF."
- DISCONNECT EBTCM.
- INSTALL J 39700 UNIVERSAL BREAKOUT BOX TO EBTCM HARNESS CONNECTOR ONLY.
- USING J 39200, MEASURE RESISTANCE BETWEEN J 39700 TERMINALS "17" AND "2".

IS RESISTANCE 1.7-2.4 OHMS?

YES → (2) / NO → (7)

(2)
- CONNECT J 39700 TO EBTCM; LEAVE CONNECTED TO HARNESS.
- DISCONNECT BPM VALVE CONNECTOR.
- CONNECT J 39200 BETWEEN J 39700 TERMINALS "2" AND "3" (USE DC VOLTS).
- IGNITION "ON."

IS VOLTAGE GREATER THAN ONE VOLT?

NO → (3) / YES → REPAIR SHORT TO B + IN CKT 828.

(7)
- DISCONNECT BPMV CONNECTOR.
- USING J 39200, MEASURE RESISTANCE BETWEEN TERMINALS "4" AND "13" ON BPMV.

IS RESISTANCE 0.9-1.6 OHMS?

YES → REPAIR OPEN IN CKT 828. / NO → REPLACE BPMV.

(3)
- IGNITION "OFF."
- DISCONNECT EBTCM.
- USING J 39200, MEASURE RESISTANCE BETWEEN J 39700 TERMINALS "2" AND "3".

DOES J 39200 DISPLAY "OL" (INFINITE)?

YES → (4) / NO → REPAIR SHORT TO GROUND IN CKT 828

(4)
- INSPECT ALL WIRING AND CONNECTORS FOR CONDITIONS WHICH COULD CAUSE INTERMITTENTS, SUCH AS CHAFING, CORROSION, OR POOR TERMINAL CONTACT.

ARE ALL CONNECTIONS AND WIRING GOOD?

YES → (5) / NO → REPAIR WIRING, CONNECTORS OR TERMINALS AS NEEDED.

(5)
- REMOVE J 39700.
- RECONNECT EBTCM AND BPMV CONNECTORS.
- INSTALL TECH 1 WITH MSC.
- IGNITION "ON."
- PERFORM TECH 1 LR SOLENOID TEST.

DO BOTH MODES OPERATE?

YES → (6) / NO → REPLACE BPMV.

(6)
- CLEAR DTCs, IGNITION "OFF."
- DISCONNECT TECH 1, IF CONNECTED.
- DRIVE VEHICLE AT LEAST 7 km/h (4 MPH) TO ACTIVATE AUTO TEST SEQUENCE.

DOES DTC 55 RESET?

NO → NO TROUBLE FOUND; SEE "DIAGNOSTIC AIDS" ON FACING PAGE.
YES → REPLACE EBTCM.

WHEN ALL DIAGNOSIS AND REPAIRS ARE COMPLETED, CLEAR DTC(s) AND VERIFY PROPER OPERATION.

DIAGNOSTIC CHART DTC 58 — 1994 FLEETWOOD

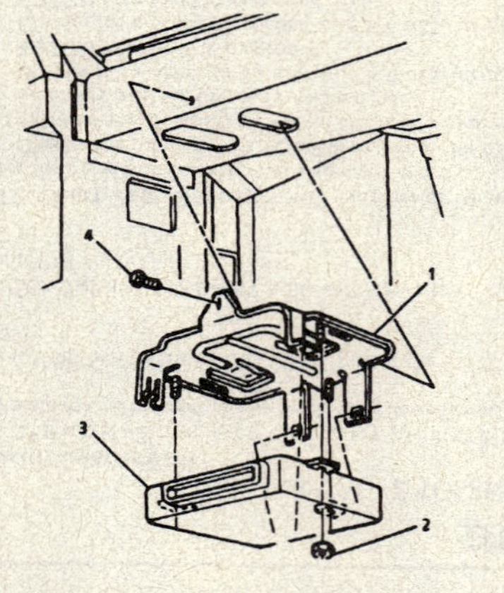

1 BRACKET ASSEMBLY, ELECTRONIC BRAKE AND TRACTION CONTROL MODULE
2 NUT, ELECTRONIC BRAKE AND TRACTION CONTROL MODULE, 1.9 N·m (17 lb. in.)
3 MODULE ASSEMBLY, ELECTRONIC BRAKE AND TRACTION CONTROL
4 BOLT/SCREW, ELECTRONIC BRAKE AND TRACTION CONTROL BRACKET, 6 N·m (53 lb. in.)

DTC 58
EBTCM INTERNAL ADJUSTER ASSEMBLY MALFUNCTION

Circuit Description:
The EBTCM contains a microprocessor for ABS/TCS calculations and functions, as well as a microprocessor for adjuster assembly calculations and functions.

DTC Will Set When: DTC 58 will set if the two microprocessors cannot communicate with each other properly.

DTC Chart Test Description: Number(s) below refer to circled number(s) on the diagnostic chart.
1. Clears DTCs and checks for DTC 58 reset. If the DTC resets, the microprocessor communication problem does exist within the EBTCM.

DIAGNOSTIC CHART DTC 58 — 1994 FLEETWOOD, CONTINUED

DTC 58
EBTCM INTERNAL ADJUSTER ASSEMBLY MALFUNCTION

(1)
- CLEAR DTCs, IGNITION "OFF."
- DISCONNECT TECH 1, IF CONNECTED.
- IGNITION "ON."

DOES DTC 58 RESET?

YES → REPLACE EBTCM.

NO → NO TROUBLE FOUND.

WHEN ALL DIAGNOSIS AND REPAIRS ARE COMPLETED, CLEAR DTC(s) AND VERIFY PROPER OPERATION.

DIAGNOSTIC CHART DTC 61 — 1994 FLEETWOOD

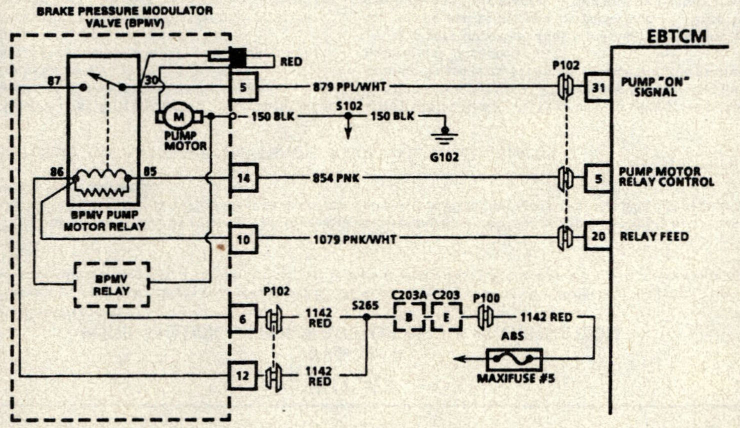

DTC 61
(Page 1 of 4)

PUMP MOTOR OR PUMP MOTOR RELAY MALFUNCTION

Circuit Description:

When the pump motor relay is grounded by the EBTCM, it closes and provides B+ to operate the pump. Once the motor circuit is energized, a "Pump ON" signal is sensed by the EBTCM through CKT 879 to verify pump operation.

DTC Will Set When: DTC 61 will set if B+ is present at the pump motor without pump motor relay activation from the EBTCM, or if B+ is NOT present at the pump motor within 60 milliseconds after the EBTCM requests pump motor relay activation.

Action Taken: The "ANTI-LOCK" and "TRACTION CONTROL" lamps will be "ON."

DTC Chart Test Description: Number(s) below refer to circled number(s) on the diagnostic chart.

1. Checks the pump motor relay coil for proper resistance.
2. Checks for an open in power feed CKT 1142.
3. Determines if the pump motor relay contacts are stuck closed.
4. Determines if the pump motor relay contacts are stuck open.
5. Verifies the integrity of the pump motor and pump motor relay circuitry internal to the BPMV assembly.
6. Checks for an open in CKT 879.
7. Checks for a short to ground in CKT 879.
8. Checks for a short to B+ in CKT 879.

Diagnostic Aids: All tests using J 39700 Universal Breakout Box terminal "3" are using that terminal as ground. These tests, of course, assume that the ground at terminal "3" is good.

It is very important that a thorough inspection of the wiring and connectors be performed. Failure to carefully and fully inspect wiring and connectors may result in misdiagnosis, causing part replacement with reappearance of the malfunction.

DIAGNOSTIC CHART DTC 61 — 1994 FLEETWOOD, CONTINUED

DTC 61
(Page 2 of 4)

PUMP MOTOR OR PUMP MOTOR RELAY MALFUNCTION

1. • IGNITION "OFF."
   • REMOVE BPMV PUMP MOTOR RELAY (4-PIN).
   • USING J 39200, MEASURE RESISTANCE BETWEEN RELAY TERMINALS "85" AND "86".
   IS RESISTANCE 45-55 OHMS?
   - YES → go to step 2
   - NO → REPLACE BPMV PUMP MOTOR RELAY.

2. • CONNECT J 39200 BETWEEN BPMV HARNESS CONNECTOR TERMINAL "12" AND A GOOD CHASSIS GROUND.
   • IGNITION "ON."
   • NOTE VOLTAGE.
   IS VOLTAGE WITHIN 2 VOLTS OF B+?
   - YES → go to step 3
   - NO → REPAIR OPEN IN CKT 1142.

3. • USING J 39200, MEASURE RESISTANCE BETWEEN RELAY TERMINALS "30" AND "87".
   DOES J 39200 DISPLAY "OL" (INFINITE)?
   - YES → go to step 4
   - NO → REPLACE BPMV PUMP MOTOR RELAY.

4. • CONNECT RELAY TERMINAL "85" TO GROUND.
   • USING A FUSED JUMPER WIRE (SUCH AS J 36169) WITH A 3 AMP FUSE, CONNECT RELAY TERMINAL "86" TO B+.
   • USING J 39200, MEASURE RESISTANCE BETWEEN RELAY TERMINALS "30" AND "87".
   IS RESISTANCE 5 OHMS OR LESS?
   - YES → go to step 5
   - NO → REPLACE BPMV PUMP MOTOR RELAY.

5. • IGNITION "OFF."
   • DISCONNECT BPMV CONNECTOR.
   • USING J 39200, MEASURE RESISTANCE BETWEEN THE FOLLOWING TERMINALS ON THE BPMV.

BPM VALVE CONNECTOR		RELAY CONNECTOR
5	AND	30
10	AND	86
14	AND	85
12	AND	87
30	AND	PUMP MOTOR GROUND STUD

   IS THE READING BETWEEN ANY OF THE SETS OF TERMINAL MEASUREMENTS MORE THAN 5 OHMS?
   - YES → REPLACE BPMV.
   - NO → go to step 6

6. • INSTALL PUMP MOTOR RELAY ON BPMV.
   • DISCONNECT EBTCM.
   • INSTALL J 39700 UNIVERSAL BREAKOUT BOX TO EBTCM HARNESS CONNECTOR ONLY.
   • USING J 39200, MEASURE RESISTANCE BETWEEN J 39700 TERMINAL "31" AND BPMV CONNECTOR TERMINAL "5".
   IS RESISTANCE 5 OHMS OR LESS?
   - YES → go to step 7
   - NO → REPAIR OPEN IN CKT 879.

7. • USING J 39200, MEASURE RESISTANCE BETWEEN J 39700 TERMINALS "3" AND "31".
   DOES J 39200 DISPLAY "OL" (INFINITE)?
   - YES → go to step 8
   - NO → REPAIR SHORT TO GROUND IN CKT 879.

8. • CONNECT J 39200 BETWEEN J 39700 TERMINALS "31" AND "3" (USE DC VOLTS).
   • IGNITION "ON."
   • NOTE VOLTAGE.
   IS VOLTAGE ONE VOLT OR LESS?
   - YES → CONTINUE DIAGNOSIS ON NEXT PAGE.
   - NO → REPAIR SHORT TO B+ IN CKT 879.

WHEN ALL DIAGNOSIS AND REPAIRS ARE COMPLETED, CLEAR DTC(S) AND VERIFY PROPER OPERATION.

DIAGNOSTIC CHART DTC 61 — 1994 FLEETWOOD, CONTINUED

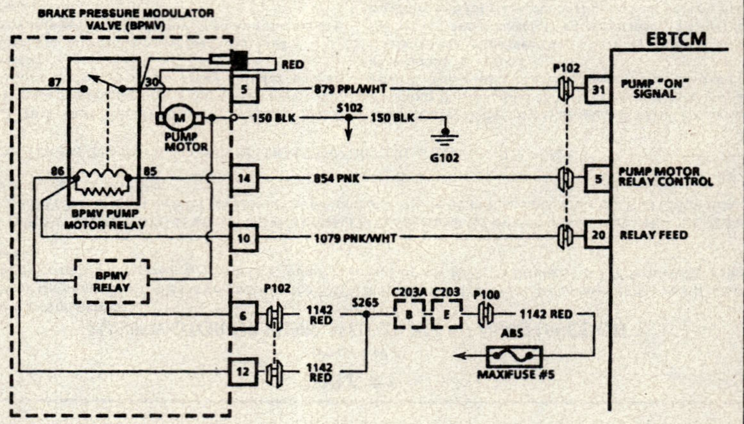

DTC 61
(Page 3 of 4)

PUMP MOTOR OR PUMP MOTOR RELAY MALFUNCTION

Circuit Description:

When the pump motor relay is grounded by the EBTCM, it closes and provides B+ to operate the pump. Once the motor circuit is energized, a "Pump ON" signal is sensed by the EBTCM through CKT 879 to verify pump operation.

DTC Will Set When: DTC 61 will set if B+ is present at the pump motor without pump motor relay activation from the EBTCM, or if B+ is NOT present at the pump motor within 60 milliseconds after the EBTCM requests pump motor relay activation.

Action Taken: The "ANTI-LOCK" and "TRACTION CONTROL" lamps will be "ON."

DTC Chart Test Description: Number(s) below refer to circled number(s) on the diagnostic chart.

9. Checks for an open in CKT 854.
10. Checks for a short to ground in CKT 854.
11. Checks for a short to B+ in CKT 854.
12. Checks wiring and connectors for intermittents.
13. Uses the Tech 1 to check for proper operation of the pump motor and associated circuitry.
14. Determines whether the problem found in Step 13 was due to a malfunctioning EBTCM or an open in pump motor ground CKT 150.

Diagnostic Aids: All tests using J 39700 Universal Breakout Box terminal "3" are using that terminal as ground. These tests, of course, assume that the ground at terminal "3" is good.

It is very important that a thorough inspection of the wiring and connectors be performed. Failure to carefully and fully inspect wiring and connectors may result in misdiagnosis, causing part replacement with reappearance of the malfunction.

DIAGNOSTIC CHART DTC 61 — 1994 FLEETWOOD, CONTINUED

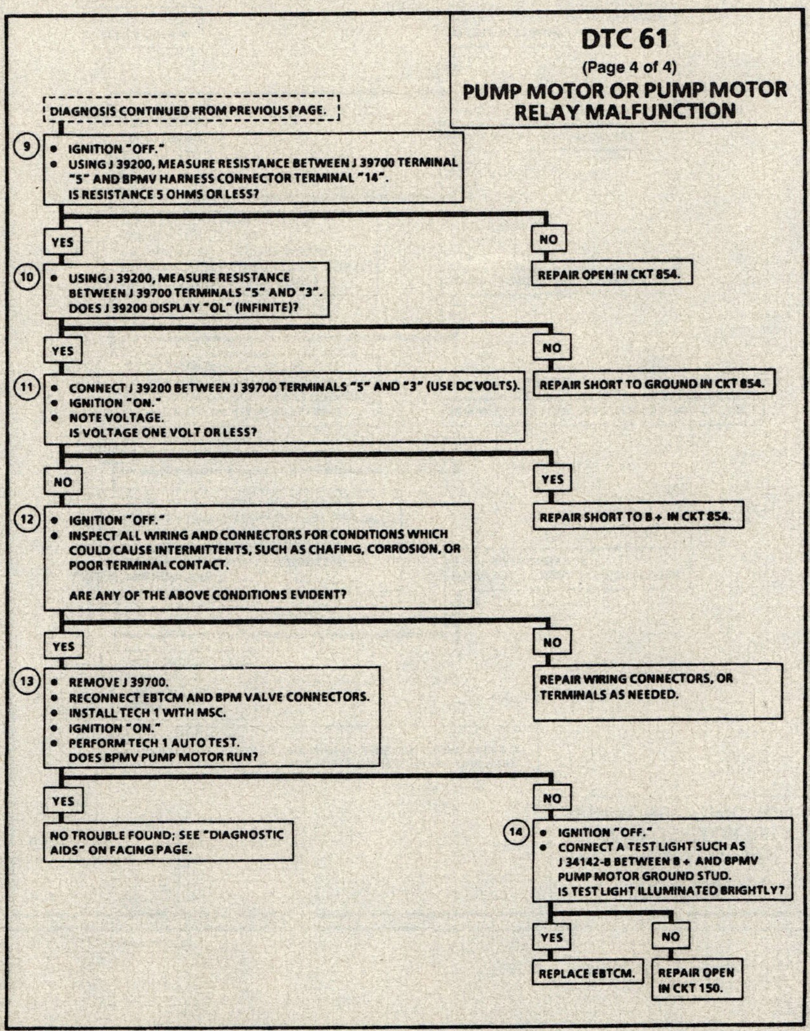

DIAGNOSTIC CHART DTC 62 — 1994 FLEETWOOD

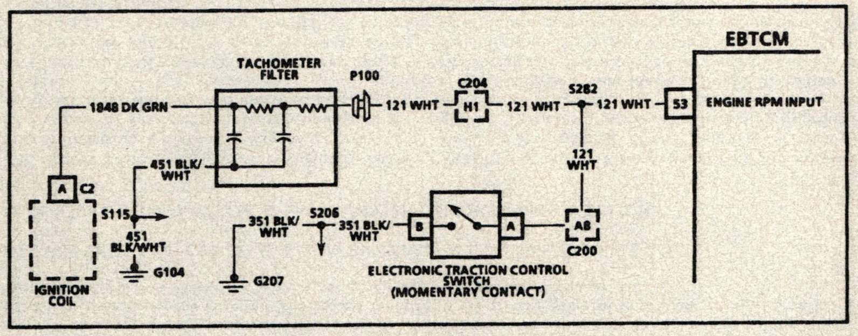

DTC 62

TRACTION CONTROL SWITCH USED OR RPM SIGNAL MALFUNCTION

Circuit Description:

The RPM signal circuit provides the EBTCM with an indication of engine RPM to help determine TCS control methods and rates when a TCS event takes place.

DTC Will Set When: DTC 62 will set if there is a short to ground, short to B +, or an open in CKT 1848 or CKT 121.

Action Taken: The "TRACTION CONTROL" lamp will be "ON."

Important

- The "Traction Control Disable" switch also can set this DTC, since it disables TCS by purposely inducing a momentary "short to ground" malfunction at EBTCM terminal "53". The traction control disable switch must be held down for at least 2 seconds to do so. Be sure to clear DTCs, turn ignition "OFF," then turn ignition "ON" again to verify the DTC was not set by the owner using the traction control disable switch. Review Tech 1 DTC History information for clues as to when the DTC set and how frequently it has set. Also, speak to the customer to determine if traction control disable switch use is the cause for this DTC.

DTC Chart Test Description: Number(s) below refer to circled number(s) on the diagnostic chart.

1. Checks for DTC reset after clearing DTCs to ensure a malfunction really exists, not that the owner used the "Traction Control Disable" switch.
2. Checks for "Traction Control Disable" switch stuck closed.
3. Checks for short to ground in CKT 121.
4. Checks for short to B+ in CKT 121.
5. Checks for normal operating voltage of about 5 volts from output side of tachometer filter. If no voltage, CKT 121 is open.

Diagnostic Aids:

Important

- The "Traction Control Disable" switch also can set this DTC, since it disables TCS by purposely inducing a momentary "short to ground" malfunction at EBTCM terminal "53". The traction control disable switch must be held down for at least 2 seconds to do so. Be sure to clear DTCs, turn ignition "OFF," then turn ignition "ON" again to verify the DTC was not set by the owner using the "Traction Control Disable" switch. Review Tech 1 DTC History information for clues as to when the DTC set and how frequently it has set. Also, speak to the customer to determine if "Traction Control Disable" switch use is the cause for this DTC.

A short to ground in CKT 1848 on the ignition system side of the tach filter may cause a no-start or extremely poor driveability condition.

DIAGNOSTIC CHART DTC 62 — 1994 FLEETWOOD, CONTINUED

DTC 62

TRACTION CONTROL SWITCH USED OR RPM SIGNAL MALFUNCTION

THIS DTC WILL BE SET IF THE OWNER USES THE ELECTRONIC TRACTION CONTROL SWITCH (REFER TO "DIAGNOSTIC AIDS" ON FACING PAGE FOR MORE INFORMATION).

(1)
- CLEAR DTCs.
- IGNITION "OFF."
- IGNITION "ON."
  DOES DTC 62 RESET?

NO: NO TROUBLE FOUND; SEE "DIAGNOSTIC AIDS" ON FACING PAGE.

YES:

(2)
- IGNITION "OFF."
- REMOVE ELECTRONIC TRACTION CONTROL SWITCH AND DISCONNECT.
- USING J 39200, MEASURE RESISTANCE BETWEEN SWITCH TERMINALS "A" AND "B". DOES J 39200 DISPLAY "OL" (INFINITE)?

NO: REPLACE ELECTRONIC TRACTION CONTROL SWITCH.

YES:

(3)
- USING J 39200, MEASURE RESISTANCE BETWEEN SWITCH CONNECTOR TERMINAL "A" AND A GOOD CHASSIS GROUND. DOES J 39200 DISPLAY "OL" (INFINITE)?

NO: REPAIR SHORT TO GROUND IN CKT 121.

YES:

(4)
- CONNECT J 39200 (USE DC VOLTS) BETWEEN SWITCH CONNECTOR TERMINAL "A" AND A GOOD CHASSIS GROUND.
- IGNITION "ON."
  IS VOLTAGE ONE VOLT OR LESS?

NO: REPAIR SHORT TO B+ IN CKT 121.

YES:

(5)
- START ENGINE.
- NOTE VOLTAGE.

0V: REPAIR OPEN IN CKT 121.

5V: NO TROUBLE FOUND; SEE "DIAGNOSTIC AIDS" ON FACING PAGE.

WHEN ALL DIAGNOSIS AND REPAIRS ARE COMPLETED, CLEAR DTC(s) AND VERIFY PROPER OPERATION.

DIAGNOSTIC CHART DTC 63 — 1994 FLEETWOOD

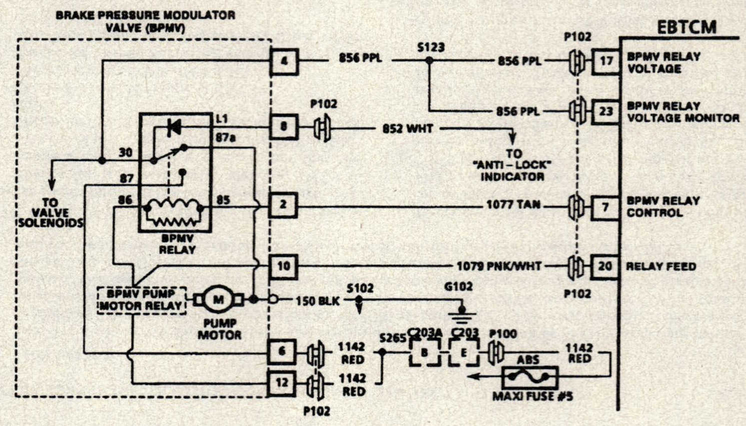

DTC 63
(Page 1 of 4)

BPMV RELAY MALFUNCTION

Circuit Description:

The BPMV relay has two functions. When the ignition is "ON" and the EBTCM is operating, the BPMV relay provides voltage to actuate the valve solenoids. However, the valve solenoids do not use this voltage unless the EBTCM provides the ground. The second function of the BPMV relay is to provide a ground path for illumination of the "ANTI-LOCK" lamp if the BPMV relay loses power or ground.

DTC Will Set When: DTC 63 will set if the BPMV relay voltage falls to less than 5 volts.

Action Taken: The "ANTI-LOCK" and "TRACTION CONTROL" lamps will be "ON."

DTC Chart Test Description: Number(s) below refer to circled number(s) on the diagnostic chart.

1. Verifies the availability of battery voltage to the BPMV relay through ABS maxifuse #5 and CKT 1142.
2. Checks the BPMV relay coil for proper resistance.
3. Checks for the BPMV relay contacts being internally open when in the BPMV relay-unenergized position.
4. Checks for the BPMV relay contacts being stuck open when in the BPMV relay-energized position.
5. Checks the integrity of the BPMV relay circuitry internal to the BPMV.
6. Checks for an open in ground CKT 150.
7. Checks for an open in CKT 856.

Diagnostic Aids: All tests using J 39700 Universal Breakout Box terminal "3" are using that terminal as ground. These tests, of course, assume that the ground at terminal "3" is good.

It is very important that a thorough inspection of the wiring and connectors be performed. Failure to carefully and fully inspect wiring and connectors may result in misdiagnosis, causing part replacement with reappearance of the malfunction.

DIAGNOSTIC CHART DTC 63 — 1994 FLEETWOOD, CONTINUED

DTC 63
(Page 2 of 4)
BPMV RELAY MALFUNCTION

1. • IGNITION "OFF."
   • DISCONNECT BPMV CONNECTOR.
   • CONNECT J 39200 BETWEEN BPMV HARNESS CONNECTOR TERMINAL "6" AND A GOOD CHASSIS GROUND.
   • NOTE VOLTAGE.
   IS VOLTAGE WITHIN 2 VOLTS OF B+?
   - NO: REPAIR OPEN IN CKT 1142 OR ABS/TC MAXIFUSE #5.
   - YES: go to 2.
2. • REMOVE BPMV RELAY (6 PIN).
   • USING J 39200, MEASURE RESISTANCE BETWEEN BPMV RELAY TERMINALS "85" AND "86".
   IS RESISTANCE BETWEEN 45-55 OHMS?
   - NO: REPLACE BPMV RELAY.
   - YES: go to 3.
3. • USING J 39200, MEASURE RESISTANCE BETWEEN BPMV RELAY TERMINALS "30" AND "87a".
   IS RESISTANCE 5 OHMS OR LESS?
   - NO: REPLACE BPMV RELAY.
   - YES: go to 4.
4. • CONNECT BPMV RELAY TERMINAL "85" TO GROUND.
   • USING A FUSED JUMPER WIRE (SUCH AS J 36169) WITH A 3 AMP FUSE, CONNECT BPMV RELAY TERMINAL "86" TO B+.
   • USING J 39200, MEASURE RESISTANCE BETWEEN BPMV RELAY TERMINALS "30" AND "87".
   IS RESISTANCE 5 OHMS OR LESS?
   - NO: REPLACE BPMV RELAY.
   - YES: go to 5.
5. • USING J 39200, MEASURE RESISTANCE BETWEEN THE FOLLOWING TERMINALS ON THE BPMV.

BPMV CONNECTOR		RELAY SOCKET TERMINALS
2	AND	85
4	AND	30
6	AND	87
MOTOR GROUND STUD	AND	87a
10	AND	86

   IS THE READING BETWEEN ANY OF THE SETS OF TERMINAL MEASUREMENTS MORE THAN 5 OHMS?
   - YES: REPLACE BPMV.
   - NO: go to 6.
6. • DISCONNECT EBTCM.
   • INSTALL J 39700 UNIVERSAL BREAKOUT BOX TO EBTCM HARNESS CONNECTOR ONLY.
   • USING J 39200, MEASURE RESISTANCE BETWEEN BPMV PUMP MOTOR GROUND STUD AND J 39700 TERMINAL "3".
   IS IT 5 OHMS OR LESS?
   - NO: REPAIR OPEN IN CKT 150 TO BPMV PUMP MOTOR GROUND STUD.
   - YES: go to 7.
7. • INSTALL BPMV RELAY.
   • USING J 39200, MEASURE RESISTANCE BETWEEN BPMV CONNECTOR TERMINAL "4" AND J 39700 TERMINAL "17".
   IS IT 5 OHMS OR LESS?
   - NO: REPAIR OPEN IN CKT 856.
   - YES: CONTINUE DIAGNOSIS ON NEXT PAGE.

WHEN ALL DIAGNOSIS AND REPAIRS ARE COMPLETED, CLEAR DTC(s) AND VERIFY PROPER OPERATION.

DIAGNOSTIC CHART DTC 63 — 1994 FLEETWOOD, CONTINUED

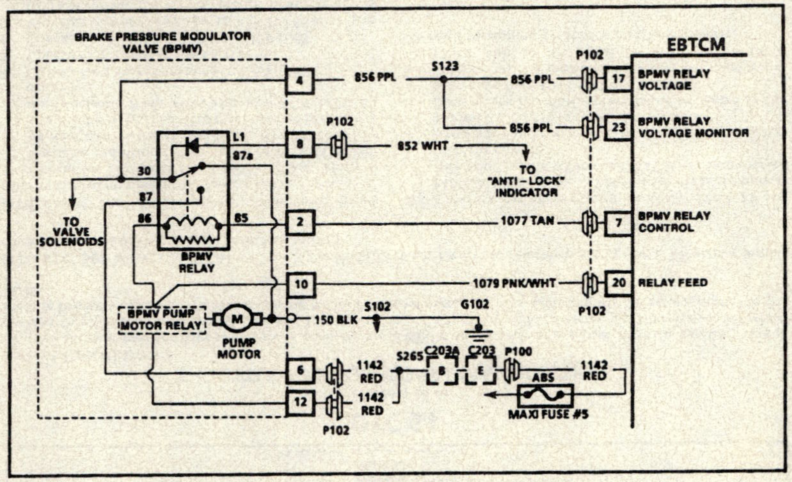

DTC 63
(Page 3 of 4)
BPMV RELAY MALFUNCTION

Circuit Description:
The BPMV relay has two functions. When the ignition is "ON" and the EBTCM is operating, the BPMV relay provides voltage to actuate the valve solenoids. However, the valve solenoids do not use this voltage unless the EBTCM provides the ground. The second function of the BPMV relay is to provide a ground path for illumination of the "ANTI-LOCK" lamp if the BPMV relay loses power or ground.

DTC Will Set When: DTC 63 will set if the BPMV relay voltage falls to less than 5 volts.

Action Taken: The "ANTI-LOCK" and "TRACTION CONTROL" lamps will be "ON."

DTC Chart Test Description: Number(s) below refer to circled number(s) on the diagnostic chart.
8. Checks for a short to ground in CKT 856.
9. Checks for a short to B+ in CKT 856.
10. Checks for an open in CKT 1079.
11. Checks for a short to ground in CKT 1079.
12. Checks for an open in CKT 1077.
13. Checks for a short to ground in CKT 1077.
14. Checks wiring and connectors for intermittents.
15. Determines whether the DTC was set due to an intermittent condition or an EBTCM malfunction.

Diagnostic Aids: All tests using J 39700 Universal Breakout Box terminal "3" are using that terminal as ground. These tests, of course, assume that the ground at terminal "3" is good.

It is very important that a thorough inspection of the wiring and connectors be performed. Failure to carefully and fully inspect wiring and connectors may result in misdiagnosis, causing part replacement with reappearance of the malfunction.

DIAGNOSTIC CHART DTC 63 — 1994 FLEETWOOD, CONTINUED

DTC 63
(Page 4 of 4)
BPMV RELAY MALFUNCTION

DIAGNOSIS CONTINUED FROM PREVIOUS PAGE.

(8)
- USING J 39200, MEASURE RESISTANCE BETWEEN J 39700 TERMINALS "17" AND "3". DOES J 39200 DISPLAY "OL" (INFINITE)?

NO → REPAIR SHORT TO GROUND IN CKT 856.

YES ↓

(9)
- CONNECT J 39200 BETWEEN J 39700 TERMINALS "17" AND "3" (USE DC VOLTS).
- IGNITION "ON."
- NOTE VOLTAGE.

IS VOLTAGE ONE VOLT OR LESS?

NO → REPAIR SHORT TO B+ IN CKT 856.

YES ↓

(10)
- IGNITION "OFF."
- USING J 39200, MEASURE RESISTANCE BETWEEN BPMV CONNECTOR TERMINAL "10" AND J 39700 TERMINAL "20". IS RESISTANCE 5 OHMS OR LESS?

NO → REPAIR OPEN IN CKT 1079.

YES ↓

(11)
- USING J 39200, MEASURE RESISTANCE BETWEEN J 39700 TERMINALS "20" AND "3". DOES J 39200 DISPLAY "OL" (INFINITE)?

NO → REPAIR SHORT TO GROUND IN CKT 1079.

YES ↓

(12)
- USING J 39200, MEASURE RESISTANCE BETWEEN BPMV CONNECTOR TERMINAL "2" AND J 39700 TERMINAL "7". IS RESISTANCE 5 OHMS OR LESS?

NO → REPAIR OPEN IN CKT 1077.

YES ↓

(13)
- USING J 39200, MEASURE RESISTANCE BETWEEN J 39700 TERMINALS "3" AND "7". DOES J 39200 DISPLAY "OL" (INFINITE)?

NO → REPAIR SHORT TO GROUND IN CKT 1077.

YES ↓

(14)
- IGNITION "OFF."
- INSPECT ALL WIRING AND CONNECTORS FOR CONDITIONS WHICH COULD CAUSE INTERMITTENTS, SUCH AS CHAFING, CORROSION, OR POOR TERMINAL CONTACT.

ARE ALL CONNECTIONS AND WIRING GOOD?

NO → REPAIR WIRING AND/OR CONNECTIONS AS NEEDED.

YES ↓

(15)
- REMOVE J 39700.
- RECONNECT EBTCM AND BPMV CONNECTORS.
- CLEAR DTCs, IGNITION "OFF."
- DISCONNECT TECH 1, IF CONNECTED.
- DRIVE VEHICLE AT LEAST 7 km/h (4 MPH) TO ACTIVATE AUTO TEST SEQUENCE. DOES DTC 63 RESET?

NO → NO TROUBLE FOUND; SEE DIAGNOSTIC AIDS ON FACING PAGE.

YES → REPLACE EBTCM

WHEN ALL DIAGNOSIS AND REPAIRS ARE COMPLETED, CLEAR DTC(s) AND VERIFY PROPER OPERATION.

DIAGNOSTIC CHART DTC 64 — 1994 FLEETWOOD

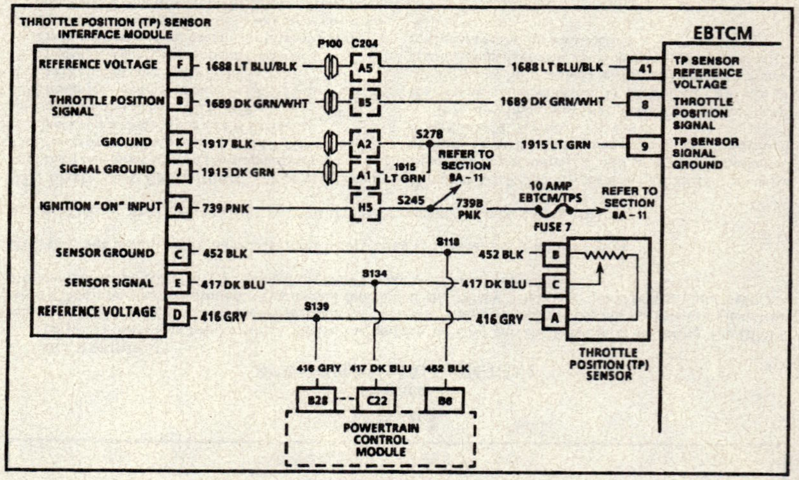

DTC 64
(Page 1 of 4)
THROTTLE POSITION SIGNAL MALFUNCTION

Circuit Description:
The throttle position signal circuit is used by the EBTCM to monitor the actual throttle position versus desired throttle position to control engine torque during a TCS event. The throttle position sensor interface module is used to allow EBTCM monitoring of the TP sensor signal without affecting the TP sensor signal to the PCM.

DTC Will Set When: DTC 64 will set if the EBTCM is not receiving reference voltage or a TP sensor signal from the throttle position sensor interface module.

DTC Chart Test Description: Number(s) below refer to circled number(s) on the diagnostic chart.

1. Checks for TP sensor-related PCM DTCs. Many of the same conditions which could cause a TCS DTC 64 will also cause PCM DTCs. The PCM DTCs should be repaired first to eliminate those malfunctions from affecting the TCS DTC 64 condition.
2. Checks for a short to ground in CKT 1689.
3. Checks for an open in CKT 1915.
4. Checks for a short to ground in CKT 1688.
5. Verifies proper voltage in CKT 1688 at the EBTCM connection. (This voltage measurement is only appropriate when the EBTCM is disconnected from the circuit.)
6. Verifies proper voltage in CKT 1689 at the EBTCM connection. (This voltage measurement is only appropriate when the EBTCM is disconnected from the circuit.)
7. Checks for an additional DTC 64 malfunction after ground CKT 1915 is repaired.
8. Determines if a malfunction found in Step 5 is due to an open fuse, CKT 739, or CKT 1688.
9. Determines if a malfunction found in Steps 5 and 13 is due to an open in CKT 1688 or a malfunctioning TP sensor interface module.

DIAGNOSTIC CHART DTC 64 — 1994 FLEETWOOD, CONTINUED

DTC 64
(Page 2 of 4)
THROTTLE POSITION SIGNAL MALFUNCTION

1. IF PCM DTCs 21 OR 22 ARE SET, USE THE PCM DIAGNOSTIC CHARTS TO REPAIR THOSE DTCs. THEN, CLEAR PCM AND TCS DTCs. TURN IGNITION "OFF." DISCONNECT TECH 1, IF CONNECTED. TURN IGNITION "ON." IF TCS DTC 64 RESETS, CONTINUE WITH THIS DIAGNOSTIC CHART.

2.
- IGNITION "OFF."
- DISCONNECT EBTCM.
- INSTALL J 39700 UNIVERSAL BREAKOUT BOX TO EBTCM HARNESS CONNECTOR ONLY.
- USING J 39200, MEASURE RESISTANCE BETWEEN J 39700 TERMINALS "8" AND "3". DOES J 39200 DISPLAY "OL" (INFINITE)?

   - YES → go to 3.
   - NO → REPAIR SHORT TO GROUND IN CKT 1689.

3.
- USING J 39200, MEASURE RESISTANCE BETWEEN J 39700 TERMINALS "9" AND "3". IS RESISTANCE 5 OHMS OR LESS?

   - YES → go to 4.
   - NO → go to 7.

4.
- USING J 39200, MEASURE RESISTANCE BETWEEN J 39700 TERMINALS "41" AND "3". DOES J 39200 DISPLAY "OL" (INFINITE)?

   - YES → go to 5.
   - NO → REPAIR SHORT TO GROUND IN CKT 1688.

5.
- CONNECT J 39200 BETWEEN J 39700 TERMINALS "41" AND "3" (USE DC VOLTS).
- IGNITION "ON."
- NOTE VOLTAGE. IS VOLTAGE APPROXIMATELY 10 VOLTS?

   - YES → go to 6.
   - NO → go to 8.

6.
- CONNECT J 39200 BETWEEN J 39700 TERMINALS "8" AND "3".
- NOTE VOLTAGE. IS VOLTAGE APPROXIMATELY 9.5 TO 10.5 VOLTS?

   - YES →
     - IGNITION "OFF."
     - DISCONNECT J 39700 BREAKOUT BOX.
     - RECONNECT EBTCM CONNECTOR.
     - IGNITION "ON." DOES DTC 64 SET?
       - YES → REPLACE EBTCM.
       - NO → MALFUNCTION IS INTERMITTENT. REFER TO "DIAGNOSTIC AIDS" ON FACING PAGE.
   - NO → CONTINUE DIAGNOSIS ON NEXT PAGE.

7.
- REPAIR OPEN IN CKT 1915.
- CLEAR DTCs, IGNITION "OFF."
- DISCONNECT TECH 1, IF CONNECTED.
- IGNITION "ON." DOES DTC 64 RESET?

   - YES → CONTINUE DIAGNOSTICS AT STEP 4.
   - NO → SYSTEM OK.

8.
- IGNITION "OFF."
- DISCONNECT TP SENSOR INTERFACE MODULE.
- CONNECT J 39200 BETWEEN TP SENSOR INTERFACE MODULE CONNECTOR TERMINAL "A" AND J 39700 TERMINAL "3".
- IGNITION "ON."
- NOTE VOLTAGE. IS VOLTAGE WITHIN 2 VOLTS OF B+?

   - YES → go to 9.
   - NO → REPLACE EBTCM/TPS FUSE #7 OR REPAIR OPEN IN CKT 739.

9.
- IGNITION "OFF."
- USING J 39200, MEASURE RESISTANCE BETWEEN J 39700 TERMINAL "41" AND TP SENSOR INTERFACE MODULE CONNECTOR TERMINAL "F". IS IT 5 OHMS OR LESS?

   - YES → REPLACE TP SENSOR INTERFACE MODULE.
   - NO → REPAIR OPEN IN CKT 1688.

WHEN ALL DIAGNOSIS AND REPAIRS ARE COMPLETED, CLEAR DTC(s) AND VERIFY PROPER OPERATION.

DIAGNOSTIC CHART DTC 64 — 1994 FLEETWOOD, CONTINUED

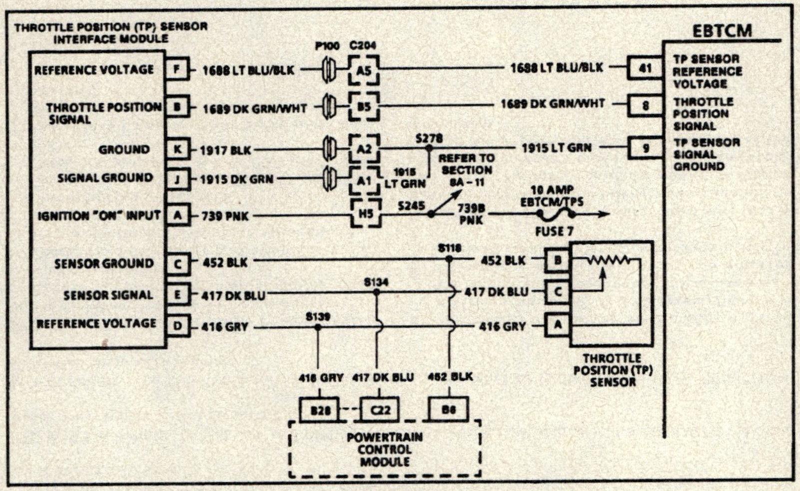

DTC 64
(Page 3 of 4)
THROTTLE POSITION SIGNAL MALFUNCTION

Circuit Description:
The throttle position signal circuit is used by the EBTCM to monitor the actual throttle position versus desired throttle position to control engine torque during a TCS event. The throttle position sensor interface module is used to allow EBTCM monitoring of the TP sensor signal without affecting the TP sensor signal to the PCM.

DTC Will Set When: DTC 64 will set if the EBTCM is not receiving reference voltage or a TP sensor signal from the throttle position sensor interface module.

DTC Chart Test Description: Number(s) below refer to circled number(s) on the diagnostic chart.

10. Checks for an open in CKT 1689.
11. Checks for TP sensor reference voltage from the PCM at the TP sensor interface module connector.
12. Checks for TP sensor signal voltage on CKT 417 at the TP sensor interface module connector.
13. Checks for an open in TP sensor signal ground CKT 452 to the TP sensor interface module connector.
14. Checks for an open in signal ground CKT 1917 to the TP sensor interface module connector.
15. Checks for DTC 64 reset after most system components have checked OK.
16. Replaces TP sensor interface module as likely cause of DTC reset in Step 15. If the DTC resets again, the EBTCM must be concluding there is a malfunction present when there is not.

Diagnostic Aids: It is very important that a thorough inspection of the wiring and connectors be performed. Failure to carefully and fully inspect wiring and connectors may result in misdiagnosis, causing part replacement with reappearance of the malfunction.

DIAGNOSTIC CHART DTC 64 — 1994 FLEETWOOD, CONTINUED

DTC 64
(Page 4 of 4)

THROTTLE POSITION SIGNAL MALFUNCTION

DIAGNOSIS CONTINUED FROM PREVIOUS PAGE.

(10)
- IGNITION "OFF."
- DISCONNECT TP SENSOR INTERFACE MODULE.
- USING J 39200, MEASURE RESISTANCE BETWEEN J 39700 TERMINAL "8" AND TP SENSOR INTERFACE MODULE CONNECTOR TERMINAL "B", USING ADAPTER FROM J 35616-A.
  IS RESISTANCE 5 OHMS OR LESS?

YES → (11) · NO → REPAIR OPEN IN CKT 1689.

(11)
- CONNECT J 39200 BETWEEN TP SENSOR INTERFACE MODULE CONNECTOR TERMINAL "D", USING ADAPTER FROM J 35616-A, AND J 39700 TERMINAL "3" (USE DC VOLTS).
- IGNITION "ON."
- NOTE VOLTAGE.
  IS VOLTAGE APPROXIMATELY 5 VOLTS?

YES → (12) · NO → REPAIR OPEN IN CKT 416.

(12)
- CONNECT J 39200 BETWEEN MODULE CONNECTOR TERMINAL "E" AND J 39700 TERMINAL "3".
- NOTE VOLTAGE.
  IS VOLTAGE APPROXIMATELY 0.5 VOLTS?

YES → (13) · NO → REPAIR OPEN IN CKT 417.

(13)
- IGNITION "OFF."
- CONNECT J 39200 BETWEEN TP SENSOR INTERFACE MODULE CONNECTOR TERMINALS "E" AND "C".
- IGNITION "ON."
- NOTE VOLTAGE.
  IS VOLTAGE APPROXIMATELY 0.5 VOLTS?

YES → (14) · NO → REPAIR OPEN IN CKT 452.

(14)
- IGNITION "OFF."
- USING J 39200, MEASURE RESISTANCE BETWEEN J 39700 TERMINAL "9" AND TP SENSOR INTERFACE MODULE CONNECTOR TERMINAL "K".
  IS RESISTANCE 5 OHMS OR LESS?

YES → (15) · NO → REPAIR OPEN IN CKT 1917.

(15)
- IGNITION "OFF."
- REMOVE J 39700; RECONNECT TP SENSOR INTERFACE MODULE.
- CLEAR DTCs, IGNITION "OFF."
- DISCONNECT TECH 1, IF CONNECTED.
- IGNITION "ON."
  DOES DTC 64 RESET?

YES → (16) · NO → NO TROUBLE FOUND; SEE "DIAGNOSTIC AIDS" ON FACING PAGE.

(16)
- IGNITION "OFF."
- REPLACE TP SENSOR INTERFACE MODULE.
- CLEAR DTCs, IGNITION "OFF."
- DISCONNECT TECH 1, IF CONNECTED.
- IGNITION "ON."
  DOES DTC 64 RESET?

YES → REPLACE EBTCM. · NO → SYSTEM OK.

WHEN ALL DIAGNOSIS AND REPAIRS ARE COMPLETED, CLEAR DTC(s) AND VERIFY PROPER OPERATION.

DIAGNOSTIC CHART DTC 65 — 1994 FLEETWOOD

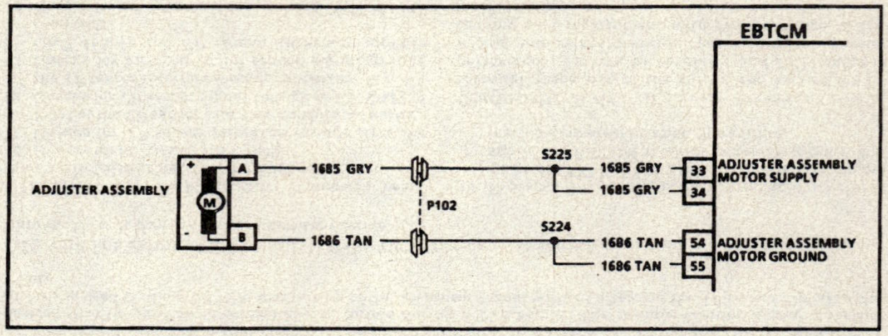

DTC 65
(Page 1 of 4)

ADJUSTER ASSEMBLY CIRCUIT MALFUNCTION

Circuit Description:
The adjuster assembly circuit provides both power and ground to the adjuster assembly to help control engine torque during a TCS event.

DTC Will Set When: DTC 65 will set if CKT 1685 or CKT 1686 are shorted to ground, shorted to B+, shorted to each other, or both open at the same time.

Action Taken: The TCS system will be disabled for the remainder of the ignition cycle and the "TRACTION CONTROL" lamp will illuminate.

DTC Chart Test Description: Number(s) below refer to circled number(s) on the diagnostic chart.

1. Checks for proper adjuster assembly motor winding resistance.
2. Checks for a poor power feed condition in CKT 1440, which the EBTCM might sense as an open.
3. Checks for open in CKT 1685.
4. Checks for open in CKT 1686.
5. Checks for a short between CKTs 1685 and 1686.
6. Checks for short(s) to B+ in CKT 1685 and/or 1685.
7. Repeat Step 1 tests to ensure motor is in "home" position and a correct resistance value can be measured.
8. Sets throttle to a position which enables a special EBTCM monitor. If DTC 65 resets, there is a malfunction which requires continuing diagnosis.
9. Determines whether wiring or the adjuster assembly are at fault for problems found in Step 7.

Diagnostic Aids: It is very important that a thorough inspection of the wiring and connectors be performed. Failure to carefully and fully inspect wiring and connectors may result in misdiagnosis, causing part replacement with reappearance of the malfunction.

DIAGNOSTIC CHART DTC 65 — 1994 FLEETWOOD, CONTINUED

DTC 65
(Page 2 of 4)
ADJUSTER ASSEMBLY CIRCUIT MALFUNCTION

1. - IGNITION "OFF."
   - DISCONNECT ADJUSTER ASSEMBLY CONNECTOR.
   - DISCONNECT ALL CABLES FROM ADJUSTER ASSEMBLY.
   - ROTATE ADJUSTER ASSEMBLY (BY HAND) FULLY, THEN RELEASE AND ALLOW TO RETURN TO THE STOP.
   - USING J 39200, MEASURE RESISTANCE BETWEEN ADJUSTER ASSEMBLY PIGTAIL CONNECTOR TERMINALS "A" AND "B", USING ADAPTERS FROM J 35616-A.

   IS RESISTANCE 0.5-10 OHMS?

   YES → 2; NO → 7

2. - DISCONNECT EBTCM AND ADJUSTER ASSEMBLY CONNECTOR (ONLY IF SENT HERE FROM STEP 15).
   - INSTALL J 39700 UNIVERSAL BREAKOUT BOX TO EBTCM HARNESS CONNECTOR ONLY (UNLESS SENT HERE FROM STEP 15).
   - CONNECT UNPOWERED TEST LIGHT BETWEEN J 39700 TERMINALS "3" AND "38", AND "3" AND "39".

   DOES BULB GLOW BRIGHTLY BOTH TIMES?

   YES → 3; NO → REPAIR POOR POWER FEED CONDITION IN CKT 1440.

3. - USING J 39200, MEASURE RESISTANCE BETWEEN ADJUSTER HARNESS CONNECTOR TERMINAL "A" AND J 39700 TERMINAL "33" OR "34".

   IS RESISTANCE 5 OHMS OR LESS?

   YES → 4; NO → REPAIR OPEN IN CKT 1685.

4. - USING J 39200, MEASURE RESISTANCE BETWEEN ADJUSTER HARNESS CONNECTOR TERMINAL "B" AND J 39700 TERMINAL "54" OR "55".

   IS RESISTANCE 5 OHMS OR LESS?

   YES → 5; NO → REPAIR OPEN IN CKT 1686.

5. - USING J 39200, MEASURE RESISTANCE BETWEEN ADJUSTER ASSEMBLY HARNESS CONNECTOR TERMINAL "A" AND "B".

   DOES J 39200 DISPLAY "OL" (INFINITE)?

   YES → 6; NO → REPAIR SHORT BETWEEN CKTs 1685 AND 1686.

6. - CONNECT J 39200 BETWEEN J 39700 TERMINALS "33" AND "3" (USE DC VOLTS).
   - IGNITION "ON."
   - NOTE VOLTAGE.
   - IS VOLTAGE ONE VOLT OR LESS?

   YES → CONTINUE DIAGNOSIS ON NEXT PAGE.; NO → REPAIR SHORT TO B+ IN CKT 1685.

7. - REPEAT "ROTATE AND RELEASE" IN STEP 1 THREE OR FOUR MORE TIMES. REPEAT RESISTANCE MEASUREMENT EACH TIME AFTER ADJUSTER RETURNS TO STOP.

   IS RESISTANCE EVER 0.5-10 OHMS?

   YES → 8; NO → 9

8. - RECONNECT ADJUSTER ASSEMBLY CABLES AND CONNECTORS.
   - CLEAR DTCs, IGNITION "OFF."
   - CONNECT TECH 1 WITH MASS STORAGE CARTRIDGE.
   - IGNITION "ON."
   - SELECT DATA LIST AND MONITOR THROTTLE ANGLE.
   - SLOWLY DEPRESS ACCELERATOR PEDAL UNTIL THROTTLE ANGLE HAS PASSED 30 DEGREES.

   DOES DTC 65 RESET?

   NO → SYSTEM OK; RECONNECT AND ADJUST CABLES.; YES → CONTINUE DIAGNOSIS AT STEP 2.

9. - THOROUGHLY INSPECT ADJUSTER ASSEMBLY WIRING PIGTAIL, CONNECTOR AND TERMINALS FOR SHORTS OR OPENS. REPAIR AS NEEDED, FOLLOWING SPECIAL REPAIR PROCEDURES SHOWN IN "ADJUSTER ASSEMBLY WIRING REPAIR" IN THIS SECTION.
   - IF NO WIRING OR CONNECTOR PROBLEMS ARE FOUND, REPLACE ADJUSTER ASSEMBLY.

WHEN ALL DIAGNOSIS AND REPAIRS ARE COMPLETED, CLEAR DTC(s) AND VERIFY PROPER OPERATION.

DIAGNOSTIC CHART DTC 65 — 1994 FLEETWOOD, CONTINUED

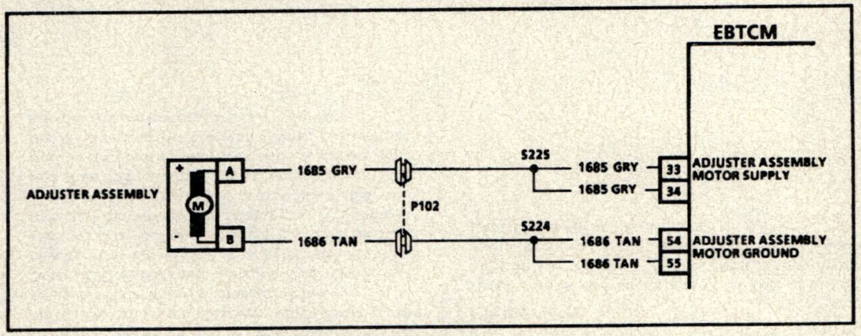

DTC 65
(Page 3 of 4)
ADJUSTER ASSEMBLY CIRCUIT MALFUNCTION

Circuit Description:
The adjuster assembly circuit provides both power and ground to the adjuster assembly to help control engine torque during a TCS event.

DTC Will Set When: DTC 65 will set if CKT 1685 or CKT 1686 are shorted to ground, shorted to B+, shorted to each other, or both open at the same time.

Action Taken: The TCS system will be disabled for the remainder of the ignition cycle and the "TRACTION CONTROL" lamp will illuminate.

DTC Chart Test Description: Number(s) below refer to circled number(s) on the diagnostic chart.

10. Checks for short to B + in CKT 1686.
11. Checks for short to ground in CKT 1685.
12. Checks for continued DTC setting after wiring and adjuster assembly inspections.
13. Sets throttle to a position which enables one of the EBTCM's special monitoring systems of the adjuster assembly circuits, checking for a DTC reset. If the DTC resets, the EBTCM is likely malfunctioning.
14. Checks for shorts to B+ or ground that are EBTCM-switched. If you have reached Step 12 for a second time after performing checks for EBTCM-switched malfunctions, the EBTCM is likely malfunctioning for the DTC reset.

Diagnostic Aids: It is very important that a thorough inspection of the wiring and connectors be performed. Failure to carefully and fully inspect wiring and connectors may result in misdiagnosis, causing part replacement with reappearance of the malfunction.

DIAGNOSTIC CHART DTC 65 — 1994 FLEETWOOD, CONTINUED

DTC 65
(Page 4 of 4)
ADJUSTER ASSEMBLY CIRCUIT MALFUNCTION

DIAGNOSIS CONTINUED FROM PREVIOUS PAGE

(10)
- CONNECT J 39200 BETWEEN J 39700 TERMINALS "54" AND "3".
- NOTE VOLTAGE.
  IS VOLTAGE ONE VOLT OR LESS?

NO: REPAIR SHORT TO B + IN CKT 1686.

YES:

(11)
- IGNITION "OFF."
- USING J 39200, MEASURE RESISTANCE BETWEEN J 39700 TERMINALS "3" AND "33".
  DOES J 39200 DISPLAY "OL" (INFINITE)?

NO: REPAIR SHORT TO GROUND IN CKT 1685.

YES:

(12)
- USING J 39200, MEASURE RESISTANCE BETWEEN J 39700 TERMINALS "3" AND "54".
  DOES J 39200 DISPLAY "OL" (INFINITE)?

NO: REPAIR SHORT TO GROUND IN CKT 1686 AND/OR CKT 1685.

YES:

(13)
- IGNITION "OFF."
- REMOVE J 39700.
- RECONNECT ADJUSTER ASSEMBLY AND EBTCM CONNECTORS.
- CLEAR DTCs, IGNITION "OFF."
- DISCONNECT TECH 1, IF CONNECTED.
- IGNITION "ON."
  DOES DTC 65 RESET?

NO:

(14)
- RECONNECT ADJUSTER ASSEMBLY CABLES AND CONNECTORS.
- CONNECT TECH 1 WITH MASS STORAGE CARTRIDGE.
- IGNITION "ON."
- SELECT DATA LIST, AND MONITOR THROTTLE ANGLE.
- SLOWLY DEPRESS ACCELERATOR PEDAL UNTIL THROTTLE ANGLE HAS PASSED 30 DEGREES.
  DOES DTC 65 RESET?

NO: NO TROUBLE FOUND, SEE "DIAGNOSTIC AIDS" ON FACING PAGE.

YES: REPLACE EBTCM.

YES:

(15)
- REPEAT DIAGNOSTICS BEGINNING AT STEP 2, EXCEPT CONNECT J 39700 TO EBTCM HARNESS CONNECTOR AND EBTCM; THIS WILL CHECK FOR EBTCM-SWITCHED SHORTS TO B + OR GROUND. REFER TO "EBTCM-SWITCHED FAULTS" IN THIS SECTION.
- HAVE YOU REPEATED STEPS 2 THROUGH 13 AS DIRECTED ABOVE, AND DID DTC 65 RESET?

YES: REPLACE EBTCM.

NO: SEE "DIAGNOSTIC AIDS" ON FACING PAGE.

WHEN ALL DIAGNOSIS AND REPAIRS ARE COMPLETED, CLEAR DTC(s) AND VERIFY PROPER OPERATION.

DIAGNOSTIC CHART DTC 66 — 1994 FLEETWOOD

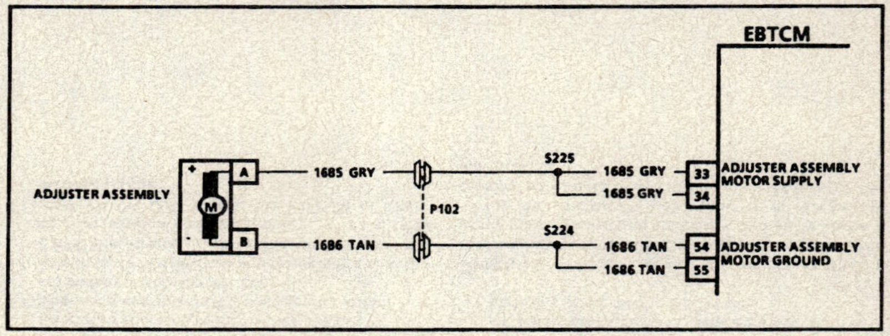

DTC 66
ADJUSTER ASSEMBLY CONTROL MALFUNCTION

Circuit Description:
The adjuster assembly circuit provides both power and ground to the adjuster assembly to help control engine torque during a TCS event.

DTC Will Set When: DTC 66 will set if, during a TCS event, the EBTCM delivers current greater than 16 amps to the adjuster assembly for more than three seconds without the adjuster assembly reaching its desired position.

DTC Chart Test Description: Number(s) below refer to circled number(s) on the diagnostic chart.
1. Checks to see if TCS DTC 65 is set, since DTC 65 conditions might affect DTC 66.
2. Checks for any mechanical binding of the adjuster assembly, attached cables, and components to which the cables are attached, which could cause excessive load on the adjuster assembly.
3. Since DTC 66 only sets during a TCS event, this check raises the vehicle on a hoist such that the rear wheels will spin while the fronts do not, inducing the system into attempting TCS control. The throttle angle must be greater than 15%, and less than 50%, and the system must attempt TCS control for greater than 3 seconds to ensure that the DTC-setting throttle and timing parameters are met when checking for a DTC 66 reset.

Diagnostic Aids: It may be helpful to disconnect all cables from the adjuster assembly and move them manually throughout their entire operating range. Also check the components to which the cables are attached for binding.

Important
- After cables are reconnected to the adjuster assembly, perform "Adjustment Procedure-Adjuster Assembly - Accelerator and Cruise Control Cables" in this section.

DIAGNOSTIC CHART DTC 66 — 1994 FLEETWOOD, CONTINUED

DTC 66
ADJUSTER ASSEMBLY CONTROL MALFUNCTION

BEFORE BEGINNING TEST, INSPECT ALL ADJUSTER ASSEMBLY CABLES AND LINKAGES TO WHICH THEY ATTACH TO ENSURE NO OBVIOUS PROBLEMS THAT COULD CAUSE BINDING OR STICKING EXIST.

(1) IS TCS DTC 65 SET?

- YES → REFER TO DTC 65 CHART.
- NO →

(2)
- IGNITION "OFF."
- HAVE ASSISTANT DEPRESS ACCELERATOR PEDAL SMOOTHLY AND FULLY TO WIDE OPEN THROTTLE (WOT), THEN SLOWLY RETURN TO IDLE POSITION WHILE INSPECTING CABLES AND LINKAGES TO WHICH THEY ARE ATTACHED FOR SMOOTH OPERATION.

DO ALL CABLES AND LINKAGES MOVE SMOOTHLY AND EVENLY WITH NO BINDING?

- NO → REPAIR LINKAGES AND/OR REPLACE CABLES, AND/OR REPAIR OR REPLACE COMPONENTS AS NEEDED TO RESTORE SMOOTH OPERATION.

  Important
  - AFTER CABLES ARE REATTACHED TO THE ADJUSTER ASSEMBLY OR COMPONENTS, OR AFTER CABLES ARE REPLACED, PERFORM "ADJUSTMENT PROCEDURE - ADJUSTER ASSEMBLY-THROTTLE AND CRUISE CONTROL CABLES," IN THIS SECTION.

  IF A NEW TP SENSOR OR THROTTLE BODY IS INSTALLED, BE SURE TO PERFORM "TP SENSOR LEARN PROCEDURE" IN THIS SECTION.
- YES →

(3)
- CLEAR DTCs, WAIT 15 SECONDS, IGNITION "OFF."
- RAISE REAR OF VEHICLE UNTIL REAR WHEELS ARE APPROXIMATELY 6" OFF FLOOR.
- CONNECT TECH 1 WITH MSC.
- START ENGINE.
- SELECT TECH 1 ECM DATA LIST AND MONITOR THROTTLE ANGLE.
- APPLY BRAKES.
- PUT VEHICLE IN GEAR.
- RELEASE BRAKES.
- DEPRESS AND HOLD ACCELERATOR PEDAL AT A THROTTLE ANGLE BETWEEN 15% AND 45% FOR NO LESS THAN FOUR SECONDS, THEN RELEASE.

DOES DTC 66 RESET?

- YES → REPLACE ADJUSTER ASSEMBLY.
- NO → NO TROUBLE FOUND, SEE "DIAGNOSTIC AIDS" ON FACING PAGE.

WHEN ALL DIAGNOSIS AND REPAIRS ARE COMPLETED, CLEAR DTC(s) AND VERIFY PROPER OPERATION.

DIAGNOSTIC CHART DTC 71 — 1994 FLEETWOOD

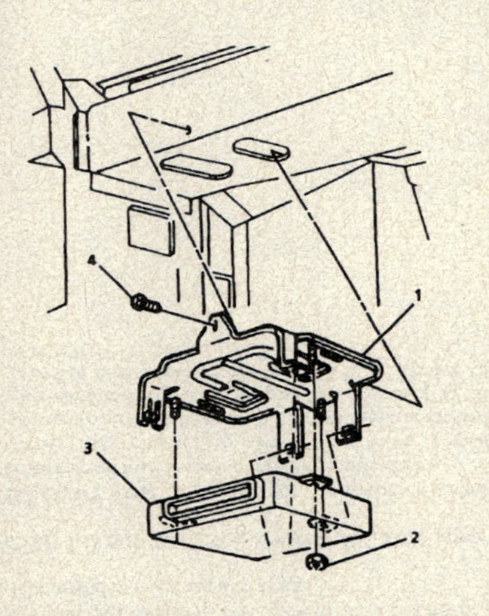

1 BRACKET ASSEMBLY, ELECTRONIC BRAKE AND TRACTION CONTROL MODULE

2 NUT, ELECTRONIC BRAKE AND TRACTION CONTROL MODULE, 1.9 N·m (17 lb. in.)

3 MODULE ASSEMBLY, ELECTRONIC BRAKE AND TRACTION CONTROL

4 BOLT/SCREW, ELECTRONIC BRAKE AND TRACTION CONTROL BRACKET, 6 N·m (53 lb. in.)

DTC 71
EBTCM INTERNAL MALFUNCTION

Circuit Description:
The EBTCM performs various diagnostic checks on itself. If it finds a malfunction, DTC 71 will set.

DTC Chart Test Description: Number(s) below refer to circled number(s) on the diagnostic chart.
1. Checks for good connections from the harness to the EBTCM itself.
2. Checks to see if the malfunction was false. If DTC 71 resets, replace the EBTCM.

DIAGNOSTIC CHART DTC 71 — 1994 FLEETWOOD, CONTINUED

DTC 71
EBTCM INTERNAL MALFUNCTION

(1)
- IGNITION "OFF."
- CHECK EBTCM CONNECTOR CAREFULLY TO ENSURE GOOD ELECTRICAL CONTACT AT ALL TERMINALS, AND THAT CONNECTOR IS PROPERLY RETAINED WHEN CONNECTED.

IS POOR CONNECTION EVIDENT?

- YES → REPAIR TERMINAL(S), CONNECTOR OR CONNECTOR RETENTION AS NECESSARY.
- NO → (2)
  - CLEAR DTCs, IGNITION "OFF."
  - DISCONNECT TECH 1, IF CONNECTED.
  - IGNITION "ON."

  DOES DTC 71 RESET?
  - YES → REPLACE EBTCM.
  - NO → NO TROUBLE FOUND; SEE "DIAGNOSTIC AIDS" ON FACING PAGE.

WHEN ALL DIAGNOSIS AND REPAIRS ARE COMPLETED, CLEAR DTC(s) AND VERIFY PROPER OPERATION.

DIAGNOSTIC CHART DTC 72 — 1994 FLEETWOOD

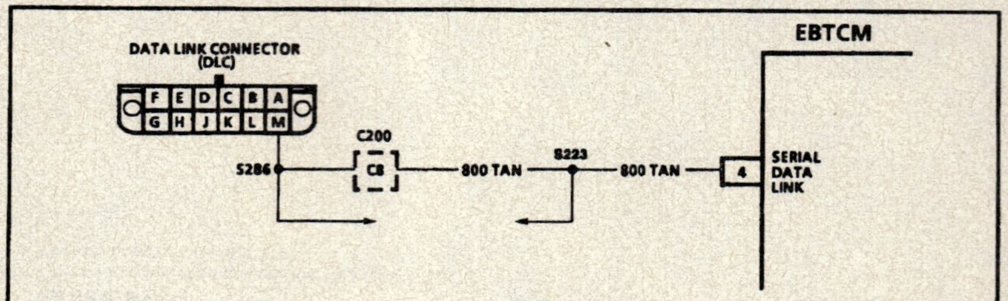

DTC 72
SERIAL DATA LINE MALFUNCTION

Circuit Description:
The serial data link is an asynchronous link operating at 8192 bits per second.

DTC Will Set When: DTC 72 will set if the EBTCM detects three consecutive serial data link messages that are ignored due to errors in transmission.

Action Taken: The "TRACTION CONTROL" lamp will illuminate.

DTC Chart Test Description: Number(s) below refer to circled number(s) on the diagnostic chart.

1. This test checks to see if the Tech 1 can communicate with the EBTCM. Automatic control air conditioning DTC display mode can also be used to communicate the the EBTCM. Refer to SECTION 1C for details.

Diagnostic Aids: The malfunction may be intermittent. Try performing the tests shown while "wiggling" wiring and connectors; this can often cause the fault to appear.

Automatic control air conditioning DTC 32 will likely be set along with this DTC if the malfunction is a short to ground or B+ on the serial data circuit.

If ABS/TCS DTC 72 is set and automatic control

DIAGNOSTIC CHART DTC 72 — 1994 FLEETWOOD, CONTINUED

DTC 72
SERIAL DATA LINE MALFUNCTION

(1)
- IGNITION "OFF."
- INSTALL TECH 1.
- IGNITION "ON."
- ATTEMPT TO ESTABLISH COMMUNICATIONS WITH THE EBTCM. ARE COMMUNICATIONS ESTABLISHED?

NO: ENSURE THERE ARE NO OPEN CIRCUITS IN SERIAL DATA LINE BETWEEN DLC TERMINAL "M" AND CCM AND PCM. IF NO OPEN CIRCUITS ARE EVIDENT,

YES: NO TROUBLE FOUND; SEE "DIAGNOSTIC AIDS" ON FACING PAGE.

WHEN ALL DIAGNOSIS AND REPAIRS ARE COMPLETED, CLEAR DTCs AND VERIFY OPERATION.

DIAGNOSTIC CHART DTC 83 — 1994 FLEETWOOD

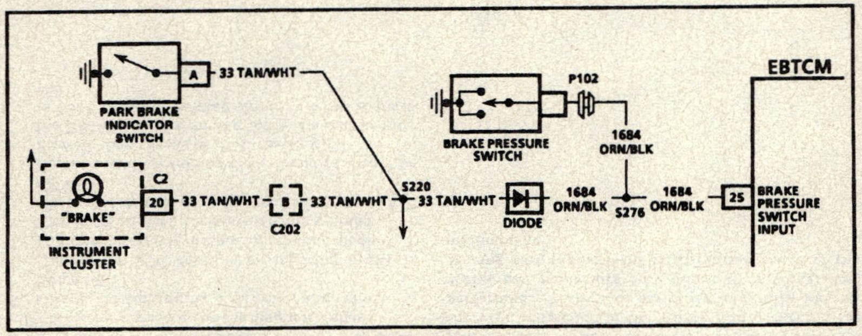

DTC 83
(Page 1 of 4)

BRAKE FLUID PRESSURE DIFFERENTIAL MALFUNCTION

Circuit Description:
The brake fluid differential pressure circuit monitors the pressure in the brake system. If the pressure becomes significantly different between the two brake hydraulic circuits, as indicated by the brake differential pressure switch, the ABS/TCS system is disabled. This is done to prevent introduction of air into the brake system. The diode is used to prevent disabling ABS/TCS if illumination of the red "BRAKE" warning lamp occurs because of application of the parking brake.

DTC Will Set When: DTC 83 will set if CKT 1684 is low (grounded).

Action Taken: The ABS/TCS system is disabled and the "ANTI-LOCK" and "TRACTION CONTROL" lamps will illuminate.

DTC Chart Test Description: Number(s) below refer to circled number(s) on the diagnostic chart.

1. Checks for proper brake fluid level in the master cylinder reservoir.
2. Checks for illumination of the red "BRAKE" warning lamp for reasons other than parking brake engagement.
3. Checks for brake pressure switch stuck closed.
4. Checks for short from CKT 1684 to chassis ground.
5. Adds fluid, noted to be low in Step 1, then checks for DTC reset.

Diagnostic Aids: It is very important that a thorough inspection of the wiring and connectors be performed. Failure to carefully and fully inspect wiring and connectors may result in misdiagnosis, causing part replacement with reappearance of the malfunction.

DIAGNOSTIC CHART DTC 83 — 1994 FLEETWOOD, CONTINUED

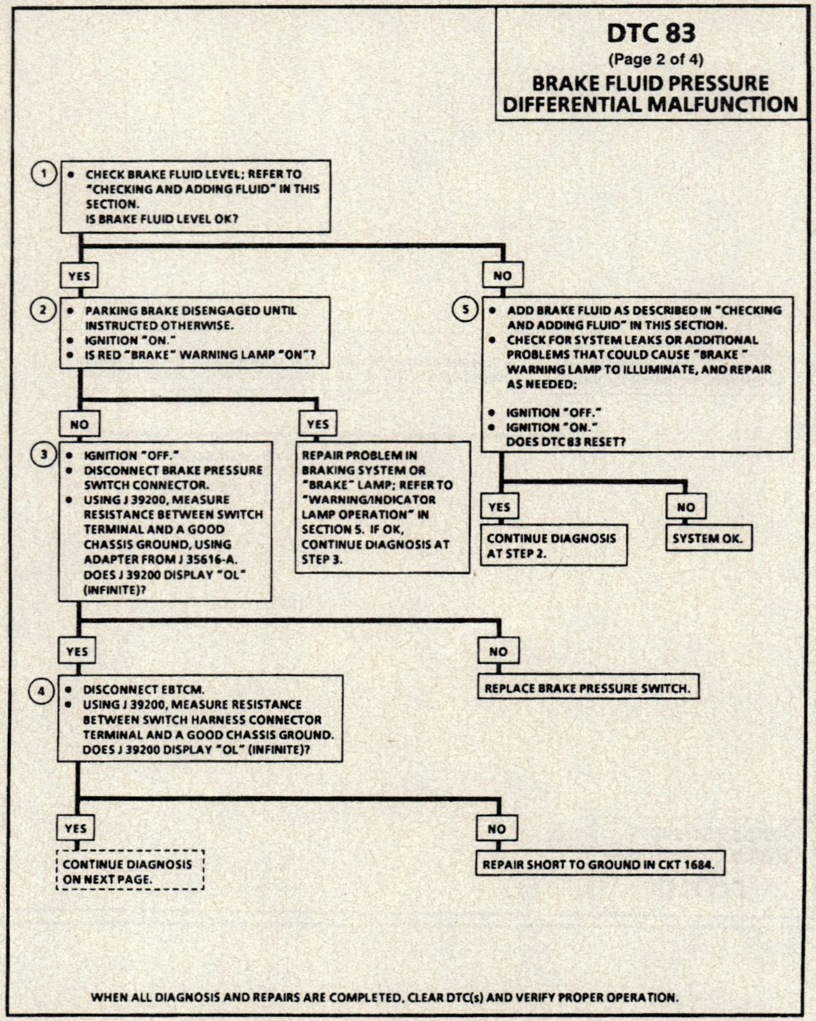

DIAGNOSTIC CHART DTC 83 — 1994 FLEETWOOD, CONTINUED

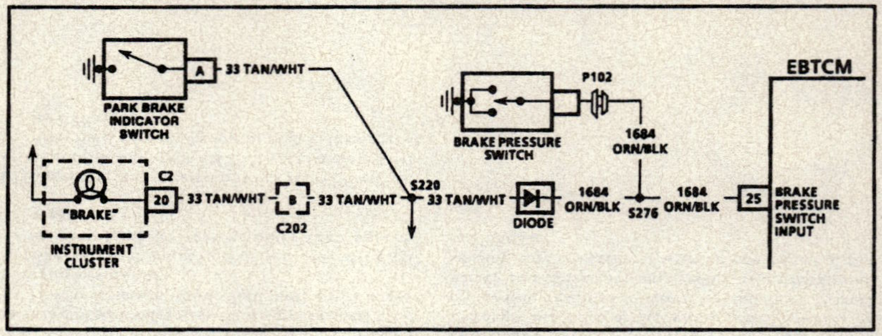

DTC 83
(Page 3 of 4)

BRAKE FLUID PRESSURE DIFFERENTIAL MALFUNCTION

Circuit Description:
The brake fluid differential pressure circuit monitors the pressure in the brake system. If the pressure becomes significantly different between the two brake hydraulic circuits, as indicated by the brake differential pressure switch, the ABS/TCS system is disabled. This is done to prevent introduction of air into the brake system. The diode is used to prevent disabling ABS/TCS if illumination of the red "BRAKE" warning lamp occurs because of application of the parking brake.

DTC Will Set When: DTC 83 will set if CKT 1684 is low (grounded).

Action Taken: The ABS/TCS system is disabled and the "ANTI-LOCK" and "TRACTION CONTROL" lamps will illuminate.

DTC Chart Test Description: Number(s) below refer to circled number(s) on the diagnostic chart.

6. Checks for continuing DTC reset after checks in Steps 1-4.
7. Checks for DTC reset when parking brake is engaged, likely due to a malfunctioning diode.
8. Checks for parking brake switch stuck closed.
9. Checks for short to ground in parking brake switch CKT 33.
10. Replaces malfunctioning diode found in Step 7 or Step 10, then checks for DTC reset. If DTC resets, the EBTCM is determining there is a malfunction in CKT 1684 and related circuitry when there is not.

Diagnostic Aids: It is very important that a thorough inspection of the wiring and connectors be performed. Failure to carefully and fully inspect wiring and connectors may result in misdiagnosis, causing part replacement with reappearance of the malfunction.

DIAGNOSTIC CHART DTC 83 — 1994 FLEETWOOD, CONTINUED

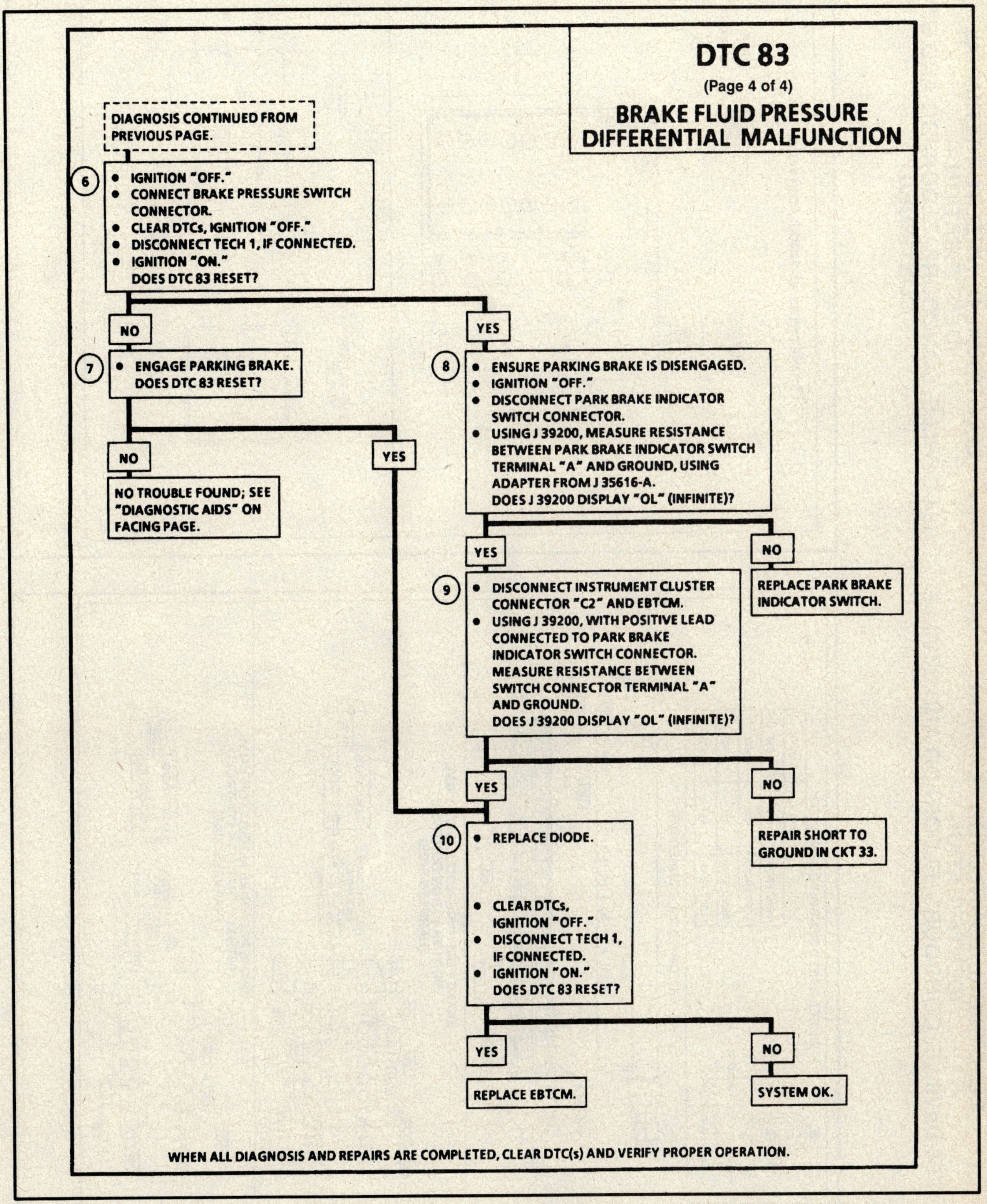

ELECTRICAL CONNECTOR TERMINAL IDENTIFICATION — 1994 DeVILLE, CONCOURS, ELDORADO AND SEVILLE

WHEEL SPEED SENSOR CONNECTOR

- WHEEL SPEED SENSOR
- C110 AND C111 BODY MAIN HARNESS SIDE

	PIN A	PIN B
RIGHT FRONT	872	833
LEFT FRONT	830	873
RIGHT REAR	882	883
LEFT REAR	884	885

EBTCM/EBCM CONNECTOR

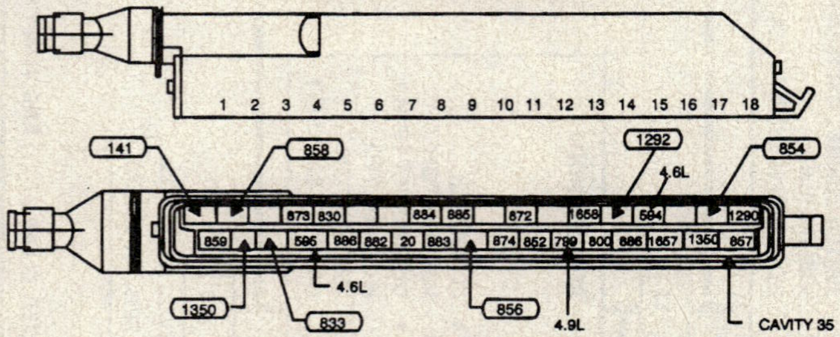

BRAKE PRESSURE MODULATOR VALVE (WITH TRACTION CONTROL) (HARNESS SIDE)

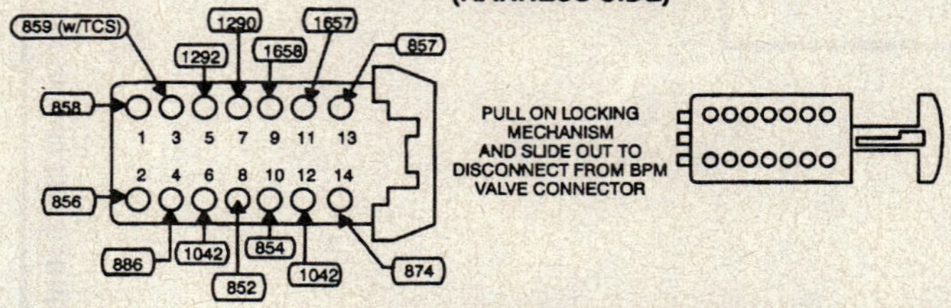

BRAKE PRESSURE MODULATOR VALVE (WITHOUT TRACTION CONTROL) (HARNESS SIDE)

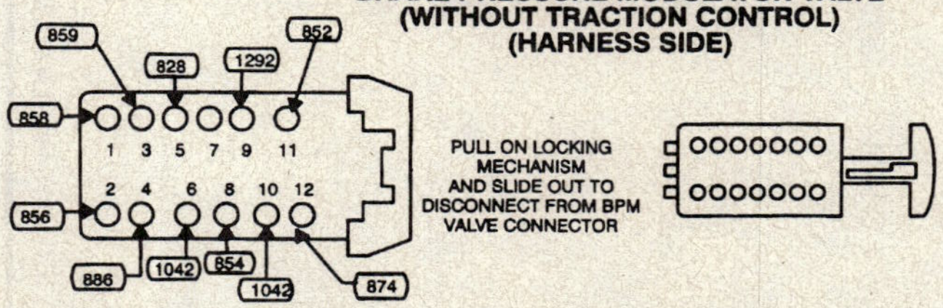

ELECTRICAL CONNECTOR TERMINAL IDENTIFICATION — 1994 DeVILLE, CONCOURS, ELDORADO AND SEVILLE, CONTINUED

BRAKE PRESSURE MODULATOR VALVE

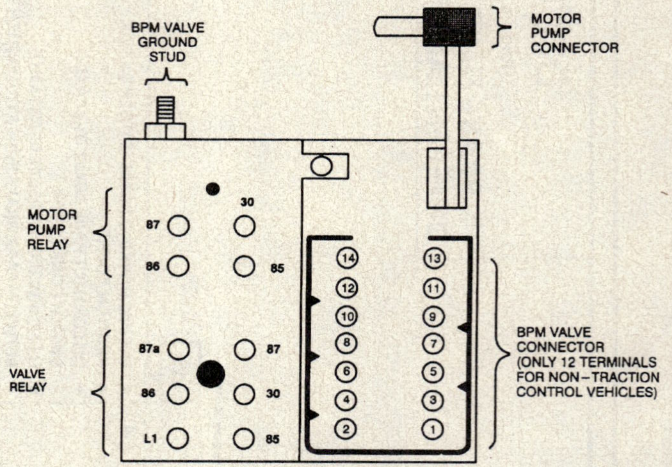

PUMP MOTOR RELAY

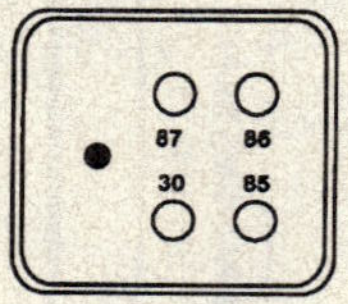

VALVE RELAY

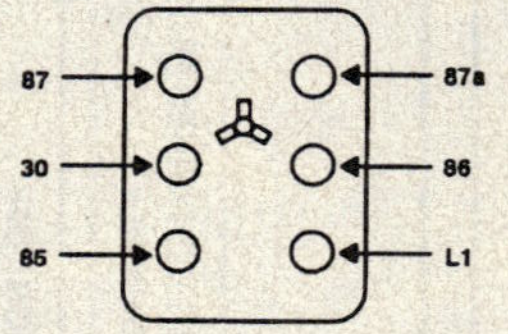

ELECTRICAL CONNECTOR TERMINAL IDENTIFICATION — 1994 DeVILLE, CONCOURS, ELDORADO AND SEVILLE, CONTINUED

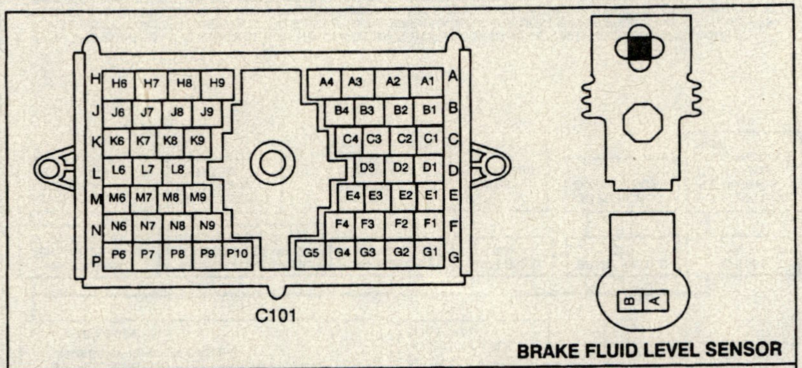

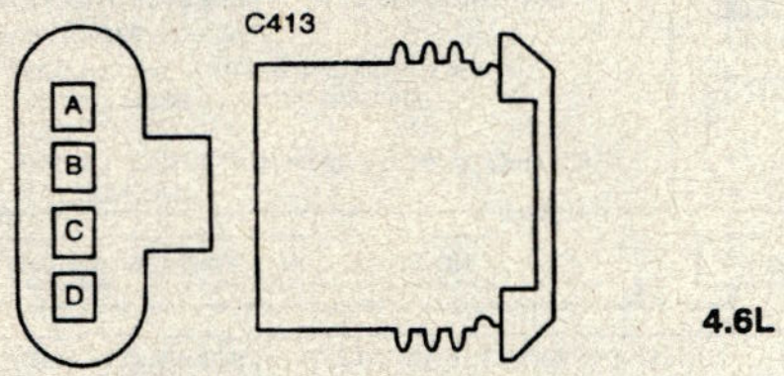

CAV	WIRE
A	884
B	885
C	882
D	883

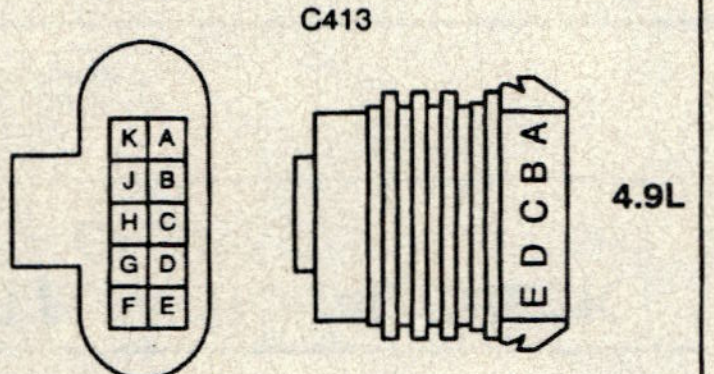

CAV	WIRE
A	884
B	885
C	882
D	883
E	1014
F	1016
G	1017
H	1010
J	1012
K	1013

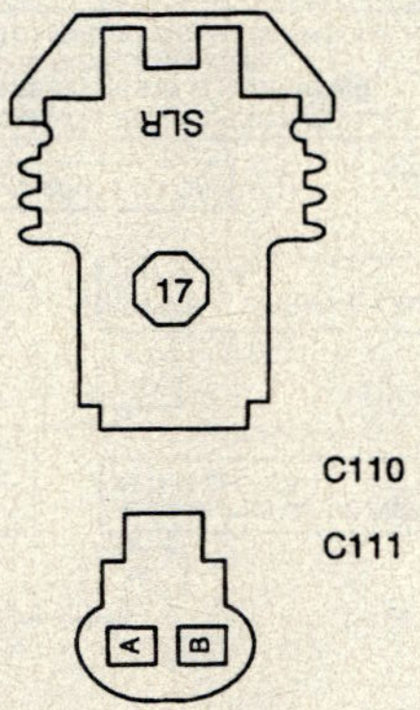

CAV	WIRE
A	830
B	873

DIAGNOSTIC CHART ABS/ASR VISUAL ILNSPECTION — 1994 DeVILLE, CONCOURS, ELDORADO AND SEVILLE

ABS/TCS VISUAL INSPECTION

ITEM	INSPECT FOR	CORRECTIVE ACTION
PARKING BRAKE	– FULL RELEASE – PROPER SWITCH FUNCTION IF NECESSARY, UNPLUG SWITCH CONNECTOR FROM PEDAL ASSEMBLY TO VERIFY – PROPER VACUUM RELEASE SYSTEM OPERATION	– OPERATE MANUAL RELEASE LEVER TO VERIFY RELEASE – ADJUST CABLE OR REPAIR RELEASE SYSTEM AS REQUIRED – REPAIR SWITCH AS REQUIRED
BRAKE LIGHT SWITCH	– IMPROPER ADJUSTMENT	– PULL UP ON BRAKE PEDAL TO ADJUST
MAJOR COMPONENTS • BRAKE FLUID RESERVOIR • BRAKE PRESSURE MODULATOR VALVE • MASTER CYLINDER • TCS PRIME PIPE • PROPORTIONING VALVES	– LOW FLUID LEVEL – EXTERNAL LEAKS – PROPER ASSEMBLY – EXTERNAL LEAKS	– ADD FLUID AS DETERMINE CAUSE OF FLUID LOSS AND REPAIR – REPAIR LEAKS AS REQUIRED – INSTALL OR POSITION COMPONENTS PROPERLY – REPAIR LEAKS AS REQUIRED
• FUSE A1 (10 A) (TRUNK COMPARTMENT FUSE BLOCK) • FUSE B3 (20 A) (ENGINE COMPARTMENT FUSE BLOCK) • MAXI® FUSE 5 (50 A) (ENGINE COMPARTMENT RH MAXI® FUSE BLOCK) FUSIBLE LINKS ON POSITIVE JUNCTION BLOCK	– BLOWN FUSE – OPEN IN FUSIBLE LINK	– REPLACE FUSE – REPAIR OPEN IN FUSIBLE LINK – VERIFY OPERATION – INSPECT FOR CAUSE OF FUSE OR FUSIBLE LINK FAILURE
CONNECTORS/WIRING UNDERHOOD • BRAKE PRESSURE MODULATOR VALVE • GROUND CONNECTION G406 • WHEEL SPEED SENSORS • C110 • C111 • BRAKE PRESSURE MODULATOR VALVE GROUND STUD IN PASSENGER COMPARTMENT OR IN TRUNK • C101 • EBTCM/EBCM • GROUND CONNECTION G406 REAR OF VEHICLE • WHEEL SPEED SENSORS • C413	– PROPER ENGAGEMENT – LOOSE WIRES OR TERMINALS – CORRODED OR BROKEN EYELETS – PAINT/PRIMER ON GROUNDS – WIRING NOT PROPERLY ROUTED OR SECURED – WIRING NEAR POSSIBLE SOURCE OF ELECTRICAL NOISE	– ADD STAR WASHER – PROPERLY ENGAGE CONNECTOR – REPAIR AS REQUIRED – REROUTE OR PROPERLY ROUTE WIRING

IF NO PROBLEM IS NOTED IN VISUAL INSPECTION, PROCEED TO ABS FUNCTIONAL CHECK

DIAGNOSTIC CHART ABS/ASR VISUAL ILNSPECTION CHART I — 1994 DeVILLE, CONCOURS, ELDORADO AND SEVILLE

ABS/TCS FUNCTIONAL CHECK CHART I

HAS VISUAL INSPECTION OF THE SYSTEM BEEN PERFORMED?

- **NO:** SEE 'ABS/TCS VISUAL INSPECTION' IN THIS SECTION
- **YES:**
  - IGNITION 'ON', START VEHICLE
  - DOES 'ANTILOCK' INDICATOR COME ON WHEN STARTING THE VEHICLE?
    - **NO:** INSPECT WARNING LAMP CIRCUITS
    - **YES:**
      - DOES 'ANTILOCK' INDICATOR STAY ON AND ARE THE 'ANTILOCK DISABLED' AND/OR 'TRACTION DISABLED' MESSAGE(S) ON?
        - **LAMP AND/OR MESSAGE(S) STAY ON:**
          - ENTER ON–BOARD DIAGNOSTICS (TCS VEHICLES)
          - TRY FLASHING ABS CODES (IF NO ON–BOARD DIAGNOSTICS OR TECH 1 AVAILABLE)
          - CHECK FOR STORED TCS DIAGNOSTIC TROUBLE CODES
            - **CODE(S) PRESENT:** GO TO ABS/TCS DIAGNOSTIC TROUBLE CODE TABLE
            - **NO CODES PRESENT OR NO TCS DATA:** GO TO ABS/TCS SYMPTOM TABLE
        - **LAMP AND MESSAGE(S) NEVER STAY ON:**
          - * IGNITION 'OFF'
          - INSTALL TECH 1 WITH MASS STORAGE CARTRIDGE
          - IGNITION 'ON'
          - SELECT MODE F1: CODE HISTORY
          - RECORD DTC (CODES) AND CODE DATA
          - IF ANY CODES HAVE SET:
            - WHICH CODES SET? (IF MULTIPLE CODES ARE STORED, BEGIN WITH THE CODE WITH THE LOWEST NUMBER OF IGNITION CYCLES)
            - CLEAR CODES

WHEEL SPEED SENSOR (21, 23, 25, 27, 28, 31, 33, 35, 37)	ABS VALVE SOLENOID OR TCS PILOT VALVE (41, 44, 45, 48, 51, 55)	PUMP MOTOR (61)	VALVE RELAY (63)	BRAKE SWITCH (67)	OTHER (71, 72, 73, 83)
GO TO ABS/TCS FUNCTIONAL CHECK CHART II	GO TO ABS/TCS FUNCTIONAL CHART III	GO TO ABS/TCS FUNCTIONAL CHART IV	GO TO ABS/TCS FUNCTIONAL CHART V	GO TO PCM DIAGNOSTICS	GO TO APPLICABLE DIAGNOSTIC TROUBLE CODE TROUBLE TREE

* FROM THIS POINT ON, IT IS ASSUMED THAT THE CONDITION(S) RESPONSIBLE FOR SETTING THE DIAGNOSTIC TROUBLE CODE(S) IS INTERMITTENT AND THE CODE IS NEVER SET CURRENT IN THE SHOP. MOST OF THE FOLLOWING TESTS ARE PERFORMED IN ORDER TO EXERCISE THE ABS/TCS CIRCUITS AND GET A CODE TO SET WITHOUT HAVING TO LEAVE THE BAY. IF AT ANYTIME A DIAGNOSTIC TROUBLE CODE IS SET 'HARD', REFER IMMEDIATELY TO THE 'ABS/TCS DIAGNOSTIC TROUBLE CODE TABLE'

DIAGNOSTIC CHART ABS/ASR VISUAL ILNSPECTION CHART II — 1994 DeVILLE, CONCOURS, ELDORADO AND SEVILLE

ABS/TCS FUNCTIONAL CHECK CHART II

WHEEL SPEED SENSORS

- USING THE TECH 1, SELECT MODE F3: SNAPSHOT
- SELECT MODE F3: AUTO TRIG
- TEST DRIVE VEHICLE AS DESCRIBED IN THIS SECTION WHILE MONITORING THE WHEEL SPEED SCREEN OF THE TECH 1 SNAPSHOT MODE. ATTEMPT TO DUPLICATE CONDITIONS AT WHICH CODE(S) SET PER OWNER DESCRIPTION AND CODE HISTORY INFORMATION:
  - SPEED AT WHICH CODE(S) SET
  - BRAKE SWITCH 'ON' OR 'OFF'
  - ABS OR TRACTION ACTIVE OR INACTIVE
- TRY DRIVING OVER GRAVEL, BUMPY ROAD, RAILROAD TRACKS AND MAKING FREQUENT TURNS. THESE MANEUVERS MAY HELP INDUCE A WIRING ASSOCIATED FAULT
- MANUALLY TRIGGER A SNAPSHOT BY SELECTING F9 IF THE WHEEL SPEED SIGNAL APPEARS ERRONEOUS
- IS A SNAPSHOT AUTOMATICALLY OR MANUALLY TRIGGERED?

NO

- CONDITION NOT PRESENT AT THIS TIME
- CHECK WHEEL SPEED SENSOR MOUNTING
- CHECK WHEEL SPEED SENSOR WIRING CONNECTIONS AND ROUTING
- IF ORIGINAL DIAGNOSTIC TROUBLE CODE WAS A CONTINUITY FAULT, WIGGLE AND FLEX SPEED SENSOR CABLE WITH IGNITION 'ON' TO SEE IF CODE RESETS WITH TECH 1 DISCONNECTED. ALSO CHECK FOR EXCESSIVE WHEEL BEARING PLAY
- SEE 'NOTE ON INTERMITTENTS'

YES

- NOTE DESCRIPTION OF WHEN SNAPSHOT OCCURRED
- SELECT MODE F0: REPLAY DATA
- RECORD ALL CRITERIA AT TRIGGER POINT
- IF FAULT ONLY OCCURS MOMENTARILY, CONDITION IS INTERMITTENT
  - INSPECT WHEEL SPEED SENSOR WIRING CONNECTIONS AND ROUTING
  - INSPECT WHEEL SPEED SENSOR CABLE FOR CHAFFING AND OTHER EVIDENCE OF DAMAGE
  - MAKE CERTAIN CONNECTIONS ARE TIGHT AND THE SENSOR CABLE IS PROPERLY ROUTED AND SECURED IN ITS CLIPS
  - LOOSE SUSPENSION COMPONENTS (I.E., STRUT, CONTROL ARM, ETC.) MAY CAUSE UNUSUAL SENSOR SIGNALS. THESE ARE REFERRED TO AS SENSOR SHOCK AND ARE THE RESULT OF SEVERE VIBRATION AT THE SENSOR PICK–UP CAUSING A SIGNAL SPIKE
- IF NO OBVIOUS PROBLEMS, REFER TO APPLICABLE DIAGNOSTIC TROUBLE CODE TROUBLE TREE

DIAGNOSTIC CHART ABS/ASR VISUAL ILNSPECTION CHART III — 1994 DeVILLE, CONCOURS, ELDORADO AND SEVILLE

ABS/TCS FUNCTIONAL CHECK CHART III
ABS VALVE SOLENOID OR TCS PILOT VALVE

- USING THE TECH 1, SELECT MODE F4: ABS/TCS TEST
- SELECT THE APPROPRIATE TEST:
  - – AUTO TEST (CODES 41, 45, 51, 55)
  - – TCS TEST (CODES 44, 48)
- PERFORM TEST
- DOES SYSTEM PASS TEST?

TCS PASSES TEST

- CONDITION MAY BE INTERMITTENT
- SELECT F3: PILOT VALVE TEST
- SELECT APPROPRIATE PILOT VALVE TO BE TESTED:
  F0: LF
  F1: RF
- DOES SYSTEM PASS TEST?

NO

- CONDITION INTERMITTENT BUT CAN BE DUPLICATED
- REFER TO APPROPRIATE DIAGNOSTIC TROUBLE CODE TROUBLE TREE

YES

- CONDITION INTERMITTENT BUT CANNOT BE DUPLICATED
- CHECK HARNESS CONNECTION AT BPM VALVE
- CHECK BPM VALVE GROUND CONNECTION
- CHECK FOR INTERMITTENT CONDITIONS

AUTO TEST PASSES

- CONDITION MAY BE INTERMITTENT
- SELECT F0: SOLENOID TEST
- SELECT APPROPRIATE SOLENOID TO BE TESTED
  F0: LF F4: LR
  F1: RF F5: RR
- PERFORM PRESSURE RELEASE TEST OF THAT VALVE SOLENOID
- DOES SYSTEM PASS TEST?

NO

- INSPECT BASE BRAKE SYSTEM FOR LEAKS
- INSPECT FOR LEAKS AT BPM VALVE
- REFER TO APPROPRIATE DIAGNOSTIC TROUBLE CODE TROUBLE TREE

YES

- CONDITION INTERMITTENT
- CHECK HARNESS CONNECTION AT BPM VALVE
- CHECK BPM VALVE GROUND CONNECTION
- CHECK FOR INTERMITTENT CONDITIONS

TCS TEST FAILS

- REFER TO APPLICABLE DIAGNOSTIC TROUBLE CODE TROUBLE TREE

AUTO TEST FAILS

- REFER TO APPLICABLE DIAGNOSTIC TROUBLE CODE TROUBLE TREE

DIAGNOSTIC CHART ABS/ASR VISUAL ILNSPECTION CHART IV — 1994 DeVILLE, CONCOURS, ELDORADO AND SEVILLE

ABS/TCS FUNCTIONAL CHECK CHART IV
PUMP MOTOR OR PUMP MOTOR RELAY

- USING THE TECH 1, SELECT MODE F4: ABS/TCS TEST
- SELECT F1: AUTO TEST
- PERFORM TEST
- DOES SYSTEM PASS TEST?

NO

- REFER TO TROUBLE TREE FOR DIAGNOSTIC TROUBLE CODE 61

YES

- CONDITION NOT PRESENT AT THIS TIME
- CHECK BPM VALVE HARNESS CONNECTION
- CHECK FOR CLEAN AND TIGHT BPM VALVE GROUND CONNECTIONS AT GROUND STUD AND ON FRAME
- CHECK PUMP RELAY CONNECTION
- CHECK PUMP RELAY FOR POOR/LOOSE TERMINALS
- CHECK FOR INTERMITTENT CONDITIONS

DIAGNOSTIC CHART ABS/ASR VISUAL ILNSPECTION CHART V — 1994 DeVILLE, CONCOURS, ELDORADO AND SEVILLE

ABS/TCS FUNCTIONAL CHECK
CHART V
VALVE RELAY

- USING THE TECH 1, SELECT MODE F4: ABS/TCS TEST
- SELECT F1: AUTO TEST
- PERFORM TEST
- DOES SYSTEM PASS TEST?

NO

- REFER TO APPLICABLE DIAGNOSTIC TROUBLE CODE TROUBLE TREE

YES

- CONDITION NOT PRESENT AT THIS TIME
- CHECK BPM VALVE HARNESS CONNECTION
- WIGGLE HARNESS CONNECTOR WHILE CONNECTED AND SEE IF DIAGNOSTIC TROUBLE CODE IS SET (TECH 1 DISCONNECTED)
- IGNITION 'OFF'
- DISCONNECT HARNESS FROM BPM VALVE. CHECK FOR ADEQUATE TERMINAL TENSION AT HARNESS CONNECTOR TERMINALS '2, 4, 6, 10, AND 12'
- REMOVE VALVE RELAY. CHECK FOR LOOSE RELAY TERMINALS
- CHECK FOR INTERMITTENT CONDITIONS

DIAGNOSTIC FAULT CODE CHART — 1994 DeVILLE, CONCOURS, ELDORADO AND SEVILLE

ABS/TCS FAULT CODE TABLE

FAULT CODE	DESCRIPTION
21	RF WHEEL SPEED SENSOR FAULT
22	RF WHEEL SPEED SENSOR FREQUENCY ERROR
23	RF WHEEL SPEED SENSOR CONTINUITY FAULT
25	LF WHEEL SPEED SENSOR FAULT
26	LF WHEEL SPEED SENSOR FREQUENCY ERROR
27	LF WHEEL SPEED SENSOR CONTINUITY FAULT
28	WHEEL SPEED SENSOR FREQUENCY ERROR
31	RR WHEEL SPEED SENSOR FAULT
32	RR WHEEL SPEED SENSOR FREQUENCY ERROR
33	RR WHEEL SPEED SENSOR CONTINUITY FAULT
35	LR WHEEL SPEED SENSOR FAULT
36	LR WHEEL SPEED SENSOR FREQUENCY ERROR
37	LR WHEEL SPEED SENSOR CONTINUITY FAULT
41	RF ABS VALVE SOLENOID FAULT
44	RF TCS PILOT VALVE FAULT
45	LF ABS VALVE SOLENOID FAULT
48	LF TCS PILOT VALVE FAULT
51	RR ABS VALVE SOLENOID FAULT
55	LR VALVE SOLENOID FAULT (TCS)
55	REAR VALVE SOLENOID FAULT (NON–TCS)
61	PUMP MOTOR OR PUMP MOTOR RELAY FAULT
63	VALVE RELAY CIRCUIT FAULT
67	BRAKE LIGHT SWITCH FAULT
71	EBTCM/EBCM INTERNAL FAULT
72	SERIAL DATA LINK FAULT
73	PCM–EBTCM/EBCM PWM SIGNAL FAULT (4.6L)
83	BRAKE FLUID LEVEL LOW

ABS/TCS SYMPTOM TABLE

CHART	DESCRIPTION
A	'ANTILOCK' INDICATOR AND 'TRACTION DISABLED' MESSAGE 'ON' – NO TCS CODES SET / NO TCS DATA
B	'ANTILOCK' INDICATOR 'ON'
C	VALVE CYCLING DURING NORMAL STOPS (UNWANTED ABS MODULATION)
D	TRACTION CONTROL OPERATION CHECK
•	'ANTILOCK' INDICATOR INOPERATIVE

DIAGNOSTIC FAULT CHART DTC 21 — 1994 DeVILLE, CONCOURS, ELDORADO AND SEVILLE

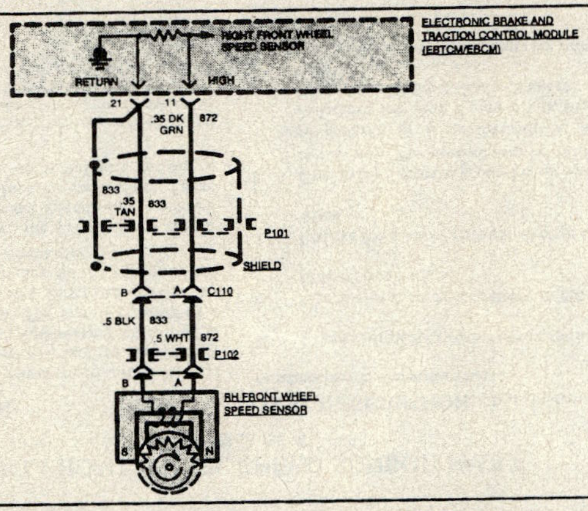

DIAGNOSTIC TROUBLE CODE 21
RIGHT FRONT WHEEL SPEED SENSOR FAULT
(Page 1 of 4)

CIRCUIT DESCRIPTION:

As each wheel turns, the wheel speed sensor for that wheel creates a small AC voltage as a toothed sensor ring, mounted in the front integral hub/bearing assembly, is passed by the stationary sensor. The sensor consists of a permanent magnet and coil and generates voltage by magnetic induction. The frequency of the wheel speed sensor signal is proportional to wheel speed.

The EBTCM/EBCM uses the wheel speed sensor signal to calculate vehicle reference speeds and individual speed, acceleration and slip values for each wheel. These values are used to determine when antilock control is required.

The EBTCM/EBCM will only set one wheel speed sensor fault code at a time, even if multiple faults exist. Detection is prioritized accordingly: LF, RR, LR, RF. Always check for additional stored codes after making repairs.

TEST CONDITION(S):

1. Tested continuously.

FAILURE CONDITION(S):

Sensor Output

1. Drive–away: Speed signal is not detected when vehicle reaches 12 kph or 7.5 mph.
2. While Driving: Speed signal is lost while vehicle speed is greater than 40 kph or 25 mph.

ACTION TAKEN:

1. Code set.
2. ABS/TCS switched 'OFF'.
3. 'ANTILOCK' indicator 'ON'.
4. 'TRACTION DISABLED' message displayed.

Code 31 is set if there is a short to voltage or ground in CKT 882 or 883, or faulty speed sensor. The testing for this fault occurs when the vehicle is in motion, it will not set with the ignition 'ON' and the vehicle at rest.

NOTICE: A Code 21 may set if CKT 872 or 833 lose continuity and the WSS signal is lost while driving. A Code 23 should accompany this code on the subsequent ignition cycle. If both Code 21 and Code 23 are set concurrently, diagnose Code 23 first.

TEST DESCRIPTION: Test numbers refer to circled numbers on the diagnostic chart.

1. Checks if the code resets during a road test.
2. Use the TECH 1 to check for sources of induced electrical noise. Some sources of induced electrical noise which may cause intermittent wheel speed sensor codes to set are: improper routing or sensor cables next to spark plug wires; cables not retained in brackets; loose fitting or improperly mounted sensors; damaged sensors or toothed rings; poor terminal engagement of connectors; or intermittent shorts or opens in wiring; or a loss of sensor circuit shield.
3. Checks for damaged wheel speed sensor circuit shield. The shield is a foil insulator that is wrapped around the wheel speed sensor circuits. Inside the EBTCM/EBCM harness connecter there is an uninsulated drain wire that is attached to the shield and jumped to the wheel speed sensor return line and taken to ground through the EBTCM/EBCM. A shield which has lost its ground through the EBTCM/EBCM leaves the sensor signal circuit susceptible to noise. It may be necessary to disassemble the EBTCM/EBCM connector to check if the drain wire has a poor connection with the wheel speed sensor return connector terminal. The shield can be repaired by redetermining the sensor return line circuit, including the drain wire.
4. Checks for intermittent opens/shorts in wheel speed sensor circuits and cables.

DIAGNOSTIC FAULT CHART DTC 21 — 1994 DeVILLE, CONCOURS, ELDORADO AND SEVILLE, CONTINUED

DIAGNOSTIC TROUBLE CODE 21
RIGHT FRONT WHEEL SPEED SENSOR FAULT
(Page 2 of 4)

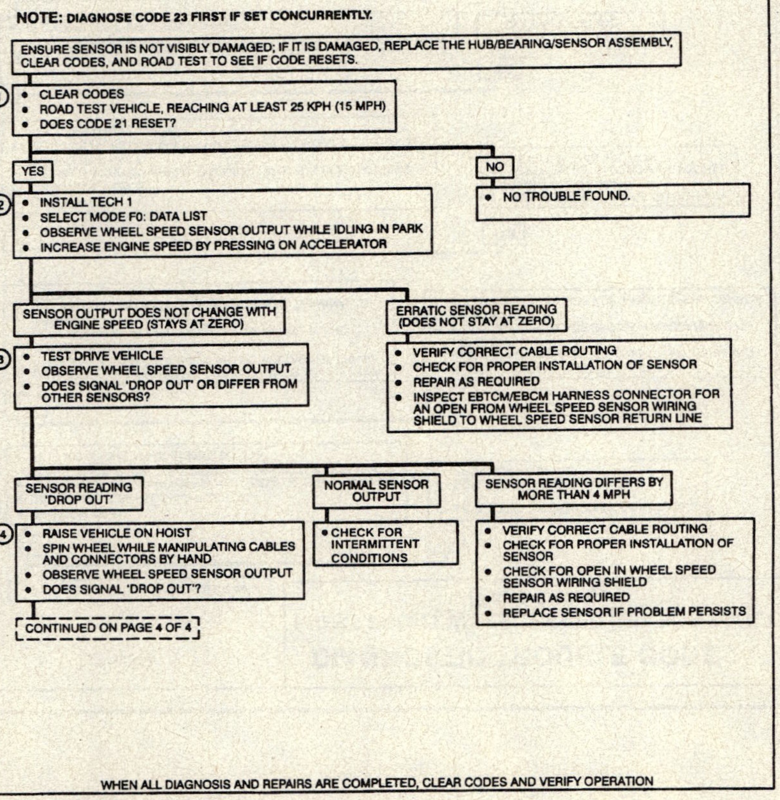

DIAGNOSTIC FAULT CHART DTC 21 — 1994 DeVILLE, CONCOURS, ELDORADO AND SEVILLE, CONTINUED

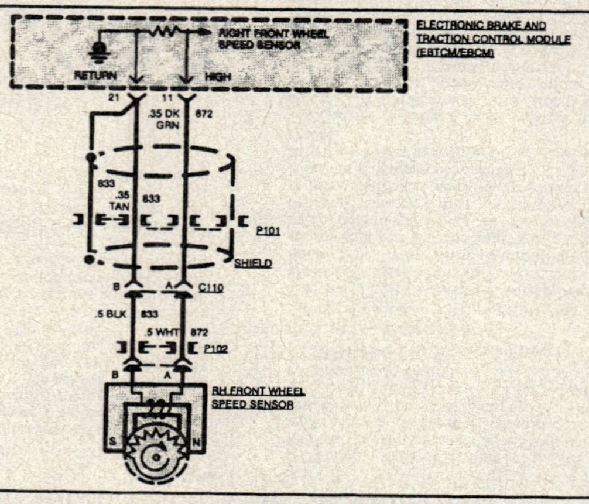

DIAGNOSTIC TROUBLE CODE 21

RIGHT FRONT WHEEL SPEED SENSOR FAULT (Page 3 of 4)

CIRCUIT DESCRIPTION:

As each wheel turns, the wheel speed sensor for that wheel creates a small AC voltage as a toothed sensor ring, mounted in the front integral hub/bearing assembly, is passed by the stationary sensor. The sensor consists of a permanent magnet and coil and generates voltage by magnetic induction. The frequency of the wheel speed sensor signal is proportional to wheel speed.

The EBTCM/EBCM uses the wheel speed sensor signal to calculate vehicle reference speeds and individual speed, acceleration and slip values for each wheel. These values are used to determine when antilock control is required.

The EBTCM/EBCM will only set one wheel speed sensor fault code at a time, even if multiple faults exist. Detection is prioritized accordingly: LF, RR, LR, RF. Always check for additional stored codes after making repairs.

TEST DESCRIPTION: Test numbers refer to circled numbers on the diagnostic chart.

5. Checks for proper resistance in the sensor itself.
6. Checks for a short to battery in either of the speed sensor circuit wires.
7. Checks for a short between the speed sensor circuit wires.
8. Checks for a short to ground in the speed sensor input circuit wire. The ground may be either a 'hard' short to ground, or a resistive short. A short with a resistance less than 2 megaohms, though not a hard short, can still cause Code 21 to set.
9. Checks wiring and connectors for intermittents.

DIAGNOSTIC FAULT CHART DTC 21 — 1994 DeVILLE, CONCOURS, ELDORADO AND SEVILLE, CONTINUED

DIAGNOSTIC TROUBLE CODE 21

RIGHT FRONT WHEEL SPEED SENSOR FAULT (Page 4 of 4)

CONTINUED CODE 21 FROM PAGE 2 OF 4

YES →
(5)
- IGNITION 'OFF'
- DISCONNECT SENSOR CONNECTOR
- USING DVM AND ADAPTER FROM J 35616–A, CHECK SENSOR RESISTANCE BY PROBING SENSOR CONNECTOR TERMINALS
- IS IT 900 – 1300 OHMS AT 25°C (77°F)?

NO →
- CHECK FOR INTERMITTENT CONDITIONS
- REPLACE EBTCM/EBCM/EBCM IF PROBLEM PERSISTS

(5) YES →
(6)
- CONNECT DVM BETWEEN SENSOR HARNESS CONNECTOR TERMINAL 'A' AND GROUND (USED DC VOLTAGE SETTING)
- IGNITION 'ON'
- NOTE VOLTAGE: REPEAT BETWEEN TERMINAL 'B' AND GROUND.
- IS IT AROUND SYSTEM VOLTAGE ON EITHER TERMINAL 'A' OR 'B'?

(5) NO →
- REMOVE WHEEL SPEED RING SENSOR
- INSPECT TOOTHED SPEED RING ON HUB/BEARING ASSEMBLY
- IF DAMAGED, REPLACE ENTIRE HUB/BEARING AND SENSOR ASSEMBLY
- IF NOT DAMAGED, REPLACE WHEEL SPEED SENSOR

(6) NO →
(7)
- IGNITION 'OFF'
- DISCONNECT EBCM
- INSTALL J 39700 UNIVERSAL BREAKOUT BOX TO EBCM HARNESS CONNECTOR ONLY
- MEASURE RESISTANCE BETWEEN J 39700 TERMINALS '11' AND '21'
- IS IT LESS THAN 5 OHMS?

(6) YES →
- REPAIR SHORT TO SYSTEM VOLTAGE IN CKT 872 OR 833
- REFER TO EBTCM/EBCM–SWITCHED FAULTS IN THIS SECTION

(7) NO →
(8)
- MEASURE RESISTANCE BETWEEN J 39700 TERMINAL '11' AND GROUND
- IS IT GREATER THAN 2 MEGAOHMS?

(7) YES →
- REPAIR SHORT BETWEEN CKT'S 872 AND 833

(8) YES →
(9)
- MEASURE RESISTANCE BETWEEN J 39700 TERMINAL '21' AND GROUND
- IS IT GREATER THAN 2 MEGAOHMS?

(8) NO →
- IF 5 OHMS OR LESS TO GROUND, REPAIR HARD SHORT IN HARNESS CKT 872
- IF BETWEEN 5 OHMS AND 2 MEGAOHMS, REPAIR RESISTIVE SHORT IN HARNESS CKT 872

(9) YES →
- CHECK ALL WIRING, CONNECTORS AND TERMINALS THOROUGHLY FOR INTERMITTENTS OR CORROSION
- CHECK FOR INTERMITTENTS AND POOR CONDITIONS
- REPAIR HARNESS OR CONNECTOR
- REPLACE SENSOR IS PROBLEM PERSISTS

(9) NO →
- IF 5 OHMS OR LESS TO GROUND, REPAIR HARD SHORT IN HARNESS CKT 833
- IF BETWEEN 5 OHMS AND 2 MEGAOHMS REPAIR RESISTIVE SHORT IN HARNESS CKT 833

WHEN ALL DIAGNOSIS AND REPAIRS ARE COMPLETED, CLEAR CODES AND VERIFY OPERATION

DIAGNOSTIC FAULT CHART DTC 22 — 1994 DeVILLE, CONCOURS, ELDORADO AND SEVILLE

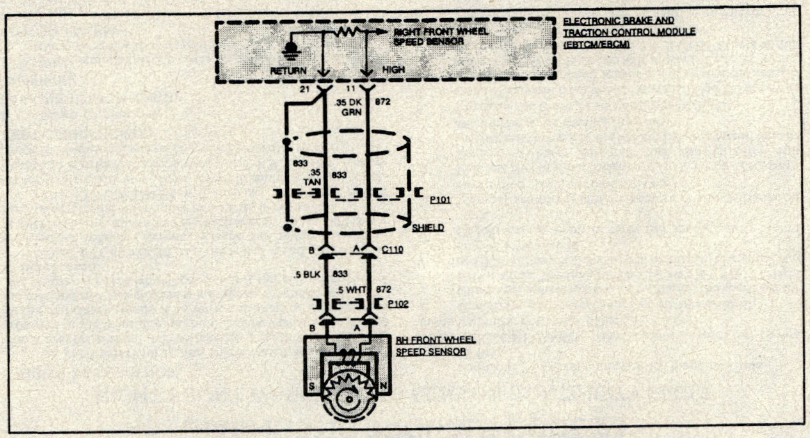

DIAGNOSTIC TROUBLE CODE 22
RIGHT FRONT WHEEL SPEED SENSOR FREQUENCY ERROR

TEST CONDITION(S):

1. Tested continuously.

FAILURE CONDITION(S):

1. While Driving: The RF wheel is detected at least 30 kph or 19 mph faster than the remaining wheels for an extended period of time.

   OR

   The RF wheel is at least 6 kph or 4 mph less than or greater than the remaining wheels for an extended period of time.

ACTION TAKEN:

1. Code set.
2. ABS/TCS switched 'OFF'.
3. 'ANTILOCK' indicator 'ON'.

TEST DESCRIPTION: Test numbers refer to circled numbers on the diagnostic chart.

1. Checks wiring and connections for problems. It is critical that this step is performed thoroughly, as wiring/connector problems are the most likely cause for this code.
2. Use the TECH 1 to monitor for vehicle electrical/electronic system 'noise' being picked up by the speed sensor circuits.
3. Use the TECH 1 to monitor the wheel speed sensors while the vehicle is operating. If the TECH 1 Auto–Trigger Snapshot mode triggers, the speed sensor or the wiring and/or connectors are going intermittent during the road test.
4. Checks for proper resistance in the sensor itself.
5. Replaces front wheel speed sensor as the likely cause of the triggering intermittent.
6. Use the TECH 1 to monitor the wheel speed sensors while the vehicle is operating. If the TECH 1 Auto–Snapshot mode triggers, the EBCM must be concluding there is a problem with wheel speed sensor frequencies when there is not.
7. Checks for a short to ground in the speed sensor input circuit wires. The circuit to be tested depends on which speed sensor triggered the TECH 1. The ground may be either a 'hard' short to ground, or a resistive short. A short with a resistance less that 2 megaohms, though not a hard short, can still cause Code 22 to set.

DIAGNOSTIC AIDS:

Be sure the speed sensor wiring is properly routed and retained. This will help prevent false signals due to electrical noise being picked up by the wiring.

It is very important that a thorough inspection of the wiring and connectors be performed. Failure to carefully and fully inspect wiring and connectors may result in misdiagnosis, causing part replacement with reappearance of the fault.

Check the toothed wheel for any large grooves, gouges, marks, etc. that might influence the tooth's signal at the wheel speed sensor. Also check for a buildup of foreign material in the gaps between teeth in the toothed wheel; this material may cause this fault.

A worn hub/bearing assembly may cause this fault in extreme cases; the bearing play allows the sensor–to–toothed–ring gap to change excessively. Never reuse a wheel speed sensor once it has been separated from the front hub/bearing assembly.

DIAGNOSTIC FAULT CHART DTC 22 — 1994 DeVILLE, CONCOURS, ELDORADO AND SEVILLE, CONTINUED

DIAGNOSTIC TROUBLE CODE 22
RIGHT FRONT WHEEL SPEED SENSOR FREQUENCY ERROR

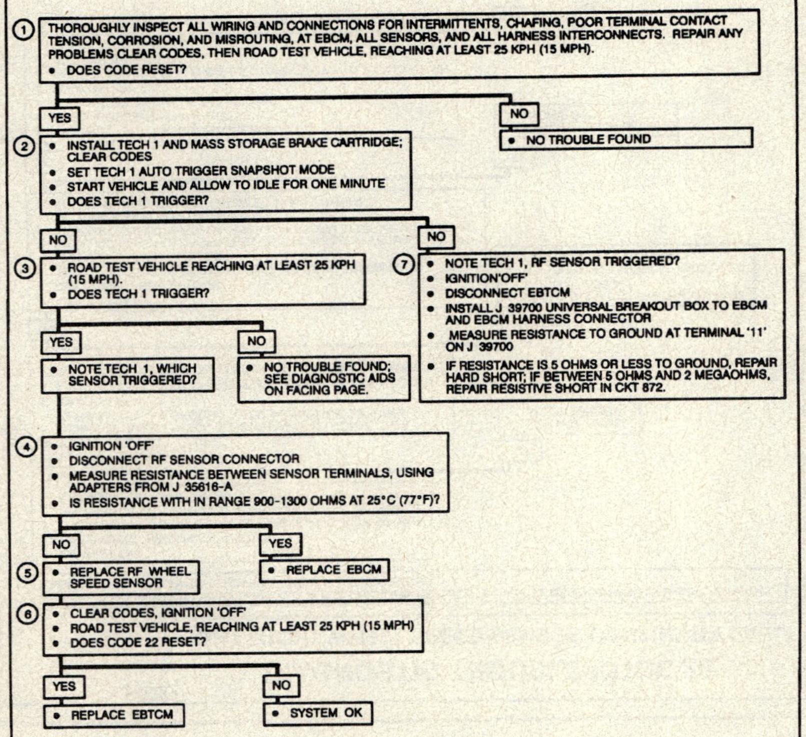

WHEN ALL DIAGNOSIS AND REPAIRS ARE COMPLETED, CLEAR CODES AND VERIFY OPERATION

DIAGNOSTIC FAULT CHART DTC 23 — 1994 DeVILLE, CONCOURS, ELDORADO AND SEVILLE

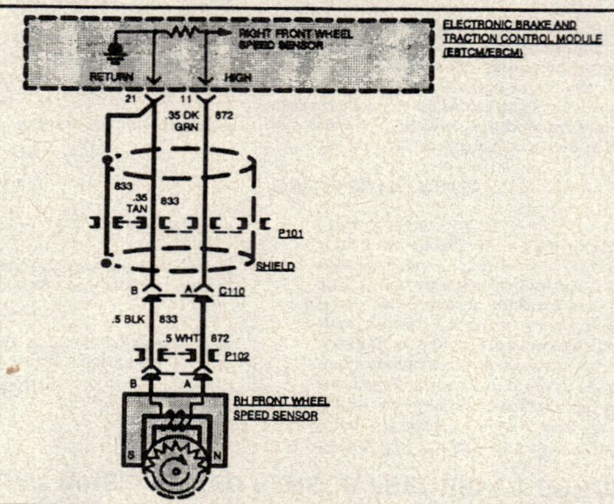

DIAGNOSTIC TROUBLE CODE 23
RIGHT FRONT WHEEL SPEED SENSOR CONTINUITY FAULT

CIRCUIT DESCRIPTION:

As each wheel turns, the wheel speed sensor for that wheel creates a small AC voltage as a toothed sensor ring, mounted in the front integral hub/bearing assembly, is passed by the stationary sensor. The sensor consists of a permanent magnet and coil and generates voltage by magnetic induction. The frequency of the wheel speed sensor signal is proportional to wheel speed.

The EBTCM/EBCM uses the wheel speed sensor signal to calculate vehicle reference speeds and individual speed, acceleration and slip values for each wheel. These values are used to determine when antilock control is required.

The EBTCM/EBCM will only set one wheel speed sensor fault code at a time, even if multiple faults exist. Detection is prioritized accordingly: LF, RR, LR, RF. Always check for additional stored codes after making repairs.

TEST CONDITION(S):

1. Tested continuously.

FAILURE CONDITION(S):

Continuity

1. While standing: Open circuit is seen while the vehicle is standing still with the ignition 'ON'

ACTION TAKEN:

1. Code set.
2. ABS/TCS switched 'OFF'.
3. 'ANTILOCK' indicator 'ON'.
4. 'TRACTION DISABLED' message displayed. (vehicles with TCS)

Code 23 is set if there is a short to voltage or ground in CKT 872 or 883, or faulty speed sensor. The testing for this fault occurs with the ignition 'ON' and the vehicle at rest.

NOTICE: A Code 21 may set if CKT 872 or 833 lose continuity and the WSS signal is lost while driving. A Code 23 should accompany this code on the subsequent ignition cycle. If both Code 21 and Code 23 are set concurrently, diagnose Code 23 first.

TEST DESCRIPTION: Test numbers refer to circled numbers on the diagnostic chart.

1. Checks for proper resistance in the sensor itself.
2. Checks for a short to battery in either of the speed sensor circuit wires. There should be 5V on CKT 872. If there is, CKT 872 has no hard opens or shorts. The condition may be intermittent.
3. Checks for an open in both the speed sensor circuit wires.
4. Manipulates wiring and connectors, trying to induce an intermittent fault not currently present.
5. This test checks if the code resets on key–up. If so, since Tests 1–4 have validated that the circuitry and components are good, the Wheel Bearing/Speed Sensor assembly may be causing the fault.
6. Checks if the Code 21 sets during a road test.
7. Code 23 sets when the vehicle is at rest: Code 21 sets with a problem in the speed sensor circuitry with the vehicle in motion. If problems are still present, Code 21 would set during the road test, not a 23. If Code 21 sets at this point, Code 21 diagnostics should be used.
8. The wheel speed sensor may be causing an intermittent fault. If Code 23 resets after sensor replacement, the EBTCM/EBCM must be concluding there is a problem present in the speed sensor circuit when there is not.

DIAGNOSTIC AIDS:

Be sure the speed sensor wiring is properly routed and retained. This will help prevent false signals due to electrical noise being picked up by the wiring.

It is very important that a thorough inspection of the wiring and connectors be performed. Failure to carefully and fully inspect wiring and connectors may result in misdiagnosis, causing part replacement with reappearance of the fault.

DIAGNOSTIC FAULT CHART DTC 23 — 1994 DeVILLE, CONCOURS, ELDORADO AND SEVILLE, CONTINUED

DIAGNOSTIC TROUBLE CODE 23
RIGHT FRONT WHEEL SPEED SENSOR CONTINUITY FAULT

ENSURE SENSOR IS NOT VISIBLY DAMAGED; IF IT IS DAMAGED, REPLACE THE WHEEL BEARING/SPEED/SENSOR ASSEMBLY. CLEAR CODES, AN TURN IGNITION 'ON' TO SEE IF CODE RESETS.

(1)
- IGNITION 'OFF'
- DISCONNECT SENSOR CONNECTOR
- USING DVM AND ADAPTER FROM J 35616–A, CHECK SENSOR RESISTANCE BY PROBING SENSOR CONNECTOR TERMINALS
- IS IT 900 – 1300 OHMS AT 25°C (77°F)?

NO:
- REPLACE WHEEL SPEED SENSOR

YES:

(2)
- CONNECT DVM BETWEEN SENSOR HARNESS CONNECTOR TERMINAL 'A' AND GROUND (USED DC VOLTAGE SETTING)
- IGNITION 'ON'
- NOTE VOLTAGE: REPEAT BETWEEN TERMINAL 'B' AND GROUND
- IS IT AROUND SYSTEM VOLTAGE ON EITHER TERMINAL 'A' OR 'B'?

YES:
- REPAIR SHORT TO SYSTEM VOLTAGE IN CKT 872 OR 883
- REFER TO EBTCM/EBCM–SWITCHED FAULTS IN THIS SECTION

NO:

(3)
- IGNITION 'OFF'
- DISCONNECT EBTCM/EBCM
- INSTALL J 39700 UNIVERSAL BREAKOUT BOX TO EBTCM/EBCM HARNESS CONNECTOR ONLY
- MEASURE RESISTANCE BETWEEN PIN 11 ON J 39700 TERMINAL 'A' OF SENSOR CONNECTOR ON HARNESS. PERFORM SAME MEASUREMENT BETWEEN PIN 21 AND TERMINAL 'B'
- IS EITHER MEASUREMENT GREATER THAN 5 OHMS?

YES:
- REPAIR OPEN IN CKT 872 AND/OR 833

NO:

(4)
- REMOVE J 39700; RECONNECT EBCM
- CLEAR CODES; IGNITION 'OFF'
- DISCONNECT TECH 1, IF CONNECTED
- IGNITION 'ON'
- WIGGLE ALL WIRING AND CONNECTORS TO TRY TO DUPLICATE AN INTERMITTENT FAULT
- DOES CODE 23 RESET?

YES:
- REPAIR WIRING AND/OR CONNECTIONS
- CHECK FOR INTERMITTENTS AND POOR CONNECTIONS

NO:

(5)
- CLEAR CODES; IGNITION 'OFF'
- DISCONNECT TECH 1, IF CONNECTED
- IGNITION 'ON'
- DOES CODE 23 RESET?

NO:

(6)
- ROAD TEST VEHICLE, REACHING AT LEAST 25 KPH (15 MPH).
- DOES CODE 21 SET?

YES: (7) GO TO CODE 21 DIAGNOSTICS

NO:
- NO TROUBLE FOUND. SEE 'DIAGNOSTIC AIDS' ON FACING PAGE.

YES:

(8)
- REPLACE WHEEL SPEED SENSOR
- CLEAR CODES; IGNITION 'OFF'
- DISCONNECT TECH 1, IF CONNECTED
- IGNITION 'ON'
- DOES CODE 23 RESET?

YES:
- REPLACE EBTCM/EBCM

NO:
- SYSTEM OK

WHEN ALL DIAGNOSIS AND REPAIRS ARE COMPLETED, CLEAR CODES AND VERIFY OPERATION

DIAGNOSTIC FAULT CHART DTC 25 — 1994 DeVILLE, CONCOURS, ELDORADO AND SEVILLE

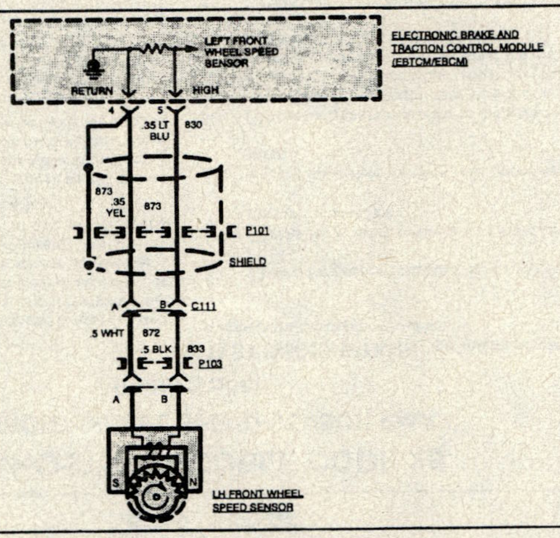

DIAGNOSTIC TROUBLE CODE 25
LEFT FRONT WHEEL SPEED SENSOR FAULT (Page 1 of 4)

CIRCUIT DESCRIPTION:

As each wheel turns, the wheel speed sensor for that wheel creates a small AC voltage as a toothed sensor ring, mounted in the front integral hub/bearing assembly, is passed by the stationary sensor. The sensor consists of a permanent magnet and coil and generates voltage by magnetic induction. The frequency of the wheel speed sensor signal is proportional to wheel speed.

The EBTCM/EBCM uses the wheel speed sensor signal to calculate vehicle reference speeds and individual speed, acceleration and slip values for each wheel. These values are used to determine when antilock control is required.

The EBTCM/EBCM will only set one wheel speed sensor fault code at a time, even if multiple faults exist. Detection is prioritized accordingly: LF, RR, LR, RF. Always check for additional stored codes after making repairs.

TEST CONDITION(S):
1. Tested continuously.

FAILURE CONDITION(S):

Sensor Output
1. Drive–away: Speed signal is not detected when vehicle reaches 12 kph or 7.5 mph.
   OR
2. While Driving: Speed signal is lost while vehicle speed is greater than 40 kph or 25 mph

ACTION TAKEN:
1. Code set.
2. ABS/TCS switched 'OFF'.
3. 'ANTILOCK' indicator 'ON'.
4. 'TRACTION DISABLED' message displayed. (vehicles with TCS)

Code 25 is set if there is a short to voltage or ground in CKT 830 or 873, or faulty speed sensor. The testing for this fault occurs when vehicle is in motion; it will not set with the ignition 'ON' and the vehicle at rest.

NOTICE: A Code 25 may set if CKT 830 or 873 lose continuity and the WSS signal is lost while driving. A Code 27 should accompany this code on the subsequent ignition cycle. If both Code 25 and Code 27 are set concurrently, diagnose Code 27 first.

TEST DESCRIPTION: Test numbers refer to circled numbers on the diagnostic chart.

1. Checks if the code resets during a road test.
2. Use the TECH 1 to check for sources of induced electrical noise. Some sources of induced electrical noise which may cause intermittent wheel speed sensor codes to set are: improper routing or sensor cables next to spark plug wires; cables not retained in brackets; loose fitting or improperly mounted sensors; damaged sensors or toothed rings; poor terminal engagement of connectors; or intermittent shorts or opens in wiring; or a loss of sensor circuit shield.
3. Checks for damaged wheel speed sensor circuit shield. The shield is a foil insulator that is wrapped around the wheel speed sensor circuits. Inside the EBTCM/EBCM harness connecter there is an uninsulated drain wire that is attached to the shield and jumped to the wheel speed sensor return line and taken to ground through the EBTCM/EBCM. A shield which has lost its ground through the EBTCM/EBCM leaves the sensor signal circuit susceptible to noise. It may be necessary to disassemble the EBTCM/EBCM connector to check if the drain wire has a poor connection with the wheel speed sensor return connector terminal. The shield can be repaired by reterminating the sensor return line circuit, including the drain wire.
4. Checks for intermittent opens/shorts in wheel speed sensor circuits and cables.

DIAGNOSTIC FAULT CHART DTC 25 — 1994 DeVILLE, CONCOURS, ELDORADO AND SEVILLE, CONTINUED

DIAGNOSTIC TROUBLE CODE 25
LEFT FRONT WHEEL SPEED SENSOR FAULT (Page 2 of 4)

NOTE: DIAGNOSE CODE 27 FIRST IF SET CONCURRENTLY.

ENSURE SENSOR IS NOT VISIBLY DAMAGED; IF IT IS DAMAGED, REPLACE THE HUB/BEARING/SENSOR ASSEMBLY, CLEAR CODES, AND ROAD TEST TO SEE IF CODE RESETS.

(1)
- CLEAR CODES
- ROAD TEST VEHICLE, REACHING AT LEAST 25 KPH (15 MPH)
- DOES CODE 25 RESET?

NO:
- NO TROUBLE FOUND
- CHECK FOR INTERMITTENTS AND LOOSE CONNECTIONS

YES:

(2)
- INSTALL TECH 1
- SELECT MODE F0: DATA LIST
- OBSERVE WHEEL SPEED SENSOR OUTPUT WHILE IDLING IN PARK
- INCREASE ENGINE SPEED BY PRESSING ON ACCELERATOR

ERRATIC SENSOR READING (DOES NOT STAY AT ZERO):
- VERIFY CORRECT CABLE ROUTING
- CHECK FOR PROPER INSTALLATION OF SENSOR
- REPAIR AS REQUIRED
- INSPECT EBTCM/EBCM HARNESS CONNECTOR FOR AN OPEN FROM THE WHEEL SPEED SENSOR WIRING SHIELD TO THE WHEEL SPEED SENSOR RETURN LINE

SENSOR OUTPUT DOES NOT CHANGE WITH ENGINE SPEED (STAYS AT ZERO):

(3)
- TEST DRIVE VEHICLE
- OBSERVE WHEEL SPEED SENSOR OUTPUT
- DOES SIGNAL 'DROP OUT' OR DIFFER FROM OTHER SENSORS?

NORMAL SENSOR OUTPUT:
- CHECK FOR INTERMITTENT CONDITIONS

SENSOR READING DIFFERS BY MORE THAN 4 MPH:
- VERIFY CORRECT CABLE ROUTING
- CHECK FOR PROPER INSTALLATION OF SENSOR
- CHECK FOR OPEN IN WHEEL SPEED SENSOR WIRING SHIELD
- REPAIR AS REQUIRED
- REPLACE SENSOR IF PROBLEM PERSISTS

SENSOR READING 'DROP OUT':

(4)
- RAISE VEHICLE ON HOIST
- SPIN WHEEL WHILE MANIPULATING CABLES AND CONNECTORS BY HAND
- OBSERVE WHEEL SPEED SENSOR OUTPUT
- DOES SIGNAL 'DROP OUT'?

CODE 25 CONTINUED PAGE 4 OF 4

WHEN ALL DIAGNOSIS AND REPAIRS ARE COMPLETED, CLEAR CODES AND VERIFY OPERATION

DIAGNOSTIC FAULT CHART DTC 25 — 1994 DeVILLE, CONCOURS, ELDORADO AND SEVILLE, CONTINUED

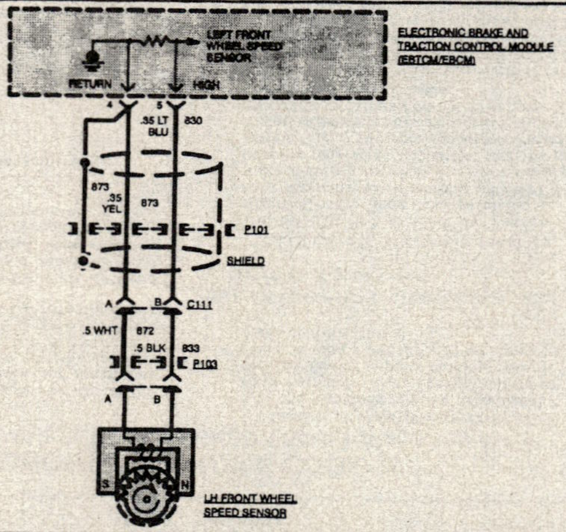

DIAGNOSTIC TROUBLE CODE 25
LEFT FRONT WHEEL SPEED SENSOR FAULT
(Page 3 of 4)

CIRCUIT DESCRIPTION:

As each wheel turns, the wheel speed sensor for that wheel creates a small AC voltage as a toothed sensor ring, mounted in the front integral hub/bearing assembly, is passed by the stationary sensor. The sensor consists of a permanent magnet and coil and generates voltage by magnetic induction. The frequency of the wheel speed sensor signal is proportional to wheel speed.

The EBTCM/EBCM uses the wheel speed sensor signal to calculate vehicle reference speeds and individual speed, acceleration and slip values for each wheel. These values are used to determine when antilock control is required.

The EBTCM/EBCM will only set one wheel speed sensor fault code at a time, even if multiple faults exist. Detection is prioritized accordingly: LF, RR, LR, RF. Always check for additional stored codes after making repairs.

TEST DESCRIPTION: Test numbers refer to circled numbers on the diagnostic chart.

5. Checks for proper resistance in the sensor itself.
6. Checks for a short to battery in either of the speed sensor circuit wires.
7. Checks for a short between the speed sensor circuit wires.
8. Checks for a short to ground in the speed sensor input circuit wire. The ground may be either a 'hard' short to ground, or a resistive short. A short with a resistance less than 2 megaohms, though not a hard short, can still cause Code 21 to set.
9. Checks wiring and connectors for intermittents.

DIAGNOSTIC FAULT CHART DTC 25 — 1994 DeVILLE, CONCOURS, ELDORADO AND SEVILLE, CONTINUED

DIAGNOSTIC TROUBLE CODE 25
LEFT FRONT WHEEL SPEED SENSOR FAULT
(Page 4 of 4)

CONTINUED FROM CODE 25 PAGE 2 OF 4

YES →

(5)
- IGNITION 'OFF'
- DISCONNECT SENSOR CONNECTOR
- USING DVM AND ADAPTER FROM J 38616-A, CHECK SENSOR RESISTANCE BY PROBING SENSOR CONNECTOR TERMINALS
- IS IT 900–1300 OHMS AT 25°C (77°F).

NO →
- CHECK FOR INTERMITTENTS AND LOOSE CONNECTIONS
- REPLACE EBTCM/EBCM IF PROBLEM PERSISTS

(5) YES →

(6)
- CONNECT DVM BETWEEN SENSOR HARNESS CONNECTOR TERMINAL 'A' AND GROUND (USE DC VOLTAGE SETTING)
- IGNITION 'ON'
- NOTE VOLTAGE: REPEAT BETWEEN TERMINAL 'B' AND GROUND
- IS IT AROUND SYSTEM VOLTAGE ON EITHER TERMINAL 'A' OR 'B'?

(5) NO →
- REMOVE WHEEL SPEED RING SENSOR
- INSPECT TOOTHED SPEED RING ON HUB/BEARING ASSEMBLY
- IF DAMAGED, REPLACE ENTIRE HUB/BEARING AND SENSOR ASSEMBLY
- IF NOT DAMAGED, REPLACE WHEEL SPEED SENSOR

(6) NO →

(7)
- IGNITION 'OFF'
- DISCONNECT EBTCM/EBCM
- INSTALL J 39700 UNIVERSAL BREAKOUT BOX TO EBTCM/EBCM HARNESS CONNECTOR ONLY
- MEASURE RESISTANCE BETWEEN J 39700 TERMINALS '4' AND '5'
- IS IT LESS THAN 5 OHMS?

(6) YES →
- REPAIR SHORT TO SYSTEM VOLTAGE IN CKTS 830 AND 873
- REFER TO EBTCM/EBCM–SWITCHED FAULTS IN THIS SECTION

(7) NO →

(8)
- MEASURE RESISTANCE BETWEEN J 39700 TERMINAL '4' AND GROUND
- IS IT GREATER THAN 2 MEGAOHMS?

(7) YES →
- REPAIR SHORT BETWEEN CKTS 830 AND 873

(8) YES →

(9)
- MEASURE RESISTANCE BETWEEN J 39700 TERMINAL '5' AND GROUND
- IS IT GREATER THAN 2 MEGAOHMS?

(8) NO →
- IF 5 OHMS OR LESS TO GROUND, REPAIR HARD SHORT IN HARNESS CKT 873
- IF BETWEEN 5 OHMS AND 2 MEGAOHMS, REPAIR RESISTIVE SHORT IN HARNESS CKT 873

(9) YES →
- CHECK ALL WIRING, CONNECTORS AND TERMINALS THOROUGHLY FOR INTERMITTENTS OR CORROSION
- REFER TO 'INTERMITTENTS AND POOR CONNECTIONS'
- REPAIR HARNESS OR CONNECTOR
- REPLACE SENSOR IF PROBLEM PERSISTS

(9) NO →
- IF 5 OHMS OR LESS TO GROUND, REPAIR HARD SHORT IN HARNESS CKT 830
- IF BETWEEN 5 OHMS AND 2 MEGAOHMS, REPAIR RESISTIVE SHORT IN HARNESS CKT 830

WHEN ALL DIAGNOSIS AND REPAIRS ARE COMPLETED, CLEAR CODES AND VERIFY OPERATION

DIAGNOSTIC FAULT CHART DTC 26 — 1994 DeVILLE, CONCOURS, ELDORADO AND SEVILLE

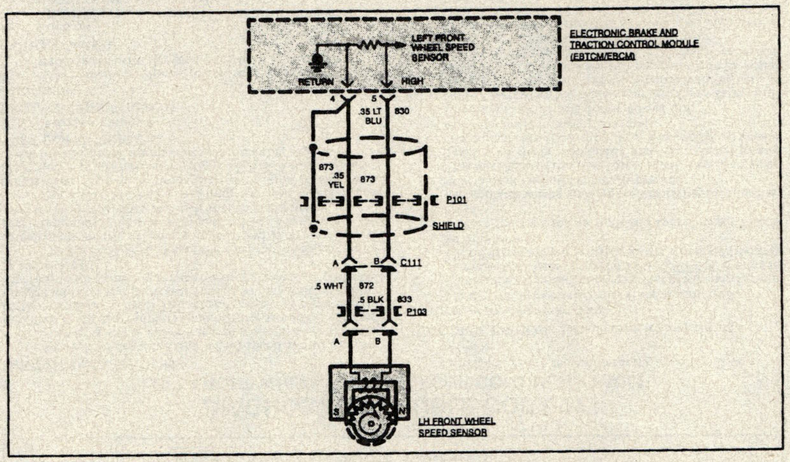

DIAGNOSTIC TROUBLE CODE 26
LEFT FRONT WHEEL SPEED SENSOR FREQUENCY ERROR

TEST CONDITION(S):

1. Tested continuously.

FAILURE CONDITION(S):

1. While Driving: The LF wheel is detected at least 30 kph or 19 mph faster than the remaining wheels for an extended period of time.

   OR

   The LF wheel is at least 6 kph or 4 mph less than or greater than the remaining wheels for an extended period of time.

ACTION TAKEN:

1. Code set.
2. ABS switched 'OFF'.
3. 'ANTILOCK' indicator 'ON'.

TEST DESCRIPTION: Test numbers refer to circled numbers on the diagnostic chart.

1. Checks wiring and connections for problems. It is critical that this step is performed thoroughly, as wiring/connector problems are the most likely cause for this code.
2. Use the TECH 1 to monitor for vehicle electrical/electronic system 'noise' being picked up by the speed sensor circuits.
3. Use the TECH 1 to monitor the wheel speed sensors while the vehicle is operating. If the TECH 1 Auto-Trigger Snapshot mode triggers, the speed sensor or the wiring and/or connectors are going intermittent during the road test.
4. Checks for proper resistance in the sensor itself.
5. Replaces front wheel speed sensor as the likely cause of the triggering intermittent.
6. Use the TECH 1 to monitor the wheel speed sensors while the vehicle is operating. If the TECH 1 Auto-Snapshot mode triggers, the EBCM must be concluding there is a problem with wheel speed sensor frequencies when there is not.
7. Checks for a short to ground in the speed sensor input circuit wires. The circuit to be tested depends on which speed sensor triggered the TECH 1. The ground may be either a 'hard' short to ground, or a resistive short. A short with a resistance less that 2 megaohms, though not a hard short, can still cause Code 26 to set.

DIAGNOSTIC AIDS:

Be sure the speed sensor wiring is properly routed and retained. This will help prevent false signals due to electrical noise being picked up by the wiring.

It is very important that a thorough inspection of the wiring and connectors be performed. Failure to carefully and fully inspect wiring and connectors may result in misdiagnosis, causing part replacement with reappearance of the fault.

Check the toothed wheel for any large grooves, gouges, marks, etc. that might influence the tooth's signal at the wheel speed sensor. Also check for a buildup of foreign material in the gaps between teeth in the toothed wheel; this material may cause this fault.

A worn hub/bearing assembly may cause this fault in extreme cases; the bearing play allows the sensor-to-toothed-ring gap to change excessively. Never reuse a wheel speed sensor once it has been separated from the front hub/bearing assembly.

DIAGNOSTIC FAULT CHART DTC 26 — 1994 DeVILLE, CONCOURS, ELDORADO AND SEVILLE, CONTINUED

DIAGNOSTIC TROUBLE CODE 26
LEFT FRONT WHEEL SPEED SENSOR FREQUENCY ERROR

(1) VERY THOROUGHLY INSPECT ALL WIRING AND CONNECTIONS FOR INTERMITTENTS, CHAFING, POOR TERMINAL CONTACT TENSION, CORROSION, AND MISROUTING, AT EBCM, ALL SENSORS, AND ALL HARNESS INTERCONNECTS. REPAIR ANY PROBLEMS CLEAR CODES, THEN ROAD TEST VEHICLE, REACHING AT LEAST 25 KPH (15 MPH).
- DOES CODE RESET?

YES → (2)
NO →
- NO TROUBLE FOUND; SEE 'DIAGNOSTIC AIDS' ON FACING PAGE

(2)
- INSTALL TECH 1 AND MASS STORAGE; CLEAR CODES
- SET TECH 1 AUTO TRIGGER SNAPSHOT MODE
- START VEHICLE AND ALLOW TO IDLE FOR ONE MINUTE
- DOES TECH 1 TRIGGER?

NO → (3)
NO → (7)

(3)
- ROAD TEST VEHICLE REACHING AT LEAST 25 KPH (15 MPH)
- DOES TECH 1 TRIGGER?

YES →
- NOTE TECH 1, WHICH SENSOR TRIGGERED?

NO →
- NO TROUBLE FOUND; SEE DIAGNOSTIC AIDS ON FACING PAGE

(7)
- NOTE TECH1, LF SENSOR TRIGGERED?
- IGNITION 'OFF'
- DISCONNECT EBCM
- INSTALL J 39700 UNIVERSAL BREAKOUT BOX TO EBCM AND EBCM HARNESS CONNECTOR
- MEASURE RESISTANCE TO GROUND AT TERMINAL '5' ON J 39700
- IF RESISTANCE IS 5 OHMS OR LESS TO GROUND, REPAIR HARD SHORT; IF BETWEEN 5 OHMS AND 2 MEGAOHMS, REPAIR RESISTIVE SHORT IN CKT 830

(4)
- IGNITION 'OFF'
- DISCONNECT SENSOR CONNECTOR
- MEASURE RESISTANCE BETWEEN SENSOR TERMINALS, USING ADAPTERS FROM J 35616-A IF FRONT SENSOR
- IS RESISTANCE WITH IN RANGE SHOWN 900-1300 OHMS AT 25°C (77°F)?

NO → (5)
YES →
- REPLACE EBCM

(5)
- REPLACE LF WHEEL SPEED SENSOR

(6)
- CLEAR CODES, IGNITION 'OFF'
- ROAD TEST VEHICLE, REACHING AT LEAST 25 KPH (15 MPH)
- DOES CODE 26 RESET?

YES →
- REPLACE EBCM

NO →
- SYSTEM OK

WHEN ALL DIAGNOSIS AND REPAIRS ARE COMPLETED, CLEAR CODES AND VERIFY OPERATION

DIAGNOSTIC FAULT CHART DTC 27 — 1994 DeVILLE, CONCOURS, ELDORADO AND SEVILLE

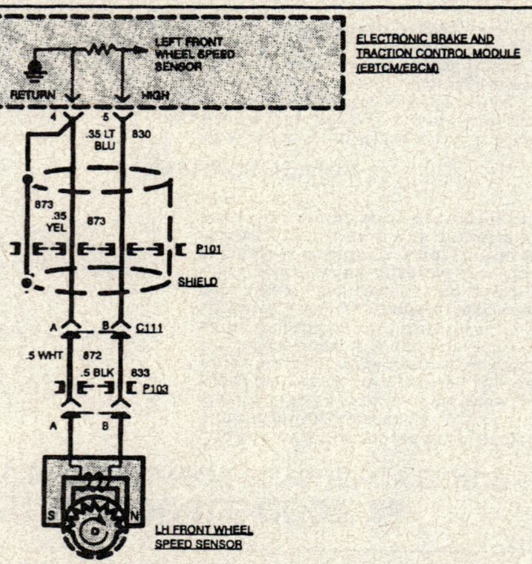

DIAGNOSTIC TROUBLE CODE 27
LEFT FRONT WHEEL SPEED SENSOR CONTINUITY FAULT

CIRCUIT DESCRIPTION:

As each wheel turns, the wheel speed sensor for that wheel creates a small AC voltage as a toothed sensor ring, mounted in the front integral hub/bearing assembly, is passed by the stationary sensor. The sensor consists of a permanent magnet and coil and generates voltage by magnetic induction. The frequency of the wheel speed sensor signal is proportional to wheel speed.

The EBTCM/EBCM uses the wheel speed sensor signal to calculate vehicle reference speeds and individual speed, acceleration and slip values for each wheel. These values are used to determine when antilock control is required.

The EBTCM/EBCM will only set one wheel speed sensor fault code at a time, even if multiple faults exist. Detection is prioritized accordingly: LF, RR, LR, RF. Always check for additional stored codes after making repairs.

TEST CONDITION(S):

1. Tested continuously.

FAILURE CONDITION(S):

Continuity

1. While Standing: Open circuit is seen while the vehicle is standing still with the ignition 'ON'.

ACTION TAKEN:

1. Code set.
2. ABS/TCS switched 'OFF'.
3. 'ANTILOCK' indicator 'ON'.
4. 'TRACTION DISABLED' message displayed. (vehicles with TCS)

Code 27 is set if there is a short to voltage or an open in CKT 830 or 873, or faulty speed sensor. The testing for this fault occurs with the ignition 'ON' and the vehicle at rest.

NOTICE: A Code 25 may set CKT 830 or 873 lose continuity and the WSS signal is lost while driving. A Code 27 should accompany this code on the subsequent ignition cycle. If both Code 25 and Code 27 are set concurrently, diagnose Code 27 first.

TEST DESCRIPTION: Test numbers refer to circled numbers on the diagnostic chart.

1. Checks for proper resistance in the sensor itself.
2. Checks for a short to battery in either of the speed sensor circuit wires. There should be 5V on CKT 830. If there is, CKT 830 has no hard opens or shorts. The condition may be intermittent.
3. Checks for an open in both the speed sensor circuit wires.
4. Manipulates wiring and connectors, trying to induce an intermittent fault not currently present.
5. This test checks if the code resets on key–up. If so, since Tests 1–4 have validated that the circuitry and components are good, the Wheel Bearing/Speed Sensor assembly may be causing the fault.
6. Checks if Code 25 sets during a road test.
7. Code 27 sets when the vehicle is at rest: Code 25 sets with a problem in the speed sensor circuitry with the vehicle in motion. If problems are still present, Code 25 would set during the road test, not a 27. If Code 25 sets at this point, Code 25 diagnostics should be used.
8. The wheel speed sensor may be causing an intermittent fault. If Code 27 resets after sensor replacement, the EBTCM/EBCM must be concluding there is a problem present in the speed sensor circuit when there is not.

DIAGNOSTIC AIDS:

Be sure the speed sensor wiring is properly routed and retained. This will help prevent false signals due to electrical noise being picked up by the wiring.

It is very important that a thorough inspection of the wiring and connectors be performed. Failure to carefully and fully inspect wiring and connectors may result in misdiagnosis, causing part replacement with reappearance of the fault.

DIAGNOSTIC FAULT CHART DTC 27 — 1994 DeVILLE, CONCOURS, ELDORADO AND SEVILLE, CONTINUED

DIAGNOSTIC TROUBLE CODE 27
LEFT FRONT WHEEL SPEED SENSOR CONTINUITY FAULT

ENSURE SENSOR IS NOT VISIBLY DAMAGED; IF IT IS DAMAGED, REPLACE THE WHEEL BEARING/SPEED/SENSOR ASSEMBLY, CLEAR CODES, AN TURN IGNITION 'ON' TO SEE IF CODE RESETS.

(1)
- IGNITION 'OFF'
- DISCONNECT SENSOR CONNECTOR
- USING DVM AND ADAPTER FROM J 35616–A, CHECK SENSOR RESISTANCE BY PROBING SENSOR CONNECTOR TERMINALS.
- IS IT 900–1300 OHMS AT 25°C (77°F)?

NO: • REPLACE WHEEL SPEED SENSOR

YES: (2)
- CONNECT DVM BETWEEN SENSOR HARNESS CONNECTOR TERMINAL 'A' AND GROUND (USED DC VOLTAGE SETTING)
- IGNITION 'ON'
- NOTE VOLTAGE: REPEAT BETWEEN TERMINAL 'B' AND GROUND.
- IS IT AROUND SYSTEM VOLTAGE ON EITHER TERMINAL 'A' OR 'B'?

YES:
- REPAIR SHORT TO SYSTEM VOLTAGE IN CKT 830 OR 873
- REFER TO EBTCM/EBCM–SWITCHED FAULTS IN THIS SECTION

NO: (3)
- IGNITION 'OFF'
- DISCONNECT EBTCM/EBCM
- INSTALL J 39700 UNIVERSAL BREAKOUT BOX TO EBTCM/EBCM HARNESS CONNECTOR ONLY
- MEASURE RESISTANCE BETWEEN PIN 5 ON J 39700 TERMINAL 'A' OF SENSOR CONNECTOR ON HARNESS. PERFORM SAME MEASUREMENT BETWEEN PIN 4 AND TERMINAL 'B'.
- IS EITHER MEASUREMENT GREATER THAN 5 OHMS?

YES: • REPAIR OPEN IN CKT 830 AND/OR 873

NO: (4)
- REMOVE J 39700; RECONNECT EBCM
- CLEAR CODES; IGNITION 'OFF'
- DISCONNECT TECH 1, IF CONNECTED
- IGNITION 'ON'
- WIGGLE ALL WIRING AND CONNECTORS TO TRY TO DUPLICATE AN INTERMITTENT FAULT
- DOES CODE 27 RESET?

YES:
- REPAIR WIRING AND/OR CONNECTIONS
- CHECK FOR INTERMITTENTS AND POOR CONNECTIONS

NO: (5)
- CLEAR CODES; IGNITION 'OFF'
- DISCONNECT TECH 1, IF CONNECTED
- IGNITION 'ON'
- DOES CODE 27 RESET?

YES: (8)
- REPLACE WHEEL SPEED SENSOR
- CLEAR CODES; IGNITION 'OFF'
- DISCONNECT TECH 1, IF CONNECTED
- IGNITION 'ON'
- DOES CODE 27 RESET?

YES: • REPLACE EBTCM/EBCM

NO: • SYSTEM OK

NO: (6)
- ROAD TEST VEHICLE, REACHING AT LEAST 25 KPH (15 MPH).
- DOES CODE 25 SET?

YES: (7) GO TO CODE 25 DIAGNOSTICS

NO: • NO TROUBLE FOUND, SEE 'DIAGNOSTIC AIDS' ON FACING PAGE.

WHEN ALL DIAGNOSIS AND REPAIRS ARE COMPLETED, CLEAR CODES AND VERIFY OPERATION

DIAGNOSTIC FAULT CHART DTC 28 — 1994 DeVILLE, CONCOURS, ELDORADO AND SEVILLE

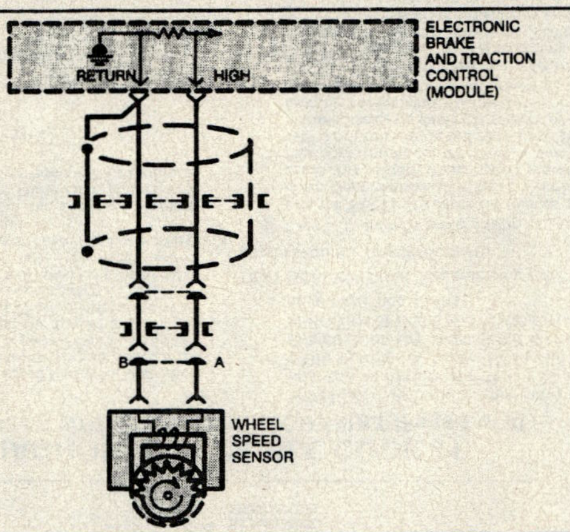

DIAGNOSTIC TROUBLE CODE 28
WHEEL SPEED SENSOR FREQUENCY ERROR

TEST CONDITION(S):

1. Tested continuously.

FAILURE CONDITION(S):

1. While Driving: One wheel is detected at least 30 kph or 19 mph faster than the remaining wheels for an extended period of time.

   OR

   One wheel is at least 6 kph or 4 mph less than or greater than the remaining wheels for an extended period of time.

ACTION TAKEN:

1. Code set.
2. ABS/TCS switched 'OFF'.
3. 'ANTILOCK' indicator 'ON'.
4. TRACTION DISABLED' message displayed.

Code 28 is set if the EBTCM cannot specifically identify which wheel speed sensor is causing the frequency error problem. If it can define the specific speed sensor causing the problem, the Sensor Fault code associated with the sensor (21, 25, 31, 35) will be set instead of Code 28.

TEST DESCRIPTION: Test numbers refer to circled numbers on the diagnostic chart.

1. Checks wiring and connections for problems. It is critical that this step is performed thoroughly, as wiring/connector problems are the most likely cause for this code.
2. Use the TECH 1 to monitor for vehicle electrical/electronic system 'noise' being picked up by the speed sensor circuits.
3. Use the TECH 1 to monitor the wheel speed sensors while the vehicle is operating. If the TECH 1 Auto–Trigger Snapshot mode triggers, the speed sensor or the wiring and/or connectors are going intermittent during the road test.
4. Checks for proper resistance in the sensor itself.

5a. Replace rear wheel speed sensor as the likely cause of the triggering intermittent.

5b. Replaces front wheel speed sensor as the likely cause of the triggering intermittent.

6. Use the TECH 1 to monitor the wheel speed sensors while the vehicle is operating. If the TECH 1 Auto–Snapshot mode triggers, the EBTCM must be concluding there is a problem with wheel speed sensor frequencies when there is not.
7. Checks for a short to ground in the speed sensor input circuit wires. The circuit to be tested depends on which speed sensor triggered the TECH 1. The ground may be either a 'hard' short to ground, or a resistive short. A short with a resistance less that 2 megaohms, though not a hard short, can still cause Code 28 to set.

DIAGNOSTIC AIDS:

Be sure the speed sensor wiring is properly routed and retained. This will help prevent false signals due to electrical noise being picked up by the wiring.

It is very important that a thorough inspection of the wiring and connectors be performed. Failure to carefully and fully inspect wiring and connectors may result in misdiagnosis, causing part replacement with reappearance of the fault.

Front Speed Sensors Only:

Check the toothed wheel for any large grooves, gouges, marks, etc. that might influence the tooth's signal at the wheel speed sensor. Also check for a buildup of foreign material in the gaps between teeth in the toothed wheel; this material may cause this fault.

A worn hub/bearing assembly may cause this fault in extreme cases; the bearing play allows the sensor–to–toothed–ring gap to change excessively. Never reuse a wheel speed sensor once it has been separated from the front hub/bearing assembly.

DIAGNOSTIC FAULT CHART DTC 28 — 1994 DeVILLE, CONCOURS, ELDORADO AND SEVILLE, CONTINUED

DIAGNOSTIC TROUBLE CODE 28
WHEEL SPEED SENSOR FREQUENCY ERROR

(1) VERY THOROUGHLY INSPECT ALL WIRING AND CONNECTIONS FOR INTERMITTENTS, CHAFING, POOR TERMINAL CONTACT TENSION, CORROSION, AND MISROUTING, AT EBTCM, ALL SENSORS, AND ALL HARNESS INTERCONNECTS. REPAIR ANY PROBLEMS CLEAR CODES, THEN ROAD TEST VEHICLE, REACHING AT LEAST 25 KPH (15 MPH).
- DOES CODE RESET?

YES → (2)

NO →
- NO TROUBLE FOUND; SEE 'DIAGNOSTIC AIDS' ON FACING PAGE.

(2)
- INSTALL TECH 1 AND MASS STORAGE; CLEAR CODES
- SET TECH 1 AUTO TRIGGER SNAPSHOT MODE
- START VEHICLE AND ALLOW TO IDLE FOR ONE MINUTE
- DOES TECH 1 TRIGGER?

NO → (3)

NO → (7)

(3)
- ROAD TEST VEHICLE REACHING AT LEAST 25 KPH (15 MPH).
- DOES TECH 1 TRIGGER?

YES →
- NOTE TECH 1, WHICH SENSOR TRIGGERED?

→ (4)

NO →
- NO TROUBLE FOUND; SEE DIAGNOSTIC AIDS ON FACING PAGE.

(4)
- IGNITION 'OFF'
- DISCONNECT SENSOR CONNECTOR
- MEASURE RESISTANCE BETWEEN SENSOR TERMINALS, USING ADAPTERS FROM J 35616-A IF FRONT SENSOR.
- IS RESISTANCE WITHIN 900-1300 OHMS AT 25°C (77°F)?

NO → FRONT → (5a)

NO → REAR → (5b)

YES →
- REPLACE EBTCM

(5a)
- REPLACE WHEEL SPEED SENSOR

(5b)
- REPLACE HUB/BEARING/SENSOR ASSEMBLY

(6)
- CLEAR CODES, IGNITION 'OFF'
- ROAD TEST VEHICLE, REACHING AT LEAST 25 KPH (15 MPH)
- DOES CODE 28 RESET?

YES →
- REPLACE EBTCM

NO →
- SYSTEM OK

(7)
- NOTE TECH1, WHICH SENSOR TRIGGERED?
- IGNITION'OFF'
- DISCONNECT EBTCM
- INSTALL J 39700 UNIVERSAL BREAKOUT BOX TO EBTCM AND EBTCM HARNESS CONNECTOR.
- USING TABLE BELOW, MEASURE RESISTANCE TO GROUND AT THE TERMINAL ON J 39700 THAT CORRESPONDS TO THE SPEED SENSOR THE TECH 1 TRIGGERED ON:

  LF-TERMINAL 5
  RF-TERMINAL 11
  LR-TERMINAL 8
  RR-TERMINAL 24

- IF RESISTANCE IS 5 OHMS OR LESS TO GROUND, REPAIR HARD SHORT; IF BETWEEN 5 OHMS AND 2 MEGAOHMS, REPAIR RESISTIVE SHORT IN CKT.

  LF-CKT 830
  RF-CKT 872
  LR-CKT 885
  RR-CKT 882

WHEN ALL DIAGNOSIS AND REPAIRS ARE COMPLETED, CLEAR CODES AND VERIFY OPERATION

DIAGNOSTIC FAULT CHART DTC 31 — 1994 DeVILLE, CONCOURS, ELDORADO AND SEVILLE

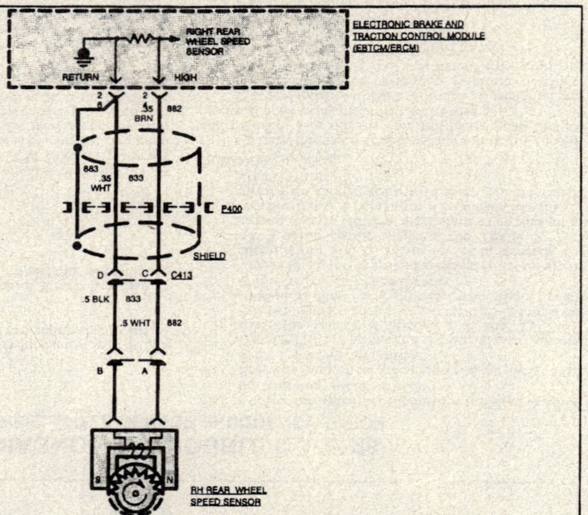

DIAGNOSTIC TROUBLE CODE 31
RIGHT REAR WHEEL SPEED SENSOR FAULT (Page 1 of 4)

CIRCUIT DESCRIPTION:

As each wheel turns, the wheel speed sensor for that wheel creates a small AC voltage as a toothed sensor ring, mounted in the front integral hub/bearing assembly, is passed by the stationary sensor. The sensor consists of a permanent magnet and coil and generates voltage by magnetic induction. The frequency of the wheel speed sensor signal is proportional to wheel speed.

The EBTCM/EBCM uses the wheel speed sensor signal to calculate vehicle reference speeds and individual speed, acceleration and slip values for each wheel. These values are used to determine when antilock control is required.

The EBTCM/EBCM will only set one wheel speed sensor fault code at a time, even if multiple faults exist. Detection is prioritized accordingly: LF, RR, LR, RF. Always check for additional stored codes after making repairs.

TEST CONDITION(S):

1. Tested continuously.

FAILURE CONDITION(S):

Sensor Output

1. Drive–away: Speed signal is not detected when vehicle reaches 12 kph or 7.5 mph.
2. While Driving: Speed signal is lost while vehicle speed is greater than 40 kph or 25 mph

ACTION TAKEN:

1. Code set.
2. ABS/TCS switched 'OFF'.
3. 'ANTILOCK' indicator 'ON'.
4. 'TRACTION DISABLED' message displayed. (vehicles with TCS)

Code 21 is set if there is a short to voltage or ground in CKT 882 or 883, or faulty speed sensor. The testing for this fault occurs when the vehicle is in motion, it will not set with the ignition 'ON' and the vehicle at rest.

NOTICE: A Code 31 may set if CKT 882 or 883 lose continuity and the WSS signal is lost while driving. A Code 33 should accompany this code on the subsequent ignition cycle. If both Code 31 and Code 33 are set concurrently, diagnose Code 33 first.

TEST DESCRIPTION: Test numbers refer to circled numbers on the diagnostic chart.

1. Checks if the code resets during a road test.
2. Use the TECH 1 to check for sources of induced electrical noise. Some sources of induced electrical noise which may cause intermittent wheel speed sensor codes to set are: improper routing or sensor cables next to spark plug wires; cables not retained in brackets; loose fitting or improperly mounted sensors; damaged sensors or toothed rings; poor terminal engagement of connectors; or intermittent shorts or opens in wiring; or a loss of sensor circuit shield.
3. Checks for damaged wheel speed sensor circuit shield. The shield is a foil insulator that is wrapped around the wheel speed sensor circuits. Inside the EBTCM/EBCM harness connecter there is an uninsulated drain wire that is attached to the shield and jumped to the wheel speed sensor return line and taken to ground through the EBTCM/EBCM. A shield which has lost its ground through the EBTCM/EBCM leaves the sensor signal circuit susceptible to noise. It may be necessary to disassemble the EBTCM/EBCM connector to check if the drain wire has a poor connection with the wheel speed sensor return connector terminal. The shield can be repaired by reterminating the sensor return line circuit, including the drain wire.
4. Checks for intermittent opens/shorts in wheel speed sensor circuits and cables.

DIAGNOSTIC FAULT CHART DTC 31 — 1994 DeVILLE, CONCOURS, ELDORADO AND SEVILLE, CONTINUED

DIAGNOSTIC TROUBLE CODE 31
RIGHT REAR WHEEL SPEED SENSOR FAULT (Page 2 of 4)

NOTE: DIAGNOSE CODE 33 FIRST IF SET CONCURRENTLY.

ENSURE SENSOR IS NOT VISIBLY DAMAGED; IF IT IS DAMAGED, REPLACE THE HUB/BEARING/SENSOR ASSEMBLY, CLEAR CODES, AND ROAD TEST TO SEE IF CODE RESETS.

(1)
- CLEAR CODES
- ROAD TEST VEHICLE, REACHING AT LEAST 25 KPH (15 MPH)
- DOES CODE 31 RESET?

NO:
- NO TROUBLE FOUND
- CHECK FOR INTERMITTENT CONDITIONS

YES:

(2)
- INSTALL TECH 1
- SELECT MODE F0: DATA LIST
- OBSERVE WHEEL SPEED SENSOR OUTPUT WHILE IDLING IN PARK
- INCREASE ENGINE SPEED BY PRESSING ON ACCELERATOR

ERRATIC SENSOR READING (DOES NOT STAY AT ZERO):
- VERIFY CORRECT CABLE ROUTING
- CHECK FOR PROPER INSTALLATION OF SENSOR
- REPAIR AS REQUIRED
- INSPECT EBTCM/EBCM HARNESS CONNECTOR FOR AN OPEN FROM THE WHEEL SPEED SENSOR WIRING SHIELD TO THE WHEEL SPEED SENSOR RETURN LINE

SENSOR OUTPUT DOES NOT CHANGE WITH ENGINE SPEED (STAYS AT ZERO):

(3)
- TEST DRIVE VEHICLE
- OBSERVE WHEEL SPEED SENSOR OUTPUT
- DOES SIGNAL 'DROP OUT' OR DIFFER FROM OTHER SENSORS?

NORMAL SENSOR OUTPUT:
- CHECK FOR INTERMITTENT CONDITIONS

SENSOR READING DIFFERS BY MORE THAN 4 MPH:
- VERIFY CORRECT CABLE ROUTING
- CHECK FOR PROPER INSTALLATION OF SENSOR
- CHECK FOR AN OPEN IN WHEEL SPEED SENSOR WIRING SHIELD
- REPAIR AS REQUIRED
- REPLACE HUB/BEARING/SENSOR ASSEMBLY IF PROBLEM PERSISTS

SENSOR READING 'DROP OUT':

(4)
- RAISE VEHICLE ON HOIST
- SPIN WHEEL WHILE MANIPULATING CABLES AND CONNECTORS BY HAND
- OBSERVE WHEEL SPEED SENSOR OUTPUT
- DOES SIGNAL 'DROP OUT'?

CONTINUED ON PAGE 4 OF 4 (CODE 31)

WHEN ALL DIAGNOSIS AND REPAIRS ARE COMPLETED, CLEAR CODES AND VERIFY OPERATION

DIAGNOSTIC FAULT CHART DTC 31 — 1994 DeVILLE, CONCOURS, ELDORADO AND SEVILLE, CONTINUED

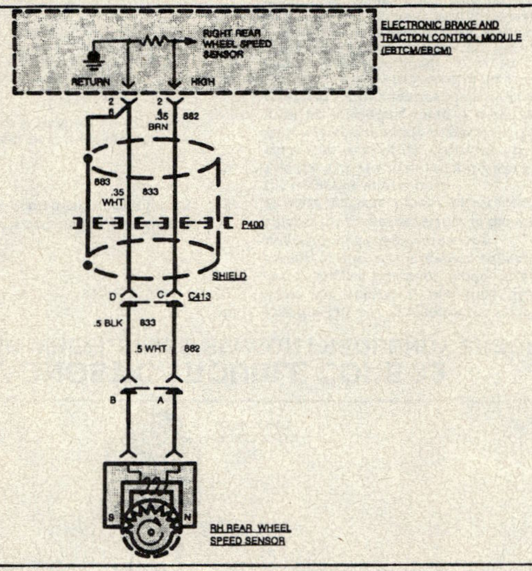

DIAGNOSTIC TROUBLE CODE 31

RIGHT REAR WHEEL SPEED SENSOR FAULT (Page 3 of 4)

CIRCUIT DESCRIPTION:

As each wheel turns, the wheel speed sensor for that wheel creates a small AC voltage as a toothed sensor ring, mounted in the front integral hub/bearing assembly, is passed by the stationary sensor. The sensor consists of a permanent magnet and coil and generates voltage by magnetic induction. The frequency of the wheel speed sensor signal is proportional to wheel speed.

The EBTCM/EBCM uses the wheel speed sensor signal to calculate vehicle reference speeds and individual speed, acceleration and slip values for each wheel. These values are used to determine when antilock control is required.

The EBTCM/EBCM will only set one wheel speed sensor fault code at a time, even if multiple faults exist. Detection is prioritized accordingly: LF, RR, LR, RF. Always check for additional stored codes after making repairs.

TEST DESCRIPTION: Test numbers refer to circled numbers on the diagnostic chart.

5. Checks for proper resistance in the sensor itself.
6. Checks for a short to battery in either of the speed sensor circuit wires.
7. Checks for a short between the speed sensor circuit wires.
8. Checks for a short to ground in the speed sensor input circuit wire. The ground may be either a 'hard' short to ground, or a resistive short. A short with a resistance less than 2 megaohms, through not a hard short, can still cause Code 21 to set.
9. Checks wiring and connectors for intermittents.

DIAGNOSTIC FAULT CHART DTC 31 — 1994 DeVILLE, CONCOURS, ELDORADO AND SEVILLE, CONTINUED

DIAGNOSTIC TROUBLE CODE 31

RIGHT REAR WHEEL SPEED SENSOR FAULT (Page 4 of 4)

CONTINUED FROM CODE 31 PAGE 2 OF 4

YES →

(5)
- IGNITION 'OFF'
- DISCONNECT SENSOR CONNECTOR
- USING DVM AND ADAPTER FROM J 35616–A, CHECK SENSOR RESISTANCE BY PROBING SENSOR CONNECTOR TERMINALS.
- IS IT 900 – 1300 OHMS AT 25°C (77°F)?

NO →
- CHECK FOR INTERMITTENTS AND LOOSE CONNECTIONS
- REPLACE EBTCM/EBCM IF PROBLEM PERSISTS

YES (from 5) →

(6)
- CONNECT DVM BETWEEN SENSOR HARNESS CONNECTOR TERMINAL 'A' AND GROUND (USE DC VOLTAGE SETTING)
- IGNITION 'ON'
- NOTE VOLTAGE: REPEAT BETWEEN TERMINAL 'B' AND GROUND.
- IS IT AROUND SYSTEM VOLTAGE ON EITHER TERMINAL 'A' OR 'D'?

NO (from 5) →
- REPLACE HUB/BEARING/SENSOR ASSEMBLY

NO (from 6) →

(7)
- IGNITION 'OFF'
- DISCONNECT EBTCM/EBCM
- INSTALL J 39700 UNIVERSAL BREAKOUT BOX TO EBTCM/EBCM HARNESS CONNECTOR ONLY
- MEASURE RESISTANCE BETWEEN J 39700 TERMINALS '26' AND '24'
- IS IT LESS THAN 5 OHMS?

YES (from 6) →
- REPAIR SHORT TO SYSTEM VOLTAGE IN CKT 882 OR 883
- REFER TO EBTCM/EBCM–SWITCHED FAULTS IN THIS SECTION

NO (from 7) →

(8)
- MEASURE RESISTANCE BETWEEN J 39700 TERMINAL '24' AND GROUND
- IS IT GREATER THAN 2 MEGAOHMS?

YES (from 7) →
- REPAIR SHORT BETWEEN CKT'S 882 AND 883

YES (from 8) →

(9)
- MEASURE RESISTANCE BETWEEN J 39700 TERMINAL '26' AND GROUND
- IS IT GREATER THAN 2 MEGAOHMS?

NO (from 8) →
- IF 5 OHMS OR LESS TO GROUND, REPAIR HARD SHORT IN HARNESS CKT 882
- IF BETWEEN 5 OHMS AND 2 MEGAOHMS, REPAIR RESISTIVE SHORT IN HARNESS CKT 882

YES (from 9) →
- CHECK ALL WIRING, CONNECTORS AND TERMINALS THOROUGHLY FOR INTERMITTENTS OR CORROSION.
- REFER TO 'INTERMITTENTS AND POOR CONNECTIONS'
- REPAIR HARNESS OR CONNECTOR
- REPLACE SENSOR IF PROBLEM PERSISTS

NO (from 9) →
- IF 5 OHMS OR LESS TO GROUND, REPAIR HARD SHORT IN HARNESS CKT 883
- IF BETWEEN 5 OHMS AND 2 MEGAOHMS REPAIR RESISTIVE SHORT IN HARNESS CKT 883

WHEN ALL DIAGNOSIS AND REPAIRS ARE COMPLETED, CLEAR CODES AND VERIFY OPERATION

DIAGNOSTIC FAULT CHART DTC 32 — 1994 DeVILLE, CONCOURS, ELDORADO AND SEVILLE

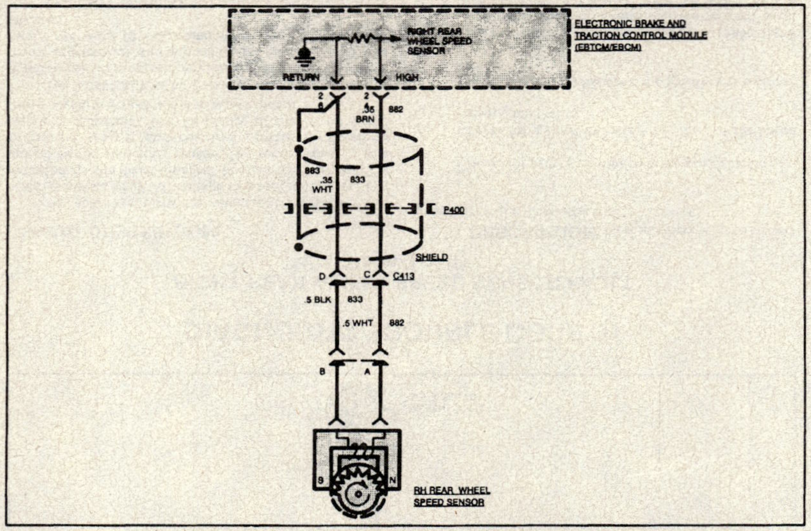

DIAGNOSTIC TROUBLE CODE 32
RIGHT REAR WHEEL SPEED SENSOR FREQUENCY ERROR

TEST CONDITION(S):

1. Tested continuously.

FAILURE CONDITION(S):

1. While Driving: One wheel is detected at least 30 kph or 19 mph faster than the remaining wheels for an extended period of time.

   OR

   One wheel is at least 6 kph or 4 mph less than or greater than the remaining wheels for an extended period of time.

ACTION TAKEN:

1. Code set.
2. ABS switched 'OFF'.
3. 'ANTILOCK' indicator 'ON'.

TEST DESCRIPTION: Test numbers refer to circled numbers on the diagnostic chart.

1. Checks wiring and connections for problems. It is critical that this step is performed thoroughly, as wiring/connector problems are the most likely cause for this code.
2. Use the TECH 1 to monitor for vehicle electrical/electronic system 'noise' being picked up by the speed sensor circuits.
3. Use the TECH 1 to monitor the wheel speed sensors while the vehicle is operating. If the TECH 1 Auto–Trigger Snapshot mode triggers, the speed sensor or the wiring and/or connectors are going intermittent during the road test.
4. Checks for proper resistance in the sensor itself.
5. Replace rear wheel speed sensor as the likely cause of the triggering intermittent.
6. Use the TECH 1 to monitor the wheel speed sensors while the vehicle is operating. If the TECH 1 Auto–Snapshot mode triggers, the EBTCM/EBCM must be concluding there is a problem with wheel speed sensor frequencies when there is not.
7. Checks for a short to ground in the speed sensor input circuit wires. The circuit to be tested depends on which speed sensor triggered the TECH 1. The ground may be either a 'hard' short to ground, or a resistive short. A short with a resistance less that 2 megaohms, though not a hard short, can still cause Code 32 to set.

DIAGNOSTIC AIDS:

Be sure the speed sensor wiring is properly routed and retained. This will help prevent false signals due to electrical noise being picked up by the wiring.

It is very important that a thorough inspection of the wiring and connectors be performed. Failure to carefully and fully inspect wiring and connectors may result in misdiagnosis, causing part replacement with reappearance of the fault.

DIAGNOSTIC FAULT CHART DTC 32 — 1994 DeVILLE, CONCOURS, ELDORADO AND SEVILLE, CONTINUED

DIAGNOSTIC TROUBLE CODE 32
RIGHT REAR WHEEL SPEED SENSOR FREQUENCY ERROR

(1) VERY THOROUGHLY INSPECT ALL WIRING AND CONNECTIONS FOR INTERMITTENTS, CHAFING, POOR TERMINAL CONTACT TENSION, CORROSION, AND MISROUTING, AT EBCM, ALL SENSORS, AND ALL HARNESS INTERCONNECTS. REPAIR ANY PROBLEMS CLEAR CODES, THEN ROAD TEST VEHICLE, REACHING AT LEAST 25 KPH (15 MPH).
- DOES CODE RESET?

YES →

(2)
- INSTALL TECH 1 AND MASS STORAGE CARTRIDGE; CLEAR CODES
- SET TECH 1 AUTO TRIGGER SNAPSHOT MODE
- START VEHICLE AND ALLOW TO IDLE FOR ONE MINUTE
- DOES TECH 1 TRIGGER?

NO →
- NO TROUBLE FOUND; SEE 'DIAGNOSTIC AIDS' ON FACING PAGE

(2) NO →

(3)
- ROAD TEST VEHICLE REACHING AT LEAST 25 KPH (15 MPH)
- DOES TECH 1 TRIGGER?

(3) YES →
- NOTE TECH 1, WHICH SENSOR TRIGGERED?

(3) NO →
- NO TROUBLE FOUND; SEE DIAGNOSTIC AIDS ON FACING PAGE

(2) NO →

(7)
- NOTE TECH1, RR SENSOR TRIGGERED?
- IGNITION 'OFF'
- DISCONNECT EBCM
- INSTALL J 39700 UNIVERSAL BREAKOUT BOX TO EBCM AND EBCM HARNESS CONNECTOR
- MEASURE RESISTANCE TO GROUND AT TERMINAL 24 ON J 39700
- IF RESISTANCE IS 5 OHMS OR LESS TO GROUND, REPAIR HARD SHORT; IF BETWEEN 5 OHMS AND 2 MEGAOHMS, REPAIR RESISTIVE SHORT IN CKT 882

(4)
- IGNITION 'OFF'
- DISCONNECT SENSOR CONNECTOR
- MEASURE RESISTANCE BETWEEN SENSOR TERMINALS
- IS RESISTANCE WITH IN RANGE 900–1300 OHMS AT 25°C (77°F)?

YES →
- REPLACE EBCM

NO →

(5)
- REPLACE HUB/BEARING/SENSOR ASSEMBLY

(6)
- CLEAR CODES, IGNITION 'OFF'
- ROAD TEST VEHICLE, REACHING AT LEAST 25 KPH (15 MPH)
- DOES CODE 32 RESET?

YES →
- REPLACE EBCM

NO →
- SYSTEM OK

WHEN ALL DIAGNOSIS AND REPAIRS ARE COMPLETED, CLEAR CODES AND VERIFY OPERATION

DIAGNOSTIC FAULT CHART DTC 33 — 1994 DeVILLE, CONCOURS, ELDORADO AND SEVILLE

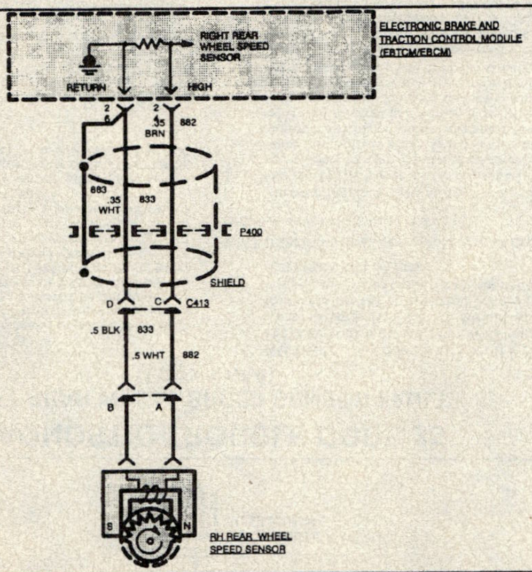

DIAGNOSTIC TROUBLE CODE 33

RIGHT REAR WHEEL SPEED SENSOR CONTINUITY FAULT

CIRCUIT DESCRIPTION:

As each wheel turns, the wheel speed sensor for that wheel creates a small AC voltage as a toothed sensor ring, mounted in the front integral hub/bearing assembly, is passed by the stationary sensor. The sensor consists of a permanent magnet and coil and generates voltage by magnetic induction. The frequency of the wheel speed sensor signal is proportional to wheel speed.

The EBTCM/EBCM uses the wheel speed sensor signal to calculate vehicle reference speeds and individual speed, acceleration and slip values for each wheel. These values are used to determine when antilock control is required.

The EBTCM/EBCM will only set one wheel speed sensor fault code at a time, even if multiple faults exist. Detection is prioritized accordingly: LF, RR, LR, RF. Always check for additional stored codes after making repairs.

TEST CONDITION(S):

1. Tested continuously.

FAILURE CONDITION(S):

Continuity

1. While standing: Open circuit is seen while the vehicle is standing still with the ignition 'ON'.

ACTION TAKEN:

1. Code set.
2. ABS/TCS switched 'OFF'.
3. 'ANTILOCK' indicator 'ON'.
4. 'TRACTION DISABLED' message displayed. (vehicles with TCS)

Code 33 is set if there is a short to voltage or an open in CKT 882 or 883, or faulty speed sensor. The testing for this fault occurs with the ignition 'ON' and the vehicle at rest.

NOTICE: A Code 31 may set if CKT 882 or 883 lose continuity and the WSS signal is lost while driving. A Code 33 should accompany this code on the subsequent ignition cycle. If both Code 31 and Code 33 are set concurrently, diagnose Code 33 first.

TEST DESCRIPTION: Test numbers refer to circled numbers on the diagnostic chart.

1. Checks for proper resistance in the sensor itself.
2. Checks for a short to battery in either of the speed sensor circuit wires. There should be 5 V on CKT 882. If there is, CKT 882 has no hard opens or shorts. The condition may be intermittent.
3. Checks for an open in both the speed sensor circuit wires.
4. Manipulates wiring and connectors, trying to induce an intermittent fault not currently present.
5. This test checks if the code resets on key–up. If so, since Tests 1–4 have validated that the circuitry and components are good, the speed sensor may be causing the fault.
6. Checks if the Code 31 sets during a road test.
7. Code 33 sets when the vehicle is at rest: Code 31 sets with a problem in the speed sensor circuitry with the vehicle in motion. If problems are still present, Code 31 would set during the road test, not a 33. If Code 33 sets at this point, Code 31 diagnostics should be used.
8. The wheel speed sensor may be causing an intermittent fault. If Code 33 resets after sensor replacement, the EBTCM/EBCM must be concluding there is a problem present in the speed sensor circuit when there is not.

DIAGNOSTIC AIDS:

Be sure the speed sensor wiring is properly routed and retained. This will help prevent false signals due to electrical noise being picked up by the wiring.

It is very important that a thorough inspection of the wiring and connectors be performed. Failure to carefully and fully inspect wiring and connectors may result in misdiagnosis, causing part replacement with reappearance of the fault.

DIAGNOSTIC FAULT CHART DTC 33 — 1994 DeVILLE, CONCOURS, ELDORADO AND SEVILLE, CONTINUED

DIAGNOSTIC TROUBLE CODE 33

RIGHT REAR WHEEL SPEED SENSOR CONTINUITY FAULT

ENSURE SENSOR IS NOT VISIBLY DAMAGED; IF IT IS DAMAGED, REPLACE THE HUB/BEARING/SPEED SENSOR ASSEMBLY. CLEAR CODES, AN TURN IGNITION 'ON' TO SEE IF CODE RESETS.

(1)
- IGNITION 'OFF'
- DISCONNECT SENSOR CONNECTOR
- USING DVM AND ADAPTER FROM J 35616–A, CHECK SENSOR RESISTANCE BY PROBING SENSOR CONNECTOR TERMINALS.
- IS IT 900 – 1300 OHMS AT 25°C (77°F)?

NO
- REPLACE HUB/BEARING/SENSOR ASSEMBLY

YES

(2)
- CONNECT DVM BETWEEN SENSOR HARNESS CONNECTOR TERMINAL 'C' AND GROUND (USED DC VOLTAGE SETTING)
- IGNITION 'ON'
- NOTE VOLTAGE: REPEAT BETWEEN TERMINAL 'D' AND GROUND.
- IS IT AROUND SYSTEM VOLTAGE ON EITHER TERMINAL 'C' OR 'D'?

YES
- REPAIR SHORT TO SYSTEM VOLTAGE IN CKT 882 OR 883
- REFER TO EBTCM/EBCM–SWITCHED FAULTS IN THIS SECTION

NO

(3)
- IGNITION 'OFF'
- DISCONNECT EBTCM/EBCM
- INSTALL J 39700 UNIVERSAL BREAKOUT BOX TO EBTCM/EBCM HARNESS CONNECTOR ONLY
- MEASURE RESISTANCE BETWEEN PIN 24 ON J 39700 TERMINAL 'A' OF SENSOR CONNECTOR ON HARNESS. PERFORM SAME MEASUREMENT BETWEEN TERMINAL 26 AND TERMINAL 'B'.
- IS EITHER MEASUREMENT GREATER THAN 5 OHMS?

YES
- REPAIR OPEN IN CKT 882 AND/OR 883

NO

(4)
- REMOVE J 39700; RECONNECT EBTCM/EBCM
- CLEAR CODES; IGNITION 'OFF'
- DISCONNECT TECH 1, IF CONNECTED
- IGNITION 'ON'
- WIGGLE ALL WIRING AND CONNECTORS TO TRY TO DUPLICATE AN INTERMITTENT FAULT
- DOES CODE 33 RESET?

YES
- REPAIR WIRING AND/OR CONNECTIONS

NO

(5)
- CLEAR CODES; IGNITION 'OFF'
- DISCONNECT TECH 1, IF CONNECTED
- IGNITION 'ON'
- DOES CODE 33 RESET?

YES

(8)
- REPLACE HUB/BEARING/SENSOR ASSEMBLY
- CLEAR CODES; IGNITION 'OFF'
- DISCONNECT TECH 1, IF CONNECTED
- IGNITION 'ON'
- DOES CODE 33 RESET?

YES
- REPLACE EBTCM/EBCM

NO
- SYSTEM OK

NO

(6)
- ROAD TEST VEHICLE, REACHING AT LEAST 25 KPH (15 MPH).
- DOES CODE 31 SET?

YES

(7) GO TO CODE 31 DIAGNOSTICS

NO
- NO TROUBLE FOUND. SEE 'DIAGNOSTIC AIDS' ON FACING PAGE.

WHEN ALL DIAGNOSIS AND REPAIRS ARE COMPLETED, CLEAR CODES AND VERIFY OPERATION

DIAGNOSTIC FAULT CHART DTC 35 — 1994 DeVILLE, CONCOURS, ELDORADO AND SEVILLE

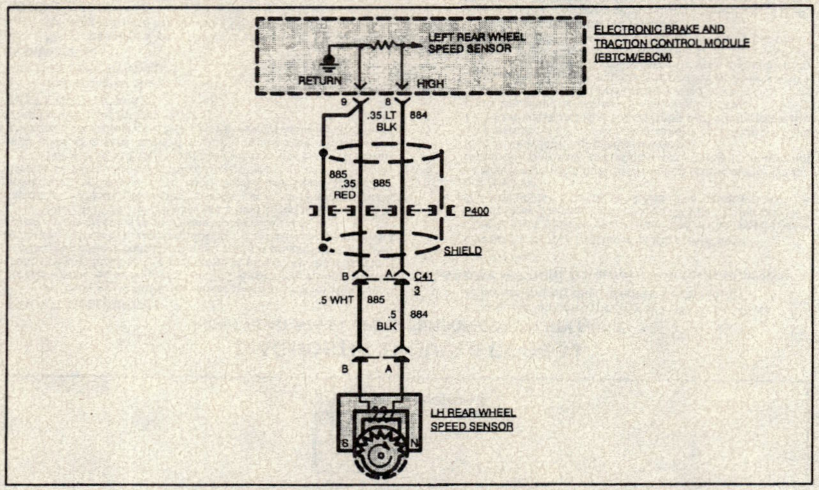

DIAGNOSTIC TROUBLE CODE 35

LEFT REAR WHEEL SPEED SENSOR FAULT
(Page 1 of 4)

CIRCUIT DESCRIPTION:

As each wheel turns, the wheel speed sensor for that wheel creates a small AC voltage as a toothed sensor ring, mounted in the front integral hub/bearing assembly, is passed by the stationary sensor. The sensor consists of a permanent magnet and coil and generates voltage by magnetic induction. The frequency of the wheel speed sensor signal is proportional to wheel speed.

The EBTCM/EBCM uses the wheel speed sensor signal to calculate vehicle reference speeds and individual speed, acceleration and slip values for each wheel. These values are used to determine when antilock control is required.

The EBTCM/EBCM will only set one wheel speed sensor fault code at a time, even if multiple faults exist. Detection is prioritized accordingly: LF, RR, LR, RF. Always check for additional stored codes after making repairs.

TEST CONDITION(S):

1. Tested continuously.

FAILURE CONDITION(S):

Sensor Output

1. Drive–away: Speed signal is not detected when vehicle reaches 12 kph or 7.5 mph.
2. While Driving: Speed signal is lost while vehicle speed is greater than 40 kph or 25 mph.

ACTION TAKEN:

1. Code set.
2. ABS/TCS switched 'OFF'.
3. 'ANTILOCK' indicator 'ON'.
4. 'TRACTION DISABLED' message displayed. (vehicles with TCS)

Code 35 is set if there is a short to voltage or ground in CKT 885 or 884, or faulty speed sensor. The testing for this fault occurs when the vehicle is in motion, it will not set with the ignition 'ON' and the vehicle at rest.

NOTICE: A Code 35 may set if CKT 884 or 885 lose continuity and the WSS signal is lost while driving. A Code 37 should accompany this code on the subsequent ignition cycle. If both Code 35 and Code 37 are set concurrently, diagnose Code 37 first.

TEST DESCRIPTION: Test numbers refer to circled numbers on the diagnostic chart.

1. Checks if the code resets during a road test.
2. Use the TECH 1 to check for sources of induced electrical noise. Some sources of induced electrical noise which may cause intermittent wheel speed sensor codes to set are: improper routing or sensor cables next to spark plug wires; cables not retained in brackets; loose fitting or improperly mounted sensors; damaged sensors or toothed rings; poor terminal engagement of connectors; or intermittent shorts or opens in wiring; or a loss of sensor circuit shield.
3. Checks for damaged wheel speed sensor circuit shield. The shield is a foil insulator that is wrapped around the wheel speed sensor circuits. Inside the EBTCM/EBCM harness connecter there is an uninsulated drain wire that is attached to the shield and jumped to the wheel speed sensor return line and taken to ground through the EBTCM/EBCM. A shield which has lost its ground through the EBTCM/EBCM leaves the sensor signal circuit susceptible to noise. It may be necessary to disassemble the EBTCM/EBCM connector to check if the drain wire has a poor connection with the wheel speed sensor return connector terminal. The shield can be repaired by reterminating the sensor return line circuit, including the drain wire.
4. Checks for intermittent opens/shorts in wheel speed sensor circuits and cables.

DIAGNOSTIC FAULT CHART DTC 35 — 1994 DeVILLE, CONCOURS, ELDORADO AND SEVILLE, CONTINUED

DIAGNOSTIC TROUBLE CODE 35

LEFT REAR WHEEL SPEED SENSOR FAULT
(Page 2 of 4)

NOTE: DIAGNOSE CODE 37 FIRST IF SET CONCURRENTLY.

(1)
- CLEAR CODES
- ROAD TEST VEHICLE, REACHING AT LEAST 25 KPH (15 MPH)
- DOES CODE 35 RESET?

NO:
- NO TROUBLE FOUND
- CHECK FOR INTERMITTENTS AND LOOSE CONNECTIONS

YES:

(2)
- INSTALL TECH 1
- SELECT MODE F0: DATA LIST
- OBSERVE WHEEL SPEED SENSOR OUTPUT WHILE IDLING IN PARK
- INCREASE ENGINE SPEED BY PRESSING ON ACCELERATOR

ERRATIC SENSOR READING (DOES NOT STAY AT ZERO):
- VERIFY CORRECT CABLE ROUTING
- CHECK FOR PROPER INSTALLATION OF SENSOR
- REPAIR AS REQUIRED
- INSPECT EBTCM/EBCM HARNESS CONNECTOR FOR AN OPEN FROM THE WHEEL SPEED SENSOR WIRING SHIELD TO THE WHEEL SPEED SENSOR RETURN LINE

SENSOR OUTPUT DOES NOT CHANGE WITH ENGINE SPEED (STAYS AT ZERO):

(3)
- TEST DRIVE VEHICLE
- OBSERVE WHEEL SPEED SENSOR OUTPUT
- DOES SIGNAL 'DROP OUT' OR DIFFER FROM OTHER SENSORS?

SENSOR READING 'DROP OUT':

(4)
- RAISE VEHICLE ON HOIST
- SPIN WHEEL WHILE MANIPULATING CABLES AND CONNECTORS BY HAND
- OBSERVE WHEEL SPEED SENSOR OUTPUT
- DOES SIGNAL 'DROP OUT'?

CONTINUED ON PAGE 4 OF 4 (CODE 35)

NORMAL SENSOR OUTPUT:
- CHECK FOR INTERMITTENT CONDITIONS

SENSOR READING DIFFERS BY MORE THAN 4 MPH:
- VERIFY CORRECT CABLE ROUTING
- CHECK FOR PROPER INSTALLATION OF SENSOR
- CHECK FOR AN OPEN IN WHEEL SPEED SENSOR WIRING SHIELD
- REPAIR AS REQUIRED
- REPLACE HUB/BEARING/SENSOR ASSEMBLY IF PROBLEM PERSISTS

WHEN ALL DIAGNOSIS AND REPAIRS ARE COMPLETED, CLEAR CODES AND VERIFY OPERATION

DIAGNOSTIC FAULT CHART DTC 35 — 1994 DeVILLE, CONCOURS, ELDORADO AND SEVILLE, CONTINUED

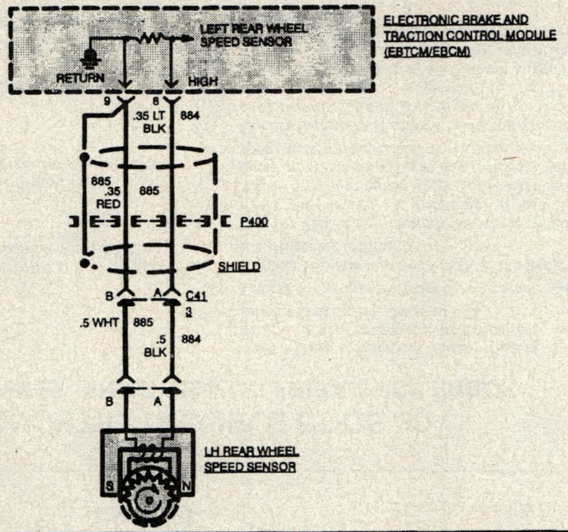

DIAGNOSTIC TROUBLE CODE 35

LEFT REAR WHEEL SPEED SENSOR FAULT (Page 3 of 4)

CIRCUIT DESCRIPTION:

As each wheel turns, the wheel speed sensor for that wheel creates a small AC voltage as a toothed sensor ring, mounted in the front integral hub/bearing assembly, is passed by the stationary sensor. The sensor consists of a permanent magnet and coil and generates voltage by magnetic induction. The frequency of the wheel speed sensor signal is proportional to wheel speed.

The EBTCM/EBCM uses the wheel speed sensor signal to calculate vehicle reference speeds and individual speed, acceleration and slip values for each wheel. These values are used to determine when antilock control is required.

The EBTCM/EBCM will only set one wheel speed sensor fault code at a time, even if multiple faults exist. Detection is prioritized accordingly: LF, RR, LR, RF. Always check for additional stored codes after making repairs.

TEST DESCRIPTION: Test numbers refer to circled numbers on the diagnostic chart.

5. Checks for proper resistance in the sensor itself.
6. Checks for a short to battery in either of the speed sensor circuit wires.
7. Checks for a short between the speed sensor circuit wires.
8. Checks for a short to ground in the speed sensor input circuit wire. The ground may be either a 'hard' short to ground, or a resistive short. A short with a resistance less than 2 megaohms, through not a hard short, can still cause Code 21 to set.
9. Checks wiring and connectors for intermittents.

DIAGNOSTIC FAULT CHART DTC 35 — 1994 DeVILLE, CONCOURS, ELDORADO AND SEVILLE, CONTINUED

DIAGNOSTIC TROUBLE CODE 35

LEFT REAR WHEEL SPEED SENSOR FAULT (Page 4 of 4)

CONTINUED FROM CODE 35 PAGE 2 OF 4

YES

(5)
- IGNITION 'OFF'
- DISCONNECT SENSOR CONNECTOR
- USING DVM AND ADAPTER FROM J 35616–A, CHECK SENSOR RESISTANCE BY PROBING SENSOR CONNECTOR TERMINALS.
- IS IT 900 – 1300 OHMS at 25°C (77°F)?

NO
- CHECK FOR INTERMITTENTS AND LOOSE CONNECTIONS
- REPLACE EBTCM/EBCM IF PROBLEM PERSISTS

YES

(6)
- CONNECT DVM BETWEEN SENSOR HARNESS CONNECTOR TERMINAL 'A' AND GROUND (USE DC VOLTAGE SETTING)
- IGNITION 'ON'
- NOTE VOLTAGE: REPEAT BETWEEN TERMINAL 'B' AND GROUND
- IS IT AROUND SYSTEM VOLTAGE ON EITHER TERMINAL 'A' OR 'B'?

NO
- REPLACE HUB/BEARING/SENSOR ASSEMBLY

YES
- REPAIR SHORT TO SYSTEM VOLTAGE IN CKT 884 OR 885
- REFER TO EBTCM/EBCM–SWITCHED FAULTS IN THIS SECTION

NO

(7)
- IGNITION 'OFF'
- DISCONNECT EBTCM/EBCM
- INSTALL J 39700 UNIVERSAL BREAKOUT BOX TO EBTCM/EBCM HARNESS CONNECTOR ONLY
- MEASURE RESISTANCE BETWEEN J 39700 TERMINALS '8' AND '9'
- IS IT LESS THAN 5 OHMS?

YES
- REPAIR SHORT BETWEEN CKT'S 884 AND 885

NO

(8)
- MEASURE RESISTANCE BETWEEN J 39700 TERMINAL '8' AND GROUND
- IS IT GREATER THAN 2 MEGAOHMS?

NO
- IF 5 OHMS OR LESS TO GROUND, REPAIR HARD SHORT IN HARNESS CKT 884
- IF BETWEEN 5 OHMS AND 2 MEGAOHMS, REPAIR RESISTIVE SHORT IN HARNESS CKT 884

YES

(9)
- MEASURE RESISTANCE BETWEEN J 39700 TERMINAL '9' AND GROUND
- IS IT GREATER THAN 2 MEGAOHMS?

NO
- IF 5 OHMS OR LESS TO GROUND, REPAIR HARD SHORT IN HARNESS CKT 885
- IF BETWEEN 5 OHMS AND 2 MEGAOHMS REPAIR RESISTIVE SHORT IN HARNESS CKT 885

YES
- CHECK ALL WIRING, CONNECTORS AND TERMINALS THOROUGHLY FOR INTERMITTENTS OR CORROSION.
- REFER TO 'INTERMITTENTS AND POOR CONNECTIONS'
- REPAIR HARNESS OR CONNECTOR
- REPLACE SENSOR IF PROBLEM PERSISTS

WHEN ALL DIAGNOSIS AND REPAIRS ARE COMPLETED, CLEAR CODES AND VERIFY OPERATION

DIAGNOSTIC FAULT CHART DTC 36 — 1994 DeVILLE, CONCOURS, ELDORADO AND SEVILLE

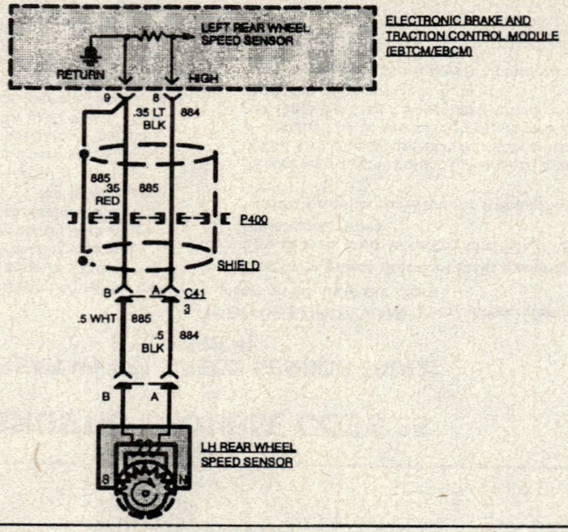

DIAGNOSTIC TROUBLE CODE 36
LEFT REAR WHEEL SPEED SENSOR FREQUENCY ERROR

TEST CONDITION(S):

1. Tested continuously.

FAILURE CONDITION(S):

1. While Driving: One wheel is detected at least 30 kph or 19 mph faster than the remaining wheels for an extended period of time.

   OR

   One wheel is at least 6 kph or 4 mph less than or greater than the remaining wheels for an extended period of time.

ACTION TAKEN:

1. Code set.
2. ABS switched 'OFF'.
3. 'ANTILOCK' indicator 'ON'.

TEST DESCRIPTION: Test numbers refer to circled numbers on the diagnostic chart.

1. Checks wiring and connections for problems. It is critical that this step is performed thoroughly, as wiring/connector problems are the most likely cause for this code.
2. Use the TECH 1 to monitor for vehicle electrical/electronic system 'noise' being picked up by the speed sensor circuits.
3. Use the TECH 1 to monitor the wheel speed sensors while the vehicle is operating. If the TECH 1 Auto–Trigger Snapshot mode triggers, the speed sensor or the wiring and/or connectors are going intermittent during the road test.
4. Checks for proper resistance in the sensor itself.
5. Replace rear wheel speed sensor as the likely cause of the triggering intermittent.
6. Use the TECH 1 to monitor the wheel speed sensors while the vehicle is operating. If the TECH 1 Auto–Snapshot mode triggers, the EBTCM/EBCM must be concluding there is a problem with wheel speed sensor frequencies when there is not.
7. Checks for a short to ground in the speed sensor input circuit wires. The circuit to be tested depends on which speed sensor triggered the TECH 1. The ground may be either a 'hard' short to ground, or a resistive short. A short with a resistance less that 2 megaohms, though not a hard short, can still cause Code 36 to set.

DIAGNOSTIC AIDS:

Be sure the speed sensor wiring is properly routed and retained. This will help prevent false signals due to electrical noise being picked up by the wiring.

It is very important that a thorough inspection of the wiring and connectors be performed. Failure to carefully and fully inspect wiring and connectors may result in misdiagnosis, causing part replacement with reappearance of the fault.

DIAGNOSTIC FAULT CHART DTC 36 — 1994 DeVILLE, CONCOURS, ELDORADO AND SEVILLE, CONTINUED

DIAGNOSTIC TROUBLE CODE 36
LEFT REAR WHEEL SPEED SENSOR FREQUENCY ERROR

(1) VERY THOROUGHLY INSPECT ALL WIRING AND CONNECTIONS FOR INTERMITTENTS, CHAFING, POOR TERMINAL CONTACT TENSION, CORROSION, AND MISROUTING, AT EBCM, ALL SENSORS, AND ALL HARNESS INTERCONNECTS. REPAIR ANY PROBLEMS CLEAR CODES, THEN ROAD TEST VEHICLE, REACHING AT LEAST 25 KPH (15 MPH).
- DOES CODE RESET?

YES → (2)

NO →
- NO TROUBLE FOUND; SEE 'DIAGNOSTIC AIDS' ON FACING PAGE.

(2)
- INSTALL TECH 1 AND MASS STORAGE CARTRIDGE; CLEAR CODES
- SET TECH 1 AUTO TRIGGER SNAPSHOT MODE
- START VEHICLE AND ALLOW TO IDLE FOR ONE MINUTE
- DOES TECH 1 TRIGGER?

NO → (3)

NO → (7)

(3)
- ROAD TEST VEHICLE REACHING AT LEAST 25 KPH (15 MPH).
- DOES TECH 1 TRIGGER?

YES →
- NOTE TECH 1, WHICH SENSOR TRIGGERED?

NO →
- NO TROUBLE FOUND; SEE DIAGNOSTIC AIDS ON FACING PAGE.

(7)
- NOTE TECH1, LR SENSOR TRIGGERED?
- IGNITION'OFF'
- DISCONNECT EBCM
- INSTALL J 39700 UNIVERSAL BREAKOUT BOX TO EBTCM/EBCM AND EBCM HARNESS CONNECTOR.
- MEASURE RESISTANCE TO GROUND AT TERMINAL 8 ON J 39700.
- IF RESISTANCE IS 5 OHMS OR LESS TO GROUND, REPAIR HARD SHORT; IF BETWEEN 5 OHMS AND 2 MEGAOHMS, REPAIR RESISTIVE SHORT IN CKT 885.

(4)
- IGNITION 'OFF'
- DISCONNECT SENSOR CONNECTOR
- MEASURE RESISTANCE BETWEEN SENSOR TERMINALS.
- IS RESISTANCE WITH IN RANGE 990-1210 OHMS?

NO → (5)

YES →
- REPLACE EBCM

(5)
- REPLACE HUB/BEARING/SENSOR ASSEMBLY

(6)
- CLEAR CODES, IGNITION 'OFF'
- ROAD TEST VEHICLE, REACHING AT LEAST 25 KPH (15 MPH)
- DOES CODE 36 RESET?

YES →
- REPLACE EBCM

NO →
- SYSTEM OK

WHEN ALL DIAGNOSIS AND REPAIRS ARE COMPLETED, CLEAR CODES AND VERIFY OPERATION

DIAGNOSTIC FAULT CHART DTC 37 — 1994 DeVILLE, CONCOURS, ELDORADO AND SEVILLE

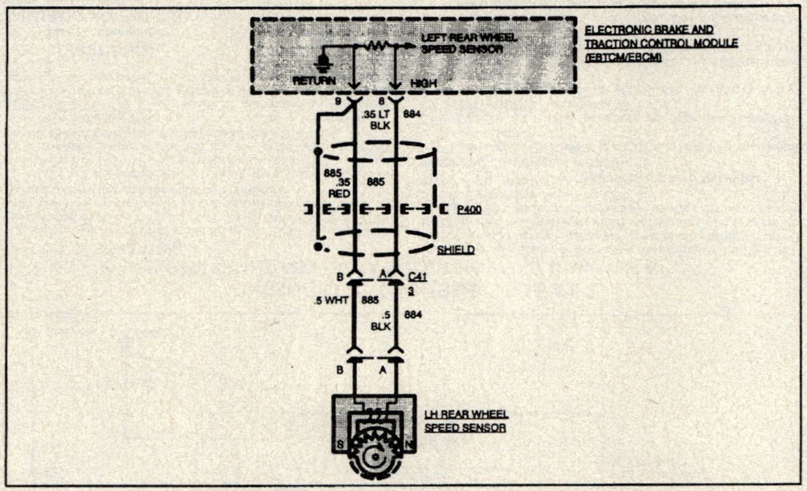

DIAGNOSTIC TROUBLE CODE 37

LEFT REAR WHEEL SPEED SENSOR CONTINUITY FAULT

CIRCUIT DESCRIPTION:

As each wheel turns, the wheel speed sensor for that wheel creates a small AC voltage as a toothed sensor ring, mounted in the front integral hub/bearing assembly, is passed by the stationary sensor. The sensor consists of a permanent magnet and coil and generates voltage by magnetic induction. The frequency of the wheel speed sensor signal is proportional to wheel speed.

The EBTCM/EBCM uses the wheel speed sensor signal to calculate vehicle reference speeds and individual speed, acceleration and slip values for each wheel. These values are used to determine when antilock control is required.

The EBTCM/EBCM will only set one wheel speed sensor fault code at a time, even if multiple faults exist. Detection is prioritized accordingly: LF, RR, LR, RF. Always check for additional stored codes after making repairs.

TEST CONDITION(S):

1. Tested continuously.

FAILURE CONDITION(S):

Continuity

1. While standing: Open circuit is seen while the vehicle is standing still with the ignition 'ON'

ACTION TAKEN:

1. Code set.
2. ABS/TCS switched 'OFF'.
3. 'ANTILOCK' indicator 'ON'.
4. 'TRACTION DISABLED' message displayed. (vehicles with TCS)

Code 37 is set if there is a short to voltage or an open in CKT 885 or 884, or faulty speed sensor. The testing for this fault occurs with the ignition 'ON' and the vehicle at rest.

NOTICE: A Code 35 may set if CKT 884 or 885 lose continuity and the WSS signal is lost while driving. A Code 37 should accompany this code on the subsequent ignition cycle. If both Code 35 and Code 37 are set concurrently, diagnose Code 37 first.

TEST DESCRIPTION: Test numbers refer to circled numbers on the diagnostic chart.

1. Checks for proper resistance in the sensor itself.
2. Checks for a short to battery in either of the speed sensor circuit wires. There should be 5V on CKT 884. If there is, CKT 884 has no hard opens or shorts. The condition may be intermittent.
3. Checks for an open in both the speed sensor circuit wires.
4. Manipulates wiring and connectors, trying to induce an intermittent fault not currently present.
5. This test checks if the code resets on key-up. If so, since Tests 1–4 have validated that the circuitry and components are good, the Speed Sensor may be causing the fault.
6. Checks if the Code 35 sets during a road test.
7. Code 37 sets when the vehicle is at rest: Code 35 sets with a problem in the speed sensor circuitry with the vehicle in motion. If problems are still present, Code 35 would set during the road test, not a 37. If Code 35 sets at this point, Code 35 diagnostics should be used.
8. The wheel speed sensor may be causing an intermittent fault. If Code 37 resets after sensor replacement, the EBTCM/EBCM must be concluding there is a problem present in the speed sensor circuit when there is not.

DIAGNOSTIC AIDS:

Be sure the speed sensor wiring is properly routed and retained. This will help prevent false signals due to electrical noise being picked up by the wiring.

It is very important that a thorough inspection of the wiring and connectors be performed. Failure to carefully and fully inspect wiring and connectors may result in misdiagnosis, causing part replacement with reappearance of the fault.

DIAGNOSTIC FAULT CHART DTC 37 — 1994 DeVILLE, CONCOURS, ELDORADO AND SEVILLE, CONTINUED

DIAGNOSTIC TROUBLE CODE 37

LR WHEEL SPEED SENSOR CONTINUITY FAULT

ENSURE SENSOR IS NOT VISIBLY DAMAGED; IF IT IS DAMAGED, REPLACE THE HUB/BEARING/SPEED SENSOR ASSEMBLY. CLEAR CODES, AN TURN IGNITION 'ON' TO SEE IF CODE RESETS.

(1)
- IGNITION 'OFF'
- DISCONNECT SENSOR CONNECTOR
- USING DVM AND ADAPTER FROM J 35616–A, CHECK SENSOR RESISTANCE BY PROBING SENSOR CONNECTOR TERMINALS
- IS IT 900 – 1300 OHMS AT 25°C (77°F)?

NO:
- REPLACE HUB/BEARING/SENSOR ASSEMBLY

YES:

(2)
- CONNECT DVM BETWEEN SENSOR HARNESS CONNECTOR TERMINAL 'A' AND GROUND (USED DC VOLTAGE SETTING)
- IGNITION 'ON'
- NOTE VOLTAGE: REPEAT BETWEEN TERMINAL 'B' AND GROUND
- IS IT AROUND SYSTEM VOLTAGE ON EITHER TERMINAL 'A' OR 'B'?

YES:
- REPAIR SHORT TO SYSTEM VOLTAGE IN CKT 884 OR 885
- REFER TO EBTCM/EBCM–SWITCHED FAULTS IN THIS SECTION

NO:

(3)
- IGNITION 'OFF'
- DISCONNECT EBTCM/EBCM
- INSTALL J 39700 UNIVERSAL BREAKOUT BOX TO EBTCM/EBCM HARNESS CONNECTOR ONLY
- MEASURE RESISTANCE BETWEEN PIN 45 ON J 39700 TERMINAL 'A' OF SENSOR CONNECTOR ON HARNESS. PERFORM SAME MEASUREMENT BETWEEN TERMINAL 44 AND TERMINAL 'B'
- IS EITHER MEASUREMENT GREATER THAN 5 OHMS?

YES:
- REPAIR OPEN IN CKT 884 AND/OR 885

NO:

(4)
- REMOVE J 39700; RECONNECT EBTCM/EBCM
- CLEAR CODES; IGNITION 'OFF'
- DISCONNECT TECH 1, IF CONNECTED
- IGNITION 'ON'
- WIGGLE ALL WIRING AND CONNECTORS TO TRY TO DUPLICATE AN INTERMITTENT FAULT
- DOES CODE 37 RESET?

YES:
- REPAIR WIRING AND/OR CONNECTIONS

NO:

(5)
- CLEAR CODES; IGNITION 'OFF'
- DISCONNECT TECH 1, IF CONNECTED
- IGNITION 'ON'
- DOES CODE 37 RESET?

YES:

(8)
- REPLACE HUB/BEARING/SENSOR ASSEMBLY
- CLEAR CODES; IGNITION 'OFF'
- DISCONNECT TECH 1, IF CONNECTED
- IGNITION 'ON'
- DOES CODE 37 RESET?

YES:
- REPLACE EBTCM/EBCM

NO:
- SYSTEM OK

NO:

(6)
- ROAD TEST VEHICLE, REACHING AT LEAST 25 KPH (15 MPH)
- DOES CODE 35 SET?

YES:

(7) GO TO CODE 35 DIAGNOSTICS

NO:
- NO TROUBLE FOUND, SEE 'DIAGNOSTIC AIDS' ON FACING PAGE

WHEN ALL DIAGNOSIS AND REPAIRS ARE COMPLETED, CLEAR CODES AND VERIFY OPERATION

DIAGNOSTIC FAULT CHART DTC 41 — 1994 DeVILLE, CONCOURS, ELDORADO AND SEVILLE

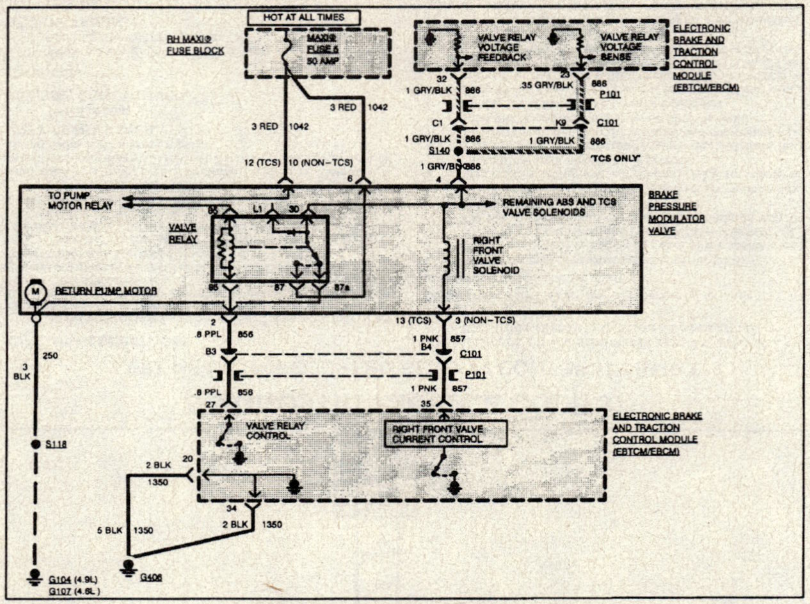

DIAGNOSTIC TROUBLE CODE 41
RIGHT FRONT ABS VALVE SOLENOID FAULT (Page 1 of 4)

CIRCUIT DESCRIPTION:

The right front valve solenoid in the BPM Valve may be commanded to three different positions by the EBTCM/EBCM. Valve position is determined by the amount of current which is allowed to flow through the valve solenoid coil.

The valve solenoid circuits receive power through the valve relay on the BPM Valve. The valve relay is engaged at KEY–ON and remains engaged throughout the ignition cycle. Valve solenoid circuits should have battery voltage available at all times.

When the key is turned to RUN, power is supplied to the EBTCM/EBCM. The EBTCM/EBCM will complete an internal self–check before providing ground to the valve relay. When the valve relay is energized, battery power is supplied to the valve solenoids.

TEST CONDITION(S):

1. Tested continuously.

FAILURE CONDITION(S):

While standing or driving:

1. Valve solenoid cannot be activated.

OR

2. Valve solenoid is activated in a defective way.

ACTION TAKEN:

1. Code set.
2. ABS/TCS switched 'OFF'.
3. 'ANTILOCK' indicator 'ON'.
4. 'TRACTION DISABLED' message displayed. (vehicles with TCS)

Code 41 will set when the expected position of the right front valve solenoid does not match the commanded position from the EBTCM/EBCM. Conditions which could cause Code 41 to set are: damage to the right front valve solenoid; an open circuit; or an open circuit to ground or battery on CKT 857.

TEST DESCRIPTION: Test numbers refer to circled numbers on the diagnostic chart.

1. Damage to CKT 886 may result in multiple valve solenoid codes setting.
2. Checks CKT 886 for opens or high resistance from EBTCM/EBCM terminal '23' to '32'.
3. Determines which branch of CKT 886 has an open or high resistance.
4. Checks the integrity of CKTs 886 and 857, the RF ABS Valve Solenoid circuitry internal to the BPM Modulator Valve, and the RF ABS Valve Solenoid coil.
5. Determines whether a problem found in Step 4 is due to an open in CKT 857, 886, or BPM Modulator Valve.
6. Determines whether the problem found in Step 4 is due to an open in CKT 857 or 886.

DIAGNOSTIC AIDS:

All tests using J 39700 Universal Breakout Box terminal '20' are using the terminal as ground. These tests, of course, assume that the ground CKT 1350 terminal '20' is good.

If Codes 41 and 45 are both set, the fault is likely to be an open or high resistance on CKT 886.

If Codes 41 and 55 are both set, the fault is likely to be a short to B+ on CKT 1290.

DIAGNOSTIC FAULT CHART DTC 41 — 1994 DeVILLE, CONCOURS, ELDORADO AND SEVILLE, CONTINUED

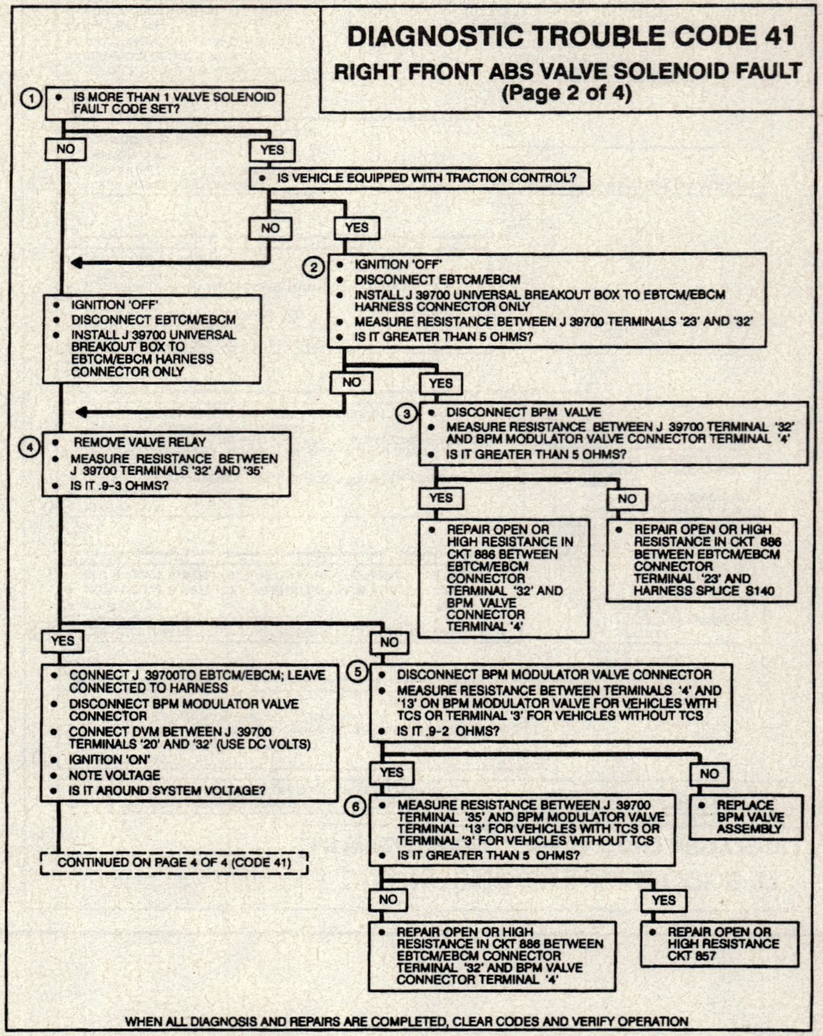

DIAGNOSTIC FAULT CHART DTC 41 — 1994 DeVILLE, CONCOURS, ELDORADO AND SEVILLE, CONTINUED

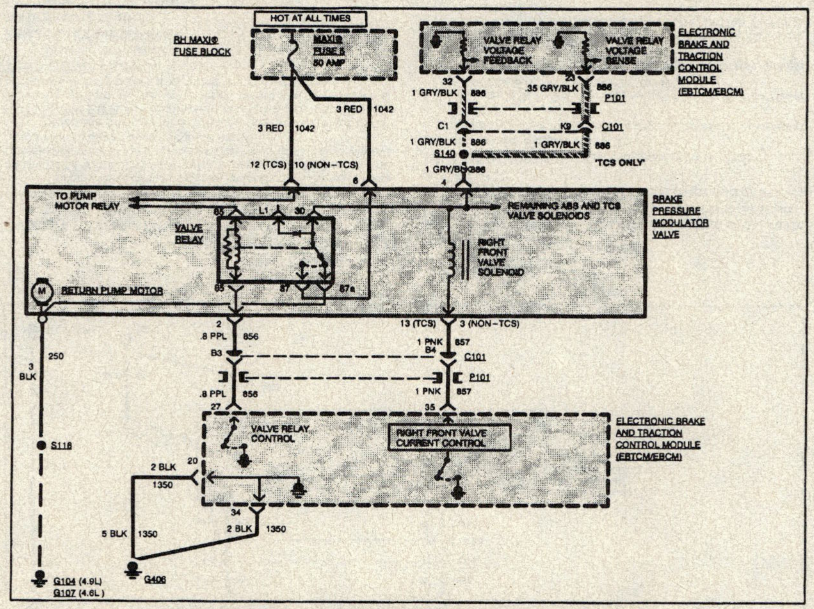

DIAGNOSTIC TROUBLE CODE 41
RIGHT FRONT ABS VALVE SOLENOID FAULT
(Page 3 of 4)

CIRCUIT DESCRIPTION:

The right front valve solenoid in the BPM Valve may be commanded to three different positions by the EBTCM/EBCM. Valve position is determined by the amount of current which is allowed to flow through the valve solenoid coil.

The valve solenoid circuits receive power through the valve relay on the BPM Valve. The valve relay is engaged at KEY–ON and remains engaged throughout the ignition cycle. Valve solenoid circuits should have battery voltage available at all times.

When the key is turned to RUN, power is supplied to the EBTCM/EBCM. The EBTCM/EBCM will complete an internal self–check before providing ground to the valve relay. When the valve relay is energized, battery power is supplied to the valve solenoids.

TEST DESCRIPTION: Test numbers refer to circled numbers on the diagnostic chart.

8. Checks for a short to battery in CKT 857 using EBTCM/EBCM terminal '20' as ground.
9. Checks for a short to ground in CKT 886 using EBTCM/EBCM terminal '20' as ground.
10. Checks for a short to ground in CKT 857 using EBTCM/EBCM terminal '20' as ground.
11. Checking wiring and connectors for intermittents.
12. Use the TECH 1 to exercise the RF ABS Valve Solenoid and check it for proper operation.
13. Determines whether the code was set by an intermittent condition or an EBTCM/EBCM fault.

DIAGNOSTIC AIDS:

All tests using J 39700 Universal Breakout Box terminal '20' are using the terminal as ground. These tests, of course, assume that the ground CKT 1350 terminal '20' is good.

If Codes 41 and 45 are both set, the fault is likely to be an open or high resistance on CKT 886.

If Codes 41 and 55 are both set, the fault is likely to be a short to B+ on CKT 1290.

DIAGNOSTIC FAULT CHART DTC 41 — 1994 DeVILLE, CONCOURS, ELDORADO AND SEVILLE, CONTINUED

DIAGNOSTIC TROUBLE CODE 41
RIGHT FRONT ABS SOLENOID VALVE FAULT
(Page 4 of 4)

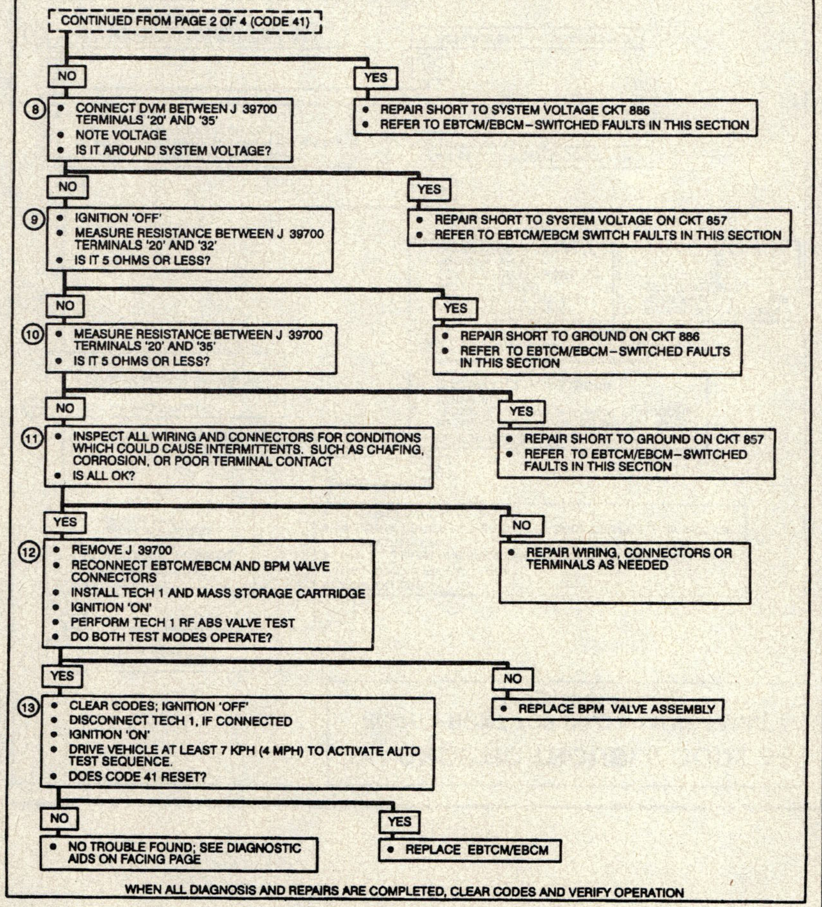

DIAGNOSTIC FAULT CHART DTC 44 — 1994 DeVILLE, CONCOURS, ELDORADO AND SEVILLE

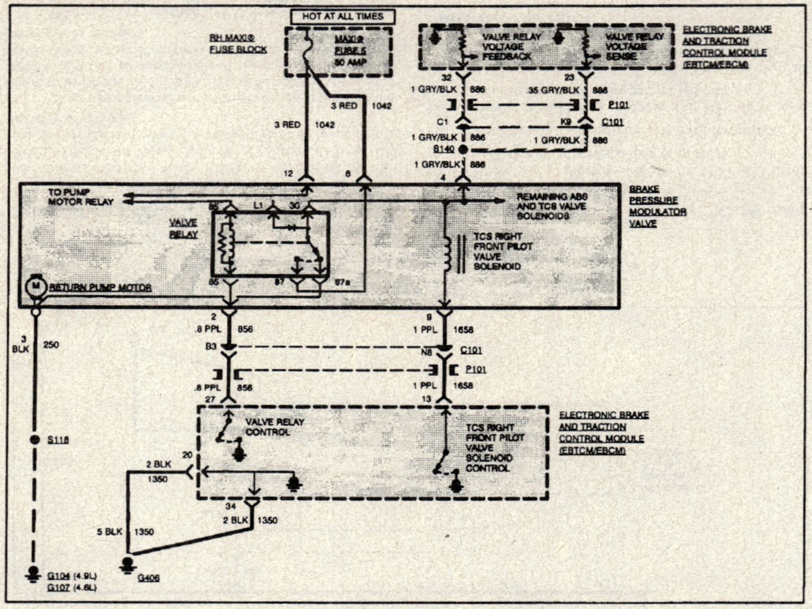

DIAGNOSTIC TROUBLE CODE 44
RIGHT FRONT TCS PILOT VALVE FAULT (Page 1 of 4)

CIRCUIT DESCRIPTION:

The right front TCS pilot valve in the BPM VALVE may be commanded to close, isolating the master cylinder.

The pilot valve circuits receive power through the valve relay on the BPM Valve. The valve relay is engaged at KEY–ON and remains engaged throughout the ignition cycle. Valve circuits should have battery voltage available at all times.

When the key is turned to RUN, power is supplied to the EBTCM. The EBTCM will complete an internal self–check before providing ground to the valve relay. When the valve relay is energized, battery power is supplied to the pilot valves.

TEST CONDITION(S):

1. Tested continuously.

FAILURE CONDITION(S):

While standing or driving:

1. Valve cannot be activated.

OR

2. Valve is activated in a defective way.

ACTION TAKEN:

1. Code set.
2. ABS/TCS switched 'OFF'.
3. 'ANTILOCK' indicator 'ON'.
4. 'TRACTION DISABLED' message displayed.

Code 44 will set when the expected position of the right front TCS pilot valve does not match the commanded position from the EBTCM. Conditions which could cause Code 44 to set are; damage to the right front TCS pilot valve, an open circuit or a short circuit to ground or battery on CKT 1658.

TEST DESCRIPTION: Test numbers refer to circled numbers on the diagnostic chart.

1. Damage to CKT 886 may result in multiple pilot valve codes setting.
2. Checks CKT 886 for opens or high resistance from EBTCM/EBCM terminal '23' to '32'.
3. Determines which branch of CKT 886 has an open or high resistance.
4. Checks the integrity of CKTs 1658 and 886, the TCS Pilot Valve circuitry internal to the BPM Valve, and the TCS Pilot Valve coil.
5. Determines whether a problem found in Step 4 is due to an open in CKTs 886, 1658 or a faulty BPM Valve.
6. Determines whether the problem found in Step 4 is due to an open in CKT 886 or 1658.

DIAGNOSTIC AIDS:

All tests using J 39700 Universal Breakout Box terminal '20' are using the terminal as ground. These tests, of course, assume that the ground CKT 1350 terminal '20' is good.

If Codes 44 and 48 are both set, the fault is likely to be an open or short in CKT 886.

DIAGNOSTIC FAULT CHART DTC 44 — 1994 DeVILLE, CONCOURS, ELDORADO AND SEVILLE, CONTINUED

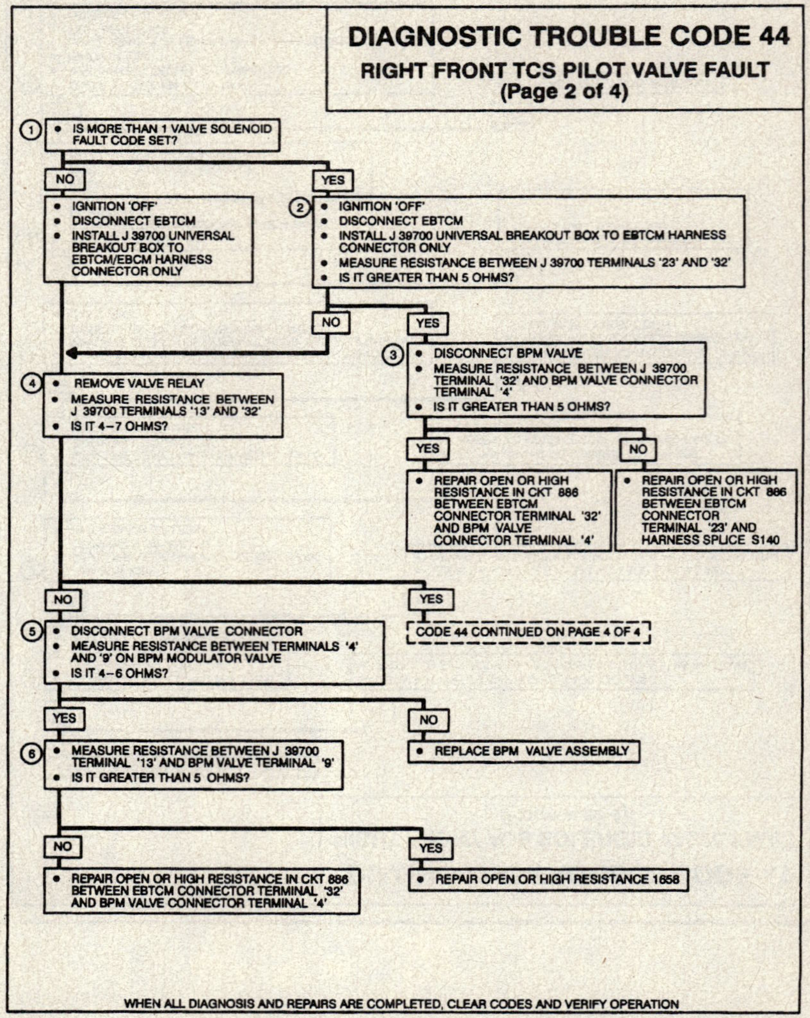

DIAGNOSTIC FAULT CHART DTC 44 — 1994 DeVILLE, CONCOURS, ELDORADO AND SEVILLE, CONTINUED

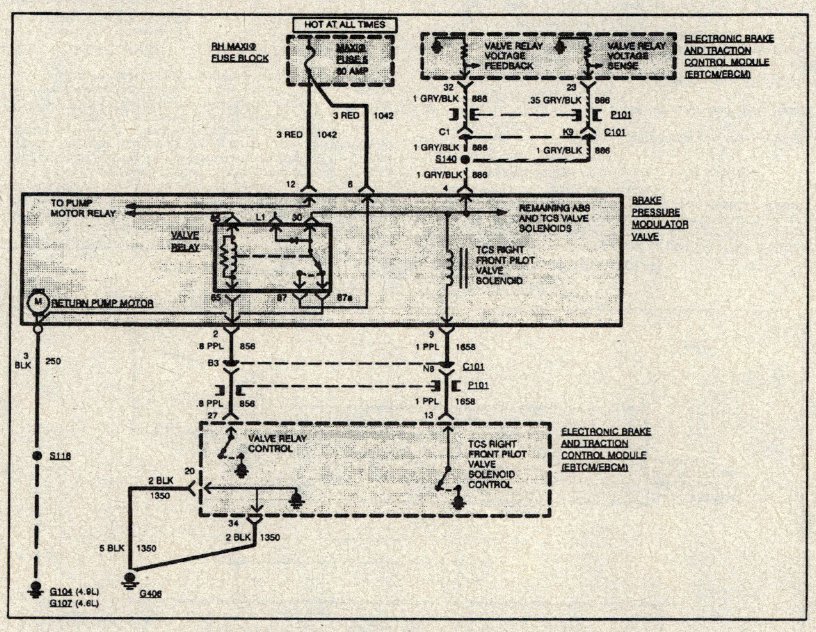

DIAGNOSTIC TROUBLE CODE 44
RIGHT FRONT TCS PILOT VALVE FAULT
(Page 3 of 4)

CIRCUIT DESCRIPTION:

The right front TCS pilot valve in the BPM Valve may be commanded to close, isolating the master cylinder.

The pilot valve circuits receive power through the valve relay on the BPM Valve. The valve relay is engaged at KEY–ON and remains engaged throughout the ignition cycle. Valve circuits should have battery voltage available at all times.

When the key is turned to RUN, power is supplied to the EBTCM. The EBTCM will complete an internal self–check before providing ground to the valve relay. When the valve relay is energized, battery power is supplied to the pilot valves.

TEST DESCRIPTION: Test numbers refer to circled numbers on the diagnostic chart.

7. Checks for a short to battery in CKT 886 using EBTCM terminal '20' as ground.
8. Checks for a short to battery in CKT 1658, using EBTCM terminal '20' as ground.
9. Checks for a short to ground in CKT 886 using EBTCM terminal '20' as ground.
10. Checks for a short to ground in CKT 1658, using EBTCM terminal '20' as ground.
11. Checks for connector or wiring problems which could cause intermittents.
12. Uses the TECH 1 to exercise the TCS pilot valve and check it for proper operation.
13. Determines whether the code was set by an intermittent conditions or an EBTCM fault.

DIAGNOSTIC AIDS:

All tests using J 39700 Universal Breakout Box terminal '20' are using the terminal as ground. These tests, of course, assume that the ground CKT 1350 terminal '20' is good.

If Codes 44 and 48 are both set, the fault is likely to be an open or short in CKT 886.

DIAGNOSTIC FAULT CHART DTC 44 — 1994 DeVILLE, CONCOURS, ELDORADO AND SEVILLE, CONTINUED

DIAGNOSTIC TROUBLE CODE 44
RIGHT FRONT TCS PILOT VALVE FAULT
(Page 4 of 4)

CODE 44 CONTINUED FROM PAGE 2 OF 4

(7)
- CONNECT J 39700TO EBTCM LEAVE CONNECTED TO HARNESS
- DISCONNECT BPM VALVE CONNECTOR
- CONNECT DVM BETWEEN J 39700 TERMINALS '20' AND '32' (USE DC VOLTS)
- IGNITION 'ON'
- NOTE VOLTAGE
- IS IT AROUND SYSTEM VOLTAGE?

YES:
- REPAIR SHORT TO SYSTEM VOLTAGE CKT 886
- REFER TO EBTCM–SWITCHED FAULTS IN THIS SECTION

NO:

(8)
- CONNECT DVM BETWEEN J 39700 TERMINALS '13' AND '20'
- NOTE VOLTAGE
- IS IT AROUND SYSTEM VOLTAGE?

YES:
- REPAIR SHORT TO SYSTEM VOLTAGE ON CKT 1658
- REFER TO EBTCM SWITCH FAULTS IN THIS SECTION

NO:

(9)
- IGNITION 'OFF'
- MEASURE RESISTANCE BETWEEN J 39700 TERMINALS '20' AND '32'
- IS IT 5 OHMS OR LESS?

YES:
- REPAIR SHORT TO GROUND ON CKT 886
- REFER TO EBTCM–SWITCHED FAULTS IN THIS SECTION

NO:

(10)
- MEASURE RESISTANCE BETWEEN J 39700 TERMINALS '13' AND '20'
- IS IT 5 OHMS OR LESS?

YES:
- REPAIR SHORT TO GROUND ON CKT 1658
- REFER TO EBTCM SWITCHED FAULTS IN THIS SECTION

NO:

(11)
- INSPECT ALL WIRING AND CONNECTORS FOR CONDITIONS WHICH COULD CAUSE INTERMITTENTS. SUCH AS CHAFING, CORROSION, OR POOR TERMINAL CONTACT
- IS ALL OK?

NO:
- REPAIR WIRING, CONNECTORS OR TERMINALS AS NEEDED

YES:

(12)
- REMOVE J 39700
- RECONNECT EBTCM AND BPM VALVE CONNECTORS
- INSTALL TECH 1 AND MASS STORAGE CARTRIDGE
- IGNITION 'ON'
- PERFORM TECH 1 RF TCS VALVE TEST
- DOES TEST MODE OPERATE?

NO:
- REPLACE BPM VALVE ASSEMBLY

YES:

(13)
- CLEAR CODES, IGNITION 'OFF'
- DISCONNECT TECH 1, IF CONNECTED
- IGNITION 'ON'
- DRIVE VEHICLE AT LEAST 7 KPH (4 MPH) TO ACTIVATE AUTO TEST SEQUENCE
- DOES CODE 44 RESET?

NO:
- NO TROUBLE FOUND; SEE DIAGNOSTIC AIDS ON FACING PAGE

YES:
- REPLACE EBTCM

WHEN ALL DIAGNOSIS AND REPAIRS ARE COMPLETED, CLEAR CODES AND VERIFY OPERATION

DIAGNOSTIC FAULT CHART DTC 45 — 1994 DeVILLE, CONCOURS, ELDORADO AND SEVILLE

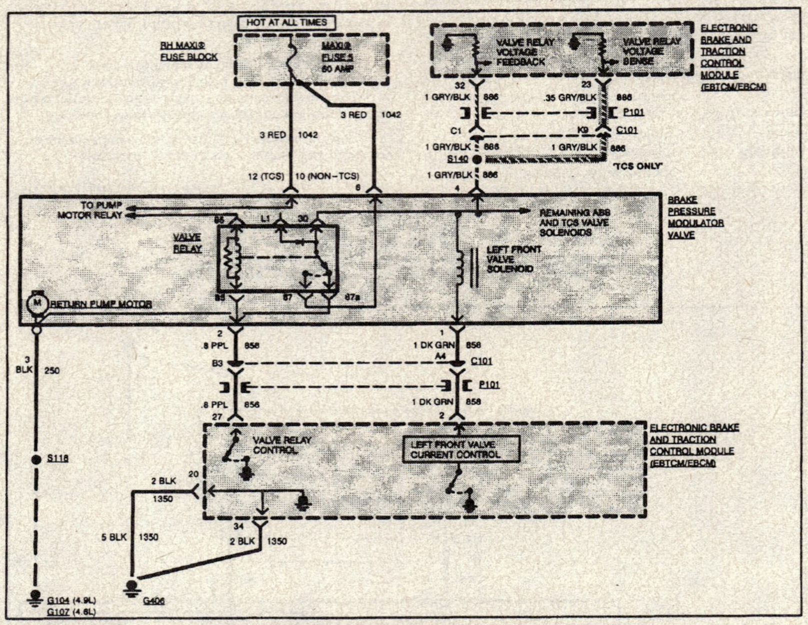

DIAGNOSTIC TROUBLE CODE 45
LEFT FRONT ABS VALVE SOLENOID FAULT (Page 1 of 4)

CIRCUIT DESCRIPTION:

The left front valve solenoid in the BPM Valve may be commanded to three different positions by the EBTCM/EBCM. Valve position is determined by the amount of current which is allowed to flow through the valve solenoid coil.

The valve solenoid circuits receive power through the valve relay on the BPM Valve. The valve relay is engaged at KEY–ON and remains engaged throughout the ignition cycle. Valve solenoid circuits should have battery voltage available at all times.

When the key is turned to RUN, power is supplied to the EBTCM/EBCM. The EBTCM/EBCM will complete an internal self–check before providing ground to the valve relay. When the valve relay is energized, battery power is supplied to the valve solenoids.

TEST CONDITION(S):

1. Tested continuously.

FAILURE CONDITION(S):

While standing or driving:

1. Valve solenoid cannot be activated.

OR

2. Valve solenoid is activated in a defective way.

ACTION TAKEN:

1. Code set.
2. ABS/TCS switched 'OFF'.
3. 'ANTILOCK' indicator 'ON'.
4. 'TRACTION DISABLED' message displayed. (vehicles with TCS)

Code 45 will set when the expected position of the left front solenoid valve does not match the commanded position from the EBTCM/EBCM. Conditions which could cause Code 45 to set are: damage to the left front valve solenoid; an open circuit; or a short circuit to ground or battery on CKT 858.

TEST DESCRIPTION: Test numbers refer to circled numbers on the diagnostic chart.

1. Damage to CKT 886 may result in multiple valve solenoid codes setting.
2. Checks CKT 886 for opens or high resistance from EBTCM/EBCM terminal '23' to '32'.
3. Determines which branch of CKT 886 has an open or high resistance.
4. Checks the integrity of CKTs 826 and 886, the LF ABS Valve Solenoid circuitry internal to the BPM Modulator Valve, and the LF ABS Valve Solenoid coil.
5. Determines whether a problem found in Step 4 is due to an open in CKT 858, 886 or a faulty BPM Modulator Valve.
6. Determines whether the problem found in Step 4 is due to an open in CKT 858 or 886.

DIAGNOSTIC AIDS:

All tests using J 39700 Universal Breakout Box terminal '20' are using the terminal as ground. These tests, of course, assume that the ground CKT 1350 terminal '20' is good.

If Codes 41 and 45 are both set, the fault is likely to be an open or high resistance on CKT 886.

DIAGNOSTIC FAULT CHART DTC 45 — 1994 DeVILLE, CONCOURS, ELDORADO AND SEVILLE, CONTINUED

DIAGNOSTIC TROUBLE CODE 45
LEFT FRONT ABS VALVE SOLENOID FAULT (Page 2 of 4)

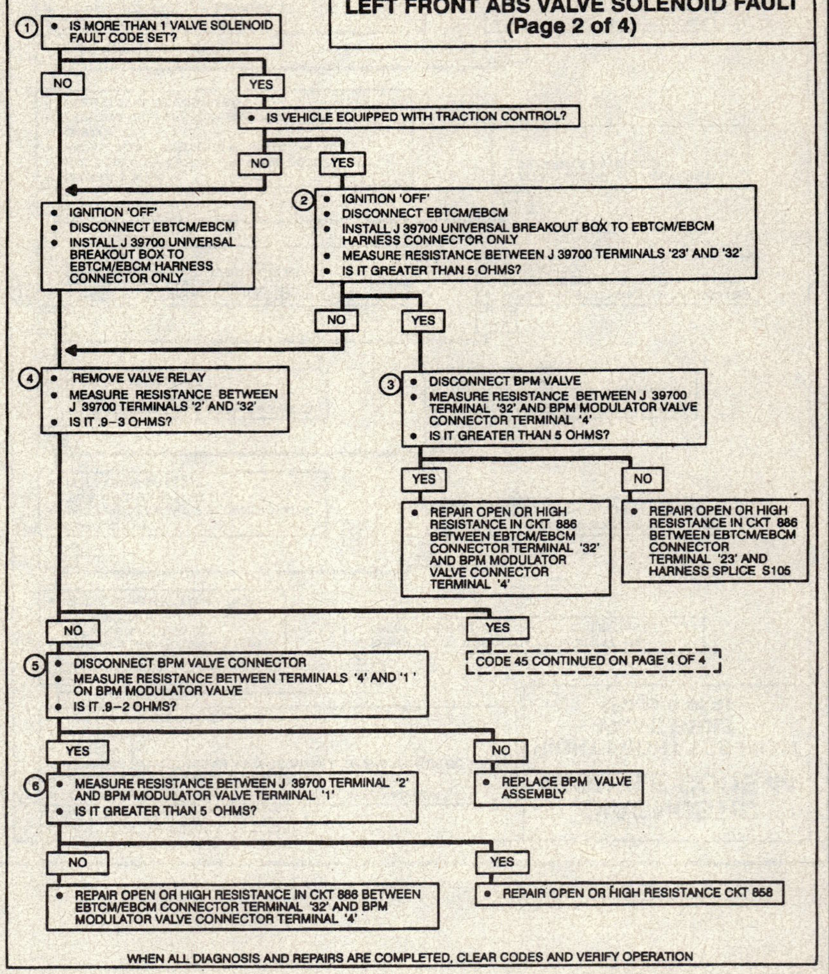

DIAGNOSTIC FAULT CHART DTC 45 — 1994 DeVILLE, CONCOURS, ELDORADO AND SEVILLE, CONTINUED

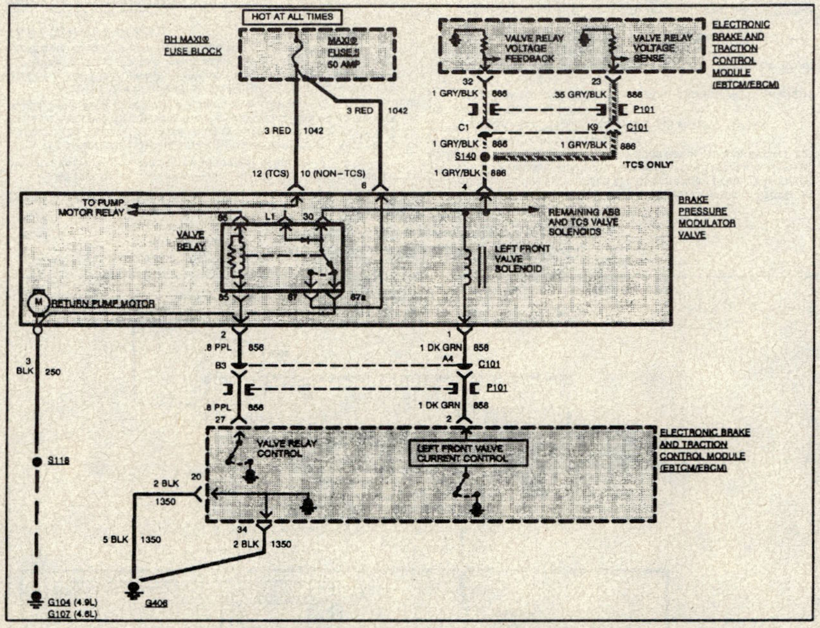

DIAGNOSTIC TROUBLE CODE 45

LEFT ABS FRONT VALVE SOLENOID FAULT (Page 3 of 4)

CIRCUIT DESCRIPTION:

The left front valve solenoid in the BPM Valve may be commanded to three different positions by the EBTCM/EBCM. Valve position is determined by the amount of current which is allowed to flow through the solenoid valve coil.

The valve solenoid circuits receive power through the valve relay on the BPM Valve. The valve relay is engaged at KEY-ON and remains engaged throughout the ignition cycle. Valve solenoid circuits should have battery voltage available at all times.

When the key is turned to RUN, power is supplied to the EBTCM/EBCM. The EBTCM/EBCM will complete an internal self-check before providing ground to the valve relay. When the valve relay is energized, battery power is supplied to the valve solenoids.

TEST DESCRIPTION: Test numbers refer to circled numbers on the diagnostic chart.

7. Checks for a short to battery in CKT 886 using EBTCM/EBCM terminal '20' as ground.
8. Checks for a short to battery in CKT 858, using EBTCM/EBCM terminal '20' as ground.
9. Checks for a short to ground in CKT 886 using EBTCM/EBCM terminal '20' as ground.
10. Checks for a short to ground in CKT 858, using EBTCM/EBCM terminal '20' as ground.
11. Checks wiring and connectors for intermittents.
12. Uses the TECH 1 to exercise the LF ABS valve solenoid and check it for proper operation.
13. Determines whether the code was set by an intermittent condition or an EBTCM/EBCM fault.

DIAGNOSTIC AIDS:

All tests using J 39700 Universal Breakout Box terminal '20' are using the terminal as ground. These tests, of course, assume that the ground CKT 1350 terminal '20' is good.

If Codes 41 and 45 are both set, the fault is likely to be an open or high resistance on CKT 886.

DIAGNOSTIC FAULT CHART DTC 45 — 1994 DeVILLE, CONCOURS, ELDORADO AND SEVILLE, CONTINUED

DIAGNOSTIC TROUBLE CODE 45
LEFT FRONT ABS VALVE SOLENOID FAULT (Page 4 of 4)

CODE 45 CONTINUED ON PAGE 4 OF 4

(7)
- CONNECT J 39700TO EBTCM/EBCM; LEAVE CONNECTED TO HARNESS
- DISCONNECT BPM MODULATOR VALVE CONNECTOR
- CONNECT DVM BETWEEN J 39700 TERMINALS '20' AND '32' (USE DC VOLTS)
- IGNITION 'ON'
- NOTE VOLTAGE
- IS IT AROUND SYSTEM VOLTAGE?

YES:
- REPAIR SHORT TO SYSTEM VOLTAGE CKT 886
- REFER TO EBTCM/EBCM-SWITCHED FAULTS IN THIS SECTION

NO → (8)
- CONNECT DVM BETWEEN J 39700 TERMINALS '2' AND '20'
- NOTE VOLTAGE
- IS IT AROUND SYSTEM VOLTAGE?

YES:
- REPAIR SHORT TO SYSTEM VOLTAGE ON CKT 858
- REFER TO 'EBTCM/EBCM SWITCH FAULTS' IN THIS SECTION

NO → (9)
- IGNITION 'OFF'
- MEASURE RESISTANCE BETWEEN J 39700 TERMINALS '20' AND '32'
- IS IT 5 OHMS OR LESS?

YES:
- REPAIR SHORT TO GROUND ON CKT 886
- REFER TO EBTCM/EBCM-SWITCHED FAULTS IN THIS SECTION

NO → (10)
- MEASURE RESISTANCE BETWEEN J 39700 TERMINALS '2' AND '20'
- IS IT 5 OHMS OR LESS?

YES:
- REPAIR SHORT TO GROUND ON CKT 858
- REFER TO EBTCM/EBCM-SWITCHED FAULTS IN THIS SECTION

NO → (11)
- INSPECT ALL WIRING AND CONNECTORS FOR CONDITIONS WHICH COULD CAUSE INTERMITTENTS. SUCH AS CHAFING, CORROSION, OR POOR TERMINAL CONTACT
- IS ALL OK?

NO:
- REPAIR WIRING, CONNECTORS OR TERMINALS AS NEEDED

YES → (12)
- REMOVE J 39700
- RECONNECT EBTCM/EBCM AND BPM MODULATOR VALVE CONNECTORS
- INSTALL TECH 1 AND MASS STORAGE CARTRIDGE
- IGNITION 'ON'
- PERFORM TECH 1 LF ABS VALVE TEST
- DOES TEST MODE OPERATE?

NO:
- REPLACE BPM MODULATOR VALVE ASSEMBLY

YES → (13)
- CLEAR CODES, IGNITION 'OFF'
- DISCONNECT TECH 1, IF CONNECTED
- IGNITION 'ON'
- DRIVE VEHICLE AT LEAST 7 KPH (4 MPH) TO ACTIVATE AUTO TEST SEQUENCE.
- DOES CODE 45 RESET?

NO:
- NO TROUBLE FOUND; SEE 'DIAGNOSTIC AIDS' ON FACING PAGE

YES:
- REPLACE EBTCM/EBCM

WHEN ALL DIAGNOSIS AND REPAIRS ARE COMPLETED, CLEAR CODES AND VERIFY OPERATION

DIAGNOSTIC FAULT CHART DTC 48 — 1994 DeVILLE, CONCOURS, ELDORADO AND SEVILLE

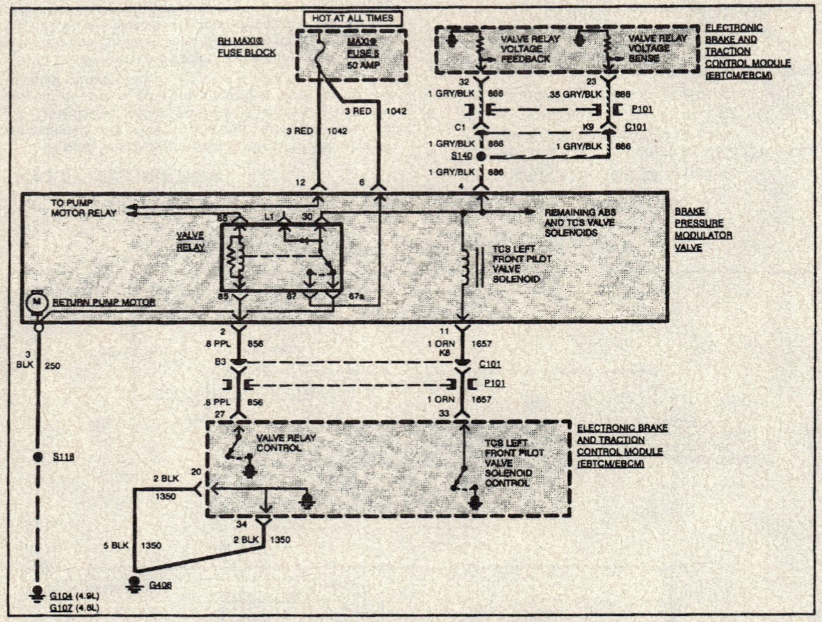

DIAGNOSTIC TROUBLE CODE 48
LEFT FRONT TCS PILOT VALVE FAULT
(Page 1 of 4)

CIRCUIT DESCRIPTION:

The left front TCS pilot valve in the BPM Valve may be commanded to close, isolating the master cylinder.

The pilot valve circuits receive power through the valve relay on the BPM Valve. The valve relay is engaged at KEY – ON and remains engaged throughout the ignition cycle. Pilot valve circuits should have battery voltage available at all times.

When the key is turned to RUN, power is supplied to the EBTCM. The EBTCM will complete an internal self – check before providing ground to the valve relay. When the valve relay is energized, battery power is supplied to the pilot valves.

TEST CONDITION(S):

1. Tested continuously.

FAILURE CONDITION(S):

While standing or driving:

1. Valve cannot be activated.

OR

2. Valve is activated in a defective way.

ACTION TAKEN:

1. Code set.
2. ABS/TCS switched 'OFF'.
3. 'ANTILOCK' indicator 'ON'.
4. 'TRACTION DISABLED' message displayed.

Code 48 will set when the expected position of the right front TCS pilot valve does not match the commanded position from the EBTCM. Conditions which could cause Code 48 to set are: damage to the right front TCS pilot valve; an open circuit; or a short circuit to ground or battery on CKT 1657.

TEST DESCRIPTION: Test numbers refer to circled numbers on the diagnostic chart.

1. Damage to CKT 886 may result in multiple valve solenoid codes setting.
2. Checks CKT 886 for opens or high resistance from EBTCM/EBCM terminal '23' to '32'.
3. Determines which branch of CKT 886 has an open or high resistance.
4. Checks the integrity of CKTs 1657 and 886, the TCS Pilot Valve circuitry internal to the BPM Valve, and the TCS Pilot Valve Solenoid coil.
5. Determines whether a problem found in Step 4 is due to an open in CKTs 886, 1657 or a faulty BPM Modulator Valve.
6. Determines whether the problem found in Step 4 is due to an open in CKT 886 or 1657.

DIAGNOSTIC AIDS:

All tests using J 39700 Universal Breakout Box terminal '20' are using the terminal as ground. These tests, of course, assume that the ground CKT 1350 terminal '20' is good.

If Codes 44 and 48 are both set, the fault is likely to be an open or short in CKT 886.

DIAGNOSTIC FAULT CHART DTC 48 — 1994 DeVILLE, CONCOURS, ELDORADO AND SEVILLE, CONTINUED

DIAGNOSTIC TROUBLE CODE 48
LEFT FRONT TCS PILOT VALVE FAULT
(Page 2 of 4)

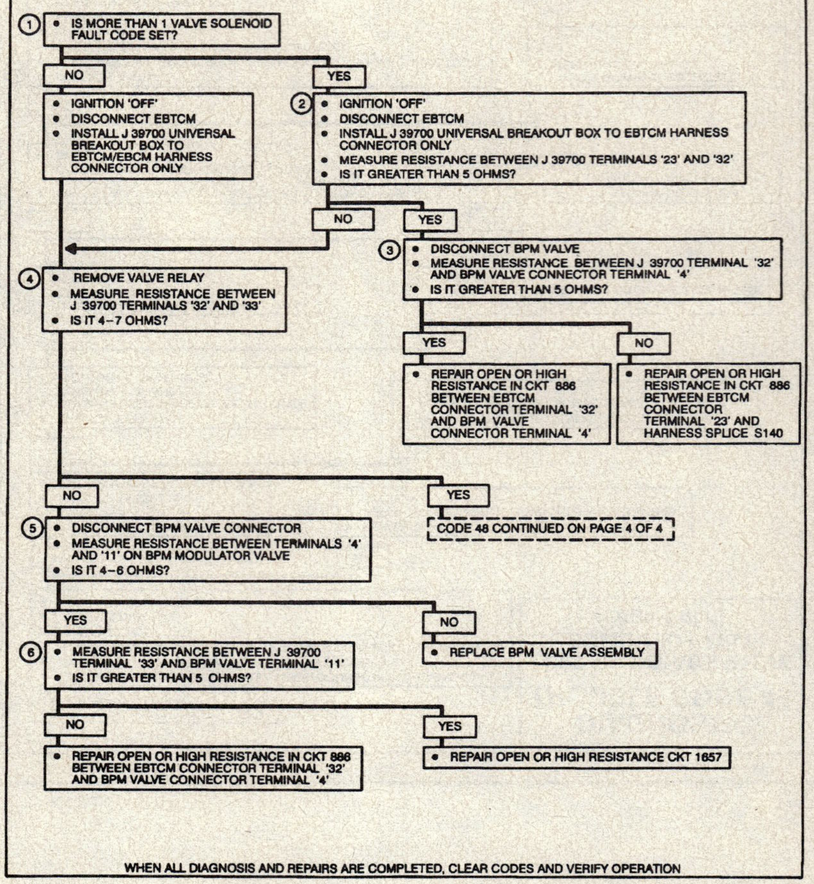

DIAGNOSTIC FAULT CHART DTC 48 — 1994 DeVILLE, CONCOURS, ELDORADO AND SEVILLE, CONTINUED

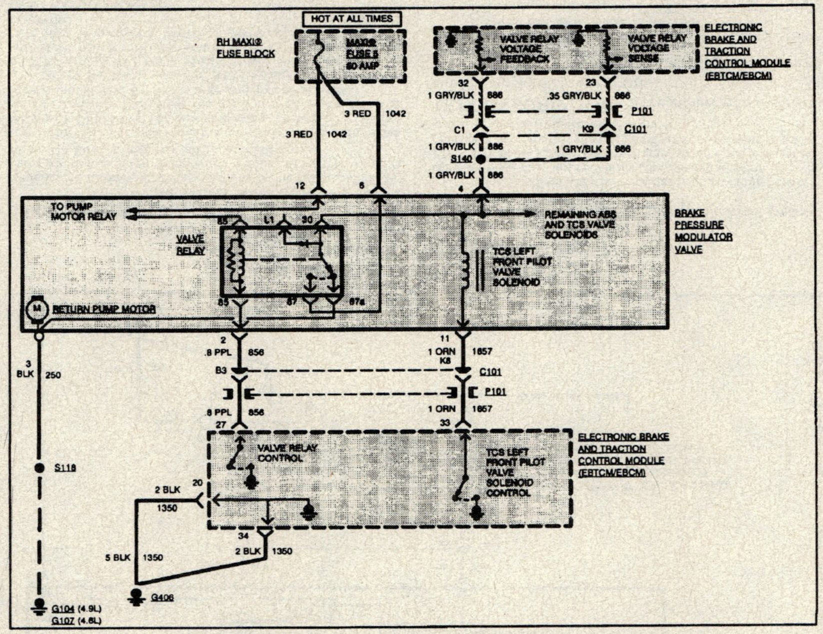

DIAGNOSTIC TROUBLE CODE 48
LEFT FRONT TCS PILOT VALVE FAULT
(Page 3 of 4)

CIRCUIT DESCRIPTION:

The left front TCS pilot valve in the BPM VALVE may be commanded to close, isolating the master cylinder.

The pilot valve circuits receive power through the valve relay on the BPM VALVE. The valve relay is engaged at KEY—ON and remains engaged throughout the ignition cycle. Pilot valve circuits should have battery voltage available at all times.

When the key is turned to RUN, power is supplied to the EBTCM. The EBTCM will complete an internal self-check before providing ground to the valve relay. When the valve relay is energized, battery power is supplied to the pilot valves.

TEST DESCRIPTION: Test numbers refer to circled numbers on the diagnostic chart.

7. Checks for a short to battery in CKT 886 using EBTCM/EBCM terminal '20' as ground.
8. Checks for a short to battery in CKT 1657, using EBTCM terminal '20' as ground.
9. Checks for a short to ground in CKT 886 using EBTCM terminal '20' as ground.
10. Checks for a short to ground in CKT 1657, using EBTCM terminal '20' as ground.
11. Checks for connector or wiring problems which could cause intermittents.
12. Uses the TECH 1 to exercise the TCS pilot valve solenoid and check it for proper operation.
13. Determines whether the code was set by an intermittent conditions or an EBTCM fault.

DIAGNOSTIC AIDS:

All tests using J 39700 Universal Breakout Box terminal '20' are using the terminal as ground. These tests, of course, assume that the ground CKT 1350 terminal '20' is good.

If Codes 44 and 48 are both set, the fault is likely to be an open or short in CKT 886.

DIAGNOSTIC FAULT CHART DTC 48 — 1994 DeVILLE, CONCOURS, ELDORADO AND SEVILLE, CONTINUED

DIAGNOSTIC TROUBLE CODE 48
LEFT FRONT TCS PILOT VALVE FAULT
(Page 4 of 4)

CONTINUED FROM CODE 48, PAGE 2 OF 4

(7)
- CONNECT J 39700TO EBTCM LEAVE CONNECTED TO HARNESS
- DISCONNECT BPM VALVE CONNECTOR
- CONNECT DVM BETWEEN J 39700 TERMINALS '20' AND '32' (USE DC VOLTS)
- IGNITION 'ON'
- NOTE VOLTAGE
- IS IT AROUND SYSTEM VOLTAGE?

YES:
- REPAIR SHORT TO SYSTEM VOLTAGE CKT 886
- REFER TO EBTCM–SWITCHED FAULTS IN THIS SECTION

NO:

(8)
- CONNECT DVM BETWEEN J 39700 TERMINALS '20' AND '33'
- NOTE VOLTAGE
- IS IT AROUND SYSTEM VOLTAGE?

YES:
- REPAIR SHORT TO SYSTEM VOLTAGE ON CKT 1657
- REFER TO EBTCM–SWITCHED FAULTS IN THIS SECTION

NO:

(9)
- IGNITION 'OFF'
- MEASURE RESISTANCE BETWEEN J 39700 TERMINALS '20' AND '32'
- IS IT 5 OHMS OR LESS?

YES:
- REPAIR SHORT TO GROUND ON CKT 886
- REFER TO EBTCM–SWITCHED FAULTS IN THIS SECTION

NO:

(10)
- MEASURE RESISTANCE BETWEEN J 39700 TERMINALS '20' AND '33'
- IS IT 5 OHMS OR LESS?

YES:
- REPAIR SHORT TO GROUND ON CKT 1657
- REFER TO EBTCM–SWITCHED FAULTS IN THIS SECTION

NO:

(11)
- INSPECT ALL WIRING AND CONNECTORS FOR CONDITIONS WHICH COULD CAUSE INTERMITTENTS. SUCH AS CHAFING, CORROSION, OR POOR TERMINAL CONTACT
- IS ALL OK?

NO:
- REPAIR WIRING, CONNECTORS OR TERMINALS AS NEEDED.

YES:

(12)
- REMOVE J 39700
- RECONNECT EBTCM AND BPM VALVE CONNECTORS
- INSTALL TECH 1 AND MASS STORAGE CARTRIDGE
- IGNITION 'ON'
- PERFORM TECH 1 LF TCS VALVE TEST
- DOES TEST MODE OPERATE?

NO:
- REPLACE BPM VALVE ASSEMBLY

YES:

(13)
- CLEAR CODES IGNITION 'OFF'
- DISCONNECT TECH 1, IF CONNECTED
- IGNITION 'ON'
- DRIVE VEHICLE AT LEAST 7 KPH (4 MPH) TO ACTIVATE AUTO TEST SEQUENCE
- DOES CODE 48 RESET?

YES:
- REPLACE EBTCM

NO:
- NO TROUBLE FOUND; SEE 'DIAGNOSTIC AIDS' ON FACING PAGE

WHEN ALL DIAGNOSIS AND REPAIRS ARE COMPLETED, CLEAR CODES AND VERIFY OPERATION

DIAGNOSTIC FAULT CHART DTC 51 — 1994 DeVILLE, CONCOURS, ELDORADO AND SEVILLE

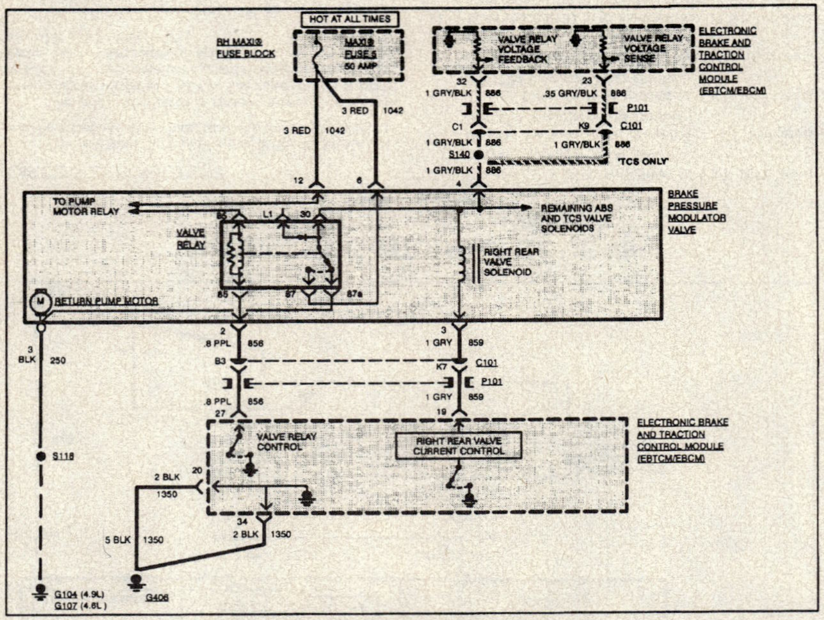

DIAGNOSTIC TROUBLE CODE 51
RIGHT REAR ABS VALVE SOLENOID FAULT
(Page 1 of 4)

CIRCUIT DESCRIPTION:

The right rear valve solenoid in the BPM Valve may be commanded to three different positions by the EBTCM/EBCM. Valve position is determined by the amount of current which is allowed to flow through the valve solenoid coil.

The valve solenoid circuits receive power through the valve relay on the BPM Valve. The valve relay is engaged at KEY – ON and remains engaged throughout the ignition cycle. Valve solenoid circuits should have battery voltage available at all times.

When the key is turned to RUN, power is supplied to the EBTCM/EBCM. The EBTCM/EBCM will complete an internal self–check before providing ground to the valve relay. When the valve relay is energized, battery power is supplied to the valve solenoids.

TEST CONDITION(S):

1. Tested continuously.

FAILURE CONDITION(S):

While standing or driving:

1. Valve solenoid cannot be activated.

OR

2. Valve solenoid is activated in a defective way.

ACTION TAKEN:

1. Code set.
2. ABS/TCS switched 'OFF'.
3. 'ANTILOCK' indicator 'ON'.
4. 'TRACTION DISABLED' message displayed. (vehicles with TCS)

Code 51 will set when the expected position of the right rear valve solenoid does not match the commanded position from the EBTCM/EBCM. Conditions which could cause Code 51 to set are: damage to the right rear valve solenoid; and open circuit; or a short circuit to ground or battery on CKT 859.

TEST DESCRIPTION: Test numbers refer to circled numbers on the diagnostic chart.

1. Damage to CKT 886 may result in multiple valve solenoid codes setting.
2. Checks CKT 886 for opens or high resistance from EBTCM/EBCM terminal '23' to '32'.
3. Determines which branch of CKT 886 has an open or high resistance.
4. Checks the integrity of CKTs 859 and 886, the RR ABS Valve Solenoid circuitry internal to the BPM Valve, and the RR ABS Valve Solenoid coil.
5. Determines whether a problem found in Step 4 is due to an open in CKT 859, 886, or a faulty BPM Valve.
6. Determines whether the problem found in Step 4 is due to an open in CKT 859 or 886.

DIAGNOSTIC AIDS:

All tests using J 39700 Universal Breakout Box terminal '20' are using the terminal as ground. These tests, of course, assume that the ground CKT 1350 terminal '20' is good.

DIAGNOSTIC FAULT CHART DTC 51 — 1994 DeVILLE, CONCOURS, ELDORADO AND SEVILLE, CONTINUED

DIAGNOSTIC TROUBLE CODE 51
RIGHT REAR ABS VALVE SOLENOID FAULT
(Page 2 of 4)

(1)
- IS MORE THAN 1 VALVE SOLENOID FAULT CODE SET?

NO:
- IGNITION 'OFF'
- DISCONNECT EBTCM/EBCM
- INSTALL J 39700 UNIVERSAL BREAKOUT BOX TO EBTCM/EBCM HARNESS CONNECTOR ONLY

YES: (2)
- IGNITION 'OFF'
- DISCONNECT EBTCM/EBCM
- INSTALL J 39700 UNIVERSAL BREAKOUT BOX TO EBTCM/EBCM HARNESS CONNECTOR ONLY
- MEASURE RESISTANCE BETWEEN J 39700 TERMINALS '23' AND '32'
- IS IT GREATER THAN 5 OHMS?

NO: (go to 4)

YES: (3)
- DISCONNECT BPM VALVE
- MEASURE RESISTANCE BETWEEN J 39700 TERMINAL '32' AND BPM VALVE CONNECTOR TERMINAL '4'
- IS IT GREATER THAN 5 OHMS?

YES:
- REPAIR OPEN OR HIGH RESISTANCE IN CKT 886 BETWEEN EBTCM/EBCM CONNECTOR TERMINAL '32' AND BPM VALVE CONNECTOR TERMINAL '4'

NO:
- REPAIR OPEN OR HIGH RESISTANCE IN CKT 886 BETWEEN EBTCM/EBCM CONNECTOR TERMINAL '23' AND HARNESS SPLICE S140

(4)
- REMOVE VALVE RELAY
- MEASURE RESISTANCE BETWEEN J 39700 TERMINALS '19' AND '32'
- IS IT .9–3 OHMS?

YES:
CODE 51 CONTINUED ON PAGE 4 OF 4

NO: (5)
- DISCONNECT BPM VALVE CONNECTOR
- MEASURE RESISTANCE BETWEEN TERMINALS '3' AND '4' ON BPM VALVE
- IS IT .9–2 OHMS?

NO:
- REPLACE BPM VALVE ASSEMBLY

YES: (6)
- MEASURE RESISTANCE BETWEEN J 39700 TERMINAL '19' AND BPM VALVE TERMINAL '3'
- IS IT GREATER THAN 5 OHMS?

NO:
- REPAIR OPEN OR HIGH RESISTANCE IN CKT 886 BETWEEN EBTCM/EBCM CONNECTOR TERMINAL '32' AND BPM VALVE CONNECTOR TERMINAL '4'

YES:
- REPAIR OPEN OR HIGH RESISTANCE CKT 859

WHEN ALL DIAGNOSIS AND REPAIRS ARE COMPLETED, CLEAR CODES AND VERIFY OPERATION

DIAGNOSTIC FAULT CHART DTC 51 — 1994 DeVILLE, CONCOURS, ELDORADO AND SEVILLE, CONTINUED

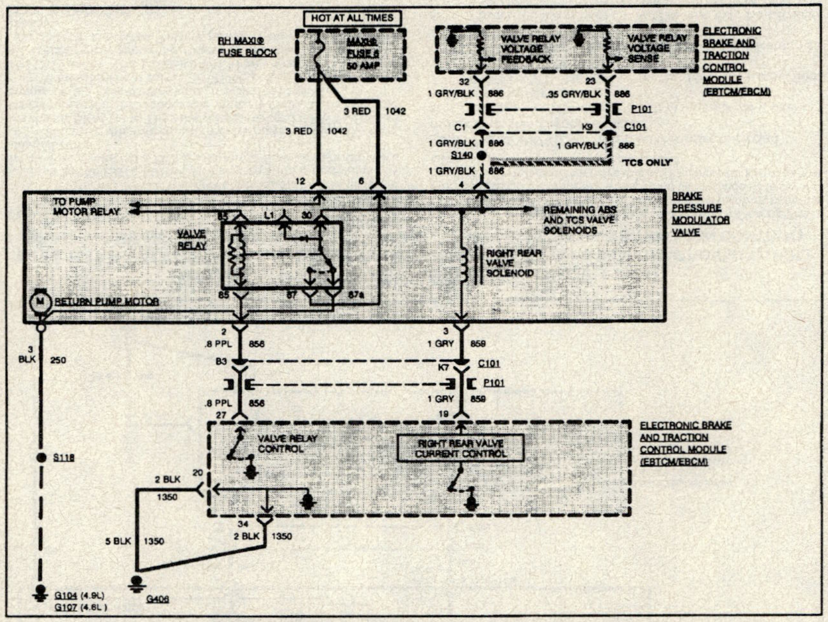

DIAGNOSTIC TROUBLE CODE 51

RIGHT REAR ABS VALVE SOLENOID FAULT (Page 3 of 4)

CIRCUIT DESCRIPTION:

The right rear valve solenoid in the BPM Valve may be commanded to three different positions by the EBTCM/EBCM. Valve position is determined by the amount of current which is allowed to flow through the valve solenoid coil.

The valve solenoid circuits receive power through the valve relay on the BPM Valve. The valve relay is engaged at KEY–ON and remains engaged throughout the ignition cycle. Valve solenoid circuits should have battery voltage available at all times.

When the key is turned to RUN, power is supplied to the EBTCM/EBCM. The EBTCM/EBCM will complete an internal self–check before providing ground to the valve relay. When the valve relay is energized, battery power is supplied to the valve solenoids.

TEST DESCRIPTION: Test numbers refer to circled numbers on the diagnostic chart.

7. Checks for a short to battery in CKT 886 using EBTCM/EBCM terminal '20' as ground.
8. Checks for a short to battery in CKT 859, using EBTCM/EBCM terminal '20' as ground.
9. Checks for a short to ground in CKT 886 using EBTCM/EBCM terminal '20' as ground.
10. Checks for a short to ground in CKT 859, using EBTCM/EBCM terminal '20' as ground.
11. Checks wiring and connectors for intermittents.
12. Uses the TECH 1 to exercise the RR ABS valve solenoid and check it for proper operation.
13. Determines whether the code was set by an intermittent conditions or an EBTCM/EBCM fault.

DIAGNOSTIC AIDS:

All tests using J 39700 Universal Breakout Box terminal '20' are using the terminal as ground. These tests, of course, assume that the ground CKT 1350 terminal '20' is good.

DIAGNOSTIC FAULT CHART DTC 51 — 1994 DeVILLE, CONCOURS, ELDORADO AND SEVILLE, CONTINUED

DIAGNOSTIC TROUBLE CODE 51

RIGHT REAR ABS VALVE SOLENOID FAULT (Page 4 of 4)

CODE 51 CONTINUED FROM PAGE 2 OF 4

(7)
- CONNECT J 39700TO EBTCM/EBCM; LEAVE CONNECTED TO HARNESS
- DISCONNECT BPM VALVE CONNECTOR
- CONNECT DVM BETWEEN J 39700 TERMINALS '20' AND '32' (USE DC VOLTS)
- IGNITION 'ON'
- NOTE VOLTAGE
- IS IT AROUND SYSTEM VOLTAGE?

YES:
- REPAIR SHORT TO SYSTEM VOLTAGE CKT 886
- REFER TO EBTCM/EBCM–SWITCHED FAULTS IN THIS SECTION

NO:

(8)
- CONNECT DVM BETWEEN J 39700 TERMINALS '19' AND '20'
- NOTE VOLTAGE
- IS IT AROUND SYSTEM VOLTAGE?

YES:
- REPAIR SHORT TO SYSTEM VOLTAGE ON CKT 859
- REFER TO EBTCM/EBCM–SWITCHED FAULTS IN THIS SECTION

NO:

(9)
- IGNITION 'OFF'
- MEASURE RESISTANCE BETWEEN J 39700 TERMINALS '20' AND '32'
- IS IT 5 OHMS OR LESS?

YES:
- REPAIR SHORT TO GROUND ON CKT 886
- REFER TO EBTCM/EBCM–SWITCHED FAULTS IN THIS SECTION

NO:

(10)
- MEASURE RESISTANCE BETWEEN J 39700 TERMINALS '19' AND '20'
- IS IT 5 OHMS OR LESS?

YES:
- REPAIR SHORT TO GROUND ON CKT 859
- REFER TO EBTCM/EBCM–SWITCHED FAULTS IN THIS SECTION

NO:

(11)
- INSPECT ALL WIRING AND CONNECTORS FOR CONDITIONS WHICH COULD CAUSE INTERMITTENTS. SUCH AS CHAFING, CORROSION, OR POOR TERMINAL CONTACT
- IS ALL OK?

NO:
- REPAIR WIRING, CONNECTORS OR TERMINALS AS NEEDED

YES:

(12)
- REMOVE J 39700
- RECONNECT EBTCM/EBCM AND BPM VALVE CONNECTORS
- INSTALL TECH 1 AND MASS STORAGE CARTRIDGE
- IGNITION 'ON'
- PERFORM TECH 1 RR ABS VALVE TEST
- DO BOTH TEST MODES OPERATE?

NO:
- REPLACE BPM VALVE ASSEMBLY

YES:

(13)
- CLEAR CODES; IGNITION 'OFF'
- DISCONNECT TECH 1, IF CONNECTED
- IGNITION 'ON'
- DRIVE VEHICLE AT LEAST 7 KPH (4 MPH) TO ACTIVATE AUTO TEST SEQUENCE
- DOES CODE 51 RESET?

NO:
- NO TROUBLE FOUND; SEE DIAGNOSTIC AIDS ON FACING PAGE

YES:
- REPLACE EBTCM/EBCM

WHEN ALL DIAGNOSIS AND REPAIRS ARE COMPLETED, CLEAR CODES AND VERIFY OPERATION

DIAGNOSTIC FAULT CHART DTC 55 — 1994 DeVILLE, CONCOURS, ELDORADO AND SEVILLE

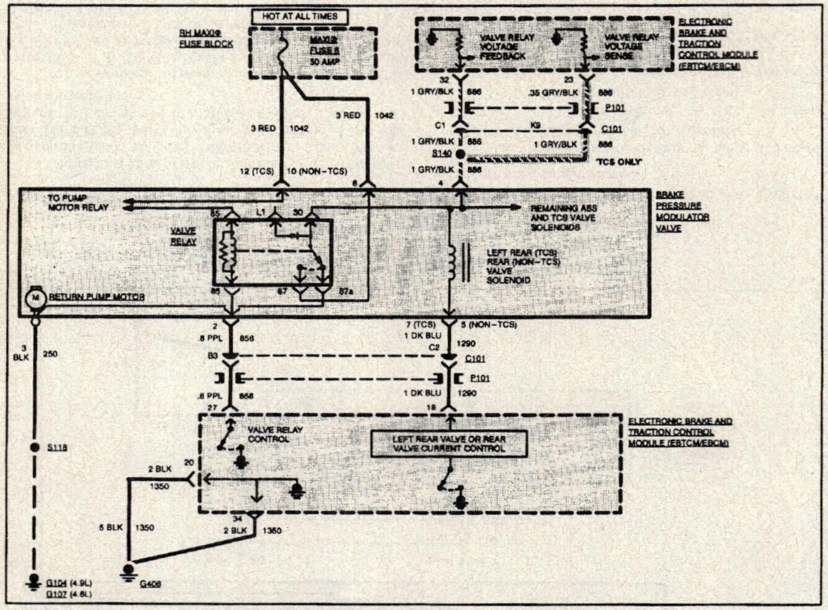

DIAGNOSTIC TROUBLE CODE 55 (Page 1 of 4)
LEFT REAR VALVE SOLENOID VALVE FAULT (VEHICLES EQUIPPED WITH TRACTION CONTROL)
REAR VALVE SOLENOID VALVE FAULT (VEHICLES NOT EQUIPPED WITH TRACTION CONTROL)

CIRCUIT DESCRIPTION:

The left rear valve solenoid (TCS) or rear valve solenoid (non–TCS) in the BPM Valve may be commanded to three different positions by the EBTCM/EBCM. Valve position is determined by the amount of current which is allowed to flow through the valve solenoid coil.

The valve solenoid circuits receive power through the valve relay on the BPM Valve. The valve relay is engaged at KEY–ON and remains engaged throughout the ignition cycle. Valve solenoid circuits should have battery voltage available at all times.

When the key is turned to RUN, power is supplied to the EBTCM/EBCM. The EBTCM/EBCM will complete an internal self–check before providing ground to the valve relay. When the valve relay is energized, battery power is supplied to the valve solenoids.

TEST CONDITION(S):

1. Tested continuously.

FAILURE CONDITION(S):

While standing or driving:

1. Valve solenoid cannot be activated.

OR

2. Valve solenoid is activated in a defective way.

ACTION TAKEN:

1. Code set.
2. ABS/TCS switched 'OFF'.
3. 'ANTILOCK' indicator 'ON'.
4. 'TRACTION DISABLED' message displayed (vehicles with TCS)

Code 55 will set when the expected position of the right rear valve solenoid does not match the commanded position from the EBTCM/EBCM. Conditions which could cause Code 55 to set are: damage to the left rear/rear valve solenoid; an open circuit; or a short circuit to ground or battery on CKT 1290.

TEST DESCRIPTION: Test numbers refer to circled numbers on the diagnostic chart.

1. Damage to CKT 886 may result in multiple valve solenoid codes setting.
2. Checks CKT 886 for opens or high resistance from EBTCM/EBCM terminal '23' to '32'.
3. Determines which branch of CKT 886 has an open or high resistance.
4. Checks the integrity of CKTs 1290 and 886, the LR/Rear ABS Valve Solenoid circuitry internal to the BPM Valve, and the LR/Rear ABS Valve Solenoid coil.
5. Determines whether a problem found in Step 4 is due to an open in CKT 1290, 886, or a faulty BPM Valve.
6. Determines whether the problem found in Step 4 is due to an open in CKT 1290 or 886.

DIAGNOSTIC AIDS:

All tests using J 39700 Universal Breakout Box terminal '20' are using the terminal as ground. These tests, of course, assume that the ground CKT 1350 terminal '20' is good.

If Codes 41 and 55 are both set, the fault is likely to be a short to B+ on 1290.

DIAGNOSTIC FAULT CHART DTC 55 — 1994 DeVILLE, CONCOURS, ELDORADO AND SEVILLE, CONTINUED

DIAGNOSTIC TROUBLE CODE 55
LEFT REAR VALVE SOLENOID VALVE FAULT (TCS)
REAR VALVE SOLENOID VALVE FAULT (NON–TCS)
(Page 2 of 4)

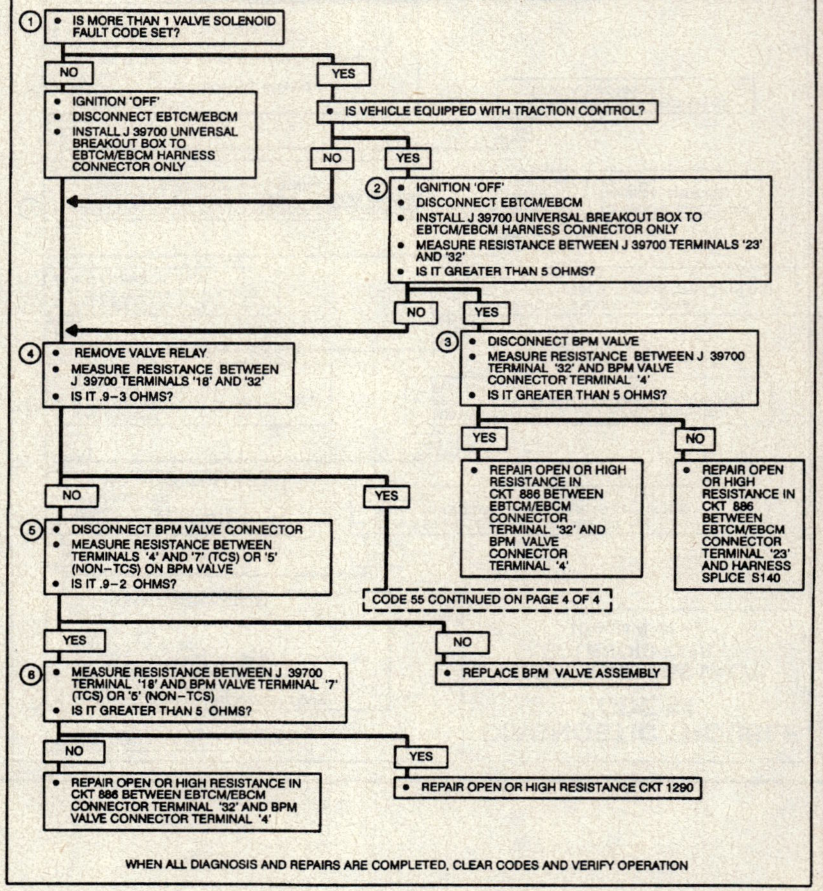

DIAGNOSTIC FAULT CHART DTC 55 — 1994 DeVILLE, CONCOURS, ELDORADO AND SEVILLE, CONTINUED

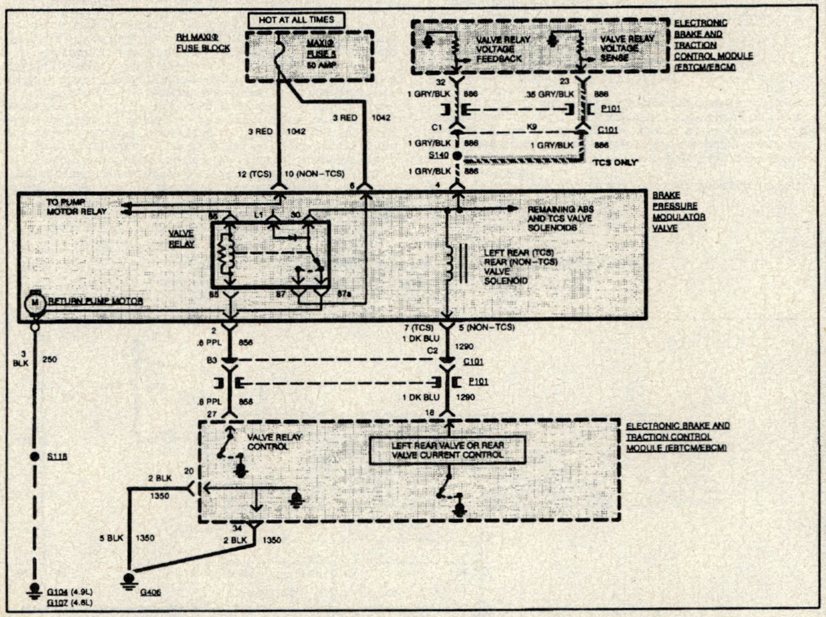

DIAGNOSTIC TROUBLE CODE 55 (Page 3 of 4)

LEFT REAR VALVE SOLENOID VALVE FAULT (VEHICLES EQUIPPED WITH TRACTION CONTROL) REAR VALVE SOLENOID VALVE FAULT (VEHICLES NOT EQUIPPED WITH TRACTION CONTROL)

CIRCUIT DESCRIPTION:

The left rear valve solenoid (TCS) or rear valve solenoid (non−TCS) in the BPM Valve may be commanded to three different positions by the EBTCM/EBCM. Valve position is determined by the amount of current which is allowed to flow through the valve solenoid coil.

The valve solenoid circuits receive power through the valve relay on the BPM valve. The valve relay is engaged at KEY−ON and remains engaged throughout the ignition cycle. Valve solenoid circuits should have battery voltage available at all times.

When the key is turned to RUN, power is supplied to the EBTCM/EBCM. The EBTCM/EBCM will complete an internal self−check before providing ground to the valve relay. When the valve relay is energized, battery power is supplied to the valve solenoids.

TEST DESCRIPTION: Test numbers refer to circled numbers on the diagnostic chart.

7. Checks for a short to battery in CKT 886 using EBTCM/EBCM terminal '20' as ground.
8. Checks for a short to battery in CKT 1290, using EBTCM/EBCM terminal '20' as ground.
9. Checks for a short to ground in CKT 886 using EBTCM/EBCM terminal '20' as ground.
10. Checks for a short to ground in CKT 1290, using EBTCM/EBCM terminal '20' as ground.
11. Checks wiring and connectors for intermittents.
12. Uses the TECH 1 to exercise the LR (TCS) or rear (non−TCS) ABS valve solenoid and check it for proper operation.
13. Determines whether the code was set by an intermittent conditions or an EBTCM/EBCM fault.

DIAGNOSTIC AIDS:

All tests using J 39700 Universal Breakout Box terminal '20' are using the terminal as ground. These tests, of course, assume that the ground CKT 1350 terminal '20' is good.

If Codes 41 and 55 are both set, the fault is likely to be a short to B+ on CKT 1290.

DIAGNOSTIC FAULT CHART DTC 55 — 1994 DeVILLE, CONCOURS, ELDORADO AND SEVILLE, CONTINUED

DIAGNOSTIC TROUBLE CODE 55

LEFT REAR VALVE SOLENOID VALVE FAULT (TCS) REAR VALVE SOLENOID VLAVE FAULT (NON−TCS) (Page 4 of 4)

CODE 55 CONTINUED FROM PAGE 4 OF 4

(7)
- CONNECT J 39700TO EBTCM/EBCM; LEAVE CONNECTED TO HARNESS
- DISCONNECT BPM VALVE CONNECTOR
- CONNECT DVM BETWEEN J 39700 TERMINALS '20' AND '32' (USE DC VOLTS)
- IGNITION 'ON'
- NOTE VOLTAGE
- IS IT AROUND SYSTEM VOLTAGE?

YES:
- REPAIR SHORT TO SYSTEM VOLTAGE CKT 886
- REFER TO EBTCM/EBCM−SWITCHED FAULTS IN THIS SECTION

NO:

(8)
- CONNECT DVM BETWEEN J 39700 TERMINALS '18' AND '20'
- NOTE VOLTAGE
- IS IT AROUND SYSTEM VOLTAGE?

YES:
- REPAIR SHORT TO SYSTEM VOLTAGE ON CKT 1290
- REFER TO EBTCM/EBCM−SWITCHED FAULTS IN THIS SECTION

NO:

(9)
- IGNITION 'OFF'
- MEASURE RESISTANCE BETWEEN J 39700 TERMINALS '20' AND '32'
- IS IT 5 OHMS OR LESS?

YES:
- REPAIR SHORT TO GROUND ON CKT 886
- REFER TO EBTCM/EBCM−SWITCHED FAULTS IN THIS SECTION

NO:

(10)
- MEASURE RESISTANCE BETWEEN J 39700 TERMINALS '18' AND '20'
- IS IT 5 OHMS OR LESS?

YES:
- REPAIR SHORT TO GROUND ON CKT 1290
- REFER TO EBTCM/EBCM−SWITCHED FAULTS IN THIS SECTION

NO:

(11)
- INSPECT ALL WIRING AND CONNECTORS FOR CONDITIONS WHICH COULD CAUSE INTERMITTENTS. SUCH AS CHAFING, CORROSION, OR POOR TERMINAL CONTACT
- IS ALL OK?

NO:
- REPAIR WIRING, CONNECTORS OR TERMINALS AS NEEDED

YES:

(12)
- REMOVE J 39700
- RECONNECT EBTCM/EBCM AND BPM VALVE CONNECTORS
- INSTALL TECH 1 AND MASS STORAGE CARTRIDGE
- IGNITION 'ON'
- PERFORM TECH 1 LR (TCS) OR REAR (NON−TCS) ABS VALVE TEST
- DO BOTH TEST MODES OPERATE?

NO:
- REPLACE BPM VALVE ASSEMBLY

YES:

(13)
- CLEAR CODES, IGNITION 'OFF'
- DISCONNECT TECH 1, IF CONNECTED
- IGNITION 'ON'
- DRIVE VEHICLE AT LEAST 7 KPH (4 MPH) TO ACTIVATE AUTO TEST SEQUENCE
- DOES CODE 55 RESET?

YES:
- REPLACE EBTCM/EBCM

NO:
- NO TROUBLE FOUND; SEE DIAGNOSTIC AIDS ON FACING PAGE

WHEN ALL DIAGNOSIS AND REPAIRS ARE COMPLETED, CLEAR CODES AND VERIFY OPERATION

DIAGNOSTIC FAULT CHART DTC 61 — 1994 DeVILLE, CONCOURS, ELDORADO AND SEVILLE

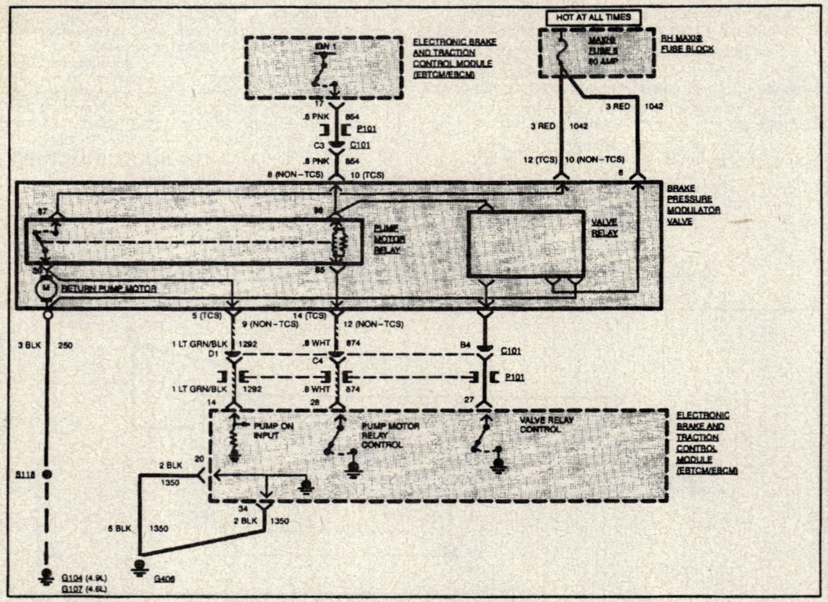

DIAGNOSTIC TROUBLE CODE 61
PUMP MOTOR OR PUMP MOTOR RELAY FAULT (Page 1 of 6)

CIRCUIT DESCRIPTION:

The pump motor returns brake fluid to the master cylinder brake circuit at the BPM Valve during antilock braking and provides brake fluid pressure to the front wheels during traction control. During normal braking, the pump does not operate. When the vehicle begins to move after start-up, the EBTCM/EBCM will turn on the pump motor and perform a self-check of the pump motor and pump motor circuit. This check may be felt and heard by the driver when the vehicle begins tc move. The self-check should be considered normal operation. The pump motor is an integral component of the BPM Valve and cannot be serviced separately.

The pump motor relay provides power to the pump motor in the BPM Valve. The pump motor relay is located on the BPM Valve and may be replaced if it is found to be defective.

The pump motor relay is not engaged during normal system operation. When antilock or traction control operation is required, the motor relay actuation line on CKT 874 (EBTCM/EBCM pin '28') is pulled to ground. The pump motor relay is engaged by the relay supply voltage line on CKT 854 (EBTCM/EBCM pin '17'). When the relay switches, battery power is provided to the pump motor.

TEST CONDITION(S):

1. Tested continuously.

FAILURE CONDITION(S):

1. Drive Away: Motor voltage is not detected during first drive away after engine start.
2. While Driving: Motor relay is energized and motor voltage is not detected.
3. Voltage at EBTCM/EBCM pin '14' below 9 volts.
4. Pump runs continuously.

ACTION TAKEN:

1. Code set.
2. ABS/TCS switched 'OFF'.
3. 'ANTILOCK' indicator 'ON'.
4. 'TRACTION DISABLED' message displayed. (vehicles with TCS)

When the pump motor relay is engaged, the motor monitor line should be at battery voltage. With the pump motor relay disengaged, the monitoring line should be at ground. If the commanded position of the pump motor relay and motor monitor line do not agree, Code 61 will set. Some conditions which will cause Code 61 to set are:

- An open circuit on CKTs 874 or 1292
- A short to voltage on CKTs 874 or 1292
- A short to ground on CKTs 874 or 1292
- A defective pump motor relay or pump motor

TEST DESCRIPTION: Test numbers refer to circled numbers on the diagnostic chart.

1. Checks the pump motor relay coil for proper resistance.
2. Determines if the pump motor relay contacts are stuck closed.
3. Determines if the pump motor relay contacts are stuck open.

DIAGNOSTIC AIDS:

All tests using J 39700 Universal Breakout Box terminal '20' are using the terminal as ground. These tests, of course, assume that the ground CKT 1350 at terminal '20' is good.

It is very important that a through inspection of the wiring and connectors be performed. Failure to carefully and fully inspect wiring and connectors may result in misdiagnosis, causing part replacement with reappearance of the fault.

DIAGNOSTIC FAULT CHART DTC 61 — 1994 DeVILLE, CONCOURS, ELDORADO AND SEVILLE, CONTINUED

DIAGNOSTIC TROUBLE CODE 61
MOTOR PUMP CIRCUIT FAULT (Page 2 of 6)

(1)
- IGNITION 'OFF'
- REMOVE BPM COVER AND PUMP MOTOR RELAY (4-PIN)
- MEASURE RESISTANCE BETWEEN PUMP MOTOR RELAY PINS 85 AND 86

45–55 Ω:

(2)
- MEASURE RESISTANCE BETWEEN PUMP MOTOR RELAY PINS 30 AND 87

OPEN CIRCUIT:

(3)
- USING A FUSED JUMPER, APPLY SYSTEM VOLTAGE TO PUMP MOTOR RELAY PIN 86 AND ATTACH A GROUND WIRE TO PIN 85
- MEASURE RESISTANCE BETWEEN PUMP MOTOR RELAY PINS 30 AND 87

CODE 61 CONTINUED ON PAGE 4 OF 6

CONTINUITY:
- REPLACE PUMP MOTOR RELAY

OUTSIDE 45–55 Ω:
- REPLACE PUMP MOTOR RELAY

WHEN ALL DIAGNOSIS AND REPAIRS ARE COMPLETED, CLEAR CODES AND VERIFY OPERATION

DIAGNOSTIC FAULT CHART DTC 61 — 1994 DeVILLE, CONCOURS, ELDORADO AND SEVILLE, CONTINUED

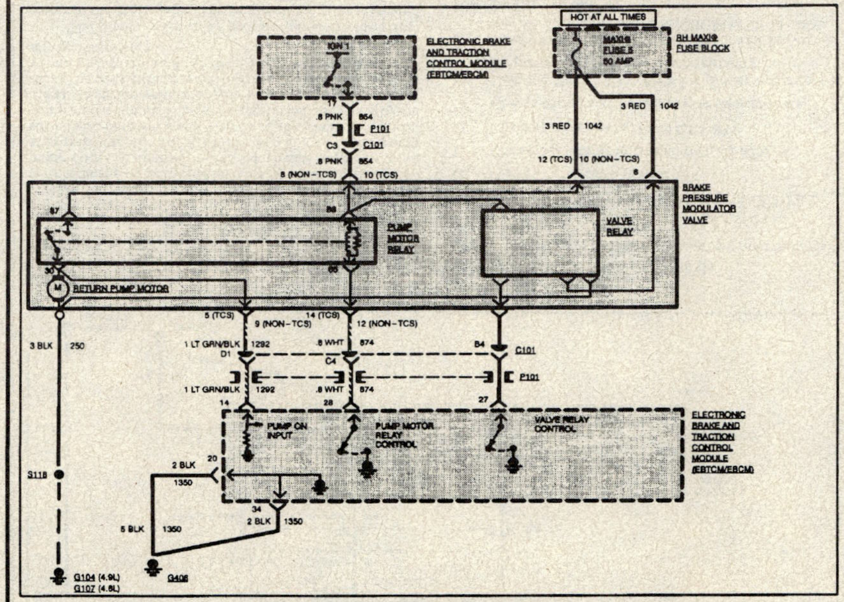

DIAGNOSTIC TROUBLE CODE 61
PUMP MOTOR OR PUMP MOTOR RELAY FAULT (Page 3 of 6)

CIRCUIT DESCRIPTION:

The pump motor returns brake fluid to the master cylinder brake circuit at the BPM valve during antilock braking and provides brake fluid pressure to the front wheels during traction control. During normal braking, the pump does not operate. When the vehicle begins to move after start–up, the EBTCM/EBCM will turn on the pump motor and perform a self–check of the pump motor and pump motor circuit. This check may be felt and heard by the driver when the vehicle begins to move. The self–check should be considered normal operation. The pump motor is an integral component of the BPM valve and cannot be serviced separately.

The pump motor relay provides power to the pump motor in the BPM valve. The pump motor relay is located on the BPM valve and may be replaced if it is found to be defective.

The pump motor relay is not engaged during normal system operation. When antilock or traction control operation is required, the motor relay actuation line on CKT 874 (EBTCM/EBCM pin '28') is pulled to ground. The pump motor relay is engaged by the relay supply voltage line on CKT 854 (EBTCM/EBCM pin '17'). When the relay switches, battery power is provided to the pump motor.

TEST DESCRIPTION: Test numbers refer to circled numbers on the diagnostic chart.

4. Checks for an open in power feed CKT 1042.
5. Checks the integrity of the pump motor and pump motor relay circuitry internal to the BPM valve.
6. Checks for a good ground to the BPM valve (CKT 250).
7. Checks for an open in CKT 1292.
8. Checks for a short to ground in CKT 1292.

DIAGNOSTIC AIDS:

All tests using J 39700 Universal Breakout Box terminal '20' are using the terminal as ground. These tests, of course, assume that the ground CKT 1350 at terminal '20' is good.

It is very important that a thorough inspection of the wiring and connectors be performed. Failure to carefully and fully inspect wiring and connectors may result in misdiagnosis, causing part replacement with reappearance of the fault.

DIAGNOSTIC FAULT CHART DTC 61 — 1994 DeVILLE, CONCOURS, ELDORADO AND SEVILLE, CONTINUED

DIAGNOSTIC TROUBLE CODE 61
MOTOR PUMP CIRCUIT FAULT (Page 4 of 6)

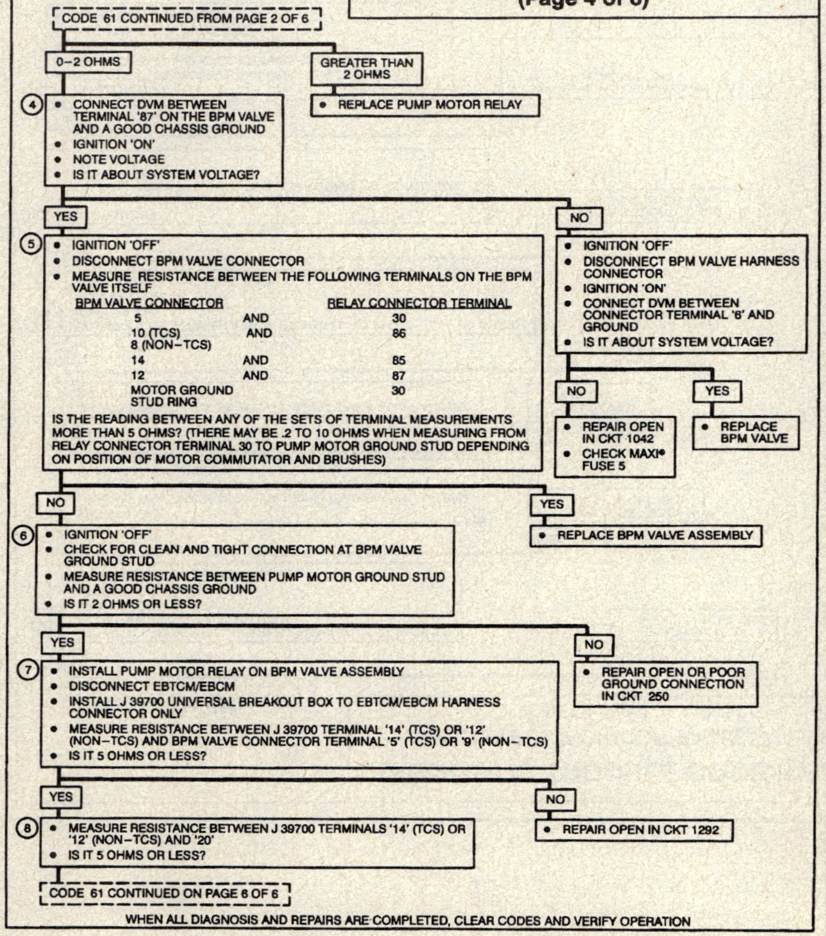

DIAGNOSTIC FAULT CHART DTC 61 — 1994 DeVILLE, CONCOURS, ELDORADO AND SEVILLE, CONTINUED

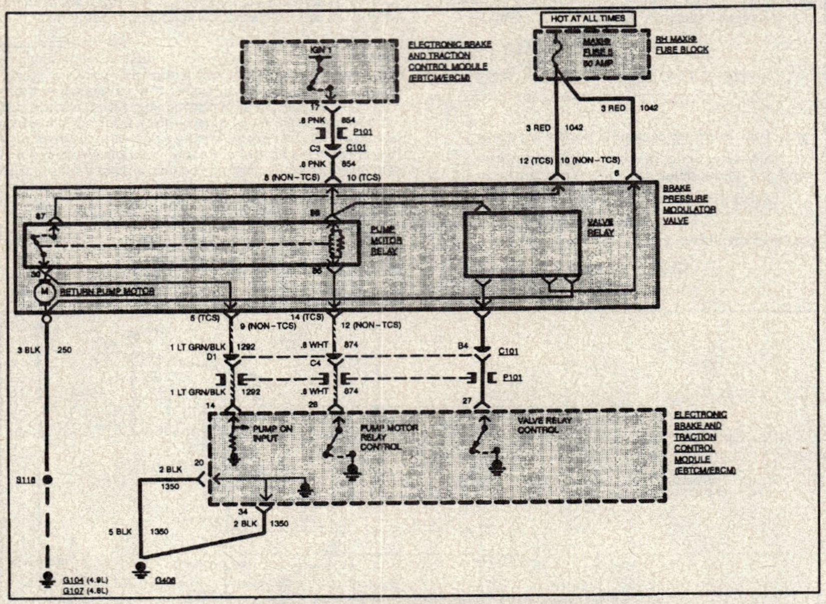

DIAGNOSTIC TROUBLE CODE 61
PUMP MOTOR OR PUMP MOTOR RELAY FAULT (Page 5 of 6)

CIRCUIT DESCRIPTION:

The pump motor returns brake fluid to the master cylinder brake circuit at the BPM valve during antilock braking and provides brake fluid pressure to the front wheels during traction control. During normal braking, the pump does not operate. When the vehicle begins to move after start–up, the EBTCM/EBCM will turn on the pump motor and perform a self–check of the pump motor and pump motor circuit. This check may be felt and heard by the driver when the vehicle begins to move. The self–check should be considered normal operation. The pump motor is an integral component of the BPM valve and cannot be serviced separately.

The pump motor relay provides power to the pump motor in the BPM valve. The pump motor relay is located on the BPM valve and may be replaced if it is found to be defective.

The pump motor relay is not engaged during normal system operation. When antilock or traction control operation is required, the motor relay actuation line on CKT 874 (EBTCM/EBCM pin '28') is pulled to ground. The pump motor relay is engaged by the relay supply voltage line on CKT 854 (EBTCM/EBCM pin '17'). When the relay switches, battery power is provided to the pump motor.

TEST DESCRIPTION: Test numbers refer to circled numbers on the diagnostic chart.

9. Checks for a short to system voltage in CKT 1292.
10. Checks for an open in CKT 874.
11. Checks for a short to ground on CKT 874.
12. Checks for a short to B+ in CKT 874.
13. Checks wiring and connectors for intermittents.
14. Uses the TECH 1 to check for proper operation of the pump motor and associated circuitry.
15. Determines whether the problem found in Step 13 was due to a faulty EBTCM/EBCM or BPM valve.

DIAGNOSTIC AIDS:

All tests using J 39700 Universal Breakout Box terminal '20' are using the terminal as ground. These tests, of course, assume that the ground CKT 1350 at terminal '20' is good.

It is very important that a thorough inspection of the wiring and connectors be performed. Failure to carefully and fully inspect wiring and connectors may result in misdiagnosis, causing part replacement with reappearance of the fault.

DIAGNOSTIC FAULT CHART DTC 61 — 1994 DeVILLE, CONCOURS, ELDORADO AND SEVILLE, CONTINUED

DIAGNOSTIC TROUBLE CODE 61
MOTOR PUMP CIRCUIT FAULT (Page 6 of 6)

CODE 61 CONTINUED FROM PAGE 4 OF 6

NO

(9)
- CONNECT DVM BETWEEN J 39700 TERMINALS '14' (TCS) OR '12' (NON–TCS) AND '20' (USE DC VOLTS)
- IGNITION 'ON'
- NOTE VOLTAGE
- IS IT AROUND SYSTEM VOLTAGE?

YES
- REPAIR SHORT TO GROUND IN CKT 1292

NO

(10)
- IGNITION 'OFF'
- MEASURE RESISTANCE BETWEEN J 39700 TERMINAL '28' AND BPM VALVE CONNECTOR TERMINAL '14' (TCS) OR '12' (NON–TCS)
- IS IT 5 OHMS OR LESS?

YES
- REPAIR SHORT SYSTEM VOLTAGE IN CKT 1292

YES

(11)
- MEASURE RESISTANCE BETWEEN J 39700 TERMINALS '20' AND '28'
- IS IT 5 OHMS OR LESS?

NO
- REPAIR OPEN IN CKT 874

NO

(12)
- CONNECT DVM BETWEEN J 39700 TERMINALS '20' AND '28' (USE DC VOLTS)
- IGNITION 'ON'
- NOTE VOLTAGE
- IS IT AROUND SYSTEM VOLTAGE?

YES
- REPAIR SHORT TO GROUND IN CKT 874

NO

(13)
- IGNITION 'OFF'
- INSPECT ALL WIRING AND CONNECTORS FOR CONDITIONS WHICH COULD CAUSE INTERMITTENTS, SUCH AS CHAFING, CORROSION, OR POOR TERMINAL CONTACT.
- IS ALL OK?

YES
- REPAIR SHORT TO SYSTEM VOLTAGE IN CKT 874

YES

(14)
- REMOVE J 39700
- RECONNECT EBTCM/EBCM AND BPM VALVE CONNECTORS
- INSTALL TECH 1 AND MASS STORAGE CARTRIDGE
- IGNITION 'ON'
- PERFORM TECH 1 AUTO TEST
- DOES PUMP MOTOR RUN?

NO
- REPAIR WIRING CONNECTORS, OR TERMINALS AS NEEDED

NO

(15)
- MONITOR VOLTAGE AT BPM VALVE TERMINAL '10' (TCS) OR '8' (NON–TCS) TO GROUND WHILE REPERFORMING TECH 1 AUTO TEST
- IS THERE SYSTEM VOLTAGE?

YES
- NO TROUBLE FOUND; SEE 'DIAGNOSTIC AIDS' ON FACING PAGE

NO
- INSPECT BPM VALVE AND EBTCM/EBCM CONNECTORS AND TERMINALS FOR DAMAGE, POOR CRIMPS, ETC.
- IF OK, REPLACE EBTCM/EBCM

YES
- INSPECT BPM VALVE CONNECTOR AND TERMINALS FOR DAMAGE, POOR CRIMPS ETC.
- IF OK, REPLACE BPM VALVE

WHEN ALL DIAGNOSIS AND REPAIRS ARE COMPLETED, CLEAR CODES AND VERIFY OPERATION

DIAGNOSTIC FAULT CHART DTC 63 — 1994 DeVILLE, CONCOURS, ELDORADO AND SEVILLE

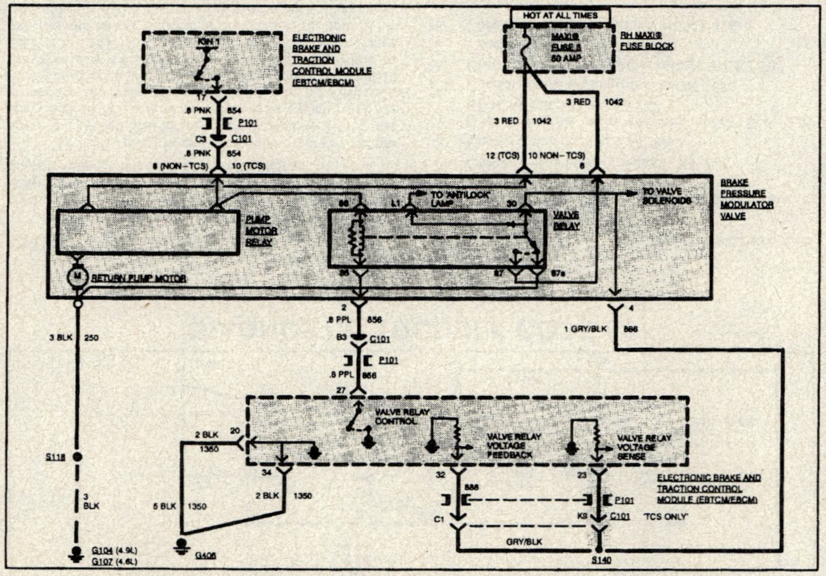

DIAGNOSTIC TROUBLE CODE 63
VALVE RELAY FAULT (Page 1 of 6)

CIRCUIT DESCRIPTION:

The valve relay provides power to the four ABS valve solenoids and two TCS Pilot Valves (if applicable) in the BPM valve. The valve relay is located on the BPM valve an may be replaced if it is found to be defective.

The valve relay is engaged during normal system operation. When the ignition is switched to RUN, the EBTCM/EBCM commands the valve relay ON by grounding the valve relay actuation line on CKT 856 (EBTCM/EBCM pin '27'). When this ground is provided, the valve relay is energized from the voltage supply line on CKT 854 (EBTCM/EBCM pin '17'). The valve relay switches and battery voltage is provided to the four valve solenoids, the pilot valves and the valve relay voltage sense line on CKT 886 (EBTCM/EBCM pin '23'). The valve relay remains engaged until the ignition is turned OFF or a failure is detected.

The valve relay voltage feedback (EBTCM/EBCM pin '32') provides suppression of voltage spikes created during operation of both front ABS valve solenoids during antilock and traction control.

TEST CONDITION(S):

1. Tested continuously.

FAILURE CONDITION(S):

1. Less than 5 volts at EBTCM/EBCM pin '23', valve relay sense;

OR

2. Voltage at EBTCM/EBCM pin '23', valve relay sense remains high, although voltage at EBTCM/EBCM pin '27', valve relay control, is low (valve relay sticks).

ACTION TAKEN:

1. Code set.
2. ABS/ TCS switched 'OFF'.
3. 'ANTILOCK' indicator 'ON'.
4. 'TRACTION DISABLED' message displayed.

When the valve relay is engaged, the valve relay sense line should be at battery voltage. With the valve relay disengaged, the sense line should be at ground. If the commanded position of the valve relay and the position indicated by the sense line do not agree, Code 63 will set. Some conditions which will cause Code 63 to set are:

- An open circuit on CKT 886 or 856.
- A short to voltage on CKT 886 or 856.
- A short to ground on CKT 886 or 856.
- A defective solenoid valve relay.
- Open circuit on CKT 854.
- Loss of voltage to BPM valve terminals '6' and '12' during pump run.

TEST DESCRIPTION: Test numbers refer to circled numbers on the diagnostic chart.

1. Checks for availability of battery voltage to the Valve Relay through Maxi® fuse 5 and CKT 1042.
2. Checks for possible loss of voltage current when system is under load. At Key–ON, system performs the power on Reset Test and cycles the valve relay, checking the current draw of valves.
3. Checks the Valve Relay coil for proper resistance.
4. Checks for the Valve Relay contacts being internally open when in the relay unenergized position.
5. Checks for the Valve Relay contacts being stuck open when in the relay–energized position.

DIAGNOSTIC AIDS:

All tests using J 39700 Universal Breakout Box terminal '20' are using the terminal as ground. These tests, of course, assume that the ground CKT 1350 at terminal '20' is good.

It is very important that a thorough inspection of the wiring and connectors be performed. Failure to carefully and fully inspect wiring and connectors may result in misdiagnosis, causing part replacement with reappearance of the fault.

DIAGNOSTIC FAULT CHART DTC 63 — 1994 DeVILLE, CONCOURS, ELDORADO AND SEVILLE, CONTINUED

DIAGNOSTIC TROUBLE CODE 63
VALVE RELAY FAULT (Page 2 of 6)

①
- IGNITION 'OFF'
- DISCONNECT BPM VALVE CONNECTOR
- CONNECT DVM BETWEEN BPM VALVE CONNECTOR TERMINAL 6 AND A GOOD CHASSIS GROUND
- NOTE VOLTAGE
- IS IT AROUND SYSTEM VOLTAGE?

NO:
- REPAIR OPEN CKT 1042
- CHECK MAX1® FUSE 5

YES:

②
- INSERT POSITIVE VOLTMETER LEAD AT REAR OF BPM VALVE CONNECTOR TERMINAL '6' AND NEGATIVE VOLTMETER LEAD TO A GOOD CHASSIS GROUND
- MONITOR THE VOLTAGE AS THE IGNITION IS TURNED 'ON'
- DOES THE VOLTAGE EVER VARY FROM SYSTEM VOLTAGE?

NO:
- IGNITION 'OFF'
- CHECK FOR POOR CRIMPING OF TERMINALS AT BPM VALVE CONNECTOR AND AT FUSE BLOCK
- CHECK FOR PROPER CRIMPING OR PARTIAL OPEN AT UNDERHOOD POWER JUNCTION BLOCK

YES:

③
- REMOVE VALVE RELAY (6–PIN)
- MEASURE RESISTANCE BETWEEN RELAY PINS 85 AND 86
- IS IT 52–64 OHMS?

NO:
- REPLACE VALVE SOLENOID RELAY

YES:

④
- MEASURE RESISTANCE BETWEEN RELAY PINS 30 AND 87A
- IS IT 2 OHMS OR LESS?

NO:
- REPLACE VALVE SOLENOID RELAY

YES:

⑤
- CONNECT RELAY PIN 85 TO GROUND
- USING A FUSED JUMPER WIRE (SUCH AS J 36169) WITH A 3 AMP FUSE CONNECT RELAY PIN 86 TO SYSTEM VOLTAGE
- MEASURE RESISTANCE BETWEEN RELAY PINS 30 AND 87
- IS IT 2 OHMS OR LESS?

CODE 63 CONTINUED ON PAGE 4 OF 6

WHEN ALL DIAGNOSIS AND REPAIRS ARE COMPLETED, CLEAR CODES AND VERIFY OPERATION

DIAGNOSTIC FAULT CHART DTC 63 — 1994 DeVILLE, CONCOURS, ELDORADO AND SEVILLE, CONTINUED

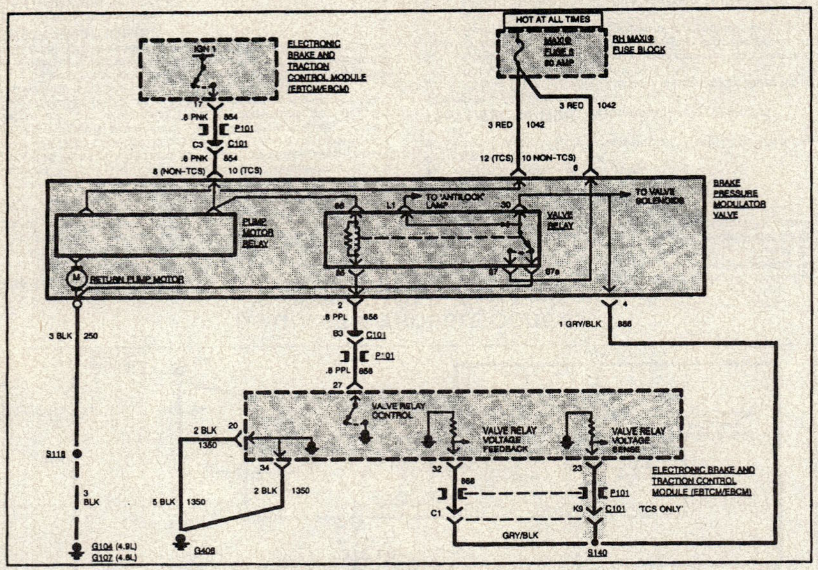

DIAGNOSTIC TROUBLE CODE 63

VALVE RELAY FAULT
(Page 3 of 6)

CIRCUIT DESCRIPTION:

The valve relay provides power to the four ABS valve solenoids and two TCS Pilot Valves (if applicable) in the BPM valve. The valve relay is located on the BPM valve an may be replaced if it is found to be defective.

The valve relay is engaged during normal system operation. When the ignition is switched to RUN, the EBTCM/EBCM commands the valve relay ON by grounding the valve relay actuation line on CKT 856 (EBTCM/EBCM pin '27'). When this ground is provided, the valve relay is energized from the voltage supply line on CKT 854 (EBTCM/EBCM pin '17'). The valve relay switches and battery voltage is provided to the four valve solenoids, the pilot valves and the valve relay voltage sense line on CKT 886 (EBTCM/EBCM pin '23'). The valve relay remains engaged until the ignition is turned 'OFF' or a failure is detected.

The valve relay voltage feedback (EBTCM/EBCM pin '32') provides suppression of voltage spikes created during operation of both front ABS valve solenoids during antilock and traction control.

TEST DESCRIPTION: Test numbers refer to circled numbers on the diagnostic chart.

6. Checks the integrity of the Valve Relay circuitry internal to the BPM valve.
7. Checks for an open in ground CKT 250.
8. Checks for an open in CKT 886.
9. Checks for an open between splice S140 and EBTCM/EBCM terminal '23'.
10. Checks for a short to ground in CKT 886.
11. Checks for a short to system voltage in CKT 886.
12. Checks for an open in CKT 854.
13. Checks for a short to ground in CKT 854.

DIAGNOSTIC AIDS:

All tests using J 39700 Universal Breakout Box terminal '20' are using the terminal as ground. These tests, of course, assume that the ground CKT 1350 at terminal '20' is good.

It is very important that a thorough inspection of the wiring and connectors be performed. Failure to carefully and fully inspect wiring and connectors may result in misdiagnosis, causing part replacement with reappearance of the fault.

DIAGNOSTIC FAULT CHART DTC 63 — 1994 DeVILLE, CONCOURS, ELDORADO AND SEVILLE, CONTINUED

DIAGNOSTIC TROUBLE CODE 63
VALVE RELAY FAULT
(Page 4 of 6)

CODE 63 CONTINUED FROM PAGE 2 OF 6

- YES → (6)
- NO → REPLACE VALVE RELAY

(6)
- MEASURE RESISTANCE BETWEEN THE FOLLOWING TERMINALS ON THE BPM VALVE ASSEMBLY ITSELF.

HYDRAULIC BPM VALVE CONNECTOR		RELAY CONNECTOR TERMINAL
2	AND	85
4	AND	30
6	AND	87
MOTOR GROUND STUD RING	AND	87A
10 (TCS) 8 (NON-TCS)	AND	86

IS THE READING BETWEEN ANY OF THE SETS OF TERMINAL MEASUREMENTS MORE THAN 2 OHMS? (THERE MAY BE .2 TO 10 OHMS WHEN MEASURING FROM RELAY CONNECTOR TERMINAL '30' TO PUMP MOTOR GROUND STUD DEPENDING ON POSITION OF MOTOR COMMUTATOR AND BRUSHES)

- NO → (7)
- YES → REPLACE BPM VALVE

(7)
- DISCONNECT EBTCM/EBCM
- INSTALL J 39700 UNIVERSAL BREAKOUT BOX TO EBTCM/EBCM HARNESS CONNECTOR ONLY
- MEASURE RESISTANCE BETWEEN PUMP MOTOR GROUND STUD AND J 39700 TERMINAL '20'
- IS IT 2 OHMS OR LESS?

- YES → (8)
- NO → REPAIR OPEN IN CKT 250 TO PUMP MOTOR GROUND STUD

(8)
- INSTALL VALVE SOLENOID RELAY
- MEASURE RESISTANCE BETWEEN BPM VALVE CONNECTOR TERMINAL '4' AND J 39700 TERMINAL '32'
- IS IT 2 OHMS OR LESS?

- YES → (9)
- NO → REPAIR OPEN IN CKT 886

(9)
- MEASURE RESISTANCE BETWEEN BPM VALVE CONNECTOR TERMINAL '4' AND J 39700 TERMINAL '23'
- IS IT 2 OHMS OR LESS?

- YES → (10)
- NO → REPAIR OPEN IN CKT 886 BETWEEN TERMINAL '23' SPLICE S140 AND THE EBTCM/EBCM

(10)
- MEASURE RESISTANCE BETWEEN J 39700 TERMINALS '20' AND '32'
- IS IT 5 OHMS OR LESS?

- NO → (11)
- YES → REPAIR SHORT GROUND IN CKT 886

(11)
- CONNECT DVM BETWEEN J 39700 TERMINALS '20' AND '32' (USE DC VOLTS)
- IGNITION 'ON'
- NOTE VOLTAGE
- IS IT AROUND SYSTEM VOLTAGE?

- YES → REPAIR SHORT TO SYSTEM VOLTAGE ON CKT 886
- NO → (12)

(12)
- IGNITION 'OFF'
- MEASURE RESISTANCE BETWEEN BPM VALVE CONNECTOR TERMINAL '10' (TCS) OR 8 (NON-TCS) AND J 39700 TERMINAL '17'
- IS IT 5 OHMS OR LESS?

- NO → REPAIR OPEN IN CKT 854
- YES → (13)

(13)
- MEASURE RESISTANCE BETWEEN J 39700 TERMINALS '17' AND '20'
- IS IT 5 OHMS OR LESS?

CODE 63 CONTINUED ON PAGE 6 OF 6

WHEN ALL DIAGNOSIS AND REPAIRS ARE COMPLETED, CLEAR CODES AND VERIFY OPERATION

DIAGNOSTIC FAULT CHART DTC 63 — 1994 DeVILLE, CONCOURS, ELDORADO AND SEVILLE, CONTINUED

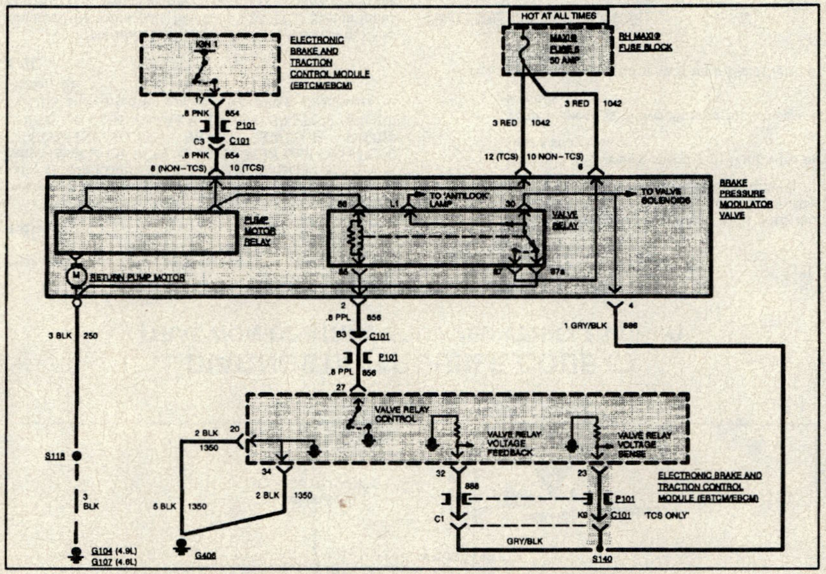

DIAGNOSTIC TROUBLE CODE 63

VALVE RELAY FAULT
(Page 5 of 6)

CIRCUIT DESCRIPTION:

The valve relay provides power to the four ABS valve solenoids and two TCS Pilot Valves (if applicable) in the BPM valve. The valve relay is located on the BPM valve an may be replaced if it is found to be defective.

The valve relay is engaged during normal system operation. When the ignition is switched to RUN, the EBTCM/EBCM commands the valve relay ON by grounding the valve relay actuation line on CKT 856 (EBTCM/EBCM pin '27'). When this ground is provided, the valve relay is energized from the voltage supply line on CKT 854 (EBTCM/EBCM pin '17'). The valve relay switches and battery voltage is provided to the four valve solenoids, the pilot valves and the valve relay voltage sense line on CKT 886 (EBTCM/EBCM pin '23'). The valve relay remains engaged until the ignition is turned 'OFF' or a failure is detected.

The valve relay voltage feedback (EBTCM/EBCM pin '32') provides suppression of voltage spikes created during operation of both front ABS valve solenoids during antilock and traction control.

TEST DESCRIPTION: Test numbers refer to circled numbers on the diagnostic chart.

13. Checks for a short to system voltage in CKT 854.
14. Checks for an open in CKT 856.
15. Checks for a short to ground in CKT 856.
16. Checks for a short to system voltage in CKT 856.
17. Checks wiring and connectors for intermittents.
18. Determines whether the code was set due to an intermittent condition or an EBTCM/EBCM fault.
19. Checks if problem found in step 13 is due to a short to system voltage in CKT 874, 856, or 854.

DIAGNOSTIC AIDS:

All tests using J 39700 Universal Breakout Box terminal '20' are using the terminal as ground. These tests, of course, assume that the ground CKT 1350 at terminal '20' is good.

It is very important that a thorough inspection of the wiring and connectors be performed. Failure to carefully and fully inspect wiring and connectors may result in misdiagnosis, causing part replacement with reappearance of the fault.

DIAGNOSTIC FAULT CHART DTC 63 — 1994 DeVILLE, CONCOURS, ELDORADO AND SEVILLE, CONTINUED

DIAGNOSTIC TROUBLE CODE 63

VALVE RELAY FAULT
(Page 6 of 6)

CODE 63 CONTINUED FROM PAGE 4 OF 6

- NO → (13)
- YES → • REPAIR SHORT TO GROUND IN CKT 854

(13)
- CONNECT DVM BETWEEN J 39700 TERMINALS '17' AND '20' (USE DC VOLTS)
- IGNITION 'ON'
- NOTE VOLTAGE
- IS IT AROUND SYSTEM VOLTAGE?

- NO → (14)
- YES → (19)

(14)
- IGNITION 'OFF'
- MEASURE RESISTANCE BETWEEN BPM VALVE CONNECTOR TERMINAL '2' AND J 39700 TERMINAL '27'
- IS IT 5 OHMS OR LESS?

(19)
- DISCONNECT VALVE RELAY
- DOES VOLTAGE MEASUREMENT DROP TO 0?

- NO →
  - DISCONNECT PUMP RELAY
  - DOES VOLTAGE MEASUREMENT DROP TO 0?
    - NO → • REPAIR SHORT TO VOLTAGE ON CKT 856
    - YES → • REPAIR SHORT TO VOLTAGE ON CKT 856
- YES → • REPAIR SHORT TO VOLTAGE ON CKT 874

- YES → (15)
- NO → • REPAIR OPEN IN CKT 856

(15)
- MEASURE RESISTANCE BETWEEN J 39700 TERMINALS '20' AND '27'
- IS IT 5 OHMS OR LESS?

- NO → (16)
- YES → • REPAIR SHORT TO GROUND IN CKT 856

(16)
- CONNECT DVM BETWEEN J 39700 TERMINALS '20' AND '27'
- IGNITION 'ON'
- NOTE VOLTAGE
- IS IT AROUND SYSTEM VOLTAGE?

- NO → (17)
- YES → • REPAIR SHORT SYSTEM VOLTAGE ON CKT 856

(17)
- IGNITION 'OFF'
- INSPECT ALL WIRING AND CONNECTORS FOR CONDITIONS WHICH COULD CAUSE INTERMITTENTS, SUCH AS CHAFING, CORROSION, OR POOR TERMINAL CONTACT
- IS ALL OK WITH CONNECTIONS AND WIRING?

- YES → (18)
- NO → • REPAIR WIRING AND/OR CONNECTIONS AS NEEDED

(18)
- REMOVE J 39700
- RECONNECT EBTCM/EBCM AND BPM VALVE CONNECTORS
- CLEAR CODES, IGNITION 'OFF'
- DISCONNECT TECH 1, IF CONNECTED
- DRIVE VEHICLE AT LEAST 7 KPH (4 MPH) TO ACTIVATE AUTO TEST SEQUENCE
- DOES CODE 63 RESET?

- NO → • NO TROUBLE FOUND; SEE 'DIAGNOSTIC AIDS' ON FACING PAGE
- YES → • REPLACE EBTCM/EBCM

WHEN ALL DIAGNOSIS AND REPAIRS ARE COMPLETED, CLEAR CODES AND VERIFY OPERATION

DIAGNOSTIC FAULT CHART DTC 67 — 1994 DeVILLE, CONCOURS, ELDORADO AND SEVILLE

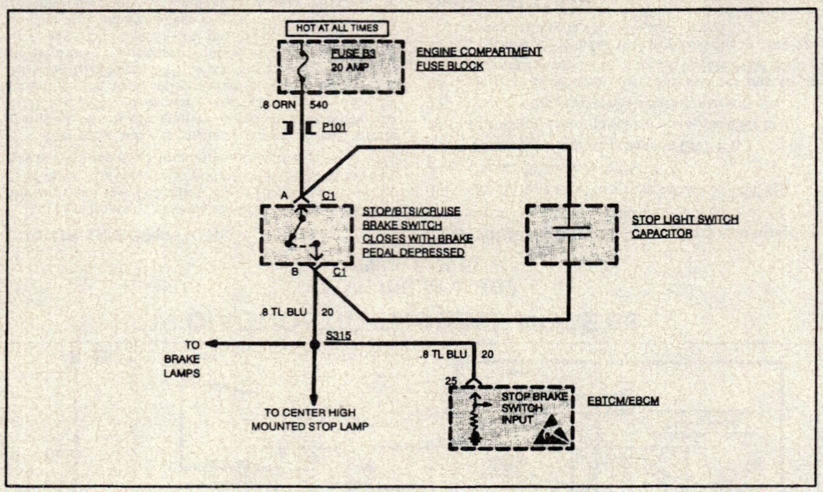

DIAGNOSTIC TROUBLE CODE 67
(TRACTION CONTROL EQUIPPED VEHICLES ONLY)

BRAKE LIGHT SWITCH FAULT

CIRCUIT DESCRIPTION:

4.6L

The electronic brake and traction control module receives Brake Light Switch status from the PCM via serial data. If fault Code 67 sets, it may be indicative of a loss of switch input to the PCM, check for PCM fault Code PO90 (STOP/BITSI/CRUISE BRAKE SWITCH FAULT CODE) or P106 (BRAKE LIGHT SWITCH FAULT CODE) and refer to appropriate PCM diagnostics in Section 6E.

4.9L

The Electronic Brake and Traction Control module monitors the serial data line for PCM TCC Switch input PI71 for TCC switch status. The EBTCM then compares this input with the voltage at terminal 25 of the EBTCM connector (system voltage indicates switch is depressed). Any discrepancy will cause the code to set. The EBTCM only performs this comparison after a certain drive pattern has been followed (see 'TEST CONDITION' for the 4.9L).

TEST CONDITION(S):

4.6L

1. Tested continuously.

4.9L

1. The EBTCM compares PCM input PI71 with the voltage at terminal 25 of the EBTCM connector after the following drive sequence has been performed:
   - Engine 'ON', vehicle speed is at 0 mph for at least 4 seconds.
   - Vehicle speed is above 30 mph for at least 4 seconds.
   - Vehicle speed returns to 0 mph for at least 4 seconds.

FAILURE CONDITION(S):

1. EBTCM receives failure status of Brake Light Switch from PCM.

ACTION TAKEN:

1. Code set.
2. TCS disabled.
3. 'TRACTION DISABLED' message ON.

DIAGNOSTIC FAULT CHART DTC 71 — 1994 DeVILLE, CONCOURS, ELDORADO AND SEVILLE

DIAGNOSTIC TROUBLE CODE 71

ELECTRONIC BRAKE AND TRACTION CONTROL (EBTCM/EBCM) INTERNAL FAULT

CIRCUIT DESCRIPTION:

The EBTCM/EBCM performs various diagnostic checks on itself. If it finds a problem, Code 71 will set.

TEST CONDITION(S):

1. Tested continuously.

FAILURE CONDITION(S):

1. While Standing or Driving: EBTCM/EBCM internal error is detected.

ACTION TAKEN:

1. Code set.
2. ABS and TCS switched 'OFF'.
3. 'ANTILOCK' indicator ON.
4. 'TRACTION DISABLED' message ON. (vehicles with TCS)

TEST DESCRIPTION: Test numbers refer to circled numbers on the diagnostic chart.

1. Checks for good connections from the harness to the EBTCM/EBCM itself.
2. Checks to see if the fault was false. If Code 71 resets, replace the EBTCM/EBCM.

DIAGNOSTIC CHART DTC 71 — 1994 DEVILLE, CONCOURS, ELDORADO AND SEVILLE, CONTINUED

DIAGNOSTIC TROUBLE CODE 71

ELECTRONIC BRAKE AND TRACTION CONTROL (EBTCM/EBCM) INTERNAL FAULT

(1)
- IGNITION 'OFF'
- CHECK EBTCM/EBCM CONNECTOR CAREFULLY TO ENSURE GOOD ELECTRICAL CONTACT AT ALL TERMINALS AND THAT CONNECTOR IS PROPERLY RETAINED WHEN CONNECTED
- IS ALL OK?

YES → (2)

YES →
- REPAIR TERMINAL(S), CONNECTOR OR CONNECTOR RETENTION AS NECESSARY
- IDENTIFY POOR CONNECTIONS

(2)
- CLEAR CODES, IGNITION 'OFF'
- DISCONNECT TECH 1, IF CONNECTED
- IGNITION 'ON'
- DOES CODE 71 RESET?

YES → REPLACE EBTCM/EBCM

YES →
- NO TROUBLE FOUND
- CHECK FOR INTERMITTENTS AND LOOSE WIRING/CONNECTORS

WHEN ALL DIAGNOSIS AND REPAIRS ARE COMPLETED, CLEAR CODES AND VERIFY OPERATION

DIAGNOSTIC CHART DTC 72 — 1994 DEVILLE, CONCOURS, ELDORADO AND SEVILLE

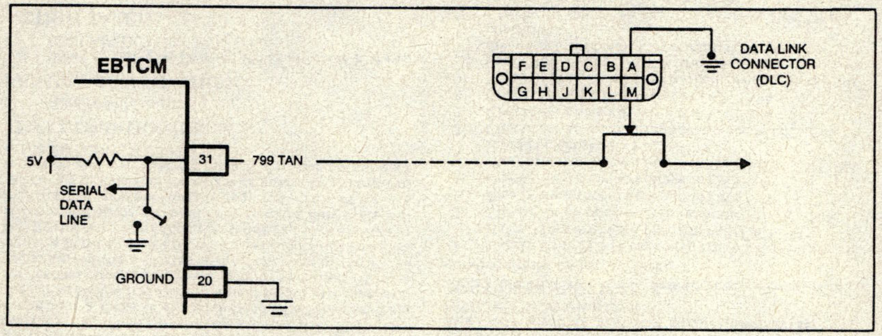

DIAGNOSTIC TROUBLE CODE 72

SERIAL DATA LINK FAULT

CIRCUIT DESCRIPTION:

The Serial Data Link is an asynchronous link operating at 8192 bits per second.

Code 72 will set and the 'ANTILOCK' indicator will illuminate if the EBTCM/EBCM does not get polled by the IPC for information, or if it does not see PCM information on the data link.

TEST CONDITION(S):

1. Tested continuously.

FAILURE CONDITION(S):

1. While Standing or Driving: Serial Data communications are lost.

ACTION TAKEN:

1. Code set.
2. ABS and TCS switched 'OFF'.
3. 'ANTILOCK' indicator ON.
4. 'TRACTION DISABLED' message ON (vehicles with TCS).

TEST DESCRIPTION: Test numbers refer to circled numbers on the diagnostic chart.

1. This test checks to see if the off–board device can communicate with the EBTCM/EBCM.
2. This test checks for a short to battery in the serial data circuit.
3. This test checks for a short to ground in the serial data circuit.
4. This test checks for an open in the serial data circuit.
5. This test checks to see if the off–board device can communicate with the EBTCM/EBCM.
6. If the connections at the EBTCM/EBCM and DLC connector are OK, replace the EBTCM/EBCM.

DIAGNOSTIC AIDS:

The problem may be intermittent. Try performing the tests shown while 'wiggling' wiring and connectors; this can often cause the fault to appear.

NOTICE: Diagnostic Trouble Code 72 will never be displayed while serial data fault is current. A current Code 72 will result in a 'NO TCS DATA', and 'TRACTION DISABLED' messages as well as IPC Diagnostic Trouble Code B332 and PCM Code P133.

DIAGNOSTIC CHART DTC 72 — 1994 DEVILLE, CONCOURS, ELDORADO AND SEVILLE, CONTINUED

DIAGNOSTIC TROUBLE CODE 72
SERIAL DATA LINK FAULT

(1)
- IGNITION 'OFF'
- INSTALL TECH 1
- IGNITION 'ON'
- ATTEMPT TO ESTABLISH COMMUNICATIONS WITH THE EBTCM/EBCM; CAN YOU COMMUNICATE WITH IT?

OTHER CODES:
- NO TROUBLE FOUND; SEE 'DIAGNOSTIC AIDS' ON FACING PAGE

NO:

(2)
- IGNITION 'OFF'
- DISCONNECT EBTCM/EBCM
- DISCONNECT TECH 1
- INSTALL J 39700 UNIVERSAL BREAKOUT BOX TO EBTCM/EBCM HARNESS CONNECTOR ONLY
- IGNITION 'ON'
- CONNECT DVM BETWEEN J 39700 TERMINALS '20' AND '30'
- NOTE VOLTAGE
- IS IT AROUND SYSTEM VOLTAGE?

YES:
- REPAIR SHORT TO SYSTEM VOLTAGE IN CKT 799

NO:

(3)
- IGNITION 'OFF'
- CHECK RESISTANCE BETWEEN J 39700 TERMINALS '20' AND '30'
- IS IT LESS THAN 5 Ω?

YES:
- REPAIR SHORT TO GROUND IN CKT 799

NO:

(4)
- CHECK RESISTANCE BETWEEN J 39700 TERMINAL '30' AND DLC TERMINAL 'M' USING ADAPTER FROM J 35616–A
- IS IT LESS THAN 5 Ω?

YES:
- REPAIR OPEN IN CKT 799

NO:

(5)
- IGNITION 'OFF'
- REMOVE J 39700
- RECONNECT EBTCM/EBCM
- INSTALL TECH 1
- IGNITION 'ON'
- ATTEMPT TO ESTABLISH COMMUNICATIONS WITH THE EBTCM/EBCM; CAN YOU COMMUNICATE WITH IT?

YES:
- NO TROUBLE FOUND; REFER TO 'DIAGNOSTIC AIDS' ON FACING PAGE

NO:

(6)
- CHECK FOR POOR CONNECTIONS AT EBTCM/EBCM AND DLC CONNECTORS; IF OK, REPLACE EBTCM/EBCM

WHEN ALL DIAGNOSIS AND REPAIRS ARE COMPLETED, CLEAR CODES AND VERIFY OPERATION

DIAGNOSTIC CHART DTC 73 — 1994 DEVILLE, CONCOURS, ELDORADO AND SEVILLE

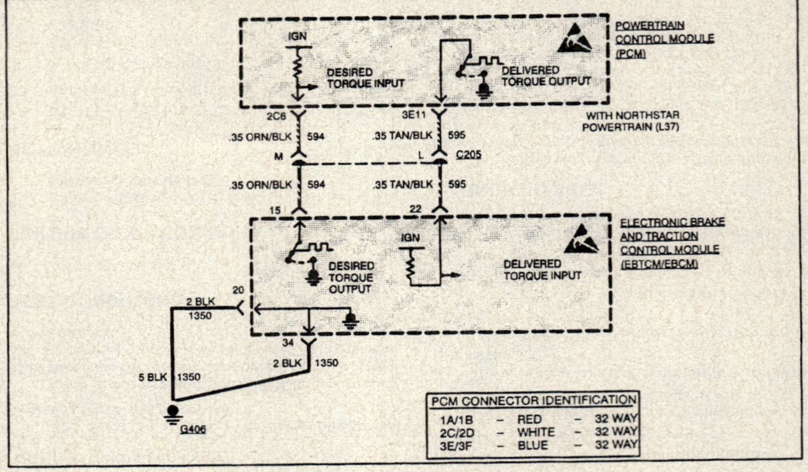

DIAGNOSTIC TROUBLE CODE 73
PCM–EBTCM/EBCM PWM SIGNAL FAULT (4.6L ONLY)

CIRCUIT DESCRIPTION:

Traction control is simultaneously controlled by the EBTCM on the vehicles equipped with the 4.6L Northstar, and the PCM. The PCM receives a 'DESIRED TORQUE' request via a Pulse Width Modulated (PWM) signal from the EBTCM (CKT 594) requesting the desired torque level for proper traction control system operation. Torque reduction will be indicated if the duty cycle of the PWM signal is between 10% (100 % torque reduction) and 90% (no torque reduction) and at least one (up to five) cylinder is requested to be disabled. The percent duty cycle will indicate the percent torque required. The PCM will selectively disable fuel to certain cylinders to achieve this reduction and communicate the actual torque output back to the EBTCM (CKT 595).

TEST CONDITION(S):

1. Tested continuously.

FAILURE CONDITION(S):

1. While Standing or Driving: PWM interface between EBTCM/EBCM and PCM is lost.

ACTION TAKEN:

1. Code set.
2. 'TRACTION DISABLED' message 'ON'.

A Code 73 may set if either CKT 594 or 595 is open, shorted to ground or shorted to voltage.

If CKT 595, Delivered Torque, is damaged and the EBTCM loses this input, a TCS Diagnostic Code 73 will be set. The PCM monitors the serial data line to see if the traction control has been disabled. If it has, the PCM will not set Code 93.

If CKT 594, Desired Torque is damaged and the PCM loses this input, the PCM monitors the serial data line to see if the traction control has been disabled. If not, the PCM sets a Code 93 and then tells the EBTCM/EBCM, via serial data, that there is a PWM circuit problem and the EBTCM/EBCM sets a Code 73.

TEST DESCRIPTION: Test numbers refer to circled numbers on the diagnostic chart.

1. Checks if the PCM Diagnostic Trouble Code 93 is set. If it is, there is likely a problem with CKT 594.
2. Checks for an open or high resistance in CKT 595.
3. Checks for a short to ground in CKT 595.
4. Checks for a short to voltage in CKT 595.
5. If no problem is found up until this point, CKT 594 should be checked.
6. Checks for an open or high resistance in CKT 594.
7. Checks for a short to ground in CKT 594.
8. Checks for a short to voltage in CKT 594.
9. If CKT 595 has not been checked, proceed to the left hand side of the trouble tree. If it has, continue downward. Problem is intermittent or in one of the controllers.
10. Checks if Code 73 and/or 93 are intermittent. If not, trouble is in either PCM or EBTCM.

DIAGNOSTIC AIDS:

- It is theoretically impossible for only a PCM Code 93 to set. If only a PCM Code 93 is present, either the TCS Code 73 has been cleared or there is a problem with the PCM.
- A Code 73 may set by itself if the PCM PROM has been improperly installed.

DIAGNOSTIC CHART DTC 73 — 1994 DEVILLE, CONCOURS, ELDORADO AND SEVILLE, CONTINUED

DIAGNOSTIC TROUBLE CODE 73
PCM–EBTCM/EBCM PWM SIGNAL FAULT

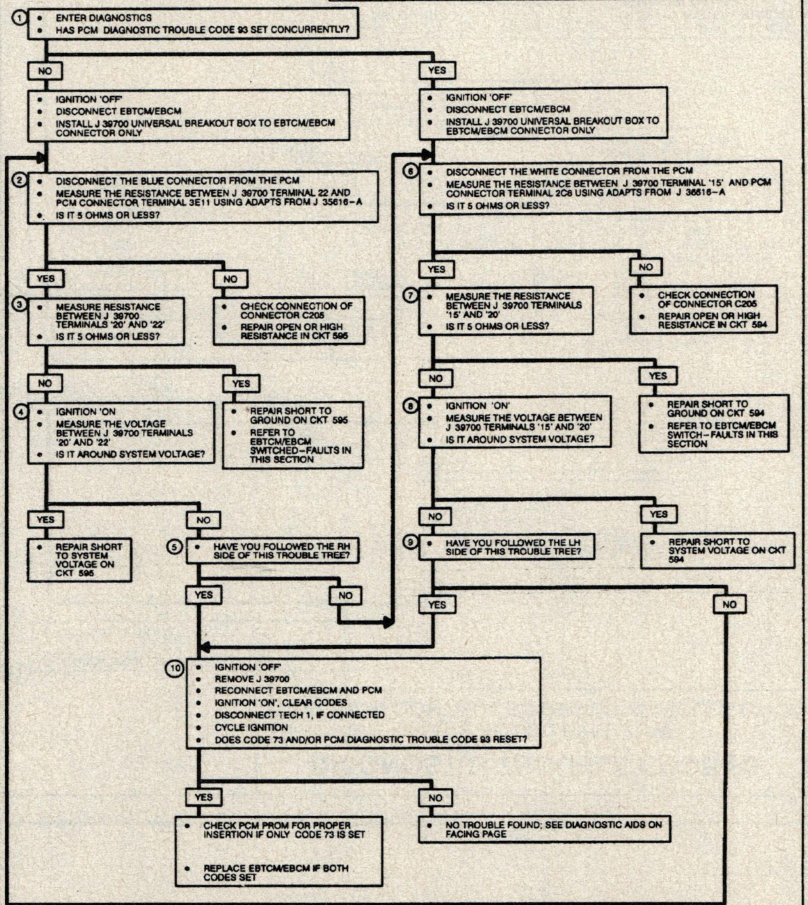

WHEN ALL DIAGNOSIS AND REPAIRS ARE COMPLETED, CLEAR CODES AND VERIFY OPERATION

DIAGNOSTIC CHART DTC 83 — 1994 DEVILLE, CONCOURS, ELDORADO AND SEVILLE

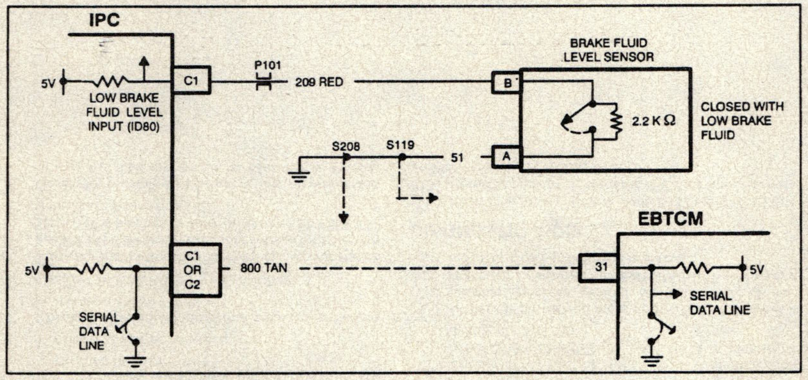

DIAGNOSTIC TROUBLE CODE 83
(TRACTION CONTROL EQUIPPED VEHICLES ONLY)

LOW BRAKE FLUID LEVEL

CIRCUIT DESCRIPTION:

The IPC outputs 5 volts on CKT 209 to the brake fluid level sensor. The IPC determines the level of brake fluid by the voltage drop it detects across the sensor circuit. This information is delivered to the EBTCM via the serial data line.

TEST CONDITION(S):

1. Tested continuously.

FAILURE CONDITION(S):

1. While Standing or Driving: Brake fluid falls below acceptable level.
2. Brake fluid level sensor disconnected.
3. Open, short to ground or short to battery on CKT 209.

ACTION TAKEN:

1. Code set.
2. ABS and TCS switched 'OFF'.
3. Red BRAKE warning lamp and 'ANTILOCK' indicator ON.
4. 'TRACTION DISABLED' message ON.

TEST DESCRIPTION: Test numbers refer to circled numbers on the diagnostic chart.

1. Check the brake fluid level in master cylinder reservoir.
2. Check brake fluid level as detected by IPC.
3. Brake fluid level not acceptable. Fill and recheck.
4. Brake fluid level as detected by IPC. See 'DIAGNOSTIC AIDS'.
5. Checks IPC input parameter ID80 with sensor disconnected for a short to ground in CKT 209, or bad sensor.
6. Checks IPC input parameter ID80 with sensor disconnected and connector shorted.
7. IPC detects acceptable brake fluid level. Problem not present.
8. Checks sensor resistor and checks for a short between CKTs 901 and 51.
9. Checks for open in CKTs 209 and 51.

DIAGNOSTIC AIDS:

IPC data parameter ID80 will display the following:

- 3.6–5V – open in sensor circuit.
- 1.5–3.5V – circuit OK. Problem not present.
- 0–1.4V – short between CKTs 209 and 51.

DIAGNOSTIC CHART DTC 83 — 1994 DEVILLE, CONCOURS, ELDORADO AND SEVILLE, CONTINUED

DIAGNOSTIC TROUBLE CODE 83

BRAKE FLUID LEVEL LOW (TRACTION CONTROL EQUIPPED VEHICLES ONLY)

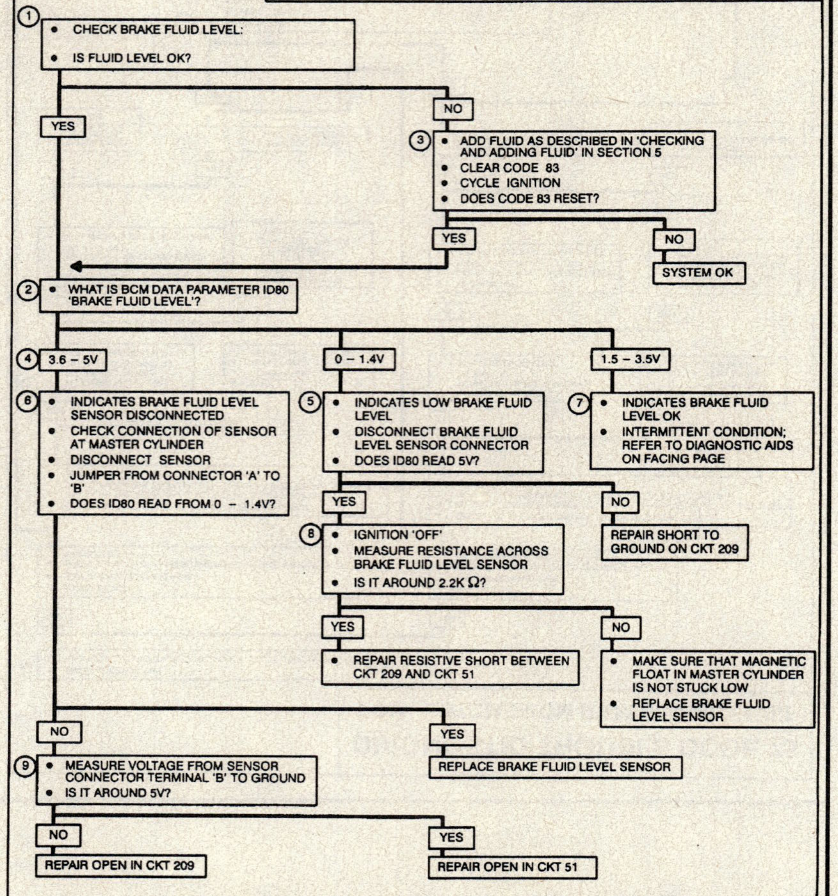

WHEN ALL DIAGNOSIS AND REPAIRS ARE COMPLETED, CLEAR CODES AND VERIFY OPERATION

DIAGNOSTIC CHART A — 1994 DEVILLE, CONCOURS, ELDORADO AND SEVILLE

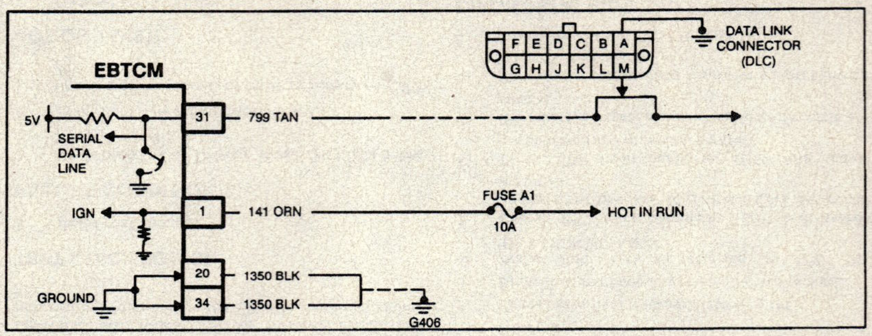

CHART A

'ANTILOCK' INDICATOR AND 'TRACTION DISABLED' MESSAGE 'ON', (IF APPLICABLE) NO CODES SET / NO TCS DATA

TEST DESCRIPTION: Test numbers refer to circled numbers on the diagnostic chart.

1. Checks if condition is due to a poor connection at the EBTCM/EBCM. Make sure the connecter is not 'cocked' on module.
2. Verifies that fuse A1 is intact.
3. Checks EBTCM/EBCM ground resistances. Should be less than 1 ohm.
4. Checks EBTCM/EBCM ignition voltage.
5. This step checks for possible open or high resistance in CKT 141 if there is insufficient voltage measured in Step 4. It is assumed a short to ground on CKT 141 would have blown fuse 'A1'.
6. Checks for possible insufficient voltage on power side of fuse 'A1'. This may be due to Maxi® fuse 6 being damaged or an intermittent connection in the ignition switch.
7. Checks for EBTCM/EBCM terminal '1' having a poor solder joint or for being partially backed out.
8. Checks if a loss of serial data communication between EBTCM/EBCM and BCM is due to a damaged CKT 799.
9. If there is no voltage measured in Step 8, there is a possible short to ground on CKT 799. If not, regulated 5V supply from EBTCM/EBCM is lost and EBTCM/EBCM should be replaced.

DIAGNOSTIC AIDS:

BCM Diagnostic Trouble Code B332 and PCM Code P137 should be present. If all of the IPC data line codes are present, refer to 'SYSTEM CHECK'.

DIAGNOSTIC CHART A — 1994 DEVILLE, CONCOURS, ELDORADO AND SEVILLE, CONTINUED

CHART A

'ANTILOCK' INDICATOR AND 'TRACTION DISABLED' MESSAGE 'ON', NO CODES SET / NO TCS DATA

- IPC CODE I332 SHOULD BE PRESENT
- IF ALL OF THE IPC DATA LINE CODES ARE PRESENT, REFER TO 'SYSTEM CHECK'

①
- VERIFY THAT THE EBTCM/EBCM CONNECTOR IS SECURELY ATTACHED TO THE EBCM. IF DISCONNECTED, CONNECT AND VERIFY OPERATION

②
- IF FUSES HAVE NOT ALREADY BEEN CHECKED, INSPECT FUSE A1 IN THE CONSOLE FUSE BLOCK

FUSE OK → ③
- INSTALL J 39700 UNIVERSAL BREAKOUT BOX TO BOTH EBTCM/EBCM CONNECTOR AND EBTCM/EBCM HARNESS CONNECTOR
- MEASURE RESISTANCE BETWEEN PIN 20 AND BODY GROUND
- MEASURE RESISTANCE BETWEEN PIN 34 AND BODY GROUND

FUSE BLOWN →
- REPLACE FUSE AND VERIFY OPERATION

BOTH 0 – 1 OHM → ④
- IGNITION SWITCH IN RUN
- MEASURE VOLTAGE FROM PIN 1 AND PIN 20

EITHER OR BOTH GREATER THAN 1 OHM →
- REPAIR OPEN OR HIGH RESISTANCE IN CKT 1350 BETWEEN EBTCM/EBCM AND NEGATIVE JUNCTION BLOCK.
- CLEAN AND TIGHTEN EBCM/EBTCM GROUND G408. SAND EYELETS ON HARNESS

LESS THAN 9 VOLTS → ⑤
- IGNITION SWITCH OFF
- FUSE A1 REMOVED
- MEASURE RESISTANCE FROM GROUND SIDE OF FUSE TO PIN 1

GREATER THAN 9 VOLTS → ⑦
- CHECK TERMINAL 1 OF EBTCM/EBCM
- CHECK PIN 1 OF EBTCM/EBCM CONNECTOR FOR INTERMITTENT OPEN OR POOR CRIMP
- WIGGLE CKT 141 FOR POSSIBLE INTERMITTENT OPEN

⑤ LESS THAN 1 OHM → ⑥
- IGNITION SWITCH IN RUN
- MEASURE VOLTAGE ON POWER SIDE OF FUSE A1

⑤ GREATER THAN 1 OHM →
- REPAIR OPEN OR PARTIAL OPEN IN CKT 141 BETWEEN TRUNK FUSE BLOCK AND EBTCM/EBCM CONNECTOR

⑥ LESS THAN 9 VOLTS →
- VERIFY MAXI® FUSE 6 (60A) IS OK (LOCATED IN LH MAXI® FUSE BLOCK)
- DISCONNECT NEGATIVE BATTERY CABLE
- INSPECT INTERIOR RELAY CENTER SPADE TERMINALS FOR POOR CRIMP
- INSPECT WIRING AT CONSOLE FUSE BLOCK FOR OPENS
- REPAIR OPEN IN CIRCUIT BETWEEN IGNITION SWITCH AND FUSEBLOCK

⑥ GREATER THAN 9 VOLTS →
- DISCONNECT NEGATIVE BATTERY CABLE
- INSPECT FUSE BLOCK SPADE TERMINALS FOR POOR CRIMP
- INSPECT WIRING AT CONSOLE FUSE BLOCK FOR OPENS
- REPLACE FUSE

⑧
- NO TROUBLE FOUND
- POSSIBLE OPEN IN CKT 799
- IGNITION 'ON'
- MEASURE VOLTAGE FROM BREAKOUT BOX TERMINAL '31' TO TERMINAL '20'

0V → ⑨
- IGNITION OFF
- MEASURE RESISTANCE FROM BREAKOUT BOX TERMINAL '31' TO TERMINAL '20'
- IS IT 5 OHMS OR LESS?

5V →
- OPEN IN CKT 799

YES →
INSPECT WIRING AND COMPONENTS

NO →
- REPLACE EBTCM/EBCM

WHEN ALL DIAGNOSIS AND REPAIRS ARE COMPLETED, CLEAR CODES AND VERIFY OPERATION

DIAGNOSTIC CHART B — 1994 DEVILLE, CONCOURS, ELDORADO AND SEVILLE

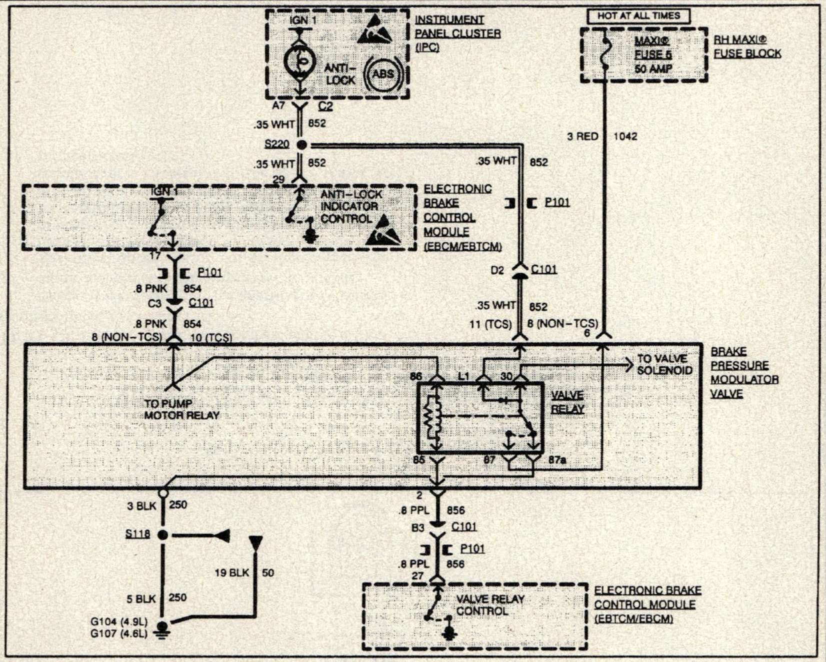

CHART B

'ANTILOCK' INDICATOR 'ON', NO CODES SET

TEST DESCRIPTION: Test numbers refer to circled numbers on the diagnostic chart.

1. Checks if 'ANTILOCK' indicator is ON due to a loss of communication between the IPC and EBTCM/EBCM.
2. Checks for a short to ground in CKT 852.
3. Final check for diagnostic trouble codes.

DIAGNOSTIC AIDS:

The EBTCM/EBCM controls the 'ANTILOCK' indicator only in instances where serial data communication with the IPC is lost. In all other instances the IPC controls this lamp.

DIAGNOSTIC CHART B — 1994 DEVILLE, CONCOURS, ELDORADO AND SEVILLE, CONTINUED

CHART B

'ANTILOCK' INDICATOR 'ON', NO TCS CODES SET

(1)
- IF 'ANTILOCK' INDICATOR OR 'TRACTION DISABLED' MESSAGE ARE 'ON' AND IPC DIAGNOSTIC TROUBLE CODE 1332 IS SET, REFER TO CHART A

(2)
- IGNITION 'OFF'
- DISCONNECT EBTCM/EBCM AND REMOVE VALVE SOLENOID RELAY
- IGNITION 'ON'
- DOES 'ANTILOCK' INDICATOR TURN OFF?

YES:

(3)
- IGNITION 'OFF'
- RECONNECT EBTCM/EBCM AND REINSTALL VALVE SOLENOID RELAY
- IGNITION 'ON'
- IS 'ANTILOCK' INDICATOR 'ON' AND/OR THE 'ANTILOCK' INDICATOR OR 'TRACTION DISABLED' MESSAGE 'ON' WITH NO IPC CODE 1332 OR DIAGNOSTIC TROUBLE CODES?

NO:
- IF CODES ARE SET, REFER TO APPROPRIATE TROUBLE TREE
- IF THERE ARE NO MESSAGES OR CODES, PERFORM IPC DIAGNOSIS

YES:
- REPLACE EBTCM/EBCM

NO:
- IGNITION 'OFF'
- INSTALL J 39700 UNIVERSAL BREAKOUT BOX TO EBTCM/EBCM HARNESS CONNECTOR ONLY
- MEASURE RESISTANCE FROM PIN 29 TO PIN 20
- IS IT LESS THAN 5 Ω?

NO:
- NO TROUBLE FOUND
- CHECK BPM VALVE AND EBTCM/EBCM CONNECTOR TERMINAL INTEGRITY
- CHECK VALVE SOLENOID RELAY TERMINAL INSERTION

YES:
REPAIR SHORT TO GROUND ON CKT 852

WHEN ALL DIAGNOSIS AND REPAIRS ARE COMPLETED, CLEAR CODES AND VERIFY OPERATION

DIAGNOSTIC CHART C — 1994 DEVILLE, CONCOURS, ELDORADO AND SEVILLE

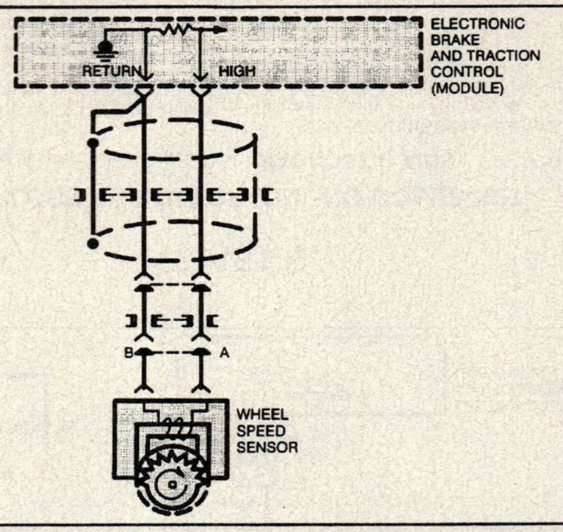

CHART C

VALVE CYCLING (CHATTER) DURING NORMAL STOPS (UNWANTED ABS MODULATION)

TEST DESCRIPTION: Test numbers refer to circled numbers on the diagnostic chart.

1. Checks if valve activation due to erroneous electrical 'noise' induced onto wheel speed sensor circuits.
2. The wheel speed sensor output should gradually increase with vehicle speed. There should be no skips or jumps. All four wheel speed sensor outputs should be identical. On a slow down, one of the signals may drop off faster than the others, indicating a lock–up condition or a problem with the speed sensor or wiring (corrosion etc.).

DIAGNOSTIC CHART C — 1994 DEVILLE, CONCOURS, ELDORADO AND SEVILLE, CONTINUED

CHART C

VALVE CYCLING (CHATTER) DURING NORMAL STOPS (UNWANTED ABS MODULATION)

①
- VERIFY CORRECT WSS CABLE ROUTING
- CHECK SENSOR MOUNTING AND CONNECTION
- ROAD TEST VEHICLE

VALVE CHATTER

②
- INSTALL TECH 1
- OBSERVE WHEEL SPEED SENSOR OUTPUT WHILE TEST DRIVING VEHICLE

RF SENSOR OUTPUT
- ERRATIC OR
- DROPS OUT OR
- DIFFERS FROM OTHER OUTPUTS

→
- GO TO CODE 21 CHART
- BEGIN ANALYSIS AT NOTE 2 OF CODE 21 CHART

LF SENSOR OUTPUT
- ERRATIC OR
- DROPS OUT OR
- DIFFERS FROM OTHER OUTPUTS

→
- GO TO CODE 25 CHART
- BEGIN ANALYSIS AT NOTE 2 OF CODE 25 CHART

RR SENSOR OUTPUT
- ERRATIC OR
- DROPS OUT OR
- DIFFERS FROM OTHER OUTPUTS

→
- GO TO CODE 31 CHART
- BEGIN ANALYSIS AT NOTE 2 OF CODE 31 CHART

LR SENSOR OUTPUT
- ERRATIC OR
- DROPS OUT OR
- DIFFERS FROM OTHER OUTPUTS

→
- GO TO CODE 35 CHART
- BEGIN ANALYSIS AT NOTE 2 OF CODE 35 CHART

NORMAL SENSOR OUTPUT

→
- REPLACE EBTCM /EBCM

NO VALVE CHATTER

- CONDITION CORRECTED

WHEN ALL DIAGNOSIS AND REPAIRS ARE COMPLETED, CLEAR CODES AND VERIFY OPERATION

DIAGNOSTIC CHART D — 1994 DEVILLE, CONCOURS, ELDORADO AND SEVILLE

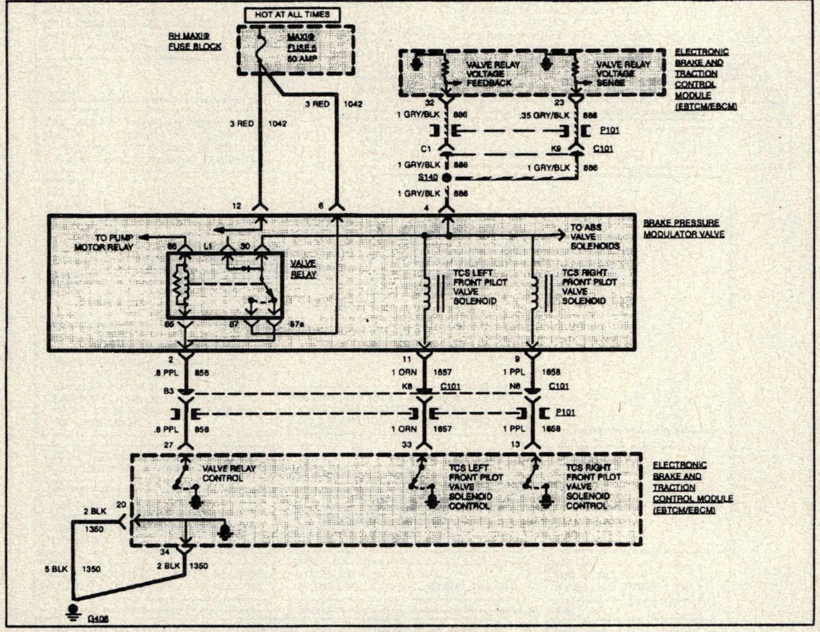

CHART D

TRACTION CONTROL OPERATION CHECK (NO DIAGNOSTIC TROUBLE CODES SET)

TEST DESCRIPTION: Test numbers refer to circled numbers on the diagnostic chart.

1. Traction control system should interpret the elevated front wheels as an icy or free spin condition. At $^1/_2$ throttle a properly operating system will allow both front wheels to spin up quickly to about 20 – 25 MPH and then slow down (but not stop) to about 5 – 10 MPH and have a jerky appearance. The traction control 'ACTIVE' message should display on the DIC. The pump motor should also be heard.
2. System is building up pressure to the front brake and operating correctly.
3. System is not building up pressure to the front brakes. Possible air or dirt in the system.
4. No dirt in reservoir and the fluid level is OK. Air may still be trapped in system. The TCS hydraulic circuit should be bled using TECH 1 TCS test and bleed procedure outlined in Section 5. One of the front caliper bleed screws must be open to allow air to escape.
5. Fluid level OK. Air may still be trapped in the system. The TCS hydraulic circuit should be bled using TECH 1 TCS test and bleed procedure outlined in Section 5. One of the front caliper bleed screws must be open to allow air to escape.
6. Reservoir TCS prime pipe may be contaminated. Flush pipe and reservoir to remove any debris. Air may still be trapped in system. The TCS hydraulic circuit should be bled using TECH 1 TCS test and bleed procedure outlined in Section 5. One of the front caliper bleed screws must be open to allow air to escape.

DIAGNOSTIC AIDS:

- The Brake Light Switch must be properly adjusted to prevent an erroneous brake application input to the EBTCM/EBCM. Misadjustment may prevent brake intervention during traction control operation. Also, the EBTCM/EBCM receives TCC switch status over serial data from the PCM. If serial data is lost or the switch remains engaged, brake intervention during traction control may be lost. Monitor PCM TCC switch input PI71 while cycling brake switch. The input should cycle 'HI' and 'LO'.
- The PCM shall not allow traction control fuel shutoff if:
  - Engine Coolant Temperature less than −40°C
  - Engine Coolant Temperature greater than or equal to 131°C
  - Low Coolant Level Detected
- Cylinders shall be disabled if the engine speed greater than or equal to 600 RPM.
- The PCM can disable traction control for up to 11 ignition cycles if override PS24, Traction Control Disable, is activated. When PS24 is selected, the display will alternate between '– –' for one second and the current state of Traction Control Disable override ('99' for override active and '00' for override not active). Pressing the WARMER button will disable traction control for the next eleven ignition cycles and display '99'. The 'TRACTION DISABLED' message will be displayed for the duration of this override and no TCS diagnostic trouble codes will be set. Pressing the COOLER button will allow normal program control of traction control and display '00'.

DIAGNOSTIC CHART D — 1994 DEVILLE, CONCOURS, ELDORADO AND SEVILLE, CONTINUED

CHART D

TRACTION CONTROL OPERATION CHECK (NO DIAGNOSTIC TROUBLE CODES SET)

NOTE: SOME OF THESE CHECKS WILL CAUSE DIAGNOSTIC TROUBLE CODES TO SET. BE SURE TO VERIFY AUTHENTICITY OF ANY CODES AND CORRECT IF NECESSARY. BE SURE TO CLEAR ANY ERRONEOUS OR CORRECTED CODES.

(1)
- CLEAR CODES
- IGNITION 'OFF'
- DISCONNECT TECH 1
- RAISE FRONT WHEELS 4" TO 5" ON HOIST
- IGNITION 'ON', ENGINE 'ON'
- WITH TRANSMISSION IN DRIVE, ACCELERATE TO $^1/_2$ THROTTLE WHILE AN OBSERVER MONITORS THE DRIVE WHEELS
- ARE THE FRONT BRAKES MODULATED?

YES → (2)
- SYSTEM IS OPERATING CORRECTLY

NO → (3)
- CHECK FLUID LEVEL IN MASTER CYLINDER RESERVOIR
- CHECK FOR DIRT OR DEBRIS IN RESERVOIR FLUID

LEVEL OK, FLUID NOT DIRTY → (4)
- POSSIBLE AIR IN SYSTEM
- BLEED TCS PRIME PIPE AT BPM VALVE
- PERFORM THE TCS TEST USING THE TECH 1 PER THE BLEED PROCEDURE.
  ONE OF THE FRONT CALIPER BLEED SCREWS MUST BE 'OPENED'
- RETEST TRACTION CONTROL OPERATION ON HOIST

LOW → (6)
- ADD FLUID
- POSSIBLE AIR IN SYSTEM
- BLEED TCS PRIME PIPE AT BPM VALVE
- PERFORM THE TCS TEST USING THE TECH 1 PER THE BLEED PROCEDURE.
  ONE OF THE FRONT CALIPER BLEED SCREWS MUST BE 'OPENED'
- RETEST TRACTION CONTROL OPERATION ON HOIST

DIRTY → (5)
- DISCONNECT TCS PRIME PIPE AT BPM VALVE
- ALLOW FLUID TO DRAIN FROM RESERVOIR
- FLUSH RESERVOIR AND TCS PRIME PIPE WITH FRESH BRAKE FLUID
- RECONNECT TCS PRIME PIPE
- REFILL
- POSSIBLE AIR IN SYSTEM
- BLEED TCS PRIME PIPE AT BPM VALVE
- PERFORM THE TCS TEST USING THE TECH 1 PER THE BLEED PROCEDURE.
  ONE OF THE FRONT CALIPER BLEED SCREWS MUST BE 'OPENED'
- RETEST TRACTION CONTROL OPERATION ON HOIST

WHEN ALL DIAGNOSIS AND REPAIRS ARE COMPLETED, CLEAR CODES AND VERIFY OPERATION

DIAGNOSTIC CHART ABS/ASR DIAGNOSTIC SYSTEM CHECK — 1995 FLEETWOOD

ABS/TCS DIAGNOSTIC SYSTEM CHECK

- ENGAGE PARKING BRAKE.
- IGNITION TO "RUN."
  DOES THE "BRAKE" WARNING LAMP COME "ON" STEADY?

NO → INSPECT BRAKE WARNING LAMP CIRCUIT.

YES →
- DISENGAGE PARKING BRAKE.
  DOES THE BRAKE WARNING LAMP GO "OFF"?

NO → CHECK PARKING BRAKE SWITCH.

YES →
- IGNITION "OFF."
- WHILE OBSERVING THE "ANTI-LOCK" INDICATOR TURN IGNITION TO "RUN."

INDICATOR DOES NOT COME "ON" OR FLICKERS VERY BRIEFLY. → GO TO THE APPROPRIATE CHART AS INDICATED BY THE "ABS/TCS SYMPTOM TABLE" IN THIS SECTION.

INDICATOR COMES "ON" AND STAYS "ON." → (to INSTALL TECH 1 step)

INDICATOR COMES "ON" STEADY FOR ABOUT 2 SECONDS, THEN GOES "OFF." →
- IGNITION "OFF."
- WHILE OBSERVING THE "TRACTION CONTROL" INDICATOR TURN IGNITION TO "RUN."

INDICATOR DOES NOT COME "ON." → GO TO THE APPROPRIATE CHART AS INDICATED BY THE "ABS/TCS SYMPTOM TABLE" IN THIS SECTION.

INDICATOR COMES "ON" AND STAYS "ON." → (to INSTALL TECH 1 step)

INDICATOR COMES "ON" STEADY FOR ABOUT 2 SECONDS, THEN GOES "OFF." →
- IGNITION "OFF."
- WHILE OBSERVING THE "TRACTION ENGAGED" INDICATOR, TURN IGNITION TO "RUN."
- DOES THE "TRACTION ENGAGED" INDICATOR COME "ON" STEADY FOR ABOUT 2 SECONDS, THEN GO "OFF"?

NO → GO TO THE APPROPRIATE CHART AS INDICATED BY THE "ABS/TCS SYMPTOM TABLE" IN THIS SECTION.

YES →
- ROAD TEST VEHICLE FOR AT LEAST TWO MINUTES, REACHING AT LEAST 25 km/h (15 MPH).
  DID THE "ANTI-LOCK" AND/OR "TRACTION CONTROL" INDICATOR(S) COME "ON"?

YES →
- INSTALL TECH 1 WITH MASS STORAGE CARTRIDGE.
- IGNITION TO "RUN."
- REQUEST ABS/TCS DTC DISPLAY, RECORD ALL DTCs ON REPAIR ORDER.

DTCs PRESENT. → GO TO "ABS/TCS DTC TABLE" IN THIS SECTION.

NO DTCs PRESENT AND/OR TECH 1 DISPLAYS "NO DATA RECEIVED FROM EBCM/EBTCM." → GO TO "ABS/TCS SYMPTOM TABLE" IN THIS SECTION FOR APPROPRIATE CHART TO DIAGNOSE INDICATOR "ON" WITH NO DTCs SET.

TECH 1 CANNOT COMMUNICATE WITH EBTCM.

NO →
DID THE VEHICLE BRAKES EXHIBIT ANY SYMPTOMS OF IMPROPER OPERATION THAT MAY BE MECHANICAL IN NATURE AND/OR DID THE "BRAKE" WARNING LAMP COME "ON"?

YES → CHECK BRAKE SYSTEM.

NO → SYMPTOM IS FUNCTIONAL AND FREE OF MALFUNCTIONS. NO FURTHER DIAGNOSIS IS REQUIRED.

DIAGNOSTIC TROUBLE CODE CHART — 1995 FLEETWOOD

ABS/TCS SYMPTOM TABLE	
CHART	SYMPTOM
A	"ANTI-LOCK" Indicator "ON" With No DTCs Set
B	"ANTI-LOCK" Indicator Inoperative or Flickers Briefly at Ingition "RUN"
C	"TRACTION CONTROL" Indicator "ON" With No DTCs Set
D	"TRACTION CONTROL" Indicator Inoperative
E	"TRACTION ENGAGED" Indicator "ON" With No DTCs Set
F	"TRACTION ENGAGED" Indicator Inoperative With No DTCs Set
	"BRAKE" Warning Lamp "ON"
	"BRAKE" Warning Lamp Inoperative

DTC	DESCRIPTION
21	RF Wheel Speed Sensor Circuit Malfunction
23	RF Wheel Speed Sensor Circuit Continuity Malfunction
25	LF Wheel Speed Sensor Circuit Malfunction
27	LF Wheel Speed Sensor Circuit Continuity Malfunction
28	Wheel Speed Sensor Circuit Frequency Malfunction
31	RR Wheel Speed Sensor Circuit Malfunction
33	RR Wheel Speed Sensor Circuit Continuity Malfunction
35	LR Wheel Speed Sensor Circuit Malfunction
37	LR Wheel Speed Sensor Circuit Continuity Malfunction
41	RF Inlet Valve Solenoid Malfunction
42	RF Outlet Valve Solenoid Malfunction
45	LF Inlet Valve Solenoid Malfunction
46	LF Outlet Valve Solenoid Malfunction
47	TCS Prime Valve (ASV) Malfunction
48	TCS Master Cylinder Isolation Valve (USV) Malfunction
51	RR Inlet Valve Solenoid Malfunction
52	RR Outlet Valve Solenoid Malfunction
55	LR Inlet Valve Solenoid Malfunction
56	LR Outlet Valve Solenoid Malfunction
58	EBTCM Internal Adjuster Assembly Malfunction
61	BPM Valve Pump Motor Malfunction
62	RPM Signal Malfunction
63	BPM Valve Power Supply Malfunction
64	Throttle Position Signal Malfunction
65	Adjuster Assembly Circuit Malfunction
66	Adjuster Assembly Control Malfunction
67	Throttle Position Sensor Comparison Malfunction
71	EBTCM Internal Malfunction
72	Serial Data Line Malfunction
73	Spark Retard Monitoring Malfunction
85	Low Voltage

ELECTRICAL CONNECTOR TERMINAL IDENTIFICATION — 1995 FLEETWOOD

C200

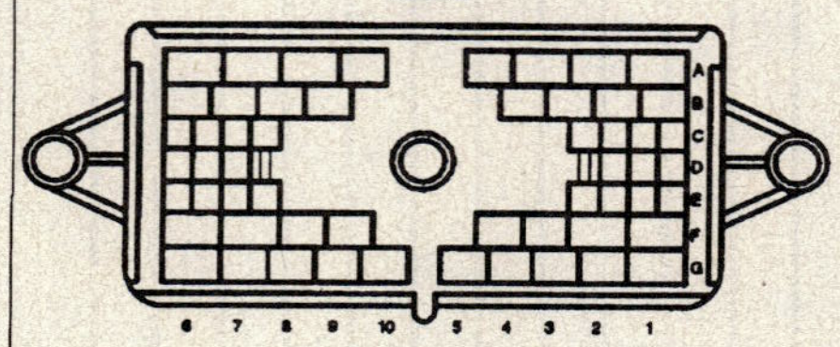

12034257
56 – WAY F METRI – PACK MIXED SERIES
BLK

C202

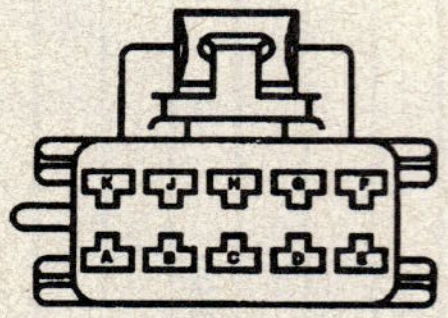

12064769
10 – WAY F METRI – PACK 150 SERIES
NAT
I/P HARNESS TO BODY HARNESS

C204

BODY HARNESS TO ENGINE HARNESS

12020916
37 – WAY F METRI – PACK MIXED SERIES
BLK

C402

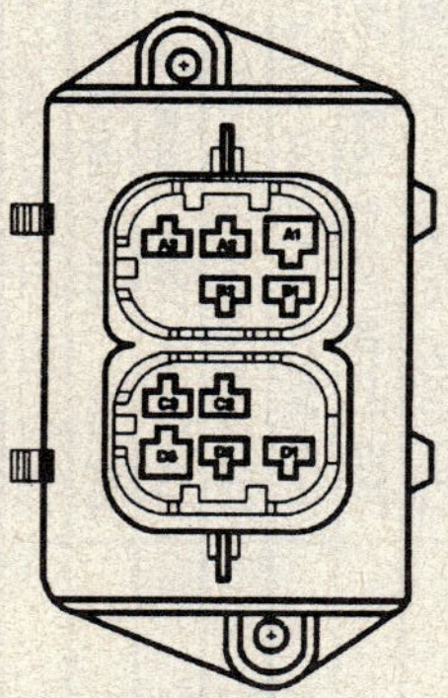

12129525
10 – WAY M METRI – PACK 150, 280 SERIES
BLK
ANTILOCK BRAKE SYSTEM PASS THRU BODY HARNESS TO ELECTRONIC BRAKE CONTROL HARNESS

ELECTRICAL CONNECTOR TERMINAL IDENTIFICATION — 1995 FLEETWOOD, CONTINUED

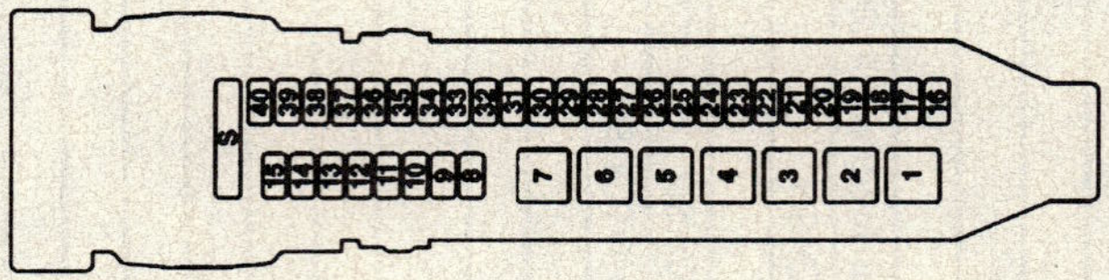

12124389
40 – WAY F CONNECTOR BODY
BLK
ELECTRONIC BRAKE AND TRACTION CONTROL MODULE (EBTCM)

CAVITIES NOT LISTED ARE NOT USED

CAVITY	WIRE COLOR	CKT	DESCRIPTION
1	BLK/WHT	51	GROUND
2	RED	1142	VOLTAGE FEED TO BPM PUMP MOTOR RELAY
3	RED	142	VOLTAGE FEED TO BPM SOLENOID VALVE RELAY
4	PNK	539	IGNITION FEED
5	RED	1142	BATTERY INPUT
6	WHT	560	THROTTLE RELAXER +
7	GRY	561	THROTTLE RELAXER -
10	GRY	416	TPS SENSOR 5 VOLT REFERENCE
12	BLK	452	SIGNAL GROUND
15	PPL/WHT	1572	TCS INDICATOR LIGHT
16	LT BLU	830	LEFT FRONT WHEEL SPEED SENSOR (HI)
17	YEL	873	LEFT FRONT WHEEL SPEED SENSOR (LOW)
18	BLK	884	LEFT REAR WHEEL SPEED SENSOR (HI)
19	RED	885	LEFT REAR WHEEL SPEED SENSOR (LOW)
20	BRN	882	RIGHT REAR WHEEL SPEED SENSOR (HI)
21	WHT	883	RIGHT REAR WHEEL SPEED SENSOR (LOW)
23	DK GRN	872	RIGHT FRONT WHEEL SPEED SENSOR (HI)
24	TAN	833	RIGHT FRONT WHEEL SPEED SENSOR (LOW)
25	WHT	627	TACH INPUT (ENGINE RPM)
28	YEL	820	BRAKE SWITCH INPUT
29	TAN	800	SERIAL DATA LINE
31	BRN/WHT	1571	TRACTION CONTROL SWITCH INPUT
34	DK BLU	417	TP SENSOR SIGNAL
35	GRY/BLK	1687	SPARK RETARD SIGNAL
36	LT GRN	1656	TCS ENGAGED INDICATOR LIGHT
40	WHT	852	ABS INDICATOR LIGHT
S	BLK/WHT	451	SHORTING BAR TO GROUND

DIAGNOSTIC CHART A — 1995 FLEETWOOD

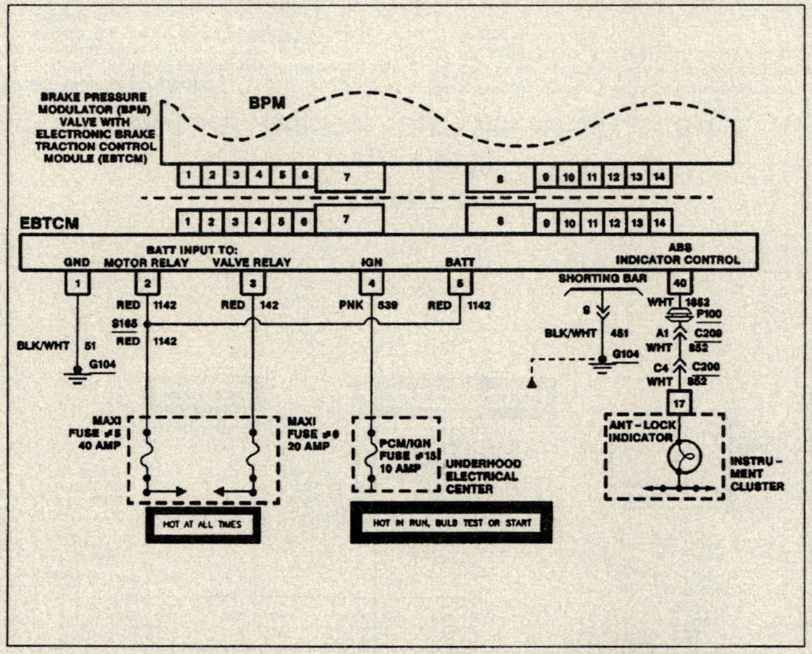

CHART A
(Page 1 of 2)
"Anti-Lock" Indicator "ON" With No DTC(s) Set

Chart Test Description:

Number(s) below refer to circled number(s) on the diagnostic chart.

1. Confirms that a malfunction condition exists.
2. Determines whether the malfunction is due to a DTC setting condition, an indicator low side circuit malfunction, or an EBTCM supply power malfunction.
3. Checks whether the malfunction is due to an open fuse.
4. Checks for an open power feed circuit.
5. Checks for an open in the ground feed to the EBTCM.
6. Verifies proper connection of the EBTCM ground CKT 51.
7. Checks whether the fuse failure is due to a short to ground.
8. Identifies if the short to ground found in Step 7 is in CKT 539 or the EBTCM.

DIAGNOSTIC CHART A — 1995 FLEETWOOD, CONTINUED

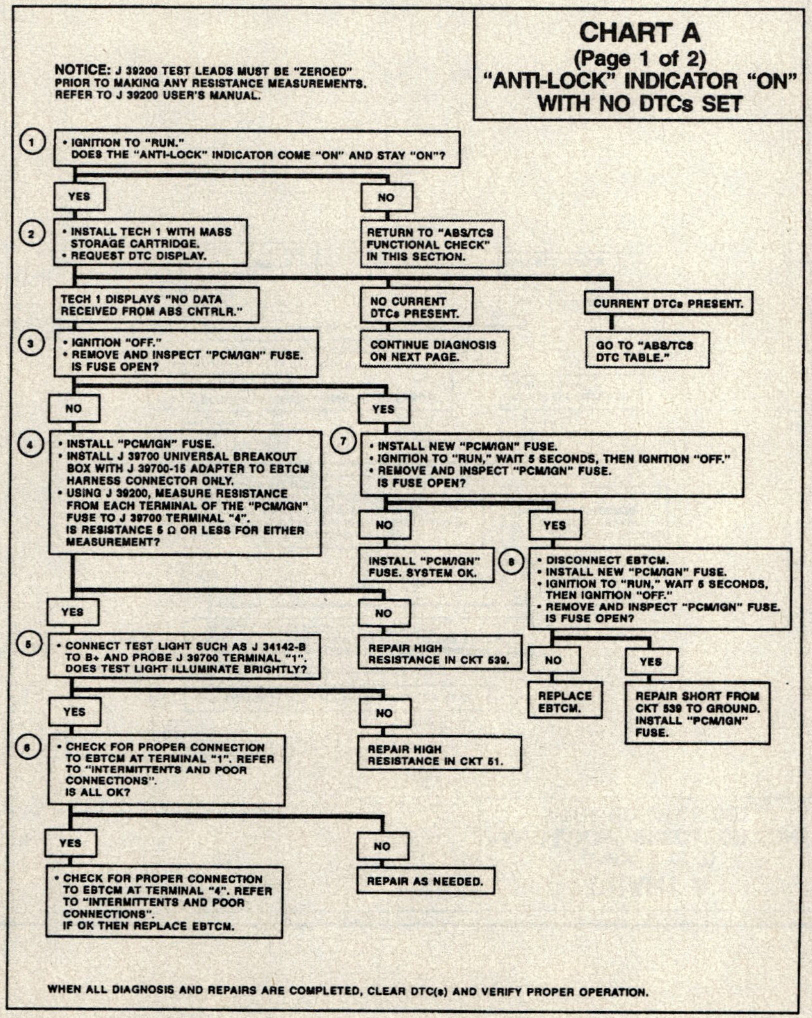

DIAGNOSTIC CHART A — 1995 FLEETWOOD, CONTINUED

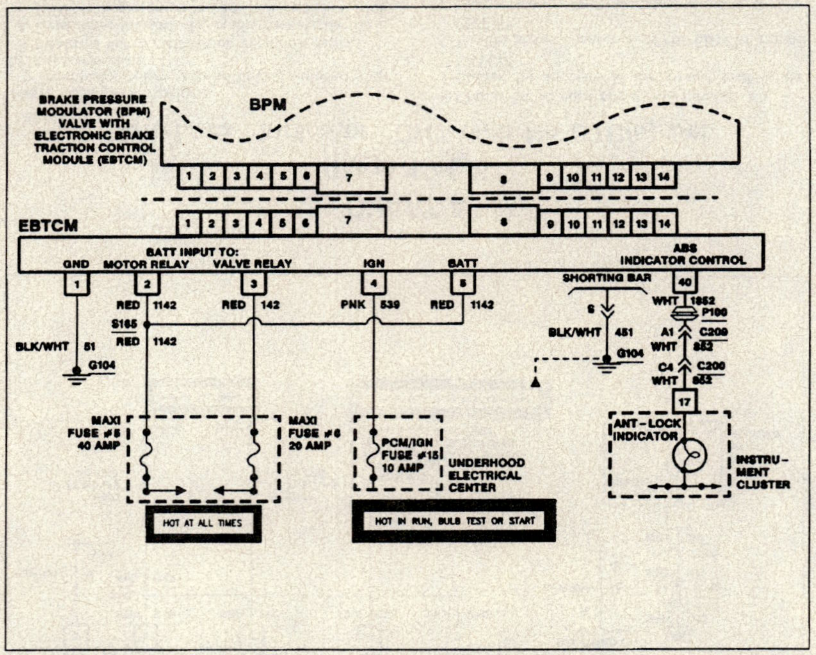

CHART A
(Page 2 of 2)
"Anti-Lock" Indicator "ON" With No DTC(s) Set

Chart Test Description:

Number(s) below refer to circled number(s) on the diagnostic chart.

9. Checks for proper retention of EBTCM harness connector.
10. Checks for damaged EBTCM harness connector.
11. Checks for a damaged connector (broken/missing lift finger) on the EBTCM itself, or for a short to ground in CKT 852.

DIAGNOSTIC CHART A — 1995 FLEETWOOD, CONTINUED

CHART A
(Page 2 of 2)
"ANTI-LOCK" INDICATOR "ON" WITH NO DTCs SET

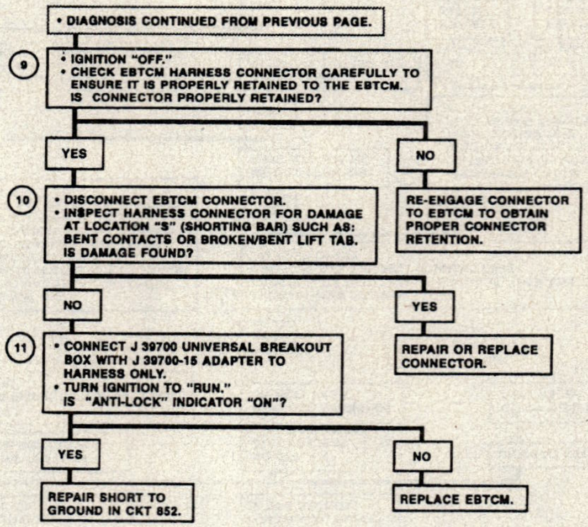

WHEN ALL DIAGNOSIS AND REPAIRS ARE COMPLETED, CLEAR DTC(s) AND VERIFY PROPER OPERATION.

DIAGNOSTIC CHART B — 1995 FLEETWOOD

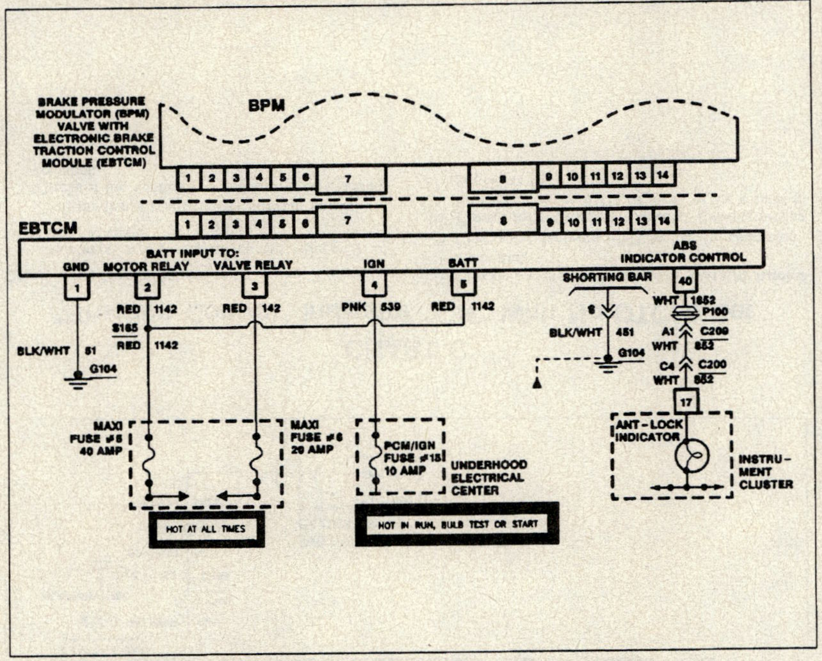

CHART B
"Anti-Lock" Indicator Inoperative Or Flickers Briefly At Ignition "RUN"

Chart Test Description:
Number(s) below refer to circled number(s) on the diagnostic chart.
1. Confirms that a malfunction condition exists.
2. Checks to see if the indicator can be turned "ON" manually.
3. Verifies the integrity of the indicator bulb.
4. Checks for an indicator low side circuit open.
5. Checks for an indicator low side circuit short to voltage.

DIAGNOSTIC CHART B — 1995 FLEETWOOD, CONTINUED

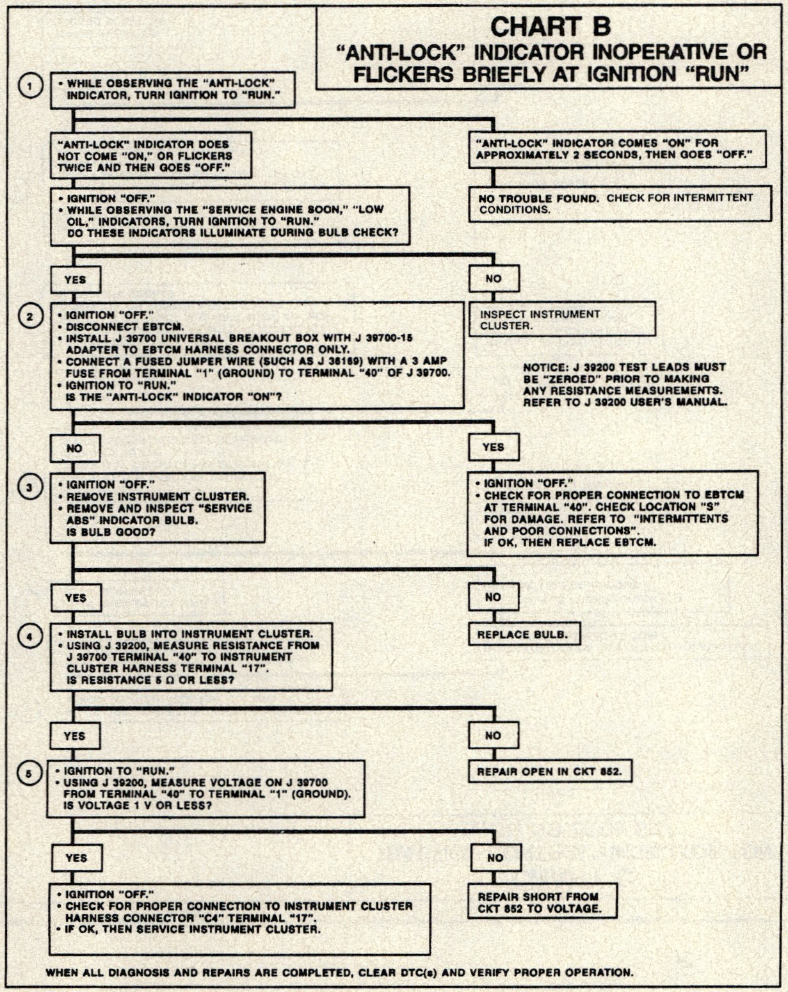

DIAGNOSTIC CHART C — 1995 FLEETWOOD

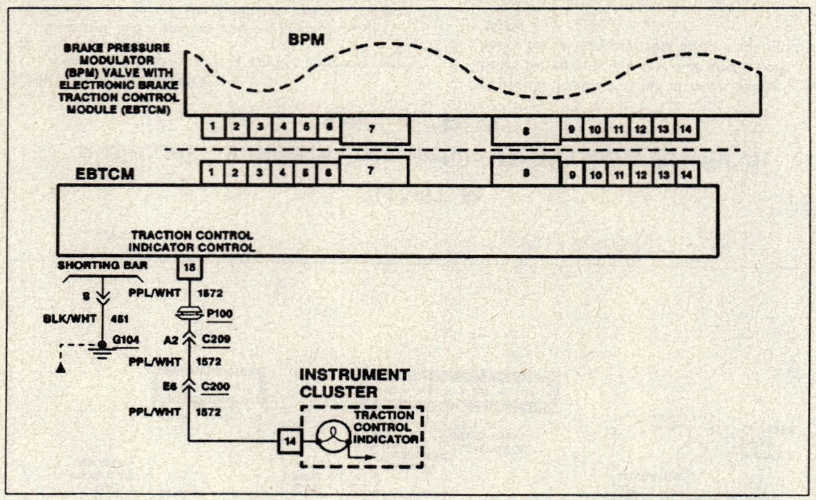

CHART C

"Traction Control" Indicator "ON" With No DTC(s) Set

Chart Test Description:

Number(s) below refer to circled number(s) on the diagnostic chart.

1. Confirms that a malfunction condition exists.
2. Checks if the malfunction is due to a DTC-setting condition.
3. Checks for proper retention of EBTCM harness connector.
4. Checks for damaged EBTCM harness connector.
5. Checks for a damaged connector (broken/missing lift finger) on the EBTCM itself, or for a short to ground in CKT 1572.

DIAGNOSTIC CHART C — 1995 FLEETWOOD, CONTINUED

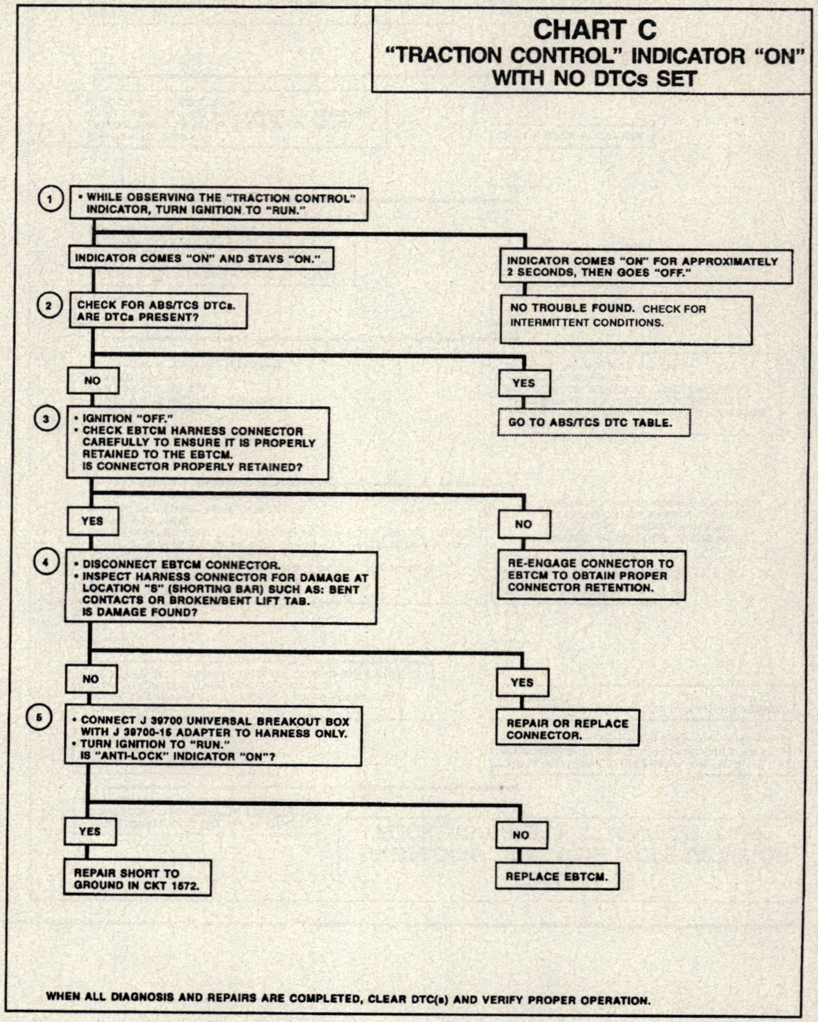

DIAGNOSTIC CHART D — 1995 FLEETWOOD

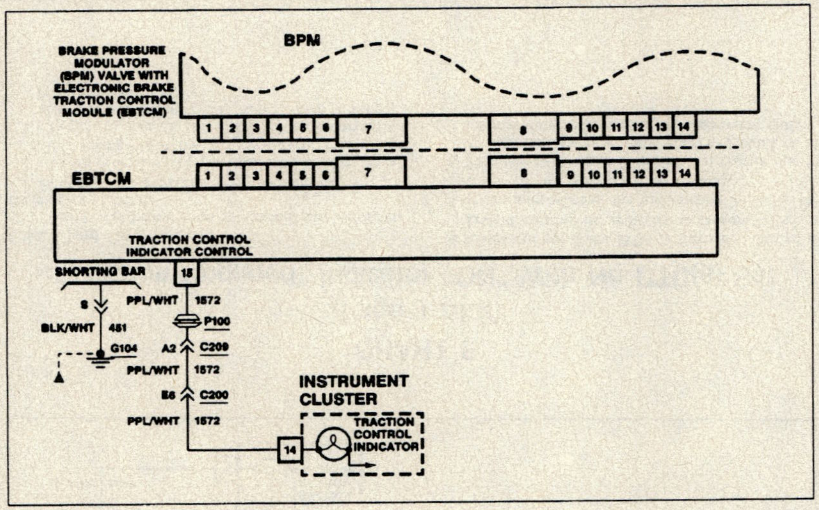

CHART D
"Traction Control" Indicator Inoperative

Chart Test Description:
Number(s) below refer to circled number(s) on the diagnostic chart.

1. Confirms that a malfunction condition exists.
2. Checks to see if the indicator can be turned "ON" manually.
3. Verifies the integrity of the indicator bulb.
4. Checks for an indicator low side circuit open.
5. Checks for an indicator low side circuit short to voltage.

DIAGNOSTIC CHART D — 1995 FLEETWOOD, CONTINUED

CHART D
"TRACTION CONTROL" INDICATOR INOPERATIVE

(1)
- WHILE OBSERVING THE "TRACTION CONTROL" INDICATOR, TURN THE IGNITION TO "RUN."

INDICATOR DOES NOT COME "ON." → go to (2)

INDICATOR COMES "ON" FOR APPROXIMATELY 2 SECONDS, THEN GOES "OFF." → NO TROUBLE FOUND. CHECK FOR INTERMITTENT CONDITIONS.

(2)
- IGNITION "OFF."
- DISCONNECT EBTCM HARNESS CONNECTOR.
- INSTALL J 39700 UNIVERSAL BREAKOUT BOX WITH J 39700-15 ADAPTER TO EBTCM HARNESS CONNECTOR ONLY.
- CONNECT A FUSED JUMPER WIRE (SUCH AS J 36169) WITH A 3 AMP FUSE FROM J 39700 TERMINAL "1" (GROUND) TO TERMINAL "15".
- IGNITION TO "RUN."

IS "TRACTION CONTROL" INDICATOR "ON"?

NO → go to (3)

YES →
- IGNITION "OFF."
- CHECK FOR PROPER CONNECTION TO EBTCM AT TERMINAL "15"

IF OK, THEN REPLACE EBTCM.

(3)
- IGNITION "OFF."
- REMOVE INSTRUMENT CLUSTER
- REMOVE AND INSPECT THE "TRACTION CONTROL" INDICATOR BULB.

IS BULB GOOD?

YES → go to (4)

NO → REPLACE BULB.

(4)
- REINSTALL BULB INTO INSTRUMENT CLUSTER.
- USING J 39200, MEASURE RESISTANCE FROM J 39700 TERMINAL "15" TO INSTRUMENT CLUSTER HARNESS CONNECTOR TERMINAL "14".

IS IT 5 Ω OR LESS?

YES → go to (5)

NO → REPAIR OPEN IN CKT 1572.

(5)
- IGNITION TO "RUN."
- USING J 39200, MEASURE VOLTAGE ON J 39700 FROM TERMINAL "15" TO TERMINAL "1" (GROUND).

IS IT 1 V OR LESS?

YES → REFER TO INSTRUMENT CLUSTER DIAGNOSIS.

NO → REPAIR SHORT TO VOLTAGE IN CKT 1572.

WHEN ALL DIAGNOSIS AND REPAIRS ARE COMPLETED, CLEAR DTC(s) AND VERIFY PROPER OPERATION.

DIAGNOSTIC CHART E — 1995 FLEETWOOD

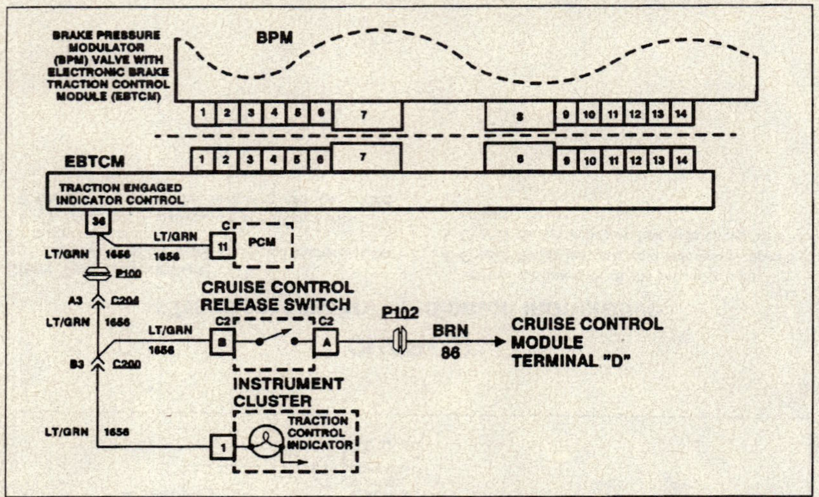

CHART E
(Page 1 of 2)
"Traction Engaged" Indicator "ON" With No DTC(s) Set

Chart Test Description:

Number(s) below refer to circled number(s) on the diagnostic chart.

1. Confirms that a malfunction condition exists.
2. Checks whether the malfunction is due to a DTC-setting condition.
3. Checks to see if the EBTCM is incorrectly commanding the indicator "ON" when it should not.
4. Checks to see if the PCM is internally shorted to ground, causing the indicator to remain "ON."
5. Determines whether the malfunction is due to the cruise control portion of the circuitry.
6. This test determines if the malfunction is due to an indicator low side circuit short to ground, or an internal short to ground in the instrument cluster.

DIAGNOSTIC CHART E — 1995 FLEETWOOD, CONTINUED

CHART E
(Page 1 of 2)
"TRACTION ENGAGED" INDICATOR "ON" WITH NO DTCs SET

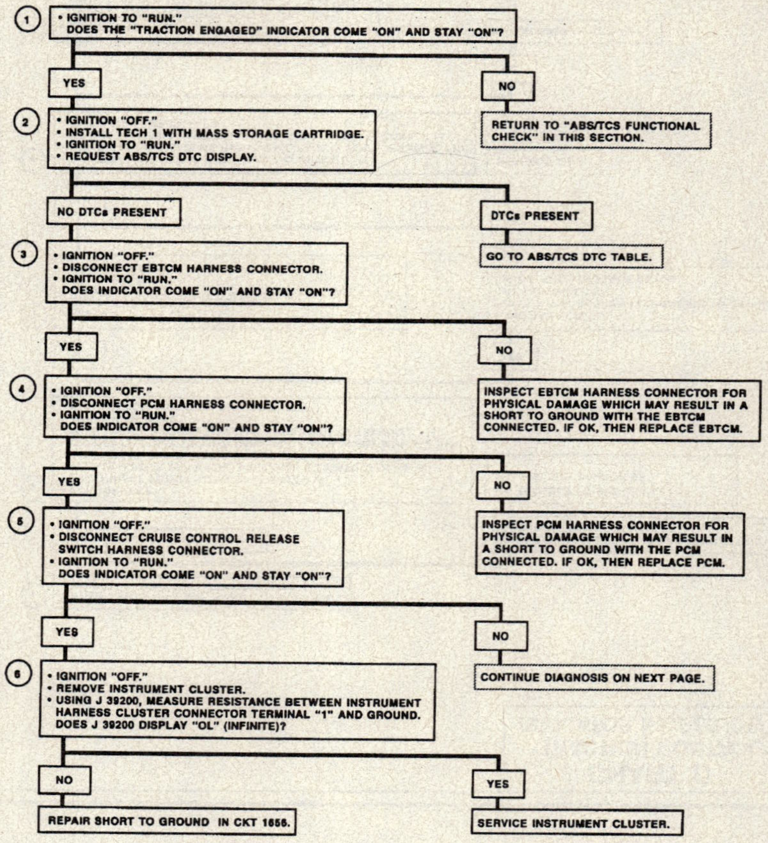

WHEN ALL DIAGNOSIS AND REPAIRS ARE COMPLETED, CLEAR DTC(s) AND VERIFY PROPER OPERATION.

DIAGNOSTIC CHART E — 1995 FLEETWOOD, CONTINUED

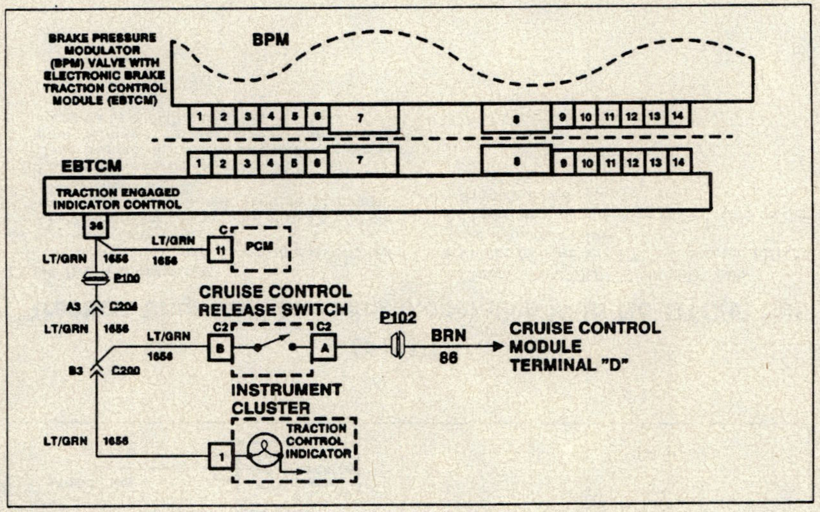

CHART E
(Page 2 of 2)
"Traction Engaged" Indicator "ON" With No DTC(s) Set

Chart Test Description:

Number(s) below refer to circled number(s) on the diagnostic chart.

7. Ensures the cruise control release switch is not shorted to ground.
8. Checks CKT 86 for short to ground.
9. Checks to see if the cruise control module is internally shorted to ground, causing the indicator to remain "ON."

DIAGNOSTIC CHART E — 1995 FLEETWOOD, CONTINUED

CHART E
(Page 2 of 2)
"TRACTION ENGAGED" INDICATOR "ON" WITH NO DTCs SET

CONTINUED FROM PREVIOUS PAGE.

(7)
- IGNITION "OFF."
- USING J 39200, MEASURE RESISTANCE BETWEEN CRUISE CONTROL RELEASE SWITCH TERMINAL "A" AND GROUND. REPEAT BETWEEN TERMINAL "B" AND GROUND. DOES J 39200 DISPLAY "OL" (INFINITE) FOR BOTH MEASUREMENTS?

NO → REPLACE CRUISE CONTROL RELEASE SWITCH.

YES →

(8)
- DISCONNECT CRUISE CONTROL MODULE.
- USING J 39200, MEASURE RESISTANCE BETWEEN CRUISE CONTROL MODULE HARNESS CONNECTOR TERMINAL "D" AND GROUND. DOES J 39200 DISPLAY "OL" (INFINITE)?

NO → REPAIR SHORT TO GROUND IN CKT 86.

YES →

(9)
- RECONNECT CRUISE CONTROL MODULE.
- RECONNECT CRUISE CONTROL RELEASE SWITCH.
- IGNITION TO "RUN." DOES "TRACTION ENGAGED" INDICATOR COME "ON" AND STAY "ON"?

YES → REPLACE CRUISE CONTROL MODULE.

NO → THOROUGHLY INSPECT ALL WIRING AND CONNECTIONS FOR INTERMITTENTS, CHAFING, AND MISROUTING AT EBTCM, INSTRUMENT CLUSTER, CRUISE CONTROL MODULE AND ALL INTERCONNECTS. REPAIR ANY PROBLEMS FOUND.

DIAGNOSTIC CHART F — 1995 FLEETWOOD

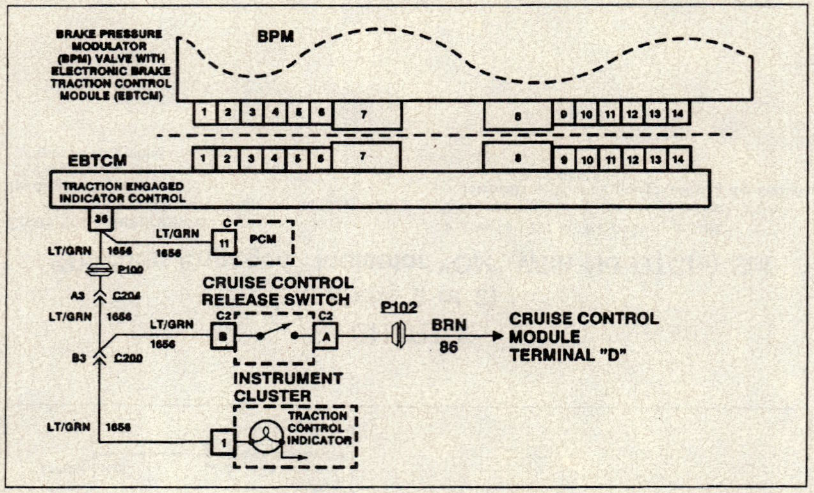

CHART F

"Traction Engaged" Indicator Inoperative With No DTC(s) Set

Chart Test Description:

Number(s) below refer to circled number(s) on the diagnostic chart.

1. Checks that other indicators fed from CKT 39 are working properly. If not, refer to Instrument Cluster Diagnosis.
2. Ensures indicator can be turned "ON" manually. If the indicator illuminates, malfunction is due to poor terminal contact at EBTCM harness connector or EBTCM.
3. Checks "TRACTION ENGAGED" bulb.
4. Checks for open in CKT 1656 between EBTCM and instrument cluster.
5. Checks for short to voltage in CKT 1656 causing inoperative bulb.

DIAGNOSTIC CHART F — 1995 FLEETWOOD, CONTINUED

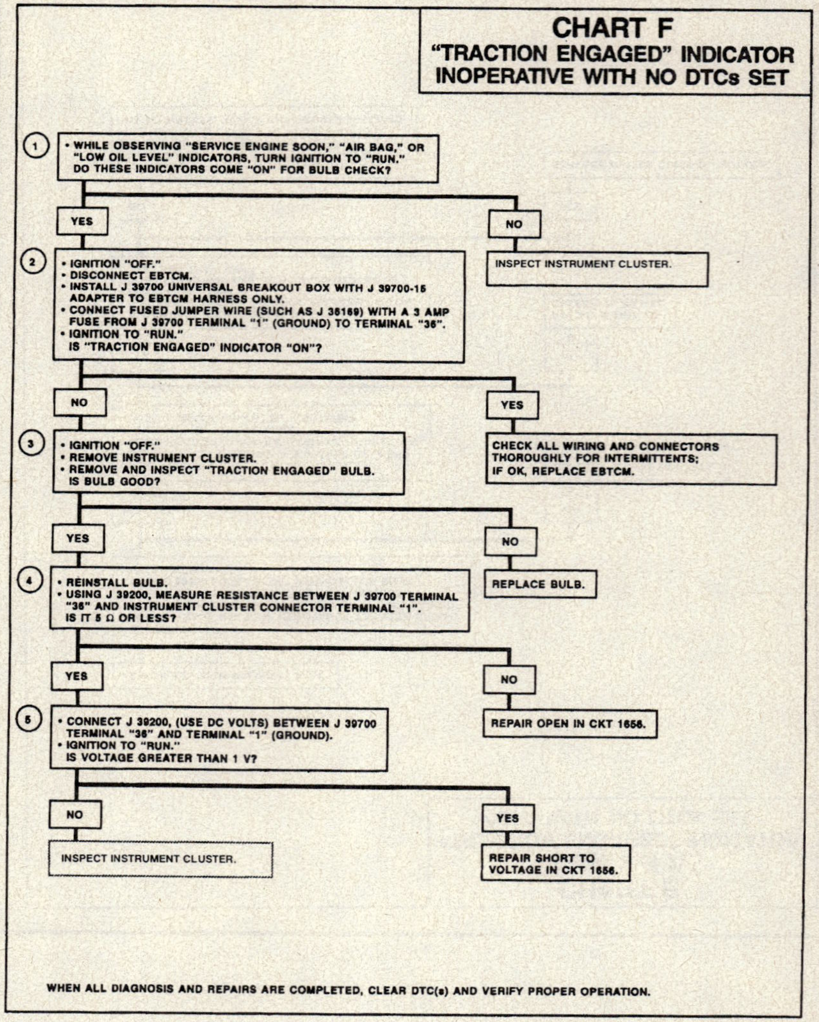

DIAGNOSTIC CHART DTC 21 AND 25 — 1995 FLEETWOOD

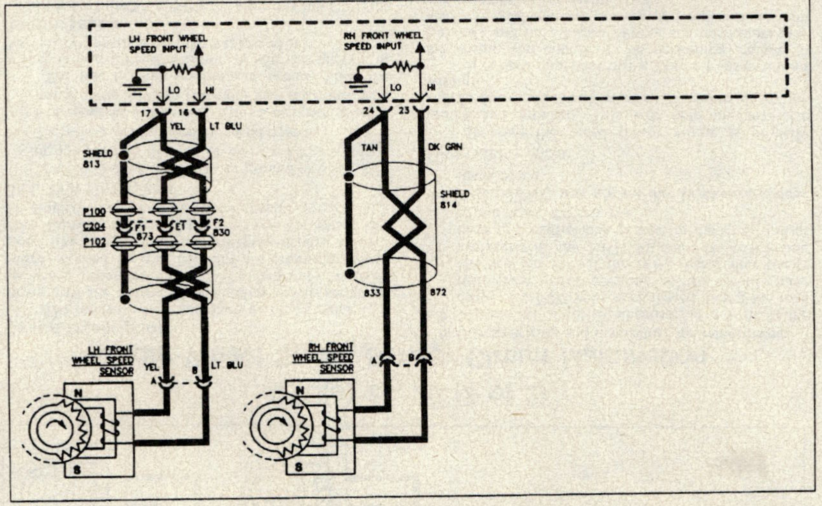

DTC(s) 21, 25 (1 of 2)

Front Wheel Speed Sensor Circuit Malfunction

Circuit Description:

The toothed wheel generates a voltage pulse as it moves past the sensor; each tooth-gap-tooth series on the wheel generates a pulse. The frequency of these pulses is used by the EBTCM to determine wheel speed. The amplitude of the voltage generated in each pulse depends on the air gap between the sensor and the toothed wheel, and on wheel speed.

DTC Will Set When:

- There is a short to voltage or ground in (CKT 872 or 833 for RF) or (873 or 830 for LF).
- There is a malfunctioning speed sensor.
- It is possible that driving with a compact spare at speeds above 40 km/h (24 mph) may cause this fault.

The test for this malfunction occurs when the vehicle is in motion; it will not set with the ignition in the "RUN" position and the vehicle at rest.

Action Taken:

The "ANTI-LOCK" and "TRACTION CONTROL" lamps will be "ON." ABS and TCS are disabled.

DTC Chart Test Description:

Number(s) below refer to circled number(s) on the diagnostic chart.

1. Checks for proper resistance in the sensor itself.
2. Checks for a short to voltage in either of the speed sensor circuit wires.
3. Checks for a short between the speed sensor circuit wires.
4. Checks for a short to ground in the speed sensor input circuit wire. The ground may be either a "hard" short to ground, or a resistive (partial) short. A short with a resistance less than 2 megohms, though not a hard short, can still cause the DTC to set.

Diagnostic Aids:

Be sure the speed sensor wiring is properly routed and retained. This will help prevent false signals due to electrical noise being picked up by the wiring.

It is very important that a thorough inspection of the wiring and connectors be performed. Failure to carefully and fully inspect wiring and connectors may result in misdiagnosis, causing part replacement with reappearance of the malfunction.

The Tech 1 can also be used to monitor the wheel speeds during a road test. Watch the wheel speeds being displayed on the Tech 1 to see if any of the readings are unusual, such as one sensor varying in speed from the other three, a signal going intermittently high or low, etc. If this does not identify the intermittent, wet the speed sensor harness on the underside of the vehicle and road test monitoring wheel speeds with the Tech 1.

DIAGNOSTIC CHART DTC 21 AND 25 — 1995 FLEETWOOD, CONTINUED

DTC(s) 21, 25
(Page 1 of 2)
FRONT WHEEL SPEED SENSOR CIRCUIT MALFUNCTION

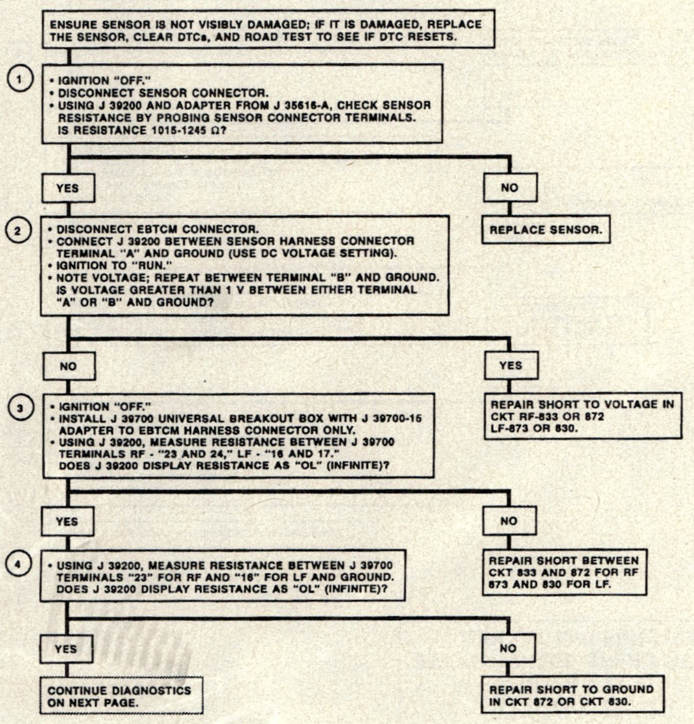

WHEN ALL DIAGNOSIS AND REPAIRS ARE COMPLETED, CLEAR DTC(s) AND VERIFY PROPER OPERATION.

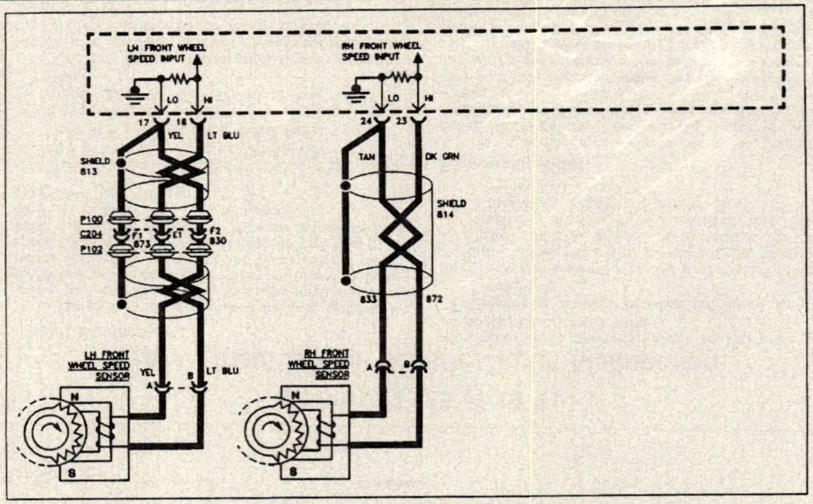

DTC(s) 21, 25 (2 of 2)

Front Wheel Speed Sensor Circuit Malfunction

Circuit Description:

The toothed wheel generates a voltage pulse as it moves past the sensor; each tooth-gap-tooth series on the wheel generates a pulse. The frequency of these pulses is used by the EBTCM to determine wheel speed. The amplitude of the voltage generated in each pulse depends on the air gap between the sensor and the toothed wheel, and on wheel speed.

DTC Will Set When:

- There is a short to voltage or ground in (CKT 872 or 833 for RF) or (873 or 830 for LF).
- There is a malfunctioning speed sensor.
- It is possible that driving with a compact spare at speeds above 40 km/h (24 mph) may cause this fault.

The test for this malfunction occurs when the vehicle is in motion; it will not set with the ignition in the "RUN" position and the vehicle at rest.

Action Taken:

The "ANTI-LOCK" and "TRACTION CONTROL" lamps will be "ON." ABS and TCS are disabled.

DTC Chart Test Description:

Number(s) below refer to circled number(s) on the diagnostic chart.

5. Checks if the DTC resets during a road test. If so, since tests 1-4 have validated that the circuitry and components are good, intermittent malfunctions are suspected.
6. Checks wiring and connectors for intermittents.
7. Replace the wheel speed sensor, as it is likely the cause of a DTC reset experienced during the road test, since other portions of the circuit have checked out OK. If the DTC resets after sensor replacement, the EBTCM must be concluding there is a malfunction present in the speed sensor circuit when there is not.
8. Ensures DTC was not set due to electromagnetic interference.

Diagnostic Aids:

Be sure the speed sensor wiring is properly routed and retained. This will help prevent false signals due to electrical noise being picked up by the wiring.

It is very important that a thorough inspection of the wiring and connectors be performed. Failure to carefully and fully inspect wiring and connectors may result in misdiagnosis, causing part replacement with reappearance of the malfunction.

The Tech 1 can also be used to monitor the wheel speeds during a road test. Watch the wheel speeds being displayed on the Tech 1 to see if any of the readings are unusual, such as one sensor varying in speed from the other three, a signal going intermittently high or low, etc. If this does not identify the intermittent, wet the speed sensor harness on the underside of the vehicle and road test monitoring wheel speeds with the Tech 1.

DIAGNOSTIC CHART DTC 21 AND 25 —
1995 FLEETWOOD, CONTINUED

DTC(s) 21, 25
(Page 2 of 2)
FRONT WHEEL SPEED SENSOR CIRCUIT MALFUNCTION

DIAGNOSTICS CONTINUED FROM PREVIOUS PAGE.

(5)
- REMOVE J 39700.
- RECONNECT ALL CONNECTORS.
- CLEAR DTCs, IGNITION "OFF."
- DISCONNECT TECH 1, IF CONNECTED.
- ROAD TEST VEHICLE, REACHING AT LEAST 25 km/h (15 MPH).

DOES DTC 21 OR 25 RESET?

- YES: go to (6)
- NO: NO TROUBLE FOUND; SEE "DIAGNOSTIC AIDS" ON FACING PAGE.

(6)
- CHECK ALL WIRING, CONNECTORS AND TERMINALS THOROUGHLY FOR INTERMITTENTS OR CORROSION.

ARE ALL CONNECTIONS AND WIRING GOOD?

- YES: go to (7)
- NO: REPAIR WIRING AND/OR CONNECTIONS AS NEEDED.

(7)
- REPLACE SENSOR.
- CLEAR DTCs, IGNITION "OFF."
- DISCONNECT TECH 1, IF CONNECTED.
- ROAD TEST VEHICLE, REACHING AT LEAST 25 km/h (15 MPH).

DOES DTC 21 OR 25 RESET?

- YES: go to (8)
- NO: SYSTEM OK; SEE "DIAGNOSTIC AIDS" ON FACING PAGE.

(8)
PERFORM EMI TEST. REFER TO EMI TEST FOUND IN "INTERMITTENTS AND POOR CONNECTIONS" IN THIS SECTION.
WERE EMI TESTS COMPLETED WITH NO TROUBLE FOUND?

- YES: REPLACE EBTCM.
- NO: PERFORM NECESSARY REPAIRS.

WHEN ALL DIAGNOSIS AND REPAIRS ARE COMPLETED, CLEAR DTC(s) AND VERIFY PROPER OPERATION.

DIAGNOSTIC CHART DTC 31 AND 35 — 1995 FLEETWOOD

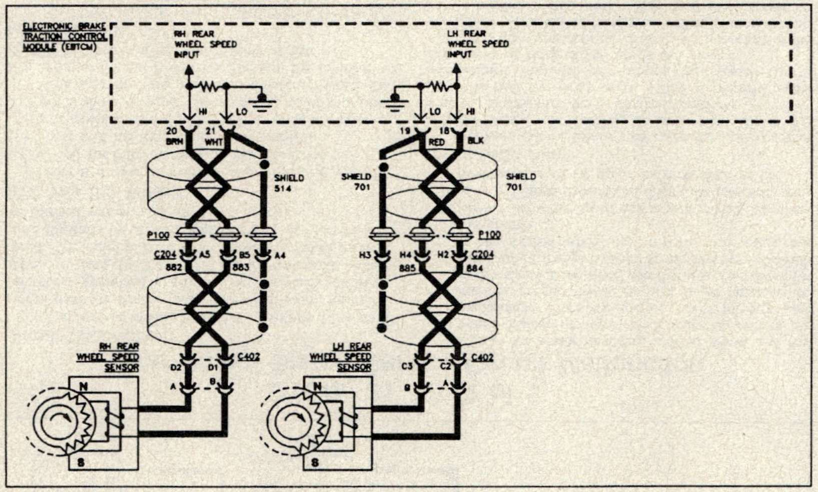

DTC(s) 31, 35 (1 of 2)

Rear Wheel Speed Sensor Circuit Malfunction

Circuit Description:

The toothed wheel generates a voltage pulse as it moves past the sensor; each tooth-gap-tooth series on the wheel generates a pulse. The frequency of these pulses is used by the EBTCM to determine wheel speed. The amplitude of the voltage generated in each pulse depends on the air gap between the sensor and the toothed wheel, and on wheel speed.

DTC Will Set When:

- There is a short to voltage or ground in (CKT 882 or 883 for RR) or (884 or 885 for LR).
- There is a malfunctioning speed sensor.
- It is possible that driving with a compact spare at speeds above 40 km/h (24 mph) may cause this fault.

The test for this malfunction occurs when the vehicle is in motion; it will not set with the ignition in the "RUN" position and the vehicle at rest.

Action Taken:

The "ANTI-LOCK" and "TRACTION CONTROL" lamps will be "ON." ABS and TCS are disabled.

DTC Chart Test Description:

Number(s) below refer to circled number(s) on the diagnostic chart.

1. Checks for proper resistance in the sensor itself.
2. Checks for a short to voltage in either of the speed sensor circuit wires.
3. Checks for a short between the speed sensor circuit wires.
4. Checks for a short to ground in the speed sensor input circuit wire. The ground may be either a "hard" short to ground, or a resistive (partial) short. A short with a resistance less than 2 megohms, though not a hard short, can still cause the DTC to set.

Diagnostic Aids:

A worn axle or bearing may cause this malfunction in extreme cases; the bearing or axle play allows the sensor-to-toothed-ring gap to change excessively.

Be sure the speed sensor wiring is properly routed and retained. This will help prevent false signals due to electrical noise being picked up by the wiring.

It is very important that a thorough inspection of the wiring and connectors be performed. Failure to carefully and fully inspect wiring and connectors may result in misdiagnosis, causing part replacement with reappearance of the malfunction.

The Tech 1 can also be used to monitor the wheel speeds during a road test. Watch the wheel speeds being displayed on the Tech 1 to see if any of the readings are unusual, such as one sensor varying in speed from the other three, a signal going intermittently high or low, etc. If this does not identify the intermittent, wet the speed sensor harness on the underside of the vehicle and road test monitoring wheel speeds with the Tech 1.

DIAGNOSTIC CHART DTC 31 AND 35 — 1995 FLEETWOOD, CONTINUED

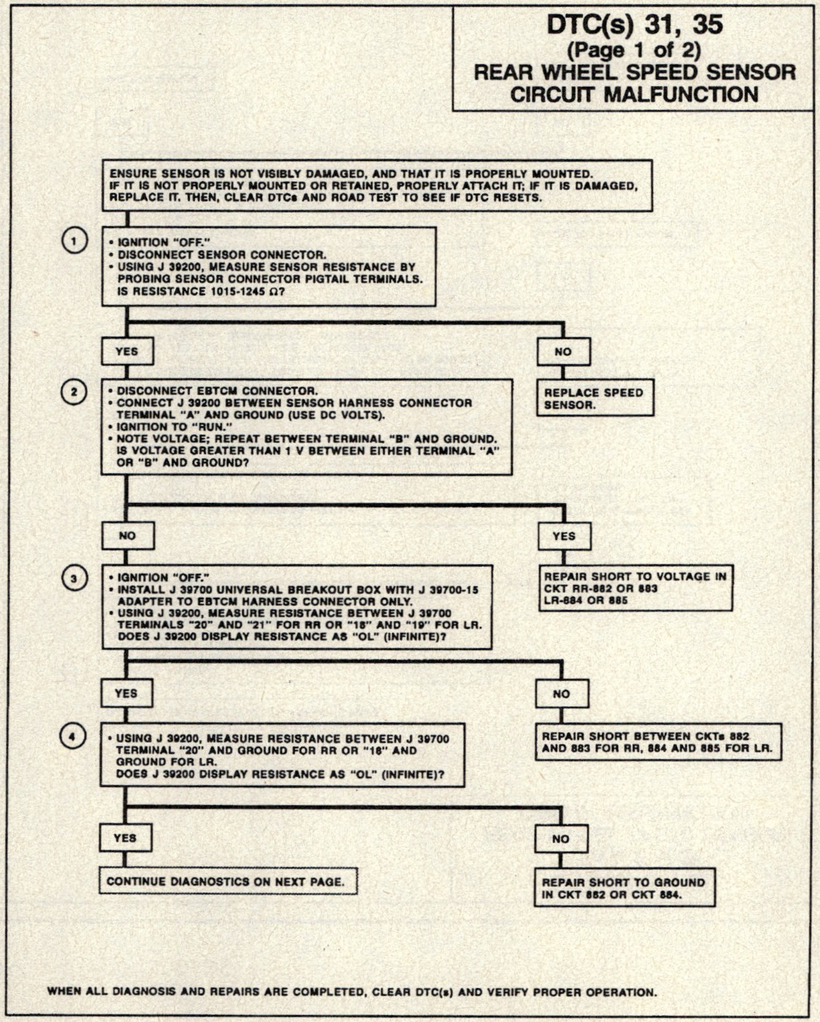

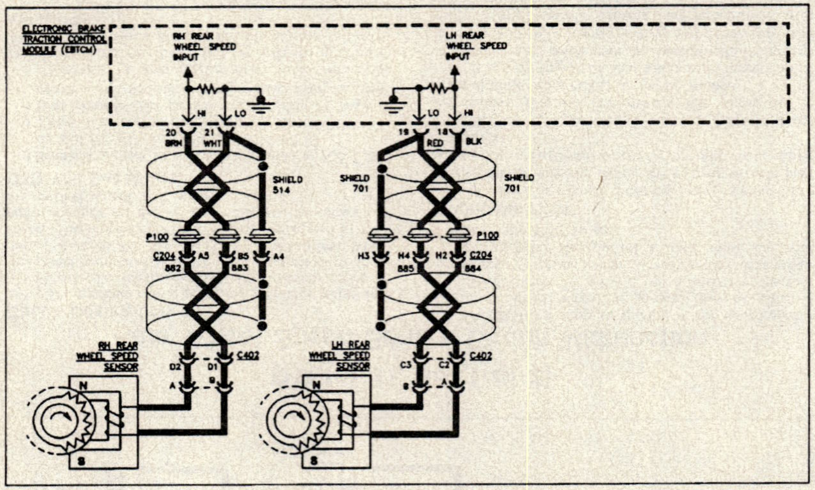

DTC(s) 31, 35 (2 of 2)
Rear Wheel Speed Sensor Circuit Malfunction

Circuit Description:

The toothed wheel generates a voltage pulse as it moves past the sensor; each tooth-gap-tooth series on the wheel generates a pulse. The frequency of these pulses is used by the EBTCM to determine wheel speed. The amplitude of the voltage generated in each pulse depends on the air gap between the sensor and the toothed wheel, and on wheel speed.

DTC Will Set When:

- There is a short to voltage or ground in (CKT 882 or 883 for RR) or (884 or 885 for LR).
- There is a malfunctioning speed sensor.
- It is possible that driving with a compact spare at speeds above 40 km/h (24 mph) may cause this fault.

The test for this malfunction occurs when the vehicle is in motion; it will not set with the ignition in the "RUN" position and the vehicle at rest.

Action Taken:

The "ANTI-LOCK" and "TRACTION CONTROL" lamps will be "ON." ABS and TCS are disabled.

DTC Chart Test Description:

Number(s) below refer to circled number(s) on the diagnostic chart.

5. Checks if the DTC resets during a road test. If so, since tests 1-4 have validated that all of the circuitry and components are good, intermittent malfunctions are suspected.
6. Checks wiring and connectors for intermittents.
7. Checks for sensor mounting, toothed wheel, and rear wheel bearing problems which may be causing the malfunction. Check the toothed wheel for any large grooves, gouges, marks, etc. that might influence the tooth's signal at the wheel speed sensor. Also check for a buildup of foreign material in the gaps between teeth in the toothed wheel; this material may cause this malfunction.
8. Replace the wheel speed sensor, as it is likely the cause of a DTC reset experienced during the road test, since other portions of the circuit have checked out OK.

Diagnostic Aids:

A worn axle or bearing may cause this malfunction in extreme cases; the bearing or axle play allows the sensor-to-toothed-ring gap to change excessively.

Be sure the speed sensor wiring is properly routed and retained. This will help prevent false signals due to electrical noise being picked up by the wiring.

It is very important that a thorough inspection of the wiring and connectors be performed. Failure to carefully and fully inspect wiring and connectors may result in misdiagnosis, causing part replacement with reappearance of the malfunction.

The Tech 1 can also be used to monitor the wheel speeds during a road test. Watch the wheel speeds being displayed on the Tech 1 to see if any of the readings are unusual, such as one sensor varying in speed from the other three, a signal going intermittently high or low, etc. If this does not identify the intermittent, wet the speed sensor harness on the underside of the vehicle and road test monitoring wheel speeds with the Tech 1.

DTC(s) 31, 35
(Page 2 of 2)
REAR WHEEL SPEED SENSOR CIRCUIT MALFUNCTION

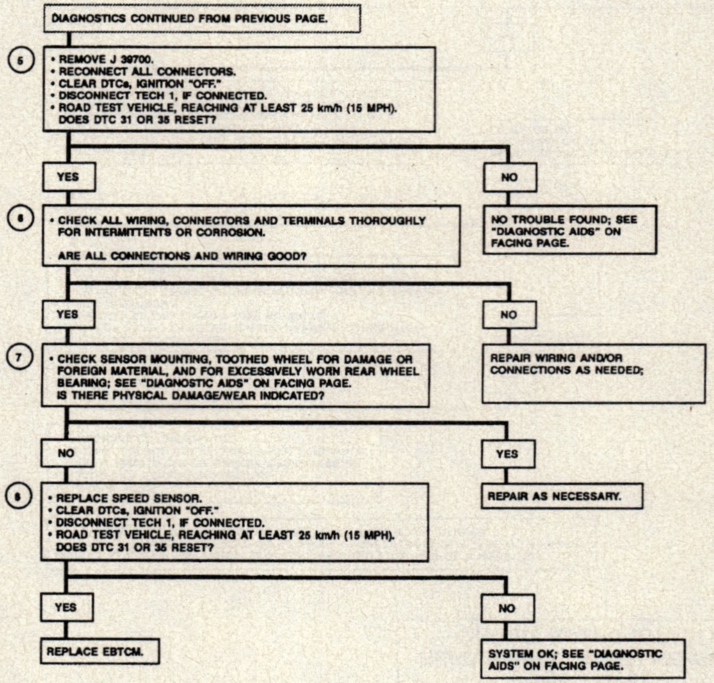

WHEN ALL DIAGNOSIS AND REPAIRS ARE COMPLETED, CLEAR DTC(s) AND VERIFY PROPER OPERATION.

DIAGNOSTIC CHART DTC 23, 27, 33 AND 37 — 1995 FLEETWOOD

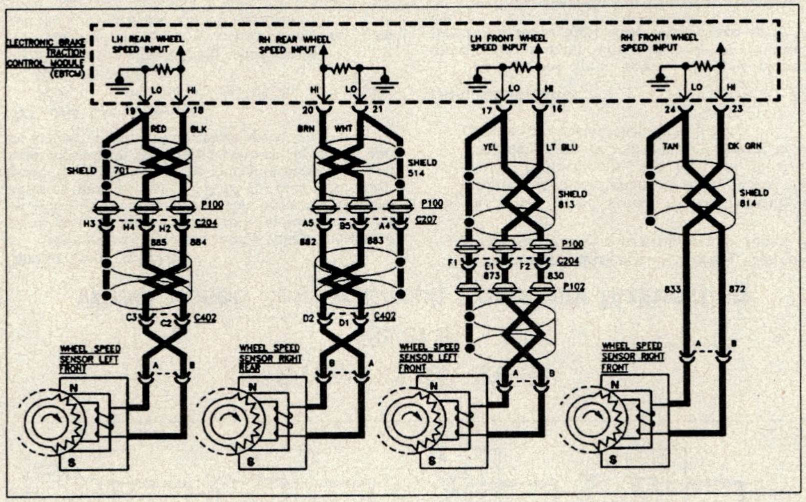

DTC(s) 23, 27, 33, 37
(1 of 2)
Wheel Speed Sensor Circuit Continuity Malfunction

Circuit Description:

The toothed wheel generates a voltage pulse as it moves past the sensor; each tooth-gap-tooth series on the wheel generates a pulse. The frequency of these pulses is used by the EBTCM to determine wheel speed. The amplitude of the voltage generated in each pulse depends on the air gap between the sensor and the toothed wheel, and on wheel speed.

DTC Will Set When:

- There is a short to voltage or an open in the affected circuit.
- There is a malfunctioning speed sensor.

The test for this malfunction occurs any time with the ignition in the "RUN" position.

Action Taken:

The "ANTI-LOCK" and "TRACTION CONTROL" lamps will be "ON." ABS and TCS are disabled.

DTC Chart Test Description:

Number(s) below refer to circled number(s) on the diagnostic chart.

1. Checks for proper resistance in the sensor itself.
2. Checks for a short to voltage in either of the speed sensor circuit wires.
3. Checks for an open in both the speed sensor circuit wires.
4. Manipulates wiring and connectors, trying to induce an intermittent malfunction not currently present.
5. This test checks if the DTC resets when the key is turned to "RUN". If so, since tests 1-4 have validated that the circuitry and components are good, the wheel speed sensor may be causing the malfunction.

Diagnostic Aids:

Be sure the speed sensor wiring is properly routed and retained. This will help prevent false signals due to electrical noise being picked up by the wiring.

It is very important that a thorough inspection of the wiring and connectors be performed. Failure to carefully and fully inspect wiring and connectors may result in misdiagnosis, causing part replacement with reappearance of the malfunction.

DIAGNOSTIC CHART DTC 23, 27, 33 AND 37 — 1995 FLEETWOOD, CONTINUED

DTC(s) 23, 27, 33, 37
(Page 1 of 2)
WHEEL SPEED SENSOR CIRCUIT CONTINUITY MALFUNCTION

ENSURE SENSOR IS NOT VISIBLY DAMAGED; IF IT IS DAMAGED, REPLACE THE SENSOR, CLEAR DTCs, AND TURN IGNITION TO "RUN" TO SEE IF DTC RESETS.

(1)
- IGNITION "OFF."
- DISCONNECT SENSOR CONNECTOR.
- USING J 39200 AND ADAPTERS FROM J 35616-A, CHECK SENSOR RESISTANCE BY PROBING SENSOR CONNECTOR TERMINALS.
- IS RESISTANCE 1015-1245 Ω?

NO → REPLACE SENSOR.

YES ↓

(2)
- DISCONNECT EBTCM CONNECTOR.
- CONNECT J 39200 BETWEEN SENSOR HARNESS CONNECTOR TERMINAL "A" AND GROUND (USE DC VOLTS).
- IGNITION TO "RUN."
- NOTE VOLTAGE; REPEAT BETWEEN TERMINAL "B" AND GROUND. IS VOLTAGE GREATER THAN 1 V BETWEEN EITHER TERMINAL "A" OR "B" AND GROUND?

YES → REPAIR SHORT TO VOLTAGE IN THE AFFECTED CIRCUIT.

NO ↓

(3)
- IGNITION "OFF."
- INSTALL J 39700 UNIVERSAL BREAKOUT BOX WITH J 39700-15 ADAPTER TO EBTCM HARNESS CONNECTOR ONLY.
- USING J 39200, MEASURE RESISTANCE BETWEEN:

J 39700 TERMINAL		SENSOR AND HARNESS		CKT	FOR DTC
24	AND	A	FOR	833	23
23		B		872	23
17		A		873	27
16		B		830	27
20		A		882	33
21		B		883	33
18		A		884	37
19		B		885	37

IS MEASUREMENT GREATER THAN 5 Ω?

YES → REPAIR OPEN OR HIGH RESISTANCE IN THE AFFECTED CIRCUIT.

NO ↓

(4)
- REMOVE J 39700.
- RECONNECT ALL CONNECTORS.
- CLEAR DTCs; IGNITION "OFF."
- DISCONNECT TECH 1, IF CONNECTED.
- IGNITION TO "RUN."
- WIGGLE ALL WIRING AND CONNECTORS TO TRY TO INDUCE AN INTERMITTENT FAULT. DOES DTC RESET?

YES → REPAIR INTERMITTENT IN CONNECTOR OR WIRING.

NO ↓

(5)
- CLEAR DTCs; IGNITION "OFF."
- DISCONNECT TECH 1, IF CONNECTED.
- IGNITION TO "RUN." DOES DTC RESET?

DIAGNOSTICS CONTINUED ON NEXT PAGE.

WHEN ALL DIAGNOSIS AND REPAIRS ARE COMPLETED, CLEAR DTC(s) AND VERIFY PROPER OPERATION.

DIAGNOSTIC CHART DTC 23, 27, 33 AND 37 — 1995 FLEETWOOD, CONTINUED

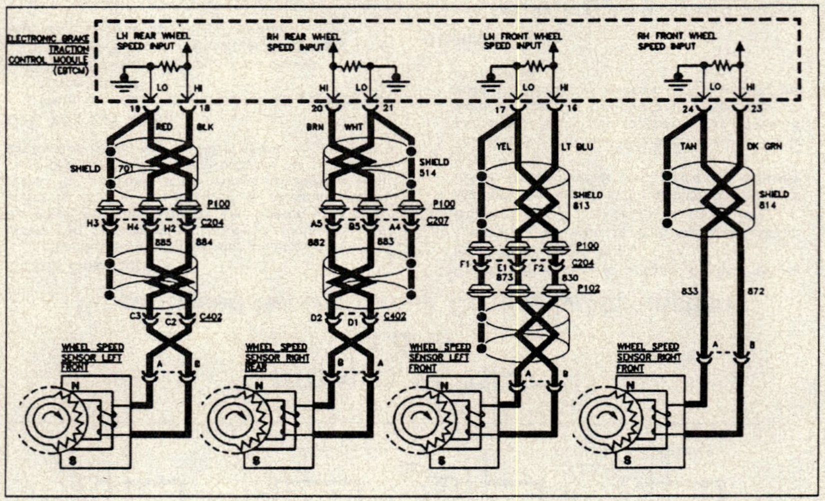

DTC(s) 23, 27, 33, 37
(2 of 2)
Wheel Speed Sensor Circuit Continuity Malfunction

Circuit Description:

The toothed wheel generates a voltage pulse as it moves past the sensor; each tooth-gap-tooth series on the wheel generates a pulse. The frequency of these pulses is used by the EBTCM to determine wheel speed. The amplitude of the voltage generated in each pulse depends on the air gap between the sensor and the toothed wheel, and on wheel speed.

DTC Will Set When:

- There is a short to voltage or an open in the affected circuit.
- There is a malfunctioning speed sensor.

The test for this malfunction occurs any time with the ignition in the "RUN" position.

Action Taken:

The "ANTI-LOCK" and "TRACTION CONTROL" lamps will be "ON." ABS and TCS are disabled.

DTC Chart Test Description:

Number(s) below refer to circled number(s) on the diagnostic chart.

6. Checks if DTCs(21 RF), (25 LF), (31 RR) (35 LR) sets during a road test.
7. If the circuit malfunction DTC sets at this point, diagnostics for circuit malfunctions should be used.
8. The wheel speed sensor may be causing an intermittent malfunction. If DTC resets after sensor replacement, the EBTCM must be concluding there is a malfunction present in the speed sensor circuit when there is not.

Diagnostic Aids:

Be sure the speed sensor wiring is properly routed and retained. This will help prevent false signals due to electrical noise being picked up by the wiring.

It is very important that a thorough inspection of the wiring and connectors be performed. Failure to carefully and fully inspect wiring and connectors may result in misdiagnosis, causing part replacement with reappearance of the malfunction.

DIAGNOSTIC CHART DTC 23, 27, 33 AND 37 — 1995 FLEETWOOD, CONTINUED

DTC(s) 23, 27, 33, 37
(Page 2 of 2)
WHEEL SPEED SENSOR CIRCUIT CONTINUITY MALFUNCTION

DIAGNOSTICS CONTINUED FROM PREVIOUS PAGE.

- NO → (6)
  - ROAD TEST VEHICLE, REACHING AT LEAST 25 km/h (15 MPH).
  - DOES DTC 21, 25, 31, OR 35 SET?
    - YES → (7) GO TO DTC 21, 25, 31, OR 35 DIAGNOSTICS.
    - NO → NO TROUBLE FOUND, SEE "DIAGNOSTIC AIDS" ON FACING PAGE.
- YES → (8)
  - REPLACE SENSOR.
  - CLEAR DTCs; IGNITION "OFF."
  - DISCONNECT TECH 1, IF CONNECTED.
  - IGNITION TO "RUN."
  - DOES DTC RESET?
    - YES → REPLACE EBTCM.
    - NO → SYSTEM OK.

WHEN ALL DIAGNOSIS AND REPAIRS ARE COMPLETED, CLEAR DTC(s) AND VERIFY PROPER OPERATION.

DIAGNOSTIC CHART DTC 28 — 1995 FLEETWOOD

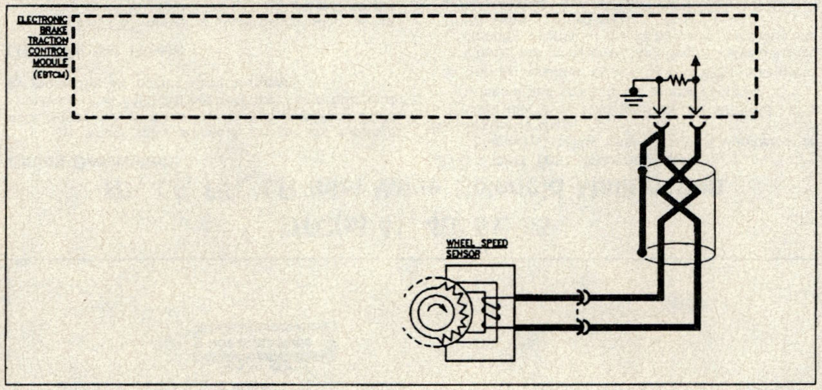

DTC 28
Wheel Speed Sensor Circuit Frequency Malfunction

Circuit Description:

The toothed wheel generates a voltage pulse as it moves past the sensor; each tooth-gap-tooth series on the wheel generates a pulse. The frequency of these pulses is used by the EBTCM to determine wheel speed. The amplitude of the voltage generated in each pulse depends on the air gap between the sensor and the toothed wheel, and on wheel speed.

DTC Will Set When:

- The EBTCM cannot specifically identify which wheel speed sensor is causing the frequency error malfunction. If the EBTCM can define the specific speed sensor causing the malfunction, the DTC associated with the sensor (21,25,31,35) will be set instead of DTC 28.
- It is possible that driving with a compact spare at speeds above 40 km/h (24 mph) may cause this fault.

Action Taken:

The "ANTI-LOCK" and "TRACTION CONTROL" lamps will be "ON." ABS and TCS are disabled.

DTC Chart Test Description:

Number(s) below refer to circled number(s) on the diagnostic chart.

1. Checks wiring and connections for malfunctions. It is critical that this step is performed thoroughly, as wiring/connector malfunctions are the most likely cause for this DTC being set.
2. Uses the Tech 1 to monitor for vehicle electrical/electronic system "noise" being picked up by the speed sensor circuits.
3. Uses the Tech 1 to monitor the wheel speed sensors while the vehicle is operating. If the Tech 1 auto-trigger snapshot mode triggers, the speed sensor or the wiring and/or connectors are intermittent during the road test.
4. Checks for proper resistance in the sensor itself.
5. 5. Replaces wheel speed sensor as the likely cause of the triggering intermittent. If DTC resets, replace EBTCM.
6. Checks for a short to ground in the speed sensor input circuit wires. The circuit to be tested depends on which speed sensor triggered the Tech 1. The ground may be either a "hard" short to ground, or a resistive (partial) short. A short with a resistance less than 2 megohms, though not a hard short, can still cause DTC 28 to set.
7. Ensures DTC was not set due to electromagnetic interference.

Diagnostic Aids:

DTC 28 may be set by running the Tech 1 "Auto Test" if the throttle angle readings are not updating while in the data list mode. If this is the case, clear DTCs, disconnect Tech 1, and road test to at least 25 km/h (15 mph) to see if DTC resets.

Check the toothed wheel for any large grooves, gouges, marks, etc. that might influence the tooth's signal at the wheel speed sensor. Also check for a buildup of foreign material in the gaps between teeth in the toothed wheel, as the material may cause this malfunction.

A worn hub/bearing assembly may cause this fault in extreme cases; the bearing play allows the sensor-to-toothed-ring gap to change excessively.

DIAGNOSTIC CHART DTC 28 — 1995 FLEETWOOD, CONTINUED

DTC 28
WHEEL SPEED SENSOR CIRCUIT FREQUENCY MALFUNCTION

(1) THOROUGHLY INSPECT ALL WIRING AND CONNECTIONS FOR INTERMITTENTS, CHAFING, POOR TERMINAL CONTACT TENSION, CORROSION, AND MISROUTING AT EBTCM, ALL SENSORS, AND ALL HARNESS INTERCONNECTS. REPAIR ANY PROBLEMS, CLEAR DTCs, THEN ROAD TEST VEHICLE, REACHING AT LEAST 25 km/h (15 MPH). DOES DTC RESET?

- YES → (2)
- NO → NO TROUBLE FOUND. SEE "DIAGNOSTIC AIDS" ON FACING PAGE.

(2)
- INSTALL TECH 1 AND MASS STORAGE CARTRIDGE. CLEAR DTCs.
- SET TECH 1 AUTO TRIGGER SNAPSHOT MODE.
- START VEHICLE AND ALLOW TO IDLE FOR ONE MINUTE. DOES TECH 1 TRIGGER?

- NO → (3)
- YES → (6)

(3)
- ROAD TEST VEHICLE REACHING AT LEAST 25 km/h (15 MPH). DOES TECH 1 TRIGGER?

- YES → NOTE TECH 1. PERFORM STEP 4 ON TRIGGERED SENSOR(s).
- NO → NO TROUBLE FOUND; SEE "DIAGNOSTIC AIDS" ON FACING PAGE.

(4)
- IGNITION "OFF."
- DISCONNECT SENSOR CONNECTOR.
- USING J 39200, MEASURE RESISTANCE BETWEEN SENSOR TERMINALS, USING ADAPTERS FROM J 35616-A IF MEASURING A FRONT SENSOR. IS RESISTANCE 1015-1245 Ω?

- NO → (5)
- YES → (7)

(5)
- REPLACE SPEED SENSOR.
- CLEAR DTCs, IGNITION "OFF."
- DISCONNECT TECH 1, IF CONNECTED.
- ROAD TEST VEHICLE, REACHING AT LEAST 25 km/h (15 MPH). DOES DTC 28 RESET?

- YES → REPLACE EBTCM.
- NO → SYSTEM OK.

(6)
- NOTE WITH TECH 1 WHICH SENSOR TRIGGERED.
- IGNITION "OFF."
- DISCONNECT EBTCM.
- INSTALL J 39700 UNIVERSAL BREAKOUT BOX WITH J 39700-15 ADAPTER TO EBTCM HARNESS CONNECTOR ONLY.
- USING TABLE BELOW, WITH J 39200, MEASURE RESISTANCE TO GROUND AT THE TERMINAL ON J 39700 THAT CORRESPONDS TO THE SPEED SENSOR THE TECH 1 TRIGGERED ON:

  LF-TERMINAL "16"
  RF-TERMINAL "23"
  LR-TERMINAL "18"
  RR-TERMINAL "20"

- IF RESISTANCE IS 5 Ω OR LESS TO GROUND, REPAIR HARD SHORT; IF BETWEEN 5 Ω AND 2 MEGOHMS, REPAIR RESISTIVE (PARTIAL) SHORT IN CKT:

  LF-CKT 830
  RF-CKT 872
  LR-CKT 884
  RR-CKT 882

(7)
- PERFORM EMI TEST. REFER TO EMI TEST FOUND IN "INTERMITTENTS AND POOR CONNECTIONS" IN THIS SECTION. WERE EMI TESTS COMPLETED WITH NO TROUBLE FOUND?

- YES → REPLACE EBTCM.
- NO → PERFORM NECESSARY REPAIRS.

WHEN ALL DIAGNOSIS AND REPAIRS ARE COMPLETED, CLEAR DTC(s) AND VERIFY PROPER OPERATION.

DIAGNOSTIC CHART DTC 41, 45, 51 AND 55 — 1995 FLEETWOOD

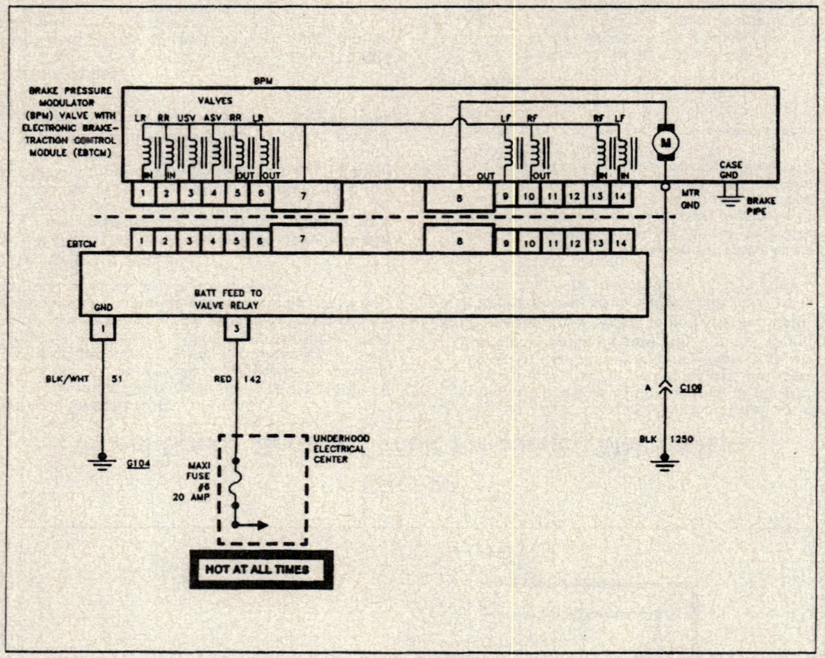

DTC(s) 41, 45, 51, 55

RF, LF, RR, LR Inlet Valve Solenoid Malfunction

Circuit Description:

The wheel valve solenoid circuits are supplied with battery power when the ignition is in the "RUN "position. The EBTCM controls the valve functions by grounding the circuit when necessary.

DTC Will Set When:

- The EBTCM senses a discrepancy such as an open, short to ground or voltage internal to the EBTCM and BPM.
- There is poor connection in the circuit internal to the EBTCM and BPM.
- There is a malfunctioning solenoid valve.

Action Taken:

The valve relay will turn "OFF," the "ANTI-LOCK" and "TRACTION CONTROL" lamps will come "ON." ABS and TCS are disabled.

DTC Chart Test Description:

Number(s) below refer to circled number(s) on the diagnostic chart.

1. Uses the Tech 1 to exercise the affected valve solenoid and check it for proper operation.
2. Checks condition of EBTCM to BPM connector.
3. Checks the integrity of the affected valve solenoid circuitry internal to the BPM valve assembly, and the valve solenoid coil.
4. Determines whether the DTC was set by an intermittent condition or an EBTCM malfunction.
5. Checks if DTC 71 was set concurrently with valve DTC.

Diagnostic Aids:

It is very important that a thorough inspection of the wiring and connectors be performed. Failure to carefully and fully inspect wiring and connectors may result in misdiagnosis, causing part replacement with reappearance of the malfunction.

DIAGNOSTIC CHART DTC 41, 45, 51 AND 55 — 1995 FLEETWOOD, CONTINUED

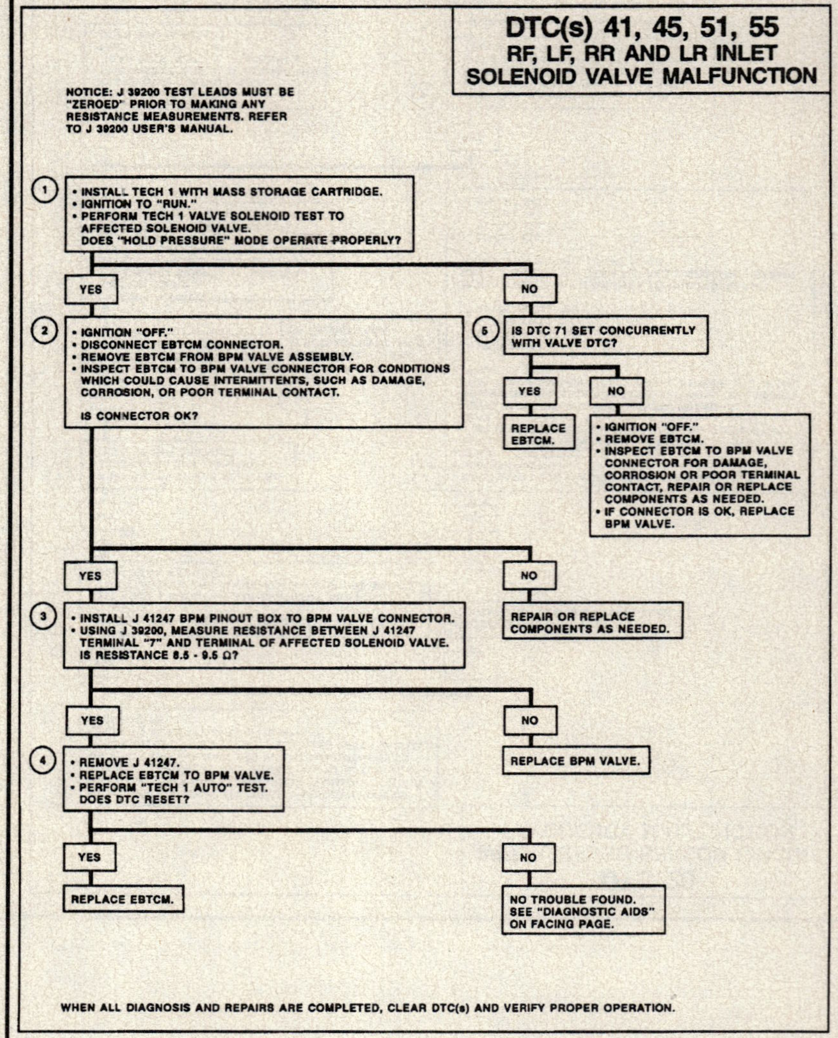

DIAGNOSTIC CHART DTC 42, 46, 52 AND 56 — 1995 FLEETWOOD

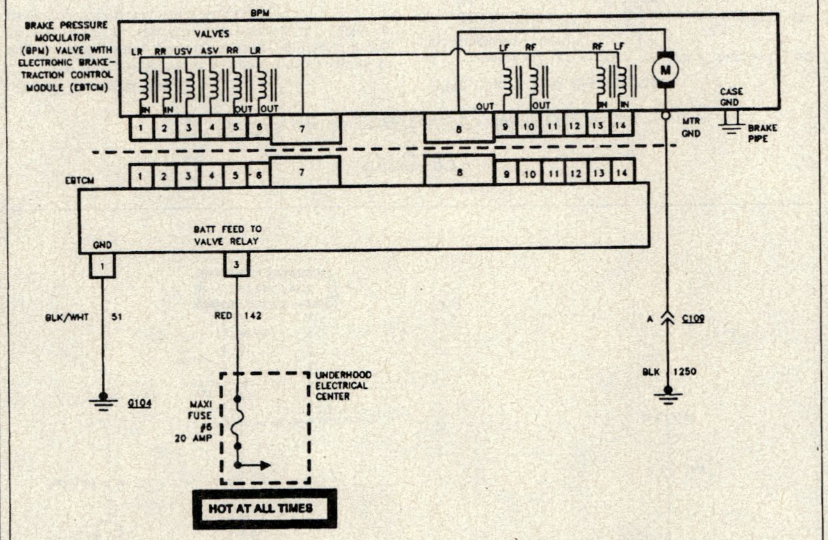

DTC(s) 42, 46, 52, 56
RF, LF, RR, LR Outlet Valve Solenoid Malfunction

Circuit Description:

The wheel valve solenoid circuits are supplied with battery power when the ignition is in the "RUN" position. The EBTCM controls the valve functions by grounding the circuit when necessary.

DTC Will Set When:

- The EBTCM senses a discrepancy such as an open, short to ground or voltage internal to the EBTCM and BPM.
- There is poor connection in the circuit internal to the EBTCM and BPM.
- There is a malfunctioning solenoid valve.

Action Taken:

The valve relay will turn "OFF," the "ANTI-LOCK" and "TRACTION CONTROL" lamps will come "ON". ABS and TCS are disabled.

DTC Chart Test Description:

Number(s) below refer to circled number(s) on the diagnostic chart.

1. Uses the Tech 1 to exercise the affected valve solenoid and check it for proper operation.
2. Checks condition of EBTCM to BPM connector.
3. Checks the integrity of the affected valve solenoid circuitry internal to the BPM valve assembly, and the valve solenoid coil.
4. Determines whether the DTC was set by an intermittent condition or an EBTCM malfunction.
5. Checks if DTC 71 was set concurrently with valve DTC.

Diagnostic Aids:

It is very important that a thorough inspection of the wiring and connectors be performed. Failure to carefully and fully inspect wiring and connectors may result in misdiagnosis, causing part replacement with reappearance of the malfunction.

DIAGNOSTIC CHART DTC 42, 46, 52 AND 56 — 1995 FLEETWOOD, CONTINUED

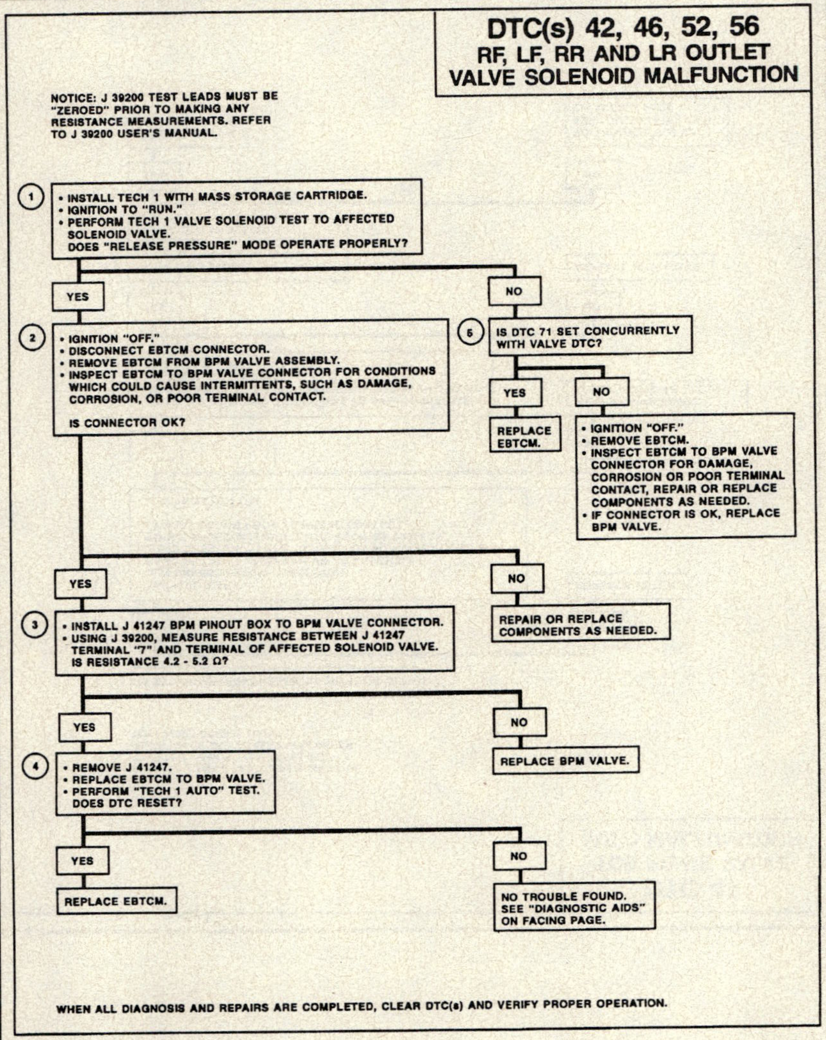

DIAGNOSTIC CHART DTC 47 — 1995 FLEETWOOD

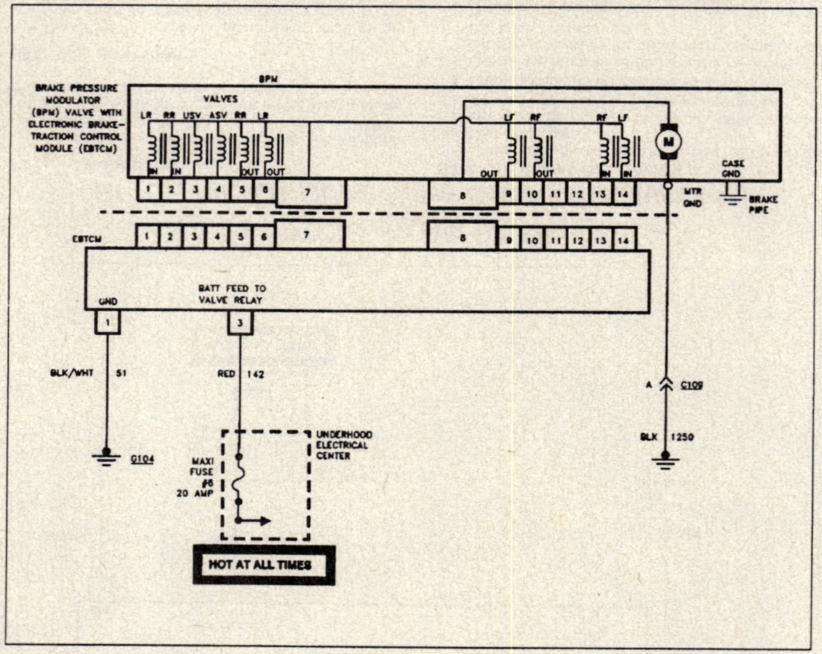

DTC 47
TCS Prime Valve (ASV) Malfunction

Circuit Description:

The TCS Prime Valve solenoid (ASV) circuit is supplied with battery power when the ignition is in the "RUN" position. The EBTCM controls the valve by grounding the circuit when necessary.

DTC Will Set When:

- The EBTCM senses a discrepancy such as an open, short to ground or voltage internal to the EBTCM and BPM.
- There is poor connection in the circuit internal to the EBTCM and BPM.
- There is a malfunctioning solenoid valve.

Action Taken:

The valve relay will turn "OFF," the "ANTI-LOCK" and "TRACTION CONTROL" lamps will come "ON." ABS and TCS are disabled.

DTC Chart Test Description:

Number(s) below refer to circled number(s) on the diagnostic chart.

1. Checks if DTC 71 was set concurrently with DTC 47.
2. Checks condition of EBTCM to BPM connector.
3. Checks the integrity of the affected valve solenoid circuitry internal to the BPM valve assembly, and the valve solenoid coil.
4. Determines whether the DTC was set by an intermittent condition or an EBTCM malfunction.

Diagnostic Aids:

It is very important that a thorough inspection of the wiring and connectors be performed. Failure to carefully and fully inspect wiring and connectors may result in misdiagnosis, causing part replacement with reappearance of the malfunction.

DIAGNOSTIC CHART DTC 47 — 1995 FLEETWOOD, CONTINUED

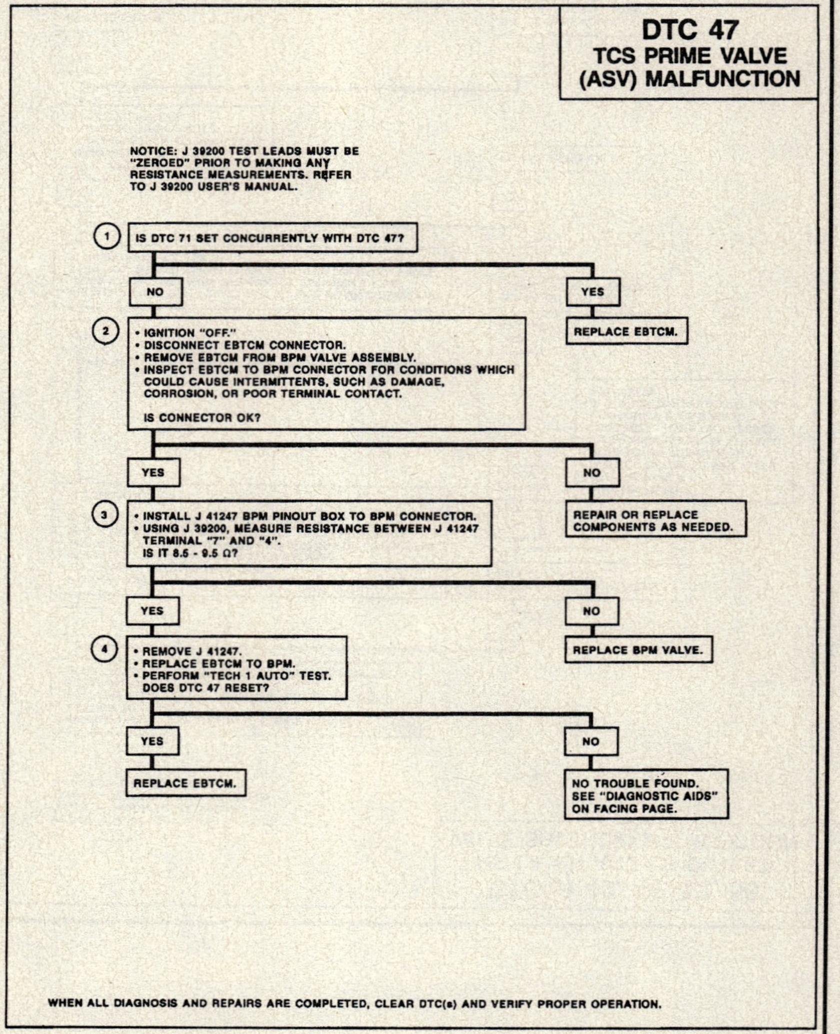

DIAGNOSTIC CHART DTC 48 — 1995 FLEETWOOD

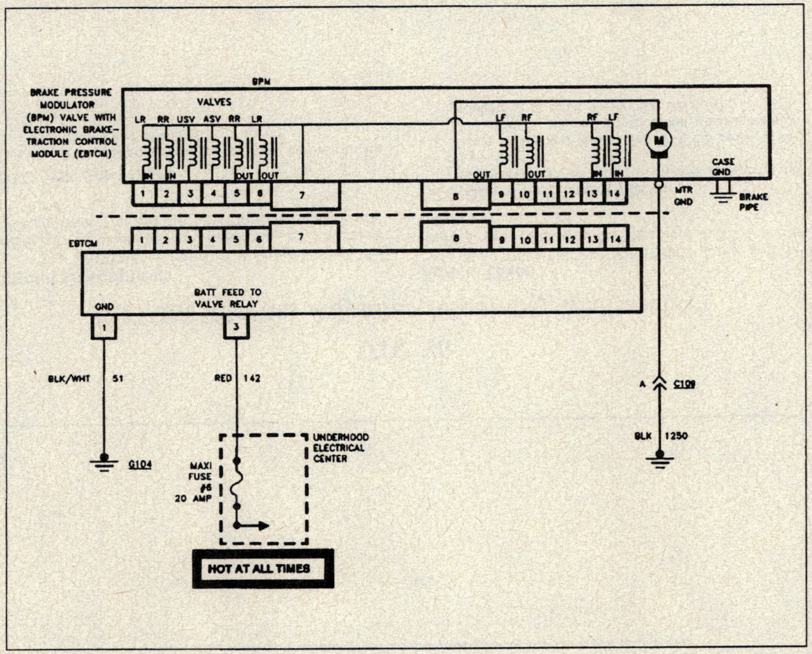

DTC 48
TCS Master Cylinder Isolation Valve (USV) Malfunction

Circuit Description:

The Master Cylinder Isolation Valve Solenoid (USV) circuit is supplied with battery power when the ignition is in the "RUN" position. The EBTCM controls the valve by grounding the circuit when necessary.

DTC Will Set When:

- The EBTCM senses a discrepancy such as an open, short to ground or voltage internal to the EBTCM and BPM.
- There is poor connection in the circuit internal to the EBTCM and BPM.
- There is a malfunctioning solenoid valve.

Action Taken:

The valve relay will turn "OFF", the "ANTI-LOCK" and "TRACTION CONTROL" lamps will come "ON", ABS and TCS are disabled.

DTC Chart Test Description:

Number(s) below refer to circled number(s) on the diagnostic chart.

1. Checks if DTC 71 was set concurrently with DTC 48.
2. Checks condition of EBTCM to BPM connector.
3. Checks the integrity of the affected valve solenoid circuitry internal to the BPM valve assembly, and the valve solenoid coil.
4. Determines whether the DTC was set by an intermittent condition or an EBTCM malfunction.

Diagnostic Aids:

It is very important that a thorough inspection of the wiring and connectors be performed. Failure to carefully and fully inspect wiring and connectors may result in misdiagnosis, causing part replacement with reappearance of the malfunction.

DIAGNOSTIC CHART DTC 48 — 1995 FLEETWOOD, CONTINUED

DTC 48
TCS MASTER CYLINDER ISOLATION VALVE (USV) MALFUNCTION

NOTICE: J 39200 TEST LEADS MUST BE "ZEROED" PRIOR TO MAKING ANY RESISTANCE MEASUREMENTS. REFER TO J 39200 USER'S MANUAL.

(1) IS DTC 71 SET CONCURRENTLY WITH DTC 48?

- YES: REPLACE EBTCM.
- NO: go to (2)

(2)
- IGNITION "OFF."
- DISCONNECT EBTCM CONNECTOR.
- REMOVE EBTCM FROM BPM VALVE ASSEMBLY.
- INSPECT EBTCM TO BPM CONNECTOR FOR CONDITIONS WHICH COULD CAUSE INTERMITTENTS, SUCH AS DAMAGE, CORROSION, OR POOR TERMINAL CONTACT.

IS CONNECTOR OK?

- NO: REPAIR OR REPLACE COMPONENTS AS NEEDED.
- YES: go to (3)

(3)
- INSTALL J 41247 BPM PINOUT BOX TO BPM CONNECTOR.
- USING J 39200, MEASURE RESISTANCE BETWEEN J 41247 TERMINALS "7" AND "3".

IS IT 8.5 - 9.5 Ω?

- NO: REPLACE BPM VALVE.
- YES: go to (4)

(4)
- REMOVE J 41247.
- REPLACE EBTCM TO BPM.
- PERFORM "TECH 1 AUTO" TEST.

DOES DTC 48 RESET?

- YES: REPLACE EBTCM.
- NO: NO TROUBLE FOUND. SEE "DIAGNOSTIC AIDS" ON FACING PAGE.

WHEN ALL DIAGNOSIS AND REPAIRS ARE COMPLETED, CLEAR DTC(s) AND VERIFY PROPER OPERATION.

DIAGNOSTIC CHART DTC 58 — 1995 FLEETWOOD

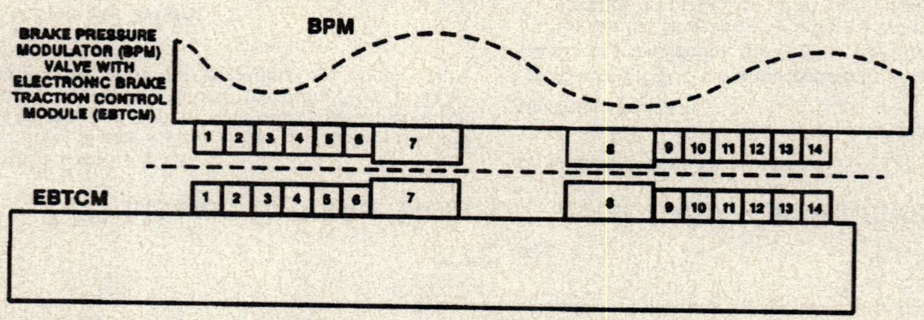

DTC 58
EBTCM Internal Adjuster Assembly Malfunction

Circuit Description:

The EBTCM contains a microprocessor for ABS/TCS calculations and functions, as well as a microprocessor for adjuster assembly calculations and functions.

DTC Will Set When:

- The two microprocessors cannot communicate with each other properly.

Action Taken:

The "TRACTION CONTROL" lamp will be "ON", and TCS will be disabled. ABS remains functional.

DTC Chart Test Description:

Number(s) below refer to circled number(s) on the diagnostic chart.

1. Clears DTCs and checks for DTC 58 reset. If the DTC resets, the microprocessor communication problem does exist within the EBTCM.

DIAGNOSTIC CHART DTC 58 — 1995 FLEETWOOD, CONTINUED

DTC 58
EBTCM INTERNAL ADJUSTER ASSEMBLY MALFUNCTION

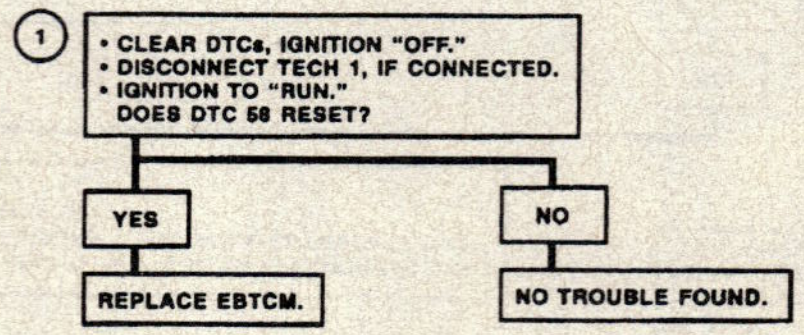

WHEN ALL DIAGNOSIS AND REPAIRS ARE COMPLETED, CLEAR DTC(s) AND VERIFY PROPER OPERATION.

DIAGNOSTIC CHART DTC 61 — 1995 FLEETWOOD

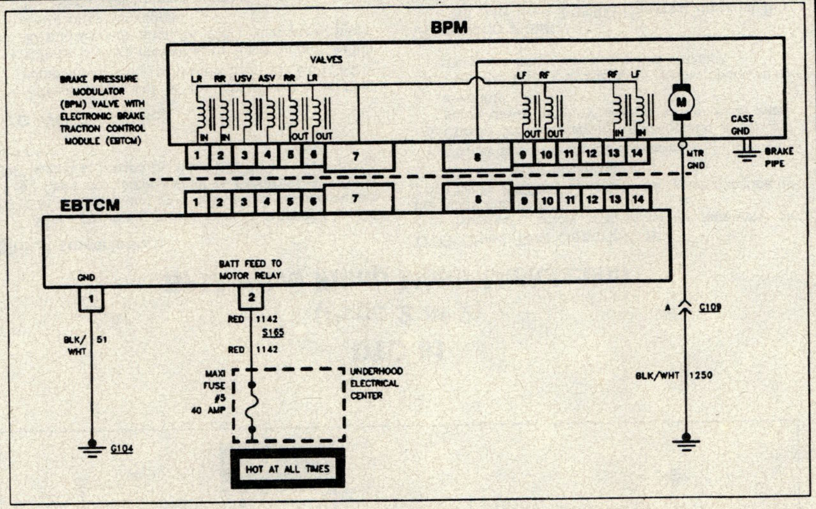

DTC 61
(Page 1 of 2)
BPM Valve Pump Motor Malfunction

Circuit Description:

When the pump motor relay is grounded by the EBTCM, it closes and provides voltage to operate the pump. Once the motor circuit is energized, a "Pump ON" signal is sensed by the EBTCM to verify pump operation.

DTC Will Set When:

- Voltage is present at the pump motor without pump motor relay activation from the EBTCM.
- Voltage is NOT present at the pump motor within 40 milliseconds after the EBTCM requests pump motor relay activation.
- The motor GND circuit is open.
- There is a pump motor malfunction.

Action Taken:

The "ANTI-LOCK" and "TRACTION CONTROL" lamps will be "ON." ABS and TCS are disabled.

DTC Chart Test Description:

Number(s) below refer to circled number(s) on the diagnostic chart.

1. Checks that Power Distribution Maxi Fuse feeding the EBTCM Motor Relay circuit is good.
2. Checks for an open in Pump Motor ground CKT 1250.
3. Measures for appropriate voltage from the EBTCM Motor Relay circuit, using the EBTCM ground line to complete the circuit.
4. Checks EBTCM connector terminals for good contact, i.e., not spread open, backed out of connector, corroded, etc.
5. Checks if fuse failure was due to a short to ground.

Diagnostic Aids:

All tests using J 39700 Universal Breakout Box terminal "1" are using that terminal as ground. These tests assume that the ground CKT 51 at terminal "1" is good.

It is very important that a thorough inspection of the wiring and connectors be performed. Failure to carefully and fully inspect wiring and connectors may result in misdiagnosis, causing part replacement with reappearance of the malfunction.

DIAGNOSTIC CHART DTC 61 — 1995 FLEETWOOD, CONTINUED

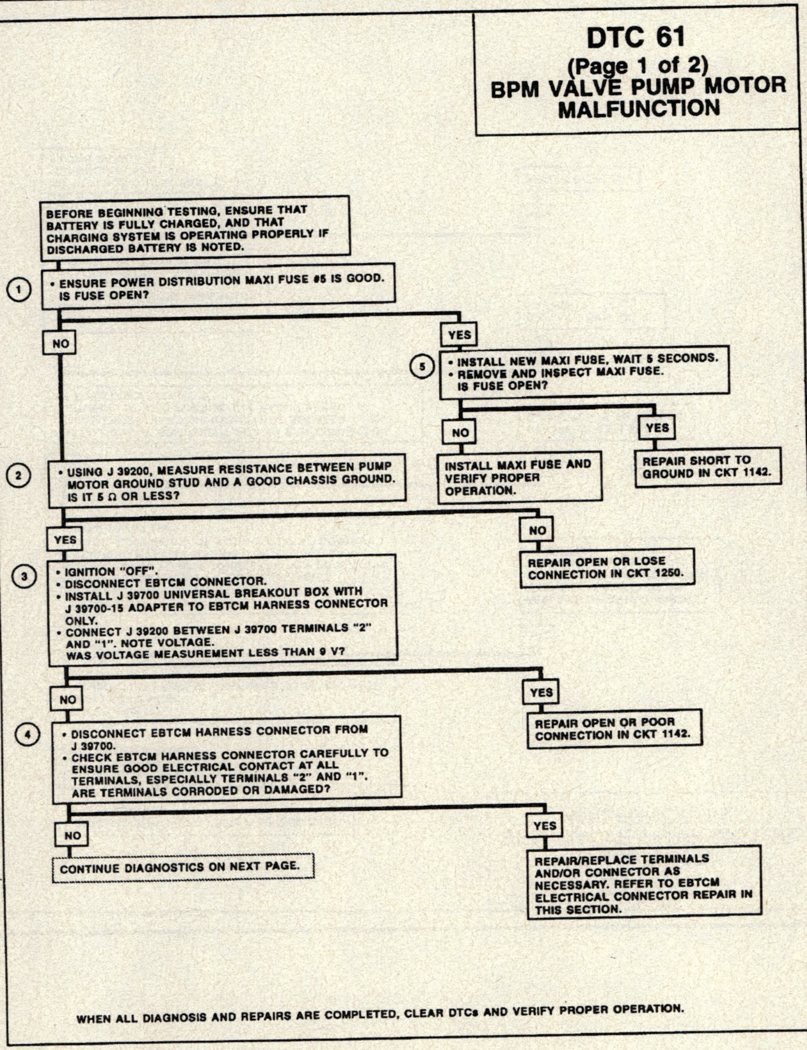

DIAGNOSTIC CHART DTC 61 — 1995 FLEETWOOD, CONTINUED

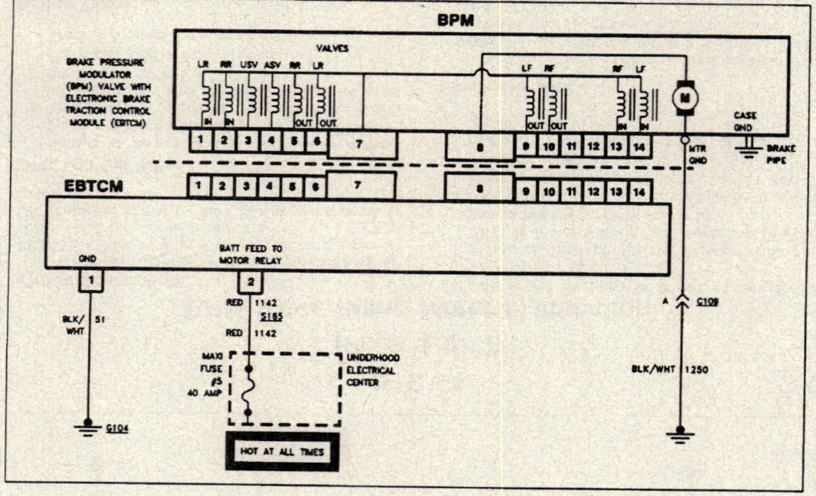

DTC 61
(Page 2 of 2)
BPM Valve Pump Motor Malfunction

Circuit Description:

When the pump motor relay is grounded by the EBTCM, it closes and provides voltage to operate the pump. Once the motor circuit is energized, a "Pump ON" signal is sensed by the EBTCM to verify pump operation.

DTC Will Set When:

- Voltage is present at the pump motor without pump motor relay activation from the EBTCM.
- Voltage is NOT present at the pump motor within 40 milliseconds after the EBTCM requests pump motor relay activation.
- The motor GND circuit is open.
- There is a pump motor malfunction.

Action Taken:

The "ANTI-LOCK" and "TRACTION CONTROL" lamps will be "ON." ABS and TCS are disabled.

DTC Chart Test Description:

Number(s) below refer to circled number(s) on the diagnostic chart.

6. Checks EBTCM connector for good retention to EBTCM when installed.
7. Checks for condition of connector.
8. Checks for the integrity of the pump motor and pump motor circuitry internal to the BPM valve assembly.
9. Uses the Tech 1 to check for proper operation of the pump motor and associated circuitry.

Diagnostic Aids:

All tests using J 39700 Universal Breakout Box terminal "1" are using that terminal as ground. These tests assume that the ground CKT 51 at terminal "1" is good.

It is very important that a thorough inspection of the wiring and connectors be performed. Failure to carefully and fully inspect wiring and connectors may result in misdiagnosis, causing part replacement with reappearance of the malfunction.

DIAGNOSTIC CHART DTC 61 — 1995 FLEETWOOD, CONTINUED

DTC 61
(Page 2 of 2)
BPM VALVE PUMP MOTOR MALFUNCTION

DIAGNOSTICS CONTINUED FROM PREVIOUS PAGE.

NOTICE: J 39200 TEST LEADS MUST BE "ZEROED" PRIOR TO MAKING ANY RESISTANCE MEASUREMENTS. REFER TO J 39200 USER'S MANUAL.

(6)
- REMOVE J 39700.
- RECONNECT EBTCM.

IS CONNECTOR PROPERLY RETAINED TO EBTCM WHEN CONNECTED?

NO: REPLACE OR REPAIR WIRING HARNESS CONNECTOR OR EBTCM AS NEEDED TO OBTAIN PROPER CONNECTOR RETENTION.

YES:

(7)
- IGNITION "OFF".
- DISCONNECT EBTCM CONNECTOR.
- REMOVE EBTCM FROM BPM VALVE ASSEMBLY.
- INSPECT EBTCM TO BPM CONNECTOR FOR CONDITIONS WHICH COULD CAUSE INTERMITTENTS, SUCH AS DAMAGE, CORROSION, OR POOR TERMINAL CONTACT.

IS CONNECTOR OK?

NO: REPAIR OR REPLACE COMPONENTS AS NEEDED.

YES:

(8)
- INSTALL J 41247 BPM PINOUT BOX TO BPM CONNECTOR.
- USING J 39200, MEASURE RESISTANCE BETWEEN J 41247 TERMINAL "8" AND PUMP MOTOR GROUND STUD.

IS RESISTANCE .05-10 Ω?

NO: REPLACE BPM VALVE ASSEMBLY.

YES:

(9)
- REMOVE J 41247.
- RE-INSTALL EBTCM TO BPM.
- RE-CONNECT EBTCM CONNECTOR.
- IGNITION TO "RUN".
- PERFORM "TECH 1 AUTO" TEST.

DOES PUMP MOTOR RUN?

YES: NO TROUBLE FOUND. SEE "DIAGNOSTIC AIDS" ON FACING PAGE.

NO: REPLACE EBTCM.

WHEN ALL DIAGNOSIS AND REPAIRS ARE COMPLETED, CLEAR DTCs AND VERIFY PROPER OPERATION.

DIAGNOSTIC CHART DTC 62 — 1995 FLEETWOOD

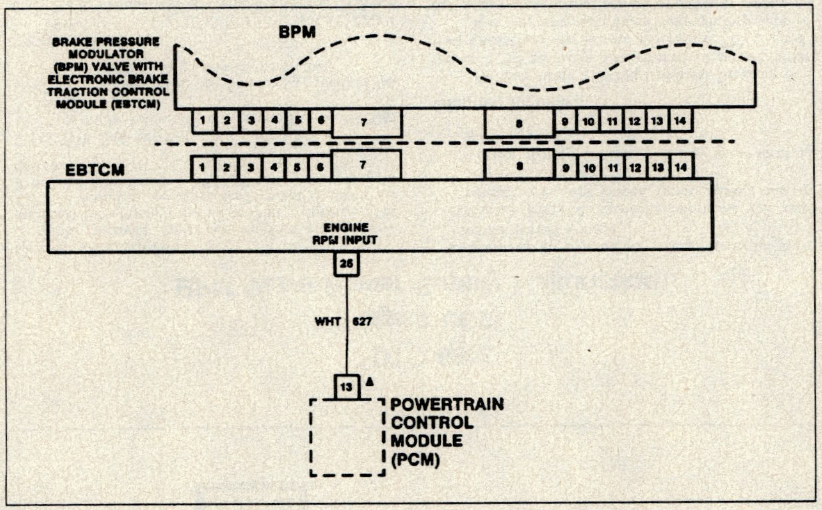

DTC 62
RPM Signal Malfunction

Circuit Description:

The RPM signal circuit provides the EBTCM with an indication of engine RPM to help determine TCS control methods and rates when a TCS event takes place.

DTC Will Set When:

- There is a short to ground, short to voltage, or an open in CKT 627.

Action Taken:

The "TRACTION CONTROL" lamp will be "ON", and TCS is disabled. ABS remains functional.

DTC Chart Test Description:

Number(s) below refer to circled number(s) on the diagnostic chart.

1. Checks for DTC reset after clearing DTCs to ensure a malfunction really exists.
2. Checks for short to ground in CKT 627.
3. Checks for short to voltage in CKT 627.
4. Checks for normal operating voltage of about 5 volts from output side of the PCM. If no voltage is found, CKT 627 is open.

Diagnostic Aids:

All tests using J 39700 Universal Breakout Box terminal "1" are using that terminal as ground. These tests assume that the ground at terminal "1" is good.

It is very important that a thorough inspection of the wiring and connectors be performed. Failure to carefully and fully inspect wiring and connectors may result in misdiagnosis, causing part replacement with reappearance of the malfunction.

DIAGNOSTIC CHART DTC 62 — 1995 FLEETWOOD, CONTINUED

DTC 62
RPM SIGNAL MALFUNCTION

NOTICE: J 39200 TEST LEADS MUST BE "ZEROED" PRIOR TO MAKING ANY RESISTANCE MEASUREMENTS. REFER TO J 39200 USER'S MANUAL.

(1)
- CLEAR DTCs.
- IGNITION "OFF".
- START ENGINE.
- DOES DTC 62 RESET?

NO → NO TROUBLE FOUND; SEE "DIAGNOSTIC AIDS" ON FACING PAGE.

YES ↓

(2)
- IGNITION "OFF".
- DISCONNECT EBTCM.
- INSTALL J 39700 UNIVERSAL BREAKOUT BOX WITH ADAPTER J 39700-15 TO EBTCM HARNESS CONNECTOR ONLY.
- USING J 39200 MEASURE RESISTANCE BETWEEN J 39700 TERMINALS "25" AND "1". DOES J 39200 DISPLAY "OL" (INFINITE)?

NO → REPAIR SHORT TO GROUND IN CKT 627.

YES ↓

(3)
- CONNECT J 39200 (USE DC VOLTS) BETWEEN J 39700 TERMINAL "25" AND "1".
- IGNITION TO "RUN". IS VOLTAGE ABOVE 1 V?

YES → REPAIR SHORT TO VOLTAGE IN CKT 627.

NO ↓

(4)
- START ENGINE.
- NOTE VOLTAGE. IS IT ABOUT 5 V?

YES → NO TROUBLE FOUND; SEE "DIAGNOSTIC AIDS" ON FACING PAGE.

NO → REPAIR OPEN IN CKT 627.

WHEN ALL DIAGNOSIS AND REPAIRS ARE COMPLETED, CLEAR DTCs AND VERIFY PROPER OPERATION.

DIAGNOSTIC CHART DTC 63 — 1995 FLEETWOOD

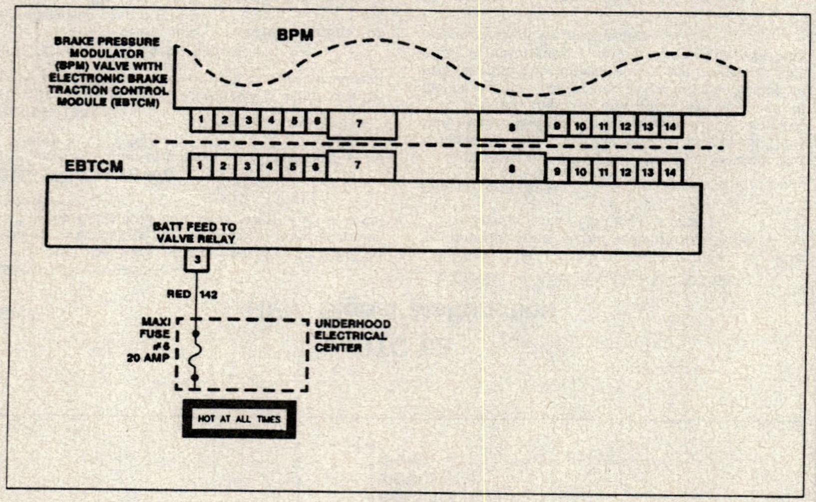

DTC 63
(Page 1 of 2)
BPM Valve Power Supply Malfunction

Circuit Description:

The solenoid valves are supplied voltage from the maxi fuse through the valve relay to operate. The EBTCM monitors the valve feed circuit to determine if a valve response to a EBTCM activation request is detected.

DTC Will Set When:

- 2 or more valves do not respond to a EBTCM request.
- The EBTCM senses a discrepancy such as an open or a short to ground in the circuit.

Action Taken:

The "ANTI-LOCK" and "TRACTION CONTROL" lamps will be "ON." ABS and TCS are disabled.

DTC Chart Test Description:

Number(s) below refer to circled number(s) on the diagnostic chart.

1. Checks that Power Distribution Fuse feeding the Valve Relay circuit is good.
2. Measures for appropriate voltage from the valve circuit supply voltage.
3. Checks EBTCM connector terminals for good contact, i.e., not spread open, backed out of connector, corroded, etc.
4. Checks if fuse failure was due to a short to ground.

Diagnostic Aids:

All tests using J 39700 Universal Breakout Box terminal "1" are using that terminal as ground. These tests assume that the ground at terminal "1" is good.

It is very important that a thorough inspection of the wiring and connectors be performed. Failure to carefully and fully inspect wiring and connectors may result in misdiagnosis, causing part replacement with reappearance of the malfunction.

DIAGNOSTIC CHART DTC 63 — 1995 FLEETWOOD, CONTINUED

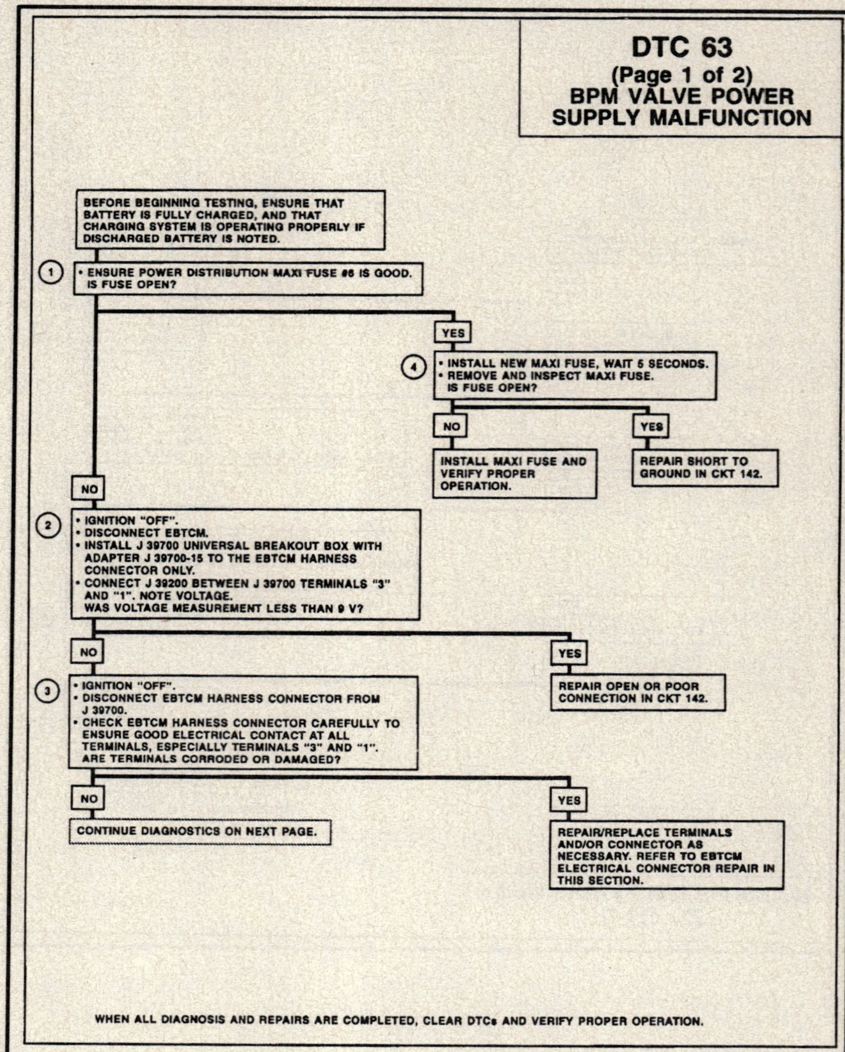

DIAGNOSTIC CHART DTC 63 — 1995 FLEETWOOD, CONTINUED

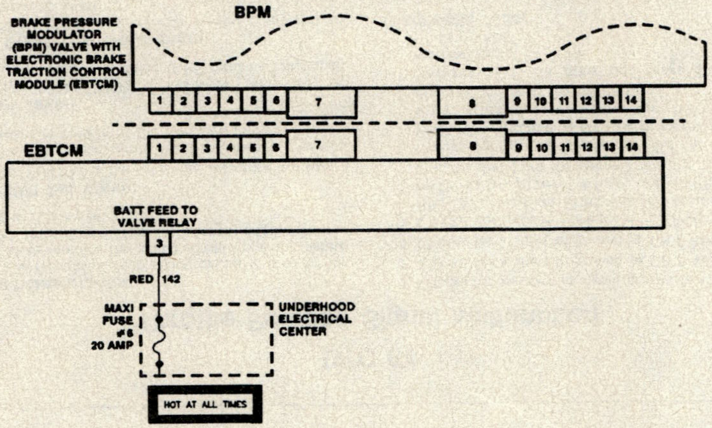

DTC 63
(Page 2 of 2)
BPM Valve Power Supply Malfunction

Circuit Description:

The solenoid valves are supplied voltage from the maxi fuse through the valve relay to operate. The EBTCM monitors the valve feed circuit to determine if a valve response to a EBTCM activation request is detected.

DTC Will Set When:

- 2 or more valves do not respond to a EBTCM request.
- The EBTCM senses a discrepancy such as an open or a short to ground in the circuit.

Action Taken:

The "ANTI-LOCK" and "TRACTION CONTROL" lamps will be "ON." ABS and TCS are disabled.

DTC Chart Test Description:

Number(s) below refer to circled number(s) on the diagnostic chart.

5. Checks condition of EBTCM to BPM connector.
6. Checks EBTCM connector for good retention to EBTCM when installed.
7. Ensures DTC 63 does not reset due to a malfunctioning EBTCM.

Diagnostic Aids:

All tests using J 39700 Universal Breakout Box terminal "1" are using that terminal as ground. These tests assume that the ground at terminal "1" is good.

It is very important that a thorough inspection of the wiring and connectors be performed. Failure to carefully and fully inspect wiring and connectors may result in misdiagnosis, causing part replacement with reappearance of the malfunction.

DIAGNOSTIC CHART DTC 63 — 1995 FLEETWOOD, CONTINUED

DTC 63
(Page 2 of 2)
BPM VALVE POWER SUPPLY MALFUNCTION

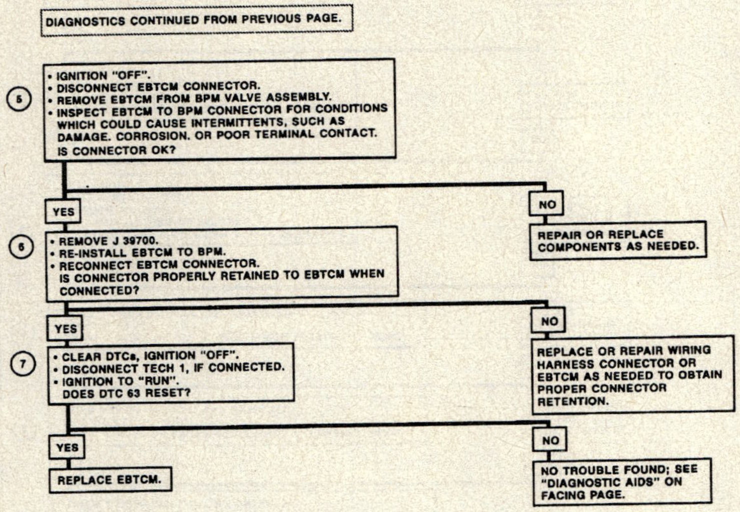

WHEN ALL DIAGNOSIS AND REPAIRS ARE COMPLETED, CLEAR DTCs AND VERIFY PROPER OPERATION.

DIAGNOSTIC CHART DTC 64 — 1995 FLEETWOOD

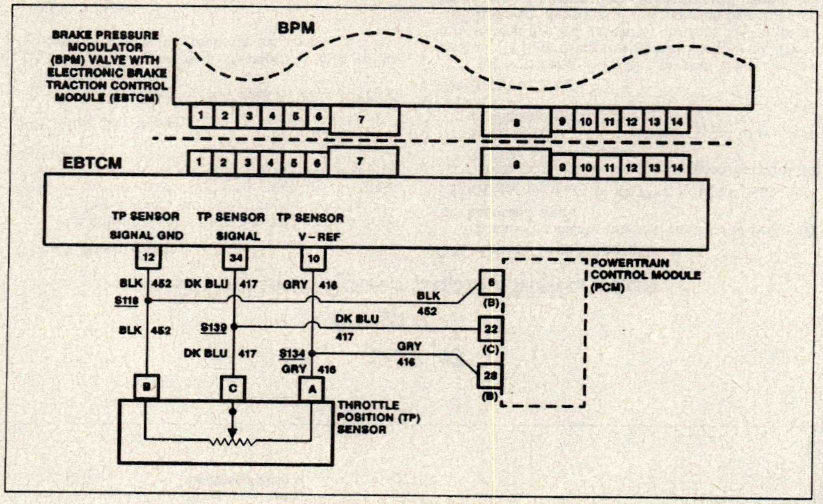

DTC 64

Throttle Position Signal Malfunction

Circuit Description:

The throttle position signal circuit is used by the EBTCM to monitor the actual throttle position versus desired throttle position to control engine torque during a TCS event.

DTC Will Set When:

- The EBTCM is not receiving reference voltage.
- The TP sensor signal from the throttle position sensor is outside of the specified range.

Action Taken:

The "TRACTION CONTROL" lamp will be "ON", and TCS is disabled. ABS remains functional.

DTC Chart Test Description:

Number(s) below refer to circled number(s) on the diagnostic chart.

1. Checks for TP sensor-related PCM DTCs. Many of the same conditions which could cause a TCS DTC 64 will also cause PCM DTCs. The PCM DTCs should be repaired first to eliminate those malfunctions from affecting the TCS DTC 64 condition.
2. Checks for an open in signal ground in CKT 452.
3. Checks for a short to ground in CKT 417.
4. Checks for a short to ground in CKT 416.
5. Checks for proper voltage in CKT 416 at the EBTCM connection. (This voltage measurement is only appropriate when the EBTCM is disconnected from the circuit.)
6. Checks for proper voltage in CKT 417 at the EBTCM connection. (This voltage measurement is only appropriate when the EBTCM is disconnected from the circuit.)
7. Ensures DTC 64 does not reset due to a malfunctioning EBTCM.

Diagnostic Aids:

It is very important that a thorough inspection of the wiring and connectors be performed. Failure to carefully and fully inspect wiring and connectors may result in misdiagnosis, causing part replacement with reappearance of the malfunction.

DIAGNOSTIC CHART DTC 64 — 1995 FLEETWOOD, CONTINUED

DTC 64
THROTTLE POSITION SIGNAL MALFUNCTION

NOTICE: THROTTLE PLATE MUST BE FULLY CLOSED WHILE DIAGNOSING THIS DTC.

(1) IF PCM DTC 21 OR 22 IS SET, USE THE PCM DIAGNOSTIC CHARTS TO REPAIR THOSE DTCs. THEN, CLEAR PCM AND TCS DTCs. TURN IGNITION "OFF". DISCONNECT TECH 1, IF CONNECTED. TURN IGNITION TO "RUN". IF TCS DTC 64 RESETS, CONTINUE WITH THIS DIAGNOSTIC CHART.

(2)
- IGNITION "OFF".
- DISCONNECT EBTCM.
- INSTALL J 39700 UNIVERSAL BREAKOUT BOX WITH ADAPTER J 39700-15 TO EBTCM HARNESS CONNECTOR ONLY.
- USING J 39700, MEASURE RESISTANCE BETWEEN J 39700 TERMINALS "12" AND "1".

DOES J 39200 DISPLAY "OL" (INFINITE)?

- YES → REPAIR OPEN IN CKT 452.
- NO → (3)

(3)
- USING J 39200, MEASURE RESISTANCE BETWEEN J 39700 TERMINALS "34" AND "1".

DOES J 39200 DISPLAY "OL" (INFINITE)?

- NO → REPAIR SHORT TO GROUND IN CKT 417.
- YES → (4)

(4)
- USING J 39200, MEASURE RESISTANCE BETWEEN J 39700 TERMINALS "10" AND "1".

DOES J 39200 DISPLAY "OL" (INFINITE)?

- NO → REPAIR SHORT TO GROUND IN CKT 416.
- YES → (5)

(5)
- CONNECT J 39200 BETWEEN J 39700 TERMINALS "10" AND "1" (USE DC VOLTS).
- IGNITION TO "RUN".
- NOTE VOLTAGE.

IS IT ABOUT 5 V?

- NO → REPAIR OPEN IN CKT 416.
- YES → (6)

(6)
- CONNECT J 39200 BETWEEN J 39700 TERMINALS "34" AND "1" (USE DC VOLTS).
- NOTE VOLTAGE.

IS IT ABOUT 1 V?

- NO → REPAIR OPEN IN CKT 417.
- YES → (7)

(7)
- IGNITION "OFF".
- DISCONNECT J 39700 BREAKOUT BOX.
- RECONNECT EBTCM CONNECTOR.
- IGNITION TO "RUN".

DOES DTC 64 SET?

- YES → REPLACE EBTCM.
- NO → MALFUNCTION IS INTERMITTENT. REFER TO "DIAGNOSTIC AIDS" ON FACING PAGE.

WHEN ALL DIAGNOSIS AND REPAIRS ARE COMPLETED, CLEAR DTCs AND VERIFY PROPER OPERATION.

DIAGNOSTIC CHART DTC 65 — 1995 FLEETWOOD

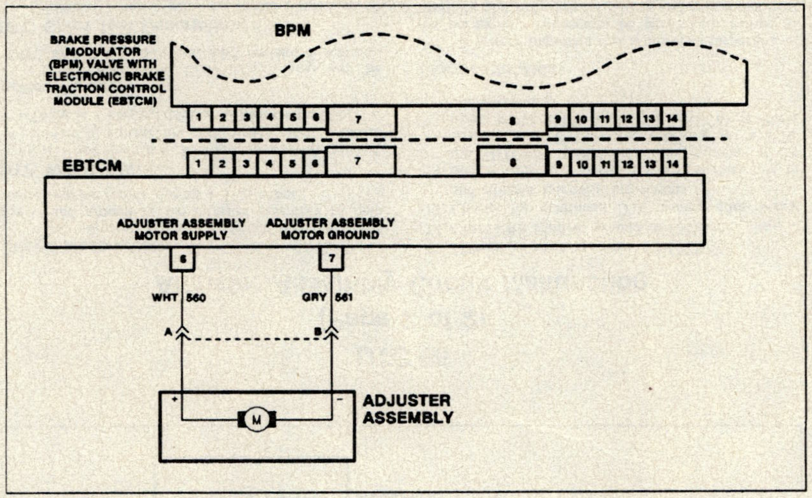

DTC 65
(Page 1 of 2)
Adjuster Assembly Circuit Malfunction

Circuit Description:

The adjuster assembly circuit provides both power and ground to the adjuster assembly to help control engine torque during a TCS event.

DTC Will Set When:

- CKT 560 or CKT 561 are shorted to ground, shorted to voltage, shorted to each other, or open.
- There is a malfunctioning Adjuster Assembly.

Action Taken:

The "TRACTION CONTROL" lamp will be "ON", and TCS is disabled. ABS remains functional.

DTC Chart Test Description:

Number(s) below refer to circled number(s) on the diagnostic chart.

1. Checks for proper adjuster assembly motor winding resistance.
2. Checks for a poor power feed condition in CKT 1142, which the EBTCM might sense as an open.
3. Checks for open in CKT 560.
4. Checks for open in CKT 561.
5. Checks for a short between CKTs 560 and 561.
6. Checks for short to voltage in CKT 560.
7. Repeat Step 1 tests to ensure motor is in "home" position and a correct resistance value can be measured.
8. Sets throttle to a position which enables a special EBTCM monitor. If DTC 65 resets, there is a malfunction which requires continuing diagnosis.
9. Determines whether wiring or the adjuster assembly are at fault for problems found in Step 7.

Diagnostic Aids:

It is very important that a thorough inspection of the wiring and connectors be performed. Failure to carefully and fully inspect wiring and connectors may result in misdiagnosis, causing part replacement with reappearance of the malfunction.

DIAGNOSTIC CHART DTC 65 — 1995 FLEETWOOD, CONTINUED

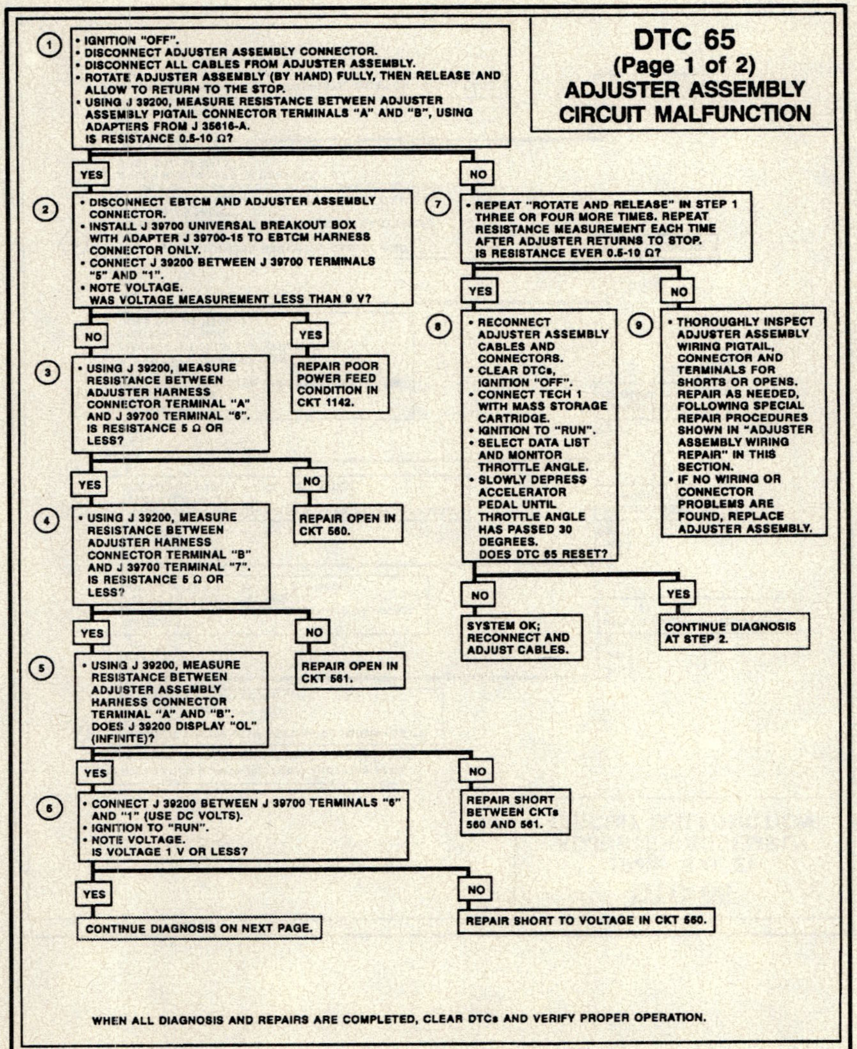

DIAGNOSTIC CHART DTC 65 — 1995 FLEETWOOD, CONTINUED

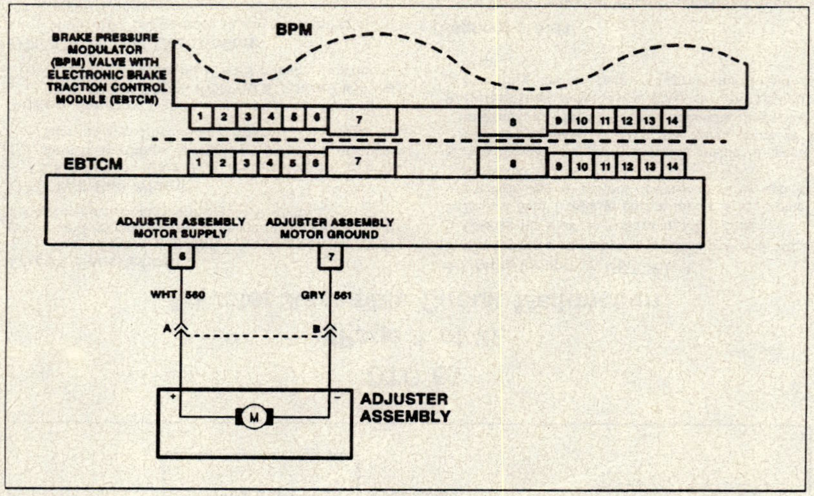

DTC 65

(Page 2 of 2)

Adjuster Assembly Circuit Malfunction

Circuit Description:

The adjuster assembly circuit provides both power and ground to the adjuster assembly to help control engine torque during a TCS event.

DTC Will Set When:

- CKT 560 or CKT 561 are shorted to ground, shorted to voltage, shorted to each other, or open.
- There is a malfunctioning Adjuster Assembly.

Action Taken:

The "TRACTION CONTROL" lamp will be "ON", and TCS is disabled. ABS remains functional.

DTC Chart Test Description:

Number(s) below refer to circled number(s) on the diagnostic chart.

10. Checks for short to voltage in CKT 561.
11. Checks for short to ground in CKT 560.
12. Checks for short to ground in CKT 561.
13. Checks for continued DTC setting after wiring and adjuster assembly inspections.
14. Sets throttle to a position which enables one of the EBTCM's special monitoring systems of the adjuster assembly circuits, checking for a DTC reset. If the DTC resets, the EBTCM is likely malfunctioning.

Diagnostic Aids:

It is very important that a thorough inspection of the wiring and connectors be performed. Failure to carefully and fully inspect wiring and connectors may result in misdiagnosis, causing part replacement with reappearance of the malfunction.

DIAGNOSTIC CHART DTC 65 — 1995 FLEETWOOD, CONTINUED

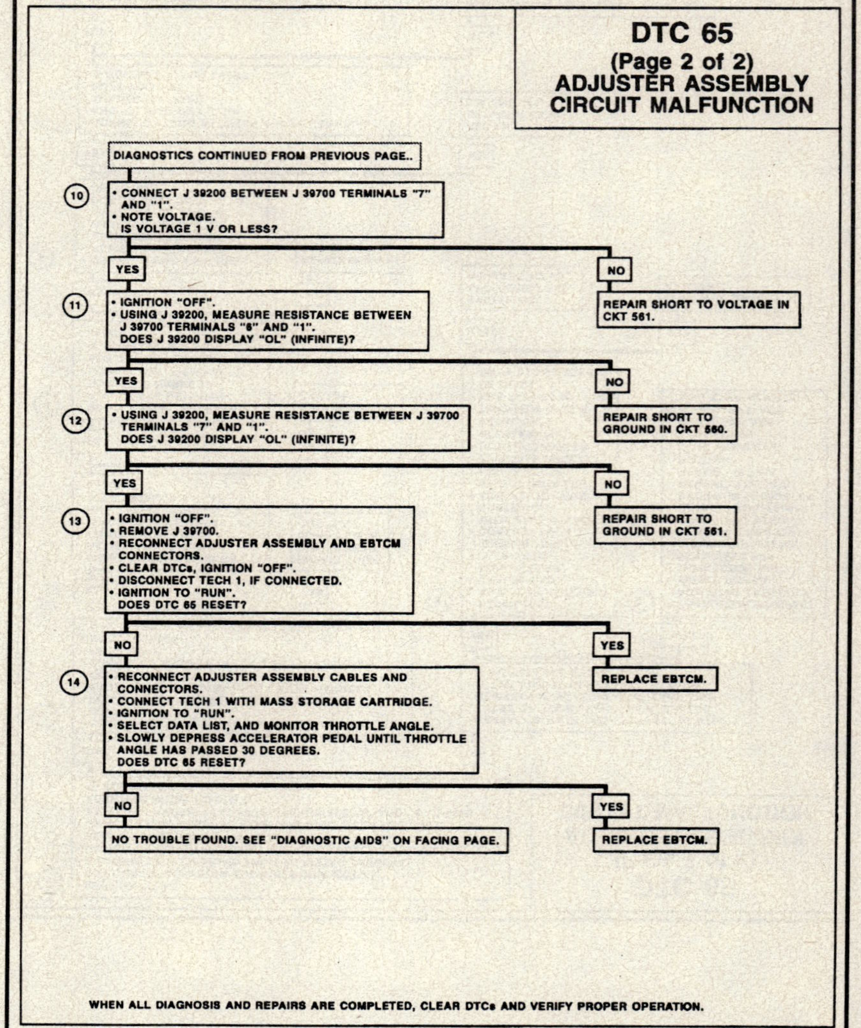

DIAGNOSTIC CHART DTC 66 — 1995 FLEETWOOD

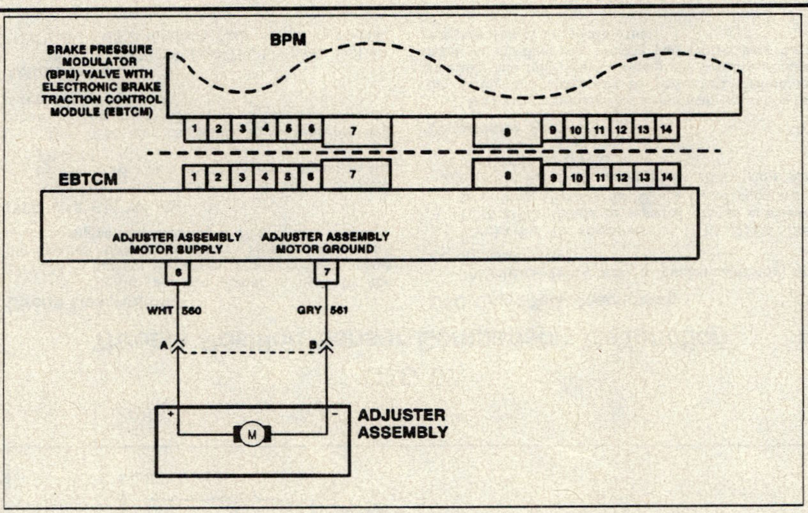

DTC 66
Adjuster Assembly Control Malfunction

Circuit Description:

The adjuster assembly circuit provides both power and ground to the adjuster assembly to help control engine torque during a TCS event.

DTC Will Set When:

- During a TCS event, the EBTCM delivers current greater than 18 amps to the adjuster assembly for more than three seconds without the adjuster assembly reaching its desired position.

Action Taken:

The "TRACTION CONTROL" lamp will be "ON", and TCS is disabled. ABS remains functional.

DTC Chart Test Description:

Number(s) below refer to circled number(s) on the diagnostic chart.

1. Checks to see if DTC 65 is set concurrently with DTC 66, since DTC 65 conditions might affect DTC 66.
2. Checks for any mechanical binding of the adjuster assembly, attached cables, and components to which the cables are attached, which could cause excessive load on the adjuster assembly.
3. Since DTC 66 only sets during a TCS event, this check raises the vehicle on a hoist such that the rear wheels will spin while the fronts do not, inducing the system into attempting TCS control. The throttle angle must be greater than 15%, and less than 50%, and the system must attempt TCS control for greater than 3 seconds to ensure that the DTC-setting throttle and timing parameters are met when checking for a DTC 66 reset.

Diagnostic Aids:

It may be helpful to disconnect all cables from the adjuster assembly and move them manually throughout their entire operating range. Also check the components to which the cables are attached for binding.

Important

- After cables are reconnected to the adjuster assembly, perform "Adjustment Procedure-Adjuster Assembly-Accelerator and Cruise Control Cables" in this section.

DIAGNOSTIC CHART DTC 66 — 1995 FLEETWOOD, CONTINUED

DTC 66
ADJUSTER ASSEMBLY CONTROL MALFUNCTION

BEFORE BEGINNING TEST, INSPECT ALL ADJUSTER ASSEMBLY CABLES AND LINKAGES TO WHICH THEY ATTACH TO ENSURE NO OBVIOUS PROBLEMS THAT COULD CAUSE BINDING OR STICKING EXIST.

(1) IS TCS DTC 65 SET CONCURRENTLY WITH DTC 66?

- YES: REFER TO DTC 65 CHART.
- NO:

(2)
- IGNITION "OFF".
- HAVE ASSISTANT DEPRESS ACCELERATOR PEDAL SMOOTHLY AND FULLY TO WIDE OPEN THROTTLE (WOT), THEN SLOWLY RETURN TO IDLE POSITION WHILE INSPECTING CABLES AND LINKAGES TO WHICH THEY ARE ATTACHED FOR SMOOTH OPERATION.

DO ALL CABLES AND LINKAGES MOVE SMOOTHLY AND EVENLY WITH NO BINDING?

- NO: REPAIR LINKAGES AND/OR REPLACE CABLES, AND/OR REPAIR OR REPLACE COMPONENTS AS NEEDED TO RESTORE SMOOTH OPERATION.

  IMPORTANT:

  - AFTER CABLES ARE REATTACHED TO THE ADJUSTER ASSEMBLY OR COMPONENTS, OR AFTER CABLES ARE REPLACED, PERFORM "ADJUSTMENT PROCEDURE-ADJUSTER ASSEMBLY - THROTTLE AND CRUISE CONTROL CABLES," IN THIS SECTION.

  IF A NEW TP SENSOR OR THROTTLE BODY IS INSTALLED, BE SURE TO PERFORM "TP SENSOR LEARN PROCEDURE" IN THIS SECTION.
- YES:

(3)
- CLEAR DTCs, IGNITION "OFF".
- RAISE REAR OF VEHICLE UNTIL REAR WHEELS ARE APPROXIMATELY 6" OFF FLOOR.
- CONNECT TECH 1 WITH MASS STORAGE CARTRIDGE.
- START ENGINE.
- SELECT TECH 1 PCM DATA LIST AND MONITOR THROTTLE ANGLE.
- APPLY BRAKES.
- PUT VEHICLE IN GEAR.
- RELEASE BRAKES.
- DEPRESS AND HOLD ACCELERATOR PEDAL AT A THROTTLE ANGLE BETWEEN 15% AND 45% FOR NO LESS THAN FOUR SECONDS, THEN RELEASE.

DOES DTC 66 RESET?

- YES: REPLACE ADJUSTER ASSEMBLY.
- NO: NO TROUBLE FOUND SEE "DIAGNOSTIC AIDS" ON FACING PAGE.

WHEN ALL DIAGNOSIS AND REPAIRS ARE COMPLETED, CLEAR DTCs AND VERIFY PROPER OPERATION.

DIAGNOSTIC CHART DTC 67 — 1995 FLEETWOOD

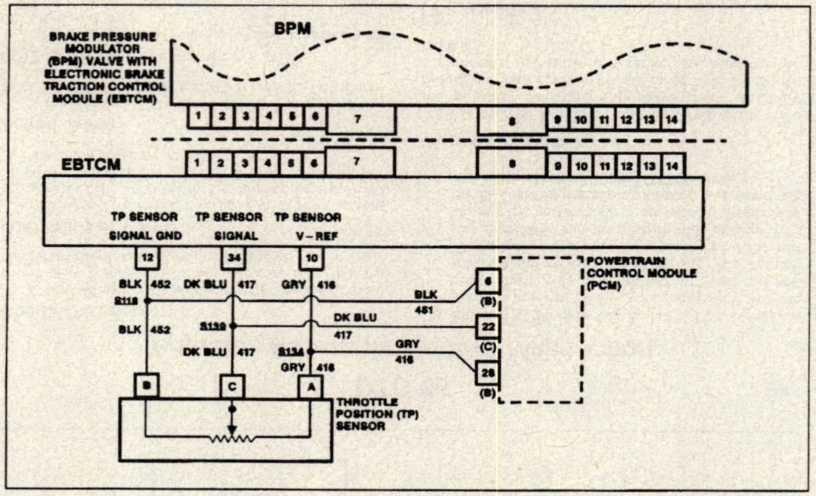

DTC 67

Throttle Position Sensor Comparison Malfunction

Circuit Description:

The Throttle position signal is used by the EBTCM to monitor the actual throttle position versus the desired throttle position. This monitors and controls engine torque during a TCS event.

DTC Will Set When:

- A throttle position sensor circuit fault is detected by the PCM which sends an error signal to the EBTCM.

The PCM diagnostics are more sensitive that those of the EBTCM, so a DTC 64 will not necessarily be set.

Action Taken:

The "TRACTION CONTROL" lamp will be "ON", and TCS is disabled. ABS remains functional.

DTC Chart Test Description:

Number(s) below refer to circled number(s) on the diagnostic chart.

1. Checks for TP sensor-related PCM DTCs. The PCM DTCs should be repaired first to eliminate those from affecting the TCS DTC 67 condition.
2. Checks for DTC reset after addressing for possible PCM DTCs.

Diagnostic Aids:

It is very important that a thorough inspection of the wiring and connectors be performed. Failure to carefully and fully inspect wiring and connectors may result in misdiagnosis, causing part replacement with reappearance of the malfunction.

DIAGNOSTIC CHART DTC 67 — 1995 FLEETWOOD, CONTIN UED

DTC 67
THROTTLE POSITION SENSOR COMPARISON MALFUNCTION

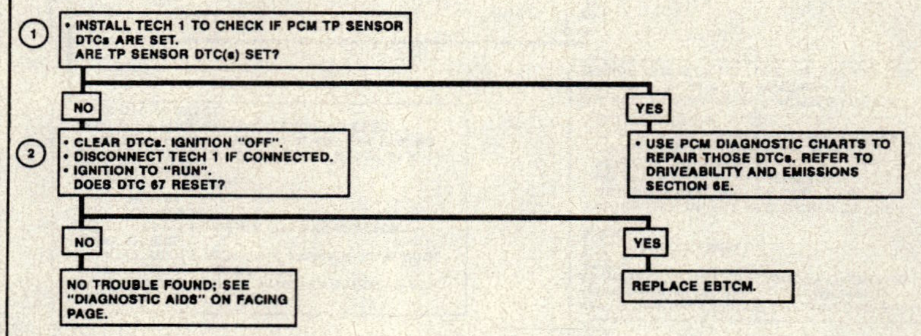

WHEN ALL DIAGNOSIS AND REPAIRS ARE COMPLETED, CLEAR DTCs AND VERIFY PROPER OPERATION.

DIAGNOSTIC CHART DTC 71 — 1995 FLEETWOOD

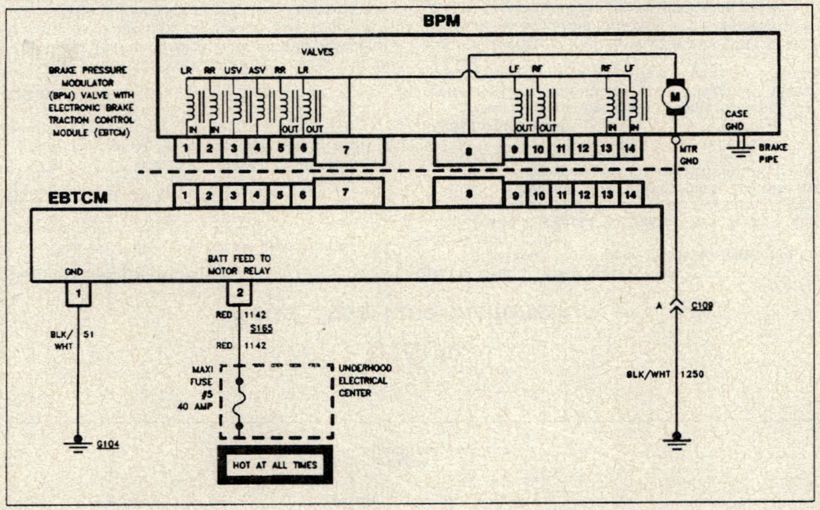

DTC 71
EBTCM Internal Malfunction

Circuit Description:

The EBTCM performs various diagnostic checks on itself. If it finds a malfunction, DTC 71 will set.

DTC Will Set When:

- The EBTCM finds an internal malfunction during self diagnosis.
- The pump motor supply line circuit is open or shorted to ground (open fuse).

Action Taken:

The "ABS" and "TRACTION CONTROL" lamps will be "ON", and ABS and TCS are disabled.

DTC Chart Test Description:

Number(s) below refer to circled number(s) on the diagnostic chart.

1. Checks that Power Distribution MAXI Fuse feeding the pump motor relay is good.
2. Measures for appropriate voltage from the pump motor supply circuit using the EBTCM ground line to complete the circuit.
3. Checks for good connections from the harness to the EBTCM itself.
4. Checks to see if the malfunction was false. If DTC 71 resets, replace the EBTCM.
5. Checks if fuse failure was due to a short to ground.

Diagnostic Aids:

All tests using J 39700 Universal Breakout Box terminal "1" are using that terminal as ground. These tests assume that the ground at terminal "1" is good.

It is very important that a thorough inspection of the wiring and connectors be performed. Failure to carefully and fully inspect wiring and connectors may result in misdiagnosis, causing part replacement with reappearance of the malfunction.

DIAGNOSTIC CHART DTC 71 — 1995 FLEETWOOD, CONTINUED

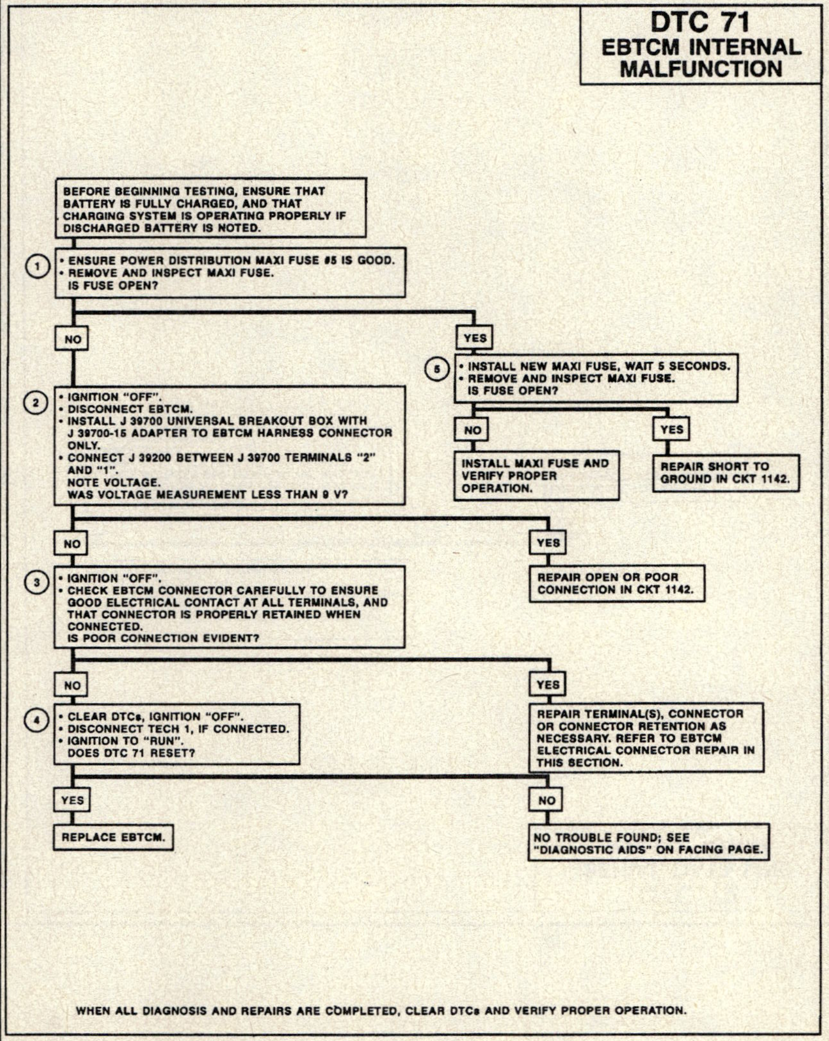

DIAGNOSTIC CHART DTC 72 — 1995 FLEETWOOD

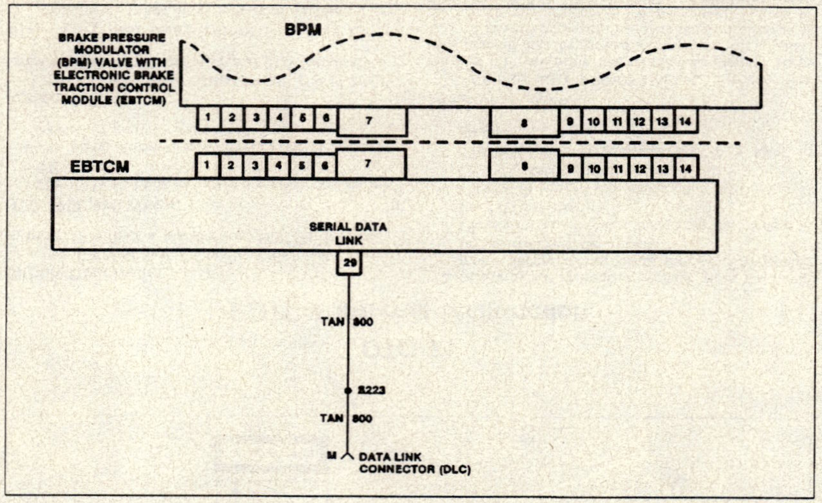

DTC 72
Serial Data Line Malfunction

Circuit Description:

The serial data link is an asynchronous link operating at 8192 bits per second.

DTC Will Set When:

- The EBTCM does not receive a valid serial data link message within 5.6 seconds due to errors in transmission.

Action Taken:

The "TRACTION CONTROL" lamp will be "ON", and TCS is disabled. ABS remains functional. When a valid message is received five consecutive times, TCS is re-enabled and the lamp will turn "OFF".

DTC Chart Test Description:

Number(s) below refer to circled number(s) on the diagnostic chart.

1. This test checks to see if the Tech 1 can communicate with the EBTCM. Automatic control air conditioning DTC display mode can also be used to communicate to the EBTCM. Refer to SECTION 1C for details.

Diagnostic Aids:

The malfunction may be intermittent. Try performing the tests shown while "wiggling" wiring and connectors; this can often cause the fault to appear.

Automatic control air conditioning DTC 32 will likely be set along with this DTC if the malfunction is likely to be set along with this DTC if the malfunction is a short to ground or B+ on the serial data circuit.

If ABS/TCS DTC 72 is set and automatic control air conditioning DTC 41 is not, the malfunction is likely to be an open between the EBTCM and the automatic control air conditioning.

DTC 72 can set by pulling the Tech 1 connector from the DLC prior to exiting from the ABS applications.

DIAGNOSTIC CHART DTC 72 — 1995 FLEETWOOD, CONTINUED

DTC 72
SERIAL DATA LINE MALFUNCTION

(1)
- IGNITION "OFF".
- INSTALL TECH 1 WITH MASS STORAGE CARTRIDGE.
- IGNITION TO "RUN".
- ATTEMPT TO ESTABLISH COMMUNICATIONS WITH THE EBTCM.

ARE COMMUNICATIONS ESTABLISHED?

NO → ENSURE THERE ARE NO OPEN CIRCUITS IN SERIAL DATA LINE BETWEEN DLC TERMINAL "M" AND CCM AND PCM. IF NO OPEN CIRCUITS ARE EVIDENT, DIAGNOSE SERIAL DATA LINE MALFUNCTION.

YES → NO TROUBLE FOUND; SEE "DIAGNOSTIC AIDS" ON FACING PAGE.

WHEN ALL DIAGNOSIS AND REPAIRS ARE COMPLETED, CLEAR DTCs AND VERIFY PROPER OPERATION.

DIAGNOSTIC CHART DTC 73 — 1995 FLEETWOOD

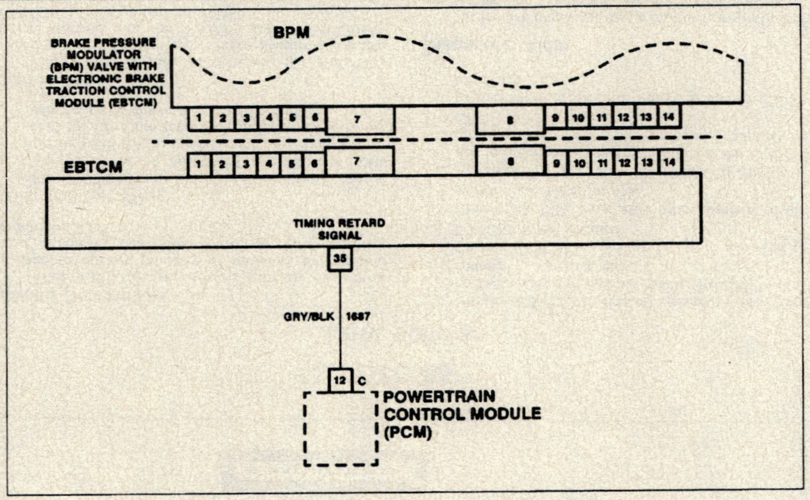

DTC 73
Spark Retard Monitoring Malfunction

Circuit Description:
The EBTCM, upon sensing the need for ASR intervention, requests spark retard from the PCM by supplying B+ to CKT 1687.

DTC Will Set When:
- The EBTCM senses an open, short to voltage, or short to ground on CKT 1687.

Action Taken:
The "TRACTION CONTROL" lamp will be "ON" and TCS is disabled. ABS remains functional.

DTC Chart Test Description:
Number(s) below refer to circled number(s) on the diagnostic chart.
1. Checks for a short to voltage on CKT 1687.
2. Checks for a short to ground on CKT 1687.
3. Checks CKT 1687 and internal PCM circuitry for proper resistance.
4. Checks for DTC reset after the rest of the system has checked OK.
5. Checks of a malfunction found in Step 3 is due to an open in CKT 1687 or an internally malfunctioning PCM.

Diagnostic Aids:

It is very important that a thorough inspection of the wiring and connectors be performed. Failure to carefully and fully inspect wiring and connectors may result in misdiagnosis, causing part replacement with reappearance of the malfunction.

DIAGNOSTIC CHART DTC 73 — 1995 FLEETWOOD, CONTINUED

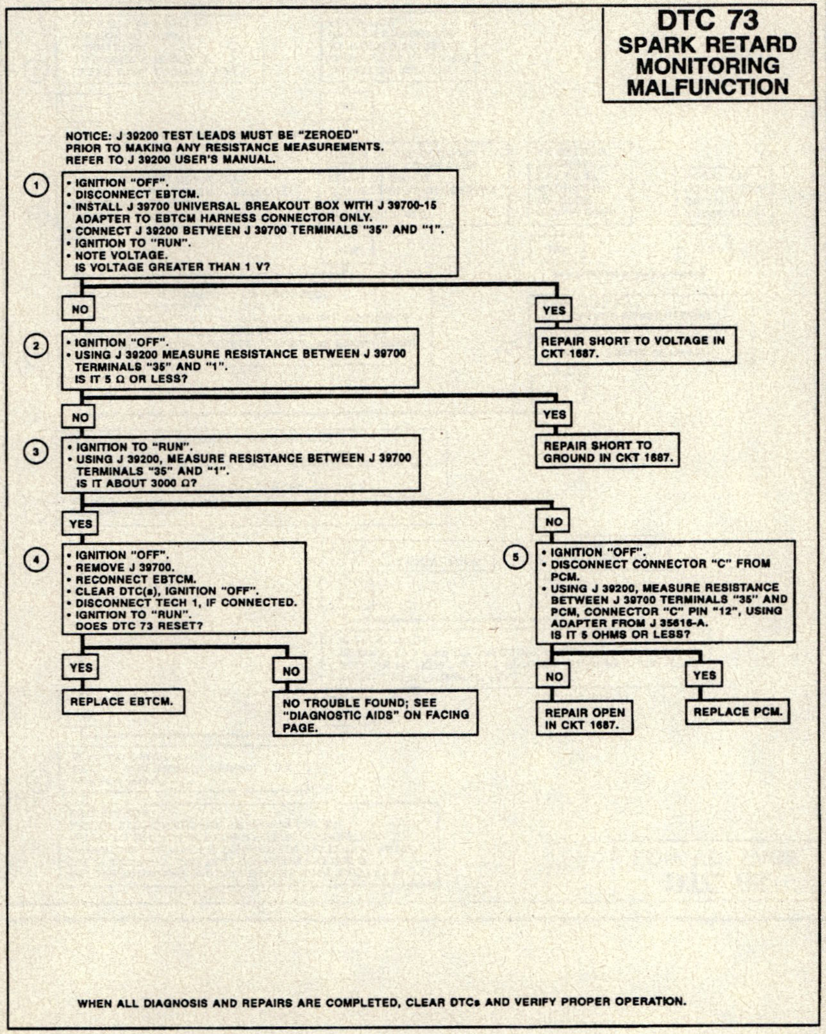

DIAGNOSTIC CHART DTC 85 — 1995 FLEETWOOD

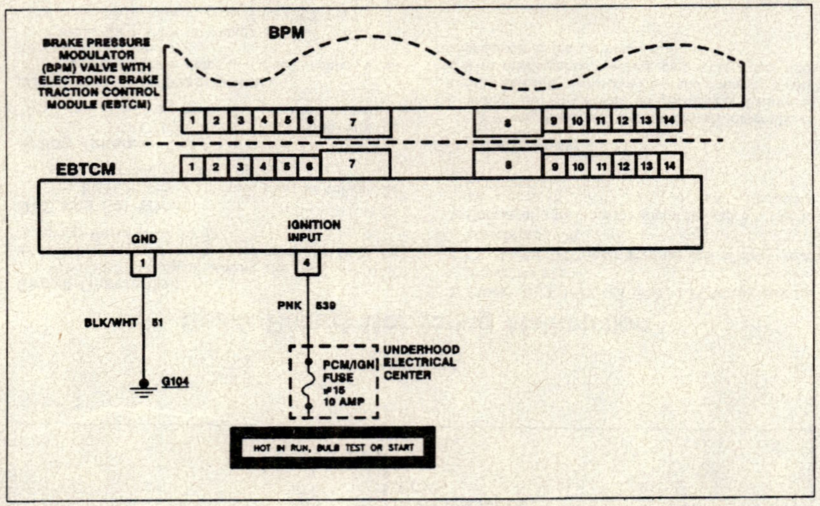

DTC 85
Low Voltage

Circuit Description:

The EBTCM, requires a certain minimum voltage to operate properly. It monitors the ignition feed circuit to determine if voltage falls below a minimum level.

DTC Will Set When:

- The EBTCM operating voltage falls below 9.4 volts out of ABS/TCS control, 8.8 volts during ABS/TCS control and low voltage is detected for more than 500 milliseconds with the vehicle speed above 6 km/h (3.6 mph).

Action Taken:

When low voltage is detected, the ABS/TCS valves are switched passive, which prevents a possible pump start-up and the "ABS" and "TRACTION CONTROL" lamps will be "ON".

DTC Chart Test Description:

Number(s) below refer to circled number(s) on the diagnostic chart.

1. Checks that PCM/IGN fuse feeding the EBTCM ignition circuit is good.
2. Measures for appropriate voltage from the EBTCM ignition circuit.
3. Checks EBTCM connector terminals for good contact, i.e., not spread open, backed out of connector, corroded, etc.
4. Checks EBTCM connector for good retention to EBTCM when installed.
5. Checks for DTC reset after other portions of the system have tested as OK.
6. Determines if a malfunction found in Step 2 is due to an open or poor connection in the EBTCM ignition feed CKT 539, or poor ground(s) in CKT 51.
7. Checks whether the fuse failure is due to a short to ground.

Diagnostic Aids:

It is very important that a thorough inspection of the wiring and connectors be performed. Failure to carefully and fully inspect wiring and connectors may result in misdiagnosis, causing part replacement with reappearance of the malfunction.

DIAGNOSTIC CHART DTC 85 — 1995 FLEETWOOD, CONTINUED

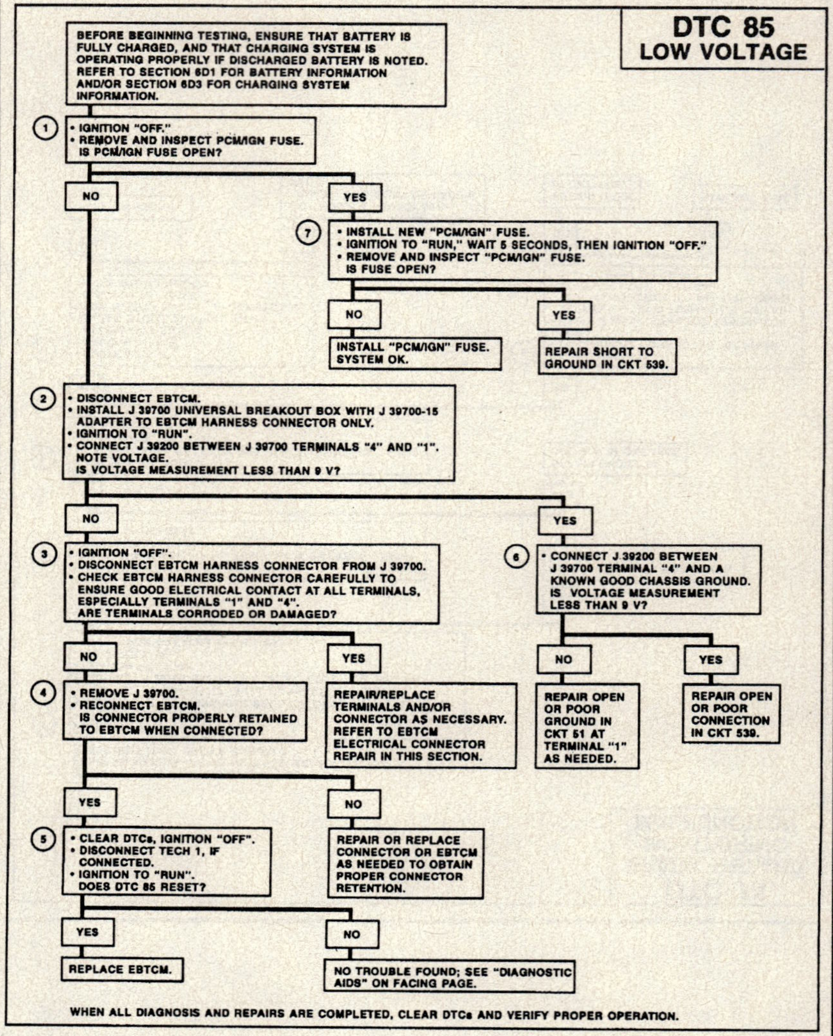

ELECTRICAL CONNECTOR TERMINAL IDENTIFICATION — 1995 DEVILLE, CONCOURS, ELDORADO AND SEVILLE

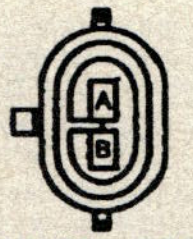

• WHEEL SPEED SENSOR
• C110 AND C111 BODY MAIN HARNESS SIDE

WHEEL SPEED SENSOR CONNECTOR

	PIN A	PIN B
RIGHT FRONT	872	833
LEFT FRONT	830	873
RIGHT REAR	882	883
LEFT REAR	884	885

EBTCM CONNECTOR

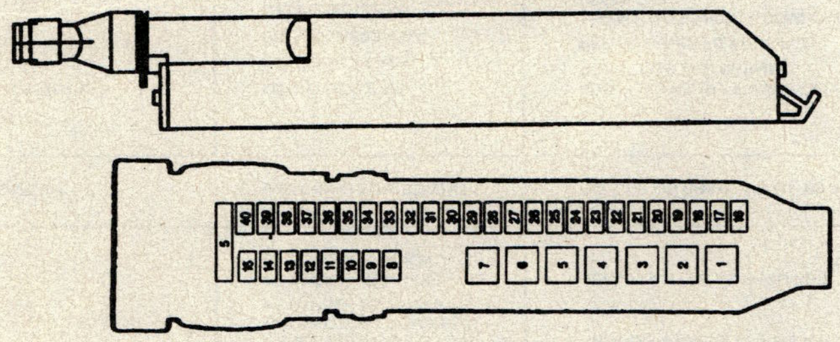

ELECTRICAL CONNECTOR TERMINAL IDENTIFICATION — 1995 DEVILLE, CONCOURS, ELDORADO AND SEVILLE, CONTINUED

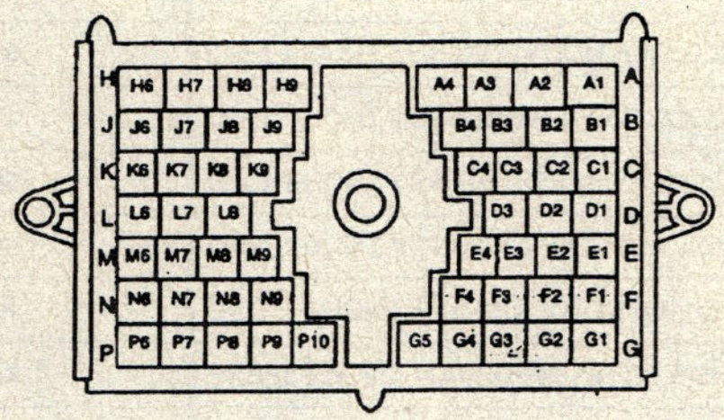

C101

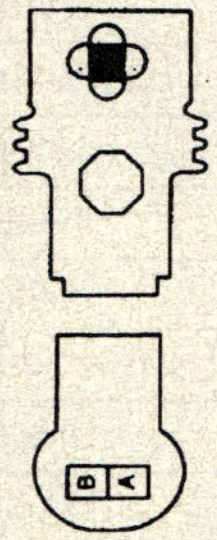

BRAKE FLUID LEVEL SENSOR

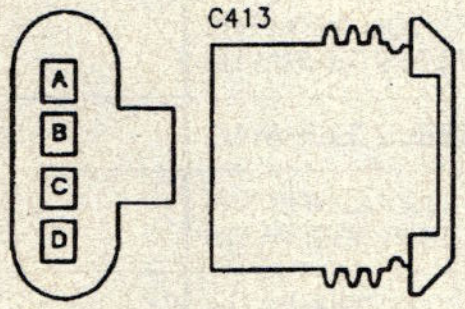

C413

4.6L

CAV	WIRE
A	884
B	885
C	882
D	883

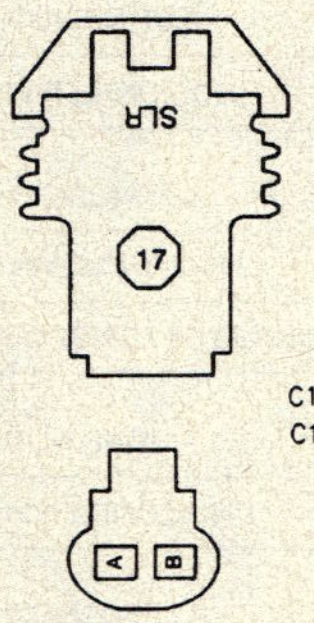

C110
C111

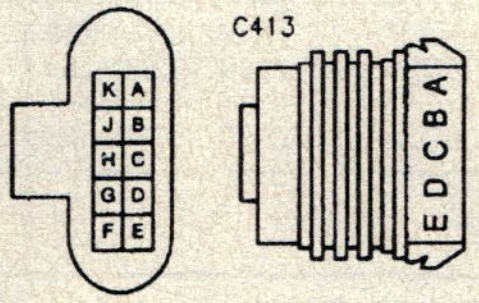

C413

4.9L

CAV	WIRE
A	884
B	885
C	882
D	883
E	1014
F	1016
G	1017
H	1010
J	1012
K	1013

CAV	WIRE
A	830
B	873

DIAGNOSTIC CHART ABS/ASR VISUAL INSPECTION — 1995 DEVILLE, CONCOURS, ELDORADO AND SEVILLE

ABS/TCS VISUAL INSPECTION		
ITEM	**INSPECT FOR**	**CORRECTIVE ACTION**
PARKING BRAKE	- FULL RELEASE - PROPER SWITCH FUNCTION IF NECESSARY, UNPLUG SWITCH CONNECTOR FROM PEDAL ASSEMBLY TO VERIFY - PROPER VACUUM RELEASE SYSTEM OPERATION	- OPERATE MANUAL RELEASE LEVER TO VERIFY RELEASE - ADJUST CABLE OR REPAIR RELEASE SYSTEM AS REQUIRED - REPAIR SWITCH AS REQUIRED
BRAKE LIGHT SWITCH	- IMPROPER ADJUSTMENT	- PULL UP ON BRAKE PEDAL TO ADJUST
MAJOR COMPONENTS • BRAKE FLUID RESERVOIR • BRAKE PRESSURE MODULATOR VALVE • MASTER CYLINDER • PROPORTIONING VALVES	- LOW FLUID LEVEL - EXTERNAL LEAKS - PROPER ASSEMBLY - EXTERNAL LEAKS	- ADD FLUID AS DETERMINE CAUSE OF FLUID LOSS AND REPAIR - REPAIR LEAKS AS REQUIRED - INSTALL OR POSITION COMPONENTS PROPERLY - REPAIR LEAKS AS REQUIRED
• FUSE A1 (10 A) (TRUNK COMPARTMENT FUSE BLOCK) • FUSE B3 (20 A) (ENGINE COMPARTMENT FUSE BLOCK) • MAXI® FUSE 5 (50 A) (ENGINE COMPARTMENT RH MAXI® FUSE BLOCK)	- BLOWN FUSE	- REPLACE FUSE - REPAIR OPEN IN FUSIBLE LINK - VERIFY OPERATION
CONNECTORS/WIRING UNDERHOOD • BRAKE PRESSURE MODULATOR VALVE • GROUND CONNECTION G303 • WHEEL SPEED SENSORS • C110 • C111 • BRAKE PRESSURE MODULATOR VALVE GROUND STUD IN PASSENGER COMPARTMENT OR IN TRUNK • C101 • EBTCM/EBCM REAR OF VEHICLE • WHEEL SPEED SENSORS • C413	- PROPER ENGAGEMENT - LOOSE WIRES OR TERMINALS - CORRODED OR BROKEN EYELETS - PAINT/PRIMER ON GROUNDS - WIRING NOT PROPERLY ROUTED OR SECURED - WIRING NEAR POSSIBLE SOURCE OF ELECTRICAL NOISE	- ADD STAR WASHER - PROPERLY ENGAGE CONNECTOR - REPAIR AS REQUIRED - REROUTE OR PROPERLY ROUTE WIRING
IF NO PROBLEM IS NOTED IN VISUAL INSPECTION, PROCEED TO ABS FUNCTIONAL CHECK		

DIAGNOSTIC TROUBLE CODE CHART — 1995 DEVILLE, CONCOURS, ELDORADO AND SEVILLE

ABS/TCS DTC TABLE	
DIAGNOSTIC TROUBLE CODE	**DESCRIPTION**
21	RF WHEEL SPEED SENSOR CIRCUIT MALFUNCTION
23	RF WHEEL SPEED SENSOR CIRCUIT CONTINUITY MALFUNCTION
25	LF WHEEL SPEED SENSOR CIRCUIT MALFUNCTION
27	LF WHEEL SPEED SENSOR CIRCUIT CONTINUITY MALFUNCTION
28	WHEEL SPEED SENSOR CIRCUIT FREQUENCY MALFUNCTION
31	RR WHEEL SPEED SENSOR CIRUCIT MALFUNCTION
33	RR WHEEL SPEED SENSOR CIRCUIT CONTINUITY MALFUNCTION
35	LR WHEEL SPEED SENSOR CIRCUIT MALFUNCTION
37	LR WHEEL SPEED SENSOR CIRCUIT CONTINUITY MALFUNCTION
41	RIGHT FRONT ABS INLET VALVE SOLENOID FAULT
42	RIGHT FRONT ABS OUTLET VALVE SOLENOID FAULT
43	RIGHT FRONT TCS PRIME VALVE SOLENOID FAULT
44	RIGHT FRONT TCS ISOLATION VALVE SOLENOID FAULT
45	LEFT FRONT ABS INLET VALVE SOLENOID FAULT
46	LEFT FRONT ABS QUTLET VALVE SOLENOID FAULT
47	LEFT FRONT TCS PRIME VALVE SOLENOID FAULT
48	LEFT FRONT TCS ISOLATION VALVE SOLENOID FAULT
51	RIGHT REAR ABS INLET VALVE SOLENOID FAULT
52	RIGHT REAR ABS OUTLET VALVE SOLENOID FAULT
55	LEFT REAR ABS INLET VALVE SOLENOID FAULT
56	LEFT REAR ABS OUTLET VALVE SOLENOID FAULT
61	PUMP MOTOR OR PUMP MOTOR CIRCUIT FAULT
63	VALVE RELAY CIRCUIT FAULT
67	TCC/BLS BRAKE SWITCH FAULT
71	ELECTRONIC BRAKE/TRACTION CONTROL MODULE INTERNAL ERROR
72	IPC/SERIAL DATA LINK FAULT

DIAGNOSTIC TROUBLE CODE CHART — 1995 DEVILLE, CONCOURS, ELDORADO AND SEVILLE, CONTINUED

ABS/TCS DTC TABLE

DIAGNOSTIC TROUBLE CODE	DESCRIPTION
73	PCM-TCS INTERFACE FAULT (4.6L ONLY)
77	PCM/SERIAL DATA LINK FAULT
83	LOW BRAKE FLUID
84	PCM TRACTION CONTROL NOT ALLOWED (4.9L)
85	LOW VOLTAGE CONDITION

ABS/TCS DTC SYMPTOM TABLE

CHART	DESCRIPTION
A	'ANTILOCK' INDICATOR AND TRACTION 'DISABLED' MESSAGE 'ON' - NO TCS CODES SET/NO TCS DATA
B	'ANTILOCK' INDICATOR 'ON'
C	VALVE CYCLING DURING NORMAL STOPS (UNWANTED ABS MODULATION)
D	TRACTION CONTROL OPERATION CHECK
•	'ANTILOCK' INDICATOR INOPERATIVE

DIAGNOSTIC CHART ABS/ASR FUNCTIONAL CHECK CHART I — 1995 DEVILLE, CONCOURS, ELDORADO AND SEVILLE

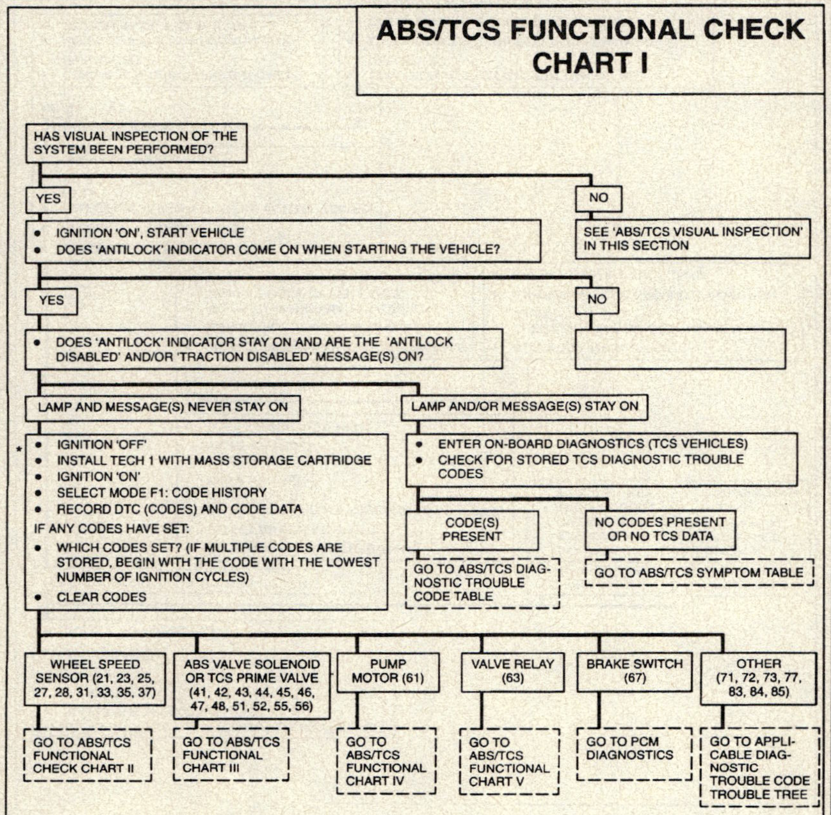

* FROM THIS POINT ON, IT IS ASSUMED THAT THE CONDITION(S) RESPONSIBLE FOR SETTING THE DIAGNOSTIC TROUBLE CODE(S) IS INTERMITTENT AND THE CODE IS NEVER SET CURRENT IN THE SHOP. MOST OF THE FOLLOWING TESTS ARE PERFORMED IN ORDER TO EXERCISE THE ABS/TCS CIRCUITS AND GET A CODE TO SET WITHOUT HAVING TO LEAVE THE BAY. IF AT ANYTIME A DIAGNOSTIC TROUBLE CODE IS SET 'HARD', REFER IMMEDIATELY TO THE 'ABS/TCS DIAGNOSTIC TROUBLE CODE TABLE'

DIAGNOSTIC CHART ABS/ASR FUNCTIONAL CHECK CHART II — 1995 DEVILLE, CONCOURS, ELDORADO AND SEVILLE

ABS/TCS FUNCTIONAL CHECK CHART II
WHEEL SPEED SENSORS

- USING THE TECH 1, SELECT MODE F3: SNAPSHOT
- SELECT MODE F2: AUTO TRIG
- TEST DRIVE VEHICLE AS DESCRIBED IN THIS SECTION WHILE MONITORING THE WHEEL SPEED SCREEN OF THE TECH 1 SNAPSHOT MODE. ATTEMPT TO DUPLICATE CONDITIONS AT WHICH CODE(S) SET PER OWNER DESCRIPTION AND CODE HISTORY INFORMATION:
  - SPEED AT WHICH CODE(S) SET
  - BRAKE SWITCH 'ON' OR 'OFF'
  - ABS OR TRACTION ACTIVE OR INACTIVE
- TRY DRIVING OVER GRAVEL, BUMPY ROAD, RAILROAD TRACKS AND MAKING FREQUENT TURNS. THESE MANEUVERS MAY HELP INDUCE A WIRING ASSOCIATED FAULT
- MANUALLY TRIGGER A SNAPSHOT BY SELECTING F1 IF THE WHEEL SPEED SIGNAL APPEARS ERRONEOUS
- IS A SNAPSHOT AUTOMATICALLY OR MANUALLY TRIGGERED?

NO

- CONDITION NOT PRESENT AT THIS TIME
- CHECK WHEEL SPEED SENSOR MOUNTING
- CHECK WHEEL SPEED SENSOR WIRING CONNECTIONS AND ROUTING
- IF ORIGINAL DIAGNOSTIC TROUBLE CODE WAS A CONTINUITY FAULT, WIGGLE AND FLEX SPEED SENSOR CABLE WITH IGNITION 'ON' TO SEE IF CODE RESETS WITH TECH 1 DISCONNECTED. ALSO CHECK FOR EXCESSIVE WHEEL BEARING PLAY
- SEE 'NOTE ON INTERMITTENTS'

YES

- NOTE DESCRIPTION OF WHEN SNAPSHOT OCCURRED
- SELECT MODE F0: REPLAY DATA
- RECORD ALL CRITERIA AT TRIGGER POINT
- IF FAULT ONLY OCCURS MOMENTARILY, CONDITION IS INTERMITTENT
  - INSPECT WHEEL SPEED SENSOR WIRING CONNECTIONS AND ROUTING
  - INSPECT WHEEL SPEED SENSOR CABLE FOR CHAFFING AND OTHER EVIDENCE OF DAMAGE
  - MAKE CERTAIN CONNECTIONS ARE TIGHT AND THE SENSOR CABLE IS PROPERLY ROUTED AND SECURED IN ITS CLIPS
  - LOOSE SUSPENSION COMPONENTS (I.E., STRUT, CONTROL ARM, ETC.) MAY CAUSE UNUSUAL SENSOR SIGNALS. THESE ARE REFERRED TO AS SENSOR SHOCK AND ARE THE RESULT OF SEVERE VIBRATION AT THE SENSOR PICK-UP CAUSING A SIGNAL SPIKE
- IF NO OBVIOUS PROBLEMS, REFER TO APPLICABLE DIAGNOSTIC TROUBLE CODE TROUBLE TREE

DIAGNOSTIC CHART ABS/ASR FUNCTIONAL CHECK CHART III — 1995 DEVILLE, CONCOURS, ELDORADO AND SEVILLE

ABS/TCS FUNCTIONAL CHECK CHART III
ABS VALVE SOLENOID OR TCS PILOT VALVE

- USING THE TECH 1, SELECT MODE F4: ABS/TCS TEST
- SELECT THE APPROPRIATE TEST:
  - AUTO TEST
  - TCS TEST
- PERFORM TEST
- DOES SYSTEM PASS TEST?

AUTO TEST FAILS

- REFER TO APPLICABLE DIAGNOSTIC TROUBLE CODE TROUBLE TREE

TCS TEST FAILS

- REFER TO APPLICABLE DIAGNOSTIC TROUBLE CODE TROUBLE TREE

AUTO TEST PASSES

- CONDITION MAY BE INTERMITTENT
- SELECT F0: SOLENOID TEST
- SELECT APPROPRIATE SOLENOID TO BE TESTED
  F0: LF F4: LR
  F1: RF F5: RR
- PERFORM PRESSURE RELEASE TEST OF THAT VALVE SOLENOID
- DOES SYSTEM PASS TEST?

NO

- INSPECT BASE BRAKE SYSTEM FOR LEAKS
- INSPECT FOR LEAKS AT BPM VALVE
- REFER TO APPROPRIATE DIAGNOSTIC TROUBLE CODE TROUBLE TREE

YES

- CONDITION INTERMITTENT
- CHECK HARNESS CONNECTION AT BPM VALVE
- CHECK BPM VALVE GROUND CONNECTION
- SEE 'NOTE ON INTERMITTENTS'

TCS PASSES TEST

- CONDITION MAY BE INTERMITTENT
- SELECT F3: PILOT VALVE TEST
- SELECT APPROPRIATE PILOT VALVE TO BE TESTED:
  F0: LF
  F1: RF
- DOES SYSTEM PASS TEST?

NO

- CONDITION INTERMITTENT BUT CAN BE DUPLICATED
- REFER TO APPROPRIATE DIAGNOSTIC TROUBLE CODE TROUBLE TREE

YES

- CONDITION INTERMITTENT BUT CANNOT BE DUPLICATED
- CHECK HARNESS CONNECTION AT BPM VALVE
- CHECK BPM VALVE GROUND CONNECTION
- SEE 'NOTE ON INTERMITTENTS'

DIAGNOSTIC CHART ABS/ASR FUNCTIONAL CHECK CHART IV — 1995 DEVILLE, CONCOURS, ELDORADO AND SEVILLE

ABS/TCS FUNCTIONAL CHECK
CHART IV
PUMP MOTOR OR PUMP MOTOR RELAY

- USING THE TECH 1, SELECT MODE F4: ABS/TCS TEST
- SELECT F1: AUTO TEST
- PERFORM TEST
- DOES SYSTEM PASS TEST?

NO

- REFER TO TROUBLE TREE FOR DIAGNOSTIC TROUBLE CODE 61

YES

- CHECK TERMINAL CONNECTIONS BETWEEN ELECTRONIC BRAKE AND TRACTION CONTROL MODULE (EBTCM) AND BRAKE PRESSURE MODULATOR VALVE (BPMV)
- CONDITION NOT PRESENT AT THIS TIME
- CHECK FOR CLEAN AND TIGHT BPM VALVE GROUND CONNECTIONS AT GROUND STUD AND ON FRAME
- SEE 'NOTE ON INTERMITTENTS'

DIAGNOSTIC CHART ABS/ASR FUNCTIONAL CHECK CHART V — 1995 DEVILLE, CONCOURS, ELDORADO AND SEVILLE

ABS/TCS FUNCTIONAL CHECK
CHART V
VALVE RELAY

- USING THE TECH 1, SELECT MODE F4: ABS/TCS TEST
- SELECT F1: AUTO TEST
- PERFORM TEST
- DOES SYSTEM PASS TEST?

NO

- REFER TO APPLICABLE DIAGNOSTIC TROUBLE CODE TROUBLE TREE

YES

- CHECK TERMINAL CONNECTIONS BETWEEN ELECTRONIC BRAKE AND TRACTION CONTROL MODULE (EBTCM) AND BRAKE PRESSURE MODULATOR VALVE (BPMV)
- CONDITION NOT PRESENT AT THIS TIME
- WIGGLE HARNESS CONNECTOR WHILE CONNECTED AND SEE IF DIAGNOSTIC TROUBLE CODE IS SET (TECH 1 DISCONNECTED)
- IGNITION 'OFF'
- DISCONNECT HARNESS CONNECTOR FROM EBTCM. CHECK FOR ADEQUATE TERMINAL TENSION AT HARNESS CONNECTOR
- SEE 'NOTE ON INTERMITTENTS'

DIAGNOSTIC TROUBLE CODE CHART DTC 21 — 1995 DEVILLE, CONCOURS, ELDORADO AND SEVILLE

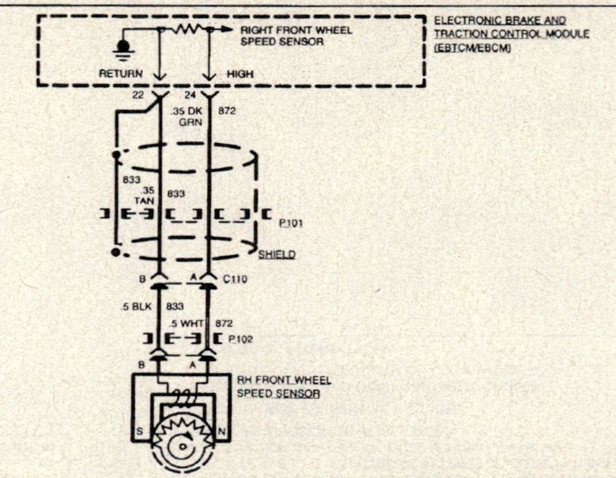

DIAGNOSTIC TROUBLE CODE 21
RIGHT FRONT WHEEL SPEED SENSOR FAULT
(Page 1 of 4)

CIRCUIT DESCRIPTION:

As each wheel turns, the wheel speed sensor for that wheel creates a small AC voltage as a toothed sensor ring, mounted in the front integral hub/bearing assembly, is passed by the stationary sensor. The sensor consists of a permanent magnet and coil and generates voltage by magnetic induction. The frequency of the wheel speed sensor signal is proportional to wheel speed.

The EBTCM uses the wheel speed sensor signal to calculate vehicle reference speeds and individual speed, acceleration and slip values for each wheel. These values are used to determine when antilock control is required.

The EBTCM will only set one wheel speed sensor fault code at a time, even if multiple faults exist. Detection is prioritized accordingly: LF, RR, LR, RF. Always check for additional stored codes after making repairs.

TEST CONDITION(S):

1. Tested continuously.

FAILURE CONDITION(S):

Sensor Output

1. Drive-away: Speed signal is not detected when vehicle reaches 12 kph or 7.5 mph.
2. While Driving: Speed signal is lost while vehicle speed is greater than 40 kph or 25 mph.

ACTION TAKEN:

1. Code set.
2. ABS/TCS switched 'OFF'.
3. 'ANTILOCK' indicator 'ON'.
4. 'TRACTION DISABLED' message displayed.

Code 21 is set if there is a short to voltage or ground in CKT 872 or 883, or faulty speed sensor. The testing for this fault occurs when the vehicle is in motion, it will not set with the ignition 'ON' and the vehicle at rest.

NOTICE: A Code 21 may set if CKT 872 or 833 lose continuity and the WSS signal is lost while driving. A Code 23 should accompany this code on the subsequent ignition cycle. If both Code 21 and Code 23 are set concurrently, diagnose Code 23 first.

TEST DESCRIPTION: Test numbers refer to circled numbers on the diagnostic chart.

1. Checks if the code resets during a road test.
2. Use the TECH 1 to check for sources of induced electrical noise. Some sources of induced electrical noise which may cause intermittent wheel speed sensor codes to set are: improper routing or sensor cables next to spark plug wires; cables not retained in brackets; loose fitting or improperly mounted sensors; damaged sensors or toothed rings; poor terminal engagement of connectors; or intermittent shorts or opens in wiring; or a loss of sensor circuit shield.
3. Checks for damaged wheel speed sensor circuit shield. The shield is a foil insulator that is wrapped around the wheel speed sensor circuits. Inside the EBTCM harness connecter there is an uninsulated drain wire that is attached to the shield and jumped to the wheel speed sensor return line and taken to ground through the EBTCM. A shield which has lost its ground through the EBTCM leaves the sensor signal circuit susceptible to noise. It may be necessary to disassemble the EBTCM connector to check if the drain wire has a poor connection with the wheel speed sensor return connector terminal. The shield can be repaired by redetermining the sensor return line circuit, including the drain wire.
4. Checks for intermittent opens/shorts in wheel speed sensor circuits and cables.

DIAGNOSTIC TROUBLE CODE CHART DTC 21 — 1995 DEVILLE, CONCOURS, ELDORADO AND SEVILLE, CONTINUED

DIAGNOSTIC TROUBLE CODE 21
RIGHT FRONT WHEEL SPEED SENSOR FAULT
(Page 2 of 4)

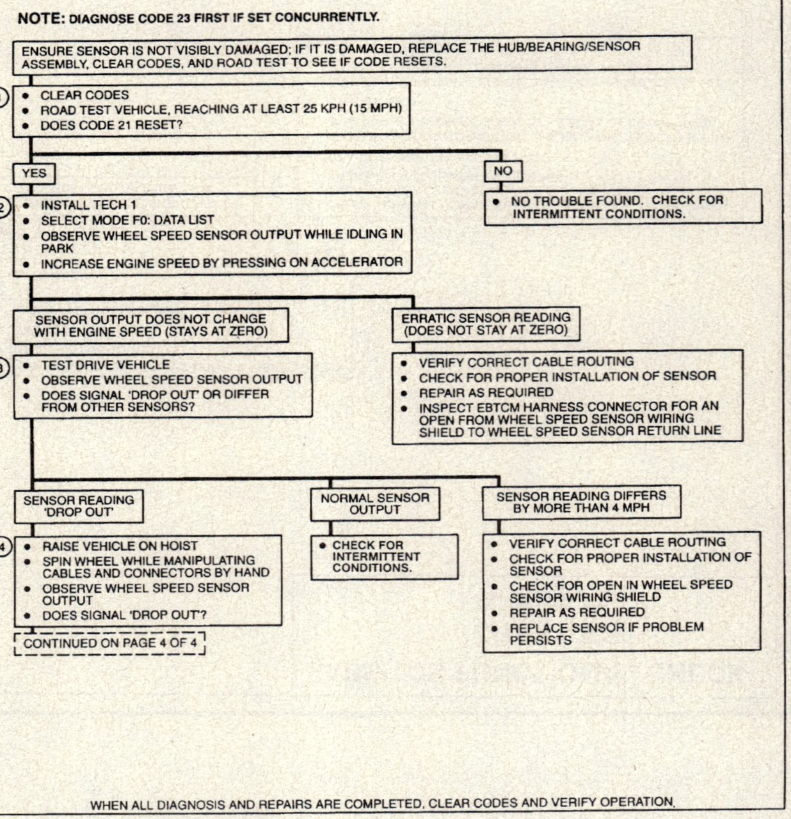

DIAGNOSTIC TROUBLE CODE CHART DTC 21 — 1995 DEVILLE, CONCOURS, ELDORADO AND SEVILLE, CONTINUED

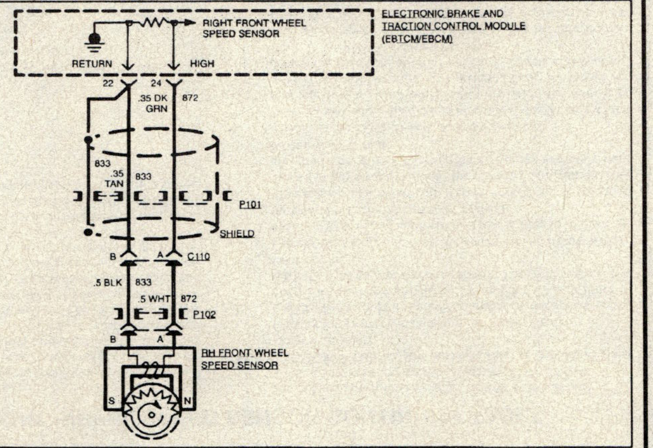

DIAGNOSTIC TROUBLE CODE 21
RIGHT FRONT WHEEL SPEED SENSOR FAULT
(Page 3 of 4)

CIRCUIT DESCRIPTION:

As each wheel turns, the wheel speed sensor for that wheel creates a small AC voltage as a toothed sensor ring, mounted in the front integral hub/bearing assembly, is passed by the stationary sensor. The sensor consists of a permanent magnet and coil and generates voltage by magnetic induction. The frequency of the wheel speed sensor signal is proportional to wheel speed.

The EBTCM uses the wheel speed sensor signal to calculate vehicle reference speeds and individual speed, acceleration and slip values for each wheel. These values are used to determine when antilock control is required.

The EBTCM will only set one wheel speed sensor fault code at a time, even if multiple faults exist. Detection is prioritized accordingly: LF, RR, LR, RF. Always check for additional stored codes after making repairs.

TEST DESCRIPTION: Test numbers refer to circled numbers on the diagnostic chart.

5. Checks for proper resistance in the sensor itself.
6. Checks for a short to battery in either of the speed sensor circuit wires.
7. Checks for a short between the speed sensor circuit wires.
8. Checks for a short to ground in the speed sensor input circuit wire. The ground may be either a 'hard' short to ground, or a resistive short. A short with a resistance less than 2 megaohms, though not a hard short, can still cause Code 21 to set.
9. Checks wiring and connectors for intermittents.

DIAGNOSTIC TROUBLE CODE CHART DTC 21 — 1995 DEVILLE, CONCOURS, ELDORADO AND SEVILLE, CONTINUED

DIAGNOSTIC TROUBLE CODE 21
RIGHT FRONT WHEEL SPEED SENSOR FAULT
(Page 4 of 4)

CONTINUED CODE 21 FROM PAGE 2 OF 4

YES

(5)
- IGNITION 'OFF'
- DISCONNECT SENSOR CONNECTOR
- USING DVM AND ADAPTER FROM J 35616-A, CHECK SENSOR RESISTANCE BY PROBING SENSOR CONNECTOR TERMINALS
- IS IT 900 - 1300 OHMS AT 25°C (77°F)?

NO
- CHECK FOR INTERMITTENT CONDITIONS.
- REPLACE EBTCM IF PROBLEM PERSISTS

YES

(6)
- CONNECT DVM BETWEEN SENSOR HARNESS CONNECTOR TERMINAL 'A' AND GROUND (USED DC VOLTAGE SETTING)
- IGNITION 'ON'
- NOTE VOLTAGE: REPEAT BETWEEN TERMINAL 'B' AND GROUND.
- IS IT AROUND SYSTEM VOLTAGE ON EITHER TERMINAL 'A' OR 'B'?

NO
- REMOVE WHEEL SPEED RING SENSOR
- INSPECT TOOTHED SPEED RING ON HUB/BEARING ASSEMBLY
- IF DAMAGED, REPLACE ENTIRE HUB/BEARING AND SENSOR ASSEMBLY
- IF NOT DAMAGED, REPLACE WHEEL SPEED SENSOR

NO

(7)
- IGNITION 'OFF'
- DISCONNECT EBCM
- INSTALL J 39700 UNIVERSAL BREAKOUT BOX TO EBTCM HARNESS CONNECTOR ONLY
- MEASURE RESISTANCE BETWEEN J 39700 TERMINALS '22' AND '24'
- IS IT LESS THAN 5 OHMS?

YES
- REPAIR SHORT TO SYSTEM VOLTAGE IN CKT 872 OR 833
- REFER TO EBTCM-SWITCHED FAULTS IN THIS SECTION

NO

(8)
- MEASURE RESISTANCE BETWEEN J 39700 TERMINAL '24' AND GROUND
- IS IT GREATER THAN 2 MEGAOHMS?

YES
- REPAIR SHORT BETWEEN CKT'S 872 AND 833

YES

(9)
- MEASURE RESISTANCE BETWEEN J 39700 TERMINAL '21' AND GROUND
- IS IT GREATER THAN 2 MEGAOHMS?

NO
- IF 5 OHMS OR LESS TO GROUND, REPAIR HARD SHORT IN HARNESS CKT 872
- IF BETWEEN 5 OHMS AND 2 MEGAOHMS, REPAIR RESISTIVE SHORT IN HARNESS CKT 872

YES
- CHECK ALL WIRING, CONNECTORS AND TERMINALS THOROUGHLY FOR INTERMITTENTS OR CORROSION
-
- REPAIR HARNESS OR CONNECTOR
- REPLACE SENSOR IS PROBLEM PERSISTS

NO
- IF 5 OHMS OR LESS TO GROUND, REPAIR HARD SHORT IN HARNESS CKT 833
- IF BETWEEN 5 OHMS AND 2 MEGAOHMS REPAIR RESISTIVE SHORT IN HARNESS CKT 833

WHEN ALL DIAGNOSIS AND REPAIRS ARE COMPLETED, CLEAR CODES AND VERIFY OPERATION

DIAGNOSTIC TROUBLE CODE CHART DTC 23 — 1995 DEVILLE, CONCOURS, ELDORADO AND SEVILLE

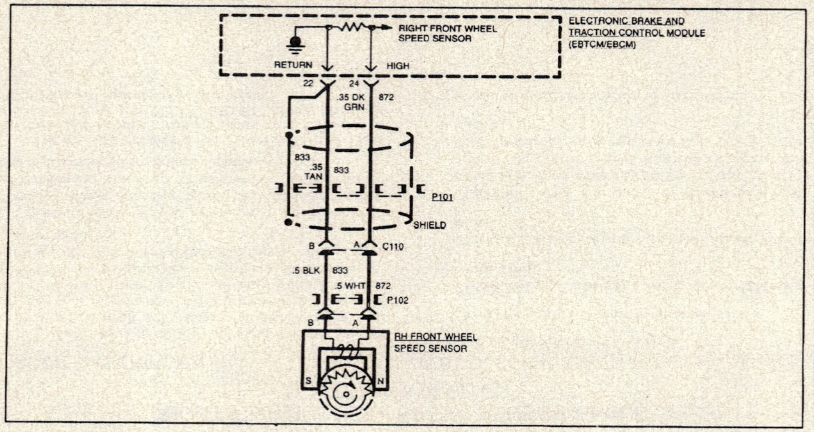

DIAGNOSTIC TROUBLE CODE 23

RIGHT FRONT WHEEL SPEED SENSOR CONTINUITY FAULT

CIRCUIT DESCRIPTION:

As each wheel turns, the wheel speed sensor for that wheel creates a small AC voltage as a toothed sensor ring, mounted in the front integral hub/bearing assembly, is passed by the stationary sensor. The sensor consists of a permanent magnet and coil and generates voltage by magnetic induction. The frequency of the wheel speed sensor signal is proportional to wheel speed.

The EBTCM uses the wheel speed sensor signal to calculate vehicle reference speeds and individual speed, acceleration and slip values for each wheel. These values are used to determine when antilock control is required.

The EBTCM will only set one wheel speed sensor fault code at a time, even if multiple faults exist. Detection is prioritized accordingly: LF, RR, LR, RF. Always check for additional stored codes after making repairs.

TEST CONDITION(S):

1. Tested continuously.

FAILURE CONDITION(S):

Continuity

1. While standing: Open circuit is seen while the vehicle is standing still with the ignition 'ON'

ACTION TAKEN:

1. Code set.
2. ABS/TCS switched 'OFF'.
3. 'ANTILOCK' indicator 'ON'.
4. 'TRACTION DISABLED' message displayed. (vehicles with TCS)

Code 23 is set if there is a short to voltage or ground in CKT 872 or 883, or faulty speed sensor. The testing for this fault occurs with the ignition 'ON' and the vehicle at rest.

NOTICE: A Code 21 may set if CKT 872 or 833 lose continuity and the WSS signal is lost while driving. A Code 23 should accompany this code on the subsequent ignition cycle. If both Code 21 and Code 23 are set concurrently, diagnose Code 23 first.

TEST DESCRIPTION: Test numbers refer to circled numbers on the diagnostic chart.

1. Checks for proper resistance in the sensor itself.
2. Checks for a short to battery in either of the speed sensor circuit wires. There should be 5V on CKT 872. If there is, CKT 872 has no hard opens or shorts. The condition may be intermittent.
3. Checks for an open in both the speed sensor circuit wires.
4. Manipulates wiring and connectors, trying to induce an intermittent fault not currently present.
5. This test checks if the code resets on key-up. If so, since Tests 1-4 have validated that the circuitry and components are good, the Wheel Bearing/Speed Sensor assembly may be causing the fault.
6. Checks if the Code 21 sets during a road test.
7. Code 23 sets when the vehicle is at rest: Code 21 sets with a problem in the speed sensor circuitry with the vehicle in motion. If problems are still present, Code 21 would set during the road test, not a 23. If Code 21 sets at this point, Code 21 diagnostics should be used.
8. The wheel speed sensor may be causing an intermittent fault. If Code 23 resets after sensor replacement, the EBTCM/EBCM must be concluding there is a problem present in the speed sensor circuit when there is not.

DIAGNOSTIC AIDS:

Be sure the speed sensor wiring is properly routed and retained. This will help prevent false signals due to electrical noise being picked up by the wiring.

It is very important that a thorough inspection of the wiring and connectors be performed. Failure to carefully and fully inspect wiring and connectors may result in misdiagnosis, causing part replacement with reappearance of the fault.

DIAGNOSTIC TROUBLE CODE CHART DTC 23 — 1995 DEVILLE, CONCOURS, ELDORADO AND SEVILLE, CONTINUED

DIAGNOSTIC TROUBLE CODE 23

RIGHT FRONT WHEEL SPEED SENSOR CONTINUITY FAULT

ENSURE SENSOR IS NOT VISIBLY DAMAGED; IF IT IS DAMAGED, REPLACE THE WHEEL BEARING/SPEED/SENSOR ASSEMBLY, CLEAR CODES, AN TURN IGNITION 'ON' TO SEE IF CODE RESETS.

(1)
- IGNITION 'OFF'
- DISCONNECT SENSOR CONNECTOR
- USING DVM AND ADAPTER FROM J 35616-A, CHECK SENSOR RESISTANCE BY PROBING SENSOR CONNECTOR TERMINALS
- IS IT 900 - 1300 OHMS AT 25°C (77°F)?

NO:
- REPLACE WHEEL SPEED SENSOR

YES:

(2)
- CONNECT DVM BETWEEN SENSOR HARNESS CONNECTOR TERMINAL 'A' AND GROUND (USED DC VOLTAGE SETTING)
- IGNITION 'ON'
- NOTE VOLTAGE: REPEAT BETWEEN TERMINAL 'B' AND GROUND
- IS IT AROUND SYSTEM VOLTAGE ON EITHER TERMINAL 'A' OR 'B'?

YES:
- REPAIR SHORT TO SYSTEM VOLTAGE IN CKT 872 OR 883
- REFER TO EBTCM-SWITCHED FAULTS IN THIS SECTION

NO:

(3)
- IGNITION 'OFF'
- DISCONNECT EBTCM
- INSTALL J 39700 UNIVERSAL BREAKOUT BOX TO EBTCM HARNESS CONNECTOR ONLY
- MEASURE RESISTANCE BETWEEN PIN 24 ON J 39700 TERMINAL 'A' OF SENSOR CONNECTOR ON HARNESS. PERFORM SAME MEASUREMENT BETWEEN PIN 22 AND TERMINAL 'B'
- IS EITHER MEASUREMENT GREATER THAN 5 OHMS?

YES:
- REPAIR OPEN IN CKT 872 AND/OR 833

NO:

(4)
- REMOVE J 39700; RECONNECT EBTCM
- CLEAR CODES; IGNITION 'OFF'
- DISCONNECT TECH 1, IF CONNECTED
- IGNITION 'ON'
- WIGGLE ALL WIRING AND CONNECTORS TO TRY TO DUPLICATE AN INTERMITTENT FAULT
- DOES CODE 23 RESET?

YES:
- REPAIR WIRING AND/OR CONNECTIONS
- REFER TO INTERMITTENTS AND POOR CONNECTIONS

NO:

(5)
- CLEAR CODES; IGNITION 'OFF'
- DISCONNECT TECH 1, IF CONNECTED
- IGNITION 'ON'
- DOES CODE 23 RESET?

YES:

(8)
- REPLACE WHEEL SPEED SENSOR
- CLEAR CODES; IGNITION 'OFF'
- DISCONNECT TECH 1, IF CONNECTED
- IGNITION 'ON'
- DOES CODE 23 RESET?

YES:
- REPLACE EBTCM

NO:
- SYSTEM OK

NO:

(6)
- ROAD TEST VEHICLE, REACHING AT LEAST 25 KPH (15 MPH).
- DOES CODE 21 SET?

YES:

(7) GO TO CODE 21 DIAGNOSTICS

NO:
- NO TROUBLE FOUND, SEE 'DIAGNOSTIC AIDS' ON FACING PAGE.

WHEN ALL DIAGNOSIS AND REPAIRS ARE COMPLETED, CLEAR CODES AND VERIFY OPERATION.

DIAGNOSTIC TROUBLE CODE CHART DTC 25 — 1995 DEVILLE, CONCOURS, ELDORADO AND SEVILLE

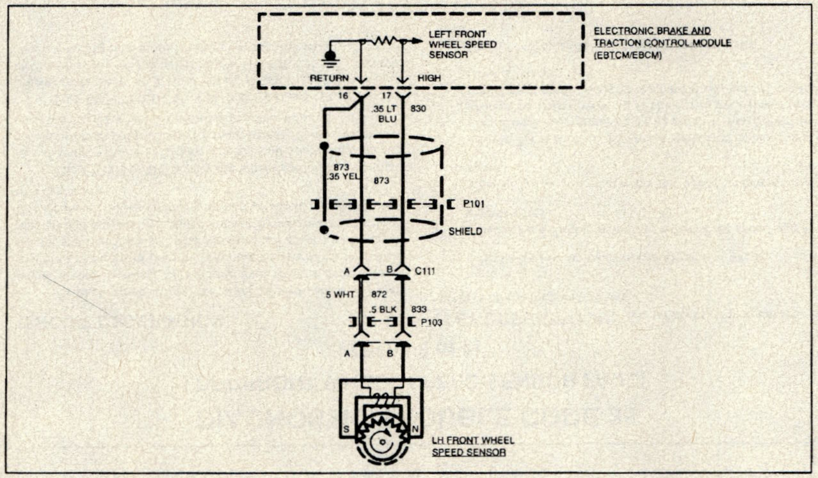

DIAGNOSTIC TROUBLE CODE 25

LEFT FRONT WHEEL SPEED SENSOR FAULT (Page 1 of 4)

CIRCUIT DESCRIPTION:

As each wheel turns, the wheel speed sensor for that wheel creates a small AC voltage as a toothed sensor ring, mounted in the front integral hub/bearing assembly, is passed by the stationary sensor. The sensor consists of a permanent magnet and coil and generates voltage by magnetic induction. The frequency of the wheel speed sensor signal is proportional to wheel speed.

The EBTCM uses the wheel speed sensor signal to calculate vehicle reference speeds and individual speed, acceleration and slip values for each wheel. These values are used to determine when antilock control is required.

The EBTCM will only set one wheel speed sensor fault code at a time, even if multiple faults exist. Detection is prioritized accordingly: LF, RR, LR, RF. Always check for additional stored codes after making repairs.

TEST CONDITION(S):

1. Tested continuously.

FAILURE CONDITION(S):

Sensor Output

1. Drive-away: Speed signal is not detected when vehicle reaches 12 kph or 7.5 mph.

OR

2. While Driving: Speed signal is lost while vehicle speed is greater than 40 kph or 25 mph

ACTION TAKEN:

1. Code set.
2. ABS/TCS switched 'OFF'.
3. 'ANTILOCK' indicator 'ON'.
4. 'TRACTION DISABLED' message displayed. (vehicles with TCS)

Code 25 is set if there is a short to voltage or ground in CKT 830 or 873, or faulty speed sensor. The testing for this fault occurs when vehicle is in motion; it will not set with the ignition 'ON' and the vehicle at rest.

NOTICE: A Code 25 may set if CKT 830 or 873 lose continuity and the WSS signal is lost while driving. A Code 27 should accompany this code on the subsequent ignition cycle. If both Code 25 and Code 27 are set concurrently, diagnose Code 27 first.

TEST DESCRIPTION: Test numbers refer to circled numbers on the diagnostic chart.

1. Checks if the code resets during a road test.
2. Use the TECH 1 to check for sources of induced electrical noise. Some sources of induced electrical noise which may cause intermittent wheel speed sensor codes to set are: improper routing or sensor cables next to spark plug wires; cables not retained in brackets; loose fitting or improperly mounted sensors; damaged sensors or toothed rings; poor terminal engagement of connectors; or intermittent shorts or opens in wiring; or a loss of sensor circuit shield.
3. Checks for damaged wheel speed sensor circuit shield. The shield is a foil insulator that is wrapped around the wheel speed sensor circuits. Inside the EBTCM harness connecter there is an uninsulated drain wire that is attached to the shield and jumped to the wheel speed sensor return line and taken to ground through the EBTCM. A shield which has lost its ground through the EBTCM leaves the sensor signal circuit susceptible to noise. It may be necessary to disassemble the EBTCM connector to check if the drain wire has a poor connection with the wheel speed sensor return connector terminal. The shield can be repaired by reterminating the sensor return line circuit, including the drain wire.
4. Checks for intermittent opens/shorts in wheel speed sensor circuits and cables.

DIAGNOSTIC TROUBLE CODE CHART DTC 25 — 1995 DEVILLE, CONCOURS, ELDORADO AND SEVILLE, CONTINUED

DIAGNOSTIC TROUBLE CODE 25

LEFT FRONT WHEEL SPEED SENSOR FAULT (Page 2 of 4)

NOTE: DIAGNOSE CODE 27 FIRST IF SET CONCURRENTLY.

ENSURE SENSOR IS NOT VISIBLY DAMAGED; IF IT IS DAMAGED, REPLACE THE HUB/BEARING/SENSOR ASSEMBLY, CLEAR CODES, AND ROAD TEST TO SEE IF CODE RESETS.

(1)
- CLEAR CODES
- ROAD TEST VEHICLE, REACHING AT LEAST 25 KPH (15 MPH)
- DOES CODE 25 RESET?

NO
- NO TROUBLE FOUND
- CHECK FOR INTERMITTENT CONDITIONS.

YES

(2)
- INSTALL TECH 1
- SELECT MODE F0: DATA LIST
- OBSERVE WHEEL SPEED SENSOR OUTPUT WHILE IDLING IN PARK
- INCREASE ENGINE SPEED BY PRESSING ON ACCELERATOR

ERRATIC SENSOR READING (DOES NOT STAY AT ZERO)
- VERIFY CORRECT CABLE ROUTING
- CHECK FOR PROPER INSTALLATION OF SENSOR
- REPAIR AS REQUIRED
- INSPECT EBTCM HARNESS CONNECTOR FOR AN OPEN FROM THE WHEEL SPEED SENSOR WIRING SHIELD TO THE WHEEL SPEED SENSOR RETURN LINE

SENSOR OUTPUT DOES NOT CHANGE WITH ENGINE SPEED (STAYS AT ZERO)

(3)
- TEST DRIVE VEHICLE
- OBSERVE WHEEL SPEED SENSOR OUTPUT
- DOES SIGNAL 'DROP OUT' OR DIFFER FROM OTHER SENSORS?

NORMAL SENSOR OUTPUT
- SEE CHECK FOR INTERMITTENTS.

SENSOR READING DIFFERS BY MORE THAN 4 MPH
- VERIFY CORRECT CABLE ROUTING
- CHECK FOR PROPER INSTALLATION OF SENSOR
- CHECK FOR OPEN IN WHEEL SPEED SENSOR WIRING SHIELD
- REPAIR AS REQUIRED
- REPLACE SENSOR IF PROBLEM PERSISTS

SENSOR READING 'DROP OUT'

(4)
- RAISE VEHICLE ON HOIST
- SPIN WHEEL WHILE MANIPULATING CABLES AND CONNECTORS BY HAND
- OBSERVE WHEEL SPEED SENSOR OUTPUT
- DOES SIGNAL 'DROP OUT'?

CODE 25 CONTINUED PAGE 4 OF 4

WHEN ALL DIAGNOSIS AND REPAIRS ARE COMPLETED, CLEAR CODES AND VERIFY OPERATION.

DIAGNOSTIC TROUBLE CODE CHART DTC 25 — 1995 DEVILLE, CONCOURS, ELDORADO AND SEVILLE, CONTINUED

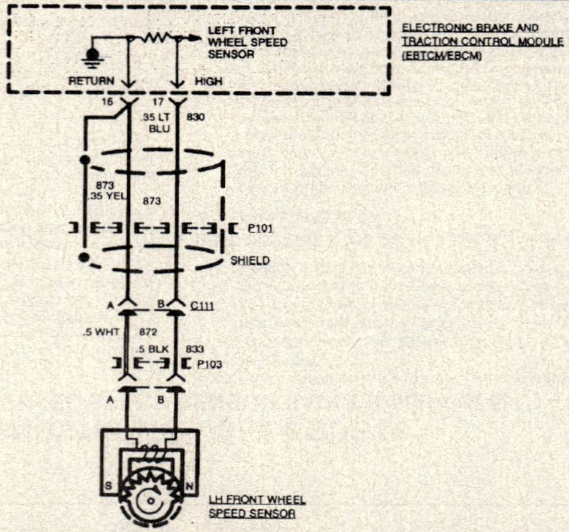

DIAGNOSTIC TROUBLE CODE 25

LEFT FRONT WHEEL SPEED SENSOR FAULT

(Page 3 of 4)

CIRCUIT DESCRIPTION:

As each wheel turns, the wheel speed sensor for that wheel creates a small AC voltage as a toothed sensor ring, mounted in the front integral hub/bearing assembly, is passed by the stationary sensor. The sensor consists of a permanent magnet and coil and generates voltage by magnetic induction. The frequency of the wheel speed sensor signal is proportional to wheel speed.

The EBTCM/EBCM uses the wheel speed sensor signal to calculate vehicle reference speeds and individual speed, acceleration and slip values for each wheel. These values are used to determine when antilock control is required.

The EBTCM/EBCM will only set one wheel speed sensor fault code at a time, even if multiple faults exist. Detection is prioritized accordingly: LF, RR, LR, RF. Always check for additional stored codes after making repairs.

TEST DESCRIPTION: Test numbers refer to circled numbers on the diagnostic chart.

5. Checks for proper resistance in the sensor itself.
6. Checks for a short to battery in either of the speed sensor circuit wires.
7. Checks for a short between the speed sensor circuit wires.
8. Checks for a short to ground in the speed sensor input circuit wire. The ground may be either a 'hard' short to ground, or a resistive short. A short with a resistance less than 2 megaohms, though not a hard short, can still cause Code 21 to set.
9. Checks wiring and connectors for intermittents.

DIAGNOSTIC TROUBLE CODE CHART DTC 25 — 1995 DEVILLE, CONCOURS, ELDORADO AND SEVILLE, CONTINUED

DIAGNOSTIC TROUBLE CODE 25

LEFT FRONT WHEEL SPEED SENSOR FAULT

(Page 4 of 4)

CONTINUED FROM CODE 25 PAGE 2 OF 4

YES →

(5)
- IGNITION 'OFF'
- DISCONNECT SENSOR CONNECTOR
- USING DVM AND ADAPTER FROM J 36616-A, CHECK SENSOR RESISTANCE BY PROBING SENSOR CONNECTOR TERMINALS
- IS IT 900-1300 OHMS AT 25°C (77°F).

NO →
- CHECK FOR INTERMITTENTS
- REPLACE EBTCM IF PROBLEM PERSISTS

(5) YES →

(6)
- CONNECT DVM BETWEEN SENSOR HARNESS CONNECTOR TERMINAL 'A' AND GROUND (USE DC VOLTAGE SETTING)
- IGNITION 'ON'
- NOTE VOLTAGE: REPEAT BETWEEN TERMINAL 'B' AND GROUND
- IS IT AROUND SYSTEM VOLTAGE ON EITHER TERMINAL 'A' OR 'B'?

(5) NO →
- REMOVE WHEEL SPEED RING SENSOR
- INSPECT TOOTHED SPEED RING ON HUB/BEARING ASSEMBLY
- IF DAMAGED, REPLACE ENTIRE HUB/BEARING AND SENSOR ASSEMBLY
- IF NOT DAMAGED, REPLACE WHEEL SPEED SENSOR

(6) NO →

(7)
- IGNITION 'OFF'
- DISCONNECT EBTCM
- INSTALL J 39700 UNIVERSAL BREAKOUT BOX TO EBTCM HARNESS CONNECTOR ONLY
- MEASURE RESISTANCE BETWEEN J 39700 TERMINALS '16' AND '17'
- IS IT LESS THAN 5 OHMS?

(6) YES →
- REPAIR SHORT TO SYSTEM VOLTAGE IN CKTS 830 AND 873
- REFER TO EBTCM-SWITCHED FAULTS IN THIS SECTION

(7) NO →

(8)
- MEASURE RESISTANCE BETWEEN J 39700 TERMINAL '16' AND GROUND
- IS IT GREATER THAN 2 MEGAOHMS?

(7) YES →
- REPAIR SHORT BETWEEN CKTS 830 AND 873

(8) YES →

(9)
- MEASURE RESISTANCE BETWEEN J 39700 TERMINAL '17' AND GROUND
- IS IT GREATER THAN 2 MEGAOHMS?

(8) NO →
- IF 5 OHMS OR LESS TO GROUND, REPAIR HARD SHORT IN HARNESS CKT 873
- IF BETWEEN 5 OHMS AND 2 MEGAOHMS, REPAIR RESISTIVE SHORT IN HARNESS CKT 873

(9) YES →
- CHECK ALL WIRING, CONNECTORS AND TERMINALS THOROUGHLY FOR INTERMITTENTS OR CORROSION
- REPAIR HARNESS OR CONNECTOR
- REPLACE SENSOR IF PROBLEM PERSISTS

(9) NO →
- IF 5 OHMS OR LESS TO GROUND, REPAIR HARD SHORT IN HARNESS CKT 830
- IF BETWEEN 5 OHMS AND 2 MEGAOHMS, REPAIR RESISTIVE SHORT IN HARNESS CKT 830

WHEN ALL DIAGNOSIS AND REPAIRS ARE COMPLETED, CLEAR CODES AND VERIFY OPERATION

DIAGNOSTIC TROUBLE CODE CHART DTC 27 — 1995 DEVILLE, CONCOURS, ELDORADO AND SEVILLE

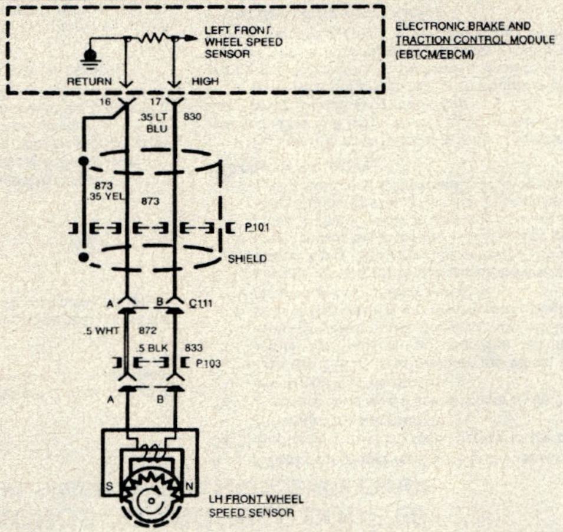

DIAGNOSTIC TROUBLE CODE 27

LEFT FRONT WHEEL SPEED SENSOR CONTINUITY FAULT

CIRCUIT DESCRIPTION:

As each wheel turns, the wheel speed sensor for that wheel creates a small AC voltage as a toothed sensor ring, mounted in the front integral hub/bearing assembly, is passed by the stationary sensor. The sensor consists of a permanent magnet and coil and generates voltage by magnetic induction. The frequency of the wheel speed sensor signal is proportional to wheel speed.

The EBTCM uses the wheel speed sensor signal to calculate vehicle reference speeds and individual speed, acceleration and slip values for each wheel. These values are used to determine when antilock control is required.

The EBTCM will only set one wheel speed sensor fault code at a time, even if multiple faults exist. Detection is prioritized accordingly: LF, RR, LR, RF. Always check for additional stored codes after making repairs.

TEST CONDITION(S):

1. Tested continuously.

FAILURE CONDITION(S):

Continuity

1. While Standing: Open circuit is seen while the vehicle is standing still with the ignition 'ON'.

ACTION TAKEN:

1. Code set.
2. ABS/TCS switched 'OFF'.
3. 'ANTILOCK' indicator 'ON'.
4. 'TRACTION DISABLED' message displayed. (vehicles with TCS)

Code 27 is set if there is a short to voltage or an open in CKT 830 or 873, or faulty speed sensor. The testing for this fault occurs with the ignition 'ON' and the vehicle at rest.

NOTICE: A Code 25 may set CKT 830 or 873 lose continuity and the WSS signal is lost while driving. A Code 27 should accompany this code on the subsequent ignition cycle. If both Code 25 and Code 27 are set concurrently, diagnose Code 27 first.

TEST DESCRIPTION: Test numbers refer to circled numbers on the diagnostic chart.

1. Checks for proper resistance in the sensor itself.
2. Checks for a short to battery in either of the speed sensor circuit wires. There should be 5V on CKT 830. If there is, CKT 830 has no hard opens or shorts. The condition may be intermittent.
3. Checks for an open in both the speed sensor circuit wires.
4. Manipulates wiring and connectors, trying to induce an intermittent fault not currently present.
5. This test checks if the code resets on key-up. If so, since Tests 1-4 have validated that the circuitry and components are good, the Wheel Bearing/Speed Sensor assembly may be causing the fault.
6. Checks if Code 25 sets during a road test.
7. Code 27 sets when the vehicle is at rest; Code 25 sets with a problem in the speed sensor circuitry with the vehicle in motion. If problems are still present, Code 25 would set during the road test, not a 27. If Code 25 sets at this point, Code 25 diagnostics should be used.
8. The wheel speed sensor may be causing an intermittent fault. If Code 27 resets after sensor replacement, the EBTCM/EBCM must be concluding there is a problem present in the speed sensor circuit when there is not.

DIAGNOSTIC AIDS:

Be sure the speed sensor wiring is properly routed and retained. This will help prevent false signals due to electrical noise being picked up by the wiring.

It is very important that a thorough inspection of the wiring and connectors be performed. Failure to carefully and fully inspect wiring and connectors may result in misdiagnosis, causing part replacement with reappearance of the fault.

DIAGNOSTIC TROUBLE CODE CHART DTC 27 — 1995 DEVILLE, CONCOURS, ELDORADO AND SEVILLE, CONTINUED

DIAGNOSTIC TROUBLE CODE 27

LEFT FRONT WHEEL SPEED SENSOR CONTINUITY FAULT

ENSURE SENSOR IS NOT VISIBLY DAMAGED; IF IT IS DAMAGED, REPLACE THE WHEEL BEARING/SPEED/SENSOR ASSEMBLY, CLEAR CODES, AN TURN IGNITION 'ON' TO SEE IF CODE RESETS.

(1)
- IGNITION 'OFF'
- DISCONNECT SENSOR CONNECTOR
- USING DVM AND ADAPTER FROM J 35616-A, CHECK SENSOR RESISTANCE BY PROBING SENSOR CONNECTOR TERMINALS.
- IS IT 900-1300 OHMS AT 25°C (77°F)?

NO
- REPLACE WHEEL SPEED SENSOR

YES

(2)
- CONNECT DVM BETWEEN SENSOR HARNESS CONNECTOR TERMINAL 'A' AND GROUND (USED DC VOLTAGE SETTING)
- IGNITION 'ON'
- NOTE VOLTAGE: REPEAT BETWEEN TERMINAL 'B' AND GROUND.
- IS IT AROUND SYSTEM VOLTAGE ON EITHER TERMINAL 'A' OR 'B'?

YES
- REPAIR SHORT TO SYSTEM VOLTAGE IN CKT 830 OR 873
- REFER TO EBTCM-SWITCHED FAULTS IN THIS SECTION

NO

(3)
- IGNITION 'OFF'
- DISCONNECT EBTCM
- INSTALL J 39700 UNIVERSAL BREAKOUT BOX TO EBTCM HARNESS CONNECTOR ONLY
- MEASURE RESISTANCE BETWEEN PIN 17 ON J 39700 TERMINAL 'A' OF SENSOR CONNECTOR ON HARNESS. PERFORM SAME MEASUREMENT BETWEEN PIN 16 AND TERMINAL 'B'.
- IS EITHER MEASUREMENT GREATER THAN 5 OHMS?

YES
- REPAIR OPEN IN CKT 830 AND/OR 873

NO

(4)
- REMOVE J 39700; RECONNECT EBTCM
- CLEAR CODES; IGNITION 'OFF'
- DISCONNECT TECH 1, IF CONNECTED
- IGNITION 'ON'
- WIGGLE ALL WIRING AND CONNECTORS TO TRY TO DUPLICATE AN INTERMITTENT FAULT
- DOES CODE 27 RESET?

YES
- REPAIR WIRING AND/OR CONNECTIONS
-

NO

(5)
- CLEAR CODES; IGNITION 'OFF'
- DISCONNECT TECH 1, IF CONNECTED
- IGNITION 'ON'
- DOES CODE 27 RESET?

NO

(6)
- ROAD TEST VEHICLE, REACHING AT LEAST 25 KPH (15 MPH).
- DOES CODE 25 SET?

YES
(7) GO TO CODE 25 DIAGNOSTICS

NO
- NO TROUBLE FOUND. SEE 'DIAGNOSTIC AIDS' ON FACING PAGE.

YES

(8)
- REPLACE WHEEL SPEED SENSOR
- CLEAR CODES; IGNITION 'OFF'
- DISCONNECT TECH 1, IF CONNECTED
- IGNITION 'ON'
- DOES CODE 27 RESET?

YES
- REPLACE EBTCM

NO
- SYSTEM OK

WHEN ALL DIAGNOSIS AND REPAIRS ARE COMPLETED, CLEAR CODES AND VERIFY OPERATION

DIAGNOSTIC TROUBLE CODE CHART DTC 28 — 1995 DEVILLE, CONCOURS, ELDORADO AND SEVILLE

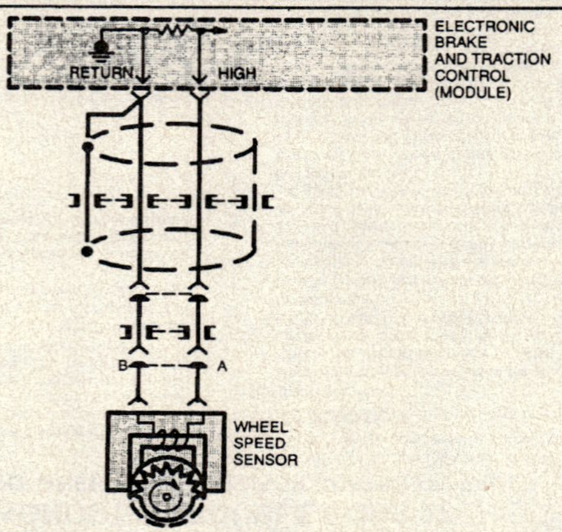

DIAGNOSTIC TROUBLE CODE 28
WHEEL SPEED SENSOR FREQUENCY ERROR

TEST CONDITION(S):
1. Tested continuously.

FAILURE CONDITION(S):
1. While Driving: One wheel is detected at least 30 kph or 19 mph faster than the remaining wheels for an extended period of time.

OR

One wheel is at least 6 kph or 4 mph less than or greater than the remaining wheels for an extended period of time.

ACTION TAKEN:
1. Code set.
2. ABS/TCS switched 'OFF'.
3. 'ANTILOCK' indicator 'ON'.
4. TRACTION DISABLED' message displayed.

Code 28 is set if the EBTCM cannot specifically identify which wheel speed sensor is causing the frequency error problem. If it can define the specific speed sensor causing the problem, the Sensor Fault code associated with the sensor (21, 25, 31, 35) will be set instead of Code 28.

TEST DESCRIPTION: Test numbers refer to circled numbers on the diagnostic chart.

1. Checks wiring and connections for problems. It is critical that this step is performed thoroughly, as wiring/connector problems are the most likely cause for this code.
2. Use the TECH 1 to monitor for vehicle electrical/electronic system 'noise' being picked up by the speed sensor circuits.
3. Use the TECH 1 to monitor the wheel speed sensors while the vehicle is operating. If the TECH 1 Auto-Trigger Snapshot mode triggers, the speed sensor or the wiring and/or connectors are going intermittent during the road test.
4. Checks for proper resistance in the sensor itself.

5a. Replaces rear wheel speed sensor as the likely cause of the triggering intermittent.

5b. Replaces front wheel speed sensor as the likely cause of the triggering intermittent.

6. Use the TECH 1 to monitor the wheel speed sensors while the vehicle is operating. If the TECH 1 Auto-Snapshot mode triggers, the EBTCM must be concluding there is a problem with wheel speed sensor frequencies when there is not.
7. Checks for a short to ground in the speed sensor input circuit wires. The circuit to be tested depends on which speed sensor triggered the TECH 1. The ground may be either a 'hard' short to ground, or a resistive short. A short with a resistance less that 2 megaohms, though not a hard short, can still cause Code 28 to set.

DIAGNOSTIC AIDS:

Be sure the speed sensor wiring is properly routed and retained. This will help prevent false signals due to electrical noise being picked up by the wiring.

It is very important that a thorough inspection of the wiring and connectors be performed. Failure to carefully and fully inspect wiring and connectors may result in misdiagnosis, causing part replacement with reappearance of the fault.

Front Speed Sensors Only:

Check the toothed wheel for any large grooves, gouges, marks, etc. that might influence the tooth's signal at the wheel speed sensor. Also check for a buildup of foreign material in the gaps between teeth in the toothed wheel; this material may cause this fault.

A worn hub/bearing assembly may cause this fault in extreme cases; the bearing play allows the sensor-to-toothed-ring gap to change excessively. Never reuse a wheel speed sensor once it has been separated from the front hub/bearing assembly.

DIAGNOSTIC TROUBLE CODE CHART DTC 28 — 1995 DEVILLE, CONCOURS, ELDORADO AND SEVILLE, CONTINUED

DIAGNOSTIC TROUBLE CODE 28
WHEEL SPEED SENSOR FREQUENCY ERROR

(1) VERY THOROUGHLY INSPECT ALL WIRING AND CONNECTIONS FOR INTERMITTENTS, CHAFING, POOR TERMINAL CONTACT TENSION, CORROSION, AND MISROUTING, AT EBTCM, ALL SENSORS, AND ALL HARNESS INTERCONNECTS. REPAIR ANY PROBLEMS CLEAR CODES, THEN ROAD TEST VEHICLE, REACHING AT LEAST 25 KPH (15 MPH).
- DOES CODE RESET?

YES → (2)

NO →
- NO TROUBLE FOUND; SEE 'DIAGNOSTIC AIDS' ON FACING PAGE.

(2)
- INSTALL TECH 1 AND MASS STORAGE; CLEAR CODES
- SET TECH 1 AUTO TRIGGER SNAPSHOT MODE
- START VEHICLE AND ALLOW TO IDLE FOR ONE MINUTE
- DOES TECH 1 TRIGGER?

NO → (3)

YES/NO →
- NOTE TECH1, WHICH SENSOR TRIGGERED?
- IGNITION'OFF'
- DISCONNECT EBTCM
- INSTALL J 39700 UNIVERSAL BREAKOUT BOX TO EBTCM AND EBTCM HARNESS CONNECTOR.
- USING TABLE BELOW, MEASURE RESISTANCE TO GROUND AT THE TERMINAL ON J 39700 THAT CORRESPONDS TO THE SPEED SENSOR THE TECH 1 TRIGGERED ON:

LF-TERMINAL 17
RF-TERMINAL 24
LR-TERMINAL 19
RR-TERMINAL 21

- IF RESISTANCE IS 5 OHMS OR LESS TO GROUND, REPAIR HARD SHORT; IF BETWEEN 5 OHMS AND 2 MEGAOHMS, REPAIR RESISTIVE SHORT IN CKT.

LF-CKT 830
RF-CKT 872
LR-CKT 885
RR-CKT 882

(3)
- ROAD TEST VEHICLE REACHING AT LEAST 25 KPH (15 MPH).
- DOES TECH 1 TRIGGER?

YES →
- NOTE TECH 1, WHICH SENSOR TRIGGERED?

NO →
- NO TROUBLE FOUND; SEE DIAGNOSTIC AIDS ON FACING PAGE.

(4)
- IGNITION 'OFF'
- DISCONNECT SENSOR CONNECTOR
- MEASURE RESISTANCE BETWEEN SENSOR TERMINALS, USING ADAPTERS FROM J 35616-A IF FRONT SENSOR.
- IS RESISTANCE WITHIN 900-1300 OHMS AT 25°C (77°F)?

YES →
- REPLACE EBTCM

NO → FRONT / REAR

FRONT: (5a)
- REPLACE WHEEL SPEED SENSOR

REAR: (5b)
- REPLACE HUB/BEARING/SENSOR ASSEMBLY

(6)
- CLEAR CODES, IGNITION 'OFF'
- DISCONNECT TECH 1, IF CONNECTED
- ROAD TEST VEHICLE, REACHING AT LEAST 25 KPH (15 MPH)
- DOES CODE 28 RESET?

YES →
- REPLACE EBTCM

NO →
- SYSTEM OK

WHEN ALL DIAGNOSIS AND REPAIRS ARE COMPLETED, CLEAR CODES AND VERIFY OPERATION

DIAGNOSTIC TROUBLE CODE CHART DTC 31 — 1995 DEVILLE, CONCOURS, ELDORADO AND SEVILLE

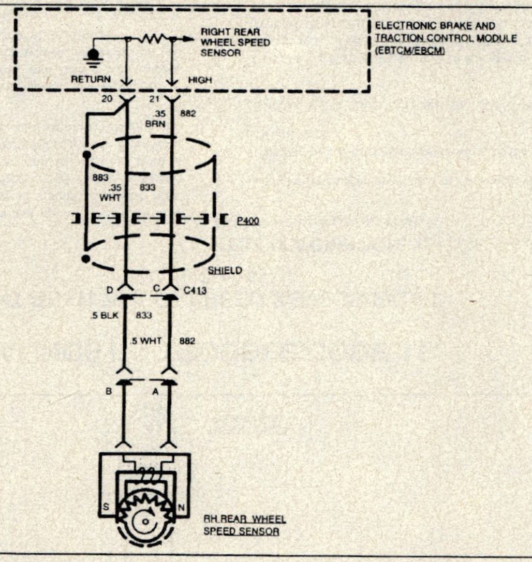

DIAGNOSTIC TROUBLE CODE 31
RIGHT REAR WHEEL SPEED SENSOR FAULT (Page 1 of 4)

CIRCUIT DESCRIPTION:

As each wheel turns, the wheel speed sensor for that wheel creates a small AC voltage as a toothed sensor ring, mounted in the front integral hub/bearing assembly, is passed by the stationary sensor. The sensor consists of a permanent magnet and coil and generates voltage by magnetic induction. The frequency of the wheel speed sensor signal is proportional to wheel speed.

The EBTCM uses the wheel speed sensor signal to calculate vehicle reference speeds and individual speed, acceleration and slip values for each wheel. These values are used to determine when antilock control is required.

The EBTCM will only set one wheel speed sensor fault code at a time, even if multiple faults exist. Detection is prioritized accordingly: LF, RR, LR, RF. Always check for additional stored codes after making repairs.

TEST CONDITION(S):

1. Tested continuously.

FAILURE CONDITION(S):

Sensor Output

1. Drive-away: Speed signal is not detected when vehicle reaches 12 kph or 7.5 mph.
2. While Driving: Speed signal is lost while vehicle speed is greater than 40 kph or 25 mph

ACTION TAKEN:

1. Code set.
2. ABS/TCS switched 'OFF'.
3. 'ANTILOCK' indicator 'ON'.
4. 'TRACTION DISABLED' message displayed. (vehicles with TCS)

Code 21 is set if there is a short to voltage or ground in CKT 882 or 883, or faulty speed sensor. The testing for this fault occurs when the vehicle is in motion, it will not set with the ignition 'ON' and the vehicle at rest.

NOTICE: A Code 31 may set if CKT 882 or 883 lose continuity and the WSS signal is lost while driving. A Code 33 should accompany this code on the subsequent ignition cycle. If both Code 31 and Code 33 are set concurrently, diagnose Code 33 first.

TEST DESCRIPTION: Test numbers refer to circled numbers on the diagnostic chart.

1. Checks if the code resets during a road test.
2. Use the TECH 1 to check for sources of induced electrical noise. Some sources of induced electrical noise which may cause intermittent wheel speed sensor codes to set are: improper routing or sensor cables next to spark plug wires; cables not retained in brackets; loose fitting or improperly mounted sensors; damaged sensors or toothed rings; poor terminal engagement of connectors; or intermittent shorts or opens in wiring; or a loss of sensor circuit shield.
3. Checks for damaged wheel speed sensor circuit shield. The shield is a foil insulator that is wrapped around the wheel speed sensor circuits. Inside the EBTCM harness connecter there is an uninsulated drain wire that is attached to the shield and jumped to the wheel speed sensor return line and taken to ground through the EBTCM. A shield which has lost its ground through the EBTCM leaves the sensor signal circuit susceptible to noise. It may be necessary to disassemble the EBTCM connector to check if the drain wire has a poor connection with the wheel speed sensor return connector terminal. The shield can be repaired by reterminating the sensor return line circuit, including the drain wire.
4. Checks for intermittent opens/shorts in wheel speed sensor circuits and cables.

DIAGNOSTIC TROUBLE CODE CHART DTC 31 — 1995 DEVILLE, CONCOURS, ELDORADO AND SEVILLE, CONTINUED

DIAGNOSTIC TROUBLE CODE 31
RIGHT REAR WHEEL SPEED SENSOR FAULT (Page 2 of 4)

NOTE: DIAGNOSE CODE 33 FIRST IF SET CONCURRENTLY.

ENSURE SENSOR IS NOT VISIBLY DAMAGED; IF IT IS DAMAGED, REPLACE THE HUB/BEARING/SENSOR ASSEMBLY, CLEAR CODES, AND ROAD TEST TO SEE IF CODE RESETS.

(1)
- CLEAR CODES
- ROAD TEST VEHICLE, REACHING AT LEAST 25 KPH (15 MPH)
- DOES CODE 31 RESET?

NO:
- NO TROUBLE FOUND
- REFER TO 'NOTE ON INTERMITTENTS'

YES:

(2)
- INSTALL TECH 1
- SELECT MODE F0: DATA LIST
- OBSERVE WHEEL SPEED SENSOR OUTPUT WHILE IDLING IN PARK
- INCREASE ENGINE SPEED BY PRESSING ON ACCELERATOR

ERRATIC SENSOR READING (DOES NOT STAY AT ZERO):
- VERIFY CORRECT CABLE ROUTING
- CHECK FOR PROPER INSTALLATION OF SENSOR
- REPAIR AS REQUIRED
- INSPECT EBTCM HARNESS CONNECTOR FOR AN OPEN FROM THE WHEEL SPEED SENSOR WIRING SHIELD TO THE WHEEL SPEED SENSOR RETURN LINE

SENSOR OUTPUT DOES NOT CHANGE WITH ENGINE SPEED (STAYS AT ZERO):

(3)
- TEST DRIVE VEHICLE
- OBSERVE WHEEL SPEED SENSOR OUTPUT
- DOES SIGNAL 'DROP OUT' OR DIFFER FROM OTHER SENSORS?

NORMAL SENSOR OUTPUT:
- SEE 'NOTE ON INTERMITTENTS'

SENSOR READING DIFFERS BY MORE THAN 4 MPH:
- VERIFY CORRECT CABLE ROUTING
- CHECK FOR PROPER INSTALLATION OF SENSOR
- CHECK FOR AN OPEN IN WHEEL SPEED SENSOR WIRING SHIELD
- REPAIR AS REQUIRED
- REPLACE HUB/BEARING/SENSOR ASSEMBLY IF PROBLEM PERSISTS

SENSOR READING 'DROP OUT':

(4)
- RAISE VEHICLE ON HOIST
- SPIN WHEEL WHILE MANIPULATING CABLES AND CONNECTORS BY HAND
- OBSERVE WHEEL SPEED SENSOR OUTPUT
- DOES SIGNAL 'DROP OUT'?

CONTINUED ON PAGE 4 OF 4 (CODE 31)

WHEN ALL DIAGNOSIS AND REPAIRS ARE COMPLETED, CLEAR CODES AND VERIFY OPERATION

DIAGNOSTIC TROUBLE CODE CHART DTC 31 — 1995 DEVILLE, CONCOURS, ELDORADO AND SEVILLE, CONTINUED

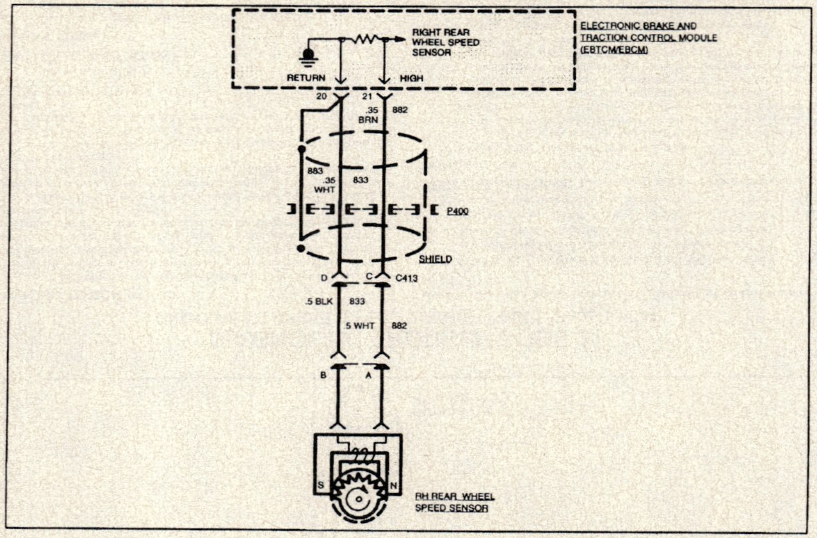

DIAGNOSTIC TROUBLE CODE 31

RIGHT REAR WHEEL SPEED SENSOR FAULT (Page 3 of 4)

CIRCUIT DESCRIPTION:

As each wheel turns, the wheel speed sensor for that wheel creates a small AC voltage as a toothed sensor ring, mounted in the front integral hub/bearing assembly, is passed by the stationary sensor. The sensor consists of a permanent magnet and coil and generates voltage by magnetic induction. The frequency of the wheel speed sensor signal is proportional to wheel speed.

The EBTCM uses the wheel speed sensor signal to calculate vehicle reference speeds and individual speed, acceleration and slip values for each wheel. These values are used to determine when antilock control is required.

The EBTCM will only set one wheel speed sensor fault code at a time, even if multiple faults exist. Detection is prioritized accordingly: LF, RR, LR, RF. Always check for additional stored codes after making repairs.

TEST DESCRIPTION: Test numbers refer to circled numbers on the diagnostic chart.

5. Checks for proper resistance in the sensor itself.
6. Checks for a short to battery in either of the speed sensor circuit wires.
7. Checks for a short between the speed sensor circuit wires.
8. Checks for a short to ground in the speed sensor input circuit wire. The ground may be either a 'hard' short to ground, or a resistive short. A short with a resistance less than 2 megaohms, through not a hard short, can still cause Code 21 to set.
9. Checks wiring and connectors for intermittents.

DIAGNOSTIC TROUBLE CODE CHART DTC 31 — 1995 DEVILLE, CONCOURS, ELDORADO AND SEVILLE, CONTINUED

DIAGNOSTIC TROUBLE CODE 31

RIGHT REAR WHEEL SPEED SENSOR FAULT (Page 4 of 4)

CONTINUED FROM CODE 31 PAGE 2 OF 4

YES →

(5)
- IGNITION 'OFF'
- DISCONNECT SENSOR CONNECTOR
- USING DVM AND ADAPTER FROM J 35616-A, CHECK SENSOR RESISTANCE BY PROBING SENSOR CONNECTOR TERMINALS.
- IS IT 900 - 1300 OHMS AT 25°C (77°F)?

NO →
- CHECK FOR INTERMITTENTS
- REPLACE EBTCM IF PROBLEM PERSISTS

YES →

(6)
- CONNECT DVM BETWEEN SENSOR HARNESS CONNECTOR TERMINAL 'A' AND GROUND (USE DC VOLTAGE SETTING)
- IGNITION 'ON'
- NOTE VOLTAGE: REPEAT BETWEEN TERMINAL 'B' AND GROUND.
- IS IT AROUND SYSTEM VOLTAGE ON EITHER TERMINAL 'A' OR 'B'?

NO →
- REPLACE HUB/BEARING/SENSOR ASSEMBLY

NO →

(7)
- IGNITION 'OFF'
- DISCONNECT EBTCM/EBCM
- INSTALL J 39700 UNIVERSAL BREAKOUT BOX TO EBTCM/EBCM HARNESS CONNECTOR ONLY
- MEASURE RESISTANCE BETWEEN J 39700 TERMINALS '20' AND '21'
- IS IT LESS THAN 5 OHMS?

YES →
- REPAIR SHORT TO SYSTEM VOLTAGE IN CKT 882 OR 883
- REFER TO EBTCM-SWITCHED FAULTS IN THIS SECTION

NO →

(8)
- MEASURE RESISTANCE BETWEEN J 39700 TERMINAL '21' AND GROUND
- IS IT GREATER THAN 2 MEGAOHMS?

YES →
- REPAIR SHORT BETWEEN CKT'S 882 AND 883

YES →

(9)
- MEASURE RESISTANCE BETWEEN J 39700 TERMINAL '20' AND GROUND
- IS IT GREATER THAN 2 MEGAOHMS?

NO →
- IF 5 OHMS OR LESS TO GROUND, REPAIR HARD SHORT IN HARNESS CKT 882
- IF BETWEEN 5 OHMS AND 2 MEGAOHMS, REPAIR RESISTIVE SHORT IN HARNESS CKT 882

YES →
- CHECK ALL WIRING, CONNECTORS AND TERMINALS THOROUGHLY FOR INTERMITTENTS OR CORROSION.
-
- REPAIR HARNESS OR CONNECTOR
- REPLACE SENSOR IF PROBLEM PERSISTS

NO →
- IF 5 OHMS OR LESS TO GROUND, REPAIR HARD SHORT IN HARNESS CKT 883
- IF BETWEEN 5 OHMS AND 2 MEGAOHMS REPAIR RESISTIVE SHORT IN HARNESS CKT 883

WHEN ALL DIAGNOSIS AND REPAIRS ARE COMPLETED, CLEAR CODES AND VERIFY OPERATION

DIAGNOSTIC TROUBLE CODE CHART DTC 33 — 1995 DEVILLE, CONCOURS, ELDORADO AND SEVILLE

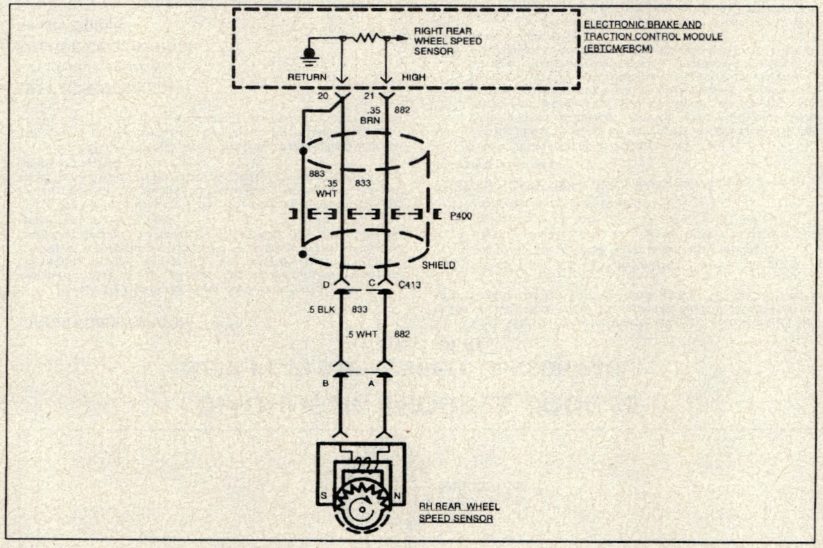

DIAGNOSTIC TROUBLE CODE 33

RIGHT REAR WHEEL SPEED SENSOR CONTINUITY FAULT

CIRCUIT DESCRIPTION:

As each wheel turns, the wheel speed sensor for that wheel creates a small AC voltage as a toothed sensor ring, mounted in the front integral hub/bearing assembly, is passed by the stationary sensor. The sensor consists of a permanent magnet and coil and generates voltage by magnetic induction. The frequency of the wheel speed sensor signal is proportional to wheel speed.

The EBTCM uses the wheel speed sensor signal to calculate vehicle reference speeds and individual speed, acceleration and slip values for each wheel. These values are used to determine when antilock control is required.

The EBTCM will only set one wheel speed sensor fault code at a time, even if multiple faults exist. Detection is prioritized accordingly: LF, RR, LR, RF. Always check for additional stored codes after making repairs.

TEST CONDITION(S):

1. Tested continuously.

FAILURE CONDITION(S):

Continuity

1. While standing: Open circuit is seen while the vehicle is standing still with the ignition 'ON'.

ACTION TAKEN:

1. Code set.
2. ABS/TCS switched 'OFF'.
3. 'ANTILOCK' indicator 'ON'.
4. 'TRACTION DISABLED' message displayed. (vehicles with TCS)

Code 33 is set if there is a short to voltage or an open in CKT 882 or 883, or faulty speed sensor. The testing for this fault occurs with the ignition 'ON' and the vehicle at rest.

NOTICE: A Code 31 may set if CKT 882 or 883 lose continuity and the WSS signal is lost while driving. A Code 33 should accompany this code on the subsequent ignition cycle. If both Code 31 and Code 33 are set concurrently, diagnose Code 33 first.

TEST DESCRIPTION: Test numbers refer to circled numbers on the diagnostic chart.

1. Checks for proper resistance in the sensor itself.
2. Checks for a short to battery in either of the speed sensor circuit wires. There should be 5V on CKT 882. If there is, CKT 882 has no hard opens or shorts. The condition may be intermittent.
3. Checks for an open in both the speed sensor circuit wires.
4. Manipulates wiring and connectors, trying to induce an intermittent fault not currently present.
5. This test checks if the code resets on key-up. If so, since Tests 1-4 have validated that the circuitry and components are good, the speed sensor may be causing the fault.
6. Checks if the Code 31 sets during a road test.
7. Code 33 sets when the vehicle is at rest: Code 31 sets with a problem in the speed sensor circuitry with the vehicle in motion. If problems are still present, Code 31 would set during the road test, not a 33. If Code 33 sets at this point, Code 31 diagnostics should be used.
8. The wheel speed sensor may be causing an intermittent fault. If Code 33 resets after sensor replacement, the EBTCM must be concluding there is a problem present in the speed sensor circuit when there is not.

DIAGNOSTIC AIDS:

Be sure the speed sensor wiring is properly routed and retained. This will help prevent false signals due to electrical noise being picked up by the wiring.

It is very important that a thorough inspection of the wiring and connectors be performed. Failure to carefully and fully inspect wiring and connectors may result in misdiagnosis, causing part replacement with reappearance of the fault.

DIAGNOSTIC TROUBLE CODE CHART DTC 33 — 1995 DEVILLE, CONCOURS, ELDORADO AND SEVILLE, CONTINUED

DIAGNOSTIC TROUBLE CODE 33

RIGHT REAR WHEEL SPEED SENSOR CONTINUITY FAULT

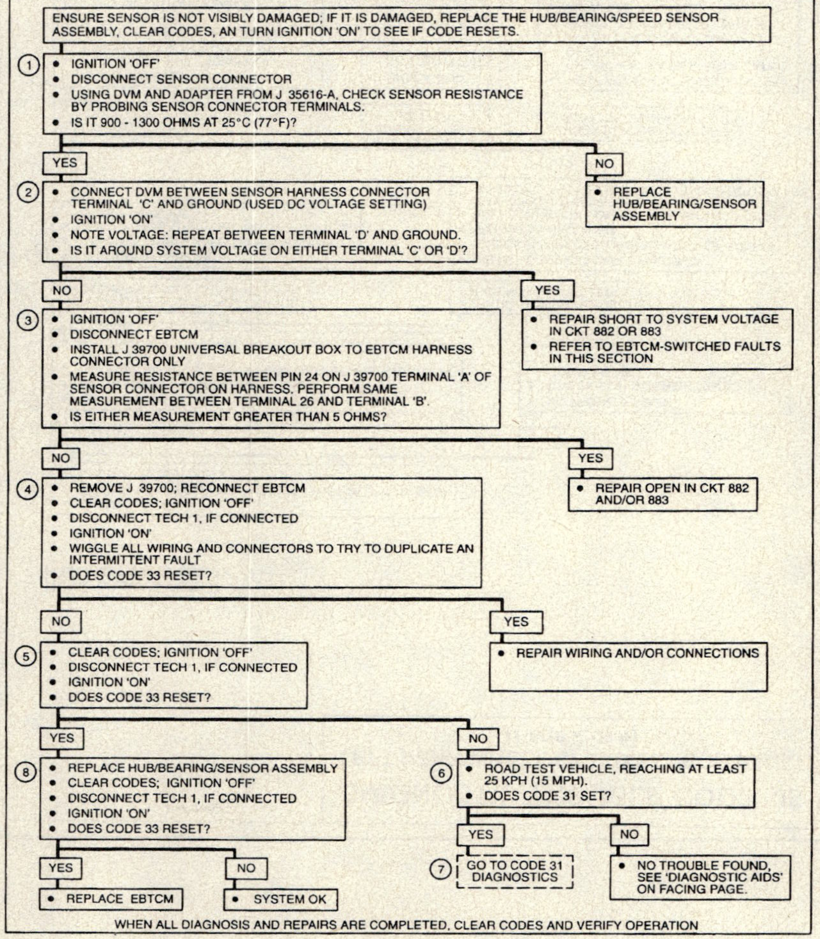

DIAGNOSTIC TROUBLE CODE CHART DTC 35 — 1995 DEVILLE, CONCOURS, ELDORADO AND SEVILLE

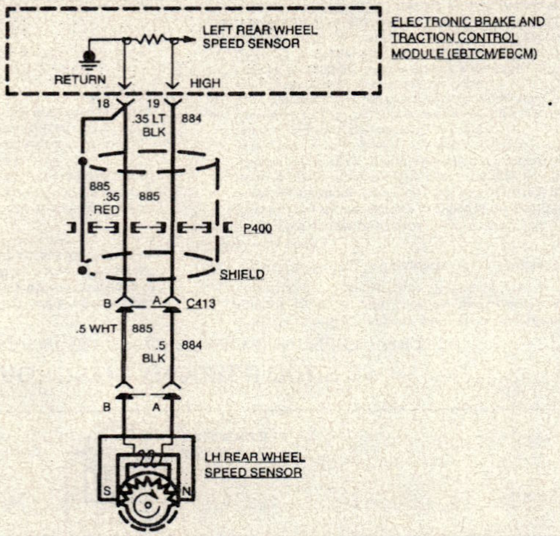

DIAGNOSTIC TROUBLE CODE 35

LEFT REAR WHEEL SPEED SENSOR FAULT (Page 1 of 4)

CIRCUIT DESCRIPTION:

As each wheel turns, the wheel speed sensor for that wheel creates a small AC voltage as a toothed sensor ring, mounted in the front integral hub/bearing assembly, is passed by the stationary sensor. The sensor consists of a permanent magnet and coil and generates voltage by magnetic induction. The frequency of the wheel speed sensor signal is proportional to wheel speed.

The EBTCM uses the wheel speed sensor signal to calculate vehicle reference speeds and individual speed, acceleration and slip values for each wheel. These values are used to determine when antilock control is required.

The EBTCM will only set one wheel speed sensor fault code at a time, even if multiple faults exist. Detection is prioritized accordingly: LF, RR, LR, RF. Always check for additional stored codes after making repairs.

TEST CONDITION(S):

1. Tested continuously.

FAILURE CONDITION(S):

Sensor Output

1. Drive-away: Speed signal is not detected when vehicle reaches 12 kph or 7.5 mph.
2. While Driving: Speed signal is lost while vehicle speed is greater than 40 kph or 25 mph.

ACTION TAKEN:

1. Code set.
2. ABS/TCS switched 'OFF'.
3. 'ANTILOCK' indicator 'ON'.
4. 'TRACTION DISABLED' message displayed. (vehicles with TCS)

Code 35 is set if there is a short to voltage or ground in CKT 885 or 884, or faulty speed sensor. The testing for this fault occurs when the vehicle is in motion, it will not set with the ignition 'ON' and the vehicle at rest.

NOTICE: A Code 35 may set if CKT 884 or 885 lose continuity and the WSS signal is lost while driving. A Code 37 should accompany this code on the subsequent ignition cycle. If both Code 35 and Code 37 are set concurrently, diagnose Code 37 first.

TEST DESCRIPTION: Test numbers refer to circled numbers on the diagnostic chart.

1. Checks if the code resets during a road test.
2. Use the TECH 1 to check for sources of induced electrical noise. Some sources of induced electrical noise which may cause intermittent wheel speed sensor codes to set are: improper routing or sensor cables next to spark plug wires; cables not retained in brackets; loose fitting or improperly mounted sensors; damaged sensors or toothed rings; poor terminal engagement of connectors; or intermittent shorts or opens in wiring; or a loss of sensor circuit shield.
3. Checks for damaged wheel speed sensor circuit shield. The shield is a foil insulator that is wrapped around the wheel speed sensor circuits. Inside the EBTCM harness connecter there is an uninsulated drain wire that is attached to the shield and jumped to the wheel speed sensor return line and taken to ground through the EBTCM. A shield which has lost its ground through the EBTCM leaves the sensor signal circuit susceptible to noise. It may be necessary to disassemble the EBTCM connector to check if the drain wire has a poor connection with the wheel speed sensor return connector terminal. The shield can be repaired by reterminating the sensor return line circuit, including the drain wire.
4. Checks for intermittent opens/shorts in wheel speed sensor circuits and cables.

DIAGNOSTIC TROUBLE CODE CHART DTC 35 — 1995 DEVILLE, CONCOURS, ELDORADO AND SEVILLE, CONTINUED

DIAGNOSTIC TROUBLE CODE 35

LEFT REAR WHEEL SPEED SENSOR FAULT (Page 2 of 4)

NOTE: DIAGNOSE CODE 37 FIRST IF SET CONCURRENTLY.

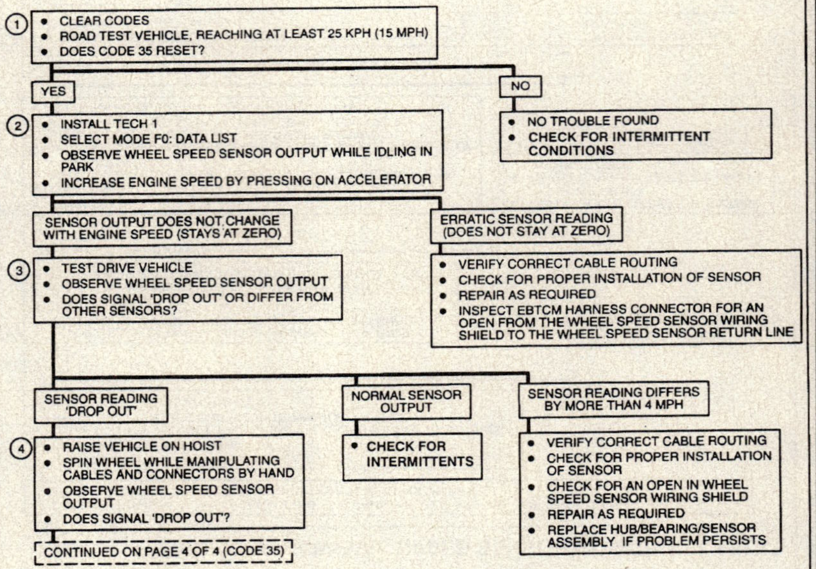

WHEN ALL DIAGNOSIS AND REPAIRS ARE COMPLETED, CLEAR CODES AND VERIFY OPERATION

DIAGNOSTIC TROUBLE CODE CHART DTC 35 — 1995 DEVILLE, CONCOURS, ELDORADO AND SEVILLE, CONTINUED

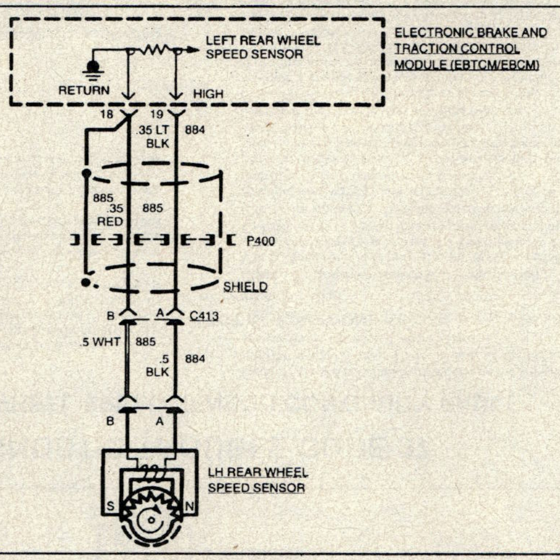

DIAGNOSTIC TROUBLE CODE 35

LEFT REAR WHEEL SPEED SENSOR FAULT
(Page 3 of 4)

CIRCUIT DESCRIPTION:

As each wheel turns, the wheel speed sensor for that wheel creates a small AC voltage as a toothed sensor ring, mounted in the front integral hub/bearing assembly, is passed by the stationary sensor. The sensor consists of a permanent magnet and coil and generates voltage by magnetic induction. The frequency of the wheel speed sensor signal is proportional to wheel speed.

The EBTCM uses the wheel speed sensor signal to calculate vehicle reference speeds and individual speed, acceleration and slip values for each wheel. These values are used to determine when antilock control is required.

The EBTCM will only set one wheel speed sensor fault code at a time, even if multiple faults exist. Detection is prioritized accordingly: LF, RR, LR, RF. Always check for additional stored codes after making repairs.

TEST DESCRIPTION: Test numbers refer to circled numbers on the diagnostic chart.

5. Checks for proper resistance in the sensor itself.
6. Checks for a short to battery in either of the speed sensor circuit wires.
7. Checks for a short between the speed sensor circuit wires.
8. Checks for a short to ground in the speed sensor input circuit wire. The ground may be either a 'hard' short to ground, or a resistive short. A short with a resistance less than 2 megaohms, through not a hard short, can still cause Code 21 to set.
9. Checks wiring and connectors for intermittents.

DIAGNOSTIC TROUBLE CODE CHART DTC 35 — 1995 DEVILLE, CONCOURS, ELDORADO AND SEVILLE, CONTINUED

CHECK FOR INTERMITTENT CONDITIONS

DIAGNOSTIC TROUBLE CODE 35

LEFT REAR WHEEL SPEED SENSOR FAULT
(Page 4 of 4)

CONTINUED FROM CODE 35 PAGE 2 OF 4

YES

(5)
- IGNITION 'OFF'
- DISCONNECT SENSOR CONNECTOR
- USING DVM AND ADAPTER FROM J 35616-A, CHECK SENSOR RESISTANCE BY PROBING SENSOR CONNECTOR TERMINALS.
- IS IT 900 - 1300 OHMS at 25°C (77°F)?

NO
-
- REPLACE EBTCM IF PROBLEM PERSISTS

YES

(6)
- CONNECT DVM BETWEEN SENSOR HARNESS CONNECTOR TERMINAL 'A' AND GROUND (USE DC VOLTAGE SETTING)
- IGNITION 'ON'
- NOTE VOLTAGE: REPEAT BETWEEN TERMINAL 'B' AND GROUND
- IS IT AROUND SYSTEM VOLTAGE ON EITHER TERMINAL 'A' OR 'B'?

NO
- REPLACE HUB/BEARING/SENSOR ASSEMBLY

NO

(7)
- IGNITION 'OFF'
- DISCONNECT EBTCM
- INSTALL J 39700 UNIVERSAL BREAKOUT BOX TO EBTCM HARNESS CONNECTOR ONLY
- MEASURE RESISTANCE BETWEEN J 39700 TERMINALS '19' AND '18'
- IS IT LESS THAN 5 OHMS?

YES
- REPAIR SHORT TO SYSTEM VOLTAGE IN CKT 884 OR 885
- REFER TO EBTCM-SWITCHED FAULTS IN THIS SECTION

NO

(8)
- MEASURE RESISTANCE BETWEEN J 39700 TERMINAL '19' AND GROUND
- IS IT GREATER THAN 2 MEGAOHMS?

YES
- REPAIR SHORT BETWEEN CKT'S 884 AND 885

YES

(9)
- MEASURE RESISTANCE BETWEEN J 39700 TERMINAL '18' AND GROUND
- IS IT GREATER THAN 2 MEGAOHMS?

NO
- IF 5 OHMS OR LESS TO GROUND, REPAIR HARD SHORT IN HARNESS CKT 884
- IF BETWEEN 5 OHMS AND 2 MEGAOHMS, REPAIR RESISTIVE SHORT IN HARNESS CKT 884

YES
- CHECK ALL WIRING, CONNECTORS AND TERMINALS THOROUGHLY FOR INTERMITTENTS OR CORROSION.
- REFER TO 'INTERMITTENTS AND POOR CONNECTIONS'
- REPAIR HARNESS OR CONNECTOR
- REPLACE SENSOR IF PROBLEM PERSISTS

NO
- IF 5 OHMS OR LESS TO GROUND, REPAIR HARD SHORT IN HARNESS CKT 885
- IF BETWEEN 5 OHMS AND 2 MEGAOHMS REPAIR RESISTIVE SHORT IN HARNESS CKT 885

WHEN ALL DIAGNOSIS AND REPAIRS ARE COMPLETED, CLEAR CODES AND VERIFY OPERATION

DIAGNOSTIC TROUBLE CODE CHART DTC 37 — 1995 DEVILLE, CONCOURS, ELDORADO AND SEVILLE

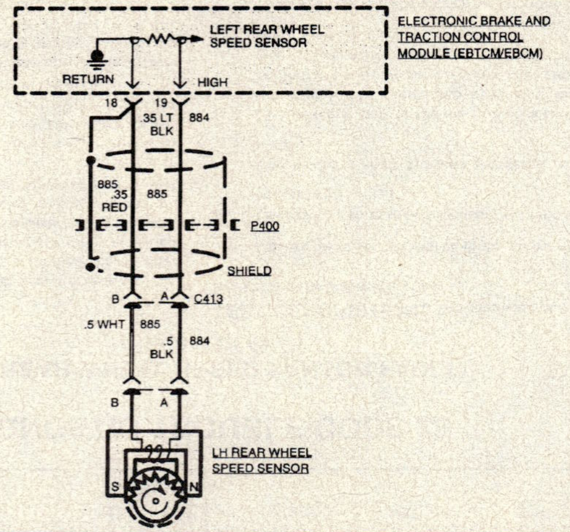

DIAGNOSTIC TROUBLE CODE 37

LEFT REAR WHEEL SPEED SENSOR CONTINUITY FAULT

CIRCUIT DESCRIPTION:

As each wheel turns, the wheel speed sensor for that wheel creates a small AC voltage as a toothed sensor ring, mounted in the front integral hub/bearing assembly, is passed by the stationary sensor. The sensor consists of a permanent magnet and coil and generates voltage by magnetic induction. The frequency of the wheel speed sensor signal is proportional to wheel speed.

The EBTCM uses the wheel speed sensor signal to calculate vehicle reference speeds and individual speed, acceleration and slip values for each wheel. These values are used to determine when antilock control is required.

The EBTCM will only set one wheel speed sensor fault code at a time, even if multiple faults exist. Detection is prioritized accordingly: LF, RR, LR, RF. Always check for additional stored codes after making repairs.

TEST CONDITION(S):

1. Tested continuously.

FAILURE CONDITION(S):

Continuity

1. While standing: Open circuit is seen while the vehicle is standing still with the ignition 'ON'

ACTION TAKEN:

1. Code set.
2. ABS/TCS switched 'OFF'.
3. 'ANTILOCK' indicator 'ON'.
4. 'TRACTION DISABLED' message displayed. (vehicles with TCS)

Code 37 is set if there is a short to voltage or an open in CKT 885 or 884, or faulty speed sensor. The testing for this fault occurs with the ignition 'ON' and the vehicle at rest.

NOTICE: A Code 35 may set if CKT 884 or 885 lose continuity and the WSS signal is lost while driving. A Code 37 should accompany this code on the subsequent ignition cycle. If both Code 35 and Code 37 are set concurrently, diagnose Code 37 first.

TEST DESCRIPTION: Test numbers refer to circled numbers on the diagnostic chart.

1. Checks for proper resistance in the sensor itself.
2. Checks for a short to battery in either of the speed sensor circuit wires. There should be 5V on CKT 884. If there is, CKT 884 has no hard opens or shorts. The condition may be intermittent.
3. Checks for an open in both the speed sensor circuit wires.
4. Manipulates wiring and connectors, trying to induce an intermittent fault not currently present.
5. This test checks if the code resets on key-up. If so, since Tests 1-4 have validated that the circuitry and components are good, the Speed Sensor may be causing the fault.
6. Checks if the Code 35 sets during a road test.
7. Code 37 sets when the vehicle is at rest: Code 35 sets with a problem in the speed sensor circuitry with the vehicle in motion. If problems are still present, Code 35 would set during the road test, not a 37. If Code 35 sets at this point, Code 35 diagnostics should be used.
8. The wheel speed sensor may be causing an intermittent fault. If Code 37 resets after sensor replacement, the EBTCM must be concluding there is a problem present in the speed sensor circuit when there is not.

DIAGNOSTIC AIDS:

Be sure the speed sensor wiring is properly routed and retained. This will help prevent false signals due to electrical noise being picked up by the wiring.

It is very important that a thorough inspection of the wiring and connectors be performed. Failure to carefully and fully inspect wiring and connectors may result in misdiagnosis, causing part replacement with reappearance of the fault.

DIAGNOSTIC TROUBLE CODE CHART DTC 37 — 1995 DEVILLE, CONCOURS, ELDORADO AND SEVILLE, CONTINUED

DIAGNOSTIC TROUBLE CODE 37

LR WHEEL SPEED SENSOR CONTINUITY FAULT

ENSURE SENSOR IS NOT VISIBLY DAMAGED; IF IT IS DAMAGED, REPLACE THE HUB/BEARING/SPEED SENSOR ASSEMBLY, CLEAR CODES, AN TURN IGNITION 'ON' TO SEE IF CODE RESETS.

(1)
- IGNITION 'OFF'
- DISCONNECT SENSOR CONNECTOR
- USING DVM AND ADAPTER FROM J 35616-A, CHECK SENSOR RESISTANCE BY PROBING SENSOR CONNECTOR TERMINALS
- IS IT 900 - 1300 OHMS AT 25°C (77°F)?

NO:
- REPLACE HUB/BEARING/SENSOR ASSEMBLY

YES:

(2)
- CONNECT DVM BETWEEN SENSOR HARNESS CONNECTOR TERMINAL 'A' AND GROUND (USED DC VOLTAGE SETTING)
- IGNITION 'ON'
- NOTE VOLTAGE: REPEAT BETWEEN TERMINAL 'B' AND GROUND
- IS IT AROUND SYSTEM VOLTAGE ON EITHER TERMINAL 'A' OR 'B'?

YES:
- REPAIR SHORT TO SYSTEM VOLTAGE IN CKT 884 OR 885
- REFER TO EBTCM-SWITCHED FAULTS IN THIS SECTION

NO:

(3)
- IGNITION 'OFF'
- DISCONNECT EBTCM
- INSTALL J 39700 UNIVERSAL BREAKOUT BOX TO EBTCM HARNESS CONNECTOR ONLY
- MEASURE RESISTANCE BETWEEN PIN 45 ON J 39700 TERMINAL 'A' OF SENSOR CONNECTOR ON HARNESS. PERFORM SAME MEASUREMENT BETWEEN TERMINAL 44 AND TERMINAL 'B'
- IS EITHER MEASUREMENT GREATER THAN 5 OHMS?

YES:
- REPAIR OPEN IN CKT 884 AND/OR 885

NO:

(4)
- REMOVE J 39700; RECONNECT EBTCM
- CLEAR CODES; IGNITION 'OFF'
- DISCONNECT TECH 1, IF CONNECTED
- IGNITION 'ON'
- WIGGLE ALL WIRING AND CONNECTORS TO TRY TO DUPLICATE AN INTERMITTENT FAULT
- DOES CODE 37 RESET?

YES:
- REPAIR WIRING AND/OR CONNECTIONS

NO:

(5)
- CLEAR CODES; IGNITION 'OFF'
- DISCONNECT TECH 1, IF CONNECTED
- IGNITION 'ON'
- DOES CODE 37 RESET?

NO:

(6)
- ROAD TEST VEHICLE, REACHING AT LEAST 25 KPH (15 MPH)
- DOES CODE 35 SET?

YES: (7) GO TO CODE 35 DIAGNOSTICS

NO:
- NO TROUBLE FOUND. SEE 'DIAGNOSTIC AIDS' ON FACING PAGE

YES:

(8)
- REPLACE HUB/BEARING/SENSOR ASSEMBLY
- CLEAR CODES; IGNITION 'OFF'
- DISCONNECT TECH 1, IF CONNECTED
- IGNITION 'ON'
- DOES CODE 37 RESET?

YES:
- REPLACE EBTCM

NO:
- SYSTEM OK

WHEN ALL DIAGNOSIS AND REPAIRS ARE COMPLETED, CLEAR CODES AND VERIFY OPERATION

DIAGNOSTIC TROUBLE CODE CHART DTC 41 — 1995 DEVILLE, CONCOURS, ELDORADO AND SEVILLE

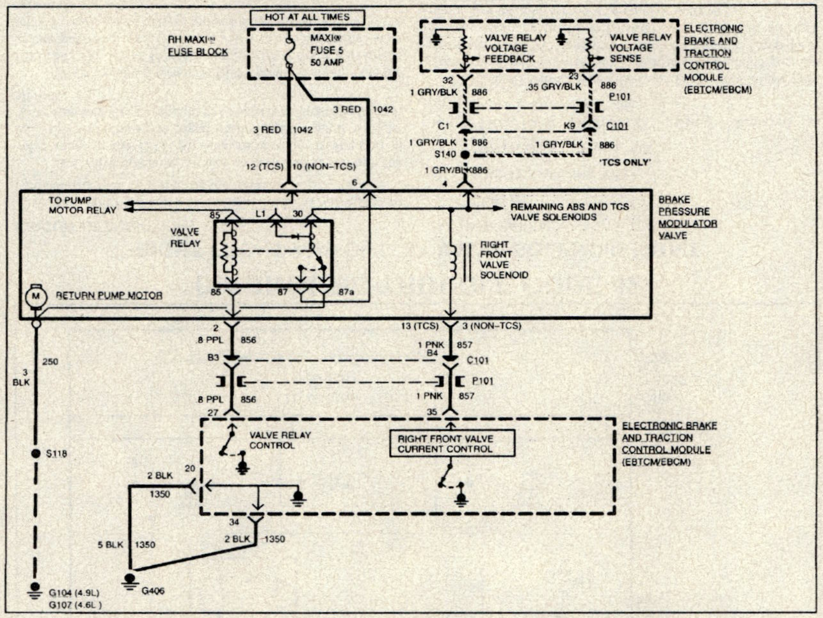

DIAGNOSTIC TROUBLE CODE 41

RIGHT FRONT ABS VALVE SOLENOID FAULT

CIRCUIT DESCRIPTION:

The right front valve solenoid in the BPM Valve may be commanded 'ON' or 'OFF' by the EBTCM.

The valve solenoid circuits receive power through the valve relay in the EBTCM. The valve relay is engaged at KEY-ON and remains engaged throughout the ignition cycle. Valve solenoid circuits should have battery voltage available at all times.

When the key is turned to RUN, power is supplied to the EBTCM. The EBTCM will complete an internal self-check before providing ground to the valve relay. When the valve relay is energized, battery power is supplied to the valve solenoids.

TEST CONDITION(S):

1. Tested continuously.

FAILURE CONDITION(S):

While standing or driving:

1. Valve solenoid cannot be activated.

OR

2. Valve solenoid is activated in a defective way.

ACTION TAKEN:

1. Code set.
2. ABS/TCS switched 'OFF'.
3. 'ANTILOCK' indicator 'ON'.
4. 'TRACTION DISABLED' message displayed on Driver Information Center.

Code 41 will set when the expected position of the right front valve solenoid does not match the commanded position from the EBTCM. Conditions which could cause Code 41 to set are: damage to the right front valve solenoid; an open circuit; or an open circuit to ground or battery.

DIAGNOSTIC AIDS:

- If measured resistance is less than 8.5 ohms or greater than 9.5 ohms, the Brake Pressure Modulator Valve (BPMV) is at fault.
- Before proceeding with any resistance tests, check the Electronic Brake and Traction Control Module (EBTCM) ground paths at terminals 1, 15 and G303 for clean, tight connections.

DIAGNOSTIC TROUBLE CODE CHART DTC 41 — 1995 DEVILLE, CONCOURS, ELDORADO AND SEVILLE, CONTINUED

DIAGNOSTIC TROUBLE CODE 41

RIGHT FRONT ABS INLET VALVE SOLENOID FAULT

NOTICE: J 39200 TEST LEADS MUST BE "ZEROED" PRIOR TO MAKING ANY RESISTANCE MEASUREMENTS. REFER TO J 39200 USER'S MANUAL.

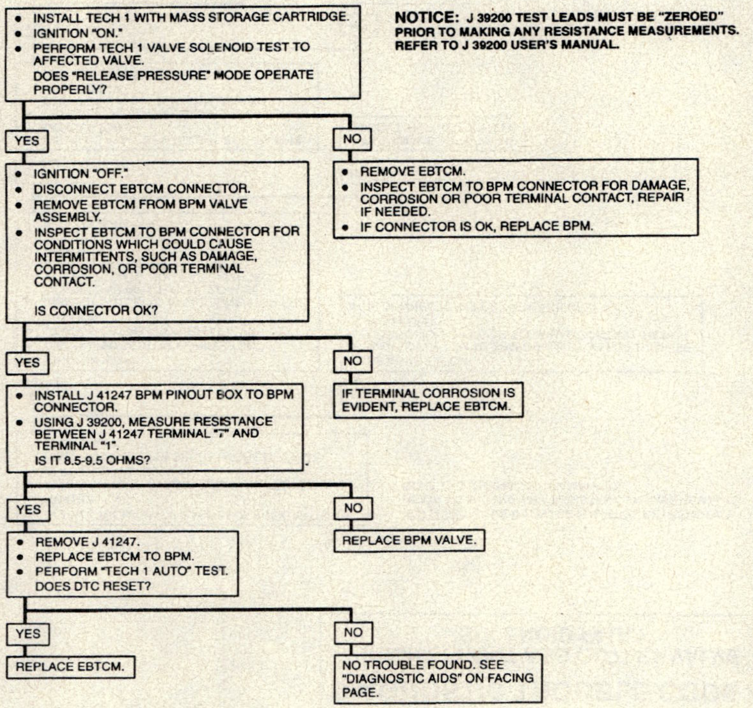

DIAGNOSTIC TROUBLE CODE CHART DTC 42 — 1995 DEVILLE, CONCOURS, ELDORADO AND SEVILLE

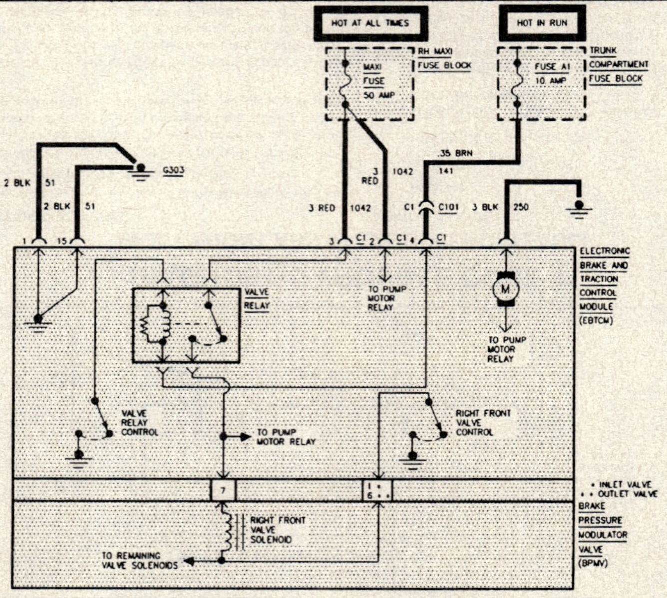

DIAGNOSTIC TROUBLE CODE 42

RIGHT FRONT ABS OUTLET VALVE SOLENOID FAULT

CIRCUIT DESCRIPTION:

The right front valve solenoid in the BPM Valve may be commanded 'ON' or 'OFF' by the EBTCM.

The valve solenoid circuits receive power through the valve relay in the EBTCM. The valve relay is engaged at KEY-ON and remains engaged throughout the ignition cycle. Valve solenoid circuits should have battery voltage available at all times.

When the key is turned to RUN, power is supplied to the EBTCM. The EBTCM will complete an internal self-check before providing ground to the valve relay. When the valve relay is energized, battery power is supplied to the valve solenoids.

TEST CONDITION(S):

1. Tested continuously.

FAILURE CONDITION(S):

While standing or driving:

1. Valve solenoid cannot be activated.

OR

2. Valve solenoid is activated in a defective way.

ACTION TAKEN:

1. Code set.
2. ABS/TCS switched 'OFF'.
3. 'ANTILOCK' indicator 'ON'.
4. 'TRACTION DISABLED' message displayed on Driver Information Center.

Code 41 will set when the expected position of the right front valve solenoid does not match the commanded position from the EBTCM. Conditions which could cause Code 41 to set are: damage to the right front valve solenoid; an open circuit; or an open circuit to ground or battery.

DIAGNOSTIC AIDS:

- If measured resistance is less than 4.5 ohms or greater than 5.0 ohms, the Brake Pressure Modulator Valve (BPMV) is at fault.
- Before proceeding with any resistance tests, check the Electronic Brake and Traction Control Module (EBTCM) ground paths at terminals 1, 15 and G303 for clean, tight connections.

DIAGNOSTIC TROUBLE CODE CHART DTC 42 — 1995 DEVILLE, CONCOURS, ELDORADO AND SEVILLE, CONTINUED

DIAGNOSTIC TROUBLE CODE 42

RIGHT FRONT ABS OUTLET VALVE SOLENOID FAULT

NOTICE: J 39200 TEST LEADS MUST BE "ZEROED" PRIOR TO MAKING ANY RESISTANCE MEASUREMENTS. REFER TO J 39200 USER'S MANUAL.

- INSTALL TECH 1 WITH MASS STORAGE CARTRIDGE.
- IGNITION "ON."
- PERFORM TECH 1 VALVE SOLENOID TEST TO AFFECTED VALVE.

DOES "RELEASE PRESSURE" MODE OPERATE PROPERLY?

NO:
- REMOVE EBTCM.
- INSPECT EBTCM TO BPM CONNECTOR FOR DAMAGE, CORROSION OR POOR TERMINAL CONTACT, REPAIR IF NEEDED.
- IF CONNECTOR IS OK, REPLACE BPM.

YES:
- IGNITION "OFF."
- DISCONNECT EBTCM CONNECTOR.
- REMOVE EBTCM FROM BPM VALVE ASSEMBLY.
- INSPECT EBTCM TO BPM CONNECTOR FOR CONDITIONS WHICH COULD CAUSE INTERMITTENTS, SUCH AS DAMAGE, CORROSION, OR POOR TERMINAL CONTACT.

IS CONNECTOR OK?

NO: IF TERMINAL CORROSION IS EVIDENT, REPLACE EBTCM.

YES:
- INSTALL J 41247 BPM PINOUT BOX TO BPM CONNECTOR.
- USING J 39200, MEASURE RESISTANCE BETWEEN J 41247 TERMINAL "7" AND TERMINAL "6".

IS IT 4.5-5.0 OHMS?

NO: REPLACE BPM VALVE.

YES:
- REMOVE J 41247.
- REPLACE EBTCM TO BPM.
- PERFORM "TECH 1 AUTO" TEST.

DOES DTC RESET?

YES: REPLACE EBTCM.

NO: NO TROUBLE FOUND. SEE "DIAGNOSTIC AIDS" ON FACING PAGE.

DIAGNOSTIC TROUBLE CODE CHART DTC 43 — 1995 DEVILLE, CONCOURS, ELDORADO AND SEVILLE

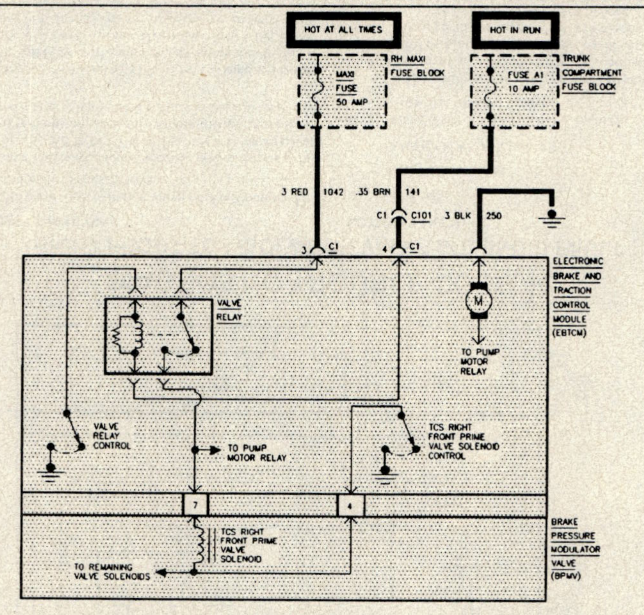

DIAGNOSTIC TROUBLE CODE 43

RIGHT FRONT TCS PRIME VALVE SOLENOID FAULT

CIRCUIT DESCRIPTION:

The right front valve solenoid in the BPM Valve may be commanded 'ON' or 'OFF' by the EBTCM.

The valve solenoid circuits receive power through the valve relay in the EBTCM. The valve relay is engaged at KEY-ON and remains engaged throughout the ignition cycle. Valve solenoid circuits should have battery voltage available at all times.

When the key is turned to RUN, power is supplied to the EBTCM. The EBTCM will complete an internal self-check before providing ground to the valve relay. When the valve relay is energized, battery power is supplied to the valve solenoids.

TEST CONDITION(S):

1. Tested continuously.

FAILURE CONDITION(S):

While standing or driving:

1. Valve solenoid cannot be activated.

OR

2. Valve solenoid is activated in a defective way.

ACTION TAKEN:

1. Code set.
2. ABS/TCS switched 'OFF'.
3. 'ANTILOCK' indicator 'ON'.
4. 'TRACTION DISABLED' message displayed on Driver Information Center.

Code 41 will set when the expected position of the right front valve solenoid does not match the commanded position from the EBTCM. Conditions which could cause Code 41 to set are: damage to the right front valve solenoid; an open circuit; or an open circuit to ground or battery.

DIAGNOSTIC AIDS:

- If measured resistance is less than 8.5 ohms or greater than 9.5 ohms, the Brake Pressure Modulator Valve (BPMV) is at fault.
- Before proceeding with any resistance tests, check the Electronic Brake and Traction Control Module (EBTCM) ground paths at terminals 1, 15 and G303 for clean, tight connections.

DIAGNOSTIC TROUBLE CODE CHART DTC 43 — 1995 DEVILLE, CONCOURS, ELDORADO AND SEVILLE, CONTINUED

DIAGNOSTIC TROUBLE CODE 43

RIGHT FRONT TCS PRIME VALVE SOLENOID FAULT

NOTICE: J 39200 TEST LEADS MUST BE "ZEROED" PRIOR TO MAKING ANY RESISTANCE MEASUREMENTS. REFER TO J 39200 USER'S MANUAL.

- INSTALL TECH 1 WITH MASS STORAGE CARTRIDGE.
- IGNITION "ON."
- PERFORM TECH 1 VALVE SOLENOID TEST TO AFFECTED VALVE.

DOES "RELEASE PRESSURE" MODE OPERATE PROPERLY?

NO:

- REMOVE EBTCM.
- INSPECT EBTCM TO BPM CONNECTOR FOR DAMAGE, CORROSION OR POOR TERMINAL CONTACT, REPAIR IF NEEDED.
- IF CONNECTOR IS OK, REPLACE BPM.

YES:

- IGNITION "OFF."
- DISCONNECT EBTCM CONNECTOR.
- REMOVE EBTCM FROM BPM VALVE ASSEMBLY.
- INSPECT EBTCM TO BPM CONNECTOR FOR CONDITIONS WHICH COULD CAUSE INTERMITTENTS, SUCH AS DAMAGE, CORROSION, OR POOR TERMINAL CONTACT.

IS CONNECTOR OK?

NO: IF TERMINAL CORROSION IS EVIDENT, REPLACE EBTCM.

YES:

- INSTALL J 41247 BPM PINOUT BOX TO BPM CONNECTOR.
- USING J 39200, MEASURE RESISTANCE BETWEEN J 41247 TERMINAL "7" AND TERMINAL "7".

IS IT 8.5-9.5 OHMS?

NO: REPLACE BPM VALVE.

YES:

- REMOVE J 41247.
- REPLACE EBTCM TO BPM.
- PERFORM "TECH 1 AUTO" TEST.

DOES DTC RESET?

YES: REPLACE EBTCM.

NO: NO TROUBLE FOUND. SEE "DIAGNOSTIC AIDS" ON FACING PAGE.

DIAGNOSTIC TROUBLE CODE CHART DTC 44 — 1995 DEVILLE, CONCOURS, ELDORADO AND SEVILLE

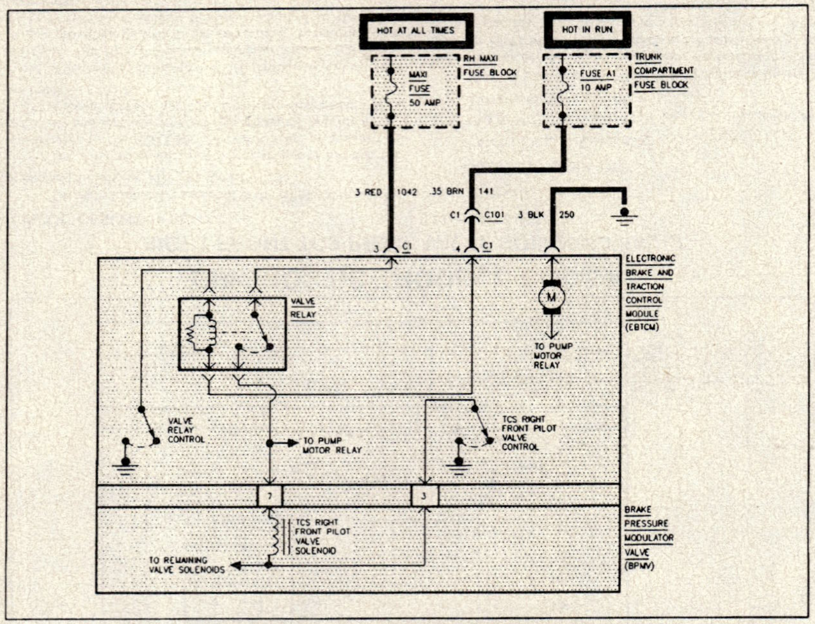

DIAGNOSTIC TROUBLE CODE 44

RIGHT FRONT TCS ISOLATION VALVE SOLENOID FAULT

CIRCUIT DESCRIPTION:

The right front TCS isolation valve may be commanded to close, isolating the master cylinder.

The valve solenoid circuits receive power through the valve relay in the EBTCM. The valve relay is engaged at KEY-ON and remains engaged throughout the ignition cycle. Valve solenoid circuits should have battery voltage available at all times.

When the key is turned to RUN, power is supplied to the EBTCM. The EBTCM will complete an internal self-check before providing ground to the valve relay. When the valve relay is energized, battery power is supplied to the valve solenoids.

TEST CONDITION(S):

1. Tested continuously.

FAILURE CONDITION(S):

While standing or driving:

1. Valve solenoid cannot be activated.

OR

2. Valve solenoid is activated in a defective way.

ACTION TAKEN:

1. Code set.
2. ABS/TCS switched 'OFF'.
3. 'ANTILOCK' indicator 'ON'.
4. 'TRACTION DISABLED' message displayed on Driver Information Center.

Code 41 will set when the expected position of the right front valve solenoid does not match the commanded position from the EBTCM. Conditions which could cause Code 41 to set are: damage to the right front valve solenoid; an open circuit; or an open circuit to ground or battery.

DIAGNOSTIC AIDS:

- If measured resistance is less than 8.5 ohms or greater than 9.5 ohms, the Brake Pressure Modulator Valve (BPMV) is at fault.
- Before proceeding with any resistance tests, check the Electronic Brake and Traction Control Module (EBTCM) ground paths at terminals 1, 15 and G303 for clean, tight connections.

DIAGNOSTIC TROUBLE CODE CHART DTC 44 — 1995 DEVILLE, CONCOURS, ELDORADO AND SEVILLE, CONTINUED

DIAGNOSTIC TROUBLE CODE 44

RIGHT FRONT TCS ISOLATION VALVE SOLENOID FAULT

NOTICE: J 39200 TEST LEADS MUST BE "ZEROED" PRIOR TO MAKING ANY RESISTANCE MEASUREMENTS. REFER TO J 39200 USER'S MANUAL.

- INSTALL TECH 1 WITH MASS STORAGE CARTRIDGE.
- IGNITION "ON."
- PERFORM TECH 1 VALVE SOLENOID TEST TO AFFECTED VALVE.

DOES "RELEASE PRESSURE" MODE OPERATE PROPERLY?

YES

- IGNITION "OFF."
- DISCONNECT EBTCM CONNECTOR.
- REMOVE EBTCM FROM BPM VALVE ASSEMBLY.
- INSPECT EBTCM TO BPM CONNECTOR FOR CONDITIONS WHICH COULD CAUSE INTERMITTENTS, SUCH AS DAMAGE, CORROSION, OR POOR TERMINAL CONTACT.

IS CONNECTOR OK?

NO

- REMOVE EBTCM.
- INSPECT EBTCM TO BPM CONNECTOR FOR DAMAGE, CORROSION OR POOR TERMINAL CONTACT, REPAIR IF NEEDED.
- IF CONNECTOR IS OK, REPLACE BPM.

(IS CONNECTOR OK?) **YES**

- INSTALL J 41247 BPM PINOUT BOX TO BPM CONNECTOR.
- USING J 39200, MEASURE RESISTANCE BETWEEN J 41247 TERMINAL "7" AND TERMINAL "3".

IS IT 8.5-9.5 OHMS?

(IS CONNECTOR OK?) **NO**

IF TERMINAL CORROSION IS EVIDENT, REPLACE EBTCM.

(IS IT 8.5-9.5 OHMS?) **YES**

- REMOVE J 41247.
- REPLACE EBTCM TO BPM.
- PERFORM "TECH 1 AUTO" TEST.

DOES DTC RESET?

(IS IT 8.5-9.5 OHMS?) **NO**

REPLACE BPM VALVE.

(DOES DTC RESET?) **YES**

REPLACE EBTCM.

(DOES DTC RESET?) **NO**

NO TROUBLE FOUND. SEE "DIAGNOSTIC AIDS" ON FACING PAGE.

DIAGNOSTIC TROUBLE CODE CHART DTC 45 — 1995 DEVILLE, CONCOURS, ELDORADO AND SEVILLE

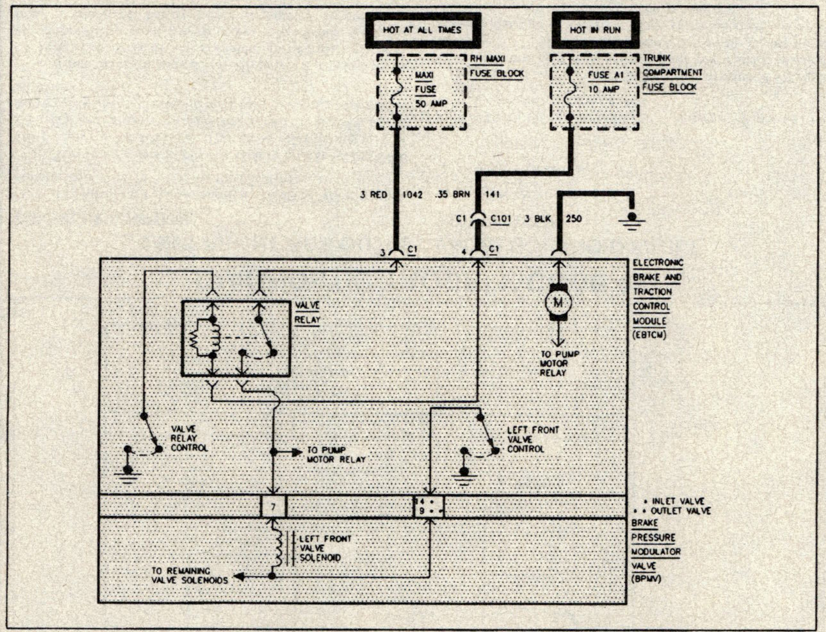

DIAGNOSTIC TROUBLE CODE 45

LEFT FRONT ABS INLET VALVE SOLENOID FAULT

CIRCUIT DESCRIPTION:

The left front valve solenoid in the BPM Valve may be commanded 'ON' or 'OFF' by the EBTCM.

The valve solenoid circuits receive power through the valve relay in the EBTCM. The valve relay is engaged at KEY-ON and remains engaged throughout the ignition cycle. Valve solenoid circuits should have battery voltage available at all times.

When the key is turned to RUN, power is supplied to the EBTCM. The EBTCM will complete an internal self-check before providing ground to the valve relay. When the valve relay is energized, battery power is supplied to the valve solenoids.

TEST CONDITION(S):

1. Tested continuously.

FAILURE CONDITION(S):

While standing or driving:

1. Valve solenoid cannot be activated.

OR

2. Valve solenoid is activated in a defective way.

ACTION TAKEN:

1. Code set.
2. ABS/TCS switched 'OFF'.
3. 'ANTILOCK' indicator 'ON'.
4. 'TRACTION DISABLED' message displayed on Driver Information Center.

Code 41 will set when the expected position of the right front valve solenoid does not match the commanded position from the EBTCM. Conditions which could cause Code 41 to set are: damage to the right front valve solenoid; an open circuit; or an open circuit to ground or battery.

DIAGNOSTIC AIDS:

- If measured resistance is less than 8.5 ohms or greater than 9.5 ohms, the Brake Pressure Modulator Valve (BPMV) is at fault.
- Before proceeding with any resistance tests, check the Electronic Brake and Traction Control Module (EBTCM) ground paths at terminals 1, 15 and G303 for clean, tight connections.

DIAGNOSTIC TROUBLE CODE CHART DTC 45 — 1995 DEVILLE, CONCOURS, ELDORADO AND SEVILLE, CONTINUED

DIAGNOSTIC TROUBLE CODE 45

LEFT FRONT ABS INLET VALVE SOLENOID FAULT

NOTICE: J 39200 TEST LEADS MUST BE "ZEROED" PRIOR TO MAKING ANY RESISTANCE MEASUREMENTS. REFER TO J 39200 USER'S MANUAL.

- INSTALL TECH 1 WITH MASS STORAGE CARTRIDGE.
- IGNITION "ON."
- PERFORM TECH 1 VALVE SOLENOID TEST TO AFFECTED VALVE.

DOES "RELEASE PRESSURE" MODE OPERATE PROPERLY?

NO:

- REMOVE EBTCM.
- INSPECT EBTCM TO BPM CONNECTOR FOR DAMAGE, CORROSION OR POOR TERMINAL CONTACT, REPAIR IF NEEDED.
- IF CONNECTOR IS OK, REPLACE BPM.

YES:

- IGNITION "OFF."
- DISCONNECT EBTCM CONNECTOR.
- REMOVE EBTCM FROM BPM VALVE ASSEMBLY.
- INSPECT EBTCM TO BPM CONNECTOR FOR CONDITIONS WHICH COULD CAUSE INTERMITTENTS, SUCH AS DAMAGE, CORROSION, OR POOR TERMINAL CONTACT.

IS CONNECTOR OK?

NO: IF TERMINAL CORROSION IS EVIDENT, REPLACE EBTCM.

YES:

- INSTALL J 41247 BPM PINOUT BOX TO BPM CONNECTOR.
- USING J 39200, MEASURE RESISTANCE BETWEEN J 41247 TERMINAL "7" AND TERMINAL "14".

IS IT 8.5-9.5 OHMS?

NO: REPLACE BPM VALVE.

YES:

- REMOVE J 41247.
- REPLACE EBTCM TO BPM.
- PERFORM "TECH 1 AUTO" TEST.

DOES DTC RESET?

YES: REPLACE EBTCM.

NO: NO TROUBLE FOUND. SEE "DIAGNOSTIC AIDS" ON FACING PAGE.

DIAGNOSTIC TROUBLE CODE CHART DTC 46 — 1995 DEVILLE, CONCOURS, ELDORADO AND SEVILLE

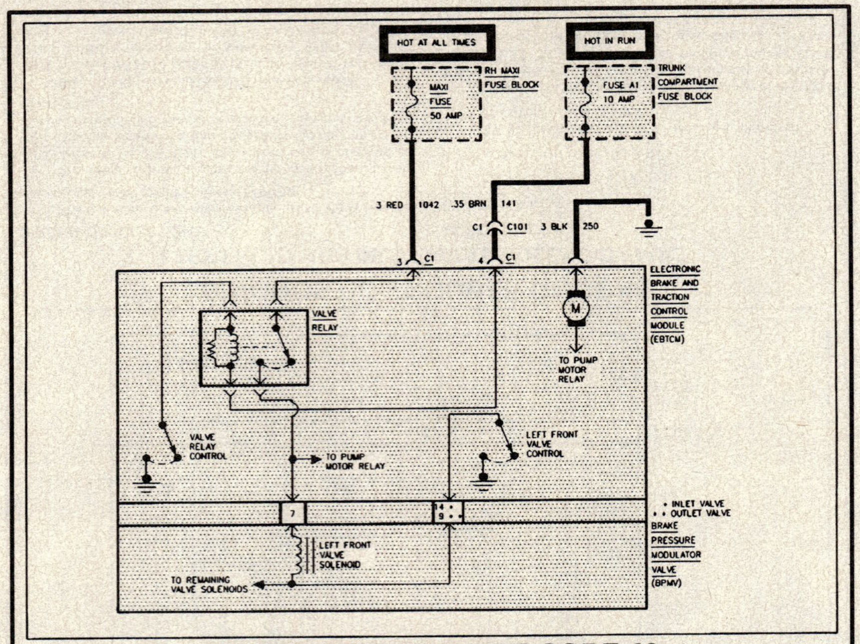

DIAGNOSTIC TROUBLE CODE 46

LEFT FRONT ABS OUTLET VALVE SOLENOID FAULT

CIRCUIT DESCRIPTION:

The left front valve solenoid in the BPM Valve may be commanded 'ON' or 'OFF' by the EBTCM.

The valve solenoid circuits receive power through the valve relay in the EBTCM. The valve relay is engaged at KEY-ON and remains engaged throughout the ignition cycle. Valve solenoid circuits should have battery voltage available at all times.

When the key is turned to RUN, power is supplied to the EBTCM. The EBTCM will complete an internal self-check before providing ground to the valve relay. When the valve relay is energized, battery power is supplied to the valve solenoids.

TEST CONDITION(S):

1. Tested continuously.

FAILURE CONDITION(S):

While standing or driving:

1. Valve solenoid cannot be activated.

OR

2. Valve solenoid is activated in a defective way.

ACTION TAKEN:

1. Code set.
2. ABS/TCS switched 'OFF'.
3. 'ANTILOCK' indicator 'ON'.
4. 'TRACTION DISABLED' message displayed on Driver Information Center.

Code 41 will set when the expected position of the right front valve solenoid does not match the commanded position from the EBTCM. Conditions which could cause Code 41 to set are: damage to the right front valve solenoid; an open circuit; or an open circuit to ground or battery.

DIAGNOSTIC AIDS:

- If measured resistance is less than 4.5 ohms <u>or</u> greater than 5.0 ohms, the Brake Pressure Modulator Valve (BPMV) is at fault.
- Before proceeding with any resistance tests, check the Electronic Brake and Traction Control Module (EBTCM) ground paths at terminals 1, 15 and G303 for clean, tight connections.

DIAGNOSTIC TROUBLE CODE CHART DTC 46 — 1995 DEVILLE, CONCOURS, ELDORADO AND SEVILLE, CONTINUED

DIAGNOSTIC TROUBLE CODE 46

LEFT FRONT ABS OUTLET VALVE SOLENOID FAULT

NOTICE: J 39200 TEST LEADS MUST BE "ZEROED" PRIOR TO MAKING ANY RESISTANCE MEASUREMENTS. REFER TO J 39200 USER'S MANUAL.

- INSTALL TECH 1 WITH MASS STORAGE CARTRIDGE.
- IGNITION "ON."
- PERFORM TECH 1 VALVE SOLENOID TEST TO AFFECTED VALVE.

DOES "RELEASE PRESSURE" MODE OPERATE PROPERLY?

NO:

- REMOVE EBTCM.
- INSPECT EBTCM TO BPM CONNECTOR FOR DAMAGE, CORROSION OR POOR TERMINAL CONTACT, REPAIR IF NEEDED.
- IF CONNECTOR IS OK, REPLACE BPM.

YES:

- IGNITION "OFF."
- DISCONNECT EBTCM CONNECTOR.
- REMOVE EBTCM FROM BPM VALVE ASSEMBLY.
- INSPECT EBTCM TO BPM CONNECTOR FOR CONDITIONS WHICH COULD CAUSE INTERMITTENTS, SUCH AS DAMAGE, CORROSION, OR POOR TERMINAL CONTACT.

IS CONNECTOR OK?

NO: IF TERMINAL CORROSION IS EVIDENT, REPLACE EBTCM.

YES:

- INSTALL J 41247 BPM PINOUT BOX TO BPM CONNECTOR.
- USING J 39200, MEASURE RESISTANCE BETWEEN J 41247 TERMINAL "7" AND TERMINAL "7".

IS IT 4.5-5.0 OHMS?

NO: REPLACE BPM VALVE.

YES:

- REMOVE J 41247.
- REPLACE EBTCM TO BPM.
- PERFORM "TECH 1 AUTO" TEST.

DOES DTC RESET?

YES: REPLACE EBTCM.

NO: NO TROUBLE FOUND. SEE "DIAGNOSTIC AIDS" ON FACING PAGE.

DIAGNOSTIC TROUBLE CODE CHART DTC 47 — 1995 DEVILLE, CONCOURS, ELDORADO AND SEVILLE

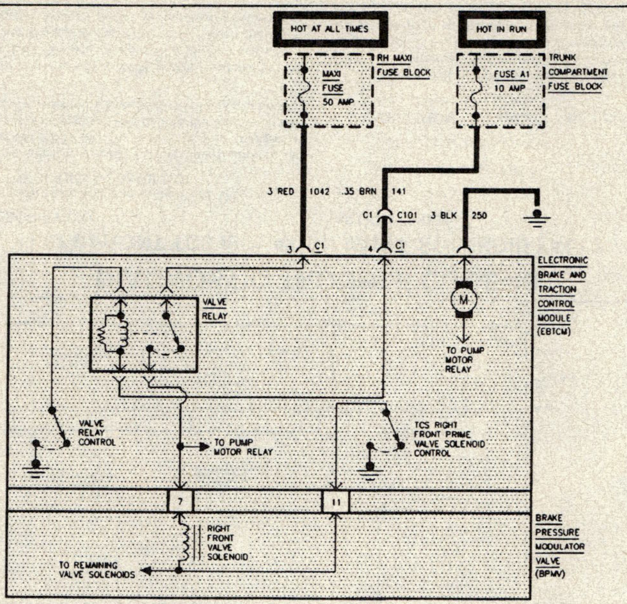

DIAGNOSTIC TROUBLE CODE 47

LEFT FRONT TCS PRIME VALVE SOLENOID FAULT

CIRCUIT DESCRIPTION:

The left front valve solenoid in the BPM Valve may be commanded 'ON' or 'OFF' by the EBTCM.

The valve solenoid circuits receive power through the valve relay in the EBTCM. The valve relay is engaged at KEY-ON and remains engaged throughout the ignition cycle. Valve solenoid circuits should have battery voltage available at all times.

When the key is turned to RUN, power is supplied to the EBTCM. The EBTCM will complete an internal self-check before providing ground to the valve relay. When the valve relay is energized, battery power is supplied to the valve solenoids.

TEST CONDITION(S):

1. Tested continuously.

FAILURE CONDITION(S):

While standing or driving:

1. Valve solenoid cannot be activated.

OR

2. Valve solenoid is activated in a defective way.

ACTION TAKEN:

1. Code set.
2. ABS/TCS switched 'OFF'.
3. 'ANTILOCK' indicator 'ON'.
4. 'TRACTION DISABLED' message displayed on Driver Information Center.

Code 41 will set when the expected position of the right front valve solenoid does not match the commanded position from the EBTCM. Conditions which could cause Code 41 to set are: damage to the right front valve solenoid; an open circuit; or an open circuit to ground or battery.

DIAGNOSTIC AIDS:

- If measured resistance is less than 8.5 ohms or greater than 9.5 ohms, the Brake Pressure Modulator Valve (BPMV) is at fault.
- Before proceeding with any resistance tests, check the Electronic Brake and Traction Control Module (EBTCM) ground paths at terminals 1, 15 and G303 for clean, tight connections.

DIAGNOSTIC TROUBLE CODE CHART DTC 47 — 1995 DEVILLE, CONCOURS, ELDORADO AND SEVILLE, CONTINUED

DIAGNOSTIC TROUBLE CODE 47

LEFT FRONT TCS PRIME VALVE SOLENOID FAULT

NOTICE: J 39200 TEST LEADS MUST BE "ZEROED" PRIOR TO MAKING ANY RESISTANCE MEASUREMENTS. REFER TO J 39200 USER'S MANUAL.

- INSTALL TECH 1 WITH MASS STORAGE CARTRIDGE.
- IGNITION "ON."
- PERFORM TECH 1 VALVE SOLENOID TEST TO AFFECTED VALVE.

DOES "RELEASE PRESSURE" MODE OPERATE PROPERLY?

NO:
- REMOVE EBTCM.
- INSPECT EBTCM TO BPM CONNECTOR FOR DAMAGE, CORROSION OR POOR TERMINAL CONTACT, REPAIR IF NEEDED.
- IF CONNECTOR IS OK, REPLACE BPM.

YES:
- IGNITION "OFF."
- DISCONNECT EBTCM CONNECTOR.
- REMOVE EBTCM FROM BPM VALVE ASSEMBLY.
- INSPECT EBTCM TO BPM CONNECTOR FOR CONDITIONS WHICH COULD CAUSE INTERMITTENTS, SUCH AS DAMAGE, CORROSION, OR POOR TERMINAL CONTACT.

IS CONNECTOR OK?

NO: IF TERMINAL CORROSION IS EVIDENT, REPLACE EBTCM.

YES:
- INSTALL J 41247 BPM PINOUT BOX TO BPM CONNECTOR.
- USING J 39200, MEASURE RESISTANCE BETWEEN J 41247 TERMINAL "7" AND TERMINAL "11".

IS IT 8.5-9.5 OHMS?

NO: REPLACE BPM VALVE.

YES:
- REMOVE J 41247.
- REPLACE EBTCM TO BPM.
- PERFORM "TECH 1 AUTO" TEST.

DOES DTC RESET?

YES: REPLACE EBTCM.

NO: NO TROUBLE FOUND. SEE "DIAGNOSTIC AIDS" ON FACING PAGE.

DIAGNOSTIC TROUBLE CODE CHART DTC 48 — 1995 DEVILLE, CONCOURS, ELDORADO AND SEVILLE

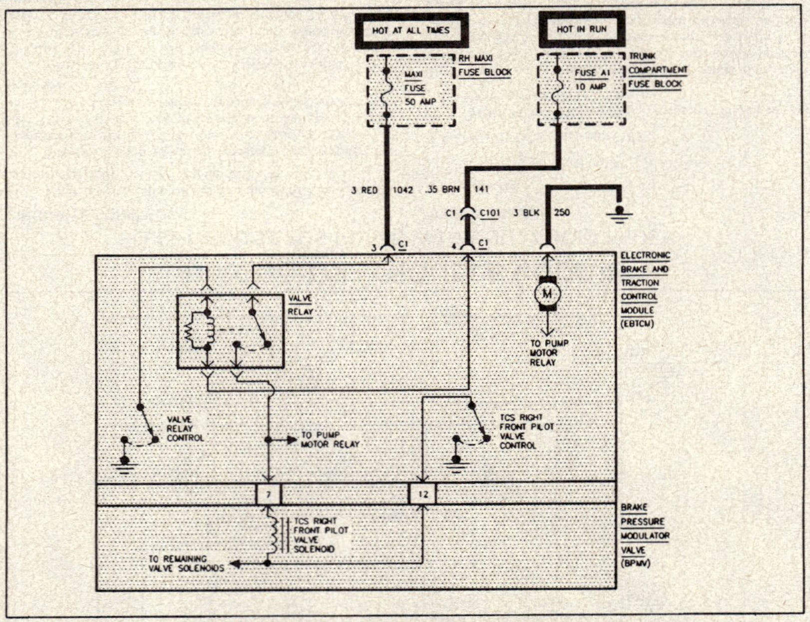

DIAGNOSTIC TROUBLE CODE 48

LEFT FRONT TCS ISOLATION VALVE SOLENOID FAULT

CIRCUIT DESCRIPTION:

The left front valve solenoid in the BPM Valve may be commanded 'ON' or 'OFF' by the EBTCM.

The valve solenoid circuits receive power through the valve relay in the EBTCM. The valve relay is engaged at KEY-ON and remains engaged throughout the ignition cycle. Valve solenoid circuits should have battery voltage available at all times.

When the key is turned to RUN, power is supplied to the EBTCM. The EBTCM will complete an internal self-check before providing ground to the valve relay. When the valve relay is energized, battery power is supplied to the valve solenoids.

TEST CONDITION(S):

1. Tested continuously.

FAILURE CONDITION(S):

While standing or driving:

1. Valve solenoid cannot be activated.

OR

2. Valve solenoid is activated in a defective way.

ACTION TAKEN:

1. Code set.
2. ABS/TCS switched 'OFF'.
3. 'ANTILOCK' indicator 'ON'.
4. 'TRACTION DISABLED' message displayed on Driver Information Center.

Code 41 will set when the expected position of the right front valve solenoid does not match the commanded position from the EBTCM. Conditions which could cause Code 41 to set are: damage to the right front valve solenoid; an open circuit; or an open circuit to ground or battery.

DIAGNOSTIC AIDS:

- If measured resistance is less than 8.5 ohms or greater than 9.5 ohms, the Brake Pressure Modulator Valve (BPMV) is at fault.
- Before proceeding with any resistance tests, check the Electronic Brake and Traction Control Module (EBTCM) ground paths at terminals 1, 15 and G303 for clean, tight connections.

DIAGNOSTIC TROUBLE CODE CHART DTC 48 — 1995 DEVILLE, CONCOURS, ELDORADO AND SEVILLE, CONTINUED

DIAGNOSTIC TROUBLE CODE 48

LEFT FRONT TCS ISOLATION VALVE SOLENOID FAULT

NOTICE: J 39200 TEST LEADS MUST BE "ZEROED" PRIOR TO MAKING ANY RESISTANCE MEASUREMENTS. REFER TO J 39200 USER'S MANUAL.

- INSTALL TECH 1 WITH MASS STORAGE CARTRIDGE.
- IGNITION "ON."
- PERFORM TECH 1 VALVE SOLENOID TEST TO AFFECTED VALVE.

DOES "RELEASE PRESSURE" MODE OPERATE PROPERLY?

NO:

- REMOVE EBTCM.
- INSPECT EBTCM TO BPM CONNECTOR FOR DAMAGE, CORROSION OR POOR TERMINAL CONTACT, REPAIR IF NEEDED.
- IF CONNECTOR IS OK, REPLACE BPM.

YES:

- IGNITION "OFF."
- DISCONNECT EBTCM CONNECTOR.
- REMOVE EBTCM FROM BPM VALVE ASSEMBLY.
- INSPECT EBTCM TO BPM CONNECTOR FOR CONDITIONS WHICH COULD CAUSE INTERMITTENTS, SUCH AS DAMAGE, CORROSION, OR POOR TERMINAL CONTACT.

IS CONNECTOR OK?

NO: IF TERMINAL CORROSION IS EVIDENT, REPLACE EBTCM.

YES:

- INSTALL J 41247 BPM PINOUT BOX TO BPM CONNECTOR.
- USING J 39200, MEASURE RESISTANCE BETWEEN J 41247 TERMINAL "7" AND TERMINAL "12".

IS IT 8.5-9.5 OHMS?

NO: REPLACE BPM VALVE.

YES:

- REMOVE J 41247.
- REPLACE EBTCM TO BPM.
- PERFORM "TECH 1 AUTO" TEST.

DOES DTC RESET?

YES: REPLACE EBTCM.

NO: NO TROUBLE FOUND. SEE "DIAGNOSTIC AIDS" ON FACING PAGE.

DIAGNOSTIC TROUBLE CODE CHART DTC 51 — 1995 DEVILLE, CONCOURS, ELDORADO AND SEVILLE

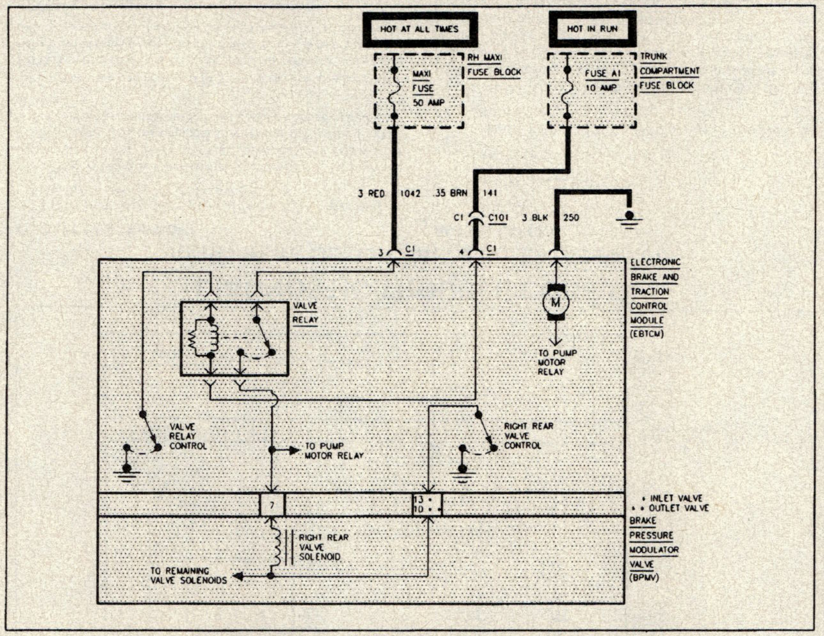

DIAGNOSTIC TROUBLE CODE 51

RIGHT REAR ABS INLET VALVE SOLENOID FAULT

CIRCUIT DESCRIPTION:

The right rear valve solenoid in the BPM Valve may be commanded 'ON' or 'OFF' by the EBTCM.

The valve solenoid circuits receive power through the valve relay in the EBTCM. The valve relay is engaged at KEY-ON and remains engaged throughout the ignition cycle. Valve solenoid circuits should have battery voltage available at all times.

When the key is turned to RUN, power is supplied to the EBTCM. The EBTCM will complete an internal self-check before providing ground to the valve relay. When the valve relay is energized, battery power is supplied to the valve solenoids.

TEST CONDITION(S):

1. Tested continuously.

FAILURE CONDITION(S):

While standing or driving:

1. Valve solenoid cannot be activated.

OR

2. Valve solenoid is activated in a defective way.

ACTION TAKEN:

1. Code set.
2. ABS/TCS switched 'OFF'.
3. 'ANTILOCK' indicator 'ON'.
4. 'TRACTION DISABLED' message displayed on Driver Information Center.

Code 41 will set when the expected position of the right front valve solenoid does not match the commanded position from the EBTCM. Conditions which could cause Code 41 to set are: damage to the right front valve solenoid; an open circuit; or an open circuit to ground or battery.

DIAGNOSTIC AIDS:

- If measured resistance is less than 8.5 ohms or greater than 9.5 ohms, the Brake Pressure Modulator Valve (BPMV) is at fault.
- Before proceeding with any resistance tests, check the Electronic Brake and Traction Control Module (EBTCM) ground paths at terminals 1, 15 and G303 for clean, tight connections.

DIAGNOSTIC TROUBLE CODE CHART DTC 51 — 1995 DEVILLE, CONCOURS, ELDORADO AND SEVILLE, CONTINUED

DIAGNOSTIC TROUBLE CODE 51

RIGHT REAR ABS INLET VALVE SOLENOID FAULT

NOTICE: J 39200 TEST LEADS MUST BE "ZEROED" PRIOR TO MAKING ANY RESISTANCE MEASUREMENTS. REFER TO J 39200 USER'S MANUAL.

- INSTALL TECH 1 WITH MASS STORAGE CARTRIDGE.
- IGNITION "ON."
- PERFORM TECH 1 VALVE SOLENOID TEST TO AFFECTED VALVE.

DOES "RELEASE PRESSURE" MODE OPERATE PROPERLY?

NO:
- REMOVE EBTCM.
- INSPECT EBTCM TO BPM CONNECTOR FOR DAMAGE, CORROSION OR POOR TERMINAL CONTACT, REPAIR IF NEEDED.
- IF CONNECTOR IS OK, REPLACE BPM.

YES:
- IGNITION "OFF."
- DISCONNECT EBTCM CONNECTOR.
- REMOVE EBTCM FROM BPM VALVE ASSEMBLY.
- INSPECT EBTCM TO BPM CONNECTOR FOR CONDITIONS WHICH COULD CAUSE INTERMITTENTS, SUCH AS DAMAGE, CORROSION, OR POOR TERMINAL CONTACT.

IS CONNECTOR OK?

NO: IF TERMINAL CORROSION IS EVIDENT, REPLACE EBTCM.

YES:
- INSTALL J 41247 BPM PINOUT BOX TO BPM CONNECTOR.
- USING J 39200, MEASURE RESISTANCE BETWEEN J 41247 TERMINAL "7" AND TERMINAL "3".

IS IT 8.5-9.5 OHMS?

NO: REPLACE BPM VALVE.

YES:
- REMOVE J 41247.
- REPLACE EBTCM TO BPM.
- PERFORM "TECH 1 AUTO" TEST.

DOES DTC RESET?

YES: REPLACE EBTCM.

NO: NO TROUBLE FOUND. SEE "DIAGNOSTIC AIDS" ON FACING PAGE.

DIAGNOSTIC TROUBLE CODE CHART DTC 52 — 1995 DEVILLE, CONCOURS, ELDORADO AND SEVILLE

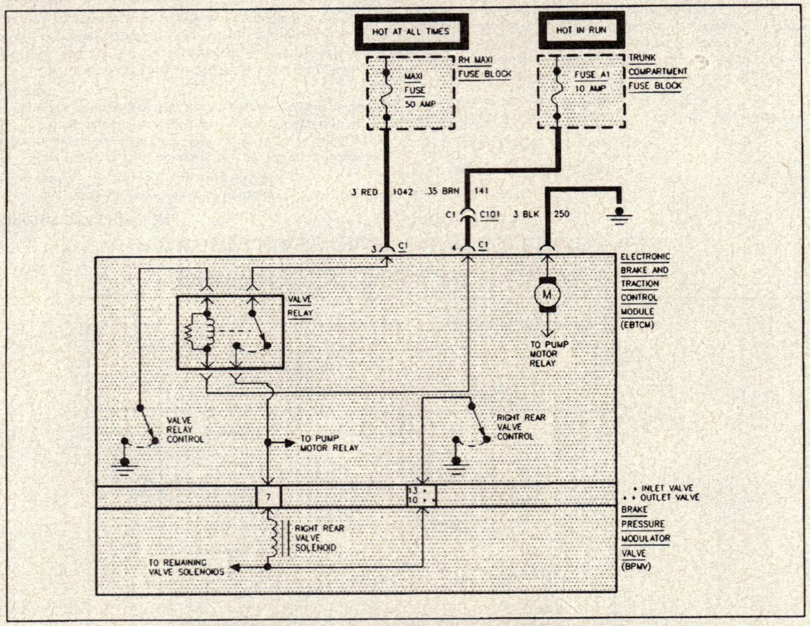

DIAGNOSTIC TROUBLE CODE 52

RIGHT REAR OUTLET VALVE SOLENOID FAULT

CIRCUIT DESCRIPTION:

The right rear valve solenoid in the BPM Valve may be commanded 'ON' or 'OFF' by the EBTCM.

The valve solenoid circuits receive power through the valve relay in the EBTCM. The valve relay is engaged at KEY-ON and remains engaged throughout the ignition cycle. Valve solenoid circuits should have battery voltage available at all times.

When the key is turned to RUN, power is supplied to the EBTCM. The EBTCM will complete an internal self-check before providing ground to the valve relay. When the valve relay is energized, battery power is supplied to the valve solenoids.

TEST CONDITION(S):

1. Tested continuously.

FAILURE CONDITION(S):

While standing or driving:

1. Valve solenoid cannot be activated.

OR

2. Valve solenoid is activated in a defective way.

ACTION TAKEN:

1. Code set.
2. ABS/TCS switched 'OFF'.
3. 'ANTILOCK' indicator 'ON'.
4. 'TRACTION DISABLED' message displayed on Driver Information Center.

Code 41 will set when the expected position of the right front valve solenoid does not match the commanded position from the EBTCM. Conditions which could cause Code 41 to set are: damage to the right front valve solenoid; an open circuit; or an open circuit to ground or battery.

DIAGNOSTIC AIDS:

- If measured resistance is less than 4.5 ohms or greater than 5.0 ohms, the Brake Pressure Modulator Valve (BPMV) is at fault.
- Before proceeding with any resistance tests, check the Electronic Brake and Traction Control Module (EBTCM) ground paths at terminals 1, 15 and G303 for clean, tight connections.

DIAGNOSTIC TROUBLE CODE CHART DTC 52 — 1995 DEVILLE, CONCOURS, ELDORADO AND SEVILLE, CONTINUED

DIAGNOSTIC TROUBLE CODE 52

RIGHT REAR OUTLET VALVE SOLENOID FAULT

NOTICE: J 39200 TEST LEADS MUST BE "ZEROED" PRIOR TO MAKING ANY RESISTANCE MEASUREMENTS. REFER TO J 39200 USER'S MANUAL.

- INSTALL TECH 1 WITH MASS STORAGE CARTRIDGE.
- IGNITION "ON."
- PERFORM TECH 1 VALVE SOLENOID TEST TO AFFECTED VALVE.

DOES "RELEASE PRESSURE" MODE OPERATE PROPERLY?

NO:

- REMOVE EBTCM.
- INSPECT EBTCM TO BPM CONNECTOR FOR DAMAGE, CORROSION OR POOR TERMINAL CONTACT, REPAIR IF NEEDED.
- IF CONNECTOR IS OK, REPLACE BPM.

YES:

- IGNITION "OFF."
- DISCONNECT EBTCM CONNECTOR.
- REMOVE EBTCM FROM BPM VALVE ASSEMBLY.
- INSPECT EBTCM TO BPM CONNECTOR FOR CONDITIONS WHICH COULD CAUSE INTERMITTENTS, SUCH AS DAMAGE, CORROSION, OR POOR TERMINAL CONTACT.

IS CONNECTOR OK?

NO: IF TERMINAL CORROSION IS EVIDENT, REPLACE EBTCM.

YES:

- INSTALL J 41247 BPM PINOUT BOX TO BPM CONNECTOR.
- USING J 39200, MEASURE RESISTANCE BETWEEN J 41247 TERMINAL "7" AND TERMINAL "10".

IS IT 4.5-5.0 OHMS?

NO: REPLACE BPM VALVE.

YES:

- REMOVE J 41247.
- REPLACE EBTCM TO BPM..
- PERFORM "TECH 1 AUTO" TEST.

DOES DTC RESET?

YES: REPLACE EBTCM.

NO: NO TROUBLE FOUND. SEE "DIAGNOSTIC AIDS" ON FACING PAGE.

DIAGNOSTIC TROUBLE CODE CHART DTC 55 — 1995 DEVILLE, CONCOURS, ELDORADO AND SEVILLE

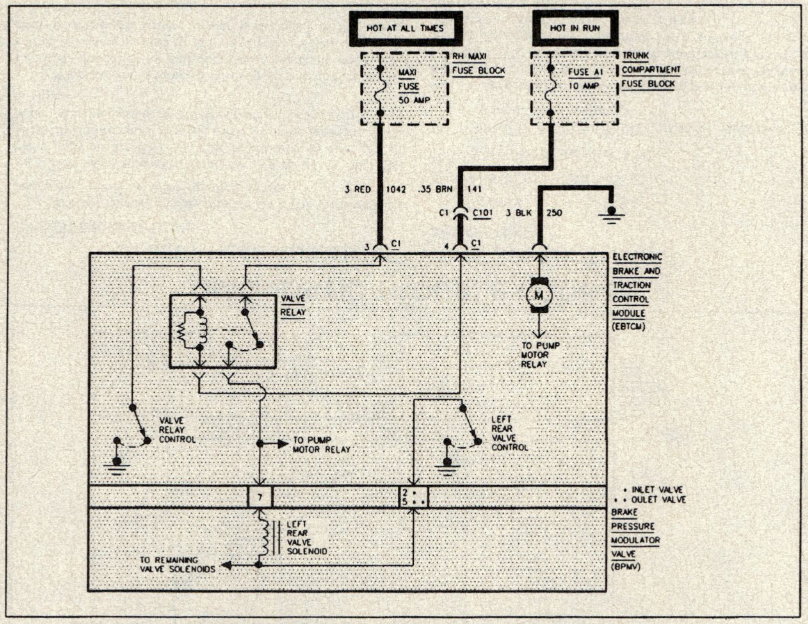

DIAGNOSTIC TROUBLE CODE 55

LEFT REAR ABS INLET VALVE SOLENOID FAULT

CIRCUIT DESCRIPTION:

The left rear valve solenoid in the BPM Valve may be commanded 'ON' or 'OFF' by the EBTCM.

The valve solenoid circuits receive power through the valve relay in the EBTCM. The valve relay is engaged at KEY-ON and remains engaged throughout the ignition cycle. Valve solenoid circuits should have battery voltage available at all times.

When the key is turned to RUN, power is supplied to the EBTCM. The EBTCM will complete an internal self-check before providing ground to the valve relay. When the valve relay is energized, battery power is supplied to the valve solenoids.

TEST CONDITION(S):

1. Tested continuously.

FAILURE CONDITION(S):

While standing or driving:

1. Valve solenoid cannot be activated.

OR

2. Valve solenoid is activated in a defective way.

ACTION TAKEN:

1. Code set.
2. ABS/TCS switched 'OFF'.
3. 'ANTILOCK' indicator 'ON'.
4. 'TRACTION DISABLED' message displayed on Driver Information Center.

Code 41 will set when the expected position of the right front valve solenoid does not match the commanded position from the EBTCM. Conditions which could cause Code 41 to set are: damage to the right front valve solenoid; an open circuit; or an open circuit to ground or battery.

DIAGNOSTIC AIDS:

- If measured resistance is less than 8.5 ohms or greater than 9.5 ohms, the Brake Pressure Modulator Valve (BPMV) is at fault.
- Before proceeding with any resistance tests, check the Electronic Brake and Traction Control Module (EBTCM) ground paths at terminals 1, 15 and G303 for clean, tight connections.

DIAGNOSTIC TROUBLE CODE CHART DTC 55 — 1995 DEVILLE, CONCOURS, ELDORADO AND SEVILLE, CONTINUED

DIAGNOSTIC TROUBLE CODE 55

LEFT REAR ABS INLET VALVE SOLENOID FAULT

NOTICE: J 39200 TEST LEADS MUST BE "ZEROED" PRIOR TO MAKING ANY RESISTANCE MEASUREMENTS. REFER TO J 39200 USER'S MANUAL.

- INSTALL TECH 1 WITH MASS STORAGE CARTRIDGE.
- IGNITION "ON."
- PERFORM TECH 1 VALVE SOLENOID TEST TO AFFECTED VALVE.

DOES "RELEASE PRESSURE" MODE OPERATE PROPERLY?

NO:
- REMOVE EBTCM.
- INSPECT EBTCM TO BPM CONNECTOR FOR DAMAGE, CORROSION OR POOR TERMINAL CONTACT, REPAIR IF NEEDED.
- IF CONNECTOR IS OK, REPLACE BPM.

YES:
- IGNITION "OFF."
- DISCONNECT EBTCM CONNECTOR.
- REMOVE EBTCM FROM BPM VALVE ASSEMBLY.
- INSPECT EBTCM TO BPM CONNECTOR FOR CONDITIONS WHICH COULD CAUSE INTERMITTENTS, SUCH AS DAMAGE, CORROSION, OR POOR TERMINAL CONTACT.

IS CONNECTOR OK?

NO: IF TERMINAL CORROSION IS EVIDENT, REPLACE EBTCM.

YES:
- INSTALL J 41247 BPM PINOUT BOX TO BPM CONNECTOR.
- USING J 39200, MEASURE RESISTANCE BETWEEN J 41247 TERMINAL "7" AND TERMINAL "2".

IS IT 8.5-9.5 OHMS?

NO: REPLACE BPM VALVE.

YES:
- REMOVE J 41247.
- REPLACE EBTCM TO BPM.
- PERFORM "TECH 1 AUTO" TEST.

DOES DTC RESET?

YES: REPLACE EBTCM.

NO: NO TROUBLE FOUND. SEE "DIAGNOSTIC AIDS" ON FACING PAGE.

DIAGNOSTIC TROUBLE CODE CHART DTC 56 — 1995 DEVILLE, CONCOURS, ELDORADO AND SEVILLE

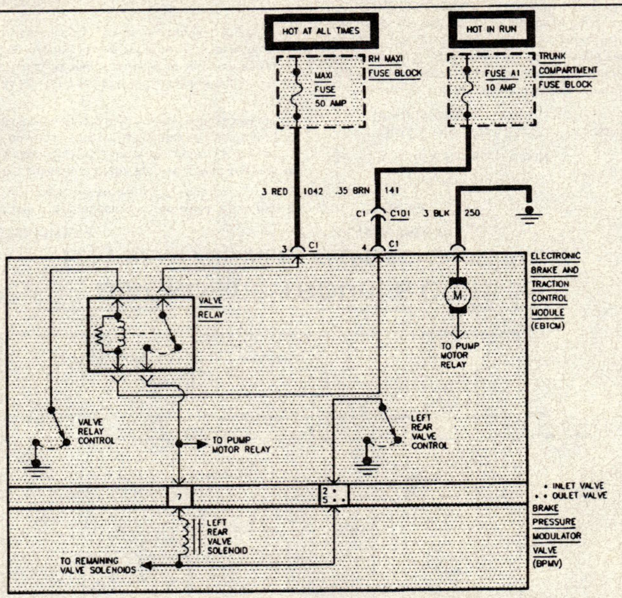

DIAGNOSTIC TROUBLE CODE 56

LEFT REAR ABS OUTLET VALVE SOLENOID FAULT

CIRCUIT DESCRIPTION:

The right front valve solenoid in the BPM Valve may be commanded 'ON' or 'OFF' by the EBTCM.

The valve solenoid circuits receive power through the valve relay in the EBTCM. The valve relay is engaged at KEY-ON and remains engaged throughout the ignition cycle. Valve solenoid circuits should have battery voltage available at all times.

When the key is turned to RUN, power is supplied to the EBTCM. The EBTCM will complete an internal self-check before providing ground to the valve relay. When the valve relay is energized, battery power is supplied to the valve solenoids.

TEST CONDITION(S):

1. Tested continuously.

FAILURE CONDITION(S):

While standing or driving:

1. Valve solenoid cannot be activated.

OR

2. Valve solenoid is activated in a defective way.

ACTION TAKEN:

1. Code set.
2. ABS/TCS switched 'OFF'.
3. 'ANTILOCK' indicator 'ON'.
4. 'TRACTION DISABLED' message displayed on Driver Information Center.

Code 41 will set when the expected position of the right front valve solenoid does not match the commanded position from the EBTCM. Conditions which could cause Code 41 to set are: damage to the right front valve solenoid; an open circuit; or an open circuit to ground or battery.

DIAGNOSTIC AIDS:

- If measured resistance is less than 4.5 ohms or greater than 5.0 ohms, the Brake Pressure Modulator Valve (BPMV) is at fault.
- Before proceeding with any resistance tests, check the Electronic Brake and Traction Control Module (EBTCM) ground paths at terminals 1, 15 and G303 for clean, tight connections.

DIAGNOSTIC TROUBLE CODE CHART DTC 56 — 1995 DEVILLE, CONCOURS, ELDORADO AND SEVILLE, CONTINUED

DIAGNOSTIC TROUBLE CODE 56

LEFT REAR ABS OUTLET VALVE SOLENOID FAULT

NOTICE: J 39200 TEST LEADS MUST BE "ZEROED" PRIOR TO MAKING ANY RESISTANCE MEASUREMENTS. REFER TO J 39200 USER'S MANUAL.

- INSTALL TECH 1 WITH MASS STORAGE CARTRIDGE.
- IGNITION "ON."
- PERFORM TECH 1 VALVE SOLENOID TEST TO AFFECTED VALVE.

DOES "RELEASE PRESSURE" MODE OPERATE PROPERLY?

NO:
- REMOVE EBTCM.
- INSPECT EBTCM TO BPM CONNECTOR FOR DAMAGE, CORROSION OR POOR TERMINAL CONTACT, REPAIR IF NEEDED.
- IF CONNECTOR IS OK, REPLACE BPM.

YES:
- IGNITION "OFF."
- DISCONNECT EBTCM CONNECTOR.
- REMOVE EBTCM FROM BPM VALVE ASSEMBLY.
- INSPECT EBTCM TO BPM CONNECTOR FOR CONDITIONS WHICH COULD CAUSE INTERMITTENTS, SUCH AS DAMAGE, CORROSION, OR POOR TERMINAL CONTACT.

IS CONNECTOR OK?

NO: IF TERMINAL CORROSION IS EVIDENT, REPLACE EBTCM.

YES:
- INSTALL J 41247 BPM PINOUT BOX TO BPM CONNECTOR.
- USING J 39200, MEASURE RESISTANCE BETWEEN J 41247 TERMINAL "7" AND TERMINAL "5".

IS IT 4.5-5.0 OHMS?

NO: REPLACE BPM VALVE.

YES:
- REMOVE J 41247.
- REPLACE EBTCM TO BPM.
- PERFORM "TECH 1 AUTO" TEST.

DOES DTC RESET?

YES: REPLACE EBTCM.

NO: NO TROUBLE FOUND. SEE "DIAGNOSTIC AIDS" ON FACING PAGE.

DIAGNOSTIC TROUBLE CODE CHART DTC 61 — 1995 DEVILLE, CONCOURS, ELDORADO AND SEVILLE

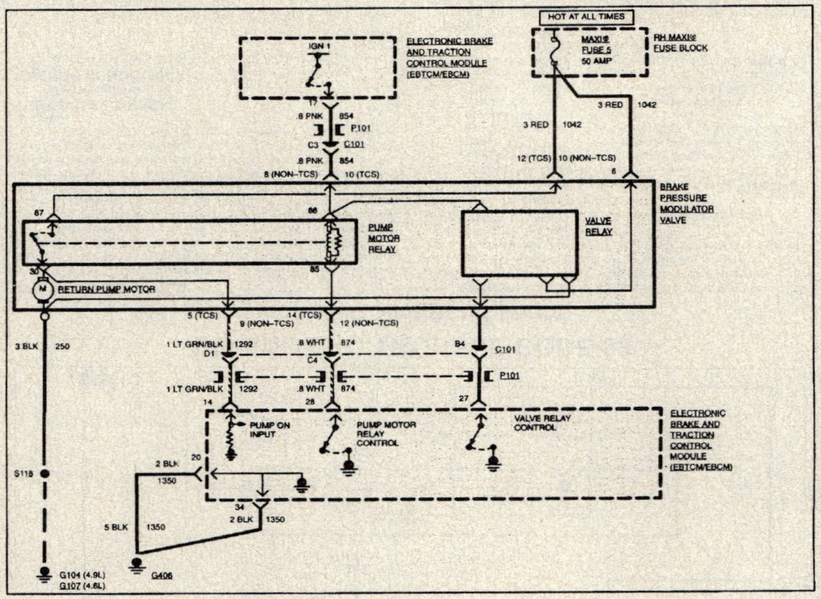

DIAGNOSTIC TROUBLE CODE 61

PUMP MOTOR OR PUMP MOTOR CIRCUIT FAULT

CIRCUIT DESCRIPTION:

The pump motor returns brake fluid to the master cylinder brake circuit at the BPM Valve during antilock braking and provides brake fluid pressure to the front wheels during traction control. During normal braking, the pump does not operate. When the vehicle begins to move after start-up, the EBTCM/EBCM will turn on the pump motor and perform a self-check of the pump motor and pump motor circuit. This check may be felt and heard by the driver when the vehicle begins to move. The self-check should be considered normal operation. The pump motor is an integral component of the BPM Valve and cannot be serviced separately.

The pump motor relay provides power to the pump motor in the BPM Valve. The pump motor relay is located in the EBTCM and cannot be replaced seperately from the EBTCM.

The pump motor relay is not engaged during normal system operation. When antilock or traction control operation is required, battery power is provided to the pump motor via the relay in the EBTCM.

TEST CONDITION(S):

1. Tested continuously.

FAILURE CONDITION(S):

1. Before Drive Away: Loss of ground on motor stud.
2. Drive Away: Motor voltage is not detected during first drive away after engine start.
3. While Driving: Motor relay is energized and motor voltage is not detected.
4. Pump runs continuously.

ACTION TAKEN:

1. Code set.
2. ABS/TCS switched 'OFF'.
3. 'ANTILOCK' indicator 'ON'.
4. 'TRACTION DISABLED' message displayed.

DIAGNOSTIC AIDS:

It is very important that a through inspection of the wiring and connectors be performed. Failure to carefully and fully inspect wiring and connectors may result in misdiagnosis, causing part replacement with reappearance of the fault.

DIAGNOSTIC TROUBLE CODE CHART DTC 61 — 1995 DEVILLE, CONCOURS, ELDORADO AND SEVILLE, CONTINUED

DIAGNOSTIC TROUBLE CODE 61

PUMP MOTOR OR PUMP MOTOR CIRCUIT FAULT

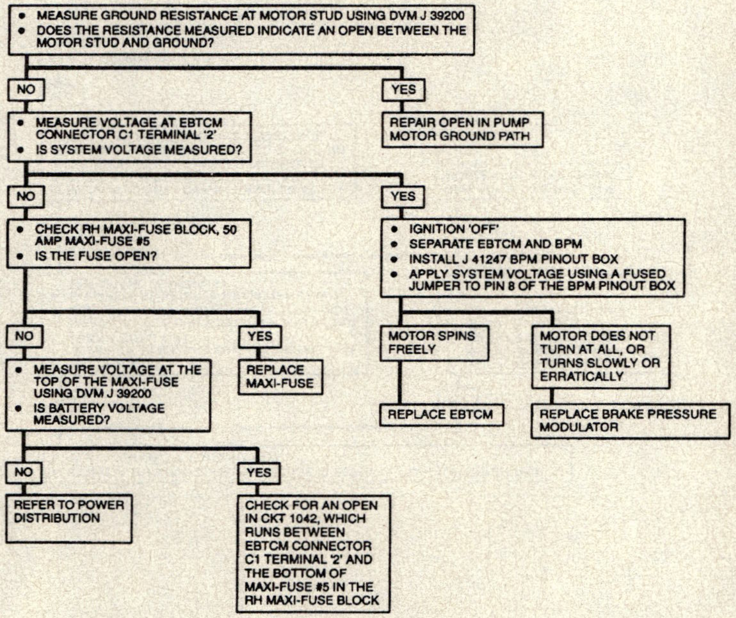

DIAGNOSTIC TROUBLE CODE CHART DTC 63 — 1995 DEVILLE, CONCOURS, ELDORADO AND SEVILLE

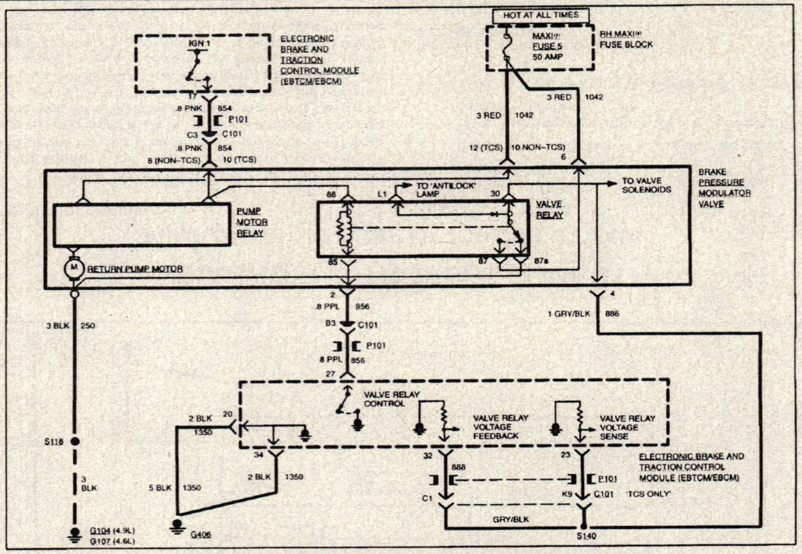

DIAGNOSTIC TROUBLE CODE 63

VALVE RELAY CIRCUIT FAULT

CIRCUIT DESCRIPTION:

The valve relay provides power to the ABS valve solenoids and TCS Pilot Valves in the BPM valve. The valve relay is located in the EBTCM an may be replaced if it is found to be defective.

The valve relay is engaged during normal system operation. When the ignition is switched to RUN, the EBTCM commands the valve relay ON by grounding the valve relay control. When this ground is provided, the valve relay is energized from the voltage supply line on CKT 1042 (EBTCM pin '3'). The valve relay switches and battery voltage is provided to the valve solenoids, the pilot valves and the valve relay voltage sense line. The valve relay remains engaged until the ignition is turned OFF or a failure is detected.

TEST CONDITION(S):

1. Tested continuously.

FAILURE CONDITION(S):

1. Less than 5 volts at EBTCM, valve relay sense;

   OR

2. Voltage at EBTCM, valve relay sense remains high, although voltage at EBTCM, valve relay control, is low (valve relay sticks).

ACTION TAKEN:

1. Code set.
2. ABS/TCS switched 'OFF'.
3. 'ANTILOCK' indicator 'ON'.
4. 'TRACTION DISABLED' message displayed.

When the valve relay is engaged, the valve relay sense line should be at battery voltage. With the valve relay disengaged, the sense line should be at ground. If the commanded position of the valve relay and the position indicated by the sense line do not agree, Code 63 will set.

TEST DESCRIPTION: Test numbers refer to circled numbers on the diagnostic chart.

1. Checks for availability of battery voltage to the Valve Relay through Maxi® fuse 5 and CKT 1042.
2. Checks for possible loss of voltage current when system is under load. At Key-ON, system performs the power on Reset Test and cycles the valve relay, checking the current draw of valves.
3. Checks the Valve Relay coil for proper resistance.
4. Checks for the Valve Relay contacts being internally open when in the relay unenergized position.
5. Checks for the Valve Relay contacts being stuck open when in the relay-energized position.

DIAGNOSTIC AIDS:

All tests using J 39700 Universal Breakout Box terminal '20' are using the terminal as ground. These tests, of course, assume that the ground CKT 1350 at terminal '20' is good.

It is very important that a thorough inspection of the wiring and connectors be performed. Failure to carefully and fully inspect wiring and connectors may result in misdiagnosis, causing parts replacement with reappearance of the fault.

DIAGNOSTIC TROUBLE CODE CHART DTC 63 — 1995 DEVILLE, CONCOURS, ELDORADO AND SEVILLE, CONTINUED

DIAGNOSTIC TROUBLE CODE 63

VALVE RELAY CIRCUIT FAULT

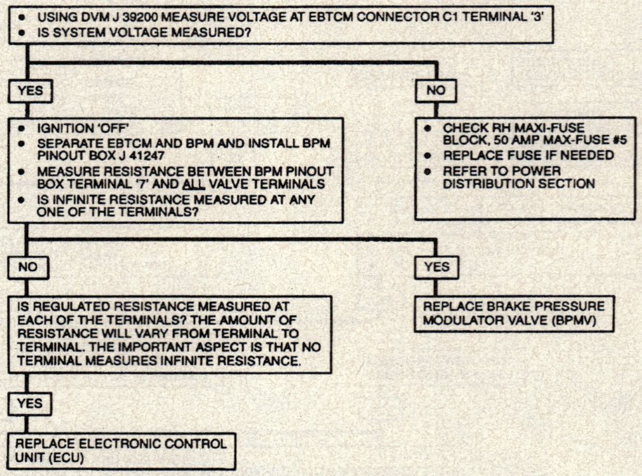

DIAGNOSTIC TROUBLE CODE CHART DTC 67 — 1995 DEVILLE, CONCOURS, ELDORADO AND SEVILLE

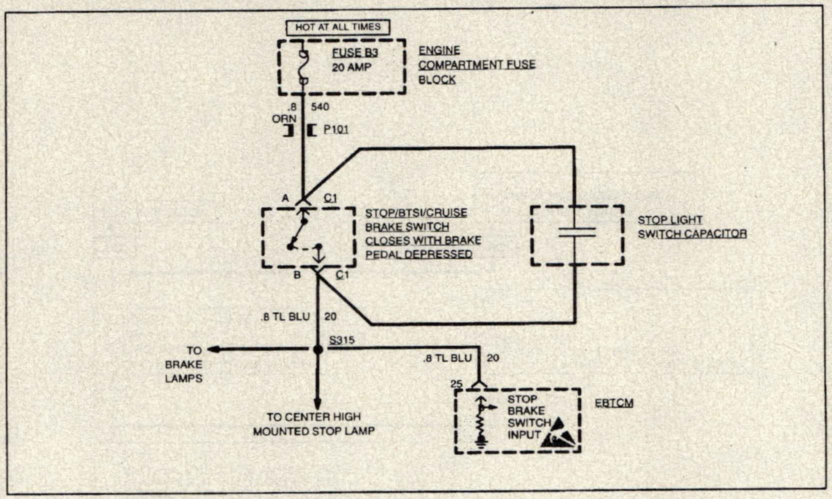

DIAGNOSTIC TROUBLE CODE 67

BRAKE LIGHT SWITCH FAULT

CIRCUIT DESCRIPTION:

The Electronic Brake and Traction Control module monitors the serial data line for PCM TCC Switch input PI71 for TCC switch status. The EBTCM then compares this input with the voltage at terminal 28 of the EBTCM connector (system voltage indicates switch is depressed). Any discrepancy will cause the code to set. The EBTCM only performs this comparison after a certain drive pattern has been followed.

TEST CONDITION(S):

1. Tested continuously.
2. The EBTCM compares PCM input PI71 with the voltage at terminal 28 of the EBTCM connector after the following drive sequence has been performed:
   - Engine 'ON', vehicle speed is at 0 mph for at least 4 seconds.
   - Vehicle speed is above 30 mph for at least 4 seconds.
   - Vehicle speed returns to 0 mph for at least 4 seconds.

FAILURE CONDITION(S):

1. From PCM.

ACTION TAKEN:

1. Code set.
2. TCS disabled.
3. 'TRACTION DISABLED' message ON.

DIAGNOSTIC TROUBLE CODE CHART DTC 71 — 1995 DEVILLE, CONCOURS, ELDORADO AND SEVILLE

DIAGNOSTIC TROUBLE CODE 71

ELECTRONIC BRAKE AND TRACTION CONTROL (EBTCM) INTERNAL FAULT

CIRCUIT DESCRIPTION:

The EBTCM performs various diagnostic checks on itself. If it finds a problem, Code 71 will set.

TEST CONDITION(S):

1. Tested continuously.

FAILURE CONDITION(S):

1. While Standing or Driving: EBTCM internal error is detected.

ACTION TAKEN:

1. Code set.
2. ABS and TCS switched 'OFF'.
3. 'ANTILOCK' indicator ON.
4. 'TRACTION DISABLED' message ON. (vehicles with TCS)

TEST DESCRIPTION: Test numbers refer to circled numbers on the diagnostic chart.

1. Checks for good connections from the harness to the EBTCM itself.
2. Checks to see if the fault was false. If Code 71 resets, replace the EBTCM.

DIAGNOSTIC TROUBLE CODE CHART DTC 71 — 1995 DEVILLE, CONCOURS, ELDORADO AND SEVILLE, CONTINUED

DIAGNOSTIC TROUBLE CODE 71

ELECTRONIC BRAKE AND TRACTION CONTROL (EBTCM) INTERNAL FAULT

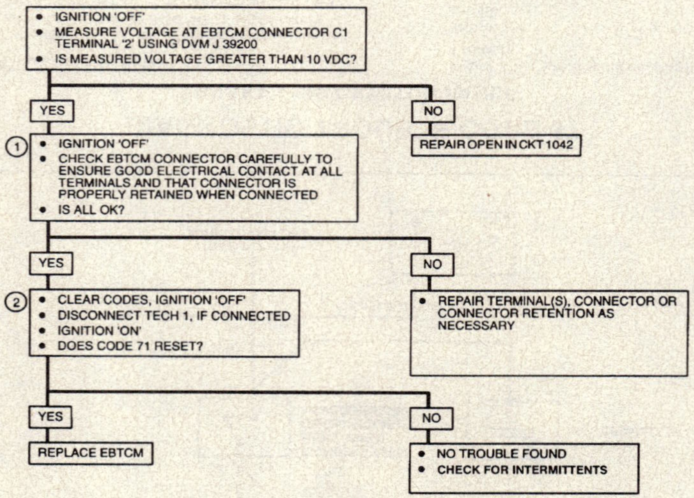

WHEN ALL DIAGNOSIS AND REPAIRS ARE COMPLETED, CLEAR CODES AND VERIFY OPERATION

DIAGNOSTIC TROUBLE CODE CHART DTC 72 — 1995 DEVILLE, CONCOURS, ELDORADO AND SEVILLE

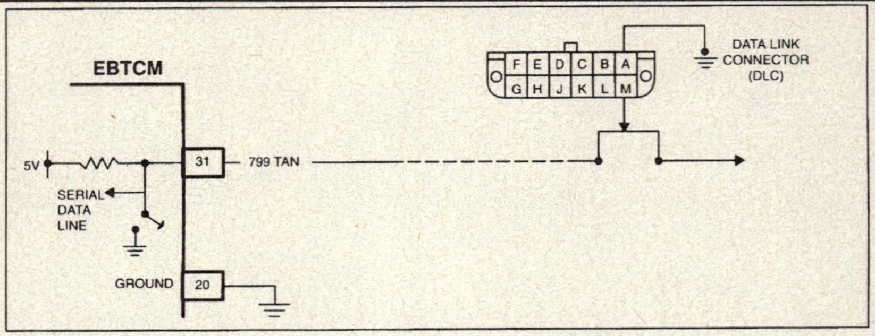

DIAGNOSTIC TROUBLE CODE 72

IPC SERIAL DATA LINK FAULT

CIRCUIT DESCRIPTION:

The Serial Data Link is an asynchronous link operating at 8192 bits per second.

Code 72 will set and the 'ANTILOCK' indicator will illuminate if the EBTCM does not get polled by the IPC for information.

TEST CONDITION(S):

1. Tested continuously.

FAILURE CONDITION(S):

1. While Standing or Driving: Serial Data communications are lost.

ACTION TAKEN:

1. Code set.
2. TCS switched 'OFF'.
3. 'TRACTION DISABLED' message ON.

TEST DESCRIPTION: Test numbers refer to circled numbers on the diagnostic chart.

1. This test checks to see if the off-board device can communicate with the EBTCM.
2. This test checks for a short to battery in the serial data circuit.
3. This test checks for a short to ground in the serial data circuit.
4. This test checks for an open in the serial data circuit.
5. This test checks to see if the off-board device can communicate with the EBTCM.
6. If the connections at the EBTCM and DLC connector are OK, replace the EBTCM.

DIAGNOSTIC AIDS:

The problem may be intermittent. Try performing the tests shown while 'wiggling' wiring and connectors; this can often cause the fault to appear.

NOTICE: Diagnostic Trouble Code 72 will never be displayed while serial data fault is current. A current Code 72 will result in a 'NO TCS DATA', and 'TRACTION DISABLED' messages as well as IPC Diagnostic Trouble Code B332.

DIAGNOSTIC TROUBLE CODE CHART DTC 72 — 1995 DEVILLE, CONCOURS, ELDORADO AND SEVILLE, CONTINUED

DIAGNOSTIC TROUBLE CODE 72
IPC SERIAL DATA LINK FAULT

①
- IGNITION 'OFF'
- INSTALL TECH 1
- IGNITION 'ON'
- ATTEMPT TO ESTABLISH COMMUNICATIONS WITH THE EBTCM; CAN YOU COMMUNICATE WITH IT?

NO → ②; OTHER CODES →
- NO TROUBLE FOUND; SEE 'DIAGNOSTIC AIDS' ON FACING PAGE

②
- IGNITION 'OFF'
- DISCONNECT EBTCM
- DISCONNECT TECH 1
- INSTALL J 39700 UNIVERSAL BREAKOUT BOX TO EBTCM HARNESS CONNECTOR ONLY
- IGNITION 'ON'
- CONNECT DVM BETWEEN J 39700 TERMINALS '42' AND '1'
- NOTE VOLTAGE
- IS IT AROUND SYSTEM VOLTAGE?

NO → ③; YES →
- REPAIR SHORT TO SYSTEM VOLTAGE IN CKT 800

③
- IGNITION 'OFF'
- CHECK RESISTANCE BETWEEN J 39700 TERMINALS '42' AND '1'
- IS IT LESS THAN 5 Ω?

NO → ④; YES →
- REPAIR SHORT TO GROUND IN CKT 800

④
- CHECK RESISTANCE BETWEEN J 39700 TERMINAL '42' AND DLC TERMINAL '9' USING ADAPTER FROM J 35616-A
- IS IT LESS THAN 5 Ω?

NO →
- REPAIR OPEN IN CKT 800

YES → ⑤

⑤
- IGNITION 'OFF'
- REMOVE J 39700
- RECONNECT EBTCM
- INSTALL TECH 1
- IGNITION 'ON'
- ATTEMPT TO ESTABLISH COMMUNICATIONS WITH THE EBTCM; CAN YOU COMMUNICATE WITH IT?

NO → ⑥; YES →
- NO TROUBLE FOUND; REFER TO 'DIAGNOSTIC AIDS' ON FACING PAGE

⑥
- CHECK FOR POOR CONNECTIONS AT EBTCM/EBCM AND DLC CONNECTORS; IF OK, REPLACE EBTCM

WHEN ALL DIAGNOSIS AND REPAIRS ARE COMPLETED, CLEAR CODES AND VERIFY OPERATION

DIAGNOSTIC TROUBLE CODE CHART DTC 73 — 1995 DEVILLE, CONCOURS, ELDORADO AND SEVILLE

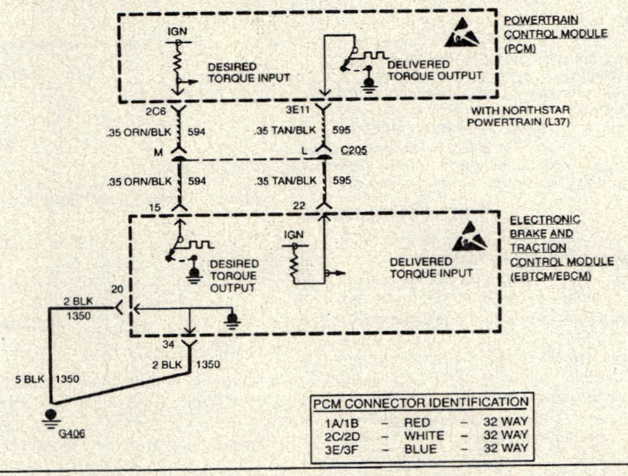

DIAGNOSTIC TROUBLE CODE 73
PCM-EBTCM PWM SIGNAL FAULT (4.6L ONLY)

CIRCUIT DESCRIPTION:

Traction control is simultaneously controlled by the EBTCM on vehicles equipped with the 4.6L Northstar and the PCM. The PCM receives a 'DESIRED TORQUE' request via a Pulse Width Modulated (PWM) signal from the EBTCM requesting the desired torque level for proper traction control system operation. Torque reduction will be indicated if the duty cycle of the PWM signal is between 10% (100 % torque reduction) and 90% (no torque reduction) and at least one (up to five) cylinder is requested to be disabled. The percent duty cycle will indicate the percent torque required. The PCM will selectively disable fuel to certain cylinders to achieve this reduction and communicate the actual torque output back to the EBTCM.

TEST CONDITION(S):

1. Tested continuously.

FAILURE CONDITION(S):

1. While Standing or Driving: PWM interface between EBTCM and PCM is lost.

ACTION TAKEN:

1. Code set.
2. 'TRACTION DISABLED' message 'ON'.

A Code 73 may set if either the torque requested or torque delivered circuits are shorted to ground or shorted to voltage.

If Delivered Torque is damaged and the EBTCM loses this input, a TCS Diagnostic Code 73 will be set. The PCM monitors the serial data line to see if the traction control has been disabled. If it has, the PCM will not set Code 93.

If Desired Torque is damaged and the PCM loses this input, the PCM monitors the serial data line to see if the traction control has been disabled. If not, the PCM sets a Code 93 and then tells the EBTCM, via serial data, that there is a PWM circuit problem and the EBTCM sets a Code 73.

TEST DESCRIPTION: Test numbers refer to circled numbers on the diagnostic chart.

1. Checks if the PCM Diagnostic Trouble Code 93 is set. If it is, there is likely a problem with CKT 594.
2. Checks for an open or high resistance in CKT 594.
3. Checks for a short to ground in CKT 594.
4. Checks for a short to voltage in CKT 595.
5. If no problem is found up until this point, CKT 594 should be checked.
6. Checks for an open or high resistance in CKT 594.
7. Checks for a short to ground in CKT 594.
8. Checks for a short to voltage in CKT 594.
9. If CKT 595 has not been checked, proceed to the left hand side of the trouble tree. If it has, continue downward. Problem is intermittent or in one of the controllers.
10. Checks if Code 73 and/or 93 are intermittent. If not, trouble is in either PCM or EBTCM.

DIAGNOSTIC AIDS:

- It is theoretically impossible for only a PCM Code 93 to set. If only a PCM Code 93 is present, either the TCS Code 73 has been cleared or there is a problem with the PCM. Refer to Section 6EA.
- A Code 73 may set by itself if the PCM PROM has been improperly installed.

DIAGNOSTIC TROUBLE CODE CHART DTC 73 — 1995 DEVILLE, CONCOURS, ELDORADO AND SEVILLE, CONTINUED

DIAGNOSTIC TROUBLE CODE 73

PCM-EBTCM PWM SIGNAL FAULT

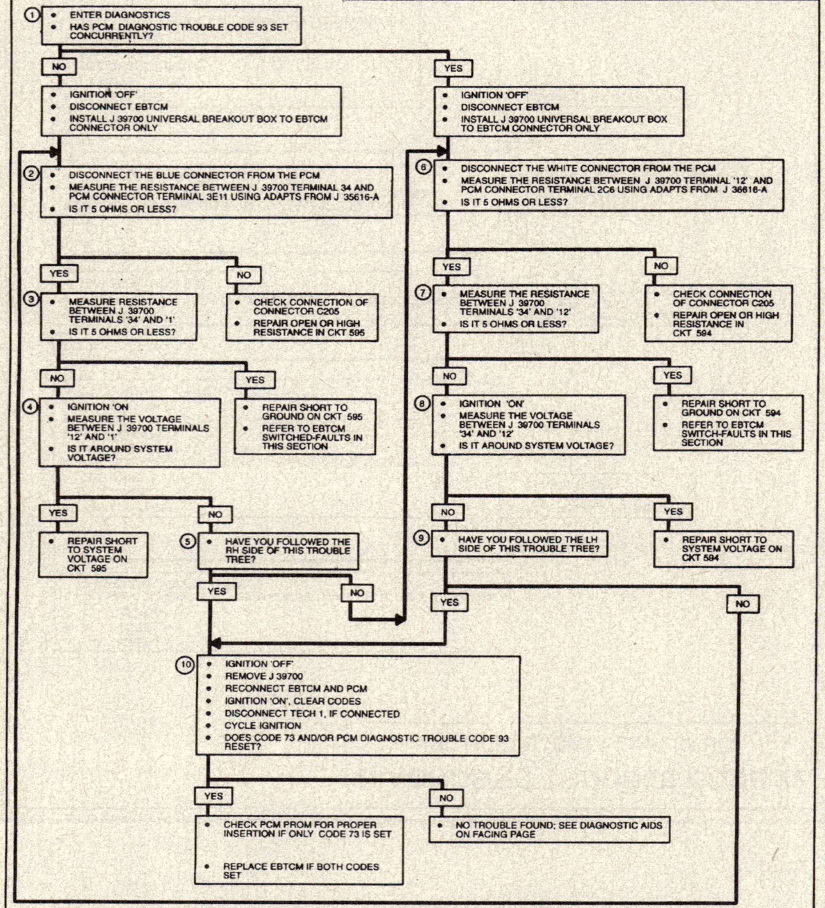

WHEN ALL DIAGNOSIS AND REPAIRS ARE COMPLETED, CLEAR CODES AND VERIFY OPERATION

DIAGNOSTIC TROUBLE CODE CHART DTC 77 — 1995 DEVILLE, CONCOURS, ELDORADO AND SEVILLE

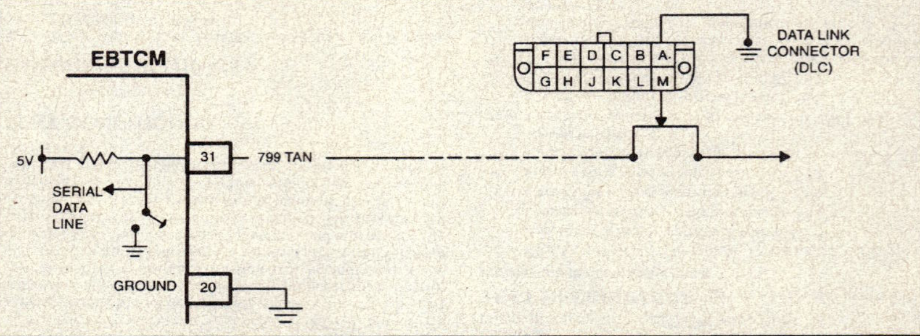

DIAGNOSTIC TROUBLE CODE 77

PCM/SERIAL DATA LINK FAULT

CIRCUIT DESCRIPTION:

The Serial Data Link is an asynchronous link operating at 8192 bits per second.

Code 77 will set and the 'ANTILOCK' indicator will illuminate if the EBTCM does not get polled by the PCM for information.

TEST CONDITION(S):

1. Tested continuously.

FAILURE CONDITION(S):

1. While Standing or Driving: Serial Data communications are lost.

ACTION TAKEN:

1. Code set.
2. TCS switched 'OFF'.
3. 'TRACTION DISABLED' message ON.

TEST DESCRIPTION: Test numbers refer to circled numbers on the diagnostic chart.

1. This test checks to see if the off-board device can communicate with the EBTCM.
2. This test checks for a short to battery in the serial data circuit.
3. This test checks for a short to ground in the serial data circuit.
4. This test checks for an open in the serial data circuit.
5. This test checks to see if the off-board device can communicate with the EBTCM.
6. If the connections at the EBTCM and DLC connector are OK, replace the EBTCM.

DIAGNOSTIC AIDS:

The problem may be intermittent. Try performing the tests shown while 'wiggling' wiring and connectors; this can often cause the fault to appear.

NOTICE: Diagnostic Trouble Code 77 will never be displayed while serial data fault is current. A current Code 77 will result in a 'NO TCS DATA', and 'TRACTION DISABLED' messages as well as PCM Diagnostic Trouble Code P133.

DIAGNOSTIC TROUBLE CODE CHART DTC 77 — 1995 DEVILLE, CONCOURS, ELDORADO AND SEVILLE, CONTINUED

DIAGNOSTIC TROUBLE CODE 77
PCM SERIAL DATA LINK FAULT

(1)
- IGNITION 'OFF'
- INSTALL TECH 1
- IGNITION 'ON'
- ATTEMPT TO ESTABLISH COMMUNICATIONS WITH THE EBTCM; CAN YOU COMMUNICATE WITH IT?

NO → (2)

OTHER CODES → • NO TROUBLE FOUND; SEE 'DIAGNOSTIC AIDS' ON FACING PAGE

(2)
- IGNITION 'OFF'
- DISCONNECT EBTCM
- DISCONNECT TECH 1
- INSTALL J 39700 UNIVERSAL BREAKOUT BOX TO EBTCM HARNESS CONNECTOR ONLY
- IGNITION 'ON'
- CONNECT DVM BETWEEN J 39700 TERMINALS '42' AND '1'
- NOTE VOLTAGE
- IS IT AROUND SYSTEM VOLTAGE?

NO → (3)

YES → • REPAIR SHORT TO SYSTEM VOLTAGE IN CKT 800

(3)
- IGNITION 'OFF'
- CHECK RESISTANCE BETWEEN J 39700 TERMINALS '42' AND '1'
- IS IT LESS THAN 5 Ω?

NO → (4)

YES → • REPAIR SHORT TO GROUND IN CKT 800

(4)
- CHECK RESISTANCE BETWEEN J 39700 TERMINAL '42' AND DLC TERMINAL '9' USING ADAPTER FROM J 35616-A
- IS IT LESS THAN 5 Ω?

NO → (5)

YES → • REPAIR OPEN IN CKT 800

(5)
- IGNITION 'OFF'
- REMOVE J 39700
- RECONNECT EBTCM
- INSTALL TECH 1
- IGNITION 'ON'
- ATTEMPT TO ESTABLISH COMMUNICATIONS WITH THE EBTCM; CAN YOU COMMUNICATE WITH IT?

NO → (6)

YES → • NO TROUBLE FOUND; REFER TO 'DIAGNOSTIC AIDS' ON FACING PAGE

(6)
- CHECK FOR POOR CONNECTIONS AT EBTCM/EBCM AND DLC CONNECTORS; IF OK, REPLACE EBTCM

WHEN ALL DIAGNOSIS AND REPAIRS ARE COMPLETED, CLEAR CODES AND VERIFY OPERATION

DIAGNOSTIC TROUBLE CODE CHART DTC 83 — 1995 DEVILLE, CONCOURS, ELDORADO AND SEVILLE

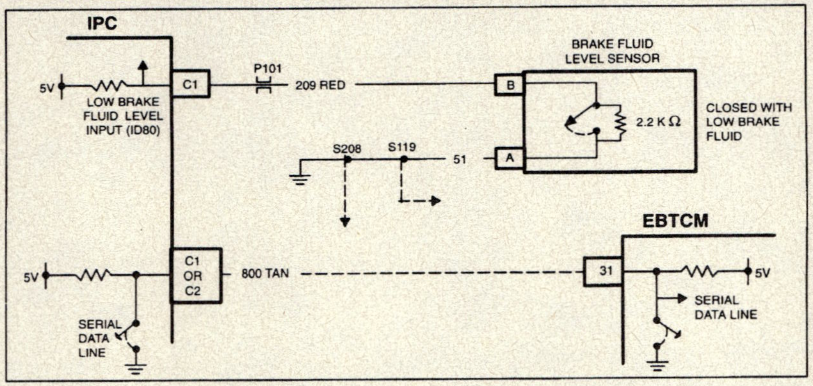

DIAGNOSTIC TROUBLE CODE 83
LOW BRAKE FLUID LEVEL

CIRCUIT DESCRIPTION:

The IPC outputs 5 volts on CKT 209 to the brake fluid level sensor. The IPC determines the level of brake fluid by the voltage drop it detects across the sensor circuit. This information is delivered to the EBTCM via the serial data line.

TEST CONDITION(S):

1. Tested continuously.

FAILURE CONDITION(S):

1. While Standing or Driving: Brake fluid falls below acceptable level.
2. Brake fluid level sensor disconnected.
3. Open, short to ground or short to battery on CKT 209.

ACTION TAKEN:

1. Code set.
2. ABS and TCS switched 'OFF'.
3. Red BRAKE warning lamp and 'ANTILOCK' indicator ON.
4. 'TRACTION DISABLED' message ON.

TEST DESCRIPTION: Test numbers refer to circled numbers on the diagnostic chart.

1. Check the brake fluid level in master cylinder reservoir.
2. Check brake fluid level as detected by IPC.
3. Brake fluid level not acceptable. Fill and recheck.
4. Brake fluid level as detected by IPC. See 'DIAGNOSTIC AIDS'.
5. Checks IPC input parameter ID80 with sensor disconnected for a short to ground in CKT 209, or bad sensor.
6. Checks IPC input parameter ID80 with sensor disconnected and connector shorted.
7. IPC detects acceptable brake fluid level. Problem not present.
8. Checks sensor resistor and checks for a short between CKTs 209 and 51.
9. Checks for open in CKTs 209 and 51.

DIAGNOSTIC AIDS:

IPC data parameter ID80 will display the following:

- 3.6-5V - open in sensor circuit.
- 1.5-3.5V - circuit OK. Problem not present.
- 0-1.4V - short between CKTs 209 and 51.

DIAGNOSTIC TROUBLE CODE CHART DTC 83 — 1995 DEVILLE, CONCOURS, ELDORADO AND SEVILLE, CONTINUED

DIAGNOSTIC TROUBLE CODE 83

BRAKE FLUID LEVEL LOW

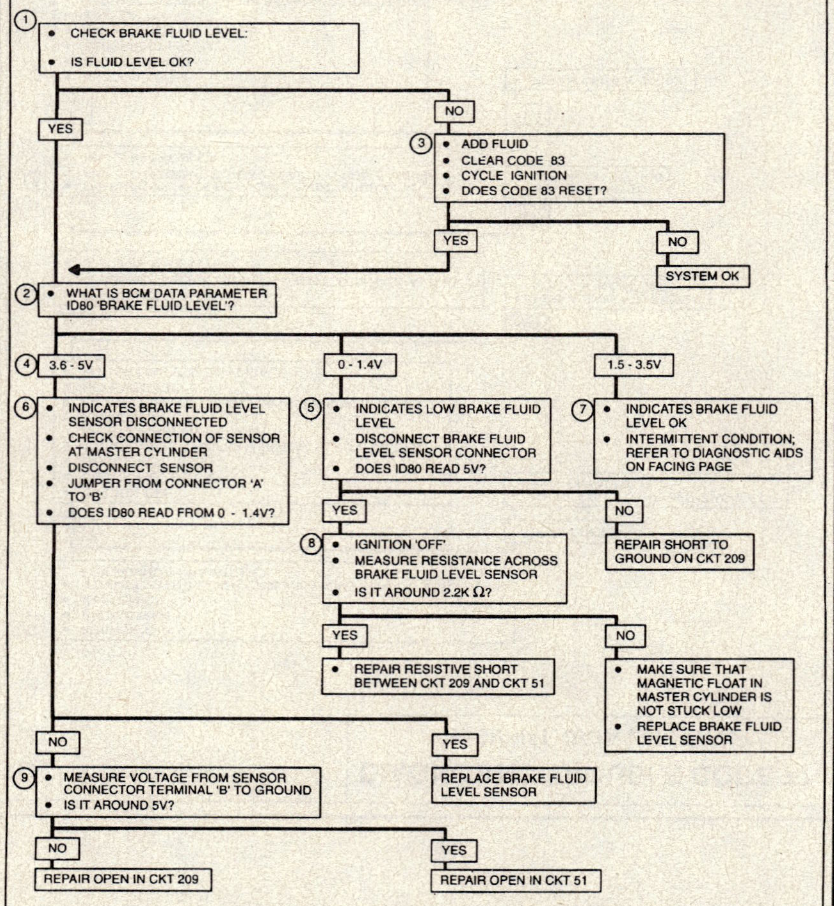

WHEN ALL DIAGNOSIS AND REPAIRS ARE COMPLETED, CLEAR CODES AND VERIFY OPERATION.

DIAGNOSTIC TROUBLE CODE CHART DTC 84 — 1995 DEVILLE, CONCOURS, ELDORADO AND SEVILLE

DIAGNOSTIC TROUBLE CODE 84

PCM TRACTION CONTROL NOT ALLOWED (4.9L ONLY)

CIRCUIT DESCRIPTION:

The PCM monitors various parameters and will not allow Traction Control operation if any of these parameters fall below a predetermined value.

DTC WILL SET WHEN:

1. Tested continuously.

ACTION TAKEN:

1. Code set.
2. TCS switched 'OFF'.
3. 'TRACTION DISABLED' message displayed on Driver Information Center.

DIAGNOSTIC AIDS:

Traction Control will not be allowed if any of the following conditions are observed.

- Engine Coolant Temperature less than -40°C.
- Engine Coolant Temperature greater than or equal to 131°C.
- Low coolant level detected.

DIAGNOSTIC TROUBLE CODE CHART DTC 84 — 1995 DEVILLE, CONCOURS, ELDORADO AND SEVILLE, CONTINUED

DIAGNOSTIC TROUBLE CODE 84
PCM TRACTION CONTROL NOT ALLOWED (4.9L ONLY)

- CLEAR CODES
- IGNITION 'OFF'
- DISCONNECT TECH 1
- RAISE FRONT WHEELS 4" TO 5" ON HOIST
- IGNITION 'ON', ENGINE 'ON'
- WITH TRANSMISSION IN DRIVE, ACCELERATE TO 1/2 THROTTLE WHILE AN OBSERVER MONITORS THE DRIVE WHEELS
- ARE THE FRONT BRAKES MODULATED?

NO

- CHECK FLUID LEVEL IN MASTER CYLINDER RESERVOIR
- CHECK FOR DIRT OR DEBRIS IN RESERVOIR FLUID

LEVEL OK, FLUID NOT DIRTY

- POSSIBLE AIR IN SYSTEM
- BLEED TCS PRIME PIPE AT BPM VALVE
- PERFORM THE TCS TEST USING THE TECH 1 PER THE BLEED PROCEDURE. ONE OF THE FRONT CALIPER BLEED SCREWS MUST BE 'OPENED'
- RETEST TRACTION CONTROL OPERATION ON HOIST

LOW

- ADD FLUID
- POSSIBLE AIR IN SYSTEM
- BLEED TCS PRIME PIPE AT BPM VALVE
- PERFORM THE TCS TEST USING THE TECH 1 PER THE BLEED PROCEDURE ONE OF THE FRONT CALIPER BLEED SCREWS MUST BE 'OPENED'
- RETEST TRACTION CONTROL OPERATION ON HOIST

DIRTY

- DISCONNECT TCS PRIME PIPE AT BPM VALVE
- ALLOW FLUID TO DRAIN FROM RESERVOIR
- FLUSH RESERVOIR AND TCS PRIME PIPE WITH FRESH BRAKE FLUID
- RECONNECT TCS PRIME PIPE
- REFILL
- POSSIBLE AIR IN SYSTEM
- BLEED TCS PRIME PIPE AT BPM VALVE
- PERFORM THE TCS TEST USING THE TECH 1 PER THE BLEED PROCEDURE ONE OF THE FRONT CALIPER BLEED SCREWS MUST BE OPENED
- RETEST TRACTION CONTROL 'OPERATION ON HOIST

YES

SYSTEM IS OPERATING CORRECTLY

DIAGNOSTIC TROUBLE CODE CHART DTC 85 — 1995 DEVILLE, CONCOURS, ELDORADO AND SEVILLE

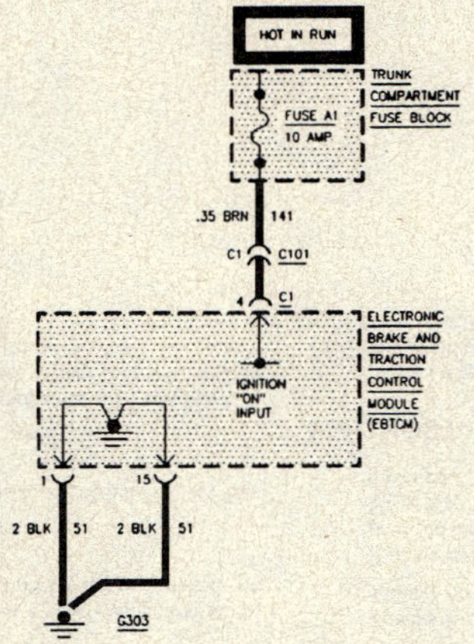

DIAGNOSTIC TROUBLE CODE 85
LOW VOLTAGE

CIRCUIT DESCRIPTION:

The EBTCM, requires a certain minimum voltage to operate properly. It monitors the ignition feed circuit to determine if voltage falls below a minimum level.

DTC WILL SET WHEN:

The EBTCM operating voltage falls below 9.0 volts.

TEST DESCRIPTION: Number(s) below refer to circled number(s) on the diagnostic chart.

1. Checks that fuse A1 feeding the EBTCM ignition circuit is good.
2. Measures for appropriate voltage from the EBTCM ignition circuit, using each of the two EBTCM ground lines to complete the circuit.
3. Checks EBTCM connector terminals for good contact, i.e. not spread open, backed out of connector, corroded, etc.
4. Checks EBTCM connector for good retention to EBTCM when installed.
5. Checks for DTC reset after other portions of the system have tested as OK.
6. Determines if a malfunction found in Step 2 is due to an open or poor connection in the EBTCM ignition feed CKT 141 or poor ground(s) in CKT 51.

DIAGNOSTIC AIDS:

It is very important that a thorough inspection of the wiring and connectors be performed. Failure to carefully and fully inspect wiring and connectors may result in misdiagnosis, causing part replacement with reappearance of the malfunction.

DIAGNOSTIC TROUBLE CODE CHART DTC 85 — 1995 DEVILLE, CONCOURS, ELDORADO AND SEVILLE, CONTINUED

DIAGNOSTIC TROUBLE CODE 85
LOW VOLTAGE

BEFORE TESTING, ENSURE THAT BATTERY IS FULLY CHARGED, AND THAT CHARGING SYSTEM IS OPERATING PROPERLY IF DISCHARGED BATTERY NOTED.

(1)
- ENSURE FUSE A1 IS GOOD.
- IGNITION "ON."
DOES FUSE A1 OPEN WHEN VOLTAGE IS APPLIED?

YES: REPAIR SHORT TO GROUND IN CKT 141.

NO:

(2)
- IGNITION "OFF."
- DISCONNECT EBTCM.
- INSTALL J 39700 UNIVERSAL BREAKOUT BOX TO EBTCM AND EBTCM HARNESS CONNECTOR.
- IGNITION "ON."
- CONNECT J 39200 BETWEEN J 39700 TERMINAL PAIRS LISTED BELOW. NOTE VOLTAGE FOR EACH CHECK

J 39700 TERMINALS
1 AND 4
15 AND 4

WAS ANY VOLTAGE MEASUREMENT LESS THAN 9 VOLTS?

YES:

(6)
- CONNECT J 39200 BETWEEN J 39700 TERMINAL "4" AND A KNOWN GOOD CHASSIS GROUND.
IS VOLTAGE MEASUREMENT LESS THAN 9 VOLTS?

NO: REPAIR OPEN OR POOR GROUND IN CKT 51 AT TERMINAL(S) "1" OR "15", AS NEEDED.

YES: REPAIR OPEN OR POOR CONNECTION IN CKT 141.

NO:

(3)
- IGNITION "OFF."
- DISCONNECT EBTCM HARNESS CONNECTOR FROM J 39700.
- CHECK EBTCM HARNESS CONNECTOR CAREFULLY TO ENSURE GOOD ELECTRICAL CONTACT AT ALL TERMINALS, ESPECIALLY TERMINALS "1", "15" AND "4".
ARE ALL TERMINALS OK?

NO: REPLACE TERMINALS AND/OR CONNECTORS AS NECESSARY.

YES:

(4)
- REMOVE J 39700.
- RECONNECT EBTCM.
DOES CONNECTOR SEEM TO BE PROPERLY RETAINED TO EBTCM WHEN CONNECTED?

NO: REPLACE WIRING HARNESS OR EBTCM AS NEEDED TO OBTAIN PROPER CONNECTOR RETENTION.

YES:

(5)
- CLEAR DTCs, IGNITION "OFF."
- DISCONNECT TECH 1, IF CONNECTED.
- IGNITION "ON."
DOES DTC 85 RESET?

YES: REPLACE EBTCM.

NO: NO TROUBLE FOUND; SEE "DIAGNOSTIC AIDS" ON FACING PAGE.

WHEN ALL DIAGNOSIS AND REPAIRS ARE COMPLETED, CLEAR DTCs AND VERIFY PROPER OPERATION.

DIAGNOSTIC TROUBLE CODE CHART A — 1995 DEVILLE, CONCOURS, ELDORADO AND SEVILLE

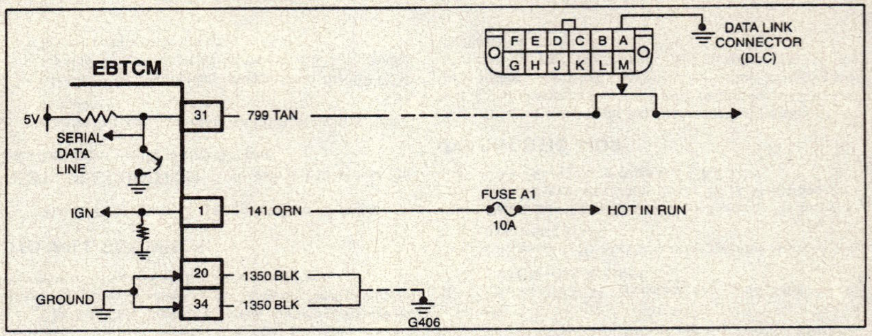

CHART A

'ANTILOCK' INDICATOR AND 'TRACTION DISABLED' MESSAGE 'ON', NO CODES SET / NO TCS DATA

TEST DESCRIPTION: Number(s) below refer to circled number(s) on the diagnostic chart.

1. Checks if condition is due to a poor connection at the EBTCM. Make sure the connecter is not 'cocked' on module.
2. Verifies that fuse A1 is intact.
3. Checks EBTCM ground resistances. Should be less than 1 ohm.
4. Checks EBTCM ignition voltage.
5. This step checks for possible open or high resistance in CKT 141 if there is insufficient voltage measured in Step 4. It is assumed a short to ground on CKT 141 would have blown fuse 'A1'.
6. Checks for EBTCM terminal '1' having a poor solder joint or for being partially backed out.
7. Checks if a loss of serial data communication is due to damage on CKT 800.
8. If there is no voltage measured in Step 8, there is a possible short to ground on CKT 800. If not, regulated 5V supply from EBTCM is lost and EBTCM should be replaced.

DIAGNOSTIC AIDS:

BCM Diagnostic Trouble Code B332 and PCM Code P137 should be present. If all of the IPC data line codes are present, refer to Section 8C1 'SYSTEM CHECK'.

DIAGNOSTIC CHART A — 1995 DEVILLE, CONCOURS, ELDORADO AND SEVILLE

CHART A

'ANTILOCK' INDICATOR AND 'TRACTION DISABLED' MESSAGE 'ON', NO CODES SET / NO TCS DATA

- IPC CODE I332 SHOULD BE PRESENT.
- IF ALL OF THE IPC DATA LINE CODES ARE PRESENT, REFER TO "SYSTEM CHECK."

(1) VERIFY THAT THE EBTCM CONNECTOR IS SECURELY ATTACHED TO THE EBTCM. IF DISCONNECTED, CONNECT AND VERIFY OPERATION.

(2) IF FUSES HAVE NOT ALREADY BEEN CHECKED, INSPECT FUSE A1 IN THE CONSOLE FUSE BLOCK.

- FUSE BLOWN
  - REPLACE FUSE AND VERIFY OPERATION
- FUSE OK

(3)
- INSTALL J 39700 UNIVERSAL BREAKOUT BOX TO BOTH EBTCM CONNECTOR AND EBTCM HARNESS CONNECTOR.
- MEASURE RESISTANCE BETWEEN PIN 1 AND BODY GROUND.
- MEASURE RESISTANCE BETWEEN PIN 15 AND BODY GROUND.

EITHER OR BOTH GREATER THAN 1 OHM
- REPAIR OPEN OR HIGH RESISTANCE IN CKT 51 BETWEEN EBTCM AND NEGATIVE JUNCTION BLOCK
- CLEAN AND TIGHTEN EBTCM GROUND G303. SAND EYELETS ON HARNESS

BOTH 0 - 1 OHM

(4)
- IGNITION SWITCH IN RUN.
- MEASURE VOLTAGE FROM PIN 1 AND PIN 4.

GREATER THAN 9 VOLTS

(6)
- CHECK TERMINAL 4 OF EBTCM.
- CHECK PIN 4 OF EBTCM CONNECTOR FOR INTERMITTENT OPEN OR POOR CRIMP

(7)
- NO TROUBLE FOUND
- POSSIBLE OPEN IN CKT 800
- IGNITION 'ON'
- MEASURE VOLTAGE FROM BREAKOUT BOX TERMINAL "42" TO TERMINAL "1"

5V
- OPEN IN CKT 800.

0V

(8)
- IGNITION OFF.
- MEASURE RESISTANCE FROM BREAKOUT BOX TERMINAL "42" TO TERMINAL "1"

IS IT 5 OHMS OR LESS?

YES — INSPECT SYSTEM WIRING

NO — REPLACE EBTCM

LESS THAN 9 VOLTS

(5)
- IGNITION SWITCH OFF
- FUSE A1 REMOVED
- MEASURE RESISTANCE FROM GROUND SIDE OF FUSE TO PIN 1

GREATER THAN 1 OHM
- REPAIR OPEN OR PARTIAL OPEN IN CKT 141 BETWEEN TRUNK FUSE BLOCK AND EBTCM CONNECTOR.

LESS THAN 1 OHM
- IGNITION SWITCH IN RUN.
- MEASURE VOLTAGE ON POWER SIDE OF FUSE A1

GREATER THAN 9 VOLTS
- DISCONNECT NEGATIVE BATTERY CABLE
- INSPECT FUSE BLOCK SPADE TERMINALS FOR POOR CRIMP
- INSPECT WIRING AT CONSOLE FUSE BLOCK FOR OPENS
- REPLACE FUSE

LESS THAN 9 VOLTS
- VERIFY MAXI® FUSE 6 (60A) IS OK (LOCATED IN LH MAXI® FUSE BLOCK)
- DISCONNECT NEGATIVE BATTERY CABLE
- INSPECT INTERIOR RELAY CENTER SPADE TERMINALS FOR POOR CRIMP
- INSPECT WIRING AT CONSOLE FUSE BLOCK FOR OPENS
- REPAIR OPEN IN CIRCUIT BETWEEN IGNITION SWITCH AND FUSE BLOCK

WHEN ALL DIAGNOSIS AND REPAIRS ARE COMPLETED, CLEAR CODES AND VERIFY OPERATION

DIAGNOSTIC CHART B — 1995 DEVILLE, CONCOURS, ELDORADO AND SEVILLE

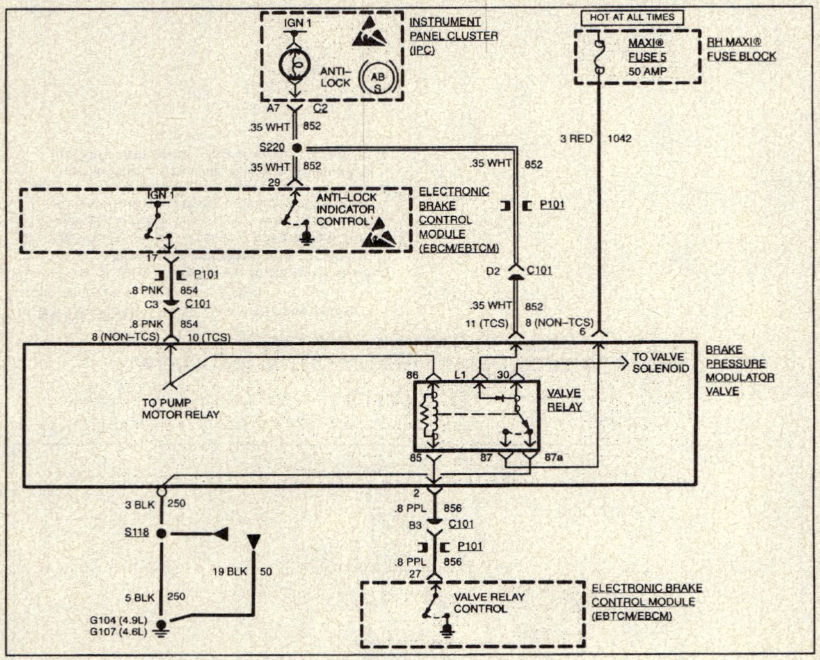

CHART B

'ANTILOCK' INDICATOR 'ON', NO CODES SET

TEST DESCRIPTION: Number(s) below refer to circled number(s) on the diagnostic chart.

1. Checks if 'ANTILOCK' indicator is ON due to a loss of communication between the IPC and EBTCM.
2. Checks for a short to ground in CKT 852.
3. Final check for diagnostic trouble codes.

DIAGNOSTIC AIDS:

The EBTCM controls the 'ANTILOCK' indicator only in instances where serial data communication with the IPC is lost. In all other instances the IPC controls this lamp.

Inspect the shorting bar which is located in the EBTCM connector. Its purpose is to provide a ground path to turn "ON" the antilock indicator, if the EBTCM connector becomes unplugged. If the shorting bar, which is molded plastic should break off, the antilock indicator could remain on at all times.

DIAGNOSTIC CHART B — 1995 DEVILLE, CONCOURS, ELDORADO AND SEVILLE, CONTINUED

CHART B
'ANTILOCK' INDICATOR 'ON', NO TCS CODES SET

(1)
- IF "ANTILOCK" INDICATOR OR "TRACTION DISABLED" MESSAGE ARE "ON" AND IPC DIAGNOSTIC TROUBLE CODE I332 IS SET, REFER TO CHART A

(2)
- IGNITION "OFF"
- DISCONNECT EBTCM AND REMOVE VALVE SOLENOID RELAY
- IGNITION "ON"

DOES "ANTILOCK" INDICATOR TURN OFF?

YES:

(3)
- IGNITION "OFF"
- RECONNECT EBTCM AND REINSTALL VALVE SOLENOID RELAY.
- IGNITION "ON"

IS "ANTILOCK" INDICATOR "ON" AND/OR THE "ANTILOCK" INDICATOR OR "TRACTION DISABLED" MESSAGE "ON" WITH NO IPC CODE I332 OR DIAGNOSTIC TROUBLE CODES?

- **NO:**
  - IF CODES ARE SET, REFER TO APPROPRIATE TROUBLE TREE
  - IF THERE ARE NO MESSAGES OR CODES, REFER TO IPC DIAGNOSIS
- **YES:**
  - REPLACE EBTCM

NO:
- IGNITION 'OFF'
- INSTALL J 39700 UNIVERSAL BREAKOUT BOX TO EBTCM/EBCM HARNESS CONNECTOR ONLY
- MEASURE RESISTANCE FROM PIN 29 TO PIN 20
- IS IT LESS THAN 5 Ω?

- **NO:**
  - NO TROUBLE FOUND
  - CHECK BPM VALVE AND EBTCM CONNECTOR TERMINAL INTEGRITY
  - CHECK VALVE SOLENOID RELAY TERMINAL INSERTION
- **YES:**
  - REPAIR SHORT TO GROUND ON CKT 852

WHEN ALL DIAGNOSIS AND REPAIRS ARE COMPLETED, CLEAR CODES AND VERIFY OPERATION

DIAGNOSTIC CHART C — 1995 DEVILLE, CONCOURS, ELDORADO AND SEVILLE

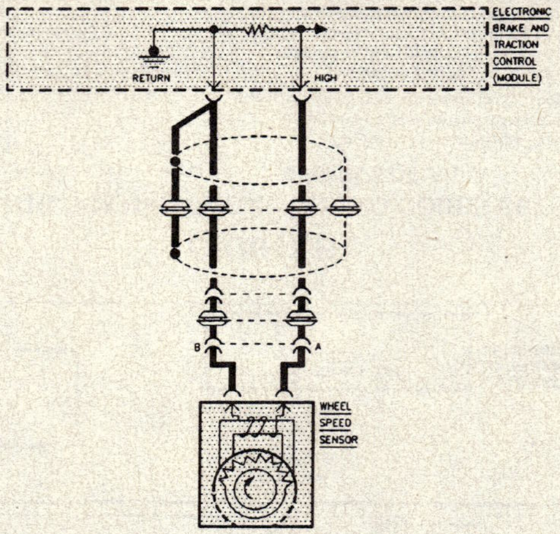

CHART C
VALVE CYCLING (CHATTER) DURING NORMAL STOPS (UNWANTED ABS MODULATION)

TEST DESCRIPTION: Number(s) below refer to circled number(s) on the diagnostic chart.

1. Checks if valve activation due to erroneous electrical 'noise' induced onto wheel speed sensor circuits.
2. The wheel speed sensor output should gradually increase with vehicle speed. There should be no skips or jumps. All four wheel speed sensor outputs should be identical. On a slow down, one of the signals may drop off faster than the others, indicating a lock-up condition or a problem with the speed sensor or wiring (corrosion etc.).

DIAGNOSTIC CHART C — 1995 DEVILLE, CONCOURS, ELDORADO AND SEVILLE, CONTINUED

CHART C

VALVE CYCLING (CHATTER) DURING NORMAL STOPS (UNWANTED ABS MODULATION)

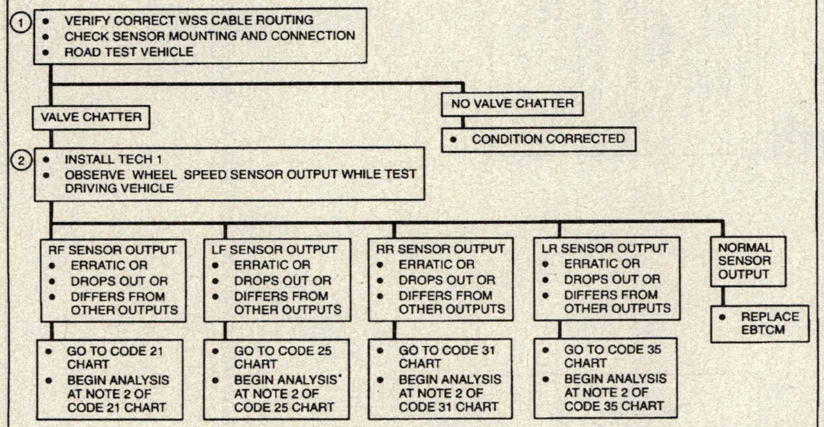

WHEN ALL DIAGNOSIS AND REPAIRS ARE COMPLETED, CLEAR CODES AND VERIFY OPERATION

DIAGNOSTIC CHART D — 1995 DEVILLE, CONCOURS, ELDORADO AND SEVILLE

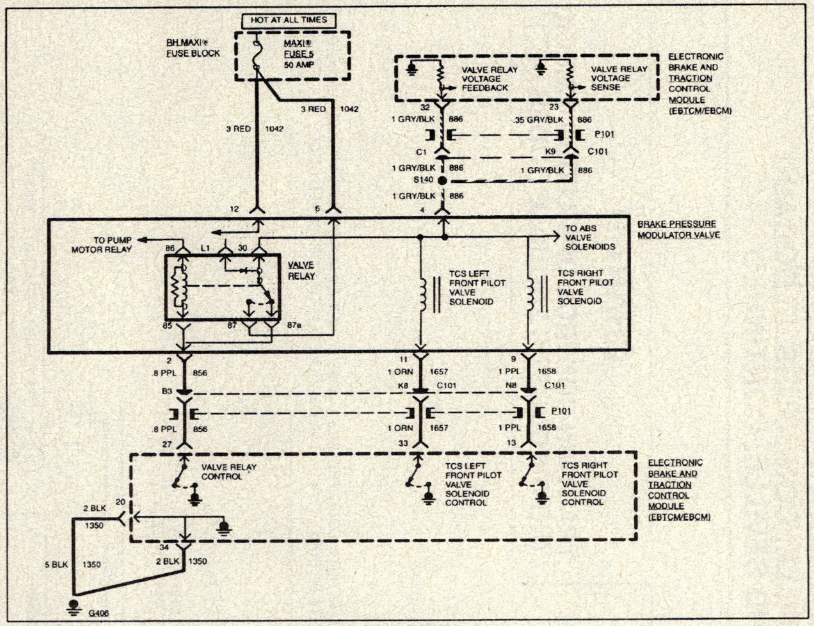

CHART D

TRACTION CONTROL OPERATION CHECK (NO DIAGNOSTIC TROUBLE CODES SET)

TEST DESCRIPTION: Test numbers refer to circled numbers on the diagnostic chart.

1. Traction control system should interpret the elevated front wheels as an icy or free spin condition. At 1/2 throttle a properly operating system will allow both front wheels to spin up quickly to about 20-25 MPH and then slow down (but not stop) to about 5-10 MPH and have a jerky appearance. The traction control 'ACTIVE' message should display on the DIC. The pump motor should also be heard.
2. System is building up pressure to the front brake and operating correctly.
3. System is not building up pressure to the front brakes. Possible air or dirt in the system.
4. No dirt in reservoir and the fluid level is OK. Air may still be trapped in system. The TCS hydraulic circuit should be bled using TECH 1 TCS test and bleed procedure outlined in Section 5. One of the front caliper bleed screws must be open to allow air to escape.
5. Fluid level OK. Air may still be trapped in the system. The TCS hydraulic circuit should be bled using TECH 1 TCS test and bleed procedure. One of the front caliper bleed screws must be open to allow air to escape.

DIAGNOSTIC AIDS:

- The Brake Light Switch must be properly adjusted to prevent an erroneous brake application input to the EBTCM. Misadjustment may prevent brake intervention during traction control operation. Also, the EBTCM receives TCC switch status over serial data from the PCM. If serial data is lost or the switch remains engaged, brake intervention during traction control may be lost. Monitor PCM TCC switch input PI71 while cycling brake switch. The input should cycle 'HI' and 'LO'.
- Cylinders shall be disabled if the engine speed greater than or equal to 600 RPM.
- The PCM can disable traction control for up to 11 ignition cycles if override PS24, Traction Control Disable, is activated. When PS24 is selected, the display will alternate between '- -' for one second and the current state of Traction Control Disable override ('99' for override active and '00' for override not active). Pressing the WARMER button will disable traction control for the next eleven ignition cycles and display '99'. The 'TRACTION DISABLED' message will be displayed for the duration of this override and no TCS diagnostic trouble codes will be set. Pressing the COOLER button will allow normal program control of traction control and display '00'.

DIAGNOSTIC CHART D — 1995 DEVILLE, CONCOURS, ELDORADO AND SEVILLE, CONTINUED

CHART D

TRACTION CONTROL OPERATION CHECK (NO DIAGNOSTIC TROUBLE CODES SET)

NOTE: SOME OF THESE CHECKS WILL CAUSE DIAGNOSTIC TROUBLE CODES TO SET. BE SURE TO VERIFY AUTHENTICITY OF ANY CODES AND CORRECT IF NECESSARY. BE SURE TO CLEAR ANY ERRONEOUS OR CORRECTED CODES.

(1)
- CLEAR CODES
- IGNITION 'OFF'
- DISCONNECT TECH 1
- RAISE FRONT WHEELS 4" TO 5" ON HOIST
- IGNITION 'ON', ENGINE 'ON'
- WITH TRANSMISSION IN DRIVE, ACCELERATE TO 1/2 THROTTLE WHILE AN OBSERVER MONITORS THE DRIVE WHEELS
- ARE THE FRONT BRAKES MODULATED?

NO

(3)
- CHECK FLUID LEVEL IN MASTER CYLINDER RESERVOIR
- CHECK FOR DIRT OR DEBRIS IN RESERVOIR FLUID

LEVEL OK, FLUID NOT DIRTY

(4)
- POSSIBLE AIR IN SYSTEM
- BLEED TCS PRIME PIPE AT BPM VALVE
- PERFORM THE TCS TEST USING THE TECH 1 PER THE BLEED PROCEDURE

  ONE OF THE FRONT CALIPER BLEED SCREWS MUST BE 'OPENED'
- RETEST TRACTION CONTROL OPERATION ON HOIST

LOW

(7)
- ADD FLUID
- POSSIBLE AIR IN SYSTEM
- BLEED TCS PRIME PIPE AT BPM VALVE
- PERFORM THE TCS TEST USING THE TECH 1 PER THE BLEED PROCEDURE

  ONE OF THE FRONT CALIPER BLEED SCREWS MUST BE 'OPENED'
- RETEST TRACTION CONTROL OPERATION ON HOIST

YES

(2)
- SYSTEM IS OPERATING CORRECTLY

WHEN ALL DIAGNOSIS AND REPAIRS ARE COMPLETED, CLEAR CODES AND VERIFY OPERATION

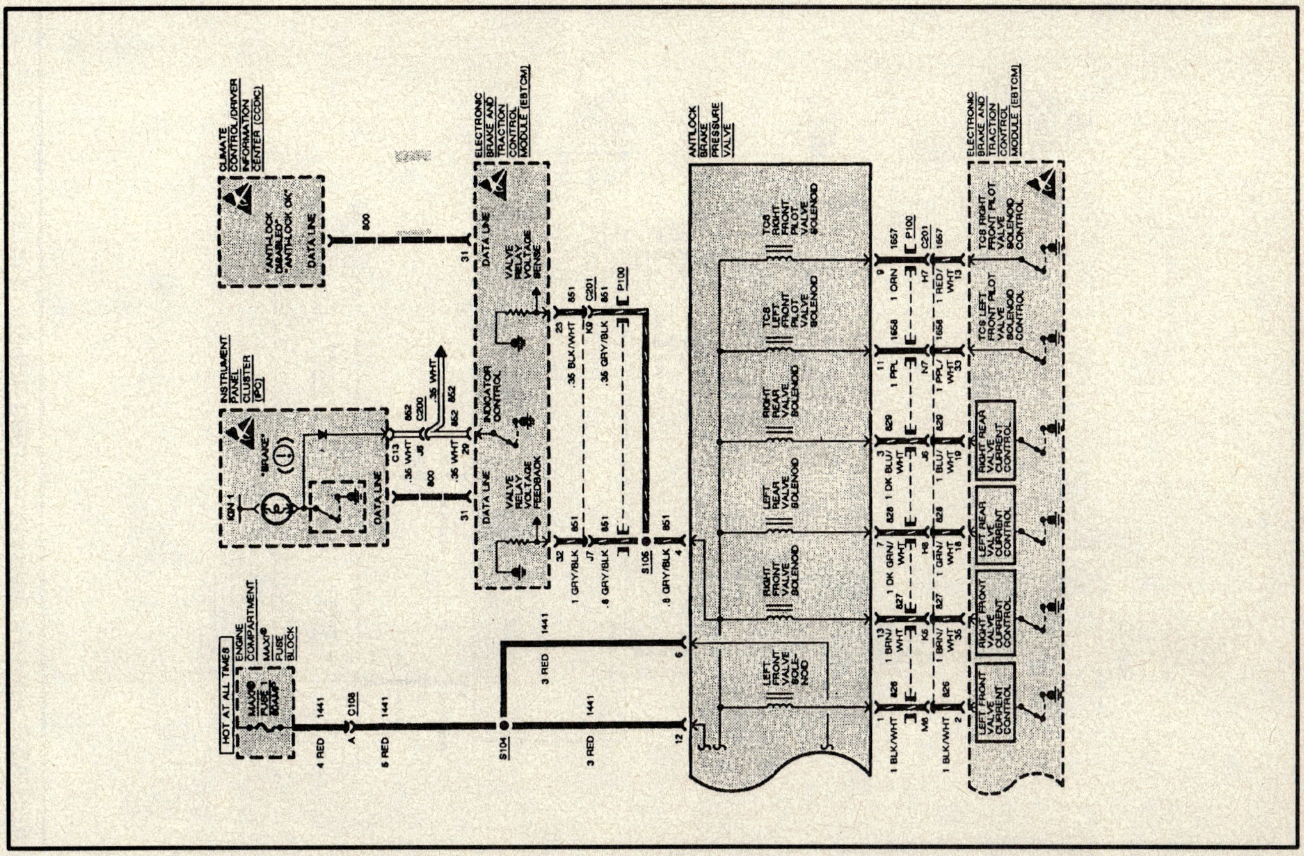

ABS/ASR wiring schematic — 1993 Allante, 2 of 4

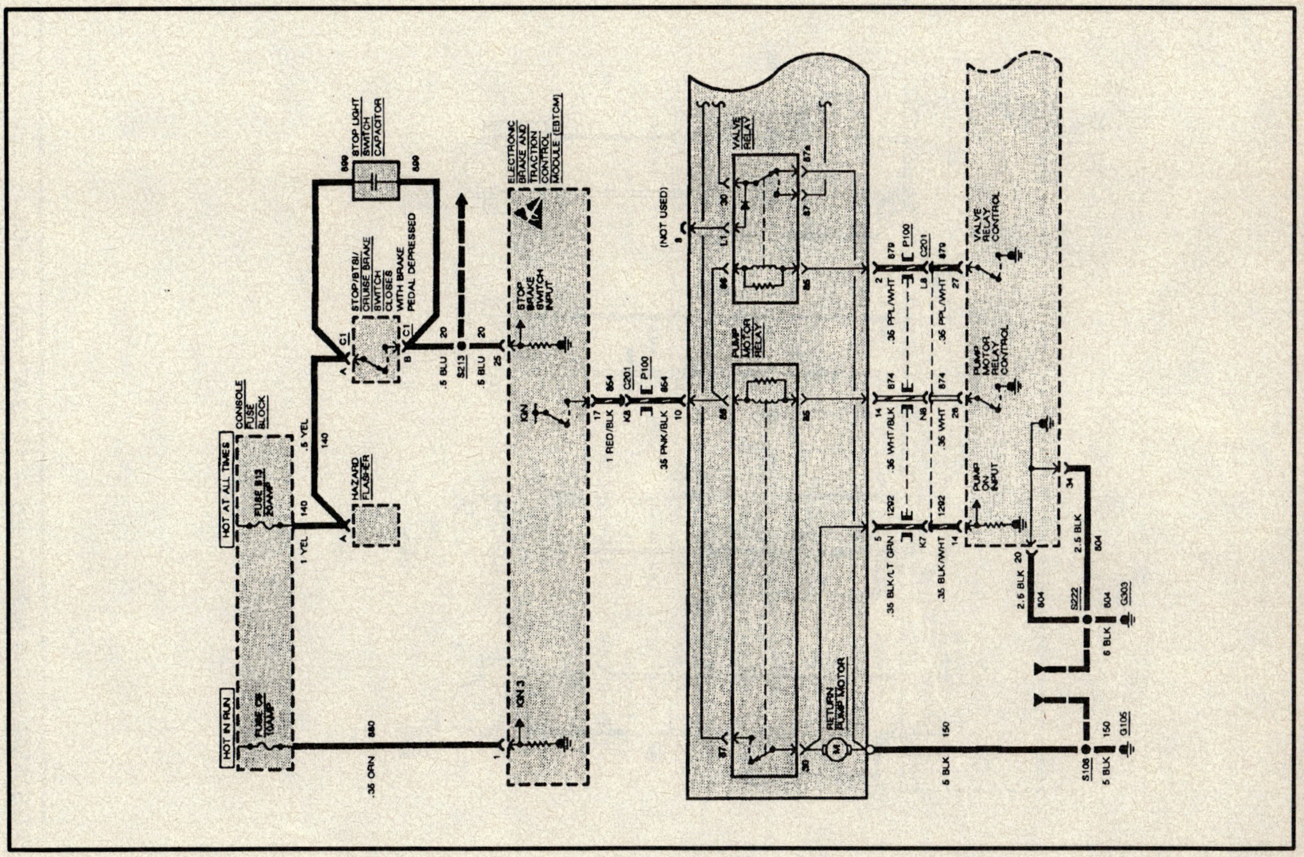

ABS/ASR wiring schematic — 1993 Allante, 1 of 4

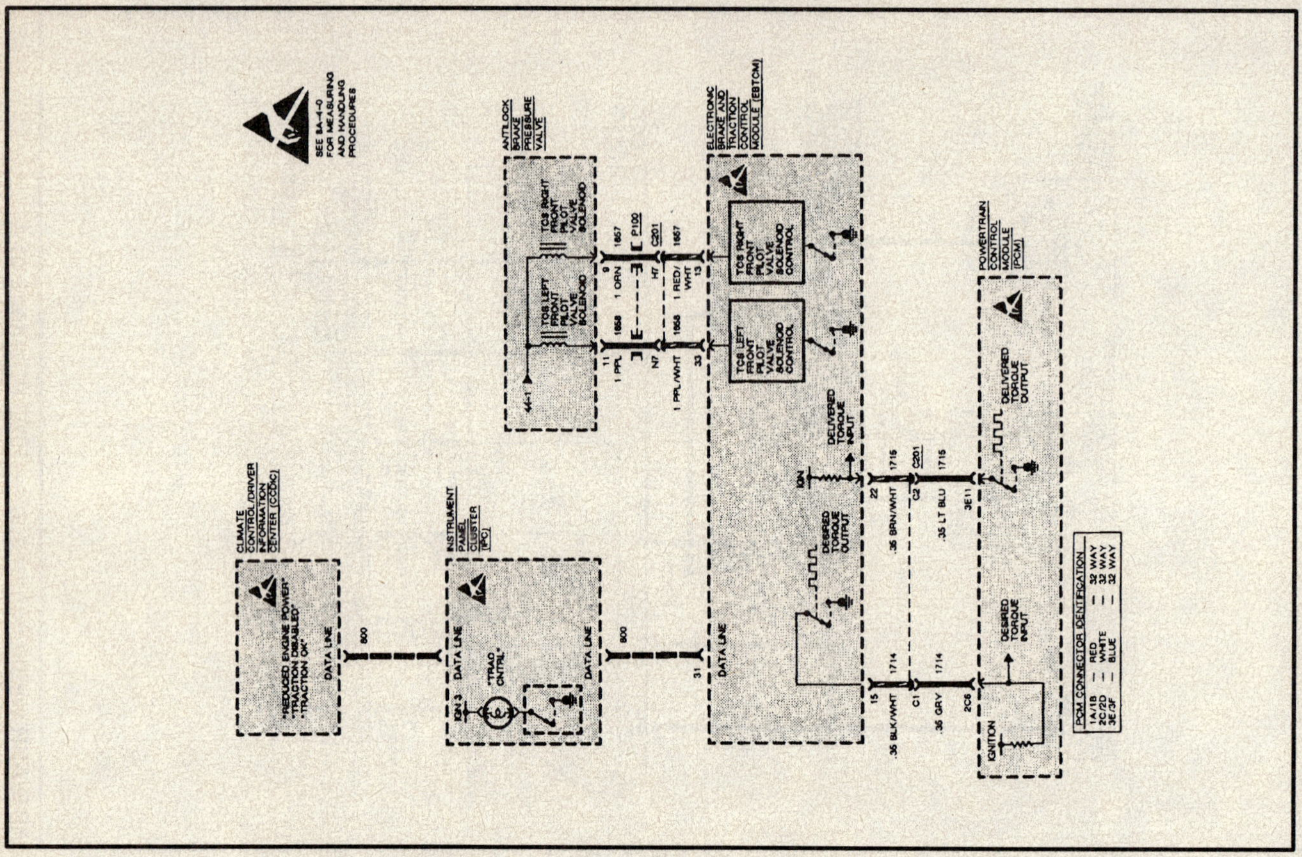

ABS/ASR wiring schematic — 1993 Allante, 4 of 4

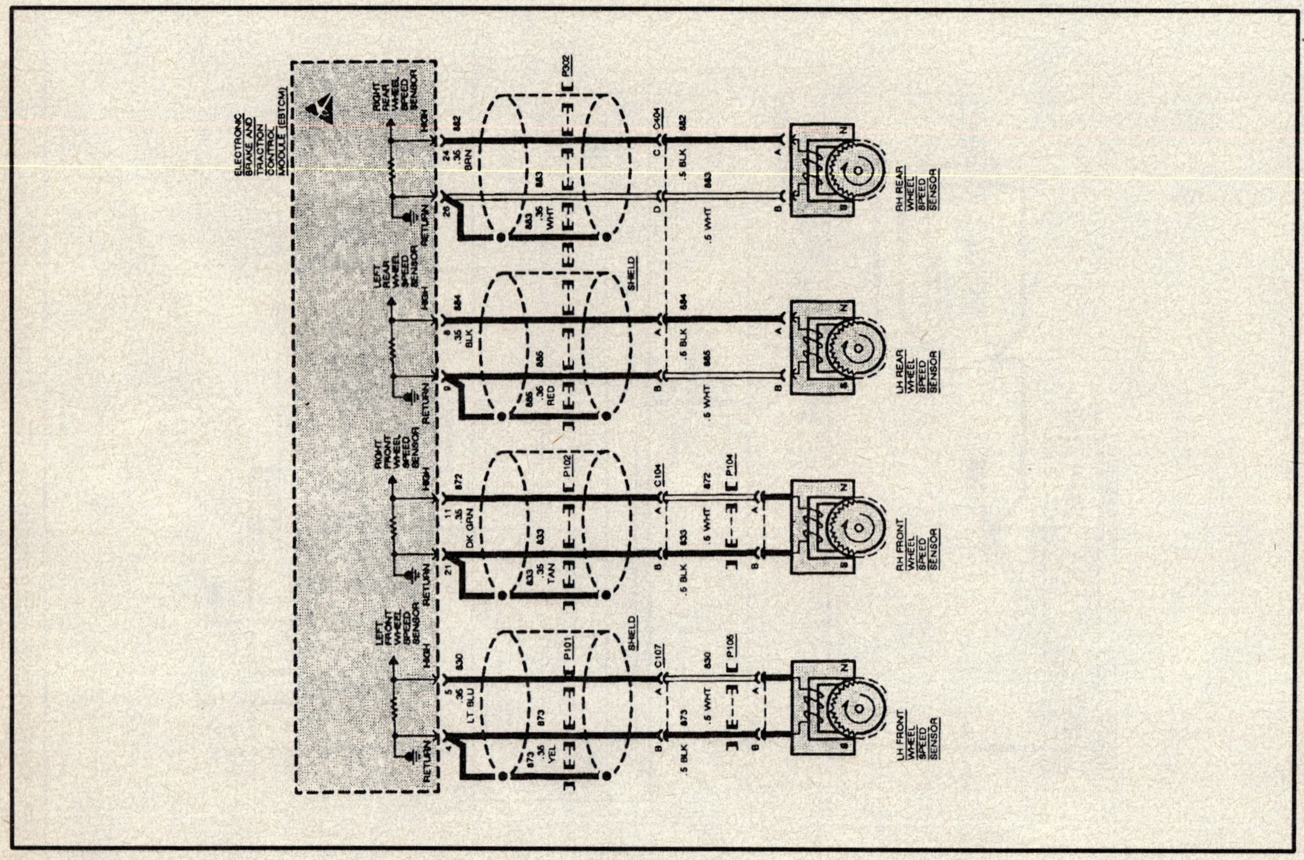

ABS/ASR wiring schematic — 1993 Allante,, 3 of 4

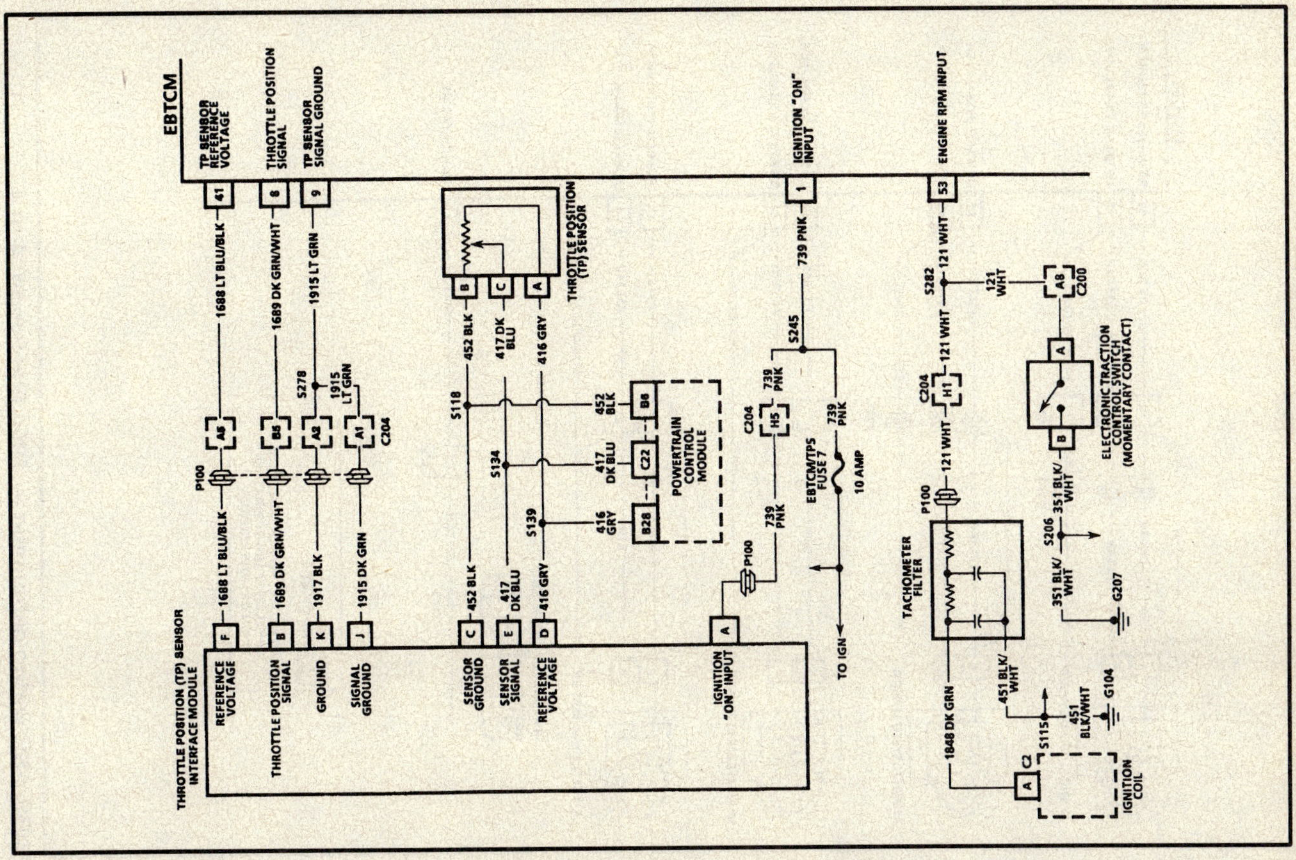

ABS/ASR wiring schematic — 1994 Fleetwood, 2 of 4

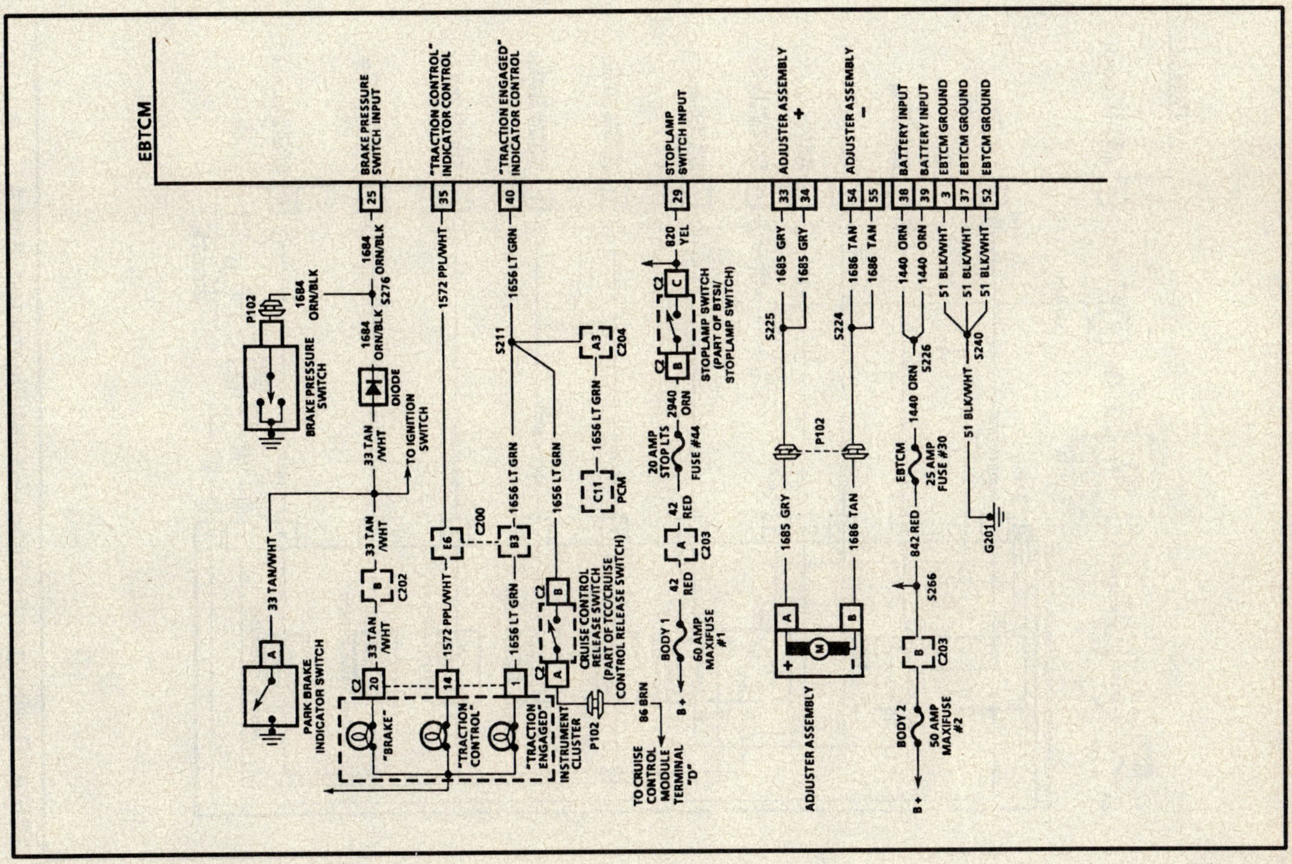

ABS/ASR wiring schematic — 1994 Fleetwood, 1 of 4

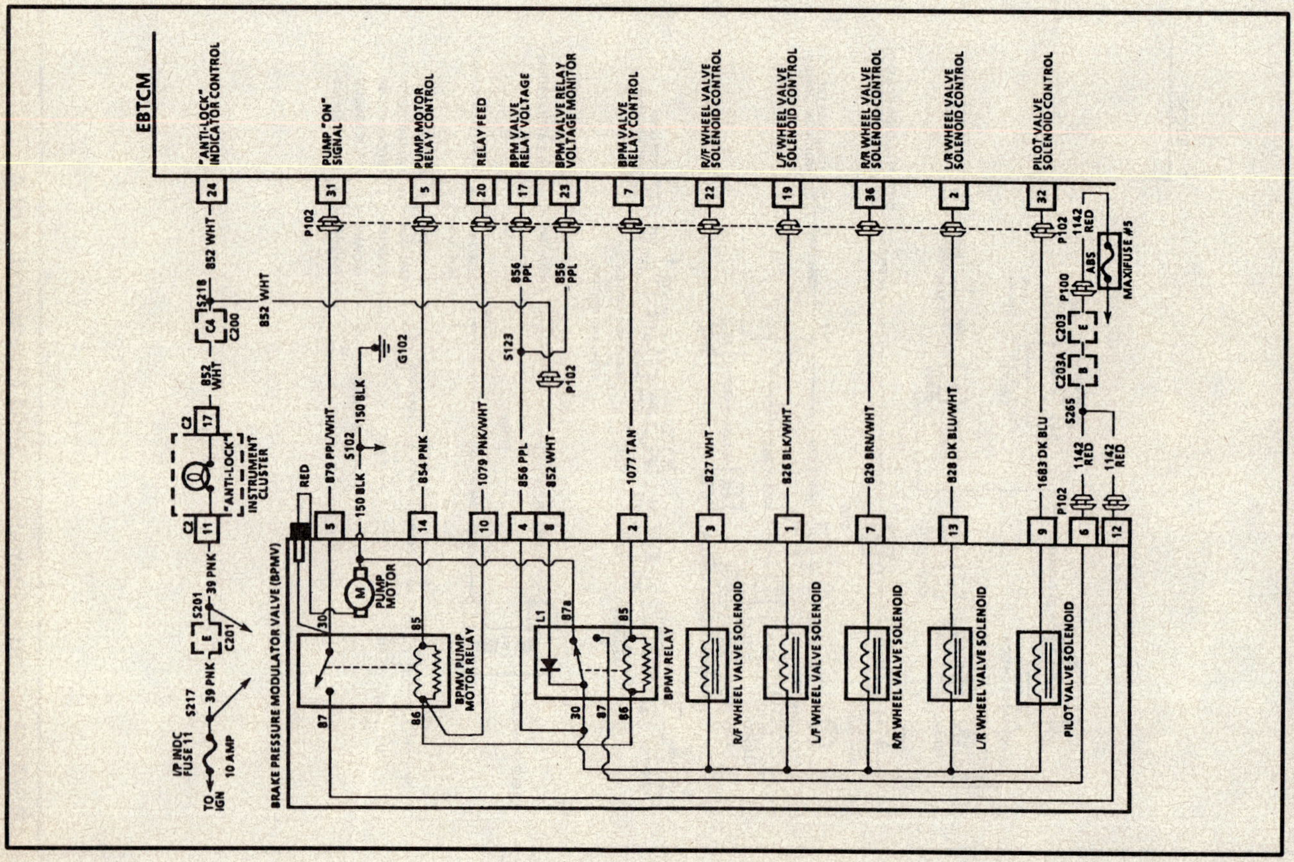

ABS/ASR wiring schematic — 1994 Fleetwood, 3 of 4

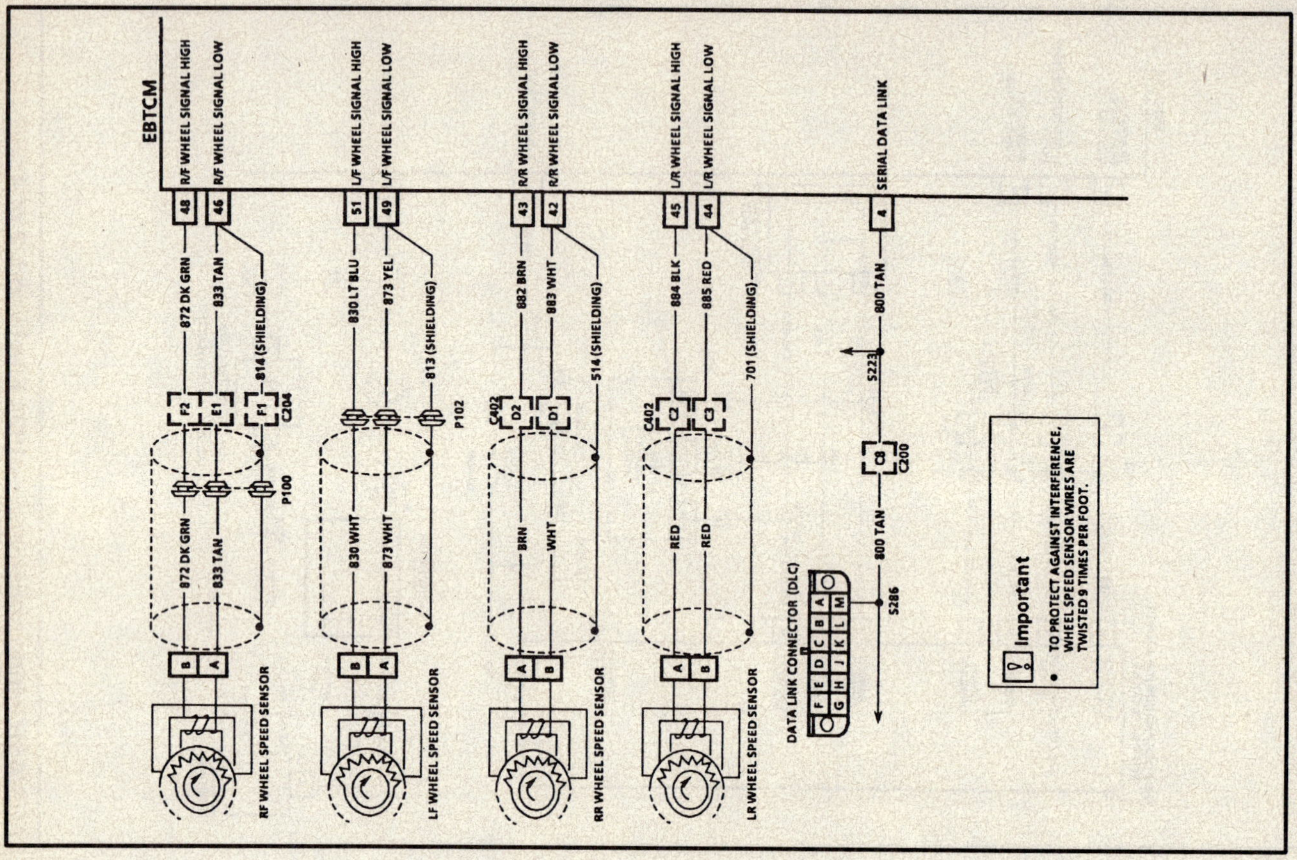

ABS/ASR wiring schematic — 1994 Fleetwood, 4 of 4

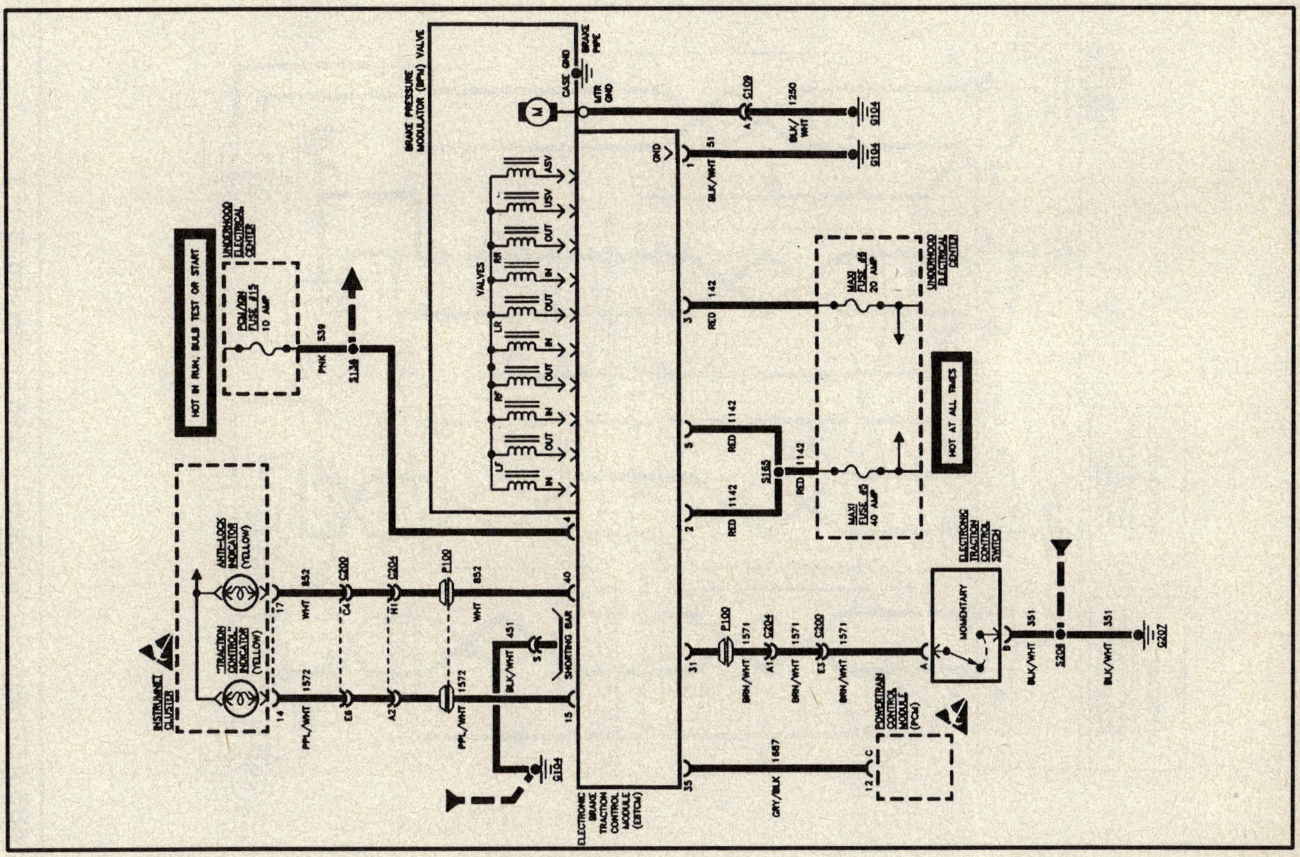

ABS/ASR wiring schematic — 1995 Fleetwood, 1 of 3

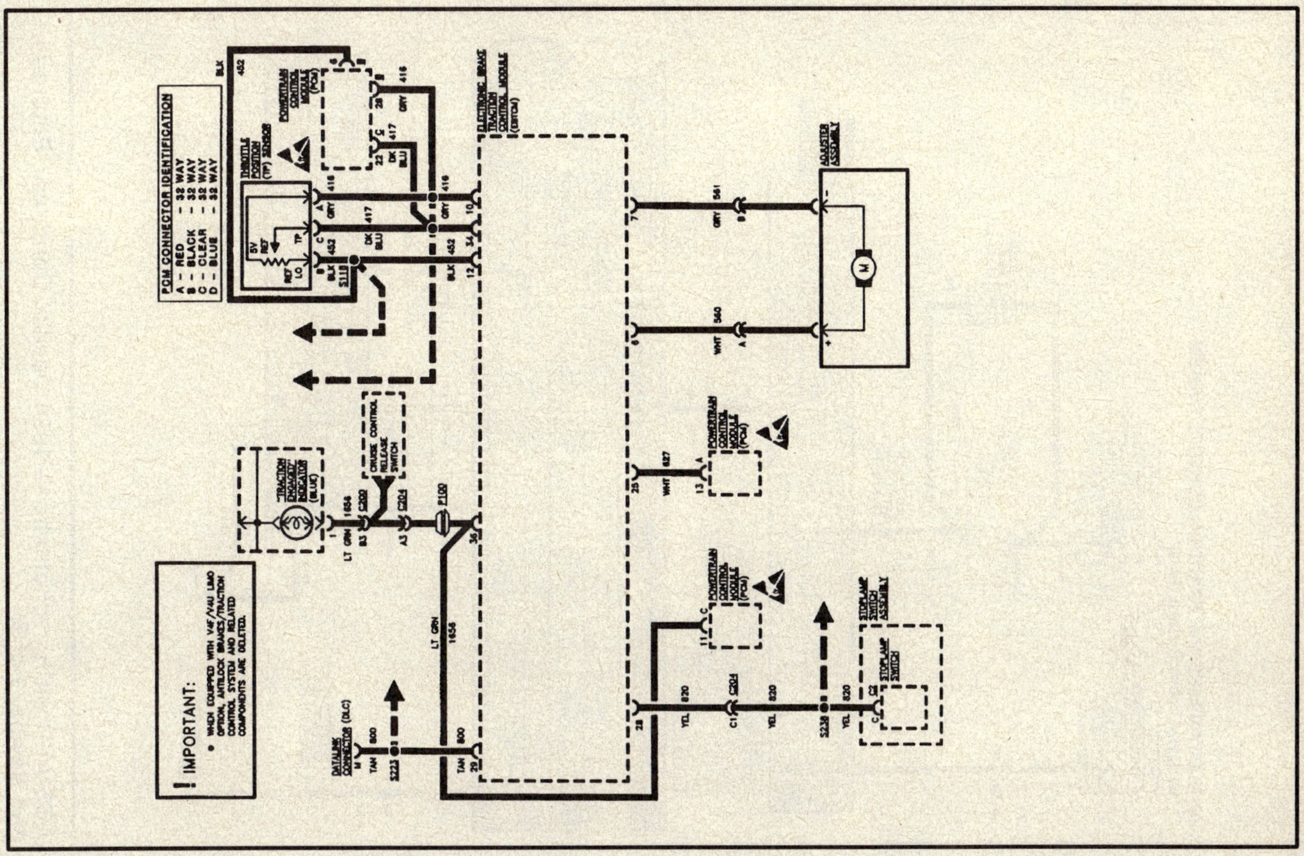

ABS/ASR wiring schematic — 1995 Fleetwood, 2 of 3

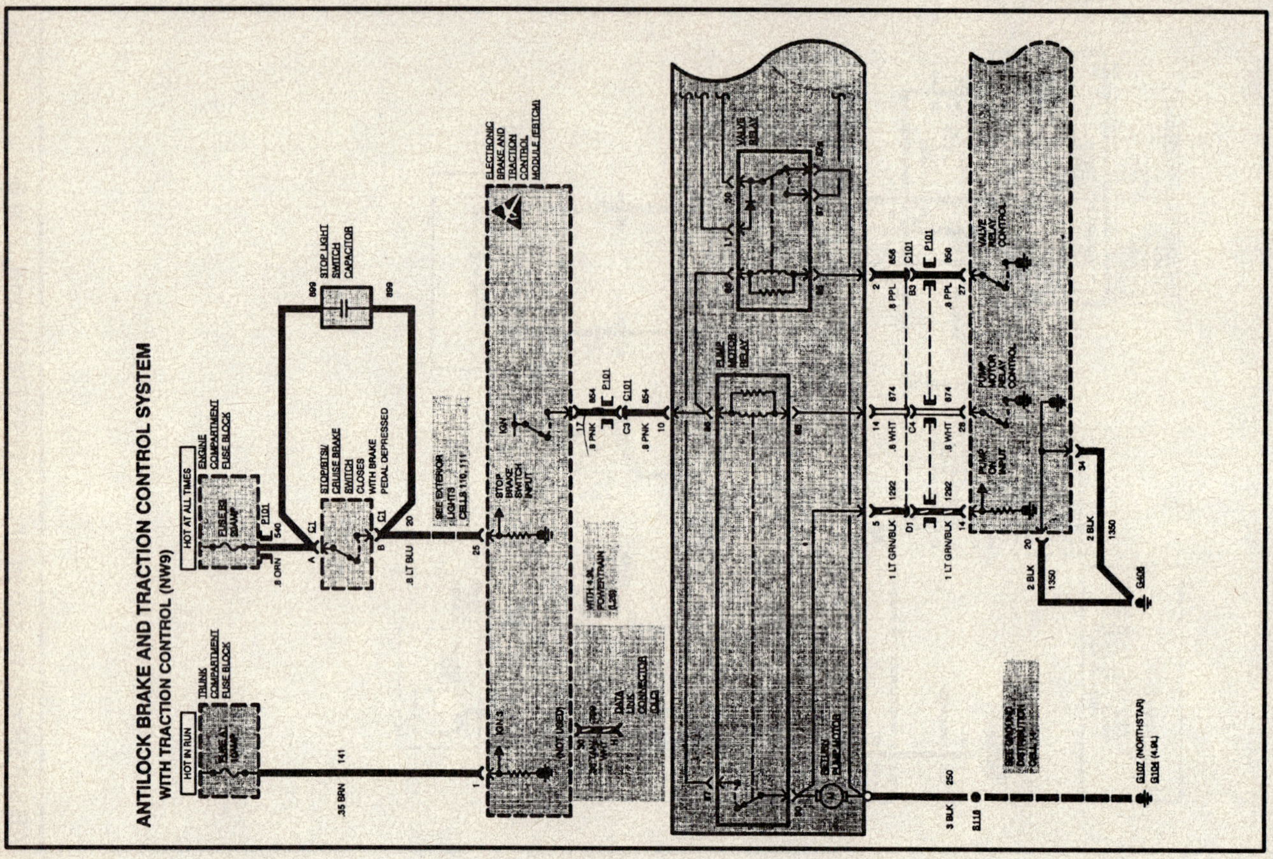

ABS/ASR wiring schematic — 1994 Deville, Concours, Eldorado and Seville, 1 of 5

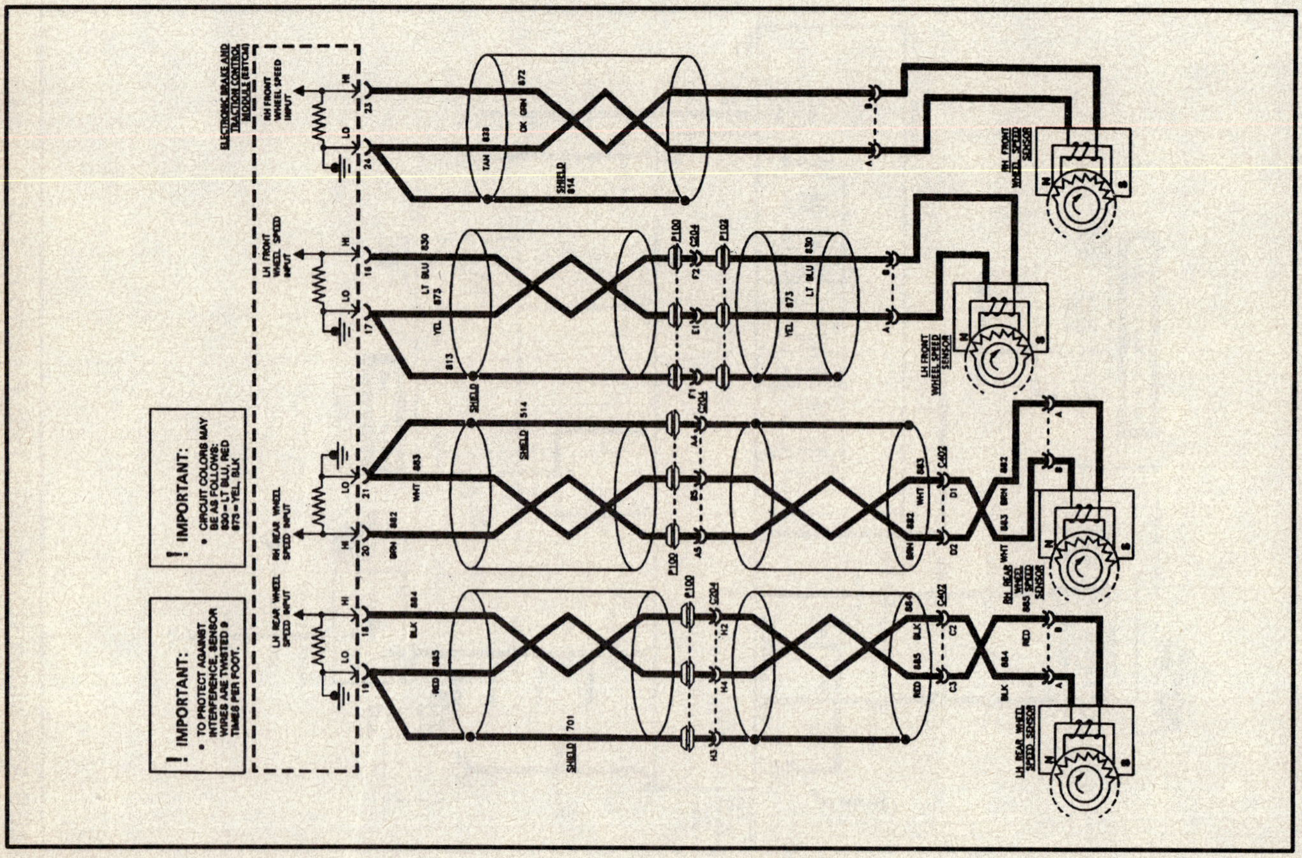

ABS/ASR wiring schematic — 1995 Fleetwood, 3 of 3

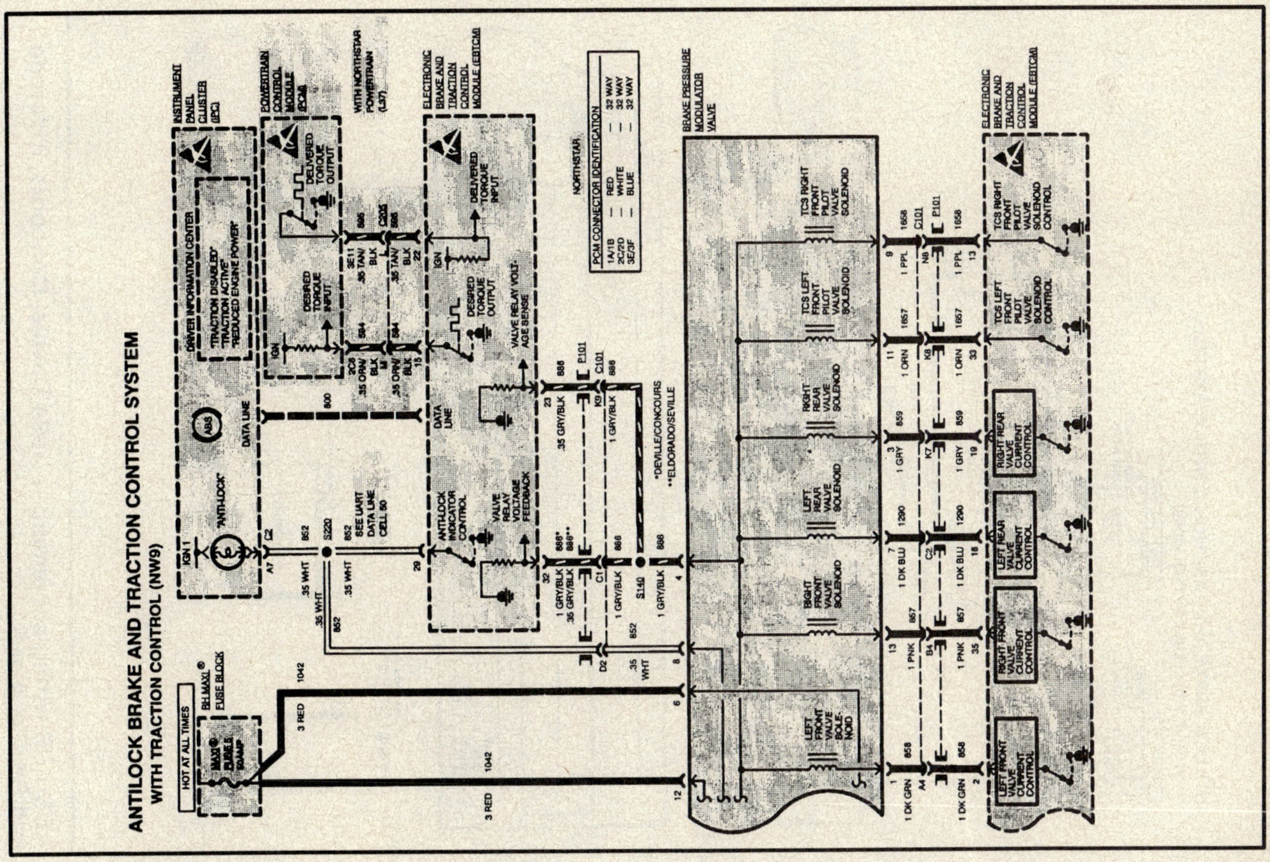

ABS/ASR wiring schematic — 1994 Deville, Concours, Eldorado and Seville, 2 of 5

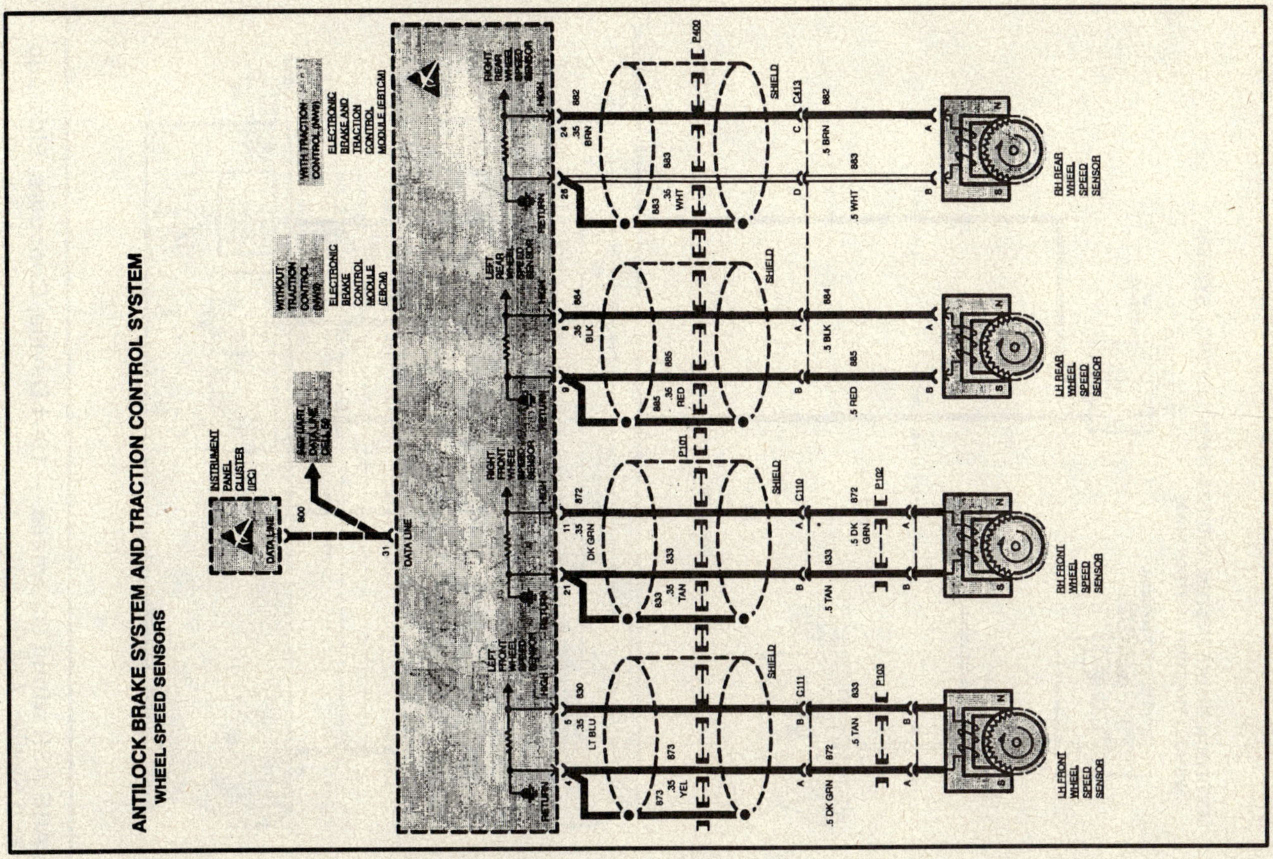

ABS/ASR wiring schematic — 1994 Deville, Concours, Eldorado and Seville, 3 of 5

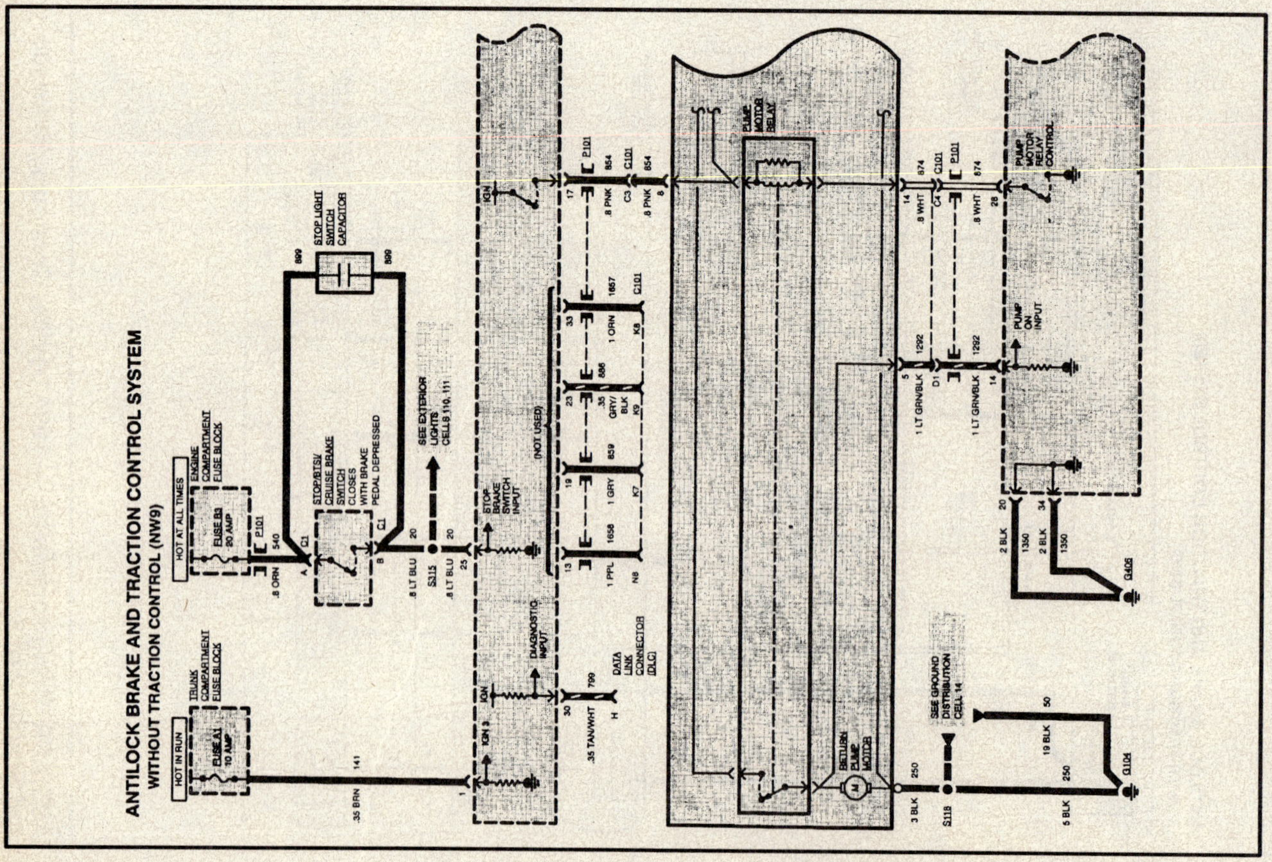

ABS/ASR wiring schematic — 1994 Deville, Concours, Eldorado and Seville, 4 of 5

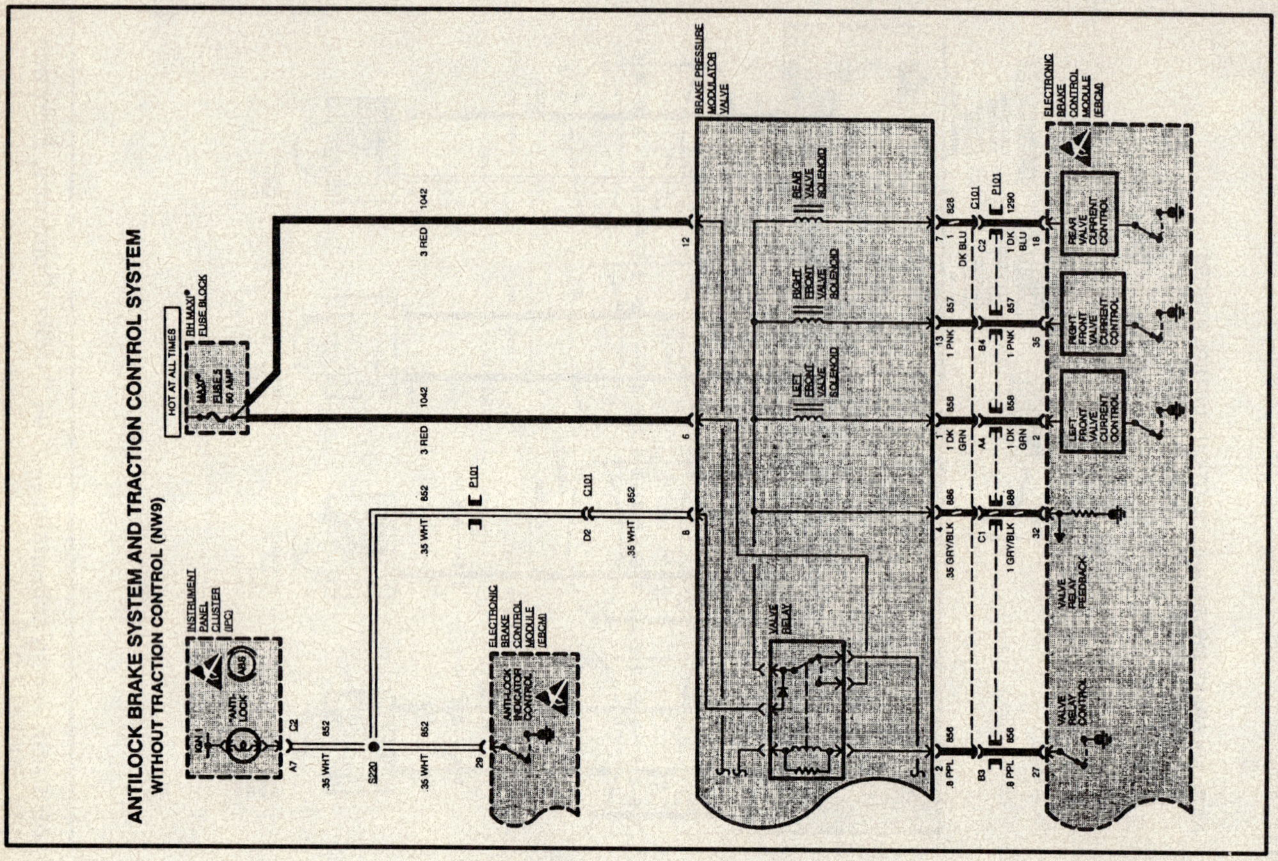

ABS/ASR wiring schematic — 1994 Deville, Concours, Eldorado and Seville, 5 of 5

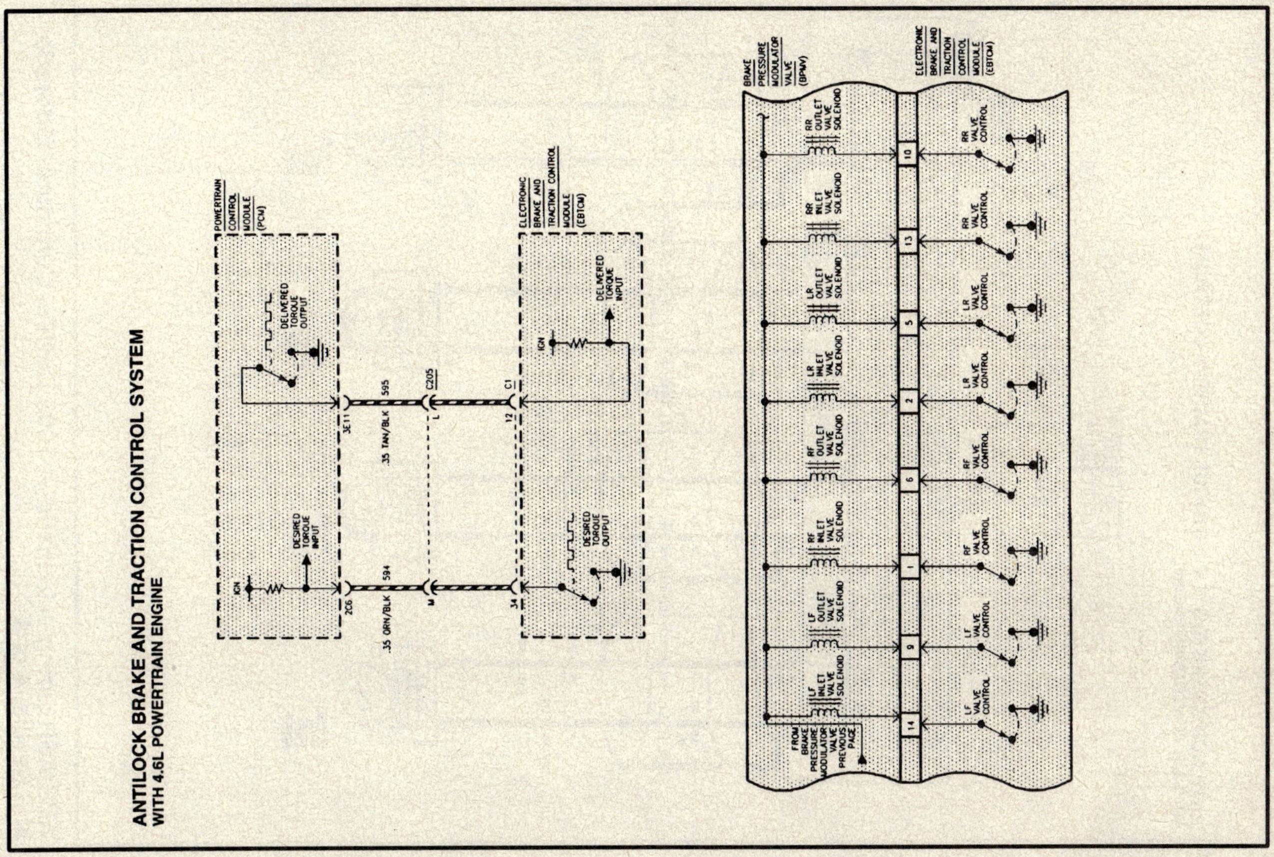

ABS/ASR wiring schematic — 1995 Deville, Concours, Eldorado and Seville, 2 of 4

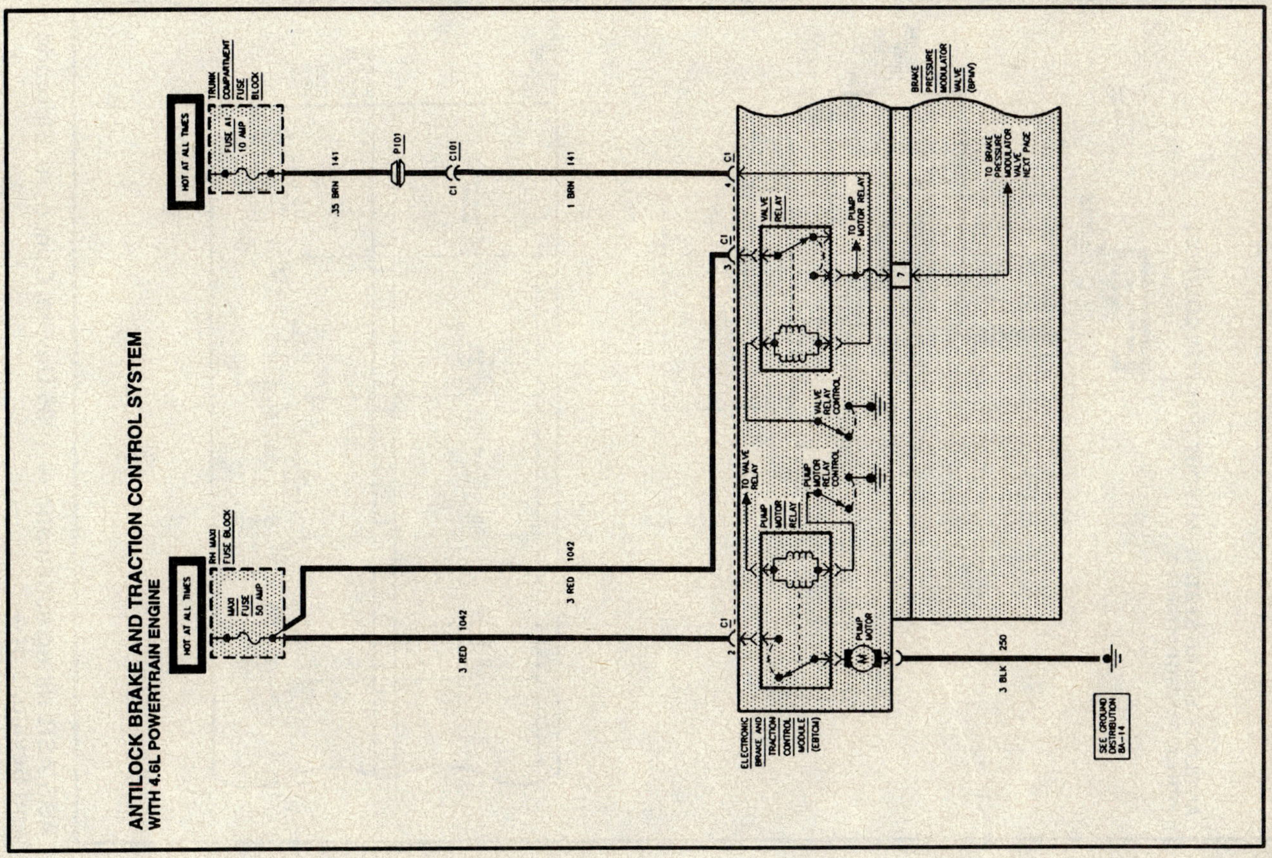

ABS/ASR wiring schematic — 1995 Deville, Concours, Eldorado and Seville, 1 of 4

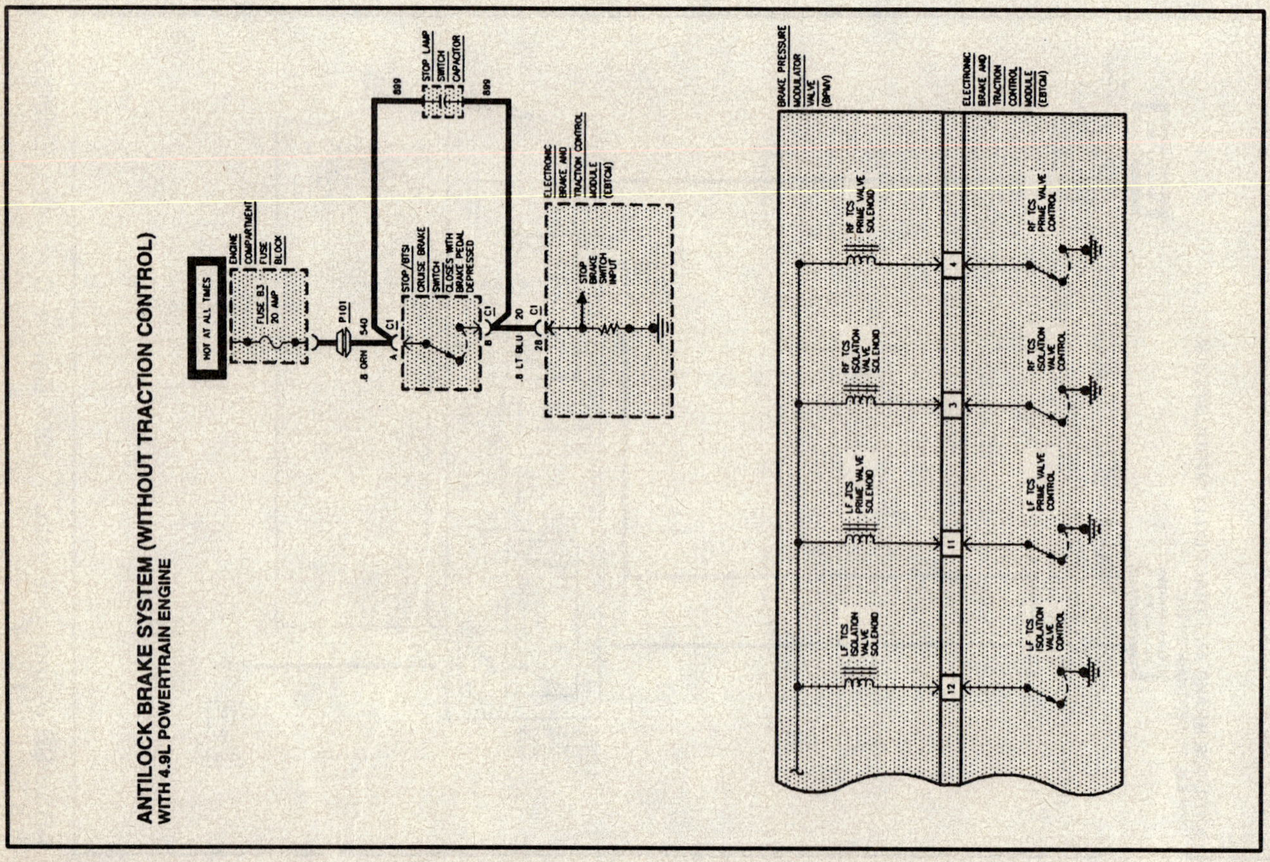

ABS/ASR wiring schematic — 1995 Deville, Concours, Eldorado and Seville, 3 of 4

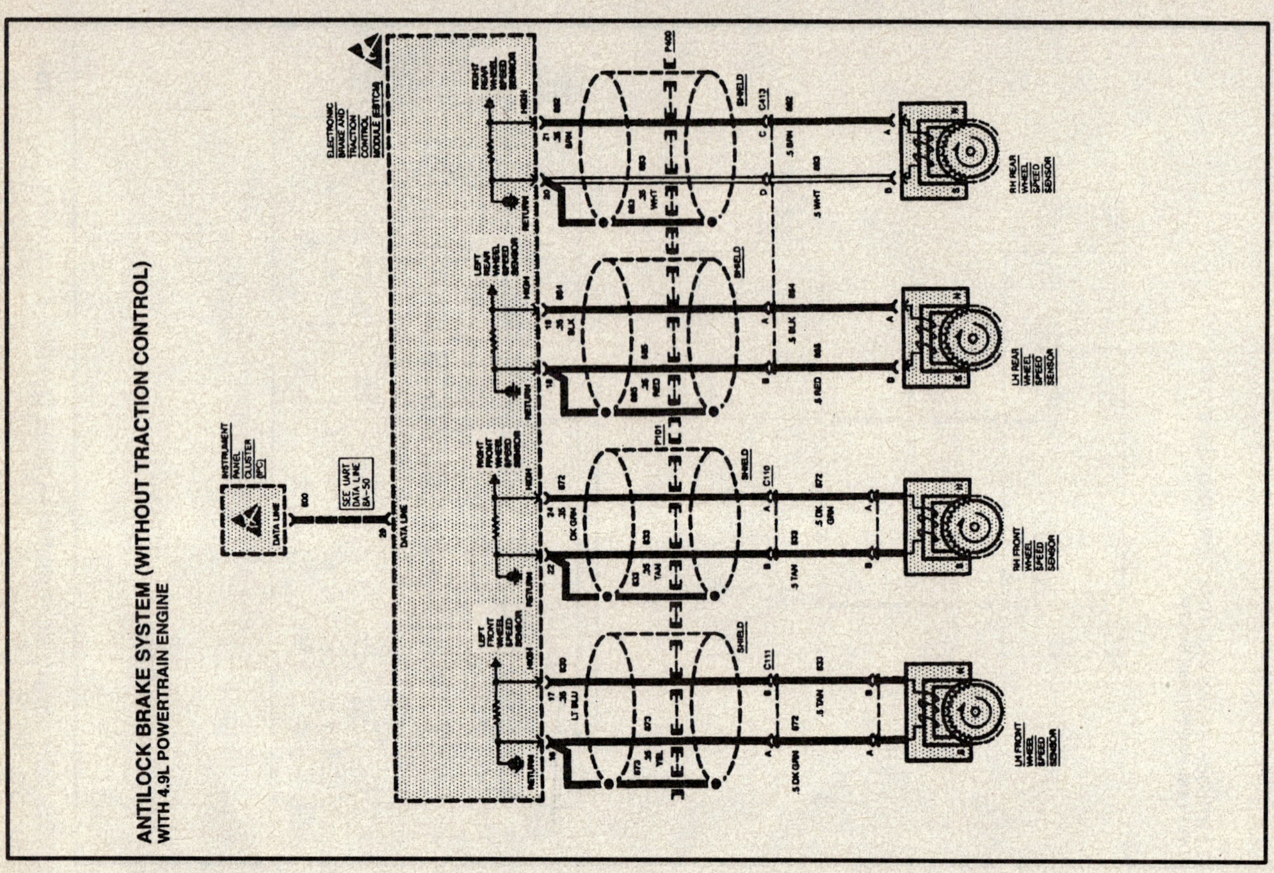

ABS/ASR wiring schematic — 1995 Deville, Concours, Eldorado and Seville, 4 of 4

Component Replacement

SYSTEM REFILL

The master cylinder brake fluid reservoir is filled in the usual manner with no special procedures being necessary. With the ignition **OFF**, wipe all dirt and debris from the cap and reservoir before removing the cap. Add DOT 3 fluid as necessary; never use DOT 5 or silicone fluid in this system. Fill the reservoir only to the **MAX** line. Do not overfill the reservoir. Replace the cap and tighten it.

SYSTEM BLEEDING

Bleeding the entire system is required if air has been introduced by a low fluid condition or by disconnecting the brake lines at the master cylinder or ABP valve. Repairs at a wheel require only bleeding of the affected circuit.

Manual Bleeding

1. If the entire system is to be bled, the ABP valve prime pipe circuit must be bled first. The pipe must be bled at the bleed fitting on the ABP valve. To bleed the prime pipe:
   a. Fill the reservoir to the correct level; leave the cap off during bleeding operations.
   b. Remove the close-out panel above the radiator.
   c. Remove the air cleaner box.
   d. Remove the cap from the bleed screw on the ABP valve; place a wrench of the correct size on the bleed fitting.
   e. Install a piece of clear tubing over the fitting; place the other end of the hose into a container partly filled with brake fluid. The end of the tube must be submerged in the fluid.
   f. Open the bleed screw and allow fluid to flow through the hose. Allow the fluid to flow until no air bubbles are visible in the hose.
   g. Tighten the bleed screw to 9 ft. lbs. (12Nm). Make sure the bleed screw seals.
   h. Reinstall the air cleaner and the close-out panel.

2. Remove the booster vacuum reserve by applying the brake pedal several times with the ignition **OFF**. The master cylinder must be free of air before bleeding the system. If the master cylinder contains air, it must be bled.
   a. Loosen the forward brake line connection at the master cylinder. Allow fluid to flow from the connection; maintain the fluid level in the reservoir.
   b. Tighten the fitting at the master cylinder, stopping the flow of fluid.
   c. Have an assistant depress the brake pedal slowly one time and hold the pedal down. Loosen the forward fitting again, purging air from the cylinder bore. Tighten the fitting, then release the pedal slowly.
   d. Wait 15 seconds. Repeat Steps **c** and **d**, including the 15-second pause, until all air is removed from the bore.
   e. Tighten the fitting to 24 ft. lbs. (32Nm) and check for leakage.
   f. Repeat the entire procedure, steps a through f, at the rearward brake line fitting.

3. Maintain the correct fluid level in the reservoir during the procedure.
4. If more than one circuit is to be bled, do so in this order: left front, right front, left rear, right rear.
5. Remove the bleeder cap at the wheel to be bled. Fit a box-end wrench of the proper size over the bleed fitting. Attach a transparent hose to the fitting and immerse the other end of the hose in a clear container half-filled with clean brake fluid. The end of the hose must be submerged in the fluid at all times during the bleeding.
6. Depress the brake pedal slowly one time and hold it down. Loosen the bleeder valve to allow fluid and air into the hose and container. Tighten the bleeder, then slowly release the pedal.
7. Wait 15 seconds, then repeat Steps 6 and 7 until all air is removed. The sequence may need to be repeated many times to clear the entire circuit.

NOTE: Rapid pumping of the pedal will prevent satisfactory bleeding.

8. When the circuit is free of all air, tighten the wheel cylinder bleed valve to 5 ft. lbs. (7Nm) or tighten the caliper bleed valve to 10 ft. lbs. (13Nm). reinstall the cap on the bleed valve.
9. Repeat Steps 5–8 for each wheel to be bled.
10. Refill the master cylinder reservoir to the **MAX** line only.
11. Check the brake pedal for poor or spongy feel. Repeat the entire bleeding procedure if pedal feel is not correct.

Pressure Bleeding Procedure

NOTE: Use of a diaphragm pressure bleeder is required. The air chamber must be separated from the brake fluid to prevent entry of dirt and moisture. The correct bleeder adapter must be used to prevent damage to the reservoir.

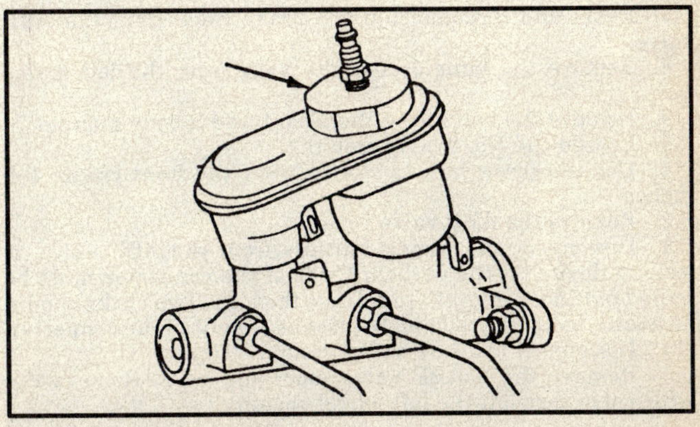

1993 Vehicles require an adapter cap before pressure bleeding

1. Fill the master cylinder reservoir to the proper level.
2. Install the bleeder adapter, J–5589 or equivalent cap to the reservoir.
3. Pressurize the bleeder at 20–25 psi (140–172 kPa). Connect the bleeder pressure hose to the adapter on the reservoir; open the line valve on the bleeder.
4. If more than one circuit is to be bled, do so in the order left front, right front, left rear, right rear.
5. Remove the bleeder cap at the wheel to be bled. Fit a box-end wrench of the proper size over the bleed fitting. Attach a transparent hose to the fitting and immerse the other end of the hose in a clear container half-filled with clean brake fluid. The end of the hose must be submerged in the fluid at all times during the bleeding.

6. Loosen the bleeder valve approximately 3/4 turn to allow fluid and air into the hose and container. Allow the fluid to flow until no air is seen in the fluid.
7. When the circuit is free of all air, tighten the wheel cylinder bleed valve to 10 ft. lbs. (13Nm). Reinstall the cover on the bleed valve.
8. Repeat steps 5–7 for each wheel to be bled.
9. After all circuits are bled, close the pressure line valve on the bleeder. Disconnect the pressure bleeding equipment. Refill the master cylinder reservoir to the **MAX** line only.
10. Check the brake pedal for poor or spongy feel. Repeat the entire bleeding procedure if pedal feel is not correct.

ANTI-LOCK BRAKE PRESSURE (ABP) UNIT

1993

Removal and Installation

NOTE: The ABP valve must not be disassembled or repaired. Only the external relays are replaceable. No screws on the ABP valve should be loosened; if any are loosened, it will be impossible to obtain leak-free brake circuits.

1. Disconnect the negative battery cable.
2. Elevate and safely support the vehicle.
3. Remove the left front wheel.
4. Use a screw jack to support the front cradle assembly.
5. Remove the exhaust Y-pipe from the catalytic converter.
6. Remove the splash shields from the left front wheel house.
7. Label and disconnect the 6 brake lines from the ABP valve.
8. Remove the bundle of brake lines from the left frame rail.
9. Remove the bolts from the Nos. 1 and 2 body mounts.
10. Loosen the No. 3 body mount.
11. Use the screw jack to slowly lower the front cradle 4–5 inches.
12. Remove the ABP valve cover.
13. Disconnect the wiring harness from the ABP valve. A locking device on the outboard side of the connector must be pulled out (not removed) to release the lock. Two clicks should be heard, indicating complete disengagement of the connector.
14. Disconnect the ABP valve ground strap.
15. Remove the 3 ABP valve mounting nuts. Remove the ABP valve through the left wheel opening.

To install:

16. Install the ABP valve to its bracket; tighten the nuts to 10 ft. lbs. (14Nm).
17. Connect the ground strap.
18. Connect the wire harness connector; make certain the locking mechanism is properly engaged.
19. Install the ABP valve cover.
20. Connect the 6 brake lines to the ABP valve. The lines must be installed in their original locations. Tighten the brake lines to 13 ft. lbs. (18Nm).
21. Use the screw jack to raise the cradle into position; do not raise above the correct position.
22. Install the 3 body mounts. Tighten the bolts to 50 ft. lbs. (68Nm).
23. Install the brake line bundle to the left frame rail.
24. Install the splash shields in the left wheel house.
25. Install the Y-pipe to the converter. Tighten the bolts 9 ft. lbs. (12Nm) at the converter.
26. Install the left front wheel.
27. Remove the screw jack; lower the vehicle.
28. Connect the negative battery cable.
29. Fill the brake fluid reservoir as necessary. Bleed the master cylinder, prime pipe and hydraulic circuits.
30. Test drive the vehicle, checking for proper system function. Check for any stored codes.

BRAKE PRESSURE MODULATOR (BPM) VALVE

Removal and Installation

1994–95 DEVILLE, CONCOURS, ELDORADO AND SEVILLE

1. Disconnect the negative battery cable.
2. Safely raise the vehicle off the ground and remove the left front wheel.
3. Support the front cradle with a screw jack.
4. Remove the exhaust Y-pipe from the catalytic converter.
5. Remove the splash shield from the left inner wheel.
6. Remove the brake lines from the BPM valve, making note of their correct location for installation.
7. Disconnect the brake line bundle from the left frame rail.
8. Remove the bolts from the No. 1 and 2 body mounts.
9. Loosen the No. 3 body mount.
10. Using the screw jack, carefully lower the front cradle about 5 inches.
11. Remove the three BPM valve mounting nuts along with the ground wire and remove the BPM through the left wheel well.

To install:

12. Install the BPM valve through the left wheel well and install the three mounting nuts and ground wire. Tighten the nuts to 86 inch lbs. (10Nm).
13. Carefully attach the brake lines to the BPM. Tighten the line fittings to 13 ft. lbs. (18Nm).
14. Using the screw jack, raise the front cradle. Install and tighten the body bolts to 50 ft. lbs. (68Nm).
15. Install the brake line bundle onto the frame rail.
16. Install the inner splash shield.
17. Connect the Y-pipe to the catalytic converter. Tighten the bolts to 9 ft. lbs. (12Nm).
18. Install the left front wheel.
19. Install the negative battery cable.
20. Bleed the brake system and road test the vehicle to insure correct operation. Add DOT 3 fluid as necessary; never use DOT 5 or silicone fluid in this system. Fill the reservoir only to the **MAX** line. Do not overfill the reservoir. Replace the cap and tighten it. Verify that no new codes have been retained in the EBTCM memory.

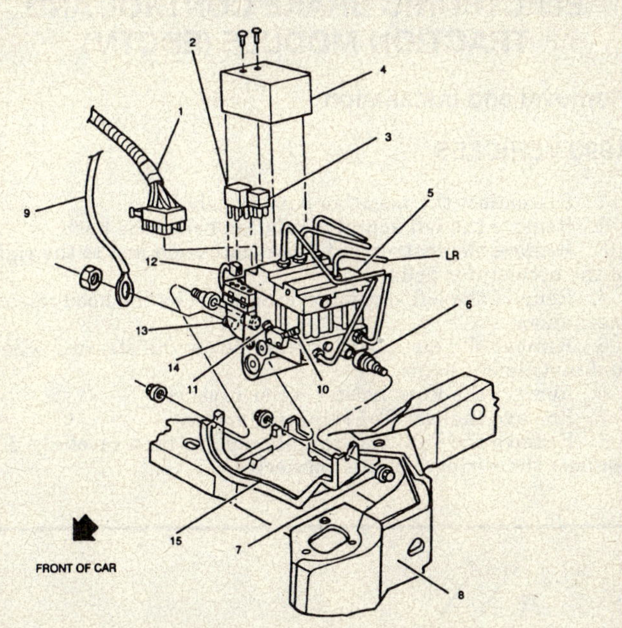

1 ABS VALVE HARNESS CONNECTOR
2 PUMP RELAY
3 VALVE RELAY
4 BPM VALVE COVER
5 BRAKE PRESSURE MODULATOR VALVE (BPM)
6 BUSHING (3)
7 MOUNTING NUT (3)
8 FRAME
9 PUMP/MOTOR GROUND STOP
10 TCS PRIME PIPE NOZZLE
11 TCS PRIME PIPE BLEED SCREW
12 PUMP/MOTOR POWER FEED
13 PUMP/MOTOR GROUND STUD
14 PUMP/MOTOR
15 BPM VALVE MOUNTING BRACKET

Brake pressure modulator valve components — Allante, DeVille, Concours, Eldorado and Seville

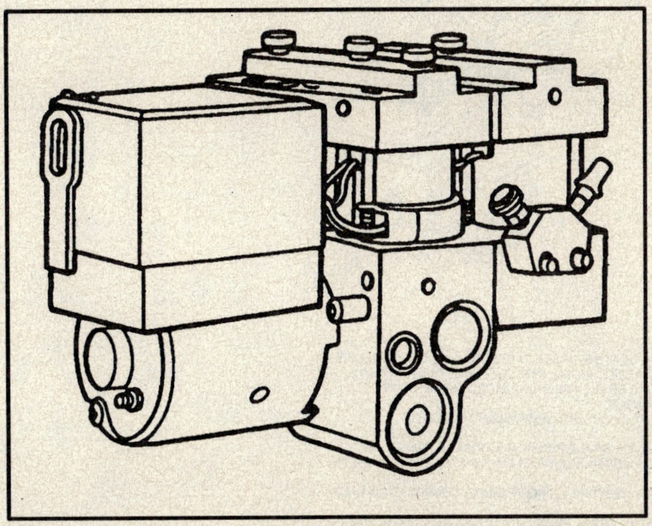

Brake pressure modulator valve — 1994–95 DeVille, Concours, Eldorado and Seville

1994–95 FLEETWOOD

1. Disconnect the negative battery cable.
2. Remove the air intake duct and the resonator assembly.
3. It may be necessary to move the radiator upper coolant hose and the canister purge hose assembly out of the way.
4. Loosen bolt/screw and remove the cover from the brake pressure modulator valve assembly on 1994 models.
5. Disconnect the wiring harness from the connector on the brake pressure modulator.
6. Remove the ground cable connector.
7. Remove the three bracket nuts and insulators.

NOTE: Mark location of brake lines for aid during installation.

8. Carefully loosen and remove the brake lines.
9. Lift the brake pressure modulator off the bracket and remove it from the vehicle.

To install:

10. Install the brake pressure modulator valve into the bracket.
11. Install nuts and insulators on the brake pressure modulator and bracket.
12. Carefully install the brake lines in their proper locations.
13. Install the ground cable connector.
14. Install the wiring harness into the brake pressure modulator.
15. Install the bolt/screw and the cover on the modulator valve.
16. If removed, install the radiator upper coolant hose and the canister purge hose assembly.
17. Install the air intake duct and the resonator assembly.
18. Connect the negative battery cable.
19. Bleed the brake system and road test the vehicle to insure correct operation. Add DOT 3 fluid as necessary; never use DOT 5 or silicone fluid in this system. Fill the reservoir only to the **MAX** line. Do not overfill the reservoir. Replace the cap and tighten it. Verify that no new codes have been retained in the EBTCM memory.

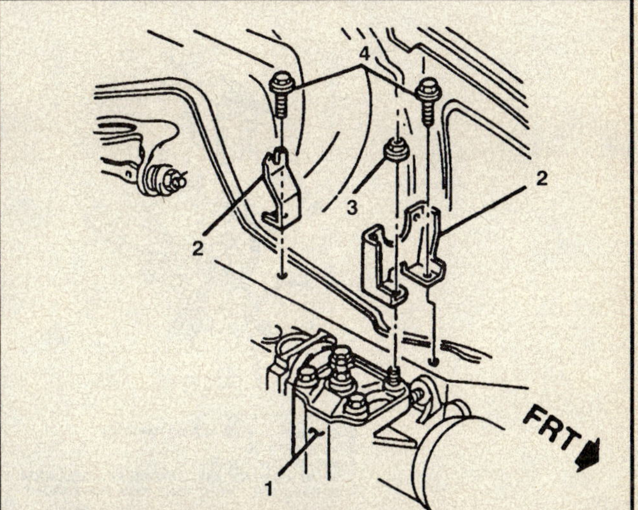

1 GEAR ASSEMBLY, STEERING
2 BRACKET, BRAKE PRESSURE MODULATOR VALVE
3 NUT, BRAKE PRESSURE MODULATOR VALVE BRACKET
4 BOLT/SCREW, BRAKE PRESSURE MODULATOR VALVE BRACKET

Brake pressure modulator valve bracket — 1994 Fleetwood

FLUID RESERVOIR

Removal and Installation

The master cylinder reservoir may be replaced after removing the master cylinder in the traditional fashion. No special procedures are necessary.

FLUID LEVEL SENSOR

Removal and Installation

The fluid level sensor switch may be removed from the fluid reservoir. Disconnect the fluid level sensor wing. Use narrow pliers to compress the switch locking tabs at the inboard side of the master cylinder, then remove the switch. Reinstall the switch, making sure the locking tabs engage securely. Connect the wiring to the switch.

ELECTRONIC BRAKE CONTROL AND TRACTION MODULE (EBCTM)

Removal and Installation

1993 VEHICLES

1. Disconnect the negative battery cable.
2. Remove the left sound insulation below the dash.
3. Remove the instrument panel extension just to the right of the accelerator pedal.
4. Remove the left shroud or kick panel by the hood release mechanism.
5. Remove the knee bolster from below the steering wheel and instrument panel.
6. Remove the knee bolster inner bracket.
7. Remove the EBCTM retaining screw.
8. Remove the EBCTM from its mount, then carefully disconnect the wiring harness connector.

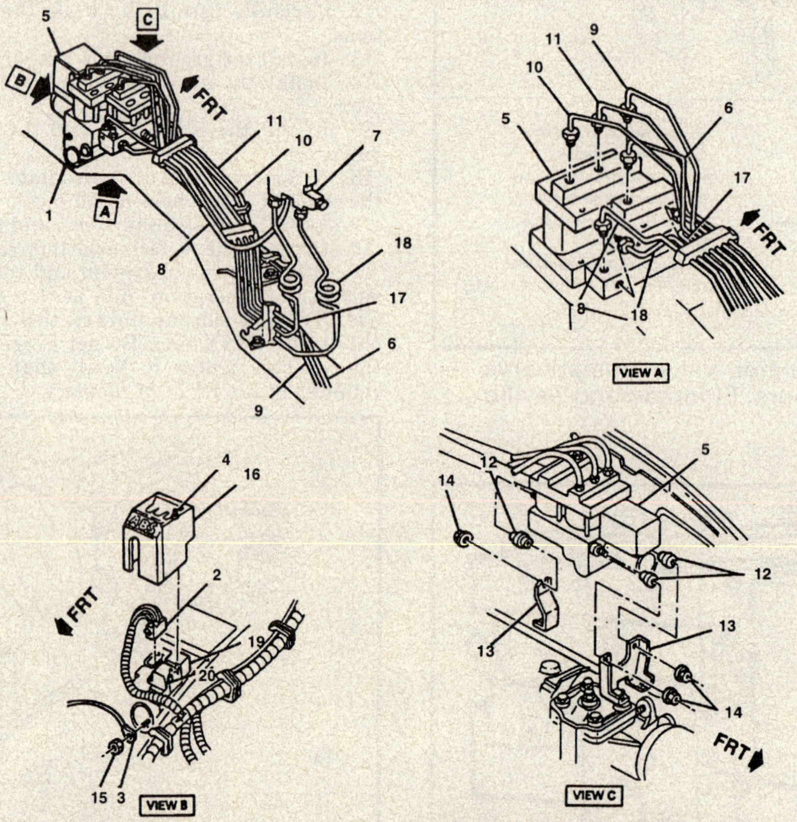

1 FRAME ASSEMBLY
2 CONNECTOR, BODY WIRING HARNESS
3 WIRE, GROUND
4 BOLT/SCREW, COVER
5 VALVE ASSEMBLY, BRAKE PRESSURE MODULATOR
6 PIPE ASSEMBLY, RIGHT-HAND REAR BRAKE FRONT
7 VALVE ASSEMBLY, BRAKE COMBINATION
8 HOSE ASSEMBLY, BRAKE MASTER CYLINDER RESERVOIR
9 PIPE ASSEMBLY, LEFT-HAND REAR BRAKE FRONT,
10 PIPE ASSEMBLY, LEFT-HAND FRONT BRAKE
11 PIPE ASSEMBLY, RIGHT-HAND FRONT BRAKE
12 INSULATOR, BRAKE PRESSURE MODULATOR VALVE
13 BRACKET, BRAKE PRESSURE MODULATOR VALVE
14 NUT, BRAKE PRESSURE MODULATOR VALVE BRACKET
15 NUT, FORWARD LAMP WIRING HARNESS
16 COVER, BRAKE PRESSURE MODULATOR VALVE
17 PIPE ASSEMBLY, REAR BRAKE COMBINATION VALVE
18 PIPE ASSEMBLY, FRONT BRAKE COMBINATION VALVE
19 RELAY, BRAKE PRESSURE MODULATOR VALVE PUMP MOTOR
20 RELAY, BRAKE PRESSURE MODULATOR VALVE

Brake pressure modulator valve assembly — 1994 Fleetwood

To install:

9. Connect the wiring harness to the EBCTM; make certain the connector is squarely placed and that the connector lock is securely engaged.
10. Mount the unit and install the retaining nut.
11. Install the bolster inner bracket and knee bolster.
12. Install the shroud or kick panel.
13. Install the instrument panel extension.
14. Install the left sound insulator.
15. Connect the negative battery cable.
16. Test drive the vehicle, checking for proper system operation. Check for any stored codes.

1994 FLEETWOOD

1. Disconnect the negative battery cable.
2. Remove the left hand instrument panel sound insulator assembly.
3. Disconnect the electrical connector to the EBTCM.
4. Remove the attaching bolts for the EBTCM and remove the unit from the vehicle.

To install:

5. Install the EBTCM into the vehicle and install the mounting bolts.
6. Install the electrical connector to the EBTCM.
7. Install the left hand instrument panel sound insulator assembly.
8. Connect the negative battery cable.

1994 DEVILLE, CONCOURS, ELDORADO AND SEVILLE

1. Remove the electronics bay carpet panel at the forward part of the trunk.
2. Remove the three mounting screws that hold the EBTCM/EBCM in the cradle.
3. Disconnect the harness connector from the EBTCM/EBCM and remove it from the vehicle.

To Install:

4. Install the harness connector to the EBTCM/EBCM.
5. Install the EBTCM/EBCM into the cradle and install the mounting screws.
6. Install the electronics bay carpet panel.

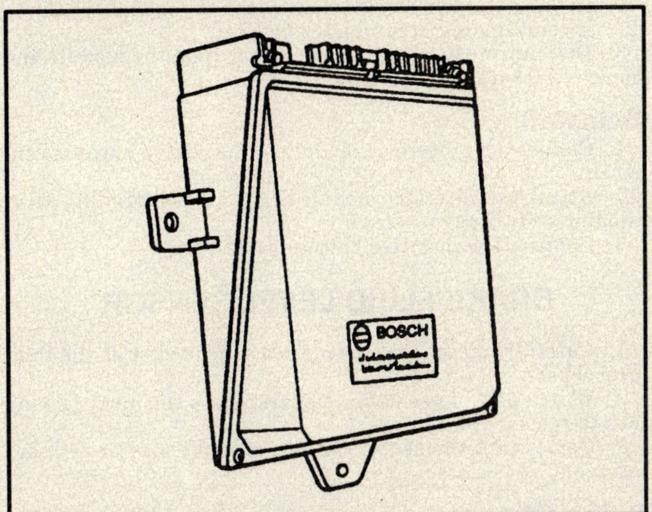

EBTCM — 1995 DeVille, Concours, Eldorado and Seville

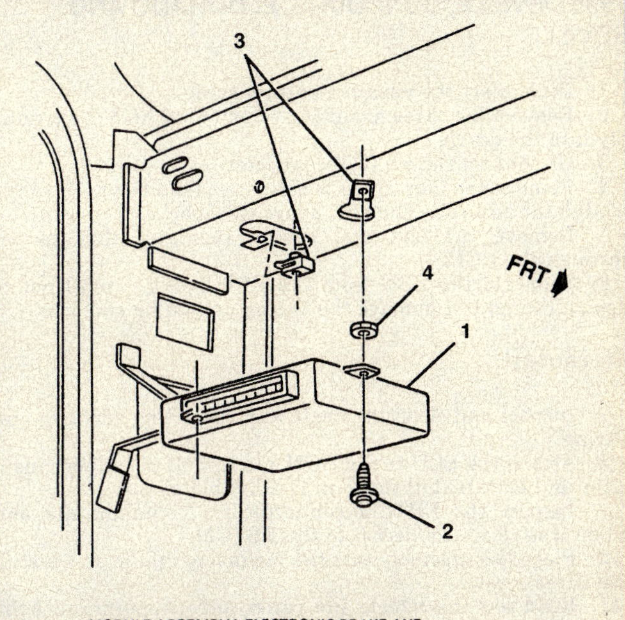

EBTCM — 1994 Fleetwood

1995 FLEETWOOD

NOTE: The EBTCM and BPM are an integrated assembly.

1. Disconnect the negative battery cable.
2. Remove the air intake duct and resonator assembly.
3. Carefully clean the EBTCM/BPM assembly to prevent the entry of dirt into the brake system.
4. Loosen and remove the wire harness from the frame rail bracket.
5. Disconnect and remove the electrical connector.
6. Remove the EBTCM/BPM mounting nuts and insulators.
7. Loosen and remove the brake pipe clip bolt that is downstream from EBTCM/BPM unit.
8. Prop up the EBTCM/BPM assembly with a block of wood.
9. Loosen and remove the four EBTCM to BPM mounting bolts.
10. Remove the front insulator bolt.
11. Carefully remove the EBTCM from the BPM valve.

To install:

12. Carefully place the EBTCM on the BPM valve, being careful not to damage the O-ring or mounting surfaces.
13. Install the front insulator bolt.
14. Install the four EBTCM to BPM mounting bolts.
15. Remove the wood block and position the EBTCM/BPM into the mounting bracket.
16. Install the brake clip bolt into the frame rail.
17. Install the mounting insulators.
18. Tighten the mounting nuts and bolts to 89 inch lbs. (10Nm).
19. Install the air duct and resonator assembly.
20. Connect the negative battery cable.

1995 DEVILLE, CONCOURS, ELDORADO AND SEVILLE

1. Disconnect the negative battery cable.
2. Remove the three mounting screws that hold the assembly into the cradle.
3. Disconnect the EBTCM connector.
4. Remove the four TORX® bolts that hold the ECU to the modulator and raise the unit above the bracket.
5. Remove the bolt and rubber insulator that passes through the ECU.
6. Separate the ECU from the HCU being careful not to pry on the unit or damage the O-ring or sealing surfaces.

To install:

7. Inspect and clean the modulator mounting surfaces and O-ring.
8. Attach the ECU to the HCU and install the rubber insulator and bolt. Install the four TORX® bolts.
9. Inspect the EBTCM connector for contaminants and when it is clean, connect it to the EBTCM.
10. Place the assembly into the mounting cradle and install the three screws.
11. Road test the vehicle and verify correct operation of the ABS/ASR system. Make sure that no new codes have been retained by the system.

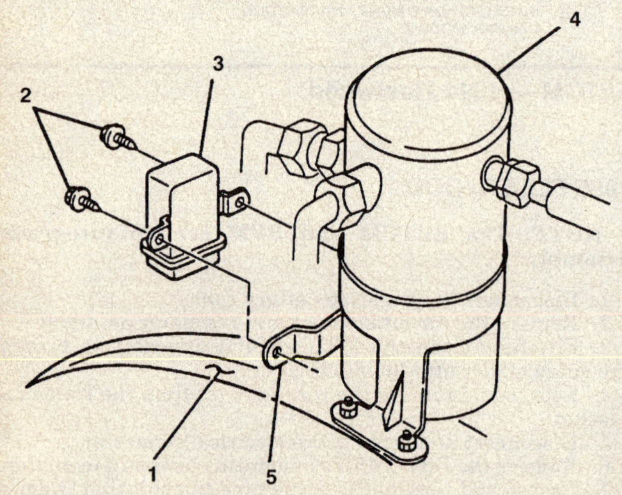

1. PANEL, WHEELHOUSE
2. BOLT/SCREW, THROTTLE POSITION SENSOR INTERFACE MODULE
3. MODULE ASSEMBLY, THROTTLE POSITION SENSOR INTERFACE
4. ACCUMULATOR ASSEMBLY, AIR CONDITIONING
5. BRACKET, THROTTLE POSITION SENSOR INTERFACE MODULE

Throttle position sensor interface module assembly — 1994 Fleetwood

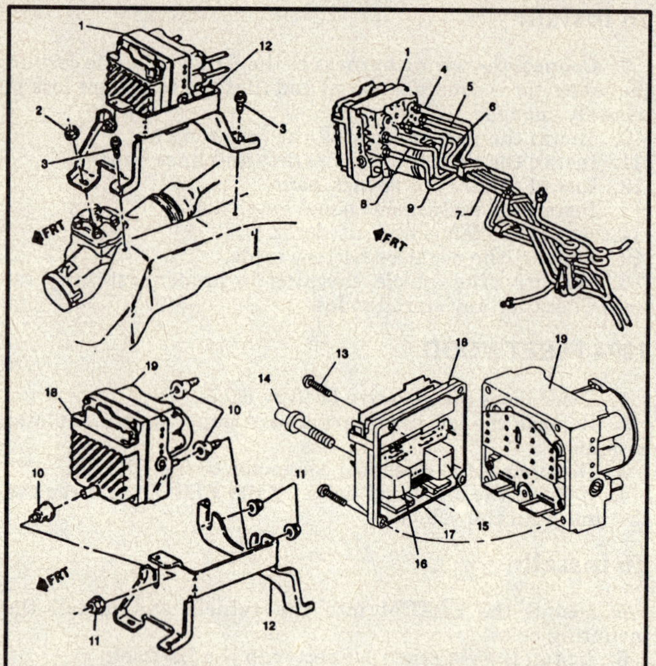

1. EBTCM/BPM VALVE ASSEMBLY
2. BRACKET TO STEERING GEAR NUT
3. BRACKET MOUNTING BOLT
4. REAR BRAKE PIPE (TO COMBINATION VALVE)
5. RIGHT REAR BRAKE PIPE
6. LEFT REAR BRAKE PIPE
7. FRONT BRAKE PIPE (TO COMBINATION VALVE)
8. RIGHT FRONT BRAKE PIPE
9. LEFT FRONT BRAKE PIPE
10. EBTCM/BPM VALVE MOUNTING INSULATION
11. EBTCM/BPM VALVE BRACKET NUT
12. EBTCM/BPM VALVE MOUNTING BRACKET
13. EBTCM TO BPM VALVE BOLT
14. INSULATOR MOUNTING BOLT
15. VALVE RELAY (NOT SERVICEABLE, INTEGRAL TO EBTCM)
16. PUMP MOTOR RELAY (NOT SERVICEABLE, INTEGRAL TO EBTCM)
17. EBTCM SEAL
18. ELECTRONIC BRAKE TRACTION CONTROL MODULE (EBTCM)
19. BRAKE PRESSURE MODULATOR (BPM) VALVE

EBTCM and BPM assembly — 1995 Fleetwood

THROTTLE POSITION SENSOR INTERFACE MODULE ASSEMBLY

Removal and Installation

1. Disconnect the negative battery cable.
2. Remove the two screws that attach the module to the to the air conditioning accumulator bracket.
3. Disconnect the electrical connector from the module and remove it from the vehicle.

To install:

4. Connect the interface module to the wiring harness connector.
5. Attach the interface module to the air conditioning accumulator with the two screws
6. Connect the negative battery cable.

BRAKE FLUID LEVEL SENSOR

1. Carefully disconnect the electrical connector at the fluid level sensor.
2. With a pair of needle nose pliers, grasp the inner locking tabs to release the sensor
3. Gently pull the sensor out of the brake master cylinder reservoir.

To install:

4. Gently install the sensor into the brake master cylinder reservoir.
5. Connect the electrical wire connector to the sensor.

FRONT WHEEL SPEED SENSORS

Removal and Installation

The front wheel speed sensors on these vehicles are integrated with the hub and bearing assembly. The sensor may be replaced and is serviced separately from the hub and bearing assembly.

NOTE: The sensor must be replaced whenever it is separated from the hub or bearing. Do not attempt to reuse the sensor once it is removed.

1. With the ignition **OFF**, disconnect the wheel speed sensor connector.
2. Remove the hub and bearing assembly. Clean all dirt from the sensor housing area.
3. Gently pry the slinger from the speed sensor assembly. Discard the used slinger.
4. Carefully pry the speed sensor from the bearing assembly. Do not allow dirt to enter the bearing when the sensor is removed. Do not add lubricant through the sensor opening; the bearing is permanently lubricated.
5. Inspect the bearing and sensor for any signs of rust or water intrusion; if such damage is found, replace the hub and bearing assembly. Do not clean the grease from the toothed sensor ring; grease does not affect operation of the sensor.
6. Use a clean, lint-free cloth to remove sealant from the outer diameter of the bearing hub.

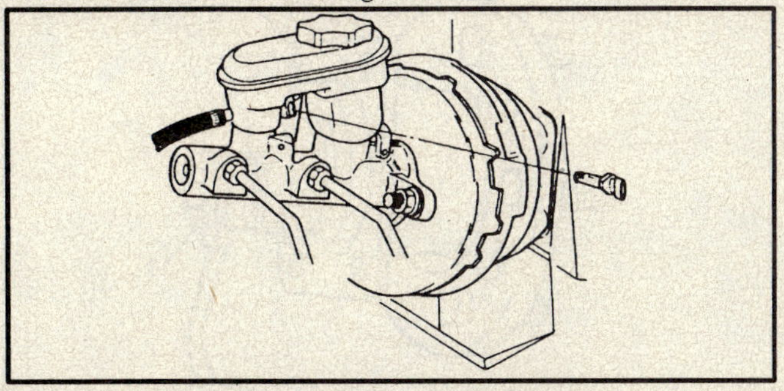

The brake fluid level sensor is removable

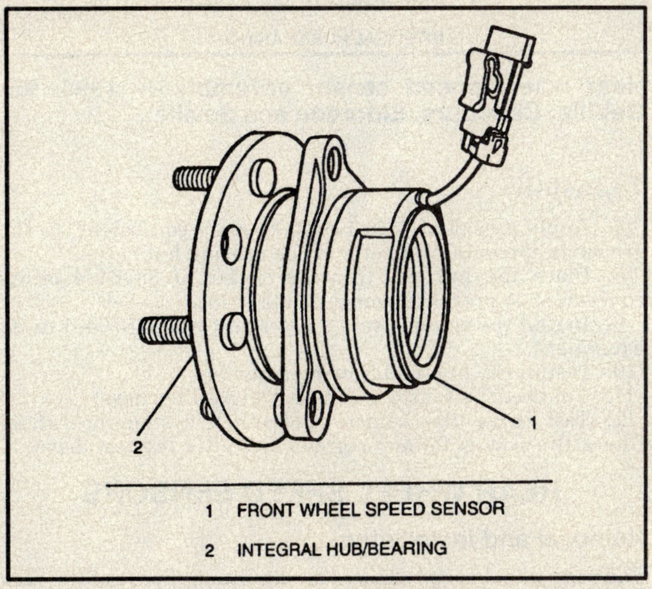

Front wheel speed sensor assembly — 1994–95 DeVille, Concours, Eldorado and Seville

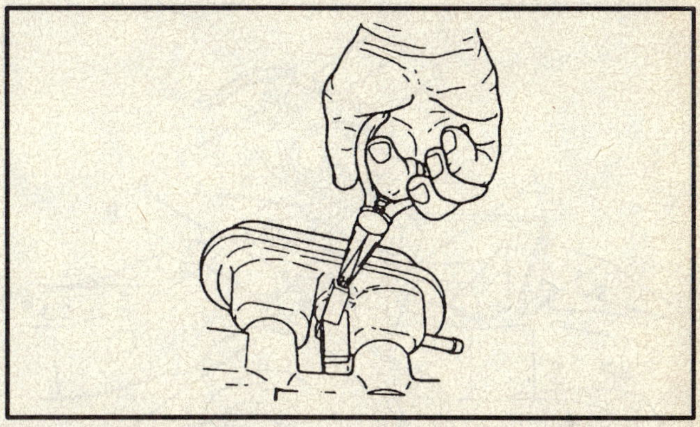

Carefully release the locking tabs before removing the sensor

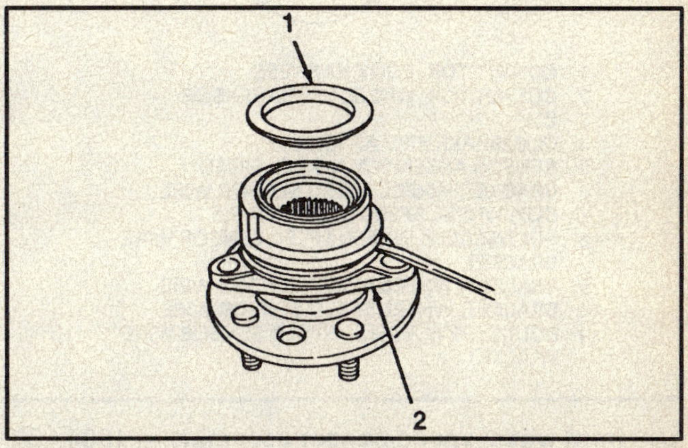

After the slinger (1) is removed, carefully pry the sensor from the hub and bearing assembly (2)

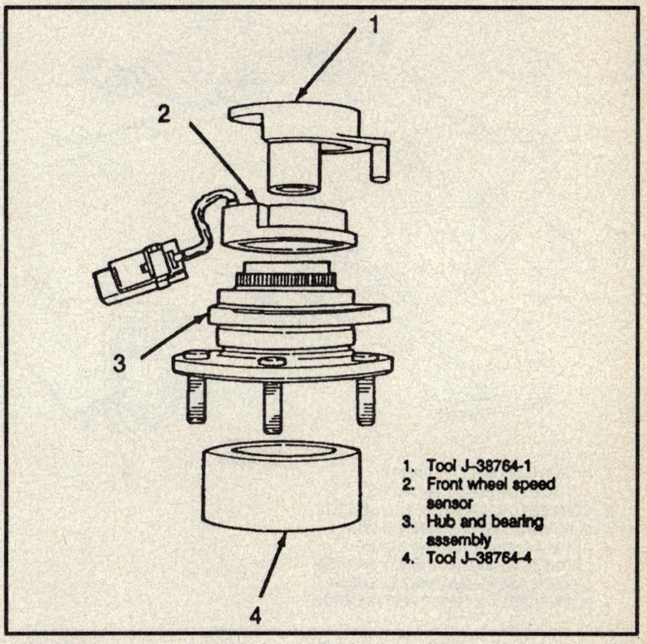

Installation of wheel speed sensor must be performed with a press

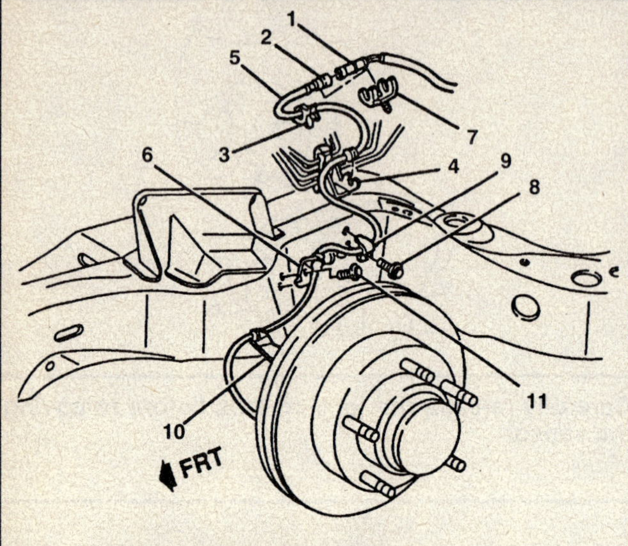

1 CONNECTOR, BODY HARNESS
2 CONNECTOR, WHEEL SPEED SENSOR
3 CLIP
4 CLIP, BRAKE PIPE ASSEMBLY
5 SENSOR ASSEMBLY, WHEEL SPEED
6 BRACKET, WHEEL SPEED SENSOR WIRE
7 CLIP, WHEEL SPEED SENSOR WIRE
8 BOLT/SCREW, WHEEL SPEED SENSOR WIRE BRACKET
9 BRACKET, WHEEL SPEED SENSOR WIRE
10 BRACKET, WHEEL SPEED SENSOR WIRE
11 BOLT/SCREW, WHEEL SPEED SENSOR WIRE BRACKET

Left front wheel speed sensor assembly — 1994–95 Fleetwood

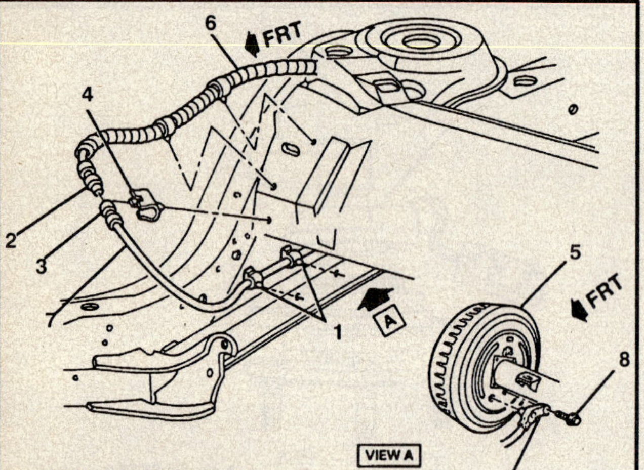

1 CLIP, WHEEL SPEED SENSOR
2 CONNECTOR, BODY HARNESS
3 CONNECTOR, WHEEL SPEED SENSOR
4 CLIP, WHEEL SPEED SENSOR WIRE
5 DRUM ASSEMBLY, REAR BRAKE
6 HARNESS ASSEMBLY, BODY WIRING
7 SENSOR ASSEMBLY, WHEEL SPEED
8 BOLT/SCREW, WHEEL SPEED SENSOR

Rear wheel speed sensor assembly — 1994–95 Fleetwood

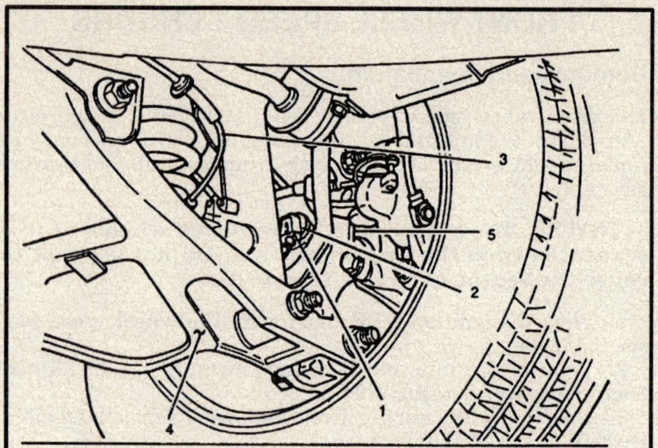

1 WHEEL SPEED SENSOR CONNECTOR—RR
2 WHEEL SPEED SENSOR—RR
3 WHEEL SPEED SENSOR CABLE
4 LOWER CONTROL ARM
5 INTEGRAL HUB/BEARING

Rear wheel speed sensor and suspension assembly — 1994–95 DeVille, Concours, Eldorado and Seville

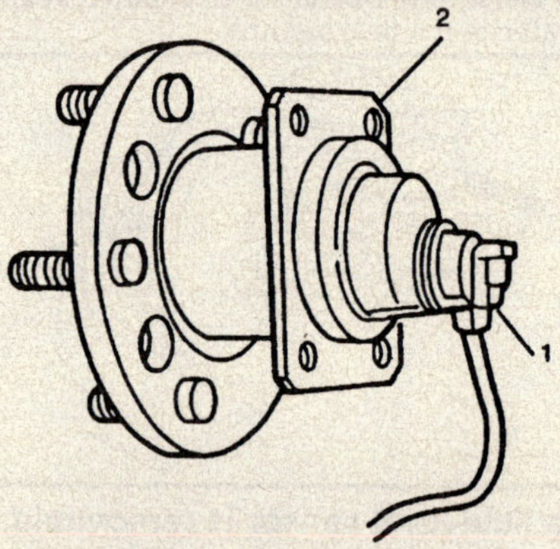

1 REAR WHEEL SPEED SENSOR
2 INTEGRAL HUB/BEARING

Rear wheel speed sensor assembly — 1994–95 DeVille, Concours, Eldorado and Seville

To install:

7. Apply a sealant, 620 Loctite® or its equivalent, to the groove in the outer diameter of the bearing hub.
8. Place the hub and bearing on tool J–38764–4 or its equivalent to prevent damage to the studs.
9. Install the sensor using a press and tool J38764–1 or its equivalent.
10. Install the hub and bearing assembly.
11. Connect the sensor wiring to the ABS harness.
12. Test drive the vehicle and check system operation. Check the system for any stored codes after the test drive.

REAR WHEEL SPEED SENSORS

Removal and Installation

The rear wheel speed sensors are an integral part of the rear hub and bearing assembly and are not removable. If the sensor fails, the hub and bearing assembly must be replaced. Do not attempt to remove the sensor from the bearing.

DELCO 6 ANTI-LOCK BRAKE SYSTEM WITH TRACTION CONTROL SYSTEM (TCS)

General Description

The Delco 6 anti-lock brake system is used on A, F, J, L, N, U and W bodied vehicles. In the Delco 6 system, quick-response electric motors change the position of displacement pistons. The pistons serve as moveable valves, altering the line pressure in small increments. The motor-operated valves compare to an analog circuit with continuous modulation. This continuous modulation of brake line pressure yields much smoother operation with less pedal vibration during operation.

The system uses input signals from each of 4 wheel sensors. Three output channels (2 front, 1 rear) are individually controlled by the control module. When locking is detected at one wheel, a solenoid closes the hydraulic path from the master cylinder. The electric motor for that brake circuit cycles the piston up and down to modulate braking force. The Delco 6 cannot increase the brake pressure above that developed in the master cylinder and the system cannot apply the brakes by itself.

A Traction Control System (TCS) is available on F, J and U body vehicles. The system utilizes the same basic components as other vehicles, but it has the ability to control wheel slip with a combination of spark retard and transmission upshift, controlled with signals from the Powertrain Control Module (PCM) as directed by the Electronic Brake Traction Control Module (EBTCM).

GM BODY/MODEL CODE DESIGNATION TABLE

BODY	MODEL
A	Century, Cutlass Ciera, Cutlass Cruiser
F	Camaro, Firebird
J	Cavalier, Sunbird, Sunfire
L	Beretta, Corsica
N	Achieva, Grand Am, Skylark
U	Lumina APV, Silhouette, Trans Sport
W	Cutlass Supreme, Grand Prix, Lumina, Monte Carlo, Regal

SYSTEM OPERATION

During normal braking, the ABS system is in a non-operational mode. Internally, each control piston is in the uppermost or home position, allowing brake fluid pressure to pass to the wheels unrestricted. A small internal brake device is applied to each piston, preventing it from being forced downward by the pressurized fluid passing above it.

When impending wheel lock is noted at one or more wheels, the Electronic Brake Control Module (EBCM) commands the system into ABS mode. Solenoids in each front wheel circuit close, shutting off the brake circuit(s) from the master cylinder. The electric brakes on the front circuit pistons are released and the pistons are driven downward by the electric motors through a system of driven gears. The amount of current applied to the motors controls the speed and distance traveled. As the motors move backwards, the piston moves downward, allowing a check valve to seat. The brake pressure to the wheel is now a function of the controlled volume within the piston chamber.

To reduce pressure, the motor continues to drive the piston downward. If an increase in pressure is necessary, the piston is driven upward. Total pressure available is limited to the amount present in the brake system when the Anti-lock Brake System was activated.

The rear brakes are controlled in similar fashion but without the solenoid to isolate the master cylinder. Wheel speed signals are received from each rear wheel. The EBCM uses a Select Low strategy, controlling a single output to the rear brakes based on the wheel with the greatest tendency to lock. During normal brake operation, the rear piston is held by a mechanical spring brake rather than an electro-magnetic brake. The spring expands outward within the housing, holding the piston. With the drive motor **ON** during ABS, the direction of spring expansion is reversed, allowing it to contract and the piston to move freely.

1. RIGHT FRONT WHEEL SPEED SENSOR
2. LEFT FRONT WHEEL SPEED SENSOR
3. ABS HYDRAULIC MODULATOR
4. ABS SOLENOIDS
5. MASTER CYLINDER
6. ENGINE COMPARTMENT FUSE/RELAY CENTER (ABS FUSES, ELECTRONIC BRAKE CONTROL RELAY)
7. DATA LINK CONNECTOR
8. ELECTRONIC INSTRUMENT PANEL CLUSTER
9. ELECTRONIC BRAKE CONTROL MODULE
10. DASH TO BODY HARNESS CONNECTOR
11. BODY TO ABS AND FUEL SENDER HARNESS CONNECTOR
12. RIGHT REAR WHEEL SPEED SENSOR
13. ABS REAR AXLE HARNESS
14. LEFT REAR WHEEL SPEED SENSOR
15. DASH TO ENGINE HARNESS CONNECTOR
16. DASH HARNESS
17. BODY HARNESS
18. ENGINE HARNESS

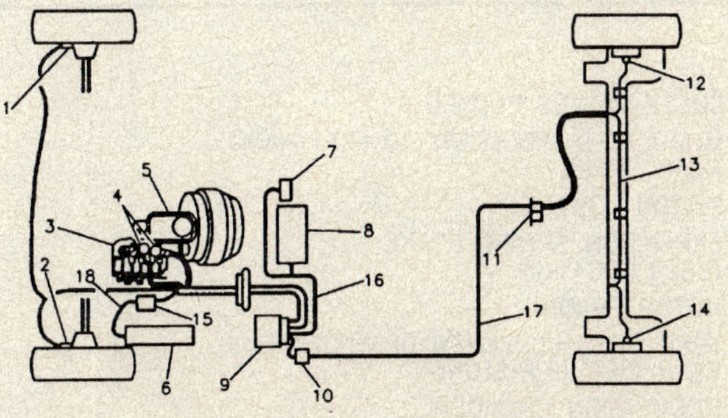

ABS component location and harness routing varies by model

1 CHECK VALVE OPEN
2 INLET FROM MASTER CYLINDER
3 NORMALLY OPEN SOLENOID
4 PISTON FULLY RAISED
5 NUT
6 BALL SCREW SPINDLE
7 MOTOR PACK
8 ELECTROMAGNETIC BRAKE (EMB)
9 BRAKE PRESSURE TO FRONT BRAKE

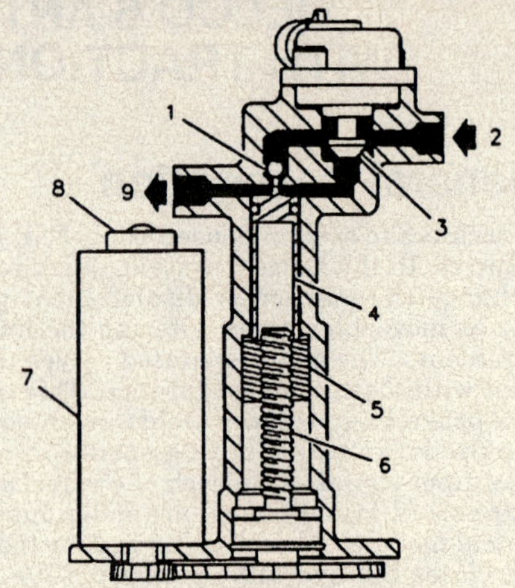

Front control circuit in normal operating mode

1 CHECK VALVE CLOSED
2 SOLENOID VALVE CLOSED
3 INLET FROM MASTER CYLINDER
4 MODULATION CHAMBER
5 PISTON IN "MODULATION" POSITION
6 BALL SCREW NUT "MODULATION" POSITION
7 FRONT MOTOR
8 ELECTROMAGNETIC BRAKE (EMB)
9 MODULATED ABS PRESSURE TO FRONT BRAKE

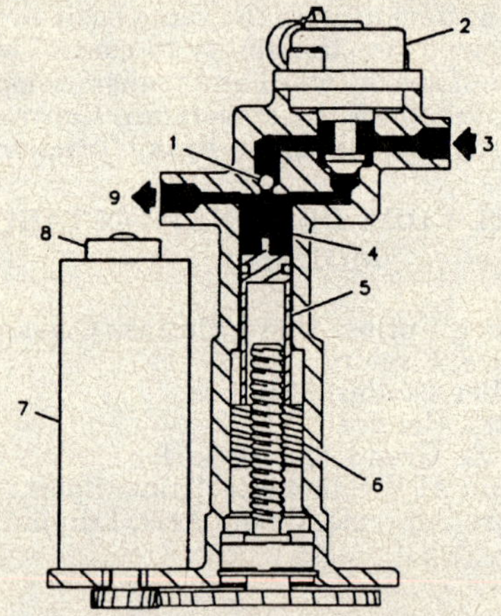

Front control circuit during ABS operation

1 CHECK VALVES CLOSED
2 MODULATED PRESSURE TO LEFT HAND REAR BRAKE
3 PISTON LOWERED
4 EXPANSION SPRING BRAKE (ESB) LOCATION
5 MOTOR PINION
6 YOKE ON BALL SCREW DRIVES BOTH REAR CIRCUIT PISTONS
7 MODULATION CHAMBER
8 MODULATED PRESSURE TO RIGHT HAND REAR BRAKE

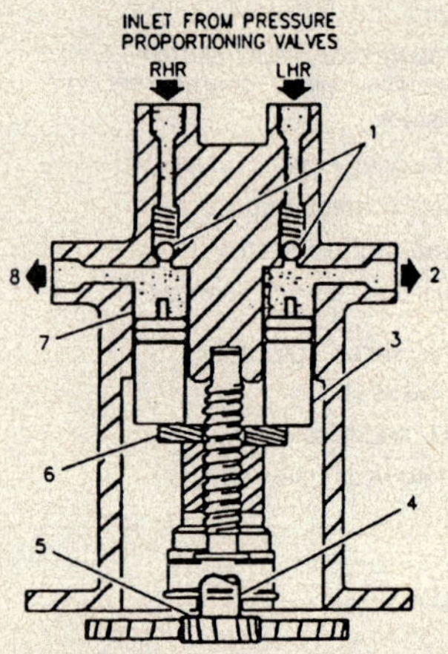

Rear control circuit during ABS operation

SYSTEM COMPONENTS

Hydraulic Modulator Assembly

Located next to the master cylinder assembly, the hydraulic modulator contains the motors, drive gears, pistons and check valve assemblies. The unit controls hydraulic brake pressure to the wheel circuits during an ABS stop. The electrical components of the hydraulic assembly function at the command of the EBCM.

Electronic Brake Control Module (EBCM)/ Electronic Brake Traction Control Module (EBTCM)

The control module is located above the left kick panel (F body), on the right firewall (A, J, L or M bodies), under the left dash near the steering column (U body) or inside the left front fender (W body).

The EBCM receives the input signals from the wheel speed sensors, brake switch, brake fluid level switch and unswitched battery voltage. Outputs include signals to three bi-directional motor controls, two Electromagnetic Brake (EMB) controls, two solenoid controls and the electronic brake control relay (enable relay). Additionally, the unit monitors itself, each input and each output for proper function. If an electrical fault is noted, the EBCM will assign a fault code and store it in memory for diagnostic assistance. A serial data link is provided to pin **M** of the ALDL so that scan tools can communicate with the system.

There are no serviceable or replaceable parts within the EBCM. It is a replaceable only component.

Wheel Speed Sensors

The wheel speed sensors generate an AC voltage as a magnetic toothed ring passes a stationary coil. The frequency of this voltage increases with wheel speed and is used by the EBCM to determine wheel speed. By comparing wheel speeds during braking, the EBCM determines impending wheel lock. The front sensors may be removed or replaced individually; however, on U vans, the front hub and bearing assembly must be removed for access to the sensor.

At the front, the speed sensor rings are located on the outer CV-joints, directly below the wheel speed sensor. No repair or air gap adjustment to the speed sensor units is possible or permissible.

The rear wheel speed sensors and rings are integral parts of the rear wheel hub and bearing assemblies. Each sensor and ring is self-contained and sealed from the environment. If a sensor or ring is damaged or malfunctioning, the entire rear bearing and sensor assembly for that side must be replaced. No repair or air gap adjustment is possible or permissible.

Speed Sensor Jumper Harnesses

Each wheel speed sensor is connected to the main harness by a jumper harness of flexible, twisted-pair wiring. Rear circuits join at some point in the harness routing. The jumpers must be flexible to allow for suspension motion as the vehicle moves over the road. No repair is allowed to the jumper harnesses; any repair would stiffen that section of wire, inducing later failure due to cable flexing. Any wiring damage to the jumpers requires the jumper to be replaced with a new harness. Never attempt to splice, solder or crimp the jumper harnesses.

Electronic Brake Control Relay (Enable Relay)

The electronic brake control relay (enable relay) is a normally open contract type relay. It has special contact material to handle the high currents required by the ABS system. The relay switches battery power to the EBM's and EBCM, which allows the EBCM to supply to the motors and solenoids. It is energized when the ignition is **ON**.

ETS Switch

Located in the gear shift handle, the ETS switch allows the operator to turn the traction control system **ON** and **OFF**.

ABS Lamp Driver Module

Located in the dashboard, the lamp driver module controls operation of the ABS warning lamp based on signals from the EBCM.

Warning Lamps

BRAKE

The red BRAKE warning lamp operates in traditional fashion, warning of parking brake application or fluid loss in the reservoir. The lamp also illuminates as a bulb test during start-up. The BRAKE warning lamp indicates a problem in the conventional brake system; when this lamp is lit, braking may be impaired. Whenever this lamp is lit, ABS is disabled and the ABS lamp may be lit.

ABS

The amber ABS warning lamp in the instrument panel advises of problems within the anti-lock system. If the EBCM detects a problem within the system, it will command the lamp **ON**.

If the lamp is flashing, the fault does not immediately impair ABS function but must be attended to as soon as possible. If the lamp is constantly lit, the fault affects operation of the ABS system. The EBCM will disable the ABS function until the fault is corrected. If the ABS warning lamp is **ON** constantly, ABS is disabled but normal braking remains available.

ABS INOP

A solid ABS INOP lamp indicates that a problem has been detected that affects the operation of the ABS system. A flashing ABS INOP lamp indicates that the EBCM has attempted to turn **ON** the red BRAKE warning lamp, but the lamp did not activate. If the lamp is flashing, the vehicle should not be driven, as a serious malfunction in the brake system has been detected.

ETS OFF

The amber ETS OFF warning lamp will activate if the EBTCM detects a malfunction in the traction control system.

LOW TRAC

The LOW TRAC lamp is used to alert the driver when the ABS is active. Anytime the EBCM detects a braking event where ABS is required, the lamp will activate and remain lit for until several seconds after the ABS system has been deactivated.

TCS Shutdown Mode

During prolonged Traction Control System operation, the brake lining temperature may rise to a higher than desired level. If that occurs, a brake thermal sensor will send a signal to the Electronic Brake Traction Control Module (EBTCM) which will deactivate the traction control system, once the event has been completed. The TCS warning lamp will come **ON** and the TCS will remain disabled until the brakes cool to the proper temperature. The TCS system will automatically re-activate at that time.

Diagnosis and Testing

SERVICE PRECAUTIONS

- If the vehicle is equipped with an air bag (SIR) system, always properly disable this system before working on or around its components. Failure to disarm the system may result in unintentional deployment of the air bag, personal injury, property damage and/or additional repairs.
- Certain components within the ABS system are not intended to be serviced or repaired individually. Only those components with removal and installation procedures should be serviced.
- Do not use rubber hoses or other parts not specifically specified for the ABS system. When using repair kits, replace all parts included in the kit. Partial or incorrect repair may lead to functional problems and require the replacement of additional components.
- Lubricate rubber parts with clean, fresh brake fluid to ease assembly. Do not use lubricated shop air to clean parts; damage to rubber components may result.
- Use only DOT 3 brake fluid from an unopened container. Use of DOT 5 silicone brake fluid is specifically prohibited.
- If any hydraulic component or line is removed or replaced, it may be necessary to bleed the entire system.
- A clean repair area is essential. Perform repairs after components have been thoroughly cleaned; use only denatured alcohol to clean components. Do not allow ABS components to come into contact with any substance containing mineral oil; this includes used shop rags.
- Remove the lock pin before disconnecting CPA connectors.
- The anti-lock brake controller is a microprocessor similar to other computer units in the vehicle. Ensure that the ignition switch is **OFF** before removing or installing controller harnesses. Avoid static electricity discharge at or near the controller.
- Fault codes stored within the system can only be read with a bi-directional scanner such as the GM Tech 1 scanner or its equivalent.
- The brake controller is equipped with on-board diagnostic programs beyond the limits of readable codes. Not all scanners are capable of using these other features. Consult the scanner manufacturer's manual for directions and capabilities.

READING CODES

Stored fault codes may only be read with a scan tool; there are no provisions for a flashing code display from the dash lamps. Once connected, the scan tool becomes part of the vehicle's electrical system and can perform many functions. Properly used, the scan tool can display ABS system data, display and clear fault codes, test or control ABS components, aid in diagnosis and help track intermittent faults.

The EBCM can store several codes simultaneously. In the CODE HISTORY mode, the scanner will note the very last code stored (most recent occurrence) as well as the first 5 faults, in order of occurrence. This is particularly helpful in tracking intermittent conditions; one fault can trigger several others. Always begin with the most recently occurring fault.

INTERMITTENT FAULTS

The diagnostic procedures may or may not be helpful in determining the cause of intermittent problems in the system electrical components. In most cases, the fault must be present to locate the problem using the fault charts.

Most intermittent problems are caused by faulty electrical connections or wiring. When an intermittent failure is encountered:

1. Check for history codes which may be stored in the anti-lock brake controller. If a history code is stored, this may indicate the circuitry which has the intermittent condition. Move the related connectors, harness and components in an effort to induce the failure.
2. Enter the Enhanced Diagnostic feature. This feature will help determine how intermittent the fault is and may help determine certain conditions that cause the fault.
3. Set the ABS snapshot to trigger **ON** the intermittent trouble code and use the enhanced diagnostic feature to recreate the conditions that cause the intermittent code to set.
4. Check for poor mating of connector halves or terminals not fully seated in the connector body.
5. Inspect for improperly formed or damaged terminals. All connector terminals in a problem circuit should be carefully reformed to increase contact tension.
6. Check for poor terminal-to-wire connection. This requires removing the terminal from the connector body to inspect.

NOTE: The diagnostic charts assume that a current code is stored, indicating that the fault is present. When using the diagnostic charts, always refer to the circuit diagram for the particular vehicle being serviced; wire colors and terminal designations may vary between vehicles.

CLEARING CODES

Stored fault codes may be cleared with a scan tool or through the ignition cycle default. The control module will not permit any codes to be cleared until all codes have been displayed. Codes can not be cleared by disconnecting the battery or EBCM.

If no diagnostic fault code has occurred for 100 drive cycles, all stored codes will be cleared. A drive cycle occurs when the ignition is **ON** and vehicle speed exceeds 10 mph.

After codes are cleared, always verify proper system operation and the absence of any new codes.

ELECTRICAL CONNECTOR TERMINAL IDENTIFICATION — 1993 CAVALIER AND SUNBIRD

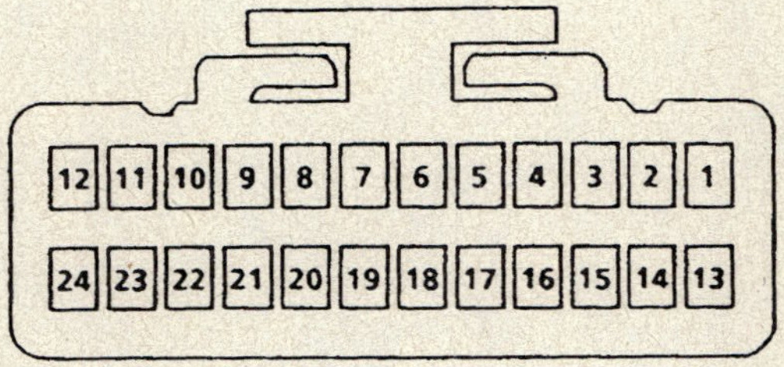

EBCM 24 PIN CONNECTOR

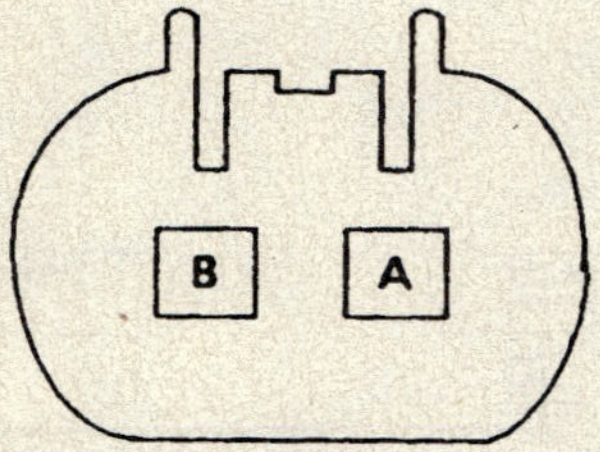

2 WAY EBCM CONNECTOR

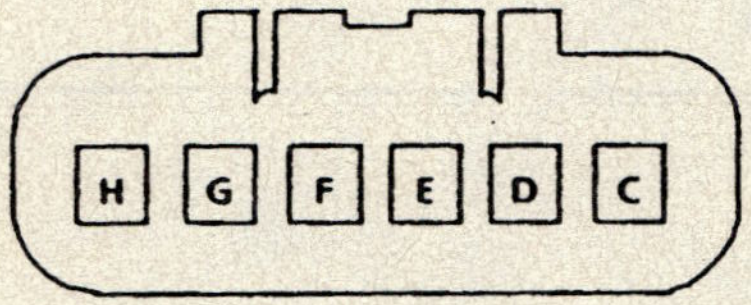

6 WAY EBCM CONNECTOR

DIAGNOSTIC CHART CIRCUIT CHECK — 1993 CAVALIER AND SUNBIRD

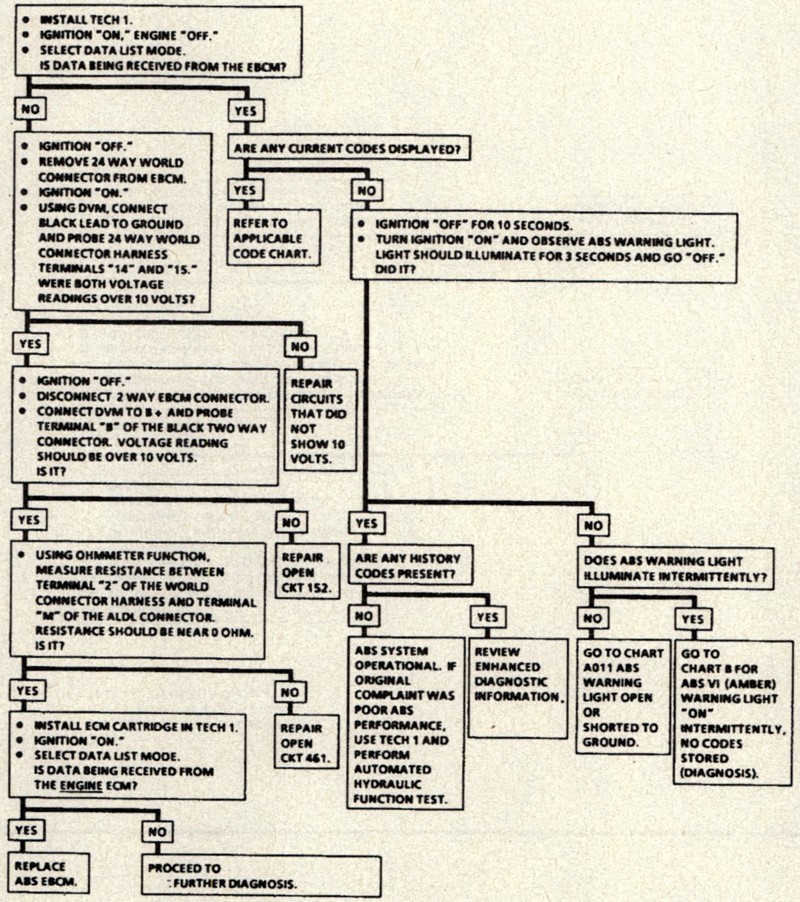

DIAGNOSTIC CHART ABS WARNING LAMP — 1993 CAVALIER AND SUNBIRD

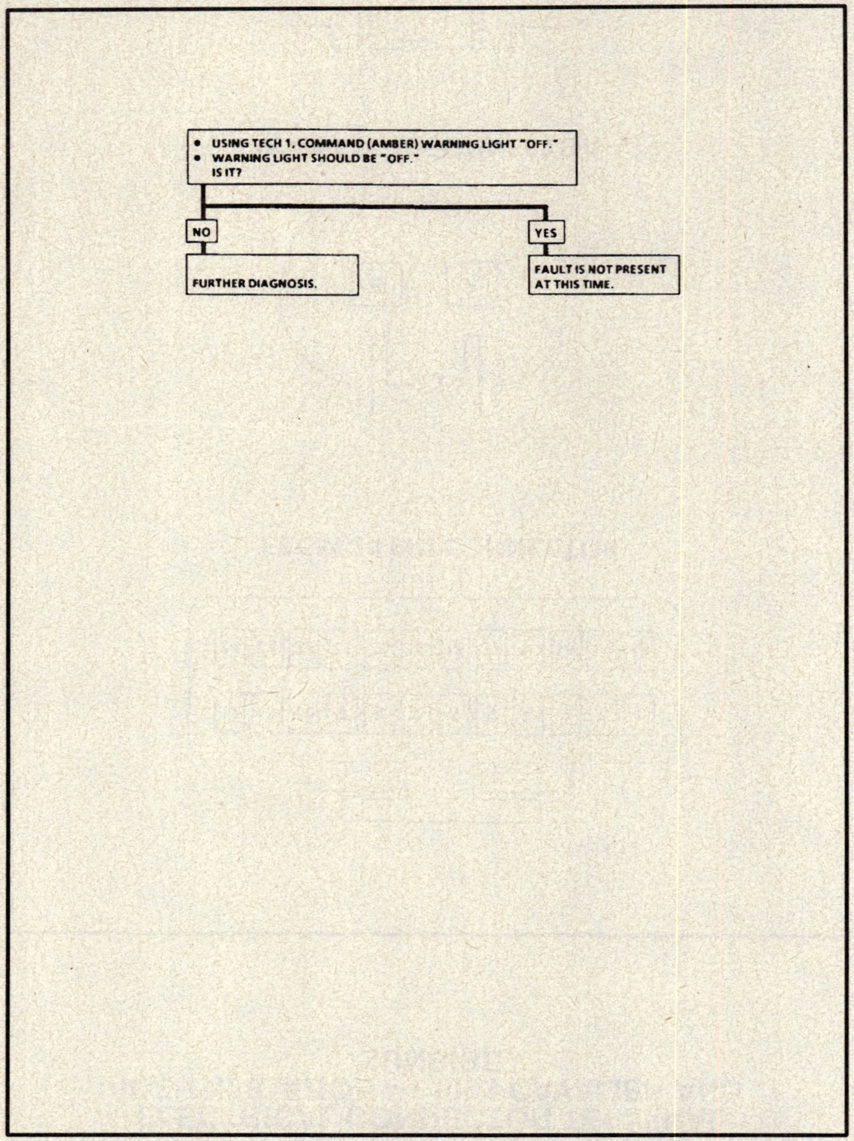

DIAGNOSTIC CHART ABS WARNING LAMP — 1993 CAVALIER AND SUNBIRD

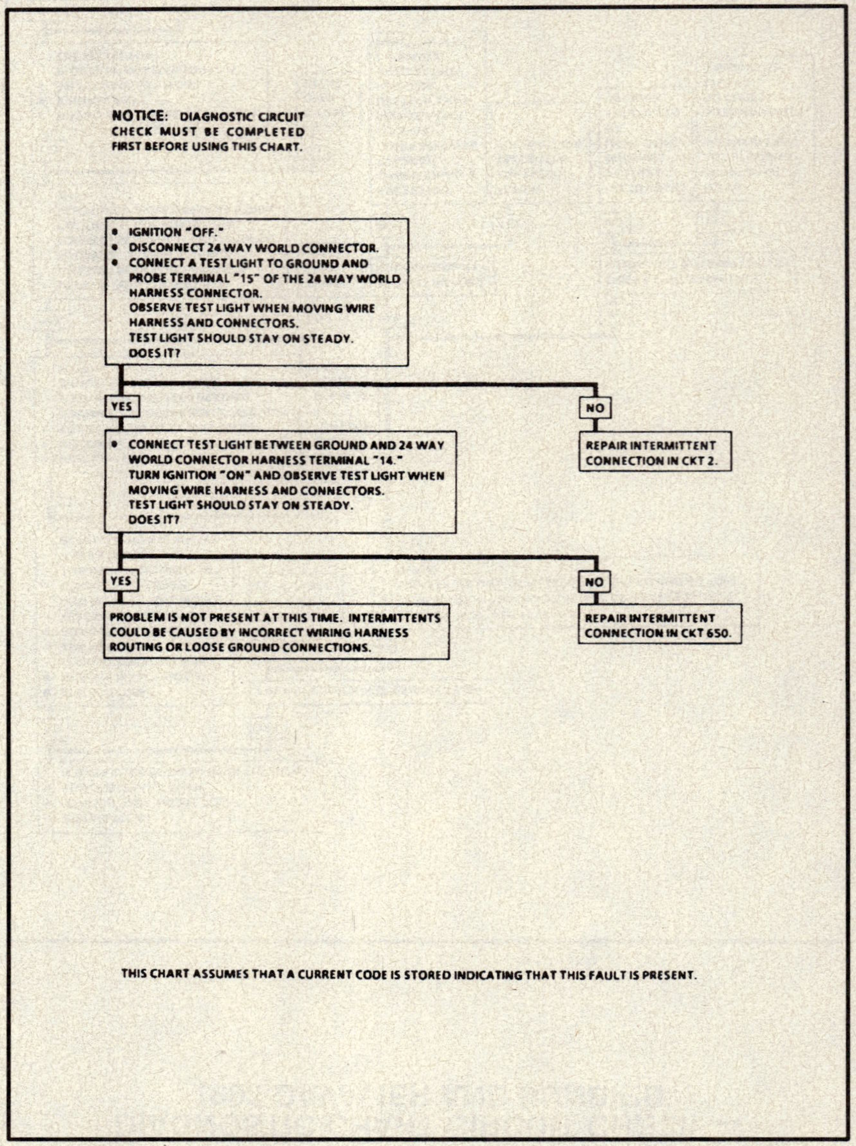

DIAGNOSTIC CHART DTC A1011 — 1993 CAVALIER AND SUNBIRD

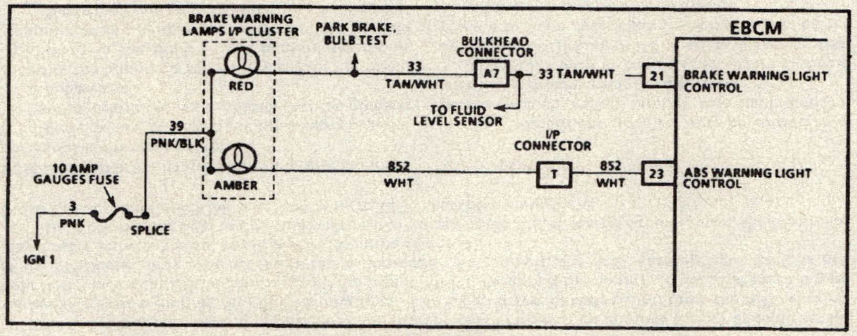

CODE A011

ABS WARNING LIGHT OPEN OR SHORTED TO GROUND

Circuit Description:

This code checks the state of the ABS warning light to identify a situation in which the driver could not be warned of a system fault by the ABS warning light, or the warning light is always "ON." The Lamp Driver Module (LDM) turns the ABS warning light "ON," unless the EBCM provides ground to turn it "OFF." Because of the circuitry in the LDM is part of the I/P cluster, only external faults can be detected. As a result, the ABS warning light itself is not diagnosable, only the control line to the EBCM can be diagnosed. In the event of an open CKT 852, the ABS warning light will be on at all times, due to the loss of ground at the LDM input. If the control line is shorted to ground, the ABS warning light is kept "OFF," due to the LDM input being grounded.

Test Description: Number(s) below refer to circled number(s) on the diagnostic chart.

1. Checks to see if ABS light functions properly. Normal operation would indicate that the problem is not present.
2. This step determines if any other codes are present.
3. This step insures that the circuitry from the EBCM to the lamp driver module is complete.
4. This step insures that the ignition circuit to the lamp driver module is functional. An open gauges fuse or open CKT 3 or 39 would result in inoperative indicator lights.
5. Isolates the cause of the ABS lamp remaining "OFF," by disconnecting the 24 way world connector, CKT 852 should be open and cause the ABS lamp to be "ON."

Diagnostic Aids:

An "Intermittent" problem may be caused by a poor connection, rubbed through wire insulation, or a wire that is broken inside the insulation.

The light test function of the "Scan" tool may be used to command the light "ON" while looking for an intermittent problem in the warning light circuitry.

Any circuitry that is suspected as causing the intermittent complaint should be thoroughly checked for backed out terminals, improper mating, broken locks, improperly formed or damaged terminals, poor terminal to wiring connections or physical damage to the wiring harness.

DIAGNOSTIC CHART DTC A1011 — 1993 CAVALIER AND SUNBIRD

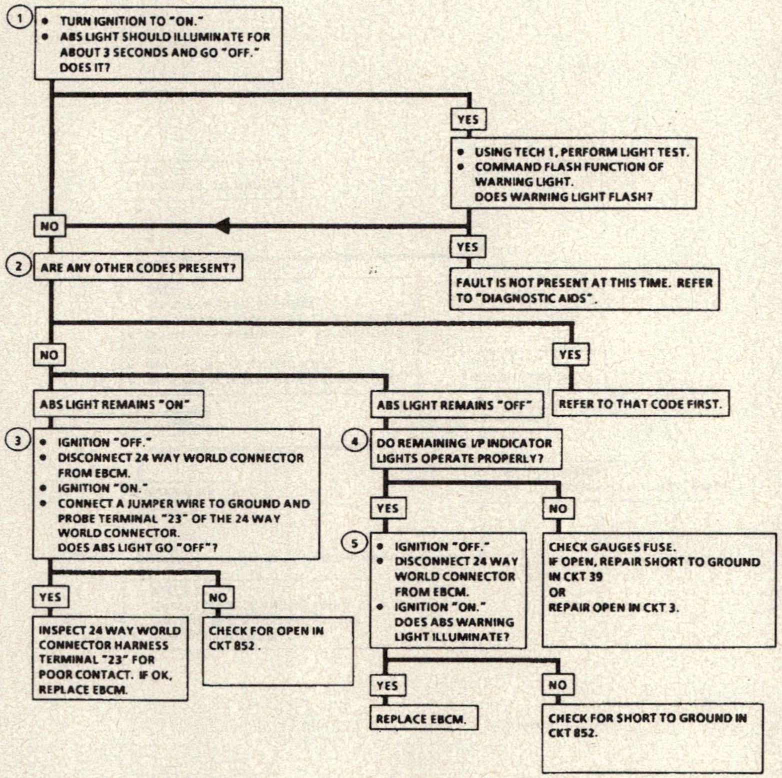

DIAGNOSTIC CHART A013 — 1993 CAVALIER AND SUNBIRD

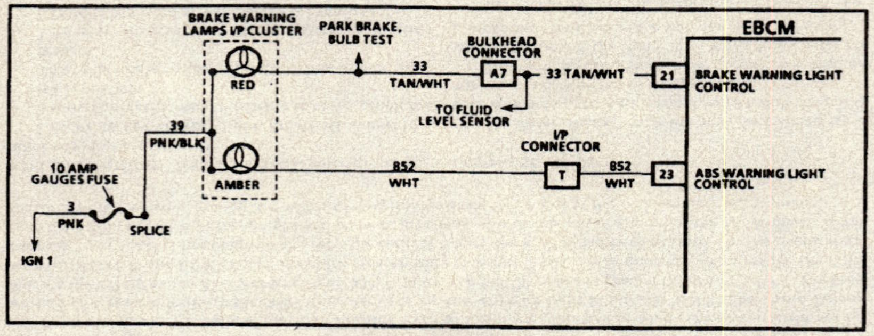

CODE A013

ABS WARNING LIGHT CIRCUIT SHORTED TO BATTERY

Circuit Description:

This code checks the state of the ABS warning light to identify a situation in which the driver could not be warned of a system fault by the ABS warning light. The Lamp Driver Module (LDM) turns the ABS warning light "ON," unless the EBCM provides ground to turn it "OFF." Because of the circuitry in the LDM which is part of the I/P cluster, only external faults can be detected. As a result, the ABS warning light itself is not diagnosable, only the control line to the EBCM can be diagnosed.

If the ABS warning light control line is shorted to battery or the EBCM control circuit is open, the Electronic Brake Control Module (EBCM) is not able to turn "OFF" the ABS warning light.

Test Description: Number(s) below refer to circled number(s) on the diagnostic chart.

1. Checks to see if ABS light functions properly. Normal operation would indicate that the problem is not present.
2. After the ignition is turned to the "OFF" position, voltage to the lamp driver module should be interrupted. A light that remains "ON" would indicate that voltage to this circuit is from a point other than ignition.
3. After removing the 10 amp gauges fuse, voltage on CKT 852 will be eliminated. If voltage is still present, it would indicate that a short to voltage is present between the lamp driver module and EBCM.

Diagnostic Aids:

An "Intermittent" problem may be caused by a poor connection, rubbed through wire insulation or a wire that is broken inside the insulation.

Code A013 must be currently set before it should be addressed. If Code A013 is set as a history code, proceed to those codes that are currently set before attempting to repair a Code A013 fault.

The frequency of the problem can be checked by using the enhanced diagnostic function of the "Scan" tool

Any circuitry that is suspected as causing the "Intermittent" complaint should be thoroughly checked for backed out terminals, improper mating, broken locks, improperly formed or damaged terminals, poor terminal to wiring connections or physical damage to the wiring harness.

DIAGNOSTIC CHART A013 — 1993 CAVALIER AND SUNBIRD

(1)
- TURN IGNITION TO "ON."
- ABS LIGHT SHOULD ILLUMINATE FOR ABOUT 3 SECONDS AND GO "OFF." DOES IT?

NO → TURN IGNITION "OFF."

YES → FAULT IS NOT PRESENT AT THIS TIME. REFER TO "DIAGNOSTIC AIDS".

After TURN IGNITION "OFF.":

LIGHT "OFF" →

(3)
- DISCONNECT 24 WAY WORLD CONNECTOR.
- REMOVE 10 AMP GAUGES FUSE.
- IGNITION "ON."
- INSTALL DVM BETWEEN PIN "23" OF 24 WAY WORLD CONNECTOR HARNESS AND GROUND.
- VOLTAGE SHOULD BE 0 VOLT. IS IT?

NO → REPAIR SHORT TO IGNITION IN CKT 852.

YES → REPLACE EBCM.

(2) LIGHT "ON" → REPAIR SHORT TO B + IN CKT 852.

THIS CHART ASSUMES THAT A CURRENT CODE IS STORED INDICATING THAT THIS FAULT IS PRESENT.

DIAGNOSTIC CHART A014 — 1993
CAVALIER AND SUNBIRD

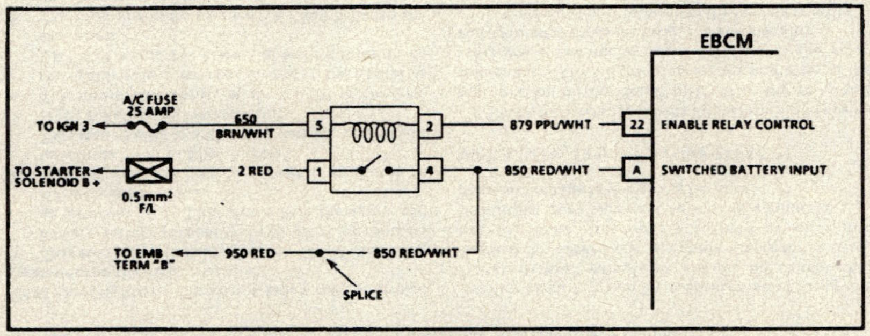

CODE A014

ENABLE RELAY CONTACTS OR FUSE OPEN

Circuit Description:

Ignition voltage is supplied through terminal "5" of the ABS enable relay. The EBCM then is able to energize the pull-in coil by completing the ground circuit at Pin "22" of the EBCM. The magnetic field created closes the enable relay contacts and allows battery voltage and current to be supplied to the EMB's and EBCM, which supplies power to the motors and solenoids.

This code monitors the availability of current/voltage to the motors, solenoids and EMB's. This fault indicates voltage is not available and would therefore not allow ABS operation if required.

Test Description: Number(s) below refer to circled number(s) on the diagnostic chart.

1. This step would indicate that the EBCM is sensing battery voltage present through the relay contacts at terminal "A." This shows that the relay and circuitry is functional.
2. This insures that battery voltage is available to terminals "1" and "5" of the relay.
3. This step checks to make sure that the pull-in coil of the enable relay has continuity.
4. This checks the ability of the EBCM to energize the enable relay. An open circuit in the EBCM or CKT 879 would not allow the enable relay to be energized.
5. This step checks for an open CKT 850 between the enable relay and EBCM.

Diagnostic Aids:

An "Intermittent" problem may be caused by a poor connection, rubbed through wire insulation, or a wire that is broken inside the insulation.

The frequency of the problem can be checked by using the enhanced diagnostic function of the "Scan" tool.

Any circuitry that is suspected as causing the "Intermittent" complaint should be thoroughly checked for backed out terminals, improper mating, broken locks, improperly formed or damaged terminals, poor terminal to wiring connections or physical damage to the wiring harness.

VIBRATION, TEMPERATURE EFFECTS:

Check for vibration effects by performing the relay test function of the "Scan" tool. With the relay test commanded "ON," lightly tap the top and sides of the relay while monitoring relay voltage. If the relay voltage changes significantly, replace the relay.

If Code A014 only fails when the vehicle is initially started in cold ambient conditions (temperature less than 32°F - 0°C), replace the relay.

DIAGNOSTIC CHART A014 — 1993
CAVALIER AND SUNBIRD

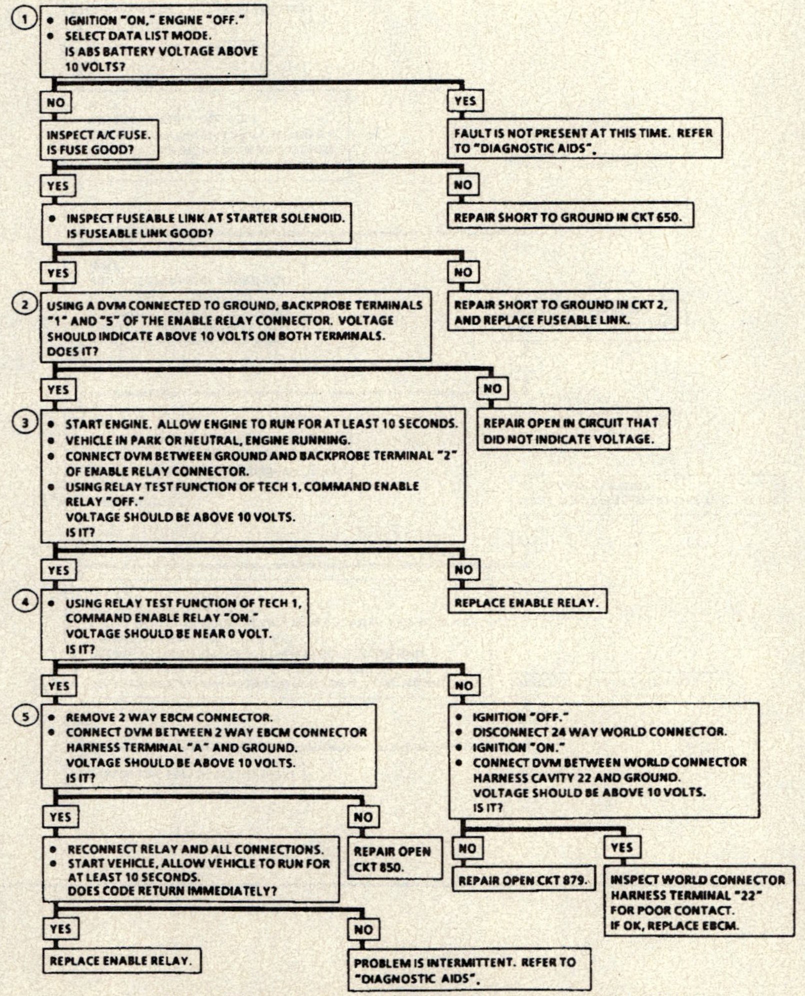

THIS CHART ASSUMES THAT A CURRENT CODE IS STORED INDICATING THAT THIS FAULT IS PRESENT.

DIAGNOSTIC CHART A014 — 1993 CAVALIER AND SUNBIRD

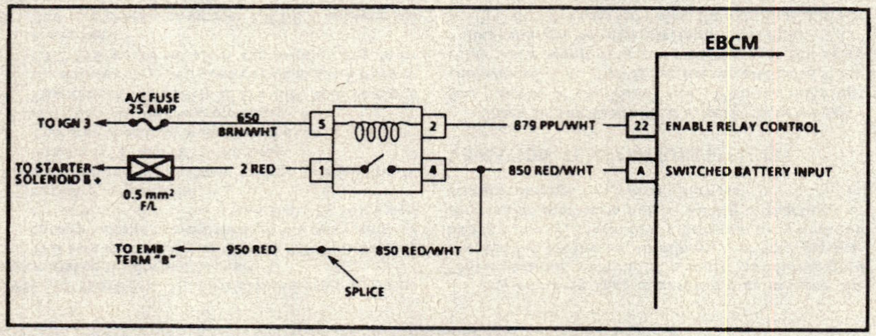

CODE A014

ENABLE RELAY CONTACTS OR FUSE OPEN

Circuit Description:

Ignition voltage is supplied through terminal "5" of the ABS enable relay. The EBCM then is able to energize the pull-in coil by completing the ground circuit at Pin "22" of the EBCM. The magnetic field created closes the enable relay contacts and allows battery voltage and current to be supplied to the EMB's and EBCM, which supplies power to the motors and solenoids.

This code monitors the availability of current/voltage to the motors, solenoids and EMB's. This fault indicates voltage is not available and would therefore not allow ABS operation if required.

Test Description: Number(s) below refer to circled number(s) on the diagnostic chart.

1. This step would indicate that the EBCM is sensing battery voltage present through the relay contacts at terminal "A." This shows that the relay and circuitry is functional.
2. This insures that battery voltage is available to terminals "1" and "5" of the relay.
3. This step checks to make sure that the pull-in coil of the enable relay has continuity.
4. This checks the ability of the EBCM to energize the enable relay. An open circuit in the EBCM or CKT 879 would not allow the enable relay to be energized.
5. This step checks for an open CKT 850 between the enable relay and EBCM.

Diagnostic Aids:

An "Intermittent" problem may be caused by a poor connection, rubbed through wire insulation, or a wire that is broken inside the insulation.

The frequency of the problem can be checked by using the enhanced diagnostic function of the "Scan" tool.

Any circuitry that is suspected as causing the "Intermittent" complaint should be thoroughly checked for backed out terminals, improper mating, broken locks, improperly formed or damaged terminals, poor terminal to wiring connections or physical damage to the wiring harness.

VIBRATION, TEMPERATURE EFFECTS:

Check for vibration effects by performing the relay test function of the "Scan" tool. With the relay test commanded "ON," lightly tap the top and sides of the relay while monitoring relay voltage. If the relay voltage changes significantly, replace the relay.

If Code A014 only fails when the vehicle is initially started in cold ambient conditions (temperature less than 32°F - 0°C), replace the relay.

DIAGNOSTIC CHART A014 — 1993 CAVALIER AND SUNBIRD

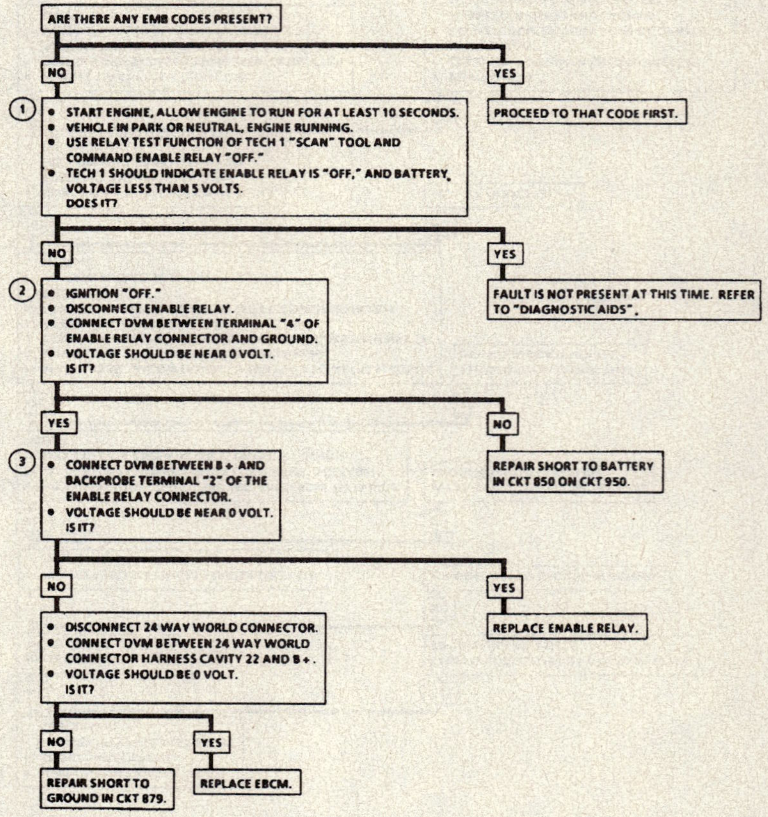

THIS CHART ASSUMES THAT A CURRENT CODE IS STORED INDICATING THAT THIS FAULT IS PRESENT.

DIAGNOSTIC CHART A016 — 1993
CAVALIER AND SUNBIRD

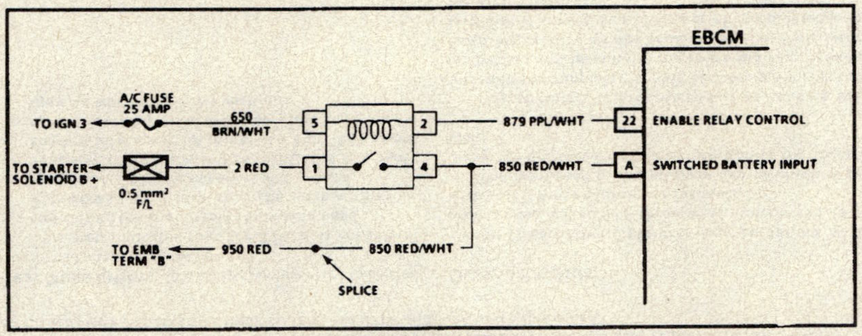

CODE A016

ENABLE RELAY COIL CIRCUIT OPEN

Circuit Description:

Ignition voltage is supplied through terminal "5" of the ABS enable relay. The EBCM then is able to energize the pull-in coil by completing the ground circuit at Pin "22" of the EBCM. The magnetic field created closes the enable relay contacts and allows battery voltage and current to be supplied to the EMB's and EBCM, which supplies power to the motors and solenoids.

This code detects opens in the enable relay coil circuit. An open in this circuit will not allow the enable relay to be energized, thus preventing voltage/current to the motors, EMB's and solenoids.

Test Description: Number(s) below refer to circled number(s) on the diagnostic chart.

1. This step indicates the EBCM is capable of controlling the enable relay as commanded.
2. This step checks to insure that voltage is available to the pull-in coil of the enable relay.
3. This step insures that there is continuity through the pull-in coil of the enable relay.

Diagnostic Aids:

An "Intermittent" problem may be caused by a poor connection, rubbed through wire insulation, or a wire that is broken inside the insulation.

The frequency of the problem can be checked by using the enhanced diagnostic function of the "Scan" tool.

If the frequency of the problem is high, but is currently intermittent, check for high coil resistance by measuring between relay terminals 5 and 2 using a DVM. If resistance shows greater than 100 ohms, replace the relay.

Any circuitry, that is suspected as causing the intermittent complaint, should be thoroughly checked for backed out terminals, improper mating, broken locks, improperly formed or damaged terminals, poor terminal to wiring connections or physical damage to the wiring harness.

DIAGNOSTIC CHART A016 — 1993
CAVALIER AND SUNBIRD

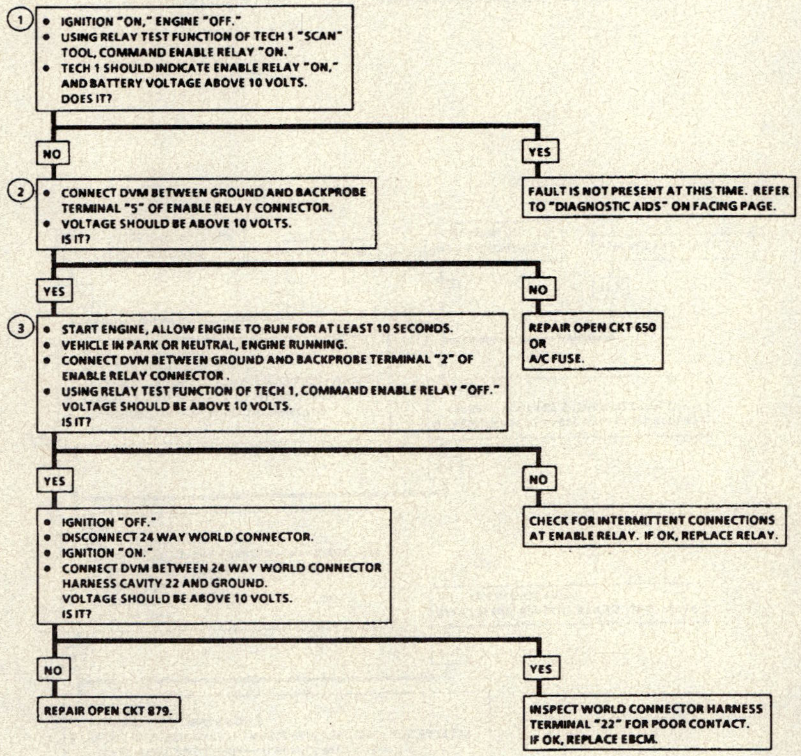

THIS CHART ASSUMES THAT A CURRENT CODE IS STORED INDICATING THAT THIS FAULT IS PRESENT.

DIAGNOSTIC CHART A017 — 1993 CAVALIER AND SUNBIRD

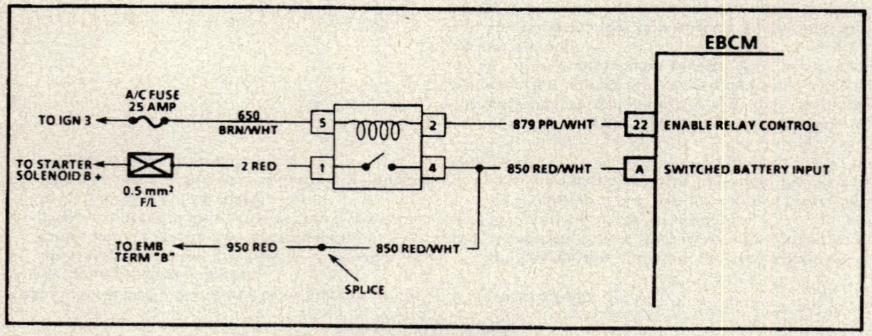

CODE A017

ENABLE RELAY COIL CIRCUIT SHORTED TO GROUND

Circuit Description:

Ignition voltage is supplied through terminal "5" of the ABS enable relay. The EBCM then is able to energize the pull-in coil by completing the ground circuit at Pin "22" of the EBCM. The magnetic field created closes the enable relay contacts and allows battery voltage and current to be supplied to the EMB's and EBCM, which supplies voltage to the motors and solenoids.

This code determines if the enable relay is energized when it should not be. This fault would not allow the enable relay to remove power to the ABS system. If a second fault were to occur that requires the enable relay to be turned "OFF," that fault cannot be removed if the relay cannot be controlled.

Test Description: Number(s) below refer to circled number(s) on the diagnostic chart.

1. This step indicates that the EBCM is capable of controlling the enable relay as commanded.
2. This checks to insure that the enable relay or control CKT 879 are not shorted to ground.
3. This step checks to see if the EBCM is internally shorted to ground. A grounded Pin "22" would cause the enable relay to be energized any time the ignition was in the "ON" position.

Diagnostic Aids:

An "Intermittent" problem may be caused by a poor connection, rubbed through wire insulation, or a wire that is broken inside the insulation.

The frequency of the problem can be checked by using the enhanced diagnostic function of the "Scan" tool,

Any circuitry, that is suspected as causing the intermittent complaint, should be thoroughly checked for backed out terminals, improper mating, broken locks, improperly formed or damaged terminals, poor terminal to wiring connections or physical damage to the wiring harness.

DIAGNOSTIC CHART A017 — 1993 CAVALIER AND SUNBIRD

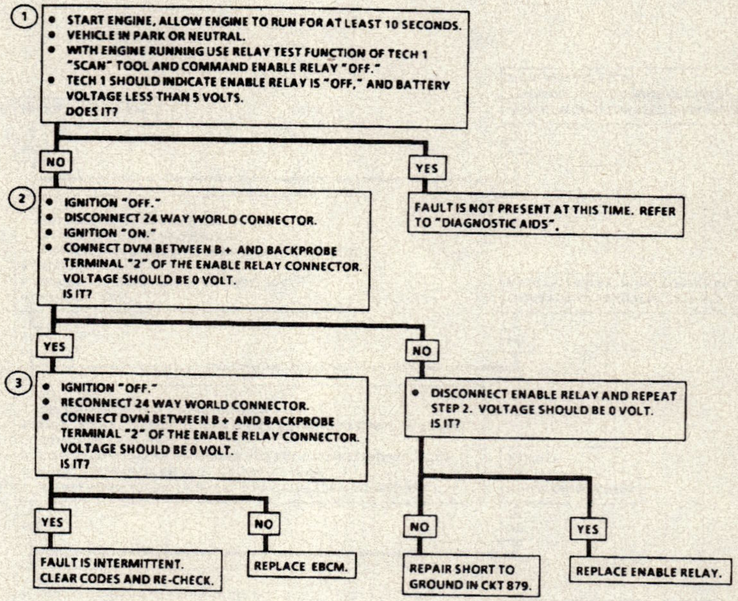

THIS CHART ASSUMES THAT A CURRENT CODE IS STORED INDICATING THAT THIS FAULT IS PRESENT.

DIAGNOSTIC CHART A018 — 1993 CAVALIER AND SUNBIRD

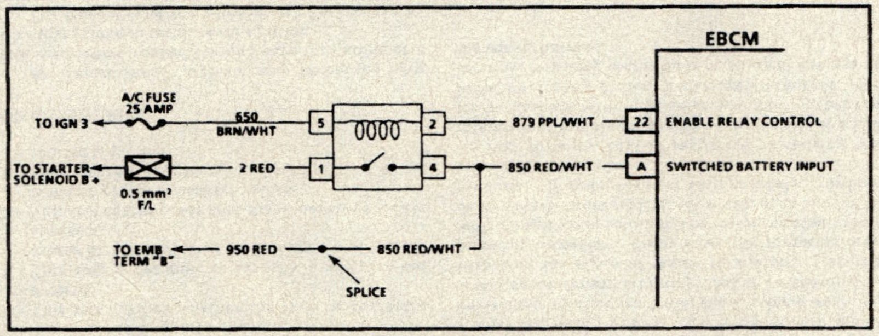

CODE A018

ENABLE RELAY COIL CIRCUIT SHORTED TO BATTERY

Circuit Description:

Ignition voltage is supplied through terminal "5" of the ABS enable relay. The EBCM then is able to energize the pull-in coil by completing the ground circuit at Pin "22" of the EBCM. The magnetic field created closes the enable relay contacts and allows battery voltage and current to be supplied to the EMB's and EBCM, which supplies voltage to the motors and solenoids.

This code monitors the availability of current/voltage to the motors, solenoids and EMB's. A fault of the type specified above would not allow voltage to be available to the motors, solenoids and EMB's and would, therefore, not allow ABS operation if required.

Test Description: Number(s) below refer to circled number(s) on the diagnostic chart.

1. This step indicates that the EBCM is capable of controlling the enable relay as commanded.
2. With the enable relay removed, voltage should not be available at cavity 2. Any voltage at this point would indicate that CKT 879 was shorted to a voltage source.
3. This step checks for an open Quad-Driver in the EBCM.

Diagnostic Aids:

An "Intermittent" problem may be caused by a poor connection, rubbed through wire insulation, or a wire that is broken inside the insulation.

The frequency of the problem can be checked by using the enhanced diagnostic function of the "Scan" tool

Any circuitry, that is suspected as causing the intermittent complaint, should be thoroughly checked for backed out terminals, improper mating, broken locks, improperly formed or damaged terminals, poor terminal to wiring connections or physical damage to the wiring harness.

DIAGNOSTIC CHART A018 — 1993 CAVALIER AND SUNBIRD

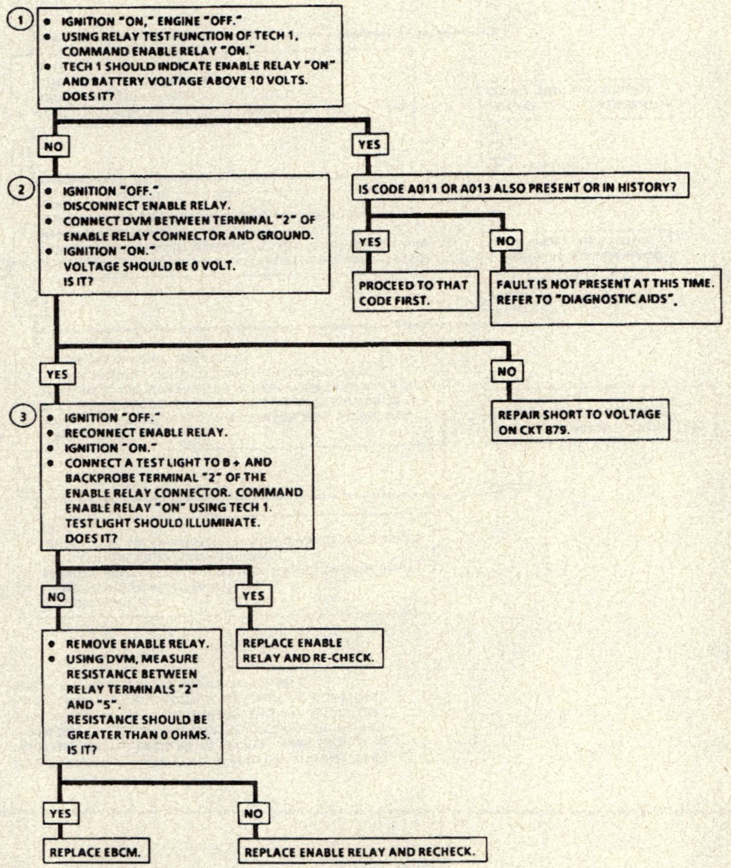

THIS CHART ASSUMES THAT A CURRENT CODE IS STORED INDICATING THAT THIS FAULT IS PRESENT.

DIAGNOSTIC CHART A021 — 1993 CAVALIER AND SUNBIRD

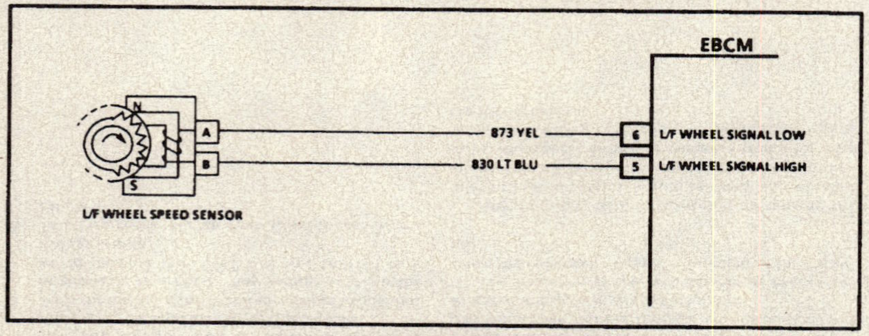

CODE A021

LEFT FRONT WHEEL SPEED = 0

Circuit Description:

As a toothed ring passes by the wheel speed sensor, changes in the electromagnetic field cause the wheel speed sensor to produce a sinusoidal (AC) voltage signal whose frequency is proportional to wheel speed. The magnitude of this signal is directly related to wheel speed and the proximity of the wheel speed sensor to the toothed ring, often referred to as the air gap.

The code detects opens, shorts to ground, and shorts to battery that cause a wheel speed of 0 mph to be calculated by the ABS EBCM. The conditions present to fail this test are: (1) Three (3) wheels are rotating at least 5 mph, and (2) The fourth wheel is 0 mph.

Test Description: Number(s) below refer to circled number(s) on the diagnostic chart.

1. This test verifies whether the fault is currently present.
2. This step is designed to identify a wheel speed sensor or circuitry that is damaged and visibly apparent.
3. This test checks to see that the wheel speed sensor is within proper resistance values.
4. This insures that the wheel speed sensor is not shorted to ground.

Diagnostic Aids:

An "Intermittent" problem may be caused by a poor connection, rubbed through wire insulation, or a wire that is broken inside the insulation.

The frequency of the problem can be checked by using the enhanced diagnostic function of the "Scan" tool,

If the customers comments reflect that the ABS Amber warning light is "ON" only during moist environmental changes, (rain, snow, vehicle wash) all wheel speed sensor circuitry should be thoroughly inspected for signs of water intrusion. Use the following procedure. Spray down the suspected area with 5% salt water solution (two teaspoons of salt to 12 oz. of water). Start vehicle, allow vehicle to run for 10 seconds. If code returns immediately, replace suspected harness.

Any circuitry, that is suspected as causing the intermittent complaint, should be thoroughly checked for backed out terminals, improper mating, broken locks, improperly formed or damaged terminals, poor terminal to wiring connections or physical damage to the wiring harness.

DIAGNOSTIC CHART A021 — 1993 CAVALIER AND SUNBIRD

IMPORTANT: WHEEL SPEED SENSOR INTERMITTENT PROBLEMS MAY BE DIFFICULT TO LOCATE. CARE SHOULD BE TAKEN NOT TO DISTURB ANY ELECTRICAL CONNECTIONS PRIOR TO AN INDICATED STEP OF THIS CHART. THIS WILL INSURE THAT AN INTERMITTENT CONNECTION WILL NOT BE CORRECTED BEFORE THE SOURCE OF THE PROBLEM IS FOUND.

(1)
- TEST DRIVE VEHICLE.
- USING TECH 1, SELECT DATA LIST.
- MONITOR WHEEL SPEED ON LEFT FRONT WHEEL WHILE SLOWLY DECELERATING FROM 35 MPH TO 0 MPH.
- WHEEL SPEED SHOULD MATCH REMAINING WHEEL SPEED INPUTS. DOES IT?

YES: FAULT IS NOT PRESENT AT THIS TIME. REFER TO "DIAGNOSTIC AIDS".

NO:

(2)
- IGNITION "OFF."
- PHYSICALLY INSPECT WHEEL SPEED SENSOR, RING, WIRING AND CONNECTORS FOR DAMAGE. PAY PARTICULAR ATTENTION TO THE SPEED RING. ANY SIGNIFICANT DAMAGE (OTHER THAN NICKS FROM STONES ETC.) WILL AFFECT THE WHEEL SPEED INPUT SIGNAL. IS PHYSICAL DAMAGE INDICATED?

YES: CORRECT PHYSICAL DAMAGE. CLEAR CODES AND RECHECK.

NO:

(3)
- DISCONNECT CONNECTOR FROM LEFT FRONT WHEEL SPEED SENSOR.
- CONNECT DVM BETWEEN TERMINAL "A" AND "B" OF SENSOR, AND MEASURE RESISTANCE. RESISTANCE SHOULD BE BETWEEN 1650 AND 1800 OHMS. IS IT?

NO: REPLACE WHEEL SPEED SENSOR. CLEAR CODES AND RECHECK.

YES:

(4)
- CONNECT DVM BETWEEN SENSOR TERMINAL "A" AND GROUND AND MEASURE RESISTANCE. RESISTANCE SHOULD BE INFINITE. IS IT?

NO: REPLACE WHEEL SPEED SENSOR. CLEAR CODES AND RECHECK.

YES: GO TO FOLLOWING PAGE OF THIS CODE CHART

THIS CHART ASSUMES THAT A CURRENT CODE IS STORED INDICATING THAT THIS FAULT IS PRESENT.

DIAGNOSTIC CHART A021 — 1993
CAVALIER AND SUNBIRD

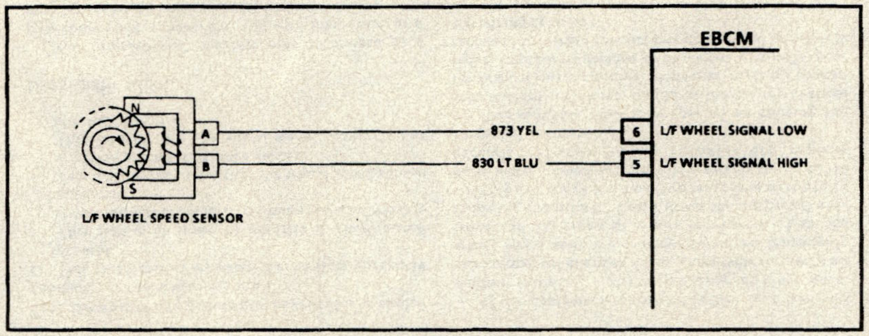

CODE A021

LEFT FRONT WHEEL SPEED = 0

Circuit Description:

As a toothed ring passes by the wheel speed sensor, changes in the electromagnetic field cause the wheel speed sensor to produce a sinusoidal (AC) voltage signal whose frequency is proportional to wheel speed. The magnitude of this signal is directly related to wheel speed and the proximity of the wheel speed sensor to the toothed ring often referred to as the air gap.

This code detects opens, shorts to ground, and shorts to battery that cause a wheel speed of 0 mph to be calculated by the ABS EBCM. The conditions present to fail this test are: (1) Three (3) wheels are rotating at least 5 mph, and (2) The fourth wheel is 0 mph.

Test Description: Number(s) below refer to circled number(s) on the diagnostic chart.

5. This step checks to see if the wheel speed sensor circuitry is shorted to voltage.
6. This step insures that the wheel speed sensor circuitry resistance is within the proper value.
7. This checks for an open in the wheel signal low circuit.
8. This checks for an open in the wheel signal high circuit.
9. This step insures that Code A021 was not set due to a poor connection between the world connector and EBCM.

Diagnostic Aids:

An "Intermittent" problem may be caused by a poor connection, rubbed through wire insulation, or a wire that is broken inside the insulation.

The frequency of the problem can be checked by using the enhanced diagnostic function of the "Scan" tool,

If the customers comments reflect that the ABS Amber warning light is "ON" only during moist environmental changes, (rain, snow, vehicle wash) all wheel speed sensor circuitry should be thoroughly inspected for signs of water intrusion. Use the following procedure. Spray down the suspected area with 5% salt water solution (two teaspoons of salt to 12 oz. of water). Start vehicle, allow vehicle to run for 10 seconds. If code returns immediately, replace suspected harness.

Any circuitry, that is suspected as causing the intermittent complaint, should be thoroughly checked for backed out terminals, improper mating, broken locks, improperly formed or damaged terminals, poor terminal to wiring connections or physical damage to the wiring harness.

DIAGNOSTIC CHART A021 — 1993
CAVALIER AND SUNBIRD

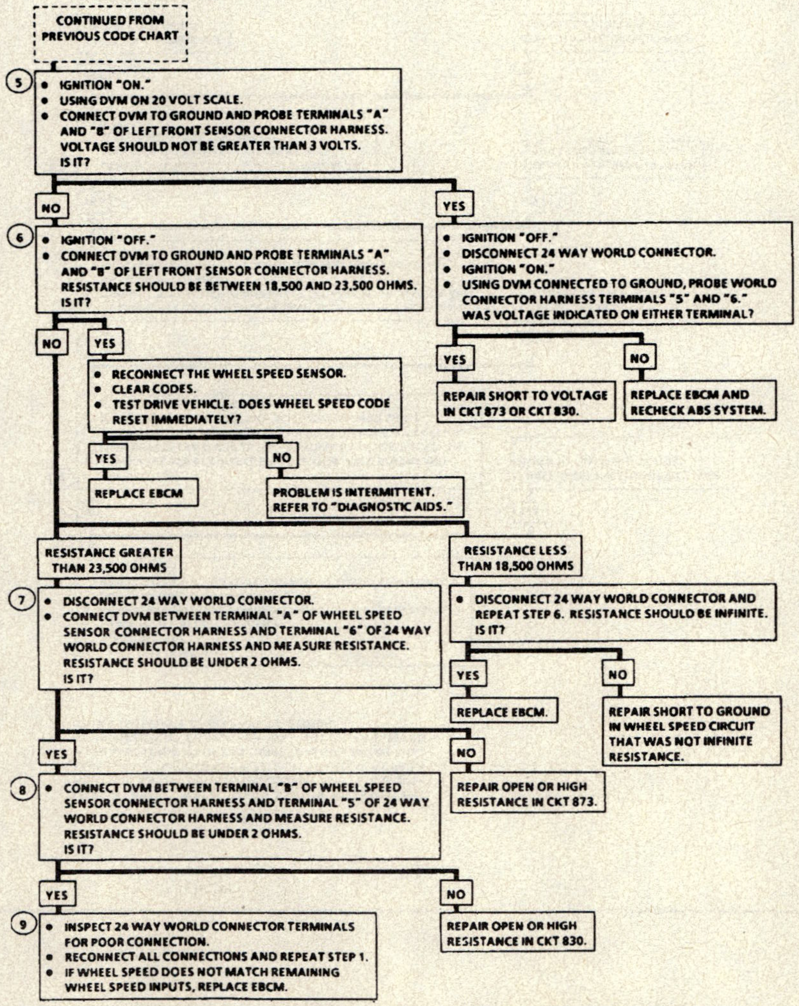

THIS CHART ASSUMES THAT A CURRENT CODE IS STORED INDICATING THAT THIS FAULT IS PRESENT.

DIAGNOSTIC CHART A022 — 1993
CAVALIER AND SUNBIRD

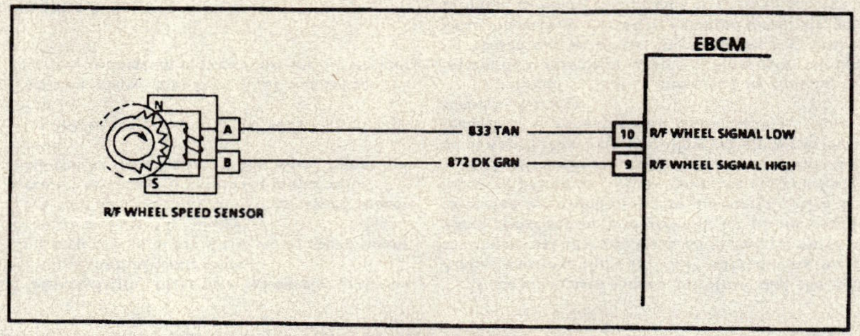

CODE A022

RIGHT FRONT WHEEL SPEED = 0

Circuit Description:

As a toothed ring passes by the wheel speed sensor, changes in the electromagnetic field cause the wheel speed sensor to produce a sinusoidal (AC) voltage signal whose frequency is proportional to wheel speed. The magnitude of this signal is directly related to wheel speed and the proximity of the wheel speed sensor to the toothed ring, often referred to as the air gap.

This code detects opens, shorts to ground, and shorts to battery that cause a wheel speed of 0 mph to be calculated by the ABS EBCM. The conditions present to fail this test are: (1) Three (3) wheels are rotating at least 5 mph, and (2) The fourth wheel is 0 mph.

Test Description: Number(s) below refer to circled number(s) on the diagnostic chart.

1. This test verifies whether the fault is currently present.
2. This step is designed to identify a wheel speed sensor or circuitry that is damaged and visibly apparent.
3. This test checks to see that the wheel speed sensor is within proper resistance values.
4. This insures that the wheel speed sensor is not shorted to ground.

Diagnostic Aids:

An "Intermittent" problem may be caused by a poor connection, rubbed through wire insulation, or a wire that is broken inside the insulation.

The frequency of the problem can be checked by using the enhanced diagnostic function of the "Scan" tool.

If the customers comments reflect that the ABS Amber warning light is on only during moist environmental changes, (rain, snow, vehicle wash) all wheel speed sensor circuitry should be thoroughly inspected for signs of water intrusion. Use the following procedure. Spray down the suspected area with 5% salt water solution (two teaspoons of salt to 12 oz. of water). Start vehicle, allow vehicle to run for 10 seconds. If code returns immediately, replace suspected harness.

Any circuitry, that is suspected as causing the intermittent complaint, should be thoroughly checked for backed out terminals, improper mating, broken locks, improperly formed or damaged terminals, poor terminal to wiring connections or physical damage to the wiring harness.

DIAGNOSTIC CHART A022 — 1993
CAVALIER AND SUNBIRD

IMPORTANT: WHEEL SPEED SENSOR INTERMITTENT PROBLEMS MAY BE DIFFICULT TO LOCATE. CARE SHOULD BE TAKEN NOT TO DISTURB ANY ELECTRICAL CONNECTIONS PRIOR TO AN INDICATED STEP OF THIS CHART. THIS WILL INSURE THAT AN INTERMITTENT CONNECTION WILL NOT BE CORRECTED BEFORE THE SOURCE OF THE PROBLEM IS FOUND.

(1)
- TEST DRIVE VEHICLE.
- IGNITION "ON."
- USING TECH 1, SELECT DATA LIST.
- MONITOR WHEEL SPEED ON RIGHT FRONT WHEEL WHILE SLOWLY DECELERATING FROM 35 MPH TO 0 MPH.

WHEEL SPEED SHOULD MATCH REMAINING WHEEL SPEED INPUTS. DOES IT?

YES → FAULT IS NOT PRESENT AT THIS TIME, REFER TO "DIAGNOSTIC AIDS".

NO ↓

(2)
- IGNITION "OFF."
- PHYSICALLY INSPECT WHEEL SPEED SENSOR, RING, WIRING AND CONNECTORS FOR DAMAGE. PAY PARTICULAR ATTENTION TO THE SPEED RING. ANY SIGNIFICANT DAMAGE (OTHER THAN NICKS FROM STONES ETC.) WILL AFFECT THE WHEEL SPEED INPUT SIGNAL.

IS PHYSICAL DAMAGE INDICATED?

YES → CORRECT PHYSICAL DAMAGE, CLEAR CODES AND RECHECK.

NO ↓

(3)
- DISCONNECT CONNECTOR FROM RIGHT FRONT WHEEL SPEED SENSOR.
- CONNECT DVM BETWEEN TERMINALS "A" AND "B" OF SENSOR, AND MEASURE RESISTANCE.

RESISTANCE SHOULD BE BETWEEN 1650 AND 1800 OHMS. IS IT?

NO → REPLACE WHEEL SPEED SENSOR, CLEAR CODES AND RECHECK.

YES ↓

(4)
- CONNECT DVM BETWEEN SENSOR TERMINAL "A" AND GROUND AND MEASURE RESISTANCE.

RESISTANCE SHOULD BE INFINITE. IS IT?

NO → REPLACE WHEEL SPEED SENSOR, CLEAR CODES AND RECHECK.

YES → GO TO FOLLOWING PAGE OF THIS CODE CHART

THIS CHART ASSUMES THAT A CURRENT CODE IS STORED INDICATING THAT THIS FAULT IS PRESENT.

DIAGNOSTIC CHART A022 — 1993 CAVALIER AND SUNBIRD

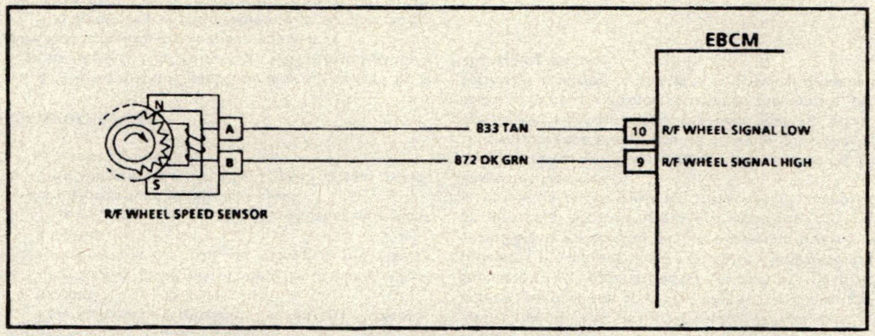

CODE A022

RIGHT FRONT WHEEL SPEED = 0

Circuit Description:

As a toothed ring passes by the wheel speed sensor, changes in the electromagnetic field cause the wheel speed sensor to produce a sinusoidal (AC) voltage signal whose frequency is proportional to wheel speed. The magnitude of this signal is directly related to wheel speed and the proximity of the wheel speed sensor to the toothed ring, often referred to as the air gap.

This code detects opens, shorts to ground, and shorts to battery that cause a wheel speed of 0 mph to be calculated by the ABS EBCM. The conditions present to fail this test are: (1) Three (3) wheels are rotating at least 5 mph, and (2) The fourth wheel is 0 mph.

Test Description: Number(s) below refer to circled number(s) on the diagnostic chart.

5. This step checks to see if the wheel speed sensor circuitry is shorted to voltage.
6. This step insures that the wheel speed sensor circuitry resistance is within the proper value.
7. This checks for an open in the wheel signal low circuit.
8. This checks for an open in the wheel signal high circuit.
9. This step insures that Code A022 was not set due to a poor connection between the world connector and EBCM.

Diagnostic Aids:

An "Intermittent" problem may be caused by a poor connection, rubbed through wire insulation, or a wire that is broken inside the insulation.

The frequency of the problem can be checked by using the enhanced diagnostic function of the "Scan" tool,

If the customers comments reflect that the ABS Amber warning light is "ON" only during moist environmental changes, (rain, snow, vehicle wash) all wheel speed sensor circuitry should be thoroughly inspected for signs of water intrusion. Use the following procedure. Spray down the suspected area with 5% salt water solution (two teaspoons of salt to 12 oz. of water). Start vehicle, allow vehicle to run for 10 seconds. If code returns immediately, replace suspected harness.

Any circuitry, that is suspected as causing the intermittent complaint, should be thoroughly checked for backed out terminals, improper mating, broken locks, improperly formed or damaged terminals, poor terminal to wiring connections or physical damage to the wiring harness.

DIAGNOSTIC CHART A022 — 1993 CAVALIER AND SUNBIRD

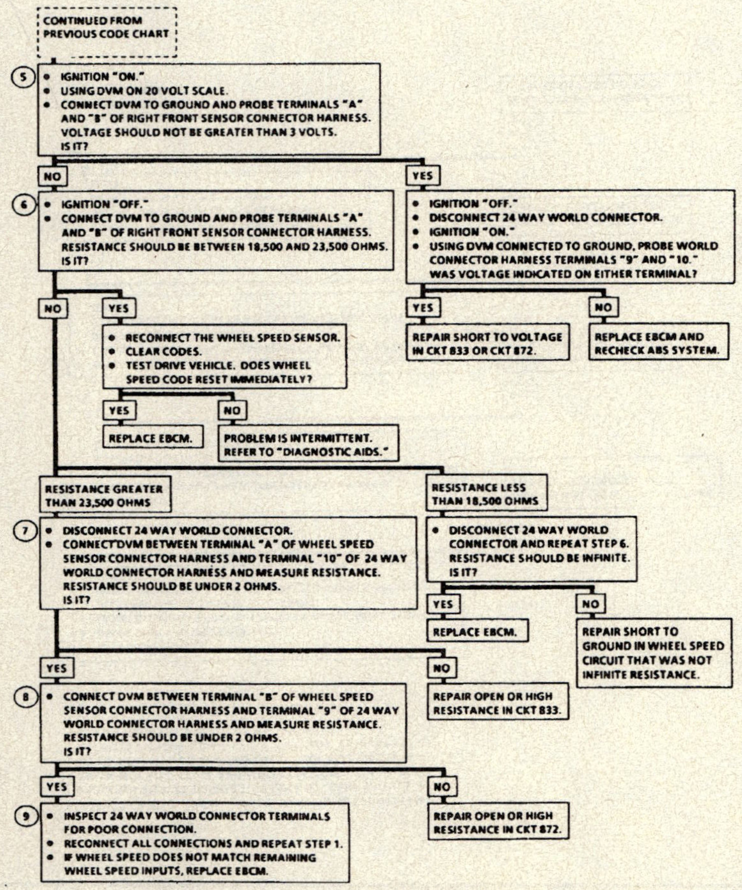

THIS CHART ASSUMES THAT A CURRENT CODE IS STORED INDICATING THAT THIS FAULT IS PRESENT.

DIAGNOSTIC CHART A023 — 1993 CAVALIER AND SUNBIRD

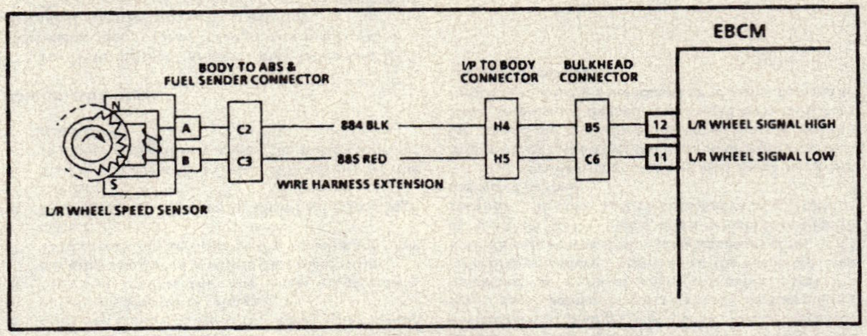

CODE A023

LEFT REAR WHEEL SPEED = 0

Circuit Description:

As a toothed ring passes by the wheel speed sensor, changes in the electromagnetic field cause the wheel speed sensor to produce a sinusoidal (AC) voltage signal whose frequency is proportional to wheel speed. The magnitude of this signal is directly related to wheel speed and the proximity of the wheel speed sensor to the toothed ring, often referred to as the air gap.

This code detects opens, shorts to ground, and shorts to battery that cause a wheel speed of 0 mph to be calculated by the ABS EBCM. The conditions present to fail this test are: (1) Three (3) wheels are rotating at least 5 mph, (2) The fourth wheel is 0 mph.

Test Description: Number(s) below refer to circled number(s) on the diagnostic chart.

1. This test verifies whether the fault is currently present.
2. This step is designed to identify a wheel speed sensor or circuitry that is damaged and visibly apparent.
3. This test checks to see that the wheel speed sensor is within proper resistance values.
4. This insures that the wheel speed sensor is not shorted to ground.

Diagnostic Aids:

An "Intermittent" problem may be caused by a poor connection, rubbed through wire insulation, or a wire that is broken inside the insulation.

The frequency of the problem can be checked by using the enhanced diagnostic function of the "Scan" tool.

If the customers comments reflect that the ABS Amber warning light is "ON" only during moist environmental changes, (rain, snow, vehicle wash) all wheel speed senor circuitry should be thoroughly inspected for signs of water intrusion. Use the following procedure. Spray down the suspected area with 5% salt water solution (two teaspoons of salt to 12 oz. of water). Start vehicle, allow vehicle to run for 10 seconds. If code returns immediately, replace suspected harness.

Any circuitry, that is suspected as causing the intermittent complaint, should be thoroughly checked for backed out terminals, improper mating, broken locks, improperly formed or damaged terminals, poor terminal to wiring connections or physical damage to the wiring harness.

DIAGNOSTIC CHART A023 — 1993 CAVALIER AND SUNBIRD

IMPORTANT: WHEEL SPEED SENSOR INTERMITTENT PROBLEMS MAY BE DIFFICULT TO LOCATE. CARE SHOULD BE TAKEN NOT TO DISTURB ANY ELECTRICAL CONNECTIONS PRIOR TO AN INDICATED STEP OF THIS CHART. THIS WILL INSURE THAT AN INTERMITTENT CONNECTION WILL NOT BE CORRECTED BEFORE THE SOURCE OF THE PROBLEM IS FOUND.

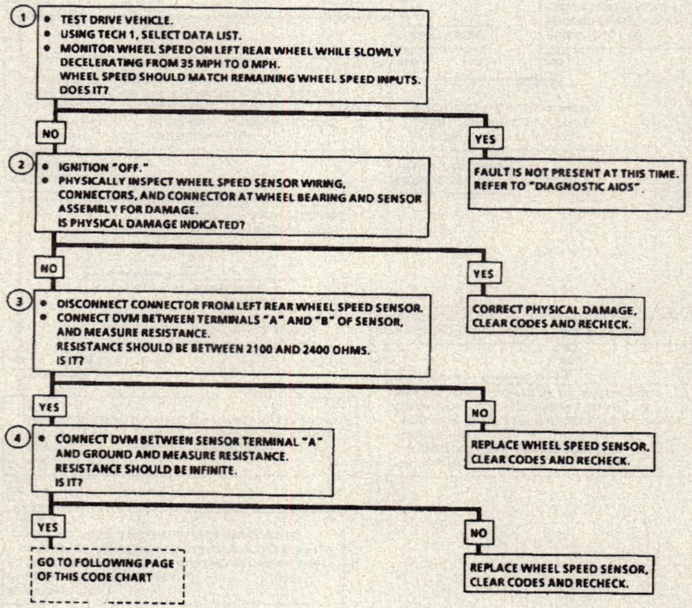

THIS CHART ASSUMES THAT A CURRENT CODE IS STORED INDICATING THAT THIS FAULT IS PRESENT.

DIAGNOSTIC CHART A023 — 1993 CAVALIER AND SUNBIRD

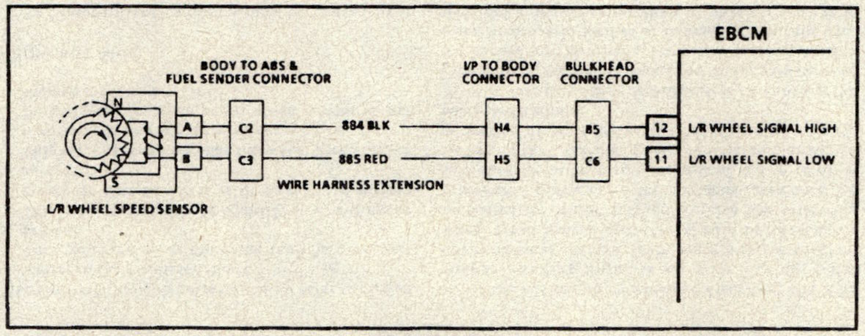

CODE A023

LEFT REAR WHEEL SPEED = 0

Circuit Description:

As a toothed ring passes by the wheel speed sensor, changes in the electromagnetic field cause the wheel speed sensor to produce a sinusoidal (AC) voltage signal whose frequency is proportional to wheel speed. The magnitude of this signal is directly related to wheel speed and the proximity of the wheel speed sensor to the toothed ring, often referred to as the air gap.

This code detects opens, shorts to ground, and shorts to battery that cause a wheel speed of 0 mph to be calculated by the ABS EBCM. The conditions present to fail this test are: (1) Three (3) wheels are rotating at least 5 mph, and (2) The fourth wheel is 0 mph.

Test Description: Number(s) below refer to circled number(s) on the diagnostic chart.

5. This step checks to see if the wheel speed sensor circuitry is shorted to voltage.
6. This step insures that the wheel speed sensor circuitry resistance is within the proper value.
7. This checks for an open in the wheel signal high circuit.
8. This checks for an open in the wheel signal low circuit.
9. This step insures that Code A023 was not set due to a poor connection between the world connector and EBCM.

Diagnostic Aids:

An "Intermittent" problem may be caused by a poor connection, rubbed through wire insulation, or a wire that is broken inside the insulation.

The frequency of the problem can be checked by using the enhanced diagnostic function of the "Scan" tool.

If the customers comments reflect that the ABS Amber warning light is "ON" only during moist environmental changes, (rain, snow, vehicle wash) all wheel speed sensor circuitry should be thoroughly inspected for signs of water intrusion. Use the following procedure. Spray down the suspected area with 5% salt water solution (two teaspoons of salt to 12 oz. of water). Start vehicle, allow vehicle to run for 10 seconds. If code returns immediately, replace suspected harness.

Any circuitry, that is suspected as causing the intermittent complaint, should be thoroughly checked for backed out terminals, improper mating, broken locks, improperly formed or damaged terminals, poor terminal to wiring connections or physical damage to the wiring harness.

DIAGNOSTIC CHART A023 — 1993 CAVALIER AND SUNBIRD

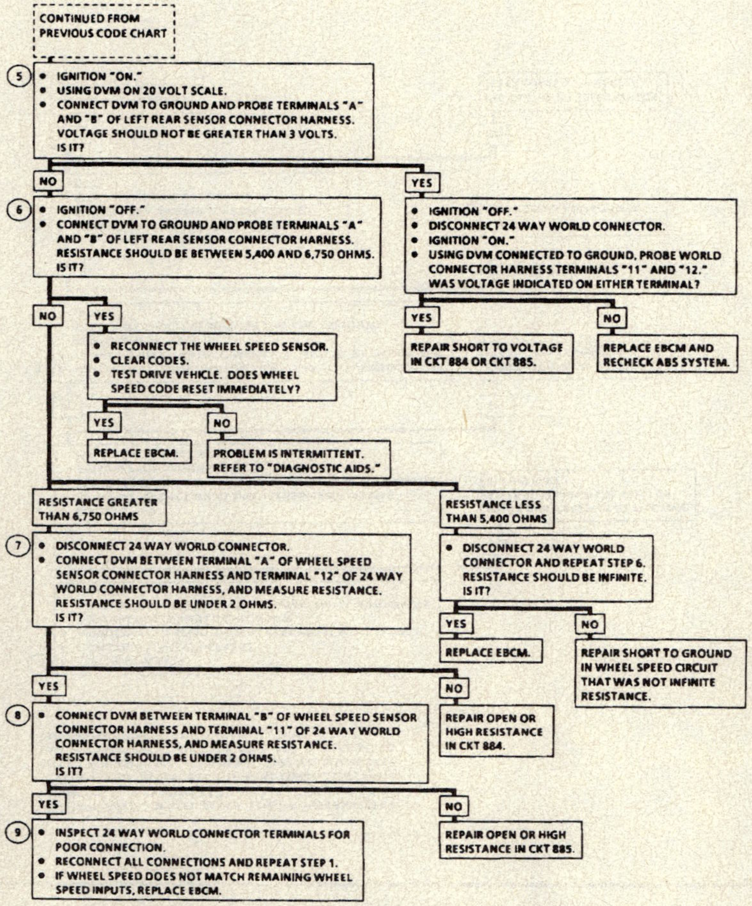

THIS CHART ASSUMES THAT A CURRENT CODE IS STORED INDICATING THAT THIS FAULT IS PRESENT.

DIAGNOSTIC CHART A024 — 1993 CAVALIER AND SUNBIRD

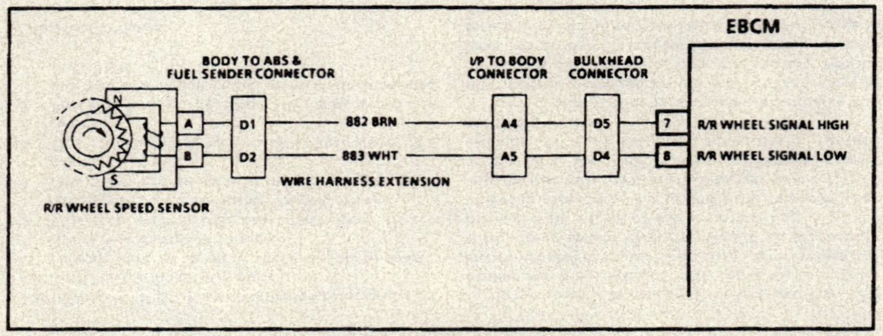

CODE A024

RIGHT REAR WHEEL SPEED = 0

Circuit Description:

As a toothed ring passes by the wheel speed sensor, changes in the electromagnetic field cause the wheel speed sensor to produce a sinusoidal (AC) voltage signal whose frequency is proportional to wheel speed. The magnitude of this signal is directly related to wheel speed and the proximity of the wheel speed sensor to the toothed ring, often referred to as the air gap.

This code detects opens, shorts to ground, and shorts to battery that cause a wheel speed of 0 mph to be calculated by the ABS EBCM. The conditions present to fail this test are: (1) Three (3) wheels are rotating at least 5 mph, and (2) The fourth wheel is 0 mph.

Test Description: Number(s) below refer to circled number(s) on the diagnostic chart.

1. This test verifies whether the fault is currently present.
2. This step is designed to identify a wheel speed sensor or circuitry that is damaged and visibly apparent.
3. This test checks to see that the wheel speed sensor is within proper resistance values.
4. This insures that the wheel speed sensor is not shorted to ground.

Diagnostic Aids:

An "Intermittent" problem may be caused by a poor connection, rubbed through wire insulation, or a wire that is broken inside the insulation.

The frequency of the problem can be checked by using the enhanced diagnostic function of the "Scan" tool

If the customers comments reflect that the ABS Amber warning light is on only during moist environmental changes, (rain, snow, vehicle wash) all wheel speed sensor circuitry should be thoroughly inspected for water intrusion. Use the following procedure. Spray down the suspected area with 5% salt water solution (two teaspoons of salt to 12 oz. of water). Start vehicle, allow vehicle to run for 10 seconds. If code returns immediately, replace suspected harness.

Any circuitry, that is suspected as causing the intermittent complaint, should be thoroughly checked for backed out terminals, improper mating, broken locks, improperly formed or damaged terminals, poor terminal to wiring connections or physical damage to the wiring harness.

DIAGNOSTIC CHART A024 — 1993 CAVALIER AND SUNBIRD

IMPORTANT: WHEEL SPEED SENSOR INTERMITTENT PROBLEMS MAY BE DIFFICULT TO LOCATE. CARE SHOULD BE TAKEN NOT TO DISTURB ANY ELECTRICAL CONNECTIONS PRIOR TO AN INDICATED STEP OF THIS CHART. THIS WILL INSURE THAT AN INTERMITTENT CONNECTION WILL NOT BE CORRECTED BEFORE THE SOURCE OF THE PROBLEM IS FOUND.

(1)
- TEST DRIVE VEHICLE.
- USING TECH 1, SELECT DATA LIST.
- MONITOR WHEEL SPEED ON RIGHT REAR WHEEL WHILE SLOWLY DECELERATING FROM 35 MPH TO 0 MPH.

WHEEL SPEED SHOULD MATCH REMAINING WHEEL SPEED INPUTS. DOES IT?

- YES: FAULT IS NOT PRESENT AT THIS TIME. REFER TO "DIAGNOSTIC AIDS" ON FACING PAGE.
- NO: go to (2)

(2)
- IGNITION "OFF."
- PHYSICALLY INSPECT WHEEL SPEED SENSOR, RING, WIRING, CONNECTORS, AND CONNECTOR AT WHEEL BEARING AND SENSOR ASSEMBLY FOR DAMAGE.

IS PHYSICAL DAMAGE INDICATED?

- YES: CORRECT PHYSICAL DAMAGE, CLEAR CODES AND RECHECK.
- NO: go to (3)

(3)
- DISCONNECT CONNECTOR FROM RIGHT REAR WHEEL SPEED SENSOR.
- CONNECT DVM BETWEEN TERMINALS "A" AND "B" OF SENSOR, AND MEASURE RESISTANCE.

RESISTANCE SHOULD BE BETWEEN 2100 AND 2400 OHMS. IS IT?

- NO: REPLACE WHEEL SPEED SENSOR, CLEAR CODES AND RECHECK.
- YES: go to (4)

(4)
- CONNECT DVM BETWEEN SENSOR TERMINAL "A" AND GROUND, AND MEASURE RESISTANCE.

RESISTANCE SHOULD BE INFINITE. IS IT?

- NO: REPLACE WHEEL SPEED SENSOR, CLEAR CODES AND RECHECK.
- YES: GO TO FOLLOWING PAGE OF THIS CODE CHART

THIS CHART ASSUMES THAT A CURRENT CODE IS STORED INDICATING THAT THIS FAULT IS PRESENT.

DIAGNOSTIC CHART A024 — 1993 CAVALIER AND SUNBIRD

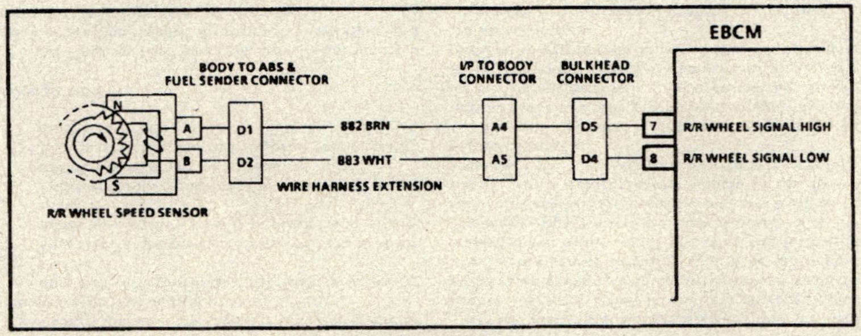

CODE A024

RIGHT REAR WHEEL SPEED = 0

Circuit Description:

As a toothed ring passes by the wheel speed sensor, changes in the electromagnetic field cause the wheel speed sensor to produce a sinusoidal (AC) voltage signal whose frequency is proportional to wheel speed. The magnitude of this signal is directly related to wheel speed and the proximity of the wheel speed sensor to the toothed ring, often referred to as the air gap.

This code detects opens, shorts to ground, and shorts to battery that cause a wheel speed of 0 mph to be calculated by the ABS EBCM. The conditions present to fail this test are: (1) Three (3) wheels are rotating at least 5 mph, and (2) The fourth wheel is 0 mph.

Test Description: Number(s) below refer to circled number(s) on the diagnostic chart.

5. This step checks to see if the wheel speed sensor circuitry is shorted to voltage.
6. This step insures that the wheel speed sensor circuitry resistance is within the proper value.
7. This checks for an open in the wheel signal high circuit.
8. This checks for an open in the wheel signal low circuit.
9. This step insures that Code A024 was not set due to a poor connection between the world connector and EBCM.

Diagnostic Aids:

An "Intermittent" problem may be caused by a poor connection, rubbed through wire insulation, or a wire that is broken inside the insulation.

If the customers comments reflect that the ABS Amber warning light is on only during moist environmental changes, (rain, snow, vehicle wash) all wheel speed sensor circuitry should be thoroughly inspected for water intrusion. Use the following procedure. Spray down the suspected area with 5% salt water solution (two teaspoons of salt to 12 oz. of water). Start vehicle, allow vehicle to run for 10 seconds. If code returns immediately, replace suspected harness.

Any circuitry, that is suspected as causing the intermittent complaint, should be thoroughly checked for backed out terminals, improper mating, broken locks, improperly formed or damaged terminals, poor terminal to wiring connections or physical damage to the wiring harness.

DIAGNOSTIC CHART A024 — 1993 CAVALIER AND SUNBIRD

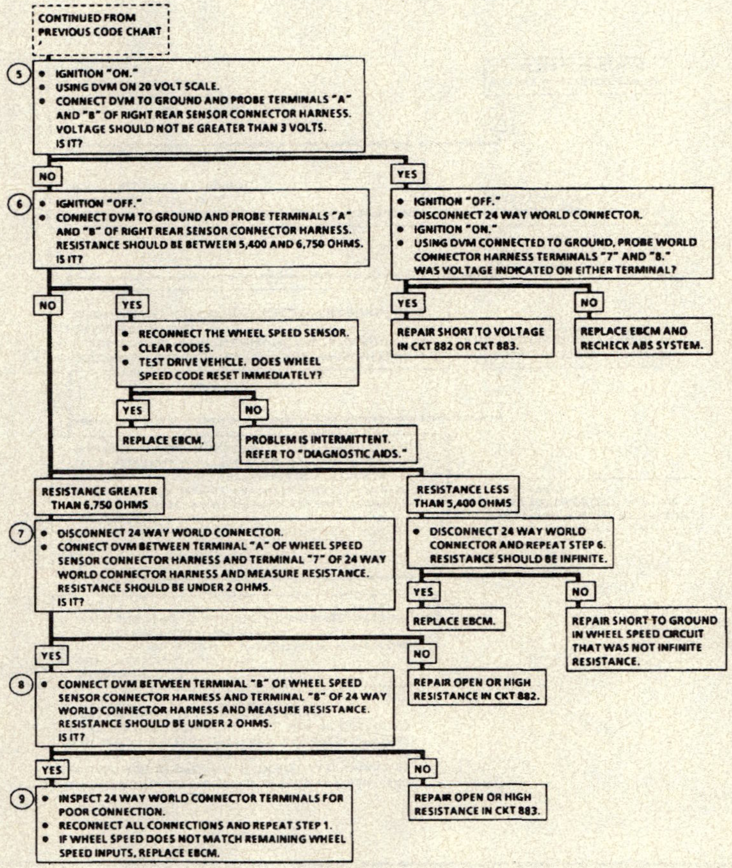

THIS CHART ASSUMES THAT A CURRENT CODE IS STORED INDICATING THAT THIS FAULT IS PRESENT.

DIAGNOSTIC CHART A025 — 1993 CAVALIER AND SUNBIRD

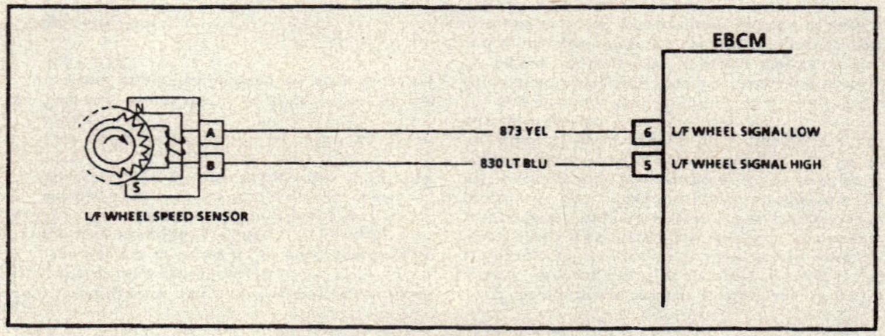

CODE A025

LEFT FRONT EXCESSIVE WHEEL SPEED VARIATION

Circuit Description:

As a toothed ring passes by the wheel speed sensor, changes in the electromagnetic field cause the wheel speed sensor to produce a sinusoidal (AC) voltage signal whose frequency is proportional to wheel speed. The magnitude of this signal is directly related to wheel speed and the proximity of the wheel speed sensor to the toothed ring, often referred to as the air gap.

This code detects opens, shorts to ground, and shorts to battery that cause intermittent wheel speed operation. Also detected is any sudden change in wheel speed determined to the unreasonable.

Test Description: Number(s) below refer to circled number(s) on the diagnostic chart.

1. This test verifies whether the fault is currently present.
2. This step is designed to identify a wheel speed sensor or circuitry that is damaged and visibly apparent.
3. This test checks to see that the wheel speed sensor circuitry is within proper resistance value.
4. This insure that the wheel speed sensor is not shorted to ground.

Diagnostic Aids:

An "Intermittent" problem may be caused by a poor connection, rubbed through wire insulation, or a wire that is broken inside the insulation.

If the customer's comments reflect that the ABS Amber warning light is on only during moist environmental changes, (rain, snow, vehicle wash) all wheel speed sensor circuitry should be thoroughly inspected for water intrusion. Use the following procedure. Spray down the suspected area with 5% salt water solution (two teaspoons of salt to 12 oz. of water). Start vehicle, allow vehicle to run for 10 seconds. If code returns immediately, replace suspected harness.

Any circuitry, that is suspected as causing the intermittent complaint, should be thoroughly checked for backed out terminals, improper mating, broken locks, improperly formed or damaged terminals, poor terminal to wiring connections or physical damage to the wiring harness.

A careful visual inspection of the toothed ring, CV joint, bearing and wheel speed sensor for physical damage should also be performed. If code sets at the same mph every time and wheel speed variation is noted above this mph on the Tech 1, the speed ring is most likely damaged. This can occur at any speed.

DIAGNOSTIC CHART A025 — 1993 CAVALIER AND SUNBIRD

IMPORTANT: WHEEL SPEED SENSOR INTERMITTENT PROBLEMS MAY BE DIFFICULT TO LOCATE. CARE SHOULD BE TAKEN NOT TO DISTURB ANY ELECTRICAL CONNECTIONS PRIOR TO AN INDICATED STEP OF THIS CHART. THIS WILL INSURE THAT AN INTERMITTENT CONNECTION WILL NOT BE CORRECTED BEFORE THE SOURCE OF THE PROBLEM IS FOUND.

(1)
- TEST DRIVE VEHICLE.
- USING TECH 1, SELECT DATA LIST.
- MONITOR WHEEL SPEED ON LEFT FRONT WHEEL WHILE SLOWLY DECELERATING FROM 35 MPH TO 0 MPH.
- WHEEL SPEED SHOULD MATCH REMAINING WHEEL SPEED INPUTS. DOES IT?

YES: FAULT IS NOT PRESENT AT THIS TIME. REFER TO "DIAGNOSTIC AIDS" ON FACING PAGE.

NO:

(2)
- IGNITION "OFF."
- PHYSICALLY INSPECT WHEEL SPEED SENSOR, RING, WIRING AND CONNECTORS FOR DAMAGE. PAY PARTICULAR ATTENTION TO THE SPEED RING. ANY SIGNIFICANT DAMAGE (OTHER THAN NICKS FROM STONES ETC.) WILL AFFECT THE WHEEL SPEED INPUT SIGNAL. IS PHYSICAL DAMAGE INDICATED?

YES: CORRECT PHYSICAL DAMAGE. CLEAR CODES AND RECHECK.

NO:

(3)
- DISCONNECT CONNECTOR FROM LEFT FRONT WHEEL SPEED SENSOR.
- CONNECT DVM BETWEEN TERMINAL "A" AND "B" OF SENSOR, AND MEASURE RESISTANCE. RESISTANCE SHOULD BE BETWEEN 1650 AND 1800 OHMS. IS IT?

NO: REPLACE WHEEL SPEED SENSOR, CLEAR CODES AND RECHECK.

YES:

(4)
- CONNECT DVM BETWEEN SENSOR TERMINAL "A" AND GROUND AND MEASURE RESISTANCE. RESISTANCE SHOULD BE INFINITE. IS IT?

NO: REPLACE WHEEL SPEED SENSOR, CLEAR CODES AND RECHECK.

YES: GO TO FOLLOWING PAGE OF THIS CODE CHART

THIS CHART ASSUMES THAT A CURRENT CODE IS STORED INDICATING THAT THIS FAULT IS PRESENT.

DIAGNOSTIC CHART A025 — 1993 CAVALIER AND SUNBIRD

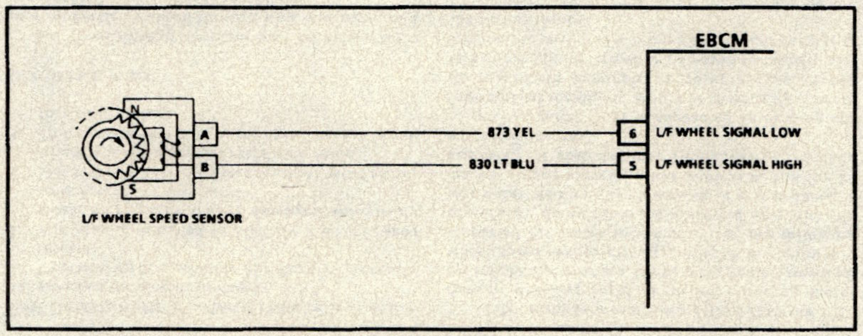

CODE A025

LEFT FRONT EXCESSIVE WHEEL SPEED VARIATION

Circuit Description:

As a toothed ring passes by the wheel speed sensor, changes in the electromagnetic field cause the wheel speed sensor to produce a sinusoidal (AC) voltage signal whose frequency is proportional to wheel speed. The magnitude of this signal is directly related to wheel speed and the proximity of the wheel speed sensor to the toothed ring, often referred to as the air gap.

This code detects opens, shorts to ground, and shorts to battery that cause intermittent wheel speed operational. Also detected is any sudden change in wheel speed determined to be unreasonable.

Test Description: Number(s) below refer to circled number(s) on the diagnostic chart.

5. This step checks to see if the wheel speed sensor circuitry is shorted to voltage.
6. This step insures that the wheel speed sensor circuitry resistance is within the proper value.
7. This checks for an open in the wheel signal low circuit.
8. This checks for an open in the wheel signal high circuit.
9. This step insures that Code A025 was not set due to a poor connection between the world connector and EBCM.

Diagnostic Aids:

An "Intermittent" problem may be caused by a poor connection, rubbed through wire insulation, or a wire that is broken inside the insulation.

If the customers comments reflect that the ABS Amber warning light is on only during moist environmental changes, (rain, snow, vehicle wash) all wheel speed sensor circuitry should be thoroughly inspected for water intrusion. Use the following procedure. Spray down the suspected area with 5% salt water solution (two teaspoons of salt to 12 oz. of water). Start vehicle, allow vehicle to run for 10 seconds. If code returns immediately, replace suspected harness.

Any circuitry, that is suspected as causing the intermittent complaint, should be thoroughly checked for backed out terminals, improper mating, broken locks, improperly formed or damaged terminals, poor terminal to wiring connections or physical damage to the wiring harness.

A careful visual inspection of the toothed ring, CV joint, bearing and wheel speed sensor for physical damage should also be performed. If code sets at the same mph every time and wheel speed variation is noted above this mph on the Tech 1, the speed ring is most likely damaged. This can occur at any speed.

DIAGNOSTIC CHART A025 — 1993 CAVALIER AND SUNBIRD

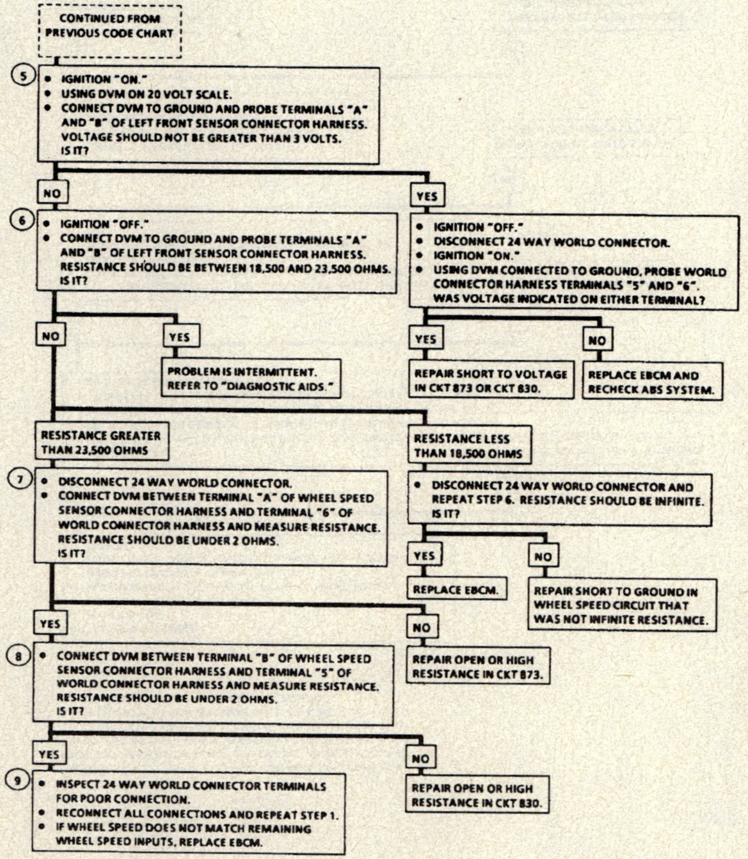

THIS CHART ASSUMES THAT A CURRENT CODE IS STORED INDICATING THAT THIS FAULT IS PRESENT.

DIAGNOSTIC CHART A026 — 1993 CAVALIER AND SUNBIRD

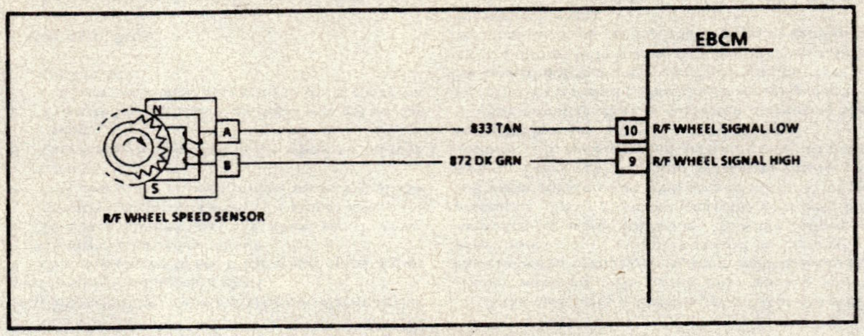

CODE A026

RIGHT FRONT EXCESSIVE WHEEL SPEED VARIATION

Circuit Description:

As a toothed ring passes by the wheel speed sensor, changes in the electromagnetic field cause the wheel speed sensor to produce a sinusoidal (AC) voltage signal whose frequency is proportional to wheel speed. The magnitude of this signal is directly related to wheel speed and the proximity of the wheel speed sensor to the toothed ring, often referred to as the air gap.

This code detects opens, shorts to ground, and shorts to battery that cause intermittent wheel speed operation. Also detected is any sudden change in wheel speed determined to be unreasonable.

Test Description: Number(s) below refer to circled number(s) on the diagnostic chart.

1. This test verifies whether the fault is currently present.
2. This step is designed to identify a wheel speed sensor or circuitry that is damaged and visibly apparent.
3. This test checks to see that the wheel speed sensor circuitry is within proper resistance value.
4. This insures that the wheel speed sensor is not shorted to ground.

Diagnostic Aids:

An "Intermittent" problem may be caused by a poor connection, rubbed through wire insulation, or a wire that is broken inside the insulation.

If the customers comments reflect that the ABS Amber warning light is on only during moist environmental changes, (rain, snow, vehicle wash) all wheel speed sensor circuitry should be thoroughly inspected for water intrusion. Use the following procedure. Spray down the suspected area with 5% salt water solution (two teaspoons of salt to 12 oz. of water). Start vehicle, allow vehicle to run for 10 seconds. If code returns immediately, replace suspected harness.

Any circuitry, that is suspected as causing the intermittent complaint, should be thoroughly checked for backed out terminals, improper mating, broken locks, improperly formed or damaged terminals, poor terminal to wiring connections or physical damage to the wiring harness.

A careful visual inspection of the toothed ring, CV joint, bearing and wheel speed sensor for physical damage should also be performed. If code sets at the same mph every time and wheel speed variation is noted above this mph on the Tech 1, the speed ring is most likely damaged. This can occur at any speed.

DIAGNOSTIC CHART A026 — 1993 CAVALIER AND SUNBIRD

IMPORTANT: WHEEL SPEED SENSOR INTERMITTENT PROBLEMS MAY BE DIFFICULT TO LOCATE. CARE SHOULD BE TAKEN NOT TO DISTURB ANY ELECTRICAL CONNECTIONS PRIOR TO AN INDICATED STEP OF THIS CHART. THIS WILL INSURE THAT AN INTERMITTENT CONNECTION WILL NOT BE CORRECTED BEFORE THE SOURCE OF THE PROBLEM IS FOUND.

(1)
- TEST DRIVE VEHICLE.
- IGNITION "ON."
- USING TECH 1, SELECT DATA LIST.
- MONITOR WHEEL SPEED ON RIGHT FRONT WHEEL WHILE SLOWLY DECELERATING FROM 35 MPH TO 0 MPH.

WHEEL SPEED SHOULD MATCH REMAINING WHEEL SPEED INPUTS. DOES IT?

- YES: FAULT IS NOT PRESENT AT THIS TIME, REFER TO "DIAGNOSTIC AIDS".
- NO: go to (2)

(2)
- IGNITION "OFF."
- PHYSICALLY INSPECT WHEEL SPEED SENSOR, RING, WIRING AND CONNECTORS FOR DAMAGE. PAY PARTICULAR ATTENTION TO THE SPEED RING. ANY SIGNIFICANT DAMAGE (OTHER THAN NICKS FROM STONES ETC.) WILL AFFECT THE WHEEL SPEED INPUT SIGNAL.

IS PHYSICAL DAMAGE INDICATED?

- YES: CORRECT PHYSICAL DAMAGE, CLEAR CODES AND RECHECK.
- NO: go to (3)

(3)
- DISCONNECT CONNECTOR FROM RIGHT FRONT WHEEL SPEED SENSOR.
- CONNECT DVM BETWEEN TERMINALS "A" AND "B" OF SENSOR, AND MEASURE RESISTANCE.

RESISTANCE SHOULD BE BETWEEN 1650 AND 1800 OHMS. IS IT?

- NO: REPLACE WHEEL SPEED SENSOR, CLEAR CODES AND RECHECK.
- YES: go to (4)

(4)
- CONNECT DVM BETWEEN SENSOR TERMINAL "A" AND GROUND AND MEASURE RESISTANCE.

RESISTANCE SHOULD BE INFINITE. IS IT?

- NO: REPLACE WHEEL SPEED SENSOR, CLEAR CODES AND RECHECK.
- YES: GO TO FOLLOWING PAGE OF THIS CODE CHART

THIS CHART ASSUMES THAT A CURRENT CODE IS STORED INDICATING THAT THIS FAULT IS PRESENT.

DIAGNOSTIC CHART A026 — 1993 CAVALIER AND SUNBIRD

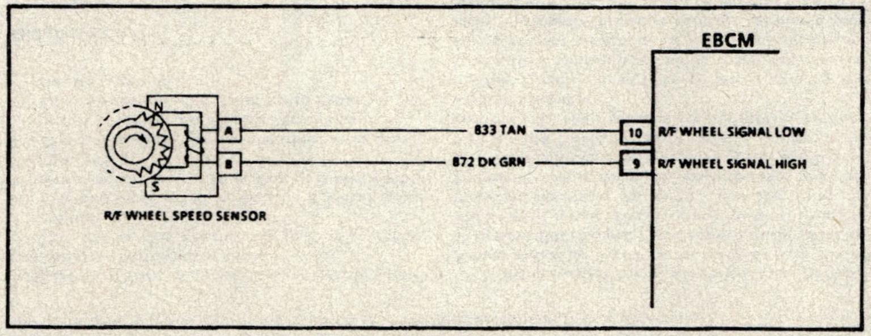

CODE A026

RIGHT FRONT EXCESSIVE WHEEL SPEED VARIATION

Circuit Description:

As a toothed ring passes by the wheel speed sensor, changes in the electromagnetic field cause the wheel speed sensor to produce a sinusoidal (AC) voltage signal whose frequency is proportional to wheel speed. The magnitude of this signal is directly related to wheel speed and the proximity of the wheel speed sensor to the toothed ring, often referred to as the air gap.

This code detects opens, shorts to ground, and shorts to battery that cause intermittent wheel speed operation. Also detected is any sudden change in wheel speed determined to be unreasonable.

Test Description: Number(s) below refer to circled number(s) on the diagnostic chart.

5. This step checks to see if the wheel speed sensor circuitry is shorted to voltage.
6. This step insures that the wheel speed sensor circuitry resistance is within the proper value.
7. This checks for an open in the wheel signal low circuit.
8. This checks for an open in the wheel signal high circuit.
9. This step insures that Code A026 was not set due to a poor connection between the world connector and EBCM.

Diagnostic Aids:

An "Intermittent" problem may be caused by a poor connection, rubbed through wire insulation, or a wire that is broken inside the insulation.

If the customers comments reflect that the ABS Amber warning light is on only during moist environmental changes, (rain, snow, vehicle wash) all wheel speed sensor circuitry should be thoroughly inspected for water intrusion. Use the following procedure. Spray down the suspected area with 5% salt water solution (two teaspoons of salt to 12 oz. of water). Start vehicle, allow vehicle to run for 10 seconds. If code returns immediately, replace suspected harness.

Any circuitry, that is suspected as causing the intermittent complaint, should be thoroughly checked for backed out terminals, improper mating, broken locks, improperly formed or damaged terminals, poor terminal to wiring connections or physical damage to the wiring harness.

A careful visual inspection of the toothed ring, CV joint, bearing and wheel speed sensor for physical damage should also be performed. If code sets at the same mph every time and wheel speed variation is noted above this mph on the Tech 1, the speed ring is most likely damaged. This can occur at any speed.

DIAGNOSTIC CHART A026 — 1993 CAVALIER AND SUNBIRD

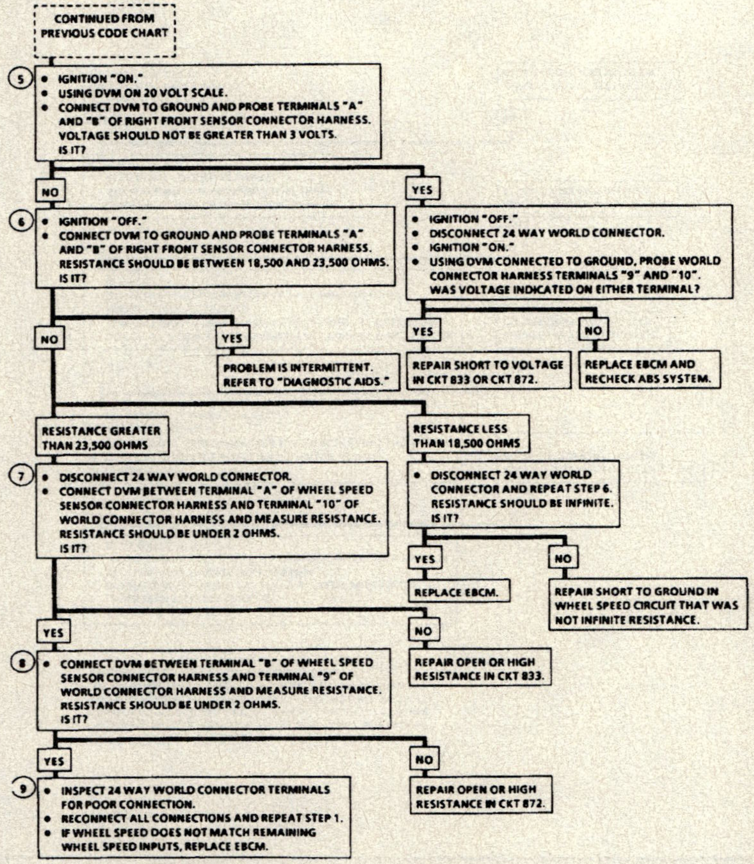

THIS CHART ASSUMES THAT A CURRENT CODE IS STORED INDICATING THAT THIS FAULT IS PRESENT.

DIAGNOSTIC CHART A027 — 1993 CAVALIER AND SUNBIRD

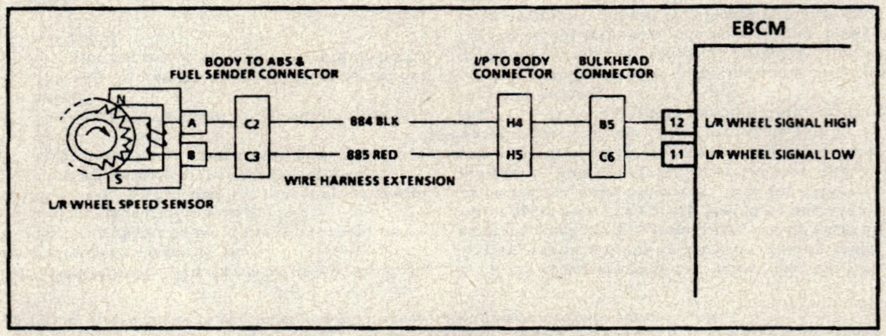

CODE A027

LEFT REAR EXCESSIVE WHEEL SPEED VARIATION

Circuit Description:

As a toothed ring passes by the wheel speed sensor, changes in the electromagnetic field cause the wheel speed sensor to produce a sinusoidal (AC) voltage signal whose frequency is proportional to wheel speed. The magnitude of this signal is directly related to wheel speed and the proximity of the wheel speed sensor to the toothed ring, often referred to as the air gap.

This code detects opens, shorts to ground, and shorts to battery that cause intermittent wheel speed operation. Also detected is any sudden change in wheel speed determined to be unreasonable.

Test Description: Number(s) below refer to circled number(s) on the diagnostic chart.

1. This test verifies whether the fault is currently present.
2. This step is designed to identify a wheel speed sensor or circuitry that is damaged and visibly apparent.
3. This test checks to see that the wheel speed sensor circuitry is within proper resistance value.
4. This insures that the wheel speed sensor is not shorted to ground.

Diagnostic Aids:

An "Intermittent" problem may be caused by a poor connection, rubbed through wire insulation, or a wire that is broken inside the insulation.

If the customers comments reflect that the ABS Amber warning light is on only during moist environmental changes, (rain, snow, vehicle wash) all wheel speed sensor circuitry should be thoroughly inspected for water intrusion. Use the following procedure. Spray down the suspected area with 5% salt water solution (two teaspoons of salt to 12 oz. of water). Start vehicle, allow vehicle to run for 10 seconds. If code returns immediately, replace suspected harness.

Any circuitry, that is suspected as causing the intermittent complaint, should be thoroughly checked for backed out terminals, improper mating, broken locks, improperly formed or damaged terminals, poor terminal to wiring connections or physical damage to the wiring harness.

Excessive end play can cause this code to set. Check integral bearing for excessive axial end play.

DIAGNOSTIC CHART A027 — 1993 CAVALIER AND SUNBIRD

IMPORTANT: WHEEL SPEED SENSOR INTERMITTENT PROBLEMS MAY BE DIFFICULT TO LOCATE. CARE SHOULD BE TAKEN NOT TO DISTURB ANY ELECTRICAL CONNECTIONS PRIOR TO AN INDICATED STEP OF THIS CHART. THIS WILL INSURE THAT AN INTERMITTENT CONNECTION WILL NOT BE CORRECTED BEFORE THE SOURCE OF THE PROBLEM IS FOUND.

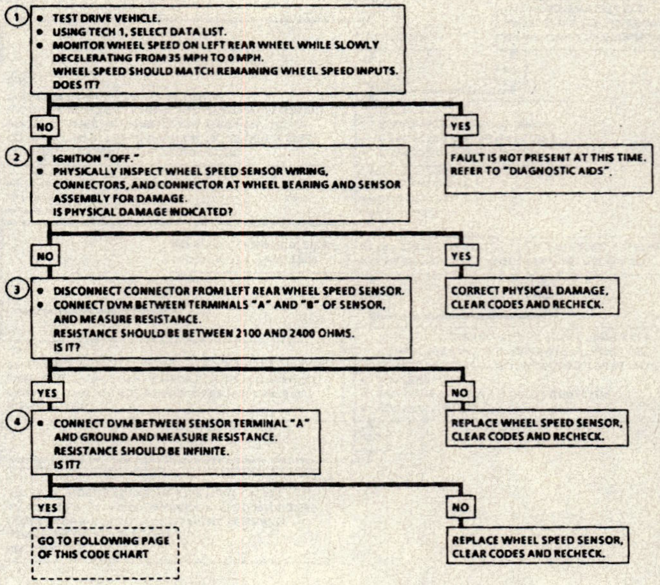

THIS CHART ASSUMES THAT A CURRENT CODE IS STORED INDICATING THAT THIS FAULT IS PRESENT.

DIAGNOSTIC CHART A027 — 1993
CAVALIER AND SUNBIRD

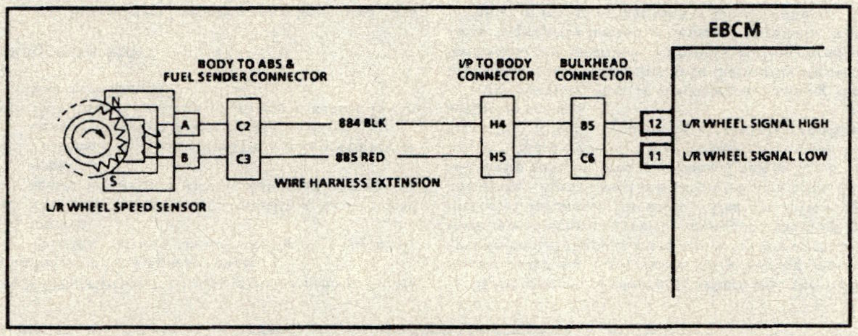

CODE A027

LEFT REAR EXCESSIVE WHEEL SPEED VARIATION

Circuit Description:

As a toothed ring passes by the wheel speed sensor, changes in the electromagnetic field cause the wheel speed sensor to produce a sinusoidal (AC) voltage signal whose frequency is proportional to wheel speed. The magnitude of this signal is directly related to wheel speed and the proximity of the wheel speed sensor to the toothed ring, often referred to as the air gap.

This code detects opens, shorts to ground, and shorts to battery that cause intermittent wheel speed operation. Also detected is any sudden change in wheel speed determined to be unreasonable.

Test Description: Number(s) below refer to circled number(s) on the diagnostic chart.

5. This step checks to see if the wheel speed sensor circuitry is shorted to voltage.
6. This step insures that the wheel speed sensor circuitry resistance is within the proper value.
7. This checks for an open in the wheel signal high circuit.
8. This checks for an open in the wheel signal low circuit.
9. This step insures that Code A027 was not set due to a poor connection between the world connector and EBCM.

Diagnostic Aids:

An "Intermittent" problem may be caused by a poor connection, rubbed through wire insulation, or a wire that is broken inside the insulation.

If the customers comments reflect that the ABS Amber warning light is on only during moist environmental changes, (rain, snow, vehicle wash) all wheel speed sensor circuitry should be thoroughly inspected for water intrusion. Use the following procedure. Spray down the suspected area with 5% salt water solution (two teaspoons of salt to 12 oz. of water). Start vehicle, allow vehicle to run for 10 seconds. If code returns immediately, replace suspected harness.

Any circuitry, that is suspected as causing the intermittent complaint, should be thoroughly checked for backed out terminals, improper mating, broken locks, improperly formed or damaged terminals, poor terminal to wiring connections or physical damage to the wiring harness.

Excessive end play can cause this code to set. Check integral bearing for excessive axial end play.

DIAGNOSTIC CHART A027 — 1993
CAVALIER AND SUNBIRD

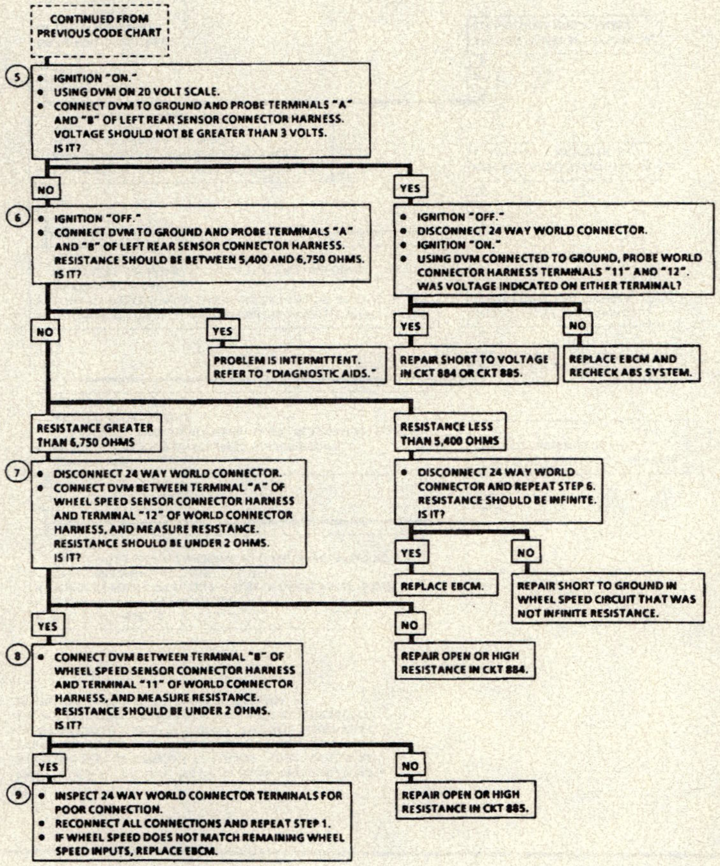

THIS CHART ASSUMES THAT A CURRENT CODE IS STORED INDICATING THAT THIS FAULT IS PRESENT.

DIAGNOSTIC CHART A028 — 1993 CAVALIER AND SUNBIRD

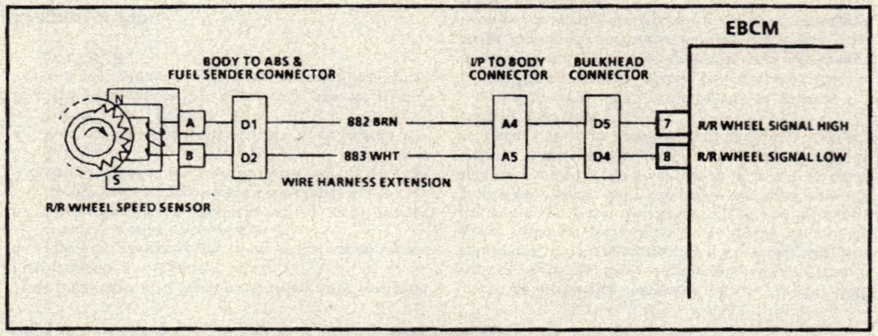

CODE A028

RIGHT REAR EXCESSIVE WHEEL SPEED VARIATION

Circuit Description:

As a toothed ring passes by the wheel speed sensor, changes in the electromagnetic field cause the wheel speed sensor to produce a sinusoidal (AC) voltage signal whose frequency is proportional to wheel speed. The magnitude of this signal is directly related to wheel speed and the proximity of the wheel speed sensor to the toothed ring, often referred to as the air gap.

This code detects opens, shorts to ground, and shorts to battery that cause intermittent wheel speed operation. Also detected is any sudden change in wheel speed determined to be unreasonable.

Test Description: Number(s) below refer to circled number(s) on the diagnostic chart.

1. This test verifies whether the fault is currently present.
2. This step is designed to identify a wheel speed sensor or circuitry that is damaged and visibly apparent.
3. This test checks to see that the wheel speed sensor circuitry is within proper resistance value.
4. This insures that the wheel speed sensor is not shorted to ground.

Diagnostic Aids:

An "Intermittent" problem may be caused by a poor connection, rubbed through wire insulation, or a wire that is broken inside the insulation.

If the customers comments reflect that the ABS Amber warning light is on only during moist environmental changes, (rain, snow, vehicle wash) all wheel speed sensor circuitry should be thoroughly inspected for water intrusion. Use the following procedure. Spray down the suspected area with 5% salt water solution (two teaspoons of salt to 12 oz. of water). Start vehicle, allow vehicle to run for 10 seconds. If code returns immediately, replace suspected harness.

Any circuitry, that is suspected as causing the intermittent complaint, should be thoroughly checked for backed out terminals, improper mating, broken locks, improperly formed or damaged terminals, poor terminal to wiring connections or physical damage to the wiring harness.

Excessive end play can cause this code to set. Check integral bearing for excessive axial end play,

DIAGNOSTIC CHART A028 — 1993 CAVALIER AND SUNBIRD

IMPORTANT: WHEEL SPEED SENSOR INTERMITTENT PROBLEMS MAY BE DIFFICULT TO LOCATE. CARE SHOULD BE TAKEN NOT TO DISTURB ANY ELECTRICAL CONNECTIONS PRIOR TO AN INDICATED STEP OF THIS CHART. THIS WILL INSURE THAT AN INTERMITTENT CONNECTION WILL NOT BE CORRECTED BEFORE THE SOURCE OF THE PROBLEM IS FOUND.

(1)
- TEST DRIVE VEHICLE.
- USING TECH 1, SELECT DATA LIST.
- MONITOR WHEEL SPEED ON RIGHT REAR WHEEL WHILE SLOWLY DECELERATING FROM 35 MPH TO 0 MPH.

WHEEL SPEED SHOULD MATCH REMAINING WHEEL SPEED INPUTS.
DOES IT?

- YES: FAULT IS NOT PRESENT AT THIS TIME. REFER TO "DIAGNOSTIC AIDS".
- NO: go to step 2.

(2)
- IGNITION "OFF."
- PHYSICALLY INSPECT WHEEL SPEED SENSOR WIRING, CONNECTORS, AND CONNECTOR AT WHEEL BEARING AND SENSOR ASSEMBLY FOR DAMAGE.

IS PHYSICAL DAMAGE INDICATED?

- YES: CORRECT PHYSICAL DAMAGE, CLEAR CODES AND RECHECK.
- NO: go to step 3.

(3)
- DISCONNECT CONNECTOR FROM RIGHT REAR WHEEL SPEED SENSOR.
- CONNECT DVM BETWEEN TERMINALS "A" AND "B" OF SENSOR, AND MEASURE RESISTANCE.

RESISTANCE SHOULD BE BETWEEN 2100 AND 2400 OHMS.
IS IT?

- NO: REPLACE WHEEL SPEED SENSOR, CLEAR CODES AND RECHECK.
- YES: go to step 4.

(4)
- CONNECT DVM BETWEEN SENSOR TERMINAL "A" AND GROUND, AND MEASURE RESISTANCE.

RESISTANCE SHOULD BE INFINITE.
IS IT?

- NO: REPLACE WHEEL SPEED SENSOR, CLEAR CODES AND RECHECK.
- YES: GO TO FOLLOWING PAGE OF THIS CODE CHART

THIS CHART ASSUMES THAT A CURRENT CODE IS STORED INDICATING THAT THIS FAULT IS PRESENT.

DIAGNOSTIC CHART A028 — 1993 CAVALIER AND SUNBIRD

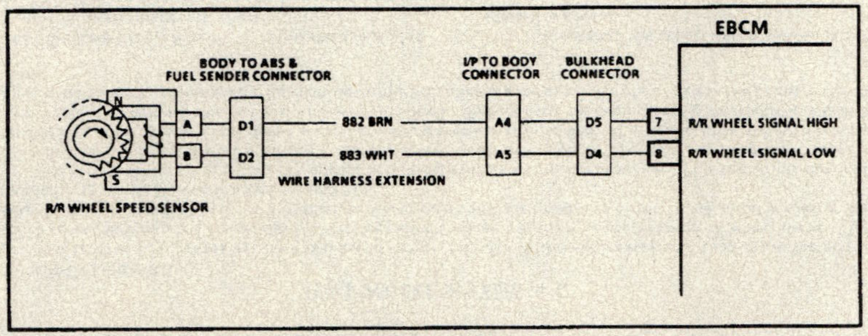

CODE A028

RIGHT REAR EXCESSIVE WHEEL SPEED VARIATION

Circuit Description:

As a toothed ring passes by the wheel speed sensor, changes in the electromagnetic field cause the wheel speed sensor to produce a sinusoidal (AC) voltage signal whose frequency is proportional to wheel speed. The magnitude of this signal is directly related to wheel speed and the proximity of the wheel speed sensor to the toothed ring, often referred to as the air gap.

This code detects opens, shorts to ground, and shorts to battery that cause intermittent wheel speed operation. Also detected is any sudden change in wheel speed determined to be unreasonable.

Test Description: Number(s) below refer to circled number(s) on the diagnostic chart.

5. This step checks to see if the wheel speed sensor circuitry is shorted to voltage.
6. This step insures that the wheel speed sensor circuitry resistance is within proper value.
7. This checks for an open in the wheel signal high circuit.
8. This checks for an open in the wheel signal low circuit.
9. This step insures that Code A028 was not set due to a poor connection between the world connector and EBCM.

Diagnostic Aids:

An "Intermittent" problem may be caused by a poor connection, rubbed through wire insulation, or a wire that is broken inside the insulation.

If the customers comments reflect that the ABS Amber warning light is on only during moist environmental changes, (rain, snow, vehicle wash) all wheel speed sensor circuitry should be thoroughly inspected for water intrusion. Use the following procedure. Spray down the suspected area with 5% salt water solution (two teaspoons of salt to 12 oz. of water). Start vehicle, allow vehicle to run for 10 seconds. If code returns immediately, replace suspected harness.

Any circuitry, that is suspected as causing the intermittent complaint, should be thoroughly checked for backed out terminals, improper mating, broken locks, improperly formed or damaged terminals, poor terminal to wiring connections or physical damage to the wiring harness.

Excessive end play can cause this code to set. Check integral bearing for excessive axial end play.

DIAGNOSTIC CHART A028 — 1993 CAVALIER AND SUNBIRD

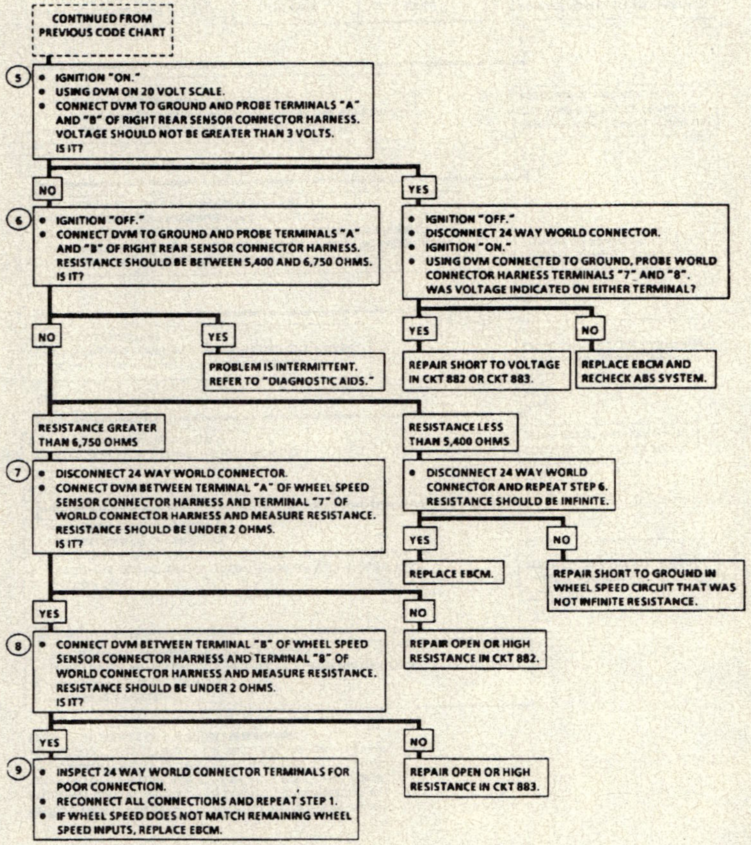

THIS CHART ASSUMES THAT A CURRENT CODE IS STORED INDICATING THAT THIS FAULT IS PRESENT.

DIAGNOSTIC CHART A031 — 1993 CAVALIER AND SUNBIRD

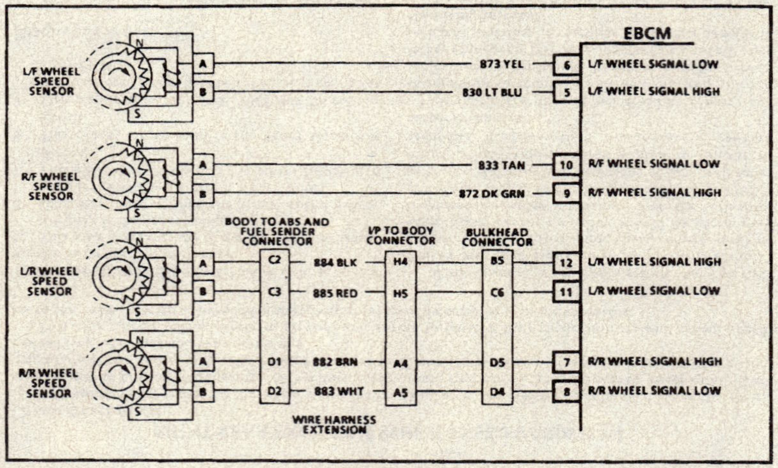

CODE A031

TWO WHEEL SPEEDS = 0

Circuit Description:

As a toothed ring passes by the wheel speed sensor, changes in the electromagnetic field cause the wheel speed sensor to produce a sinusoidal (AC) voltage signal whose frequency is proportional to wheel speed. The magnitude of this signal is directly related to wheel speed and the proximity of the wheel speed sensor to the toothed ring, often referred to as the air gap.

This code is designed to detect two open wheel speed sensors in any combination. This test exists due to a condition that may exist that does not allow the driver to be informed by the ABS warning light of non-functioning ABS. If one wheel speed sensor is open, the driver is informed. If a second fails, the driver is no longer informed without this test and ABS is not available. By requiring the two operating wheels to be within 7 mph, a false code is prevented during spin-ups while still allowing for detection when the stowaway spare tire is used.

Test Description: Number(s) below refer to circled number(s) on the diagnostic chart.

1. This test verifies whether the fault is currently present.
2. This step is designed to identify a wheel speed sensor or circuitry that is damaged and visibly apparent.
3. This step identifies if the cause of the problem is common to the rear wheel sensor circuitry.
4. Because both rear sensors are common to the extension harness, the 4 way connectors and harness should be inspected.
5. This test checks to see that the wheel speed sensor is within proper resistance values.
6. This insures that the wheel speed sensor is not shorted to ground.

Diagnostic Aids:

A false Code A031 may be set as a result of vehicle operation while on a hoist or a lifted condition or stuck in mud or snow, when the vehicle is in gear.

An "Intermittent" problem may be caused by a poor connection, rubbed through wire insulation, or a wire that is broken inside the insulation.

DIAGNOSTIC CHART A031 — 1993 CAVALIER AND SUNBIRD

IMPORTANT: WHEEL SPEED SENSOR INTERMITTENT PROBLEMS MAY BE DIFFICULT TO LOCATE. CARE SHOULD BE TAKEN NOT TO DISTURB ANY ELECTRICAL CONNECTIONS PRIOR TO AN INDICATED STEP OF THIS CHART. THIS WILL INSURE THAT AN INTERMITTENT CONNECTION WILL NOT BE CORRECTED BEFORE THE SOURCE OF THE PROBLEM IS FOUND.

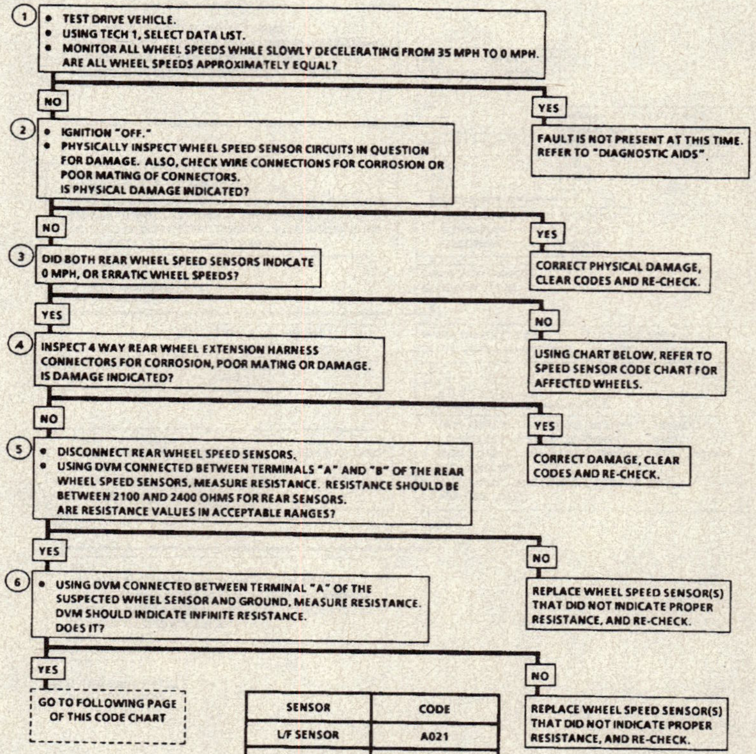

SENSOR	CODE
L/F SENSOR	A021
R/F SENSOR	A022
L/R SENSOR	A023
R/R SENSOR	A024

DIAGNOSTIC CHART A031 — 1993 CAVALIER AND SUNBIRD

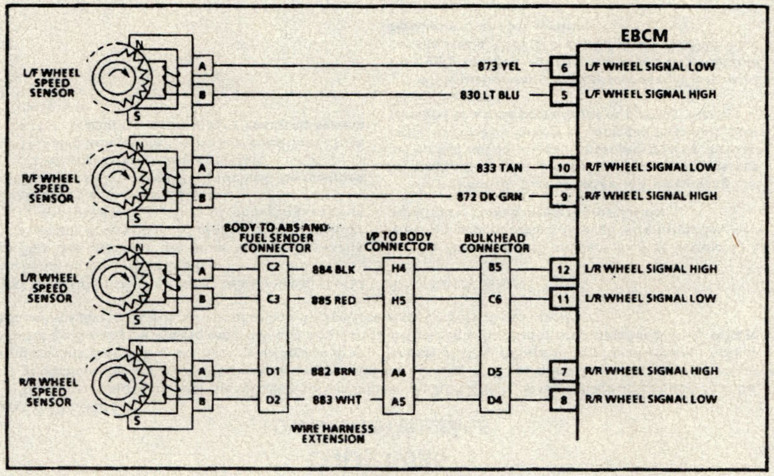

CODE A031

TWO WHEEL SPEEDS = 0

Circuit Description:

As a toothed ring passes by the wheel speed sensor, changes in the electromagnetic field cause the wheel speed sensor to produce a sinusoidal (AC) voltage signal whose frequency is proportional to wheel speed. The magnitude of this signal is directly related to wheel speed and the proximity of the wheel speed sensor to the toothed ring, often referred to as the air gap.

This code is designed to detect two open wheel speed sensors in any combination. This test exists due to a condition that may exist that does not allow the driver to be informed by the ABS warning light of non-functioning ABS. If one wheel speed sensor is open, the driver is informed. If a second fails, the driver is no longer informed without this test and ABS is not available. By requiring the two operating wheels to be within 7 mph, a false code is prevented during spin-ups while still allowing for detection when the stowaway spare tire is used.

Test Description: Number(s) below refer to circled number(s) on the diagnostic chart.

7. This step checks for an open in the wheel speed sensor circuitry of one side of the sensor.
8. This step checks for an open on the other side of the sensor in question.
9. This step checks for a short to ground in the wheel speed sensor circuitry.
10. This insures that the circuitry is not shorted to voltage.

Diagnostic Aids:

A false Code A031 may be set as a result of vehicle operation while on a hoist or a lifted condition, or stuck in mud or snow when the vehicle is in gear.

An "Intermittent" problem may be caused by a poor connection, rubbed through wire insulation, or a wire that is broken inside the insulation.

DIAGNOSTIC CHART A031 — 1993 CAVALIER AND SUNBIRD

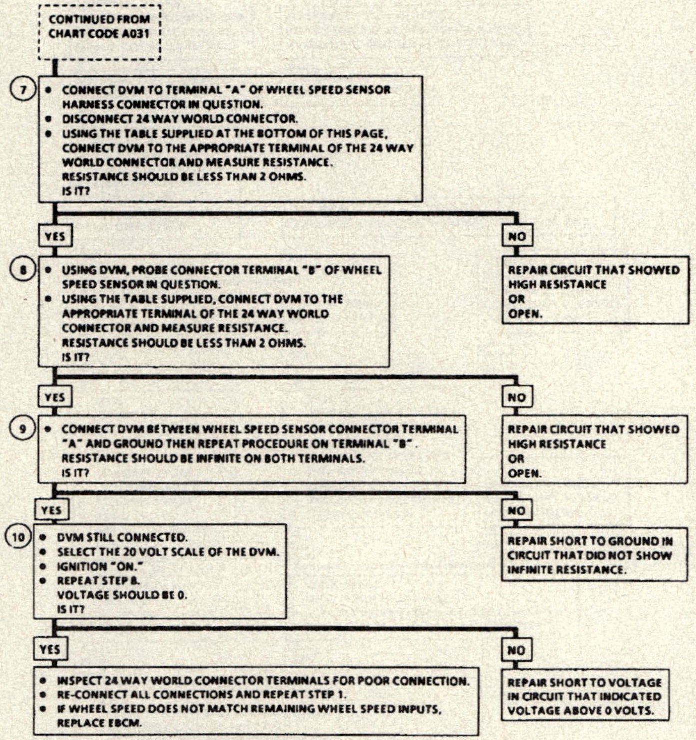

SENSOR TERMINAL	24 WAY WORLD CONNECTOR	CIRCUIT NUMBER
L/R WHEEL A	PIN 12	884
L/R WHEEL B	PIN 11	885
R/R WHEEL A	PIN 7	882
R/R WHEEL B	PIN 8	883

THIS CHART ASSUMES THAT A CURRENT CODE IS STORED INDICATING THAT THIS FAULT IS PRESENT.

DIAGNOSTIC CHART A036 — 1993
CAVALIER AND SUNBIRD

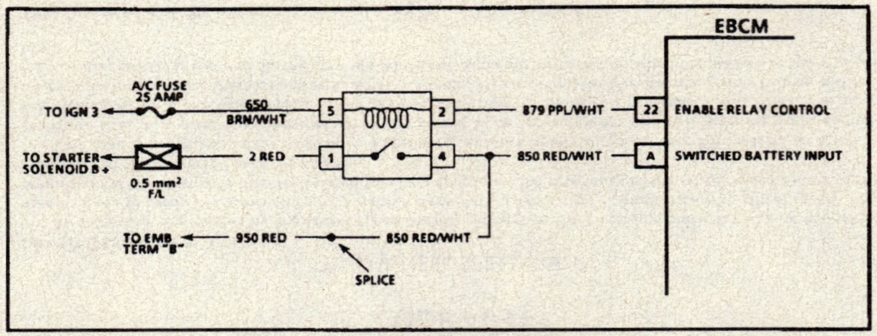

CODE A036
LOW SYSTEM VOLTAGE

Circuit Description:

This code is used to monitor the voltage level available to the ABS EBCM. If the voltage drops below 11 volts, full performance of the ABS system cannot be guaranteed. During ABS operation, there are several current requirements that will cause battery voltage to drop. Because of this, voltage is monitored prior to ABS to indicate good charging system condition and also during ABS when voltage may drop significantly. If voltage falls below a level that will provide adequate ABS performance, Code A036 will result.

Test Description: Number(s) below refer to circled number(s) on the diagnostic chart.

1. This test checks to see what voltage is readily available at terminal "A" of the EBCM. If voltage is representative of good charging system condition, a fault is not present.
2. This step isolates the EBCM to check the condition of the circuitry and charging system.
3. This step isolates the low voltage condition to high circuit resistance, or improper charging system operation.

Diagnostic Aids:

An "Intermittent" problem may be caused by a poor connection, rubbed through wire insulation, or a wire that is broken inside the insulation.

Any circuitry, that is suspected as causing the intermittent complaint, should be thoroughly checked for backed out terminals, improper mating, broken locks, improperly formed or damaged terminals, poor terminal to wiring connections or physical damage to the wiring harness.

While performing a voltage load test, if it is noted that only ignition voltage drops below acceptable voltage levels, CKT 650 should be checked for high resistance or an open condition.

DIAGNOSTIC CHART A036 — 1993
CAVALIER AND SUNBIRD

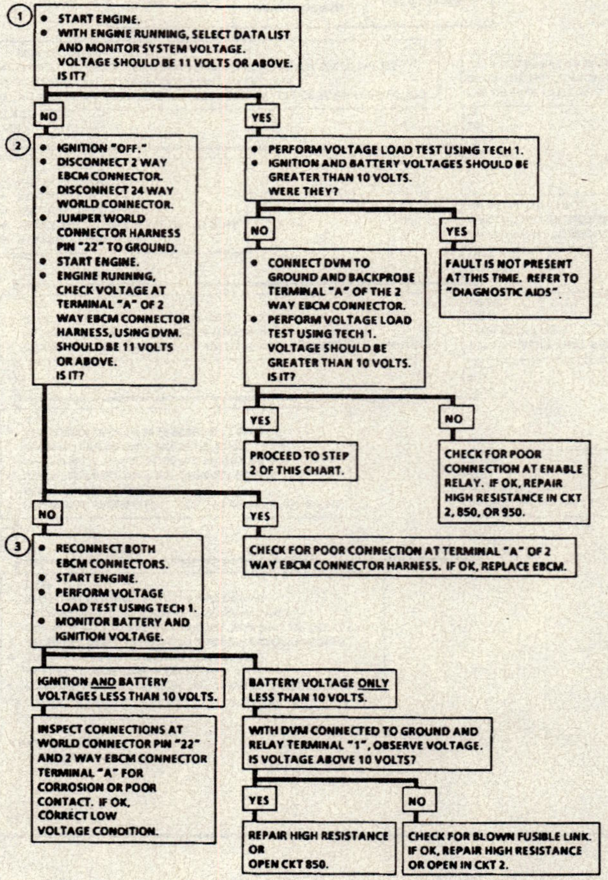

DIAGNOSTIC CHART A037 — 1993 CAVALIER AND SUNBIRD

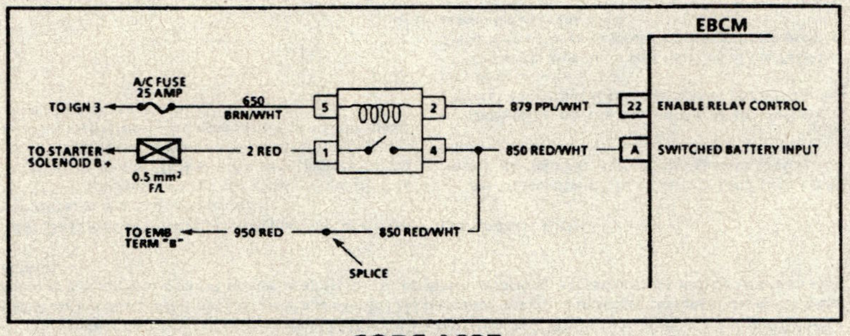

CODE A037

HIGH SYSTEM VOLTAGE

Circuit Description:

This code is designed to detect high vehicle voltage levels prior to any required motor movement (initialization or ABS). If excessive voltage exists, demagnetization of the motor magnets may occur, which would eventually affect or eliminate ABS performance.

Test Description: Number(s) below refer to circled number(s) on the diagnostic chart.

1. This step checks the voltage level being received by the EBCM. If high voltage is present, a charging system problem is indicated.

Diagnostic Aids:

An "Intermittent" problem may be caused by a poor connection, rubbed through wire insulation, or a wire that is broken inside the insulation.

Any circuitry, that is suspected as causing the intermittent complaint, should be thoroughly checked for backed out terminals, improper mating, broken locks, improperly formed or damaged terminals, poor terminal to wiring connections or physical damage to the wiring harness.

DIAGNOSTIC CHART A037 — 1993 CAVALIER AND SUNBIRD

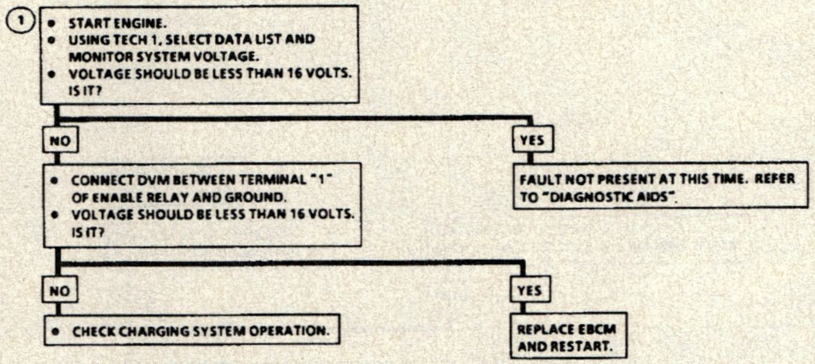

THIS CHART ASSUMES THAT A CURRENT CODE IS STORED INDICATING THAT THIS FAULT IS PRESENT.

DIAGNOSTIC CHART A038 — 1993
CAVALIER AND SUNBIRD

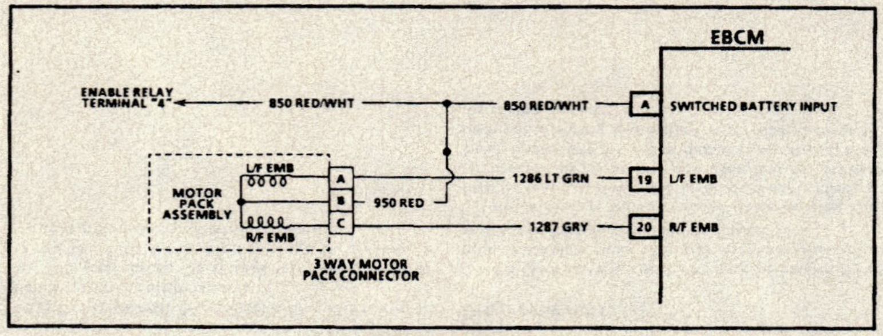

CODE A038

LEFT FRONT EMB WILL NOT HOLD MOTOR

Circuit Description:

Current is supplied to the EMB's through terminal "B" of the three way motor pack connector. When a release of the EMB is commanded, the EBCM grounds the EMB input and allows the EMB to be energized to release the motor.

This code is designed to detect slipping EMB's during initialization. When the release is commanded during initialization, the EMB is not energized and should hold the motor, resulting in sensed current of the motor being equal to commanded current (motor is at stall). If the motor is moving, sensed current will be less than stall current.

Test Description: Number(s) below refer to circled number(s) on the diagnostic chart.

1. This functionally checks the ability of the EMB to hold the motor. If the brake pedal moves during this test, it would indicate the EMB is slipping.
2. This step is used to release the tension of the motor pack prior to removal.

Diagnostic Aids:

An "Intermittent" fault in this code may result from a mechanical part of the system that sticks, binds, or slips.

Depending on the frequency of the failure, a physical inspection of the mechanical parts suspected may be necessary.

The static modulator test function of the Tech 1 or T-100 may be used to locate an intermittent problem associated with the EMB.

NOTICE: This test may fail if low voltage (10 volts or less) was present when initialization occurred. Ensure battery/charging system is functioning properly.

DIAGNOSTIC CHART A038 — 1993
CAVALIER AND SUNBIRD

NOTICE: TESTING EMB WITH JUMPER WIRES TO VOLTAGE OR GROUND WILL DESTROY THE EMB.

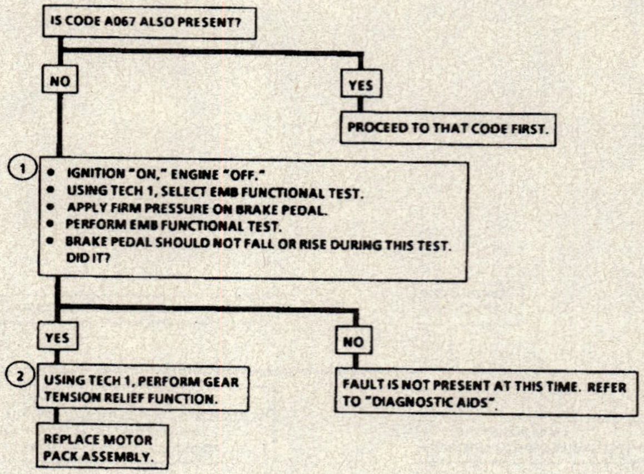

DIAGNOSTIC CHART A041 — 1993 CAVALIER AND SUNBIRD

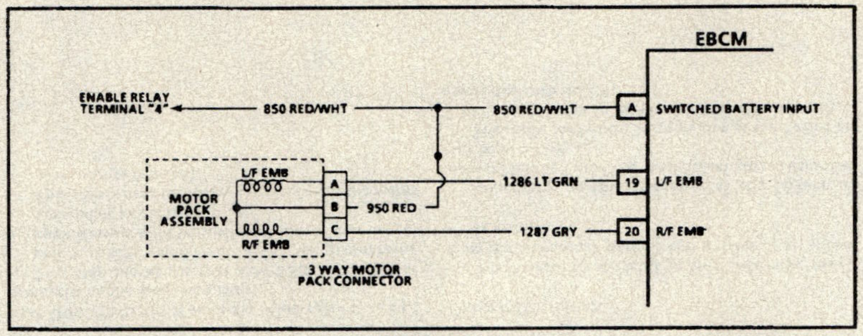

CODE A041

RIGHT FRONT EMB WILL NOT HOLD MOTOR

Circuit Description:

Current is supplied to the EMB's through terminal "B" of the three way motor pack connector. When a release of the EMB is commanded, the EBCM grounds the EMB input and allows the EMB to be energized to release the motor.

This code is designed to detect slipping EMB's during initialization. When the release is commanded during initialization, the EMB is not energized and should hold the motor, resulting in sensed current of the motor being equal to commanded current (motor is at stall). If the motor is moving, sensed current will be less than stall current.

Test Description: Number(s) below refer to circled number(s) on the diagnostic chart.

1. This functionally checks the ability of the EMB to hold the motor. If the brake pedal moves during this test, it would indicate the EMB is slipping.
2. This step is used to release the tension of the motor pack prior to removal.

Diagnostic Aids:

An "Intermittent" fault in this code may result from a mechanical part of the system that sticks, binds, or slips.

Depending on the frequency of the failure, a physical inspection of the mechanical parts suspected may be necessary.

The static modulator test function of the Tech 1 or T-100 may be used to locate an intermittent problem associated with the EMB.

NOTICE: This code may fail if low voltage (10 volts or less) was present when initialization occurred. Ensure battery/charging system is functioning properly.

DIAGNOSTIC CHART A041 — 1993 CAVALIER AND SUNBIRD

NOTICE: TESTING END WITH JUMPER WIRES TO VOLTAGE OR GROUND WILL DESTROY THE EMB.

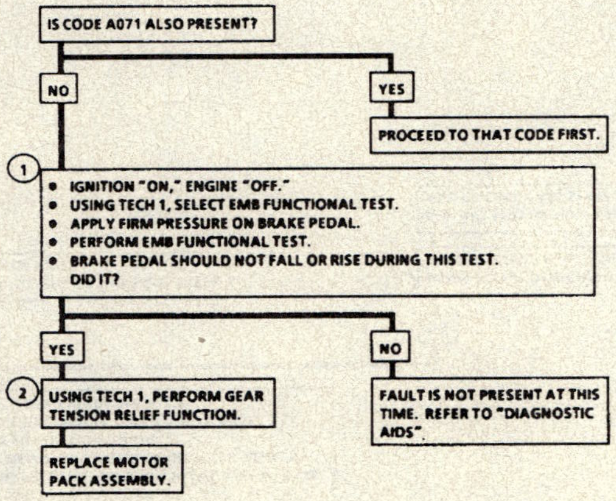

DIAGNOSTIC CHART A042 — 1993 CAVALIER AND SUNBIRD

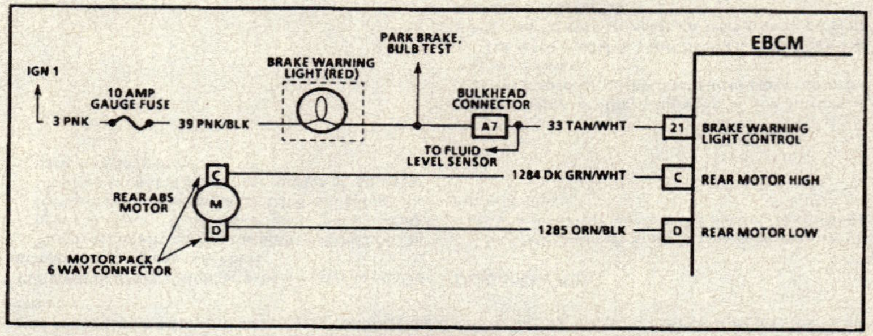

CODE A042

REAR AXLE ESB WILL NOT HOLD MOTOR

Circuit Description:

This code is designed to detect a slipping rear axle ESB. During initialization and braking, the rear motor is re-homed. If the ESB slips, the motor/piston will move. During the next key on initialization, a re-home of the motor verifies the motor/piston remained at the home position. If motor movement is detected, the ESB must be slipping.

Test Description: Number(s) below refer to circled number(s) on the diagnostic chart.

1. This step checks the rear axle ESB. A broken or defective ESB would result in the rear piston being back driven by hydraulic pressure, and wheel movement would result.
2. This step is used to release the tension of the motor pack prior to removal.

Diagnostic Aids:

An "Intermittent" fault in this code may result from a mechanical part of the system that sticks, binds, or slips.

Depending on the frequency of the failure, a physical inspection of the mechanical parts suspected may be necessary.

The static modulator test function of the Tech 1 or T-100 may be used to locate an intermittent problem associated with the ESB.

DIAGNOSTIC CHART A042 — 1993 CAVALIER AND SUNBIRD

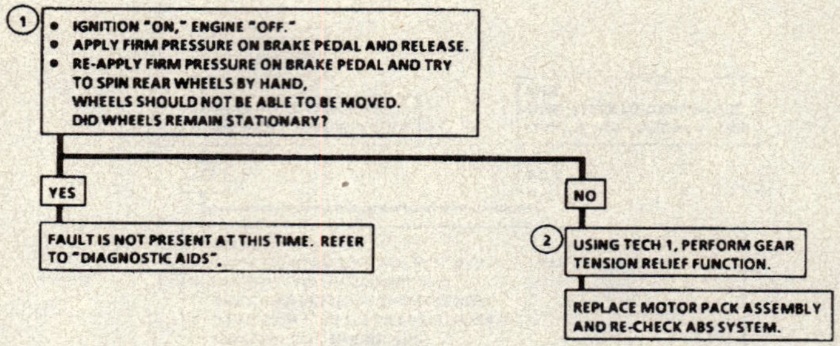

DIAGNOSTIC CHART A044 — 1993 CAVALIER AND SUNBIRD

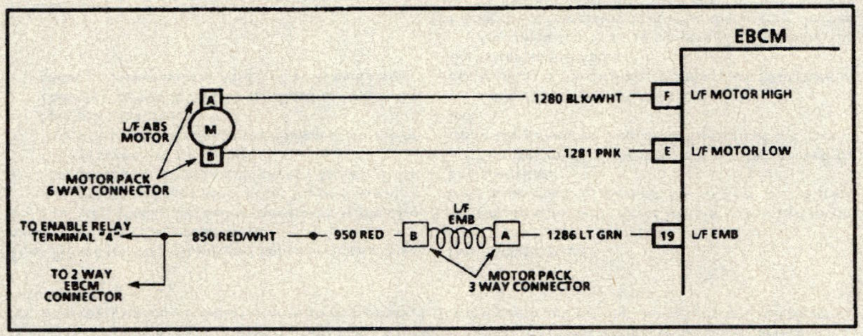

CODE A044

LEFT FRONT CHANNEL WILL NOT MOVE

Circuit Description:

Current is supplied to the EMB through terminal "B" of the three way motor pack connector. When a release of the EMB is commanded, the EBCM grounds the EMB input and allows the EMB to be energized to release the motor.

This code is designed to detect non-actuating EMB, stuck motor or a seized hydraulic modulator. When the release is commanded during initialization, the EMB should release the motor, resulting in sensed current being less than commanded current (motor is spinning freely). If the motor is not moving, sensed current will be equal to stall current.

Test Description: Number(s) below refer to circled number(s) on the diagnostic chart.

1. This step checks for motor movement while also positioning the brake pedal at the top of its travel.
2. This procedure functionally checks to see if the EMB is capable of releasing.
3. This step identifies a short to voltage in the EMB control circuitry.
4. This step is used to release the tension of the motor pack prior to removal.
5. This step serves to isolate the fault to either the motor pack assembly or ABS hydraulic modulator.

Diagnostic Aids:

An "Intermittent" fault in this code may result from a mechanical part of the system that sticks, binds, or slips.

The frequency of the problem can be checked by using the enhanced diagnostic function of the "Scan" tool

Code A044 may set after modulator disassembly if the modulator pistons are positioned at the bottom of their bore.

Any circuitry, that is suspected as causing the intermittent complaint, should be thoroughly checked for backed out terminals, improper mating, broken locks, improperly formed or damaged terminals, poor terminal to wiring connections or physical damage to the wiring harness.

NOTICE: This code may fail if low voltage (10 volts or less) was present when initialization occurred. Ensure battery/charging system is functioning properly.

DIAGNOSTIC CHART A044 — 1993 CAVALIER AND SUNBIRD

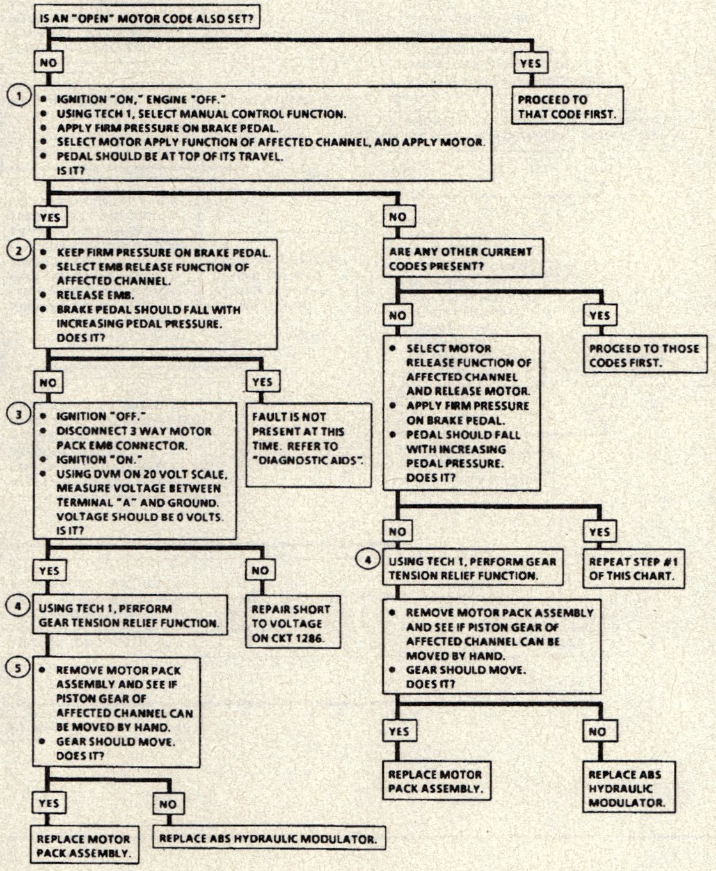

DIAGNOSTIC CHART A045 — 1993 CAVALIER AND SUNBIRD

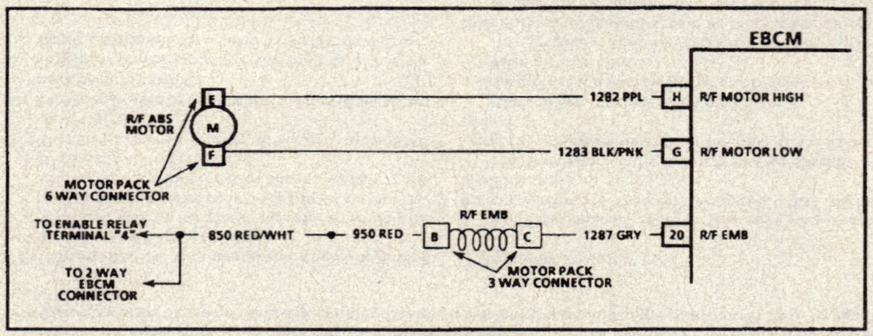

CODE A045

RIGHT FRONT CHANNEL WILL NOT MOVE

Circuit Description:

Current is supplied to the EMB through terminal "B" of the three way motor pack connector. When a release of the EMB is commanded, the EBCM grounds the EMB input and allows the EMB to be energized to release the motor.

This code is designed to detect non-actuating EMB, stuck motor or a seized hydraulic modulator. When the release is commanded during initialization, the EMB should release the motor, resulting in sensed current being less than commanded current (motor is spinning freely). If the motor is not moving, sensed current will be equal to stall current.

Test Description: Number(s) below refer to circled number(s) on the diagnostic chart.

1. This step checks for motor movement while also positioning the brake pedal at the top of its travel.
2. This procedure functionally checks to see if the EMB is capable of releasing.
3. This step is used to release the tension of the motor pack prior to removal.
4. This step serves to isolate the fault to either the motor pack assembly or ABS hydraulic modulator.

Diagnostic Aids:

An "Intermittent" fault in this code may result from a mechanical part of the system that sticks, binds, or slips.

The frequency of the problem can be checked by using the enhanced diagnostic function of the "Scan" tool

Code A045 may set after modulator disassembly if the modulator pistons are positioned at the bottom of their bore.

Any circuitry, that is suspected as causing the intermittent complaint, should be thoroughly checked for backed out terminals, improper mating, broken locks, improperly formed or damaged terminals, poor terminal to wiring connections or physical damage to the wiring harness.

NOTICE: This code may fail if low voltage (10 volts or less) was present when initialization occurred. Ensure battery/charging system is functioning properly.

DIAGNOSTIC CHART A045 — 1993 CAVALIER AND SUNBIRD

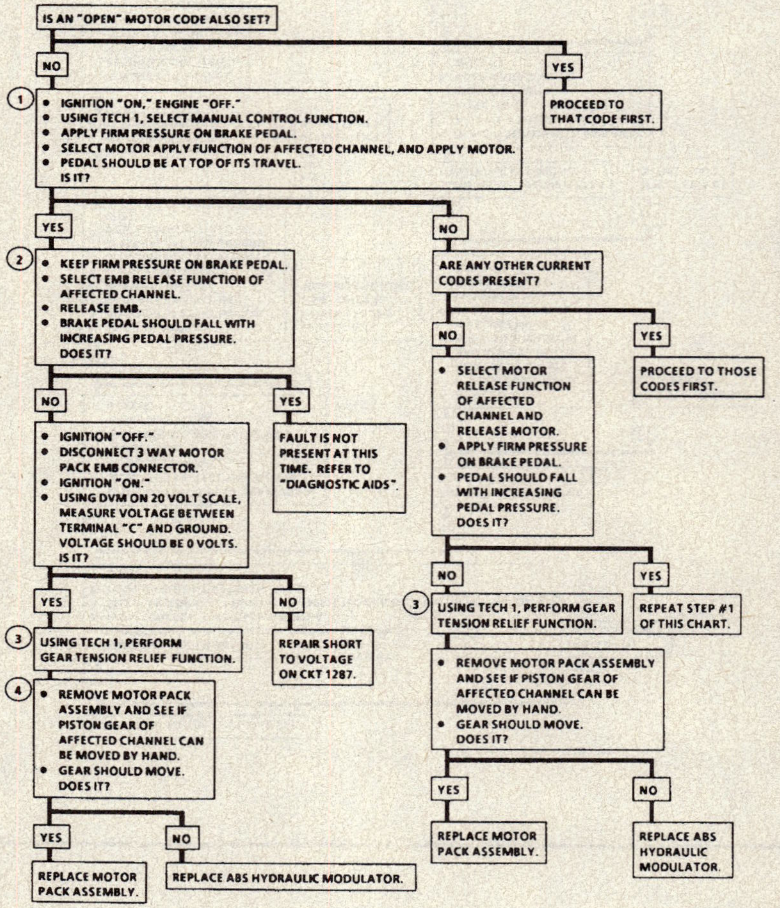

DIAGNOSTIC CHART A046 — 1993 CAVALIER AND SUNBIRD

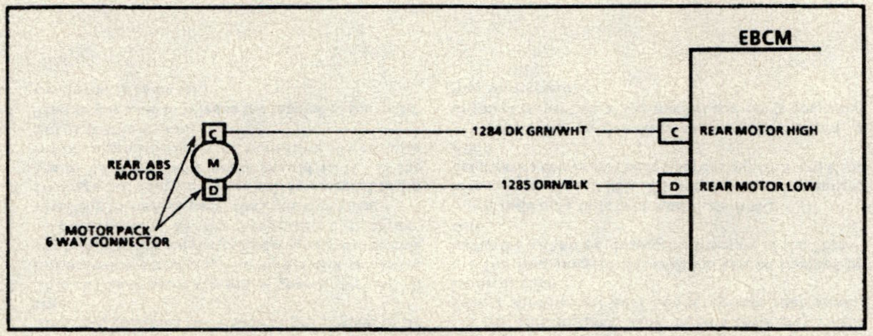

CODE A046

REAR AXLE CHANNEL WILL NOT MOVE

Circuit Description:

This code is designed to detect bound-up ESB's, stuck motor or a seized hydraulic modulator. When the release is commanded during initialization, the ESB should release the motor, resulting in sensed current being less than commanded current (motor is spinning freely). If the motor is not moving, sensed current will be equal to stall current.

Test Description: Number(s) below refer to circled number(s) on the diagnostic chart.

1. This checks the ability of the motor to move, applying hydraulic pressure to the wheel cylinders. A frozen piston or gear would result in no apply of rear brakes and a wheel that freely spins.
2. This step checks to see that the motor can also move to the release position. If a motor were not able to release hydraulic pressure, the wheel would not spin.
3. This step is used to release the tension of the motor pack prior to removal.
4. Isolation of the faulty component is identified in this step.

Diagnostic Aids:

An "Intermittent" fault in this code may result from a mechanical part of the system that sticks or binds.

The frequency of the problem can be checked by using the enhanced diagnostic function of the Tech 1 "Scan" tool

Code A046 may set after modulator disassembly if the modulator pistons are positioned at the bottom of their bore.

Depending on the frequency of the failure, a physical inspection of the mechanical parts suspected may be necessary.

Any circuitry, that is suspected as causing the intermittent complaint, should be thoroughly checked for backed out terminals, improper mating, broken locks, improperly formed or damaged terminals, poor terminal to wiring connections or physical damage to the wiring harness.

DIAGNOSTIC CHART A046 — 1993 CAVALIER AND SUNBIRD

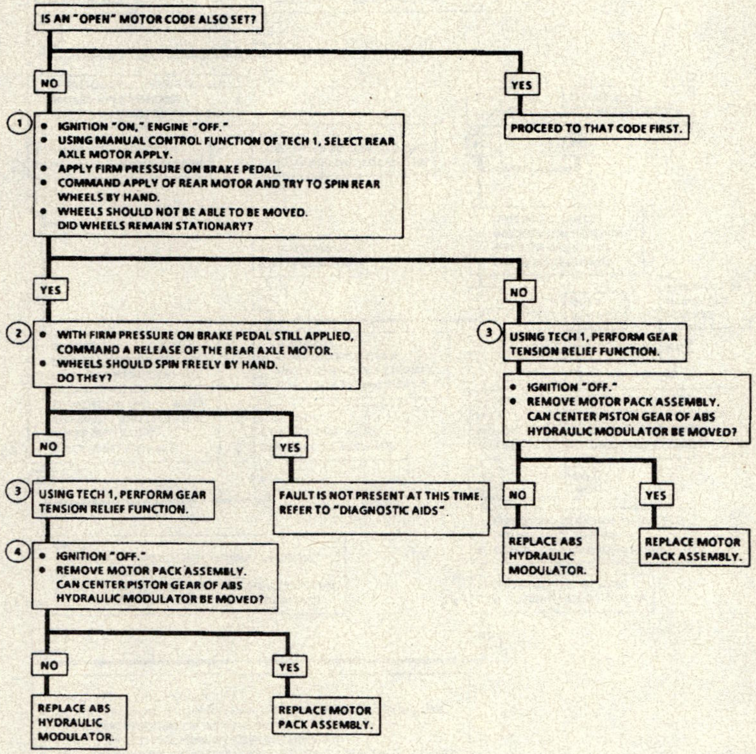

DIAGNOSTIC CHART A047 — 1993 CAVALIER AND SUNBIRD

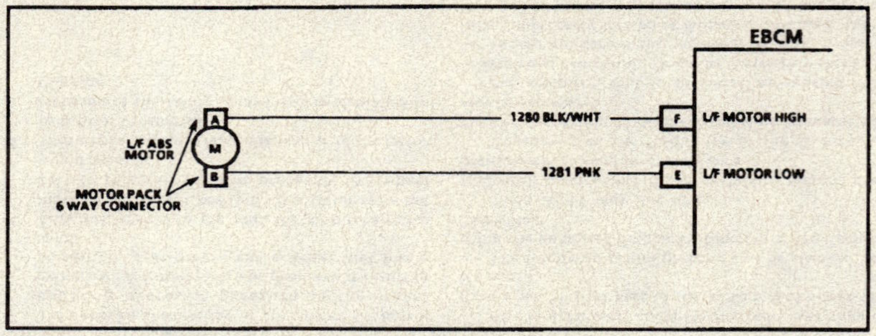

CODE A047

LEFT FRONT MOTOR FREE SPINS

Circuit Description:

This code is designed to detect a stripped nut or gear assembly during initialization. During the homing sequence, the piston should reach the top of the bore resulting in a stalled motor. If this does not occur, the motor must be spinning with little or no resistance, indicating a nut/screw or gear failure.

Test Description: Number(s) below refer to circled number(s) on the diagnostic chart.

1. This step positions the motor at the bottom of its travel.
2. This checks to see if the nut or gear is stripped. If either were stripped, the hydraulic pressure applied from the master cylinder would not be able to be overcome by the hydraulic ball screw, resulting in a brake pedal that would not rise.
3. During a full apply, once the ball screw reaches the top of its travel, the motor should stall. If the motor continues to run, an indication of a stripped nut or gear is present.
4. This step is used to release the tension of the motor pack prior to removal.

Diagnostic Aids:

An "Intermittent" fault in this code may result from a mechanical part of the system that sticks, binds, or slips.

The frequency of the problem can be checked by using the enhanced diagnostic function of the "Scan" tool

If code only fails once and Code A056 also fails see Code A056. If intermittent and enhanced diagnostics shows this code fails during ABS, see Code A056.

Depending on the frequency of the failure, a physical inspection of the mechanical parts suspected may be necessary.

DIAGNOSTIC CHART A047 — 1993 CAVALIER AND SUNBIRD

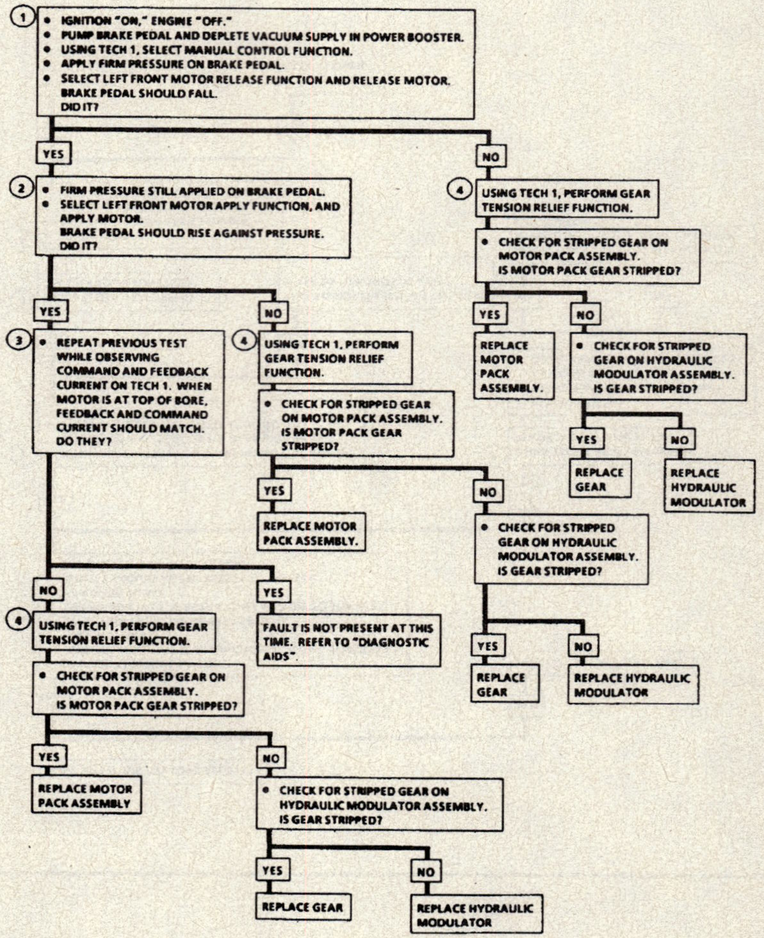

DIAGNOSTIC CHART A048 — 1993 CAVALIER AND SUNBIRD

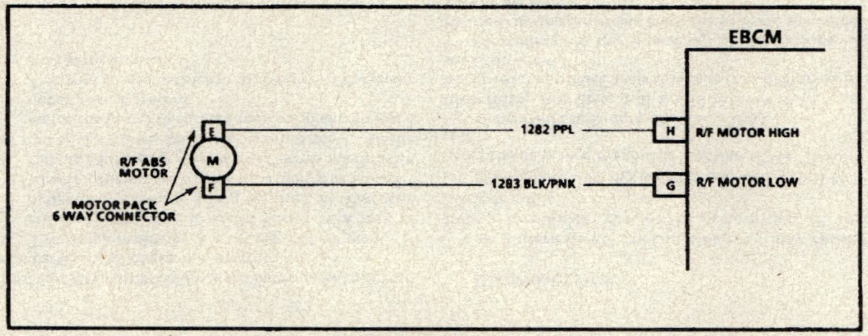

CODE A048

RIGHT FRONT MOTOR FREE SPINS

Circuit Description:

This code is designed to detect a stripped nut or gear assembly during initialization. During the homing sequence, the piston should reach the top of the bore resulting in a stalled motor. If this does not occur, the motor must be spinning with little or no resistance, indicating a nut/screw or gear failure.

Test Description: Number(s) below refer to circled number(s) on the diagnostic chart.

1. This step positions the motor at the bottom of its travel.
2. This checks to see if the nut or gear is stripped. If either were stripped, the hydraulic pressure applied from the master cylinder would not be able to be overcome by the hydraulic ball screw, resulting in a bake pedal that would not rise.
3. During a full apply, once the ball screw reaches the top of its travel, the motor should stall. If the motor continues to run, an indication of a stripped nut or gear is present.
4. This step is used to release the tension of the motor pack prior to removal.

Diagnostic Aids:

An "Intermittent" fault in this code may result from a mechanical part of the system that sticks, binds, or slips.

The frequency of the problem can be checked by using the enhanced diagnostic function of the "Scan"

If code only fails once and Code A061 also fails see Code A061. If intermittent and enhanced diagnostics shows this code fails during ABS, see Code A061.

Depending on the frequency of the failure, a physical inspection of the mechanical parts suspected may be necessary.

DIAGNOSTIC CHART A048 — 1993 CAVALIER AND SUNBIRD

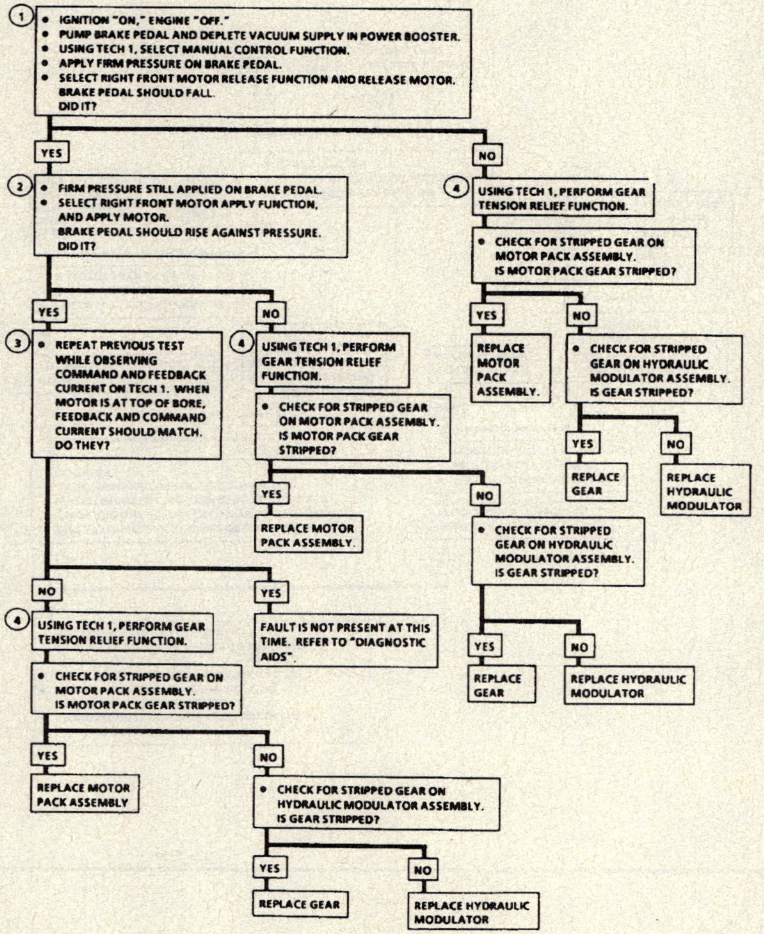

DIAGNOSTIC CHART A051 — 1993 CAVALIER AND SUNBIRD

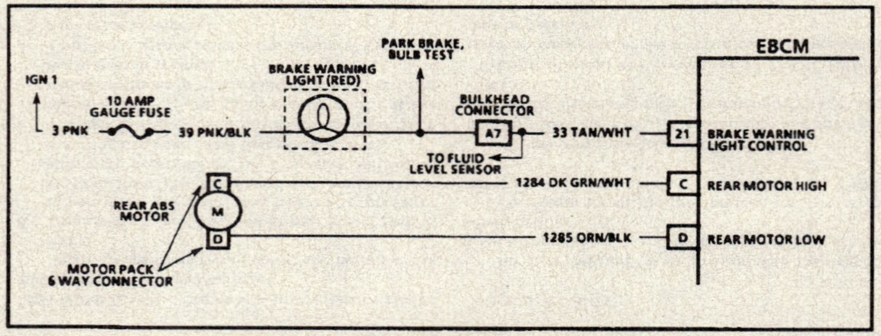

CODE A051

REAR AXLE MOTOR FREE SPINS

Circuit Description:

This code is designed to detect a stripped nut or gear assembly during initialization. During the homing sequence, the piston should reach the top of the bore resulting in a stalled motor. If this does not occur, the motor must be spinning with little or no resistance, indicating a nut/screw or gear failure.

Test Description: Number(s) below refer to circled number(s) on the diagnostic chart.

1. This step checks for a slipping nut or gear. If either were to slip, it would result in a lack of hydraulic pressure being applied at the rear wheels, causing the wheels to be able to be moved.
2. During a full apply, once the ball screw reaches the top of its travel, the motor should stall. If the motor continues to run, an indication of a stripped nut or gear is present.
3. This step is used to release the tension of the motor pack prior to removal.

Diagnostic Aids:

An "Intermittent" fault in this code may result from a mechanical part of the system that sticks, binds, or slips.

The frequency of the problem can be checked by using the enhanced diagnostic function of the "Scan" tool

If code only fails once and Code A064 also fails, see Code A064. If intermittent and enhanced diagnostics show this code fails during ABS, see Code A064.

Depending on the frequency of the failure, a physical inspection of the mechanical parts suspected may be necessary.

DIAGNOSTIC CHART A051 — 1993 CAVALIER AND SUNBIRD

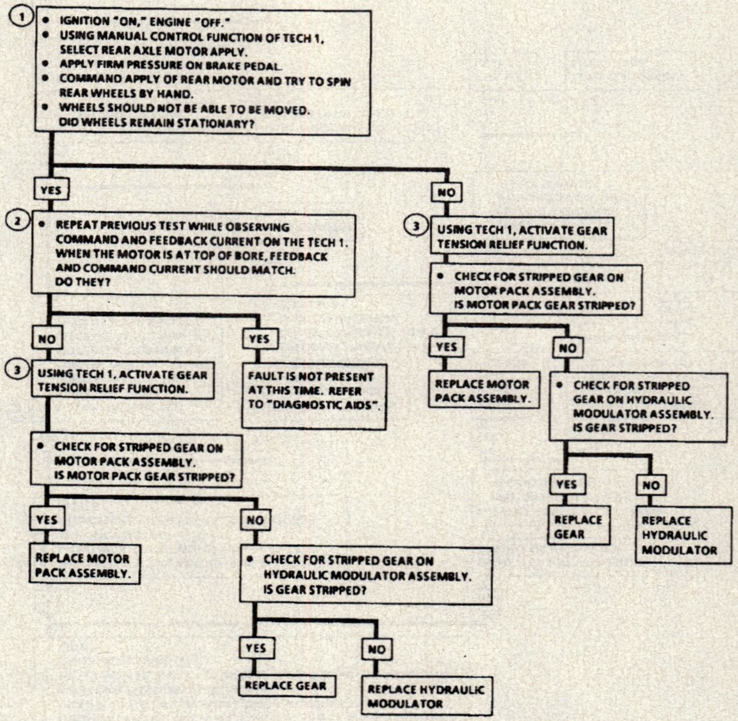

DIAGNOSTIC CHART A052 — 1993 CAVALIER AND SUNBIRD

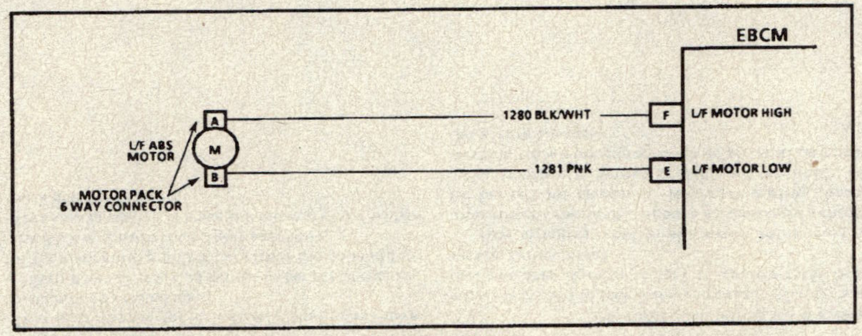

CODE A052

LEFT FRONT CHANNEL IN RELEASE TOO LONG

Circuit Description:

This code will diagnose a motor that is energized longer than expected. This could occur if a wheel speed sensor is faulty, the motor does not turn, the left front solenoid mechanically fails open, or the motor wires are crossed.

Test Description: Number(s) below refer to circled number(s) on the diagnostic chart.

1. This step identifies a problem in a wheel speed sensor that may cause the system to be in release too long.
2. This step identifies a motor fault as being failed or wired incorrectly.
3. This step checks for a solenoid that may have mechanically failed open.
4. If Step 3 has failed, this serves to isolate the cause of the hydraulic problem, to either the solenoid or the ABS hydraulic modulator assembly.
5. This step is used to release the tension of the motor pack.

Diagnostic Aids:

An "Intermittent" problem may be caused by a mechanical part of the system that sticks or binds.

The frequency of the problem can be checked by using the enhanced diagnostic function of the "Scan" tool

Code A052 may fail if on ice and steering wheel is turned to lock during braking. Using "scan" tool perform Hydraulic test to ensure total brake system is functional.

Any circuitry, that is suspected as causing the intermittent complaint, should be thoroughly checked for backed out terminals, improper mating, broken locks, improperly formed or damaged terminals, poor terminal to wiring connections or physical damage to the wiring harness.

DIAGNOSTIC CHART A052 — 1993 CAVALIER AND SUNBIRD

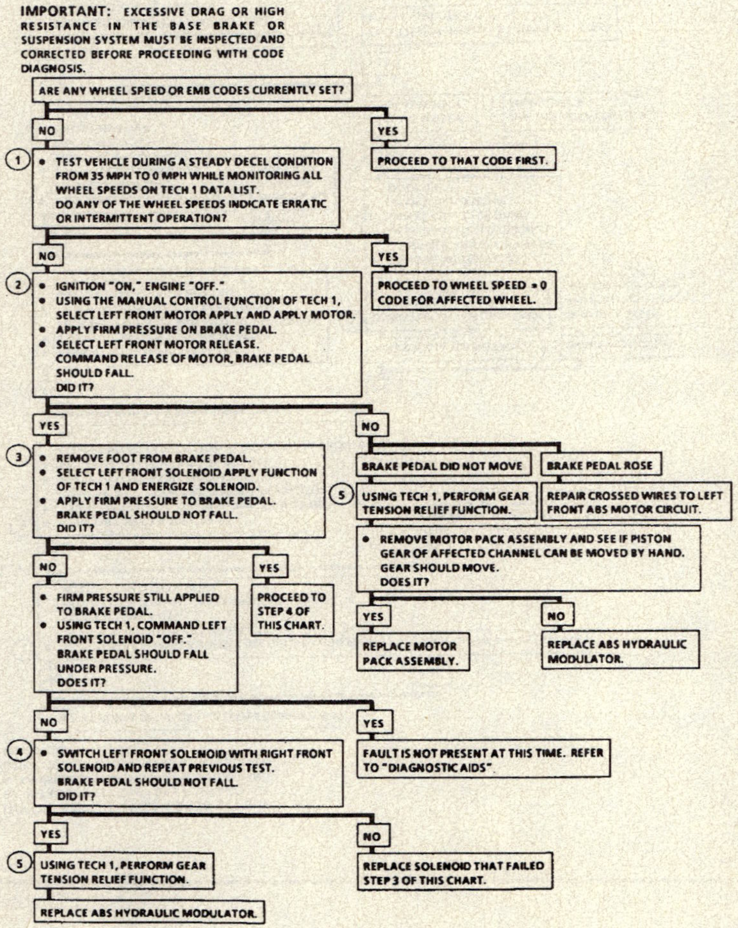

DIAGNOSTIC CHART A053 — 1993 CAVALIER AND SUNBIRD

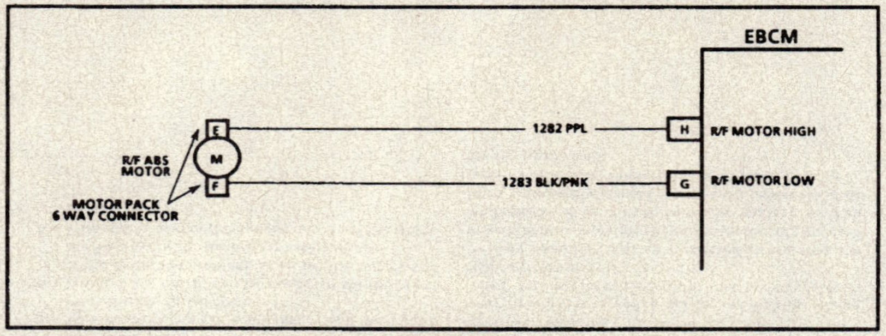

CODE A053

RIGHT FRONT CHANNEL IN RELEASE TOO LONG

Circuit Description:

This code will diagnose a motor that is energized longer than expected. This could occur if a wheel speed sensor is faulty, the motor does not turn, the right front solenoid mechanically fails open, or the motor wires are crossed.

Test Description: Number(s) below refer to circled number(s) on the diagnostic chart.

1. This step identifies a problem in a wheel speed sensor that may cause the system to be in release too long.
2. This step identifies a motor fault as being failed or wired incorrectly.
3. This step checks for a solenoid that may have mechanically failed open.
4. If Step 3 has failed, this serves to isolate the cause of the hydraulic problem, to either the solenoid or the ABS hydraulic modulator assembly.
5. This step is used to release the tension of the motor pack prior to removal.

Diagnostic Aids:

An "Intermittent" problem may be caused by a mechanical part of the system that sticks or binds.

The frequency of the problem can be checked by using the enhanced diagnostic function of the "Scan" tool,

Code A053 may fail on ice if steering wheel is turned to lock during braking. Using "Scan" tool perform Hydraulic test to ensure total brake system is functional.

Any circuitry, that is suspected as causing the intermittent complaint, should be thoroughly checked for backed out terminals, improper mating, broken locks, improperly formed or damaged terminals, poor terminal to wiring connections or physical damage to the wiring harness.

DIAGNOSTIC CHART A053 — 1993 CAVALIER AND SUNBIRD

IMPORTANT: EXCESSIVE DRAG OR HIGH RESISTANCE IN THE BASE BRAKE OR SUSPENSION SYSTEM MUST BE INSPECTED AND CORRECTED BEFORE PROCEEDING WITH CODE DIAGNOSIS.

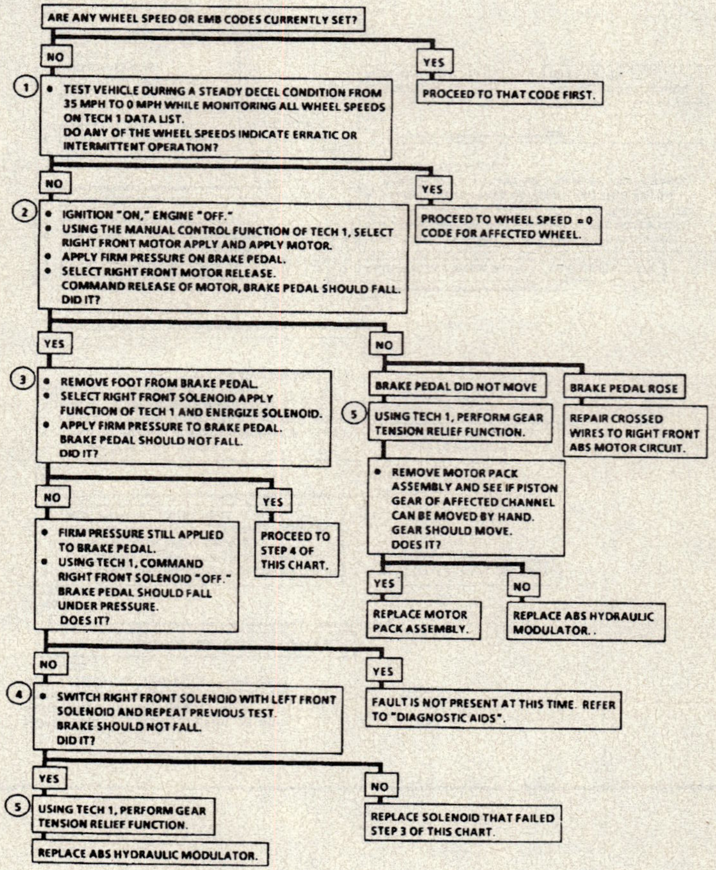

DIAGNOSTIC CHART A054 — 1993 CAVALIER AND SUNBIRD

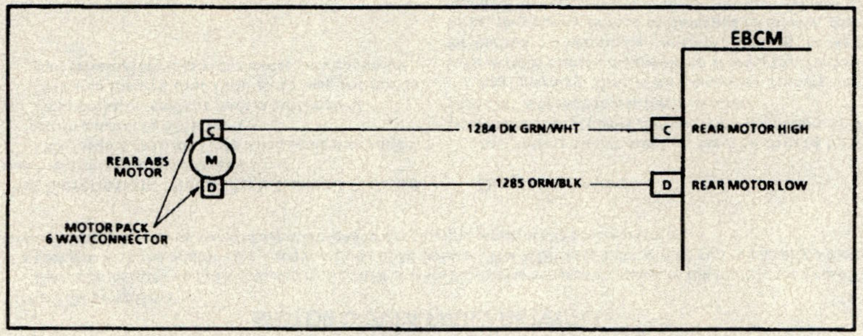

CODE A054

REAR AXLE IN RELEASE TOO LONG

Circuit Description:

This code will diagnose a motor that is energized longer than expected. This could occur if a wheel speed sensor is faulty, the motor does not turn, or the motor wires are crossed.

Test Description: Numbers below refer to circled numbers on the diagnostic chart.

1. This step identifies a problem in a wheel speed sensor that may cause the system to be in release too long.
2. This checks for a wheel that may stick or bind because of a mechanical fault.
3. This checks to see if the motor is capable of moving and applying the hydraulic piston for the rear wheels.
4. This step insures that the motor wiring is not crossed.
5. This step is used to release the tension of the motor pack prior to removal.
6. This isolates the fault of a "no-apply" situation to either the motor pack or ABS hydraulic modulator assembly.

Diagnostic Aids:

An "Intermittent" problem may be caused by a mechanical part of the system that sticks or binds.

The frequency of the problem can be checked by using the enhanced diagnostic function of the "Scan" tool.

Using "Scan" tool perform hydraulic test to ensure total brake system is functional.

Any circuitry, that is suspected as causing the intermittent complaint, should be thoroughly checked for backed out terminals, improper mating, broken locks, improperly formed or damaged terminals, poor terminal to wiring connections or physical damage to the wiring harness.

DIAGNOSTIC CHART A054 — 1993 CAVALIER AND SUNBIRD

IMPORTANT: EXCESSIVE DRAG OR HIGH RESISTANCE IN THE BASE BRAKE OR SUSPENSION SYSTEM MUST BE INSPECTED AND CORRECTED BEFORE PROCEEDING WITH CODE DIAGNOSIS.

ARE ANY WHEEL SPEED CODES CURRENTLY SET?

- YES: PROCEED TO THAT CODE FIRST.
- NO:

(1)
- TEST VEHICLE DURING A STEADY DECEL CONDITION FROM 35 MPH TO 0 MPH WHILE MONITORING ALL WHEEL SPEEDS ON TECH 1 DATA LIST. DO ANY OF THE WHEEL SPEEDS INDICATE ERRATIC OR INTERMITTENT OPERATION?

- YES: PROCEED TO WHEEL SPEED = 0 CODE FOR AFFECTED WHEEL.
- NO:

(2)
- RAISE REAR OF VEHICLE AND SUPPORT.
- SPIN BOTH REAR WHEELS AND CONFORM PHYSICAL CONDITION OF BASE BRAKE SYSTEM. BOTH WHEELS SHOULD SPIN FREELY BY HAND. DO THEY?

- NO: REPAIR SOURCE OF RESISTANCE IN BASE BRAKE SYSTEM.
- YES:

(3)
- IGNITION "ON," ENGINE "OFF."
- USING MANUAL CONTROL FUNCTION OF TECH 1, SELECT REAR AXLE MOTOR APPLY.
- APPLY FIRM PRESSURE ON BRAKE PEDAL.
- COMMAND APPLY OF REAR MOTOR AND TRY AND SPIN REAR WHEELS BY HAND.
- WHEELS SHOULD NOT BE ABLE TO BE MOVED. DID WHEELS REMAIN STATIONARY?

- YES: FAULT IS NOT PRESENT AT THIS TIME. REFER TO "DIAGNOSTIC AIDS".
- NO:

(4)
- WITH FIRM PRESSURE ON BRAKE PEDAL STILL APPLIED, COMMAND A RELEASE OF THE REAR AXLE MOTOR. CAN WHEELS BE MOVED?

- NO: REPAIR CROSSED WIRES AT REAR MOTOR CIRCUIT.
- YES:

(5) USING TECH 1, PERFORM GEAR TENSION RELIEF FUNCTION.

(6)
- REMOVE MOTOR PACK ASSEMBLY.
- CONNECT MOTOR PACK WIRING.
- USING TECH 1, PERFORM MOTOR TEST. ALL THREE MOTORS SHOULD SPIN FREELY IN BOTH DIRECTIONS. DO THEY?

- NO: REPLACE MOTOR PACK ASSEMBLY.
- YES: REPLACE ABS HYDRAULIC MODULATOR.

DIAGNOSTIC CHART A055 — 1993 CAVALIER AND SUNBIRD

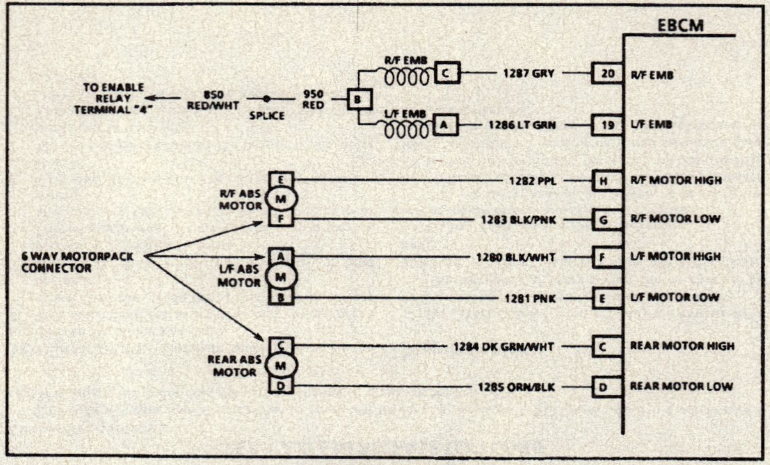

CODE A055

MOTOR DRIVER FAULT DETECTED

Circuit Description:

This code identifies a fault detected by the MDI custom IC. It also ensures the cause of the fault is not a result of a problem with the enable relay under a Code A055 failure. The MDI fault tests will be run to identify which circuit has failed, and may also cause additional codes to set to pinpoint the problem circuit.

Test Description: Number(s) below refer to circled number(s) on the diagnostic chart.

1. This insures that the Code A055 fault was not set due to a motor or EMB fault.
2. This checks to see if the fault is still present.
3. This step insures that Code A055 was not due to poor terminal contact at the motor pack connector.

Diagnostic Aids:

An "Intermittent" problem may be caused by a poor connection, rubbed through wire insulation, or a wire that is broken inside the insulation.

Any circuitry, that is suspected as causing the intermittent complaint, should be thoroughly checked for backed out terminals, improper mating, broken locks, improperly formed or damaged terminals, poor terminal to wiring connections or physical damage to the wiring harness.

DIAGNOSTIC CHART A055 — 1993 CAVALIER AND SUNBIRD

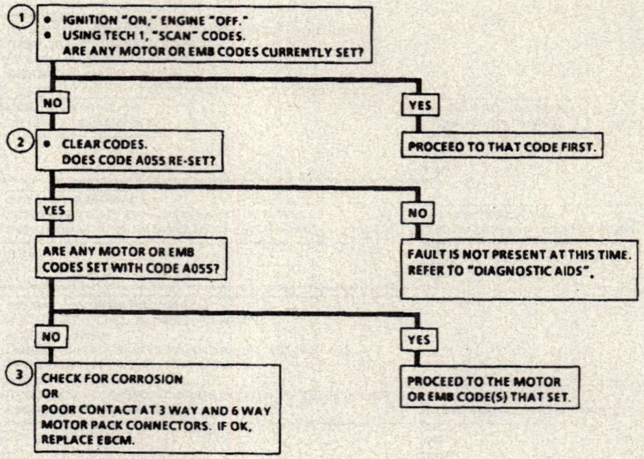

THIS CHART ASSUMES THAT A CURRENT CODE IS STORED INDICATING THAT THIS FAULT IS PRESENT.

DIAGNOSTIC CHART A056 — 1993
CAVALIER AND SUNBIRD

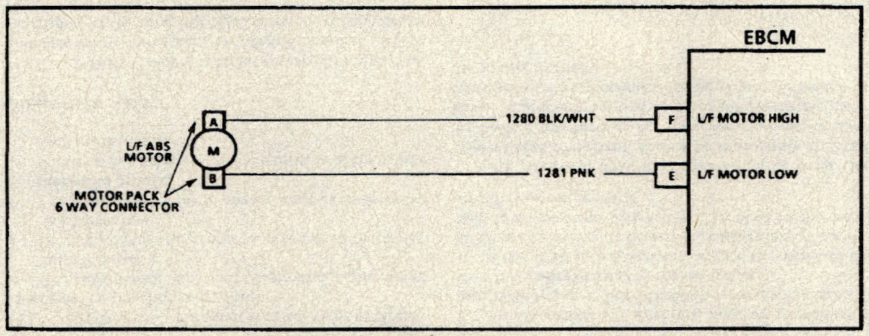

CODE A056
LEFT FRONT MOTOR CIRCUIT OPEN

Circuit Description:

This code identifies a motor that cannot be energized due to an open in its circuitry. This fault will not allow proper front ABS operation.

Test Description: Number(s) below refer to circled number(s) on the diagnostic chart.

1. This step checks for proper resistance of the motor.
2. This step is used to release the tension of the motor pack prior to removal.
3. This step checks for an open in the motor "high" circuitry.
4. This step checks for an open in the motor "low" circuitry.

Diagnostic Aids:

Using Tech 1, select manual control function, and exercise motor movement of affected channel in both directions while applying light pressure on the brake pedal.

If erratic or "jumpy" brake pedal movement is detected while performing an "apply" or "release" function of the motor, an intermittent fault may be indicated.

An "Intermittent" problem may be caused by a poor connection, rubbed through wire insulation, or a wire that is broken inside the insulation.

If the fault is not current, wiggle the wires of the affected channel and check if the code resets. This will help to pinpoint an intermittent problem in the motor circuitry or connections.

Any circuitry, that is suspected as causing the intermittent complaint, should be thoroughly checked for backed out terminals, improper mating, broken locks, improperly formed or damaged terminals, poor terminal to wiring connections or physical damage to the wiring harness.

DIAGNOSTIC CHART A056 — 1993
CAVALIER AND SUNBIRD

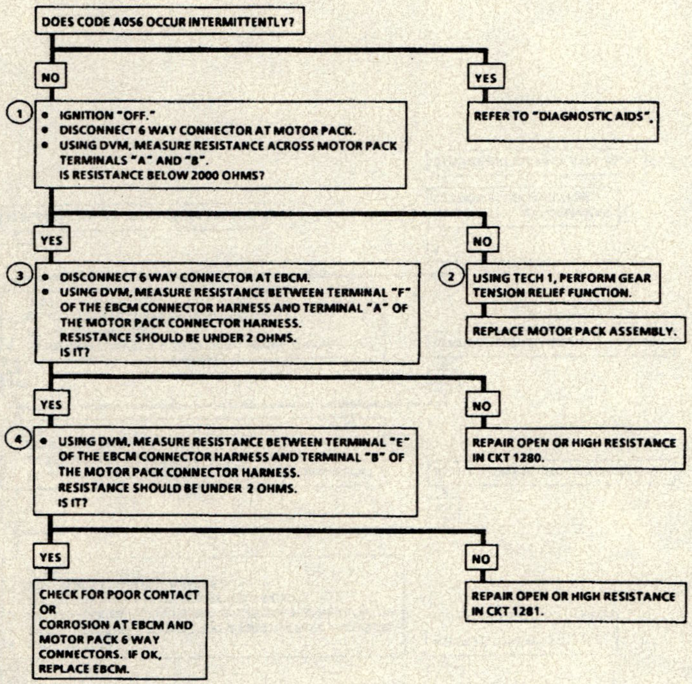

THIS CHART ASSUMES THAT A CURRENT CODE IS STORED INDICATING THAT THIS FAULT IS PRESENT.

DIAGNOSTIC CHART A057 — 1993 CAVALIER AND SUNBIRD

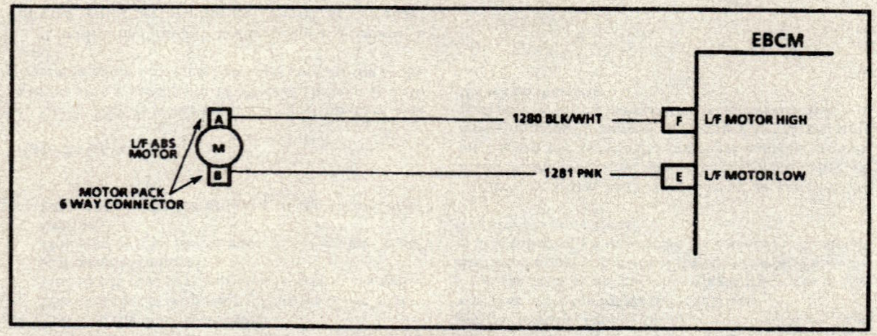

CODE A057

LEFT FRONT MOTOR CIRCUIT SHORTED TO GROUND

Circuit Description:

This code identifies a motor circuit that is shorted to ground. This fault will not allow the motor to be controlled at the commanded current rate or will cause the driver circuit to allow current directly to ground.

Test Description: Number(s) below refer to circled number(s) on the diagnostic chart.

1. This step checks for a short to ground in the motor "high" circuitry.
2. This step checks for a short to ground in the motor "low" circuitry.
3. This step checks for a motor that is internally shorted to ground.
4. This step is used to release the tension of the motor pack prior to removal.

Diagnostic Aids:

Using Tech 1, select manual control function, and exercise motor movement of affected channel in both directions while applying light pressure on the brake pedal.

If erratic or "jumpy" brake pedal movement is detected while performing an "apply" or "release" function of the motor, an intermittent fault may be indicated.

An "Intermittent" problem may be caused by a poor connection, rubbed through wire insulation, or a wire that is broken inside the insulation.

If the fault is not current, wiggle the wires of the affected channel and check if the code resets. This will help to pinpoint an intermittent problem in the motor circuitry or connections.

Any circuitry, that is suspected as causing the intermittent complaint, should be thoroughly checked for backed out terminals, improper mating, broken locks, improperly formed or damaged terminals, poor terminal to wiring connections or physical damage to the wiring harness.

DIAGNOSTIC CHART A057 — 1993 CAVALIER AND SUNBIRD

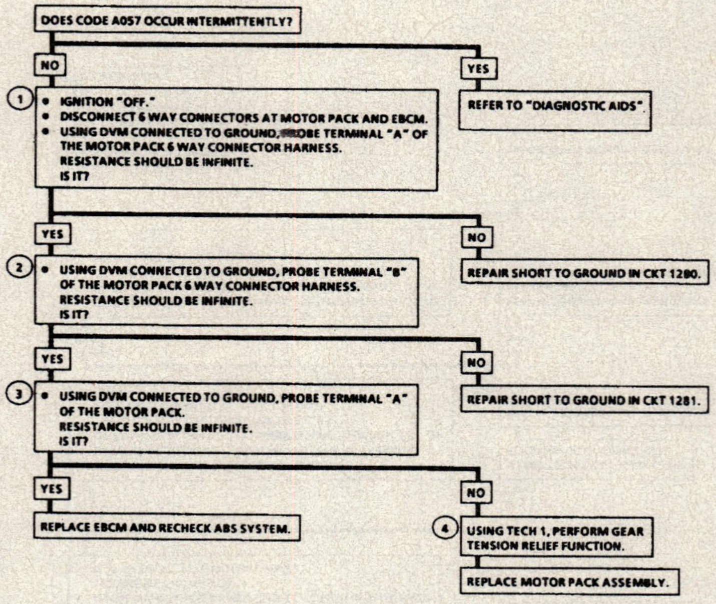

THIS CHART ASSUMES THAT A CURRENT CODE IS STORED INDICATING THAT THIS FAULT IS PRESENT.

DIAGNOSTIC CHART A058 — 1993 CAVALIER AND SUNBIRD

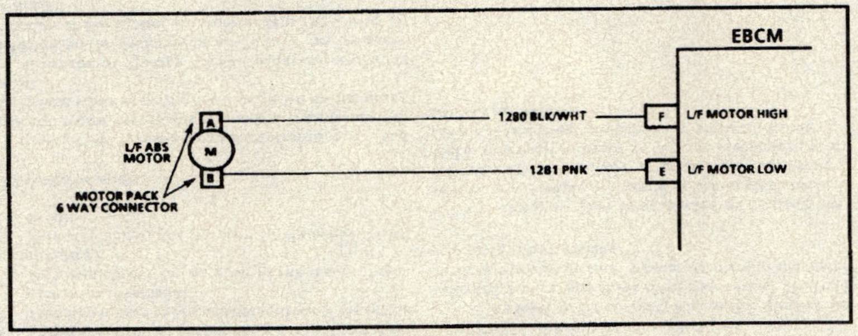

CODE A058

LEFT FRONT MOTOR CIRCUIT SHORTED TO BATTERY OR MOTOR SHORTED

Circuit Description:

This code identifies a motor circuit that is shorted to battery or a motor that has low or no resistance. This fault will not allow the motor to be controlled at the commanded current rate or will cause the motor to turn in the opposite direction, or not at all.

Test Description: Number(s) below refer to circled number(s) on the diagnostic chart.

1. This step checks for a short to voltage in the motor "high" circuitry.
2. This step checks for a short to voltage in the motor "low" circuitry.
3. This step checks for a motor that is internally shorted.
4. This step is used to release the tension of the motor pack prior to removal.

Diagnostic Aids:

Using Tech 1, select manual control function, and exercise motor movement of affected channel in both directions while applying light pressure on the brake pedal.

If erratic or "jumpy" brake pedal movement is detected while performing an "apply" or "release" function of the motor, an intermittent fault may be indicated.

An "Intermittent" problem may be caused by a poor connection, rubbed through wire insulation, or a wire that is broken inside the insulation.

If the fault is not current, wiggle the wires of the affected channel and check if the code resets. This will help to pinpoint an intermittent problem in the motor circuitry or connections.

Code A038 sets as a current code and A058 is an infrequent history code, see Code A038 diagnostics.

Any circuitry, that is suspected as causing the intermittent complaint, should be thoroughly checked for backed out terminals, improper mating, broken locks, improperly formed or damaged terminals, poor terminal to wiring connections or physical damage to the wiring harness.

DIAGNOSTIC CHART A058 — 1993 CAVALIER AND SUNBIRD

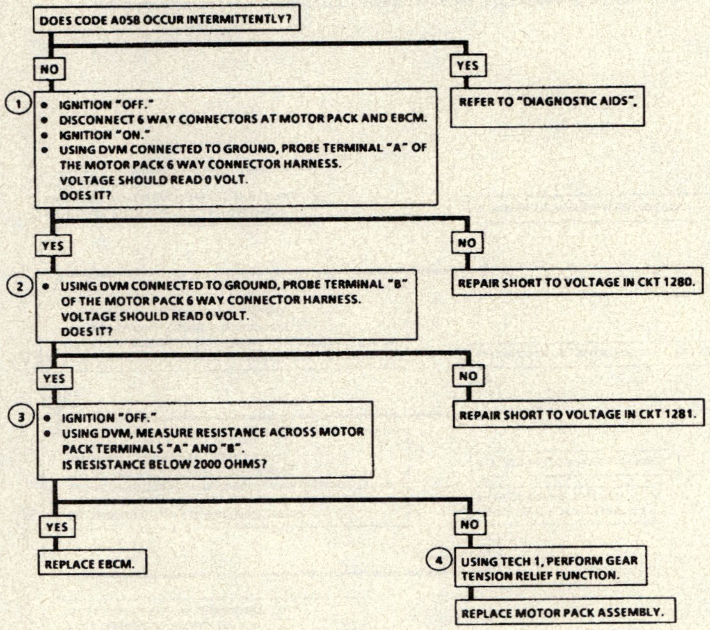

THIS CHART ASSUMES THAT A CURRENT CODE IS STORED INDICATING THAT THIS FAULT IS PRESENT.

DIAGNOSTIC CHART A061 — 1993 CAVALIER AND SUNBIRD

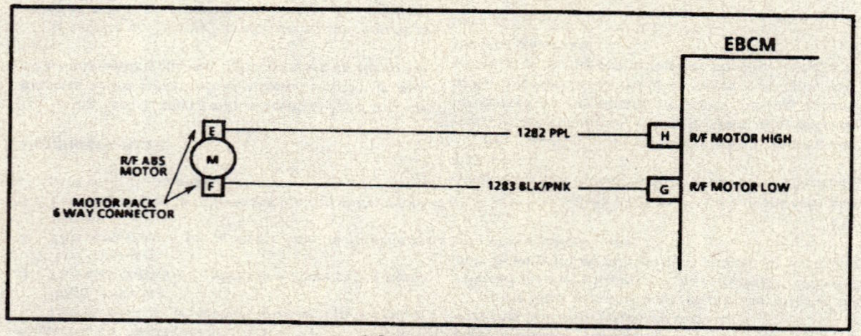

CODE A061

RIGHT FRONT MOTOR CIRCUIT OPEN

Circuit Description:

This code identifies a motor that cannot be energized due to an open in its circuitry. This fault will not allow proper front ABS operation.

Test Description: Number(s) below refer to circled number(s) on the diagnostic chart.

1. This step checks for proper resistance of the motor.
2. This step is used to release the tension of the motor pack prior to removal.
3. This step checks for an open in the motor "high" circuitry.
4. This step checks for an open in the motor "low" circuitry.

Diagnostic Aids:

Using Tech 1, select manual control function, and exercise motor movement of affected channel in both directions while applying light pressure on the brake pedal.

If erratic or "jumpy" brake pedal movement is detected while performing an "apply" or "release" function of the motor, an intermittent fault may be indicated.

An "Intermittent" problem may be caused by a poor connection, rubbed through wire insulation, or a wire that is broken inside the insulation.

If the fault is not current, wiggle the wires of the affected channel and check if the code resets. This will help to pinpoint an intermittent problem in the motor circuitry or connections.

Any circuitry, that is suspected as causing the intermittent complaint, should be thoroughly checked for backed out terminals, improper mating, broken locks, improperly formed or damaged terminals, poor terminal to wiring connections or physical damage to the wiring harness.

DIAGNOSTIC CHART A061 — 1993 CAVALIER AND SUNBIRD

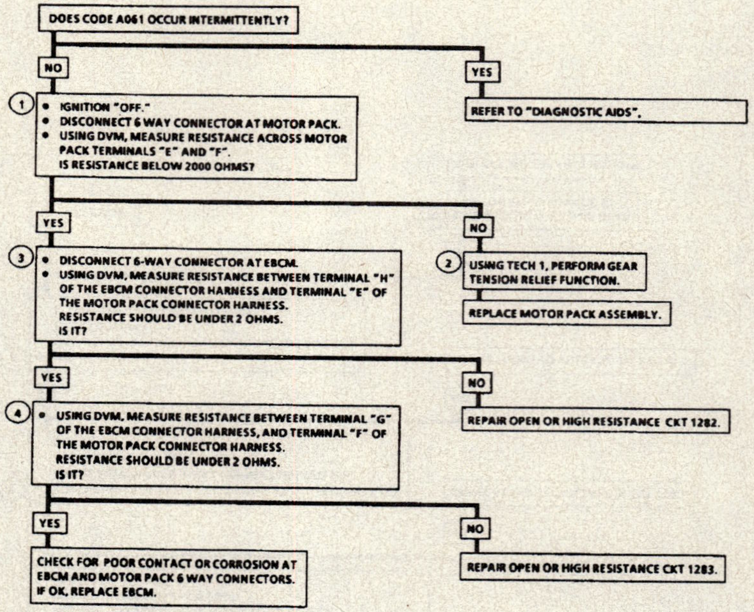

THIS CHART ASSUMES THAT A CURRENT CODE IS STORED INDICATING THAT THIS FAULT IS PRESENT.

DIAGNOSTIC CHART A062 — 1993 CAVALIER AND SUNBIRD

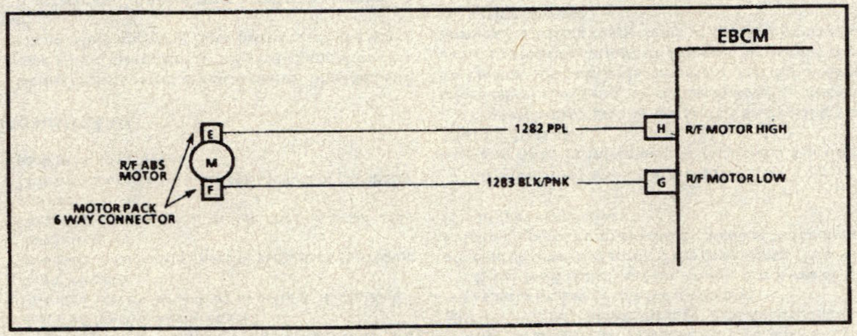

CODE A062

RIGHT FRONT MOTOR CIRCUIT SHORTED TO GROUND

Circuit Description:

This code identifies a motor circuit that is shorted to ground. This fault will not allow the motor to be controlled at the commanded current rate or will cause the motor driver circuit to allow current directly to ground.

Test Description: Number(s) below refer to circled number(s) on the diagnostic chart.

1. This step checks for a short to ground in the motor "high" circuitry.
2. This step checks for a short to ground in the motor "low" circuitry.
3. This step checks for a motor that is internally shorted to ground.
4. This step is used to release the tension of the motor pack prior to removal.

Diagnostic Aids:

Using Tech 1, select manual control function, and exercise motor movement of affected channel in both directions while applying light pressure on the brake pedal.

If erratic or "jumpy" brake pedal movement is detected while performing an "apply" or "release" function of the motor, an intermittent fault may be indicated.

An "Intermittent" problem may be caused by a poor connection, rubbed through wire insulation, or a wire that is broken inside the insulation.

If the fault is not current, wiggle the wires of the affected channel and check if the code resets. This will help to pinpoint an intermittent problem in the motor circuitry or connections.

Any circuitry, that is suspected as causing the intermittent complaint, should be thoroughly checked for backed out terminals, improper mating, broken locks, improperly formed or damaged terminals, poor terminal to wiring connections or physical damage to the wiring harness.

DIAGNOSTIC CHART A062 — 1993 CAVALIER AND SUNBIRD

DOES CODE A062 OCCUR INTERMITTENTLY?

- YES → REFER TO "DIAGNOSTIC AIDS".
- NO →

(1)
- IGNITION "OFF."
- DISCONNECT 6 WAY CONNECTORS AT MOTOR PACK AND EBCM.
- USING DVM CONNECTED TO GROUND, PROBE TERMINAL "E" OF THE MOTOR PACK 6 WAY CONNECTOR HARNESS. RESISTANCE SHOULD BE INFINITE. IS IT?

- NO → REPAIR SHORT TO GROUND IN CKT 1282.
- YES →

(2)
- USING DVM CONNECTED TO GROUND, PROBE TERMINAL "F" OF THE MOTOR PACK 6 WAY CONNECTOR HARNESS. RESISTANCE SHOULD BE INFINITE. IS IT?

- NO → REPAIR SHORT TO GROUND IN CKT 1283.
- YES →

(3)
- USING DVM CONNECTED TO GROUND, PROBE TERMINAL "E" OF THE MOTOR PACK. RESISTANCE SHOULD BE INFINITE. IS IT?

- YES → REPLACE EBCM AND RE-CHECK ABS SYSTEM.
- NO → (4) USING TECH 1, PERFORM GEAR TENSION RELIEF FUNCTION. → REPLACE MOTOR PACK ASSEMBLY.

THIS CHART ASSUMES THAT A CURRENT CODE IS STORED INDICATING THAT THIS FAULT IS PRESENT.

DIAGNOSTIC CHART A063 — 1993 CAVALIER AND SUNBIRD

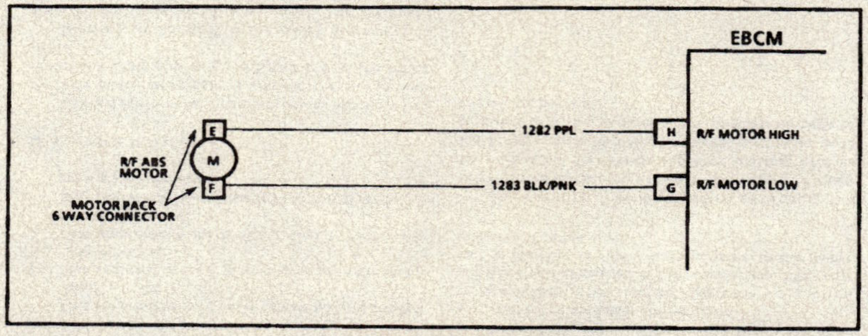

CODE A063

RIGHT FRONT MOTOR CIRCUIT SHORTED TO BATTERY OR MOTOR SHORTED

Circuit Description:

This code identifies a motor circuit that is shorted to battery or a motor that has low or no resistance. This fault will not allow the motor to be controlled at the commanded current rate or will cause the motor to turn in the opposite direction or not at all.

Test Description: Number(s) below refer to circled number(s) on the diagnostic chart.

1. This step checks for a short to voltage in the motor "high" circuitry.
2. This step checks for a short to voltage in the motor "low" circuitry.
3. This step checks for a motor that is internally shorted.
4. This test is used to release the tension of the motor pack prior to removal.

Diagnostic Aids:

Using Tech 1, select manual control function, and exercise motor movement of affected channel in both directions while applying light pressure on the brake pedal.

If erratic or "jumpy" brake pedal movement is detected while performing an "apply" or "release" function of the motor, an intermittent fault may be indicated.

An "Intermittent" problem may be caused by a poor connection, rubbed through wire insulation, or a wire that is broken inside the insulation.

If the fault is not current, wiggle the wires of the affected channel and check if the code resets. This will help to pinpoint an intermittent problem in the motor circuitry or connections.

If Code A041 sets as a current code and A063 is infrequent history code, see Code A041 for diagnosis.

Any circuitry, that is suspected as causing the intermittent complaint, should be thoroughly checked for backed out terminals, improper mating, broken locks, improperly formed or damaged terminals, poor terminal to wiring connections or physical damage to the wiring harness.

DIAGNOSTIC CHART A063 — 1993 CAVALIER AND SUNBIRD

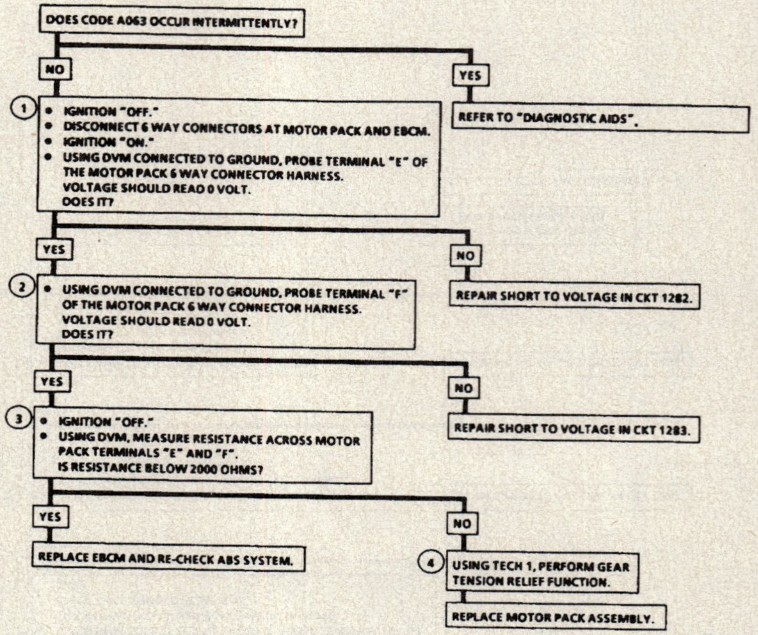

THIS CHART ASSUMES THAT A CURRENT CODE IS STORED INDICATING THAT THIS FAULT IS PRESENT.

DIAGNOSTIC CHART A064 — 1993 CAVALIER AND SUNBIRD

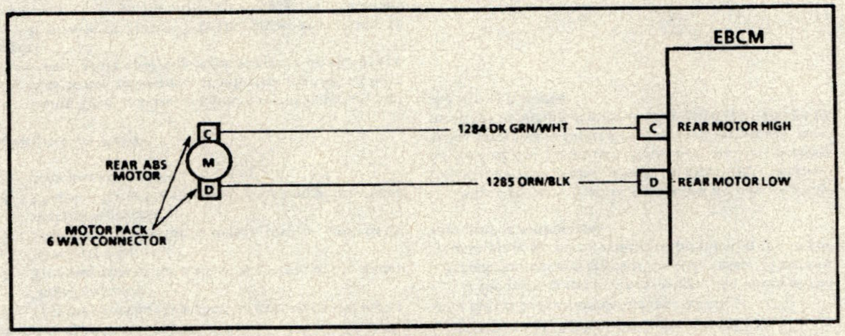

CODE A064

REAR AXLE MOTOR CIRCUIT OPEN

Circuit Description:

This code identifies a motor that cannot be energized due to an open in its circuitry. This fault will not allow proper rear ABS operation.

Test Description: Number(s) below refer to circled number(s) on the diagnostic chart.

1. This checks for an open circuit in the motor.
2. This step is used to release the tension of the motor pack prior to removal.
3. This step checks for an open in the motor "high" circuitry.
4. This step checks for an open in the motor "low" circuitry.

Diagnostic Aids:

Using Tech 1, select manual control function, and exercise motor movement of affected channel in both directions while applying light pressure on the brake pedal.

If erratic or "jumpy" brake pedal movement is detected while performing an "apply" or "release" function of the motor, an intermittent fault may be indicated.

An "Intermittent" problem may be caused by a poor connection, rubbed through wire insulation, or a wire that is broken inside the insulation.

If the fault is not current, wiggle the wires of the affected channel and check if the code resets. This will help to pinpoint an intermittent problem in the motor circuitry or connections.

Any circuitry, that is suspected as causing the intermittent complaint, should be thoroughly checked for backed out terminals, improper mating, broken locks, improperly formed or damaged terminals, poor terminal to wiring connections or physical damage to the wiring harness.

DIAGNOSTIC CHART A064 — 1993 CAVALIER AND SUNBIRD

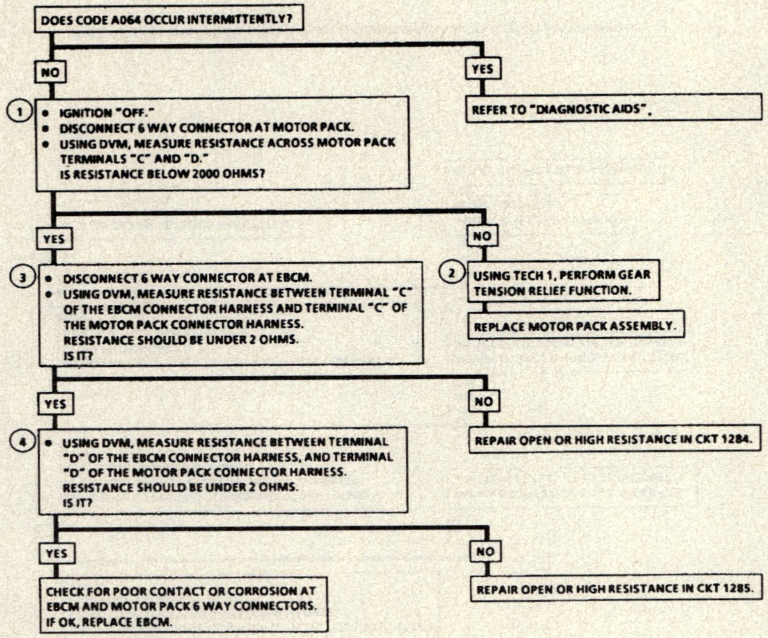

THIS CHART ASSUMES THAT A CURRENT CODE IS STORED INDICATING THAT THIS FAULT IS PRESENT.

DIAGNOSTIC CHART A065 — 1993
CAVALIER AND SUNBIRD

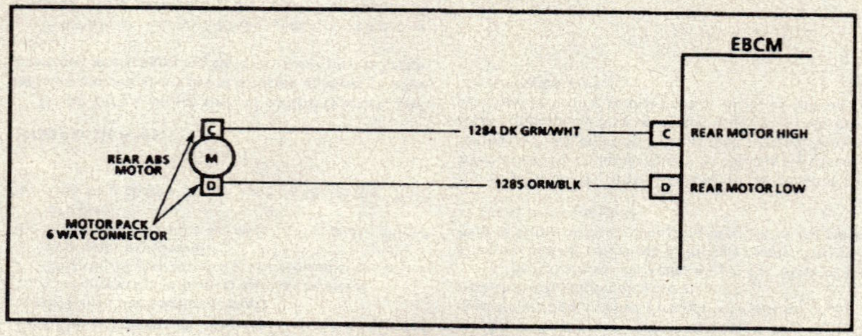

CODE A065

REAR AXLE MOTOR CIRCUIT SHORTED TO GROUND

Circuit Description:

This code identifies a motor circuit that is shorted to ground. This fault will not allow the motor to be controlled at the commanded current rate or will cause the driver circuit to allow current directly to ground.

Test Description: Number(s) below refer to circled number(s) on the diagnostic chart.

1. This step checks for a short to ground in the motor "high" circuitry.
2. This step checks for a short to ground in the motor "low" circuitry.
3. This step checks for a motor that is internally shorted to ground.
4. This step is used to release the tension of the motor pack prior to removal.

Diagnostic Aids:

Using Tech 1, select manual control function, and exercise motor movement of affected channel in both directions while applying light pressure on the brake pedal.

If erratic or "jumpy" brake pedal movement is detected while performing an "apply" or "release" function of the motor, an intermittent fault may be indicated.

An "Intermittent" problem may be caused by a poor connection, rubbed through wire insulation, or a wire that is broken inside the insulation.

If the fault is not current, wiggle the wires of the affected channel and check if the code resets. This will help to pinpoint an intermittent problem in the motor circuitry or connections.

Any circuitry, that is suspected as causing the intermittent complaint, should be thoroughly checked for backed out terminals, improper mating, broken locks, improperly formed or damaged terminals, poor terminal to wiring connections or physical damage to the wiring harness.

DIAGNOSTIC CHART A065 — 1993
CAVALIER AND SUNBIRD

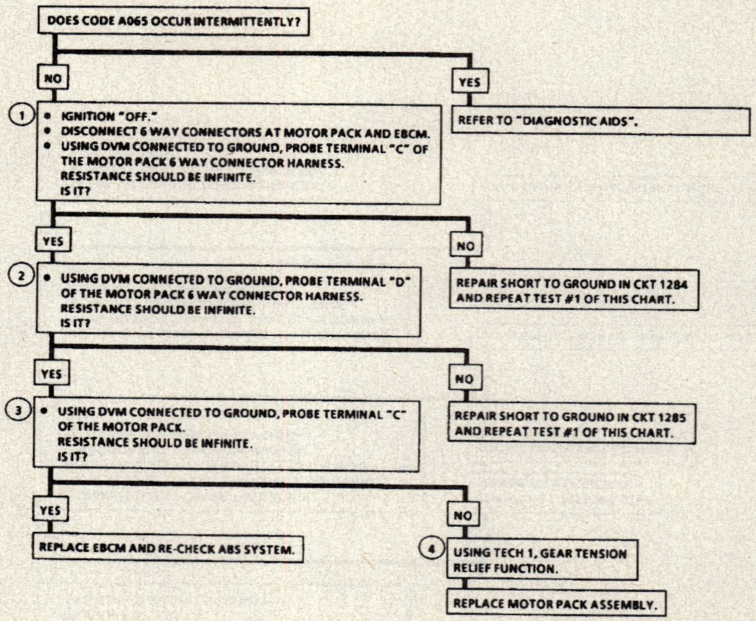

THIS CHART ASSUMES THAT A CURRENT CODE IS STORED INDICATING THAT THIS FAULT IS PRESENT.

DIAGNOSTIC CHART A066 — 1993 CAVALIER AND SUNBIRD

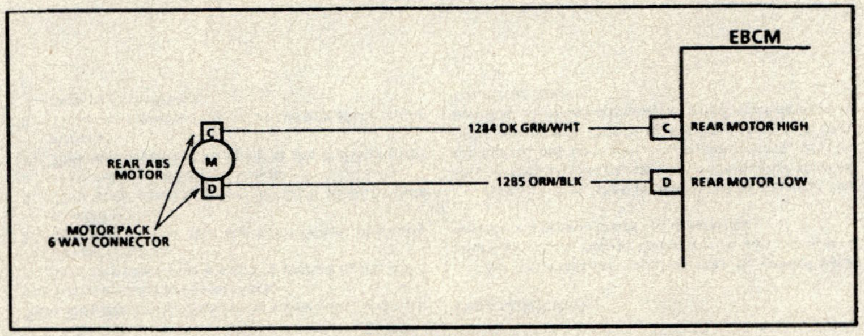

CODE A066

REAR AXLE MOTOR CIRCUIT SHORTED TO BATTERY OR MOTOR SHORTED

Circuit Description:

This code identifies a motor circuit that is shorted to battery or a motor that has low or zero resistance. This fault will not allow the motor to be controlled at the commanded current rate or will cause the motor to turn in the opposite direction or not turn at all.

Test Description: Number(s) below refer to circled number(s) on the diagnostic chart.

1. This step checks for a short to voltage in the motor "high" circuitry.
2. This step checks for a short to voltage in the motor "low" circuitry.
3. This step checks for a motor that is internally shorted.
4. This step is used to release the tension of the motor pack prior to removal.

Diagnostic Aids:

Using Tech 1, select manual control function, and exercise motor movement of affected channel in both directions while applying light pressure on the brake pedal.

If erratic or "jumpy" brake pedal movement is detected while performing an "apply" or "release" function of the motor, an intermittent fault may be indicated.

An "Intermittent" problem may be caused by a poor connection, rubbed through wire insulation, or a wire that is broken inside the insulation.

If the fault is not current, wiggle the wires of the affected channel and check if the code resets. This will help to pinpoint an intermittent problem in the motor circuitry or connections.

The frequency of the problem can be checked by using the enhanced diagnostic function of the "Scan" tool,

If Code A042 sets as current code and A066 is infrequent history code, see Code A042 diagnostics.

Any circuitry, that is suspected as causing the intermittent complaint, should be thoroughly checked for backed out terminals, improper mating, broken locks, improperly formed or damaged terminals, poor terminal to wiring connections or physical damage to the wiring harness.

DIAGNOSTIC CHART A066 — 1993 CAVALIER AND SUNBIRD

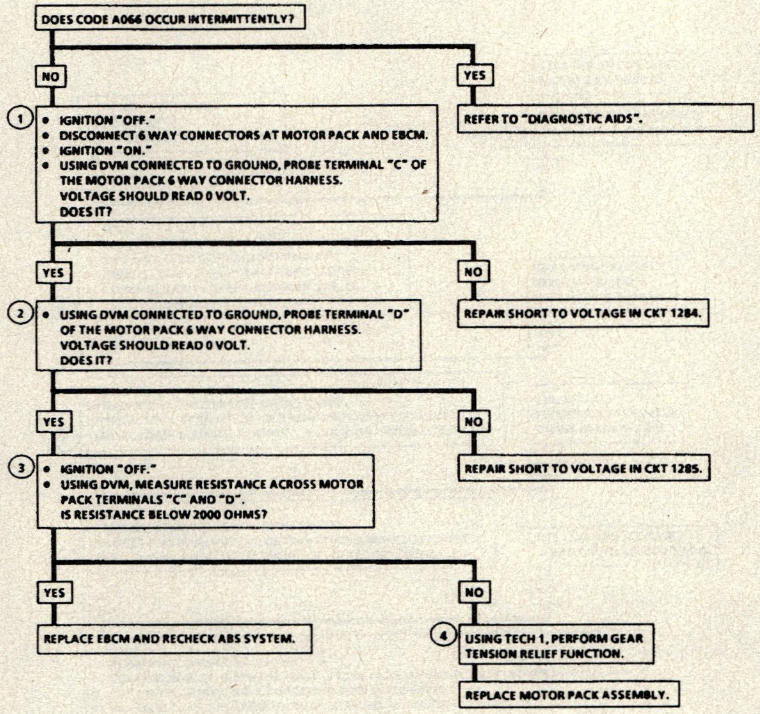

THIS CHART ASSUMES THAT A CURRENT CODE IS STORED INDICATING THAT THIS FAULT IS PRESENT.

DIAGNOSTIC CHART A067 — 1993 CAVALIER AND SUNBIRD

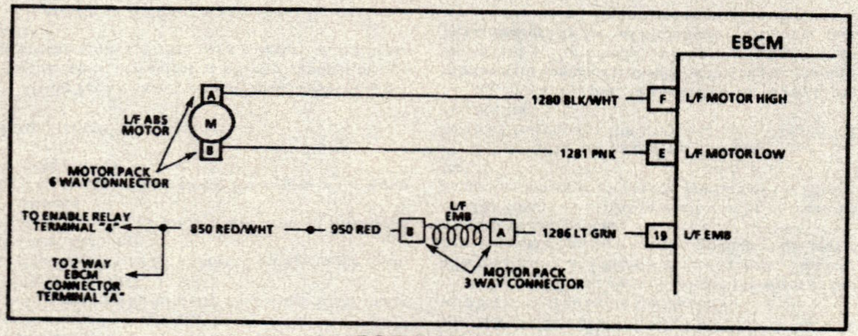

CODE A067

LEFT FRONT EMB RELEASE CIRCUIT OPEN OR SHORTED TO GROUND

Circuit Description:

This code is designed to detect shorts to ground and opens in the EMB release circuit.

A short to ground between the EMB and the ABS EBCM will cause the EMB to be energized indefinitely at an uncontrolled current, causing the EMB to damage itself. To remove the effects of the fault, the enable relay must be turned "OFF," eliminating power to the EMB as well as the ABS system.

Test Description: Number(s) below refer to circled number(s) on the diagnostic chart.

1. This step checks for a short to ground in the EMB control circuitry.
2. This step insures that the EMB is not internally shorted or open.
3. This step checks to see if CKT 850/950 has continuity to the EMB connector.
4. This step looks for an open in the EMB control circuitry.
5. This step is used to release the tension of the motor pack prior to removal.

Diagnostic Aids:

An "Intermittent" problem may be caused by a poor connection, rubbed through wire insulation, or a wire that is broken inside the insulation.

Any circuitry, that is suspected as causing the intermittent complaint, should be thoroughly checked for backed out terminals, improper mating, broken locks, improperly formed or damaged terminals, poor terminal to wiring connections or physical damage to the wiring harness.

DIAGNOSTIC CHART A067 — 1993 CAVALIER AND SUNBIRD

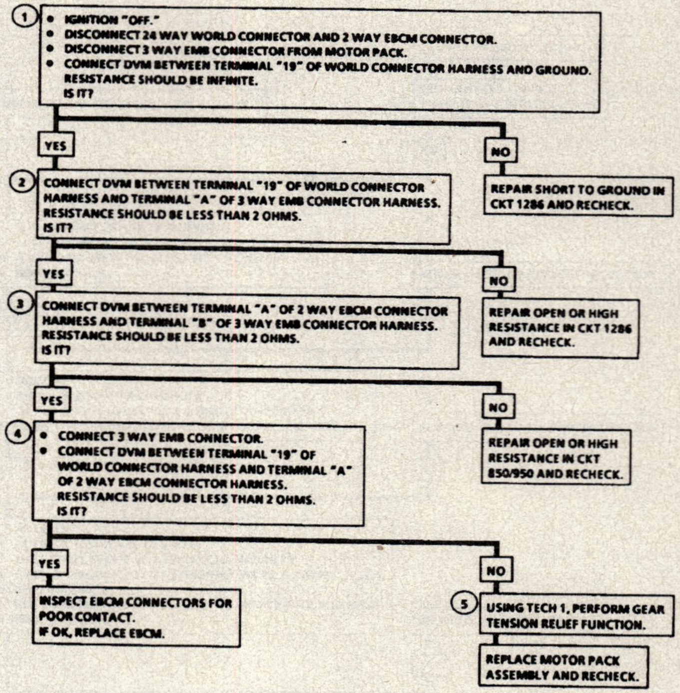

THIS CHART ASSUMES THAT A CURRENT CODE IS STORED INDICATING THAT THIS FAULT IS PRESENT.

DIAGNOSTIC CHART A068 — 1993
CAVALIER AND SUNBIRD

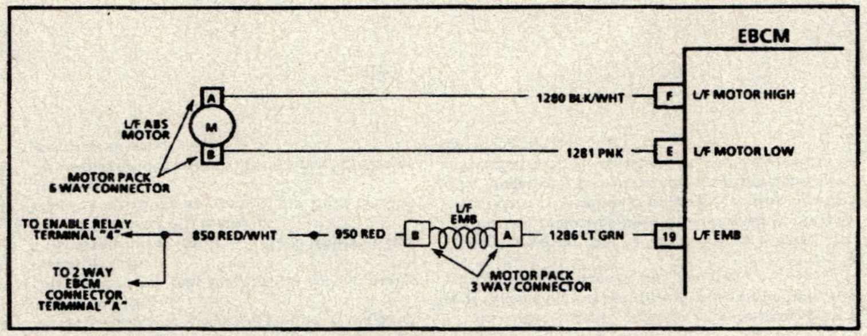

CODE A068

LEFT FRONT EMB CIRCUIT SHORTED TO BATTERY OR EMB SHORTED

Circuit Description:

This code is designed to detect shorts to battery and shorts in the EMB release circuit. These faults will not allow the EMB to be energized when needed.

Test Description: Number(s) below refer to circled number(s) on the diagnostic chart.

1. This step checks to see if the fault is currently present.
2. This step identifies a short to voltage in the EMB control circuitry.
3. This step checks for an EMB that has lower than normal resistance.
4. This step is used to release the tension of the motor pack prior to removal.

Diagnostic Aids:

An "Intermittent" problem may be caused by a poor connection, rubbed through wire insulation, or a wire that is broken inside the insulation.

Any circuitry, that is suspected as causing the intermittent complaint, should be thoroughly checked for backed out terminals, improper mating, broken locks, improperly formed or damaged terminals, poor terminal to wiring connections or physical damage to the wiring harness.

DIAGNOSTIC CHART A068 — 1993
CAVALIER AND SUNBIRD

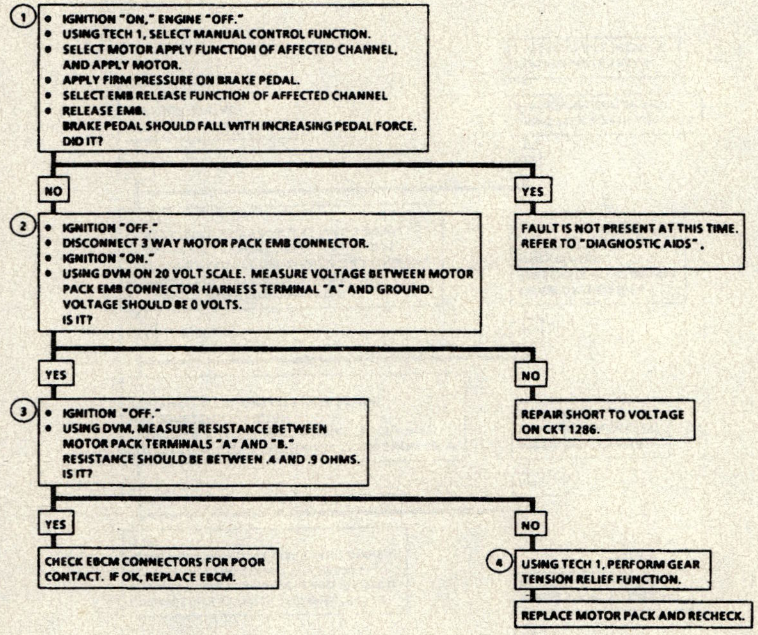

THIS CHART ASSUMES THAT A CURRENT CODE IS STORED INDICATING THAT THIS FAULT IS PRESENT.

DIAGNOSTIC CHART A071 — 1993 CAVALIER AND SUNBIRD

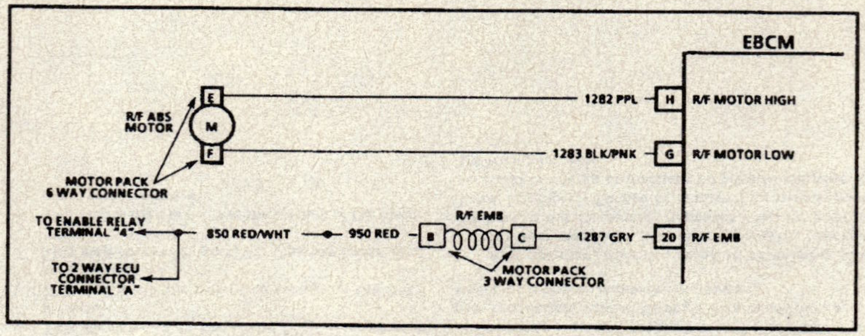

CODE A071

RIGHT FRONT EMB CIRCUIT OPEN OR SHORTED TO GROUND

Circuit Description:

This code is designed to detect shorts to ground and opens in the EMB release circuit.

A short to ground between the EMB and the ABS EBCM will cause the EMB to be energized indefinitely at an uncontrolled current causing the EMB to damage itself.

Test Description: Number(s) below refer to circled number(s) on the diagnostic chart.

1. This step checks for a short to ground in the EMB control circuitry.
2. This step insures that the EMB is not internally shorted or open.
3. This step checks to see if CKT 850/950 has continuity to the EMB connector.
4. This step looks for an open in the EMB control circuitry.
5. This step is used to release the tension of the motor pack prior to removal.

Diagnostic Aids:

An "Intermittent" problem may be caused by a poor connection, rubbed through wire insulation, or a wire that is broken inside the insulation.

Any circuitry, that is suspected as causing the intermittent complaint, should be thoroughly checked for backed out terminals, improper mating, broken locks, improperly formed or damaged terminals, poor terminal to wiring connections or physical damage to the wiring harness.

DIAGNOSTIC CHART A071 — 1993 CAVALIER AND SUNBIRD

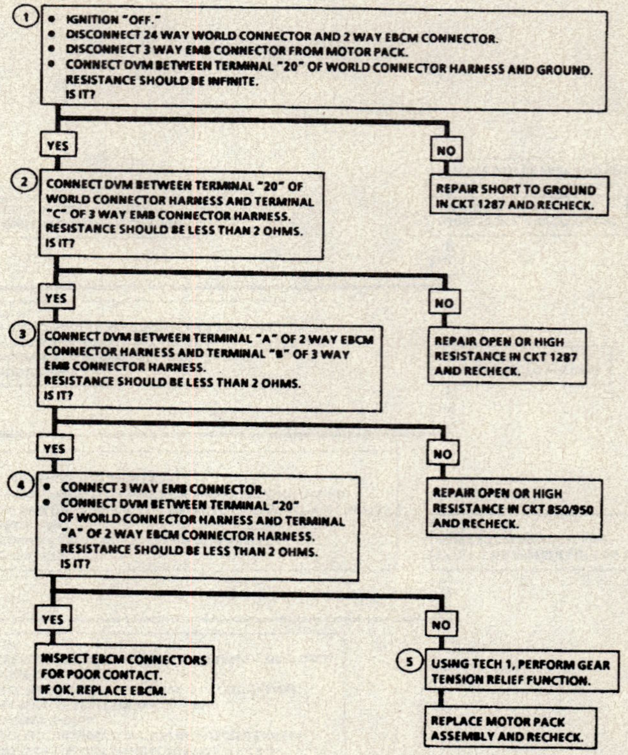

THIS CHART ASSUMES THAT A CURRENT CODE IS STORED INDICATING THAT THIS FAULT IS PRESENT.

DIAGNOSTIC CHART A072 — 1993
CAVALIER AND SUNBIRD

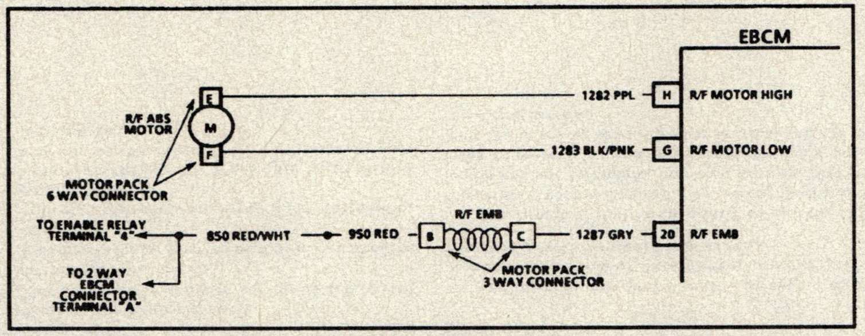

CODE A072

RIGHT FRONT EMB CIRCUIT SHORTED TO BATTERY OR EMB SHORTED

Circuit Description:

This code is designed to detect shorts to battery and shorts in the EMB release circuit. These faults will not allow the EMB to be energized when needed.

Test Description: Number(s) below refer to circled number(s) on the diagnostic chart.

1. This step checks to see if the fault is currently present.
2. This step identifies a short to voltage in the EMB control circuitry.
3. This step checks for an EMB that has lower than normal resistance.
4. This step is used to release the tension of the motor pack prior to removal.

Diagnostic Aids:

An "Intermittent" problem may be caused by a poor connection, rubbed through wire insulation, or a wire that is broken inside the insulation.

Any circuitry, that is suspected as causing the intermittent complaint, should be thoroughly checked for backed out terminals, improper mating, broken locks, improperly formed or damaged terminals, poor terminal to wiring connections or physical damage to the wiring harness.

DIAGNOSTIC CHART A072 — 1993
CAVALIER AND SUNBIRD

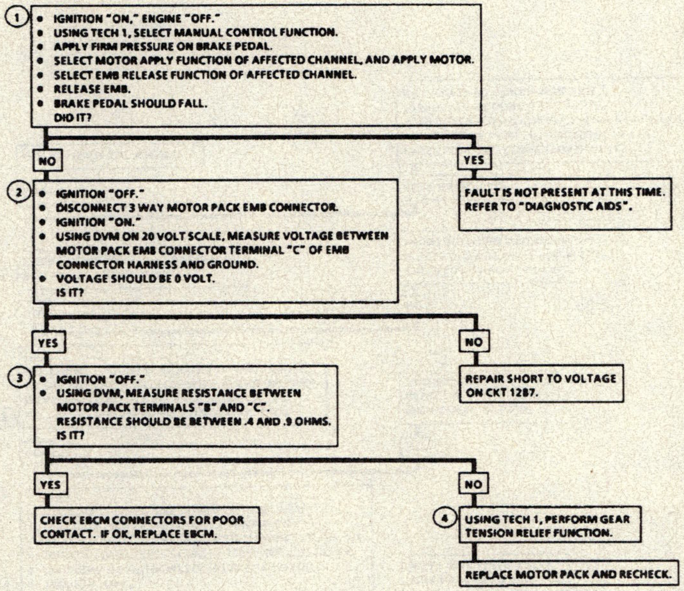

THIS CHART ASSUMES THAT A CURRENT CODE IS STORED INDICATING THAT THIS FAULT IS PRESENT.

DIAGNOSTIC CHART A076 — 1993
CAVALIER AND SUNBIRD

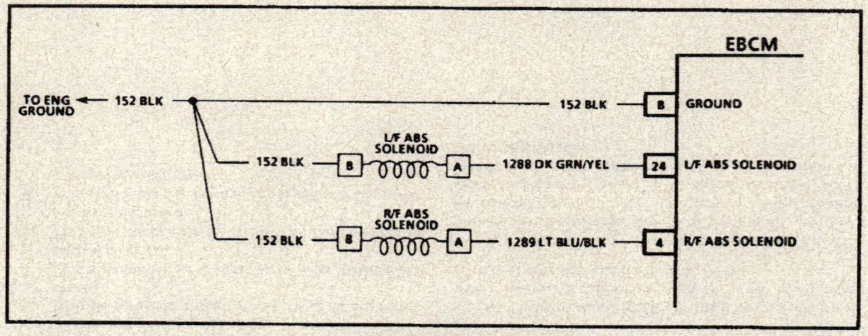

CODE A076

LEFT FRONT SOLENOID CIRCUIT SHORTED TO BATTERY OR OPEN

Circuit Description:

This code identifies a solenoid that cannot be energized due to an open in its circuitry, or a solenoid that is always energized due to a short to battery in its circuitry between the driver and the solenoid. An open will not allow proper ABS operation, but the short to battery simply turns "ON" the solenoid. A path for base brakes is still allowed once the motor re-homes and the check valve is lifted off its seat during key on initialization.

Test Description: Number(s) below refer to circled number(s) on the diagnostic chart.

1. This checks for a short to voltage in the control circuitry of the solenoid.
2. This checks for an open in the control circuit of the solenoid.
3. This step identifies an open in the solenoid or ground circuit.
4. This step checks for a possible intermittent problem in the solenoid circuitry due to poor terminal contact.

Diagnostic Aids:

An "Intermittent" problem may be caused by a poor connection, rubbed through wire insulation, or a wire that is broken inside the insulation.

Any circuitry, that is suspected as causing the intermittent complaint, should be thoroughly checked for backed out terminals, improper mating, broken locks, improperly formed or damaged terminals, poor terminal to wiring connections or physical damage to the wiring harness.

DIAGNOSTIC CHART A076 — 1993
CAVALIER AND SUNBIRD

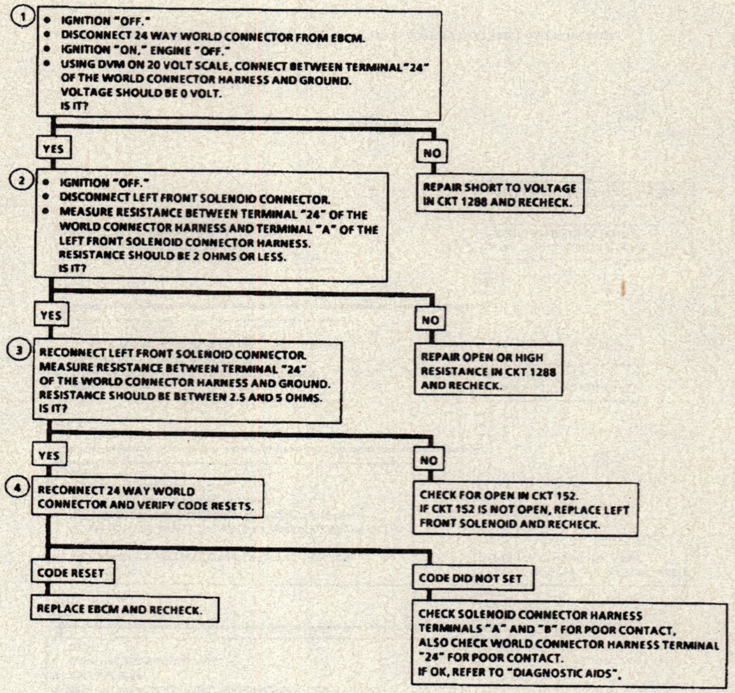

THIS CHART ASSUMES THAT A CURRENT CODE IS STORED INDICATING THAT THIS FAULT IS PRESENT.

DIAGNOSTIC CHART A077 — 1993 CAVALIER AND SUNBIRD

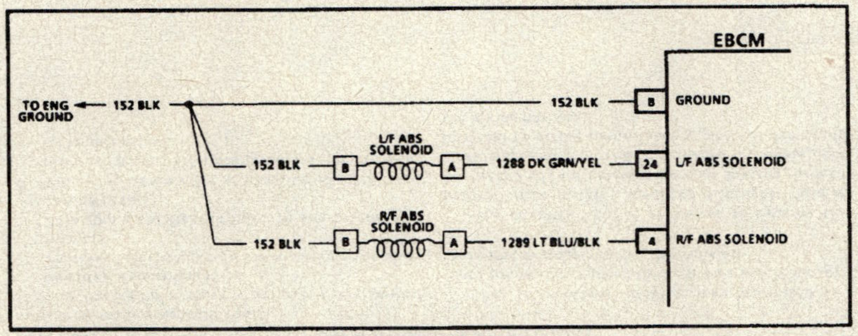

CODE A077

LEFT FRONT SOLENOID CIRCUIT SHORTED TO GROUND OR DRIVER OPEN

Circuit Description:

This code identifies a solenoid that cannot be energized due to an open in its driver circuitry, or a short to ground between the solenoid driver and the solenoid. These faults can affect ABS operation since the flow of brake fluid to the wheel cylinder cannot be stopped, making ABS operation for that channel impossible.

Test Description: Number(s) below refer to circled number(s) on the diagnostic chart.

1. This step checks for a solenoid that is internally shorted to ground.
2. This step checks for a solenoid that is not within proper resistance values.
3. This step indicates if a short to ground exists in the solenoid circuitry.
4. This checks for an open control circuit to the EBCM.

Diagnostic Aids:

An "Intermittent" problem may be caused by a poor connection, rubbed through wire insulation, or a wire that is broken inside the insulation.

Any circuitry, that is suspected as causing the intermittent complaint, should be thoroughly checked for backed out terminals; improper mating, broken locks, improperly formed or damaged terminals, poor terminal to wiring connections or physical damage to the wiring harness.

DIAGNOSTIC CHART A077 — 1993 CAVALIER AND SUNBIRD

(1)
- IGNITION "OFF."
- DISCONNECT LEFT FRONT SOLENOID CONNECTOR.
- USING DVM CONNECTED TO GROUND, PROBE TERMINAL "A" OF LEFT FRONT SOLENOID. RESISTANCE SHOULD BE INFINITE. IS IT?

NO → REPLACE SHORTED SOLENOID AND RE-TEST.

YES ↓

(2)
- CONNECT DVM BETWEEN SOLENOID TERMINALS "A" AND "B," AND MEASURE RESISTANCE. RESISTANCE SHOULD BE BETWEEN 2.5 AND 5 OHMS. IS IT?

NO → REPLACE SOLENOID AND RE-TEST.

YES ↓

(3)
- DISCONNECT 24 WAY WORLD CONNECTOR.
- USING DVM CONNECTED TO GROUND, PROBE TERMINAL 24 OF WORLD CONNECTOR HARNESS. RESISTANCE SHOULD BE INFINITE. IS IT?

NO → REPAIR SHORT TO GROUND IN CKT 1288 AND RE-TEST.

YES ↓

(4)
- CONNECT DVM BETWEEN TERMINAL "24" OF WORLD CONNECTOR HARNESS AND TERMINAL "A" OF THE LEFT FRONT SOLENOID CONNECTOR HARNESS. RESISTANCE SHOULD BE 2 OHMS OR LESS. IS IT?

NO → REPAIR OPEN CKT 1288 AND RE-TEST.

YES → CHECK EBCM 24 WAY WORLD CONNECTOR FOR POOR CONTACT. IF OK, REPLACE EBCM.

THIS CHART ASSUMES THAT A CURRENT CODE IS STORED INDICATING THAT THIS FAULT IS PRESENT.

DIAGNOSTIC CHART A078 — 1993 CAVALIER AND SUNBIRD

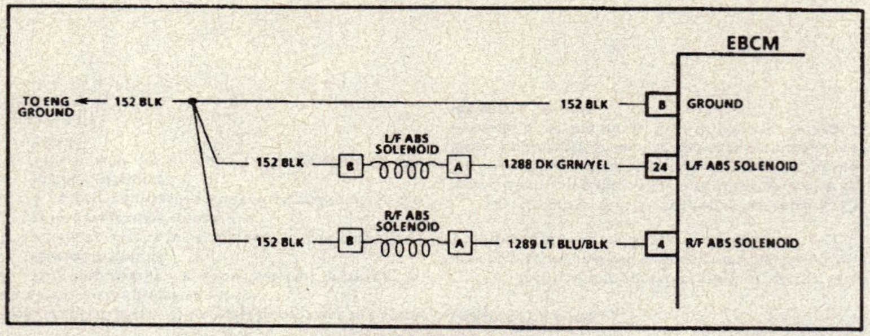

CODE A078

RIGHT FRONT SOLENOID CIRCUIT SHORTED TO BATTERY OR OPEN

Circuit Description:

This code identifies a solenoid that cannot be energized due to an open in its circuitry, or a solenoid that is always energized due to a short to battery in its circuitry between the driver and the solenoid. An open will not allow proper ABS operation, but the short to battery simply turns "ON" the solenoid. A path for base brakes is still allowed once the motor re-homes and the check valve is lifted off its seat during key on initialization.

Test Description: Number(s) below refer to circled number(s) on the diagnostic chart.

1. This checks for a short to voltage in the control circuitry of the solenoid.
2. This checks for an open in the control circuit of the solenoid.
3. This step identifies an open in the solenoid or ground circuit.
4. This step checks for a possible intermittent problem in the solenoid circuitry due to poor terminal contact.

Diagnostic Aids:

An "Intermittent" problem may be caused by a poor connection, rubbed through wire insulation, or a wire that is broken inside the insulation.

Any circuitry, that is suspected as causing the intermittent complaint, should be thoroughly checked for backed out terminals, improper mating, broken locks, improperly formed or damaged terminals, poor terminal to wiring connections or physical damage to the wiring harness.

DIAGNOSTIC CHART A078 — 1993 CAVALIER AND SUNBIRD

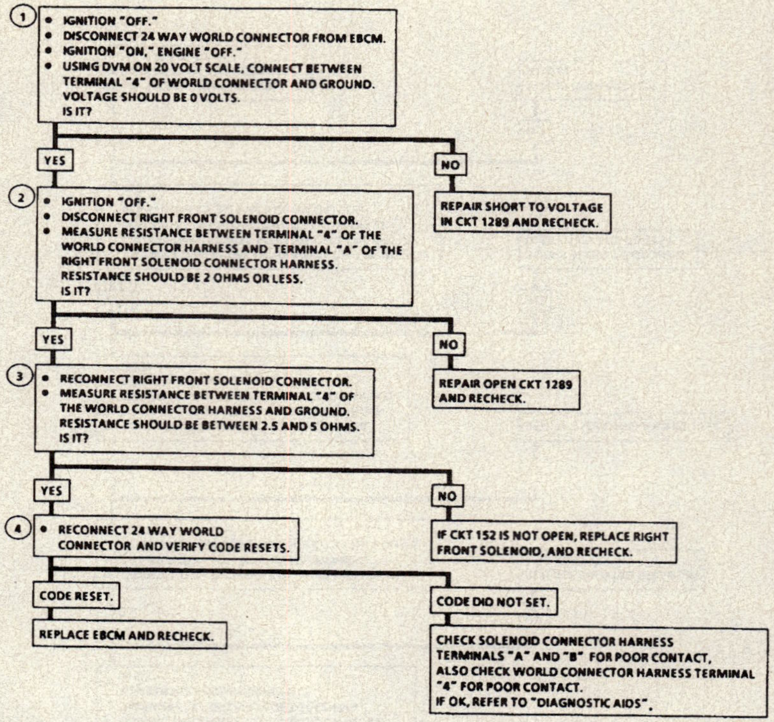

THIS CHART ASSUMES THAT A CURRENT CODE IS STORED INDICATING THAT THIS FAULT IS PRESENT.

DIAGNOSTIC CHART A081 — 1993
CAVALIER AND SUNBIRD

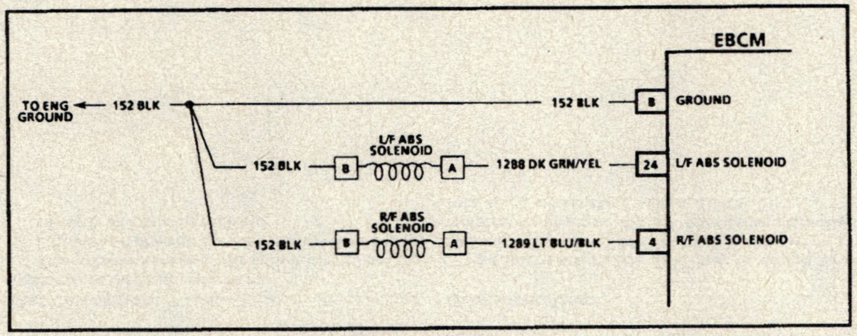

CODE A081

RIGHT FRONT SOLENOID CIRCUIT SHORTED TO GROUND OR DRIVER OPEN

Circuit Description:

This code identifies a solenoid that cannot be energized due to an open in its driver circuitry, or a short to ground between the solenoid driver and the solenoid. These faults can affect ABS operation since the flow of brake fluid to the wheel cylinder cannot be stopped, making ABS operation for that channel impossible.

Test Description: Number(s) below refer to circled number(s) on the diagnostic chart.

1. This step checks for a solenoid that is internally shorted to ground.
2. This step checks for a solenoid that is not within proper resistance values.
3. This step indicates if a short to ground exists in the solenoid circuitry.
4. This checks for an open control circuit to the EBCM.

Diagnostic Aids:

An "Intermittent" problem may be caused by a poor connection, rubbed through wire insulation, or a wire that is broken inside the insulation.

Any circuitry, that is suspected as causing the intermittent complaint, should be thoroughly checked for backed out terminals, improper mating, broken locks, improperly formed or damaged terminals, poor terminal to wiring connections or physical damage to the wiring harness.

DIAGNOSTIC CHART A081 — 1993
CAVALIER AND SUNBIRD

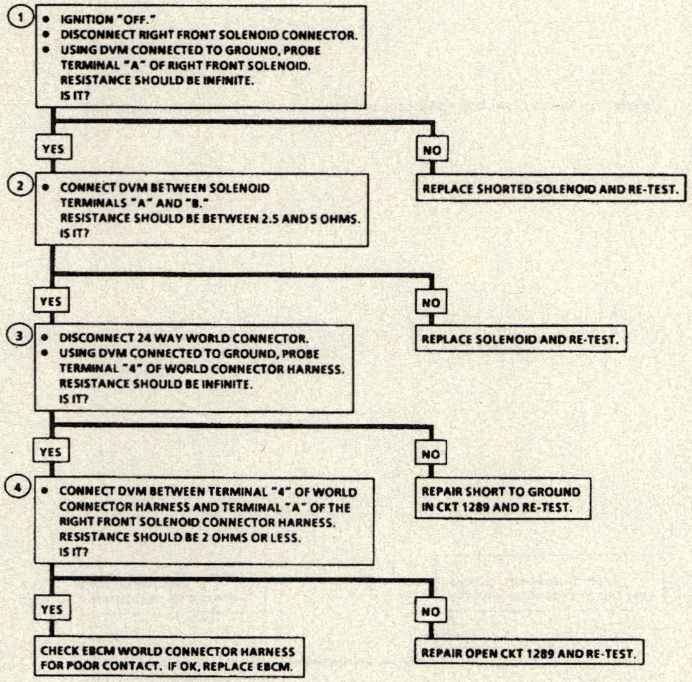

THIS CHART ASSUMES THAT A CURRENT CODE IS STORED INDICATING THAT THIS FAULT IS PRESENT.

DIAGNOSTIC CHART A082 — 1993 CAVALIER AND SUNBIRD

D.M. A.B.S. VI
ELECTRONIC CONTROL UNIT
(TYPICAL)

CODE A082

CALIBRATION MEMORY FAILURE

Circuit Description:

This code allows the EBCM to check for a calibration failure by comparing the calibration value to a known value stored in the EEPROM.

This test is also used as a security measure to prevent improper use of calibrations or changes to these calibrations that may alter the designed function of ABS.

Test Description: Number(s) below refer to circled number(s) on the diagnostic chart.

1. This step checks to see if the fault is present during diagnosis. If present, the EBCM is not functioning correctly and must be replaced.

Diagnostic Aids:

An intermittent Code A082 may be caused by a bad cell in the EEPROM that is sensitive to temperature changes. If Code A082 failed more than once, but is intermittent, replace EBCM.

DIAGNOSTIC CHART A082 — 1993 CAVALIER AND SUNBIRD

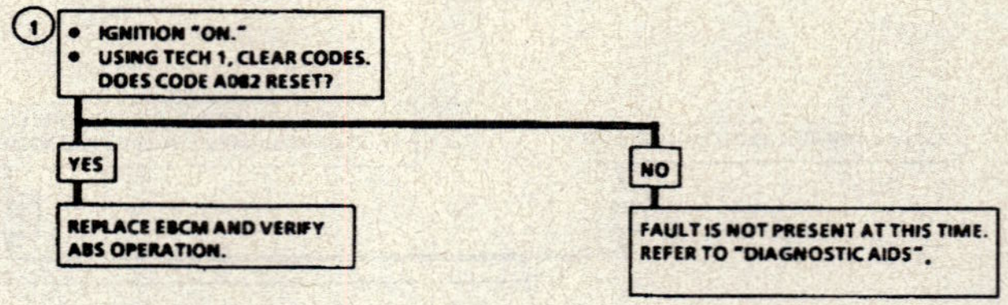

THIS CHART ASSUMES THAT A CURRENT CODE IS STORED INDICATING THAT THIS FAULT IS PRESENT.

DIAGNOSTIC CHART A086 — 1993
CAVALIER AND SUNBIRD

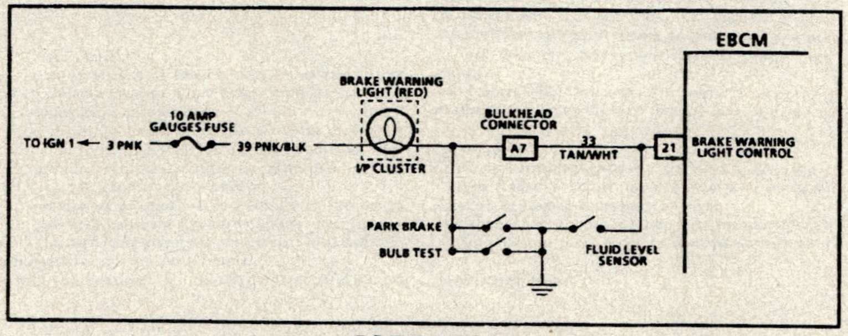

CODE A086

RED BRAKE WARNING LIGHT ACTIVATED BY ABS

Circuit Description:

This code is provided as an information only test, and reflects the status of the command issued by the ABS EBCM to illuminate the red brake warning light. If another diagnostic code issues a command to illuminate the red brake warning light, Code A086 will be stored in EEPROM as a history code at the conclusion of the ignition cycle.

Test Description: Number(s) below refer to circled number(s) on the diagnostic chart.

1. This step serves to identify if a code other than Code A086 commanded the red brake warning light "ON."

Diagnostic Aids:

Any ABS mechanical code that issues a command to illuminate the red brake warning light will also result in Code A086 being stored in EEPROM during shut down. These codes are: A038, A041, A042, A045, A046, A047, A048 and A051. If the motors are not in their home position, certain electrical codes will also command the red brake warning light "ON." These codes are: A014, A018, A055, A056, A057, A0587, A061, A062, A063, A064, A065, A066, A067, A068, A071 and A072.

If any of these codes are indicated along with Cod A086, they must be corrected prior to addressing a Code A086 fault.

DIAGNOSTIC CHART A086 — 1993
CAVALIER AND SUNBIRD

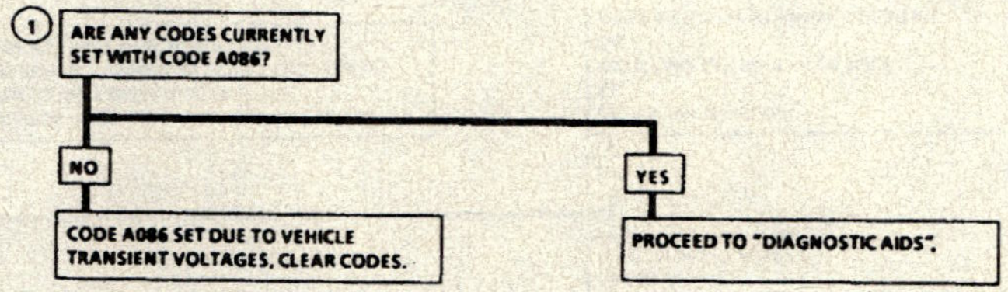

DIAGNOSTIC CHART A087 — 1993 CAVALIER AND SUNBIRD

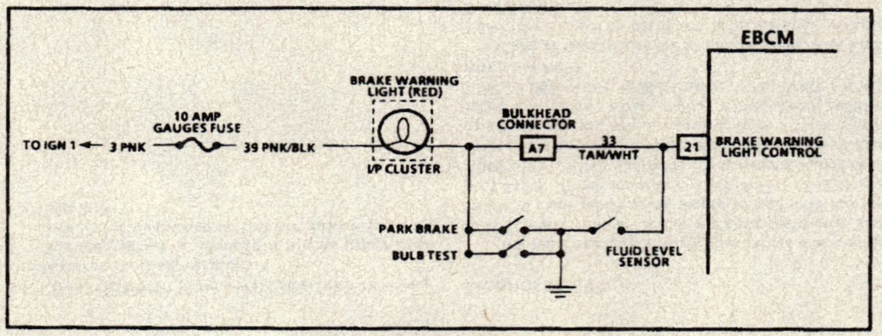

CODE A087

RED BRAKE WARNING LIGHT CIRCUIT OPEN

Circuit Description:

This code is used to verify the EBCM has continuity to the brake warning light in case the EBCM must turn it "ON." This will only occur if an ABS fault is detected that may degrade base brake operation.

Because ABS is not the only device controlling the brake warning light (parking brake and low fluid sensor may also turn "ON" the light) a short to ground in this circuit cannot be detected.

Test Description: Number(s) below refer to circled number(s) on the diagnostic chart.

1. This step identifies if the fault is currently present.
2. This step indicates if the EBCM and circuitry has the ability to complete the ground to the brake warning light and illuminate it.
3. Because the brake warning light has current supplied to it through the 10 amp gauge fuse, this step would indicate if the ignition circuit is complete to the I/P cluster.
4. This step isolates if the open circuit is due to a EBCM failure or open circuitry to the brake warning light.

Diagnostic Aids:

An "Intermittent" problem may be caused by a poor connection, rubbed through wire insulation, or a wire that is broken inside the insulation.

The frequency of the problem can be checked by using the enhanced diagnostic function of the Tech 1 "Scan" tool,

If a Tech 1 or T-100 are not available the parking brake handle may be lifted to verify proper light operation and continuity of CKTs 3 and 39.

Any circuitry, that is suspected as causing the intermittent complaint, should be thoroughly checked for backed out terminals, improper mating, broken locks, improperly formed or damaged terminals, poor terminal to wiring connections or physical damage to the wiring harness.

DIAGNOSTIC CHART A087 — 1993 CAVALIER AND SUNBIRD

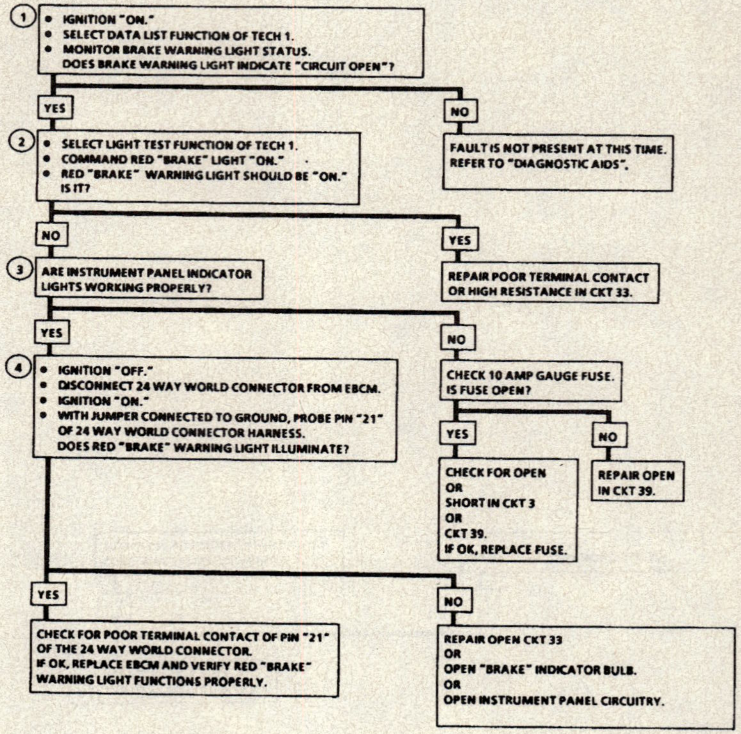

THIS CHART ASSUMES THAT A CURRENT CODE IS STORED INDICATING THAT THIS FAULT IS PRESENT.

DIAGNOSTIC CHART A088 — 1993 CAVALIER AND SUNBIRD

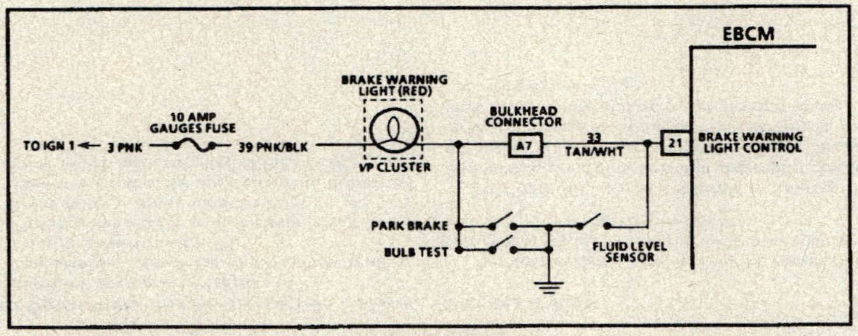

CODE A088

RED BRAKE WARNING LIGHT CIRCUIT SHORTED TO BATTERY

Circuit Description:

This code identifies a short to battery between the EBCM and the RED brake warning light, or an open driver that does not allow the RED brake warning light to be illuminated by the EBCM. This will only occur if an ABS fault is detected that may degrade base brake operation.

Test Description: Number(s) below refer to circled number(s) on the diagnostic chart.

1. This step identifies if the ground circuit to the Red brake warning light is being completed by a source other than the EBCM.
2. This step identifies if the fault is currently present.
3. By removing the 10 amp gauge fuse, the voltage source is eliminated. This test indicates if voltage is being supplied from a source other than the gauge fuse.

Diagnostic Aids:

An "Intermittent" problem may be caused by a poor connection, rubbed through wire insulation, or a wire that is broken inside the insulation.

Any circuitry, that is suspected as causing the intermittent complaint, should be thoroughly checked for backed out terminals, improper mating, brokep locks, improperly formed or damaged terminals, po terminal to wiring connections or physical damage to the wiring harness.

DIAGNOSTIC CHART A088 — 1993 CAVALIER AND SUNBIRD

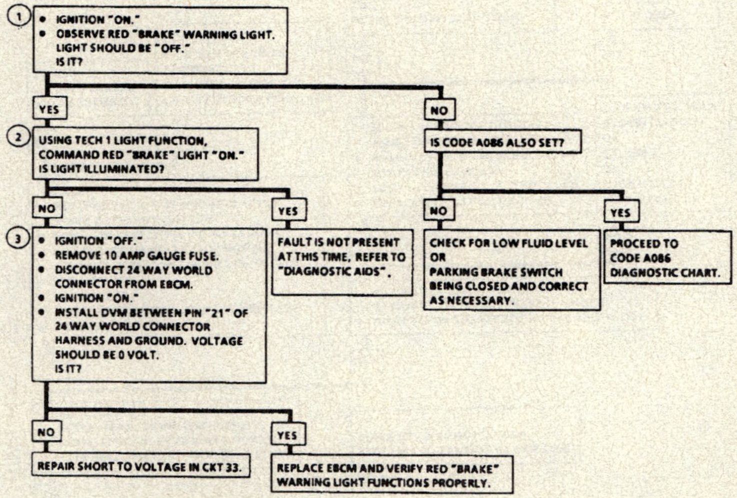

THIS CHART ASSUMES THAT A CURRENT CODE IS STORED INDICATING THAT THIS FAULT IS PRESENT.

DIAGNOSTIC CHART A091 — 1993 CAVALIER AND SUNBIRD

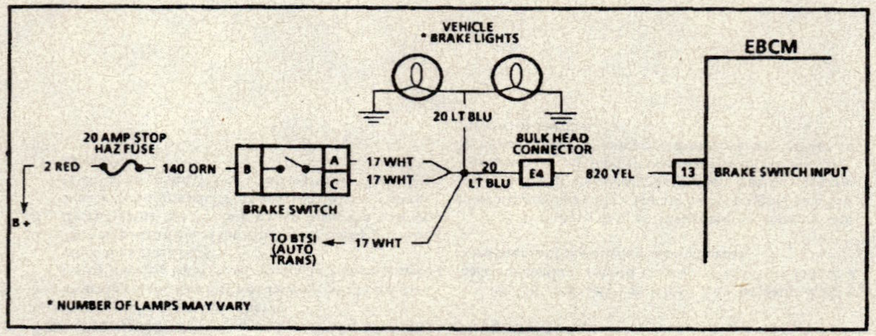

CODE A091

OPEN BRAKE SWITCH CONTACTS DURING DECELERATION

Circuit Description:

This code is used to detect an open brake switch in the non-ABS mode. The EBCM looks for decel rates that would indicate braking action and verifies this assumption by requiring several repeats of this detection method. In each case, ABS will not be available since no brake switch is seen.

Test Description: Number(s) below refer to circled number(s) on the diagnostic chart.

1. This step indicates if the brake switch signal is being received by the EBCM.
2. This step indicates if an open circuit exists in the brake switch or brake lamp circuitry.
3. This step isolates the open circuit to either the brake switch input circuitry, or the EBCM.

Diagnostic Aids:

An "Intermittent" problem may be caused by a poor connection, rubbed through wire insulation, or a wire that is broken inside the insulation.

Any circuitry, that is suspected as causing the intermittent complaint, should be thoroughly checked for backed out terminals, improper mating, broken locks, improperly formed or damaged terminals, poor terminal to wiring connections or physical damage to the wiring harness.

DIAGNOSTIC CHART A091 — 1993 CAVALIER AND SUNBIRD

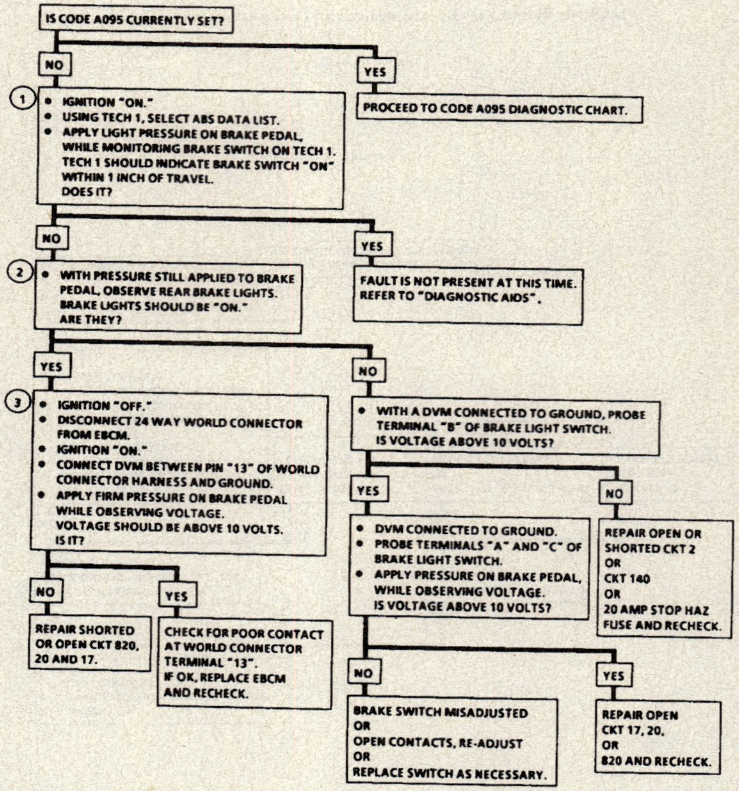

DIAGNOSTIC CHART A092 — 1993 CAVALIER AND SUNBIRD

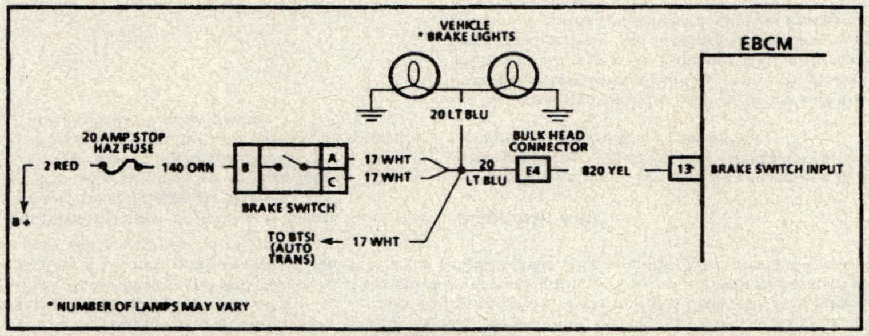

CODE A092

OPEN BRAKE SWITCH CONTACTS WHEN ABS WAS REQUIRED

Circuit Description:

This code is run to determine the proper operation of the brake switch. This is important because ABS is activated when the brake switch is "ON." If the brake switch is "OFF," ABS will never be activated. Since this fault is difficult to detect under normal braking conditions, this fault is only detected when ABS is required.

Test Description: Number(s) below refer to circled number(s) on the diagnostic chart.

1. This step indicates if the brake switch signal is being received by the EBCM.
2. This step indicates if an open circuit exists in the brake switch or brake lamp circuitry.
3. This step isolates the open circuit to either the brake switch input circuitry, or the EBCM.

Diagnostic Aids:

An "Intermittent" problem may be caused by a poor connection, rubbed through wire insulation, or a wire that is broken inside the insulation.

Any circuitry, that is suspected as causing the intermittent complaint, should be thoroughly checked for backed out terminals, improper mating, broken locks, improperly formed or damaged terminals, poor terminal to wiring connections or physical damage to the wiring harness.

DIAGNOSTIC CHART A092 — 1993 CAVALIER AND SUNBIRD

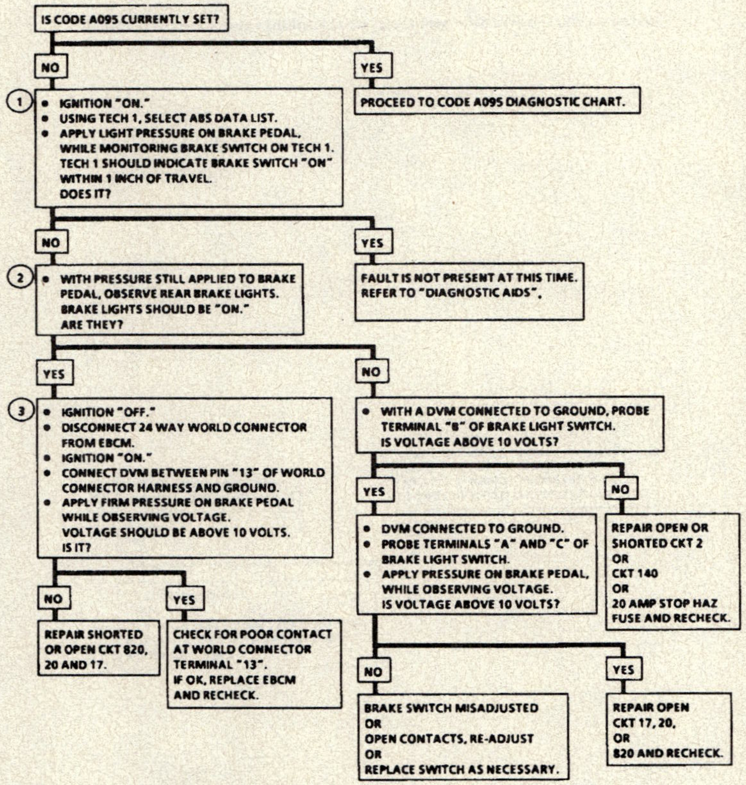

DIAGNOSTIC CHART A093 — 1993 CAVALIER AND SUNBIRD

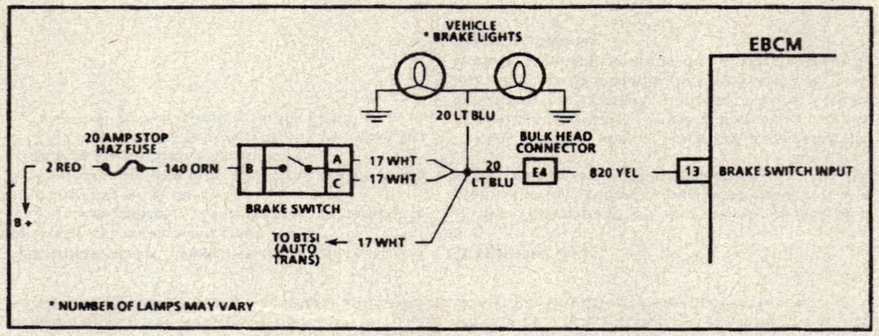

CODE A093

CODES A091 OR A092 SET IN CURRENT OR PREVIOUS IGNITION CYCLE

Circuit Description:

This code is the second portion of Codes A091 and A092. If Codes A091 or A092 failed during the last ignition cycle, Code A093 becomes a current failure during the next ignition cycle, keeping ABS disabled until a brake switch "ON" state is seen. When a change is seen during an ignition cycle in which Code A093 is a current failure, Code A091 or A092 will clear itself at the end of the current ignition cycle, and ABS will enable itself at the start of the next ignition cycle. Code A093 alone indicates Codes A091 or A092 failed previously, but is intermittent, or has been corrected.

Test Description: Number(s) below refer to circled number(s) on the diagnostic chart.

1. This step indicates which code (either A091 or A092) caused Code A093 to set.
2. This insures that the code that set is repaired so that Code A093 can be cleared.

Diagnostic Aids:

An "Intermittent" problem may be caused by a poor connection, rubbed through wire insulation, or a wire that is broken inside the insulation.

Any circuitry, that is suspected as causing the intermittent complaint, should be thoroughly checked for backed out terminals, improper mating, broken locks, improperly formed or damaged terminals, poor terminal to wiring connections or physical damage to the wiring harness.

Also, verify proper brake switch operation using the data list of the "Scan" tool. As the brake is applied, the data list should display the brake switch "ON" within 1 inch of travel.

DIAGNOSTIC CHART A093 — 1993 CAVALIER AND SUNBIRD

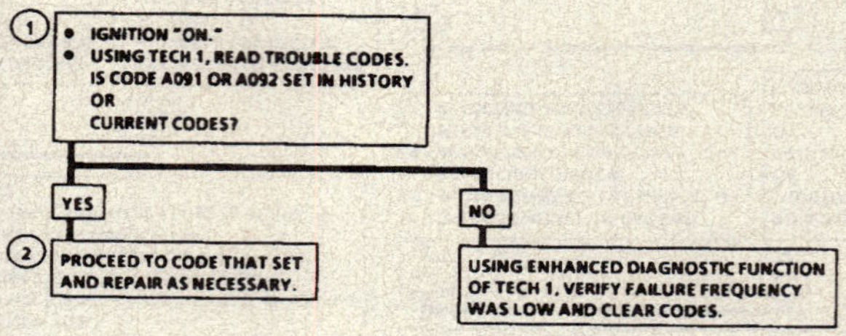

THIS CHART ASSUMES THAT A CURRENT CODE IS STORED INDICATING THAT THIS FAULT IS PRESENT.

DIAGNOSTIC CHART A094 — 1993
CAVALIER AND SUNBIRD

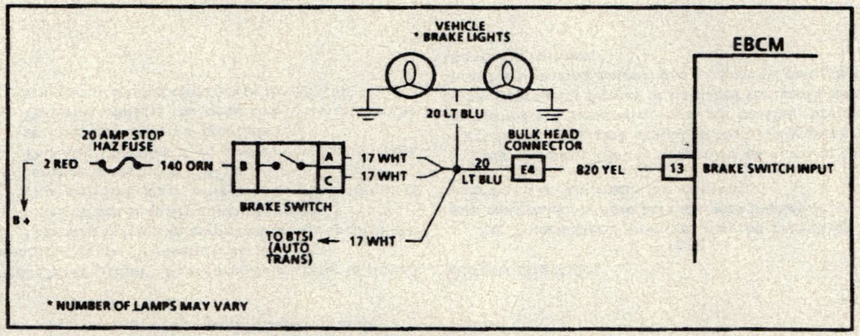

CODE A094

BRAKE SWITCH CONTACTS ALWAYS CLOSED

Circuit Description:

This code is run to determine the proper operation of the brake switch. This is important because ABS is activated when the brake switch is "ON" and turned "OFF" when the brake is "OFF." If the brake switch is always "ON," ABS operation will always be requested resulting in potential modulator cycling on rough roads. Additionally, this failure will most likely result in a dead battery (due to the brake lights staying "ON") if the driver is not informed of this fault.

Test Description: Number(s) below refer to circled number(s) on the diagnostic chart.

1. This checks if the fault is currently present.
2. This step isolates the cause of the fault to either a faulty or misadjusted brake switch, or a short to voltage in the brake switch circuitry.

Diagnostic Aids:

An "Intermittent" problem may be caused by a poor connection, rubbed through wire insulation, or a wire that is broken inside the insulation.

Any circuitry, that is suspected as causing the intermittent complaint, should be thoroughly checked for backed out terminals, improper mating, broken locks, improperly formed or damaged terminals, poor terminal to wiring connections or physical damage to the wiring harness.

DIAGNOSTIC CHART A094 — 1993
CAVALIER AND SUNBIRD

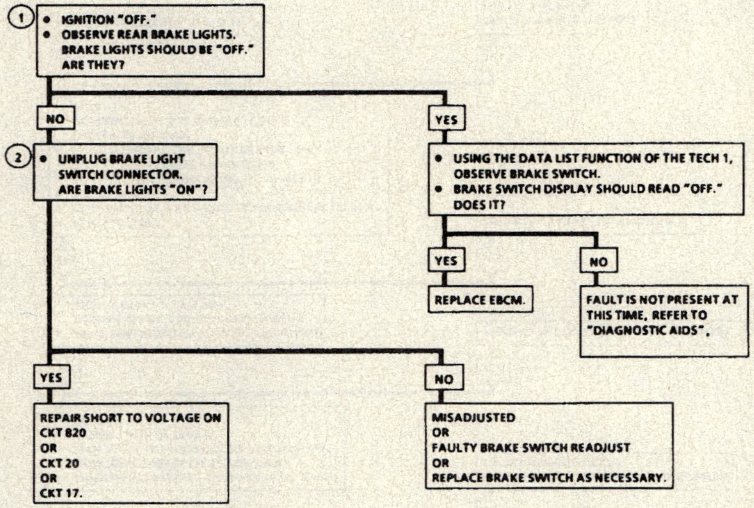

THIS CHART ASSUMES THAT A CURRENT CODE IS STORED INDICATING THAT THIS FAULT IS PRESENT.

DIAGNOSTIC CHART A095 — 1993 CAVALIER AND SUNBIRD

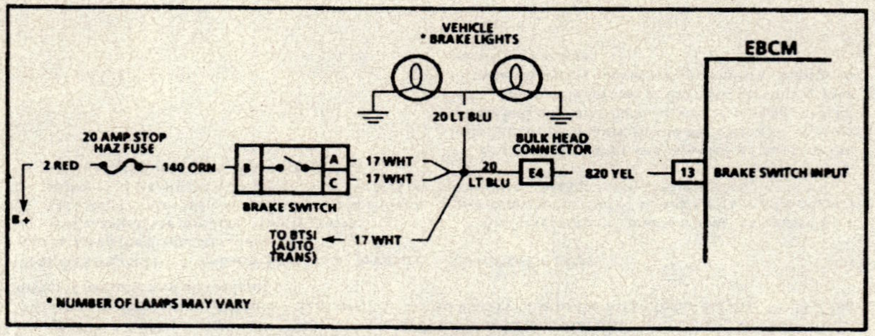

CODE A095

BRAKE SWITCH CIRCUIT OPEN

Circuit Description:

This code is used to identify open brake switch circuitry that prevents the brake switch input to the EBCM from changing states when the brake is applied. This code is used in conjunction with Codes A091 and A092 to determine the cause of an open brake switch fault.

Test Description: Number(s) below refer to circled number(s) on the diagnostic chart.

1. This step is used to confirm that an open in the brake switch circuitry currently exists.
2. This step indicates if the brake switch signal is being received by the EBCM.
3. This step indicates if an open circuit exists in the brake switch or brake light circuitry.
4. This step isolates the open circuit to either the brake switch input circuitry,or the EBCM.

Diagnostic Aids:

An "Intermittent" problem may be caused by a poor connection, rubbed through wire insulation, or a wire that is broken inside the insulation.

Any circuitry, that is suspected as causing the intermittent complaint, should be thoroughly checked for backed out terminals, improper mating, broken locks, improperly formed or damaged terminals, poor terminal to wiring connections or physical damage to the wiring harness.

DIAGNOSTIC CHART A095 — 1993
CAVALIER AND SUNBIRD

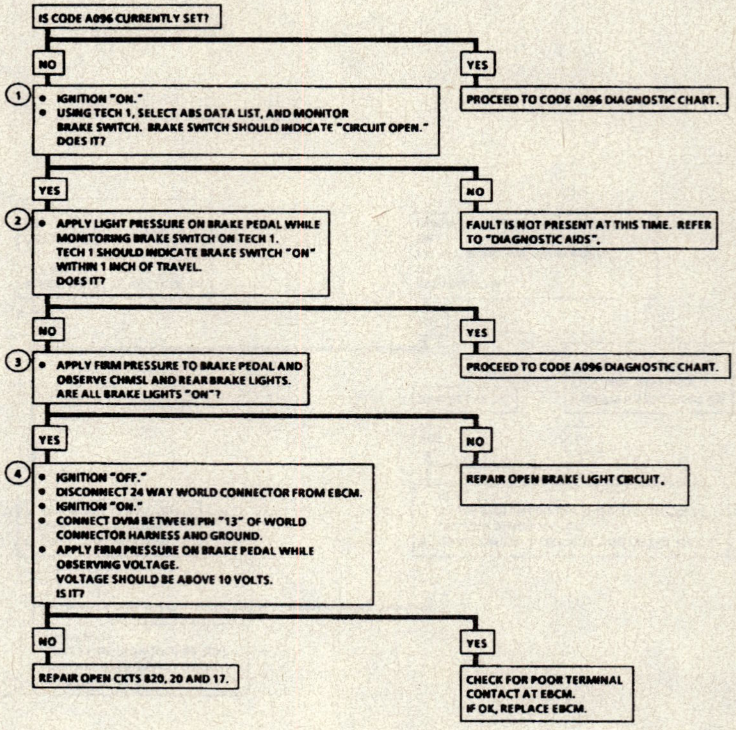

THIS CHART ASSUMES THAT A CURRENT CODE IS STORED INDICATING THAT THIS FAULT IS PRESENT.

DIAGNOSTIC CHART A096 — 1993 CAVALIER AND SUNBIRD

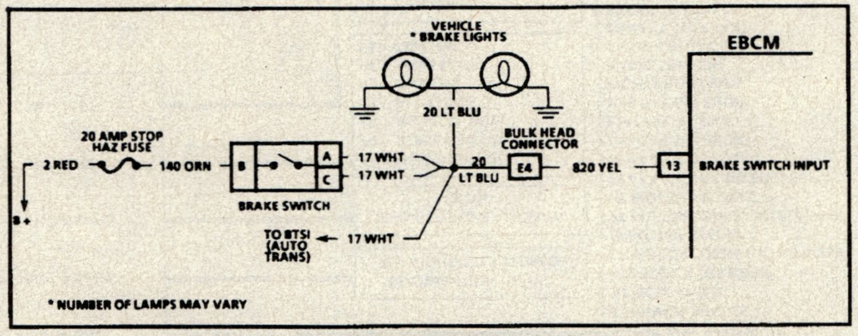

CODE A096

BRAKE LIGHTS CIRCUIT OPEN

Circuit Description:

This code is designed to isolate the cause of a Code A095 failure and indicate to the driver ABS is still available. If Code A095 fails with Code A096, the brake lights are open, the brake light grounds are open, or the Center High Mounted Stop Light (CHMSL) circuit is open during 4 way flasher use. The presence of battery voltage at the brake switch input indicates a valid brake switch input is still available.

Test Description: Number(s) below refer to circled number(s) on the diagnostic chart.

1. As a result of a failure of an additional brake switch circuit code, this code may be set. To insure proper diagnosis, any additional brake switch codes must be repaired first.
2. This step identifies if the fault is currently present in the brake circuit.

Diagnostic Aids:

An "Intermittent" problem may be caused by a poor connection, rubbed through wire insulation, or a wire that is broken inside the insulation.

Any circuitry, that is suspected as causing the intermittent complaint, should be thoroughly checked for backed out terminals, improper mating, broken locks, improperly formed or damaged terminals, poor terminal to wiring connections or physical damage to the wiring harness.

DIAGNOSTIC CHART A096 — 1993 CAVALIER AND SUNBIRD

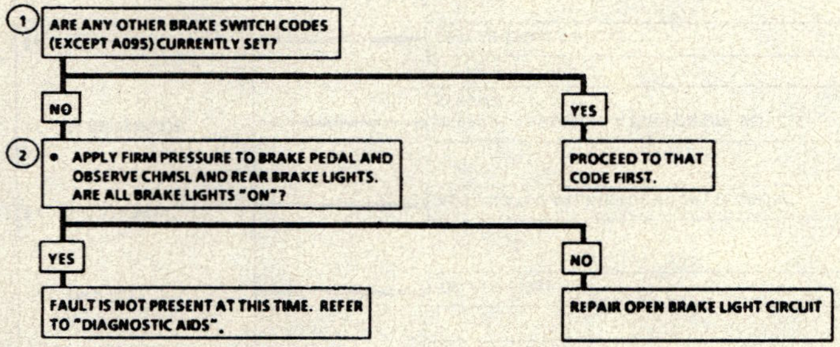

THIS CHART ASSUMES THAT A CURRENT CODE IS STORED INDICATING THAT THIS FAULT IS PRESENT.

DIAGNOSTIC PROCESS CHART — 1994 CAVALIER AND SUNBIRD

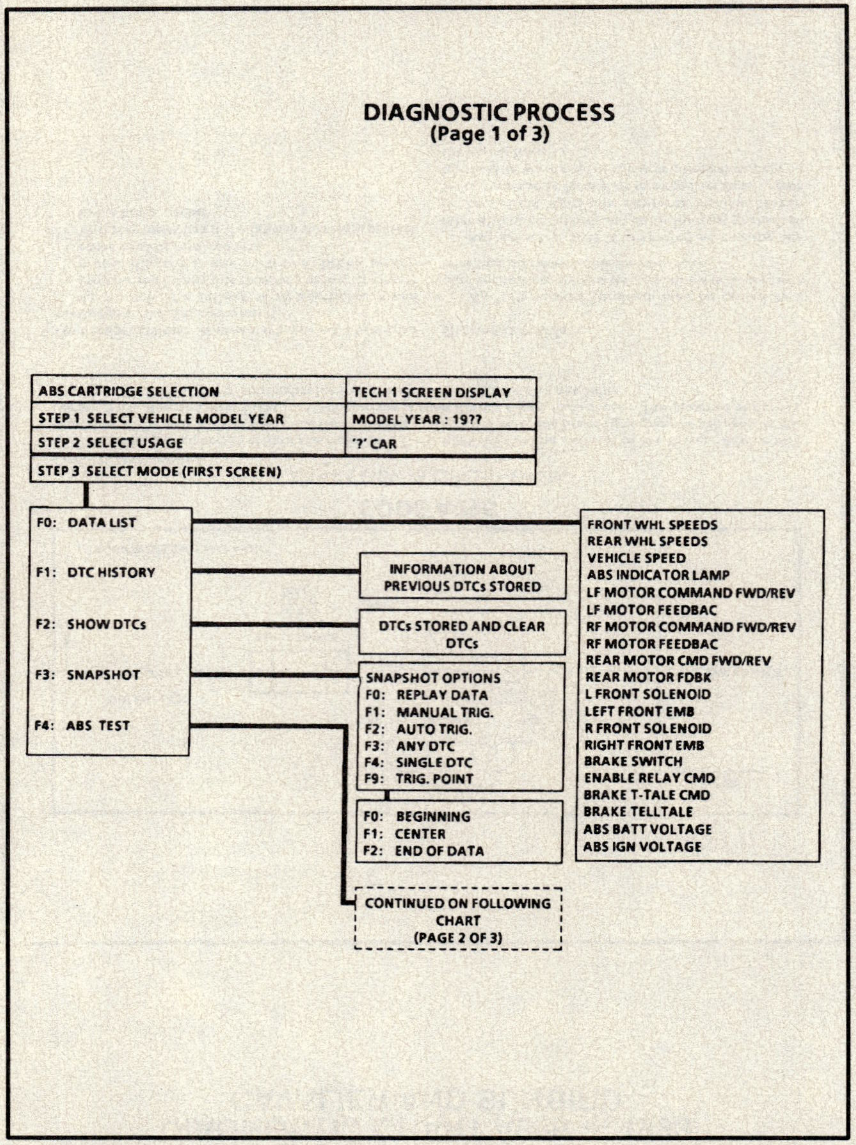

DIAGNOSTIC PROCESS CHART — 1994 CAVALIER AND SUNBIRD

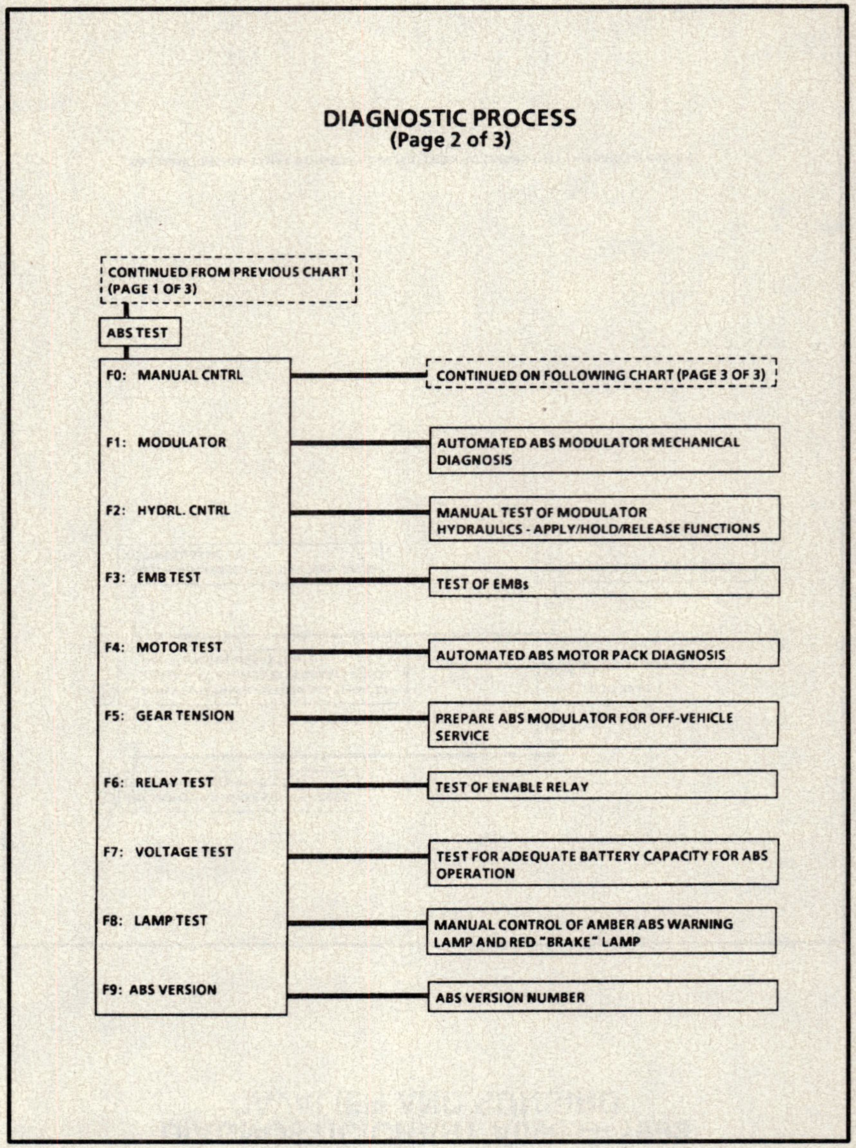

DIAGNOSTIC PROCESS CHART — 1994 CAVALIER AND SUNBIRD

DIAGNOSTIC PROCESS (PAGE 3 OF 3)

CONTINUED FROM PREVIOUS CHART (PAGE 2 OF 3)

MANUAL CONTROL RELAY, EMBS, SOLENOIDS AND MOTORS.

↑ = TURN ON
↓ = TURN OFF
ENTER = RELAY "ON/OFF"
YES TO CONTINUE

SELECT
F0: LF APPLY
F1: LF RELEASE
F2: LF SOLENOID
F3: LF EMB
F4: RF APPLY
F5: RF RELEASE
F6: RF SOLENOID
F7: RF EMB
F8: REAR APPLY
F9: REAR RELEASE

MOTOR APPLY
COMMAND: X AMPS
FEEDBACK: X AMPS
ENABLE RELAY: "ON/OFF"

MOTOR RELEASE
COMMAND: X AMPS
FEEDBACK: X AMPS
ENABLE RELAY: "ON/OFF"

SOLENOID
COMMAND: "ON/OFF"
FEEDBACK: "ON/OFF"
ENABLE RELAY: "ON/OFF"

EMB
COMMAND: HOLD/RELEASE
ENABLE RELAY: "ON/OFF"

DIAGNOSTIC CHART SYSTEM CHECK — 1994 CAVALIER AND SUNBIRD

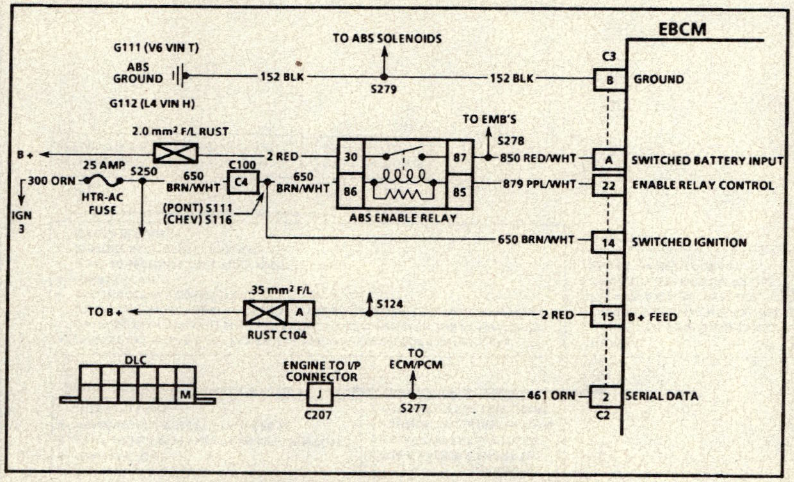

DIAGNOSTIC SYSTEM CHECK

System Description:

The diagnostic system check is an organized approach to identifying an Antilock Brake System (ABS) malfunction. It must be the starting point for any ABS complaint diagnosis because it directs the service technician to the next logical step in diagnosing the complaint.

Serial data is transmitted by the EBCM through terminal "2". The EBCM is supplied switched ignition voltage through terminal "14", battery voltage through terminal "15" and switched battery voltage through terminal "A". The EBCM ground is provided through terminal "B".

Chart Test Description: Number(s) below refer to circled number(s) on the diagnostic chart.

1. Indicates if EBCM is transmitting data.
2. Checks for a serial data link malfunction.
3. Checks for proper Tech 1 connection to DLC.
4. Checks for open heater-A/C fuse.
5. Checks for high resistance in the EBCM ground circuit.
6. Checks for current DTCs.
7. Checks for proper ABS warning lamp operation.
8. Checks for history DTCs.
9. Checks for possible short to ground in the switched ignition circuitry.
10. Isolates the short to ground to either the circuitry or the EBCM.
11. Ensures short to ground is not due to physical damage of the circuitry.
12. Checks for proper battery voltage at the EBCM.
13. Checks for high resistance in the B + feed circuitry.
14. Checks for high resistance in the switched ignition circuitry.
15. Ensures malfunction is not due to poor terminal contact.
16. Checks for high resistance in the switched ignition circuit between the fuse block and the EBCM.

Diagnostic Aids: Excessive resistance in the ground or power supply circuits will not allow communication with the EBCM. If communication with the EBCM is not possible, ensure the ABS ground connection is good and that there is no excessive resistance in any of the power supply circuits.

DIAGNOSTIC CHART SYSTEM CHECK — 1994
CAVALIER AND SUNBIRD

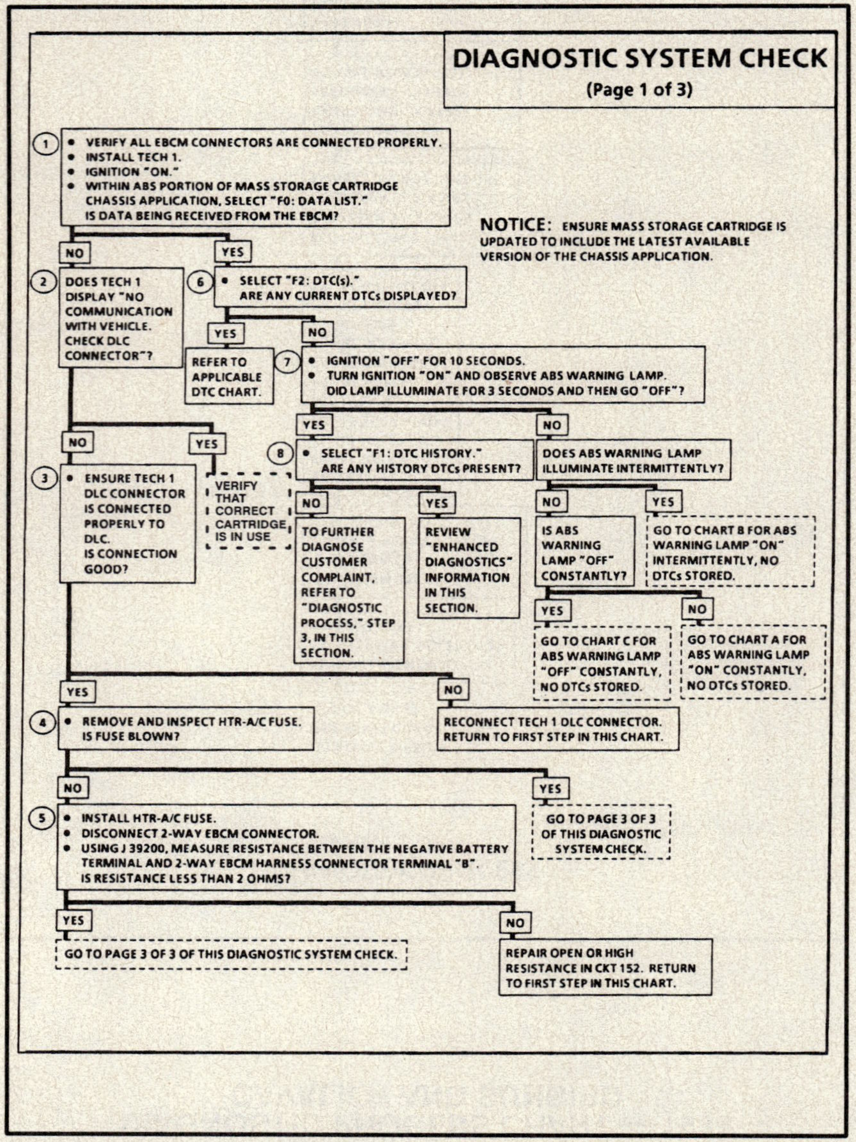

DIAGNOSTIC CHART SYSTEM CHECK — 1994
CAVALIER AND SUNBIRD

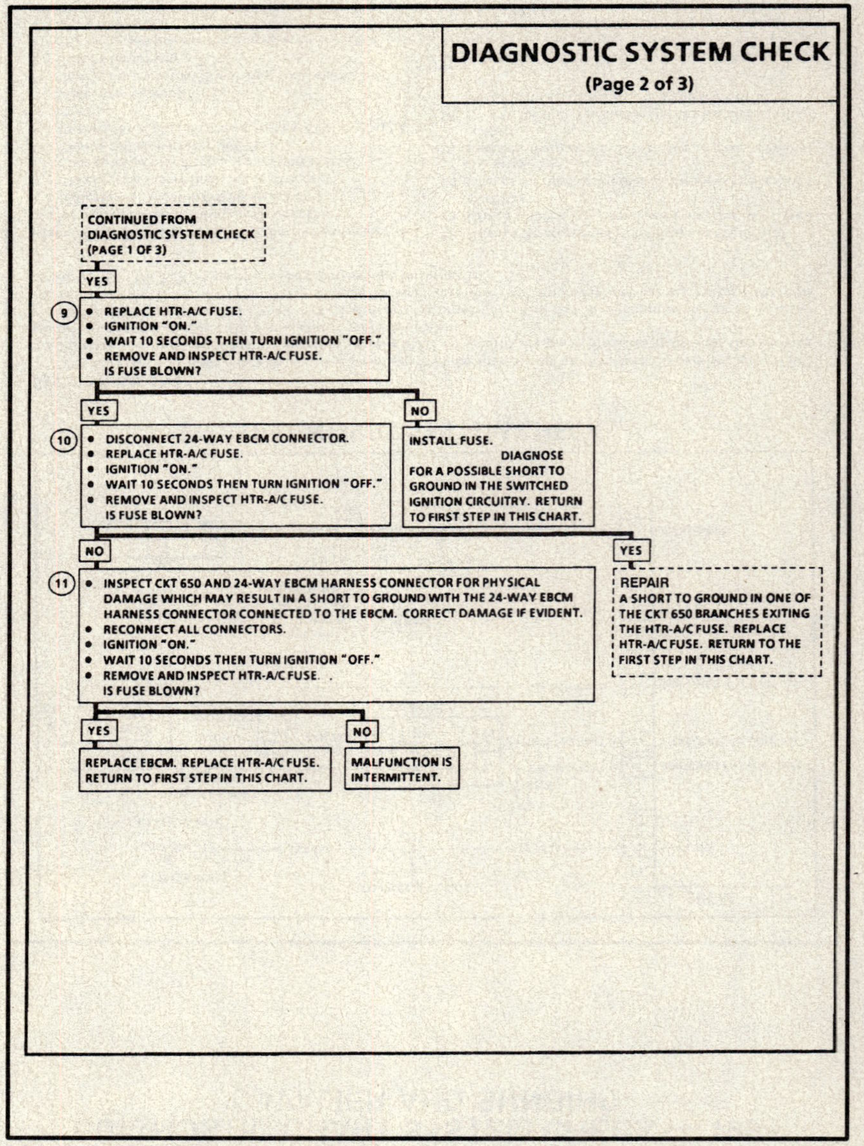

DIAGNOSTIC CHART SYSTEM CHECK — 1994 CAVALIER AND SUNBIRD

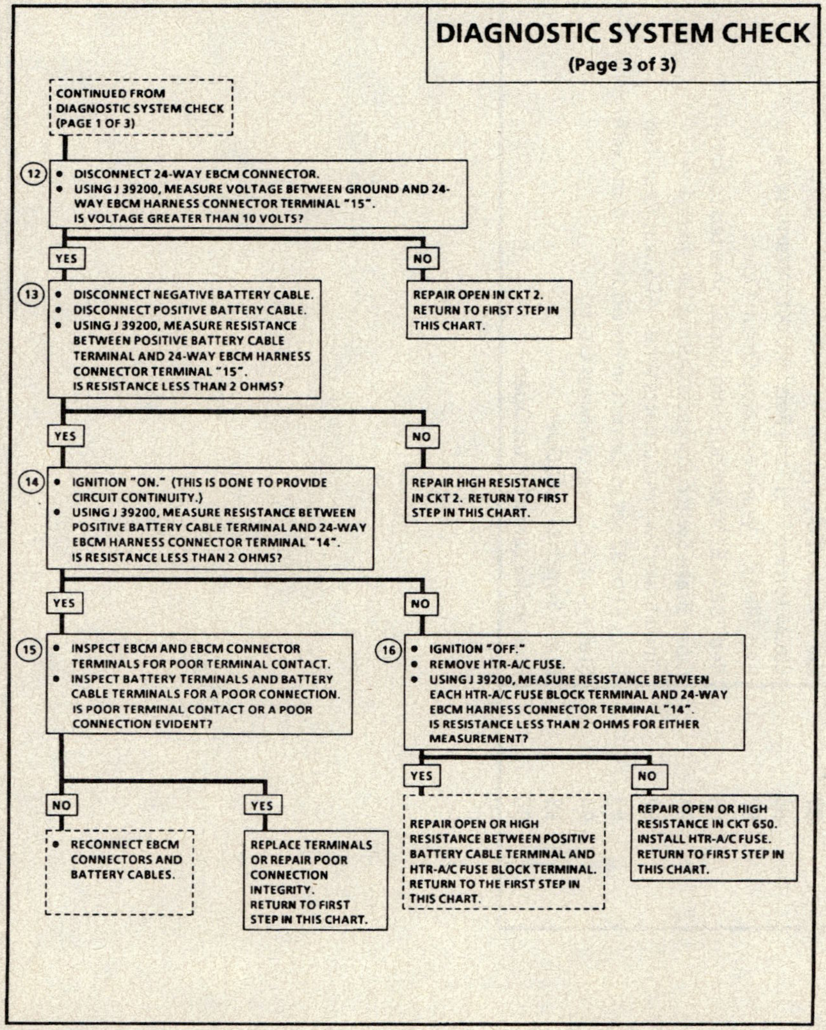

DIAGNOSTIC TROUBLE CODE CHART — 1994 CAVALIER AND SUNBIRD

DIAGNOSTIC TROUBLE CODE AND SYMPTOM TABLE

CHART	SYMPTOM
A	ABS (Amber) Warning Lamp "ON" Constantly, No DTCs Stored
B	ABS (Amber) Warning Lamp "ON" Intermittently, No DTCs Stored
C	ABS (Amber) Warning Lamp "OFF" Constantly, No DTCs Stored
D	Tech 1 Displays Undefined DTCs

DIAGNOSTIC TROUBLE CODE	DESCRIPTION
11	ABS Warning Lamp Circuit Malfunction (Pontiac Only)
11	ABS Warning Lamp Circuit Malfunction (Chevrolet Only)
13	ABS Warning Lamp Circuit Shorted to Battery (Pontiac Only)
13	ABS Warning Lamp Circuit Shorted to Battery (Chevrolet Only)
14	ABS Enable Relay Contact Circuit Open (1 of 3)
15	ABS Enable Relay Contact Circuit Shorted to Battery or Always Closed
16	ABS Enable Relay Coil Circuit Open
17	ABS Enable Relay Coil Circuit Shorted to Ground
18	ABS Enable Relay Coil Circuit Shorted to Battery or Coil Shorted
21	Left Front Wheel Speed = 0 or Unreasonable (1 of 2)
22	Right Front Wheel Speed = 0 or Unreasonable (1 of 2)
23	Left Rear Wheel Speed = 0 or Unreasonable (1 of 4)
24	Right Rear Wheel Speed = 0 or Unreasonable (1 of 4)
25	Left Front Excessive Wheel Speed Variation (1 of 2)
26	Right Front Excessive Wheel Speed Variation (1 of 2)
27	Left Rear Excessive Wheel Speed Variation (1 of 4)
28	Right Rear Excessive Wheel Speed Variation (1 of 4)
36	Low System Voltage
37	High System Voltage
38	Left Front EMB Will Not Hold Motor
41	Right Front EMB Will Not Hold Motor
42	Rear ESB Will Not Hold Motor
44	Left Front ABS Channel Will Not Move (1 of 2)
45	Right Front ABS Channel Will Not Move (1 of 2)
46	Rear ABS Channel Will Not Move
47	Left Front ABS Motor Free Spins (1 of 2)
48	Right Front ABS Motor Free Spins (1 of 2)
51	Rear ABS Motor Free Spins (1 of 2)
52	Left Front ABS Channel in Release Too Long
53	Right Front ABS Channel In Release Too Long
54	Rear ABS Channel in Release Too Long
55	EBCM Malfunction
56	Left Front ABS Motor Circuit Open
57	Left Front ABS Motor Circuit Shorted to Ground
58	Left Front ABS Motor Circuit Shorted to Battery or Motor Shorted

DIAGNOSTIC TROUBLE CODE CHART — 1994
CAVALIER AND SUNBIRD

DIAGNOSTIC TROUBLE CODE AND SYMPTOM TABLE

DIAGNOSTIC TROUBLE CODE	DESCRIPTION
61	Right Front ABS Motor Circuit Open
62	Right Front ABS Motor Circuit Shorted to Ground
63	Right Front ABS Motor Circuit Shorted to Battery or Motor Shorted
64	Rear ABS Motor Circuit Open
65	Rear ABS Motor Circuit Shorted to Ground
66	Rear ABS Motor Circuit Shorted to Battery or Motor Shorted
67	Left Front EMB Circuit Open or Shorted to Ground
68	Left Front EMB Circuit Shorted to Battery or Driver Open
71	Right Front EMB Circuit Open or Shorted to Ground
72	Right Front EMB Circuit Shorted to Battery or Driver Open
76	Left Front Solenoid Circuit Open or Shorted to Battery
77	Left Front Solenoid Circuit Shorted to Ground or Driver Open
78	Right Front Solenoid Circuit Open or Shorted to Battery
81	Right Front Solenoid Circuit Shorted to Ground or Driver Open
82	Calibration Malfunction
86	EBCM Turned "ON" the Red "BRAKE" Warning Lamp
87	Red "BRAKE" Warning Lamp Circuit Open
88	Red "BRAKE" Warning Lamp Circuit Shorted to Battery
91	Open Brake Switch Contacts During Deceleration
92	Open Brake Switch Contacts When ABS was Required
93	DTCs 91 or 92 Set in Current or Previous Ignition Cycle
94	Brake Switch Contacts Always Closed
95	Brake Switch Circuit Open
96	Rear Brake Lamp Circuit Open

ELECTRICAL CONNECTOR TERMINAL IDENTIFICATION — 1994 CAVALIER AND SUNBIRD

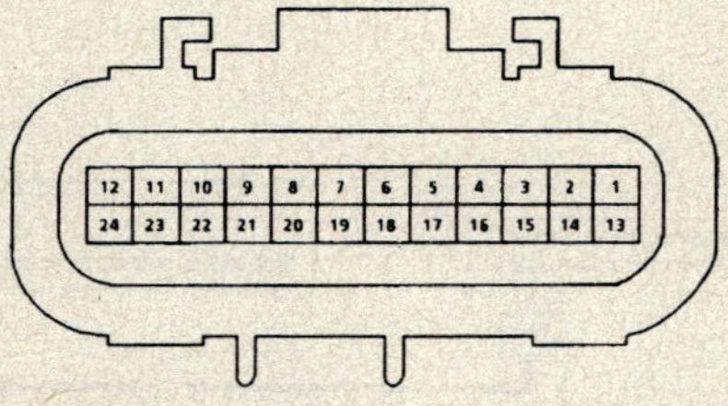

24-WAY EBCM HARNESS CONNECTOR

PIN	CIRCUIT NO.	COLOR	CIRCUIT
1		TAN/WHT	NOT USED
2	461	ORN	SERIAL DATA
3	OPEN		NOT USED
4	1289	LT BLU/BLK	R/F ABS SOLENOID CONTROL
5	830	LT BLU	L/F WHEEL SIGNAL HIGH
6	873	YEL	L/F WHEEL SIGNAL LOW
7	882	BRN	R/R WHEEL SIGNAL HIGH
8	883	WHT	R/R WHEEL SIGNAL LOW
9	872	DK GRN	R/F WHEEL SIGNAL HIGH
10	833	TAN	R/F WHEEL SIGNAL LOW
11	885	RED	L/R WHEEL SIGNAL LOW
12	884	BLK	L/R WHEEL SIGNAL HIGH
13	820	YEL	BRAKE SWITCH INPUT
14	650	BRN/WHT	SWITCH IGNITION
15	2	RED	B + FEED
16	OPEN		NOT USED
17	OPEN		NOT USED
18	VENT TUBE	BLK	VENT TUBE
19	1286	LT GRN	L/F EMB CONTROL
20	1287	GRY	R/F EMB CONTROL
21	33	TAN/WHT	BRAKE WARNING LAMP CONTROL
22	879	PPL/WHT	ENABLE RELAY CONTROL
23	852	WHT	ABS INDICATOR LAMP CONTROL
24	1288	DK GRN/YEL	L/F ABS SOLENOID CONTROL

ELECTRICAL CONNECTOR TERMINAL IDENTIFICATION — 1994 CAVALIER AND SUNBIRD

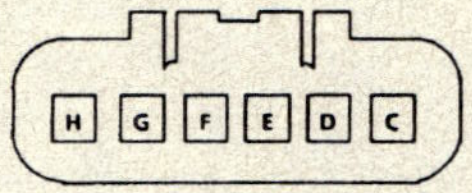

6-WAY EBCM HARNESS CONNECTOR

PIN	CIRCUIT NO.	COLOR	CIRCUIT
C	1284	DK GRN/WHT	REAR MOTOR HIGH
D	1285	ORN/BLK	REAR MOTOR LOW
E	1281	PNK	L/F MOTOR LOW
F	1280	BLK/WHT	L/F MOTOR HIGH
G	1283	BLK/PNK	R/F MOTOR LOW
H	1282	PPL	R/F MOTOR HIGH

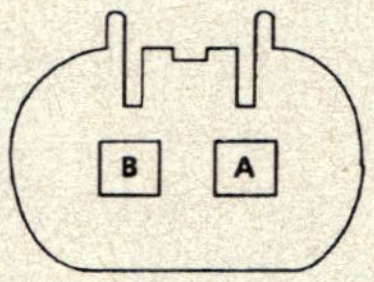

2-WAY EBCM HARNESS CONNECTOR

PIN	CIRCUIT NO.	COLOR	CIRCUIT
A	650	RED/WHT	SWITCHED BATTERY INPUT
B	152	BLK	GROUND

DIAGNOSTIC CHART A — 1994 CAVALIER AND SUNBIRD

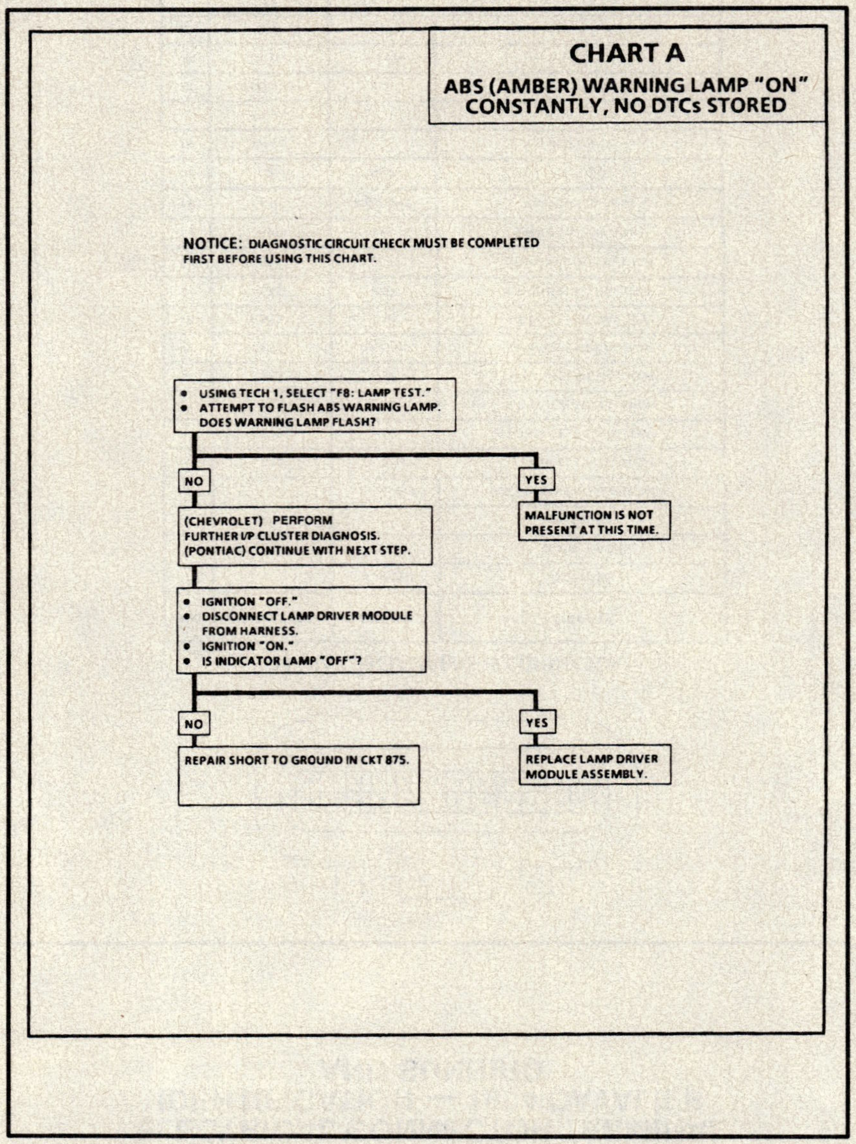

DIAGNOSTIC CHART B — 1994 CAVALIER AND SUNBIRD

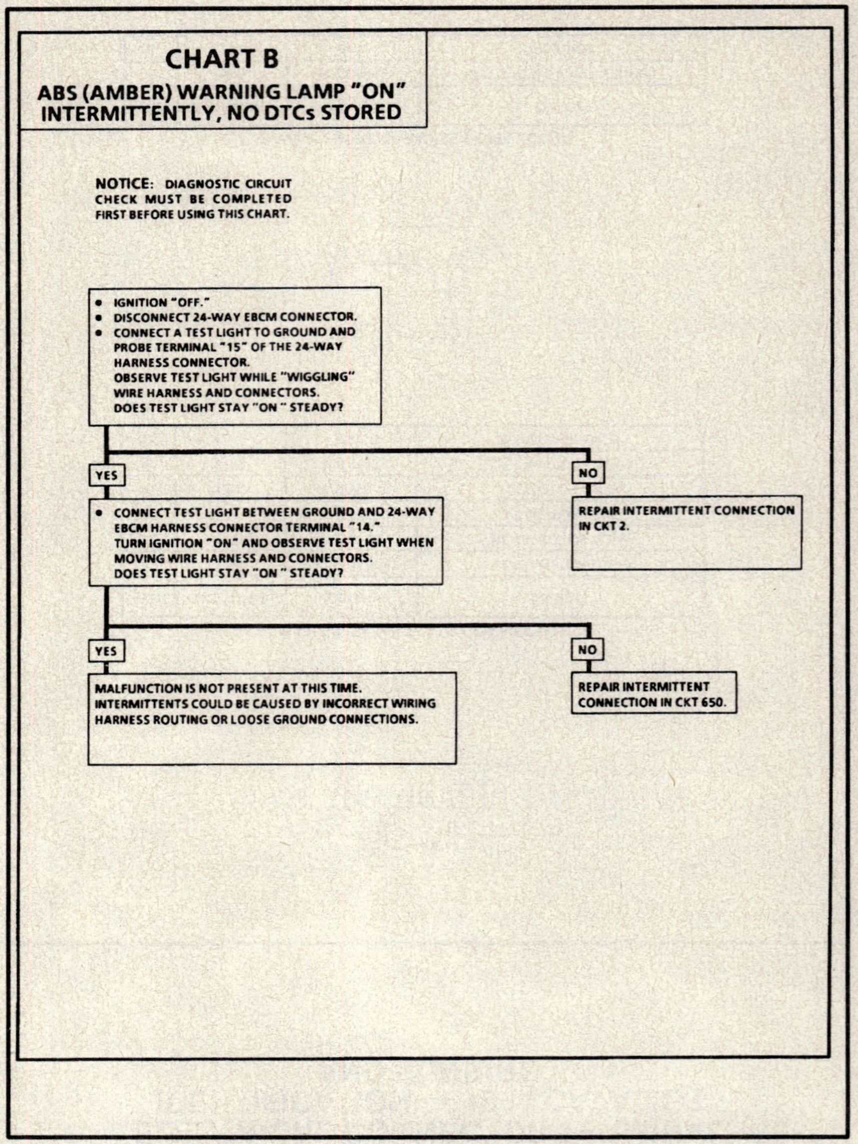

DIAGNOSTIC CHART C — 1994 CAVALIER AND SUNBIRD

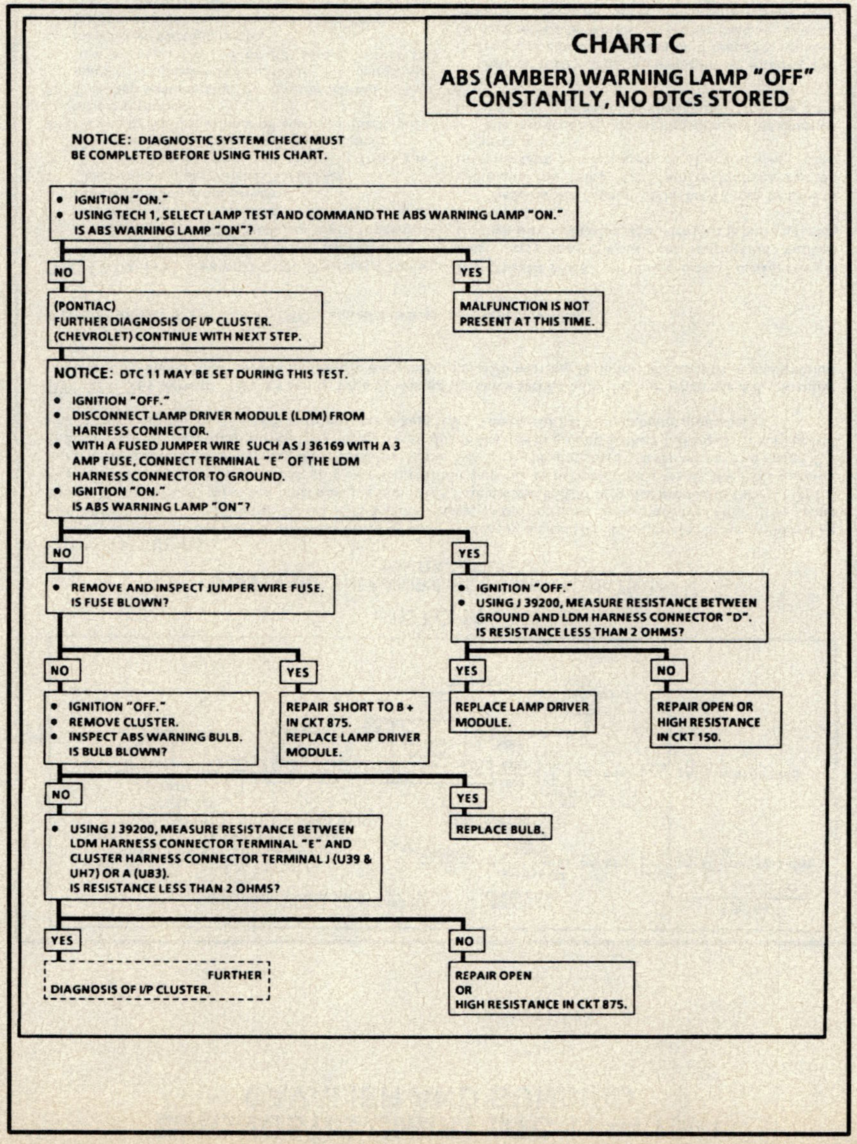

DIAGNOSTIC CHART D — 1994 CAVALIER AND SUNBIRD

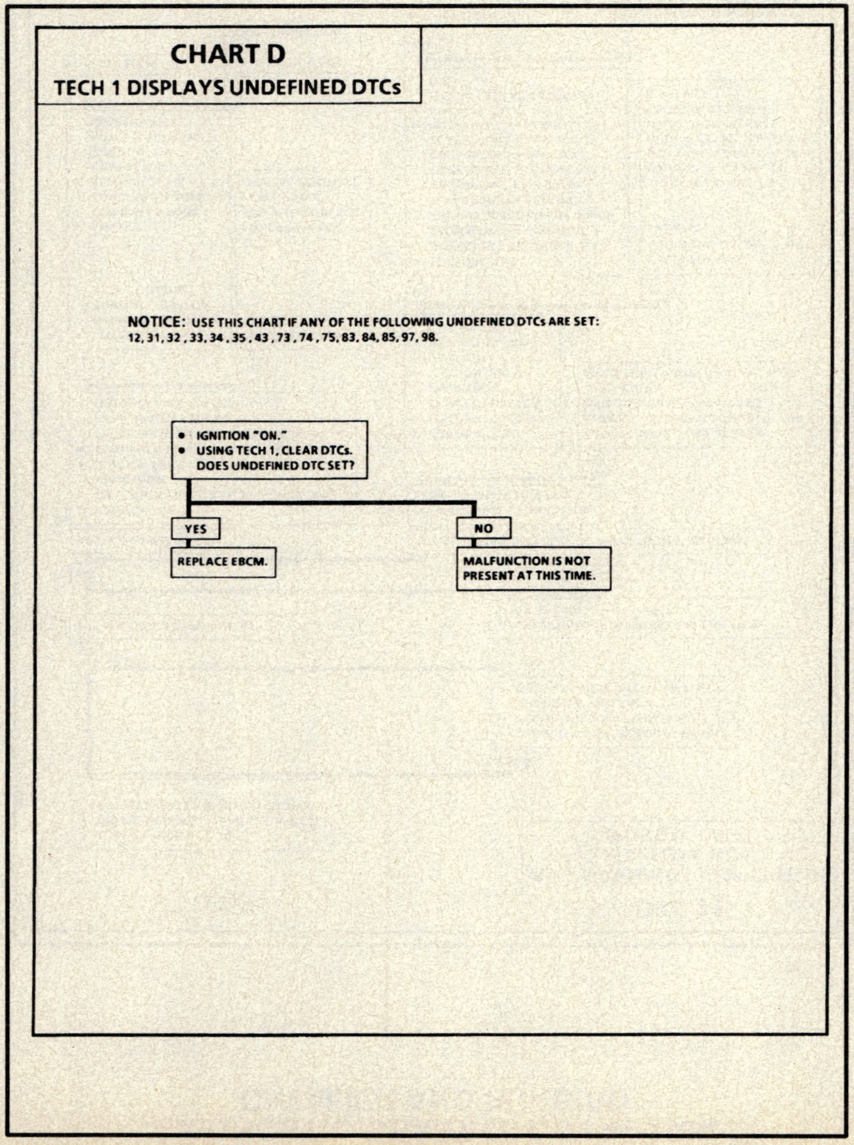

DIAGNOSTIC CHART DTC 11 — 1994 CAVALIER AND SUNBIRD

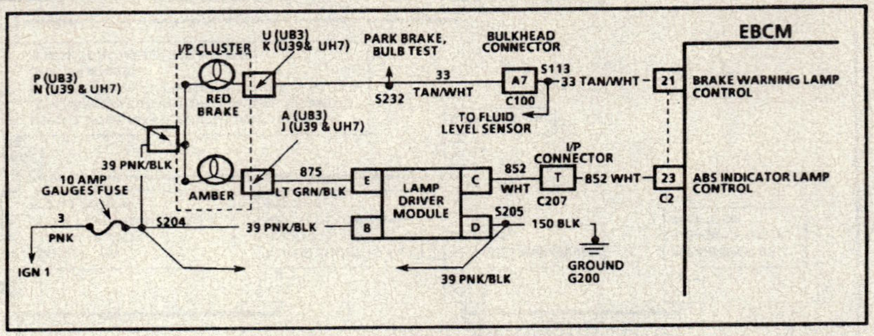

DTC 11

ABS WARNING LAMP CIRCUIT MALFUNCTION (PONTIAC ONLY)

Circuit Description:

This test checks the state of the ABS warning lamp to identify a situation in which the driver could not be warned of a system malfunction by the ABS warning lamp, or the warning lamp is always "ON." The Lamp Driver Module (LDM) turns the ABS warning lamp "ON," unless the EBCM provides ground to turn it "OFF." Because of the circuitry in the LDM, only external malfunctions can be detected. As a result, the ABS warning lamp itself is not diagnosable, only the control line to the EBCM can be diagnosed. In the event of an open CKT 852, the ABS warning lamp will be "ON" at all times, due to the loss of ground at the LDM input. If the control line is shorted to ground, the ABS warning lamp is kept "OFF," due to the LDM input being grounded.

DTC Will Set When: DTC 11 can be set only during the three second bulb check or when the ABS warning lamp is commanded "ON." If the EBCM can not control the ABS warning lamp for two seconds, a malfunction exists.

Action Taken: A malfunction DTC is stored; however, ABS is not disabled.

DTC Chart Test Description: Number(s) below refer to circled number(s) on the diagnostic chart.

1. Checks to see if ABS warning lamp functions properly. Normal operation would indicate that the malfunction is not present.
2. Checks to see if any other DTCs are set.
3. This step ensures that the circuitry from the EBCM to the lamp driver module is complete.
4. Checks to see if malfunction occurred due to poor terminal contact.
5. This step ensures that the ignition circuit to the lamp driver module is functional. An open gauge fuse or open CKT 3 or 39 would result in inoperative warning lamps.
6. Isolates the cause of the ABS lamp remaining "OFF," by disconnecting the 24-way EBCM connector, CKT 852 should be open and cause the ABS lamp to be "ON."
7. Ensures the malfunction is not due to a short in the ABS indicator lamp control circuit with EBCM connected.

Diagnostic Aids: An "Intermittent" malfunction is most likely caused by a poor connection, rubbed through wire insulation, or a wire that is broken inside the insulation.

The lamp test function of the Tech 1 may be used to command the lamp "ON" while looking for an intermittent malfunction in the warning lamp circuitry.

The frequency of the malfunction can be checked by using the enhanced diagnostic function of the Tech 1.

Any circuitry that is suspected as causing the intermittent complaint should be thoroughly checked for backed out terminals, improper mating, broken locks, improperly formed or damaged terminals, poor terminal to wiring connections or physical damage to the wiring harness.

DIAGNOSTIC CHART DTC 11 — 1994 CAVALIER AND SUNBIRD

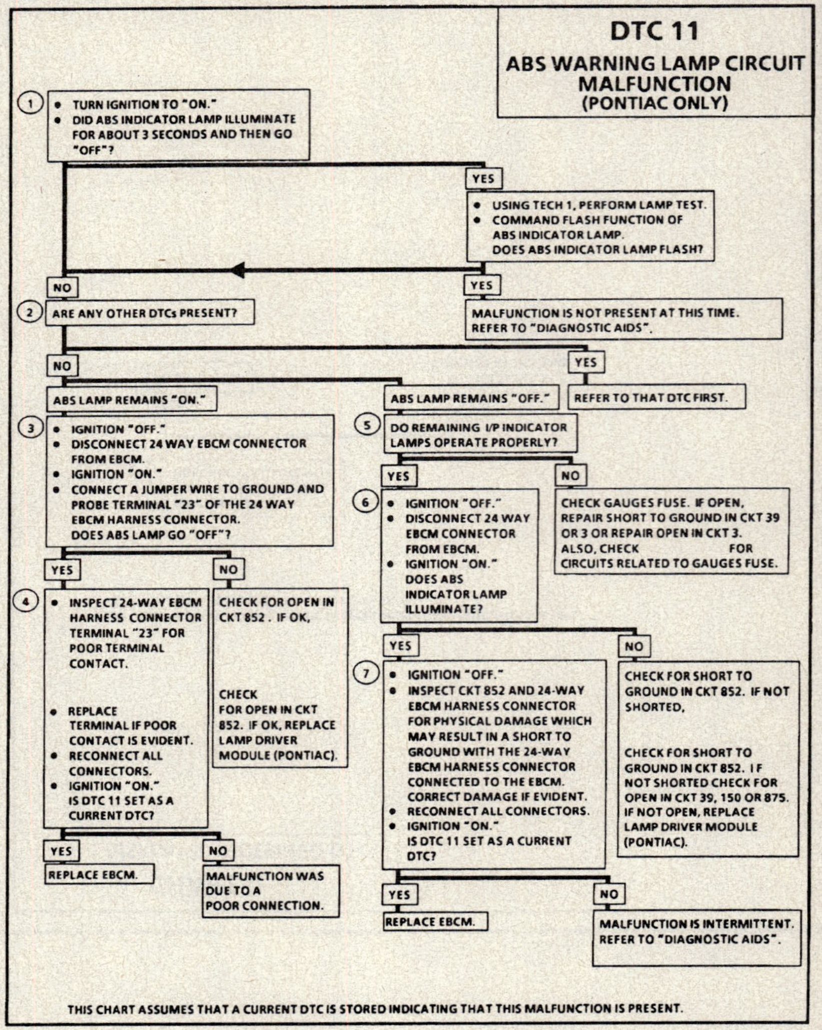

DIAGNOSTIC CHART DTC 11 — 1994 CAVALIER AND SUNBIRD

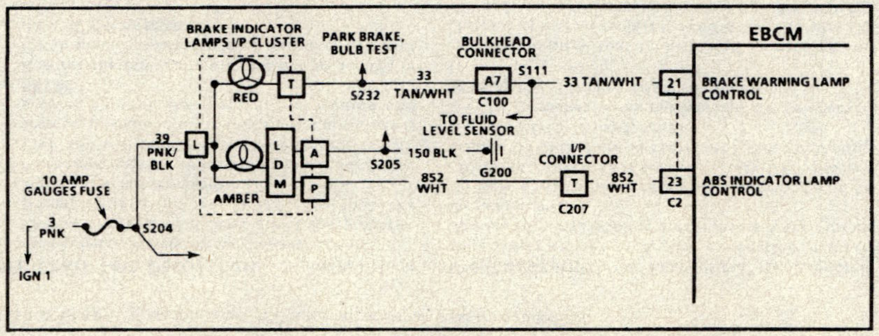

DTC 11

ABS WARNING LAMP CIRCUIT MALFUNCTION
(CHEVROLET ONLY)

Circuit Description:

This test checks the state of the ABS warning lamp to identify a situation in which the driver could not be warned of a system malfunction by the ABS warning lamp, or the warning lamp is always "ON." The Lamp Driver Module (LDM) turns the ABS warning lamp "ON," unless the EBCM provides ground to turn it "OFF." Because of the circuitry in the LDM, only external malfunctions can be detected. As a result, the ABS warning lamp itself is not diagnosable, only the control line to the EBCM can be diagnosed. In the event of an open CKT 852, the ABS warning lamp will be "ON" at all times, due to the loss of ground at the LDM input. If the control line is shorted to ground, the ABS warning lamp is kept "OFF," due to the LDM input being grounded.

DTC Will Set When: DTC 11 can be set only during the three second bulb check or when the ABS warning lamp is commanded "ON." If the EBCM can not control the ABS warning lamp for two seconds, a malfunction exists.

Action Taken: A malfunction DTC is stored; however, ABS is not disabled.

DTC Chart Test Description: Number(s) below refer to circled number(s) on the diagnostic chart.

1. Checks to see if ABS warning lamp functions properly. Normal operation would indicate that the malfunction is not present.
2. Checks to see if any other DTCs are set.
3. This step ensures that the circuitry from the EBCM to the lamp driver module is complete.
4. Checks to see if malfunction occurred due to poor terminal contact.
5. This step ensures that the ignition circuit to the lamp driver module is functional. An open gauge fuse or open CKT 3 or 39 would result in inoperative warning lamps.
6. Isolates the cause of the ABS lamp remaining "OFF," by disconnecting the 24-way EBCM connector, CKT 852 should be open and cause the ABS lamp to be "ON."
7. Ensures the malfunction is not due to a short in the ABS indicator lamp control circuit with EBCM connected.

Diagnostic Aids: An "Intermittent" malfunction is most likely caused by a poor connection, rubbed through wire insulation, or a wire that is broken inside the insulation.

The lamp test function of the Tech 1 may be used to command the lamp "ON" while looking for an intermittent malfunction in the warning lamp circuitry.

The frequency of the malfunction can be checked by using the enhanced diagnostic function of the Tech 1.

Any circuitry that is suspected as causing the intermittent complaint should be thoroughly checked for backed out terminals, improper mating, broken locks, improperly formed or damaged terminals, poor terminal to wiring connections or physical damage to the wiring harness.

DIAGNOSTIC CHART DTC 11 — 1994 CAVALIER AND SUNBIRD

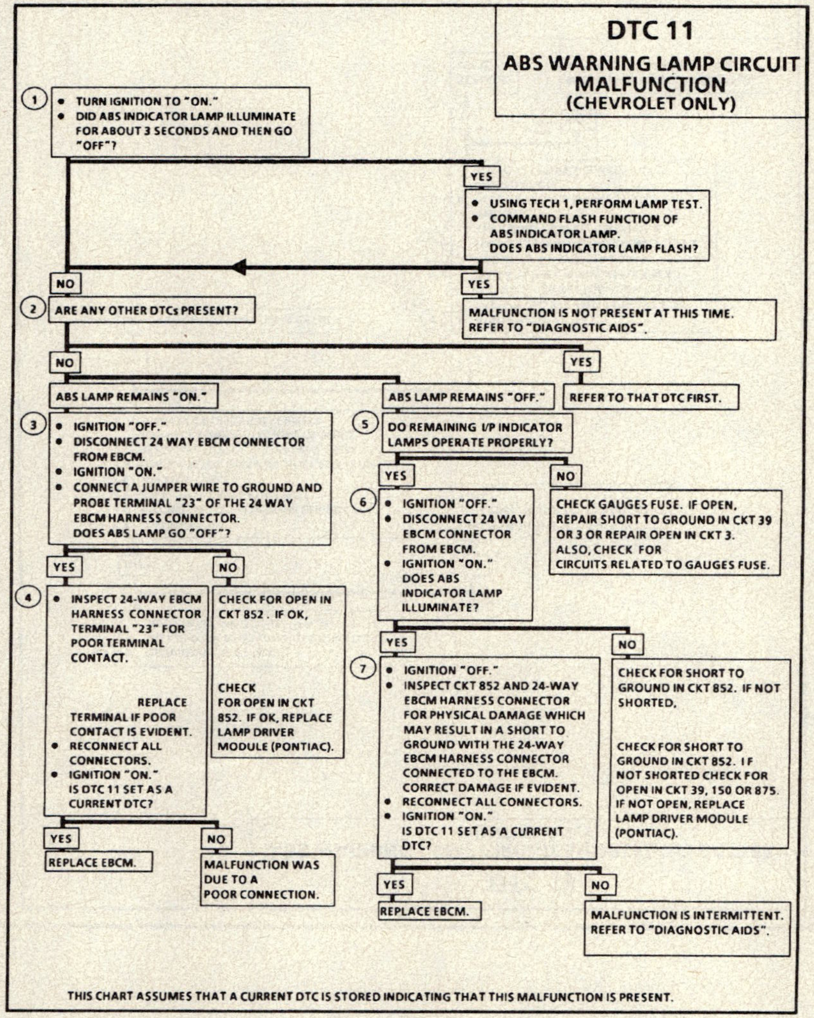

DIAGNOSTIC CHART DTC 13 — 1994 SUNBIRD

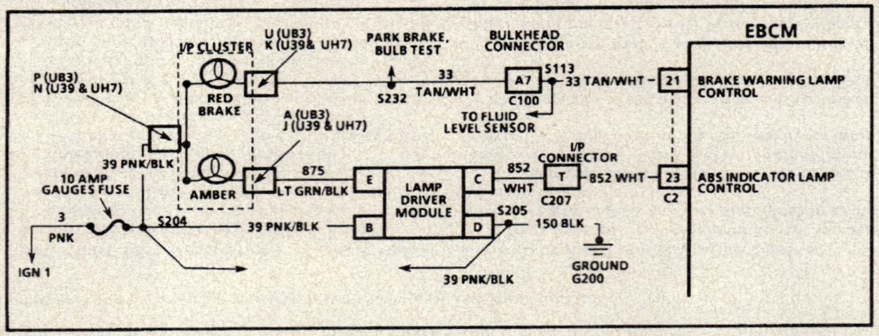

DTC 13

ABS WARNING LAMP CIRCUIT SHORTED TO BATTERY (PONTIAC ONLY)

Circuit Description:

This test checks the state of the ABS warning lamp to identify a situation in which the driver could not be warned of a system malfunction by the ABS warning lamp. The Lamp Driver Module (LDM) turns the ABS warning lamp "ON," unless the EBCM provides ground to turn it "OFF." Because of the circuitry internal to the LDM, only external malfunctions can be detected. As a result, the ABS warning lamp itself is not diagnosable, only the control line to the EBCM can be diagnosed.

If the ABS warning lamp control line is shorted to battery or the EBCM control circuit is open, the Electronic Brake Control Module (EBCM) is not able to turn "OFF" the ABS warning lamp.

DTC Will Set When: DTC 13 can be set anytime the EBCM commands the ABS warning lamp "OFF." If the EBCM can not control the ABS warning lamp for two seconds, a malfunction exists.

Action Taken: A malfunction DTC is stored; however, ABS is not disabled.

DTC Chart Test Description: Number(s) below refer to circled number(s) on the diagnostic chart.

1. Checks to see if ABS lamp functions properly. Normal operation would indicate that the malfunction is not present.
2. After removing the 10 amp gauge fuse, voltage on CKT 852 will be eliminated. If voltage is still present, it would indicate that a short to voltage is present between the lamp driver module and EBCM.
3. Ensures the malfunction is not due to a short in B+ in the ABS indicator lamp control circuit with the EBCM connected.

Diagnostic Aids: An "Intermittent" malfunction is most likely caused by a poor connection, rubbed through wire insulation or a wire that is broken inside the insulation.

DTC 13 must be currently set before it should be addressed. If DTC 13 is set as a history DTC, proceed to those DTCs that are currently set before attempting to repair a DTC 13 malfunction.

The frequency of the malfunction can be checked by using the enhanced diagnostic function of the Tech 1.

Any circuitry that is suspected as causing the "Intermittent" complaint should be thoroughly checked for backed out terminals, improper mating, broken locks, improperly formed or damaged terminals, poor terminal to wiring connections or physical damage to the wiring harness.

DIAGNOSTIC CHART DTC 13 — 1994 SUNBIRD

DTC 13

ABS WARNING LAMP CIRCUIT SHORTED TO BATTERY (PONTIAC ONLY)

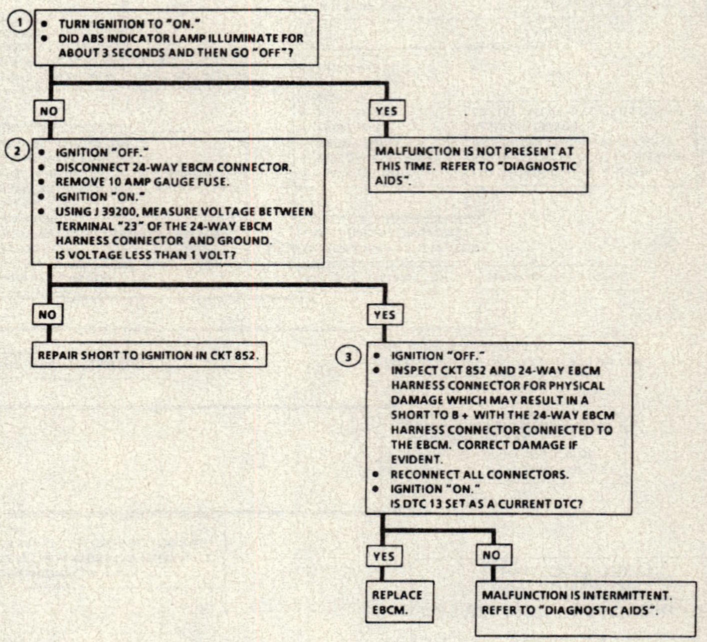

THIS CHART ASSUMES THAT A CURRENT DTC IS STORED INDICATING THAT THIS MALFUNCTION IS PRESENT.

DIAGNOSTIC CHART DTC 13 — 1994
CAVALIER

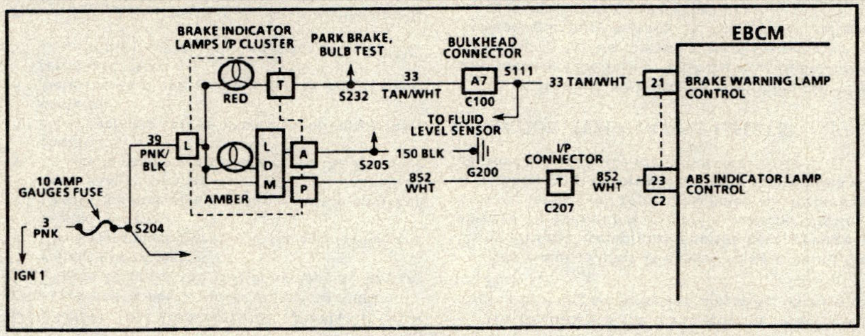

DTC 13

ABS WARNING LAMP CIRCUIT SHORTED TO BATTERY
(CHEVROLET ONLY)

Circuit Description:

This test checks the state of the ABS warning lamp to identify a situation in which the driver could not be warned of a system malfunction by the ABS warning lamp. The Lamp Driver Module (LDM) turns the ABS warning lamp "ON," unless the EBCM provides ground to turn it "OFF." Because of the circuitry internal to the LDM, only external malfunctions can be detected. As a result, the ABS warning lamp itself is not diagnosable, only the control line to the EBCM can be diagnosed.

If the ABS warning lamp control line is shorted to battery or the EBCM control circuit is open, the Electronic Brake Control Module (EBCM) is not able to turn "OFF" the ABS warning lamp.

DTC Will Set When: DTC 13 can be set anytime the EBCM commands the ABS warning lamp "OFF." If the EBCM can not control the ABS warning lamp for two seconds, a malfunction exists.

Action Taken: A malfunction DTC is stored, however, ABS is not disabled.

DTC Chart Test Description: Number(s) below refer to circled number(s) on the diagnostic chart.

1. Checks to see if ABS lamp functions properly. Normal operation would indicate that the malfunction is not present.
2. After removing the 10 amp gauge fuse, voltage on CKT 852 will be eliminated. If voltage is still present, it would indicate that a short to voltage is present between the lamp driver module and EBCM.
3. Ensures the malfunction is not due to a short in B+ in the ABS indicator lamp control circuit with the EBCM connected.

Diagnostic Aids: An "Intermittent" malfunction is most likely caused by a poor connection, rubbed through wire insulation or a wire that is broken inside the insulation.

DTC 13 must be currently set before it should be addressed. If DTC 13 is set as a history DTC, proceed to those DTCs that are currently set before attempting to repair a DTC 13 malfunction.

The frequency of the malfunction can be checked by using the enhanced diagnostic function of the Tech 1.

Any circuitry that is suspected as causing the "Intermittent" complaint should be thoroughly checked for backed out terminals, improper mating, broken locks, improperly formed or damaged terminals, poor terminal to wiring connections or physical damage to the wiring harness.

DIAGNOSTIC CHART DTC 13 — 1994
CAVALIER

DTC 13

ABS WARNING LAMP CIRCUIT SHORTED TO BATTERY
(CHEVROLET ONLY)

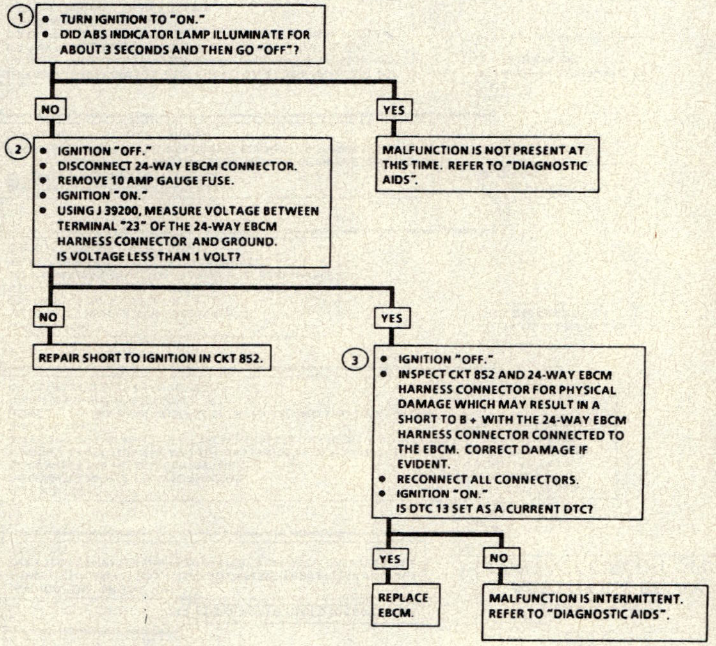

THIS CHART ASSUMES THAT A CURRENT DTC IS STORED INDICATING THAT THIS MALFUNCTION IS PRESENT.

DIAGNOSTIC CHART DTC 14 — 1994
CAVALIER AND SUNBIRD

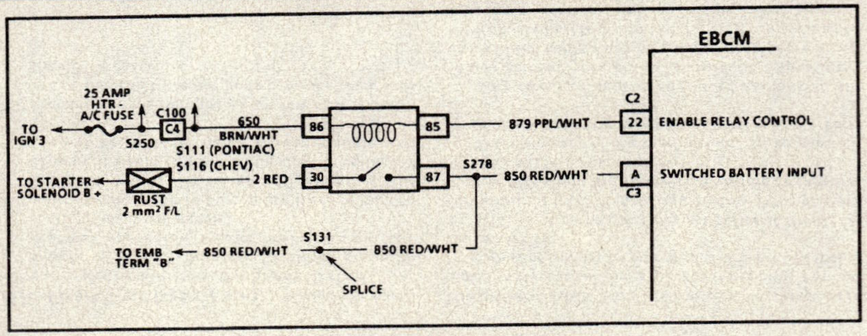

DTC 14

ABS ENABLE RELAY CONTACT CIRCUIT OPEN

Circuit Description:
Ignition voltage is supplied through terminal "86" of the ABS enable relay. The EBCM is then able to energize the pull-in coil by completing the ground circuit at EBCM terminal "22". The magnetic field created closes the enable relay contacts and allows battery voltage and current to be supplied to the EMBs and EBCM, which supplies power to the motors and solenoids.

DTC Will Set When: DTC 14 can be set anytime after the EBCM commands the enable relay "ON" (the relay is first commanded "ON" during the three second bulb check). This test monitors the availability of current/voltage to the motors, solenoids, and EMBs. This malfunction indicates voltage is not available and would therefore not allow ABS operation if required.

Action Taken: A malfunction DTC is stored, ABS is disabled and the ABS warning lamp is turned "ON."

DTC Chart Test Description: Number(s) below refer to circled number(s) on the diagnostic chart.

1. Checks to see if the EBCM is sensing battery voltage at terminal "A".
2. Checks for proper operation of the enable relay and related circuitry.
3. Checks for excessive resistance in the switched battery input circuitry.
4. Checks for excessive resistance in the B+ feed circuit.
5. Checks for excessive resistance in the enable relay contacts.
6. Determines if the malfunction is due to poor terminal contact.
7. Determines if the malfunction is due to the EBCM.

Diagnostic Aids: An "Intermittent" malfunction is most likely caused by a poor connection, rubbed through wire insulation, or a wire that is broken inside the insulation.

The frequency of the malfunction is most likely checked by using the enhanced diagnostic function of the Tech 1.

Any circuitry that is suspected as causing the "Intermittent" complaint should be thoroughly checked for backed out terminals, improper mating, broken locks, improperly formed or damaged terminals, poor terminal to wiring connections or physical damage to the wiring harness.

VIBRATION, TEMPERATURE EFFECTS:

Check for vibration effects by performing the relay test function of the Tech 1. With the relay commanded "ON," lightly tap the top and sides of the relay while monitoring battery voltage. If the battery voltage changes significantly, replace the relay.

If test 14 only fails when the vehicle is initially started in cold ambient conditions (temperature less than 32°F-0°C), replace the relay.

DIAGNOSTIC CHART DTC 14 — 1994
CAVALIER AND SUNBIRD

DTC 14
(Page 1 of 3)
ABS ENABLE RELAY CONTACT CIRCUIT OPEN

IS DTC 16 ALSO PRESENT OR IN HISTORY?
- YES → PROCEED TO THAT DTC CHART FIRST.
- NO ↓

(1)
- IGNITION "ON," ENGINE "OFF."
- USING TECH 1, SELECT DATA LIST AND OBSERVE ABS BATTERY VOLTAGE. DOES TECH 1 INDICATE THAT ABS BATTERY VOLTAGE IS GREATER THAN 10 VOLTS?

- YES → MALFUNCTION IS NOT PRESENT AT THIS TIME. REFER TO "DIAGNOSTIC AIDS" ON FACING PAGE.
- NO ↓

(2)
- IGNITION "OFF."
- DISCONNECT 24-WAY EBCM CONNECTOR.
- DISCONNECT 2-WAY EBCM CONNECTOR.
- CONNECT A FUSED JUMPER WIRE SUCH AS J 36169 WITH A 3 AMP FUSE BETWEEN GROUND AND 24-WAY EBCM HARNESS CONNECTOR TERMINAL "22."
- IGNITION "ON."
- USING J 39200, MEASURE VOLTAGE BETWEEN GROUND AND 2-WAY EBCM HARNESS CONNECTOR TERMINAL "A". IS VOLTAGE GREATER THAN 10 VOLTS?

- NO → REFER TO PAGE 2 OF 3 OF THIS DTC CHART.
- YES ↓

(3)
- IGNITION "OFF."
- DISCONNECT JUMPER WIRE USED IN STEP 2.
- DISCONNECT ENABLE RELAY.
- USING J 39200, MEASURE RESISTANCE BETWEEN 2-WAY EBCM HARNESS CONNECTOR TERMINAL "A" AND ENABLE RELAY HARNESS CONNECTOR TERMINAL "87". IS RESISTANCE LESS THAN 2 OHMS?

- NO → REPAIR HIGH RESISTANCE IN CKT 850.
- YES ↓

(4)
- DISCONNECT NEGATIVE BATTERY CABLE.
- DISCONNECT POSITIVE BATTERY CABLE.
- USING J 39200, MEASURE RESISTANCE BETWEEN POSITIVE BATTERY CABLE TERMINAL AND ENABLE RELAY HARNESS CONNECTOR TERMINAL "30". IS RESISTANCE LESS THAN 2 OHMS?

- NO → REPAIR HIGH RESISTANCE IN B+ FEED CIRCUIT.
- YES ↓

(5)
- USING A JUMPER WIRE, CONNECT ENABLE RELAY TERMINAL "85" TO GROUND. THEN, USING A FUSED JUMPER WIRE SUCH AS J 36169 WITH A 3 AMP FUSE, CONNECT ENABLE RELAY TERMINAL "86" TO B+.
- USING J 39200, MEASURE THE RESISTANCE BETWEEN ENABLE RELAY TERMINALS "30" AND "87". IS RESISTANCE LESS THAN 2 OHMS?

- NO → REPLACE ENABLE RELAY.
- YES ↓

(6)
- INSPECT ENABLE RELAY AND ENABLE RELAY HARNESS CONNECTOR FOR A POOR CONNECTION OR POOR TERMINAL CONTACT. INSPECT EBCM AND EBCM CONNECTOR TERMINALS "22" AND "A" FOR POOR TERMINAL CONTACT. INSPECT POSITIVE BATTERY TERMINAL AND POSITIVE BATTERY CABLE TERMINAL FOR A POOR CONNECTION.

- YES → REPLACE TERMINALS OR REPAIR CONNECTION INTEGRITY (IF POOR CONTACT EXISTS AT BATTERY TERMINAL) THAT EXHIBIT(S) POOR CONTACT.
- NO ↓

(7)
- RECONNECT ALL CONNECTORS.
- IGNITION "ON." IS DTC 14 SET AS A CURRENT DTC?

- YES → REPLACE EBCM.
- NO → MALFUNCTION IS INTERMITTENT. REFER TO "DIAGNOSTIC AIDS".

THIS CHART ASSUMES THAT A CURRENT DTC IS STORED INDICATING THAT THIS MALFUNCTION IS PRESENT.

DIAGNOSTIC CHART DTC 14 — 1994
CAVALIER AND SUNBIRD

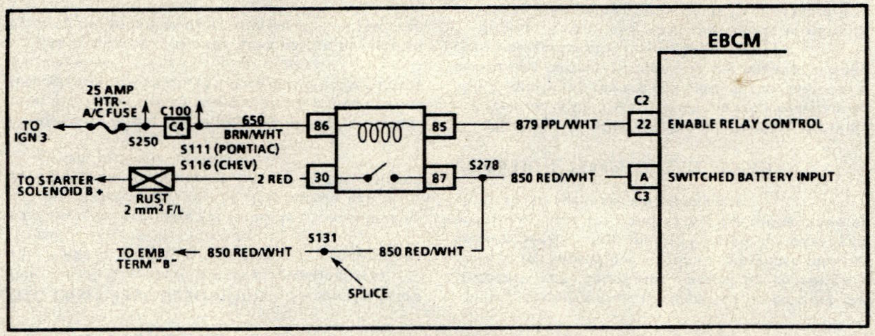

DTC 14

(Page 2 of 3)

ABS ENABLE RELAY CONTACT CIRCUIT OPEN

Circuit Description:

Ignition voltage is supplied through terminal "86" of the ABS enable relay. The EBCM is then able to energize the pull-in coil by completing the ground circuit at EBCM terminal "22". The magnetic field created closes the enable relay contacts and allows battery voltage and current to be supplied to the EMBs and EBCM, which supplies power to the motors and solenoids.

DTC Will Set When: DTC 14 can be set anytime after the EBCM commands the enable relay "ON" (the relay is first commanded "ON" during the three second bulb check). This test monitors the availability of current/voltage to the motors, solenoids, and EMBs. This malfunction indicates voltage is not available and would therefore not allow ABS operation if required.

Action Taken: A malfunction DTC is stored, ABS is disabled and the ABS warning lamp is turned "ON."

DTC Chart Test Description: Number(s) below refer to circled number(s) on the diagnostic chart.

8. Checks for excessive resistance in the enable relay control circuitry.
9. Checks for proper battery voltage.
10. Checks for an open in the B+ feed circuit.
11. Checks for an open in the switched battery input circuit.
12. Checks for proper operation of the enable relay.

Diagnostic Aids: An "Intermittent" malfunction is most likely caused by a poor connection, rubbed through wire insulation, or a wire that is broken inside the insulation.

The frequency of the malfunction can be checked by using the enhanced diagnostic function of the Tech 1.

Any circuitry that is suspected as causing the "Intermittent" complaint should be thoroughly checked for backed out terminals, improper mating, broken locks, improperly formed or damaged terminals, poor terminal to wiring connections or physical damage to the wiring harness.

VIBRATION, TEMPERATURE EFFECTS:

Check for vibration effects by performing the relay test function of the Tech 1. With the relay commanded "ON," lightly tap the top and sides of the relay while monitoring battery voltage. If the battery voltage changes significantly, replace the relay.

If test 14 only fails when the vehicle is initially started in cold ambient conditions (temperature less than 32°F-0°C), replace the relay.

DIAGNOSTIC CHART DTC 14 — 1994
CAVALIER AND SUNBIRD

DTC 14

(Page 2 of 3)

ABS ENABLE RELAY CONTACT CIRCUIT OPEN

CONTINUED FROM PAGE 1 OF 3 OF THIS CHART.

(8)
- IGNITION "OFF."
- DISCONNECT JUMPER WIRE USED IN STEP 2.
- DISCONNECT NEGATIVE BATTERY CABLE FROM BATTERY.
- DISCONNECT POSITIVE BATTERY CABLE FROM BATTERY.
- TURN IGNITION "ON" (THIS IS DONE TO PROVIDE CIRCUIT CONTINUITY).
- USING J 39200, MEASURE RESISTANCE BETWEEN POSITIVE BATTERY CABLE TERMINAL AND 24-WAY EBCM HARNESS CONNECTOR TERMINAL "22".

IS RESISTANCE LESS THAN 95 OHMS WHEN RELAY IS APPROXIMATELY 25°C (77°F)?

NO: REFER TO PAGE 3 OF 3 OF THIS DTC CHART.

YES:

(9)
- USING J 39200, MEASURE VOLTAGE BETWEEN THE NEGATIVE AND POSITIVE BATTERY TERMINALS.

IS VOLTAGE GREATER THAN 10 VOLTS?

NO: CORRECT LOW VOLTAGE CONDITION.

YES:

(10)
- IGNITION "OFF."
- RECONNECT BATTERY CABLES.
- DISCONNECT ENABLE RELAY.
- USING J 39200, MEASURE VOLTAGE BETWEEN GROUND AND ENABLE RELAY HARNESS CONNECTOR TERMINAL "30".

IS VOLTAGE GREATER THAN 10 VOLTS?

NO: REPAIR OPEN IN B+ FEED CIRCUIT.

YES:

(11)
- USING J 39200, MEASURE RESISTANCE BETWEEN 2-WAY EBCM HARNESS CONNECTOR TERMINAL "A" AND ENABLE RELAY HARNESS CONNECTOR TERMINAL "87".

IS RESISTANCE LESS THAN 2 OHMS?

NO: REPAIR OPEN IN CKT 850.

YES:

(12)
- USING A JUMPER WIRE, CONNECT ENABLE RELAY TERMINAL "85" TO GROUND. THEN, USING A FUSED JUMPER WIRE SUCH AS J 36169 WITH A 3 AMP FUSE, CONNECT ENABLE RELAY TERMINAL "86" TO B+.
- USING J 39200, MEASURE RESISTANCE BETWEEN ENABLE RELAY TERMINALS "30" AND "87".

IS RESISTANCE LESS THAN 2 OHMS?

NO: REPLACE ENABLE RELAY.

YES: MALFUNCTION WAS DUE TO A POOR CONNECTION.

REPLACE OR REPAIR TERMINALS THAT EXHIBIT POOR TERMINAL CONTACT.

THIS CHART ASSUMES THAT A CURRENT DTC IS STORED INDICATING THAT THIS MALFUNCTION IS PRESENT.

DIAGNOSTIC CHART DTC 14 — 1994
CAVALIER AND SUNBIRD

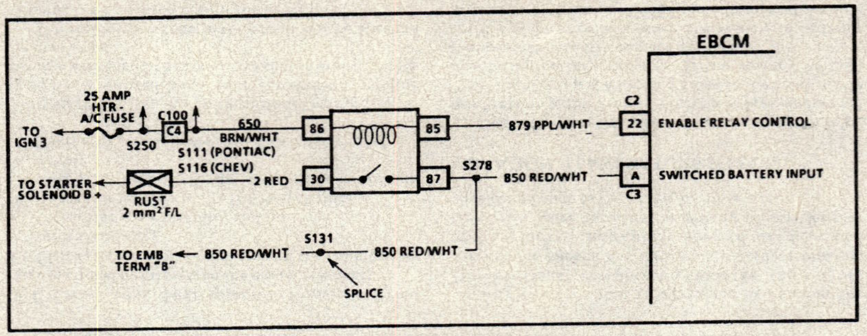

DTC 14
(Page 3 of 3)
ABS ENABLE RELAY CONTACT CIRCUIT OPEN

Circuit Description:
Ignition voltage is supplied through terminal "86" of the ABS enable relay. The EBCM is then able to energize the pull-in coil by completing the ground circuit at EBCM terminal "22". The magnetic field created closes the enable relay contacts and allows battery voltage and current to be supplied to the EMBs and EBCM, which supplies power to the motors and solenoids.

DTC Will Set When: DTC 14 can be set anytime after the EBCM commands the enable relay "ON" (the relay is first commanded "ON" during the three second bulb check). This test monitors the availability of current/voltage to the motors, solenoids, and EMBs. This malfunction indicates voltage is not available and would therefore not allow ABS operation if required.

Action Taken: A malfunction DTC is stored, ABS is disabled and the ABS warning lamp is turned "ON."

DTC Chart Test Description: Number(s) below refer to circled number(s) on the diagnostic chart.
13. Checks for excessive resistance in the enable relay coil.
14. Checks for excessive resistance in the enable relay control circuit between the relay and the EBCM.
15. Checks for excessive resistance in the ignition switch "run" fuse feed circuit.

Diagnostic Aids: An "Intermittent" malfunction is most likely caused by a poor connection, rubbed through wire insulation, or a wire that is broken inside the insulation.

The frequency of the malfunction can be checked by using the enhanced diagnostic function of the Tech 1.

Any circuitry that is suspected as causing the "Intermittent" complaint should be thoroughly checked for backed out terminals, improper mating, broken locks, improperly formed or damaged terminals, poor terminal to wiring connections or physical damage to the wiring harness.

VIBRATION, TEMPERATURE EFFECTS:

Check for vibration effects by performing the relay test function of the Tech 1. With the relay commanded "ON," lightly tap the top and sides of the relay while monitoring battery voltage. If the battery voltage changes significantly, replace the relay.

If test 14 only fails when the vehicle is initially started in cold ambient conditions (temperature less than 32°F-0°C), replace the relay.

DIAGNOSTIC CHART DTC 14 — 1994
CAVALIER AND SUNBIRD

DTC 14
(Page 3 of 3)
ABS ENABLE RELAY CONTACT CIRCUIT OPEN

CONTINUED FROM PAGE 2 OF 3 OF THIS CHART.

(13)
- USING J 39200, MEASURE RESISTANCE BETWEEN ENABLE RELAY TERMINALS "85" AND "86".
- IS RESISTANCE LESS THAN 95 OHMS WHEN RELAY IS APPROXIMATELY 25°C (77°F)?

NO: REPLACE ENABLE RELAY.

YES:

(14)
- USING J 39200, MEASURE RESISTANCE BETWEEN ENABLE RELAY HARNESS CONNECTOR TERMINAL "85" AND 24-WAY EBCM HARNESS CONNECTOR TERMINAL "22".
- IS RESISTANCE LESS THAN 2 OHMS?

NO: REPAIR HIGH RESISTANCE IN CKT 879.

YES:

(15)
- DISCONNECT ENABLE RELAY.
- USING J 39200, MEASURE RESISTANCE BETWEEN POSITIVE BATTERY CABLE TERMINAL AND ENABLE RELAY HARNESS CONNECTOR TERMINAL "86".
- IS RESISTANCE LESS THAN 2 OHMS?

YES: MALFUNCTION WAS DUE TO A POOR CONNECTION. TEST FOR POOR TERMINAL CONTACT BETWEEN ENABLE RELAY AND ENABLE RELAY HARNESS CONNECTOR. REPLACE TERMINALS THAT EXHIBIT POOR TERMINAL CONTACT.

NO: REPAIR HIGH RESISTANCE IN CIRCUIT BETWEEN POSITIVE BATTERY CABLE TERMINAL AND ENABLE RELAY HARNESS CONNECTOR TERMINAL "86".

THIS CHART ASSUMES THAT A CURRENT DTC IS STORED INDICATING THAT THIS MALFUNCTION IS PRESENT.

DIAGNOSTIC CHART DTC 15 — 1994 CAVALIER AND SUNBIRD

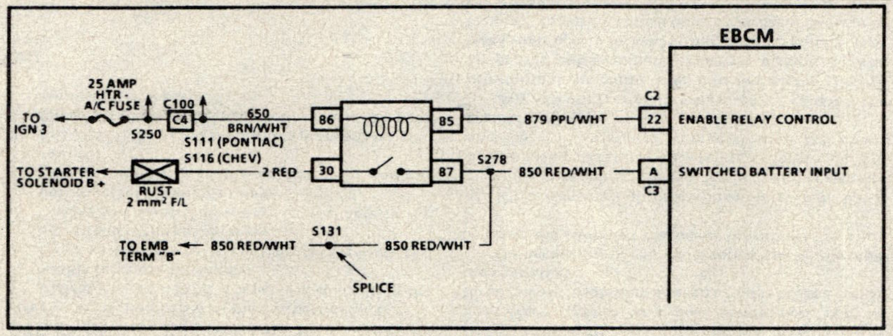

DTC 15

ABS ENABLE RELAY CONTACT CIRCUIT SHORTED TO BATTERY OR ALWAYS CLOSED

Circuit Description:

Ignition voltage is supplied through terminal "86" of the ABS enable relay. The EBCM is then able to energize the pull-in coil by completing the ground circuit at terminal "22" of the EBCM. The magnetic field created closes the enable relay contacts and allows battery voltage and current to be supplied to the EMBs and EBCM, which supplies power to the motors and solenoids.

DTC Will Set When: DTC 15 can be set only before the EBCM commands the relay "ON." This test determines if the enable relay is energized when it should not be. This malfunction would not allow the enable relay to remove power to the ABS system. If a second malfunction were to occur that requires the enable relay to be turned "OFF," that malfunction can not be removed if the relay can not be controlled. The malfunction must be present for three consecutive ignition cycles before the DTC is set.

Action Taken: A malfunction DTC is stored. ABS is not disabled; however, the ABS indicator lamp will flash to indicate a malfunction exists.

DTC Chart Test Description: Number(s) below refer to circled number(s) on the diagnostic chart.

1. Indicates if the EBCM is capable of controlling the enable relay as commanded.
2. Checks for voltage at terminal "A" of the 2-way EBCM harness connector. If voltage is present, the malfunction exists in the enable relay and/or its circuitry.
3. Checks for a short to voltage in the switched battery input circuit.
4. Ensures the malfunction is not due to a short to B+ in the switched battery input circuit with EBCM connected.

Diagnostic Aids: An "Intermittent" malfunction is most likely caused by a poor connection, rubbed through wire insulation, or a wire that is broken inside the insulation.

The frequency of the malfunction can be checked by using the enhanced diagnostic function of the Tech 1

Any circuitry, that is suspected as causing the intermittent complaint, should be thoroughly checked for backed out terminals, improper mating, broken locks, improperly formed or damaged terminals, poor terminal to wiring connections or physical damage to the wiring harness.

DIAGNOSTIC CHART DTC 15 — 1994 CAVALIER AND SUNBIRD

DTC 15

ABS ENABLE RELAY CONTACT CIRCUIT SHORTED TO BATTERY OR ALWAYS CLOSED

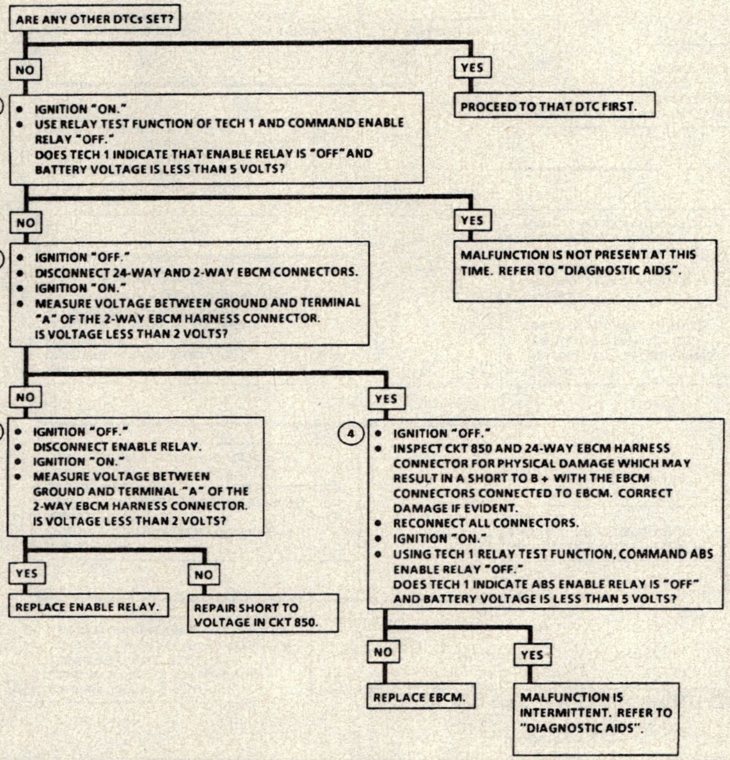

- AFTER DIAGNOSIS IS COMPLETE, CLEAR DTCs AND TEST DRIVE VEHICLE FOR THREE (3) DRIVE CYCLES TO VERIFY DTC DOES NOT RESET. A DRIVE CYCLE CONSISTS OF STARTING THE VEHICLE, DRIVING ABOVE 16 km/h (10 MPH) AND THEN KEYING DOWN.

THIS CHART ASSUMES THAT A CURRENT DTC IS STORED INDICATING THAT THIS MALFUNCTION IS PRESENT.

DIAGNOSTIC CHART DTC 16 — 1994 CAVALIER AND SUNBIRD

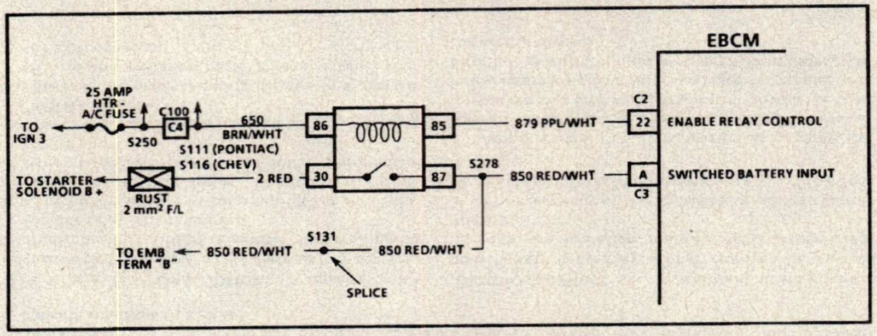

DTC 16

ABS ENABLE RELAY COIL CIRCUIT OPEN

Circuit Description:
Ignition voltage is supplied through terminal "86" of the ABS enable relay. The EBCM is then able to energize the pull-in coil by completing the ground circuit at terminal "22" of the EBCM. The magnetic field created closes the enable relay contacts and allows battery voltage and current to be supplied to the EMBs and EBCM, which supplies power to the motors and solenoids.

DTC Will Set When: DTC 16 can be set only after DTC 14 has been set. This test detects an open in the enable relay coil circuit. An open in this circuit will not allow the enable relay to be energized thus preventing voltage/current to the motors, EMBs and solenoids. If this malfunction is present and the ignition is turned "OFF" before 5 km/h (3 mph) is reached, DTC 14 is set alone.

Action Taken: A malfunction DTC is stored, ABS is disabled and the ABS warning lamp is turned "ON."

DTC Chart Test Description: Number(s) below refer to circled number(s) on the diagnostic chart.

1. Indicates if the EBCM is capable of controlling the enable relay as commanded.
2. Ensures that there is continuity through the pull-in coil of the enable relay.
3. Checks to ensure that voltage is available to the pull-in coil of the enable relay.
4. Checks continuity in the enable relay control circuit.
5. Ensures DTC was not set due to a poor connection.
6. Identifies whether the malfunction is due to a poor connection or the EBCM.

Diagnostic Aids: An "Intermittent" malfunction is most likely caused by a poor connection, rubbed through wire insulation, or a wire that is broken inside the insulation.

The frequency of the malfunction can be checked by using the enhanced diagnostic function of the Tech 1.

If the frequency of the malfunction is high, but is currently intermittent, check for high coil resistance by measuring resistance between relay terminals "86" and "85". If resistance is greater than 100 ohms, replace the relay.

Any circuitry, that is suspected as causing the intermittent complaint, should be thoroughly checked for backed out terminals, improper mating, broken locks, improperly formed or damaged terminals, poor terminal to wiring connections or physical damage to the wiring harness.

DIAGNOSTIC CHART DTC 16 — 1994 CAVALIER AND SUNBIRD

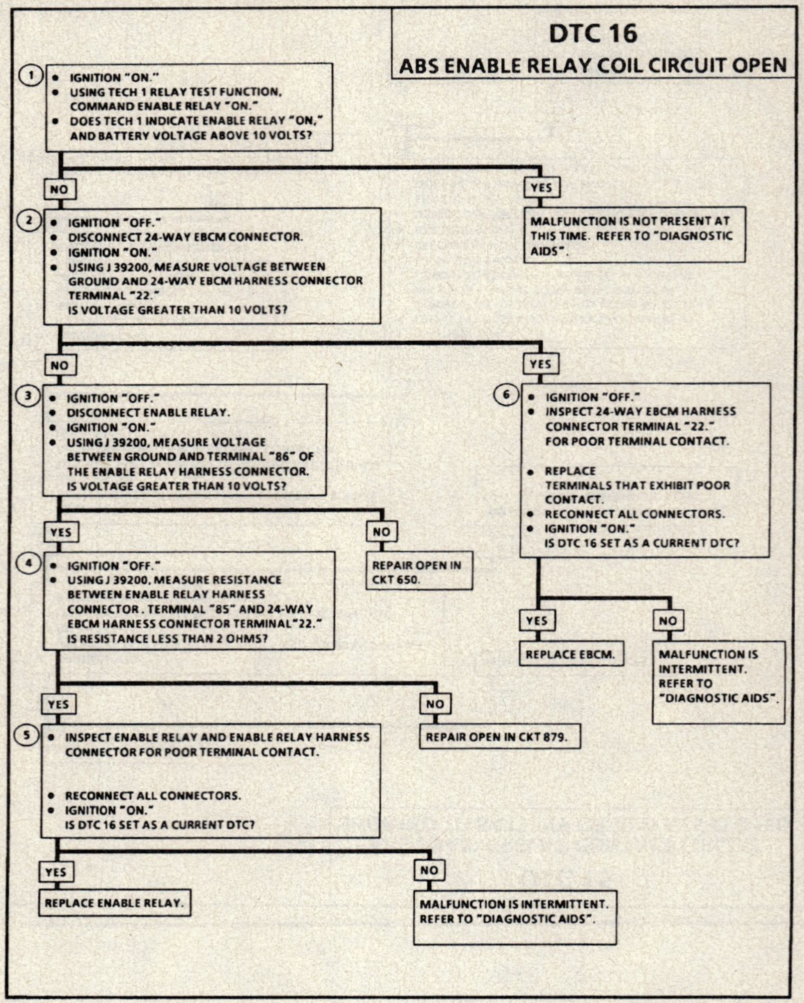

DIAGNOSTIC CHART DTC 17 — 1994
CAVALIER AND SUNBIRD

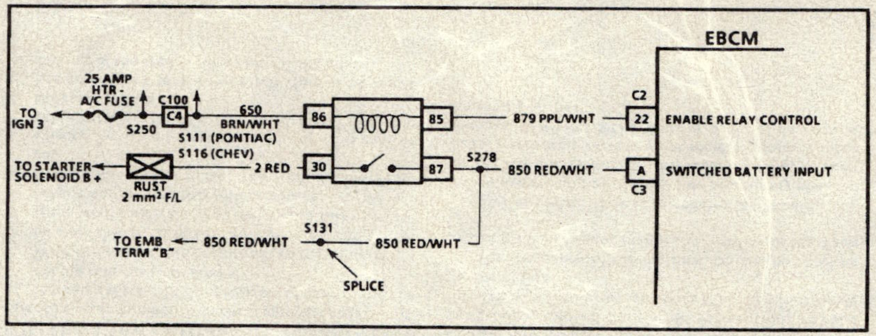

DTC 17

ABS ENABLE RELAY COIL CIRCUIT SHORTED TO GROUND

Circuit Description:

Ignition voltage is supplied through terminal "86" of the ABS enable relay. The EBCM is then able to energize the pull-in coil by completing the ground circuit at terminal "22" of the EBCM. The magnetic field created closes the enable relay contacts and allows battery voltage and current to be supplied to the EMBs and EBCM, which supplies voltage to the motors and solenoids.

DTC Will Set When: DTC 17 can be set before the EBCM commands the enable relay "ON." This test determines if the enable relay is energized when it should not be. This malfunction would not allow the enable relay to remove power to the ABS system. If a second malfunction were to occur that requires the enable relay to be turned "OFF," that malfunction can not be removed if the relay can not be controlled.

Action Taken: A malfunction DTC is stored. ABS is not disabled; however the ABS warning lamp flashes to indicate a malfunction exists.

DTC Chart Test Description: Number(s) below refer to circled number(s) on the diagnostic chart.

1. Indicates if the EBCM is capable of controlling the enable relay as commanded.
2. Ensures the enable relay or control circuit are not shorted to ground.
3. This step isolates the malfunction to either the enable relay or the enable relay control circuit.
4. Ensures short to ground is not due to physical damage of the circuitry.

Diagnostic Aids: An "Intermittent" malfunction is most likely caused by a poor connection, rubbed through wire insulation, or a wire that is broken inside the insulation.

The frequency of the malfunction can be checked by using the enhanced diagnostic function of the Tech 1.

Any circuitry, that is suspected as causing the intermittent complaint, should be thoroughly checked for backed out terminals, improper mating, broken locks, improperly formed or damaged terminals, poor terminal to wiring connections or physical damage to the wiring harness.

DIAGNOSTIC CHART DTC 17 — 1994
CAVALIER AND SUNBIRD

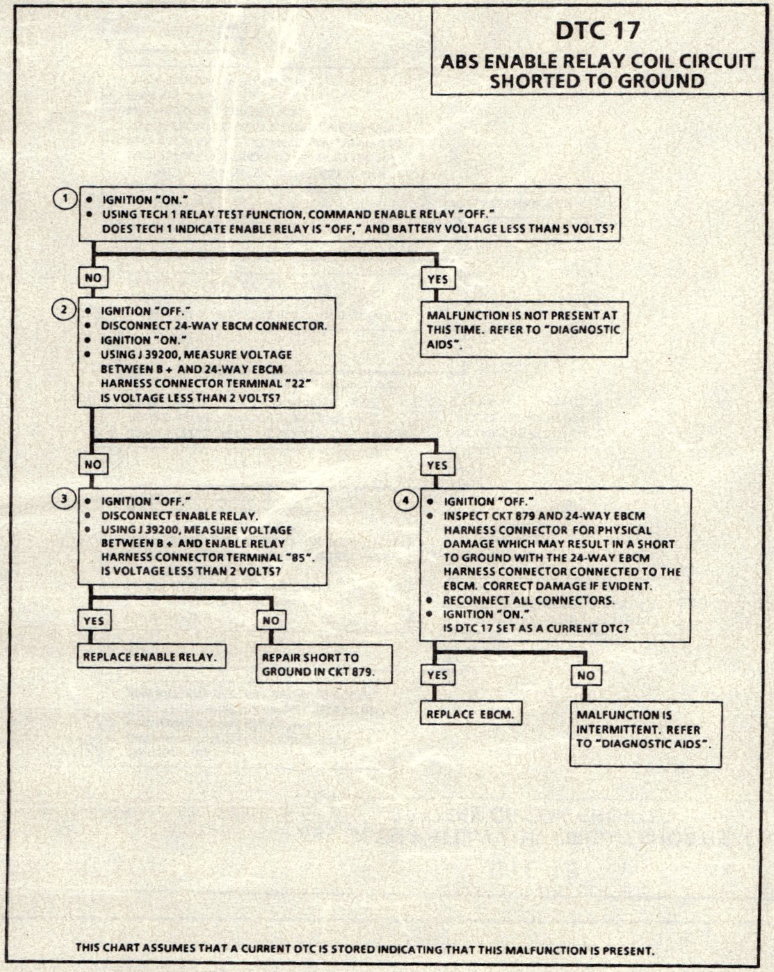

DIAGNOSTIC CHART DTC 18 — 1994 CAVALIER AND SUNBIRD

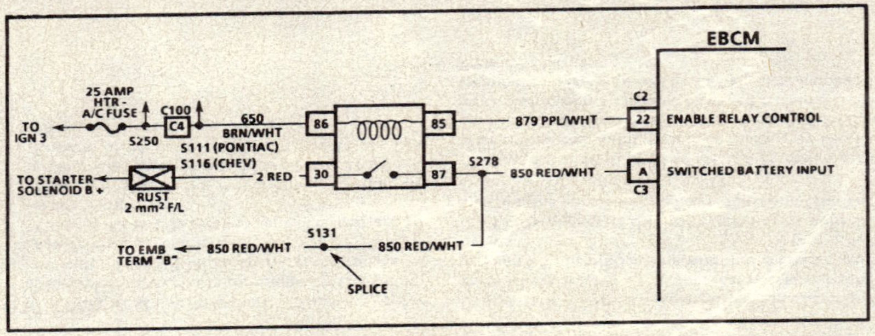

DTC 18

ABS ENABLE RELAY COIL CIRCUIT SHORTED TO BATTERY OR COIL SHORTED

Circuit Description:
Ignition voltage is supplied through terminal "86" of the ABS enable relay. The EBCM is then able to energize the pull-in coil by completing the ground circuit at terminal "22" of the EBCM. The magnetic field created closes the enable relay contacts and allows battery voltage and current to be supplied to the EMBs and EBCM, which supplies voltage to the motors and solenoids.

DTC Will Set When: DTC 18 can be set after the EBCM commands the enable relay "ON." This test monitors the availability of current/voltage to the motors, solenoids and EMBs. A malfunction of this type will not allow voltage to be available to the motors, solenoids and EMBs; therefore, ABS operation would not be allowed if required.

Action Taken: A malfunction DTC is stored, ABS is disabled and the ABS warning lamp is turned "ON."

DTC Chart Test Description: Number(s) below refer to circled number(s) on the diagnostic chart.

1. Indicates if the EBCM is capable of controlling the enable relay as commanded.
2. With the enable relay removed, voltage should not be available at terminal "85". Any voltage at this point would indicate that the enable relay control circuit was shorted to a voltage source.
3. Checks for a short across the ABS enable relay coil.
4. Checks for an internal short in the ABS enable relay between the B+ feed at terminal "30" and the relay pull-in coil.
5. Ensures short to B+ was not due to physical damage of the enable relay control circuit.

Diagnostic Aids: An "Intermittent" malfunction is most likely caused by a poor connection, rubbed through wire insulation, or a wire that is broken inside the insulation.

The frequency of the malfunction can be checked by using the enhanced diagnostic function of the Tech 1.

Any circuitry, that is suspected as causing the intermittent complaint, should be thoroughly checked for backed out terminals, improper mating, broken locks, improperly formed or damaged terminals, poor terminal to wiring connections or physical damage to the wiring harness.

DIAGNOSTIC CHART DTC 18 — 1994 CAVALIER AND SUNBIRD

DTC 18

ABS ENABLE RELAY COIL CIRCUIT SHORTED TO BATTERY OR COIL SHORTED

(1)
- IGNITION "ON."
- USING RELAY TEST FUNCTION OF TECH 1, COMMAND ENABLE RELAY "ON."

DOES TECH 1 INDICATE ENABLE RELAY "ON" AND BATTERY VOLTAGE ABOVE 10 VOLTS?

- YES: MALFUNCTION IS NOT PRESENT AT THIS TIME. REFER TO "DIAGNOSTIC AIDS".
- NO: go to (2)

(2)
- IGNITION "OFF."
- DISCONNECT ENABLE RELAY.
- DISCONNECT 24-WAY EBCM CONNECTOR.
- IGNITION "ON."
- USING J 39200, MEASURE VOLTAGE BETWEEN GROUND AND ENABLE RELAY HARNESS CONNECTOR TERMINAL "85".

IS VOLTAGE LESS THAN 2 VOLTS?

- NO: REPAIR SHORT TO VOLTAGE ON CKT 879.
- YES: go to (3)

(3)
- USING J 39200, MEASURE RESISTANCE BETWEEN ENABLE RELAY TERMINALS "85" AND "86".

IS RESISTANCE GREATER THAN 40 OHMS?

- NO: REPLACE ENABLE RELAY.
- YES: go to (4)

(4)
- USING J 39200, MEASURE RESISTANCE BETWEEN ENABLE RELAY TERMINALS "85" AND "30".

DOES J 39200 DISPLAY "OL" (INFINITE)?

- NO: REPLACE ENABLE RELAY.
- YES: go to (5)

(5)
- IGNITION "OFF."
- *INSPECT CKT 879 AND 24-WAY EBCM HARNESS* CONNECTOR FOR PHYSICAL DAMAGE WHICH MAY RESULT IN A SHORT TO B+ WITH THE 24-WAY EBCM HARNESS CONNECTOR CONNECTED TO THE EBCM. CORRECT DAMAGE IF EVIDENT.
- RECONNECT ALL CONNECTORS.
- IGNITION "ON."

IS DTC 18 SET AS A CURRENT DTC?

- YES: REPLACE EBCM.
- NO: MALFUNCTION IS INTERMITTENT. REFER TO "DIAGNOSTIC AIDS".

THIS CHART ASSUMES THAT A CURRENT DTC IS STORED INDICATING THAT THIS MALFUNCTION IS PRESENT.

DIAGNOSTIC CHART DTC 21 — 1994 CAVALIER AND SUNBIRD

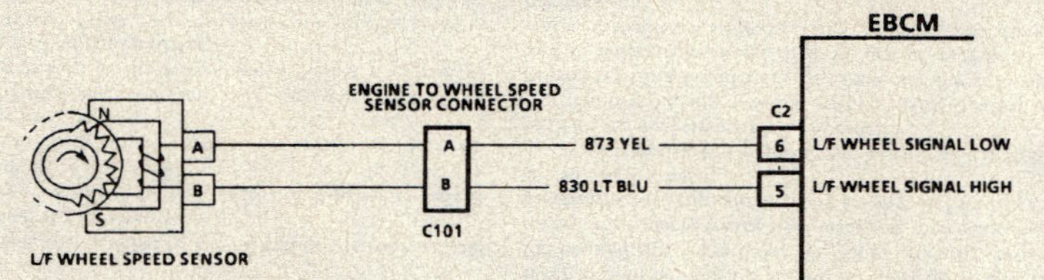

DTC 21

(Page 1 of 2)
LEFT FRONT WHEEL SPEED = 0
OR UNREASONABLE

Circuit Description:

As a toothed ring passes by the wheel speed sensor, changes in the electromagnetic field cause the wheel speed sensor to produce a sinusoidal (AC) voltage signal whose frequency is proportional to wheel speed. The magnitude of this signal is directly related to wheel speed and the proximity of the wheel speed sensor to the toothed ring, often referred to as the air gap.

DTC Will Set When: DTC 21 can be set when the vehicle is not in an ABS stop. If the left front wheel speed is less than 1/2 of the vehicle speed and the vehicle speed is greater than 8 km/h (5 mph), a malfunction exists.

Action Taken: A malfunction DTC is stored, ABS is disabled and the ABS warning lamp is turned "ON."

DTC Chart Test Description: Number(s) below refer to circled number(s) on the diagnostic chart.

1. Identifies if a wheel speed sensor or circuitry is damaged and visibly apparent.
2. Checks wheel speed sensor for proper resistance values.
3. Ensures the wheel speed sensor and toothed ring generate the proper voltage.
4. Ensures the wheel speed sensor is not shorted to ground.

Diagnostic Aids: An "Intermittent" malfunction is most likely caused by a poor connection, rubbed through wire insulation, or a wire that is broken inside the insulation.

The frequency of the malfunction can be checked by using the enhanced diagnostic function of the Tech 1.

If the customer's comments reflect that the ABS amber warning lamp is "ON" only during moist environmental changes, (rain, snow, vehicle wash) all wheel speed sensor circuitry should be thoroughly inspected for signs of water intrusion.

Use the following procedure. Spray down the suspected area with 5% salt water solution (two teaspoons of salt to 12 oz. of water). Drive vehicle above 24 km/h (15 mph) for at least 30 seconds. If DTC returns immediately, replace suspected harness.

Any circuitry, that is suspected as causing the intermittent complaint, should be thoroughly checked for backed out terminals, improper mating, broken locks, improperly formed or damaged terminals, poor terminal to wiring connections or physical damage to the wiring harness.

Resistance of the wheel speed sensor will increase with an increase in sensor temperature.

When replacing a wheel speed sensor, inspect the sensor terminals and harness connector for corrosion and/or water intrusion. If evidence of corrosion or water intrusion exists, replace wheel speed sensor harness.

If wheel speed sensor harness shows evidence of water intrusion, inspect wheel speed sensor for water intrusion. If water intrusion is evident, replace wheel speed sensor.

DIAGNOSTIC CHART DTC 21 — 1994 CAVALIER AND SUNBIRD

DTC 21

(Page 1 of 2)
LEFT FRONT WHEEL SPEED = 0
OR UNREASONABLE

IMPORTANT:

- WHEEL SPEED SENSOR INTERMITTENT MALFUNCTIONS MAY BE DIFFICULT TO LOCATE. CARE SHOULD BE TAKEN NOT TO DISTURB ANY ELECTRICAL CONNECTIONS PRIOR TO AN INDICATED STEP OF THIS CHART. THIS WILL ENSURE THAT AN INTERMITTENT CONNECTION WILL NOT BE CORRECTED BEFORE THE SOURCE OF THE MALFUNCTION IS FOUND.

(1)
- IGNITION "OFF."
- INSPECT WHEEL SPEED SENSOR WIRING AND CONNECTORS FOR DAMAGE. INSPECT WHEEL SPEED SENSOR AND TOOTHED RING FOR LOOSENESS OR DAMAGE. PAY PARTICULAR ATTENTION TO THE TOOTHED RING. ANY SIGNIFICANT DAMAGE WILL AFFECT THE WHEEL SPEED INPUT SIGNAL.
  IS DAMAGE OR SENSOR/TOOTHED RING LOOSENESS EVIDENT?

YES: REPAIR DAMAGE OR SENSOR/TOOTHED RING LOOSENESS. DRIVE VEHICLE ABOVE 24 km/h (15 MPH) FOR AT LEAST 30 SECONDS TO ENSURE DTC 21 DOES NOT SET.

NO:

(2)
- DISCONNECT CONNECTOR FROM LEFT FRONT WHEEL SPEED SENSOR.
- USING J 39200, MEASURE RESISTANCE BETWEEN SENSOR TERMINALS "A" AND "B".
  IS RESISTANCE BETWEEN 1628 AND 1889 OHMS WHEN SENSOR IS 25°C (77°F)? *[IF TEMPERATURE OF THE SENSOR IS SUSPECTED TO BE OTHER THAN 25°C (77°F), REFER TO TABLE BELOW FOR APPLICABLE SENSOR RESISTANCE VALUES; THE VALUES IN THE TABLE ARE FOR TEMPERATURE OF THE SENSOR NOT AIR TEMPERATURE.]*

NO: REPLACE WHEEL SPEED SENSOR.

YES:

(3)
- WITH J 39200 STILL CONNECTED, SELECT AC VOLTAGE SCALE.
- SPIN WHEEL BY HAND WHILE MONITORING VOLTAGE.
  IS VOLTAGE AT LEAST 100 mV (VOLTAGE WILL INCREASE AS WHEEL SPEED INCREASES)?

NO: REPLACE WHEEL SPEED SENSOR.

YES:

(4)
- USING J 39200, MEASURE RESISTANCE BETWEEN SENSOR TERMINAL "A" AND GROUND.
  DOES J 39200 DISPLAY "OL" (INFINITE)?

NO: REPLACE WHEEL SPEED SENSOR.

YES: GO TO PAGE 2 OF 2 OF THIS DTC CHART.

FRONT WHEEL SPEED SENSOR

TEMPERATURE OF SENSOR VS. SENSOR RESISTANCE (APPROXIMATE)

TEMP. (°C)	TEMP. (°F)	RESISTANCE (OHMS)
-34 TO 4	-30 TO 40	1258 TO 1739
5 TO 43	41 TO 110	1503 TO 2023
44 TO 93	111 TO 200	1747 TO 2384

THIS CHART ASSUMES THAT A CURRENT DTC IS STORED INDICATING THAT THIS MALFUNCTION IS PRESENT.

DIAGNOSTIC CHART DTC 21 — 1994 CAVALIER AND SUNBIRD

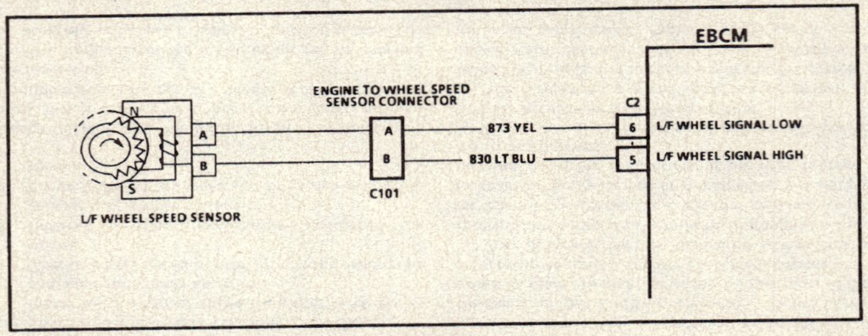

DTC 21

(Page 2 of 2)

LEFT FRONT WHEEL SPEED = 0 OR UNREASONABLE

Circuit Description:

As a toothed ring passes by the wheel speed sensor, changes in the electromagnetic field cause the wheel speed sensor to produce a sinusoidal (AC) voltage signal whose frequency is proportional to wheel speed. The magnitude of this signal is directly related to wheel speed and the proximity of the wheel speed sensor to the toothed ring often referred to as the air gap.

DTC Will Set When: DTC 21 can be set when the vehicle is not in an ABS stop. If the left front wheel speed is less than 1/2 of the vehicle speed and the vehicle speed is greater than 8 km/h (5 mph), a malfunction exists.

Action Taken: A malfunction DTC is stored, ABS is disabled and the ABS warning lamp is turned "ON."

DTC Chart Test Description: Number(s) below refer to circled number(s) on the diagnostic chart.

5. Checks for proper voltages at speed sensor harness connector.
6. Ensures that the wheel speed sensor circuitry is not shorted to ground.
7. Checks for an open in the wheel signal low circuit.
8. Checks for an open in the wheel signal high circuit.
9. Ensures that DTC 21 was not set due to a poor connection at the EBCM or physical damage of the circuitry.
10. Checks for a short to B+ in either the high or low wheel signal circuits.
11. Ensures short to B+ was not due to physical damage of the circuitry.
12. Tests for a short between the wheel signal high and low circuits.
13. Ensures malfunction was not due to a poor connection at the wheel speed sensor or physical damage of the circuitry.

Diagnostic Aids: An "Intermittent" malfunction is most likely caused by a poor connection, rubbed through wire insulation, or a wire that is broken inside the insulation.

The frequency of the malfunction can be checked by using the enhanced diagnostic function of the Tech 1.

If the customer's comments reflect that the ABS amber warning lamp is "ON" only during moist environmental changes, (rain, snow, vehicle wash) all wheel speed sensor circuitry should be thoroughly inspected for signs of water intrusion. Use the following procedure. Spray down the suspected area with 5% salt water solution (two teaspoons of salt to 12 oz. of water). Drive vehicle above 24 km/h (15 mph) for at least 30 seconds. If DTC returns immediately, replace suspected harness.

Any circuitry, that is suspected as causing the intermittent complaint, should be thoroughly checked for backed out terminals, improper mating, broken locks, improperly formed or damaged terminals, poor terminal to wiring connections or physical damage to the wiring harness.

Resistance of the wheel speed sensor will increase with an increase in sensor temperature.

When replacing a wheel speed sensor, inspect the sensor terminals and harness connector for corrosion and/or water intrusion. If evidence of corrosion or water intrusion exists, replace wheel speed sensor harness.

If wheel speed sensor harness shows evidence of water intrusion, inspect wheel speed sensor for water intrusion. If water intrusion is evident, replace wheel speed sensor.

DIAGNOSTIC CHART DTC 21 — 1994 CAVALIER AND SUNBIRD

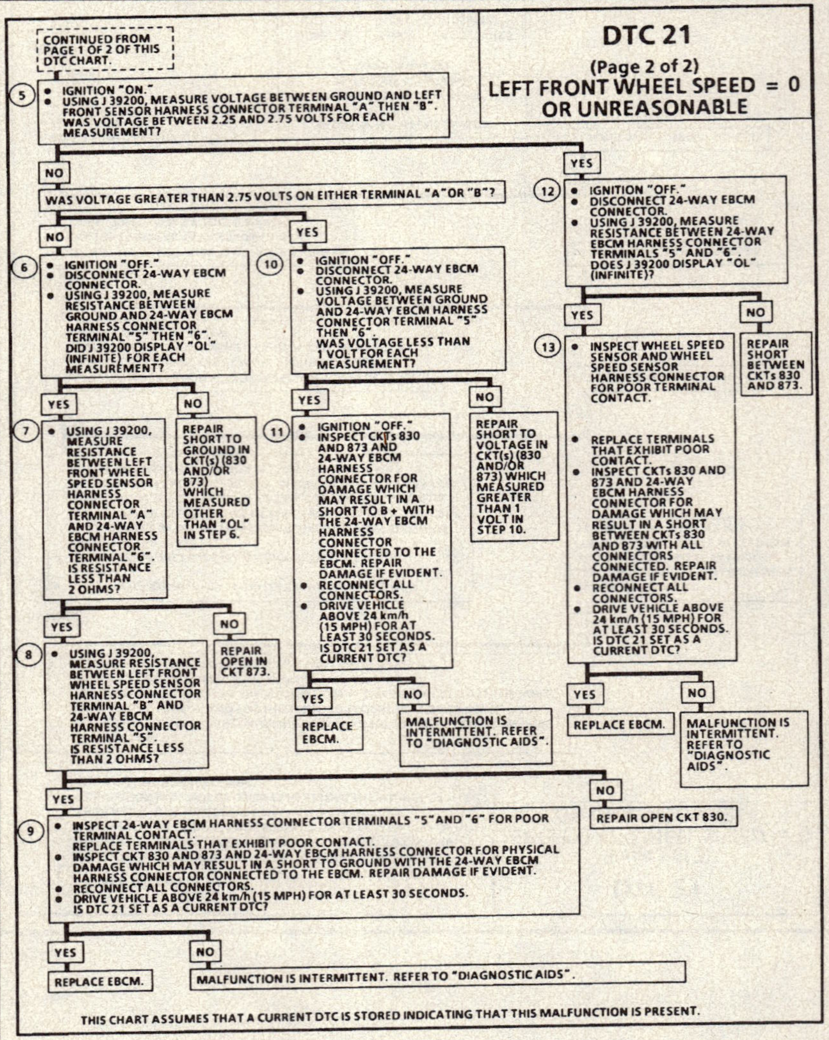

DIAGNOSTIC CHART DTC 22 — 1994 CAVALIER AND SUNBIRD

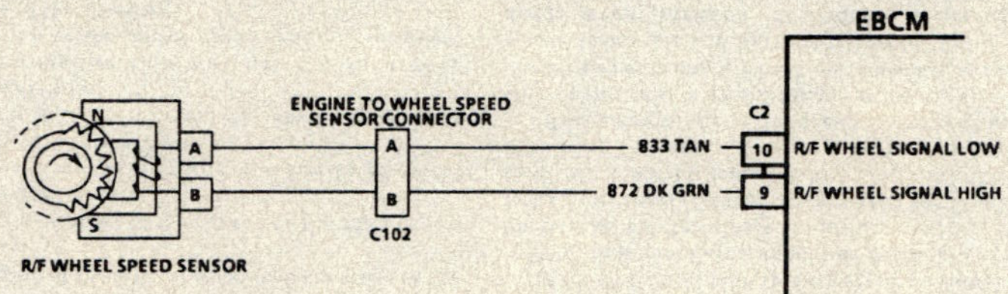

DTC 22

RIGHT FRONT WHEEL SPEED = 0 OR UNREASONABLE

Circuit Description:

As a toothed ring passes by the wheel speed sensor, changes in the electromagnetic field cause the wheel speed sensor to produce a sinusoidal (AC) voltage signal whose frequency is proportional to wheel speed. The magnitude of this signal is directly related to wheel speed and the proximity of the wheel speed sensor to the toothed ring, often referred to as the air gap.

DTC Will Set When: DTC 22 can be set when the vehicle is not in an ABS stop. If the right front wheel speed is less than 1/2 of the vehicle speed and the vehicle speed is greater than 8 km/h (5 mph), a malfunction exists.

Action Taken: A malfunction DTC is stored, ABS is disabled and the ABS warning lamp is turned "ON."

DTC Chart Test Description: Number(s) below refer to circled number(s) on the diagnostic chart.

1. Identifies a wheel speed sensor or circuitry that is damaged and visibly apparent.
2. Checks wheel speed sensor for proper resistance values.
3. Ensures the wheel speed sensor and sensor ring generate the proper voltage.
4. Ensures the wheel speed sensor is not shorted to ground.

Diagnostic Aids: An "Intermittent" malfunction is most likely caused by a poor connection, rubbed through wire insulation, or a wire that is broken inside the insulation.

The frequency of the malfunction can be checked by using the enhanced diagnostic function of the Tech 1.

If the customer's comments reflect that the ABS amber warning lamp is "ON" only during moist environmental changes, (rain, snow, vehicle wash) all wheel speed sensor circuitry should be thoroughly inspected for signs of water intrusion. Use the following procedure. Spray down the suspected area with 5% salt water solution (two teaspoons of salt to 12 oz. of water). Drive vehicle above 24 km/h (15 mph) for at least 30 seconds. If DTC returns immediately, replace suspected harness.

Any circuitry, that is suspected as causing the intermittent complaint, should be thoroughly checked for backed out terminals, improper mating, broken locks, improperly formed or damaged terminals, poor terminal to wiring connections or physical damage to the wiring harness.

Resistance of the wheel speed sensor will increase with an increase in sensor temperature.

When replacing a wheel speed sensor, inspect the sensor terminals and harness connector for corrosion and/or water intrusion. If evidence of corrosion or water intrusion exists, replace wheel speed sensor harness.

If wheel speed sensor harness shows evidence of water intrusion, inspect wheel speed sensor for water intrusion. If water intrusion is evident, replace wheel speed sensor.

DIAGNOSTIC CHART DTC 22 — 1994 CAVALIER AND SUNBIRD

DTC 22

(Page 1 of 2)

RIGHT FRONT WHEEL SPEED = 0 OR UNREASONABLE

IMPORTANT:

- WHEEL SPEED SENSOR INTERMITTENT MALFUNCTIONS MAY BE DIFFICULT TO LOCATE. CARE SHOULD BE TAKEN NOT TO DISTURB ANY ELECTRICAL CONNECTIONS PRIOR TO AN INDICATED STEP OF THIS CHART. THIS WILL ENSURE THAT AN INTERMITTENT CONNECTION WILL NOT BE CORRECTED BEFORE THE SOURCE OF THE MALFUNCTION IS FOUND.

(1)
- IGNITION "OFF."
- INSPECT WHEEL SPEED SENSOR WIRING AND CONNECTORS FOR DAMAGE. INSPECT WHEEL SPEED SENSOR AND TOOTHED RING FOR LOOSENESS OR DAMAGE. PAY PARTICULAR ATTENTION TO THE TOOTHED RING. ANY SIGNIFICANT DAMAGE WILL AFFECT THE WHEEL SPEED INPUT SIGNAL.

IS DAMAGE OR SENSOR/TOOTHED RING LOOSENESS EVIDENT?

YES → REPAIR DAMAGE OR SENSOR/TOOTHED RING LOOSENESS. DRIVE VEHICLE ABOVE 24 km/h (15 MPH) FOR AT LEAST 30 SECONDS TO ENSURE DTC 22 DOES NOT SET.

NO ↓

(2)
- DISCONNECT CONNECTOR FROM RIGHT FRONT WHEEL SPEED SENSOR.
- USING J 39200, MEASURE RESISTANCE BETWEEN SENSOR TERMINALS "A" AND "B".

IS RESISTANCE BETWEEN 1628 AND 1889 OHMS WHEN SENSOR IS 25°C (77°F)? *[IF TEMPERATURE OF THE SENSOR IS SUSPECTED TO BE OTHER THAN 25°C (77°F), REFER TO TABLE BELOW FOR APPLICABLE SENSOR RESISTANCE VALUES; THE VALUES IN THE TABLE ARE FOR TEMPERATURE OF THE SENSOR NOT AIR TEMPERATURE.]*

NO → REPLACE WHEEL SPEED SENSOR.

YES ↓

(3)
- WITH J 39200 STILL CONNECTED, SELECT AC VOLTAGE SCALE.
- SPIN WHEEL BY HAND WHILE MONITORING VOLTAGE.

IS VOLTAGE AT LEAST 100 mV (VOLTAGE WILL INCREASE AS WHEEL SPEED INCREASES)?

NO → REPLACE WHEEL SPEED SENSOR.

YES ↓

(4)
- USING J 39200, MEASURE RESISTANCE BETWEEN SENSOR TERMINAL "A" AND GROUND.

DOES J 39200 DISPLAY "OL" (INFINITE)?

NO → REPLACE WHEEL SPEED SENSOR.

YES → GO TO PAGE 2 OF 2 OF THIS DTC CHART.

FRONT WHEEL SPEED SENSOR

TEMPERATURE OF SENSOR VS. SENSOR RESISTANCE (APPROXIMATE)

TEMP. (°C)	TEMP. (°F)	RESISTANCE (OHMS)
-34 TO 4	-30 TO 40	1258 TO 1739
5 TO 43	41 TO 110	1503 TO 2023
44 TO 93	111 TO 200	1747 TO 2384

THIS CHART ASSUMES THAT A CURRENT DTC IS STORED INDICATING THAT THIS MALFUNCTION IS PRESENT.

DIAGNOSTIC CHART DTC 22 — 1994 CAVALIER AND SUNBIRD

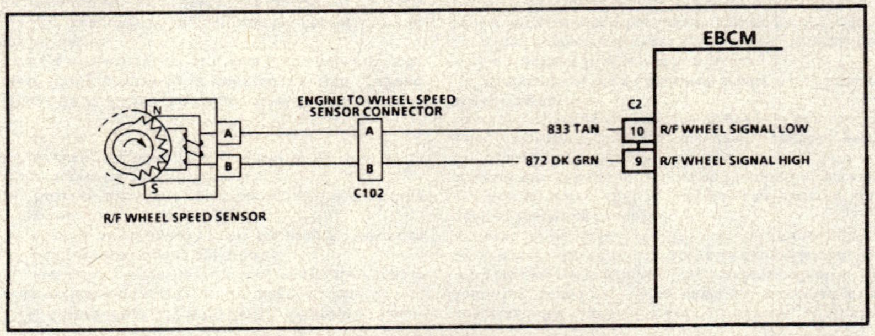

DTC 22
(Page 2 of 2)
RIGHT FRONT WHEEL SPEED = 0 OR UNREASONABLE

Circuit Description:

As a toothed ring passes by the wheel speed sensor, changes in the electromagnetic field cause the wheel speed sensor to produce a sinusoidal (AC) voltage signal whose frequency is proportional to wheel speed. The magnitude of this signal is directly related to wheel speed and the proximity of the wheel speed sensor to the toothed ring, often referred to as the air gap.

DTC Will Set When: DTC 22 can be set when the vehicle is not in an ABS stop. If the right front wheel speed is less than 1/2 of the vehicle speed and the vehicle speed is greater than 8 km/h (5 mph), a malfunction exists.

Action Taken: A malfunction DTC is stored, ABS is disabled and the ABS warning lamp is turned "ON."

DTC Chart Test Description: Number(s) below refer to circled number(s) on the diagnostic chart.

5. Checks for proper voltages at speed sensor harness connector.
6. Ensures that the wheel speed sensor circuitry is not shorted to ground.
7. Checks for an open in the wheel signal low circuit.
8. Checks for an open in the wheel signal high circuit.
9. Ensures that DTC 22 was not set due to a poor connection at the EBCM or physical damage of the circuitry.
10. Checks for a short to B+ in either the high or low wheel signal circuits.
11. Ensures short to B+ was not due to physical damage of the circuitry.
12. Tests for a short between the wheel signal high and low circuits.
13. Ensures malfunction was not due to a poor connection at the wheel speed sensor or physical damage of the circuitry.

Diagnostic Aids: An "Intermittent" malfunction is most likely caused by a poor connection, rubbed through wire insulation, or a wire that is broken inside the insulation.

The frequency of the malfunction can be checked by using the enhanced diagnostic function of the Tech 1.

If the customer's comments reflect that the ABS amber warning lamp is "ON" only during moist environmental changes, (rain, snow, vehicle wash) all wheel speed sensor circuitry should be thoroughly inspected for signs of water intrusion. Use the following procedure. Spray down the suspected area with 5% salt water solution (two teaspoons of salt to 12 oz. of water). Drive vehicle above 24 km/h (15 mph) for at least 30 seconds. If DTC returns immediately, replace suspected harness.

Any circuitry, that is suspected as causing the intermittent complaint, should be thoroughly checked for backed out terminals, improper mating, broken locks, improperly formed or damaged terminals, poor terminal to wiring connections or physical damage to the wiring harness.

Resistance of the wheel speed sensor will increase with an increase in sensor temperature.

When replacing a wheel speed sensor, inspect the sensor terminals and harness connector for corrosion and/or water intrusion. If evidence of corrosion or water intrusion exists, replace wheel speed sensor harness.

If wheel speed sensor harness shows evidence of water intrusion, inspect wheel speed sensor for water intrusion. If water intrusion is evident, replace wheel speed sensor.

DIAGNOSTIC CHART DTC 22 — 1994 CAVALIER AND SUNBIRD

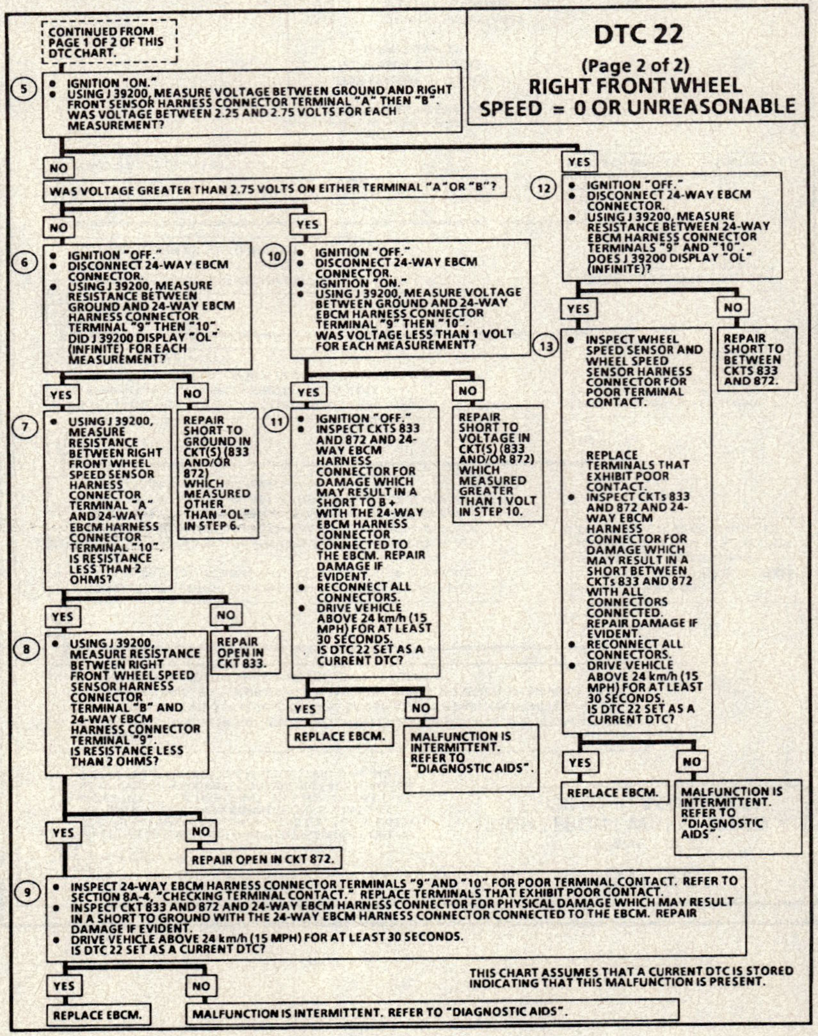

DIAGNOSTIC CHART DTC 23 — 1994
CAVALIER AND SUNBIRD

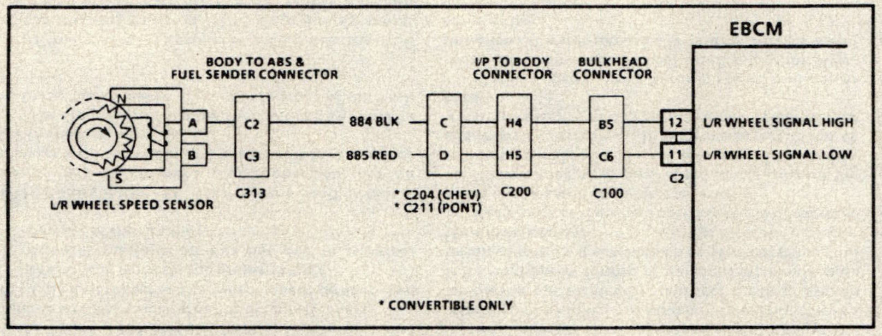

DTC 23

LEFT REAR WHEEL SPEED = 0 OR UNREASONABLE

Circuit Description:
As a toothed ring passes by the wheel speed sensor, changes in the electromagnetic field cause the wheel speed sensor to produce a sinusoidal (AC) voltage signal whose frequency is proportional to wheel speed. The magnitude of this signal is directly related to wheel speed and the proximity of the wheel speed sensor to the toothed ring, often referred to as the air gap.

DTC Will Set When: DTC 23 can be set when the vehicle is not in an ABS stop. If the left rear wheel speed is less than 1/2 of the vehicle speed and the vehicle speed is greater than 8 km/h (5 mph), a malfunction exists.

Action Taken: A malfunction DTC is stored, ABS is disabled and the ABS warning lamp is turned "ON."

DTC Chart Test Description: Number(s) below refer to circled number(s) on the diagnostic chart.

1. This step will identify circuitry that is damaged and visibly apparent.
2. Checks for proper voltages at the rear sensor harness connector.
3. Checks for a short between the wheel signal high and low circuits.
4. This test checks wheel speed sensor for proper resistance values.
5. This step ensures the wheel speed sensor is not shorted to ground.

Diagnostic Aids: An "Intermittent" malfunction is most likely caused by a poor connection, rubbed through wire insulation, or a wire that is broken inside the insulation.

The frequency of the malfunction can be checked by using the enhanced diagnostic function of the Tech 1.

If the customer's comments reflect that the ABS amber warning lamp is "ON" only during moist environmental changes, (rain, snow, vehicle wash) all wheel speed sensor circuitry should be thoroughly inspected for signs of water intrusion. Use the following procedure. Spray down the suspected area with 5% salt water solution (two teaspoons of salt to 12 oz. of water). Drive vehicle above 24 km/h (15 mph) for at least 30 seconds. If DTC returns immediately, replace suspected harness.

Any circuitry, that is suspected as causing the intermittent complaint, should be thoroughly checked for backed out terminals, improper mating, broken locks, improperly formed or damaged terminals, poor terminal to wiring connections or physical damage to the wiring harness.

Resistance of the wheel speed sensor will increase with an increase in sensor temperature.

When replacing a wheel speed sensor, inspect the sensor terminals and harness connector for corrosion and/or water intrusion. If evidence or corrosion or water intrusion exists, replace wheel speed sensor harness.

If wheel speed sensor harness shows evidence of water intrusion, inspect wheel speed sensor for water intrusion. If water intrusion is evident, replace wheel speed sensor.

DIAGNOSTIC CHART DTC 23 — 1994
CAVALIER AND SUNBIRD

DTC 23
(Page 1 of 4)
LEFT REAR WHEEL SPEED = 0 OR UNREASONABLE

Important
- WHEEL SPEED SENSOR INTERMITTENT MALFUNCTIONS MAY BE DIFFICULT TO LOCATE. CARE SHOULD BE TAKEN NOT TO DISTURB ANY ELECTRICAL CONNECTIONS PRIOR TO AN INDICATED STEP OF THIS CHART. THIS WILL ENSURE THAT AN INTERMITTENT CONNECTION WILL NOT BE CORRECTED BEFORE THE SOURCE OF THE MALFUNCTION IS FOUND.

(1)
- IGNITION "OFF."
- INSPECT WHEEL SPEED SENSOR WIRING AND CONNECTORS FOR DAMAGE. IS DAMAGE EVIDENT?

YES: REPAIR DAMAGE. DRIVE VEHICLE ABOVE 24 km/h (15 MPH) FOR AT LEAST 30 SECONDS TO ENSURE DTC 23 DOES NOT SET.

NO: (2)
- DISCONNECT CONNECTOR FROM LEFT REAR WHEEL SPEED SENSOR.
- IGNITION "ON."
- USING J 39200, MEASURE VOLTAGE BETWEEN GROUND AND LEFT REAR SENSOR HARNESS CONNECTOR TERMINAL "A" THEN "B". WAS VOLTAGE BETWEEN 2.25 AND 2.75 VOLTS FOR EACH MEASUREMENT?

NO: GO TO PAGE 3 OF 4 OF THIS DTC CHART.

YES: (3)
- IGNITION "OFF."
- DISCONNECT 24-WAY EBCM CONNECTOR.
- USING J 39200, MEASURE RESISTANCE BETWEEN 24-WAY EBCM HARNESS CONNECTOR TERMINALS "11" AND "12." DOES J 39200 DISPLAY "OL" (INFINITE)?

NO: REPAIR SHORT BETWEEN CKTs 884 AND 885.

YES: (4)
- USING J 39200, MEASURE THE RESISTANCE BETWEEN LEFT REAR WHEEL SPEED SENSOR TERMINALS "A" AND "B". IS RESISTANCE BETWEEN 2100 AND 2400 OHMS WHEN SENSOR IS 25°C (77°F)? (IF TEMPERATURE OF THE SENSOR IS SUSPECTED TO BE OTHER THAN 25°C (77°F). REFER TO THE TABLE BELOW FOR APPLICABLE SENSOR RESISTANCE VALUES; THE VALUES IN THE TABLE ARE FOR TEMPERATURE OF THE SENSOR NOT AIR TEMPERATURE).

NO: REPLACE INTERGRAL BEARING AND SPEED SENSOR ASSEMBLY.

YES: (5)
- USING J 39200, MEASURE RESISTANCE BETWEEN SENSOR TERMINAL "A" AND INTEGRAL BEARING AND SPEED SENSOR ASSEMBLY MOUNTING FLANGE. DOES J 39200 DISPLAY "OL" (INFINITE)?

YES: CONTINUE DIAGNOSIS ON NEXT PAGE.

NO: REPLACE INTEGRAL BEARING AND SPEED SENSOR ASSEMBLY.

REAR WHEEL SPEED SENSOR
TEMPERATURE OF SENSOR VS. SENSOR RESISTANCE (APPROXIMATE)

TEMP. (°C)	TEMP. (°F)	RESISTANCE (OHMS)
-34 TO 4	-30 TO 40	1622 TO 2205
5 TO 43	41 TO 110	1930 TO 2566
44 TO 93	111 TO 200	2245 TO 3028

THIS CHART ASSUMES THAT A CURRENT DTC IS STORED INDICATING THAT THIS MALFUNCTION IS PRESENT.

DIAGNOSTIC CHART DTC 23 — 1994 CAVALIER AND SUNBIRD

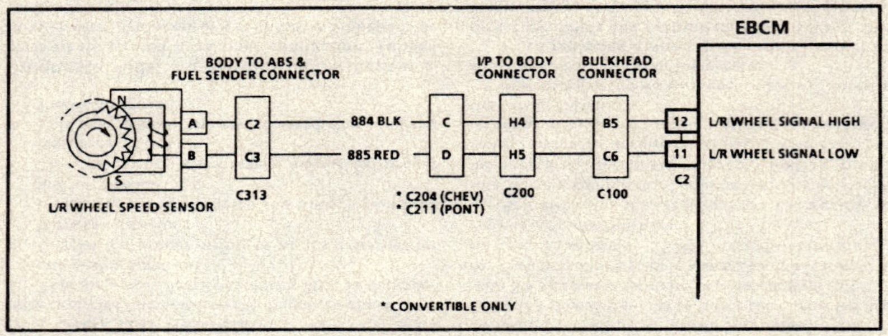

DTC 23

(Page 2 of 4)
LEFT REAR WHEEL SPEED = 0
OR UNREASONABLE

Circuit Description:

As a toothed ring passes by the wheel speed sensor, changes in the electromagnetic field cause the wheel speed sensor to produce a sinusoidal (AC) voltage signal whose frequency is proportional to wheel speed. The magnitude of this signal is directly related to wheel speed and the proximity of the wheel speed sensor to the toothed ring, often referred to as the air gap.

DTC Will Set When: DTC 23 can be set when the vehicle is not in an ABS stop. If the left rear wheel speed is less than 1/2 of the vehicle speed and the vehicle speed is greater than 8 km/h (5 mph), a malfunction exists.

Action Taken: A malfunction DTC is stored, ABS is disabled and the ABS warning lamp is turned "ON."

DTC Chart Test Description: Number(s) below refer to circled number(s) on the diagnostic chart.

6. This test ensures the wheel speed sensor and sensor ring generate the proper voltage.
7. Ensures malfunction was not due to physical damage of the circuitry.

Diagnostic Aids: An "Intermittent" malfunction is most likely caused by a poor connection, rubbed through wire insulation, or a wire that is broken inside the insulation.

The frequency of the malfunction can be checked by using the enhanced diagnostic function of the Tech 1.

If the customer's comments reflect that the ABS amber warning lamp is "ON" only during moist environmental changes, (rain, snow, vehicle wash) all wheel speed sensor circuitry should be thoroughly inspected for signs of water intrusion. Use the following procedure. Spray down the suspected area with 5% salt water solution (two teaspoons of salt to 12 oz. of water). Drive vehicle above 24 km/h (15 mph) for at least 30 seconds. If DTC returns immediately, replace suspected harness.

Any circuitry, that is suspected as causing the intermittent complaint, should be thoroughly checked for backed out terminals, improper mating, broken locks, improperly formed or damaged terminals, poor terminal to wiring connections or physical damage to the wiring harness.

Resistance of the wheel speed sensor will increase with an increase in sensor temperature.

When replacing a wheel speed sensor, inspect the sensor terminals and harness connector for corrosion and/or water intrusion. If evidence or corrosion or water intrusion exists, replace wheel speed sensor harness.

If wheel speed sensor harness shows evidence of water intrusion, inspect wheel speed sensor for water intrusion. If water intrusion is evident, replace wheel speed sensor.

DIAGNOSTIC CHART DTC 23 — 1994 CAVALIER AND SUNBIRD

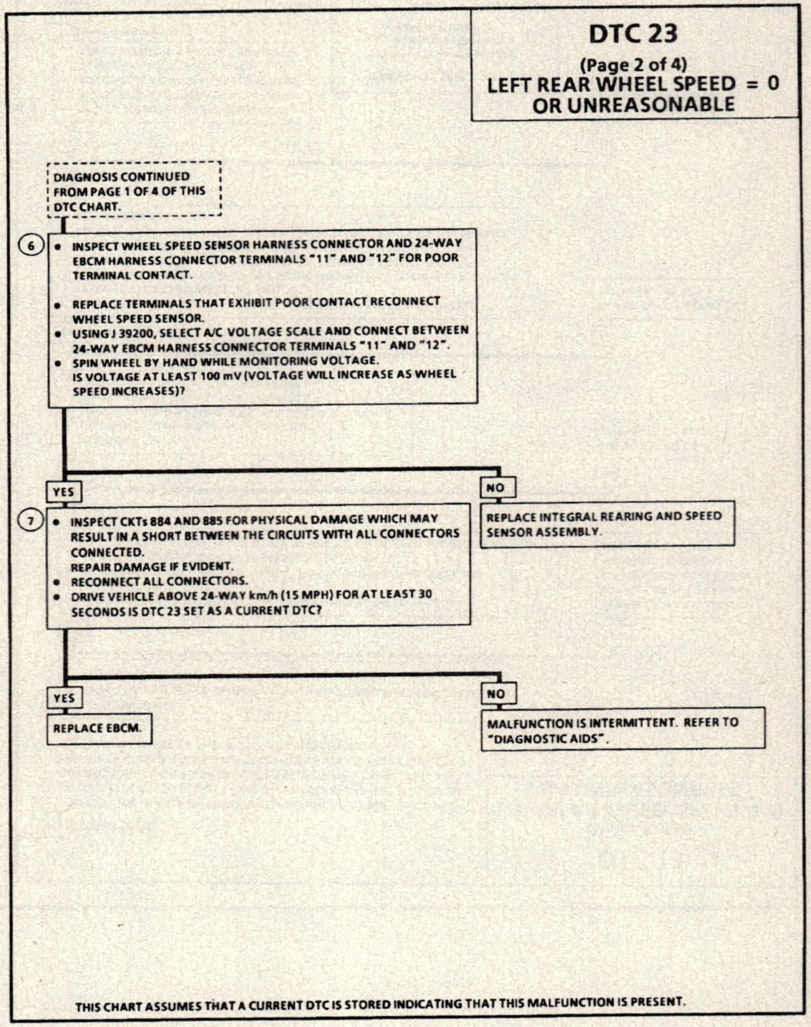

DIAGNOSTIC CHART DTC 23 — 1994
CAVALIER AND SUNBIRD

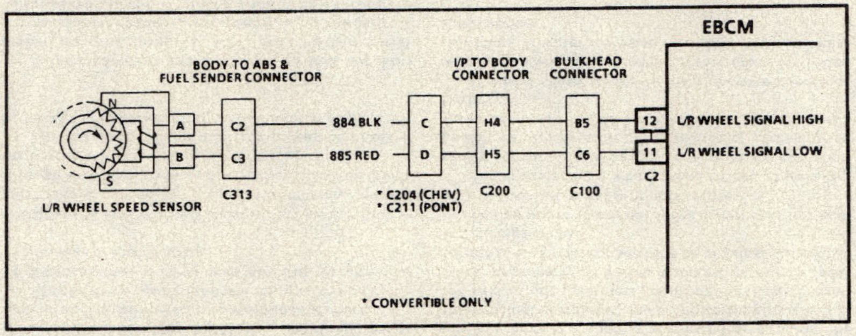

DTC 23
(Page 3 of 4)
LEFT REAR WHEEL SPEED = 0 OR UNREASONABLE

Circuit Description:
As a toothed ring passes by the wheel speed sensor, changes in the electromagnetic field cause the wheel speed sensor to produce a sinusoidal (AC) voltage signal whose frequency is proportional to wheel speed. The magnitude of this signal is directly related to wheel speed and the proximity of the wheel speed sensor to the toothed ring, often referred to as the air gap.

DTC Will Set When: DTC 23 can be set when the vehicle is not in an ABS stop. If the left rear wheel speed is less than 1/2 of the vehicle speed and the vehicle speed is greater than 8 km/h (5 mph), a malfunction exists.

Action Taken: A malfunction DTC is stored, ABS is disabled and the ABS warning lamp is turned "ON."

DTC Chart Test Description: Number(s) below refer to circled number(s) on the diagnostic chart.

8. This step checks for proper voltages at body to ABS and fuel sender harness connector.
9. Checks for proper voltages at the engine to body harness connector.
10. Checks for a short to ground in the wheel signal circuit.
11. This checks for an open in the wheel signal high circuit.
12. Checks wheel speed circuits for short to B+ at engine harness.

Diagnostic Aids: An "Intermittent" malfunction is most likely caused by a poor connection, rubbed through wire insulation, or a wire that is broken inside the insulation.

The frequency of the malfunction can be checked by using the enhanced diagnostic function of the Tech 1.

If the customer's comments reflect that the ABS amber warning lamp is "ON" only during moist environmental changes, (rain, snow, vehicle wash) all wheel speed sensor circuitry should be thoroughly inspected for signs of water intrusion. Use the following procedure. Spray down the suspected area with 5% salt water solution (two teaspoons of salt to 12 oz. of water). Drive vehicle above 24 km/h (15 mph) for at least 30 seconds. If DTC returns immediately, replace suspected harness.

Any circuitry, that is suspected as causing the intermittent complaint, should be thoroughly checked for backed out terminals, improper mating, broken locks, improperly formed or damaged terminals, poor terminal to wiring connections or physical damage to the wiring harness.

Resistance of the wheel speed sensor will increase with an increase in sensor temperature.

When replacing a wheel speed sensor, inspect the sensor terminals and harness connector for corrosion and/or water intrusion. If evidence of corrosion or water intrusion exists, replace wheel speed sensor harness.

If wheel speed sensor harness shows evidence of water intrusion, inspect wheel speed sensor for water intrusion. If water intrusion is evident, replace wheel speed sensor.

DIAGNOSTIC CHART DTC 23 — 1994
CAVALIER AND SUNBIRD

DTC 23
(Page 3 of 4)
LEFT REAR WHEEL SPEED = 0 OR UNREASONABLE

CONTINUED FROM PAGE 1 OF 4 OF THIS DTC CHART.

(8)
- IGNITION "OFF."
- DISCONNECT REAR ABS HARNESS.
- IGNITION "ON."
- USING J 39200 MEASURE VOLTAGE BETWEEN GROUND AND BODY HARNESS CONNECTOR TERMINAL "C2" THE "C3". WAS VOLTAGE BETWEEN 2.25 AND 2.75 VOLTS FOR EACH MEASUREMENT?

- YES → INSPECT BODY TO REAR ABS HARNESS CONNECTOR FOR POOR TERMINAL CONTACT. REPLACE TERMINALS THAT EXHIBIT POOR CONTACT. IF POOR TERMINAL CONTACT IS NOT EVIDENT, REPLACE REAR ABS HARNESS.
- NO → (9)

(9)
- IGNITION "OFF."
- DISCONNECT BODY HARNESS FROM ENGINE HARNESS.
- IGNITION "ON."
- USING J 39200, MEASURE VOLTAGE BETWEEN GROUND AND ENGINE HARNESS CONNECTOR TERMINAL "B5" THEN "C6". WAS VOLTAGE BETWEEN 2.25 AND 2.75 VOLTS FOR EACH MEASUREMENT?

- YES → WAS VOLTAGE GREATER THAN 2.75 FOR ONE OR BOTH OF THE MEASUREMENTS IN STEP 8?
  - NO → REPAIR OPEN OR SHORT TO GROUND IN BODY HARNESS OR I/P HARNESS CKT(s) (884 AND/OR 885) WHICH MEASURED LESS THAN 2.25 VOLTS IN STEP 8.
  - YES → REPAIR SHORT TO VOLTAGE IN ENGINE HARNESS CKT(s) (884 AND/OR 885) WHICH MEASURED GREATER THAN 2.75 VOLTS IN STEP 8.
- NO → IS VOLTAGE GREATER THAN 2.75 VOLTS ON EITHER TERMINAL "D" OR "E"?
  - YES → (12)
  - NO → (10)

(10)
- IGNITION "OFF."
- DISCONNECT 24-WAY EBCM CONNECTOR.
- USING J 39200, MEASURE RESISTANCE BETWEEN GROUND AND 24-WAY EBCM HARNESS CONNECTOR TERMINAL "11" THEN "12". DID J 39200 DISPLAY "OL" (INFINITE) FOR EACH MEASUREMENT?

- YES → (11)
- NO → REPAIR SHORT TO GROUND IN ENGINE HARNESS CKT(s) (884 AND/OR 885) WHICH MEASURED OTHER THAN "OL" (INFINITE).

(12)
- IGNITION "OFF."
- DISCONNECT 24-WAY EBCM CONNECTOR.
- IGNITION "ON."
- USING J 39200 MEASURE VOLTAGE BETWEEN GROUND AND 24-WAY EBCM HARNESS CONNECTOR TERMINAL "11" THEN "12". WAS VOLTAGE LESS THAN 1 VOLT FOR EACH MEASUREMENT?

- YES → CONTINUE DIAGNOSIS AT STEP 13 ON PAGE 4 OF 4 OF THIS DTC CHART.
- NO → REPAIR SHORT TO VOLTAGE IN ENGINE HARNESS CKT(s) (884 AND/OR 885 WHICH MEASURED GREATER THAN 1 VOLT.

(11)
- USING J 39200 MEASURE RESISTANCE BETWEEN ENGINE HARNESS CONNECTOR TERMINAL "B5" AND 24-WAY EBCM HARNESS CONNECTOR TERMINAL "12". IS RESISTANCE LESS THAN 2 OHMS?

- YES → CONTINUE DIAGNOSIS ON NEXT PAGE.
- NO → REPAIR OPEN IN ENGINE HARNESS CKT 884.

THIS CHART ASSUMES THAT A CURRENT DTC IS STORED INDICATING THAT THIS MALFUNCTION IS PRESENT.

DIAGNOSTIC CHART DTC 23 — 1994 CAVALIER AND SUNBIRD

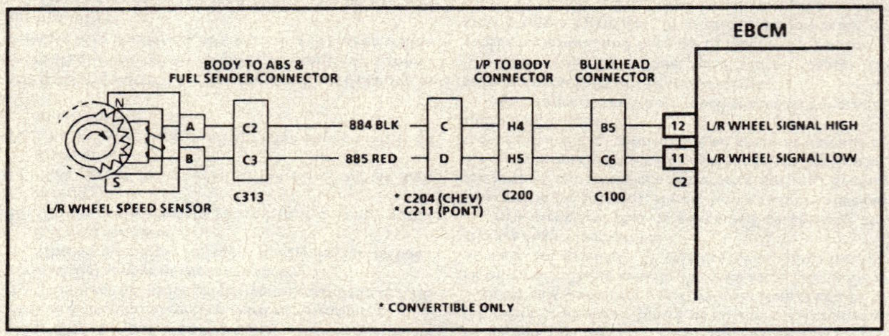

DTC 23

(Page 4 of 4)
LEFT REAR WHEEL SPEED = 0
OR UNREASONABLE

Circuit Description:

As a toothed ring passes by the wheel speed sensor, changes in the electromagnetic field cause the wheel speed sensor to produce a sinusoidal (AC) voltage signal whose frequency is proportional to wheel speed. The magnitude of this signal is directly related to wheel speed and the proximity of the wheel speed sensor to the toothed ring, often referred to as the air gap.

DTC Will Set When: DTC 23 can be set when the vehicle is not in an ABS stop. If the left rear wheel speed is less than 1/2 of the vehicle speed and the vehicle speed is greater than 8 km/h (5 mph), a malfunction exists.

Action Taken: A malfunction DTC is stored, ABS is disabled and the ABS warning lamp is turned "ON."

DTC Chart Test Description: Number(s) below refer to circled number(s) on the diagnostic chart.

13. Checks for an open in the wheel signal low circuit.
14. Ensures malfunction was not due to physical damage of the circuitry.

Diagnostic Aids: An "Intermittent" malfunction is most likely caused by a poor connection, rubbed through wire insulation, or a wire that is broken inside the insulation.

The frequency of the malfunction can be checked by using the enhanced diagnostic function of the Tech 1.

If the customer's comments reflect that the ABS amber warning lamp is "ON" only during moist environmental changes, (rain, snow, vehicle wash) all wheel speed sensor circuitry should be thoroughly inspected for signs of water intrusion. Use the following procedure. Spray down the suspected area with 5% salt water solution (two teaspoons of salt to 12 oz. of water). Drive vehicle above 24 km/h (15 mph) for at least 30 seconds. If DTC returns immediately, replace suspected harness.

Any circuitry, that is suspected as causing the intermittent complaint, should be thoroughly checked for backed out terminals, improper mating, broken locks, improperly formed or damaged terminals, poor terminal to wiring connections or physical damage to the wiring harness.

Resistance of the wheel speed sensor will increase with an increase in sensor temperature.

When replacing a wheel speed sensor, inspect the sensor terminals and harness connector for corrosion and/or water intrusion. If evidence of corrosion or water intrusion exists, replace wheel speed sensor harness.

If wheel speed sensor harness shows evidence of water intrusion, inspect wheel speed sensor for water intrusion. If water intrusion is evident, replace wheel speed sensor.

DIAGNOSTIC CHART DTC 23 — 1994 CAVALIER AND SUNBIRD

DTC 23

(Page 4 of 4)
LEFT REAR WHEEL SPEED = 0
OR UNREASONABLE

CONTINUED FROM PAGE 3 OF 4 OF THIS DTC CHART.

(13)
- USING J 39200, MEASURE RESISTANCE BETWEEN ENGINE HARNESS CONNECTOR TERMINAL "C6" AND 24-WAY EBCM HARNESS CONNECTOR TERMINAL "11". IS RESISTANCE LESS THAN 2 OHMS?

NO → REPAIR OPEN IN ENGINE HARNESS CKT 885.

YES →

(14)
- IGNITION "OFF."
- INSPECT ALL WHEEL SPEED SENSOR HARNESS CONNECTORS FOR POOR TERMINAL CONTACT OR CORROSION. REPLACE TERMINALS THAT EXHIBIT POOR CONTACT OR CORROSION.
- INSPECT ALL WHEEL SPEED SENSOR CONNECTORS AND HARNESSES FOR DAMAGE WHICH MAY RESULT IN AN OPEN OR SHORT TO GROUND/BATTERY WHEN ALL COMPONENTS ARE CONNECTED. REPAIR DAMAGE IF EVIDENT.
- RECONNECT ALL CONNECTORS.
- DRIVE VEHICLE ABOVE 24 km/h (15 MPH) FOR AT LEAST 30 SECONDS. IS DTC 23 SET AS A CURRENT DTC?

YES → REPLACE EBCM.

NO → MALFUNCTION IS INTERMITTENT. REFER TO "DIAGNOSTIC AIDS".

THIS CHART ASSUMES THAT A CURRENT DTC IS STORED INDICATING THAT THIS MALFUNCTION IS PRESENT.

DIAGNOSTIC CHART DTC 24 — 1994 CAVALIER AND SUNBIRD

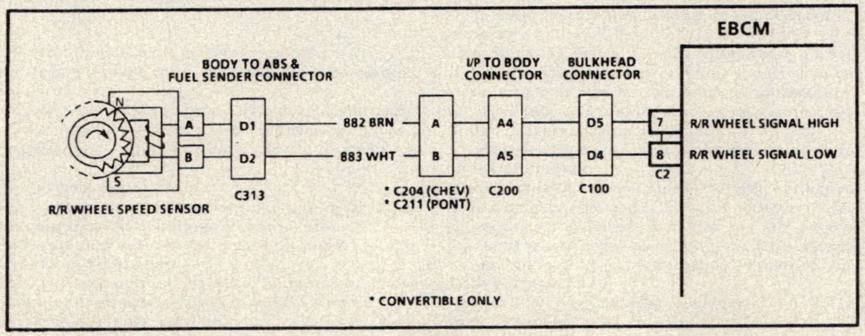

DTC 24

(Page 1 of 4)
RIGHT REAR WHEEL SPEED = 0
OR UNREASONABLE

Circuit Description:

As a toothed ring passes by the wheel speed sensor, changes in the electromagnetic field cause the wheel speed sensor to produce a sinusoidal (AC) voltage signal whose frequency is proportional to wheel speed. The magnitude of this signal is directly related to wheel speed and the proximity of the wheel speed sensor to the toothed ring, often referred to as the air gap.

DTC Will Set When: DTC 24 can be set when the vehicle is not in an ABS stop. If the right rear wheel speed is less than 1/2 of the vehicle speed and the vehicle speed is greater than 8 km/h (5 mph), a malfunction exists.

Action Taken: A malfunction DTC is stored, ABS is disabled and the ABS warning lamp is turned "ON."

DTC Chart Test Description: Number(s) below refer to circled number(s) on the diagnostic chart.

1. This step will identify circuitry that is damaged and visibly apparent.
2. Checks for proper voltages at the rear sensor harness connector.
3. Checks for a short between the wheel signal high and low circuits.
4. This test checks wheel speed sensor for proper resistance values.

Diagnostic Aids: An "Intermittent" malfunction is most likely caused by a poor connection, rubbed through wire insulation, or a wire that is broken inside the insulation.

The frequency of the malfunction can be checked by using the enhanced diagnostic function of the Tech 1.

If the customer's comments reflect that the ABS amber warning lamp is "ON" only during moist environmental changes, (rain, snow, vehicle wash) all wheel speed sensor circuitry should be thoroughly inspected for water intrusion.

Use the following procedure. Spray down the suspected area with 5% salt water solution (two teaspoons of salt to 12 oz. of water). Drive vehicle above 24 km/h (15 mph) for at least 30 seconds. If DTC returns immediately, replace suspected harness.

Any circuitry, that is suspected as causing the intermittent complaint, should be thoroughly checked for backed out terminals, improper mating, broken locks, improperly formed or damaged terminals, poor terminal to wiring connections or physical damage to the wiring harness.

Resistance of the wheel speed sensor will increase with an increase in sensor temperature.

When replacing a wheel speed sensor, inspect the sensor terminals and harness connector for corrosion and/or water intrusion. If evidence of corrosion or water intrusion exists, replace wheel speed sensor harness.

If wheel speed sensor harness shows evidence of water intrusion, inspect wheel speed sensor for water intrusion. If water intrusion is evident, replace wheel speed sensor.

DIAGNOSTIC CHART DTC 24 — 1994 CAVALIER AND SUNBIRD

DTC 24
(Page 1 of 4)
RIGHT REAR WHEEL SPEED = 0
OR UNREASONABLE

Important

- WHEEL SPEED SENSOR INTERMITTENT MALFUNCTIONS MAY BE DIFFICULT TO LOCATE. CARE SHOULD BE TAKEN NOT TO DISTURB ANY ELECTRICAL CONNECTIONS PRIOR TO AN INDICATED STEP OF THIS CHART. THIS WILL ENSURE THAT AN INTERMITTENT CONNECTION WILL NOT BE CORRECTED BEFORE THE SOURCE OF THE MALFUNCTION IS FOUND.

(1)
- IGNITION "OFF."
- INSPECT WHEEL SPEED SENSOR WIRING AND CONNECTORS FOR DAMAGE. IS DAMAGE EVIDENT?

YES: REPAIR DAMAGE. DRIVE VEHICLE ABOVE 24 km/h (15MPH) FOR AT LEAST 30 SECONDS TO ENSURE DTC 24 DOES NOT SET.

NO:

(2)
- DISCONNECT CONNECTOR FROM RIGHT REAR WHEEL SPEED SENSOR.
- IGNITION "ON."
- USING J 39200, MEASURE VOLTAGE BETWEEN GROUND AND RIGHT REAR SENSOR HARNESS CONNECTOR TERMINAL "A" THEN "B". WAS VOLTAGE BETWEEN 2.25 AND 2.75 VOLTS FOR EACH MEASUREMENT?

YES: GO TO PAGE 3 OF 4 OF THIS CHART.

NO:

(3)
- IGNITION "OFF."
- DISCONNECT 24-WAY EBCM CONNECTOR.
- USING J39200, MEASURE RESISTANCE BETWEEN 24-WAY EBCM HARNESS CONNECTOR TERMINALS "8" AND "7." DOES J 39200 DISPLAY "OL" (INFINITE).

NO: REPAIR SHORT BETWEEN CKT(s) 882 AND 883.

YES:

(4)
- USING J 39200, MEASURE RESISTANCE BETWEEN RIGHT REAR WHEEL SPEED SENSOR TERMINALS "A" AND "B." IS RESISTANCE BETWEEN 2100 AND 2400 OHMS WHEN SENSOR IS 25°C (77°F)? (IF TEMPERATURE OF THE SENSOR IS SUSPECTED TO BE OTHER THAN 25°C (77°F). REFER TO THE TABLE BELOW FOR APPLICABLE SENSOR RESISTANCE VALUES; THE VALUES IN THE TABLE ARE FOR TEMPERATURE OF THE SENSOR NOT AIR TEMPERATURE).

YES: CONTINUE DIAGNOSIS ON NEXT PAGE.

NO: REPLACE INTEGRAL BEARING AND SPEED SENSOR ASSEMBLY.

REAR WHEEL SPEED SENSOR
TEMPERATURE OF SENSOR VS. SENSOR RESISTANCE (APPROXIMATE)

TEMP. (°C)	TEMP. (°F)	RESISTANCE (OHMS)
-34 TO 4	-30 TO 40	1623 TO 2210
5 TO 43	41 TO 110	1939 TO 2570
44 TO 93	111 TO 200	2254 TO 3029

THIS CHART ASSUMES THAT A CURRENT DTC IS STORED INDICATING THAT THIS MALFUNCTION IS PRESENT.

DIAGNOSTIC CHART DTC 24 — 1994 CAVALIER AND SUNBIRD

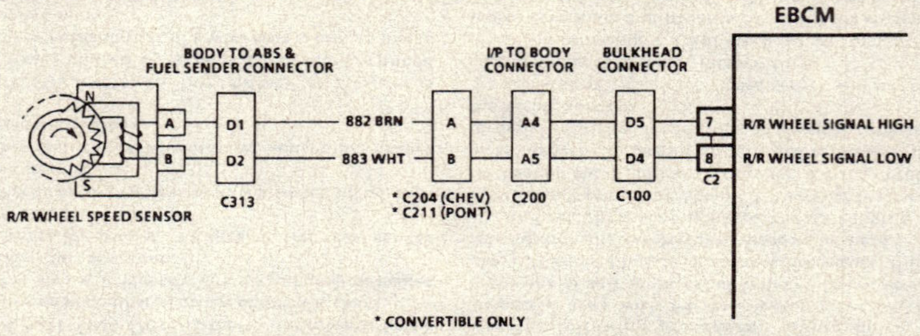

DTC 24

(Page 2 of 4)
RIGHT REAR WHEEL SPEED = 0
OR UNREASONABLE

Circuit Description:

As a toothed ring passes by the wheel speed sensor, changes in the electromagnetic field cause the wheel speed sensor to produce a sinusoidal (AC) voltage signal whose frequency is proportional to wheel speed. The magnitude of this signal is directly related to wheel speed and the proximity of the wheel speed sensor to the toothed ring, often referred to as the air gap.

DTC Will Set When: DTC 24 can be set when the vehicle is not in an ABS stop. If the right rear wheel speed is less than 1/2 of the vehicle speed and the vehicle speed is greater than 8 km/h (5 mph), a malfunction exists.

Action Taken: A malfunction DTC is stored, ABS is disabled and the ABS warning lamp is turned "ON."

DTC Chart Test Description: Number(s) below refer to circled number(s) on the diagnostic chart.

5. This step ensures the wheel speed sensor is not shorted to ground.
6. This test ensures the wheel speed sensor and sensor ring generate the proper voltage.
7. Ensures malfunction was not due to physical damage of the circuitry.

Diagnostic Aids: An "Intermittent" malfunction is most likely caused by a poor connection, rubbed through wire insulation, or a wire that is broken inside the insulation.

The frequency of the malfunction can be checked by using the enhanced diagnostic function of the Tech 1.

If the customer's comments reflect that the ABS amber warning lamp is "ON" only during moist environmental changes, (rain, snow, vehicle wash) all wheel speed sensor circuitry should be thoroughly inspected for water intrusion. Use the following procedure. Spray down the suspected area with 5% salt water solution (two teaspoons of salt to 12 oz. of water). Drive vehicle above 24 km/h (15 mph) for at least 30 seconds. If DTC returns immediately, replace suspected harness.

Any circuitry, that is suspected as causing the intermittent complaint, should be thoroughly checked for backed out terminals, improper mating, broken locks, improperly formed or damaged terminals, poor terminal to wiring connections or physical damage to the wiring harness.

Resistance of the wheel speed sensor will increase with an increase in sensor temperature.

When replacing a wheel speed sensor, inspect the sensor terminals and harness connector for corrosion and/or water intrusion. If evidence of corrosion or water intrusion exists, replace wheel speed sensor harness.

If wheel speed sensor harness shows evidence of water intrusion, inspect wheel speed sensor for water intrusion. If water intrusion is evident, replace wheel speed sensor.

DIAGNOSTIC CHART DTC 24 — 1994 CAVALIER AND SUNBIRD

DTC 24

(Page 2 of 4)
RIGHT REAR WHEEL SPEED = 0
OR UNREASONABLE

CONTINUED FROM PAGE 1 OF 4 OF THIS DTC CHART.

(5)
- USING J 39200, MEASURE RESISTANCE BETWEEN SENSOR TERMINAL "A" AND INTEGRAL BEARING AND SPEED SENSOR ASSEMBLY MOUNTING FLANGE.

DOES J 39200 DISPLAY "OL" (INFINITE)?

NO: REPLACE INTEGRAL BEARING AND SPEED SENSOR ASSEMBLY.

YES:

(6)
- INSPECT WHEEL SPEED SENSOR HARNESS CONNECTOR AND 24-WAY EBCM HARNESS CONNECTOR TERMINALS "8" AND "7" FOR POOR TERMINAL CONTACT.

REPLACE TERMINALS THAT EXHIBIT POOR CONTACT.
- RECONNECT WHEEL SPEED SENSOR.
- USING J 39200, SELECT A/C VOLTAGE SCALE AND CONNECT CONNECTOR TERMINALS "8" AND "7".
- SPIN WHEEL BY HAND WHILE MONITORING VOLTAGE.

IS VOLTAGE AT LEAST 100 mV (VOLTAGE WILL INCREASE AS WHEEL SPEED INCREASES)?

NO: REPLACE INTEGRAL BEARING AND SPEED SENSOR ASSEMBLY.

YES:

(7)
- INSPECT CKTs 882 AND 883 FOR PHYSICAL DAMAGE WHICH MAY RESULT IN A SHORT BETWEEN THE CIRCUITS WITH ALL CONNECTORS CONNECTED REPAIR DAMAGE IF EVIDENT.
- RECONNECT ALL CONNECTORS.
- DRIVE VEHICLE ABOVE 24 km/h (15 MPH) FOR AT LEAST 30 SECONDS.

IS DTC 24 SET AS A CURRENT DTC?

YES: REPLACE EBCM.

NO: MALFUNCTION IS INTERMITTENT. REFER TO "DIAGNOSTIC AIDS".

THIS CHART ASSUMES THAT A CURRENT DTC IS STORED INDICATING THAT THIS MALFUNCTION IS PRESENT.

DIAGNOSTIC CHART DTC 24 — 1994
CAVALIER AND SUNBIRD

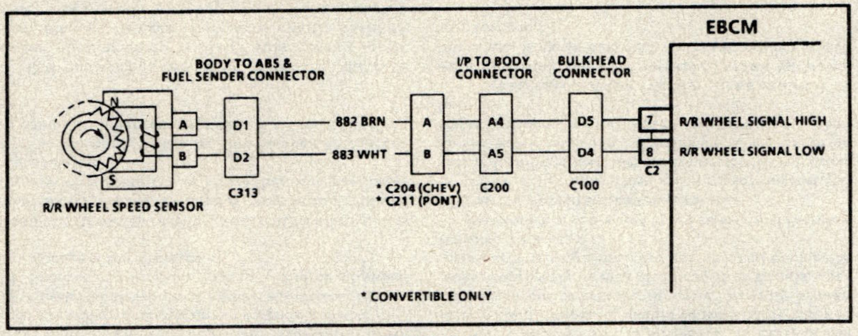

DTC 24
(Page 3 of 4)
RIGHT REAR WHEEL SPEED = 0 OR UNREASONABLE

Circuit Description:
As a toothed ring passes by the wheel speed sensor, changes in the electromagnetic field cause the wheel speed sensor to produce a sinusoidal (AC) voltage signal whose frequency is proportional to wheel speed. The magnitude of this signal is directly related to wheel speed and the proximity of the wheel speed sensor to the toothed ring, often referred to as the air gap.

DTC Will Set When: DTC 24 can be set when the vehicle is not in an ABS stop. If the right rear wheel speed is less than 1/2 of the vehicle speed and the vehicle speed is greater than 8 km/h (5 mph), a malfunction exists.

Action Taken: A malfunction DTC is stored, ABS is disabled and the ABS warning lamp is turned "ON."

DTC Chart Test Description: Number(s) below refer to circled number(s) on the diagnostic chart.

8. This step checks for proper voltages at body to ABS and fuel sender harness connector.
9. Checks for proper voltages at the engine to body harness connector.
10. Checks for a short to ground in the wheel signal circuits.
11. This checks for an open in the wheel signal high circuit.
12. Checks wheel speed circuits for short to B+ at engine harness.

Diagnostic Aids: An "Intermittent" malfunction is most likely caused by a poor connection, rubbed through wire insulation, or a wire that is broken inside the insulation.

The frequency of the malfunction can be checked by using the enhanced diagnostic function of the Tech 1.

If the customer's comments reflect that the ABS amber warning lamp is "ON" only during moist environmental changes, (rain, snow, vehicle wash) all wheel speed sensor circuitry should be thoroughly inspected for water intrusion. Use the following procedure. Spray down the suspected area with 5% salt water solution (two teaspoons of salt to 12 oz. of water). Drive vehicle above 24 km/h (15 mph) for at least 30 seconds. If DTC returns immediately, replace suspected harness.

Any circuitry, that is suspected as causing the intermittent complaint, should be thoroughly checked for backed out terminals, improper mating, broken locks, improperly formed or damaged terminals, poor terminal to wiring connections or physical damage to the wiring harness.

Resistance of the wheel speed sensor will increase with an increase in sensor temperature.

When replacing a wheel speed sensor, inspect the sensor terminals and harness connector for corrosion and/or water intrusion. If evidence of corrosion or water intrusion exists, replace wheel speed sensor harness.

If wheel speed sensor harness shows evidence of water intrusion, inspect wheel speed sensor for water intrusion. If water intrusion is evident, replace wheel speed sensor.

DIAGNOSTIC CHART DTC 24 — 1994
CAVALIER AND SUNBIRD

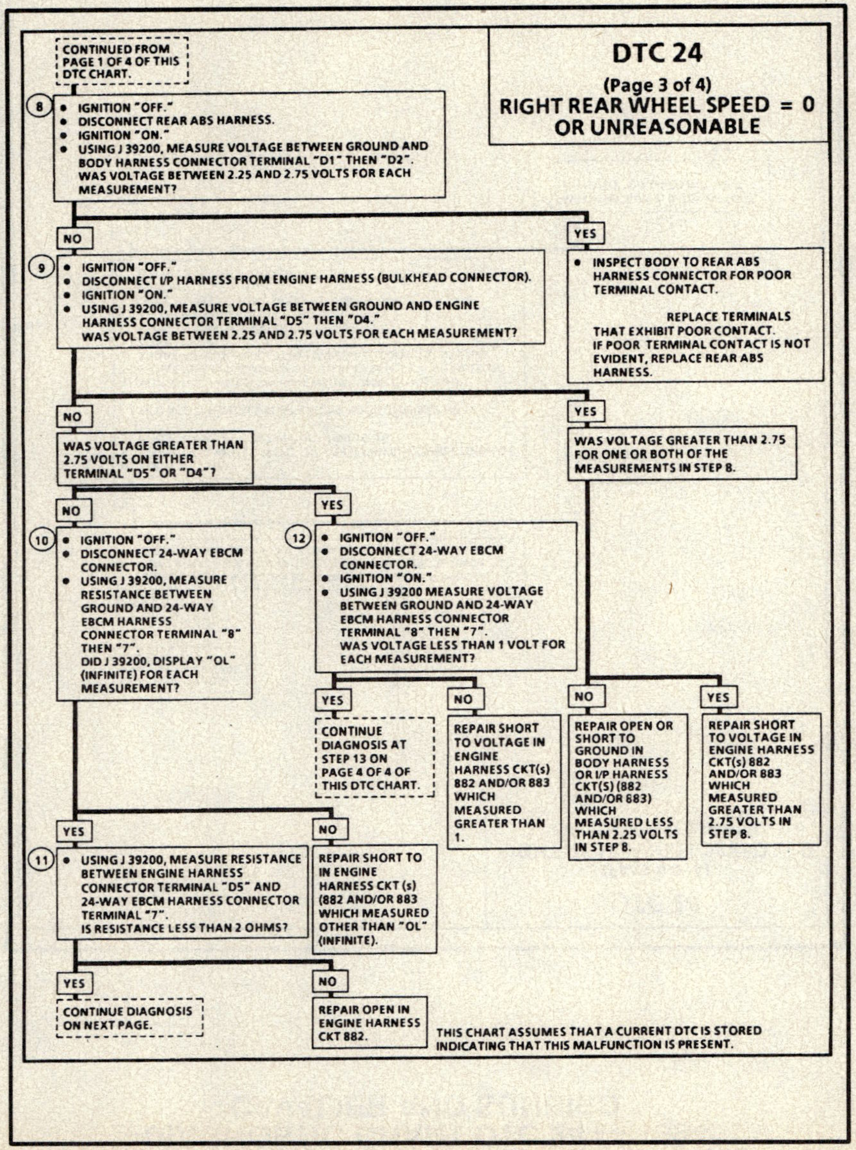

DIAGNOSTIC CHART DTC 24 — 1994 CAVALIER AND SUNBIRD

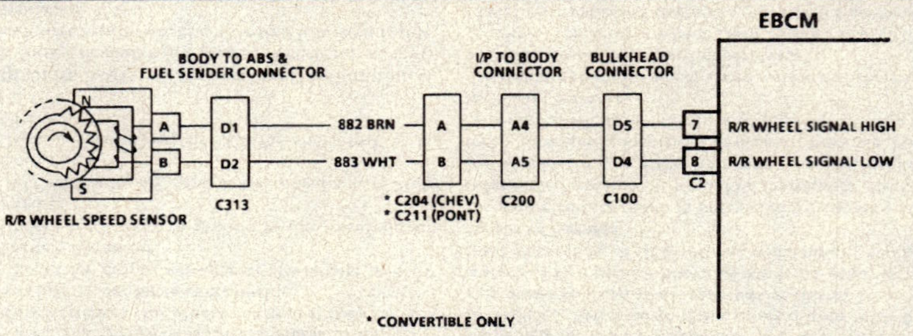

DTC 24
(Page 4 of 4)
RIGHT REAR WHEEL SPEED = 0 OR UNREASONABLE

Circuit Description:
As a toothed ring passes by the wheel speed sensor, changes in the electromagnetic field cause the wheel speed sensor to produce a sinusoidal (AC) voltage signal whose frequency is proportional to wheel speed. The magnitude of this signal is directly related to wheel speed and the proximity of the wheel speed sensor to the toothed ring, often referred to as the air gap.

DTC Will Set When: DTC 24 can be set when the vehicle is not in an ABS stop. If the right rear wheel speed is less than 1/2 of the vehicle speed and the vehicle speed is greater than 8 km/h (5 mph), a malfunction exists.

Action Taken: A malfunction DTC is stored, ABS is disabled and the ABS warning lamp is turned "ON."

DTC Chart Test Description: Number(s) below refer to circled number(s) on the diagnostic chart.
13. Checks for an open in the wheel signal low circuit.
14. Ensures malfunction was not due to physical damage of the circuitry.

Diagnostic Aids: An "Intermittent" malfunction is most likely caused by a poor connection, rubbed through wire insulation, or a wire that is broken inside the insulation.

The frequency of the malfunction can be checked by using the enhanced diagnostic function of the Tech 1.

If the customer's comments reflect that the ABS amber warning lamp is "ON" only during moist environmental changes, (rain, snow, vehicle wash) all wheel speed sensor circuitry should be thoroughly inspected for water intrusion. Use the following procedure. Spray down the suspected area with 5% salt water solution (two teaspoons of salt to 12 oz. of water). Drive vehicle above 24 km/h (15 mph) for at least 30 seconds. If DTC returns immediately, replace suspected harness.

Any circuitry, that is suspected as causing the intermittent complaint, should be thoroughly checked for backed out terminals, improper mating, broken locks, improperly formed or damaged terminals, poor terminal to wiring connections or physical damage to the wiring harness.

Resistance of the wheel speed sensor will increase with an increase in sensor temperature.

When replacing a wheel speed sensor, inspect the sensor terminals and harness connector for corrosion and/or water intrusion. If evidence of corrosion or water intrusion exists, replace wheel speed sensor harness.

If wheel speed sensor harness shows evidence of water intrusion, inspect wheel speed sensor for water intrusion. If water intrusion is evident, replace wheel speed sensor.

DIAGNOSTIC CHART DTC 24 — 1994 CAVALIER AND SUNBIRD

DTC 24
(Page 4 of 4)
RIGHT REAR WHEEL SPEED = 0 OR UNREASONABLE

CONTINUED FROM PAGE 3 OF 4 OF THIS DTC CHART.

YES

(13)
- USING J 39200, MEASURE RESISTANCE BETWEEN TERMINAL "D4" AND 24-WAY EBCM HARNESS CONNECTOR TERMINAL "8". IS RESISTANCE LESS THAN 2 OHMS?

NO → REPAIR OPEN IN ENGINE HARNESS CKT 883.

YES

(14)
- IGNITION "OFF."
- INSPECT ALL WHEEL SPEED SENSOR HARNESS CONNECTORS FOR POOR TERMINAL CONTACT OR CORROSION.

  REPLACE TERMINALS THAT EXHIBIT POOR CONTACT OR CORROSION.
- INSPECT ALL WHEEL SPEED SENSOR CONNECTORS AND HARNESSES FOR DAMAGE WHICH MAY RESULT IN AN OPEN OR SHORT TO GROUND/BATTERY WHEN ALL COMPONENTS ARE CONNECTED.
  REPAIR DAMAGE IF EVIDENT.
- RECONNECT ALL CONNECTORS.
- DRIVE VEHICLE ABOVE 24 km/h (15 MPH) FOR AT LEAST 30 SECONDS.
  IS DTC 24 SET AS A CURRENT DTC?

YES → REPLACE EBCM.

NO → MALFUNCTION IS INTERMITTENT. REFER TO "DIAGNOSTIC AIDS".

THIS CHART ASSUMES THAT A CURRENT DTC IS STORED INDICATING THAT THIS MALFUNCTION IS PRESENT.

DIAGNOSTIC CHART DTC 25 — 1994 CAVALIER AND SUNBIRD

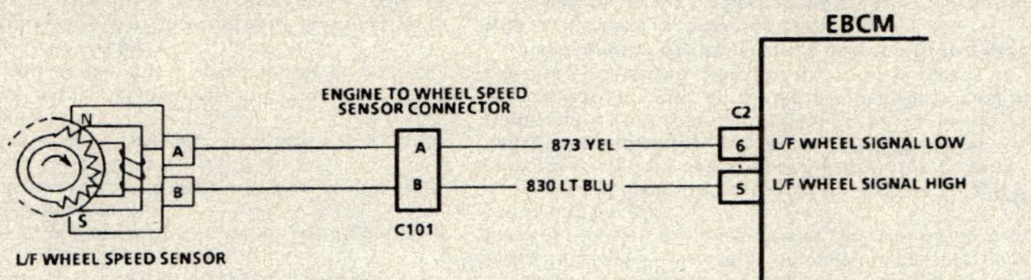

DTC 25

LEFT FRONT EXCESSIVE WHEEL SPEED VARIATION

Circuit Description:

As a toothed ring passes by the wheel speed sensor, changes in the electromagnetic field cause the wheel speed sensor to produce a sinusoidal (AC) voltage signal whose frequency is proportional to wheel speed. The magnitude of this signal is directly related to wheel speed and the proximity of the wheel speed sensor to the toothed ring, often referred to as the air gap.

DTC Will Set When: DTC 25 can be set when the brake is "OFF." The purpose of this test is to detect a situation in which the left front wheel acceleration or deceleration is beyond specified limits.

Action Taken: A malfunction DTC is stored, ABS is disabled and the ABS warning lamp is turned "ON."

DTC Chart Test Description: Number(s) below refer to circled number(s) on the diagnostic chart.

1. Identifies if a wheel speed sensor or circuitry is damaged and visibly apparent.
2. Checks wheel speed sensor for proper resistance values.
3. Ensures the wheel speed sensor and toothed ring generate the proper voltage.
4. Ensures the wheel speed sensor is not shorted to ground.

Diagnostic Aids: An "Intermittent" malfunction is most likely caused by a poor connection, rubbed through wire insulation, or a wire that is broken inside the insulation.

The frequency of the malfunction can be checked by using the enhanced diagnostic function of the Tech 1.

If the customer's comments reflect that the ABS amber warning lamp is "ON" only during moist environmental changes, (rain, snow, vehicle wash) all wheel speed sensor circuitry should be thoroughly inspected for water intrusion. Use the following procedure. Spray down the suspected area with 5% salt water solution (two teaspoons of salt to 12 oz. of water). Drive vehicle above 24 km/h (15 mph) for at least 30 seconds. If DTC returns immediately, replace suspected harness.

Any circuitry, that is suspected as causing the intermittent complaint, should be thoroughly checked for backed out terminals, improper mating, broken locks, improperly formed or damaged terminals, poor terminal to wiring connections or physical damage to the wiring harness.

A careful visual inspection of the toothed ring, CV joint, bearing and wheel speed sensor for physical damage should also be performed. If DTC sets at the same mph every time and wheel speed variation is noted above this mph on the Tech 1, the speed ring is most likely damaged. This can occur at any speed.

Resistance of the wheel speed sensor will increase with an increase in sensor temperature.

When replacing a wheel speed sensor, inspect the sensor terminals and harness connector for corrosion and/or water intrusion. If evidence of corrosion or water intrusion exists, replace wheel speed sensor harness.

If wheel speed sensor harness shows evidence of water intrusion, inspect wheel speed sensor for water intrusion. If water intrusion is evident, replace wheel speed sensor.

DIAGNOSTIC CHART DTC 25 — 1994 CAVALIER AND SUNBIRD

DTC 25
(Page 1 of 2)
LEFT FRONT EXCESSIVE WHEEL SPEED VARIATION

IMPORTANT:

- WHEEL SPEED SENSOR INTERMITTENT MALFUNCTIONS MAY BE DIFFICULT TO LOCATE. CARE SHOULD BE TAKEN NOT TO DISTURB ANY ELECTRICAL CONNECTIONS PRIOR TO AN INDICATED STEP OF THIS CHART. THIS WILL ENSURE THAT AN INTERMITTENT CONNECTION WILL NOT BE CORRECTED BEFORE THE SOURCE OF THE MALFUNCTION IS FOUND.

(1)
- IGNITION "OFF."
- INSPECT WHEEL SPEED SENSOR WIRING AND CONNECTORS FOR DAMAGE. INSPECT WHEEL SPEED SENSOR AND TOOTHED RING FOR LOOSENESS OR DAMAGE. PAY PARTICULAR ATTENTION TO THE TOOTHED RING. ANY SIGNIFICANT DAMAGE WILL AFFECT THE WHEEL SPEED INPUT SIGNAL.

IS DAMAGE OR SENSOR/TOOTHED RING LOOSENESS EVIDENT?

- YES: REPAIR DAMAGE OR SENSOR/TOOTHED RING LOOSENESS. DRIVE VEHICLE ABOVE 24 km/h (15 MPH) FOR AT LEAST 30 SECONDS TO ENSURE DTC 25 DOES NOT SET.
- NO: go to (2)

(2)
- DISCONNECT CONNECTOR FROM LEFT FRONT WHEEL SPEED SENSOR.
- USING J 39200, MEASURE RESISTANCE BETWEEN SENSOR TERMINALS "A" AND "B".

IS RESISTANCE BETWEEN 1628 AND 1889 OHMS WHEN SENSOR IS 25°C (77°F)? *[IF TEMPERATURE OF THE SENSOR IS SUSPECTED TO BE OTHER THAN 25°C (77°F), REFER TO TABLE BELOW FOR APPLICABLE SENSOR RESISTANCE VALUES; THE VALUES IN THE TABLE ARE FOR TEMPERATURE OF THE SENSOR NOT AIR TEMPERATURE.]*

- NO: REPLACE WHEEL SPEED SENSOR.
- YES: go to (3)

(3)
- WITH J 39200 STILL CONNECTED, SELECT AC VOLTAGE SCALE.
- SPIN WHEEL BY HAND WHILE MONITORING VOLTAGE.

IS VOLTAGE AT LEAST 100 mV (VOLTAGE WILL INCREASE AS WHEEL SPEED INCREASES)?

- NO: REPLACE WHEEL SPEED SENSOR.
- YES: go to (4)

(4)
- USING J 39200, MEASURE RESISTANCE BETWEEN SENSOR TERMINAL "A" AND GROUND.

DOES J 39200 DISPLAY "OL" (INFINITE)?

- NO: REPLACE WHEEL SPEED SENSOR.
- YES: GO TO PAGE 2 OF 2 OF THIS DTC CHART.

FRONT WHEEL SPEED SENSOR
TEMPERATURE OF SENSOR VS. SENSOR RESISTANCE (APPROXIMATE)

TEMP. (°C)	TEMP. (°F)	RESISTANCE (OHMS)
-34 TO 4	-30 TO 40	1258 TO 1739
5 TO 43	41 TO 110	1503 TO 2023
44 TO 93	111 TO 200	1747 TO 2384

THIS CHART ASSUMES THAT A CURRENT DTC IS STORED INDICATING THAT THIS MALFUNCTION IS PRESENT.

DIAGNOSTIC CHART DTC 25 — 1994 CAVALIER AND SUNBIRD

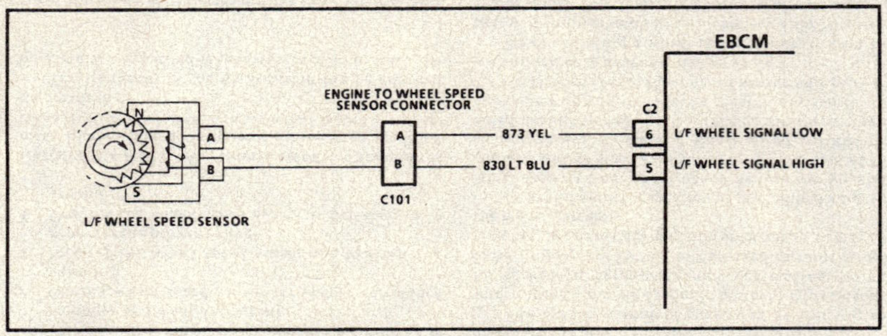

DTC 25

(Page 2 of 2)

LEFT FRONT EXCESSIVE WHEEL SPEED VARIATION

Circuit Description:

As a toothed ring passes by the wheel speed sensor, changes in the electromagnetic field cause the wheel speed sensor to produce a sinusoidal (AC) voltage signal whose frequency is proportional to wheel speed. The magnitude of this signal is directly related to wheel speed and the proximity of the wheel speed sensor to the toothed ring, often referred to as the air gap.

DTC Will Set When: DTC 25 can be set when the brake is "OFF." The purpose of this test is to detect a situation in which the left front wheel acceleration or deceleration is beyond specified limits.

Action Taken: A malfunction DTC is stored, ABS is disabled and the ABS warning lamp is turned "ON."

DTC Chart Test Description: Number(s) below refer to circled number(s) on the diagnostic chart.

5. Checks for proper voltages at speed sensor harness connector.
6. Ensures that the wheel speed sensor circuitry is not shorted to ground.
7. Checks for an open in the wheel signal low circuit.
8. Checks for an open in the wheel signal high circuit.
9. Ensures that DTC 25 was not set due to a poor connection at the EBCM or physical damage of the circuitry.
10. Checks for a short to B+ in either the high or low wheel signal circuits.
11. Ensures short to B+ was not due to physical damage of the circuitry.
12. Tests for a short between the wheel signal high and low circuits.
13. Ensures malfunction was not due to a poor connection at the wheel speed sensor or physical damage of the circuitry.

Diagnostic Aids: An "Intermittent" malfunction is most likely caused by a poor connection, rubbed through wire insulation, or a wire that is broken inside the insulation.

The frequency of the malfunction can be checked by using the enhanced diagnostic function of the Tech 1.

If the customer's comments reflect that the ABS amber warning lamp is "ON" only during moist environmental changes, (rain, snow, vehicle wash) all wheel speed sensor circuitry should be thoroughly inspected for water intrusion. Use the following procedure. Spray down the suspected area with 5% salt water solution (two teaspoons of salt to 12 oz. of water). Drive vehicle above 24 km/h (15 mph) for at least 30 seconds. If DTC returns immediately, replace suspected harness.

Any circuitry, that is suspected as causing the intermittent complaint, should be thoroughly checked for backed out terminals, improper mating, broken locks, improperly formed or damaged terminals, poor terminal to wiring connections or physical damage to the wiring harness.

A careful visual inspection of the toothed ring, CV joint, bearing and wheel speed sensor for physical damage should also be performed. If DTC sets at the same mph every time and wheel speed variation is noted above this mph on the Tech 1, the speed ring is most likely damaged. This can occur at any speed.

Resistance of the wheel speed sensor will increase with an increase in sensor temperature.

When replacing a wheel speed sensor, inspect the sensor terminals and harness connector for corrosion and/or water intrusion. If evidence of corrosion or water intrusion exists, replace wheel speed sensor harness.

If wheel speed sensor harness shows evidence of water intrusion, inspect wheel speed sensor for water intrusion. If water intrusion is evident, replace wheel speed sensor.

DIAGNOSTIC CHART DTC 25 — 1994 CAVALIER AND SUNBIRD

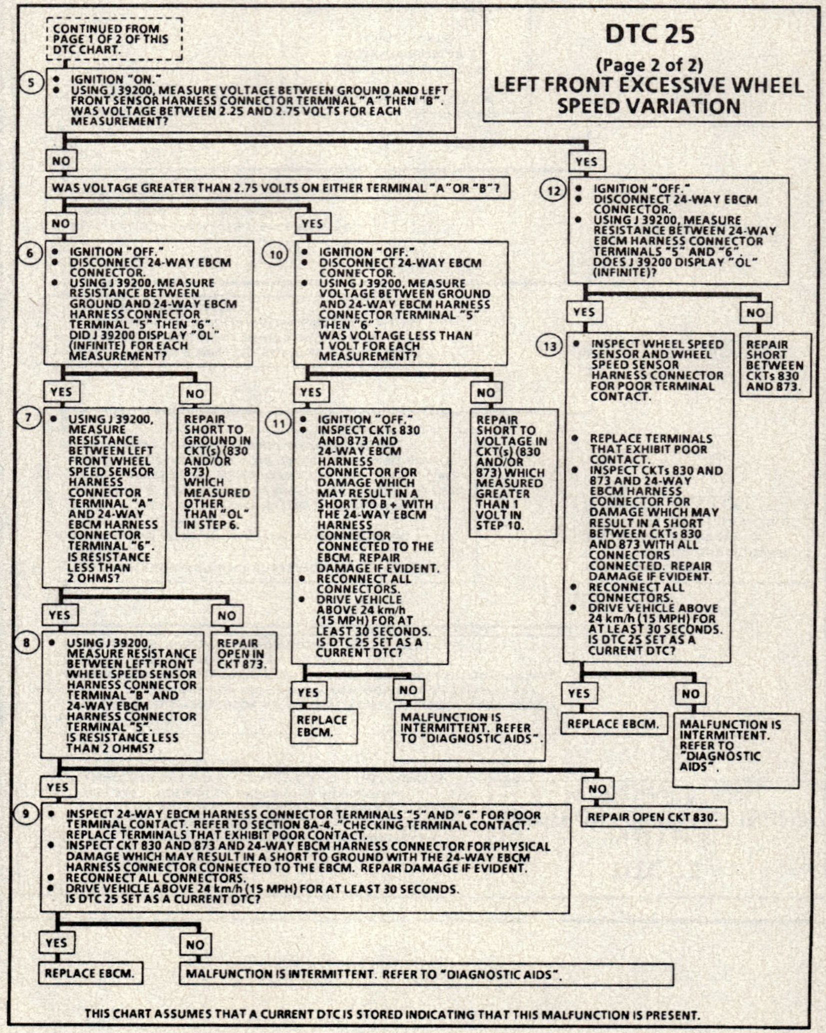

DIAGNOSTIC CHART DTC 26 — 1994 CAVALIER AND SUNBIRD

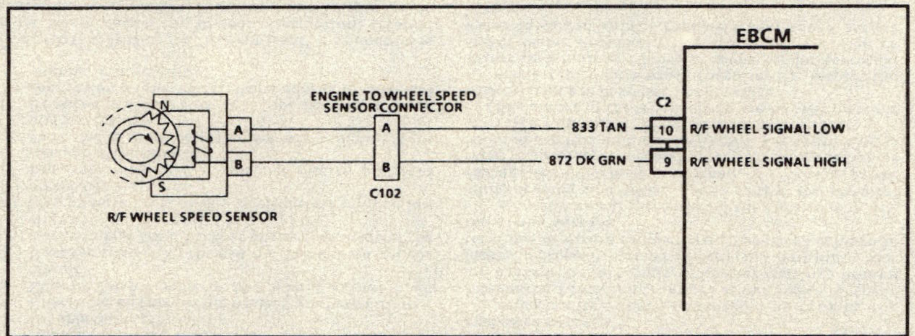

DTC 26

(Page 1 of 2)

RIGHT FRONT EXCESSIVE WHEEL SPEED VARIATION

Circuit Description:

As a toothed ring passes by the wheel speed sensor, changes in the electromagnetic field cause the wheel speed sensor to produce a sinusoidal (AC) voltage signal whose frequency is proportional to wheel speed. The magnitude of this signal is directly related to wheel speed and the proximity of the wheel speed sensor to the toothed ring, often referred to as the air gap.

DTC Will Set When: DTC 26 can be set when the brake is "OFF." The purpose of this test is to detect a situation in which the right front wheel acceleration or deceleration is beyond specified limits.

Action Taken: A malfunction DTC is stored, ABS is disabled and the ABS warning lamp is turned "ON."

DTC Chart Test Description: Number(s) below refer to circled number(s) on the diagnostic chart.

1. Identifies a wheel speed sensor or circuitry that is damaged and visibly apparent.
2. Checks wheel speed sensor for proper resistance values.
3. Ensures the wheel speed sensor and sensor ring generate the proper voltage.
4. Ensures the wheel speed sensor is not shorted to ground.

Diagnostic Aids: An "Intermittent" malfunction is most likely caused by a poor connection, rubbed through wire insulation, or a wire that is broken inside the insulation.

The frequency of the malfunction can be checked by using the enhanced diagnostic function of the Tech 1.

If the customer's comments reflect that the ABS amber warning lamp is "ON" only during moist environmental changes, (rain, snow, vehicle wash) all wheel speed sensor circuitry should be thoroughly inspected for water intrusion. Use the following procedure. Spray down the suspected area with 5% salt water solution (two teaspoons of salt to 12 oz. of water). Drive vehicle above 24 km/h (15 mph) for at least 30 seconds. If DTC returns immediately, replace suspected harness.

Any circuitry, that is suspected as causing the intermittent complaint, should be thoroughly checked for backed out terminals, improper mating, broken locks, improperly formed or damaged terminals, poor terminal to wiring connections or physical damage to the wiring harness.

A careful visual inspection of the toothed ring, CV joint, bearing and wheel speed sensor for physical damage should also be performed. If DTC sets at the same mph every time and wheel speed variation is noted above this mph on the Tech 1, the speed ring is most likely damaged. This can occur at any speed.

Resistance of the wheel speed sensor will increase with an increase in sensor temperature.

When replacing a wheel speed sensor, inspect the sensor terminals and harness connector for corrosion and/or water intrusion. If evidence of corrosion or water intrusion exists, replace wheel speed sensor harness.

If wheel speed sensor harness shows evidence of water intrusion, inspect wheel speed sensor for water intrusion. If water intrusion is evident, replace wheel speed sensor.

DIAGNOSTIC CHART DTC 26 — 1994 CAVALIER AND SUNBIRD

DTC 26

(Page 1 of 2)

RIGHT FRONT EXCESSIVE WHEEL SPEED VARIATION

IMPORTANT:

- WHEEL SPEED SENSOR INTERMITTENT MALFUNCTIONS MAY BE DIFFICULT TO LOCATE. CARE SHOULD BE TAKEN NOT TO DISTURB ANY ELECTRICAL CONNECTIONS PRIOR TO AN INDICATED STEP OF THIS CHART. THIS WILL ENSURE THAT AN INTERMITTENT CONNECTION WILL NOT BE CORRECTED BEFORE THE SOURCE OF THE MALFUNCTION IS FOUND.

(1)
- IGNITION "OFF."
- INSPECT WHEEL SPEED SENSOR WIRING AND CONNECTORS FOR DAMAGE. INSPECT WHEEL SPEED SENSOR AND TOOTHED RING FOR LOOSENESS OR DAMAGE. PAY PARTICULAR ATTENTION TO THE TOOTHED RING. ANY SIGNIFICANT DAMAGE WILL AFFECT THE WHEEL SPEED INPUT SIGNAL. IS DAMAGE OR SENSOR/TOOTHED RING LOOSENESS EVIDENT?

YES: REPAIR DAMAGE OR SENSOR/TOOTHED RING LOOSENESS. DRIVE VEHICLE ABOVE 24 km/h (15 MPH) FOR AT LEAST 30 SECONDS TO ENSURE DTC 26 DOES NOT SET.

NO:

(2)
- DISCONNECT CONNECTOR FROM RIGHT FRONT WHEEL SPEED SENSOR.
- USING J 39200, MEASURE RESISTANCE BETWEEN SENSOR TERMINALS "A" AND "B".

IS RESISTANCE BETWEEN 1628 AND 1889 OHMS WHEN SENSOR IS 25°C (77°F)? *[IF TEMPERATURE OF THE SENSOR IS SUSPECTED TO BE OTHER THAN 25°C (77°F), REFER TO TABLE BELOW FOR APPLICABLE SENSOR RESISTANCE VALUES; THE VALUES IN THE TABLE ARE FOR TEMPERATURE OF THE SENSOR NOT AIR TEMPERATURE.]*

NO: REPLACE WHEEL SPEED SENSOR.

YES:

(3)
- WITH J 39200 STILL CONNECTED, SELECT AC VOLTAGE SCALE.
- SPIN WHEEL BY HAND WHILE MONITORING VOLTAGE.

IS VOLTAGE AT LEAST 100 mV (VOLTAGE WILL INCREASE AS WHEEL SPEED INCREASES)?

NO: REPLACE WHEEL SPEED SENSOR.

YES:

(4)
- USING J 39200, MEASURE RESISTANCE BETWEEN SENSOR TERMINAL "A" AND GROUND.

DOES J 39200 DISPLAY "OL" (INFINITE)?

NO: REPLACE WHEEL SPEED SENSOR.

YES: GO TO PAGE 2 OF 2 OF THIS DTC CHART.

FRONT WHEEL SPEED SENSOR

TEMPERATURE OF SENSOR VS. SENSOR RESISTANCE (APPROXIMATE)

TEMP. (°C)	TEMP. (°F)	RESISTANCE (OHMS)
-34 TO 4	-30 TO 40	1258 TO 1739
5 TO 43	41 TO 110	1503 TO 2023
44 TO 93	111 TO 200	1747 TO 2384

THIS CHART ASSUMES THAT A CURRENT DTC IS STORED INDICATING THAT THIS MALFUNCTION IS PRESENT.

DIAGNOSTIC CHART DTC 26 — 1994
CAVALIER AND SUNBIRD

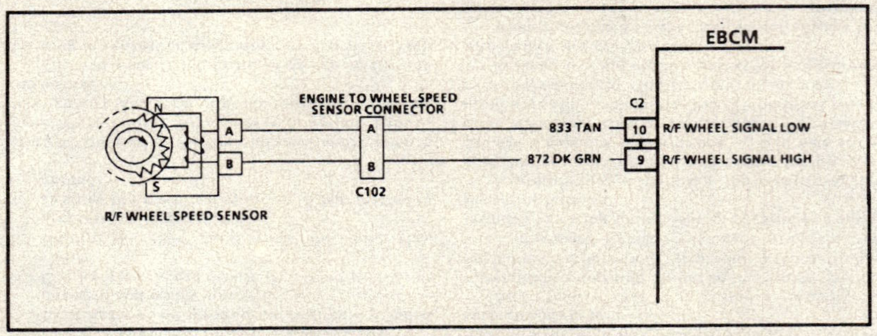

DTC 26

(Page 2 of 2)

RIGHT FRONT EXCESSIVE WHEEL SPEED VARIATION

Circuit Description:

As a toothed ring passes by the wheel speed sensor, changes in the electromagnetic field cause the wheel speed sensor to produce a sinusoidal (AC) voltage signal whose frequency is proportional to wheel speed. The magnitude of this signal is directly related to wheel speed and the proximity of the wheel speed sensor to the toothed ring, often referred to as the air gap.

DTC Will Set When: DTC 26 can be set when the brake is "OFF." The purpose of this test is to detect a situation in which the right front wheel acceleration or deceleration is beyond specified limits.

Action Taken: A malfunction DTC is stored, ABS is disabled and the ABS warning lamp is turned "ON."

DTC Chart Test Description: Number(s) below refer to circled number(s) on the diagnostic chart.

5. Checks for proper voltages at speed sensor harness connector.
6. Ensures that the wheel speed sensor circuitry is not shorted to ground.
7. Checks for an open in the wheel signal low circuit.
8. Checks for an open in the wheel signal high circuit.
9. Ensures that DTC 26 was not set due to a poor connection at the EBCM or physical damage of the circuitry.
10. Checks for a short to B+ in either the high or low wheel signal circuits.
11. Ensures short to B+ was not due to physical damage of the circuitry.
12. Tests for a short between the wheel signal high and low circuits.
13. Ensures malfunction was not due to a poor connection at the wheel speed sensor or physical damage of the circuitry.

Diagnostic Aids: An "Intermittent" malfunction is most likely caused by a poor connection, rubbed through wire insulation, or a wire that is broken inside the insulation.

The frequency of the malfunction can be checked by using the enhanced diagnostic function of the Tech 1.

If the customer's comments reflect that the ABS amber warning lamp is "ON" only during moist environmental changes, (rain, snow, vehicle wash) all wheel speed sensor circuitry should be thoroughly inspected for water intrusion. Use the following procedure. Spray down the suspected area with 5% salt water solution (two teaspoons of salt to 12 oz. of water). Drive vehicle above 24 km/h (15 mph) for at least 30 seconds. If DTC returns immediately, replace suspected harness.

Any circuitry, that is suspected as causing the intermittent complaint, should be thoroughly checked for backed out terminals, improper mating, broken locks, improperly formed or damaged terminals, poor terminal to wiring connections or physical damage to the wiring harness.

A careful visual inspection of the toothed ring, CV joint, bearing and wheel speed sensor for physical damage should also be performed. If DTC sets at the same mph every time and wheel speed variation is noted above this mph on the Tech 1, the speed ring is most likely damaged. This can occur at any speed.

Resistance of the wheel speed sensor will increase with an increase in sensor temperature.

When replacing a wheel speed sensor, inspect the sensor terminals and harness connector for corrosion and/or water intrusion. If evidence of corrosion or water intrusion exists, replace wheel speed sensor harness.

If wheel speed sensor harness shows evidence of water intrusion, inspect wheel speed sensor for water intrusion. If water intrusion is evident, replace wheel speed sensor.

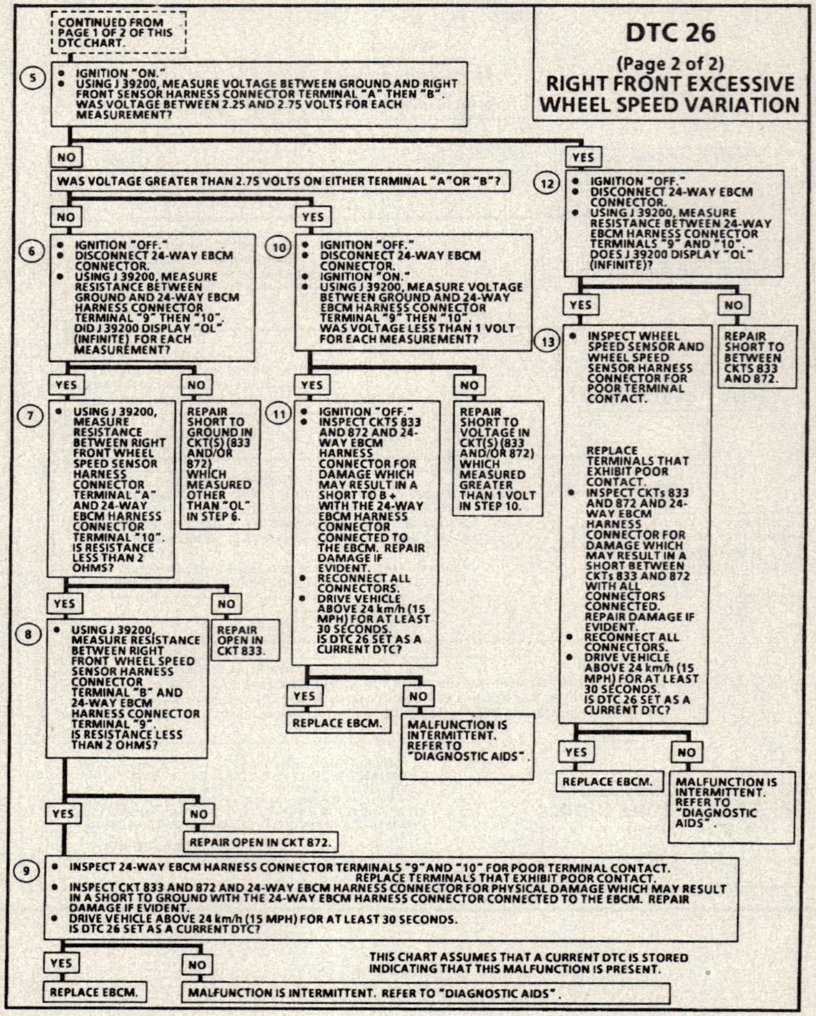

DIAGNOSTIC CHART DTC 27 — 1994 CAVALIER AND SUNBIRD

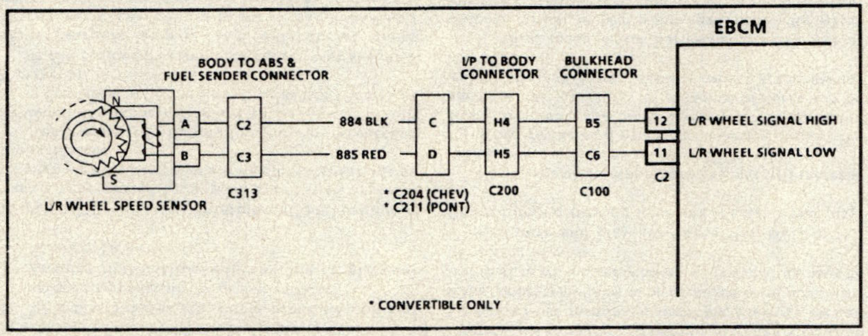

DTC 27

LEFT REAR EXCESSIVE WHEEL SPEED VARIATION

Circuit Description:

As a toothed ring passes by the wheel speed sensor, changes in the electromagnetic field cause the wheel speed sensor to produce a sinusoidal (AC) voltage signal whose frequency is proportional to wheel speed. The magnitude of this signal is directly related to wheel speed and the proximity of the wheel speed sensor to the toothed ring, often referred to as the air gap.

DTC Will Set When: DTC 27 can be set when the brake is "OFF." The purpose of this test is to detect a situation in which the left rear wheel acceleration or deceleration is beyond specified limits.

Action Taken: A malfunction DTC is stored, ABS is disabled and the ABS warning lamp is turned "ON."

DTC Chart Test Description: Number(s) below refer to circled number(s) on the diagnostic chart.

1. This step will identify circuitry that is damaged and visibly apparent.
2. Checks for proper voltages at the rear sensor harness connector.
3. Checks for a short between the wheel signal high and low circuits.
4. This test checks wheel speed sensor for proper resistance values.
5. This step ensures the wheel speed sensor is not shorted to ground.

Diagnostic Aids: An "Intermittent" malfunction is most likely caused by a poor connection, rubbed through wire insulation, or a wire that is broken inside the insulation.

The frequency of the malfunction can be checked by using the enhanced diagnostic function of the Tech 1.

If the customer's comments reflect that the ABS amber warning lamp is "ON" only during moist environmental changes, (rain, snow, vehicle wash) all wheel speed sensor circuitry should be thoroughly inspected for water intrusion.

Use the following procedure. Spray down the suspected area with 5% salt water solution (two teaspoons of salt to 12 oz. of water). Drive vehicle above 24 km/h (15 mph) for at least 30 seconds. If DTC returns immediately, replace suspected harness.

Any circuitry, that is suspected as causing the intermittent complaint, should be thoroughly checked for backed out terminals, improper mating, broken locks, improperly formed or damaged terminals, poor terminal to wiring connections or physical damage to the wiring harness.

Excessive end play can cause this DTC to set. Check integral bearing for excessive axial end play.

Resistance of the wheel speed sensor will increase with an increase in sensor temperature.

When replacing a wheel speed sensor, inspect the sensor terminals and harness connector for corrosion and/or water intrusion. If evidence of corrosion or water intrusion exists, replace wheel speed sensor harness.

If wheel speed sensor harness shows evidence of water intrusion, inspect wheel speed sensor for water intrusion. If water intrusion is evident, replace wheel speed sensor.

DIAGNOSTIC CHART DTC 27 — 1994 CAVALIER AND SUNBIRD

DTC 27
(Page 1 of 4)
LEFT REAR EXCESSIVE WHEEL SPEED VARIATION

Important

- WHEEL SPEED SENSOR INTERMITTENT MALFUNCTIONS MAY BE DIFFICULT TO LOCATE. CARE SHOULD BE TAKEN NOT TO DISTURB ANY ELECTRICAL CONNECTIONS PRIOR TO AN INDICATED STEP OF THIS CHART. THIS WILL ENSURE THAT AN INTERMITTENT CONNECTION WILL NOT BE CORRECTED BEFORE THE SOURCE OF THE MALFUNCTION IS FOUND.

(1)
- IGNITION "OFF."
- INSPECT WHEEL SPEED SENSOR WIRING AND CONNECTORS FOR DAMAGE. IS DAMAGE EVIDENT?

YES: REPAIR DAMAGE. DRIVE VEHICLE ABOVE 24 km/h (15 MPH) FOR AT LEAST 30 SECONDS TO ENSURE DTC 23 DOES NOT SET.

NO:

(2)
- DISCONNECT CONNECTOR FROM LEFT REAR WHEEL SPEED SENSOR.
- IGNITION "ON."
- USING J 39200, MEASURE VOLTAGE BETWEEN GROUND AND LEFT REAR SENSOR HARNESS CONNECTOR TERMINAL "A" THEN "B". WAS VOLTAGE BETWEEN 2.25 AND 2.75 VOLTS FOR EACH MEASUREMENT?

NO: GO TO PAGE 3 OF 4 OF THIS DTC CHART.

YES:

(3)
- IGNITION "OFF."
- DISCONNECT 24-WAY EBCM CONNECTOR.
- USING J 39200, MEASURE RESISTANCE BETWEEN 24-WAY EBCM HARNESS CONNECTOR TERMINALS "11" AND "12." DOES J 39200 DISPLAY "OL" (INFINITE)?

NO: REPAIR SHORT BETWEEN CKTs 884 AND 885.

YES:

(4)
- USING J 39200, MEASURE THE RESISTANCE BETWEEN LEFT REAR WHEEL SPEED SENSOR TERMINALS "A" AND "B". IS RESISTANCE BETWEEN 2100 AND 2400 OHMS WHEN SENSOR IS 25°C (77°F)? (IF TEMPERATURE OF THE SENSOR IS SUSPECTED TO BE OTHER THAN 25°C (77°F). REFER TO THE TABLE BELOW FOR APPLICABLE SENSOR RESISTANCE VALUES; THE VALUES IN THE TABLE ARE FOR TEMPERATURE OF THE SENSOR NOT AIR TEMPERATURE).

NO: REPLACE INTERGRAL BEARING AND SPEED SENSOR ASSEMBLY.

YES:

(5)
- USING J 39200, MEASURE RESISTANCE BETWEEN SENSOR TERMINAL "A" AND INTEGRAL BEARING AND SPEED SENSOR ASSEMBLY MOUNTING FLANGE. DOES J 39200 DISPLAY "OL" (INFINITE)?

YES: CONTINUE DIAGNOSIS ON NEXT PAGE.

NO: REPLACE INTEGRAL BEARING AND SPEED SENSOR ASSEMBLY.

REAR WHEEL SPEED SENSOR
TEMPERATURE OF SENSOR VS. SENSOR RESISTANCE (APPROXIMATE)

TEMP. (°C)	TEMP. (°F)	RESISTANCE (OHMS)
-34 TO 4	-30 TO 40	1622 TO 2205
5 TO 43	41 TO 110	1930 TO 2566
44 TO 93	111 TO 200	2245 TO 3028

THIS CHART ASSUMES THAT A CURRENT DTC IS STORED INDICATING THAT THIS MALFUNCTION IS PRESENT.

DIAGNOSTIC CHART DTC 27 — 1994 CAVALIER AND SUNBIRD

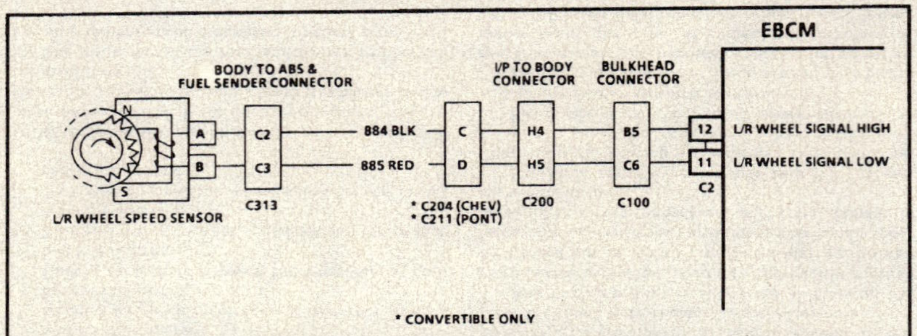

DTC 27

(Page 2 of 4)

LEFT REAR EXCESSIVE WHEEL SPEED VARIATION

Circuit Description:

As a toothed ring passes by the wheel speed sensor, changes in the electromagnetic field cause the wheel speed sensor to produce a sinusoidal (AC) voltage signal whose frequency is proportional to wheel speed. The magnitude of this signal is directly related to wheel speed and the proximity of the wheel speed sensor to the toothed ring, often referred to as the air gap.

DTC Will Set When: DTC 27 can be set when the brake is "OFF." The purpose of this test is to detect a situation in which the left rear wheel acceleration or deceleration is beyond specified limits.

Action Taken: A malfunction DTC is stored, ABS is disabled and the ABS warning lamp is turned "ON."

DTC Chart Test Description: Number(s) below refer to circled number(s) on the diagnostic chart.

6. This test ensures the wheel speed sensor and sensor ring generate the proper voltage.
7. Ensures malfunction was not due to physical damage of the circuitry.

Diagnostic Aids: An "Intermittent" malfunction is most likely caused by a poor connection, rubbed through wire insulation, or a wire that is broken inside the insulation.

The frequency of the malfunction can be checked by using the enhanced diagnostic function of the Tech 1, as described in "Enhanced Diagnostics," found in this section.

If the customer's comments reflect that the ABS amber warning lamp is "ON" only during moist environmental changes, (rain, snow, vehicle wash) all wheel speed sensor circuitry should be thoroughly inspected for water intrusion. Use the following procedure. Spray down the suspected area with 5% salt water solution (two teaspoons of salt to 12 oz. of water). Drive vehicle above 24 km/h (15 mph) for at least 30 seconds. If DTC returns immediately, replace suspected harness.

Any circuitry, that is suspected as causing the intermittent complaint, should be thoroughly checked for backed out terminals, improper mating, broken locks, improperly formed or damaged terminals, poor terminal to wiring connections or physical damage to the wiring harness.

Excessive end play can cause this DTC to set. Check integral bearing for excessive axial end play.

Resistance of the wheel speed sensor will increase with an increase in sensor temperature.

When replacing a wheel speed sensor, inspect the sensor terminals and harness connector for corrosion and/or water intrusion. If evidence of corrosion or water intrusion exists, replace wheel speed sensor harness.

If wheel speed sensor harness shows evidence of water intrusion, inspect wheel speed sensor for water intrusion. If water intrusion is evident, replace wheel speed sensor.

DIAGNOSTIC CHART DTC 27 — 1994 CAVALIER AND SUNBIRD

DTC 27

(Page 2 of 4)

LEFT REAR EXCESSIVE WHEEL SPEED VARIATION

DIAGNOSIS CONTINUED FROM PAGE 1 OF 4 OF THIS DTC CHART.

(6)
- INSPECT WHEEL SPEED SENSOR HARNESS CONNECTOR AND 24-WAY EBCM HARNESS CONNECTOR TERMINALS "11" AND "12" FOR POOR TERMINAL CONTACT.

  REPLACE TERMINALS THAT EXHIBIT POOR CONTACT RECONNECT WHEEL SPEED SENSOR.
- USING J 39200, SELECT A/C VOLTAGE SCALE AND CONNECT BETWEEN 24-WAY EBCM HARNESS CONNECTOR TERMINALS "11" AND "12".
- SPIN WHEEL BY HAND WHILE MONITORING VOLTAGE.

  IS VOLTAGE AT LEAST 100 mV (VOLTAGE WILL INCREASE AS WHEEL SPEED INCREASES)?

YES →

(7)
- INSPECT CKTs 884 AND 885 FOR PHYSICAL DAMAGE WHICH MAY RESULT IN A SHORT BETWEEN THE CIRCUITS WITH ALL CONNECTORS CONNECTED.

  REPAIR DAMAGE IF EVIDENT.
- RECONNECT ALL CONNECTORS.
- DRIVE VEHICLE ABOVE 24-WAY km/h (15 MPH) FOR AT LEAST 30 SECONDS IS DTC 23 SET AS A CURRENT DTC?

  YES → REPLACE EBCM.

  NO → MALFUNCTION IS INTERMITTENT. REFER TO "DIAGNOSTIC AIDS".

NO → REPLACE INTEGRAL REARING AND SPEED SENSOR ASSEMBLY.

THIS CHART ASSUMES THAT A CURRENT DTC IS STORED INDICATING THAT THIS MALFUNCTION IS PRESENT.

DIAGNOSTIC CHART DTC 27 — 1994 CAVALIER AND SUNBIRD

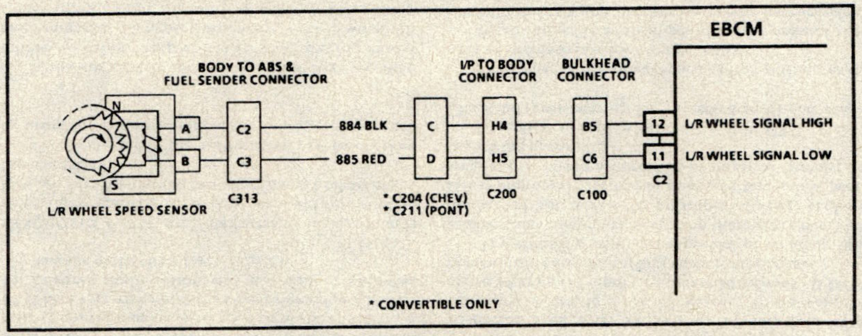

DTC 27
(Page 3 of 4)
LEFT REAR EXCESSIVE WHEEL SPEED VARIATION

Circuit Description:
As a toothed ring passes by the wheel speed sensor, changes in the electromagnetic field cause the wheel speed sensor to produce a sinusoidal (AC) voltage signal whose frequency is proportional to wheel speed. The magnitude of this signal is directly related to wheel speed and the proximity of the wheel speed sensor to the toothed ring, often referred to as the air gap.

DTC Will Set When: DTC 27 can be set when the brake is "OFF." The purpose of this test is to detect a situation in which the left rear wheel acceleration or deceleration is beyond specified limits.

Action Taken: A malfunction DTC is stored, ABS is disabled and the ABS warning lamp is turned "ON."

DTC Chart Test Description: Number(s) below refer to circled number(s) on the diagnostic chart.

8. This step checks for proper voltages at body to ABS and fuel sender harness connector.
9. Checks for proper voltages at the engine to body harness connector.
10. Checks for a short to ground in the wheel signal circuit.
11. This checks for an open in the wheel signal high circuit.
12. Checks wheel speed circuits for short to B+ at engine harness.

Diagnostic Aids: An "Intermittent" malfunction is most likely caused by a poor connection, rubbed through wire insulation, or a wire that is broken inside the insulation.

The frequency of the malfunction can be checked by using the enhanced diagnostic function of the Tech 1.

If the customer's comments reflect that the ABS amber warning lamp is "ON" only during moist environmental changes, (rain, snow, vehicle wash) all wheel speed sensor circuitry should be thoroughly inspected for water intrusion.

Use the following procedure. Spray down the suspected area with 5% salt water solution (two teaspoons of salt to 12 oz. of water). Drive vehicle above 24 km/h (15 mph) for at least 30 seconds. If DTC returns immediately, replace suspected harness.

Any circuitry, that is suspected as causing the intermittent complaint, should be thoroughly checked for backed out terminals, improper mating, broken locks, improperly formed or damaged terminals, poor terminal to wiring connections or physical damage to the wiring harness.

Excessive end play can cause this DTC to set. Check integral bearing for excessive axial end play.

Resistance of the wheel speed sensor will increase with an increase in sensor temperature.

When replacing a wheel speed sensor, inspect the sensor terminals and harness connector for corrosion and/or water intrusion. If evidence of corrosion or water intrusion exists, replace wheel speed sensor harness.

If wheel speed sensor harness shows evidence of water intrusion, inspect wheel speed sensor for water intrusion. If water intrusion is evident, replace wheel speed sensor.

DIAGNOSTIC CHART DTC 27 — 1994 CAVALIER AND SUNBIRD

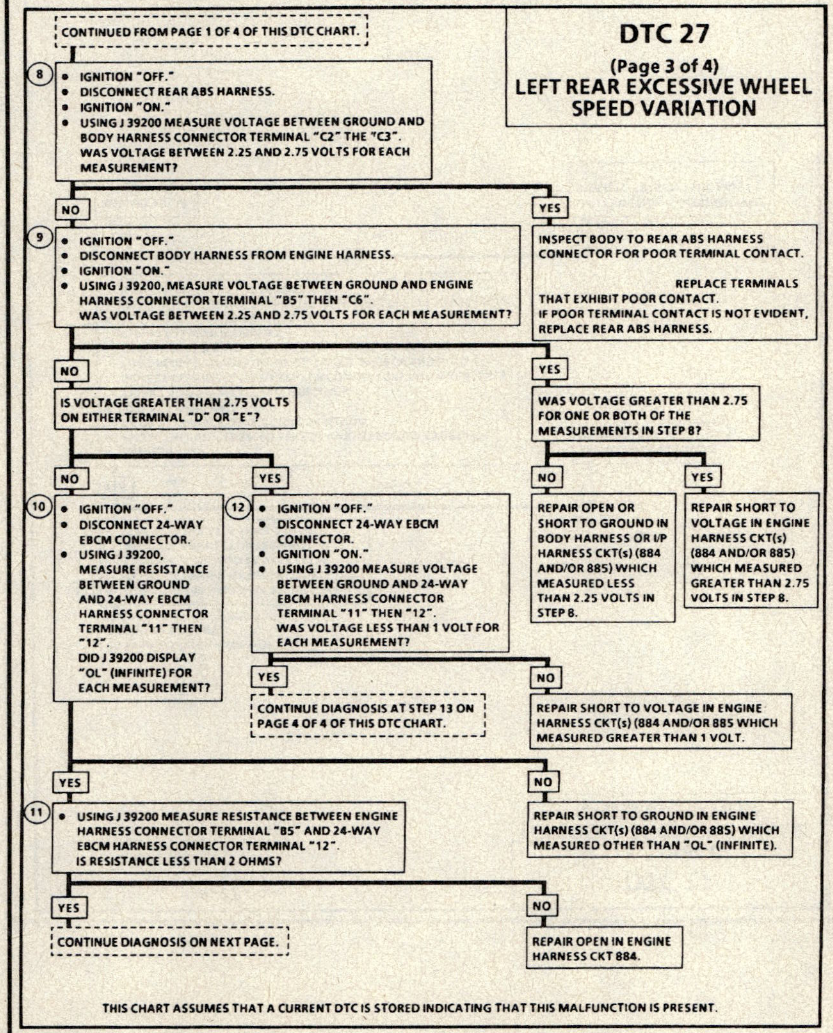

DIAGNOSTIC CHART DTC 27 — 1994 CAVALIER AND SUNBIRD

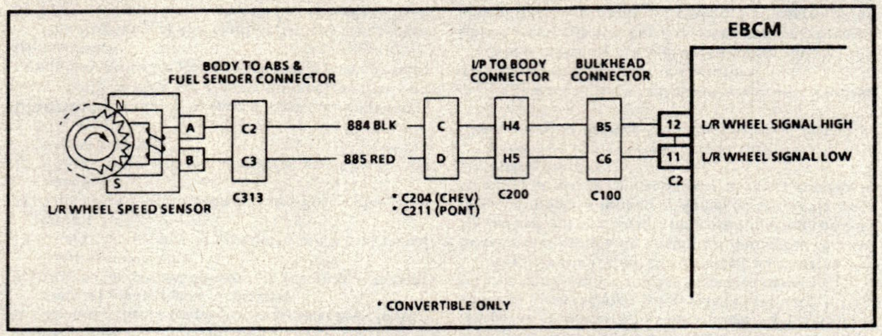

DTC 27

(Page 4 of 4)

LEFT REAR EXCESSIVE WHEEL SPEED VARIATION

Circuit Description:

As a toothed ring passes by the wheel speed sensor, changes in the electromagnetic field cause the wheel speed sensor to produce a sinusoidal (AC) voltage signal whose frequency is proportional to wheel speed. The magnitude of this signal is directly related to wheel speed and the proximity of the wheel speed sensor to the toothed ring, often referred to as the air gap.

DTC Will Set When: DTC 27 can be set when the brake is "OFF." The purpose of this test is to detect a situation in which the left rear wheel acceleration or deceleration is beyond specified limits.

Action Taken: A malfunction DTC is stored, ABS is disabled and the ABS warning lamp is turned "ON."

DTC Chart Test Description: Number(s) below refer to circled number(s) on the diagnostic chart.

13. Checks for an open in the wheel signal low circuit.
14. Ensures malfunction was not due to physical damage of the circuitry.

Diagnostic Aids: An "Intermittent" malfunction is most likely caused by a poor connection, rubbed through wire insulation, or a wire that is broken inside the insulation.

The frequency of the malfunction can be checked by using the enhanced diagnostic function of the Tech 1.

If the customer's comments reflect that the ABS amber warning lamp is "ON" only during moist environmental changes, (rain, snow, vehicle wash) all wheel speed sensor circuitry should be thoroughly inspected for water intrusion.

Use the following procedure. Spray down the suspected area with 5% salt water solution (two teaspoons of salt to 12 oz. of water). Drive vehicle above 24 km/h (15 mph) for at least 30 seconds. If DTC returns immediately, replace suspected harness.

Any circuitry, that is suspected as causing the intermittent complaint, should be thoroughly checked for backed out terminals, improper mating, broken locks, improperly formed or damaged terminals, poor terminal to wiring connections or physical damage to the wiring harness.

Excessive end play can cause this DTC to set. Check integral bearing for excessive axial end play.

Resistance of the wheel speed sensor will increase with an increase in sensor temperature.

When replacing a wheel speed sensor, inspect the sensor terminals and harness connector for corrosion and/or water intrusion. If evidence of corrosion or water intrusion exists, replace wheel speed sensor harness.

If wheel speed sensor harness shows evidence of water intrusion, inspect wheel speed sensor for water intrusion. If water intrusion is evident, replace wheel speed sensor.

DIAGNOSTIC CHART DTC 27 — 1994 CAVALIER AND SUNBIRD

DTC 27

(Page 4 of 4)

LEFT REAR EXCESSIVE WHEEL SPEED VARIATION

CONTINUED FROM PAGE 3 OF 4 OF THIS DTC CHART.

(13)
- USING J 39200, MEASURE RESISTANCE BETWEEN ENGINE HARNESS CONNECTOR TERMINAL "C6" AND 24-WAY EBCM HARNESS CONNECTOR TERMINAL "11". IS RESISTANCE LESS THAN 2 OHMS?

NO: REPAIR OPEN IN ENGINE HARNESS CKT 885.

YES:

(14)
- IGNITION "OFF."
- INSPECT ALL WHEEL SPEED SENSOR HARNESS CONNECTORS FOR POOR TERMINAL CONTACT OR CORROSION.
- REPLACE TERMINALS THAT EXHIBIT POOR CONTACT OR CORROSION.
- INSPECT ALL WHEEL SPEED SENSOR CONNECTORS AND HARNESSES FOR DAMAGE WHICH MAY RESULT IN AN OPEN OR SHORT TO GROUND/BATTERY WHEN ALL COMPONENTS ARE CONNECTED. REPAIR DAMAGE IF EVIDENT.
- RECONNECT ALL CONNECTORS.
- DRIVE VEHICLE ABOVE 24 km/h (15 MPH) FOR AT LEAST 30 SECONDS. IS DTC 23 SET AS A CURRENT DTC?

YES: REPLACE EBCM.

NO: MALFUNCTION IS INTERMITTENT. REFER TO "DIAGNOSTIC AIDS".

THIS CHART ASSUMES THAT A CURRENT DTC IS STORED INDICATING THAT THIS MALFUNCTION IS PRESENT.

DIAGNOSTIC CHART DTC 28 — 1994 CAVALIER AND SUNBIRD

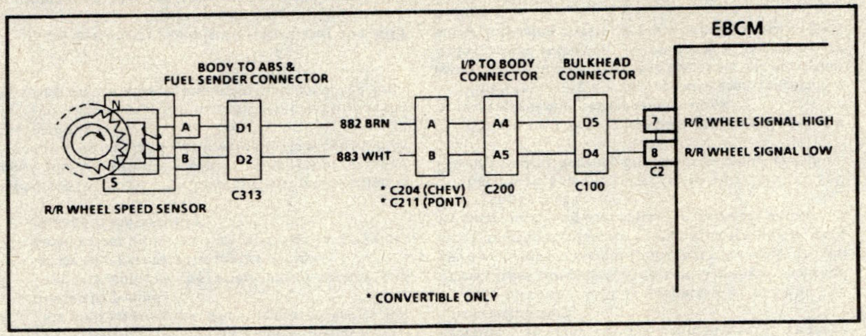

DTC 28
(Page 1 of 4)
RIGHT REAR EXCESSIVE WHEEL SPEED VARIATION

Circuit Description:

As a toothed ring passes by the wheel speed sensor, changes in the electromagnetic field cause the wheel speed sensor to produce a sinusoidal (AC) voltage signal whose frequency is proportional to wheel speed. The magnitude of this signal is directly related to wheel speed and the proximity of the wheel speed sensor to the toothed ring, often referred to as the air gap.

DTC Will Set When: DTC 28 can be set when the brake is "OFF." The purpose of this test is to detect a situation in which the right rear wheel acceleration or deceleration is beyond specified limits.

Action Taken: A malfunction DTC is stored, ABS is disabled and the ABS warning lamp is turned "ON."

DTC Chart Test Description: Number(s) below refer to circled number(s) on the diagnostic chart.

1. This step will identify circuitry that is damaged and visibly apparent.
2. Checks for proper voltages at the rear sensor harness connector.
3. Checks for a short between the wheel signal high and low circuits.
4. This test checks wheel speed sensor for proper resistance values.

Diagnostic Aids: An "Intermittent" malfunction is most likely caused by a poor connection, rubbed through wire insulation, or a wire that is broken inside the insulation.

The frequency of the malfunction can be checked by using the enhanced diagnostic function of the Tech 1.

If the customer's comments reflect that the ABS amber warning lamp is "ON" only during moist environmental changes, (rain, snow, vehicle wash) all wheel speed sensor circuitry should be thoroughly inspected for water intrusion. Use the following procedure.

Spray down the suspected area with 5% salt water solution (two teaspoons of salt to 12 oz. of water). Drive vehicle above 24 km/h (15 mph) for at least 30 seconds. If DTC returns immediately, replace suspected harness.

Any circuitry, that is suspected as causing the intermittent complaint, should be thoroughly checked for backed out terminals, improper mating, broken locks, improperly formed or damaged terminals, poor terminal to wiring connections or physical damage to the wiring harness.

Excessive end play can cause this DTC to set. Check integral bearing for excessive axial end play.

Resistance of the wheel speed sensor will increase with an increase in sensor temperature.

When replacing a wheel speed sensor, inspect the sensor terminals and harness connector for corrosion and/or water intrusion. If evidence of corrosion or water intrusion exists, replace wheel speed sensor harness.

If wheel speed sensor harness shows evidence of water intrusion, inspect wheel speed sensor for water intrusion. If water intrusion is evident, replace wheel speed sensor.

DIAGNOSTIC CHART DTC 28 — 1994 CAVALIER AND SUNBIRD

DTC 28
(Page 1 of 4)
RIGHT REAR EXCESSIVE WHEEL SPEED VARIATION

Important

- WHEEL SPEED SENSOR INTERMITTENT MALFUNCTIONS MAY BE DIFFICULT TO LOCATE. CARE SHOULD BE TAKEN NOT TO DISTURB ANY ELECTRICAL CONNECTIONS PRIOR TO AN INDICATED STEP OF THIS CHART. THIS WILL ENSURE THAT AN INTERMITTENT CONNECTION WILL NOT BE CORRECTED BEFORE THE SOURCE OF THE MALFUNCTION IS FOUND.

(1)
- IGNITION "OFF."
- INSPECT WHEEL SPEED SENSOR WIRING AND CONNECTORS FOR DAMAGE.
IS DAMAGE EVIDENT?

NO → go to step 2
YES → REPAIR DAMAGE. DRIVE VEHICLE ABOVE 24 km/h (15MPH) FOR AT LEAST 30 SECONDS TO ENSURE DTC 24 DOES NOT SET.

(2)
- DISCONNECT CONNECTOR FROM RIGHT REAR WHEEL SPEED SENSOR.
- IGNITION "ON."
- USING J 39200, MEASURE VOLTAGE BETWEEN GROUND AND RIGHT REAR SENSOR HARNESS CONNECTOR TERMINAL "A" THEN "B".
WAS VOLTAGE BETWEEN 2.25 AND 2.75 VOLTS FOR EACH MEASUREMENT?

NO → go to step 3
YES → GO TO PAGE 3 OF 4 OF THIS CHART.

(3)
- IGNITION "OFF."
- DISCONNECT 24-WAY EBCM CONNECTOR.
- USING J39200, MEASURE RESISTANCE BETWEEN 24-WAY EBCM HARNESS CONNECTOR TERMINALS "B" AND "7."
DOES J 39200 DISPLAY "OL" (INFINITE).

YES → go to step 4
NO → REPAIR SHORT BETWEEN CKT(s) 882 AND 883.

(4)
- USING J 39200, MEASURE RESISTANCE BETWEEN RIGHT REAR WHEEL SPEED SENSOR TERMINALS "A" AND "B."
IS RESISTANCE BETWEEN 2100 AND 2400 OHMS WHEN SENSOR IS 25°C (77°F)? (IF TEMPERATURE OF THE SENSOR IS SUSPECTED TO BE OTHER THAN 25°C (77°F). REFER TO THE TABLE BELOW FOR APPLICABLE SENSOR RESISTANCE VALUES; THE VALUES IN THE TABLE ARE FOR TEMPERATURE OF THE SENSOR NOT AIR TEMPERATURE).

YES → CONTINUE DIAGNOSIS ON NEXT PAGE.
NO → REPLACE INTEGRAL BEARING AND SPEED SENSOR ASSEMBLY.

REAR WHEEL SPEED SENSOR
TEMPERATURE OF SENSOR VS. SENSOR RESISTANCE (APPROXIMATE)

TEMP. (°C)	TEMP. (°F)	RESISTANCE (OHMS)
-34 TO 4	-30 TO 40	1623 TO 2210
5 TO 43	41 TO 110	1939 TO 2570
44 TO 93	111 TO 200	2254 TO 3029

THIS CHART ASSUMES THAT A CURRENT DTC IS STORED INDICATING THAT THIS MALFUNCTION IS PRESENT.

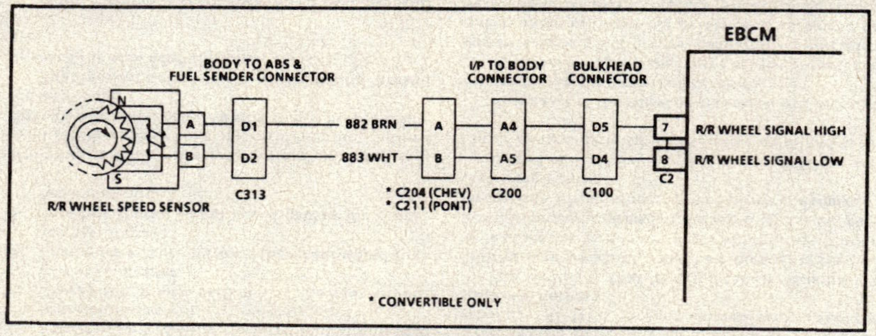

DTC 28

(Page 2 of 4)

RIGHT REAR EXCESSIVE WHEEL SPEED VARIATION

Circuit Description:

As a toothed ring passes by the wheel speed sensor, changes in the electromagnetic field cause the wheel speed sensor to produce a sinusoidal (AC) voltage signal whose frequency is proportional to wheel speed. The magnitude of this signal is directly related to wheel speed and the proximity of the wheel speed sensor to the toothed ring, often referred to as the air gap.

DTC Will Set When: DTC 28 can be set when the brake is "OFF." The purpose of this test is to detect a situation in which the right rear wheel acceleration or deceleration is beyond specified limits.

Action Taken: A malfunction DTC is stored, ABS is disabled and the ABS warning lamp is turned "ON."

DTC Chart Test Description: Number(s) below refer to circled number(s) on the diagnostic chart.

5. This step ensures the wheel speed sensor is not shorted to ground.
6. This test ensures the wheel speed sensor and sensor ring generate the proper voltage.
7. Ensures malfunction was not due to physical damage of the circuitry.

Diagnostic Aids: An "Intermittent" malfunction is most likely caused by a poor connection, rubbed through wire insulation, or a wire that is broken inside the insulation.

The frequency of the malfunction can be checked by using the enhanced diagnostic function of the Tech 1.

If the customer's comments reflect that the ABS amber warning lamp is "ON" only during moist environmental changes, (rain, snow, vehicle wash) all wheel speed sensor circuitry should be thoroughly inspected for water intrusion. Use the following procedure. Spray down the suspected area with 5% salt water solution (two teaspoons of salt to 12 oz. of water). Drive vehicle above 24 km/h (15 mph) for at least 30 seconds. If DTC returns immediately, replace suspected harness.

Any circuitry, that is suspected as causing the intermittent complaint, should be thoroughly checked for backed out terminals, improper mating, broken locks, improperly formed or damaged terminals, poor terminal to wiring connections or physical damage to the wiring harness.

Excessive end play can cause this DTC to set. Check integral bearing for excessive axial end play.

Resistance of the wheel speed sensor will increase with an increase in sensor temperature.

When replacing a wheel speed sensor, inspect the sensor terminals and harness connector for corrosion and/or water intrusion. If evidence of corrosion or water intrusion exists, replace wheel speed sensor harness.

If wheel speed sensor harness shows evidence of water intrusion, inspect wheel speed sensor for water intrusion. If water intrusion is evident, replace wheel speed sensor.

DTC 28

(Page 2 of 4)

RIGHT REAR EXCESSIVE WHEEL SPEED VARIATION

CONTINUED FROM PAGE 1 OF 4 OF THIS DTC CHART.

(5)
- USING J 39200, MEASURE RESISTANCE BETWEEN SENSOR TERMINAL "A" AND INTEGRAL BEARING AND SPEED SENSOR ASSEMBLY MOUNTING FLANGE.

DOES J 39200 DISPLAY "OL" (INFINITE)?

NO: REPLACE INTEGRAL BEARING AND SPEED SENSOR ASSEMBLY.

YES:

(6)
- INSPECT WHEEL SPEED SENSOR HARNESS CONNECTOR AND 24-WAY EBCM HARNESS CONNECTOR TERMINALS "8" AND "7" FOR POOR TERMINAL CONTACT.
- REPLACE TERMINALS THAT EXHIBIT POOR CONTACT.
- RECONNECT WHEEL SPEED SENSOR.
- USING J 39200, SELECT A/C VOLTAGE SCALE AND CONNECT CONNECTOR TERMINALS "8" AND "7".
- SPIN WHEEL BY HAND WHILE MONITORING VOLTAGE.

IS VOLTAGE AT LEAST 100 mV (VOLTAGE WILL INCREASE AS WHEEL SPEED INCREASES)?

NO: REPLACE INTEGRAL BEARING AND SPEED SENSOR ASSEMBLY.

YES:

(7)
- INSPECT CKTs 882 AND 883 FOR PHYSICAL DAMAGE WHICH MAY RESULT IN A SHORT BETWEEN THE CIRCUITS WITH ALL CONNECTORS CONNECTED REPAIR DAMAGE IF EVIDENT.
- RECONNECT ALL CONNECTORS.
- DRIVE VEHICLE ABOVE 24 km/h (15 MPH) FOR AT LEAST 30 SECONDS.

IS DTC 24 SET AS A CURRENT DTC?

YES: REPLACE EBCM.

NO: MALFUNCTION IS INTERMITTENT. REFER TO "DIAGNOSTIC AIDS".

THIS CHART ASSUMES THAT A CURRENT DTC IS STORED INDICATING THAT THIS MALFUNCTION IS PRESENT.

DIAGNOSTIC CHART DTC 28 — 1994 CAVALIER AND SUNBIRD

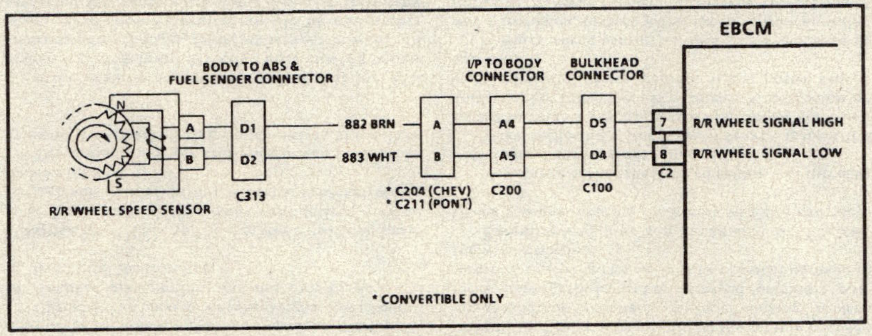

DTC 28

(Page 3 of 4)

RIGHT REAR EXCESSIVE WHEEL SPEED VARIATION

Circuit Description:

As a toothed ring passes by the wheel speed sensor, changes in the electromagnetic field cause the wheel speed sensor to produce a sinusoidal (AC) voltage signal whose frequency is proportional to wheel speed. The magnitude of this signal is directly related to wheel speed and the proximity of the wheel speed sensor to the toothed ring, often referred to as the air gap.

DTC Will Set When: DTC 28 can be set when the brake is "OFF." The purpose of this test is to detect a situation in which the right rear wheel acceleration or deceleration is beyond specified limits.

Action Taken: A malfunction DTC is stored, ABS is disabled and the ABS warning lamp is turned "ON."

DTC Chart Test Description: Number(s) below refer to circled number(s) on the diagnostic chart.

8. This step checks for proper voltages at body to ABS and fuel sender harness connector.
9. Checks for proper voltages at the engine to body harness connector.
10. Checks for a short to ground in the wheel signal circuits.
11. This checks for an open in the wheel signal high circuit.
12. Checks wheel speed circuits for short to B+ at engine harness.

Diagnostic Aids: An "Intermittent" malfunction is most likely caused by a poor connection, rubbed through wire insulation, or a wire that is broken inside the insulation.

The frequency of the malfunction can be checked by using the enhanced diagnostic function of the Tech 1.

If the customer's comments reflect that the ABS amber warning lamp is "ON" only during moist environmental changes, (rain, snow, vehicle wash) all wheel speed sensor circuitry should be thoroughly inspected for water intrusion.

Use the following procedure. Spray down the suspected area with 5% salt water solution (two teaspoons of salt to 12 oz. of water). Drive vehicle above 24 km/h (15 mph) for at least 30 seconds. If DTC returns immediately, replace suspected harness.

Any circuitry, that is suspected as causing the intermittent complaint, should be thoroughly checked for backed out terminals, improper mating, broken locks, improperly formed or damaged terminals, poor terminal to wiring connections or physical damage to the wiring harness.

Excessive end play can cause this DTC to set. Check integral bearing for excessive axial end play.

Resistance of the wheel speed sensor will increase with an increase in sensor temperature.

When replacing a wheel speed sensor, inspect the sensor terminals and harness connector for corrosion and/or water intrusion. If evidence of corrosion or water intrusion exists, replace wheel speed sensor harness.

If wheel speed sensor harness shows evidence of water intrusion, inspect wheel speed sensor for water intrusion. If water intrusion is evident, replace wheel speed sensor.

DIAGNOSTIC CHART DTC 28 — 1994 CAVALIER AND SUNBIRD

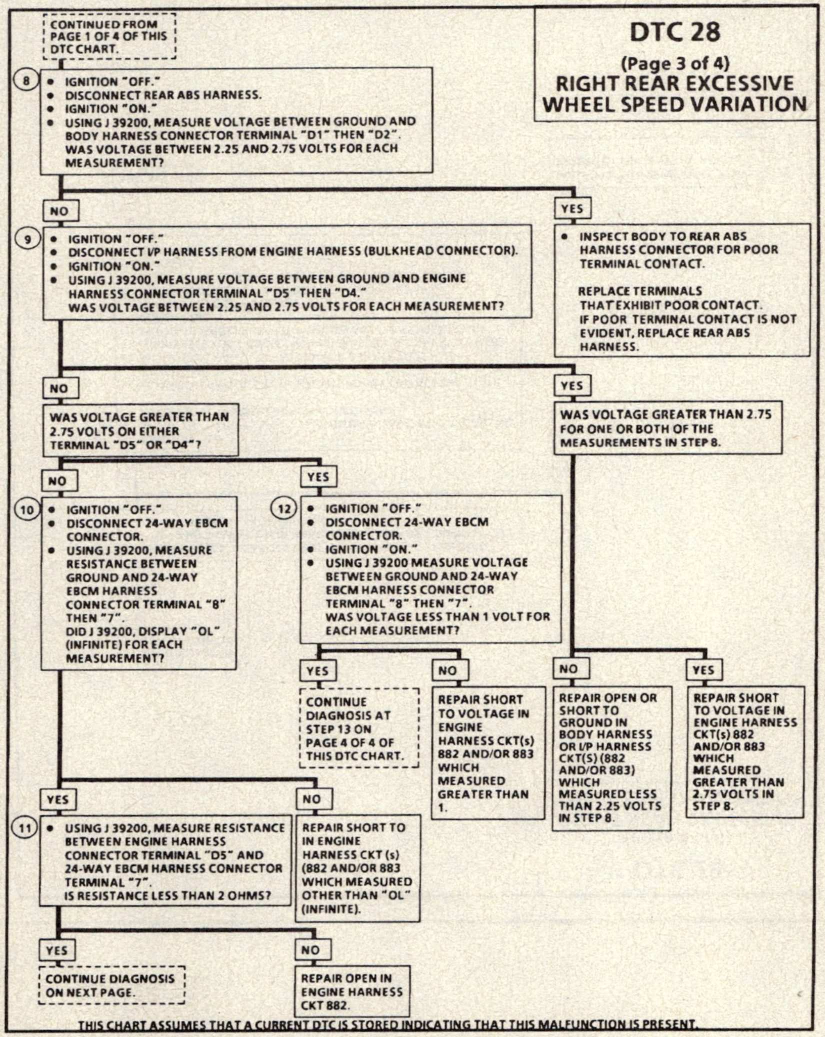

DIAGNOSTIC CHART DTC 28 — 1994 CAVALIER AND SUNBIRD

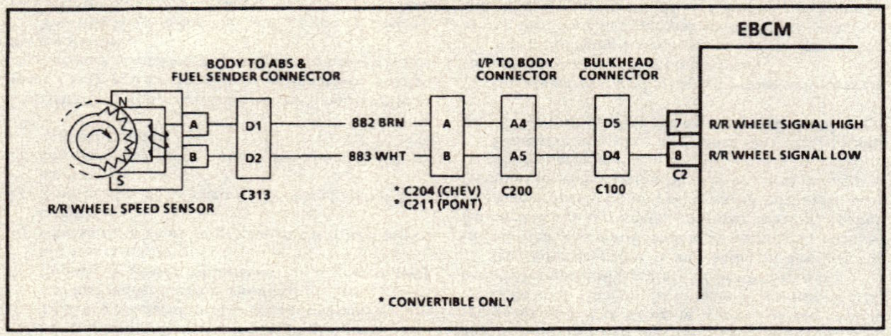

DTC 28

(Page 4 of 4)

RIGHT REAR EXCESSIVE WHEEL SPEED VARIATION

Circuit Description:

As a toothed ring passes by the wheel speed sensor, changes in the electromagnetic field cause the wheel speed sensor to produce a sinusoidal (AC) voltage signal whose frequency is proportional to wheel speed. The magnitude of this signal is directly related to wheel speed and the proximity of the wheel speed sensor to the toothed ring, often referred to as the air gap.

DTC Will Set When: DTC 28 can be set when the brake is "OFF." The purpose of this test is to detect a situation in which the right rear wheel acceleration or deceleration is beyond specified limits.

Action Taken: A malfunction DTC is stored, ABS is disabled and the ABS warning lamp is turned "ON."

DTC Chart Test Description: Number(s) below refer to circled number(s) on the diagnostic chart.

13. Checks for an open in the wheel signal low circuit.
14. Ensures malfunction was not due to physical damage of the circuitry.

Diagnostic Aids: An "Intermittent" malfunction is most likely caused by a poor connection, rubbed through wire insulation, or a wire that is broken inside the insulation.

The frequency of the malfunction can be checked by using the enhanced diagnostic function of the Tech 1.

If the customer's comments reflect that the ABS amber warning lamp is "ON" only during moist environmental changes, (rain, snow, vehicle wash) all wheel speed sensor circuitry should be thoroughly inspected for water intrusion. Use the following procedure. Spray down the suspected area with 5% salt water solution (two teaspoons of salt to 12 oz. of water). Drive vehicle above 24 km/h (15 mph) for at least 30 seconds. If DTC returns immediately, replace suspected harness.

Any circuitry, that is suspected as causing the intermittent complaint, should be thoroughly checked for backed out terminals, improper mating, broken locks, improperly formed or damaged terminals, poor terminal to wiring connections or physical damage to the wiring harness.

Excessive end play can cause this DTC to set. Check integral bearing for excessive axial end play.

Resistance of the wheel speed sensor will increase with an increase in sensor temperature.

When replacing a wheel speed sensor, inspect the sensor terminals and harness connector for corrosion and/or water intrusion. If evidence of corrosion or water intrusion exists, replace wheel speed sensor harness.

If wheel speed sensor harness shows evidence of water intrusion, inspect wheel speed sensor for water intrusion. If water intrusion is evident, replace wheel speed sensor.

DIAGNOSTIC CHART DTC 28 — 1994 CAVALIER AND SUNBIRD

DTC 28

(Page 4 of 4)

RIGHT REAR EXCESSIVE WHEEL SPEED VARIATION

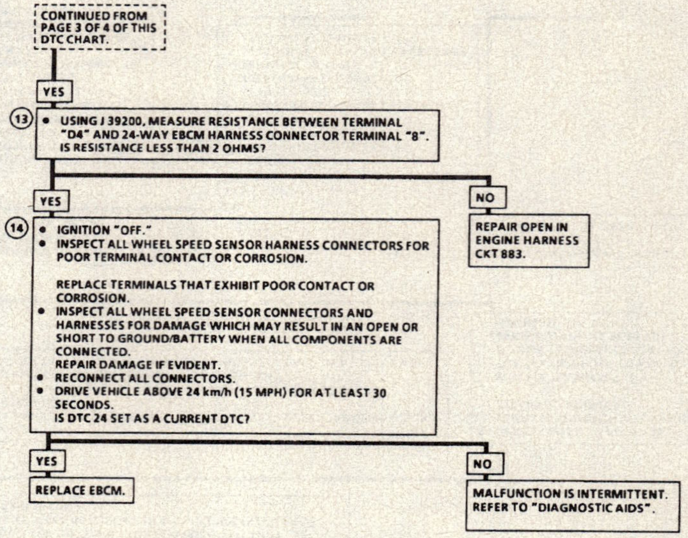

THIS CHART ASSUMES THAT A CURRENT DTC IS STORED INDICATING THAT THIS MALFUNCTION IS PRESENT.

DIAGNOSTIC CHART DTC 36 — 1994 CAVALIER AND SUNBIRD

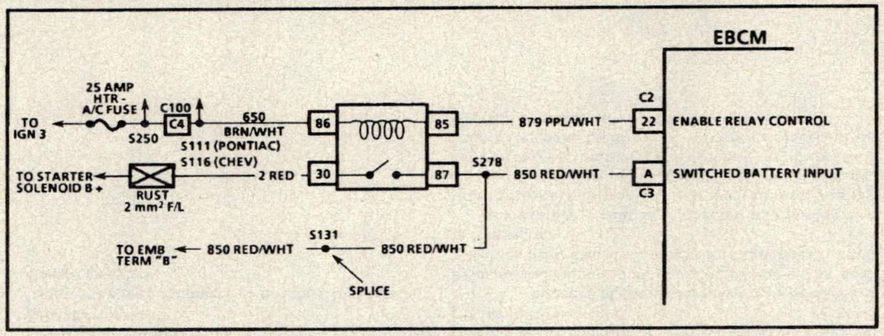

DTC 36
LOW SYSTEM VOLTAGE

Circuit Description:
This DTC is used to monitor the voltage level available to the ABS EBCM. If the voltage drops below 11 volts, full performance of the ABS system cannot be guaranteed. During ABS operation, there are several current requirements that will cause battery voltage to drop. Because of this, voltage is monitored prior to ABS operation to indicate good charging system condition and also during ABS operation when voltage may drop significantly.

DTC Will Set When: DTC 36 can only be set if the vehicle's speed is greater than 5 km/h (3 mph). If both the switched battery and switched ignition voltage is less than 11.2 volts non-ABS or less than 9.6 volts ABS, a malfunction exists.

Action Taken: A malfunction DTC is stored, ABS is disabled and the ABS warning lamp is turned "ON."

DTC Chart Test Description: Number(s) below refer to circled number(s) on the diagnostic chart.
1. Tests to see if malfunction is present.
2. Checks charging system integrity.
3. Checks for excessive resistance in the switched battery input circuit.
4. Tests for high resistance in the enable relay control circuit.
5. Ensures low voltage condition was not due to a poor connection.

Diagnostic Aids: An "Intermittent" malfunction is most likely caused by a poor connection, rubbed through wire insulation, or a wire that is broken inside the insulation.

The frequency of the malfunction can be checked by using the enhanced diagnostic function of the Tech 1.

Any circuitry, that is suspected as causing the intermittent complaint, should be thoroughly checked for backed out terminals, improper mating, broken locks, improperly formed or damaged terminals, poor terminal to wiring connections or physical damage to the wiring harness.

While performing a voltage load test, if it is noted that only ignition voltage drops below acceptable voltage levels, CKT 650 should be checked for high resistance or an open condition.

DIAGNOSTIC CHART DTC 36 — 1994 CAVALIER AND SUNBIRD

DTC 36
LOW SYSTEM VOLTAGE

1.
- START ENGINE.
- USING TECH 1, SELECT "F3: SNAPSHOT."
- SELECT "F4: SINGLE DTC"; THEN ENTER "36".
- DRIVE VEHICLE ABOVE 5 km/h (3 MPH).

DID TECH 1 TRIGGER?

YES → go to step 2.

NO →
- WITH ENGINE RUNNING, USE TECH 1 AND SELECT "F4: MISC. TESTS."
- PERFORM "VOLTAGE LOAD" TEST.

ARE IGNITION AND BATTERY VOLTAGES GREATER THAN 10 VOLTS?

- NO → go to step 2.
- YES → MALFUNCTION IS NOT PRESENT AT THIS TIME. REFER TO "DIAGNOSTIC AIDS".

2.
- DIAGNOSE CHARGING SYSTEM.

IS CHARGING SYSTEM OK?

- YES → go to step 3.
- NO → CORRECT LOW VOLTAGE CONDITION.

3.
- IGNITION "OFF."
- DISCONNECT 2-WAY AND 24-WAY EBCM CONNECTORS.
- DISCONNECT ENABLE RELAY.
- USING J 39200, MEASURE RESISTANCE BETWEEN ENABLE RELAY HARNESS CONNECTOR TERMINAL "87" AND 2-WAY EBCM HARNESS CONNECTOR TERMINAL "A".

IS RESISTANCE LESS THAN 2 OHMS?

YES →
- DISCONNECT NEGATIVE BATTERY CABLE.
- DISCONNECT POSITIVE BATTERY CABLE.
- USING J 39200 MEASURE RESISTANCE BETWEEN THE POSITIVE BATTERY CABLE TERMINAL AND ENABLE RELAY HARNESS CONNECTOR TERMINAL "30".

IS RESISTANCE LESS THAN 2 OHMS?

- YES → go to step 4.
- NO →
  - REPAIR HIGH RESISTANCE IN CKT 2.
  - PROCEED TO STEP 4.

NO →
- REPAIR HIGH RESISTANCE IN CKT 850.
- PROCEED TO STEP 4.

4.
- REMOVE HTR-A/C FUSE.
- USING J 39200 MEASURE RESISTANCE BETWEEN EACH HTR-A/C FUSE BLOCK TERMINAL AND 24-WAY EBCM HARNESS CONNECTOR TERMINAL "14".

IS EITHER MEASUREMENT LESS THAN 2 OHMS?

NO → REPAIR HIGH RESISTANCE IN CKT 650.

YES →
- WITH BATTERY CABLES DISCONNECTED, TURN IGNITION "ON" (THIS PROVIDES CIRCUIT CONTINUITY).
- USING J 39200, MEASURE RESISTANCE BETWEEN POSITIVE BATTERY CABLE TERMINAL AND EACH HTR-A/C FUSE BLOCK TERMINAL.

IS EITHER MEASUREMENT LESS THAN 2 OHMS?

- NO → REPAIR HIGH RESISTANCE IN THE "IGNITION SWITCH 'RUN' FUSE FEED" CIRCUIT.
- YES → go to step 5.

5.
- IGNITION "OFF."
- INSPECT ENABLE RELAY AND ENABLE RELAY HARNESS CONNECTOR FOR POOR TERMINAL CONTACT. INSPECT 2-WAY AND 24-WAY EBCM HARNESS CONNECTORS FOR A POOR CONNECTION TO THE EBCM. INSPECT POSITIVE BATTERY CABLE TERMINAL FOR POOR CONTACT TO THE BATTERY. REPLACE TERMINALS IF POOR CONTACT IS EVIDENT.
- RECONNECT ALL CONNECTORS.
- DRIVE VEHICLE ABOVE 5 km/h (3 MPH).

IS DTC 36 SET AS A CURRENT DTC?

- YES → REPLACE EBCM.
- NO → MALFUNCTION IS INTERMITTENT. REFER TO "DIAGNOSTIC AIDS" ON FACING PAGE.

THIS CHART ASSUMES THAT A CURRENT DTC IS STORED INDICATING THAT THIS MALFUNCTION IS PRESENT.

DIAGNOSTIC CHART DTC 37 — 1994 CAVALIER AND SUNBIRD

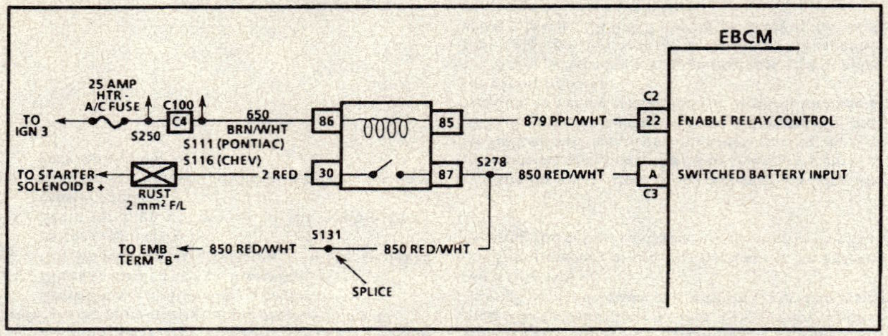

DTC 37

HIGH SYSTEM VOLTAGE

Circuit Description:
This DTC is designed to detect high vehicle voltage levels prior to any required motor movement (initialization or ABS). If excessive voltage exists, demagnetization of the motor magnets may occur, which would eventually affect or eliminate ABS performance.

DTC Will Set When: DTC 37 can only be set if the vehicle's speed is greater than 5 km/h (3 mph). If the switched battery voltage is greater than 17 volts, a malfunction exists.

Action Taken: A malfunction DTC is stored, ABS is disabled and the ABS warning lamp is turned "ON."

DTC Chart Test Description: Number(s) below refer to circled number(s) on the diagnostic chart.

1. Checks the voltage level being received by the EBCM.
2. Isolates the malfunction to either the EBCM or the charging system.

Diagnostic Aids: An "Intermittent" malfunction is most likely caused by a poor connection, rubbed through wire insulation, or a wire that is broken inside the insulation.

The frequency of the malfunction can be checked by using the enhanced diagnostic function of the Tech 1 tool, as described in "Enhanced Diagnostics," found in this section.

Any circuitry, that is suspected as causing the intermittent malfunction, should be thoroughly checked for backed out terminals, improper mating, broken locks, improperly formed or damaged terminals, poor terminal to wiring connections or physical damage to the wiring harness.

DIAGNOSTIC CHART DTC 37 — 1994
CAVALIER AND SUNBIRD

DTC 37
HIGH SYSTEM VOLTAGE

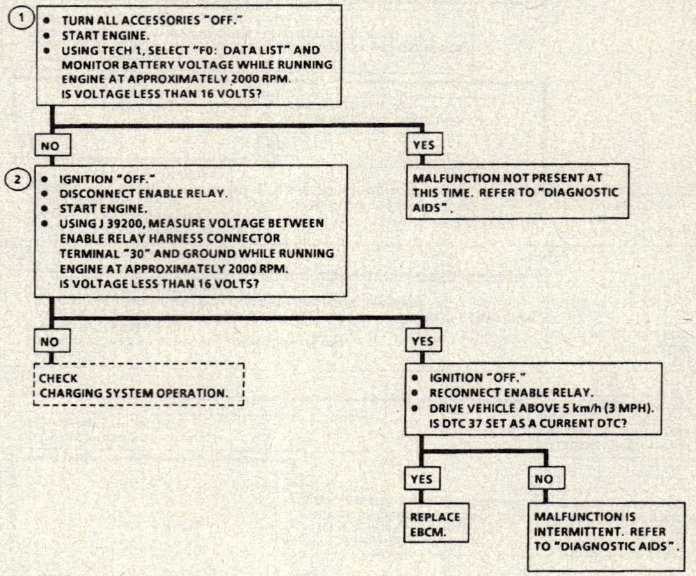

THIS CHART ASSUMES THAT A CURRENT DTC IS STORED INDICATING THAT THIS MALFUNCTION IS PRESENT.

DIAGNOSTIC CHART DTC 38 — 1994
CAVALIER AND SUNBIRD

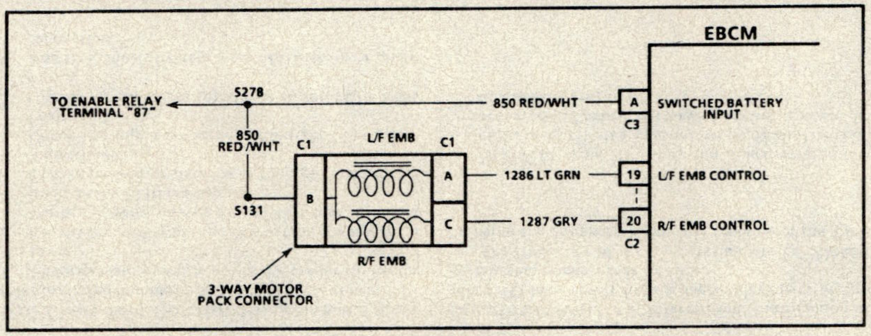

DTC 38

LEFT FRONT EMB WILL NOT HOLD MOTOR

Circuit Description:

Current is supplied to the EMBs through terminal "B" of the three-way motor pack connector. When a release of the EMB is commanded, the EBCM grounds the EMB input and allows the EMB to be energized to release the motor.

The left front motor is commanded in the forward direction while the EMB is released (energized).

DTC Will Set When: DTC 38 can be set during initialization. A malfunction exists if the feedback current is less than the command current and the left front channel was expected to be in the home position. DTC 86 is always set with DTC 38.

Action Taken: If an EMB can not hold a motor in the home position, the motor may be backdriven when the brake pedal is applied causing the brake pedal to drop. A malfunction DTC is stored, ABS is disabled and the ABS warning lamp is turned "ON."

DTC Chart Test Description: Number(s) below refer to circled number(s) on the diagnostic chart.

1. Ensures motor circuits are in the proper connector cavity.
2. Checks the ability of the EMB to hold the motor. If the brake pedal moves during this test, it would indicate the EMB is slipping.
3. Checks for high resistance in the left front motor high circuit.
4. Checks for high resistance in the left front motor low circuit.
5. Checks for excessive resistance in the left front motor.
6. Ensures malfunction was not due to a poor connection.

Diagnostic Aids: An "Intermittent" malfunction in this DTC may result from a mechanical part of the system that sticks, binds, or slips.

The frequency of the malfunction can be checked by using the enhanced diagnostic function of the Tech 1.

NOTICE: This test may fail if low voltage (10 volts or less) was present when initialization occurred. Ensure battery/charging system is functioning properly.

DIAGNOSTIC CHART DTC 38 — 1994
CAVALIER AND SUNBIRD

NOTICE: TESTING EMB WITH JUMPER WIRES TO VOLTAGE OR GROUND WILL DESTROY THE EMB.

DTC 38

LEFT FRONT EMB WILL NOT HOLD MOTOR

NOTICE: J 39200 TEST LEADS MUST BE "ZEROED" PRIOR TO MAKING ANY RESISTANCE MEASUREMENTS. REFER TO J 39200 USER'S MANUAL.

(1)
- INSPECT THE 6-WAY MOTOR PACK AND EBCM HARNESS CONNECTORS FOR PROPER WIRE COLOR/CONNECTOR CAVITY COMBINATION.
ARE THE PROPER WIRES LOCATED IN THE PROPER CONNECTOR CAVITY?

NO: USING THE CIRCUIT DIAGRAMS LOCATED IN THIS SECTION, PLACE THE PROPER WIRES IN THE PROPER CONNECTOR CAVITIES. CLEAR DTCs AND RECHECK.

YES: IS DTC 67 ALSO PRESENT?

YES: PROCEED TO DTC 67 CHART.

NO:

(2)
- IGNITION "ON."
- USING TECH 1, SELECT EMB FUNCTIONAL TEST.
- APPLY FIRM PRESSURE ON BRAKE PEDAL.
- PERFORM EMB FUNCTIONAL TEST.
- DOES BRAKE PEDAL FALL OR RISE DURING THIS TEST?

YES: REPLACE MOTOR PACK.

NO:

(3)
- IGNITION "OFF."
- DISCONNECT 6-WAY MOTOR PACK AND EBCM CONNECTORS.
- USING J 39200, MEASURE THE RESISTANCE BETWEEN 6-WAY EBCM HARNESS CONNECTOR TERMINAL "F" AND 6-WAY MOTOR PACK HARNESS CONNECTOR TERMINAL "A".
IS RESISTANCE LESS THAN 1.5Ω?

NO: REPAIR HIGH RESISTANCE IN CKT 1280.

YES:

(4)
- USING J 39200, MEASURE RESISTANCE BETWEEN 6-WAY EBCM HARNESS CONNECTOR TERMINAL "E" AND 6-WAY MOTOR PACK HARNESS CONNECTOR TERMINAL "B".
IS RESISTANCE LESS THAN 1.5Ω?

NO: REPAIR HIGH RESISTANCE IN CKT 1281.

YES:

(5)
- USING J 39200, MEASURE RESISTANCE BETWEEN MOTOR PACK CONNECTOR TERMINALS "A" AND "B".
IS RESISTANCE LESS THAN 1.5Ω?

NO: REPLACE MOTOR PACK.

YES:

(6)
- INSPECT 6-WAY EBCM AND MOTOR PACK CONNECTORS FOR POOR TERMINAL CONTACT OR CORROSION.
- REPLACE TERMINALS THAT EXHIBIT POOR CONTACT OR CORROSION.
- RECONNECT ALL CONNECTORS.
- DRIVE VEHICLE FOR THREE DRIVE CYCLES (A DRIVE CYCLE CONSISTS OF STARTING ENGINE, DRIVING OVER 16 km/h (10 MPH) AND THEN TURNING IGNITION "OFF").
DID DTC 38 SET IN THE LAST THREE DRIVE CYCLES?

YES: REPLACE EBCM.

NO: MALFUNCTION IS INTERMITTENT. REFER TO "DIAGNOSTIC AIDS".

DIAGNOSTIC CHART DTC 41 — 1994 CAVALIER AND SUNBIRD

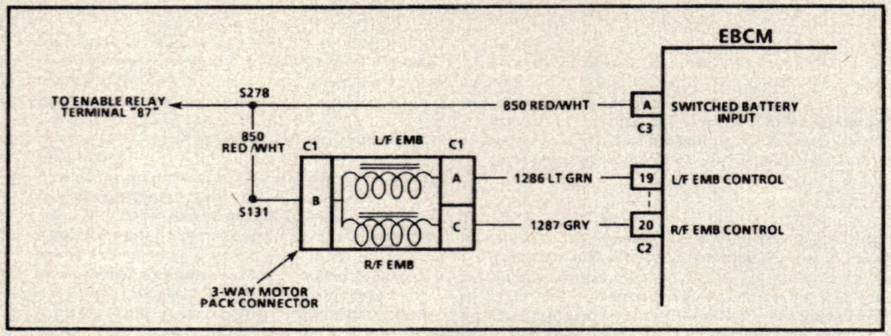

DTC 41

RIGHT FRONT EMB WILL NOT HOLD MOTOR

Circuit Description:
Current is supplied to the EMBs through terminal "B" of the three-way motor pack connector. When a release of the EMB is commanded, the EBCM grounds the EMB input and allows the EMB to be energized to release the motor.

The right front motor is commanded in the forward direction while the EMB is released (energized).

DTC Will Set When: DTC 41 can be set during initialization. A malfunction exists if the feedback current is less than the command current and the right front channel was expected to be in the home position. DTC 86 is always set with DTC 41.

Action Taken: If an EMB can not hold a motor in the home position, the motor may be backdriven when the brake pedal is applied causing the brake pedal to drop. A malfunction DTC is stored, ABS is disabled and the ABS warning lamp is turned "ON."

DTC Chart Test Description: Number(s) below refer to circled number(s) on the diagnostic chart.
1. Ensures motor circuits are in the proper connector cavity.
2. Checks the ability of the EMB to hold the motor. If the brake pedal moves during this test, it would indicate the EMB is slipping.
3. Checks for high resistance in the right front motor high circuit.
4. Checks for high resistance in the right front motor low circuit.
5. Checks for excessive resistance in the right front motor.
6. Ensures malfunction was not due to a poor connection.

Diagnostic Aids: An "Intermittent" malfunction in this DTC may result from a mechanical part of the system that sticks, binds, or slips.

The frequency of the malfunction can be checked by using the enhanced diagnostic function of the Tech 1.

NOTICE: This DTC may fail if low voltage (10 volts or less) was present when initialization occurred. Ensure battery/charging system is functioning properly.

DIAGNOSTIC CHART DTC 41 — 1994 CAVALIER AND SUNBIRD

NOTICE: TESTING EMB WITH JUMPER WIRES TO VOLTAGE OR GROUND WILL DESTROY THE EMB.

DTC 41

RIGHT FRONT EMB WILL NOT HOLD MOTOR

NOTICE: J 39200 TEST LEADS MUST BE "ZEROED" PRIOR TO MAKING ANY RESISTANCE MEASUREMENTS. REFER TO J 39200 USER'S MANUAL.

(1)
- INSPECT THE 6-WAY MOTOR PACK AND EBCM HARNESS CONNECTORS FOR PROPER WIRE COLOR/CONNECTOR CAVITY COMBINATION.
ARE THE PROPER WIRES LOCATED IN THE PROPER CONNECTOR CAVITY?

NO → USING THE CIRCUIT DIAGRAMS LOCATED IN THIS SECTION, PLACE THE PROPER WIRES IN THE PROPER CONNECTOR CAVITIES. CLEAR DTCs AND RECHECK.

YES → IS DTC 71 ALSO PRESENT?

YES → PROCEED TO DTC 71 CHART.

NO →

(2)
- IGNITION "ON."
- USING TECH 1, SELECT EMB FUNCTIONAL TEST.
- APPLY FIRM PRESSURE ON BRAKE PEDAL.
- PERFORM EMB FUNCTIONAL TEST.
DOES BRAKE PEDAL FALL OR RISE DURING THIS TEST?

YES → REPLACE MOTOR PACK.

NO →

(3)
- IGNITION "OFF."
- DISCONNECT 6-WAY MOTOR PACK AND EBCM CONNECTORS.
- USING J 39200, MEASURE THE RESISTANCE BETWEEN 6-WAY EBCM HARNESS CONNECTOR TERMINAL "H" AND 6-WAY MOTOR PACK HARNESS CONNECTOR TERMINAL "E".
IS RESISTANCE LESS THAN 1.5Ω?

NO → REPAIR HIGH RESISTANCE IN CKT 1282.

YES →

(4)
- USING J 39200, MEASURE RESISTANCE BETWEEN 6-WAY EBCM HARNESS CONNECTOR TERMINAL "G" AND 6-WAY MOTOR PACK HARNESS CONNECTOR TERMINAL "F".
IS RESISTANCE LESS THAN 1.5Ω?

NO → REPAIR HIGH RESISTANCE IN CKT 1283.

YES →

(5)
- USING J 39200, MEASURE RESISTANCE BETWEEN MOTOR PACK CONNECTOR TERMINALS "E" AND "F".
IS RESISTANCE LESS THAN 1.5Ω?

NO → REPLACE MOTOR PACK.

YES →

(6)
- INSPECT 6-WAY EBCM AND MOTOR PACK CONNECTORS FOR POOR TERMINAL CONTACT OR CORROSION. REPLACE TERMINALS THAT EXHIBIT POOR CONTACT OR CORROSION.
- RECONNECT ALL CONNECTORS.
- DRIVE VEHICLE FOR THREE DRIVE CYCLES (A DRIVE CYCLE CONSISTS OF STARTING ENGINE, DRIVING OVER 16 km/h (10 MPH) AND THEN TURNING IGNITION "OFF").
DID DTC 41 SET IN THE LAST THREE DRIVE CYCLES?

YES → REPLACE EBCM.

NO → MALFUNCTION IS INTERMITTENT. REFER TO "DIAGNOSTIC AIDS".

DIAGNOSTIC CHART DTC 42 — 1994 CAVALIER AND SUNBIRD

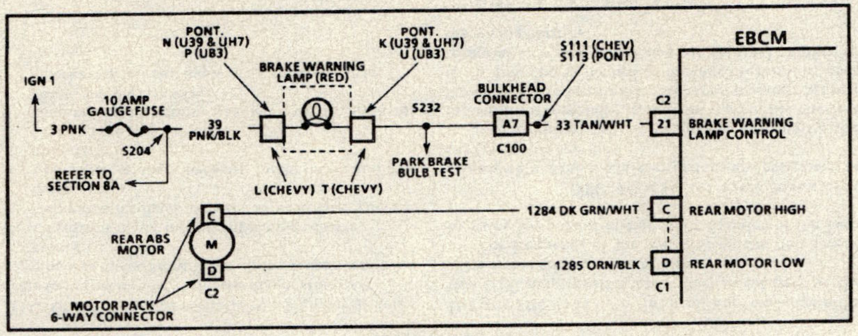

DTC 42

REAR ESB WILL NOT HOLD MOTOR

Circuit Description:
This DTC is designed to detect a slipping rear ESB. During initialization and braking, the rear motor is re-homed. If the ESB slips, the motor/piston will move. During the next key "ON" initialization, a re-home of the motor verifies the motor/piston remained at the home position. If motor movement is detected, the ESB must be slipping.

DTC Will Set When: DTC 42 can be set during initialization. DTC 86 is always set with DTC 42.

Action Taken: If an ESB can not hold a motor in the home position, the motor may be backdriven when the brake pedal is applied causing the brake pedal to drop. A malfunction DTC is stored, ABS is disabled and the ABS warning lamp is turned "ON."

DTC Chart Test Description: Number(s) below refer to circled number(s) on the diagnostic chart.

1. This step ensures motor circuits are in the proper connector cavity.
2. This step checks the rear ESB. A broken or defective ESB would result in the rear piston being back driven by hydraulic pressure, and wheel movement would result.
3. Checks for high resistance in the rear motor high circuit.
4. Checks for high resistance in the rear motor low circuit.
5. Checks for excessive resistance in the rear motor.
6. Ensures malfunction was not due to a poor connection.

Diagnostic Aids: An "Intermittent" malfunction in this DTC may result from a mechanical part of the system that sticks, binds, or slips.
The frequency of the malfunction can be checked by using the enhanced diagnostic function of the Tech 1.

DIAGNOSTIC CHART DTC 42 — 1994 CAVALIER AND SUNBIRD

DTC 42

REAR ESB WILL NOT HOLD MOTOR

NOTICE: J 39200 TEST LEADS MUST BE "ZEROED" PRIOR TO MAKING ANY RESISTANCE MEASUREMENTS. REFER TO J 39200 USER'S MANUAL.

(1)
- INSPECT THE 6-WAY EBCM AND MOTOR PACK CONNECTORS FOR PROPER WIRE COLOR/CONNECTOR CAVITY COMBINATION.
- ARE THE PROPER WIRES LOCATED IN THE PROPER CONNECTOR CAVITIES?

NO: USING THE CIRCUIT DIAGRAMS LOCATED IN THE FRONT OF THIS SECTION, PLACE THE PROPER WIRES INTO THE PROPER CONNECTOR CAVITIES. CLEAR DTCs AND RECHECK SYSTEM.

YES:

(2)
- RAISE THE VEHICLE SUCH THAT THE REAR WHEELS ARE APPROXIMATELY SIX INCHES OFF THE FLOOR.
- IGNITION "ON."
- APPLY FIRM PRESSURE ON BRAKE PEDAL AND RELEASE.
- REAPPLY FIRM PRESSURE ON BRAKE PEDAL AND HAVE AN ASSISTANT TRY TO SPIN THE REAR WHEEL BY HAND.
COULD ASSISTANT SPIN THE REAR WHEELS?

YES: REPLACE MOTOR PACK.

NO:

(3)
- IGNITION "OFF."
- DISCONNECT 6-WAY MOTOR PACK AND EBCM CONNECTORS.
- USING J 39200, MEASURE THE RESISTANCE BETWEEN 6-WAY EBCM HARNESS CONNECTOR TERMINAL "C" AND 6-WAY MOTOR PACK HARNESS CONNECTOR TERMINAL "C".
IS RESISTANCE LESS THAN 1.5Ω?

NO: REPAIR HIGH RESISTANCE IN CKT 1284.

YES:

(4)
- USING J 39200, MEASURE RESISTANCE BETWEEN 6-WAY EBCM HARNESS CONNECTOR TERMINAL "D" AND 6-WAY MOTOR PACK HARNESS CONNECTOR TERMINAL "D".
IS RESISTANCE LESS THAN 1.5Ω?

NO: REPAIR HIGH RESISTANCE IN CKT 1285.

YES:

(5)
- USING J 39200, MEASURE RESISTANCE BETWEEN MOTOR PACK CONNECTOR TERMINALS "C" AND "D".
IS RESISTANCE LESS THAN 1.5Ω?

NO: REPLACE MOTOR PACK.

YES:

(6)
- INSPECT 6-WAY EBCM AND MOTOR PACK CONNECTORS FOR POOR TERMINAL CONTACT OR CORROSION.
- REPLACE TERMINALS THAT EXHIBIT POOR CONTACT OR CORROSION.
- RECONNECT ALL CONNECTORS.
- DRIVE VEHICLE FOR THREE DRIVE CYCLES (A DRIVE CYCLE CONSISTS OF STARTING ENGINE, DRIVING OVER 16 km/h (10 MPH) AND THEN TURNING IGNITION "OFF").
DID DTC 42 SET IN THE LAST THREE DRIVE CYCLES?

YES: REPLACE EBCM.

NO: MALFUNCTION IS INTERMITTENT. REFER TO "DIAGNOSTIC AIDS".

DIAGNOSTIC CHART DTC 44 — 1994 CAVALIER AND SUNBIRD

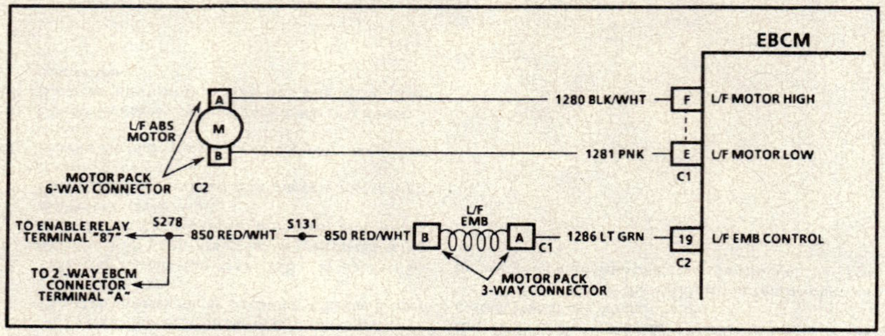

DTC 44

LEFT FRONT ABS CHANNEL WILL NOT MOVE

Circuit Description:
Current is supplied to the EMB through terminal "B" of the three-way motor pack connector. When a release of the EMB is commanded, the EBCM grounds the EMB input and allows the EMB to be energized to release the motor. This DTC is designed to detect non-actuating EMB, stuck motor or a seized hydraulic modulator.

DTC Will Set When: DTC 44 can be set at initialization. The EBCM commands the left front motor in the reverse and forward direction to ensure the motor moves properly. If the EBCM detects a condition in which it cannot move the motor in either direction, a malfunction exists.

Action Taken: This malfunction indicates the channel can not be moved properly. A malfunction DTC is stored, ABS is disabled and the ABS warning lamp is turned "ON."

DTC Chart Test Description: Number(s) below refer to circled number(s) on the diagnostic chart.

1. Ensures motor circuits are in the proper connector cavity.
2. Checks for motor movement during release.
3. Compares feedback current to command current during apply.
4. Checks for a short between motor high and low circuits.
5. Checks for a shorted motor.
6. Ensures malfunction was not due to physical damage of the circuitry.
7. Checks if motor can apply properly under load.

Diagnostic Aids: An "Intermittent" malfunction in this DTC may result from a mechanical part of the system that sticks, binds, or slips.

The frequency of the malfunction can be checked by using the enhanced diagnostic function of the Tech 1.

DTC 44 may set after modulator disassembly if the modulator pistons are positioned at the bottom of their bore.

Any circuitry, that is suspected as causing the intermittent complaint, should be thoroughly checked for backed out terminals, improper mating, broken locks, improperly formed or damaged terminals, poor terminal to wiring connections or physical damage to the wiring harness.

NOTICE: This DTC may fail if low voltage (10 volts or less) was present when initialization occurred. Ensure battery/charging system is functioning properly.

DIAGNOSTIC CHART DTC 44 — 1994 CAVALIER AND SUNBIRD

DTC 44
(Page 1 of 2)
LEFT FRONT ABS CHANNEL WILL NOT MOVE

NOTICE: J 39200 TEST LEADS MUST BE "ZEROED" PRIOR TO MAKING ANY RESISTANCE MEASUREMENTS. REFER TO J 39200 USER'S MANUAL.

(1)
- VISUALLY INSPECT THE 6-WAY EBCM AND MOTOR PACK CONNECTORS FOR PROPER WIRE COLOR/CONNECTOR CAVITY COMBINATION.
ARE THE PROPER WIRES LOCATED IN THE PROPER CONNECTOR CAVITIES?

YES → go to (2)
NO → USING THE CIRCUIT DIAGRAMS LOCATED IN THE FRONT OF THIS SECTION, PLACE THE PROPER WIRES IN THE PROPER CONNECTOR CAVITIES. CLEAR DTCs AND RECHECK SYSTEM.

(2)
- IGNITION "ON."
- USING TECH 1 ABS TESTS, SELECT MANUAL CONTROL FUNCTION.
- SELECT LEFT FRONT MOTOR APPLY FUNCTION AND APPLY THE MOTOR.
- APPLY FIRM PRESSURE ON BRAKE PEDAL.
- SELECT LEFT FRONT MOTOR RELEASE FUNCTION AND RELEASE MOTOR.
DID BRAKE PEDAL FALL?

NO → go to (3)
YES → go to (7)

(7)
- FIRM PRESSURE STILL ON BRAKE PEDAL.
- SELECT LEFT FRONT MOTOR APPLY FUNCTION AND APPLY THE MOTOR.
DID BRAKE PEDAL RISE?

NO → go to (3)
YES → MALFUNCTION IS INTERMITTENT. REFER TO "DIAGNOSTIC AIDS".

(3)
- REMOVE FOOT FROM BRAKE PEDAL.
- SELECT LEFT FRONT MOTOR APPLY FUNCTION AND APPLY THE MOTOR WHILE CAREFULLY OBSERVING THE COMMANDED CURRENT AND THE FEEDBACK CURRENT ON THE TECH 1.
WAS THE FEEDBACK CURRENT HIGHER THAN THE COMMANDED CURRENT?

YES → go to (4)
NO → GO TO PAGE 2 OF 2 OF THIS DTC CHART.

(4)
- IGNITION "OFF."
- DISCONNECT 6-WAY EBCM CONNECTOR AND THE 6-WAY MOTOR PACK CONNECTOR.
- USING J 39200, MEASURE RESISTANCE BETWEEN TERMINAL "E" AND TERMINAL "F" OF THE 6-WAY EBCM HARNESS CONNECTOR.
DOES J 39200 DISPLAY RESISTANCE AS "OL" (INFINITE)?

YES → go to (5)
NO → REPAIR SHORT BETWEEN CKT 1280 AND CKT 1281.

(5)
- USING J 39200, MEASURE RESISTANCE BETWEEN MOTOR PACK TERMINALS "A" AND "B".
IS RESISTANCE GREATER THAN .4 OHMS?

YES → go to (6)
NO → REPLACE MOTOR PACK.

(6)
- INSPECT CKTs 1280 AND 1281 FOR DAMAGE WHICH MAY RESULT IN A SHORT BETWEEN THE TWO CIRCUITS. REPAIR DAMAGE IF EVIDENT.
- RECONNECT ALL CONNECTORS.
- DRIVE VEHICLE FOR THREE DRIVE CYCLES (A DRIVE CYCLE CONSISTS OF STARTING ENGINE, DRIVING OVER 16 KM/H (10 MPH) AND THEN TURNING IGNITION "OFF").
DID DTC 44 SET IN THE LAST THREE DRIVE CYCLES?

YES → REPLACE EBCM.
NO → MALFUNCTION IS INTERMITTENT. REFER TO "DIAGNOSTIC AIDS".

- AFTER DIAGNOSIS IS COMPLETE, CLEAR DTCs AND TEST DRIVE VEHICLE FOR THREE (3) DRIVE CYCLES TO VERIFY DTC DOES NOT RESET. A DRIVE CYCLE CONSISTS OF STARTING THE VEHICLE, DRIVING OVER 10 MPH AND THEN KEYING DOWN.

DIAGNOSTIC CHART DTC 44 — 1994 CAVALIER AND SUNBIRD

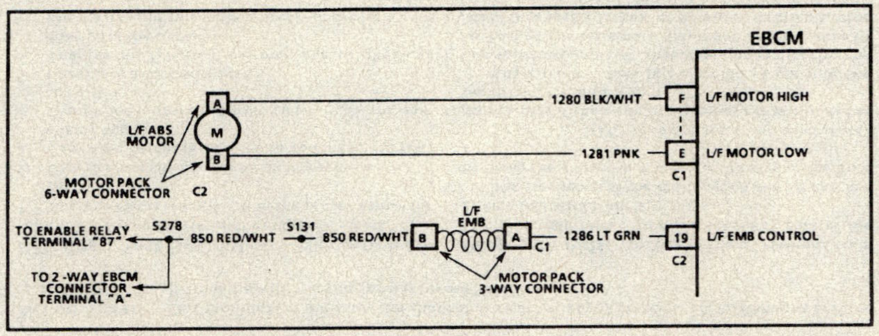

DTC 44
(Page 2 of 2)
LEFT FRONT ABS CHANNEL WILL NOT MOVE

Circuit Description:
Current is supplied to the EMB through terminal "B" of the three-way motor pack connector. When a release of the EMB is commanded, the EBCM grounds the EMB input and allows the EMB to be energized to release the motor. This DTC is designed to detect non-actuating EMB, stuck motor or a seized hydraulic modulator.

DTC Will Set When: DTC 44 can be set at initialization. The EBCM commands the left front motor in the reverse and forward direction to ensure the motor moves properly. If the EBCM detects a condition in which it cannot move the motor in either direction, a malfunction exists.

Action Taken: This malfunction indicates the channel can not be moved properly. A malfunction DTC is stored, ABS is disabled and the ABS warning lamp is turned "ON."

DTC Chart Test Description: Number(s) below refer to circled number(s) on the diagnostic chart.
8. Checks for a short to B+ in the left front EMB control circuit.
9. Checks for high resistance in the left front EMB control circuit.
10. Checks for high resistance in the EMB power feed circuit.
11. Checks for high resistance in the left front EMB.
12. Checks for proper operation of the hydraulic modulator.
13. Ensures malfunction was not due to poor terminal contact.
14. Isolates malfunction to either the left front EMB circuitry or the EBCM.
15. Ensures malfunction was not due to physical damage of the EMB circuit.

Diagnostic Aids: An "Intermittent" malfunction in this DTC may result from a mechanical part of the system that sticks, binds, or slips.

The frequency of the malfunction can be checked by using the enhanced diagnostic function of the Tech 1.

DTC 44 may set after modulator disassembly if the modulator pistons are positioned at the bottom of their bore.

Any circuitry, that is suspected as causing the intermittent complaint, should be thoroughly checked for backed out terminals, improper mating, broken locks, improperly formed or damaged terminals, poor terminal to wiring connections or physical damage to the wiring harness.

NOTICE: This DTC may fail if low voltage (10 volts or less) was present when initialization occurred. Ensure battery/charging system is functioning properly.

DIAGNOSTIC CHART DTC 44 — 1994 CAVALIER AND SUNBIRD

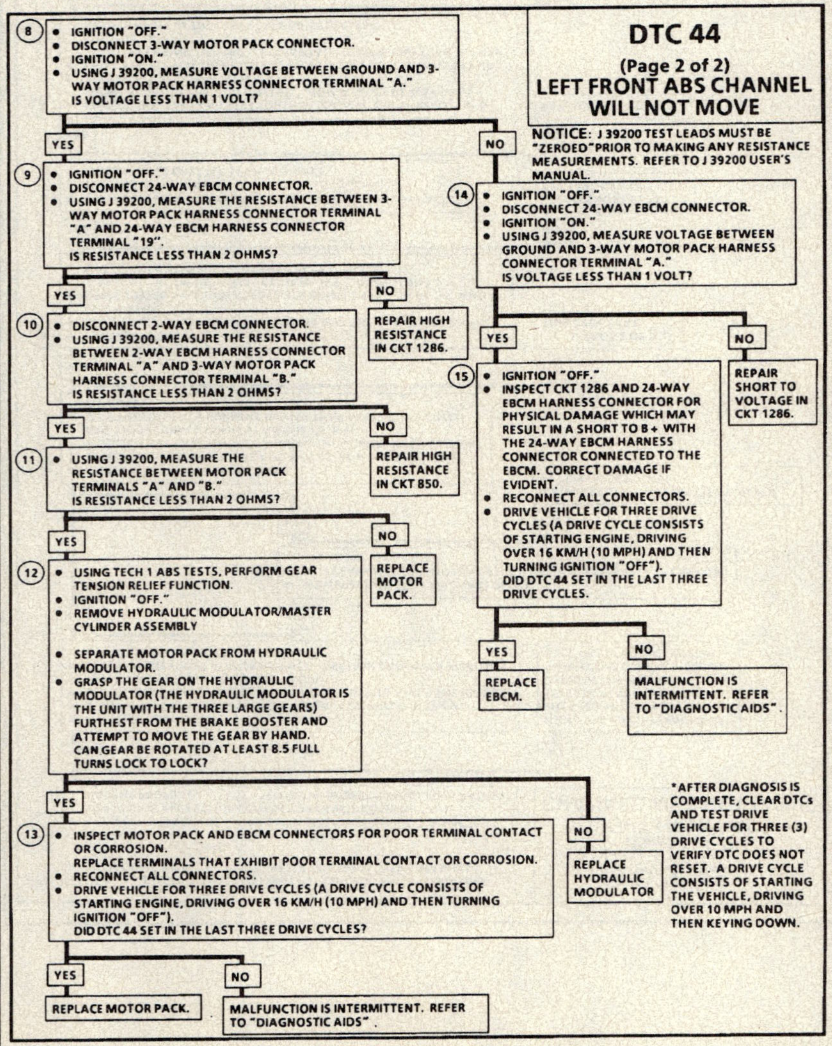

DIAGNOSTIC CHART DTC 45 — 1994 CAVALIER AND SUNBIRD

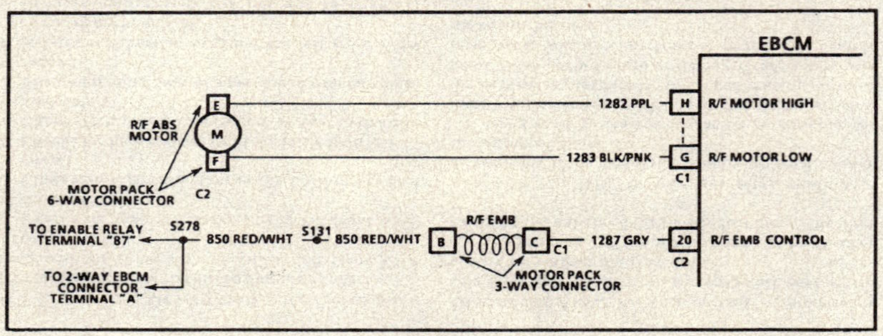

DTC 45

(Page 1 of 2)

RIGHT FRONT ABS CHANNEL WILL NOT MOVE

Circuit Description:

Current is supplied to the EMB through terminal "B" of the three-way motor pack connector. When a release of the EMB is commanded, the EBCM grounds the EMB input and allows the EMB to be energized to release the motor.

This DTC is designed to detect non-actuating EMB, stuck motor or a seized hydraulic modulator. When the release is commanded during initialization, the EMB should release the motor, resulting in sensed current being less than commanded current (motor is spinning freely). If the motor is not moving, sensed current will be equal to stall current.

DTC Will Set When: DTC 45 can be set during initialization. The EBCM commands the right front motor in the reverse and forward direction to ensure the motor moves properly. If the EBCM detects a condition in which it cannot move the motor in either direction, a malfunction exists.

Action Taken: This malfunction indicates the channel can not be moved properly. A malfunction DTC is stored, ABS is disabled and the ABS warning lamp is turned "ON."

DTC Chart Test Description: Number(s) below refer to circled number(s) on the diagnostic chart.

1. Ensures motor circuits are in the proper connector cavity.
2. Checks for motor movement during release.
3. Compares feedback current to command current during apply.
4. Checks for a short between motor high and low circuits.
5. Checks for a shorted motor.
6. Ensures malfunction was not due to physical damage of the circuitry.
7. Checks if motor can apply properly under load.

Diagnostic Aids: An "Intermittent" malfunction in this DTC may result from a mechanical part of the system that sticks, binds, or slips.

The frequency of the malfunction can be checked by using the enhanced diagnostic function of the Tech 1.

DTC 45 may set after modulator disassembly if the modulator pistons are positioned at the bottom of their bore.

Any circuitry, that is suspected as causing the intermittent complaint, should be thoroughly checked for backed out terminals, improper mating, broken locks, improperly formed or damaged terminals, poor terminal to wiring connections or physical damage to the wiring harness.

NOTICE: This DTC may fail if low voltage (10 volts or less) was present when initialization occurred. Ensure battery/charging system is functioning properly.

DIAGNOSTIC CHART DTC 45 — 1994 CAVALIER AND SUNBIRD

DTC 45
(Page 1 of 2)
RIGHT FRONT ABS CHANNEL WILL NOT MOVE

NOTICE: J 39200 TEST LEADS MUST BE "ZEROED" PRIOR TO MAKING ANY RESISTANCE MEASUREMENTS. REFER TO J 39200 USER'S MANUAL.

(1)
- VISUALLY INSPECT THE 6-WAY EBCM AND MOTOR PACK CONNECTORS FOR PROPER WIRE COLOR/CONNECTOR CAVITY COMBINATION. ARE THE PROPER WIRES LOCATED IN THE PROPER CONNECTOR CAVITIES?

NO: USING THE CIRCUIT DIAGRAMS LOCATED IN THE FRONT OF THIS SECTION, PLACE THE PROPER WIRES IN THE PROPER CONNECTOR CAVITIES. CLEAR DTCs AND RECHECK SYSTEM.

YES:

(2)
- IGNITION "ON."
- USING TECH 1 ABS TESTS, SELECT MANUAL CONTROL FUNCTION.
- SELECT RIGHT FRONT MOTOR APPLY FUNCTION AND APPLY THE MOTOR.
- APPLY FIRM PRESSURE ON BRAKE PEDAL.
- SELECT RIGHT FRONT MOTOR RELEASE FUNCTION AND RELEASE MOTOR. DID BRAKE PEDAL FALL?

YES:

(7)
- FIRM PRESSURE STILL ON BRAKE PEDAL.
- SELECT RIGHT FRONT MOTOR APPLY FUNCTION AND APPLY THE MOTOR. DID BRAKE PEDAL RISE?

YES: MALFUNCTION IS INTERMITTENT. REFER TO "DIAGNOSTIC AIDS".

NO (from 2 and 7):

(3)
- REMOVE FOOT FROM BRAKE PEDAL.
- SELECT RIGHT FRONT MOTOR APPLY FUNCTION AND APPLY THE MOTOR WHILE CAREFULLY OBSERVING THE COMMANDED CURRENT AND THE FEEDBACK CURRENT ON THE TECH 1. WAS THE FEEDBACK CURRENT HIGHER THAN THE COMMANDED CURRENT?

NO: GO TO PAGE 2 OF 2 OF THIS DTC CHART.

YES:

(4)
- IGNITION "OFF."
- DISCONNECT 6-WAY EBCM CONNECTOR AND THE 6-WAY MOTOR PACK CONNECTOR.
- USING J 39200, MEASURE RESISTANCE BETWEEN TERMINAL "G" AND TERMINAL "H" OF THE 6-WAY EBCM HARNESS CONNECTOR. DOES J 39200 DISPLAY RESISTANCE AS "OL" (INFINITE)?

NO: REPAIR SHORT BETWEEN CKT 1282 AND CKT 1283.

YES:

(5)
- USING J 39200, MEASURE RESISTANCE BETWEEN MOTOR PACK TERMINALS "E" AND "F". IS RESISTANCE GREATER THAN .4 OHMS?

NO: REPLACE MOTOR PACK.

YES:

(6)
- INSPECT CKTs 1282 AND 1283 FOR DAMAGE WHICH MAY RESULT IN A SHORT BETWEEN THE TWO CIRCUITS. REPAIR DAMAGE IF EVIDENT.
- RECONNECT ALL CONNECTORS.
- DRIVE VEHICLE FOR THREE DRIVE CYCLES (A DRIVE CYCLE CONSISTS OF STARTING ENGINE, DRIVING OVER 16 KM/H (10 MPH) AND THEN TURNING IGNITION "OFF"). DID DTC 45 SET IN THE LAST THREE DRIVE CYCLES?

YES: REPLACE EBCM.

NO: MALFUNCTION IS INTERMITTENT. REFER TO "DIAGNOSTIC AIDS".

- AFTER DIAGNOSIS IS COMPLETE, CLEAR DTCs AND TEST DRIVE VEHICLE FOR THREE (3) DRIVE CYCLES TO VERIFY DTC DOES NOT RESET. A DRIVE CYCLE CONSISTS OF STARTING THE VEHICLE, DRIVING OVER 10 MPH AND THEN KEYING DOWN.

DIAGNOSTIC CHART DTC 45 — 1994 CAVALIER AND SUNBIRD

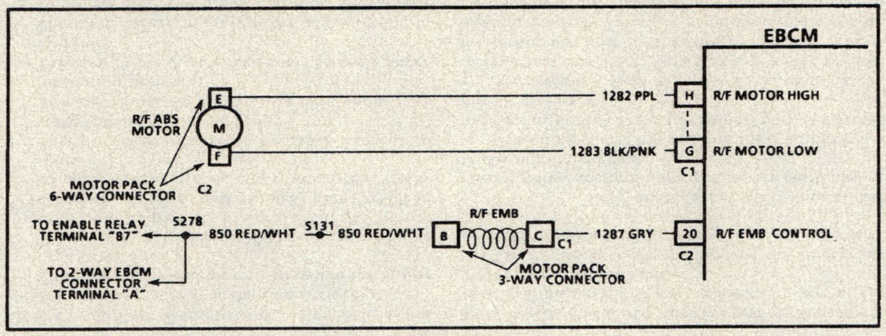

DTC 45

(Page 2 of 2)

RIGHT FRONT ABS CHANNEL WILL NOT MOVE

Circuit Description:

Current is supplied to the EMB through terminal "B" of the three-way motor pack connector. When a release of the EMB is commanded, the EBCM grounds the EMB input and allows the EMB to be energized to release the motor. This DTC is designed to detect non-actuating EMB, stuck motor or a seized hydraulic modulator.

DTC Will Set When: DTC 45 can be set during initialization. The EBCM commands the right front motor in the reverse and forward direction to ensure the motor moves properly. If the EBCM detects a condition in which it cannot move the motor in either direction, a malfunction exists.

Action Taken: This malfunction indicates the channel cannot be moved properly. A malfunction DTC is stored, ABS is disabled and the ABS warning lamp is turned "ON."

DTC Chart Test Description: Number(s) below refer to circled number(s) on the diagnostic chart.

8. Checks for a short to B+ in the right front EMB control circuit.
9. Checks for high resistance in the right front EMB control circuit.
10. Checks for high resistance in the EMB power feed circuit.
11. Checks for high resistance in the right front EMB.
12. Checks for proper operation of the hydraulic modulator.
13. Ensures malfunction was not due to poor terminal contact.
14. Isolates malfunction to either the right front EMB circuitry or the EBCM.
15. Ensures malfunction was not due to physical damage of the EMB circuitry.

Diagnostic Aids: An "Intermittent" malfunction in this DTC may result from a mechanical part of the system that sticks, binds, or slips.

The frequency of the malfunction can be checked by using the enhanced diagnostic function of the Tech 1, as described in "Enhanced Diagnostics," found in this section. DTC 45 may set after modulator disassembly if the modulator pistons are positioned at the bottom of their bore.

Any circuitry, that is suspected as causing the intermittent complaint, should be thoroughly checked for backed out terminals, improper mating, broken locks, improperly formed or damaged terminals, poor terminal to wiring connections or physical damage to the wiring harness.

NOTICE: This DTC may fail if low voltage (10 volts or less) was present when initialization occurred. Ensure battery/charging system is functioning properly.

DIAGNOSTIC CHART DTC 45 — 1994 CAVALIER AND SUNBIRD

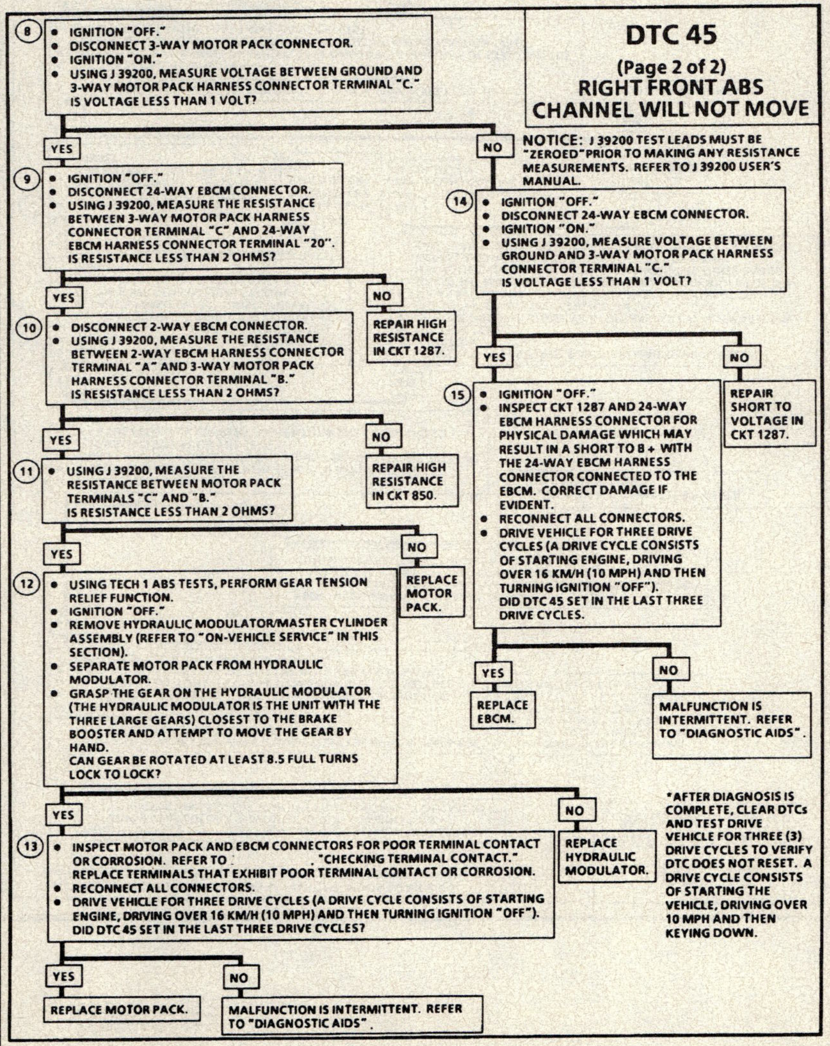

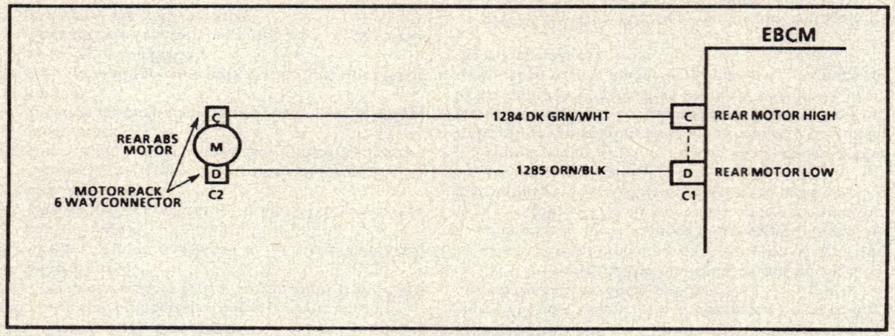

DTC 46

REAR ABS CHANNEL WILL NOT MOVE

Circuit Description:
This DTC is designed to detect bound-up ESBs, stuck motor or a seized hydraulic modulator. Release is commanded during initialization, the ESB should release the motor, resulting in sensed current being less than commanded current (motor is spinning freely). If the motor is not moving, sensed current will be equal to stall current.

DTC Will Set When: DTC 46 can be set during initialization. The EBCM commands the rear motor in the reverse and forward direction to ensure the motor moves properly. If the EBCM detects a condition in which it cannot move the motor in either direction, a malfunction exists.

Action Taken: This malfunction indicates the channel can not be moved properly. A malfunction DTC is stored, ABS is disabled, and the ABS warning lamp is turned "ON."

DTC Chart Test Description: Number(s) below refer to circled number(s) on the diagnostic chart.

1. This step ensures motor circuits are in the proper connector cavity.
2. This checks the ability of the motor to move applying hydraulic pressure to the wheel cylinders. A frozen piston or gear would result in no apply of rear brakes and a wheel that freely spins.
3. This procedure compares feedback current to command current during apply.
4. This step checks to see if hydraulic modulator is functioning properly.
5. This step checks if motor can apply properly under load.
6. This step serves to isolate the malfunction to either the EBCM or a short between CKTs 1284 and 1285.
7. Checks for a shorted motor.
8. Ensures the malfunction was not due to physical damage of the circuitry.

Diagnostic Aids: An "Intermittent" malfunction in this DTC may result from a mechanical part of the system that sticks or binds.

The frequency of the malfunction can be checked by using the enhanced diagnostic function of the Tech 1.

DTC 46 may set after modulator disassembly if the modulator pistons are positioned at the bottom of their bore.

Depending on the frequency of the malfunction, a physical inspection of the mechanical parts suspected may be necessary.

Any circuitry, that is suspected as causing the intermittent complaint, should be thoroughly checked for backed out terminals, improper mating, broken locks, improperly formed or damaged terminals, poor terminal to wiring connections or physical damage to the wiring harness.

DIAGNOSTIC CHART DTC 46 — 1994
CAVALIER AND SUNBIRD

DTC 46
REAR ABS CHANNEL WILL NOT MOVE

NOTICE: J 39200 TEST LEADS MUST BE "ZEROED" PRIOR TO MAKING ANY RESISTANCE MEASUREMENTS. REFER TO J 39200 USER'S MANUAL.

(1)
- VISUALLY INSPECT THE 6-WAY EBCM AND MOTOR PACK CONNECTORS FOR PROPER WIRE COLOR/CONNECTOR CAVITY COMBINATION. ARE THE PROPER WIRES LOCATED IN THE PROPER CONNECTOR CAVITIES?

NO: USING THE CIRCUIT DIAGRAMS LOCATED IN THE FRONT OF THIS SECTION, PLACE THE PROPER WIRES IN THE PROPER CONNECTOR CAVITIES. CLEAR DTCs AND RECHECK SYSTEM.

YES:

(2)
- RAISE AND SUPPORT THE VEHICLE SUCH THAT THE REAR WHEELS ARE APPROXIMATELY SIX INCHES OFF THE FLOOR.
- IGNITION "ON."
- APPLY FIRM PRESSURE ON BRAKE PEDAL.
- USING TECH 1 ABS TESTS, SELECT MANUAL CONTROL FUNCTION.
- SELECT REAR MOTOR APPLY FUNCTION AND APPLY THE MOTOR.
- HAVE AN ASSISTANT TRY TO SPIN THE REAR WHEELS BY HAND. COULD ASSISTANT SPIN THE REAR WHEELS?

YES: (go to 3)

NO:

(7)
- FIRM PRESSURE STILL ON BRAKE PEDAL.
- SELECT REAR MOTOR RELEASE FUNCTION AND RELEASE THE MOTOR.
- HAVE AN ASSISTANT TRY TO SPIN THE REAR WHEELS BY HAND. COULD ASSISTANT SPIN THE REAR WHEELS?

NO: (go to 3)

YES: MALFUNCTION IS INTERMITTENT. REFER TO "DIAGNOSTIC AIDS".

(3)
- REMOVE FOOT FROM BRAKE PEDAL.
- SELECT REAR MOTOR APPLY FUNCTION AND APPLY THE MOTOR WHILE CAREFULLY OBSERVING THE COMMANDED CURRENT AND THE FEEDBACK CURRENT ON THE TECH 1. WAS THE FEEDBACK CURRENT HIGHER THAN THE COMMANDED CURRENT?

YES:

(4)
- IGNITION "OFF."
- DISCONNECT THE 6-WAY EBCM CONNECTOR AND THE 6-WAY MOTOR PACK CONNECTOR.
- USING A J 39200, MEASURE THE RESISTANCE BETWEEN TERMINAL "C" AND TERMINAL "D" OF THE 6-WAY EBCM HARNESS CONNECTOR. DOES J 39200 DISPLAY RESISTANCE AS "OL" (INFINITE)?

YES:

(5)
- USING J 39200, MEASURE RESISTANCE BETWEEN MOTOR PACK TERMINALS "C" AND "D". IS RESISTANCE GREATER THAN .4 OHMS?

YES:

(6)
- INSPECT CKTs 1284 AND 1285 FOR DAMAGE WHICH MAY RESULT IN A SHORT BETWEEN THE TWO CIRCUITS. REPAIR DAMAGE IF EVIDENT.
- RECONNECT ALL CONNECTOR.
- DRIVE VEHICLE FOR THREE DRIVE CYCLES (A DRIVE CYCLE CONSISTS OF STARTING THE ENGINE, DRIVING OVER 16 km/h (10 MPH) AND THEN TURNING IGNITION "OFF"). DID DTC 46 SET IN THE LAST THREE DRIVE CYCLES?

YES: REPLACE EBCM.

NO: MALFUNCTION IS INTERMITTENT. REFER TO "DIAGNOSTIC AIDS".

NO (5): REPLACE MOTOR PACK.

NO (4): REPAIR SHORT BETWEEN CKT 1284 AND CKT 1285.

NO (3):

(8)
- USING TECH 1 ABS TESTS, PERFORM GEAR TENSION RELIEF FUNCTION.
- IGNITION "OFF."
- REMOVE HYDRAULIC MODULATOR/MASTER CYLINDER ASSEMBLY (REFER TO "ON-VEHICLE SERVICE" IN THIS SECTION).
- SEPARATE MOTOR PACK FROM HYDRAULIC MODULATOR.
- GRASP THE MIDDLE GEAR ON THE HYDRAULIC MODULATOR (THE HYDRAULIC MODULATOR IS THE UNIT WITH THE THREE LARGE GEARS) AND ATTEMPT TO MOVE THE GEAR BY HAND. CAN GEAR BE ROTATED AT LEAST 3.5 FULL TURNS LOCK TO LOCK?

NO: REPLACE HYDRAULIC MODULATOR.

YES: REPLACE MOTOR PACK.

* AFTER ALL DIAGNOSIS IS COMPLETE, CLEAR DTCs AND TEST DRIVE VEHICLE FOR THREE (3) DRIVE CYCLES TO VERIFY THAT DTC DOES NOT RESET. A DRIVE CYCLE CONSISTS OF STARTING THE VEHICLE, DRIVING VEHICLE OVER 10 MPH AND THEN KEYING DOWN.

DIAGNOSTIC CHART DTC 47 — 1994 CAVALIER AND SUNBIRD

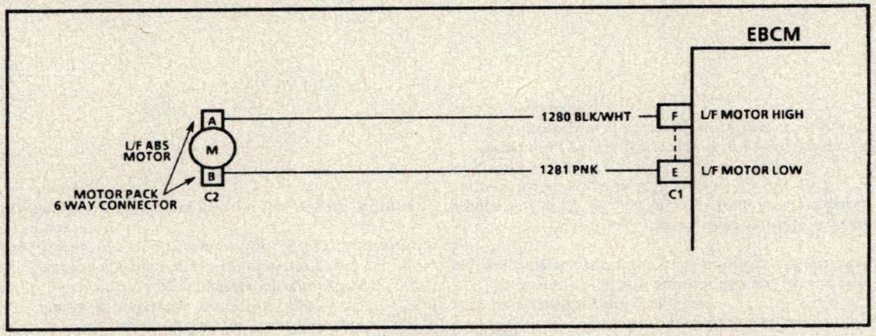

DTC 47

LEFT FRONT ABS MOTOR FREE SPINS

Circuit Description:
This DTC is designed to detect a stripped nut or gear assembly during initialization. During the homing sequence, the piston should reach the top of the bore resulting in a stalled motor. If this does not occur, the motor must be spinning with little or no resistance, indicating a nut/screw or gear malfunction.

DTC Will Set When: DTC 47 can only be set during initialization. If the feedback current is less than the command current for a specified period to time, a malfunction exists.

Action Taken: A malfunction DTC is stored, ABS is disabled, and the ABS warning lamp is turned "ON."

DTC Chart Test Description: Number(s) below refer to circled number(s) on the diagnostic chart.
1. This step compares feedback current to command current during apply.
2. This step compares feedback current to command current during release.
3. Steps 1, 2 and 3 verify malfunction exists. This step also moves the position to the top of the bore.
4. Checks for excessive resistance in the left front motor high circuit.
5. Checks for excessive resistance in the left front motor low circuit.
6. Checks for resistance in the left front motor.

Diagnostic Aids: An "Intermittent" malfunction in this DTC may result from a mechanical part of the system that sticks, binds, or slips.
The frequency of the malfunction can be checked by using the enhanced diagnostic function of the Tech 1.
If DTC only occurs once and DTC 56 also occurs see DTC 56. If intermittent and enhanced diagnostics shows this DTC occurs during ABS, see DTC 56.
Depending on the frequency of the malfunction, a physical inspection of the mechanical parts suspected may be necessary.

DIAGNOSTIC CHART DTC 47 — 1994 CAVALIER AND SUNBIRD

DTC 47
(Page 1 of 2)
LEFT FRONT ABS MOTOR FREE SPINS

AFTER DIAGNOSIS IS COMPLETE, CLEAR DTCs AND TEST DRIVE VEHICLE FOR THREE(3) DRIVE CYCLES TO VERIFY THAT DTC DOES NOT RESET. A DRIVE CYCLE CONSISTS OF STARTING THE ENGINE, DRIVING THE VEHICLE ABOVE 16 km/h (10 MPH) AND THEN KEYING DOWN.

NOTICE: J 39200 TEST LEADS MUST BE "ZEROED" PRIOR TO MAKING ANY RESISTANCE MEASUREMENTS. REFER TO J 39200 USER'S MANUAL.

IS DTC 56 ALSO PRESENT?

YES: PROCEED TO DTC 56 CHART FIRST.

NO:

(1)
- IGNITION "ON," ENGINE "OFF."
- PUMP BRAKE PEDAL TO DEPLETE VACUUM POWER BOOSTER.
- USING TECH 1 ABS TESTS, SELECT MANUAL CONTROL FUNCTION.
- SELECT LEFT FRONT MOTOR APPLY FUNCTION AND APPLY THE MOTOR WHILE CAREFULLY OBSERVING THE COMMAND CURRENT AND FEEDBACK CURRENT ON THE TECH 1. WERE BOTH COMMAND AND FEEDBACK CURRENT 10 AMPS?

YES:

(2)
- APPLY FIRM PRESSURE ON BRAKE PEDAL.
- SELECT LEFT FRONT MOTOR RELEASE FUNCTION AND RELEASE MOTOR. DID THE BRAKE PEDAL FALL AND WERE BOTH COMMAND AND FEEDBACK CURRENT 6 AMPS?

NO: GO TO STEP 4.

YES:

(3)
- FIRM PRESSURE STILL ON BRAKE PEDAL.
- SELECT LEFT FRONT MOTOR APPLY FUNCTION AND APPLY THE MOTOR (THIS MOVES THE HYDRAULIC MODULATOR PISTON TO THE TOP OF ITS BORE). DID BRAKE PEDAL RISE?

YES: MALFUNCTION IS INTERMITTENT. REFER TO "DIAGNOSTIC AIDS".

NO:

(4)
- IGNITION "OFF."
- DISCONNECT 6-WAY EBCM AND MOTOR PACK HARNESS CONNECTORS.
- USING J 39200, MEASURE RESISTANCE BETWEEN 6-WAY EBCM HARNESS CONNECTOR TERMINAL "F" AND 6-WAY MOTOR PACK HARNESS CONNECTOR TERMINAL "A". IS RESISTANCE LESS THAN 1.5 OHMS?

NO: REPAIR HIGH RESISTANCE IN CKT 1280.

YES:

(5)
- USING J 39200, MEASURE RESISTANCE BETWEEN 6-WAY EBCM HARNESS CONNECTOR TERMINAL "E" AND 6-WAY MOTOR PACK HARNESS CONNECTOR TERMINAL "B". IS RESISTANCE LESS THAN 1.5 OHMS?

NO: REPAIR HIGH RESISTANCE IN CKT 1281.

YES:

(6)
- USING J 39200, MEASURE RESISTANCE BETWEEN 6-WAY MOTOR PACK CONNECTOR TERMINALS "A" AND "B". IS RESISTANCE LESS THAN 1.5 OHMS?

NO: REPLACE MOTOR PACK.

YES: GO TO PAGE 2 OF 2 OF THIS CHART.

DIAGNOSTIC CHART DTC 47 — 1994 CAVALIER AND SUNBIRD

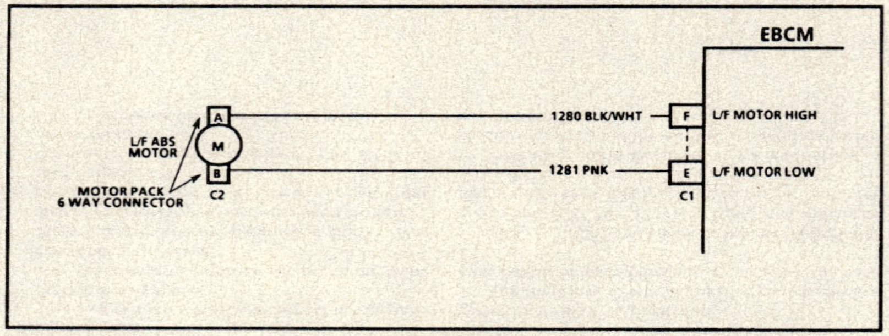

DTC 47
(Page 2 of 2)
LEFT FRONT ABS MOTOR FREE SPINS

Circuit Description:
This DTC is designed to detect a stripped nut or gear assembly during initialization. During the homing sequence, the piston should reach the top of the bore resulting in a stalled motor. If this does not occur, the motor must be spinning with little or no resistance, indicating a nut/screw or gear malfunction.

DTC Will Set When: DTC 47 can only be set during initialization. If the feedback current is less than the command current for a specified period to time, a malfunction exists.

Action Taken: A malfunction DTC is stored, ABS is disabled, and the ABS warning lamp is turned "ON."

DTC Chart Test Description: Number(s) below refer to circled number(s) on the diagnostic chart.
7. Checks for a stripped motor pack gear.
8. Checks for a stripped hydraulic modulator gear.
9. Checks for proper operation of motor pack.
10. Checks for proper operation of the hydraulic modulator.
11. Ensures malfunction was not due to poor terminal contact.

Diagnostic Aids: An "Intermittent" malfunction in this DTC may result from a mechanical part of the system that sticks, binds, or slips.

The frequency of the malfunction can be checked by using the enhanced diagnostic function of the Tech 1.

If DTC only occurs once and DTC 56 also occurs, see DTC 56. If intermittent and enhanced diagnostics shows this DTC occurs during ABS, see DTC 56.

Depending on the frequency of the malfunction, a physical inspection of the mechanical parts suspected may be necessary.

DIAGNOSTIC CHART DTC 47 — 1994 CAVALIER AND SUNBIRD

DTC 47
(Page 2 of 2)
LEFT FRONT ABS MOTOR FREE SPINS

CONTINUED FROM PAGE 1 OF 2 OF THIS CHART.

(7)
- REMOVE HYDRAULIC MODULATOR/MASTER CYLINDER ASSEMBLY.
- REMOVE GEAR COVER.
- CHECK FOR STRIPPED GEAR ON MOTOR PACK; THE MOTOR PACK IS THE UNIT WITH THE THREE SMALL GEARS. THE LEFT FRONT GEAR IS FURTHEST FROM THE BRAKE BOOSTER WHEN THE ASSEMBLY IS INSTALLED IN THE VEHICLE. IS THE MOTOR PACK GEAR STRIPPED?

YES: REPLACE MOTOR PACK.
NO: go to (8)

(8)
- CHECK FOR A STRIPPED GEAR ON THE HYDRAULIC MODULATOR; THE HYDRAULIC MODULATOR IS THE UNIT WITH THE THREE LARGE GEARS. THE LEFT FRONT GEAR IS FURTHEST FROM THE BRAKE BOOSTER WHEN THE ASSEMBLY IS INSTALLED IN THE VEHICLE. IS THE HYDRAULIC MODULATOR GEAR STRIPPED?

YES: REPLACE HYDRAULIC MODULATOR GEAR.
NO: go to (9)

(9)
- SECURELY POSITION HYDRAULIC MODULATOR/MASTER CYLINDER ASSEMBLY IN VEHICLE ALLOWING LEFT FRONT GEAR SET OBSERVATION. BE CAREFUL NOT TO DAMAGE GEARS OR ALLOW THE HYDRAULIC MODULATOR/MASTER CYLINDER ASSEMBLY TO MOVE DURING TESTING.
- RECONNECT ELECTRICAL CONNECTORS.
- IGNITION "ON."
- USING TECH 1 ABS TESTS, SELECT MANUAL CONTROL FUNCTION.
- SELECT LEFT FRONT MOTOR APPLY AND APPLY THE MOTOR WHILE OBSERVING THE GEAR SET FURTHEST FORWARD FROM THE BOOSTER WHEN THE ASSEMBLY IS INSTALLED IN THE VEHICLE.
- SELECT LEFT FRONT MOTOR RELEASE FUNCTION AND RELEASE THE MOTOR WHILE STILL OBSERVING THE GEAR SET.
- SELECT LEFT FRONT MOTOR APPLY FUNCTION AND APPLY THE MOTOR. DID GEAR SET MOVE IN BOTH DIRECTIONS FOR AT LEAST ONE REVOLUTION?

NO: REPLACE MOTOR PACK.
YES: go to (10)

(10)
- USING TECH 1, SELECT "F4: ABS TESTS" AND PERFORM GEAR TENSION RELIEF FUNCTION.
- IGNITION "OFF."
- SEPARATE MOTOR PACK FROM HYDRAULIC MODULATOR.
- GRASP THE LEFT FRONT GEAR ON THE HYDRAULIC MODULATOR; THE HYDRAULIC MODULATOR IS THE UNIT WITH THE THREE LARGE GEARS. THE LEFT FRONT GEAR IS FURTHEST FROM THE BRAKE BOOSTER WHEN THE ASSEMBLY IS INSTALLED IN THE VEHICLE.
- ATTEMPT TO ROTATE THE GEAR. CAN THE GEAR BE ROTATED MORE THAN 9 TURNS LOCK TO LOCK?

YES: REPLACE HYDRAULIC MODULATOR.
NO: go to (11)

(11)
- INSPECT 6-WAY MOTOR PACK AND EBCM CONNECTORS FOR CORROSION OR POOR TERMINAL CONTACT. REFER TO "CHECKING TERMINAL CONTACT." REPLACE TERMINALS THAT EXHIBIT POOR TERMINAL CONTACT OR CORROSION.
- INSTALL HYDRAULIC MODULATOR/MASTER CYLINDER ASSEMBLY.
- RECONNECT ALL CONNECTORS.
- DRIVE VEHICLE FOR THREE DRIVE CYCLES (A DRIVE CYCLE CONSISTS OF STARTING ENGINE, DRIVING OVER 16 km/h (10 MPH) AND THEN TURNING IGNITION "OFF"). DID DTC 47 SET IN THE LAST THREE DRIVE CYCLES?

YES: REPLACE EBCM.
NO: MALFUNCTION WAS DUE TO A POOR CONNECTION.

DIAGNOSTIC CHART DTC 48 — 1994 CAVALIER AND SUNBIRD

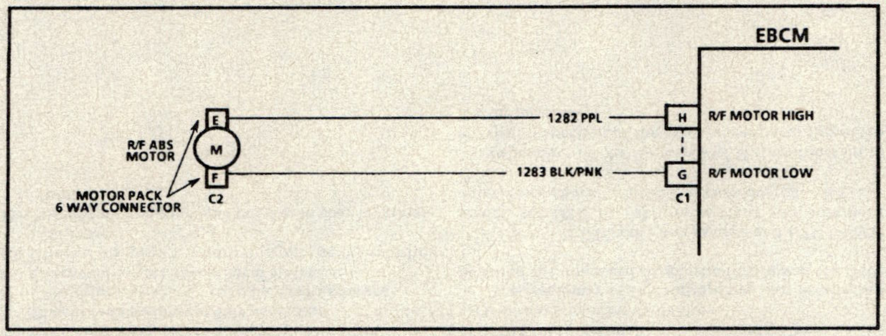

DTC 48

(Page 1 of 2)

RIGHT FRONT ABS MOTOR FREE SPINS

Circuit Description:

This DTC is designed to detect a stripped nut or gear assembly during initialization. During the homing sequence, the piston should reach the top of the bore resulting in a stalled motor. If this does not occur, the motor must be spinning with little or no resistance, indicating a nut/screw or gear malfunction.

DTC Will Set When: DTC 48 can only be set during initialization. If the feedback current is less than the command current for a specified period of time, a malfunction exists.

Action Taken: A malfunction DTC is stored, ABS is disabled, and the ABS warning lamp is turned "ON."

DTC Chart Test Description: Number(s) below refer to circled number(s) on the diagnostic chart.

1. This step compares feedback current to command current during apply.
2. This step compares feedback current to command current during release.
3. Steps 1, 2 and 3 verify malfunction exists. This step also moves the piston to the top of the bore.
4. Checks for excessive resistance in the right front motor high circuit.
5. Checks for excessive resistance in the right front motor low circuit.
6. Checks for resistance in the right front motor.

Diagnostic Aids: An "Intermittent" malfunction in this DTC may result from a mechanical part of the system that sticks, binds, or slips.

The frequency of the malfunction can be checked by using the enhanced diagnostic function of the Tech 1.

If DTC only occurs once and DTC 61 also occurs see DTC 61. If intermittent and enhanced diagnostics shows this DTC occurs during ABS, see DTC 61.

Depending on the frequency of the malfunction, a physical inspection of the mechanical parts suspected may be necessary.

DIAGNOSTIC CHART DTC 48 — 1994 CAVALIER AND SUNBIRD

DTC 48

(Page 1 of 2)

RIGHT FRONT ABS MOTOR FREE SPINS

AFTER DIAGNOSIS IS COMPLETE, CLEAR DTCs AND TEST DRIVE VEHICLE FOR THREE(3) DRIVE CYCLES TO VERIFY THAT DTC DOES NOT RESET. A DRIVE CYCLE CONSISTS OF STARTING THE ENGINE, DRIVING THE VEHICLE ABOVE 16 km/h (10 MPH) AND THEN KEYING DOWN.

NOTICE: J 39200 TEST LEADS MUST BE "ZEROED" PRIOR TO MAKING ANY RESISTANCE MEASUREMENTS. REFER TO J 39200 USER'S MANUAL.

IS DTC 61 ALSO PRESENT?

- YES → PROCEED TO DTC 61 CHART FIRST.
- NO ↓

(1)
- IGNITION "ON," ENGINE "OFF."
- PUMP BRAKE PEDAL TO DEPLETE VACUUM POWER BOOSTER.
- USING TECH 1 ABS TESTS, SELECT MANUAL CONTROL FUNCTION.
- SELECT RIGHT FRONT MOTOR APPLY FUNCTION AND APPLY THE MOTOR WHILE CAREFULLY OBSERVING THE COMMAND CURRENT AND FEEDBACK CURRENT ON THE TECH 1. WERE BOTH COMMAND AND FEEDBACK CURRENT 10 AMPS?

- YES ↓
- NO → GO TO STEP 4.

(2)
- APPLY FIRM PRESSURE ON BRAKE PEDAL.
- SELECT RIGHT FRONT MOTOR RELEASE FUNCTION AND RELEASE MOTOR. DID THE BRAKE PEDAL FALL AND WERE BOTH COMMAND AND FEEDBACK CURRENT 6 AMPS?

- YES ↓
- NO → GO TO STEP 4.

(3)
- FIRM PRESSURE STILL ON BRAKE PEDAL.
- SELECT RIGHT FRONT MOTOR APPLY FUNCTION AND APPLY THE MOTOR (THIS MOVES THE HYDRAULIC MODULATOR PISTON TO THE TOP OF ITS BORE). DID BRAKE PEDAL RISE?

- YES → MALFUNCTION IS INTERMITTENT. REFER TO "DIAGNOSTIC AIDS".
- NO ↓

(4)
- IGNITION "OFF."
- DISCONNECT 6-WAY EBCM AND MOTOR PACK HARNESS CONNECTORS.
- USING J 39200, MEASURE RESISTANCE BETWEEN 6-WAY EBCM HARNESS CONNECTOR TERMINAL "H" AND 6-WAY MOTOR PACK HARNESS CONNECTOR TERMINAL "E". IS RESISTANCE LESS THAN 1.5 OHMS?

- YES ↓
- NO → REPAIR HIGH RESISTANCE IN CKT 1282.

(5)
- USING J 39200, MEASURE RESISTANCE BETWEEN 6-WAY EBCM HARNESS CONNECTOR TERMINAL "G" AND 6-WAY MOTOR PACK HARNESS CONNECTOR TERMINAL "F". IS RESISTANCE LESS THAN 1.5 OHMS?

- YES ↓
- NO → REPAIR HIGH RESISTANCE IN CKT 1283.

(6)
- USING J 39200, MEASURE RESISTANCE BETWEEN 6-WAY MOTOR PACK CONNECTOR TERMINALS "E" AND "F". IS RESISTANCE LESS THAN 1.5 OHMS?

- YES → GO TO PAGE 2 OF 2 OF THIS CHART.
- NO → REPLACE MOTOR PACK.

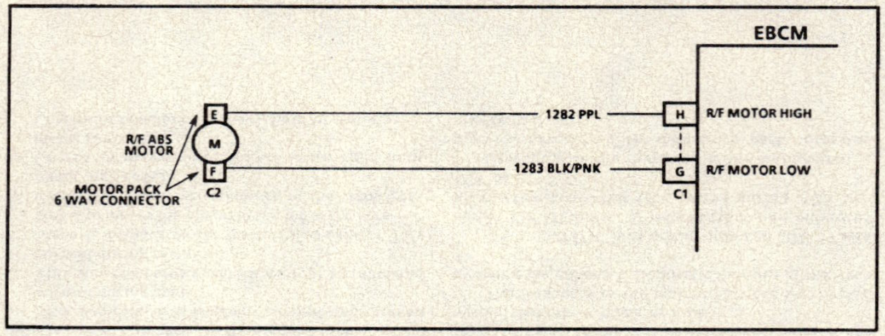

DTC 48
(Page 2 of 2)
RIGHT FRONT ABS MOTOR FREE SPINS

Circuit Description:
This DTC is designed to detect a stripped nut or gear assembly during initialization. During the homing sequence, the piston should reach the top of the bore resulting in a stalled motor. If this does not occur, the motor must be spinning with little or no resistance, indicating a nut/screw or gear malfunction.

DTC Will Set When: DTC 48 can only be set during initialization. If the feedback current is less than the command current for a specified period of time, a malfunction exists.

Action Taken: A malfunction DTC is stored, ABS is disabled, and the ABS warning lamp is turned "ON."

DTC Chart Test Description: Number(s) below refer to circled number(s) on the diagnostic chart.
7. Checks for a stripped motor pack gear.
8. Checks for a stripped hydraulic modulator gear.
9. Checks for proper operation of motor pack.
10. Checks for proper operation of the hydraulic modulator.
11. Ensures malfunction was not due to poor terminal contact.

Diagnostic Aids: An "Intermittent" malfunction in this DTC may result from a mechanical part of the system that sticks, binds, or slips.

The frequency of the malfunction can be checked by using the enhanced diagnostic function of the Tech 1.

If DTC only occurs once and DTC 61 also occurs, see DTC 61. If intermittent and enhanced diagnostics shows this DTC occurs during ABS, see DTC 61.

Depending on the frequency of the malfunction, a physical inspection of the mechanical parts suspected may be necessary.

DIAGNOSTIC CHART DTC 48 — 1994
CAVALIER AND SUNBIRD

DTC 48
(Page 2 of 2)
RIGHT FRONT ABS MOTOR FREE SPINS

CONTINUED FROM PAGE 1 OF 2 OF THIS CHART.

(7)
- REMOVE HYDRAULIC MODULATOR/MASTER CYLINDER ASSEMBLY.
- REMOVE GEAR COVER.
- CHECK FOR STRIPPED GEAR ON MOTOR PACK; THE MOTOR PACK IS THE UNIT WITH THE THREE SMALL GEARS. THE RIGHT FRONT GEAR IS CLOSEST TO THE BRAKE BOOSTER WHEN THE ASSEMBLY IS INSTALLED IN THE VEHICLE. IS THE MOTOR PACK GEAR STRIPPED?

YES → REPLACE MOTOR PACK.

NO ↓

(8)
- CHECK FOR A STRIPPED GEAR ON THE HYDRAULIC MODULATOR; THE HYDRAULIC MODULATOR IS THE UNIT WITH THE THREE LARGE GEARS. THE RIGHT FRONT GEAR IS CLOSEST TO THE BRAKE BOOSTER WHEN THE ASSEMBLY IS INSTALLED IN THE VEHICLE. IS THE HYDRAULIC MODULATOR GEAR STRIPPED?

YES → REPLACE HYDRAULIC MODULATOR GEAR.

NO ↓

(9)
- SECURELY POSITION HYDRAULIC MODULATOR/MASTER CYLINDER ASSEMBLY IN VEHICLE ALLOWING RIGHT FRONT GEAR SET OBSERVATION. BE CAREFUL NOT TO DAMAGE GEARS OR ALLOW THE HYDRAULIC MODULATOR/MASTER CYLINDER ASSEMBLY TO MOVE DURING TESTING.
- RECONNECT ELECTRICAL CONNECTORS.
- IGNITION "ON."
- USING TECH 1 ABS TESTS, SELECT MANUAL CONTROL FUNCTION.
- SELECT RIGHT FRONT MOTOR APPLY AND APPLY THE MOTOR WHILE OBSERVING THE GEAR SET CLOSEST TO THE BOOSTER WHEN THE ASSEMBLY IS INSTALLED IN THE VEHICLE.
- SELECT RIGHT FRONT MOTOR RELEASE FUNCTION AND RELEASE THE MOTOR WHILE STILL OBSERVING THE GEAR SET.
- SELECT RIGHT FRONT MOTOR APPLY FUNCTION AND APPLY THE MOTOR. DID GEAR SET MOVE IN BOTH DIRECTIONS FOR AT LEAST ONE REVOLUTION?

NO → REPLACE MOTOR PACK.

YES ↓

(10)
- USING TECH 1, SELECT "F4: ABS TESTS" AND PERFORM GEAR TENSION RELIEF FUNCTION.
- IGNITION "OFF."
- SEPARATE MOTOR PACK FROM HYDRAULIC MODULATOR.
- GRASP THE RIGHT FRONT GEAR ON THE HYDRAULIC MODULATOR; THE HYDRAULIC MODULATOR IS THE UNIT WITH THE THREE LARGE GEARS. THE RIGHT FRONT GEAR IS CLOSEST TO THE BRAKE BOOSTER WHEN THE ASSEMBLY IS INSTALLED IN THE VEHICLE.
- ATTEMPT TO ROTATE THE GEAR. CAN THE GEAR BE ROTATED MORE THAN 9 TURNS LOCK TO LOCK?

YES → REPLACE HYDRAULIC MODULATOR.

NO ↓

(11)
- INSPECT 6-WAY MOTOR PACK AND EBCM CONNECTORS FOR CORROSION OR POOR TERMINAL CONTACT. REPLACE TERMINALS THAT EXHIBIT POOR TERMINAL CONTACT OR CORROSION.
- INSTALL HYDRAULIC MODULATOR/MASTER CYLINDER ASSEMBLY.
- RECONNECT ALL CONNECTORS.
- DRIVE VEHICLE FOR THREE DRIVE CYCLES (A DRIVE CYCLE CONSISTS OF STARTING ENGINE, DRIVING OVER 16 km/h (10 MPH) AND THEN TURNING IGNITION "OFF"). DID DTC 48 SET IN THE LAST THREE DRIVE CYCLES?

YES → REPLACE EBCM.

NO → MALFUNCTION WAS DUE TO A POOR CONNECTION.

DIAGNOSTIC CHART DTC 51 — 1994 CAVALIER AND SUNBIRD

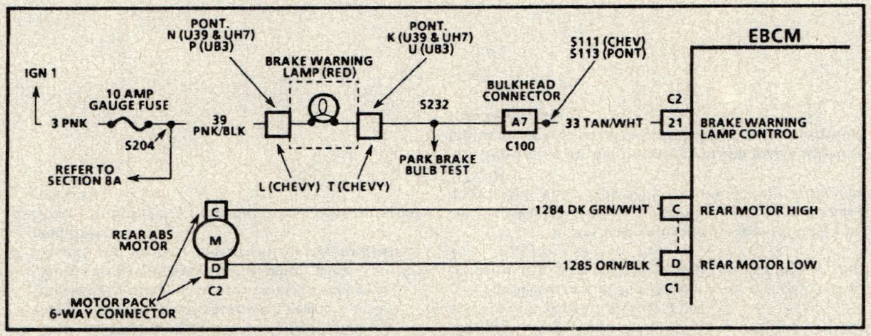

DTC 51

REAR ABS MOTOR FREE SPINS

Circuit Description:
This DTC is designed to detect a stripped nut or gear assembly during initialization. During the homing sequence, the piston should reach the top of the bore resulting in a stalled motor. If this does not occur, the motor must be spinning with little or no resistance, indicating a nut/screw or gear malfunction.

DTC Will Set When: DTC 51 can be set during initialization. If the feedback current is less than the command current for a specified period of time, a malfunction exists.

Action Taken: A malfunction DTC is stored, ABS is disabled, and the ABS warning lamp is turned "ON."

DTC Chart Test Description: Number(s) below refer to circled number(s) on the diagnostic chart.

1. This step compares feedback current to command current during apply.
2. This step compares feedback current to command current during release.
3. Steps 1, 2 and 3 verify malfunction exists. This step also moves the piston to the top of the bore.
4. Checks for excessive resistance in the rear motor high circuit.
5. Checks for excessive resistance in the rear motor low circuit.
6. Checks for resistance in the rear motor.

Diagnostic Aids: An "Intermittent" malfunction in this DTC may result from a mechanical part of the system that sticks, binds, or slips.

The frequency of the malfunction can be checked by using the enhanced diagnostic function of the Tech 1.

If DTC only occurs once and DTC 64 also occurs, see DTC 64. If intermittent and enhanced diagnostics show this DTC occurs during ABS, see DTC 64.

Depending on the frequency of the failure, a physical inspection of the mechanical parts suspected may be necessary.

DIAGNOSTIC CHART DTC 51 — 1994 CAVALIER AND SUNBIRD

DTC 51
(Page 1 of 2)
REAR ABS MOTOR FREE SPINS

AFTER DIAGNOSIS IS COMPLETE, CLEAR DTCs AND TEST DRIVE VEHICLE FOR THREE(3) DRIVE CYCLES TO VERIFY THAT DTC DOES NOT RESET. A DRIVE CYCLE CONSISTS OF STARTING THE ENGINE, DRIVING THE VEHICLE ABOVE 16 km/h (10 MPH) AND THEN KEYING DOWN.

NOTICE: J 39200 TEST LEADS MUST BE "ZEROED" PRIOR TO MAKING ANY RESISTANCE MEASUREMENTS. REFER TO J 39200 USER'S MANUAL.

IS DTC 64 ALSO PRESENT?

- YES: PROCEED TO DTC 64 CHART FIRST.
- NO:

(1)
- IGNITION "ON," ENGINE "OFF."
- PUMP BRAKE PEDAL TO DEPLETE VACUUM POWER BOOSTER.
- USING TECH 1 ABS TESTS, SELECT MANUAL CONTROL FUNCTION.
- SELECT REAR MOTOR APPLY FUNCTION AND APPLY THE MOTOR WHILE CAREFULLY OBSERVING THE COMMAND CURRENT AND FEEDBACK CURRENT ON THE TECH 1.

WERE BOTH COMMAND AND FEEDBACK CURRENT 10 AMPS?

- NO: GO TO STEP 4.
- YES:

(2)
- APPLY FIRM PRESSURE ON BRAKE PEDAL.
- SELECT REAR MOTOR RELEASE FUNCTION AND RELEASE MOTOR.

DID THE BRAKE PEDAL FALL AND WERE BOTH COMMAND AND FEEDBACK CURRENT 6 AMPS?

- NO: GO TO STEP 4.
- YES:

(3)
- FIRM PRESSURE STILL ON BRAKE PEDAL.
- SELECT REAR MOTOR APPLY FUNCTION AND APPLY THE MOTOR (THIS MOVES THE HYDRAULIC MODULATOR PISTON TO THE TOP OF ITS BORE).

DID BRAKE PEDAL RISE?

- YES: MALFUNCTION IS INTERMITTENT. REFER TO "DIAGNOSTIC AIDS".
- NO:

(4)
- IGNITION "OFF."
- DISCONNECT 6-WAY EBCM AND MOTOR PACK HARNESS CONNECTORS.
- USING J 39200, MEASURE RESISTANCE BETWEEN 6-WAY EBCM HARNESS CONNECTOR TERMINAL "C" AND 6-WAY MOTOR PACK HARNESS CONNECTOR TERMINAL "C".

IS RESISTANCE LESS THAN 1.5 OHMS?

- NO: REPAIR HIGH RESISTANCE IN CKT 1284.
- YES:

(5)
- USING J 39200, MEASURE RESISTANCE BETWEEN 6-WAY EBCM HARNESS CONNECTOR TERMINAL "D" AND 6-WAY MOTOR PACK HARNESS CONNECTOR TERMINAL "D".

IS RESISTANCE LESS THAN 1.5 OHMS?

- NO: REPAIR HIGH RESISTANCE IN CKT 1285.
- YES:

(6)
- USING J 39200, MEASURE RESISTANCE BETWEEN 6-WAY MOTOR PACK CONNECTOR TERMINALS "C" AND "D".

IS RESISTANCE LESS THAN 1.5 OHMS?

- NO: REPLACE MOTOR PACK.
- YES: GO TO PAGE 2 OF 2 OF THIS CHART.

DIAGNOSTIC CHART DTC 51 — 1994 CAVALIER AND SUNBIRD

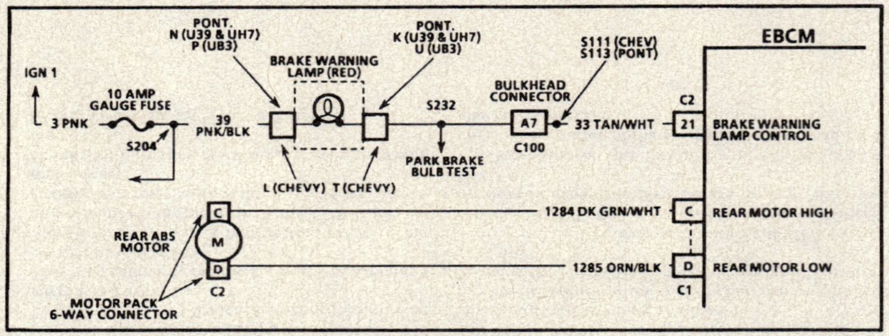

DTC 51

(Page 2 of 2)

REAR ABS MOTOR FREE SPINS

Circuit Description:

This DTC is designed to detect a stripped nut or gear assembly during initialization. During the homing sequence, the piston should reach the top of the bore resulting in a stalled motor. If this does not occur, the motor must be spinning with little or no resistance, indicating a nut/screw or gear malfunction.

DTC Will Set When: DTC 51 can be set during initialization. If the feedback current is less than the command current for a specified period of time, a malfunction exists.

Action Taken: A malfunction DTC is stored, ABS is disabled, and the ABS warning lamp is turned "ON."

DTC Chart Test Description: Number(s) below refer to circled number(s) on the diagnostic chart.

7. Checks for a stripped motor pack gear.
8. Checks for a stripped hydraulic modulator gear.
9. Checks for proper operation of motor pack.
10. Checks for proper operation of the hydraulic modulator.
11. Ensures malfunction was not due to poor terminal contact.

Diagnostic Aids: An "Intermittent" malfunction in this DTC may result from a mechanical part of the system that sticks, binds, or slips.

The frequency of the malfunction can be checked by using the enhanced diagnostic function of the Tech 1.

If DTC only occurs once and DTC 64 also occurs, see DTC 64. If intermittent and enhanced diagnostics show this DTC occurs during ABS, see DTC 64.

Depending on the frequency of the malfunction, a physical inspection of the mechanical parts suspected may be necessary.

DIAGNOSTIC CHART DTC 51 — 1994 CAVALIER AND SUNBIRD

DTC 51

(Page 2 of 2)

REAR ABS MOTOR FREE SPINS

CONTINUED FROM PAGE 1 OF 2 OF THIS CHART.

(7)
- REMOVE HYDRAULIC MODULATOR/MASTER CYLINDER ASSEMBLY.
- REMOVE GEAR COVER.
- CHECK FOR STRIPPED GEAR ON MOTOR PACK; THE MOTOR PACK IS THE UNIT WITH THE THREE SMALL GEARS. THE REAR CHANNEL GEAR IS THE MIDDLE GEAR OF THE THREE.

IS THE MOTOR PACK GEAR STRIPPED?

YES → REPLACE MOTOR PACK.

NO →

(8)
- CHECK FOR A STRIPPED GEAR ON THE HYDRAULIC MODULATOR; THE HYDRAULIC MODULATOR IS THE UNIT WITH THE THREE LARGE GEARS. THE REAR CHANNEL GEAR IS THE MIDDLE GEAR OF THE THREE.

IS THE HYDRAULIC MODULATOR GEAR STRIPPED?

YES → REPLACE HYDRAULIC MODULATOR GEAR.

NO →

(9)
- SECURELY POSITION HYDRAULIC MODULATOR/MASTER CYLINDER ASSEMBLY IN VEHICLE ALLOWING REAR CHANNEL GEAR SET OBSERVATION. BE CAREFUL NOT TO DAMAGE GEARS OR ALLOW THE HYDRAULIC MODULATOR/MASTER CYLINDER ASSEMBLY TO MOVE DURING TESTING.
- RECONNECT ELECTRICAL CONNECTORS.
- IGNITION "ON."
- USING TECH 1 ABS TESTS, SELECT MANUAL CONTROL FUNCTION.
- SELECT REAR MOTOR APPLY AND APPLY THE MOTOR WHILE OBSERVING THE MIDDLE GEAR SET.
- SELECT REAR MOTOR RELEASE FUNCTION AND RELEASE THE MOTOR WHILE STILL OBSERVING THE GEAR SET.
- SELECT REAR MOTOR APPLY FUNCTION AND APPLY THE MOTOR.

DID GEAR SET MOVE IN BOTH DIRECTIONS FOR AT LEAST ONE REVOLUTION?

NO → REPLACE MOTOR PACK.

YES →

(10)
- USING TECH 1, SELECT "F4: ABS TESTS" AND PERFORM GEAR TENSION RELIEF FUNCTION.
- IGNITION "OFF."
- SEPARATE MOTOR PACK FROM HYDRAULIC MODULATOR.
- GRASP THE REAR CHANNEL GEAR ON THE HYDRAULIC MODULATOR; THE HYDRAULIC MODULATOR IS THE UNIT WITH THE THREE LARGE GEARS. THE REAR CHANNEL GEAR IS THE MIDDLE GEAR OF THE THREE.
- ATTEMPT TO ROTATE THE GEAR.

CAN THE GEAR BE ROTATED MORE THAN 4 TURNS LOCK TO LOCK?

YES → REPLACE HYDRAULIC MODULATOR.

NO →

(11)
- INSPECT 6-WAY MOTOR PACK AND EBCM CONNECTORS FOR CORROSION OR POOR TERMINAL CONTACT. REFER TO "CHECKING TERMINAL CONTACT." REPLACE TERMINALS THAT EXHIBIT POOR TERMINAL CONTACT OR CORROSION.
- INSTALL HYDRAULIC MODULATOR/MASTER CYLINDER ASSEMBLY.
- RECONNECT ALL CONNECTORS.
- DRIVE VEHICLE FOR THREE DRIVE CYCLES (A DRIVE CYCLE CONSISTS OF STARTING ENGINE, DRIVING OVER 16 km/h (10 MPH) AND THEN TURNING IGNITION "OFF").

DID DTC 51 SET IN THE LAST THREE DRIVE CYCLES?

YES → REPLACE EBCM.

NO → MALFUNCTION WAS DUE TO A POOR CONNECTION.

DIAGNOSTIC CHART DTC 52 — 1994 CAVALIER AND SUNBIRD

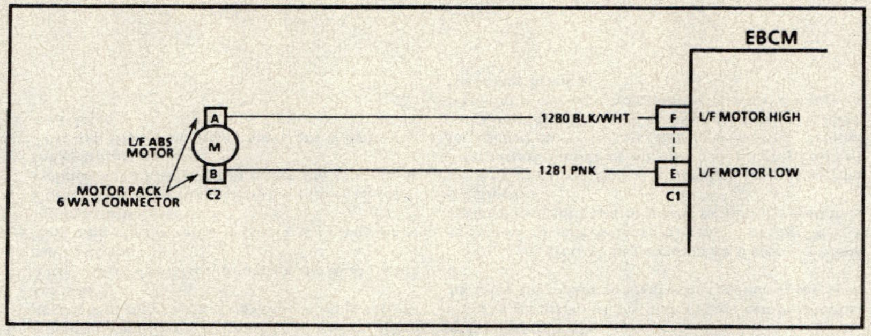

DTC 52

LEFT FRONT ABS CHANNEL IN RELEASE TOO LONG

Circuit Description:

This DTC will diagnose a motor that is energized longer than expected. This could occur if a wheel speed sensor is faulty, the motor does not turn, the left front solenoid mechanically fails open, the brakes are dragging, or the motor wires are crossed.

DTC Will Set When: DTC 52 can be set only during an ABS stop. If the EBCM commands the left front channel in release for three seconds, a malfunction exists.

Action Taken: A malfunction DTC is stored, ABS is disabled, and the ABS warning lamp is turned "ON."

DTC Chart Test Description: Number(s) below refer to circled number(s) on the diagnostic chart.

1. This step identifies a malfunction in a wheel speed sensor that may cause the system to be in release too long.
2. This step identifies a motor or modulator malfunction.
3. This step checks for a solenoid that may have mechanically failed open.
4. Checks for a solenoid that may have failed closed.
5. Isolates the malfunction to either the solenoid or the hydraulic modulator.
6. This step checks for proper operation of hydraulic modulator.

Diagnostic Aids: An "Intermittent" malfunction is most likely caused by a mechanical part of the system that sticks or binds.

The frequency of the malfunction can be checked by using the enhanced diagnostic function of the "Tech 1.

DTC 52 may occur if on ice and steering wheel is turned to lock during braking. Using Tech 1 perform hydraulic test to ensure total brake system is functional.

Any circuitry, that is suspected as causing the intermittent complaint, should be thoroughly checked for backed out terminals, improper mating, broken locks, improperly formed or damaged terminals, poor terminal to wiring connections or physical damage to the wiring harness.

DIAGNOSTIC CHART DTC 52 — 1994 CAVALIER AND SUNBIRD

DTC 52
LEFT FRONT ABS CHANNEL IN RELEASE TOO LONG

IMPORTANT:
- EXCESSIVE DRAG OR HIGH RESISTANCE IN THE BASE BRAKE OR SUSPENSION SYSTEM MUST BE INSPECTED AND CORRECTED BEFORE PROCEEDING WITH DTC DIAGNOSIS.

ARE ANY WHEEL SPEED DTCs CURRENTLY SET?
- YES: PROCEED TO THAT DTC CHART FIRST.
- NO: go to step 1.

(1)
- USING TECH 1, SELECT "F0: DATA LIST" AND TEST VEHICLE DURING A STEADY DECEL CONDITION FROM 35 MPH TO 0 MPH WHILE MONITORING ALL WHEEL SPEEDS. DO ANY OF THE WHEEL SPEEDS INDICATE ERRATIC OR INTERMITTENT OPERATION?
  - YES: PROCEED TO EXCESSIVE WHEEL SPEED VARIATION DTC FOR AFFECTED WHEEL.
  - NO: go to step 2.

(2)
- IGNITION "ON."
- USING TECH 1 MANUAL CONTROL FUNCTION, SELECT LEFT FRONT MOTOR APPLY AND APPLY MOTOR.
- APPLY FIRM PRESSURE ON BRAKE PEDAL.
- SELECT LEFT FRONT MOTOR RELEASE AND COMMAND RELEASE OF MOTOR. DID BRAKE PEDAL FALL?
  - YES: go to step 3.
  - NO:
    - BRAKE PEDAL ROSE: REPAIR CROSSED WIRES (CKTs 1280 AND 1281) TO LEFT FRONT ABS MOTOR.
    - BRAKE PEDAL DID NOT MOVE: go to step 6.

(6)
- USING TECH 1, PERFORM GEAR TENSION RELIEF FUNCTION.
- SEPARATE MOTOR PACK FROM HYDRAULIC MODULATOR.
- GRASP THE GEAR ON THE HYDRAULIC MODULATOR (THE HYDRAULIC MODULATOR IS THE UNIT WITH THE THREE LARGE GEARS) FURTHEST FROM THE BRAKE BOOSTER AND ATTEMPT TO MOVE THE GEAR IN EITHER DIRECTION. CAN GEAR BE MOVED BY HAND?
  - YES: REPLACE MOTOR PACK.
  - NO: REPLACE ABS HYDRAULIC MODULATOR.

(3)
- REMOVE FOOT FROM BRAKE PEDAL.
- USING TECH 1 MANUAL CONTROL FUNCTION, SELECT LEFT FRONT SOLENOID AND COMMAND THE SOLENOID "ON."
- APPLY FIRM PRESSURE TO BRAKE PEDAL. DID BRAKE PEDAL FALL?
  - YES: PROCEED TO STEP 5 OF THIS CHART.
  - NO: go to step 4.

(4)
- FIRM PRESSURE STILL APPLIED TO BRAKE PEDAL.
- USING TECH 1, COMMAND LEFT FRONT SOLENOID "OFF." DID BRAKE PEDAL FALL?
  - YES: MALFUNCTION IS NOT PRESENT AT THIS TIME. REFER TO "DIAGNOSTIC AIDS".
  - NO: go to step 5.

(5)
- REMOVE FOOT FROM BRAKE PEDAL.
- IGNITION "OFF."
- SWITCH LEFT FRONT SOLENOID WITH RIGHT FRONT SOLENOID.
- IGNITION "ON."
- USING TECH 1 MANUAL CONTROL FUNCTION, SELECT LEFT FRONT SOLENOID AND COMMAND THE SOLENOID "ON."
- APPLY FIRM PRESSURE TO BRAKE PEDAL. DID BRAKE PEDAL FALL?
  - YES: REPLACE HYDRAULIC MODULATOR.
  - NO: REPLACE THE SOLENOID WHICH WAS ORIGINALLY IN THE LEFT FRONT CHANNEL.

* AFTER DIAGNOSIS IS COMPLETE, CLEAR DTCs AND TEST DRIVE VEHICLE FOR THREE (3) DRIVE CYCLES TO VERIFY DTC DOES NOT RESET. A DRIVE CYCLE CONSISTS OF STARTING THE VEHICLE, DRIVING OVER 16 km/h (10 MPH) AND THEN KEYING DOWN.

DIAGNOSTIC CHART DTC 53 — 1994 CAVALIER AND SUNBIRD

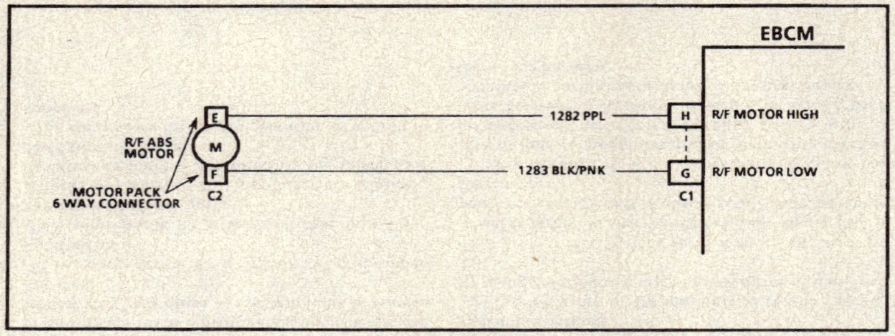

DTC 53

RIGHT FRONT ABS CHANNEL IN RELEASE TOO LONG

Circuit Description:

This DTC will diagnose a motor that is energized longer than expected. This could occur if a wheel speed sensor is faulty, the motor does not turn, the right front solenoid mechanically fails open, the brakes are dragging, or the motor wires are crossed.

DTC Will Set When: DTC 53 can be set only during an ABS stop. If the EBCM commands the right front channel in release for three seconds, a malfunction exists.

Action Taken: A malfunction DTC is stored, ABS is disabled, and the ABS warning lamp is turned "ON."

DTC Chart Test Description: Number(s) below refer to circled number(s) on the diagnostic chart.

1. This step identifies a malfunction in a wheel speed sensor that may cause the system to be in release too long.
2. This step identifies a motor or modulator malfunction.
3. This step checks for a solenoid that may have mechanically failed open.
4. Checks for a solenoid that may have failed closed.
5. Isolates the malfunction to either the solenoid or the hydraulic modular.
6. This step checks for proper operation of hydraulic modulator.

Diagnostic Aids: An "Intermittent" malfunction is most likely caused by a mechanical part of the system that sticks or binds.

The frequency of the malfunction can be checked by using the enhanced diagnostic function of the Tech 1.

DTC 53 may occur on ice if steering wheel is turned to lock during braking. Using Tech 1 perform hydraulic test to ensure total brake system is functional.

Any circuitry, that is suspected as causing the intermittent complaint, should be thoroughly checked for backed out terminals, improper mating, broken locks, improperly formed or damaged terminals, poor terminal to wiring connections or physical damage to the wiring harness.

DIAGNOSTIC CHART DTC 53 — 1994 CAVALIER AND SUNBIRD

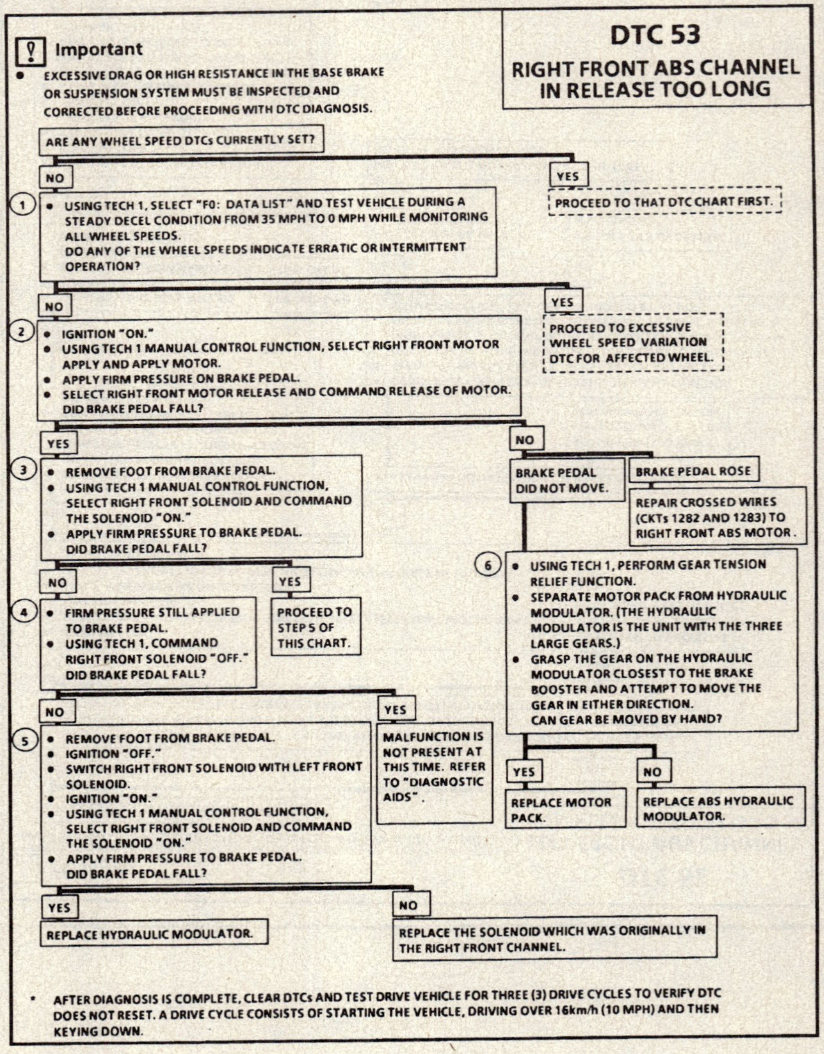

DIAGNOSTIC CHART DTC 54 — 1994 CAVALIER AND SUNBIRD

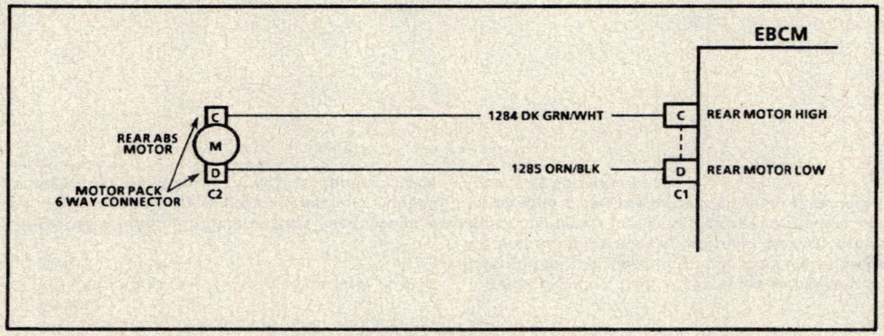

DTC 54

REAR ABS CHANNEL IN RELEASE TOO LONG

Circuit Description:
This DTC will diagnose a motor that is energized longer than expected. This could occur if a wheel speed sensor is faulty, the motor does not turn, the brakes are dragging, or the motor wires are crossed.

DTC Will Set When: DTC 54 can be set only during an ABS stop. If the EBCM commands the rear ABS channel in release for three seconds, a malfunction exists.

Action Taken: A malfunction DTC is stored, ABS is disabled, and the ABS warning lamp is turned "ON."

DTC Chart Test Description: Number(s) below refer to circled number(s) on the diagnostic chart.

1. This step identifies a problem in a wheel speed sensor that may cause the system to be in release too long.
2. This checks for a wheel that may stick or bind because of a mechanical fault.
3. This checks to see if the motor is capable of moving and applying the hydraulic piston for the rear wheels.
4. This step ensures that the motor wiring is not crossed.
5. This isolates the malfunction of a "no-apply" situation to either the motor pack or ABS hydraulic modulator.

Diagnostic Aids: An "Intermittent" malfunction is most likely caused by a mechanical part of the system that sticks or binds.

The frequency of the malfunction can be checked by using the enhanced diagnostic function of the Tech 1.

Using Tech 1 perform hydraulic test to ensure total brake system is functional.

Any circuitry, that is suspected as causing the intermittent complaint, should be thoroughly checked for backed out terminals, improper mating, broken locks, improperly formed or damaged terminals, poor terminal to wiring connections or physical damage to the wiring harness.

DIAGNOSTIC CHART DTC 54 CAVALIER AND SUNBIRD

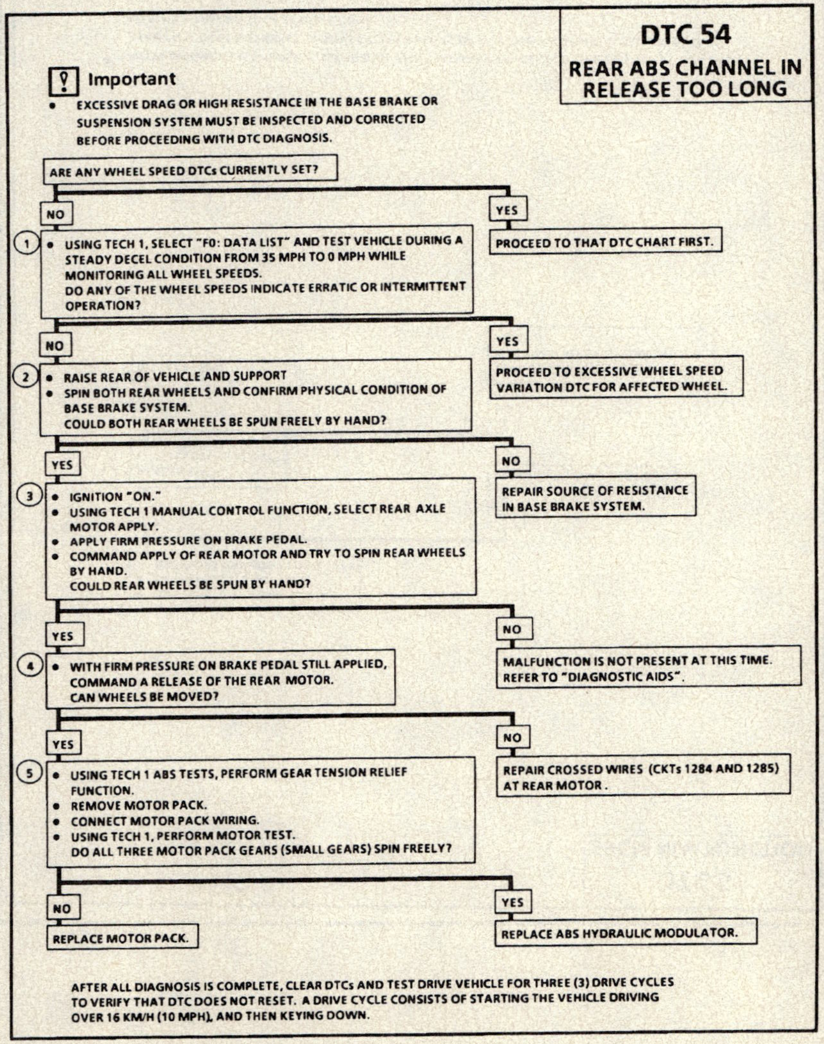

DIAGNOSTIC CHART DTC 55 — 1994 CAVALIER AND SUNBIRD

DTC 55
EBCM MALFUNCTION

Circuit Description:
This DTC identifies a malfunction detected by the MDI custom IC. It also ensures the cause of the malfunction is not a result of a malfunction with the enable relay under a DTC 55 failure.

DTC Will Set When: DTC 55 is set when an internal EBCM malfunction exists.

Action Taken: A malfunction DTC is stored, ABS is disabled and the ABS warning lamp is turned "ON."

DTC Chart Test Description: Number(s) below refer to circled number(s) on the diagnostic chart.
1. Checks to see if there are any other malfunctions present.
2. This checks to see if the malfunction is still present.

Diagnostic Aids: An "Intermittent" malfunction is most likely caused by a poor connection, rubbed through wire insulation, or a wire that is broken inside the insulation.

The frequency of the malfunction can be checked by using the enhanced diagnostic function of the Tech 1.

Any circuitry, that is suspected as causing the intermittent complaint, should be thoroughly checked for backed out terminals, improper mating, broken locks, improperly formed or damaged terminals, poor terminal to wiring connections or physical damage to the wiring harness.

DIAGNOSTIC CHART DTC 55 — 1994 CAVALIER AND SUNBIRD

DTC 55
EBCM MALFUNCTION

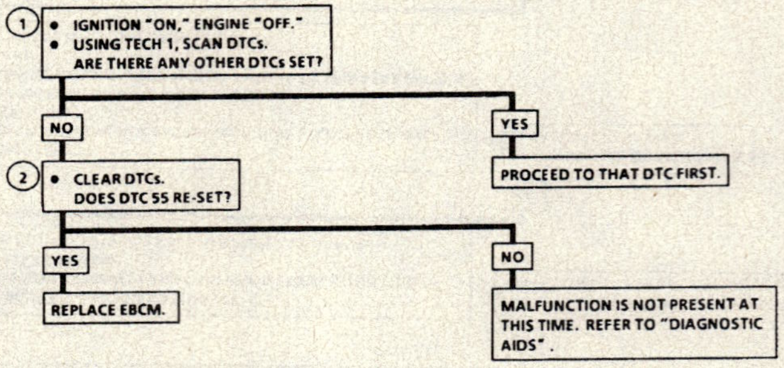

* AFTER DIAGNOSIS IS COMPLETE, CLEAR DTCs AND TEST DRIVE VEHICLE FOR THREE (3) DRIVE CYCLES TO VERIFY DTC DOES NOT RESET. A DRIVE CYCLE CONSISTS OF STARTING THE VEHICLE, DRIVING OVER 16 km/h (10 MPH) AND THEN KEYING DOWN.

THIS CHART ASSUMES THAT A CURRENT DTC IS STORED INDICATING THAT THIS MALFUNCTION IS PRESENT.

DIAGNOSTIC CHART DTC 56 — 1994 CAVALIER AND SUNBIRD

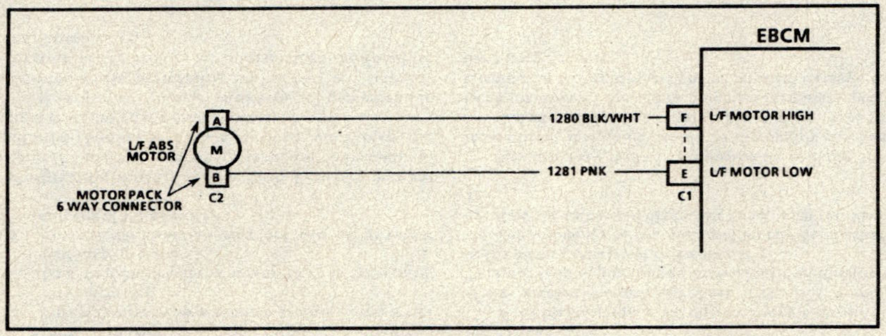

DTC 56

LEFT FRONT ABS MOTOR CIRCUIT OPEN

Circuit Description:
This DTC identifies a motor that cannot be energized due to an open in its circuitry. This malfunction will not allow proper front ABS operation.

DTC Will Set When: DTC 56 can be set only when the motor is commanded "OFF." If the EBCM detects an out of range voltage in either of the left front motor circuits indicating an open circuit, a malfunction exists.

Action Taken: An open motor will not activate when requested. A malfunction DTC is stored, ABS is disabled and the ABS warning lamp is turned "ON."

DTC Chart Test Description: Number(s) below refer to circled number(s) on the diagnostic chart.

1. This step checks for proper resistance of the motor.
2. This step checks for an open in the motor "high" circuitry.
3. This step checks for an open in the motor "low" circuitry.
4. Ensures malfunction was not due to poor terminal contact.

Diagnostic Aids: Using Tech 1, select manual control function, and exercise motor movement of affected channel in both directions while applying light pressure on the brake pedal.

If erratic or "jumpy" brake pedal movement is detected while performing an "apply" or "release" function of the motor, an intermittent malfunction may be indicated.

An "Intermittent" malfunction is most likely caused by a poor connection, rubbed through wire insulation, or a wire that is broken inside the insulation.

If the malfunction is not current, wiggle the wires of the affected channel and check if the DTC resets. This will help to pinpoint an intermittent malfunction in the motor circuitry or connections.

The frequency of the malfunction can be checked by using the enhanced diagnostic function of the Tech 1.

Any circuitry, that is suspected as causing the intermittent complaint, should be thoroughly checked for backed out terminals, improper mating, broken locks, improperly formed or damaged terminals, poor terminal to wiring connections or physical damage to the wiring harness.

DIAGNOSTIC CHART DTC 56 — 1994 CAVALIER AND SUNBIRD

DTC 56
LEFT FRONT ABS MOTOR CIRCUIT OPEN

DOES DTC 56 OCCUR INTERMITTENTLY?

- YES: REFER TO "DIAGNOSTIC AIDS".
- NO:

(1)
- IGNITION "OFF."
- DISCONNECT 6-WAY MOTOR PACK CONNECTOR.
- USING J 39200, MEASURE RESISTANCE ACROSS MOTOR PACK TERMINALS "A" AND "B".

IS RESISTANCE BELOW 2000 OHMS?

- NO: REPLACE MOTOR PACK.
- YES:

(2)
- DISCONNECT 6-WAY EBCM CONNECTOR.
- USING J 39200, MEASURE RESISTANCE BETWEEN 6-WAY EBCM HARNESS CONNECTOR TERMINAL "F" AND 6-WAY MOTOR PACK HARNESS CONNECTOR TERMINAL "A".

IS RESISTANCE LESS THAN 2 OHMS?

- NO: REPAIR OPEN OR HIGH RESISTANCE IN CKT 1280.
- YES:

(3)
- USING J 39200, MEASURE RESISTANCE BETWEEN 6-WAY EBCM HARNESS CONNECTOR TERMINAL "E" AND 6-WAY MOTOR PACK HARNESS CONNECTOR TERMINAL "B".

IS RESISTANCE LESS THAN 2 OHMS?

- NO: REPAIR OPEN OR HIGH RESISTANCE IN CKT 1281.
- YES:

(4)
- INSPECT 6-WAY MOTOR PACK AND EBCM HARNESS CONNECTORS FOR EVIDENCE OF POOR TERMINAL CONTACT OR CORROSION.
- RECONNECT ALL CONNECTORS.
- DRIVE VEHICLE FOR THREE DRIVE CYCLES (A DRIVE CYCLE CONSISTS OF STARTING ENGINE, DRIVING ABOVE 16 km/h (10 MPH) AND THEN TURNING IGNITION "OFF").

DID DTC 56 SET IN THE LAST THREE DRIVE CYCLES?

- YES: REPLACE EBCM.
- NO: REFER TO "DIAGNOSTIC AIDS".

* AFTER DIAGNOSIS IS COMPLETE, CLEAR DTCs AND TEST DRIVE VEHICLE FOR THREE (3) DRIVE CYCLES TO VERIFY DTC DOES NOT RESET. A DRIVE CYCLE CONSISTS OF STARTING THE VEHICLE, DRIVING OVER 16 km/h (10 MPH) AND THEN KEYING DOWN.

THIS CHART ASSUMES THAT A CURRENT DTC IS STORED INDICATING THAT THIS MALFUNCTION IS PRESENT.

DIAGNOSTIC CHART DTC 57 — 1994 CAVALIER AND SUNBIRD

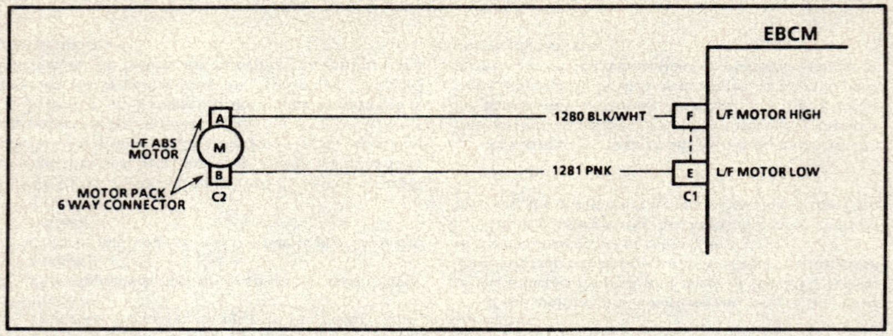

DTC 57

LEFT FRONT ABS MOTOR CIRCUIT SHORTED TO GROUND

Circuit Description:

This DTC identifies a motor circuit that is shorted to ground. This malfunction will not allow the motor to be controlled at the commanded current rate or will cause the driver circuit to allow current directly to ground.

DTC Will Set When: DTC 57 can be set anytime. If the EBCM detects an out of range voltage in either of the left front motor circuits indicating a circuit shorted to ground, a malfunction exists.

Action Taken: A malfunction DTC is stored, ABS is disabled and the ABS warning lamp is turned "ON."

DTC Chart Test Description: Number(s) below refer to circled number(s) on the diagnostic chart.

1. This step checks for a short to ground in the motor "high" circuitry.
2. This step checks for a short to ground in the motor "low" circuitry.
3. This step checks for a motor that is internally shorted to ground.
4. Ensures malfunction was not due to physical damage of the circuitry.

Diagnostic Aids: Using Tech 1, select manual control function, and exercise motor movement of affected channel in both directions while applying light pressure on the brake pedal.

If erratic or "jumpy" brake pedal movement is detected while performing an "apply" or "release" function of the motor, an intermittent malfunction may be indicated.

An "Intermittent" malfunction is most likely caused by a poor connection, rubbed through wire insulation, or a wire that is broken inside the insulation.

If the malfunction is not current, wiggle the wires of the affected channel and check if the DTC resets. This will help to pinpoint an intermittent malfunction in the motor circuitry or connections.

The frequency of the malfunction can be checked by using the enhanced diagnostic function of the Tech 1.

Any circuitry, that is suspected as causing the intermittent complaint, should be thoroughly checked for backed out terminals, improper mating, broken locks, improperly formed or damaged terminals, poor terminal to wiring connections or physical damage to the wiring harness.

DIAGNOSTIC CHART DTC 57 — 1994 CAVALIER AND SUNBIRD

DTC 57

LEFT FRONT ABS MOTOR CIRCUIT SHORTED TO GROUND

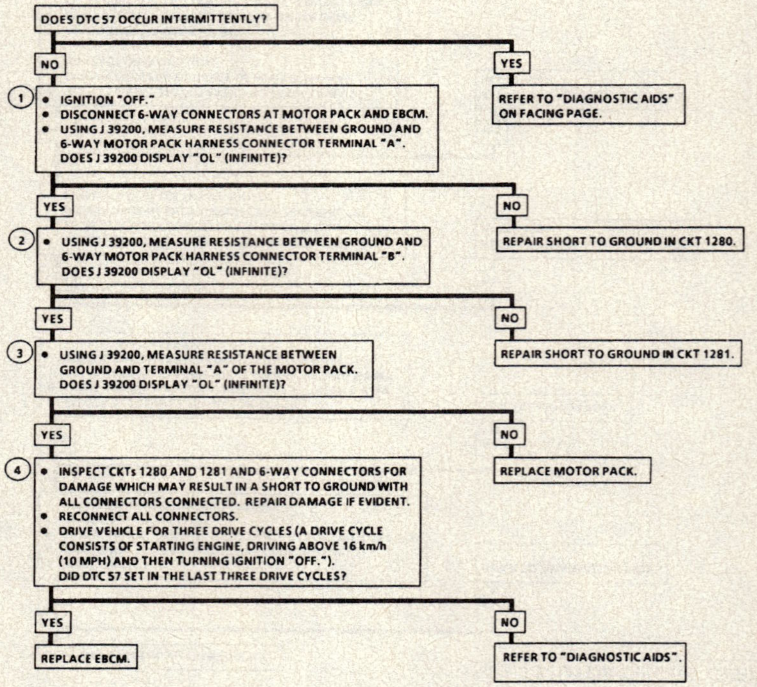

* AFTER DIAGNOSIS IS COMPLETE, CLEAR DTCs AND TEST DRIVE VEHICLE FOR THREE (3) DRIVE CYCLES TO VERIFY DTC DOES NOT RESET. A DRIVE CYCLE CONSISTS OF STARTING THE VEHICLE, DRIVING OVER 16 km/h (10 MPH) AND THEN KEYING DOWN.

THIS CHART ASSUMES THAT A CURRENT DTC IS STORED INDICATING THAT THIS MALFUNCTION IS PRESENT.

DIAGNOSTIC CHART DTC 58 — 1994 CAVALIER AND SUNBIRD

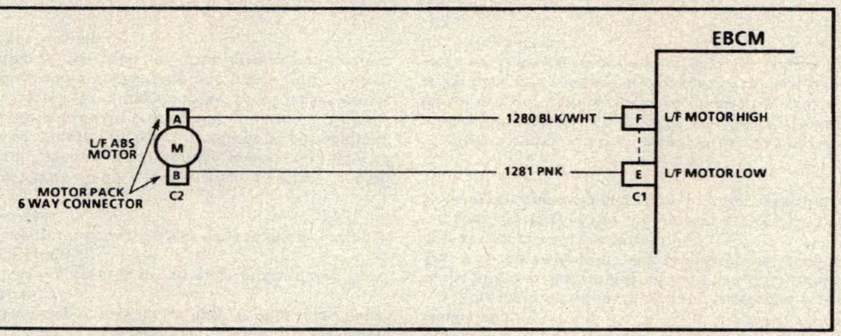

DTC 58

LEFT FRONT ABS MOTOR CIRCUIT SHORTED TO BATTERY OR MOTOR SHORTED

Circuit Description:

This DTC identifies a motor circuit that is shorted to battery or a motor that has low or no resistance. This malfunction will not allow the motor to be controlled at the commanded current rate or will cause the motor to turn in the opposite direction, or not at all.

DTC Will Set When: DTC 58 can be set anytime. If the EBCM detects an out of range voltage in either of the left front motor circuits indicating a circuit shorted to battery or a motor shorted, a malfunction exists.

Action Taken: A malfunction DTC is stored, ABS is disabled and the ABS warning lamp is turned "ON."

DTC Chart Test Description: Number(s) below refer to circled number(s) on the diagnostic chart.

1. This step checks for a short to voltage in the motor "high" circuitry.
2. This step checks for a short to voltage in the motor "low" circuitry.
3. Checks for a short between "high" and "low" motor circuits.
4. This step checks for a motor that is internally shorted.
5. Checks for a short between the EMB and motor circuits.
6. Checks for an internal motor pack short between the EMB and motor.
7. Ensures malfunction was not due to physical damage of the circuitry.

Diagnostic Aids: Using Tech 1, select manual control function, and exercise motor movement of affected channel in both directions while applying light pressure on the brake pedal.

If erratic or "jumpy" brake pedal movement is detected while performing an "apply" or "release" function of the motor, an intermittent malfunction may be indicated.

An "Intermittent" malfunction is most likely caused by a poor connection, rubbed through wire insulation, or a wire that is broken inside the insulation.

If the malfunction is not current, wiggle the wires of the affected channel and check if the DTC resets. This will help to pinpoint an intermittent malfunction in the motor circuitry or connections.

The frequency of the malfunction can be checked by using the enhanced diagnostic function of the Tech 1.

If DTC 38 sets as a current DTC and DTC 58 is an infrequent history DTC, see DTC 38 diagnostics.

Any circuitry, that is suspected as causing the intermittent complaint, should be thoroughly checked for backed out terminals, improper mating, broken locks, improperly formed or damaged terminals, poor terminal to wiring connections or physical damage to the wiring harness.

DIAGNOSTIC CHART DTC 58 — 1994 CAVALIER AND SUNBIRD

DTC 58

LEFT FRONT ABS MOTOR CIRCUIT SHORTED TO BATTERY OR MOTOR SHORTED

NOTICE: J 39200 TEST LEADS MUST BE "ZEROED" PRIOR TO MAKING ANY RESISTANCE MEASUREMENTS. REFER TO J 39200 USER'S MANUAL.

DOES DTC 58 OCCUR INTERMITTENTLY?

- YES: REFER TO "DIAGNOSTIC AIDS".
- NO:

(1) **NOTICE:** THIS STEP MAY CAUSE ADDITIONAL DTCs TO SET.
- IGNITION "OFF."
- DISCONNECT 6-WAY CONNECTORS AT MOTOR PACK AND EBCM.
- IGNITION "ON."
- USING J 39200, MEASURE VOLTAGE BETWEEN GROUND AND 6-WAY MOTOR PACK HARNESS CONNECTOR TERMINAL "A". IS VOLTAGE LESS THAN 1 VOLT?

- NO: REPAIR SHORT TO VOLTAGE IN CKT 1280.
- YES:

(2)
- USING J 39200, MEASURE VOLTAGE BETWEEN GROUND AND 6-WAY MOTOR PACK HARNESS CONNECTOR TERMINAL "B". IS VOLTAGE LESS THAN 1 VOLT?

- NO: REPAIR SHORT TO VOLTAGE IN CKT 1281.
- YES:

(3)
- IGNITION "OFF."
- USING J 39200, MEASURE RESISTANCE BETWEEN 6-WAY MOTOR PACK HARNESS CONNECTOR TERMINALS "A" AND "B". DOES J 39200 DISPLAY "OL" (INFINITE)?

- NO: REPAIR SHORT BETWEEN CKTs 1280 AND 1281.
- YES:

(4)
- USING J 39200, MEASURE RESISTANCE BETWEEN MOTOR PACK TERMINALS "A" AND "B". IS RESISTANCE GREATER THAN .4 OHMS?

- NO: REPLACE MOTOR PACK.
- YES:

(5)
- DISCONNECT 3-WAY MOTOR PACK CONNECTOR.
- USING J 39200, MEASURE RESISTANCE BETWEEN 3-WAY MOTOR PACK HARNESS CONNECTOR TERMINAL "B" AND 6-WAY MOTOR PACK HARNESS CONNECTOR TERMINAL "A" THEN "B". DID J 39200 DISPLAY "OL" (INFINITE) FOR EACH MEASUREMENT?

- NO: REPAIR SHORT BETWEEN CKT 850 AND MOTOR CIRCUIT WHICH MEASURED OTHER THAN "OL" (INFINITE).
- YES:

(6)
- USING J 39200, MEASURE RESISTANCE BETWEEN 3-WAY MOTOR PACK CONNECTOR TERMINAL "B" AND 6-WAY MOTOR PACK CONNECTOR TERMINAL "A". DID J 39200 DISPLAY "OL" (INFINITE)?

- NO: REPLACE MOTOR PACK.
- YES:

(7)
- INSPECT CKTs 1280 AND 1281 AND 6-WAY CONNECTORS FOR DAMAGE WHICH MAY RESULT IN A SHORT TO B + WITH ALL CONNECTORS CONNECTED.
- REPAIR DAMAGE IF EVIDENT.
- RECONNECT ALL CONNECTORS.
- DRIVE VEHICLE FOR THREE DRIVE CYCLES (A DRIVE CYCLE CONSISTS OF STARTING ENGINE, DRIVE ABOVE 16 km/h (10 MPH) AND THEN TURNING IGNITION "OFF"). DID DTC 58 SET IN THE LAST THREE DRIVE CYCLES?

- YES: REPLACE EBCM.
- NO: REFER TO "DIAGNOSTIC AIDS".

* AFTER DIAGNOSIS IS COMPLETE, CLEAR DTCs AND TEST DRIVE VEHICLE FOR THREE (3) DRIVE CYCLES TO VERIFY DTC DOES NOT RESET. A DRIVE CYCLE CONSISTS OF STARTING THE VEHICLE, DRIVING OVER 16 km/h (10 MPH) AND THEN KEYING DOWN.

THIS CHART ASSUMES THAT A CURRENT DTC IS STORED INDICATING THAT THIS MALFUNCTION IS PRESENT.

DIAGNOSTIC CHART DTC 61 — 1994 CAVALIER AND SUNBIRD

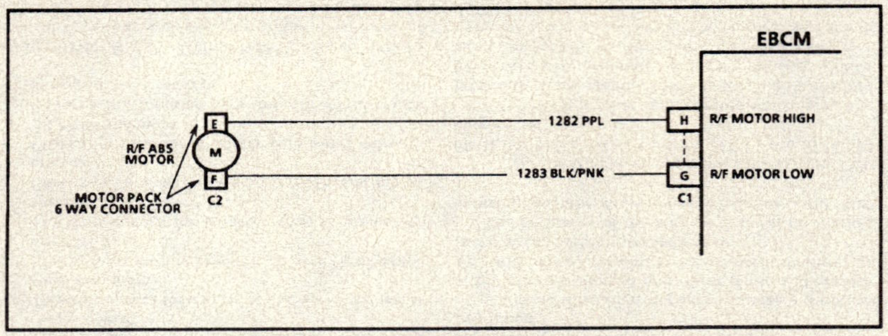

DTC 61

RIGHT FRONT ABS MOTOR CIRCUIT OPEN

Circuit Description:
This DTC identifies a motor that cannot be energized due to an open in its circuitry. This malfunction will not allow proper front ABS operation.

DTC Will Set When: DTC 61 can be set only when the motor is commanded "OFF." If the EBCM detects an out of range voltage in either of the right front motor circuits indicating an open circuit, a malfunction exists.

Action Taken: An open motor will not activate when requested. A malfunction DTC is stored, ABS is disabled and the ABS warning lamp is turned "ON."

DTC Chart Test Description: Number(s) below refer to circled number(s) on the diagnostic chart.

1. This step checks for proper resistance of the motor.
2. This step checks for an open in the motor "high" circuitry.
3. This step checks for an open in the motor "low" circuitry.
4. Ensures malfunction was not due to poor terminal contact.

Diagnostic Aids: Using Tech 1, select manual control function, and exercise motor movement of affected channel in both directions while applying light pressure on the brake pedal.

If erratic or "jumpy" brake pedal movement is detected while performing an "apply" or "release" function of the motor, an intermittent malfunction may be indicated.

An "Intermittent" malfunction is most likely caused by a poor connection, rubbed through wire insulation, or a wire that is broken inside the insulation.

If the malfunction is not current, wiggle the wires of the affected channel and check if the DTC resets. This will help to pinpoint an intermittent malfunction in the motor circuitry or connections.

The frequency of the malfunction can be checked by using the enhanced diagnostic function of the Tech 1.

Any circuitry, that is suspected as causing the intermittent complaint, should be thoroughly checked for backed out terminals, improper mating, broken locks, improperly formed or damaged terminals, poor terminal to wiring connections or physical damage to the wiring harness.

DIAGNOSTIC CHART DTC 61 — 1994 CAVALIER AND SUNBIRD

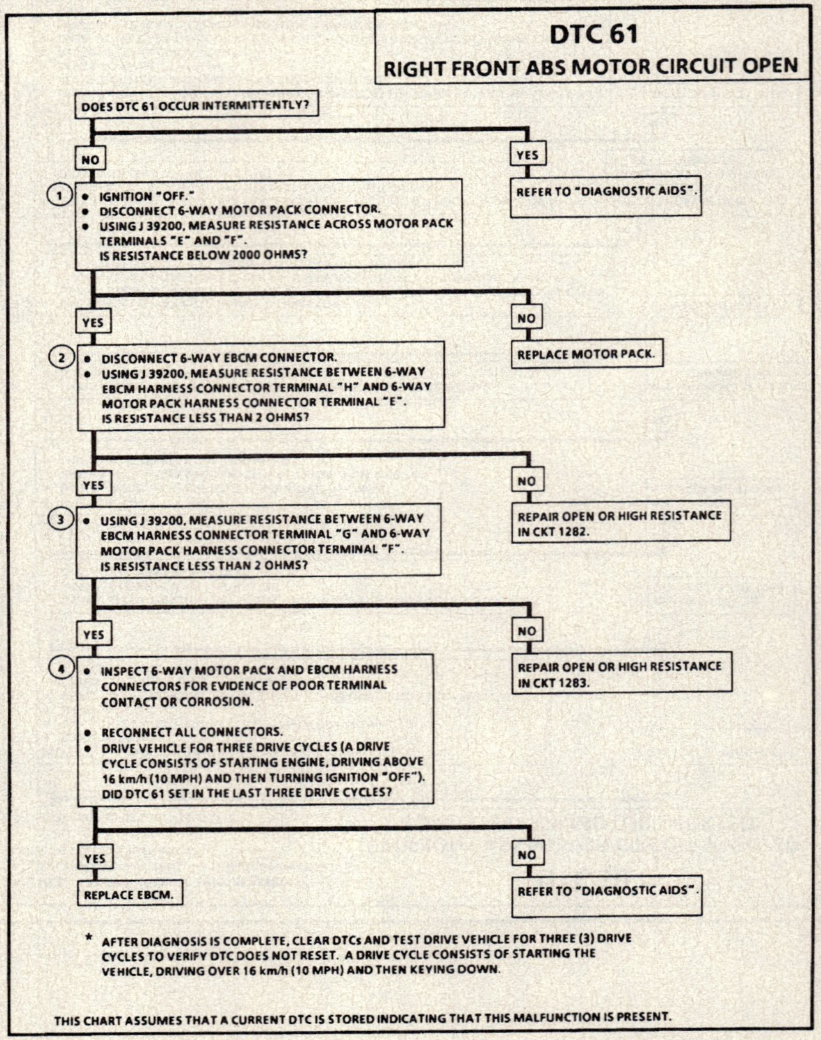

DIAGNOSTIC CHART DTC 62 — 1994 CAVALIER AND SUNBIRD

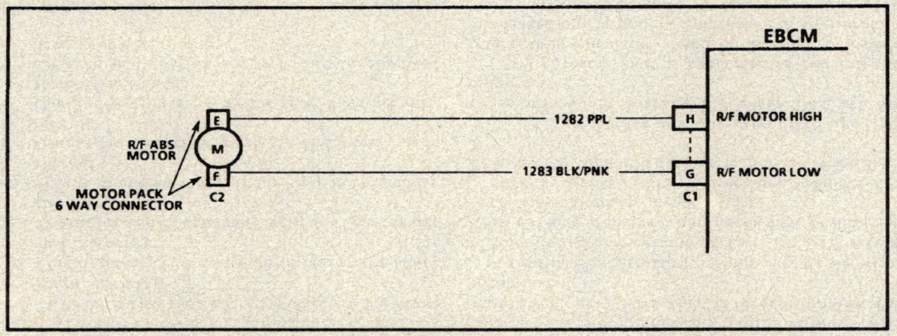

DTC 62

RIGHT FRONT ABS MOTOR CIRCUIT SHORTED TO GROUND

Circuit Description:
This DTC identifies a motor circuit that is shorted to ground. This malfunction will not allow the motor to be controlled at the commanded current rate or will cause the motor driver circuit to allow current directly to ground.

DTC Will Set When: DTC 62 can be set anytime. If the EBCM detects an out of range voltage in either of the right front motor circuits indicating a circuit shorted to ground, a malfunction exists.

Action Taken: A malfunction DTC is stored, ABS is disabled and the ABS warning lamp is turned "ON."

DTC Chart Test Description: Number(s) below refer to circled number(s) on the diagnostic chart.

1. This step checks for a short to ground in the motor "high" circuitry.
2. This step checks for a short to ground in the motor "low" circuitry.
3. This step checks for a motor that is internally shorted to ground.
4. Ensures malfunction was not due to physical damage of the circuitry.

Diagnostic Aids: Using Tech 1, select manual control function, and exercise motor movement of affected channel in both directions while applying light pressure on the brake pedal.

If erratic or "jumpy" brake pedal movement is detected while performing an "apply" or "release" function of the motor, an intermittent malfunction may be indicated.

An "Intermittent" malfunction is most likely caused by a poor connection, rubbed through wire insulation, or a wire that is broken inside the insulation.

If the malfunction is not current, wiggle the wires of the affected channel and check if the DTC resets. This will help to pinpoint an intermittent malfunction in the motor circuitry or connections.

The frequency of the malfunction can be checked by using the enhanced diagnostic function of the Tech 1.

Any circuitry, that is suspected as causing the intermittent complaint, should be thoroughly checked for backed out terminals, improper mating, broken locks, improperly formed or damaged terminals, poor terminal to wiring connections or physical damage to the wiring harness.

DIAGNOSTIC CHART DTC 62 — 1994 CAVALIER AND SUNBIRD

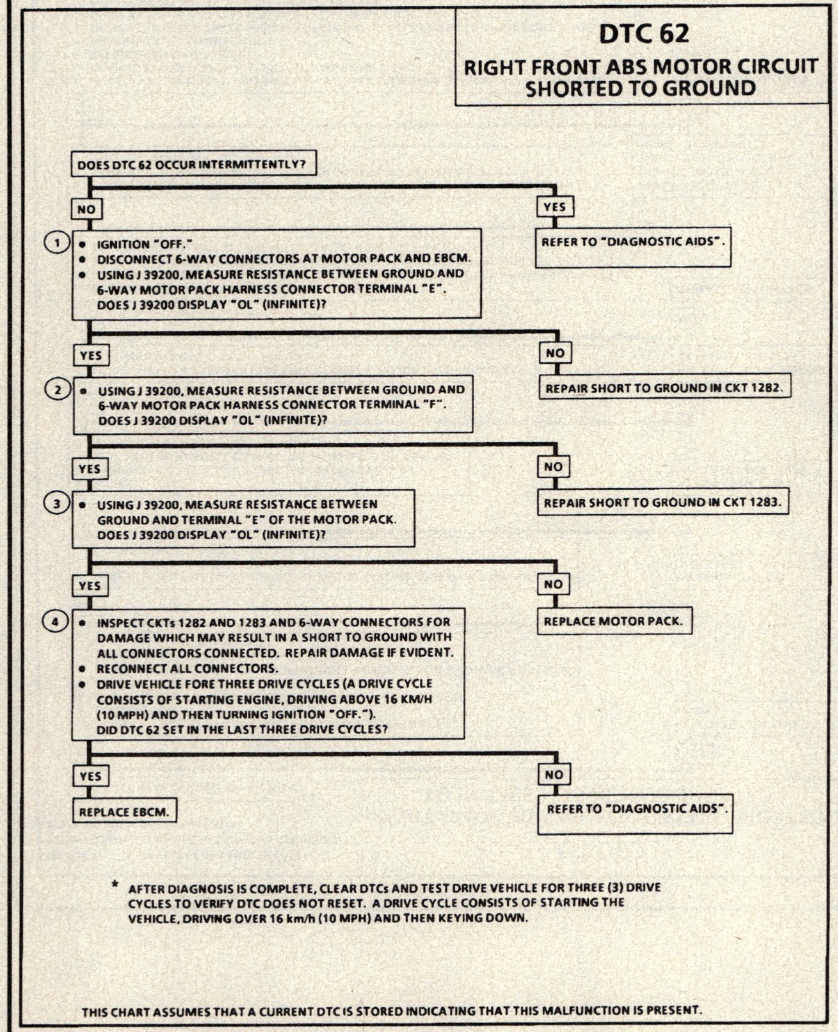

DIAGNOSTIC CHART DTC 63 — 1994 CAVALIER AND SUNBIRD

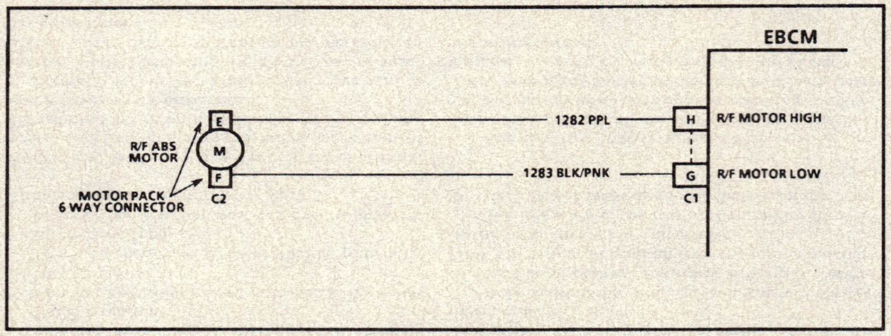

DTC 63

RIGHT FRONT ABS MOTOR CIRCUIT SHORTED TO BATTERY OR MOTOR SHORTED

Circuit Description:

This DTC identifies a motor circuit that is shorted to battery or a motor that has low or no resistance. This malfunction will not allow the motor to be controlled at the commanded current rate or will cause the motor to turn in the opposite direction or not at all.

DTC Will Set When: DTC 63 can be set anytime. If the EBCM detects an out of range voltage in either of the right front motor circuits indicating a circuit shorted to battery or a motor shorted, a malfunction exists.

Action Taken: A malfunction DTC is stored, ABS is disabled and the ABS warning lamp is turned "ON."

DTC Chart Test Description: Number(s) below refer to circled number(s) on the diagnostic chart.

1. This step checks for a short to voltage in the motor "high" circuitry.
2. This step checks for a short to voltage in the motor "low" circuitry.
3. Checks for a short between "high" and "low" motor circuits.
4. This step checks for a motor that is internally shorted.
5. Checks for a short between the EMB and motor circuits.
6. Checks for an internal motor pack short between the EMB and motor.
7. Ensures malfunction was not due to physical damage of the circuitry.

Diagnostic Aids: Using Tech 1, select manual control function, and exercise motor movement of affected channel in both directions while applying light pressure on the brake pedal.

If erratic or "jumpy" brake pedal movement is detected while performing an "apply" or "release" function of the motor, an intermittent malfunction may be indicated.

An "Intermittent" malfunction is most likely caused by a poor connection, rubbed through wire insulation, or a wire that is broken inside the insulation.

If the malfunction is not current, wiggle the wires of the affected channel and check if the DTC resets. This will help to pinpoint an intermittent malfunction in the motor circuitry or connections.

The frequency of the malfunction can be checked by using the enhanced diagnostic function of the Tech 1.

If DTC 41 sets as a current DTC and DTC 63 is infrequent history DTC, see DTC 41 for diagnosis.

Any circuitry, that is suspected as causing the intermittent complaint, should be thoroughly checked for backed out terminals, improper mating, broken locks, improperly formed or damaged terminals, poor terminal to wiring connections or physical damage to the wiring harness.

DIAGNOSTIC CHART DTC 63 — 1994 CAVALIER AND SUNBIRD

NOTICE: J 39200 TEST LEADS MUST BE "ZEROED" PRIOR TO MAKING ANY RESISTANCE MEASUREMENTS. REFER TO J 39200 USER'S MANUAL.

DTC 63

RIGHT FRONT ABS MOTOR CIRCUIT SHORTED TO BATTERY OR MOTOR SHORTED

DOES DTC 63 OCCUR INTERMITTENTLY?

- YES: REFER TO "DIAGNOSTIC AIDS".
- NO:

(1) NOTICE: THIS STEP MAY CAUSE ADDITIONAL DTCs TO SET.
- IGNITION "OFF."
- DISCONNECT 6-WAY CONNECTORS AT MOTOR PACK AND EBCM.
- IGNITION "ON."
- USING J 39200, MEASURE VOLTAGE BETWEEN GROUND AND 6-WAY MOTOR PACK HARNESS CONNECTOR TERMINAL "E".

IS VOLTAGE LESS THAN 1 VOLT?

- NO: REPAIR SHORT TO VOLTAGE IN CKT 1282.
- YES:

(2)
- USING J 39200, MEASURE VOLTAGE BETWEEN GROUND AND 6-WAY MOTOR PACK HARNESS CONNECTOR TERMINAL "F".

IS VOLTAGE LESS THAN 1 VOLT?

- NO: REPAIR SHORT TO VOLTAGE IN CKT 1283.
- YES:

(3)
- IGNITION "OFF."
- USING J 39200, MEASURE RESISTANCE BETWEEN 6-WAY MOTOR PACK HARNESS CONNECTOR TERMINALS "E" AND "F".

DOES J 39200 DISPLAY "OL" (INFINITE)?

- NO: REPAIR SHORT BETWEEN CKTs 1282 AND 1283.
- YES:

(4)
- USING J 39200, MEASURE RESISTANCE BETWEEN MOTOR PACK TERMINALS "E" AND "F".

IS RESISTANCE GREATER THAN .4 OHMS?

- NO: REPLACE MOTOR PACK.
- YES:

(5)
- DISCONNECT 3-WAY MOTOR PACK CONNECTOR.
- USING J 39200, MEASURE RESISTANCE BETWEEN 3-WAY MOTOR PACK HARNESS CONNECTOR TERMINAL "B" AND 6-WAY MOTOR PACK HARNESS CONNECTOR TERMINAL "E" THEN "F".

DID J 39200 DISPLAY "OL" (INFINITE) FOR EACH MEASUREMENT?

- NO: REPAIR SHORT BETWEEN CKT 850 AND MOTOR CIRCUIT WHICH MEASURED OTHER THAN "OL" (INFINITE).
- YES:

(6)
- USING J 39200, MEASURE RESISTANCE BETWEEN 3-WAY MOTOR PACK CONNECTOR TERMINAL "B" AND 6-WAY MOTOR PACK CONNECTOR TERMINAL "E".

DID J 39200 DISPLAY "OL" (INFINITE)?

- NO: REPLACE MOTOR PACK.
- YES:

(7)
- INSPECT CKTs 1282 AND 1283 AND 6-WAY CONNECTORS FOR DAMAGE WHICH MAY RESULT IN A SHORT TO B+ WITH ALL CONNECTORS CONNECTED.
- REPAIR DAMAGE IF EVIDENT.
- RECONNECT ALL CONNECTORS.
- DRIVE VEHICLE FOR THREE DRIVE CYCLES (A DRIVE CYCLE CONSISTS OF STARTING ENGINE, DRIVE ABOVE 16 km/h (10 MPH) AND THEN TURNING IGNITION "OFF").

DID DTC 63 SET IN THE LAST THREE DRIVE CYCLES?

- YES: REPLACE EBCM.
- NO: REFER TO "DIAGNOSTIC AIDS".

* AFTER DIAGNOSIS IS COMPLETE, CLEAR DTCs AND TEST DRIVE VEHICLE FOR THREE (3) DRIVE CYCLES TO VERIFY DTC DOES NOT RESET. A DRIVE CYCLE CONSISTS OF STARTING THE VEHICLE, DRIVING OVER 16 km/h (10 MPH) AND THEN KEYING DOWN.

THIS CHART ASSUMES THAT A CURRENT DTC IS STORED INDICATING THAT THIS MALFUNCTION IS PRESENT.

DIAGNOSTIC CHART DTC 64 — 1994 CAVALIER AND SUNBIRD

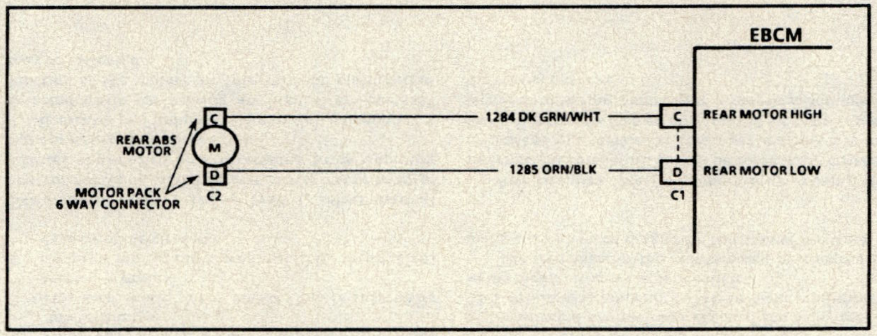

DTC 64

REAR ABS MOTOR CIRCUIT OPEN

Circuit Description:
This DTC identifies a motor that cannot be energized due to an open in its circuitry. This malfunction will not allow proper rear ABS operation.

DTC Will Set When: DTC 64 can be set only when the motor is commanded "OFF." If the EBCM detects an out of range voltage in either of the rear ABS motor circuits indicating an open circuit, a malfunction exists.

Action Taken: An open motor will not activate when requested. A malfunction DTC is stored, ABS is disabled and the ABS warning lamp is turned "ON."

DTC Chart Test Description: Number(s) below refer to circled number(s) on the diagnostic chart.

1. This checks for proper resistance of the motor.
2. This step checks for an open in the motor "high" circuitry.
3. This step checks for an open in the motor "low" circuitry.
4. Ensures malfunction was not due to poor terminal contact.

Diagnostic Aids: Using Tech 1, select manual control function, and exercise motor movement of affected channel in both directions while applying light pressure on the brake pedal.

If erratic or "jumpy" brake pedal movement is detected while performing an "apply" or "release" function of the motor, an intermittent malfunction may be indicated.

An "Intermittent" malfunction is most likely caused by a poor connection, rubbed through wire insulation, or a wire that is broken inside the insulation.

If the malfunction is not current, wiggle the wires of the affected channel and check if the DTC resets. This will help to pinpoint an intermittent malfunction in the motor circuitry or connections.

The frequency of the malfunction can be checked by using the enhanced diagnostic function of the Tech 1.

Any circuitry, that is suspected as causing the intermittent complaint, should be thoroughly checked for backed out terminals, improper mating, broken locks, improperly formed or damaged terminals, poor terminal to wiring connections or physical damage to the wiring harness.

DIAGNOSTIC CHART DTC 64 — 1994 CAVALIER AND SUNBIRD

DTC 64

REAR ABS MOTOR CIRCUIT OPEN

DOES DTC 64 OCCUR INTERMITTENTLY?

- YES: REFER TO "DIAGNOSTIC AIDS".
- NO:

(1)
- IGNITION "OFF."
- DISCONNECT 6- WAY MOTOR PACK CONNECTOR.
- USING J 39200, MEASURE RESISTANCE ACROSS MOTOR PACK TERMINALS "C" AND "D".

IS RESISTANCE BELOW 2000 OHMS?

- NO: REPLACE MOTOR PACK.
- YES:

(2)
- DISCONNECT 6-WAY EBCM CONNECTOR.
- USING J 39200, MEASURE RESISTANCE BETWEEN 6-WAY EBCM HARNESS CONNECTOR TERMINAL "C" AND 6-WAY MOTOR PACK HARNESS CONNECTOR TERMINAL "C".

IS RESISTANCE LESS THAN 2 OHMS?

- NO: REPAIR OPEN OR HIGH RESISTANCE IN CKT 1284.
- YES:

(3)
- USING J 39200, MEASURE RESISTANCE BETWEEN 6-WAY EBCM HARNESS CONNECTOR TERMINAL "D" AND 6-WAY MOTOR PACK HARNESS CONNECTOR TERMINAL "D".

IS RESISTANCE LESS THAN 2 OHMS?

- NO: REPAIR OPEN OR HIGH RESISTANCE IN CKT 1285.
- YES:

(4)
- INSPECT 6-WAY MOTOR PACK AND EBCM HARNESS CONNECTORS FOR EVIDENCE OF POOR TERMINAL CONTACT OR CORROSION.
- RECONNECT ALL CONNECTORS.
- DRIVE VEHICLE FOR THREE DRIVE CYCLES (A DRIVE CYCLE CONSISTS OF STARTING ENGINE, DRIVING ABOVE 16 km/h (10 MPH) AND THEN TURNING IGNITION "OFF").

DID DTC 64 SET IN THE LAST THREE DRIVE CYCLES?

- YES: REPLACE EBCM.
- NO: REFER TO "DIAGNOSTIC AIDS".

* AFTER DIAGNOSIS IS COMPLETE, CLEAR DTCs AND TEST DRIVE VEHICLE FOR THREE (3) DRIVE CYCLES TO VERIFY DTC DOES NOT RESET. A DRIVE CYCLE CONSISTS OF STARTING THE VEHICLE, DRIVING OVER 16 km/h (10 MPH) AND THEN KEYING DOWN.

THIS CHART ASSUMES THAT A CURRENT DTC IS STORED INDICATING THAT THIS MALFUNCTION IS PRESENT.

DIAGNOSTIC CHART DTC 65 — 1994 CAVALIER AND SUNBIRD

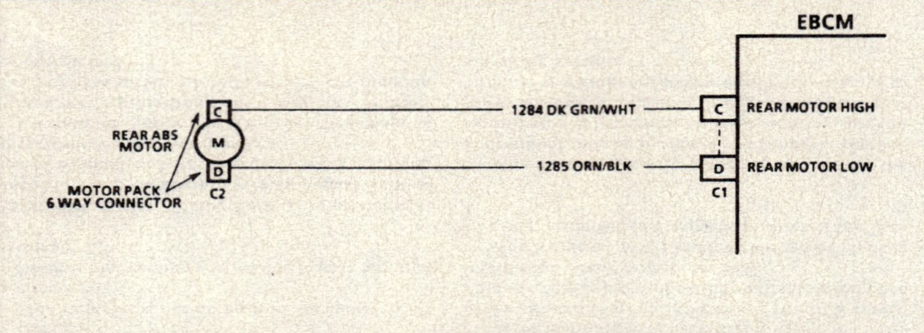

DTC 65

REAR ABS MOTOR CIRCUIT SHORTED TO GROUND

Circuit Description:
This DTC identifies a motor circuit that is shorted to ground. This malfunction will not allow the motor to be controlled at the commanded current rate or will cause the driver circuit to allow current directly to ground.

DTC Will Set When: DTC 65 can be set anytime. If the EBCM detects an out of range voltage in either of the rear ABS motor circuits indicating a circuit shorted to ground, a malfunction exists.

Action Taken: A malfunction DTC is stored, ABS is disabled and the ABS warning lamp is turned "ON."

DTC Chart Test Description: Number(s) below refer to circled number(s) on the diagnostic chart.

1. This step checks for a short to ground in the motor "high" circuitry.
2. This step checks for a short to ground in the motor "low" circuitry.
3. This step checks for a motor that is internally shorted to ground.
4. Ensures malfunction was not due to physical damage of the circuitry.

Diagnostic Aids: Using Tech 1, select manual control function, and exercise motor movement of affected channel in both directions while applying light pressure on the brake pedal.

If erratic or "jumpy" brake pedal movement is detected while performing an "apply" or "release" function of the motor, an intermittent malfunction may be indicated.

An "Intermittent" malfunction is most likely caused by a poor connection, rubbed through wire insulation, or a wire that is broken inside the insulation.

If the malfunction is not current, wiggle the wires of the affected channel and check if the DTC resets. This will help to pinpoint an intermittent malfunction in the motor circuitry or connections.

The frequency of the problem can be checked by using the enhanced diagnostic function of the Tech 1.

Any circuitry, that is suspected as causing the intermittent complaint, should be thoroughly checked for backed out terminals, improper mating, broken locks, improperly formed or damaged terminals, poor terminal to wiring connections or physical damage to the wiring harness.

DIAGNOSTIC CHART DTC 65 — 1994 CAVALIER AND SUNBIRD

DTC 65

REAR ABS MOTOR CIRCUIT SHORTED TO GROUND

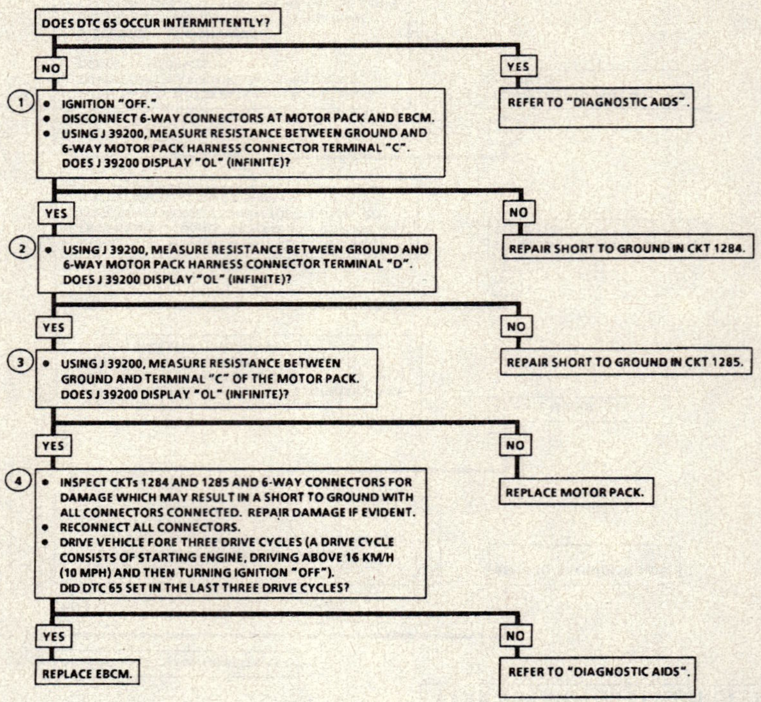

* AFTER DIAGNOSIS IS COMPLETE, CLEAR DTCs AND TEST DRIVE VEHICLE FOR THREE (3) DRIVE CYCLES TO VERIFY DTC DOES NOT RESET. A DRIVE CYCLE CONSISTS OF STARTING THE VEHICLE, DRIVING OVER 16 km/h (10 MPH) AND THEN KEYING DOWN.

THIS CHART ASSUMES THAT A CURRENT DTC IS STORED INDICATING THAT THIS MALFUNCTION IS PRESENT.

DIAGNOSTIC CHART DTC 66 — 1994
CAVALIER AND SUNBIRD

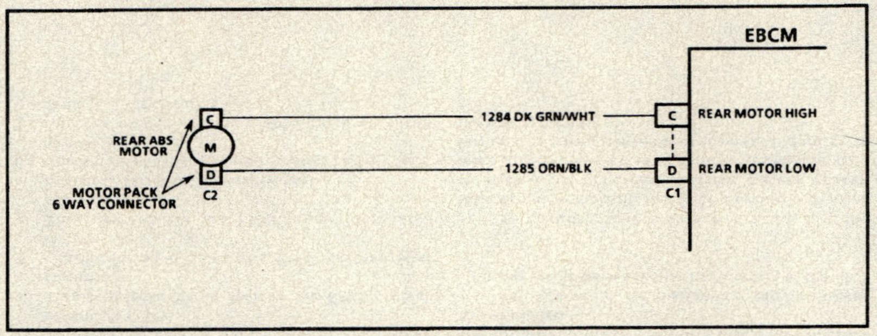

DTC 66

REAR ABS MOTOR CIRCUIT SHORTED TO BATTERY OR MOTOR SHORTED

Circuit Description:

This DTC identifies a motor circuit that is shorted to battery or a motor that has low or zero resistance. This malfunction will not allow the motor to be controlled at the commanded current rate or will cause the motor to turn in the opposite direction or not turn at all.

DTC Will Set When: DTC 66 can be set anytime. If the EBCM detects an out of range voltage in either of the rear ABS motor circuits indicating a circuit shorted to battery or a motor shorted, a malfunction exists.

Action Taken: A malfunction DTC is stored, ABS is disabled and the ABS warning lamp is turned "ON."

DTC Chart Test Description: Number(s) below refer to circled number(s) on the diagnostic chart.

1. This step checks for a short to voltage in the motor "high" circuitry.
2. This step checks for a short to voltage in the motor "low" circuitry.
3. Checks for a short between "high" and "low" motor circuits.
4. This step checks for a motor that is internally shorted.
5. Checks for a short between the EMB and motor circuits.
6. Checks for an internal motor pack short between the EMB and motor.
7. Ensures malfunction was not due to physical damage of the circuitry.

Diagnostic Aids: Using Tech 1, select manual control function, and exercise motor movement of affected channel in both directions while applying light pressure on the brake pedal.

If erratic or "jumpy" brake pedal movement is detected while performing an "apply" or "release" function of the motor, an intermittent malfunction may be indicated.

An "Intermittent" malfunction is most likely caused by a poor connection, rubbed through wire insulation, or a wire that is broken inside the insulation.

If the malfunction is not current, wiggle the wires of the affected channel and check if the DTC resets. This will help to pinpoint an intermittent malfunction in the motor circuitry or connections.

The frequency of the malfunction can be checked by using the enhanced diagnostic function of the Tech 1.

If DTC 42 sets as current DTC and DTC 66 is infrequent history DTC, see DTC 42 diagnostics.

Any circuitry, that is suspected as causing the intermittent complaint, should be thoroughly checked for backed out terminals, improper mating, broken locks, improperly formed or damaged terminals, poor terminal to wiring connections or physical damage to the wiring harness.

DIAGNOSTIC CHART DTC 66 — 1994
CAVALIER AND SUNBIRD

DTC 66

REAR ABS MOTOR CIRCUIT SHORTED TO BATTERY OR MOTOR SHORTED

NOTICE: J 39200 TEST LEADS MUST BE "ZEROED" PRIOR TO MAKING ANY RESISTANCE MEASUREMENTS. REFER TO J 39200 USER'S MANUAL.

DOES DTC 66 OCCUR INTERMITTENTLY?

- YES: REFER TO "DIAGNOSTIC AIDS".
- NO: go to step 1.

(1) NOTICE: THIS STEP MAY CAUSE ADDITIONAL DTCs TO SET.
- IGNITION "OFF."
- DISCONNECT 6-WAY CONNECTORS AT MOTOR PACK AND EBCM.
- IGNITION "ON."
- USING J 39200, MEASURE VOLTAGE BETWEEN GROUND AND 6-WAY MOTOR PACK HARNESS CONNECTOR TERMINAL "C".

IS VOLTAGE LESS THAN 1 VOLT?

- NO: REPAIR SHORT TO VOLTAGE IN CKT 1284.
- YES: go to step 2.

(2)
- USING J 39200, MEASURE VOLTAGE BETWEEN GROUND AND 6-WAY MOTOR PACK HARNESS CONNECTOR TERMINAL "D".

IS VOLTAGE LESS THAN 1 VOLT?

- NO: REPAIR SHORT TO VOLTAGE IN CKT 1285.
- YES: go to step 3.

(3)
- IGNITION "OFF."
- USING J 39200, MEASURE RESISTANCE BETWEEN 6-WAY MOTOR PACK HARNESS CONNECTOR TERMINALS "C" AND "D".

DOES J 39200 DISPLAY "OL" (INFINITE)?

- NO: REPAIR SHORT BETWEEN CKTs 1284 AND 1285.
- YES: go to step 4.

(4)
- USING J 39200, MEASURE RESISTANCE BETWEEN MOTOR PACK TERMINALS "C" AND "D".

IS RESISTANCE GREATER THAN .4 OHMS?

- NO: REPLACE MOTOR PACK.
- YES: go to step 5.

(5)
- DISCONNECT 3-WAY MOTOR PACK CONNECTOR.
- USING J 39200, MEASURE RESISTANCE BETWEEN 3-WAY MOTOR PACK HARNESS CONNECTOR TERMINAL "B" AND 6-WAY MOTOR PACK HARNESS CONNECTOR TERMINAL "C" THEN "D".

DID J 39200 DISPLAY "OL" (INFINITE) FOR EACH MEASUREMENT?

- NO: REPAIR SHORT BETWEEN CIRCUIT 850 AND MOTOR CIRCUIT WHICH MEASURED OTHER THEN "OL" (INFINITE).
- YES: go to step 6.

(6)
- USING J 39200, MEASURE RESISTANCE BETWEEN 3-WAY MOTOR PACK CONNECTOR TERMINAL "B" AND 6-WAY MOTOR PACK CONNECTOR TERMINAL "C".

DOES J 39200 DISPLAY "OL" (INFINITE)?

- NO: REPLACE MOTOR PACK.
- YES: go to step 7.

(7)
- INSPECT CKTs 1284 AND 1285 AND 6-WAY CONNECTORS FOR DAMAGE WHICH MAY RESULT IN A SHORT TO B + WITH ALL CONNECTORS CONNECTED.
- REPAIR DAMAGE IF EVIDENT.
- RECONNECT ALL CONNECTORS.
- DRIVE VEHICLE FOR THREE DRIVE CYCLES (A DRIVE CYCLE CONSISTS OF STARTING ENGINE, DRIVE ABOVE 16 km/h (10 MPH) AND THEN TURNING IGNITION "OFF").

DID DTC 66 SET IN THE LAST THREE DRIVE CYCLES?

- YES: REPLACE EBCM.
- NO: REFER TO "DIAGNOSTIC AIDS".

* AFTER DIAGNOSIS IS COMPLETE, CLEAR DTCs AND TEST DRIVE VEHICLE FOR THREE (3) DRIVE CYCLES TO VERIFY DTC DOES NOT RESET. A DRIVE CYCLE CONSISTS OF STARTING THE VEHICLE, DRIVING OVER 16 km/h (10 MPH) AND THEN KEYING DOWN

THIS CHART ASSUMES THAT A CURRENT DTC IS STORED INDICATING THAT THIS MALFUNCTION IS PRESENT.

DIAGNOSTIC CHART DTC 67 — 1994 CAVALIER AND SUNBIRD

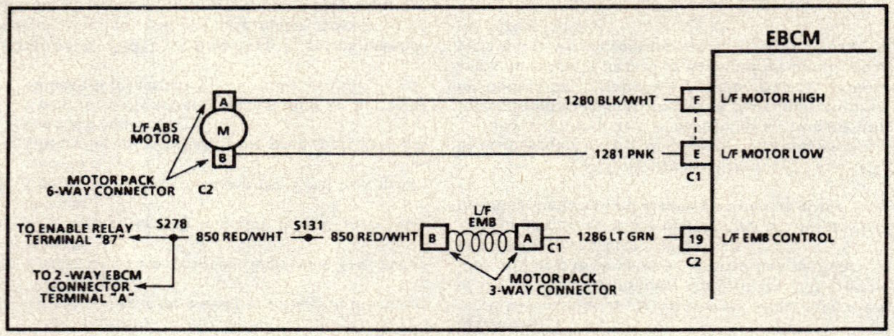

DTC 67

LEFT FRONT EMB CIRCUIT OPEN OR SHORTED TO GROUND

Circuit Description:

This DTC is designed to detect shorts to ground and opens in the EMB release circuit.

A short to ground between the EMB and the ABS EBCM will cause the EMB to be energized indefinitely at an uncontrolled current, causing the EMB to be destroyed.

DTC Will Set When: DTC 67 can be set only when the left front EMB is commanded "OFF" (not released). If the EBCM detects the left front EMB control circuit is constantly at ground, a malfunction exists.

Action Taken: A malfunction DTC is stored, ABS is disabled and the ABS warning lamp is turned "ON."

DTC Chart Test Description: Number(s) below refer to circled number(s) on the diagnostic chart.

1. This step checks for a short to ground in the EMB control circuitry.
2. This step looks for an open in the EMB control circuitry.
3. Checks for an open in the EMB battery feed circuit.
4. This step ensures the EMB is not internally shorted to ground.
5. Ensures EMB is not internally open.
6. Checks for short to ground in EMB battery feed circuit.
7. Ensures malfunction was not due to physical damage of the circuitry.

Diagnostic Aids: An "Intermittent" malfunction is most likely caused by a poor connection, rubbed through wire insulation, or a wire that is broken inside the insulation.

The frequency of the malfunction can be checked by using the enhanced diagnostic function of the Tech 1.

Any circuitry, that is suspected as causing the intermittent complaint, should be thoroughly checked for backed out terminals, improper mating, broken locks, improperly formed or damaged terminals, poor terminal to wiring connections or physical damage to the wiring harness.

DIAGNOSTIC CHART DTC 67 — 1994 CAVALIER AND SUNBIRD

DTC 67
LEFT FRONT EMB CIRCUIT OPEN OR SHORTED TO GROUND

(1)
- IGNITION "OFF."
- DISCONNECT 24-WAY AND 2-WAY EBCM HARNESS CONNECTORS.
- DISCONNECT 3-WAY MOTOR PACK HARNESS CONNECTOR.
- USING J 39200, MEASURE RESISTANCE BETWEEN GROUND AND 24-WAY EBCM HARNESS CONNECTOR TERMINAL "19". DOES J 39200 DISPLAY "OL" (INFINITE)?

NO: REPAIR SHORT TO GROUND IN CKT 1286.

YES:

(2)
- USING J 39200, MEASURE RESISTANCE BETWEEN 3-WAY MOTOR PACK HARNESS CONNECTOR TERMINAL "A" AND 24-WAY EBCM HARNESS CONNECTOR TERMINAL "19". IS RESISTANCE LESS THAN 2 OHMS?

NO: REPAIR OPEN OR HIGH RESISTANCE IN CKT 1286.

YES:

(3)
- USING J 39200, MEASURE RESISTANCE BETWEEN 2-WAY EBCM HARNESS CONNECTOR TERMINAL "A" AND 3-WAY MOTOR PACK HARNESS CONNECTOR TERMINAL "B". IS RESISTANCE LESS THAN 2 OHMS?

NO: REPAIR OPEN OR HIGH RESISTANCE IN CKT 850.

YES:

(4)
- USING J 39200, MEASURE RESISTANCE BETWEEN MOTOR PACK TERMINAL "A" AND GROUND. DOES J 39200 DISPLAY "OL" (INFINITE)?

NO: REPLACE MOTOR PACK.

YES:

(5)
- USING J 39200, MEASURE RESISTANCE BETWEEN MOTOR PACK TERMINALS "A" AND "B". IS RESISTANCE LESS THAN 2 OHMS?

NO: REPLACE MOTOR PACK.

YES:

(6)
- USING J 39200, MEASURE RESISTANCE BETWEEN GROUND AND 2-WAY EBCM HARNESS CONNECTOR TERMINAL "A". DOES J 39200 DISPLAY "OL" (INFINITE)?

NO: REPAIR SHORT TO GROUND IN CKT 850.

YES:

(7)
- INSPECT 2-WAY AND 24-WAY EBCM AND 3-WAY MOTOR PACK CONNECTIONS FOR POOR TERMINAL CONTACT. REPLACE TERMINALS THAT EXHIBIT POOR CONTACT.
- INSPECT CKT 850, CKT 1286, 3-WAY MOTOR PACK AND EBCM HARNESS CONNECTORS FOR PHYSICAL DAMAGE WHICH MAY RESULT IN A SHORT TO GROUND WITH ALL CONNECTORS CONNECTED. REPAIR DAMAGE IF EVIDENT.
- RECONNECT ALL CONNECTORS.
- TEST DRIVE VEHICLE FOR THREE DRIVE CYCLES. A DRIVE CYCLE CONSISTS OF STARTING THE ENGINE, DRIVING OVER 16 km/h (10 MPH) AND THEN KEYING DOWN. DID DTC 67 SET IN THE LAST THREE DRIVE CYCLES?

YES: REPLACE EBCM.

NO: MALFUNCTION IS INTERMITTENT. REFER TO "DIAGNOSTIC AIDS".

* AFTER DIAGNOSIS IS COMPLETE, CLEAR DTCs AND TEST DRIVE VEHICLE FOR THREE (3) DRIVE CYCLES TO VERIFY DTC DOES NOT RESET. A DRIVE CYCLE CONSISTS OF STARTING THE VEHICLE, DRIVING OVER 16km/h (10 MPH) AND THEN KEYING DOWN.

THIS CHART ASSUMES THAT A CURRENT DTC IS STORED INDICATING THAT THIS MALFUNCTION IS PRESENT.

DIAGNOSTIC CHART DTC 68 — 1994
CAVALIER AND SUNBIRD

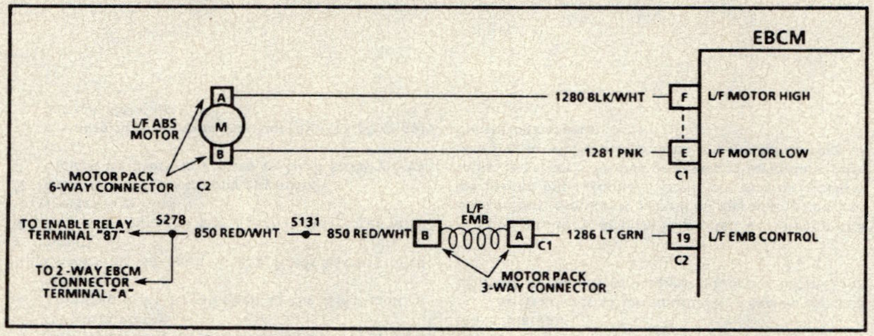

DTC 68

LEFT FRONT EMB CIRCUIT SHORTED TO BATTERY OR DRIVER OPEN

Circuit Description:

This DTC is designed to detect shorts to battery, shorts in the EMB release circuit, or an open EMB driver. These malfunctions will not allow the EMB to be energized (released) when needed.

DTC Will Set When: DTC 68 can be set only when the left front EMB is commanded "ON" (released). If an excessive voltage drop is across the motor driver for the left front EMB, a malfunction exists.

Action Taken: A malfunction DTC is stored, ABS is disabled and the ABS warning lamp is turned "ON."

DTC Chart Test Description: Number(s) below refer to circled number(s) on the diagnostic chart.

1. This step checks to see if the malfunction is currently present.
2. This step identifies a short to voltage in the EMB control circuitry.
3. This step checks for an EMB that has lower than normal resistance.
4. Checks for a short between the EMB battery feed and control circuits.
5. Ensures malfunction was not due to physical damage of the circuitry.

Diagnostic Aids: An "Intermittent" malfunction is most likely caused by a poor connection, rubbed through wire insulation, or a wire that is broken inside the insulation.

The frequency of the malfunction can be checked by using the enhanced diagnostic function of the Tech 1.

Any circuitry, that is suspected as causing the intermittent complaint, should be thoroughly checked for backed out terminals, improper mating, broken locks, improperly formed or damaged terminals, poor terminal to wiring connections or physical damage to the wiring harness.

DIAGNOSTIC CHART DTC 68 — 1994
CAVALIER AND SUNBIRD

DTC 68

LEFT FRONT EMB CIRCUIT SHORTED TO BATTERY OR DRIVER OPEN

*NOTICE: J 39200 TEST LEADS MUST BE "ZEROED" PRIOR TO MAKING ANY RESISTANCE MEASUREMENTS. REFER TO J 39200 USER'S MANUAL.

(1)
- IGNITION "ON."
- USING TECH 1 ABS TESTS, SELECT MANUAL CONTROL FUNCTION.
- SELECT LEFT FRONT MOTOR APPLY FUNCTION AND APPLY MOTOR.
- APPLY FIRM PRESSURE ON BRAKE PEDAL.
- SELECT LEFT FRONT EMB RELEASE FUNCTION AND RELEASE EMB.

DID BRAKE PEDAL FALL WITH INCREASING PEDAL FORCE?

- YES → MALFUNCTION IS NOT PRESENT AT THIS TIME. REFER TO "DIAGNOSTIC AIDS".
- NO → go to step 2.

(2) NOTICE: THIS STEP MAY CAUSE ADDITIONAL DTCs TO SET.
- IGNITION "OFF."
- DISCONNECT 3-WAY MOTOR PACK CONNECTOR.
- DISCONNECT 24-WAY EBCM CONNECTOR.
- IGNITION "ON."
- USING J 39200, MEASURE VOLTAGE BETWEEN GROUND AND 3-WAY MOTOR PACK HARNESS CONNECTOR TERMINAL "A".

IS VOLTAGE LESS THAN 1 VOLT?

- NO → REPAIR SHORT TO VOLTAGE IN CKT 1286.
- YES → go to step 3.

(3)
- IGNITION "OFF."
- USING J 39200, MEASURE RESISTANCE BETWEEN MOTOR PACK TERMINALS "A" AND "B".

IS RESISTANCE GREATER THAN .4 OHMS?

- NO → REPLACE MOTOR PACK.
- YES → go to step 4.

(4)
- USING J 39200, MEASURE THE RESISTANCE BETWEEN 3-WAY MOTOR PACK HARNESS CONNECTOR TERMINALS "B" AND "A".

DOES J 39200 DISPLAY "OL" (INFINITE)?

- NO → REPAIR SHORT BETWEEN CKT 850.
- YES → go to step 5.

(5)
- INSPECT CKT 1286, 3-WAY MOTOR PACK AND 24-WAY EBCM CONNECTORS FOR DAMAGE WHICH MAY RESULT IN A SHORT TO B+ WITH ALL CONNECTORS CONNECTED. REPAIR DAMAGE IF EVIDENT.
- RECONNECT ALL CONNECTORS.
- DRIVE VEHICLE FOR THREE DRIVE CYCLES (A DRIVE CYCLE CONSISTS OF STARTING ENGINE, DRIVING ABOVE 16 km/h (10 MPH) AND THEN TURNING IGNITION "OFF").
- IGNITION "ON."
- USING TECH 1, SELECT "F1: DTC HISTORY."

DID DTC 68 SET IN LAST THREE DRIVE CYCLES?

- YES → REPLACE EBCM.
- NO → MALFUNCTION IS INTERMITTENT. REFER TO "DIAGNOSTIC AIDS".

* AFTER DIAGNOSIS IS COMPLETE, CLEAR DTCs AND TEST DRIVE VEHICLE FOR THREE (3) DRIVE CYCLES TO VERIFY DTC DOES NOT RESET. A DRIVE CYCLE CONSISTS OF STARTING THE VEHICLE, DRIVING OVER 16 km/h (10 MPH) AND THEN KEYING DOWN.

THIS CHART ASSUMES THAT A CURRENT DTC IS STORED INDICATING THAT THIS MALFUNCTION IS PRESENT.

DIAGNOSTIC CHART DTC 71 — 1994 CAVALIER AND SUNBIRD

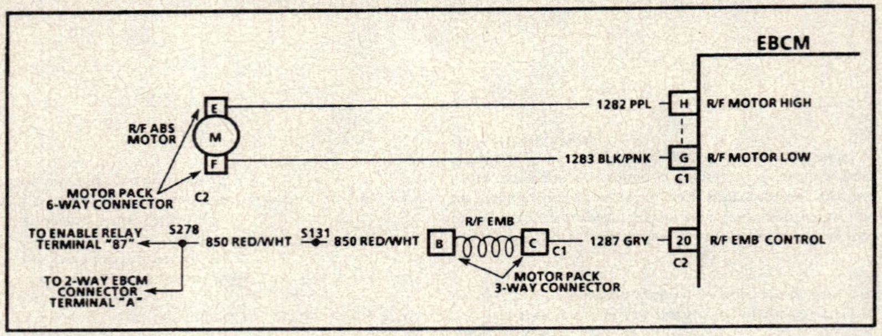

DTC 71

RIGHT FRONT EMB CIRCUIT OPEN OR SHORTED TO GROUND

Circuit Description:

This DTC is designed to detect shorts to ground and opens in the EMB release circuit.

A short to ground between the EMB and the ABS EBCM will cause the EMB to be energized indefinitely at an uncontrolled current causing the EMB to be destroyed.

DTC Will Set When: DTC 71 can be set only when the right front EMB is commanded "OFF" (not released). If the EBCM detects the right front EMB control circuit is constantly at ground, a malfunction exists.

Action Taken: A malfunction DTC is stored, ABS is disabled and the ABS warning lamp is turned "ON."

DTC Chart Test Description: Number(s) below refer to circled number(s) on the diagnostic chart.

1. This step checks for a short to ground in the EMB control circuitry.
2. This step looks for an open in the EMB control circuitry.
3. Checks for an open in the EMB battery feed circuit.
4. This step ensures the EMB is not internally shorted to ground.
5. Ensures EMB is not internally open.
6. Checks for short to ground in EMB battery feed circuit.
7. Ensures malfunction was not due to physical damage of the circuitry.

Diagnostic Aids: An "Intermittent" malfunction is most likely caused by a poor connection, rubbed through wire insulation, or a wire that is broken inside the insulation.

The frequency of the malfunction can be checked by using the enhanced diagnostic function of the Tech 1.

Any circuitry, that is suspected as causing the intermittent complaint, should be thoroughly checked for backed out terminals, improper mating, broken locks, improperly formed or damaged terminals, poor terminal to wiring connections or physical damage to the wiring harness.

DIAGNOSTIC CHART DTC 71 — 1994 CAVALIER AND SUNBIRD

DTC 71

RIGHT FRONT EMB CIRCUIT OPEN OR SHORTED TO GROUND

(1)
- IGNITION "OFF."
- DISCONNECT 24-WAY AND 2-WAY EBCM HARNESS CONNECTORS.
- DISCONNECT 3-WAY MOTOR PACK HARNESS CONNECTOR.
- USING J 39200, MEASURE RESISTANCE BETWEEN GROUND AND 24-WAY EBCM HARNESS CONNECTOR TERMINAL "20". DOES J 39200 DISPLAY "OL" (INFINITE)?

NO: REPAIR SHORT TO GROUND IN CKT 1287.

YES:

(2)
- USING J 39200, MEASURE RESISTANCE BETWEEN 3-WAY MOTOR PACK HARNESS CONNECTOR TERMINAL "A" AND 24-WAY EBCM HARNESS CONNECTOR TERMINAL "20". IS RESISTANCE LESS THAN 2 OHMS?

NO: REPAIR OPEN OR HIGH RESISTANCE IN CKT 1287.

YES:

(3)
- USING J 39200, MEASURE RESISTANCE BETWEEN 2-WAY EBCM HARNESS CONNECTOR TERMINAL "A" AND 3-WAY MOTOR PACK HARNESS CONNECTOR TERMINAL "B". IS RESISTANCE LESS THAN 2 OHMS?

NO: REPAIR OPEN OR HIGH RESISTANCE IN CKT 850.

YES:

(4)
- USING J 39200, MEASURE RESISTANCE BETWEEN MOTOR PACK TERMINAL "C" AND GROUND. DOES J 39200 DISPLAY "OL" (INFINITE)?

NO: REPLACE MOTOR PACK.

YES:

(5)
- USING J 39200, MEASURE RESISTANCE BETWEEN MOTOR PACK TERMINALS "C" AND "B". IS RESISTANCE LESS THAN 2 OHMS?

NO: REPLACE MOTOR PACK.

YES:

(6)
- USING J 39200, MEASURE RESISTANCE BETWEEN GROUND AND 2-WAY EBCM HARNESS CONNECTOR TERMINAL "A". DOES J 39200 DISPLAY "OL" (INFINITE)?

NO: REPAIR SHORT TO GROUND IN CKT 850.

YES:

(7)
- INSPECT 2-WAY AND 24-WAY EBCM AND 3-WAY MOTOR PACK CONNECTIONS FOR POOR TERMINAL CONTACT. REPLACE TERMINALS THAT EXHIBIT POOR CONTACT.
- INSPECT CKT 850, CKT 1287, 3-WAY MOTOR PACK AND EBCM HARNESS CONNECTORS FOR PHYSICAL DAMAGE WHICH MAY RESULT IN A SHORT TO GROUND WITH ALL CONNECTORS CONNECTED. REPAIR DAMAGE IF EVIDENT.
- RECONNECT ALL CONNECTORS.
- TEST DRIVE VEHICLE FOR THREE DRIVE CYCLES. A DRIVE CYCLE CONSISTS OF STARTING THE ENGINE, DRIVING OVER 16 km/h (10 MPH) AND THEN KEYING DOWN. DID DTC 71 SET THE LAST THREE DRIVE CYCLES?

YES: REPLACE EBCM.

NO: MALFUNCTION IS INTERMITTENT. REFER TO "DIAGNOSTIC AIDS".

- AFTER DIAGNOSIS IS COMPLETE, CLEAR DTCs AND TEST DRIVE VEHICLE FOR THREE (3) DRIVE CYCLES TO VERIFY DTC DOES NOT RESET. A DRIVE CYCLE CONSISTS OF STARTING THE VEHICLE, DRIVING OVER 16km/h (10 MPH) AND THEN KEYING DOWN.

THIS CHART ASSUMES THAT A CURRENT DTC IS STORED INDICATING THAT THIS MALFUNCTION IS PRESENT.

DIAGNOSTIC CHART DTC 72 — 1994 CAVALIER AND SUNBIRD

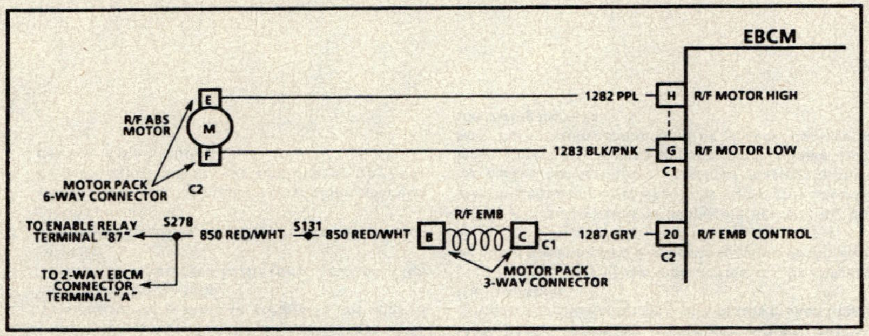

DTC 72

RIGHT FRONT EMB CIRCUIT SHORTED TO BATTERY OR DRIVER OPEN

Circuit Description:

This DTC is designed to detect shorts to battery, shorts in the EMB release circuit, or an open EMB driver. These malfunctions will not allow the EMB to be energized (released) when needed.

DTC Will Set When: DTC 72 can be set only when the right front EMB is commanded "ON" (released). If an excessive voltage drop is across the motor driver for the right front EMB, a malfunction exists.

Action Taken: A malfunction DTC is stored, ABS is disabled and the ABS warning lamp is turned "ON."

DTC Chart Test Description: Number(s) below refer to circled number(s) on the diagnostic chart.

1. This step checks to see if the malfunction is currently present.
2. This step identifies a short to voltage in the EMB control circuitry.
3. This step checks for an EMB that has lower than normal resistance.
4. Checks for a short between the EMB battery feed and control circuits.
5. Ensures malfunction was not due to physical damage of the circuitry.

Diagnostic Aids: An "Intermittent" malfunction is most likely caused by a poor connection, rubbed through wire insulation, or a wire that is broken inside the insulation.

The frequency of the malfunction can be checked by using the enhanced diagnostic function of the Tech 1.

Any circuitry, that is suspected as causing the intermittent complaint, should be thoroughly checked for backed out terminals, improper mating, broken locks, improperly formed or damaged terminals, poor terminal to wiring connections or physical damage to the wiring harness.

DIAGNOSTIC CHART DTC 72 — 1994 CAVALIER AND SUNBIRD

DTC 72

RIGHT FRONT EMB CIRCUIT SHORTED TO BATTERY OR DRIVER OPEN

*NOTICE: J 39200 TEST LEADS MUST BE "ZEROED" PRIOR TO MAKING ANY RESISTANCE MEASUREMENTS. REFER TO J 39200 USER'S MANUAL.

(1)
- IGNITION "ON."
- USING TECH 1 ABS TESTS, SELECT MANUAL CONTROL FUNCTION.
- SELECT RIGHT FRONT MOTOR APPLY FUNCTION AND APPLY MOTOR.
- APPLY FIRM PRESSURE ON BRAKE PEDAL.
- SELECT RIGHT FRONT EMB RELEASE FUNCTION AND RELEASE EMB. DID BRAKE PEDAL FALL WITH INCREASING PEDAL FORCE?

YES: MALFUNCTION IS NOT PRESENT AT THIS TIME. REFER TO "DIAGNOSTIC AIDS".

NO: (2) NOTICE: THIS STEP MAY CAUSE ADDITIONAL DTCs TO SET.
- IGNITION "OFF."
- DISCONNECT 3-WAY MOTOR PACK CONNECTOR.
- DISCONNECT 24-WAY EBCM CONNECTOR.
- IGNITION "ON."
- USING J 39200, MEASURE VOLTAGE BETWEEN GROUND AND 3-WAY MOTOR PACK HARNESS CONNECTOR TERMINAL "C". IS VOLTAGE LESS THAN 1 VOLT?

NO: REPAIR SHORT TO VOLTAGE IN CKT 1287.

YES: (3)
- IGNITION "OFF."
- USING J 39200, MEASURE RESISTANCE BETWEEN MOTOR PACK TERMINALS "C" AND "B". IS RESISTANCE GREATER THAN .4 OHMS?

NO: REPLACE MOTOR PACK.

YES: (4)
- USING J 39200, MEASURE RESISTANCE BETWEEN 3-WAY MOTOR PACK HARNESS CONNECTOR TERMINALS "B" AND "C". DOES J 39200 DISPLAY "OL" (INFINITE)?

NO: REPAIR SHORT BETWEEN CKT 850.

YES: (5)
- INSPECT CKT 1287, 3-WAY MOTOR PACK AND 24-WAY EBCM CONNECTORS FOR DAMAGE WHICH MAY RESULT IN A SHORT TO B + WITH ALL CONNECTORS CONNECTED. REPAIR DAMAGE IF EVIDENT.
- RECONNECT ALL CONNECTORS.
- DRIVE VEHICLE FOR THREE DRIVE CYCLES (A DRIVE CYCLE CONSISTS OF STARTING ENGINE, DRIVING ABOVE 16 KM/H (10 MPH) AND THEN TURNING IGNITION "OFF").
- IGNITION "ON."
- USING TECH 1, SELECT "F1: DTC HISTORY." DID DTC 72 SET IN LAST THREE DRIVE CYCLES?

YES: REPLACE EBCM.

NO: MALFUNCTION IS INTERMITTENT. REFER TO "DIAGNOSTIC AIDS".

* AFTER DIAGNOSIS IS COMPLETE, CLEAR DTCs AND TEST DRIVE VEHICLE FOR THREE (3) DRIVE CYCLES TO VERIFY DTC DOES NOT RESET. A DRIVE CYCLE CONSISTS OF STARTING THE VEHICLE, DRIVING OVER 16 km/h (10 MPH) AND THEN KEYING DOWN.

THIS CHART ASSUMES THAT A CURRENT DTC IS STORED INDICATING THAT THIS MALFUNCTION IS PRESENT.

DIAGNOSTIC CHART DTC 76 — 1994 CAVALIER AND SUNBIRD

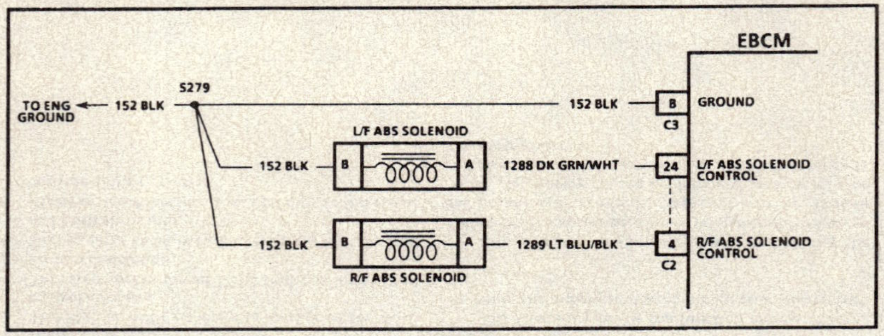

DTC 76

LEFT FRONT SOLENOID CIRCUIT OPEN OR SHORTED TO BATTERY

Circuit Description:

This DTC identifies a solenoid that cannot be energized due to an open in its circuitry, or a solenoid that is always energized due to a short to battery in its circuitry between the driver and the solenoid. An open will not allow proper ABS operation, but the short to battery simply turns "ON" the solenoid. A path for base brakes is still allowed once the motor re-homes and the check valve is lifted off its seat during key "ON" initialization.

DTC Will Set When: DTC 76 can be set only when the solenoid is commanded "OFF." If the EBCM detects an excessive voltage in the left front solenoid control circuit, a malfunction exists.

Action Taken: A malfunction DTC is stored, ABS is disabled and the ABS warning lamp is turned "ON."

DTC Chart Test Description: Number(s) below refer to circled number(s) on the diagnostic chart.

1. This checks for a short to voltage in the control circuitry of the solenoid.
2. This checks for an open in the control circuit of the solenoid.
3. This step identifies an open in the solenoid or ground circuit.
4. Checks for proper resistance values of the solenoid.
5. Ensures malfunction was not due to physical damage of the circuitry.

Diagnostic Aids: An "Intermittent" malfunction is most likely caused by a poor connection, rubbed through wire insulation, or a wire that is broken inside the insulation.

The frequency of the malfunction can be checked by using the enhanced diagnostic function of the Tech 1.

Any circuitry, that is suspected as causing the intermittent complaint, should be thoroughly checked for backed out terminals, improper mating, broken locks, improperly formed or damaged terminals, poor terminal to wiring connections or physical damage to the wiring harness.

DIAGNOSTIC CHART DTC 76 — 1994 CAVALIER AND SUNBIRD

DTC 76

LEFT FRONT SOLENOID CIRCUIT OPEN OR SHORTED TO BATTERY

*NOTICE: J 39200 TEST LEADS MUST BE "ZEROED" PRIOR TO MAKING ANY RESISTANCE MEASUREMENTS. REFER TO J 39200 USER'S MANUAL.

(1)
- IGNITION "OFF."
- DISCONNECT 24-WAY EBCM CONNECTOR.
- IGNITION "ON."
- USING J 39200, MEASURE VOLTAGE BETWEEN GROUND AND 24-WAY EBCM HARNESS CONNECTOR TERMINAL "24".

IS VOLTAGE LESS THAN 1 VOLT?

NO: REPAIR SHORT TO VOLTAGE IN CKT 1288.

YES:

(2)
- IGNITION "OFF."
- DISCONNECT LEFT FRONT SOLENOID CONNECTOR.
- USING J 39200, MEASURE RESISTANCE BETWEEN LEFT FRONT SOLENOID HARNESS CONNECTOR TERMINAL "A" AND 24-WAY EBCM HARNESS CONNECTOR TERMINAL "24".

IS RESISTANCE LESS THAN 2 OHMS?

NO: REPAIR OPEN OR HIGH RESISTANCE IN CKT 1288.

YES:

(3)
- USING J 39200, MEASURE RESISTANCE BETWEEN GROUND AND LEFT FRONT SOLENOID HARNESS CONNECTOR TERMINAL "B".

IS RESISTANCE LESS THAN 2 OHMS?

NO: REPAIR OPEN OR HIGH RESISTANCE IN CKT 152.

YES:

(4)
- USING J 39200, MEASURE RESISTANCE BETWEEN LEFT FRONT SOLENOID TERMINALS "A" AND "B".

IS RESISTANCE BETWEEN 2.5 AND 5 OHMS?

NO: REPLACE LEFT FRONT SOLENOID.

YES:

(5)
- INSPECT LEFT FRONT SOLENOID AND 24-WAY EBCM HARNESS CONNECTORS FOR POOR TERMINAL CONTACT OR CORROSION. REPLACE TERMINALS THAT EXHIBIT POOR CONTACT OR CORROSION.
- INSPECT CKT 1288 FOR DAMAGE WHICH MAY RESULT IN A SHORT TO B+ WITH ALL CONNECTORS CONNECTED. REPAIR DAMAGE IF EVIDENT.
- RECONNECT ALL CONNECTORS.
- IGNITION "ON."

IS DTC 76 SET AS A CURRENT DTC?

YES: REPLACE EBCM.

NO: REFER TO "DIAGNOSTIC AIDS"

THIS CHART ASSUMES THAT A CURRENT DTC IS STORED INDICATING THAT THIS MALFUNCTION IS PRESENT.

DIAGNOSTIC CHART DTC 77 — 1994 CAVALIER AND SUNBIRD

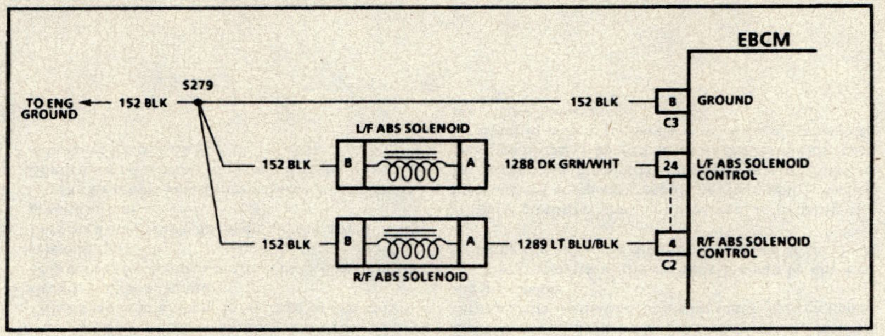

DTC 77

LEFT FRONT SOLENOID CIRCUIT SHORTED TO GROUND OR DRIVER OPEN

Circuit Description:

This DTC identifies a solenoid that cannot be energized due to an open in its driver circuitry, or a short to ground between the solenoid driver and the solenoid. These malfunctions can affect ABS operation since the flow of brake fluid to the wheel cylinder cannot be stopped, making ABS operation for that channel impossible.

DTC Will Set When: DTC 77 can be set only when the solenoid is commanded "ON." If the EBCM detects the left front solenoid control circuit voltage is out of specification, a malfunction exists.

Action Taken: A malfunction DTC is stored, ABS is disabled and the ABS warning lamp is turned "ON."

DTC Chart Test Description: Number(s) below refer to circled number(s) on the diagnostic chart.

1. This step checks for a solenoid that is internally shorted to ground.
2. This step checks for a solenoid that is not within proper resistance values.
3. This step indicates if a short to ground exists in the solenoid control circuitry.
4. Ensures malfunction was not due to physical damage of the circuitry.

Diagnostic Aids: An "Intermittent" malfunction is most likely caused by a poor connection, rubbed through wire insulation, or a wire that is broken inside the insulation.

The frequency of the malfunction can be checked by using the enhanced diagnostic function of the Tech 1.

Any circuitry, that is suspected as causing the intermittent complaint, should be thoroughly checked for backed out terminals, improper mating, broken locks, improperly formed or damaged terminals, poor terminal to wiring connections or physical damage to the wiring harness.

DIAGNOSTIC CHART DTC 77 — 1994 CAVALIER AND SUNBIRD

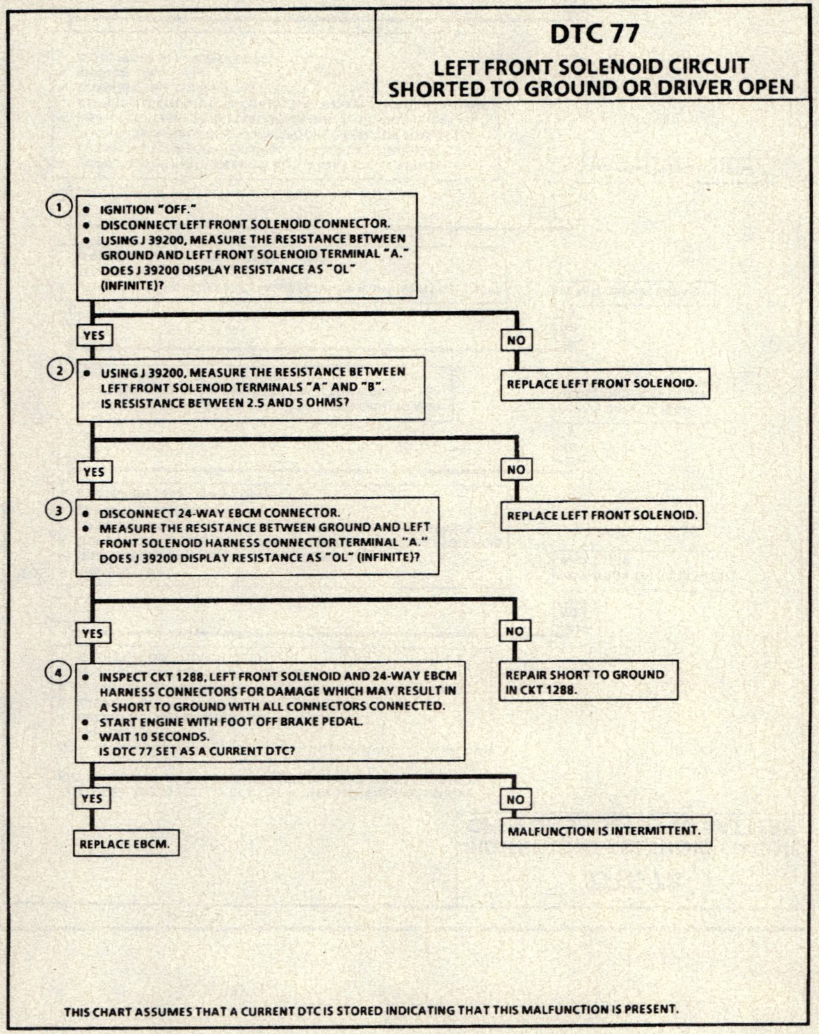

DIAGNOSTIC CHART DTC 78 — 1994
CAVALIER AND SUNBIRD

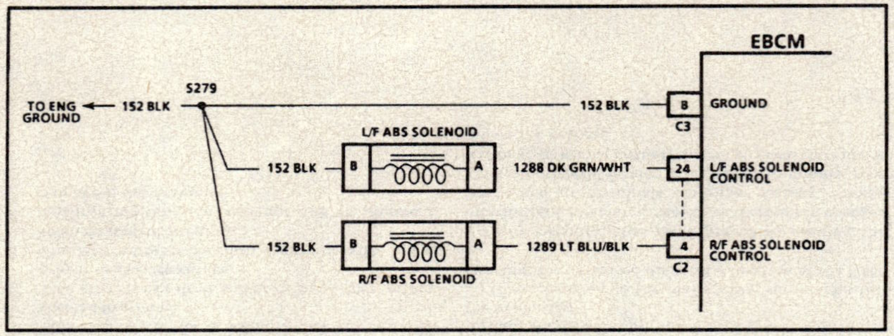

DTC 78

RIGHT FRONT SOLENOID CIRCUIT OPEN OR SHORTED TO BATTERY

Circuit Description:

This DTC identifies a solenoid that cannot be energized due to an open in its circuitry, or a solenoid that is always energized due to a short to battery in its circuitry between the driver and the solenoid. An open will not allow proper ABS operation but the short to battery simply turns "ON" the solenoid. A path for base brakes is still allowed once the motor re-homes and the check valve is lifted off its seat during key "ON" initialization.

DTC Will Set When: DTC 78 can be set only when the solenoid is commanded "OFF." If the EBCM detects an excessive voltage in the right front solenoid control circuit, a malfunction exists.

Action Taken: A malfunction DTC is stored, ABS is disabled and the ABS warning lamp is turned "ON."

DTC Chart Test Description: Number(s) below refer to circled number(s) on the diagnostic chart.

1. This checks for a short to voltage in the control circuitry of the solenoid.
2. This checks for an open in the control circuit of the solenoid.
3. This step identifies an open in the solenoid or ground circuit.
4. Checks for proper resistance values of the solenoid.
5. Ensures malfunction was not due to physical damage of the circuitry.

Diagnostic Aids: An "Intermittent" malfunction is most likely caused by a poor connection, rubbed through wire insulation, or a wire that is broken inside the insulation.

The frequency of the malfunction can be checked by using the enhanced diagnostic function of the Tech 1.

Any circuitry, that is suspected as causing the intermittent complaint, should be thoroughly checked for backed out terminals, improper mating, broken locks, improperly formed or damaged terminals, poor terminal to wiring connections or physical damage to the wiring harness.

DIAGNOSTIC CHART DTC 78 — 1994
CAVALIER AND SUNBIRD

DTC 78
RIGHT FRONT SOLENOID CIRCUIT OPEN OR SHORTED TO BATTERY

*NOTICE: J39200 TEST LEADS MUST BE "ZEROED" PRIOR TO MAKING ANY RESISTANCE MEASUREMENTS. REFER TO J39200 USER'S MANUAL.

(1)
- IGNITION "OFF."
- DISCONNECT 24-WAY EBCM CONNECTOR.
- IGNITION "ON."
- USING J 39200, MEASURE VOLTAGE BETWEEN GROUND AND 24-WAY EBCM HARNESS CONNECTOR TERMINAL "4".

IS VOLTAGE LESS THAN 1 VOLT?

YES → (2)
NO → REPAIR SHORT TO VOLTAGE IN CKT 1289.

(2)
- IGNITION "OFF."
- DISCONNECT RIGHT FRONT SOLENOID CONNECTOR.
- USING J 39200, MEASURE RESISTANCE BETWEEN RIGHT FRONT SOLENOID HARNESS CONNECTOR TERMINAL "A" AND 24-WAY EBCM HARNESS CONNECTOR TERMINAL "4".

IS RESISTANCE LESS THAN 2 OHMS?

YES → (3)
NO → REPAIR OPEN OR HIGH RESISTANCE IN CKT 1289.

(3)
- USING J39200, MEASURE RESISTANCE BETWEEN GROUND AND RIGHT FRONT SOLENOID HARNESS CONNECTOR TERMINAL "B".

IS RESISTANCE LESS THAN 2 OHMS?

YES → (4)
NO → REPAIR OPEN IN CKT 152.

(4)
- USING J39200, MEASURE RESISTANCE BETWEEN RIGHT FRONT SOLENOID TERMINALS "A" AND "B".

IS RESISTANCE BETWEEN 2.5 AND 5 OHMS?

YES → (5)
NO → REPLACE RIGHT FRONT SOLENOID.

(5)
- INSPECT RIGHT FRONT SOLENOID AND 24-WAY EBCM HARNESS CONNECTORS FOR POOR TERMINAL CONTACT OR CORROSION. REPLACE TERMINALS THAT EXHIBIT POOR CONTACT OR CORROSION.
- INSPECT CKT 1289 FOR DAMAGE WHICH MAY RESULT IN A SHORT TO B+ WITH ALL CONNECTORS CONNECTED. REPAIR DAMAGE IF EVIDENT.
- RECONNECT ALL CONNECTORS.
- IGNITION "ON."

IS DTC 78 SET AS A CURRENT DTC?

YES → REPLACE EBCM.
NO → MALFUNCTION IS INTERMITTENT.

THIS CHART ASSUMES THAT A CURRENT DTC IS STORED INDICATING THAT THIS MALFUNCTION IS PRESENT.

DIAGNOSTIC CHART DTC 81 — 1994 CAVALIER AND SUNBIRD

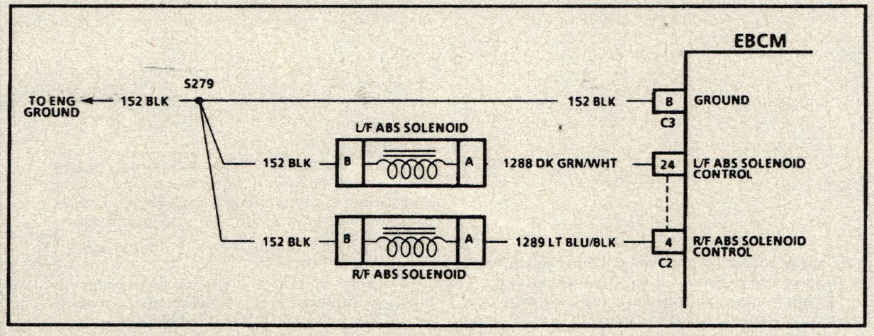

DTC 81

RIGHT FRONT SOLENOID CIRCUIT SHORTED TO GROUND OR DRIVER OPEN

Circuit Description:
This DTC identifies a solenoid that cannot be energized due to an open in its driver circuitry, or a short to ground between the solenoid driver and the solenoid. These malfunctions can affect ABS operation since the flow of brake fluid to the wheel cylinder cannot be stopped, making ABS operation for that channel impossible.

DTC Will Set When: DTC 81 can be set only when the solenoid is commanded "ON." If the EBCM detects the right front solenoid control circuit voltage is out of specification, a malfunction exists.

Action Taken: A malfunction DTC is stored, ABS is disabled and the ABS warning lamp is turned "ON."

DTC Chart Test Description: Number(s) below refer to circled number(s) on the diagnostic chart.

1. This step checks for a solenoid that is internally shorted to ground.
2. This step checks for a solenoid that is not within proper resistance values.
3. This step indicates if a short to ground exists in the solenoid control circuitry.
4. Ensures malfunction was not due to physical damage of the circuitry.

Diagnostic Aids: An "Intermittent" malfunction is most likely caused by a poor connection, rubbed through wire insulation, or a wire that is broken inside the insulation.

The frequency of the malfunction can be checked by using the enhanced diagnostic function of the Tech 1.

Any circuitry, that is suspected as causing the intermittent complaint, should be thoroughly checked for backed out terminals, improper mating, broken locks, improperly formed or damaged terminals, poor terminal to wiring connections or physical damage to the wiring harness.

DIAGNOSTIC CHART DTC 81 — 1994 CAVALIER AND SUNBIRD

DTC 81

RIGHT FRONT SOLENOID CIRCUIT SHORTED TO GROUND OR DRIVER OPEN

(1)
- IGNITION "OFF."
- DISCONNECT RIGHT FRONT SOLENOID CONNECTOR.
- USING J 39200, MEASURE THE RESISTANCE BETWEEN GROUND AND RIGHT FRONT SOLENOID TERMINAL "A".
  DOES J 39200 DISPLAY RESISTANCE AS "OL" (INFINITE)?

NO → REPLACE RIGHT FRONT SOLENOID.

YES ↓

(2)
- USING J 39200, MEASURE RESISTANCE BETWEEN RIGHT FRONT SOLENOID TERMINALS "A" AND "B".
  IS RESISTANCE BETWEEN 2.5 AND 5 OHMS?

NO → REPLACE RIGHT FRONT SOLENOID.

YES ↓

(3)
- DISCONNECT 24-WAY EBCM CONNECTOR.
- MEASURE THE RESISTANCE BETWEEN GROUND AND RIGHT FRONT SOLENOID HARNESS CONNECTOR TERMINAL "A".
  DOES J 39200 DISPLAY RESISTANCE AS "OL" (INFINITE)?

NO → REPAIR SHORT TO GROUND IN CKT 1289.

YES ↓

(4)
- INSPECT CKT 1289, RIGHT FRONT SOLENOID AND 24-WAY EBCM HARNESS CONNECTORS FOR DAMAGE WHICH MAY RESULT IN A SHORT TO GROUND WITH ALL CONNECTORS CONNECTED.
- START ENGINE WITH FOOT OFF BRAKE PEDAL.
- WAIT 10 SECONDS.
  IS DTC 81 SET AS A CURRENT DTC?

YES → REPLACE EBCM.

NO → MALFUNCTION IS INTERMITTENT.

THIS CHART ASSUMES THAT A CURRENT DTC IS STORED INDICATING THAT THIS MALFUNCTION IS PRESENT.

DIAGNOSTIC CHART DTC 82 — 1994 CAVALIER AND SUNBIRD

D.M. A.B.S. VI MODULE ASSEMBLY - ELECTRONIC BRAKE CONTROL (TYPICAL)

DTC 82
CALIBRATION MALFUNCTION

Circuit Description:

This DTC allows the EBCM to check for a calibration failure by comparing the calibration value to a known value stored in the EEPROM.

This test is also used as a security measure to prevent improper use of calibrations or changes to these calibrations that may alter the designed function of ABS.

DTC Will Set When: DTC 82 can be set at key-up. If the program identifier is incorrect or the memory checksum is incorrect, a malfunction exists.

Action Taken: A malfunction DTC is stored, ABS is disabled and the ABS warning lamp is turned "ON."

DTC Chart Test Description: Number(s) below refer to circled number(s) on the diagnostic chart.

1. This step checks to see if the malfunction is present during diagnosis. If present, the EBCM is not functioning correctly and must be replaced.

Important:

- At the time of printing, 1994 vehicles with ABS VI were not being field reprogrammed to correct this failure. A service bulletin will be issued when EEPROM reprogramming is authorized.

Diagnostic Aids: An intermittent DTC 82 is most likely caused by a bad cell in the EEPROM that is sensitive to temperature changes. If DTC 82 occurred more than once, but is intermittent, replace EBCM.

The frequency of the malfunction can be checked by using the enhanced diagnostic function of the Tech 1.

DIAGNOSTIC CHART DTC 82 — 1994 CAVALIER AND SUNBIRD

DTC 82
CALIBRATION MALFUNCTION

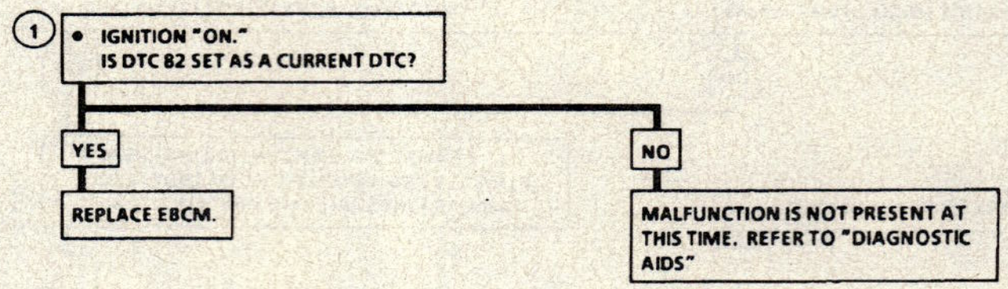

THIS CHART ASSUMES THAT A CURRENT DTC IS STORED INDICATING THAT THIS MALFUNCTION IS PRESENT.

DIAGNOSTIC CHART DTC 86 — 1994 CAVALIER AND SUNBIRD

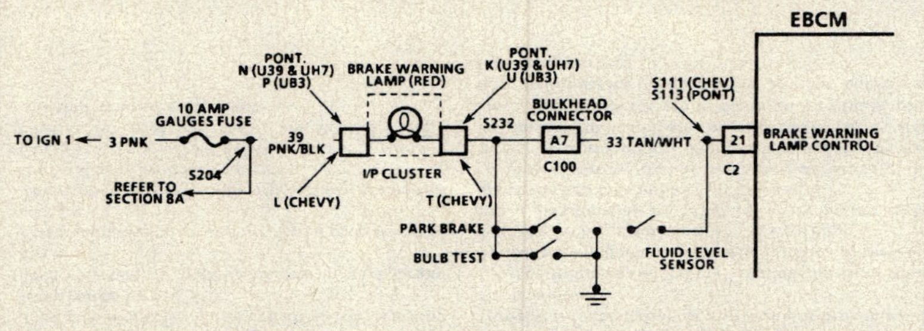

DTC 86

EBCM TURNED "ON" THE RED "BRAKE" WARNING LAMP

Circuit Description:

This DTC is provided as an information only test, and reflects the status of the command issued by the ABS EBCM to illuminate the red "BRAKE" warning lamp. If another diagnostic DTC issues a command to illuminate the red "BRAKE" warning lamp, DTC 86 will be stored in EEPROM as a history DTC at the conclusion of the ignition cycle.

DTC Chart Test Description: Number(s) below refer to circled number(s) on the diagnostic chart.

1. This step serves to identify if a DTC other than DTC 86 commanded the red "BRAKE" warning lamp "ON."

Diagnostic Aids: Any ABS mechanical DTC that issues a command to illuminate the red "BRAKE" warning lamp will also result in DTC 86 being stored in EEPROM during shut down. These DTCs are: 38, 41, 42, 45, 46, 47, 48 and 51. If the motors are not in their home position, certain electrical DTCs will also command the red "BRAKE" warning lamp "ON." These DTCs are: 14, 18, 55, 56, 57, 58, 61, 62, 63, 64, 65, 66, 67, 68, 71 and 72.

If any of these DTCs are indicated along with DTC 86, they must be corrected prior to addressing a DTC 86 malfunction.

DIAGNOSTIC CHART DTC 86 — 1994 CAVALIER AND SUNBIRD

DTC 86

EBCM TURNED "ON" THE RED "BRAKE" WARNING LAMP

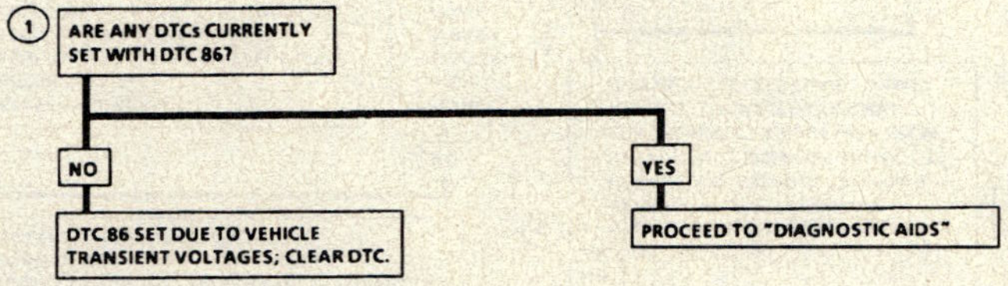

DIAGNOSTIC CHART DTC 87 — 1994 CAVALIER AND SUNBIRD

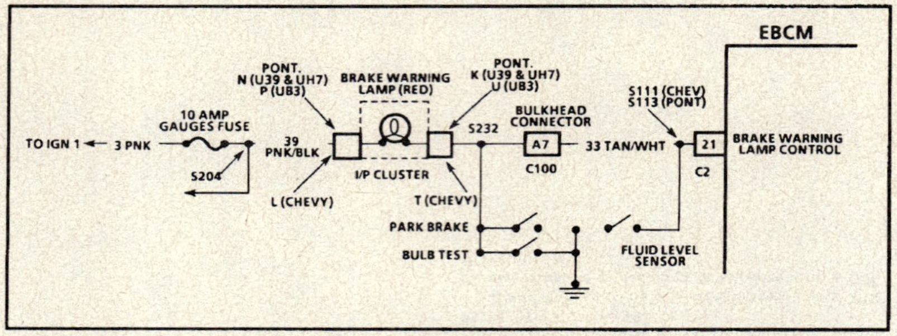

DTC 87

RED "BRAKE" WARNING LAMP CIRCUIT OPEN

Circuit Description:

This DTC is used to verify the EBCM has continuity to the "BRAKE" warning lamp in case the EBCM must turn it "ON." This will only occur if an ABS malfunction is detected that may degrade base brake operation.

Because ABS is not the only device controlling the "BRAKE" warning lamp (parking brake and low fluid sensor may also turn "ON" the lamp) a short to ground in this circuit can not necessarily be detected. Excessive resistance in the fluid level switch or parking brake circuitry may set this DTC.

DTC Will Set When: DTC 87 can be set after initialization is completed. If the red "BRAKE" warning lamp circuit voltage is out of specification indicating an open circuit, a malfunction exists.

Action Taken: A malfunction DTC is stored, however ABS is not disabled.

DTC Chart Test Description: Number(s) below refer to circled number(s) on the diagnostic chart.

1. This step identifies if the red "BRAKE" warning lamp can be turned "ON."
2. This step test for high resistance in park brake circuitry.
3. Ensures malfunction was not due to poor terminal contact.
4. This step identifies if high resistance is present in the fluid level switch circuitry.
5. Checks integrity of red "BRAKE" warning bulb.
6. Checks for high resistance in the "BRAKE" warning lamp control circuit.

Diagnostic Aids: An "Intermittent" malfunction is most likely caused by a poor connection, rubbed through wire insulation, or a wire that is broken inside the insulation.

The frequency of the malfunction can be checked by using the enhanced diagnostic function of the Tech 1. If the Tech 1 is not available, the parking brake handle may be lifted to verify proper lamp operation and continuity of CKTs 3 and 39.

Any circuitry, that is suspected as causing the intermittent complaint, should be thoroughly checked for backed out terminals, improper mating, broken locks, improperly formed or damaged terminals, poor terminal to wiring connections or physicai damage to the wiring harness.

DIAGNOSTIC CHART DTC 87 — 1994 CAVALIER AND SUNBIRD

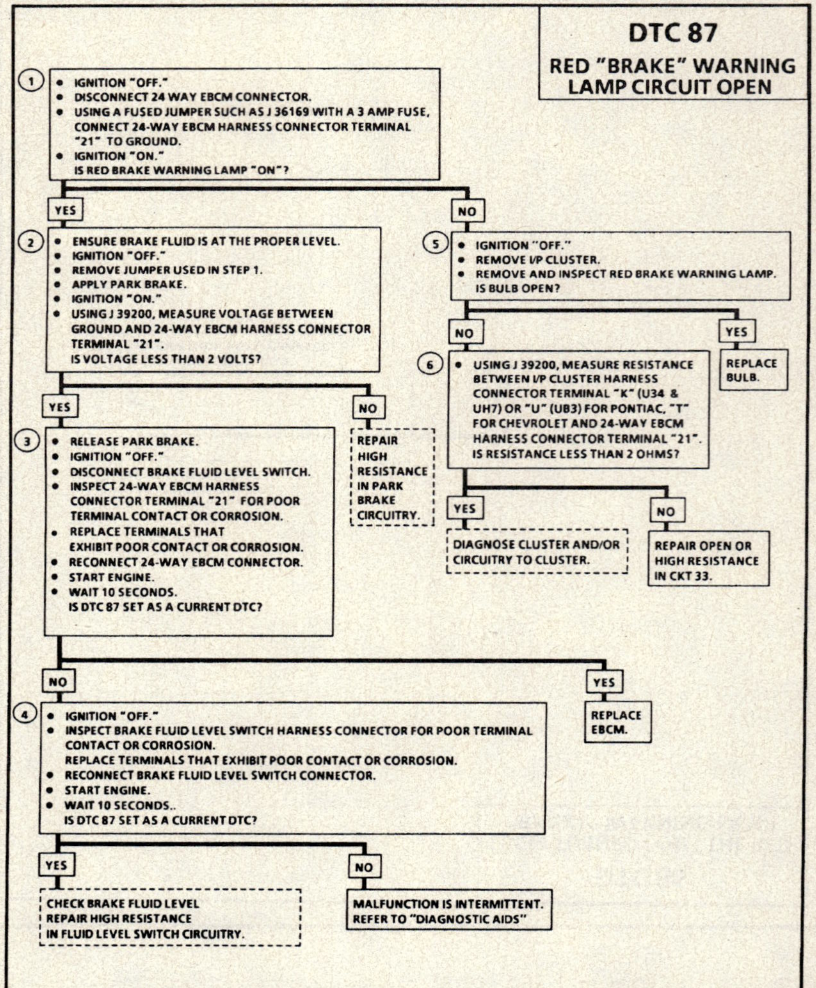

THIS CHART ASSUMES THAT A CURRENT DTC IS STORED INDICATING THAT THIS MALFUNCTION IS PRESENT.

DIAGNOSTIC CHART DTC 88 — 1994 CAVALIER AND SUNBIRD

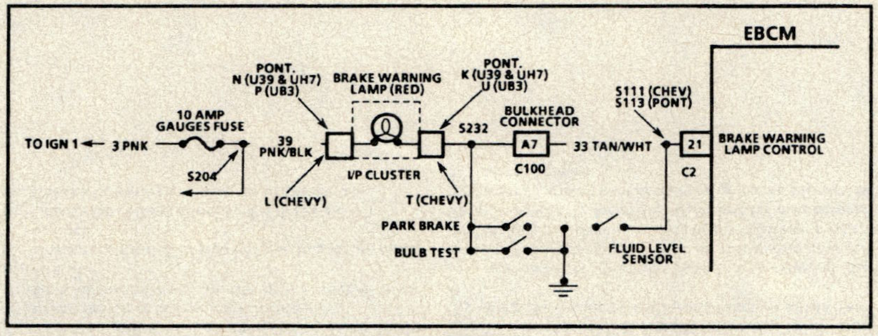

DTC 88

RED "BRAKE" WARNING LAMP CIRCUIT SHORTED TO BATTERY

Circuit Description:

This DTC identifies a short to battery between the EBCM and the red "BRAKE" warning lamp, or an open driver that does not allow the red "BRAKE" warning lamp to be illuminated by the EBCM. This will only occur if an ABS malfunction is detected that may degrade base brake operation.

DTC Will Set When: DTC 88 can be set only at the beginning of the three second bulb check. If the EBCM detects battery voltage on the red "BRAKE" warning lamp control circuit, a malfunction exists.

Action Taken: A malfunction DTC is stored. ABS is not disabled; however, the ABS warning lamp will flash to indicate a malfunction exists.

DTC Chart Test Description: Number(s) below refer to circled number(s) on the diagnostic chart.

1. This step identifies if the ground circuit to the red "BRAKE" warning lamp is being completed by a source other than the EBCM.
2. This step identifies if the malfunction is currently present.
3. By removing the 10 amp gauge fuse, the voltage source is eliminated. This test indicates if voltage is being supplied from a source other than the gauge fuse.
4. Ensures malfunction was not due to physical damage of the circuitry.

Diagnostic Aids: An "Intermittent" malfunction is most likely caused by a poor connection, rubbed through wire insulation, or a wire that is broken inside the insulation.

The frequency of the malfunction can be checked by using the enhanced diagnostic function of the Tech 1.

Any circuitry, that is suspected as causing the intermittent complaint, should be thoroughly checked for backed out terminals, improper mating, broken locks, improperly formed or damaged terminals, poor terminal to wiring connections or physical damage to the wiring harness.

DIAGNOSTIC CHART DTC 88 — 994 CAVALIER AND SUNBIRD

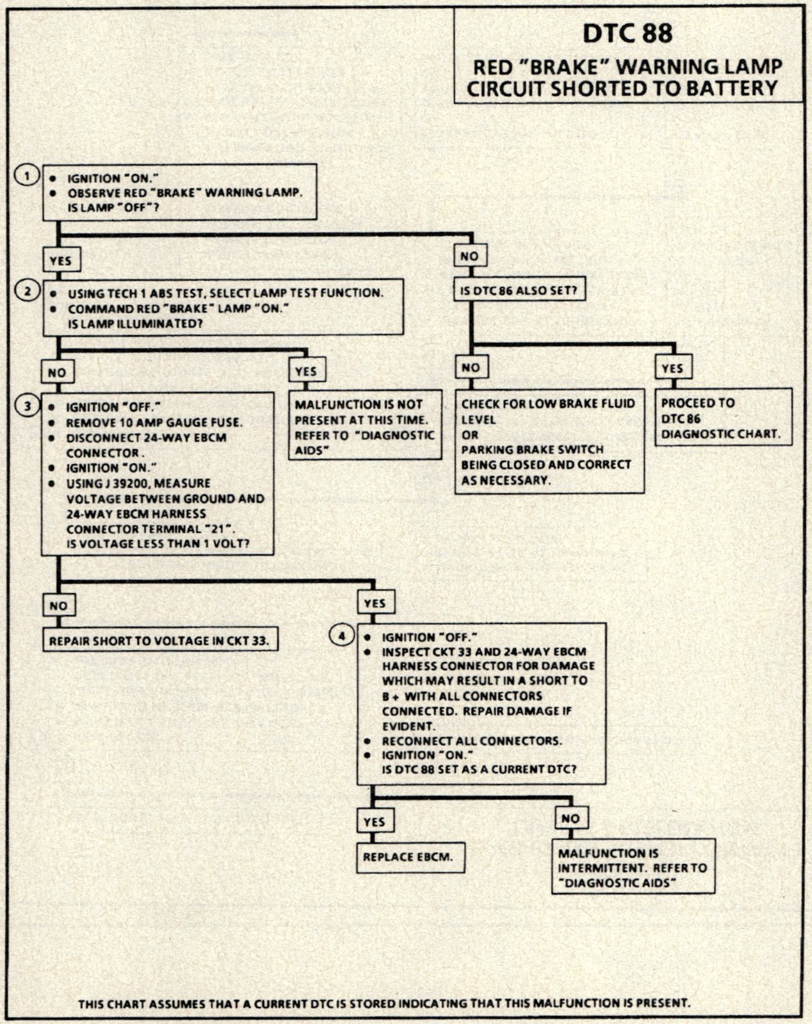

DIAGNOSTIC CHART DTC 91 — 1994 CAVALIER AND SUNBIRD

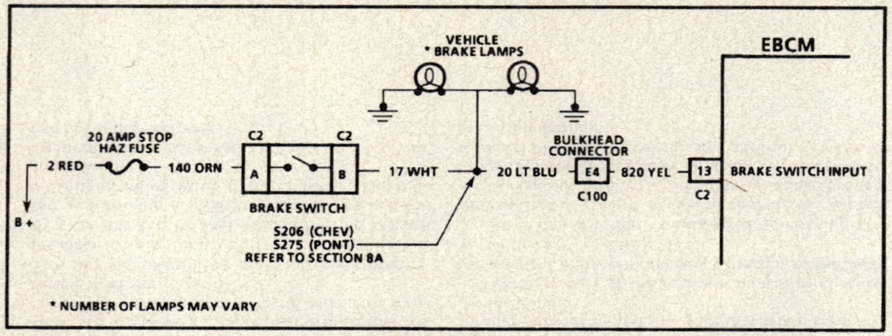

DTC 91

OPEN BRAKE SWITCH CONTACTS DURING DECELERATION

Circuit Description:

This DTC is used to detect an open brake switch in the non-ABS mode. The EBCM looks for decel rates that would indicate braking action and verifies this assumption by requiring several repeats of this detection method. In each case, ABS will not be available since no brake switch is seen.

DTC Will Set When: DTC 91 can be set if three deceleration cycles occur with the brake switch "OFF."

Action Taken: A malfunction DTC is stored, ABS is disabled and the ABS warning lamp is turned "ON."

DTC Chart Test Description: Number(s) below refer to circled number(s) on the diagnostic chart.

1. This step indicates if the brake switch signal is being received by the EBCM.
2. This step indicates if an open circuit exists in the brake switch or brake lamp circuitry.
3. Checks for an open in the brake switch input circuit.
4. Ensures malfunction was not due to poor terminal contact.
5. Checks for voltage entering the brake switch.
6. Checks for voltage exiting the brake switch.

Diagnostic Aids: An "Intermittent" malfunction is most likely caused by a poor connection, rubbed through wire insulation, or a wire that is broken inside the insulation.

The frequency of the malfunction can be checked by using the enhanced diagnostic function of the Tech 1.

Any circuitry, that is suspected as causing the intermittent complaint, should be thoroughly checked for backed out terminals, improper mating, broken locks, improperly formed or damaged terminals, poor terminal to wiring connections or physical damage to the wiring harness.

DIAGNOSTIC CHART DTC 91 — 1994 CAVALIER AND SUNBIRD

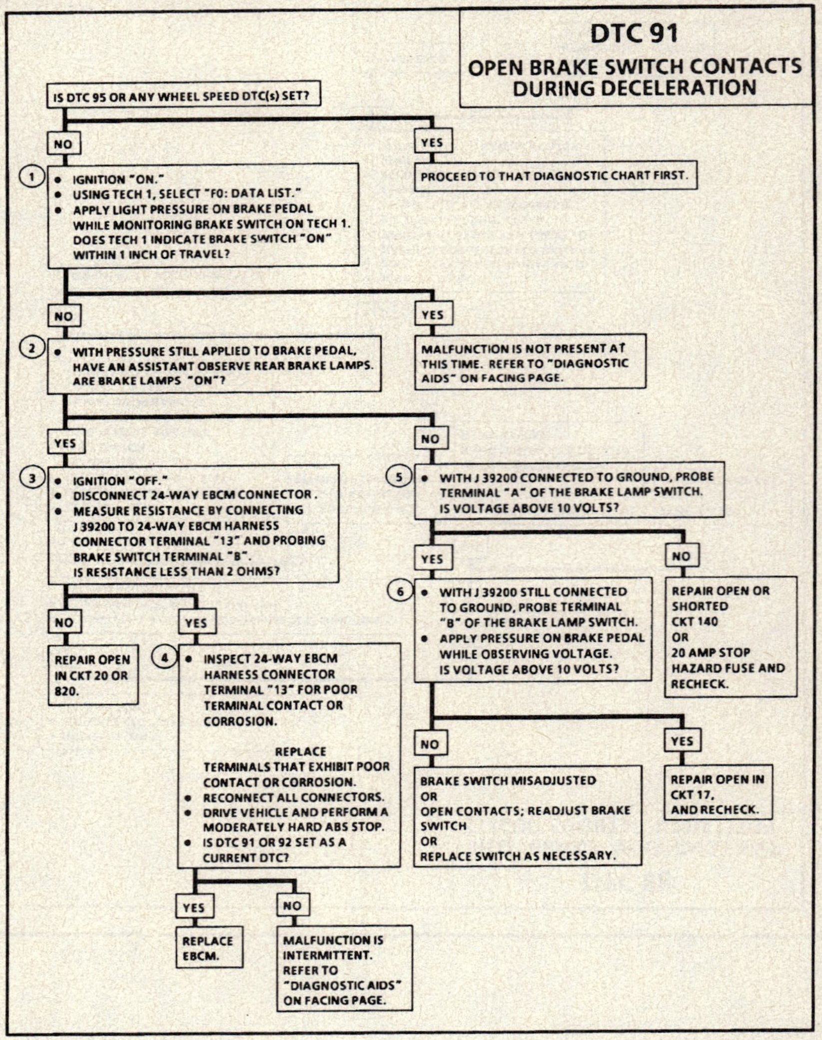

DIAGNOSTIC CHART DTC 92 — 1994 CAVALIER AND SUNBIRD

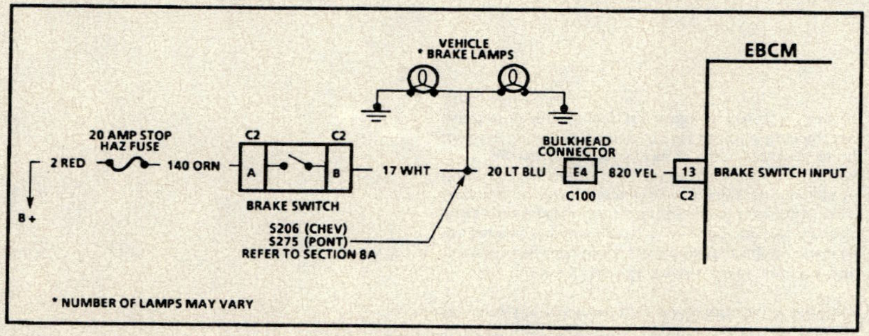

DTC 92

OPEN BRAKE SWITCH CONTACTS WHEN ABS WAS REQUIRED

Circuit Description:

This DTC is run to determine the proper operation of the brake switch. This is important because ABS is activated when the brake switch is "ON." If the brake switch is "OFF," ABS will never be activated. Since this malfunction is difficult to detect under normal braking conditions, this malfunction is only detected when ABS is required.

DTC Will Set When: DTC 92 can be set if the vehicle's speed is greater than 8 km/h (5 mph). If the brake was not "ON" and a release was required on two channels for .7 second, a malfunction exists.

Action Taken: A malfunction DTC is stored, ABS is disabled and the ABS warning lamp is turned "ON."

DTC Chart Test Description: Number(s) below refer to circled number(s) on the diagnostic chart.

1. This step indicates if the brake switch signal is being received by the EBCM.
2. This step indicates if an open circuit exists in the brake switch or brake lamp circuitry.
3. Checks for an open in the brake switch input circuit.
4. Ensures malfunction was not due to poor terminal contact.
5. Checks for voltage entering the brake switch.
6. Checks for voltage exiting the brake switch.

Diagnostic Aids: An "Intermittent" malfunction is most likely caused by a poor connection, rubbed through wire insulation, or a wire that is broken inside the insulation.

The frequency of the malfunction can be checked by using the enhanced diagnostic function of the Tech 1.

Any circuitry, that is suspected as causing the intermittent complaint, should be thoroughly checked for backed out terminals, improper mating, broken locks, improperly formed or damaged terminals, poor terminal to wiring connections or physical damage to the wiring harness.

DIAGNOSTIC CHART DTC 92 — 1994 CAVALIER AND SUNBIRD

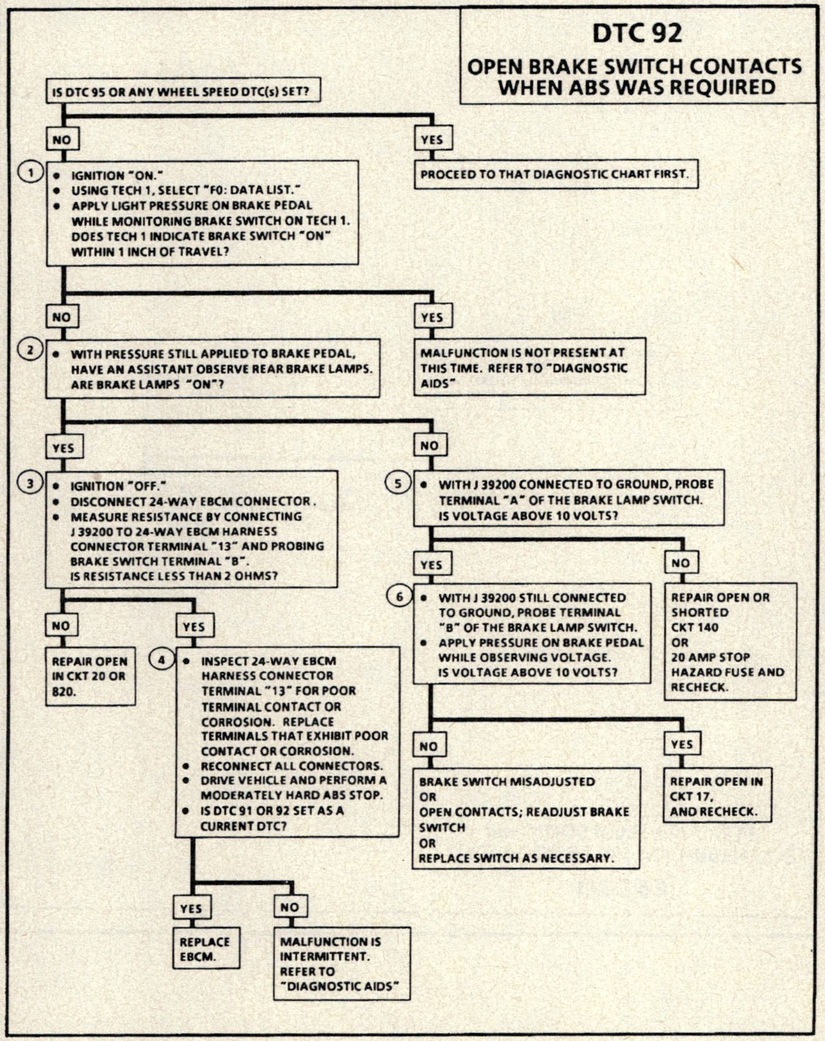

DIAGNOSTIC CHART DTC 93 — 1994 CAVALIER AND SUNBIRD

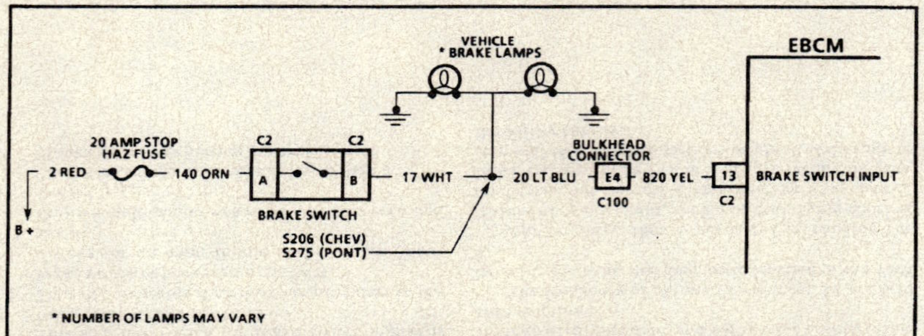

DTC 93

DTCs 91 OR 92 SET IN CURRENT OR PREVIOUS IGNITION CYCLE

Circuit Description:

This DTC is the second portion of DTCs 91 and 92. If DTCs 91 or 92 occurred during the last ignition cycle, DTC 93 becomes a current malfunction during the next ignition cycle, keeping ABS disabled until a brake switch "ON" state is seen. When a change is seen during an ignition cycle in which DTC 93 is a current malfunction, DTC 91 or 92 will clear itself at the end of the current ignition cycle, and ABS will enable itself at the start of the next ignition cycle. DTC 93 alone indicates DTCs 91 or 92 occurred previously, but is intermittent, or has been corrected.

DTC Chart Test Description: Number(s) below refer to circled number(s) on the diagnostic chart.

1. This step indicates which DTC (either 91 or 92) caused DTC 93 to set.

Diagnostic Aids: An "Intermittent" malfunction is most likely caused by a poor connection, rubbed through wire insulation, or a wire that is broken inside the insulation.

The frequency of the malfunction can be checked by using the enhanced diagnostic function of the Tech 1.

Any circuitry, that is suspected as causing the intermittent complaint, should be thoroughly checked for backed out terminals, improper mating, broken locks, improperly formed or damaged terminals, poor terminal to wiring connections or physical damage to the wiring harness.

Also, verify proper brake switch operation using the data list of the Tech 1. As the brake is applied, the data list should display the brake switch "ON" within 1 inch of travel.

DIAGNOSTIC CHART DTC 93 — 1994 CAVALIER AND SUNBIRD

DTC 93

DTCs 91 OR 92 SET IN CURRENT OR PREVIOUS IGNITION CYCLE

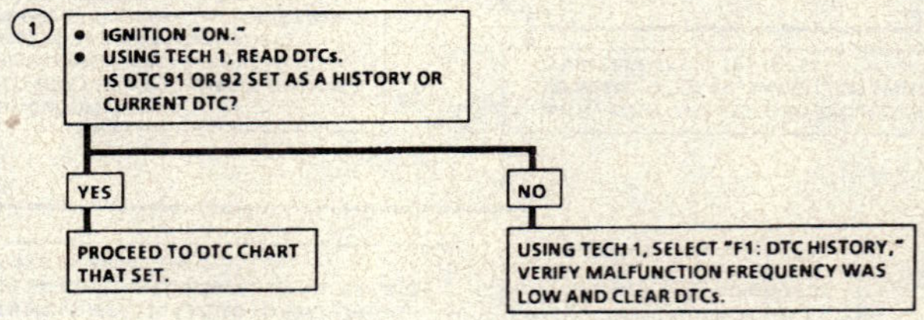

THIS CHART ASSUMES THAT A CURRENT DTC IS STORED INDICATING THAT THIS MALFUNCTION IS PRESENT.

DIAGNOSTIC CHART DTC 94 — 1994 CAVALIER AND SUNBIRD

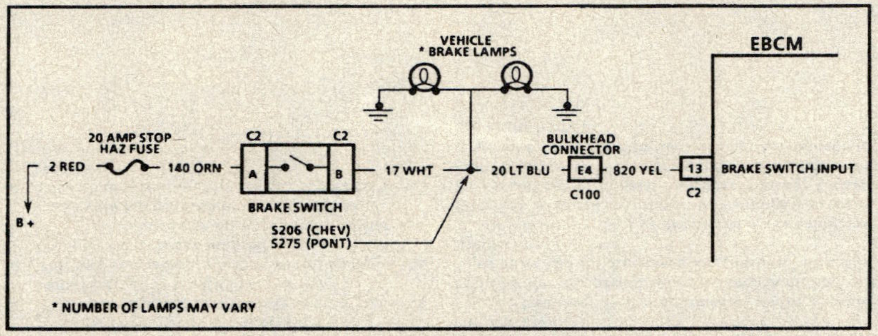

DTC 94

BRAKE SWITCH CONTACTS ALWAYS CLOSED

Circuit Description:

This DTC is run to determine the proper operation of the brake switch. This is important because ABS is activated when the brake switch is "ON" and turned "OFF" when the brake is "OFF." If the brake switch is always "ON," ABS operation will always be requested resulting in potential modulator cycling on rough roads. Additionally, this malfunction will most likely result in a dead battery (due to the brake lamps staying "ON") if the driver is not informed of this malfunction.

DTC Will Set When: DTC 94 can be set if the vehicle speed reaches at least 40 km/h (25 mph). If the brake was never "OFF" during two consecutive drive cycles, a malfunction exists.

Action Taken: A malfunction DTC is stored, however ABS is not disabled.

DTC Chart Test Description: Number(s) below refer to circled number(s) on the diagnostic chart.

1. This step checks to see if brake lamps are "ON."
2. This step isolates the cause of the malfunction to either a faulty or misadjusted brake switch, or a short to voltage in the brake switch circuitry.
3. This step checks for a resistive short to voltage in the brake switch circuitry.
4. Ensures malfunction was not due to a physical damage of the circuitry.

Diagnostic Aids: An "Intermittent" malfunction is most likely caused by a poor connection, rubbed through wire insulation, or a wire that is broken inside the insulation.

The frequency of the malfunction can be checked by using the enhanced diagnostic function of the Tech 1.

Any circuitry, that is suspected as causing the intermittent complaint, should be thoroughly checked for backed out terminals, improper mating, broken locks, improperly formed or damaged terminals, poor terminal to wiring connections or physical damage to the wiring harness.

DIAGNOSTIC CHART DTC 94 — 1994 CAVALIER AND SUNBIRD

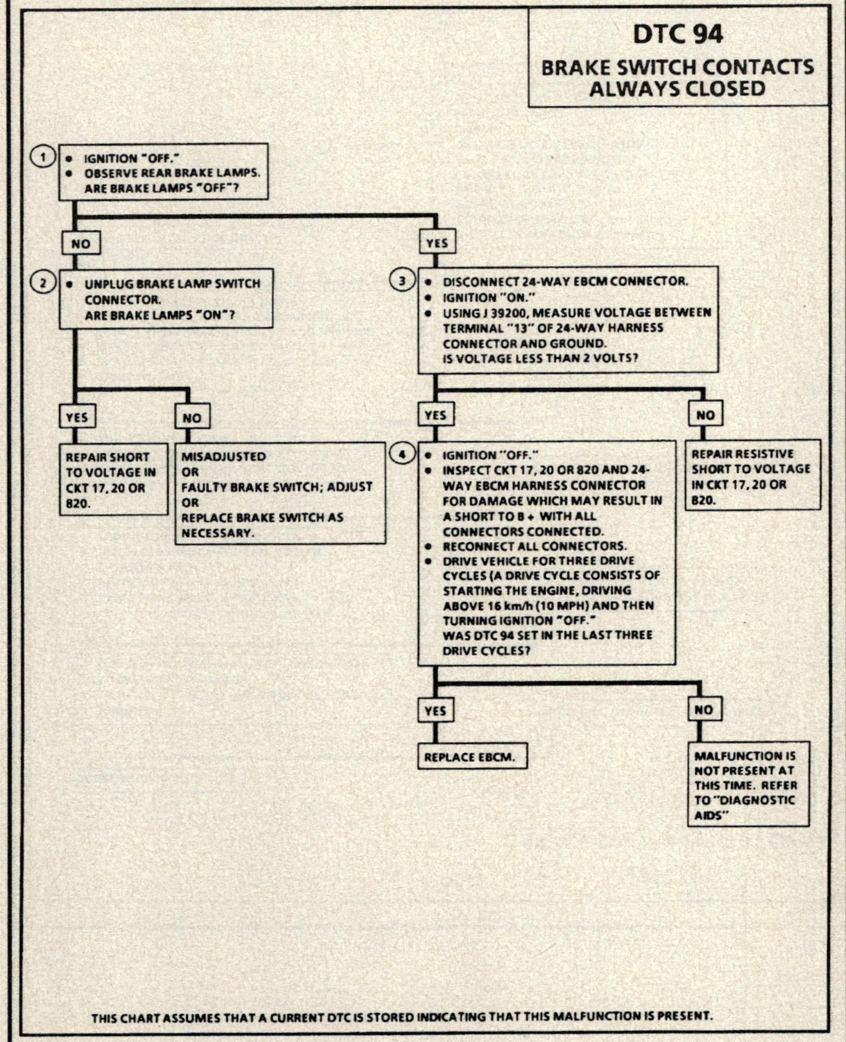

DIAGNOSTIC CHART DTC 95 — 1994 CAVALIER AND SUNBIRD

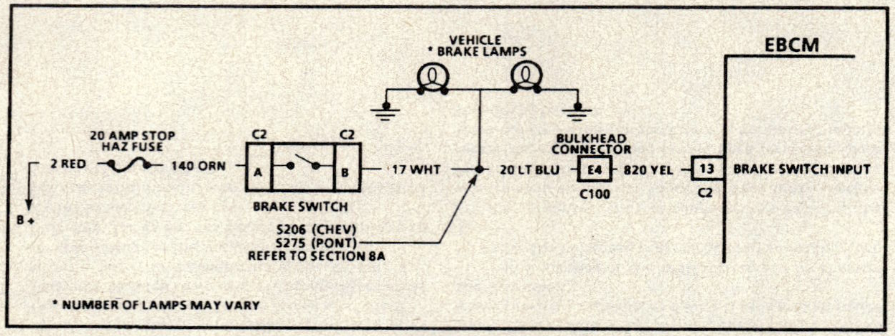

DTC 95

BRAKE SWITCH CIRCUIT OPEN

Circuit Description:

This DTC is used to identify open brake switch circuitry that prevents the brake switch input to the EBCM from changing states when the brake is applied. This DTC is used in conjunction with DTCs 91 and 92 to determine the cause of an open brake switch malfunction.

DTC Will Set When: DTC 95 can be set after initialization is completed. If the brake switch input voltage is out of specification for one second indicating an open circuit, a malfunction exists.

Action Taken: A malfunction DTC is stored, ABS is disabled and the ABS warning lamp is turned "ON."

DTC Chart Test Description: Number(s) below refer to circled number(s) on the diagnostic chart.

1. This step is used to confirm that an open in the brake switch circuitry currently exists.
2. This step indicates if the brake switch signal is being received by the EBCM.
3. This step indicates if an open circuit exists in the brake switch or brake lamp circuitry.
4. This step isolates the open circuit to either the brake switch input circuitry, or the EBCM.
5. Ensures malfunction was not due to poor terminal contact.

Diagnostic Aids: An "Intermittent" malfunction is most likely caused by a poor connection, rubbed through wire insulation, or a wire that is broken inside the insulation.

The frequency of the malfunction can be checked by using the enhanced diagnostic function of the Tech 1, as described in "Enhanced Diagnostics," found in this section.

Any circuitry, that is suspected as causing the intermittent complaint, should be thoroughly checked for backed out terminals, improper mating, broken locks, improperly formed or damaged terminals, poor terminal to wiring connections or physical damage to the wiring harness.

DIAGNOSTIC CHART DTC 95 — 1994 CAVALIER AND SUNBIRD

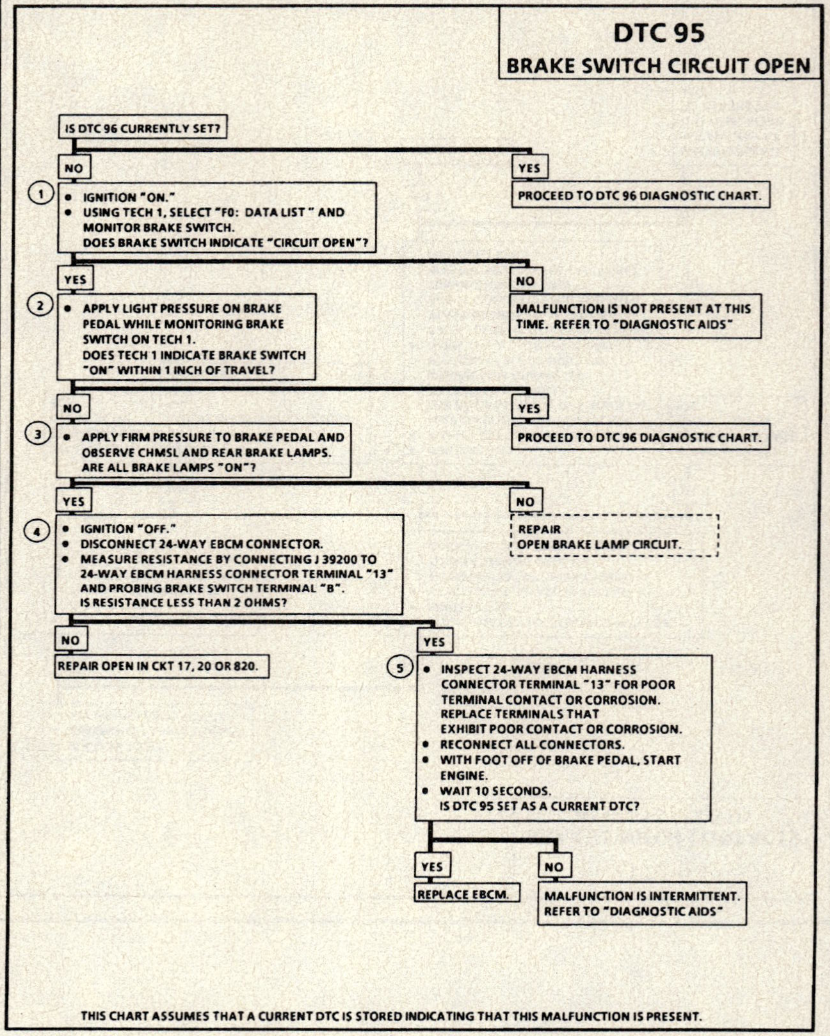

DIAGNOSTIC CHART DTC 96 — 1994 CAVALIER AND SUNBIRD

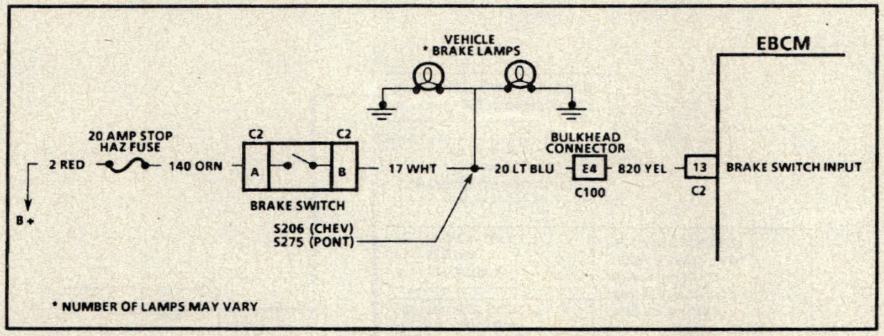

DTC 96

REAR BRAKE LAMP CIRCUIT OPEN

Circuit Description:

This DTC is designed to isolate the cause of a DTC 95 failure and indicate to the driver ABS is still available. If DTC 95 occurs with DTC 96, the brake lamps are open, the brake lamp grounds are open, or the Center High Mounted Stop Lamp (CHMSL) circuit is open during 4-way flasher use. The presence of battery voltage at the brake switch input indicates a valid brake switch input is still available.

DTC Will Set When: DTC 96 can be set only after DTC 95 has been set. If the EBCM detects battery voltage on the brake switch input circuit for one second, a malfunction exists.

Action Taken: A malfunction DTC is stored. ABS is not disabled; however, the ABS warning lamp will flash indicating a malfunction exists.

DTC Chart Test Description: Number(s) below refer to circled number(s) on the diagnostic chart.

1. As a result of a malfunction of an additional brake switch circuit DTC, this DTC may be set. To insure proper diagnosis, any additional brake switch DTCs must be repaired first.
2. This step identifies if the malfunction is currently present in the brake circuit.

Diagnostic Aids: An "Intermittent" malfunction is most likely caused by a poor connection, rubbed through wire insulation, or a wire that is broken inside the insulation.

The frequency of the malfunction can be checked by using the enhanced diagnostic function of the Tech 1.

Any circuitry, that is suspected as causing the intermittent complaint, should be thoroughly checked for backed out terminals, improper mating, broken locks, improperly formed or damaged terminals, poor terminal to wiring connections or physical damage to the wiring harness.

DIAGNOSTIC CHART DTC 96 — 1994 CAVALIER AND SUNBIRD

DTC 96

REAR BRAKE LAMP CIRCUIT OPEN

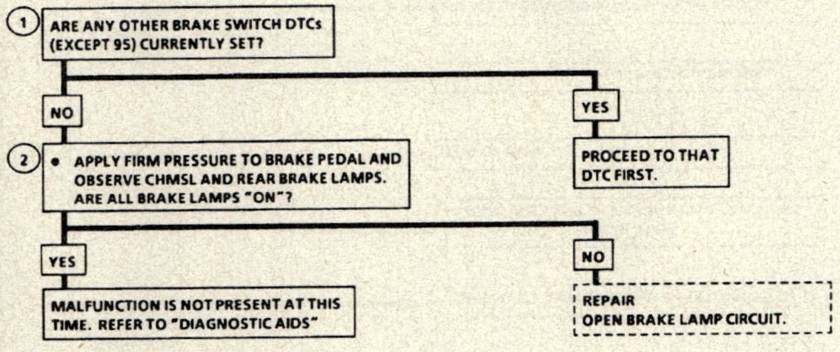

THIS CHART ASSUMES THAT A CURRENT DTC IS STORED INDICATING THAT THIS MALFUNCTION IS PRESENT.

DIAGNOSTIC PROCESS CHART — 1995 CAVALIER AND SUNFIRE

TECH 1 DIAGNOSTIC PROCESS
(Page 1 of 3)

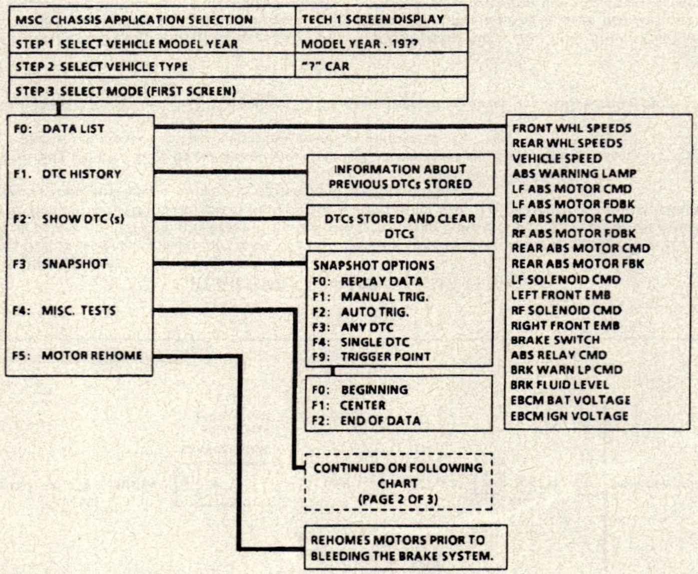

DIAGNOSTIC PROCESS CHART — 1995 CAVALIER AND SUNFIRE

TECH 1 DIAGNOSTIC PROCESS
(Page 2 of 3)

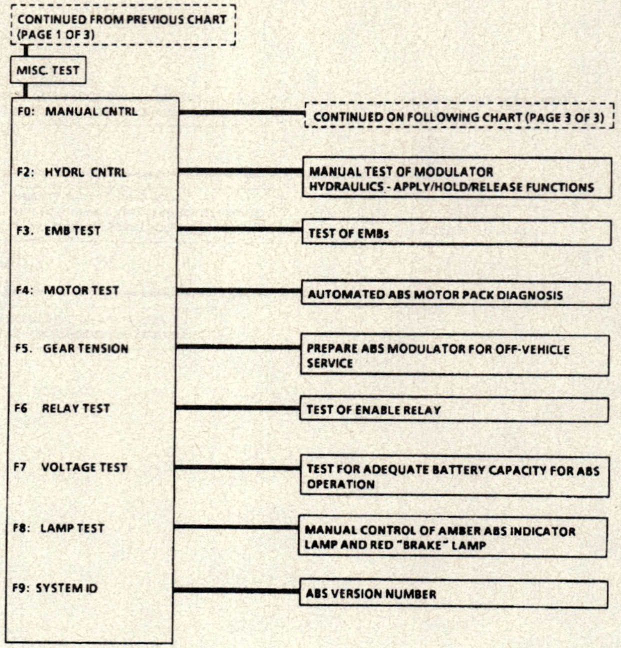

DIAGNOSTIC PROCESS CHART — 1995 CAVALIER AND SUNFIRE

TECH 1 DIAGNOSTIC PROCESS (PAGE 3 OF 3)

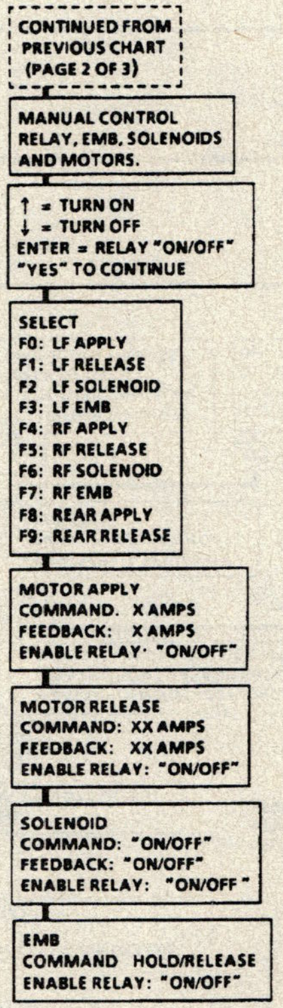

DIAGNOSTIC CHART SYSTEM CHECK — 1995 CAVALIER AND SUNFIRE

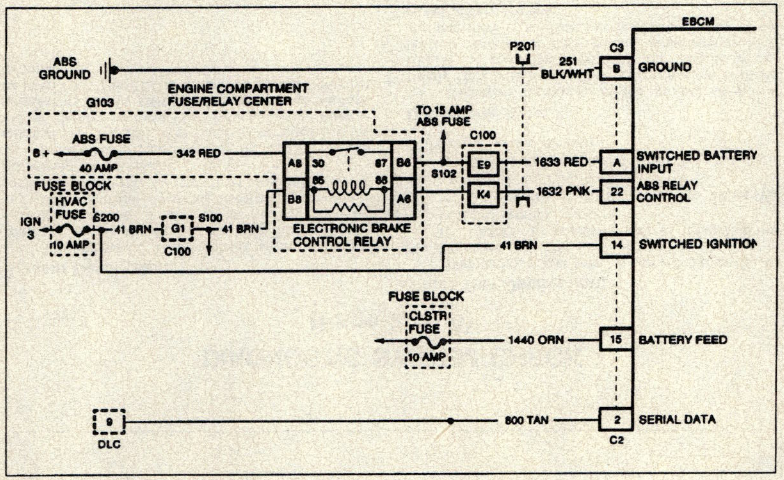

DIAGNOSTIC SYSTEM CHECK

System Description:

The diagnostic system check is an organized approach to identifying an Antilock Brake System (ABS) malfunction. It must be the starting point for any ABS complaint diagnosis because it directs the service technician to the next logical step in diagnosing the complaint.

Serial data is transmitted by the EBCM through terminal "2." The EBCM is supplied switched ignition voltage through terminal "14," battery voltage through terminal "15" and switched battery voltage through terminal "A." The EBCM ground is provided through terminal "B."

Chart Test Description:

Number(s) below refer to circled number(s) on the diagnostic chart.

1. Indicates if EBCM is transmitting data.
2. Checks for a serial data link malfunction.
3. Checks for proper Tech 1 connection to DLC.
4. Checks for open HVAC fuse.
5. Checks for open CLSTR fuse.
6. Checks for high resistance in the EBCM battery feed circuit.
7. Checks for current DTC's.
8. Checks for proper "ABS" warning lamp operation.
9. Checks for history DTC's.

Diagnostic Aids:

Excessive resistance in the ground or power supply circuits will not allow communication with the EBCM. If communication with the EBCM is not possible, ensure the ABS ground connection is good and that there is no excessive resistance in any of the power supply circuits.

DIAGNOSTIC CHART SYSTEM CHECK — 1995 CAVALIER AND SUNFIRE

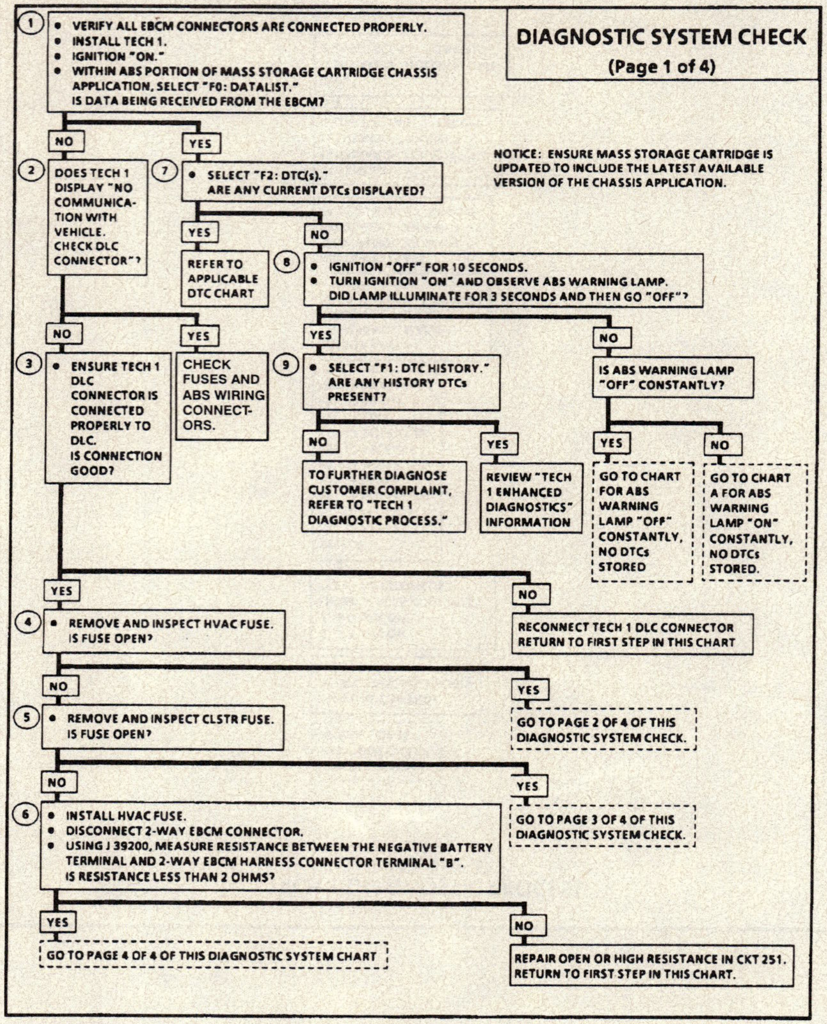

DIAGNOSTIC CHART SYSTEM CHECK — 1995 CAVALIER AND SUNFIRE

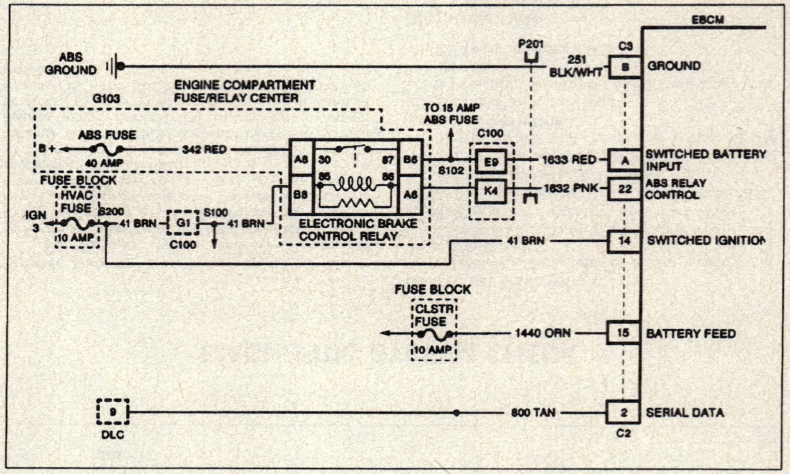

DIAGNOSTIC SYSTEM CHECK
(Page 2 of 4)

System Description:

The diagnostic system check is an organized approach to identifying an Antilock Brake System (ABS) malfunction. It must be the starting point for any ABS complaint diagnosis because it directs the service technician to the next logical step in diagnosing the complaint.

Serial data is transmitted by the EBCM through terminal "2." The EBCM is supplied switched ignition voltage through terminal "14," battery voltage through terminal "15" and switched battery voltage through terminal "A." The EBCM ground is provided through terminal "B."

Chart Test Description:

Number(s) below refer to circled number(s) on the diagnostic chart.

10. Checks for possible short to ground in the switched ignition circuitry.

11. Isolates the short to ground to either the circuitry or the EBCM.

12. Ensures short to ground is not due to physical damage of the circuitry.

Diagnostic Aids:

Excessive resistance in the ground or power supply circuits will not allow communication with the EBCM. If communication with the EBCM is not possible, ensure the ABS ground connection is good and that there is no excessive resistance in any of the power supply circuits.

DIAGNOSTIC CHART SYSTEM CHECK — 1995 CAVALIER AND SUNFIRE

DIAGNOSTIC SYSTEM CHECK (Page 2 of 4)

CONTINUED FROM DIAGNOSTIC SYSTEM CHECK (PAGE 1 OF 3)

YES

(10)
- REPLACE HVAC FUSE.
- IGNITION "ON."
- WAIT 10 SECONDS THEN TURN IGNITION "OFF."
- REMOVE AND INSPECT HVAC FUSE.

IS FUSE BLOWN?

YES → (11)
- DISCONNECT 24-WAY EBCM CONNECTOR.
- REPLACE HVAC FUSE
- IGNITION "ON."
- WAIT 10 SECONDS THEN TURN IGNITION "OFF."
- REMOVE AND INSPECT HVAC FUSE

IS FUSE BLOWN?

NO → INSTALL FUSE. DIAGNOSE FOR A POSSIBLE SHORT TO GROUND IN THE SWITCHED IGNITION CIRCUITRY. RETURN TO FIRST STEP IN THIS CHART.

(11) NO → (12)
- INSPECT CKT 41 AND 24-WAY EBCM HARNESS CONNECTOR FOR PHYSICAL DAMAGE WHICH MAY RESULT IN A SHORT TO GROUND WITH THE 24-WAY EBCM HARNESS CONNECTOR CONNECTED TO THE EBCM. CORRECT DAMAGE IF EVIDENT.
- RECONNECT ALL CONNECTORS.
- IGNITION "ON."
- WAIT 10 SECONDS THEN TURN IGNITION "OFF."
- REMOVE AND INSPECT HVAC FUSE.

IS FUSE OPEN?

(11) YES → REPAIR A SHORT TO GROUND IN ONE OF THE CKT 41 BRANCHES EXITING THE HVAC FUSE. REPLACE HVAC FUSE. RETURN TO THE FIRST STEP IN THIS CHART.

(12) YES → REPLACE EBCM. REPLACE HVAC FUSE. RETURN TO FIRST STEP IN THIS CHART.

(12) NO → MALFUNCTION IS INTERMITTENT.

DIAGNOSTIC CHART SYSTEM CHECK — 1995 CAVALIER AND SUNFIRE

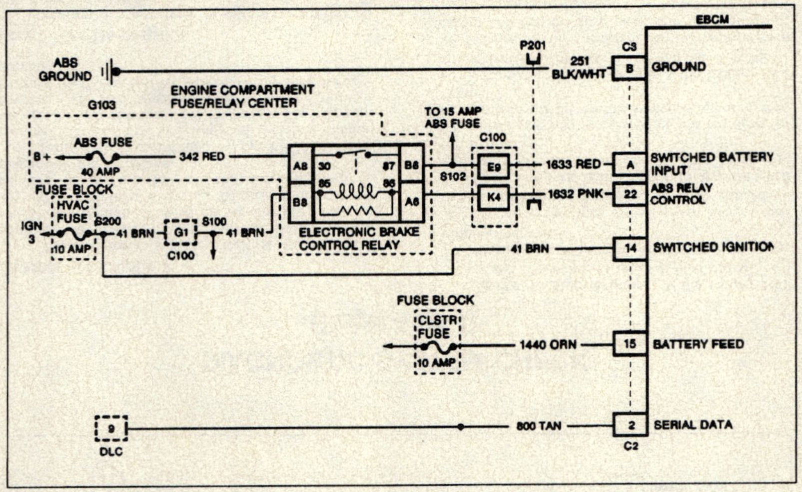

DIAGNOSTIC SYSTEM CHECK (Page 3 of 4)

System Description:

The diagnostic system check is an organized approach to identifying an Antilock Brake System (ABS) malfunction. It must be the starting point for any ABS complaint diagnosis because it directs the service technician to the next logical step in diagnosing the complaint.

Serial data is transmitted by the EBCM through terminal "2." The EBCM is supplied switched ignition voltage through terminal "14," battery voltage through terminal "15" and switched battery voltage through terminal "A." The EBCM ground is provided through terminal "B."

Chart Test Description:

Number(s) below refer to circled number(s) on the diagnostic chart.

13. Checks for possible short to ground in the battery feed circuitry.
14. Isolates the short to ground to either the circuitry or the EBCM.
15. Ensures short to ground is not due to physical damage of the circuitry.

Diagnostic Aids:

Excessive resistance in the ground or power supply circuits will not allow communication with the EBCM. If communication with the EBCM is not possible, ensure the ABS ground connection is good and that there is no excessive resistance in any of the power supply circuits.

DIAGNOSTIC CHART SYSTEM CHECK — 1995 CAVALIER AND SUNFIRE

DIAGNOSTIC SYSTEM CHECK
(Page 3 of 4)

CONTINUED FROM DIAGNOSTIC SYSTEM CHECK (PAGE 1 OF 4)

YES

(13)
- REPLACE CLSTR FUSE.
- WAIT 10 SECONDS.
- REMOVE AND INSPECT CLSTR FUSE. IS FUSE OPEN?

NO: INSTALL FUSE. DIAGNOSE FOR A POSSIBLE SHORT TO GROUND IN THE BATTERY FEED CIRCUITRY. RETURN TO FIRST STEP IN THIS CHART.

YES:

(14)
- DISCONNECT 24-WAY EBCM CONNECTOR.
- REPLACE CLSTR FUSE.
- WAIT 10 SECONDS.
- REMOVE AND INSPECT CLSTR FUSE. IS FUSE OPEN?

YES: REPAIR A SHORT TO GROUND IN ONE OF THE CKT 1440 BRANCHES EXITING THE CLSTR FUSE. REPLACE CLSTR FUSE. RETURN TO THE FIRST STEP IN THIS CHART

NO:

(15)
- INSPECT CKT 1440 AND 24-WAY EBCM HARNESS CONNECTOR FOR PHYSICAL DAMAGE WHICH MAY RESULT IN A SHORT TO GROUND WITH THE 24-WAY EBCM HARNESS CONNECTOR CONNECTED TO THE EBCM. CORRECT DAMAGE IF EVIDENT
- RECONNECT ALL CONNECTORS.
- WAIT 10 SECONDS.
- REMOVE AND INSPECT CLSTR FUSE. IS FUSE OPEN?

YES: REPLACE EBCM. REPLACE CLSTR FUSE. RETURN TO FIRST STEP IN THIS CHART.

NO: MALFUNCTION IS INTERMITTENT.

DIAGNOSTIC CHART SYSTEM CHECK — 1995 CAVALIER AND SUNFIRE

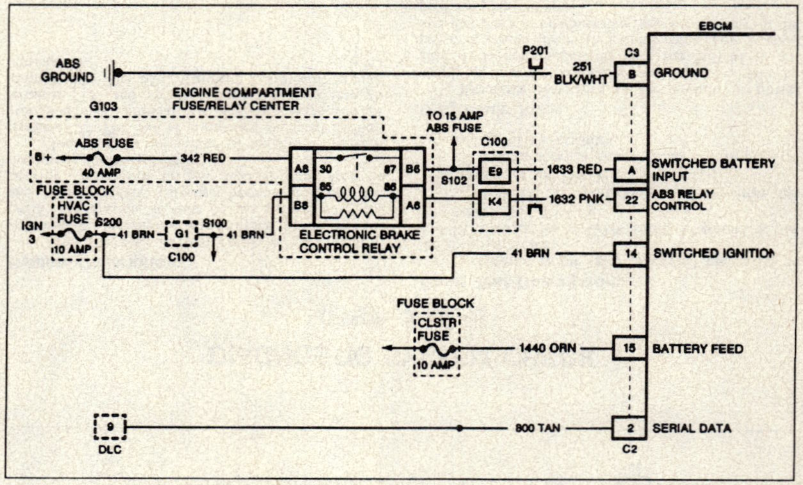

DIAGNOSTIC SYSTEM CHECK
(Page 4 of 4)

System Description:

The diagnostic system check is an organized approach to identifying an Antilock Brake System (ABS) malfunction. It must be the starting point for any ABS complaint diagnosis because it directs the service technician to the next logical step in diagnosing the complaint.

Serial data is transmitted by the EBCM through terminal "2." The EBCM is supplied switched ignition voltage through terminal "14," battery voltage through terminal "15" and switched battery voltage through terminal "A." The EBCM ground is provided through terminal "B."

Chart Test Description:

Number(s) below refer to circled number(s) on the diagnostic chart.

16. Checks for proper battery feed voltage at the EBCM.
17. Checks for proper ignition feed voltage at the EBCM.
18. Checks for high resistance in the battery feed circuit and the switched ignition feed circuit.
19. Ensures malfunction is not due to poor terminal contact.
20. Checks for an open circuit in the battery feed circuit between the fuse block and the EBCM.
21. Checks for an open circuit in the switched ignition feed circuit between the fuse block and the EBCM.
22. Isolates the high resistance circuit to either the battery feed circuit or the switched ignition feed circuit.
23. Isolates the high resistance in the battery feed circuit between the fuse block and the EBCM.
24. Isolates the high resistance in the ignition feed circuit between the fuse block and the EBCM.

Diagnostic Aids:

Excessive resistance in the ground or power supply circuits will not allow communication with the EBCM. If communication with the EBCM is not possible, ensure the ABS ground connection is good and that there is no excessive resistance in any of the power supply circuits.

DIAGNOSTIC CHART SYSTEM CHECK — 1995 CAVALIER AND SUNFIRE

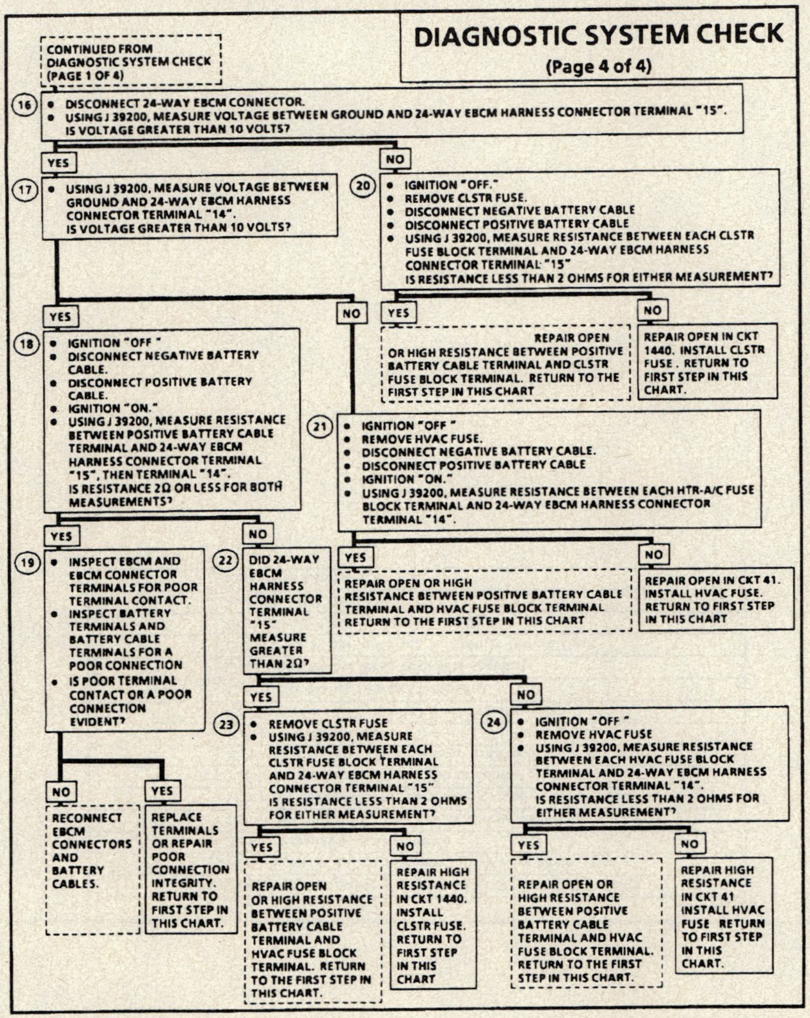

DIAGNOSTIC TROUBLE CODE CHART — 1995 CAVALIER AND SUNFIRE

DIAGNOSTIC TROUBLE CODE AND SYMPTOM TABLE

CHART	SYMPTOM
A	ABS (Amber) Indicator Lamp "ON" Constantly, No DTCs Stored
B	Tech 1 Displays Undefined DTCs
C	ABS (Amber) Indicator Lamp "OFF" Constantly, No DTCs Stored
DTC	**DESCRIPTION**
14	Electronic Brake Control Relay Contacts Circuit Open
15	Electronic Brake Control Relay Contacts Circuit Shorted to Battery
16	Electronic Brake Control Relay Coil Circuit Open
17	Electronic Brake Control Relay Coil Circuit Shorted to Ground
18	Electronic Brake Control Relay Coil Circuit Shorted to Battery or Coil Shorted
21	Left Front Wheel Speed = 0 or Unreasonable
22	Right Front Wheel Speed = 0 or Unreasonable
23	Left Rear Wheel Speed = 0 or Unreasonable
24	Right Rear Wheel Speed = 0 or Unreasonable
25	Left Front Excessive Wheel Speed Variation
26	Right Front Excessive Wheel Speed Variation
27	Left Rear Excessive Wheel Speed Variation
28	Right Rear Excessive Wheel Speed Variation
36	Low System Voltage
37	High System Voltage
38	Left Front EMB Will Not Hold Motor
41	Right Front EMB Will Not Hold Motor
42	Rear ESB Will Not Hold Motor
44	Left Front Channel Will Not Move
45	Right Front Channel Will Not Move
46	Rear Axle Channel Will Not Move
47	Left Front Motor Free Spins
48	Right Front Motor Free Spins
51	Rear Motor Free Spins
52	Left Front Channel in Release Too Long
53	Right Front Channel in Release Too Long
54	Rear Channel in Release Too Long
55	EBCM Malfunction
56	Left Front Motor Circuit Open
57	Left Front Motor Circuit Shorted to Ground
58	Left Front Motor Circuit Shorted to Battery or Motor Shorted
61	Right Front Motor Circuit Open
62	Right Front Motor Circuit Shorted to Ground
63	Right Front Motor Circuit Shorted to Battery or Motor Shorted
64	Rear Motor Circuit Open
65	Rear Motor Circuit Shorted to Ground
66	Rear Motor Circuit Shorted to Battery or Motor Shorted
67	Left Front EMB Circuit Open or Shorted to Ground
68	Left Front EMB Circuit Shorted to Battery or Driver Open
71	Right Front EMB Circuit Open or Shorted to Ground

DIAGNOSTIC TROUBLE CODE CHART — 1995 CAVALIER AND SUNFIRE

DIAGNOSTIC TROUBLE CODE AND SYMPTOM TABLE	
72	Right Front EMB Circuit Shorted to Battery or Driver Open
75	Serial Communication Malfunction
76	Left Front Solenoid Circuit Open or Shorted to Battery
77	Left Front Solenoid Circuit Shorted to Ground or Driver Open
78	Right Front Solenoid Circuit Open or Shorted to Battery
81	Right Front Solenoid Circuit Shorted to Ground or Driver Open
82	Calibration Malfunction
86	EBCM Turned "ON" the Red Brake Warning Lamp
91	Open Brake Switch During Deceleration
92	Open Brake Switch When ABS Was Required
93	DTCs 91 or 92 Set in Current or Previous Ignition Cycle
94	Brake Switch Contacts Always Closed
95	Brake Switch Circuit Open
96	Rear Brake Lamp Circuit Open

ELECTRICAL CONNECTOR TERMINAL IDENTIFICATION — 1995 CAVALIER AND SUNFIRE

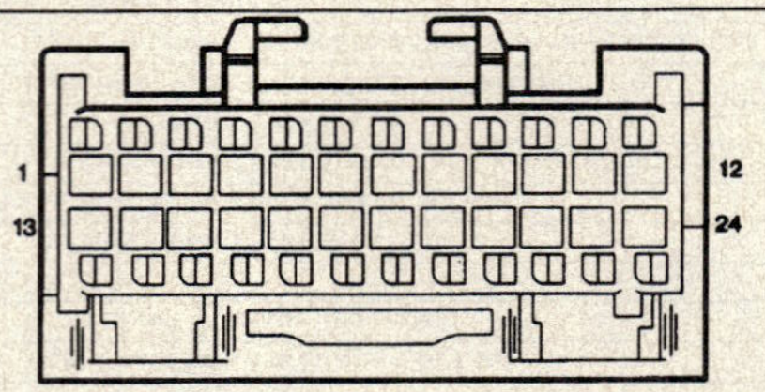

PIN	CIRCUIT NO.	COLOR	CIRCUIT
1	OPEN		NOT USED
2	800	TAN	SERIAL DATA LINK
3	OPEN		NOT USED
4	1289	LT BLU	R/F ABS SOLENOID
5	830	LT BLU	L/F WHEEL SIGNAL HIGH
6	873	YEL	L/F WHEEL SIGNAL LOW
7	882	BRN	R/R WHEEL SIGNAL HIGH
8	883	WHT	R/R WHEEL SIGNAL LOW
9	872	DK GRN	R/F WHEEL SIGNAL HIGH
10	833	TAN	R/F WHEEL SIGNAL LOW
11	885	RED	L/R WHEEL SIGNAL LOW
12	884	BLK	L/R WHEEL SIGNAL HIGH
13	17	WHT	BRAKE SWITCH INPUT
14	41	BRN	SWITCH IGNITION
15	1440	ORN	B+ FEED
16	OPEN		NOT USED
17	OPEN		NOT USED
18	OPEN		NOT USED
19	1286	LT GRN	L/F EMB
20	1287	GRY	R/F EMB
21	849	PNK	BRAKE FLUID LEVEL SWITCH INPUT
22	1632	PNK	ENABLE RELAY CONTROL
23	OPEN		NOT USED
24	1288	DK GRN	L/F ABS SOLENOID

ELECTRICAL CONNECTOR TERMINAL IDENTIFICATION — 1995 CAVALIER AND SUNFIRE

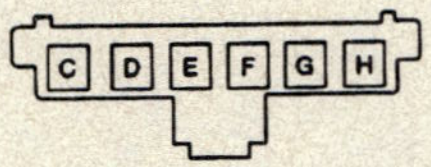

6 WAY EBCM HARNESS CONNECTOR

PIN	CIRCUIT NO.	COLOR	CIRCUIT
C	1284	DK GRN	REAR MOTOR HIGH
D	1285	ORN	REAR MOTOR LOW
E	1281	PNK	L/F MOTOR LOW
F	1280	BLK	L/F MOTOR HIGH
G	1283	BLK	R/F MOTOR LOW
H	1282	PPL	R/F MOTOR HIGH

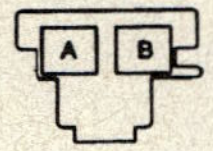

2 WAY EBCM HARNESS CONNECTOR

PIN	CIRCUIT NO.	COLOR	CIRCUIT
A	1633	RED	SWITCHED BATTERY INPUT
B	251	BLK/WHT	GROUND

DIAGNOSTIC CHART A — 1995 CAVALIER AND SUNFIRE

CHART A
ABS (AMBER) INDICATOR LAMP "ON" CONSTANTLY, NO DTCs STORED

NOTICE: DIAGNOSTIC SYSTEM CHECK MUST BE COMPLETED FIRST BEFORE USING THIS CHART.

- USING TECH 1, EXIT BACK TO MSC APPLICATION MENU.
- SELECT BODY PORTION OF MASS STORAGE CARTRIDGE. ENTER VEHICLE INFORMATION. SELECT IPC SYSTEM. SELECT SDL MONITOR MODE. SELECT ABS TO IPC SDL MODE.

IS TECH 1 DISPLAYING ABS T/T "ON"?

- NO → DIAGNOSE INSTRUMENT CLUSTER.
- YES → REPLACE EBCM.

DIAGNOSTIC CHART B — 1995 CAVALIER AND SUNFIRE

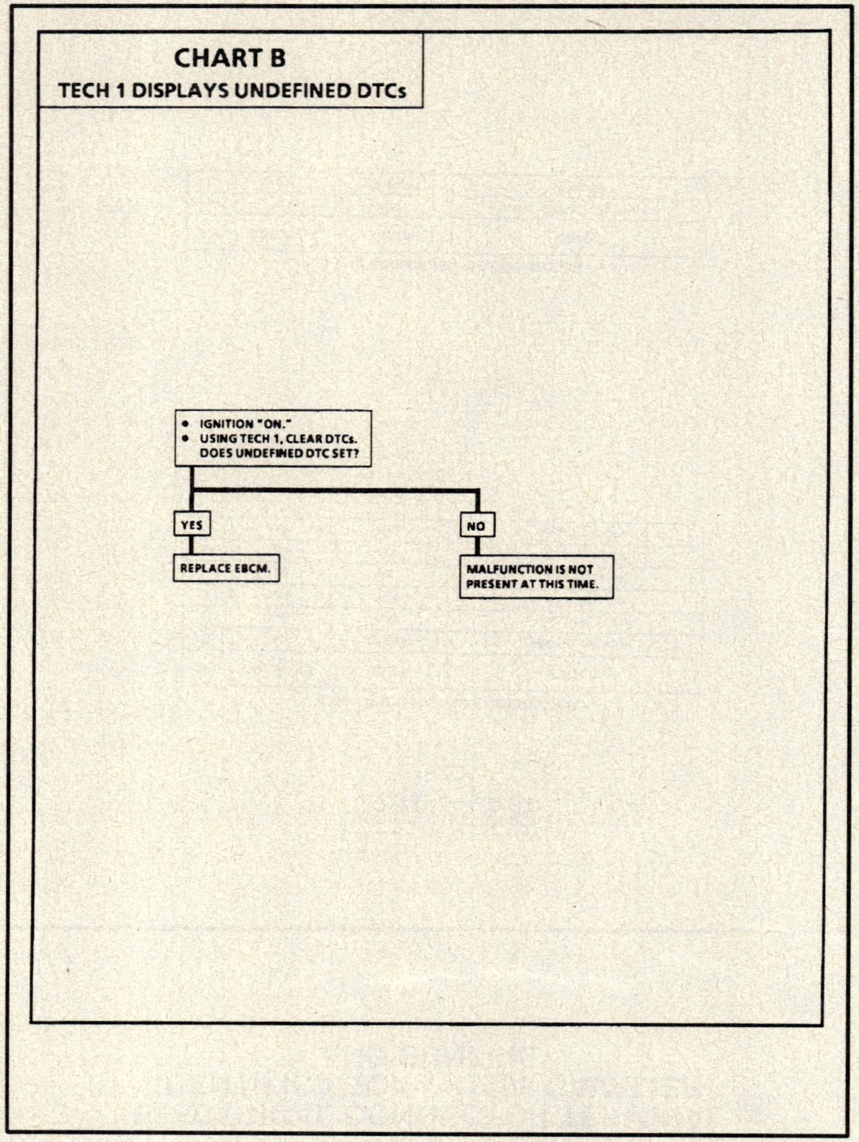

DIAGNOSTIC CHART C — 1995 CAVALIER AND SUNFIRE

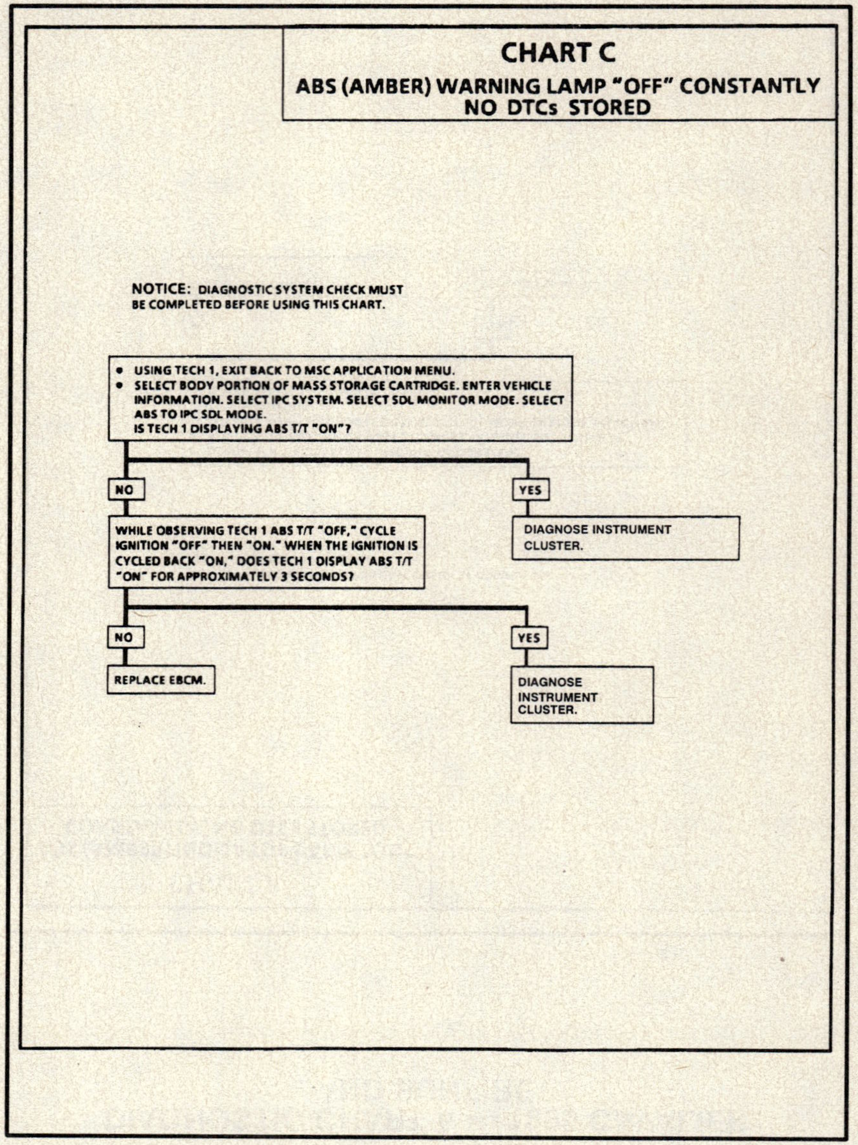

DIAGNOSTIC CHART DTC 14 — 1995 CAVALIER AND SUNFIRE

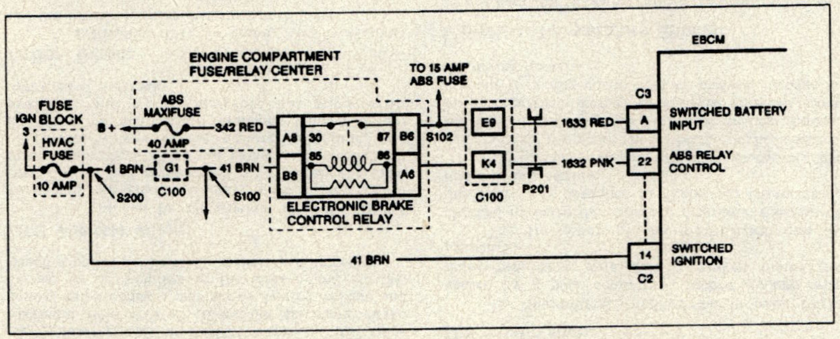

DTC 14

ELECTRONIC BRAKE CONTROL RELAY CONTACT CIRCUIT OPEN

Circuit Description:

Ignition voltage is supplied through terminal "85" of the electronic brake control relay. The EBCM is then able to energize the pull-in coil by completing the ground circuit at EBCM terminal "22". The magnetic field created closes the electronic brake control relay contacts and allows battery voltage and current to be supplied to the EMB's and EBCM, which supplies power to the motors and solenoids.

DTC Will Set When:

DTC 14 can be set anytime after the EBCM commands the electronic brake control relay "ON" (the relay is first commanded "ON" during the three second bulb check). This test monitors the availability of current/voltage to the motors, solenoids, and EMB's. This malfunction indicates voltage is not available and would therefore not allow ABS operation if required.

Action Taken:

A malfunction DTC is stored, ABS is disabled and the "ABS" warning lamp is turned "ON."

DTC Chart Test Description: Number(s) below refer to circled number(s) on the diagnostic chart.

1. Checks to see if the EBCM is sensing battery voltage at terminal "A".
2. Checks for proper operation of the electronic brake control relay and related circuitry.
3. Checks for excessive resistance in the switched battery input circuitry.
4. Checks for excessive resistance in the battery feed circuit.
5. Checks for excessive resistance in the electronic brake control relay contacts.
6. Determines if the malfunction is due to poor terminal contact.
7. Determines if the malfunction is due to the EBCM.

Diagnostic Aids:

An "intermittent" malfunction is most likely caused by a poor connection, rubbed through wire insulation, or a wire that is broken inside the insulation. The frequency of the malfunction is most likely checked by using the enhanced diagnostic function of the Tech 1, as described in "Enhanced Diagnostics," found in this section. Any circuitry that is suspected as causing the "intermittent" complaint should be thoroughly checked for backed out terminals, improper mating, broken locks, improperly formed or damaged terminals, poor terminal to wiring connections or physical damage to the wiring harness.

Vibration, Temperature Effects:

Check for vibration effects by performing the relay test function of the Tech 1. With the relay commanded "ON," lightly tap the top and sides of the relay while monitoring battery voltage. If the battery voltage changes significantly, replace the relay.

If DTC 14 only sets when the vehicle is initially started in cold ambient conditions (temperature less than 32°-0°C),replace the electronic brake control relay.

DIAGNOSTIC CHART DTC 14 — 1995 CAVALIER AND SUNFIRE

DTC 14
(Page 1 of 3)
ELECTRONIC BRAKE CONTROL RELAY CONTACTS CIRCUIT OPEN

IS DTC 16 ALSO PRESENT OR IN HISTORY?
- YES → PROCEED TO THAT DTC CHART FIRST.
- NO ↓

(1)
- IGNITION "ON," ENGINE "OFF."
- USING TECH 1, SELECT DATA LIST AND OBSERVE ABS BATTERY VOLTAGE.

DOES TECH 1 INDICATE THAT ABS BATTERY VOLTAGE IS GREATER THAN 10 VOLTS?
- YES → MALFUNCTION IS NOT PRESENT AT THIS TIME. REFER TO "DIAGNOSTIC AIDS" ON PREVIOUS PAGE.
- NO ↓

(2)
- IGNITION "OFF."
- DISCONNECT 24-WAY EBCM CONNECTOR.
- DISCONNECT 2-WAY EBCM CONNECTOR.
- CONNECT A FUSED JUMPER WIRE SUCH AS J 36169 WITH A 3 AMP FUSE BETWEEN GROUND AND 24-WAY EBCM HARNESS CONNECTOR TERMINAL "22".
- IGNITION "ON."
- USING J 39200, MEASURE VOLTAGE BETWEEN GROUND AND 2-WAY EBCM HARNESS CONNECTOR TERMINAL "A".

IS VOLTAGE GREATER THAN 10 VOLTS?
- NO → REFER TO PAGE 2 OF 3 OF THIS DTC CHART.
- YES ↓

(3)
- IGNITION "OFF."
- DISCONNECT JUMPER WIRE USED IN STEP 2.
- DISCONNECT ELECTRONIC BRAKE CONTROL RELAY.
- USING J 39200, MEASURE RESISTANCE BETWEEN 2-WAY EBCM HARNESS CONNECTOR TERMINAL "A" AND ENGINE COMPARTMENT FUSE/RELAY CENTER TERMINAL "B6".

IS RESISTANCE LESS THAN 2 OHMS?
- NO → REPAIR HIGH RESISTANCE IN CKT 1633.
- YES ↓

(4)
- DISCONNECT NEGATIVE BATTERY CABLE.
- DISCONNECT POSITIVE BATTERY CABLE.
- USING J 39200, MEASURE RESISTANCE BETWEEN POSITIVE BATTERY CABLE TERMINAL AND ENGINE COMPARTMENT FUSE/RELAY CENTER TERMINAL "A8".

IS RESISTANCE LESS THAN 2 OHMS?
- NO → REPAIR HIGH RESISTANCE IN B+ FEED CIRCUIT.
- YES ↓

(5)
- USING A JUMPER WIRE, CONNECT ELECTRONIC BRAKE CONTROL RELAY TERMINAL "85" TO GROUND. THEN, USING A FUSED JUMPER WIRE SUCH AS J 36169 WITH A 3 AMP FUSE, CONNECT ELECTRONIC BRAKE CONTROL RELAY TERMINAL "86" TO B+.
- USING J 39200, MEASURE THE RESISTANCE BETWEEN ENABLE RELAY TERMINALS "30" AND "87".

IS RESISTANCE LESS THAN 2 OHMS?
- NO → REPLACE ELECTRONIC BRAKE CONTROL RELAY.
- YES ↓

(6)
- INSPECT ELECTRONIC BRAKE CONTROL RELAY AND ENGINE COMPARTMENT FUSE/RELAY CENTER FOR A POOR CONNECTION OR POOR TERMINAL CONTACT. INSPECT EBCM AND EBCM CONNECTOR TERMINALS "22" AND "A" FOR POOR TERMINAL CONTACT. INSPECT POSITIVE BATTERY TERMINAL AND POSITIVE BATTERY CABLE TERMINAL FOR A POOR CONNECTION.

IS POOR TERMINAL CONTACT OR A POOR CONNECTION EVIDENT?
- YES → REPLACE TERMINALS OR REPAIR CONNECTION INTEGRITY (IF POOR CONTACT EXISTS AT BATTERY TERMINAL) THAT EXHIBIT(S) POOR CONTACT.
- NO ↓

(7)
- RECONNECT ALL CONNECTORS.
- IGNITION "ON."

IS DTC 14 SET AS A CURRENT DTC?
- YES → REPLACE EBCM.
- NO → MALFUNCTION IS INTERMITTENT. REFER TO "DIAGNOSTIC AIDS" ON PREVIOUS PAGE.

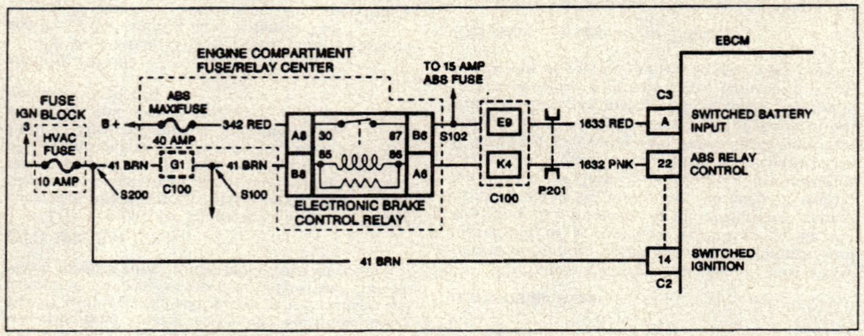

DTC 14
(Page 2 of 3)
ELECTRONIC BRAKE CONTROL RELAY CONTACT CIRCUIT OPEN

Circuit Description:

Ignition voltage is supplied through terminal "85" of the electronic brake control relay. The EBCM is then able to energize the pull-in coil by completing the ground circuit at EBCM terminal "22". The magnetic field created closes the electronic brake control relay contacts and allows battery voltage and current to be supplied to the EMB's and EBCM, which supplies power to the motors and solenoids.

DTC Will Set When:

DTC 14 can be set anytime after the EBCM commands the electronic brake control relay "ON" (the relay is first commanded "ON" during the three second bulb check). This test monitors the availability of current/voltage to the motors, solenoids, and EMB's. This malfunction indicates voltage is not available and would therefore not allow ABS operation if required.

Action Taken:

A malfunction DTC is stored, ABS is disabled and the "ABS" warning lamp is turned "ON."

DTC Chart Test Description:

Number(s) below refer to circled number(s) on the diagnostic chart.

8. Checks for excessive resistance in the electronic brake control relay control circuitry.
9. Checks for proper battery voltage.
10. Checks for an open in the BATTERY feed circuit.
11. Checks for an open in the switched battery input circuit.
12. Checks for proper operation of the electronic brake control relay.

Diagnostic Aids:

An "intermittent" malfunction is most likely caused by a poor connection, rubbed through wire insulation, or a wire that is broken inside the insulation.

The frequency of the malfunction can be checked by using the enhanced diagnostic function of the Tech 1, as described in "Enhanced Diagnostics," found in this section.

Any circuitry that is suspected as causing the "intermittent" complaint should be thoroughly checked for backed out terminals, improper mating, broken locks, improperly formed or damaged terminals, poor terminal to wiring connections or physical damage to the wiring harness.

Vibration, Temperature Effects:

Check for vibration effects by performing the relay test function of the Tech 1. With the relay commanded "ON," lightly tap the top and sides of the relay while monitoring battery voltage. If the battery voltage changes significantly, replace the relay.

If DTC 14 only sets when the vehicle is initially started in cold ambient conditions (temperature less than 32°F-0°C),replace the electronic brake control relay.

DTC 14
(Page 2 of 3)
ELECTRONIC BRAKE CONTROL RELAY CONTACT CIRCUITS OPEN

CONTINUED FROM PAGE 1 OF 3 OF THIS CHART.

(8)
- IGNITION "OFF."
- DISCONNECT JUMPER WIRE USED IN STEP 2.
- DISCONNECT NEGATIVE BATTERY CABLE FROM BATTERY.
- DISCONNECT POSITIVE BATTERY CABLE FROM BATTERY.
- TURN IGNITION "ON" (THIS IS DONE TO PROVIDE CIRCUIT CONTINUITY).
- USING J 39200, MEASURE RESISTANCE BETWEEN POSITIVE BATTERY CABLE TERMINAL AND 24-WAY EBCM HARNESS CONNECTOR TERMINAL "22". IS RESISTANCE LESS THAN 95 OHMS WHEN RELAY IS APPROXIMATELY 25°C (77°F)?

NO: REFER TO PAGE 3 OF 3 OF THIS DTC CHART.

YES:

(9)
- USING J 39200, MEASURE VOLTAGE BETWEEN THE NEGATIVE AND POSITIVE BATTERY TERMINALS. IS VOLTAGE GREATER THAN 10 VOLTS?

NO: CORRECT LOW VOLTAGE CONDITION.

YES:

(10)
- IGNITION "OFF."
- RECONNECT BATTERY CABLES.
- DISCONNECT ELECTRONIC BRAKE CONTROL RELAY.
- USING J 39200, MEASURE VOLTAGE BETWEEN GROUND AND ENGINE COMPARTMENT FUSE/RELAY CENTER TERMINAL "A8". IS VOLTAGE GREATER THAN 10 VOLTS?

NO: REPAIR OPEN IN B+ FEED CIRCUIT.

YES:

(11)
- USING J 39200, MEASURE RESISTANCE BETWEEN 2-WAY EBCM HARNESS CONNECTOR TERMINAL "A" AND ENGINE COMPARTMENT FUSE/RELAY CENTER TERMINAL "B6". IS RESISTANCE LESS THAN 2 OHMS?

NO: INSPECT IP TO ENGINE HARNESS CONNECTOR FOR A POOR CONNECTION OR POOR TERMINAL CONTACT. IS POOR TERMINAL CONTACT OR A POOR CONNECTION EVIDENT?
- NO: REPAIR OPEN IN CKT 1633.
- YES: REPLACE TERMINALS OR REPAIR CONNECTION INTEGRITY THAT EXHIBITS POOR CONTACT.

YES:

(12)
- USING A JUMPER WIRE, CONNECT ELECTRONIC BRAKE CONTROL RELAY TERMINAL "86" TO GROUND. THEN, USING A FUSED JUMPER WIRE SUCH AS J 36169 WITH A 3 AMP FUSE, CONNECT ENABLE RELAY TERMINAL "85" TO B+.
- USING J 39200, MEASURE RESISTANCE BETWEEN ENABLE RELAY TERMINALS "30" AND "87". IS RESISTANCE LESS THAN 2 OHMS?

NO: REPLACE ELECTRONIC BRAKE CONTROL RELAY.

YES: MALFUNCTION WAS DUE TO A POOR CONNECTION.

BETWEEN ELECTRONIC BRAKE CONTROL RELAY AND ENGINE COMPARTMENT FUSE/RELAY CENTER AND/OR AT POSITIVE BATTERY TERMINAL. REPLACE OR REPAIR TERMINALS THAT EXHIBIT POOR TERMINAL CONTACT.

DIAGNOSTIC CHART DTC 14 — 1995 CAVALIER AND SUNFIRE

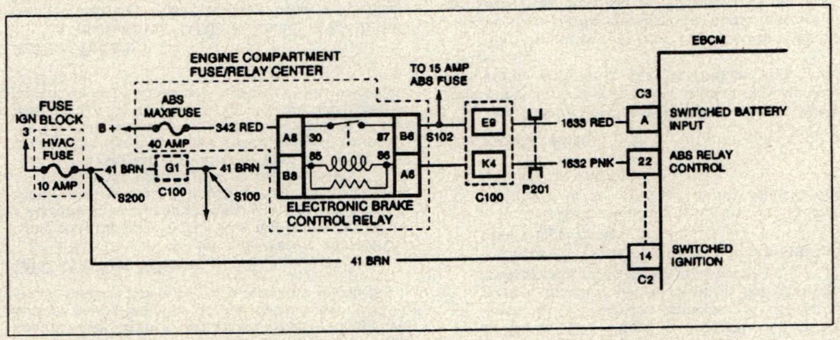

DTC 14
(Page 3 of 3)
ELECTRONIC BRAKE CONTROL RELAY CONTACT CIRCUIT OPEN

Circuit Description:
Ignition voltage is supplied through terminal "85" of the electronic brake control relay. The EBCM is then able to energize the pull-in coil by completing the ground circuit at EBCM terminal "22". The magnetic field created closes the electronic brake control relay contacts and allows battery voltage and current to be supplied to the EMB's and EBCM, which supplies power to the motors and solenoids.

DTC Will Set When:
DTC 14 can be set anytime after the EBCM commands the electronic brake control relay "ON" (the relay is first commanded "ON" during the three second bulb check). This test monitors the availability of current/voltage to the motors, solenoids, and EMB's. This malfunction indicates voltage is not available and would therefore not allow ABS operation if required.

Action Taken:
A malfunction DTC is stored, ABS is disabled and the "ABS" warning lamp is turned "ON."

DTC Chart Test Description:
Number(s) below refer to circled number(s) on the diagnostic chart.
13. Checks for excessive resistance in the electronic brake control relay coil.
14. Checks for excessive resistance in the electronic brake control relay control circuit between the relay and the EBCM.
15. Checks for excessive resistance in the ignition switch "run" fuse feed circuit.

Diagnostic Aids:
An "intermittent" malfunction is most likely caused by a poor connection, rubbed through wire insulation, or a wire that is broken inside the insulation.

The frequency of the malfunction can be checked by using the enhanced diagnostic function of the Tech 1, as described in "Enhanced Diagnostics," found in this section.

Any circuitry that is suspected as causing the "intermittent" complaint should be thoroughly checked for backed out terminals, improper mating, broken locks, improperly formed or damaged terminals, poor terminal to wiring connections or physical damage to the wiring harness.

Vibration, Temperature Effects:
Check for vibration effects by performing the relay test function of the Tech 1. With the relay commanded "ON," lightly tap the top and sides of the relay while monitoring battery voltage. If the battery voltage changes significantly, replace the relay.

If DTC 14 only sets when the vehicle is initially started in cold ambient conditions (temperature less than 32°F-0°C),replace the electronic brake control relay.

DIAGNOSTIC CHART DTC 14 — 1995 CAVALIER AND SUNFIRE

DTC 14
(Page 3 of 3)
ELECTRONIC BRAKE CONTROL RELAY CONTACTS CIRCUIT OPEN

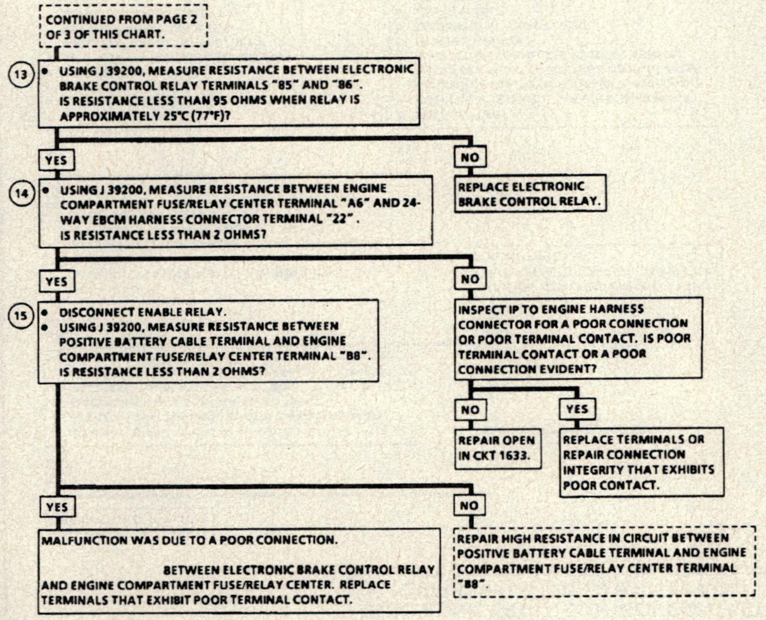

DIAGNOSTIC CHART DTC 15 — 1995 CAVALIER AND SUNFIRE

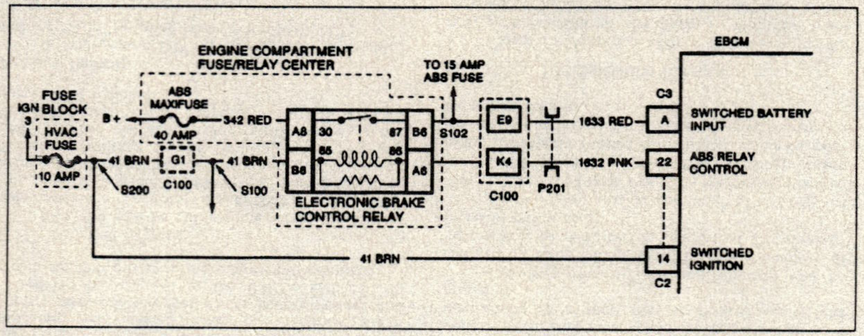

DTC 15

ELECTRONIC BRAKE CONTROL RELAY CONTACT CIRCUIT SHORTED TO BATTERY OR ALWAYS CLOSED

Circuit Description:

Ignition voltage is supplied through terminal "85" of the electronic brake control relay. The EBCM is then able to energize the pull-in coil by completing the ground circuit at terminal "22" of the EBCM. The magnetic field created closes the electronic brake control relay contacts and allows battery voltage and current to be supplied to the EMB's and EBCM, which supplies power to the motors and solenoids.

DTC Will Set When:

DTC 15 can be set only before the EBCM commands the relay "ON." This test determines if the electronic brake control relay is energized when it should not be. This malfunction would not allow the electronic brake control relay to remove power to the ABS. If a second malfunction were to occur that requires the electronic brake control relay to be turned "OFF," that malfunction cannot be removed if the relay cannot be controlled. The malfunction must be present for three consecutive ignition cycles before the DTC is set.

Action Taken:

A malfunction DTC is stored. ABS is not disabled and the "ABS" warning lamp will remain "OFF."

DTC Chart Test Description:

Number(s) below refer to circled number(s) on the diagnostic chart.

1. Indicates if the EBCM is capable of controlling the electronic brake control relay as commanded.
2. Checks for voltage at terminal "A" of the 2-way EBCM harness connector. If voltage is present, the malfunction exists in the electronic brake control relay and/or its circuitry.
3. Checks for a short to voltage in the switched battery input circuit.
4. Ensures the malfunction is not due to a short to BATTERY in the switched battery input circuit with EBCM connected.

Diagnostic Aids:

An "intermittent" malfunction is most likely caused by a poor connection, rubbed through wire insulation, or a wire that is broken inside the insulation.

The frequency of the malfunction can be checked by using the enhanced diagnostic function of the Tech 1, as described in "Enhanced Diagnostics," found in this section. Any circuitry, that is suspected as causing the intermittent complaint, should be thoroughly checked for backed out terminals, improper mating, broken locks, improperly formed or damaged terminals, poor terminal to wiring connections or physical damage to the wiring harness.

DIAGNOSTIC CHART DTC 15 — 1995 CAVALIER AND SUNFIRE

DTC 15
ELECTRONIC BRAKE CONTROL CONTACTS CIRCUIT SHORTED TO BATTERY OR

ARE ANY OTHER DTCs SET?

- YES: PROCEED TO THAT DTC FIRST.
- NO:
  - (1) • IGNITION "ON." • USE RELAY TEST FUNCTION OF TECH 1 AND COMMAND ELECTRONIC BRAKE CONTROL RELAY "OFF." DOES TECH 1 INDICATE THAT ELECTRONIC BRAKE CONTROL RELAY IS "OFF" AND BATTERY VOLTAGE IS 5.0V OR LESS?
    - YES: MALFUNCTION IS NOT PRESENT AT THIS TIME. REFER TO "DIAGNOSTIC AIDS" ON FACING PAGE.
    - NO:
      - (2) • IGNITION "OFF." • DISCONNECT 24-WAY AND 2-WAY EBCM CONNECTORS. • IGNITION "ON." • MEASURE VOLTAGE BETWEEN GROUND AND TERMINAL "A" OF THE 2-WAY EBCM HARNESS CONNECTOR. IS VOLTAGE 2.0V OR LESS?
        - NO:
          - (3) • IGNITION "OFF." • DISCONNECT ELECTRONIC BRAKE CONTROL RELAY. • IGNITION "ON." • MEASURE VOLTAGE BETWEEN GROUND AND TERMINAL "A" OF THE 2-WAY EBCM HARNESS CONNECTOR. IS VOLTAGE 2.0V OR LESS?
            - YES: REPLACE ELECTRONIC BRAKE CONTROL RELAY.
            - NO: REPAIR SHORT TO VOLTAGE IN CKT 1633.
        - YES:
          - (4) • IGNITION "OFF." • INSPECT CKT 1633 AND 24-WAY EBCM HARNESS CONNECTOR FOR PHYSICAL DAMAGE WHICH MAY RESULT IN A SHORT TO VOLTAGE WITH THE EBCM CONNECTORS CONNECTED TO EBCM. CORRECT DAMAGE IF EVIDENT. • RECONNECT ALL CONNECTORS. • IGNITION "ON." • USING TECH 1 RELAY TEST FUNCTION, COMMAND ELECTRONIC BRAKE CONTROL RELAY "OFF." DOES TECH 1 INDICATE ELECTRONIC BRAKE CONTROL ENABLE RELAY IS "OFF" AND BATTERY VOLTAGE IS LESS THAN 5 VOLTS?
            - NO: REPLACE EBCM.
            - YES: MALFUNCTION IS INTERMITTENT. REFER TO "DIAGNOSTIC AIDS" ON FACING PAGE.

• AFTER DIAGNOSIS IS COMPLETE, CLEAR DTCs AND TEST DRIVE VEHICLE FOR THREE (3) DRIVE CYCLES TO VERIFY DTC DOES NOT RESET. A DRIVE CYCLE CONSISTS OF STARTING THE VEHICLE, DRIVING ABOVE 16 km/h (10 MPH) AND THEN TURNING IGNITION "OFF."

DIAGNOSTIC CHART DTC 16 — 1995
CAVALIER AND SUNFIRE

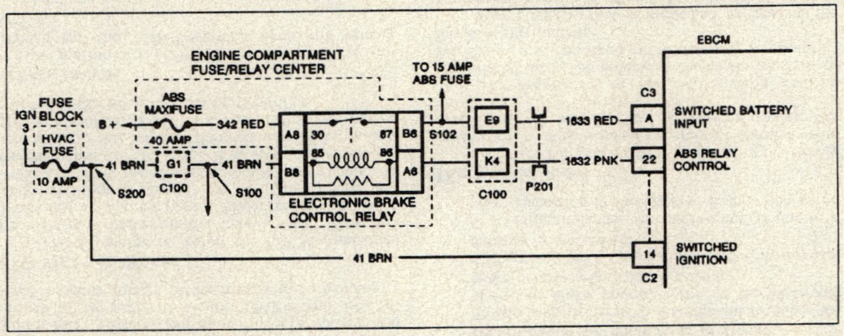

DTC 16
ELECTRONIC BRAKE CONTROL RELAY COIL CIRCUIT OPEN

Circuit Description:

Ignition voltage is supplied through terminal "85" of the electronic brake control relay. The EBCM is then able to energize the pull-in coil by completing the ground circuit at terminal "22" of the EBCM. The magnetic field created closes the electronic brake control relay contacts and allows battery voltage and current to be supplied to the EMB's and EBCM, which supplies power to the motors and solenoids.

DTC Will Set When:

DTC 16 can be set anytime after the electronic brake control relay is commanded "ON." This test detects an open in the electronic brake control relay coil circuit. An open in this circuit will not allow the electronic brake control relay to be energized thus preventing power to the motors, EMB's and solenoids. A code 86 will be set with code 16 if the rear channel is not expected to be in the home position.

Action Taken:

A malfunction DTC is stored, ABS is disabled and the "ABS" warning lamp is turned "ON."

DTC Chart Test Description:

Number(s) below refer to circled number(s) on the diagnostic chart.

1. Indicates if the EBCM is capable of controlling the electronic brake control relay as commanded.
2. Ensures that there is continuity through the pull-in coil of the electronic brake control relay.
3. Checks to ensure that voltage is available to the pull-in coil of the electronic brake control relay.
4. Checks continuity in the electronic brake control relay control circuit.
5. Ensures DTC was not set due to a poor connection.
6. Identifies whether the malfunction is due to a poor connection or the EBCM.
7. Identifies whether the malfunction is due to a poor connection in the Dash to Engine connector.

Diagnostic Aids:

An "intermittent" malfunction is most likely caused by a poor connection, rubbed through wire insulation, or a wire that is broken inside the insulation.

The frequency of the malfunction can be checked by using the enhanced diagnostic function of the Tech 1, as described in "Enhanced Diagnostics," found in this section.

If the frequency of the malfunction is high, but is currently intermittent, check for high coil resistance by measuring resistance between relay terminals "86" and "85". If resistance is greater than 100 ohms, replace the relay.

Any circuitry, that is suspected as causing the intermittent complaint, should be thoroughly checked for backed out terminals, improper mating, broken locks, improperly formed or damaged terminals, poor terminal to wiring connections or physical damage to the wiring harness.

DIAGNOSTIC CHART DTC 16 — 1995
CAVALIER AND SUNFIRE

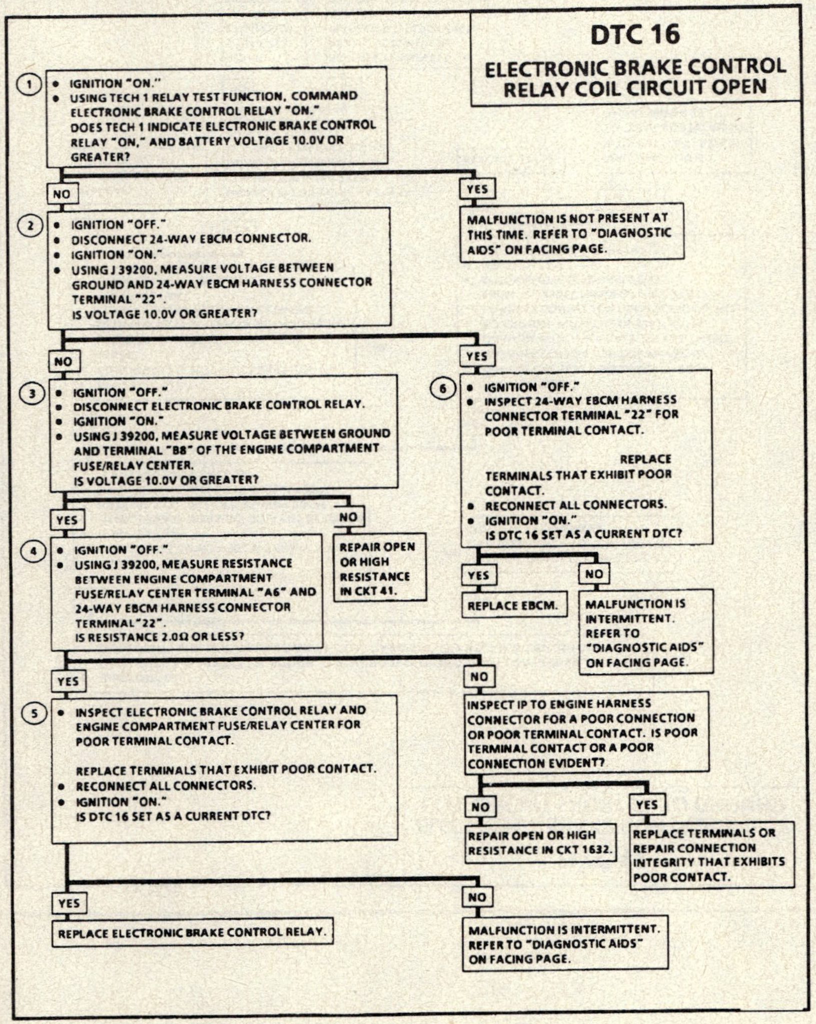

DIAGNOSTIC CHART DTC 17 — 1995 CAVALIER AND SUNFIRE

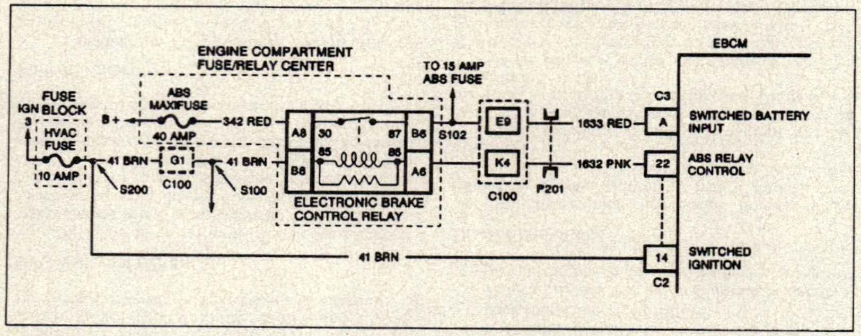

DTC 17
ELECTRONIC BRAKE CONTROL RELAY COIL CIRCUIT SHORTED TO GROUND

Circuit Description:

Ignition voltage is supplied through terminal "85" of the electronic brake control relay. The EBCM is then able to energize the pull-in coil by completing the ground circuit at terminal "22" of the EBCM. The magnetic field created closes the electronic brake control relay contacts and allows battery voltage and current to be supplied to the EMB's and EBCM, which supplies voltage to the motors and solenoids.

DTC Will Set When:

DTC 17 can be set before the EBCM commands the electronic brake control relay "ON." This test determines if the electronic brake control relay is energized when it should not be. This malfunction would not allow the electronic brake control relay to remove power to the ABS. If a second malfunction were to occur that requires the electronic brake control relay to be turned "OFF," that malfunction can not be removed if the relay can not be controlled.

Action Taken:

A malfunction DTC is stored. ABS is not disabled and the "ABS" warning lamp will remain "OFF."

DTC Chart Test Description:

Number(s) below refer to circled number(s) on the diagnostic chart.

1. Indicates if the EBCM is capable of controlling the electronic brake control relay as commanded.
2. Ensures the electronic brake control relay or control circuit are not shorted to ground.
3. This step isolates the malfunction to either the electronic brake control relay or the electronic brake control relay control circuit.
4. Ensures short to ground is not due to physical damage of the circuitry.
5. Identifies whether the malfunction is due to a poor connection in the Dash to Engine connector.

Diagnostic Aids:

An "intermittent" malfunction is most likely caused by a poor connection, rubbed through wire insulation, or a wire that is broken inside the insulation.

The frequency of the malfunction can be checked by using the enhanced diagnostic function of the Tech 1, as described in "Enhanced Diagnostics," found in this section.

Any circuitry, that is suspected as causing the intermittent complaint, should be thoroughly checked for backed out terminals, improper mating, broken locks, improperly formed or damaged terminals, poor terminal to wiring connections or physical damage to the wiring harness.

DIAGNOSTIC CHART DTC 17 — 1995 CAVALIER AND SUNFIRE

DTC 17
ELECTRONIC BRAKE CONTROL RELAY COIL CIRCUIT SHORTED TO GROUND

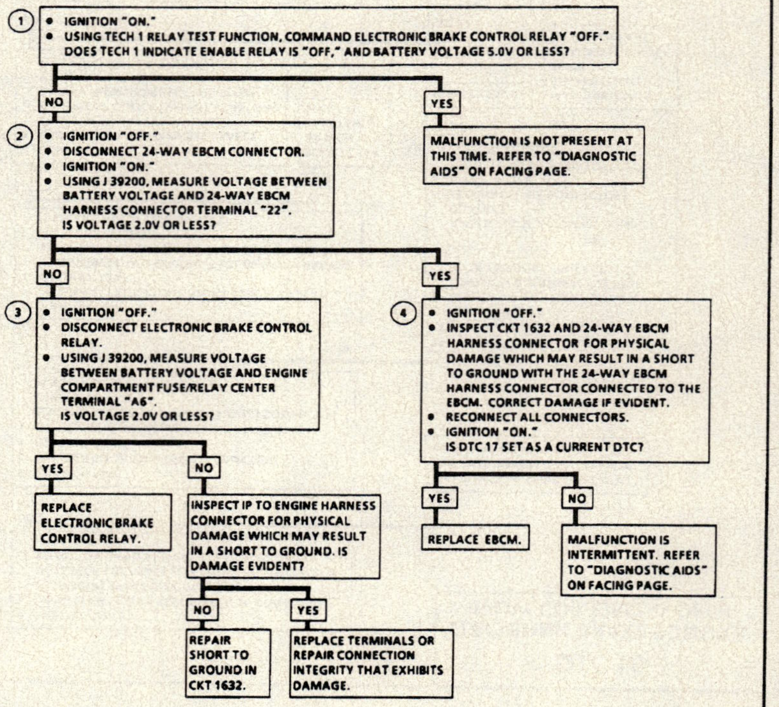

DIAGNOSTIC CHART DTC 18 — 1995 CAVALIER AND SUNFIRE

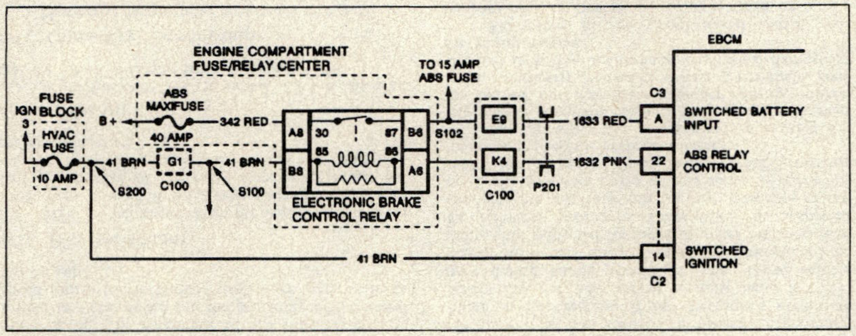

DTC 18

ELECTRONIC BRAKE CONTROL RELAY COIL CIRCUIT SHORTED TO BATTERY OR COIL SHORTED

Circuit Description:

Ignition voltage is supplied through terminal "85" of the electronic brake control relay. The EBCM is then able to energize the pull-in coil by completing the ground circuit at terminal "22" of the EBCM. The magnetic field created closes the electronic brake control relay contacts and allows battery voltage and current to be supplied to the EMB's and EBCM, which supplies voltage to the motors and solenoids.

DTC Will Set When:

DTC 18 can be set after the EBCM commands the electronic brake control relay "ON." This test monitors the availability of current/voltage to the motors, solenoids and EMB's. A malfunction of this type will not allow voltage to be available to the motors, solenoids and EMB's; therefore, ABS operation would not be allowed if required. DTC 86 is set with DTC 18 if the rear channel is not expected to be in the home position.

Action Taken:

A malfunction DTC is stored, ABS is disabled and the "ABS" warning lamp is turned "ON."

DTC Chart Test Description:

Number(s) below refer to circled number(s) on the diagnostic chart.

1. Indicates if the EBCM is capable of controlling the electronic brake control relay as commanded.
2. With the electronic brake control relay removed, voltage should not be available at terminal "86". Any voltage at this point would indicate that the electronic brake control relay circuit is shorted to a voltage source.
3. Checks for a short across the electronic brake control relay coil.
4. Checks for an internal short in the electronic brake control relay between the BATTERY feed terminal "30" and the relay pull-in coil.
5. Ensures short to BATTERY was not due to physical damage of the electronic brake control relay control circuit.
6. Identifies whether the malfunction is due to a shorted connection in the Dash to Engine connector.

Diagnostic Aids:

An "intermittent" malfunction is most likely caused by a poor connection, rubbed through wire insulation, or a wire that is broken inside the insulation.

The frequency of the malfunction can be checked by using the enhanced diagnostic function of the Tech 1, as described in "Enhanced Diagnostics," found in this section.

Any circuitry, that is suspected as causing the intermittent complaint, should be thoroughly checked for backed out terminals, improper mating, broken locks, improperly formed or damaged terminals, poor terminal to wiring connections or physical damage to the wiring harness.

DIAGNOSTIC CHART DTC 18 — 1995 CAVALIER AND SUNFIRE

DTC 18

ELECTRONIC BRAKE CONTROL RELAY COIL CIRCUIT SHORTED TO BATTERY OR COIL SHORTED

(1)
- IGNITION "ON."
- USING RELAY TEST FUNCTION OF TECH 1, COMMAND ELECTRONIC BRAKE CONTROL RELAY "ON." DOES TECH 1 INDICATE ELECTRONIC BRAKE CONTROL RELAY "ON" AND BATTERY VOLTAGE 10.0V OR GREATER?

  - YES → MALFUNCTION IS NOT PRESENT AT THIS TIME. REFER TO "DIAGNOSTIC AIDS" ON FACING PAGE.
  - NO → (2)

(2)
- IGNITION "OFF."
- DISCONNECT ELECTRONIC BRAKE CONTROL RELAY.
- DISCONNECT 24-WAY EBTCM CONNECTOR.
- IGNITION "ON."
- USING J 39200, MEASURE VOLTAGE BETWEEN GROUND AND ENGINE COMPARTMENT FUSE/RELAY CENTER TERMINAL "A6". IS VOLTAGE 2.0V OR LESS?

  - NO → INSPECT IP TO ENGINE HARNESS CONNECTOR FOR PHYSICAL DAMAGE WHICH MAY RESULT IN A SHORT TO BATTERY. IS DAMAGE EVIDENT?
    - NO → REPAIR SHORT TO GROUND IN CKT 1632.
    - YES → REPLACE TERMINALS OR REPAIR CONNECTION INTEGRITY THAT EXHIBITS DAMAGE.
  - YES → (3)

(3)
- USING J 39200, MEASURE RESISTANCE BETWEEN ELECTRONIC BRAKE CONTROL RELAY TERMINALS "85" AND "86". IS RESISTANCE 40.0Ω OR GREATER?

  - NO → REPLACE ELECTRONIC BRAKE CONTROL RELAY.
  - YES → (4)

(4)
- USING J 39200, MEASURE RESISTANCE BETWEEN ELECTRONIC BRAKE CONTROL RELAY TERMINALS "86" AND "30". DOES J 39200 DISPLAY RESISTANCE AS "OL" (INFINITE)?

  - NO → REPLACE ELECTRONIC BRAKE CONTROL RELAY.
  - YES → (5)

(5)
- IGNITION "OFF."
- INSPECT CKT 1632 AND 24-WAY EBTCM HARNESS CONNECTOR FOR PHYSICAL DAMAGE WHICH MAY RESULT IN A SHORT TO BATTERY WITH THE 24-WAY EBTCM HARNESS CONNECTOR CONNECTED TO THE EBTCM. CORRECT DAMAGE IF EVIDENT.
- RECONNECT ALL CONNECTORS.
- IGNITION "ON."
- IS DTC 18 SET AS A CURRENT DTC?

  - YES → REPLACE EBTCM.
  - NO → MALFUNCTION IS INTERMITTENT. REFER TO "DIAGNOSTIC AIDS" ON FACING PAGE.

DIAGNOSTIC CHART DTC 21 — 1995 CAVALIER AND SUNFIRE

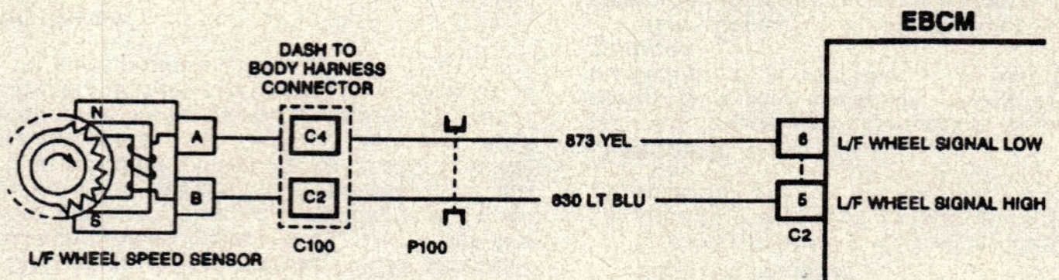

DTC 21

LEFT FRONT WHEEL SPEED = 0 OR UNREASONABLE

Circuit Description:

As a toothed ring passes by the wheel speed sensor, changes in the electromagnetic field cause the wheel speed sensor to produce a sinusoidal (AC) voltage signal whose frequency is proportional to wheel speed. The magnitude of this signal is directly related to wheel speed and the proximity of the wheel speed sensor to the toothed ring, often referred to as the air gap.

DTC Will Set When:

DTC 21 can be set when the vehicle is not in an ABS stop. If the left front wheel speed is less than half the average of the other three wheels and the vehicle speed is greater than 8 km/h (5 mph), a malfunction exists.

Action Taken:

A malfunction DTC is stored, ABS is disabled and the "ABS" warning lamp is turned "ON."

DTC Chart Test Description:

Number(s) below refer to circled number(s) on the diagnostic chart.

1. Identifies if a wheel speed sensor or circuitry is damaged and visibly apparent.
2. Checks wheel speed sensor for proper resistance value.
3. Ensures the wheel speed sensor and toothed ring generate the proper voltage.
4. Ensures the wheel speed sensor is not shorted to ground.

Diagnostic Aids:

An "intermittent" malfunction is most likely caused by a poor connection, rubbed through wire insulation, or a wire that is broken inside the insulation. The frequency of the malfunction can be checked by using the enhanced diagnostic function of the Tech 1, as described in "Enhanced Diagnostics," found in this section. If the customer's comments reflect that the ABS amber warning lamp is "ON" only during humid conditions (rain, snow, vehicle wash), all wheel speed sensor circuitry should be thoroughly inspected for signs of water intrusion. Use the following procedure. Spray down the suspected area with 5% salt water solution (two teaspoons of salt to 12 oz. of water). Drive vehicle above 24 km/h (15 mph) for at least 30 seconds. If DTC returns immediately, replace suspected harness.

Any circuitry, that is suspected as causing the intermittent complaint, should be thoroughly checked for backed out terminals, improper mating, broken locks, improperly formed or damaged terminals, poor terminal to wiring connections or physical damage to the wiring harness.

Resistance of the wheel speed sensor will increase with an increase in sensor temperature.

When replacing a wheel speed sensor, inspect the sensor terminals and harness connector for corrosion and/or water intrusion. If evidence of corrosion or water intrusion exists, replace wheel speed sensor harness. Refer to "On-Vehicle Service" in this section.

If wheel speed sensor harness shows evidence of water intrusion, inspect wheel speed sensor for water intrusion. If water intrusion is evident, replace wheel speed sensor.

DIAGNOSTIC CHART DTC 21 — 1995 CAVALIER AND SUNFIRE

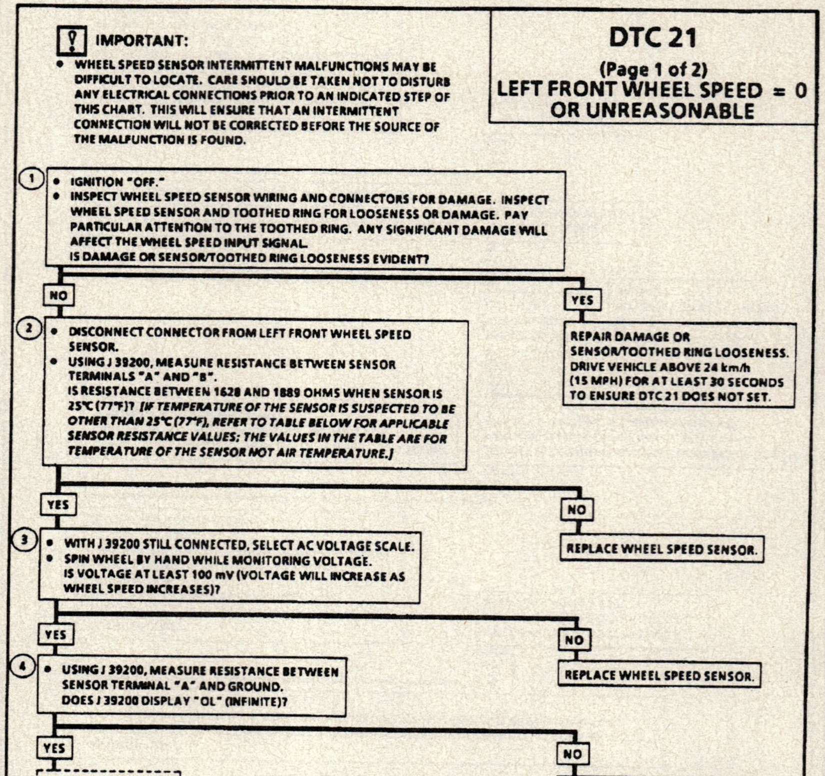

FRONT WHEEL SPEED SENSOR

TEMPERATURE OF SENSOR VS. SENSOR RESISTANCE (APPROXIMATE)

TEMP. (°C)	TEMP. (°F)	RESISTANCE (OHMS)
-34 TO 4	-30 TO 40	1258 TO 1739
5 TO 43	41 TO 110	1503 TO 2023
44 TO 93	111 TO 200	1747 TO 2384

DIAGNOSTIC CHART DTC 21 — 1995 CAVALIER AND SUNFIRE

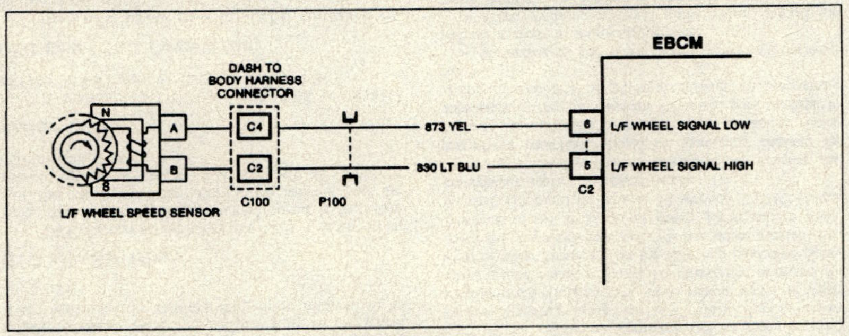

DTC 21
(Page 2 of 2)
LEFT FRONT WHEEL SPEED = 0 OR UNREASONABLE

Circuit Description:

As a toothed ring passes by the wheel speed sensor, changes in the electromagnetic field cause the wheel speed sensor to produce a sinusoidal (AC) voltage signal whose frequency is proportional to wheel speed. The magnitude of this signal is directly related to wheel speed and the proximity of the wheel speed sensor to the toothed ring often referred to as the air gap.

DTC Will Set When:

DTC 21 can be set when the vehicle is not in an ABS stop If the left front wheel speed is less than half the average of the other three wheels and the vehicle speed is greater than 8 km/h (5 mph), a malfunction exists.

Action Taken:

A malfunction DTC is stored, ABS is disabled and the "ABS" warning lamp is turned "ON."

DTC Chart Test Description:

Number(s) below refer to circled number(s) on the diagnostic chart.

5. Checks for proper voltages at speed sensor harness connector.
6. Ensures that the wheel speed sensor circuitry is not shorted to ground.
7. Checks for an open in the wheel signal low circuit.
8. Checks for an open in the wheel signal high circuit.
9. Ensures that DTC 21 was not set due to a poor connection at the EBCM or physical damage of the circuitry.
10. Checks for a short to BATTERY in either the high or low wheel signal circuits.
11. Ensures short to BATTERY was not due to physical damage of the circuitry.
12. Tests for a short between the wheel signal high and low circuits.
13. Ensures malfunction was not due to a poor connection at the wheel speed sensor or physical damage of the circuitry.

Diagnostic Aids:

An "intermittent" malfunction is most likely caused by a poor connection, rubbed through wire insulation, or a wire that is broken inside the insulation.

The frequency of the malfunction can be checked by using the enhanced diagnostic function of the Tech 1, as described in "Enhanced Diagnostics," found in this section.

If the customer's comments reflect that the ABS amber warning lamp is "ON" only during moist environmental changes, (rain, snow, vehicle wash) all wheel speed sensor circuitry should be thoroughly inspected for signs of water intrusion. Use the following procedure. Spray down the suspected area with 5% salt water solution (two teaspoons of salt to 12 oz. of water). Drive vehicle above 24 km/h (15 mph) for at least 30 seconds. If DTC returns immediately, replace suspected harness.

Any circuitry, that is suspected as causing the intermittent complaint, should be thoroughly checked for backed out terminals, improper mating, broken locks, improperly formed or damaged terminals, poor terminal to wiring connections or physical damage to the wiring harness. Resistance of the wheel speed sensor will increase with an increase in sensor temperature. When replacing a wheel speed sensor, inspect the sensor terminals and harness connector for corrosion and/or water intrusion. If evidence of corrosion or water intrusion exists, replace wheel speed sensor harness. Refer to "On-Vehicle Service" in this section.

If wheel speed sensor harness shows evidence of water intrusion, inspect wheel speed sensor for water intrusion. If water intrusion is evident, replace wheel speed sensor.

DIAGNOSTIC CHART DTC 21 — 1995 CAVALIER AND SUNFIRE

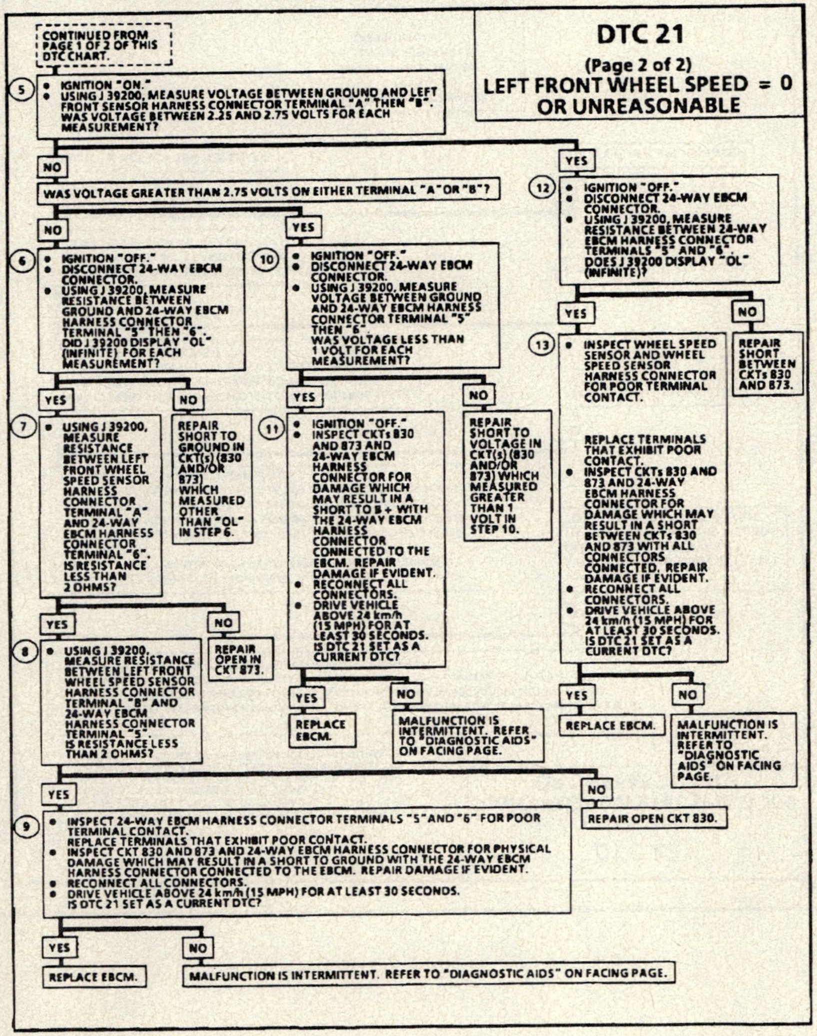

DIAGNOSTIC CHART DTC 22 — 1995 CAVALIER AND SUNFIRE

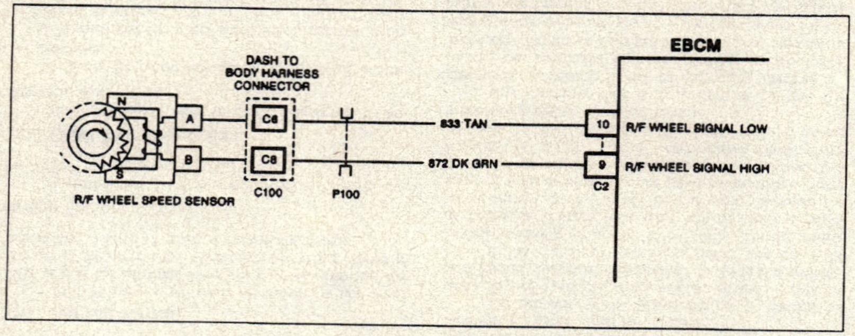

DTC 22
(Page 1 of 2)
RIGHT FRONT WHEEL SPEED = 0 OR UNREASONABLE

Circuit Description:

As a toothed ring passes by the wheel speed sensor, changes in the electromagnetic field cause the wheel speed sensor to produce a sinusoidal (AC) voltage signal whose frequency is proportional to wheel speed. The magnitude of this signal is directly related to wheel speed and the proximity of the wheel speed sensor to the toothed ring, often referred to as the air gap.

DTC Will Set When:

DTC 22 can be set when the vehicle is not in an ABS stop. If the right front wheel speed is less than half the average of the other three wheels and the vehicle speed is greater than 8 km/h (5 mph), a malfunction exists.

Action Taken:

A malfunction DTC is stored, ABS is disabled and the "ABS" warning lamp is turned "ON."

DTC Chart Test Description:

Number(s) below refer to circled number(s) on the diagnostic chart.

1. Identifies a wheel speed sensor or circuitry that is damaged and visibly apparent.
2. Checks wheel speed sensor for proper resistance value.
3. Ensures the wheel speed sensor and sensor ring generate the proper voltage.
4. Ensures the wheel speed sensor is not shorted to ground.

Diagnostic Aids:

An "intermittent" malfunction is most likely caused by a poor connection, rubbed through wire insulation, or a wire that is broken inside the insulation.

The frequency of the malfunction can be checked by using the enhanced diagnostic function of the Tech 1, as described in "Diagnosis Electrical," found in this section.

If the customer's comments reflect that the ABS amber warning lamp is "ON" only during moist environmental changes, (rain, snow, vehicle wash) all wheel speed sensor circuitry should be thoroughly inspected for signs of water intrusion. Use the following procedure. Spray down the suspected area with 5% salt water solution (two teaspoons of salt to 12 oz. of water). Drive vehicle above 24 km/h (15 mph) for at least 30 seconds. If DTC returns immediately, replace suspected harness.

Any circuitry, that is suspected as causing the intermittent complaint, should be thoroughly checked for backed out terminals, improper mating, broken locks, improperly formed or damaged terminals, poor terminal to wiring connections or physical damage to the wiring harness.

Resistance of the wheel speed sensor will increase with an increase in sensor temperature.

When replacing a wheel speed sensor, inspect the sensor terminals and harness connector for corrosion and/or water intrusion. If evidence of corrosion or water intrusion exists, replace wheel speed sensor harness. Refer to "On-Vehicle Service" in this section.

If wheel speed sensor harness shows evidence of water intrusion, inspect wheel speed sensor for water intrusion. If water intrusion is evident, replace wheel speed sensor.

DIAGNOSTIC CHART DTC 22 — 1995 CAVALIER AND SUNFIRE

DTC 22
(Page 1 of 2)
RIGHT FRONT WHEEL SPEED = 0 OR UNREASONABLE

IMPORTANT:
- WHEEL SPEED SENSOR INTERMITTENT MALFUNCTIONS MAY BE DIFFICULT TO LOCATE. CARE SHOULD BE TAKEN NOT TO DISTURB ANY ELECTRICAL CONNECTIONS PRIOR TO AN INDICATED STEP OF THIS CHART. THIS WILL ENSURE THAT AN INTERMITTENT CONNECTION WILL NOT BE CORRECTED BEFORE THE SOURCE OF THE MALFUNCTION IS FOUND.

(1)
- IGNITION "OFF."
- INSPECT WHEEL SPEED SENSOR WIRING AND CONNECTORS FOR DAMAGE. INSPECT WHEEL SPEED SENSOR AND TOOTHED RING FOR LOOSENESS OR DAMAGE. PAY PARTICULAR ATTENTION TO THE TOOTHED RING. ANY SIGNIFICANT DAMAGE WILL AFFECT THE WHEEL SPEED INPUT SIGNAL.

IS DAMAGE OR SENSOR/TOOTHED RING LOOSENESS EVIDENT?

- YES: REPAIR DAMAGE OR SENSOR/TOOTHED RING LOOSENESS. DRIVE VEHICLE ABOVE 24 km/h (15 MPH) FOR AT LEAST 30 SECONDS TO ENSURE DTC 22 DOES NOT SET.
- NO: go to step 2.

(2)
- DISCONNECT CONNECTOR FROM RIGHT FRONT WHEEL SPEED SENSOR.
- USING J 39200, MEASURE RESISTANCE BETWEEN SENSOR TERMINALS "A" AND "B".

IS RESISTANCE BETWEEN 1628 AND 1889 OHMS WHEN SENSOR IS 25°C (77°F)? *[IF TEMPERATURE OF THE SENSOR IS SUSPECTED TO BE OTHER THAN 25°C (77°F), REFER TO TABLE BELOW FOR APPLICABLE SENSOR RESISTANCE VALUES; THE VALUES IN THE TABLE ARE FOR TEMPERATURE OF THE SENSOR NOT AIR TEMPERATURE.]*

- NO: REPLACE WHEEL SPEED SENSOR.
- YES: go to step 3.

(3)
- WITH J 39200 STILL CONNECTED, SELECT AC VOLTAGE SCALE.
- SPIN WHEEL BY HAND WHILE MONITORING VOLTAGE.

IS VOLTAGE AT LEAST 100 mV (VOLTAGE WILL INCREASE AS WHEEL SPEED INCREASES)?

- NO: REPLACE WHEEL SPEED SENSOR.
- YES: go to step 4.

(4)
- USING J 39200, MEASURE RESISTANCE BETWEEN SENSOR TERMINAL "A" AND GROUND.

DOES J 39200 DISPLAY "OL" (INFINITE)?

- NO: REPLACE WHEEL SPEED SENSOR.
- YES: GO TO PAGE 2 OF 2 OF THIS DTC CHART.

FRONT WHEEL SPEED SENSOR

TEMPERATURE OF SENSOR VS. SENSOR RESISTANCE (APPROXIMATE)

TEMP. (°C)	TEMP. (°F)	RESISTANCE (OHMS)
-34 TO 4	-30 TO 40	1258 TO 1739
5 TO 43	41 TO 110	1503 TO 2023
44 TO 93	111 TO 200	1747 TO 2384

DIAGNOSTIC CHART DTC 22 — 1995 CAVALIER AND SUNFIRE

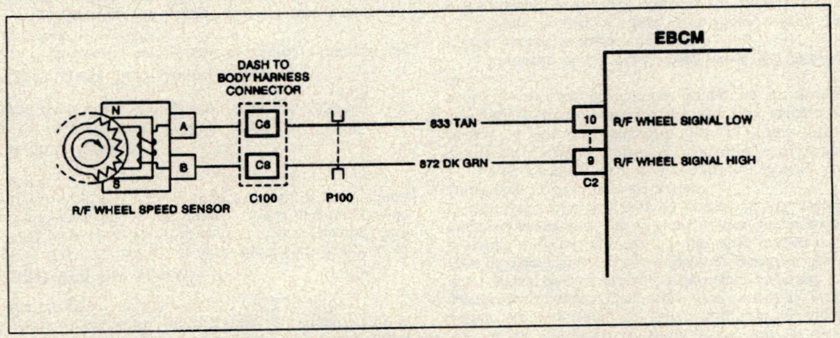

DTC 22
(Page 2 of 2)
RIGHT FRONT WHEEL SPEED = 0 OR UNREASONABLE

Circuit Description:

As a toothed ring passes by the wheel speed sensor, changes in the electromagnetic field cause the wheel speed sensor to produce a sinusoidal (AC) voltage signal whose frequency is proportional to wheel speed. The magnitude of this signal is directly related to wheel speed and the proximity of the wheel speed sensor to the toothed ring, often referred to as the air gap.

DTC Will Set When:

DTC 22 can be set when the vehicle is not in an ABS stop. If the right front wheel speed is less than half the average of the other three wheels and the vehicle speed is greater than 8 km/h (5 mph), a malfunction exists.

Action Taken:

A malfunction DTC is stored, ABS is disabled and the "ABS" warning lamp is turned "ON."

DTC Chart Test Description:

Number(s) below refer to circled number(s) on the diagnostic chart.

5. Checks for proper voltages at speed sensor harness connector.
6. Ensures that the wheel speed sensor circuitry is not shorted to ground.
7. Checks for an open in the wheel signal low circuit.
8. Checks for an open in the wheel signal high circuit.
9. Ensures that DTC 22 was not set due to a poor connection at the EBCM or physical damage of the circuitry.
10. Checks for a short to BATTERY in either the high or low wheel signal circuits.
11. Ensures short to BATTERY was not due to physical damage of the circuitry.
12. Tests for a short between the wheel signal high and low circuits.
13. Ensures malfunction was not due to a poor connection at the wheel speed sensor or physical damage of the circuitry.

Diagnostic Aids:

An "intermittent" malfunction is most likely caused by a poor connection, rubbed through wire insulation, or a wire that is broken inside the insulation.

The frequency of the malfunction can be checked by using the enhanced diagnostic function of the Tech 1, as described in "Enhanced Diagnostics," found in this section.

If the customer's comments reflect that the ABS amber warning lamp is "ON" only during moist environmental changes, (rain, snow, vehicle wash) all wheel speed sensor circuitry should be thoroughly inspected for signs of water intrusion. Use the following procedure. Spray down the suspected area with 5% salt water solution (two teaspoons of salt to 12 oz. of water). Drive vehicle above 24 km/h (15 mph) for at least 30 seconds. If DTC returns immediately, replace suspected harness.

Any circuitry, that is suspected as causing the intermittent complaint, should be thoroughly checked for backed out terminals, improper mating, broken locks, improperly formed or damaged terminals, poor terminal to wiring connections or physical damage to the wiring harness.

Resistance of the wheel speed sensor will increase with an increase in sensor temperature. When replacing a wheel speed sensor, inspect the sensor terminals and harness connector for corrosion and/or water intrusion. If evidence of corrosion or water intrusion exists, replace wheel speed sensor harness. Refer to "On-Vehicle Service" in this section.

If wheel speed sensor harness shows evidence of water intrusion, inspect wheel speed sensor for water intrusion. If water intrusion is evident, replace wheel speed sensor.

DIAGNOSTIC CHART DTC 22 — 1995 CAVALIER AND SUNFIRE

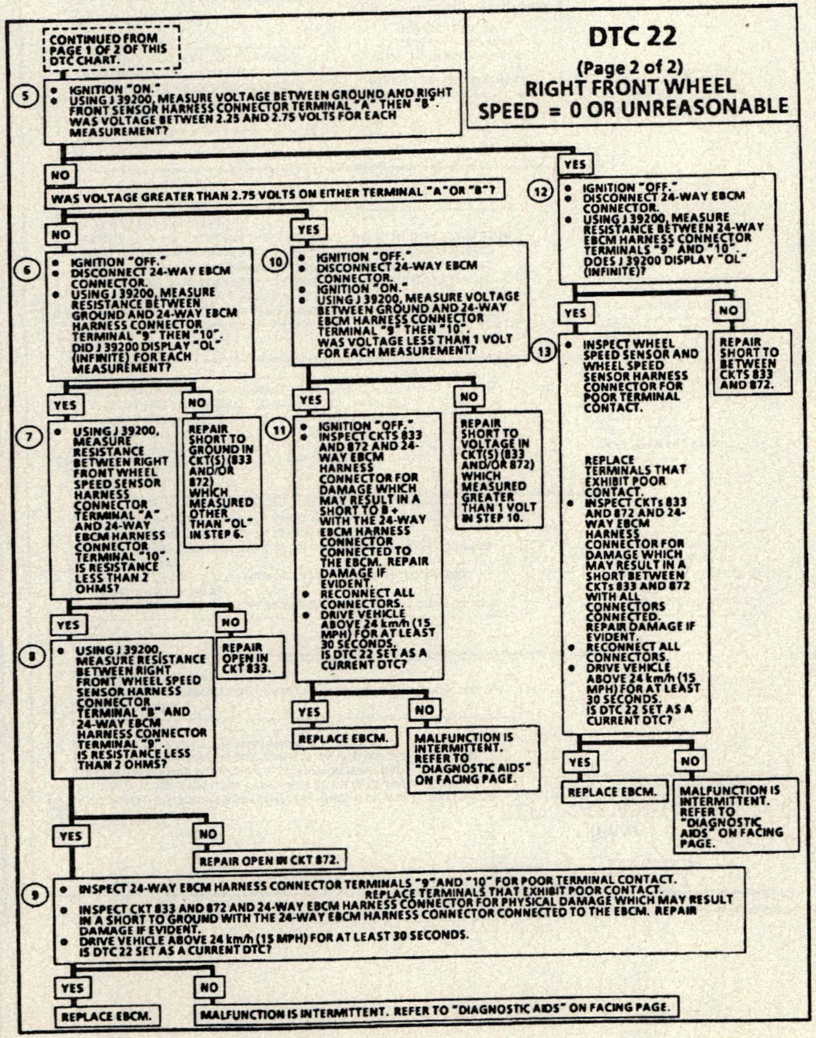

DIAGNOSTIC CHART DTC 23 — 1995 CAVALIER AND SUNFIRE

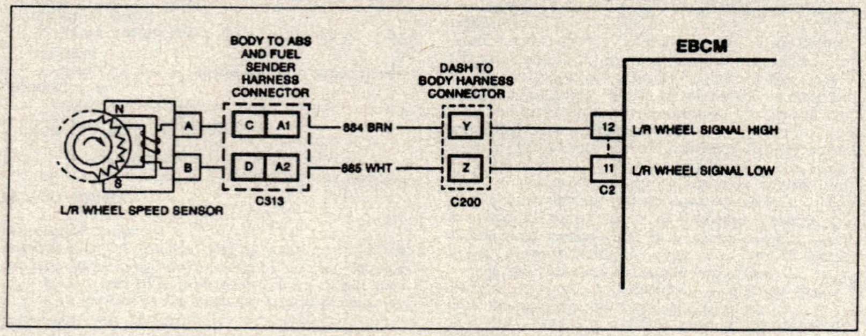

DTC 23
(Page 1 of 4)
LEFT REAR WHEEL SPEED = 0 OR UNREASONABLE

Circuit Description:

As a toothed ring passes by the wheel speed sensor, changes in the electromagnetic field cause the wheel speed sensor to produce a sinusoidal (AC) voltage signal whose frequency is proportional to wheel speed. The magnitude of this signal is directly related to wheel speed and the proximity of the wheel speed sensor to the toothed ring, often referred to as the air gap.

DTC Will Set When:

DTC 23 can be set when the vehicle is not in an ABS stop. If the left rear wheel speed is less than half the average of the other three wheels and the vehicle speed is greater than 8 km/h (5 mph), a malfunction exists.

Action Taken:

A malfunction DTC is stored, ABS is disabled and the "ABS" warning lamp is turned "ON."

DTC Chart Test Description:

Number(s) below refer to circled number(s) on the diagnostic chart.

1. This step will identify circuitry that is damaged and visibly apparent.
2. Checks for proper voltages at the rear sensor harness connector.
3. Checks for a short between the wheel signal high and low circuits.
4. This test checks wheel speed sensor for proper resistance value.
5. This step ensures the wheel speed sensor is not shorted to ground.

Diagnostic Aids:

An "intermittent" malfunction is most likely caused by a poor connection, rubbed through wire insulation, or a wire that is broken inside the insulation.

The frequency of the malfunction can be checked by using the enhanced diagnostic function of the Tech 1, as described in "Enhanced Diagnostics," found in this section.

If the customer's comments reflect that the ABS amber warning lamp is "ON" only during moist environmental changes, (rain, snow, vehicle wash) all wheel speed sensor circuitry should be thoroughly inspected for signs of water intrusion. Use the following procedure. Spray down the suspected area with 5% salt water solution (two teaspoons of salt to 12 oz. of water). Drive vehicle above 24 km/h (15 mph) for at least 30 seconds. If DTC returns immediately, replace suspected harness.

Any circuitry, that is suspected as causing the intermittent complaint, should be thoroughly checked for backed out terminals, improper mating, broken locks, improperly formed or damaged terminals, poor terminal to wiring connections or physical damage to the wiring harness.

Resistance of the wheel speed sensor will increase with an increase in sensor temperature.

When replacing a wheel speed sensor, inspect the sensor terminals and harness connector for corrosion and/or water intrusion. If evidence or corrosion or water intrusion exists, replace wheel speed sensor harness.

If wheel speed sensor harness shows evidence of water intrusion, inspect wheel speed sensor for water intrusion. If water intrusion is evident, replace wheel speed sensor.

DIAGNOSTIC CHART DTC 23 — 1995 CAVALIER AND SUNFIRE

DTC 23
(Page 1 of 4)
LEFT REAR WHEEL SPEED = 0 OR UNREASONABLE

Important

- WHEEL SPEED SENSOR INTERMITTENT MALFUNCTIONS MAY BE DIFFICULT TO LOCATE. CARE SHOULD BE TAKEN NOT TO DISTURB ANY ELECTRICAL CONNECTIONS PRIOR TO AN INDICATED STEP OF THIS CHART. THIS WILL ENSURE THAT AN INTERMITTENT CONNECTION WILL NOT BE CORRECTED BEFORE THE SOURCE OF THE MALFUNCTION IS FOUND.

(1)
- IGNITION "OFF."
- INSPECT WHEEL SPEED SENSOR WIRING AND CONNECTORS FOR DAMAGE. IS DAMAGE EVIDENT?

YES: REPAIR DAMAGE. DRIVE VEHICLE ABOVE 24 km/h (15 MPH) FOR AT LEAST 30 SECONDS TO ENSURE DTC 23 DOES NOT SET.

NO:

(2)
- REMOVE LEFT REAR INTEGRAL BEARING AND SPEED SENSOR ASSEMBLY (TUBULAR AXLE ONLY).
- DISCONNECT CONNECTOR FROM LEFT REAR WHEEL SPEED SENSOR.
- IGNITION "ON."
- USING J 39200, MEASURE VOLTAGE BETWEEN GROUND AND LEFT REAR SENSOR HARNESS CONNECTOR TERMINAL "A" THEN "B". WAS VOLTAGE BETWEEN 2.25V AND 2.75 V FOR EACH MEASUREMENT?

NO: GO TO PAGE 3 OF 4 OF THIS DTC CHART.

YES:

(3)
- IGNITION "OFF."
- DISCONNECT 24-WAY EBCM CONNECTOR.
- USING J 39200, MEASURE RESISTANCE BETWEEN 24-WAY EBCM HARNESS CONNECTOR TERMINALS "11" AND "12". DOES J 39200 DISPLAY RESISTANCE AS "OL" (INFINITE)?

NO: REPAIR SHORT BETWEEN CKTs 884 AND 885.

YES:

(4)
- USING J 39200, MEASURE THE RESISTANCE BETWEEN LEFT REAR WHEEL SPEED SENSOR TERMINALS "A" AND "B". IS RESISTANCE BETWEEN 2100 AND 2400 Ω WHEN SENSOR IS 25°C (77°F)? (IF TEMPERATURE OF THE SENSOR IS SUSPECTED TO BE OTHER THAN 25°C (77°F). REFER TO THE TABLE BELOW FOR APPLICABLE SENSOR RESISTANCE VALUES; THE VALUES IN THE TABLE ARE FOR TEMPERATURE OF THE SENSOR NOT AIR TEMPERATURE.)

NO: REPLACE INTEGRAL BEARING AND SPEED SENSOR ASSEMBLY.

YES:

(5)
- USING J 39200, MEASURE RESISTANCE BETWEEN SENSOR TERMINAL "A" AND INTEGRAL BEARING AND SPEED SENSOR ASSEMBLY MOUNTING FLANGE. DOES J 39200 DISPLAY "OL" (INFINITE)?

YES: CONTINUE DIAGNOSIS ON NEXT PAGE.

NO: REPLACE INTEGRAL BEARING AND SPEED SENSOR ASSEMBLY.

REAR WHEEL SPEED SENSOR

TEMPERATURE OF SENSOR VS. SENSOR RESISTANCE (APPROXIMATE)

TEMP. (°C)	TEMP. (°F)	RESISTANCE (OHMS)
-34 TO 4	-30 TO 40	1622 TO 2205
5 TO 43	41 TO 110	1930 TO 2566
44 TO 93	111 TO 200	2245 TO 3028

DIAGNOSTIC CHART DTC 23 — 1995 CAVALIER AND SUNFIRE

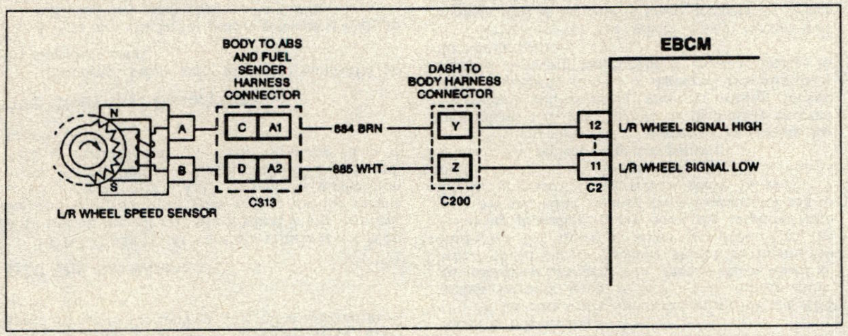

DTC 23
(Page 2 of 4)
LEFT REAR WHEEL SPEED = 0 OR UNREASONABLE

Circuit Description:

As a toothed ring passes by the wheel speed sensor, changes in the electromagnetic field cause the wheel speed sensor to produce a sinusoidal (AC) voltage signal whose frequency is proportional to wheel speed. The magnitude of this signal is directly related to wheel speed and the proximity of the wheel speed sensor to the toothed ring, often referred to as the air gap.

DTC Will Set When:

DTC 23 can be set when the vehicle is not in an ABS stop. If the left rear wheel speed is less than half the average of the other three wheels and the vehicle speed is greater than 8 km/h (5 mph), a malfunction exists.

Action Taken:

A malfunction DTC is stored, ABS is disabled and the "ABS" warning lamp is turned "ON."

DTC Chart Test Description:

Number(s) below refer to circled number(s) on the diagnostic chart.

6. This test ensures the wheel speed sensor and sensor ring generate the proper voltage.
7. Ensures malfunction was not due to physical damage of the circuitry.

Diagnostic Aids:

An "intermittent" malfunction is most likely caused by a poor connection, rubbed through wire insulation, or a wire that is broken inside the insulation.

The frequency of the malfunction can be checked by using the enhanced diagnostic function of the Tech 1, as described in "Enhanced Diagnostics," found in this section.

If the customer's comments reflect that the ABS amber warning lamp is "ON" only during moist environmental changes, (rain, snow, vehicle wash) all wheel speed sensor circuitry should be thoroughly inspected for signs of water intrusion. Use the following procedure. Spray down the suspected area with 5% salt water solution (two teaspoons of salt to 12 oz. of water). Drive vehicle above 24 km/h (15 mph) for at least 30 seconds. If DTC returns immediately, replace suspected harness.

Any circuitry, that is suspected as causing the intermittent complaint, should be thoroughly checked for backed out terminals, improper mating, broken locks, improperly formed or damaged terminals, poor terminal to wiring connections or physical damage to the wiring harness.

Resistance of the wheel speed sensor will increase with an increase in sensor temperature. When replacing a wheel speed sensor, inspect the sensor terminals and harness connector for corrosion and/or water intrusion. If evidence or corrosion or water intrusion exists, replace wheel speed sensor harness. Refer to "On-Vehicle Service" in this section.

If wheel speed sensor harness shows evidence of water intrusion, inspect wheel speed sensor for water intrusion. If water intrusion is evident, replace wheel speed sensor.

DIAGNOSTIC CHART DTC 23 — 1995 CAVALIER AND SUNFIRE

DTC 23
(Page 2 of 4)
LEFT REAR WHEEL SPEED = 0 OR UNREASONABLE

DIAGNOSIS CONTINUED FROM PAGE 1 OF 4 OF THIS DTC CHART.

(6)
- INSPECT WHEEL SPEED SENSOR HARNESS CONNECTOR AND 24-WAY EBCM HARNESS CONNECTOR TERMINALS "11" AND "12" FOR POOR TERMINAL CONTACT.

  REPLACE TERMINALS THAT EXHIBIT POOR CONTACT RECONNECT WHEEL SPEED SENSOR.
- INSTALL INTEGRAL BEARING AND SPEED SENSOR ASSEMBLY (TUBULAR AXLE ONLY).
- USING J 39200, SELECT A/C VOLTAGE SCALE AND CONNECT BETWEEN 24-WAY EBCM HARNESS CONNECTOR TERMINALS "11" AND "12".
- SPIN WHEEL BY HAND WHILE MONITORING VOLTAGE.

  IS VOLTAGE AT LEAST 100 mV OR GREATER (VOLTAGE WILL INCREASE AS WHEEL SPEED INCREASES)?

YES → (7)
- INSPECT CKTs 884 AND 885 FOR PHYSICAL DAMAGE WHICH MAY RESULT IN A SHORT BETWEEN THE CIRCUITS WITH ALL CONNECTORS CONNECTED.

  REPAIR DAMAGE IF EVIDENT.
- RECONNECT ALL CONNECTORS.
- DRIVE VEHICLE ABOVE 24-WAY km/h (15 MPH) FOR AT LEAST 30 SECONDS IS DTC 23 SET AS A CURRENT DTC?

  YES → REPLACE EBCM.

  NO → MALFUNCTION IS INTERMITTENT. REFER TO "DIAGNOSTIC AIDS" ON FACING PAGE.

NO → REPLACE INTEGRAL REARING AND SPEED SENSOR ASSEMBLY.

DIAGNOSTIC CHART DTC 23 — 1995 CAVALIER AND SUNFIRE

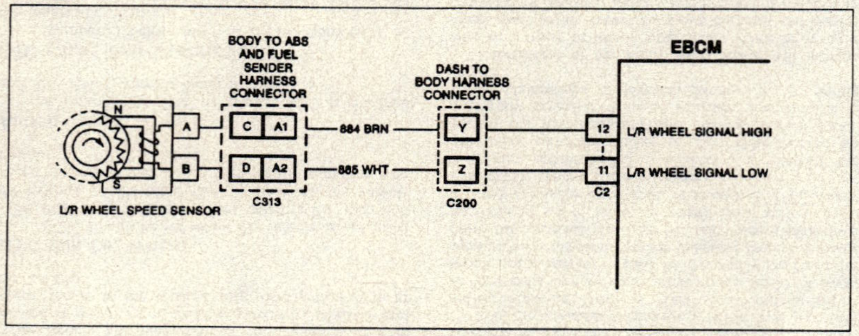

DTC 23
(Page 3 of 4)
LEFT REAR WHEEL SPEED = 0 OR UNREASONABLE

Circuit Description:

As a toothed ring passes by the wheel speed sensor, changes in the electromagnetic field cause the wheel speed sensor to produce a sinusoidal (AC) voltage signal whose frequency is proportional to wheel speed. The magnitude of this signal is directly related to wheel speed and the proximity of the wheel speed sensor to the toothed ring, often referred to as the air gap.

DTC Will Set When:

DTC 23 can be set when the vehicle is not in an ABS stop. If the left rear wheel speed is less than half the average of the other three wheels and the vehicle speed is greater than 8 km/h (5 mph), a malfunction exists. Action Taken: A malfunction DTC is stored, ABS is disabled and the "ABS" warning lamp is turned "ON."

DTC Chart Test Description:

Number(s) below refer to circled number(s) on the diagnostic chart.

8. This step checks for proper voltages at body to ABS and fuel sender harness connector.
9. Checks for proper voltages at the engine to dash harness connector.
10. Checks for a short to ground in the wheel signal circuit.
11. This checks for an open in the wheel signal high circuit.
12. Checks wheel speed circuits for short to BATTERY at engine harness.

Diagnostic Aids:

An "intermittent" malfunction is most likely caused by a poor connection, rubbed through wire insulation, or a wire that is broken inside the insulation.

The frequency of the malfunction can be checked by using the enhanced diagnostic function of the Tech 1, as described in "Enhanced Diagnostics," found in this section.

If the customer's comments reflect that the ABS amber warning lamp is "ON" only during moist environmental changes, (rain, snow, vehicle wash) all wheel speed sensor circuitry should be thoroughly inspected for signs of water intrusion. Use the following procedure. Spray down the suspected area with 5% salt water solution (two teaspoons of salt to 12 oz. of water). Drive vehicle above 24 km/h (15 mph) for at least 30 seconds. If DTC returns immediately, replace suspected harness.

Any circuitry, that is suspected as causing the intermittent complaint, should be thoroughly checked for backed out terminals, improper mating, broken locks, improperly formed or damaged terminals, poor terminal to wiring connections or physical damage to the wiring harness.

Resistance of the wheel speed sensor will increase with an increase in sensor temperature. When replacing a wheel speed sensor, inspect the sensor terminals and harness connector for corrosion and/or water intrusion. If evidence of corrosion or water intrusion exists, replace wheel speed sensor harness. Refer to "On-Vehicle Service" in this section.

If wheel speed sensor harness shows evidence of water intrusion, inspect wheel speed sensor for water intrusion. If water intrusion is evident, replace wheel speed sensor.

DIAGNOSTIC CHART DTC 23 — 1995 CAVALIER AND SUNFIRE

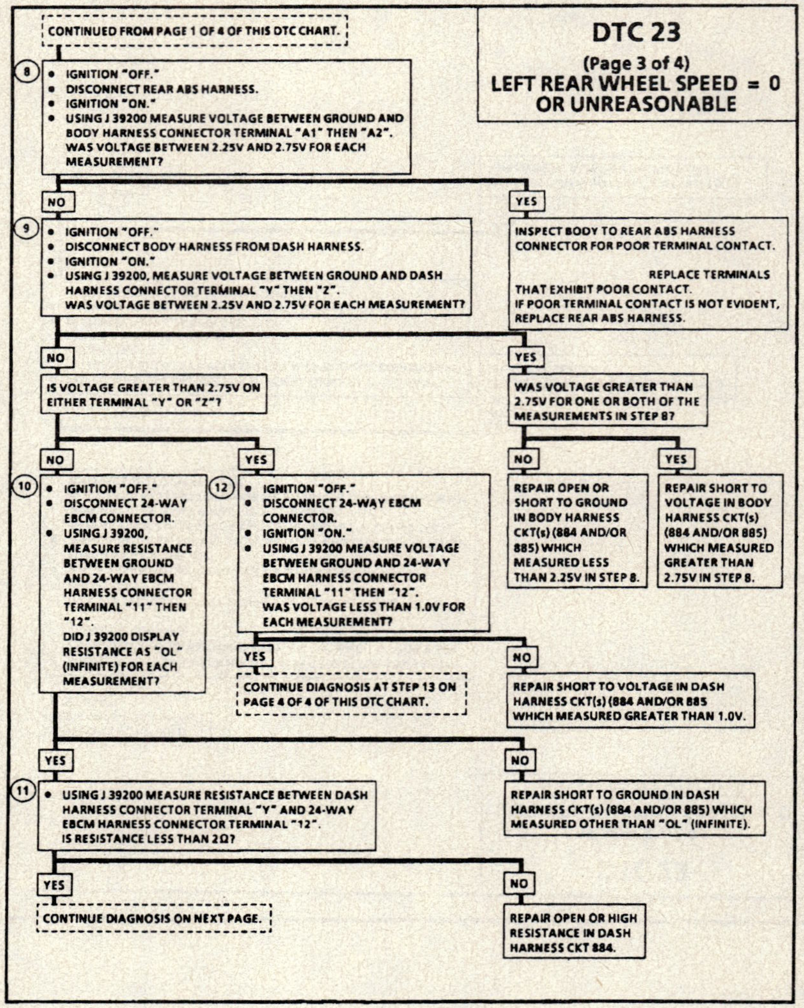

DIAGNOSTIC CHART DTC 23 — 1995 CAVALIER AND SUNFIRE

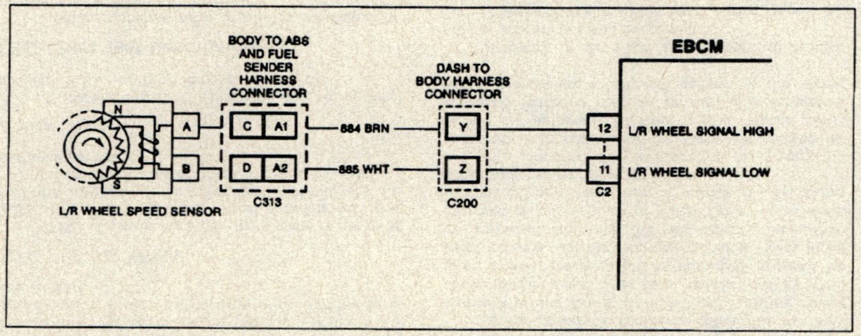

DTC 23
(Page 4 of 4)
LEFT REAR WHEEL SPEED = 0 OR UNREASONABLE

Circuit Description:

As a toothed ring passes by the wheel speed sensor, changes in the electromagnetic field cause the wheel speed sensor to produce a sinusoidal (AC) voltage signal whose frequency is proportional to wheel speed. The magnitude of this signal is directly related to wheel speed and the proximity of the wheel speed sensor to the toothed ring, often referred to as the air gap.

DTC Will Set When:

DTC 23 can be set when the vehicle is not in an ABS stop. If the left rear wheel speed is less than half the average of the other three wheels and the vehicle speed is greater than 8 km/h (5 mph), a malfunction exists.

Action Taken:

A malfunction DTC is stored, ABS is disabled and the "ABS" warning lamp is turned "ON."

DTC Chart Test Description:

Number(s) below refer to circled number(s) on the diagnostic chart.

13. Checks for an open in the wheel signal low circuit.

14. Ensures malfunction was not due to physical damage of the circuitry.

Diagnostic Aids:

An "intermittent" malfunction is most likely caused by a poor connection, rubbed through wire insulation, or a wire that is broken inside the insulation.

The frequency of the malfunction can be checked by using the enhanced diagnostic function of the Tech 1, as described in "Enhanced Diagnostics," found in this section.

If the customer's comments reflect that the ABS amber warning lamp is "ON" only during moist environmental changes, (rain, snow, vehicle wash) all wheel speed sensor circuitry should be thoroughly inspected for signs of water intrusion. Use the following procedure. Spray down the suspected area with 5% salt water solution (two teaspoons of salt to 12 oz. of water). Drive vehicle above 24 km/h (15 mph) for at least 30 seconds. If DTC returns immediately, replace suspected harness.

Any circuitry, that is suspected as causing the intermittent complaint, should be thoroughly checked for backed out terminals, improper mating, broken locks, improperly formed or damaged terminals, poor terminal to wiring connections or physical damage to the wiring harness.

Resistance of the wheel speed sensor will increase with an increase in sensor temperature. When replacing a wheel speed sensor, inspect the sensor terminals and harness connector for corrosion and/or water intrusion. If evidence of corrosion or water intrusion exists, replace wheel speed sensor harness. Refer to "On-Vehicle Service" in this section.

If wheel speed sensor harness shows evidence of water intrusion, inspect wheel speed sensor for water intrusion. If water intrusion is evident, replace wheel speed sensor.

DIAGNOSTIC CHART DTC 23 — 1995 CAVALIER AND SUNFIRE

DTC 23
(Page 4 of 4)
LEFT FRONT WHEEL SPEED = 0 OR UNREASONABLE

CONTINUED FROM PAGE 3 OF 4 OF THIS DTC CHART.

(13)
- USING J 39200, MEASURE RESISTANCE BETWEEN DASH HARNESS CONNECTOR TERMINAL "Z" AND 24-WAY EBCM HARNESS CONNECTOR TERMINAL "11". IS RESISTANCE 2Ω OR LESS?

NO → REPAIR OPEN OR HIGH RESISTANCE IN DASH HARNESS CKT 885.

YES ↓

(14)
- IGNITION "OFF."
- INSPECT ALL WHEEL SPEED SENSOR HARNESS CONNECTORS FOR POOR TERMINAL CONTACT OR CORROSION. REPLACE TERMINALS THAT EXHIBIT POOR CONTACT OR CORROSION.
- INSPECT ALL WHEEL SPEED SENSOR CONNECTORS AND HARNESSES FOR DAMAGE WHICH MAY RESULT IN AN OPEN OR SHORT TO GROUND/BATTERY WHEN ALL COMPONENTS ARE CONNECTED. REPAIR DAMAGE IF EVIDENT.
- RECONNECT ALL CONNECTORS.
- DRIVE VEHICLE ABOVE 24 km/h (15 MPH) FOR AT LEAST 30 SECONDS. IS DTC 23 SET AS A CURRENT DTC?

YES → REPLACE EBCM.

NO → MALFUNCTION IS INTERMITTENT. REFER TO "DIAGNOSTIC AIDS" ON FACING PAGE.

DIAGNOSTIC CHART DTC 24 — 1995 CAVALIER AND SUNFIRE

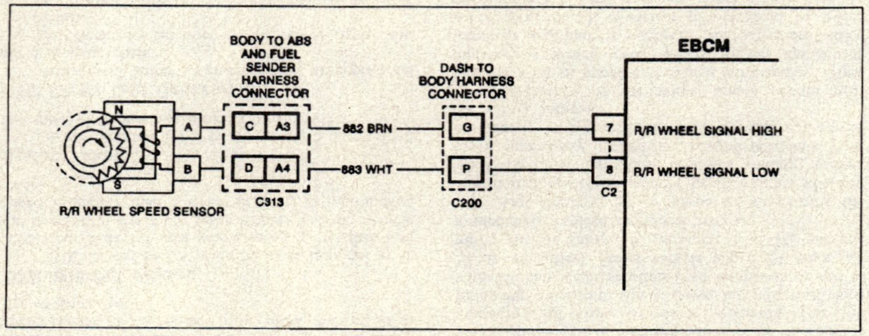

DTC 24
(Page 1 of 4)
RIGHT REAR WHEEL SPEED = 0 OR UNREASONABLE

Circuit Description:

As a toothed ring passes by the wheel speed sensor, changes in the electromagnetic field cause the wheel speed sensor to produce a sinusoidal (AC) voltage signal whose frequency is proportional to wheel speed. The magnitude of this signal is directly related to wheel speed and the proximity of the wheel speed sensor to the toothed ring, often referred to as the air gap.

DTC Will Set When:

DTC 24 can be set when the vehicle is not in an ABS stop. If the right rear wheel speed is less than half the average of the other three wheels and the vehicle speed is greater than 8 km/h (5 mph), a malfunction exists.

Action Taken:

A malfunction DTC is stored, ABS is disabled and the "ABS" warning lamp is turned "ON."

DTC Chart Test Description:

Number(s) below refer to circled number(s) on the diagnostic chart.

1. This step will identify circuitry that is damaged and visibly apparent.
2. Checks for proper voltages at the rear sensor harness connector.
3. Checks for a short between the wheel signal high and low circuits.
4. This test checks wheel speed sensor for proper resistance value.

Diagnostic Aids:

An "intermittent" malfunction is most likely caused by a poor connection, rubbed through wire insulation, or a wire that is broken inside the insulation.

The frequency of the malfunction can be checked by using the enhanced diagnostic function of the Tech 1, as described in "Enhanced Diagnostics," found in this section.

If the customer's comments reflect that the ABS amber warning lamp is "ON" only during moist environmental changes, (rain, snow, vehicle wash) all wheel speed sensor circuitry should be thoroughly inspected for water intrusion. Use the following procedure. Spray down the suspected area with 5% salt water solution (two teaspoons of salt to 12 oz. of water). Drive vehicle above 24 km/h (15 mph) for at least 30 seconds. If DTC returns immediately, replace suspected harness.

Any circuitry, that is suspected as causing the intermittent complaint, should be thoroughly checked for backed out terminals, improper mating, broken locks, improperly formed or damaged terminals, poor terminal to wiring connections or physical damage to the wiring harness.

Resistance of the wheel speed sensor will increase with an increase in sensor temperature.

When replacing a wheel speed sensor, inspect the sensor terminals and harness connector for corrosion and/or water intrusion. If evidence of corrosion or water intrusion exists, replace wheel speed sensor harness. Refer to "On-Vehicle Service" in this section.

If wheel speed sensor harness shows evidence of water intrusion, inspect wheel speed sensor for water intrusion. If water intrusion is evident, replace wheel speed sensor.

DIAGNOSTIC CHART DTC 24 — 1995 CAVALIER AND SUNFIRE

DTC 24
(Page 1 of 4)
RIGHT REAR WHEEL SPEED = 0 OR UNREASONABLE

Important
- WHEEL SPEED SENSOR INTERMITTENT MALFUNCTIONS MAY BE DIFFICULT TO LOCATE. CARE SHOULD BE TAKEN NOT TO DISTURB ANY ELECTRICAL CONNECTIONS PRIOR TO AN INDICATED STEP OF THIS CHART. THIS WILL ENSURE THAT AN INTERMITTENT CONNECTION WILL NOT BE CORRECTED BEFORE THE SOURCE OF THE MALFUNCTION IS FOUND.

(1)
- IGNITION "OFF."
- INSPECT WHEEL SPEED SENSOR WIRING AND CONNECTORS FOR DAMAGE. IS DAMAGE EVIDENT?

YES: REPAIR DAMAGE. DRIVE VEHICLE ABOVE 24 km/h (15 MPH) FOR AT LEAST 30 SECONDS TO ENSURE DTC 24 DOES NOT SET.

NO:

(2)
- REMOVE RIGHT REAR INTEGRAL BEARING AND SPEED SENSOR ASSEMBLY (TUBULAR AXLE ONLY).
- DISCONNECT CONNECTOR FROM RIGHT REAR WHEEL SPEED SENSOR.
- IGNITION "ON."
- USING J 39200, MEASURE VOLTAGE BETWEEN GROUND AND RIGHT REAR SENSOR HARNESS CONNECTOR TERMINAL "A" THEN "B". WAS VOLTAGE BETWEEN 2.25V AND 2.75V FOR EACH MEASUREMENT?

NO: GO TO PAGE 3 OF 4 OF THIS CHART.

YES:

(3)
- IGNITION "OFF."
- DISCONNECT 24-WAY EBCM CONNECTOR.
- USING J 39200, MEASURE RESISTANCE BETWEEN 24-WAY EBCM HARNESS CONNECTOR TERMINALS "8" AND "7." DOES J 39200 DISPLAY RESISTANCE AS "OL" (INFINITE)?

NO: REPAIR SHORT BETWEEN CKTs 882 AND 883.

YES:

(4)
- USING J 39200, MEASURE RESISTANCE BETWEEN RIGHT REAR WHEEL SPEED SENSOR TERMINALS "A" AND "B." IS RESISTANCE BETWEEN 2100 AND 2400Ω WHEN SENSOR IS 25°C (77°F)? (IF TEMPERATURE OF THE SENSOR IS SUSPECTED TO BE OTHER THAN 25°C (77°F). REFER TO THE TABLE BELOW FOR APPLICABLE SENSOR RESISTANCE VALUES; THE VALUES IN THE TABLE ARE FOR TEMPERATURE OF THE SENSOR NOT AIR TEMPERATURE.)

NO: REPLACE INTEGRAL BEARING AND SPEED SENSOR ASSEMBLY.

YES: CONTINUE DIAGNOSIS ON NEXT PAGE.

REAR WHEEL SPEED SENSOR
TEMPERATURE OF SENSOR VS. SENSOR RESISTANCE (APPROXIMATE)

TEMP. (°C)	TEMP. (°F)	RESISTANCE (OHMS)
-34 TO 4	-30 TO 40	1623 TO 2210
5 TO 43	41 TO 110	1939 TO 2570
44 TO 93	111 TO 200	2254 TO 3029

DIAGNOSTIC CHART DTC 24 — 1995 CAVALIER AND SUNFIRE

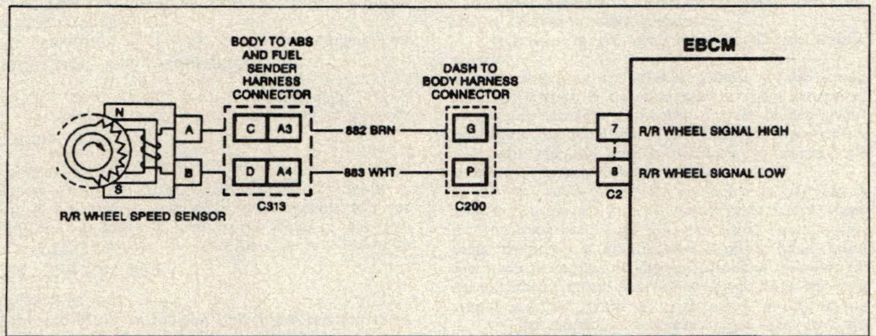

DTC 24
(Page 2 of 4)
RIGHT REAR WHEEL SPEED = 0 OR UNREASONABLE

Circuit Description:

As a toothed ring passes by the wheel speed sensor, changes in the electromagnetic field cause the wheel speed sensor to produce a sinusoidal (AC) voltage signal whose frequency is proportional to wheel speed. The magnitude of this signal is directly related to wheel speed and the proximity of the wheel speed sensor to the toothed ring, often referred to as the air gap.

DTC Will Set When:

DTC 24 can be set when the vehicle is not in an ABS stop. If the right rear wheel speed is less than half the average of the other three wheels and the vehicle speed is greater than 8 km/h (5 mph), a malfunction exists.

Action Taken:

A malfunction DTC is stored, ABS is disabled and the "ABS" warning lamp is turned "ON."

DTC Chart Test Description:

Number(s) below refer to circled number(s) on the diagnostic chart.

5. This step ensures the wheel speed sensor is not shorted to ground.
6. This test ensures the wheel speed sensor and sensor ring generate the proper voltage.
7. Ensures malfunction was not due to physical damage of the circuitry.

Diagnostic Aids:

An "intermittent" malfunction is most likely caused by a poor connection, rubbed through wire insulation, or a wire that is broken inside the insulation.

The frequency of the malfunction can be checked by using the enhanced diagnostic function of the Tech 1, as described in "Enhanced Diagnostics," found in this section.

If the customer's comments reflect that the ABS amber warning lamp is "ON" only during moist environmental changes, (rain, snow, vehicle wash) all wheel speed sensor circuitry should be thoroughly inspected for water intrusion. Use the following procedure. Spray down the suspected area with 5% salt water solution (two teaspoons of salt to 12 oz. of water). Drive vehicle above 24 km/h (15 mph) for at least 30 seconds. If DTC returns immediately, replace suspected harness.

Any circuitry, that is suspected as causing the intermittent complaint, should be thoroughly checked for backed out terminals, improper mating, broken locks, improperly formed or damaged terminals, poor terminal to wiring connections or physical damage to the wiring harness.

Resistance of the wheel speed sensor will increase with an increase in sensor temperature.

When replacing a wheel speed sensor, inspect the sensor terminals and harness connector for corrosion and/or water intrusion. If evidence of corrosion or water intrusion exists, replace wheel speed sensor harness. Refer to "On-Vehicle Service" in this section.

If wheel speed sensor harness shows evidence of water intrusion, inspect wheel speed sensor for water intrusion. If water intrusion is evident, replace wheel speed sensor.

DIAGNOSTIC CHART DTC 24 — 1995 CAVALIER AND SUNFIRE

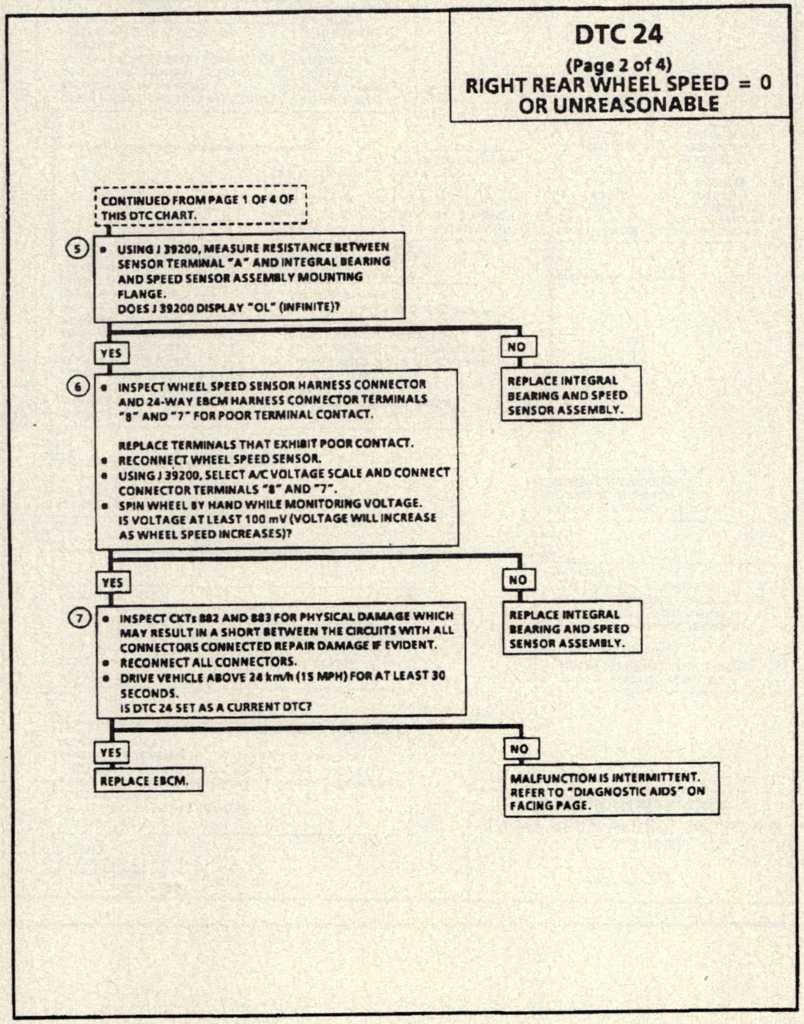

DIAGNOSTIC CHART DTC 24 — 1995 CAVALIER AND SUNFIRE

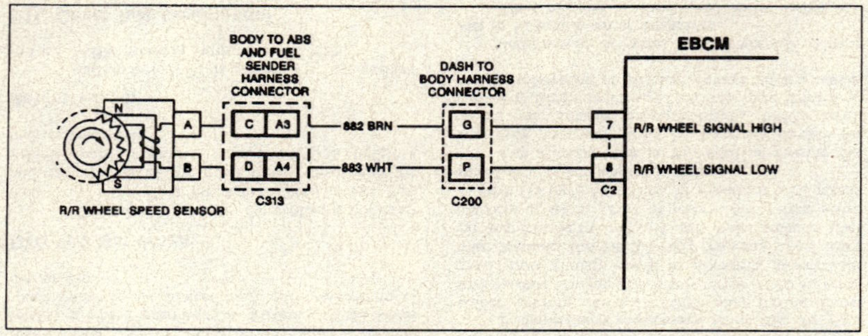

DTC 24
(Page 3 of 4)
RIGHT REAR WHEEL SPEED = 0 OR UNREASONABLE

Circuit Description:

As a toothed ring passes by the wheel speed sensor, changes in the electromagnetic field cause the wheel speed sensor to produce a sinusoidal (AC) voltage signal whose frequency is proportional to wheel speed. The magnitude of this signal is directly related to wheel speed and the proximity of the wheel speed sensor to the toothed ring, often referred to as the air gap.

DTC Will Set When:

DTC 24 can be set when the vehicle is not in an ABS stop. If the right rear wheel speed is less than half the average of the other three wheels and the vehicle speed is greater than 8 km/h (5 mph), a malfunction exists.

Action Taken:

A malfunction DTC is stored, ABS is disabled and the "ABS" warning lamp is turned "ON."

DTC Chart Test Description:

Number(s) below refer to circled number(s) on the diagnostic chart.

8. This step checks for proper voltages at body to ABS and fuel sender harness connector.
9. Checks for proper voltages at the body harness to dash harness connector.
10. Checks for a short to ground in the wheel signal circuits.
11. This checks for an open in the wheel signal high circuit.
12 Checks wheel speed circuits for short to BATTERY at engine harness.

Diagnostic Aids:

An "intermittent" malfunction is most likely caused by a poor connection, rubbed through wire insulation, or a wire that is broken inside the insulation.

The frequency of the malfunction can be checked by using the enhanced diagnostic function of the Tech 1, as described in "Enhanced Diagnostics," found in this section.

If the customer's comments reflect that the ABS amber warning lamp is "ON" only during moist environmental changes, (rain, snow, vehicle wash) all wheel speed sensor circuitry should be thoroughly inspected for water intrusion. Use the following procedure. Spray down the suspected area with 5% salt water solution (two teaspoons of salt to 12 oz. of water). Drive vehicle above 24 km/h (15 mph) for at least 30 seconds. If DTC returns immediately, replace suspected harness.

Any circuitry, that is suspected as causing the intermittent complaint, should be thoroughly checked for backed out terminals, improper mating, broken locks, improperly formed or damaged terminals, poor terminal to wiring connections or physical damage to the wiring harness.

Resistance of the wheel speed sensor will increase with an increase in sensor temperature.

When replacing a wheel speed sensor, inspect the sensor terminals and harness connector for corrosion and/or water intrusion. If evidence of corrosion or water intrusion exists, replace wheel speed sensor harness. Refer to "On-Vehicle Service" in this section.

If wheel speed sensor harness shows evidence of water intrusion, inspect wheel speed sensor for water intrusion. If water intrusion is evident, replace wheel speed sensor.

DIAGNOSTIC CHART DTC 24 — 1995 CAVALIER AND SUNFIRE

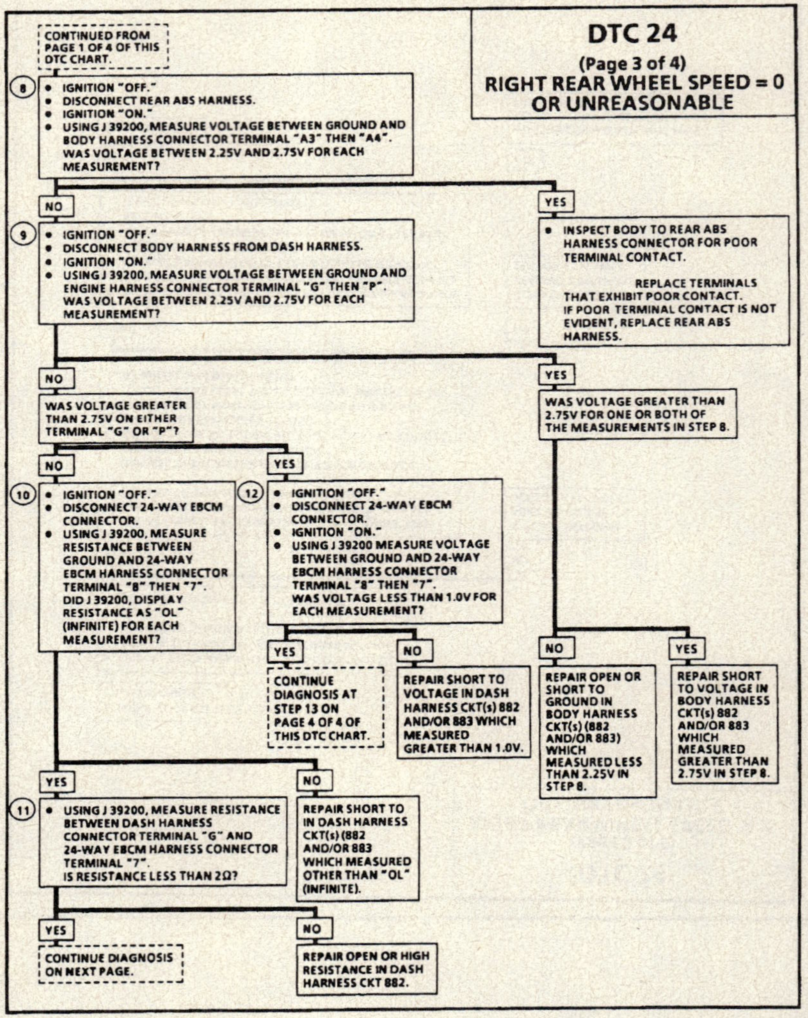

DIAGNOSTIC CHART DTC 24 — 1995 CAVALIER AND SUNFIRE

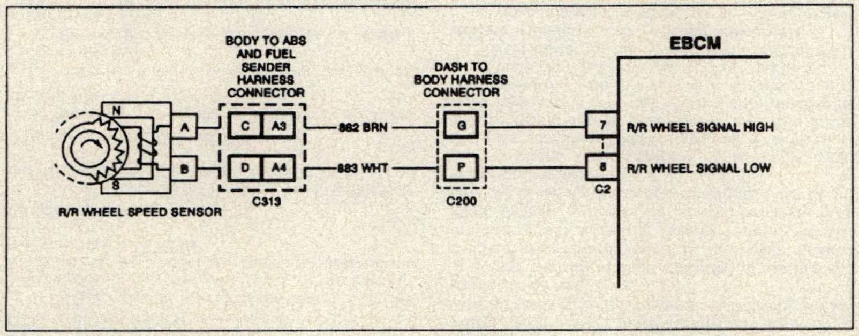

DTC 24
(Page 4 of 4)
RIGHT REAR WHEEL SPEED = 0 OR UNREASONABLE

Circuit Description:

As a toothed ring passes by the wheel speed sensor, changes in the electromagnetic field cause the wheel speed sensor to produce a sinusoidal (AC) voltage signal whose frequency is proportional to wheel speed. The magnitude of this signal is directly related to wheel speed and the proximity of the wheel speed sensor to the toothed ring, often referred to as the air gap.

DTC Will Set When:

DTC 24 can be set when the vehicle is not in an ABS stop. If the right rear wheel speed is less than half the average of the other three wheels and the vehicle speed is greater than 8 km/h (5 mph), a malfunction exists.

Action Taken:

A malfunction DTC is stored, ABS is disabled and the "ABS" warning lamp is turned "ON."

DTC Chart Test Description:

Number(s) below refer to circled number(s) on the diagnostic chart.

13. Checks for an open in the wheel signal low circuit.

14. Ensures malfunction was not due to physical damage of the circuitry.

Diagnostic Aids:

An "intermittent" malfunction is most likely caused by a poor connection, rubbed through wire insulation, or a wire that is broken inside the insulation.

The frequency of the malfunction can be checked by using the enhanced diagnostic function of the Tech 1, as described in "Enhanced Diagnostics," found in this section.

If the customer's comments reflect that the ABS amber warning lamp is "ON" only during moist environmental changes, (rain, snow, vehicle wash) all wheel speed sensor circuitry should be thoroughly inspected for water intrusion. Use the following procedure. Spray down the suspected area with 5% salt water solution (two teaspoons of salt to 12 oz. of water). Drive vehicle above 24 km/h (15 mph) for at least 30 seconds. If DTC returns immediately, replace suspected harness.

Any circuitry, that is suspected as causing the intermittent complaint, should be thoroughly checked for backed out terminals, improper mating, broken locks, improperly formed or damaged terminals, poor terminal to wiring connections or physical damage to the wiring harness.

Resistance of the wheel speed sensor will increase with an increase in sensor temperature.

When replacing a wheel speed sensor, inspect the sensor terminals and harness connector for corrosion and/or water intrusion. If evidence of corrosion or water intrusion exists, replace wheel speed sensor harness. Refer to "On-Vehicle Service" in this section.

If wheel speed sensor harness shows evidence of water intrusion, inspect wheel speed sensor for water intrusion. If water intrusion is evident, replace wheel speed sensor.

DIAGNOSTIC CHART DTC 24 — 1995 CAVALIER AND SUNFIRE

DTC 24
(Page 4 of 4)
RIGHT REAR WHEEL SPEED = 0 OR UNREASONABLE

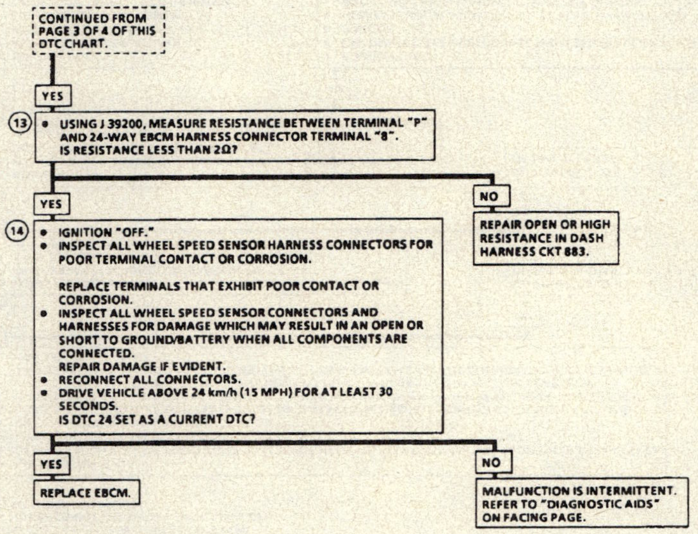

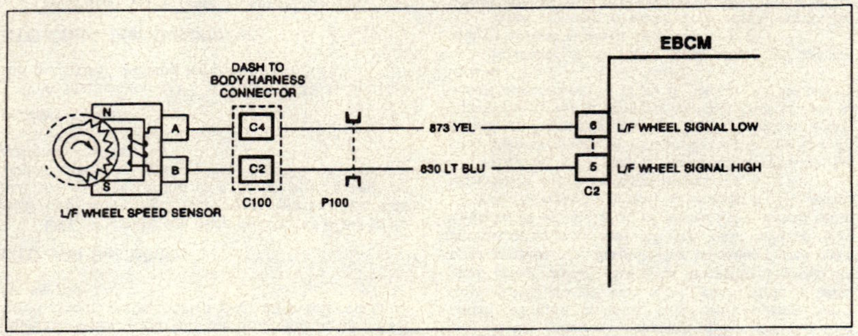

DTC 25
(Page 1 of 3)
LEFT FRONT EXCESSIVE WHEEL SPEED VARIATION

Circuit Description:

As a toothed ring passes by the wheel speed sensor, changes in the electromagnetic field cause the wheel speed sensor to produce a sinusoidal (AC) voltage signal whose frequency is proportional to wheel speed. The magnitude of this signal is directly related to wheel speed and the proximity of the wheel speed sensor to the toothed ring, often referred to as the air gap.

DTC Will Set When:

DTC 25 can be set when the brake is "OFF." The purpose of this test is to detect a situation in which the left front wheel acceleration or deceleration is beyond specified limits.

Action Taken:

A malfunction DTC is stored, ABS is disabled and the "ABS" warning lamp is turned "ON."

DTC Chart Test Description:

Number(s) below refer to circled number(s) on the diagnostic chart.

1. Checks wheel speed sensor for proper resistance value.
2. Ensures the wheel speed sensor is not shorted to ground.
3. Ensures the wheel speed sensor is not shorted to voltage.
4. Ensures that DTC 25 was not set due to a poor connection at the EBCM or physical damage of the circuitry.
5. Checks for proper voltage at the wheel signal low speed sensor harness connector.
6. Checks for proper voltage at the wheel signal high speed sensor harness connector.

The frequency of the malfunction can be checked by using the enhanced diagnostic function of the Tech 1, as described in "Enhanced Diagnostics," found in this section.

If the customer's comments reflect that the ABS amber warning lamp is "ON" only during moist environmental changes, (rain, snow, vehicle wash) all wheel speed sensor circuitry should be thoroughly inspected for water intrusion. Use the following procedure. Spray down the suspected area with 5% salt water solution (two teaspoons of salt to 12 oz. of water). Drive vehicle above 24 km/h (15 mph) for at least 30 seconds. If DTC returns immediately, replace suspected harness.

Any circuitry, that is suspected as causing the intermittent complaint, should be thoroughly checked for backed out terminals, improper mating, broken locks, improperly formed or damaged terminals, poor terminal to wiring connections or physical damage to the wiring harness.

A careful visual inspection of the toothed ring, CV joint, bearing and wheel speed sensor for physical damage should also be performed. If DTC sets at the same mph every time and wheel speed variation is noted above this mph on the Tech 1, the speed ring is most likely damaged. This can occur at any speed.

Resistance of the wheel speed sensor will increase with an increase in sensor temperature.

When replacing a wheel speed sensor, inspect the sensor terminals and harness connector for corrosion and/or water intrusion. If evidence of corrosion or water intrusion exists, replace wheel speed sensor harness. Refer to "On-Vehicle Service" in this section.

If wheel speed sensor harness shows evidence of water intrusion, inspect wheel speed sensor for water intrusion. If water intrusion is evident, replace wheel speed sensor.

DIAGNOSTIC CHART DTC 25 — 1995
CAVALIER AND SUNFIRE

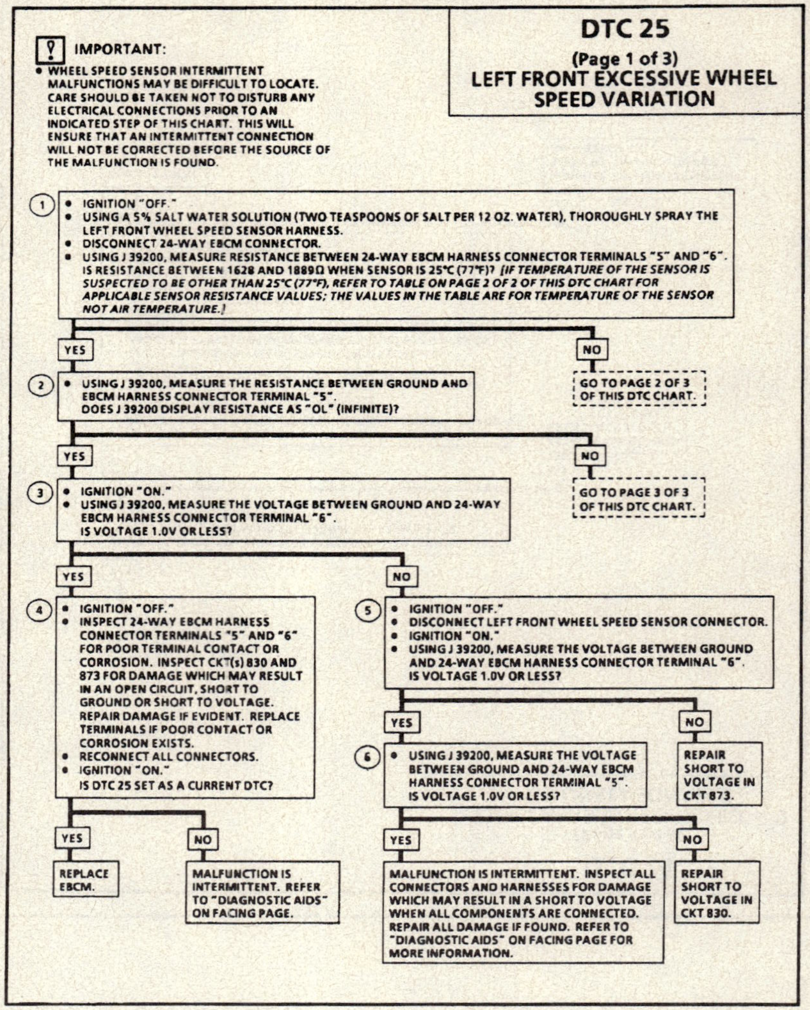

DIAGNOSTIC CHART DTC 25 — 1995 CAVALIER AND SUNFIRE

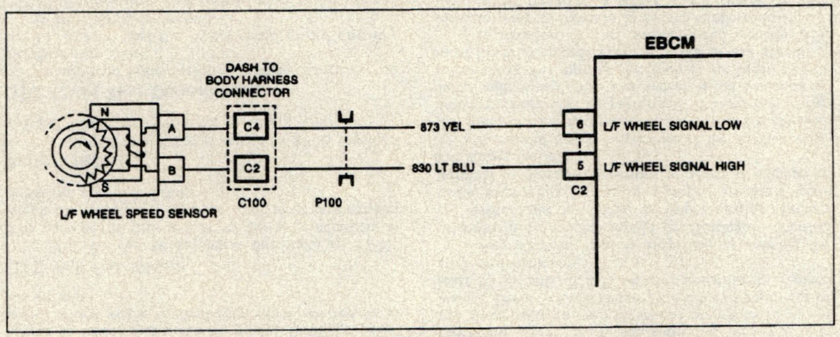

DTC 25
(Page 2 of 3)
LEFT FRONT EXCESSIVE WHEEL SPEED VARIATION

Circuit Description:

As a toothed ring passes by the wheel speed sensor, changes in the electromagnetic field cause the wheel speed sensor to produce a sinusoidal (AC) voltage signal whose frequency is proportional to wheel speed. The magnitude of this signal is directly related to wheel speed and the proximity of the wheel speed sensor to the toothed ring, often referred to as the air gap.

DTC Will Set When:

DTC 25 can be set when the brake is "OFF." The purpose of this test is to detect a situation in which the left front wheel acceleration or deceleration is beyond specified limits.

Action Taken:

A malfunction DTC is stored, ABS is disabled and the "ABS" warning lamp is turned "ON."

DTC Chart Test Description:

Number(s) below refer to circled number(s) on the diagnostic chart.

7. Tests for a short between the wheel signal high and wheel signal low circuits.
8. Checks for an open in the wheel signal low circuit.
9. Checks for an open in the wheel signal high circuit.
10. Checks for proper resistance of the wheel speed sensor.

Diagnostic Aids:

An "intermittent" malfunction is most likely caused by a poor connection, rubbed through wire insulation, or a wire that is broken inside the insulation.

The frequency of the malfunction can be checked by using the enhanced diagnostic function of the Tech 1, as described in "Enhanced Diagnostics," found in this section.

If the customer's comments reflect that the ABS amber warning lamp is "ON" only during moist environmental changes, (rain, snow, vehicle wash) all wheel speed sensor circuitry should be thoroughly inspected for water intrusion. Use the following procedure. Spray down the suspected area with 5% salt water solution (two teaspoons of salt to 12 oz. of water). Drive vehicle above 24 km/h (15 mph) for at least 30 seconds. If DTC returns immediately, replace suspected harness.

Any circuitry, that is suspected as causing the intermittent complaint, should be thoroughly checked for backed out terminals, improper mating, broken locks, improperly formed or damaged terminals, poor terminal to wiring connections or physical damage to the wiring harness.

A careful visual inspection of the toothed ring, CV joint, bearing and wheel speed sensor for physical damage should also be performed. If DTC sets at the same mph every time and wheel speed variation is noted above this mph on the Tech 1, the speed ring is most likely damaged. This can occur at any speed.

Resistance of the wheel speed sensor will increase with an increase in sensor temperature.

When replacing a wheel speed sensor, inspect the sensor terminals and harness connector for corrosion and/or water intrusion. If evidence of corrosion or water intrusion exists, replace wheel speed sensor harness. Refer to "On-Vehicle Service" in this section.

If wheel speed sensor harness shows evidence of water intrusion, inspect wheel speed sensor for water intrusion. If water intrusion is evident, replace wheel speed sensor.

DIAGNOSTIC CHART DTC 25 — 1995 CAVALIER AND SUNFIRE

DTC 25
(Page 2 of 3)
LEFT FRONT EXCESSIVE WHEEL SPEED VARIATION

CONTINUED FROM PAGE 1 OF 3 OF THIS DTC CHART.

(7)
- DISCONNECT LEFT FRONT WHEEL SPEED SENSOR CONNECTOR.
- USING J 39200, MEASURE THE RESISTANCE BETWEEN 24-WAY EBCM HARNESS CONNECTOR TERMINALS "5" AND "6".

DOES J 39200 DISPLAY RESISTANCE AS "OL" (INFINITE)?

NO → REPAIR SHORT BETWEEN CKTs 830 AND 873.

YES ↓

(8)
- USING J 39200, MEASURE THE RESISTANCE BETWEEN 24-WAY EBCM HARNESS CONNECTOR TERMINAL "6" AND LEFT FRONT WHEEL SPEED SENSOR HARNESS CONNECTOR TERMINAL "A."

IS RESISTANCE 2.0Ω OR LESS?

NO → REPAIR OPEN OR HIGH RESISTANCE IN CKT 873.

YES ↓

(9)
- USING J 39200, MEASURE THE RESISTANCE BETWEEN 24-WAY EBCM HARNESS CONNECTOR TERMINAL "5" AND LEFT FRONT WHEEL SPEED SENSOR HARNESS CONNECTOR TERMINAL "B."

IS RESISTANCE 2.0Ω OR LESS?

NO → REPAIR OPEN OR HIGH RESISTANCE IN CKT 830.

YES ↓

(10)
- USING J 39200, MEASURE THE RESISTANCE BETWEEN LEFT FRONT WHEEL SPEED SENSOR TERMINALS "A" AND "B."

IS RESISTANCE BETWEEN 1628 AND 1889Ω WHEN SENSOR IS 25°C (77°F)?

[IF TEMPERATURE OF THE SENSOR IS SUSPECTED TO BE OTHER THAN 25°C (77°F), REFER TO THE TABLE BELOW FOR APPLICABLE SENSOR RESISTANCE VALUES; THE VALUES IN THE TABLE ARE FOR TEMPERATURE OF THE SENSOR NOT AIR TEMPERATURE.]

NO → REPLACE WHEEL SPEED SENSOR.

YES → MALFUNCTION IS INTERMITTENT. INSPECT ALL CONNECTORS AND HARNESSES FOR DAMAGES WHICH MAY RESULT IN HIGH RESISTANCE WHEN ALL COMPONENTS ARE CONNECTED. REFER TO "DIAGNOSTIC AIDS" ON FACING PAGE FOR MORE INFORMATION. REPAIR ALL DAMAGE IF FOUND.

FRONT WHEEL SPEED SENSOR

TEMPERATURE OF SENSOR VS. SENSOR RESISTANCE (APPROXIMATE)

TEMP. (°C)	TEMP. (°F)	RESISTANCE (OHMS)
-34 TO 4	-30 TO 40	1258 TO 1739
5 TO 43	41 TO 110	1503 TO 2023
44 TO 93	111 TO 200	1747 TO 2384

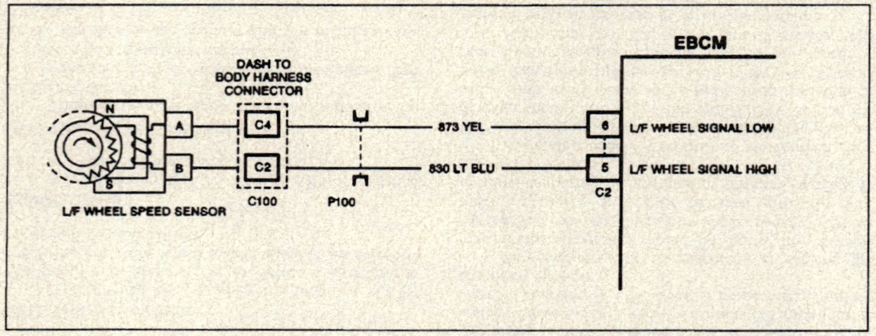

DTC 25
(Page 3 of 3)
LEFT FRONT EXCESSIVE WHEEL SPEED VARIATION

Circuit Description:

As a toothed ring passes by the wheel speed sensor, changes in the electromagnetic field cause the wheel speed sensor to produce a sinusoidal (AC) voltage signal whose frequency is proportional to wheel speed. The magnitude of this signal is directly related to wheel speed and the proximity of the wheel speed sensor to the toothed ring, often referred to as the air gap.

DTC Will Set When:

DTC 25 can be set when the brake is "OFF." The purpose of this test is to detect a situation in which the left front wheel acceleration or deceleration is beyond specified limits.

Action Taken:

A malfunction DTC is stored, ABS is disabled and the "ABS" warning lamp is turned "ON."

DTC Chart Test Description:

Number(s) below refer to circled number(s) on the diagnostic chart.

11. Ensures that the wheel speed sensor circuitry is not shorted to ground.
12. Ensures the wheel speed sensor is not shorted to ground.

Diagnostic Aids:

An "intermittent" malfunction is most likely caused by a poor connection, rubbed through wire insulation, or a wire that is broken inside the insulation.

The frequency of the malfunction can be checked by using the enhanced diagnostic function of the Tech 1, as described in "Enhanced Diagnostics," found in this section.

If the customer's comments reflect that the ABS amber warning lamp is "ON" only during moist environmental changes, (rain, snow, vehicle wash) all wheel speed sensor circuitry should be thoroughly inspected for water intrusion. Use the following procedure. Spray down the suspected area with 5% salt water solution (two teaspoons of salt to 12 oz. of water). Drive vehicle above 24 km/h (15 mph) for at least 30 seconds. If DTC returns immediately, replace suspected harness.

Any circuitry, that is suspected as causing the intermittent complaint, should be thoroughly checked for backed out terminals, improper mating, broken locks, improperly formed or damaged terminals, poor terminal to wiring connections or physical damage to the wiring harness.

A careful visual inspection of the toothed ring, CV joint, bearing and wheel speed sensor for physical damage should also be performed. If DTC sets at the same mph every time and wheel speed variation is noted above this mph on the Tech 1, the speed ring is most likely damaged. This can occur at any speed.

Resistance of the wheel speed sensor will increase with an increase in sensor temperature.

When replacing a wheel speed sensor, inspect the sensor terminals and harness connector for corrosion and/or water intrusion. If evidence of corrosion or water intrusion exists, replace wheel speed sensor harness. Refer to "On-Vehicle Service" in this section.

If wheel speed sensor harness shows evidence of water intrusion, inspect wheel speed sensor for water intrusion. If water intrusion is evident, replace wheel speed sensor.

DIAGNOSTIC CHART DTC 25 — 1995
CAVALIER AND SUNFIRE

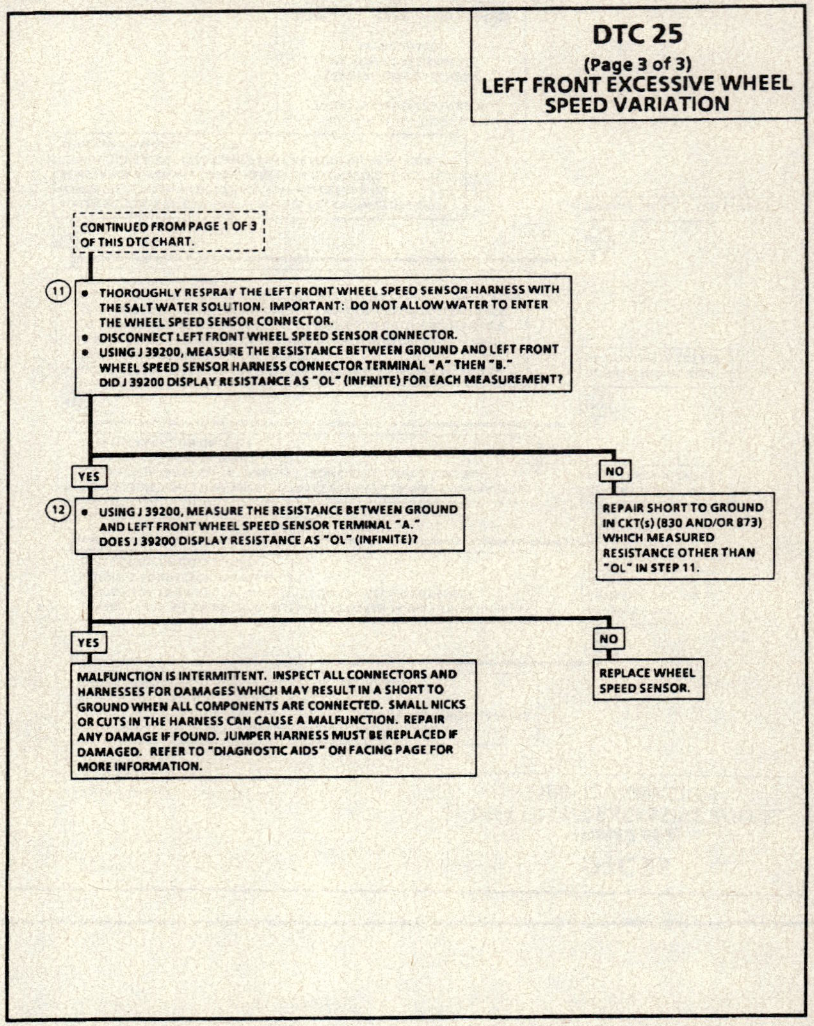

DIAGNOSTIC CHART DTC 26 — 1995 CAVALIER AND SUNFIRE

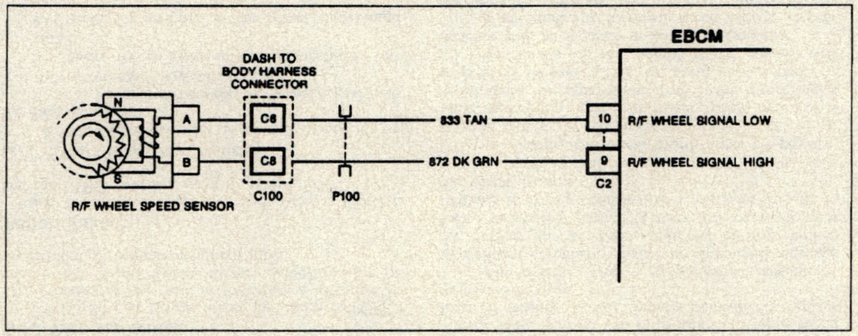

DTC 26

RIGHT FRONT EXCESSIVE WHEEL SPEED VARIATION

Circuit Description:

As a toothed ring passes by the wheel speed sensor, changes in the electromagnetic field cause the wheel speed sensor to produce a sinusoidal (AC) voltage signal whose frequency is proportional to wheel speed. The magnitude of this signal is directly related to wheel speed and the proximity of the wheel speed sensor to the toothed ring, often referred to as the air gap.

DTC Will Set When:

DTC 26 can be set when the brake is "OFF." The purpose of this test is to detect a situation in which the right front wheel acceleration or deceleration is beyond specified limits.

Action Taken:

A malfunction DTC is stored, ABS is disabled and the "ABS" warning lamp is turned "ON."

DTC Chart Test Description:

Number(s) below refer to circled number(s) on the diagnostic chart.

1. Checks wheel speed sensor for proper resistance value.
2. Ensures the wheel speed sensor is not shorted to ground.
3. Ensures the wheel speed sensor is not shorted to voltage.
4. Ensures that DTC 26 was not set due to a poor connection at the EBCM or physical damage of the circuitry.
5. Checks for proper voltage at the wheel signal low speed sensor harness connector.
6. Checks for proper voltage at the wheel signal high speed sensor harness connector.

The frequency of the malfunction can be checked by using the enhanced diagnostic function of the Tech 1, as described in "Enhanced Diagnostics," found in this section.

If the customer's comments reflect that the ABS amber warning lamp is "ON" only during moist environmental changes, (rain, snow, vehicle wash) all wheel speed sensor circuitry should be thoroughly inspected for water intrusion. Use the following procedure. Spray down the suspected area with 5% salt water solution (two teaspoons of salt to 12 oz. of water). Drive vehicle above 24 km/h (15 mph) for at least 30 seconds. If DTC returns immediately, replace suspected harness.

Any circuitry, that is suspected as causing the intermittent complaint, should be thoroughly checked for backed out terminals, improper mating, broken locks, improperly formed or damaged terminals, poor terminal to wiring connections or physical damage to the wiring harness.

A careful visual inspection of the toothed ring, CV joint, bearing and wheel speed sensor for physical damage should also be performed. If DTC sets at the same mph every time and wheel speed variation is noted above this mph on the Tech 1, the speed ring is most likely damaged. This can occur at any speed.

Resistance of the wheel speed sensor will increase with an increase in sensor temperature.

When replacing a wheel speed sensor, inspect the sensor terminals and harness connector for corrosion and/or water intrusion. If evidence of corrosion or water intrusion exists, replace wheel speed sensor harness. Refer to "On-Vehicle Service" in this section.

If wheel speed sensor harness shows evidence of water intrusion, inspect wheel speed sensor for water intrusion. If water intrusion is evident, replace wheel speed sensor.

DIAGNOSTIC CHART DTC 26 — 1995 CAVALIER AND SUNFIRE

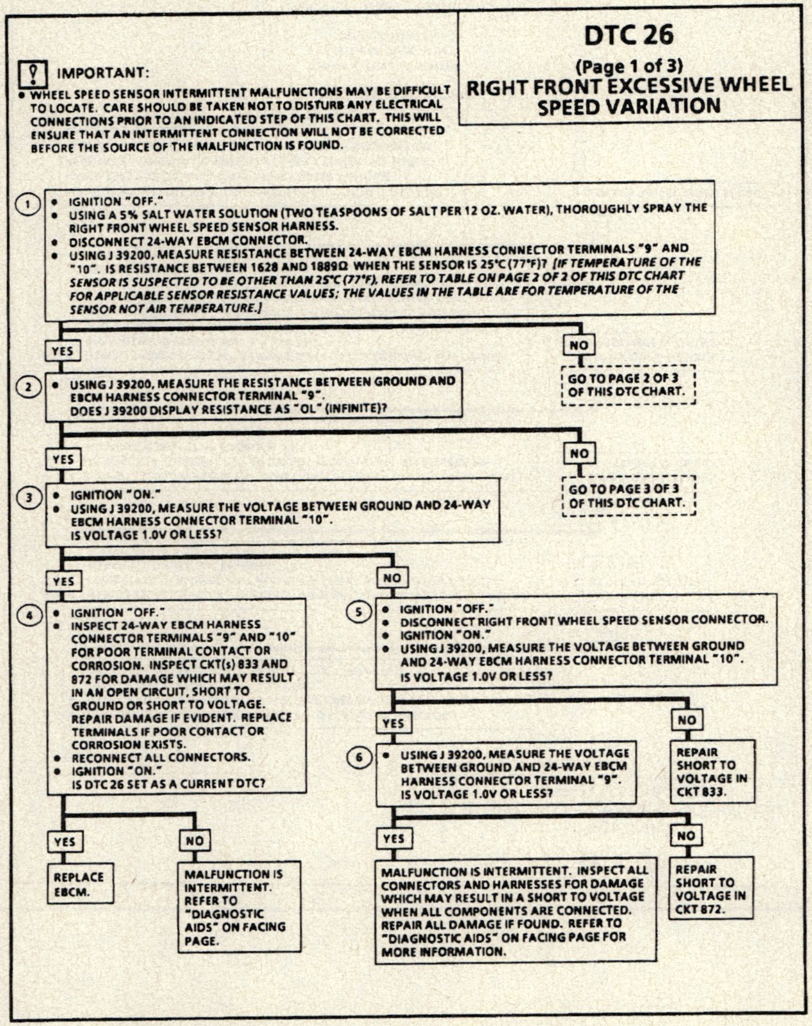

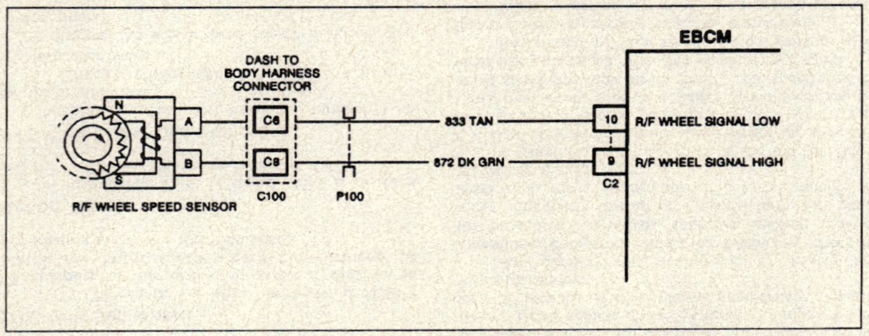

DTC 26 (Page 2 of 3)
RIGHT FRONT EXCESSIVE WHEEL SPEED VARIATION

Circuit Description:
As a toothed ring passes by the wheel speed sensor, changes in the electromagnetic field cause the wheel speed sensor to produce a sinusoidal (AC) voltage signal whose frequency is proportional to wheel speed. The magnitude of this signal is directly related to wheel speed and the proximity of the wheel speed sensor to the toothed ring, often referred to as the air gap.

DTC Will Set When:
DTC 26 can be set when the brake is "OFF." The purpose of this test is to detect a situation in which the right front wheel acceleration or deceleration is beyond specified limits.

Action Taken:
A malfunction DTC is stored, ABS is disabled and the "ABS" warning lamp is turned "ON."

DTC Chart Test Description:
Number(s) below refer to circled number(s) on the diagnostic chart.
7. Tests for a short between the wheel signal high and wheel signal low circuits.
8. Checks for an open in the wheel signal low circuit.
9. Checks for an open in the wheel signal high circuit.
10. Checks for proper resistance of the wheel speed sensor.

Diagnostic Aids:
An "intermittent" malfunction is most likely caused by a poor connection, rubbed through wire insulation, or a wire that is broken inside the insulation.

The frequency of the malfunction can be checked by using the enhanced diagnostic function of the Tech 1, as described in "Enhanced Diagnostics," found in this section.

If the customer's comments reflect that the ABS amber warning lamp is "ON" only during moist environmental changes, (rain, snow, vehicle wash) all wheel speed sensor circuitry should be thoroughly inspected for water intrusion. Use the following procedure. Spray down the suspected area with 5% salt water solution (two teaspoons of salt to 12 oz. of water). Drive vehicle above 24 km/h (15 mph) for at least 30 seconds. If DTC returns immediately, replace suspected harness.

Any circuitry, that is suspected as causing the intermittent complaint, should be thoroughly checked for backed out terminals, improper mating, broken locks, improperly formed or damaged terminals, poor terminal to wiring connections or physical damage to the wiring harness.

A careful visual inspection of the toothed ring, CV joint, bearing and wheel speed sensor for physical damage should also be performed. If DTC sets at the same mph every time and wheel speed variation is noted above this mph on the Tech 1, the speed ring is most likely damaged. This can occur at any speed.

Resistance of the wheel speed sensor will increase with an increase in sensor temperature.

When replacing a wheel speed sensor, inspect the sensor terminals and harness connector for corrosion and/or water intrusion. If evidence of corrosion or water intrusion exists, replace wheel speed sensor harness. Refer to "On-Vehicle Service" in this section.

If wheel speed sensor harness shows evidence of water intrusion, inspect wheel speed sensor for water intrusion. If water intrusion is evident, replace wheel speed sensor.

DTC 26
(Page 2 of 3)
RIGHT FRONT EXCESSIVE WHEEL SPEED VARIATION

CONTINUED FROM PAGE 1 OF 3 OF THIS DTC CHART.

(7)
- DISCONNECT RIGHT FRONT WHEEL SPEED SENSOR CONNECTOR.
- USING J 39200, MEASURE THE RESISTANCE BETWEEN 24-WAY EBCM HARNESS CONNECTOR TERMINALS "9" AND "10".

DOES J 39200 DISPLAY RESISTANCE AS "OL" (INFINITE)?

NO → REPAIR SHORT BETWEEN CKTs 872 AND 833.

YES ↓

(8)
- USING J 39200, MEASURE THE RESISTANCE BETWEEN 24-WAY EBCM HARNESS CONNECTOR TERMINAL "10" AND RIGHT FRONT WHEEL SPEED SENSOR HARNESS CONNECTOR TERMINAL "A."

IS RESISTANCE 2.0Ω OR LESS?

NO → REPAIR OPEN OR HIGH RESISTANCE IN CKT 833.

YES ↓

(9)
- USING J 39200, MEASURE THE RESISTANCE BETWEEN 24-WAY EBCM HARNESS CONNECTOR TERMINAL "9" AND RIGHT FRONT WHEEL SPEED SENSOR HARNESS CONNECTOR TERMINAL "B."

IS RESISTANCE 2.0Ω OR LESS?

NO → REPAIR OPEN OR HIGH RESISTANCE IN CKT 872.

YES ↓

(10)
- USING J 39200, MEASURE THE RESISTANCE BETWEEN RIGHT FRONT WHEEL SPEED SENSOR TERMINALS "A" AND "B."

IS RESISTANCE BETWEEN 1628 AND 1889Ω WHEN SENSOR IS 25°C (77°F)?
[IF TEMPERATURE OF THE SENSOR IS SUSPECTED TO BE OTHER THAN 25°C (77°F), REFER TO THE TABLE BELOW FOR APPLICABLE SENSOR RESISTANCE VALUES: THE VALUES IN THE TABLE ARE FOR TEMPERATURE OF THE SENSOR NOT AIR TEMPERATURE.]

NO → REPLACE WHEEL SPEED SENSOR.

YES ↓

MALFUNCTION IS INTERMITTENT. INSPECT ALL CONNECTORS AND HARNESSES FOR DAMAGES WHICH MAY RESULT IN HIGH RESISTANCE WHEN ALL COMPONENTS ARE CONNECTED. REFER TO "DIAGNOSTIC AIDS" ON FACING PAGE FOR MORE INFORMATION. REPAIR ALL DAMAGE IF FOUND.

FRONT WHEEL SPEED SENSOR

TEMPERATURE OF SENSOR VS. SENSOR RESISTANCE (APPROXIMATE)

TEMP. (°C)	TEMP. (°F)	RESISTANCE (OHMS)
-34 TO 4	-30 TO 40	1258 TO 1739
5 TO 43	41 TO 110	1503 TO 2023
44 TO 93	111 TO 200	1747 TO 2384

DIAGNOSTIC CHART DTC 26 — 1995 CAVALIER AND SUNFIRE

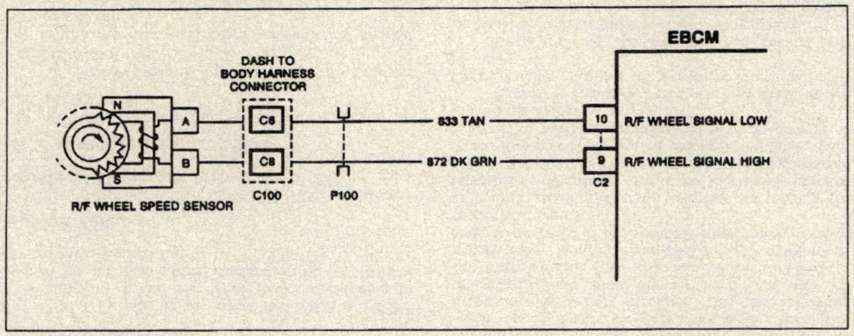

DTC 26
(Page 3 of 3)
RIGHT FRONT EXCESSIVE WHEEL SPEED VARIATION

Circuit Description:

As a toothed ring passes by the wheel speed sensor, changes in the electromagnetic field cause the wheel speed sensor to produce a sinusoidal (AC) voltage signal whose frequency is proportional to wheel speed. The magnitude of this signal is directly related to wheel speed and the proximity of the wheel speed sensor to the toothed ring, often referred to as the air gap.

DTC Will Set When:

DTC 26 can be set when the brake is "OFF." The purpose of this test is to detect a situation in which the right front wheel acceleration or deceleration is beyond specified limits.

Action Taken:

A malfunction DTC is stored, ABS is disabled and the "ABS" warning lamp is turned "ON."

DTC Chart Test Description:

Number(s) below refer to circled number(s) on the diagnostic chart.

11. Ensures that the wheel speed sensor circuitry is not shorted to ground.
12. Ensures the wheel speed sensor is not shorted to ground.

Diagnostic Aids:

An "intermittent" malfunction is most likely caused by a poor connection, rubbed through wire insulation, or a wire that is broken inside the insulation.

The frequency of the malfunction can be checked by using the enhanced diagnostic function of the Tech 1, as described in "Enhanced Diagnostics," found in this section.

If the customer's comments reflect that the ABS amber warning lamp is "ON" only during moist environmental changes, (rain, snow, vehicle wash) all wheel speed sensor circuitry should be thoroughly inspected for water intrusion. Use the following procedure. Spray down the suspected area with 5% salt water solution (two teaspoons of salt to 12 oz. of water). Drive vehicle above 24 km/h (15 mph) for at least 30 seconds. If DTC returns immediately, replace suspected harness.

Any circuitry, that is suspected as causing the intermittent complaint, should be thoroughly checked for backed out terminals, improper mating, broken locks, improperly formed or damaged terminals, poor terminal to wiring connections or physical damage to the wiring harness.

A careful visual inspection of the toothed ring, CV joint, bearing and wheel speed sensor for physical damage should also be performed. If DTC sets at the same mph every time and wheel speed variation is noted above this mph on the Tech 1, the speed ring is most likely damaged. This can occur at any speed.

Resistance of the wheel speed sensor will increase with an increase in sensor temperature.

When replacing a wheel speed sensor, inspect the sensor terminals and harness connector for corrosion and/or water intrusion. If evidence of corrosion or water intrusion exists, replace wheel speed sensor harness. Refer to "On-Vehicle Service" in this section.

If wheel speed sensor harness shows evidence of water intrusion, inspect wheel speed sensor for water intrusion. If water intrusion is evident, replace wheel speed sensor.

DIAGNOSTIC CHART DTC 26 — 1995 CAVALIER AND SUNFIRE

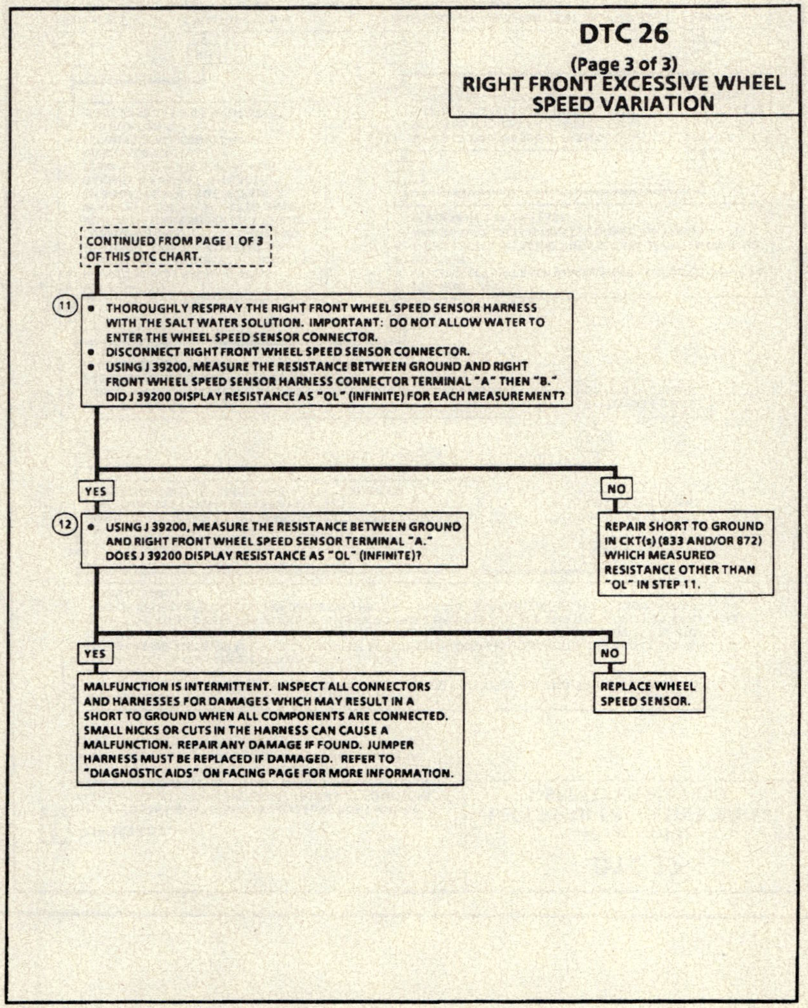

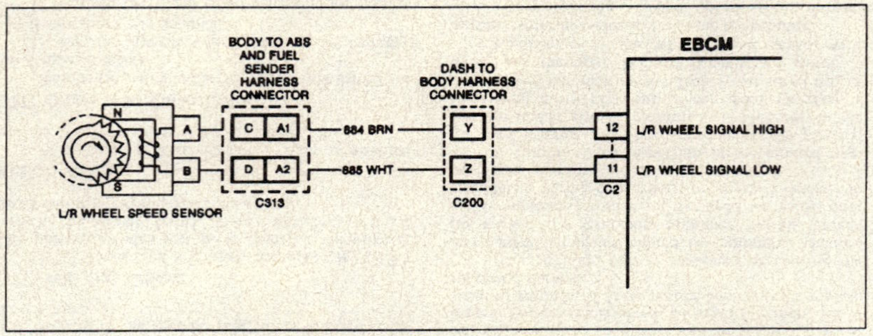

DTC 27

LEFT REAR EXCESSIVE WHEEL SPEED VARIATION

Circuit Description:

As a toothed ring passes by the wheel speed sensor, changes in the electromagnetic field cause the wheel speed sensor to produce a sinusoidal (AC) voltage signal whose frequency is proportional to wheel speed. The magnitude of this signal is directly related to wheel speed and the proximity of the wheel speed sensor to the toothed ring, often referred to as the air gap.

DTC Will Set When:

DTC 27 can be set when the brake is "OFF." The purpose of this test is to detect a situation in which the left rear wheel acceleration or deceleration is beyond specified limits.

Action Taken:

A malfunction DTC is stored, ABS is disabled and the "ABS" warning lamp is turned "ON."

DTC Chart Test Description:

Number(s) below refer to circled number(s) on the diagnostic chart.

1. Checks wheel speed sensor for proper resistance value.
2. Ensures the wheel speed sensor is not shorted to ground.
3. Ensures the wheel speed sensor is not shorted to voltage.
4. Ensures that DTC 27 was not set due to a poor connection at the EBCM or physical damage of the circuitry.
5. Checks for proper voltage at the wheel signal low speed sensor harness connector.
6. Checks for proper voltage at the wheel signal high speed sensor harness connector.

Diagnostic Aids:

An "intermittent" malfunction is most likely caused by a poor connection, rubbed through wire insulation, or a wire that is broken inside the insulation.

The frequency of the malfunction can be checked by using the enhanced diagnostic function of the Tech 1, as described in "Enhanced Diagnostics," found in this section.

If the customer's comments reflect that the ABS amber warning lamp is "ON" only during moist environmental changes, (rain, snow, vehicle wash) all wheel speed sensor circuitry should be thoroughly inspected for water intrusion. Use the following procedure. Spray down the suspected area with 5% salt water solution (two teaspoons of salt to 12 oz. of water). Drive vehicle above 24 km/h (15 mph) for at least 30 seconds. If DTC returns immediately, replace suspected harness.

Any circuitry, that is suspected as causing the intermittent complaint, should be thoroughly checked for backed out terminals, improper mating, broken locks, improperly formed or damaged terminals, poor terminal to wiring connections or physical damage to the wiring harness.

Excessive end play can cause this DTC to set. Check integral bearing for excessive axial end play.

Resistance of the wheel speed sensor will increase with an increase in sensor temperature.

When replacing a wheel speed sensor, inspect the sensor terminals and harness connector for corrosion and/or water intrusion. If evidence of corrosion or water intrusion exists, replace wheel speed sensor harness.

If wheel speed sensor harness shows evidence of water intrusion, inspect wheel speed sensor for water intrusion. If water intrusion is evident, replace wheel speed sensor.

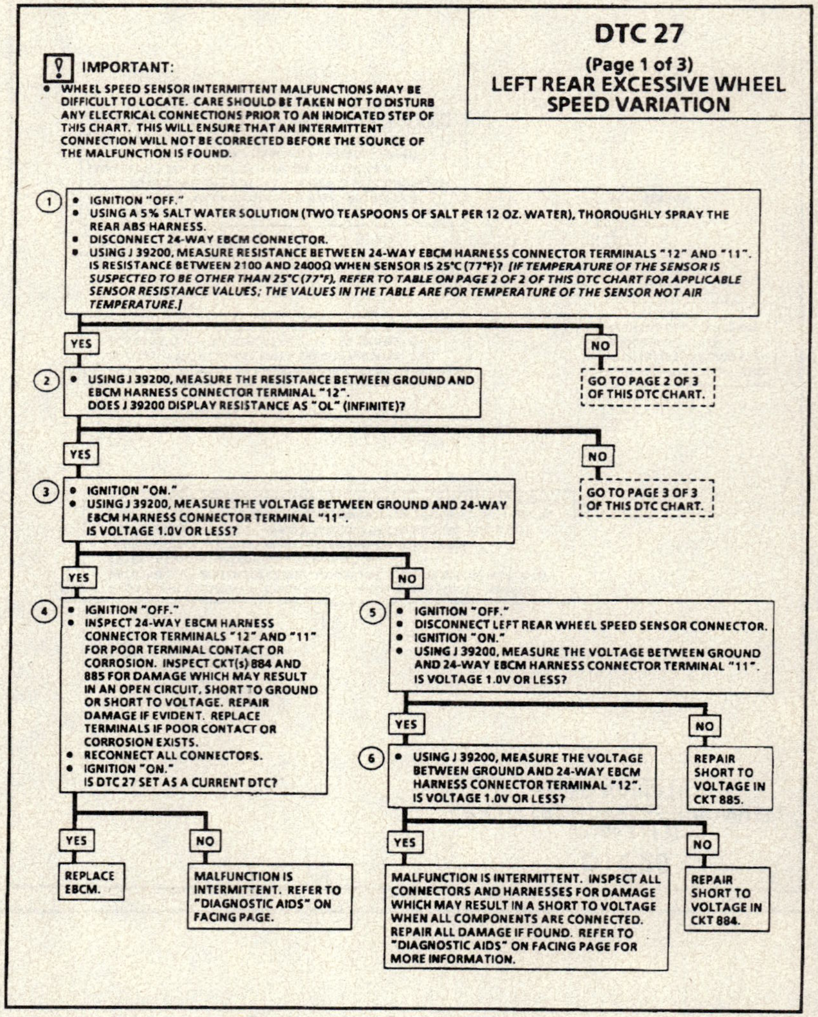

DIAGNOSTIC CHART DTC 27 — 1995 CAVALIER AND SUNFIRE

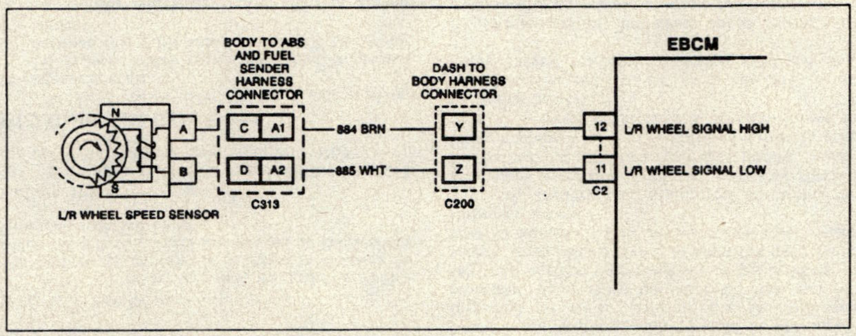

DTC 27
(Page 2 of 3)
LEFT REAR EXCESSIVE WHEEL SPEED VARIATION

Circuit Description:

As a toothed ring passes by the wheel speed sensor, changes in the electromagnetic field cause the wheel speed sensor to produce a sinusoidal (AC) voltage signal whose frequency is proportional to wheel speed. The magnitude of this signal is directly related to wheel speed and the proximity of the wheel speed sensor to the toothed ring, often referred to as the air gap.

DTC Will Set When:

DTC 27 can be set when the brake is "OFF." The purpose of this test is to detect a situation in which the left rear wheel acceleration or deceleration is beyond specified limits.

Action Taken:

A malfunction DTC is stored, ABS is disabled and the "ABS" warning lamp is turned "ON."

DTC Chart Test Description:

Number(s) below refer to circled number(s) on the diagnostic chart.

7. Tests for a short between the wheel signal high and wheel signal low circuits.
8. Checks for an open in the wheel signal low circuit.
9. Checks for an open in the wheel signal high circuit.
10. Checks for proper resistance of the wheel speed sensor.

Diagnostic Aids:

An "intermittent" malfunction is most likely caused by a poor connection, rubbed through wire insulation, or a wire that is broken inside the insulation.

The frequency of the malfunction can be checked by using the enhanced diagnostic function of the Tech 1, as described in "Enhanced Diagnostics," found in this section. If the customer's comments reflect that the ABS amber warning lamp is "ON" only during moist environmental changes, (rain, snow, vehicle wash) all wheel speed sensor circuitry should be thoroughly inspected for water intrusion. Use the following procedure. Spray down the suspected area with 5% salt water solution (two teaspoons of salt to 12 oz. of water). Drive vehicle above 24 km/h (15 mph) for at least 30 seconds. If DTC returns immediately, replace suspected harness.

Any circuitry, that is suspected as causing the intermittent complaint, should be thoroughly checked for backed out terminals, improper mating, broken locks, improperly formed or damaged terminals, poor terminal to wiring connections or physical damage to the wiring harness.

Excessive end play can cause this DTC to set. Check integral bearing for excessive axial end play, refer to SECTION 3.

Resistance of the wheel speed sensor will increase with an increase in sensor temperature.

When replacing a wheel speed sensor, inspect the sensor terminals and harness connector for corrosion and/or water intrusion. If evidence of corrosion or water intrusion exists, replace wheel speed sensor harness. Refer to "On-Vehicle Service" in this section.

If wheel speed sensor harness shows evidence of water intrusion, inspect wheel speed sensor for water intrusion. If water intrusion is evident, replace wheel speed sensor.

DIAGNOSTIC CHART DTC 27 — 1995 CAVALIER AND SUNFIRE

DTC 27
(Page 2 of 3)
LEFT REAR EXCESSIVE WHEEL SPEED VARIATION

CONTINUED FROM PAGE 1 OF 3 OF THIS DTC CHART.

(7)
- DISCONNECT LEFT REAR WHEEL SPEED SENSOR CONNECTOR.
- USING J 39200, MEASURE THE RESISTANCE BETWEEN 24-WAY EBCM HARNESS CONNECTOR TERMINALS "11" AND "12". DOES J 39200 DISPLAY RESISTANCE AS "OL" (INFINITE)?

NO: REPAIR SHORT BETWEEN CKTs 884 AND 885.

YES:

(8)
- USING J 39200, MEASURE THE RESISTANCE BETWEEN 24-WAY EBCM HARNESS CONNECTOR TERMINAL "12" AND LEFT REAR WHEEL SPEED SENSOR HARNESS CONNECTOR TERMINAL "A." IS RESISTANCE 2.0Ω OR LESS?

NO: REPAIR OPEN OR HIGH RESISTANCE IN CKT 884.

YES:

(9)
- USING J 39200, MEASURE THE RESISTANCE BETWEEN 24-WAY EBCM HARNESS CONNECTOR TERMINAL "11" AND LEFT REAR WHEEL SPEED SENSOR HARNESS CONNECTOR TERMINAL "B." IS RESISTANCE 2.0Ω OR LESS?

NO: REPAIR OPEN OR HIGH RESISTANCE IN CKT 885.

YES:

(10)
- USING J 39200, MEASURE THE RESISTANCE BETWEEN LEFT REAR WHEEL SPEED SENSOR TERMINALS "A" AND "B." IS RESISTANCE BETWEEN 2100 AND 2400Ω WHEN SENSOR IS 25°C (77°F)? *[IF TEMPERATURE OF THE SENSOR IS SUSPECTED TO BE OTHER THAN 25°C (77°F), REFER TO THE TABLE BELOW FOR APPLICABLE SENSOR RESISTANCE VALUES; THE VALUES IN THE TABLE ARE FOR TEMPERATURE OF THE SENSOR NOT AIR TEMPERATURE.]*

NO: REPLACE INTEGRAL BEARING AND SPEED SENSOR ASSEMBLY.

YES: MALFUNCTION IS INTERMITTENT. INSPECT ALL CONNECTORS AND HARNESSES FOR DAMAGES WHICH MAY RESULT IN HIGH RESISTANCE WHEN ALL COMPONENTS ARE CONNECTED. REFER TO "DIAGNOSTIC AIDS" ON FACING PAGE FOR MORE INFORMATION. REPAIR ALL DAMAGE IF FOUND.

REAR WHEEL SPEED SENSOR

TEMPERATURE OF SENSOR VS. SENSOR RESISTANCE (APPROXIMATE)

TEMP. (°C)	TEMP. (°F)	RESISTANCE (OHMS)
-34 TO 4	-30 TO 40	1623 TO 2210
5 TO 43	41 TO 110	1939 TO 2570
44 TO 93	111 TO 200	2254 TO 3029

DIAGNOSTIC CHART DTC 27 — 1995 CAVALIER AND SUNFIRE

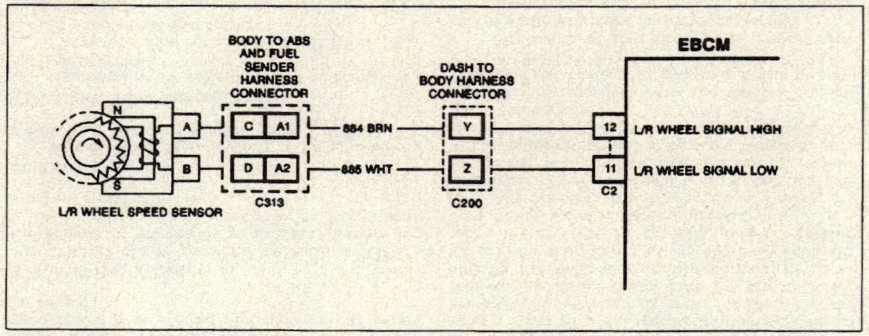

DTC 27 (Page 3 of 3)
LEFT REAR EXCESSIVE WHEEL SPEED VARIATION

Circuit Description:

As a toothed ring passes by the wheel speed sensor, changes in the electromagnetic field cause the wheel speed sensor to produce a sinusoidal (AC) voltage signal whose frequency is proportional to wheel speed. The magnitude of this signal is directly related to wheel speed and the proximity of the wheel speed sensor to the toothed ring, often referred to as the air gap.

DTC Will Set When:

DTC 27 can be set when the brake is "OFF." The purpose of this test is to detect a situation in which the left rear wheel acceleration or deceleration is beyond specified limits.

Action Taken:

A malfunction DTC is stored, ABS is disabled and the "ABS" warning lamp is turned "ON."

DTC Chart Test Description:

Number(s) below refer to circled number(s) on the diagnostic chart.

11. Isolates wheel signal high short to ground between rear ABS axle harness and dash/body harnesses.
12. Isolates wheel signal low short to ground between rear ABS axle harness and dash/body harnesses.
13. Ensures that the wheel speed sensor harness is not shorted to ground.
14. Ensures that the wheel speed sensor is not shorted to ground.

Diagnostic Aids:

An "intermittent" malfunction is most likely caused by a poor connection, rubbed through wire insulation, or a wire that is broken inside the insulation.

The frequency of the malfunction can be checked by using the enhanced diagnostic function of the Tech 1, as described in "Enhanced Diagnostics," found in this section.

If the customer's comments reflect that the ABS amber warning lamp is "ON" only during moist environmental changes, (rain, snow, vehicle wash) all wheel speed sensor circuitry should be thoroughly inspected for water intrusion. Use the following procedure. Spray down the suspected area with 5% salt water solution (two teaspoons of salt to 12 oz. of water). Drive vehicle above 24 km/h (15 mph) for at least 30 seconds. If DTC returns immediately, replace suspected harness.

Any circuitry, that is suspected as causing the intermittent complaint, should be thoroughly checked for backed out terminals, improper mating, broken locks, improperly formed or damaged terminals, poor terminal to wiring connections or physical damage to the wiring harness.

Excessive end play can cause this DTC to set. Check integral bearing for excessive axial end play.

Resistance of the wheel speed sensor will increase with an increase in sensor temperature.

When replacing a wheel speed sensor, inspect the sensor terminals and harness connector for corrosion and/or water intrusion. If evidence of corrosion or water intrusion exists, replace wheel speed sensor harness.

If wheel speed sensor harness shows evidence of water intrusion, inspect wheel speed sensor for water intrusion. If water intrusion is evident, replace wheel speed sensor.

DIAGNOSTIC CHART DTC 27 — 1995 CAVALIER AND SUNFIRE

DTC 27
(Page 3 of 3)
LEFT REAR EXCESSIVE WHEEL SPEED VARIATION

CONTINUED FROM PAGE 1 OF 3 OF THIS DTC CHART.

(11)
- DISCONNECT REAR ABS HARNESS FROM BODY TO ABS AND FUEL SENDER HARNESS CONNECTOR.
- USING J 39200, MEASURE THE RESISTANCE BETWEEN GROUND AND 24-WAY EBCM HARNESS CONNECTOR TERMINAL "12".

DOES J 39200 DISPLAY RESISTANCE AS "OL" (INFINITE)?

NO → REPAIR SHORT TO GROUND IN CKT 884.

YES ↓

(12)
- USING J 39200, MEASURE THE RESISTANCE BETWEEN GROUND AND 24-WAY EBCM HARNESS CONNECTOR TERMINAL "11".

DOES J 39200 DISPLAY RESISTANCE AS "OL" (INFINITE)?

NO → REPAIR SHORT TO GROUND IN CKT 885.

YES ↓

(13)
- THOROUGHLY RESPRAY THE REAR ABS HARNESS WITH THE SALT WATER SOLUTION. IMPORTANT: DO NOT ALLOW WATER TO ENTER THE WHEEL SPEED SENSOR CONNECTOR.
- DISCONNECT LEFT REAR WHEEL SPEED SENSOR CONNECTOR.
- USING J 39200, MEASURE THE RESISTANCE BETWEEN GROUND AND LEFT REAR WHEEL SPEED SENSOR HARNESS CONNECTOR TERMINAL "A" THEN "B."

DID J 39200 DISPLAY RESISTANCE AS "OL" (INFINITE) FOR EACH MEASUREMENT?

NO → REPLACE REAR ABS HARNESS.

YES ↓

(14)
- USING J 39200, MEASURE THE RESISTANCE BETWEEN SENSOR TERMINAL "A" AND INTEGRAL BEARING AND SPEED SENSOR ASSEMBLY MOUNTING FLANGE.

DOES J 39200 DISPLAY RESISTANCE AS "OL" (INFINITE)?

NO → REPLACE INTEGRAL BEARING AND SPEED SENSOR ASSEMBLY.

YES ↓

MALFUNCTION IS INTERMITTENT. INSPECT ALL CONNECTORS AND HARNESSES FOR DAMAGES WHICH MAY RESULT IN A SHORT TO GROUND WHEN ALL COMPONENTS ARE CONNECTED. SMALL NICKS OR CUTS IN THE HARNESS CAN CAUSE A MALFUNCTION. REPAIR ANY DAMAGE IF FOUND. JUMPER HARNESS MUST BE REPLACED IF DAMAGED. REFER TO "DIAGNOSTIC AIDS" ON FACING PAGE FOR MORE INFORMATION.

DIAGNOSTIC CHART DTC 28 — 1995 CAVALIER AND SUNFIRE

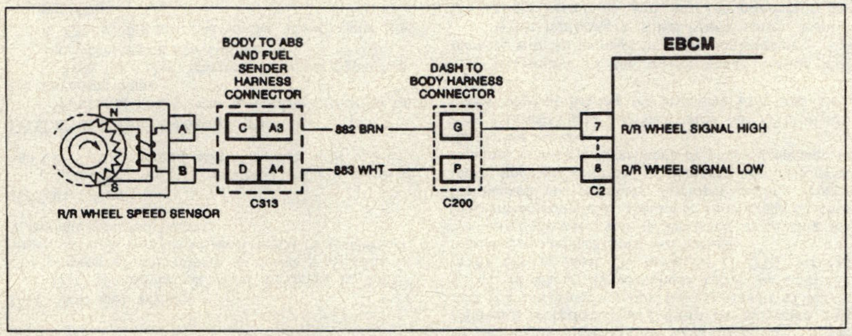

DTC 28 (Page 1 of 3)

RIGHT REAR EXCESSIVE WHEEL SPEED VARIATION

Circuit Description:

As a toothed ring passes by the wheel speed sensor, changes in the electromagnetic field cause the wheel speed sensor to produce a sinusoidal (AC) voltage signal whose frequency is proportional to wheel speed. The magnitude of this signal is directly related to wheel speed and the proximity of the wheel speed sensor to the toothed ring, often referred to as the air gap.

DTC Will Set When:

DTC 28 can be set when the brake is "OFF." The purpose of this test is to detect a situation in which the right rear wheel acceleration or deceleration is beyond specified limits.

Action Taken:

A malfunction DTC is stored, ABS is disabled and the "ABS" warning lamp is turned "ON."

DTC Chart Test Description:

Number(s) below refer to circled number(s) on the diagnostic chart.

1. Checks wheel speed sensor for proper resistance value.
2. Ensures the wheel speed sensor is not shorted to ground.
3. Ensures the wheel speed sensor is not shorted to voltage.
4. Ensures that DTC 28 was not set due to a poor connection at the EBCM or physical damage of the circuitry.
5. Checks for proper voltage at the wheel signal low speed sensor harness connector.
6. Checks for proper voltage at the wheel signal high speed sensor harness connector.

Diagnostic Aids:

An "intermittent" malfunction is most likely caused by a poor connection, rubbed through wire insulation, or a wire that is broken inside the insulation.

The frequency of the malfunction can be checked by using the enhanced diagnostic function of the Tech 1, as described in "Enhanced Diagnostics," found in this section.

If the customer's comments reflect that the ABS amber warning lamp is "ON" only during moist environmental changes, (rain, snow, vehicle wash) all wheel speed sensor circuitry should be thoroughly inspected for water intrusion. Use the following procedure. Spray down the suspected area with 5% salt water solution (two teaspoons of salt to 12 oz. of water). Drive vehicle above 24 km/h (15 mph) for at least 30 seconds. If DTC returns immediately, replace suspected harness.

Any circuitry, that is suspected as causing the intermittent complaint, should be thoroughly checked for backed out terminals, improper mating, broken locks, improperly formed or damaged terminals, poor terminal to wiring connections or physical damage to the wiring harness.

Excessive end play can cause this DTC to set. Check integral bearing for excessive axial end play.

Resistance of the wheel speed sensor will increase with an increase in sensor temperature.

When replacing a wheel speed sensor, inspect the sensor terminals and harness connector for corrosion and/or water intrusion. If evidence of corrosion or water intrusion exists, replace wheel speed sensor harness.

If wheel speed sensor harness shows evidence of water intrusion, inspect wheel speed sensor for water intrusion. If water intrusion is evident, replace wheel speed sensor.

DIAGNOSTIC CHART DTC 28 — 1995 CAVALIER AND SUNFIRE

DTC 28 (Page 1 of 3) RIGHT REAR EXCESSIVE WHEEL SPEED VARIATION

IMPORTANT:

- WHEEL SPEED SENSOR INTERMITTENT MALFUNCTIONS MAY BE DIFFICULT TO LOCATE. CARE SHOULD BE TAKEN NOT TO DISTURB ANY ELECTRICAL CONNECTIONS PRIOR TO AN INDICATED STEP OF THIS CHART. THIS WILL ENSURE THAT AN INTERMITTENT CONNECTION WILL NOT BE CORRECTED BEFORE THE SOURCE OF THE MALFUNCTION IS FOUND.

(1)
- IGNITION "OFF."
- USING A 5% SALT WATER SOLUTION (TWO TEASPOONS OF SALT PER 12 OZ. WATER), THOROUGHLY SPRAY THE REAR ABS HARNESS.
- DISCONNECT 24-WAY EBCM CONNECTOR.
- USING J 39200, MEASURE RESISTANCE BETWEEN 24-WAY EBCM HARNESS CONNECTOR TERMINALS "7" AND "8". IS RESISTANCE BETWEEN 2100 AND 2400 Ω WHEN SENSOR IS 25°C (77°F)? *[IF TEMPERATURE OF THE SENSOR IS SUSPECTED TO BE OTHER THAN 25°C (77°F), REFER TO TABLE ON PAGE 2 OF 2 OF THIS DTC CHART FOR APPLICABLE SENSOR RESISTANCE VALUES; THE VALUES IN THE TABLE ARE FOR TEMPERATURE OF THE SENSOR NOT AIR TEMPERATURE.]*

NO → GO TO PAGE 2 OF 3 OF THIS DTC CHART.

YES → (2)
- USING J 39200, MEASURE THE RESISTANCE BETWEEN GROUND AND EBCM HARNESS CONNECTOR TERMINAL "7". DOES J 39200 DISPLAY RESISTANCE AS "OL" (INFINITE)?

NO → GO TO PAGE 3 OF 3 OF THIS DTC CHART.

YES → (3)
- IGNITION "ON."
- USING J 39200, MEASURE THE VOLTAGE BETWEEN GROUND AND 24-WAY EBCM HARNESS CONNECTOR TERMINAL "8". IS VOLTAGE 1.0V OR LESS?

YES → (4)
- IGNITION "OFF."
- INSPECT 24-WAY EBCM HARNESS CONNECTOR TERMINALS "7" AND "8" FOR POOR TERMINAL CONTACT OR CORROSION. INSPECT CKT(s) 882 AND 883 FOR DAMAGE WHICH MAY RESULT IN AN OPEN CIRCUIT, SHORT TO GROUND OR SHORT TO VOLTAGE. REPAIR DAMAGE IF EVIDENT. REPLACE TERMINALS IF POOR CONTACT OR CORROSION EXISTS.
- RECONNECT ALL CONNECTORS.
- IGNITION "ON."
- IS DTC 28 SET AS A CURRENT DTC?

  - YES → REPLACE EBCM.
  - NO → MALFUNCTION IS INTERMITTENT. REFER TO "DIAGNOSTIC AIDS" ON FACING PAGE.

NO → (5)
- IGNITION "OFF."
- DISCONNECT RIGHT REAR WHEEL SPEED SENSOR CONNECTOR.
- IGNITION "ON."
- USING J 39200, MEASURE THE VOLTAGE BETWEEN GROUND AND 24-WAY EBCM HARNESS CONNECTOR TERMINAL "8". IS VOLTAGE 1.0V OR LESS?

  - NO → REPAIR SHORT TO VOLTAGE IN CKT 883.
  - YES → (6)
    - USING J 39200, MEASURE THE VOLTAGE BETWEEN GROUND AND 24-WAY EBCM HARNESS CONNECTOR TERMINAL "7". IS VOLTAGE 1.0V OR LESS?
    - YES → MALFUNCTION IS INTERMITTENT. INSPECT ALL CONNECTORS AND HARNESSES FOR DAMAGE WHICH MAY RESULT IN A SHORT TO VOLTAGE WHEN ALL COMPONENTS ARE CONNECTED. REPAIR ALL DAMAGE IF FOUND. REFER TO "DIAGNOSTIC AIDS" ON FACING PAGE FOR MORE INFORMATION.
    - NO → REPAIR SHORT TO VOLTAGE IN CKT 882.

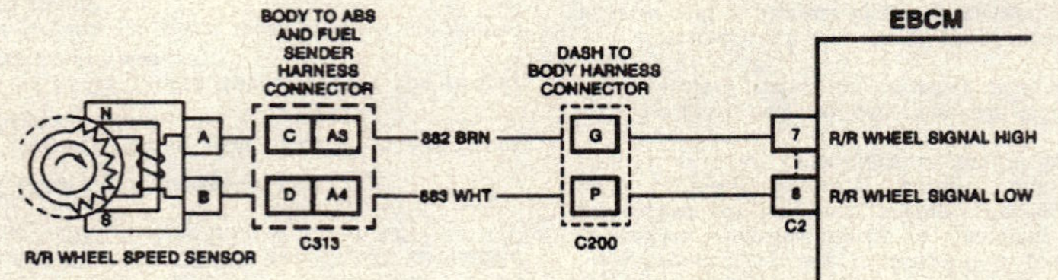

DTC 28
(Page 2 of 3)
RIGHT REAR EXCESSIVE WHEEL SPEED VARIATION

Circuit Description:

As a toothed ring passes by the wheel speed sensor, changes in the electromagnetic field cause the wheel speed sensor to produce a sinusoidal (AC) voltage signal whose frequency is proportional to wheel speed. The magnitude of this signal is directly related to wheel speed and the proximity of the wheel speed sensor to the toothed ring, often referred to as the air gap.

DTC Will Set When:

DTC 28 can be set when the brake is "OFF." The purpose of this test is to detect a situation in which the right rear wheel acceleration or deceleration is beyond specified limits.

Action Taken:

A malfunction DTC is stored, ABS is disabled and the "ABS" warning lamp is turned "ON."

DTC Chart Test Description:

Number(s) below refer to circled number(s) on the diagnostic chart.

7. Tests for a short between the wheel signal high and wheel signal low circuits.
8. Checks for an open in the wheel signal low circuit.
9. Checks for an open in the wheel signal high circuit.
10. Checks for proper resistance of the wheel speed sensor.

Diagnostic Aids:

An "intermittent" malfunction is most likely caused by a poor connection, rubbed through wire insulation, or a wire that is broken inside the insulation.

The frequency of the malfunction can be checked by using the enhanced diagnostic function of the Tech 1, as described in "Enhanced Diagnostics," found in this section. If the customer's comments reflect that the ABS amber warning lamp is "ON" only during moist environmental changes, (rain, snow, vehicle wash) all wheel speed sensor circuitry should be thoroughly inspected for water intrusion. Use the following procedure. Spray down the suspected area with 5% salt water solution (two teaspoons of salt to 12 oz. of water). Drive vehicle above 24 km/h (15 mph) for at least 30 seconds. If DTC returns immediately, replace suspected harness.

Any circuitry, that is suspected as causing the intermittent complaint, should be thoroughly checked for backed out terminals, improper mating, broken locks, improperly formed or damaged terminals, poor terminal to wiring connections or physical damage to the wiring harness.

Excessive end play can cause this DTC to set. Check integral bearing for excessive axial end play.

Resistance of the wheel speed sensor will increase with an increase in sensor temperature.

When replacing a wheel speed sensor, inspect the sensor terminals and harness connector for corrosion and/or water intrusion. If evidence of corrosion or water intrusion exists, replace wheel speed sensor harness.

If wheel speed sensor harness shows evidence of water intrusion, inspect wheel speed sensor for water intrusion. If water intrusion is evident, replace wheel speed sensor.

DIAGNOSTIC CHART DTC 28 — 1995
CAVALIER AND SUNFIRE

DTC 28
(Page 2 of 3)
RIGHT REAR EXCESSIVE WHEEL SPEED VARIATION

CONTINUED FROM PAGE 1 OF 3 OF THIS DTC CHART.

(7)
- DISCONNECT RIGHT REAR WHEEL SPEED SENSOR CONNECTOR.
- USING J 39200, MEASURE THE RESISTANCE BETWEEN 24-WAY EBCM HARNESS CONNECTOR TERMINALS "7" AND "8". DOES J 39200 DISPLAY RESISTANCE AS "OL" (INFINITE)?

NO → REPAIR SHORT BETWEEN CKTs 882 AND 883.

YES ↓

(8)
- USING J 39200, MEASURE THE RESISTANCE BETWEEN 24-WAY EBCM HARNESS CONNECTOR TERMINAL "7" AND RIGHT REAR WHEEL SPEED SENSOR HARNESS CONNECTOR TERMINAL "A." IS RESISTANCE 2.0Ω OR LESS?

NO → REPAIR OPEN OR HIGH RESISTANCE IN CKT 882.

YES ↓

(9)
- USING J 39200, MEASURE THE RESISTANCE BETWEEN 24-WAY EBCM HARNESS CONNECTOR TERMINAL "8" AND RIGHT REAR WHEEL SPEED SENSOR HARNESS CONNECTOR TERMINAL "B." IS RESISTANCE 2.0Ω OR LESS?

NO → REPAIR OPEN OR HIGH RESISTANCE IN CKT 883.

YES ↓

(10)
- USING J 39200, MEASURE THE RESISTANCE BETWEEN RIGHT REAR WHEEL SPEED SENSOR TERMINALS "A" AND "B." IS RESISTANCE BETWEEN 2100 AND 2400Ω WHEN SENSOR IS 25°C (77°F)? *[IF TEMPERATURE OF THE SENSOR IS SUSPECTED TO BE OTHER THAN 25°C (77°F), REFER TO THE TABLE BELOW FOR APPLICABLE SENSOR RESISTANCE VALUES; THE VALUES IN THE TABLE ARE FOR TEMPERATURE OF THE SENSOR NOT AIR TEMPERATURE.]*

NO → REPLACE INTEGRAL BEARING AND SPEED SENSOR ASSEMBLY.

YES → MALFUNCTION IS INTERMITTENT. INSPECT ALL CONNECTORS AND HARNESSES FOR DAMAGES WHICH MAY RESULT IN HIGH RESISTANCE WHEN ALL COMPONENTS ARE CONNECTED. REFER TO "DIAGNOSTIC AIDS" ON FACING PAGE FOR MORE INFORMATION. REPAIR ALL DAMAGE IF FOUND.

REAR WHEEL SPEED SENSOR

TEMPERATURE OF SENSOR VS. SENSOR RESISTANCE (APPROXIMATE)

TEMP. (°C)	TEMP. (°F)	RESISTANCE (OHMS)
-34 TO 4	-30 TO 40	1623 TO 2210
5 TO 43	41 TO 110	1939 TO 2570
44 TO 93	111 TO 200	2254 TO 3029

DIAGNOSTIC CHART DTC 28 — 1995 CAVALIER AND SUNFIRE

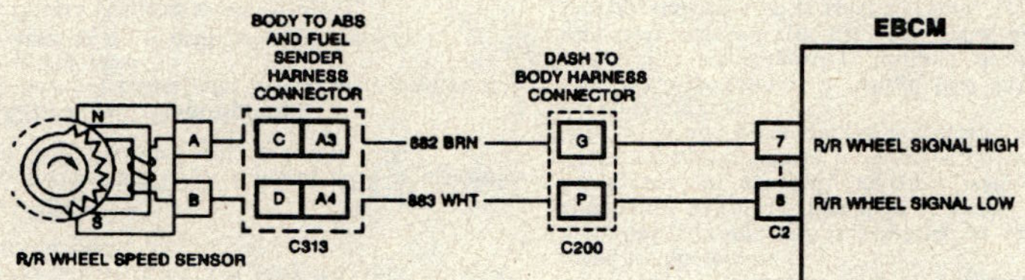

DTC 28 (Page 3 of 3)
RIGHT REAR EXCESSIVE WHEEL SPEED VARIATION

Circuit Description:

As a toothed ring passes by the wheel speed sensor, changes in the electromagnetic field cause the wheel speed sensor to produce a sinusoidal (AC) voltage signal whose frequency is proportional to wheel speed. The magnitude of this signal is directly related to wheel speed and the proximity of the wheel speed sensor to the toothed ring, often referred to as the air gap.

DTC Will Set When:

DTC 28 can be set when the brake is "OFF." The purpose of this test is to detect a situation in which the right rear wheel acceleration or deceleration is beyond specified limits.

Action Taken:

A malfunction DTC is stored, ABS is disabled and the "ABS" warning lamp is turned "ON."

DTC Chart Test Description:

Number(s) below refer to circled number(s) on the diagnostic chart.

11. Isolates wheel signal high short to ground between rear ABS axle harness and dash/body harnesses.
12. Isolates wheel signal low short to ground between rear ABS axle harness and dash/body harnesses.
13. Ensures that the wheel speed sensor harness is not shorted to ground.
14. Ensures that the wheel speed sensor is not shorted to ground.

Diagnostic Aids:

An "intermittent" malfunction is most likely caused by a poor connection, rubbed through wire insulation, or a wire that is broken inside the insulation.

The frequency of the malfunction can be checked by using the enhanced diagnostic function of the Tech 1, as described in "Enhanced Diagnostics," found in this section.

If the customer's comments reflect that the ABS amber warning lamp is "ON" only during moist environmental changes, (rain, snow, vehicle wash) all wheel speed sensor circuitry should be thoroughly inspected for water intrusion. Use the following procedure. Spray down the suspected area with 5% salt water solution (two teaspoons of salt to 12 oz. of water). Drive vehicle above 24 km/h (15 mph) for at least 30 seconds. If DTC returns immediately, replace suspected harness.

Any circuitry, that is suspected as causing the intermittent complaint, should be thoroughly checked for backed out terminals, improper mating, broken locks, improperly formed or damaged terminals, poor terminal to wiring connections or physical damage to the wiring harness.

Excessive end play can cause this DTC to set. Check integral bearing for excessive axial end play.

Resistance of the wheel speed sensor will increase with an increase in sensor temperature.

When replacing a wheel speed sensor, inspect the sensor terminals and harness connector for corrosion and/or water intrusion. If evidence of corrosion or water intrusion exists, replace wheel speed sensor harness.

If wheel speed sensor harness shows evidence of water intrusion, inspect wheel speed sensor for water intrusion. If water intrusion is evident, replace wheel speed sensor.

DIAGNOSTIC CHART DTC 28 — 1995 CAVALIER AND SUNFIRE

DTC 28
(Page 3 of 3)
RIGHT REAR EXCESSIVE WHEEL SPEED VARIATION

CONTINUED FROM PAGE 1 OF 3 OF THIS DTC CHART.

(11)
- DISCONNECT REAR ABS HARNESS FROM BODY TO ABS AND FUEL SENDER HARNESS CONNECTOR.
- USING J 39200, MEASURE THE RESISTANCE BETWEEN GROUND AND 24-WAY EBCM HARNESS CONNECTOR TERMINAL "7".

DOES J 39200 DISPLAY RESISTANCE AS "OL" (INFINITE)?

NO → REPAIR SHORT TO GROUND IN CKT 882.

YES ↓

(12)
- USING J 39200, MEASURE THE RESISTANCE BETWEEN GROUND AND 24-WAY EBCM HARNESS CONNECTOR TERMINAL "8".

DOES J 39200 DISPLAY RESISTANCE AS "OL" (INFINITE)?

NO → REPAIR SHORT TO GROUND IN CKT 883.

YES ↓

(13)
- THOROUGHLY RESPRAY THE REAR ABS HARNESS WITH THE SALT WATER SOLUTION. IMPORTANT: DO NOT ALLOW WATER TO ENTER THE WHEEL SPEED SENSOR CONNECTOR.
- DISCONNECT RIGHT REAR WHEEL SPEED SENSOR CONNECTOR.
- USING J 39200, MEASURE THE RESISTANCE BETWEEN GROUND AND RIGHT REAR WHEEL SPEED SENSOR HARNESS CONNECTOR TERMINAL "A" THEN "B."

DID J 39200 DISPLAY RESISTANCE AS "OL" (INFINITE) FOR EACH MEASUREMENT?

NO → REPLACE REAR ABS HARNESS.

YES ↓

(14)
- USING J 39200, MEASURE THE RESISTANCE BETWEEN SENSOR TERMINAL "A" AND INTEGRAL BEARING AND SPEED SENSOR ASSEMBLY MOUNTING FLANGE.

DOES J 39200 DISPLAY RESISTANCE AS "OL" (INFINITE)?

NO → REPLACE INTEGRAL BEARING AND SPEED SENSOR ASSEMBLY.

YES → MALFUNCTION IS INTERMITTENT. INSPECT ALL CONNECTORS AND HARNESSES FOR DAMAGES WHICH MAY RESULT IN A SHORT TO GROUND WHEN ALL COMPONENTS ARE CONNECTED. SMALL NICKS OR CUTS IN THE HARNESS CAN CAUSE A MALFUNCTION. REPAIR ANY DAMAGE IF FOUND. JUMPER HARNESS MUST BE REPLACED IF DAMAGED. REFER TO "DIAGNOSTIC AIDS" ON FACING PAGE FOR MORE INFORMATION.

DIAGNOSTIC CHART DTC 36 — 1995 CAVALIER AND SUNFIRE

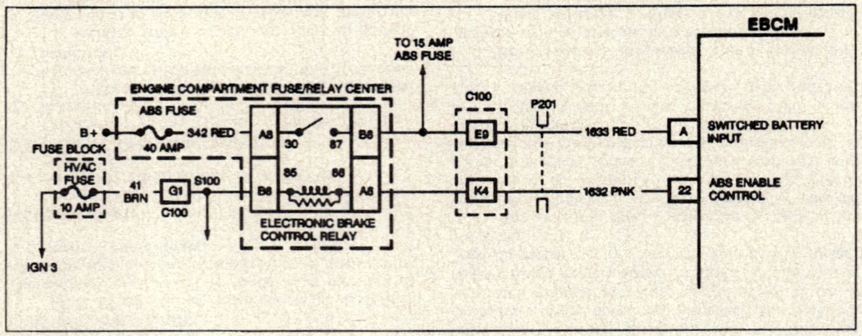

DTC 36
LOW SYSTEM VOLTAGE

Circuit Description:

This DTC is used to monitor the voltage level available to the EBCM. If the voltage drops below 11.0 volts, full performance of the ABS cannot be guaranteed. During ABS operation, there are several current requirements that will cause battery voltage to drop. Because of this, voltage is monitored prior to ABS operation to indicate good charging system condition and also during ABS operation when voltage may drop significantly.

DTC Will Set When:

DTC 36 can only be set if the vehicle's speed is greater than 5 km/h (3 mph). If both the switched battery and switched ignition voltages are less than 11.2 volts non-ABS or less than 9.6 volts ABS, a malfunction exists.

Action Taken:

A malfunction DTC is stored, ABS is disabled and the "ABS" warning lamp is turned "ON."

DTC Chart Test Description:

Number(s) below refer to circled number(s) on the diagnostic chart.

1. Tests to see if malfunction is present.
2. Checks charging system integrity.
3. Checks for excessive resistance in the switched battery input circuit.
4. Tests for high resistance in the electronic brake control relay circuit.
5. Ensures low voltage condition was not due to a poor connection.

Diagnostic Aids:

An "intermittent" malfunction is most likely caused by a poor connection, rubbed through wire insulation, or a wire that is broken inside the insulation.

The frequency of the malfunction can be checked by using the enhanced diagnostic function of the Tech 1, as described in "Enhanced Diagnostics," found in this section.

Any circuitry, that is suspected as causing the intermittent complaint, should be thoroughly checked for backed out terminals, improper mating, broken locks, improperly formed or damaged terminals, poor terminal to wiring connections or physical damage to the wiring harness.

While performing a voltage load test, if it is noted that only ignition voltage drops below acceptable voltage levels, CKT 650 should be checked for high resistance or an open condition.

DIAGNOSTIC CHART DTC 36 — 1995 CAVALIER AND SUNFIRE

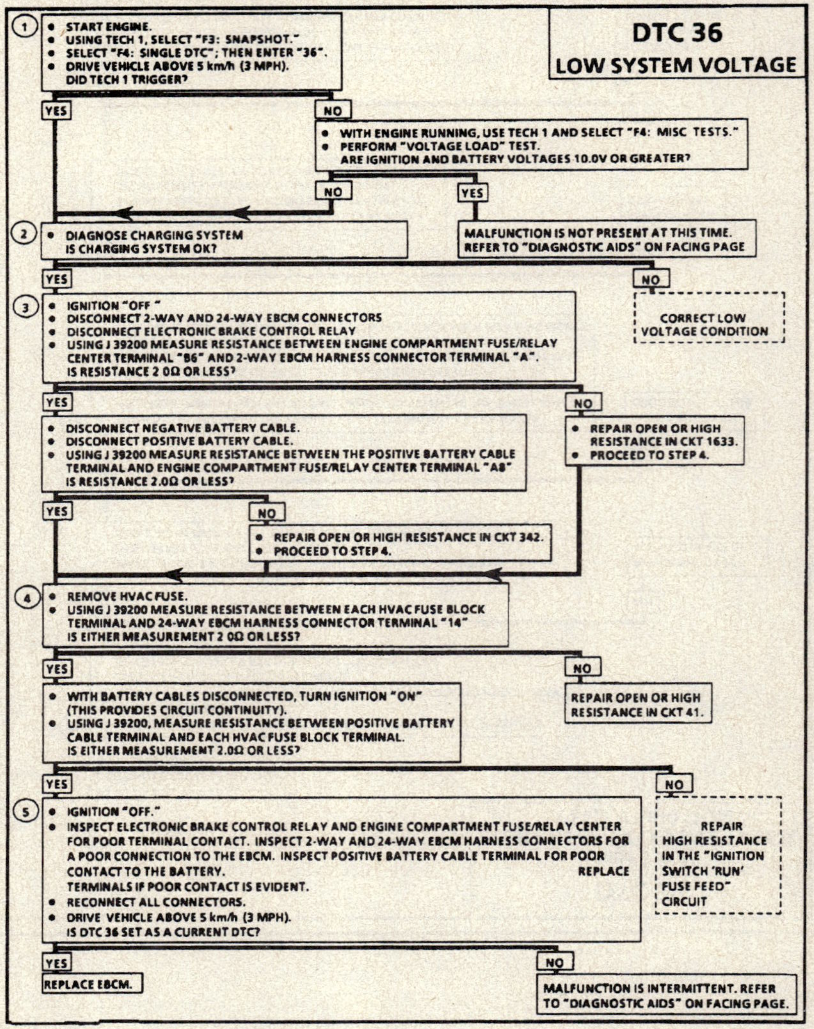

DIAGNOSTIC CHART DTC 37 — 1995 CAVALIER AND SUNFIRE

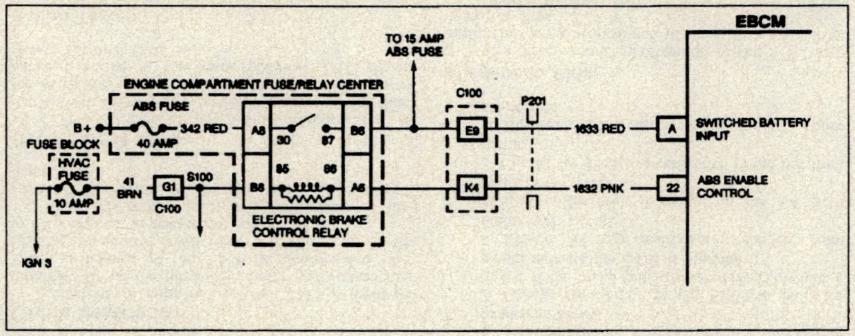

DTC 37
HIGH SYSTEM VOLTAGE

Circuit Description:
This DTC is designed to detect high vehicle voltage levels prior to any required motor movement (initialization or ABS). If excessive voltage exists, demagnetization of the motor magnets may occur, which would eventually affect or eliminate ABS performance.

DTC Will Set When:
DTC 37 can only be set if the vehicle's speed is greater than 5 km/h (3 mph). If the switched battery voltage is greater than 17 volts, a malfunction exists.

Action Taken:
A malfunction DTC is stored, ABS is disabled and the "ABS" warning lamp is turned "ON."

DTC Chart Test Description:
Number(s) below refer to circled number(s) on the diagnostic chart.
1. Checks the voltage level being received by the EBCM.
2. Isolates the malfunction to either the EBCM or the charging system.

Diagnostic Aids:
An "intermittent" malfunction is most likely caused by a poor connection, rubbed through wire insulation, or a wire that is broken inside the insulation.

The frequency of the malfunction can be checked by using the enhanced diagnostic function of the Tech 1 tool, as described in "Enhanced Diagnostics," found in this section.

Any circuitry, that is suspected as causing the intermittent malfunction, should be thoroughly checked for backed out terminals, improper mating, broken locks, improperly formed or damaged terminals, poor terminal to wiring connections or physical damage to the wiring harness.

DIAGNOSTIC CHART DTC 37 — 1995 CAVALIER AND SUNFIRE

DTC 37
HIGH SYSTEM VOLTAGE

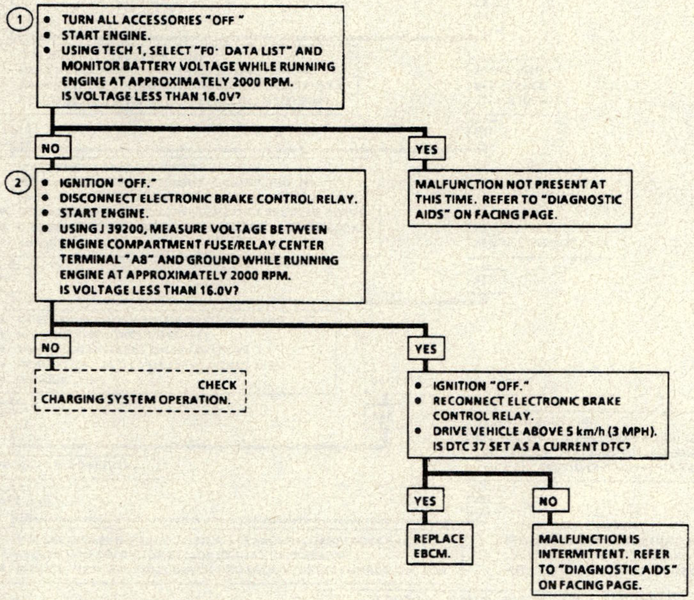

DIAGNOSTIC CHART DTC 38 — 1995 CAVALIER AND SUNFIRE

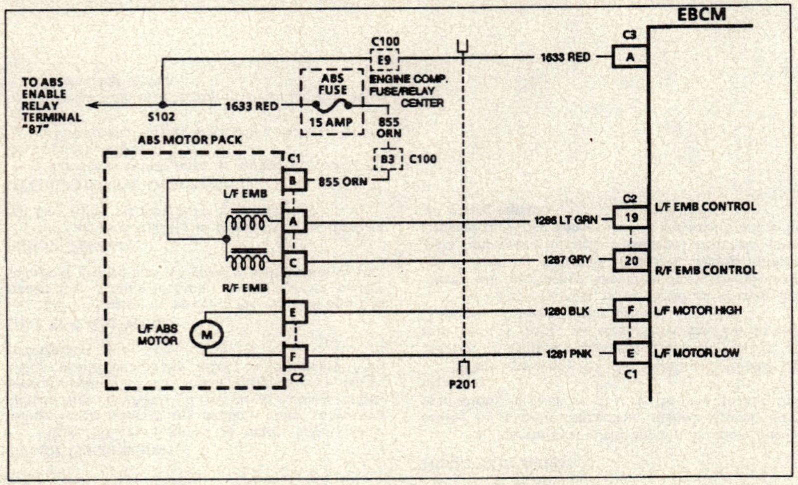

DTC 38

LEFT FRONT EMB WILL NOT HOLD MOTOR

Circuit Description:

Current is supplied to the EMB's through terminal "B" of the three-way motor pack connector. When a release of the EMB is commanded, the EBCM grounds the EMB input and allows the EMB to be energized to release the motor.

The left front motor is commanded in the forward direction while the EMB is released (energized).

DTC Will Set When:

DTC 38 can be set during initialization. A malfunction exists if the feedback current is less than the command current and the left front channel was expected to be in the home position. DTC 86 is always set with DTC 38.

Action Taken:

If an EMB can not hold a motor in the home position, the motor may be back driven when the brake pedal is applied causing the brake pedal to drop. A malfunction DTC is stored, ABS is disabled and the "ABS" warning lamp is turned "ON."

DTC Chart Test Description:

Number(s) below refer to circled number(s) on the diagnostic chart.

1. Ensures motor circuits are in the proper connector cavity.
2. Checks the ability of the EMB to hold the motor. If the brake pedal moves during this test, it would indicate the EMB is slipping.
3. Checks for high resistance in the left front motor high circuit.
4. Checks for high resistance in the left front motor low circuit.
5. Checks for excessive resistance in the left front motor.
6. Ensures malfunction was not due to a poor connection.

Diagnostic Aids:

An "intermittent" malfunction in this DTC may result from a mechanical part of the system that sticks, binds, or slips.

The frequency of the malfunction can be checked by using the enhanced diagnostic function of the Tech 1 as described in "Enhanced Diagnostics" found in this section.

NOTICE: This test may fail if low voltage (10 volts or less) was present when initialization occurred. Ensure battery/charging system is functioning properly.

DIAGNOSTIC CHART DTC 38 — 1995 CAVALIER AND SUNFIRE

DTC 38

LEFT FRONT EMB WILL NOT HOLD MOTOR

NOTICE: TESTING EMB WITH JUMPER WIRES TO VOLTAGE OR GROUND WILL DESTROY THE EMB.

NOTICE: J 39200 TEST LEADS MUST BE "ZEROED" PRIOR TO MAKING ANY RESISTANCE MEASUREMENTS. REFER TO J 39200 USER'S MANUAL.

(1)
- INSPECT THE 6-WAY MOTOR PACK AND EBCM HARNESS CONNECTORS FOR PROPER WIRE COLOR/CONNECTOR CAVITY COMBINATION.
- ARE THE PROPER WIRES LOCATED IN THE PROPER CONNECTOR CAVITY?

NO → USING THE CIRCUIT DIAGRAMS LOCATED IN THIS SECTION, PLACE THE PROPER WIRES IN THE PROPER CONNECTOR CAVITIES CLEAR DTCs AND RECHECK.

YES → IS DTC 67 ALSO PRESENT?

YES → PROCEED TO DTC 67 CHART.

NO ↓

(2)
- IGNITION "ON"
- USING TECH 1, SELECT EMB FUNCTIONAL TEST.
- APPLY FIRM PRESSURE ON BRAKE PEDAL.
- PERFORM EMB FUNCTIONAL TEST.
- DOES BRAKE PEDAL FALL OR RISE DURING THIS TEST?

YES → REPLACE MOTOR PACK.

NO ↓

(3)
- IGNITION "OFF."
- DISCONNECT 6-WAY MOTOR PACK AND EBCM CONNECTORS.
- USING J 39200, MEASURE THE RESISTANCE BETWEEN 6-WAY EBCM HARNESS CONNECTOR TERMINAL "F" AND 6-WAY MOTOR PACK HARNESS CONNECTOR TERMINAL "A" IS RESISTANCE LESS THAN 1.5Ω?

NO → REPAIR HIGH RESISTANCE IN CKT 1280

YES ↓

(4)
- USING J 39200, MEASURE RESISTANCE BETWEEN 6-WAY EBCM HARNESS CONNECTOR TERMINAL "E" AND 6-WAY MOTOR PACK HARNESS CONNECTOR TERMINAL "B". IS RESISTANCE LESS THAN 1.5Ω?

NO → REPAIR HIGH RESISTANCE IN CKT 1281.

YES ↓

(5)
- USING J 39200, MEASURE RESISTANCE BETWEEN MOTOR PACK CONNECTOR TERMINALS "A" AND "B" IS RESISTANCE LESS THAN 1.5Ω?

NO → REPLACE MOTOR PACK.

YES ↓

(6)
- INSPECT 6-WAY EBCM AND MOTOR PACK CONNECTORS FOR POOR TERMINAL CONTACT OR CORROSION.

REPLACE TERMINALS THAT EXHIBIT POOR CONTACT OR CORROSION.
- RECONNECT ALL CONNECTORS.
- DRIVE VEHICLE FOR THREE DRIVE CYCLES (A DRIVE CYCLE CONSISTS OF STARTING ENGINE, DRIVING OVER 16 km/h (10 MPH) AND THEN TURNING IGNITION "OFF").

DID DTC 38 SET IN THE LAST THREE DRIVE CYCLES?

YES → REPLACE EBCM.

NO → MALFUNCTION IS INTERMITTENT. REFER TO "DIAGNOSTIC AIDS" ON FACING PAGE

DIAGNOSTIC CHART DTC 41 — 1995 CAVALIER AND SUNFIRE

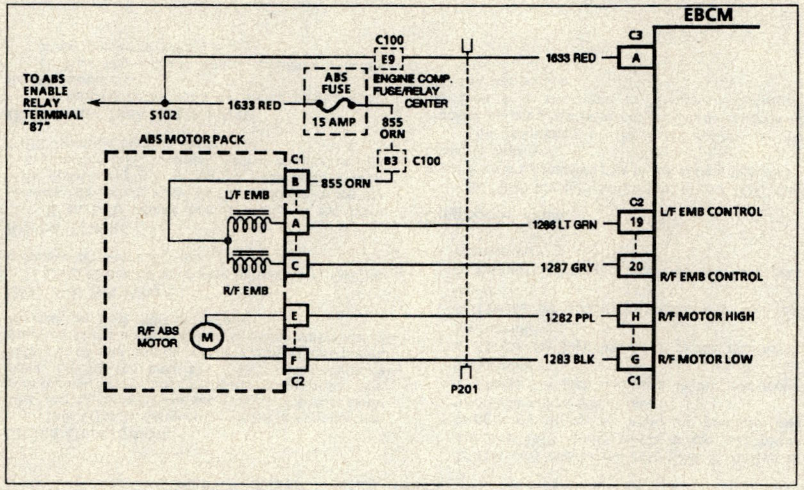

DTC 41
RIGHT FRONT EMB WILL NOT HOLD MOTOR

Circuit Description:

Current is supplied to the EMB's through terminal "B" of the three-way motor pack connector. When a release of the EMB is commanded, the EBCM grounds the EMB input and allows the EMB to be energized to release the motor.

The right front motor is commanded in the forward direction while the EMB is released (energized).

DTC Will Set When:

DTC 41 can be set during initialization. A malfunction exists if the feedback current is less than the command current and the right front channel was expected to be in the home position. DTC 86 is always set with DTC 41.

Action Taken:

If an EMB can not hold a motor in the home position, the motor may be back driven when the brake pedal is applied causing the brake pedal to drop. A malfunction DTC is stored, ABS is disabled and the "ABS" warning lamp is turned "ON."

DTC Chart Test Description:

Number(s) below refer to circled number(s) on the diagnostic chart.

1. Ensures motor circuits are in the proper connector cavity.
2. Checks the ability of the EMB to hold the motor. If the brake pedal moves during this test, it would indicate the EMB is slipping.
3. Checks for high resistance in the right front motor high circuit.
4. Checks for high resistance in the right front motor low circuit.
5. Checks for excessive resistance in the right front motor.
6. Ensures malfunction was not due to a poor connection.

Diagnostic Aids:

An "intermittent" malfunction in this DTC may result from a mechanical part of the system that sticks, binds, or slips.

The frequency of the malfunction can be checked by using the enhanced diagnostic function of the Tech 1 tool as described in "Enhanced Diagnostics" found in this section.

NOTICE: This DTC may fail if low voltage (10 volts or less) was present when initialization occurred. Ensure battery/charging system is functioning properly.

DIAGNOSTIC CHART DTC 41 — 1995 CAVALIER AND SUNFIRE

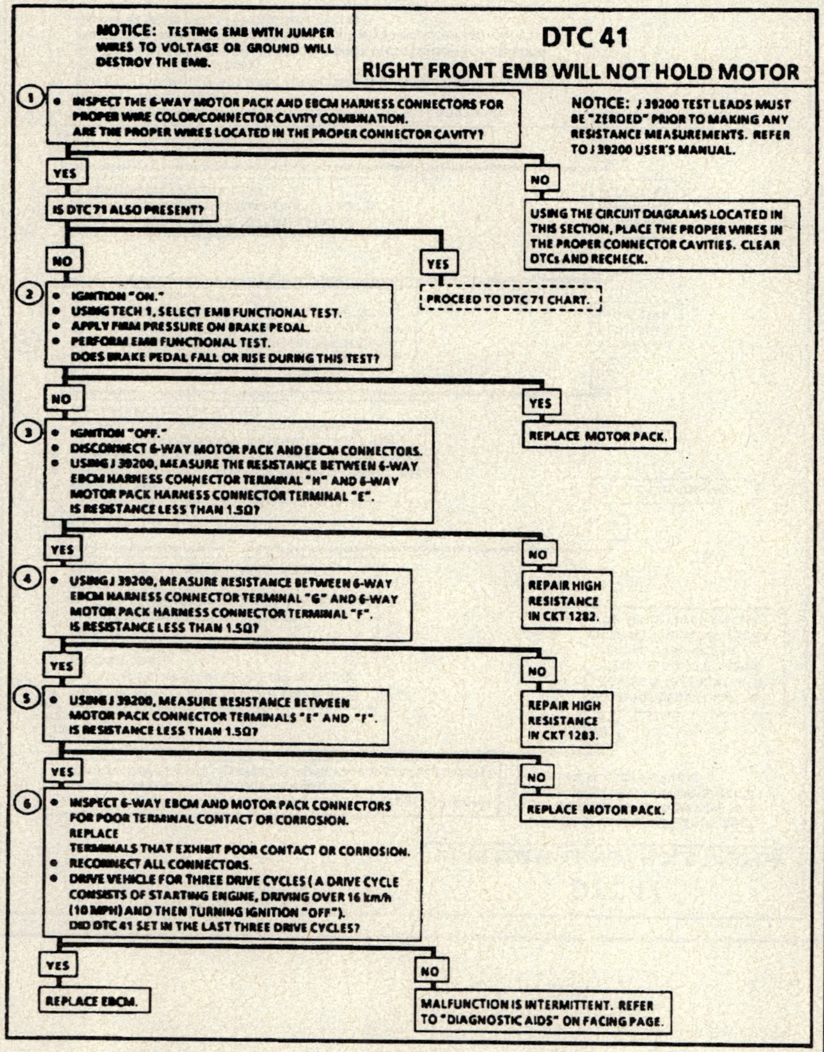

DIAGNOSTIC CHART DTC 42 — 1995 CAVALIER AND SUNFIRE

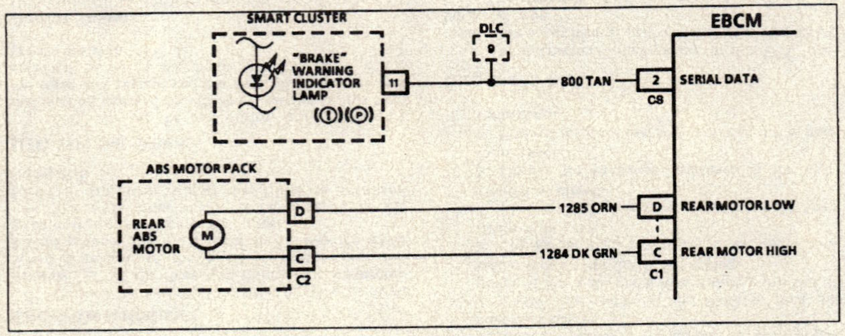

DTC 42
REAR ESB WILL NOT HOLD MOTOR

Circuit Description:

This DTC is designed to detect a slipping rear ESB. During initialization and braking, the rear motor is re-homed. If the ESB slips, the motor/piston will move. During the next key "ON" initialization, a re-home of the motor verifies the motor/piston remained at the home position. If motor movement is detected, the ESB must be slipping.

DTC Will Set When:

DTC 42 can be set during initialization. DTC 86 is always set with DTC 42.

Action Taken:

If an ESB cannot hold a motor in the home position, the motor may be back driven when the brake pedal is applied causing the brake pedal to drop. A malfunction DTC is stored, ABS is disabled and the "ABS" warning lamp is turned "ON."

DTC Chart Test Description:

Number(s) below refer to circled number(s) on the diagnostic chart.

1. This step ensures motor circuits are in the proper connector cavity.
2. This step checks the rear ESB. A broken or defective ESB would result in the rear piston being back driven by hydraulic pressure, and wheel movement would result.
3. Checks for high resistance in the rear motor high circuit.
4. Checks for high resistance in the rear motor low circuit.
5. Checks for excessive resistance in the rear motor.
6. Ensures malfunction was not due to a poor connection.

Diagnostic Aids:

An "intermittent" malfunction in this DTC may result from a mechanical part of the system that sticks, binds, or slips.

The frequency of the malfunction can be checked by using the enhanced diagnostic function of the Tech 1 as described in "Diagnosis-Electrical" found in this section.

DIAGNOSTIC CHART DTC 42 — 1995 CAVALIER AND SUNFIRE

DTC 42
REAR ESB WILL NOT HOLD MOTOR

NOTICE: J 39200 TEST LEADS MUST BE "ZEROED" PRIOR TO MAKING ANY RESISTANCE MEASUREMENTS. REFER TO J 39200 USER'S MANUAL.

(1)
- INSPECT THE 6-WAY EBCM AND MOTOR PACK CONNECTORS FOR PROPER WIRE COLOR/CONNECTOR CAVITY COMBINATION.

ARE THE PROPER WIRES LOCATED IN THE PROPER CONNECTOR CAVITIES?

- YES: go to 2.
- NO: USING THE CIRCUIT DIAGRAMS LOCATED IN THE FRONT OF THIS SECTION, PLACE THE PROPER WIRES INTO THE PROPER CONNECTOR CAVITIES. CLEAR DTCs AND RECHECK SYSTEM.

(2)
- RAISE THE VEHICLE SUCH THAT THE REAR WHEELS ARE APPROXIMATELY SIX INCHES OFF THE FLOOR.
- IGNITION "ON."
- APPLY FIRM PRESSURE ON BRAKE PEDAL AND RELEASE.
- REAPPLY FIRM PRESSURE ON BRAKE PEDAL AND HAVE AN ASSISTANT TRY TO SPIN THE REAR WHEEL BY HAND.

COULD ASSISTANT SPIN THE REAR WHEELS?

- NO: go to 3.
- YES: REPLACE MOTOR PACK.

(3)
- IGNITION "OFF."
- DISCONNECT 6-WAY MOTOR PACK AND EBCM CONNECTORS.
- USING J 39200, MEASURE THE RESISTANCE BETWEEN 6-WAY EBCM HARNESS CONNECTOR TERMINAL "C" AND 6-WAY MOTOR PACK HARNESS CONNECTOR TERMINAL "C".

IS RESISTANCE LESS THAN 1.5Ω?

- YES: go to 4.
- NO: REPAIR HIGH RESISTANCE IN CKT 1284.

(4)
- USING J 39200, MEASURE RESISTANCE BETWEEN 6-WAY EBCM HARNESS CONNECTOR TERMINAL "D" AND 6-WAY MOTOR PACK HARNESS CONNECTOR TERMINAL "D".

IS RESISTANCE LESS THAN 1.5Ω?

- YES: go to 5.
- NO: REPAIR HIGH RESISTANCE IN CKT 1285.

(5)
- USING J 39200, MEASURE RESISTANCE BETWEEN MOTOR PACK CONNECTOR TERMINALS "C" AND "D".

IS RESISTANCE LESS THAN 1.5Ω?

- YES: go to 6.
- NO: REPLACE MOTOR PACK.

(6)
- INSPECT 6-WAY EBCM AND MOTOR PACK CONNECTORS FOR POOR TERMINAL CONTACT OR CORROSION.
- REPLACE TERMINALS THAT EXHIBIT POOR CONTACT OR CORROSION.
- RECONNECT ALL CONNECTORS.
- DRIVE VEHICLE FOR THREE DRIVE CYCLES (A DRIVE CYCLE CONSISTS OF STARTING ENGINE, DRIVING OVER 16 km/h (10 MPH) AND THEN TURNING IGNITION "OFF").

DID DTC 42 SET IN THE LAST THREE DRIVE CYCLES?

- YES: REPLACE EBCM.
- NO: MALFUNCTION IS INTERMITTENT. REFER TO "DIAGNOSTIC AIDS" ON FACING PAGE.

DIAGNOSTIC CHART DTC 44 — 1995 CAVALIER AND SUNFIRE

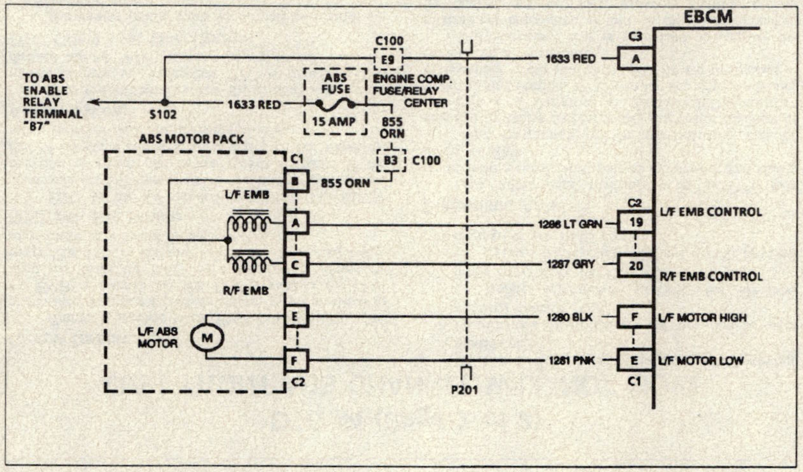

DTC 44
LEFT FRONT ABS CHANNEL WILL NOT MOVE

Circuit Description:

Current is supplied to the EMB through terminal "B" of the three-way motor pack connector. When a release of the EMB is commanded, the EBCM grounds the EMB input and allows the EMB to be energized to release the motor. This DTC is designed to detect non-actuating EMB, stuck motor or a seized hydraulic modulator.

DTC Will Set When:

DTC 44 can be set at initialization. The EBCM commands the left front motor in the reverse and forward direction to ensure the motor moves properly. If the EBCM detects a condition in which it cannot move the motor in either direction, a malfunction exists.

Action Taken:

This malfunction indicates the channel can not be moved properly. A malfunction DTC is stored, ABS is disabled and the "ABS" warning lamp is turned "ON."

DTC Chart Test Description:

Number(s) below refer to circled number(s) on the diagnostic chart.

1. Ensures motor circuits are in the proper connector cavity.
2. Checks for motor movement during release.
3. Compares feedback current to command current during apply.
4. Checks for a short between motor high and low circuits.
5. Checks for a shorted motor.
6. Ensures malfunction was not due to physical damage of the circuitry.
7. Checks if motor can apply properly under load.

Diagnostic Aids:

An "intermittent" malfunction in this DTC may result from a mechanical part of the system that sticks, binds, or slips.

The frequency of the malfunction can be checked by using the enhanced diagnostic function of the Tech 1 as described in "Enhanced Diagnostics," found in this section. DTC 44 may set after modulator disassembly if the modulator pistons are positioned at the bottom of their bore.

Any circuitry, that is suspected as causing the intermittent complaint, should be thoroughly checked for backed out terminals, improper mating, broken locks, improperly formed or damaged terminals, poor terminal to wiring connections or physical damage to the wiring harness.

NOTICE: This DTC may fail if low voltage (10 volts or less) was present when initialization occurred. Ensure battery/charging system is functioning properly.

DIAGNOSTIC CHART DTC 44 — 1995 CAVALIER AND SUNFIRE

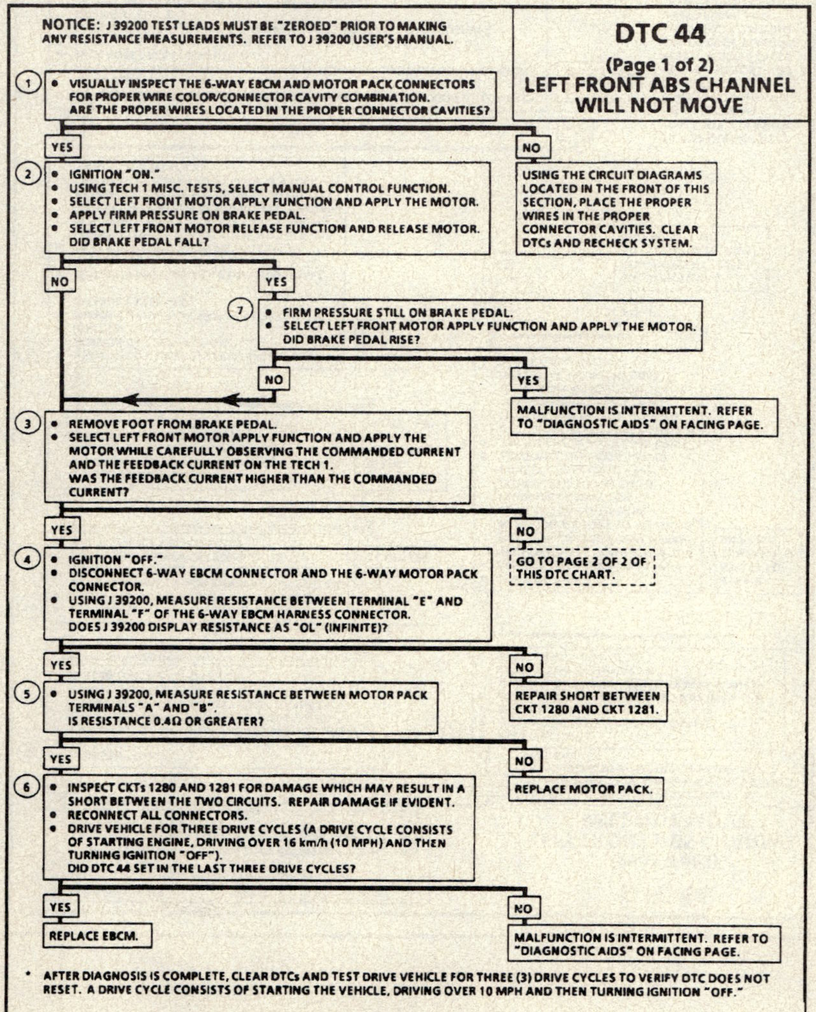

DIAGNOSTIC CHART DTC 44 — 1995 CAVALIER AND SUNFIRE

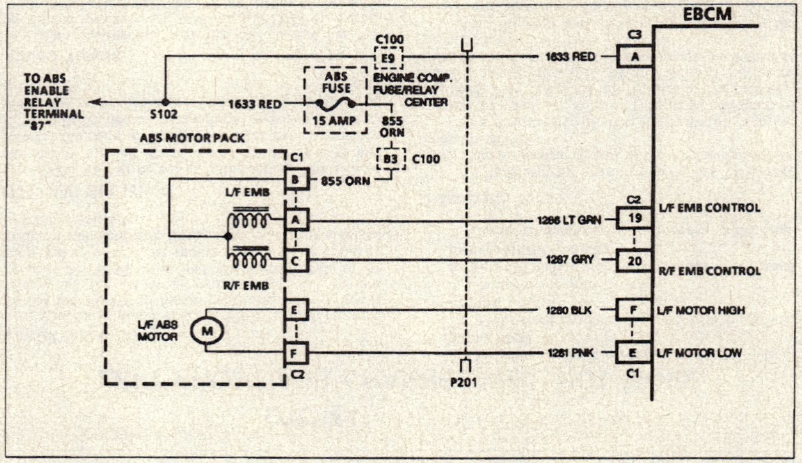

DTC 44 (Page 2 of 2)

LEFT FRONT ABS CHANNEL WILL NOT MOVE

Circuit Description:

Current is supplied to the EMB through terminal "B" of the three-way motor pack connector. When a release of the EMB is commanded, the EBCM grounds the EMB input and allows the EMB to be energized to release the motor. This DTC is designed to detect non-actuating EMB, stuck motor or a seized hydraulic modulator.

DTC Will Set When:

DTC 44 can be set at initialization. The EBCM commands the left front motor in the reverse and forward direction to ensure the motor moves properly. If the EBCM detects a condition in which it cannot move the motor in either direction, a malfunction exists.

Action Taken:

This malfunction indicates the channel can not be moved properly. A malfunction DTC is stored, ABS is disabled and the "ABS" warning lamp is turned "ON."

DTC Chart Test Description:

Number(s) below refer to circled number(s) on the diagnostic chart.

8. Checks for a short to BATTERY in the left front EMB control circuit.
9. Checks for high resistance in the left front EMB control circuit.
10. Checks for high resistance in the EMB power feed circuit.
11. Checks for high resistance in the left front EMB.
12. Checks for proper operation of the hydraulic modulator.
13. Ensures malfunction was not due to poor terminal contact.
14. Isolates malfunction to either the left front EMB circuitry or the EBCM.
15. Ensures malfunction was not due to physical damage of the EMB circuit.

Diagnostic Aids:

An "intermittent" malfunction in this DTC may result from a mechanical part of the system that sticks, binds, or slips.

The frequency of the malfunction can be checked by using the enhanced diagnostic function of the Tech 1 as described in "Enhanced Diagnostics," found in this section. DTC 44 may set after modulator disassembly if the modulator pistons are positioned at the bottom of their bore.

Any circuitry, that is suspected as causing the intermittent complaint, should be thoroughly checked for backed out terminals, improper mating, broken locks, improperly formed or damaged terminals, poor terminal to wiring connections or physical damage to the wiring harness.

NOTICE: This DTC may fail if low voltage (10 volts or less) was present when initialization occurred. Ensure battery/charging system is functioning properly.

DIAGNOSTIC CHART DTC 44 — 1995 CAVALIER AND SUNFIRE

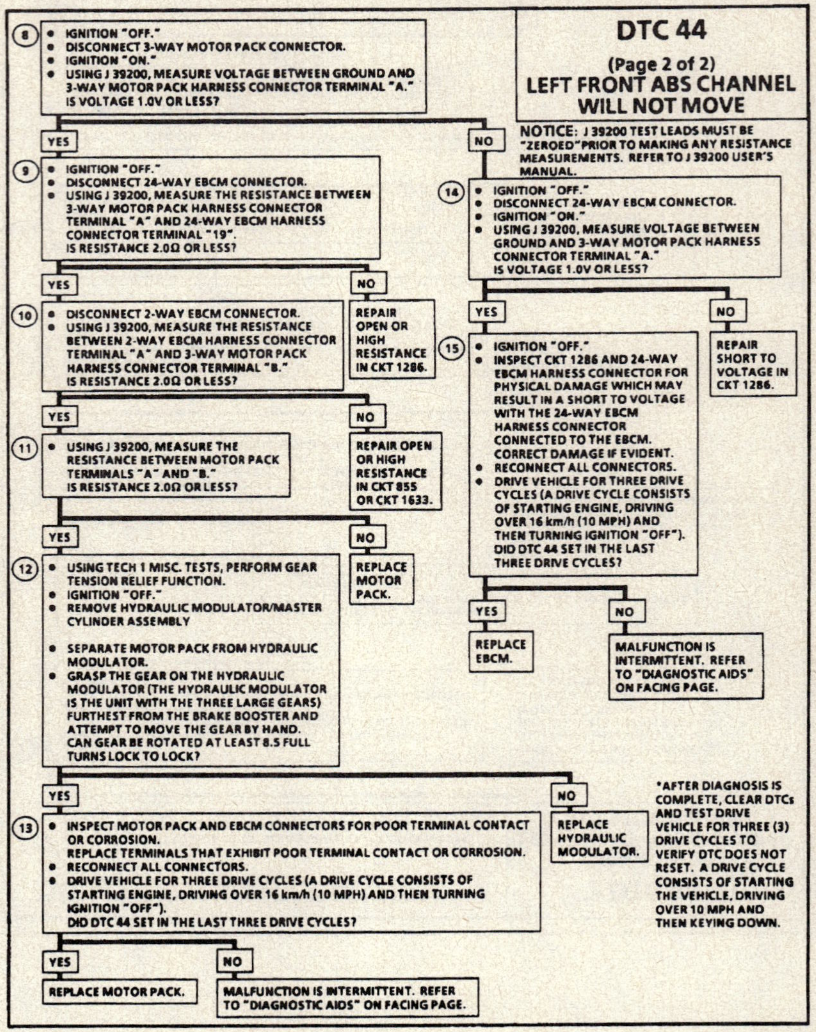

DIAGNOSTIC CHART DTC 45 — 1995 CAVALIER AND SUNFIRE

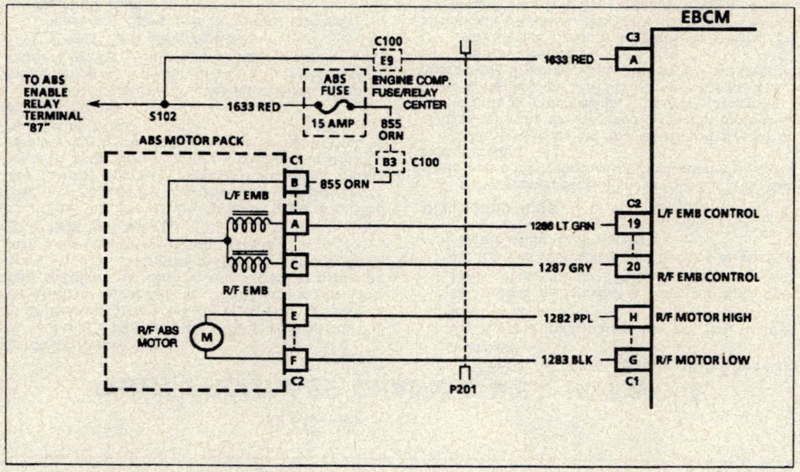

DTC 45
RIGHT FRONT ABS CHANNEL WILL NOT MOVE

Circuit Description:

Current is supplied to the EMB through terminal "B" of the three-way motor pack connector. When a release of the EMB is commanded, the EBCM grounds the EMB input and allows the EMB to be energized to release the motor.

This DTC is designed to detect non-actuating EMB, stuck motor or a seized hydraulic modulator. When the release is commanded during initialization, the EMB should release the motor, resulting in sensed current being less than commanded current (motor is spinning freely). If the motor is not moving, sensed current will be equal to stall current.

DTC Will Set When:

DTC 45 can be set during initialization. The EBCM commands the right front motor in the reverse and forward direction to ensure the motor moves properly. If the EBCM detects a condition in which it cannot move the motor in either direction, a malfunction exists.

Action Taken:

This malfunction indicates the channel can not be moved properly. A malfunction DTC is stored, ABS is disabled and the "ABS" warning lamp is turned "ON."

DTC Chart Test Description:

Number(s) below refer to circled number(s) on the diagnostic chart.

1. Ensures motor circuits are in the proper connector cavity.
2. Checks for motor movement during release.
3. Compares feedback current to command current during apply.
4. Checks for a short between motor high and low circuits.
5. Checks for a shorted motor.
6. Ensures malfunction was not due to physical damage of the circuitry.
7. Checks if motor can apply properly under load.

Diagnostic Aids:

An "intermittent" malfunction in this DTC may result from a mechanical part of the system that sticks, binds, or slips.

The frequency of the malfunction can be checked by using the enhanced diagnostic function of the Tech 1 as described in "Enhanced Diagnostics," found in this section. DTC 45 may set after modulator disassembly if the modulator pistons are positioned at the bottom of their bore.

Any circuitry, that is suspected as causing the intermittent complaint, should be thoroughly checked for backed out terminals, improper mating, broken locks, improperly formed or damaged terminals, poor terminal to wiring connections or physical damage to the wiring harness.

NOTICE: This DTC may fail if low voltage (10 volts or less) was present when initialization occurred. Ensure battery/charging system is functioning properly.

DIAGNOSTIC CHART DTC 45 — 1995 CAVALIER AND SUNFIRE

DTC 45
(Page 1 of 2)
RIGHT FRONT ABS CHANNEL WILL NOT MOVE

NOTICE: J 39200 TEST LEADS MUST BE "ZEROED" PRIOR TO MAKING ANY RESISTANCE MEASUREMENTS. REFER TO J 39200 USER'S MANUAL.

(1)
- VISUALLY INSPECT THE 6-WAY EBCM AND MOTOR PACK CONNECTORS FOR PROPER WIRE COLOR/CONNECTOR CAVITY COMBINATION. ARE THE PROPER WIRES LOCATED IN THE PROPER CONNECTOR CAVITIES?

NO: USING THE CIRCUIT DIAGRAMS LOCATED IN THE FRONT OF THIS SECTION, PLACE THE PROPER WIRES IN THE PROPER CONNECTOR CAVITIES. CLEAR DTCs AND RECHECK SYSTEM.

YES: (2)
- IGNITION "ON."
- USING TECH 1 MISC. TESTS, SELECT MANUAL CONTROL FUNCTION.
- SELECT RIGHT FRONT MOTOR APPLY FUNCTION AND APPLY THE MOTOR.
- APPLY FIRM PRESSURE ON BRAKE PEDAL.
- SELECT RIGHT FRONT MOTOR RELEASE FUNCTION AND RELEASE MOTOR. DID BRAKE PEDAL FALL?

YES: (7)
- FIRM PRESSURE STILL ON BRAKE PEDAL.
- SELECT RIGHT FRONT MOTOR APPLY FUNCTION AND APPLY THE MOTOR. DID BRAKE PEDAL RISE?

YES: MALFUNCTION IS INTERMITTENT. REFER TO "DIAGNOSTIC AIDS" ON FACING PAGE.

NO (from 2 or 7): (3)
- REMOVE FOOT FROM BRAKE PEDAL.
- SELECT RIGHT FRONT MOTOR APPLY FUNCTION AND APPLY THE MOTOR WHILE CAREFULLY OBSERVING THE COMMANDED CURRENT AND THE FEEDBACK CURRENT ON THE TECH 1. WAS THE FEEDBACK CURRENT HIGHER THAN THE COMMANDED CURRENT?

NO: GO TO PAGE 2 OF 2 OF THIS DTC CHART.

YES: (4)
- IGNITION "OFF."
- DISCONNECT 6-WAY EBCM CONNECTOR AND THE 6-WAY MOTOR PACK CONNECTOR.
- USING J 39200, MEASURE RESISTANCE BETWEEN TERMINAL "G" AND TERMINAL "H" OF THE 6-WAY EBCM HARNESS CONNECTOR. DOES J 39200 DISPLAY RESISTANCE AS "OL" (INFINITE)?

NO: REPAIR SHORT BETWEEN CKT 1282 AND CKT 1283.

YES: (5)
- USING J 39200, MEASURE RESISTANCE BETWEEN MOTOR PACK TERMINALS "E" AND "F". IS RESISTANCE .4Ω OR GREATER?

NO: REPLACE MOTOR PACK.

YES: (6)
- INSPECT CKTs 1282 AND 1283 FOR DAMAGE WHICH MAY RESULT IN A SHORT BETWEEN THE TWO CIRCUITS. REPAIR DAMAGE IF EVIDENT.
- RECONNECT ALL CONNECTORS.
- DRIVE VEHICLE FOR THREE DRIVE CYCLES (A DRIVE CYCLE CONSISTS OF STARTING ENGINE, DRIVING OVER 16 km/h (10 MPH) AND THEN TURNING IGNITION "OFF"). DID DTC 45 SET IN THE LAST THREE DRIVE CYCLES?

YES: REPLACE EBCM.

NO: MALFUNCTION IS INTERMITTENT. REFER TO "DIAGNOSTIC AIDS" ON FACING PAGE.

- AFTER DIAGNOSIS IS COMPLETE, CLEAR DTCs AND TEST DRIVE VEHICLE FOR THREE (3) DRIVE CYCLES TO VERIFY DTC DOES NOT RESET. A DRIVE CYCLE CONSISTS OF STARTING THE VEHICLE, DRIVING OVER 10 MPH AND THEN TURNING IGNITION "OFF."

DIAGNOSTIC CHART DTC 45 — 1995 CAVALIER AND SUNFIRE

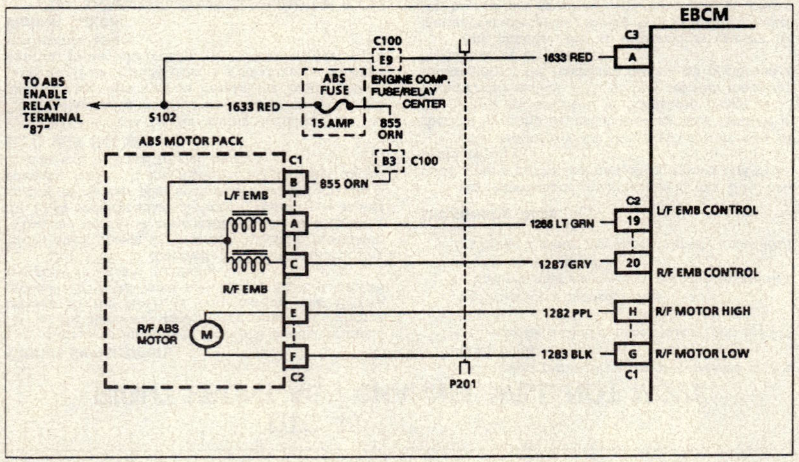

DTC 45

RIGHT FRONT ABS CHANNEL WILL NOT MOVE

Circuit Description:

Current is supplied to the EMB through terminal "B" of the three-way motor pack connector. When a release of the EMB is commanded, the EBCM grounds the EMB input and allows the EMB to be energized to release the motor. This DTC is designed to detect non-actuating EMB, stuck motor or a seized hydraulic modulator.

DTC Will Set When:

DTC 45 can be set during initialization. The EBCM commands the right front motor in the reverse and forward direction to ensure the motor moves properly. If the EBCM detects a condition in which it cannot move the motor in either direction, a malfunction exists.

Action Taken:

This malfunction indicates the channel cannot be moved properly. A malfunction DTC is stored, ABS is disabled and the "ABS" warning lamp is turned "ON."

DTC Chart Test Description:

Number(s) below refer to circled number(s) on the diagnostic chart.

8. Checks for a short to BATTERY in the right front EMB control circuit.
9. Checks for high resistance in the right front EMB control circuit.
10. Checks for high resistance in the EMB power feed circuit.
11. Checks for high resistance in the right front EMB.
12. Checks for proper operation of the hydraulic modulator.
13. Ensures malfunction was not due to poor terminal contact.
14. Isolates malfunction to either the right front EMB circuitry or the EBCM.
15. Ensures malfunction was not due to physical damage of the EMB circuitry.

Diagnostic Aids:

An "intermittent" malfunction in this DTC may result from a mechanical part of the system that sticks, binds, or slips.

The frequency of the malfunction can be checked by using the enhanced diagnostic function of the Tech 1, as described in "Enhanced Diagnostics," found in this section. DTC 45 may set after modulator disassembly if the modulator pistons are positioned at the bottom of their bore.

Any circuitry, that is suspected as causing the intermittent complaint, should be thoroughly checked for backed out terminals, improper mating, broken locks, improperly formed or damaged terminals, poor terminal to wiring connections or physical damage to the wiring harness.

NOTICE: This DTC may fail if low voltage (10 volts or less) was present when initialization occurred. Ensure battery/charging system is functioning properly.

DIAGNOSTIC CHART DTC 45 — 1995 CAVALIER AND SUNFIRE

DTC 45 (Page 2 of 2) RIGHT FRONT ABS CHANNEL WILL NOT MOVE

NOTICE: J 39200 TEST LEADS MUST BE "ZEROED" PRIOR TO MAKING ANY RESISTANCE MEASUREMENTS. REFER TO J 39200 USER'S MANUAL.

(8)
- IGNITION "OFF."
- DISCONNECT 3-WAY MOTOR PACK CONNECTOR.
- IGNITION "ON."
- USING J 39200, MEASURE VOLTAGE BETWEEN GROUND AND 3-WAY MOTOR PACK HARNESS CONNECTOR TERMINAL "C." IS VOLTAGE 1.0V OR LESS?

YES → (9); NO → (14)

(9)
- IGNITION "OFF."
- DISCONNECT 24-WAY EBCM CONNECTOR.
- USING J 39200, MEASURE THE RESISTANCE BETWEEN 3-WAY MOTOR PACK HARNESS CONNECTOR TERMINAL "A" AND 24-WAY EBCM HARNESS CONNECTOR TERMINAL "20". IS RESISTANCE 2.0Ω OR LESS?

YES → (10); NO → REPAIR OPEN OR HIGH RESISTANCE IN CKT 1287.

(10)
- DISCONNECT 2-WAY EBCM CONNECTOR.
- USING J 39200, MEASURE THE RESISTANCE BETWEEN 2-WAY EBCM HARNESS CONNECTOR TERMINAL "A" AND 3-WAY MOTOR PACK HARNESS CONNECTOR TERMINAL "B." IS RESISTANCE 2.0Ω OR LESS?

YES → (11); NO → REPAIR OPEN OR HIGH RESISTANCE IN CKT 855 OR CKT 1633.

(11)
- USING J 39200, MEASURE THE RESISTANCE BETWEEN MOTOR PACK TERMINALS "C" AND "B." IS RESISTANCE 2.0Ω OR LESS?

YES → (12); NO → REPLACE MOTOR PACK.

(12)
- USING TECH 1 MISC. TESTS, PERFORM GEAR TENSION RELIEF FUNCTION.
- IGNITION "OFF."
- REMOVE HYDRAULIC MODULATOR/MASTER CYLINDER ASSEMBLY (REFER TO "ON-VEHICLE SERVICE" IN THIS SECTION).
- SEPARATE MOTOR PACK FROM HYDRAULIC MODULATOR.
- GRASP THE GEAR ON THE HYDRAULIC MODULATOR (THE HYDRAULIC MODULATOR IS THE UNIT WITH THE THREE LARGE GEARS) FURTHEST FROM THE BRAKE BOOSTER AND ATTEMPT TO MOVE THE GEAR BY HAND. CAN GEAR BE ROTATED AT LEAST 8.5 FULL TURNS LOCK TO LOCK?

YES → (13); NO → REPLACE HYDRAULIC MODULATOR.

(13)
- INSPECT MOTOR PACK AND EBCM CONNECTORS FOR POOR TERMINAL CONTACT OR CORROSION. REPLACE TERMINALS THAT EXHIBIT POOR TERMINAL CONTACT OR CORROSION.
- RECONNECT ALL CONNECTORS.
- DRIVE VEHICLE FOR THREE DRIVE CYCLES (A DRIVE CYCLE CONSISTS OF STARTING ENGINE, DRIVING OVER 16 km/h (10 MPH) AND THEN TURNING IGNITION "OFF").

DID DTC 45 SET IN THE LAST THREE DRIVE CYCLES?

YES → REPLACE MOTOR PACK.; NO → MALFUNCTION IS INTERMITTENT. REFER TO "DIAGNOSTIC AIDS" ON FACING PAGE.

(14)
- IGNITION "OFF."
- DISCONNECT 24-WAY EBCM CONNECTOR.
- IGNITION "ON."
- USING J 39200, MEASURE VOLTAGE BETWEEN GROUND AND 3-WAY MOTOR PACK HARNESS CONNECTOR TERMINAL "C." IS VOLTAGE 1.0V OR LESS?

YES → (15); NO → REPAIR SHORT TO VOLTAGE IN CKT 1286.

(15)
- IGNITION "OFF."
- INSPECT CKT 1287 AND 24-WAY EBCM HARNESS CONNECTOR FOR PHYSICAL DAMAGE WHICH MAY RESULT IN A SHORT TO VOLTAGE WITH THE 24-WAY EBCM HARNESS CONNECTOR CONNECTED TO THE EBCM. CORRECT DAMAGE IF EVIDENT.
- RECONNECT ALL CONNECTORS.
- DRIVE VEHICLE FOR THREE DRIVE CYCLES (A DRIVE CYCLE CONSISTS OF STARTING ENGINE, DRIVING OVER 16 km/h (10 MPH) AND THEN TURNING IGNITION "OFF"). DID DTC 45 SET IN THE LAST THREE DRIVE CYCLES?

YES → REPLACE EBCM.; NO → MALFUNCTION IS INTERMITTENT. REFER TO "DIAGNOSTIC AIDS" ON FACING PAGE.

*AFTER DIAGNOSIS IS COMPLETE, CLEAR DTCs AND TEST DRIVE VEHICLE FOR THREE (3) DRIVE CYCLES TO VERIFY DTC DOES NOT RESET. A DRIVE CYCLE CONSISTS OF STARTING THE VEHICLE, DRIVING OVER 10 MPH AND THEN TURNING IGNITION "OFF."

DIAGNOSTIC CHART DTC 46 — 1995 CAVALIER AND SUNFIRE

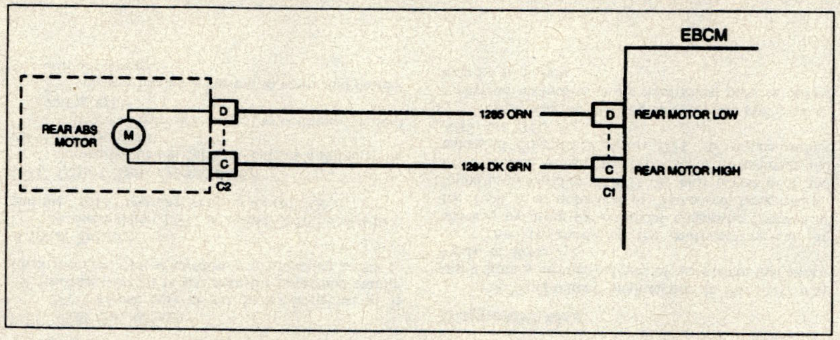

DTC 46
REAR ABS CHANNEL WILL NOT MOVE

Circuit Description:

This DTC is designed to detect bound-up ESB's, stuck motor or a seized hydraulic modulator. Release is commanded during initialization, the ESB should release the motor, resulting in sensed current being less than commanded current (motor is spinning freely). If the motor is not moving, sensed current will be equal to stall current.

DTC Will Set When:

DTC 46 can be set during initialization. The EBCM commands the rear motor in the reverse and forward direction to ensure the motor moves properly. If the EBCM detects a condition in which it cannot move the motor in either direction, a malfunction exists.

Action Taken:

This malfunction indicates the channel can not be moved properly. A malfunction DTC is stored, ABS is disabled, and the "ABS" warning lamp is turned "ON."

DTC Chart Test Description:

Number(s) below refer to circled number(s) on the diagnostic chart.

1. Ensures motor circuits are in the proper connector cavity.
2. Checks the ability of the motor to move applying hydraulic pressure to the wheel cylinders. A frozen piston or gear would result in no apply of rear brakes and a wheel that freely spins.
3. Compares feedback current to command current during apply.
4. Checks to see if hydraulic modulator is functioning properly.
5. Checks if motor can apply properly under load.
6. Isolates the malfunction to either the EBCM or a short between CKTs 1284 and 1285.
7. Checks for a shorted motor.
8. Ensures the malfunction was not due to physical damage of the circuitry.

Diagnostic Aids:

An "intermittent" malfunction in this DTC may result from a mechanical part of the system that sticks or binds.

The frequency of the malfunction can be checked by using the enhanced diagnostic function of the Tech 1, as described in "Enhanced Diagnostics," found in this section. DTC 46 may set after modulator disassembly if the modulator pistons are positioned at the bottom of their bore.

Depending on the frequency of the malfunction, a physical inspection of the mechanical parts suspected may be necessary.

Any circuitry, that is suspected as causing the intermittent complaint, should be thoroughly checked for backed out terminals, improper mating, broken locks, improperly formed or damaged terminals, poor terminal to wiring connections or physical damage to the wiring harness.

DIAGNOSTIC CHART DTC 46 — 1995 CAVALIER AND SUNFIRE

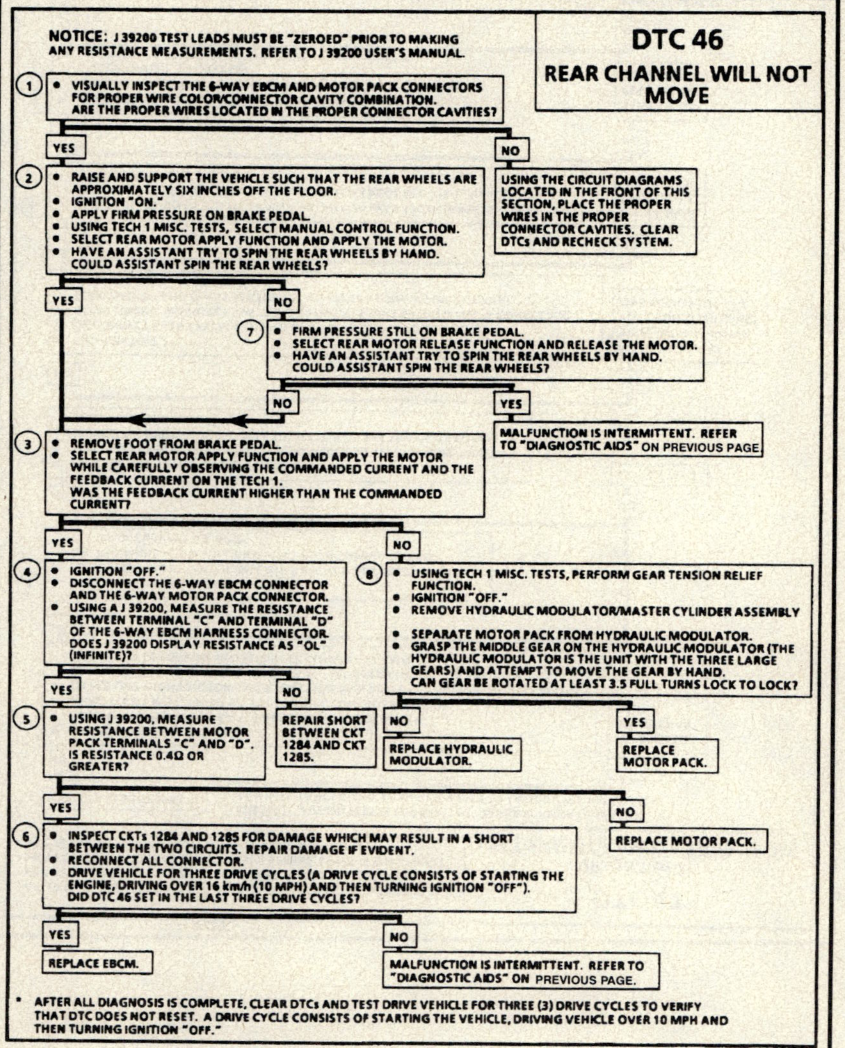

DIAGNOSTIC CHART DTC 47 — 1995 CAVALIER AND SUNFIRE

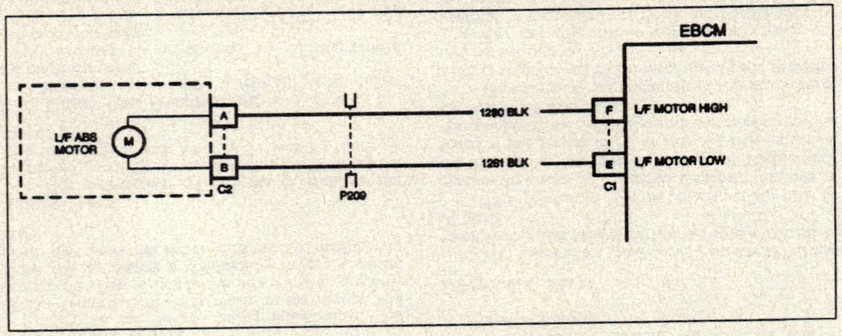

DTC 47

LEFT FRONT ABS MOTOR FREE SPINS

Circuit Description:
This DTC is designed to detect a stripped nut or gear assembly during initialization. During the homing sequence, the piston should reach the top of the bore resulting in a stalled motor. If this does not occur, the motor must be spinning with little or no resistance, indicating a nut/screw or gear malfunction.

DTC Will Set When:
DTC 47 can only be set during initialization. If the feedback current is less than the command current for a specified period to time, a malfunction exists.

Action Taken:
A malfunction DTC is stored, ABS is disabled, and the "ABS" warning lamp is turned "ON."

DTC Chart Test Description:
Number(s) below refer to circled number(s) on the diagnostic chart.
1. Compares feedback current to command current during apply.
2. Compares feedback current to command current during release.
3. Moves the position to the top of the bore.
4. Checks for excessive resistance in the left front motor high circuit.
5. Checks for excessive resistance in the left front motor low circuit.
6. Checks for high resistance in the left front motor.

Diagnostic Aids:
An "intermittent" malfunction in this DTC may result from a mechanical part of the system that sticks, binds, or slips.
The frequency of the malfunction can be checked by using the enhanced diagnostic function of the Tech 1 as described in "Enhanced Diagnostics," found in this section. If DTC 47 only occurs once and DTC 56 also occurs, see DTC 56. If intermittent and enhanced diagnostics shows DTC 47 occurs during ABS, see DTC 56.
Depending on the frequency of the malfunction, a physical inspection of the mechanical parts suspected may be necessary.

DIAGNOSTIC CHART DTC 47 — 1995 CAVALIER AND SUNFIRE

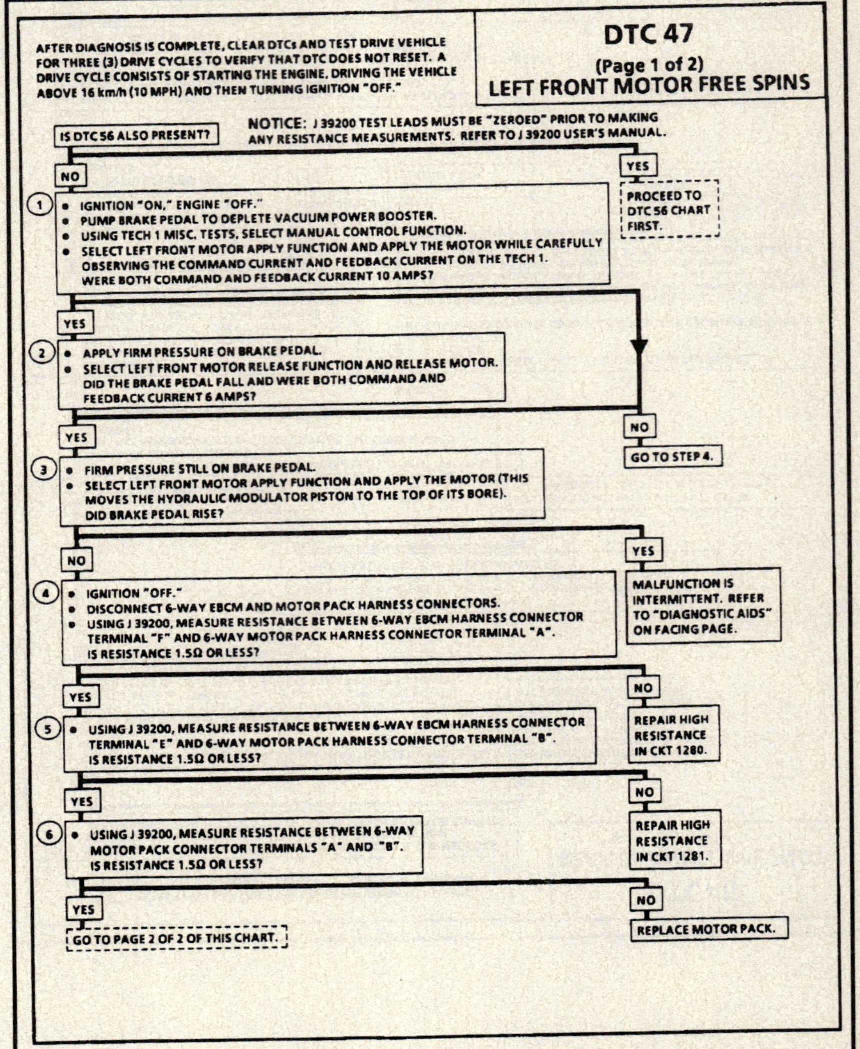

DIAGNOSTIC CHART DTC 47 — 1995 CAVALIER AND SUNFIRE

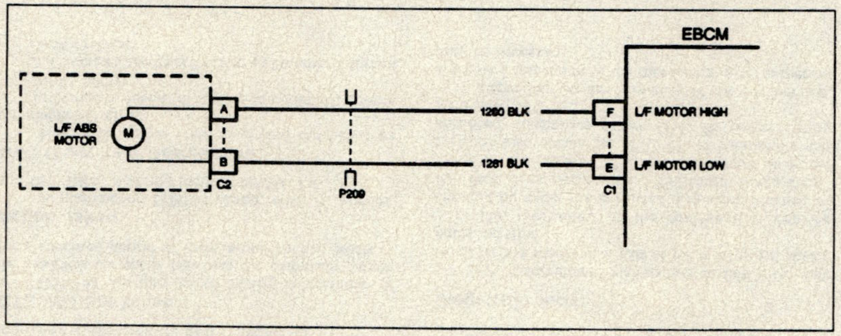

DTC 47
(Page 2 of 2)
LEFT FRONT ABS MOTOR FREE SPINS

Circuit Description:

This DTC is designed to detect a stripped nut or gear assembly during initialization. During the homing sequence, the piston should reach the top of the bore resulting in a stalled motor. If this does not occur, the motor must be spinning with little or no resistance, indicating a nut/screw or gear malfunction.

DTC Will Set When:

DTC 47 can only be set during initialization. If the feedback current is less than the command current for a specified period to time, a malfunction exists.

Action Taken:

A malfunction DTC is stored, ABS is disabled, and the "ABS" warning lamp is turned "ON."

DTC Chart Test Description:

Number(s) below refer to circled number(s) on the diagnostic chart.

7. Checks for a stripped motor pack gear.
8. Checks for a stripped hydraulic modulator gear.
9. Checks for proper operation of motor pack.
10. Checks for proper operation of the hydraulic modulator.
11. Ensures malfunction was not due to poor terminal contact.

Diagnostic Aids:

An "intermittent" malfunction in this DTC may result from a mechanical part of the system that sticks, binds, or slips.

The frequency of the malfunction can be checked by using the enhanced diagnostic function of the Tech 1, as described in "Enhanced Diagnostics," found in this section. If DTC 47 only occurs once and DTC 56 also occurs, see DTC 56. If intermittent and enhanced diagnostics shows DTC 47 occurs during ABS, see DTC 56.

Depending on the frequency of the malfunction, a physical inspection of the mechanical parts suspected may be necessary.

DIAGNOSTIC CHART DTC 47 — 1995 CAVALIER AND SUNFIRE

DTC 47
(Page 2 of 2)
LEFT FRONT MOTOR FREE SPINS

CONTINUED FROM PAGE 1 OF 2 OF THIS CHART.

(7)
- REMOVE HYDRAULIC MODULATOR/MASTER CYLINDER ASSEMBLY.
- REMOVE GEAR COVER.
- CHECK FOR STRIPPED GEAR ON MOTOR PACK; THE MOTOR PACK IS THE UNIT WITH THE THREE SMALL GEARS. THE LEFT FRONT GEAR IS FURTHEST FROM THE BRAKE BOOSTER WHEN THE ASSEMBLY IS INSTALLED IN THE VEHICLE. IS THE MOTOR PACK GEAR STRIPPED?

YES → REPLACE MOTOR PACK.

NO ↓

(8)
- CHECK FOR A STRIPPED GEAR ON THE HYDRAULIC MODULATOR; THE HYDRAULIC MODULATOR IS THE UNIT WITH THE THREE LARGE GEARS. THE LEFT FRONT GEAR IS FURTHEST FROM THE BRAKE BOOSTER WHEN THE ASSEMBLY IS INSTALLED IN THE VEHICLE. IS THE HYDRAULIC MODULATOR GEAR STRIPPED?

YES → REPLACE HYDRAULIC MODULATOR GEAR.

NO ↓

(9)
- SECURELY POSITION HYDRAULIC MODULATOR/MASTER CYLINDER ASSEMBLY IN VEHICLE ALLOWING LEFT FRONT GEAR SET OBSERVATION. BE CAREFUL NOT TO DAMAGE GEARS OR ALLOW THE HYDRAULIC MODULATOR/MASTER CYLINDER ASSEMBLY TO MOVE DURING TESTING.
- RECONNECT ELECTRICAL CONNECTORS.
- IGNITION "ON."
- USING TECH 1 MISC. TESTS, SELECT MANUAL CONTROL FUNCTION.
- SELECT LEFT FRONT MOTOR APPLY AND APPLY THE MOTOR WHILE OBSERVING THE GEAR SET FURTHEST FORWARD FROM THE BOOSTER WHEN THE ASSEMBLY IS INSTALLED IN THE VEHICLE.
- SELECT LEFT FRONT MOTOR RELEASE FUNCTION AND RELEASE THE MOTOR WHILE STILL OBSERVING THE GEAR SET.
- SELECT LEFT FRONT MOTOR APPLY FUNCTION AND APPLY THE MOTOR. DID GEAR SET MOVE IN BOTH DIRECTIONS FOR AT LEAST ONE REVOLUTION?

NO → REPLACE MOTOR PACK.

YES ↓

(10)
- USING TECH 1, SELECT "F4: MISC. TESTS" AND PERFORM GEAR TENSION RELIEF FUNCTION.
- IGNITION "OFF."
- SEPARATE MOTOR PACK FROM HYDRAULIC MODULATOR.
- GRASP THE LEFT FRONT GEAR ON THE HYDRAULIC MODULATOR; THE HYDRAULIC MODULATOR IS THE UNIT WITH THE THREE LARGE GEARS. THE LEFT FRONT GEAR IS FURTHEST FROM THE BRAKE BOOSTER WHEN THE ASSEMBLY IS INSTALLED IN THE VEHICLE.
- ATTEMPT TO ROTATE THE GEAR. CAN THE GEAR BE ROTATED AT LEAST 8.5 FULL TURNS LOCK TO LOCK?

YES → REPLACE HYDRAULIC MODULATOR.

NO ↓

(11)
- INSPECT 6-WAY MOTOR PACK AND EBCM CONNECTORS FOR CORROSION OR POOR TERMINAL CONTACT. REPLACE TERMINALS THAT EXHIBIT POOR TERMINAL CONTACT OR CORROSION.
- INSTALL HYDRAULIC MODULATOR/MASTER CYLINDER ASSEMBLY.
- RECONNECT ALL CONNECTORS.
- DRIVE VEHICLE FOR THREE DRIVE CYCLES (A DRIVE CYCLE CONSISTS OF STARTING ENGINE, DRIVING OVER 16 km/h (10 MPH) AND THEN TURNING IGNITION "OFF"). DID DTC 47 SET IN THE LAST THREE DRIVE CYCLES?

YES → REPLACE EBCM.

NO → MALFUNCTION WAS DUE TO A POOR CONNECTION.

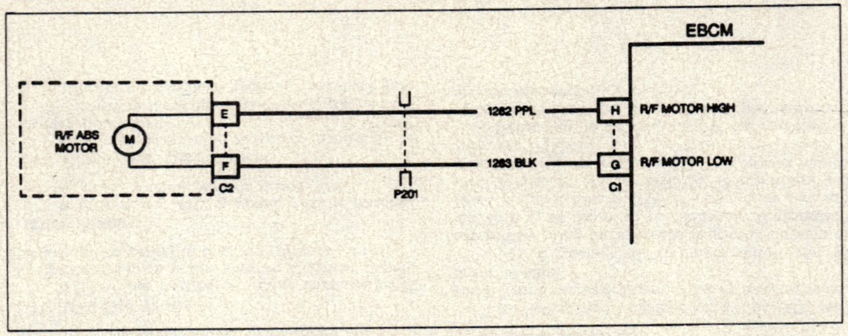

DTC 48

RIGHT FRONT ABS MOTOR FREE SPINS

Circuit Description:

This DTC is designed to detect a stripped nut or gear assembly during initialization. During the homing sequence, the piston should reach the top of the bore resulting in a stalled motor. If this does not occur, the motor must be spinning with little or no resistance, indicating a nut/screw or gear malfunction.

DTC Will Set When:

DTC 48 can only be set during initialization. If the feedback current is less than the command current for a specified period of time, a malfunction exists.

Action Taken:

A malfunction DTC is stored, ABS is disabled, and the "ABS" warning lamp is turned "ON."

DTC Chart Test Description:

Number(s) below refer to circled number(s) on the diagnostic chart.

1. Compares feedback current to command current during apply.
2. Compares feedback current to command current during release.
3. Moves the position to the top of the bore.
4. Checks for excessive resistance in the right front motor high circuit.
5. Checks for excessive resistance in the right front motor low circuit.
6. Checks for high resistance in the right front motor.

Diagnostic Aids:

An "intermittent" malfunction in this DTC may result from a mechanical part of the system that sticks, binds, or slips.

The frequency of the malfunction can be checked by using the enhanced diagnostic function of the Tech 1 as described in "Enhanced Diagnostics," found in this section. If DTC only occurs once and DTC 61 also occurs see DTC 61. If intermittent and enhanced diagnostics shows DTC 48 occurs during ABS, see DTC 61.

Depending on the frequency of the malfunction, a physical inspection of the mechanical parts suspected may be necessary.

DIAGNOSTIC CHART DTC 48 — 1995
CAVALIER AND SUNFIRE

DTC 48
(Page 1 of 2)
RIGHT FRONT MOTOR FREE SPINS

AFTER DIAGNOSIS IS COMPLETE, CLEAR DTCs AND TEST DRIVE VEHICLE FOR THREE(3) DRIVE CYCLES TO VERIFY THAT DTC DOES NOT RESET. A DRIVE CYCLE CONSISTS OF STARTING THE ENGINE, DRIVING THE VEHICLE ABOVE 16 km/h (10 MPH) AND THEN TURNING IGNITION "OFF."

NOTICE: J 39200 TEST LEADS MUST BE "ZEROED" PRIOR TO MAKING ANY RESISTANCE MEASUREMENTS. REFER TO J 39200 USER'S MANUAL.

IS DTC 61 ALSO PRESENT?
- YES: PROCEED TO DTC 61 CHART FIRST.
- NO: go to 1.

1.
- IGNITION "ON," ENGINE "OFF."
- PUMP BRAKE PEDAL TO DEPLETE VACUUM POWER BOOSTER.
- USING TECH 1 MISC. TESTS, SELECT MANUAL CONTROL FUNCTION.
- SELECT RIGHT FRONT MOTOR APPLY FUNCTION AND APPLY THE MOTOR WHILE CAREFULLY OBSERVING THE COMMAND CURRENT AND FEEDBACK CURRENT ON THE TECH 1.

WERE BOTH COMMAND AND FEEDBACK CURRENT 10 AMPS?
- YES: go to 2.
- NO: GO TO STEP 4.

2.
- APPLY FIRM PRESSURE ON BRAKE PEDAL.
- SELECT RIGHT FRONT MOTOR RELEASE FUNCTION AND RELEASE MOTOR.

DID THE BRAKE PEDAL FALL AND WERE BOTH COMMAND AND FEEDBACK CURRENT 6 AMPS?
- YES: go to 3.
- NO: GO TO STEP 4.

3.
- FIRM PRESSURE STILL ON BRAKE PEDAL.
- SELECT RIGHT FRONT MOTOR APPLY FUNCTION AND APPLY THE MOTOR (THIS MOVES THE HYDRAULIC MODULATOR PISTON TO THE TOP OF ITS BORE).

DID BRAKE PEDAL RISE?
- YES: MALFUNCTION IS INTERMITTENT. REFER TO "DIAGNOSTIC AIDS" ON FACING PAGE.
- NO: go to 4.

4.
- IGNITION "OFF."
- DISCONNECT 6-WAY EBCM AND MOTOR PACK HARNESS CONNECTORS.
- USING J 39200, MEASURE RESISTANCE BETWEEN 6-WAY EBCM HARNESS CONNECTOR TERMINAL "H" AND 6-WAY MOTOR PACK HARNESS CONNECTOR TERMINAL "E".

IS RESISTANCE 1.5Ω OR LESS?
- YES: go to 5.
- NO: REPAIR OPEN OR HIGH RESISTANCE IN CKT 1282.

5.
- USING J 39200, MEASURE RESISTANCE BETWEEN 6-WAY EBCM HARNESS CONNECTOR TERMINAL "G" AND 6-WAY MOTOR PACK HARNESS CONNECTOR TERMINAL "F".

IS RESISTANCE 1.5Ω OR LESS?
- YES: go to 6.
- NO: REPAIR OPEN OR HIGH RESISTANCE IN CKT 1283.

6.
- USING J 39200, MEASURE RESISTANCE BETWEEN 6-WAY MOTOR PACK CONNECTOR TERMINALS "E" AND "F".

IS RESISTANCE 1.5Ω OR LESS?
- YES: GO TO PAGE 2 OF 2 OF THIS CHART.
- NO: REPLACE MOTOR PACK.

DIAGNOSTIC CHART DTC 48 — 1995 CAVALIER AND SUNFIRE

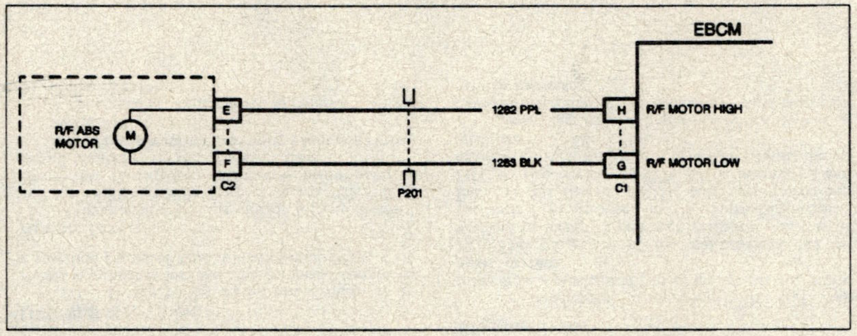

DTC 48
(Page 2 of 2)
RIGHT FRONT ABS MOTOR FREE SPINS

Circuit Description:

This DTC is designed to detect a stripped nut or gear assembly during initialization. During the homing sequence, the piston should reach the top of the bore resulting in a stalled motor. If this does not occur, the motor must be spinning with little or no resistance, indicating a nut/screw or gear malfunction.

DTC Will Set When:

DTC 48 can only be set during initialization. If the feedback current is less than the command current for a specified period of time, a malfunction exists.

Action Taken:

A malfunction DTC is stored, ABS is disabled, and the "ABS" warning lamp is turned "ON."

DTC Chart Test Description:

Number(s) below refer to circled number(s) on the diagnostic chart.

7. Checks for a stripped motor pack gear.
8. Checks for a stripped hydraulic modulator gear.
9. Checks for proper operation of motor pack.
10. Checks for proper operation of the hydraulic modulator.
11. Ensures malfunction was not due to poor terminal contact.

Diagnostic Aids:

An "intermittent" malfunction in this DTC may result from a mechanical part of the system that sticks, binds, or slips.

The frequency of the malfunction can be checked by using the enhanced diagnostic function of the Tech 1, as described in "Enhanced Diagnostics," found in this section. If DTC only occurs once and DTC 61 also occurs, see DTC 61. If intermittent and enhanced diagnostics shows DTC 48 occurs during ABS, see DTC 61.

Depending on the frequency of the malfunction, a physical inspection of the mechanical parts suspected may be necessary.

DIAGNOSTIC CHART DTC 48 — 1995 CAVALIER AND SUNFIRE

DTC 48
(Page 2 of 2)
RIGHT FRONT MOTOR FREE SPINS

CONTINUED FROM PAGE 1 OF 2 OF THIS CHART.

(7)
- REMOVE HYDRAULIC MODULATOR/MASTER CYLINDER ASSEMBLY.
- REMOVE GEAR COVER.
- CHECK FOR STRIPPED GEAR ON MOTOR PACK; THE MOTOR PACK IS THE UNIT WITH THE THREE SMALL GEARS. THE RIGHT FRONT GEAR IS CLOSEST TO THE BRAKE BOOSTER WHEN THE ASSEMBLY IS INSTALLED IN THE VEHICLE.

IS THE MOTOR PACK GEAR STRIPPED?

YES → REPLACE MOTOR PACK.

NO ↓

(8)
- CHECK FOR A STRIPPED GEAR ON THE HYDRAULIC MODULATOR; THE HYDRAULIC MODULATOR IS THE UNIT WITH THE THREE LARGE GEARS. THE RIGHT FRONT GEAR IS CLOSEST TO THE BRAKE BOOSTER WHEN THE ASSEMBLY IS INSTALLED IN THE VEHICLE.

IS THE HYDRAULIC MODULATOR GEAR STRIPPED?

YES → REPLACE HYDRAULIC MODULATOR GEAR.

NO ↓

(9)
- SECURELY POSITION HYDRAULIC MODULATOR/MASTER CYLINDER ASSEMBLY IN VEHICLE ALLOWING RIGHT FRONT GEAR SET OBSERVATION. BE CAREFUL NOT TO DAMAGE GEARS OR ALLOW THE HYDRAULIC MODULATOR/MASTER CYLINDER ASSEMBLY TO MOVE DURING TESTING.
- RECONNECT ELECTRICAL CONNECTORS.
- IGNITION "ON."
- USING TECH 1 MISC. TESTS, SELECT MANUAL CONTROL FUNCTION.
- SELECT RIGHT FRONT MOTOR APPLY AND APPLY THE MOTOR WHILE OBSERVING THE GEAR SET CLOSEST TO THE BOOSTER WHEN THE ASSEMBLY IS INSTALLED IN THE VEHICLE.
- SELECT RIGHT FRONT MOTOR RELEASE FUNCTION AND RELEASE THE MOTOR WHILE STILL OBSERVING THE GEAR SET.
- SELECT RIGHT FRONT MOTOR APPLY FUNCTION AND APPLY THE MOTOR.

DID GEAR SET MOVE IN BOTH DIRECTIONS FOR AT LEAST ONE REVOLUTION?

NO → REPLACE MOTOR PACK.

YES ↓

(10)
- USING TECH 1, SELECT "F4: MISC. TESTS" AND PERFORM GEAR TENSION RELIEF FUNCTION.
- IGNITION "OFF."
- SEPARATE MOTOR PACK FROM HYDRAULIC MODULATOR.
- GRASP THE RIGHT FRONT GEAR ON THE HYDRAULIC MODULATOR; THE HYDRAULIC MODULATOR IS THE UNIT WITH THE THREE LARGE GEARS. THE RIGHT FRONT GEAR IS CLOSEST TO THE BRAKE BOOSTER WHEN THE ASSEMBLY IS INSTALLED IN THE VEHICLE.
- ATTEMPT TO ROTATE THE GEAR.

CAN THE GEAR BE ROTATED AT LEAST 8.5 FULL TURNS LOCK TO LOCK?

YES → REPLACE HYDRAULIC MODULATOR.

NO ↓

(11)
- INSPECT 6-WAY MOTOR PACK AND EBCM CONNECTORS FOR CORROSION OR POOR TERMINAL CONTACT.

  REPLACE TERMINALS THAT EXHIBIT POOR TERMINAL CONTACT OR CORROSION.
- INSTALL HYDRAULIC MODULATOR/MASTER CYLINDER ASSEMBLY.
- RECONNECT ALL CONNECTORS.
- DRIVE VEHICLE FOR THREE DRIVE CYCLES (A DRIVE CYCLE CONSISTS OF STARTING ENGINE, DRIVING OVER 16 km/h (10 MPH) AND THEN TURNING IGNITION "OFF").

DID DTC 48 SET IN THE LAST THREE DRIVE CYCLES?

YES → REPLACE EBCM.

NO → MALFUNCTION WAS DUE TO A POOR CONNECTION.

DIAGNOSTIC CHART DTC 51 — 1995 CAVALIER AND SUNFIRE

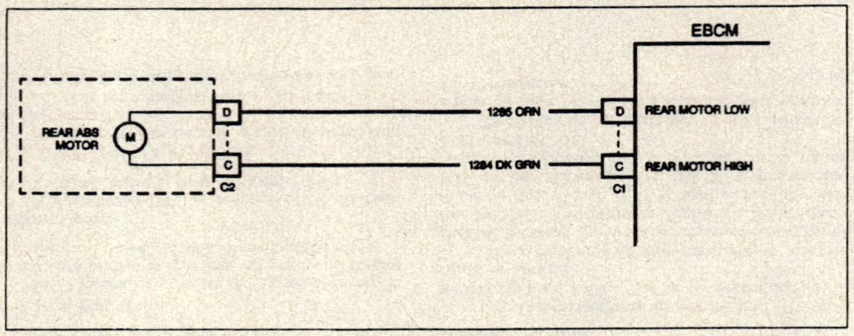

DTC 51

REAR ABS MOTOR FREE SPINS

Circuit Description:

This DTC is designed to detect a stripped nut or gear assembly during initialization. During the homing sequence, the piston should reach the top of the bore resulting in a stalled motor. If this does not occur, the motor must be spinning with little or no resistance, indicating a nut/screw or gear malfunction.

DTC Will Set When:

DTC 51 can be set during initialization. If the feedback current is less than the command current for a specified period of time, a malfunction exists.

Action Taken:

A malfunction DTC is stored, ABS is disabled, and the "ABS" warning lamp is turned "ON." DTC Chart Test Description: Number(s) below refer to circled number(s) on the diagnostic chart.

1. Compares feedback current to command current during apply.
2. Compares feedback current to command current during release.
3. Moves the piston to the top of the bore.
4. Checks for excessive resistance in the rear motor high circuit.
5. Checks for excessive resistance in the rear motor low circuit.
6. Checks for high resistance in the rear motor.

Diagnostic Aids:

An "intermittent" malfunction in this DTC may result from a mechanical part of the system that sticks, binds, or slips.

The frequency of the malfunction can be checked by using the enhanced diagnostic function of the Tech 1 as described in "Enhanced Diagnostics," found in this section. If DTC only occurs once and DTC 64 also occurs, see DTC 64. If intermittent and enhanced diagnostics show DTC 51 occurs during ABS, see DTC 64.

Depending on the frequency of the failure, a physical inspection of the mechanical parts suspected may be necessary.

DIAGNOSTIC CHART DTC 51 — 1995 CAVALIER AND SUNFIRE

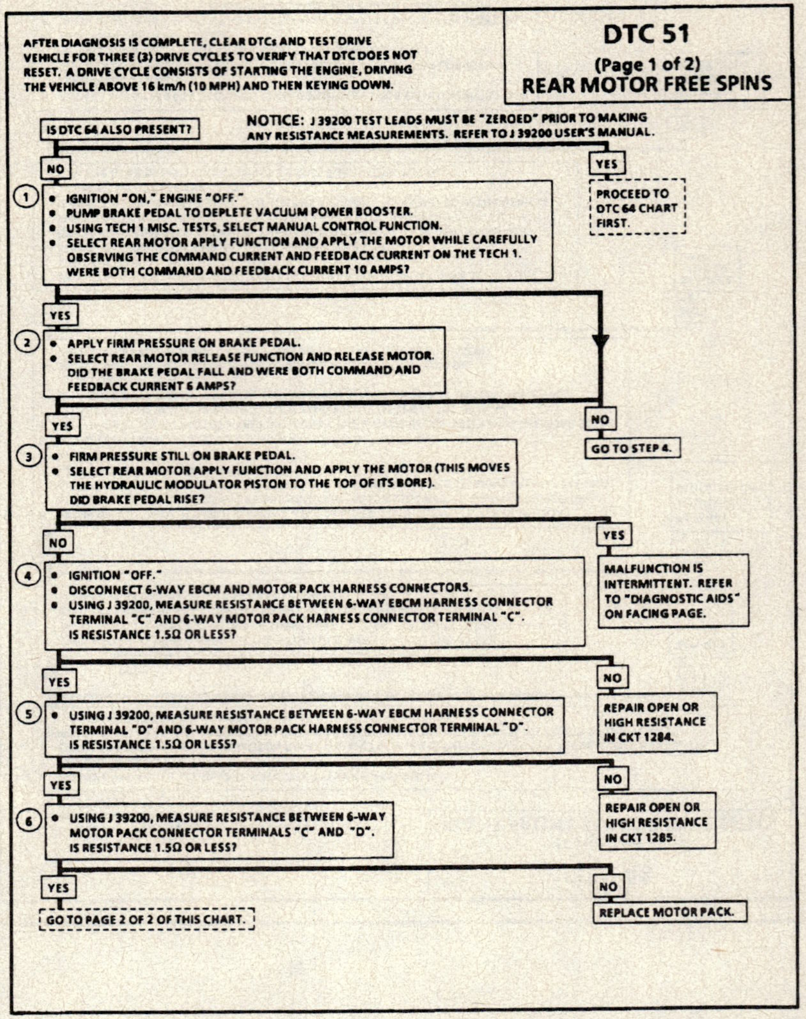

DIAGNOSTIC CHART DTC 51 — 1995 CAVALIER AND SUNFIRE

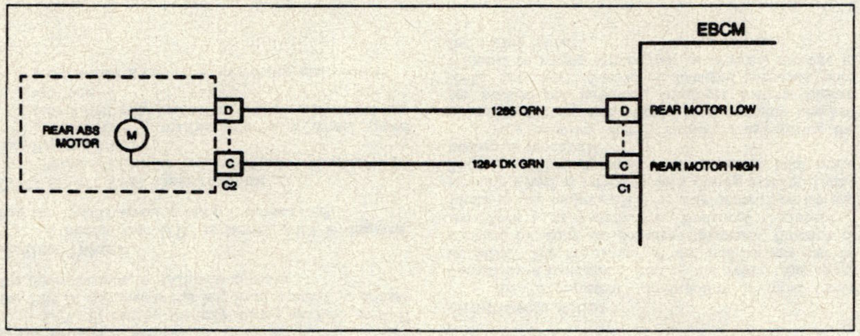

DTC 51
(Page 2 of 2)
REAR ABS MOTOR FREE SPINS

Circuit Description:
This DTC is designed to detect a stripped nut or gear assembly during initialization. During the homing sequence, the piston should reach the top of the bore resulting in a stalled motor. If this does not occur, the motor must be spinning with little or no resistance, indicating a nut/screw or gear malfunction.

DTC Will Set When:
DTC 51 can be set during initialization. If the feedback current is less than the command current for a specified period of time, a malfunction exists.

Action Taken:
A malfunction DTC is stored, ABS is disabled, and the "ABS" warning lamp is turned "ON."

DTC Chart Test Description:
Number(s) below refer to circled number(s) on the diagnostic chart.
7. Checks for a stripped motor pack gear.
8. Checks for a stripped hydraulic modulator gear.
9. Checks for proper operation of motor pack.
10. Checks for proper operation of the hydraulic modulator.
11. Ensures malfunction was not due to poor terminal contact.

Diagnostic Aids:
An "intermittent" malfunction in this DTC may result from a mechanical part of the system that sticks, binds, or slips.

The frequency of the malfunction can be checked by using the enhanced diagnostic function of the Tech 1, as described in "Enhanced Diagnostics," found in this section. If DTC only occurs once and DTC 64 also occurs, see DTC 64. If intermittent and enhanced diagnostics show DTC 51 occurs during ABS, see DTC 64.

Depending on the frequency of the malfunction, a physical inspection of the mechanical parts suspected may be necessary.

DIAGNOSTIC CHART DTC 51 — 1995 CAVALIER AND SUNFIRE

DTC 51
(Page 2 of 2)
REAR MOTOR FREE SPINS

CONTINUED FROM PAGE 1 OF 2 OF THIS CHART.

(7)
- REMOVE HYDRAULIC MODULATOR/MASTER CYLINDER ASSEMBLY.
- REMOVE GEAR COVER.
- CHECK FOR STRIPPED GEAR ON MOTOR PACK; THE MOTOR PACK IS THE UNIT WITH THE THREE SMALL GEARS. THE REAR CHANNEL GEAR IS THE MIDDLE GEAR OF THE THREE.

IS THE MOTOR PACK GEAR STRIPPED?

YES: REPLACE MOTOR PACK.
NO: go to (8)

(8)
- CHECK FOR A STRIPPED GEAR ON THE HYDRAULIC MODULATOR; THE HYDRAULIC MODULATOR IS THE UNIT WITH THE THREE LARGE GEARS. THE REAR CHANNEL GEAR IS THE MIDDLE GEAR OF THE THREE.

IS THE HYDRAULIC MODULATOR GEAR STRIPPED?

YES: REPLACE HYDRAULIC MODULATOR GEAR.
NO: go to (9)

(9)
- SECURELY POSITION HYDRAULIC MODULATOR/MASTER CYLINDER ASSEMBLY IN VEHICLE ALLOWING REAR CHANNEL GEAR SET OBSERVATION. BE CAREFUL NOT TO DAMAGE GEARS OR ALLOW THE HYDRAULIC MODULATOR/MASTER CYLINDER ASSEMBLY TO MOVE DURING TESTING.
- RECONNECT ELECTRICAL CONNECTORS.
- IGNITION "ON."
- USING TECH 1 MISC. TESTS, SELECT MANUAL CONTROL FUNCTION.
- SELECT REAR MOTOR APPLY AND APPLY THE MOTOR WHILE OBSERVING THE MIDDLE GEAR SET.
- SELECT REAR MOTOR RELEASE FUNCTION AND RELEASE THE MOTOR WHILE STILL OBSERVING THE GEAR SET.
- SELECT REAR MOTOR APPLY FUNCTION AND APPLY THE MOTOR.

DID GEAR SET MOVE IN BOTH DIRECTIONS FOR AT LEAST ONE REVOLUTION?

NO: REPLACE MOTOR PACK.
YES: go to (10)

(10)
- USING TECH 1, SELECT "F4: MISC. TESTS" AND PERFORM GEAR TENSION RELIEF FUNCTION.
- IGNITION "OFF."
- SEPARATE MOTOR PACK FROM HYDRAULIC MODULATOR.
- GRASP THE REAR CHANNEL GEAR ON THE HYDRAULIC MODULATOR; THE HYDRAULIC MODULATOR IS THE UNIT WITH THE THREE LARGE GEARS. THE REAR CHANNEL GEAR IS THE MIDDLE GEAR OF THE THREE.
- ATTEMPT TO ROTATE THE GEAR.

CAN THE GEAR BE ROTATED AT LEAST 3.5 FULL TURNS LOCK TO LOCK?

YES: REPLACE HYDRAULIC MODULATOR.
NO: go to (11)

(11)
- INSPECT 6-WAY MOTOR PACK AND EBCM CONNECTORS FOR CORROSION OR POOR TERMINAL CONTACT. REFER TO SECTION 8A, "CHECKING TERMINAL CONTACT." REPLACE TERMINALS THAT EXHIBIT POOR TERMINAL CONTACT OR CORROSION.
- INSTALL HYDRAULIC MODULATOR/MASTER CYLINDER ASSEMBLY.
- RECONNECT ALL CONNECTORS.
- DRIVE VEHICLE FOR THREE DRIVE CYCLES (A DRIVE CYCLE CONSISTS OF STARTING ENGINE, DRIVING OVER 16 km/h (10 MPH) AND THEN TURNING IGNITION "OFF").

DID DTC 51 SET IN THE LAST THREE DRIVE CYCLES?

YES: REPLACE EBCM.
NO: MALFUNCTION WAS DUE TO A POOR CONNECTION.

DIAGNOSTIC CHART DTC 52 — 1995 CAVALIER AND SUNFIRE

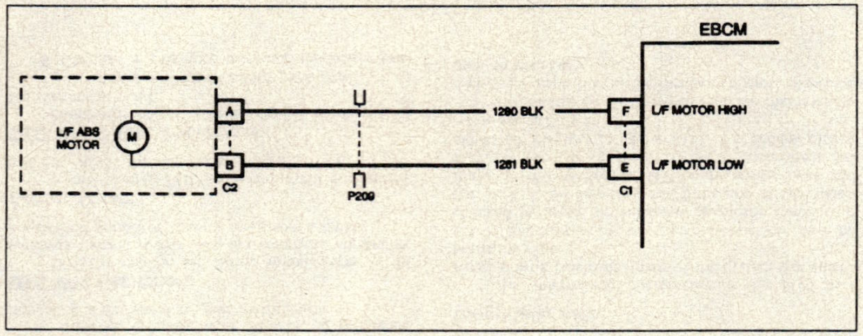

DTC 52
LEFT FRONT ABS CHANNEL IN RELEASE TOO LONG

Circuit Description:
This DTC will diagnose a motor that is energized longer than expected. This could occur if a wheel speed sensor is faulty, the motor does not turn, the left front solenoid mechanically fails open, the brakes are dragging, or the motor wires are crossed.

DTC Will Set When:
DTC 52 can be set only during an ABS stop. If the EBCM commands the left front channel in release for three seconds, a malfunction exists.

Action Taken:
A malfunction DTC is stored, ABS is disabled, and the "ABS" warning lamp is turned "ON."

DTC Chart Test Description:
Number(s) below refer to circled number(s) on the diagnostic chart.
1. Identifies a malfunction in a wheel speed sensor that may cause the system to be in release too long.
2. Identifies a motor or modulator malfunction.
3. Checks for a solenoid that may have mechanically failed open.
4. Checks for a solenoid that may have failed closed.
5. Isolates the malfunction to either the solenoid or the hydraulic modulator.
6. Checks for proper operation of hydraulic modulator.

Diagnostic Aids:
An "intermittent" malfunction is most likely caused by a mechanical part of the system that sticks or binds. The frequency of the malfunction can be checked by using the enhanced diagnostic function of the "Tech 1, as described in "Enhanced Diagnostics," found in this section. DTC 52 may occur if on ice and steering wheel is turned to lock during braking. Using Tech 1 perform hydraulic test to ensure total brake system is functional.

Any circuitry, that is suspected as causing the intermittent complaint, should be thoroughly checked for backed out terminals, improper mating, broken locks, improperly formed or damaged terminals, poor terminal to wiring connections or physical damage to the wiring harness.

DIAGNOSTIC CHART DTC 52 — 1995 CAVALIER AND SUNFIRE

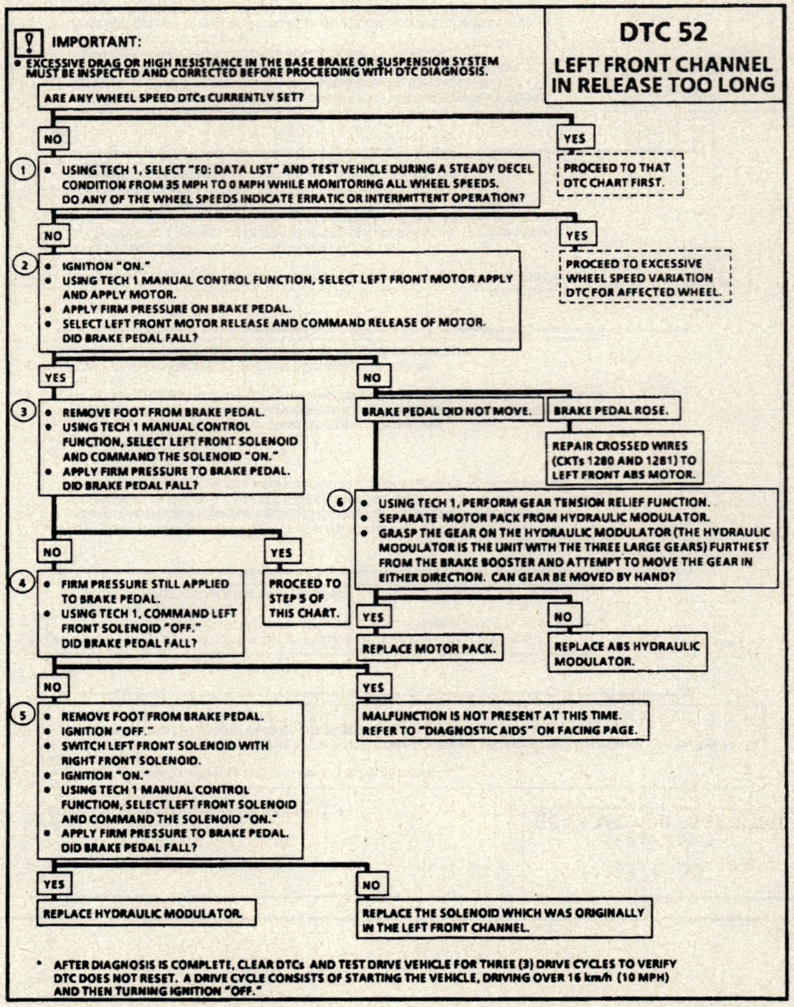

DIAGNOSTIC CHART DTC 53 — 1995 CAVALIER AND SUNFIRE

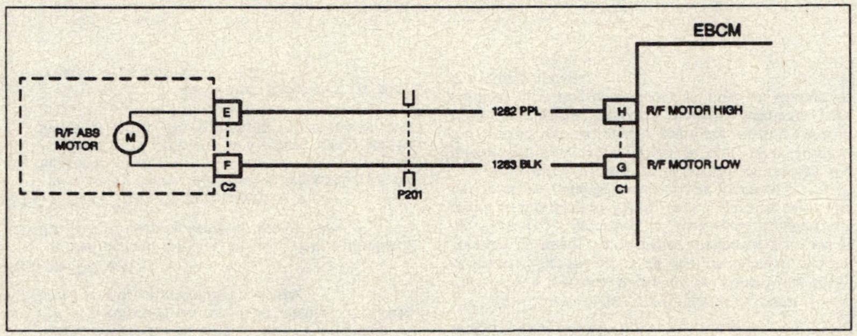

DTC 53
RIGHT FRONT ABS CHANNEL IN RELEASE TOO LONG

Circuit Description:

This DTC will diagnose a motor that is energized longer than expected. This could occur if a wheel speed sensor is faulty, the motor does not turn, the right front solenoid mechanically fails open, the brakes are dragging, or the motor wires are crossed.

DTC Will Set When:

DTC 53 can be set only during an ABS stop. If the EBCM commands the right front channel in release for three seconds, a malfunction exists.

Action Taken:

A malfunction DTC is stored, ABS is disabled, and the "ABS" warning lamp is turned "ON."

DTC Chart Test Description:

Number(s) below refer to circled number(s) on the diagnostic chart.

1. Identifies a malfunction in a wheel speed sensor that may cause the system to be in release too long.
2. Identifies a motor or modulator malfunction.
3. Checks for a solenoid that may have mechanically failed open.
4. Checks for a solenoid that may have failed closed.
5. Isolates the malfunction to either the solenoid or the hydraulic modular.
6. Checks for proper operation of hydraulic modulator.

Diagnostic Aids:

An "intermittent" malfunction is most likely caused by a mechanical part of the system that sticks or binds. The frequency of the malfunction can be checked by using the enhanced diagnostic function of the Tech 1, as described in "Enhanced Diagnostics," found in this section. DTC 53 may occur on ice if steering wheel is turned to lock during braking. Using Tech 1 perform hydraulic test to ensure total brake system is functional.

Any circuitry, that is suspected as causing the intermittent complaint, should be thoroughly checked for backed out terminals, improper mating, broken locks, improperly formed or damaged terminals, poor terminal to wiring connections or physical damage to the wiring harness.

DIAGNOSTIC CHART DTC 53 — 1995 CAVALIER AND SUNFIRE

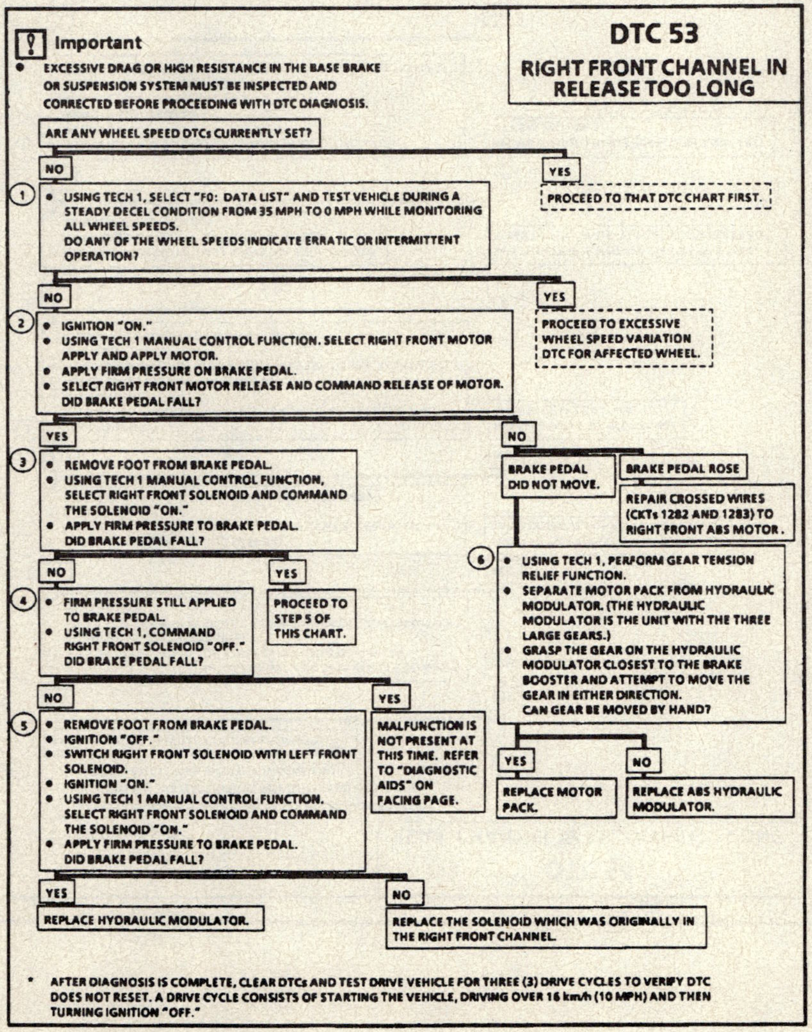

DIAGNOSTIC CHART DTC 54 — 1995 CAVALIER AND SUNFIRE

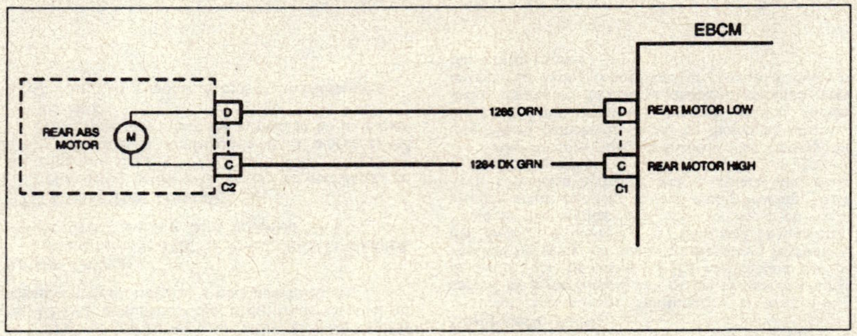

DTC 54
REAR ABS CHANNEL IN RELEASE TOO LONG

Circuit Description:
This DTC will diagnose a motor that is energized longer than expected. This could occur if a wheel speed sensor is faulty, the motor does not turn, the brakes are dragging, or the motor wires are crossed.

DTC Will Set When:
DTC 54 can be set only during an ABS stop. If the EBCM commands the rear ABS channel in release for three seconds, a malfunction exists.

Action Taken:
A malfunction DTC is stored, ABS is disabled, and the "ABS" warning lamp is turned "ON."

DTC Chart Test Description:
Number(s) below refer to circled number(s) on the diagnostic chart.
1. Identifies a problem in a wheel speed sensor that may cause the system to be in release too long.
2. Checks for a wheel that may stick or bind because of a mechanical fault.
3. Checks to see if the motor is capable of moving and applying the hydraulic piston for the rear wheels.
4. Ensures that the motor wiring is not crossed.
5. Isolates the malfunction of a "no-apply" situation to either the motor pack or ABS hydraulic modulator.

Diagnostic Aids:
An "intermittent" malfunction is most likely caused by a mechanical part of the system that sticks or binds. The frequency of the malfunction can be checked by using the enhanced diagnostic function of the Tech 1, as described in "Enhanced Diagnostics," found in this section. Using Tech 1 perform hydraulic test to ensure total brake system is functional.
Any circuitry, that is suspected as causing the intermittent complaint, should be thoroughly checked for backed out terminals, improper mating, broken locks, improperly formed or damaged terminals, poor terminal to wiring connections or physical damage to the wiring harness.

DIAGNOSTIC CHART DTC 54 — 1995 CAVALIER AND SUNFIRE

DTC 54
REAR CHANNEL IN RELEASE TOO LONG

Important
- EXCESSIVE DRAG OR HIGH RESISTANCE IN THE BASE BRAKE OR SUSPENSION SYSTEM MUST BE INSPECTED AND CORRECTED BEFORE PROCEEDING WITH DTC DIAGNOSIS.

ARE ANY WHEEL SPEED DTCs CURRENTLY SET?
- YES: PROCEED TO THAT DTC CHART FIRST.
- NO:

(1)
- USING TECH 1, SELECT "F0: DATA LIST" AND TEST VEHICLE DURING A STEADY DECEL CONDITION FROM 35 MPH TO 0 MPH WHILE MONITORING ALL WHEEL SPEEDS.
DO ANY OF THE WHEEL SPEEDS INDICATE ERRATIC OR INTERMITTENT OPERATION?
  - YES: PROCEED TO EXCESSIVE WHEEL SPEED VARIATION DTC FOR AFFECTED WHEEL.
  - NO:

(2)
- RAISE REAR OF VEHICLE AND SUPPORT.
- SPIN BOTH REAR WHEELS AND CONFIRM PHYSICAL CONDITION OF BASE BRAKE SYSTEM.
COULD BOTH REAR WHEELS BE SPUN FREELY BY HAND?
  - NO: REPAIR SOURCE OF RESISTANCE IN BASE BRAKE SYSTEM.
  - YES:

(3)
- IGNITION "ON."
- USING TECH 1 MANUAL CONTROL FUNCTION, SELECT REAR AXLE MOTOR APPLY.
- APPLY FIRM PRESSURE ON BRAKE PEDAL.
- COMMAND APPLY OF REAR MOTOR AND TRY TO SPIN REAR WHEELS BY HAND.
COULD REAR WHEELS BE SPUN BY HAND?
  - NO: MALFUNCTION IS NOT PRESENT AT THIS TIME. REFER TO "DIAGNOSTIC AIDS" ON FACING PAGE.
  - YES:

(4)
- WITH FIRM PRESSURE ON BRAKE PEDAL STILL APPLIED, COMMAND A RELEASE OF THE REAR MOTOR.
CAN WHEELS BE MOVED?
  - NO: REPAIR CROSSED WIRES (CKTs 1284 AND 1285) AT REAR MOTOR.
  - YES:

(5)
- USING TECH 1 MISC. TESTS, PERFORM GEAR TENSION RELIEF FUNCTION.
- REMOVE MOTOR PACK.
- CONNECT MOTOR PACK WIRING.
- USING TECH 1, PERFORM MOTOR TEST.
DO ALL THREE MOTOR PACK GEARS (SMALL GEARS) SPIN FREELY?
  - NO: REPLACE MOTOR PACK.
  - YES: REPLACE ABS HYDRAULIC MODULATOR.

AFTER ALL DIAGNOSIS IS COMPLETE, CLEAR DTCs AND TEST DRIVE VEHICLE FOR THREE (3) DRIVE CYCLES TO VERIFY THAT DTC DOES NOT RESET. A DRIVE CYCLE CONSISTS OF STARTING THE VEHICLE DRIVING OVER 16 km/h (10 MPH), AND THEN TURNING IGNITION "OFF."

DIAGNOSTIC CHART DTC 55 — 1995 CAVALIER AND SUNFIRE

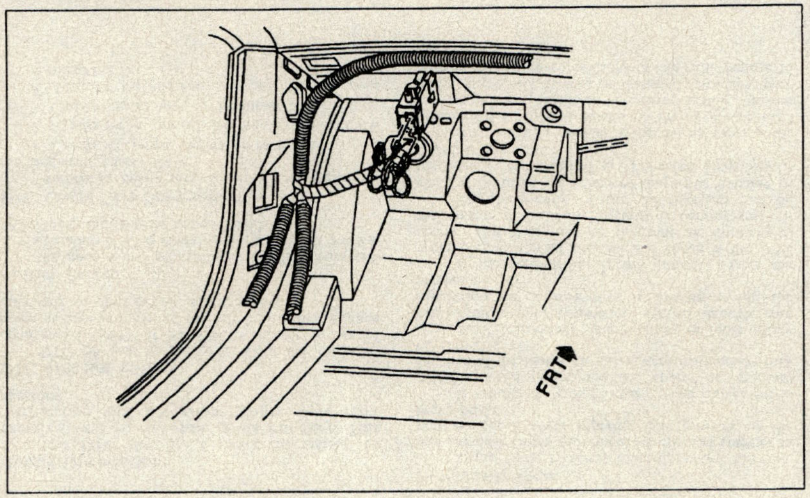

DTC 55
EBCM MALFUNCTION

Circuit Description:
This DTC identifies a malfunction within the EBCM. It also ensures the cause of the malfunction is not a result of a malfunction with the electronic brake control relay.

DTC Will Set When:
DTC 55 is set when an internal EBCM malfunction exists.

Action Taken:
A malfunction DTC is stored, ABS is disabled and the "ABS" warning lamp is turned "ON."

DTC Chart Test Description:
Number(s) below refer to circled number(s) on the diagnostic chart.
1. Checks to see if there are any other malfunctions present.
2. This checks to see if the malfunction is still present.

Diagnostic Aids:
An "intermittent" malfunction is most likely caused by a poor connection, rubbed through wire insulation, or a wire that is broken inside the insulation.
The frequency of the malfunction can be checked by using the enhanced diagnostic function of the Tech 1, as described in "Enhanced Diagnostics," found in this section.
Any circuitry, that is suspected as causing the intermittent complaint, should be thoroughly checked for backed out terminals, improper mating, broken locks, improperly formed or damaged terminals, poor terminal to wiring connections or physical damage to the wiring harness.

DIAGNOSTIC CHART DTC 55 — 1995 CAVALIER AND SUNFIRE

DTC 55
EBCM MALFUNCTION

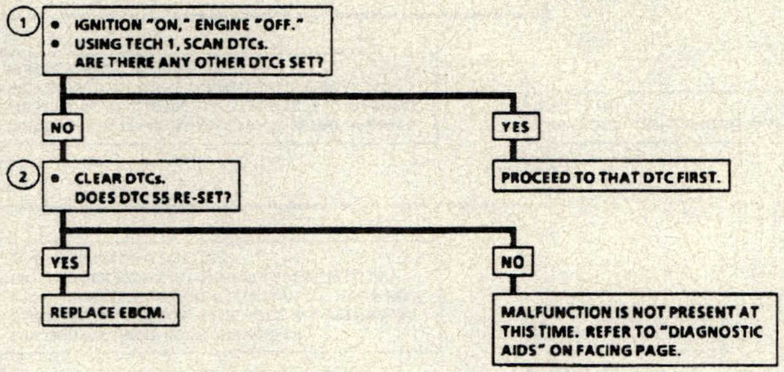

AFTER DIAGNOSIS IS COMPLETE, CLEAR DTCs AND TEST DRIVE VEHICLE FOR THREE (3) DRIVE CYCLES TO VERIFY DTC DOES NOT RESET. A DRIVE CYCLE CONSISTS OF STARTING THE VEHICLE, DRIVING OVER 16 km/h (10 MPH) AND THEN TURNING IGNITION "OFF."

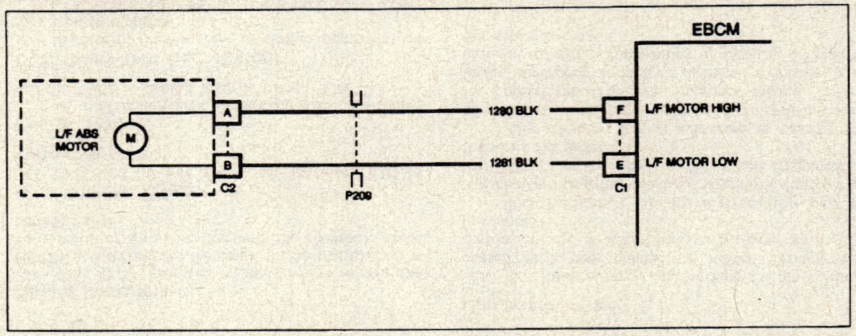

DTC 56
LEFT FRONT ABS MOTOR CIRCUIT OPEN

Circuit Description:
This DTC identifies a motor that cannot be energized due to an open in its circuitry. This malfunction will not allow proper front ABS operation.

DTC Will Set When:
DTC 56 can be set only when the motor is commanded "OFF." If the EBCM detects an out of range voltage in either of the left front motor circuits indicating an open circuit, a malfunction exists.

Action Taken:
An open motor will not activate when requested. A malfunction DTC is stored, ABS is disabled and the "ABS" warning lamp is turned "ON."

DTC Chart Test Description:
Number(s) below refer to circled number(s) on the diagnostic chart.

1. Checks for proper resistance of the motor.
2. Checks for an open in the motor high circuitry.
3. Checks for an open in the motor low circuitry.
4. Ensures malfunction was not due to poor terminal contact.

Diagnostic Aids:
Using Tech 1, select manual control function, and exercise motor movement of effected channel in both directions while applying light pressure on the brake pedal.

If erratic or "jumpy" brake pedal movement is detected while performing an "apply" or "release" function of the motor, an intermittent malfunction may be indicated.

An "intermittent" malfunction is most likely caused by a poor connection, rubbed through wire insulation, or a wire that is broken inside the insulation.

If the malfunction is not current, wiggle the wires of the effected channel and check if the DTC resets. This will help to pinpoint an intermittent malfunction in the motor circuitry or connections.

The frequency of the malfunction can be checked by using the enhanced diagnostic function of the Tech 1, as described in "Enhanced Diagnostics," found in this section.

Any circuitry, that is suspected as causing the intermittent complaint, should be thoroughly checked for backed out terminals, improper mating, broken locks, improperly formed or damaged terminals, poor terminal to wiring connections or physical damage to the wiring harness.

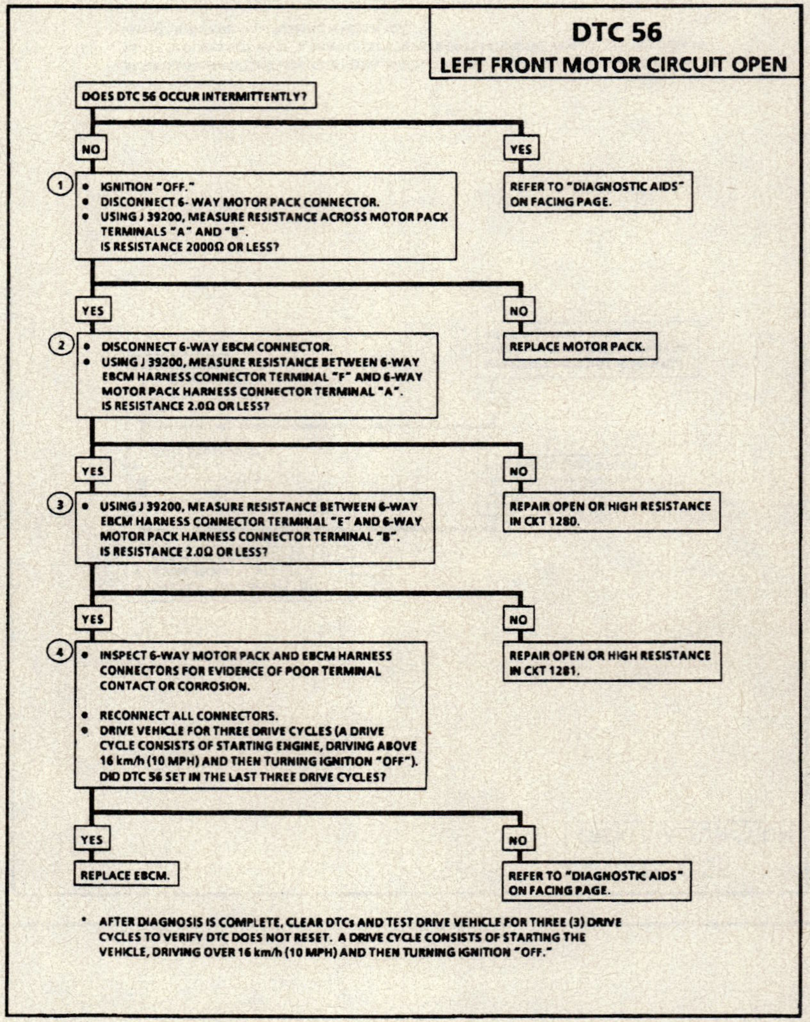

DIAGNOSTIC CHART DTC 57 — 1995
CAVALIER AND SUNFIRE

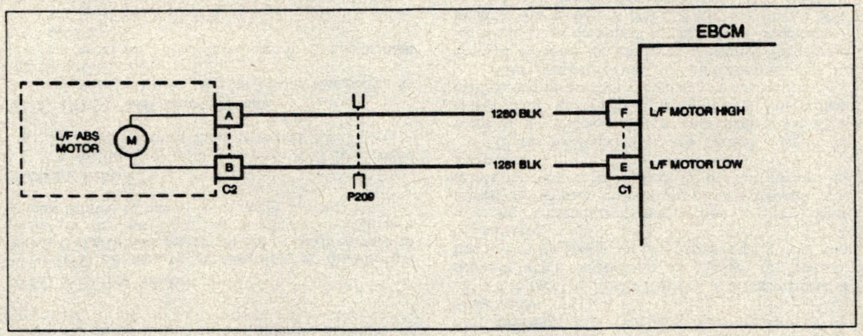

DTC 57
LEFT FRONT ABS MOTOR CIRCUIT SHORTED TO GROUND

Circuit Description:

This DTC identifies a motor circuit that is shorted to ground. This malfunction will not allow the motor to be controlled at the commanded current rate or will cause the motor driver circuit to allow current directly to ground.

DTC Will Set When:

DTC 57 can be set anytime. A malfunction exists if the EBCM detects an out of range voltage in either of the left front motor circuits indicating a circuit shorted to ground.

Action Taken:

A malfunction DTC is stored, ABS is disabled and the "ABS" warning lamp is turned "ON."

DTC Chart Test Description:

Number(s) below refer to circled number(s) on the diagnostic chart.

1. Checks for a short to ground in the motor high circuitry.
2. Checks for a short to ground in the motor low circuitry.
3. Checks for a motor that is internally shorted to ground.
4. Ensures malfunction was not due to physical damage of the circuitry.

Diagnostic Aids:

Using Tech 1, select manual control function, and exercise motor movement of effected channel in both directions while applying light pressure on the brake pedal.

If erratic or "jumpy" brake pedal movement is detected while performing an "apply" or "release" function of the motor, an intermittent malfunction may be indicated.

An "intermittent" malfunction is most likely caused by a poor connection, rubbed through wire insulation, or a wire that is broken inside the insulation.

If the malfunction is not current, wiggle the wires of the effected channel and check if the DTC resets. This will help to pinpoint an intermittent malfunction in the motor circuitry or connections.

The frequency of the malfunction can be checked by using the enhanced diagnostic function of the Tech 1, as described in "Enhanced Diagnostics," found in this section.

Any circuitry, that is suspected as causing the intermittent complaint, should be thoroughly checked for backed out terminals, improper mating, broken locks, improperly formed or damaged terminals, poor terminal to wiring connections or physical damage to the wiring harness.

DIAGNOSTIC CHART DTC 57 — 1995
CAVALIER AND SUNFIRE

DTC 57
LEFT FRONT MOTOR CIRCUIT SHORTED TO GROUND

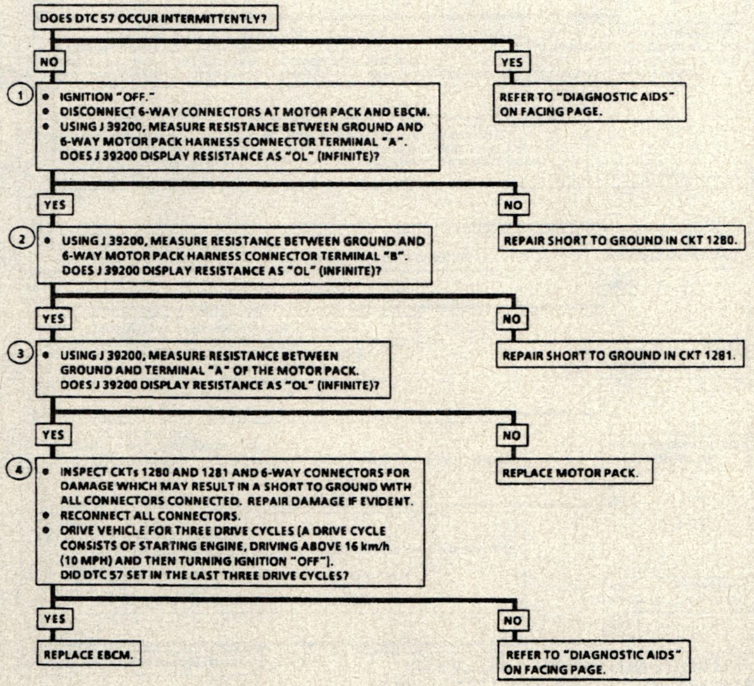

- AFTER DIAGNOSIS IS COMPLETE, CLEAR DTCs AND TEST DRIVE VEHICLE FOR THREE (3) DRIVE CYCLES TO VERIFY DTC DOES NOT RESET. A DRIVE CYCLE CONSISTS OF STARTING THE VEHICLE, DRIVING OVER 16 km/h (10 MPH) AND THEN KEYING DOWN.

DIAGNOSTIC CHART DTC 58 — 1995 CAVALIER AND SUNFIRE

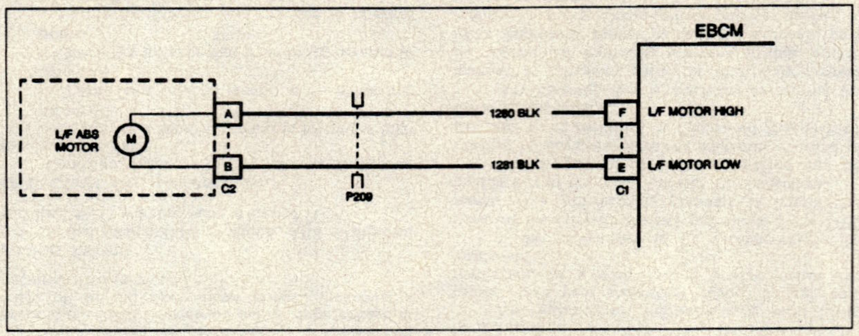

DTC 58
LEFT FRONT ABS MOTOR CIRCUIT SHORTED TO BATTERY OR MOTOR SHORTED

Circuit Description:

This DTC identifies a motor circuit that is shorted to battery or a motor that has low or no resistance. This malfunction will not allow the motor to be controlled at the commanded current rate or will cause the motor to turn in the opposite direction, or not at all.

DTC Will Set When:

DTC 58 can be set anytime. A malfunction exists if the EBCM detects an out of range voltage in either of the left front motor circuits indicating a circuit shorted to battery or a motor shorted.

Action Taken:

A malfunction DTC is stored, ABS is disabled and the "ABS" warning lamp is turned "ON."

DTC Chart Test Description:

Number(s) below refer to circled number(s) on the diagnostic chart.

1. Checks for a short to voltage in the motor high circuitry.
2. Checks for a short to voltage in the motor low circuitry.
3. Checks for a short between high and low motor circuits.
4. Checks for a motor that is internally shorted.
5. Checks for a short between the EMB and motor circuits.
6. Checks for an internal motor pack short between the EMB and motor.
7. Ensures malfunction was not due to physical damage of the circuitry.

Diagnostic Aids:

Using Tech 1, select manual control function, and exercise motor movement of effected channel in both directions while applying light pressure on the brake pedal.

If erratic or "jumpy" brake pedal movement is detected while performing an "apply" or "release" function of the motor, an intermittent malfunction may be indicated.

An "intermittent" malfunction is most likely caused by a poor connection, rubbed through wire insulation, or a wire that is broken inside the insulation.

If the malfunction is not current, wiggle the wires of the effected channel and check if the DTC resets. This will help to pinpoint an intermittent malfunction in the motor circuitry or connections.

The frequency of the malfunction can be checked by using the enhanced diagnostic function of the Tech 1, as described in "Enhanced Diagnostics," found in this section. If DTC 38 sets as a current DTC and DTC 58 is an infrequent history DTC, see DTC 38 diagnostics.

Any circuitry, that is suspected as causing the intermittent complaint, should be thoroughly checked for backed out terminals, improper mating, broken locks, improperly formed or damaged terminals, poor terminal to wiring connections or physical damage to the wiring harness.

DIAGNOSTIC CHART DTC 58 — 1995 CAVALIER AND SUNFIRE

DTC 58
LEFT FRONT MOTOR CIRCUIT SHORTED TO BATTERY OR MOTOR SHORTED

NOTICE: J 39200 TEST LEADS MUST BE "ZEROED" PRIOR TO MAKING ANY RESISTANCE MEASUREMENTS. REFER TO J 39200 USER'S MANUAL.

DOES DTC 58 OCCUR INTERMITTENTLY?

- YES → REFER TO "DIAGNOSTIC AIDS" ON FACING PAGE.
- NO →

(1) NOTICE: THIS STEP MAY CAUSE ADDITIONAL DTCs TO SET.
- IGNITION "OFF."
- DISCONNECT 6-WAY CONNECTORS AT MOTOR PACK AND EBCM.
- IGNITION "ON."
- USING J 39200, MEASURE VOLTAGE BETWEEN GROUND AND 6-WAY MOTOR PACK HARNESS CONNECTOR TERMINAL "A".

IS VOLTAGE 1.0V OR LESS?

- NO → REPAIR SHORT TO VOLTAGE IN CKT 1280.
- YES →

(2)
- USING J 39200, MEASURE VOLTAGE BETWEEN GROUND AND 6-WAY MOTOR PACK HARNESS CONNECTOR TERMINAL "B".

IS VOLTAGE 1.0V OR LESS?

- NO → REPAIR SHORT TO VOLTAGE IN CKT 1281.
- YES →

(3)
- IGNITION "OFF."
- USING J 39200, MEASURE RESISTANCE BETWEEN 6-WAY MOTOR PACK HARNESS CONNECTOR TERMINALS "A" AND "B".

DOES J 39200 DISPLAY RESISTANCE AS "OL" (INFINITE)?

- NO → REPAIR SHORT BETWEEN CKTs 1280 AND 1281.
- YES →

(4)
- USING J 39200, MEASURE RESISTANCE BETWEEN MOTOR PACK TERMINALS "A" AND "B".

IS RESISTANCE 0.4Ω OR GREATER?

- NO → REPLACE MOTOR PACK.
- YES →

(5)
- DISCONNECT 3-WAY MOTOR PACK CONNECTOR.
- USING J 39200, MEASURE RESISTANCE BETWEEN 3-WAY MOTOR PACK HARNESS CONNECTOR TERMINAL "B" AND 6-WAY MOTOR PACK HARNESS CONNECTOR TERMINAL "A" THEN "B".

DID J 39200 DISPLAY RESISTANCE AS "OL" (INFINITE) FOR EACH MEASUREMENT?

- NO → REPAIR SHORT BETWEEN CKT 855/1633 AND MOTOR CIRCUIT WHICH MEASURED RESISTANCE OTHER THAN "OL" (INFINITE).
- YES →

(6)
- USING J 39200, MEASURE RESISTANCE BETWEEN 3-WAY MOTOR PACK CONNECTOR TERMINAL "B" AND 6-WAY MOTOR PACK CONNECTOR TERMINAL "A".

DID J 39200 DISPLAY "OL" (INFINITE)?

- NO → REPLACE MOTOR PACK.
- YES →

(7)
- INSPECT CKTs 1280 AND 1281 AND 6-WAY CONNECTORS FOR DAMAGE WHICH MAY RESULT IN A SHORT TO VOLTAGE WITH ALL CONNECTORS CONNECTED.
- REPAIR DAMAGE IF EVIDENT.
- RECONNECT ALL CONNECTORS.
- DRIVE VEHICLE FOR THREE DRIVE CYCLES (A DRIVE CYCLE CONSISTS OF STARTING ENGINE, DRIVE ABOVE 16 km/h (10 MPH) AND THEN TURNING IGNITION "OFF").

DID DTC 58 SET IN THE LAST THREE DRIVE CYCLES?

- YES → REPLACE EBCM.
- NO → REFER TO "DIAGNOSTIC AIDS" ON FACING PAGE.

• AFTER DIAGNOSIS IS COMPLETE, CLEAR DTCs AND TEST DRIVE VEHICLE FOR THREE (3) DRIVE CYCLES TO VERIFY DTC DOES NOT RESET. A DRIVE CYCLE CONSISTS OF STARTING THE VEHICLE, DRIVING OVER 16 km/h (10 MPH) AND THEN TURNING IGNITION "OFF."

DIAGNOSTIC CHART DTC 61 — 1995 CAVALIER AND SUNFIRE

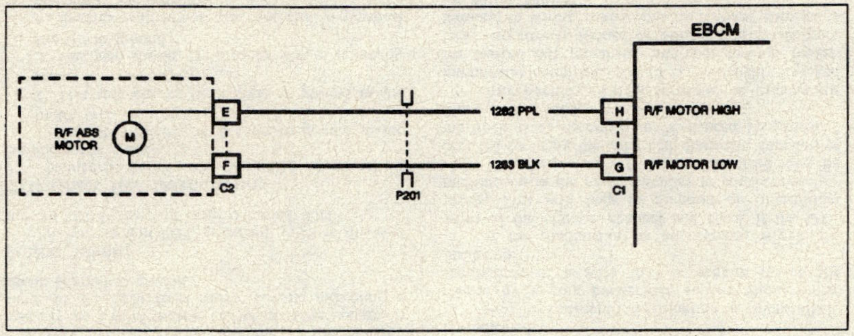

DTC 61
RIGHT FRONT ABS MOTOR CIRCUIT OPEN

Circuit Description:
This DTC identifies a motor that cannot be energized due to an open in its circuitry. This malfunction will not allow proper front ABS operation.

DTC Will Set When:
DTC 61 can be set only when the motor is commanded "OFF." A malfunction exists if the EBCM detects an out of range voltage in either of the right front motor circuits indicating an open circuit.

Action Taken:
An open motor will not activate when requested. A malfunction DTC is stored, ABS is disabled and the "ABS" warning lamp is turned "ON."

DTC Chart Test Description:
Number(s) below refer to circled number(s) on the diagnostic chart.

1. Checks for proper resistance of the motor.
2. Checks for an open in the motor high circuitry.
3. Checks for an open in the motor low circuitry.
4. Ensures malfunction was not due to poor terminal contact.

Diagnostic Aids:
Using Tech 1, select manual control function, and exercise motor movement of effected channel in both directions while applying light pressure on the brake pedal.

If erratic or "jumpy" brake pedal movement is detected while performing an "apply" or "release" function of the motor, an intermittent malfunction may be indicated.

An "intermittent" malfunction is most likely caused by a poor connection, rubbed through wire insulation, or a wire that is broken inside the insulation.

If the malfunction is not current, wiggle the wires of the effected channel and check if the DTC resets. This will help to pinpoint an intermittent malfunction in the motor circuitry or connections.

The frequency of the malfunction can be checked by using the enhanced diagnostic function of the Tech 1, as described in "Enhanced Diagnostics," found in this section.

Any circuitry, that is suspected as causing the intermittent complaint, should be thoroughly checked for backed out terminals, improper mating, broken locks, improperly formed or damaged terminals, poor terminal to wiring connections or physical damage to the wiring harness.

DIAGNOSTIC CHART DTC 61 — 1995 CAVALIER AND SUNFIRE

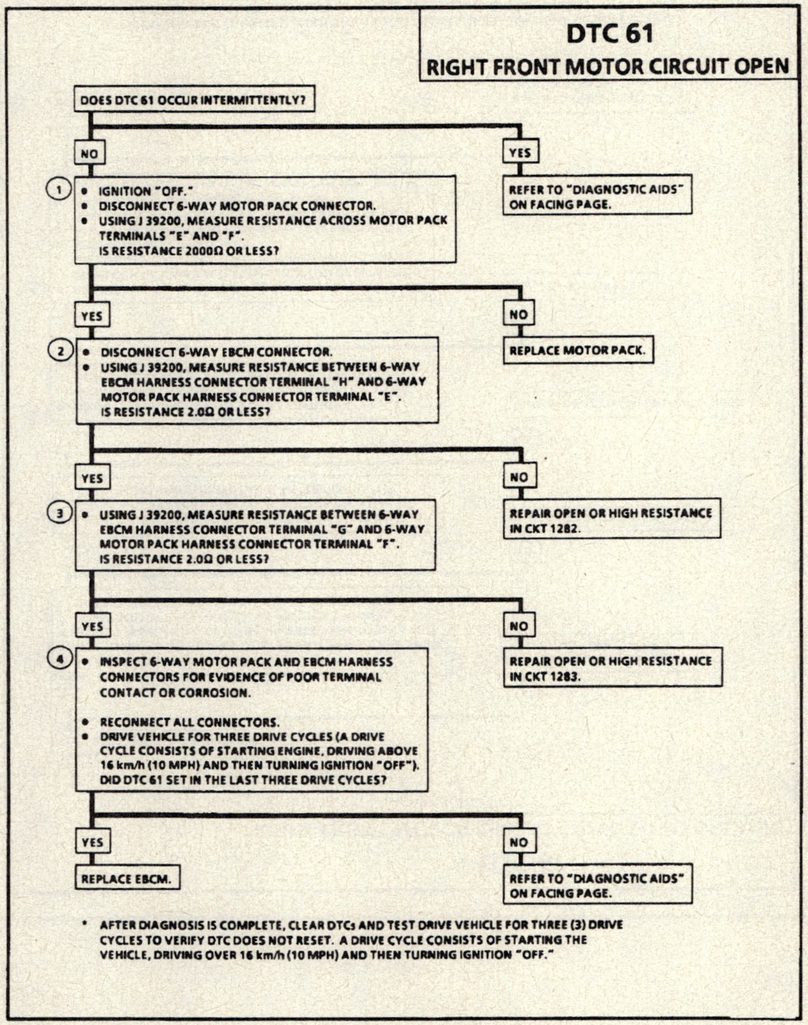

DIAGNOSTIC CHART DTC 62 — 1995 CAVALIER AND SUNFIRE

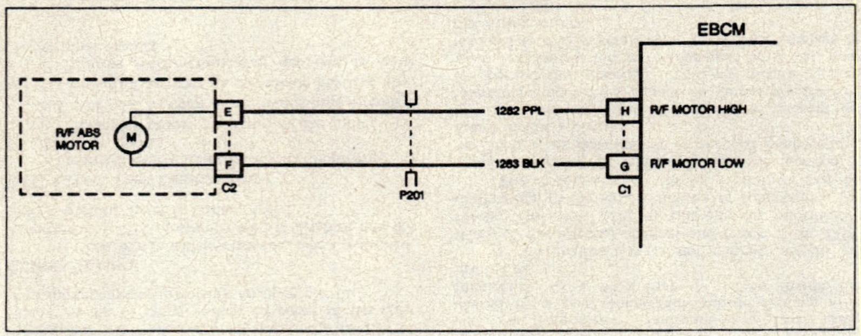

DTC 62
RIGHT FRONT ABS MOTOR CIRCUIT SHORTED TO GROUND

Circuit Description:

This DTC identifies a motor circuit that is shorted to ground. This malfunction will not allow the motor to be controlled at the commanded current rate or will cause the motor driver circuit to allow current directly to ground.

DTC Will Set When:

DTC 62 can be set anytime. A malfunction exists if the EBCM detects an out of range voltage in either of the right front motor circuits indicating a circuit shorted to ground.

Action Taken:

A malfunction DTC is stored, ABS is disabled and the "ABS" warning lamp is turned "ON."

DTC Chart Test Description:

Number(s) below refer to circled number(s) on the diagnostic chart.

1. This step checks for a short to ground in the motor high circuitry.
2. This step checks for a short to ground in the motor low circuitry.
3. This step checks for a motor that is internally shorted to ground.
4. Ensures malfunction was not due to physical damage of the circuitry.

Diagnostic Aids:

Using Tech 1, select manual control function, and exercise motor movement of effected channel in both directions while applying light pressure on the brake pedal.

If erratic or "jumpy" brake pedal movement is detected while performing an "apply" or "release" function of the motor, an intermittent malfunction may be indicated.

An "intermittent" malfunction is most likely caused by a poor connection, rubbed through wire insulation, or a wire that is broken inside the insulation.

If the malfunction is not current, wiggle the wires of the effected channel and check if the DTC resets. This will help to pinpoint an intermittent malfunction in the motor circuitry or connections.

The frequency of the malfunction can be checked by using the enhanced diagnostic function of the Tech 1, as described in "Enhanced Diagnostics," found in this section.

Any circuitry, that is suspected as causing the intermittent complaint, should be thoroughly checked for backed out terminals, improper mating, broken locks, improperly formed or damaged terminals, poor terminal to wiring connections or physical damage to the wiring harness.

DIAGNOSTIC CHART DTC 62 — 1995 CAVALIER AND SUNFIRE

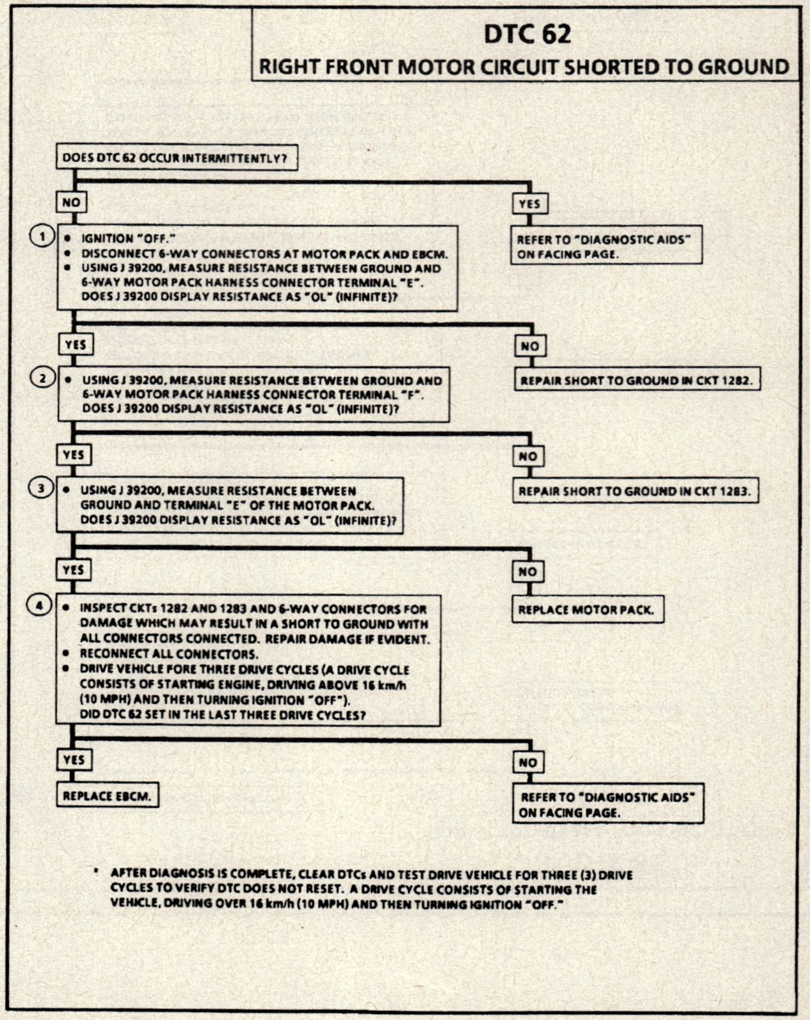

DIAGNOSTIC CHART DTC 63 — 1995
CAVALIER AND SUNFIRE

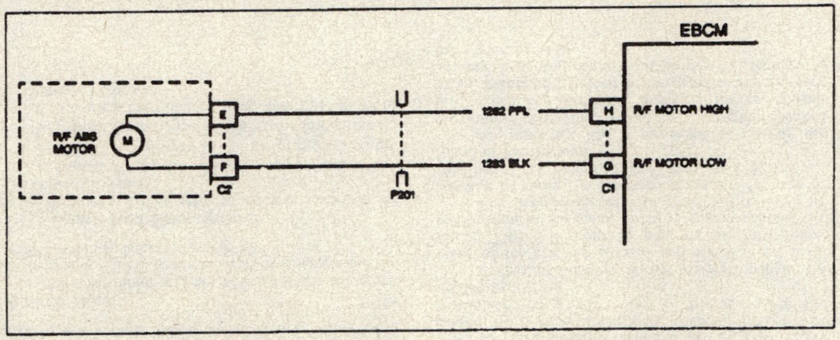

DTC 63
RIGHT FRONT ABS MOTOR CIRCUIT SHORTED TO BATTERY OR MOTOR SHORTED

Circuit Description:
This DTC identifies a motor circuit that is shorted to battery or a motor that has low or no resistance. This malfunction will not allow the motor to be controlled at the commanded current rate or will cause the motor to turn in the opposite direction or not at all.

DTC Will Set When:
DTC 63 can be set anytime. A malfunction exists if the EBCM detects an out of range voltage in either of the right front motor circuits indicating a circuit shorted to battery or a motor shorted.

Action Taken:
A malfunction DTC is stored. ABS is disabled and the "ABS" warning lamp is turned "ON."

DTC Chart Test Description:
Number(s) below refer to circled number(s) on the diagnostic chart.
1. Checks for a short to voltage in the motor high circuitry.
2. Checks for a short to voltage in the motor low circuitry.
3. Checks for a short between high and low motor circuits.
4. Checks for a motor that is internally shorted.
5. Checks for a short between the EMB and motor circuits.
6. Checks for an internal motor pack short between the EMB and motor.
7. Ensures malfunction was not due to physical damage of the circuitry.

Diagnostic Aids:
Using Tech 1, select manual control function, and exercise motor movement of effected channel in both directions while applying light pressure on the brake pedal.

If erratic or "jumpy" brake pedal movement is detected while performing an "apply" or "release" function of the motor, an intermittent malfunction may be indicated.

An "intermittent" malfunction is most likely caused by a poor connection, rubbed through wire insulation, or a wire that is broken inside the insulation.

If the malfunction is not current, wiggle the wires of the effected channel and check if the DTC resets. This will help to pinpoint an intermittent malfunction in the motor circuitry or connections.

The frequency of the malfunction can be checked by using the enhanced diagnostic function of the Tech 1, as described in "Enhanced Diagnostics," found in this section. If DTC 41 sets as a current DTC and DTC 63 is infrequent history DTC, see DTC 41 for diagnosis.

Any circuitry, that is suspected as causing the intermittent complaint, should be thoroughly checked for backed out terminals, improper mating, broken locks, improperly formed or damaged terminals, poor terminal to wiring connections or physical damage to the wiring harness.

DIAGNOSTIC CHART DTC 63 — 1995
CAVALIER AND SUNFIRE

DTC 63
RIGHT FRONT MOTOR CIRCUIT SHORTED TO BATTERY OR MOTOR SHORTED

NOTICE: J 39200 TEST LEADS MUST BE "ZEROED" PRIOR TO MAKING ANY RESISTANCE MEASUREMENTS. REFER TO J 39200 USER'S MANUAL.

DOES DTC 63 OCCUR INTERMITTENTLY?
- YES → REFER TO "DIAGNOSTIC AIDS" ON FACING PAGE.
- NO ↓

(1)
NOTICE: THIS STEP MAY CAUSE ADDITIONAL DTCs TO SET.
- IGNITION "OFF."
- DISCONNECT 6-WAY CONNECTORS AT MOTOR PACK AND EBCM.
- IGNITION "ON."
- USING J 39200, MEASURE VOLTAGE BETWEEN GROUND AND 6-WAY MOTOR PACK HARNESS CONNECTOR TERMINAL "E". IS VOLTAGE 1.0V OR LESS?

- NO → REPAIR SHORT TO VOLTAGE IN CKT 1282.
- YES ↓

(2)
- USING J 39200, MEASURE VOLTAGE BETWEEN GROUND AND 6-WAY MOTOR PACK HARNESS CONNECTOR TERMINAL "F". IS VOLTAGE 1.0V OR LESS?

- NO → REPAIR SHORT TO VOLTAGE IN CKT 1283.
- YES ↓

(3)
- IGNITION "OFF."
- USING J 39200, MEASURE RESISTANCE BETWEEN 6-WAY MOTOR PACK HARNESS CONNECTOR TERMINALS "E" AND "F". DOES J 39200 DISPLAY RESISTANCE AS "OL" (INFINITE)?

- NO → REPAIR SHORT BETWEEN CKTs 1282 AND 1283.
- YES ↓

(4)
- USING J 39200, MEASURE RESISTANCE BETWEEN MOTOR PACK TERMINALS "E" AND "F". IS RESISTANCE 0.4Ω OR GREATER?

- NO → REPLACE MOTOR PACK.
- YES ↓

(5)
- DISCONNECT 3-WAY MOTOR PACK CONNECTOR.
- USING J 39200, MEASURE RESISTANCE BETWEEN 3-WAY MOTOR PACK HARNESS CONNECTOR TERMINAL "B" AND 6-WAY MOTOR PACK HARNESS CONNECTOR TERMINAL "E" THEN "F". DID J 39200 DISPLAY "OL" (INFINITE) FOR EACH MEASUREMENT?

- NO → REPAIR SHORT BETWEEN CKT 850 AND MOTOR CIRCUIT WHICH MEASURED RESISTANCE OTHER THAN "OL" (INFINITE).
- YES ↓

(6)
- USING J 39200, MEASURE RESISTANCE BETWEEN 3-WAY MOTOR PACK CONNECTOR TERMINAL "B" AND 6-WAY MOTOR PACK CONNECTOR TERMINAL "E". DID J 39200 DISPLAY RESISTANCE AS "OL" (INFINITE)?

- NO → REPLACE MOTOR PACK.
- YES ↓

(7)
- INSPECT CKTs 1282 AND 1283 AND 6-WAY CONNECTORS FOR DAMAGE WHICH MAY RESULT IN A SHORT TO VOLTAGE WITH ALL CONNECTORS CONNECTED.
- REPAIR DAMAGE IF EVIDENT.
- RECONNECT ALL CONNECTORS.
- DRIVE VEHICLE FOR THREE DRIVE CYCLES (A DRIVE CYCLE CONSISTS OF STARTING ENGINE, DRIVE ABOVE 16 km/h (10 MPH) AND THEN TURNING IGNITION "OFF"). DID DTC 63 SET IN THE LAST THREE DRIVE CYCLES?

- YES → REPLACE EBCM.
- NO → REFER TO "DIAGNOSTIC AIDS" ON FACING PAGE.

- AFTER DIAGNOSIS IS COMPLETE, CLEAR DTCs AND TEST DRIVE VEHICLE FOR THREE (3) DRIVE CYCLES TO VERIFY DTC DOES NOT RESET. A DRIVE CYCLE CONSISTS OF STARTING THE VEHICLE, DRIVING OVER 16 km/h (10 MPH) AND THEN TURNING IGNITION "OFF."

DIAGNOSTIC CHART DTC 64 — 1995
CAVALIER AND SUNFIRE

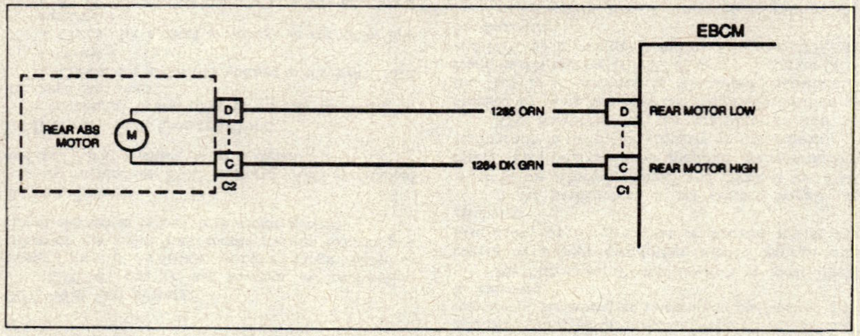

DTC 64
REAR ABS MOTOR CIRCUIT OPEN

Circuit Description:
This DTC identifies a motor that cannot be energized due to an open in its circuitry. This malfunction will not allow proper rear ABS operation.

DTC Will Set When:
DTC 64 can be set only when the motor is commanded "OFF." A malfunction exists if the EBCM detects an out of range voltage in either of the rear ABS motor circuits indicating an open circuit.

Action Taken:
An open motor will not activate when requested. A malfunction DTC is stored, ABS is disabled and the "ABS" warning lamp is turned "ON."

DTC Chart Test Description:
Number(s) below refer to circled number(s) on the diagnostic chart.
1. Checks for proper resistance of the motor.
2. Checks for an open in the motor high circuitry.
3. Checks for an open in the motor low circuitry.
4. Ensures malfunction was not due to poor terminal contact.

Diagnostic Aids:
Using Tech 1, select manual control function, and exercise motor movement of effected channel in both directions while applying light pressure on the brake pedal.

If erratic or "jumpy" brake pedal movement is detected while performing an "apply" or "release" function of the motor, an intermittent malfunction may be indicated.

An "intermittent" malfunction is most likely caused by a poor connection, rubbed through wire insulation, or a wire that is broken inside the insulation.

If the malfunction is not current, wiggle the wires of the effected channel and check if the DTC resets. This will help to pinpoint an intermittent malfunction in the motor circuitry or connections.

The frequency of the malfunction can be checked by using the enhanced diagnostic function of the Tech 1, as described in "Enhanced Diagnostics," found in this section.

Any circuitry, that is suspected as causing the intermittent complaint, should be thoroughly checked for backed out terminals, improper mating, broken locks, improperly formed or damaged terminals, poor terminal to wiring connections or physical damage to the wiring harness.

DIAGNOSTIC CHART DTC 64 — 1995
CAVALIER AND SUNFIRE

DTC 64
LEFT FRONT MOTOR CIRCUIT OPEN

DOES DTC 64 OCCUR INTERMITTENTLY?

- YES: REFER TO "DIAGNOSTIC AIDS" ON FACING PAGE.
- NO:

(1)
- IGNITION "OFF."
- DISCONNECT 6- WAY MOTOR PACK CONNECTOR.
- USING J 39200, MEASURE RESISTANCE ACROSS MOTOR PACK TERMINALS "C" AND "D".

IS RESISTANCE 2000Ω OR LESS?

- NO: REPLACE MOTOR PACK.
- YES:

(2)
- DISCONNECT 6-WAY EBCM CONNECTOR.
- USING J 39200, MEASURE RESISTANCE BETWEEN 6-WAY EBCM HARNESS CONNECTOR TERMINAL "C" AND 6-WAY MOTOR PACK HARNESS CONNECTOR TERMINAL "C".

IS RESISTANCE 2.0Ω OR LESS?

- NO: REPAIR OPEN OR HIGH RESISTANCE IN CKT 1284.
- YES:

(3)
- USING J 39200, MEASURE RESISTANCE BETWEEN 6-WAY EBCM HARNESS CONNECTOR TERMINAL "D" AND 6-WAY MOTOR PACK HARNESS CONNECTOR TERMINAL "D".

IS RESISTANCE 2.0Ω OR LESS?

- NO: REPAIR OPEN OR HIGH RESISTANCE IN CKT 1285.
- YES:

(4)
- INSPECT 6-WAY MOTOR PACK AND EBCM HARNESS CONNECTORS FOR EVIDENCE OF POOR TERMINAL CONTACT OR CORROSION.
- RECONNECT ALL CONNECTORS.
- DRIVE VEHICLE FOR THREE DRIVE CYCLES (A DRIVE CYCLE CONSISTS OF STARTING ENGINE, DRIVING ABOVE 16 km/h (10 MPH) AND THEN TURNING IGNITION "OFF").

DID DTC 64 SET IN THE LAST THREE DRIVE CYCLES?

- YES: REPLACE EBCM.
- NO: REFER TO "DIAGNOSTIC AIDS" ON FACING PAGE.

AFTER DIAGNOSIS IS COMPLETE, CLEAR DTCs AND TEST DRIVE VEHICLE FOR THREE (3) DRIVE CYCLES TO VERIFY DTC DOES NOT RESET. A DRIVE CYCLE CONSISTS OF STARTING THE VEHICLE, DRIVING OVER 16 km/h (10 MPH) AND THEN TURNING IGNITION "OFF."

DIAGNOSTIC CHART DTC 65 — 1995
CAVALIER AND SUNFIRE

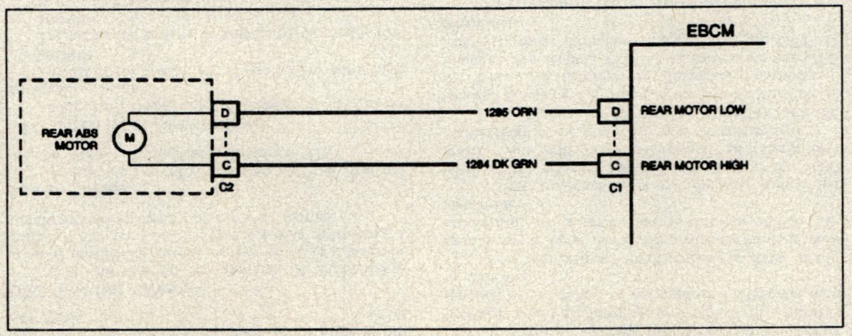

DTC 65
REAR ABS MOTOR CIRCUIT SHORTED TO GROUND

Circuit Description:

This DTC identifies a motor circuit that is shorted to ground. This malfunction will not allow the motor to be controlled at the commanded current rate or will cause the driver circuit to allow current directly to ground.

DTC Will Set When:

DTC 65 can be set anytime. A malfunction exists if the EBCM detects an out of range voltage in either of the rear ABS motor circuits indicating a circuit shorted to ground.

Action Taken:

A malfunction DTC is stored, ABS is disabled and the "ABS" warning lamp is turned "ON."

DTC Chart Test Description:

Number(s) below refer to circled number(s) on the diagnostic chart.

1. Checks for a short to ground in the motor high circuitry.
2. Checks for a short to ground in the motor low circuitry.
3. Checks for a motor that is internally shorted to ground.
4. Ensures malfunction was not due to physical damage of the circuitry.

Diagnostic Aids:

Using Tech 1, select manual control function, and exercise motor movement of effected channel in both directions while applying light pressure on the brake pedal.

If erratic or "jumpy" brake pedal movement is detected while performing an "apply" or "release" function of the motor, an intermittent malfunction may be indicated.

An "intermittent" malfunction is most likely caused by a poor connection, rubbed through wire insulation, or a wire that is broken inside the insulation.

If the malfunction is not current, wiggle the wires of the effected channel and check if the DTC resets. This will help to pinpoint an intermittent malfunction in the motor circuitry or connections.

The frequency of the problem can be checked by using the enhanced diagnostic function of the Tech 1, as described in "Enhanced Diagnostics," found in this section.

Any circuitry, that is suspected as causing the intermittent complaint, should be thoroughly checked for backed out terminals, improper mating, broken locks, improperly formed or damaged terminals, poor terminal to wiring connections or physical damage to the wiring harness.

DIAGNOSTIC CHART DTC 65 — 1995
CAVALIER AND SUNFIRE

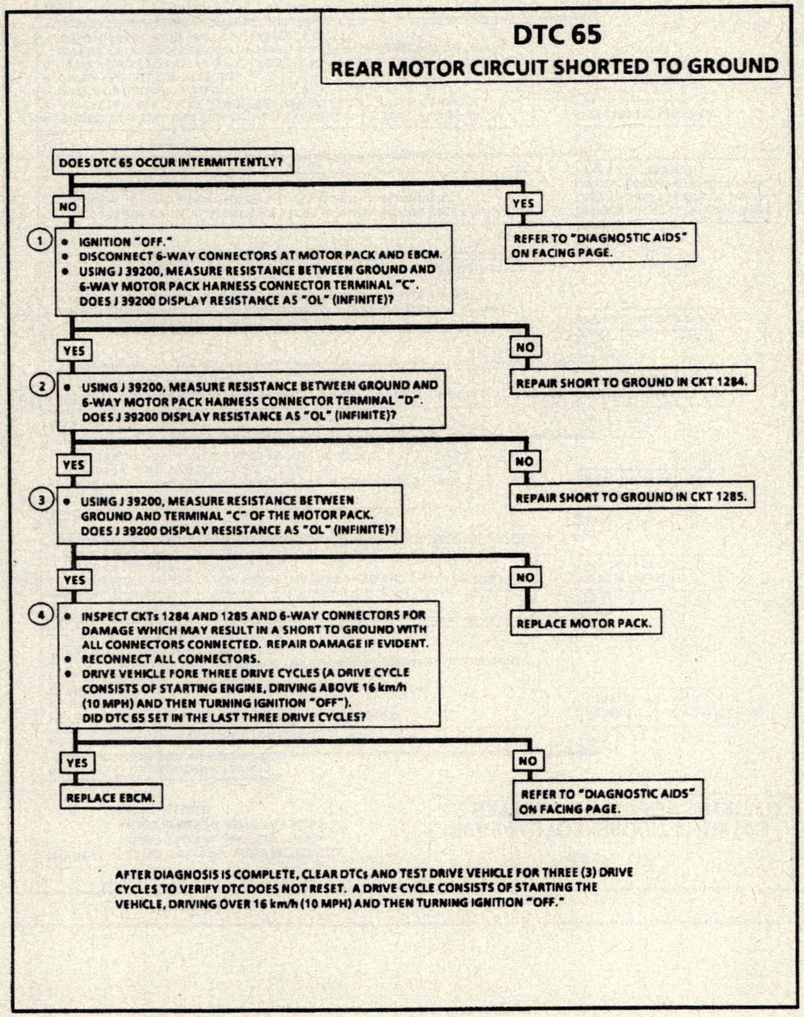

DIAGNOSTIC CHART DTC 66 — 1995 CAVALIER AND SUNFIRE

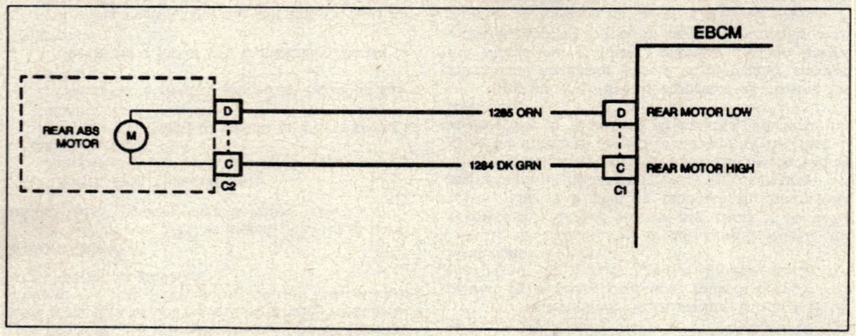

DTC 66
REAR ABS MOTOR CIRCUIT SHORTED TO BATTERY OR MOTOR SHORTED

Circuit Description:
This DTC identifies a motor circuit that is shorted to battery or a motor that has low or zero resistance. This malfunction will not allow the motor to be controlled at the commanded current rate or will cause the motor to turn in the opposite direction or not turn at all.

DTC Will Set When:
DTC 66 can be set anytime. A malfunction exists if the EBCM detects an out of range voltage in either of the rear ABS motor circuits indicating a circuit shorted to battery or a motor shorted.

Action Taken:
A malfunction DTC is stored, ABS is disabled and the "ABS" warning lamp is turned "ON."

DTC Chart Test Description:
Number(s) below refer to circled number(s) on the diagnostic chart.
1. Checks for a short to voltage in the motor high circuitry.
2. Checks for a short to voltage in the motor low circuitry.
3. Checks for a short between high and low motor circuits.
4. Checks for a motor that is internally shorted.
5. Checks for a short between the EMB and motor circuits.
6. Checks for an internal motor pack short between the EMB and motor.
7. Ensures malfunction was not due to physical damage of the circuitry.

Diagnostic Aids:
Using Tech 1, select manual control function, and exercise motor movement of effected channel in both directions while applying light pressure on the brake pedal.

If erratic or "jumpy" brake pedal movement is detected while performing an "apply" or "release" function of the motor, an intermittent malfunction may be indicated.

An "intermittent" malfunction is most likely caused by a poor connection, rubbed through wire insulation, or a wire that is broken inside the insulation.

If the malfunction is not current, wiggle the wires of the effected channel and check if the DTC resets. This will help to pinpoint an intermittent malfunction in the motor circuitry or connections.

The frequency of the malfunction can be checked by using the enhanced diagnostic function of the Tech 1, as described in "Enhanced Diagnostics," found in this section. If DTC 42 sets as current DTC and DTC 66 is infrequent history DTC, see DTC 42 diagnostics.

Any circuitry, that is suspected as causing the intermittent complaint, should be thoroughly checked for backed out terminals, improper mating, broken locks, improperly formed or damaged terminals, poor terminal to wiring connections or physical damage to the wiring harness.

DIAGNOSTIC CHART DTC 66 — 1995 CAVALIER AND SUNFIRE

DTC 66
REAR MOTOR CIRCUIT SHORTED TO BATTERY OR MOTOR SHORTED

NOTICE: J 39200 TEST LEADS MUST BE "ZEROED" PRIOR TO MAKING ANY RESISTANCE MEASUREMENTS. REFER TO J 39200 USER'S MANUAL.

DOES DTC 66 OCCUR INTERMITTENTLY?
- YES → REFER TO "DIAGNOSTIC AIDS" ON FACING PAGE.
- NO →

(1) NOTICE: THIS STEP MAY CAUSE ADDITIONAL DTCs TO SET.
- IGNITION "OFF."
- DISCONNECT 6-WAY CONNECTORS AT MOTOR PACK AND EBCM.
- IGNITION "ON."
- USING J 39200, MEASURE VOLTAGE BETWEEN GROUND AND 6-WAY MOTOR PACK HARNESS CONNECTOR TERMINAL "C".

IS VOLTAGE 1.0 VOLT OR LESS?
- NO → REPAIR SHORT TO VOLTAGE IN CKT 1284.
- YES →

(2)
- USING J 39200, MEASURE VOLTAGE BETWEEN GROUND AND 6-WAY MOTOR PACK HARNESS CONNECTOR TERMINAL "D".

IS VOLTAGE 1.0 VOLT OR LESS?
- NO → REPAIR SHORT TO VOLTAGE IN CKT 1285.
- YES →

(3)
- IGNITION "OFF."
- USING J 39200, MEASURE RESISTANCE BETWEEN 6-WAY MOTOR PACK HARNESS CONNECTOR TERMINALS "C" AND "D".

DOES J 39200 DISPLAY RESISTANCE AS "OL" (INFINITE)?
- NO → REPAIR SHORT BETWEEN CKTs 1284 AND 1285.
- YES →

(4)
- USING J 39200, MEASURE RESISTANCE BETWEEN MOTOR PACK TERMINALS "C" AND "D".

IS RESISTANCE 0.4Ω OR GREATER?
- NO → REPLACE MOTOR PACK.
- YES →

(5)
- DISCONNECT 3-WAY MOTOR PACK CONNECTOR.
- USING J 39200, MEASURE RESISTANCE BETWEEN 3-WAY MOTOR PACK HARNESS CONNECTOR TERMINAL "B" AND 6-WAY MOTOR PACK HARNESS CONNECTOR TERMINAL "C" THEN "D".

DID J 39200 DISPLAY RESISTANCE AS "OL" (INFINITE) FOR EACH MEASUREMENT?
- NO → REPAIR SHORT BETWEEN CIRCUIT 850 AND MOTOR CIRCUIT WHICH MEASURED RESISTANCE OTHER THEN "OL" (INFINITE).
- YES →

(6)
- USING J 39200, MEASURE RESISTANCE BETWEEN 3-WAY MOTOR PACK CONNECTOR TERMINAL "B" AND 6-WAY MOTOR PACK CONNECTOR TERMINAL "C".

DOES J 39200 DISPLAY "OL" (INFINITE)?
- NO → REPLACE MOTOR PACK.
- YES →

(7)
- INSPECT CKTs 1284 AND 1285 AND 6-WAY CONNECTORS FOR DAMAGE WHICH MAY RESULT IN A SHORT TO VOLTAGE WITH ALL CONNECTORS CONNECTED.
- REPAIR DAMAGE IF EVIDENT.
- RECONNECT ALL CONNECTORS.
- DRIVE VEHICLE FOR THREE DRIVE CYCLES (A DRIVE CYCLE CONSISTS OF STARTING ENGINE, DRIVE ABOVE 16 km/h (10 MPH) AND THEN TURNING IGNITION "OFF").

DID DTC 66 SET IN THE LAST THREE DRIVE CYCLES?
- YES → REPLACE EBCM.
- NO → REFER TO "DIAGNOSTIC AIDS" ON FACING PAGE.

* AFTER DIAGNOSIS IS COMPLETE, CLEAR DTCs AND TEST DRIVE VEHICLE FOR THREE (3) DRIVE CYCLES TO VERIFY DTC DOES NOT RESET. A DRIVE CYCLE CONSISTS OF STARTING THE VEHICLE, DRIVING OVER 16 km/h (10 MPH) AND THEN TURNING IGNITION "OFF."

DIAGNOSTIC CHART DTC 67 — 1995 CAVALIER AND SUNFIRE

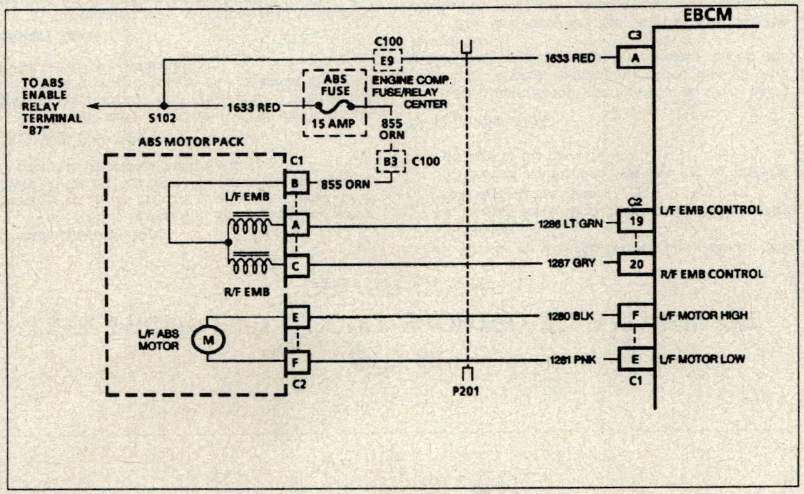

DTC 67

LEFT FRONT EMB CIRCUIT OPEN OR SHORTED TO GROUND

Circuit Description:

This DTC is designed to detect shorts to ground and opens in the EMB release circuit. A short to ground between the EMB and the ABS EBCM will cause the EMB to be energized indefinitely at an uncontrolled current, causing the EMB to be destroyed.

DTC Will Set When:

DTC 67 can be set only when the left front EMB is commanded "OFF" (not released). A malfunction exists if the EBCM detects the left front EMB control circuit is constantly at ground.

Action Taken:

A malfunction DTC is stored, ABS is disabled and the "ABS" warning lamp is turned "ON."

DTC Chart Test Description:

Number(s) below refer to circled number(s) on the diagnostic chart.

1. Checks for a short to ground in the EMB control circuitry.
2. Checks for an open in the EMB control circuitry.
3. Checks for an open in the EMB battery feed circuit.
4. Ensures the EMB is not internally shorted to ground.
5. Ensures EMB is not internally open.
6. Checks for short to ground in EMB battery feed circuit.
7. Ensures malfunction was not due to physical damage of the circuitry.

Diagnostic Aids:

An "intermittent" malfunction is most likely caused by a poor connection, rubbed through wire insulation, or a wire that is broken inside the insulation.

The frequency of the malfunction can be checked by using the enhanced diagnostic function of the Tech 1, as described in "Enhanced Diagnostics," found in this section.

Any circuitry, that is suspected as causing the intermittent complaint, should be thoroughly checked for backed out terminals, improper mating, broken locks, improperly formed or damaged terminals, poor terminal to wiring connections or physical damage to the wiring harness.

DIAGNOSTIC CHART DTC 67 — 1995 CAVALIER AND SUNFIRE

DTC 67
LEFT FRONT EMB CIRCUIT OPEN OR SHORTED TO GROUND

(1)
- IGNITION "OFF."
- DISCONNECT 24-WAY AND 2-WAY EBCM HARNESS CONNECTORS.
- DISCONNECT 3-WAY MOTOR PACK HARNESS CONNECTOR.
- USING J 39200, MEASURE RESISTANCE BETWEEN GROUND AND 24-WAY EBCM HARNESS CONNECTOR TERMINAL "19". DOES J 39200 DISPLAY RESISTANCE AS "OL" (INFINITE)?

YES → (2) | NO → REPAIR SHORT TO GROUND IN CKT 1286.

(2)
- USING J 39200, MEASURE RESISTANCE BETWEEN 3-WAY MOTOR PACK HARNESS CONNECTOR TERMINAL "A" AND 24-WAY EBCM HARNESS CONNECTOR TERMINAL "19". IS RESISTANCE 2.0Ω OR LESS?

YES → (3) | NO → REPAIR OPEN OR HIGH RESISTANCE IN CKT 1286.

(3)
- USING J 39200, MEASURE RESISTANCE BETWEEN 2-WAY EBCM HARNESS CONNECTOR TERMINAL "A" AND 3-WAY MOTOR PACK HARNESS CONNECTOR TERMINAL "B". IS RESISTANCE 2.0Ω OR LESS?

YES → (4) | NO → REPAIR OPEN OR HIGH RESISTANCE IN CKT 850.

(4)
- USING J 39200, MEASURE RESISTANCE BETWEEN MOTOR PACK TERMINAL "A" AND GROUND. DOES J 39200 DISPLAY RESISTANCE AS "OL" (INFINITE)?

YES → (5) | NO → REPLACE MOTOR PACK.

(5)
- USING J 39200, MEASURE RESISTANCE BETWEEN MOTOR PACK TERMINALS "A" AND "B". IS RESISTANCE 2.0Ω OR LESS?

YES → (6) | NO → REPLACE MOTOR PACK.

(6)
- USING J 39200, MEASURE RESISTANCE BETWEEN GROUND AND 2-WAY EBCM HARNESS CONNECTOR TERMINAL "A". DOES J 39200 DISPLAY RESISTANCE AS "OL" (INFINITE)?

YES → (7) | NO → REPAIR SHORT TO GROUND IN CKT 850.

(7)
- INSPECT 2-WAY AND 24-WAY EBCM AND 3-WAY MOTOR PACK CONNECTIONS FOR POOR TERMINAL CONTACT. REPLACE TERMINALS THAT EXHIBIT POOR CONTACT.
- INSPECT CKT 850, CKT 1286, 3-WAY MOTOR PACK AND EBCM HARNESS CONNECTORS FOR PHYSICAL DAMAGE WHICH MAY RESULT IN A SHORT TO GROUND WITH ALL CONNECTORS CONNECTED. REPAIR DAMAGE IF EVIDENT.
- RECONNECT ALL CONNECTORS.
- TEST DRIVE VEHICLE FOR THREE DRIVE CYCLES. A DRIVE CYCLE CONSISTS OF STARTING THE ENGINE, DRIVING OVER 16 km/h (10 MPH) AND THEN KEYING DOWN. DID DTC 67 SET IN THE LAST THREE DRIVE CYCLES?

YES → REPLACE EBCM. | NO → MALFUNCTION IS INTERMITTENT. REFER TO "DIAGNOSTIC AIDS" ON FACING PAGE.

- AFTER DIAGNOSIS IS COMPLETE, CLEAR DTCs AND TEST DRIVE VEHICLE FOR THREE (3) DRIVE CYCLES TO VERIFY DTC DOES NOT RESET. A DRIVE CYCLE CONSISTS OF STARTING THE VEHICLE, DRIVING OVER 16km/h (10 MPH) AND THEN KEYING DOWN.

DIAGNOSTIC CHART DTC 68 — 1995 CAVALIER AND SUNFIRE

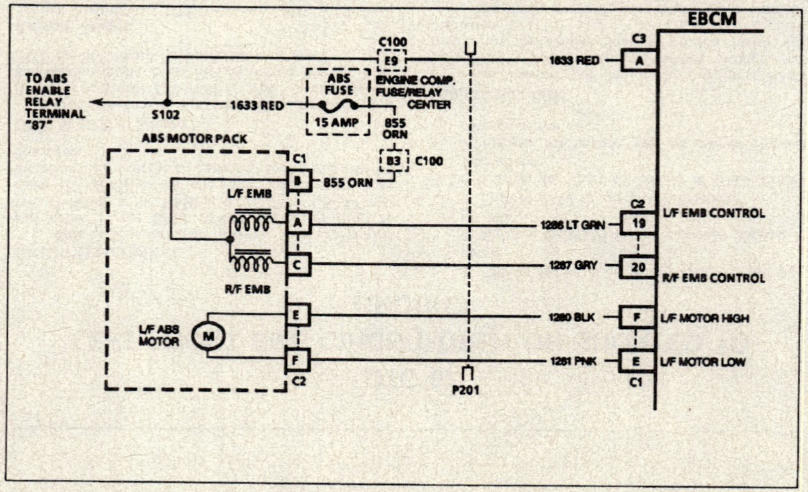

DTC 68
LEFT FRONT EMB CIRCUIT SHORTED TO BATTERY OR DRIVER OPEN

Circuit Description:
This DTC is designed to detect shorts to battery, shorts in the EMB release circuit, or an open EMB driver. These malfunctions will not allow the EMB to be energized (released) when needed.

DTC Will Set When:
DTC 68 can be set only when the left front EMB is commanded "ON" (released). A malfunction exists if an excessive voltage drop is across the motor driver for the left front EMB.

Action Taken:
A malfunction DTC is stored, ABS is disabled and the "ABS" warning lamp is turned "ON."

DTC Chart Test Description:
Number(s) below refer to circled number(s) on the diagnostic chart.
1. Checks to see if the malfunction is currently present.
2. Identifies a short to voltage in the EMB control circuitry.
3. Checks for an EMB that has lower than normal resistance.
4. Checks for a short between the EMB battery feed and control circuits.
5. Ensures malfunction was not due to physical damage of the circuitry.

Diagnostic Aids:
An "intermittent" malfunction is most likely caused by a poor connection, rubbed through wire insulation, or a wire that is broken inside the insulation.

The frequency of the malfunction can be checked by using the enhanced diagnostic function of the Tech 1, as described in "Enhanced Diagnostics," found in this section.

Any circuitry, that is suspected as causing the intermittent complaint, should be thoroughly checked for backed out terminals, improper mating, broken locks, improperly formed or damaged terminals, poor terminal to wiring connections or physical damage to the wiring harness.

DIAGNOSTIC CHART DTC 68 — 1995 CAVALIER AND SUNFIRE

DTC 68
LEFT FRONT EMB CIRCUIT SHORTED TO BATTERY OR DRIVER OPEN

*NOTICE: J 39200 TEST LEADS MUST BE "ZEROED" PRIOR TO MAKING ANY RESISTANCE MEASUREMENTS. REFER TO J 39200 USER'S MANUAL.

(1)
- IGNITION "ON."
- USING TECH 1 ABS TESTS, SELECT MANUAL CONTROL FUNCTION.
- SELECT LEFT FRONT MOTOR APPLY FUNCTION AND APPLY MOTOR.
- APPLY FIRM PRESSURE ON BRAKE PEDAL.
- SELECT LEFT FRONT EMB RELEASE FUNCTION AND RELEASE EMB. DID BRAKE PEDAL FALL WITH INCREASING PEDAL FORCE?

YES: MALFUNCTION IS NOT PRESENT AT THIS TIME. REFER TO "DIAGNOSTIC AIDS" ON FACING PAGE.

NO:

(2) NOTICE: THIS STEP MAY CAUSE ADDITIONAL DTCs TO SET.
- IGNITION "OFF."
- DISCONNECT 3-WAY MOTOR PACK CONNECTOR.
- DISCONNECT 24-WAY EBCM CONNECTOR.
- IGNITION "ON."
- USING J 39200, MEASURE VOLTAGE BETWEEN GROUND AND 3-WAY MOTOR PACK HARNESS CONNECTOR TERMINAL "A". IS VOLTAGE 1.0V OR LESS?

NO: REPAIR SHORT TO VOLTAGE IN CKT 1286.

YES:

(3)
- IGNITION "OFF."
- USING J 39200, MEASURE RESISTANCE BETWEEN MOTOR PACK TERMINALS "A" AND "B". IS RESISTANCE 0.4Ω OR GREATER?

NO: REPLACE MOTOR PACK.

YES:

(4)
- USING J 39200, MEASURE THE RESISTANCE BETWEEN 3-WAY MOTOR PACK HARNESS CONNECTOR TERMINALS "B" AND "A". DOES J 39200 DISPLAY RESISTANCE AS "OL" (INFINITE)?

NO: REPAIR SHORT BETWEEN CKT(s) 1286 AND 850.

YES:

(5)
- INSPECT CKT 1286, 3-WAY MOTOR PACK AND 24-WAY EBCM CONNECTORS FOR DAMAGE WHICH MAY RESULT IN A SHORT TO VOLTAGE WITH ALL CONNECTORS CONNECTED. REPAIR DAMAGE IF EVIDENT.
- RECONNECT ALL CONNECTORS.
- DRIVE VEHICLE FOR THREE DRIVE CYCLES (A DRIVE CYCLE CONSISTS OF STARTING ENGINE, DRIVING ABOVE 16 km/h (10 MPH) AND THEN TURNING IGNITION "OFF").
- IGNITION "ON."
- USING TECH 1, SELECT "F1: DTC HISTORY." DID DTC 68 SET IN LAST THREE DRIVE CYCLES?

YES: REPLACE EBCM.

NO: MALFUNCTION IS INTERMITTENT. REFER TO "DIAGNOSTIC AIDS" ON FACING PAGE.

- AFTER DIAGNOSIS IS COMPLETE, CLEAR DTCs AND TEST DRIVE VEHICLE FOR THREE (3) DRIVE CYCLES TO VERIFY DTC DOES NOT RESET. A DRIVE CYCLE CONSISTS OF STARTING THE VEHICLE, DRIVING OVER 16 km/h (10 MPH) AND THEN KEYING DOWN.

DIAGNOSTIC CHART DTC 71 — 1995 CAVALIER AND SUNFIRE

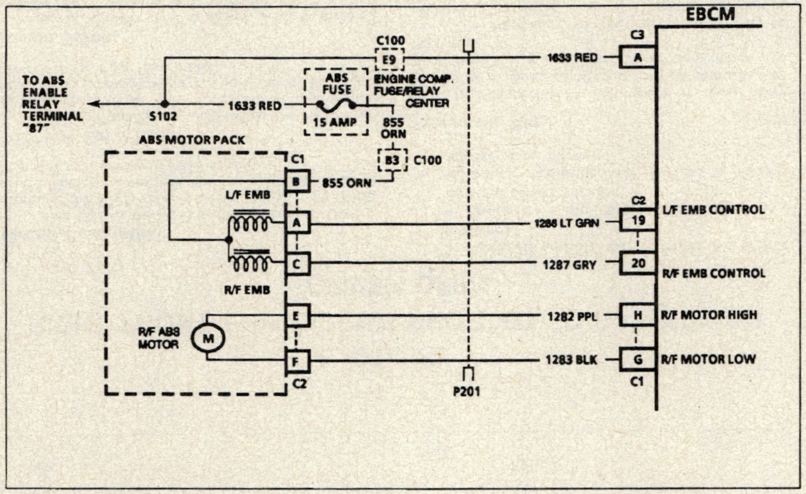

DTC 71
RIGHT FRONT EMB CIRCUIT OPEN OR SHORTED TO GROUND

Circuit Description:

This DTC is designed to detect shorts to ground and opens in the EMB release circuit. A short to ground between the EMB and the ABS EBCM will cause the EMB to be energized indefinitely at an uncontrolled current causing the EMB to be destroyed.

DTC Will Set When:

DTC 71 can be set only when the right front EMB is commanded "OFF" (not released). A malfunction exists if the EBCM detects the right front EMB control circuit is constantly at ground.

Action Taken:

A malfunction DTC is stored, ABS is disabled and the "ABS" warning lamp is turned "ON."

DTC Chart Test Description:

Number(s) below refer to circled number(s) on the diagnostic chart

1. Checks for a short to ground in the EMB control circuitry.
2. Checks for an open in the EMB control circuitry.
3. Checks for an open in the EMB battery feed circuit.
4. Ensures the EMB is not internally shorted to ground.
5. Ensures EMB is not internally open.
6. Checks for short to ground in EMB battery feed circuit.
7. Ensures malfunction was not due to physical damage of the circuitry.

Diagnostic Aids:

An "intermittent" malfunction is most likely caused by a poor connection, rubbed through wire insulation, or a wire that is broken inside the insulation.

The frequency of the malfunction can be checked by using the enhanced diagnostic function of the Tech 1, as described in "Enhanced Diagnostics," found in this section.

Any circuitry, that is suspected as causing the intermittent complaint, should be thoroughly checked for backed out terminals, improper mating, broken locks, improperly formed or damaged terminals, poor terminal to wiring connections or physical damage to the wiring harness.

DIAGNOSTIC CHART DTC 71 — 1995 CAVALIER AND SUNFIRE

DTC 71
RIGHT FRONT EMB CIRCUIT OPEN OR SHORTED TO GROUND

(1)
- IGNITION "OFF."
- DISCONNECT 24-WAY AND 2-WAY EBCM HARNESS CONNECTORS.
- DISCONNECT 3-WAY MOTOR PACK HARNESS CONNECTOR.
- USING J 39200, MEASURE RESISTANCE BETWEEN GROUND AND 24-WAY EBCM HARNESS CONNECTOR TERMINAL "20". DOES J 39200 DISPLAY RESISTANCE AS "OL" (INFINITE)?

NO: REPAIR SHORT TO GROUND IN CKT 1287.

YES:

(2)
- USING J 39200, MEASURE RESISTANCE BETWEEN 3-WAY MOTOR PACK HARNESS CONNECTOR TERMINAL "C" AND 24-WAY EBCM HARNESS CONNECTOR TERMINAL "20". IS RESISTANCE 2.0Ω OR LESS?

NO: REPAIR OPEN OR HIGH RESISTANCE IN CKT 1287.

YES:

(3)
- USING J 39200, MEASURE RESISTANCE BETWEEN 2-WAY EBCM HARNESS CONNECTOR TERMINAL "A" AND 3-WAY MOTOR PACK HARNESS CONNECTOR TERMINAL "B". IS RESISTANCE 2.0Ω OR LESS?

NO: REPAIR OPEN OR HIGH RESISTANCE IN CKT 850.

YES:

(4)
- USING J 39200, MEASURE RESISTANCE BETWEEN MOTOR PACK TERMINAL "C" AND GROUND. DOES J 39200 DISPLAY RESISTANCE AS "OL" (INFINITE)?

NO: REPLACE MOTOR PACK.

YES:

(5)
- USING J 39200, MEASURE RESISTANCE BETWEEN MOTOR PACK TERMINALS "C" AND "B". IS RESISTANCE 2.0Ω OR LESS?

NO: REPLACE MOTOR PACK.

YES:

(6)
- USING J 39200, MEASURE RESISTANCE BETWEEN GROUND AND 2-WAY EBCM HARNESS CONNECTOR TERMINAL "A". DOES J 39200 DISPLAY RESISTANCE AS "OL" (INFINITE)?

NO: REPAIR SHORT TO GROUND IN CKT 850.

YES:

(7)
- INSPECT 2-WAY AND 24-WAY EBCM AND 3-WAY MOTOR PACK CONNECTIONS FOR POOR TERMINAL CONTACT. REPLACE TERMINALS THAT EXHIBIT POOR CONTACT.
- INSPECT CKT 850, CKT 1287, 3-WAY MOTOR PACK AND EBCM HARNESS CONNECTORS FOR PHYSICAL DAMAGE WHICH MAY RESULT IN A SHORT TO GROUND WITH ALL CONNECTORS CONNECTED. REPAIR DAMAGE IF EVIDENT.
- RECONNECT ALL CONNECTORS.
- TEST DRIVE VEHICLE FOR THREE DRIVE CYCLES. A DRIVE CYCLE CONSISTS OF STARTING THE ENGINE, DRIVING OVER 16 KM/H (10 MPH) AND THEN KEYING DOWN. DID DTC 71 SET THE LAST THREE DRIVE CYCLES?

YES: REPLACE EBCM.

NO: MALFUNCTION IS INTERMITTENT. REFER TO "DIAGNOSTIC AIDS" ON FACING PAGE.

- AFTER DIAGNOSIS IS COMPLETE, CLEAR DTCs AND TEST DRIVE VEHICLE FOR THREE (3) DRIVE CYCLES TO VERIFY DTC DOES NOT RESET. A DRIVE CYCLE CONSISTS OF STARTING THE VEHICLE, DRIVING OVER 16km/h (10 MPH) AND THEN KEYING DOWN.

DIAGNOSTIC CHART DTC 72 — 1995 CAVALIER AND SUNFIRE

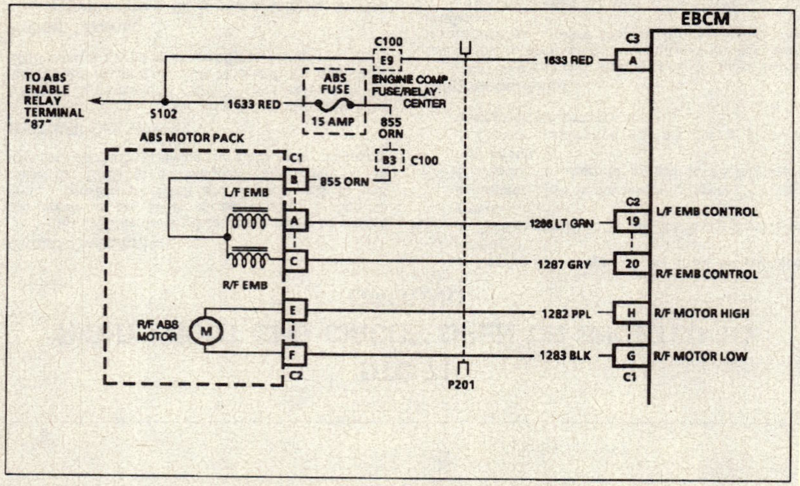

DTC 72
RIGHT FRONT EMB CIRCUIT SHORTED TO BATTERY OR DRIVER OPEN

Circuit Description:

This DTC is designed to detect shorts to battery, shorts in the EMB release circuit, or an open EMB driver. These malfunctions will not allow the EMB to be energized (released) when needed.

DTC Will Set When:

DTC 72 can be set only when the right front EMB is commanded "ON" (released). A malfunction exists if an excessive voltage drop is across the motor driver for the right front EMB.

Action Taken:

A malfunction DTC is stored, ABS is disabled and the "ABS" warning lamp is turned "ON."

DTC Chart Test Description:

Number(s) below refer to circled number(s) on the diagnostic chart.

1. Checks to see if the malfunction is currently present.
2. Identifies a short to voltage in the EMB control circuitry.
3. Checks for an EMB that has lower than normal resistance.
4. Checks for a short between the EMB battery feed and control circuits.
5. Ensures malfunction was not due to physical damage of the circuitry.

Diagnostic Aids:

An "intermittent" malfunction is most likely caused by a poor connection, rubbed through wire insulation, or a wire that is broken inside the insulation.

The frequency of the malfunction can be checked by using the enhanced diagnostic function of the Tech 1, as described in "Enhanced Diagnostics," found in this section.

Any circuitry, that is suspected as causing the intermittent complaint, should be thoroughly checked for backed out terminals, improper mating, broken locks, improperly formed or damaged terminals, poor terminal to wiring connections or physical damage to the wiring harness.

DIAGNOSTIC CHART DTC 72 — 1995 CAVALIER AND SUNFIRE

DTC 72
RIGHT FRONT EMB CIRCUIT SHORTED TO BATTERY OR DRIVER OPEN

*NOTICE: J 39200 TEST LEADS MUST BE "ZEROED" PRIOR TO MAKING ANY RESISTANCE MEASUREMENTS. REFER TO J 39200 USER'S MANUAL.

(1)
- IGNITION "ON."
- USING TECH 1 ABS TESTS, SELECT MANUAL CONTROL FUNCTION.
- SELECT RIGHT FRONT MOTOR APPLY FUNCTION AND APPLY MOTOR.
- APPLY FIRM PRESSURE ON BRAKE PEDAL.
- SELECT RIGHT FRONT EMB RELEASE FUNCTION AND RELEASE EMB. DID BRAKE PEDAL FALL WITH INCREASING PEDAL FORCE?

YES: MALFUNCTION IS NOT PRESENT AT THIS TIME. REFER TO "DIAGNOSTIC AIDS" ON FACING PAGE.

NO: go to (2)

(2) NOTICE: THIS STEP MAY CAUSE ADDITIONAL DTCs TO SET.
- IGNITION "OFF."
- DISCONNECT 3-WAY MOTOR PACK CONNECTOR.
- DISCONNECT 24-WAY EBCM CONNECTOR.
- IGNITION "ON."
- USING J 39200, MEASURE VOLTAGE BETWEEN GROUND AND 3-WAY MOTOR PACK HARNESS CONNECTOR TERMINAL "C". IS VOLTAGE 1.0V OR LESS?

NO: REPAIR SHORT TO VOLTAGE IN CKT 1287.

YES: go to (3)

(3)
- IGNITION "OFF."
- USING J 39200, MEASURE RESISTANCE BETWEEN MOTOR PACK TERMINALS "C" AND "B". IS RESISTANCE 0.4Ω OR GREATER?

NO: REPLACE MOTOR PACK.

YES: go to (4)

(4)
- USING J 39200, MEASURE RESISTANCE BETWEEN 3-WAY MOTOR PACK HARNESS CONNECTOR TERMINALS "B" AND "C". DOES J 39200 DISPLAY RESISTANCE AS "OL" (INFINITE)?

NO: REPAIR SHORT BETWEEN CKT(s) 1287 and 850.

YES: go to (5)

(5)
- INSPECT CKT 1287, 3-WAY MOTOR PACK AND 24-WAY EBCM CONNECTORS FOR DAMAGE WHICH MAY RESULT IN A SHORT TO VOLTAGE WITH ALL CONNECTORS CONNECTED. REPAIR DAMAGE IF EVIDENT.
- RECONNECT ALL CONNECTORS.
- DRIVE VEHICLE FOR THREE DRIVE CYCLES (A DRIVE CYCLE CONSISTS OF STARTING ENGINE, DRIVING ABOVE 16 KM/H (10 MPH) AND THEN TURNING IGNITION "OFF").
- IGNITION "ON."
- USING TECH 1, SELECT "F1: DTC HISTORY." DID DTC 72 SET IN LAST THREE DRIVE CYCLES?

YES: REPLACE EBCM.

NO: MALFUNCTION IS INTERMITTENT. REFER TO "DIAGNOSTIC AIDS" ON FACING PAGE.

- AFTER DIAGNOSIS IS COMPLETE, CLEAR DTCs AND TEST DRIVE VEHICLE FOR THREE (3) DRIVE CYCLES TO VERIFY DTC DOES NOT RESET. A DRIVE CYCLE CONSISTS OF STARTING THE VEHICLE, DRIVING OVER 16 km/h (10 MPH) AND THEN KEYING DOWN.

DIAGNOSTIC CHART DTC 75 — 1995 CAVALIER AND SUNFIRE

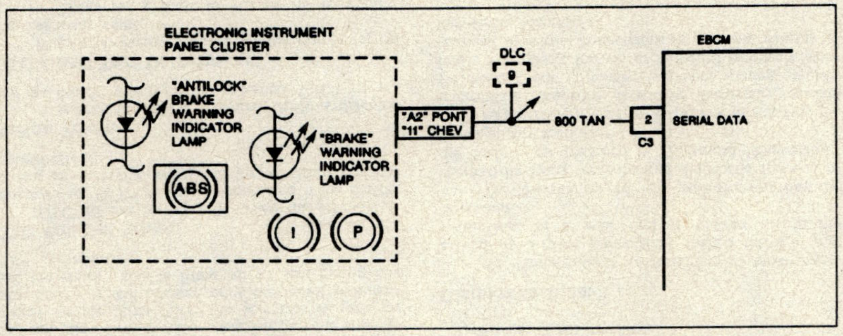

DTC 75
SERIAL COMMUNICATION FAILURE

Circuit Description:
This DTC is designed to detect no communication with the Electronic Instrument Panel Cluster (IPC) via the serial data link.

DTC Will Set When:
DTC 75 can be set anytime after ABS initialization. A malfunction exists if a message from the IPC is not received within 7 seconds.

Action Taken:
A serial communication failure does not allow the proper lamp commands to be sent to the IPC. A malfunction DTC is stored.

DTC Chart Test Description:
Number(s) below refer to circled number(s) on the diagnostic chart.
1. Checks to see if the IPC is operating properly.
2. Checks to see if communication between the EBCM and the IPC is possible.

Diagnostic Aids:
The Tech 1 serial data link (SDL) monitor used in this diagnostic is within the body portion of the Tech 1 mass storage cartridge. This requires exiting from the chassis portion of the Tech 1 to the main menu and entering into the body portion of the Tech 1 and selecting "SDL monitor."

With the SDL monitor (ABS to IPC mode), any message that is being transmitted on the serial data link can be observed (see section 8C for further serial data link information).

The frequency of the malfunction can be checked by using the enhanced diagnostic function of the Tech 1, as described in "Enhanced Diagnostics," found in this section.

Any circuitry, that is suspected as causing the intermittent complaint, should be thoroughly checked for backed out terminals, improper mating, broken locks, improperly formed or damaged terminals, poor terminal to wiring connections or physical damage to the wiring harness.

DIAGNOSTIC CHART DTC 75 — 1995 CAVALIER AND SUNFIRE

DTC 75
SERIAL COMMUNICATION FAILURE

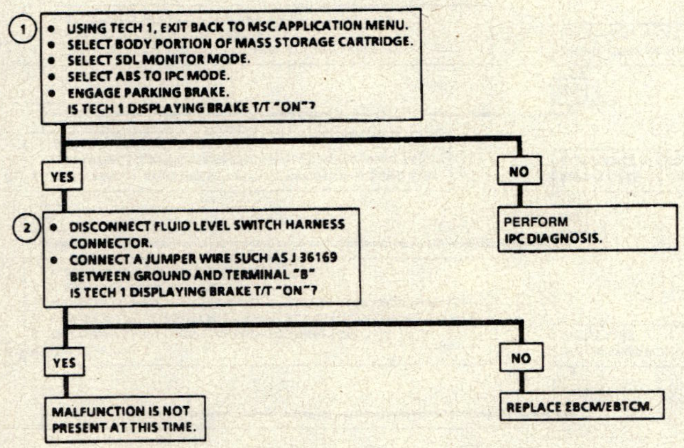

DIAGNOSTIC CHART DTC 76 — 1995 CAVALIER AND SUNFIRE

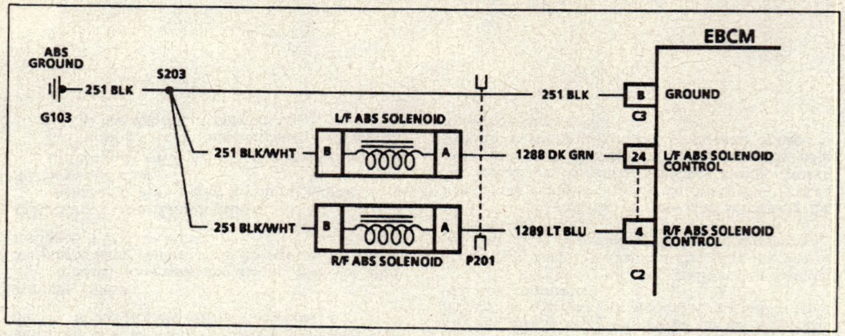

DTC 76
LEFT FRONT SOLENOID CIRCUIT OPEN OR SHORTED TO BATTERY

Circuit Description:
This DTC identifies a solenoid that cannot be energized due to an open in its circuitry, or a solenoid that is always energized due to a short to battery in its circuitry between the driver and the solenoid. An open will not allow proper ABS operation, but the short to battery simply turns "ON" the solenoid. A path for base brakes is still allowed once the motor re-homes and the check valve is lifted off its seat during key "ON" initialization.

DTC Will Set When:
DTC 76 can be set only when the solenoid is commanded "OFF." A malfunction exists if the EBCM detects an excessive voltage in the left front solenoid control circuit.

Action Taken:
A malfunction DTC is stored, ABS is disabled and the "ABS" warning lamp is turned "ON."

DTC Chart Test Description:
Number(s) below refer to circled number(s) on the diagnostic chart.
1. Checks for a short to voltage in the control circuitry of the solenoid.
2. Checks for an open in the control circuit of the solenoid.
3. Identifies an open in the solenoid or ground circuit.
4. Checks for proper resistance value of the solenoid.
5. Ensures malfunction was not due to physical damage of the circuitry.

Diagnostic Aids:
An "intermittent" malfunction is most likely caused by a poor connection, rubbed through wire insulation, or a wire that is broken inside the insulation.

The frequency of the malfunction can be checked by using the enhanced diagnostic function of the Tech 1, as described in "Enhanced Diagnostics," found in this section.

Any circuitry, that is suspected as causing the intermittent complaint, should be thoroughly checked for backed out terminals, improper mating, broken locks, improperly formed or damaged terminals, poor terminal to wiring connections or physical damage to the wiring harness.

DIAGNOSTIC CHART DTC 76 — 1995 CAVALIER AND SUNFIRE

DTC 76
LEFT FRONT SOLENOID CIRCUIT OPEN OR SHORTED TO BATTERY

*NOTICE: J 39200 TEST LEADS MUST BE "ZEROED" PRIOR TO MAKING ANY RESISTANCE MEASUREMENTS. REFER TO J 39200 USER'S MANUAL.

(1)
- IGNITION "OFF."
- DISCONNECT 24-WAY EBCM CONNECTOR.
- IGNITION "ON."
- USING J 39200, MEASURE VOLTAGE BETWEEN GROUND AND 24-WAY EBCM HARNESS CONNECTOR TERMINAL "24".

IS VOLTAGE 1.0V OR LESS?

- NO: REPAIR SHORT TO VOLTAGE IN CKT 1288.
- YES: go to 2.

(2)
- IGNITION "OFF."
- DISCONNECT LEFT FRONT SOLENOID CONNECTOR.
- USING J 39200, MEASURE RESISTANCE BETWEEN LEFT FRONT SOLENOID HARNESS CONNECTOR TERMINAL "A" AND 24-WAY EBCM HARNESS CONNECTOR TERMINAL "24".

IS RESISTANCE 2.0Ω OR LESS?

- NO: REPAIR OPEN OR HIGH RESISTANCE IN CKT 1288.
- YES: go to 3.

(3)
- USING J 39200, MEASURE RESISTANCE BETWEEN GROUND AND LEFT FRONT SOLENOID HARNESS CONNECTOR TERMINAL "B".

IS RESISTANCE 2.0Ω OR LESS?

- NO: REPAIR OPEN OR HIGH RESISTANCE IN CKT 152.
- YES: go to 4.

(4)
- USING J 39200, MEASURE RESISTANCE BETWEEN LEFT FRONT SOLENOID TERMINALS "A" AND "B".

IS RESISTANCE BETWEEN 2.5Ω AND 5.0Ω?

- NO: REPLACE LEFT FRONT SOLENOID.
- YES: go to 5.

(5)
- INSPECT LEFT FRONT SOLENOID AND 24-WAY EBCM HARNESS CONNECTORS FOR POOR TERMINAL CONTACT OR CORROSION. REPLACE TERMINALS THAT EXHIBIT POOR CONTACT OR CORROSION.
- INSPECT CKT 1288 FOR DAMAGE WHICH MAY RESULT IN A SHORT TO VOLTAGE WITH ALL CONNECTORS CONNECTED. REPAIR DAMAGE IF EVIDENT.
- RECONNECT ALL CONNECTORS.
- IGNITION "ON."

IS DTC 76 SET AS A CURRENT DTC?

- YES: REPLACE EBCM.
- NO: REFER TO "DIAGNOSTIC AIDS" ON FACING PAGE.

DIAGNOSTIC CHART DTC 77 — 1995 CAVALIER AND SUNFIRE

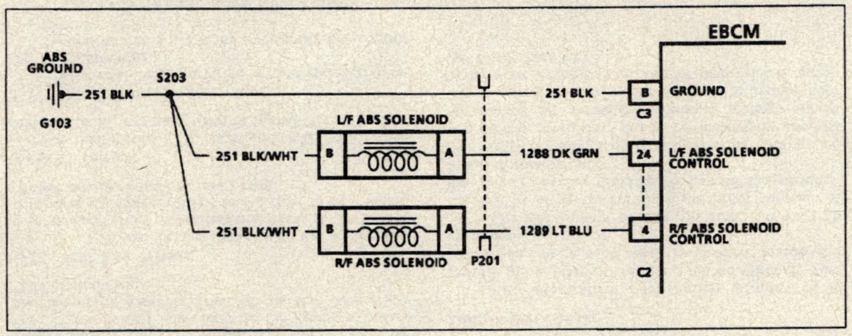

DTC 77
LEFT FRONT SOLENOID CIRCUIT SHORTED TO GROUND OR DRIVER OPEN

Circuit Description:

This DTC identifies a solenoid that cannot be energized due to an open in its driver circuitry, or a short to ground between the solenoid driver and the solenoid. These malfunctions can affect ABS operation since the flow of brake fluid to the wheel cylinder cannot be stopped, making ABS operation for that channel impossible.

DTC Will Set When:

DTC 77 can be set only when the solenoid is commanded "ON." A malfunction exists if the EBCM detects the left front solenoid control circuit voltage is out of specification.

Action Taken:

A malfunction DTC is stored, ABS is disabled and the "ABS" warning lamp is turned "ON."

DTC Chart Test Description:

Number(s) below refer to circled number(s) on the diagnostic chart.

1. Checks for a solenoid that is internally shorted to ground.
2. Checks for a solenoid that is not within proper resistance value.
3. Indicates if a short to ground exists in the solenoid control circuitry.
4. Ensures malfunction was not due to physical damage of the circuitry.

Diagnostic Aids:

An "intermittent" malfunction is most likely caused by a poor connection, rubbed through wire insulation, or a wire that is broken inside the insulation.

The frequency of the malfunction can be checked by using the enhanced diagnostic function of the Tech 1, as described in "Enhanced Diagnostics," found in this section.

Any circuitry, that is suspected as causing the intermittent complaint, should be thoroughly checked for backed out terminals, improper mating, broken locks, improperly formed or damaged terminals, poor terminal to wiring connections or physical damage to the wiring harness.

DIAGNOSTIC CHART DTC 77 — 1995 CAVALIER AND SUNFIRE

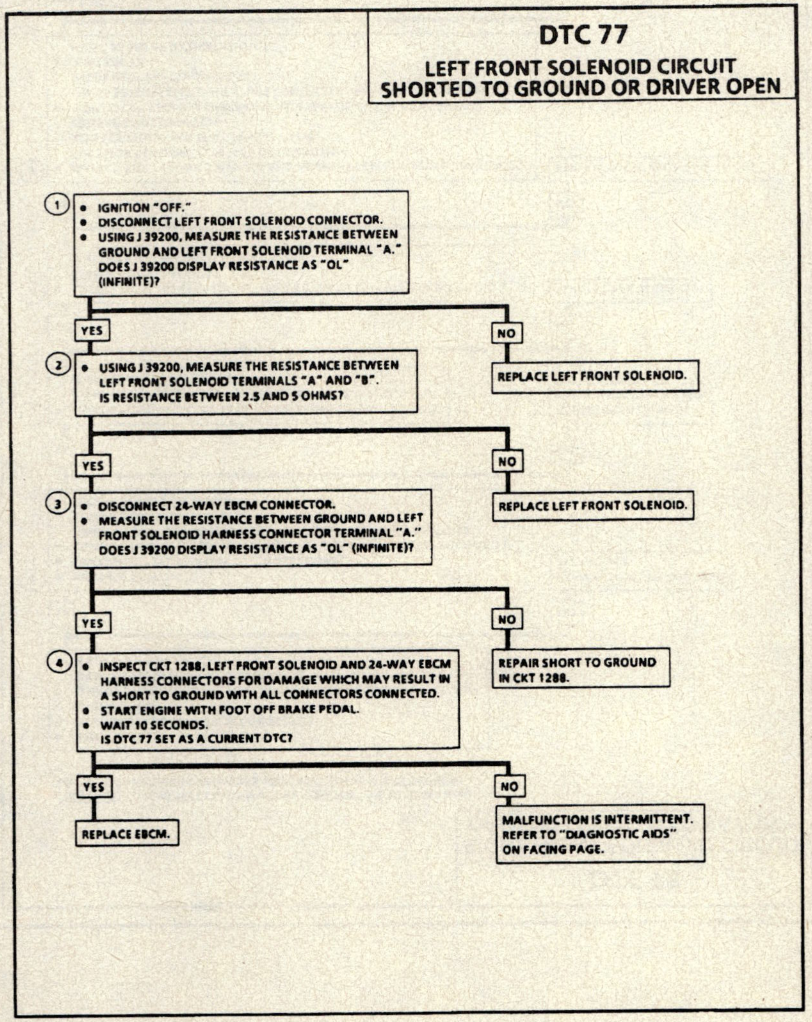

DIAGNOSTIC CHART DTC 78 — 1995 CAVALIER AND SUNFIRE

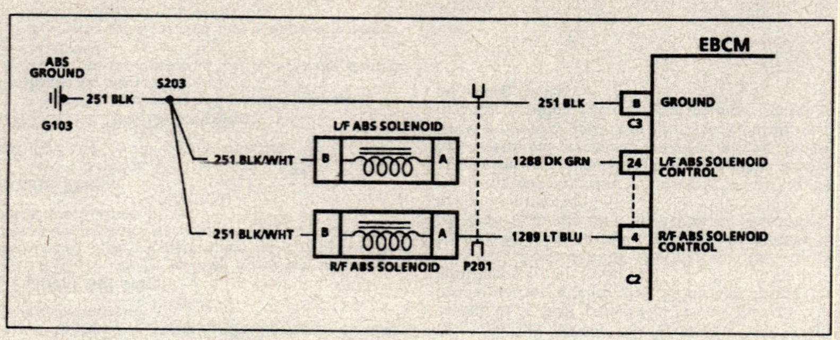

DTC 78
RIGHT FRONT SOLENOID CIRCUIT OPEN OR SHORTED TO BATTERY

Circuit Description:

This DTC identifies a solenoid that cannot be energized due to an open in its circuitry, or a solenoid that is always energized due to a short to battery in its circuitry between the driver and the solenoid. An open will not allow proper ABS operation but the short to battery simply turns "ON" the solenoid. A path for base brakes is still allowed once the motor re-homes and the check valve is lifted off its seat during key "ON" initialization.

DTC Will Set When:

DTC 78 can be set only when the solenoid is commanded "OFF." A malfunction exists if the EBCM detects an excessive voltage in the right front solenoid control circuit, a malfunction exists.

Action Taken:

A malfunction DTC is stored, ABS is disabled and the "ABS" warning lamp is turned "ON."

DTC Chart Test Description:

Number(s) below refer to circled number(s) on the diagnostic chart.

1. Checks for a short to voltage in the control circuitry of the solenoid.
2. Checks for an open in the control circuit of the solenoid.
3. Identifies an open in the solenoid or ground circuit.
4. Checks for proper resistance value of the solenoid.
5. Ensures malfunction was not due to physical damage of the circuitry.

Diagnostic Aids:

An "intermittent" malfunction is most likely caused by a poor connection, rubbed through wire insulation, or a wire that is broken inside the insulation.

The frequency of the malfunction can be checked by using the enhanced diagnostic function of the Tech 1, as described in "Enhanced Diagnostics," found in this section.

Any circuitry, that is suspected as causing the intermittent complaint, should be thoroughly checked for backed out terminals, improper mating, broken locks, improperly formed or damaged terminals, poor terminal to wiring connections or physical damage to the wiring harness.

DIAGNOSTIC CHART DTC 78 — 1995 CAVALIER AND SUNFIRE

DTC 78
RIGHT FRONT SOLENOID CIRCUIT OPEN OR SHORTED TO BATTERY

*NOTICE: J39200 TEST LEADS MUST BE "ZEROED" PRIOR TO MAKING ANY RESISTANCE MEASUREMENTS. REFER TO J39200 USER'S MANUAL.

(1)
- IGNITION "OFF."
- DISCONNECT 24-WAY EBCM CONNECTOR.
- IGNITION "ON."
- USING J 39200, MEASURE VOLTAGE BETWEEN GROUND AND 24-WAY EBCM HARNESS CONNECTOR TERMINAL "4".

IS VOLTAGE 1.0V OR LESS?

YES → go to (2). NO → REPAIR SHORT TO VOLTAGE IN CKT 1289.

(2)
- IGNITION "OFF."
- DISCONNECT RIGHT FRONT SOLENOID CONNECTOR.
- USING J 39200, MEASURE RESISTANCE BETWEEN RIGHT FRONT SOLENOID HARNESS CONNECTOR TERMINAL "A" AND 24-WAY EBCM HARNESS CONNECTOR TERMINAL "4".

IS RESISTANCE 2.0Ω OR LESS?

YES → go to (3). NO → REPAIR OPEN OR HIGH RESISTANCE IN CKT 1289.

(3)
- USING J39200, MEASURE RESISTANCE BETWEEN GROUND AND RIGHT FRONT SOLENOID HARNESS CONNECTOR TERMINAL "B".

IS RESISTANCE 2.0Ω OR LESS?

YES → go to (4). NO → REPAIR OPEN IN CKT 152.

(4)
- USING J39200, MEASURE RESISTANCE BETWEEN RIGHT FRONT SOLENOID TERMINALS "A" AND "B".

IS RESISTANCE BETWEEN 2.5Ω AND 5Ω?

YES → go to (5). NO → REPLACE RIGHT FRONT SOLENOID.

(5)
- INSPECT RIGHT FRONT SOLENOID AND 24-WAY EBCM HARNESS CONNECTORS FOR POOR TERMINAL CONTACT OR CORROSION. REPLACE TERMINALS THAT EXHIBIT POOR CONTACT OR CORROSION.
- INSPECT CKT 1289 FOR DAMAGE WHICH MAY RESULT IN A SHORT TO VOLTAGE WITH ALL CONNECTORS CONNECTED. REPAIR DAMAGE IF EVIDENT.
- RECONNECT ALL CONNECTORS.
- IGNITION "ON."

IS DTC 78 SET AS A CURRENT DTC?

YES → REPLACE EBCM. NO → MALFUNCTION IS INTERMITTENT. REFER TO "DIAGNOSTIC AIDS" ON FACING PAGE.

DIAGNOSTIC CHART DTC 81 — 1995
CAVALIER AND SUNFIRE

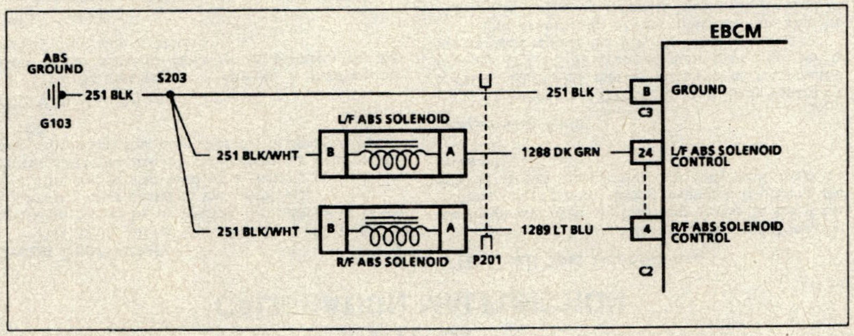

DTC 81
RIGHT FRONT SOLENOID CIRCUIT SHORTED TO GROUND OR DRIVER OPEN

Circuit Description:
This DTC identifies a solenoid that cannot be energized due to an open in its driver circuitry, or a short to ground between the solenoid driver and the solenoid. These malfunctions can affect ABS operation since the flow of brake fluid to the wheel cylinder cannot be stopped, making ABS operation for that channel impossible.

DTC Will Set When:
DTC 81 can be set only when the solenoid is commanded "ON." A malfunction exists if the EBCM detects the right front solenoid control circuit voltage is out of specification.

Action Taken:
A malfunction DTC is stored, ABS is disabled and the "ABS" warning lamp is turned "ON."

DTC Chart Test Description:
Number(s) below refer to circled number(s) on the diagnostic chart.

1. Checks for a solenoid that is internally shorted to ground.
2. Checks for a solenoid that is not within proper resistance values.
3. Indicates if a short to ground exists in the solenoid control circuitry.
4. Ensures malfunction was not due to physical damage of the circuitry.

Diagnostic Aids:
An "intermittent" malfunction is most likely caused by a poor connection, rubbed through wire insulation, or a wire that is broken inside the insulation.

The frequency of the malfunction can be checked by using the enhanced diagnostic function of the Tech 1, as described in "Enhanced Diagnostics," found in this section.

Any circuitry, that is suspected as causing the intermittent complaint, should be thoroughly checked for backed out terminals, improper mating, broken locks, improperly formed or damaged terminals, poor terminal to wiring connections or physical damage to the wiring harness.

DIAGNOSTIC CHART DTC 81 — 1995
CAVALIER AND SUNFIRE

DTC 81
RIGHT FRONT SOLENOID CIRCUIT SHORTED TO GROUND OR DRIVER OPEN

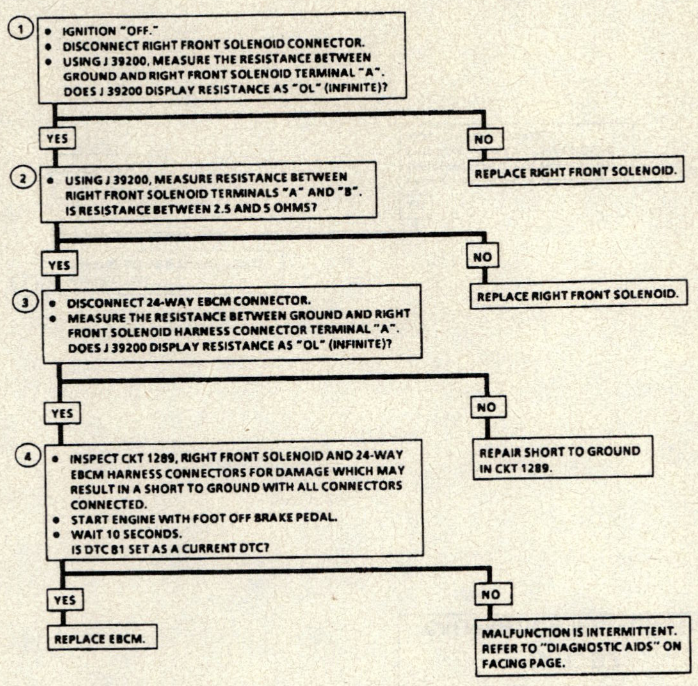

DIAGNOSTIC CHART DTC 82 — 1995 CAVALIER AND SUNFIRE

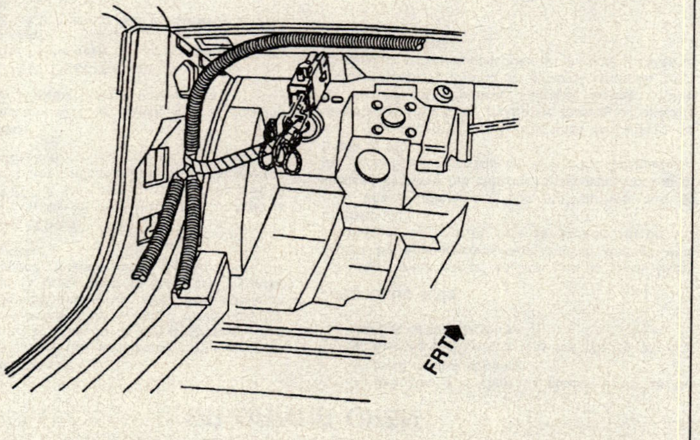

DTC 82
CALIBRATION MALFUNCTION

Circuit Description:
This DTC allows the EBCM to check for a calibration failure by comparing the calibration value to a known value stored in the EEPROM.
This test is also used as a security measure to prevent improper use of calibrations or changes to these calibrations that may alter the designed function of ABS.

DTC Will Set When:
DTC 82 can be set at key-up. A malfunction exists if the program identifier is incorrect or the memory checksum is incorrect.

Action Taken:
A malfunction DTC is stored, ABS is disabled and the "ABS" warning lamp is turned "ON."

DTC Chart Test Description:
Number(s) below refer to circled number(s) on the diagnostic chart. 1. This step checks to see if the malfunction is present during diagnosis. If present, the EBCM is not functioning correctly and must be replaced.

Diagnostic Aids:
An intermittent DTC 82 is most likely caused by a faulty EEPROM that is sensitive to temperature changes. If DTC 82 occurred more than once, but is intermittent, replace the EBCM.
The frequency of the malfunction can be checked by using the enhanced diagnostic function of the Tech 1, as described in "Enhanced Diagnostics," found in this section.

DIAGNOSTIC CHART DTC 82 — 1995 CAVALIER AND SUNFIRE

DTC 82
CALIBRATION MALFUNCTION

(1) • IGNITION "ON."
IS DTC 82 SET AS A CURRENT DTC?

- YES → REPLACE EBCM.
- NO → MALFUNCTION IS NOT PRESENT AT THIS TIME. REFER TO "DIAGNOSTIC AIDS" ON FACING PAGE.

DIAGNOSTIC CHART DTC 86 — 1995 CAVALIER AND SUNFIRE

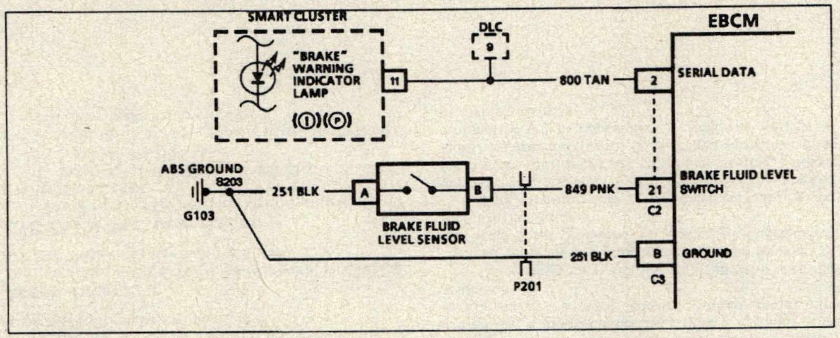

DTC 86
EBCM TURNED "ON" THE RED "BRAKE" WARNING LAMP

Circuit Description:

This DTC is provided as an "information only" test and reflects the status of the command issued by the ABS EBCM to illuminate the red "BRAKE" warning lamp. If another diagnostic DTC issues a command to illuminate the red "BRAKE" warning lamp, DTC 86 will be stored in EEPROM as a history DTC at the conclusion of the ignition cycle.

DTC Chart Test Description:

Number(s) below refer to circled number(s) on the diagnostic chart.

1. This step serves to identify if a DTC other than DTC 86 commanded the red "BRAKE" warning lamp "ON."

Diagnostic Aids:

Any ABS mechanical DTC that issues a command to illuminate the red "BRAKE" warning lamp will also result in DTC 86 being stored in EEPROM during shut down. These DTC's are: 38, 41, 42, 45, 46, 47, 48 and 51. If the motors are not in their home position, certain electrical DTC's will also command the red "BRAKE" warning lamp "ON." These DTC's are: 14, 18, 55, 56, 57, 58, 61, 62, 63, 64, 65, 66, 67, 68, 71 and 72.

If any of these DTC's are indicated along with DTC 86, they must be corrected prior to addressing a DTC 86 malfunction.

DIAGNOSTIC CHART DTC 86 — 1995 CAVALIER AND SUNFIRE

DTC 86
EBCM TURNED ON THE RED BRAKE WARNING LAMP

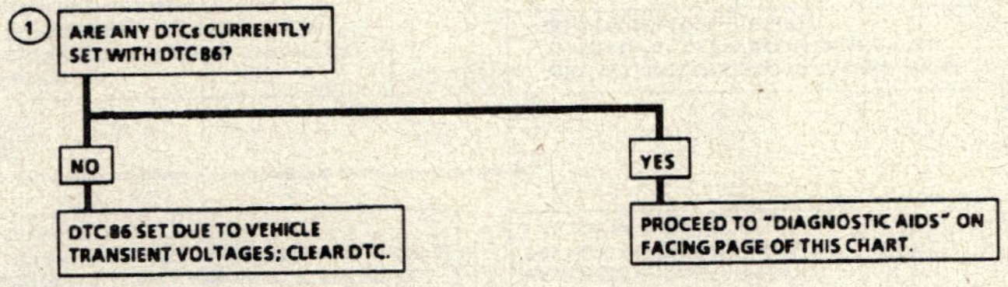

DIAGNOSTIC CHART DTC 91 — 1995 CAVALIER AND SUNFIRE

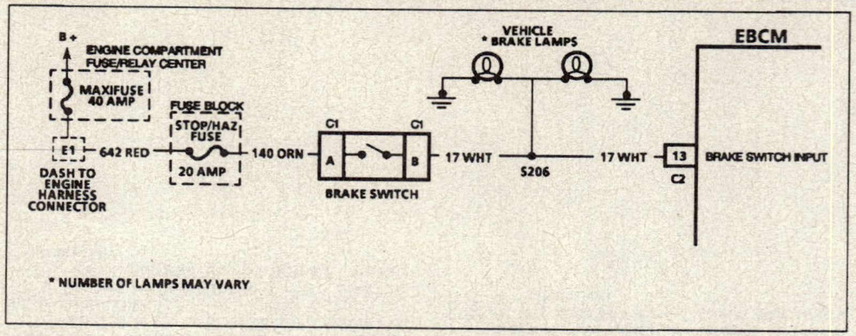

DTC 91
OPEN BRAKE SWITCH CONTACTS DURING DECELERATION

Circuit Description:

This DTC is used to detect an open brake switch in the non-ABS mode. The EBCM looks for a deceleration rate that would indicate braking action and verifies this assumption by requiring several repeats of this detection method. In each case, ABS will not be available since no brake switch is seen.

DTC Will Set When:

DTC 91 can be set if three deceleration cycles occur with the brake switch "OFF."

Action Taken:

A malfunction DTC is stored, ABS is disabled and the "ABS" warning lamp is turned "ON."

DTC Chart Test Description:

Number(s) below refer to circled number(s) on the diagnostic chart.

1. Indicates if the brake switch signal is being received by the EBCM.
2. Indicates if an open circuit exists in the brake switch or brake lamp circuitry.
3. Checks for an open in the brake switch input circuit.
4. Ensures malfunction was not due to poor terminal contact.
5. Checks for voltage entering the brake switch.
6. Checks for voltage exiting the brake switch.

Diagnostic Aids:

An "intermittent" malfunction is most likely caused by a poor connection, rubbed through wire insulation, or a wire that is broken inside the insulation.

The frequency of the malfunction can be checked by using the enhanced diagnostic function of the Tech 1, as described in "Enhanced Diagnostics" found in this section.

Any circuitry, that is suspected as causing the intermittent complaint, should be thoroughly checked for backed out terminals, improper mating, broken locks, improperly formed or damaged terminals, poor terminal to wiring connections or physical damage to the wiring harness.

DIAGNOSTIC CHART DTC 91 — 1995 CAVALIER AND SUNFIRE

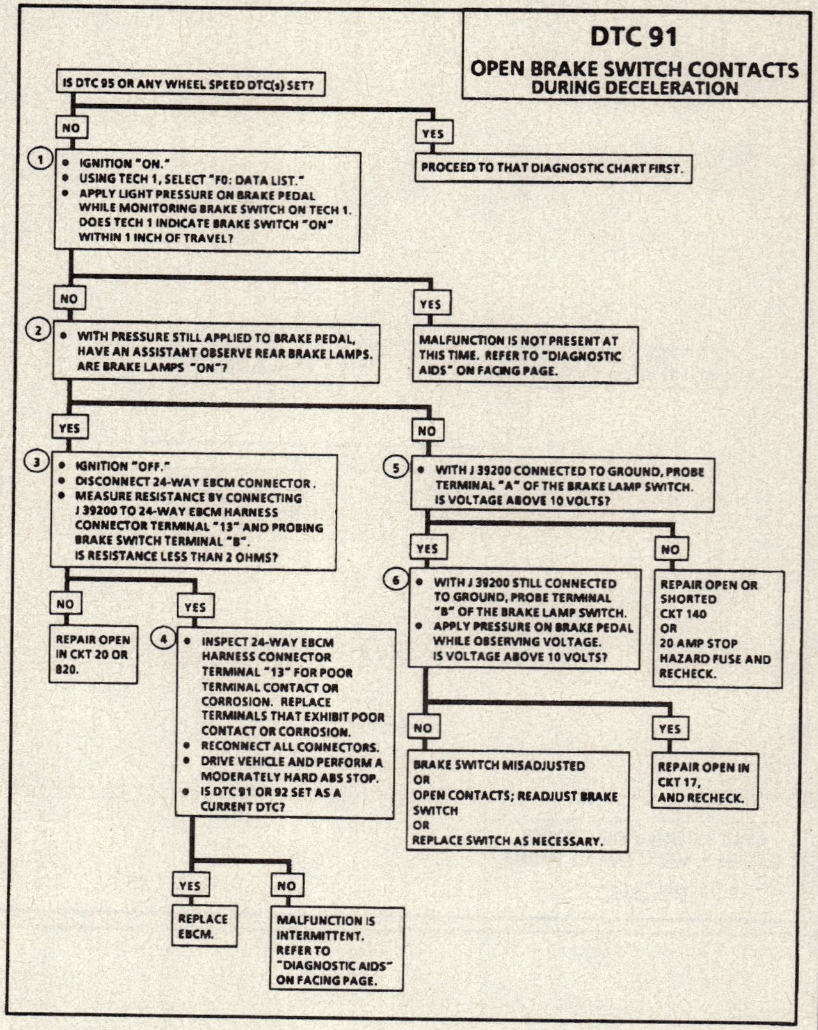

DIAGNOSTIC CHART DTC 92 — 1995 CAVALIER AND SUNFIRE

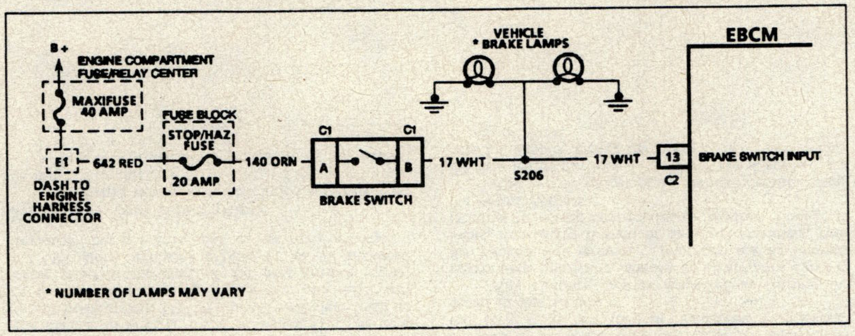

DTC 92

OPEN BRAKE SWITCH CONTACTS WHEN ABS WAS REQUIRED

Circuit Description:

This DTC is run to determine the proper operation of the brake switch. This is important because ABS is activated when the brake switch is "ON." If the brake switch is "OFF," ABS will never be activated. Since this malfunction is difficult to detect under normal braking conditions, this malfunction is only detected when ABS is required.

DTC Will Set When:

DTC 92 can be set if the vehicle's speed is greater than 8 km/h (5 mph). If the brake was not "ON" and a release was required on two channels for.7 second, a malfunction exists.

Action Taken:

A malfunction DTC is stored, ABS is disabled and the "ABS" warning lamp is turned "ON."

DTC Chart Test Description:

Number(s) below refer to circled number(s) on the diagnostic chart.

1. Indicates if the brake switch signal is being received by the EBCM.
2. Indicates if an open circuit exists in the brake switch or brake lamp circuitry.
3. Checks for an open in the brake switch input circuit.
4. Ensures malfunction was not due to poor terminal contact.
5. Checks for voltage entering the brake switch.
6. Checks for voltage exiting the brake switch.

Diagnostic Aids:

An "intermittent" malfunction is most likely caused by a poor connection, rubbed through wire insulation, or a wire that is broken inside the insulation.

The frequency of the malfunction can be checked by using the enhanced diagnostic function of the Tech 1, as described in "Enhanced Diagnostics" found in this section.

Any circuitry, that is suspected as causing the intermittent complaint, should be thoroughly checked for backed out terminals, improper mating, broken locks, improperly formed or damaged terminals, poor terminal to wiring connections or physical damage to the wiring harness.

DIAGNOSTIC CHART DTC 92 — 1995 CAVALIER AND SUNFIRE

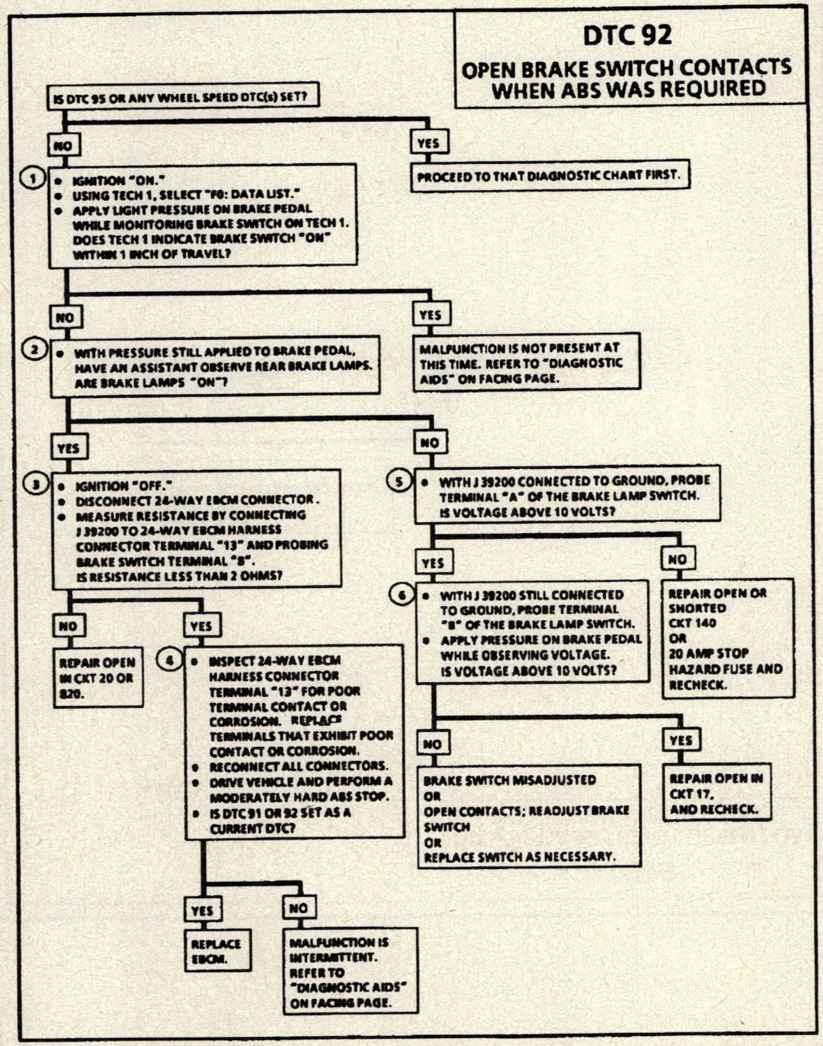

DIAGNOSTIC CHART DTC 93 — 1995 CAVALIER AND SUNFIRE

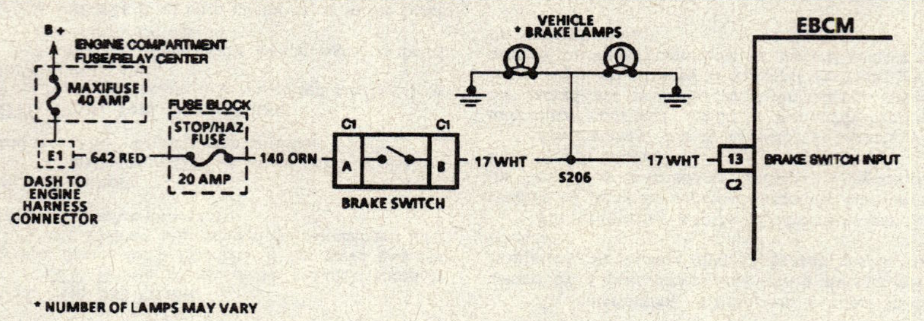

DTC 93

DTC's 91 OR 92 SET IN CURRENT OR PREVIOUS IGNITION CYCLE

Circuit Description:

This DTC is the second portion of DTC's 91 and 92. If DTC's 91 or 92 occurred during the last ignition cycle, DTC 93 becomes a current malfunction during the next ignition cycle, keeping ABS disabled until a brake switch "ON" state is seen. When a change is seen during an ignition cycle in which DTC 93 is a current malfunction, DTC 91 or 92 will clear itself at the end of the current ignition cycle, and ABS will enable itself at the start of the next ignition cycle. DTC 93 alone indicates DTC's 91 or 92 occurred previously, but is intermittent, or has been corrected.

DTC Chart Test Description:

Number(s) below refer to circled number(s) on the diagnostic chart.

1. Indicates which DTC (either 91 or 92) caused DTC 93 to set.

Diagnostic Aids:

An "intermittent" malfunction is most likely caused by a poor connection, rubbed through wire insulation, or a wire that is broken inside the insulation.

The frequency of the malfunction can be checked by using the enhanced diagnostic function of the Tech 1, as described in "Enhanced Diagnostics" found in this section.

Any circuitry, that is suspected as causing the intermittent complaint, should be thoroughly checked for backed out terminals, improper mating, broken locks, improperly formed or damaged terminals, poor terminal to wiring connections or physical damage to the wiring harness.

Also, verify proper brake switch operation using the data list of the Tech 1. As the brake is applied, the data list should display the brake switch "ON" within 1 inch of travel.

DIAGNOSTIC CHART DTC 93 — 1995 CAVALIER AND SUNFIRE

DTC 93

DTCs 91 OR 92 SET IN CURRENT OR PREVIOUS IGNITION CYCLE

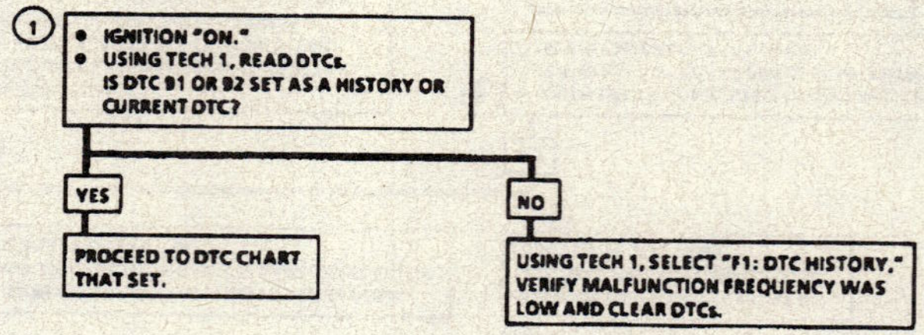

DIAGNOSTIC CHART DTC 94 — 1995 CAVALIER AND SUNFIRE

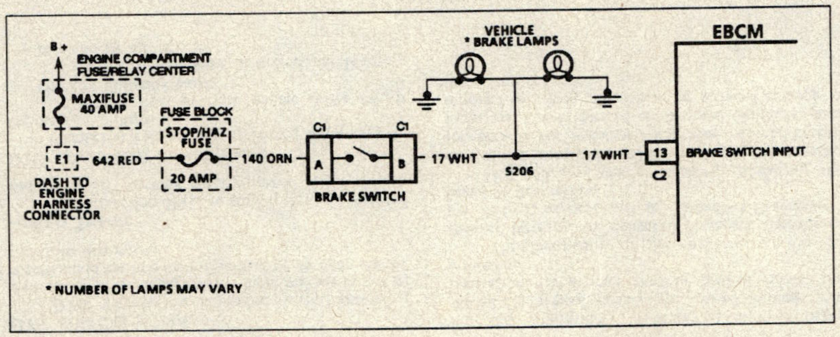

DTC 94
BRAKE SWITCH CONTACTS ALWAYS CLOSED

Circuit Description:

This DTC is run to determine the proper operation of the brake switch. This is important because ABS is activated when the brake switch is "ON" and turned "OFF" when the brake is "OFF." If the brake switch is always "ON," ABS operation will always be requested resulting in potential modulator cycling on rough roads. Additionally, this malfunction will most likely result in a dead battery (due to the brake lamps staying "ON") if the driver is not informed of this malfunction.

DTC Will Set When:

DTC 94 can be set if the vehicle speed reaches at least 40 km/h (25 mph). If the brake was never "OFF" during two consecutive drive cycles, a malfunction exists.

Action Taken:

A malfunction DTC is stored, however, ABS is not disabled.

DTC Chart Test Description:

Number(s) below refer to circled number(s) on the diagnostic chart.

1. Checks to see if brake lamps are "ON."
2. Isolates the cause of the malfunction to either a faulty or misadjusted brake switch, or a short to voltage in the brake switch circuitry.
3. Checks for a resistive short to voltage in the brake switch circuitry.
4. Ensures malfunction was not due to a physical damage of the circuitry.

Diagnostic Aids:

An "intermittent" malfunction is most likely caused by a poor connection, rubbed through wire insulation, or a wire that is broken inside the insulation.

The frequency of the malfunction can be checked by using the enhanced diagnostic function of the Tech 1, as described in "Enhanced Diagnostics" found in this section.

Any circuitry that is suspected as causing the intermittent complaint should be thoroughly checked for backed out terminals, improper mating, broken locks, improperly formed or damaged terminals, poor terminal to wiring connections or physical damage to the wiring harness.

DIAGNOSTIC CHART DTC 94 — 1995 CAVALIER AND SUNFIRE

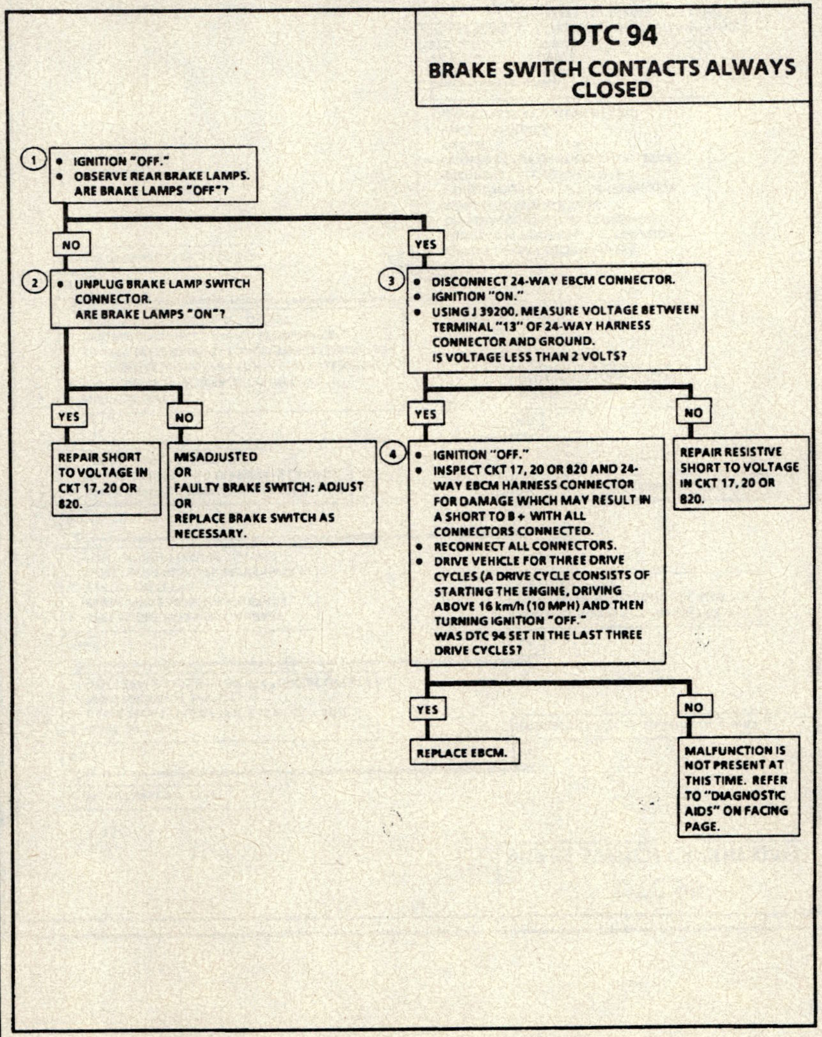

DIAGNOSTIC CHART DTC 95 — 1995 CAVALIER AND SUNFIRE

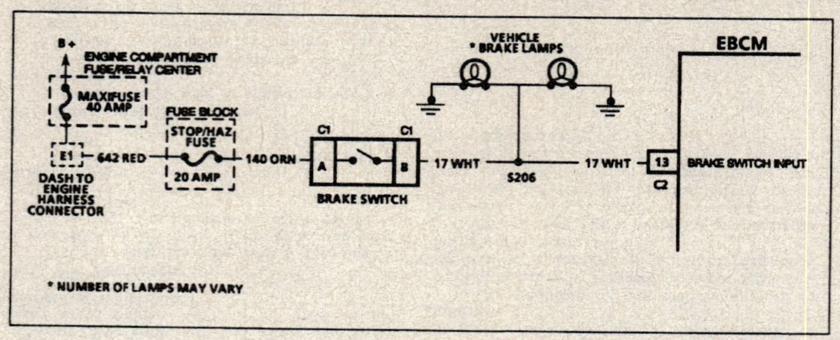

DTC 95
BRAKE SWITCH CIRCUIT OPEN

Circuit Description:

This DTC is used to identify open brake switch circuitry that prevents the brake switch input to the EBCM from changing states when the brake is applied.

This DTC is used in conjunction with DTC's 91 and 92 to determine the cause of an open brake switch malfunction.

DTC Will Set When:

DTC 95 can be set after initialization is completed. If the brake switch input voltage is out of specification for one second indicating an open circuit, a malfunction exists.

Action Taken:

A malfunction DTC is stored, ABS is disabled and the "ABS" warning lamp is turned "ON."

DTC Chart Test Description:

Number(s) below refer to circled number(s) on the diagnostic chart.

1. Confirms that an open in the brake switch circuitry currently exists.
2. Indicates if the brake switch signal is being received by the EBCM.
3. Indicates if an open circuit exists in the brake switch or brake lamp circuitry.
4. Isolates the open circuit to either the brake switch input circuitry, or the EBCM.
5. Ensures malfunction was not due to poor terminal contact.

Diagnostic Aids:

An "intermittent" malfunction is most likely caused by a poor connection, rubbed through wire insulation, or a wire that is broken inside the insulation.

The frequency of the malfunction can be checked by using the enhanced diagnostic function of the Tech 1, as described in "Enhanced Diagnostics" found in this section.

Any circuitry, that is suspected as causing the intermittent complaint, should be thoroughly checked for backed out terminals, improper mating, broken locks, improperly formed or damaged terminals, poor terminal to wiring connections or physical damage to the wiring harness.

DIAGNOSTIC CHART DTC 95 — 1995 CAVALIER AND SUNFIRE

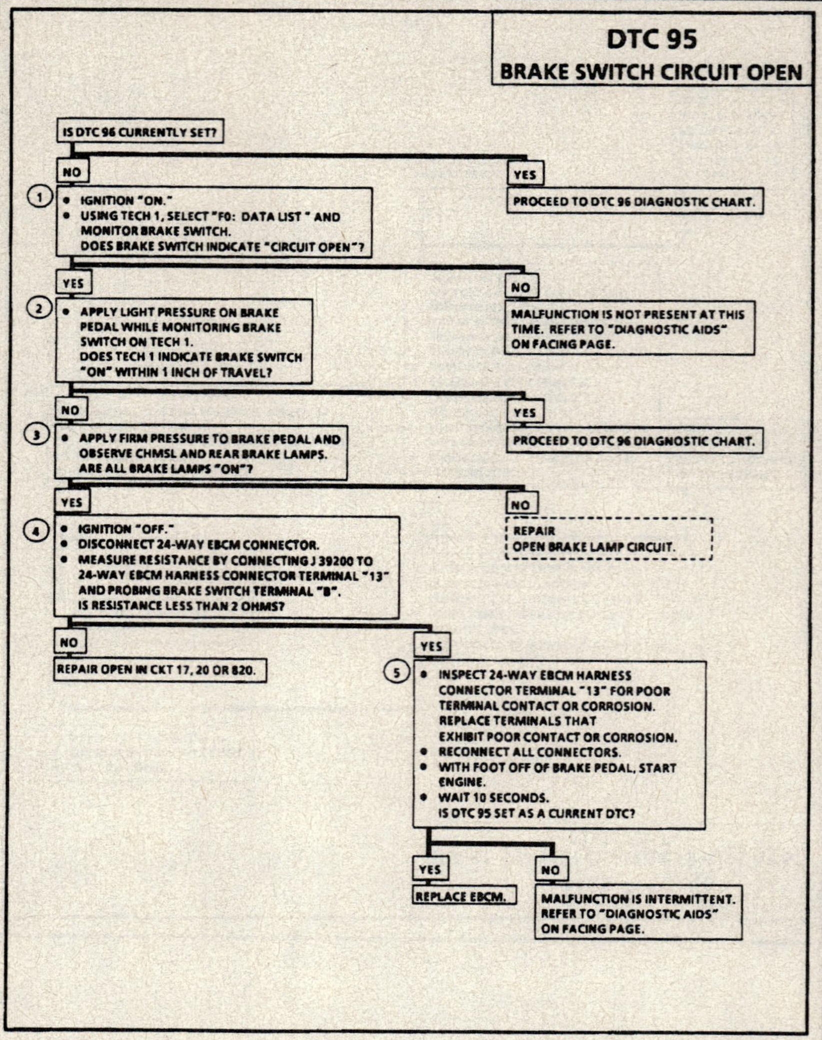

DIAGNOSTIC CHART DTC 96 — 1995
CAVALIER AND SUNFIRE

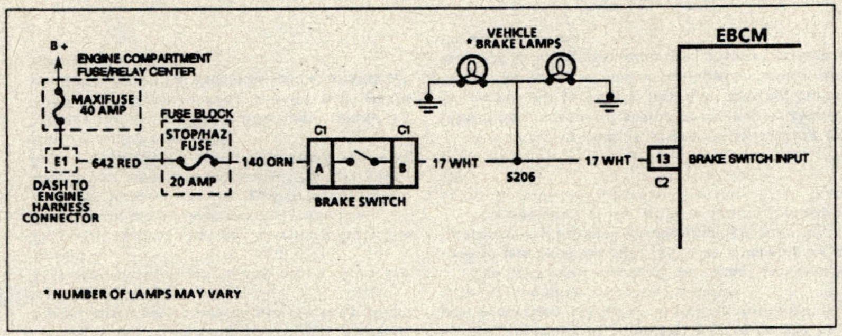

DTC 96
REAR BRAKE LAMP CIRCUIT OPEN

Circuit Description:

This DTC is designed to isolate the cause of a DTC 95 failure and indicate to the driver ABS is still available. If DTC 95 occurs with DTC 96, the brake lamps are open, the brake lamp grounds are open, or the Center High Mounted Stop Lamp (CHMSL) circuit is open during 4-way flasher use. The presence of battery voltage at the brake switch input indicates a valid brake switch input is still available.

DTC Will Set When:

DTC 96 can be set only after DTC 95 has been set. If the EBCM detects battery voltage on the brake switch input circuit for one second, a malfunction exists.

Action Taken:

A malfunction DTC is stored. ABS is not disabled; however, the "ABS" warning lamp will flash indicating a malfunction exists.

DTC Chart Test Description:

Number(s) below refer to circled number(s) on the diagnostic chart.

1. Checks for additional brake switch DTC's set.
2. Identifies if the malfunction is currently present in the brake circuit.

Diagnostic Aids:

An "intermittent" malfunction is most likely caused by a poor connection, rubbed through wire insulation, or a wire that is broken inside the insulation.

The frequency of the malfunction can be checked by using the enhanced diagnostic function of the Tech 1, as described in "Enhanced Diagnostics" found in this section.

Any circuitry, that is suspected as causing the intermittent complaint, should be thoroughly checked for backed out terminals, improper mating, broken locks, improperly formed or damaged terminals, poor terminal to wiring connections or physical damage to the wiring harness.

DIAGNOSTIC CHART DTC 96 — 1995
CAVALIER AND SUNFIRE

DTC 96
REAR BRAKE LAMP CIRCUIT OPEN

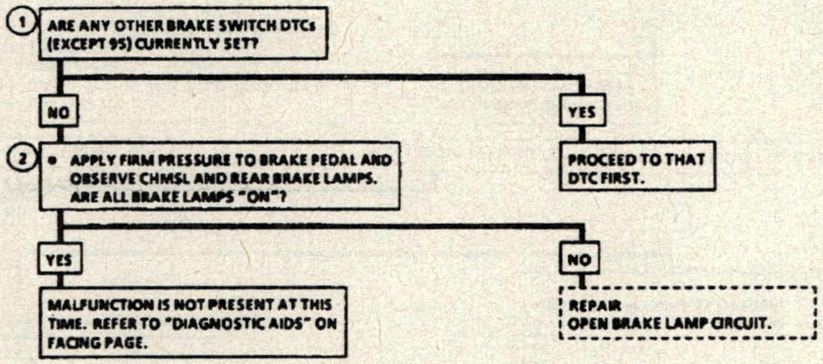

DIAGNOSTIC CHART DTC A011 — 1993 BERETTA AND CORSICA

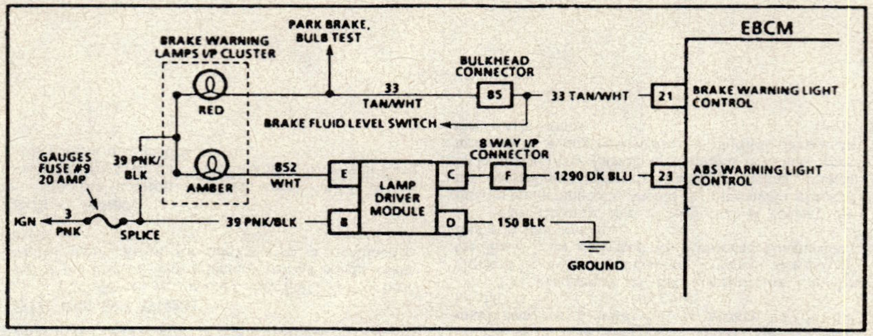

Circuit Description:

This code checks the state of the ABS warning light to identify a situation in which the driver could not be warned of a system fault by the ABS warning light, or the warning light is always "ON." The Lamp Driver Module (LDM) turns the ABS warning light "ON," unless the EBCM provides ground to turn it "OFF." Because of the circuitry in the LDM, only external faults can be detected. As a result, the ABS warning light itself is not diagnosable, only the control line to the EBCM can be diagnosed. In the event of an open CKT 1290, the ABS warning light will be on at all times, due to the loss of ground at the LDM input. If the control line is shorted to ground, the ABS warning light is kept "OFF," due to the LDM input being grounded.

Test Description:

Number(s) below refer to circled number(s) on the diagnostic chart.

1. Checks to see if ABS light functions properly. Normal operation would indicate that the problem is not present.
2. This step determines if any other codes are present.
3. This step insures that the circuitry from the EBCM to the lamp driver module is complete.
4. This step insures that the ignition circuit to the lamp driver module is functional. An open gauge fuse or open CKT 3, 39 or 852 would result in inoperative indicator lights.
5. Isolates the cause of the ABS lamp remaining "OFF," by disconnecting the 24 way world connector, CKT 1290 should be open and cause the ABS lamp to be "ON."

Diagnostic Aids:

An "Intermittent" problem may be caused by a poor connection, rubbed through wire insulation, or a wire that is broken inside the insulation.

The light test function of the "Scan" tool may be used to command the light "ON" while looking for an intermittent problem in the warning light circuitry.

The frequency of the problem can be checked by using the enhanced diagnostic function of the "Scan" tool.

Any circuitry that is suspected as causing the intermittent complaint should be thoroughly checked for backed out terminals, improper mating, broken locks, improperly formed or damaged terminals, poor terminal to wiring connections or physical damage to the wiring harness.

DIAGNOSTIC CHART DTC A011 — 1993 BERETTA AND CORSICA

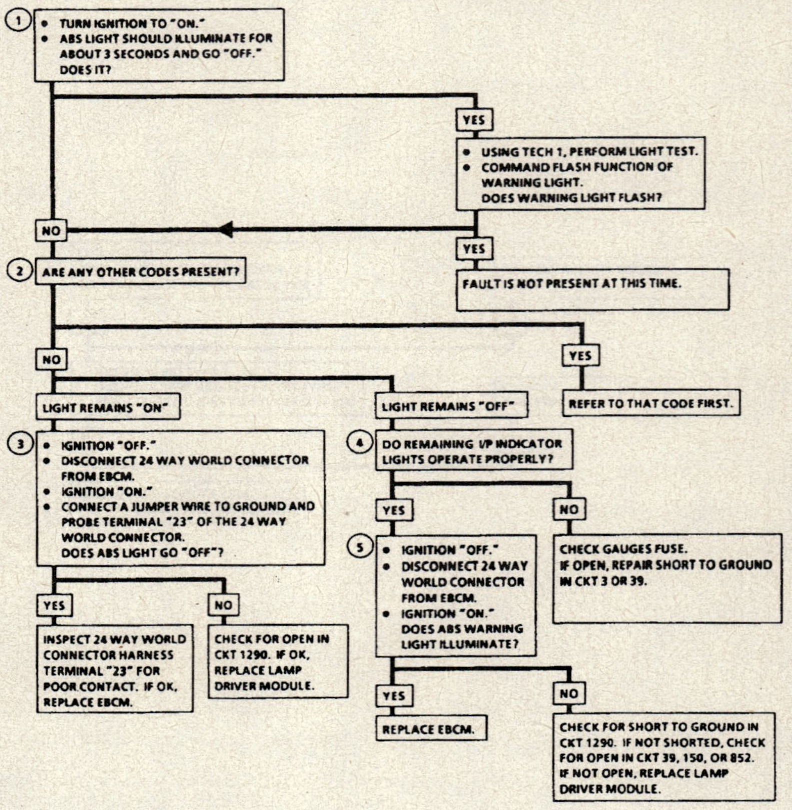

THIS CHART ASSUMES THAT A CURRENT CODE IS STORED INDICATING THAT THIS FAULT IS PRESENT.

DIAGNOSTIC CHART DTC A013 — 1993 BERETTA AND CORSICA

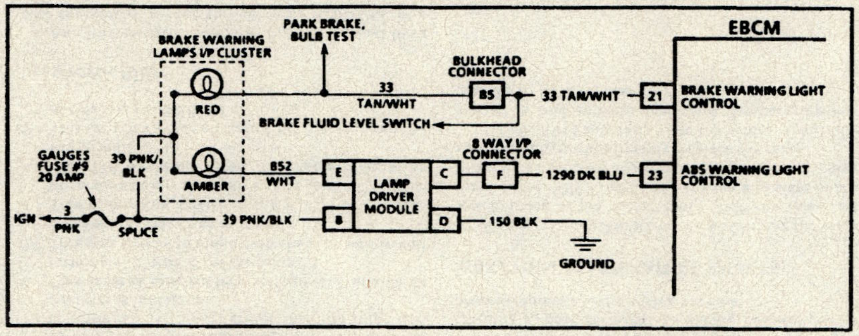

Circuit Description:

This code checks the state of the ABS warning light to identify a situation in which the driver could not be warned of a system fault by the ABS warning light. The Lamp Driver Module (LDM) turns the ABS warning light "ON," unless the EBCM provides ground to turn it "OFF." Because of the circuitry internal to the LDM, only external faults can be detected. As a result, the ABS warning light itself is not diagnosable, only the control line to the EBCM can be diagnosed.

If the ABS warning light control line is shorted to battery or the EBCM control circuit is open, the Electronic Brake Control Module (EBCM) is not able to turn "OFF" the ABS warning light.

Test Description: Number(s) below refer to circled number(s) on the diagnostic chart.

1. Checks to see if ABS light functions properly. Normal operation would indicate that the problem is not present.
2. After the ignition is turned to the "OFF" position, voltage to the lamp driver module should be interrupted. A light that remains "ON" would indicate that voltage to this circuit is from a point other than ignition.
3. After removing the 20 amp gauges fuse, voltage on CKT 1290 will be eliminated. If voltage is still present, it would indicate that a short to voltage is present between the lamp driver module and EBCM.

Diagnostic Aids:

An "Intermittent" problem may be caused by a poor connection, rubbed through wire insulation or a wire that is broken inside the insulation.

Code A013 must be currently set before it should be addressed. If Code A013 is set as a history code, proceed to those codes that are currently set before attempting to repair a Code A013 fault.

The frequency of the problem can be checked by using the enhanced diagnostic function of the "Scan" tool

Any circuitry that is suspected as causing the "Intermittent" complaint should be thoroughly checked for backed out terminals, improper mating, broken locks, improperly formed or damaged terminals, poor terminal to wiring connections or physical damage to the wiring harness.

DIAGNOSTIC CHART DTC A013 — 1993 BERETTA AND CORSICA

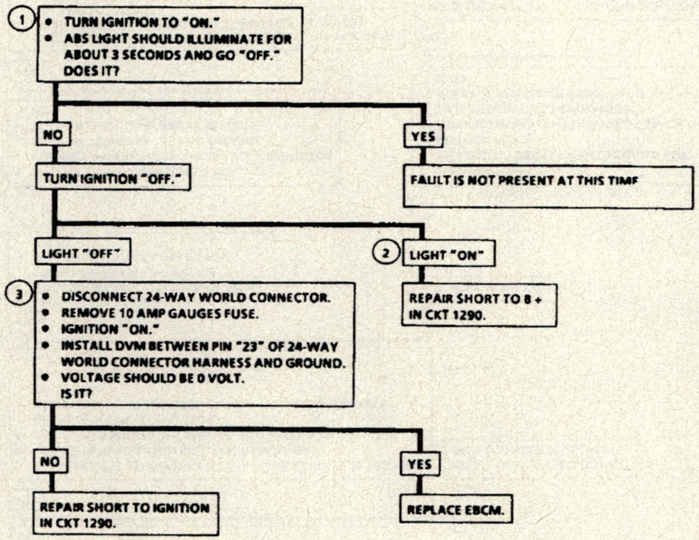

THIS CHART ASSUMES THAT A CURRENT CODE IS STORED INDICATING THAT THIS FAULT IS PRESENT.

DIAGNOSTIC CHART DTC A014 — 1993 BERETTA AND CORSICA

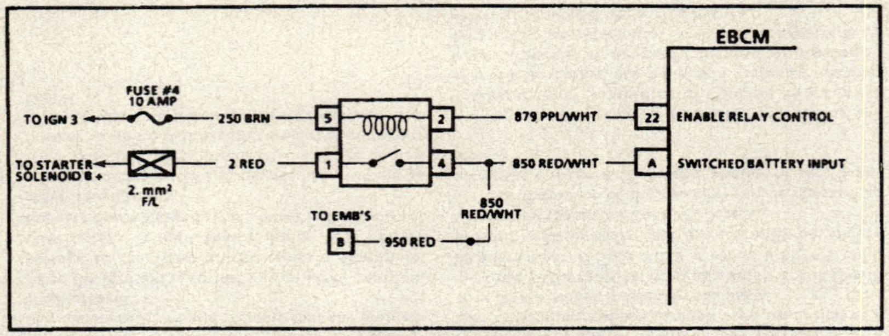

CODE A014

ENABLE RELAY CONTACTS OR FUSE OPEN

Circuit Description:

Ignition voltage is supplied through terminal "5" of the ABS enable relay. The EBCM then is able to energize the pull-in coil by completing the ground circuit at Pin "22" of the EBCM. The magnetic field created closes the enable relay contacts and allows battery voltage and current to be supplied to the EMB's and EBCM, which supplies power to the motors and solenoids.

This code monitors the availability of current/voltage to the motors, solenoids and EMB's. This fault indicates voltage is not available and would therefore not allow ABS operation if required.

Test Description: Number(s) below refer to circled number(s) on the diagnostic chart.

1. This step would indicate that the EBCM is sensing battery voltage present through the relay contacts at terminal "A." This shows that the relay and circuitry is functional.
2. This insures that battery voltage is available to terminals "1" and "5" of the relay.
3. This step checks to make sure that the pull-in coil of the enable relay has continuity.
4. This checks the ability of the EBCM to energize the enable relay. An open circuit in the EBCM or CKT 879 would not allow the enable relay to be energized.
5. This step checks for an open CKT 850 between the enable relay and EBCM.

Diagnostic Aids:

An "Intermittent" problem may be caused by a poor connection, rubbed through wire insulation, or a wire that is broken inside the insulation.

Any circuitry that is suspected as causing the "Intermittent" complaint should be thoroughly checked for backed out terminals, improper mating, broken locks, improperly formed or damaged terminals, poor terminal to wiring connections or physical damage to the wiring harness.

VIBRATION, TEMPERATURE EFFECTS:

Check for vibration effects by performing the relay test function of the "Scan" tool. With the relay test commanded "ON," lightly tap the top and sides of the relay while monitoring relay voltage. If the relay voltage changes significantly, replace the relay.

If Code A014 only fails when the vehicle is initially started in cold ambient conditions (temperature less than 32°F - 0°C), replace the relay.

DIAGNOSTIC CHART DTC A014 — 1993 BERETTA AND CORSICA

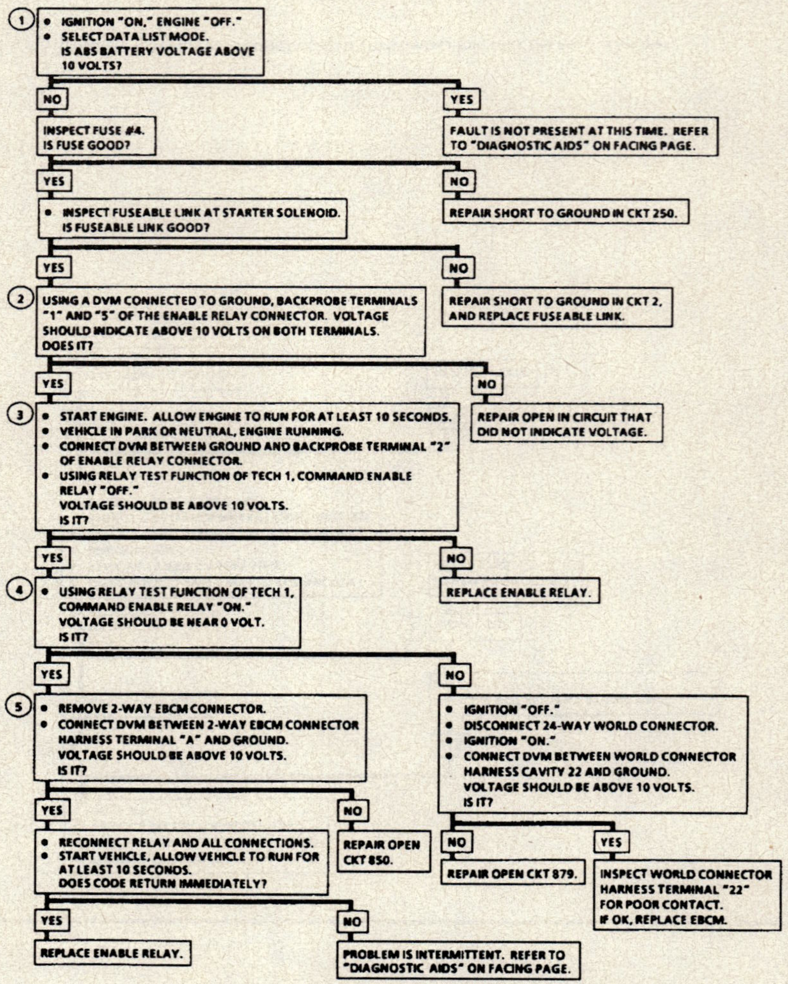

THIS CHART ASSUMES THAT A CURRENT CODE IS STORED INDICATING THAT THIS FAULT IS PRESENT.

DIAGNOSTIC CHART DTC A015 — 1993 BERETTA AND CORSICA

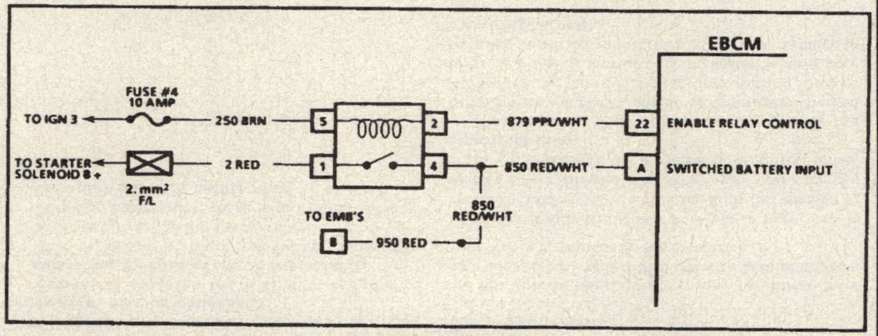

CODE A015

ENABLE RELAY CONTACTS SHORTED TO BATTERY

Circuit Description:

Ignition voltage is supplied through terminal "5" of the ABS enable relay. The EBCM then is able to energize the pull-in coil by completing the ground circuit at Pin "22" of the EBCM. The magnetic field created closes the enable relay contacts and allows battery voltage and current to be supplied to the EMB's and EBCM, which supplies power to the motors and solenoids.

This Code determines if the enable relay is energized when it should not be. This fault would not allow the enable relay to remove power to the ABS system. If a second fault were to occur that requires the enable relay to be turned "OFF," that fault cannot be removed if the relay cannot be controlled.

Test Description: Number(s) below refer to circled number(s) on the diagnostic chart.

1. This step indicates that the EBCM is capable of controlling the enable relay as commanded.
2. This step checks for a possible short to battery voltage in CKT 850 or CKT 950.
3. This step insures that the EBCM and also CKT 879 are not shorted to ground. A short to ground in either case would cause the enable relay to be energized constantly.

Diagnostic Aids:

An "Intermittent" problem may be caused by a poor connection, rubbed through wire insulation, or a wire that is broken inside the insulation.

Any circuitry, that is suspected as causing the intermittent complaint, should be thoroughly checked for backed out terminals, improper mating, broken locks, improperly formed or damaged terminals, poor terminal to wiring connections or physical damage to the wiring harness.

DIAGNOSTIC CHART DTC A015 — 1993 BERETTA AND CORSICA

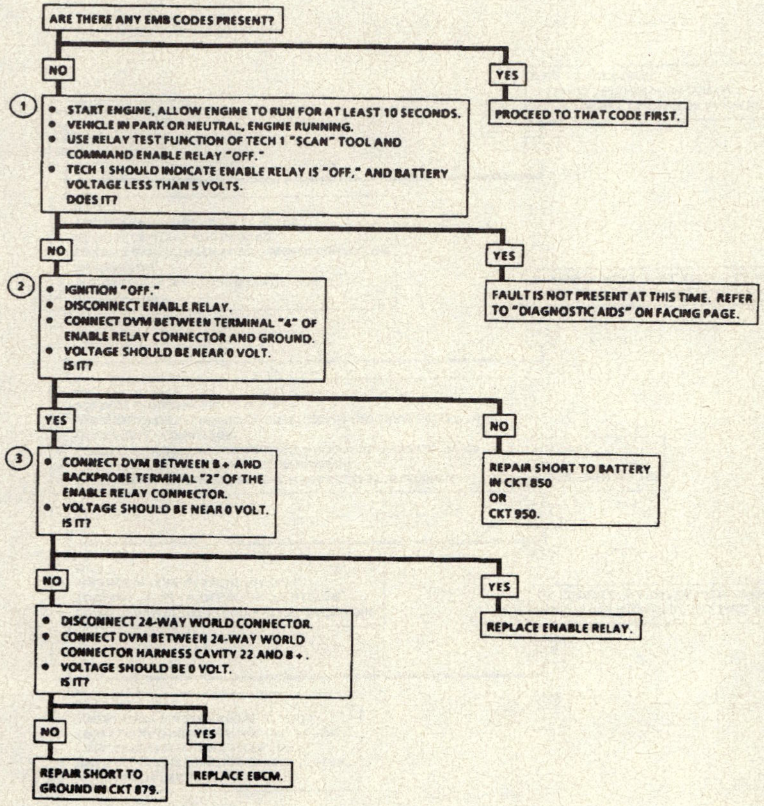

THIS CHART ASSUMES THAT A CURRENT CODE IS STORED INDICATING THAT THIS FAULT IS PRESENT.

DIAGNOSTIC CHART DTC A016 — 1993 BERETTA AND CORSICA

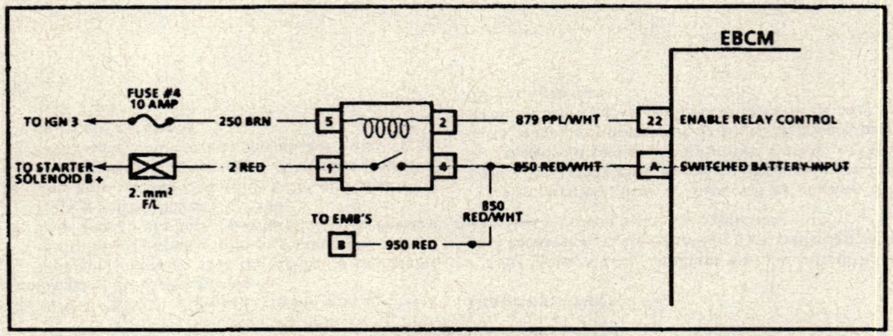

CODE A016

ENABLE RELAY COIL CIRCUIT OPEN

Circuit Description:

Ignition voltage is supplied through terminal "5" of the ABS enable relay. The EBCM then is able to energize the pull-in coil by completing the ground circuit at Pin "22" of the EBCM. The magnetic field created closes the enable relay contacts and allows battery voltage and current to be supplied to the EMB's and EBCM, which supplies power to the motors and solenoids.

This code detects opens in the enable relay coil circuit. An open in this circuit will not allow the enable relay to be energized, thus preventing voltage/current to the motors, EMB's and solenoids.

Test Description: Number(s) below refer to circled number(s) on the diagnostic chart.

1. This step indicates the EBCM is capable of controlling the enable relay as commanded.
2. This step checks to insure that voltage is available to the pull-in coil of the enable relay.
3. This step insures that there is continuity through the pull-in coil of the enable relay.

Diagnostic Aids:

An "Intermittent" problem may be caused by a poor connection, rubbed through wire insulation, or a wire that is broken inside the insulation.

If the frequency of the problem is high, but is currently intermittent, check for high coil resistance by measuring between relay terminals 5 and 2 using a DVM. If resistance shows greater than 100 ohms, replace the relay.

Any circuitry, that is suspected as causing the intermittent complaint, should be thoroughly checked for backed out terminals, improper mating, broken locks, improperly formed or damaged terminals, poor terminal to wiring connections or physical damage to the wiring harness.

DIAGNOSTIC CHART DTC A016 — 1993 BERETTA AND CORSICA

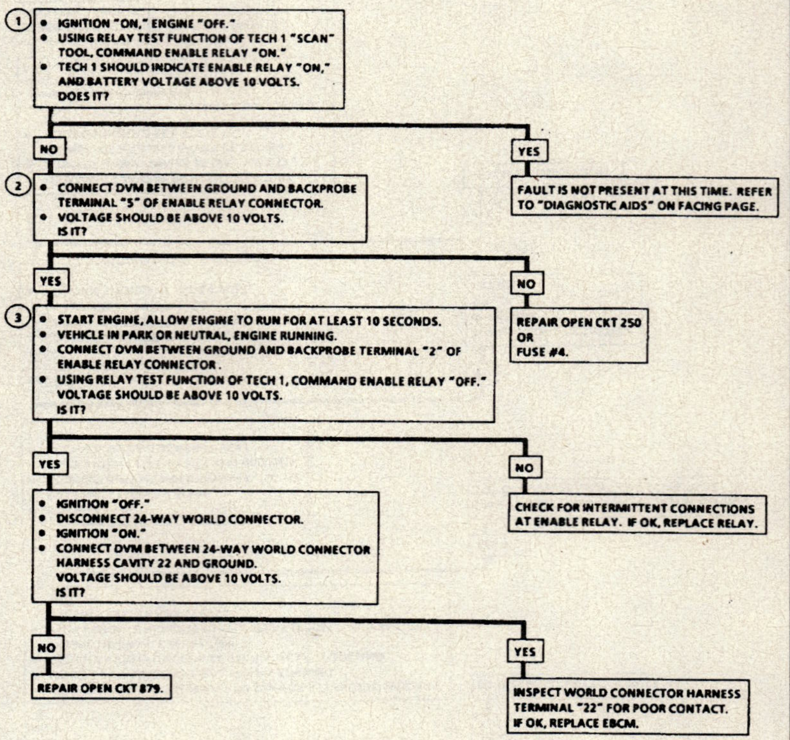

THIS CHART ASSUMES THAT A CURRENT CODE IS STORED INDICATING THAT THIS FAULT IS PRESENT

DIAGNOSTIC CHART DTC A017 — 1993 BERETTA AND CORSICA

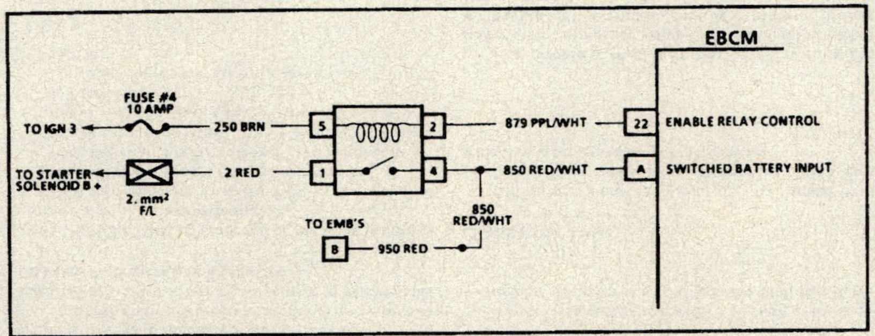

CODE A017

ENABLE RELAY COIL CIRCUIT SHORTED TO GROUND

Circuit Description:

Ignition voltage is supplied through terminal "5" of the ABS enable relay. The EBCM then is able to energize the pull-in coil by completing the ground circuit at Pin "22" of the EBCM. The magnetic field created closes the enable relay contacts and allows battery voltage and current to be supplied to the EMB's and EBCM, which supplies voltage to the motors and solenoids.

This code determines if the enable relay is energized when it should not be. This fault would not allow the enable relay to remove power to the ABS system. If a second fault were to occur that requires the enable relay to be turned "OFF," that fault cannot be removed if the relay cannot be controlled.

Test Description: Number(s) below refer to circled number(s) on the diagnostic chart.

1. This step indicates that the EBCM is capable of controlling the enable relay as commanded.
2. This checks to insure that the enable relay or control CKT 879 are not shorted to ground.
3. This step checks to see if the EBCM is internally shorted to ground. A grounded Pin "22" would cause the enable relay to be energized any time the ignition was in the "ON" position.

Diagnostic Aids:

An "Intermittent" problem may be caused by a poor connection, rubbed through wire insulation, or a wire that is broken inside the insulation.

Any circuitry, that is suspected as causing the intermittent complaint, should be thoroughly checked for backed out terminals, improper mating, broken locks, improperly formed or damaged terminals, poor terminal to wiring connections or physical damage to the wiring harness.

DIAGNOSTIC CHART DTC A017 — 1993 BERETTA AND CORSICA

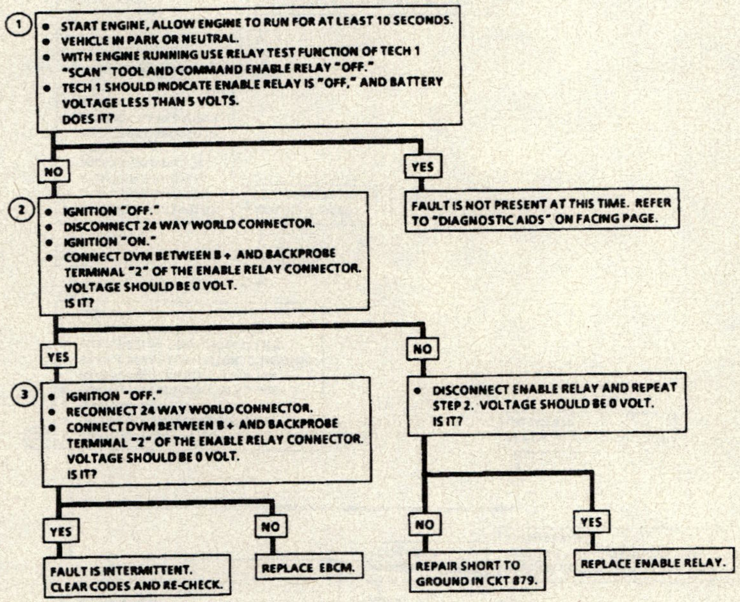

THIS CHART ASSUMES THAT A CURRENT CODE IS STORED INDICATING THAT THIS FAULT IS PRESENT.

DIAGNOSTIC CHART DTC A018 — 1993 BERETTA AND CORSICA

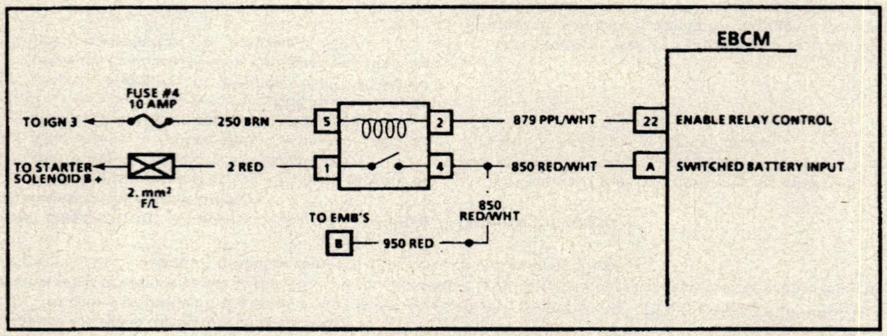

CODE A018

ENABLE RELAY COIL CIRCUIT SHORTED TO BATTERY

Circuit Description:

Ignition voltage is supplied through terminal "5" of the ABS enable relay. The EBCM then is able to energize the pull-in coil by completing the ground circuit at Pin "22" of the EBCM. The magnetic field created closes the enable relay contacts and allows battery voltage and current to be supplied to the EMB's and EBCM, which supplies voltage to the motors and solenoids.

This code monitors the availability of current/voltage to the motors, solenoids and EMB's. A fault of the type specified above would not allow voltage to be available to the motors, solenoids and EMB's and would, therefore, not allow ABS operation if required.

Test Description: Number(s) below refer to circled number(s) on the diagnostic chart.

1. This step indicates that the EBCM is capable of controlling the enable relay as commanded.
2. With the enable relay removed, voltage should not be available at cavity 2. Any voltage at this point would indicate that CKT 879 was shorted to a voltage source.
3. This step checks for an open Quad-Driver in the EBCM.

Diagnostic Aids:

An "Intermittent" problem may be caused by a poor connection, rubbed through wire insulation, or a wire that is broken inside the insulation.

Any circuitry, that is suspected as causing the intermittent complaint, should be thoroughly checked for backed out terminals, improper mating, broken locks, improperly formed or damaged terminals, poor terminal to wiring connections or physical damage to the wiring harness.

DIAGNOSTIC CHART DTC A018 — 1993 BERETTA AND CORSICA

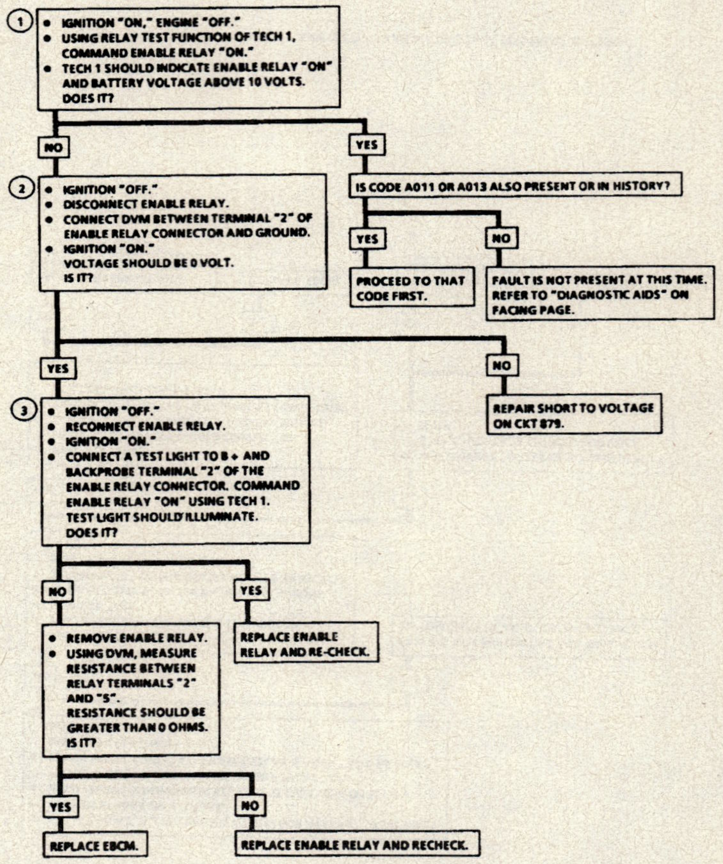

THIS CHART ASSUMES THAT A CURRENT CODE IS STORED INDICATING THAT THIS FAULT IS PRESENT.

DIAGNOSTIC CHART DTC A021 — 1993 BERETTA AND CORSICA

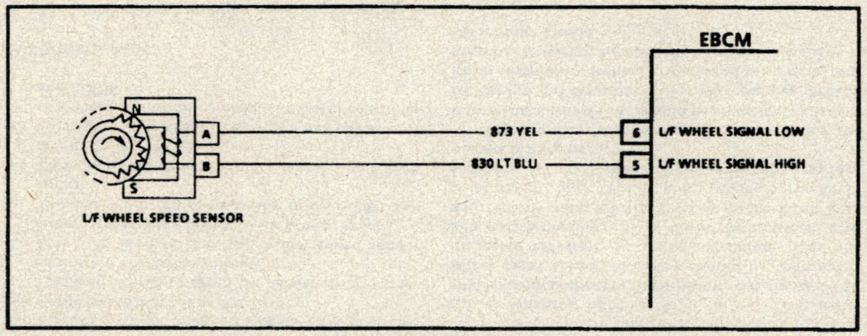

CODE A021

LEFT FRONT WHEEL SPEED = 0

Circuit Description:

As a toothed ring passes by the wheel speed sensor, changes in the electromagnetic field cause the wheel speed sensor to produce a sinusoidal (AC) voltage signal whose frequency is proportional to wheel speed. The magnitude of this signal is directly related to wheel speed and the proximity of the wheel speed sensor to the toothed ring, often referred to as the air gap.

The code detects opens, shorts to ground, and shorts to battery that cause a wheel speed of 0 mph to be calculated by the ABS EBCM. The conditions present to fail this test are: (1) Three (3) wheels are rotating at least 5 mph, and (2) The fourth wheel is 0 mph.

Test Description: Number(s) below refer to circled number(s) on the diagnostic chart.

1. This test verifies whether the fault is currently present.
2. This step is designed to identify a wheel speed sensor or circuitry that is damaged and visibly apparent.
3. This test checks to see that the wheel speed sensor is within proper resistance values.
4. This insures that the wheel speed sensor is not shorted to ground.

Diagnostic Aids:

An "Intermittent" problem may be caused by a poor connection, rubbed through wire insulation, or a wire that is broken inside the insulation.

If the customers comments reflect that the ABS Amber warning light is "ON" only during moist environmental changes, (rain, snow, vehicle wash) all wheel speed sensor circuitry should be thoroughly inspected for signs of water intrusion. Use the following procedure. Spray down the suspected area with 5% salt water solution (two teaspoons of salt to 12 oz. of water). Start vehicle, allow vehicle to run for 10 seconds. If code returns immediately, replace suspected harness.

Any circuitry, that is suspected as causing the intermittent complaint, should be thoroughly checked for backed out terminals, improper mating, broken locks, improperly formed or damaged terminals, poor terminal to wiring connections or physical damage to the wiring harness.

DIAGNOSTIC CHART DTC A021 — 1993 BERETTA AND CORSICA

IMPORTANT: WHEEL SPEED SENSOR INTERMITTENT PROBLEMS MAY BE DIFFICULT TO LOCATE. CARE SHOULD BE TAKEN NOT TO DISTURB ANY ELECTRICAL CONNECTIONS PRIOR TO AN INDICATED STEP OF THIS CHART. THIS WILL INSURE THAT AN INTERMITTENT CONNECTION WILL NOT BE CORRECTED BEFORE THE SOURCE OF THE PROBLEM IS FOUND.

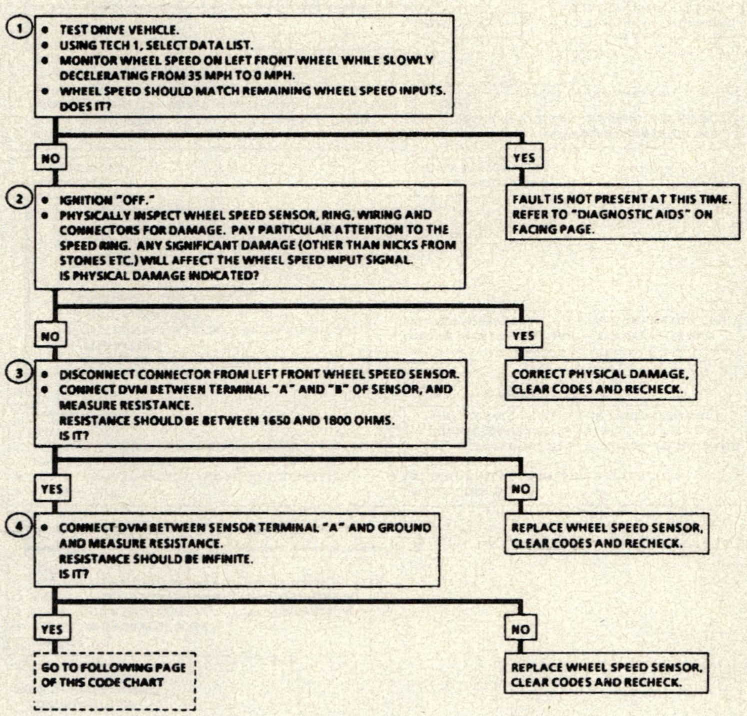

THIS CHART ASSUMES THAT A CURRENT CODE IS STORED INDICATING THAT THIS FAULT IS PRESENT.

DIAGNOSTIC CHART DTC A021 — 1993 BERETTA AND CORSICA

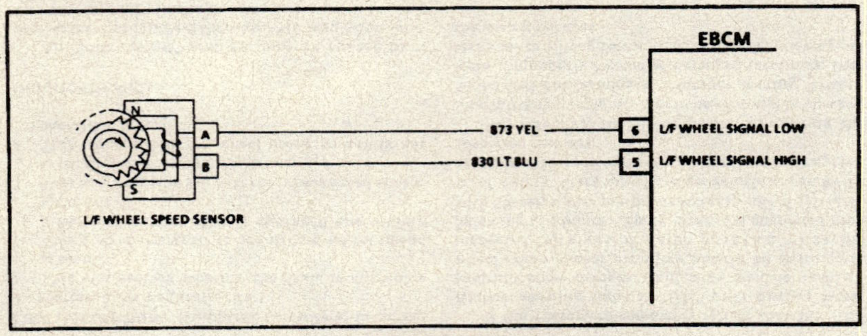

CODE A021

LEFT FRONT WHEEL SPEED = 0

Circuit Description:

As a toothed ring passes by the wheel speed sensor, changes in the electromagnetic field cause the wheel speed sensor to produce a sinusoidal (AC) voltage signal whose frequency is proportional to wheel speed. The magnitude of this signal is directly related to wheel speed and the proximity of the wheel speed sensor to the toothed ring often referred to as the air gap.

This code detects opens, shorts to ground, and shorts to battery that cause a wheel speed of 0 mph to be calculated by the ABS EBCM. The conditions present to fail this test are: (1) Three (3) wheels are rotating at least 5 mph, and (2) The fourth wheel is 0 mph.

Test Description: Number(s) below refer to circled number(s) on the diagnostic chart.

5. This step checks to see if the wheel speed sensor circuitry is shorted to voltage.
6. This step insures that the wheel speed sensor circuitry resistance is within the proper value.
7. This checks for an open in the wheel signal low circuit.
8. This checks for an open in the wheel signal high circuit.
9. This step insures that Code A021 was not set due to a poor connection between the world connector and EBCM.

Diagnostic Aids:

An "Intermittent" problem may be caused by a poor connection, rubbed through wire insulation, or a wire that is broken inside the insulation.

If the customers comments reflect that the ABS Amber warning light is "ON" only during moist environmental changes, (rain, snow, vehicle wash) all wheel speed sensor circuitry should be thoroughly inspected for signs of water intrusion. Use the following procedure. Spray down the suspected area with 5% salt water solution (two teaspoons of salt to 12 oz. of water). Start vehicle, allow vehicle to run for 10 seconds. If code returns immediately, replace suspected harness.

Any circuitry, that is suspected as causing the intermittent complaint, should be thoroughly checked for backed out terminals, improper mating, broken locks, improperly formed or damaged terminals, poor terminal to wiring connections or physical damage to the wiring harness.

DIAGNOSTIC CHART DTC A021 — 1993 BERETTA AND CORSICA

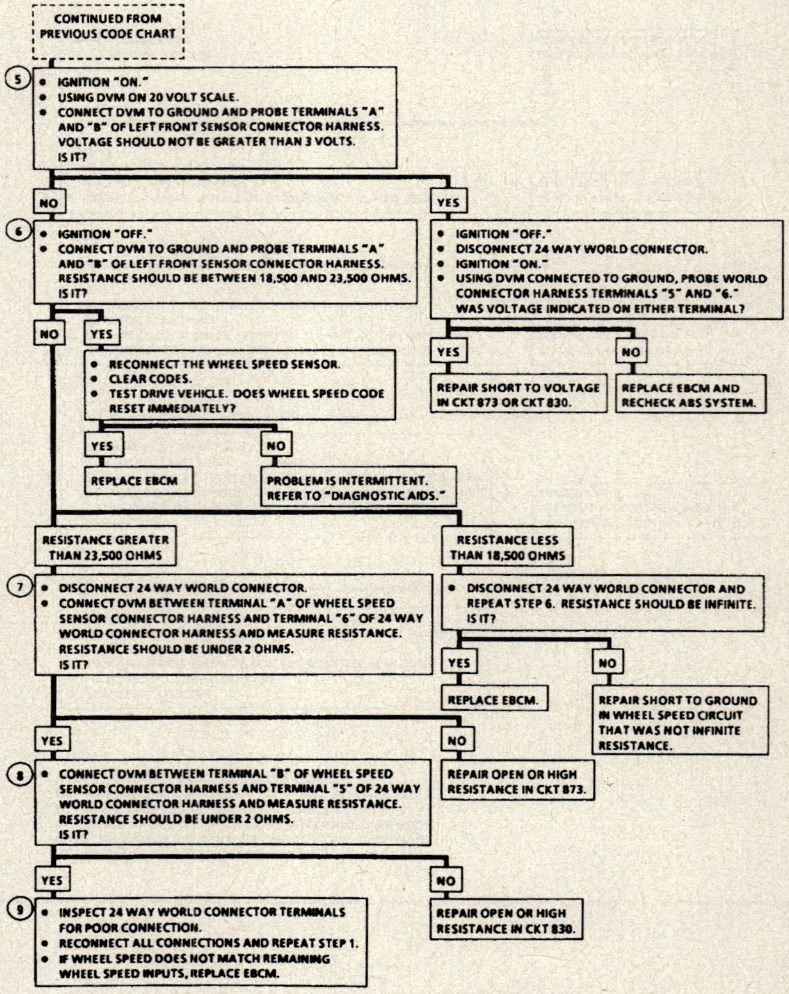

THIS CHART ASSUMES THAT A CURRENT CODE IS STORED INDICATING THAT THIS FAULT IS PRESENT.

DIAGNOSTIC CHART DTC A022 — 1993 BERETTA AND CORSICA

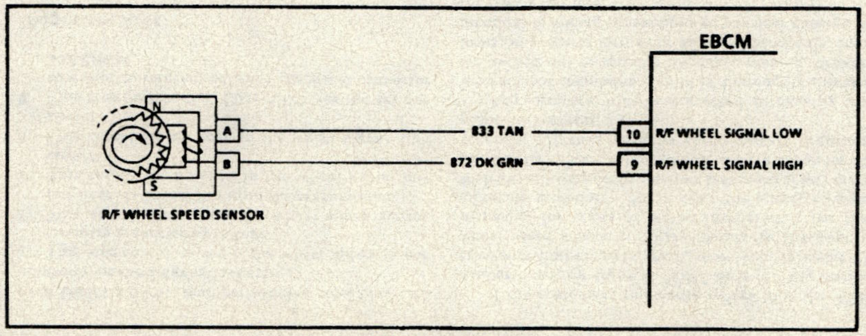

CODE A022

RIGHT FRONT WHEEL SPEED = 0

Circuit Description:

As a toothed ring passes by the wheel speed sensor, changes in the electromagnetic field cause the wheel speed sensor to produce a sinusoidal (AC) voltage signal whose frequency is proportional to wheel speed. The magnitude of this signal is directly related to wheel speed and the proximity of the wheel speed sensor to the toothed ring, often referred to as the air gap.

This code detects opens, shorts to ground, and shorts to battery that cause a wheel speed of 0 mph to be calculated by the ABS EBCM. The conditions present to fail this test are: (1) Three (3) wheels are rotating at least 5 mph, and (2) The fourth wheel is 0 mph.

Test Description: Number(s) below refer to circled number(s) on the diagnostic chart.

1. This test verifies whether the fault is currently present.
2. This step is designed to identify a wheel speed sensor or circuitry that is damaged and visibly apparent.
3. This test checks to see that the wheel speed sensor is within proper resistance values.
4. This insures that the wheel speed sensor is not shorted to ground.

Diagnostic Aids:

An "Intermittent" problem may be caused by a poor connection, rubbed through wire insulation, or a wire that is broken inside the insulation.

If the customers comments reflect that the ABS Amber warning light is on only during moist environmental changes, (rain, snow, vehicle wash) all wheel speed sensor circuitry should be thoroughly inspected for signs of water intrusion. Use the following procedure. Spray down the suspected area with 5% salt water solution (two teaspoons of salt to 12 oz. of water). Start vehicle, allow vehicle to run for 10 seconds. If code returns immediately, replace suspected harness.

Any circuitry, that is suspected as causing the intermittent complaint, should be thoroughly checked for backed out terminals, improper mating, broken locks, improperly formed or damaged terminals, poor terminal to wiring connections or physical damage to the wiring harness.

DIAGNOSTIC CHART DTC A022 — 1993 BERETTA AND CORSICA

IMPORTANT: WHEEL SPEED SENSOR INTERMITTENT PROBLEMS MAY BE DIFFICULT TO LOCATE. CARE SHOULD BE TAKEN NOT TO DISTURB ANY ELECTRICAL CONNECTIONS PRIOR TO AN INDICATED STEP OF THIS CHART. THIS WILL INSURE THAT AN INTERMITTENT CONNECTION WILL NOT BE CORRECTED BEFORE THE SOURCE OF THE PROBLEM IS FOUND.

(1)
- TEST DRIVE VEHICLE.
- IGNITION "ON."
- USING TECH 1, SELECT DATA LIST.
- MONITOR WHEEL SPEED ON RIGHT FRONT WHEEL WHILE SLOWLY DECELERATING FROM 35 MPH TO 0 MPH.

WHEEL SPEED SHOULD MATCH REMAINING WHEEL SPEED INPUTS. DOES IT?

- YES: FAULT IS NOT PRESENT AT THIS TIME. REFER TO "DIAGNOSTIC AIDS" ON FACING PAGE.
- NO: go to step 2.

(2)
- IGNITION "OFF."
- PHYSICALLY INSPECT WHEEL SPEED SENSOR, RING, WIRING AND CONNECTORS FOR DAMAGE. PAY PARTICULAR ATTENTION TO THE SPEED RING. ANY SIGNIFICANT DAMAGE (OTHER THAN NICKS FROM STONES ETC.) WILL AFFECT THE WHEEL SPEED INPUT SIGNAL.

IS PHYSICAL DAMAGE INDICATED?

- YES: CORRECT PHYSICAL DAMAGE, CLEAR CODES AND RECHECK.
- NO: go to step 3.

(3)
- DISCONNECT CONNECTOR FROM RIGHT FRONT WHEEL SPEED SENSOR.
- CONNECT DVM BETWEEN TERMINALS "A" AND "B" OF SENSOR, AND MEASURE RESISTANCE.

RESISTANCE SHOULD BE BETWEEN 1650 AND 1800 OHMS. IS IT?

- NO: REPLACE WHEEL SPEED SENSOR, CLEAR CODES AND RECHECK.
- YES: go to step 4.

(4)
- CONNECT DVM BETWEEN SENSOR TERMINAL "A" AND GROUND AND MEASURE RESISTANCE.

RESISTANCE SHOULD BE INFINITE. IS IT?

- NO: REPLACE WHEEL SPEED SENSOR, CLEAR CODES AND RECHECK.
- YES: GO TO FOLLOWING PAGE OF THIS CODE CHART

THIS CHART ASSUMES THAT A CURRENT CODE IS STORED INDICATING THAT THIS FAULT IS PRESENT.

DIAGNOSTIC CHART DTC A022 — 1993 BERETTA AND CORSICA

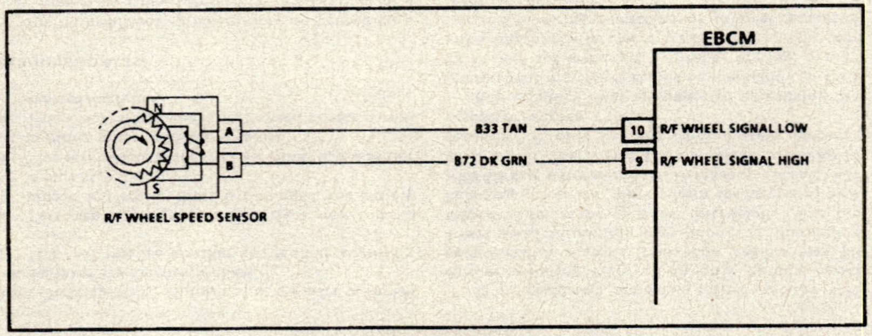

CODE A022

RIGHT FRONT WHEEL SPEED = 0

Circuit Description:

As a toothed ring passes by the wheel speed sensor, changes in the electromagnetic field cause the wheel speed sensor to produce a sinusoidal (AC) voltage signal whose frequency is proportional to wheel speed. The magnitude of this signal is directly related to wheel speed and the proximity of the wheel speed sensor to the toothed ring, often referred to as the air gap.

This code detects opens, shorts to ground, and shorts to battery that cause a wheel speed of 0 mph to be calculated by the ABS EBCM. The conditions present to fail this test are: (1) Three (3) wheels are rotating at least 5 mph, and (2) The fourth wheel is 0 mph.

Test Description: Number(s) below refer to circled number(s) on the diagnostic chart.

5. This step checks to see if the wheel speed sensor circuitry is shorted to voltage.
6. This step insures that the wheel speed sensor circuitry resistance is within the proper value.
7. This checks for an open in the wheel signal low circuit.
8. This checks for an open in the wheel signal high circuit.
9. This step insures that Code A022 was not set due to a poor connection between the world connector and EBCM.

Diagnostic Aids:

An "Intermittent" problem may be caused by a poor connection, rubbed through wire insulation, or a wire that is broken inside the insulation.

If the customers comments reflect that the ABS Amber warning light is "ON" only during moist environmental changes, (rain, snow, vehicle wash) all wheel speed sensor circuitry should be thoroughly inspected for signs of water intrusion. Use the following procedure. Spray down the suspected area with 5% salt water solution (two teaspoons of salt to 12 oz. of water). Start vehicle, allow vehicle to run for 10 seconds. If code returns immediately, replace suspected harness.

Any circuitry, that is suspected as causing the intermittent complaint, should be thoroughly checked for backed out terminals, improper mating, broken locks, improperly formed or damaged terminals, poor terminal to wiring connections or physical damage to the wiring harness.

DIAGNOSTIC CHART DTC A022 — 1993 BERETTA AND CORSICA

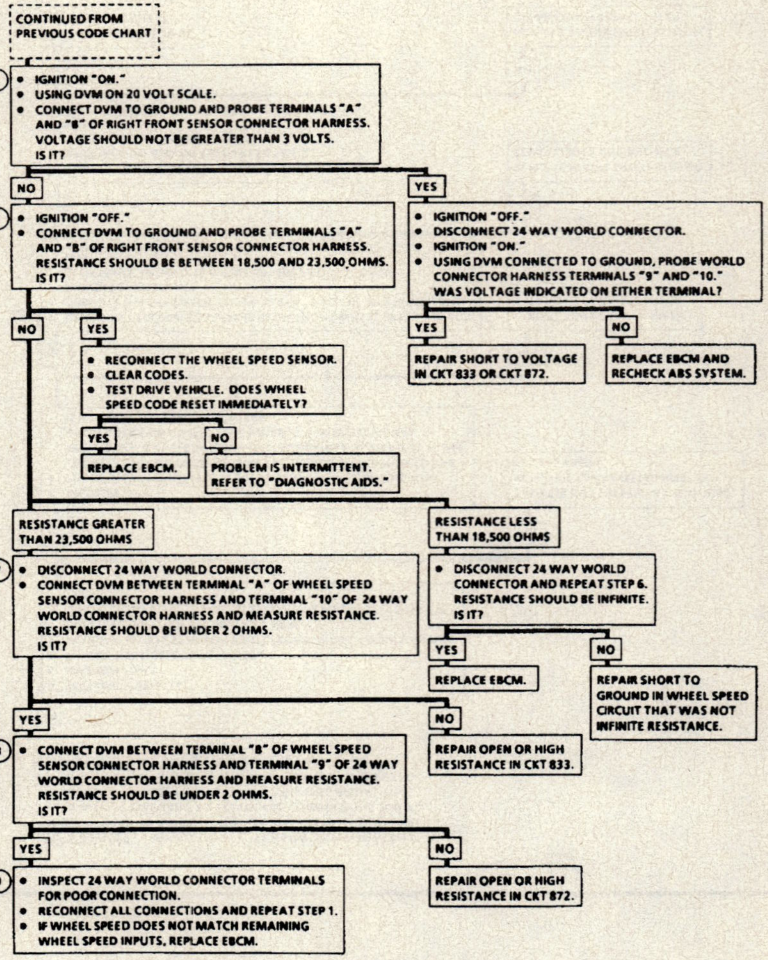

THIS CHART ASSUMES THAT A CURRENT CODE IS STORED INDICATING THAT THIS FAULT IS PRESENT.

DIAGNOSTIC CHART DTC A023 — 1993 BERETTA AND CORSICA

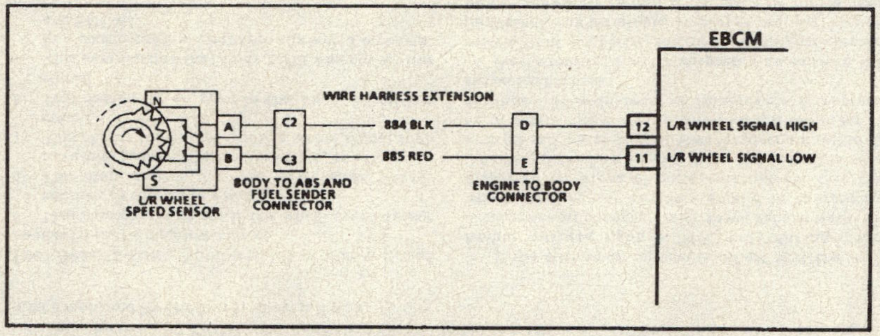

CODE A023

LEFT REAR WHEEL SPEED = 0

Circuit Description:

As a toothed ring passes by the wheel speed sensor, changes in the electromagnetic field cause the wheel speed sensor to produce a sinusoidal (AC) voltage signal whose frequency is proportional to wheel speed. The magnitude of this signal is directly related to wheel speed and the proximity of the wheel speed sensor to the toothed ring, often referred to as the air gap.

This code detects opens, shorts to ground, and shorts to battery that cause a wheel speed of 0 mph to be calculated by the ABS EBCM. The conditions present to fail this test are: (1) Three (3) wheels are rotating at least 5 mph, (2) The fourth wheel is 0 mph.

Test Description: Number(s) below refer to circled number(s) on the diagnostic chart.

1. This test verifies whether the fault is currently present.
2. This step is designed to identify a wheel speed sensor or circuitry that is damaged and visibly apparent.
3. This test checks to see that the wheel speed sensor is within proper resistance values.
4. This insures that the wheel speed sensor is not shorted to ground.

Diagnostic Aids:

An "Intermittent" problem may be caused by a poor connection, rubbed through wire insulation, or a wire that is broken inside the insulation.

If the customers comments reflect that the ABS Amber warning light is "ON" only during moist environmental changes, (rain, snow, vehicle wash) all wheel speed senor circuitry should be thoroughly inspected for signs of water intrusion. Use the following procedure. Spray down the suspected area with 5% salt water solution (two teaspoons of salt to 12 oz. of water). Start vehicle, allow vehicle to run for 10 seconds. If code returns immediately, replace suspected harness.

Any circuitry, that is suspected as causing the intermittent complaint, should be thoroughly checked for backed out terminals, improper mating, broken locks, improperly formed or damaged terminals, poor terminal to wiring connections or physical damage to the wiring harness.

DIAGNOSTIC CHART DTC A023 — 1993 BERETTA AND CORSICA

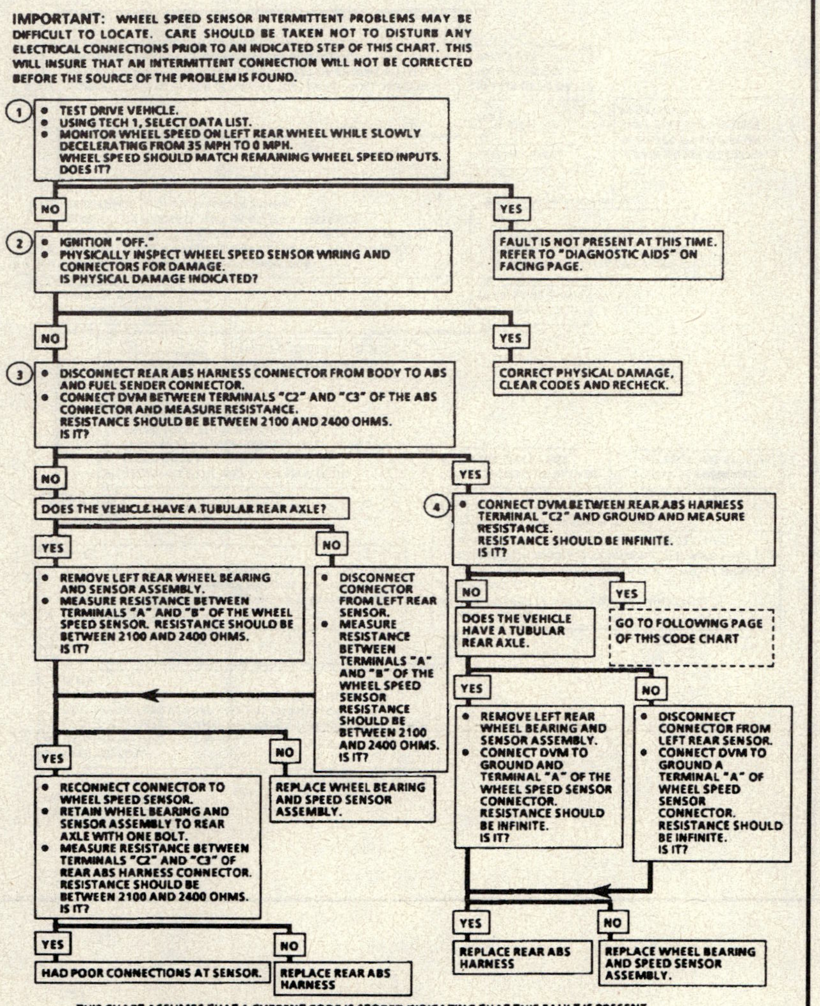

THIS CHART ASSUMES THAT A CURRENT CODE IS STORED INDICATING THAT THIS FAULT IS PRESENT.

DIAGNOSTIC CHART DTC A023 — 1993 BERETTA AND CORSICA

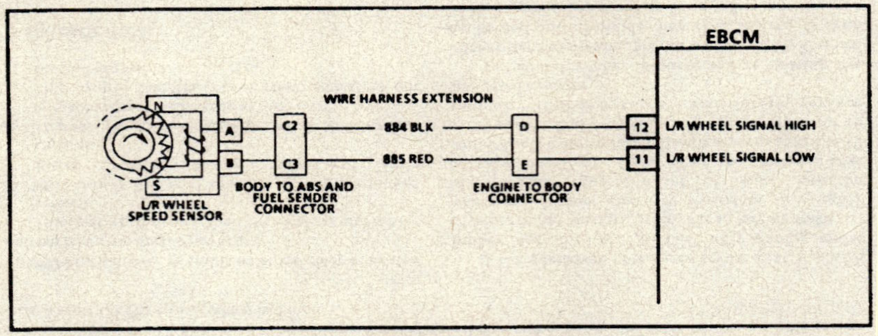

CODE A023

LEFT REAR WHEEL SPEED = 0

Circuit Description:

As a toothed ring passes by the wheel speed sensor, changes in the electromagnetic field cause the wheel speed sensor to produce a sinusoidal (AC) voltage signal whose frequency is proportional to wheel speed. The magnitude of this signal is directly related to wheel speed and the proximity of the wheel speed sensor to the toothed ring, often referred to as the air gap.

This code detects opens, shorts to ground, and shorts to battery that cause a wheel speed of 0 mph to be calculated by the ABS EBCM. The conditions present to fail this test are: (1) Three (3) wheels are rotating at least 5 mph, and (2) The fourth wheel is 0 mph.

Test Description: Number(s) below refer to circled number(s) on the diagnostic chart.

5. This step checks to see if the wheel speed sensor circuitry is shorted to voltage.
6. This step insures that the wheel speed sensor circuitry resistance is within the proper value.
7. This checks for an open in the wheel signal high circuit.
8. This checks for an open in the wheel signal low circuit.
9. This step insures that Code A023 was not set due to a poor connection between the world connector and EBCM.

Diagnostic Aids:

An "Intermittent" problem may be caused by a poor connection, rubbed through wire insulation, or a wire that is broken inside the insulation.

If the customers comments reflect that the ABS Amber warning light is "ON" only during moist environmental changes, (rain, snow, vehicle wash) all wheel speed sensor circuitry should be thoroughly inspected for signs of water intrusion. Use the following procedure. Spray down the suspected area with 5% salt water solution (two teaspoons of salt to 12 oz. of water). Start vehicle, allow vehicle to run for 10 seconds. If code returns immediately, replace suspected harness.

Any circuitry, that is suspected as causing the intermittent complaint, should be thoroughly checked for backed out terminals, improper mating, broken locks, improperly formed or damaged terminals, poor terminal to wiring connections or physical damage to the wiring harness.

DIAGNOSTIC CHART DTC A023 — 1993 BERETTA AND CORSICA

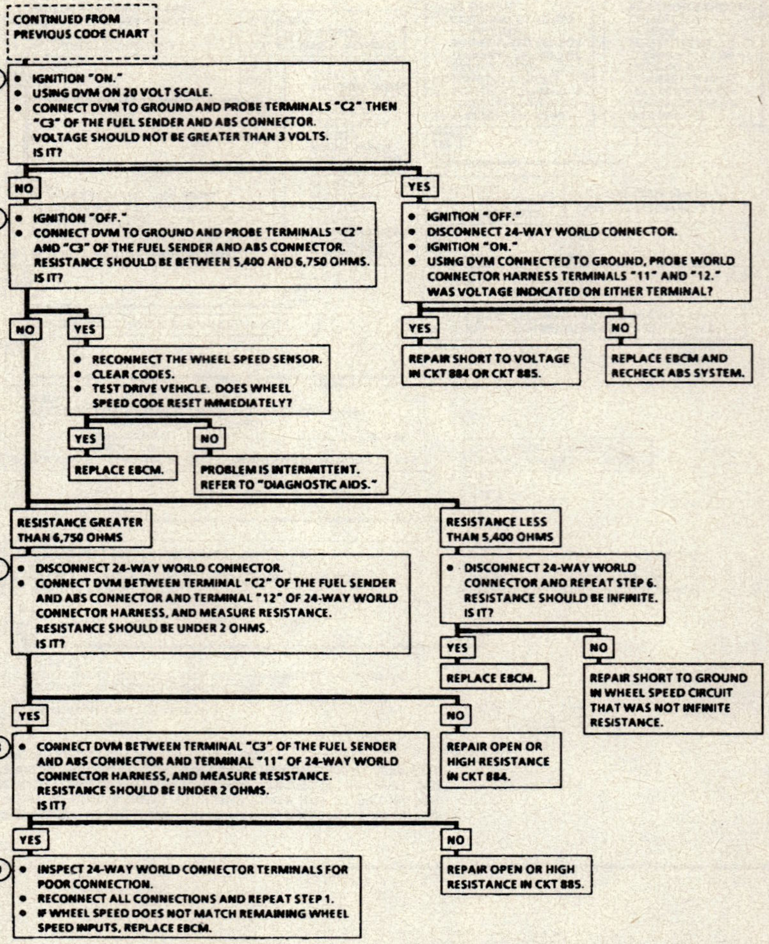

THIS CHART ASSUMES THAT A CURRENT CODE IS STORED INDICATING THAT THIS FAULT IS PRESENT.

DIAGNOSTIC CHART DTC A024 — 1993 BERETTA AND CORSICA

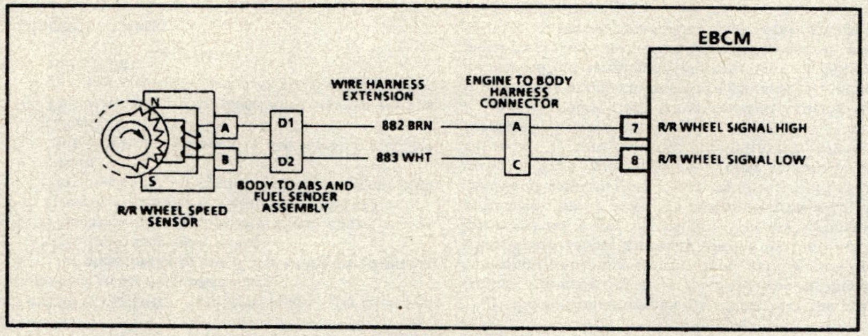

CODE A024

RIGHT REAR WHEEL SPEED = 0

Circuit Description:

As a toothed ring passes by the wheel speed sensor, changes in the electromagnetic field cause the wheel speed sensor to produce a sinusoidal (AC) voltage signal whose frequency is proportional to wheel speed. The magnitude of this signal is directly related to wheel speed and the proximity of the wheel speed sensor to the toothed ring, often referred to as the air gap.

This code detects opens, shorts to ground, and shorts to battery that cause a wheel speed of 0 mph to be calculated by the ABS EBCM. The conditions present to fail this test are: (1) Three (3) wheels are rotating at least 5 mph, and (2) The fourth wheel is 0 mph.

Test Description: Number(s) below refer to circled number(s) on the diagnostic chart.

1. This test verifies whether the fault is currently present.
2. This step is designed to identify a wheel speed sensor or circuitry that is damaged and visibly apparent.
3. This test checks to see that the wheel speed sensor is within proper resistance values.
4. This insures that the wheel speed sensor is not shorted to ground.

Diagnostic Aids:

An "Intermittent" problem may be caused by a poor connection, rubbed through wire insulation, or a wire that is broken inside the insulation.

If the customers comments reflect that the ABS Amber warning light is on only during moist environmental changes, (rain, snow, vehicle wash) all wheel speed sensor circuitry should be thoroughly inspected for water intrusion. Use the following procedure. Spray down the suspected area with 5% salt water solution (two teaspoons of salt to 12 oz. of water). Start vehicle, allow vehicle to run for 10 seconds. If code returns immediately, replace suspected harness.

Any circuitry, that is suspected as causing the intermittent complaint, should be thoroughly checked for backed out terminals, improper mating, broken locks, improperly formed or damaged terminals, poor terminal to wiring connections or physical damage to the wiring harness.

DIAGNOSTIC CHART DTC A024 — 1993 BERETTA AND CORSICA

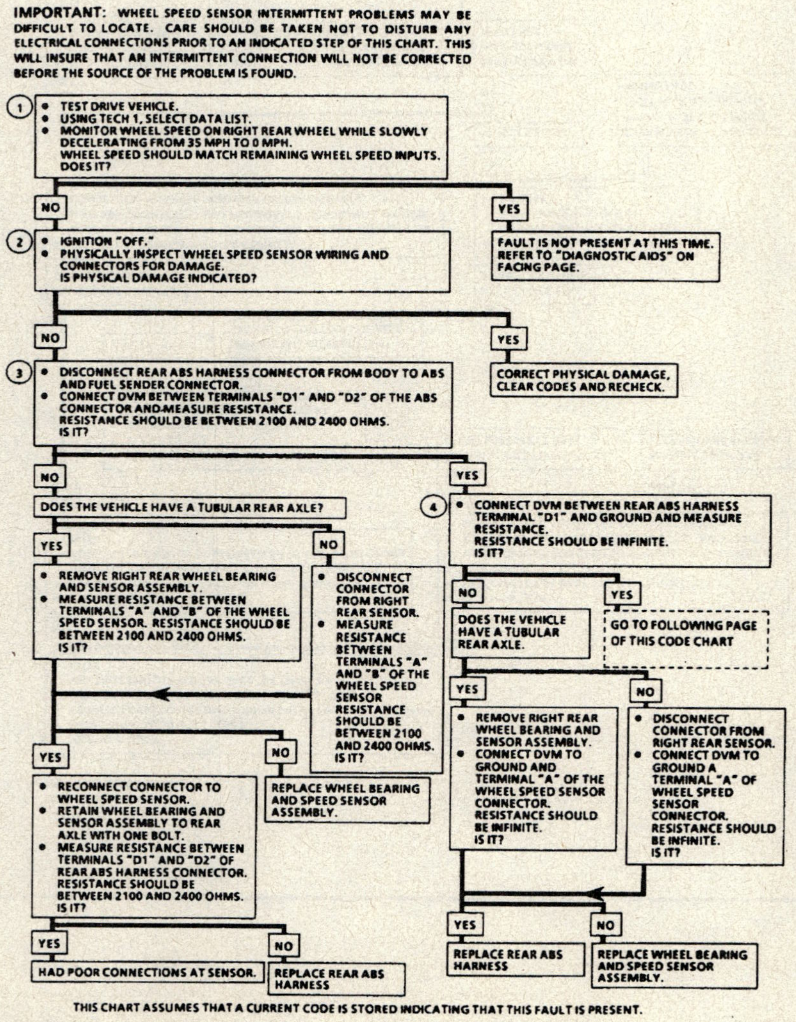

DIAGNOSTIC CHART DTC A024 — 1993 BERETTA AND CORSICA

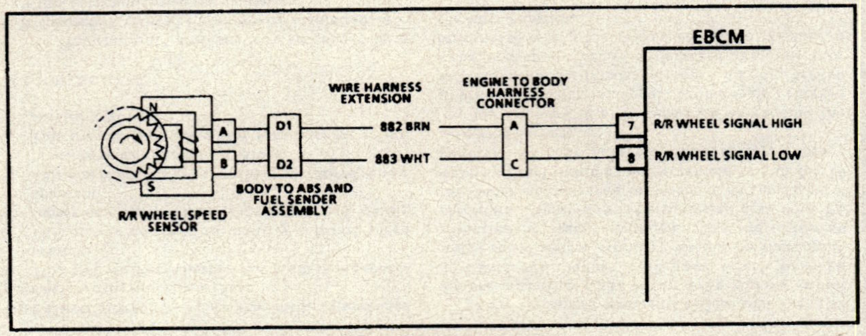

CODE A024

RIGHT REAR WHEEL SPEED = 0

Circuit Description:

As a toothed ring passes by the wheel speed sensor, changes in the electromagnetic field cause the wheel speed sensor to produce a sinusoidal (AC) voltage signal whose frequency is proportional to wheel speed. The magnitude of this signal is directly related to wheel speed and the proximity of the wheel speed sensor to the toothed ring, often referred to as the air gap.

This code detects opens, shorts to ground, and shorts to battery that cause a wheel speed of 0 mph to be calculated by the ABS EBCM. The conditions present to fail this test are: (1) Three (3) wheels are rotating at least 5 mph, and (2) The fourth wheel is 0 mph.

Test Description: Number(s) below refer to circled number(s) on the diagnostic chart.

5. This step checks to see if the wheel speed sensor circuitry is shorted to voltage.
6. This step insures that the wheel speed sensor circuitry resistance is within the proper value.
7. This checks for an open in the wheel signal high circuit.
8. This checks for an open in the wheel signal low circuit.
9. This step insures that Code A024 was not set due to a poor connection between the world connector and EBCM.

Diagnostic Aids:

An "Intermittent" problem may be caused by a poor connection, rubbed through wire insulation, or a wire that is broken inside the insulation.

If the customers comments reflect that the ABS Amber warning light is on only during moist environmental changes, (rain, snow, vehicle wash) all wheel speed sensor circuitry should be thoroughly inspected for water intrusion. Use the following procedure. Spray down the suspected area with 5% salt water solution (two teaspoons of salt to 12 oz. of water). Start vehicle, allow vehicle to run for 10 seconds. If code returns immediately, replace suspected harness.

Any circuitry, that is suspected as causing the intermittent complaint, should be thoroughly checked for backed out terminals, improper mating, broken locks, improperly formed or damaged terminals, poor terminal to wiring connections or physical damage to the wiring harness.

DIAGNOSTIC CHART DTC A024 — 1993 BERETTA AND CORSICA

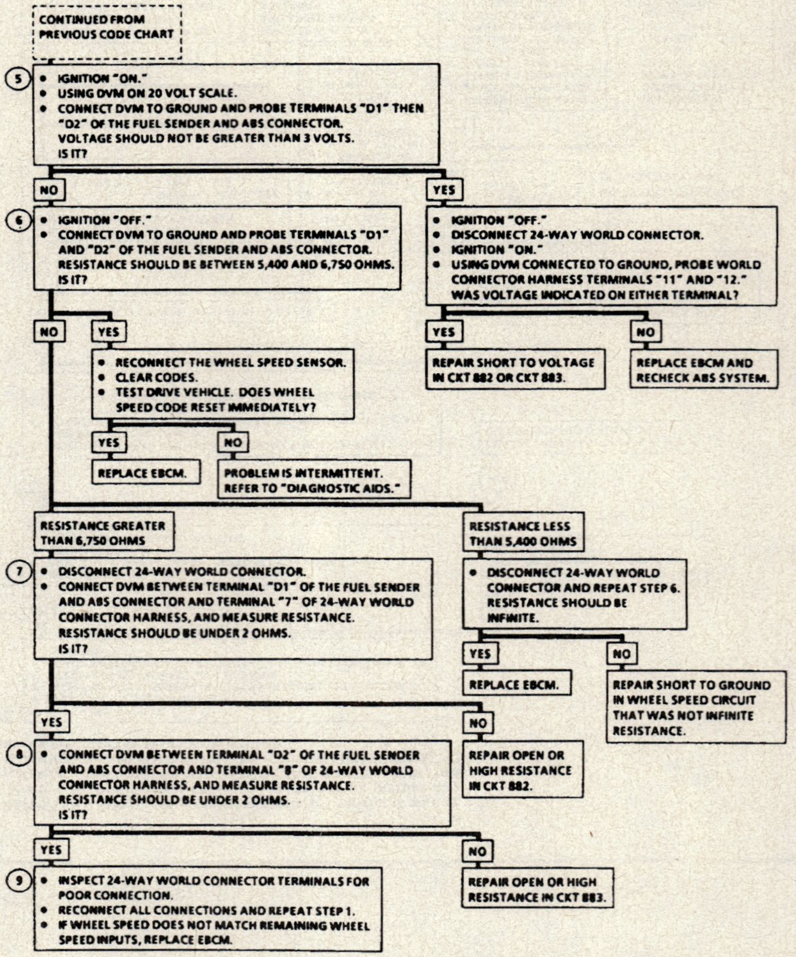

THIS CHART ASSUMES THAT A CURRENT CODE IS STORED INDICATING THAT THIS FAULT IS PRESENT.

DIAGNOSTIC CHART DTC A025 — 1993 BERETTA AND CORSICA

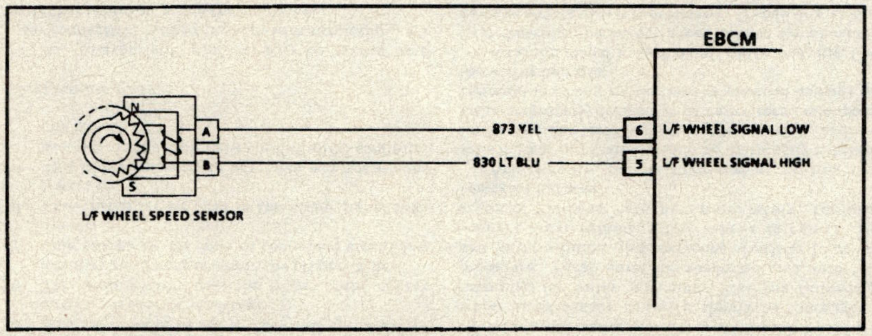

CODE A025

LEFT FRONT EXCESSIVE WHEEL SPEED VARIATION

Circuit Description:

As a toothed ring passes by the wheel speed sensor, changes in the electromagnetic field cause the wheel speed sensor to produce a sinusoidal (AC) voltage signal whose frequency is proportional to wheel speed. The magnitude of this signal is directly related to wheel speed and the proximity of the wheel speed sensor to the toothed ring, often referred to as the air gap.

This code detects opens, shorts to ground, and shorts to battery that cause intermittent wheel speed operation. Also detected is any sudden change in wheel speed determined to the unreasonable.

Test Description: Number(s) below refer to circled number(s) on the diagnostic chart.

1. This test verifies whether the fault is currently present.
2. This step is designed to identify a wheel speed sensor or circuitry that is damaged and visibly apparent.
3. This test checks to see that the wheel speed sensor circuitry is within proper resistance value.
4. This insure that the wheel speed sensor is not shorted to ground.

Diagnostic Aids:

An "Intermittent" problem may be caused by a poor connection, rubbed through wire insulation, or a wire that is broken inside the insulation.

If the customer's comments reflect that the ABS Amber warning light is on only during moist environmental changes, (rain, snow, vehicle wash) all wheel speed sensor circuitry should be thoroughly inspected for water intrusion. Use the following procedure. Spray down the suspected area with 5% salt water solution (two teaspoons of salt to 12 oz. of water). Start vehicle, allow vehicle to run for 10 seconds. If code returns immediately, replace suspected harness.

Any circuitry, that is suspected as causing the intermittent complaint, should be thoroughly checked for backed out terminals, improper mating, broken locks, improperly formed or damaged terminals, poor terminal to wiring connections or physical damage to the wiring harness.

A careful visual inspection of the toothed ring, CV joint, bearing and wheel speed sensor for physical damage should also be performed. If code sets at the same mph every time and wheel speed variation is noted above this mph on the Tech 1, the speed ring is most likely damaged. This can occur at any speed.

DIAGNOSTIC CHART DTC A025 — 1993 BERETTA AND CORSICA

IMPORTANT: WHEEL SPEED SENSOR INTERMITTENT PROBLEMS MAY BE DIFFICULT TO LOCATE. CARE SHOULD BE TAKEN NOT TO DISTURB ANY ELECTRICAL CONNECTIONS PRIOR TO AN INDICATED STEP OF THIS CHART. THIS WILL INSURE THAT AN INTERMITTENT CONNECTION WILL NOT BE CORRECTED BEFORE THE SOURCE OF THE PROBLEM IS FOUND.

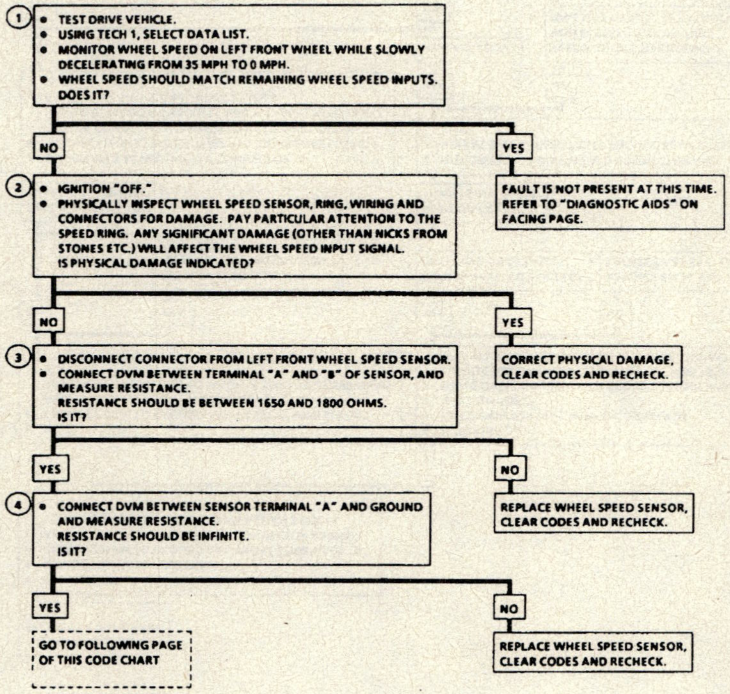

THIS CHART ASSUMES THAT A CURRENT CODE IS STORED INDICATING THAT THIS FAULT IS PRESENT.

DIAGNOSTIC CHART DTC A025 — 1993 BERETTA AND CORSICA

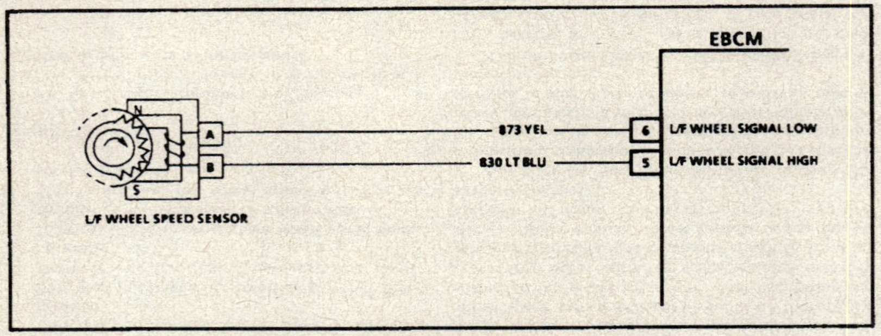

CODE A025

LEFT FRONT EXCESSIVE WHEEL SPEED VARIATION

Circuit Description:

As a toothed ring passes by the wheel speed sensor, changes in the electromagnetic field cause the wheel speed sensor to produce a sinusoidal (AC) voltage signal whose frequency is proportional to wheel speed. The magnitude of this signal is directly related to wheel speed and the proximity of the wheel speed sensor to the toothed ring, often referred to as the air gap.

This code detects opens, shorts to ground, and shorts to battery that cause intermittent wheel speed operational. Also detected is any sudden change in wheel speed determined to be unreasonable.

Test Description: Number(s) below refer to circled number(s) on the diagnostic chart.

5. This step checks to see if the wheel speed sensor circuitry is shorted to voltage.
6. This step insures that the wheel speed sensor circuitry resistance is within the proper value.
7. This checks for an open in the wheel signal low circuit.
8. This checks for an open in the wheel signal high circuit.
9. This step insures that Code A025 was not set due to a poor connection between the world connector and EBCM.

Diagnostic Aids:

An "Intermittent" problem may be caused by a poor connection, rubbed through wire insulation, or a wire that is broken inside the insulation.

If the customers comments reflect that the ABS Amber warning light is on only during moist environmental changes, (rain, snow, vehicle wash) all wheel speed sensor circuitry should be thoroughly inspected for water intrusion. Use the following procedure. Spray down the suspected area with 5% salt water solution (two teaspoons of salt to 12 oz. of water). Start vehicle, allow vehicle to run for 10 seconds. If code returns immediately, replace suspected harness.

Any circuitry, that is suspected as causing the intermittent complaint, should be thoroughly checked for backed out terminals, improper mating, broken locks, improperly formed or damaged terminals, poor terminal to wiring connections or physical damage to the wiring harness.

A careful visual inspection of the toothed ring, CV joint, bearing and wheel speed sensor for physical damage should also be performed. If code sets at the same mph every time and wheel speed variation is noted above this mph on the Tech 1, the speed ring is most likely damaged. This can occur at any speed.

DIAGNOSTIC CHART DTC A025 — 1993 BERETTA AND CORSICA

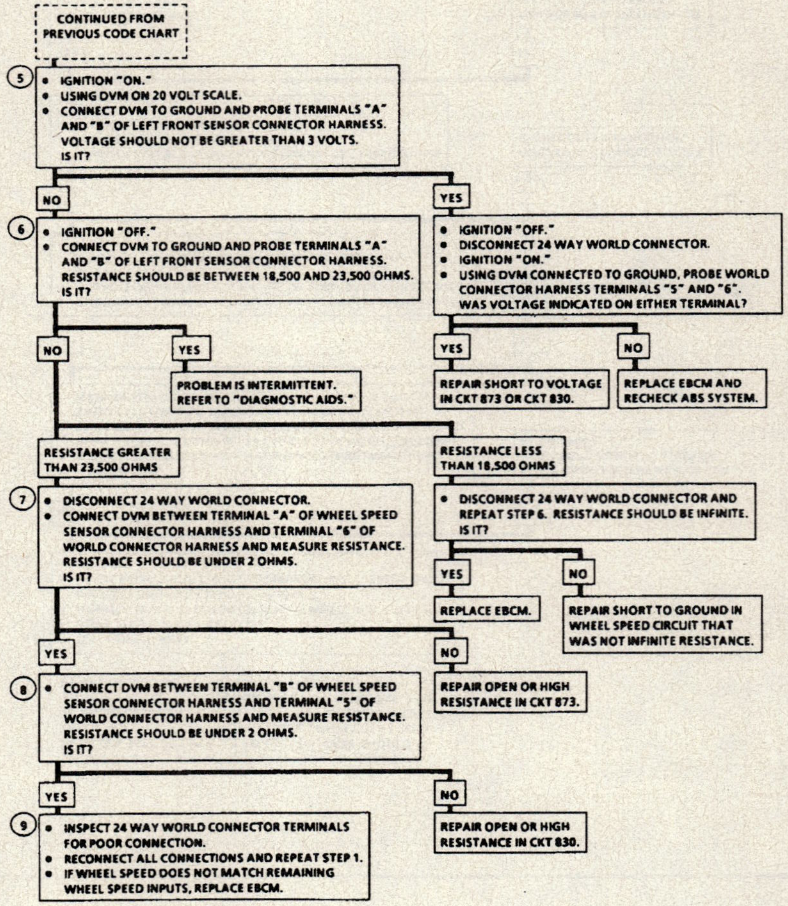

THIS CHART ASSUMES THAT A CURRENT CODE IS STORED INDICATING THAT THIS FAULT IS PRESENT.

DIAGNOSTIC CHART DTC A026 — 1993 BERETTA AND CORSICA

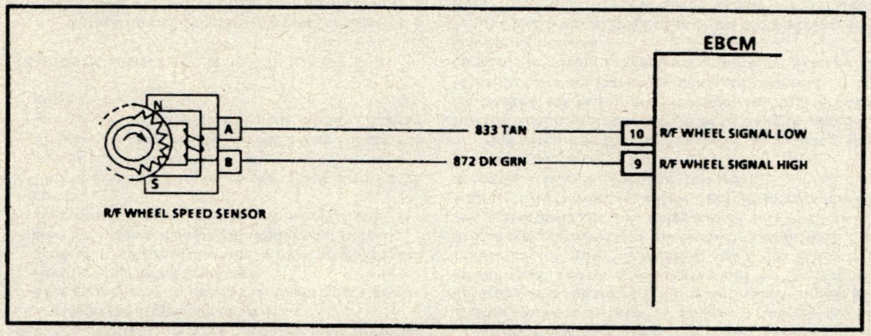

CODE A026

RIGHT FRONT EXCESSIVE WHEEL SPEED VARIATION

Circuit Description:

As a toothed ring passes by the wheel speed sensor, changes in the electromagnetic field cause the wheel speed sensor to produce a sinusoidal (AC) voltage signal whose frequency is proportional to wheel speed. The magnitude of this signal is directly related to wheel speed and the proximity of the wheel speed sensor to the toothed ring, often referred to as the air gap.

This code detects opens, shorts to ground, and shorts to battery that cause intermittent wheel speed operation. Also detected is any sudden change in wheel speed determined to be unreasonable.

Test Description: Number(s) below refer to circled number(s) on the diagnostic chart.

1. This test verifies whether the fault is currently present.
2. This step is designed to identify a wheel speed sensor or circuitry that is damaged and visibly apparent.
3. This test checks to see that the wheel speed sensor circuitry is within proper resistance value.
4. This insures that the wheel speed sensor is not shorted to ground.

Diagnostic Aids:

An "Intermittent" problem may be caused by a poor connection, rubbed through wire insulation, or a wire that is broken inside the insulation.

If the customers comments reflect that the ABS Amber warning light is on only during moist environmental changes, (rain, snow, vehicle wash) all wheel speed sensor circuitry should be thoroughly inspected for water intrusion. Use the following procedure. Spray down the suspected area with 5% salt water solution (two teaspoons of salt to 12 oz. of water). Start vehicle, allow vehicle to run for 10 seconds. If code returns immediately, replace suspected harness.

Any circuitry, that is suspected as causing the intermittent complaint, should be thoroughly checked for backed out terminals, improper mating, broken locks, improperly formed or damaged terminals, poor terminal to wiring connections or physical damage to the wiring harness.

A careful visual inspection of the toothed ring, CV joint, bearing and wheel speed sensor for physical damage should also be performed. If code sets at the same mph every time and wheel speed variation is noted above this mph on the Tech 1, the speed ring is most likely damaged. This can occur at any speed.

DIAGNOSTIC CHART DTC A026 — 1993 BERETTA AND CORSICA

IMPORTANT: WHEEL SPEED SENSOR INTERMITTENT PROBLEMS MAY BE DIFFICULT TO LOCATE. CARE SHOULD BE TAKEN NOT TO DISTURB ANY ELECTRICAL CONNECTIONS PRIOR TO AN INDICATED STEP OF THIS CHART. THIS WILL INSURE THAT AN INTERMITTENT CONNECTION WILL NOT BE CORRECTED BEFORE THE SOURCE OF THE PROBLEM IS FOUND.

(1)
- TEST DRIVE VEHICLE.
- IGNITION "ON."
- USING TECH 1, SELECT DATA LIST.
- MONITOR WHEEL SPEED ON RIGHT FRONT WHEEL WHILE SLOWLY DECELERATING FROM 35 MPH TO 0 MPH. WHEEL SPEED SHOULD MATCH REMAINING WHEEL SPEED INPUTS. DOES IT?

YES: FAULT IS NOT PRESENT AT THIS TIME. REFER TO "DIAGNOSTIC AIDS" ON FACING PAGE.

NO:

(2)
- IGNITION "OFF."
- PHYSICALLY INSPECT WHEEL SPEED SENSOR, RING, WIRING AND CONNECTORS FOR DAMAGE. PAY PARTICULAR ATTENTION TO THE SPEED RING. ANY SIGNIFICANT DAMAGE (OTHER THAN NICKS FROM STONES ETC.) WILL AFFECT THE WHEEL SPEED INPUT SIGNAL. IS PHYSICAL DAMAGE INDICATED?

YES: CORRECT PHYSICAL DAMAGE. CLEAR CODES AND RECHECK.

NO:

(3)
- DISCONNECT CONNECTOR FROM RIGHT FRONT WHEEL SPEED SENSOR.
- CONNECT DVM BETWEEN TERMINALS "A" AND "B" OF SENSOR, AND MEASURE RESISTANCE. RESISTANCE SHOULD BE BETWEEN 1650 AND 1800 OHMS. IS IT?

NO: REPLACE WHEEL SPEED SENSOR, CLEAR CODES AND RECHECK.

YES:

(4)
- CONNECT DVM BETWEEN SENSOR TERMINAL "A" AND GROUND AND MEASURE RESISTANCE. RESISTANCE SHOULD BE INFINITE. IS IT?

NO: REPLACE WHEEL SPEED SENSOR, CLEAR CODES AND RECHECK.

YES: GO TO FOLLOWING PAGE OF THIS CODE CHART

THIS CHART ASSUMES THAT A CURRENT CODE IS STORED INDICATING THAT THIS FAULT IS PRESENT.

DIAGNOSTIC CHART DTC A026 — 1993 BERETTA AND CORSICA

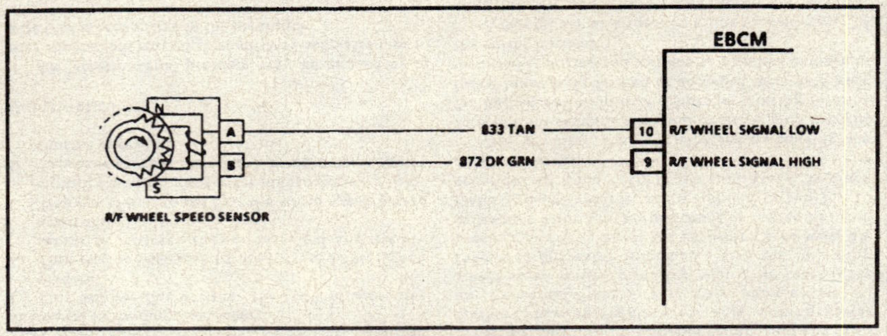

CODE A026

RIGHT FRONT EXCESSIVE WHEEL SPEED VARIATION

Circuit Description:

As a toothed ring passes by the wheel speed sensor, changes in the electromagnetic field cause the wheel speed sensor to produce a sinusoidal (AC) voltage signal whose frequency is proportional to wheel speed. The magnitude of this signal is directly related to wheel speed and the proximity of the wheel speed sensor to the toothed ring, often referred to as the air gap.

This code detects opens, shorts to ground, and shorts to battery that cause intermittent wheel speed operation. Also detected is any sudden change in wheel speed determined to be unreasonable.

Test Description: Number(s) below refer to circled number(s) on the diagnostic chart.

5. This step checks to see if the wheel speed sensor circuitry is shorted to voltage.
6. This step insures that the wheel speed sensor circuitry resistance is within the proper value.
7. This checks for an open in the wheel signal low circuit.
8. This checks for an open in the wheel signal high circuit.
9. This step insures that Code A026 was not set due to a poor connection between the world connector and EBCM.

Diagnostic Aids:

An "Intermittent" problem may be caused by a poor connection, rubbed through wire insulation, or a wire that is broken inside the insulation.

If the customers comments reflect that the ABS Amber warning light is on only during moist environmental changes, (rain, snow, vehicle wash) all wheel speed sensor circuitry should be thoroughly inspected for water intrusion. Use the following procedure. Spray down the suspected area with 5% salt water solution (two teaspoons of salt to 12 oz. of water). Start vehicle, allow vehicle to run for 10 seconds. If code returns immediately, replace suspected harness.

Any circuitry, that is suspected as causing the intermittent complaint, should be thoroughly checked for backed out terminals, improper mating, broken locks, improperly formed or damaged terminals, poor terminal to wiring connections or physical damage to the wiring harness.

A careful visual inspection of the toothed ring, CV joint, bearing and wheel speed sensor for physical damage should also be performed. If code sets at the same mph every time and wheel speed variation is noted above this mph on the Tech 1, the speed ring is most likely damaged. This can occur at any speed.

DIAGNOSTIC CHART DTC A026 — 1993 BERETTA AND CORSICA

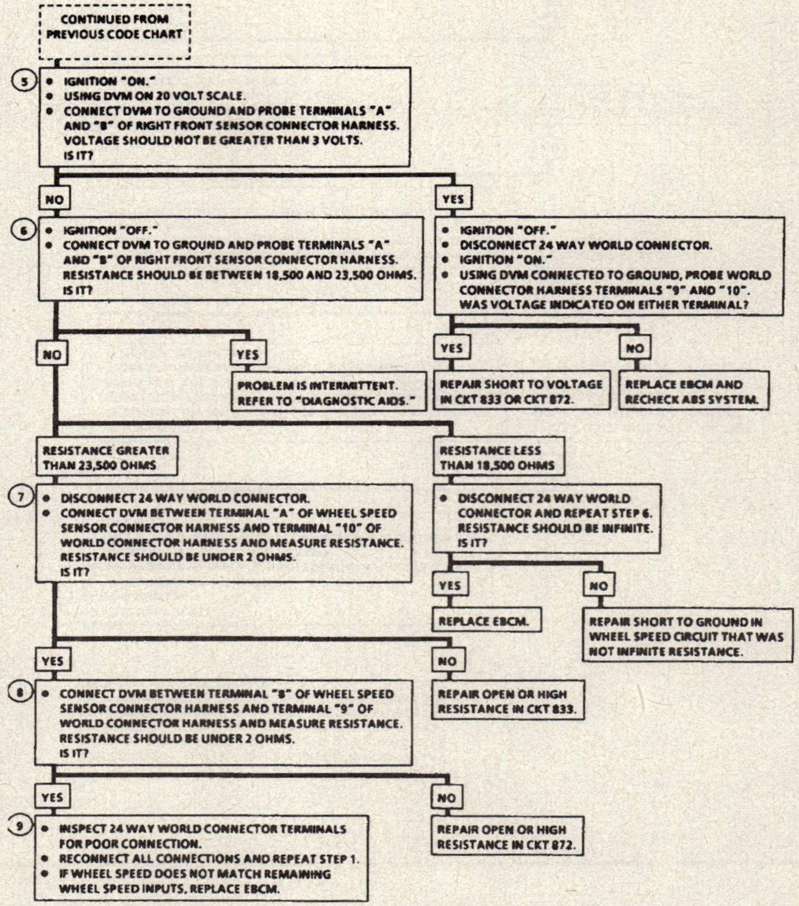

THIS CHART ASSUMES THAT A CURRENT CODE IS STORED INDICATING THAT THIS FAULT IS PRESENT.

DIAGNOSTIC CHART DTC A027 — 1993 BERETTA AND CORSICA

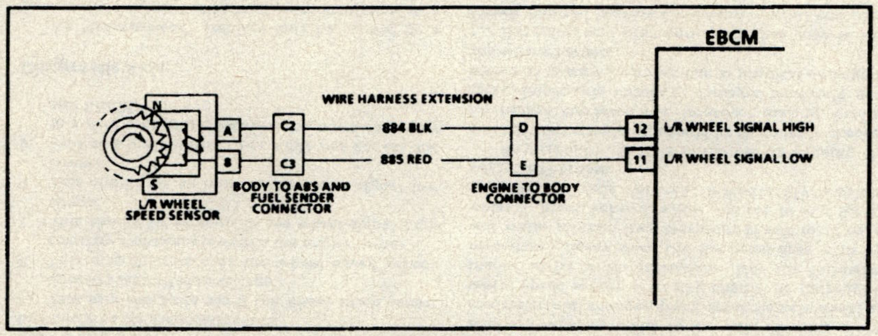

CODE A027

LEFT REAR EXCESSIVE WHEEL SPEED VARIATION

Circuit Description:

As a toothed ring passes by the wheel speed sensor, changes in the electromagnetic field cause the wheel speed sensor to produce a sinusoidal (AC) voltage signal whose frequency is proportional to wheel speed. The magnitude of this signal is directly related to wheel speed and the proximity of the wheel speed sensor to the toothed ring, often referred to as the air gap.

This code detects opens, shorts to ground, and shorts to battery that cause intermittent wheel speed operation. Also detected is any sudden change in wheel speed determined to be unreasonable.

Test Description: Number(s) below refer to circled number(s) on the diagnostic chart.

1. This test verifies whether the fault is currently present.
2. This step is designed to identify a wheel speed sensor or circuitry that is damaged and visibly apparent.
3. This test checks to see that the wheel speed sensor circuitry is within proper resistance value.
4. This insures that the wheel speed sensor is not shorted to ground.

Diagnostic Aids:

An "Intermittent" problem may be caused by a poor connection, rubbed through wire insulation, or a wire that is broken inside the insulation.

If the customers comments reflect that the ABS Amber warning light is on only during moist environmental changes, (rain, snow, vehicle wash) all wheel speed sensor circuitry should be thoroughly inspected for water intrusion. Use the following procedure. Spray down the suspected area with 5% salt water solution (two teaspoons of salt to 12 oz. of water). Start vehicle, allow vehicle to run for 10 seconds. If code returns immediately, replace suspected harness.

Any circuitry, that is suspected as causing the intermittent complaint, should be thoroughly checked for backed out terminals, improper mating, broken locks, improperly formed or damaged terminals, poor terminal to wiring connections or physical damage to the wiring harness.

Excessive end play can cause this code to set. Check integral bearing for excessive axial end play.

DIAGNOSTIC CHART DTC A027 — 1993 BERETTA AND CORSICA

IMPORTANT: WHEEL SPEED SENSOR INTERMITTENT PROBLEMS MAY BE DIFFICULT TO LOCATE. CARE SHOULD BE TAKEN NOT TO DISTURB ANY ELECTRICAL CONNECTIONS PRIOR TO AN INDICATED STEP OF THIS CHART. THIS WILL INSURE THAT AN INTERMITTENT CONNECTION WILL NOT BE CORRECTED BEFORE THE SOURCE OF THE PROBLEM IS FOUND.

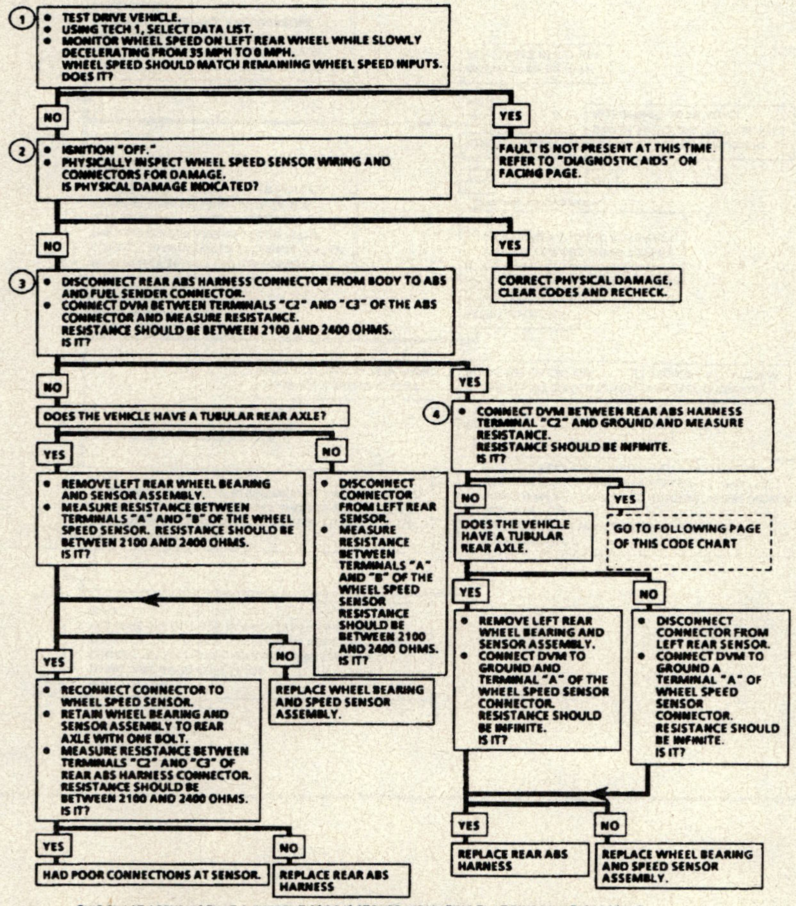

THIS CHART ASSUMES THAT A CURRENT CODE IS STORED INDICATING THAT THIS FAULT IS PRESENT.

DIAGNOSTIC CHART DTC A027 — 1993 BERETTA AND CORSICA

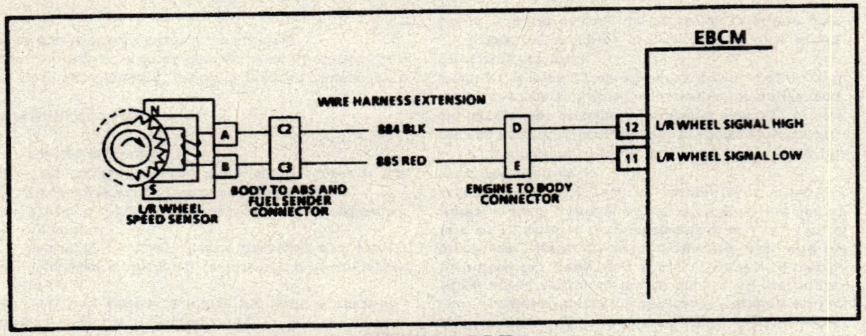

CODE A027

LEFT REAR EXCESSIVE WHEEL SPEED VARIATION

Circuit Description:

As a toothed ring passes by the wheel speed sensor, changes in the electromagnetic field cause the wheel speed sensor to produce a sinusoidal (AC) voltage signal whose frequency is proportional to wheel speed. The magnitude of this signal is directly related to wheel speed and the proximity of the wheel speed sensor to the toothed ring, often referred to as the air gap.

This code detects opens, shorts to ground, and shorts to battery that cause intermittent wheel speed operation. Also detected is any sudden change in wheel speed determined to be unreasonable.

Test Description: Number(s) below refer to circled number(s) on the diagnostic chart.

5. This step checks to see if the wheel speed sensor circuitry is shorted to voltage.
6. This step insures that the wheel speed sensor circuitry resistance is within the proper value.
7. This checks for an open in the wheel signal high circuit.
8. This checks for an open in the wheel signal low circuit.
9. This step insures that Code A027 was not set due to a poor connection between the world connector and EBCM.

Diagnostic Aids:

An "Intermittent" problem may be caused by a poor connection, rubbed through wire insulation, or a wire that is broken inside the insulation.

If the customers comments reflect that the ABS Amber warning light is on only during moist environmental changes, (rain, snow, vehicle wash) all wheel speed sensor circuitry should be thoroughly inspected for water intrusion. Use the following procedure. Spray down the suspected area with 5% salt water solution (two teaspoons of salt to 12 oz. of water). Start vehicle, allow vehicle to run for 10 seconds. If code returns immediately, replace suspected harness.

Any circuitry, that is suspected as causing the intermittent complaint, should be thoroughly checked for backed out terminals, improper mating, broken locks, improperly formed or damaged terminals, poor terminal to wiring connections or physical damage to the wiring harness.

Excessive end play can cause this code to set. Check integral bearing for excessive axial end play,

DIAGNOSTIC CHART DTC A027 — 1993 BERETTA AND CORSICA

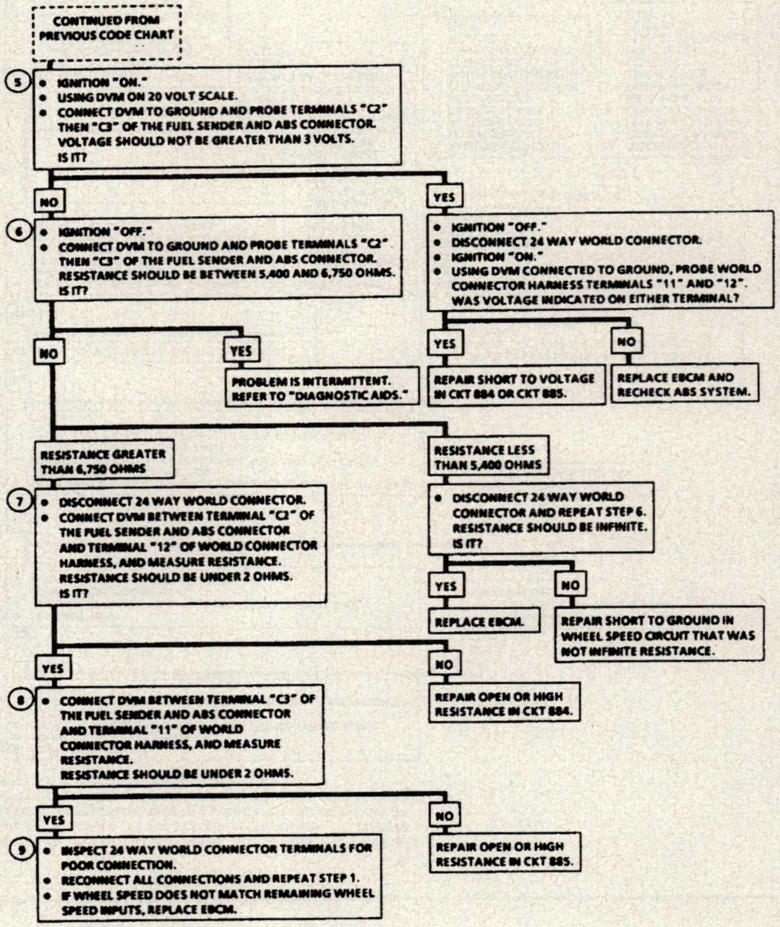

THIS CHART ASSUMES THAT A CURRENT CODE IS STORED INDICATING THAT THIS FAULT IS PRESENT.

DIAGNOSTIC CHART DTC A028 — 1993 BERETTA AND CORSICA

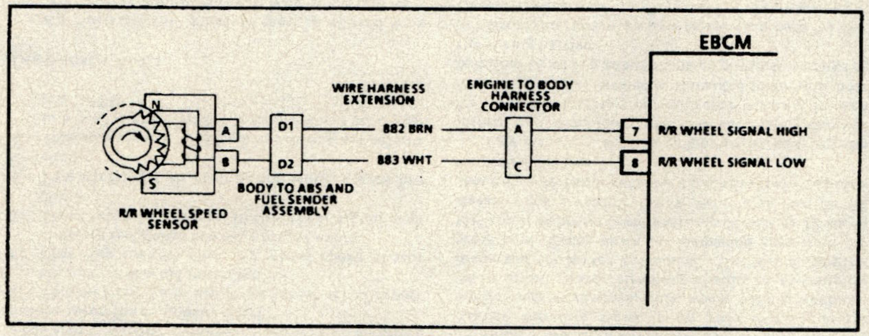

CODE A028

RIGHT REAR EXCESSIVE WHEEL SPEED VARIATION

Circuit Description:

As a toothed ring passes by the wheel speed sensor, changes in the electromagnetic field cause the wheel speed sensor to produce a sinusoidal (AC) voltage signal whose frequency is proportional to wheel speed. The magnitude of this signal is directly related to wheel speed and the proximity of the wheel speed sensor to the toothed ring, often referred to as the air gap.

This code detects opens, shorts to ground, and shorts to battery that cause intermittent wheel speed operation. Also detected is any sudden change in wheel speed determined to be unreasonable.

Test Description: Number(s) below refer to circled number(s) on the diagnostic chart.

1. This test verifies whether the fault is currently present.
2. This step is designed to identify a wheel speed sensor or circuitry that is damaged and visibly apparent.
3. This test checks to see that the wheel speed sensor circuitry is within proper resistance value.
4. This insures that the wheel speed sensor is not shorted to ground.

Diagnostic Aids:

An "Intermittent" problem may be caused by a poor connection, rubbed through wire insulation, or a wire that is broken inside the insulation.

If the customers comments reflect that the ABS Amber warning light is on only during moist environmental changes, (rain, snow, vehicle wash) all wheel speed sensor circuitry should be thoroughly inspected for water intrusion. Use the following procedure. Spray down the suspected area with 5% salt water solution (two teaspoons of salt to 12 oz. of water). Start vehicle, allow vehicle to run for 10 seconds. If code returns immediately, replace suspected harness.

Any circuitry, that is suspected as causing the intermittent complaint, should be thoroughly checked for backed out terminals, improper mating, broken locks, improperly formed or damaged terminals, poor terminal to wiring connections or physical damage to the wiring harness.

Excessive end play can cause this code to set. Check integral bearing for excessive axial end play.

DIAGNOSTIC CHART DTC A028 — 1993 BERETTA AND CORSICA

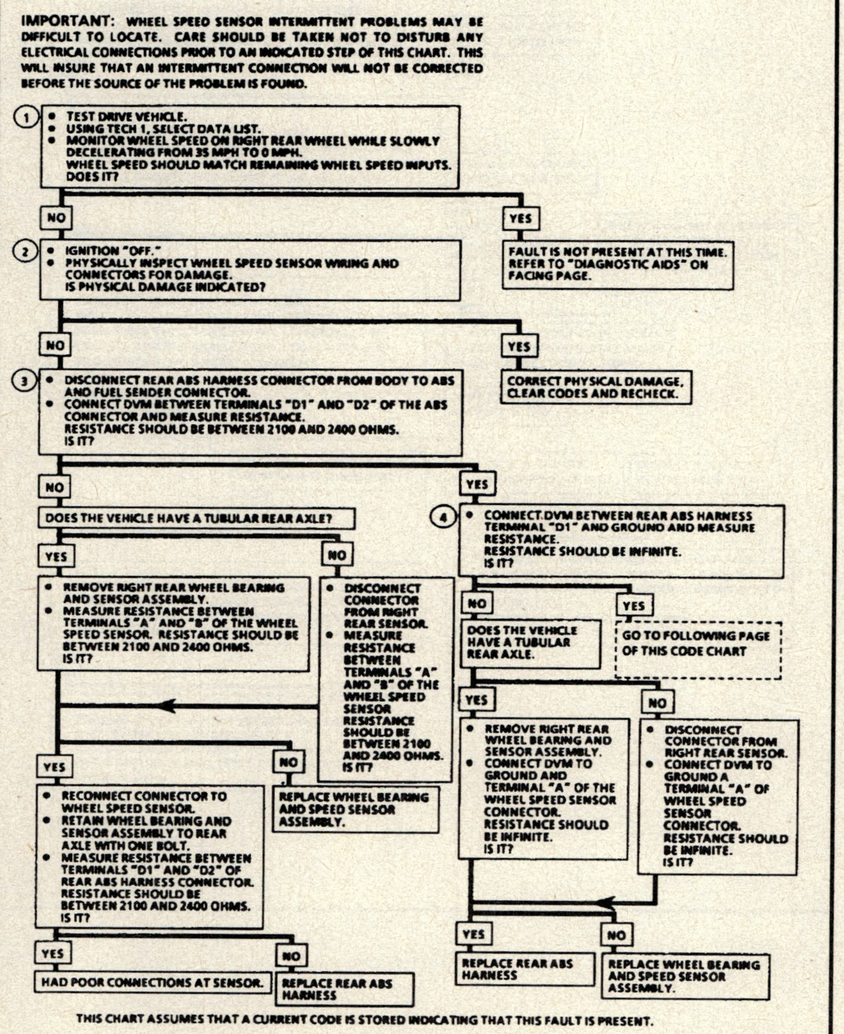

DIAGNOSTIC CHART DTC A028 — 1993 BERETTA AND CORSICA

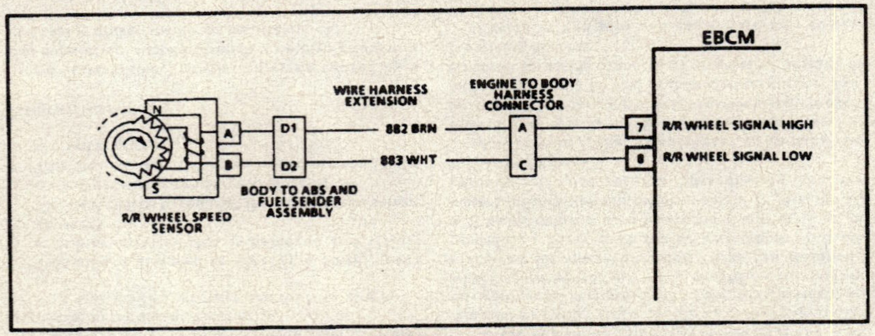

CODE A028

RIGHT REAR EXCESSIVE WHEEL SPEED VARIATION

Circuit Description:

As a toothed ring passes by the wheel speed sensor, changes in the electromagnetic field cause the wheel speed sensor to produce a sinusoidal (AC) voltage signal whose frequency is proportional to wheel speed. The magnitude of this signal is directly related to wheel speed and the proximity of the wheel speed sensor to the toothed ring, often referred to as the air gap.

This code detects opens, shorts to ground, and shorts to battery that cause intermittent wheel speed operation. Also detected is any sudden change in wheel speed determined to be unreasonable.

Test Description: Number(s) below refer to circled number(s) on the diagnostic chart.

5. This step checks to see if the wheel speed sensor circuitry is shorted to voltage.
6. This step insures that the wheel speed sensor circuitry resistance is within proper value.
7. This checks for an open in the wheel signal high circuit.
8. This checks for an open in the wheel signal low circuit.
9. This step insures that Code A028 was not set due to a poor connection between the world connector and EBCM.

Diagnostic Aids:

An "Intermittent" problem may be caused by a poor connection, rubbed through wire insulation, or a wire that is broken inside the insulation.

If the customers comments reflect that the ABS Amber warning light is on only during moist environmental changes, (rain, snow, vehicle wash) all wheel speed sensor circuitry should be thoroughly inspected for water intrusion. Use the following procedure. Spray down the suspected area with 5% salt water solution (two teaspoons of salt to 12 oz. of water). Start vehicle, allow vehicle to run for 10 seconds. If code returns immediately, replace suspected harness.

Any circuitry, that is suspected as causing the intermittent complaint, should be thoroughly checked for backed out terminals, improper mating, broken locks, improperly formed or damaged terminals, poor terminal to wiring connections or physical damage to the wiring harness.

Excessive end play can cause this code to set. Check integral bearing for excessive axial end play,

DIAGNOSTIC CHART DTC A028 — 1993 BERETTA AND CORSICA

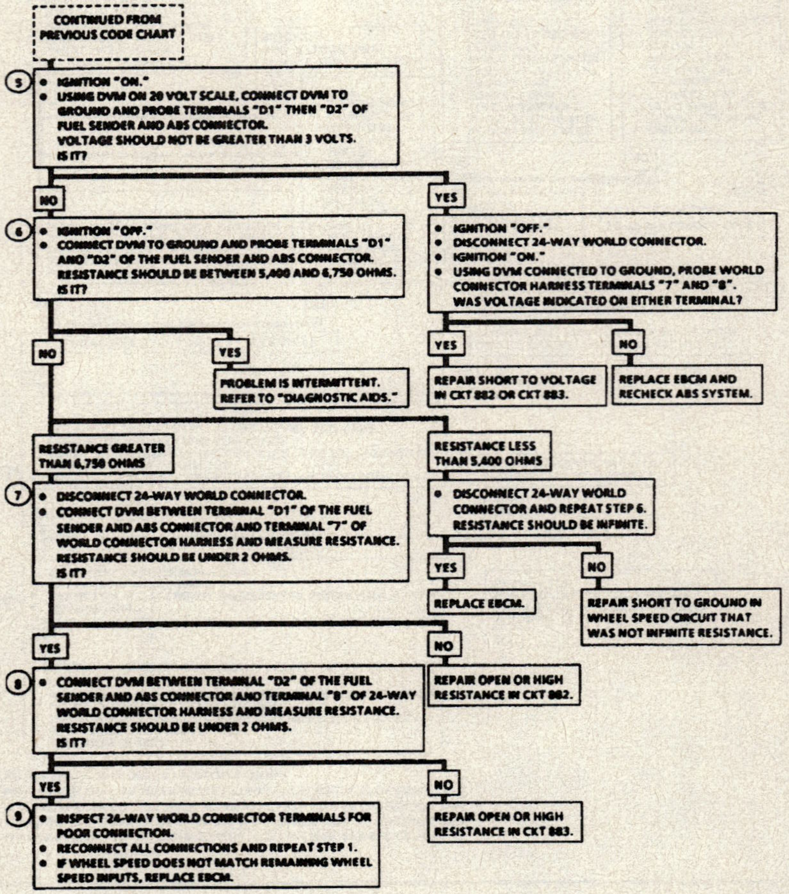

THIS CHART ASSUMES THAT A CURRENT CODE IS STORED INDICATING THAT THIS FAULT IS PRESENT.

DIAGNOSTIC CHART DTC A031 — 1993 BERETTA AND CORSICA

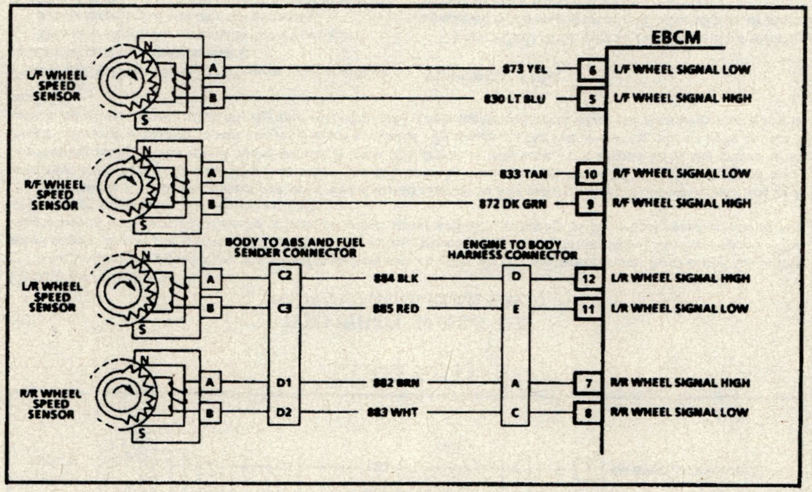

CODE A031

TWO WHEEL SPEEDS = 0 (NON-TUBULAR REAR AXLE)

Circuit Description:

As a toothed ring passes by the wheel speed sensor, changes in the electromagnetic field cause the wheel speed sensor to produce a sinusoidal (AC) voltage signal whose frequency is proportional to wheel speed. The magnitude of this signal is directly related to wheel speed and the proximity of the wheel speed sensor to the toothed ring, often referred to as the air gap.

This code is designed to detect two open wheel speed sensors in any combination. This test exists due to a condition that may exist that does not allow the driver to be informed by the ABS warning light of non-functioning ABS. If one wheel speed sensor is open, the driver is informed. If a second fails, the driver is no longer informed without this test and ABS is not available. By requiring the two operating wheels to be within 7 mph, a false code is prevented during spin-ups while still allowing for detection when the stowaway spare tire is used.

Test Description: Number(s) below refer to circled number(s) on the diagnostic chart.

1. This test verifies whether the fault is currently present.
2. This step is designed to identify a wheel speed sensor or circuitry that is damaged and visibly apparent.
3. This step identifies if the cause of the problem is common to the rear wheel sensor circuitry.
4. Because both rear sensors are common to the extension harness, the 4 way connectors and harness should be inspected.
5. This test checks to see that the wheel speed sensor is within proper resistance values.
6. This insures that the wheel speed sensor is not shorted to ground.

Diagnostic Aids:

A false Code A031 may be set as a result of vehicle operation while on a hoist or a lifted condition or stuck in mud or snow, when the vehicle is in gear.

An "Intermittent" problem may be caused by a poor connection, rubbed through wire insulation, or a wire that is broken inside the insulation.

DIAGNOSTIC CHART DTC A031 — 1993 BERETTA AND CORSICA

IMPORTANT: WHEEL SPEED SENSOR INTERMITTENT PROBLEMS MAY BE DIFFICULT TO LOCATE. CARE SHOULD BE TAKEN NOT TO DISTURB ANY ELECTRICAL CONNECTIONS PRIOR TO AN INDICATED STEP OF THIS CHART. THIS WILL INSURE THAT AN INTERMITTENT CONNECTION WILL NOT BE CORRECTED BEFORE THE SOURCE OF THE PROBLEM IS FOUND.

(1)
- TEST DRIVE VEHICLE.
- USING TECH 1, SELECT DATA LIST.
- MONITOR ALL WHEEL SPEEDS WHILE SLOWLY DECELERATING FROM 35 MPH TO 0 MPH.

ARE ALL WHEEL SPEEDS APPROXIMATELY EQUAL?

- YES: FAULT IS NOT PRESENT AT THIS TIME. REFER TO "DIAGNOSTIC AIDS" ON FACING PAGE.
- NO: go to step 2.

(2)
- IGNITION "OFF."
- PHYSICALLY INSPECT WHEEL SPEED SENSOR CIRCUITS IN QUESTION FOR DAMAGE. ALSO, CHECK WIRE CONNECTIONS FOR CORROSION OR POOR MATING OF CONNECTORS.

IS PHYSICAL DAMAGE INDICATED?

- YES: CORRECT PHYSICAL DAMAGE, CLEAR CODES AND RE-CHECK.
- NO: go to step 3.

(3) DID BOTH REAR WHEEL SPEED SENSORS INDICATE 0 MPH, OR ERRATIC WHEEL SPEEDS?

- NO: USING CHART BELOW, REFER TO SPEED SENSOR CODE CHART FOR AFFECTED WHEELS.
- YES: go to step 4.

(4) INSPECT 4 WAY REAR WHEEL EXTENSION HARNESS CONNECTORS FOR CORROSION, POOR MATING OR DAMAGE. IS DAMAGE INDICATED?

- YES: CORRECT DAMAGE, CLEAR CODES AND RE-CHECK.
- NO: go to step 5.

(5)
- DISCONNECT REAR WHEEL SPEED SENSORS.
- USING DVM CONNECTED BETWEEN TERMINALS "A" AND "B" OF THE REAR WHEEL SPEED SENSORS, MEASURE RESISTANCE. RESISTANCE SHOULD BE BETWEEN 2100 AND 2400 OHMS FOR REAR SENSORS.

ARE RESISTANCE VALUES IN ACCEPTABLE RANGES?

- NO: REPLACE WHEEL SPEED SENSOR(S) THAT DID NOT INDICATE PROPER RESISTANCE, AND RE-CHECK.
- YES: go to step 6.

(6)
- USING DVM CONNECTED BETWEEN TERMINAL "A" OF THE SUSPECTED WHEEL SENSOR AND GROUND, MEASURE RESISTANCE. DVM SHOULD INDICATE INFINITE RESISTANCE.

DOES IT?

- YES: GO TO FOLLOWING PAGE OF THIS CODE CHART
- NO: REPLACE WHEEL SPEED SENSOR(S) THAT DID NOT INDICATE PROPER RESISTANCE, AND RE-CHECK.

SENSOR	CODE
L/F SENSOR	A021
R/F SENSOR	A022
L/R SENSOR	A023
R/R SENSOR	A024

DIAGNOSTIC CHART DTC A031 — 1993 BERETTA AND CORSICA

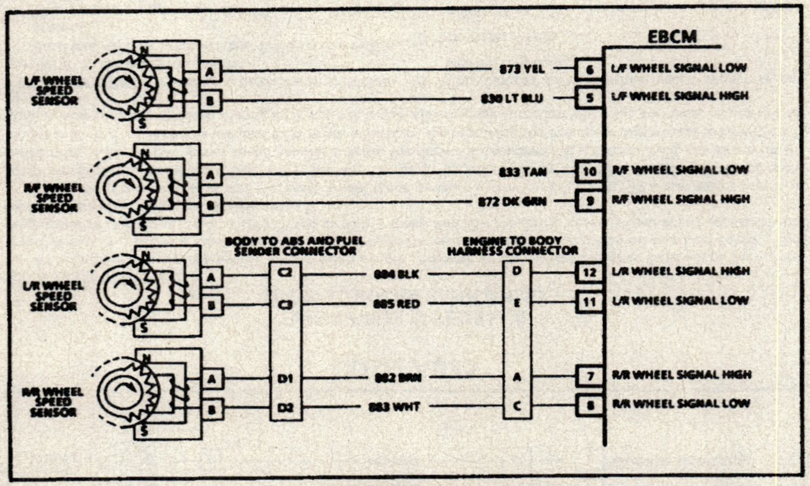

CODE A031

TWO WHEEL SPEEDS = 0 (NON-TUBULAR REAR AXLE)

Circuit Description:

As a toothed ring passes by the wheel speed sensor, changes in the electromagnetic field cause the wheel speed sensor to produce a sinusoidal (AC) voltage signal whose frequency is proportional to wheel speed. The magnitude of this signal is directly related to wheel speed and the proximity of the wheel speed sensor to the toothed ring, often referred to as the air gap.

This code is designed to detect two open wheel speed sensors in any combination. This test exists due to a condition that may exist that does not allow the driver to be informed by the ABS warning light of non-functioning ABS. If one wheel speed sensor is open, the driver is informed. If a second fails, the driver is no longer informed without this test and ABS is not available. By requiring the two operating wheels to be within 7 mph, a false code is prevented during spin-ups while still allowing for detection when the stowaway spare tire is used.

Test Description: Number(s) below refer to circled number(s) on the diagnostic chart.

7. This step checks for an open in the wheel speed sensor circuitry of one side of the sensor.
8. This step checks for an open on the other side of the sensor in question.
9. This step checks for a short to ground in the wheel speed sensor circuitry.
10. This insures that the circuitry is not shorted to voltage.

Diagnostic Aids:

A false Code A031 may be set as a result of vehicle operation while on a hoist or a lifted condition, or stuck in mud or snow when the vehicle is in gear.

An "Intermittent" problem may be caused by a poor connection, rubbed through wire insulation, or a wire that is broken inside the insulation.

DIAGNOSTIC CHART DTC A031 — 1993 BERETTA AND CORSICA

CONTINUED FROM CHART CODE A031

(7)
- CONNECT DVM TO TERMINAL "A" OF WHEEL SPEED SENSOR HARNESS CONNECTOR IN QUESTION.
- DISCONNECT 24 WAY WORLD CONNECTOR.
- USING THE TABLE SUPPLIED AT THE BOTTOM OF THIS PAGE, CONNECT DVM TO THE APPROPRIATE TERMINAL OF THE 24 WAY WORLD CONNECTOR AND MEASURE RESISTANCE.

RESISTANCE SHOULD BE LESS THAN 2 OHMS.
IS IT?

NO: REPAIR CIRCUIT THAT SHOWED HIGH RESISTANCE OR OPEN.

YES:

(8)
- USING DVM, PROBE CONNECTOR TERMINAL "B" OF WHEEL SPEED SENSOR IN QUESTION.
- USING THE TABLE SUPPLIED, CONNECT DVM TO THE APPROPRIATE TERMINAL OF THE 24 WAY WORLD CONNECTOR AND MEASURE RESISTANCE.

RESISTANCE SHOULD BE LESS THAN 2 OHMS.
IS IT?

NO: REPAIR CIRCUIT THAT SHOWED HIGH RESISTANCE OR OPEN.

YES:

(9)
- CONNECT DVM BETWEEN WHEEL SPEED SENSOR CONNECTOR TERMINAL "A" AND GROUND THEN REPEAT PROCEDURE ON TERMINAL "B".

RESISTANCE SHOULD BE INFINITE ON BOTH TERMINALS.
IS IT?

NO: REPAIR SHORT TO GROUND IN CIRCUIT THAT DID NOT SHOW INFINITE RESISTANCE.

YES:

(10)
- DVM STILL CONNECTED.
- SELECT THE 20 VOLT SCALE OF THE DVM.
- IGNITION "ON."
- REPEAT STEP 8.

VOLTAGE SHOULD BE 0.
IS IT?

NO: REPAIR SHORT TO VOLTAGE IN CIRCUIT THAT INDICATED VOLTAGE ABOVE 0 VOLTS.

YES:
- INSPECT 24 WAY WORLD CONNECTOR TERMINALS FOR POOR CONNECTION.
- RE-CONNECT ALL CONNECTIONS AND REPEAT STEP 1.
- IF WHEEL SPEED DOES NOT MATCH REMAINING WHEEL SPEED INPUTS, REPLACE EBCM.

SENSOR TERMINAL	24 WAY WORLD CONNECTOR	CIRCUIT NUMBER
L/R WHEEL A	PIN 12	884
L/R WHEEL B	PIN 11	885
R/R WHEEL A	PIN 7	882
R/R WHEEL B	PIN 8	883

THIS CHART ASSUMES THAT A CURRENT CODE IS STORED INDICATING THAT THIS FAULT IS PRESENT.

DIAGNOSTIC CHART DTC A031 — 1993 BERETTA AND CORSICA

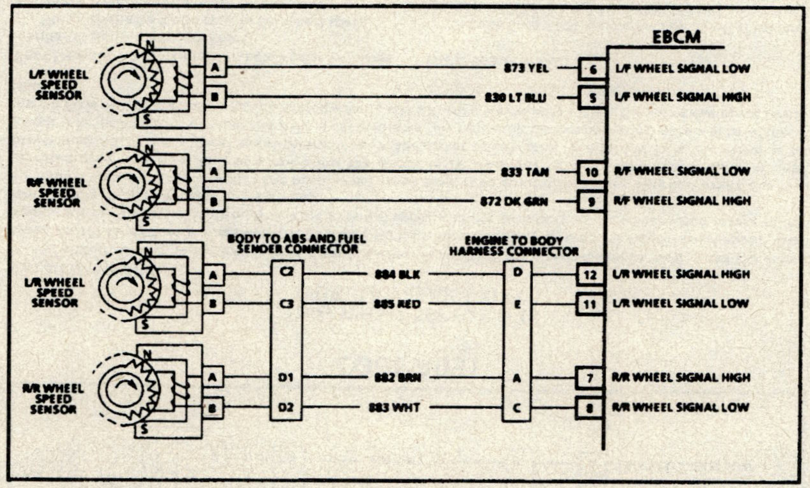

CODE A031

TWO WHEEL SPEEDS = 0 (TUBULAR REAR AXLE)

Circuit Description:

As a toothed ring passes by the wheel speed sensor, changes in the electromagnetic field cause the wheel speed sensor to produce a sinusoidal (AC) voltage signal whose frequency is proportional to wheel speed. The magnitude of this signal is directly related to wheel speed and the proximity of the wheel speed sensor to the toothed ring, often referred to as the air gap.

This code is designed to detect two open wheel speed sensors in any combination. This test exists due to a condition that may exist that does not allow the driver to be informed by the ABS warning light of non-functioning ABS. If one wheel speed sensor is open, the driver is informed. If a second fails, the driver is no longer informed without this test and ABS is not available. By requiring the two operating wheels to be within 7 mph, a false code is prevented during spin-ups while still allowing for detection when the stowaway spare tire is used.

Test Description: Number(s) below refer to circled number(s) on the diagnostic chart.

1. This test verifies whether the fault is currently present.
2. This step is designed to identify a wheel speed sensor or circuitry that is damaged and visibly apparent.
3. This step identifies if the cause of the problem is common to the rear wheel sensor circuitry.
4. Because both rear sensors are common to the extension harness, the 4 way connectors and harness should be inspected.
5. This test checks to see that the wheel speed sensor is within proper resistance values.
6. This insures that the wheel speed sensor is not shorted to ground.

Diagnostic Aids:

A false Code A031 may be set as a result of vehicle operation while on a hoist or a lifted condition or stuck in mud or snow, when the vehicle is in gear.

An "Intermittent" problem may be caused by a poor connection, rubbed through wire insulation, or a wire that is broken inside the insulation.

DIAGNOSTIC CHART DTC A031 — 1993 BERETTA AND CORSICA

IMPORTANT: WHEEL SPEED SENSOR INTERMITTENT PROBLEMS MAY BE DIFFICULT TO LOCATE. CARE SHOULD BE TAKEN NOT TO DISTURB ANY ELECTRICAL CONNECTIONS PRIOR TO AN INDICATED STEP OF THIS CHART. THIS WILL INSURE THAT AN INTERMITTENT CONNECTION WILL NOT BE CORRECTED BEFORE THE SOURCE OF THE PROBLEM IS FOUND.

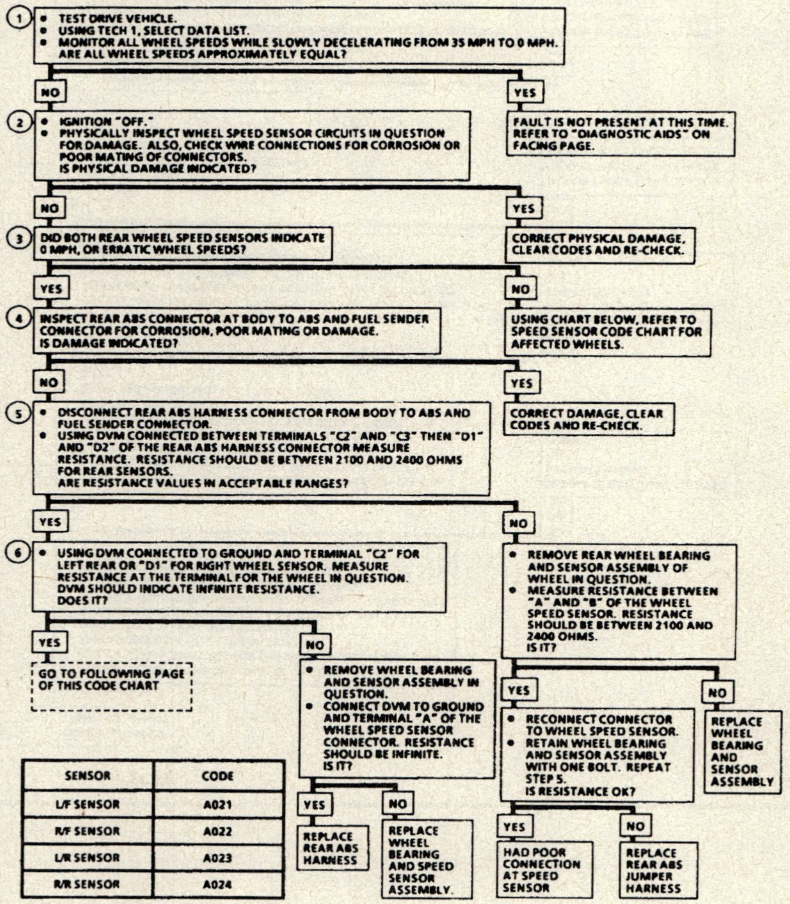

SENSOR	CODE
L/F SENSOR	A021
R/F SENSOR	A022
L/R SENSOR	A023
R/R SENSOR	A024

DIAGNOSTIC CHART DTC A031 — 1993 BERETTA AND CORSICA

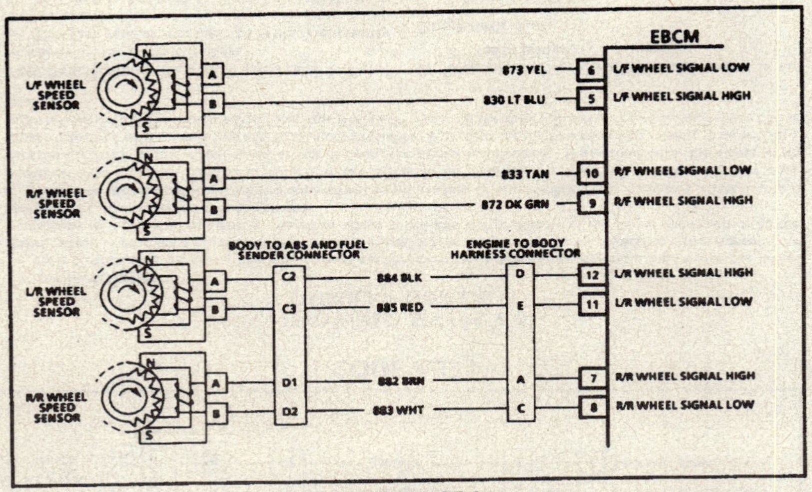

CODE A031

TWO WHEEL SPEEDS = 0
(TUBULAR REAR AXLE)

Circuit Description:

As a toothed ring passes by the wheel speed sensor, changes in the electromagnetic field cause the wheel speed sensor to produce a sinusoidal (AC) voltage signal whose frequency is proportional to wheel speed. The magnitude of this signal is directly related to wheel speed and the proximity of the wheel speed sensor to the toothed ring, often referred to as the air gap.

This code is designed to detect two open wheel speed sensors in any combination. This test exists due to a condition that may exist that does not allow the driver to be informed by the ABS warning light of non-functioning ABS. If one wheel speed sensor is open, the driver is informed. If a second fails, the driver is no longer informed without this test and ABS is not available. By requiring the two operating wheels to be within 7 mph, a false code is prevented during spin-ups while still allowing for detection when the stowaway spare tire is used.

Test Description: Number(s) below refer to circled number(s) on the diagnostic chart.

7. This step checks for an open in the wheel speed sensor circuitry of one side of the sensor.
8. This step checks for an open on the other side of the sensor in question.
9. This step checks for a short to ground in the wheel speed sensor circuitry.
10. This insures that the circuitry is not shorted to voltage.

Diagnostic Aids:

A false Code A031 may be set as a result of vehicle operation while on a hoist or a lifted condition, or stuck in mud or snow when the vehicle is in gear.

An "Intermittent" problem may be caused by a poor connection, rubbed through wire insulation, or a wire that is broken inside the insulation.

DIAGNOSTIC CHART DTC A031 — 1993 BERETTA AND CORSICA

CONTINUED FROM CHART CODE A031

(7)
- CONNECT DVM TO THE TERMINAL OF THE BODY TO ABS AND FUEL SENDER CONNECTOR FOR THE WHEEL IN QUESTION. (USE TABLE SUPPLIED AT BOTTOM OF THIS PAGE.)
- DISCONNECT 24 WAY WORLD CONNECTOR.
- USING THE TABLE SUPPLIED AT THE BOTTOM OF THIS PAGE, CONNECT DVM TO THE APPROPRIATE TERMINAL OF THE 24 WAY WORLD CONNECTOR AND MEASURE RESISTANCE.

RESISTANCE SHOULD BE LESS THAN 2 OHMS.
IS IT?

NO: REPAIR CIRCUIT THAT SHOWED HIGH RESISTANCE OR OPEN.

YES:

(8)
- CONNECT DVM TO THE OTHER TERMINAL OF THE BODY TO ABS AND FUEL SENDER CONNECTOR FOR THE WHEEL IN QUESTION. (USE TABLE SUPPLIED AT BOTTOM OF THIS PAGE.)
- USING THE TABLE SUPPLIED, CONNECT DVM TO THE APPROPRIATE TERMINAL OF THE 24 WAY WORLD CONNECTOR AND MEASURE RESISTANCE.

RESISTANCE SHOULD BE LESS THAN 2 OHMS.
IS IT?

NO: REPAIR CIRCUIT THAT SHOWED HIGH RESISTANCE OR OPEN.

YES:

(9)
- CONNECT DVM BETWEEN BODY TO ABS AND FUEL SENDER CONNECTOR TERMINAL OF THE WHEEL IN QUESTION, (USE TABLE BELOW) AND GROUND. REPEAT PROCEDURE ON THE OTHER TERMINAL.

RESISTANCE SHOULD BE INFINITE ON BOTH TERMINALS.
IS IT?

NO: REPAIR SHORT TO GROUND IN CIRCUIT THAT DID NOT SHOW INFINITE RESISTANCE.

YES:

(10)
- DVM STILL CONNECTED.
- SELECT THE 20 VOLT SCALE OF THE DVM.
- IGNITION "ON."
- REPEAT STEP 9.

VOLTAGE SHOULD BE 0.
IS IT?

NO: REPAIR SHORT TO VOLTAGE IN CIRCUIT THAT INDICATED VOLTAGE ABOVE 0 VOLTS.

YES:
- INSPECT 24 WAY WORLD CONNECTOR TERMINALS FOR POOR CONNECTION.
- RE-CONNECT ALL CONNECTIONS AND REPEAT STEP 1.
- IF WHEEL SPEED DOES NOT MATCH REMAINING WHEEL SPEED INPUTS, REPLACE EBCM.

BODY TO ABS AND FUEL SENDER CONNECTOR	24 WAY WORLD CONNECTOR	CIRCUIT NUMBER
L/R WHEEL C2	PIN 12	884
L/R WHEEL C3	PIN 11	885
R/R WHEEL D1	PIN 7	882
R/R WHEEL D2	PIN 8	883

THIS CHART ASSUMES THAT A CURRENT CODE IS STORED INDICATING THAT THIS FAULT IS PRESENT.

DIAGNOSTIC CHART DTC A036 — 1993 BERETTA AND CORSICA

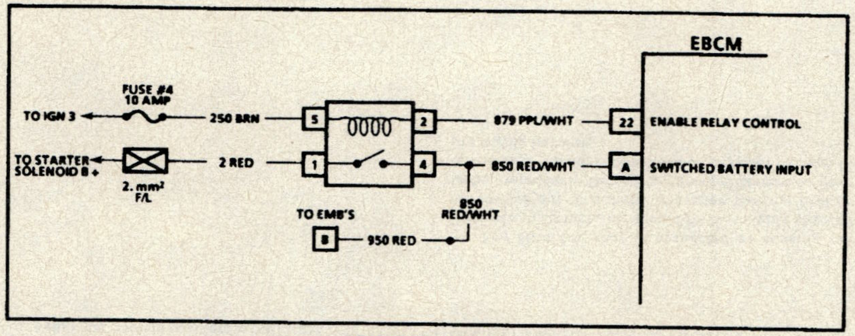

CODE A036

LOW SYSTEM VOLTAGE

Circuit Description:

This code is used to monitor the voltage level available to the ABS EBCM. If the voltage drops below 11 volts, full performance of the ABS system cannot be guaranteed. During ABS operation, there are several current requirements that will cause battery voltage to drop. Because of this, voltage is monitored prior to ABS to indicate good charging system condition and also during ABS when voltage may drop significantly. If voltage falls below a level that will provide adequate ABS performance, Code A036 will result.

Test Description: Number(s) below refer to circled number(s) on the diagnostic chart.

1. This test checks to see what voltage is readily available at terminal "A" of the EBCM. If voltage is representative of good charging system condition, a fault is not present.
2. This step isolates the EBCM to check the condition of the circuitry and charging system.
3. This step isolates the low voltage condition to high circuit resistance, or improper charging system operation.

Diagnostic Aids:

An "Intermittent" problem may be caused by a poor connection, rubbed through wire insulation, or a wire that is broken inside the insulation.

Any circuitry, that is suspected as causing the intermittent complaint, should be thoroughly checked for backed out terminals, improper mating, broken locks, improperly formed or damaged terminals, poor terminal to wiring connections or physical damage to the wiring harness.

While performing a voltage load test, if it is noted that only ignition voltage drops below acceptable voltage levels, CKT 250 should be checked for high resistance or an open condition.

DIAGNOSTIC CHART DTC A036 — 1993 BERETTA AND CORSICA

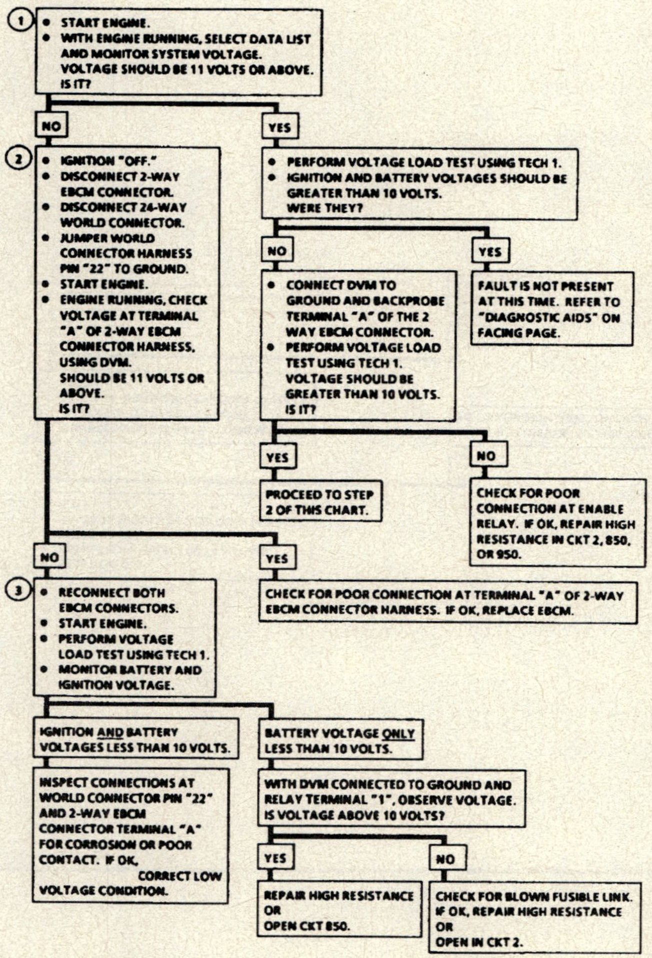

DIAGNOSTIC CHART DTC A037 — 1993 BERETTA AND CORSICA

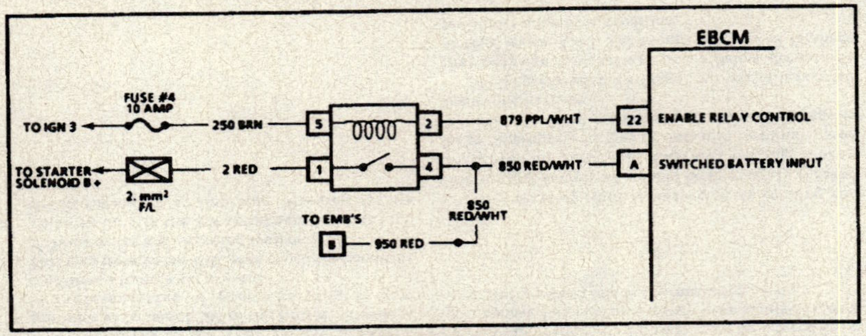

CODE A037

HIGH SYSTEM VOLTAGE

Circuit Description:

This code is designed to detect high vehicle voltage levels prior to any required motor movement (initialization or ABS). If excessive voltage exists, demagnetization of the motor magnets may occur, which would eventually affect or eliminate ABS performance.

Test Description: Number(s) below refer to circled number(s) on the diagnostic chart.

1. This step checks the voltage level being received by the EBCM. If high voltage is present, a charging system problem is indicated.

Diagnostic Aids:

An "Intermittent" problem may be caused by a poor connection, rubbed through wire insulation, or a wire that is broken inside the insulation.

Any circuitry, that is suspected as causing the intermittent complaint, should be thoroughly checked for backed out terminals, improper mating, broken locks, improperly formed or damaged terminals, poor terminal to wiring connections or physical damage to the wiring harness.

DIAGNOSTIC CHART DTC A037 — 1993 BERETTA AND CORSICA

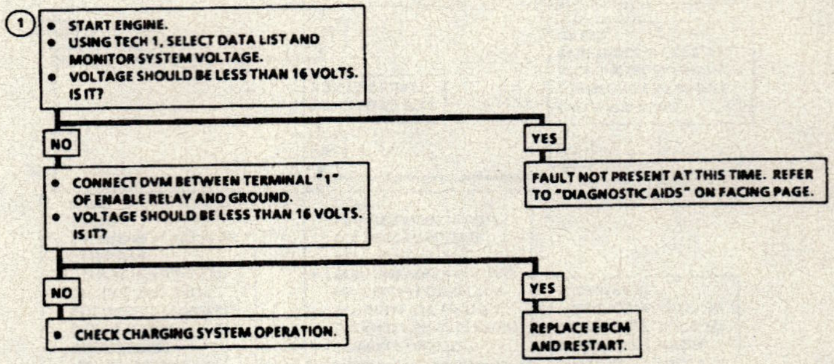

THIS CHART ASSUMES THAT A CURRENT CODE IS STORED INDICATING THAT THIS FAULT IS PRESENT.

DIAGNOSTIC CHART DTC A038 — 1993 BERETTA AND CORSICA

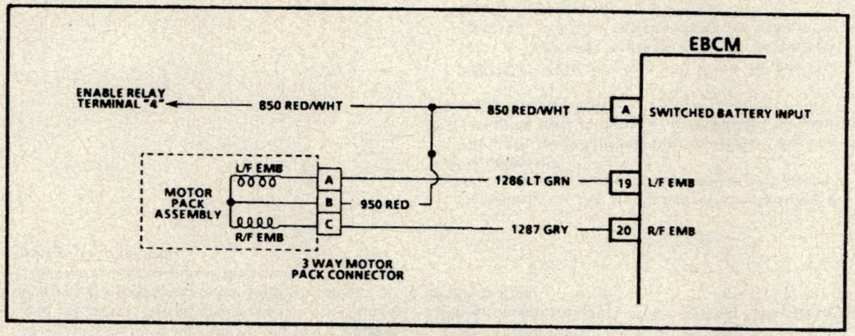

CODE A038

LEFT FRONT EMB WILL NOT HOLD MOTOR

Circuit Description:

Current is supplied to the EMB's through terminal "B" of the three way motor pack connector. When a release of the EMB is commanded, the EBCM grounds the EMB input and allows the EMB to be energized to release the motor.

This code is designed to detect slipping EMB's during initialization. When the release is commanded during initialization, the EMB is not energized and should hold the motor, resulting in sensed current of the motor being equal to commanded current (motor is at stall). If the motor is moving, sensed current will be less than stall current.

Test Description: Number(s) below refer to circled number(s) on the diagnostic chart.

1. This functionally checks the ability of the EMB to hold the motor. If the brake pedal moves during this test, it would indicate the EMB is slipping.
2. This step is used to release the tension of the motor pack prior to removal.

Diagnostic Aids:

An "Intermittent" fault in this code may result from a mechanical part of the system that sticks, binds, or slips.

Depending on the frequency of the failure, a physical inspection of the mechanical parts suspected may be necessary.

The static modulator test function of the Tech 1 or T-100 may be used to locate an intermittent problem associated with the EMB.

NOTICE: This test may fail if low voltage (10 volts or less) was present when initialization occurred. Ensure battery/charging system is functioning properly.

DIAGNOSTIC CHART DTC A038 — 1993 BERETTA AND CORSICA

NOTICE: TESTING EMB WITH JUMPER WIRES TO VOLTAGE OR GROUND WILL DESTROY THE EMB.

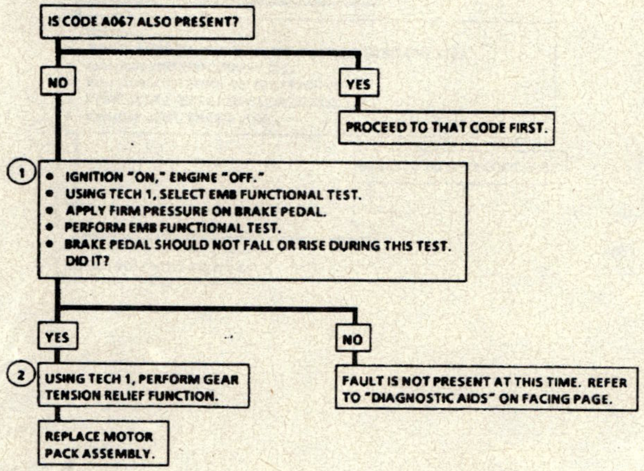

DIAGNOSTIC CHART DTC A041 — 1993 BERETTA AND CORSICA

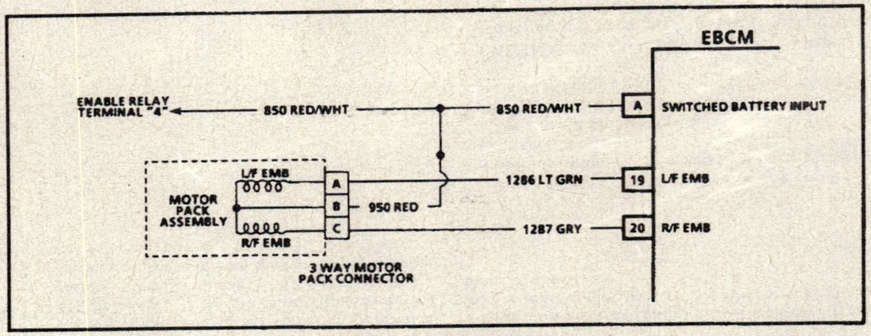

CODE A041

RIGHT FRONT EMB WILL NOT HOLD MOTOR

Circuit Description:

Current is supplied to the EMB's through terminal "B" of the three way motor pack connector. When a release of the EMB is commanded, the EBCM grounds the EMB input and allows the EMB to be energized to release the motor.

This code is designed to detect slipping EMB's during initialization. When the release is commanded during initialization, the EMB is not energized and should hold the motor, resulting in sensed current of the motor being equal to commanded current (motor is at stall). If the motor is moving, sensed current will be less than stall current.

Test Description: Number(s) below refer to circled number(s) on the diagnostic chart.

1. This functionally checks the ability of the EMB to hold the motor. If the brake pedal moves during this test, it would indicate the EMB is slipping.
2. This step is used to release the tension of the motor pack prior to removal.

Diagnostic Aids:

An "Intermittent" fault in this code may result from a mechanical part of the system that sticks, binds, or slips.

Depending on the frequency of the failure, a physical inspection of the mechanical parts suspected may be necessary.

The static modulator test function of the Tech 1 or T-100 may be used to locate an intermittent problem associated with the EMB.

NOTICE: This code may fail if low voltage (10 volts or less) was present when initialization occurred. Ensure battery/charging system is functioning properly.

DIAGNOSTIC CHART DTC A041 — 1993 BERETTA AND CORSICA

NOTICE: TESTING EMB WITH JUMPER WIRES TO VOLTAGE OR GROUND WILL DESTROY THE EMB.

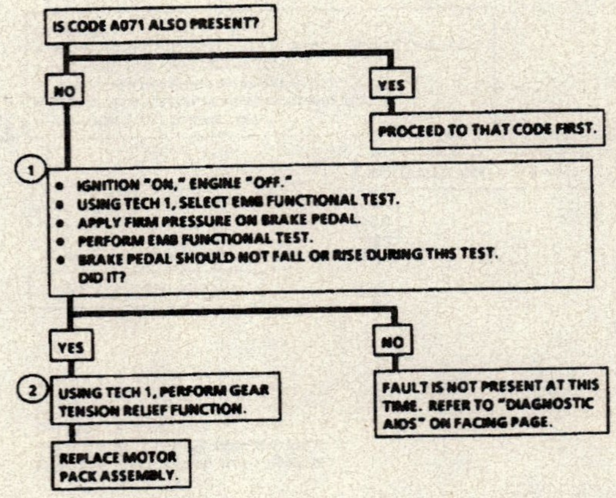

DIAGNOSTIC CHART DTC A042 — 1993 BERETTA AND CORSICA

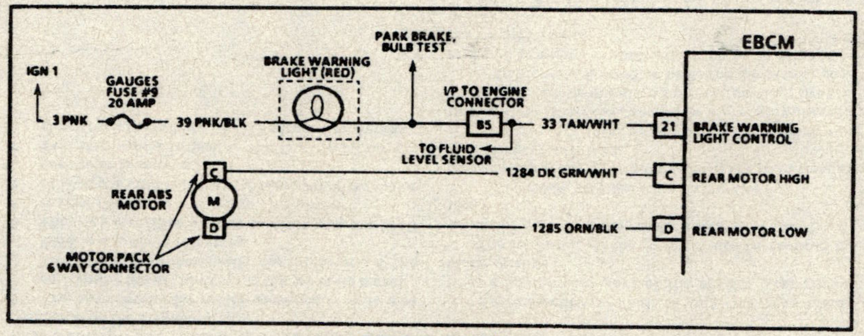

CODE A042

REAR AXLE ESB WILL NOT HOLD MOTOR

Circuit Description:

This code is designed to detect a slipping rear axle ESB. During initialization and braking, the rear motor is re-homed. If the ESB slips, the motor/piston will move. During the next key on initialization, a re-home of the motor verifies the motor/piston remained at the home position. If motor movement is detected, the ESB must be slipping.

Test Description: Number(s) below refer to circled number(s) on the diagnostic chart.

1. This step checks the rear axle ESB. A broken or defective ESB would result in the rear piston being back driven by hydraulic pressure, and wheel movement would result.
2. This step is used to release the tension of the motor pack prior to removal.

Diagnostic Aids:

An "Intermittent" fault in this code may result from a mechanical part of the system that sticks, binds, or slips.

Depending on the frequency of the failure, a physical inspection of the mechanical parts suspected may be necessary.

The static modulator test function of the Tech 1 or T-100 may be used to locate an intermittent problem associated with the ESB.

DIAGNOSTIC CHART DTC A042 — 1993 BERETTA AND CORSICA

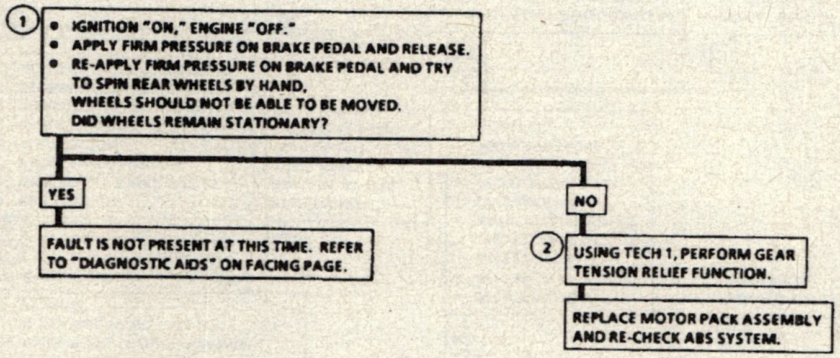

DIAGNOSTIC CHART DTC A044 — 1993 BERETTA AND CORSICA

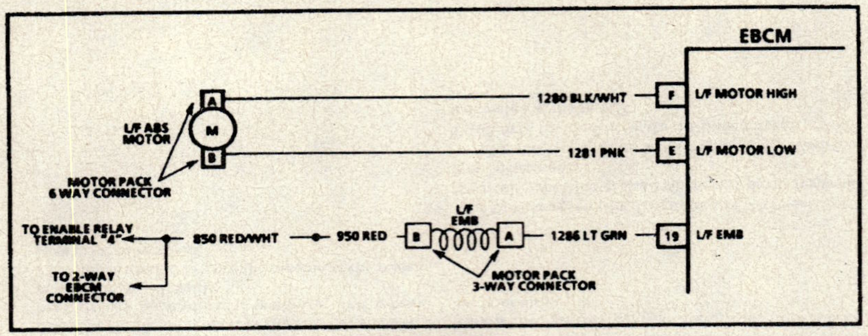

CODE A044

LEFT FRONT CHANNEL WILL NOT MOVE

Circuit Description:

Current is supplied to the EMB through terminal "B" of the three way motor pack connector. When a release of the EMB is commanded, the EBCM grounds the EMB input and allows the EMB to be energized to release the motor.

This code is designed to detect non-actuating EMB, stuck motor or a seized hydraulic modulator. When the release is commanded during initialization, the EMB should release the motor, resulting in sensed current being less than commanded current (motor is spinning freely). If the motor is not moving, sensed current will be equal to stall current.

Test Description: Number(s) below refer to circled number(s) on the diagnostic chart.

1. This step checks for motor movement while also positioning the brake pedal at the top of its travel.
2. This procedure functionally checks to see if the EMB is capable of releasing.
3. This step identifies a short to voltage in the EMB control circuitry.
4. This step is used to release the tension of the motor pack prior to removal.
5. This step serves to isolate the fault to either the motor pack assembly or ABS hydraulic modulator.

Diagnostic Aids:

An "Intermittent" fault in this code may result from a mechanical part of the system that sticks, binds, or slips.

The frequency of the problem can be checked by using the enhanced diagnostic function of the "Scan" tool

Code A044 may set after modulator disassembly if the modulator pistons are positioned at the bottom of their bore.

Any circuitry, that is suspected as causing the intermittent complaint, should be thoroughly checked for backed out terminals, improper mating, broken locks, improperly formed or damaged terminals, poor terminal to wiring connections or physical damage to the wiring harness.

NOTICE: This code may fail if low voltage (10 volts or less) was present when initialization occurred. Ensure battery/charging system is functioning properly.

DIAGNOSTIC CHART DTC A044 — 1993 BERETTA AND CORSICA

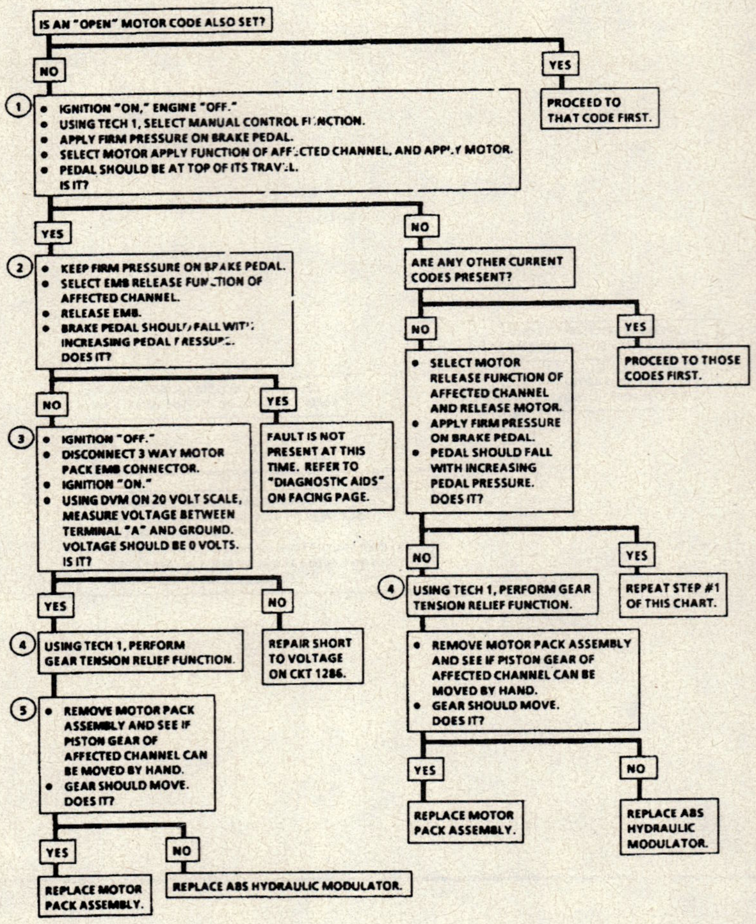

DIAGNOSTIC CHART DTC A045 — 1993 BERETTA AND CORSICA

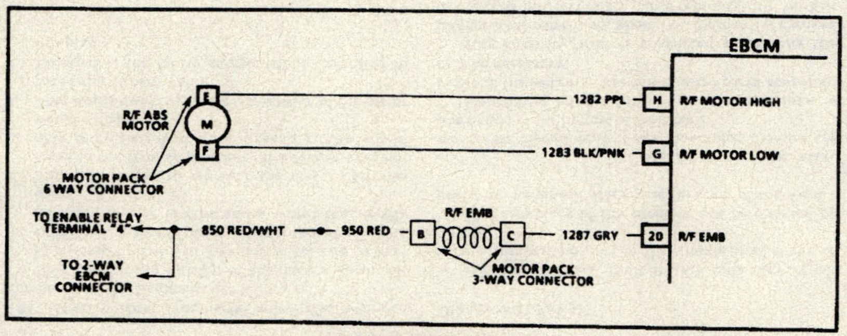

CODE A045

RIGHT FRONT CHANNEL WILL NOT MOVE

Circuit Description:

Current is supplied to the EMB through terminal "B" of the three way motor pack connector. When a release of the EMB is commanded, the EBCM grounds the EMB input and allows the EMB to be energized to release the motor.

This code is designed to detect non-actuating EMB, stuck motor or a seized hydraulic modulator. When the release is commanded during initialization, the EMB should release the motor, resulting in sensed current being less than commanded current (motor is spinning freely). If the motor is not moving, sensed current will be equal to stall current.

Test Description: Number(s) below refer to circled number(s) on the diagnostic chart.

1. This step checks for motor movement while also positioning the brake pedal at the top of its travel.
2. This procedure functionally checks to see if the EMB is capable of releasing.
3. This step is used to release the tension of the motor pack prior to removal.
4. This step serves to isolate the fault to either the motor pack assembly or ABS hydraulic modulator.

Diagnostic Aids:

An "Intermittent" fault in this code may result from a mechanical part of the system that sticks, binds, or slips.

The frequency of the problem can be checked by using the enhanced diagnostic function of the "Scan" tool

Code A045 may set after modulator disassembly if the modulator pistons are positioned at the bottom of their bore.

Any circuitry, that is suspected as causing the intermittent complaint, should be thoroughly checked for backed out terminals, improper mating, broken locks, improperly formed or damaged terminals, poor terminal to wiring connections or physical damage to the wiring harness.

NOTICE: This code may fail if low voltage (10 volts or less) was present when initialization occurred. Ensure battery/charging system is functioning properly.

DIAGNOSTIC CHART DTC A045 — 1993 BERETTA AND CORSICA

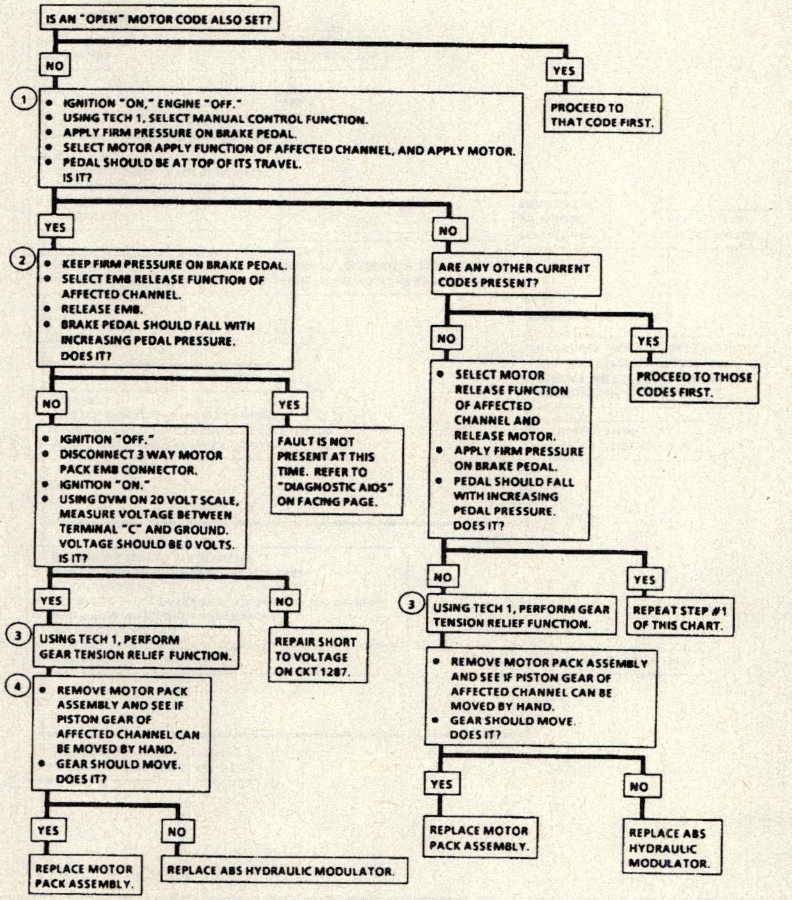

DIAGNOSTIC CHART DTC A046 — 1993 BERETTA AND CORSICA

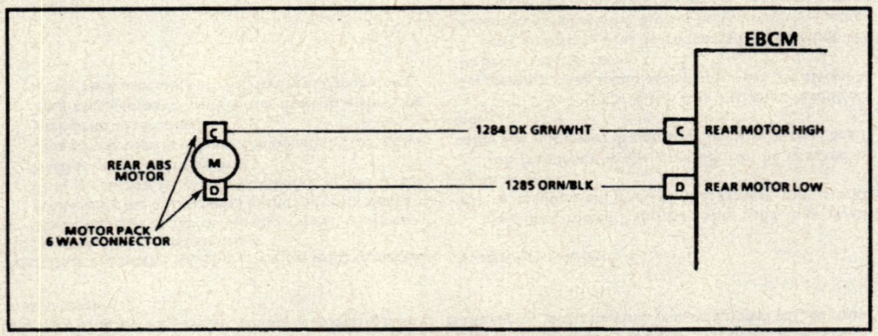

CODE A046

REAR AXLE CHANNEL WILL NOT MOVE

Circuit Description:

This code is designed to detect bound-up ESB's, stuck motor or a seized hydraulic modulator. When the release is commanded during initialization, the ESB should release the motor, resulting in sensed current being less than commanded current (motor is spinning freely). If the motor is not moving, sensed current will be equal to stall current.

Test Description: Number(s) below refer to circled number(s) on the diagnostic chart.

1. This checks the ability of the motor to move, applying hydraulic pressure to the wheel cylinders. A frozen piston or gear would result in no apply of rear brakes and a wheel that freely spins.
2. This step checks to see that the motor can also move to the release position. If a motor were not able to release hydraulic pressure, the wheel would not spin.
3. This step is used to release the tension of the motor pack prior to removal.
4. Isolation of the faulty component is identified in this step.

Diagnostic Aids:

An "Intermittent" fault in this code may result from a mechanical part of the system that sticks or binds.

The frequency of the problem can be checked by using the enhanced diagnostic function of the Tech 1

Code A046 may set after modulator disassembly if the modulator pistons are positioned at the bottom of their bore.

Depending on the frequency of the failure, a physical inspection of the mechanical parts suspected may be necessary.

Any circuitry, that is suspected as causing the intermittent complaint, should be thoroughly checked for backed out terminals, improper mating, broken locks, improperly formed or damaged terminals, poor terminal to wiring connections or physical damage to the wiring harness.

DIAGNOSTIC CHART DTC A046 — 1993 BERETTA AND CORSICA

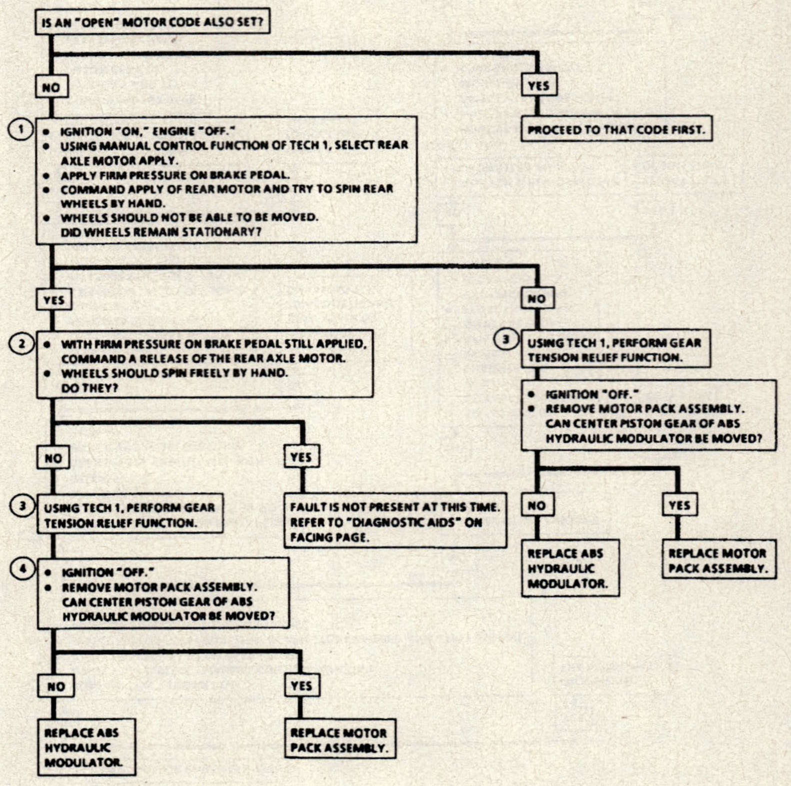

DIAGNOSTIC CHART DTC A047 — 1993 BERETTA AND CORSICA

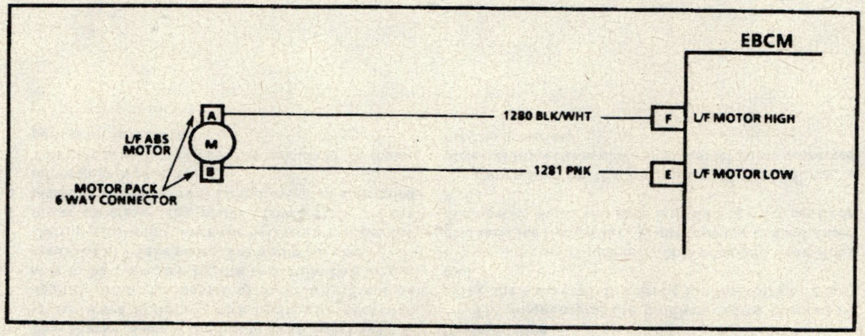

CODE A047

LEFT FRONT MOTOR FREE SPINS

Circuit Description:

This code is designed to detect a stripped nut or gear assembly during initialization. During the homing sequence, the piston should reach the top of the bore resulting in a stalled motor. If this does not occur, the motor must be spinning with little or no resistance, indicating a nut/screw or gear failure.

Test Description: Number(s) below refer to circled number(s) on the diagnostic chart.

1. This step positions the motor at the bottom of its travel.
2. This checks to see if the nut or gear is stripped. If either were stripped, the hydraulic pressure applied from the master cylinder would not be able to be overcome by the hydraulic ball screw, resulting in a brake pedal that would not rise.
3. During a full apply, once the ball screw reaches the top of its travel, the motor should stall. If the motor continues to run, an indication of a stripped nut or gear is present.
4. This step is used to release the tension of the motor pack prior to removal.

Diagnostic Aids:

An "Intermittent" fault in this code may result from a mechanical part of the system that sticks, binds, or slips.

The frequency of the problem can be checked by using the enhanced diagnostic function of the "Scan" tool

If code only fails once and Code A056 also fails see Code A056. If intermittent and enhanced diagnostics shows this code fails during ABS, see Code A056.

Depending on the frequency of the failure, a physical inspection of the mechanical parts suspected may be necessary.

DIAGNOSTIC CHART DTC A047 — 1993 BERETTA AND CORSICA

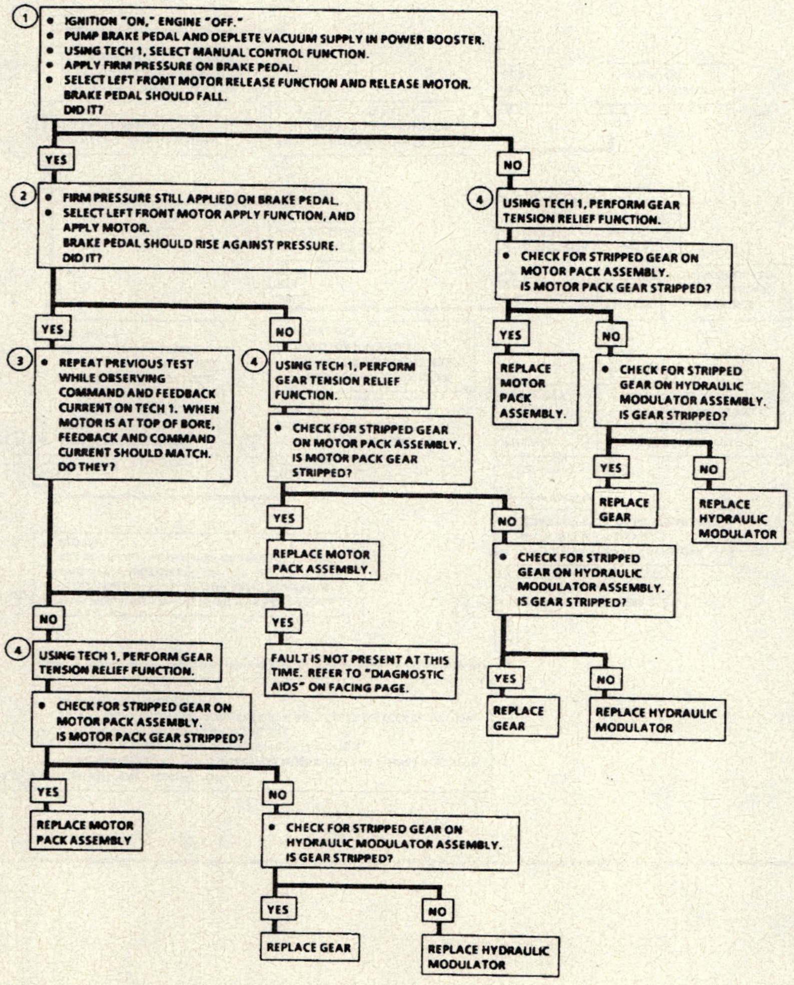

DIAGNOSTIC CHART DTC A08 — 1993 BERETTA AND CORSICA

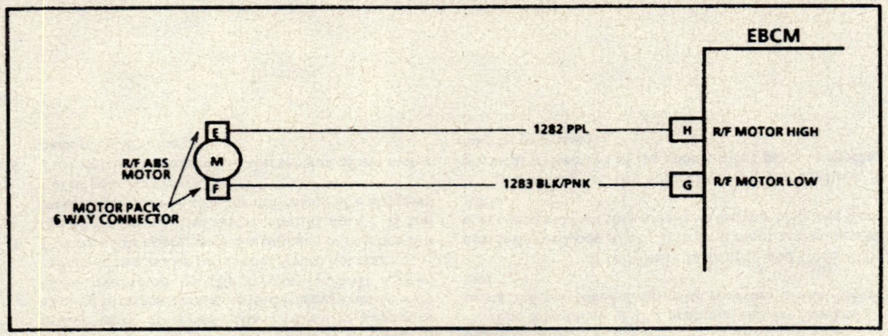

CODE A048

RIGHT FRONT MOTOR FREE SPINS

Circuit Description:

This code is designed to detect a stripped nut or gear assembly during initialization. During the homing sequence, the piston should reach the top of the bore resulting in a stalled motor. If this does not occur, the motor must be spinning with little or no resistance, indicating a nut/screw or gear failure.

Test Description: Number(s) below refer to circled number(s) on the diagnostic chart.

1. This step positions the motor at the bottom of its travel.
2. This checks to see if the nut or gear is stripped. If either were stripped, the hydraulic pressure applied from the master cylinder would not be able to be overcome by the hydraulic ball screw, resulting in a bake pedal that would not rise.
3. During a full apply, once the ball screw reaches the top of its travel, the motor should stall. If the motor continues to run, an indication of a stripped nut or gear is present.
4. This step is used to release the tension of the motor pack prior to removal.

Diagnostic Aids:

An "Intermittent" fault in this code may result from a mechanical part of the system that sticks, binds, or slips.

The frequency of the problem can be checked by using the enhanced diagnostic function of the "Scan" tool

If code only fails once and Code A061 also fails see Code A061. If intermittent and enhanced diagnostics shows this code fails during ABS, see Code A061.

Depending on the frequency of the failure, a physical inspection of the mechanical parts suspected may be necessary.

DIAGNOSTIC CHART DTC A08 — 1993 BERETTA AND CORSICA

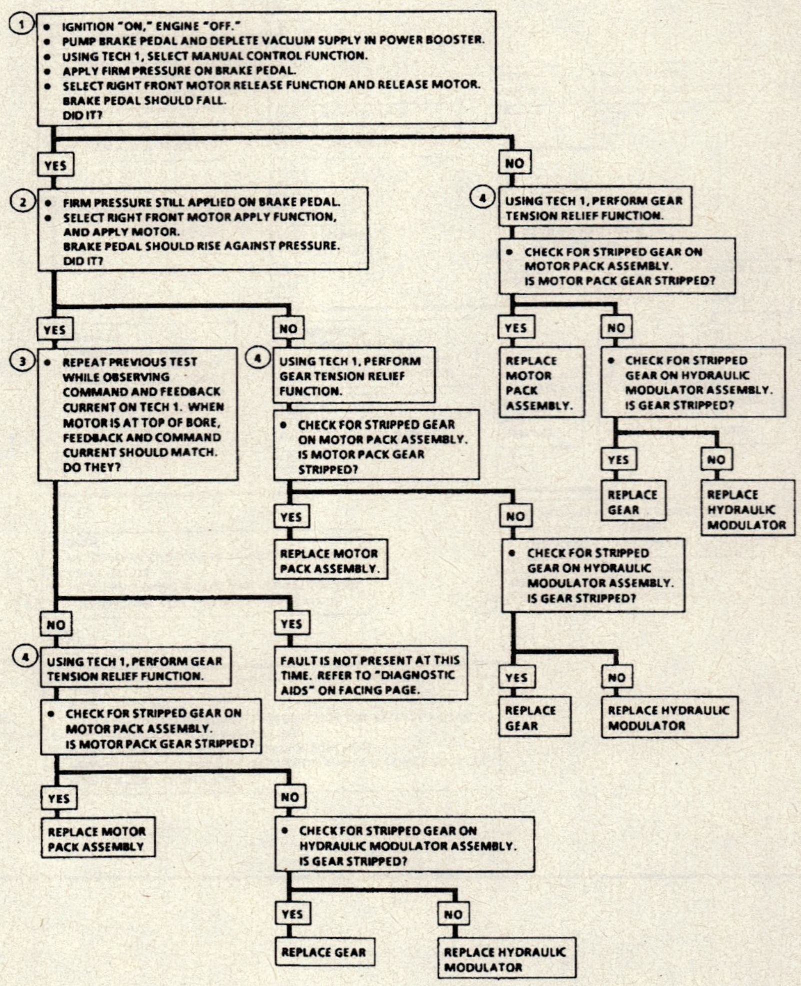

DIAGNOSTIC CHART DTC A051 — 1993 BERETTA AND CORSICA

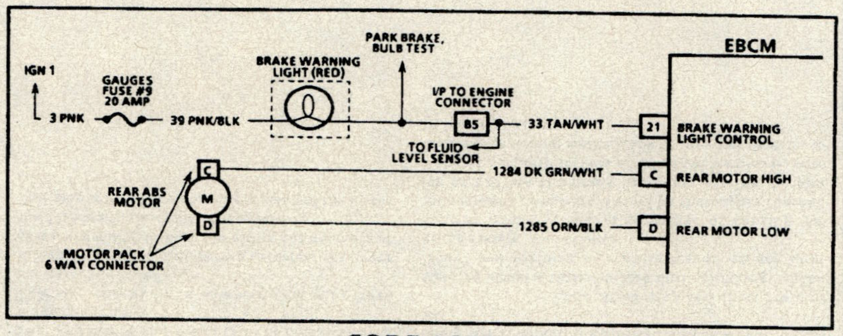

CODE A051

REAR AXLE MOTOR FREE SPINS

Circuit Description:

This code is designed to detect a stripped nut or gear assembly during initialization. During the homing sequence, the piston should reach the top of the bore resulting in a stalled motor. If this does not occur, the motor must be spinning with little or no resistance, indicating a nut/screw or gear failure.

Test Description: Number(s) below refer to circled number(s) on the diagnostic chart.

1. This step checks for a slipping nut or gear. If either were to slip, it would result in a lack of hydraulic pressure being applied at the rear wheels, causing the wheels to be able to be moved.
2. During a full apply, once the ball screw reaches the top of its travel, the motor should stall. If the motor continues to run, an indication of a stripped nut or gear is present.
3. This step is used to release the tension of the motor pack prior to removal.

Diagnostic Aids:

An "Intermittent" fault in this code may result from a mechanical part of the system that sticks, binds, or slips.

The frequency of the problem can be checked by using the enhanced diagnostic function of the "Scan" tool

If code only fails once and Code A064 also fails, see Code A064. If intermittent and enhanced diagnostics show this code fails during ABS, see Code A064.

Depending on the frequency of the failure, a physical inspection of the mechanical parts suspected may be necessary.

DIAGNOSTIC CHART DTC A051 — 1993 BERETTA AND CORSICA

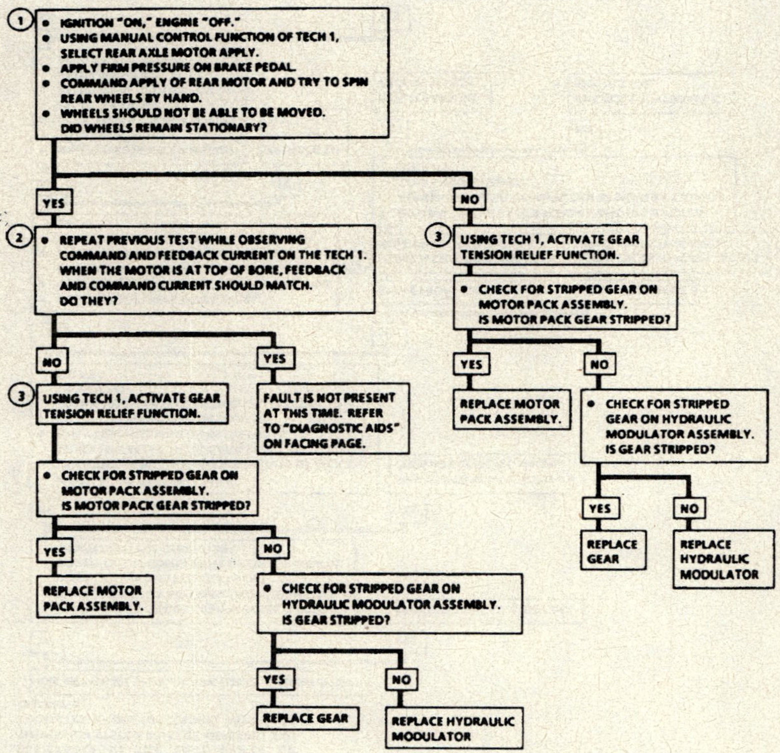

DIAGNOSTIC CHART DTC A052 — 1993 BERETTA AND CORSICA

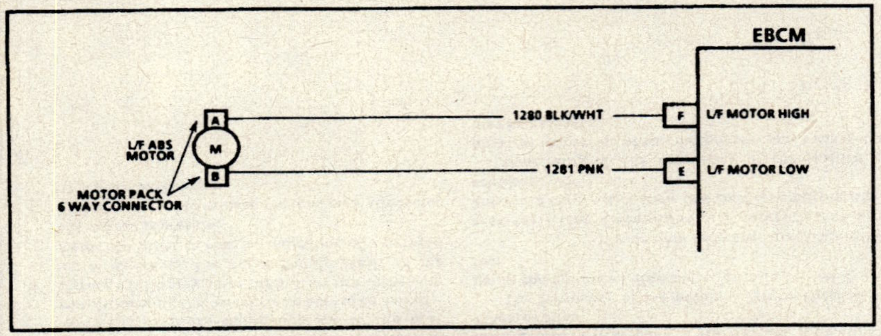

CODE A052

LEFT FRONT CHANNEL IN RELEASE TOO LONG

Circuit Description:

This code will diagnose a motor that is energized longer than expected. This could occur if a wheel speed sensor is faulty, the motor does not turn, the left front solenoid mechanically fails open, or the motor wires are crossed.

Test Description: Number(s) below refer to circled number(s) on the diagnostic chart.

1. This step identifies a problem in a wheel speed sensor that may cause the system to be in release too long.
2. This step identifies a motor fault as being failed or wired incorrectly.
3. This step checks for a solenoid that may have mechanically failed open.
4. If Step 3 has failed, this serves to isolate the cause of the hydraulic problem, to either the solenoid or the ABS hydraulic modulator assembly.
5. This step is used to release the tension of the motor pack.

Diagnostic Aids:

An "Intermittent" problem may be caused by a mechanical part of the system that sticks or binds.

The frequency of the problem can be checked by using the enhanced diagnostic function of the "Scan" tool,

Code A052 may fail if on ice and steering wheel is turned to lock during braking. Using "scan" tool perform Hydraulic test to ensure total brake system is functional.

Any circuitry, that is suspected as causing the intermittent complaint, should be thoroughly checked for backed out terminals, improper mating, broken locks, improperly formed or damaged terminals, poor terminal to wiring connections or physical damage to the wiring harness.

DIAGNOSTIC CHART DTC A052 — 1993 BERETTA AND CORSICA

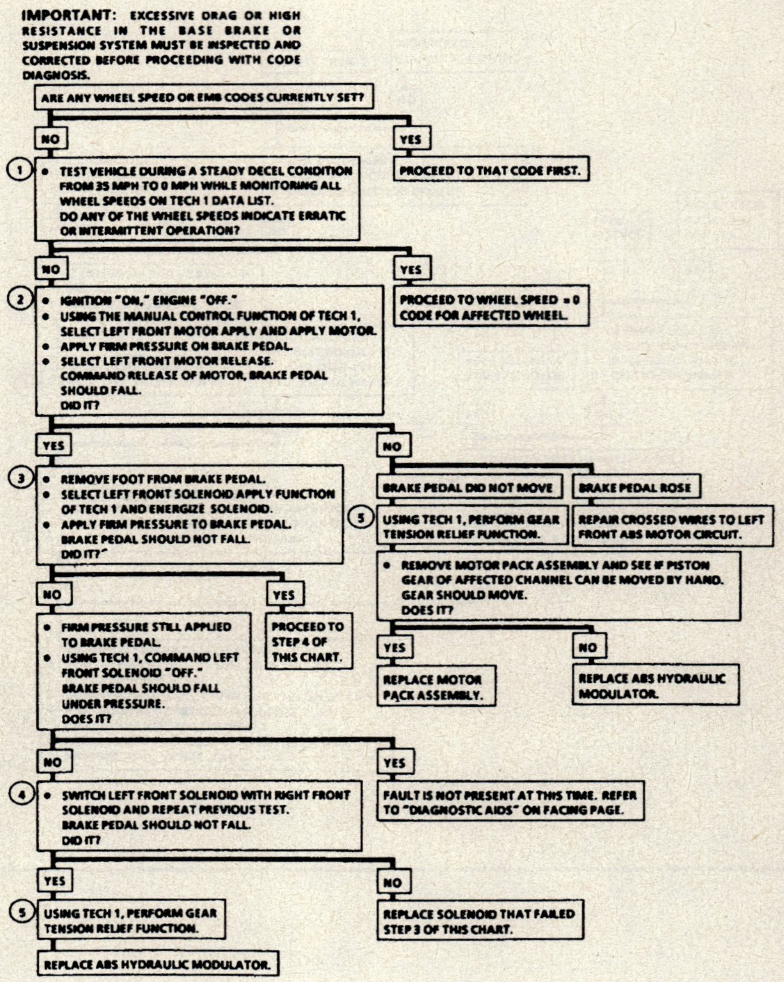

DIAGNOSTIC CHART DTC A053 — 1993 BERETTA AND CORSICA

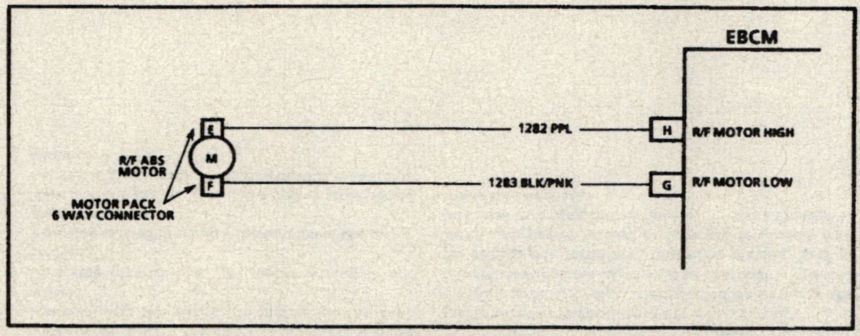

CODE A053

RIGHT FRONT CHANNEL IN RELEASE TOO LONG

Circuit Description:

This code will diagnose a motor that is energized longer than expected. This could occur if a wheel speed sensor is faulty, the motor does not turn, the right front solenoid mechanically fails open, or the motor wires are crossed.

Test Description: Number(s) below refer to circled number(s) on the diagnostic chart.

1. This step identifies a problem in a wheel speed sensor that may cause the system to be in release too long.
2. This step identifies a motor fault as being failed or wired incorrectly.
3. This step checks for a solenoid that may have mechanically failed open.
4. If Step 3 has failed, this serves to isolate the cause of the hydraulic problem, to either the solenoid or the ABS hydraulic modulator assembly.
5. This step is used to release the tension of the motor pack prior to removal.

Diagnostic Aids:

An "Intermittent" problem may be caused by a mechanical part of the system that sticks or binds.

The frequency of the problem can be checked by using the enhanced diagnostic function of the "Scan" tool,

Code A053 may fail on ice if steering wheel is turned to lock during braking. Using "Scan" tool perform Hydraulic test to ensure total brake system is functional.

Any circuitry, that is suspected as causing the intermittent complaint, should be thoroughly checked for backed out terminals, improper mating, broken locks, improperly formed or damaged terminals, poor terminal to wiring connections or physical damage to the wiring harness.

DIAGNOSTIC CHART DTC A053 — 1993 BERETTA AND CORSICA

IMPORTANT: EXCESSIVE DRAG OR HIGH RESISTANCE IN THE BASE BRAKE OR SUSPENSION SYSTEM MUST BE INSPECTED AND CORRECTED BEFORE PROCEEDING WITH CODE DIAGNOSIS.

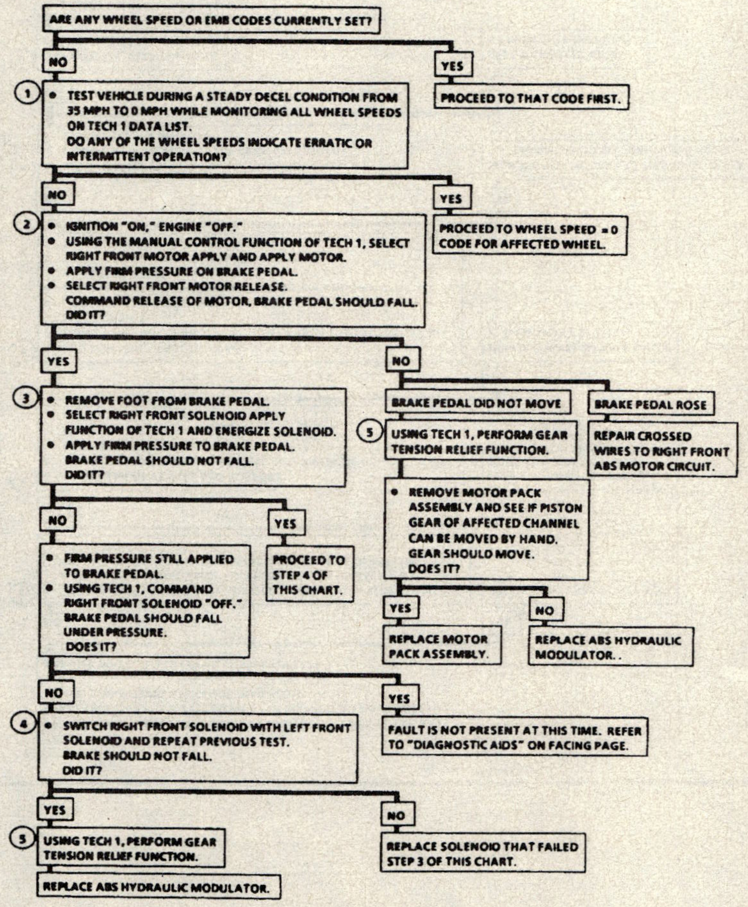

DIAGNOSTIC CHART DTC A054 — 1993 BERETTA AND CORSICA

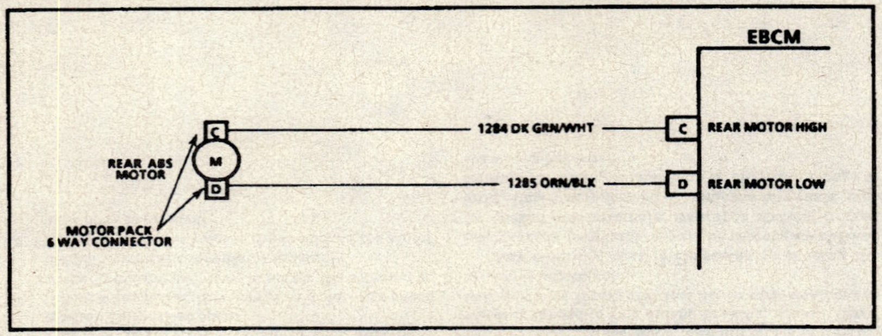

CODE A054

REAR AXLE CHANNEL IN RELEASE TOO LONG

Circuit Description:

This code will diagnose a motor that is energized longer than expected. This could occur if a wheel speed sensor is faulty, the motor does not turn, or the motor wires are crossed.

Test Description: Numbers below refer to circled numbers on the diagnostic chart.

1. This step identifies a problem in a wheel speed sensor that may cause the system to be in release too long.
2. This checks for a wheel that may stick or bind because of a mechanical fault.
3. This checks to see if the motor is capable of moving and applying the hydraulic piston for the rear wheels.
4. This step insures that the motor wiring is not crossed.
5. This step is used to release the tension of the motor pack prior to removal.
6. This isolates the fault of a "no-apply" situation to either the motor pack or ABS hydraulic modulator assembly.

Diagnostic Aids:

An "Intermittent" problem may be caused by a mechanical part of the system that sticks or binds.

The frequency of the problem can be checked by using the enhanced diagnostic function of the "Scan" tool,

Using "Scan" tool perform hydraulic test to ensure total brake system is functional.

Any circuitry, that is suspected as causing the intermittent complaint, should be thoroughly checked for backed out terminals, improper mating, broken locks, improperly formed or damaged terminals, poor terminal to wiring connections or physical damage to the wiring harness.

DIAGNOSTIC CHART DTC A054 — 1993 BERETTA AND CORSICA

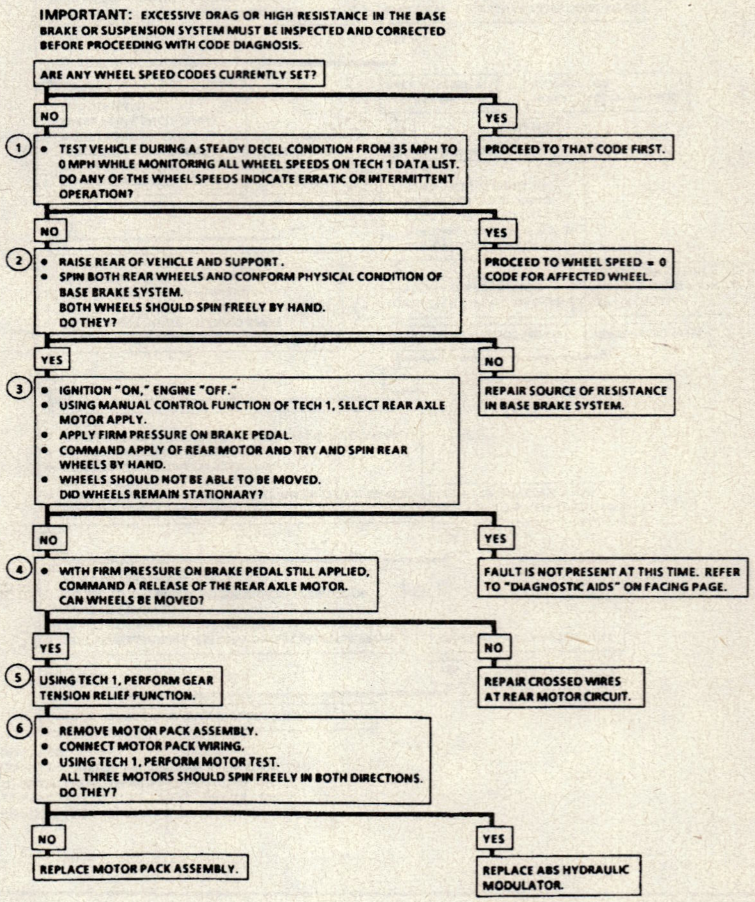

DIAGNOSTIC CHART DTC A055 — 1993 BERETTA AND CORSICA

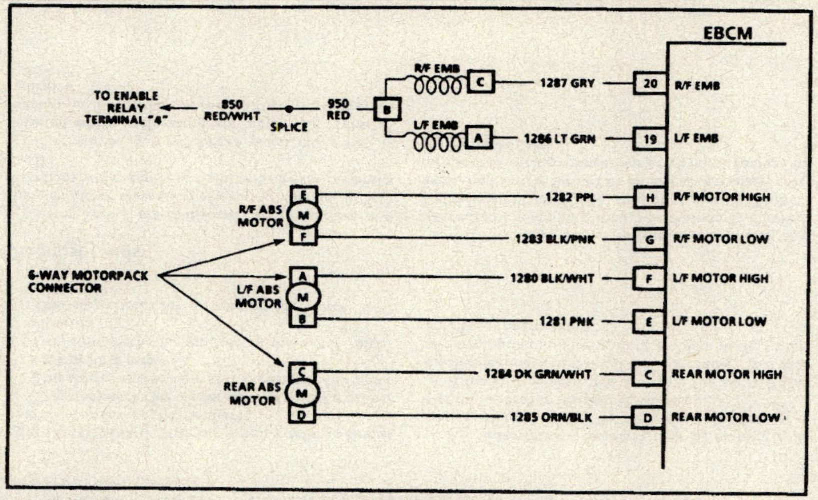

CODE A055

MOTOR DRIVER FAULT DETECTED

Circuit Description:

This code identifies a fault detected by the MDI custom IC. It also ensures the cause of the fault is not a result of a problem with the enable relay under a Code A055 failure. The MDI fault tests will be run to identify which circuit has failed, and may also cause additional codes to set to pinpoint the problem circuit.

Test Description: Number(s) below refer to circled number(s) on the diagnostic chart.

1. This insures that the Code A055 fault was not set due to a motor or EMB fault.
2. This checks to see if the fault is still present.
3. This step insures that Code A055 was not due to poor terminal contact at the motor pack connector.

Diagnostic Aids:

An "Intermittent" problem may be caused by a poor connection, rubbed through wire insulation, or a wire that is broken inside the insulation.

Any circuitry, that is suspected as causing the intermittent complaint, should be thoroughly checked for backed out terminals, improper mating, broken locks, improperly formed or damaged terminals, poor terminal to wiring connections or physical damage to the wiring harness.

DIAGNOSTIC CHART DTC A055 — 1993 BERETTA AND CORSICA

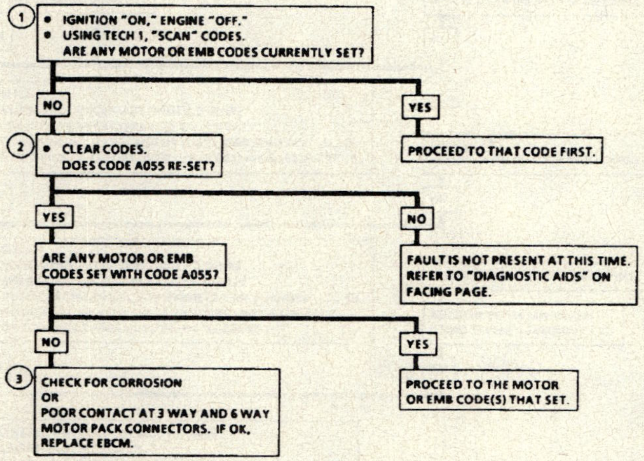

THIS CHART ASSUMES THAT A CURRENT CODE IS STORED INDICATING THAT THIS FAULT IS PRESENT.

DIAGNOSTIC CHART DTC A056 — 1993 BERETTA AND CORSICA

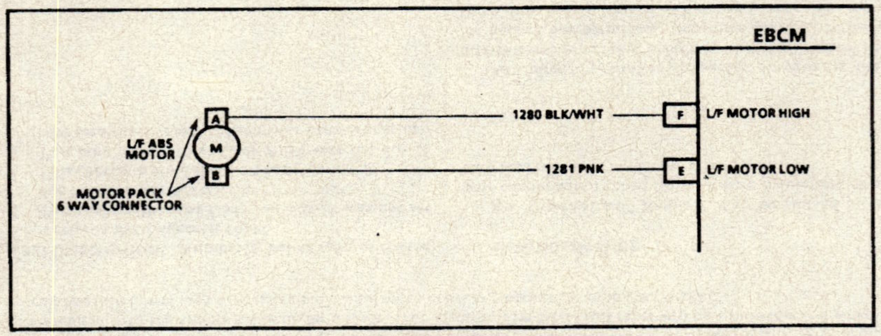

CODE A056

LEFT FRONT MOTOR CIRCUIT OPEN

Circuit Description:

This code identifies a motor that cannot be energized due to an open in its circuitry. This fault will not allow proper front ABS operation.

Test Description: Number(s) below refer to circled number(s) on the diagnostic chart.

1. This step checks for proper resistance of the motor.
2. This step is used to release the tension of the motor pack prior to removal.
3. This step checks for an open in the motor "high" circuitry.
4. This step checks for an open in the motor "low" circuitry.

An "Intermittent" problem may be caused by a poor connection, rubbed through wire insulation, or a wire that is broken inside the insulation.

If the fault is not current, wiggle the wires of the affected channel and check if the code resets. This will help to pinpoint an intermittent problem in the motor circuitry or connections.

Diagnostic Aids:

Using Tech 1, select manual control function, and exercise motor movement of affected channel in both directions while applying light pressure on the brake pedal.

If erratic or "jumpy" brake pedal movement is detected while performing an "apply" or "release" function of the motor, an intermittent fault may be indicated.

Any circuitry, that is suspected as causing the intermittent complaint, should be thoroughly checked for backed out terminals, improper mating, broken locks, improperly formed or damaged terminals, poor terminal to wiring connections or physical damage to the wiring harness.

DIAGNOSTIC CHART DTC A056 — 1993 BERETTA AND CORSICA

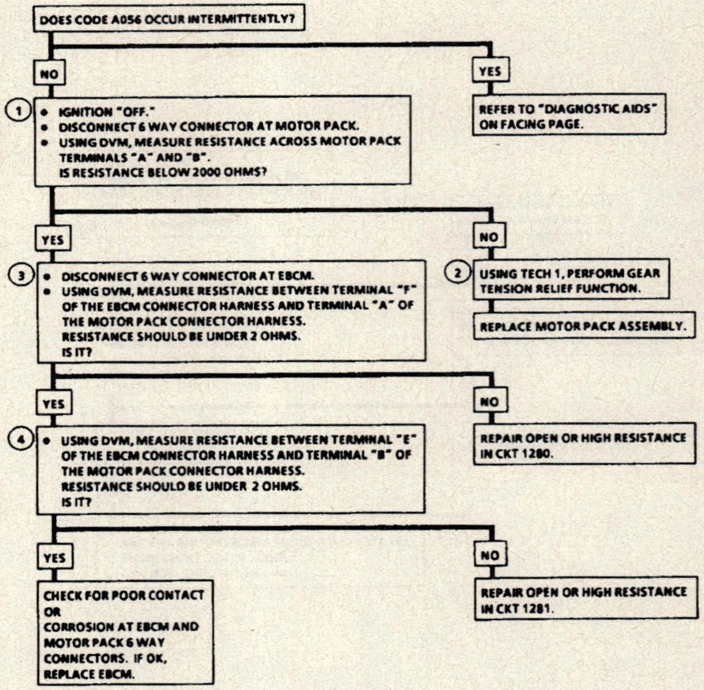

THIS CHART ASSUMES THAT A CURRENT CODE IS STORED INDICATING THAT THIS FAULT IS PRESENT.

DIAGNOSTIC CHART DTC A057 — 1993 BERETTA AND CORSICA

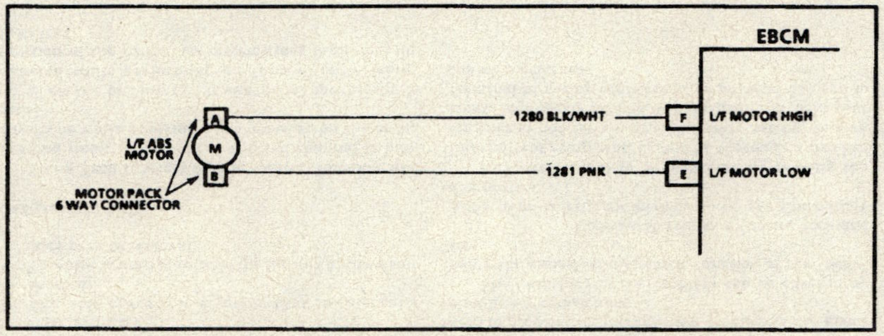

CODE A057

LEFT FRONT MOTOR CIRCUIT SHORTED TO GROUND

Circuit Description:

This code identifies a motor circuit that is shorted to ground. This fault will not allow the motor to be controlled at the commanded current rate or will cause the driver circuit to allow current directly to ground.

Test Description: Number(s) below refer to circled number(s) on the diagnostic chart.

1. This step checks for a short to ground in the motor "high" circuitry.
2. This step checks for a short to ground in the motor "low" circuitry.
3. This step checks for a motor that is internally shorted to ground.
4. This step is used to release the tension of the motor pack prior to removal.

Diagnostic Aids:

Using Tech 1, select manual control function, and exercise motor movement of affected channel in both directions while applying light pressure on the brake pedal.

If erratic or "jumpy" brake pedal movement is detected while performing an "apply" or "release" function of the motor, an intermittent fault may be indicated.

An "Intermittent" problem may be caused by a poor connection, rubbed through wire insulation, or a wire that is broken inside the insulation.

If the fault is not current, wiggle the wires of the affected channel and check if the code resets. This will help to pinpoint an intermittent problem in the motor circuitry or connections.

Any circuitry, that is suspected as causing the intermittent complaint, should be thoroughly checked for backed out terminals, improper mating, broken locks, improperly formed or damaged terminals, poor terminal to wiring connections or physical damage to the wiring harness.

DIAGNOSTIC CHART DTC A057 — 1993 BERETTA AND CORSICA

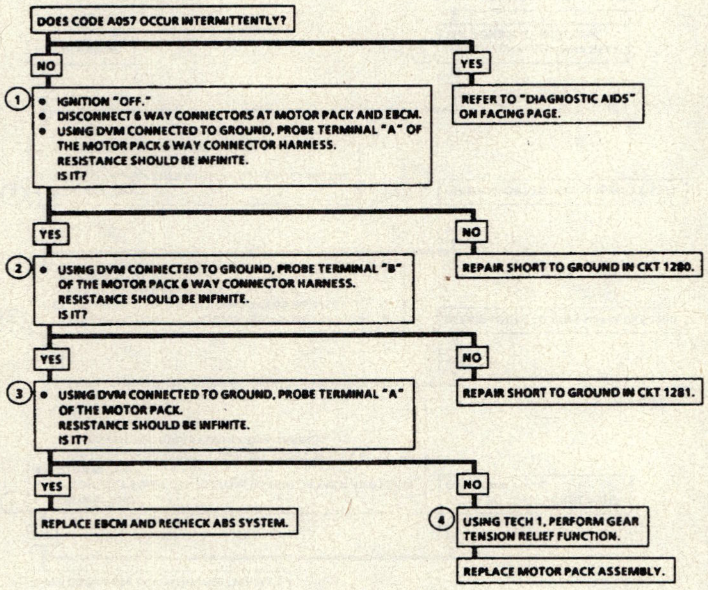

THIS CHART ASSUMES THAT A CURRENT CODE IS STORED INDICATING THAT THIS FAULT IS PRESENT.

DIAGNOSTIC CHART DTC A058 — 1993 BERETTA AND CORSICA

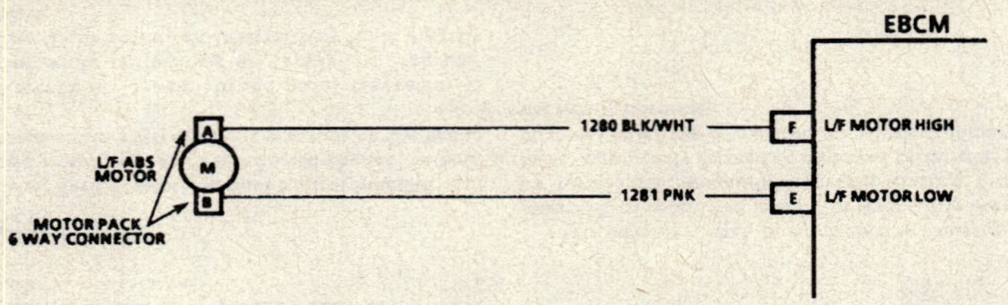

CODE A058

LEFT FRONT MOTOR CIRCUIT SHORTED TO BATTERY OR MOTOR SHORTED

Circuit Description:

This code identifies a motor circuit that is shorted to battery or a motor that has low or no resistance. This fault will not allow the motor to be controlled at the commanded current rate or will cause the motor to turn in the opposite direction, or not at all.

Test Description: Number(s) below refer to circled number(s) on the diagnostic chart.

1. This step checks for a short to voltage in the motor "high" circuitry.
2. This step checks for a short to voltage in the motor "low" circuitry.
3. This step checks for a motor that is internally shorted.
4. This step is used to release the tension of the motor pack prior to removal.

Diagnostic Aids:

Using Tech 1, select manual control function, and exercise motor movement of affected channel in both directions while applying light pressure on the brake pedal.

If erratic or "jumpy" brake pedal movement is detected while performing an "apply" or "release" function of the motor, an intermittent fault may be indicated.

An "Intermittent" problem may be caused by a poor connection, rubbed through wire insulation, or a wire that is broken inside the insulation.

If the fault is not current, wiggle the wires of the affected channel and check if the code resets. This will help to pinpoint an intermittent problem in the motor circuitry or connections.

The frequency of the problem can be checked by using the enhanced diagnostic function of the "Scan" tool,

Code A038 sets as a current code and A058 is an infrequent history code, see Code A038 diagnostics.

Any circuitry, that is suspected as causing the intermittent complaint, should be thoroughly checked for backed out terminals, improper mating, broken locks, improperly formed or damaged terminals, poor terminal to wiring connections or physical damage to the wiring harness.

DIAGNOSTIC CHART DTC A058 — 1993 BERETTA AND CORSICA

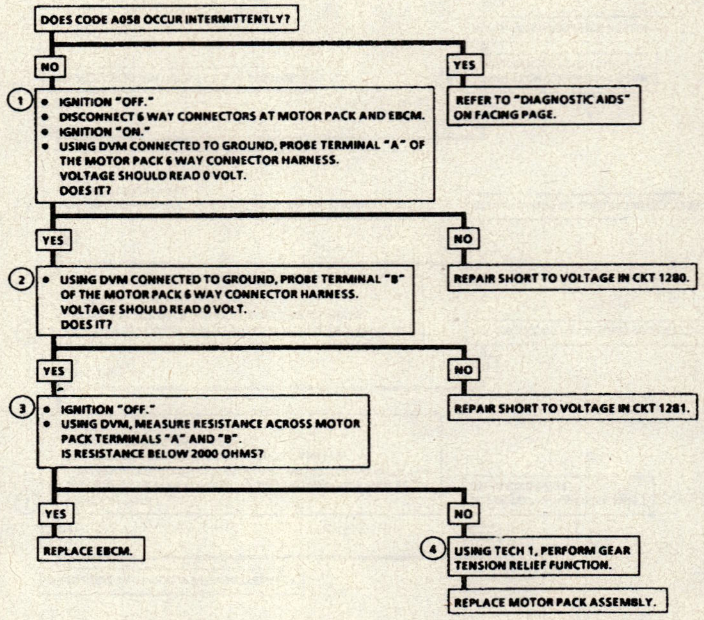

THIS CHART ASSUMES THAT A CURRENT CODE IS STORED INDICATING THAT THIS FAULT IS PRESENT.

DIAGNOSTIC CHART DTC A061 — 1993 BERETTA AND CORSICA

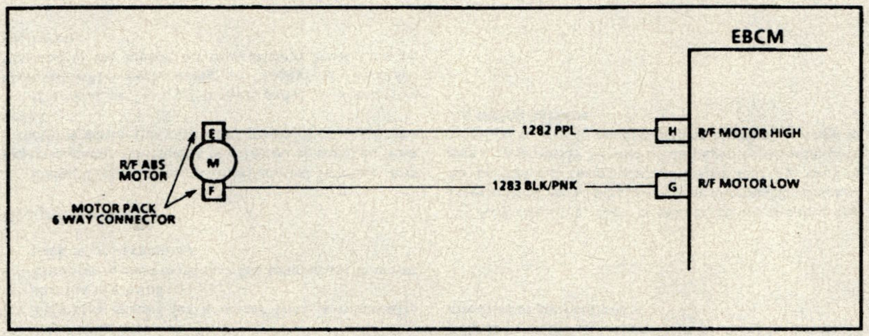

CODE A061

RIGHT FRONT MOTOR CIRCUIT OPEN

Circuit Description:

This code identifies a motor that cannot be energized due to an open in its circuitry. This fault will not allow proper front ABS operation.

Test Description: Number(s) below refer to circled number(s) on the diagnostic chart.

1. This step checks for proper resistance of the motor.
2. This step is used to release the tension of the motor pack prior to removal.
3. This step checks for an open in the motor "high" circuitry.
4. This step checks for an open in the motor "low" circuitry.

Diagnostic Aids:

Using Tech 1, select manual control function, and exercise motor movement of affected channel in both directions while applying light pressure on the brake pedal.

If erratic or "jumpy" brake pedal movement is detected while performing an "apply" or "release" function of the motor, an intermittent fault may be indicated.

An "Intermittent" problem may be caused by a poor connection, rubbed through wire insulation, or a wire that is broken inside the insulation.

If the fault is not current, wiggle the wires of the affected channel and check if the code resets. This will help to pinpoint an intermittent problem in the motor circuitry or connections.

Any circuitry, that is suspected as causing the intermittent complaint, should be thoroughly checked for backed out terminals, improper mating, broken locks, improperly formed or damaged terminals, poor terminal to wiring connections or physical damage to the wiring harness.

DIAGNOSTIC CHART DTC A061 — 1993 BERETTA AND CORSICA

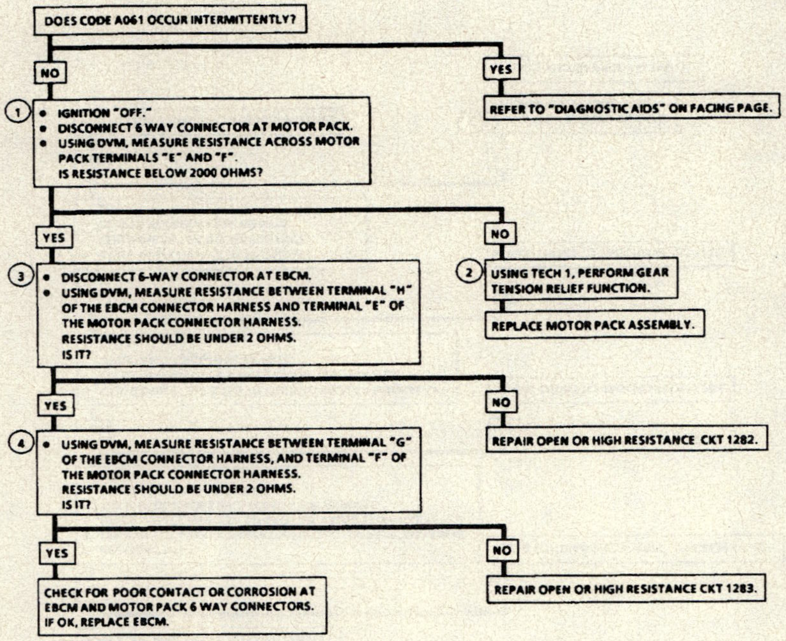

THIS CHART ASSUMES THAT A CURRENT CODE IS STORED INDICATING THAT THIS FAULT IS PRESENT.

DIAGNOSTIC CHART DTC A062 — 1993 BERETTA AND CORSICA

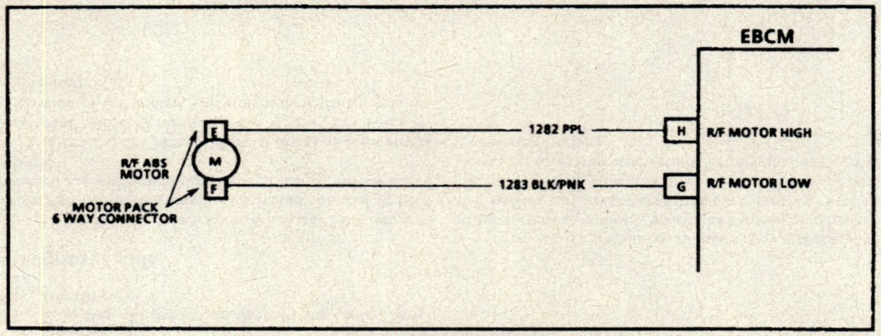

CODE A062

RIGHT FRONT MOTOR CIRCUIT SHORTED TO GROUND

Circuit Description:

This code identifies a motor circuit that is shorted to ground. This fault will not allow the motor to be controlled at the commanded current rate or will cause the motor driver circuit to allow current directly to ground.

Test Description: Number(s) below refer to circled number(s) on the diagnostic chart.

1. This step checks for a short to ground in the motor "high" circuitry.
2. This step checks for a short to ground in the motor "low" circuitry.
3. This step checks for a motor that is internally shorted to ground.
4. This step is used to release the tension of the motor pack prior to removal.

An "Intermittent" problem may be caused by a poor connection, rubbed through wire insulation, or a wire that is broken inside the insulation.

If the fault is not current, wiggle the wires of the affected channel and check if the code resets. This will help to pinpoint an intermittent problem in the motor circuitry or connections.

Diagnostic Aids:

Using Tech 1, select manual control function, and exercise motor movement of affected channel in both directions while applying light pressure on the brake pedal.

If erratic or "jumpy" brake pedal movement is detected while performing an "apply" or "release" function of the motor, an intermittent fault may be indicated.

Any circuitry, that is suspected as causing the intermittent complaint, should be thoroughly checked for backed out terminals, improper mating, broken locks, improperly formed or damaged terminals, poor terminal to wiring connections or physical damage to the wiring harness.

DIAGNOSTIC CHART DTC A062 — 1993 BERETTA AND CORSICA

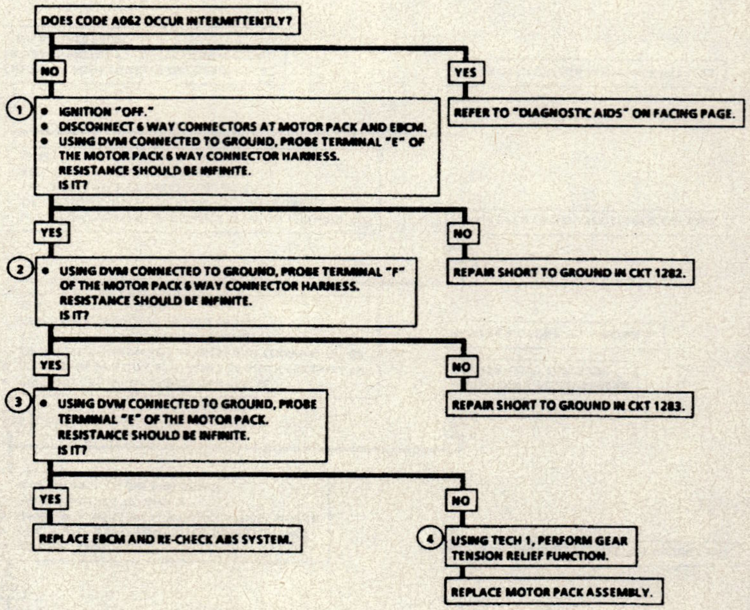

THIS CHART ASSUMES THAT A CURRENT CODE IS STORED INDICATING THAT THIS FAULT IS PRESENT.

DIAGNOSTIC CHART DTC A063 — 1993 BERETTA AND CORSICA

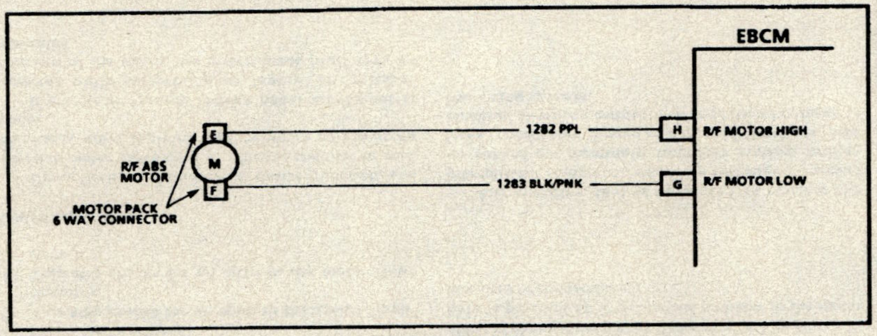

CODE A063

RIGHT FRONT MOTOR CIRCUIT SHORTED TO BATTERY OR MOTOR SHORTED

Circuit Description:

This code identifies a motor circuit that is shorted to battery or a motor that has low or no resistance. This fault will not allow the motor to be controlled at the commanded current rate or will cause the motor to turn in the opposite direction or not at all.

Test Description: Number(s) below refer to circled number(s) on the diagnostic chart.

1. This step checks for a short to voltage in the motor "high" circuitry.
2. This step checks for a short to voltage in the motor "low" circuitry.
3. This step checks for a motor that is internally shorted.
4. This test is used to release the tension of the motor pack prior to removal.

Diagnostic Aids:

Using Tech 1, select manual control function, and exercise motor movement of affected channel in both directions while applying light pressure on the brake pedal.

If erratic or "jumpy" brake pedal movement is detected while performing an "apply" or "release" function of the motor, an intermittent fault may be indicated.

An "Intermittent" problem may be caused by a poor connection, rubbed through wire insulation, or a wire that is broken inside the insulation.

If the fault is not current, wiggle the wires of the affected channel and check if the code resets. This will help to pinpoint an intermittent problem in the motor circuitry or connections.

The frequency of the problem can be checked by using the enhanced diagnostic function of the Tech 1 "Scan" tool,

If Code A041 sets as a current code and A063 is infrequent history code, see Code A041 for diagnosis.

Any circuitry, that is suspected as causing the intermittent complaint, should be thoroughly checked for backed out terminals, improper mating, broken locks, improperly formed or damaged terminals, poor terminal to wiring connections or physical damage to the wiring harness.

DIAGNOSTIC CHART DTC A063 — 1993 BERETTA AND CORSICA

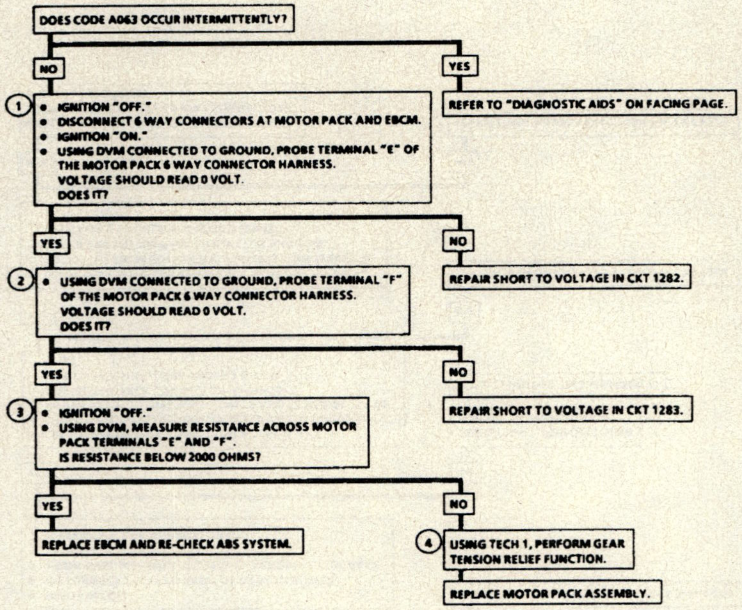

THIS CHART ASSUMES THAT A CURRENT CODE IS STORED INDICATING THAT THIS FAULT IS PRESENT.

DIAGNOSTIC CHART DTC A064 — 1993 BERETTA AND CORSICA

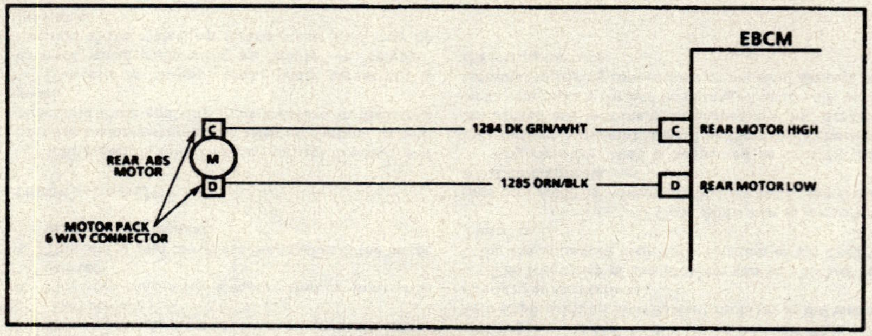

CODE A064

REAR AXLE MOTOR CIRCUIT OPEN

Circuit Description:

This code identifies a motor that cannot be energized due to an open in its circuitry. This fault will not allow proper rear ABS operation.

Test Description: Number(s) below refer to circled number(s) on the diagnostic chart.

1. This checks for an open circuit in the motor.
2. This step is used to release the tension of the motor pack prior to removal.
3. This step checks for an open in the motor "high" circuitry.
4. This step checks for an open in the motor "low" circuitry.

An "Intermittent" problem may be caused by a poor connection, rubbed through wire insulation, or a wire that is broken inside the insulation.

If the fault is not current, wiggle the wires of the affected channel and check if the code resets. This will help to pinpoint an intermittent problem in the motor circuitry or connections.

Diagnostic Aids:

Using Tech 1, select manual control function, and exercise motor movement of affected channel in both directions while applying light pressure on the brake pedal.

If erratic or "jumpy" brake pedal movement is detected while performing an "apply" or "release" function of the motor, an intermittent fault may be indicated.

Any circuitry, that is suspected as causing the intermittent complaint, should be thoroughly checked for backed out terminals, improper mating, broken locks, improperly formed or damaged terminals, poor terminal to wiring connections or physical damage to the wiring harness.

DIAGNOSTIC CHART DTC A064 — 1993 BERETTA AND CORSICA

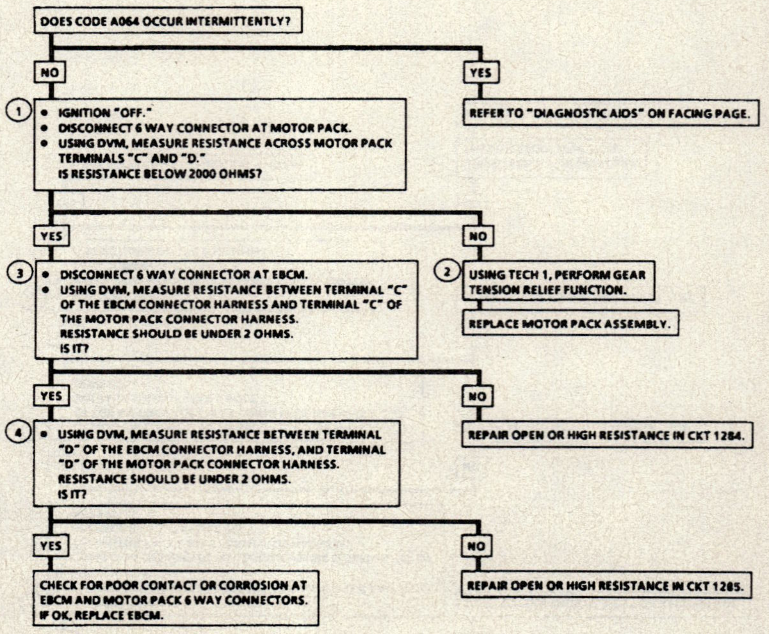

THIS CHART ASSUMES THAT A CURRENT CODE IS STORED INDICATING THAT THIS FAULT IS PRESENT.

DIAGNOSTIC CHART DTC A065 — 1993 BERETTA AND CORSICA

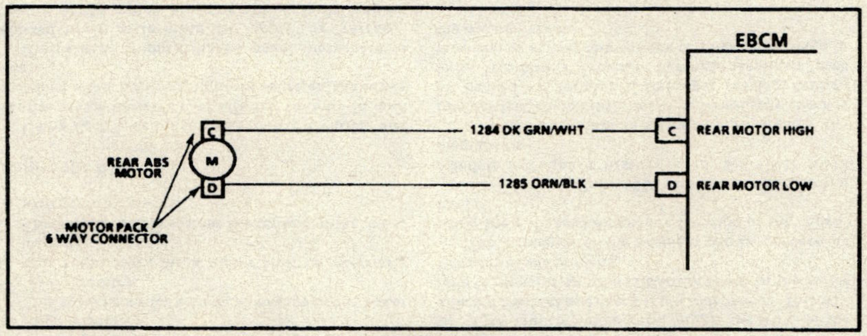

CODE A065

REAR AXLE MOTOR CIRCUIT SHORTED TO GROUND

Circuit Description:

This code identifies a motor circuit that is shorted to ground. This fault will not allow the motor to be controlled at the commanded current rate or will cause the driver circuit to allow current directly to ground.

Test Description: Number(s) below refer to circled number(s) on the diagnostic chart.

1. This step checks for a short to ground in the motor "high" circuitry.
2. This step checks for a short to ground in the motor "low" circuitry.
3. This step checks for a motor that is internally shorted to ground.
4. This step is used to release the tension of the motor pack prior to removal.

An "Intermittent" problem may be caused by a poor connection, rubbed through wire insulation, or a wire that is broken inside the insulation.

If the fault is not current, wiggle the wires of the affected channel and check if the code resets. This will help to pinpoint an intermittent problem in the motor circuitry or connections.

Diagnostic Aids:

Using Tech 1, select manual control function, and exercise motor movement of affected channel in both directions while applying light pressure on the brake pedal.

If erratic or "jumpy" brake pedal movement is detected while performing an "apply" or "release" function of the motor, an intermittent fault may be indicated.

Any circuitry, that is suspected as causing the intermittent complaint, should be thoroughly checked for backed out terminals, improper mating, broken locks, improperly formed or damaged terminals, poor terminal to wiring connections or physical damage to the wiring harness.

DIAGNOSTIC CHART DTC A065 — 1993 BERETTA AND CORSICA

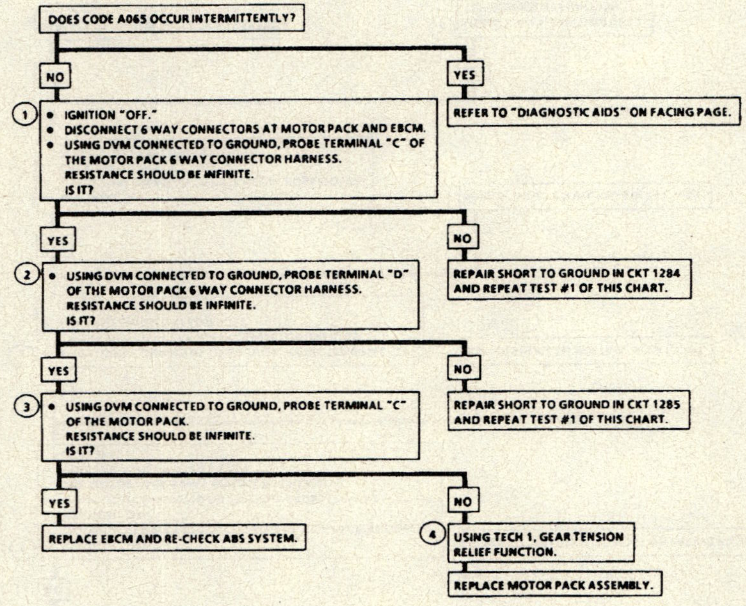

THIS CHART ASSUMES THAT A CURRENT CODE IS STORED INDICATING THAT THIS FAULT IS PRESENT.

DIAGNOSTIC CHART DTC A066 — 1993 BERETTA AND CORSICA

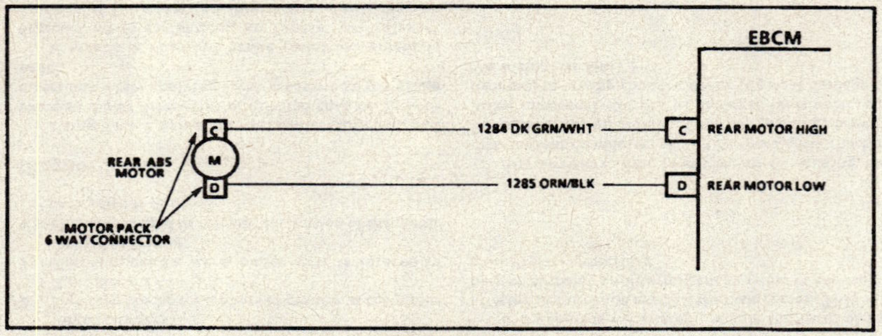

CODE A066

REAR AXLE MOTOR CIRCUIT SHORTED TO BATTERY OR MOTOR SHORTED

Circuit Description:

This code identifies a motor circuit that is shorted to battery or a motor that has low or zero resistance. This fault will not allow the motor to be controlled at the commanded current rate or will cause the motor to turn in the opposite direction or not turn at all.

Test Description: Number(s) below refer to circled number(s) on the diagnostic chart.

1. This step checks for a short to voltage in the motor "high" circuitry.
2. This step checks for a short to voltage in the motor "low" circuitry.
3. This step checks for a motor that is internally shorted.
4. This step is used to release the tension of the motor pack prior to removal.

Diagnostic Aids:

Using Tech 1, select manual control function, and exercise motor movement of affected channel in both directions while applying light pressure on the brake pedal.

If erratic or "jumpy" brake pedal movement is detected while performing an "apply" or "release" function of the motor, an intermittent fault may be indicated.

An "Intermittent" problem may be caused by a poor connection, rubbed through wire insulation, or a wire that is broken inside the insulation.

If the fault is not current, wiggle the wires of the affected channel and check if the code resets. This will help to pinpoint an intermittent problem in the motor circuitry or connections.

The frequency of the problem can be checked by using the enhanced diagnostic function of the "Scan" tool,

If Code A042 sets as current code and A066 is infrequent history code, see Code A042 diagnostics.

Any circuitry, that is suspected as causing the intermittent complaint, should be thoroughly checked for backed out terminals, improper mating, broken locks, improperly formed or damaged terminals, poor terminal to wiring connections or physical damage to the wiring harness.

DIAGNOSTIC CHART DTC A066 — 1993 BERETTA AND CORSICA

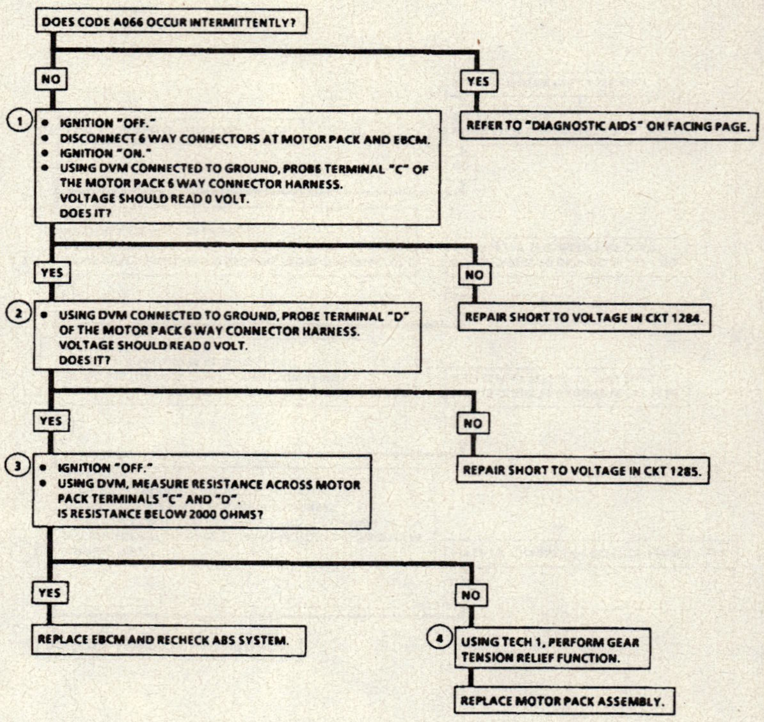

THIS CHART ASSUMES THAT A CURRENT CODE IS STORED INDICATING THAT THIS FAULT IS PRESENT.

DIAGNOSTIC CHART DTC A067 — 1993 BERETTA AND CORSICA

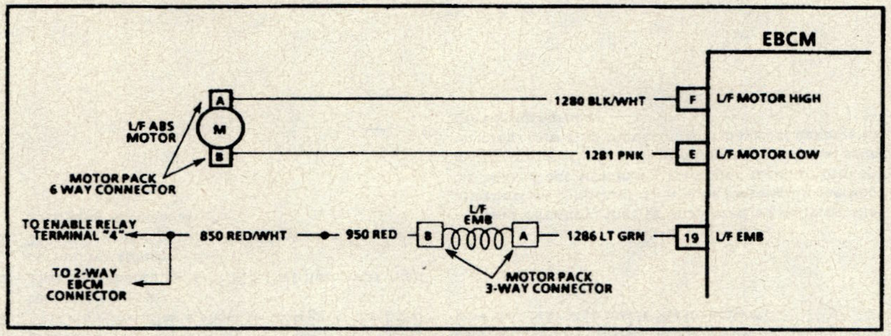

CODE A067

LEFT FRONT EMB CIRCUIT OPEN OR SHORTED TO GROUND

Circuit Description:

This code is designed to detect shorts to ground and opens in the EMB release circuit.

A short to ground between the EMB and the ABS EBCM will cause the EMB to be energized indefinitely at an uncontrolled current, causing the EMB to damage itself. To remove the effects of the fault, the enable relay must be turned "OFF," eliminating power to the EMB as well as the ABS system.

Test Description: Number(s) below refer to circled number(s) on the diagnostic chart.

1. This step checks for a short to ground in the EMB control circuitry.
2. This step insures that the EMB is not internally shorted or open.
3. This step checks to see if CKT 850/950 has continuity to the EMB connector.
4. This step looks for an open in the EMB control circuitry.
5. This step is used to release the tension of the motor pack prior to removal.

Diagnostic Aids:

An "Intermittent" problem may be caused by a poor connection, rubbed through wire insulation, or a wire that is broken inside the insulation.

Any circuitry, that is suspected as causing the intermittent complaint, should be thoroughly checked for backed out terminals, improper mating, broken locks, improperly formed or damaged terminals, poor terminal to wiring connections or physical damage to the wiring harness.

DIAGNOSTIC CHART DTC A067 — 1993 BERETTA AND CORSICA

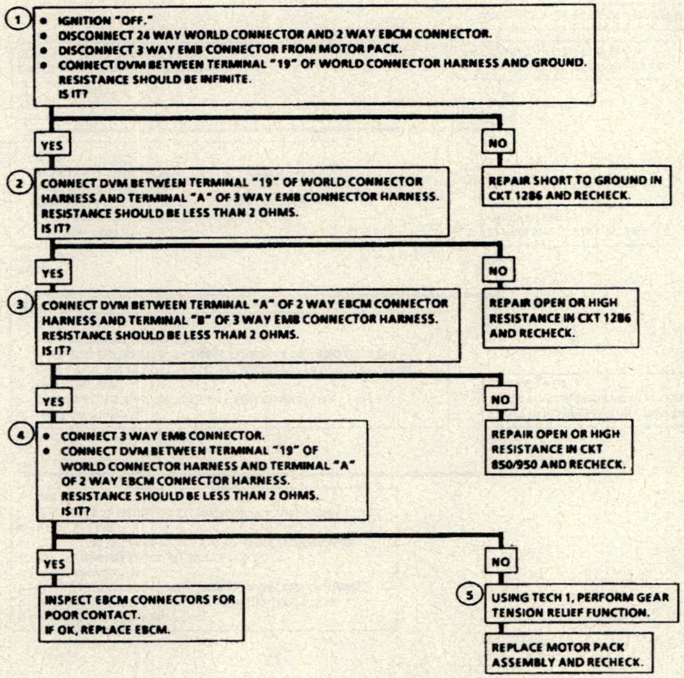

THIS CHART ASSUMES THAT A CURRENT CODE IS STORED INDICATING THAT THIS FAULT IS PRESENT.

DIAGNOSTIC CHART DTC A068 — 1993 BERETTA AND CORSICA

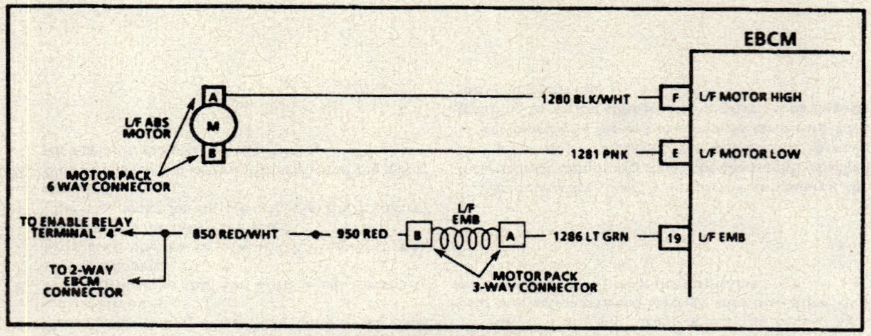

CODE A068

LEFT FRONT EMB CIRCUIT SHORTED TO BATTERY OR EMB SHORTED

Circuit Description:

This code is designed to detect shorts to battery and shorts in the EMB release circuit. These faults will not allow the EMB to be energized when needed.

Test Description: Number(s) below refer to circled number(s) on the diagnostic chart.

1. This step checks to see if the fault is currently present.
2. This step identifies a short to voltage in the EMB control circuitry.
3. This step checks for an EMB that has lower than normal resistance.
4. This step is used to release the tension of the motor pack prior to removal.

Diagnostic Aids:

An "Intermittent" problem may be caused by a poor connection, rubbed through wire insulation, or a wire that is broken inside the insulation.

Any circuitry, that is suspected as causing the intermittent complaint, should be thoroughly checked for backed out terminals, improper mating, broken locks, improperly formed or damaged terminals, poor terminal to wiring connections or physical damage to the wiring harness.

DIAGNOSTIC CHART DTC A068 — 1993 BERETTA AND CORSICA

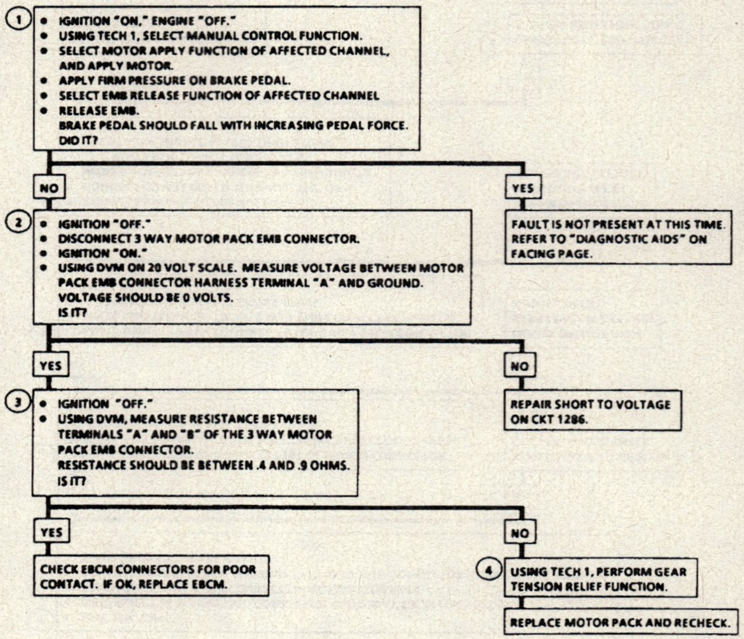

THIS CHART ASSUMES THAT A CURRENT CODE IS STORED INDICATING THAT THIS FAULT IS PRESENT.

DIAGNOSTIC CHART DTC A071 — 1993 BERETTA AND CORSICA

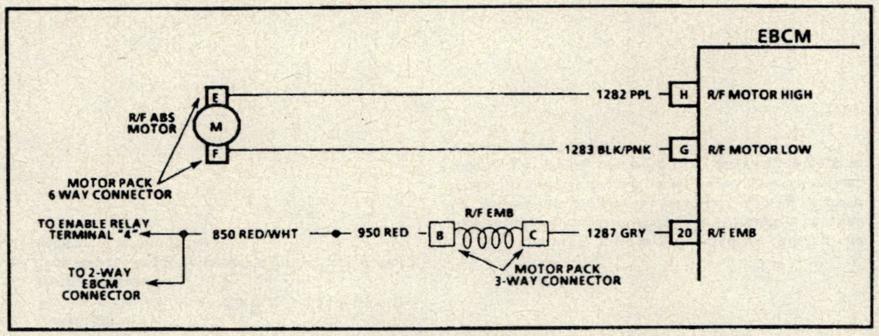

CODE A071

RIGHT FRONT EMB CIRCUIT OPEN OR SHORTED TO GROUND

Circuit Description:

This code is designed to detect shorts to ground and opens in the EMB release circuit.

A short to ground between the EMB and the ABS EBCM will cause the EMB to be energized indefinitely at an uncontrolled current causing the EMB to damage itself.

Test Description: Number(s) below refer to circled number(s) on the diagnostic chart.

1. This step checks for a short to ground in the EMB control circuitry.
2. This step insures that the EMB is not internally shorted or open.
3. This step checks to see if CKT 850/950 has continuity to the EMB connector.
4. This step looks for an open in the EMB control circuitry.
5. This step is used to release the tension of the motor pack prior to removal.

Diagnostic Aids:

An "Intermittent" problem may be caused by a poor connection, rubbed through wire insulation, or a wire that is broken inside the insulation.

Any circuitry, that is suspected as causing the intermittent complaint, should be thoroughly checked for backed out terminals, improper mating, broken locks, improperly formed or damaged terminals, poor terminal to wiring connections or physical damage to the wiring harness.

DIAGNOSTIC CHART DTC A071 — 1993 BERETTA AND CORSICA

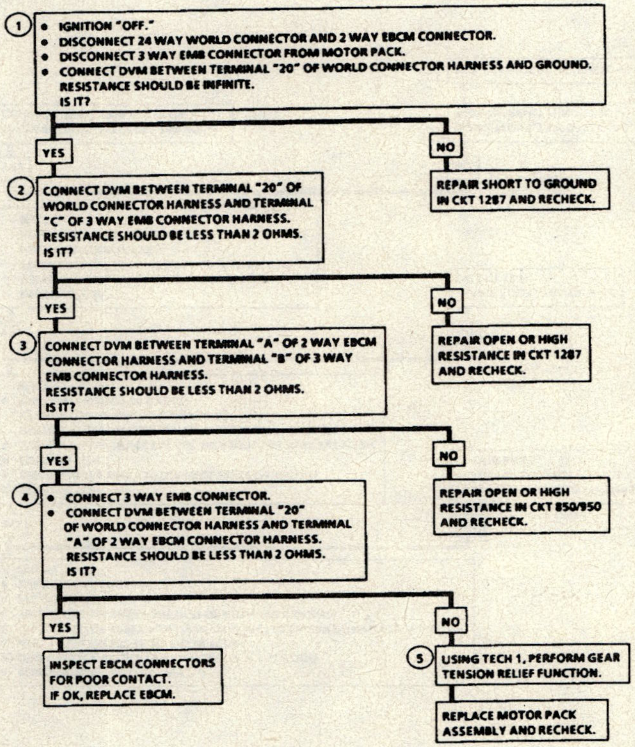

THIS CHART ASSUMES THAT A CURRENT CODE IS STORED INDICATING THAT THIS FAULT IS PRESENT.

DIAGNOSTIC CHART DTC A072 — 1993 BERETTA AND CORSICA

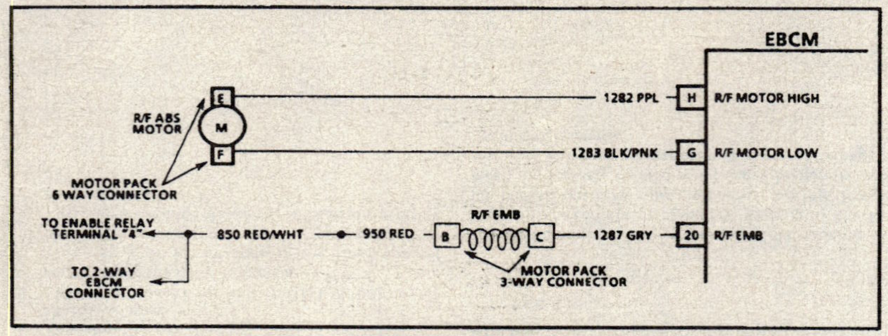

CODE A072

RIGHT FRONT EMB CIRCUIT SHORTED TO BATTERY OR EMB SHORTED

Circuit Description:

This code is designed to detect shorts to battery and shorts in the EMB release circuit. These faults will not allow the EMB to be energized when needed.

Test Description: Number(s) below refer to circled number(s) on the diagnostic chart.

1. This step checks to see if the fault is currently present.
2. This step identifies a short to voltage in the EMB control circuitry.
3. This step checks for an EMB that has lower than normal resistance.
4. This step is used to release the tension of the motor pack prior to removal.

Diagnostic Aids:

An "Intermittent" problem may be caused by a poor connection, rubbed through wire insulation, or a wire that is broken inside the insulation.

Any circuitry, that is suspected as causing the intermittent complaint, should be thoroughly checked for backed out terminals, improper mating, broken locks, improperly formed or damaged terminals, poor terminal to wiring connections or physical damage to the wiring harness.

DIAGNOSTIC CHART DTC A072 — 1993 BERETTA AND CORSICA

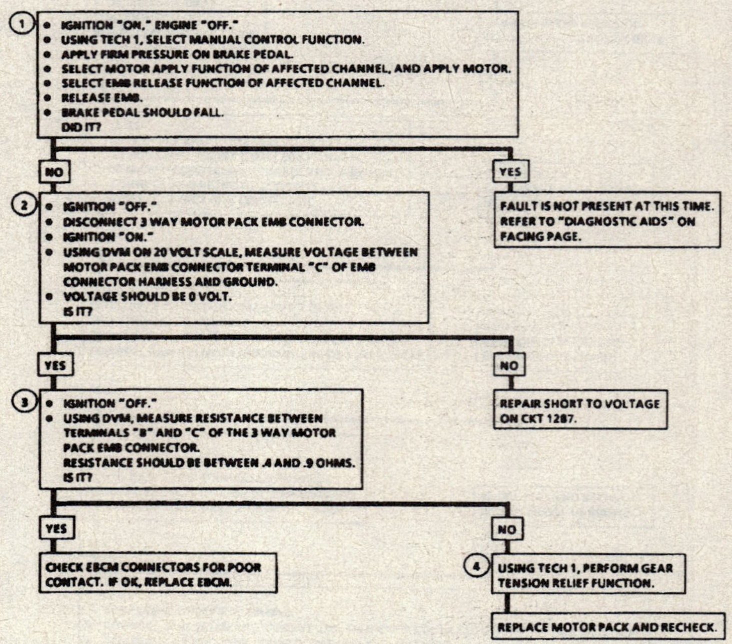

THIS CHART ASSUMES THAT A CURRENT CODE IS STORED INDICATING THAT THIS FAULT IS PRESENT.

DIAGNOSTIC CHART DTC A076 — 1993 BERETTA AND CORSICA

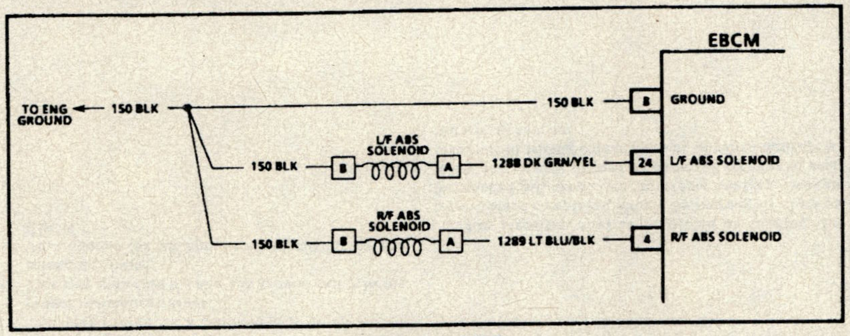

CODE A076

LEFT FRONT SOLENOID CIRCUIT SHORTED TO BATTERY OR OPEN

Circuit Description:

This code identifies a solenoid that cannot be energized due to an open in its circuitry, or a solenoid that is always energized due to a short to battery in its circuitry between the driver and the solenoid. An open will not allow proper ABS operation, but the short to battery simply turns "ON" the solenoid. A path for base brakes is still allowed once the motor re-homes and the check valve is lifted off its seat during key on initialization.

Test Description: Number(s) below refer to circled number(s) on the diagnostic chart.

1. This checks for a short to voltage in the control circuitry of the solenoid.
2. This checks for an open in the control circuit of the solenoid.
3. This step identifies an open in the solenoid or ground circuit.
4. This step checks for a possible intermittent problem in the solenoid circuitry due to poor terminal contact.

Diagnostic Aids:

An "Intermittent" problem may be caused by a poor connection, rubbed through wire insulation, or a wire that is broken inside the insulation.

Any circuitry, that is suspected as causing the intermittent complaint, should be thoroughly checked for backed out terminals, improper mating, broken locks, improperly formed or damaged terminals, poor terminal to wiring connections or physical damage to the wiring harness.

DIAGNOSTIC CHART DTC A076 — 1993 BERETTA AND CORSICA

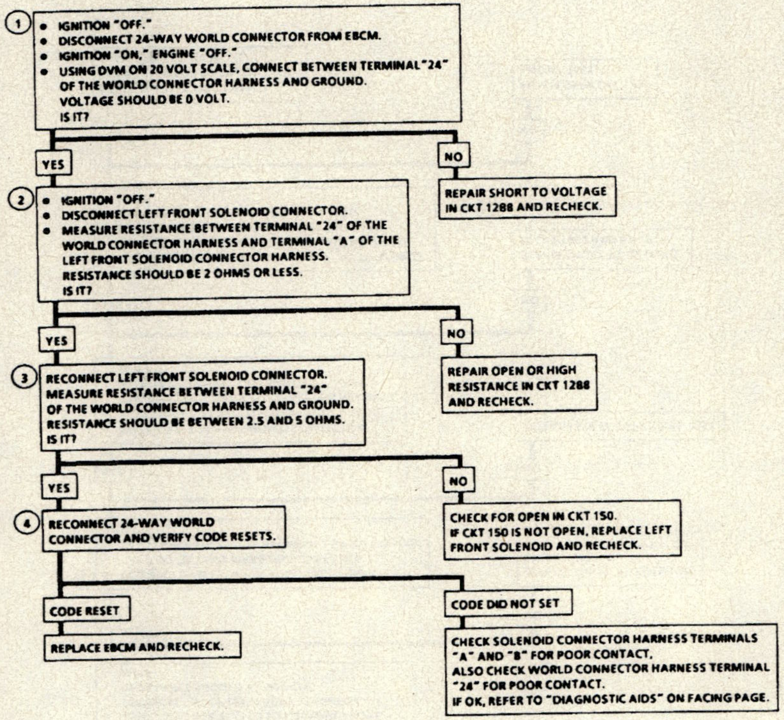

THIS CHART ASSUMES THAT A CURRENT CODE IS STORED INDICATING THAT THIS FAULT IS PRESENT.

DIAGNOSTIC CHART DTC A077 — 1993 BERETTA AND CORSICA

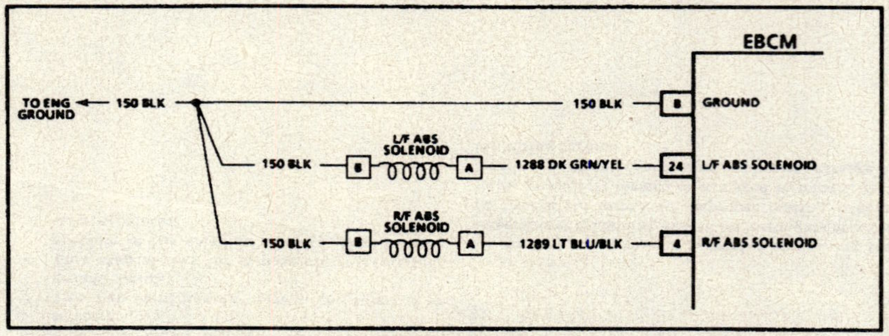

CODE A077

LEFT FRONT SOLENOID CIRCUIT SHORTED TO GROUND OR DRIVER OPEN

Circuit Description:

This code identifies a solenoid that cannot be energized due to an open in its driver circuitry, or a short to ground between the solenoid driver and the solenoid. These faults can affect ABS operation since the flow of brake fluid to the wheel cylinder cannot be stopped, making ABS operation for that channel impossible.

Test Description: Number(s) below refer to circled number(s) on the diagnostic chart.

1. This step checks for a solenoid that is internally shorted to ground.
2. This step checks for a solenoid that is not within proper resistance values.
3. This step indicates if a short to ground exists in the solenoid circuitry.
4. This checks for an open control circuit to the EBCM.

Diagnostic Aids:

An "Intermittent" problem may be caused by a poor connection, rubbed through wire insulation, or a wire that is broken inside the insulation.

Any circuitry, that is suspected as causing the intermittent complaint, should be thoroughly checked for backed out terminals, improper mating, broken locks, improperly formed or damaged terminals, poor terminal to wiring connections or physical damage to the wiring harness.

DIAGNOSTIC CHART DTC A077 — 1993 BERETTA AND CORSICA

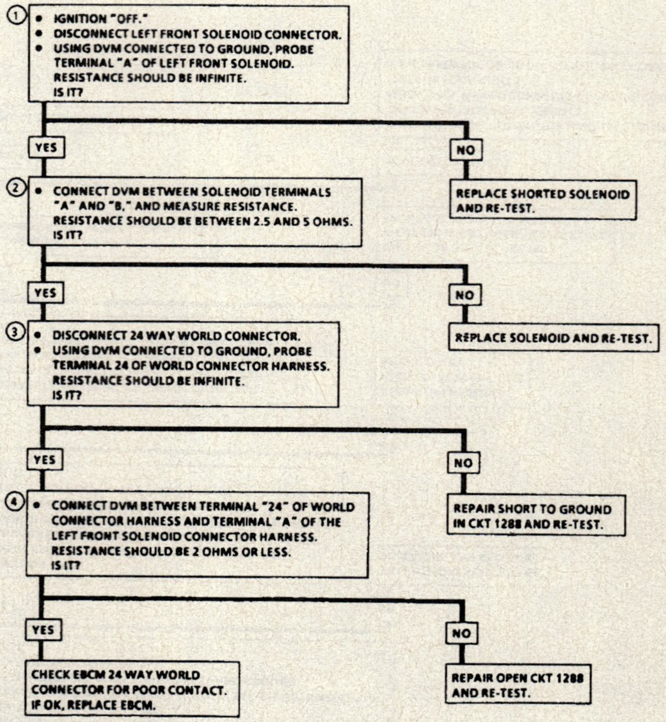

THIS CHART ASSUMES THAT A CURRENT CODE IS STORED INDICATING THAT THIS FAULT IS PRESENT.

DIAGNOSTIC CHART DTC A078 — 1993 BERETTA AND CORSICA

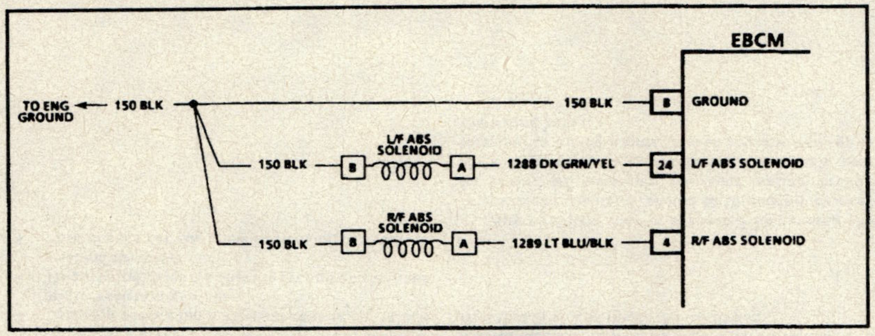

CODE A078

RIGHT FRONT SOLENOID CIRCUIT SHORTED TO BATTERY OR OPEN

Circuit Description:

This code identifies a solenoid that cannot be energized due to an open in its circuitry, or a solenoid that is always energized due to a short to battery in its circuitry between the driver and the solenoid. An open will not allow proper ABS operation, but the short to battery simply turns "ON" the solenoid. A path for base brakes is still allowed once the motor re-homes and the check valve is lifted off its seat during key on initialization.

Test Description: Number(s) below refer to circled number(s) on the diagnostic chart.

1. This checks for a short to voltage in the control circuitry of the solenoid.
2. This checks for an open in the control circuit of the solenoid.
3. This step identifies an open in the solenoid or ground circuit.
4. This step checks for a possible intermittent problem in the solenoid circuitry due to poor terminal contact.

Diagnostic Aids:

An "Intermittent" problem may be caused by a poor connection, rubbed through wire insulation, or a wire that is broken inside the insulation.

Any circuitry, that is suspected as causing the intermittent complaint, should be thoroughly checked for backed out terminals, improper mating, broken locks, improperly formed or damaged terminals, poor terminal to wiring connections or physical damage to the wiring harness.

DIAGNOSTIC CHART DTC A078 — 1993 BERETTA AND CORSICA

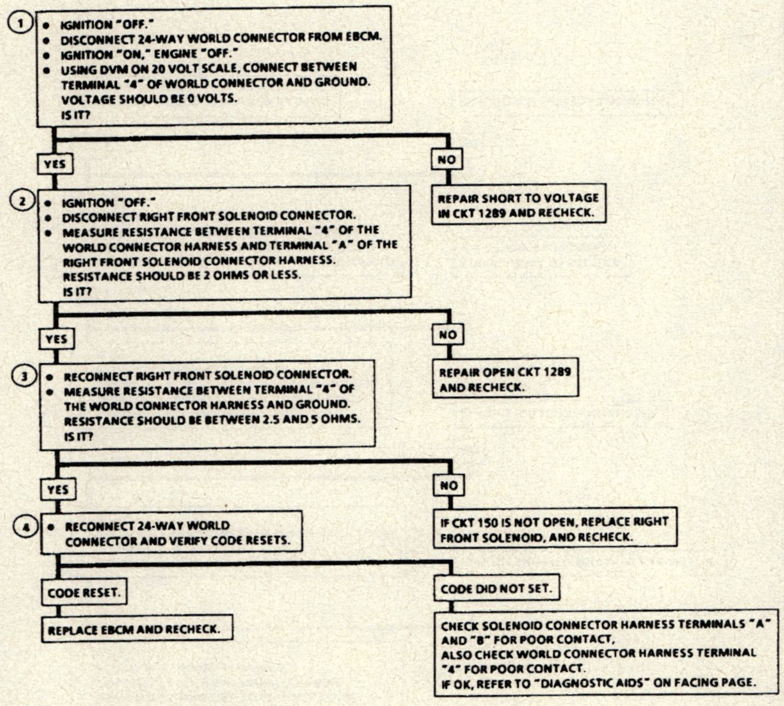

THIS CHART ASSUMES THAT A CURRENT CODE IS STORED INDICATING THAT THIS FAULT IS PRESENT.

DIAGNOSTIC CHART DTC A081 — 1993 BERETTA AND CORSICA

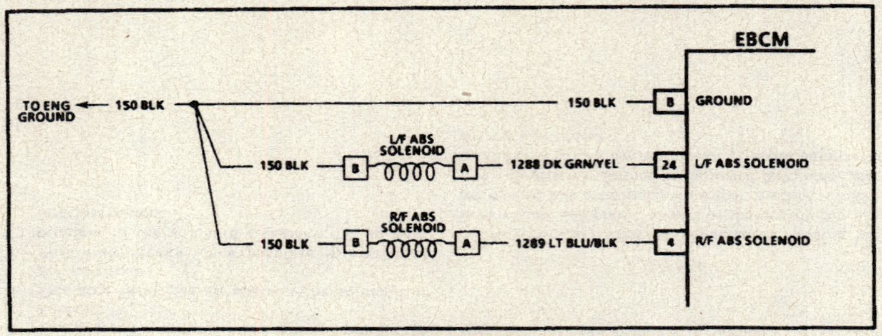

CODE A081

RIGHT FRONT SOLENOID CIRCUIT SHORTED TO GROUND OR DRIVER OPEN

Circuit Description:

This code identifies a solenoid that cannot be energized due to an open in its driver circuitry, or a short to ground between the solenoid driver and the solenoid. These faults can affect ABS operation since the flow of brake fluid to the wheel cylinder cannot be stopped, making ABS operation for that channel impossible.

Test Description: Number(s) below refer to circled number(s) on the diagnostic chart.

1. This step checks for a solenoid that is internally shorted to ground.
2. This step checks for a solenoid that is not within proper resistance values.
3. This step indicates if a short to ground exists in the solenoid circuitry.
4. This checks for an open control circuit to the EBCM.

Diagnostic Aids:

An "Intermittent" problem may be caused by a poor connection, rubbed through wire insulation, or a wire that is broken inside the insulation.

Any circuitry, that is suspected as causing the intermittent complaint, should be thoroughly checked for backed out terminals, improper mating, broken locks, improperly formed or damaged terminals, poor terminal to wiring connections or physical damage to the wiring harness.

DIAGNOSTIC CHART DTC A081 — 1993 BERETTA AND CORSICA

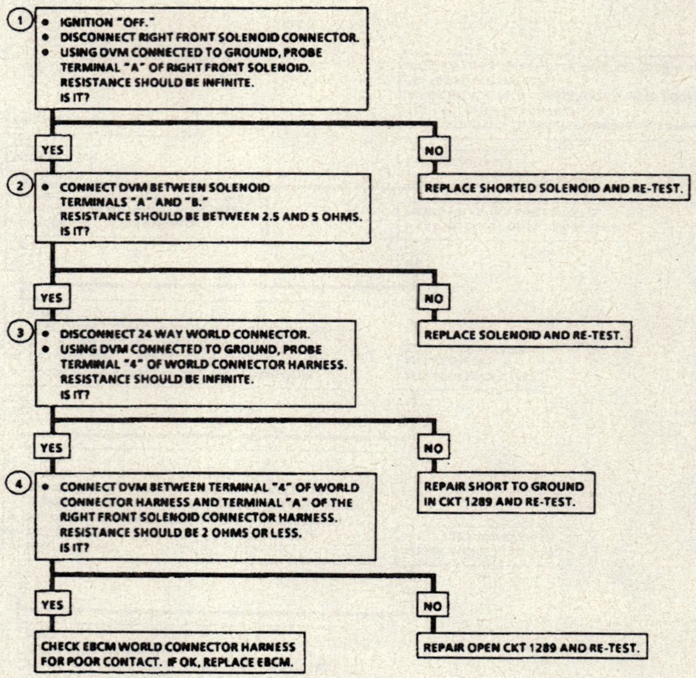

THIS CHART ASSUMES THAT A CURRENT CODE IS STORED INDICATING THAT THIS FAULT IS PRESENT.

DIAGNOSTIC CHART DTC A082 — 1993 BERETTA AND CORSICA

D.M. A.B.S. VI
ELECTRONIC CONTROL UNIT
(TYPICAL)

CODE A082

CALIBRATION MEMORY FAILURE

Circuit Description:

This code allows the EBCM to check for a calibration failure by comparing the calibration value to a known value stored in the EEPROM.

This test is also used as a security measure to prevent improper use of calibrations or changes to these calibrations that may alter the designed function of ABS.

Test Description: Number(s) below refer to circled number(s) on the diagnostic chart.

1. This step checks to see if the fault is present during diagnosis. If present, the EBCM is not functioning correctly and must be replaced.

IMPORTANT: At the time of printing, 1992 vehicles with ABS-VI were not being field reprogrammed to correct this failure. A service bulletin will be issued when EEPROM reprogramming is authorized.

Diagnostic Aids:

An intermittent Code A082 may be caused by a bad cell in the EEPROM that is sensitive to temperature changes. If Code A082 failed more than once, but is intermittent, replace EBCM.

DIAGNOSTIC CHART DTC A082 — 1993 BERETTA AND CORSICA

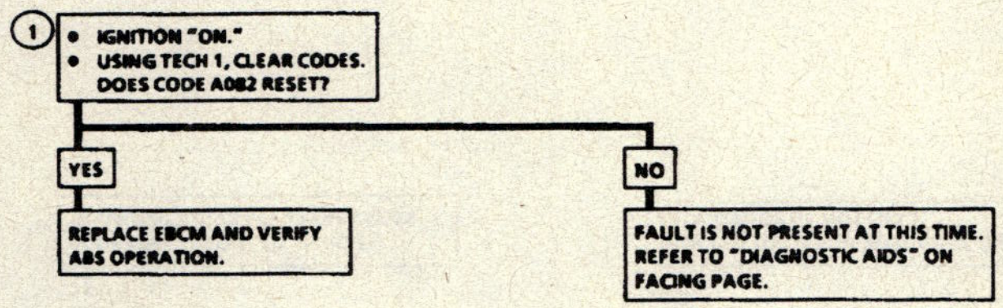

THIS CHART ASSUMES THAT A CURRENT CODE IS STORED INDICATING THAT THIS FAULT IS PRESENT.

DIAGNOSTIC CHART DTC A086 — 1993 BERETTA AND CORSICA

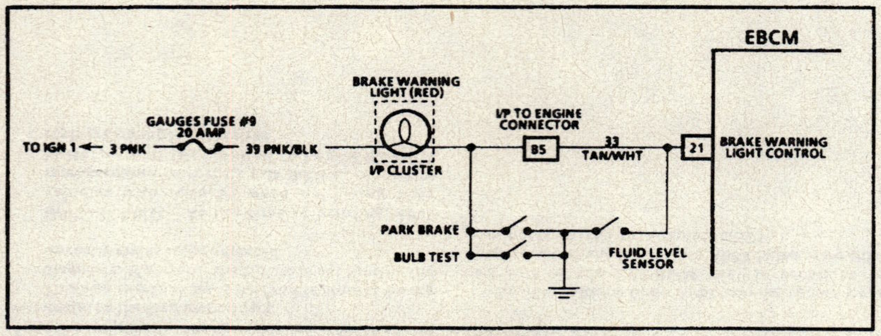

CODE A086

RED BRAKE WARNING LIGHT ACTIVATED BY ABS

Circuit Description:

This code is provided as an information only test, and reflects the status of the command issued by the ABS EBCM to illuminate the red brake warning light. If another diagnostic code issues a command to illuminate the red brake warning light, Code A086 will be stored in EEPROM as a history code at the conclusion of the ignition cycle.

Test Description: Number(s) below refer to circled number(s) on the diagnostic chart.

1. This step serves to identify if a code other than Code A086 commanded the red brake warning light "ON."

Diagnostic Aids:

Any ABS mechanical code that issues a command to illuminate the red brake warning light will also result in Code A086 being stored in EEPROM during shut down. These codes are: A038, A041, A042, A045, A046, A047, A048 and A051. If the motors are not in their home position, certain electrical codes will also command the red brake warning light "ON." These codes are: A014, A018, A055, A056, A057, A058, A061, A062, A063, A064, A065, A066, A067, A068, A071 and A072.

If any of these codes are indicated along with Code A086, they must be corrected prior to addressing a Code A086 fault.

DIAGNOSTIC CHART DTC A086 — 1993 BERETTA AND CORSICA

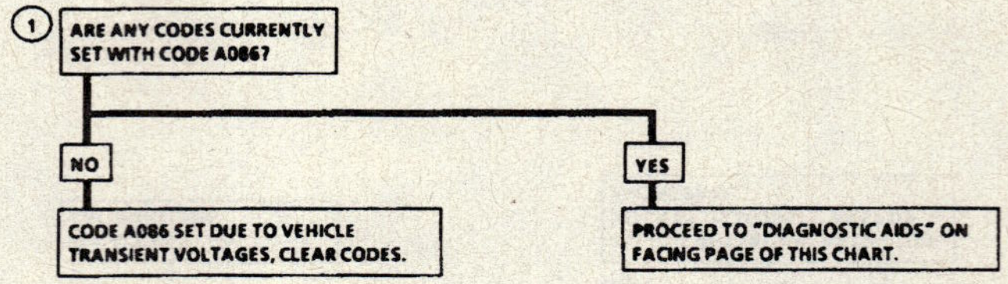

DIAGNOSTIC CHART DTC A087 — 1993 BERETTA AND CORSICA

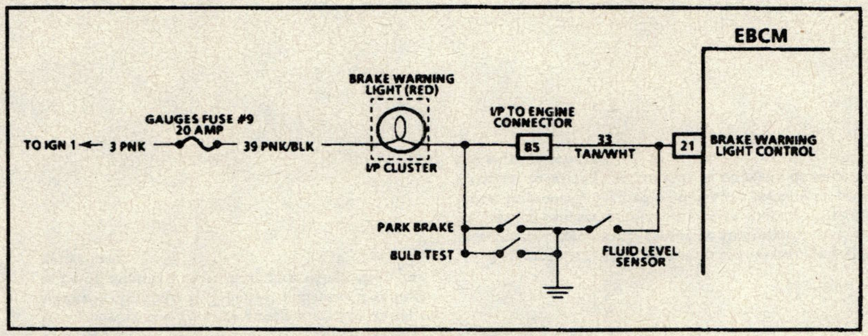

CODE A087

RED BRAKE WARNING LIGHT CIRCUIT OPEN

Circuit Description:

This code is used to verify the EBCM has continuity to the brake warning light in case the EBCM must turn it "ON." This will only occur if an ABS fault is detected that may degrade base brake operation.

Because ABS is not the only device controlling the brake warning light (parking brake and low fluid sensor may also turn "ON" the light) a short to ground in this circuit cannot be detected.

Test Description: Number(s) below refer to circled number(s) on the diagnostic chart.

1. This step identifies if the fault is currently present.
2. This step indicates if the EBCM and circuitry has the ability to complete the ground to the brake warning light and illuminate it.
3. Because the brake warning light has current supplied to it through the 20 amp gauge fuse, this step would indicate if the ignition circuit is complete to the I/P cluster.
4. This step isolates if the open circuit is due to a EBCM failure or open circuitry to the brake warning light.

Diagnostic Aids:

An "Intermittent" problem may be caused by a poor connection, rubbed through wire insulation, or a wire that is broken inside the insulation.

The frequency of the problem can be checked by using the enhanced diagnostic function of the Tech 1 "Scan" tool.

If a Tech 1 or T-100 are not available the parking brake handle may be lifted to verify proper light operation and continuity of CKTs 3 and 39.

Any circuitry, that is suspected as causing the intermittent complaint, should be thoroughly checked for backed out terminals, improper mating, broken locks, improperly formed or damaged terminals, poor terminal to wiring connections or physical damage to the wiring harness.

DIAGNOSTIC CHART DTC A087 — 1993 BERETTA AND CORSICA

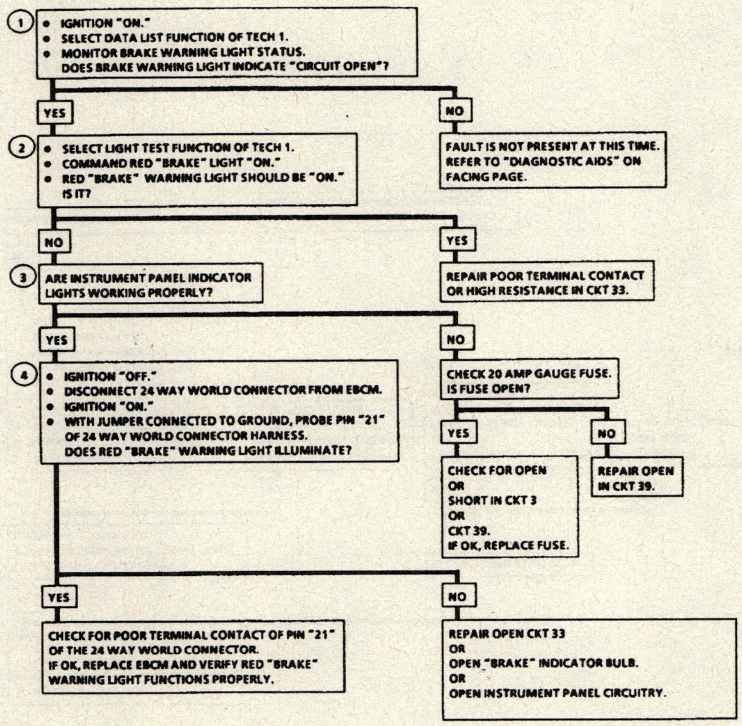

THIS CHART ASSUMES THAT A CURRENT CODE IS STORED INDICATING THAT THIS FAULT IS PRESENT.

DIAGNOSTIC CHART DTC A088 — 1993 BERETTA AND CORSICA

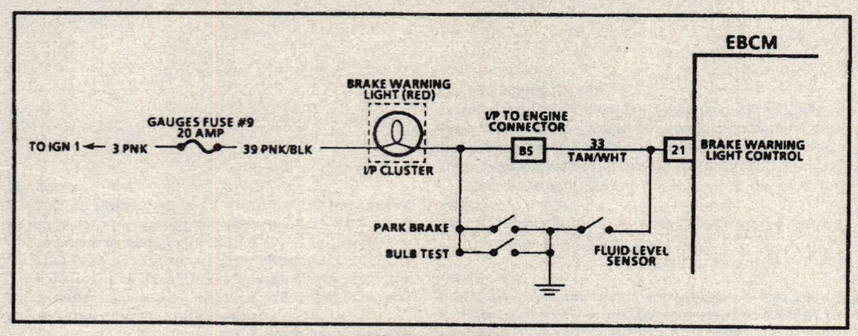

CODE A088

RED BRAKE WARNING LIGHT CIRCUIT SHORTED TO BATTERY

Circuit Description:

This code identifies a short to battery between the EBCM and the RED brake warning light, or an open driver that does not allow the RED brake warning light to be illuminated by the EBCM. This will only occur if an ABS fault is detected that may degrade base brake operation.

Test Description: Number(s) below refer to circled number(s) on the diagnostic chart.

1. This step identifies if the ground circuit to the Red brake warning light is being completed by a source other than the EBCM.
2. This step identifies if the fault is currently present.
3. By removing the 20 amp gauge fuse, the voltage source is eliminated. This test indicates if voltage is being supplied from a source other than the gauge fuse.

Diagnostic Aids:

An "Intermittent" problem may be caused by a poor connection, rubbed through wire insulation, or a wire that is broken inside the insulation.

Any circuitry, that is suspected as causing the intermittent complaint, should be thoroughly checked for backed out terminals, improper mating, broken locks, improperly formed or damaged terminals, poor terminal to wiring connections or physical damage to the wiring harness.

DIAGNOSTIC CHART DTC A088 — 1993 BERETTA AND CORSICA

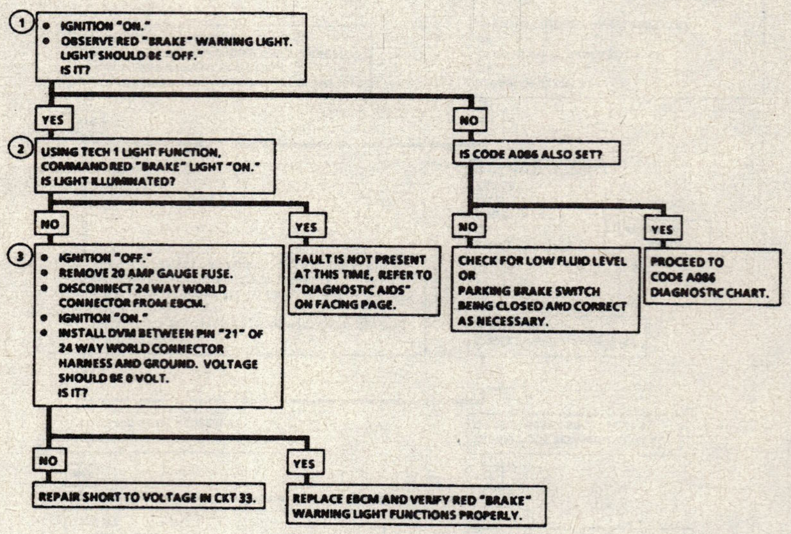

THIS CHART ASSUMES THAT A CURRENT CODE IS STORED INDICATING THAT THIS FAULT IS PRESENT.

DIAGNOSTIC CHART DTC A091 — 1993 BERETTA AND CORSICA

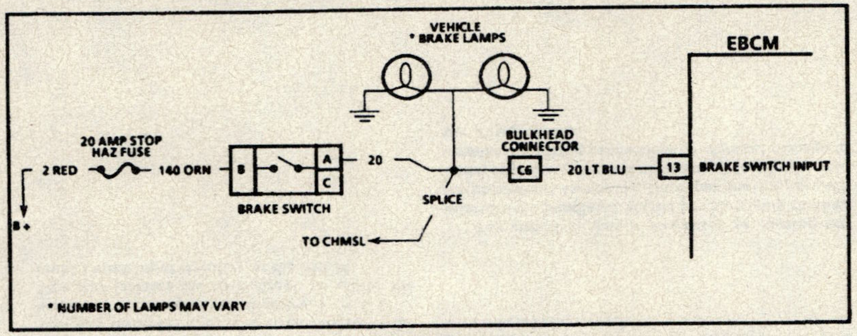

CODE A091

OPEN BRAKE SWITCH CONTACTS DURING DECELERATION

Circuit Description:

This code is used to detect an open brake switch in the non-ABS mode. The EBCM looks for decel rates that would indicate braking action and verifies this assumption by requiring several repeats of this detection method. In each case, ABS will not be available since no brake switch is seen.

Test Description: Number(s) below refer to circled number(s) on the diagnostic chart.

1. This step indicates if the brake switch signal is being received by the EBCM.
2. This step indicates if an open circuit exists in the brake switch or brake lamp circuitry.
3. This step isolates the open circuit to either the brake switch input circuitry, or the EBCM.

Diagnostic Aids:

An "Intermittent" problem may be caused by a poor connection, rubbed through wire insulation, or a wire that is broken inside the insulation.

Any circuitry, that is suspected as causing the intermittent complaint, should be thoroughly checked for backed out terminals, improper mating, broken locks, improperly formed or damaged terminals, poor terminal to wiring connections or physical damage to the wiring harness.

DIAGNOSTIC CHART DTC A091 — 1993 BERETTA AND CORSICA

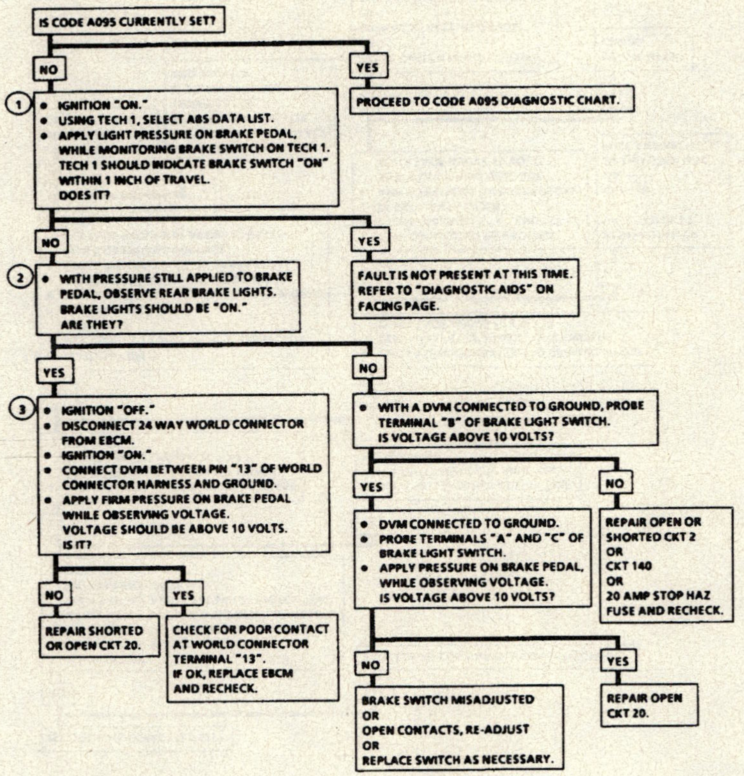

DIAGNOSTIC CHART DTC A092 — 1993 BERETTA AND CORSICA

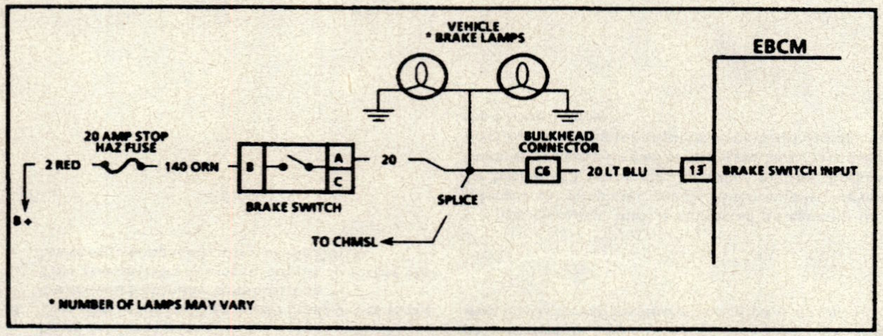

CODE A092

OPEN BRAKE SWITCH CONTACTS WHEN ABS WAS REQUIRED

Circuit Description:

This code is run to determine the proper operation of the brake switch. This is important because ABS is activated when the brake switch is "ON." If the brake switch is "OFF," ABS will never be activated. Since this fault is difficult to detect under normal braking conditions, this fault is only detected when ABS is required.

Test Description: Number(s) below refer to circled number(s) on the diagnostic chart.

1. This step indicates if the brake switch signal is being received by the EBCM.
2. This step indicates if an open circuit exists in the brake switch or brake lamp circuitry.
3. This step isolates the open circuit to either the brake switch input circuitry, or the EBCM.

Diagnostic Aids:

An "Intermittent" problem may be caused by a poor connection, rubbed through wire insulation, or a wire that is broken inside the insulation.

Any circuitry, that is suspected as causing the intermittent complaint, should be thoroughly checked for backed out terminals, improper mating, broken locks, improperly formed or damaged terminals, poor terminal to wiring connections or physical damage to the wiring harness.

DIAGNOSTIC CHART DTC A092 — 1993 BERETTA AND CORSICA

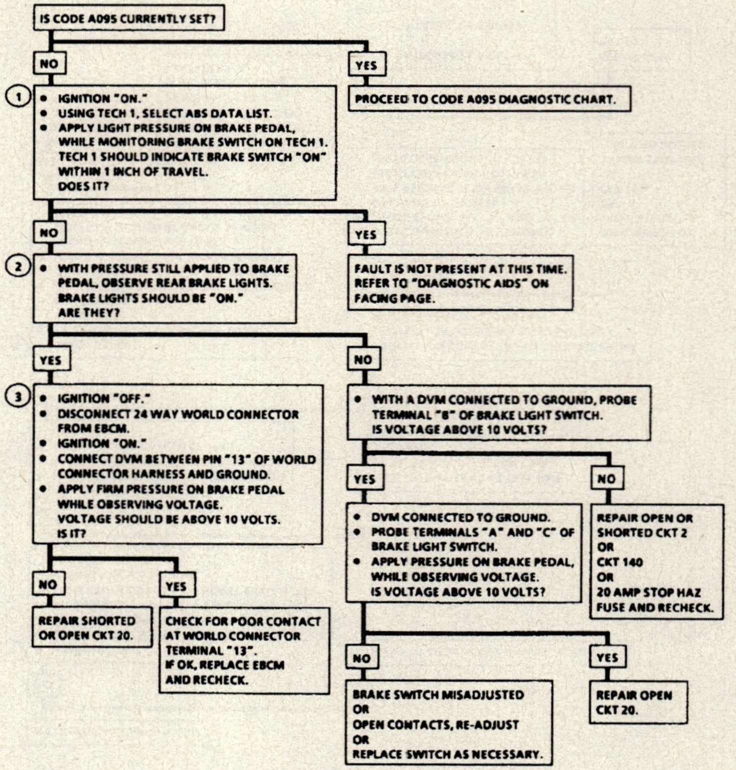

DIAGNOSTIC CHART DTC A093 — 1993 BERETTA AND CORSICA

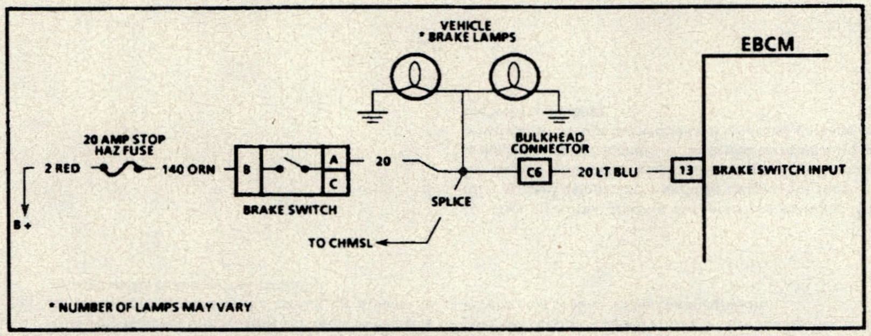

CODE A093

CODES A091 OR A092 SET IN CURRENT OR PREVIOUS IGNITION CYCLE

Circuit Description:

This code is the second portion of Codes A091 and A092. If Codes A091 or A092 failed during the last ignition cycle, Code A093 becomes a current failure during the next ignition cycle, keeping ABS disabled until a brake switch "ON" state is seen. When a change is seen during an ignition cycle in which Code A093 is a current failure, Code A091 or A092 will clear itself at the end of the current ignition cycle, and ABS will enable itself at the start of the next ignition cycle. Code A093 alone indicates Codes A091 or A092 failed previously, but is intermittent, or has been corrected.

Test Description: Number(s) below refer to circled number(s) on the diagnostic chart.

1. This step indicates which code (either A091 or A092) caused Code A093 to set.
2. This insures that the code that set is repaired so that Code A093 can be cleared.

Diagnostic Aids:

An "Intermittent" problem may be caused by a poor connection, rubbed through wire insulation, or a wire that is broken inside the insulation.

Any circuitry, that is suspected as causing the intermittent complaint, should be thoroughly checked for backed out terminals, improper mating, broken locks, improperly formed or damaged terminals, poor terminal to wiring connections or physical damage to the wiring harness.

Also, verify proper brake switch operation using the data list of the "Scan" tool. As the brake is applied, the data list should display the brake switch "ON" within 1 inch of travel.

DIAGNOSTIC CHART DTC A093 — 1993 BERETTA AND CORSICA

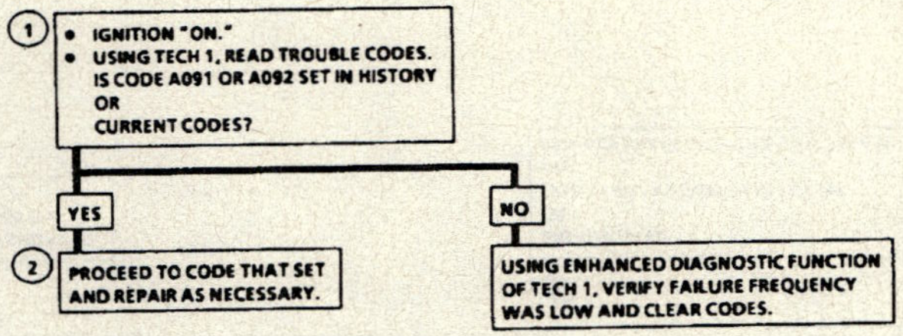

THIS CHART ASSUMES THAT A CURRENT CODE IS STORED INDICATING THAT THIS FAULT IS PRESENT.

DIAGNOSTIC CHART DTC A094 — 1993 BERETTA AND CORSICA

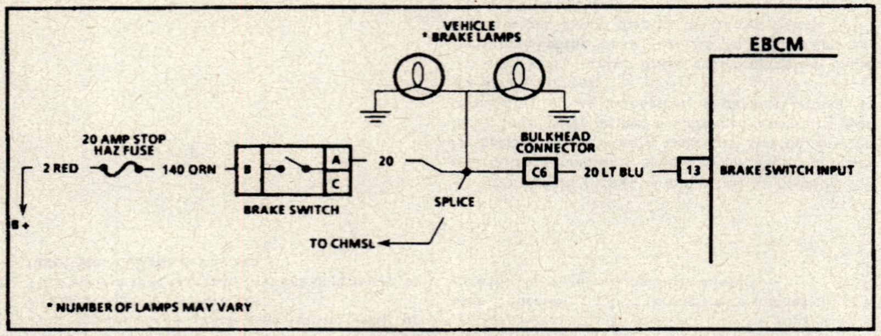

CODE A094

BRAKE SWITCH CONTACTS ALWAYS CLOSED

Circuit Description:

This code is run to determine the proper operation of the brake switch. This is important because ABS is activated when the brake switch is "ON" and turned "OFF" when the brake is "OFF." If the brake switch is always "ON," ABS operation will always be requested resulting in potential modulator cycling on rough roads. Additionally, this failure will most likely result in a dead battery (due to the brake lights staying "ON") if the driver is not informed of this fault.

Test Description: Number(s) below refer to circled number(s) on the diagnostic chart.

1. This checks if the fault is currently present.
2. This step isolates the cause of the fault to either a faulty or misadjusted brake switch, or a short to voltage in the brake switch circuitry.

Diagnostic Aids:

An "Intermittent" problem may be caused by a poor connection, rubbed through wire insulation, or a wire that is broken inside the insulation.

Any circuitry, that is suspected as causing the intermittent complaint, should be thoroughly checked for backed out terminals, improper mating, broken locks, improperly formed or damaged terminals, poor terminal to wiring connections or physical damage to the wiring harness.

DIAGNOSTIC CHART DTC A094 — 1993 BERETTA AND CORSICA

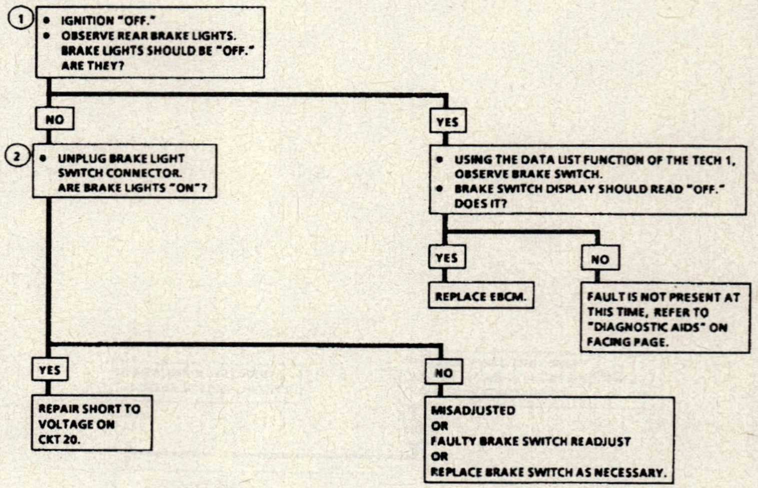

THIS CHART ASSUMES THAT A CURRENT CODE IS STORED INDICATING THAT THIS FAULT IS PRESENT.

DIAGNOSTIC CHART DTC A095 — 1993 BERETTA AND CORSICA

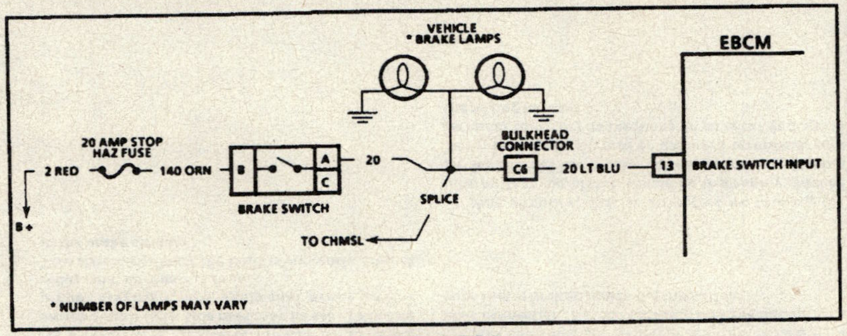

CODE A095

BRAKE SWITCH CIRCUIT OPEN

Circuit Description:

This code is used to identify open brake switch circuitry that prevents the brake switch input to the EBCM from changing states when the brake is applied. This code is used in conjunction with Codes A091 and A092 to determine the cause of an open brake switch fault.

Test Description: Number(s) below refer to circled number(s) on the diagnostic chart.

1. This step is used to confirm that an open in the brake switch circuitry currently exists.
2. This step indicates if the brake switch signal is being received by the EBCM.
3. This step indicates if an open circuit exists in the brake switch or brake light circuitry.
4. This step isolates the open circuit to either the brake switch input circuitry,or the EBCM.

Diagnostic Aids:

An "Intermittent" problem may be caused by a poor connection, rubbed through wire insulation, or a wire that is broken inside the insulation.

Any circuitry, that is suspected as causing the intermittent complaint, should be thoroughly checked for backed out terminals, improper mating, broken locks, improperly formed or damaged terminals, poor terminal to wiring connections or physical damage to the wiring harness.

DIAGNOSTIC CHART DTC A095 — 1993 BERETTA AND CORSICA

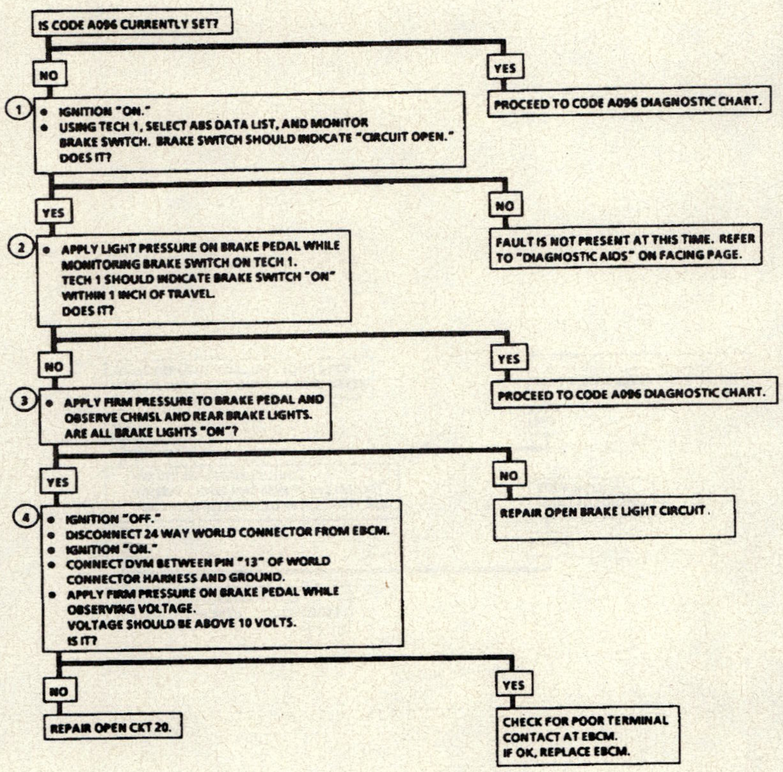

THIS CHART ASSUMES THAT A CURRENT CODE IS STORED INDICATING THAT THIS FAULT IS PRESENT.

DIAGNOSTIC CHART DTC A096 — 1993 BERETTA AND CORSICA

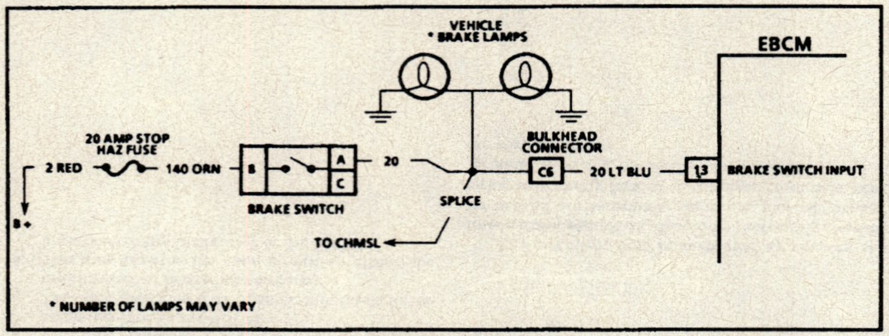

CODE A096

BRAKE LIGHTS CIRCUIT OPEN

Circuit Description:

This code is designed to isolate the cause of a Code A095 failure and indicate to the driver ABS is still available. If Code A095 fails with Code A096, the brake lights are open, the brake light grounds are open, or the Center High Mounted Stop Light (CHMSL) circuit is open during 4 way flasher use. The presence of battery voltage at the brake switch input indicates a valid brake switch input is still available.

Test Description: Number(s) below refer to circled number(s) on the diagnostic chart.

1. As a result of a failure of an additional brake switch circuit code, this code may be set. To insure proper diagnosis, any additional brake switch codes must be repaired first.
2. This step identifies if the fault is currently present in the brake circuit.

Diagnostic Aids:

An "Intermittent" problem may be caused by a poor connection, rubbed through wire insulation, or a wire that is broken inside the insulation.

Any circuitry, that is suspected as causing the intermittent complaint, should be thoroughly checked for backed out terminals, improper mating, broken locks, improperly formed or damaged terminals, poor terminal to wiring connections or physical damage to the wiring harness.

DIAGNOSTIC CHART DTC A096 — 1993 BERETTA AND CORSICA

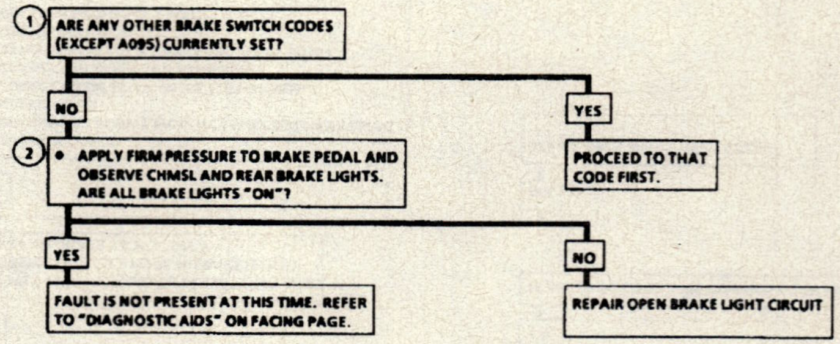

THIS CHART ASSUMES THAT A CURRENT CODE IS STORED INDICATING THAT THIS FAULT IS PRESENT.

DIAGNOSTIC PROCESS CHART — 1994–95 BERETTA AND CORSICA

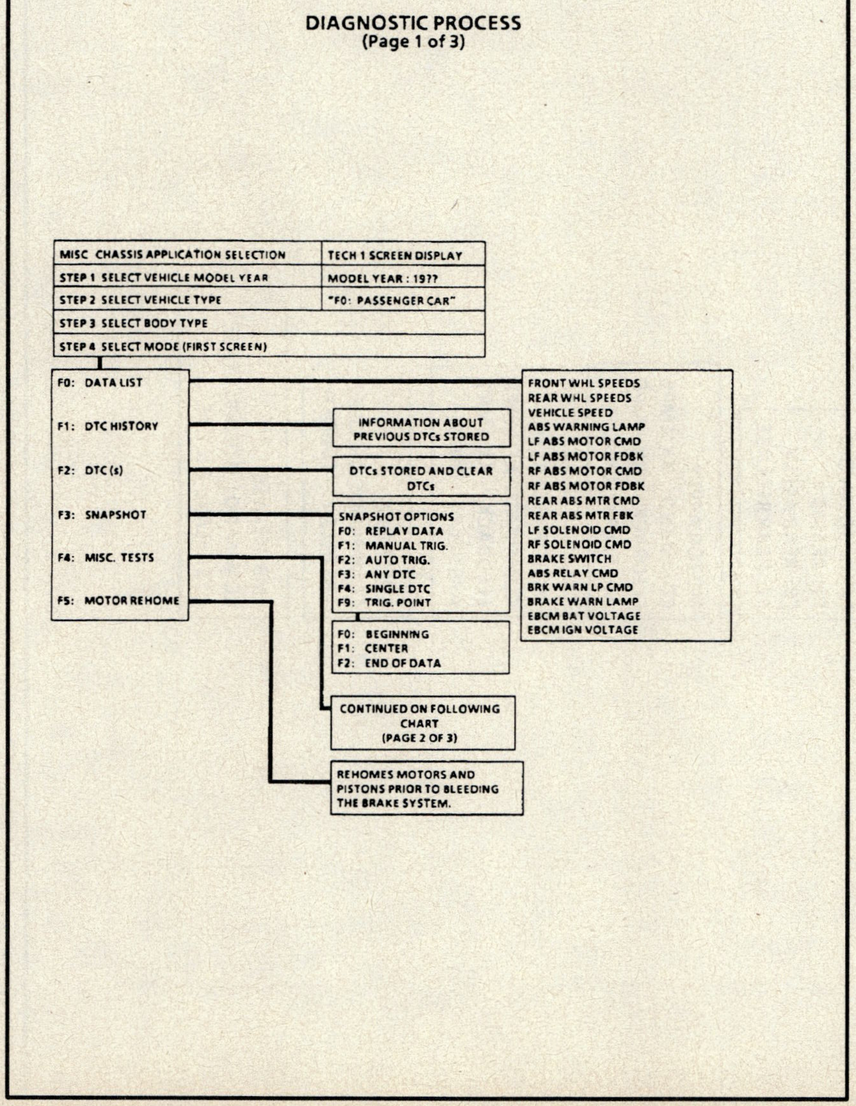

DIAGNOSTIC PROCESS CHART — 1994–95 BERETTA AND CORSICA

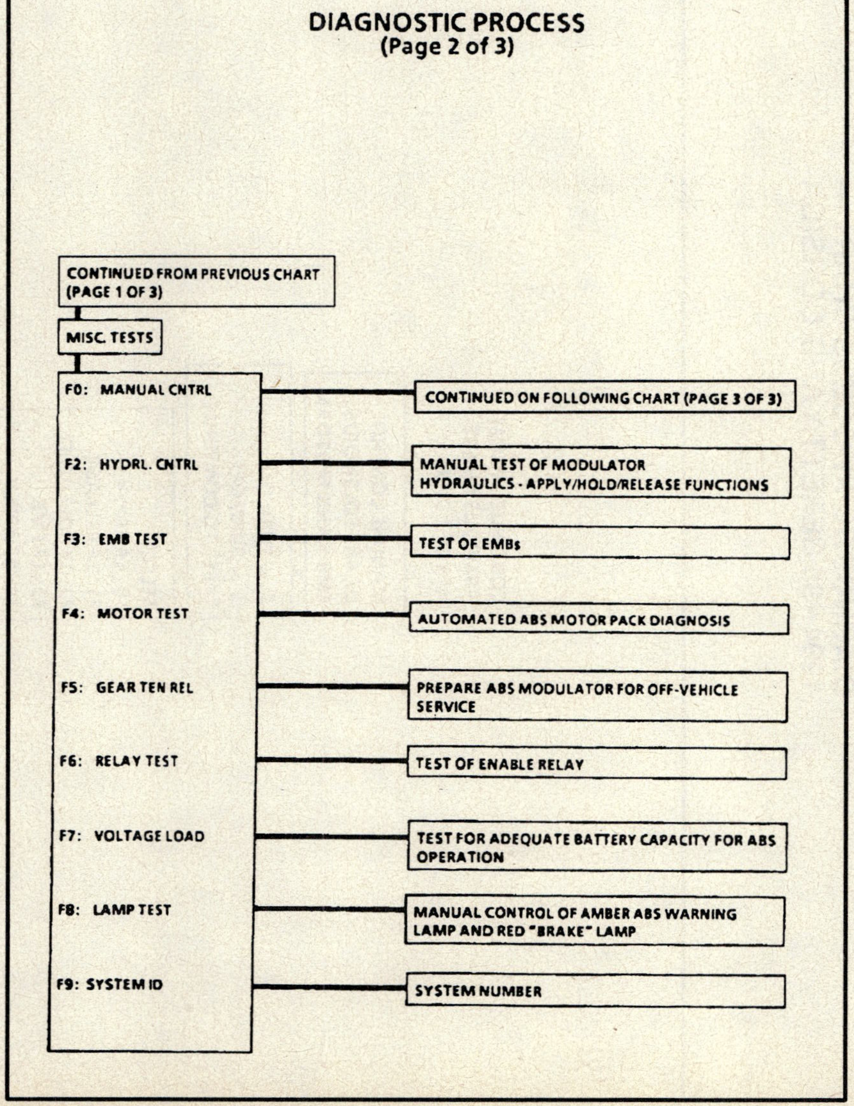

DIAGNOSTIC PROCESS CHART — 1994–95 BERETTA AND CORSICA

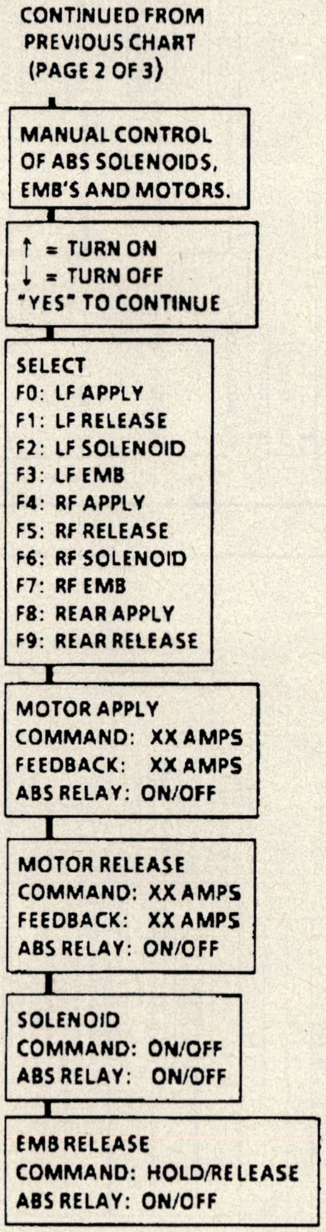

DIAGNOSTIC CHART SYSTEM CHECK — 1994 BERETTA AND CORSICA

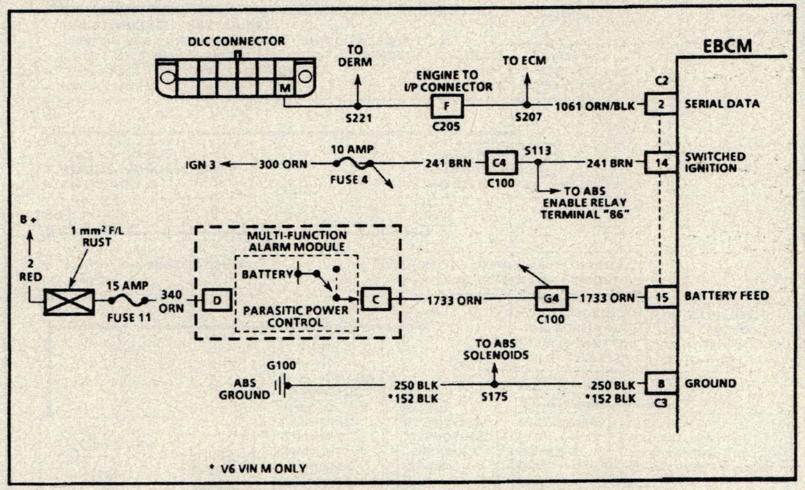

DIAGNOSTIC SYSTEM CHECK

System Description:

The diagnostic system check is an organized approach to identifying a malfunction created by an Antilock Brake System (ABS). It must be the starting point for any ABS complaint diagnosis because it directs the service technician to the next logical step in diagnosing the complaint.

Serial data is transmitted by the EBCM through terminal "2". The EBCM is supplied switched ignition voltage through terminal "14", battery voltage through terminal "15" and switched battery voltage through terminal "A". The EBCM ground is provided through terminal "B".

Chart Test Description:

Number(s) below refer to circled number(s) on the diagnostic chart.

1. Indicates if EBCM is transmitting data.
2. Checks for a serial data link malfunction.
3. Checks for proper Tech 1 connection to DLC.
4. Checks for open malfunction alarm module fuse.
5. Checks for open Fuse 4.
6. Checks for high resistance in the EBCM ground circuit.
7. Checks for current DTCs.
8. Checks for proper ABS warning lamp operation.
9. Checks for history DTCs.
10. Checks for possible short to ground in the B+ feed circuitry.
11. Checks for possible short to ground in the EBCM.
12. Checks for shor to ground in the B+ feed circuitry between the malfunction alarm module and the EBCM.
13. Ensures short to ground is not due to physical damage of the circuitry.
14. Checks for possible short to ground in the switched ignition circuitry.
15. Isolates the short to ground to either the circuitry or the EBCM.
16. Ensures short to ground is not due to physical damage of the circuitry.
17. Checks for proper battery voltage at the EBCM.
18. Checks for high resistance in the B+ feed circuitry.
19. Checks for high resistance in the switched ignition circuitry.
20. Ensures malfunction is not due to poor terminal contact.

DIAGNOSTIC CHART SYSTEM CHECK — 1994 BERETTA AND CORSICA

Chart Test Description: (Continued)

Number(s) below refer to circled number(s) on the diagnostic chart.

21. Checks for an open in the B+ feed circuit between the multifunction alarm module and the EBCM.
22. Checks for high resistance in the B+ feed circuit between the multifunction alarm module and the EBCM.
23. Checks for high resistance in the switched ignition circuit between the fuse block and the EBCM.

Diagnostic Aids:

Excessive resistance in the ground or power supply circuits will not allow communication with the EBCM. If communication with the EBCM is not possible, ensure the ABS ground connection is good and that there is no excessive resistance in any of the power supply circuits.

DIAGNOSTIC CHART SYSTEM CHECK — 1994 BERETTA AND CORSICA

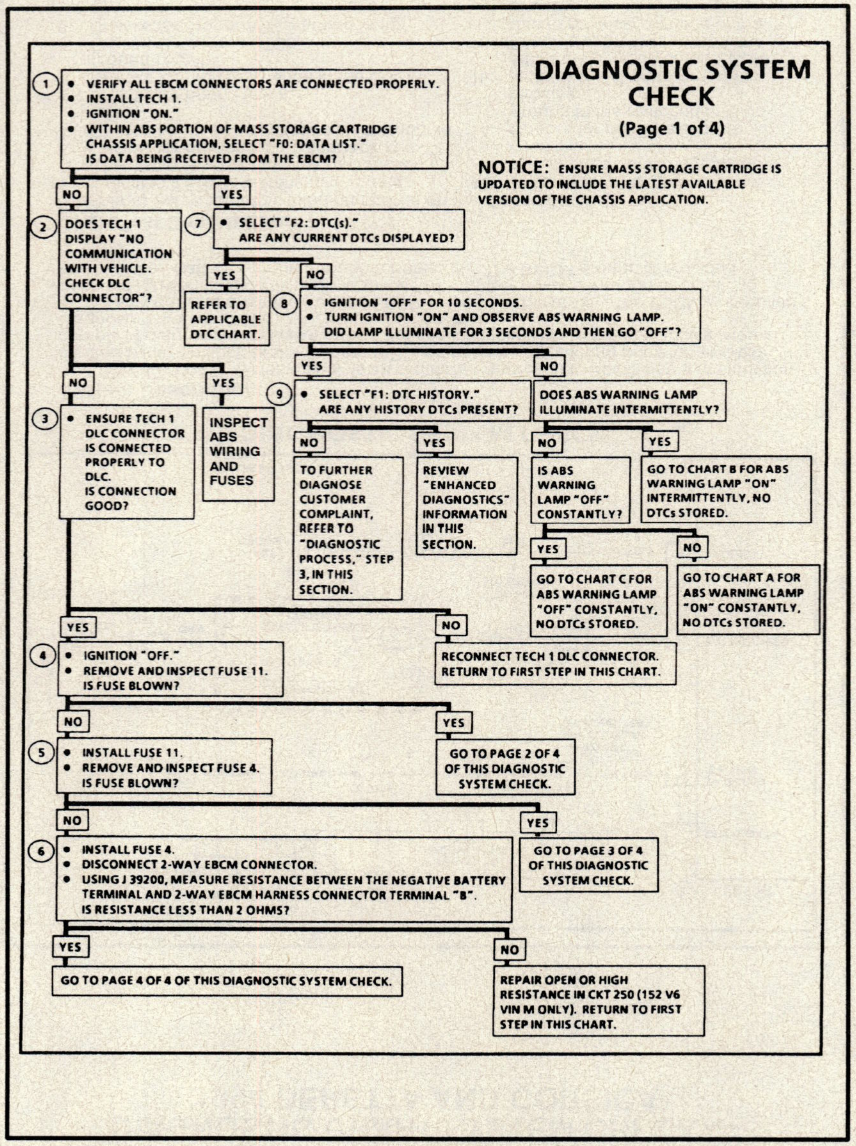

DIAGNOSTIC CHART SYSTEM CHECK — 1994 BERETTA AND CORSICA

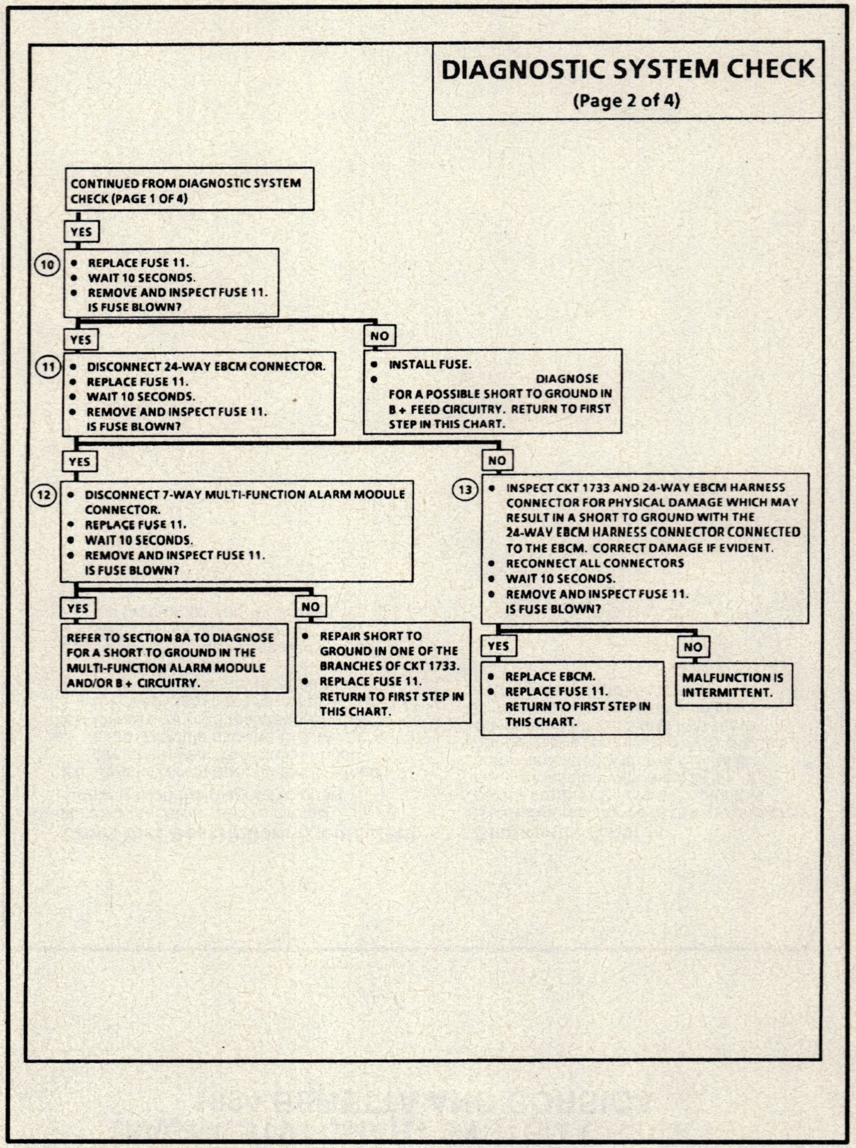

DIAGNOSTIC CHART SYSTEM CHECK — 1994 BERETTA AND CORSICA

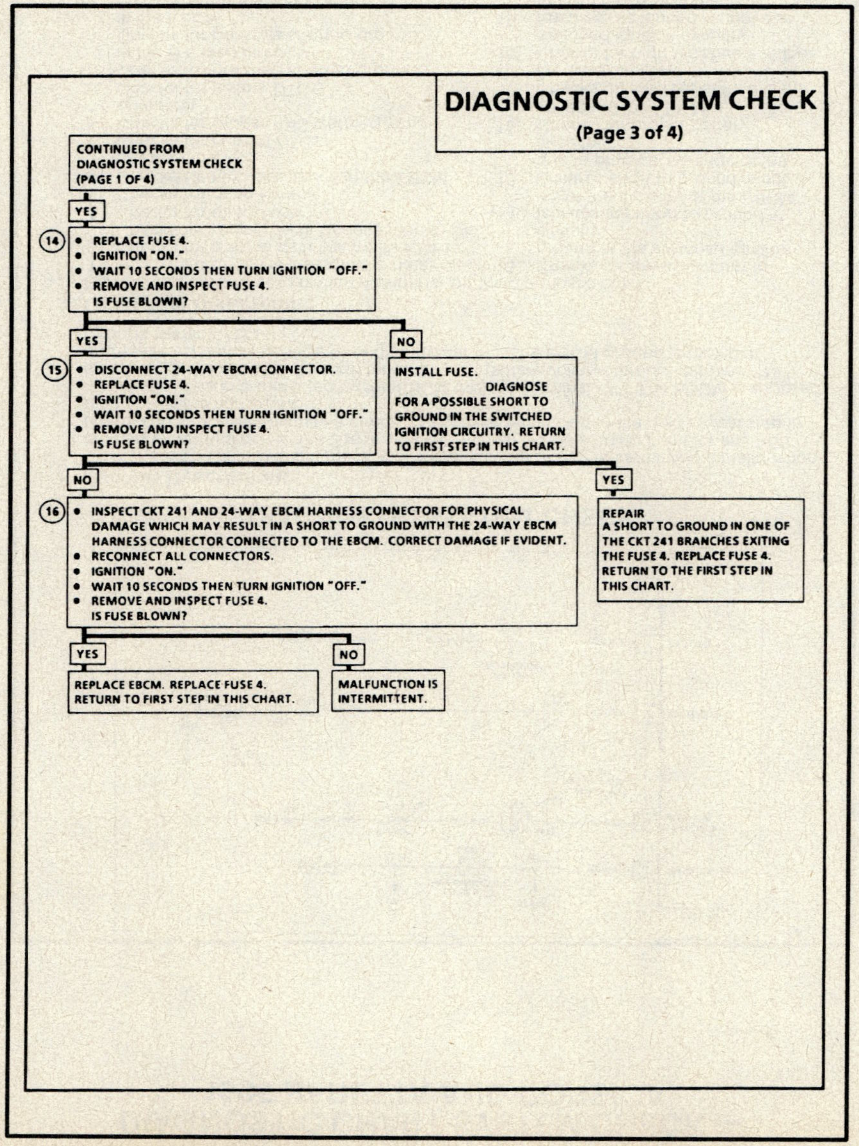

DIAGNOSTIC CHART SYSTEM CHECK — 1994 BERETTA AND CORSICA

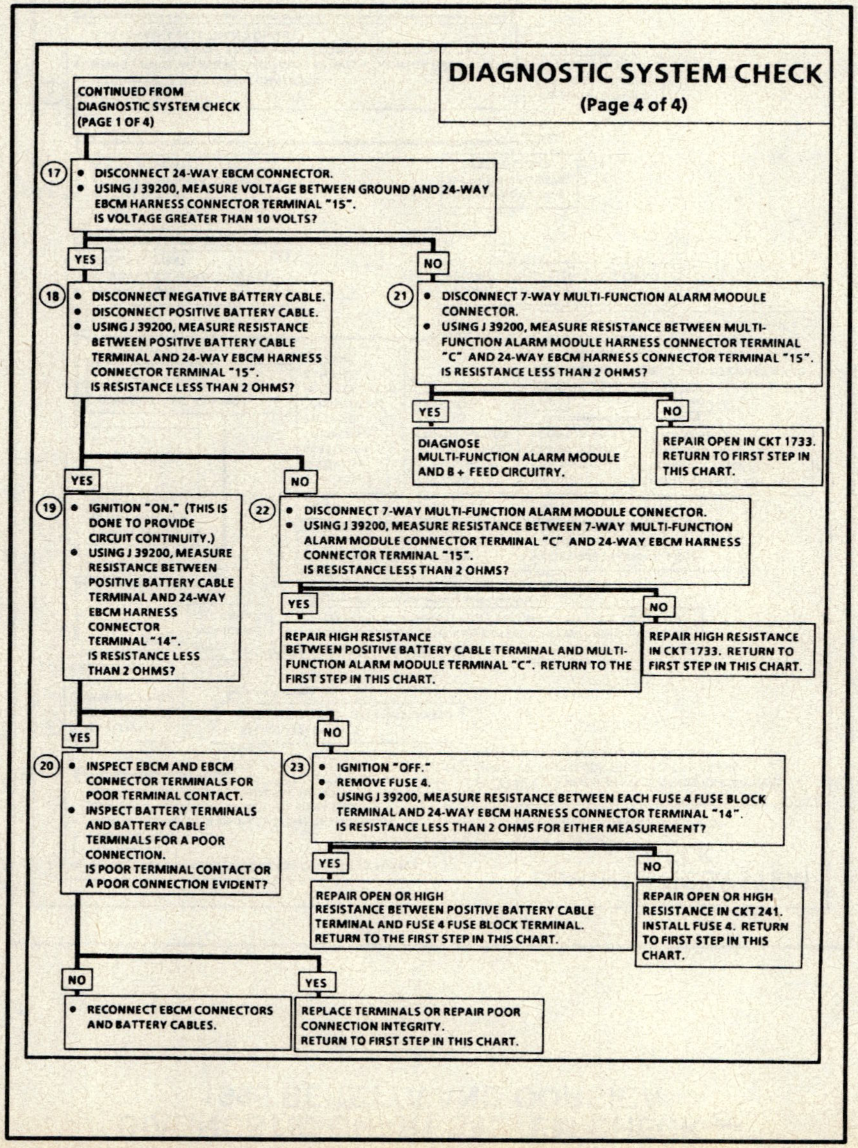

DIAGNOSTIC CHART SYSTEM CHECK — 1995 BERETTA AND CORSICA

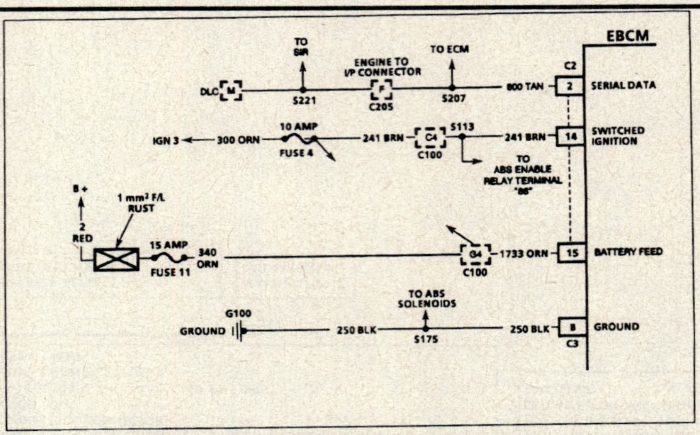

DIAGNOSTIC SYSTEM CHECK

System Description:

The diagnostic system check is an organized approach to identifying a malfunction created by an Antilock Brake System (ABS). It must be the starting point for any ABS complaint diagnosis because it directs the service technician to the next logical step in diagnosing the complaint.

Serial data is transmitted by the EBCM through terminal "2". The EBCM is supplied switched ignition voltage through terminal "14", battery voltage through terminal "15" and switched battery voltage through terminal "A". The EBCM ground is provided through terminal "B".

Chart Test Description:

Number(s) below refer to circled number(s) on the diagnostic chart.

1. Indicates if EBCM is transmitting data.
2. Checks for a serial data link malfunction.
3. Checks for proper Tech 1 connection to DLC.
4. Checks for open Fuse 11.
5. Checks for open Fuse 4.
6. Checks for high resistance in the EBCM ground circuit.
7. Checks for current DTCs.
8. Checks for proper ABS warning lamp operation.
9. Checks for history DTCs.
10. Checks for possible short to ground in the B+ feed circuitry.
11. Checks for possible short to ground in the EBCM.
12. Ensures short to ground is not due to physical damage of the circuitry.
13. Checks for possible short to ground in the switched ignition circuitry.
14. Isolates the short to ground to either the circuitry or the EBCM.
15. Ensures short to ground is not due to physical damage of the circuitry.
16. Checks for proper battery voltage at the EBCM.
17. Checks for high resistance in the B+ feed circuitry.
18. Checks for high resistance in the switched ignition circuitry.
19. Ensures malfunction is not due to poor terminal contact.
20. Checks for high resistance in the switched ignition circuit between the fuse block and the EBCM.

Diagnostic Aids:

Excessive resistance in the ground or power supply circuits will not allow communication with the EBCM. If communication with the EBCM is not possible, ensure the ABS ground connection is good and that there is no excessive resistance in any of the power supply circuits.

DIAGNOSTIC CHART SYSTEM CHECK — 1995 BERETTA AND CORSICA

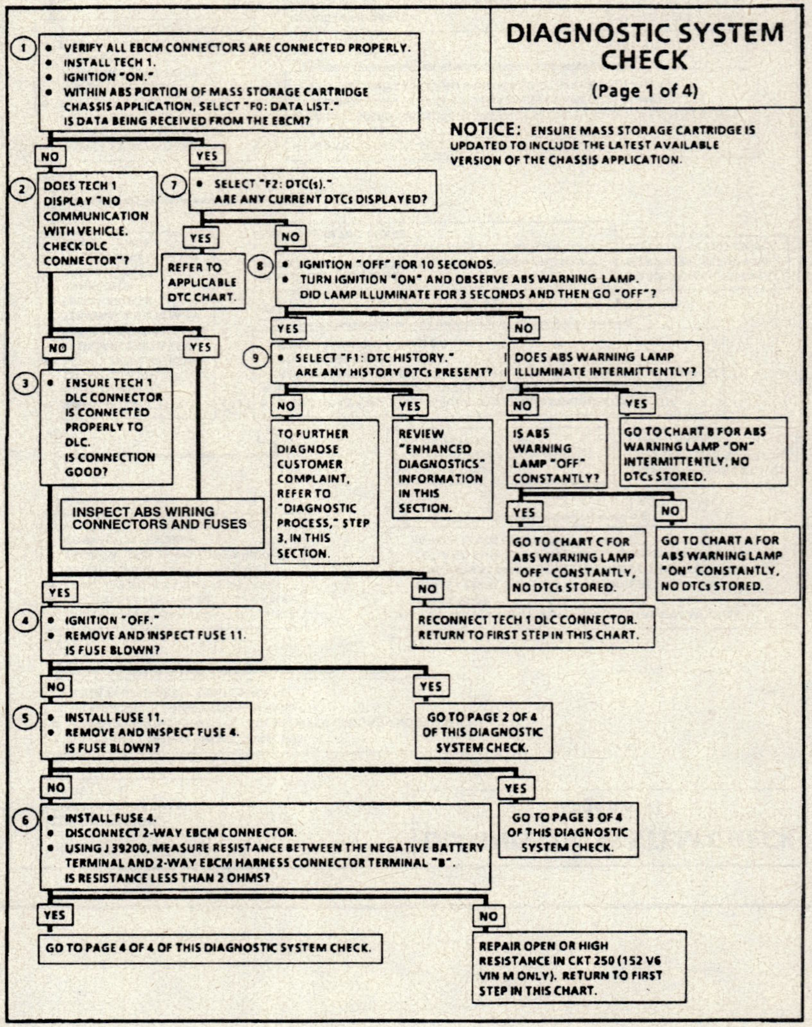

DIAGNOSTIC CHART SYSTEM CHECK — 1995 BERETTA AND CORSICA

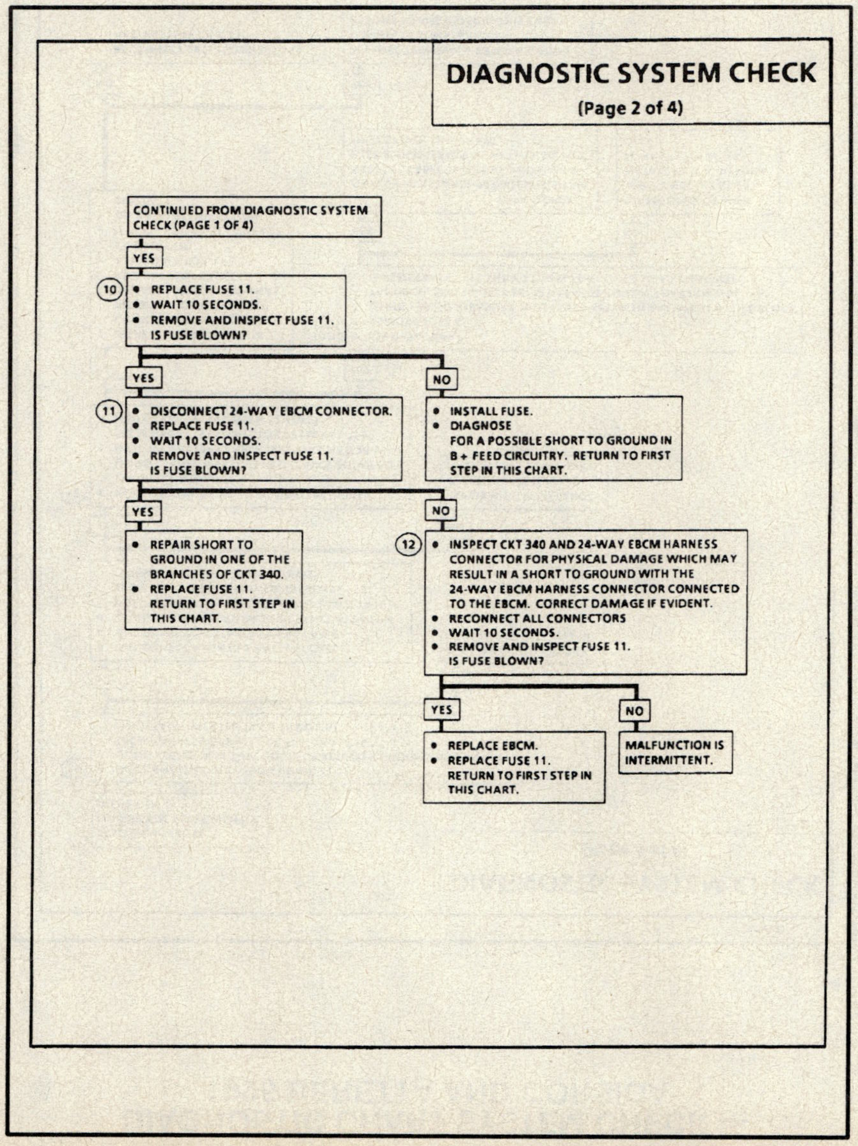

DIAGNOSTIC CHART SYSTEM CHECK — 1995 BERETTA AND CORSICA

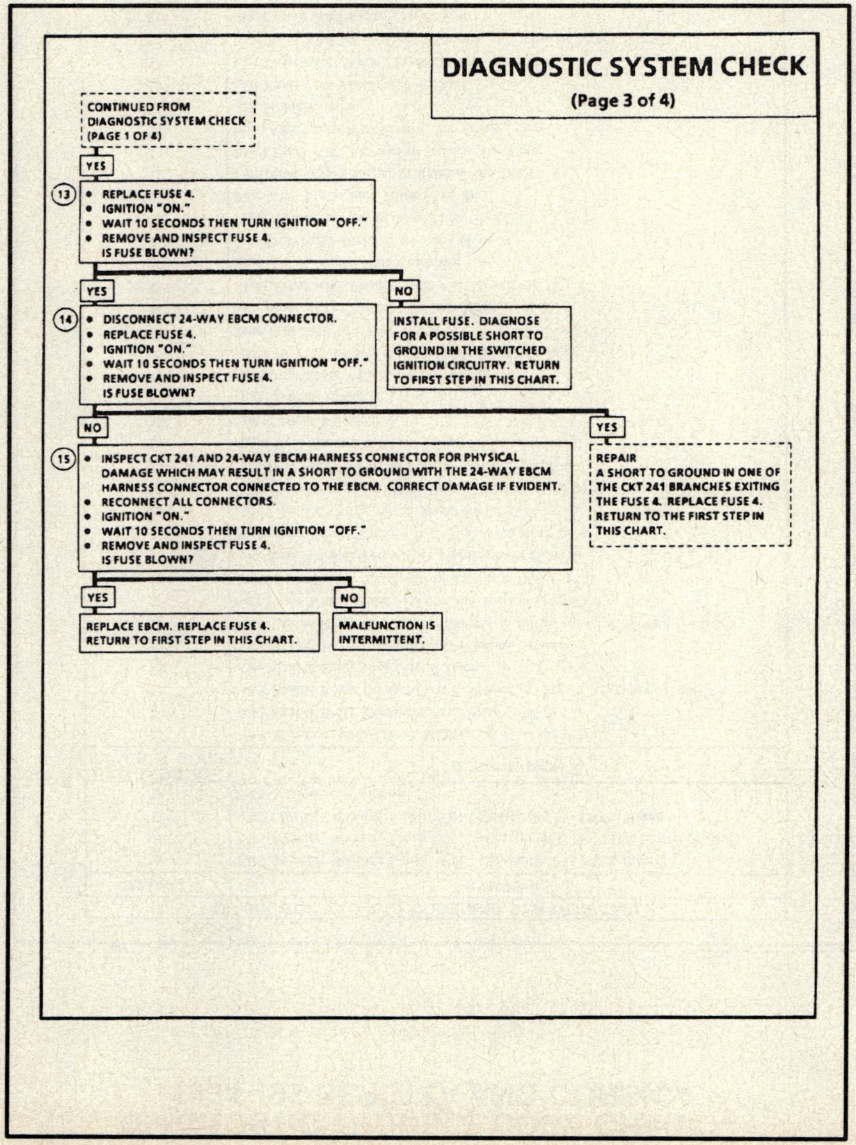

DIAGNOSTIC CHART SYSTEM CHECK — 1995 BERETTA AND CORSICA

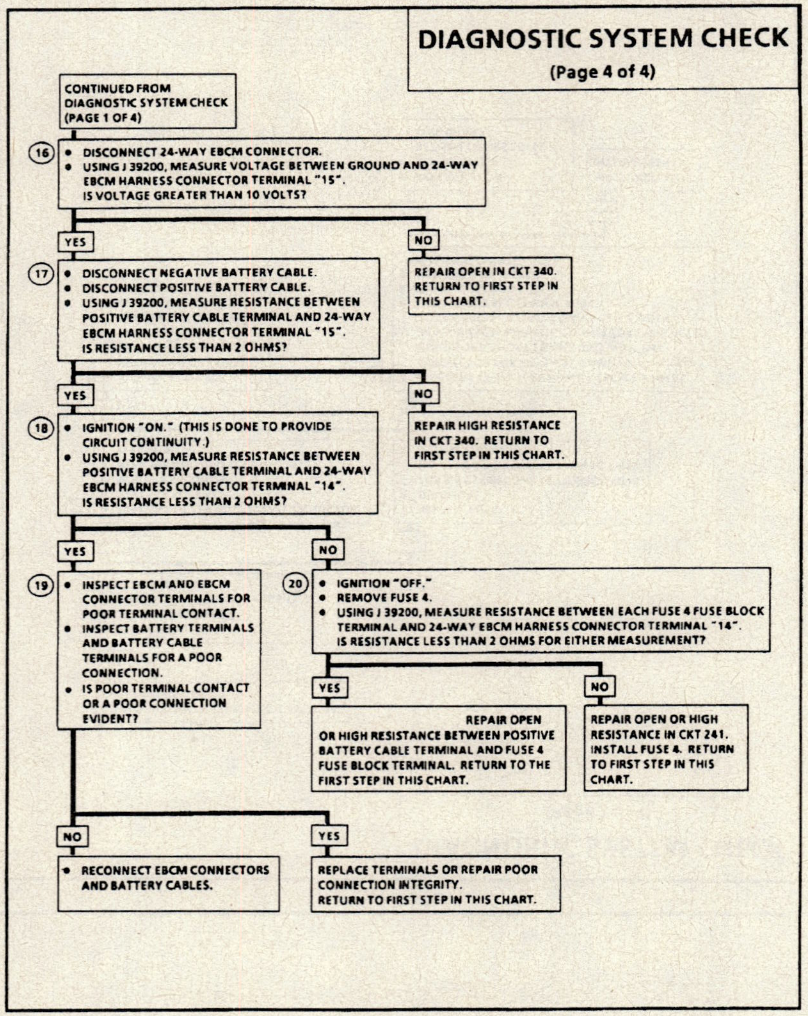

DIAGNOSTIC TROUBLE CODE CHART — 1994–95 BERETTA AND CORSICA

DIAGNOSTIC TROUBLE CODE AND SYMPTOM TABLE

CHART	SYMPTOM
A	ABS (Amber) Warning Lamp "ON" Constantly, No DTCs Stored
B	ABS (Amber) Warning Lamp "ON" Intermittently, No DTCs Stored (1 of 2)
C	ABS (Amber) Warning Lamp "ON" Constantly, No DTCs Stored
D	Tech 1 Displays Undefined DTCs
DIAGNOSTIC TROUBLE CODE	**DESCRIPTION**
11	ABS Warning Lamp Circuit Malfunction (1 of 2)
14	ABS Enable Relay Contact Circuit Open (1 of 3)
15	ABS Enable Relay Contact Circuit Shorted to Battery or Always Closed
16	ABS Enable Relay Coil Circuit Open
17	ABS Enable Relay Coil Circuit Shorted to Ground
18	ABS Enable Relay Coil Circuit Shorted to Battery or Coil Shorted
21	Left Front Wheel Speed = 0 or Unreasonable (1 of 2)
22	Right Front Wheel Speed = 0 or Unreasonable (1 of 2)
23	Left Rear Wheel Speed = 0 or Unreasonable (1 of 4)
24	Right Rear Wheel Speed = 0 or Unreasonable (1 of 4)
25	Left Front Excessive Wheel Speed Variation (1 of 2)
26	Right Front Excessive Wheel Speed Variation (1 of 2)
27	Left Rear Excessive Wheel Speed Variation (1 of 4)
28	Right Rear Excessive Wheel Speed Variation (1 of 4)
36	Low System Voltage
37	High System Voltage
38	Left Front EMB Will Not Hold Motor
41	Right Front EMB Will Not Hold Motor
42	Rear ESB Will Not Hold Motor
44	Left Front ABS Channel Will Not Move (1 of 2)
45	Right Front ABS Channel Will Not Move (1 of 2)
46	Rear ABS Channel Will Not Move
47	Left Front ABS Motor Free Spins (1 of 2)
48	Right Front ABS Motor Free Spins (1 of 2)
51	Rear ABS Motor Free Spins (1 of 2)
52	Left Front ABS Channel in Release Too Long
53	Right Front ABS Channel in Release Too Long
54	Rear ABS Channel in Release Too Long
55	EBCM Malfunction
56	Left Front ABS Motor Circuit Open
57	Left Front ABS Motor Circuit Shorted to Ground
58	Left Front ABS Motor Circuit Shorted to Battery or Motor Shorted
61	Right Front ABS Motor Circuit Open
62	Right Front ABS Motor Circuit Shorted to Ground
63	Right Front ABS Motor Circuit Shorted to Battery or Motor Shorted
64	Rear ABS Motor Circuit Open
65	Rear ABS Motor Circuit Shorted to Ground
66	Rear ABS Motor Circuit Shorted to Battery or Motor Shorted

DIAGNOSTIC TROUBLE CODE CHART — 1994–95 BERETTA AND CORSICA

DIAGNOSTIC TROUBLE CODE AND SYMPTOM TABLE	
CHART	SYMPTOM
67	Left Front EMB Circuit Open or Shorted to Ground
68	Left Front EMB Circuit Shorted to Battery or Driver Open
71	Right Front EMB Circuit Open or Shorted to Ground
72	Right Front EMB Circuit Shorted to Battery or Driver Open
76	Left Front Solenoid Circuit Open or Shorted to Battery
77	Left Front Solenoid Circuit Shorted to Ground or Driver Open
78	Right Front Solenoid Circuit Open or Shorted to Battery
81	Right Front Solenoid Circuit Shorted to Ground or Driver Open
82	Calibration Malfunction
86	EBCM Turned "ON" the Red "BRAKE" Warning Lamp
87	Red "BRAKE" Warning Lamp Circuit Open
88	Red "BRAKE" Warning Lamp Circuit Shorted to Battery
91	Open Brake Switch Contacts During Deceleration
92	Open Brake Switch Contacts When ABS Was Required
93	DTC 91 or 92 Set in Current or Previous Ignition Cycle
94	Brake Switch Contacts Always Closed
95	Brake Switch Circuit Open
96	Brake Lamp Circuit Open

ELECTRICAL CONNECTOR TERMINAL IDENTIFICATION — 1994–95 BERETTA AND CORSICA

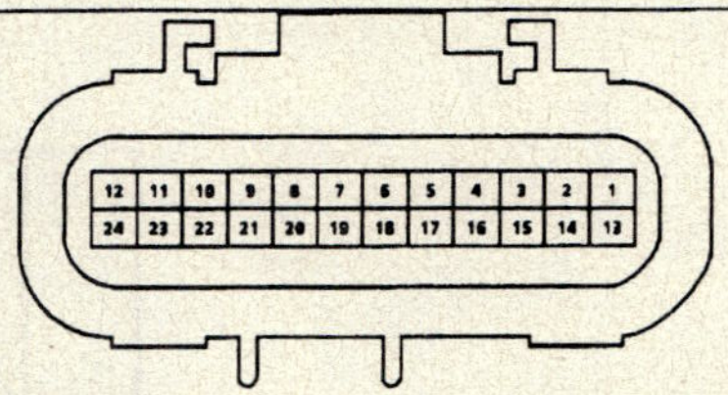

24-WAY EBCM HARNESS CONNECTOR

PIN	CIRCUIT NO.	COLOR	CIRCUIT
1	OPEN		NOT USED
2	805	TAN	SERIAL DATA
3	OPEN		NOT USED
4	1289	LT BLU	R/F ABS SOLENOID CONTROL
5	830	LT BLU	L/F WHEEL SIGNAL HIGH
6	873	YEL	L/F WHEEL SIGNAL LOW
7	882	BRN	R/R WHEEL SIGNAL HIGH
8	883	WHT	R/R WHEEL SIGNAL LOW
9	872	DK GRN	R/F WHEEL SIGNAL HIGH
10	833	TAN	R/F WHEEL SIGNAL LOW
11	885	RED	L/R WHEEL SIGNAL LOW
12	884	BLK	L/R WHEEL SIGNAL HIGH
13	17	WHT	BRAKE SWITCH INPUT
14	241	BRN	SWITCHED IGNITION
15	1733	ORN	B + FEED
16	OPEN		NOT USED
17	OPEN		NOT USED
18	VENT TUBE	BLK	VENT TUBE
19	1286	LT GRN	L/F EMB CONTROL
20	1287	GRY	R/F EMB CONTROL
21	33	TAN/WHT	BRAKE WARNING LAMP CONTROL
22	1632	PNK	ELECTRONIC BRAKE RELAY CONTROL
23	852	WHT	ABS INDICATOR LAMP CONTROL
24	1288	DK GRN	L/F ABS SOLENOID CONTROL

ELECTRICAL CONNECTOR TERMINAL IDENTIFICATION — 1994–95 BERETTA AND CORSICA

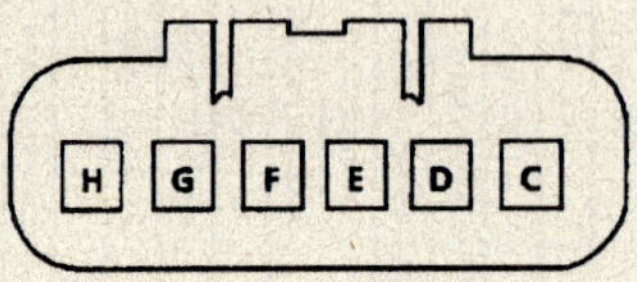

6 WAY EBCM CONNECTOR

PIN	CIRCUIT NO.	COLOR	CIRCUIT
C	1284	DK GRN	REAR MOTOR HIGH
D	1285	ORN	REAR MOTOR LOW
E	1281	PNK	L/F MOTOR LOW
F	1280	BLK	L/F MOTOR HIGH
G	1283	BLK	R/F MOTOR LOW
H	1282	PPL	R/F MOTOR HIGH

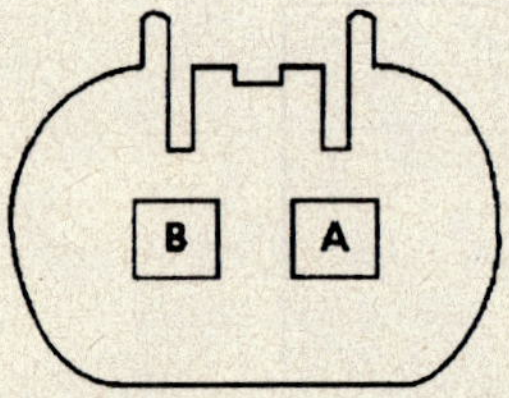

2 WAY EBCM CONNECTOR

PIN	CIRCUIT NO.	COLOR	CIRCUIT
A	1633	RED	SWITCHED BATTERY INPUT
B	250	BLK	GROUND

DIAGNOSTIC CHART A — 1994–95 BERETTA AND CORSICA

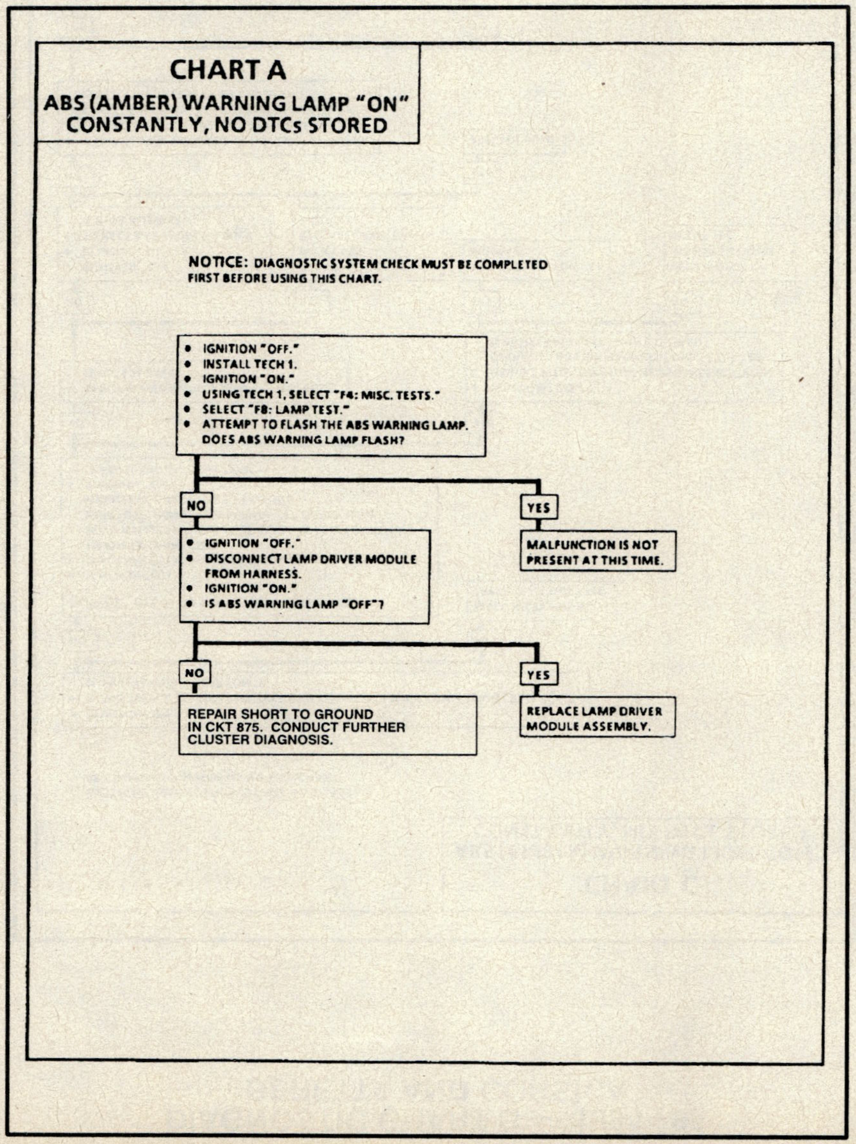

DIAGNOSTIC CHART B — 1994–95 BERETTA AND CORSICA

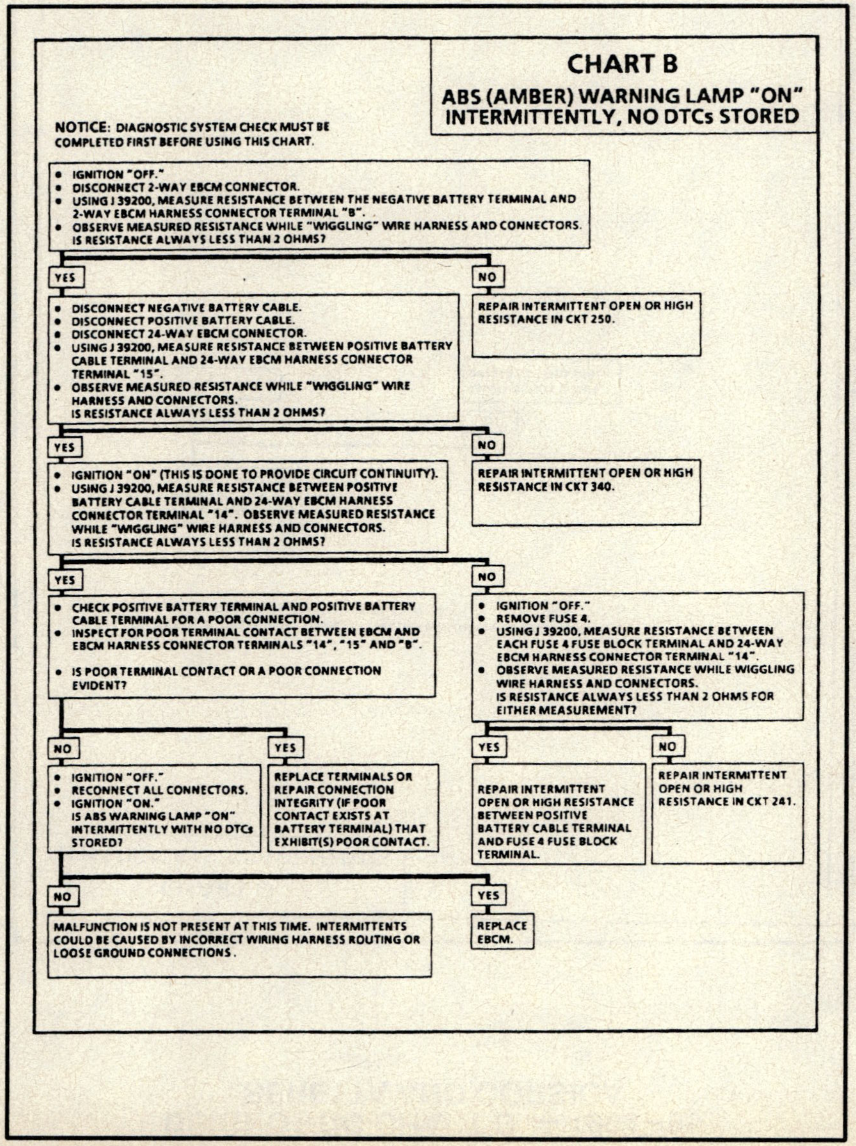

DIAGNOSTIC CHART C — 1994–95 BERETTA AND CORSICA

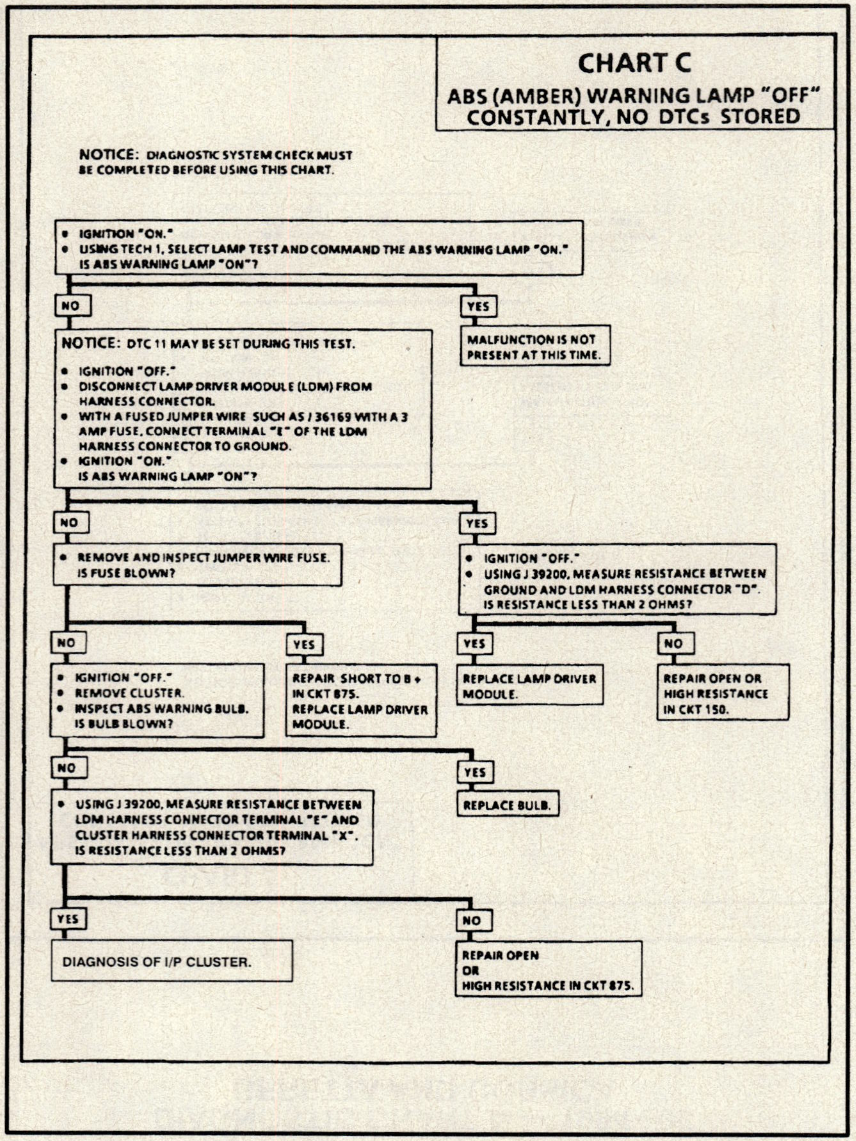

DIAGNOSTIC CHART D — 1994–95 BERETTA AND CORSICA

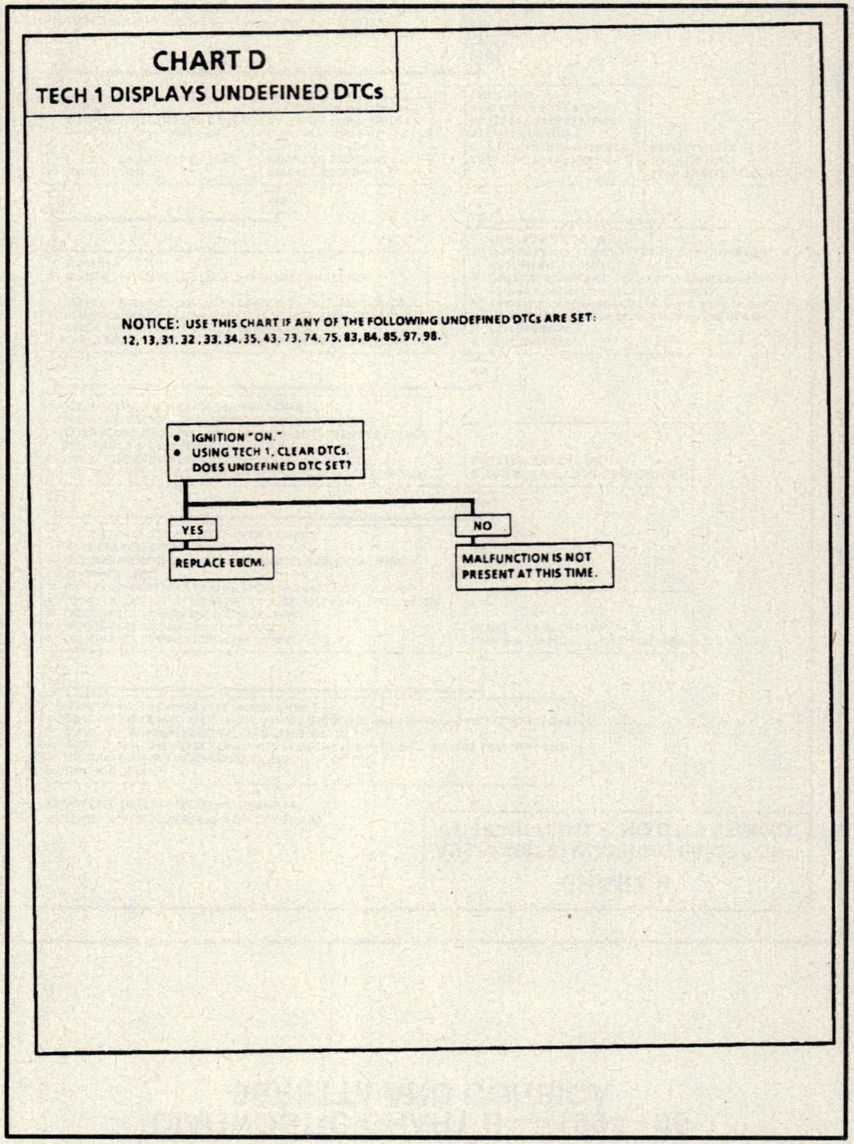

DIAGNOSTIC CHART DTC 11 — 1994–95 BERETTA AND CORSICA

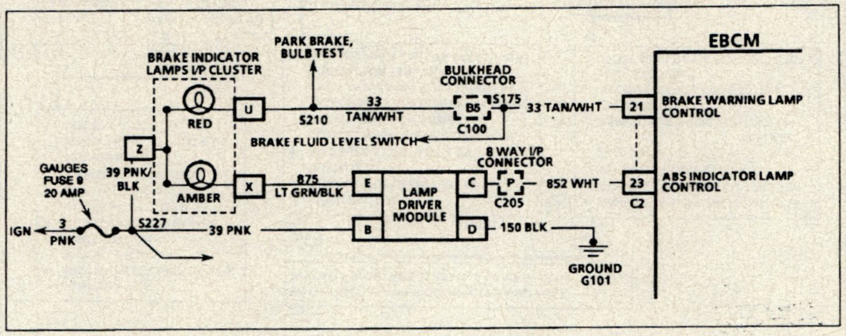

DTC 11

Circuit Description:

This test checks the state of the ABS warning lamp to identify a situation in which the driver could not be warned of a system malfunction by the ABS warning lamp, or the warning lamp is always "ON". The Lamp Driver Module (LDM) turns the ABS warning lamp "ON" unless the EBCM provides ground to turn it "OFF". Because of the circuitry in the LDM, only external malfunctions can be detected. As a result, the ABS warning lamp itself is not diagnosable, only the control line of the EBCM can be diagnosed. In the event of an open CKT 852, the ABS warning lamp will be "ON" at all times, due to the loss of ground at the LDM input. If the control line is shoreted to ground, the ABS warning lamp is kept "OFF" due to the LDM input being grounded. If the control line is shorted to B+, the EBCM cannot command the ABS warning lamp "OFF".

DTC Will Set When:

DTC 11 can be set anytime. Opens or shorts to ground are detected during the three second bulb check. Shorts to battery are detected when the EBCM commands the ABS warning lamp "OFF".

Action Taken:

A malfunction DTC is stored; however, ABS is not disabled. during the three second bulb check. Shorts to battery are detected when the

DIAGNOSTIC CHART DTC 11 — 1994–95 BERETTA AND CORSICA

DTC Chart Test Description:

Number(s) below refer to circled number(s) on the diagnostic chart.

1. Checks to see if ABS warning lamp functions properly. Normal operation would indicate that the malfunction is not present.
2. Checks for a short to B+ in the ABS warning lamp control circuit.
3. Determines whether the short to B+ is in the LDM or the ABS warning lamp control circuit.
4. Checks for an open in the ABS warning lamp control circuit.
5 Checks to see if the malfunction was due to a poor connection.
6. Checks for a possible shor to ground in the EBCM.
7. Check for a possible shor to ground in the ABS warning lamp control circuit.
8. Checks for high resistance in the LDM power feed circuit.
9. Ensures malfunction was not due to physical damage of the circuitry.
10. Checks for malfunction in the LDM power feed circuit.

Diagnostic Aids:

An "Intermittent" malfunction is most likely caused by a poor connection, rubbed through wire insulation, or a wire that is broken inside the insulation.

The lamp test funciton of the Tech 1 may be used to command the lamp "ON" while looking for an intermittent malfunction in the warning lamp circuitry.

The frequency of the malfunction can be checked by using the enhanced diagnostic function of the Tech 1 as described in "Enhanced Diagnostics," found in this section.

Any circuitry that is suspected as causing the intermittent complaint should be thoroughly checked for backed out terminals, improper mating, broken locks, improperly formed or damaged terminals, poor terminal to wiring connections or physical damage to the wiring harness.

DIAGNOSTIC CHART DTC 11 — 1994–95 BERETTA AND CORSICA

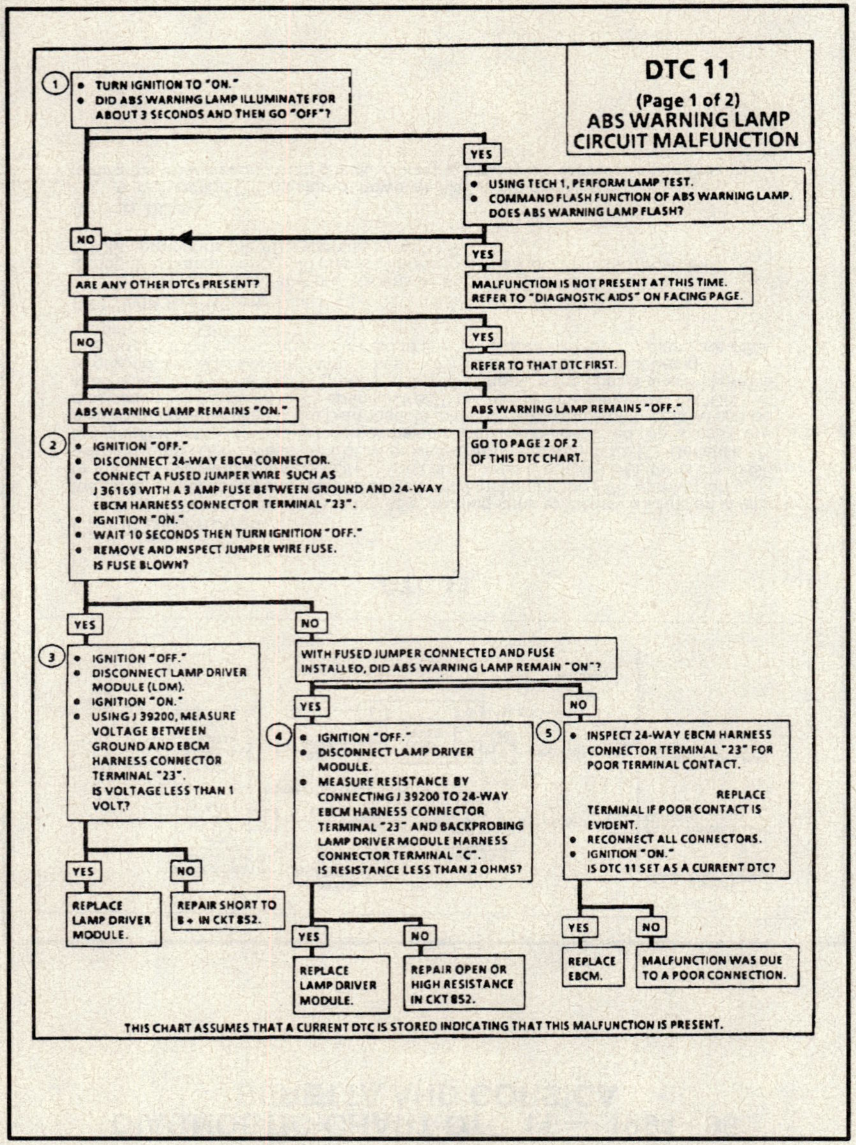

DIAGNOSTIC CHART DTC 11 — 1994–95 BERETTA AND CORSICA

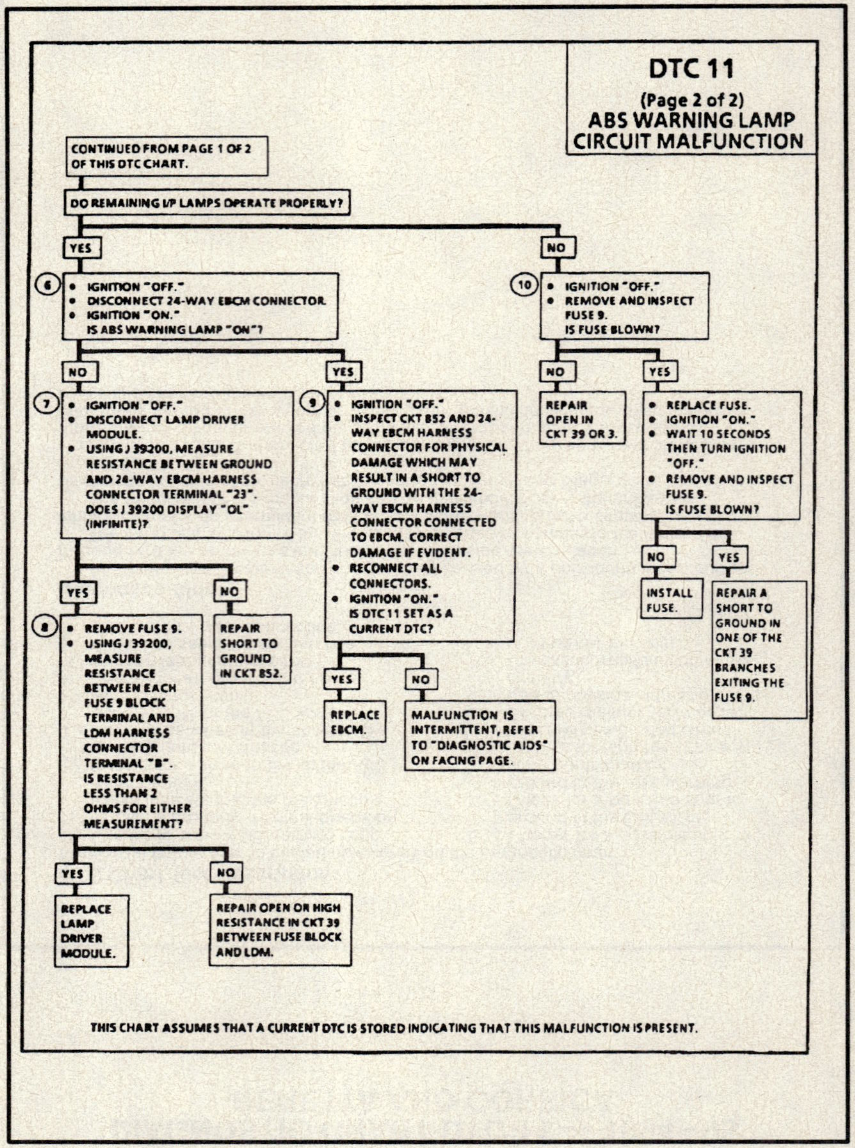

DIAGNOSTIC CHART DTC 14 — 1994–95 BERETTA AND CORSICA

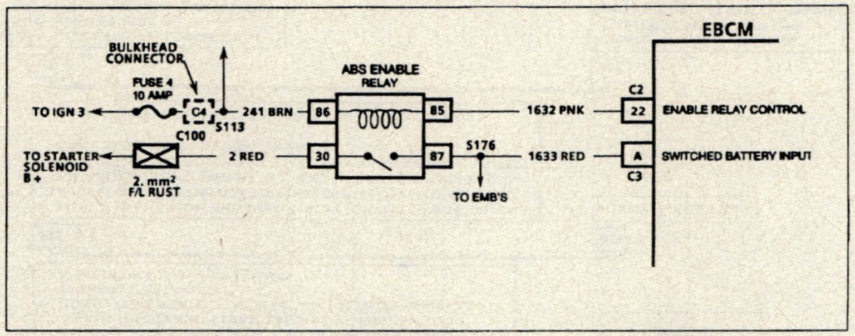

DTC 14
ABS ENABLE RELAY CONTACT CIRCUIT OPEN

Circuit Description:

Ignition voltage is supplied through terminal "86" of the ABS enable relay. The EBCM is then able to energize the pull–in coil by completing the ground circuit at EBCM terminal "22". The magnetic field created closes the enable relay contacts and allows battery voltage and current to be supplied to the EMBs and EBCM, which supplies power to the motors and solenoids.

DTC Will Set When:

DTC 14 can be set anytime after the EBCM commands the enable relay "ON" (the relay is first commanded "ON" during the three second bulb check). This test monitors the availability of current/voltage to the motors, solenoids, and EMBs. This malfunction indicates voltage is not available and would therefore not allow ABS operation if required.

Action Taken:

A malfunction DTC is stored, ABS is disabled and the ABS warning lamp is turned "ON".

DIAGNOSTIC CHART DTC 14 — 1994–95 BERETTA AND CORSICA

DTC Chart Test Description:

Number(s) below refer to circled number(s) on the diagnostic chart.

1. Checks to see if the EBCM is sensing battery voltage at terminal "A".
2. Checks for proper operation of the enable relay and related circuitry.
3. Checks for excessive resistance in the switched battery input circuitry.
4. Checks for excessive resistance in the B+ feed circuit.
5. Checks for excessive resistance in the B+ feed circuit.
6. Determines if the malfunction is due to poor terminal connection.
7. Determines if the malfunction is due to the EBCM.
8. Checks for excessive resistance in the enable relay control circuitry.
9. Checks for proper battery voltage.
10. Checks for an open in the B+ feed circuit.
11. Checks for an open in the switched battery input circuit.
12. Checks for proper operation of the enable relay.

Diagnostic Aids:

An "Intermittent" malfunction is most likely caused by a poor connection, rubbed through wire insulation, or a wire that is broken inside the insulation.

The lamp test funciton of the Tech 1 may be used to command the lamp "ON" while looking for an intermittent malfunction in the warning lamp circuitry.

The frequency of the malfunction can be checked by using the enhanced diagnostic function of the Tech 1 as described in "Enhanced Diagnostics," found in this section.

Any circuitry that is suspected as causing the intermittent complaint should be thoroughly checked for backed out terminals, improper mating, broken locks, improperly formed or damaged terminals, poor terminal to wiring connections or physical damage to the wiring harness.

VIBRATION, TEMPERATURE EFFECTS:

Check for vibration effects by performing the relay test function of the Tech 1. With the relay commended "ON", lightly tap the top and sides of the relay while monitoring battery voltage. If the battery voltage changes significantly, replace the relay.

If test 14 only fails when the vehicle is initially started in cold ambient conditions (temperature less than 325F 05C), replace the relay.

DIAGNOSTIC CHART DTC 14 — 1994–95 BERETTA AND CORSICA

DTC 14
(Page 1 of 3)
ABS ENABLE RELAY CONTACT CIRCUIT OPEN

1. • IGNITION "ON," ENGINE "OFF."
   • USING TECH 1, SELECT DATA LIST AND OBSERVE ABS BATTERY VOLTAGE.
   DOES TECH 1 INDICATE THAT ABS BATTERY VOLTAGE IS GREATER THAN 10 VOLTS?
   - YES: MALFUNCTION IS NOT PRESENT AT THIS TIME. REFER TO "DIAGNOSTIC AIDS" ON FACING PAGE.
   - NO: go to step 2.

2. • IGNITION "OFF."
   • DISCONNECT 24-WAY EBCM CONNECTOR.
   • DISCONNECT 2-WAY EBCM CONNECTOR.
   • CONNECT A FUSED JUMPER WIRE SUCH AS J 36169 WITH A 3 AMP FUSE BETWEEN GROUND AND 24-WAY EBCM HARNESS CONNECTOR TERMINAL "22".
   • IGNITION "ON."
   • USING J 39200, MEASURE VOLTAGE BETWEEN GROUND AND 2-WAY EBCM HARNESS CONNECTOR TERMINAL "A".
   IS VOLTAGE GREATER THAN 10 VOLTS?
   - NO: REFER TO PAGE 2 OF 3 OF THIS DTC CHART.
   - YES: go to step 3.

3. • IGNITION "OFF."
   • DISCONNECT JUMPER WIRE USED IN STEP 2.
   • DISCONNECT ENABLE RELAY.
   • USING J 39200, MEASURE RESISTANCE BETWEEN 2-WAY EBCM HARNESS CONNECTOR TERMINAL "A" AND ENABLE RELAY HARNESS CONNECTOR TERMINAL "87".
   IS RESISTANCE LESS THAN 2 OHMS?
   - NO: REPAIR HIGH RESISTANCE IN CKT 1633.
   - YES: go to step 4.

4. • DISCONNECT NEGATIVE BATTERY CABLE.
   • DISCONNECT POSITIVE BATTERY CABLE.
   • USING J 39200, MEASURE RESISTANCE BETWEEN POSITIVE BATTERY CABLE TERMINAL AND ENABLE RELAY HARNESS CONNECTOR TERMINAL "30".
   IS RESISTANCE LESS THAN 2 OHMS?
   - NO: REPAIR HIGH RESISTANCE IN B+ FEED CIRCUIT.
   - YES: go to step 5.

5. • USING A JUMPER WIRE, CONNECT ENABLE RELAY TERMINAL "85" TO GROUND. THEN, USING A FUSED JUMPER WIRE SUCH AS J 36169 WITH A 3 AMP FUSE, CONNECT ENABLE RELAY TERMINAL "86" TO B+.
   • USING J 39200, MEASURE THE RESISTANCE BETWEEN ENABLE RELAY TERMINALS "30" AND "87".
   IS RESISTANCE LESS THAN 2 OHMS?
   - NO: REPLACE ENABLE RELAY.
   - YES: go to step 6.

6. • INSPECT ENABLE RELAY AND ENABLE RELAY HARNESS CONNECTOR FOR A POOR CONNECTION OR POOR TERMINAL CONTACT. INSPECT EBCM AND EBCM CONNECTOR TERMINALS "22" AND "A" FOR POOR TERMINAL CONTACT. INSPECT POSITIVE BATTERY TERMINAL AND POSITIVE BATTERY CABLE TERMINAL FOR A POOR CONNECTION.
   IS POOR TERMINAL CONTACT OR A POOR CONNECTION EVIDENT?
   - YES: REPLACE TERMINALS OR REPAIR CONNECTION INTEGRITY (IF POOR CONTACT EXISTS AT BATTERY TERMINAL) THAT EXHIBIT(S) POOR CONTACT.
   - NO: go to step 7.

7. • RECONNECT ALL CONNECTORS.
   • IGNITION "ON."
   IS DTC 14 SET AS A CURRENT DTC?
   - YES: REPLACE EBCM.
   - NO: MALFUNCTION IS INTERMITTENT. REFER TO "DIAGNOSTIC AIDS" ON FACING PAGE.

DIAGNOSTIC CHART DTC 14 — 1994–95 BERETTA AND CORSICA

DTC 14
(Page 2 of 3)
ABS ENABLE RELAY CONTACT CIRCUIT OPEN

CONTINUED FROM PAGE 1 OF 3 OF THIS CHART.

8. • IGNITION "OFF."
   • DISCONNECT JUMPER WIRE USED IN STEP 2.
   • DISCONNECT NEGATIVE BATTERY CABLE FROM BATTERY.
   • DISCONNECT POSITIVE BATTERY CABLE FROM BATTERY.
   • TURN IGNITION "ON" (THIS IS DONE TO PROVIDE CIRCUIT CONTINUITY).
   • USING J 39200, MEASURE RESISTANCE BETWEEN POSITIVE BATTERY CABLE TERMINAL AND 24-WAY EBCM HARNESS CONNECTOR TERMINAL "22".
   IS RESISTANCE LESS THAN 95 OHMS WHEN RELAY IS APPROXIMATELY 20°C (68°F)?
   - NO: REFER TO PAGE 3 OF 3 OF THIS DTC CHART.
   - YES: go to step 9.

9. • USING J 39200, MEASURE VOLTAGE BETWEEN THE NEGATIVE AND POSITIVE BATTERY TERMINALS.
   IS VOLTAGE GREATER THAN 10 VOLTS?
   - NO: CORRECT LOW VOLTAGE CONDITION.
   - YES: go to step 10.

10. • IGNITION "OFF."
    • RECONNECT BATTERY CABLES.
    • DISCONNECT ENABLE RELAY.
    • USING J 39200, MEASURE VOLTAGE BETWEEN GROUND AND ENABLE RELAY HARNESS CONNECTOR TERMINAL "30".
    IS VOLTAGE GREATER THAN 10 VOLTS?
    - NO: REPAIR OPEN IN B+ FEED CIRCUIT.
    - YES: go to step 11.

11. • USING J 39200, MEASURE RESISTANCE BETWEEN 2-WAY EBCM HARNESS CONNECTOR TERMINAL "A" AND ENABLE RELAY HARNESS CONNECTOR TERMINAL "87".
    IS RESISTANCE LESS THAN 2 OHMS?
    - NO: REPAIR OPEN IN CKT 1633.
    - YES: go to step 12.

12. • USING A JUMPER WIRE, CONNECT ENABLE RELAY TERMINAL "85" TO GROUND. THEN, USING A FUSED JUMPER WIRE SUCH AS J 36169 WITH A 3 AMP FUSE, CONNECT ENABLE RELAY TERMINAL "86" TO B+.
    • USING J 39200, MEASURE RESISTANCE BETWEEN ENABLE RELAY TERMINALS "30" AND "87".
    IS RESISTANCE LESS THAN 2 OHMS?
    - NO: REPLACE ENABLE RELAY.
    - YES: MALFUNCTION WAS DUE TO A POOR CONNECTION. TEST FOR POOR TERMINAL CONTACT BETWEEN ENABLE RELAY AND ENABLE RELAY HARNESS CONNECTOR AND/OR AT POSITIVE BATTERY TERMINAL. REPLACE OR REPAIR TERMINALS THAT EXHIBIT POOR TERMINAL CONTACT.

DIAGNOSTIC CHART DTC 14 — 1994–95 BERETTA AND CORSICA

DTC 14
(Page 3 of 3)
ABS ENABLE RELAY CONTACT CIRCUIT OPEN

CONTINUED FROM PAGE 2 OF 3 OF THIS CHART.

(13)
- USING J 39200, MEASURE RESISTANCE BETWEEN ENABLE RELAY TERMINALS "85" AND "86".

IS RESISTANCE LESS THAN 95 OHMS WHEN RELAY IS APPROXIMATELY 20°C (68°F)?

- NO: REPLACE ENABLE RELAY.
- YES:

(14)
- USING J 39200, MEASURE RESISTANCE BETWEEN ENABLE RELAY HARNESS CONNECTOR TERMINAL "85" AND 24-WAY EBCM HARNESS CONNECTOR TERMINAL "22".

IS RESISTANCE LESS THAN 2 OHMS?

- NO: REPAIR HIGH RESISTANCE IN CKT 1632.
- YES:

(15)
- DISCONNECT ENABLE RELAY.
- USING J 39200, MEASURE RESISTANCE BETWEEN POSITIVE BATTERY CABLE TERMINAL AND ENABLE RELAY HARNESS CONNECTOR TERMINAL "86".

IS RESISTANCE LESS THAN 2 OHMS?

- YES: MALFUNCTION WAS DUE TO A POOR CONNECTION. TEST FOR POOR TERMINAL CONTACT BETWEEN ENABLE RELAY AND ENABLE RELAY HARNESS CONNECTOR. REPLACE TERMINALS THAT EXHIBIT POOR TERMINAL CONTACT.
- NO: REPAIR HIGH RESISTANCE IN CIRCUIT BETWEEN POSITIVE BATTERY CABLE TERMINAL AND ENABLE RELAY HARNESS CONNECTOR TERMINAL "86".

DIAGNOSTIC CHART DTC 15 — 1994–95 BERETTA AND CORSICA

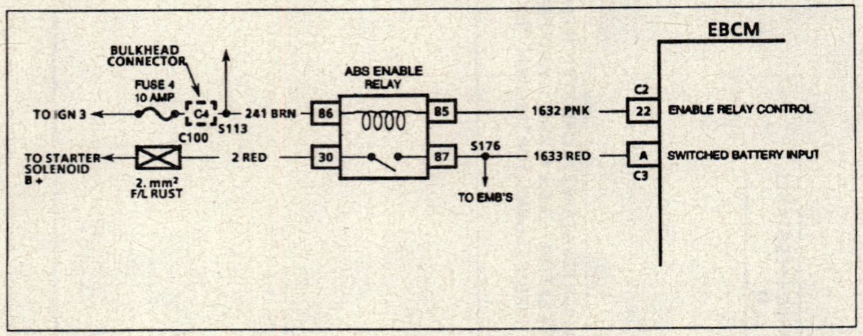

DTC 15
ABS ENABLE RELAY CONTACT CIRCUIT SHORTED TO BATTERY OR ALWAYS CLOSED

Circuit Description:

Ignition voltage is supplied through terminal "86" of the ABS enable relay. The EBCM is then able to energize the pull-in coil by completing the ground circuit at terminal "22" of the EBCM. The magnetic field created closes the enable relay contacts and allows battery voltage and current to be supplied to the EMBs and EBCM, which supplies power to the motors and solenoids.

DTC Will Set When:

DTC 15 can be set only before the EBCM commands the relay "ON." This test determines if the enable relay is energized when it should not be. This malfunction would not allow the enable relay to remove power to the ABS system. If a second malfunction were to occur that requires the enable relay to be turned "OFF," that malfunction can not be removed if the relay can not be controlled. The malfunction must be present for three consecutive ignition cycles before the DTC is set.

Action Taken:

A malfunction DTC is stored; however, ABS is not disabled.

DTC Chart Test Description:

Number(s) below refer to circled number(s) on the diagnostic chart.

1. Indicates if the EBCM is capable of controlling the enable relay as commanded.
2. Checks for voltage at terminal "A" of the 2-way EBCM harness connector. If voltage is present, the malfunction exists in the enable relay and/or its circuitry.
3. Checks for a short to voltage in the switched battery input circuit.
4. Identifies if the malfunction is due to the EBCM.

Diagnostic Aids:

An "Intermittent" malfunction is most likely caused by a poor connection, rubbed through wire insulation, or a wire that is broken inside the insulation.

The frequency of the malfunction can be checked by using the enhanced diagnostic function of the Tech 1, as described in "Enhanced Diagnostics," found in this section.

Any circuitry, that is suspected as causing the intermittent complaint, should be thoroughly checked for backed out terminals, improper mating, broken locks, improperly formed or damaged terminals, poor terminal to wiring connections or physical damage to the wiring harness.

DIAGNOSTIC CHART DTC 15 — 1994–95 BERETTA AND CORSICA

DTC 15
ABS ENABLE RELAY CONTACT CIRCUIT SHORTED TO BATTERY OR ALWAYS CLOSED

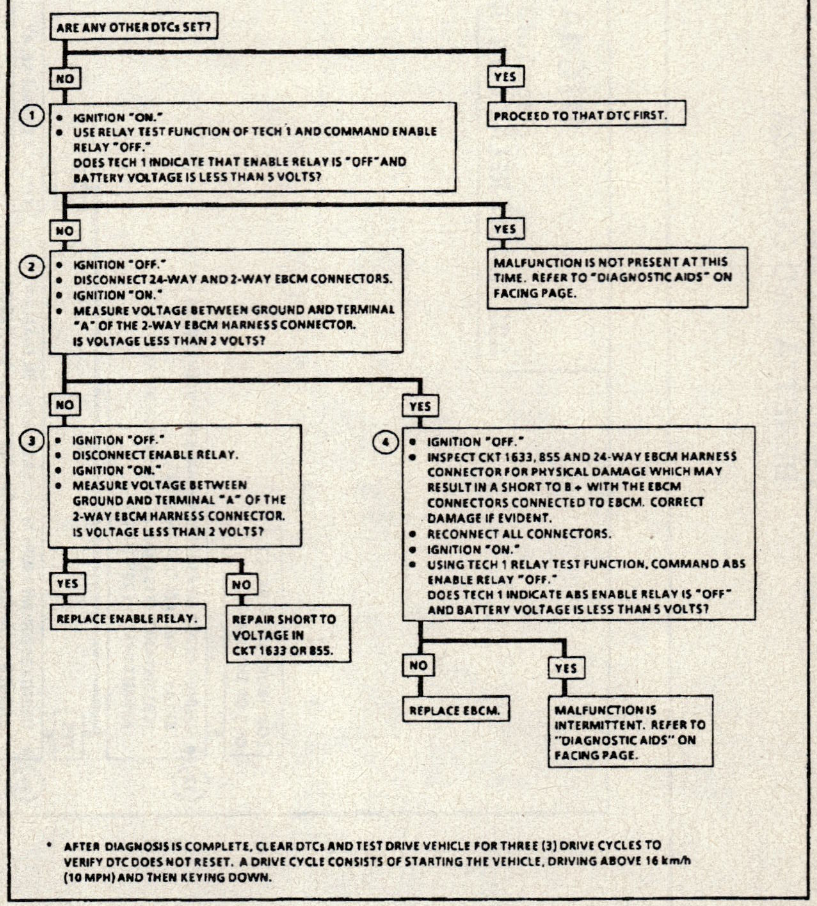

DIAGNOSTIC CHART DTC 16 — 1994–95 BERETTA AND CORSICA

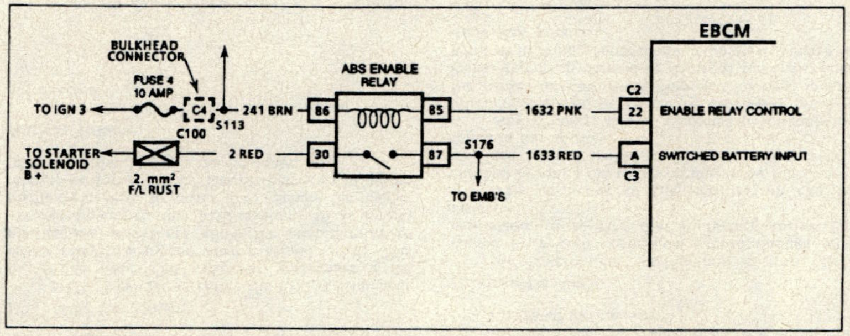

DTC 16
ABS ENABLE RELAY COIL CIRCUIT OPEN

Circuit Description:

Ignition voltage is supplied through terminal "86" of the ABS enable relay. The EBCM is then able to energize the pull-in coil by completing the ground circuit at terminal "22" of the EBCM. The magnetic field created closes the enable relay contacts and allows battery voltage and current to be supplied to the EMBs and EBCM, which supplies power to the motors and solenoids.

DTC Will Set When:

DTC 16 can be set anytime after the relay is commanded "ON". This test detects an open in the enable relay coil circuit. An open in this circuit will not allow the enable relay to be energized thus preventing voltage/current to the motors, EMBs and solenoids.

Action Taken:

A malfunction DTC is stored, ABS is disabled and the ABS warning lamp is turned "ON."

DTC Chart Test Description:

Number(s) below refer to circled number(s) on the diagnostic chart.

1. Indicates if the EBCM is capable of controlling the enable relay as commanded.
2. Ensures that there is continuity through the pull-in coil of the enable relay.
3. Checks to ensure that voltage is available to the pull-in coil of the enable relay.
4. Checks continuity in the enable relay control circuit.
5. Ensures DTC was not set due to a poor connection.
6. Identifies whether the malfunction is due to a poor connection or the EBCM.

Diagnostic Aids:

An "Intermittent" malfunction is most likely caused by a poor connection, rubbed through wire insulation, or a wire that is broken inside the insulation.

The frequency of the malfunction can be checked by using the enhanced diagnostic function of the Tech 1, as described in "Enhanced Diagnostics," found in this section.

If the frequency of the malfunction is high, but is currently intermittent, check for high coil resistance by measuring resistance between relay terminals "86" and "85". If resistance is greater than 100 ohms, replace the relay.

Any circuitry, that is suspected as causing the intermittent complaint, should be thoroughly checked for backed out terminals, improper mating, broken locks, improperly formed or damaged terminals, poor terminal to wiring connections or physical damage to the wiring harness.

DIAGNOSTIC CHART DTC 16 — 1994–95 BERETTA AND CORSICA

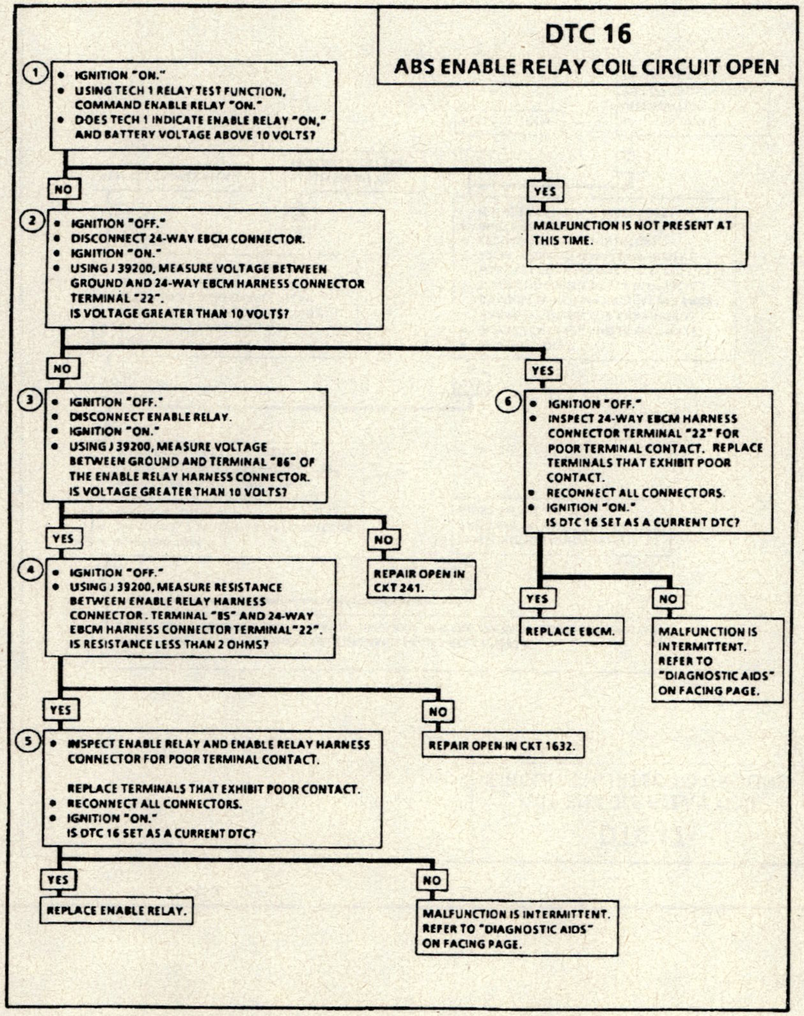

DIAGNOSTIC CHART DTC 17 — 1994–95 BERETTA AND CORSICA

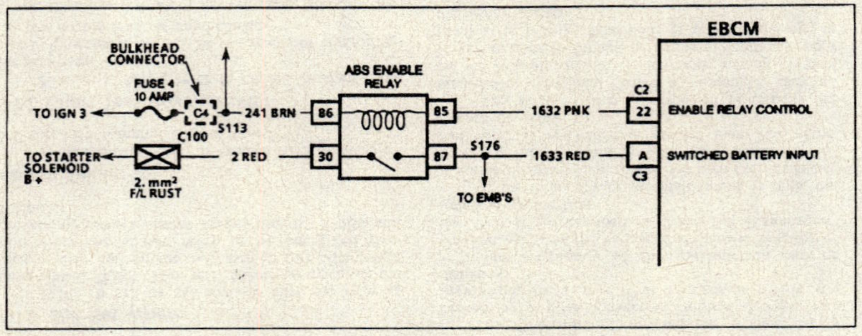

DTC 17
ABS ENABLE RELAY COIL CIRCUIT SHORTED TO GROUND

Circuit Description:

Ignition voltage is supplied through terminal "86" of the ABS enable relay. The EBCM is then able to energize the pull-in coil by completing the ground circuit at terminal "22" of the EBCM. The magnetic field created closes the enable relay contacts and allows battery voltage and current to be supplied to the EMBs and EBCM, which supplies voltage to the motors and solenoids.

DTC Will Set When:

DTC 17 can be set before the EBCM commands the enable relay "ON." This test determines if the enable relay is energized when it should not be. This malfunction would not allow the enable relay to remove power to the ABS system. If a second malfunction were to occur that requires the enable relay to be turned "OFF," that malfunction can not be removed if the relay can not be controlled.

Action Taken:

A malfunction DTC is stored; however, ABS is not disabled.

DTC Chart Test Description:

Number(s) below refer to circled number(s) on the diagnostic chart.

1. Indicates if the EBCM is capable of controlling the enable relay as commanded.
2. Ensures the enable relay or control circuit are not shorted to ground.
3. This step isolates the malfunction to either the enable relay or the enable relay control circuit.
4. Ensures short to ground is not due to physical damage of the circuitry.

Diagnostic Aids:

An "Intermittent" malfunction is most likely caused by a poor connection, rubbed through wire insulation, or a wire that is broken inside the insulation.

The frequency of the malfunction can be checked by using the enhanced diagnostic function of the Tech 1, as described in "Enhanced Diagnostics," found in this section.

Any circuitry, that is suspected as causing the intermittent complaint, should be thoroughly checked for backed out terminals, improper mating, broken locks, improperly formed or damaged terminals, poor terminal to wiring connections or physical damage to the wiring harness.

DIAGNOSTIC CHART DTC 17 — 1994–95 BERETTA AND CORSICA

DTC 17
ABS ENABLE RELAY COIL CIRCUIT SHORTED TO GROUND

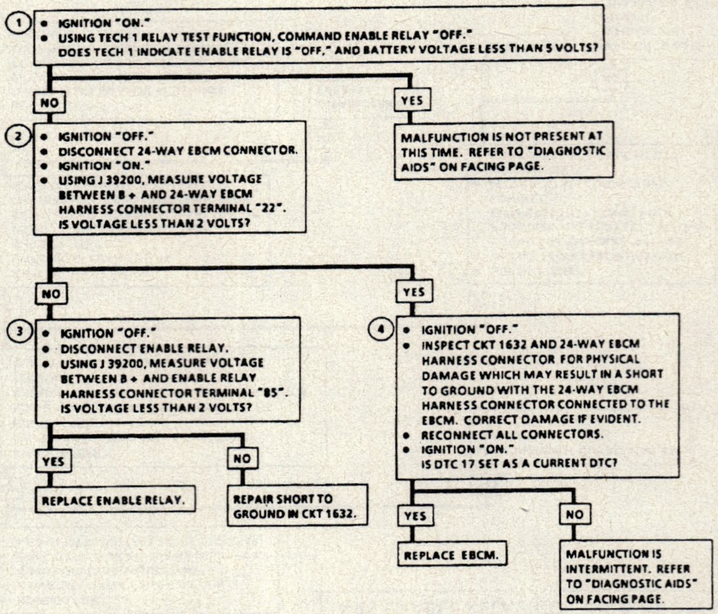

THIS CHART ASSUMES THAT A CURRENT DTC IS STORED INDICATING THAT THIS MALFUNCTION IS PRESENT.

DIAGNOSTIC CHART DTC 18 — 1994–95 BERETTA AND CORSICA

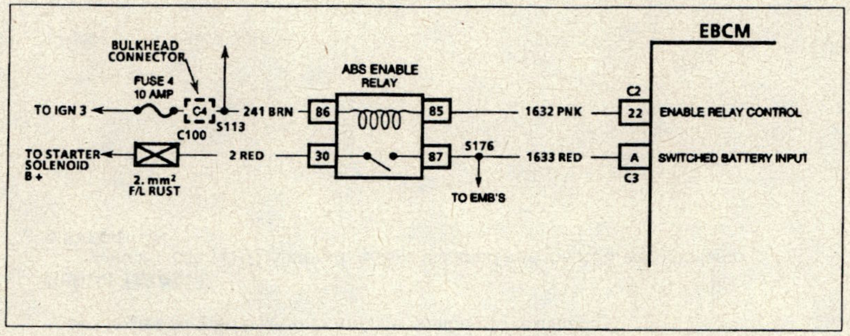

DTC 18
ABS ENABLE RELAY COIL CIRCUIT SHORTED TO BATTERY OR COIL SHORTED

Circuit Description:

Ignition voltage is supplied through terminal "86" of the ABS enable relay. The EBCM is then able to energize the pull-in coil by completing the ground circuit at terminal "22" of the EBCM. The magnetic field created closes the enable relay contacts and allows battery voltage and current to be supplied to the EMBs and EBCM, which supplies voltage to the motors and solenoids.

DTC Will Set When:

DTC 18 can be set after the EBCM commands the enable relay "ON." This test monitors the availability of current/voltage to the motors, solenoids and EMBs. A malfunction of this type will not allow voltage to be available to the motors, solenoids and EMBs; therefore, ABS operation would not be allowed if required.

Action Taken:

A malfunction DTC is stored, ABS is disabled and the ABS warning lamp is turned "ON."

DTC Chart Test Description:

Number(s) below refer to circled number(s) on the diagnostic chart.

1. Indicates if the EBCM is capable of controlling the enable relay as commanded.
2. With the enable relay removed, voltage should not be available at terminal "85". Any voltage at this point would indicate that the enable relay control circuit was shorted to a voltage source.
3. Checks for a short across the ABS enable relay coil.
4. Checks for an internal short in the ABS enable relay between the B+ feed at terminal "30" and the relay pull-in coil.
5. Ensures short to B+ was not due to physical damage of the enable relay control circuit.

Diagnostic Aids:

An "Intermittent" malfunction is most likely caused by a poor connection, rubbed through wire insulation, or a wire that is broken inside the insulation.

The frequency of the malfunction can be checked by using the enhanced diagnostic function of the Tech 1, as described in "Enhanced Diagnostics," found in this section.

Any circuitry, that is suspected as causing the intermittent complaint, should be thoroughly checked for backed out terminals, improper mating, broken locks, improperly formed or damaged terminals, poor terminal to wiring connections or physical damage to the wiring harness.

DIAGNOSTIC CHART DTC 18 — 1994–95 BERETTA AND CORSICA

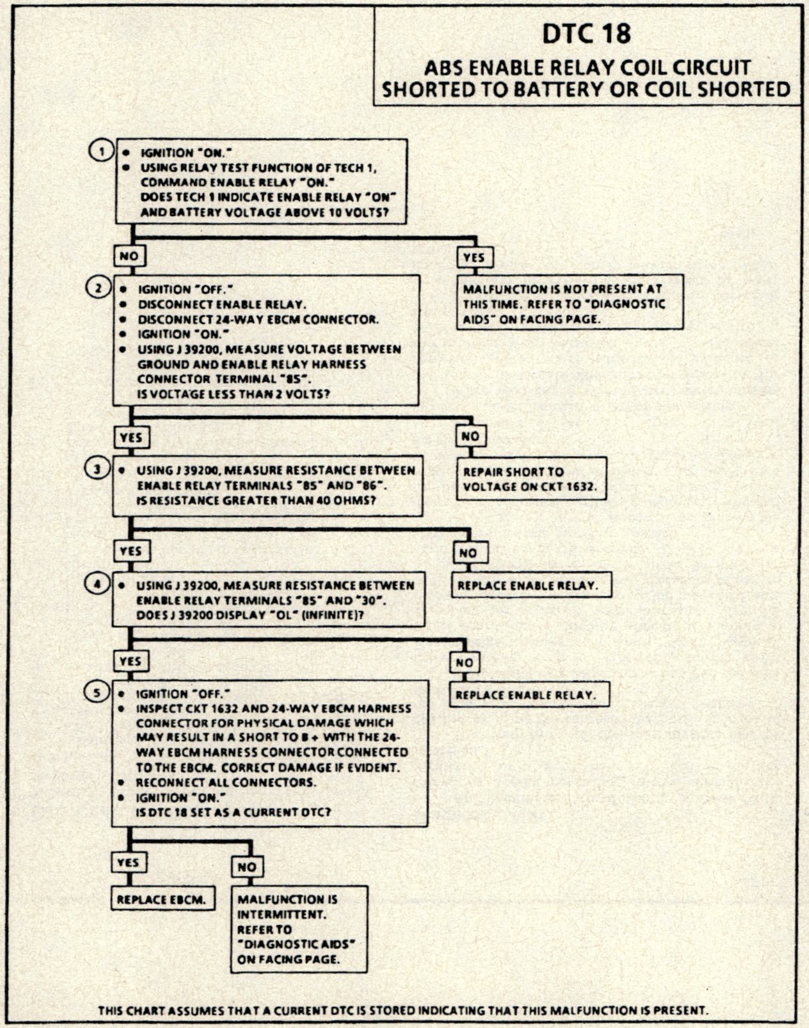

DIAGNOSTIC CHART DTC 21 — 1994–95 BERETTA AND CORSICA

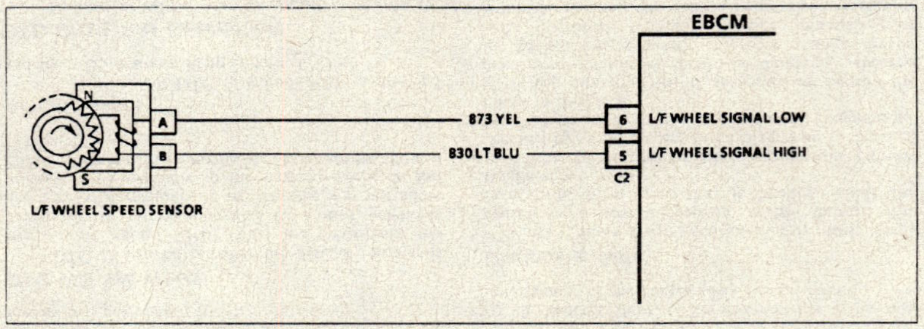

DTC 21
LEFT FRONT WHEEL SPEED = 0 OR UNREASONABLE

Circuit Description:

As a toothed ring passes by the wheel speed sensor, changes in the electromagnetic field cause the wheel speed sensor to produce a sinusoidal (AC) voltage signal whose frequency is proportional to wheel speed. The magnitude of this signal is directly related to wheel speed and the proximity of the wheel speed sensor to the toothed ring, often referred to as the air gap.

DTC Will Set When:

DTC 21 can be set when the vehicle is not in an ABS stop. If the left front wheel speed is less than one half of the vehicle speed and the vehicle speed is greater than 8 km/h (5 mph), a malfunction exists.

Action Taken:

A malfunction DTC is stored, ABS is disabled and the ABS warning lamp is turned "ON".

DIAGNOSTIC CHART DTC 21 — 1994–95 BERETTA AND CORSICA

DTC Chart Test Description:

Number(s) below refer to circled number(s) on the diagnostic chart.

1. Identifies if a wheel speed sensor or circuitry is damaged and visibly apparent.
2. Checks wheel speed sensor for proper resistance values.
3. Ensures the wheel speed sensor and toothed ring generate the proper voltage.
4. Ensures the wheel speed sensor is not shorted to ground.
5. Checks for proper voltages at speed sensor harness connector.
6. Ensures that the wheel speed sensor circuitry is not shorted to ground.
7. Checks for an open in the wheel signal low circuit.
8. Checks for an open in the wheel signal high circuit.
9. Ensures that DTC 21 was not set due to a poor connection at the EBCM or physical damage of the circuitry.
10. Checks for a short to B+ in either the high or low wheel signal circuits.
11. Ensures short to B+ was not due to physical damage of the circuitry.
12. Tests for a short between the wheel signal high and low circuits.
13. Ensures malfunction was not due to a poor connection at the wheel speed sensor or physical damage of the circuitry.

Diagnostic Aids:

An "Intermittent" malfunction is most likely caused by a poor connection, rubbed through wire insulation, or a wire that is broken inside the insulation.

The frequency of the malfunction can be checked by using the enhanced diagnostic function of the Tech 1, as described in "Enhanced Diagnostics," found in this section.

If the customer's comments reflect that the ABS amber warning lamp is "ON" only during moist environmental changes, (rain, snow, vehicle wash) all wheel speed sensor circuitry should be thoroughly inspected for signs of water intrusion. Use the following procedure. Spray down the suspected area with 5% salt water solution (two teaspoons of salt to 12 oz. of water). Drive vehicle above 24 km/h (15 mph) for at least 30 seconds. If DTC returns immediately, replace suspected harness.

Any circuitry, that is suspected as causing the intermittent complaint, should be thoroughly checked for backed out terminals, improper mating, broken locks, improperly formed or damaged terminals, poor terminal to wiring connections or physical damage to the wiring harness.

Resistance of the wheel speed sensor will increase with an increase in sensor temperature.

When replacing a wheel speed sensor, inspect the sensor terminals and harness connector for corrosion and/or water intrusion. If evidence of corrosion or water intrusion exists, replace wheel speed sensor harness. Refer to "On Vehicle Service" in this section.

If wheel speed sensor harness shows evidence of water intrusion, inspect wheel speed sensor for water intrusion. If water intrusion is evident, replace wheel speed sensor.

DIAGNOSTIC CHART DTC 21 — 1994–95 BERETTA AND CORSICA

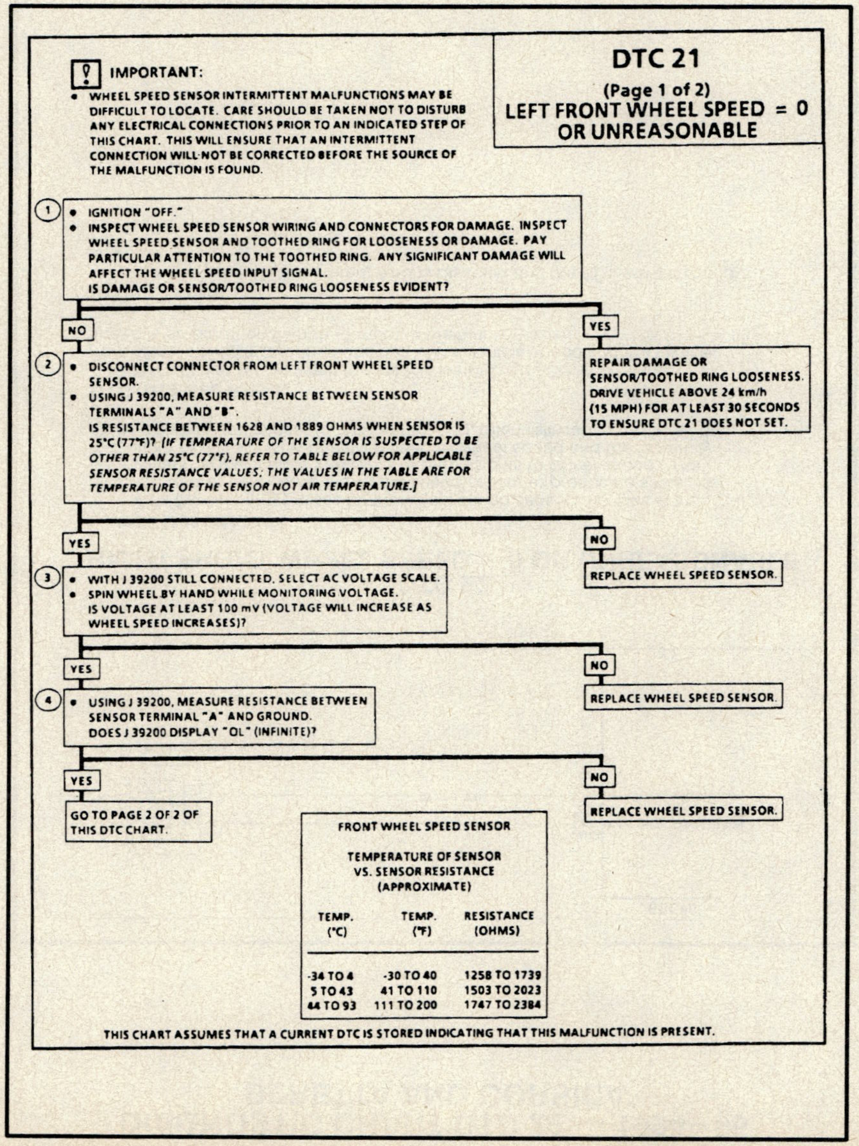

DIAGNOSTIC CHART DTC 21 — 1994–95 BERETTA AND CORSICA

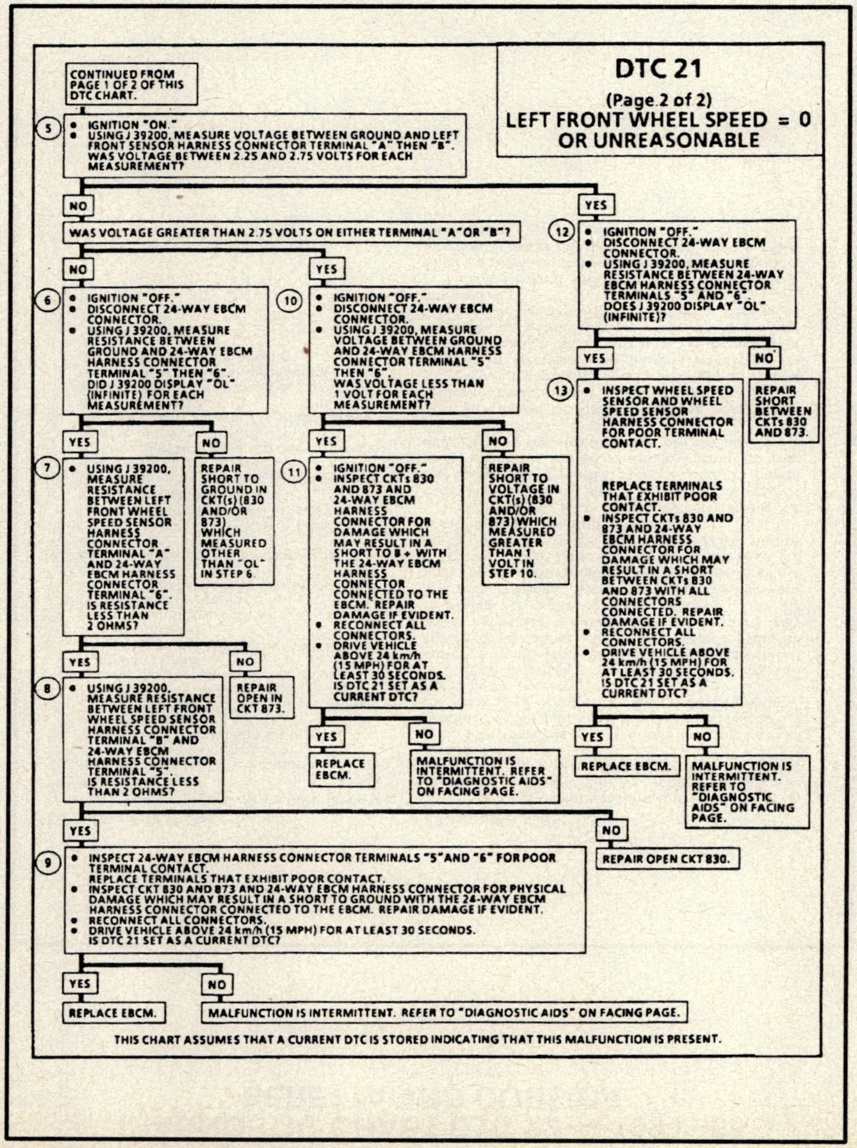

DIAGNOSTIC CHART DTC 22 — 1994–95 BERETTA AND CORSICA

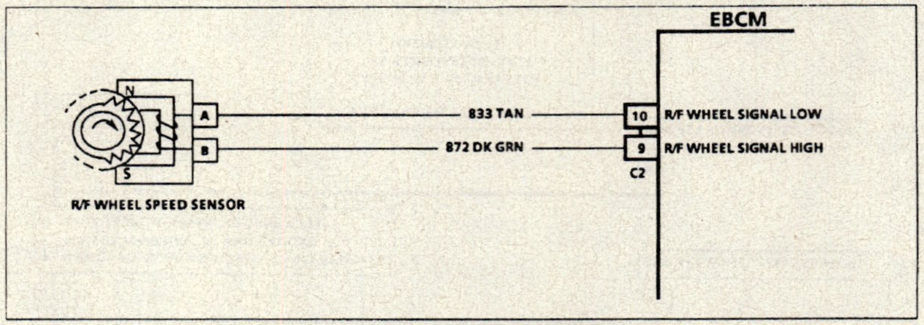

DTC 22
RIGHT FRONT WHEEL SPEED = 0 OR UNREASONABLE

Circuit Description:

As a toothed ring passes by the wheel speed sensor, changes in the electromagnetic field cause the wheel speed sensor to produce a sinusoidal (AC) voltage signal whose frequency is proportional to wheel speed. The magniture of this signal is directly related to wheel speed and the proximity of the wheel speed sensor to the toothed ring, often referred to as the air gap.

DTC Will Set When:

DTC 22 can be set when the vehicle is not in an ABS stop. If the front wheel speed is less than one half of the vehicle speed and the vehicle speed is greater than 8 km/h (5 mph), a malfunction exists.

Action Taken:

A malfunction DTC is stored, ABS is disabled and the ABS warning lamp is turned "ON".

DIAGNOSTIC CHART DTC 22 — 1994–95 BERETTA AND CORSICA

DTC Chart Test Description:

Number(s) below refer to circled number(s) on the diagnostic chart.

1. Identifies if a wheel speed sensor or circuitry is damaged and visibly apparent.
2. Checks wheel speed sensor for proper resistance values.
3. Ensures the wheel speed sensor and toothed ring generate the proper voltage.
4. Ensures the wheel speed sensor is not shorted to ground.
5. Checks for proper voltages at speed sensor harness connector.
6. Ensures that the wheel speed sensor circuitry is not shorted to ground.
7. Checks for an open in the wheel signal low circuit.
8. Checks for an open in the wheel signal high circuit.
9. Ensures that DTC 22 was not set due to a poor connection at the EBCM or physical damage of the circuitry.
10. Checks for a short to B+ in either the high or low wheel signal circuits.
11. Ensures short to B+ was not due to physical damage of the circuitry.
12. Tests for a short between the wheel signal high and low circuits.
13. Ensures malfunction was not due to a poor connection at the wheel speed sensor or physical damage of the circuitry.

Diagnostic Aids:

An "Intermittent" malfunction is most likely caused by a poor connection, rubbed through wire insulation, or a wire that is broken inside the insulation.

The frequency of the malfunction can be checked by using the enhanced diagnostic function of the Tech 1, as described in "Enhanced Diagnostics," found in this section.

If the customer's comments reflect that the ABS amber warning lamp is "ON" only during moist environmental changes, (rain, snow, vehicle wash) all wheel speed sensor circuitry should be thoroughly inspected for signs of water intrusion. Use the following procedure. Spray down the suspected area with 5% salt water solution (two teaspoons of salt to 12 oz. of water). Drive vehicle above 24 km/h (15 mph) for at least 30 seconds. If DTC returns immediately, replace suspected harness.

Any circuitry, that is suspected as causing the intermittent complaint, should be thoroughly checked for backed out terminals, improper mating, broken locks, improperly formed or damaged terminals, poor terminal to wiring connections or physical damage to the wiring harness.

Resistance of the wheel speed sensor will increase with an increase in sensor temperature.

When replacing a wheel speed sensor, inspect the sensor terminals and harness connector for corrosion and/or water intrusion. If evidence of corrosion or water intrusion exists, replace wheel speed sensor harness. Refer to "On Vehicle Service" in this section.

If wheel speed sensor harness shows evidence of water intrusion, inspect wheel speed sensor for water intrusion. If water intrusion is evident, replace wheel speed sensor.

DIAGNOSTIC CHART DTC 22 — 1994–95 BERETTA AND CORSICA

DTC 22
(Page 1 of 2)
RIGHT FRONT WHEEL SPEED = 0 OR UNREASONABLE

IMPORTANT:
- WHEEL SPEED SENSOR INTERMITTENT MALFUNCTIONS MAY BE DIFFICULT TO LOCATE. CARE SHOULD BE TAKEN NOT TO DISTURB ANY ELECTRICAL CONNECTIONS PRIOR TO AN INDICATED STEP OF THIS CHART. THIS WILL ENSURE THAT AN INTERMITTENT CONNECTION WILL NOT BE CORRECTED BEFORE THE SOURCE OF THE MALFUNCTION IS FOUND.

1.
- IGNITION "OFF."
- INSPECT WHEEL SPEED SENSOR WIRING AND CONNECTORS FOR DAMAGE. INSPECT WHEEL SPEED SENSOR AND TOOTHED RING FOR LOOSENESS OR DAMAGE. PAY PARTICULAR ATTENTION TO THE TOOTHED RING. ANY SIGNIFICANT DAMAGE WILL AFFECT THE WHEEL SPEED INPUT SIGNAL.

IS DAMAGE OR SENSOR/TOOTHED RING LOOSENESS EVIDENT?

- YES: REPAIR DAMAGE OR SENSOR/TOOTHED RING LOOSENESS. DRIVE VEHICLE ABOVE 24 km/h (15 MPH) FOR AT LEAST 30 SECONDS TO ENSURE DTC 22 DOES NOT SET.
- NO: Go to step 2.

2.
- DISCONNECT CONNECTOR FROM RIGHT FRONT WHEEL SPEED SENSOR.
- USING J 39200, MEASURE RESISTANCE BETWEEN SENSOR TERMINALS "A" AND "B".

IS RESISTANCE BETWEEN 1628 AND 1889 OHMS WHEN SENSOR IS 25°C (77°F)? *(IF TEMPERATURE OF THE SENSOR IS SUSPECTED TO BE OTHER THAN 25°C (77°F), REFER TO TABLE BELOW FOR APPLICABLE SENSOR RESISTANCE VALUES; THE VALUES IN THE TABLE ARE FOR TEMPERATURE OF THE SENSOR NOT AIR TEMPERATURE.)*

- YES: Go to step 3.
- NO: REPLACE WHEEL SPEED SENSOR.

3.
- WITH J 39200 STILL CONNECTED, SELECT AC VOLTAGE SCALE.
- SPIN WHEEL BY HAND WHILE MONITORING VOLTAGE.

IS VOLTAGE AT LEAST 100 mV (VOLTAGE WILL INCREASE AS WHEEL SPEED INCREASES)?

- YES: Go to step 4.
- NO: REPLACE WHEEL SPEED SENSOR.

4.
- USING J 39200, MEASURE RESISTANCE BETWEEN SENSOR TERMINAL "A" AND GROUND.

DOES J 39200 DISPLAY "OL" (INFINITE)?

- YES: GO TO PAGE 2 OF 2 OF THIS DTC CHART.
- NO: REPLACE WHEEL SPEED SENSOR.

FRONT WHEEL SPEED SENSOR
TEMPERATURE OF SENSOR VS. SENSOR RESISTANCE (APPROXIMATE)

TEMP. (°C)	TEMP. (°F)	RESISTANCE (OHMS)
-34 TO 4	-30 TO 40	1258 TO 1739
5 TO 43	41 TO 110	1503 TO 2023
44 TO 93	111 TO 200	1747 TO 2384

THIS CHART ASSUMES THAT A CURRENT DTC IS STORED INDICATING THAT THIS MALFUNCTION IS PRESENT.

DIAGNOSTIC CHART DTC 22 — 1994–95 BERETTA AND CORSICA

DTC 22
(Page 2 of 2)
RIGHT FRONT WHEEL SPEED = 0 OR UNREASONABLE

CONTINUED FROM PAGE 1 OF 2 OF THIS DTC CHART.

5.
- IGNITION "ON."
- USING J 39200, MEASURE VOLTAGE BETWEEN GROUND AND RIGHT FRONT SENSOR HARNESS CONNECTOR TERMINAL "A" THEN "B".

WAS VOLTAGE BETWEEN 2.25 AND 2.75 VOLTS FOR EACH MEASUREMENT?

- YES: Go to step 12.
- NO: WAS VOLTAGE GREATER THAN 2.75 VOLTS ON EITHER TERMINAL "A" OR "B"?
  - NO: Go to step 6.
  - YES: Go to step 10.

6.
- IGNITION "OFF."
- DISCONNECT 24-WAY EBCM CONNECTOR.
- USING J 39200, MEASURE RESISTANCE BETWEEN GROUND AND 24-WAY EBCM HARNESS CONNECTOR TERMINAL "9" THEN "10".

DID J 39200 DISPLAY "OL" (INFINITE) FOR EACH MEASUREMENT?

- YES: Go to step 7.
- NO: REPAIR SHORT TO GROUND IN CKT(S) (833 AND/OR 872) WHICH MEASURED OTHER THAN "OL" IN STEP 6.

7.
- USING J 39200, MEASURE RESISTANCE BETWEEN RIGHT FRONT WHEEL SPEED SENSOR HARNESS CONNECTOR TERMINAL "A" AND 24-WAY EBCM HARNESS CONNECTOR TERMINAL "10".

IS RESISTANCE LESS THAN 2 OHMS?

- YES: Go to step 8.
- NO: REPAIR OPEN IN CKT 833.

8.
- USING J 39200, MEASURE RESISTANCE BETWEEN RIGHT FRONT WHEEL SPEED SENSOR HARNESS CONNECTOR TERMINAL "B" AND 24-WAY EBCM HARNESS CONNECTOR TERMINAL "9".

IS RESISTANCE LESS THAN 2 OHMS?

- YES: Go to step 9.
- NO: REPAIR OPEN IN CKT 872.

9.
- INSPECT 24-WAY EBCM HARNESS CONNECTOR TERMINALS "9" AND "10" FOR POOR TERMINAL CONTACT. REPLACE TERMINALS THAT EXHIBIT POOR CONTACT.
- INSPECT CKT 833 AND 872 AND 24-WAY EBCM HARNESS CONNECTOR FOR PHYSICAL DAMAGE WHICH MAY RESULT IN A SHORT TO GROUND WITH THE 24-WAY EBCM HARNESS CONNECTOR CONNECTED TO THE EBCM. REPAIR DAMAGE IF EVIDENT.
- DRIVE VEHICLE ABOVE 24 km/h (15 MPH) FOR AT LEAST 30 SECONDS.

IS DTC 22 SET AS A CURRENT DTC?

- YES: REPLACE EBCM.
- NO: MALFUNCTION IS INTERMITTENT. REFER TO "DIAGNOSTIC AIDS" ON FACING PAGE.

10.
- IGNITION "OFF."
- DISCONNECT 24-WAY EBCM CONNECTOR.
- IGNITION "ON."
- USING J 39200, MEASURE VOLTAGE BETWEEN GROUND AND 24-WAY EBCM HARNESS CONNECTOR TERMINAL "9" THEN "10".

WAS VOLTAGE LESS THAN 1 VOLT FOR EACH MEASUREMENT?

- YES: Go to step 11.
- NO: REPAIR SHORT TO VOLTAGE IN CKT(S) (833 AND/OR 872) WHICH MEASURED GREATER THAN 1 VOLT IN STEP 10.

11.
- IGNITION "OFF."
- INSPECT CKTS 833 AND 872 AND 24-WAY EBCM HARNESS CONNECTOR FOR DAMAGE WHICH MAY RESULT IN A SHORT TO B+ WITH THE 24-WAY EBCM HARNESS CONNECTOR CONNECTED TO THE EBCM. REPAIR DAMAGE IF EVIDENT.
- RECONNECT ALL CONNECTORS.
- DRIVE VEHICLE ABOVE 24 km/h (15 MPH) FOR AT LEAST 30 SECONDS.

IS DTC 22 SET AS A CURRENT DTC?

- YES: REPLACE EBCM.
- NO: MALFUNCTION IS INTERMITTENT. REFER TO "DIAGNOSTIC AIDS" ON FACING PAGE.

12.
- IGNITION "OFF."
- DISCONNECT 24-WAY EBCM CONNECTOR.
- USING J 39200, MEASURE RESISTANCE BETWEEN 24-WAY EBCM HARNESS CONNECTOR TERMINALS "9" AND "10".

DOES J 39200 DISPLAY "OL" (INFINITE)?

- YES: Go to step 13.
- NO: REPAIR SHORT TO BETWEEN CKTS 833 AND 872.

13.
- INSPECT WHEEL SPEED SENSOR AND WHEEL SPEED SENSOR HARNESS CONNECTOR FOR POOR TERMINAL CONTACT.
- REPLACE TERMINALS THAT EXHIBIT POOR CONTACT.
- INSPECT CKTs 833 AND 872 AND 24-WAY EBCM HARNESS CONNECTOR FOR DAMAGE WHICH MAY RESULT IN A SHORT BETWEEN CKTs 833 AND 872 WITH ALL CONNECTORS CONNECTED. REPAIR DAMAGE IF EVIDENT.
- RECONNECT ALL CONNECTORS.
- DRIVE VEHICLE ABOVE 24 km/h (15 MPH) FOR AT LEAST 30 SECONDS.

IS DTC 22 SET AS A CURRENT DTC?

- YES: REPLACE EBCM.
- NO: MALFUNCTION IS INTERMITTENT. REFER TO "DIAGNOSTIC AIDS" ON FACING PAGE.

DIAGNOSTIC CHART DTC 23 — 1994–95 BERETTA AND CORSICA

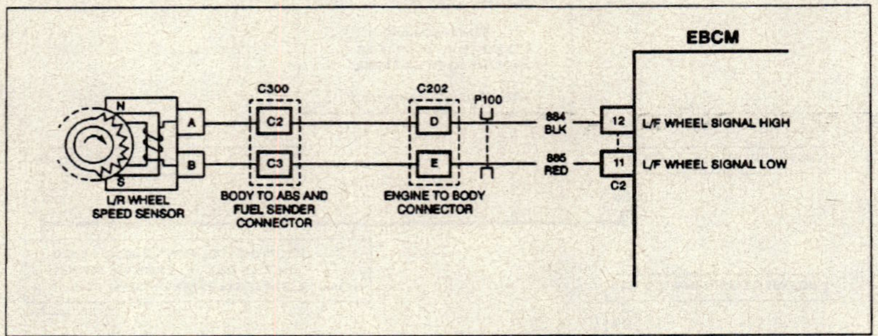

DTC 23
LEFT REAR WHEEL SPEED = 0 OR UNREASONABLE

Circuit Description:

As a toothed ring passes by the wheel speed sensor, changes in the electromagnetic field cause the wheel speed sensor to produce a sinusoidal (AC) voltage signal whose frequency is proportional to wheel speed. The magnitude of this signal is directly related to wheel speed and the proximity of the wheel speed sensor to the toothed ring, often referred to as the air gap.

DTC Will Set When:

DTC 23 can be set when the vehicle is not in an ABS stop. If the left rear wheel speed is less than one half of the vehicle speed and the vehicle speed is greater than 8 km/h (5 mph), a malfunction exists.

Action Taken:

A malfunction DTC is stored, ABS is disabled and the ABS warning lamp is turned "ON".

DIAGNOSTIC CHART DTC 23 — 1994–95 BERETTA AND CORSICA

DTC Chart Test Description:

Number(s) below refer to circled number(s) on the diagnostic chart.

1. This step will identify circuitry that is damaged and visibly apparent.
2. Checks for proper voltages at the rear sensor harness connector.
3. Checks for a short between the wheel signal high and low circuit.
4. This test checks wheel speed sensor for proper resistance values.
5. This step ensures the wheel speed sensor is not shorted to ground.
6. This test ensures the wheel speed sensor and sensor ring generate the proper voltage.
7. Ensures malfunction was not due to physical damage of the circuitry.
8. This step checks for proper voltages at body to ABS and fuel sender harness connector.
9. Checks for proper voltages at the engine to body harness connector.
10. Checks for short to ground in the wheel signal circuit.
11. This checks for an open in the wheel signal high circuit.
12. Checks wheel speed circuits for short to B+ at engine harness.
13. Checks for an open in the wheel signal low circuit.
14. Ensures malfunction was not due to physical damage of the circuitry.

Diagnostic Aids:

An "Intermittent" malfunction is most likely caused by a poor connection, rubbed through wire insulation, or a wire that is broken inside the insulation.

The frequency of the malfunction can be checked by using the enhanced diagnostic function of the Tech 1, as described in "Enhanced Diagnostics," found in this section.

If the customer's comments reflect that the ABS amber warning lamp is "ON" only during moist environmental changes, (rain, snow, vehicle wash) all wheel speed sensor circuitry should be thoroughly inspected for signs of water intrusion. Use the following procedure. Spray down the suspected area with 5% salt water solution (two teaspoons of salt to 12 oz. of water). Drive vehicle above 24 km/h (15 mph) for at least 30 seconds. If DTC returns immediately, replace suspected harness. Any circuitry, that is suspected as causing the intermittent complaint, should be thoroughly checked for backed out terminals, improper mating, broken locks, improperly formed or damaged terminals, poor terminal to wiring connections or physical damage to the wiring harness.

Resistance of the wheel speed sensor will increase with an increase in sensor temperature. When replacing a wheel speed sensor, inspect the sensor terminals and harness connector for corrosion and/or water intrusion. If evidence or corrosion or water intrusion exists, replace wheel speed sensor harness.

If wheel speed sensor harness shows evidence of water intrusion, inspect wheel speed sensor for water intrusion. If water intrusion is evident, replace wheel speed sensor.

DIAGNOSTIC CHART DTC 23 — 1994–95 BERETTA AND CORSICA

DTC 23
(Page 1 of 4)
LEFT REAR WHEEL SPEED = 0 OR UNREASONABLE

Important
- WHEEL SPEED SENSOR INTERMITTENT MALFUNCTIONS MAY BE DIFFICULT TO LOCATE. CARE SHOULD BE TAKEN NOT TO DISTURB ANY ELECTRICAL CONNECTIONS PRIOR TO AN INDICATED STEP OF THIS CHART. THIS WILL ENSURE THAT AN INTERMITTENT CONNECTION WILL NOT BE CORRECTED BEFORE THE SOURCE OF THE MALFUNCTION IS FOUND.

(1)
- IGNITION "OFF."
- INSPECT WHEEL SPEED SENSOR WIRING AND CONNECTORS FOR DAMAGE. IS DAMAGE EVIDENT?

YES: REPAIR DAMAGE. DRIVE VEHICLE ABOVE 24 km/h (15 MPH) FOR AT LEAST 30 SECONDS TO ENSURE DTC 23 DOES NOT SET.

NO: go to step 2.

(2)
- REMOVE LEFT REAR INTEGRAL BEARING AND SPEED SENSOR ASSEMBLY (TUBULAR AXLE ONLY).
- DISCONNECT CONNECTOR FROM LEFT REAR WHEEL SPEED SENSOR.
- IGNITION "ON."
- USING J 39200, MEASURE VOLTAGE BETWEEN GROUND AND LEFT REAR SENSOR HARNESS CONNECTOR TERMINAL "A" THEN "B". WAS VOLTAGE BETWEEN 2.25 AND 2.75 VOLTS FOR EACH MEASUREMENT?

NO: GO TO PAGE 3 OF 4 OF THIS DTC CHART.

YES: go to step 3.

(3)
- IGNITION "OFF."
- DISCONNECT 24-WAY EBCM CONNECTOR.
- USING J 39200, MEASURE RESISTANCE BETWEEN 24-WAY EBCM HARNESS CONNECTOR TERMINALS "11" AND "12." DOES J 39200 DISPLAY "OL" (INFINITE)?

NO: REPAIR SHORT BETWEEN CKTs 884 AND 885.

YES: go to step 4.

(4)
- USING J 39200, MEASURE THE RESISTANCE BETWEEN LEFT REAR WHEEL SPEED SENSOR TERMINALS "A" AND "B". IS RESISTANCE BETWEEN 2100 AND 2400 OHMS WHEN SENSOR IS 25°C (77°F)? (IF TEMPERATURE OF THE SENSOR IS SUSPECTED TO BE OTHER THAN 25°C (77°F). REFER TO THE TABLE BELOW FOR APPLICABLE SENSOR RESISTANCE VALUES; THE VALUES IN THE TABLE ARE FOR TEMPERATURE OF THE SENSOR NOT AIR TEMPERATURE).

NO: REPLACE INTERGRAL BEARING AND SPEED SENSOR ASSEMBLY.

YES: go to step 5.

(5)
- USING J 39200, MEASURE RESISTANCE BETWEEN SENSOR TERMINAL "A" AND INTEGRAL BEARING AND SPEED SENSOR ASSEMBLY MOUNTING FLANGE. DOES J 39200 DISPLAY "OL" (INFINITE)?

YES: CONTINUE DIAGNOSIS ON NEXT PAGE.

NO: REPLACE INTEGRAL BEARING AND SPEED SENSOR ASSEMBLY.

REAR WHEEL SPEED SENSOR
TEMPERATURE OF SENSOR VS. SENSOR RESISTANCE (APPROXIMATE)

TEMP. (°C)	TEMP. (°F)	RESISTANCE (OHMS)
-34 TO 4	-30 TO 40	1622 TO 2205
5 TO 43	41 TO 110	1930 TO 2566
44 TO 93	111 TO 200	2245 TO 3028

THIS CHART ASSUMES THAT A CURRENT DTC IS STORED INDICATING THAT THIS MALFUNCTION IS PRESENT.

DIAGNOSTIC CHART DTC 23 — 1994–95 BERETTA AND CORSICA

DTC 23
(Page 2 of 4)
LEFT REAR WHEEL SPEED = 0 OR UNREASONABLE

DIAGNOSIS CONTINUED FROM PAGE 1 OF 4 OF THIS DTC CHART.

(6)
- INSPECT WHEEL SPEED SENSOR HARNESS CONNECTOR AND 24-WAY EBCM HARNESS CONNECTOR TERMINALS "11" AND "12" FOR POOR TERMINAL CONTACT.

  REPLACE TERMINALS THAT EXHIBIT POOR CONTACT RECONNECT WHEEL SPEED SENSOR.
- INSTALL INTEGRAL BEARING AND SPEED SENSOR ASSEMBLY (TUBULAR AXLE ONLY).
- USING J 39200, SELECT A/C VOLTAGE SCALE AND CONNECT BETWEEN 24-WAY EBCM HARNESS CONNECTOR TERMINALS "11" AND "12".
- SPIN WHEEL BY HAND WHILE MONITORING VOLTAGE. IS VOLTAGE AT LEAST 100 mV (VOLTAGE WILL INCREASE AS WHEEL SPEED INCREASES)?

NO: REPLACE INTEGRAL REARING AND SPEED SENSOR ASSEMBLY.

YES: go to step 7.

(7)
- INSPECT CKTs 884 AND 885 FOR PHYSICAL DAMAGE WHICH MAY RESULT IN A SHORT BETWEEN THE CIRCUITS WITH ALL CONNECTORS CONNECTED. REPAIR DAMAGE IF EVIDENT.
- RECONNECT ALL CONNECTORS.
- DRIVE VEHICLE ABOVE 24-WAY km/h (15 MPH) FOR AT LEAST 30 SECONDS IS DTC 23 SET AS A CURRENT DTC?

YES: REPLACE EBCM.

NO: MALFUNCTION IS INTERMITTENT. REFER TO "DIAGNOSTIC AIDS" ON FACING PAGE.

THIS CHART ASSUMES THAT A CURRENT DTC IS STORED INDICATING THAT THIS MALFUNCTION IS PRESENT.

DIAGNOSTIC CHART DTC 23 — 1994–95 BERETTA AND CORSICA

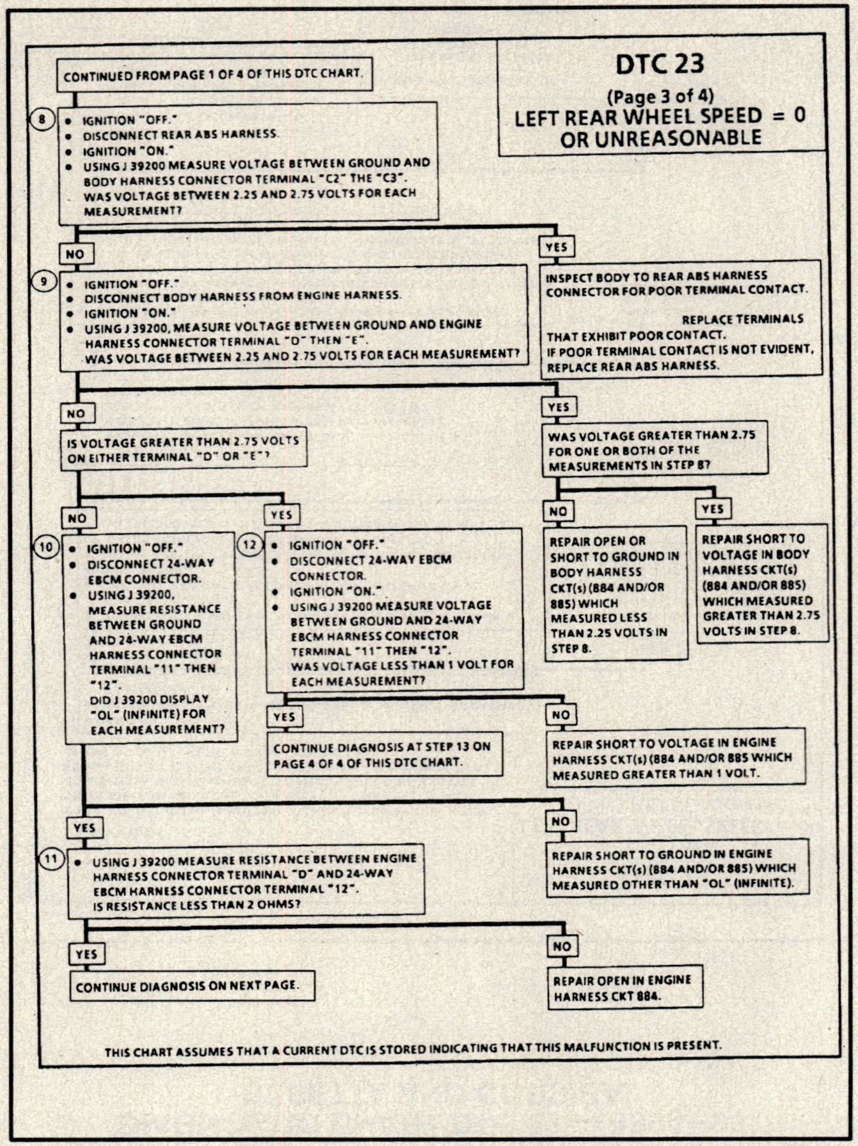

DIAGNOSTIC CHART DTC 23 — 1994–95 BERETTA AND CORSICA

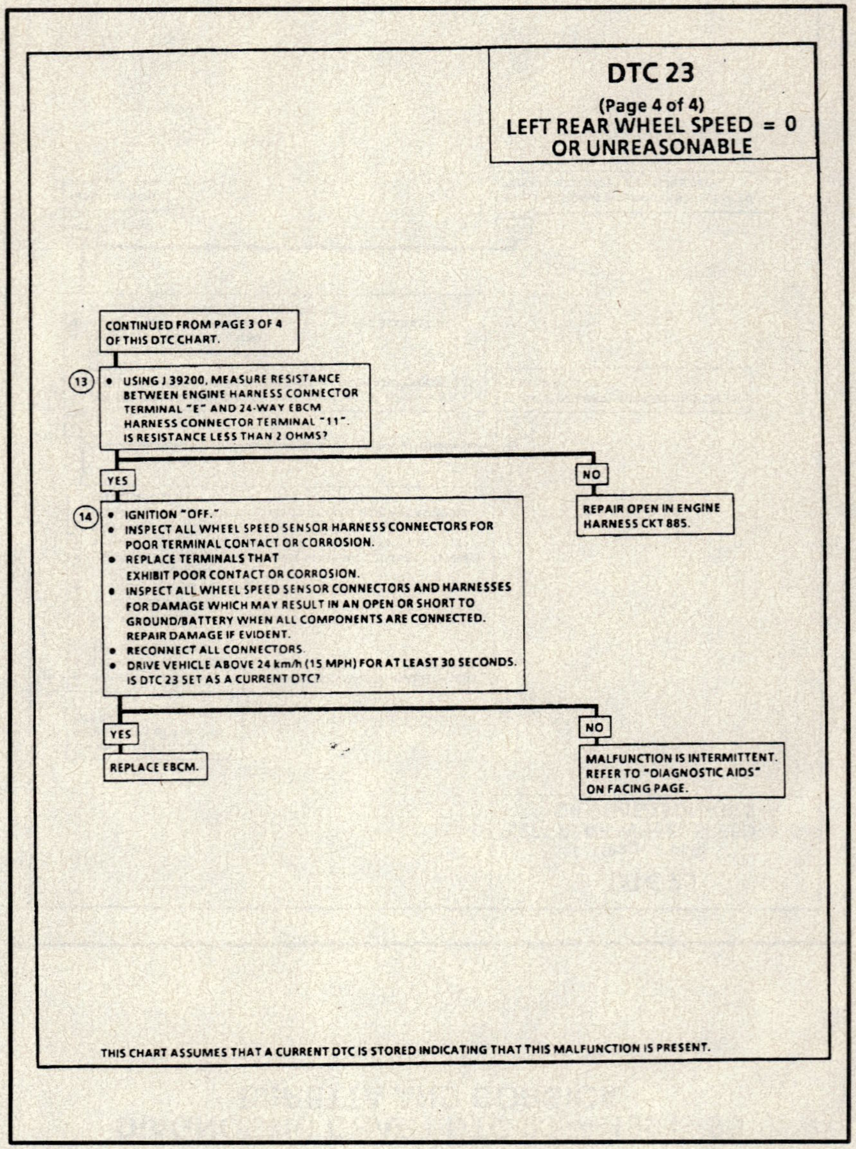

DIAGNOSTIC CHART DTC 24 — 1994–95 BERETTA AND CORSICA

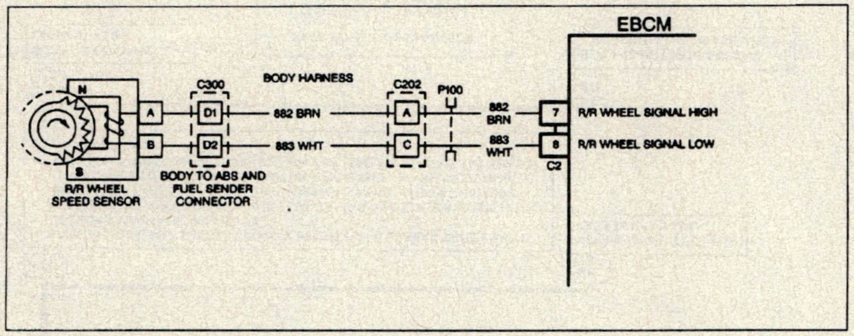

DTC 24
RIGHT REAR WHEEL SPEED = 0 OR UNREASONABLE

Circuit Description:

As a toothed ring passes by the wheel speed sensor, changes in the electromagnetic field cause the wheel speed sensor to produce a sinusoidal (AC) voltage signal whose frequency is proportional to wheel speed. The magnitude of this signal is directly related to wheel speed and the proximity of the wheel speed sensor to the toothed ring, often referred to as the air gap.

DTC Will Set When:

DTC 24 can be set when the vehicle is not in an ABS stop. If the right rear wheel speed is less than one half of the vehicle speed and the vehicle speed is greater than 8 km/h (5 mph), a malfunction exists.

Action Taken:

A malfunction DTC is stored, ABS is disabled and the ABS warning lamp is turned "ON".

DIAGNOSTIC CHART DTC 24 — 1994–95 BERETTA AND CORSICA

DTC Chart Test Description:

Number(s) below refer to circled number(s) on the diagnostic chart.

1. This step will identify circuitry that is damaged and visibly apparent.
2. Checks for proper voltages at the rear sensor harness connector.
3. Checks for a short between the wheel signal high and low circuit.
4. This test checks wheel speed sensor for proper resistance values.
5. This step ensures the wheel speed sensor is not shorted to ground.
6. This test ensures the wheel speed sensor and sensor ring generate the proper voltage.
7. Ensures malfunction was not due to physical damage of the circuitry.
8. This step checks for proper voltages at body to ABS and fuel sender harness connector.
9. Checks for proper voltages at the engine to body harness connector.
10. Checks for short to ground in the wheel signal circuit.
11. This checks for an open in the wheel signal high circuit.
12. Checks wheel speed circuits for short to B+ at engine harness.
13. Checks for an open in the wheel signal low circuit.
14. Ensures malfunction was not due to physical damage of the circuitry.

Diagnostic Aids:

An "Intermittent" malfunction is most likely caused by a poor connection, rubbed through wire insulation, or a wire that is broken inside the insulation.

The frequency of the malfunction can be checked by using the enhanced diagnostic function of the Tech 1, as described in "Enhanced Diagnostics," found in this section.

If the customer's comments reflect that the ABS amber warning lamp is "ON" only during moist environmental changes, (rain, snow, vehicle wash) all wheel speed sensor circuitry should be thoroughly inspected for water intrusion. Use the following procedure. Spray down the suspected area with 5% salt water solution (two teaspoons of salt to 12 oz. of water). Drive vehicle above 24 km/h (15 mph) for at least 30 seconds. If DTC returns immediately, replace suspected harness.

Any circuitry, that is suspected as causing the intermittent complaint, should be thoroughly checked for backed out terminals, improper mating, broken locks, improperly formed or damaged terminals, poor terminal to wiring connections or physical damage to the wiring harness.

Resistance of the wheel speed sensor will increase with an increase in sensor temperature. When replacing a wheel speed sensor, inspect the sensor terminals and harness connector for corrosion and/or water intrusion. If evidence of corrosion or water intrusion exists, replace wheel speed sensor harness. Refer to "On Vehicle Service" in this section.

If wheel speed sensor harness shows evidence of water intrusion, inspect wheel speed sensor for water intrusion. If water intrusion is evident, replace wheel speed sensor.

DIAGNOSTIC CHART DTC 24 — 1994–95 BERETTA AND CORSICA

DTC 24
(Page 1 of 4)
RIGHT REAR WHEEL SPEED = 0 OR UNREASONABLE

Important
- WHEEL SPEED SENSOR INTERMITTENT MALFUNCTIONS MAY BE DIFFICULT TO LOCATE. CARE SHOULD BE TAKEN NOT TO DISTURB ANY ELECTRICAL CONNECTIONS PRIOR TO AN INDICATED STEP OF THIS CHART. THIS WILL ENSURE THAT AN INTERMITTENT CONNECTION WILL NOT BE CORRECTED BEFORE THE SOURCE OF THE MALFUNCTION IS FOUND.

(1)
- IGNITION "OFF."
- INSPECT WHEEL SPEED SENSOR WIRING AND CONNECTORS FOR DAMAGE.

IS DAMAGE EVIDENT?

YES: REPAIR DAMAGE. DRIVE VEHICLE ABOVE 24 km/h (15MPH) FOR AT LEAST 30 SECONDS TO ENSURE DTC 24 DOES NOT SET.

NO:

(2)
- REMOVE RIGHT REAR INTEGRAL BEARING AND SPEED SENSOR ASSEMBLY (TUBULAR AXLE ONLY).
- DISCONNECT CONNECTOR FROM RIGHT REAR WHEEL SPEED SENSOR.
- IGNITION "ON."
- USING J 39200, MEASURE VOLTAGE BETWEEN GROUND AND RIGHT REAR SENSOR HARNESS CONNECTOR TERMINAL "A" THEN "B". WAS VOLTAGE BETWEEN 2.25 AND 2.75 VOLTS FOR EACH MEASUREMENT?

YES: GO TO PAGE 3 OF 4 OF THIS CHART.

NO:

(3)
- IGNITION "OFF."
- DISCONNECT 24-WAY EBCM CONNECTOR.
- USING J39200, MEASURE RESISTANCE BETWEEN 24-WAY EBCM HARNESS CONNECTOR TERMINALS "8" AND "7." DOES J 39200 DISPLAY "OL" (INFINITE).

NO: REPAIR SHORT BETWEEN CKTS 882 AND 883.

YES:

(4)
- USING J 39200, MEASURE RESISTANCE BETWEEN RIGHT REAR WHEEL SPEED SENSOR TERMINALS "A" AND "B." IS RESISTANCE BETWEEN 2100 AND 2400 OHMS WHEN SENSOR IS 25°C (77°F)? (IF TEMPERATURE OF THE SENSOR IS SUSPECTED TO BE OTHER THAN 25°C (77°F). REFER TO THE TABLE BELOW FOR APPLICABLE SENSOR RESISTANCE VALUES; THE VALUES IN THE TABLE ARE FOR TEMPERATURE OF THE SENSOR NOT AIR TEMPERATURE).

YES: CONTINUE DIAGNOSIS ON NEXT PAGE.

NO: REPLACE INTEGRAL BEARING AND SPEED SENSOR ASSEMBLY.

REAR WHEEL SPEED SENSOR
TEMPERATURE OF SENSOR VS. SENSOR RESISTANCE (APPROXIMATE)

TEMP. (°C)	TEMP (°F)	RESISTANCE (OHMS)
-34 TO 4	-30 TO 40	1623 TO 2210
5 TO 43	41 TO 110	1939 TO 2570
44 TO 93	111 TO 200	2254 TO 3029

THIS CHART ASSUMES THAT A CURRENT DTC IS STORED INDICATING THAT THIS MALFUNCTION IS PRESENT.

DIAGNOSTIC CHART DTC 24 — 1994–95 BERETTA AND CORSICA

DTC 24
(Page 2 of 4)
RIGHT REAR WHEEL SPEED = 0 OR UNREASONABLE

CONTINUED FROM PAGE 1 OF 4 OF THIS DTC CHART.

(5)
- USING J 39200, MEASURE RESISTANCE BETWEEN SENSOR TERMINAL "A" AND INTEGRAL BEARING AND SPEED SENSOR ASSEMBLY MOUNTING FLANGE. DOES J 39200 DISPLAY "OL" (INFINITE)?

NO: REPLACE INTEGRAL BEARING AND SPEED SENSOR ASSEMBLY.

YES:

(6)
- INSPECT WHEEL SPEED SENSOR HARNESS CONNECTOR AND 24-WAY EBCM HARNESS CONNECTOR TERMINALS "8" AND "7" FOR POOR TERMINAL CONTACT.

REPLACE TERMINALS THAT EXHIBIT POOR CONTACT.
- RECONNECT WHEEL SPEED SENSOR.
- INSTALL INTEGRAL BEARING AND SPEED SENSOR ASSEMBLY (TUBULAR AXLE ONLY).
- USING J 39200, SELECT A/C VOLTAGE SCALE AND CONNECT CONNECTOR TERMINALS "8" AND "7".
- SPIN WHEEL BY HAND WHILE MONITORING VOLTAGE. IS VOLTAGE AT LEAST 100 mV (VOLTAGE WILL INCREASE AS WHEEL SPEED INCREASES)?

NO: REPLACE INTEGRAL BEARING AND SPEED SENSOR ASSEMBLY.

YES:

(7)
- INSPECT CKTs 882 AND 883 FOR PHYSICAL DAMAGE WHICH MAY RESULT IN A SHORT BETWEEN THE CIRCUITS WITH ALL CONNECTORS CONNECTED REPAIR DAMAGE IF EVIDENT.
- RECONNECT ALL CONNECTORS.
- DRIVE VEHICLE ABOVE 24 km/h (15 MPH) FOR AT LEAST 30 SECONDS. IS DTC 24 SET AS A CURRENT DTC?

YES: REPLACE EBCM.

NO: MALFUNCTION IS INTERMITTENT. REFER TO "DIAGNOSTIC AIDS" ON FACING PAGE.

THIS CHART ASSUMES THAT A CURRENT DTC IS STORED INDICATING THAT THIS MALFUNCTION IS PRESENT.

DIAGNOSTIC CHART DTC 24 — 1994–95 BERETTA AND CORSICA

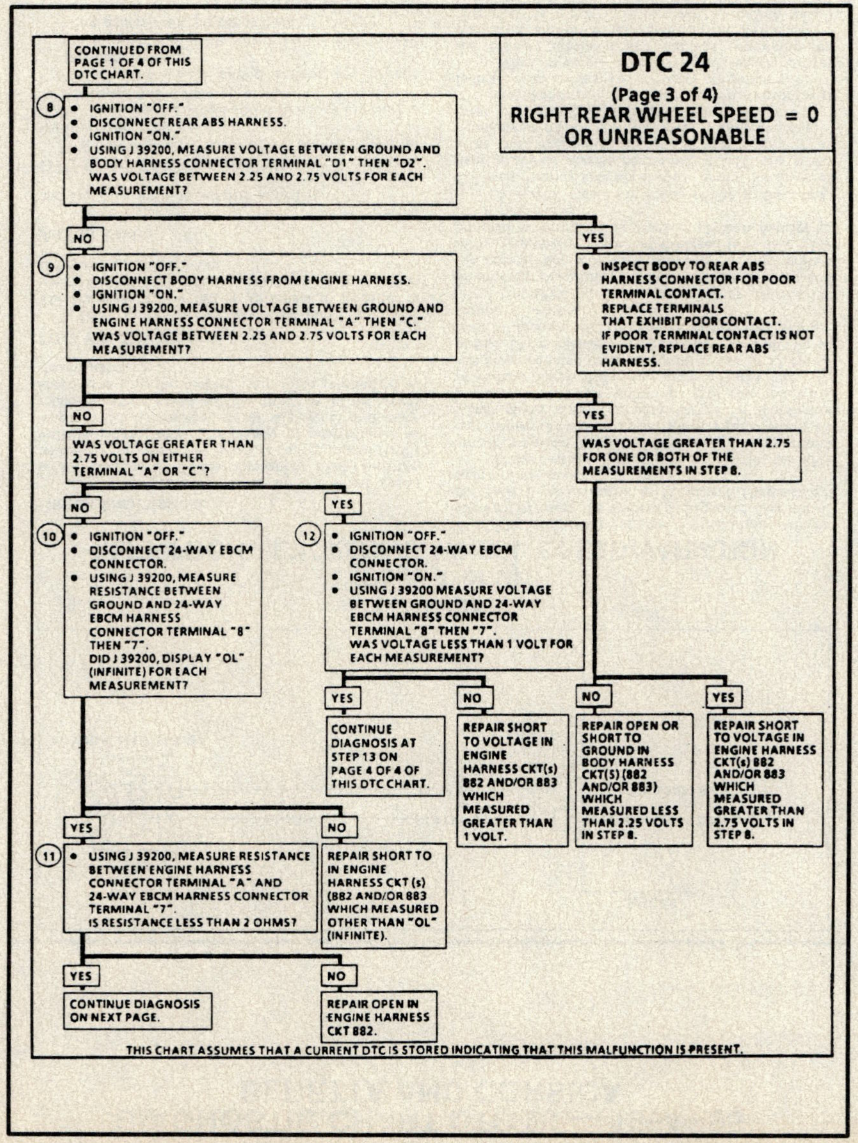

DIAGNOSTIC CHART DTC 24 — 1994–95 BERETTA AND CORSICA

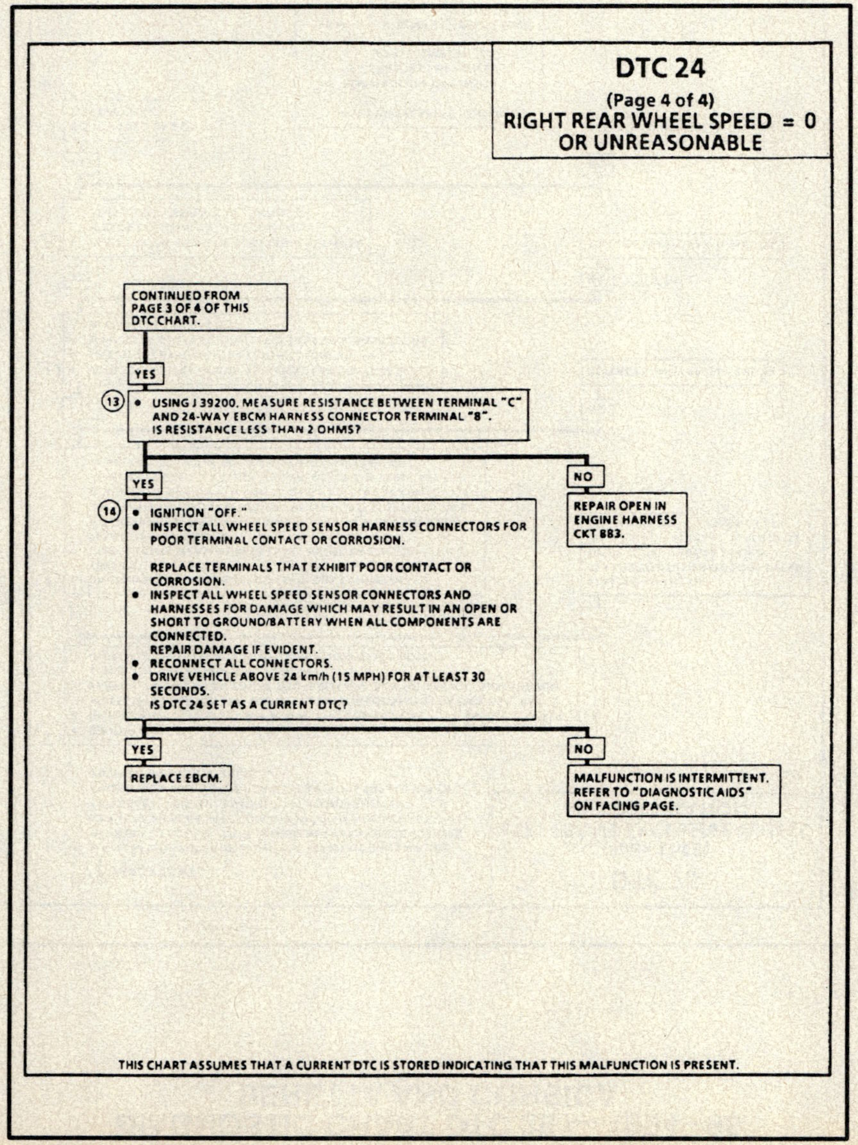

DIAGNOSTIC CHART DTC 25 — 1994–95 BERETTA AND CORSICA

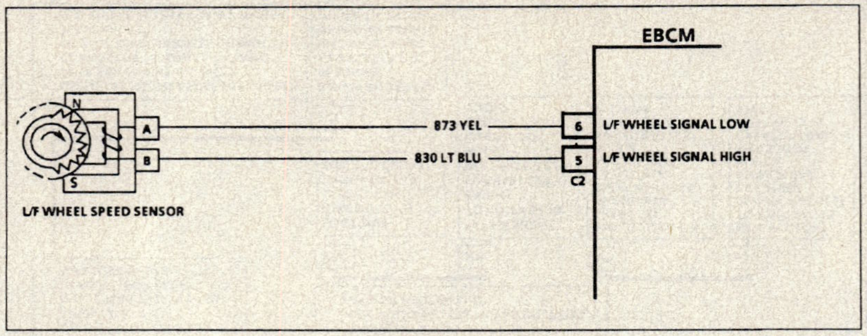

DTC 25
(Page 1 of 2)
LEFT FRONT EXCESSIVE WHEEL SPEED VARIATION

Circuit Description:

As a toothed ring passes by the wheel speed sensor, changes in the electromagnetic field cause the wheel speed sensor to produce a sinusoidal (AC) voltage signal whose frequency is proportional to wheel speed. The magnitude of this signal is directly related to wheel speed and the proximity of the wheel speed sensor to the toothed ring, often referred to as the air gap.

DTC Will Set When:

DTC 25 can be set when the brake is "OFF." The purpose of this test is to detect a situation in which the left front wheel acceleration or deceleration is beyond specified limits.

Action Taken:

A malfunction DTC is stored, ABS is disabled and the ABS warning lamp is turned "ON."

DTC Chart Test Description:

Number(s) below refer to circled number(s) on the diagnostic chart.

1. Identifies if a wheel speed sensor or circuitry is damaged and visibly apparent.
2. Checks wheel speed sensor for proper resistance values.
3. Ensures the wheel speed sensor and toothed ring generate the proper voltage.
4. Ensures the wheel speed sensor is not shorted to ground.

Diagnostic Aids:

An "Intermittent" malfunction is most likely caused by a poor connection, rubbed through wire insulation, or a wire that is broken inside the insulation.

The frequency of the malfunction can be checked by using the enhanced diagnostic function of the Tech 1, as described in "Enhanced Diagnostics," found in this section.

If the customer's comments reflect that the ABS amber warning lamp is "ON" only during moist environmental changes, (rain, snow, vehicle wash) all wheel speed sensor circuitry should be thoroughly inspected for water intrusion. Use the following procedure. Spray down the suspected area with 5% salt water solution (two teaspoons of salt to 12 oz. of water). Drive vehicle above 24 km/h (15 mph) for at least 30 seconds. If DTC returns immediately, replace suspected harness.

Any circuitry, that is suspected as causing the intermittent complaint, should be thoroughly checked for backed out terminals, improper mating, broken locks, improperly formed or damaged terminals, poor terminal to wiring connections or physical damage to the wiring harness.

A careful visual inspection of the toothed ring, CV joint, bearing and wheel speed sensor for physical damage should also be performed. If DTC sets at the same mph every time and wheel speed variation is noted above this mph on the Tech 1, the speed ring is most likely damaged. This can occur at any speed.

Resistance of the wheel speed sensor will increase with an increase in sensor temperature.

When replacing a wheel speed sensor, inspect the sensor terminals and harness connector for corrosion and/or water intrusion. If evidence of corrosion or water intrusion exists, replace wheel speed sensor harness. Refer to "On Vehicle Service" in this section.

If wheel speed sensor harness shows evidence of water intrusion, inspect wheel speed sensor for water intrusion. If water intrusion is evident, replace wheel speed sensor.

DIAGNOSTIC CHART DTC 25 — 1994–95 BERETTA AND CORSICA

DTC 25
(Page 1 of 2)
LEFT FRONT EXCESSIVE WHEEL SPEED VARIATION

IMPORTANT:
- WHEEL SPEED SENSOR INTERMITTENT MALFUNCTIONS MAY BE DIFFICULT TO LOCATE. CARE SHOULD BE TAKEN NOT TO DISTURB ANY ELECTRICAL CONNECTIONS PRIOR TO AN INDICATED STEP OF THIS CHART. THIS WILL ENSURE THAT AN INTERMITTENT CONNECTION WILL NOT BE CORRECTED BEFORE THE SOURCE OF THE MALFUNCTION IS FOUND.

(1)
- IGNITION "OFF."
- INSPECT WHEEL SPEED SENSOR WIRING AND CONNECTORS FOR DAMAGE. INSPECT WHEEL SPEED SENSOR AND TOOTHED RING FOR LOOSENESS OR DAMAGE. PAY PARTICULAR ATTENTION TO THE TOOTHED RING. ANY SIGNIFICANT DAMAGE WILL AFFECT THE WHEEL SPEED INPUT SIGNAL.

IS DAMAGE OR SENSOR/TOOTHED RING LOOSENESS EVIDENT?

YES: REPAIR DAMAGE OR SENSOR/TOOTHED RING LOOSENESS. DRIVE VEHICLE ABOVE 24 km/h (15 MPH) FOR AT LEAST 30 SECONDS TO ENSURE DTC 21 DOES NOT SET.

NO:

(2)
- DISCONNECT CONNECTOR FROM LEFT FRONT WHEEL SPEED SENSOR.
- USING J 39200, MEASURE RESISTANCE BETWEEN SENSOR TERMINALS "A" AND "B".

IS RESISTANCE BETWEEN 1628 AND 1889 OHMS WHEN SENSOR IS 25°C (77°F)? *[IF TEMPERATURE OF THE SENSOR IS SUSPECTED TO BE OTHER THAN 25°C (77°F), REFER TO TABLE BELOW FOR APPLICABLE SENSOR RESISTANCE VALUES. THE VALUES IN THE TABLE ARE FOR TEMPERATURE OF THE SENSOR NOT AIR TEMPERATURE.]*

NO: REPLACE WHEEL SPEED SENSOR.

YES:

(3)
- WITH J 39200 STILL CONNECTED, SELECT AC VOLTAGE SCALE.
- SPIN WHEEL BY HAND WHILE MONITORING VOLTAGE.

IS VOLTAGE AT LEAST 100 mV (VOLTAGE WILL INCREASE AS WHEEL SPEED INCREASES)?

NO: REPLACE WHEEL SPEED SENSOR.

YES:

(4)
- USING J 39200, MEASURE RESISTANCE BETWEEN SENSOR TERMINAL "A" AND GROUND.

DOES J 39200 DISPLAY "OL" (INFINITE)?

NO: REPLACE WHEEL SPEED SENSOR.

YES: GO TO PAGE 2 OF 2 OF THIS DTC CHART.

FRONT WHEEL SPEED SENSOR

TEMPERATURE OF SENSOR VS. SENSOR RESISTANCE (APPROXIMATE)

TEMP. (°C)	TEMP. (°F)	RESISTANCE (OHMS)
-34 TO 4	-30 TO 40	1258 TO 1739
5 TO 43	41 TO 110	1503 TO 2023
44 TO 93	111 TO 200	1747 TO 2384

THIS CHART ASSUMES THAT A CURRENT DTC IS STORED INDICATING THAT THIS MALFUNCTION IS PRESENT.

DIAGNOSTIC CHART DTC 25 — 1994–95 BERETTA AND CORSICA

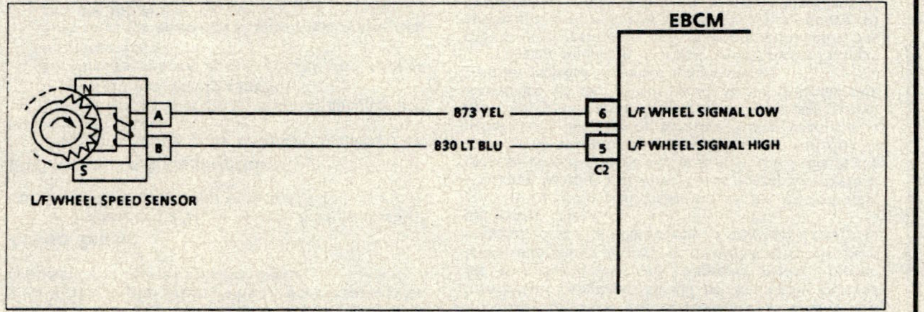

DTC 25 (Page 2 of 2)
LEFT FRONT EXCESSIVE WHEEL SPEED VARIATION

Circuit Description:

As a toothed ring passes by the wheel speed sensor, changes in the electromagnetic field cause the wheel speed sensor to produce a sinusoidal (AC) voltage signal whose frequency is proportional to wheel speed. The magnitude of this signal is directly related to wheel speed and the proximity of the wheel speed sensor to the toothed ring, often referred to as the air gap.

DTC Will Set When:

DTC 25 can be set when the brake is "OFF." The purpose of this test is to detect a situation in which the left front wheel acceleration or deceleration is beyond specified limits.

Action Taken:

A malfunction DTC is stored, ABS is disabled and the ABS warning lamp is turned "ON."

DTC Chart Test Description:

Number(s) below refer to circled number(s) on the diagnostic chart.

5. Checks for proper voltages at speed sensor harness connector.
6. Ensures that the wheel speed sensor circuitry is not shorted to ground.
7. Checks for an open in the wheel signal low circuit.
8. Checks for an open in the wheel signal high circuit.
9. Ensures that DTC 25 was not set due to a poor connection at the EBCM or physical damage of the circuitry.
10. Checks for a short to B+ in either the high or low wheel signal circuits.
11. Ensures short to B+ was not due to physical damage of the circuitry.
12. Tests for a short between the wheel signal high and low circuits.
13. Ensures malfunction was not due to a poor connection at the wheel speed sensor or physical damage of the circuitry.

Diagnostic Aids:

An "Intermittent" malfunction is most likely caused by a poor connection, rubbed through wire insulation, or a wire that is broken inside the insulation.

The frequency of the malfunction can be checked by using the enhanced diagnostic function of the Tech 1, as described in "Enhanced Diagnostics," found in this section.

If the customer's comments reflect that the ABS amber warning lamp is "ON" only during moist environmental changes, (rain, snow, vehicle wash) all wheel speed sensor circuitry should be thoroughly inspected for water intrusion. Use the following procedure. Spray down the suspected area with 5% salt water solution (two teaspoons of salt to 12 oz. of water). Drive vehicle above 24 km/h (15 mph) for at least 30 seconds. If DTC returns immediately, replace suspected harness.

Any circuitry, that is suspected as causing the intermittent complaint, should be thoroughly checked for backed out terminals, improper mating, broken locks, improperly formed or damaged terminals, poor terminal to wiring connections or physical damage to the wiring harness.

A careful visual inspection of the toothed ring, CV joint, bearing and wheel speed sensor for physical damage should also be performed. If DTC sets at the same mph every time and wheel speed variation is noted above this mph on the Tech 1, the speed ring is most likely damaged. This can occur at any speed. Resistance of the wheel speed sensor will increase with an increase in sensor temperature.

When replacing a wheel speed sensor, inspect the sensor terminals and harness connector for corrosion and/or water intrusion. If evidence of corrosion or water intrusion exists, replace wheel speed sensor harness. Refer to "On Vehicle Service" in this section.

If wheel speed sensor harness shows evidence of water intrusion, inspect wheel speed sensor for water intrusion. If water intrusion is evident, replace wheel speed sensor.

DIAGNOSTIC CHART DTC 25 — 1994–95 BERETTA AND CORSICA

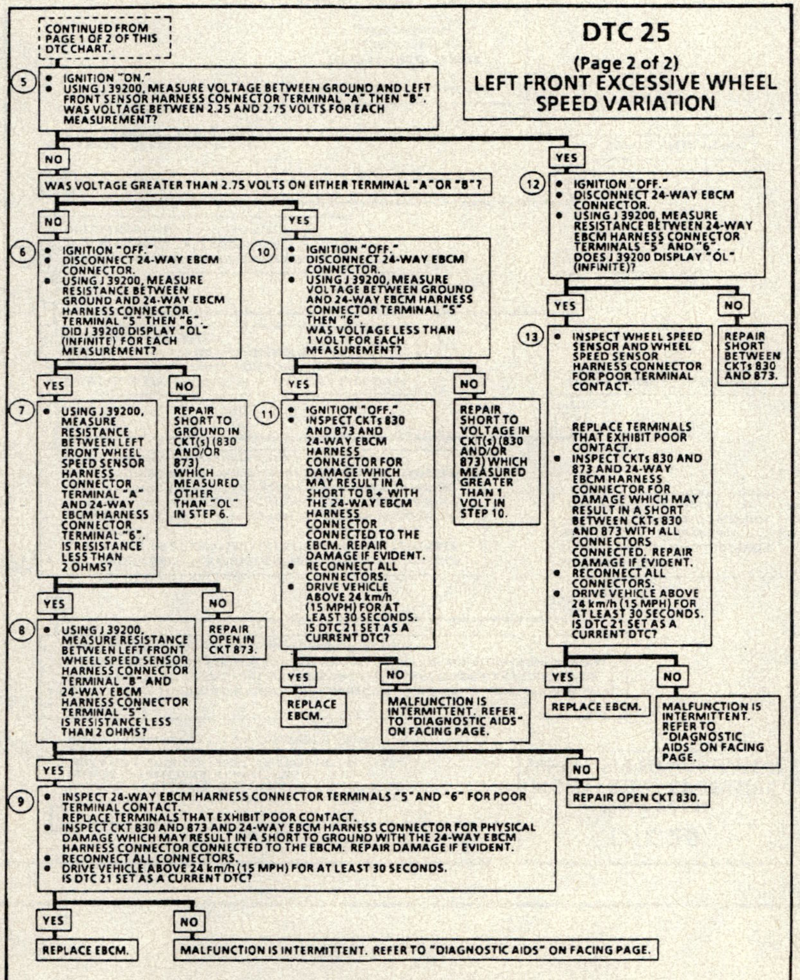

THIS CHART ASSUMES THAT A CURRENT DTC IS STORED INDICATING THAT THIS MALFUNCTION IS PRESENT.

DIAGNOSTIC CHART DTC 26 — 1994–95 BERETTA AND CORSICA

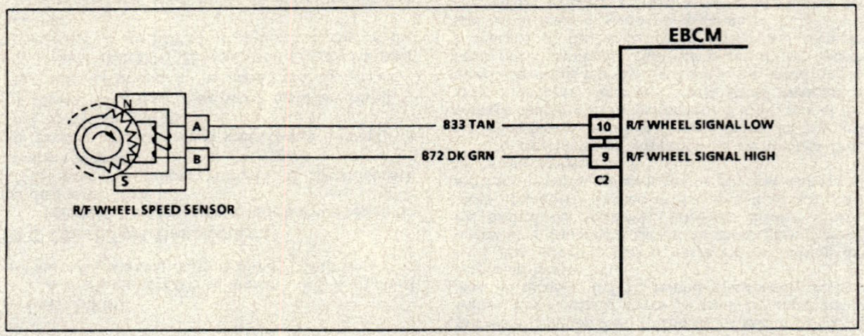

DTC 26
(Page 1 of 2)
RIGHT FRONT EXCESSIVE WHEEL SPEED VARIATION

Circuit Description:

As a toothed ring passes by the wheel speed sensor, changes in the electromagnetic field cause the wheel speed sensor to produce a sinusoidal (AC) voltage signal whose frequency is proportional to wheel speed. The magnitude of this signal is directly related to wheel speed and the proximity of the wheel speed sensor to the toothed ring, often referred to as the air gap.

DTC Will Set When:

DTC 26 can be set when the brake is "OFF." The purpose of this test is to detect a situation in which the right front wheel acceleration or deceleration is beyond specified limits.

Action Taken:

A malfunction DTC is stored, ABS is disabled and the ABS warning lamp is turned "ON."

DTC Chart Test Description:

Number(s) below refer to circled number(s) on the diagnostic chart.

1. Identifies a wheel speed sensor or circuitry that is damaged and visibly apparent.
2. Checks wheel speed sensor for proper resistance values.
3. Ensures the wheel speed sensor and sensor ring generate the proper voltage.
4. Ensures the wheel speed sensor is not shorted to ground.

Diagnostic Aids:

An "Intermittent" malfunction is most likely caused by a poor connection, rubbed through wire insulation, or a wire that is broken inside the insulation.

The frequency of the malfunction can be checked by using the enhanced diagnostic function of the Tech 1, as described in "Enhanced Diagnostics," found in this section.

If the customer's comments reflect that the ABS amber warning lamp is "ON" only during moist environmental changes, (rain, snow, vehicle wash) all wheel speed sensor circuitry should be thoroughly inspected for water intrusion. Use the following procedure. Spray down the suspected area with 5% salt water solution (two teaspoons of salt to 12 oz. of water). Drive vehicle above 24 km/h (15 mph) for at least 30 seconds. If DTC returns immediately, replace suspected harness.

Any circuitry, that is suspected as causing the intermittent complaint, should be thoroughly checked for backed out terminals, improper mating, broken locks, improperly formed or damaged terminals, poor terminal to wiring connections or physical damage to the wiring harness.

A careful visual inspection of the toothed ring, CV joint, bearing and wheel speed sensor for physical damage should also be performed. If DTC sets at the same mph every time and wheel speed variation is noted above this mph on the Tech 1, the speed ring is most likely damaged. This can occur at any speed. Resistance of the wheel speed sensor will increase with an increase in sensor temperature.

When replacing a wheel speed sensor, inspect the sensor terminals and harness connector for corrosion and/or water intrusion. If evidence of corrosion or water intrusion exists, replace wheel speed sensor harness. Refer to "On Vehicle Service" in this section.

If wheel speed sensor harness shows evidence of water intrusion, inspect wheel speed sensor for water intrusion. If water intrusion is evident, replace wheel speed sensor.

DIAGNOSTIC CHART DTC 26 — 1994–95 BERETTA AND CORSICA

DTC 26
(Page 1 of 2)
RIGHT FRONT EXCESSIVE WHEEL SPEED VARIATION

IMPORTANT:

- WHEEL SPEED SENSOR INTERMITTENT MALFUNCTIONS MAY BE DIFFICULT TO LOCATE. CARE SHOULD BE TAKEN NOT TO DISTURB ANY ELECTRICAL CONNECTIONS PRIOR TO AN INDICATED STEP OF THIS CHART. THIS WILL ENSURE THAT AN INTERMITTENT CONNECTION WILL NOT BE CORRECTED BEFORE THE SOURCE OF THE MALFUNCTION IS FOUND.

(1)
- IGNITION "OFF."
- INSPECT WHEEL SPEED SENSOR WIRING AND CONNECTORS FOR DAMAGE. INSPECT WHEEL SPEED SENSOR AND TOOTHED RING FOR LOOSENESS OR DAMAGE. PAY PARTICULAR ATTENTION TO THE TOOTHED RING. ANY SIGNIFICANT DAMAGE WILL AFFECT THE WHEEL SPEED INPUT SIGNAL.

IS DAMAGE OR SENSOR/TOOTHED RING LOOSENESS EVIDENT?

- YES: REPAIR DAMAGE OR SENSOR/TOOTHED RING LOOSENESS. DRIVE VEHICLE ABOVE 24 km/h (15 MPH) FOR AT LEAST 30 SECONDS TO ENSURE DTC 22 DOES NOT SET.
- NO: go to (2).

(2)
- DISCONNECT CONNECTOR FROM RIGHT FRONT WHEEL SPEED SENSOR.
- USING J 39200, MEASURE RESISTANCE BETWEEN SENSOR TERMINALS "A" AND "B".

IS RESISTANCE BETWEEN 1628 AND 1889 OHMS WHEN SENSOR IS 25°C (77°F)? *[IF TEMPERATURE OF THE SENSOR IS SUSPECTED TO BE OTHER THAN 25°C (77°F), REFER TO TABLE BELOW FOR APPLICABLE SENSOR RESISTANCE VALUES; THE VALUES IN THE TABLE ARE FOR TEMPERATURE OF THE SENSOR NOT AIR TEMPERATURE.]*

- NO: REPLACE WHEEL SPEED SENSOR.
- YES: go to (3).

(3)
- WITH J 39200 STILL CONNECTED, SELECT AC VOLTAGE SCALE.
- SPIN WHEEL BY HAND WHILE MONITORING VOLTAGE.

IS VOLTAGE AT LEAST 100 mV (VOLTAGE WILL INCREASE AS WHEEL SPEED INCREASES)?

- NO: REPLACE WHEEL SPEED SENSOR.
- YES: go to (4).

(4)
- USING J 39200, MEASURE RESISTANCE BETWEEN SENSOR TERMINAL "A" AND GROUND.

DOES J 39200 DISPLAY "OL" (INFINITE)?

- NO: REPLACE WHEEL SPEED SENSOR.
- YES: GO TO PAGE 2 OF 2 OF THIS DTC CHART.

FRONT WHEEL SPEED SENSOR

TEMPERATURE OF SENSOR VS. SENSOR RESISTANCE (APPROXIMATE)

TEMP. (°C)	TEMP. (°F)	RESISTANCE (OHMS)
-34 TO 4	-30 TO 40	1258 TO 1739
5 TO 43	41 TO 110	1503 TO 2023
44 TO 93	111 TO 200	1747 TO 2384

THIS CHART ASSUMES THAT A CURRENT DTC IS STORED INDICATING THAT THIS MALFUNCTION IS PRESENT.

DIAGNOSTIC CHART DTC 26 — 1994–95 BERETTA AND CORSICA

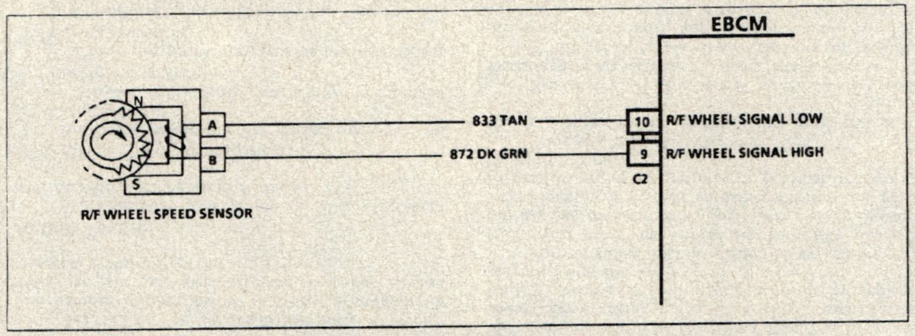

DTC 26 (Page 2 of 2)
RIGHT FRONT EXCESSIVE WHEEL SPEED VARIATION

Circuit Description:

As a toothed ring passes by the wheel speed sensor, changes in the electromagnetic field cause the wheel speed sensor to produce a sinusoidal (AC) voltage signal whose frequency is proportional to wheel speed. The magnitude of this signal is directly related to wheel speed and the proximity of the wheel speed sensor to the toothed ring, often referred to as the air gap.

DTC Will Set When:

DTC 26 can be set when the brake is "OFF." The purpose of this test is to detect a situation in which the right front wheel acceleration or deceleration is beyond specified limits.

Action Taken:

A malfunction DTC is stored, ABS is disabled and the ABS warning lamp is turned "ON."

DTC Chart Test Description:

Number(s) below refer to circled number(s) on the diagnostic chart.

5. Checks for proper voltages at speed sensor harness connector.
6. Ensures that the wheel speed sensor circuitry is not shorted to ground.
7. Checks for an open in the wheel signal low circuit.
8. Checks for an open in the wheel signal high circuit.
9. Ensures that DTC 26 was not set due to a poor connection at the EBCM or physical damage of the circuitry.
10. Checks for a short to B+ in either the high or low wheel signal circuits.
11. Ensures short to B+ was not due to physical damage of the circuitry.
12. Tests for a short between the wheel signal high and low circuits.
13. Ensures malfunction was not due to a poor connection at the wheel speed sensor or physical damage of the circuitry.

Diagnostic Aids:

An "Intermittent" malfunction is most likely caused by a poor connection, rubbed through wire insulation, or a wire that is broken inside the insulation.

The frequency of the malfunction can be checked by using the enhanced diagnostic function of the Tech 1, as described in "Enhanced Diagnostics," found in this section.

If the customer's comments reflect that the ABS amber warning lamp is "ON" only during moist environmental changes, (rain, snow, vehicle wash) all wheel speed sensor circuitry should be thoroughly inspected for water intrusion. Use the following procedure. Spray down the suspected area with 5% salt water solution (two teaspoons of salt to 12 oz. of water). Drive vehicle above 24 km/h (15 mph) for at least 30 seconds. If DTC returns immediately, replace suspected harness.

Any circuitry, that is suspected as causing the intermittent complaint, should be thoroughly checked for backed out terminals, improper mating, broken locks, improperly formed or damaged terminals, poor terminal to wiring connections or physical damage to the wiring harness.

A careful visual inspection of the toothed ring, CV joint, bearing and wheel speed sensor for physical damage should also be performed. If DTC sets at the same mph every time and wheel speed variation is noted above this mph on the Tech 1, the speed ring is most likely damaged. This can occur at any speed. Resistance of the wheel speed sensor will increase with an increase in sensor temperature.

When replacing a wheel speed sensor, inspect the sensor terminals and harness connector for corrosion and/or water intrusion. If evidence of corrosion or water intrusion exists, replace wheel speed sensor harness. Refer to "On Vehicle Service" in this section.

If wheel speed sensor harness shows evidence of water intrusion, inspect wheel speed sensor for water intrusion. If water intrusion is evident, replace wheel speed sensor.

DIAGNOSTIC CHART DTC 26 — 1994–95 BERETTA AND CORSICA

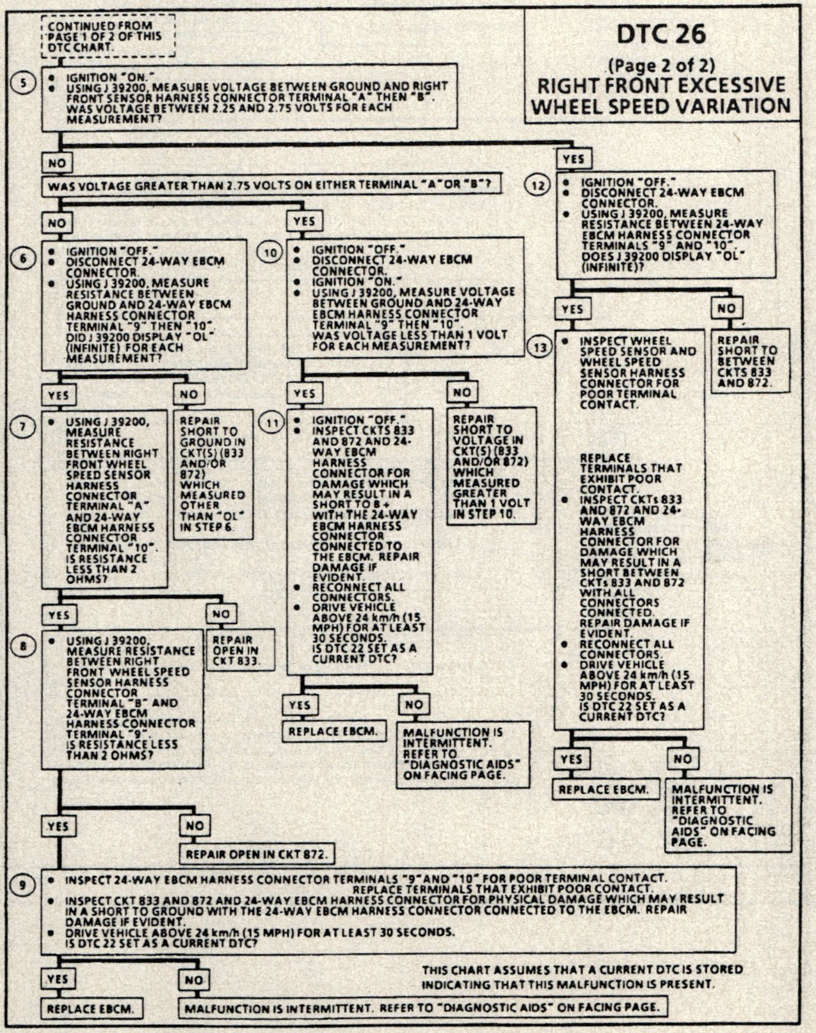

DIAGNOSTIC CHART DTC 27 — 1994–95 BERETTA AND CORSICA

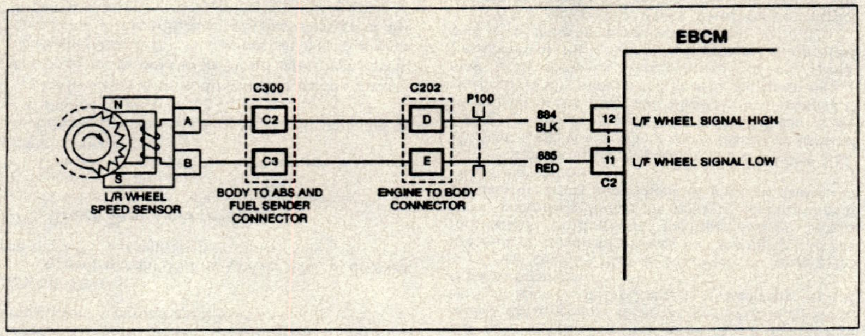

DTC 27

LEFT REAR EXCESSIVE WHEEL SPEED VARIATION

Circuit Description:

As a toothed ring passes by the wheel speed sensor, changes in the electromagnetic field cause the wheel speed sensor to produce a sinusoidal (AC) voltage signal whose frequency is proportional to wheel speed. The magnitude of this signal is directly related to wheel speed and the proximity of the wheel speed sensor to the toothed ring, often referred to as the air gap.

DTC Will Set When:

DTC 27 can be set when the brake is "OFF." The purpose of this test is to detect a situation in which the left rear wheel acceleration or deceleration is beyond specified limits.

Action Taken:

A malfunction DTC is stored, ABS is disabled and the ABS warning lamp is turned "ON."

DTC Chart Test Description:

Number(s) below refer to circled number(s) on the diagnostic chart.

1. This step will identify circuitry that is damaged and visibly apparent.
2. Checks for proper voltages at the rear sensor harness connector.
3. Checks for a short between the wheel signal high and low circuits.
4. This test checks wheel speed sensor for proper resistance values.
5. This step ensures the wheel speed sensor is not shorted to ground.

Diagnostic Aids:

An "Intermittent" malfunction is most likely caused by a poor connection, rubbed through wire insulation, or a wire that is broken inside the insulation.

The frequency of the malfunction can be checked by using the enhanced diagnostic function of the Tech 1.

If the customer's comments reflect that the ABS amber warning lamp is "ON" only during moist environmental changes, (rain, snow, vehicle wash) all wheel speed sensor circuitry should be thoroughly inspected for water intrusion. Use the following procedure. Spray down the suspected area with 5% salt water solution (two teaspoons of salt to 12 oz. of water). Drive vehicle above 24 km/h (15 mph) for at least 30 seconds. If DTC returns immediately, replace suspected harness.

Any circuitry, that is suspected as causing the intermittent complaint, should be thoroughly checked for backed out terminals, improper mating, broken locks, improperly formed or damaged terminals, poor terminal to wiring connections or physical damage to the wiring harness.

Excessive end play can cause this DTC to set. Check integral bearing for excessive axial end play.

Resistance of the wheel speed sensor will increase with an increase in sensor temperature.

When replacing a wheel speed sensor, inspect the sensor terminals and harness connector for corrosion and/or water intrusion. If evidence of corrosion or water intrusion exists, replace wheel speed sensor harness.

If wheel speed sensor harness shows evidence of water intrusion, inspect wheel speed sensor for water intrusion. If water intrusion is evident, replace wheel speed sensor.

DIAGNOSTIC CHART DTC 27 — 1994–95 BERETTA AND CORSICA

DTC 27
(Page 1 of 4)
LEFT REAR EXCESSIVE WHEEL SPEED VARIATION

Important

- WHEEL SPEED SENSOR INTERMITTENT MALFUNCTIONS MAY BE DIFFICULT TO LOCATE. CARE SHOULD BE TAKEN NOT TO DISTURB ANY ELECTRICAL CONNECTIONS PRIOR TO AN INDICATED STEP OF THIS CHART. THIS WILL ENSURE THAT AN INTERMITTENT CONNECTION WILL NOT BE CORRECTED BEFORE THE SOURCE OF THE MALFUNCTION IS FOUND.

(1)
- IGNITION "OFF."
- INSPECT WHEEL SPEED SENSOR WIRING AND CONNECTORS FOR DAMAGE. IS DAMAGE EVIDENT?

YES: REPAIR DAMAGE. DRIVE VEHICLE ABOVE 24 km/h (15 MPH) FOR AT LEAST 30 SECONDS TO ENSURE DTC 23 DOES NOT SET.

NO: go to step 2.

(2)
- REMOVE LEFT REAR INTEGRAL BEARING AND SPEED SENSOR ASSEMBLY (TUBULAR AXLE ONLY).
- DISCONNECT CONNECTOR FROM LEFT REAR WHEEL SPEED SENSOR.
- IGNITION "ON."
- USING J 39200, MEASURE VOLTAGE BETWEEN GROUND AND LEFT REAR SENSOR HARNESS CONNECTOR TERMINAL "A" THEN "B". WAS VOLTAGE BETWEEN 2.25 AND 2.75 VOLTS FOR EACH MEASUREMENT?

NO: GO TO PAGE 3 OF 4 OF THIS DTC CHART.

YES: go to step 3.

(3)
- IGNITION "OFF."
- DISCONNECT 24-WAY EBCM CONNECTOR.
- USING J 39200, MEASURE RESISTANCE BETWEEN 24-WAY EBCM HARNESS CONNECTOR TERMINALS "11" AND "12." DOES J 39200 DISPLAY "OL" (INFINITE)?

NO: REPAIR SHORT BETWEEN CKTs 884 AND 885.

YES: go to step 4.

(4)
- USING J 39200, MEASURE THE RESISTANCE BETWEEN LEFT REAR WHEEL SPEED SENSOR TERMINALS "A" AND "B". IS RESISTANCE BETWEEN 2100 AND 2400 OHMS WHEN SENSOR IS 25°C (77°F)? (IF TEMPERATURE OF THE SENSOR IS SUSPECTED TO BE OTHER THAN 25°C (77°F). REFER TO THE TABLE BELOW FOR APPLICABLE SENSOR RESISTANCE VALUES; THE VALUES IN THE TABLE ARE FOR TEMPERATURE OF THE SENSOR NOT AIR TEMPERATURE).

NO: REPLACE INTERGRAL BEARING AND SPEED SENSOR ASSEMBLY.

YES: go to step 5.

(5)
- USING J 39200, MEASURE RESISTANCE BETWEEN SENSOR TERMINAL "A" AND INTEGRAL BEARING AND SPEED SENSOR ASSEMBLY MOUNTING FLANGE. DOES J 39200 DISPLAY "OL" (INFINITE)?

YES: CONTINUE DIAGNOSIS ON NEXT PAGE.

NO: REPLACE INTEGRAL BEARING AND SPEED SENSOR ASSEMBLY.

REAR WHEEL SPEED SENSOR
TEMPERATURE OF SENSOR VS. SENSOR RESISTANCE (APPROXIMATE)

TEMP. (°C)	TEMP. (°F)	RESISTANCE (OHMS)
-34 TO 4	-30 TO 40	1622 TO 2205
5 TO 43	41 TO 110	1930 TO 2566
44 TO 93	111 TO 200	2245 TO 3028

THIS CHART ASSUMES THAT A CURRENT DTC IS STORED INDICATING THAT THIS MALFUNCTION IS PRESENT.

DIAGNOSTIC CHART DTC 27 — 1994–95 BERETTA AND CORSICA

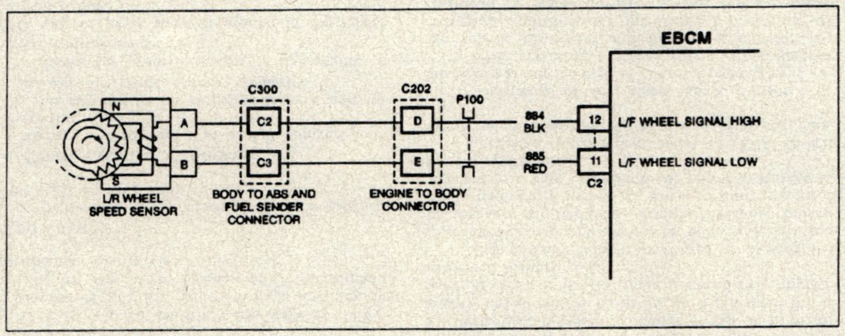

DTC 27
(Page 2 of 4)
LEFT REAR EXCESSIVE WHEEL SPEED VARIATION

Circuit Description:

As a toothed ring passes by the wheel speed sensor, changes in the electromagnetic field cause the wheel speed sensor to produce a sinusoidal (AC) voltage signal whose frequency is proportional to wheel speed. The magnitude of this signal is directly related to wheel speed and the proximity of the wheel speed sensor to the toothed ring, often referred to as the air gap.

DTC Will Set When:

DTC 27 can be set when the brake is "OFF." The purpose of this test is to detect a situation in which the left rear wheel acceleration or deceleration is beyond specified limits.

Action Taken:

A malfunction DTC is stored, ABS is disabled and the ABS warning lamp is turned "ON."

DTC Chart Test Description:

Number(s) below refer to circled number(s) on the diagnostic chart.

6. This test ensures the wheel speed sensor and sensor ring generate the proper voltage.
7. Ensures malfunction was not due to physical damage of the circuitry.

Diagnostic Aids:

An "Intermittent" malfunction is most likely caused by a poor connection, rubbed through wire insulation, or a wire that is broken inside the insulation.

The frequency of the malfunction can be checked by using the enhanced diagnostic function of the Tech 1.

If the customer's comments reflect that the ABS amber warning lamp is "ON" only during moist environmental changes, (rain, snow, vehicle wash) all wheel speed sensor circuitry should be thoroughly inspected for water intrusion. Use the following procedure. Spray down the suspected area with 5% salt water solution (two teaspoons of salt to 12 oz. of water). Drive vehicle above 24 km/h (15 mph) for at least 30 seconds. If DTC returns immediately, replace suspected harness.

Any circuitry, that is suspected as causing the intermittent complaint, should be thoroughly checked for backed out terminals, improper mating, broken locks, improperly formed or damaged terminals, poor terminal to wiring connections or physical damage to the wiring harness.

Excessive end play can cause this DTC to set. Check integral bearing for excessive axial end play.

Resistance of the wheel speed sensor will increase with an increase in sensor temperature.

When replacing a wheel speed sensor, inspect the sensor terminals and harness connector for corrosion and/or water intrusion. If evidence of corrosion or water intrusion exists, replace wheel speed sensor harness.

If wheel speed sensor harness shows evidence of water intrusion, inspect wheel speed sensor for water intrusion. If water intrusion is evident, replace wheel speed sensor.

DIAGNOSTIC CHART DTC 27 — 1994–95 BERETTA AND CORSICA

DTC 27
(Page 2 of 4)
LEFT REAR EXCESSIVE WHEEL SPEED VARIATION

DIAGNOSIS CONTINUED FROM PAGE 1 OF 4 OF THIS DTC CHART.

(6)
- INSPECT WHEEL SPEED SENSOR HARNESS CONNECTOR AND 24-WAY EBCM HARNESS CONNECTOR TERMINALS "11" AND "12" FOR POOR TERMINAL CONTACT.

  REPLACE TERMINALS THAT EXHIBIT POOR CONTACT RECONNECT WHEEL SPEED SENSOR.
- INSTALL INTEGRAL BEARING AND SPEED SENSOR ASSEMBLY (TUBULAR AXLE ONLY).
- USING J 39200, SELECT A/C VOLTAGE SCALE AND CONNECT BETWEEN 24-WAY EBCM HARNESS CONNECTOR TERMINALS "11" AND "12".
- SPIN WHEEL BY HAND WHILE MONITORING VOLTAGE.

  IS VOLTAGE AT LEAST 100 mV (VOLTAGE WILL INCREASE AS WHEEL SPEED INCREASES)?

NO: REPLACE INTEGRAL REARING AND SPEED SENSOR ASSEMBLY.

YES:

(7)
- INSPECT CKTs 884 AND 885 FOR PHYSICAL DAMAGE WHICH MAY RESULT IN A SHORT BETWEEN THE CIRCUITS WITH ALL CONNECTORS CONNECTED.

  REPAIR DAMAGE IF EVIDENT.
- RECONNECT ALL CONNECTORS.
- DRIVE VEHICLE ABOVE 24-WAY km/h (15 MPH) FOR AT LEAST 30 SECONDS IS DTC 23 SET AS A CURRENT DTC?

YES: REPLACE EBCM.

NO: MALFUNCTION IS INTERMITTENT. REFER TO "DIAGNOSTIC AIDS" ON FACING PAGE.

THIS CHART ASSUMES THAT A CURRENT DTC IS STORED INDICATING THAT THIS MALFUNCTION IS PRESENT.

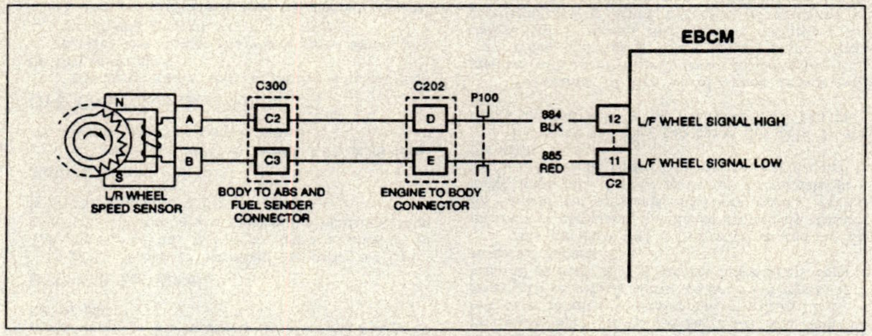

DTC 27 (Page 3 of 4)
LEFT REAR EXCESSIVE WHEEL SPEED VARIATION

Circuit Description:

As a toothed ring passes by the wheel speed sensor, changes in the electromagnetic field cause the wheel speed sensor to produce a sinusoidal (AC) voltage signal whose frequency is proportional to wheel speed. The magnitude of this signal is directly related to wheel speed and the proximity of the wheel speed sensor to the toothed ring, often referred to as the air gap.

DTC Will Set When:

DTC 27 can be set when the brake is "OFF." The purpose of this test is to detect a situation in which the left rear wheel acceleration or deceleration is beyond specified limits.

Action Taken:

A malfunction DTC is stored, ABS is disabled and the ABS warning lamp is turned "ON."

DTC Chart Test Description:

Number(s) below refer to circled number(s) on the diagnostic chart.

8. This step checks for proper voltages at body to ABS and fuel sender harness connector.
9. Checks for proper voltages at the engine to body harness connector.
10. Checks for a short to ground in the wheel signal circuit.
11. This checks for an open in the wheel signal high circuit.
12. Checks wheel speed circuits for short to B+ at engine harness.

Diagnostic Aids:

An "Intermittent" malfunction is most likely caused by a poor connection, rubbed through wire insulation, or a wire that is broken inside the insulation.

The frequency of the malfunction can be checked by using the enhanced diagnostic function of the Tech 1.

If the customer's comments reflect that the ABS amber warning lamp is "ON" only during moist environmental changes, (rain, snow, vehicle wash) all wheel speed sensor circuitry should be thoroughly inspected for water intrusion. Use the following procedure. Spray down the suspected area with 5% salt water solution (two teaspoons of salt to 12 oz. of water). Drive vehicle above 24 km/h (15 mph) for at least 30 seconds. If DTC returns immediately, replace suspected harness.

Any circuitry, that is suspected as causing the intermittent complaint, should be thoroughly checked for backed out terminals, improper mating, broken locks, improperly formed or damaged terminals, poor terminal to wiring connections or physical damage to the wiring harness.

Excessive end play can cause this DTC to set. Check integral bearing for excessive axial end play.

Resistance of the wheel speed sensor will increase with an increase in sensor temperature.

When replacing a wheel speed sensor, inspect the sensor terminals and harness connector for corrosion and/or water intrusion. If evidence of corrosion or water intrusion exists, replace wheel speed sensor harness.

If wheel speed sensor harness shows evidence of water intrusion, inspect wheel speed sensor for water intrusion. If water intrusion is evident, replace wheel speed sensor.

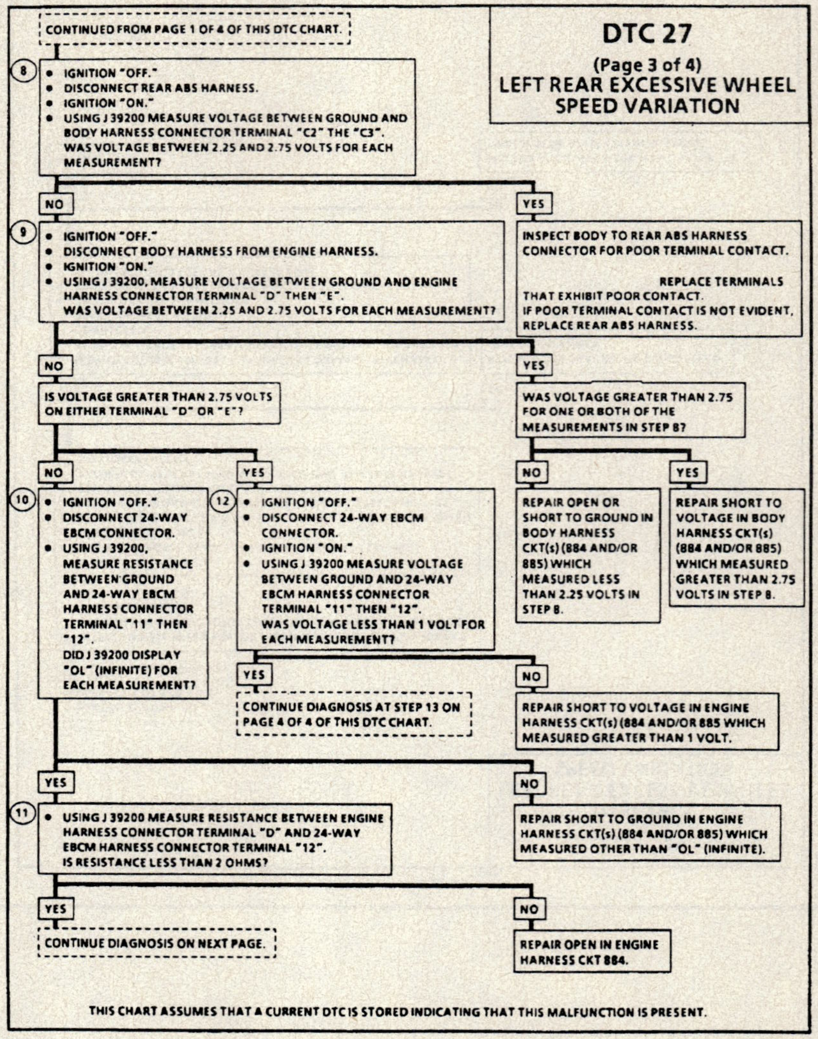

DIAGNOSTIC CHART DTC 27 — 1994–95 BERETTA AND CORSICA

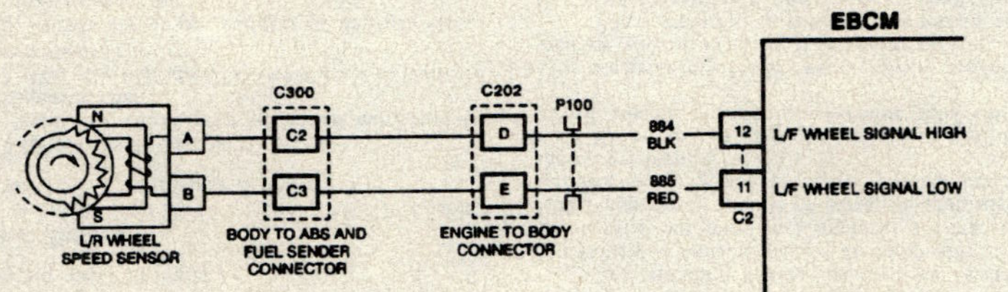

DTC 27
(Page 4 of 4)
LEFT REAR EXCESSIVE WHEEL SPEED VARIATION

Circuit Description:

As a toothed ring passes by the wheel speed sensor, changes in the electromagnetic field cause the wheel speed sensor to produce a sinusoidal (AC) voltage signal whose frequency is proportional to wheel speed. The magnitude of this signal is directly related to wheel speed and the proximity of the wheel speed sensor to the toothed ring, often referred to as the air gap.

DTC Will Set When:

DTC 27 can be set when the brake is "OFF." The purpose of this test is to detect a situation in which the left rear wheel acceleration or deceleration is beyond specified limits.

Action Taken:

A malfunction DTC is stored, ABS is disabled and the ABS warning lamp is turned "ON."

DTC Chart Test Description:

Number(s) below refer to circled number(s) on the diagnostic chart.

13. Checks for an open in the wheel signal low circuit.

14. Ensures malfunction was not due to physical damage of the circuitry.

Diagnostic Aids:

An "Intermittent" malfunction is most likely caused by a poor connection, rubbed through wire insulation, or a wire that is broken inside the insulation.

The frequency of the malfunction can be checked by using the enhanced diagnostic function of the Tech 1.

If the customer's comments reflect that the ABS amber warning lamp is "ON" only during moist environmental changes, (rain, snow, vehicle wash) all wheel speed sensor circuitry should be thoroughly inspected for water intrusion. Use the following procedure. Spray down the suspected area with 5% salt water solution (two teaspoons of salt to 12 oz. of water). Drive vehicle above 24 km/h (15 mph) for at least 30 seconds. If DTC returns immediately, replace suspected harness.

Any circuitry, that is suspected as causing the intermittent complaint, should be thoroughly checked for backed out terminals, improper mating, broken locks, improperly formed or damaged terminals, poor terminal to wiring connections or physical damage to the wiring harness.

Excessive end play can cause this DTC to set. Check integral bearing for excessive axial end play.

Resistance of the wheel speed sensor will increase with an increase in sensor temperature.

When replacing a wheel speed sensor, inspect the sensor terminals and harness connector for corrosion and/or water intrusion. If evidence of corrosion or water intrusion exists, replace wheel speed sensor harness. Refer to "On Vehicle Service" in this section.

If wheel speed sensor harness shows evidence of water intrusion, inspect wheel speed sensor for water intrusion. If water intrusion is evident, replace wheel speed sensor.

DIAGNOSTIC CHART DTC 27 — 1994–95 BERETTA AND CORSICA

DTC 27
(Page 4 of 4)
LEFT REAR EXCESSIVE WHEEL SPEED VARIATION

CONTINUED FROM PAGE 3 OF 4 OF THIS DTC CHART.

(13)
- USING J 39200, MEASURE RESISTANCE BETWEEN ENGINE HARNESS CONNECTOR TERMINAL "E" AND 24-WAY EBCM HARNESS CONNECTOR TERMINAL "11". IS RESISTANCE LESS THAN 2 OHMS?

YES → (14)

NO → REPAIR OPEN IN ENGINE HARNESS CKT 885.

(14)
- IGNITION "OFF."
- INSPECT ALL WHEEL SPEED SENSOR HARNESS CONNECTORS FOR POOR TERMINAL CONTACT OR CORROSION. REPLACE TERMINALS THAT EXHIBIT POOR CONTACT OR CORROSION.
- INSPECT ALL WHEEL SPEED SENSOR CONNECTORS AND HARNESSES FOR DAMAGE WHICH MAY RESULT IN AN OPEN OR SHORT TO GROUND/BATTERY WHEN ALL COMPONENTS ARE CONNECTED. REPAIR DAMAGE IF EVIDENT.
- RECONNECT ALL CONNECTORS.
- DRIVE VEHICLE ABOVE 24 km/h (15 MPH) FOR AT LEAST 30 SECONDS. IS DTC 23 SET AS A CURRENT DTC?

YES → REPLACE EBCM.

NO → MALFUNCTION IS INTERMITTENT. REFER TO "DIAGNOSTIC AIDS" ON FACING PAGE.

THIS CHART ASSUMES THAT A CURRENT DTC IS STORED INDICATING THAT THIS MALFUNCTION IS PRESENT.

DIAGNOSTIC CHART DTC 28 — 1994–95 BERETTA AND CORSICA

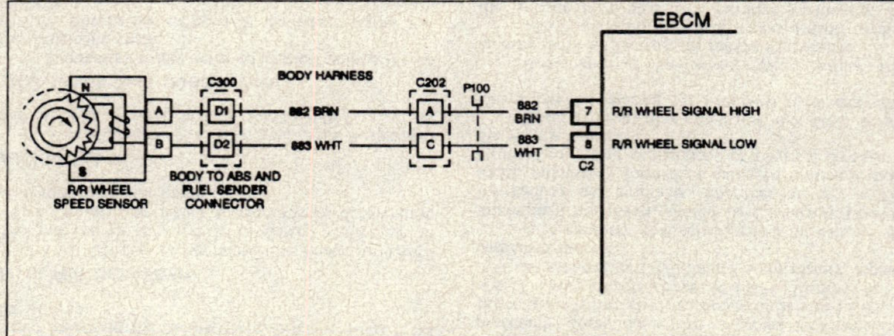

DTC 28
(Page 1 of 4)
RIGHT REAR EXCESSIVE WHEEL SPEED VARIATION

Circuit Description:

As a toothed ring passes by the wheel speed sensor, changes in the electromagnetic field cause the wheel speed sensor to produce a sinusoidal (AC) voltage signal whose frequency is proportional to wheel speed.

The magnitude of this signal is directly related to wheel speed and the proximity of the wheel speed sensor to the toothed ring, often referred to as the air gap.

DTC Will Set When:

DTC 28 can be set when the brake is "OFF." The purpose of this test is to detect a situation in which the right rear wheel acceleration or deceleration is beyond specified limits.

Action Taken:

A malfunction DTC is stored, ABS is disabled and the ABS warning lamp is turned "ON."

DTC Chart Test Description:

Number(s) below refer to circled number(s) on the diagnostic chart.

1. This step will identify circuitry that is damaged and visibly apparent.
2. Checks for proper voltages at the rear sensor harness connector.
3. Checks for a short between the wheel signal high and low circuits.
4. This test checks wheel speed sensor for proper resistance values.

Diagnostic Aids:

An "Intermittent" malfunction is most likely caused by a poor connection, rubbed through wire insulation, or a wire that is broken inside the insulation.

The frequency of the malfunction can be checked by using the enhanced diagnostic function of the Tech 1, as described in "Enhanced Diagnostics," found in this section.

If the customer's comments reflect that the ABS amber warning lamp is "ON" only during moist environmental changes, (rain, snow, vehicle wash) all wheel speed sensor circuitry should be thoroughly inspected for water intrusion. Use the following procedure. Spray down the suspected area with 5% salt water solution (two teaspoons of salt to 12 oz. of water). Drive vehicle above 24 km/h (15 mph) for at least 30 seconds. If DTC returns immediately, replace suspected harness.

Any circuitry, that is suspected as causing the intermittent complaint, should be thoroughly checked for backed out terminals, improper mating, broken locks, improperly formed or damaged terminals, poor terminal to wiring connections or physical damage to the wiring harness.

Excessive end play can cause this DTC to set. Check integral bearing for excessive axial end play.

Resistance of the wheel speed sensor will increase with an increase in sensor temperature.

When replacing a wheel speed sensor, inspect the sensor terminals and harness connector for corrosion and/or water intrusion. If evidence of corrosion or water intrusion exists, replace wheel speed sensor harness. Refer to "On Vehicle Service" in this section.

If wheel speed sensor harness shows evidence of water intrusion, inspect wheel speed sensor for water intrusion. If water intrusion is evident, replace wheel speed sensor.

DIAGNOSTIC CHART DTC 28 — 1994–95 BERETTA AND CORSICA

DTC 28
(Page 1 of 4)
RIGHT REAR EXCESSIVE WHEEL SPEED VARIATION

Important
- WHEEL SPEED SENSOR INTERMITTENT MALFUNCTIONS MAY BE DIFFICULT TO LOCATE. CARE SHOULD BE TAKEN NOT TO DISTURB ANY ELECTRICAL CONNECTIONS PRIOR TO AN INDICATED STEP OF THIS CHART. THIS WILL ENSURE THAT AN INTERMITTENT CONNECTION WILL NOT BE CORRECTED BEFORE THE SOURCE OF THE MALFUNCTION IS FOUND.

(1)
- IGNITION "OFF."
- INSPECT WHEEL SPEED SENSOR WIRING AND CONNECTORS FOR DAMAGE. IS DAMAGE EVIDENT?

YES: REPAIR DAMAGE. DRIVE VEHICLE ABOVE 24 km/h (15MPH) FOR AT LEAST 30 SECONDS TO ENSURE DTC 24 DOES NOT SET.

NO:

(2)
- REMOVE RIGHT REAR INTEGRAL BEARING AND SPEED SENSOR ASSEMBLY (TUBULAR AXLE ONLY).
- DISCONNECT CONNECTOR FROM RIGHT REAR WHEEL SPEED SENSOR.
- IGNITION "ON."
- USING J 39200, MEASURE VOLTAGE BETWEEN GROUND AND RIGHT REAR SENSOR HARNESS CONNECTOR TERMINAL "A" THEN "B". WAS VOLTAGE BETWEEN 2.25 AND 2.75 VOLTS FOR EACH MEASUREMENT?

YES: GO TO PAGE 3 OF 4 OF THIS CHART.

NO:

(3)
- IGNITION "OFF."
- DISCONNECT 24-WAY EBCM CONNECTOR.
- USING J39200, MEASURE RESISTANCE BETWEEN 24-WAY EBCM HARNESS CONNECTOR TERMINALS "8" AND "7." DOES J 39200 DISPLAY "OL" (INFINITE).

NO: REPAIR SHORT BETWEEN CKTS 882 AND 883.

YES:

(4)
- USING J 39200, MEASURE RESISTANCE BETWEEN RIGHT REAR WHEEL SPEED SENSOR TERMINALS "A" AND "B." IS RESISTANCE BETWEEN 2100 AND 2400 OHMS WHEN SENSOR IS 25°C (77°F)? (IF TEMPERATURE OF THE SENSOR IS SUSPECTED TO BE OTHER THAN 25°C (77°F). REFER TO THE TABLE BELOW FOR APPLICABLE SENSOR RESISTANCE VALUES; THE VALUES IN THE TABLE ARE FOR TEMPERATURE OF THE SENSOR NOT AIR TEMPERATURE).

YES: CONTINUE DIAGNOSIS ON NEXT PAGE.

NO: REPLACE INTEGRAL BEARING AND SPEED SENSOR ASSEMBLY.

REAR WHEEL SPEED SENSOR

TEMPERATURE OF SENSOR VS. SENSOR RESISTANCE (APPROXIMATE)

TEMP. (°C)	TEMP. (°F)	RESISTANCE (OHMS)
-34 TO 4	-30 TO 40	1623 TO 2210
5 TO 43	41 TO 110	1939 TO 2570
44 TO 93	111 TO 200	2254 TO 3029

THIS CHART ASSUMES THAT A CURRENT DTC IS STORED INDICATING THAT THIS MALFUNCTION IS PRESENT.

DIAGNOSTIC CHART DTC 28 — 1994–95 BERETTA AND CORSICA

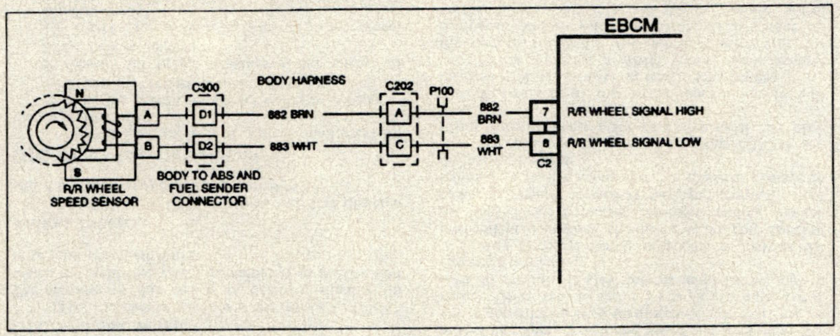

DTC 28
(Page 2 of 4)
RIGHT REAR EXCESSIVE WHEEL SPEED VARIATION

Circuit Description:

As a toothed ring passes by the wheel speed sensor, changes in the electromagnetic field cause the wheel speed sensor to produce a sinusoidal (AC) voltage signal whose frequency is proportional to wheel speed. The magnitude of this signal is directly related to wheel speed and the proximity of the wheel speed sensor to the toothed ring, often referred to as the air gap.

DTC Will Set When:

DTC 28 can be set when the brake is "OFF." The purpose of this test is to detect a situation in which the right rear wheel acceleration or deceleration is beyond specified limits.

Action Taken:

A malfunction DTC is stored, ABS is disabled and the ABS warning lamp is turned "ON."

DTC Chart Test Description:

Number(s) below refer to circled number(s) on the diagnostic chart.

5. This step ensures the wheel speed sensor is not shorted to ground.
6. This test ensures the wheel speed sensor and sensor ring generate the proper voltage.
7. Ensures malfunction was not due to physical damage of the circuitry.

Diagnostic Aids:

An "Intermittent" malfunction is most likely caused by a poor connection, rubbed through wire insulation, or a wire that is broken inside the insulation.

The frequency of the malfunction can be checked by using the enhanced diagnostic function of the Tech 1, as described in "Enhanced Diagnostics," found in this section.

If the customer's comments reflect that the ABS amber warning lamp is "ON" only during moist environmental changes, (rain, snow, vehicle wash) all wheel speed sensor circuitry should be thoroughly inspected for water intrusion. Use the following procedure. Spray down the suspected area with 5% salt water solution (two teaspoons of salt to 12 oz. of water). Drive vehicle above 24 km/h (15 mph) for at least 30 seconds. If DTC returns immediately, replace suspected harness.

Any circuitry, that is suspected as causing the intermittent complaint, should be thoroughly checked for backed out terminals, improper mating, broken locks, improperly formed or damaged terminals, poor terminal to wiring connections or physical damage to the wiring harness.

Excessive end play can cause this DTC to set. Check integral bearing for excessive axial end play.

Resistance of the wheel speed sensor will increase with an increase in sensor temperature.

When replacing a wheel speed sensor, inspect the sensor terminals and harness connector for corrosion and/or water intrusion. If evidence of corrosion or water intrusion exists, replace wheel speed sensor harness. Refer to "On Vehicle Service" in this section.

If wheel speed sensor harness shows evidence of water intrusion, inspect wheel speed sensor for water intrusion. If water intrusion is evident, replace wheel speed sensor.

DIAGNOSTIC CHART DTC 28 — 1994–95 BERETTA AND CORSICA

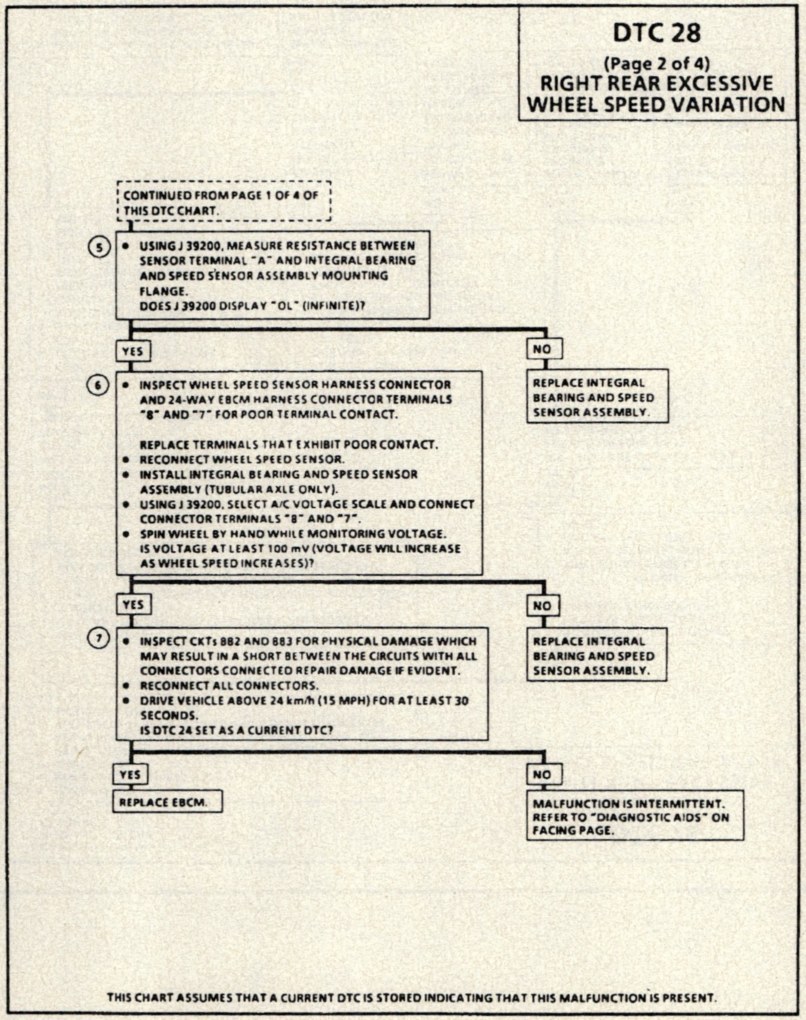

DIAGNOSTIC CHART DTC 28 — 1994–95 BERETTA AND CORSICA

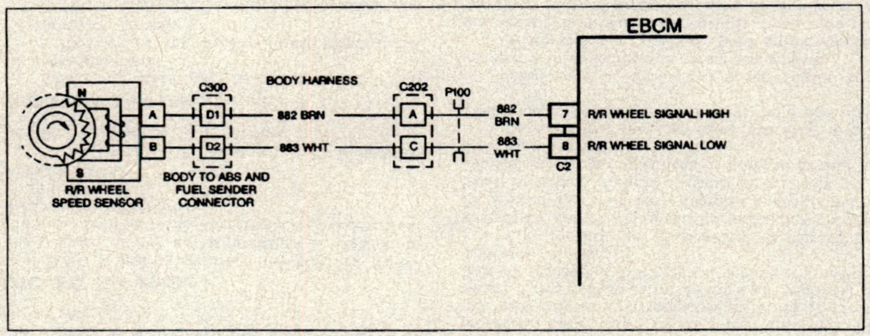

DTC 28 (Page 3 of 4)
RIGHT REAR EXCESSIVE WHEEL SPEED VARIATION

Circuit Description:

As a toothed ring passes by the wheel speed sensor, changes in the electromagnetic field cause the wheel speed sensor to produce a sinusoidal (AC) voltage signal whose frequency is proportional to wheel speed. The magnitude of this signal is directly related to wheel speed and the proximity of the wheel speed sensor to the toothed ring, often referred to as the air gap.

DTC Will Set When:

DTC 28 can be set when the brake is "OFF." The purpose of this test is to detect a situation in which the right rear wheel acceleration or deceleration is beyond specified limits.

Action Taken:

A malfunction DTC is stored, ABS is disabled and the ABS warning lamp is turned "ON."

DTC Chart Test Description:

Number(s) below refer to circled number(s) on the diagnostic chart.

8. This step checks for proper voltages at body to ABS and fuel sender harness connector.
9. Checks for proper voltages at the engine to body harness connector.
10. Checks for a short to ground in the wheel signal circuits.
11. This checks for an open in the wheel signal high circuit.
12. Checks wheel speed circuits for short to B+ at engine harness.

Diagnostic Aids:

An "Intermittent" malfunction is most likely caused by a poor connection, rubbed through wire insulation, or a wire that is broken inside the insulation.

The frequency of the malfunction can be checked by using the enhanced diagnostic function of the Tech 1, as described in "Enhanced Diagnostics," found in this section.

If the customer's comments reflect that the ABS amber warning lamp is "ON" only during moist environmental changes, (rain, snow, vehicle wash) all wheel speed sensor circuitry should be thoroughly inspected for water intrusion. Use the following procedure. Spray down the suspected area with 5% salt water solution (two teaspoons of salt to 12 oz. of water). Drive vehicle above 24 km/h (15 mph) for at least 30 seconds. If DTC returns immediately, replace suspected harness.

Any circuitry, that is suspected as causing the intermittent complaint, should be thoroughly checked for backed out terminals, improper mating, broken locks, improperly formed or damaged terminals, poor terminal to wiring connections or physical damage to the wiring harness.

Excessive end play can cause this DTC to set. Check integral bearing for excessive axial end play.

Resistance of the wheel speed sensor will increase with an increase in sensor temperature.

When replacing a wheel speed sensor, inspect the sensor terminals and harness connector for corrosion and/or water intrusion. If evidence of corrosion or water intrusion exists, replace wheel speed sensor harness. Refer to "On Vehicle Service" in this section.

If wheel speed sensor harness shows evidence of water intrusion, inspect wheel speed sensor for water intrusion. If water intrusion is evident, replace wheel speed sensor.

DIAGNOSTIC CHART DTC 28 — 1994–95 BERETTA AND CORSICA

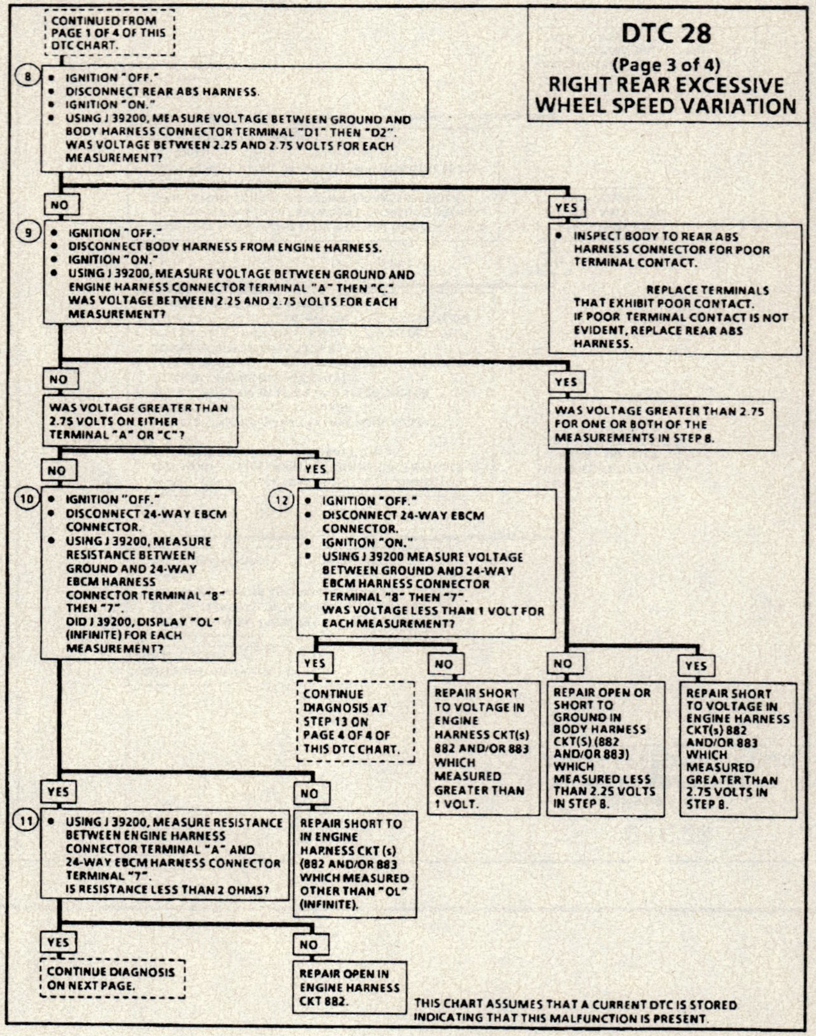

DIAGNOSTIC CHART DTC 28 — 1994–95 BERETTA AND CORSICA

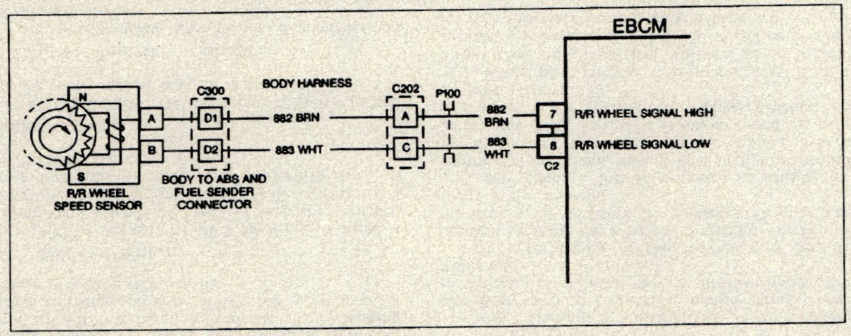

DTC 28
(Page 4 of 4)
RIGHT REAR EXCESSIVE WHEEL SPEED VARIATION

Circuit Description:

As a toothed ring passes by the wheel speed sensor, changes in the electromagnetic field cause the wheel speed sensor to produce a sinusoidal (AC) voltage signal whose frequency is proportional to wheel speed. The magnitude of this signal is directly related to wheel speed and the proximity of the wheel speed sensor to the toothed ring, often referred to as the air gap.

DTC Will Set When:

DTC 28 can be set when the brake is "OFF." The purpose of this test is to detect a situation in which the right rear wheel acceleration or deceleration is beyond specified limits.

Action Taken:

A malfunction DTC is stored, ABS is disabled and the ABS warning lamp is turned "ON."

DTC Chart Test Description:

Number(s) below refer to circled number(s) on the diagnostic chart.

13. Checks for an open in the wheel signal low circuit.

14. Ensures malfunction was not due to physical damage of the circuitry.

Diagnostic Aids:

An "Intermittent" malfunction is most likely caused by a poor connection, rubbed through wire insulation, or a wire that is broken inside the insulation.

The frequency of the malfunction can be checked by using the enhanced diagnostic function of the Tech 1, as described in "Enhanced Diagnostics," found in this section.

If the customer's comments reflect that the ABS amber warning lamp is "ON" only during moist environmental changes, (rain, snow, vehicle wash) all wheel speed sensor circuitry should be thoroughly inspected for water intrusion. Use the following procedure. Spray down the suspected area with 5% salt water solution (two teaspoons of salt to 12 oz. of water). Drive vehicle above 24 km/h (15 mph) for at least 30 seconds. If DTC returns immediately, replace suspected harness.

Any circuitry, that is suspected as causing the intermittent complaint, should be thoroughly checked for backed out terminals, improper mating, broken locks, improperly formed or damaged terminals, poor terminal to wiring connections or physical damage to the wiring harness.

Excessive end play can cause this DTC to set. Check integral bearing for excessive axial end play.

Resistance of the wheel speed sensor will increase with an increase in sensor temperature.

When replacing a wheel speed sensor, inspect the sensor terminals and harness connector for corrosion and/or water intrusion. If evidence of corrosion or water intrusion exists, replace wheel speed sensor harness. Refer to "On Vehicle Service" in this section.

If wheel speed sensor harness shows evidence of water intrusion, inspect wheel speed sensor for water intrusion. If water intrusion is evident, replace wheel speed sensor.

DIAGNOSTIC CHART DTC 28 — 1994–95 BERETTA AND CORSICA

DTC 28
(Page 4 of 4)
RIGHT REAR EXCESSIVE WHEEL SPEED VARIATION

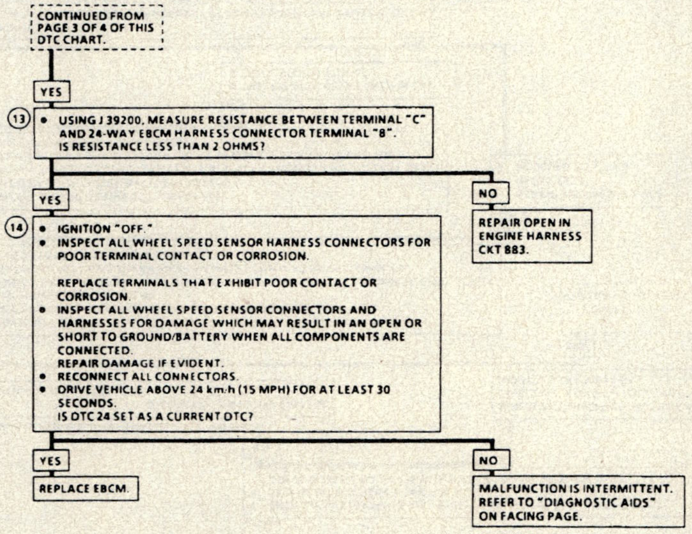

THIS CHART ASSUMES THAT A CURRENT DTC IS STORED INDICATING THAT THIS MALFUNCTION IS PRESENT.

DIAGNOSTIC CHART DTC 36 — 1994–95 BERETTA AND CORSICA

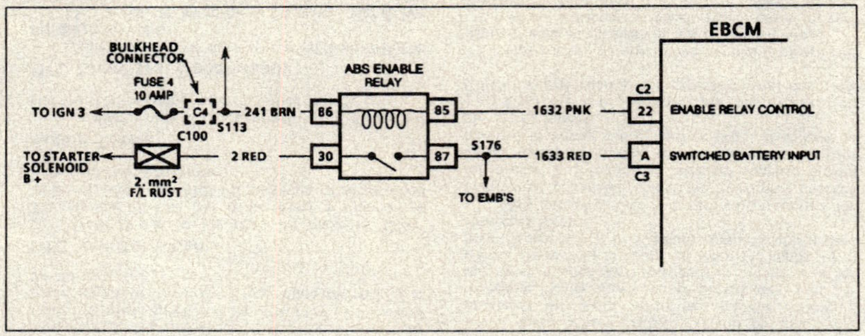

DTC 36
LOW SYSTEM VOLTAGE

Circuit Description:

This DTC is used to monitor the voltage level available to the ABS EBCM. If the voltage drops below 11 volts, full performance of the ABS system cannot be guaranteed. During ABS operation, there are several current requirements that will cause battery voltage to drop. Because of this, voltage is monitored prior to ABS operation to indicate good charging system condition and also during ABS operation when voltage may drop significantly.

DTC Will Set When:

DTC 36 can only be set if the vehicle's speed is greater than 5 km/h (3 mph). If both the switched battery and switched ignition voltage is less than 11.2 volts non-ABS or less than 9.6 volts ABS, a malfunction exists.

Action Taken:

A malfunction DTC is stored, ABS is disabled and the ABS warning lamp is turned "ON."

DTC Chart Test Description:

Number(s) below refer to circled number(s) on the diagnostic chart.

1. Tests to see if malfunction is present.
2. Checks charging system integrity.
3. Checks for excessive resistance in the switched battery input circuit.
4. Tests for high resistance in the enable relay control circuit.
5. Ensures low voltage condition was not due to a poor connection.

Diagnostic Aids:

An "Intermittent" malfunction is most likely caused by a poor connection, rubbed through wire insulation, or a wire that is broken inside the insulation.

The frequency of the malfunction can be checked by using the enhanced diagnostic function of the Tech 1, as described in "Enhanced Diagnostics," found in this section.

Any circuitry, that is suspected as causing the intermittent complaint, should be thoroughly checked for backed out terminals, improper mating, broken locks, improperly formed or damaged terminals, poor terminal to wiring connections or physical damage to the wiring harness.

While performing a voltage load test, if it is noted that only ignition voltage drops below acceptable voltage levels, CKT 650 should be checked for high resistance or an open condition.

DIAGNOSTIC CHART DTC 36 — 1994–95 BERETTA AND CORSICA

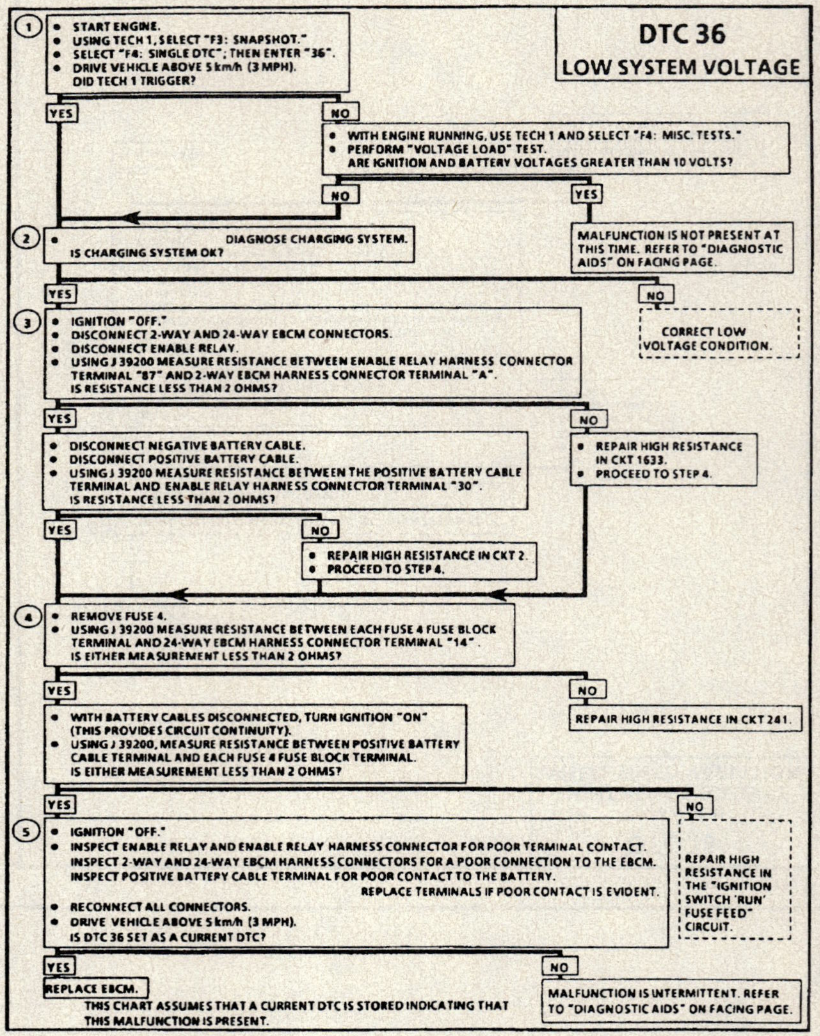

DIAGNOSTIC CHART DTC 37 — 1994–95
BERETTA AND CORSICA

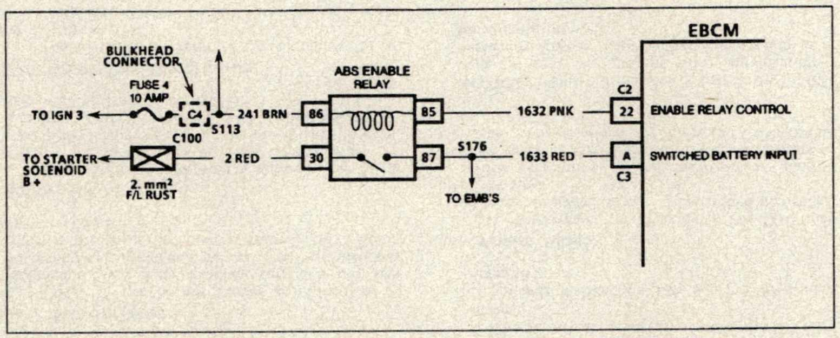

DTC 37
HIGH SYSTEM VOLTAGE

Circuit Description:

This DTC is designed to detect high vehicle voltage levels prior to any required motor movement (initialization or ABS). If excessive voltage exists, demagnetization of the motor magnets may occur, which would eventually affect or eliminate ABS performance.

DTC Will Set When:

DTC 37 can only be set if the vehicle's speed is greater than 5 km/h (3 mph). If the switched battery voltage is greater than 17 volts, a malfunction exists.

Action Taken:

A malfunction DTC is stored, ABS is disabled and the ABS warning lamp is turned "ON."

DTC Chart Test Description:

Number(s) below refer to circled number(s) on the diagnostic chart.

1. Checks the voltage level being received by the EBCM.
2. Isolates the malfunction to either the EBCM or the charging system.

Diagnostic Aids:

An "Intermittent" malfunction is most likely caused by a poor connection, rubbed through wire insulation, or a wire that is broken inside the insulation.

The frequency of the malfunction can be checked by using the enhanced diagnostic function of the Tech 1 tool, as described in "Enhanced Diagnostics," found in this section.

Any circuitry, that is suspected as causing the intermittent malfunction, should be thoroughly checked for backed out terminals, improper mating, broken locks, improperly formed or damaged terminals, poor terminal to wiring connections or physical damage to the wiring harness.

DIAGNOSTIC CHART DTC 37 — 1994–95
BERETTA AND CORSICA

DTC 37
HIGH SYSTEM VOLTAGE

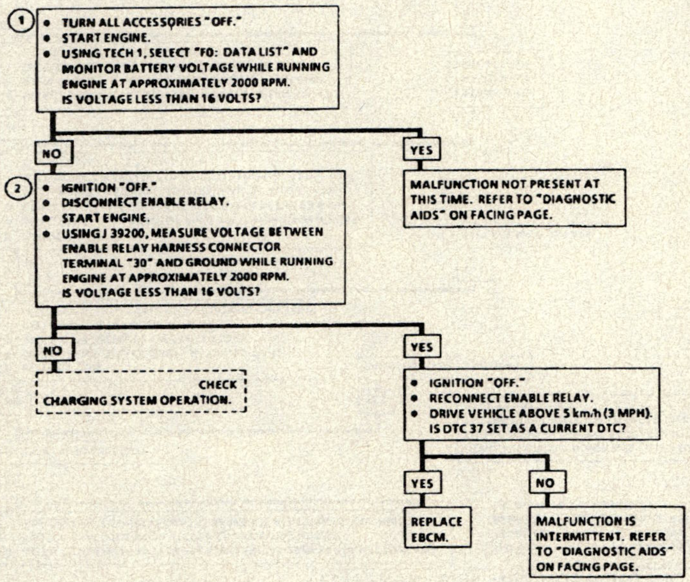

THIS CHART ASSUMES THAT A CURRENT DTC IS STORED INDICATING THAT THIS MALFUNCTION IS PRESENT.

DIAGNOSTIC CHART DTC 38 — 1994–95 BERETTA AND CORSICA

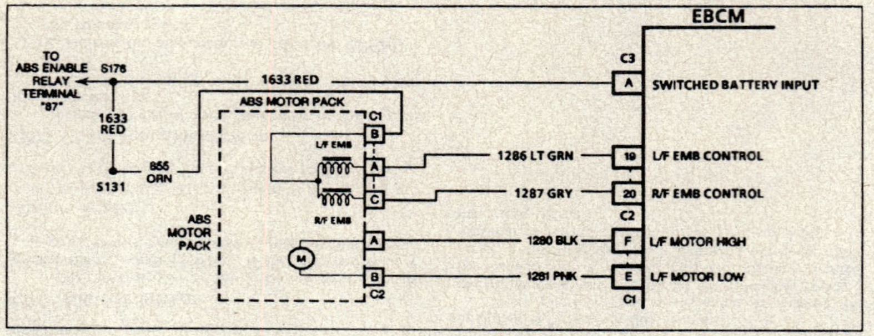

DTC 38
LEFT FRONT EMB WILL NOT HOLD MOTOR

Circuit Description:

Current is supplied to the EMBs through terminal "B" of the three-way motor pack connector. When a release of the EMB is commanded, the EBCM grounds the EMB input and allows the EMB to be energized to release the motor.

DTC Will Set When:

DTC 38 can be set during initialization. A malfunction exists if the feedback current is less than the command current and the left front channel was expected to be in the home position. DTC 86 is always set with DTC 38.

Action Taken:

If an EMB can not hold a motor in the home position, the motor may be back driven when the brake pedal is applied causing the brake pedal to drop. A malfunction DTC is stored, ABS is disabled and the ABS warning lamp is turned "ON."

DTC Chart Test Description:

Number(s) below refer to circled number(s) on the diagnostic chart.

1. Ensures motor circuits are in the proper connector cavity.
2. Checks the ability of the EMB to hold the motor. If the brake pedal moves during this test, it would indicate the EMB is slipping.
3. Checks for high resistance in the left front motor high circuit.
4. Checks for high resistance in the left front motor low circuit.
5. Checks for excessive resistance in the left front motor.
6. Ensures malfunction was not due to a poor connection.

Diagnostic Aids:

An "Intermittent" malfunction in this DTC may result from a mechanical part of the system that sticks, binds, or slips.

The frequency of the malfunction can be checked by using the enhanced diagnostic function of the Tech 1 as described in "Enhanced Diagnostics" found in this section.

NOTICE: This test may fail if low voltage (10 volts or less) was present when initialization occurred. Ensure battery/charging system is functioning properly.

DIAGNOSTIC CHART DTC 38 — 1994–95 BERETTA AND CORSICA

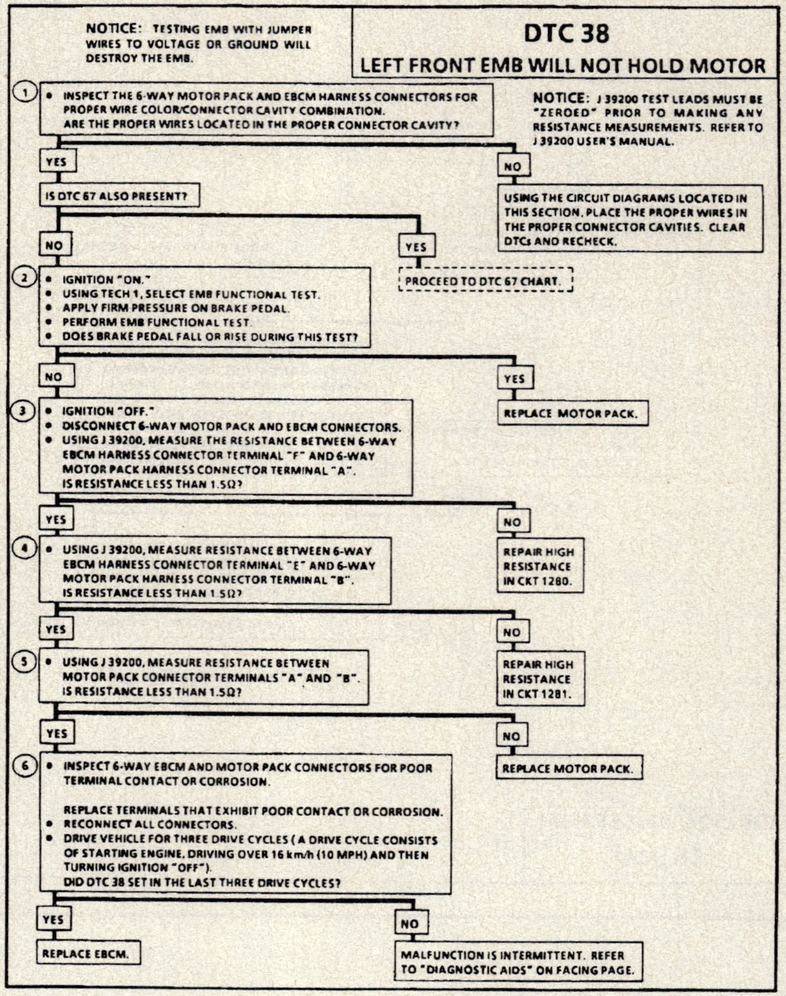

DIAGNOSTIC CHART DTC 41 — 1994–95 BERETTA AND CORSICA

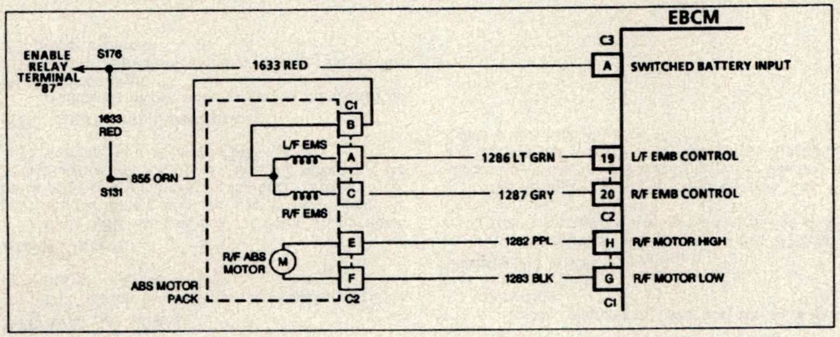

DTC 41
RIGHT FRONT EMB WILL NOT HOLD MOTOR

Circuit Description:

Current is supplied to the EMBs through terminal "B" of the three-way motor pack connector. When a release of the EMB is commanded, the EBCM grounds the EMB input and allows the EMB to be energized to release the motor. The right front motor is commanded in the forward direction while the EMB is released (energized).

DTC Will Set When:

DTC 41 can be set during initialization. A malfunction exists if the feedback current is less than the command current and the right front channel was expected to be in the home position. DTC 86 is always set with DTC 41.

Action Taken:

If an EMB can not hold a motor in the home position, the motor may be back driven when the brake pedal is applied causing the brake pedal to drop. A malfunction DTC is stored, ABS is disabled and the ABS warning lamp is turned "ON."

DTC Chart Test Description:

Number(s) below refer to circled number(s) on the diagnostic chart.

1. Ensures motor circuits are in the proper connector cavity.
2. Checks the ability of the EMB to hold the motor. If the brake pedal moves during this test, it would indicate the EMB is slipping.
3. Checks for high resistance in the right front motor high circuit.
4. Checks for high resistance in the right front motor low circuit.
5. Checks for excessive resistance in the right front motor.
6. Ensures malfunction was not due to a poor connection.

Diagnostic Aids:

An "Intermittent" malfunction in this DTC may result from a mechanical part of the system that sticks, binds, or slips.

The frequency of the malfunction can be checked by using the enhanced diagnostic function of the Tech 1 tool as described in "Enhanced Diagnostics" found in this section.

NOTICE: This DTC may fail if low voltage (10 volts or less) was present when initialization occurred. Ensure battery/charging system is functioning properly.

DIAGNOSTIC CHART DTC 41 — 1994–95 BERETTA AND CORSICA

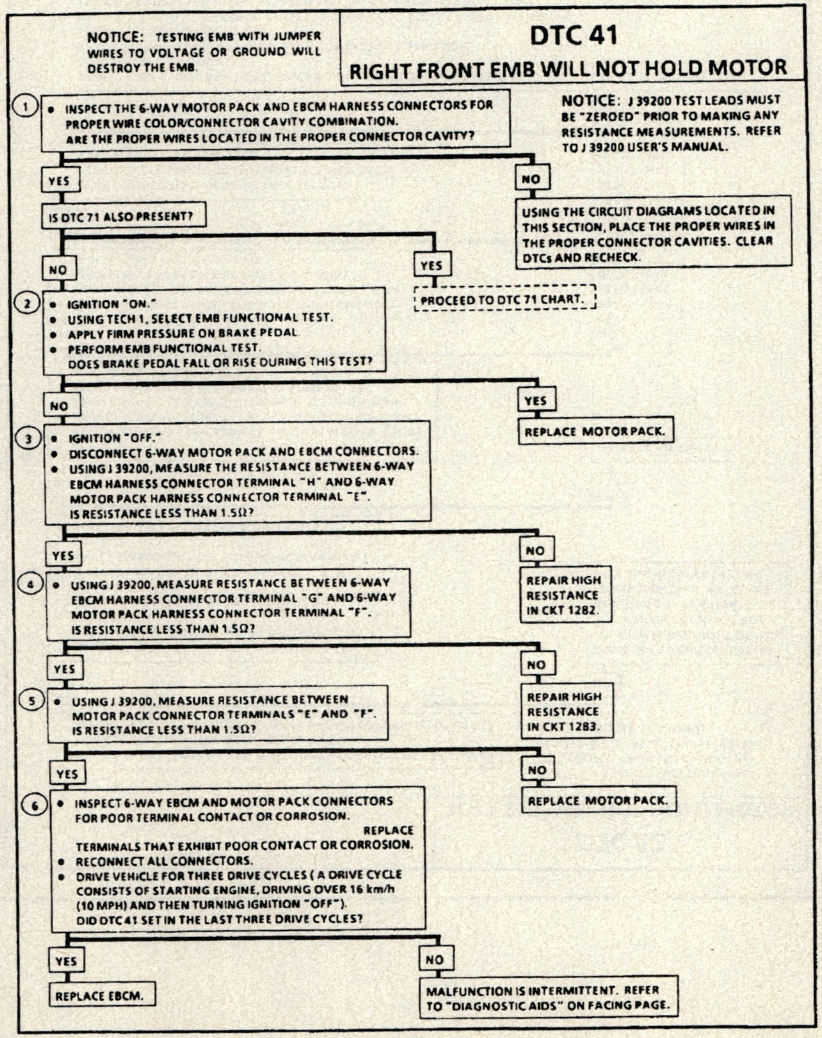

DIAGNOSTIC CHART DTC 42 — 1994–95 BERETTA AND CORSICA

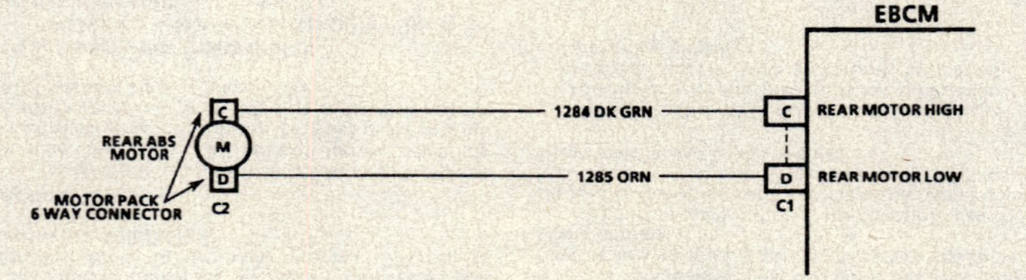

DTC 42
REAR ESB WILL NOT HOLD MOTOR

Circuit Description:

This DTC is designed to detect a slipping rear ESB. During initialization and braking, the rear motor is re homed. If the ESB slips, the motor/piston will move. During the next key "ON" initialization, a rehome of the motor verifies the motor/piston remained at the home position. If motor movement is detected, the ESB must be slipping.

DTC Will Set When:

DTC 42 can be set during initialization. DTC 86 is always set with DTC 42.

Action Taken:

If an ESB can not hold a motor in the home position, the motor may be back-driven when the brake pedal is applied causing the brake pedal to drop. A malfunction DTC is stored, ABS is disabled and the ABS warning lamp is turned "ON."

DTC Chart Test Description:

Number(s) below refer to circled number(s) on the diagnostic chart.

1. This step ensures motor circuits are in the proper connector cavity.
2. This step checks the rear ESB. A broken or defective ESB would result in the rear piston being back driven by hydraulic pressure, and wheel movement would result.
3. Checks for high resistance in the rear motor high circuit.
4. Checks for high resistance in the rear motor low circuit.
5. Checks for excessive resistance in the rear motor.
6. Ensures malfunction was not due to a poor connection.

Diagnostic Aids:

An "Intermittent" malfunction in this DTC may result from a mechanical part of the system that sticks, binds, or slips.

The frequency of the malfunction can be checked by using the enhanced diagnostic function of the Tech 1 as described in "Diagnosis-Electrical" found in this section.

DIAGNOSTIC CHART DTC 42 — 1994–95 BERETTA AND CORSICA

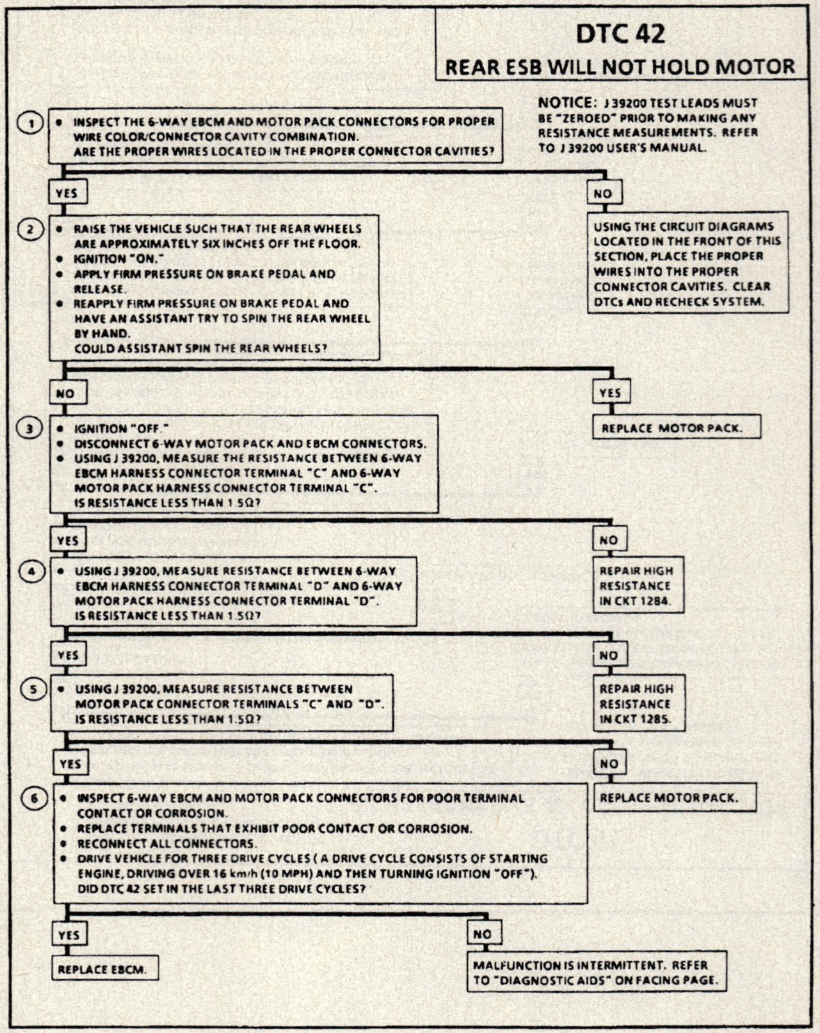

DIAGNOSTIC CHART DTC 44 — 1994–95 BERETTA AND CORSICA

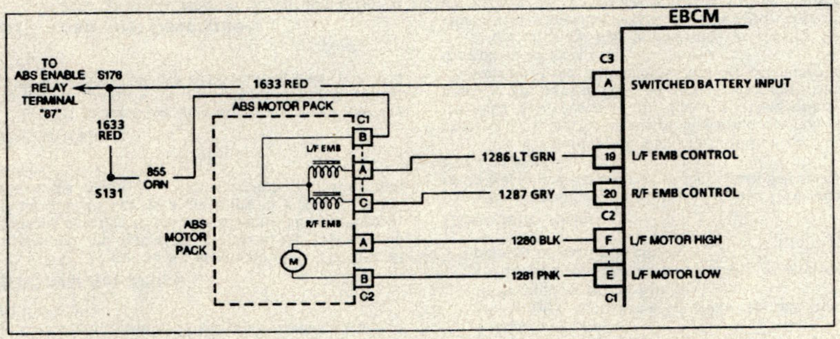

DTC 44
(Page 1 of 2)
LEFT FRONT ABS CHANNEL WILL NOT MOVE

Circuit Description:

Current is supplied to the EMB through terminal "B" of the three-way motor pack connector. When a release of the EMB is commanded, the EBCM grounds the EMB input and allows the EMB to be energized to release the motor. This DTC is designed to detect non-actuating EMB, stuck motor or a seized hydraulic modulator.

DTC Will Set When:

DTC 44 can be set at initialization. The EBCM commands the left front motor in the reverse and forward direction to ensure the motor moves properly. If the EBCM detects a condition in which it cannot move the motor in either direction, a malfunction exists.

Action Taken:

This malfunction indicates the channel can not be moved properly. A malfunction DTC is stored, ABS is disabled and the ABS warning lamp is turned "ON."

DTC Chart Test Description:

Number(s) below refer to circled number(s) on the diagnostic chart.

1. Ensures motor circuits are in the proper connector cavity.
2. Checks for motor movement during release.
3. Compares feedback current to command current during apply.
4. Checks for a short between motor high and low circuits.
5. Checks for a shorted motor.
6. Ensures malfunction was not due to physical damage of the circuitry.
7. Checks if motor can apply properly under load.

Diagnostic Aids:

An "Intermittent" malfunction in this DTC may result from a mechanical part of the system that sticks, binds, or slips.

The frequency of the malfunction can be checked by using the enhanced diagnostic function of the Tech 1 as described in "Enhanced Diagnostics," found in this section. DTC 44 may set after modulator disassembly if the modulator pistons are positioned at the bottom of their bore.

Any circuitry, that is suspected as causing the intermittent complaint, should be thoroughly checked for backed out terminals, improper mating, broken locks, improperly formed or damaged terminals, poor terminal to wiring connections or physical damage to the wiring harness.

NOTICE: This DTC may fail if low voltage (10 volts or less) was present when initialization occurred. Ensure battery/charging system is functioning properly.

DIAGNOSTIC CHART DTC 44 — 1994–95 BERETTA AND CORSICA

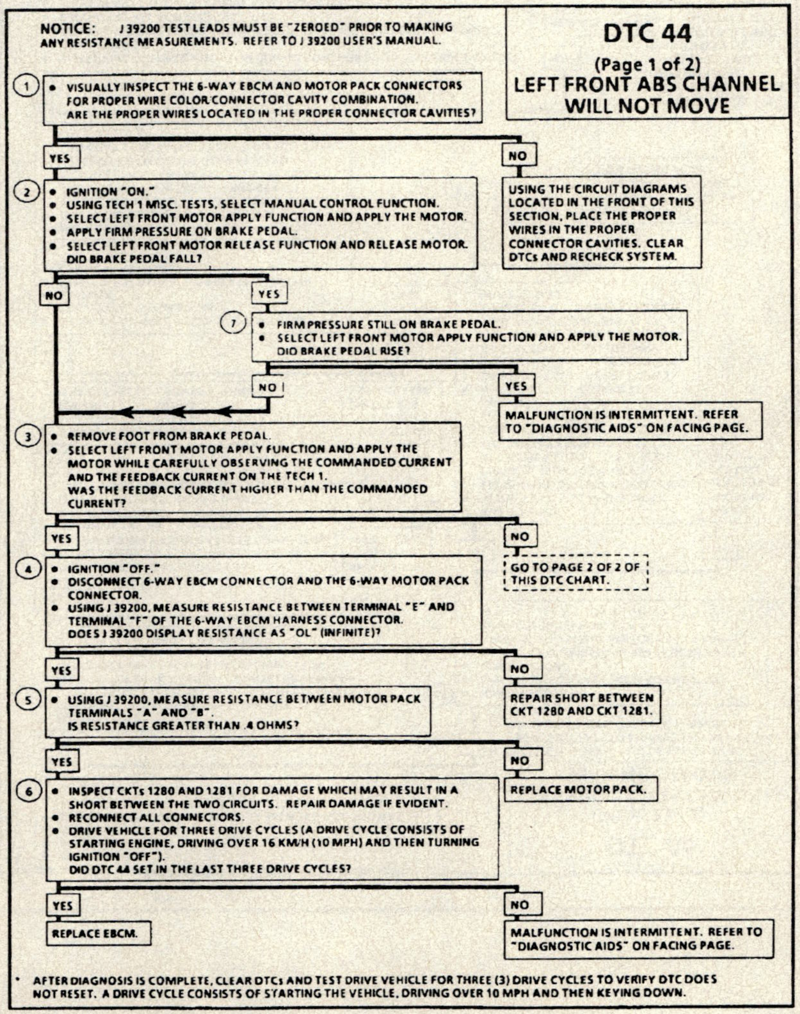

DIAGNOSTIC CHART DTC 44 — 1994–95 BERETTA AND CORSICA

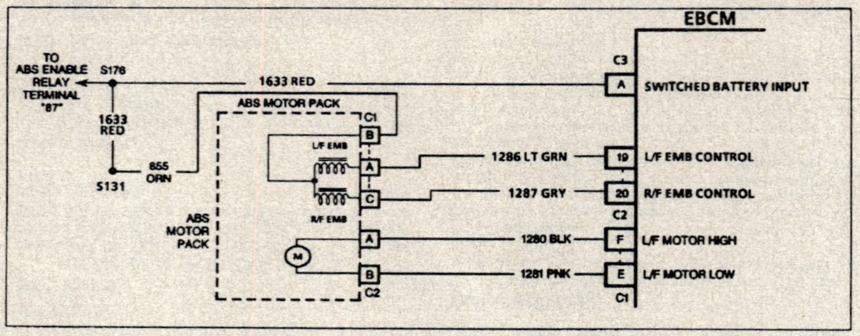

DTC 44
(Page 2 of 2)
LEFT FRONT ABS CHANNEL WILL NOT MOVE

Circuit Description:

Current is supplied to the EMB through terminal "B" of the three-way motor pack connector. When a release of the EMB is commanded, the EBCM grounds the EMB input and allows the EMB to be energized to release the motor. This DTC is designed to detect non-actuating EMB, stuck motor or a seized hydraulic modulator.

DTC Will Set When:

DTC 44 can be set at initialization. The EBCM commands the left front motor in the reverse and forward direction to ensure the motor moves properly. If the EBCM detects a condition in which it cannot move the motor in either direction, a malfunction exists.

Action Taken:

This malfunction indicates the channel can not be moved properly. A malfunction DTC is stored, ABS is disabled and the ABS warning lamp is turned "ON."

DTC Chart Test Description:

Number(s) below refer to circled number(s) on the diagnostic chart.

8. Checks for a short to B+ in the left front EMB control circuit.
9. Checks for high resistance in the left front EMB control circuit.
10. Checks for high resistance in the EMB power feed circuit.
11. Checks for high resistance in the left front EMB.
12. Checks for proper operation of the hydraulic modulator.
13. Ensures malfunction was not due to poor terminal contact.
14. Isolates malfunction to either the left front EMB circuitry or the EBCM.
15. Ensures malfunction was not due to physical damage of the EMB circuit.

Diagnostic Aids:

An "Intermittent" malfunction in this DTC may result from a mechanical part of the system that sticks, binds, or slips.

The frequency of the malfunction can be checked by using the enhanced diagnostic function of the Tech 1 as described in "Enhanced Diagnostics," found in this section. DTC 44 may set after modulator disassembly if the modulator pistons are positioned at the bottom of their bore.

Any circuitry, that is suspected as causing the intermittent complaint, should be thoroughly checked for backed out terminals, improper mating, broken locks, improperly formed or damaged terminals, poor terminal to wiring connections or physical damage to the wiring harness.

NOTICE: This DTC may fail if low voltage (10 volts or less) was present when initialization occurred. Ensure battery/charging system is functioning properly.

DIAGNOSTIC CHART DTC 44 — 1994–95 BERETTA AND CORSICA

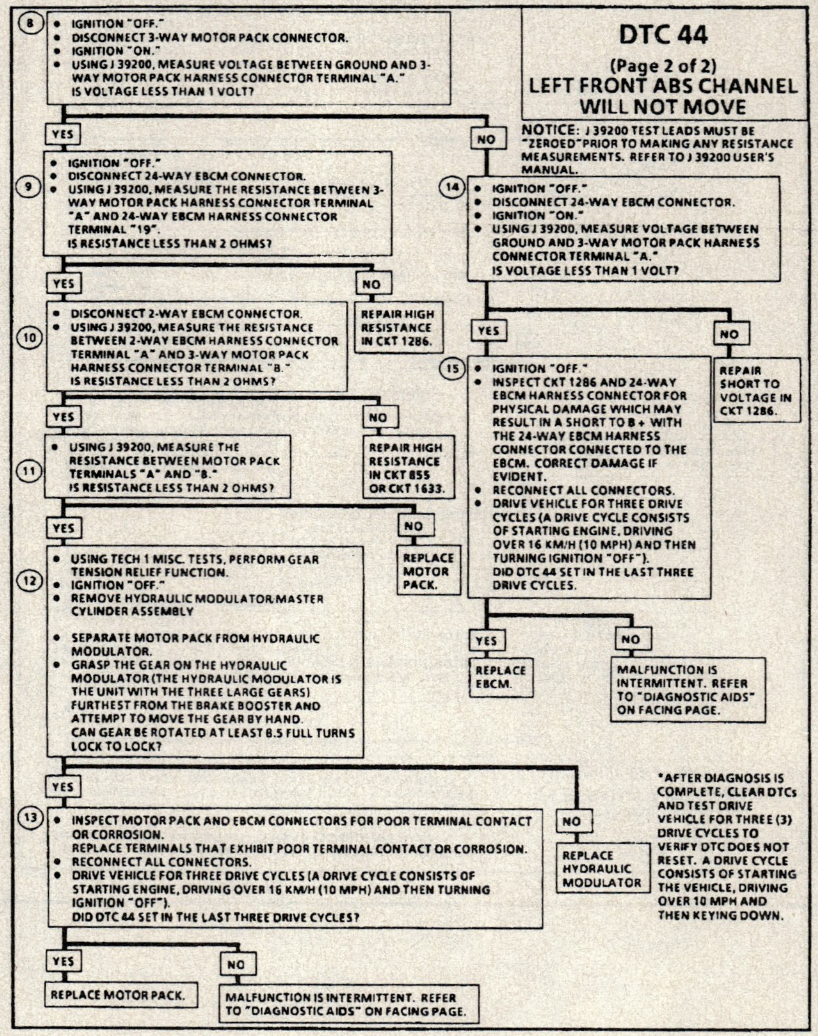

DIAGNOSTIC CHART DTC 45 — 1994–95 BERETTA AND CORSICA

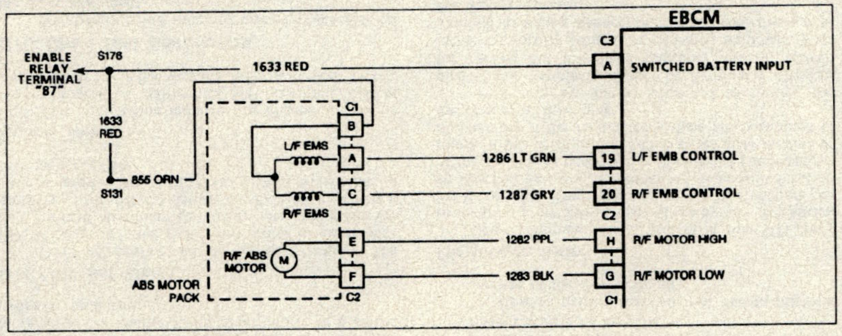

DTC 45
(Page 1 of 2)
RIGHT FRONT ABS CHANNEL WILL NOT MOVE

Circuit Description:

Current is supplied to the EMB through terminal "B" of the three-way motor pack connector. When a release of the EMB is commanded, the EBCM grounds the EMB input and allows the EMB to be energized to release the motor. This DTC is designed to detect non-actuating EMB, stuck motor or a seized hydraulic modulator. When the release is commanded during initialization, the EMB should release the motor, resulting in sensed current being less than commanded current (motor is spinning freely). If the motor is not moving, sensed current will be equal to stall current.

DTC Will Set When:

DTC 45 can be set during initialization. The EBCM commands the right front motor in the reverse and forward direction to ensure the motor moves properly. If the EBCM detects a condition in which it cannot move the motor in either direction, a malfunction exists.

Action Taken:

This malfunction indicates the channel can not be moved properly. A malfunction DTC is stored, ABS is disabled and the ABS warning lamp is turned "ON."

DTC Chart Test Description:

Number(s) below refer to circled number(s) on the diagnostic chart.

1. Ensures motor circuits are in the proper connector cavity.
2. Checks for motor movement during release.
3. Compares feedback current to command current during apply.
4. Checks for a short between motor high and low circuits.
5. Checks for a shorted motor.
6. Ensures malfunction was not due to physical damage of the circuitry.
7. Checks if motor can apply properly under load.

Diagnostic Aids:

An "Intermittent" malfunction in this DTC may result from a mechanical part of the system that sticks, binds, or slips.

The frequency of the malfunction can be checked by using the enhanced diagnostic function of the Tech 1 as described in "Enhanced Diagnostics," found in this section. DTC 45 may set after modulator disassembly if the modulator pistons are positioned at the bottom of their bore.

Any circuitry, that is suspected as causing the intermittent complaint, should be thoroughly checked for backed out terminals, improper mating, broken locks, improperly formed or damaged terminals, poor terminal to wiring connections or physical damage to the wiring harness.

NOTICE: This DTC may fail if low voltage (10 volts or less) was present when initialization occurred. Ensure battery/charging system is functioning properly.

DIAGNOSTIC CHART DTC 45 — 1994–95 BERETTA AND CORSICA

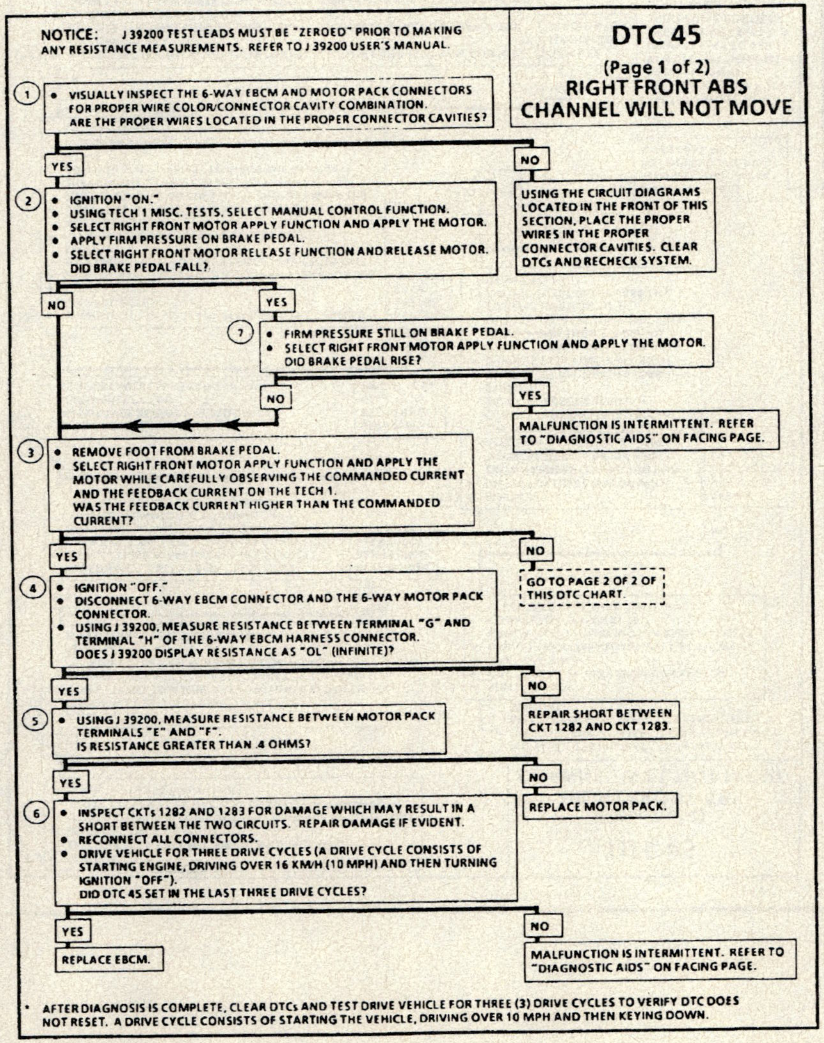

DIAGNOSTIC CHART DTC 45 — 1994–95 BERETTA AND CORSICA

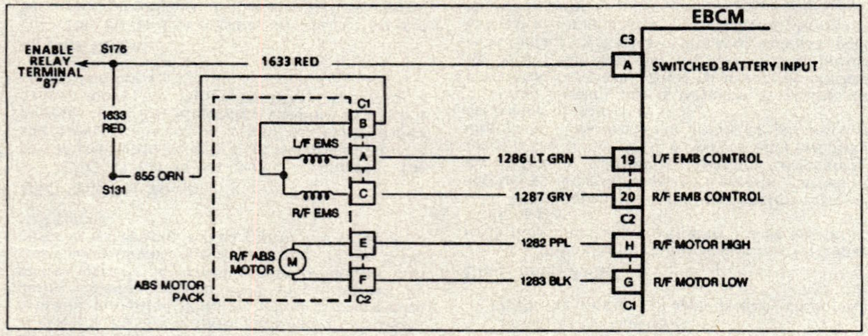

DTC 45
(Page 2 of 2)
RIGHT FRONT ABS CHANNEL WILL NOT MOVE

Circuit Description:

Current is supplied to the EMB through terminal "B" of the three-way motor pack connector. When a release of the EMB is commanded, the EBCM grounds the EMB input and allows the EMB to be energized to release the motor. This DTC is designed to detect non-actuating EMB, stuck motor or a seized hydraulic modulator.

DTC Will Set When:

DTC 45 can be set during initialization. The EBCM commands the right front motor in the reverse and forward direction to ensure the motor moves properly. If the EBCM detects a condition in which it cannot move the motor in either direction, a malfunction exists.

Action Taken:

This malfunction indicates the channel cannot be moved properly. A malfunction DTC is stored, ABS is disabled and the ABS warning lamp is turned "ON."

DTC Chart Test Description:

Number(s) below refer to circled number(s) on the diagnostic chart.

8. Checks for a short to B+ in the right front EMB control circuit.
9. Checks for high resistance in the right front EMB control circuit.
10. Checks for high resistance in the EMB power feed circuit.
11. Checks for high resistance in the right front EMB.
12. Checks for proper operation of the hydraulic modulator.
13. Ensures malfunction was not due to poor terminal contact.
14. Isolates malfunction to either the right front EMB circuitry or the EBCM.
15. Ensures malfunction was not due to physical damage of the EMB circuitry.

Diagnostic Aids:

An "Intermittent" malfunction in this DTC may result from a mechanical part of the system that sticks, binds, or slips. The frequency of the malfunction can be checked by using the enhanced diagnostic function of the Tech 1, as described in "Enhanced Diagnostics," found in this section. DTC 45 may set after modulator disassembly if the modulator pistons are positioned at the bottom of their bore.

Any circuitry, that is suspected as causing the intermittent complaint, should be thoroughly checked for backed out terminals, improper mating, broken locks, improperly formed or damaged terminals, poor terminal to wiring connections or physical damage to the wiring harness.

NOTICE: This DTC may fail if low voltage (10 volts or less) was present when initialization occurred. Ensure battery/charging system is functioning properly.

DIAGNOSTIC CHART DTC 45 — 1994–95 BERETTA AND CORSICA

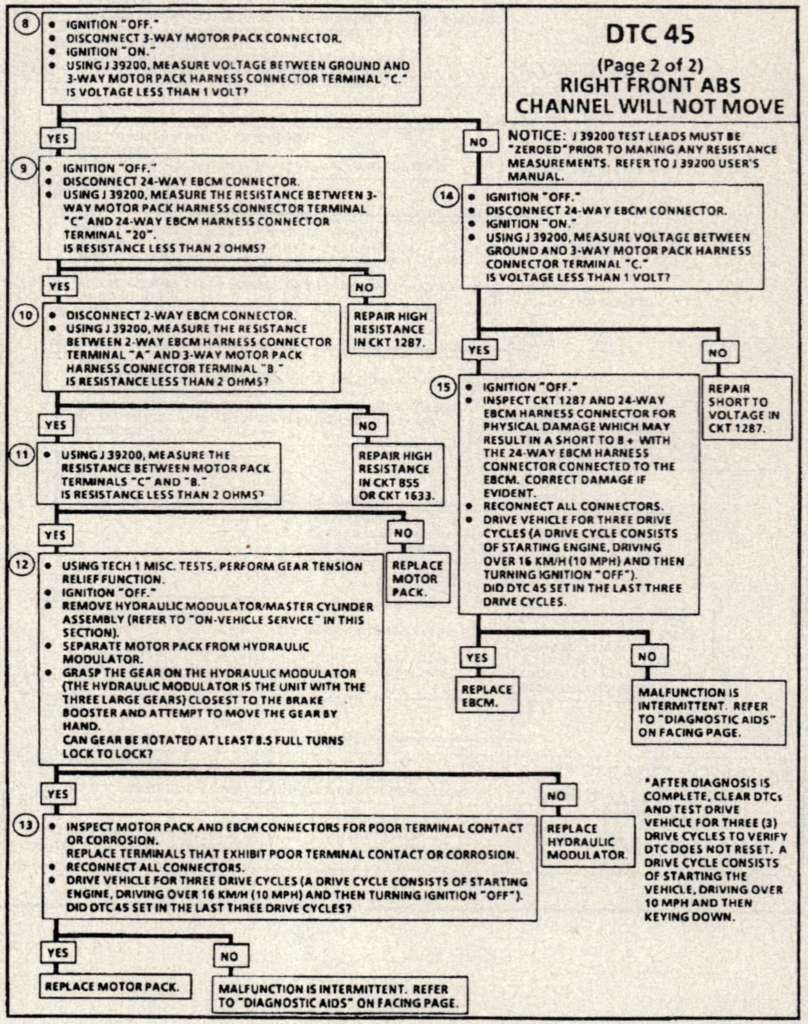

DIAGNOSTIC CHART DTC 46 — 1994–95 BERETTA AND CORSICA

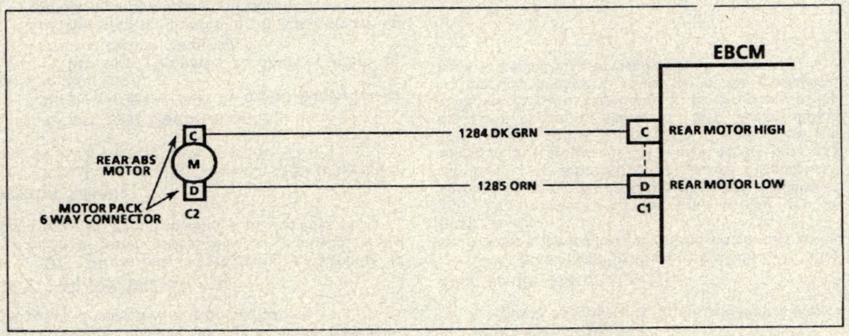

DTC 46
REAR ABS CHANNEL WILL NOT MOVE

Circuit Description:

This DTC is designed to detect bound-up ESBs, stuck motor or a seized hydraulic modulator. Release is commanded during initialization, the ESB should release the motor, resulting in sensed current being less than commanded current (motor is spinning freely). If the motor is not moving, sensed current will be equal to stall current.

DTC Will Set When:

DTC 46 can be set during initialization. The EBCM commands the rear motor in the reverse and forward direction to ensure the motor moves properly. If the EBCM detects a condition in which it cannot move the motor in either direction, a malfunction exists.

Action Taken:

This malfunction indicates the channel can not be moved properly. A malfunction DTC is stored, ABS is disabled, and the ABS warning lamp is turned "ON."

DTC Chart Test Description:

Number(s) below refer to circled number(s) on the diagnostic chart.

1. This step ensures motor circuits are in the proper connector cavity.
2. This checks the ability of the motor to move applying hydraulic pressure to the wheel cylinders. A frozen piston or gear would result in no apply of rear brakes and a wheel that freely spins.
3. This procedure compares feedback current to command current during apply.
4. This step checks to see if hydraulic modulator is functioning properly.
5. This step checks if motor can apply properly under load.
6. This step serves to isolate the malfunction to either the EBCM or a short between CKTs 1284 and 1285.
7. Checks for a shorted motor.
8. Ensures the malfunction was not due to physical damage of the circuitry.

Diagnostic Aids:

An "Intermittent" malfunction in this DTC may result from a mechanical part of the system that sticks or binds. The frequency of the malfunction can be checked by using the enhanced diagnostic function of the Tech 1, as described in "Enhanced Diagnostics," found in this section. DTC 46 may set after modulator disassembly if the modulator pistons are positioned at the bottom of their bore.

Depending on the frequency of the malfunction, a physical inspection of the mechanical parts suspected may be necessary.

Any circuitry, that is suspected as causing the intermittent complaint, should be thoroughly checked for backed out terminals, improper mating, broken locks, improperly formed or damaged terminals, poor terminal to wiring connections or physical damage to the wiring harness.

DIAGNOSTIC CHART DTC 46 — 1994–95 BERETTA AND CORSICA

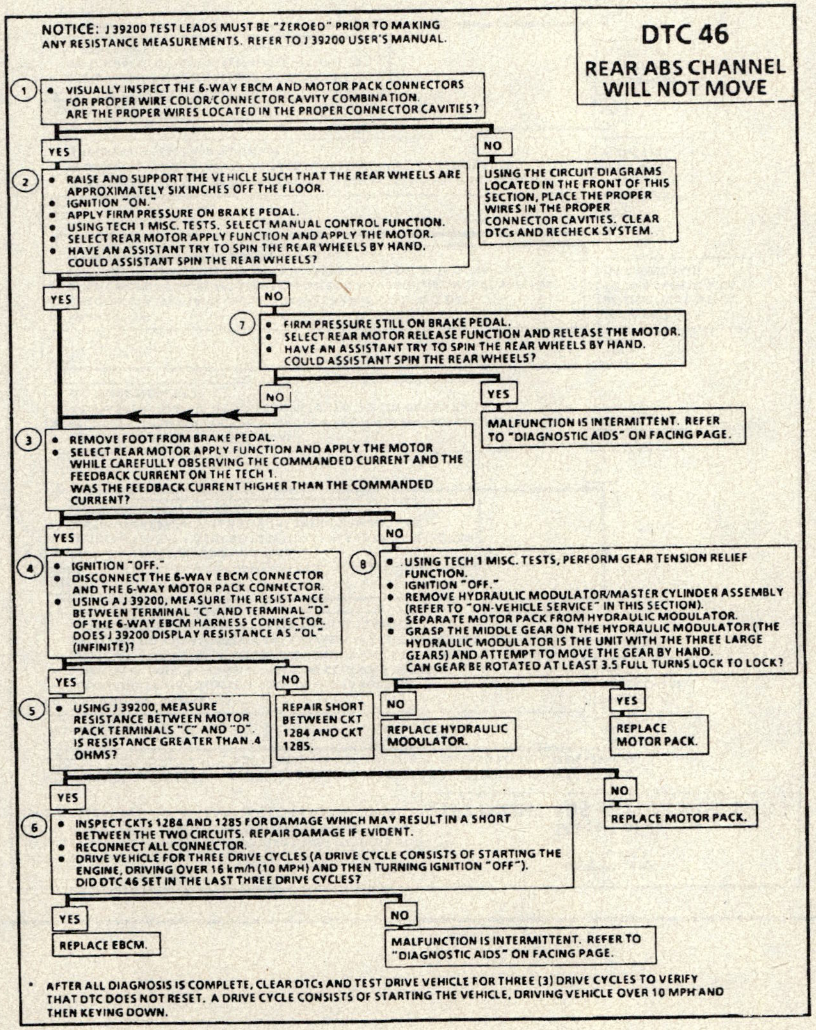

DIAGNOSTIC CHART DTC 47 — 1994–95
BERETTA AND CORSICA

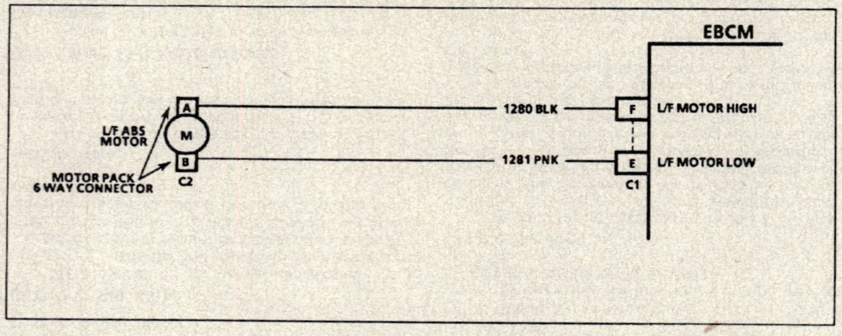

DTC 47

LEFT FRONT ABS MOTOR FREE SPINS

Circuit Description:

This DTC is designed to detect a stripped nut or gear assembly during initialization. During the homing sequence, the piston should reach the top of the bore resulting in a stalled motor. If this does not occur, the motor must be spinning with little or no resistance, indicating a nut/screw or gear malfunction.

DTC Will Set When:

DTC 47 can only be set during initialization. If the feedback current is less than the command current for a specified period to time, a malfunction exists.

Action Taken:

A malfunction DTC is stored, ABS is disabled, and the ABS warning lamp is turned "ON."

DTC Chart Test Description:

Number(s) below refer to circled number(s) on the diagnostic chart.

1. This step compares feedback current to command current during apply.
2. This step compares feedback current to command current during release.
3. Steps 1, 2 and 3 verify malfunction exists. This step also moves the position to the top of the bore.
4. Checks for excessive resistance in the left front motor high circuit.
5. Checks for excessive resistance in the left front motor low circuit.
6. Checks for resistance in the left front motor.

Diagnostic Aids:

An "Intermittent" malfunction in this DTC may result from a mechanical part of the system that sticks, binds, or slips.

The frequency of the malfunction can be checked by using the enhanced diagnostic function of the Tech 1 as described in "Enhanced Diagnostics," found in this section. If DTC only occurs once and DTC 56 also occurs see DTC 56. If intermittent and enhanced diagnostics shows this DTC occurs during ABS, see DTC 56. Depending on the frequency of the malfunction, a physical inspection of the mechanical parts suspected may be necessary.

DIAGNOSTIC CHART DTC 47 — 1994–95
BERETTA AND CORSICA

AFTER DIAGNOSIS IS COMPLETE, CLEAR DTCs AND TEST DRIVE VEHICLE FOR THREE(3) DRIVE CYCLES TO VERIFY THAT DTC DOES NOT RESET. A DRIVE CYCLE CONSISTS OF STARTING THE ENGINE, DRIVING THE VEHICLE ABOVE 16 km/h (10 MPH) AND THEN KEYING DOWN.

DTC 47
(Page 1 of 2)
LEFT FRONT ABS MOTOR FREE SPINS

NOTICE: J 39200 TEST LEADS MUST BE "ZEROED" PRIOR TO MAKING ANY RESISTANCE MEASUREMENTS. REFER TO J 39200 USER'S MANUAL.

IS DTC 56 ALSO PRESENT?

- YES: PROCEED TO DTC 56 CHART FIRST.
- NO: go to step 1.

(1)
- IGNITION "ON," ENGINE "OFF."
- PUMP BRAKE PEDAL TO DEPLETE VACUUM POWER BOOSTER.
- USING TECH 1 MISC. TESTS, SELECT MANUAL CONTROL FUNCTION.
- SELECT LEFT FRONT MOTOR APPLY FUNCTION AND APPLY THE MOTOR WHILE CAREFULLY OBSERVING THE COMMAND CURRENT AND FEEDBACK CURRENT ON THE TECH 1. WERE BOTH COMMAND AND FEEDBACK CURRENT 10 AMPS?

YES → (2); NO → GO TO STEP 4.

(2)
- APPLY FIRM PRESSURE ON BRAKE PEDAL.
- SELECT LEFT FRONT MOTOR RELEASE FUNCTION AND RELEASE MOTOR. DID THE BRAKE PEDAL FALL AND WERE BOTH COMMAND AND FEEDBACK CURRENT 6 AMPS?

YES → (3); NO → GO TO STEP 4.

(3)
- FIRM PRESSURE STILL ON BRAKE PEDAL.
- SELECT LEFT FRONT MOTOR APPLY FUNCTION AND APPLY THE MOTOR (THIS MOVES THE HYDRAULIC MODULATOR PISTON TO THE TOP OF ITS BORE). DID BRAKE PEDAL RISE?

NO → (4); YES → MALFUNCTION IS INTERMITTENT. REFER TO "DIAGNOSTIC AIDS" ON FACING PAGE.

(4)
- IGNITION "OFF."
- DISCONNECT 6-WAY EBCM AND MOTOR PACK HARNESS CONNECTORS.
- USING J 39200, MEASURE RESISTANCE BETWEEN 6-WAY EBCM HARNESS CONNECTOR TERMINAL "F" AND 6-WAY MOTOR PACK HARNESS CONNECTOR TERMINAL "A". IS RESISTANCE LESS THAN 1.5 OHMS?

YES → (5); NO → REPAIR HIGH RESISTANCE IN CKT 1280.

(5)
- USING J 39200, MEASURE RESISTANCE BETWEEN 6-WAY EBCM HARNESS CONNECTOR TERMINAL "E" AND 6-WAY MOTOR PACK HARNESS CONNECTOR TERMINAL "B". IS RESISTANCE LESS THAN 1.5 OHMS?

YES → (6); NO → REPAIR HIGH RESISTANCE IN CKT 1281.

(6)
- USING J 39200, MEASURE RESISTANCE BETWEEN 6-WAY MOTOR PACK CONNECTOR TERMINALS "A" AND "B". IS RESISTANCE LESS THAN 1.5 OHMS?

YES → GO TO PAGE 2 OF 2 OF THIS CHART.; NO → REPLACE MOTOR PACK.

DIAGNOSTIC CHART DTC 47 — 1994–95 BERETTA AND CORSICA

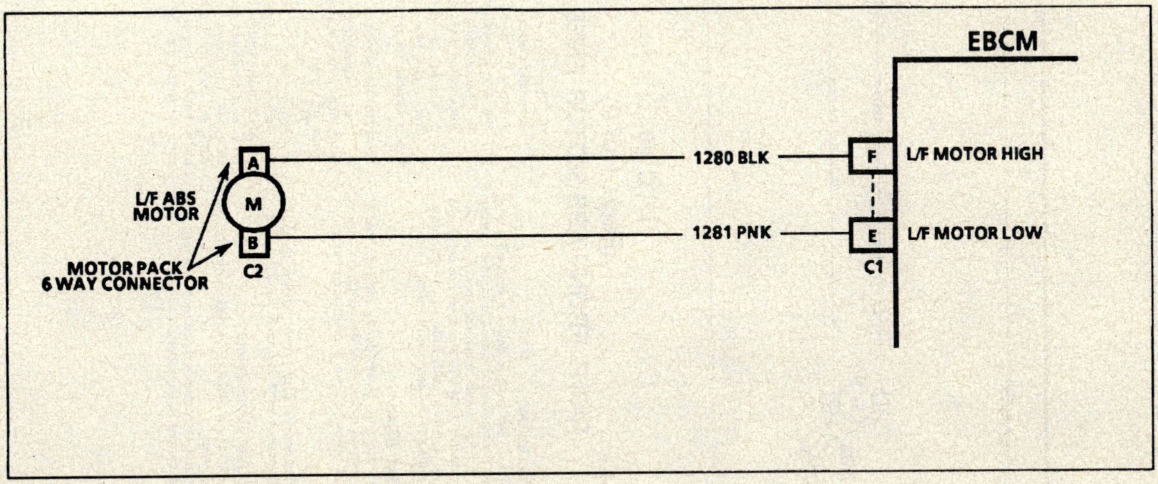

DTC 47
(Page 2 of 2)
LEFT FRONT ABS MOTOR FREE SPINS

Circuit Description:

This DTC is designed to detect a stripped nut or gear assembly during initialization. During the homing sequence, the piston should reach the top of the bore resulting in a stalled motor. If this does not occur, the motor must be spinning with little or no resistance, indicating a nut/screw or gear malfunction.

DTC Will Set When:

DTC 47 can only be set during initialization. If the feedback current is less than the command current for a specified period to time, a malfunction exists.

Action Taken:

A malfunction DTC is stored, ABS is disabled, and the ABS warning lamp is turned "ON."

DTC Chart Test Description:

Number(s) below refer to circled number(s) on the diagnostic .chart

7. Checks for a stripped motor pack gear.
8. Checks for a stripped hydraulic modulator gear.
9. Checks for proper operation of motor pack.
10. Checks for proper operation of the hydraulic modulator.
11. Ensures malfunction was not due to poor terminal contact.

Diagnostic Aids:

An "Intermittent" malfunction in this DTC may result from a mechanical part of the system that sticks, binds, or slips.

The frequency of the malfunction can be checked by using the enhanced diagnostic function of the Tech 1, as described in "Enhanced Diagnostics," found in this section. If DTC only occurs once and DTC 56 also occurs, see DTC 56. If intermittent and enhanced diagnostics shows this DTC occurs during ABS, see DTC 56.

Depending on the frequency of the malfunction, a physical inspection of the mechanical parts suspected may be necessary.

DIAGNOSTIC CHART DTC 47 — 1994–95 BERETTA AND CORSICA

DTC 47
(Page 2 of 2)
LEFT FRONT ABS MOTOR FREE SPINS

CONTINUED FROM PAGE 1 OF 2 OF THIS CHART.

(7)
- REMOVE HYDRAULIC MODULATOR/MASTER CYLINDER ASSEMBLY.
- REMOVE GEAR COVER.
- CHECK FOR STRIPPED GEAR ON MOTOR PACK; THE MOTOR PACK IS THE UNIT WITH THE THREE SMALL GEARS. THE LEFT FRONT GEAR IS FURTHEST FROM THE BRAKE BOOSTER WHEN THE ASSEMBLY IS INSTALLED IN THE VEHICLE. IS THE MOTOR PACK GEAR STRIPPED?

YES: REPLACE MOTOR PACK.

NO:

(8)
- CHECK FOR A STRIPPED GEAR ON THE HYDRAULIC MODULATOR; THE HYDRAULIC MODULATOR IS THE UNIT WITH THE THREE LARGE GEARS. THE LEFT FRONT GEAR IS FURTHEST FROM THE BRAKE BOOSTER WHEN THE ASSEMBLY IS INSTALLED IN THE VEHICLE. IS THE HYDRAULIC MODULATOR GEAR STRIPPED?

YES: REPLACE HYDRAULIC MODULATOR GEAR.

NO:

(9)
- SECURELY POSITION HYDRAULIC MODULATOR/MASTER CYLINDER ASSEMBLY IN VEHICLE ALLOWING LEFT FRONT GEAR SET OBSERVATION. BE CAREFUL NOT TO DAMAGE GEARS OR ALLOW THE HYDRAULIC MODULATOR MASTER CYLINDER ASSEMBLY TO MOVE DURING TESTING.
- RECONNECT ELECTRICAL CONNECTORS.
- IGNITION "ON."
- USING TECH 1 MISC. TESTS, SELECT MANUAL CONTROL FUNCTION.
- SELECT LEFT FRONT MOTOR APPLY AND APPLY THE MOTOR WHILE OBSERVING THE GEAR SET FURTHEST FORWARD FROM THE BOOSTER WHEN THE ASSEMBLY IS INSTALLED IN THE VEHICLE.
- SELECT LEFT FRONT MOTOR RELEASE FUNCTION AND RELEASE THE MOTOR WHILE STILL OBSERVING THE GEAR SET.
- SELECT LEFT FRONT MOTOR APPLY FUNCTION AND APPLY THE MOTOR. DID GEAR SET MOVE IN BOTH DIRECTIONS FOR AT LEAST ONE REVOLUTION?

NO: REPLACE MOTOR PACK.

YES:

(10)
- USING TECH 1, SELECT "F4: MISC. TESTS" AND PERFORM GEAR TENSION RELIEF FUNCTION.
- IGNITION "OFF."
- SEPARATE MOTOR PACK FROM HYDRAULIC MODULATOR.
- GRASP THE LEFT FRONT GEAR ON THE HYDRAULIC MODULATOR; THE HYDRAULIC MODULATOR IS THE UNIT WITH THE THREE LARGE GEARS. THE LEFT FRONT GEAR IS FURTHEST FROM THE BRAKE BOOSTER WHEN THE ASSEMBLY IS INSTALLED IN THE VEHICLE.
- ATTEMPT TO ROTATE THE GEAR. CAN THE GEAR BE ROTATED MORE THAN 9 TURNS LOCK TO LOCK?

YES: REPLACE HYDRAULIC MODULATOR.

NO:

(11)
- INSPECT 6-WAY MOTOR PACK AND EBCM CONNECTORS FOR CORROSION OR POOR TERMINAL CONTACT. REPLACE TERMINALS THAT EXHIBIT POOR TERMINAL CONTACT OR CORROSION.
- INSTALL HYDRAULIC MODULATOR/MASTER CYLINDER ASSEMBLY.
- RECONNECT ALL CONNECTORS.
- DRIVE VEHICLE FOR THREE DRIVE CYCLES (A DRIVE CYCLE CONSISTS OF STARTING ENGINE, DRIVING OVER 16 km/h (10 MPH) AND THEN TURNING IGNITION "OFF"). DID DTC 47 SET IN THE LAST THREE DRIVE CYCLES?

YES: REPLACE EBCM.

NO: MALFUNCTION WAS DUE TO A POOR CONNECTION.

DIAGNOSTIC CHART DTC 48 — 1994–95 BERETTA AND CORSICA

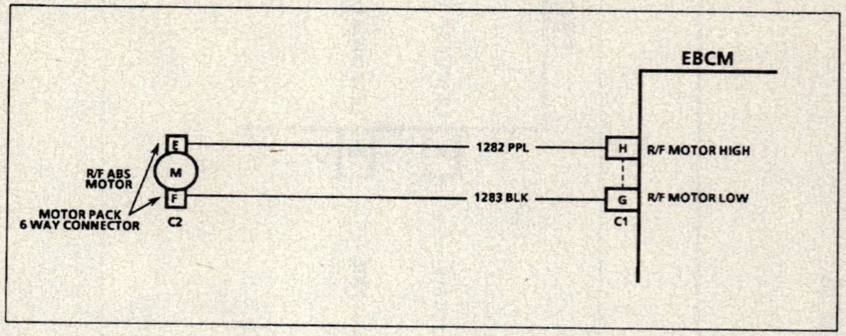

DTC 48
(Page 1 of 2)
RIGHT FRONT ABS MOTOR FREE SPINS

Circuit Description:

This DTC is designed to detect a stripped nut or gear assembly during initialization. During the homing sequence, the piston should reach the top of the bore resulting in a stalled motor. If this does not occur, the motor must be spinning with little or no resistance, indicating a nut/screw or gear malfunction.

DTC Will Set When:

DTC 48 can only be set during initialization. If the feedback current is less than the command current for a specified period of time, a malfunction exists.

Action Taken:

A malfunction DTC is stored, ABS is disabled, and the ABS warning lamp is turned "ON."

DTC Chart Test Description:

Number(s) below refer to circled number(s) on the diagnostic chart.

1. This step compares feedback current to command current during apply.
2. This step compares feedback current to command current during release.
3. Steps 1, 2 and 3 verify malfunction exists. This step also moves the piston to the top of the bore.
4. Checks for excessive resistance in the right front motor high circuit.
5. Checks for excessive resistance in the right front motor low circuit.
6. Checks for resistance in the right front motor.

Diagnostic Aids:

An "Intermittent" malfunction in this DTC may result from a mechanical part of the system that sticks, binds, or slips.

The frequency of the malfunction can be checked by using the enhanced diagnostic function of the Tech 1 as described in "Enhanced Diagnostics," found in this section. If DTC only occurs once and DTC 61 also occurs see DTC 61. If intermittent and enhanced diagnostics shows this DTC occurs during ABS, see DTC 61. Depending on the frequency of the malfunction, a physical inspection of the mechanical parts suspected may be necessary.

DIAGNOSTIC CHART DTC 48 — 1994–95 BERETTA AND CORSICA

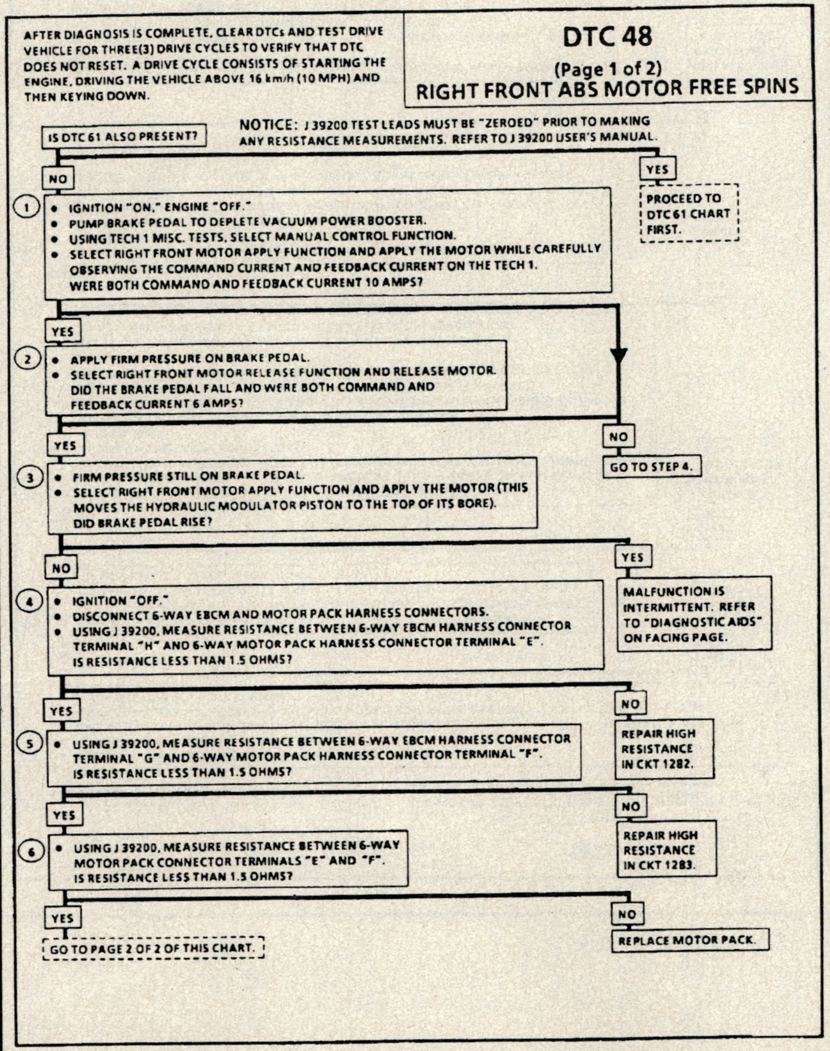

DIAGNOSTIC CHART DTC 48 — 1994–95 BERETTA AND CORSICA

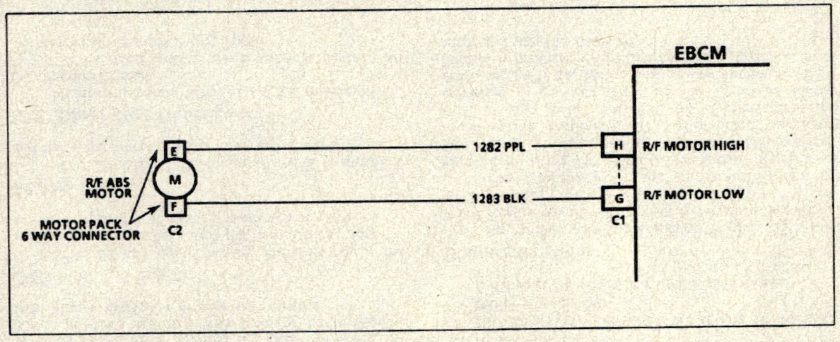

DTC 48
(Page 2 of 2)
RIGHT FRONT ABS MOTOR FREE SPINS

Circuit Description:

This DTC is designed to detect a stripped nut or gear assembly during initialization. During the homing sequence, the piston should reach the top of the bore resulting in a stalled motor. If this does not occur, the motor must be spinning with little or no resistance, indicating a nut/screw or gear malfunction.

DTC Will Set When:

DTC 48 can only be set during initialization. If the feedback current is less than the command current for a specified period of time, a malfunction exists.

Action Taken:

A malfunction DTC is stored, ABS is disabled, and the ABS warning lamp is turned "ON."

DTC Chart Test Description:

Number(s) below refer to circled number(s) on the diagnostic chart.

7. Checks for a stripped motor pack gear.
8. Checks for a stripped hydraulic modulator gear.
9. Checks for proper operation of motor pack.
10. Checks for proper operation of the hydraulic modulator.
11. Ensures malfunction was not due to poor terminal contact.

Diagnostic Aids:

An "Intermittent" malfunction in this DTC may result from a mechanical part of the system that sticks, binds, or slips.

The frequency of the malfunction can be checked by using the enhanced diagnostic function of the Tech 1, as described in "Enhanced Diagnostics," found in this section. If DTC only occurs once and DTC 61 also occurs, see DTC 61. If intermittent and enhanced diagnostics shows this DTC occurs during ABS, see DTC 61.

Depending on the frequency of the malfunction, a physical inspection of the mechanical parts suspected may be necessary.

DIAGNOSTIC CHART DTC 48 — 1994–95 BERETTA AND CORSICA

DTC 48
(Page 2 of 2)
RIGHT FRONT ABS MOTOR FREE SPINS

CONTINUED FROM PAGE 1 OF 2 OF THIS CHART.

7
- REMOVE HYDRAULIC MODULATOR/MASTER CYLINDER ASSEMBLY.
- REMOVE GEAR COVER.
- CHECK FOR STRIPPED GEAR ON MOTOR PACK; THE MOTOR PACK IS THE UNIT WITH THE THREE SMALL GEARS. THE RIGHT FRONT GEAR IS CLOSEST TO THE BRAKE BOOSTER WHEN THE ASSEMBLY IS INSTALLED IN THE VEHICLE. IS THE MOTOR PACK GEAR STRIPPED?

YES: REPLACE MOTOR PACK.

NO:

8
- CHECK FOR A STRIPPED GEAR ON THE HYDRAULIC MODULATOR; THE HYDRAULIC MODULATOR IS THE UNIT WITH THE THREE LARGE GEARS. THE RIGHT FRONT GEAR IS CLOSEST TO THE BRAKE BOOSTER WHEN THE ASSEMBLY IS INSTALLED IN THE VEHICLE. IS THE HYDRAULIC MODULATOR GEAR STRIPPED?

YES: REPLACE HYDRAULIC MODULATOR GEAR.

NO:

9
- SECURELY POSITION HYDRAULIC MODULATOR/MASTER CYLINDER ASSEMBLY IN VEHICLE ALLOWING RIGHT FRONT GEAR SET OBSERVATION. BE CAREFUL NOT TO DAMAGE GEARS OR ALLOW THE HYDRAULIC MODULATOR/MASTER CYLINDER ASSEMBLY TO MOVE DURING TESTING.
- RECONNECT ELECTRICAL CONNECTORS.
- IGNITION "ON."
- USING TECH 1 MISC. TESTS, SELECT MANUAL CONTROL FUNCTION.
- SELECT RIGHT FRONT MOTOR APPLY AND APPLY THE MOTOR WHILE OBSERVING THE GEAR SET CLOSEST TO THE BOOSTER WHEN THE ASSEMBLY IS INSTALLED IN THE VEHICLE.
- SELECT RIGHT FRONT MOTOR RELEASE FUNCTION AND RELEASE THE MOTOR WHILE STILL OBSERVING THE GEAR SET.
- SELECT RIGHT FRONT MOTOR APPLY FUNCTION AND APPLY THE MOTOR. DID GEAR SET MOVE IN BOTH DIRECTIONS FOR AT LEAST ONE REVOLUTION?

NO: REPLACE MOTOR PACK.

YES:

10
- USING TECH 1, SELECT "F4: MISC. TESTS" AND PERFORM GEAR TENSION RELIEF FUNCTION.
- IGNITION "OFF."
- SEPARATE MOTOR PACK FROM HYDRAULIC MODULATOR.
- GRASP THE RIGHT FRONT GEAR ON THE HYDRAULIC MODULATOR; THE HYDRAULIC MODULATOR IS THE UNIT WITH THE THREE LARGE GEARS. THE RIGHT FRONT GEAR IS CLOSEST TO THE BRAKE BOOSTER WHEN THE ASSEMBLY IS INSTALLED IN THE VEHICLE.
- ATTEMPT TO ROTATE THE GEAR. CAN THE GEAR BE ROTATED MORE THAN 9 TURNS LOCK TO LOCK?

YES: REPLACE HYDRAULIC MODULATOR.

NO:

11
- INSPECT 6-WAY MOTOR PACK AND EBCM CONNECTORS FOR CORROSION OR POOR TERMINAL CONTACT. REPLACE TERMINALS THAT EXHIBIT POOR TERMINAL CONTACT OR CORROSION.
- INSTALL HYDRAULIC MODULATOR/MASTER CYLINDER ASSEMBLY.
- RECONNECT ALL CONNECTORS.
- DRIVE VEHICLE FOR THREE DRIVE CYCLES (A DRIVE CYCLE CONSISTS OF STARTING ENGINE, DRIVING OVER 16 km/h (10 MPH) AND THEN TURNING IGNITION "OFF"). DID DTC 48 SET IN THE LAST THREE DRIVE CYCLES?

YES: REPLACE EBCM.

NO: MALFUNCTION WAS DUE TO A POOR CONNECTION.

DIAGNOSTIC CHART DTC 51 — 1994–95 BERETTA AND CORSICA

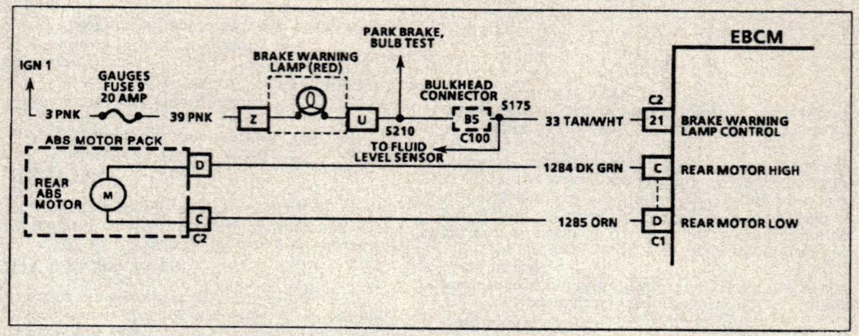

DTC 51

REAR ABS MOTOR FREE SPINS

Circuit Description:

This DTC is designed to detect a stripped nut or gear assembly during initialization. During the homing sequence, the piston should reach the top of the bore resulting in a stalled motor. If this does not occur, the motor must be spinning with little or no resistance, indicating a nut/screw or gear malfunction.

DTC Will Set When:

DTC 51 can be set during initialization. If the feedback current is less than the command current for a specified period of time, a malfunction exists.

Action Taken:

A malfunction DTC is stored, ABS is disabled, and the ABS warning lamp is turned "ON."

DTC Chart Test Description:

Number(s) below refer to circled number(s) on the diagnostic chart.

1. This step compares feedback current to command current during apply.
2. This step compares feedback current to command current during release.
3. Steps 1, 2 and 3 verify malfunction exists. This step also moves the piston to the top of the bore.
4. Checks for excessive resistance in the rear motor high circuit.
5. Checks for excessive resistance in the rear motor low circuit.
6. Checks for resistance in the rear motor.

Diagnostic Aids:

An "Intermittent" malfunction in this DTC may result from a mechanical part of the system that sticks, binds, or slips.

The frequency of the malfunction can be checked by using the enhanced diagnostic function of the Tech 1 as described in "Enhanced Diagnostics," found in this section. If DTC only occurs once and DTC 64 also occurs, see DTC 64. If intermittent and enhanced diagnostics show this DTC occurs during ABS, see DTC 64. Depending on the frequency of the failure, a physical inspection of the mechanical parts suspected may be necessary.

DIAGNOSTIC CHART DTC 51 — 1994–95 BERETTA AND CORSICA

DTC 51 (Page 1 of 2) REAR ABS MOTOR FREE SPINS

AFTER DIAGNOSIS IS COMPLETE, CLEAR DTCs AND TEST DRIVE VEHICLE FOR THREE(3) DRIVE CYCLES TO VERIFY THAT DTC DOES NOT RESET. A DRIVE CYCLE CONSISTS OF STARTING THE ENGINE, DRIVING THE VEHICLE ABOVE 16 km/h (10 MPH) AND THEN KEYING DOWN.

NOTICE: J 39200 TEST LEADS MUST BE "ZEROED" PRIOR TO MAKING ANY RESISTANCE MEASUREMENTS. REFER TO J 39200 USER'S MANUAL.

IS DTC 64 ALSO PRESENT?

- YES: PROCEED TO DTC 64 CHART FIRST.
- NO: go to step 1.

1.
   - IGNITION "ON," ENGINE "OFF."
   - PUMP BRAKE PEDAL TO DEPLETE VACUUM POWER BOOSTER.
   - USING TECH 1 MISC. TESTS, SELECT MANUAL CONTROL FUNCTION.
   - SELECT REAR MOTOR APPLY FUNCTION AND APPLY THE MOTOR WHILE CAREFULLY OBSERVING THE COMMAND CURRENT AND FEEDBACK CURRENT ON THE TECH 1. WERE BOTH COMMAND AND FEEDBACK CURRENT 10 AMPS?

   - YES: go to step 2.
   - NO: GO TO STEP 4.

2.
   - APPLY FIRM PRESSURE ON BRAKE PEDAL.
   - SELECT REAR MOTOR RELEASE FUNCTION AND RELEASE MOTOR. DID THE BRAKE PEDAL FALL AND WERE BOTH COMMAND AND FEEDBACK CURRENT 6 AMPS?

   - YES: go to step 3.
   - NO: GO TO STEP 4.

3.
   - FIRM PRESSURE STILL ON BRAKE PEDAL.
   - SELECT REAR MOTOR APPLY FUNCTION AND APPLY THE MOTOR (THIS MOVES THE HYDRAULIC MODULATOR PISTON TO THE TOP OF ITS BORE). DID BRAKE PEDAL RISE?

   - YES: MALFUNCTION IS INTERMITTENT. REFER TO "DIAGNOSTIC AIDS" ON FACING PAGE.
   - NO: go to step 4.

4.
   - IGNITION "OFF."
   - DISCONNECT 6-WAY EBCM AND MOTOR PACK HARNESS CONNECTORS.
   - USING J 39200, MEASURE RESISTANCE BETWEEN 6-WAY EBCM HARNESS CONNECTOR TERMINAL "C" AND 6-WAY MOTOR PACK HARNESS CONNECTOR TERMINAL "C". IS RESISTANCE LESS THAN 1.5 OHMS?

   - YES: go to step 5.
   - NO: REPAIR HIGH RESISTANCE IN CKT 1284.

5.
   - USING J 39200, MEASURE RESISTANCE BETWEEN 6-WAY EBCM HARNESS CONNECTOR TERMINAL "D" AND 6-WAY MOTOR PACK HARNESS CONNECTOR TERMINAL "D". IS RESISTANCE LESS THAN 1.5 OHMS?

   - YES: go to step 6.
   - NO: REPAIR HIGH RESISTANCE IN CKT 1285.

6.
   - USING J 39200, MEASURE RESISTANCE BETWEEN 6-WAY MOTOR PACK CONNECTOR TERMINALS "C" AND "D". IS RESISTANCE LESS THAN 1.5 OHMS?

   - YES: GO TO PAGE 2 OF 2 OF THIS CHART.
   - NO: REPLACE MOTOR PACK.

DIAGNOSTIC CHART DTC 51 — 1994–95 BERETTA AND CORSICA

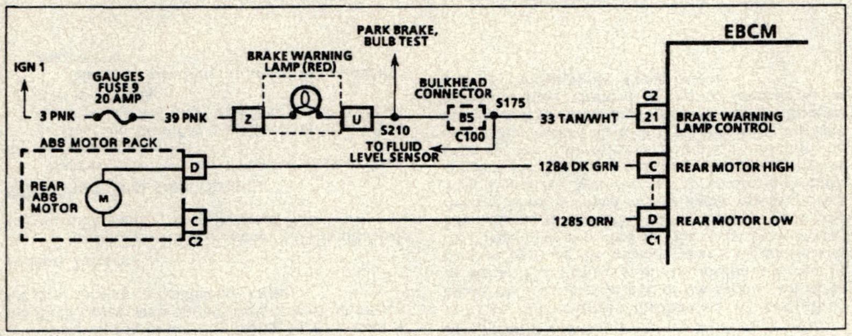

DTC 51 (Page 2 of 2) REAR ABS MOTOR FREE SPINS

Circuit Description:

This DTC is designed to detect a stripped nut or gear assembly during initialization. During the homing sequence, the piston should reach the top of the bore resulting in a stalled motor. If this does not occur, the motor must be spinning with little or no resistance, indicating a nut/screw or gear malfunction.

DTC Will Set When:

DTC 51 can be set during initialization. If the feedback current is less than the command current for a specified period of time, a malfunction exists.

Action Taken:

A malfunction DTC is stored, ABS is disabled, and the ABS warning lamp is turned "ON."

DTC Chart Test Description:

Number(s) below refer to circled number(s) on the diagnostic chart.

7. Checks for a stripped motor pack gear.
8. Checks for a stripped hydraulic modulator gear.
9. Checks for proper operation of motor pack.
10. Checks for proper operation of the hydraulic modulator.
11. Ensures malfunction was not due to poor terminal contact.

Diagnostic Aids:

An "Intermittent" malfunction in this DTC may result from a mechanical part of the system that sticks, binds, or slips.

The frequency of the malfunction can be checked by using the enhanced diagnostic function of the Tech 1, as described in "Enhanced Diagnostics," found in this section. If DTC only occurs once and DTC 64 also occurs, see DTC 64. If intermittent and enhanced diagnostics show this DTC occurs during ABS, see DTC 64. Depending on the frequency of the malfunction, a physical inspection of the mechanical parts suspected may be necessary.

DIAGNOSTIC CHART DTC 51 — 1994–95 BERETTA AND CORSICA

DTC 51
(Page 2 of 2)
REAR ABS MOTOR FREE SPINS

CONTINUED FROM PAGE 1 OF 2 OF THIS CHART.

(7)
- REMOVE HYDRAULIC MODULATOR/MASTER CYLINDER ASSEMBLY.
- REMOVE GEAR COVER.
- CHECK FOR STRIPPED GEAR ON MOTOR PACK; THE MOTOR PACK IS THE UNIT WITH THE THREE SMALL GEARS. THE REAR CHANNEL GEAR IS THE MIDDLE GEAR OF THE THREE. IS THE MOTOR PACK GEAR STRIPPED?

YES: REPLACE MOTOR PACK.

NO:

(8)
- CHECK FOR A STRIPPED GEAR ON THE HYDRAULIC MODULATOR; THE HYDRAULIC MODULATOR IS THE UNIT WITH THE THREE LARGE GEARS. THE REAR CHANNEL GEAR IS THE MIDDLE GEAR OF THE THREE IS THE HYDRAULIC MODULATOR GEAR STRIPPED?

YES: REPLACE HYDRAULIC MODULATOR GEAR.

NO:

(9)
- SECURELY POSITION HYDRAULIC MODULATOR/MASTER CYLINDER ASSEMBLY IN VEHICLE ALLOWING REAR CHANNEL GEAR SET OBSERVATION. BE CAREFUL NOT TO DAMAGE GEARS OR ALLOW THE HYDRAULIC MODULATOR MASTER CYLINDER ASSEMBLY TO MOVE DURING TESTING.
- RECONNECT ELECTRICAL CONNECTORS.
- IGNITION "ON."
- USING TECH 1 MISC. TESTS, SELECT MANUAL CONTROL FUNCTION.
- SELECT REAR MOTOR APPLY AND APPLY THE MOTOR WHILE OBSERVING THE MIDDLE GEAR SET.
- SELECT REAR MOTOR RELEASE FUNCTION AND RELEASE THE MOTOR WHILE STILL OBSERVING THE GEAR SET.
- SELECT REAR MOTOR APPLY FUNCTION AND APPLY THE MOTOR. DID GEAR SET MOVE IN BOTH DIRECTIONS FOR AT LEAST ONE REVOLUTION?

NO: REPLACE MOTOR PACK.

YES:

(10)
- USING TECH 1, SELECT "F4: MISC TESTS" AND PERFORM GEAR TENSION RELIEF FUNCTION.
- IGNITION "OFF."
- SEPARATE MOTOR PACK FROM HYDRAULIC MODULATOR.
- GRASP THE REAR CHANNEL GEAR ON THE HYDRAULIC MODULATOR; THE HYDRAULIC MODULATOR IS THE UNIT WITH THE THREE LARGE GEARS. THE REAR CHANNEL GEAR IS THE MIDDLE GEAR OF THE THREE.
- ATTEMPT TO ROTATE THE GEAR. CAN THE GEAR BE ROTATED MORE THAN 4 TURNS LOCK TO LOCK?

YES: REPLACE HYDRAULIC MODULATOR.

NO:

(11)
- INSPECT 6-WAY MOTOR PACK AND EBCM CONNECTORS FOR CORROSION OR POOR TERMINAL CONTACT. REPLACE TERMINALS THAT EXHIBIT POOR TERMINAL CONTACT OR CORROSION.
- INSTALL HYDRAULIC MODULATOR/MASTER CYLINDER ASSEMBLY.
- RECONNECT ALL CONNECTORS.
- DRIVE VEHICLE FOR THREE DRIVE CYCLES (A DRIVE CYCLE CONSISTS OF STARTING ENGINE, DRIVING OVER 16 km/h (10 MPH) AND THEN TURNING IGNITION "OFF"). DID DTC 51 SET IN THE LAST THREE DRIVE CYCLES?

YES: REPLACE EBCM.

NO: MALFUNCTION WAS DUE TO A POOR CONNECTION.

DIAGNOSTIC CHART DTC 52 — 1994–95 BERETTA AND CORSICA

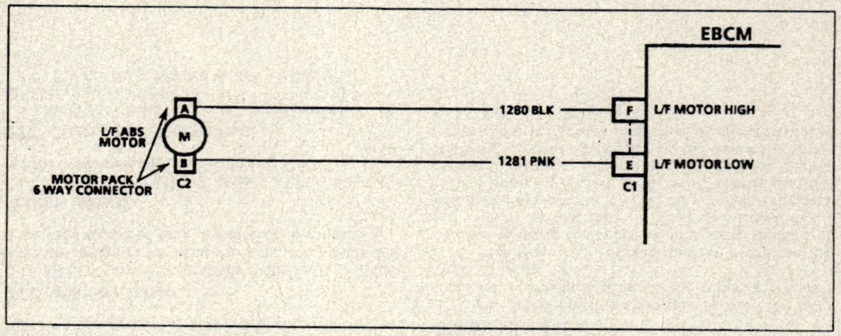

DTC 52
LEFT FRONT ABS CHANNEL IN RELEASE TOO LONG

Circuit Description:

This DTC will diagnose a motor that is energized longer than expected. This could occur if a wheel speed sensor is faulty, the motor does not turn, the left front solenoid mechanically fails open, the brakes are dragging, or the motor wires are crossed.

DTC Will Set When:

DTC 52 can be set only during an ABS stop. If the EBCM commands the left front channel in release for three seconds, a malfunction exists.

Action Taken:

A malfunction DTC is stored, ABS is disabled, and the ABS warning lamp is turned "ON."

DTC Chart Test Description:

Number(s) below refer to circled number(s) on the diagnostic chart.

1. This step identifies a malfunction in a wheel speed sensor that may cause the system to be in release too long.
2. This step identifies a motor or modulator malfunction.
3. This step checks for a solenoid that may have mechanically failed open.
4. Checks for a solenoid that may have failed closed.
5. Isolates the malfunction to either the solenoid or the hydraulic modulator.
6. This step checks for proper operation of hydraulic modulator.

Diagnostic Aids:

An "Intermittent" malfunction is most likely caused by a mechanical part of the system that sticks or binds. The frequency of the malfunction can be checked by using the enhanced diagnostic function of the "Tech 1, as described in "Enhanced Diagnostics," found in this section. DTC 52 may occur if on ice and steering wheel is turned to lock during braking. Using Tech 1 perform hydraulic test to ensure total brake system is functional. Any circuitry, that is suspected as causing the intermittent complaint, should be thoroughly checked for backed out terminals, improper mating, broken locks, improperly formed or damaged terminals, poor terminal to wiring connections or physical damage to the wiring harness.

DIAGNOSTIC CHART DTC 52 — 1994–95 BERETTA AND CORSICA

DTC 52
LEFT FRONT ABS CHANNEL IN RELEASE TOO LONG

IMPORTANT:
- EXCESSIVE DRAG OR HIGH RESISTANCE IN THE BASE BRAKE OR SUSPENSION SYSTEM MUST BE INSPECTED AND CORRECTED BEFORE PROCEEDING WITH DTC DIAGNOSIS.

ARE ANY WHEEL SPEED DTCs CURRENTLY SET?
- YES: PROCEED TO THAT DTC CHART FIRST.
- NO:

(1)
- USING TECH 1, SELECT "F0: DATA LIST" AND TEST VEHICLE DURING A STEADY DECEL CONDITION FROM 35 MPH TO 0 MPH WHILE MONITORING ALL WHEEL SPEEDS. DO ANY OF THE WHEEL SPEEDS INDICATE ERRATIC OR INTERMITTENT OPERATION?
  - YES: PROCEED TO EXCESSIVE WHEEL SPEED VARIATION DTC FOR AFFECTED WHEEL.
  - NO:

(2)
- IGNITION "ON."
- USING TECH 1 MANUAL CONTROL FUNCTION, SELECT LEFT FRONT MOTOR APPLY AND APPLY MOTOR.
- APPLY FIRM PRESSURE ON BRAKE PEDAL.
- SELECT LEFT FRONT MOTOR RELEASE AND COMMAND RELEASE OF MOTOR. DID BRAKE PEDAL FALL?
  - YES: go to (3)
  - NO: BRAKE PEDAL DID NOT MOVE. → go to (6); BRAKE PEDAL ROSE. → REPAIR CROSSED WIRES (CKTs 1280 AND 1281) TO LEFT FRONT ABS MOTOR.

(3)
- REMOVE FOOT FROM BRAKE PEDAL.
- USING TECH 1 MANUAL CONTROL FUNCTION, SELECT LEFT FRONT SOLENOID AND COMMAND THE SOLENOID "ON."
- APPLY FIRM PRESSURE TO BRAKE PEDAL. DID BRAKE PEDAL FALL?
  - YES: PROCEED TO STEP 5 OF THIS CHART.
  - NO:

(4)
- FIRM PRESSURE STILL APPLIED TO BRAKE PEDAL.
- USING TECH 1, COMMAND LEFT FRONT SOLENOID "OFF." DID BRAKE PEDAL FALL?
  - YES: MALFUNCTION IS NOT PRESENT AT THIS TIME. REFER TO "DIAGNOSTIC AIDS" ON FACING PAGE.
  - NO:

(5)
- REMOVE FOOT FROM BRAKE PEDAL.
- IGNITION "OFF."
- SWITCH LEFT FRONT SOLENOID WITH RIGHT FRONT SOLENOID.
- IGNITION "ON."
- USING TECH 1 MANUAL CONTROL FUNCTION, SELECT LEFT FRONT SOLENOID AND COMMAND THE SOLENOID "ON."
- APPLY FIRM PRESSURE TO BRAKE PEDAL. DID BRAKE PEDAL FALL?
  - YES: REPLACE HYDRAULIC MODULATOR.
  - NO: REPLACE THE SOLENOID WHICH WAS ORIGINALLY IN THE LEFT FRONT CHANNEL.

(6)
- USING TECH 1, PERFORM GEAR TENSION RELIEF FUNCTION.
- SEPARATE MOTOR PACK FROM HYDRAULIC MODULATOR.
- GRASP THE GEAR ON THE HYDRAULIC MODULATOR (THE HYDRAULIC MODULATOR IS THE UNIT WITH THE THREE LARGE GEARS) FURTHEST FROM THE BRAKE BOOSTER AND ATTEMPT TO MOVE THE GEAR IN EITHER DIRECTION. CAN GEAR BE MOVED BY HAND?
  - YES: REPLACE MOTOR PACK.
  - NO: REPLACE ABS HYDRAULIC MODULATOR.

* AFTER DIAGNOSIS IS COMPLETE, CLEAR DTCs AND TEST DRIVE VEHICLE FOR THREE (3) DRIVE CYCLES TO VERIFY DTC DOES NOT RESET. A DRIVE CYCLE CONSISTS OF STARTING THE VEHICLE, DRIVING OVER 16 km/h (10 MPH) AND THEN KEYING DOWN.

DIAGNOSTIC CHART DTC 53 — 1994–95 BERETTA AND CORSICA

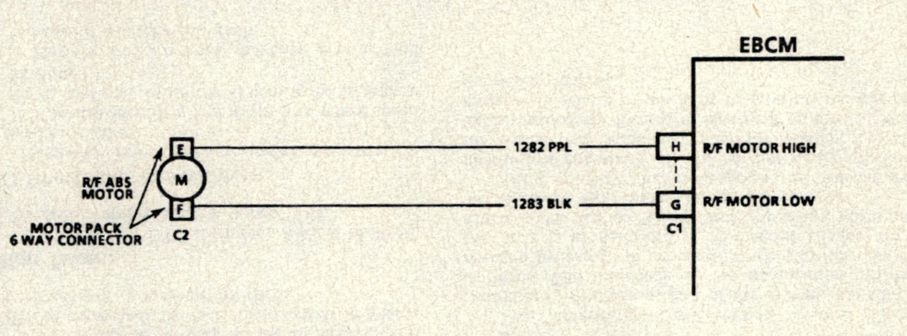

DTC 53
RIGHT FRONT ABS CHANNEL IN RELEASE TOO LONG

Circuit Description:

This DTC will diagnose a motor that is energized longer than expected. This could occur if a wheel speed sensor is faulty, the motor does not turn, the right front solenoid mechanically fails open, the brakes are dragging, or the motor wires are crossed.

DTC Will Set When:

DTC 53 can be set only during an ABS stop. If the EBCM commands the right front channel in release for three seconds, a malfunction exists.

Action Taken:

A malfunction DTC is stored, ABS is disabled, and the ABS warning lamp is turned "ON."

DTC Chart Test Description:

Number(s) below refer to circled number(s) on the diagnostic chart.

1. This step identifies a malfunction in a wheel speed sensor that may cause the system to be in release too long.
2. This step identifies a motor or modulator malfunction.
3. This step checks for a solenoid that may have mechanically failed open.
4. Checks for a solenoid that may have failed closed.
5. Isolates the malfunction to either the solenoid or the hydraulic modular.
6. This step checks for proper operation of hydraulic modulator.

Diagnostic Aids:

An "Intermittent" malfunction is most likely caused by a mechanical part of the system that sticks or binds. The frequency of the malfunction can be checked by using the enhanced diagnostic function of the Tech 1, as described in "Enhanced Diagnostics," found in this section. DTC 53 may occur on ice if steering wheel is turned to lock during braking. Using Tech 1 perform hydraulic test to ensure total brake system is functional. Any circuitry, that is suspected as causing the intermittent complaint, should be thoroughly checked for backed out terminals, improper mating, broken locks, improperly formed or damaged terminals, poor terminal to wiring connections or physical damage to the wiring harness.

DIAGNOSTIC CHART DTC 53 — 1994–95 BERETTA AND CORSICA

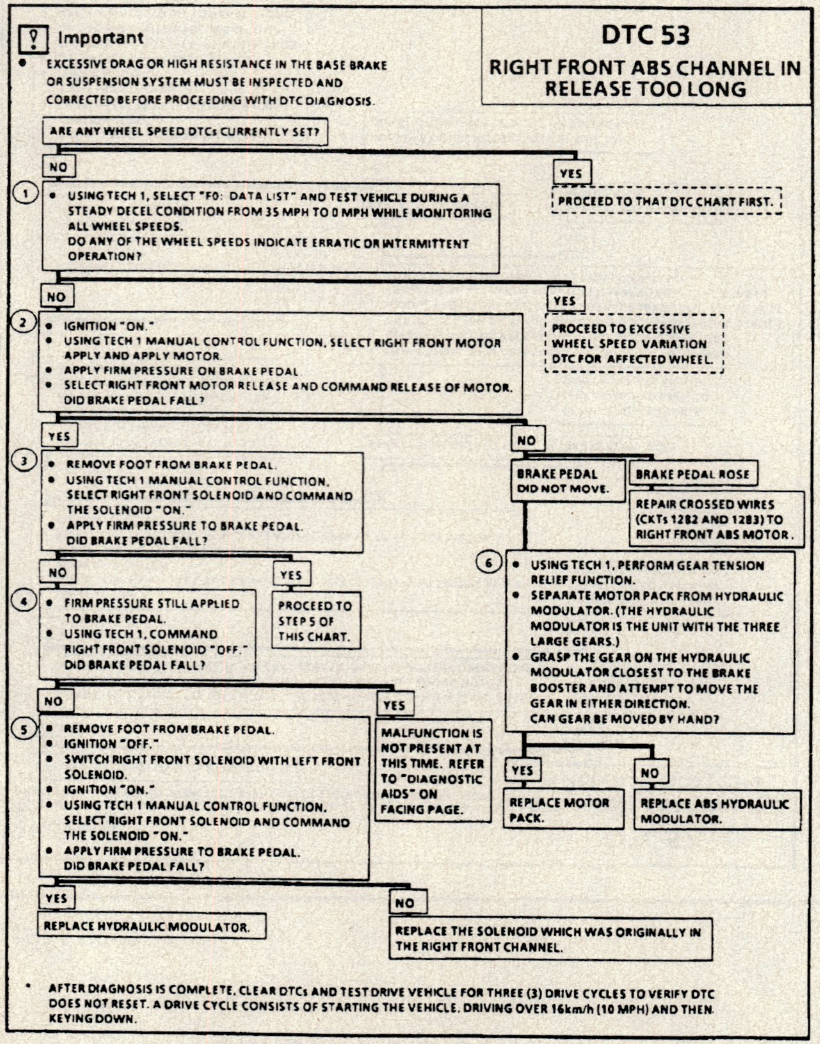

DIAGNOSTIC CHART DTC 54 — 1994–95 BERETTA AND CORSICA

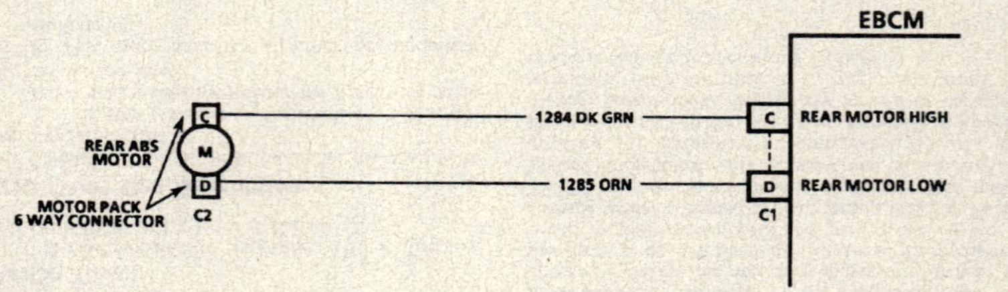

DTC 54
REAR ABS CHANNEL IN RELEASE TOO LONG

Circuit Description:

This DTC will diagnose a motor that is energized longer than expected. This could occur if a wheel speed sensor is faulty, the motor does not turn, the brakes are dragging, or the motor wires are crossed.

DTC Will Set When:

DTC 54 can be set only during an ABS stop. If the EBCM commands the rear ABS channel in release for three seconds, a malfunction exists.

Action Taken:

A malfunction DTC is stored, ABS is disabled, and the ABS warning lamp is turned "ON."

DTC Chart Test Description:

Number(s) below refer to circled number(s) on the diagnostic chart.

1. This step identifies a problem in a wheel speed sensor that may cause the system to be in release too long.
2. This checks for a wheel that may stick or bind because of a mechanical fault.
3. This checks to see if the motor is capable of moving and applying the hydraulic piston for the rear wheels.
4. This step ensures that the motor wiring is not crossed.
5. This isolates the malfunction of a "no apply" situation to either the motor pack or ABS hydraulic modulator.

Diagnostic Aids:

An "Intermittent" malfunction is most likely caused by a mechanical part of the system that sticks or binds. The frequency of the malfunction can be checked by using the enhanced diagnostic function of the Tech 1, as described in "Enhanced Diagnostics," found in this section. Using Tech 1 perform hydraulic test to ensure total brake system is functional.

Any circuitry, that is suspected as causing the intermittent complaint, should be thoroughly checked for backed out terminals, improper mating, broken locks, improperly formed or damaged terminals, poor terminal to wiring connections or physical damage to the wiring harness.

DIAGNOSTIC CHART DTC 54 — 1994–95 BERETTA AND CORSICA

DTC 54
REAR ABS CHANNEL IN RELEASE TOO LONG

Important
- EXCESSIVE DRAG OR HIGH RESISTANCE IN THE BASE BRAKE OR SUSPENSION SYSTEM MUST BE INSPECTED AND CORRECTED BEFORE PROCEEDING WITH DTC DIAGNOSIS.

ARE ANY WHEEL SPEED DTCs CURRENTLY SET?
- YES: PROCEED TO THAT DTC CHART FIRST.
- NO:

(1)
- USING TECH 1, SELECT "F0: DATA LIST" AND TEST VEHICLE DURING A STEADY DECEL CONDITION FROM 35 MPH TO 0 MPH WHILE MONITORING ALL WHEEL SPEEDS.

DO ANY OF THE WHEEL SPEEDS INDICATE ERRATIC OR INTERMITTENT OPERATION?
- YES: PROCEED TO EXCESSIVE WHEEL SPEED VARIATION DTC FOR AFFECTED WHEEL.
- NO:

(2)
- RAISE REAR OF VEHICLE AND SUPPORT
- SPIN BOTH REAR WHEELS AND CONFIRM PHYSICAL CONDITION OF BASE BRAKE SYSTEM.

COULD BOTH REAR WHEELS BE SPUN FREELY BY HAND?
- NO: REPAIR SOURCE OF RESISTANCE IN BASE BRAKE SYSTEM.
- YES:

(3)
- IGNITION "ON."
- USING TECH 1 MANUAL CONTROL FUNCTION, SELECT REAR AXLE MOTOR APPLY.
- APPLY FIRM PRESSURE ON BRAKE PEDAL.
- COMMAND APPLY OF REAR MOTOR AND TRY TO SPIN REAR WHEELS BY HAND.

COULD REAR WHEELS BE SPUN BY HAND?
- NO: MALFUNCTION IS NOT PRESENT AT THIS TIME. REFER TO "DIAGNOSTIC AIDS" ON FACING PAGE.
- YES:

(4)
- WITH FIRM PRESSURE ON BRAKE PEDAL STILL APPLIED, COMMAND A RELEASE OF THE REAR MOTOR.

CAN WHEELS BE MOVED?
- NO: REPAIR CROSSED WIRES (CKTs 1284 AND 1285) AT REAR MOTOR.
- YES:

(5)
- USING TECH 1 MISC. TESTS, PERFORM GEAR TENSION RELIEF FUNCTION.
- REMOVE MOTOR PACK.
- CONNECT MOTOR PACK WIRING.
- USING TECH 1, PERFORM MOTOR TEST.

DO ALL THREE MOTOR PACK GEARS (SMALL GEARS) SPIN FREELY?
- NO: REPLACE MOTOR PACK.
- YES: REPLACE ABS HYDRAULIC MODULATOR.

AFTER ALL DIAGNOSIS IS COMPLETE, CLEAR DTCs AND TEST DRIVE VEHICLE FOR THREE (3) DRIVE CYCLES TO VERIFY THAT DTC DOES NOT RESET. A DRIVE CYCLE CONSISTS OF STARTING THE VEHICLE DRIVING OVER 16 KM/H (10 MPH), AND THEN KEYING DOWN.

DIAGNOSTIC CHART DTC 55 — 1994–95 BERETTA AND CORSICA

D.M. ABS VI ELECTRONIC BRAKE CONTROL MODULE (TYPICAL)

DTC 55
EBCM MALFUNCTION

Circuit Description:

This DTC identifies a malfunction detected by the MDI custom IC. It also ensures the cause of the malfunction is not a result of a malfunction with the enable relay under a DTC 55 failure.

DTC Will Set When:

DTC 55 is set when an internal EBCM malfunction exists.

Action Taken:

A malfunction DTC is stored, ABS is disabled and the ABS warning lamp is turned "ON."

DTC Chart Test Description:

Number(s) below refer to circled number(s) on the diagnostic chart.

1. Checks to see if there are any other malfunctions present.
2. This checks to see if the malfunction is still present.

Diagnostic Aids:

An "Intermittent" malfunction is most likely caused by a poor connection, rubbed through wire insulation, or a wire that is broken inside the insulation.

The frequency of the malfunction can be checked by using the enhanced diagnostic function of the Tech 1, as described in "Enhanced Diagnostics," found in this section.

Any circuitry, that is suspected as causing the intermittent complaint, should be thoroughly checked for backed out terminals, improper mating, broken locks, improperly formed or damaged terminals, poor terminal to wiring connections or physical damage to the wiring harness.

DIAGNOSTIC CHART DTC 55 — 1994–95 BERETTA AND CORSICA

DTC 55
EBCM MALFUNCTION

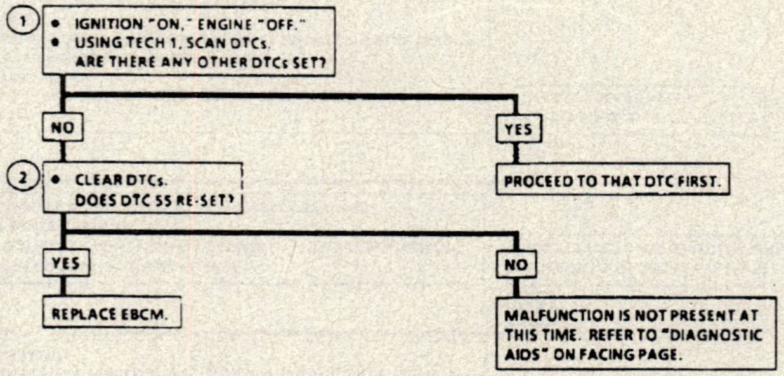

* AFTER DIAGNOSIS IS COMPLETE, CLEAR DTCs AND TEST DRIVE VEHICLE FOR THREE (3) DRIVE CYCLES TO VERIFY DTC DOES NOT RESET. A DRIVE CYCLE CONSISTS OF STARTING THE VEHICLE, DRIVING OVER 16 km/h (10 MPH) AND THEN KEYING DOWN.

THIS CHART ASSUMES THAT A CURRENT DTC IS STORED INDICATING THAT THIS MALFUNCTION IS PRESENT.

DIAGNOSTIC CHART DTC 56 — 1994–95 BERETTA AND CORSICA

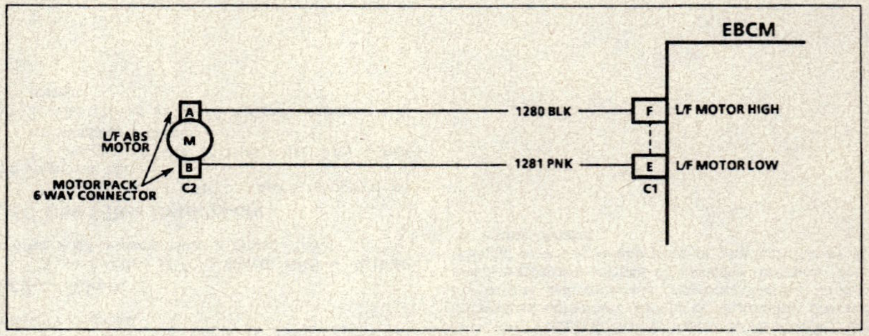

DTC 56
LEFT FRONT ABS MOTOR CIRCUIT OPEN

Circuit Description:

This DTC identifies a motor that cannot be energized due to an open in its circuitry. This malfunction will not allow proper front ABS operation.

DTC Will Set When:

DTC 56 can be set only when the motor is commanded "OFF." If the EBCM detects an out of range voltage in either of the left front motor circuits indicating an open circuit, a malfunction exists.

Action Taken:

An open motor will not activate when requested. A malfunction DTC is stored, ABS is disabled and the ABS warning lamp is turned "ON."

DTC Chart Test Description:

Number(s) below refer to circled number(s) on the diagnostic chart.

1. This step checks for proper resistance of the motor.
2. This step checks for an open in the motor "high" circuitry.
3. This step checks for an open in the motor "low" circuitry.
4. Ensures malfunction was not due to poor terminal contact.

Diagnostic Aids:

Using Tech 1, select manual control function, and exercise motor movement of affected channel in both directions while applying light pressure on the brake pedal.

If erratic or "jumpy" brake pedal movement is detected while performing an "apply" or "release" function of the motor, an intermittent malfunction may be indicated. An "Intermittent" malfunction is most likely caused by a poor connection, rubbed through wire insulation, or a wire that is broken inside the insulation.

If the malfunction is not current, wiggle the wires of the affected channel and check if the DTC resets. This will help to pinpoint an intermittent malfunction in the motor circuitry or connections.

The frequency of the malfunction can be checked by using the enhanced diagnostic function of the Tech 1, as described in "Enhanced Diagnostics," found in this section.

Any circuitry, that is suspected as causing the intermittent complaint, should be thoroughly checked for backed out terminals, improper mating, broken locks, improperly formed or damaged terminals, poor terminal to wiring connections or physical damage to the wiring harness.

DIAGNOSTIC CHART DTC 56 — 1994–95 BERETTA AND CORSICA

DTC 56
LEFT FRONT ABS MOTOR CIRCUIT OPEN

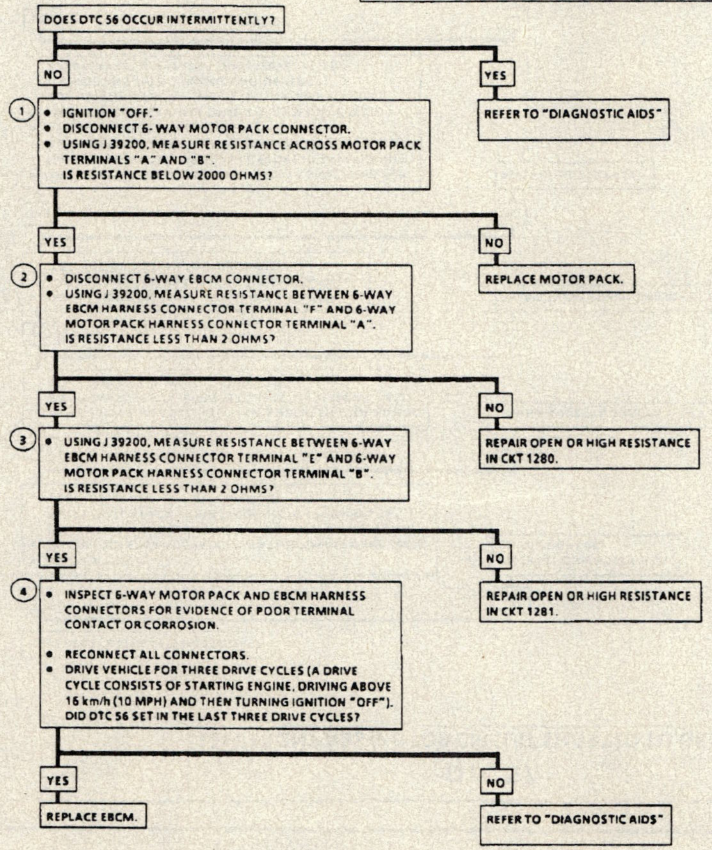

* AFTER DIAGNOSIS IS COMPLETE, CLEAR DTCs AND TEST DRIVE VEHICLE FOR THREE (3) DRIVE CYCLES TO VERIFY DTC DOES NOT RESET. A DRIVE CYCLE CONSISTS OF STARTING THE VEHICLE, DRIVING OVER 16 km/h (10 MPH) AND THEN KEYING DOWN.

THIS CHART ASSUMES THAT A CURRENT DTC IS STORED INDICATING THAT THIS MALFUNCTION IS PRESENT.

DIAGNOSTIC CHART DTC 57 — 1994–95 BERETTA AND CORSICA

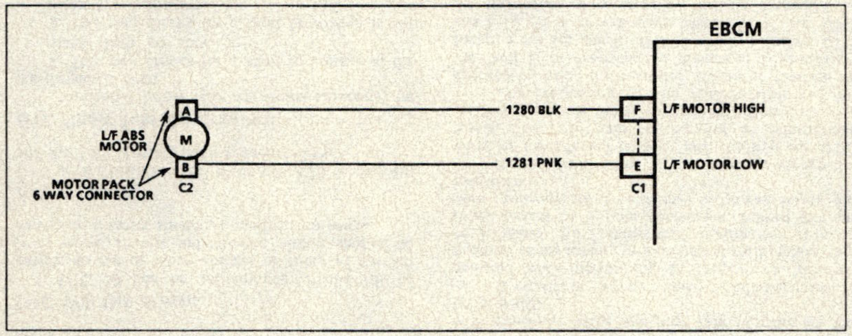

DTC 57
LEFT FRONT ABS MOTOR CIRCUIT SHORTED TO GROUND

Circuit Description:

This DTC identifies a motor circuit that is shorted to ground. This malfunction will not allow the motor to be controlled at the commanded current rate or will cause the driver circuit to allow current directly to ground.

DTC Will Set When:

DTC 57 can be set anytime. If the EBCM detects an out of range voltage in either of the left front motor circuits indicating a circuit shorted to ground, a malfunction exists.

Action Taken:

A malfunction DTC is stored, ABS is disabled and the ABS warning lamp is turned "ON."

DTC Chart Test Description:

Number(s) below refer to circled number(s) on the diagnostic chart.

1. This step checks for a short to ground in the motor "high" circuitry.
2. This step checks for a short to ground in the motor "low" circuitry.
3. This step checks for a motor that is internally shorted to ground.
4. Ensures malfunction was not due to physical damage of the circuitry.

Diagnostic Aids:

Using Tech 1, select manual control function, and exercise motor movement of affected channel in both directions while applying light pressure on the brake pedal.

If erratic or "jumpy" brake pedal movement is detected while performing an "apply" or "release" function of the motor, an intermittent malfunction may be indicated. An "Intermittent" malfunction is most likely caused by a poor connection, rubbed through wire insulation, or a wire that is broken inside the insulation.

If the malfunction is not current, wiggle the wires of the affected channel and check if the DTC resets. This will help to pinpoint an intermittent malfunction in the motor circuitry or connections.

The frequency of the malfunction can be checked by using the enhanced diagnostic function of the Tech 1, as described in "Enhanced Diagnostics," found in this section.

Any circuitry, that is suspected as causing the intermittent complaint, should be thoroughly checked for backed out terminals, improper mating, broken locks, improperly formed or damaged terminals, poor terminal to wiring connections or physical damage to the wiring harness.

DIAGNOSTIC CHART DTC 57 — 1994–95 BERETTA AND CORSICA

DTC 57
LEFT FRONT ABS MOTOR CIRCUIT SHORTED TO GROUND

DOES DTC 57 OCCUR INTERMITTENTLY?

- YES → REFER TO "DIAGNOSTIC AIDS" ON FACING PAGE.
- NO →

(1)
- IGNITION "OFF."
- DISCONNECT 6-WAY CONNECTORS AT MOTOR PACK AND EBCM.
- USING J 39200, MEASURE RESISTANCE BETWEEN GROUND AND 6-WAY MOTOR PACK HARNESS CONNECTOR TERMINAL "A". DOES J 39200 DISPLAY "OL" (INFINITE)?

- NO → REPAIR SHORT TO GROUND IN CKT 1280.
- YES →

(2)
- USING J 39200, MEASURE RESISTANCE BETWEEN GROUND AND 6-WAY MOTOR PACK HARNESS CONNECTOR TERMINAL "B". DOES J 39200 DISPLAY "OL" (INFINITE)?

- NO → REPAIR SHORT TO GROUND IN CKT 1281.
- YES →

(3)
- USING J 39200, MEASURE RESISTANCE BETWEEN GROUND AND TERMINAL "A" OF THE MOTOR PACK. DOES J 39200 DISPLAY "OL" (INFINITE)?

- NO → REPLACE MOTOR PACK.
- YES →

(4)
- INSPECT CKTs 1280 AND 1281 AND 6-WAY CONNECTORS FOR DAMAGE WHICH MAY RESULT IN A SHORT TO GROUND WITH ALL CONNECTORS CONNECTED. REPAIR DAMAGE IF EVIDENT.
- RECONNECT ALL CONNECTORS.
- DRIVE VEHICLE FOR THREE DRIVE CYCLES (A DRIVE CYCLE CONSISTS OF STARTING ENGINE, DRIVING ABOVE 16 km/h (10 MPH) AND THEN TURNING IGNITION "OFF."). DID DTC 57 SET IN THE LAST THREE DRIVE CYCLES?

- YES → REPLACE EBCM.
- NO → REFER TO "DIAGNOSTIC AIDS" ON FACING PAGE.

* AFTER DIAGNOSIS IS COMPLETE, CLEAR DTCs AND TEST DRIVE VEHICLE FOR THREE (3) DRIVE CYCLES TO VERIFY DTC DOES NOT RESET. A DRIVE CYCLE CONSISTS OF STARTING THE VEHICLE, DRIVING OVER 16 km/h (10 MPH) AND THEN KEYING DOWN.

THIS CHART ASSUMES THAT A CURRENT DTC IS STORED INDICATING THAT THIS MALFUNCTION IS PRESENT.

DIAGNOSTIC CHART DTC 58 — 1994–95 BERETTA AND CORSICA

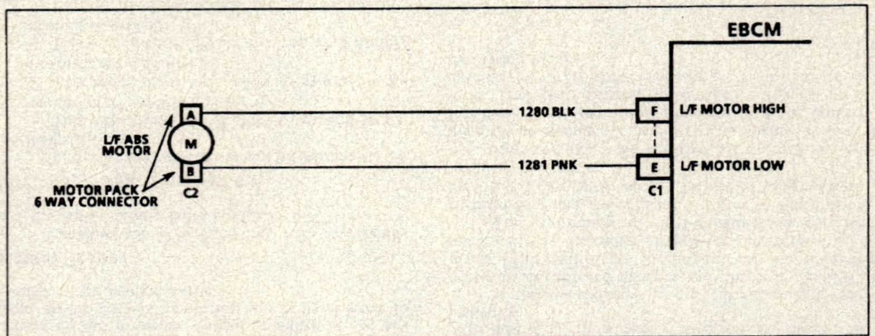

DTC 58
LEFT FRONT ABS MOTOR CIRCUIT SHORTED TO BATTERY OR MOTOR SHORTED

Circuit Description:

This DTC identifies a motor circuit that is shorted to battery or a motor that has low or no resistance. This malfunction will not allow the motor to be controlled at the commanded current rate or will cause the motor to turn in the opposite direction, or not at all.

DTC Will Set When:

DTC 58 can be set anytime. If the EBCM detects an out of range voltage in either of the left front motor circuits indicating a circuit shorted to battery or a motor shorted, a malfunction exists.

Action Taken:

A malfunction DTC is stored, ABS is disabled and the ABS warning lamp is turned "ON."

DTC Chart Test Description:

Number(s) below refer to circled number(s) on the diagnostic chart.

1. This step checks for a short to voltage in the motor "high" circuitry.
2. This step checks for a short to voltage in the motor "low" circuitry.
3. Checks for a short between "high" and "low" motor circuits.
4. This step checks for a motor that is internally shorted.
5. Checks for a short between the EMB and motor circuits.
6. Checks for an internal motor pack short between the EMB and motor.
7. Ensures malfunction was not due to physical damage of the circuitry.

Diagnostic Aids:

Using Tech 1, select manual control function, and exercise motor movement of affected channel in both directions while applying light pressure on the brake pedal.

If erratic or "jumpy" brake pedal movement is detected while performing an "apply" or "release" function of the motor, an intermittent malfunction may be indicated. An "Intermittent" malfunction is most likely caused by a poor connection, rubbed through wire insulation, or a wire that is broken inside the insulation.

If the malfunction is not current, wiggle the wires of the affected channel and check if the DTC resets. This will help to pinpoint an intermittent malfunction in the motor circuitry or connections.

The frequency of the malfunction can be checked by using the enhanced diagnostic function of the Tech 1, as described in "Enhanced Diagnostics," found in this section. If DTC 38 sets as a current DTC and DTC 58 is an infrequent history DTC, see DTC 38 diagnostics. Any circuitry, that is suspected as causing the intermittent complaint, should be thoroughly checked for backed out terminals, improper mating, broken locks, improperly formed or damaged terminals, poor terminal to wiring connections or physical damage to the wiring harness.

DIAGNOSTIC CHART DTC 58 — 1994–95 BERETTA AND CORSICA

NOTICE: J 39200 TEST LEADS MUST BE "ZEROED" PRIOR TO MAKING ANY RESISTANCE MEASUREMENTS. REFER TO J 39200 USER'S MANUAL.

DTC 58
LEFT FRONT ABS MOTOR CIRCUIT SHORTED TO BATTERY OR MOTOR SHORTED

DOES DTC 58 OCCUR INTERMITTENTLY?
- YES: REFER TO "DIAGNOSTIC AIDS" ON FACING PAGE.
- NO: go to step 1.

(1) NOTICE: THIS STEP MAY CAUSE ADDITIONAL DTCs TO SET.
- IGNITION "OFF."
- DISCONNECT 6-WAY CONNECTORS AT MOTOR PACK AND EBCM.
- IGNITION "ON."
- USING J 39200, MEASURE VOLTAGE BETWEEN GROUND AND 6-WAY MOTOR PACK HARNESS CONNECTOR TERMINAL "A". IS VOLTAGE LESS THAN 1 VOLT?

YES: go to step 2. NO: REPAIR SHORT TO VOLTAGE IN CKT 1280.

(2)
- USING J 39200, MEASURE VOLTAGE BETWEEN GROUND AND 6-WAY MOTOR PACK HARNESS CONNECTOR TERMINAL "B". IS VOLTAGE LESS THAN 1 VOLT?

YES: go to step 3. NO: REPAIR SHORT TO VOLTAGE IN CKT 1281.

(3)
- IGNITION "OFF."
- USING J 39200, MEASURE RESISTANCE BETWEEN 6-WAY MOTOR PACK HARNESS CONNECTOR TERMINALS "A" AND "B". DOES J 39200 DISPLAY "OL" (INFINITE)?

YES: go to step 4. NO: REPAIR SHORT BETWEEN CKTs 1280 AND 1281.

(4)
- USING J 39200, MEASURE RESISTANCE BETWEEN MOTOR PACK TERMINALS "A" AND "B". IS RESISTANCE GREATER THAN 4 OHMS?

YES: go to step 5. NO: REPLACE MOTOR PACK.

(5)
- DISCONNECT 3-WAY MOTOR PACK CONNECTOR.
- USING J 39200, MEASURE RESISTANCE BETWEEN 3-WAY MOTOR PACK HARNESS CONNECTOR TERMINAL "B" AND 6-WAY MOTOR PACK HARNESS CONNECTOR TERMINAL "A" THEN "B". DID J 39200 DISPLAY "OL" (INFINITE) FOR EACH MEASUREMENT?

YES: go to step 6. NO: REPAIR SHORT BETWEEN CKT 855/1633 AND MOTOR CIRCUIT WHICH MEASURED OTHER THAN "OL" (INFINITE).

(6)
- USING J 39200, MEASURE RESISTANCE BETWEEN 3-WAY MOTOR PACK CONNECTOR TERMINAL "B" AND 6-WAY MOTOR PACK CONNECTOR TERMINAL "A". DID J 39200 DISPLAY "OL" (INFINITE)?

YES: go to step 7. NO: REPLACE MOTOR PACK.

(7)
- INSPECT CKTs 1280 AND 1281 AND 6-WAY CONNECTORS FOR DAMAGE WHICH MAY RESULT IN A SHORT TO B+ WITH ALL CONNECTORS CONNECTED.
- REPAIR DAMAGE IF EVIDENT.
- RECONNECT ALL CONNECTORS.
- DRIVE VEHICLE FOR THREE DRIVE CYCLES (A DRIVE CYCLE CONSISTS OF STARTING ENGINE, DRIVE ABOVE 16 km/h (10 MPH) AND THEN TURNING IGNITION "OFF"). DID DTC 58 SET IN THE LAST THREE DRIVE CYCLES?

YES: REPLACE EBCM.* NO: REFER TO "DIAGNOSTIC AIDS" ON FACING PAGE.

* AFTER DIAGNOSIS IS COMPLETE, CLEAR DTCs AND TEST DRIVE VEHICLE FOR THREE (3) DRIVE CYCLES TO VERIFY DTC DOES NOT RESET. A DRIVE CYCLE CONSISTS OF STARTING THE VEHICLE, DRIVING OVER 16 km/h (10 MPH) AND THEN KEYING DOWN.

THIS CHART ASSUMES THAT A CURRENT DTC IS STORED INDICATING THAT THIS MALFUNCTION IS PRESENT.

DIAGNOSTIC CHART DTC 61 — 1994–95 BERETTA AND CORSICA

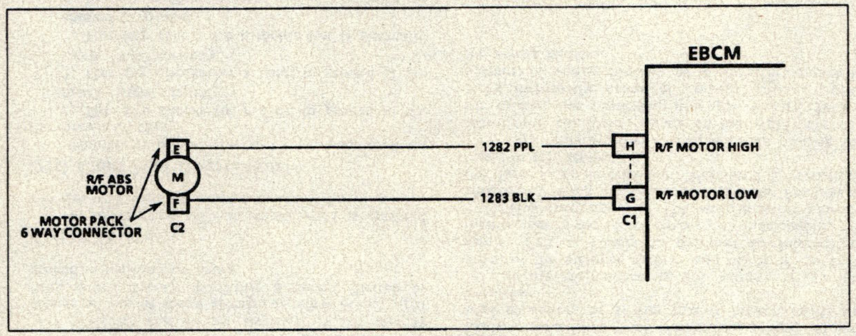

DTC 61
RIGHT FRONT ABS MOTOR CIRCUIT OPEN

Circuit Description:

This DTC identifies a motor that cannot be energized due to an open in its circuitry. This malfunction will not allow proper front ABS operation.

DTC Will Set When:

DTC 61 can be set only when the motor is commanded "OFF." If the EBCM detects an out of range voltage in either of the right front motor circuits indicating an open circuit, a malfunction exists.

Action Taken:

An open motor will not activate when requested. A malfunction DTC is stored, ABS is disabled and the ABS warning lamp is turned "ON."

DTC Chart Test Description:

Number(s) below refer to circled number(s) on the diagnostic chart.

1. This step checks for proper resistance of the motor.
2. This step checks for an open in the motor "high" circuitry.
3. This step checks for an open in the motor "low" circuitry.
4. Ensures malfunction was not due to poor terminal contact.

Diagnostic Aids:

Using Tech 1, select manual control function, and exercise motor movement of affected channel in both directions while applying light pressure on the brake pedal.

If erratic or "jumpy" brake pedal movement is detected while performing an "apply" or "release" function of the motor, an intermittent malfunction may be indicated.

An "Intermittent" malfunction is most likely caused by a poor connection, rubbed through wire insulation, or a wire that is broken inside the insulation.

If the malfunction is not current, wiggle the wires of the affected channel and check if the DTC resets. This will help to pinpoint an intermittent malfunction in the motor circuitry or connections.

The frequency of the malfunction can be checked by using the enhanced diagnostic function of the Tech 1, as described in "Enhanced Diagnostics," found in this section.

Any circuitry, that is suspected as causing the intermittent complaint, should be thoroughly checked for backed out terminals, improper mating, broken locks, improperly formed or damaged terminals, poor terminal to wiring connections or physical damage to the wiring harness.

DIAGNOSTIC CHART DTC 61 — 1994–95 BERETTA AND CORSICA

DTC 61
RIGHT FRONT ABS MOTOR CIRCUIT OPEN

DOES DTC 61 OCCUR INTERMITTENTLY?

- YES → REFER TO "DIAGNOSTIC AIDS" ON FACING PAGE.
- NO →

(1)
- IGNITION "OFF."
- DISCONNECT 6-WAY MOTOR PACK CONNECTOR.
- USING J 39200, MEASURE RESISTANCE ACROSS MOTOR PACK TERMINALS "E" AND "F".

IS RESISTANCE BELOW 2000 OHMS?

- NO → REPLACE MOTOR PACK.
- YES →

(2)
- DISCONNECT 6-WAY EBCM CONNECTOR.
- USING J 39200, MEASURE RESISTANCE BETWEEN 6-WAY EBCM HARNESS CONNECTOR TERMINAL "H" AND 6-WAY MOTOR PACK HARNESS CONNECTOR TERMINAL "E".

IS RESISTANCE LESS THAN 2 OHMS?

- NO → REPAIR OPEN OR HIGH RESISTANCE IN CKT 1282.
- YES →

(3)
- USING J 39200, MEASURE RESISTANCE BETWEEN 6-WAY EBCM HARNESS CONNECTOR TERMINAL "G" AND 6-WAY MOTOR PACK HARNESS CONNECTOR TERMINAL "F".

IS RESISTANCE LESS THAN 2 OHMS?

- NO → REPAIR OPEN OR HIGH RESISTANCE IN CKT 1283.
- YES →

(4)
- INSPECT 6-WAY MOTOR PACK AND EBCM HARNESS CONNECTORS FOR EVIDENCE OF POOR TERMINAL CONTACT OR CORROSION.
- RECONNECT ALL CONNECTORS.
- DRIVE VEHICLE FOR THREE DRIVE CYCLES (A DRIVE CYCLE CONSISTS OF STARTING ENGINE, DRIVING ABOVE 16 km/h (10 MPH) AND THEN TURNING IGNITION "OFF").

DID DTC 61 SET IN THE LAST THREE DRIVE CYCLES?

- YES → REPLACE EBCM.
- NO → REFER TO "DIAGNOSTIC AIDS" ON FACING PAGE.

* AFTER DIAGNOSIS IS COMPLETE, CLEAR DTCs AND TEST DRIVE VEHICLE FOR THREE (3) DRIVE CYCLES TO VERIFY DTC DOES NOT RESET. A DRIVE CYCLE CONSISTS OF STARTING THE VEHICLE, DRIVING OVER 16 km/h (10 MPH) AND THEN KEYING DOWN.

THIS CHART ASSUMES THAT A CURRENT DTC IS STORED INDICATING THAT THIS MALFUNCTION IS PRESENT.

DIAGNOSTIC CHART DTC 62 — 1994–95 BERETTA AND CORSICA

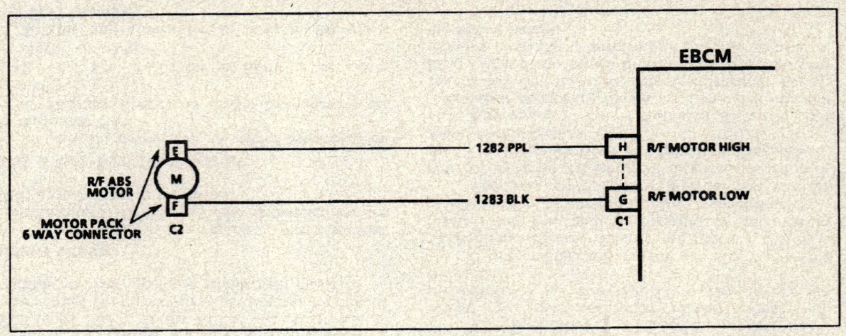

DTC 62
RIGHT FRONT ABS MOTOR CIRCUIT SHORTED TO GROUND

Circuit Description:

This DTC identifies a motor circuit that is shorted to ground. This malfunction will not allow the motor to be controlled at the commanded current rate or will cause the motor driver circuit to allow current directly to ground.

DTC Will Set When:

DTC 62 can be set anytime. If the EBCM detects an out of range voltage in either of the right front motor circuits indicating a circuit shorted to ground, a malfunction exists.

Action Taken:

A malfunction DTC is stored, ABS is disabled and the ABS warning lamp is turned "ON."

DTC Chart Test Description:

Number(s) below refer to circled number(s) on the diagnostic chart.

1. This step checks for a short to ground in the motor "high" circuitry.
2. This step checks for a short to ground in the motor "low" circuitry.
3. This step checks for a motor that is internally shorted to ground.
4. Ensures malfunction was not due to physical damage of the circuitry.

Diagnostic Aids:

Using Tech 1, select manual control function, and exercise motor movement of affected channel in both directions while applying light pressure on the brake pedal.

If erratic or "jumpy" brake pedal movement is detected while performing an "apply" or "release" function of the motor, an intermittent malfunction may be indicated. An "Intermittent" malfunction is most likely caused by a poor connection, rubbed through wire insulation, or a wire that is broken inside the insulation.

If the malfunction is not current, wiggle the wires of the affected channel and check if the DTC resets. This will help to pinpoint an intermittent malfunction in the motor circuitry or connections.

The frequency of the malfunction can be checked by using the enhanced diagnostic function of the Tech 1, as described in "Enhanced Diagnostics," found in this section.

Any circuitry, that is suspected as causing the intermittent complaint, should be thoroughly checked for backed out terminals, improper mating, broken locks, improperly formed or damaged terminals, poor terminal to wiring connections or physical damage to the wiring harness.

DIAGNOSTIC CHART DTC 62 — 1994–95 BERETTA AND CORSICA

DTC 62
RIGHT FRONT ABS MOTOR CIRCUIT SHORTED TO GROUND

DOES DTC 62 OCCUR INTERMITTENTLY?

- YES: REFER TO "DIAGNOSTIC AIDS" ON FACING PAGE.
- NO: go to step 1.

(1)
- IGNITION "OFF."
- DISCONNECT 6-WAY CONNECTORS AT MOTOR PACK AND EBCM.
- USING J 39200, MEASURE RESISTANCE BETWEEN GROUND AND 6-WAY MOTOR PACK HARNESS CONNECTOR TERMINAL "E". DOES J 39200 DISPLAY "OL" (INFINITE)?

- YES: go to step 2.
- NO: REPAIR SHORT TO GROUND IN CKT 1282.

(2)
- USING J 39200, MEASURE RESISTANCE BETWEEN GROUND AND 6-WAY MOTOR PACK HARNESS CONNECTOR TERMINAL "F". DOES J 39200 DISPLAY "OL" (INFINITE)?

- YES: go to step 3.
- NO: REPAIR SHORT TO GROUND IN CKT 1283.

(3)
- USING J 39200, MEASURE RESISTANCE BETWEEN GROUND AND TERMINAL "E" OF THE MOTOR PACK. DOES J 39200 DISPLAY "OL" (INFINITE)?

- YES: go to step 4.
- NO: REPLACE MOTOR PACK.

(4)
- INSPECT CKTs 1282 AND 1283 AND 6-WAY CONNECTORS FOR DAMAGE WHICH MAY RESULT IN A SHORT TO GROUND WITH ALL CONNECTORS CONNECTED. REPAIR DAMAGE IF EVIDENT.
- RECONNECT ALL CONNECTORS.
- DRIVE VEHICLE FORE THREE DRIVE CYCLES (A DRIVE CYCLE CONSISTS OF STARTING ENGINE, DRIVING ABOVE 16 KM/H (10 MPH) AND THEN TURNING IGNITION "OFF."). DID DTC 62 SET IN THE LAST THREE DRIVE CYCLES?

- YES: REPLACE EBCM.
- NO: REFER TO "DIAGNOSTIC AIDS" ON FACING PAGE.

* AFTER DIAGNOSIS IS COMPLETE, CLEAR DTCs AND TEST DRIVE VEHICLE FOR THREE (3) DRIVE CYCLES TO VERIFY DTC DOES NOT RESET. A DRIVE CYCLE CONSISTS OF STARTING THE VEHICLE, DRIVING OVER 16 km/h (10 MPH) AND THEN KEYING DOWN.

THIS CHART ASSUMES THAT A CURRENT DTC IS STORED INDICATING THAT THIS MALFUNCTION IS PRESENT.

DIAGNOSTIC CHART DTC 63 — 1994–95 BERETTA AND CORSICA

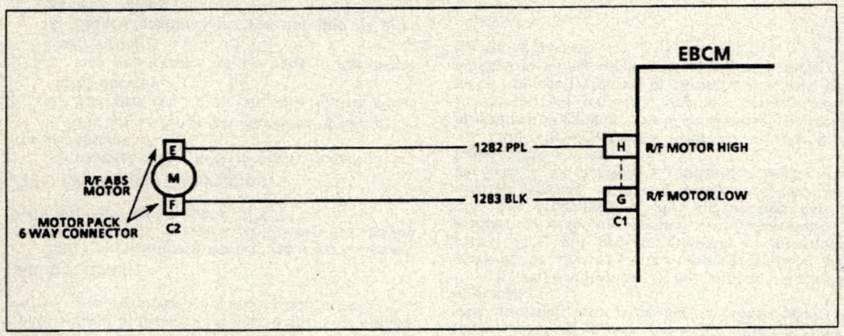

DTC 63
RIGHT FRONT ABS MOTOR CIRCUIT SHORTED TO BATTERY OR MOTOR SHORTED

Circuit Description:

This DTC identifies a motor circuit that is shorted to battery or a motor that has low or no resistance. This malfunction will not allow the motor to be controlled at the commanded current rate or will cause the motor to turn in the opposite direction or not at all.

DTC Will Set When:

DTC 63 can be set anytime. If the EBCM detects an out of range voltage in either of the right front motor circuits indicating a circuit shorted to battery or a motor shorted, a malfunction exists.

Action Taken:

A malfunction DTC is stored, ABS is disabled and the ABS warning lamp is turned "ON."

DTC Chart Test Description:

Number(s) below refer to circled number(s) on the diagnostic chart.

1. This step checks for a short to voltage in the motor "high" circuitry.
2. This step checks for a short to voltage in the motor "low" circuitry.
3. Checks for a short between "high" and "low" motor circuits.
4. This step checks for a motor that is internally shorted.
5. Checks for a short between the EMB and motor circuits.
6. Checks for an internal motor pack short between the EMB and motor.
7. Ensures malfunction was not due to physical damage of the circuitry.

Diagnostic Aids:

Using Tech 1, select manual control function, and exercise motor movement of affected channel in both directions while applying light pressure on the brake pedal.

If erratic or "jumpy" brake pedal movement is detected while performing an "apply" or "release" function of the motor, an intermittent malfunction may be indicated. An "Intermittent" malfunction is most likely caused by a poor connection, rubbed through wire insulation, or a wire that is broken inside the insulation.

If the malfunction is not current, wiggle the wires of the affected channel and check if the DTC resets. This will help to pinpoint an intermittent malfunction in the motor circuitry or connections.

The frequency of the malfunction can be checked by using the enhanced diagnostic function of the Tech 1, as described in "Enhanced Diagnostics," found in this section. If DTC 41 sets as a current DTC and DTC 63 is infrequent history DTC, see DTC 41 for diagnosis. Any circuitry, that is suspected as causing the intermittent complaint, should be thoroughly checked for backed out terminals, improper mating, broken locks, improperly formed or damaged terminals, poor terminal to wiring connections or physical damage to the wiring harness.

DIAGNOSTIC CHART DTC 63 — 1994–95 BERETTA AND CORSICA

NOTICE: J 39200 TEST LEADS MUST BE "ZEROED" PRIOR TO MAKING ANY RESISTANCE MEASUREMENTS. REFER TO J 39200 USER'S MANUAL.

DTC 63

RIGHT FRONT ABS MOTOR CIRCUIT SHORTED TO BATTERY OR MOTOR SHORTED

DOES DTC 63 OCCUR INTERMITTENTLY?

- YES → REFER TO "DIAGNOSTIC AIDS" ON FACING PAGE.
- NO →

(1) NOTICE: THIS STEP MAY CAUSE ADDITIONAL DTCs TO SET.
- IGNITION "OFF."
- DISCONNECT 6-WAY CONNECTORS AT MOTOR PACK AND EBCM.
- IGNITION "ON."
- USING J 39200, MEASURE VOLTAGE BETWEEN GROUND AND 6-WAY MOTOR PACK HARNESS CONNECTOR TERMINAL "E". IS VOLTAGE LESS THAN 1 VOLT?

  - NO → REPAIR SHORT TO VOLTAGE IN CKT 1282.
  - YES →

(2)
- USING J 39200, MEASURE VOLTAGE BETWEEN GROUND AND 6-WAY MOTOR PACK HARNESS CONNECTOR TERMINAL "F". IS VOLTAGE LESS THAN 1 VOLT?

  - NO → REPAIR SHORT TO VOLTAGE IN CKT 1283.
  - YES →

(3)
- IGNITION "OFF."
- USING J 39200, MEASURE RESISTANCE BETWEEN 6-WAY MOTOR PACK HARNESS CONNECTOR TERMINALS "E" AND "F". DOES J 39200 DISPLAY "OL" (INFINITE)?

  - NO → REPAIR SHORT BETWEEN CKTs 1282 AND 1283.
  - YES →

(4)
- USING J 39200, MEASURE RESISTANCE BETWEEN MOTOR PACK TERMINALS "E" AND "F". IS RESISTANCE GREATER THAN 4 OHMS?

  - NO → REPLACE MOTOR PACK.
  - YES →

(5)
- DISCONNECT 3-WAY MOTOR PACK CONNECTOR.
- USING J 39200, MEASURE RESISTANCE BETWEEN 3-WAY MOTOR PACK HARNESS CONNECTOR TERMINAL "B" AND 6-WAY MOTOR PACK HARNESS CONNECTOR TERMINAL "E" THEN "F". DID J 39200 DISPLAY "OL" (INFINITE) FOR EACH MEASUREMENT?

  - NO → REPAIR SHORT BETWEEN CKT 855/1633 AND MOTOR CIRCUIT WHICH MEASURED OTHER THAN "OL" (INFINITE).
  - YES →

(6)
- USING J 39200, MEASURE RESISTANCE BETWEEN 3-WAY MOTOR PACK CONNECTOR TERMINAL "B" AND 6-WAY MOTOR PACK CONNECTOR TERMINAL "E". DID J 39200 DISPLAY "OL" (INFINITE)?

  - NO → REPLACE MOTOR PACK.
  - YES →

(7)
- INSPECT CKTs 1282 AND 1283 AND 6-WAY CONNECTORS FOR DAMAGE WHICH MAY RESULT IN A SHORT TO B+ WITH ALL CONNECTORS CONNECTED.
- REPAIR DAMAGE IF EVIDENT.
- RECONNECT ALL CONNECTORS.
- DRIVE VEHICLE FOR THREE DRIVE CYCLES (A DRIVE CYCLE CONSISTS OF STARTING ENGINE, DRIVE ABOVE 16 km/h (10 MPH) AND THEN TURNING IGNITION "OFF"). DID DTC 63 SET IN THE LAST THREE DRIVE CYCLES?

  - YES → REPLACE EBCM.
  - NO → REFER TO "DIAGNOSTIC AIDS" ON FACING PAGE.

* AFTER DIAGNOSIS IS COMPLETE, CLEAR DTCs AND TEST DRIVE VEHICLE FOR THREE (3) DRIVE CYCLES TO VERIFY DTC DOES NOT RESET. A DRIVE CYCLE CONSISTS OF STARTING THE VEHICLE, DRIVING OVER 16 km/h (10 MPH) AND THEN KEYING DOWN.

THIS CHART ASSUMES THAT A CURRENT DTC IS STORED INDICATING THAT THIS MALFUNCTION IS PRESENT.

DIAGNOSTIC CHART DTC 64 — 1994–95 BERETTA AND CORSICA

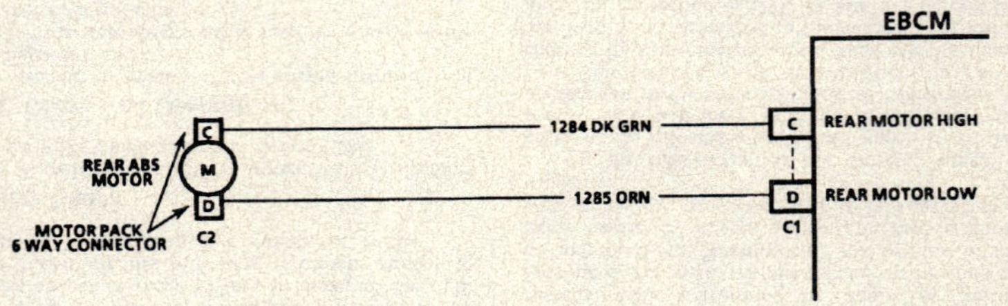

DTC 64

REAR ABS MOTOR CIRCUIT OPEN

Circuit Description:

This DTC identifies a motor that cannot be energized due to an open in its circuitry. This malfunction will not allow proper rear ABS operation.

DTC Will Set When:

DTC 64 can be set only when the motor is commanded "OFF." If the EBCM detects an out of range voltage in either of the rear ABS motor circuits indicating an open circuit, a malfunction exists.

Action Taken:

An open motor will not activate when requested. A malfunction DTC is stored, ABS is disabled and the ABS warning lamp is turned "ON."

DTC Chart Test Description:

Number(s) below refer to circled number(s) on the diagnostic chart.

1. This checks for proper resistance of the motor.
2. This step checks for an open in the motor "high" circuitry.
3. This step checks for an open in the motor "low" circuitry.
4. Ensures malfunction was not due to poor terminal contact.

Diagnostic Aids:

Using Tech 1, select manual control function, and exercise motor movement of affected channel in both directions while applying light pressure on the brake pedal.

If erratic or "jumpy" brake pedal movement is detected while performing an "apply" or "release" function of the motor, an intermittent malfunction may be indicated. An "Intermittent" malfunction is most likely caused by a poor connection, rubbed through wire insulation, or a wire that is broken inside the insulation.

If the malfunction is not current, wiggle the wires of the affected channel and check if the DTC resets. This will help to pinpoint an intermittent malfunction in the motor circuitry or connections.

The frequency of the malfunction can be checked by using the enhanced diagnostic function of the Tech 1, as described in "Enhanced Diagnostics," found in this section.

Any circuitry, that is suspected as causing the intermittent complaint, should be thoroughly checked for backed out terminals, improper mating, broken locks, improperly formed or damaged terminals, poor terminal to wiring connections or physical damage to the wiring harness.

DIAGNOSTIC CHART DTC 64 — 1994–95 BERETTA AND CORSICA

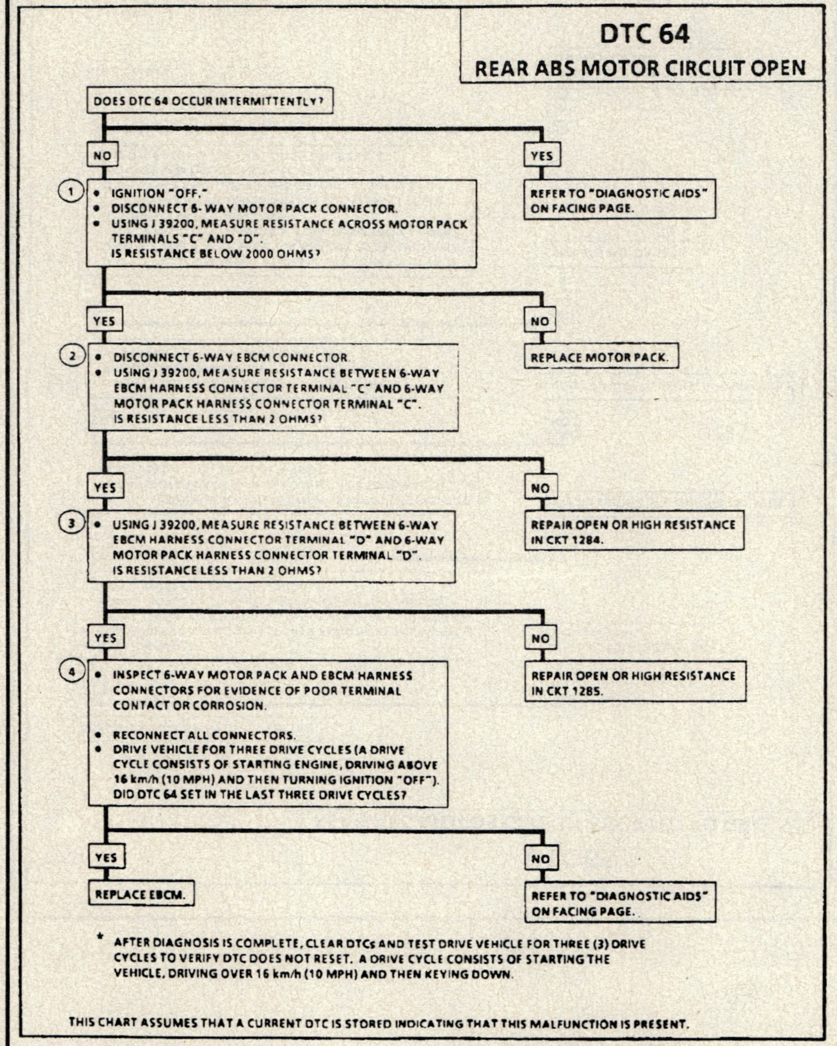

DIAGNOSTIC CHART DTC 65 — 1994–95 BERETTA AND CORSICA

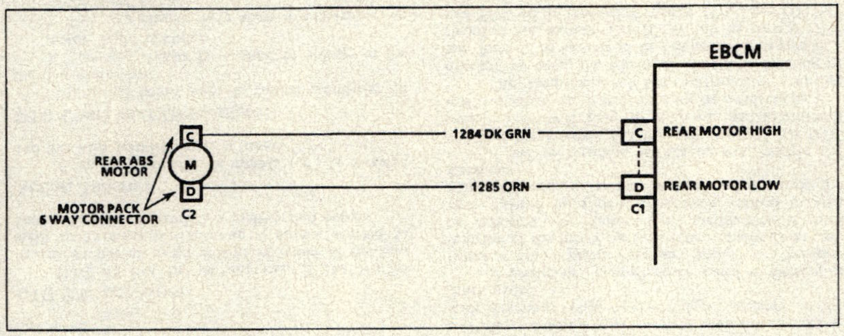

DTC 65
REAR ABS MOTOR CIRCUIT SHORTED TO GROUND

Circuit Description:

This DTC identifies a motor circuit that is shorted to ground. This malfunction will not allow the motor to be controlled at the commanded current rate or will cause the driver circuit to allow current directly to ground.

DTC Will Set When:

DTC 65 can be set anytime. If the EBCM detects an out of range voltage in either of the rear ABS motor circuits indicating a circuit shorted to ground, a malfunction exists.

Action Taken:

A malfunction DTC is stored, ABS is disabled and the ABS warning lamp is turned "ON."

DTC Chart Test Description:

Number(s) below refer to circled number(s) on the diagnostic chart.

1. This step checks for a short to ground in the motor "high" circuitry.
2. This step checks for a short to ground in the motor "low" circuitry.
3. This step checks for a motor that is internally shorted to ground.
4. Ensures malfunction was not due to physical damage of the circuitry.

Diagnostic Aids:

Using Tech 1, select manual control function, and exercise motor movement of affected channel in both directions while applying light pressure on the brake pedal.

If erratic or "jumpy" brake pedal movement is detected while performing an "apply" or "release" function of the motor, an intermittent malfunction may be indicated. An "Intermittent" malfunction is most likely caused by a poor connection, rubbed through wire insulation, or a wire that is broken inside the insulation.

If the malfunction is not current, wiggle the wires of the affected channel and check if the DTC resets. This will help to pinpoint an intermittent malfunction in the motor circuitry or connections.

The frequency of the problem can be checked by using the enhanced diagnostic function of the Tech 1, as described in "Enhanced Diagnostics," found in this section.

Any circuitry, that is suspected as causing the intermittent complaint, should be thoroughly checked for backed out terminals, improper mating, broken locks, improperly formed or damaged terminals, poor terminal to wiring connections or physical damage to the wiring harness.

DIAGNOSTIC CHART DTC 65 — 1994–95 BERETTA AND CORSICA

DTC 65
REAR ABS MOTOR CIRCUIT SHORTED TO GROUND

DOES DTC 65 OCCUR INTERMITTENTLY?

- YES → REFER TO "DIAGNOSTIC AIDS" ON FACING PAGE.
- NO →

(1)
- IGNITION "OFF."
- DISCONNECT 6-WAY CONNECTORS AT MOTOR PACK AND EBCM.
- USING J 39200, MEASURE RESISTANCE BETWEEN GROUND AND 6-WAY MOTOR PACK HARNESS CONNECTOR TERMINAL "C". DOES J 39200 DISPLAY "OL" (INFINITE)?

- NO → REPAIR SHORT TO GROUND IN CKT 1284.
- YES →

(2)
- USING J 39200, MEASURE RESISTANCE BETWEEN GROUND AND 6-WAY MOTOR PACK HARNESS CONNECTOR TERMINAL "D". DOES J 39200 DISPLAY "OL" (INFINITE)?

- NO → REPAIR SHORT TO GROUND IN CKT 1285.
- YES →

(3)
- USING J 39200, MEASURE RESISTANCE BETWEEN GROUND AND TERMINAL "C" OF THE MOTOR PACK. DOES J 39200 DISPLAY "OL" (INFINITE)?

- NO → REPLACE MOTOR PACK.
- YES →

(4)
- INSPECT CKTs 1284 AND 1285 AND 6-WAY CONNECTORS FOR DAMAGE WHICH MAY RESULT IN A SHORT TO GROUND WITH ALL CONNECTORS CONNECTED. REPAIR DAMAGE IF EVIDENT.
- RECONNECT ALL CONNECTORS.
- DRIVE VEHICLE FORE THREE DRIVE CYCLES (A DRIVE CYCLE CONSISTS OF STARTING ENGINE, DRIVING ABOVE 16 KM/H (10 MPH) AND THEN TURNING IGNITION "OFF"). DID DTC 65 SET IN THE LAST THREE DRIVE CYCLES?

- YES → REPLACE EBCM.
- NO → REFER TO "DIAGNOSTIC AIDS" ON FACING PAGE.

* AFTER DIAGNOSIS IS COMPLETE, CLEAR DTCs AND TEST DRIVE VEHICLE FOR THREE (3) DRIVE CYCLES TO VERIFY DTC DOES NOT RESET. A DRIVE CYCLE CONSISTS OF STARTING THE VEHICLE, DRIVING OVER 16 km/h (10 MPH) AND THEN KEYING DOWN.

THIS CHART ASSUMES THAT A CURRENT DTC IS STORED INDICATING THAT THIS MALFUNCTION IS PRESENT.

DIAGNOSTIC CHART DTC 66 — 1994–95 BERETTA AND CORSICA

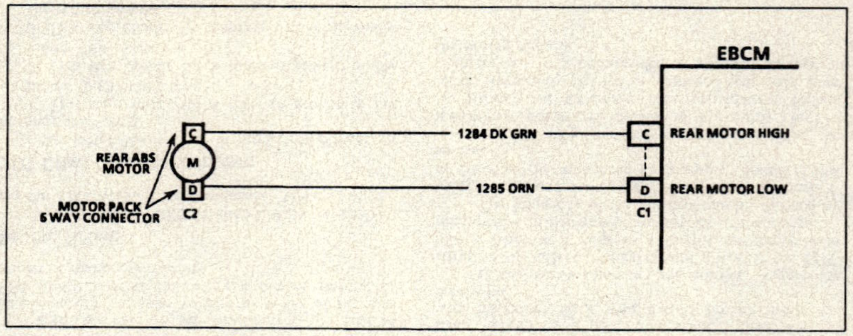

DTC 66
REAR ABS MOTOR CIRCUIT SHORTED TO BATTERY OR MOTOR SHORTED

Circuit Description:

This DTC identifies a motor circuit that is shorted to battery or a motor that has low or zero resistance. This malfunction will not allow the motor to be controlled at the commanded current rate or will cause the motor to turn in the opposite direction or not turn at all.

DTC Will Set When:

DTC 66 can be set anytime. If the EBCM detects an out of range voltage in either of the rear ABS motor circuits indicating a circuit shorted to battery or a motor shorted, a malfunction exists.

Action Taken:

A malfunction DTC is stored, ABS is disabled and the ABS warning lamp is turned "ON."

DTC Chart Test Description:

Number(s) below refer to circled number(s) on the diagnostic chart.

1. This step checks for a short to voltage in the motor "high" circuitry.
2. This step checks for a short to voltage in the motor "low" circuitry.
3. Checks for a short between "high" and "low" motor circuits.
4. This step checks for a motor that is internally shorted.
5. Checks for a short between the EMB and motor circuits.
6. Checks for an internal motor pack short between the EMB and motor.
7. Ensures malfunction was not due to physical damage of the circuitry.

Diagnostic Aids:

Using Tech 1, select manual control function, and exercise motor movement of affected channel in both directions while applying light pressure on the brake pedal.

If erratic or "jumpy" brake pedal movement is detected while performing an "apply" or "release" function of the motor, an intermittent malfunction may be indicated. An "Intermittent" malfunction is most likely caused by a poor connection, rubbed through wire insulation, or a wire that is broken inside the insulation.

If the malfunction is not current, wiggle the wires of the affected channel and check if the DTC resets. This will help to pinpoint an intermittent malfunction in the motor circuitry or connections.

The frequency of the malfunction can be checked by using the enhanced diagnostic function of the Tech 1, as described in "Enhanced Diagnostics," found in this section. If DTC 42 sets as current DTC and DTC 66 is infrequent history DTC, see DTC 42 diagnostics.

Any circuitry, that is suspected as causing the intermittent complaint, should be thoroughly checked for backed out terminals, improper mating, broken locks, improperly formed or damaged terminals, poor terminal to wiring connections or physical damage to the wiring harness.

DIAGNOSTIC CHART DTC 66 — 1994–95 BERETTA AND CORSICA

NOTICE: J 39200 TEST LEADS MUST BE "ZEROED" PRIOR TO MAKING ANY RESISTANCE MEASUREMENTS. REFER TO J 39200 USER'S MANUAL.

DTC 66
REAR ABS MOTOR CIRCUIT SHORTED TO BATTERY OR MOTOR SHORTED

DOES DTC 66 OCCUR INTERMITTENTLY?

- YES: REFER TO "DIAGNOSTIC AIDS" ON FACING PAGE.
- NO:

(1)
NOTICE: THIS STEP MAY CAUSE ADDITIONAL DTCs TO SET.
- IGNITION "OFF."
- DISCONNECT 6-WAY CONNECTORS AT MOTOR PACK AND EBCM.
- IGNITION "ON."
- USING J 39200, MEASURE VOLTAGE BETWEEN GROUND AND 6-WAY MOTOR PACK HARNESS CONNECTOR TERMINAL "C".

IS VOLTAGE LESS THAN 1 VOLT?

- NO: REPAIR SHORT TO VOLTAGE IN CKT 1284.
- YES:

(2)
- USING J 39200, MEASURE VOLTAGE BETWEEN GROUND AND 6-WAY MOTOR PACK HARNESS CONNECTOR TERMINAL "D"

IS VOLTAGE LESS THAN 1 VOLT?

- NO: REPAIR SHORT TO VOLTAGE IN CKT 1285.
- YES:

(3)
- IGNITION "OFF."
- USING J 39200, MEASURE RESISTANCE BETWEEN 6-WAY MOTOR PACK HARNESS CONNECTOR TERMINALS "C" AND "D".

DOES J 39200 DISPLAY "OL" (INFINITE)?

- NO: REPAIR SHORT BETWEEN CKTs 1284 AND 1285.
- YES:

(4)
- USING J 39200, MEASURE RESISTANCE BETWEEN MOTOR PACK TERMINALS "C" AND "D".

IS RESISTANCE GREATER THAN 4 OHMS?

- NO: REPLACE MOTOR PACK.
- YES:

(5)
- DISCONNECT 3-WAY MOTOR PACK CONNECTOR.
- USING J 39200, MEASURE RESISTANCE BETWEEN 3-WAY MOTOR PACK HARNESS CONNECTOR TERMINAL "B" AND 6-WAY MOTOR PACK HARNESS CONNECTOR TERMINAL "C" THEN "D".

DID J 39200 DISPLAY "OL" (INFINITE) FOR EACH MEASUREMENT?

- NO: REPAIR SHORT BETWEEN CIRCUIT 855/1633 AND MOTOR CIRCUIT WHICH MEASURED OTHER THEN "OL" (INFINITE).
- YES:

(6)
- USING J 39200, MEASURE RESISTANCE BETWEEN 3-WAY MOTOR PACK CONNECTOR TERMINAL "B" AND 6-WAY MOTOR PACK CONNECTOR TERMINAL "C".

DOES J 39200 DISPLAY "OL" (INFINITE)?

- NO: REPLACE MOTOR PACK.
- YES:

(7)
- INSPECT CKTs 1284 AND 1285 AND 6-WAY CONNECTORS FOR DAMAGE WHICH MAY RESULT IN A SHORT TO B+ WITH ALL CONNECTORS CONNECTED.
- REPAIR DAMAGE IF EVIDENT.
- RECONNECT ALL CONNECTORS.
- DRIVE VEHICLE FOR THREE DRIVE CYCLES (A DRIVE CYCLE CONSISTS OF STARTING ENGINE, DRIVE ABOVE 16 km/h (10 MPH) AND THEN TURNING IGNITION "OFF").

DID DTC 66 SET IN THE LAST THREE DRIVE CYCLES?

- YES: REPLACE EBCM.
- NO: REFER TO "DIAGNOSTIC AIDS" ON FACING PAGE.

* AFTER DIAGNOSIS IS COMPLETE, CLEAR DTCs AND TEST DRIVE VEHICLE FOR THREE (3) DRIVE CYCLES TO VERIFY DTC DOES NOT RESET. A DRIVE CYCLE CONSISTS OF STARTING THE VEHICLE, DRIVING OVER 16 km/h (10 MPH) AND THEN KEYING DOWN.

THIS CHART ASSUMES THAT A CURRENT DTC IS STORED INDICATING THAT THIS MALFUNCTION IS PRESENT.

DIAGNOSTIC CHART DTC 67 — 1994–95 BERETTA AND CORSICA

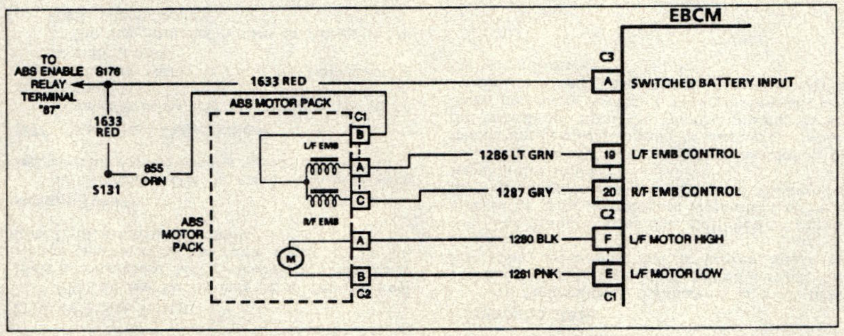

DTC 67
LEFT FRONT EMB CIRCUIT OPEN OR SHORTED TO GROUND

Circuit Description:

This DTC is designed to detect shorts to ground and opens in the EMB release circuit. A short to ground between the EMB and the ABS EBCM will cause the EMB to be energized indefinitely at an uncontrolled current, causing the EMB to be destroyed.

DTC Will Set When:

DTC 67 can be set only when the left front EMB is commanded "OFF" (not released). If the EBCM detects the left front EMB control circuit is constantly at ground, a malfunction exists.

Action Taken:

A malfunction DTC is stored, ABS is disabled and the ABS warning lamp is turned "ON."

DTC Chart Test Description:

Number(s) below refer to circled number(s) on the diagnostic chart.

1. This step checks for a short to ground in the EMB control circuitry.
2. This step looks for an open in the EMB control circuitry.
3. Checks for an open in the EMB battery feed circuit.
4. This step ensures the EMB is not internally shorted to ground.
5. Ensures EMB is not internally open.
6. Checks for short to ground in EMB battery feed circuit.
7. Ensures malfunction was not due to physical damage of the circuitry.

Diagnostic Aids:

An "Intermittent" malfunction is most likely caused by a poor connection, rubbed through wire insulation, or a wire that is broken inside the insulation.

The frequency of the malfunction can be checked by using the enhanced diagnostic function of the Tech 1, as described in "Enhanced Diagnostics," found in this section.

Any circuitry, that is suspected as causing the intermittent complaint, should be thoroughly checked for backed out terminals, improper mating, broken locks, improperly formed or damaged terminals, poor terminal to wiring connections or physical damage to the wiring harness.

DIAGNOSTIC CHART DTC 67 — 1994–95 BERETTA AND CORSICA

DTC 67
LEFT FRONT EMB CIRCUIT OPEN OR SHORTED TO GROUND

1. • IGNITION "OFF."
   • DISCONNECT 24-WAY AND 2-WAY EBCM HARNESS CONNECTORS.
   • DISCONNECT 3-WAY MOTOR PACK HARNESS CONNECTOR.
   • USING J 39200, MEASURE RESISTANCE BETWEEN GROUND AND 24-WAY EBCM HARNESS CONNECTOR TERMINAL "19".
   DOES J 39200 DISPLAY "OL" (INFINITE)?
   - NO: REPAIR SHORT TO GROUND IN CKT 1286.
   - YES: go to 2.
2. • USING J 39200, MEASURE RESISTANCE BETWEEN 3-WAY MOTOR PACK HARNESS CONNECTOR TERMINAL "A" AND 24-WAY EBCM HARNESS CONNECTOR TERMINAL "19".
   IS RESISTANCE LESS THAN 2 OHMS?
   - NO: REPAIR OPEN OR HIGH RESISTANCE IN CKT 1286.
   - YES: go to 3.
3. • USING J 39200, MEASURE RESISTANCE BETWEEN 2-WAY EBCM HARNESS CONNECTOR TERMINAL "A" AND 3-WAY MOTOR PACK HARNESS CONNECTOR TERMINAL "B".
   IS RESISTANCE LESS THAN 2 OHMS?
   - NO: REPAIR OPEN OR HIGH RESISTANCE IN CKT 1286.
   - YES: go to 4.
4. • USING J 39200, MEASURE RESISTANCE BETWEEN MOTOR PACK TERMINAL "A" AND GROUND.
   DOES J 39200 DISPLAY "OL" (INFINITE)?
   - NO: REPAIR OPEN OR HIGH RESISTANCE IN CKT 1633/855.
   - YES: go to 5.
5. • USING J 39200, MEASURE RESISTANCE BETWEEN MOTOR PACK TERMINALS "A" AND "B".
   IS RESISTANCE LESS THAN 2 OHMS?
   - NO: REPLACE MOTOR PACK.
   - YES: go to 6.
6. • USING J 39200, MEASURE RESISTANCE BETWEEN GROUND AND 2-WAY EBCM HARNESS CONNECTOR TERMINAL "A".
   DOES J 39200 DISPLAY "OL" (INFINITE)?
   - NO: REPLACE MOTOR PACK.
   - YES: go to 7.
7. • INSPECT 2-WAY AND 24-WAY EBCM AND 3-WAY MOTOR PACK CONNECTIONS FOR POOR TERMINAL CONTACT.
   REPLACE TERMINALS THAT EXHIBIT POOR CONTACT.
   • INSPECT CKT 855/1633, CKT 1286, 3-WAY MOTOR PACK AND EBCM HARNESS CONNECTORS FOR PHYSICAL DAMAGE WHICH MAY RESULT IN A SHORT TO GROUND WITH ALL CONNECTORS CONNECTED. REPAIR DAMAGE IF EVIDENT.
   • RECONNECT ALL CONNECTORS.
   • TEST DRIVE VEHICLE FOR THREE DRIVE CYCLES. A DRIVE CYCLE CONSISTS OF STARTING THE ENGINE, DRIVING OVER 16 km/h (10 MPH) AND THEN KEYING DOWN.
   DID DTC 67 SET IN THE LAST THREE DRIVE CYCLES?
   - NO: REPAIR SHORT TO GROUND IN CKT 855/1633.
   - YES: go to the next step.
   - Next step (DID DTC 67 SET…): YES: REPLACE EBCM. NO: MALFUNCTION IS INTERMITTENT. REFER TO "DIAGNOSTIC AIDS" ON FACING PAGE.

• AFTER DIAGNOSIS IS COMPLETE, CLEAR DTCs AND TEST DRIVE VEHICLE FOR THREE (3) DRIVE CYCLES TO VERIFY DTC DOES NOT RESET. A DRIVE CYCLE CONSISTS OF STARTING THE VEHICLE, DRIVING OVER 16km/h (10 MPH) AND THEN KEYING DOWN.

THIS CHART ASSUMES THAT A CURRENT DTC IS STORED INDICATING THAT THIS MALFUNCTION IS PRESENT.

DIAGNOSTIC CHART DTC 68 — 1994–95 BERETTA AND CORSICA

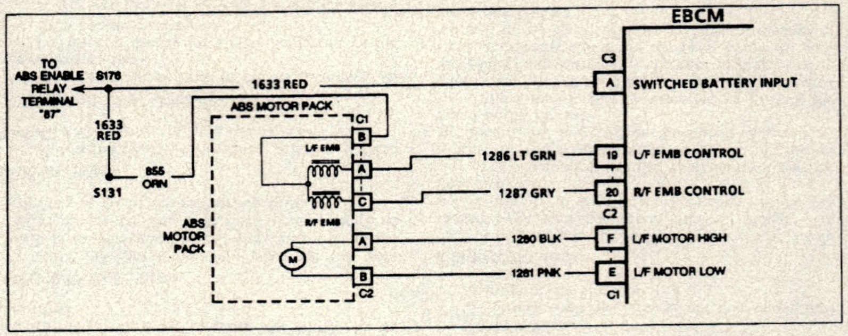

DTC 68
LEFT FRONT EMB CIRCUIT SHORTED TO BATTERY OR DRIVER OPEN

Circuit Description:

This DTC is designed to detect shorts to battery, shorts in the EMB release circuit, or an open EMB driver. These malfunctions will not allow the EMB to be energized (released) when needed.

DTC Will Set When:

DTC 68 can be set only when the left front EMB is commanded "ON" (released). If an excessive voltage drop is across the motor driver for the left front EMB, a malfunction exists.

Action Taken:

A malfunction DTC is stored, ABS is disabled and the ABS warning lamp is turned "ON."

DTC Chart Test Description:

Number(s) below refer to circled number(s) on the diagnostic chart.

1. This step checks to see if the malfunction is currently present.
2. This step identifies a short to voltage in the EMB control circuitry.
3. This step checks for an EMB that has lower than normal resistance.
4. Checks for a short between the EMB battery feed and control circuits.
5. Ensures malfunction was not due to physical damage of the circuitry.

Diagnostic Aids:

An "Intermittent" malfunction is most likely caused by a poor connection, rubbed through wire insulation, or a wire that is broken inside the insulation.

The frequency of the malfunction can be checked by using the enhanced diagnostic function of the Tech 1, as described in "Enhanced Diagnostics," found in this section.

Any circuitry, that is suspected as causing the intermittent complaint, should be thoroughly checked for backed out terminals, improper mating, broken locks, improperly formed or damaged terminals, poor terminal to wiring connections or physical damage to the wiring harness.

DIAGNOSTIC CHART DTC 68 — 1994–95 BERETTA AND CORSICA

*NOTICE: J 39200 TEST LEADS MUST BE "ZEROED" PRIOR TO MAKING ANY RESISTANCE MEASUREMENTS. REFER TO J 39200 USER'S MANUAL.

DTC 68
LEFT FRONT EMB CIRCUIT SHORTED TO BATTERY OR DRIVER OPEN

(1)
- IGNITION "ON."
- USING TECH 1 MISC. TESTS, SELECT MANUAL CONTROL FUNCTION.
- SELECT LEFT FRONT MOTOR APPLY FUNCTION AND APPLY MOTOR.
- APPLY FIRM PRESSURE ON BRAKE PEDAL.
- SELECT LEFT FRONT EMB RELEASE FUNCTION AND RELEASE EMB.

DID BRAKE PEDAL FALL WITH INCREASING PEDAL FORCE?

- YES: MALFUNCTION IS NOT PRESENT AT THIS TIME. REFER TO "DIAGNOSTIC AIDS" ON FACING PAGE.
- NO: go to step 2.

(2) NOTICE: THIS STEP MAY CAUSE ADDITIONAL DTCs TO SET.
- IGNITION "OFF."
- DISCONNECT 3-WAY MOTOR PACK CONNECTOR.
- DISCONNECT 24-WAY EBCM CONNECTOR.
- IGNITION "ON."
- USING J 39200, MEASURE VOLTAGE BETWEEN GROUND AND 3-WAY MOTOR PACK HARNESS CONNECTOR TERMINAL "A".

IS VOLTAGE LESS THAN 1 VOLT?

- NO: REPAIR SHORT TO VOLTAGE IN CKT 1286.
- YES: go to step 3.

(3)
- IGNITION "OFF."
- USING J 39200, MEASURE RESISTANCE BETWEEN MOTOR PACK TERMINALS "A" AND "B".

IS RESISTANCE GREATER THAN .4 OHMS?

- NO: REPLACE MOTOR PACK.
- YES: go to step 4.

(4)
- USING J 39200, MEASURE THE RESISTANCE BETWEEN 3-WAY MOTOR PACK HARNESS CONNECTOR TERMINALS "B" AND "A".

DOES J 39200 DISPLAY "OL" (INFINITE)?

- NO: REPAIR SHORT BETWEEN CKTs 1286 AND 855/1633.
- YES: go to step 5.

(5)
- INSPECT CKT 1286, 3-WAY MOTOR PACK AND 24-WAY EBCM CONNECTORS FOR DAMAGE WHICH MAY RESULT IN A SHORT TO B+ WITH ALL CONNECTORS CONNECTED. REPAIR DAMAGE IF EVIDENT.
- RECONNECT ALL CONNECTORS.
- DRIVE VEHICLE FOR THREE DRIVE CYCLES (A DRIVE CYCLE CONSISTS OF STARTING ENGINE, DRIVING ABOVE 16 km/h (10 MPH) AND THEN TURNING IGNITION "OFF").
- IGNITION "ON."
- USING TECH 1, SELECT "F1: DTC HISTORY."

DID DTC 68 SET IN LAST THREE DRIVE CYCLES?

- YES: REPLACE EBCM.
- NO: MALFUNCTION IS INTERMITTENT. REFER TO "DIAGNOSTIC AIDS" ON FACING PAGE.

* AFTER DIAGNOSIS IS COMPLETE, CLEAR DTCs AND TEST DRIVE VEHICLE FOR THREE (3) DRIVE CYCLES TO VERIFY DTC DOES NOT RESET. A DRIVE CYCLE CONSISTS OF STARTING THE VEHICLE, DRIVING OVER 16 km/h (10 MPH) AND THEN KEYING DOWN.

THIS CHART ASSUMES THAT A CURRENT DTC IS STORED INDICATING THAT THIS MALFUNCTION IS PRESENT.

DIAGNOSTIC CHART DTC 71 — 1994–95 BERETTA AND CORSICA

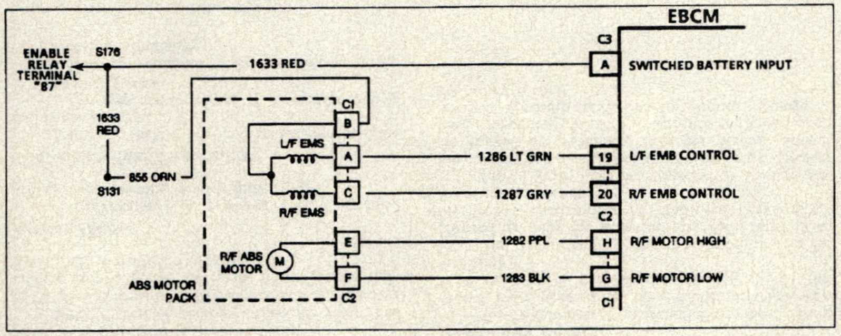

DTC 71
RIGHT FRONT EMB CIRCUIT OPEN OR SHORTED TO GROUND

Circuit Description:

This DTC is designed to detect shorts to ground and opens in the EMB release circuit. A short to ground between the EMB and the ABS EBCM will cause the EMB to be energized indefinitely at an uncontrolled current causing the EMB to be destroyed.

DTC Will Set When:

DTC 71 can be set only when the right front EMB is commanded "OFF" (not released). If the EBCM detects the right front EMB control circuit is constantly at ground, a malfunction exists.

Action Taken:

A malfunction DTC is stored, ABS is disabled and the ABS warning lamp is turned "ON."

DTC Chart Test Description:

Number(s) below refer to circled number(s) on the diagnostic chart

1. This step checks for a short to ground in the EMB control circuitry
2. This step looks for an open in the EMB control circuitry.
3. Checks for an open in the EMB battery feed circuit.
4. This step ensures the EMB is not internally shorted to ground.
5. Ensures EMB is not internally open.
6. Checks for short to ground in EMB battery feed circuit.
7. Ensures malfunction was not due to physical damage of the circuitry.

Diagnostic Aids:

An "Intermittent" malfunction is most likely caused by a poor connection, rubbed through wire insulation, or a wire that is broken inside the insulation.

The frequency of the malfunction can be checked by using the enhanced diagnostic function of the Tech 1, as described in "Enhanced Diagnostics," found in this section.

Any circuitry, that is suspected as causing the intermittent complaint, should be thoroughly checked for backed out terminals, improper mating, broken locks, improperly formed or damaged terminals, poor terminal to wiring connections or physical damage to the wiring harness.

DIAGNOSTIC CHART DTC 71 — 1994–95 BERETTA AND CORSICA

DTC 71
RIGHT FRONT EMB CIRCUIT OPEN OR SHORTED TO GROUND

1.
   - IGNITION "OFF."
   - DISCONNECT 24-WAY AND 2-WAY EBCM HARNESS CONNECTORS.
   - DISCONNECT 3-WAY MOTOR PACK HARNESS CONNECTOR.
   - USING J 39200, MEASURE RESISTANCE BETWEEN GROUND AND 24-WAY EBCM HARNESS CONNECTOR TERMINAL "20". DOES J 39200 DISPLAY "OL" (INFINITE)?

   - YES: go to step 2.
   - NO: REPAIR SHORT TO GROUND IN CKT 1287.

2.
   - USING J 39200, MEASURE RESISTANCE BETWEEN 3-WAY MOTOR PACK HARNESS CONNECTOR TERMINAL "A" AND 24-WAY EBCM HARNESS CONNECTOR TERMINAL "20". IS RESISTANCE LESS THAN 2 OHMS?

   - YES: go to step 3.
   - NO: REPAIR OPEN OR HIGH RESISTANCE IN CKT 1287.

3.
   - USING J 39200, MEASURE RESISTANCE BETWEEN 2-WAY EBCM HARNESS CONNECTOR TERMINAL "A" AND 3-WAY MOTOR PACK HARNESS CONNECTOR TERMINAL "B". IS RESISTANCE LESS THAN 2 OHMS?

   - YES: go to step 4.
   - NO: REPAIR OPEN OR HIGH RESISTANCE IN CKT 1287.

4.
   - USING J 39200, MEASURE RESISTANCE BETWEEN MOTOR PACK TERMINAL "C" AND GROUND. DOES J 39200 DISPLAY "OL" (INFINITE)?

   - YES: go to step 5.
   - NO: REPAIR OPEN OR HIGH RESISTANCE IN CKT 1633/855.

5.
   - USING J 39200, MEASURE RESISTANCE BETWEEN MOTOR PACK TERMINALS "C" AND "B". IS RESISTANCE LESS THAN 2 OHMS?

   - YES: go to step 6.
   - NO: REPLACE MOTOR PACK.

6.
   - USING J 39200, MEASURE RESISTANCE BETWEEN GROUND AND 2-WAY EBCM HARNESS CONNECTOR TERMINAL "A". DOES J 39200 DISPLAY "OL" (INFINITE)?

   - YES: go to step 7.
   - NO: REPLACE MOTOR PACK.

7.
   - INSPECT 2-WAY AND 24-WAY EBCM AND 3-WAY MOTOR PACK CONNECTIONS FOR POOR TERMINAL CONTACT. REPLACE TERMINALS THAT EXHIBIT POOR CONTACT.
   - INSPECT CKT 855/1633, CKT 1287, 3-WAY MOTOR PACK AND EBCM HARNESS CONNECTORS FOR PHYSICAL DAMAGE WHICH MAY RESULT IN A SHORT TO GROUND WITH ALL CONNECTORS CONNECTED. REPAIR DAMAGE IF EVIDENT.
   - RECONNECT ALL CONNECTORS.
   - TEST DRIVE VEHICLE FOR THREE DRIVE CYCLES. A DRIVE CYCLE CONSISTS OF STARTING THE ENGINE, DRIVING OVER 16 km/h (10 MPH) AND THEN KEYING DOWN. DID DTC 71 SET THE LAST THREE DRIVE CYCLES?

   - NO: REPAIR SHORT TO GROUND IN CKT 855/1633.
   - YES: REPLACE EBCM.
   - NO: MALFUNCTION IS INTERMITTENT. REFER TO "DIAGNOSTIC AIDS" ON FACING PAGE.

- AFTER DIAGNOSIS IS COMPLETE, CLEAR DTCs AND TEST DRIVE VEHICLE FOR THREE (3) DRIVE CYCLES TO VERIFY DTC DOES NOT RESET. A DRIVE CYCLE CONSISTS OF STARTING THE VEHICLE, DRIVING OVER 16km/h (10 MPH) AND THEN KEYING DOWN.

THIS CHART ASSUMES THAT A CURRENT DTC IS STORED INDICATING THAT THIS MALFUNCTION IS PRESENT.

DIAGNOSTIC CHART DTC 72 — 1994–95 BERETTA AND CORSICA

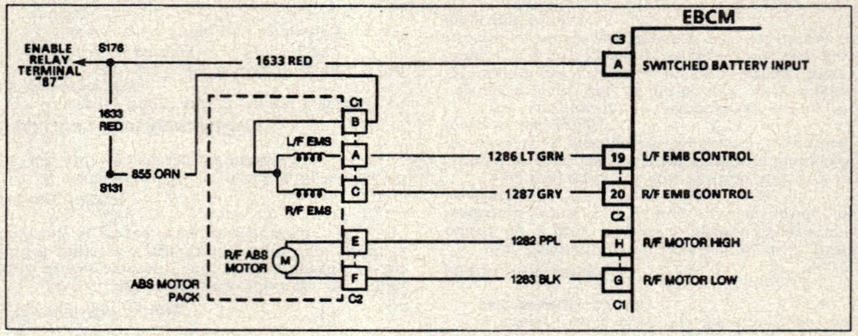

DTC 72
RIGHT FRONT EMB CIRCUIT SHORTED TO BATTERY OR DRIVER OPEN

Circuit Description:

This DTC is designed to detect shorts to battery, shorts in the EMB release circuit, or an open EMB driver. These malfunctions will not allow the EMB to be energized (released) when needed.

DTC Will Set When:

DTC 72 can be set only when the right front EMB is commanded "ON" (released). If an excessive voltage drop is across the motor driver for the right front EMB, a malfunction exists.

Action Taken:

A malfunction DTC is stored, ABS is disabled and the ABS warning lamp is turned "ON."

DTC Chart Test Description:

Number(s) below refer to circled number(s) on the diagnostic chart.

1. This step checks to see if the malfunction is currently present.
2. This step identifies a short to voltage in the EMB control circuitry.
3. This step checks for an EMB that has lower than normal resistance.
4. Checks for a short between the EMB battery feed and control circuits.
5. Ensures malfunction was not due to physical damage of the circuitry.

Diagnostic Aids:

An "Intermittent" malfunction is most likely caused by a poor connection, rubbed through wire insulation, or a wire that is broken inside the insulation.

The frequency of the malfunction can be checked by using the enhanced diagnostic function of the Tech 1, as described in "Enhanced Diagnostics," found in this section.

Any circuitry, that is suspected as causing the intermittent complaint, should be thoroughly checked for backed out terminals, improper mating, broken locks, improperly formed or damaged terminals, poor terminal to wiring connections or physical damage to the wiring harness.

DIAGNOSTIC CHART DTC 72 — 1994–95 BERETTA AND CORSICA

DTC 72
RIGHT FRONT EMB CIRCUIT SHORTED TO BATTERY OR DRIVER OPEN

*NOTICE: J 39200 TEST LEADS MUST BE "ZEROED" PRIOR TO MAKING ANY RESISTANCE MEASUREMENTS. REFER TO J 39200 USER'S MANUAL.

(1)
- IGNITION "ON."
- USING TECH 1 MISC. TESTS, SELECT MANUAL CONTROL FUNCTION.
- SELECT RIGHT FRONT MOTOR APPLY FUNCTION AND APPLY MOTOR.
- APPLY FIRM PRESSURE ON BRAKE PEDAL.
- SELECT RIGHT FRONT EMB RELEASE FUNCTION AND RELEASE EMB.

DID BRAKE PEDAL FALL WITH INCREASING PEDAL FORCE?

YES: MALFUNCTION IS NOT PRESENT AT THIS TIME. REFER TO "DIAGNOSTIC AIDS" ON FACING PAGE.

NO:

(2) NOTICE: THIS STEP MAY CAUSE ADDITIONAL DTCs TO SET.
- IGNITION "OFF."
- DISCONNECT 3-WAY MOTOR PACK CONNECTOR.
- DISCONNECT 24-WAY EBCM CONNECTOR.
- IGNITION "ON."
- USING J 39200, MEASURE VOLTAGE BETWEEN GROUND AND 3-WAY MOTOR PACK HARNESS CONNECTOR TERMINAL "C".

IS VOLTAGE LESS THAN 1 VOLT?

NO: REPAIR SHORT TO VOLTAGE IN CKT 1287.

YES:

(3)
- IGNITION "OFF."
- USING J 39200, MEASURE RESISTANCE BETWEEN MOTOR PACK TERMINALS "C" AND "B".

IS RESISTANCE GREATER THAN .4 OHMS?

NO: REPLACE MOTOR PACK.

YES:

(4)
- USING J 39200, MEASURE RESISTANCE BETWEEN 3-WAY MOTOR PACK HARNESS CONNECTOR TERMINALS "B" AND "C".

DOES J 39200 DISPLAY "OL" (INFINITE)?

NO: REPAIR SHORT BETWEEN CKT(s) 1287 AND 855/1633.

YES:

(5)
- INSPECT CKT 1287, 3-WAY MOTOR PACK AND 24-WAY EBCM CONNECTORS FOR DAMAGE WHICH MAY RESULT IN A SHORT TO B+ WITH ALL CONNECTORS CONNECTED. REPAIR DAMAGE IF EVIDENT.
- RECONNECT ALL CONNECTORS.
- DRIVE VEHICLE FOR THREE DRIVE CYCLES (A DRIVE CYCLE CONSISTS OF STARTING ENGINE, DRIVING ABOVE 16 KM/H (10 MPH) AND THEN TURNING IGNITION "OFF").
- IGNITION "ON."
- USING TECH 1, SELECT "F1: DTC HISTORY."

DID DTC 72 SET IN LAST THREE DRIVE CYCLES?

YES: REPLACE EBCM.

NO: MALFUNCTION IS INTERMITTENT. REFER TO "DIAGNOSTIC AIDS" ON FACING PAGE.

* AFTER DIAGNOSIS IS COMPLETE, CLEAR DTCs AND TEST DRIVE VEHICLE FOR THREE (3) DRIVE CYCLES TO VERIFY DTC DOES NOT RESET. A DRIVE CYCLE CONSISTS OF STARTING THE VEHICLE, DRIVING OVER 16 km/h (10 MPH) AND THEN KEYING DOWN.

THIS CHART ASSUMES THAT A CURRENT DTC IS STORED INDICATING THAT THIS MALFUNCTION IS PRESENT.

DIAGNOSTIC CHART DTC 76 — 1994–95 BERETTA AND CORSICA

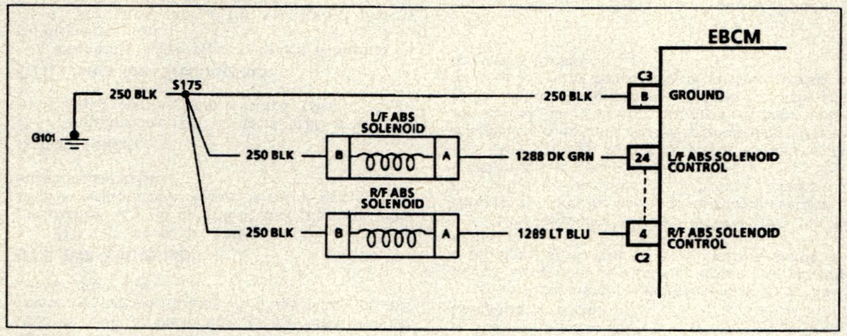

DTC 76
LEFT FRONT SOLENOID CIRCUIT OPEN OR SHORTED TO BATTERY

Circuit Description:

This DTC identifies a solenoid that cannot be energized due to an open in its circuitry, or a solenoid that is always energized due to a short to battery in its circuitry between the driver and the solenoid. An open will not allow proper ABS operation, but the short to battery simply turns "ON" the solenoid. A path for base brakes is still allowed once the motor re-homes and the check valve is lifted off its seat during key "ON" initialization.

DTC Will Set When:

DTC 76 can be set only when the solenoid is commanded "OFF." If the EBCM detects an excessive voltage in the left front solenoid control circuit, a malfunction exists.

Action Taken:

A malfunction DTC is stored, ABS is disabled and the ABS warning lamp is turned "ON."

DTC Chart Test Description:

Number(s) below refer to circled number(s) on the diagnostic chart.

1. This checks for a short to voltage in the control circuitry of the solenoid.
2. This checks for an open in the control circuit of the solenoid.
3. This step identifies an open in the solenoid or ground circuit.
4. Checks for proper resistance values of the solenoid.
5. Ensures malfunction was not due to physical damage of the circuitry.

Diagnostic Aids:

An "Intermittent" malfunction is most likely caused by a poor connection, rubbed through wire insulation, or a wire that is broken inside the insulation.

The frequency of the malfunction can be checked by using the enhanced diagnostic function of the Tech 1, as described in "Enhanced Diagnostics," found in this section.

Any circuitry, that is suspected as causing the intermittent complaint, should be thoroughly checked for backed out terminals, improper mating, broken locks, improperly formed or damaged terminals, poor terminal to wiring connections or physical damage to the wiring harness.

DIAGNOSTIC CHART DTC 76 — 1994–95 BERETTA AND CORSICA

DTC 76
LEFT FRONT SOLENOID CIRCUIT OPEN OR SHORTED TO BATTERY

*NOTICE: J 39200 TEST LEADS MUST BE "ZEROED" PRIOR TO MAKING ANY RESISTANCE MEASUREMENTS. REFER TO J 39200 USER'S MANUAL.

(1)
- IGNITION "OFF."
- DISCONNECT 24-WAY EBCM CONNECTOR.
- IGNITION "ON."
- USING J 39200, MEASURE VOLTAGE BETWEEN GROUND AND 24-WAY EBCM HARNESS CONNECTOR TERMINAL "24". IS VOLTAGE LESS THAN 1 VOLT?

YES → go to step 2. NO → REPAIR SHORT TO VOLTAGE IN CKT 1288.

(2)
- IGNITION "OFF."
- DISCONNECT LEFT FRONT SOLENOID CONNECTOR.
- USING J 39200, MEASURE RESISTANCE BETWEEN LEFT FRONT SOLENOID HARNESS CONNECTOR TERMINAL "A" AND 24-WAY EBCM HARNESS CONNECTOR TERMINAL "24". IS RESISTANCE LESS THAN 2 OHMS?

YES → go to step 3. NO → REPAIR OPEN OR HIGH RESISTANCE IN CKT 1288.

(3)
- USING J 39200, MEASURE RESISTANCE BETWEEN GROUND AND LEFT FRONT SOLENOID HARNESS CONNECTOR TERMINAL "B". IS RESISTANCE LESS THAN 2 OHMS?

YES → go to step 4. NO → REPAIR OPEN OR HIGH RESISTANCE IN CKT 250

(4)
- USING J 39200, MEASURE RESISTANCE BETWEEN LEFT FRONT SOLENOID TERMINALS "A" AND "B". IS RESISTANCE BETWEEN 2.5 AND 5 OHMS?

YES → go to step 5. NO → REPLACE LEFT FRONT SOLENOID.

(5)
- INSPECT LEFT FRONT SOLENOID AND 24-WAY EBCM HARNESS CONNECTORS FOR POOR TERMINAL CONTACT OR CORROSION. REPLACE TERMINALS THAT EXHIBIT POOR CONTACT OR CORROSION.
- INSPECT CKT 1288 FOR DAMAGE WHICH MAY RESULT IN A SHORT TO B+ WITH ALL CONNECTORS CONNECTED. REPAIR DAMAGE IF EVIDENT.
- RECONNECT ALL CONNECTORS.
- IGNITION "ON."
- IS DTC 76 SET AS A CURRENT DTC?

YES → REPLACE EBCM. NO → REFER TO "DIAGNOSTIC AIDS" ON FACING PAGE.

THIS CHART ASSUMES THAT A CURRENT DTC IS STORED INDICATING THAT THIS MALFUNCTION IS PRESENT.

DIAGNOSTIC CHART DTC 77 — 1994–95 BERETTA AND CORSICA

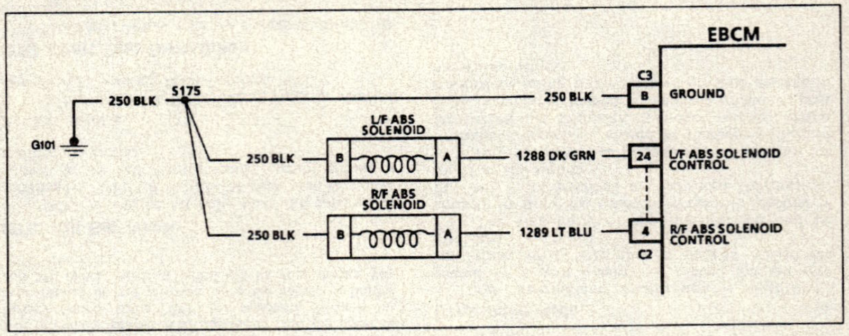

DTC 77
LEFT FRONT SOLENOID CIRCUIT SHORTED TO GROUND OR DRIVER OPEN

Circuit Description:

This DTC identifies a solenoid that cannot be energized due to an open in its driver circuitry, or a short to ground between the solenoid driver and the solenoid. These malfunctions can affect ABS operation since the flow of brake fluid to the wheel cylinder cannot be stopped, making ABS operation for that channel impossible.

DTC Will Set When:

DTC 77 can be set only when the solenoid is commanded "ON." If the EBCM detects the left front solenoid control circuit voltage is out of specification, a malfunction exists.

Action Taken:

A malfunction DTC is stored, ABS is disabled and the ABS warning lamp is turned "ON."

DTC Chart Test Description:

Number(s) below refer to circled number(s) on the diagnostic chart.

1. This step checks for a solenoid that is internally shorted to ground.
2. This step checks for a solenoid that is not within proper resistance values.
3. This step indicates if a short to ground exists in the solenoid control circuitry.
4. Ensures malfunction was not due to physical damage of the circuitry.

Diagnostic Aids:

An "Intermittent" malfunction is most likely caused by a poor connection, rubbed through wire insulation, or a wire that is broken inside the insulation.

The frequency of the malfunction can be checked by using the enhanced diagnostic function of the Tech 1, as described in "Enhanced Diagnostics," found in this section.

Any circuitry, that is suspected as causing the intermittent complaint, should be thoroughly checked for backed out terminals, improper mating, broken locks, improperly formed or damaged terminals, poor terminal to wiring connections or physical damage to the wiring harness.

DIAGNOSTIC CHART DTC 77 — 1994–95 BERETTA AND CORSICA

DTC 77
LEFT FRONT SOLENOID CIRCUIT SHORTED TO GROUND OR DRIVER OPEN

(1)
- IGNITION "OFF."
- DISCONNECT LEFT FRONT SOLENOID CONNECTOR.
- USING J 39200, MEASURE THE RESISTANCE BETWEEN GROUND AND LEFT FRONT SOLENOID TERMINAL "A." DOES J 39200 DISPLAY RESISTANCE AS "OL" (INFINITE)?

NO → REPLACE LEFT FRONT SOLENOID.

YES ↓

(2)
- USING J 39200, MEASURE THE RESISTANCE BETWEEN LEFT FRONT SOLENOID TERMINALS "A" AND "B". IS RESISTANCE BETWEEN 2.5 AND 5 OHMS?

NO → REPLACE LEFT FRONT SOLENOID.

YES ↓

(3)
- DISCONNECT 24-WAY EBCM CONNECTOR.
- MEASURE THE RESISTANCE BETWEEN GROUND AND LEFT FRONT SOLENOID HARNESS CONNECTOR TERMINAL "A." DOES J 39200 DISPLAY RESISTANCE AS "OL" (INFINITE)?

NO → REPAIR SHORT TO GROUND IN CKT 1288.

YES ↓

(4)
- INSPECT CKT 1288, LEFT FRONT SOLENOID AND 24-WAY EBCM HARNESS CONNECTORS FOR DAMAGE WHICH MAY RESULT IN A SHORT TO GROUND WITH ALL CONNECTORS CONNECTED.
- START ENGINE WITH FOOT OFF BRAKE PEDAL.
- WAIT 10 SECONDS. IS DTC 77 SET AS A CURRENT DTC?

YES → REPLACE EBCM.

NO → MALFUNCTION IS INTERMITTENT. REFER TO "DIAGNOSTIC AIDS" ON FACING PAGE.

THIS CHART ASSUMES THAT A CURRENT DTC IS STORED INDICATING THAT THIS MALFUNCTION IS PRESENT.

DIAGNOSTIC CHART DTC 78 — 1994–95 BERETTA AND CORSICA

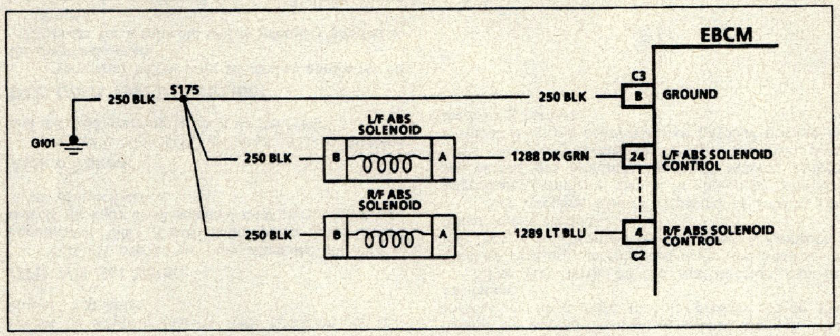

DTC 78
RIGHT FRONT SOLENOID CIRCUIT OPEN OR SHORTED TO BATTERY

Circuit Description:

This DTC identifies a solenoid that cannot be energized due to an open in its circuitry, or a solenoid that is always energized due to a short to battery in its circuitry between the driver and the solenoid. An open will not allow proper ABS operation but the short to battery simply turns "ON" the solenoid. A path for base brakes is still allowed once the motor re-homes and the check valve is lifted off its seat during key "ON" initialization.

DTC Will Set When:

DTC 78 can be set only when the solenoid is commanded "OFF." If the EBCM detects an excessive voltage in the right front solenoid control circuit, a malfunction exists.

Action Taken:

A malfunction DTC is stored, ABS is disabled and the ABS warning lamp is turned "ON."

DTC Chart Test Description:

Number(s) below refer to circled number(s) on the diagnostic chart.

1. This checks for a short to voltage in the control circuitry of the solenoid.
2. This checks for an open in the control circuit of the solenoid.
3. This step identifies an open in the solenoid or ground circuit.
4. Checks for proper resistance values of the solenoid.
5. Ensures malfunction was not due to physical damage of the circuitry.

Diagnostic Aids:

An "Intermittent" malfunction is most likely caused by a poor connection, rubbed through wire insulation, or a wire that is broken inside the insulation.

The frequency of the malfunction can be checked by using the enhanced diagnostic function of the Tech 1, as described in "Enhanced Diagnostics," found in this section.

Any circuitry, that is suspected as causing the intermittent complaint, should be thoroughly checked for backed out terminals, improper mating, broken locks, improperly formed or damaged terminals, poor terminal to wiring connections or physical damage to the wiring harness.

DIAGNOSTIC CHART DTC 78 — 1994–95 BERETTA AND CORSICA

DTC 78
RIGHT FRONT SOLENOID CIRCUIT OPEN OR SHORTED TO BATTERY

*NOTICE: J39200 TEST LEADS MUST BE "ZEROED" PRIOR TO MAKING ANY RESISTANCE MEASUREMENTS. REFER TO J39200 USER'S MANUAL.

(1)
- IGNITION "OFF."
- DISCONNECT 24-WAY EBCM CONNECTOR.
- IGNITION "ON."
- MEASURE VOLTAGE BETWEEN GROUND AND 24-WAY EBCM HARNESS CONNECTOR TERMINAL "4".

IS VOLTAGE LESS THAN 1 VOLT?

- NO: REPAIR SHORT TO VOLTAGE IN CKT 1289.
- YES: go to step 2.

(2)
- IGNITION "OFF."
- DISCONNECT RIGHT FRONT SOLENOID CONNECTOR.
- USING J 39200, MEASURE RESISTANCE BETWEEN RIGHT FRONT SOLENOID HARNESS CONNECTOR TERMINAL "A" AND 24-WAY EBCM HARNESS CONNECTOR TERMINAL "4".

IS RESISTANCE LESS THAN 2 OHMS?

- NO: REPAIR OPEN OR HIGH RESISTANCE IN CKT 1289.
- YES: go to step 3.

(3)
- USING J39200, MEASURE RESISTANCE BETWEEN GROUND AND RIGHT FRONT SOLENOID HARNESS CONNECTOR TERMINAL "B".

IS RESISTANCE LESS THAN 2 OHMS?

- NO: REPAIR OPEN IN CKT 250
- YES: go to step 4.

(4)
- USING J39200, MEASURE RESISTANCE BETWEEN RIGHT FRONT SOLENOID TERMINALS "A" AND "B".

IS RESISTANCE BETWEEN 2.5 AND 5 OHMS?

- NO: REPLACE RIGHT FRONT SOLENOID.
- YES: go to step 5.

(5)
- INSPECT RIGHT FRONT SOLENOID AND 24-WAY EBCM HARNESS CONNECTORS FOR POOR TERMINAL CONTACT OR CORROSION. REPLACE TERMINALS THAT EXHIBIT POOR CONTACT OR CORROSION.
- INSPECT CKT 1289 FOR DAMAGE WHICH MAY RESULT IN A SHORT TO B+ WITH ALL CONNECTORS CONNECTED. REPAIR DAMAGE IF EVIDENT.
- RECONNECT ALL CONNECTORS.
- IGNITION "ON."

IS DTC 78 SET AS A CURRENT DTC?

- YES: REPLACE EBCM.
- NO: MALFUNCTION IS INTERMITTENT, REFER TO "DIAGNOSTIC AIDS" ON FACING PAGE.

THIS CHART ASSUMES THAT A CURRENT DTC IS STORED INDICATING THAT THIS MALFUNCTION IS PRESENT.

DIAGNOSTIC CHART DTC 81 — 1994–95 BERETTA AND CORSICA

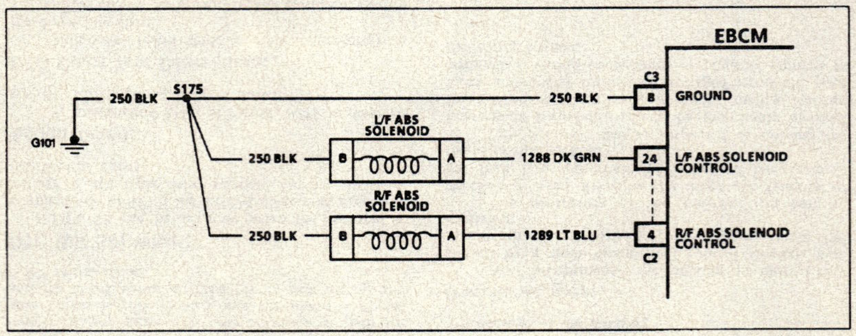

DTC 81
RIGHT FRONT SOLENOID CIRCUIT SHORTED TO GROUND OR DRIVER OPEN

Circuit Description:

This DTC identifies a solenoid that cannot be energized due to an open in its driver circuitry, or a short to ground between the solenoid driver and the solenoid. These malfunctions can affect ABS operation since the flow of brake fluid to the wheel cylinder cannot be stopped, making ABS operation for that channel impossible.

DTC Will Set When:

DTC 81 can be set only when the solenoid is commanded "ON." A malfunction exists if the EBCM detects the right front solenoid control circuit voltage is out of specification.

Action Taken:

A malfunction DTC is stored, ABS is disabled and the ABS warning lamp is turned "ON."

DTC Chart Test Description:

Number(s) below refer to circled number(s) on the diagnostic chart.

1. Checks for a solenoid that is internally shorted to ground.
2. Checks for a solenoid that is not within proper resistance values.
3. Indicates if a short to ground exists in the solenoid control circuitry.
4. Ensures malfunction was not due to physical damage of the circuitry.

Diagnostic Aids:

An "Intermittent" malfunction is most likely caused by a poor connection, rubbed through wire insulation, or a wire that is broken inside the insulation.

The frequency of the malfunction can be checked by using the enhanced diagnostic function of the Tech 1, as described in "Enhanced Diagnostics," found in this section.

Any circuitry, that is suspected as causing the intermittent complaint, should be thoroughly checked for backed out terminals, improper mating, broken locks, improperly formed or damaged terminals, poor terminal to wiring connections or physical damage to the wiring harness.

DIAGNOSTIC CHART DTC 81 — 1994–95 BERETTA AND CORSICA

DTC 81
RIGHT FRONT SOLENOID CIRCUIT SHORTED TO GROUND OR DRIVER OPEN

1.
- IGNITION "OFF."
- DISCONNECT RIGHT FRONT SOLENOID CONNECTOR.
- USING J 39200, MEASURE THE RESISTANCE BETWEEN GROUND AND RIGHT FRONT SOLENOID TERMINAL "A". DOES J 39200 DISPLAY RESISTANCE AS "OL" (INFINITE)?

NO → REPLACE RIGHT FRONT SOLENOID.
YES ↓

2.
- USING J 39200, MEASURE RESISTANCE BETWEEN RIGHT FRONT SOLENOID TERMINALS "A" AND "B". IS RESISTANCE BETWEEN 2.5 AND 5 OHMS?

NO → REPLACE RIGHT FRONT SOLENOID.
YES ↓

3.
- DISCONNECT 24-WAY EBCM CONNECTOR.
- MEASURE THE RESISTANCE BETWEEN GROUND AND RIGHT FRONT SOLENOID HARNESS CONNECTOR TERMINAL "A". DOES J 39200 DISPLAY RESISTANCE AS "OL" (INFINITE)?

NO → REPAIR SHORT TO GROUND IN CKT 1289.
YES ↓

4.
- INSPECT CKT 1289, RIGHT FRONT SOLENOID AND 24-WAY EBCM HARNESS CONNECTORS FOR DAMAGE WHICH MAY RESULT IN A SHORT TO GROUND WITH ALL CONNECTORS CONNECTED.
- START ENGINE WITH FOOT OFF BRAKE PEDAL.
- WAIT 10 SECONDS. IS DTC 81 SET AS A CURRENT DTC?

YES → REPLACE EBCM.
NO → MALFUNCTION IS INTERMITTENT. REFER TO "DIAGNOSTIC AIDS" ON FACING PAGE.

THIS CHART ASSUMES THAT A CURRENT DTC IS STORED INDICATING THAT THIS MALFUNCTION IS PRESENT.

DIAGNOSTIC CHART DTC 82 — 1994–95 BERETTA AND CORSICA

D.M. ABS VI ELECTRONIC BRAKE CONTROL MODULE (TYPICAL)

DTC 82
CALIBRATION MALFUNCTION

Circuit Description:

This DTC allows the EBCM to check for a calibration failure by comparing the calibration value to a known value stored in the EEPROM.

This test is also used as a security measure to prevent improper use of calibrations or changes to these calibrations that may alter the designed function of ABS.

DTC Will Set When:

DTC 82 can be set at key-up. A malfunction exists if the program identifier is incorrect or the memory checksum is incorrect.

Action Taken:

A malfunction DTC is stored, ABS is disabled and the ABS warning lamp is turned "ON."

DTC Chart Test Description:

Number(s) below refer to circled number(s) on the diagnostic chart.

1. This step checks to see if the malfunction is present during diagnosis. If present, the EBCM is not functioning correctly and must be replaced.

Diagnostic Aids:

An intermittent DTC 82 is most likely caused by a bad cell in the EEPROM that is sensitive to temperature changes. If DTC 82 occurred more than once, but is intermittent, replace EBCM.

The frequency of the malfunction can be checked by using the enhanced diagnostic function of the Tech 1, as described in "Enhanced Diagnostics," found in this section.

DIAGNOSTIC CHART DTC 82 — 1994–95 BERETTA AND CORSICA

DTC 82
CALIBRATION MALFUNCTION

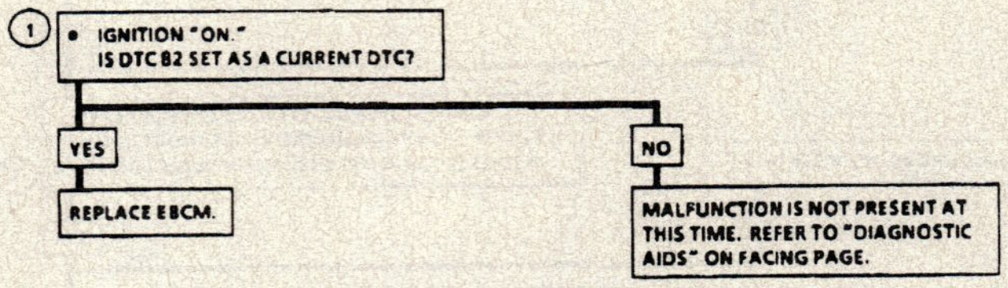

THIS CHART ASSUMES THAT A CURRENT DTC IS STORED INDICATING THAT THIS MALFUNCTION IS PRESENT.

DIAGNOSTIC CHART DTC 86 — 1994–95 BERETTA AND CORSICA

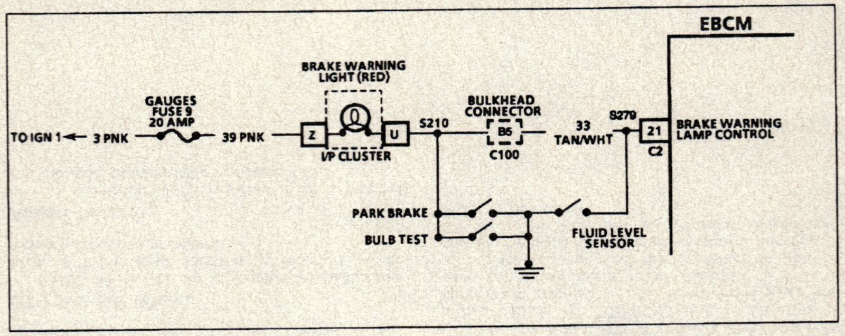

DTC 86
EBCM TURNED "ON" THE RED "BRAKE" WARNING LAMP

Circuit Description:

This DTC is provided as an "information only" test and reflects the status of the command issued by the ABS EBCM to illuminate the red "BRAKE" warning lamp. If another diagnostic DTC issues a command to illuminate the red "BRAKE" warning lamp, DTC 86 will be stored in EEPROM as a history DTC at the conclusion of the ignition cycle.

DTC Chart Test Description:

Number(s) below refer to circled number(s) on the diagnostic chart.

1. This step serves to identify if a DTC other than DTC 86 commanded the red "BRAKE" warning lamp "ON."

Diagnostic Aids:

Any ABS mechanical DTC that issues a command to illuminate the red "BRAKE" warning lamp will also result in DTC 86 being stored in EEPROM during shut down. These DTC's are: 38, 41, 42, 45, 46, 47, 48 and 51. If the motors are not in their home position, certain electrical DTC's will also command the red "BRAKE" warning lamp "ON." These DTC's are: 14, 18, 55, 56, 57, 58, 61, 62, 63, 64, 65, 66, 67, 68, 71 and 72.

If any of these DTC's are indicated along with DTC 86, they must be corrected prior to addressing a DTC 86 malfunction.

DIAGNOSTIC CHART DTC 86 — 1994–95 BERETTA AND CORSICA

DTC 86
EBCM TURNED "ON" THE RED "BRAKE" WARNING LAMP

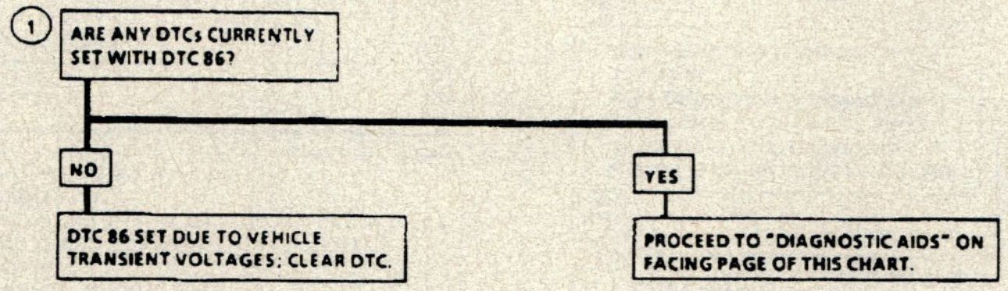

DIAGNOSTIC CHART DTC 87 — 1994–95 BERETTA AND CORSICA

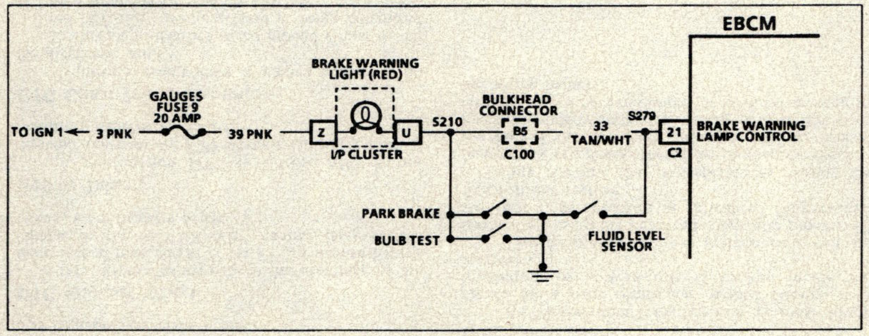

DTC 87
RED "BRAKE" WARNING LAMP CIRCUIT OPEN

Circuit Description:

This DTC is used to verify the EBCM has continuity to the "BRAKE" warning lamp in case the EBCM must turn it "ON." This will only occur if an ABS malfunction is detected that may degrade base brake operation.

Because ABS is not the only device controlling the "BRAKE" warning lamp (parking brake and low fluid sensor may also turn "ON" the lamp) a short to ground in this circuit can not necessarily be detected. Excessive resistance in the fluid level switch or parking brake circuitry may set this DTC.

DTC Will Set When:

DTC 87 can be set after initialization is completed. If the red "BRAKE" warning lamp circuit voltage is out of specification indicating an open circuit, a malfunction exists.

Action Taken:

A malfunction DTC is stored, however ABS is not disabled.

DTC Chart Test Description:

Number(s) below refer to circled number(s) on the diagnostic chart.

1. This step identifies if the red "BRAKE" warning lamp can be turned "ON."
2. This step test for high resistance in park brake circuitry.
3. Ensures malfunction was not due to poor terminal contact.
4. This step identifies if high resistance is present in the fluid level switch circuitry.
5. Checks integrity of red "BRAKE" warning bulb.
6. Checks for high resistance in the "BRAKE" warning lamp control circuit.

Diagnostic Aids:

An "Intermittent" malfunction is most likely caused by a poor connection, rubbed through wire insulation, or a wire that is broken inside the insulation.

The frequency of the malfunction can be checked by using the enhanced diagnostic function of the Tech 1, as described in "Enhanced Diagnostics," found in this section. If the Tech 1 is not available, the parking brake handle may be lifted to verify proper lamp operation and continuity of CKTs 3 and 39.

Any circuitry, that is suspected as causing the intermittent complaint, should be thoroughly checked for backed out terminals, improper mating, broken locks, improperly formed or damaged terminals, poor terminal to wiring connections or physical damage to the wiring harness.

DIAGNOSTIC CHART DTC 87 — 1994–95 BERETTA AND CORSICA

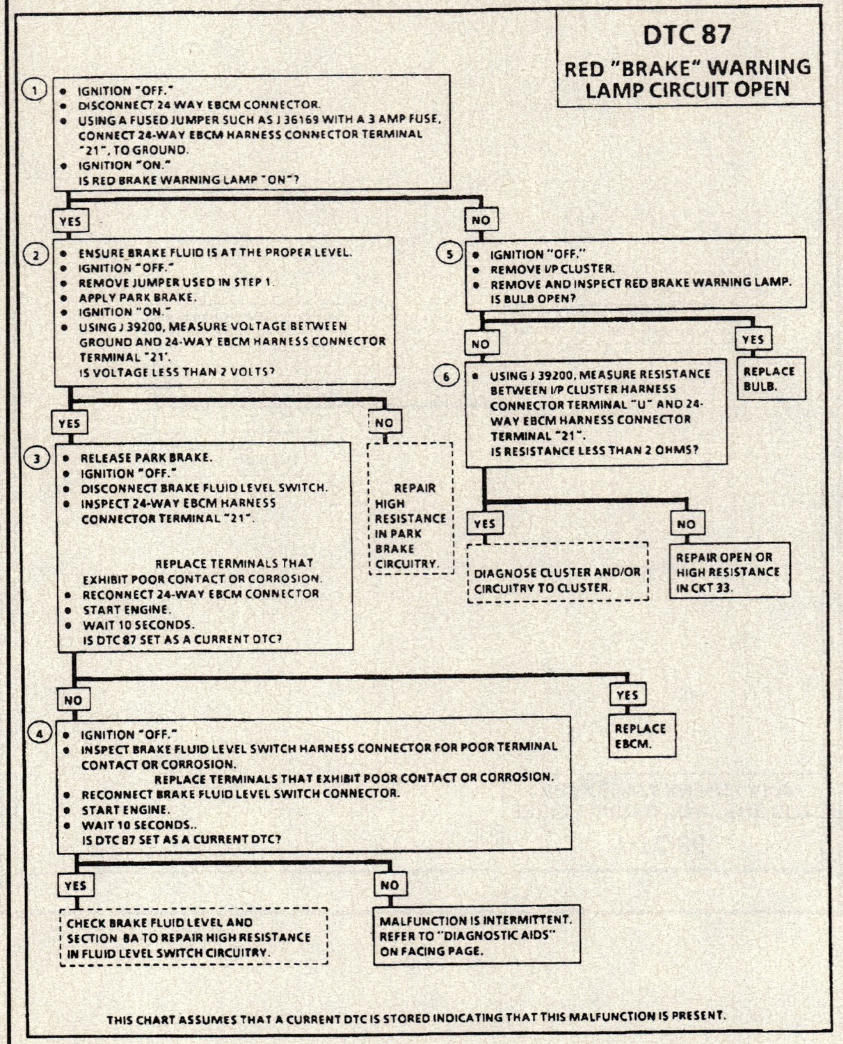

DIAGNOSTIC CHART DTC 88 — 1994–95 BERETTA AND CORSICA

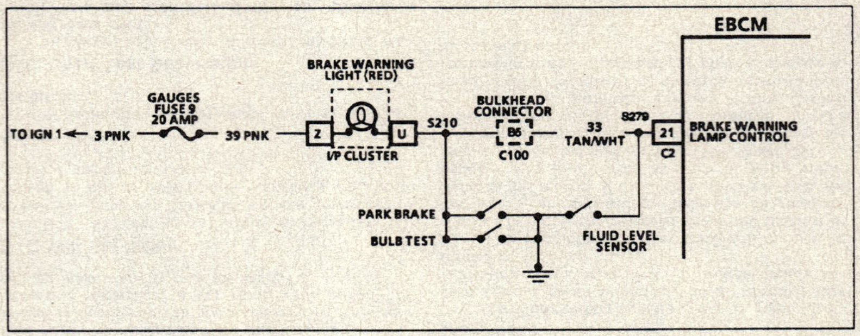

DTC 88
RED "BRAKE" WARNING LAMP CIRCUIT SHORTED TO BATTERY

Circuit Description:

This DTC identifies a short to battery between the EBCM and the red "BRAKE" warning lamp, or an open driver that does not allow the red "BRAKE" warning lamp to be illuminated by the EBCM. This will only occur if an ABS malfunction is detected that may degrade base brake operation.

DTC Will Set When:

DTC 88 can be set only at the beginning of the three second bulb check. If the EBCM detects battery voltage on the red "BRAKE" warning lamp control circuit, a malfunction exists.

Action Taken:

A malfunction DTC is stored. ABS is not disabled; however, the ABS warning lamp will flash to indicate a malfunction exists.

DTC Chart Test Description:

Number(s) below refer to circled number(s) on the diagnostic chart.

1. This step identifies if the ground circuit to the red "BRAKE" warning lamp is being completed by a source other than the EBCM.
2. This step identifies if the malfunction is currently present.
3. By removing the 20 amp Fuse 9, the voltage source is eliminated. This test indicates if voltage is being supplied from a source other than the gauge fuse.
4. Ensures malfunction was not due to physical damage of the circuitry.

Diagnostic Aids:

An "Intermittent" malfunction is most likely caused by a poor connection, rubbed through wire insulation, or a wire that is broken inside the insulation.

The frequency of the malfunction can be checked by using the enhanced diagnostic function of the Tech 1, as described in "Enhanced Diagnostics," found in this section.

Any circuitry, that is suspected as causing the intermittent complaint, should be thoroughly checked for backed out terminals, improper mating, broken locks, improperly formed or damaged terminals, poor terminal to wiring connections or physical damage to the wiring harness.

DIAGNOSTIC CHART DTC 88 — 1994–95 BERETTA AND CORSICA

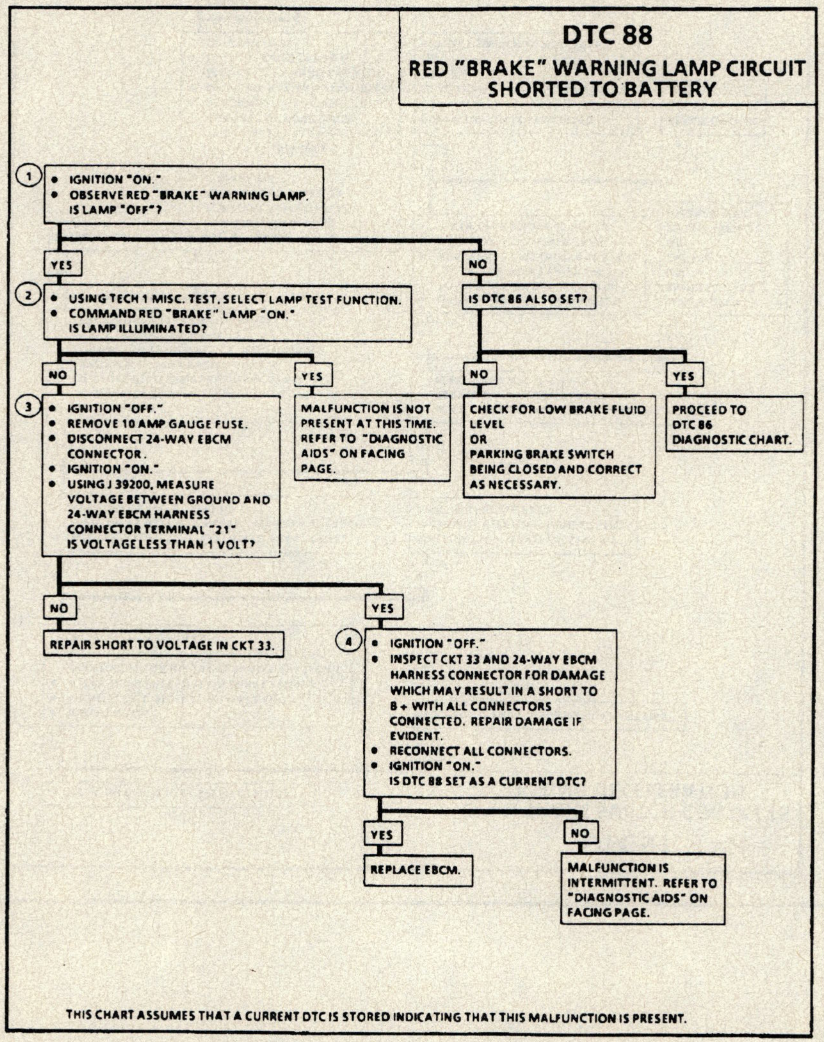

DIAGNOSTIC CHART DTC 91 — 1994–95 BERETTA AND CORSICA

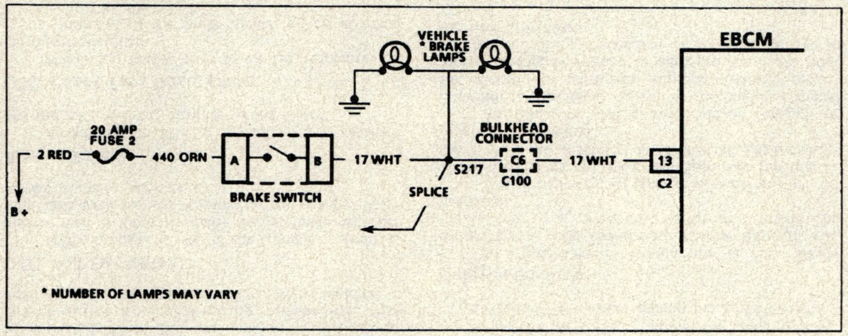

DTC 91
OPEN BRAKE SWITCH CONTACTS DURING DECELERATION

Circuit Description:

This DTC is used to detect an open brake switch in the non-ABS mode. The EBCM looks for decel rates that would indicate braking action and verifies this assumption by requiring several repeats of this detection method. In each case, ABS will not be available since no brake switch is seen.

DTC Will Set When:

DTC 91 can be set if three deceleration cycles occur with the brake switch "OFF."

Action Taken:

A malfunction DTC is stored, ABS is disabled and the ABS warning lamp is turned "ON."

DTC Chart Test Description:

Number(s) below refer to circled number(s) on the diagnostic chart.

1. Indicates if the brake switch signal is being received by the EBCM.
2. Indicates if an open circuit exists in the brake switch or brake lamp circuitry.
3. Checks for an open in the brake switch input circuit.
4. Ensures malfunction was not due to poor terminal contact.
5. Checks for voltage entering the brake switch.
6. Checks for voltage exiting the brake switch.

Diagnostic Aids:

An "Intermittent" malfunction is most likely caused by a poor connection, rubbed through wire insulation, or a wire that is broken inside the insulation.

The frequency of the malfunction can be checked by using the enhanced diagnostic function of the Tech 1, as described in "Enhanced Diagnostics," found in this section.

Any circuitry, that is suspected as causing the intermittent complaint, should be thoroughly checked for backed out terminals, improper mating, broken locks, improperly formed or damaged terminals, poor terminal to wiring connections or physical damage to the wiring harness.

DIAGNOSTIC CHART DTC 91 — 1994–95 BERETTA AND CORSICA

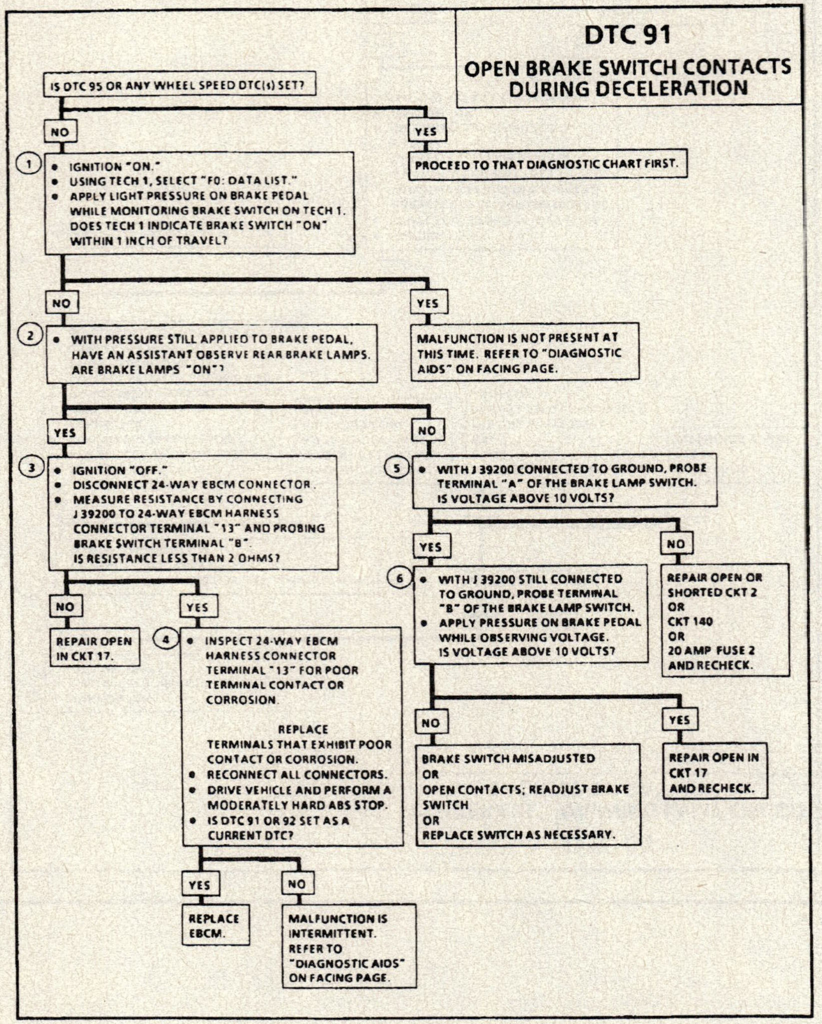

DIAGNOSTIC CHART DTC 92 — 1994–95 BERETTA AND CORSICA

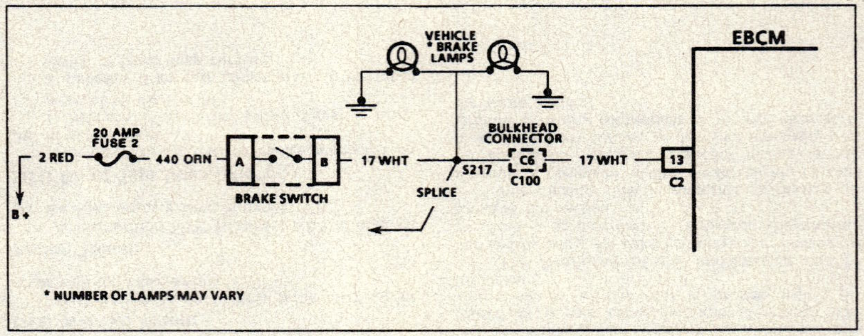

DTC 92

OPEN BRAKE SWITCH CONTACTS WHEN ABS WAS REQUIRED

Circuit Description:

This DTC is run to determine the proper operation of the brake switch. This is important because ABS is activated when the brake switch is "ON." If the brake switch is "OFF," ABS will never be activated. Since this malfunction is difficult to detect under normal braking conditions, this malfunction is only detected when ABS is required.

DTC Will Set When:

DTC 92 can be set if the vehicle's speed is greater than 8 km/h (5 mph). If the brake was not "ON" and a release was required on two channels for.7 second, a malfunction exists.

Action Taken:

A malfunction DTC is stored, ABS is disabled and the ABS warning lamp is turned "ON."

DTC Chart Test Description:

Number(s) below refer to circled number(s) on the diagnostic chart.

1. Indicates if the brake switch signal is being received by the EBCM.
2. Indicates if an open circuit exists in the brake switch or brake lamp circuitry.
3. Checks for an open in the brake switch input circuit.
4. Ensures malfunction was not due to poor terminal contact.
5. Checks for voltage entering the brake switch.
6. Checks for voltage exiting the brake switch.

Diagnostic Aids:

An "Intermittent" malfunction is most likely caused by a poor connection, rubbed through wire insulation, or a wire that is broken inside the insulation.

The frequency of the malfunction can be checked by using the enhanced diagnostic function of the Tech 1, as described in "Enhanced Diagnostics," found in this section.

Any circuitry, that is suspected as causing the intermittent complaint, should be thoroughly checked for backed out terminals, improper mating, broken locks, improperly formed or damaged terminals, poor terminal to wiring connections or physical damage to the wiring harness.

DIAGNOSTIC CHART DTC 92 — 1994–95 BERETTA AND CORSICA

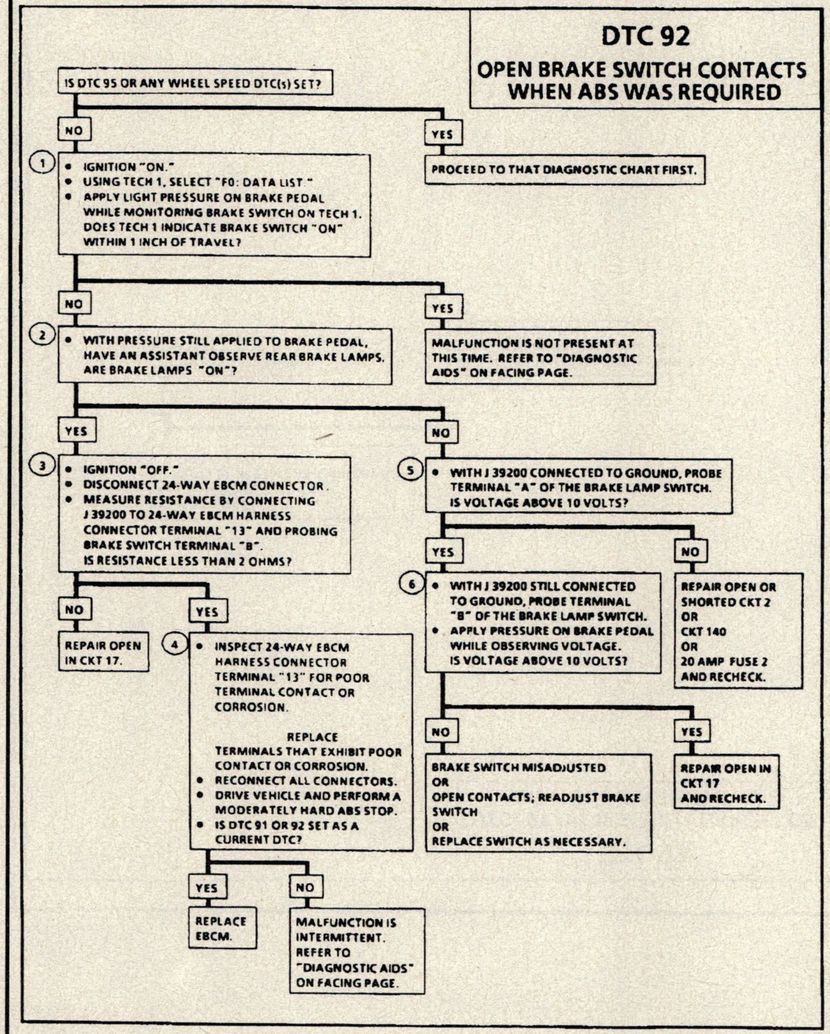

DIAGNOSTIC CHART DTC 93 — 1994–95 BERETTA AND CORSICA

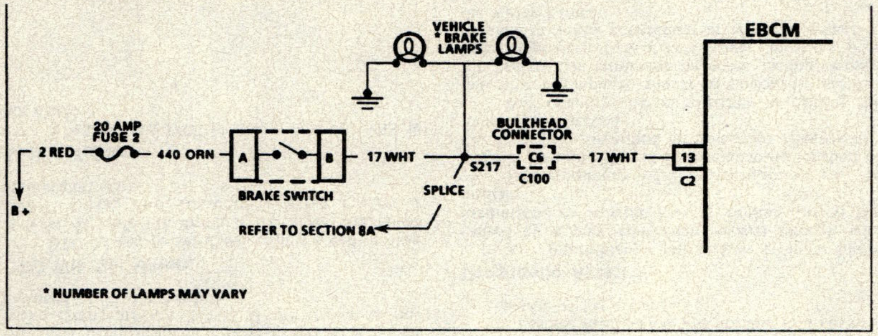

DTC 93
DTC's 91 OR 92 SET IN CURRENT OR PREVIOUS IGNITION CYCLE

Circuit Description:

This DTC is the second portion of DTC's 91 and 92. If DTC's 91 or 92 occurred during the last ignition cycle, DTC 93 becomes a current malfunction during the next ignition cycle, keeping ABS disabled until a brake switch "ON" state is seen. When a change is seen during an ignition cycle in which DTC 93 is a current malfunction, DTC 91 or 92 will clear itself at the end of the current ignition cycle, and ABS will enable itself at the start of the next ignition cycle. DTC 93 alone indicates DTC's 91 or 92 occurred previously, but is intermittent, or has been corrected.

DTC Chart Test Description:

Number(s) below refer to circled number(s) on the diagnostic chart.

1. Indicates which DTC (either 91 or 92) caused DTC 93 to set.

Diagnostic Aids:

An "Intermittent" malfunction is most likely caused by a poor connection, rubbed through wire insulation, or a wire that is broken inside the insulation.

The frequency of the malfunction can be checked by using the enhanced diagnostic function of the Tech 1, as described in "Enhanced Diagnostics," found in this section.

Any circuitry, that is suspected as causing the intermittent complaint, should be thoroughly checked for backed out terminals, improper mating, broken locks, improperly formed or damaged terminals, poor terminal to wiring connections or physical damage to the wiring harness.

Also, verify proper brake switch operation using the data list of the Tech 1. As the brake is applied, the data list should display the brake switch "ON" within 1 inch of travel.

DIAGNOSTIC CHART DTC 93 — 1994–95 BERETTA AND CORSICA

DTC 93
DTCs 91 OR 92 SET IN CURRENT OR PREVIOUS IGNITION CYCLE

(1)
- IGNITION "ON."
- USING TECH 1, READ DTCs. IS DTC 91 OR 92 SET AS A HISTORY OR CURRENT DTC?

YES: PROCEED TO DTC CHART THAT SET.

NO: USING TECH 1, SELECT "F1: DTC HISTORY." VERIFY MALFUNCTION FREQUENCY WAS LOW AND CLEAR DTCs.

THIS CHART ASSUMES THAT A CURRENT DTC IS STORED INDICATING THAT THIS MALFUNCTION IS PRESENT.

DIAGNOSTIC CHART DTC 94 — 1994–95 BERETTA AND CORSICA

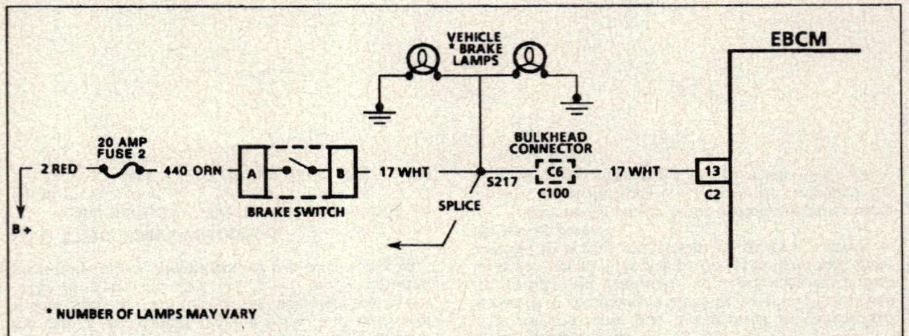

DTC 94
BRAKE SWITCH CONTACTS ALWAYS CLOSED

Circuit Description:

This DTC is run to determine the proper operation of the brake Switch. This is important because ABS is activated when the brake switch is "ON" and turned "OFF" when the brake is "OFF." If the brake switch is always "ON," ABS operation will always be requested resulting in potential modulator cycling on rough roads. Additionally, this malfunction will most likely result in a dead battery (due to the brake lamps staying "ON") if the driver is not informed of this malfunction.

DTC Will Set When:

DTC 94 can be set if the vehicle speed reaches at least 40 km/h (25 mph). If the brake was never "OFF" during two consecutive drive cycles, a malfunction exists.

Action Taken:

A malfunction DTC is stored, however ABS is not disabled.

DTC Chart Test Description:

Number(s) below refer to circled number(s) on the diagnostic chart.

1. Checks to see if brake lamps are "ON."
2. Isolates the cause of the malfunction to either a faulty or misadjusted brake switch, or a short to voltage in the brake switch circuitry.
3. Checks for a resistive short to voltage in the brake switch circuitry.
4. Ensures malfunction was not due to a physical damage of the circuitry.

Diagnostic Aids:

An "Intermittent" malfunction is most likely caused by a poor connection, rubbed through wire insulation, or a wire that is broken inside the insulation.

The frequency of the malfunction can be checked by using the enhanced diagnostic function of the Tech 1, as described in "Enhanced Diagnostics," found in this section.

Any circuitry, that is suspected as causing the intermittent complaint, should be thoroughly checked for backed out terminals, improper mating, broken locks, improperly formed or damaged terminals, poor terminal to wiring connections or physical damage to the wiring harness.

DIAGNOSTIC CHART DTC 94 — 1994–95 BERETTA AND CORSICA

DTC 94
BRAKE SWITCH CONTACTS ALWAYS CLOSED

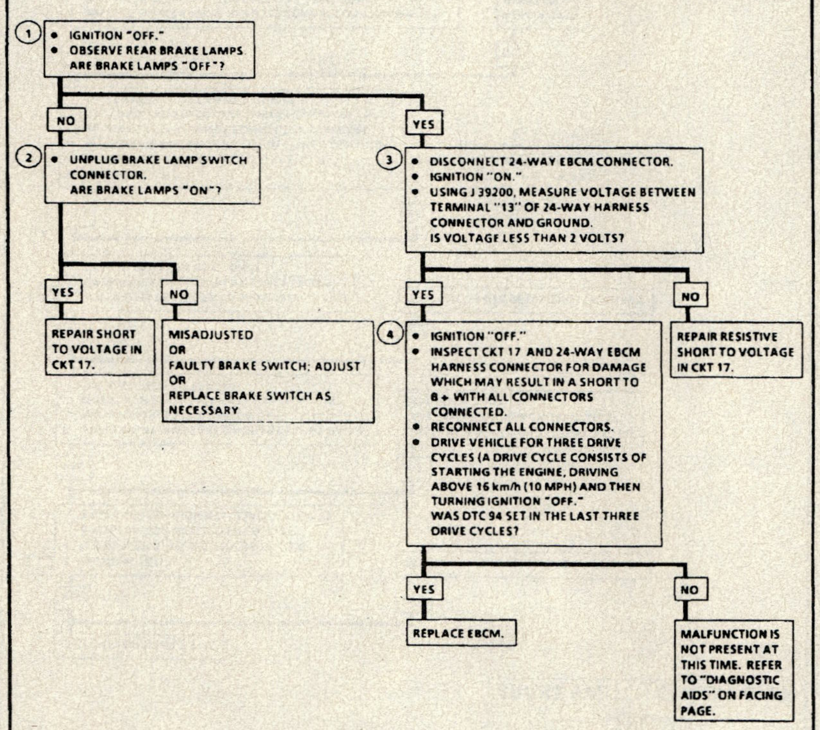

THIS CHART ASSUMES THAT A CURRENT DTC IS STORED INDICATING THAT THIS MALFUNCTION IS PRESENT.

DIAGNOSTIC CHART DTC 95 — 1994–95 BERETTA AND CORSICA

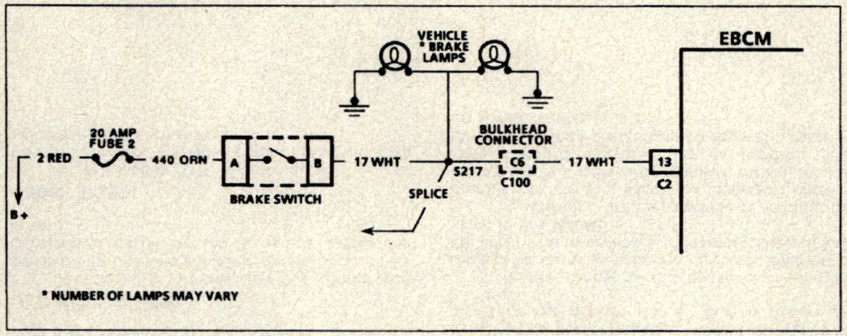

DTC 95
BRAKE SWITCH CIRCUIT OPEN

Circuit Description:

This DTC is used to identify open brake switch circuitry that prevents the brake switch input to the EBCM from changing states when the brake is applied.

This DTC is used in conjunction with DTC's 91 and 92 to determine the cause of an open brake switch malfunction.

DTC Will Set When:

DTC 95 can be set after initialization is completed. If the brake switch input voltage is out of specification for one second indicating an open circuit, a malfunction exists.

Action Taken:

A malfunction DTC is stored, ABS is disabled and the ABS warning lamp is turned "ON."

DTC Chart Test Description:

Number(s) below refer to circled number(s) on the diagnostic chart.

1. Confirms that an open in the brake switch circuitry currently exists.
2. Indicates if the brake switch signal is being received by the EBCM.
3. Indicates if an open circuit exists in the brake switch or brake lamp circuitry.
4. Isolates the open circuit to either the brake switch input circuitry, or the EBCM.
5. Ensures malfunction was not due to poor terminal contact.

Diagnostic Aids:

An "Intermittent" malfunction is most likely caused by a poor connection, rubbed through wire insulation, or a wire that is broken inside the insulation.

The frequency of the malfunction can be checked by using the enhanced diagnostic function of the Tech 1, as described in "Enhanced Diagnostics," found in this section.

Any circuitry, that is suspected as causing the intermittent complaint, should be thoroughly checked for backed out terminals, improper mating, broken locks, improperly formed or damaged terminals, poor terminal to wiring connections or physical damage to the wiring harness.

DIAGNOSTIC CHART DTC 95 — 1994–95 BERETTA AND CORSICA

DTC 95
BRAKE SWITCH CIRCUIT OPEN

IS DTC 96 CURRENTLY SET?

- YES: PROCEED TO DTC 96 DIAGNOSTIC CHART.
- NO:

1. • IGNITION "ON."
   • USING TECH 1, SELECT "F0: DATA LIST" AND MONITOR BRAKE SWITCH. DOES BRAKE SWITCH INDICATE "CIRCUIT OPEN"?
   - NO: MALFUNCTION IS NOT PRESENT AT THIS TIME. REFER TO "DIAGNOSTIC AIDS" ON FACING PAGE.
   - YES:

2. • APPLY LIGHT PRESSURE ON BRAKE PEDAL WHILE MONITORING BRAKE SWITCH ON TECH 1. DOES TECH 1 INDICATE BRAKE SWITCH "ON" WITHIN 1 INCH OF TRAVEL?
   - YES: PROCEED TO DTC 96 DIAGNOSTIC CHART.
   - NO:

3. • APPLY FIRM PRESSURE TO BRAKE PEDAL AND OBSERVE CHMSL AND REAR BRAKE LAMPS. ARE ALL BRAKE LAMPS "ON"?
   - NO: REPAIR OPEN BRAKE LAMP CIRCUIT.
   - YES:

4. • IGNITION "OFF."
   • DISCONNECT 24-WAY EBCM CONNECTOR.
   • MEASURE RESISTANCE BY CONNECTING J 39200 TO 24-WAY EBCM HARNESS CONNECTOR TERMINAL "13" AND PROBING BRAKE SWITCH TERMINAL "B". IS RESISTANCE LESS THAN 2 OHMS?
   - NO: REPAIR OPEN IN CKT 17.
   - YES:

5. • INSPECT 24-WAY EBCM HARNESS CONNECTOR TERMINAL "13" FOR POOR TERMINAL CONTACT OR CORROSION. REPLACE TERMINALS THAT EXHIBIT POOR CONTACT OR CORROSION.
   • RECONNECT ALL CONNECTORS.
   • WITH FOOT OFF OF BRAKE PEDAL, START ENGINE.
   • WAIT 10 SECONDS. IS DTC 95 SET AS A CURRENT DTC?
   - YES: REPLACE EBCM.
   - NO: MALFUNCTION IS INTERMITTENT. REFER TO "DIAGNOSTIC AIDS" ON FACING PAGE.

THIS CHART ASSUMES THAT A CURRENT DTC IS STORED INDICATING THAT THIS MALFUNCTION IS PRESENT.

DIAGNOSTIC CHART DTC 96 — 1994–95 BERETTA AND CORSICA

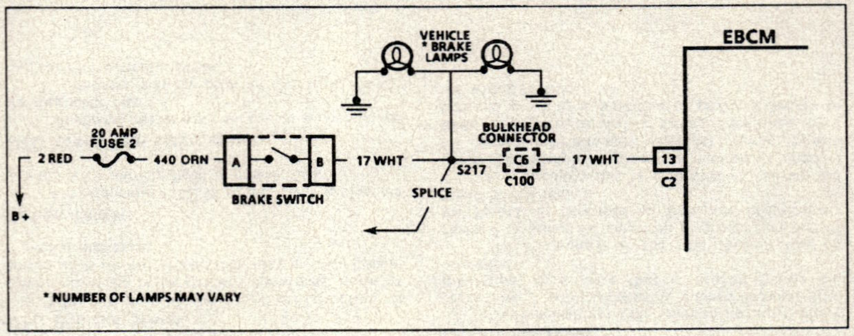

DTC 96
REAR BRAKE LAMP CIRCUIT OPEN

Circuit Description:

This DTC is designed to isolate the cause of a DTC 95 failure and indicate to the driver ABS is still available. If DTC 95 occurs with DTC 96, the brake lamps are open, the brake lamp grounds are open, or the Center High Mounted Stop Lamp (CHMSL) circuit is open during 4-way flasher use. The presence of battery voltage at the brake switch input indicates a valid brake switch input is still available.

DTC Will Set When:

DTC 96 can be set only after DTC 95 has been set. If the EBCM detects battery voltage on the brake switch input circuit for one second, a malfunction exists.

Action Taken:

A malfunction DTC is stored. ABS is not disabled; however, the ABS warning lamp will flash indicating a malfunction exists.

DTC Chart Test Description:

Number(s) below refer to circled number(s) on the diagnostic chart.

1. Checks for additional brake switch DTC's set.
2. Identifies if the malfunction is currently present in the brake circuit.

Diagnostic Aids:

An "Intermittent" malfunction is most likely caused by a poor connection, rubbed through wire insulation, or a wire that is broken inside the insulation.

The frequency of the malfunction can be checked by using the enhanced diagnostic function of the Tech 1, as described in "Enhanced Diagnostics," found in this section.

Any circuitry, that is suspected as causing the intermittent complaint, should be thoroughly checked for backed out terminals, improper mating, broken locks, improperly formed or damaged terminals, poor terminal to wiring connections or physical damage to the wiring harness.

DIAGNOSTIC CHART DTC 96 — 1994–95 BERETTA AND CORSICA

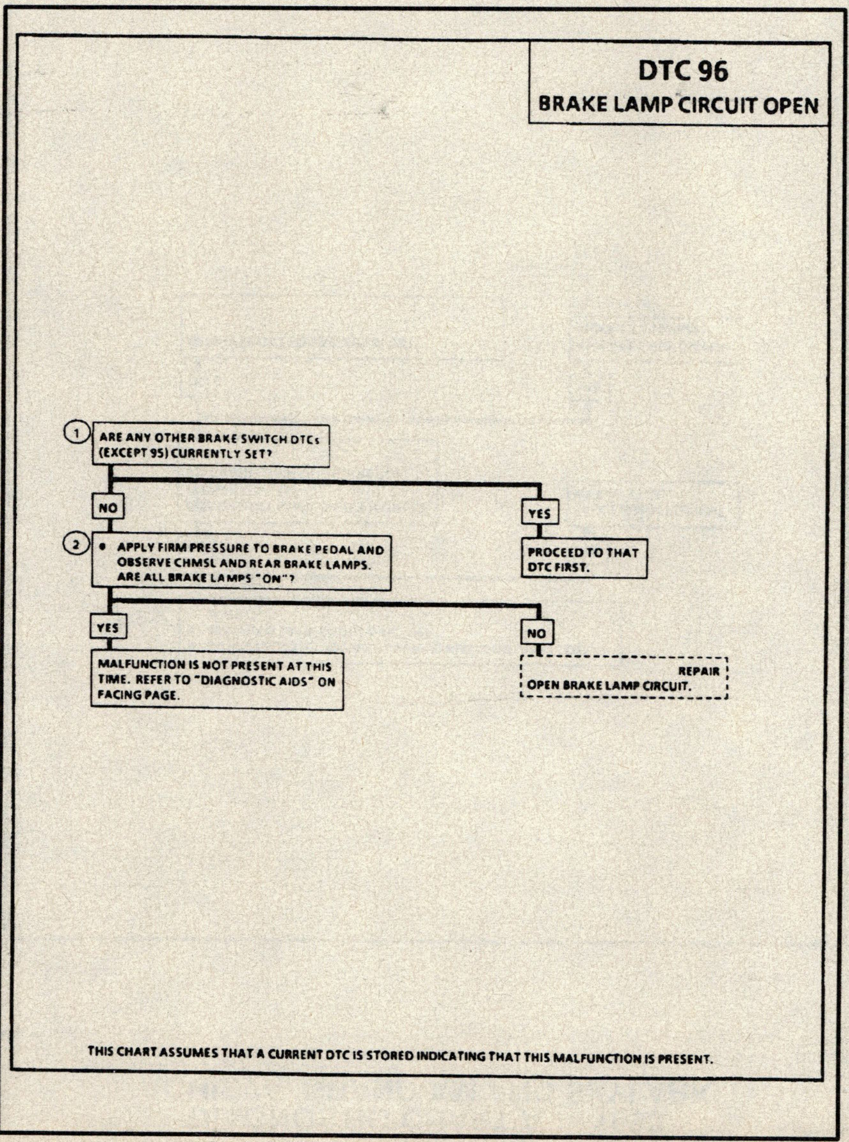

DIAGNOSTIC CIRCUIT CHECK CHART — 1993 ACHIEVA, GRAND AM AND SKYLARK

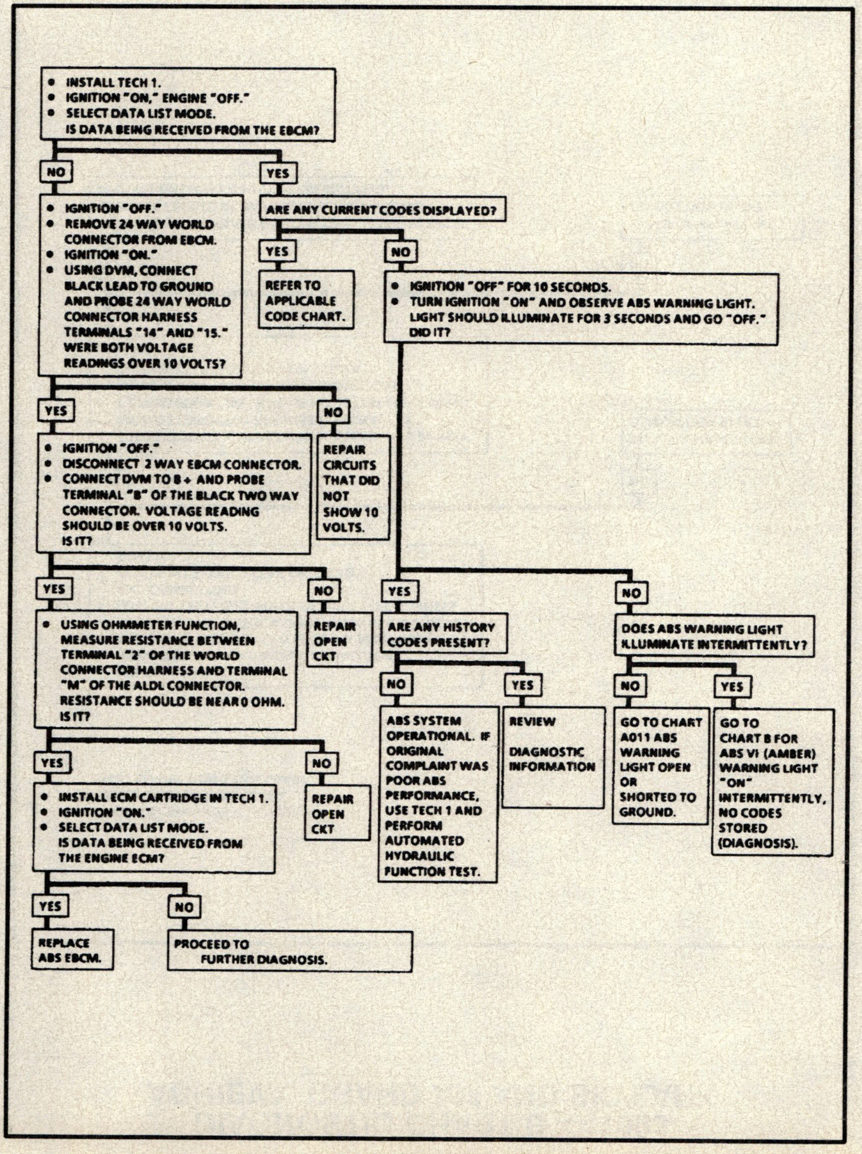

DIAGNOSTIC CHART A — 1993 ACHIEVA, GRAND AM AND SKYLARK

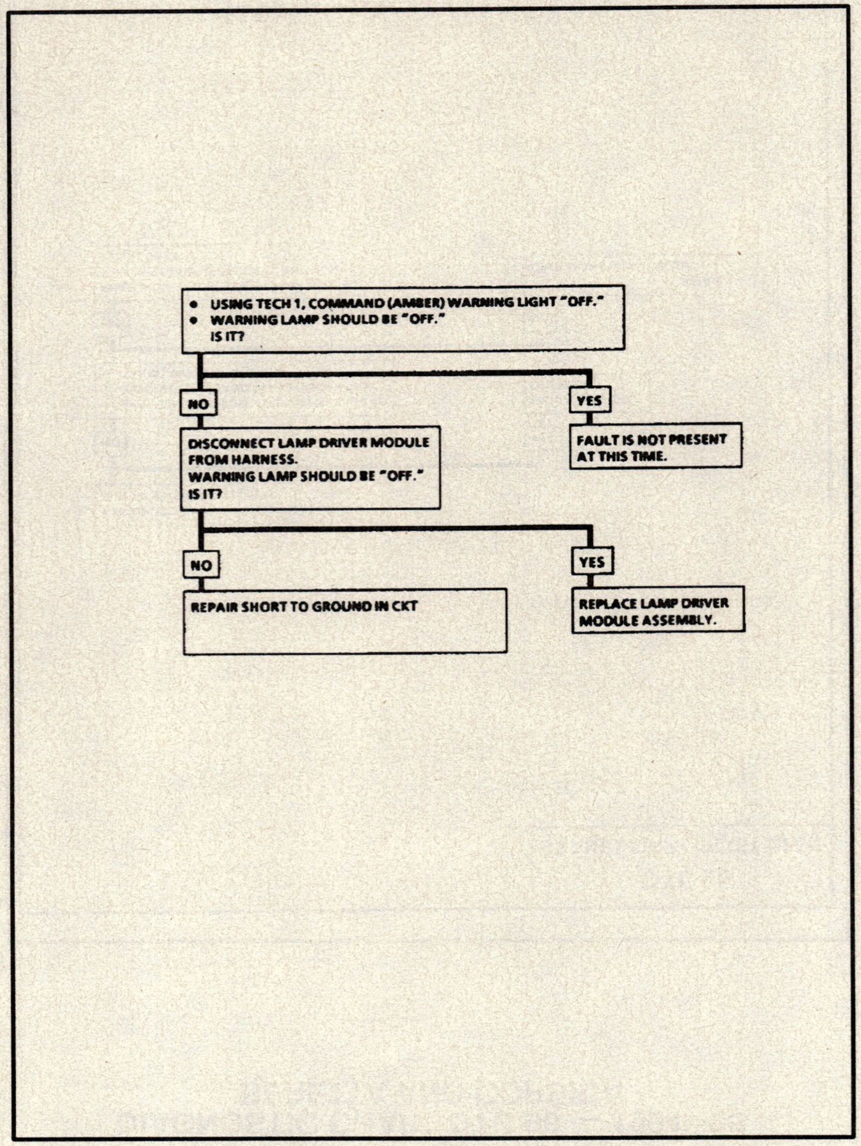

DIAGNOSTIC CHART B — 1993 ACHIEVA, GRAND AM AND SKYLARK

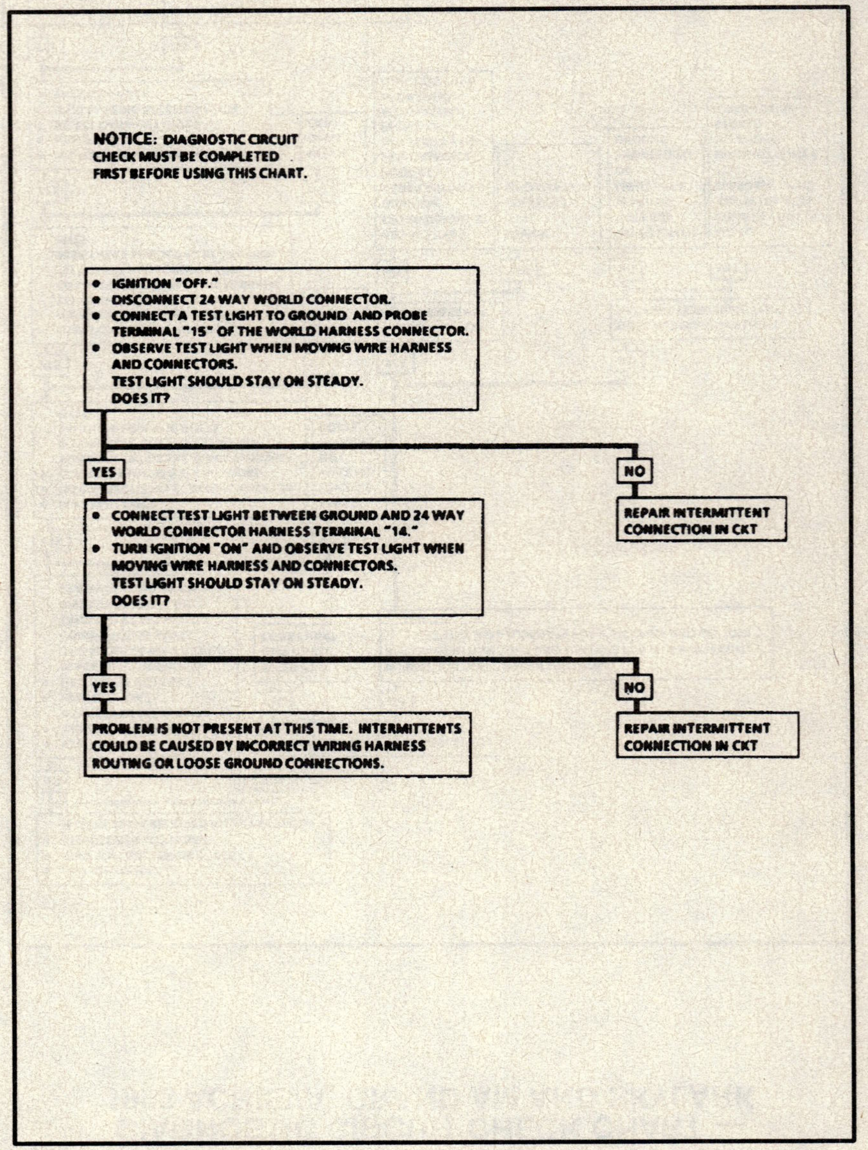

DIAGNOSTIC CHART DTC A011 — 1993
ACHIEVA, GRAND AM AND SKYLARK

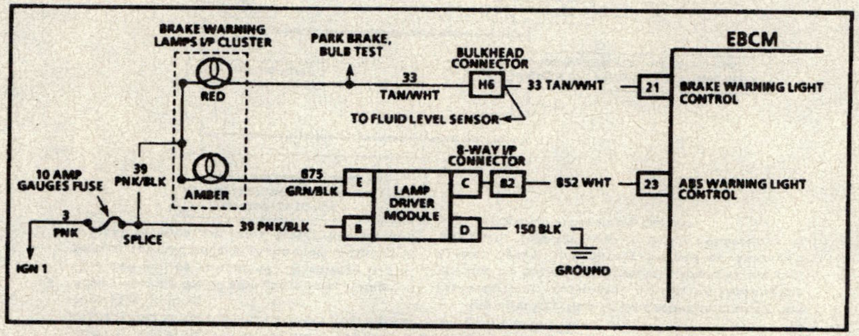

CODE A011

ABS WARNING LIGHT CIRCUIT OPEN OR SHORTED TO GROUND

Circuit Description:

This test checks the state of the ABS warning light to identify a situation in which the driver could not be warned of a system fault by the ABS warning light, or the warning light is always "ON." The Lamp Driver Module (LDM) turns the ABS warning light "ON," unless the EBCM provides ground to turn it "OFF." Because of the circuitry in the LDM, only external faults can be detected. As a result, the ABS warning light itself is not diagnosable, only the control line to the EBCM can be diagnosed. In the event of an open CKT 852, the ABS warning light will be on at all times, due to the loss of ground at the LDM input. If the control line is shorted to ground, the ABS warning light is kept "OFF," due to the LDM input being grounded.

Test Description: Number(s) below refer to circled number(s) on the diagnostic chart.

1. Checks to see if ABS light functions properly. Normal operation would indicate that the problem is not present.
2. This step insures that the circuitry from the EBCM to the lamp driver module is complete.
3. This step insures that the ignition circuit to the lamp driver module is functional. An open gauge fuse or open CKT 3, 39, or 875 would result in inoperative indicator lights.
4. Isolates the cause of the ABS lamp remaining "OFF," by disconnecting the 24 way world connector, CKT 852 should be open and cause the ABS lamp to be "ON."

Diagnostic Aids:

An "Intermittent" problem may be caused by a poor connection, rubbed through wire insulation, or a wire that is broken inside the insulation.

The light test function of the "Scan" tool may be used to command the light "ON" while looking for an intermittent problem in the warning light circuitry.

Any circuitry that is suspected as causing the intermittent complaint should be thoroughly checked for backed out terminals, improper mating, broken locks, improperly formed or damaged terminals, poor terminal to wiring connections or physical damage to the wiring harness.

DIAGNOSTIC CHART DTC A011 — 1993
ACHIEVA, GRAND AM AND SKYLARK

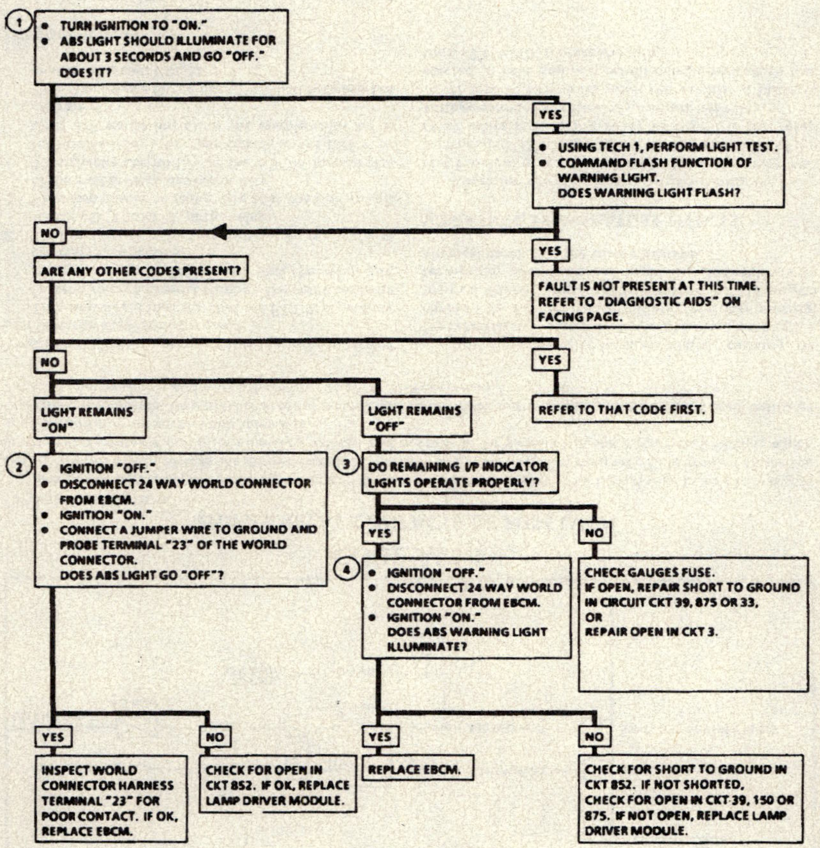

DIAGNOSTIC CHART DTC A013 — 1993 ACHIEVA, GRAND AM AND SKYLARK

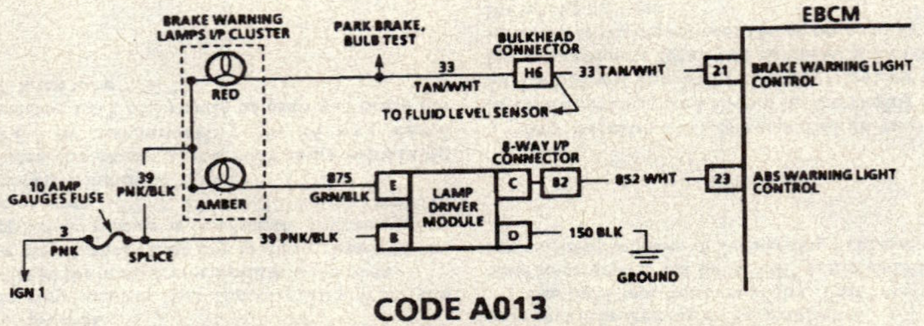

CODE A013

ABS WARNING LIGHT CIRCUIT SHORTED TO BATTERY

Circuit Description:

This test checks the state of the ABS warning light to identify a situation in which the driver could not be warned of a system fault by the ABS warning light. The Lamp Driver Module (LDM) turns the ABS warning light "ON," unless the EBCM provides ground to turn it "OFF." Because of the circuitry internal to the LDM, only external faults can be detected. As a result, the ABS warning light itself is not diagnosable, only the control line to the EBCM can be diagnosed.

If the ABS warning light control line is shorted to battery or the EBCM control circuit is open, the Electronic Brake Control Module (EBCM) is not able to turn "OFF" the ABS warning light.

Test Description: Number(s) below refer to circled number(s) on the diagnostic chart.

1. Checks to see if ABS light functions properly. Normal operation would indicate that the problem is not present.
2. After the ignition is turned to the "OFF" position, voltage to the lamp driver module should be interrupted. A light that remains "ON" would indicate that voltage to this circuit is from a point other than ignition.
3. After removing the 10 amp gauge fuse, voltage on CKT 852 will be eliminated. If voltage is still present, it would indicate that a short to voltage is present between the lamp driver module and EBCM.

Diagnostic Aids:

An "Intermittent" problem may be caused by a poor connection, rubbed through wire insulation or a wire that is broken inside the insulation.

Code 13 must be currently set before it should be addressed. If Code 13 is set as a history code, proceed to those codes that are currently set before attempting to repair a Code 13 fault.

Any circuitry that is suspected as causing the "Intermittent" complaint should be thoroughly checked for backed out terminals, improper mating, broken locks, improperly formed or damaged terminals, poor terminal to wiring connections or physical damage to the wiring harness.

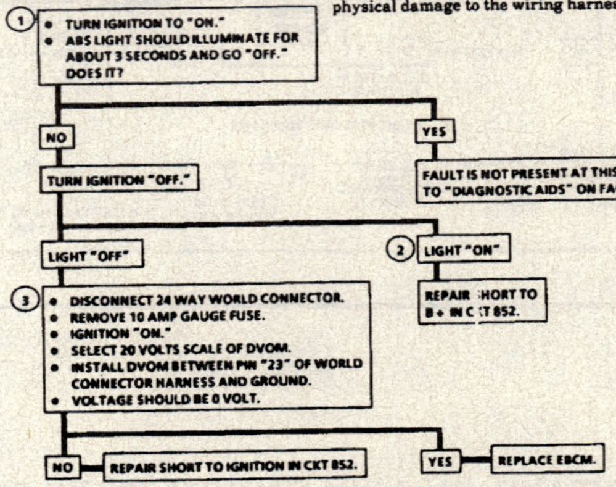

DIAGNOSTIC CHART DTC A014 — 1993 ACHIEVA, GRAND AM AND SKYLARK

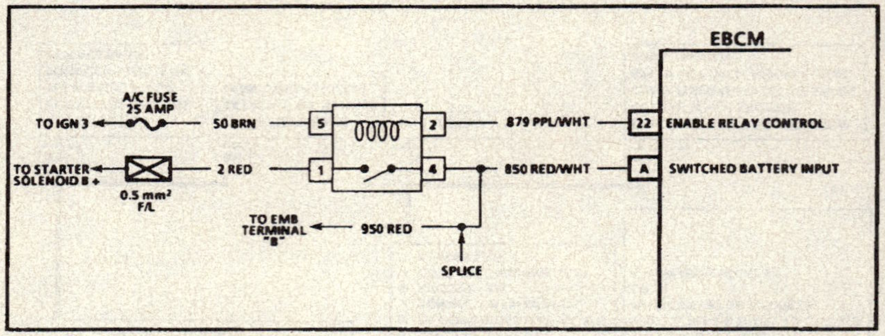

CODE A014

ENABLE RELAY CONTACTS OR FUSE OPEN

Circuit Description:

Ignition voltage is supplied through terminal "5" of the ABS enable relay. The EBCM then is able to energize the pull-in coil by completing the ground circuit at Pin "22" of the EBCM. The magnetic field created closes the enable relay contacts and allows battery voltage and current to be supplied to the EMB's and EBCM, which supplies power to the motors and solenoids.

This test monitors the availability of current/voltage to the motors, solenoids and EMB's. This fault indicates voltage is not available and would therefore not allow ABS operation if required.

Test Description: Number(s) below refer to circled number(s) on the diagnostic chart.

1. This step would indicate that the EBCM is sensing battery voltage present through the relay contacts at terminal "A." This shows that the relay and circuitry is functional.
2. This insures that battery voltage is available to terminals "1" and "5" of the relay.
3. This step checks to make sure that the pull-in coil of the enable relay has continuity.
4. This checks the ability of the EBCM to energize the enable relay. An open circuit in the EBCM or CKT 879 would not allow the enable relay to be energized.
5. This step checks for an open CKT 850 between the enable relay and EBCM.

Diagnostic Aids:

An "Intermittent" problem may be caused by a poor connection, rubbed through wire insulation, or a wire that is broken inside the insulation.

Any circuitry that is suspected as causing the "Intermittent" complaint should be thoroughly checked for backed out terminals, improper mating, broken locks, improperly formed or damaged terminals, poor terminal to wiring connections or physical damage to the wiring harness.

VIBRATION, TEMPERATURE EFFECTS:

Check for vibration effects by performing the relay test function of the "Scan" tool. With the relay test commanded "ON," lightly tap the top and sides of the relay while monitoring relay voltage. If the relay voltage changes significantly, replace the relay.

If test 14 only fails when the vehicle is initially started in cold ambient conditions (temperature less than 32°F - 0°C), replace the relay.

DIAGNOSTIC CHART DTC A014 — 1993
ACHIEVA, GRAND AM AND SKYLARK

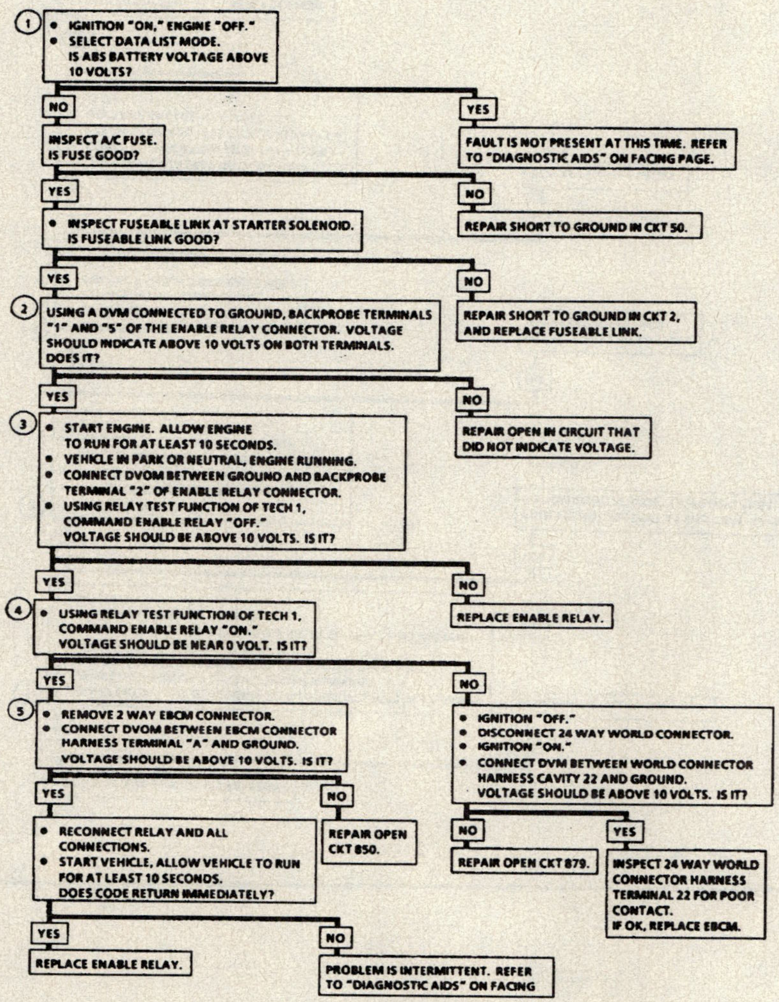

DIAGNOSTIC CHART DTC A015 — 1993
ACHIEVA, GRAND AM AND SKYLARK

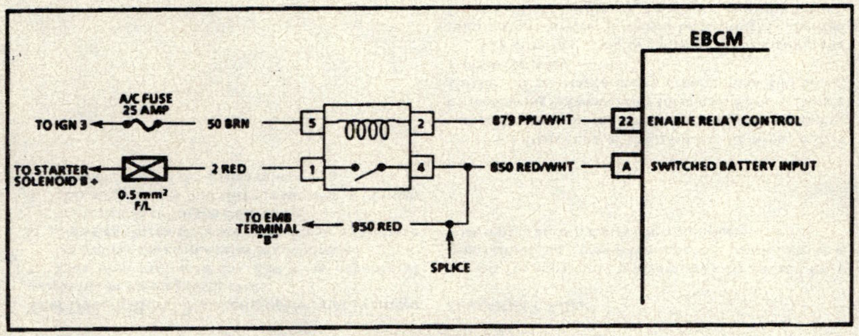

CODE A015

ENABLE RELAY CONTACTS SHORTED TO BATTERY

Circuit Description:

Ignition voltage is supplied through terminal "5" of the ABS enable relay. The EBCM then is able to energize the pull-in coil by completing the ground circuit at Pin "22" of the EBCM. The magnetic field created closes the enable relay contacts and allows battery voltage and current to be supplied to the EMB's and EBCM, which supplies power to the motors and solenoids.

This test determines if the enable relay is energized when it should not be. This fault would not allow the enable relay to remove power to the ABS system. If a second fault were to occur that requires the enable relay to be turned "OFF," that fault cannot be removed if the relay cannot be controlled.

Test Description: Number(s) below refer to circled number(s) on the diagnostic chart.

1. This step indicates that the EBCM is capable of controlling the enable relay as commanded.
2. This step checks for a possible short to battery voltage in CKT 850.
3. This step insures that the EBCM and also CKT 879 are not shorted to ground. A short to ground in either case would cause the enable relay to be energized constantly.

Diagnostic Aids:

An "Intermittent" problem may be caused by a poor connection, rubbed through wire insulation, or a wire that is broken inside the insulation.

Any circuitry, that is suspected as causing the intermittent complaint, should be thoroughly checked for backed out terminals, improper mating, broken locks, improperly formed or damaged terminals, poor terminal to wiring connections or physical damage to the wiring harness.

DIAGNOSTIC CHART DTC A015 — 1993 ACHIEVA, GRAND AM AND SKYLARK

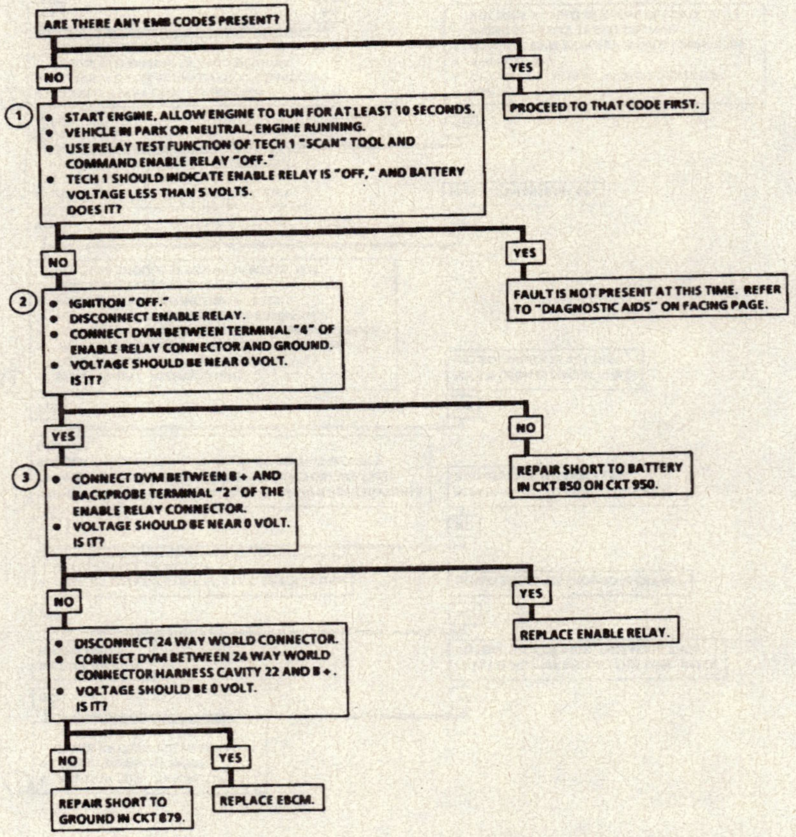

DIAGNOSTIC CHART DTC A016 — 1993 ACHIEVA, GRAND AM AND SKYLARK

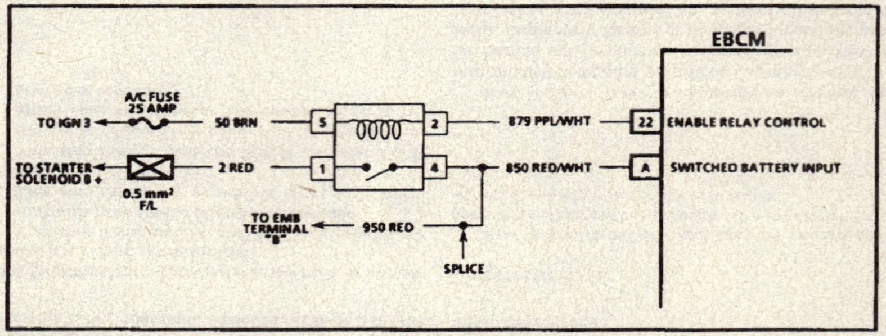

CODE A016

ENABLE RELAY COIL CIRCUIT OPEN

Circuit Description:

Ignition voltage is supplied through terminal "5" of the ABS enable relay. The EBCM then is able to energize the pull-in coil by completing the ground circuit at Pin "22" of the EBCM. The magnetic field created closes the enable relay contacts and allows battery voltage and current to be supplied to the EMB's and EBCM, which supplies power to the motors and solenoids.

This test detects opens in the enable relay coil circuit. An open in this circuit will not allow the enable relay to be energized, thus preventing voltage/current to the motors, EMB's and solenoids.

Test Description: Number(s) below refer to circled number(s) on the diagnostic chart.

1. This step indicates the EBCM is capable of controlling the enable relay as commanded.
2. This step checks to insure that voltage is available to the pull-in coil of the enable relay.
3. This step insures that there is continuity through the pull-in coil of the enable relay.

Diagnostic Aids:

An "Intermittent" problem may be caused by a poor connection, rubbed through wire insulation, or a wire that is broken inside the insulation.

If the frequency of the problem is high, but is currently intermittent, check for high coil resistance by measuring between relay terminals 5 and 2 using a DVOM. If resistance shows greater than 100 ohms, replace the relay.

Any circuitry, that is suspected as causing the intermittent complaint, should be thoroughly checked for backed out terminals, improper mating, broken locks, improperly formed or damaged terminals, poor terminal to wiring connections or physical damage to the wiring harness.

DIAGNOSTIC CHART DTC A016 — 1993 ACHIEVA, GRAND AM AND SKYLARK

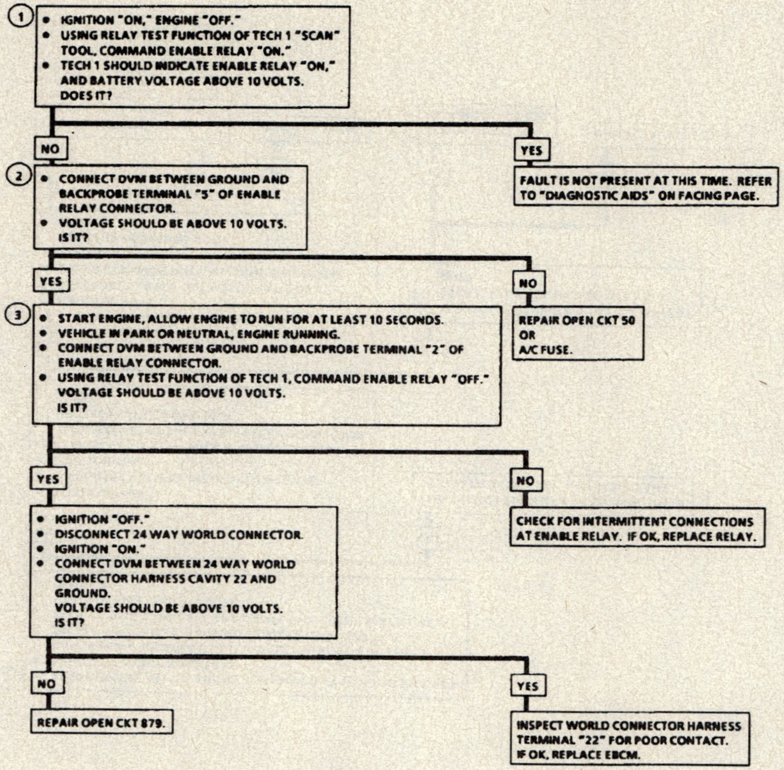

DIAGNOSTIC CHART DTC A017 — 1993 ACHIEVA, GRAND AM AND SKYLARK

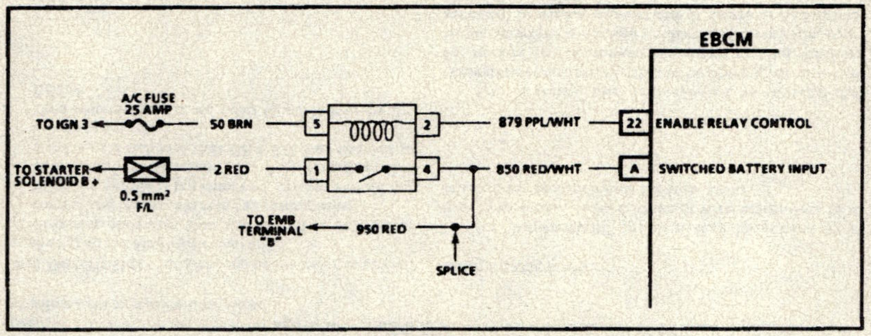

CODE A017

ENABLE RELAY COIL CIRCUIT SHORTED TO GROUND

Circuit Description:

Ignition voltage is supplied through terminal "5" of the ABS enable relay. The EBCM then is able to energize the pull-in coil by completing the ground circuit at Pin "22" of the EBCM. The magnetic field created closes the enable relay contacts and allows battery voltage and current to be supplied to the EMB's and EBCM, which supplies voltage to the motors and solenoids.

This test determines if the enable relay is energized when it should not be. This fault would not allow the enable relay to remove power to the ABS system. If a second fault were to occur that requires the enable relay to be turned "OFF," that fault cannot be removed if the relay cannot be controlled.

Test Description: Number(s) below refer to circled number(s) on the diagnostic chart.

1. This step indicates that the EBCM is capable of controlling the enable relay as commanded.
2. This checks to insure that the enable relay or control CKT 879 are not shorted to ground.
3. This step checks to see if the EBCM is internally shorted to ground. A grounded Pin "22" would cause the enable relay to be energized any time the ignition was in the "ON" position.

Diagnostic Aids:

An "Intermittent" problem may be caused by a poor connection, rubbed through wire insulation, or a wire that is broken inside the insulation.

Any circuitry, that is suspected as causing the intermittent complaint, should be thoroughly checked for backed out terminals, improper mating, broken locks, improperly formed or damaged terminals, poor terminal to wiring connections or physical damage to the wiring harness.

DIAGNOSTIC CHART DTC A017 — 1993 ACHIEVA, GRAND AM AND SKYLARK

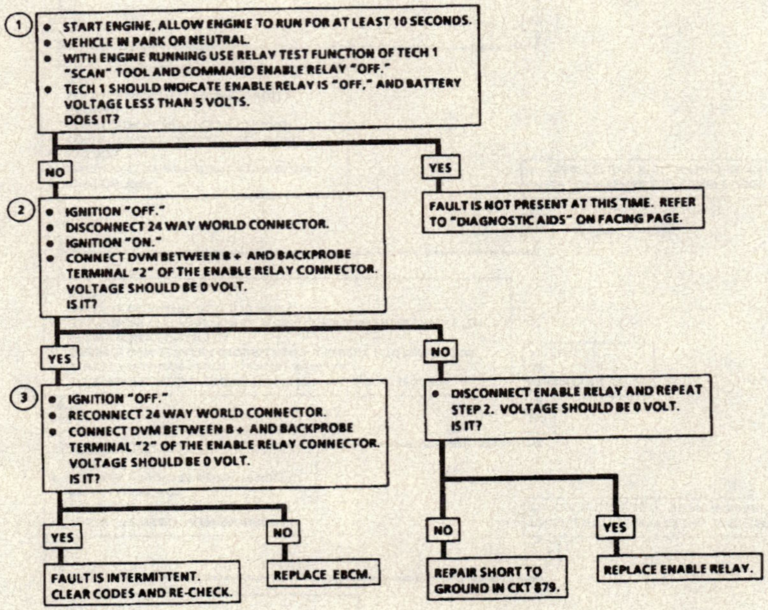

DIAGNOSTIC CHART DTC A018 — 1993 ACHIEVA, GRAND AM AND SKYLARK

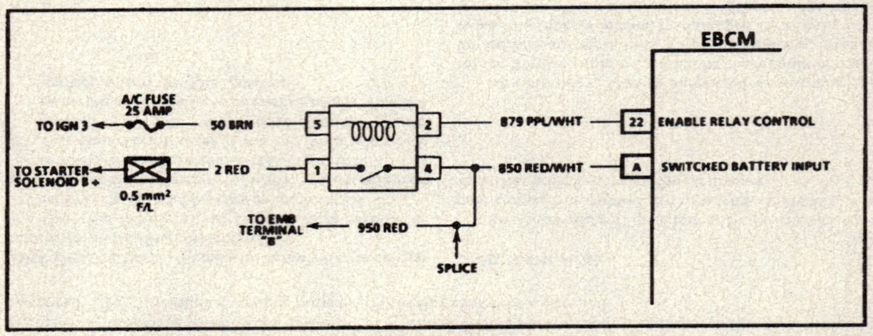

CODE A018

ENABLE RELAY COIL CIRCUIT SHORTED TO BATTERY

Circuit Description:

Ignition voltage is supplied through terminal "5" of the ABS enable relay. The EBCM then is able to energize the pull-in coil by completing the ground circuit at Pin "22" of the EBCM. The magnetic field created closes the enable relay contacts and allows battery voltage and current to be supplied to the EMB's and EBCM, which supplies voltage to the motors and solenoids.

This test monitors the availability of current/voltage to the motors, solenoids and EMB's. A fault of the type specified above would not allow voltage to be available to the motors, solenoids and EMB's and would, therefore, not allow ABS operation if required.

Test Description: Number(s) below refer to circled number(s) on the diagnostic chart.

1. This step indicates that the EBCM is capable of controlling the enable relay as commanded.
2. With the enable relay removed, voltage should not be available at terminal "2". Any voltage at this point would indicate that CKT 879 was shorted to a voltage source.
3. This step checks for an open Quad-Driver in the EBCM.

Diagnostic Aids:

An "Intermittent" problem may be caused by a poor connection, rubbed through wire insulation, or a wire that is broken inside the insulation.

Any circuitry, that is suspected as causing the intermittent complaint, should be thoroughly checked for backed out terminals, improper mating, broken locks, improperly formed or damaged terminals, poor terminal to wiring connections or physical damage to the wiring harness.

DIAGNOSTIC CHART DTC A018 — 1993
ACHIEVA, GRAND AM AND SKYLARK

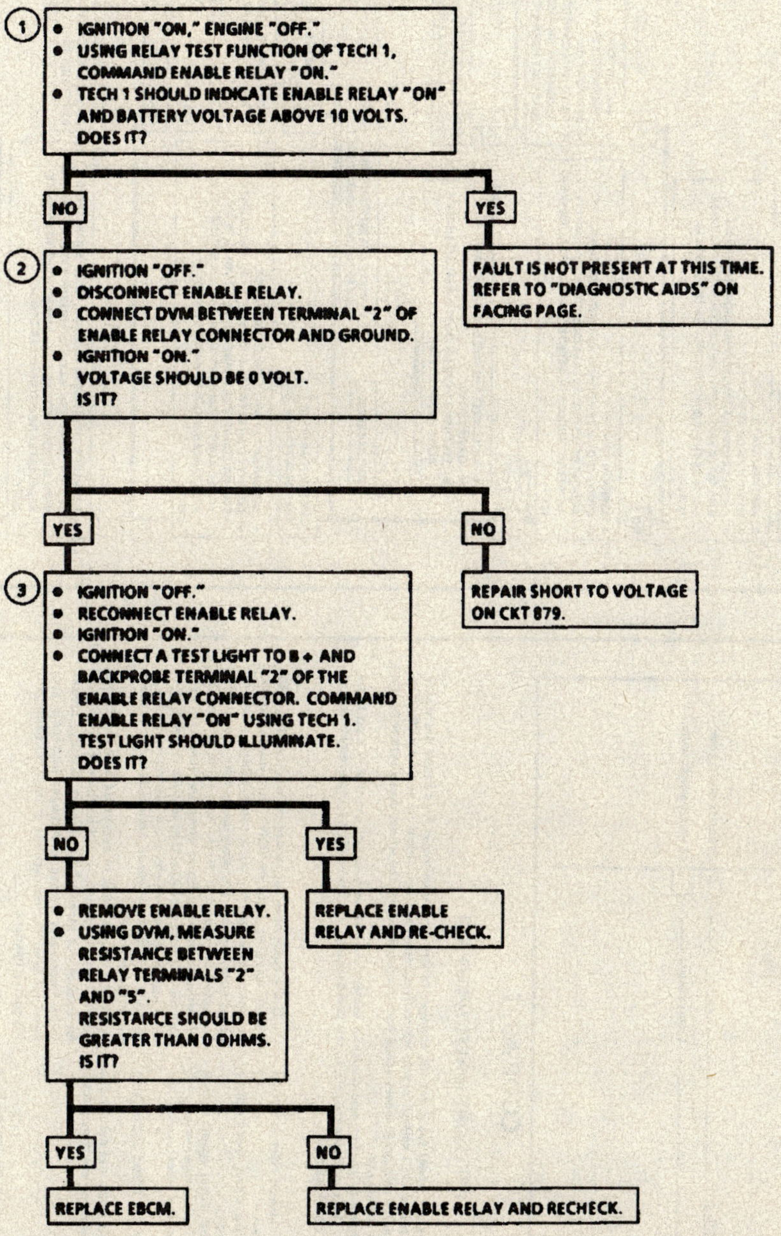

DIAGNOSTIC CHART DTC A021 — 1993 ACHIEVA GRAND AM AND SKYLARK

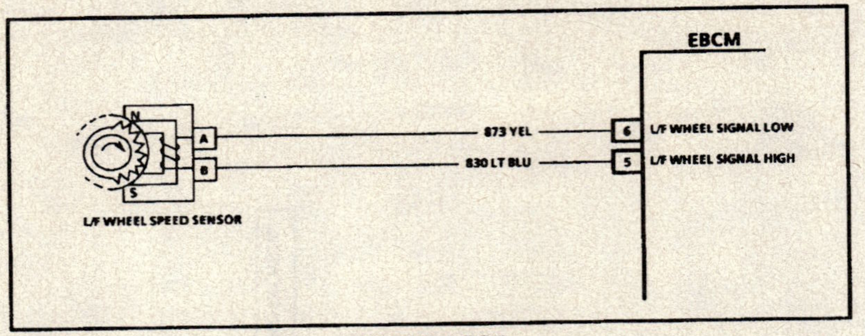

CODE A021

LEFT FRONT WHEEL SPEED = 0

Circuit Description:

As a toothed ring passes by the wheel speed sensor, changes in the electromagnetic field cause the wheel speed sensor to produce a sinusoidal (AC) voltage signal whose frequency is proportional to wheel speed. The magnitude of this signal is directly related to wheel speed and the proximity of the wheel speed sensor to the toothed ring often referred to as the air gap.

This code detects opens, shorts to ground, and shorts to battery that cause a wheel speed of 0 mph to be calculated by the ABS EBCM. The conditions present to fail this test are: (1) Three (3) wheels are rotating at least 5 mph, and (2) The fourth wheel is 0 mph.

Test Description: Number(s) below refer to circled number(s) on the diagnostic chart.

5. This step checks to see if the wheel speed sensor circuitry is shorted to voltage.
6. This step insures that the wheel speed sensor circuitry resistance is within the proper value.
7. This checks for an open in the wheel signal low circuit.
8. This checks for an open in the wheel signal high circuit.
9. This step insures that Code A021 was not set due to a poor connection between the world connector and EBCM.

Diagnostic Aids:

An "Intermittent" problem may be caused by a poor connection, rubbed through wire insulation, or a wire that is broken inside the insulation.

If the customers comments reflect that the ABS Amber warning light is "ON" only during moist environmental changes, (rain, snow, vehicle wash) all wheel speed sensor circuitry should be thoroughly inspected for signs of water intrusion. Use the following procedure. Spray down the suspected area with 5% salt water solution (two teaspoons of salt to 12 oz. of water). Start vehicle, allow vehicle to run for 10 seconds. If code returns immediately, replace suspected harness.

Any circuitry, that is suspected as causing the intermittent complaint, should be thoroughly checked for backed out terminals, improper mating, broken locks, improperly formed or damaged terminals, poor terminal to wiring connections or physical damage to the wiring harness.

DIAGNOSTIC CHART DTC A021 — 1993 ACHIEVA GRAND AM AND SKYLARK

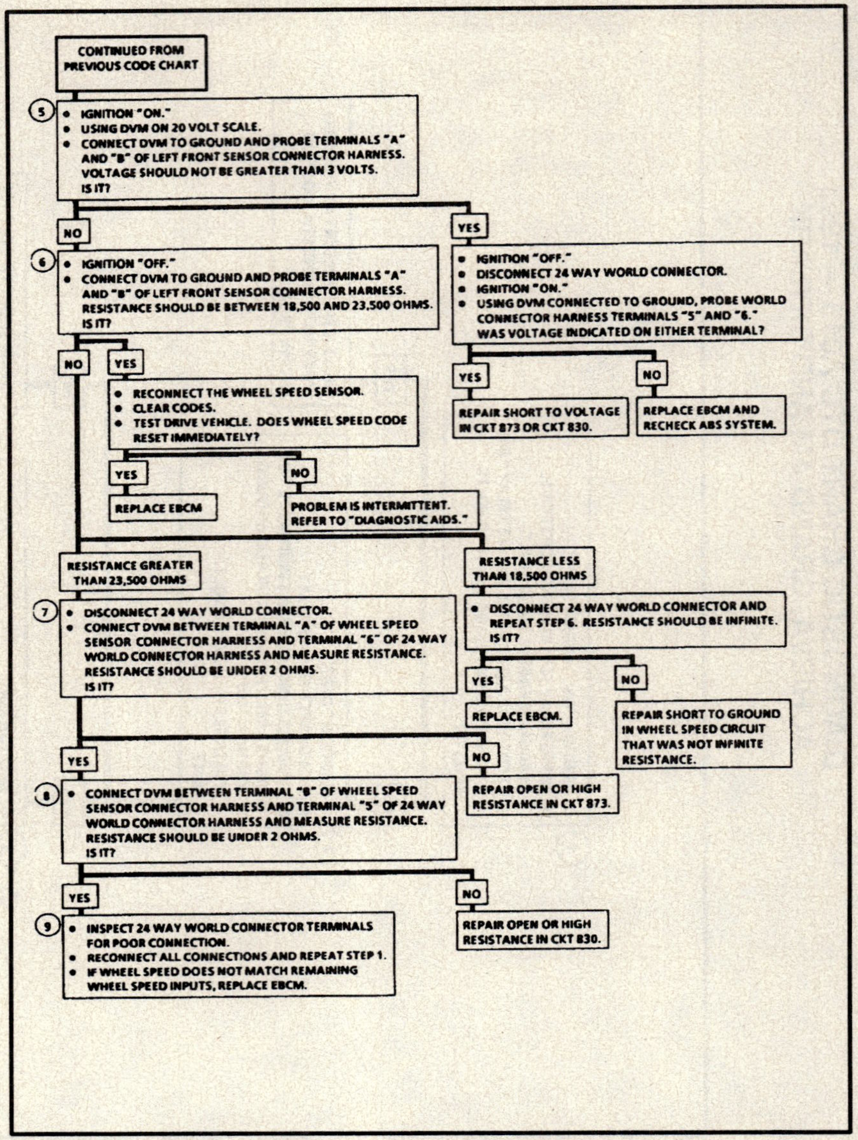

DIAGNOSTIC CHART DTC A022 — 1993 ACHIEVA GRAND AM AND SKYLARK

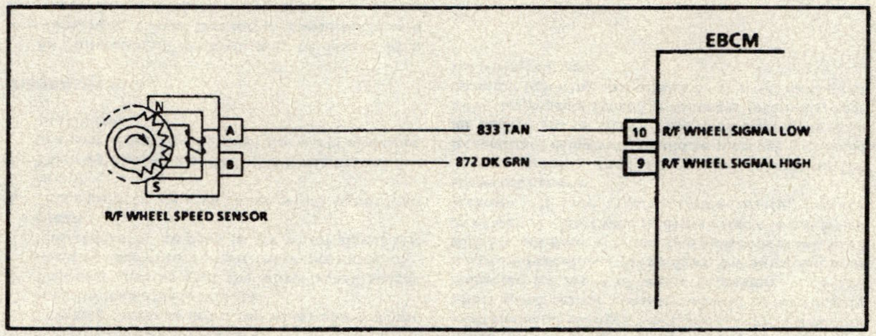

CODE A022

RIGHT FRONT WHEEL SPEED = 0

Circuit Description:

As a toothed ring passes by the wheel speed sensor, changes in the electromagnetic field cause the wheel speed sensor to produce a sinusoidal (AC) voltage signal whose frequency is proportional to wheel speed. The magnitude of this signal is directly related to wheel speed and the proximity of the wheel speed sensor to the toothed ring, often referred to as the air gap.

This code detects opens, shorts to ground, and shorts to battery that cause a wheel speed of 0 mph to be calculated by the ABS EBCM. The conditions present to fail this test are: (1) Three (3) wheels are rotating at least 5 mph, and (2) The fourth wheel is 0 mph.

Test Description: Number(s) below refer to circled number(s) on the diagnostic chart.

1. This test verifies whether the fault is currently present.
2. This step is designed to identify a wheel speed sensor or circuitry that is damaged and visibly apparent.
3. This test checks to see that the wheel speed sensor is within proper resistance values.
4. This insures that the wheel speed sensor is not shorted to ground.

Diagnostic Aids:

An "Intermittent" problem may be caused by a poor connection, rubbed through wire insulation, or a wire that is broken inside the insulation.

If the customers comments reflect that the ABS Amber warning light is on only during moist environmental changes, (rain, snow, vehicle wash) all wheel speed sensor circuitry should be thoroughly inspected for signs of water intrusion. Use the following procedure. Spray down the suspected area with 5% salt water solution (two teaspoons of salt to 12 oz. of water). Start vehicle, allow vehicle to run for 10 seconds. If code returns immediately, replace suspected harness.

Any circuitry, that is suspected as causing the intermittent complaint, should be thoroughly checked for backed out terminals, improper mating, broken locks, improperly formed or damaged terminals, poor terminal to wiring connections or physical damage to the wiring harness.

DIAGNOSTIC CHART DTC A022 — 1993 ACHIEVA GRAND AM AND SKYLARK

IMPORTANT: WHEEL SPEED SENSOR INTERMITTENT PROBLEMS MAY BE DIFFICULT TO LOCATE. CARE SHOULD BE TAKEN NOT TO DISTURB ANY ELECTRICAL CONNECTIONS PRIOR TO AN INDICATED STEP OF THIS CHART. THIS WILL INSURE THAT AN INTERMITTENT CONNECTION WILL NOT BE CORRECTED BEFORE THE SOURCE OF THE PROBLEM IS FOUND.

(1)
- TEST DRIVE VEHICLE.
- IGNITION "ON."
- USING TECH 1, SELECT DATA LIST.
- MONITOR WHEEL SPEED ON RIGHT FRONT WHEEL WHILE SLOWLY DECELERATING FROM 35 MPH TO 0 MPH.
  WHEEL SPEED SHOULD MATCH REMAINING WHEEL SPEED INPUTS.
  DOES IT?

YES → FAULT IS NOT PRESENT AT THIS TIME, REFER TO "DIAGNOSTIC AIDS" ON FACING PAGE.

NO ↓

(2)
- IGNITION "OFF."
- PHYSICALLY INSPECT WHEEL SPEED SENSOR, RING, WIRING AND CONNECTORS FOR DAMAGE. PAY PARTICULAR ATTENTION TO THE SPEED RING. ANY SIGNIFICANT DAMAGE (OTHER THAN NICKS FROM STONES ETC.) WILL AFFECT THE WHEEL SPEED INPUT SIGNAL.
  IS PHYSICAL DAMAGE INDICATED?

YES → CORRECT PHYSICAL DAMAGE, CLEAR CODES AND RECHECK.

NO ↓

(3)
- DISCONNECT CONNECTOR FROM RIGHT FRONT WHEEL SPEED SENSOR.
- CONNECT DVM BETWEEN TERMINALS "A" AND "B" OF SENSOR, AND MEASURE RESISTANCE.
  RESISTANCE SHOULD BE BETWEEN 1650 AND 1800 OHMS.
  IS IT?

NO → REPLACE WHEEL SPEED SENSOR, CLEAR CODES AND RECHECK.

YES ↓

(4)
- CONNECT DVM BETWEEN SENSOR TERMINAL "A" AND GROUND AND MEASURE RESISTANCE.
  RESISTANCE SHOULD BE INFINITE.
  IS IT?

NO → REPLACE WHEEL SPEED SENSOR, CLEAR CODES AND RECHECK.

YES → GO TO FOLLOWING PAGE OF THIS CODE CHART

DIAGNOSTIC CHART DTC A022 — 1993 ACHIEVA GRAND AM AND SKYLARK

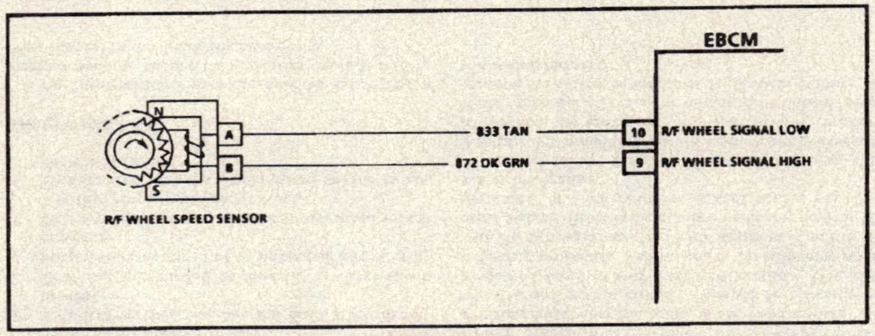

CODE A022

RIGHT FRONT WHEEL SPEED = 0

Circuit Description:

As a toothed ring passes by the wheel speed sensor, changes in the electromagnetic field cause the wheel speed sensor to produce a sinusoidal (AC) voltage signal whose frequency is proportional to wheel speed. The magnitude of this signal is directly related to wheel speed and the proximity of the wheel speed sensor to the toothed ring, often referred to as the air gap.

This code detects opens, shorts to ground, and shorts to battery that cause a wheel speed of 0 mph to be calculated by the ABS EBCM. The conditions present to fail this test are: (1) Three (3) wheels are rotating at least 5 mph, and (2) The fourth wheel is 0 mph.

Test Description: Number(s) below refer to circled number(s) on the diagnostic chart.

5. This step checks to see if the wheel speed sensor circuitry is shorted to voltage.
6. This step insures that the wheel speed sensor circuitry resistance is within the proper value.
7. This checks for an open in the wheel signal low circuit.
8. This checks for an open in the wheel signal high circuit.
9. This step insures that Code A022 was not set due to a poor connection between the world connector and EBCM.

Diagnostic Aids:

An "Intermittent" problem may be caused by a poor connection, rubbed through wire insulation, or a wire that is broken inside the insulation.

If the customers comments reflect that the ABS Amber warning light is "ON" only during moist environmental changes, (rain, snow, vehicle wash) all wheel speed sensor circuitry should be thoroughly inspected for signs of water intrusion. Use the following procedure. Spray down the suspected area with 5% salt water solution (two teaspoons of salt to 12 oz. of water). Start vehicle, allow vehicle to run for 10 seconds. If code returns immediately, replace suspected harness.

Any circuitry, that is suspected as causing the intermittent complaint, should be thoroughly checked for backed out terminals, improper mating, broken locks, improperly formed or damaged terminals, poor terminal to wiring connections or physical damage to the wiring harness.

DIAGNOSTIC CHART DTC A022 — 1993 ACHIEVA GRAND AM AND SKYLARK

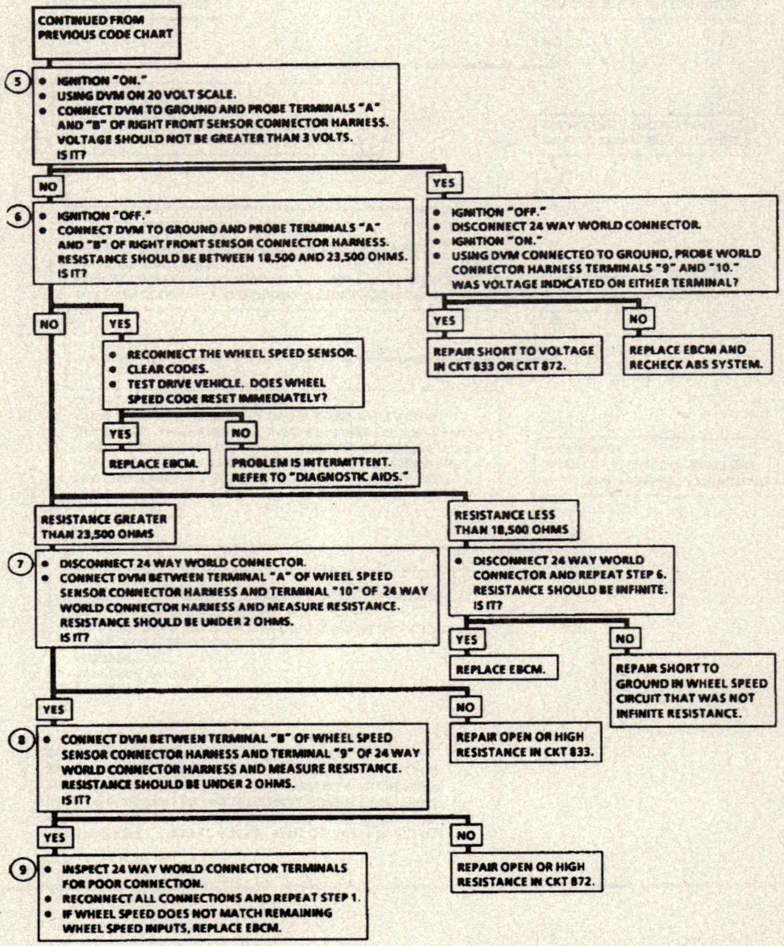

DIAGNOSTIC CHART DTC A023 — 1993 ACHIEVA GRAND AM AND SKYLARK

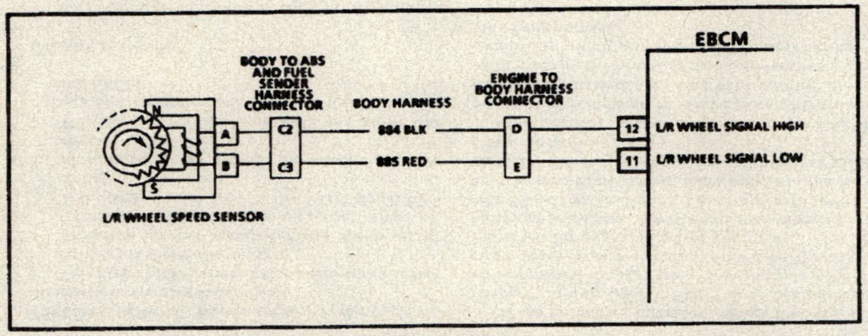

CODE A023

LEFT REAR WHEEL SPEED = 0

Circuit Description:

As a toothed ring passes by the wheel speed sensor, changes in the electromagnetic field cause the wheel speed sensor to produce a sinusoidal (AC) voltage signal whose frequency is proportional to wheel speed. The magnitude of this signal is directly related to wheel speed and the proximity of the wheel speed sensor to the toothed ring, often referred to as the air gap.

This code detects opens, shorts to ground, and shorts to battery that cause a wheel speed of 0 mph to be calculated by the ABS EBCM. The conditions present to fail this test are: (1) Three (3) wheels are rotating at least 5 mph, (2) The fourth wheel is 0 mph.

Test Description: Number(s) below refer to circled number(s) on the diagnostic chart.

1. This test verifies whether the fault is currently present.
2. This step is designed to identify a wheel speed sensor or circuitry that is damaged and visibly apparent.
3. This test checks to see that the wheel speed sensor is within proper resistance values.
4. This insures that the wheel speed sensor is not shorted to ground.

Diagnostic Aids:

An "Intermittent" problem may be caused by a poor connection, rubbed through wire insulation, or a wire that is broken inside the insulation.

If the customers comments reflect that the ABS Amber warning light is "ON" only during moist environmental changes, (rain, snow, vehicle wash) all wheel speed senor circuitry should be thoroughly inspected for signs of water intrusion. Use the following procedure. Spray down the suspected area with 5% salt water solution (two teaspoons of salt to 12 oz. of water). Start vehicle, allow vehicle to run for 10 seconds. If code returns immediately, replace suspected harness.

Any circuitry, that is suspected as causing the intermittent complaint, should be thoroughly checked for backed out terminals, improper mating, broken locks, improperly formed or damaged terminals, poor terminal to wiring connections or physical damage to the wiring harness.

DIAGNOSTIC CHART DTC A023 — 1993 ACHIEVA GRAND AM AND SKYLARK

IMPORTANT: WHEEL SPEED SENSOR INTERMITTENT PROBLEMS MAY BE DIFFICULT TO LOCATE. CARE SHOULD BE TAKEN NOT TO DISTURB ANY ELECTRICAL CONNECTIONS PRIOR TO AN INDICATED STEP OF THIS CHART. THIS WILL INSURE THAT AN INTERMITTENT CONNECTION WILL NOT BE CORRECTED BEFORE THE SOURCE OF THE PROBLEM IS FOUND.

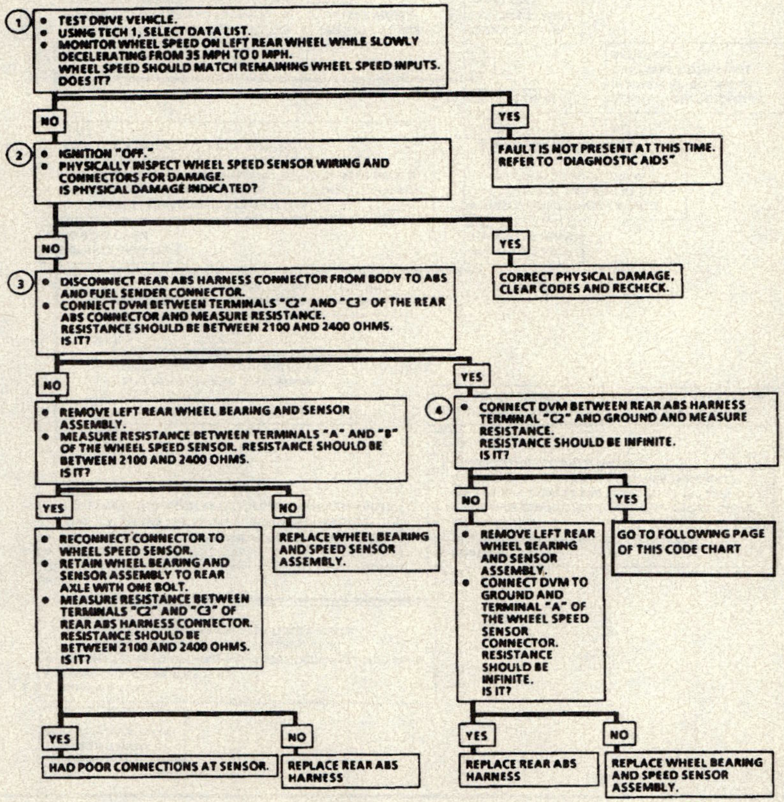

DIAGNOSTIC CHART DTC A023 — 1993 ACHIEVA GRAND AM AND SKYLARK

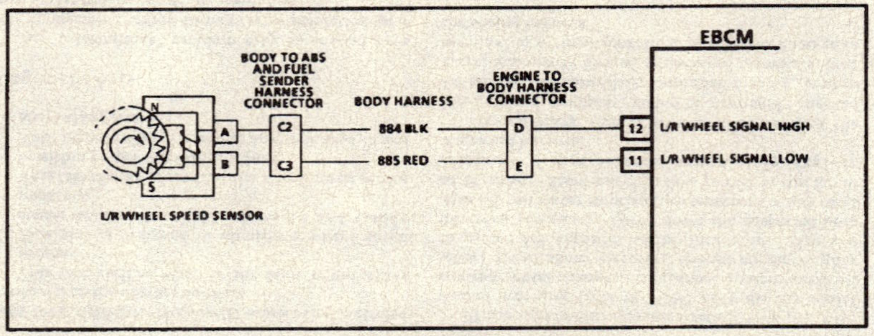

CODE A023

LEFT REAR WHEEL SPEED = 0

Circuit Description:

As a toothed ring passes by the wheel speed sensor, changes in the electromagnetic field cause the wheel speed sensor to produce a sinusoidal (AC) voltage signal whose frequency is proportional to wheel speed. The magnitude of this signal is directly related to wheel speed and the proximity of the wheel speed sensor to the toothed ring, often referred to as the air gap.

This code detects opens, shorts to ground, and shorts to battery that cause a wheel speed of 0 mph to be calculated by the ABS EBCM. The conditions present to fail this test are: (1) Three (3) wheels are rotating at least 5 mph, and (2) The fourth wheel is 0 mph.

Test Description: Number(s) below refer to circled number(s) on the diagnostic chart.

5. This step checks to see if the wheel speed sensor circuitry is shorted to voltage.
6. This step insures that the wheel speed sensor circuitry resistance is within the proper value.
7. This checks for an open in the wheel signal high circuit.
8. This checks for an open in the wheel signal low circuit.
9. This step insures that Code A023 was not set due to a poor connection between the world connector and EBCM.

Diagnostic Aids:

An "Intermittent" problem may be caused by a poor connection, rubbed through wire insulation, or a wire that is broken inside the insulation.

If the customers comments reflect that the ABS Amber warning light is "ON" only during moist environmental changes, (rain, snow, vehicle wash) all wheel speed sensor circuitry should be thoroughly inspected for signs of water intrusion. Use the following procedure. Spray down the suspected area with 5% salt water solution (two teaspoons of salt to 12 oz. of water). Start vehicle, allow vehicle to run for 10 seconds. If code returns immediately, replace suspected harness.

Any circuitry, that is suspected as causing the intermittent complaint, should be thoroughly checked for backed out terminals, improper mating, broken locks, improperly formed or damaged terminals, poor terminal to wiring connections or physical damage to the wiring harness.

DIAGNOSTIC CHART DTC A023 — 1993 ACHIEVA GRAND AM AND SKYLARK

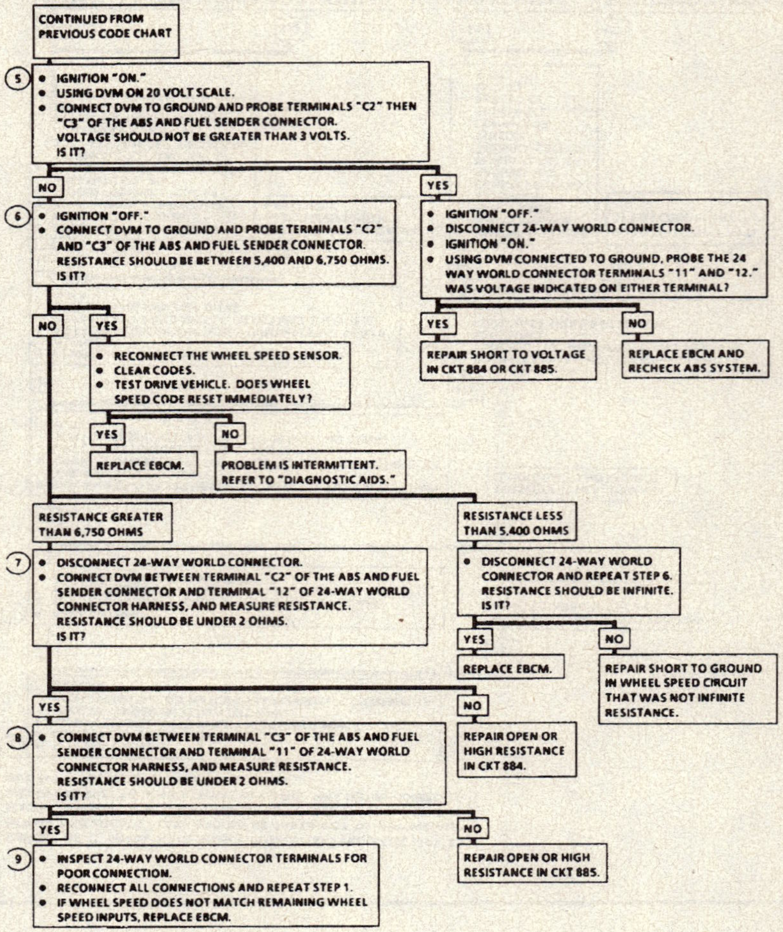

DIAGNOSTIC CHART DTC A024 — 1993 ACHIEVA GRAND AM AND SKYLARK

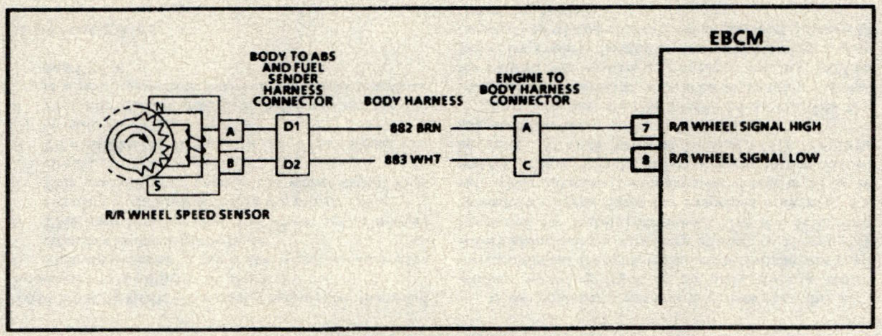

CODE A024

RIGHT REAR WHEEL SPEED = 0

Circuit Description:

As a toothed ring passes by the wheel speed sensor, changes in the electromagnetic field cause the wheel speed sensor to produce a sinusoidal (AC) voltage signal whose frequency is proportional to wheel speed. The magnitude of this signal is directly related to wheel speed and the proximity of the wheel speed sensor to the toothed ring, often referred to as the air gap.

This code detects opens, shorts to ground, and shorts to battery that cause a wheel speed of 0 mph to be calculated by the ABS EBCM. The conditions present to fail this test are: (1) Three (3) wheels are rotating at least 5 mph, and (2) The fourth wheel is 0 mph.

Test Description: Number(s) below refer to circled number(s) on the diagnostic chart.

1. This test verifies whether the fault is currently present.
2. This step is designed to identify a wheel speed sensor or circuitry that is damaged and visibly apparent.
3. This test checks to see that the wheel speed sensor is within proper resistance values.
4. This insures that the wheel speed sensor is not shorted to ground.

Diagnostic Aids:

An "Intermittent" problem may be caused by a poor connection, rubbed through wire insulation, or a wire that is broken inside the insulation.

If the customers comments reflect that the ABS Amber warning light is on only during moist environmental changes, (rain, snow, vehicle wash) all wheel speed sensor circuitry should be thoroughly inspected for water intrusion. Use the following procedure. Spray down the suspected area with 5% salt water solution (two teaspoons of salt to 12 oz. of water). Start vehicle, allow vehicle to run for 10 seconds. If code returns immediately, replace suspected harness.

Any circuitry, that is suspected as causing the intermittent complaint, should be thoroughly checked for backed out terminals, improper mating, broken locks, improperly formed or damaged terminals, poor terminal to wiring connections or physical damage to the wiring harness.

DIAGNOSTIC CHART DTC A024 — 1993 ACHIEVA GRAND AM AND SKYLARK

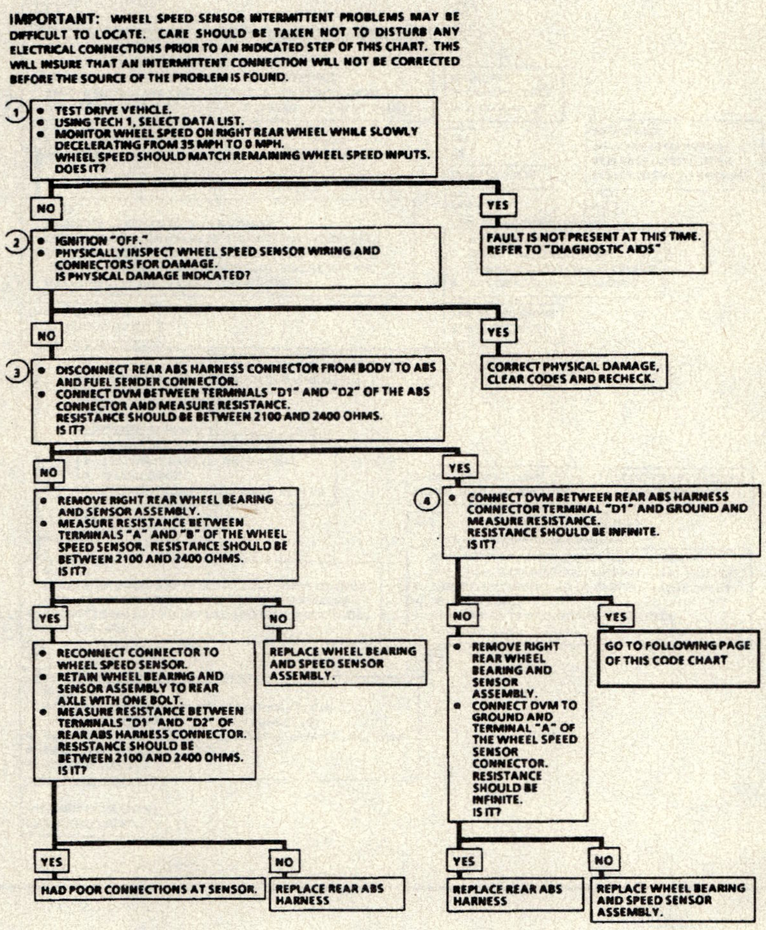

DIAGNOSTIC CHART DTC A024 — 1993 ACHIEVA GRAND AM AND SKYLARK

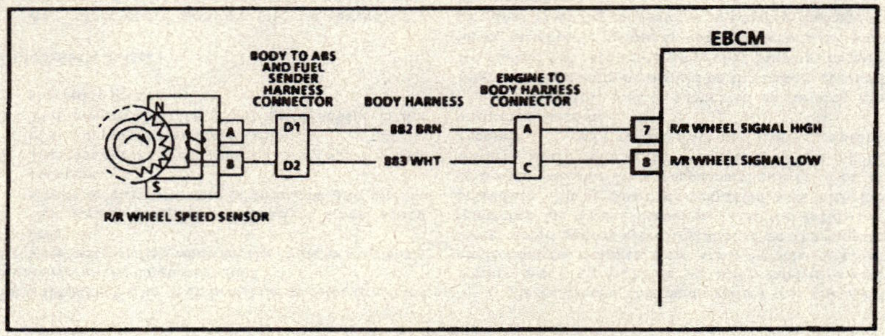

CODE A024

RIGHT REAR WHEEL SPEED = 0

Circuit Description:

As a toothed ring passes by the wheel speed sensor, changes in the electromagnetic field cause the wheel speed sensor to produce a sinusoidal (AC) voltage signal whose frequency is proportional to wheel speed. The magnitude of this signal is directly related to wheel speed and the proximity of the wheel speed sensor to the toothed ring, often referred to as the air gap.

This code detects opens, shorts to ground, and shorts to battery that cause a wheel speed of 0 mph to be calculated by the ABS EBCM. The conditions present to fail this test are: (1) Three (3) wheels are rotating at least 5 mph, and (2) The fourth wheel is 0 mph.

Test Description: Number(s) below refer to circled number(s) on the diagnostic chart.

5. This step checks to see if the wheel speed sensor circuitry is shorted to voltage.
6. This step insures that the wheel speed sensor circuitry resistance is within the proper value.
7. This checks for an open in the wheel signal high circuit.
8. This checks for an open in the wheel signal low circuit.
9. This step insures that Code A024 was not set due to a poor connection between the world connector and EBCM.

Diagnostic Aids:

An "Intermittent" problem may be caused by a poor connection, rubbed through wire insulation, or a wire that is broken inside the insulation.

If the customers comments reflect that the ABS Amber warning light is on only during moist environmental changes, (rain, snow, vehicle wash) all wheel speed sensor circuitry should be thoroughly inspected for water intrusion. Use the following procedure. Spray down the suspected area with 5% salt water solution (two teaspoons of salt to 12 oz. of water). Start vehicle, allow vehicle to run for 10 seconds. If code returns immediately, replace suspected harness.

Any circuitry, that is suspected as causing the intermittent complaint, should be thoroughly checked for backed out terminals, improper mating, broken locks, improperly formed or damaged terminals, poor terminal to wiring connections or physical damage to the wiring harness.

DIAGNOSTIC CHART DTC A024 — 1993 ACHIEVA GRAND AM AND SKYLARK

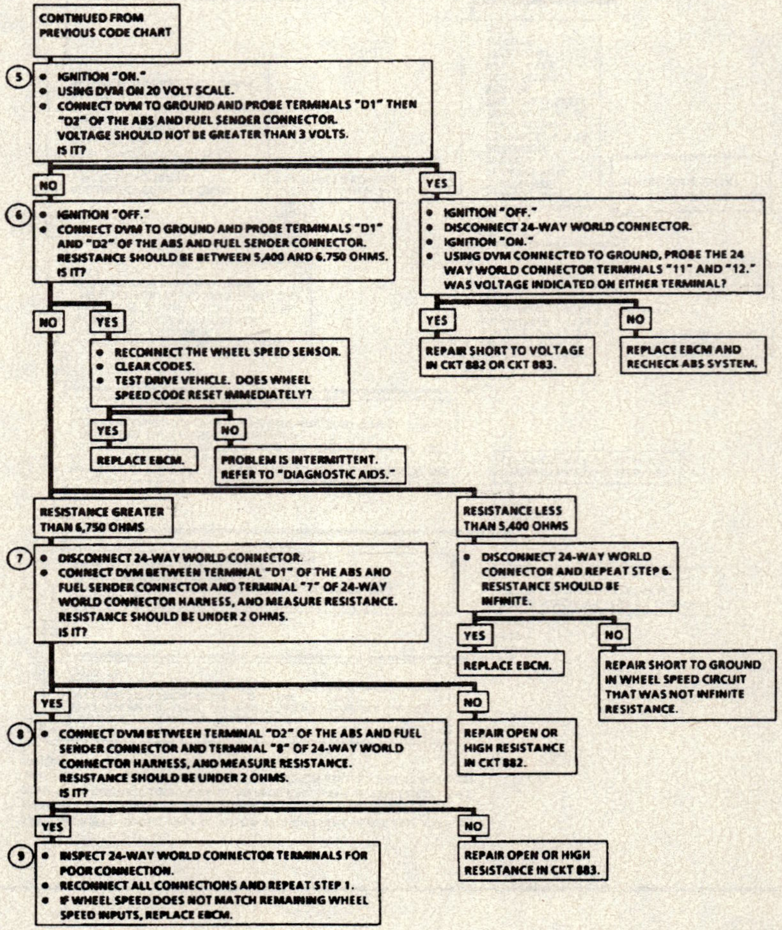

DIAGNOSTIC CHART DTC A025 — 1993 ACHIEVA GRAND AM AND SKYLARK

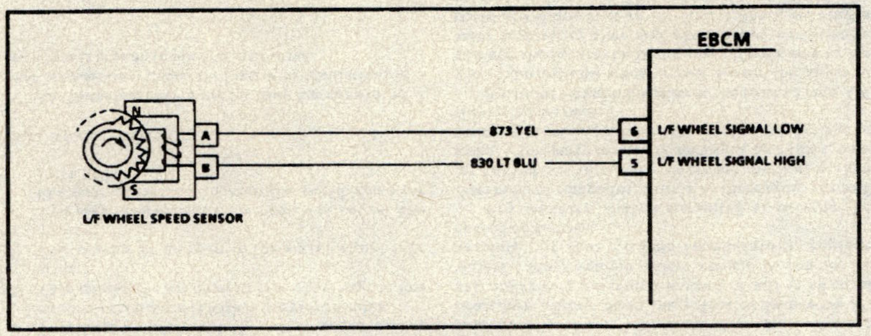

CODE A025

LEFT FRONT EXCESSIVE WHEEL SPEED VARIATION

Circuit Description:

As a toothed ring passes by the wheel speed sensor, changes in the electromagnetic field cause the wheel speed sensor to produce a sinusoidal (AC) voltage signal whose frequency is proportional to wheel speed. The magnitude of this signal is directly related to wheel speed and the proximity of the wheel speed sensor to the toothed ring, often referred to as the air gap.

This code detects opens, shorts to ground, and shorts to battery that cause intermittent wheel speed operation. Also detected is any sudden change in wheel speed determined to the unreasonable.

Test Description: Number(s) below refer to circled number(s) on the diagnostic chart.

1. This test verifies whether the fault is currently present.
2. This step is designed to identify a wheel speed sensor or circuitry that is damaged and visibly apparent.
3. This test checks to see that the wheel speed sensor circuitry is within proper resistance value.
4. This insure that the wheel speed sensor is not shorted to ground.

Diagnostic Aids:

An "Intermittent" problem may be caused by a poor connection, rubbed through wire insulation, or a wire that is broken inside the insulation.

If the customer's comments reflect that the ABS Amber warning light is on only during moist environmental changes, (rain, snow, vehicle wash) all wheel speed sensor circuitry should be thoroughly inspected for water intrusion. Use the following procedure. Spray down the suspected area with 5% salt water solution (two teaspoons of salt to 12 oz. of water). Start vehicle, allow vehicle to run for 10 seconds. If code returns immediately, replace suspected harness.

Any circuitry, that is suspected as causing the intermittent complaint, should be thoroughly checked for backed out terminals, improper mating, broken locks, improperly formed or damaged terminals, poor terminal to wiring connections or physical damage to the wiring harness.

A careful visual inspection of the toothed ring, CV joint, bearing and wheel speed sensor for physical damage should also be performed. If code sets at the same mph every time and wheel speed variation is noted above this mph on the Tech 1, the speed ring is most likely damaged. This can occur at any speed.

DIAGNOSTIC CHART DTC A025 — 1993 ACHIEVA GRAND AM AND SKYLARK

IMPORTANT: WHEEL SPEED SENSOR INTERMITTENT PROBLEMS MAY BE DIFFICULT TO LOCATE. CARE SHOULD BE TAKEN NOT TO DISTURB ANY ELECTRICAL CONNECTIONS PRIOR TO AN INDICATED STEP OF THIS CHART. THIS WILL INSURE THAT AN INTERMITTENT CONNECTION WILL NOT BE CORRECTED BEFORE THE SOURCE OF THE PROBLEM IS FOUND.

(1)
- TEST DRIVE VEHICLE.
- USING TECH 1, SELECT DATA LIST.
- MONITOR WHEEL SPEED ON LEFT FRONT WHEEL WHILE SLOWLY DECELERATING FROM 35 MPH TO 0 MPH.
- WHEEL SPEED SHOULD MATCH REMAINING WHEEL SPEED INPUTS.

DOES IT?

- YES: FAULT IS NOT PRESENT AT THIS TIME. REFER TO "DIAGNOSTIC AIDS" ON
- NO: go to (2)

(2)
- IGNITION "OFF."
- PHYSICALLY INSPECT WHEEL SPEED SENSOR, RING, WIRING AND CONNECTORS FOR DAMAGE. PAY PARTICULAR ATTENTION TO THE SPEED RING. ANY SIGNIFICANT DAMAGE (OTHER THAN NICKS FROM STONES ETC.) WILL AFFECT THE WHEEL SPEED INPUT SIGNAL.

IS PHYSICAL DAMAGE INDICATED?

- YES: CORRECT PHYSICAL DAMAGE, CLEAR CODES AND RECHECK.
- NO: go to (3)

(3)
- DISCONNECT CONNECTOR FROM LEFT FRONT WHEEL SPEED SENSOR.
- CONNECT DVM BETWEEN TERMINAL "A" AND "B" OF SENSOR, AND MEASURE RESISTANCE.

RESISTANCE SHOULD BE BETWEEN 1650 AND 1800 OHMS.
IS IT?

- NO: REPLACE WHEEL SPEED SENSOR, CLEAR CODES AND RECHECK.
- YES: go to (4)

(4)
- CONNECT DVM BETWEEN SENSOR TERMINAL "A" AND GROUND AND MEASURE RESISTANCE.

RESISTANCE SHOULD BE INFINITE.
IS IT?

- NO: REPLACE WHEEL SPEED SENSOR, CLEAR CODES AND RECHECK.
- YES: GO TO FOLLOWING PAGE OF THIS CODE CHART

DIAGNOSTIC CHART DTC A025 — 1993 ACHIEVA GRAND AM AND SKYLARK

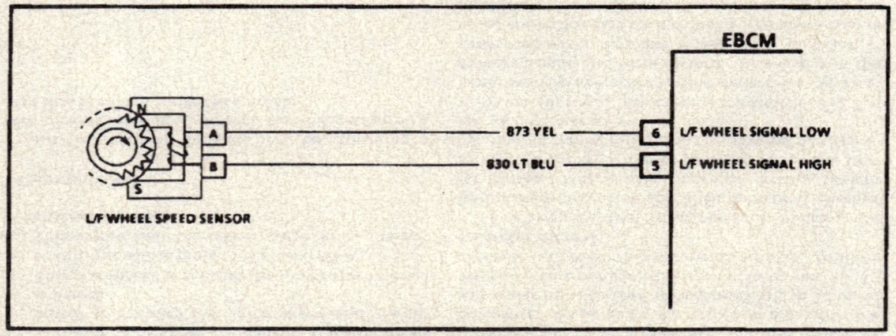

CODE A025

LEFT FRONT EXCESSIVE WHEEL SPEED VARIATION

Circuit Description:

As a toothed ring passes by the wheel speed sensor, changes in the electromagnetic field cause the wheel speed sensor to produce a sinusoidal (AC) voltage signal whose frequency is proportional to wheel speed. The magnitude of this signal is directly related to wheel speed and the proximity of the wheel speed sensor to the toothed ring, often referred to as the air gap.

This code detects opens, shorts to ground, and shorts to battery that cause intermittent wheel speed operational. Also detected is any sudden change in wheel speed determined to be unreasonable.

Test Description: Number(s) below refer to circled number(s) on the diagnostic chart.

5. This step checks to see if the wheel speed sensor circuitry is shorted to voltage.
6. This step insures that the wheel speed sensor circuitry resistance is within the proper value.
7. This checks for an open in the wheel signal low circuit.
8. This checks for an open in the wheel signal high circuit.
9. This step insures that Code A025 was not set due to a poor connection between the world connector and EBCM.

Diagnostic Aids:

An "Intermittent" problem may be caused by a poor connection, rubbed through wire insulation, or a wire that is broken inside the insulation.

If the customers comments reflect that the ABS Amber warning light is on only during moist environmental changes, (rain, snow, vehicle wash) all wheel speed sensor circuitry should be thoroughly inspected for water intrusion. Use the following procedure. Spray down the suspected area with 5% salt water solution (two teaspoons of salt to 12 oz. of water). Start vehicle, allow vehicle to run for 10 seconds. If code returns immediately, replace suspected harness.

Any circuitry, that is suspected as causing the intermittent complaint, should be thoroughly checked for backed out terminals, improper mating, broken locks, improperly formed or damaged terminals, poor terminal to wiring connections or physical damage to the wiring harness.

A careful visual inspection of the toothed ring, CV joint, bearing and wheel speed sensor for physical damage should also be performed. If code sets at the same mph every time and wheel speed variation is noted above this mph on the Tech 1, the speed ring is most likely damaged. This can occur at any speed.

DIAGNOSTIC CHART DTC A025 — 1993 ACHIEVA GRAND AM AND SKYLARK

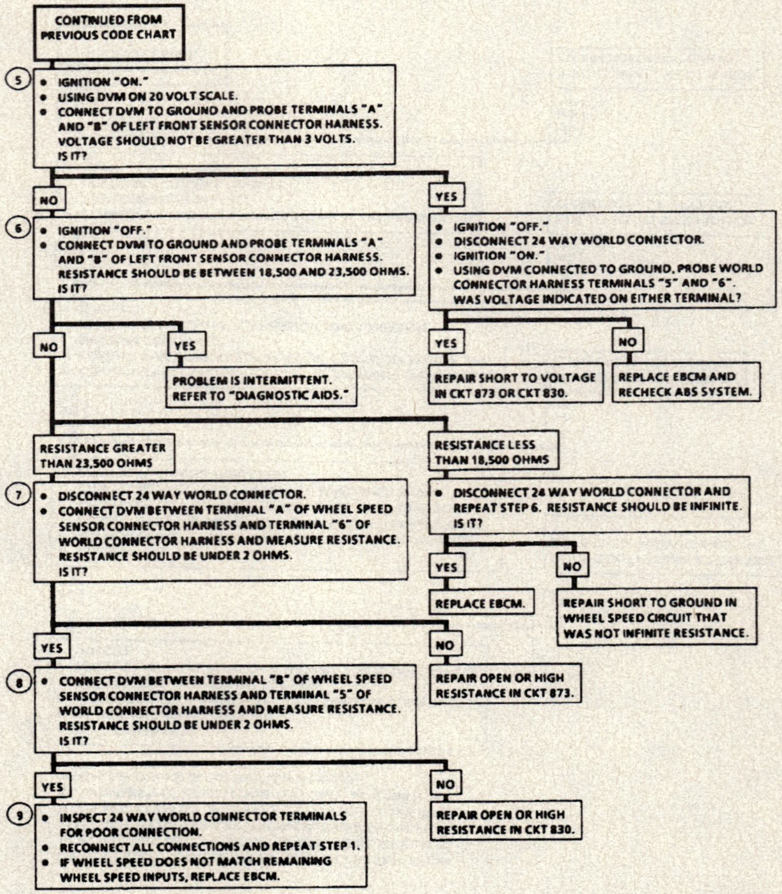

DIAGNOSTIC CHART DTC A026 — 1993 ACHIEVA GRAND AM AND SKYLARK

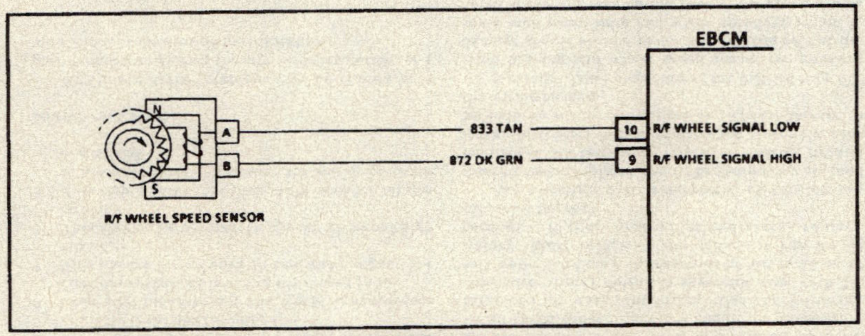

CODE A026

RIGHT FRONT EXCESSIVE WHEEL SPEED VARIATION

Circuit Description:

As a toothed ring passes by the wheel speed sensor, changes in the electromagnetic field cause the wheel speed sensor to produce a sinusoidal (AC) voltage signal whose frequency is proportional to wheel speed. The magnitude of this signal is directly related to wheel speed and the proximity of the wheel speed sensor to the toothed ring, often referred to as the air gap.

This code detects opens, shorts to ground, and shorts to battery that cause intermittent wheel speed operation. Also detected is any sudden change in wheel speed determined to be unreasonable.

Test Description: Number(s) below refer to circled number(s) on the diagnostic chart.

1. This test verifies whether the fault is currently present.
2. This step is designed to identify a wheel speed sensor or circuitry that is damaged and visibly apparent.
3. This test checks to see that the wheel speed sensor circuitry is within proper resistance value.
4. This insures that the wheel speed sensor is not shorted to ground.

Diagnostic Aids:

An "Intermittent" problem may be caused by a poor connection, rubbed through wire insulation, or a wire that is broken inside the insulation.

If the customers comments reflect that the ABS Amber warning light is on only during moist environmental changes, (rain, snow, vehicle wash) all wheel speed sensor circuitry should be thoroughly inspected for water intrusion. Use the following procedure. Spray down the suspected area with 5% salt water solution (two teaspoons of salt to 12 oz. of water). Start vehicle, allow vehicle to run for 10 seconds. If code returns immediately, replace suspected harness.

Any circuitry, that is suspected as causing the intermittent complaint, should be thoroughly checked for backed out terminals, improper mating, broken locks, improperly formed or damaged terminals, poor terminal to wiring connections or physical damage to the wiring harness.

A careful visual inspection of the toothed ring, CV joint, bearing and wheel speed sensor for physical damage should also be performed. If code sets at the same mph every time and wheel speed variation is noted above this mph on the Tech 1, the speed ring is most likely damaged. This can occur at any speed.

DIAGNOSTIC CHART DTC A026 — 1993 ACHIEVA GRAND AM AND SKYLARK

IMPORTANT: WHEEL SPEED SENSOR INTERMITTENT PROBLEMS MAY BE DIFFICULT TO LOCATE. CARE SHOULD BE TAKEN NOT TO DISTURB ANY ELECTRICAL CONNECTIONS PRIOR TO AN INDICATED STEP OF THIS CHART. THIS WILL INSURE THAT AN INTERMITTENT CONNECTION WILL NOT BE CORRECTED BEFORE THE SOURCE OF THE PROBLEM IS FOUND.

(1)
- TEST DRIVE VEHICLE.
- IGNITION "ON."
- USING TECH 1, SELECT DATA LIST.
- MONITOR WHEEL SPEED ON RIGHT FRONT WHEEL WHILE SLOWLY DECELERATING FROM 35 MPH TO 0 MPH.

WHEEL SPEED SHOULD MATCH REMAINING WHEEL SPEED INPUTS.
DOES IT?

YES: FAULT IS NOT PRESENT AT THIS TIME, REFER TO "DIAGNOSTIC AIDS"

NO:

(2)
- IGNITION "OFF."
- PHYSICALLY INSPECT WHEEL SPEED SENSOR, RING, WIRING AND CONNECTORS FOR DAMAGE. PAY PARTICULAR ATTENTION TO THE SPEED RING. ANY SIGNIFICANT DAMAGE (OTHER THAN NICKS FROM STONES ETC.) WILL AFFECT THE WHEEL SPEED INPUT SIGNAL.

IS PHYSICAL DAMAGE INDICATED?

YES: CORRECT PHYSICAL DAMAGE, CLEAR CODES AND RECHECK.

NO:

(3)
- DISCONNECT CONNECTOR FROM RIGHT FRONT WHEEL SPEED SENSOR.
- CONNECT DVM BETWEEN TERMINALS "A" AND "B" OF SENSOR, AND MEASURE RESISTANCE.

RESISTANCE SHOULD BE BETWEEN 1650 AND 1800 OHMS.
IS IT?

NO: REPLACE WHEEL SPEED SENSOR, CLEAR CODES AND RECHECK.

YES:

(4)
- CONNECT DVM BETWEEN SENSOR TERMINAL "A" AND GROUND AND MEASURE RESISTANCE.

RESISTANCE SHOULD BE INFINITE.
IS IT?

NO: REPLACE WHEEL SPEED SENSOR, CLEAR CODES AND RECHECK.

YES: GO TO FOLLOWING PAGE OF THIS CODE CHART

DIAGNOSTIC CHART DTC A026 — 1993 ACHIEVA GRAND AM AND SKYLARK

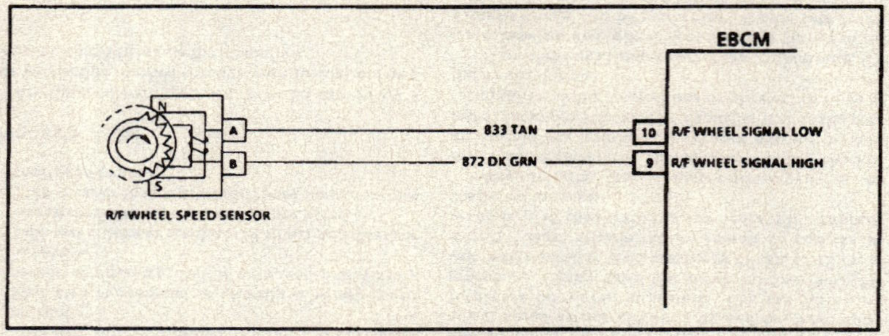

CODE A026

RIGHT FRONT EXCESSIVE WHEEL SPEED VARIATION

Circuit Description:

As a toothed ring passes by the wheel speed sensor, changes in the electromagnetic field cause the wheel speed sensor to produce a sinusoidal (AC) voltage signal whose frequency is proportional to wheel speed. The magnitude of this signal is directly related to wheel speed and the proximity of the wheel speed sensor to the toothed ring, often referred to as the air gap.

This code detects opens, shorts to ground, and shorts to battery that cause intermittent wheel speed operation. Also detected is any sudden change in wheel speed determined to be unreasonable.

Test Description: Number(s) below refer to circled number(s) on the diagnostic chart.

5. This step checks to see if the wheel speed sensor circuitry is shorted to voltage.
6. This step insures that the wheel speed sensor circuitry resistance is within the proper value.
7. This checks for an open in the wheel signal low circuit.
8. This checks for an open in the wheel signal high circuit.
9. This step insures that Code A026 was not set due to a poor connection between the world connector and EBCM.

Diagnostic Aids:

An "Intermittent" problem may be caused by a poor connection, rubbed through wire insulation, or a wire that is broken inside the insulation.

If the customers comments reflect that the ABS Amber warning light is on only during moist environmental changes, (rain, snow, vehicle wash) all wheel speed sensor circuitry should be thoroughly inspected for water intrusion. Use the following procedure. Spray down the suspected area with 5% salt water solution (two teaspoons of salt to 12 oz. of water). Start vehicle, allow vehicle to run for 10 seconds. If code returns immediately, replace suspected harness.

Any circuitry, that is suspected as causing the intermittent complaint, should be thoroughly checked for backed out terminals, improper mating, broken locks, improperly formed or damaged terminals, poor terminal to wiring connections or physical damage to the wiring harness.

A careful visual inspection of the toothed ring, CV joint, bearing and wheel speed sensor for physical damage should also be performed. If code sets at the same mph every time and wheel speed variation is noted above this mph on the Tech 1, the speed ring is most likely damaged. This can occur at any speed.

DIAGNOSTIC CHART DTC A026 — 1993 ACHIEVA GRAND AM AND SKYLARK

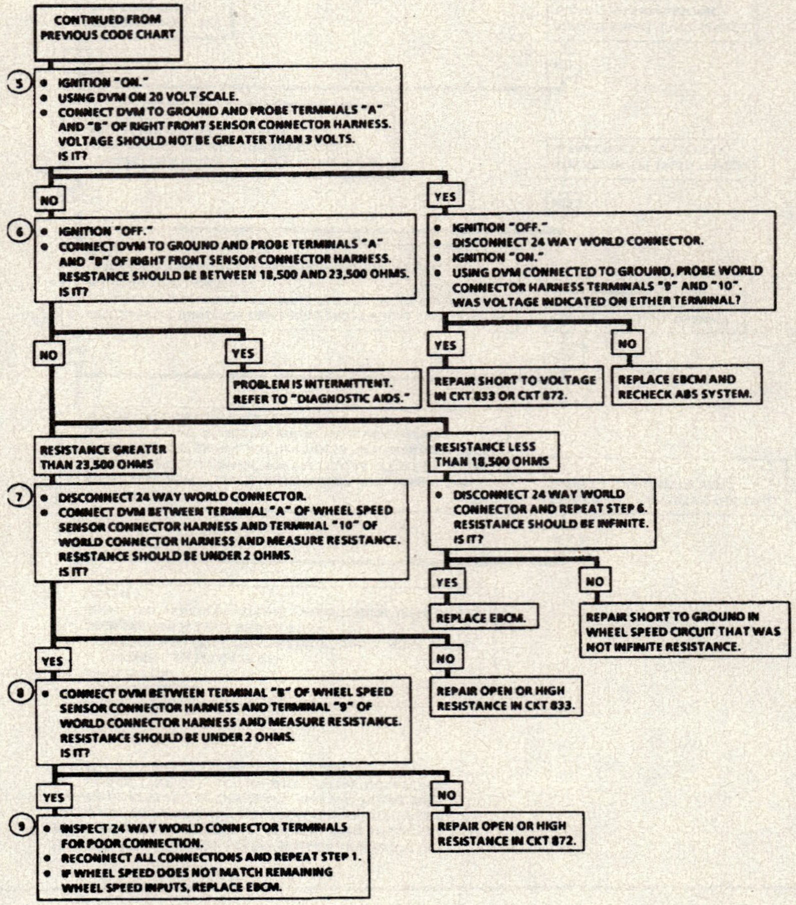

DIAGNOSTIC CHART DTC A027 — 1993 ACHIEVA GRAND AM AND SKYLARK

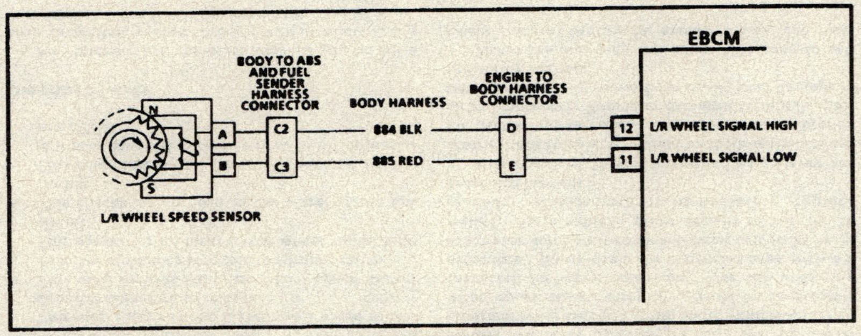

CODE A027

LEFT REAR EXCESSIVE WHEEL SPEED VARIATION

Circuit Description:

As a toothed ring passes by the wheel speed sensor, changes in the electromagnetic field cause the wheel speed sensor to produce a sinusoidal (AC) voltage signal whose frequency is proportional to wheel speed. The magnitude of this signal is directly related to wheel speed and the proximity of the wheel speed sensor to the toothed ring, often referred to as the air gap.

This code detects opens, shorts to ground, and shorts to battery that cause intermittent wheel speed operation. Also detected is any sudden change in wheel speed determined to be unreasonable.

Test Description: Number(s) below refer to circled number(s) on the diagnostic chart.

1. This test verifies whether the fault is currently present.
2. This step is designed to identify a wheel speed sensor or circuitry that is damaged and visibly apparent.
3. This test checks to see that the wheel speed sensor circuitry is within proper resistance value.
4. This insures that the wheel speed sensor is not shorted to ground.

Diagnostic Aids:

An "Intermittent" problem may be caused by a poor connection, rubbed through wire insulation, or a wire that is broken inside the insulation.

If the customers comments reflect that the ABS Amber warning light is on only during moist environmental changes, (rain, snow, vehicle wash) all wheel speed sensor circuitry should be thoroughly inspected for water intrusion. Use the following procedure. Spray down the suspected area with 5% salt water solution (two teaspoons of salt to 12 oz. of water). Start vehicle, allow vehicle to run for 10 seconds. If code returns immediately, replace suspected harness.

Any circuitry, that is suspected as causing the intermittent complaint, should be thoroughly checked for backed out terminals, improper mating, broken locks, improperly formed or damaged terminals, poor terminal to wiring connections or physical damage to the wiring harness.

Excessive end play can cause this code to set. Check integral bearing for excessive axial end play,

DIAGNOSTIC CHART DTC A027 — 1993 ACHIEVA GRAND AM AND SKYLARK

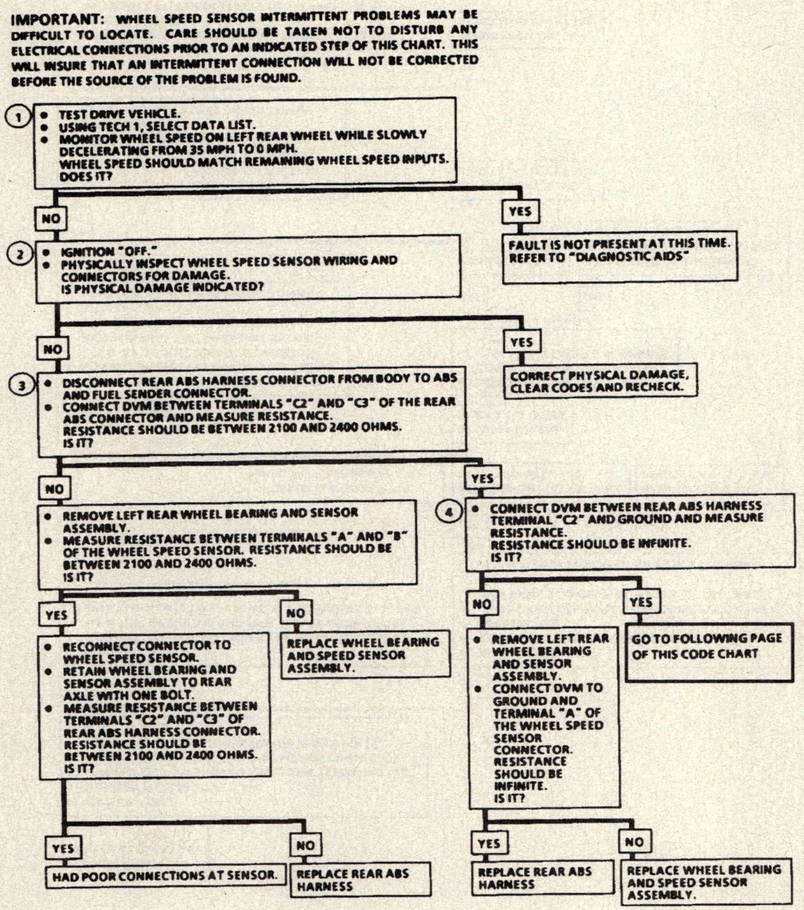

DIAGNOSTIC CHART DTC A027 — 1993 ACHIEVA GRAND AM AND SKYLARK

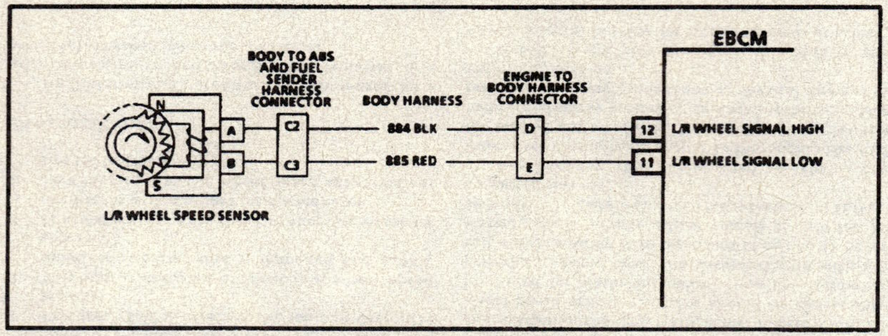

CODE A027

LEFT REAR EXCESSIVE WHEEL SPEED VARIATION

Circuit Description:

As a toothed ring passes by the wheel speed sensor, changes in the electromagnetic field cause the wheel speed sensor to produce a sinusoidal (AC) voltage signal whose frequency is proportional to wheel speed. The magnitude of this signal is directly related to wheel speed and the proximity of the wheel speed sensor to the toothed ring, often referred to as the air gap.

This code detects opens, shorts to ground, and shorts to battery that cause intermittent wheel speed operation. Also detected is any sudden change in wheel speed determined to be unreasonable.

Test Description: Number(s) below refer to circled number(s) on the diagnostic chart.

5. This step checks to see if the wheel speed sensor circuitry is shorted to voltage.
6. This step insures that the wheel speed sensor circuitry resistance is within the proper value.
7. This checks for an open in the wheel signal high circuit.
8. This checks for an open in the wheel signal low circuit.
9. This step insures that Code A027 was not set due to a poor connection between the world connector and EBCM.

Diagnostic Aids:

An "Intermittent" problem may be caused by a poor connection, rubbed through wire insulation, or a wire that is broken inside the insulation.

If the customers comments reflect that the ABS Amber warning light is on only during moist environmental changes, (rain, snow, vehicle wash) all wheel speed sensor circuitry should be thoroughly inspected for water intrusion. Use the following procedure. Spray down the suspected area with 5% salt water solution (two teaspoons of salt to 12 oz. of water). Start vehicle, allow vehicle to run for 10 seconds. If code returns immediately, replace suspected harness.

Any circuitry, that is suspected as causing the intermittent complaint, should be thoroughly checked for backed out terminals, improper mating, broken locks, improperly formed or damaged terminals, poor terminal to wiring connections or physical damage to the wiring harness.

Excessive end play can cause this code to set. Check integral bearing for excessive axial end play.

DIAGNOSTIC CHART DTC A027 — 1993 ACHIEVA GRAND AM AND SKYLARK

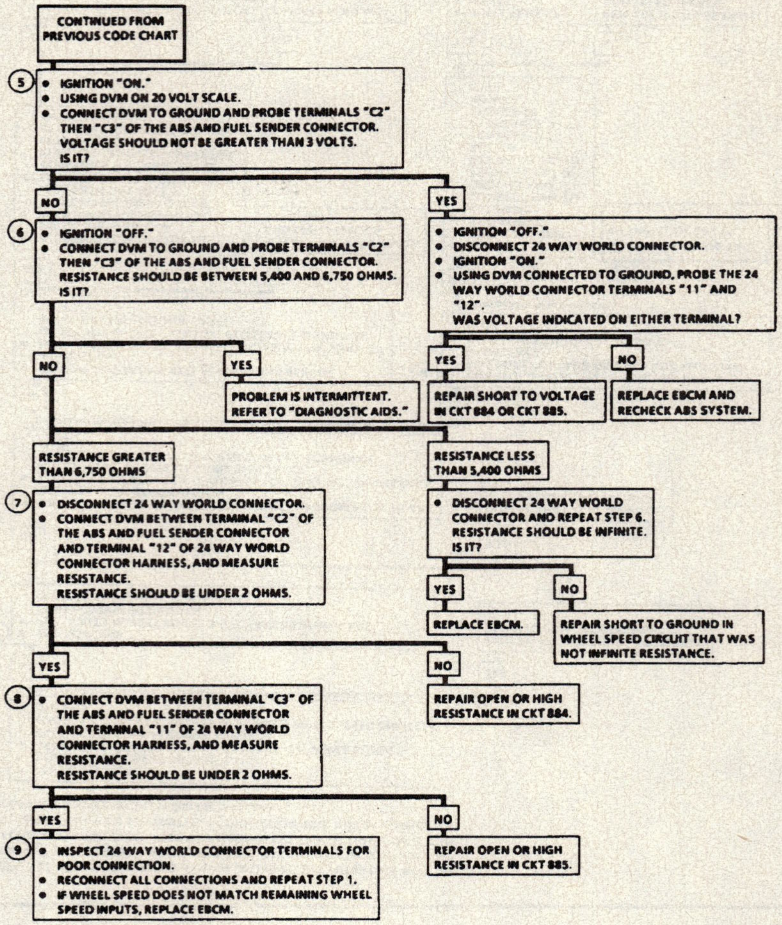

DIAGNOSTIC CHART DTC A028 — 1993 ACHIEVA GRAND AM AND SKYLARK

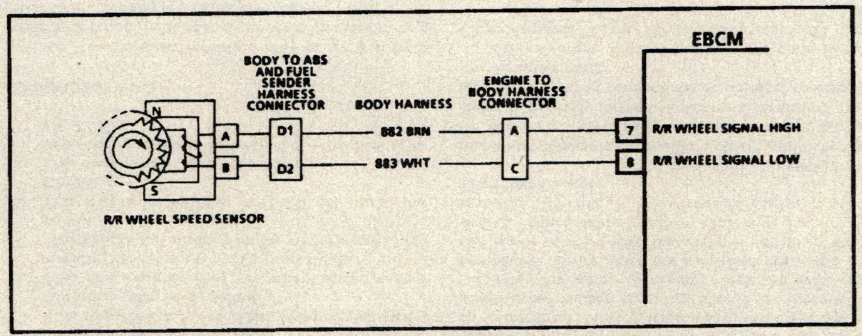

CODE A028

RIGHT REAR EXCESSIVE WHEEL SPEED VARIATION

Circuit Description:

As a toothed ring passes by the wheel speed sensor, changes in the electromagnetic field cause the wheel speed sensor to produce a sinusoidal (AC) voltage signal whose frequency is proportional to wheel speed. The magnitude of this signal is directly related to wheel speed and the proximity of the wheel speed sensor to the toothed ring, often referred to as the air gap.

This code detects opens, shorts to ground, and shorts to battery that cause intermittent wheel speed operation. Also detected is any sudden change in wheel speed determined to be unreasonable.

Test Description: Number(s) below refer to circled number(s) on the diagnostic chart.

1. This test verifies whether the fault is currently present.
2. This step is designed to identify a wheel speed sensor or circuitry that is damaged and visibly apparent.
3. This test checks to see that the wheel speed sensor circuitry is within proper resistance value.
4. This insures that the wheel speed sensor is not shorted to ground.

Diagnostic Aids:

An "Intermittent" problem may be caused by a poor connection, rubbed through wire insulation, or a wire that is broken inside the insulation.

If the customers comments reflect that the ABS Amber warning light is on only during moist environmental changes, (rain, snow, vehicle wash) all wheel speed sensor circuitry should be thoroughly inspected for water intrusion. Use the following procedure. Spray down the suspected area with 5% salt water solution (two teaspoons of salt to 12 oz. of water). Start vehicle, allow vehicle to run for 10 seconds. If code returns immediately, replace suspected harness.

Any circuitry, that is suspected as causing the intermittent complaint, should be thoroughly checked for backed out terminals, improper mating, broken locks, improperly formed or damaged terminals, poor terminal to wiring connections or physical damage to the wiring harness.

Excessive end play can cause this code to set. Check integral bearing for excessive axial end play,

DIAGNOSTIC CHART DTC A028 — 1993 ACHIEVA GRAND AM AND SKYLARK

IMPORTANT: WHEEL SPEED SENSOR INTERMITTENT PROBLEMS MAY BE DIFFICULT TO LOCATE. CARE SHOULD BE TAKEN NOT TO DISTURB ANY ELECTRICAL CONNECTIONS PRIOR TO AN INDICATED STEP OF THIS CHART. THIS WILL INSURE THAT AN INTERMITTENT CONNECTION WILL NOT BE CORRECTED BEFORE THE SOURCE OF THE PROBLEM IS FOUND.

(1)
- TEST DRIVE VEHICLE.
- USING TECH 1, SELECT DATA LIST.
- MONITOR WHEEL SPEED ON RIGHT REAR WHEEL WHILE SLOWLY DECELERATING FROM 35 MPH TO 0 MPH. WHEEL SPEED SHOULD MATCH REMAINING WHEEL SPEED INPUTS. DOES IT?

YES: FAULT IS NOT PRESENT AT THIS TIME. REFER TO "DIAGNOSTIC AIDS" ON FACING PAGE.

NO:

(2)
- IGNITION "OFF."
- PHYSICALLY INSPECT WHEEL SPEED SENSOR WIRING AND CONNECTORS FOR DAMAGE. IS PHYSICAL DAMAGE INDICATED?

YES: CORRECT PHYSICAL DAMAGE. CLEAR CODES AND RECHECK.

NO:

(3)
- DISCONNECT REAR ABS HARNESS CONNECTOR FROM BODY TO ABS AND FUEL SENDER CONNECTOR.
- CONNECT DVM BETWEEN TERMINALS "D1" AND "D2" OF THE ABS CONNECTOR AND MEASURE RESISTANCE. RESISTANCE SHOULD BE BETWEEN 2100 AND 2400 OHMS. IS IT?

NO:
- REMOVE RIGHT REAR WHEEL BEARING AND SENSOR ASSEMBLY.
- MEASURE RESISTANCE BETWEEN TERMINALS "A" AND "B" OF THE WHEEL SPEED SENSOR. RESISTANCE SHOULD BE BETWEEN 2100 AND 2400 OHMS. IS IT?

- NO: REPLACE WHEEL BEARING AND SPEED SENSOR ASSEMBLY.
- YES:
  - RECONNECT CONNECTOR TO WHEEL SPEED SENSOR.
  - RETAIN WHEEL BEARING AND SENSOR ASSEMBLY TO REAR AXLE WITH ONE BOLT.
  - MEASURE RESISTANCE BETWEEN TERMINALS "D1" AND "D2" OF REAR ABS HARNESS CONNECTOR. RESISTANCE SHOULD BE BETWEEN 2100 AND 2400 OHMS. IS IT?
    - YES: HAD POOR CONNECTIONS AT SENSOR.
    - NO: REPLACE REAR ABS HARNESS

YES:

(4)
- CONNECT DVM BETWEEN REAR ABS HARNESS CONNECTOR TERMINAL "D1" AND GROUND AND MEASURE RESISTANCE. RESISTANCE SHOULD BE INFINITE. IS IT?

- YES: GO TO FOLLOWING PAGE OF THIS CODE CHART
- NO:
  - REMOVE RIGHT REAR WHEEL BEARING AND SENSOR ASSEMBLY.
  - CONNECT DVM TO GROUND AND TERMINAL "A" OF THE WHEEL SPEED SENSOR CONNECTOR. RESISTANCE SHOULD BE INFINITE. IS IT?
    - YES: REPLACE REAR ABS HARNESS
    - NO: REPLACE WHEEL BEARING AND SPEED SENSOR ASSEMBLY.

DIAGNOSTIC CHART DTC A028 — 1993 ACHIEVA GRAND AM AND SKYLARK

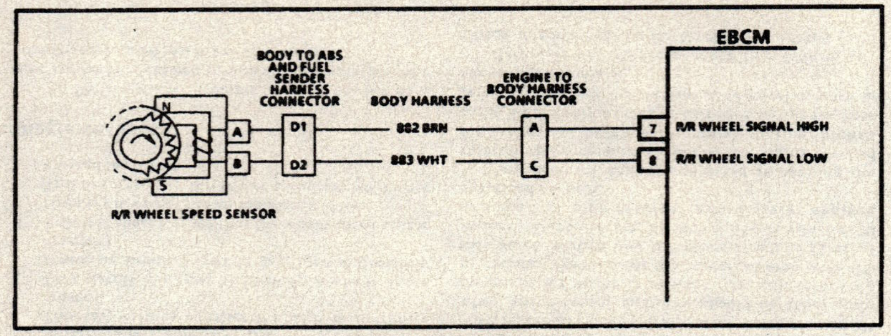

CODE A028

RIGHT REAR EXCESSIVE WHEEL SPEED VARIATION

Circuit Description:

As a toothed ring passes by the wheel speed sensor, changes in the electromagnetic field cause the wheel speed sensor to produce a sinusoidal (AC) voltage signal whose frequency is proportional to wheel speed. The magnitude of this signal is directly related to wheel speed and the proximity of the wheel speed sensor to the toothed ring, often referred to as the air gap.

This code detects opens, shorts to ground, and shorts to battery that cause intermittent wheel speed operation. Also detected is any sudden change in wheel speed determined to be unreasonable.

Test Description: Number(s) below refer to circled number(s) on the diagnostic chart.

5. This step checks to see if the wheel speed sensor circuitry is shorted to voltage.
6. This step insures that the wheel speed sensor circuitry resistance is within proper value.
7. This checks for an open in the wheel signal high circuit.
8. This checks for an open in the wheel signal low circuit.
9. This step insures that Code A028 was not set due to a poor connection between the world connector and EBCM.

Diagnostic Aids:

An "Intermittent" problem may be caused by a poor connection, rubbed through wire insulation, or a wire that is broken inside the insulation.

If the customers comments reflect that the ABS Amber warning light is on only during moist environmental changes, (rain, snow, vehicle wash) all wheel speed sensor circuitry should be thoroughly inspected for water intrusion. Use the following procedure. Spray down the suspected area with 5% salt water solution (two teaspoons of salt to 12 oz. of water). Start vehicle, allow vehicle to run for 10 seconds. If code returns immediately, replace suspected harness.

Any circuitry, that is suspected as causing the intermittent complaint, should be thoroughly checked for backed out terminals, improper mating, broken locks, improperly formed or damaged terminals, poor terminal to wiring connections or physical damage to the wiring harness.

Excessive end play can cause this code to set. Check integral bearing for excessive axial end play,

DIAGNOSTIC CHART DTC A028 — 1993 ACHIEVA GRAND AM AND SKYLARK

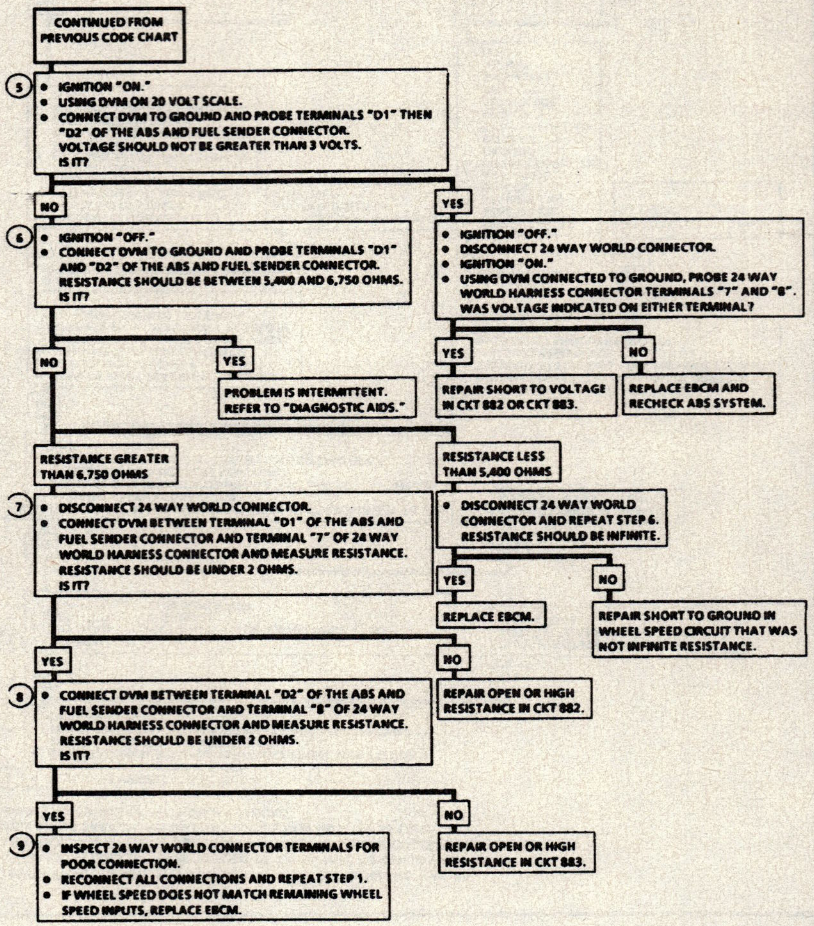

DIAGNOSTIC CHART DTC A031 — 1993 ACHIEVA GRAND AM AND SKYLARK

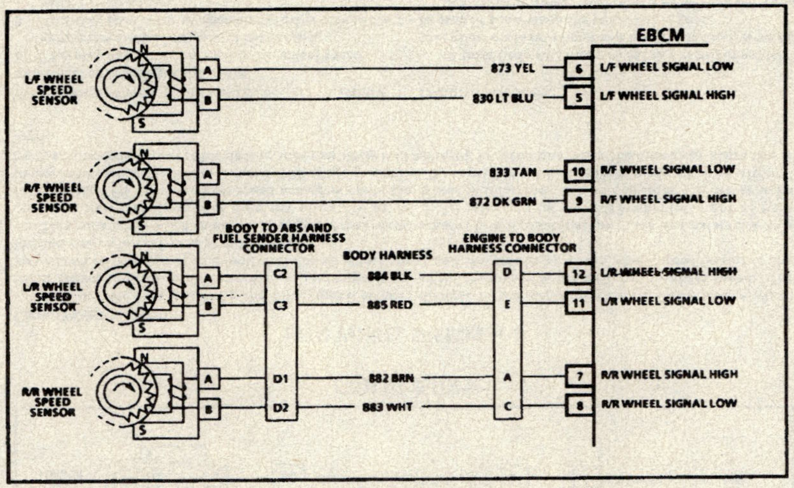

CODE A031

TWO WHEEL SPEEDS = 0

Circuit Description:

As a toothed ring passes by the wheel speed sensor, changes in the electromagnetic field cause the wheel speed sensor to produce a sinusoidal (AC) voltage signal whose frequency is proportional to wheel speed. The magnitude of this signal is directly related to wheel speed and the proximity of the wheel speed sensor to the toothed ring, often referred to as the air gap.

This code is designed to detect two open wheel speed sensors in any combination. This test exists due to a condition that may exist that does not allow the driver to be informed by the ABS warning light of non-functioning ABS. If one wheel speed sensor is open, the driver is informed. If a second fails, the driver is no longer informed without this test and ABS is not available. By requiring the two operating wheels to be within 7 mph, a false code is prevented during spin-ups while still allowing for detection when the stowaway spare tire is used.

Test Description: Number(s) below refer to circled number(s) on the diagnostic chart.

1. This test verifies whether the fault is currently present.
2. This step is designed to identify a wheel speed sensor or circuitry that is damaged and visibly apparent.
3. This step identifies if the cause of the problem is common to the rear wheel sensor circuitry.
4. Because both rear sensors are common to the extension harness, the 4 way connectors and harness should be inspected.
5. This test checks to see that the wheel speed sensor is within proper resistance values.
6. This insures that the wheel speed sensor is not shorted to ground.

Diagnostic Aids:

A false Code A031 may be set as a result of vehicle operation while on a hoist or a lifted condition or stuck in mud or snow, when the vehicle is in gear.

An "Intermittent" problem may be caused by a poor connection, rubbed through wire insulation, or a wire that is broken inside the insulation.

DIAGNOSTIC CHART DTC A031 — 1993 ACHIEVA GRAND AM AND SKYLARK

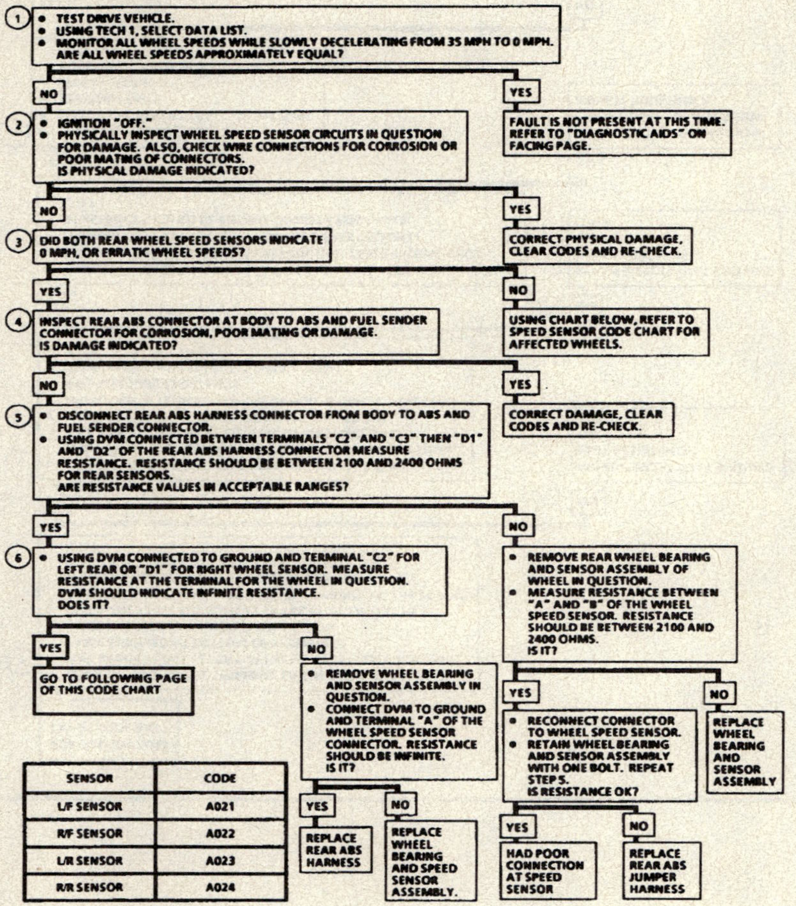

SENSOR	CODE
L/F SENSOR	A021
R/F SENSOR	A022
L/R SENSOR	A023
R/R SENSOR	A024

DIAGNOSTIC CHART DTC A031 — 1993 ACHIEVA GRAND AM AND SKYLARK

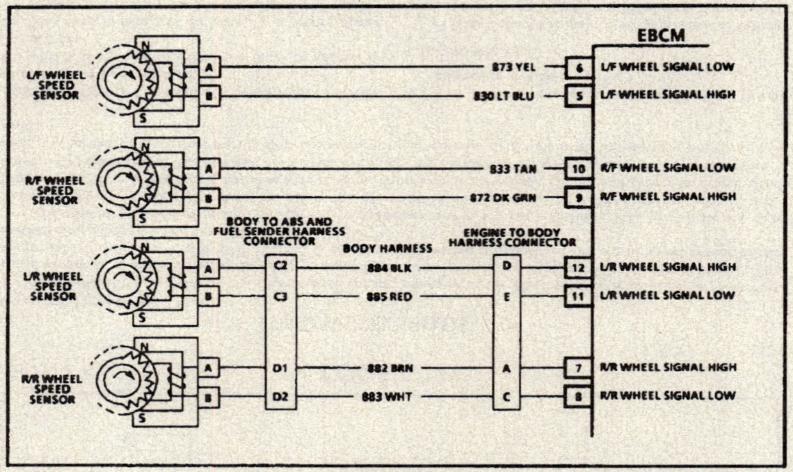

CODE A031

TWO WHEEL SPEEDS = 0

Circuit Description:

As a toothed ring passes by the wheel speed sensor, changes in the electromagnetic field cause the wheel speed sensor to produce a sinusoidal (AC) voltage signal whose frequency is proportional to wheel speed. The magnitude of this signal is directly related to wheel speed and the proximity of the wheel speed sensor to the toothed ring, often referred to as the air gap.

This code is designed to detect two open wheel speed sensors in any combination. This test exists due to a condition that may exist that does not allow the driver to be informed by the ABS warning light of non-functioning ABS. If one wheel speed sensor is open, the driver is informed. If a second fails, the driver is no longer informed without this test and ABS is not available. By requiring the two operating wheels to be within 7 mph, a false code is prevented during spin-ups while still allowing for detection when the stowaway spare tire is used.

Test Description: Number(s) below refer to circled number(s) on the diagnostic chart.

7. This step checks for an open in the wheel speed sensor circuitry of one side of the sensor.
8. This step checks for an open on the other side of the sensor in question.
9. This step checks for a short to ground in the wheel speed sensor circuitry.
10. This insures that the circuitry is not shorted to voltage.

Diagnostic Aids:

A false Code A031 may be set as a result of vehicle operation while on a hoist or a lifted condition, or stuck in mud or snow when the vehicle is in gear.

An "Intermittent" problem may be caused by a poor connection, rubbed through wire insulation, or a wire that is broken inside the insulation.

DIAGNOSTIC CHART DTC A031 — 1993 ACHIEVA GRAND AM AND SKYLARK

CONTINUED FROM CHART CODE A031

(7)
- CONNECT DVM TO THE TERMINAL OF THE BODY TO ABS AND FUEL SENDER CONNECTOR FOR THE WHEEL IN QUESTION. (USE TABLE SUPPLIED AT BOTTOM OF THIS PAGE.)
- DISCONNECT 24 WAY WORLD CONNECTOR.
- USING THE TABLE SUPPLIED AT THE BOTTOM OF THIS PAGE, CONNECT DVM TO THE APPROPRIATE TERMINAL OF THE 24 WAY WORLD CONNECTOR AND MEASURE RESISTANCE. RESISTANCE SHOULD BE LESS THAN 2 OHMS. IS IT?

NO → REPAIR CIRCUIT THAT SHOWED HIGH RESISTANCE OR OPEN.

YES ↓

(8)
- CONNECT DVM TO THE OTHER TERMINAL OF THE BODY TO ABS AND FUEL SENDER CONNECTOR FOR THE WHEEL IN QUESTION. (USE TABLE SUPPLIED AT BOTTOM OF THIS PAGE.)
- USING THE TABLE SUPPLIED, CONNECT DVM TO THE APPROPRIATE TERMINAL OF THE 24 WAY WORLD CONNECTOR AND MEASURE RESISTANCE. RESISTANCE SHOULD BE LESS THAN 2 OHMS. IS IT?

NO → REPAIR CIRCUIT THAT SHOWED HIGH RESISTANCE OR OPEN.

YES ↓

(9)
- CONNECT DVM BETWEEN BODY TO ABS AND FUEL SENDER CONNECTOR TERMINAL OF THE WHEEL IN QUESTION, (USE TABLE BELOW) AND GROUND. REPEAT PROCEDURE ON THE OTHER TERMINAL. RESISTANCE SHOULD BE INFINITE ON BOTH TERMINALS. IS IT?

NO → REPAIR CIRCUIT THAT SHOWED HIGH RESISTANCE OR OPEN.

YES ↓

(10)
- DVM STILL CONNECTED.
- SELECT THE 20 VOLT SCALE OF THE DVM.
- IGNITION "ON."
- REPEAT STEP 9. VOLTAGE SHOULD BE 0. IS IT?

NO → REPAIR SHORT TO GROUND IN CIRCUIT THAT DID NOT SHOW INFINITE RESISTANCE.

YES ↓

- INSPECT 24 WAY WORLD CONNECTOR TERMINALS FOR POOR CONNECTION.
- RE-CONNECT ALL CONNECTIONS AND REPEAT STEP 1.
- IF WHEEL SPEED DOES NOT MATCH REMAINING WHEEL SPEED INPUTS, REPLACE EBCM.

NO → REPAIR SHORT TO VOLTAGE IN CIRCUIT THAT INDICATED VOLTAGE ABOVE 0 VOLTS.

BODY TO ABS AND FUEL SENDER CONNECTOR	24 WAY WORLD CONNECTOR	CIRCUIT NUMBER
L/R WHEEL C2	PIN 12	884
L/R WHEEL C3	PIN 11	885
R/R WHEEL D1	PIN 7	882
R/R WHEEL D2	PIN 8	883

DIAGNOSTIC CHART DTC A036 — 1993 ACHIEVA GRAND AM AND SKYLARK

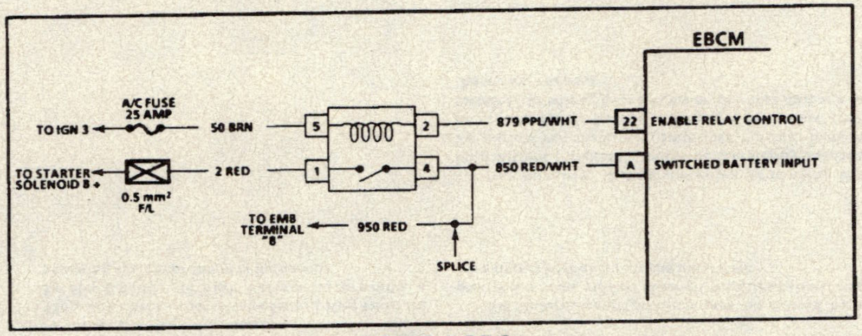

CODE A036

LOW SYSTEM VOLTAGE

Circuit Description:

This code is used to monitor the voltage level available to the ABS EBCM. If the voltage drops below 11 volts, full performance of the ABS system cannot be guaranteed. During ABS operation, there are several current requirements that will cause battery voltage to drop. Because of this, voltage is monitored prior to ABS to indicate good charging system condition and also during ABS when voltage may drop significantly. If voltage falls below a level that will provide adequate ABS performance, Code A036 will result.

Test Description: Number(s) below refer to circled number(s) on the diagnostic chart.

1. This test checks to see what voltage is readily available at terminal "A" of the EBCM. If voltage is representative of good charging system condition, a fault is not present.
2. This step isolates the EBCM to check the condition of the circuitry and charging system.
3. This step isolates the low voltage condition to high circuit resistance, or improper charging system operation.

Diagnostic Aids:

An "Intermittent" problem may be caused by a poor connection, rubbed through wire insulation, or a wire that is broken inside the insulation.

Any circuitry, that is suspected as causing the intermittent complaint, should be thoroughly checked for backed out terminals, improper mating, broken locks, improperly formed or damaged terminals, poor terminal to wiring connections or physical damage to the wiring harness.

While performing a voltage load test, if it is noted that only ignition voltage drops below acceptable voltage levels, CKT 650 should be checked for high resistance or an open condition.

DIAGNOSTIC CHART DTC A036 — 1993 ACHIEVA GRAND AM AND SKYLARK

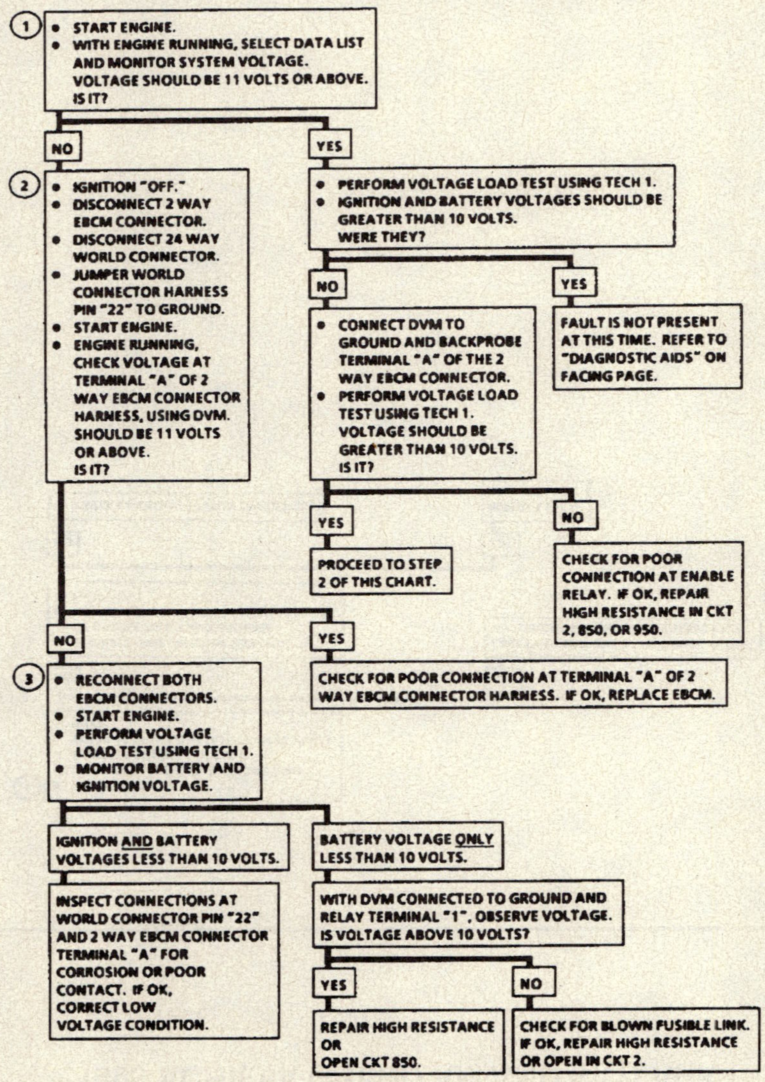

DIAGNOSTIC CHART DTC A037 — 1993 ACHIEVA GRAND AM AND SKYLARK

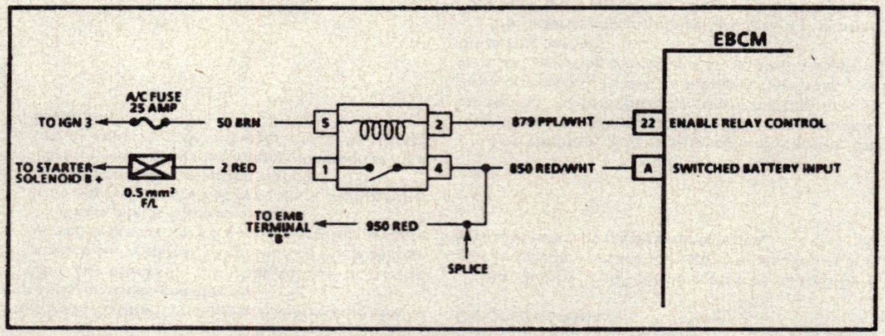

CODE A037
HIGH SYSTEM VOLTAGE

Circuit Description:

This code is designed to detect high vehicle voltage levels prior to any required motor movement (initialization or ABS). If excessive voltage exists, demagnetization of the motor magnets may occur, which would eventually affect or eliminate ABS performance.

Test Description: Number(s) below refer to circled number(s) on the diagnostic chart.

1. This step checks the voltage level being received by the EBCM. If high voltage is present, a charging system problem is indicated.

Diagnostic Aids:

An "Intermittent" problem may be caused by a poor connection, rubbed through wire insulation, or a wire that is broken inside the insulation.

Any circuitry, that is suspected as causing the intermittent complaint, should be thoroughly checked for backed out terminals, improper mating, broken locks, improperly formed or damaged terminals, poor terminal to wiring connections or physical damage to the wiring harness.

DIAGNOSTIC CHART DTC A037 — 1993 ACHIEVA GRAND AM AND SKYLARK

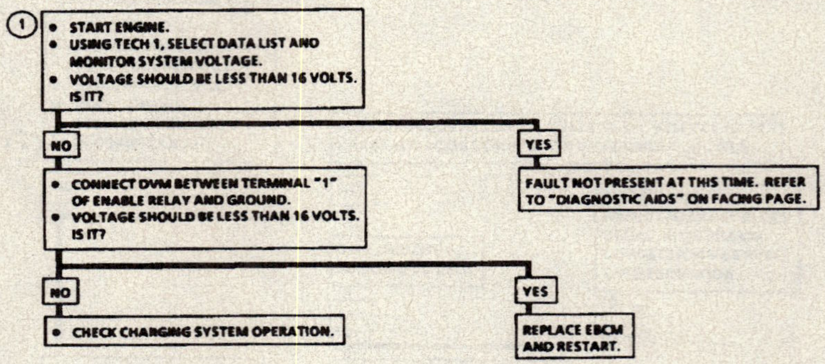

DIAGNOSTIC CHART DTC A038 — 1993 ACHIEVA GRAND AM AND SKYLARK

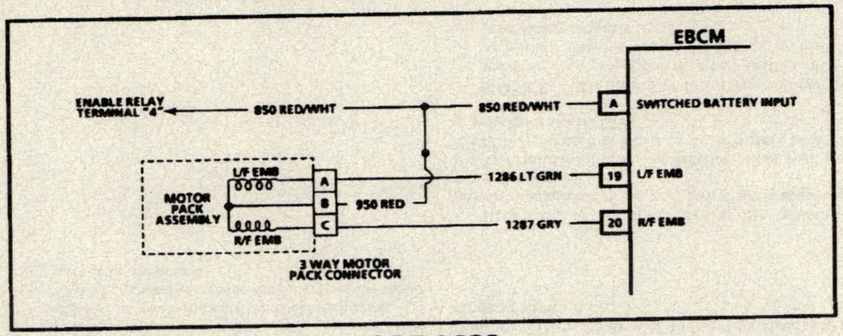

CODE A038

LEFT FRONT EMB WILL NOT HOLD MOTOR

Circuit Description:
Current is supplied to the EMB's through terminal "B" of the three way motor pack connector. When a release of the EMB is commanded, the EBCM grounds the EMB input and allows the EMB to be energized to release the motor.

This code is designed to detect slipping EMB's during initialization. When the release is commanded during initialization, the EMB is not energized and should hold the motor, resulting in sensed current of the motor being equal to commanded current (motor is at stall). If the motor is moving, sensed current will be less than stall current.

Test Description: Number(s) below refer to circled number(s) on the diagnostic chart.

1. This functionally checks the ability of the EMB to hold the motor. If the brake pedal moves during this test, it would indicate the EMB is slipping.
2. This step is used to release the tension of the motor pack prior to removal.

Diagnostic Aids:

An "Intermittent" fault in this code may result from a mechanical part of the system that sticks, binds, or slips.

Depending on the frequency of the failure, a physical inspection of the mechanical parts suspected may be necessary.

The static modulator test function of the Tech 1 or T-100 may be used to locate an intermittent problem associated with the EMB.

NOTICE: This test may fail if low voltage (10 volts or less) was present when initialization occurred. Ensure battery/charging system is functioning properly.

DIAGNOSTIC CHART DTC A038 — 1993 ACHIEVA GRAND AM AND SKYLARK

NOTICE: TESTING EMB WITH JUMPER WIRES TO VOLTAGE OR GROUND WILL DESTROY THE EMB.

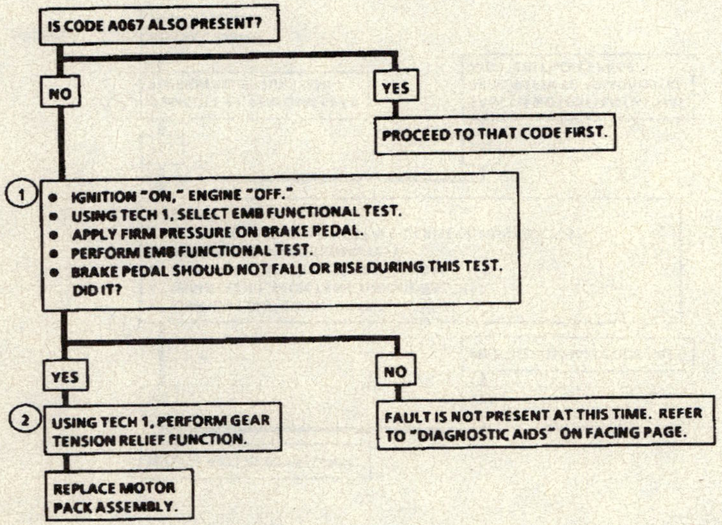

DIAGNOSTIC CHART DTC A041 — 1993 ACHIEVA GRAND AM AND SKYLARK

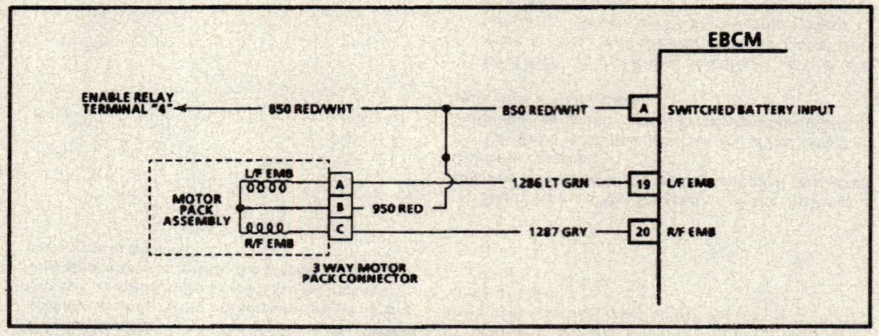

CODE A041

RIGHT FRONT EMB WILL NOT HOLD MOTOR

Circuit Description:

Current is supplied to the EMB's through terminal "B" of the three way motor pack connector. When a release of the EMB is commanded, the EBCM grounds the EMB input and allows the EMB to be energized to release the motor.

This code is designed to detect slipping EMB's during initialization. When the release is commanded during initialization, the EMB is not energized and should hold the motor, resulting in sensed current of the motor being equal to commanded current (motor is at stall). If the motor is moving, sensed current will be less than stall current.

Test Description: Number(s) below refer to circled number(s) on the diagnostic chart.

1. This functionally checks the ability of the EMB to hold the motor. If the brake pedal moves during this test, it would indicate the EMB is slipping.
2. This step is used to release the tension of the motor pack prior to removal.

Diagnostic Aids:

An "Intermittent" fault in this code may result from a mechanical part of the system that sticks, binds, or slips.

Depending on the frequency of the failure, a physical inspection of the mechanical parts suspected may be necessary.

The static modulator test function of the Tech 1 or T-100 may be used to locate an intermittent problem associated with the EMB.

NOTICE: This code may fail if low voltage (10 volts or less) was present when initialization occurred. Ensure battery/charging system is functioning properly.

DIAGNOSTIC CHART DTC A041 — 1993 ACHIEVA GRAND AM AND SKYLARK

NOTICE: TESTING END WITH JUMPER WIRES TO VOLTAGE OR GROUND WILL DESTROY THE EMB.

IS CODE A071 ALSO PRESENT?

- YES: PROCEED TO THAT CODE FIRST.
- NO: (1)
  - IGNITION "ON," ENGINE "OFF."
  - USING TECH 1, SELECT EMB FUNCTIONAL TEST.
  - APPLY FIRM PRESSURE ON BRAKE PEDAL.
  - PERFORM EMB FUNCTIONAL TEST.
  - BRAKE PEDAL SHOULD NOT FALL OR RISE DURING THIS TEST.

  DID IT?
  - YES: (2) USING TECH 1, PERFORM GEAR TENSION RELIEF FUNCTION. → REPLACE MOTOR PACK ASSEMBLY.
  - NO: FAULT IS NOT PRESENT AT THIS TIME. REFER TO "DIAGNOSTIC AIDS" ON FACING PAGE.

DIAGNOSTIC CHART DTC A042 — 1993 ACHIEVA GRAND AM AND SKYLARK

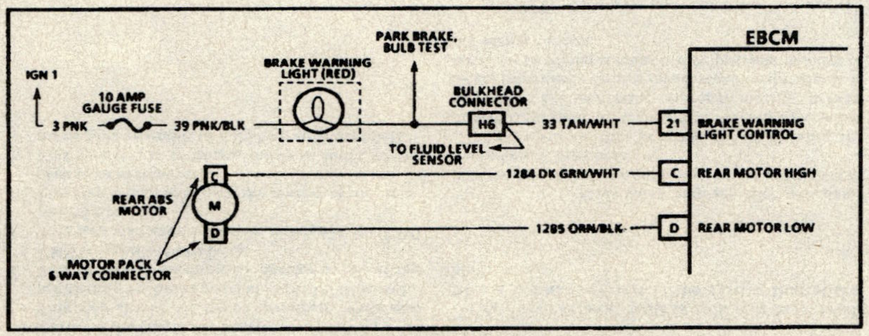

CODE A042

REAR AXLE ESB WILL NOT HOLD MOTOR

Circuit Description:

This code is designed to detect a slipping rear axle ESB. During initialization and braking, the rear motor is re-homed. If the ESB slips, the motor/piston will move. During the next key on initialization, a re-home of the motor verifies the motor/piston remained at the home position. If motor movement is detected, the ESB must be slipping.

Test Description: Number(s) below refer to circled number(s) on the diagnostic chart.

1. This step checks the rear axle ESB. A broken or defective ESB would result in the rear piston being back driven by hydraulic pressure, and wheel movement would result.
2. This step is used to release the tension of the motor pack prior to removal.

Diagnostic Aids:

An "Intermittent" fault in this code may result from a mechanical part of the system that sticks, binds, or slips.

Depending on the frequency of the failure, a physical inspection of the mechanical parts suspected may be necessary.

The static modulator test function of the Tech 1 or T-100 may be used to locate an intermittent problem associated with the ESB.

DIAGNOSTIC CHART DTC A042 — 1993 ACHIEVA GRAND AM AND SKYLARK

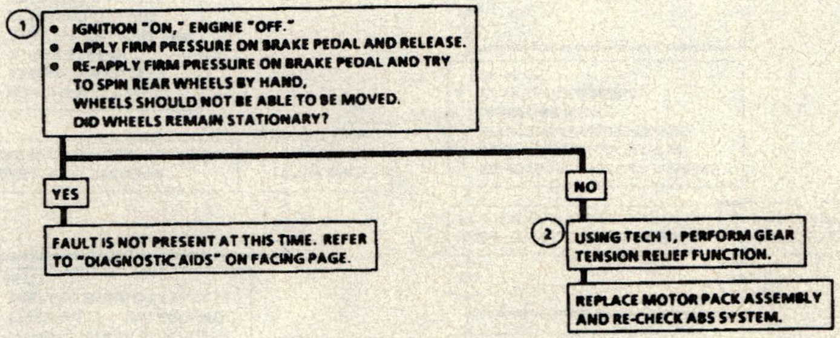

DIAGNOSTIC CHART DTC A044 — 1993 ACHIEVA GRAND AM AND SKYLARK

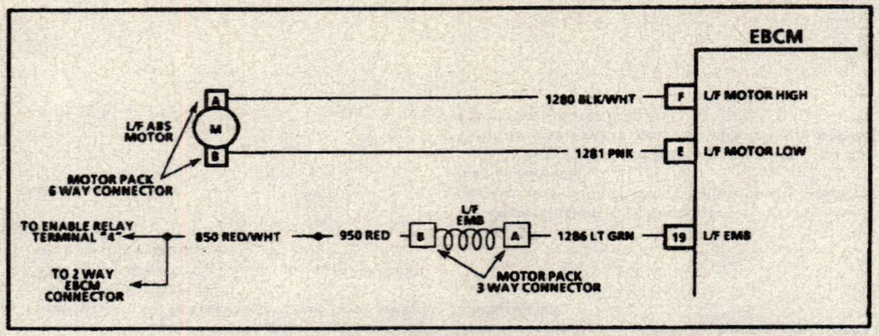

CODE A044

LEFT FRONT CHANNEL WILL NOT MOVE

Circuit Description:

Current is supplied to the EMB through terminal "B" of the three way motor pack connector. When a release of the EMB is commanded, the EBCM grounds the EMB input and allows the EMB to be energized to release the motor.

This code is designed to detect non-actuating EMB, stuck motor or a seized hydraulic modulator. When the release is commanded during initialization, the EMB should release the motor, resulting in sensed current being less than commanded current (motor is spinning freely). If the motor is not moving, sensed current will be equal to stall current.

Test Description: Number(s) below refer to circled number(s) on the diagnostic chart.

1. This step checks for motor movement while also positioning the brake pedal at the top of its travel.
2. This procedure functionally checks to see if the EMB is capable of releasing.
3. This step identifies a short to voltage in the EMB control circuitry.
4. This step is used to release the tension of the motor pack prior to removal.
5. This step serves to isolate the fault to either the motor pack assembly or ABS hydraulic modulator.

Diagnostic Aids:

An "Intermittent" fault in this code may result from a mechanical part of the system that sticks, binds, or slips.

Code A044 may set after modulator disassembly if the modulator pistons are positioned at the bottom of their bore.

Any circuitry, that is suspected as causing the intermittent complaint, should be thoroughly checked for backed out terminals, improper mating, broken locks, improperly formed or damaged terminals, poor terminal to wiring connections or physical damage to the wiring harness.

NOTICE: This code may fail if low voltage (10 volts or less) was present when initialization occurred. Ensure battery/charging system is functioning properly.

DIAGNOSTIC CHART DTC A044 — 1993 ACHIEVA GRAND AM AND SKYLARK

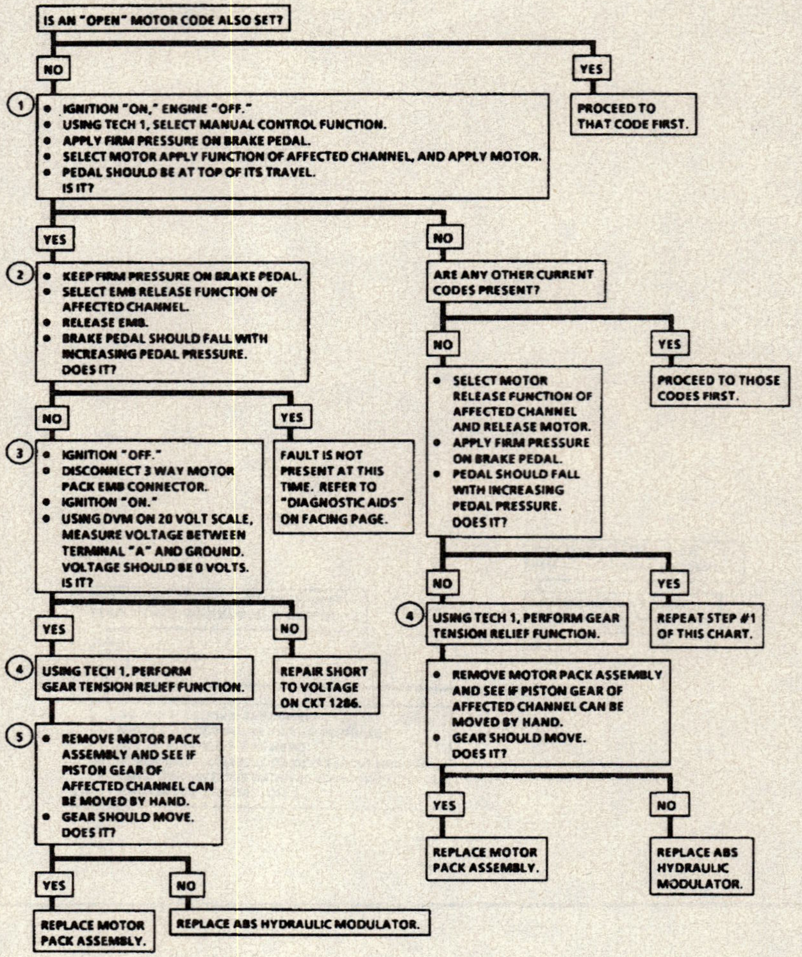

DIAGNOSTIC CHART DTC A045 — 1993 ACHIEVA GRAND AM AND SKYLARK

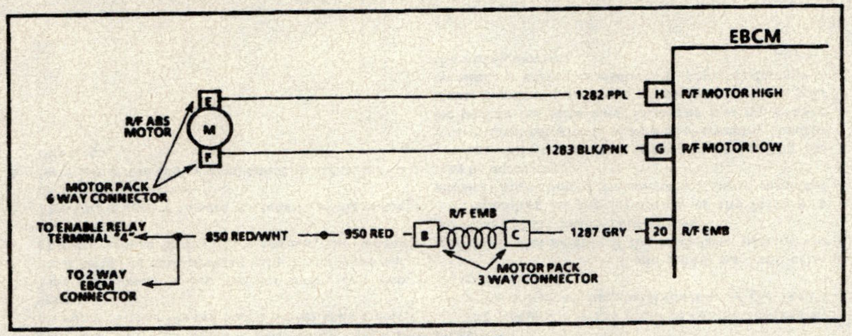

CODE A045

RIGHT FRONT CHANNEL WILL NOT MOVE

Circuit Description:

Current is supplied to the EMB through terminal "B" of the three way motor pack connector. When a release of the EMB is commanded, the EBCM grounds the EMB input and allows the EMB to be energized to release the motor.

This code is designed to detect non-actuating EMB, stuck motor or a seized hydraulic modulator. When the release is commanded during initialization, the EMB should release the motor, resulting in sensed current being less than commanded current (motor is spinning freely). If the motor is not moving, sensed current will be equal to stall current.

Test Description: Number(s) below refer to circled number(s) on the diagnostic chart.

1. This step checks for motor movement while also positioning the brake pedal at the top of its travel.
2. This procedure functionally checks to see if the EMB is capable of releasing.
3. This step is used to release the tension of the motor pack prior to removal.
4. This step serves to isolate the fault to either the motor pack assembly or ABS hydraulic modulator.

Diagnostic Aids:

An "Intermittent" fault in this code may result from a mechanical part of the system that sticks, binds, or slips.

The frequency of the problem can be checked by using the enhanced diagnostic function of the "Scan" tool

Code A045 may set after modulator disassembly if the modulator pistons are positioned at the bottom of their bore.

Any circuitry, that is suspected as causing the intermittent complaint, should be thoroughly checked for backed out terminals, improper mating, broken locks, improperly formed or damaged terminals, poor terminal to wiring connections or physical damage to the wiring harness.

NOTICE: This code may fail if low voltage (10 volts or less) was present when initialization occurred. Ensure battery/charging system is functioning properly.

DIAGNOSTIC CHART DTC A045 — 1993 ACHIEVA GRAND AM AND SKYLARK

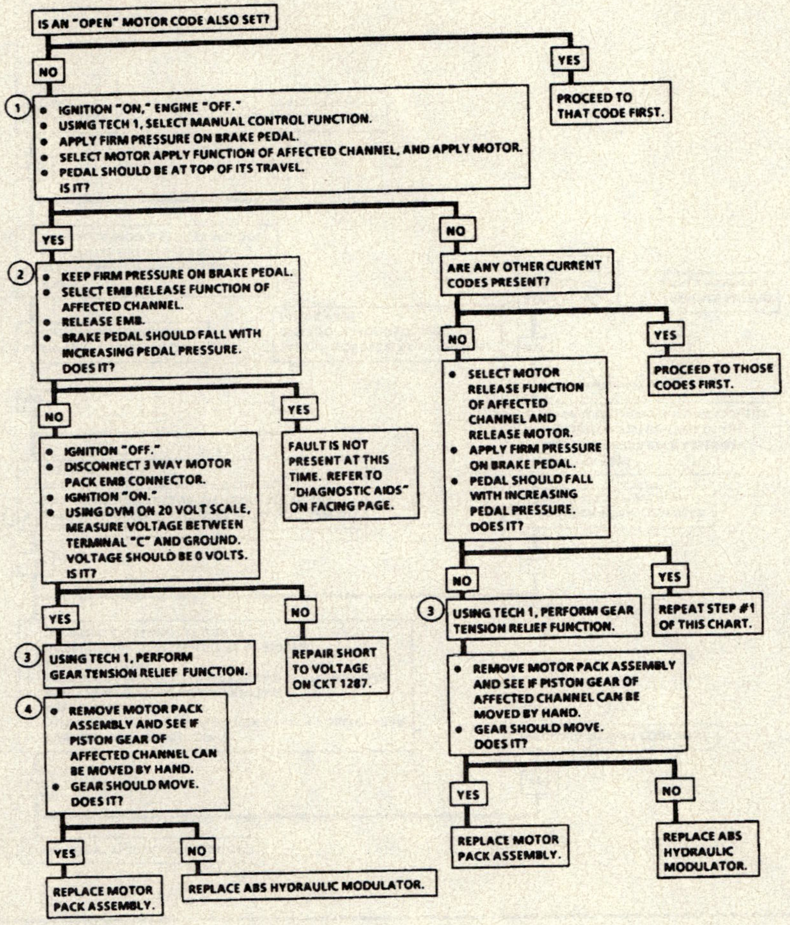

DIAGNOSTIC CHART DTC A046 — 1993 ACHIEVA GRAND AM AND SKYLARK

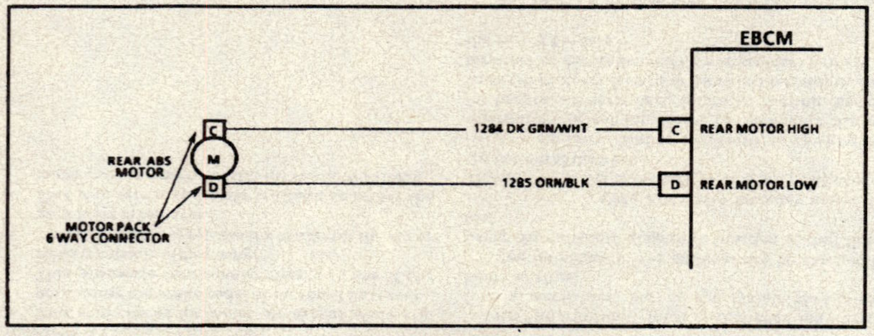

CODE A046

REAR AXLE CHANNEL WILL NOT MOVE

Circuit Description:

This code is designed to detect bound-up ESB's, stuck motor or a seized hydraulic modulator. When the release is commanded during initialization, the ESB should release the motor, resulting in sensed current being less than commanded current (motor is spinning freely). If the motor is not moving, sensed current will be equal to stall current.

Test Description: Number(s) below refer to circled number(s) on the diagnostic chart.

1. This checks the ability of the motor to move, applying hydraulic pressure to the wheel cylinders. A frozen piston or gear would result in no apply of rear brakes and a wheel that freely spins.
2. This step checks to see that the motor can also move to the release position. If a motor were not able to release hydraulic pressure, the wheel would not spin.
3. This step is used to release the tension of the motor pack prior to removal.
4. Isolation of the faulty component is identified in this step.

Diagnostic Aids:

An "Intermittent" fault in this code may result from a mechanical part of the system that sticks or binds.

The frequency of the problem can be checked by using the enhanced diagnostic function of the Tech 1 "Scan" tool.

Code A046 may set after modulator disassembly if the modulator pistons are positioned at the bottom of their bore.

Depending on the frequency of the failure, a physical inspection of the mechanical parts suspected may be necessary.

Any circuitry, that is suspected as causing the intermittent complaint, should be thoroughly checked for backed out terminals, improper mating, broken locks, improperly formed or damaged terminals, poor terminal to wiring connections or physical damage to the wiring harness.

DIAGNOSTIC CHART DTC A046 — 1993 ACHIEVA GRAND AM AND SKYLARK

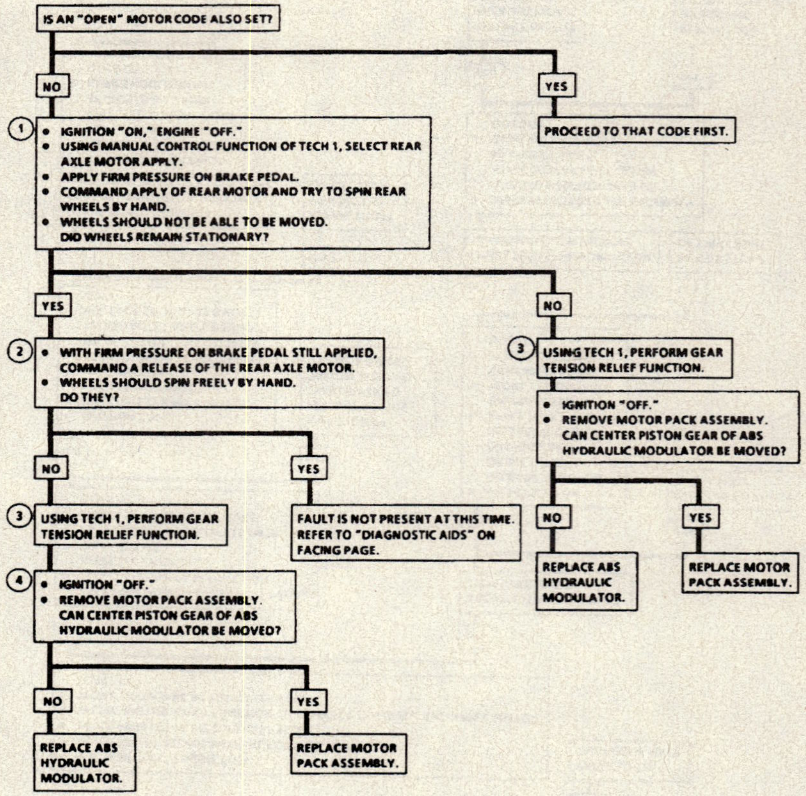

DIAGNOSTIC CHART DTC A047 — 1993 ACHIEVA GRAND AM AND SKYLARK

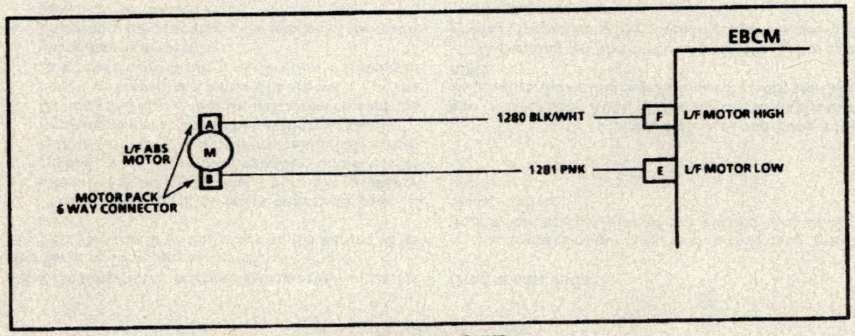

CODE A047

LEFT FRONT MOTOR FREE SPINS

Circuit Description:

This code is designed to detect a stripped nut or gear assembly during initialization. During the homing sequence, the piston should reach the top of the bore resulting in a stalled motor. If this does not occur, the motor must be spinning with little or no resistance, indicating a nut/screw or gear failure.

Test Description: Number(s) below refer to circled number(s) on the diagnostic chart.

1. This step positions the motor at the bottom of its travel.
2. This checks to see if the nut or gear is stripped. If either were stripped, the hydraulic pressure applied from the master cylinder would not be able to be overcome by the hydraulic ball screw, resulting in a brake pedal that would not rise.
3. During a full apply, once the ball screw reaches the top of its travel, the motor should stall. If the motor continues to run, an indication of a stripped nut or gear is present.
4. This step is used to release the tension of the motor pack prior to removal.

Diagnostic Aids:

An "Intermittent" fault in this code may result from a mechanical part of the system that sticks, binds, or slips.

If code only fails once and Code A056 also fails see Code A056. If intermittent and enhanced diagnostics shows this code fails during ABS, see Code A056.

Depending on the frequency of the failure, a physical inspection of the mechanical parts suspected may be necessary.

DIAGNOSTIC CHART DTC A047 — 1993 ACHIEVA GRAND AM AND SKYLARK

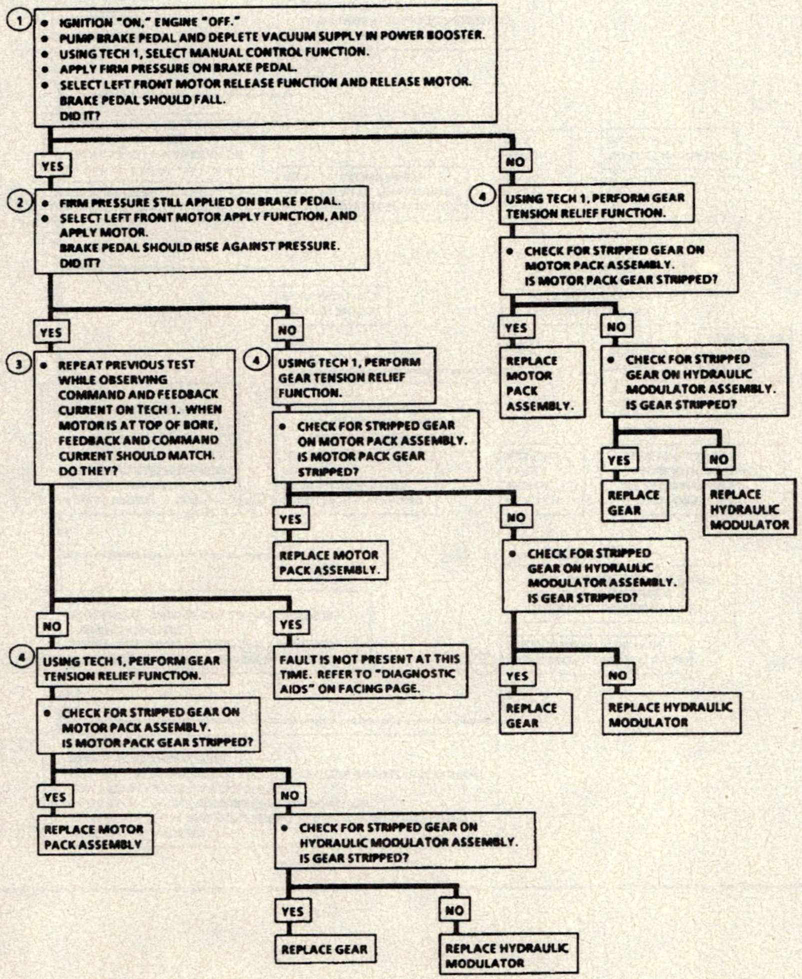

DIAGNOSTIC CHART DTC A048 — 1993 ACHIEVA GRAND AM AND SKYLARK

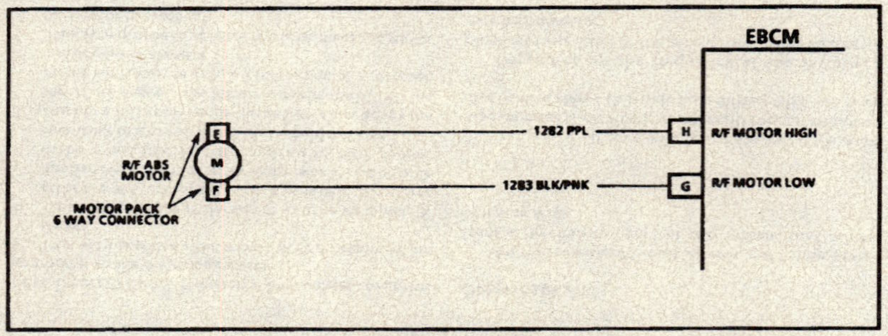

CODE A048

RIGHT FRONT MOTOR FREE SPINS

Circuit Description:

This code is designed to detect a stripped nut or gear assembly during initialization. During the homing sequence, the piston should reach the top of the bore resulting in a stalled motor. If this does not occur, the motor must be spinning with little or no resistance, indicating a nut/screw or gear failure.

Test Description: Number(s) below refer to circled number(s) on the diagnostic chart.

1. This step positions the motor at the bottom of its travel.
2. This checks to see if the nut or gear is stripped. If either were stripped, the hydraulic pressure applied from the master cylinder would not be able to be overcome by the hydraulic ball screw, resulting in a bake pedal that would not rise.
3. During a full apply, once the ball screw reaches the top of its travel, the motor should stall. If the motor continues to run, an indication of a stripped nut or gear is present.
4. This step is used to release the tension of the motor pack prior to removal.

Diagnostic Aids:

An "Intermittent" fault in this code may result from a mechanical part of the system that sticks, binds, or slips.

If code only fails once and Code A061 also fails see Code A061. If intermittent and enhanced diagnostics shows this code fails during ABS, see Code A061.

Depending on the frequency of the failure, a physical inspection of the mechanical parts suspected may be necessary.

DIAGNOSTIC CHART DTC A048 — 1993 ACHIEVA GRAND AM AND SKYLARK

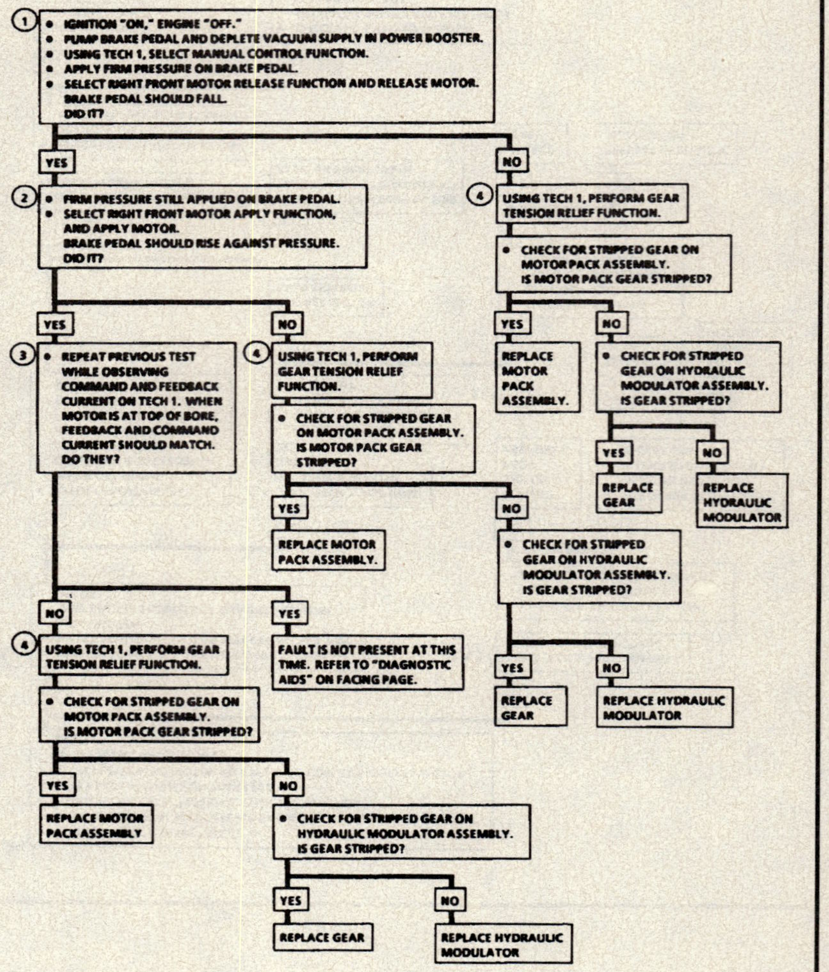

DIAGNOSTIC CHART DTC A051 — 1993 ACHIEVA GRAND AM AND SKYLARK

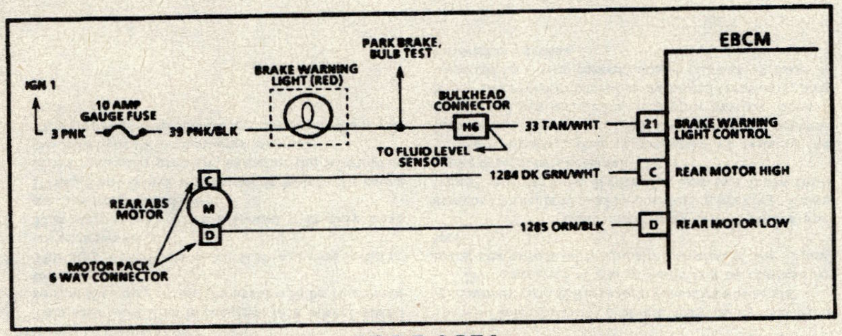

CODE A051

REAR AXLE MOTOR FREE SPINS

Circuit Description:

This code is designed to detect a stripped nut or gear assembly during initialization. During the homing sequence, the piston should reach the top of the bore resulting in a stalled motor. If this does not occur, the motor must be spinning with little or no resistance, indicating a nut/screw or gear failure.

Test Description: Number(s) below refer to circled number(s) on the diagnostic chart.

1. This step checks for a slipping nut or gear. If either were to slip, it would result in a lack of hydraulic pressure being applied at the rear wheels, causing the wheels to be able to be moved.
2. During a full apply, once the ball screw reaches the top of its travel, the motor should stall. If the motor continues to run, an indication of a stripped nut or gear is present.
3. This step is used to release the tension of the motor pack prior to removal.

Diagnostic Aids:

An "Intermittent" fault in this code may result from a mechanical part of the system that sticks, binds, or slips.

The frequency of the problem can be checked by using the enhanced diagnostic function of the "Scan" tool

If code only fails once and Code A064 also fails, see Code A064. If intermittent and enhanced diagnostics show this code fails during ABS, see Code A064.

Depending on the frequency of the failure, a physical inspection of the mechanical parts suspected may be necessary.

DIAGNOSTIC CHART DTC A051 — 1993 ACHIEVA GRAND AM AND SKYLARK

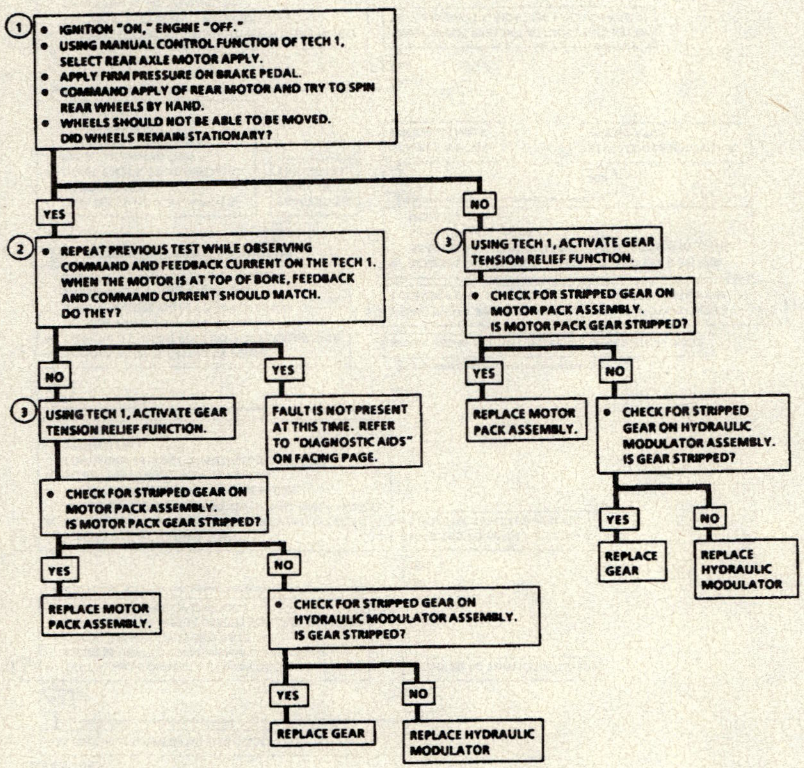

DIAGNOSTIC CHART DTC A052 — 1993 ACHIEVA GRAND AM AND SKYLARK

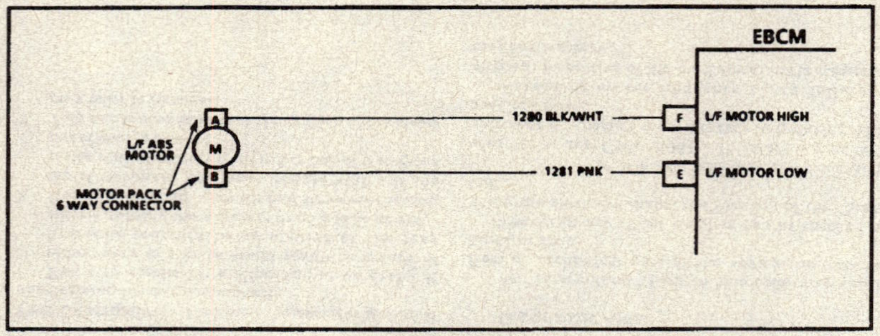

CODE A052

LEFT FRONT CHANNEL IN RELEASE TOO LONG

Circuit Description:

This code will diagnose a motor that is energized longer than expected. This could occur if a wheel speed sensor is faulty, the motor does not turn, the left front solenoid mechanically fails open, or the motor wires are crossed.

Test Description: Number(s) below refer to circled number(s) on the diagnostic chart.

1. This step identifies a problem in a wheel speed sensor that may cause the system to be in release too long.
2. This step identifies a motor fault as being failed or wired incorrectly.
3. This step checks for a solenoid that may have mechanically failed open.
4. If Step 3 has failed, this serves to isolate the cause of the hydraulic problem, to either the solenoid or the ABS hydraulic modulator assembly.
5. This step is used to release the tension of the motor pack.

Diagnostic Aids:

An "Intermittent" problem may be caused by a mechanical part of the system that sticks or binds.

The frequency of the problem can be checked by using the enhanced diagnostic function of the "Scan" tool

Code A052 may fail if on ice and steering wheel is turned to lock during braking. Using "scan" tool perform Hydraulic test to ensure total brake system is functional.

Any circuitry, that is suspected as causing the intermittent complaint, should be thoroughly checked for backed out terminals, improper mating, broken locks, improperly formed or damaged terminals, poor terminal to wiring connections or physical damage to the wiring harness.

DIAGNOSTIC CHART DTC A052 — 1993 ACHIEVA GRAND AM AND SKYLARK

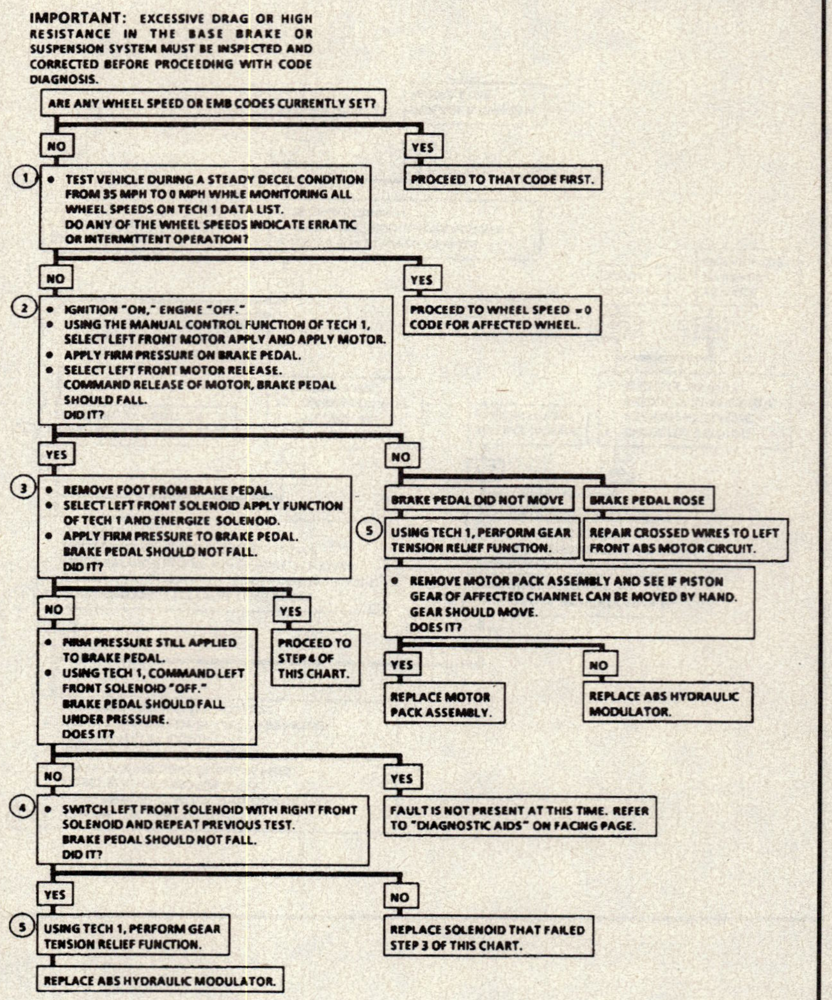

DIAGNOSTIC CHART DTC A053 — 1993 ACHIEVA GRAND AM AND SKYLARK

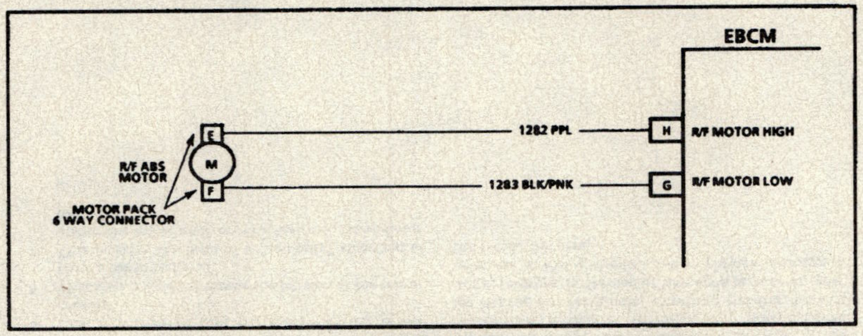

CODE A053

RIGHT FRONT CHANNEL IN RELEASE TOO LONG

Circuit Description:

This code will diagnose a motor that is energized longer than expected. This could occur if a wheel speed sensor is faulty, the motor does not turn, the right front solenoid mechanically fails open, or the motor wires are crossed.

Test Description: Number(s) below refer to circled number(s) on the diagnostic chart.

1. This step identifies a problem in a wheel speed sensor that may cause the system to be in release too long.
2. This step identifies a motor fault as being failed or wired incorrectly.
3. This step checks for a solenoid that may have mechanically failed open.
4. If Step 3 has failed, this serves to isolate the cause of the hydraulic problem, to either the solenoid or the ABS hydraulic modulator assembly.
5. This step is used to release the tension of the motor pack prior to removal.

Diagnostic Aids:

An "Intermittent" problem may be caused by a mechanical part of the system that sticks or binds.

The frequency of the problem can be checked by using the enhanced diagnostic function of the "Scan" tool

Code A053 may fail on ice if steering wheel is turned to lock during braking. Using "Scan" tool perform Hydraulic test to ensure total brake system is functional.

Any circuitry, that is suspected as causing the intermittent complaint, should be thoroughly checked for backed out terminals, improper mating, broken locks, improperly formed or damaged terminals, poor terminal to wiring connections or physical damage to the wiring harness.

DIAGNOSTIC CHART DTC A053 — 1993 ACHIEVA GRAND AM AND SKYLARK

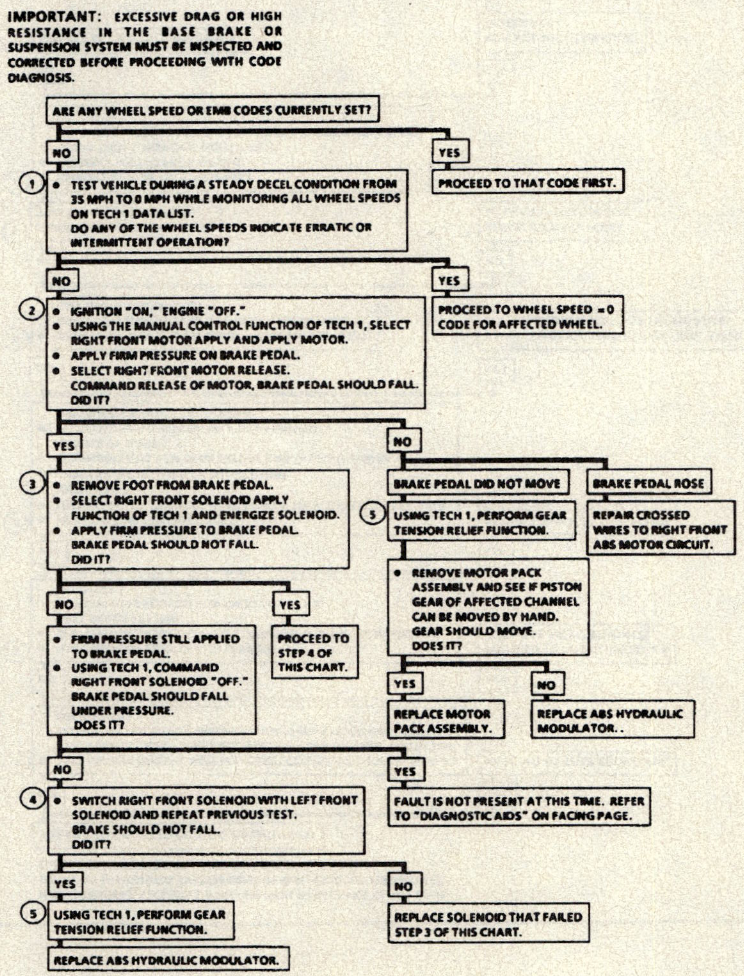

DIAGNOSTIC CHART DTC A054 — 1993 ACHIEVA GRAND AM AND SKYLARK

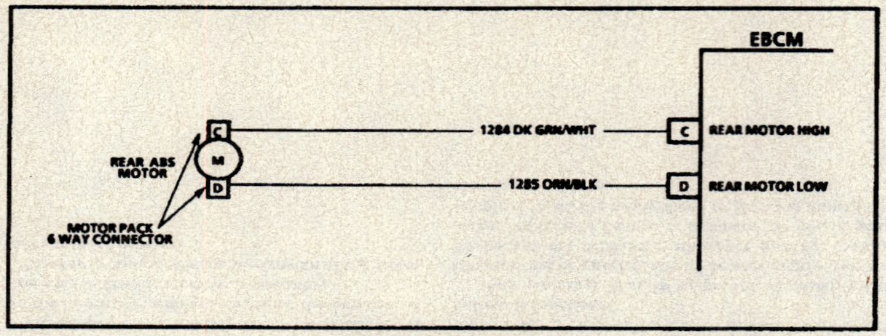

CODE A054

REAR AXLE IN RELEASE TOO LONG

Circuit Description:

This code will diagnose a motor that is energized longer than expected. This could occur if a wheel speed sensor is faulty, the motor does not turn, or the motor wires are crossed.

Test Description: Numbers below refer to circled numbers on the diagnostic chart.

1. This step identifies a problem in a wheel speed sensor that may cause the system to be in release too long.
2. This checks for a wheel that may stick or bind because of a mechanical fault.
3. This checks to see if the motor is capable of moving and applying the hydraulic piston for the rear wheels.
4. This step insures that the motor wiring is not crossed.
5. This step is used to release the tension of the motor pack prior to removal.
6. This isolates the fault of a "no-apply" situation to either the motor pack or ABS hydraulic modulator assembly.

Diagnostic Aids:

An "Intermittent" problem may be caused by a mechanical part of the system that sticks or binds.

Using "Scan" tool perform hydraulic test to ensure total brake system is functional.

Any circuitry, that is suspected as causing the intermittent complaint, should be thoroughly checked for backed out terminals, improper mating, broken locks, improperly formed or damaged terminals, poor terminal to wiring connections or physical damage to the wiring harness.

DIAGNOSTIC CHART DTC A054 — 1993 ACHIEVA GRAND AM AND SKYLARK

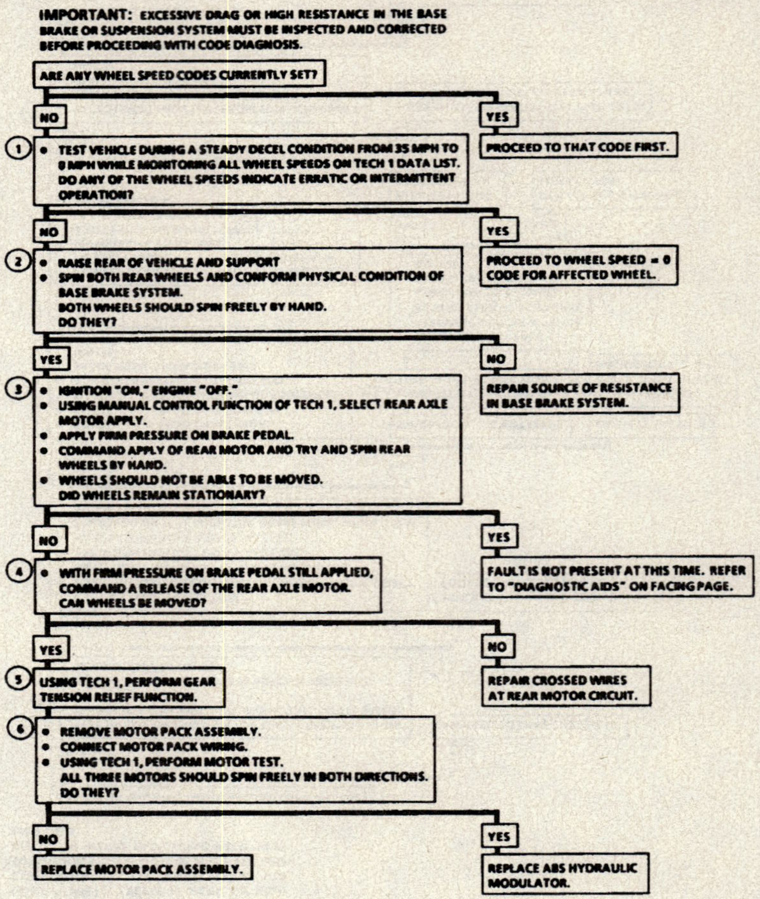

DIAGNOSTIC CHART DTC A055 — 1993 ACHIEVA GRAND AM AND SKYLARK

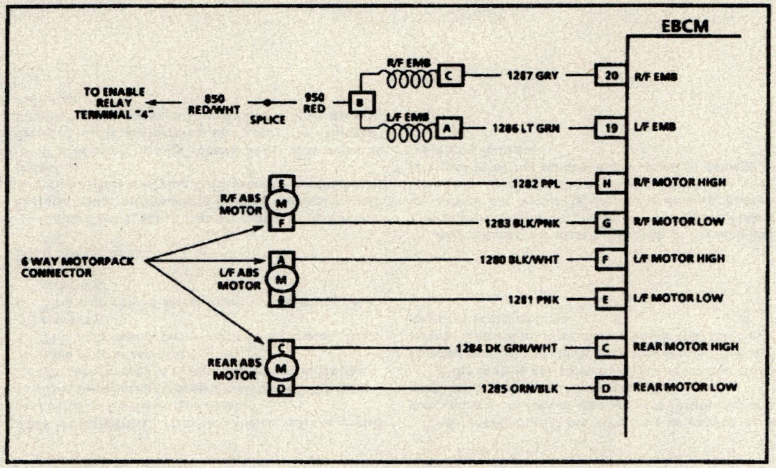

CODE A055

MOTOR DRIVER FAULT DETECTED

Circuit Description:

This code identifies a fault detected by the MDI custom IC. It also ensures the cause of the fault is not a result of a problem with the enable relay under a Code A055 failure. The MDI fault tests will be run to identify which circuit has failed, and may also cause additional codes to set to pinpoint the problem circuit.

Test Description: Number(s) below refer to circled number(s) on the diagnostic chart.

1. This insures that the Code A055 fault was not set due to a motor or EMB fault.
2. This checks to see if the fault is still present.
3. This step insures that Code A055 was not due to poor terminal contact at the motor pack connector.

Diagnostic Aids:

An "Intermittent" problem may be caused by a poor connection, rubbed through wire insulation, or a wire that is broken inside the insulation.

Any circuitry, that is suspected as causing the intermittent complaint, should be thoroughly checked for backed out terminals, improper mating, broken locks, improperly formed or damaged terminals, poor terminal to wiring connections or physical damage to the wiring harness.

DIAGNOSTIC CHART DTC A055 — 1993 ACHIEVA GRAND AM AND SKYLARK

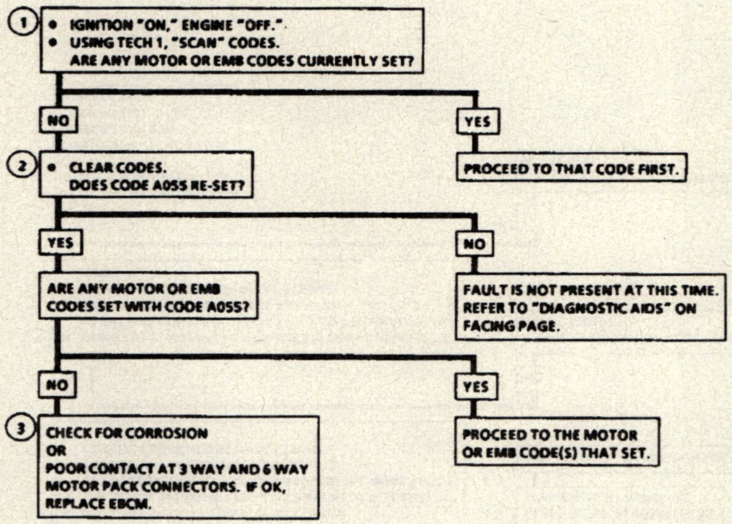

DIAGNOSTIC CHART DTC A056 — 1993 ACHIEVA GRAND AM AND SKYLARK

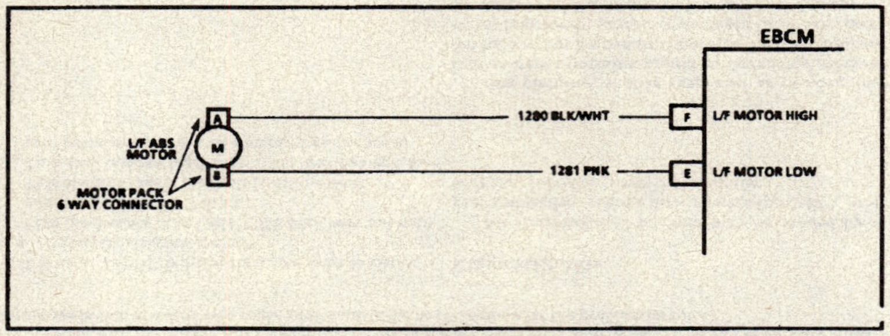

CODE A056

LEFT FRONT MOTOR CIRCUIT OPEN

Circuit Description:

This code identifies a motor that cannot be energized due to an open in its circuitry. This fault will not allow proper front ABS operation.

Test Description: Number(s) below refer to circled number(s) on the diagnostic chart.

1. This step checks for proper resistance of the motor.
2. This step is used to release the tension of the motor pack prior to removal.
3. This step checks for an open in the motor "high" circuitry.
4. This step checks for an open in the motor "low" circuitry.

An "Intermittent" problem may be caused by a poor connection, rubbed through wire insulation, or a wire that is broken inside the insulation.

If the fault is not current, wiggle the wires of the affected channel and check if the code resets. This will help to pinpoint an intermittent problem in the motor circuitry or connections.

Diagnostic Aids:

Using Tech 1, select manual control function, and exercise motor movement of affected channel in both directions while applying light pressure on the brake pedal.

If erratic or "jumpy" brake pedal movement is detected while performing an "apply" or "release" function of the motor, an intermittent fault may be indicated.

Any circuitry, that is suspected as causing the intermittent complaint, should be thoroughly checked for backed out terminals, improper mating, broken locks, improperly formed or damaged terminals, poor terminal to wiring connections or physical damage to the wiring harness.

DIAGNOSTIC CHART DTC A056 — 1993 ACHIEVA GRAND AM AND SKYLARK

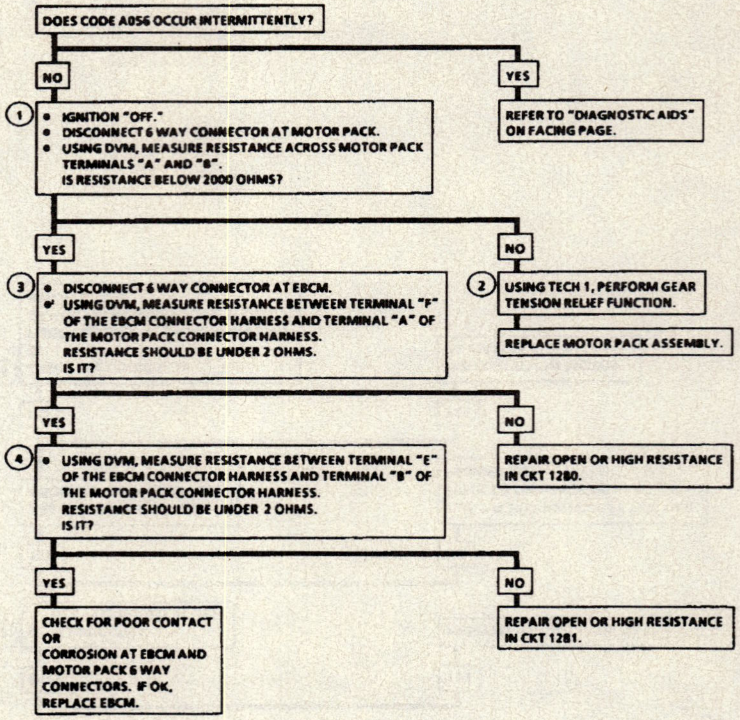

DIAGNOSTIC CHART DTC A057 — 1993 ACHIEVA GRAND AM AND SKYLARK

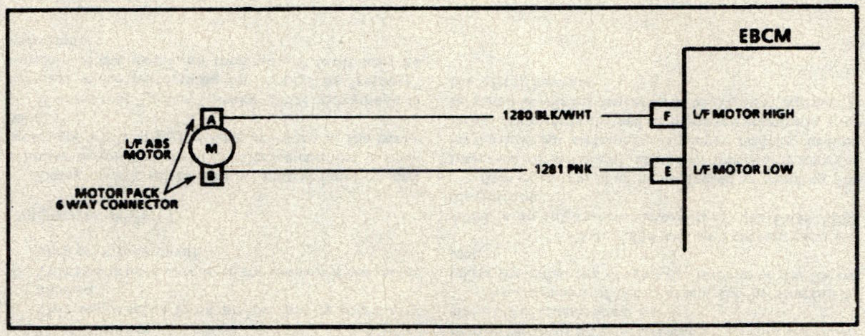

CODE A057

LEFT FRONT MOTOR CIRCUIT SHORTED TO GROUND

Circuit Description:

This code identifies a motor circuit that is shorted to ground. This fault will not allow the motor to be controlled at the commanded current rate or will cause the driver circuit to allow current directly to ground.

Test Description: Number(s) below refer to circled number(s) on the diagnostic chart.

1. This step checks for a short to ground in the motor "high" circuitry.
2. This step checks for a short to ground in the motor "low" circuitry.
3. This step checks for a motor that is internally shorted to ground.
4. This step is used to release the tension of the motor pack prior to removal.

An "Intermittent" problem may be caused by a poor connection, rubbed through wire insulation, or a wire that is broken inside the insulation.

If the fault is not current, wiggle the wires of the affected channel and check if the code resets. This will help to pinpoint an intermittent problem in the motor circuitry or connections.

Diagnostic Aids:

Using Tech 1, select manual control function, and exercise motor movement of affected channel in both directions while applying light pressure on the brake pedal.

If erratic or "jumpy" brake pedal movement is detected while performing an "apply" or "release" function of the motor, an intermittent fault may be indicated.

Any circuitry, that is suspected as causing the intermittent complaint, should be thoroughly checked for backed out terminals, improper mating, broken locks, improperly formed or damaged terminals, poor terminal to wiring connections or physical damage to the wiring harness.

DIAGNOSTIC CHART DTC A057 — 1993 ACHIEVA GRAND AM AND SKYLARK

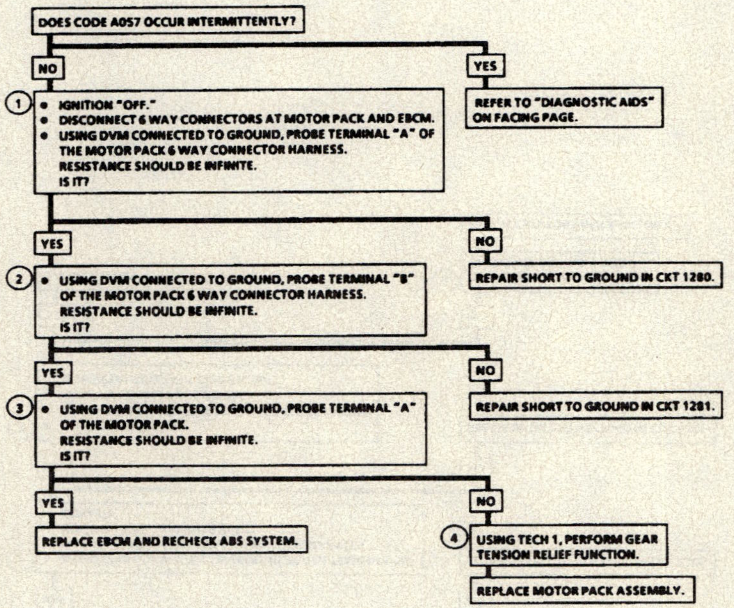

DIAGNOSTIC CHART DTC A058 — 1993 ACHIEVA GRAND AM AND SKYLARK

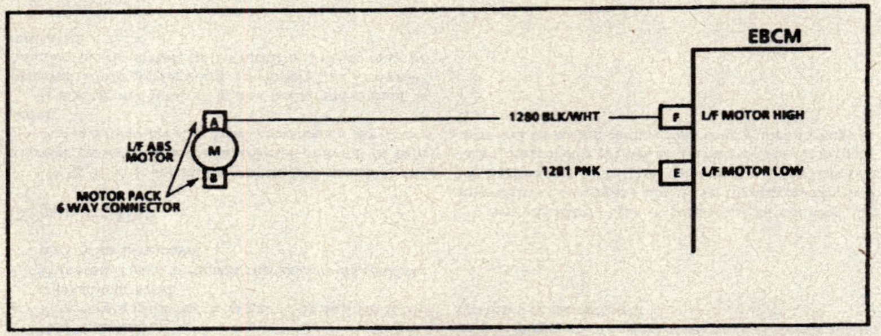

CODE A058

LEFT FRONT MOTOR CIRCUIT SHORTED TO BATTERY OR MOTOR SHORTED

Circuit Description:

This code identifies a motor circuit that is shorted to battery or a motor that has low or no resistance. This fault will not allow the motor to be controlled at the commanded current rate or will cause the motor to turn in the opposite direction, or not at all.

Test Description: Number(s) below refer to circled number(s) on the diagnostic chart.

1. This step checks for a short to voltage in the motor "high" circuitry.
2. This step checks for a short to voltage in the motor "low" circuitry.
3. This step checks for a motor that is internally shorted.
4. This step is used to release the tension of the motor pack prior to removal.

Diagnostic Aids:

Using Tech 1, select manual control function, and exercise motor movement of affected channel in both directions while applying light pressure on the brake pedal.

If erratic or "jumpy" brake pedal movement is detected while performing an "apply" or "release" function of the motor, an intermittent fault may be indicated.

An "Intermittent" problem may be caused by a poor connection, rubbed through wire insulation, or a wire that is broken inside the insulation.

If the fault is not current, wiggle the wires of the affected channel and check if the code resets. This will help to pinpoint an intermittent problem in the motor circuitry or connections.

The frequency of the problem can be checked by using the enhanced diagnostic function of the "Scan" tool.

Code A038 sets as a current code and A058 is an infrequent history code, see Code A038 diagnostics.

Any circuitry, that is suspected as causing the intermittent complaint, should be thoroughly checked for backed out terminals, improper mating, broken locks, improperly formed or damaged terminals, poor terminal to wiring connections or physical damage to the wiring harness.

DIAGNOSTIC CHART DTC A058 — 1993 ACHIEVA GRAND AM AND SKYLARK

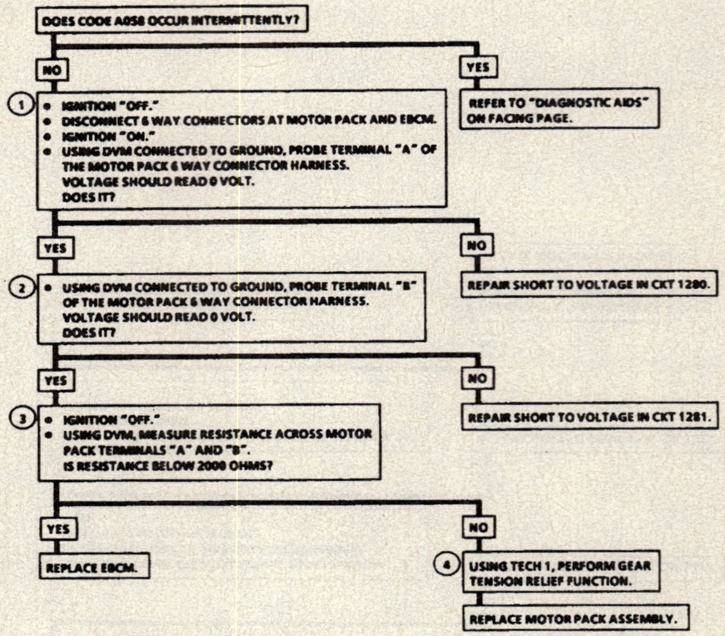

DIAGNOSTIC CHART DTC A061 — 1993 ACHIEVA GRAND AM AND SKYLARK

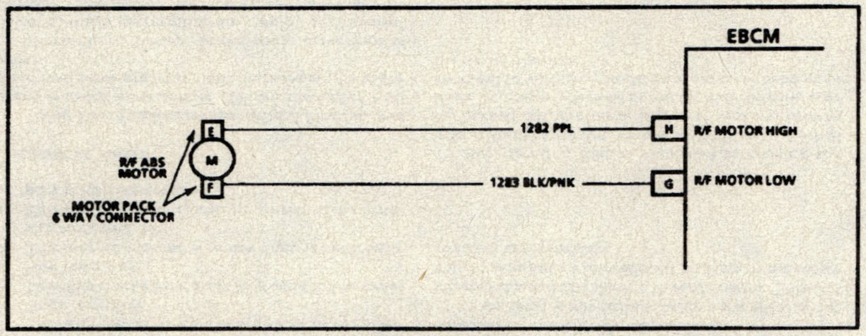

CODE A061
RIGHT FRONT MOTOR CIRCUIT OPEN

Circuit Description:

This code identifies a motor that cannot be energized due to an open in its circuitry. This fault will not allow proper front ABS operation.

Test Description: Number(s) below refer to circled number(s) on the diagnostic chart.

1. This step checks for proper resistance of the motor.
2. This step is used to release the tension of the motor pack prior to removal.
3. This step checks for an open in the motor "high" circuitry.
4. This step checks for an open in the motor "low" circuitry.

An "Intermittent" problem may be caused by a poor connection, rubbed through wire insulation, or a wire that is broken inside the insulation.

If the fault is not current, wiggle the wires of the affected channel and check if the code resets. This will help to pinpoint an intermittent problem in the motor circuitry or connections.

Diagnostic Aids:

Using Tech 1, select manual control function, and exercise motor movement of affected channel in both directions while applying light pressure on the brake pedal.

If erratic or "jumpy" brake pedal movement is detected while performing an "apply" or "release" function of the motor, an intermittent fault may be indicated.

Any circuitry, that is suspected as causing the intermittent complaint, should be thoroughly checked for backed out terminals, improper mating, broken locks, improperly formed or damaged terminals, poor terminal to wiring connections or physical damage to the wiring harness.

DIAGNOSTIC CHART DTC A061 — 1993 ACHIEVA GRAND AM AND SKYLARK

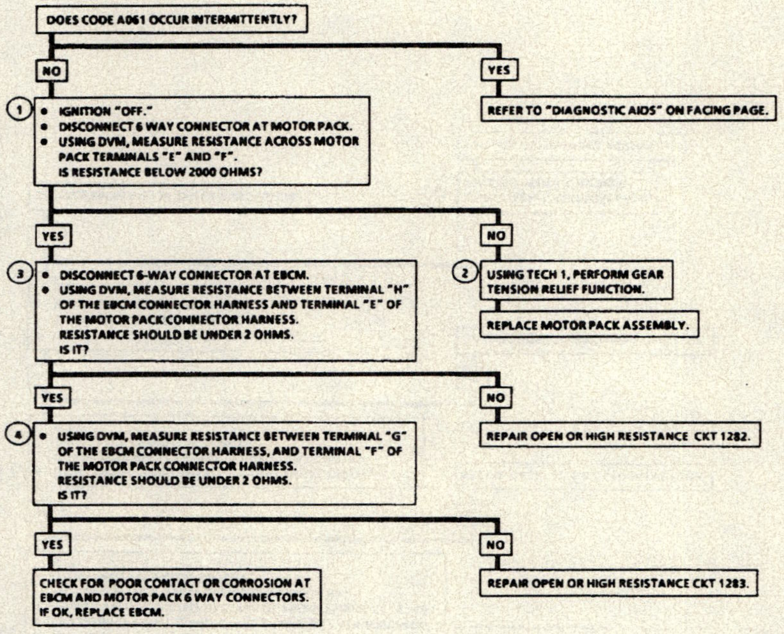

DIAGNOSTIC CHART DTC A062 — 1993 ACHIEVA GRAND AM AND SKYLARK

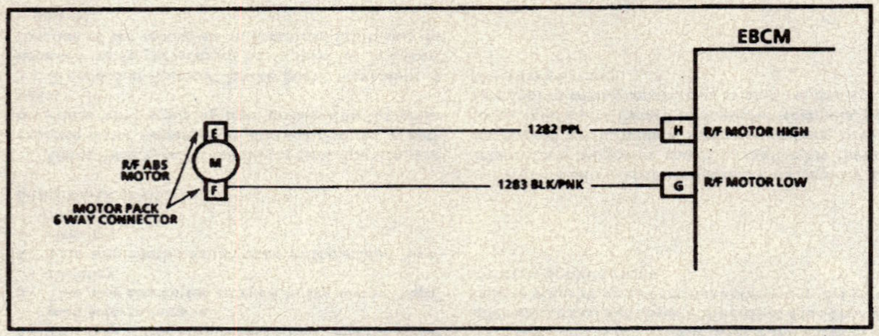

CODE A062

RIGHT FRONT MOTOR CIRCUIT SHORTED TO GROUND

Circuit Description:

This code identifies a motor circuit that is shorted to ground. This fault will not allow the motor to be controlled at the commanded current rate or will cause the motor driver circuit to allow current directly to ground.

Test Description: Number(s) below refer to circled number(s) on the diagnostic chart.

1. This step checks for a short to ground in the motor "high" circuitry.
2. This step checks for a short to ground in the motor "low" circuitry.
3. This step checks for a motor that is internally shorted to ground.
4. This step is used to release the tension of the motor pack prior to removal.

An "Intermittent" problem may be caused by a poor connection, rubbed through wire insulation, or a wire that is broken inside the insulation.

If the fault is not current, wiggle the wires of the affected channel and check if the code resets. This will help to pinpoint an intermittent problem in the motor circuitry or connections.

Diagnostic Aids:

Using Tech 1, select manual control function, and exercise motor movement of affected channel in both directions while applying light pressure on the brake pedal.

If erratic or "jumpy" brake pedal movement is detected while performing an "apply" or "release" function of the motor, an intermittent fault may be indicated.

Any circuitry, that is suspected as causing the intermittent complaint, should be thoroughly checked for backed out terminals, improper mating, broken locks, improperly formed or damaged terminals, poor terminal to wiring connections or physical damage to the wiring harness.

DIAGNOSTIC CHART DTC A062 — 1993 ACHIEVA GRAND AM AND SKYLARK

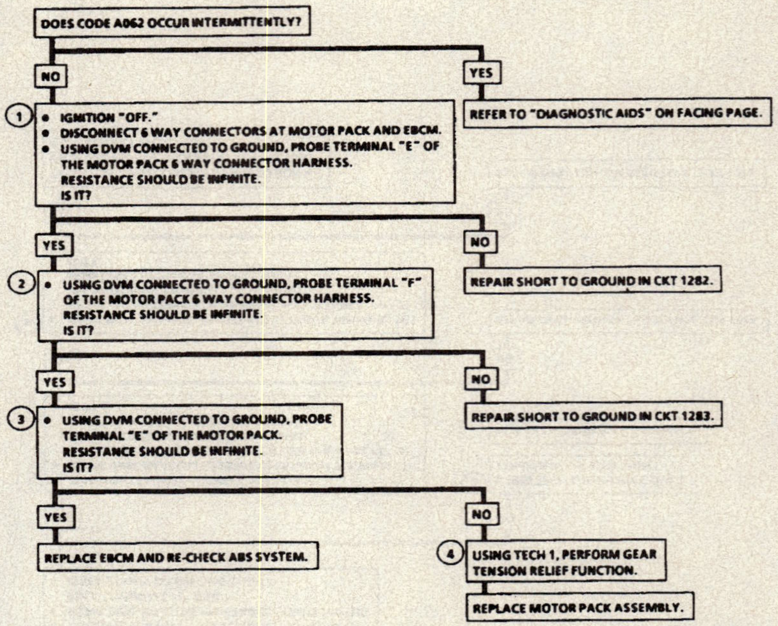

DIAGNOSTIC CHART DTC A063 — 1993 ACHIEVA GRAND AM AND SKYLARK

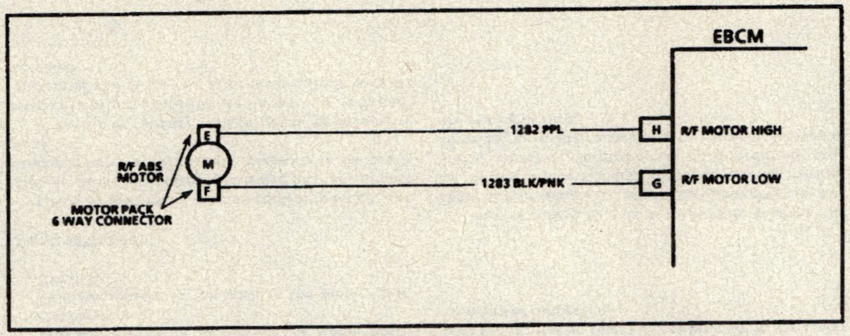

CODE A063

RIGHT FRONT MOTOR CIRCUIT SHORTED TO BATTERY OR MOTOR SHORTED

Circuit Description:

This code identifies a motor circuit that is shorted to battery or a motor that has low or no resistance. This fault will not allow the motor to be controlled at the commanded current rate or will cause the motor to turn in the opposite direction or not at all.

Test Description: Number(s) below refer to circled number(s) on the diagnostic chart.

1. This step checks for a short to voltage in the motor "high" circuitry.
2. This step checks for a short to voltage in the motor "low" circuitry.
3. This step checks for a motor that is internally shorted.
4. This test is used to release the tension of the motor pack prior to removal.

Diagnostic Aids:

Using Tech 1, select manual control function, and exercise motor movement of affected channel in both directions while applying light pressure on the brake pedal.

If erratic or "jumpy" brake pedal movement is detected while performing an "apply" or "release" function of the motor, an intermittent fault may be indicated.

An "Intermittent" problem may be caused by a poor connection, rubbed through wire insulation, or a wire that is broken inside the insulation.

If the fault is not current, wiggle the wires of the affected channel and check if the code resets. This will help to pinpoint an intermittent problem in the motor circuitry or connections.

The frequency of the problem can be checked by using the enhanced diagnostic function of the Tech 1 "Scan" tool,

If Code A041 sets as a current code and A063 is infrequent history code, see Code A041 for diagnosis.

Any circuitry, that is suspected as causing the intermittent complaint, should be thoroughly checked for backed out terminals, improper mating, broken locks, improperly formed or damaged terminals, poor terminal to wiring connections or physical damage to the wiring harness.

DIAGNOSTIC CHART DTC A063 — 1993 ACHIEVA GRAND AM AND SKYLARK

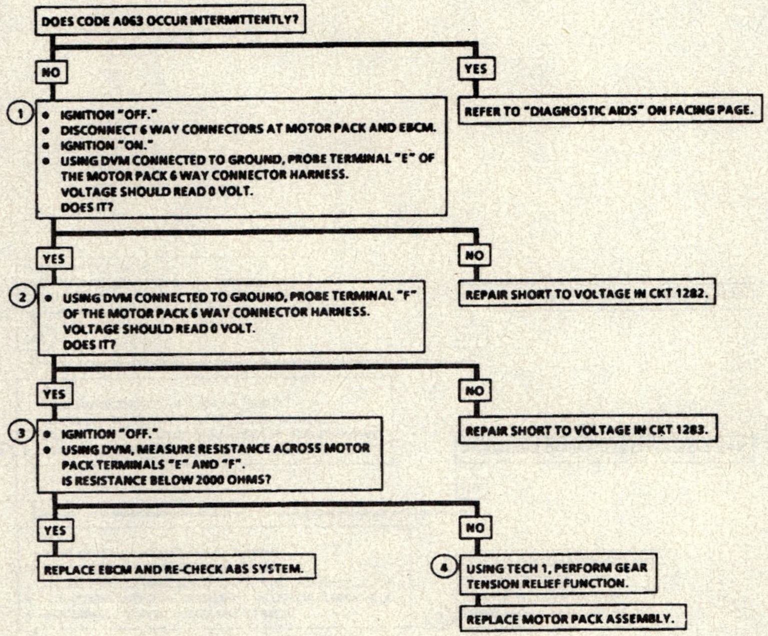

DIAGNOSTIC CHART DTC A064 — 1993 ACHIEVA GRAND AM AND SKYLARK

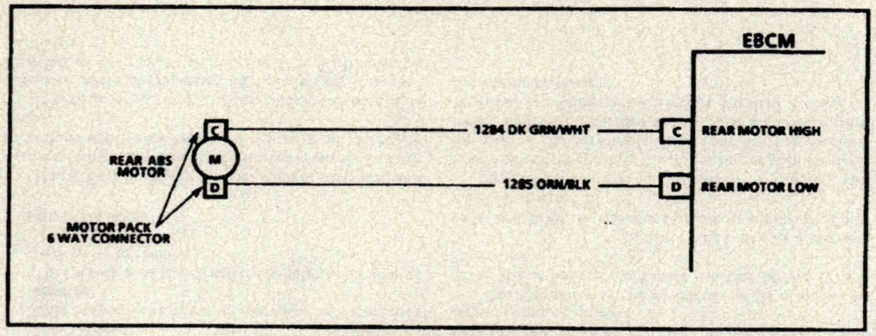

CODE A064

REAR AXLE MOTOR CIRCUIT OPEN

Circuit Description:

This code identifies a motor that cannot be energized due to an open in its circuitry. This fault will not allow proper rear ABS operation.

Test Description: Number(s) below refer to circled number(s) on the diagnostic chart.

1. This checks for an open circuit in the motor.
2. This step is used to release the tension of the motor pack prior to removal.
3. This step checks for an open in the motor "high" circuitry.
4. This step checks for an open in the motor "low" circuitry.

An "Intermittent" problem may be caused by a poor connection, rubbed through wire insulation, or a wire that is broken inside the insulation.

If the fault is not current, wiggle the wires of the affected channel and check if the code resets. This will help to pinpoint an intermittent problem in the motor circuitry or connections.

Diagnostic Aids:

Using Tech 1, select manual control function, and exercise motor movement of affected channel in both directions while applying light pressure on the brake pedal.

If erratic or "jumpy" brake pedal movement is detected while performing an "apply" or "release" function of the motor, an intermittent fault may be indicated.

Any circuitry, that is suspected as causing the intermittent complaint, should be thoroughly checked for backed out terminals, improper mating, broken locks, improperly formed or damaged terminals, poor terminal to wiring connections or physical damage to the wiring harness.

DIAGNOSTIC CHART DTC A064 — 1993 ACHIEVA GRAND AM AND SKYLARK

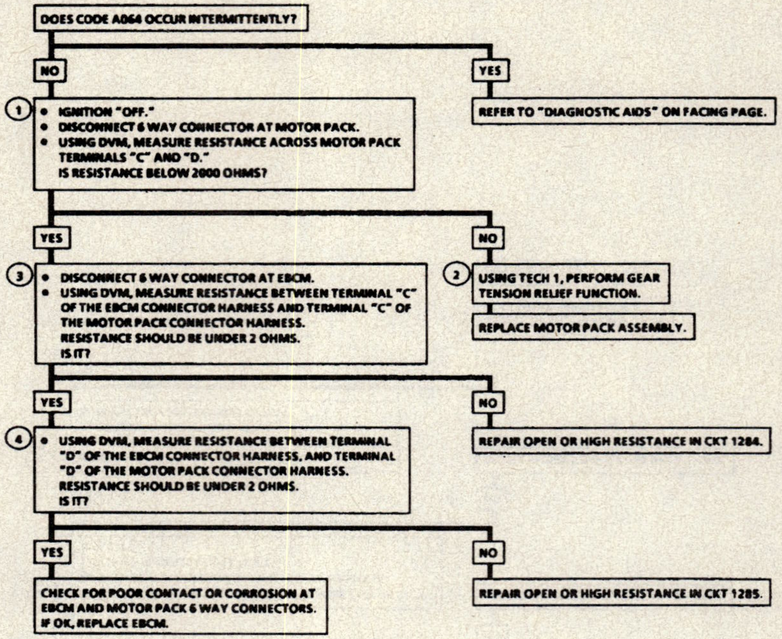

DIAGNOSTIC CHART DTC A065 — 1993 ACHIEVA GRAND AM AND SKYLARK

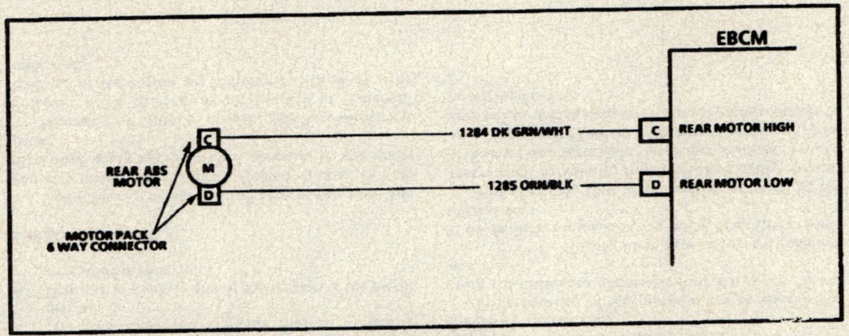

CODE A065

REAR AXLE MOTOR CIRCUIT SHORTED TO GROUND

Circuit Description:

This code identifies a motor circuit that is shorted to ground. This fault will not allow the motor to be controlled at the commanded current rate or will cause the driver circuit to allow current directly to ground.

Test Description: Number(s) below refer to circled number(s) on the diagnostic chart.

1. This step checks for a short to ground in the motor "high" circuitry.
2. This step checks for a short to ground in the motor "low" circuitry.
3. This step checks for a motor that is internally shorted to ground.
4. This step is used to release the tension of the motor pack prior to removal.

An "Intermittent" problem may be caused by a poor connection, rubbed through wire insulation, or a wire that is broken inside the insulation.

If the fault is not current, wiggle the wires of the affected channel and check if the code resets. This will help to pinpoint an intermittent problem in the motor circuitry or connections.

Diagnostic Aids:

Using Tech 1, select manual control function, and exercise motor movement of affected channel in both directions while applying light pressure on the brake pedal.

If erratic or "jumpy" brake pedal movement is detected while performing an "apply" or "release" function of the motor, an intermittent fault may be indicated.

Any circuitry, that is suspected as causing the intermittent complaint, should be thoroughly checked for backed out terminals, improper mating, broken locks, improperly formed or damaged terminals, poor terminal to wiring connections or physical damage to the wiring harness.

DIAGNOSTIC CHART DTC A065 — 1993 ACHIEVA GRAND AM AND SKYLARK

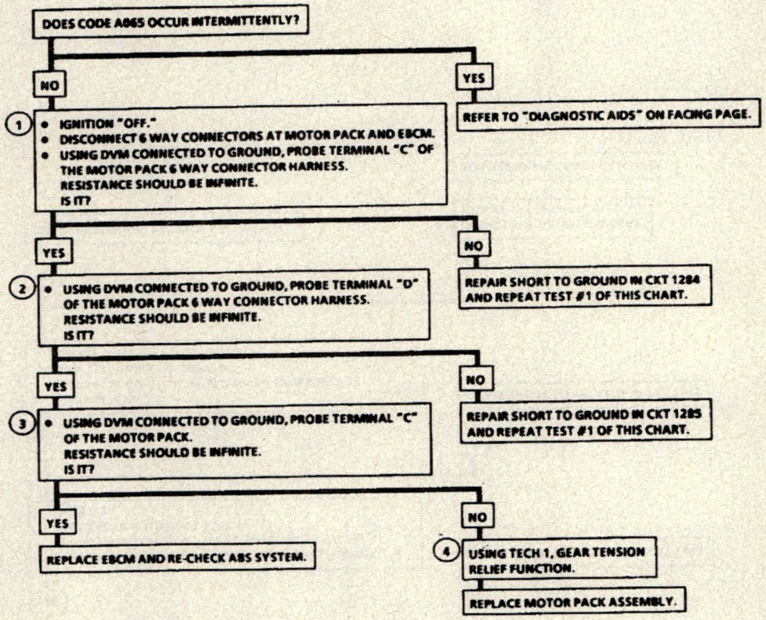

DIAGNOSTIC CHART DTC A066 — 1993 ACHIEVA GRAND AM AND SKYLARK

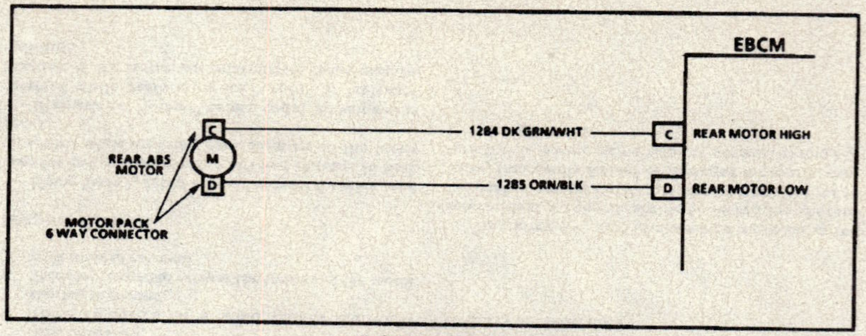

CODE A066

REAR AXLE MOTOR CIRCUIT SHORTED TO BATTERY OR MOTOR SHORTED

Circuit Description:

This code identifies a motor circuit that is shorted to battery or a motor that has low or zero resistance. This fault will not allow the motor to be controlled at the commanded current rate or will cause the motor to turn in the opposite direction or not turn at all.

Test Description: Number(s) below refer to circled number(s) on the diagnostic chart.

1. This step checks for a short to voltage in the motor "high" circuitry.
2. This step checks for a short to voltage in the motor "low" circuitry.
3. This step checks for a motor that is internally shorted.
4. This step is used to release the tension of the motor pack prior to removal.

Diagnostic Aids:

Using Tech 1, select manual control function, and exercise motor movement of affected channel in both directions while applying light pressure on the brake pedal.

If erratic or "jumpy" brake pedal movement is detected while performing an "apply" or "release" function of the motor, an intermittent fault may be indicated.

An "Intermittent" problem may be caused by a poor connection, rubbed through wire insulation, or a wire that is broken inside the insulation.

If the fault is not current, wiggle the wires of the affected channel and check if the code resets. This will help to pinpoint an intermittent problem in the motor circuitry or connections.

The frequency of the problem can be checked by using the enhanced diagnostic function of the "Scan" tool,

If Code A042 sets as current code and A066 is infrequent history code, see Code A042 diagnostics.

Any circuitry, that is suspected as causing the intermittent complaint, should be thoroughly checked for backed out terminals, improper mating, broken locks, improperly formed or damaged terminals, poor terminal to wiring connections or physical damage to the wiring harness.

DIAGNOSTIC CHART DTC A066 — 1993 ACHIEVA GRAND AM AND SKYLARK

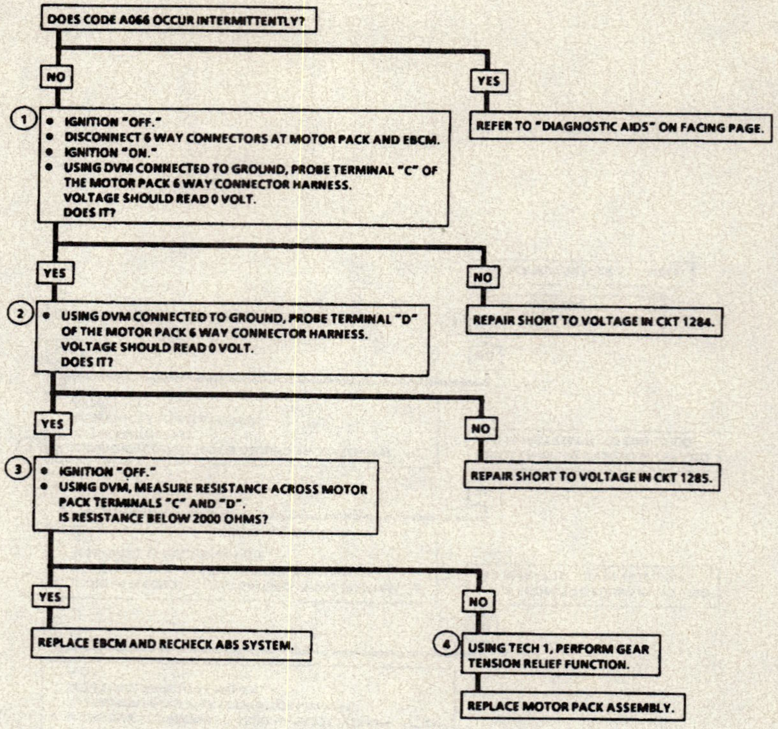

DIAGNOSTIC CHART DTC A067 — 1993 ACHIEVA GRAND AM AND SKYLARK

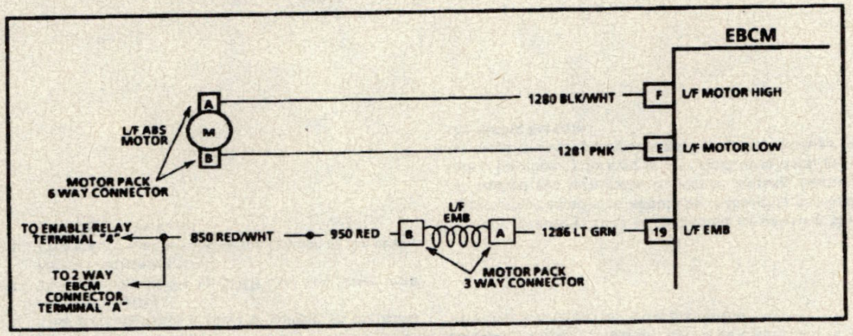

CODE A067

LEFT FRONT EMB RELEASE CIRCUIT OPEN OR SHORTED TO GROUND

Circuit Description:

This code is designed to detect shorts to ground and opens in the EMB release circuit.

A short to ground between the EMB and the ABS EBCM will cause the EMB to be energized indefinitely at an uncontrolled current, causing the EMB to damage itself. To remove the effects of the fault, the enable relay must be turned "OFF," eliminating power to the EMB as well as the ABS system.

Test Description: Number(s) below refer to circled number(s) on the diagnostic chart.

1. This step checks for a short to ground in the EMB control circuitry.
2. This step insures that the EMB is not internally shorted or open.
3. This step checks to see if CKT 850/950 has continuity to the EMB connector.
4. This step looks for an open in the EMB control circuitry.
5. This step is used to release the tension of the motor pack prior to removal.

Diagnostic Aids:

An "Intermittent" problem may be caused by a poor connection, rubbed through wire insulation, or a wire that is broken inside the insulation.

Any circuitry, that is suspected as causing the intermittent complaint, should be thoroughly checked for backed out terminals, improper mating, broken locks, improperly formed or damaged terminals, poor terminal to wiring connections or physical damage to the wiring harness.

DIAGNOSTIC CHART DTC A067 — 1993 ACHIEVA GRAND AM AND SKYLARK

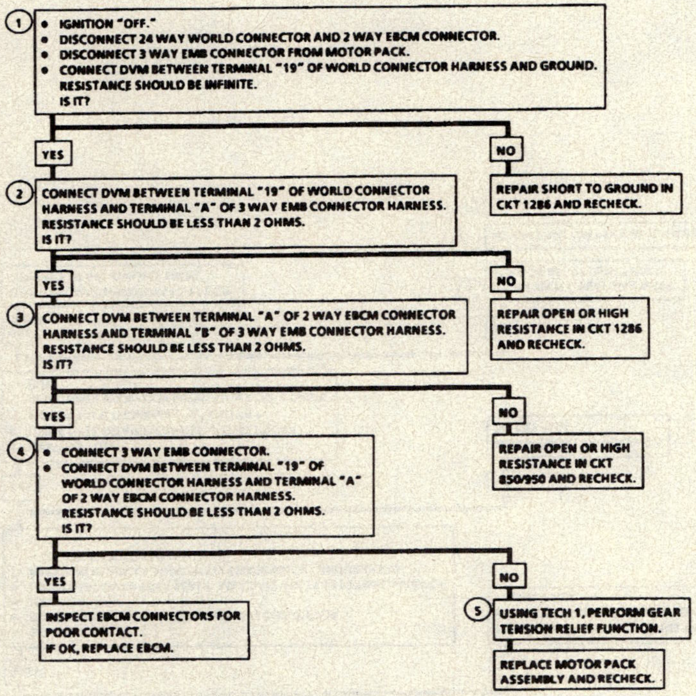

DIAGNOSTIC CHART DTC A068 — 1993 ACHIEVA GRAND AM AND SKYLARK

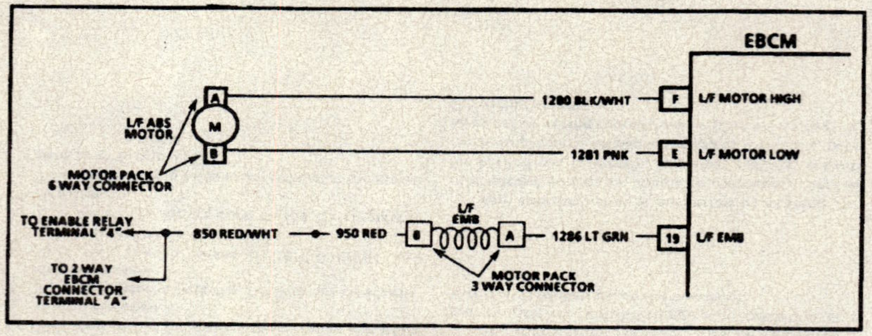

CODE A068

LEFT FRONT EMB CIRCUIT SHORTED TO BATTERY OR EMB SHORTED

Circuit Description:

This code is designed to detect shorts to battery and shorts in the EMB release circuit. These faults will not allow the EMB to be energized when needed.

Test Description: Number(s) below refer to circled number(s) on the diagnostic chart.

1. This step checks to see if the fault is currently present.
2. This step identifies a short to voltage in the EMB control circuitry.
3. This step checks for an EMB that has lower than normal resistance.
4. This step is used to release the tension of the motor pack prior to removal.

Diagnostic Aids:

An "Intermittent" problem may be caused by a poor connection, rubbed through wire insulation, or a wire that is broken inside the insulation.

Any circuitry, that is suspected as causing the intermittent complaint, should be thoroughly checked for backed out terminals, improper mating, broken locks, improperly formed or damaged terminals, poor terminal to wiring connections or physical damage to the wiring harness.

DIAGNOSTIC CHART DTC A068 — 1993 ACHIEVA GRAND AM AND SKYLARK

(1)
- IGNITION "ON," ENGINE "OFF."
- USING TECH 1, SELECT MANUAL CONTROL FUNCTION.
- SELECT MOTOR APPLY FUNCTION OF AFFECTED CHANNEL, AND APPLY MOTOR.
- APPLY FIRM PRESSURE ON BRAKE PEDAL.
- SELECT EMB RELEASE FUNCTION OF AFFECTED CHANNEL
- RELEASE EMB.

BRAKE PEDAL SHOULD FALL WITH INCREASING PEDAL FORCE. DID IT?

YES → FAULT IS NOT PRESENT AT THIS TIME. REFER TO "DIAGNOSTIC AIDS" ON FACING PAGE.

NO ↓

(2)
- IGNITION "OFF."
- DISCONNECT 3 WAY MOTOR PACK EMB CONNECTOR.
- IGNITION "ON."
- USING DVM ON 20 VOLT SCALE. MEASURE VOLTAGE BETWEEN MOTOR PACK EMB CONNECTOR HARNESS TERMINAL "A" AND GROUND.

VOLTAGE SHOULD BE 0 VOLTS. IS IT?

NO → REPAIR SHORT TO VOLTAGE ON CKT 1286.

YES ↓

(3)
- IGNITION "OFF."
- USING DVM, MEASURE RESISTANCE BETWEEN MOTOR PACK TERMINALS "A" AND "B."

RESISTANCE SHOULD BE BETWEEN .4 AND .9 OHMS. IS IT?

YES → CHECK EBCM CONNECTORS FOR POOR CONTACT. IF OK, REPLACE EBCM.

NO → (4) USING TECH 1, PERFORM GEAR TENSION RELIEF FUNCTION. → REPLACE MOTOR PACK AND RECHECK.

DIAGNOSTIC CHART DTC A071 — 1993 ACHIEVA GRAND AM AND SKYLARK

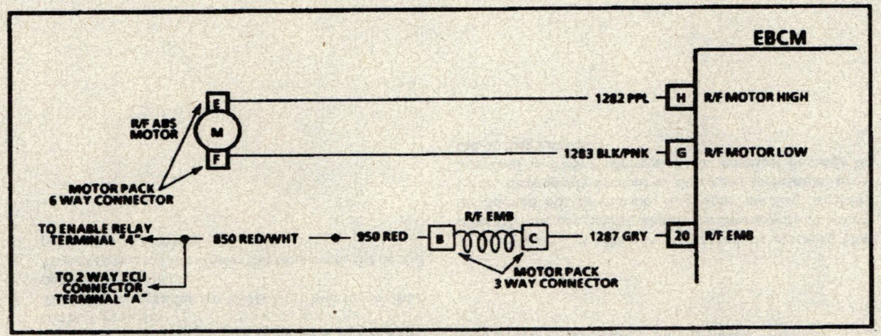

CODE A071

RIGHT FRONT EMB CIRCUIT OPEN OR SHORTED TO GROUND

Circuit Description:

This code is designed to detect shorts to ground and opens in the EMB release circuit.

A short to ground between the EMB and the ABS EBCM will cause the EMB to be energized indefinitely at an uncontrolled current causing the EMB to damage itself.

Test Description: Number(s) below refer to circled number(s) on the diagnostic chart.

1. This step checks for a short to ground in the EMB control circuitry.
2. This step insures that the EMB is not internally shorted or open.
3. This step checks to see if CKT 850/950 has continuity to the EMB connector.
4. This step looks for an open in the EMB control circuitry.
5. This step is used to release the tension of the motor pack prior to removal.

Diagnostic Aids:

An "Intermittent" problem may be caused by a poor connection, rubbed through wire insulation, or a wire that is broken inside the insulation.

Any circuitry, that is suspected as causing the intermittent complaint, should be thoroughly checked for backed out terminals, improper mating, broken locks, improperly formed or damaged terminals, poor terminal to wiring connections or physical damage to the wiring harness.

DIAGNOSTIC CHART DTC A071 — 1993 ACHIEVA GRAND AM AND SKYLARK

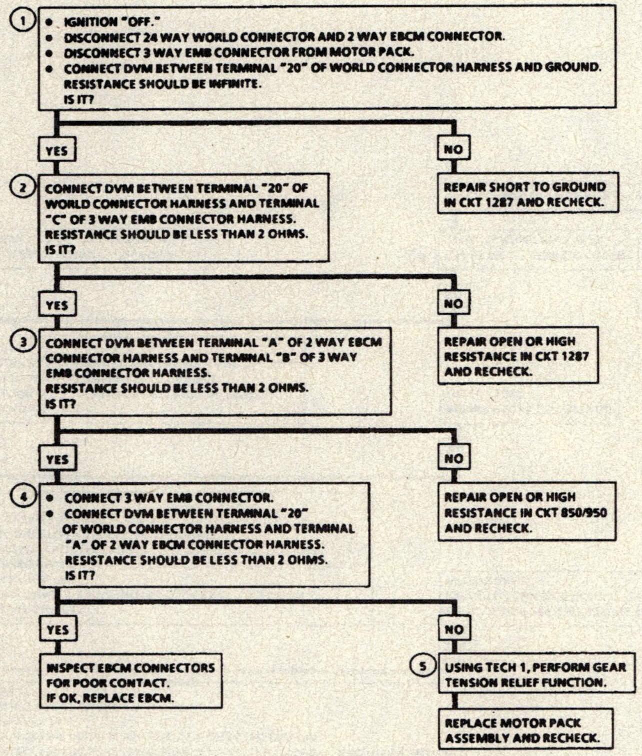

DIAGNOSTIC CHART DTC A072 — 1993 ACHIEVA GRAND AM AND SKYLARK

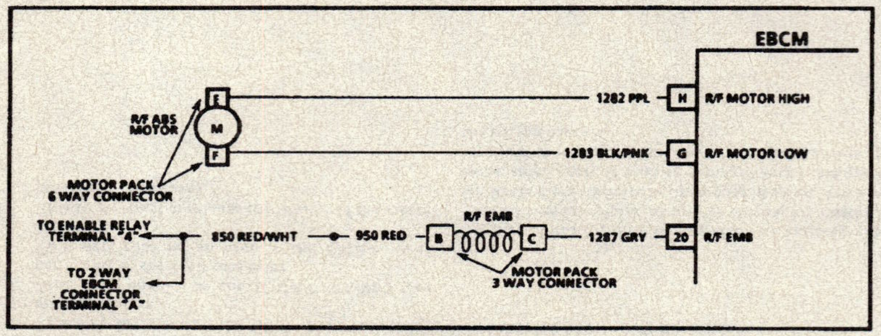

CODE A072

RIGHT FRONT EMB CIRCUIT SHORTED TO BATTERY OR EMB SHORTED

Circuit Description:

This code is designed to detect shorts to battery and shorts in the EMB release circuit. These faults will not allow the EMB to be energized when needed.

Test Description: Number(s) below refer to circled number(s) on the diagnostic chart.

1. This step checks to see if the fault is currently present.
2. This step identifies a short to voltage in the EMB control circuitry.
3. This step checks for an EMB that has lower than normal resistance.
4. This step is used to release the tension of the motor pack prior to removal.

Diagnostic Aids:

An "Intermittent" problem may be caused by a poor connection, rubbed through wire insulation, or a wire that is broken inside the insulation.

Any circuitry, that is suspected as causing the intermittent complaint, should be thoroughly checked for backed out terminals, improper mating, broken locks, improperly formed or damaged terminals, poor terminal to wiring connections or physical damage to the wiring harness.

DIAGNOSTIC CHART DTC A072 — 1993 ACHIEVA GRAND AM AND SKYLARK

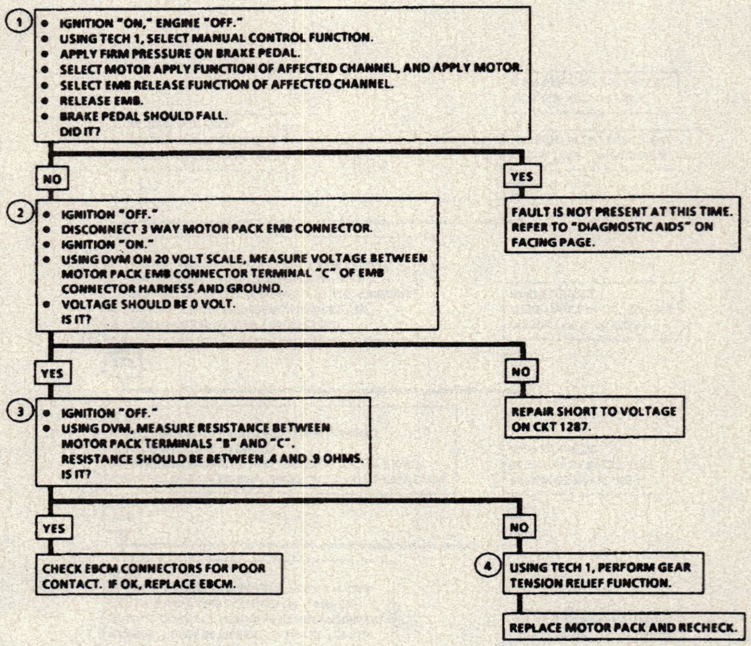

DIAGNOSTIC CHART DTC A076 — 1993 ACHIEVA GRAND AM AND SKYLARK

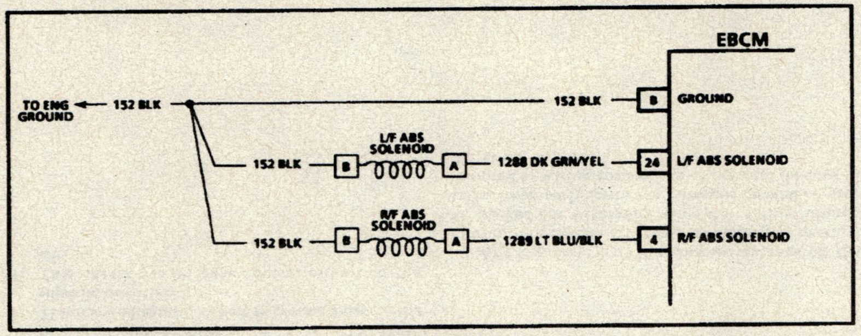

CODE A076

LEFT FRONT SOLENOID CIRCUIT SHORTED TO BATTERY OR OPEN

Circuit Description:

This code identifies a solenoid that cannot be energized due to an open in its circuitry, or a solenoid that is always energized due to a short to battery in its circuitry between the driver and the solenoid. An open will not allow proper ABS operation, but the short to battery simply turns "ON" the solenoid. A path for base brakes is still allowed once the motor re-homes and the check valve is lifted off its seat during key on initialization.

Test Description: Number(s) below refer to circled number(s) on the diagnostic chart.

1. This checks for a short to voltage in the control circuitry of the solenoid.
2. This checks for an open in the control circuit of the solenoid.
3. This step identifies an open in the solenoid or ground circuit.
4. This step checks for a possible intermittent problem in the solenoid circuitry due to poor terminal contact.

Diagnostic Aids:

An "Intermittent" problem may be caused by a poor connection, rubbed through wire insulation, or a wire that is broken inside the insulation.

Any circuitry, that is suspected as causing the intermittent complaint, should be thoroughly checked for backed out terminals, improper mating, broken locks, improperly formed or damaged terminals, poor terminal to wiring connections or physical damage to the wiring harness.

DIAGNOSTIC CHART DTC A076 — 1993 ACHIEVA GRAND AM AND SKYLARK

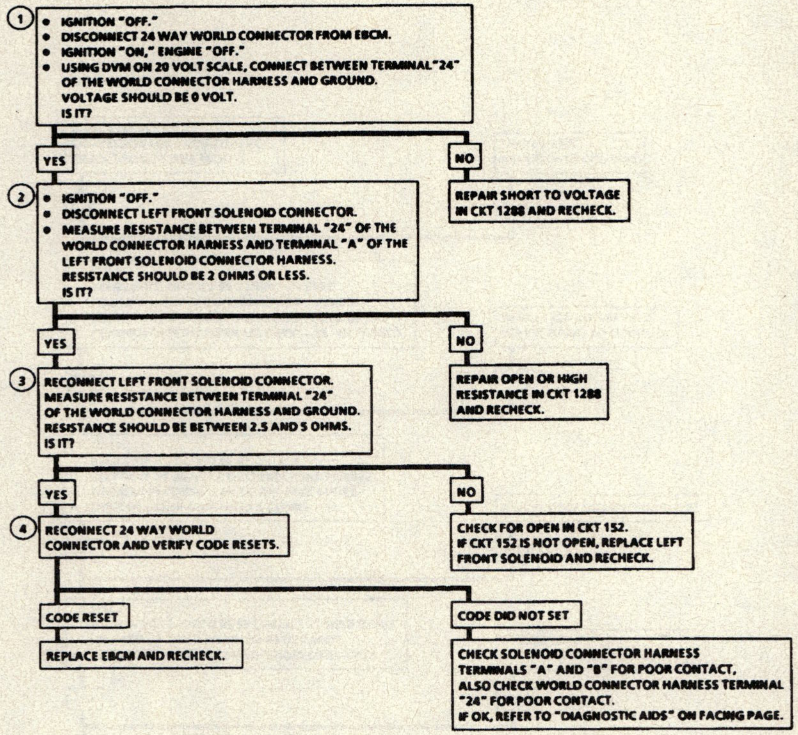

DIAGNOSTIC CHART DTC A077 — 1993 ACHIEVA GRAND AM AND SKYLARK

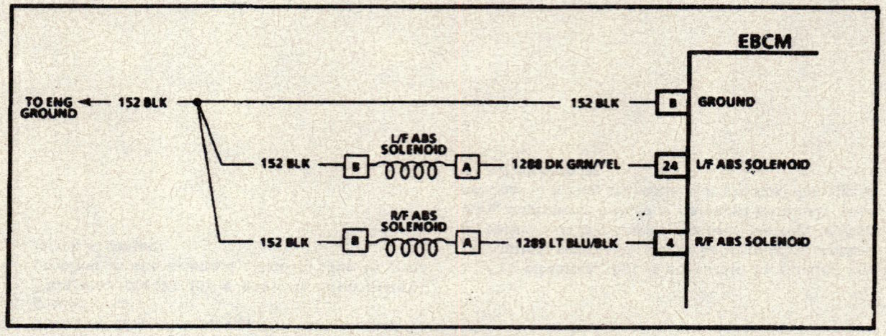

CODE A077

LEFT FRONT SOLENOID CIRCUIT SHORTED TO GROUND OR DRIVER OPEN

Circuit Description:

This code identifies a solenoid that cannot be energized due to an open in its driver circuitry, or a short to ground between the solenoid driver and the solenoid. These faults can affect ABS operation since the flow of brake fluid to the wheel cylinder cannot be stopped, making ABS operation for that channel impossible.

Test Description: Number(s) below refer to circled number(s) on the diagnostic chart.

1. This step checks for a solenoid that is internally shorted to ground.
2. This step checks for a solenoid that is not within proper resistance values.
3. This step indicates if a short to ground exists in the solenoid circuitry.
4. This checks for an open control circuit to the EBCM.

Diagnostic Aids:

An "Intermittent" problem may be caused by a poor connection, rubbed through wire insulation, or a wire that is broken inside the insulation.

Any circuitry, that is suspected as causing the intermittent complaint, should be thoroughly checked for backed out terminals, improper mating, broken locks, improperly formed or damaged terminals, poor terminal to wiring connections or physical damage to the wiring harness.

DIAGNOSTIC CHART DTC A077 — 1993 ACHIEVA GRAND AM AND SKYLARK

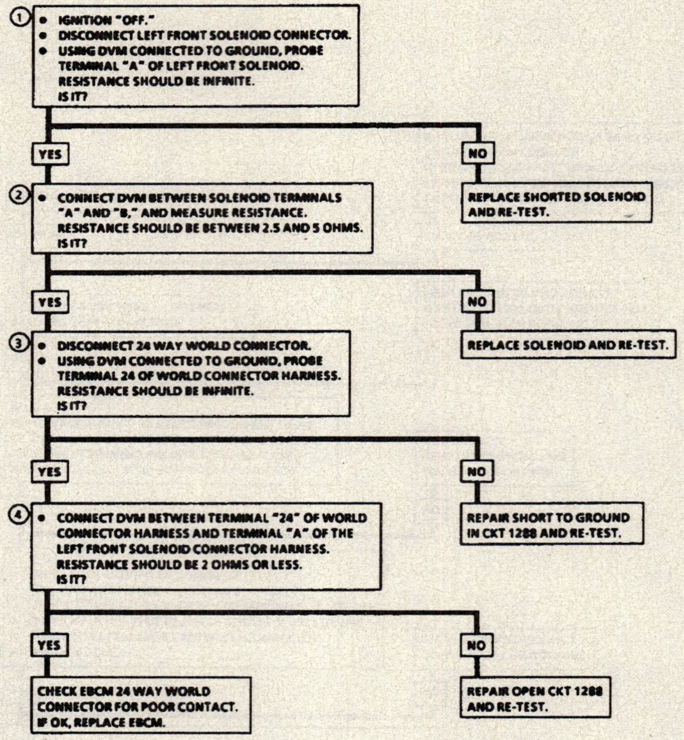

DIAGNOSTIC CHART DTC A078 — 1993 ACHIEVA GRAND AM AND SKYLARK

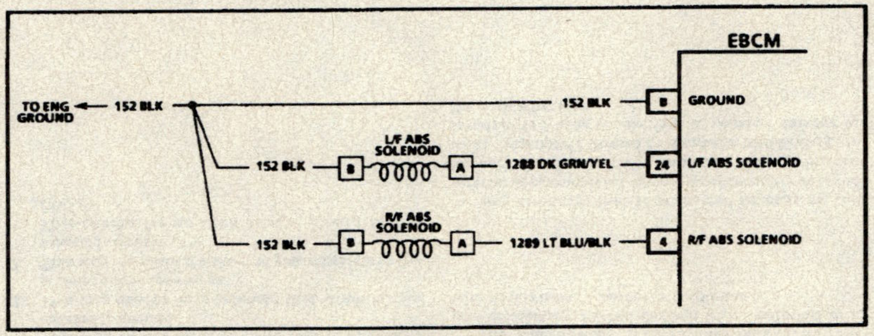

CODE A078

RIGHT FRONT SOLENOID CIRCUIT SHORTED TO BATTERY OR OPEN

Circuit Description:

This code identifies a solenoid that cannot be energized due to an open in its circuitry, or a solenoid that is always energized due to a short to battery in its circuitry between the driver and the solenoid. An open will not allow proper ABS operation, but the short to battery simply turns "ON" the solenoid. A path for base brakes is still allowed once the motor re-homes and the check valve is lifted off its seat during key on initialization.

Test Description: Number(s) below refer to circled number(s) on the diagnostic chart.

1. This checks for a short to voltage in the control circuitry of the solenoid.
2. This checks for an open in the control circuit of the solenoid.
3. This step identifies an open in the solenoid or ground circuit.
4. This step checks for a possible intermittent problem in the solenoid circuitry due to poor terminal contact.

Diagnostic Aids:

An "Intermittent" problem may be caused by a poor connection, rubbed through wire insulation, or a wire that is broken inside the insulation.

Any circuitry, that is suspected as causing the intermittent complaint, should be thoroughly checked for backed out terminals, improper mating, broken locks, improperly formed or damaged terminals, poor terminal to wiring connections or physical damage to the wiring harness.

DIAGNOSTIC CHART DTC A078 — 1993 ACHIEVA GRAND AM AND SKYLARK

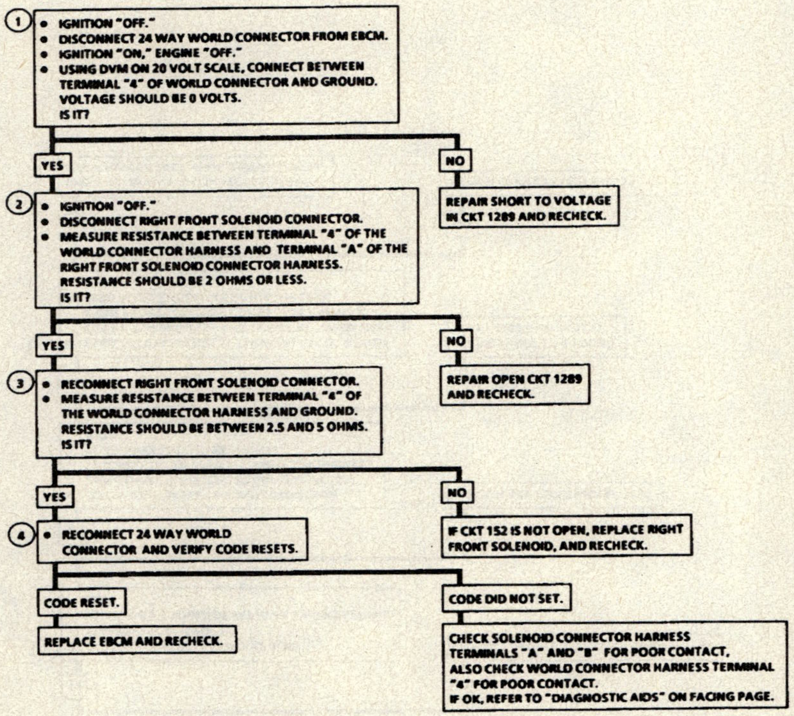

DIAGNOSTIC CHART DTC A081 — 1993 ACHIEVA GRAND AM AND SKYLARK

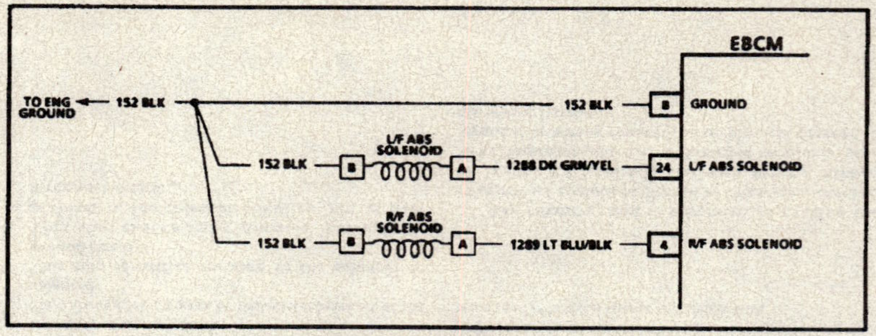

CODE A081

RIGHT FRONT SOLENOID CIRCUIT SHORTED TO GROUND OR DRIVER OPEN

Circuit Description:

This code identifies a solenoid that cannot be energized due to an open in its driver circuitry, or a short to ground between the solenoid driver and the solenoid. These faults can affect ABS operation since the flow of brake fluid to the wheel cylinder cannot be stopped, making ABS operation for that channel impossible.

Test Description: Number(s) below refer to circled number(s) on the diagnostic chart.

1. This step checks for a solenoid that is internally shorted to ground.
2. This step checks for a solenoid that is not within proper resistance values.
3. This step indicates if a short to ground exists in the solenoid circuitry.
4. This checks for an open control circuit to the EBCM.

Diagnostic Aids:

An "Intermittent" problem may be caused by a poor connection, rubbed through wire insulation, or a wire that is broken inside the insulation.

Any circuitry, that is suspected as causing the intermittent complaint, should be thoroughly checked for backed out terminals, improper mating, broken locks, improperly formed or damaged terminals, poor terminal to wiring connections or physical damage to the wiring harness.

DIAGNOSTIC CHART DTC A081 — 1993 ACHIEVA GRAND AM AND SKYLARK

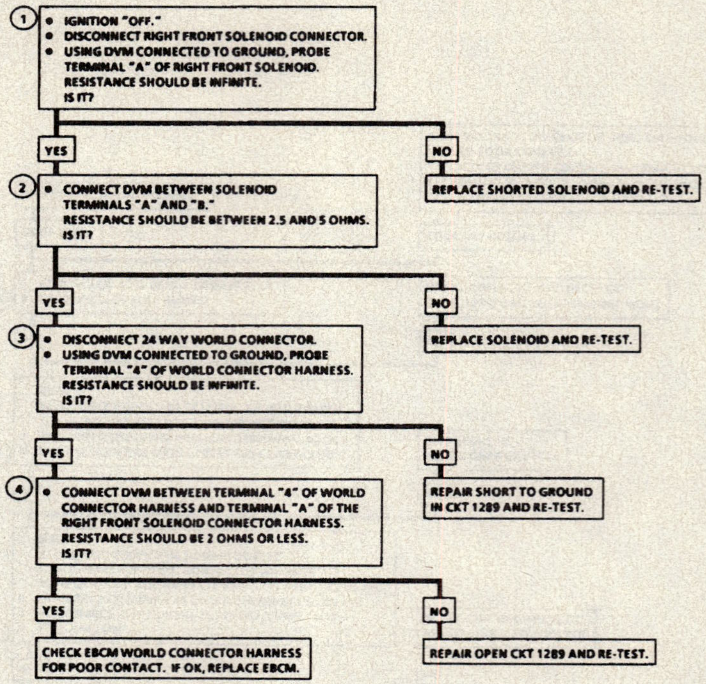

DIAGNOSTIC CHART DTC A082 — 1993 ACHIEVA GRAND AM AND SKYLARK

D.M. A.B.S. VI
ELECTRONIC CONTROL UNIT
(TYPICAL)

CODE A082

CALIBRATION MEMORY FAILURE

Circuit Description:

This code allows the EBCM to check for a calibration failure by comparing the calibration value to a known value stored in the EEPROM.

This test is also used as a security measure to prevent improper use of calibrations or changes to these calibrations that may alter the designed function of ABS.

Test Description: Number(s) below refer to circled number(s) on the diagnostic chart.

1. This step checks to see if the fault is present during diagnosis. If present, the EBCM is not functioning correctly and must be replaced.

IMPORTANT: At the time of printing, vehicles with ABS-VI were not being field reprogrammed to correct this failure. A service bulletin will be issued when EEPROM reprogramming is authorized.

Diagnostic Aids:

An intermittent Code A082 may be caused by a bad cell in the EEPROM that is sensitive to temperature changes. If Code A082 failed more than once, but is intermittent, replace EBCM.

DIAGNOSTIC CHART DTC A082 — 1993 ACHIEVA GRAND AM AND SKYLARK

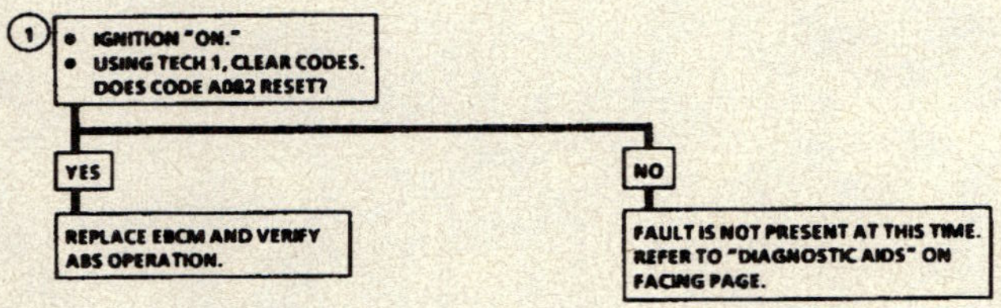

DIAGNOSTIC CHART DTC A086 — 1993 ACHIEVA GRAND AM AND SKYLARK

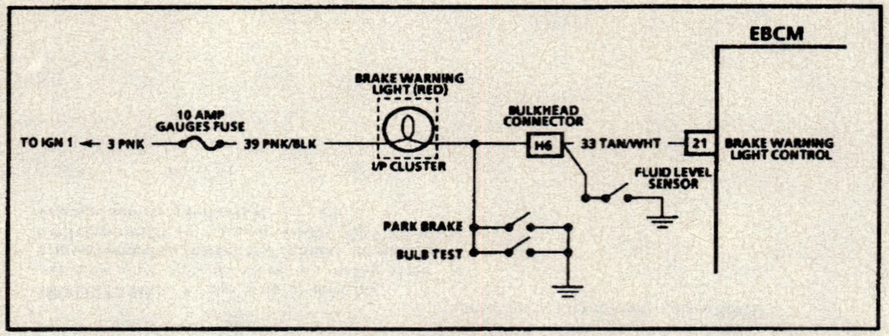

CODE A086

RED BRAKE WARNING LIGHT ACTIVATED BY ABS

Circuit Description:

This code is provided as an information only test, and reflects the status of the command issued by the ABS EBCM to illuminate the red brake warning light. If another diagnostic code issues a command to illuminate the red brake warning light, Code A086 will be stored in EEPROM as a history code at the conclusion of the ignition cycle.

Test Description: Number(s) below refer to circled number(s) on the diagnostic chart.

1. This step serves to identify if a code other than Code A086 commanded the red brake warning light "ON."

Diagnostic Aids:

Any ABS mechanical code that issues a command to illuminate the red brake warning light will also result in Code A086 being stored in EEPROM during shut down. These Codes are: A038, A041, A042, A045, A046, A047, A048 and A051. If the motors are not in their home position, certain electrical codes will also command the red brake warning light "ON." These Codes are: A014, A018, A055, A056, A057, A0587, A061, A062, A063, A064, A065, A066, A067, A068, A071 and A072.

If any of these codes are indicated along with Code A086, they must be corrected prior to addressing a Code A086 fault.

DIAGNOSTIC CHART DTC A086 — 1993 ACHIEVA GRAND AM AND SKYLARK

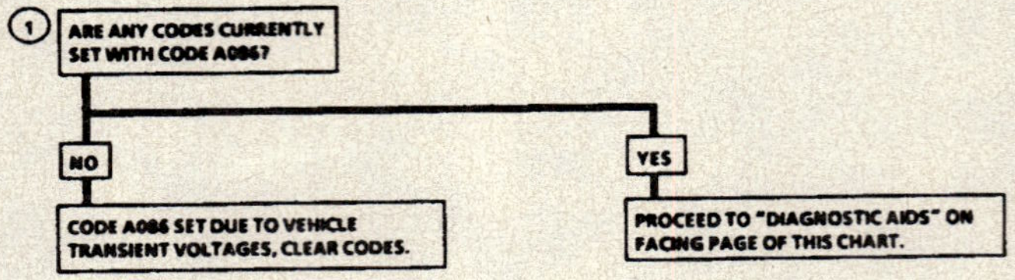

DIAGNOSTIC CHART DTC A087 — 1993 ACHIEVA GRAND AM AND SKYLARK

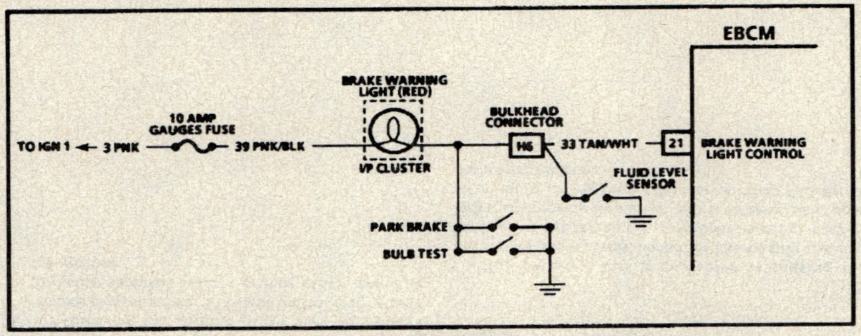

CODE A087

RED BRAKE WARNING LIGHT CIRCUIT OPEN

Circuit Description:

This code is used to verify the EBCM has continuity to the brake warning light in case the EBCM must turn it "ON." This will only occur if an ABS fault is detected that may degrade base brake operation.

Because ABS is not the only device controlling the brake warning light (parking brake and low fluid sensor may also turn "ON" the light) a short to ground in this circuit cannot be detected.

Test Description: Number(s) below refer to circled number(s) on the diagnostic chart.

1. This step identifies if the fault is currently present.
2. This step indicates if the EBCM and circuitry has the ability to complete the ground to the brake warning light and illuminate it.
3. Because the brake warning light has current supplied to it through the 10 amp gauge fuse, this step would indicate if the ignition circuit is complete to the I/P cluster.
4. This step isolates if the open circuit is due to a EBCM failure or open circuitry to the brake warning light.

Diagnostic Aids:

An "Intermittent" problem may be caused by a poor connection, rubbed through wire insulation, or a wire that is broken inside the insulation.

The frequency of the problem can be checked by using the enhanced diagnostic function of the Tech 1 "Scan" tool, If a Tech 1 or T-100 are not available the parking brake handle may be lifted to verify proper light operation and continuity of CKTs 3 and 39.

Any circuitry, that is suspected as causing the intermittent complaint, should be thoroughly checked for backed out terminals, improper mating, broken locks, improperly formed or damaged terminals, poor terminal to wiring connections or physical damage to the wiring harness.

DIAGNOSTIC CHART DTC A087 — 1993 ACHIEVA GRAND AM AND SKYLARK

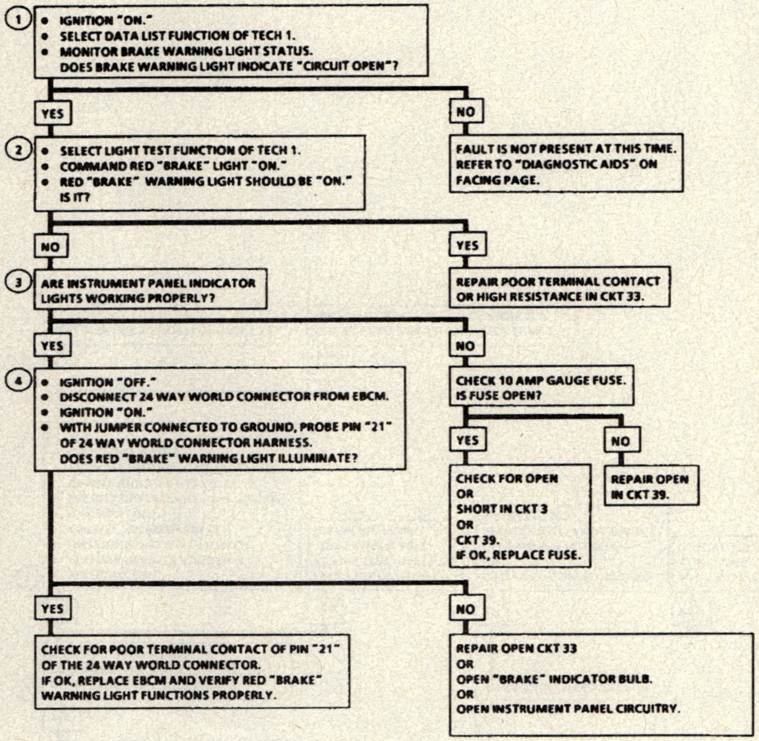

DIAGNOSTIC CHART DTC A088 — 1993 ACHIEVA GRAND AM AND SKYLARK

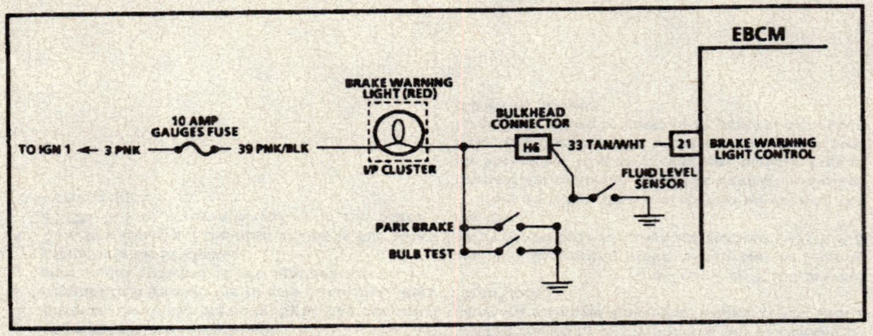

CODE A088

RED BRAKE WARNING LIGHT CIRCUIT SHORTED TO BATTERY

Circuit Description:

This code identifies a short to battery between the EBCM and the RED brake warning light, or an open driver that does not allow the RED brake warning light to be illuminated by the EBCM. This will only occur if an ABS fault is detected that may degrade base brake operation.

Test Description: Number(s) below refer to circled number(s) on the diagnostic chart.

1. This step identifies if the ground circuit to the Red brake warning light is being completed by a source other than the EBCM.
2. This step identifies if the fault is currently present.
3. By removing the 10 amp gauge fuse, the voltage source is eliminated. This test indicates if voltage is being supplied from a source other than the gauge fuse.

Diagnostic Aids:

An "Intermittent" problem may be caused by a poor connection, rubbed through wire insulation, or a wire that is broken inside the insulation.

Any circuitry, that is suspected as causing the intermittent complaint, should be thoroughly checked for backed out terminals, improper mating, broken locks, improperly formed or damaged terminals, poor terminal to wiring connections or physical damage to the wiring harness.

DIAGNOSTIC CHART DTC A088 — 1993 ACHIEVA GRAND AM AND SKYLARK

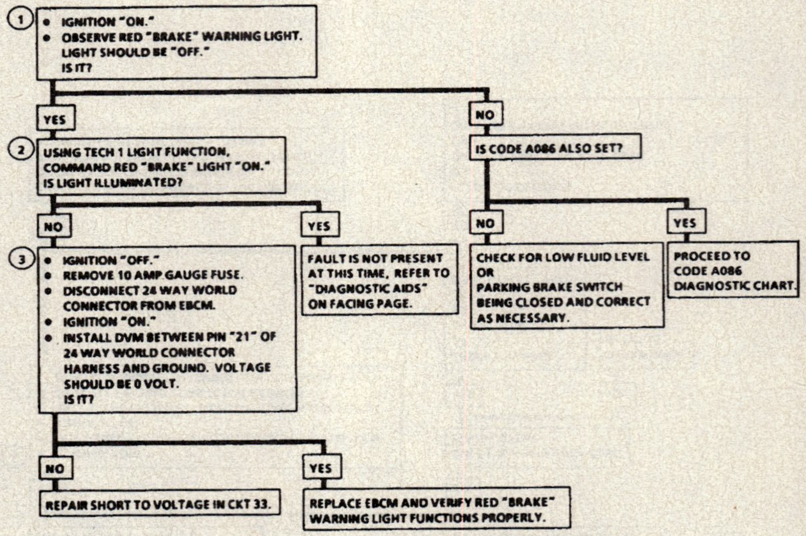

DIAGNOSTIC CHART DTC A091 — 1993 ACHIEVA GRAND AM AND SKYLARK

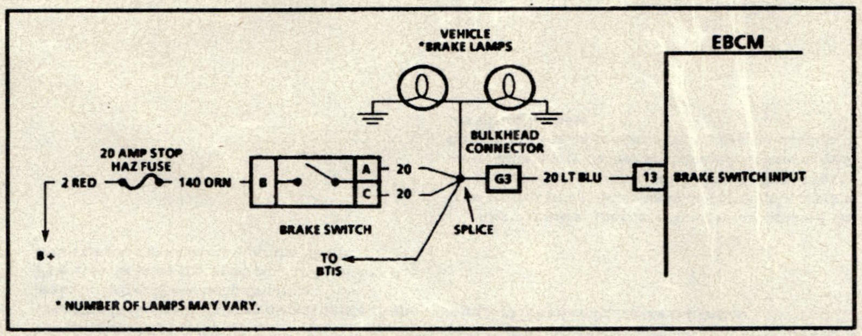

CODE A091

OPEN BRAKE SWITCH CONTACTS DURING DECELERATION

Circuit Description:

This code is used to detect an open brake switch in the non-ABS mode. The EBCM looks for decel rates that would indicate braking action and verifies this assumption by requiring several repeats of this detection method. In each case, ABS will not be available since no brake switch is seen.

Test Description: Number(s) below refer to circled number(s) on the diagnostic chart.

1. This step indicates if the brake switch signal is being received by the EBCM.
2. This step indicates if an open circuit exists in the brake switch or brake lamp circuitry.
3. This step isolates the open circuit to either the brake switch input circuitry, or the EBCM.

Diagnostic Aids:

An "Intermittent" problem may be caused by a poor connection, rubbed through wire insulation, or a wire that is broken inside the insulation.

Any circuitry, that is suspected as causing the intermittent complaint, should be thoroughly checked for backed out terminals, improper mating, broken locks, improperly formed or damaged terminals, poor terminal to wiring connections or physical damage to the wiring harness.

DIAGNOSTIC CHART DTC A091 — 1993 ACHIEVA GRAND AM AND SKYLARK

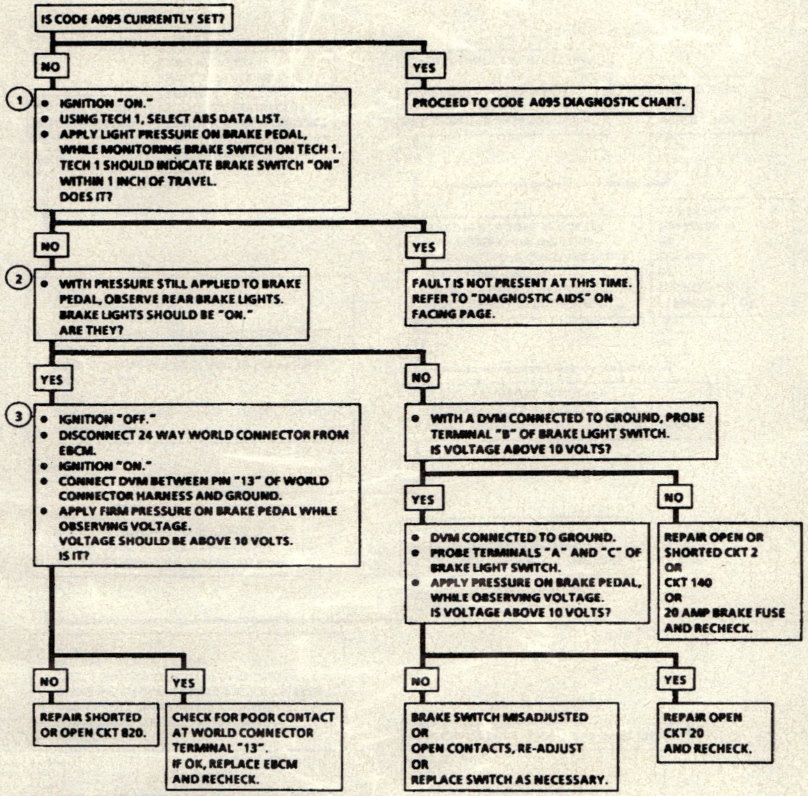

DIAGNOSTIC CHART DTC A092 — 1993 ACHIEVA GRAND AM AND SKYLARK

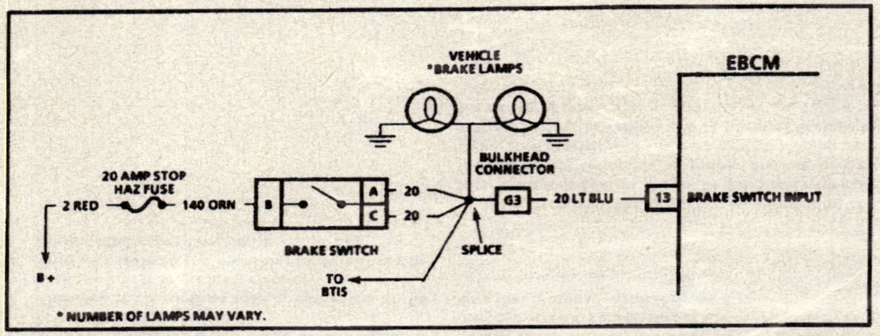

CODE A092

OPEN BRAKE SWITCH CONTACTS WHEN ABS WAS REQUIRED

Circuit Description:

This code is run to determine the proper operation of the brake switch. This is important because ABS is activated when the brake switch is "ON." If the brake switch is "OFF," ABS will never be activated. Since this fault is difficult to detect under normal braking conditions, this fault is only detected when ABS is required.

Test Description: Number(s) below refer to circled number(s) on the diagnostic chart.

1. This step indicates if the brake switch signal is being received by the EBCM.
2. This step indicates if an open circuit exists in the brake switch or brake lamp circuitry.
3. This step isolates the open circuit to either the brake switch input circuitry, or the EBCM.

Diagnostic Aids:

An "Intermittent" problem may be caused by a poor connection, rubbed through wire insulation, or a wire that is broken inside the insulation.

Any circuitry, that is suspected as causing the intermittent complaint, should be thoroughly checked for backed out terminals, improper mating, broken locks, improperly formed or damaged terminals, poor terminal to wiring connections or physical damage to the wiring harness.

DIAGNOSTIC CHART DTC A092 — 1993 ACHIEVA GRAND AM AND SKYLARK

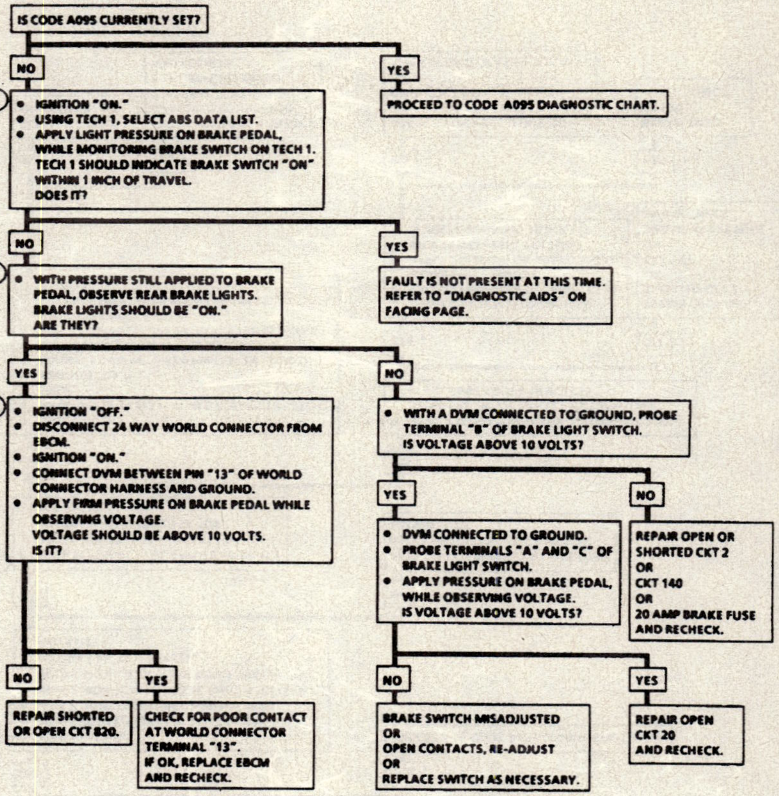

DIAGNOSTIC CHART DTC A093 — 1993 ACHIEVA GRAND AM AND SKYLARK

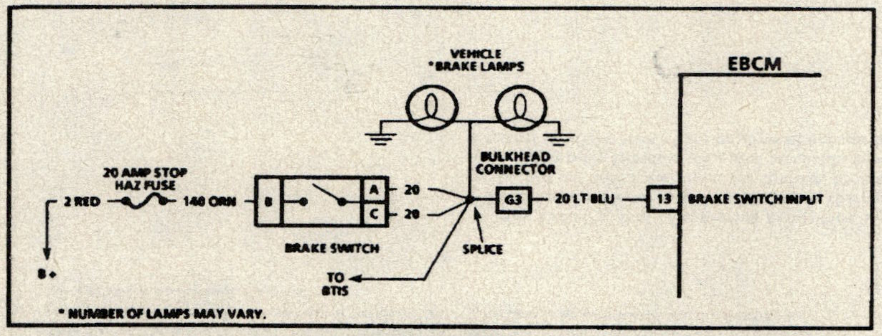

CODE A093

CODES A091 OR A092 SET IN CURRENT OR PREVIOUS IGNITION CYCLE

Circuit Description:

This code is the second portion of Codes A091 and A092. If Codes A091 or A092 failed during the last ignition cycle, Code A093 becomes a current failure during the next ignition cycle, keeping ABS disabled until a brake switch "ON" state is seen. When a change is seen during an ignition cycle in which Code A093 is a current failure, Code A091 or A092 will clear itself at the end of the current ignition cycle, and ABS will enable itself at the start of the next ignition cycle. Code A093 alone indicates Codes A091 or A092 failed previously, but is intermittent, or has been corrected.

Test Description: Number(s) below refer to circled number(s) on the diagnostic chart.

1. This step indicates which Code (either A091 or A092) caused Code A093 to set.
2. This insures that the code that set is repaired so that Code A093 can be cleared.

Diagnostic Aids:

An "Intermittent" problem may be caused by a poor connection, rubbed through wire insulation, or a wire that is broken inside the insulation.

Any circuitry, that is suspected as causing the intermittent complaint, should be thoroughly checked for backed out terminals, improper mating, broken locks, improperly formed or damaged terminals, poor terminal to wiring connections or physical damage to the wiring harness.

Also, verify proper brake switch operation using the data list of the "Scan" tool. As the brake is applied, the data list should display the brake switch "ON" within 1 inch of travel.

DIAGNOSTIC CHART DTC A093 — 1993 ACHIEVA GRAND AM AND SKYLARK

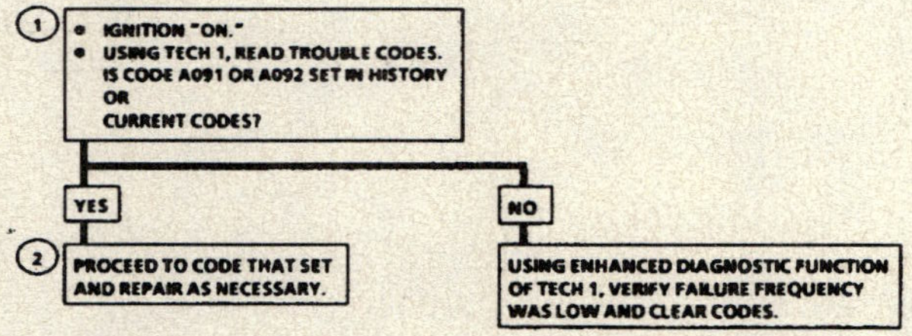

DIAGNOSTIC CHART DTC A094 — 1993 ACHIEVA GRAND AM AND SKYLARK

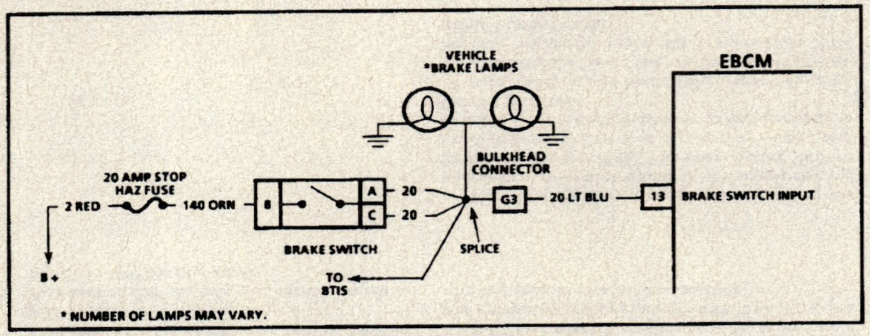

CODE A094

BRAKE SWITCH CONTACTS ALWAYS CLOSED

Circuit Description:

This code is run to determine the proper operation of the brake switch. This is important because ABS is activated when the brake switch is "ON" and turned "OFF" when the brake is "OFF." If the brake switch is always "ON," ABS operation will always be requested resulting in potential modulator cycling on rough roads. Additionally, this failure will most likely result in a dead battery (due to the brake lights staying "ON") if the driver is not informed of this fault.

Test Description: Number(s) below refer to circled number(s) on the diagnostic chart.

1. This checks if the fault is currently present.
2. This step isolates the cause of the fault to either a faulty or misadjusted brake switch, or a short to voltage in the brake switch circuitry.

Diagnostic Aids:

An "Intermittent" problem may be caused by a poor connection, rubbed through wire insulation, or a wire that is broken inside the insulation.

Any circuitry, that is suspected as causing the intermittent complaint, should be thoroughly checked for backed out terminals, improper mating, broken locks, improperly formed or damaged terminals, poor terminal to wiring connections or physical damage to the wiring harness.

DIAGNOSTIC CHART DTC A094 — 1993 ACHIEVA GRAND AM AND SKYLARK

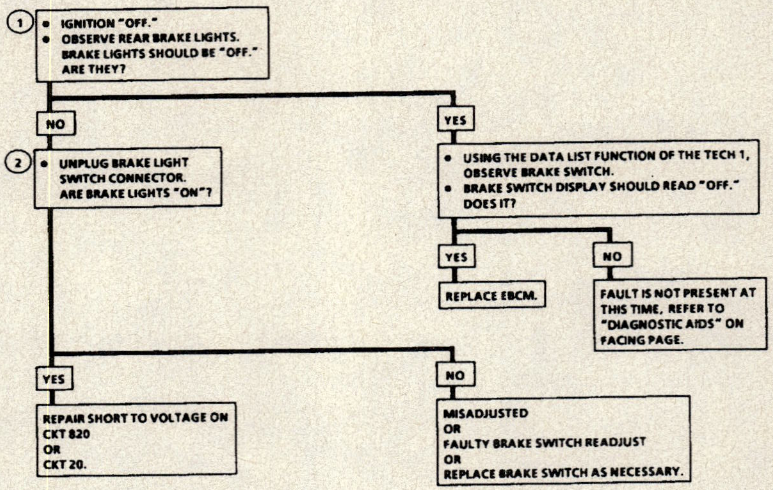

DIAGNOSTIC CHART DTC A095 — 1993 ACHIEVA GRAND AM AND SKYLARK

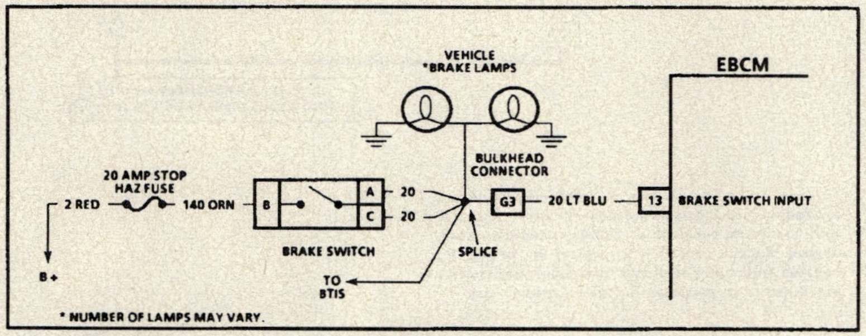

CODE A095

BRAKE SWITCH CIRCUIT OPEN

Circuit Description:

This code is used to identify open brake switch circuitry that prevents the brake switch input to the EBCM from changing states when the brake is applied. This code is used in conjunction with Codes A091 and A092 to determine the cause of an open brake switch fault.

Test Description: Number(s) below refer to circled number(s) on the diagnostic chart.

1. This step is used to confirm that an open in the brake switch circuitry currently exists.
2. This step indicates if the brake switch signal is being received by the EBCM.
3. This step indicates if an open circuit exists in the brake switch or brake light circuitry.
4. This step isolates the open circuit to either the brake switch input circuitry,or the EBCM.

Diagnostic Aids:

An "Intermittent" problem may be caused by a poor connection, rubbed through wire insulation, or a wire that is broken inside the insulation

Any circuitry, that is suspected as causing the intermittent complaint, should be thoroughly checked for backed out terminals, improper mating, broken locks, improperly formed or damaged terminals, poor terminal to wiring connections or physical damage to the wiring harness.

DIAGNOSTIC CHART DTC A095 — 1993 ACHIEVA GRAND AM AND SKYLARK

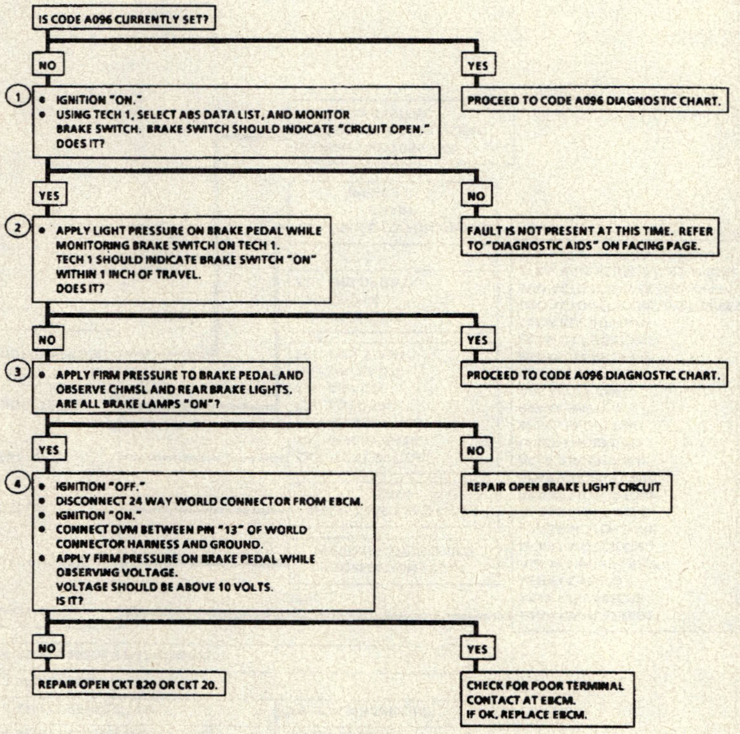

DIAGNOSTIC CHART DTC A096 — 1993 ACHIEVA GRAND AM AND SKYLARK

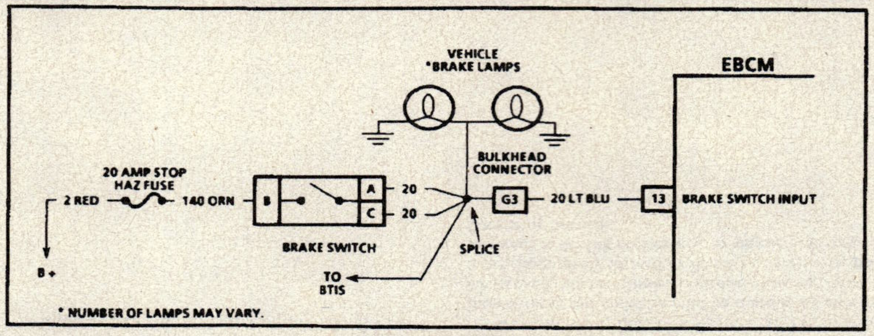

CODE A096

BRAKE LIGHTS CIRCUIT OPEN

Circuit Description:

This code is designed to isolate the cause of a Code A095 failure and indicate to the driver ABS is still available. If Code A095 fails with Code A096, the brake lights are open, the brake light grounds are open, or the Center High Mounted Stop Light (CHMSL) circuit is open during 4 way flasher use. The presence of battery voltage at the brake switch input indicates a valid brake switch input is still available.

Test Description: Number(s) below refer to circled number(s) on the diagnostic chart.

1. As a result of a failure of an additional brake switch circuit code, this code may be set. To insure proper diagnosis, any additional brake switch codes must be repaired first.
2. This step identifies if the fault is currently present in the brake circuit.

Diagnostic Aids:

An "Intermittent" problem may be caused by a poor connection, rubbed through wire insulation, or a wire that is broken inside the insulation.

Any circuitry, that is suspected as causing the intermittent complaint, should be thoroughly checked for backed out terminals, improper mating, broken locks, improperly formed or damaged terminals, poor terminal to wiring connections or physical damage to the wiring harness.

(1) ARE ANY OTHER BRAKE SWITCH CODES (EXCEPT A095) CURRENTLY SET?

- YES → PROCEED TO THAT CODE FIRST.
- NO → (2) • APPLY FIRM PRESSURE TO BRAKE PEDAL AND OBSERVE CHMSL AND REAR BRAKE LIGHTS. ARE ALL BRAKE LIGHTS "ON"?
  - YES → FAULT IS NOT PRESENT AT THIS TIME. REFER TO "DIAGNOSTIC AIDS" ON FACING PAGE.
  - NO → REPAIR OPEN BRAKE LIGHT CIRCUIT

DIAGNOSTIC PROCESS CHART 1994–95 ACHIEVA, GRAND AM AND SKYLARK (Page 1 of 3)

MISC CHASSIS APPLICATION SELECTION	TECH 1 SCREEN DISPLAY
STEP 1 SELECT VEHICLE MODEL YEAR	MODEL YEAR : 19??
STEP 2 SELECT VEHICLE TYPE	"F0: PASSENGER CAR"
STEP 3 SELECT BODY TYPE	
STEP 4 SELECT MODE (FIRST SCREEN)	

- F0: DATA LIST → FRONT WHL SPEEDS, REAR WHL SPEEDS, VEHICLE SPEED, ABS WARNING LAMP, LF ABS MOTOR CMD, LF ABS MOTOR FDBK, RF ABS MOTOR CMD, RF ABS MOTOR FDBK, REAR ABS MTR CMD, REAR ABS MTR FBK, LF SOLENOID CMD, RF SOLENOID CMD, BRAKE SWITCH, ABS RELAY CMD, BRK WARN LP CMD, BRAKE WARN LAMP, EBCM BAT VOLTAGE, EBCM IGN VOLTAGE, EVO ACTUATOR CMD (ABS W/VES), EVO ACTUATOR FBK (ABS W/VES), STRG. WHL SENSOR (ABS W/VES), STRG. WHL SENSOR (ABS W/VES)
- F1: DTC HISTORY → INFORMATION ABOUT PREVIOUS DTCs STORED
- F2: DTC(s) → DTCs STORED AND CLEAR DTCs
- F3: SNAPSHOT → SNAPSHOT OPTIONS
  - F0: REPLAY DATA
  - F1: MANUAL TRIG.
  - F2: AUTO TRIG.
  - F3: ANY DTC
  - F4: SINGLE DTC
  - F9: TRIG. POINT
  - → F0: BEGINNING / F1: CENTER / F2: END OF DATA
- F4: MISC. TESTS → CONTINUED ON FOLLOWING CHART (PAGE 2 OF 3)
- F5: MOTOR REHOME → REHOMES MOTORS AND PISTONS PRIOR TO BLEEDING THE BRAKE SYSTEM.

DIAGNOSTIC PROCESS CHART
1994–95 ACHIEVA, GRAND AM AND SKYLARK
(Page 2 of 3)

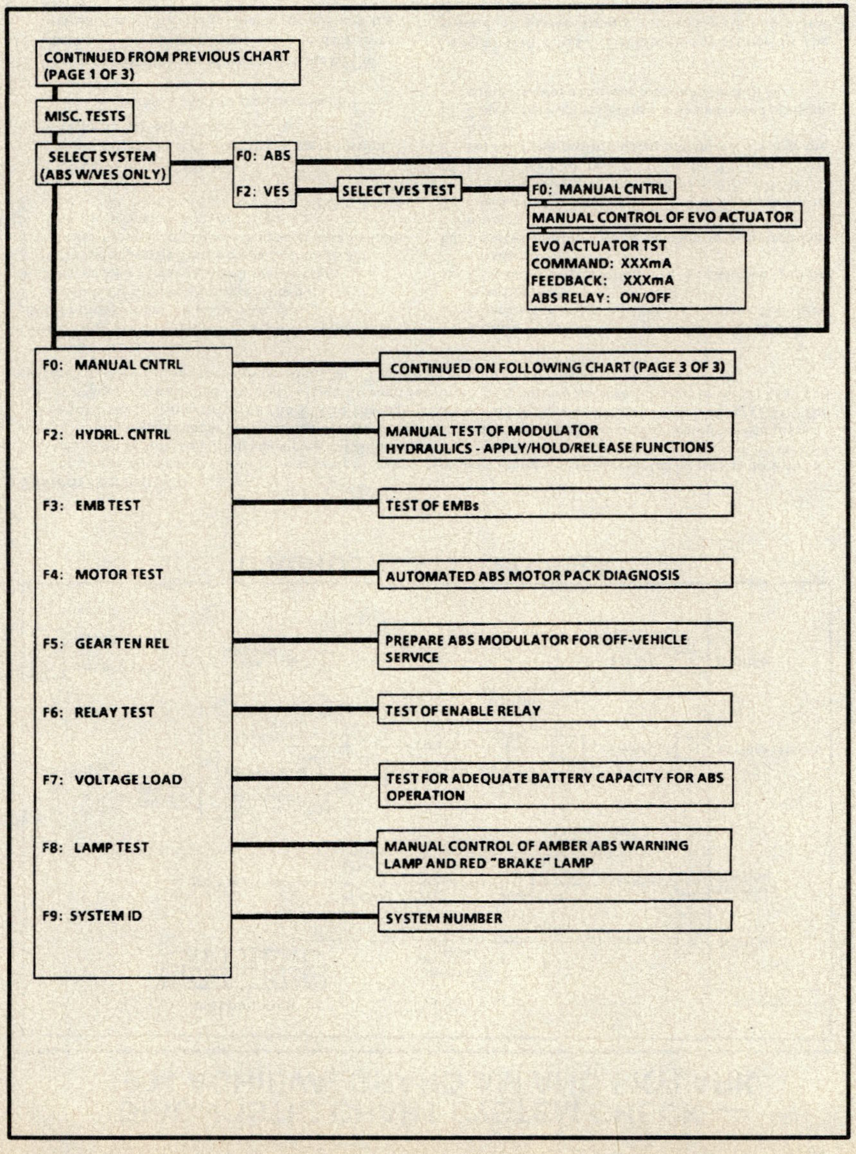

DIAGNOSTIC PROCESS CHART
1994–95 ACHIEVA, GRAND AM AND SKYLARK
(Page 3 of 3)

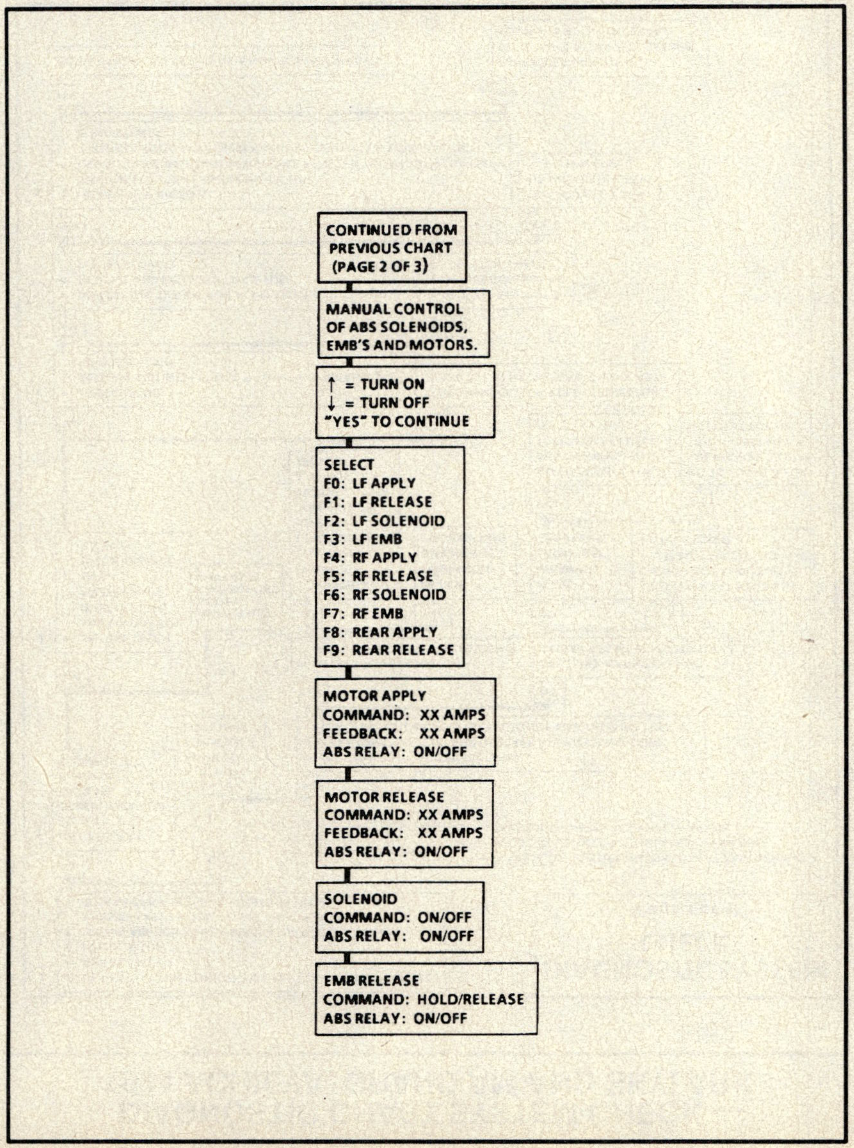

DIAGNOSTIC CHART SYSTEM CHECK — 1994 ACHIEVA, GRAND AM AND SKYLARK

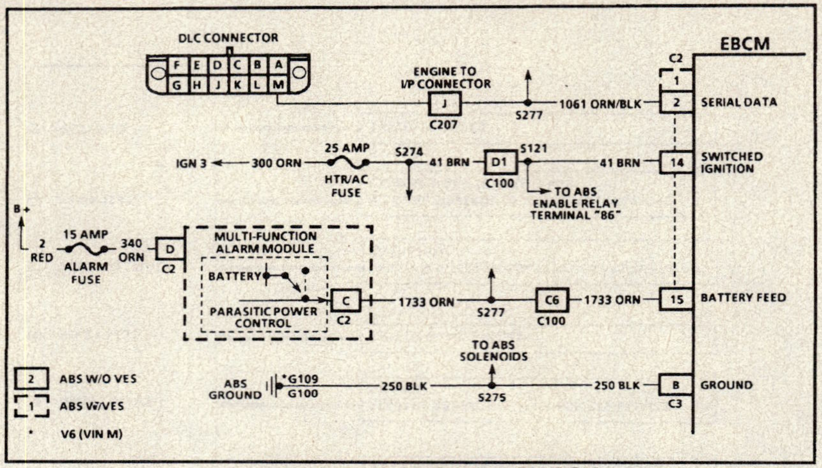

DIAGNOSTIC SYSTEM CHECK

System Description: The diagnostic system check is an organized approach to identifying a problem created by an Antilock Brake System (ABS) or Variable Effort Steering (VES) malfunction. It must be the starting point for any ABS or VES complaint diagnosis because it directs the service technician to the next logical step in diagnosing the complaint.

Serial data is transmitted by the EBCM through terminal "2" (ABS w/o VES) or "1" (ABS w/VES). The EBCM is supplied switched ignition voltage through terminal "14", battery voltage through terminal "15" and switched battery voltage through terminal "A". The EBCM ground is provided through terminal "B".

Chart Test Description: Number(s) below refer to circled number(s) on the diagnostic chart.

1. Indicates if EBCM is transmitting data.
2. Checks for a serial data link malfunction.
3. Checks for proper Tech 1 connection to DLC.
4. Checks for open multifunction alarm module fuse.
5. Checks for open heater-A/C fuse.
6. Checks for high resistance in the EBCM ground circuit.
7. Checks for current DTCs.
8. Checks for proper ABS warning lamp operation.
9. Checks for history DTCs.
10. Checks for possible short to ground in the B+ feed circuitry.
11. Checks for possible short to ground in the EBCM.
12. Checks for short to ground in the B+ feed circuitry between the multifunction alarm module and the EBCM.
13. Ensures short to ground is not due to physical damage of the circuitry.
14. Checks for possible short to ground in the switched ignition circuitry.
15. Isolates the short to ground to either the circuitry or the EBCM.
16. Ensures short to ground is not due to physical damage of the circuitry.
17. Checks for proper battery voltage at the EBCM.
18. Checks for high resistance in the B+ feed circuitry.
19. Checks for high resistance in the switched ignition circuitry.
20. Ensures malfunction is not due to poor terminal contact.
21. Checks for an open in the B+ feed circuit between the multifunction alarm module and the EBCM.
22. Checks for high resistance in the B+ feed circuit between the multifunction alarm module and the EBCM.
23. Checks for high resistance in the switched ignition circuit between the fuse block and the EBCM.

Diagnostic Aids: Excessive resistance in the ground or power supply circuits will not allow communication with the EBCM. If communication with the EBCM is not possible, ensure the ABS ground connection is good and that there is no excessive resistance in any of the power supply circuits.

DIAGNOSTIC CHART SYSTEM CHECK — 1994 ACHIEVA, GRAND AM AND SKYLARK

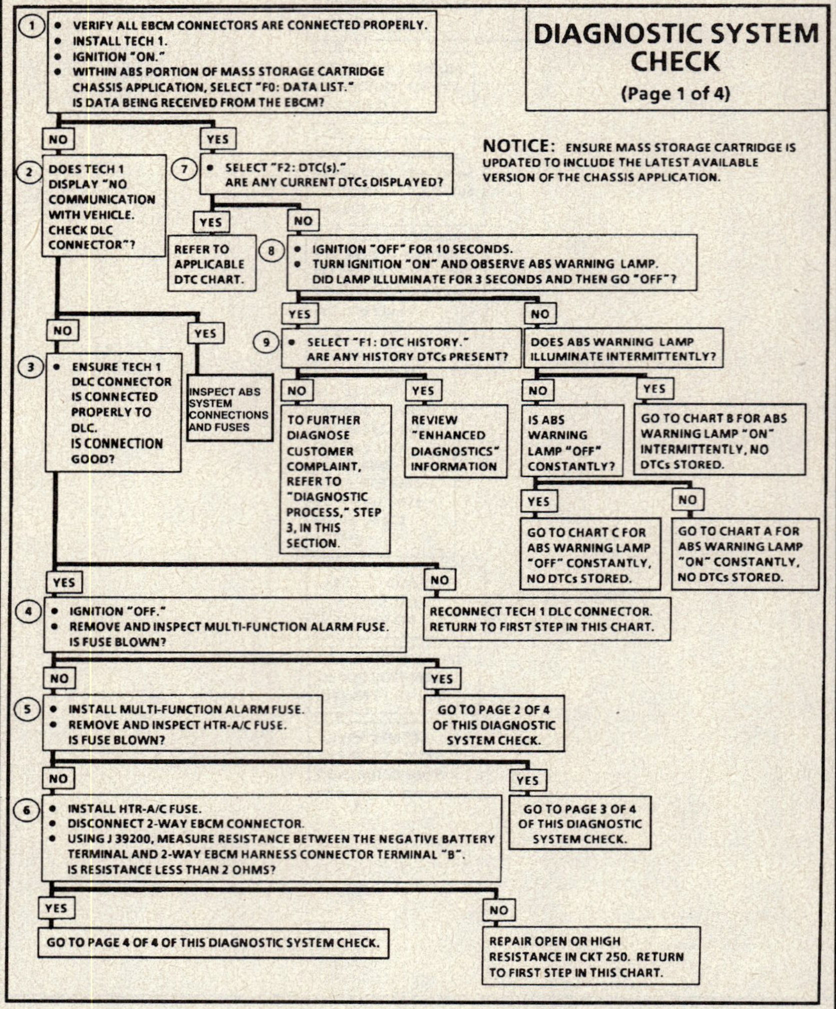

DIAGNOSTIC CHART SYSTEM CHECK — 1994 ACHIEVA, GRAND AM AND SKYLARK

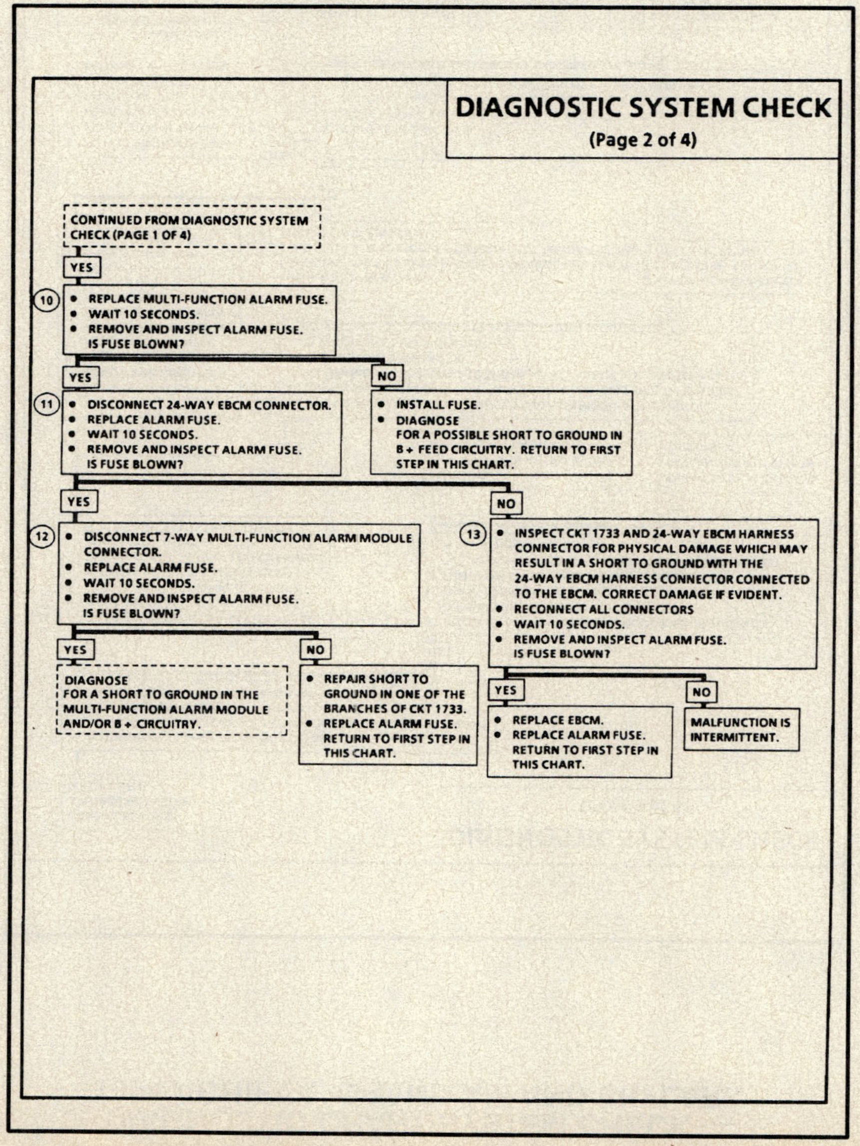

DIAGNOSTIC CHART SYSTEM CHECK — 1994 ACHIEVA, GRAND AM AND SKYLARK

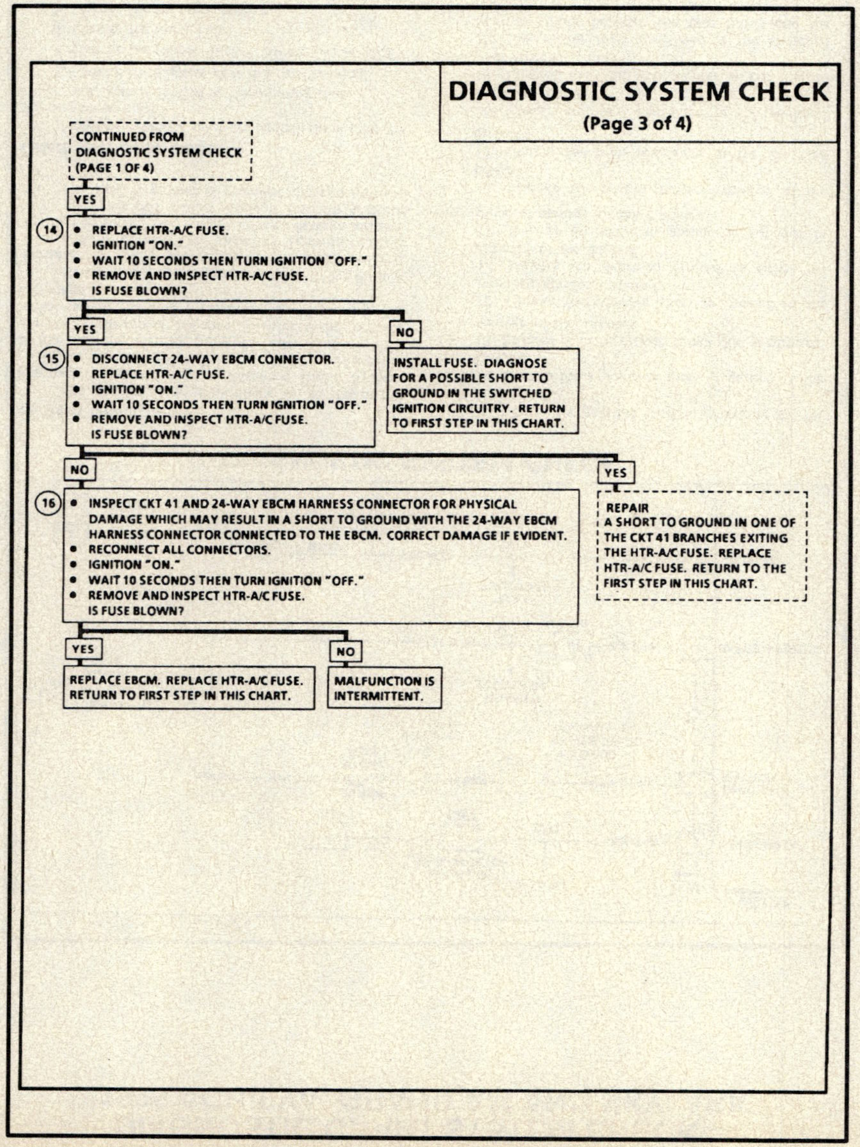

DIAGNOSTIC CHART SYSTEM CHECK — 1994 ACHIEVA, GRAND AM AND SKYLARK

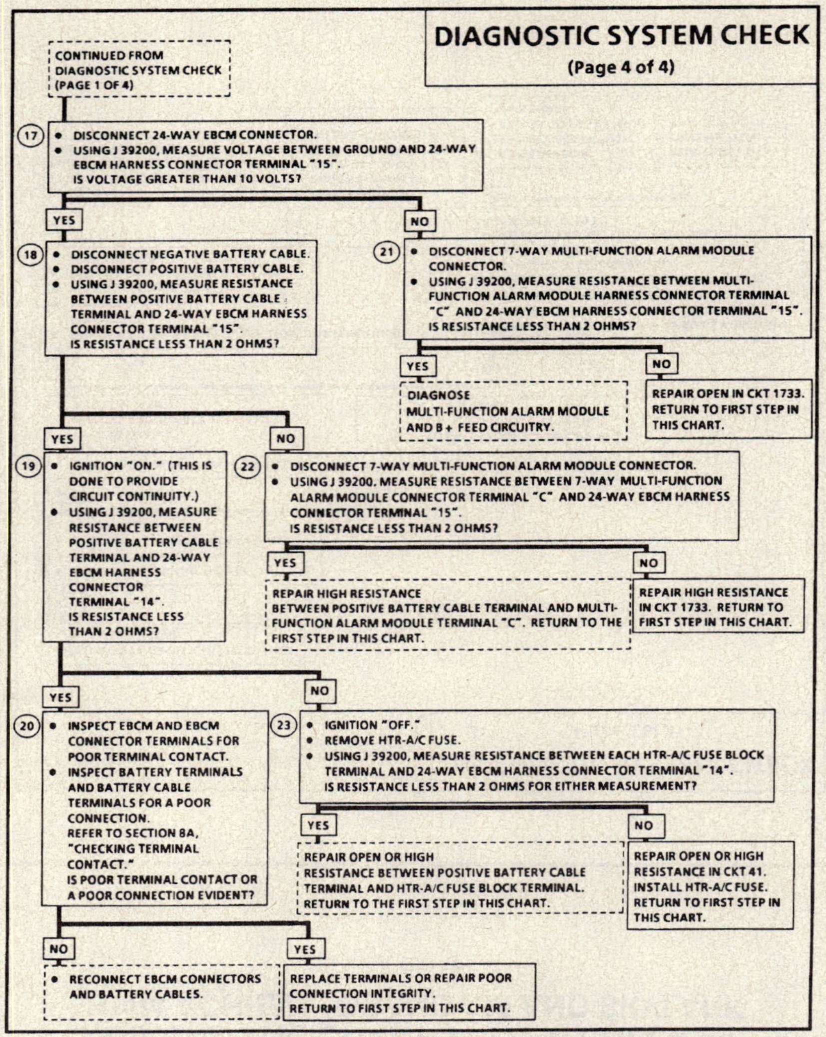

DIAGNOSTIC CHART SYSTEM CHECK — 1995 ACHIEVA, GRAND AM AND SKYLARK

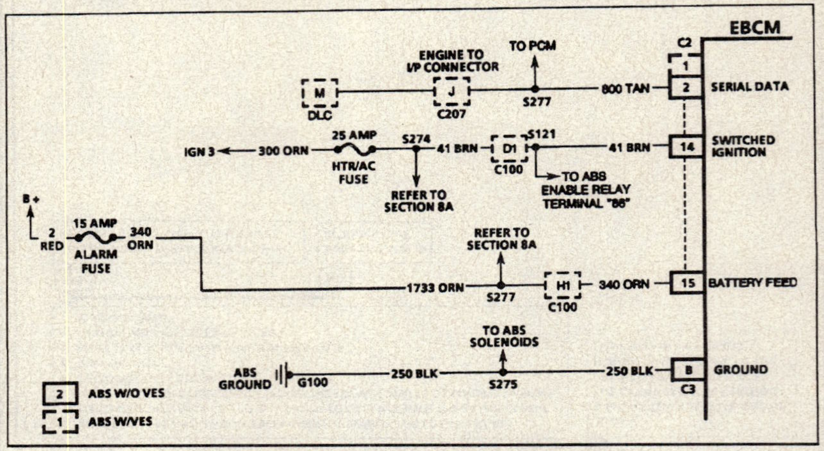

DIAGNOSTIC SYSTEM CHECK

System Description:

The diagnostic system check is an organized approach to identifying an Antilock Brake System (ABS) or Variable Effort Steering (VES) malfunction. It must be the starting point for any ABS or VES complaint diagnosis because it directs the service technician to the next logical step in diagnosing the complaint.

Serial data is transmitted by the EBCM through terminal "2" (ABS w/o VES) or "1" (ABS w/VES). The EBCM is supplied switched ignition voltage through terminal "14", battery voltage through terminal "15" and switched battery voltage through terminal "A". The EBCM ground is provided through terminal "B."

Chart Test Description:

Number(s) below refer to circled number(s) on the diagnostic chart.

1. Indicates if EBCM is transmitting data.
2. Checks for a serial data link malfunction.
3. Checks for proper Tech 1 connection to DLC.
4. Checks for open Fuse 11.
5. Checks for an open Fuse 4
6. Checks for high resistance in the EBCM ground circuit.
9. Checks for current DTCs.
8. Checks for proper ABS warning lamp operation.
9. Checks for history DTCs.
10. Checks for possible short to ground in the B+ feed circuitry.
11. Checks for a possible short to ground in the EBCM.
12. Ensures short to ground is not due to physical damage of the circuitry.
13. Checks for possible short to ground in the switched ignition circuitry.
14. Isolates the short to ground to either the circuitry or the EBCM.
15. Ensures the short to ground is not due to physical damage of the circuitry.
16. Checks for proper battery voltage at the EBCM.
17. Checks for high resistance in the B+ feed circuitry.
18. Checks for high resistance in the switched ignition circuitry.
19. Ensures that the malfunction is not due to poor terminal contact.
20. Checks for high resistance in the switched ignition circuit between the fuse block and the EBCM.

Diagnostic Aids:

Excessive resistance in the ground or power supply circuits will not allow communication with the EBCM. If communication with the EBCM is not possible, ensure the ABS ground connection is good and that there is no excessive resistance in any of the power supply circuits.

DIAGNOSTIC CHART SYSTEM CHECK — 1994 ACHIEVA, GRAND AM AND SKYLARK

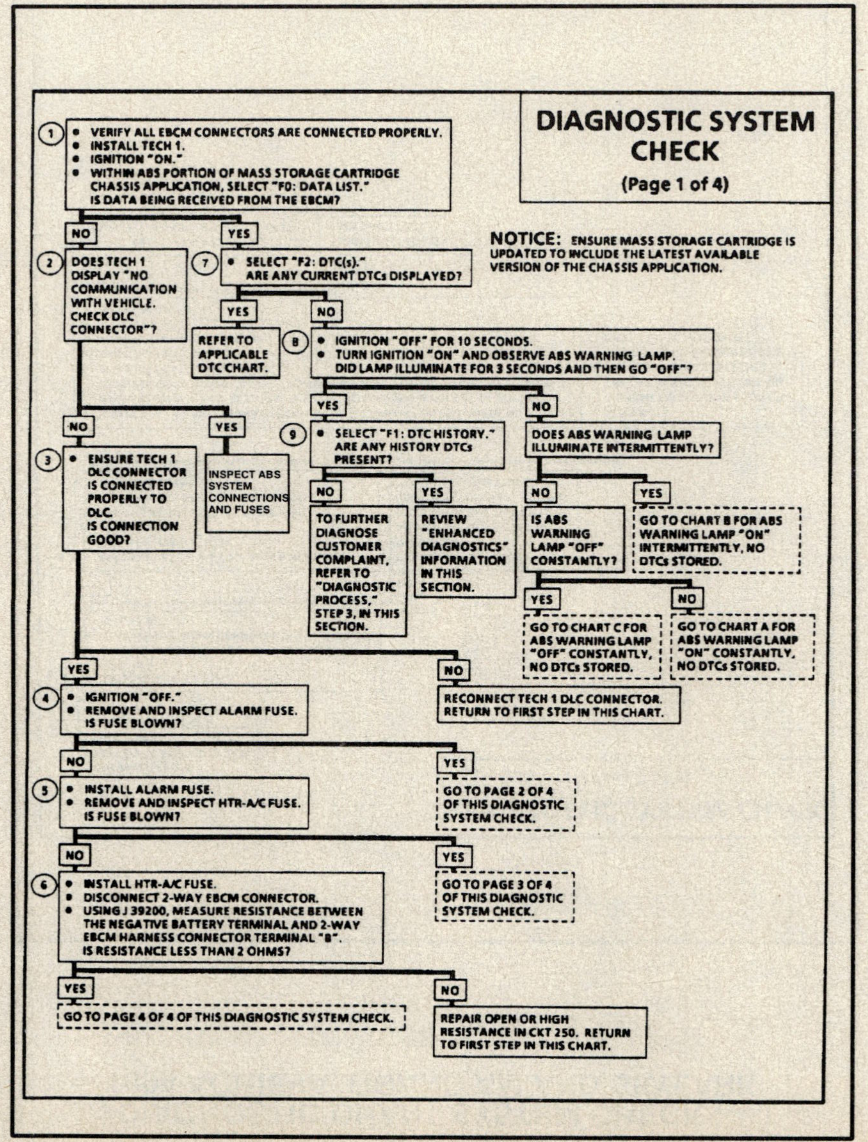

DIAGNOSTIC CHART SYSTEM CHECK — 1995 ACHIEVA, GRAND AM AND SKYLARK

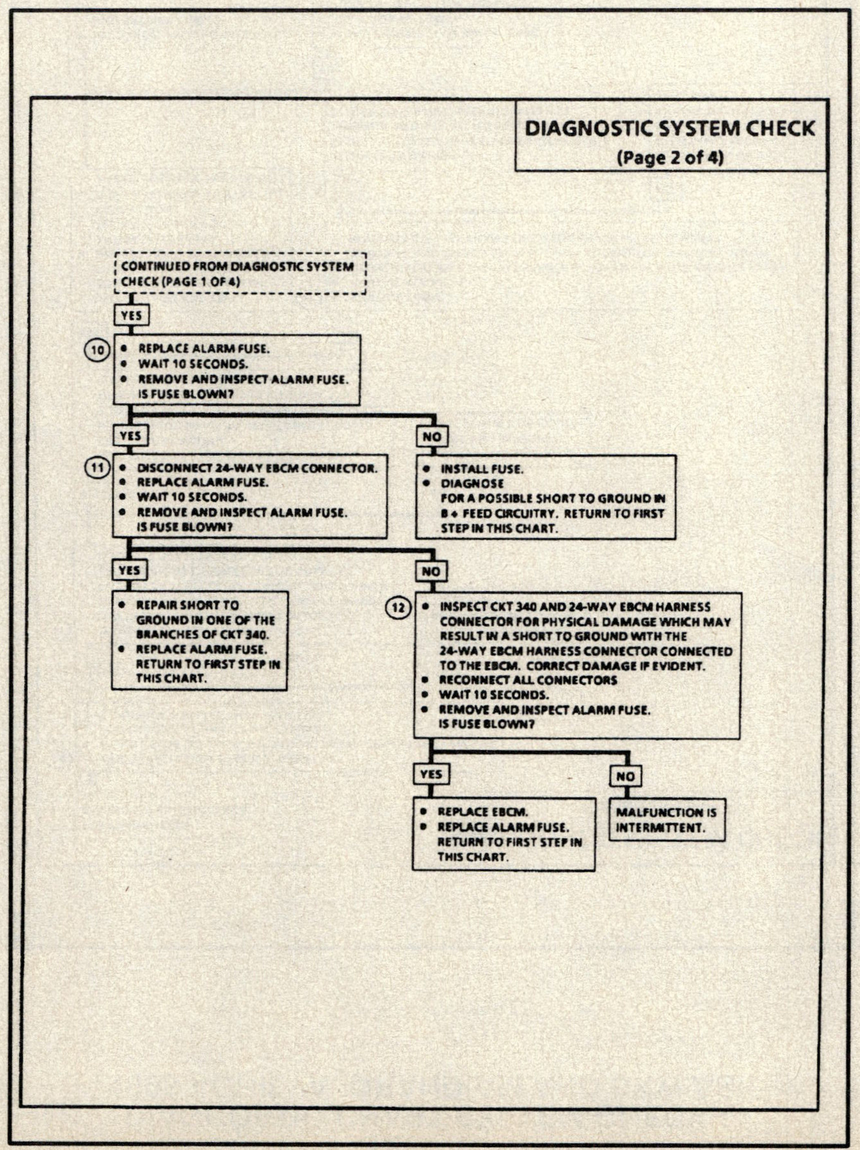

DIAGNOSTIC CHART SYSTEM CHECK — 1994 ACHIEVA, GRAND AM AND SKYLARK

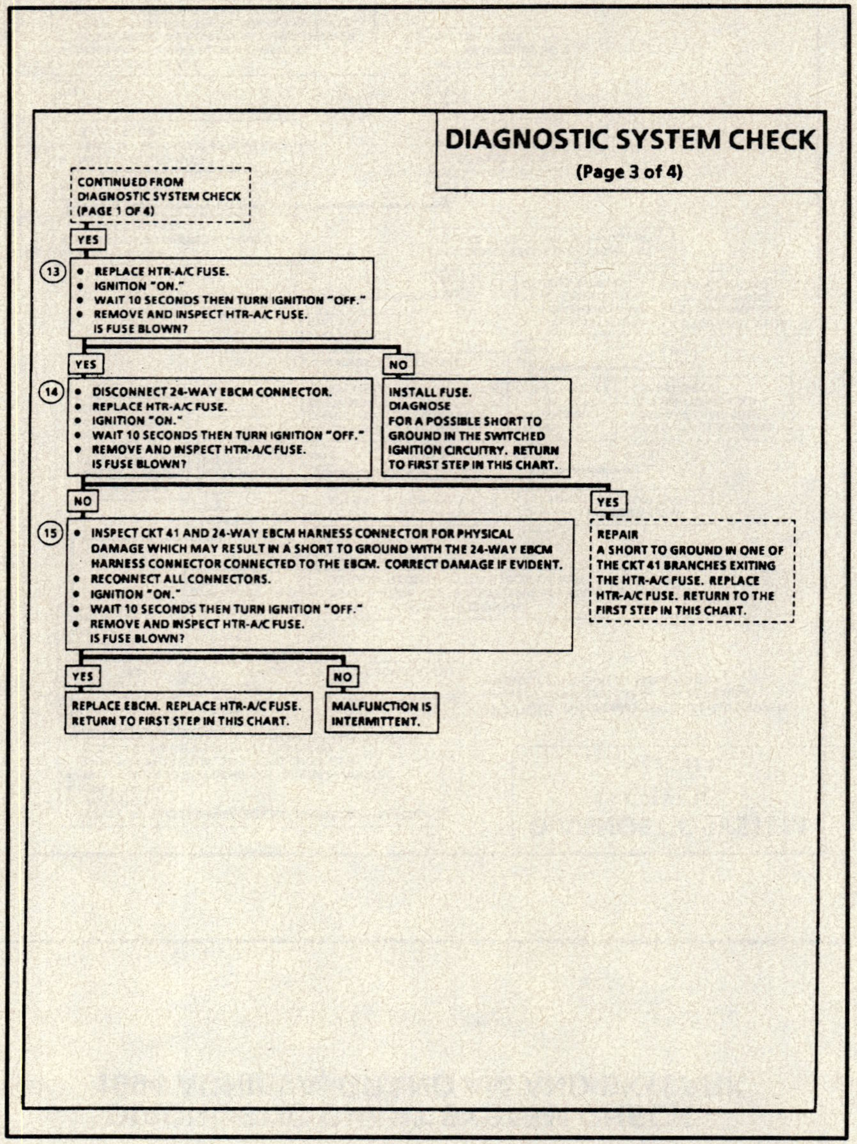

DIAGNOSTIC CHART SYSTEM CHECK — 1995 ACHIEVA, GRAND AM AND SKYLARK

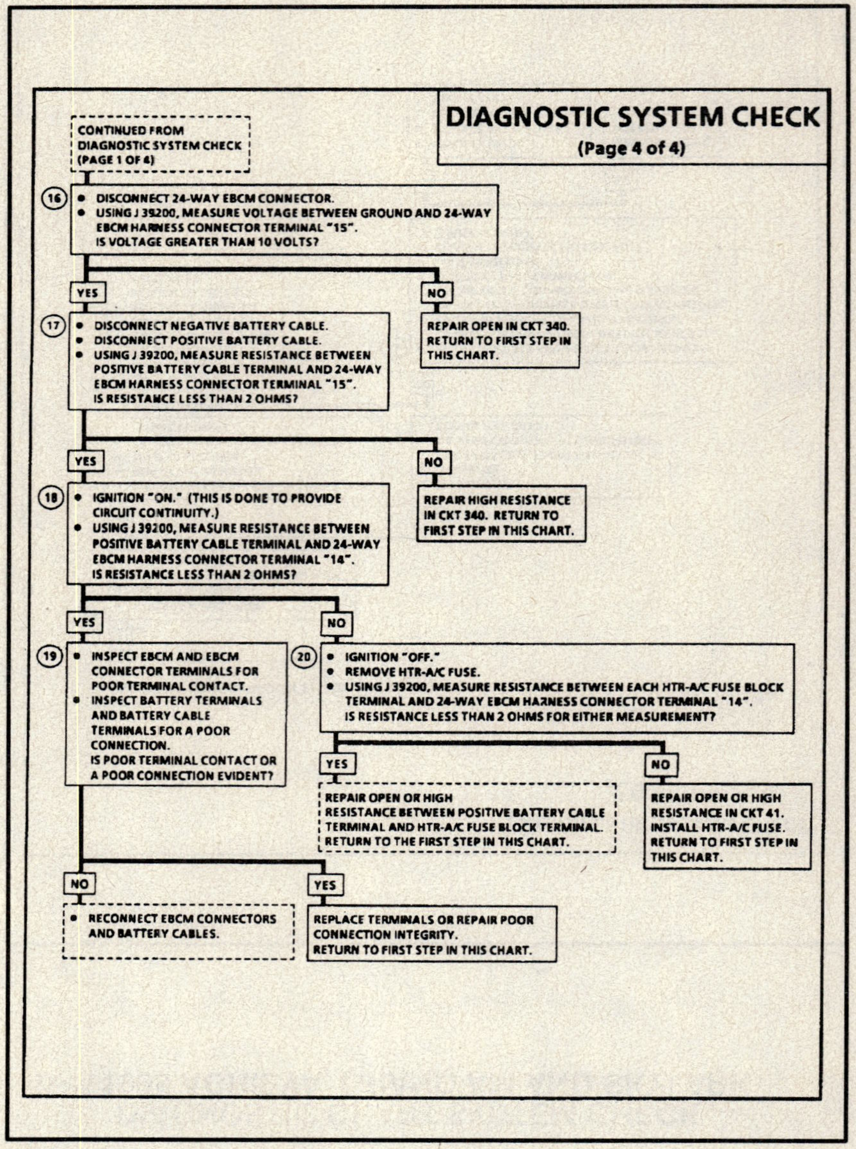

DIAGNOSTIC TROUBLE CODE CHART
1994–95 ACHIEVA, GRAND AM AND SKYLARK

DIAGNOSTIC TROUBLE CODE AND SYMPTOM TABLE

CHART	SYMPTOM
A	ABS (Amber) Warning Lamp "ON" Constantly, No DTCs Stored
B	ABS (Amber) Warning Lamp "ON" Intermittently, No DTCs Stored (1 of 2)
C	ABS (Amber) Warning Lamp "OFF" Constantly, No DTCs Stored
D	Tech 1 Displays Undefined DTCs
DIAGNOSTIC TROUBLE CODE	**DESCRIPTION**
11	ABS Warning Lamp Circuit Malfunction (1 of 2)
14	ABS Enable Relay Contact Circuit Open (1 of 3)
15	ABS Enable Relay Contact Circuit Shorted to Battery or Always Closed
16	ABS Enable Relay Coil Circuit Open
17	ABS Enable Relay Coil Circuit Shorted to Ground
18	ABS Enable Relay Coil Circuit Shorted to Battery or Coil Shorted
21	Left Front Wheel Speed = 0 or Unreasonable (For ABS W/O VES Only) (1 of 2)
21	Left Front Wheel Speed = 0 (For ABS With VES Only)
22	Right Front Wheel Speed = 0 or Unreasonable (For ABS W/O VES Only) (1 of 2)
22	Right Front Wheel Speed = 0 (For ABS With VES Only)
23	Left Rear Wheel Speed = 0 or Unreasonable (For ABS W/O VES Only) (1 of 4)
23	Left Rear Wheel Speed = 0 (For ABS With VES Only) (1 of 2)
24	Right Rear Wheel Speed = 0 or Unreasonable (For ABS W/O VES Only) (1 of 4)
24	Right Rear Wheel Speed = 0 (For ABS With VES Only) (1 of 2)
25	Left Front Excessive Wheel Speed Variation (1 of 3)
26	Right Front Excessive Wheel Speed Variation (1 of 3)
27	Left Rear Excessive Wheel Speed Variation (1 of 3)
28	Right Rear Excessive Wheel Speed Variation (1 of 3)
32	Left Front Wheel Speed Sensor Circuit Open or Shorted to Ground/Battery (For ABS With VES Only) (1 of 3)
33	Right Front Wheel Speed Sensor Circuit Open or Shorted to Ground/Battery (For ABS With VES Only) (1 of 3)
34	Left Rear Wheel Speed Sensor Circuit Open or Shorted to Ground/Battery (For ABS With VES Only) (1 of 3)
35	Right Rear Wheel Speed Sensor Circuit Open or Shorted to Ground/Battery (For ABS With VES Only) (1 of 3)
36	Low System Voltage
37	High System Voltage
38	Left Front EMB Will Not Hold Motor
41	Right Front EMB Will Not Hold Motor
42	Rear ESB Will Not Hold Motor
43	VES Steering Wheel Speed Sensor Circuit Malfunction (For ABS With VES Only) (1 of 3)

DIAGNOSTIC TROUBLE CODE CHART
1994–95 ACHIEVA, GRAND AM AND SKYLARK

DIAGNOSTIC TROUBLE CODE AND SYMPTOM TABLE

DIAGNOSTIC TROUBLE CODE	DESCRIPTION
44	Left Front ABS Channel Will Not Move (1 of 2)
45	Right Front ABS Channel Will Not Move (1 of 2)
46	Rear ABS Channel Will Not Move
47	Left Front ABS Motor Free Spins (1 of 2)
48	Right Front ABS Motor Free Spins (1 of 2)
51	Rear ABS Motor Free Spins (1 of 2)
52	Left Front ABS Channel in Release Too Long
53	Right Front ABS Channel In Release Too Long
54	Rear ABS Channel in Release Too Long
55	EBCM Malfunction
56	Left Front ABS Motor Circuit Open
57	Left Front ABS Motor Circuit Shorted to Ground
58	Left Front ABS Motor Circuit Shorted to Battery or Motor Shorted
61	Right Front ABS Motor Circuit Open
62	Right Front ABS Motor Circuit Shorted to Ground
63	Right Front ABS Motor Circuit Shorted to Battery or Motor Shorted
64	Rear ABS Motor Circuit Open
65	Rear ABS Motor Circuit Shorted to Ground
66	Rear ABS Motor Circuit Shorted to Battery or Motor Shorted
67	Left Front EMB Circuit Open or Shorted to Ground
68	Left Front EMB Circuit Shorted to Battery or Driver Open
71	Right Front EMB Circuit Open or Shorted to Ground
72	Right Front EMB Circuit Shorted to Battery or Driver Open
73	EVO Actuator Circuit Open or Shorted to Ground (For ABS With VES Only)
74	EVO Actuator Circuit Shorted to Battery or Solenoid Shorted (For ABS With VES Only)
76	Left Front Solenoid Circuit Open or Shorted to Battery
77	Left Front Solenoid Circuit Shorted to Ground or Driver Open
78	Right Front Solenoid Circuit Open or Shorted to Battery
81	Right Front Solenoid Circuit Shorted to Ground or Driver Open
82	Calibration Malfunction
86	EBCM Turned "ON" the Red "BRAKE" Warning Lamp
87	Red "BRAKE" Warning Lamp Circuit Open
88	Red "BRAKE" Warning Lamp Circuit Shorted to Battery
91	Open Brake Switch Contacts During Deceleration
92	Open Brake Switch Contacts When ABS Was Required
93	DTCs 91 or 92 Set in Current or Previous Ignition Cycle
94	Brake Switch Contacts Always Closed
95	Brake Switch Circuit Open
96	Brake Lamp Circuit Open

ELECTRICAL CONNECTOR TERMINAL IDENTIFICATION WITHOUT VES — 1994–95 ACHIEVA, GRAND AM AND SKYLARK

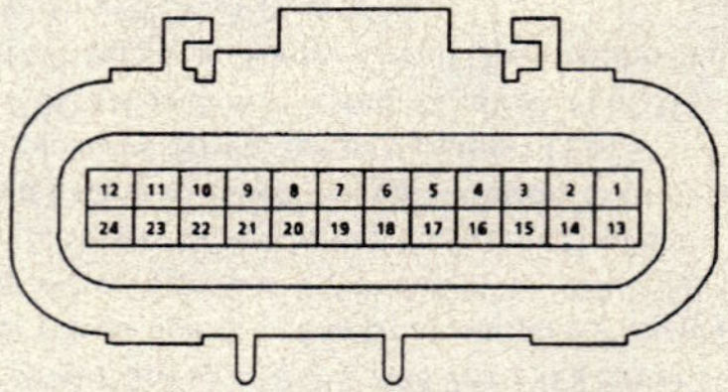

24-WAY EBCM HARNESS CONNECTOR (ABS WITHOUT VES)

PIN	CIRCUIT NO.	COLOR	CIRCUIT
1	OPEN		NOT USED
2	1061	ORN/BLK	SERIAL DATA
3	OPEN		NOT USED
4	1289	LT BLU	R/F ABS SOLENOID CONTROL
5	830	LT BLU	L/F WHEEL SIGNAL HIGH
6	873	YEL	L/F WHEEL SIGNAL LOW
7	882	BRN	R/R WHEEL SIGNAL HIGH
8	883	WHT	R/R WHEEL SIGNAL LOW
9	872	DK GRN	R/F WHEEL SIGNAL HIGH
10	833	TAN	R/F WHEEL SIGNAL LOW
11	885	RED	L/R WHEEL SIGNAL LOW
12	884	BLK	L/R WHEEL SIGNAL HIGH
13	17	WHT	BRAKE SWITCH INPUT
14	41	BRN	SWITCHED IGNITION
15	1733	ORN	BATTERY FEED
16	OPEN		NOT USED
17	OPEN		NOT USED
18	VENT TUBE	BLK	VENT TUBE
19	1286	LT GRN	L/F EMB CONTROL
20	1287	GRY	R/F EMB CONTROL
21	33	TAN/WHT	BRAKE WARNING LAMP CONTROL
22	1632	PNK	ENABLE RELAY CONTROL
23	852	WHT	ABS WARNING LAMP CONTROL
24	1288	DK GRN	L/F ABS SOLENOID CONTROL

ELECTRICAL CONNECTOR TERMINAL IDENTIFICATION WITH VES — 1994–95 ACHIEVA, GRAND AM AND SKYLARK

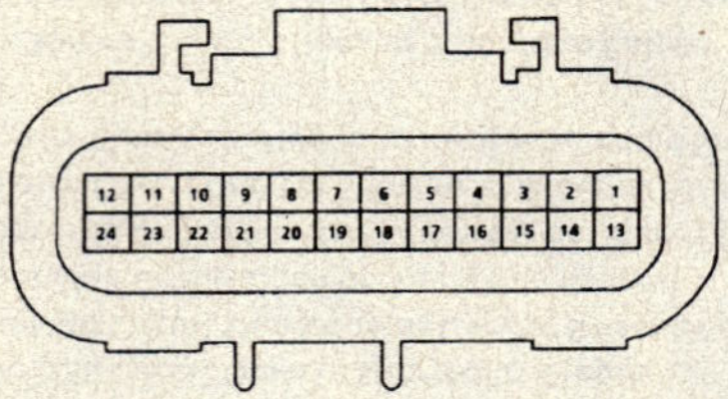

24-WAY EBCM HARNESS CONNECTOR (ABS WITH VES)

PIN	CIRCUIT NO.	COLOR	CIRCUIT
1	1061	ORN/BLK	SERIAL DATA
2	1289	LT BLU	R/F ABS SOLENOID CONTROL
3	830	LT BLU	L/F WHEEL SIGNAL HIGH
4	873	YEL	L/F WHEEL SIGNAL LOW
5	882	BRN	R/R WHEEL SIGNAL HIGH
6	883	WHT	R/R WHEEL SIGNAL LOW
7	872	DK GRN	R/F WHEEL SIGNAL HIGH
8	833	TAN	R/F WHEEL SIGNAL LOW
9	884	BLK	L/R WHEEL SIGNAL HIGH
10	885	RED	L/R WHEEL SIGNAL LOW
11	1295	BRN	EVO ACTUATOR CONTROL
12	1056	GRY	STEERING WHEEL SENSOR VCC
13	17	WHT	BRAKE SWITCH INPUT
14	41	BRN	SWITCHED IGNITION
15	1733	ORN	BATTERY FEED
16	556	ORN/BLK	STEERING WHEEL SENSOR GROUND
17	1286	LT GRN	L/F EMB CONTROL
18	1287	GRY	R/F EMB CONTROL
19	33	TAN/WHT	BRAKE WARNING LAMP CONTROL
20	1632	PNK	ENABLE RELAY CONTROL
21	852	WHT	ABS WARNING LAMP CONTROL
22	1288	DK GRN	L/F ABS SOLENOID CONTROL
23	1059	LT BLU	STEERING WHEEL SENSOR INPUT
24	VENT TUBE	BLK	VENT TUBE

ELECTRICAL CONNECTOR TERMINAL IDENTIFICATION — 1994–95 ACHIEVA, GRAND AM AND SKYLARK

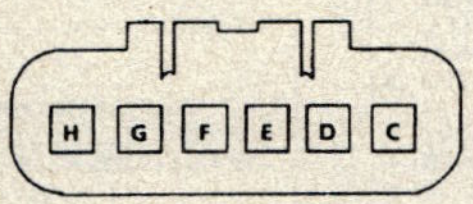

6-WAY EBCM HARNESS CONNECTOR

PIN	CIRCUIT NO.	COLOR	CIRCUIT
C	1284	DK GRN	REAR MOTOR HIGH
D	1285	ORN	REAR MOTOR LOW
E	1281	PNK	L/F MOTOR LOW
F	1280	BLK	L/F MOTOR HIGH
G	1283	BLK	R/F MOTOR LOW
H	1282	PPL	R/F MOTOR HIGH

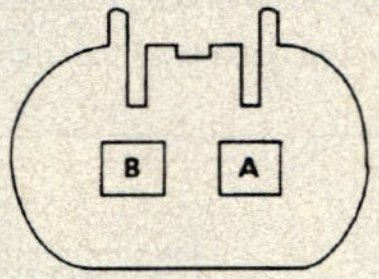

2-WAY EBCM HARNESS CONNECTOR

PIN	CIRCUIT NO.	COLOR	CIRCUIT
A	1633	RED	SWITCHED BATTERY INPUT
B	250	BLK	GROUND

DIAGNOSTIC CHART A 1994–95 ACHIEVA, GRAND AM AND SKYLARK

CHART A

ABS (AMBER) WARNING LAMP "ON" CONSTANTLY, NO DTCs STORED

NOTICE: DIAGNOSTIC SYSTEM CHECK MUST BE COMPLETED FIRST BEFORE USING THIS CHART.

- IGNITION "OFF."
- INSTALL TECH 1.
- IGNITION "ON."
- USING TECH 1, SELECT "F4: MISC. TESTS."
- SELECT "F8: LAMP TEST."
- ATTEMPT TO FLASH THE ABS WARNING LAMP. DOES ABS WARNING LAMP FLASH?

YES: MALFUNCTION IS NOT PRESENT AT THIS TIME.

NO:

- IGNITION "OFF."
- DISCONNECT LAMP DRIVER MODULE FROM HARNESS.
- IGNITION "ON."
- IS ABS WARNING LAMP "OFF"?

NO: REPAIR SHORT TO GROUND IN CKT 875. REFER TO SECTION 8A-81 FOR FURTHER CLUSTER DIAGNOSIS.

YES: REPLACE LAMP DRIVER MODULE ASSEMBLY.

DIAGNOSTIC CHART B
1994–95 ACHIEVA, GRAND AM AND SKYLARK
(Page 1 of 2)

CHART B
(Page 1 of 2)
ABS (AMBER) WARNING LAMP "ON" INTERMITTENTLY, NO DTCs STORED

NOTICE: DIAGNOSTIC SYSTEM CHECK MUST BE COMPLETED FIRST BEFORE USING THIS CHART.

- IGNITION "OFF."
- DISCONNECT 2-WAY EBCM CONNECTOR.
- USING J 39200, MEASURE RESISTANCE BETWEEN THE NEGATIVE BATTERY TERMINAL AND 2-WAY EBCM HARNESS CONNECTOR TERMINAL "B".
- OBSERVE MEASURED RESISTANCE WHILE "WIGGLING" WIRE HARNESS AND CONNECTORS.

IS RESISTANCE ALWAYS LESS THAN 2 OHMS?

NO: REPAIR INTERMITTENT OPEN OR HIGH RESISTANCE IN CKT 250.

YES:

- DISCONNECT NEGATIVE BATTERY CABLE.
- DISCONNECT POSITIVE BATTERY CABLE.
- DISCONNECT 24-WAY EBCM CONNECTOR.
- USING J 39200, MEASURE RESISTANCE BETWEEN POSITIVE BATTERY CABLE TERMINAL AND 24-WAY EBCM HARNESS CONNECTOR TERMINAL "15".
- OBSERVE MEASURED RESISTANCE WHILE "WIGGLING" WIRE HARNESS AND CONNECTORS.

IS RESISTANCE ALWAYS LESS THAN 2 OHMS?

NO: GO TO PAGE 2 OF 2 OF THIS CHART.

YES:

- IGNITION "ON" (THIS IS DONE TO PROVIDE CIRCUIT CONTINUITY).
- USING J 39200, MEASURE RESISTANCE BETWEEN POSITIVE BATTERY CABLE TERMINAL AND 24-WAY EBCM HARNESS CONNECTOR TERMINAL "14". OBSERVE MEASURED RESISTANCE WHILE "WIGGLING" WIRE HARNESS AND CONNECTORS.

IS RESISTANCE ALWAYS LESS THAN 2 OHMS?

NO:

- IGNITION "OFF."
- REMOVE HTR-A/C FUSE.
- USING J 39200, MEASURE RESISTANCE BETWEEN EACH HTR-A/C FUSE BLOCK TERMINAL AND 24-WAY EBCM HARNESS CONNECTOR TERMINAL "14".
- OBSERVE MEASURED RESISTANCE WHILE WIGGLING WIRE HARNESS AND CONNECTORS.

IS RESISTANCE ALWAYS LESS THAN 2 OHMS FOR EITHER MEASUREMENT?

- **YES:** REPAIR INTERMITTENT OPEN OR HIGH RESISTANCE BETWEEN POSITIVE BATTERY CABLE TERMINAL AND HTR-A/C FUSE BLOCK TERMINAL.
- **NO:** REPAIR INTERMITTENT OPEN OR HIGH RESISTANCE IN CKT 41. REFER TO "INTERMITTENTS AND POOR CONNECTIONS"

YES:

- CHECK POSITIVE BATTERY TERMINAL AND POSITIVE BATTERY CABLE TERMINAL FOR A POOR CONNECTION.
- INSPECT FOR POOR TERMINAL CONTACT BETWEEN EBCM AND EBCM HARNESS CONNECTOR TERMINALS "14", "15" AND "B".

IS POOR TERMINAL CONTACT OR A POOR CONNECTION EVIDENT?

YES: REPLACE TERMINALS OR REPAIR CONNECTION INTEGRITY (IF POOR CONTACT EXISTS AT BATTERY TERMINAL) THAT EXHIBIT(S) POOR CONTACT.

NO:

- IGNITION "OFF."
- RECONNECT ALL CONNECTORS.
- IGNITION "ON."

IS ABS WARNING LAMP "ON" INTERMITTENTLY WITH NO DTCs STORED?

- **NO:** MALFUNCTION IS NOT PRESENT AT THIS TIME. INTERMITTENTS COULD BE CAUSED BY INCORRECT WIRING HARNESS ROUTING OR LOOSE GROUND CONNECTIONS.
- **YES:** REPLACE EBCM.

DIAGNOSTIC CHART B
1994–95 ACHIEVA, GRAND AM AND SKYLARK
(Page 2 of 2)

CHART B
(Page 2 of 2)
ABS (AMBER) WARNING LAMP "ON" INTERMITTENTLY, NO DTCs STORED

CONTINUED FROM PAGE 1 OF 2 OF THIS CHART.

- DISCONNECT 7-WAY MALFUNCTION ALARM MODULE CONNECTOR.
- USING J 39200, MEASURE RESISTANCE BETWEEN MULTIFUNCTION ALARM MODULE CONNECTOR TERMINAL "C" AND 24-WAY EBCM HARNESS CONNECTOR TERMINAL "15".
- OBSERVE MEASURED RESISTANCE WHILE WIGGLING WIRE HARNESS AND CONNECTORS.

IS RESISTANCE ALWAYS LESS THAN 2 OHMS?

- **YES:** REPAIR INTERMITTENT OPEN OR HIGH RESISTANCE BETWEEN POSITIVE BATTERY CABLE TERMINAL AND MULTIFUNCTION 7-WAY ALARM MODULE CONNECTOR TERMINAL "C".
- **NO:** REPAIR INTERMITTENT OPEN OR HIGH RESISTANCE IN CKT 1733.

DIAGNOSTIC CHART C
1994–95 ACHIEVA, GRAND AM AND SKYLARK

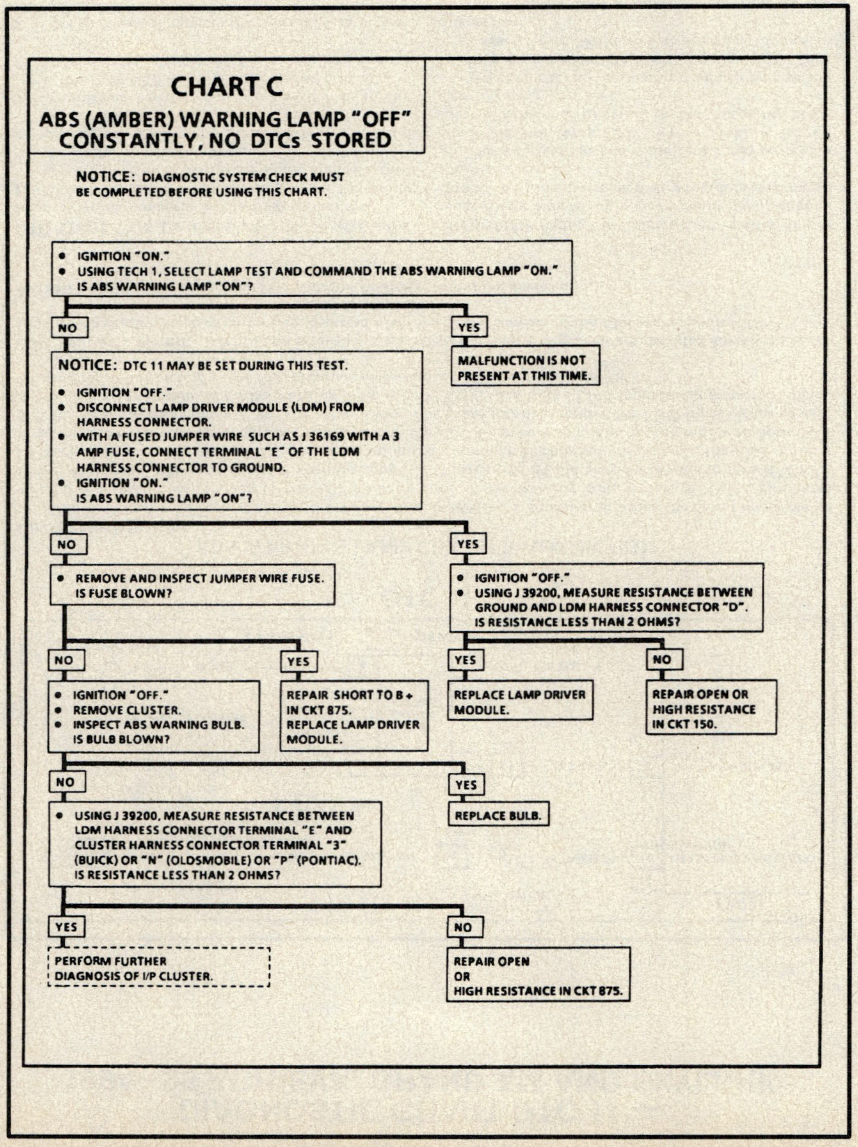

DIAGNOSTIC CHART D
1994–95 ACHIEVA, GRAND AM AND SKYLARK

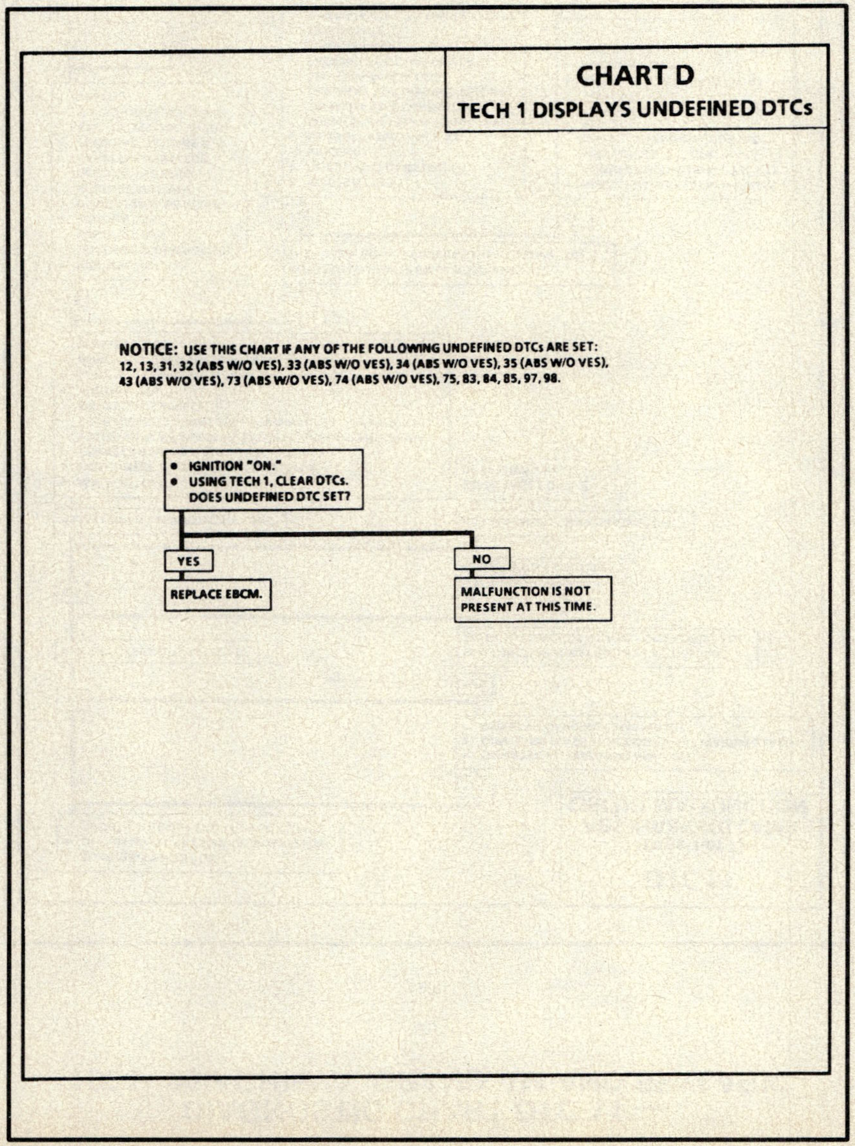

DIAGNOSTIC CHART DTC 11 — 1994–95 ACHIEVA, GRAND AM AND SKYLARK

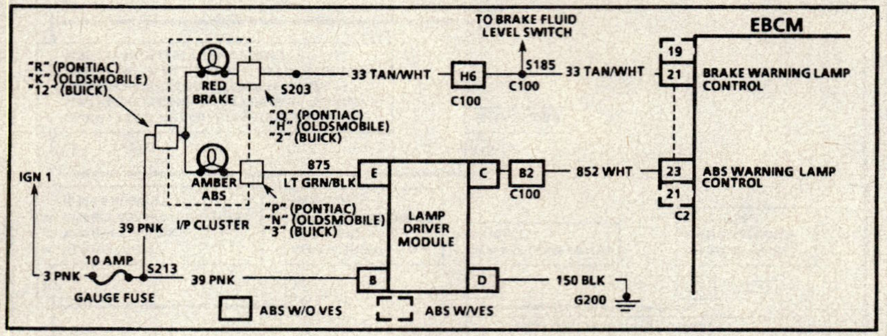

DTC 11

ABS WARNING LAMP CIRCUIT MALFUNCTION

Circuit Description:

This test checks the state of the ABS warning lamp to identify a situation in which the driver could not be warned of a system malfunction by the ABS warning lamp, or the warning lamp is always "ON." The Lamp Driver Module (LDM) turns the ABS warning lamp "ON," unless the EBCM provides ground to turn it "OFF." Because of the circuitry in the LDM, only external malfunctions can be detected. As a result, the ABS warning lamp itself is not diagnosable, only the control line to the EBCM can be diagnosed. In the event of an open CKT 852, the ABS warning lamp will be "ON" at all times, due to the loss of ground at the LDM input. If the control line is shorted to ground, the ABS warning lamp is kept "OFF," due to the LDM input being grounded. If the control line is shorted to B+, the EBCM cannot command the ABS warning lamp "OFF."

DTC Will Set When: DTC 11 can be set anytime. Opens or shorts to ground are detected during the three second bulb check. Shorts to battery are detected when the EBCM commands the ABS warning lamp "OFF."

Action Taken: A malfunction DTC is stored; however, ABS is not disabled.

DTC Chart Test Description: Number(s) below refer to circled number(s) on the diagnostic chart.

1. Checks to see if ABS warning lamp functions properly. Normal operation would indicate that the malfunction is not present.
2. Checks for a short to B+ in the ABS warning lamp control circuit.
3. Determines whether the short to B+ is in the LDM or the ABS warning lamp control circuit.
4. Checks for an open the ABS warning lamp control circuit.
5. Checks to see if malfunction was due to a poor connection.
6. Checks for a possible short to ground in the EBCM.
7. Checks for a short to ground in the ABS warning lamp control circuit.
8. Checks for high resistance in the LDM power feed circuit.
9. Ensures malfunction was not due to physical damage of the circuitry.
10. Checks for a malfunction in the LDM power feed circuit.

Diagnostic Aids: An "Intermittent" malfunction is most likely caused by a poor connection, rubbed through wire insulation, or a wire that is broken inside the insulation.

The lamp test function of the Tech 1 may be used to command the lamp "ON" while looking for an intermittent malfunction in the warning lamp circuitry.

The frequency of the malfunction can be checked by using the enhanced diagnostic function of the Tech 1, as described in "Enhanced Diagnostics," found in this section.

Any circuitry that is suspected as causing the intermittent complaint should be thoroughly checked for backed out terminals, improper mating, broken locks, improperly formed or damaged terminals, poor terminal to wiring connections or physical damage to the wiring harness.

DIAGNOSTIC CHART DTC 11 — 1994–95 ACHIEVA, GRAND AM AND SKYLARK

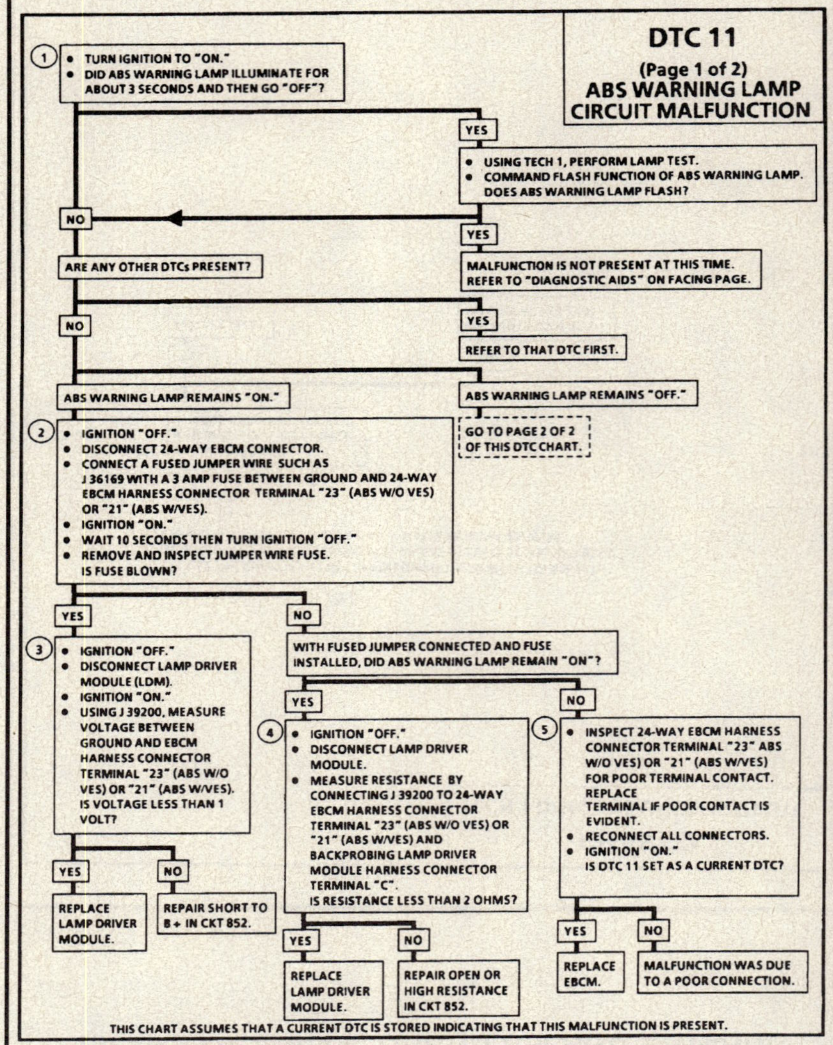

DIAGNOSTIC CHART DTC 11 — 1994–95 ACHIEVA, GRAND AM AND SKYLARK

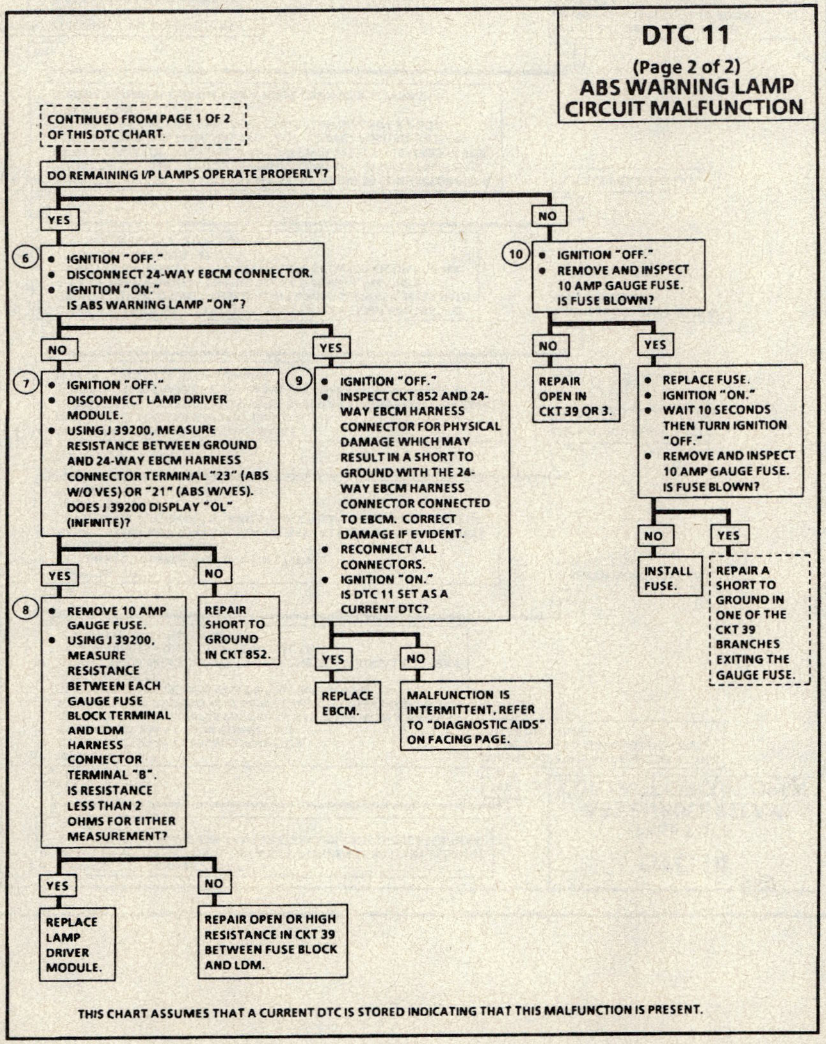

DIAGNOSTIC CHART DTC 14 — 1994–95 ACHIEVA, GRAND AM AND SKYLARK

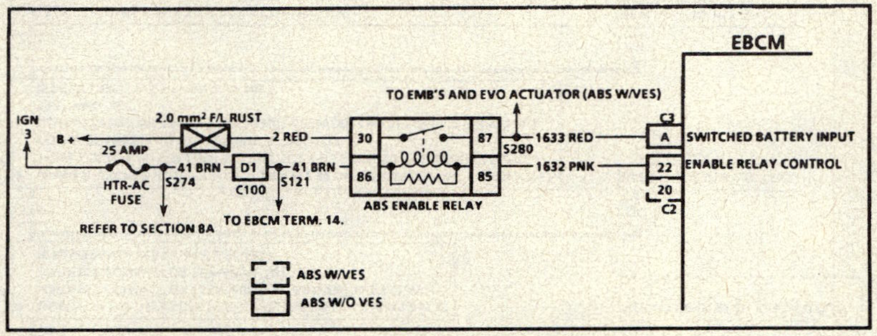

DTC 14

ABS ENABLE RELAY CONTACT CIRCUIT OPEN

Circuit Description:

Ignition voltage is supplied through terminal "86" of the ABS enable relay. The EBCM is then able to energize the pull-in coil by completing the ground circuit at EBCM terminal "22" (ABS w/o VES) or "20" (ABS w/VES). The magnetic field created closes the enable relay contacts and allows battery voltage and current to be supplied to the EMBs, EVO actuator (ABS w/VES), and EBCM, which supplies power to the motors and solenoids.

DTC Will Set When: DTC 14 can be set anytime after the EBCM commands the enable relay "ON" (the relay is first commanded "ON" during the three second bulb check). This test monitors the availability of current/voltage to the motors, solenoids, EMBs, and EVO actuator. This malfunction indicates voltage is not available and would therefore not allow ABS operation if required.

Action Taken: A malfunction DTC is stored, ABS/VES is disabled and the ABS warning lamp is turned "ON."

DTC Chart Test Description: Number(s) below refer to circled number(s) on the diagnostic chart.

1. Checks to see if the EBCM is sensing battery voltage at terminal "A".
2. Checks for proper operation of the enable relay and related circuitry.
3. Checks for excessive resistance in the switched battery input circuitry.
4. Checks for excessive resistance in the B+ feed circuit.
5. Checks for excessive resistance in the enable relay contacts.
6. Determines if the malfunction is due to poor terminal contact.
7. Determines if the malfunction is due to the EBCM.
8. Checks for excessive resistance in the enable relay control circuitry.
9. Checks for proper battery voltage.
10. Checks for an open in the B+ feed circuit.
11. Checks for an open in the switched battery input circuit.
12. Checks for proper operation of the enable relay.
13. Checks for excessive resistance in the enable relay coil.
14. Checks for excessive resistance in the enable relay control circuit between the relay and the EBCM.
15. Checks for excessive resistance in the ignition switch "run" fuse feed circuit.

Diagnostic Aids: An "Intermittent" malfunction is most likely caused by a poor connection, rubbed through wire insulation, or a wire that is broken inside the insulation.

The frequency of the malfunction can be checked by using the enhanced diagnostic function of the Tech 1, as described in "Enhanced Diagnostics," found in this section.

Any circuitry that is suspected as causing the "Intermittent" complaint should be thoroughly checked for backed out terminals, improper mating, broken locks, improperly formed or damaged terminals, poor terminal to wiring connections or physical damage to the wiring harness.

VIBRATION, TEMPERATURE EFFECTS:

Check for vibration effects by performing the relay test function of the Tech 1. With the relay commanded "ON," lightly tap the top and sides of the relay while monitoring battery voltage. If the battery voltage changes significantly, replace the relay.

If test 14 only fails when the vehicle is initially started in cold ambient conditions (temperature less than 32°F-0°C), replace the relay.

DIAGNOSTIC CHART DTC 14 — 1994–95 ACHIEVA, GRAND AM AND SKYLARK

DTC 14
(Page 1 of 3)
ABS ENABLE RELAY CONTACT CIRCUIT OPEN

1.
- IGNITION "ON," ENGINE "OFF."
- USING TECH 1, SELECT DATA LIST AND OBSERVE ABS BATTERY VOLTAGE. DOES TECH 1 INDICATE THAT ABS BATTERY VOLTAGE IS GREATER THAN 10 VOLTS?

NO → Step 2
YES → MALFUNCTION IS NOT PRESENT AT THIS TIME. REFER TO "DIAGNOSTIC AIDS" ON FACING PAGE.

2.
- IGNITION "OFF."
- DISCONNECT 24-WAY EBCM CONNECTOR.
- DISCONNECT 2-WAY EBCM CONNECTOR.
- CONNECT A FUSED JUMPER WIRE SUCH AS J 36169 WITH A 3 AMP FUSE BETWEEN GROUND AND 24-WAY EBCM HARNESS CONNECTOR TERMINAL "22" (ABS W/O VES) OR "20" (ABS W/VES).
- IGNITION "ON."
- USING J 39200, MEASURE VOLTAGE BETWEEN GROUND AND 2-WAY EBCM HARNESS CONNECTOR TERMINAL "A". IS VOLTAGE GREATER THAN 10 VOLTS?

YES → Step 3
NO → REFER TO PAGE 2 OF 3 OF THIS DTC CHART.

3.
- IGNITION "OFF."
- DISCONNECT JUMPER WIRE USED IN STEP 2.
- DISCONNECT ENABLE RELAY.
- USING J 39200, MEASURE RESISTANCE BETWEEN 2-WAY EBCM HARNESS CONNECTOR TERMINAL "A" AND ENABLE RELAY HARNESS CONNECTOR TERMINAL "87". IS RESISTANCE LESS THAN 2 OHMS?

YES → Step 4
NO → REPAIR HIGH RESISTANCE IN CKT 1633.

4.
- DISCONNECT NEGATIVE BATTERY CABLE.
- DISCONNECT POSITIVE BATTERY CABLE.
- USING J 39200, MEASURE RESISTANCE BETWEEN POSITIVE BATTERY CABLE TERMINAL AND ENABLE RELAY HARNESS CONNECTOR TERMINAL "30". IS RESISTANCE LESS THAN 2 OHMS?

YES → Step 5
NO → REPAIR HIGH RESISTANCE IN B+ FEED CIRCUIT.

5.
- USING A JUMPER WIRE, CONNECT ENABLE RELAY TERMINAL "85" TO GROUND. THEN, USING A FUSED JUMPER WIRE SUCH AS J 36169 WITH A 3 AMP FUSE, CONNECT ENABLE RELAY TERMINAL "86" TO B+.
- USING J 39200, MEASURE THE RESISTANCE BETWEEN ENABLE RELAY TERMINALS "30" AND "87". IS RESISTANCE LESS THAN 2 OHMS?

YES → Step 6
NO → REPLACE ENABLE RELAY.

6.
- INSPECT ENABLE RELAY AND ENABLE RELAY HARNESS CONNECTOR FOR A POOR CONNECTION OR POOR TERMINAL CONTACT. INSPECT EBCM AND EBCM CONNECTOR TERMINALS "22" (ABS W/O VES) OR "20" (ABS W/VES) AND "A" FOR POOR TERMINAL CONTACT. INSPECT POSITIVE BATTERY TERMINAL AND POSITIVE BATTERY CABLE TERMINAL FOR A POOR CONNECTION. IS POOR TERMINAL CONTACT OR A POOR CONNECTION EVIDENT?

NO → Step 7
YES → REPLACE TERMINALS OR REPAIR CONNECTION INTEGRITY (IF POOR CONTACT EXISTS AT BATTERY TERMINAL) THAT EXHIBIT(S) POOR CONTACT.

7.
- RECONNECT ALL CONNECTORS.
- IGNITION "ON." IS DTC 14 SET AS A CURRENT DTC?

YES → REPLACE EBCM.
NO → MALFUNCTION IS INTERMITTENT. REFER TO "DIAGNOSTIC AIDS" ON FACING PAGE.

THIS CHART ASSUMES THAT A CURRENT DTC IS STORED INDICATING THAT THIS MALFUNCTION IS PRESENT.

DIAGNOSTIC CHART DTC 14 — 1994–95 ACHIEVA, GRAND AM AND SKYLARK

DTC 14
(Page 2 of 3)
ABS ENABLE RELAY CONTACT CIRCUIT OPEN

CONTINUED FROM PAGE 1 OF 3 OF THIS CHART.

8.
- IGNITION "OFF."
- DISCONNECT JUMPER WIRE USED IN STEP 2.
- DISCONNECT NEGATIVE BATTERY CABLE FROM BATTERY.
- DISCONNECT POSITIVE BATTERY CABLE FROM BATTERY.
- TURN IGNITION "ON" (THIS IS DONE TO PROVIDE CIRCUIT CONTINUITY).
- USING J 39200, MEASURE RESISTANCE BETWEEN POSITIVE BATTERY CABLE TERMINAL AND 24-WAY EBCM HARNESS CONNECTOR TERMINAL "22" (ABS W/O VES) OR "20" (ABS W/VES). IS RESISTANCE LESS THAN 95 OHMS WHEN RELAY IS APPROXIMATELY 20°C (68°F)?

YES → Step 9
NO → REFER TO PAGE 3 OF 3 OF THIS DTC CHART.

9.
- USING J 39200, MEASURE VOLTAGE BETWEEN THE NEGATIVE AND POSITIVE BATTERY TERMINALS. IS VOLTAGE GREATER THAN 10 VOLTS?

YES → Step 10
NO → CORRECT LOW VOLTAGE CONDITION.

10.
- IGNITION "OFF."
- RECONNECT BATTERY CABLES.
- DISCONNECT ENABLE RELAY.
- USING J 39200, MEASURE VOLTAGE BETWEEN GROUND AND ENABLE RELAY HARNESS CONNECTOR TERMINAL "30". IS VOLTAGE GREATER THAN 10 VOLTS?

YES → Step 11
NO → REPAIR OPEN IN B+ FEED CIRCUIT.

11.
- USING J 39200, MEASURE RESISTANCE BETWEEN 2-WAY EBCM HARNESS CONNECTOR TERMINAL "A" AND ENABLE RELAY HARNESS CONNECTOR TERMINAL "87". IS RESISTANCE LESS THAN 2 OHMS?

YES → Step 12
NO → REPAIR OPEN IN CKT 1633.

12.
- USING A JUMPER WIRE, CONNECT ENABLE RELAY TERMINAL "85" TO GROUND. THEN, USING A FUSED JUMPER WIRE SUCH AS J 36169 WITH A 3 AMP FUSE, CONNECT ENABLE RELAY TERMINAL "86" TO B+.
- USING J 39200, MEASURE RESISTANCE BETWEEN ENABLE RELAY TERMINALS "30" AND "87". IS RESISTANCE LESS THAN 2 OHMS?

YES → MALFUNCTION WAS DUE TO A POOR CONNECTION. TEST FOR POOR TERMINAL CONTACT BETWEEN ENABLE RELAY AND ENABLE RELAY HARNESS CONNECTOR AND/OR AT POSITIVE BATTERY TERMINAL. REPLACE OR REPAIR TERMINALS THAT EXHIBIT POOR TERMINAL CONTACT.
NO → REPLACE ENABLE RELAY.

THIS CHART ASSUMES THAT A CURRENT DTC IS STORED INDICATING THAT THIS MALFUNCTION IS PRESENT.

DIAGNOSTIC CHART DTC 14 — 1994–95 ACHIEVA, GRAND AM AND SKYLARK

DTC 14

(Page 3 of 3)

ABS ENABLE RELAY CONTACT CIRCUIT OPEN

CONTINUED FROM PAGE 2 OF 3 OF THIS CHART.

(13)
- USING J 39200, MEASURE RESISTANCE BETWEEN ENABLE RELAY TERMINALS "85" AND "86".
IS RESISTANCE LESS THAN 95 OHMS WHEN RELAY IS APPROXIMATELY 20°C (68°F)?

NO: REPLACE ENABLE RELAY.

YES:

(14)
- USING J 39200, MEASURE RESISTANCE BETWEEN ENABLE RELAY HARNESS CONNECTOR TERMINAL "85" AND 24-WAY EBCM HARNESS CONNECTOR TERMINAL "22" (ABS W/O VES) OR "20" (ABS W/ VES).
IS RESISTANCE LESS THAN 2 OHMS?

NO: REPAIR HIGH RESISTANCE IN CKT 1632.

YES:

(15)
- DISCONNECT ENABLE RELAY.
- USING J 39200, MEASURE RESISTANCE BETWEEN POSITIVE BATTERY CABLE TERMINAL AND ENABLE RELAY HARNESS CONNECTOR TERMINAL "86".
IS RESISTANCE LESS THAN 2 OHMS?

YES: MALFUNCTION WAS DUE TO A POOR CONNECTION. TEST FOR POOR TERMINAL CONTACT BETWEEN ENABLE RELAY AND ENABLE RELAY HARNESS CONNECTOR. REPLACE TERMINALS THAT EXHIBIT POOR TERMINAL CONTACT.

NO: REPAIR HIGH RESISTANCE IN CIRCUIT BETWEEN POSITIVE BATTERY CABLE TERMINAL AND ENABLE RELAY HARNESS CONNECTOR TERMINAL "86".

THIS CHART ASSUMES THAT A CURRENT DTC IS STORED INDICATING THAT THIS MALFUNCTION IS PRESENT.

DIAGNOSTIC CHART DTC 15 — 1994–95 ACHIEVA, GRAND AM AND SKYLARK

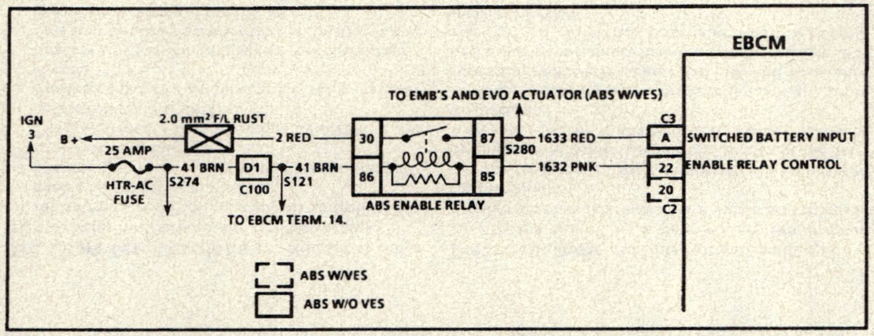

DTC 15

ABS ENABLE RELAY CONTACT CIRCUIT SHORTED TO BATTERY OR ALWAYS CLOSED

Circuit Description:

Ignition voltage is supplied through terminal "86" of the ABS enable relay. The EBCM is then able to energize the pull-in coil by completing the ground circuit at EBCM terminal "22" (ABS w/o VES), or "20" (ABS w/VES). The magnetic field created closes the enable relay contacts and allows battery voltage and current to be supplied to the EMBs, EVO actuator (ABS w/VES), and EBCM, which supplies power to the motors and solenoids.

DTC Will Set When: DTC 15 can be set only before the EBCM commands the relay "ON." This test determines if the enable relay is energized when it should not be. This malfunction would not allow the enable relay to remove power to the ABS. If a second malfunction were to occur that requires the enable relay to be turned "OFF," that malfunction can not be removed if the relay can not be controlled. The malfunction must be present for three consecutive ignition cycles before the DTC is set.

Action Taken: A malfunction DTC is stored; however, ABS/VES is not disabled.

DTC Chart Test Description: Number(s) below refer to circled number(s) on the diagnostic chart.

1. Indicates if the EBCM is capable of controlling the enable relay as commanded.
2. Checks for voltage at terminal "A" of the 2-way EBCM harness connector. If voltage is present, the malfunction exists in the enable relay and/or its circuitry.
3. Checks for a short to voltage in the switched battery input circuit.
4. Identifies if the malfunction is due to the EBCM.

Diagnostic Aids: An "Intermittent" malfunction is most likely caused by a poor connection, rubbed through wire insulation, or a wire that is broken inside the insulation.

The frequency of the malfunction can be checked by using the enhanced diagnostic function of the Tech 1, as described in "Enhanced Diagnostics," found in this section.

Any circuitry, that is suspected as causing the intermittent complaint, should be thoroughly checked for backed out terminals, improper mating, broken locks, improperly formed or damaged terminals, poor terminal to wiring connections or physical damage to the wiring harness.

DIAGNOSTIC CHART DTC 15 — 1994–95 ACHIEVA, GRAND AM AND SKYLARK

DTC 15
ABS ENABLE RELAY CONTACT CIRCUIT SHORTED TO BATTERY OR ALWAYS CLOSED

ARE ANY OTHER DTCs SET?

- YES: PROCEED TO THAT DTC FIRST.
- NO: go to step 1.

(1)
- IGNITION "ON."
- USE RELAY TEST FUNCTION OF TECH 1 AND COMMAND ENABLE RELAY "OFF."

DOES TECH 1 INDICATE THAT ENABLE RELAY IS "OFF" AND BATTERY VOLTAGE IS LESS THAN 5 VOLTS?

- YES: MALFUNCTION IS NOT PRESENT AT THIS TIME. REFER TO "DIAGNOSTIC AIDS" ON FACING PAGE.
- NO: go to step 2.

(2)
- IGNITION "OFF."
- DISCONNECT 24-WAY AND 2-WAY EBCM CONNECTORS.
- IGNITION "ON."
- MEASURE VOLTAGE BETWEEN GROUND AND TERMINAL "A" OF THE 2-WAY EBCM HARNESS CONNECTOR.

IS VOLTAGE LESS THAN 2 VOLTS?

- NO: go to step 3.
- YES: go to step 4.

(3)
- IGNITION "OFF."
- DISCONNECT ENABLE RELAY.
- IGNITION "ON."
- MEASURE VOLTAGE BETWEEN GROUND AND TERMINAL "A" OF THE 2-WAY EBCM HARNESS CONNECTOR.

IS VOLTAGE LESS THAN 2 VOLTS?

- YES: REPLACE ENABLE RELAY.
- NO: REPAIR SHORT TO VOLTAGE IN CKT 1633, 855 OR 1294 (ABS W/VES).

(4)
- IGNITION "OFF."
- INSPECT CKT 1633, 855 OR 1294 (ABS W/VES) AND 24-WAY EBCM HARNESS CONNECTOR FOR PHYSICAL DAMAGE WHICH MAY RESULT IN A SHORT TO B+ WITH THE EBCM CONNECTORS CONNECTED TO EBCM. CORRECT DAMAGE IF EVIDENT.
- RECONNECT ALL CONNECTORS.
- IGNITION "ON."
- USING TECH 1 RELAY TEST FUNCTION, COMMAND ABS ENABLE RELAY "OFF."

DOES TECH 1 INDICATE ABS ENABLE RELAY IS "OFF" AND BATTERY VOLTAGE IS LESS THAN 5 VOLTS?

- NO: REPLACE EBCM.
- YES: MALFUNCTION IS INTERMITTENT. REFER TO "DIAGNOSTIC AIDS" ON FACING PAGE.

- AFTER DIAGNOSIS IS COMPLETE, CLEAR DTCs AND TEST DRIVE VEHICLE FOR THREE (3) DRIVE CYCLES TO VERIFY DTC DOES NOT RESET. A DRIVE CYCLE CONSISTS OF STARTING THE VEHICLE, DRIVING ABOVE 16 km/h (10 MPH) AND THEN KEYING DOWN.

THIS CHART ASSUMES THAT A CURRENT DTC IS STORED INDICATING THAT THIS MALFUNCTION IS PRESENT.

DIAGNOSTIC CHART DTC 16 — 1994–95 ACHIEVA, GRAND AM AND SKYLARK

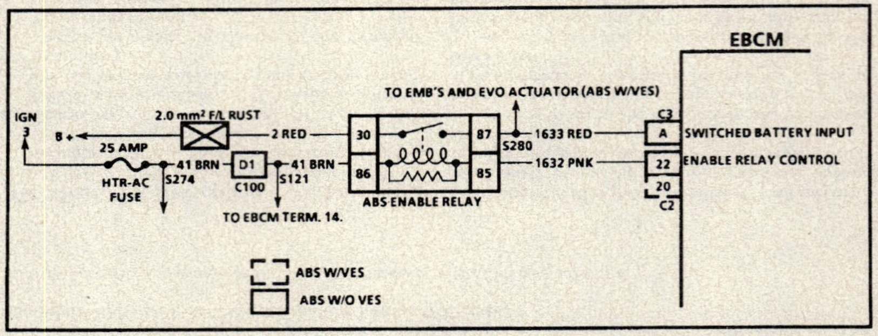

DTC 16
ABS ENABLE RELAY COIL CIRCUIT OPEN

Circuit Description:
Ignition voltage is supplied through terminal "86" of the ABS enable relay. The EBCM is then able to energize the pull-in coil by completing the ground circuit at EBCM terminal "22" (ABS w/o VES) or "20" (ABS w/VES). The magnetic field created closes the enable relay contacts and allows battery voltage and current to be supplied to the EMBs, EVO actuator (ABS w/VES), and EBCM, which supplies power to the motors and solenoids.

DTC Will Set When: DTC 16 can be set anytime after the relay is commanded "ON." This test detects an open in the enable relay coil circuit. An open in this circuit will not allow the enable relay to be energized thus preventing voltage/current to the motors, EMBs, EVO actuator and solenoids.

Action Taken: A malfunction DTC is stored, ABS/VES is disabled and the ABS warning lamp is turned "ON."

DTC Chart Test Description: Number(s) below refer to circled number(s) on the diagnostic chart.

1. Indicates if the EBCM is capable of controlling the enable relay as commanded.
2. Ensures that there is continuity through the pull-in coil of the enable relay.
3. Checks to ensure that voltage is available to the pull-in coil of the enable relay.
4. Checks continuity in the enable relay control circuit.
5. Ensures DTC was not set due to a poor connection.
6. This step identifies whether the malfunction is due to a poor connection or the EBCM.

Diagnostic Aids: An "Intermittent" malfunction is most likely caused by a poor connection, rubbed through wire insulation, or a wire that is broken inside the insulation.

The frequency of the malfunction can be checked by using the enhanced diagnostic function of the Tech 1, as described in "Enhanced Diagnostics," found in this section.

If the frequency of the malfunction is high, but is currently intermittent, check for high coil resistance by measuring resistance between relay terminals "85" and "86". If resistance is greater than 100 ohms, replace the relay.

Any circuitry, that is suspected as causing the intermittent complaint, should be thoroughly checked for backed out terminals, improper mating, broken locks, improperly formed or damaged terminals, poor terminal to wiring connections or physical damage to the wiring harness.

DIAGNOSTIC CHART DTC 16 — 1994–95 ACHIEVA, GRAND AM AND SKYLARK

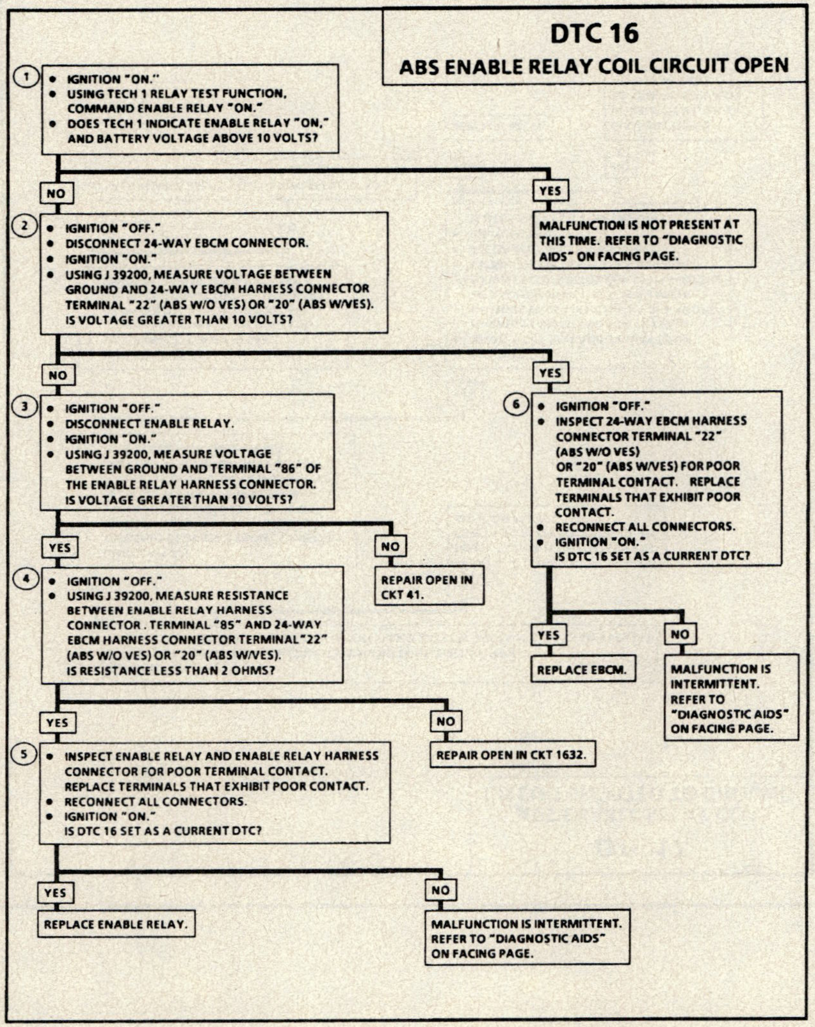

DIAGNOSTIC CHART DTC 17 — 1994–95 ACHIEVA, GRAND AM AND SKYLARK

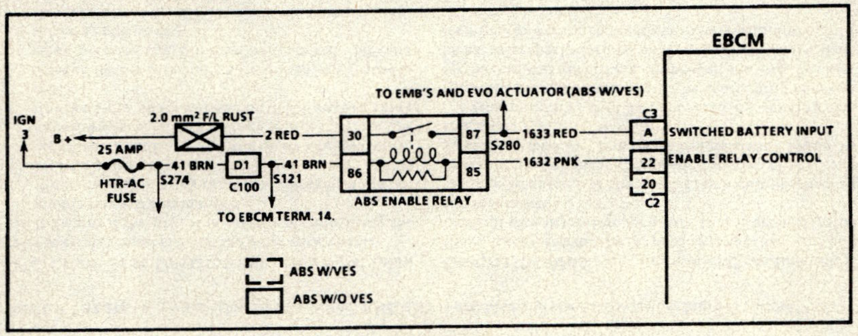

DTC 17

ABS ENABLE RELAY COIL CIRCUIT SHORTED TO GROUND

Circuit Description:

Ignition voltage is supplied through terminal "86" of the ABS enable relay. The EBCM is then able to energize the pull-in coil by completing the ground circuit at EBCM terminal "22" (ABS w/o VES) or "20" (ABS w/VES). The magnetic field created closes the enable relay contacts and allows battery voltage and current to be supplied to the EMBs, EVO actuator (ABS w/VES), and EBCM, which supplies voltage to the motors and solenoids.

DTC Will Set When: DTC 17 can be set before the EBCM commands the enable relay "ON." This test determines if the enable relay is energized when it should not be. This malfunction would not allow the enable relay to remove power to the ABS. If a second malfunction were to occur that requires the enable relay to be turned "OFF," that malfunction can not be removed if the relay can not be controlled.

Action Taken: A malfunction DTC is stored; however, ABS/VES is not disabled.

DTC Chart Test Description: Number(s) below refer to circled number(s) on the diagnostic chart.

1. Indicates if the EBCM is capable of controlling the enable relay as commanded.
2. Ensures the enable relay or control circuit are not shorted to ground.
3. This step isolates the malfunction to either the enable relay or enable relay control circuit.
4. Ensures short to ground is not due to physical damage of the circuitry.

Diagnostic Aids: An "Intermittent" malfunction is most likely caused by a poor connection, rubbed through wire insulation, or a wire that is broken inside the insulation.

The frequency of the malfunction can be checked by using the enhanced diagnostic function of the Tech 1, as described in "Enhanced Diagnostics," found in this section.

Any circuitry, that is suspected as causing the intermittent complaint, should be thoroughly checked for backed out terminals, improper mating, broken locks, improperly formed or damaged terminals, poor terminal to wiring connections or physical damage to the wiring harness.

DIAGNOSTIC CHART DTC 17 — 1994–95 ACHIEVA, GRAND AM AND SKYLARK

DTC 17
ABS ENABLE RELAY COIL CIRCUIT SHORTED TO GROUND

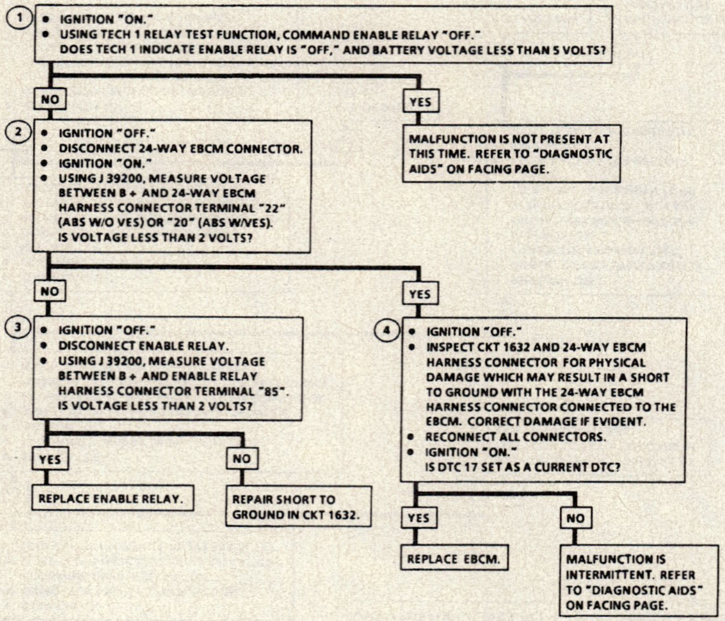

THIS CHART ASSUMES THAT A CURRENT DTC IS STORED INDICATING THAT THIS MALFUNCTION IS PRESENT.

DIAGNOSTIC CHART DTC 18 — 1994–95 ACHIEVA, GRAND AM AND SKYLARK

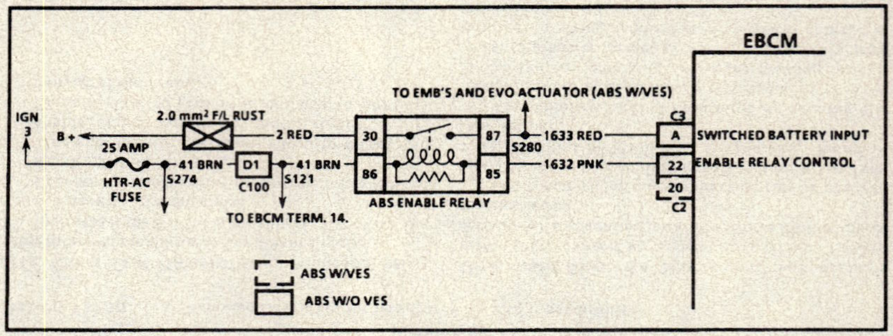

DTC 18

ABS ENABLE RELAY COIL CIRCUIT SHORTED TO BATTERY OR COIL SHORTED

Circuit Description:

Ignition voltage is supplied through terminal "86" of the ABS enable relay. The EBCM is then able to energize the pull-in coil by completing the ground circuit at EBCM terminal "22" (ABS w/o VES) or "20" (ABS w/VES). The magnetic field created closes the enable relay contacts and allows battery voltage and current to be supplied to the EMBs, EVO actuator (ABS w/VES), and EBCM, which supplies voltage to the motors and solenoids.

DTC Will Set When: DTC 18 can be set after the EBCM commands the enable relay "ON." This test monitors the availability of current/voltage to the motors, solenoids, EVO actuator, and EMBs. A malfunction of this type will not allow voltage to be available to the motors, solenoids, EVO actuator, and EMBs; therefore, ABS/VES operation would not be allowed if required.

Action Taken: A malfunction DTC is stored, ABS/VES is disabled and the ABS warning lamp is turned "ON."

DTC Chart Test Description: Number(s) below refer to circled number(s) on the diagnostic chart.

1. Indicates if the EBCM is capable of controlling the enable relay as commanded.
2. With the enable relay removed, voltage should not be available at terminal "85". Any voltage at this point would indicate that the enable relay control circuit was shorted to a voltage source.
3. Checks for a short across the ABS enable relay coil.
4. Checks for an internal short in the ABS enable relay between the B+ feed at terminal "30" and the relay pull-in coil.
5. Ensures short to B+ was not due to physical damage of the enable relay control circuit.

Diagnostic Aids: An "Intermittent" malfunction is most likely caused by a poor connection, rubbed through wire insulation, or a wire that is broken inside the insulation.

The frequency of the malfunction can be checked by using the enhanced diagnostic function of the Tech 1, as described in "Enhanced Diagnostics," found in this section.

Any circuitry, that is suspected as causing the intermittent complaint, should be thoroughly checked for backed out terminals, improper mating, broken locks, improperly formed or damaged terminals, poor terminal to wiring connections or physical damage to the wiring harness.

DIAGNOSTIC CHART DTC 18 — 1994–95 ACHIEVA, GRAND AM AND SKYLARK

DTC 18
ABS ENABLE RELAY COIL CIRCUIT SHORTED TO BATTERY OR COIL SHORTED

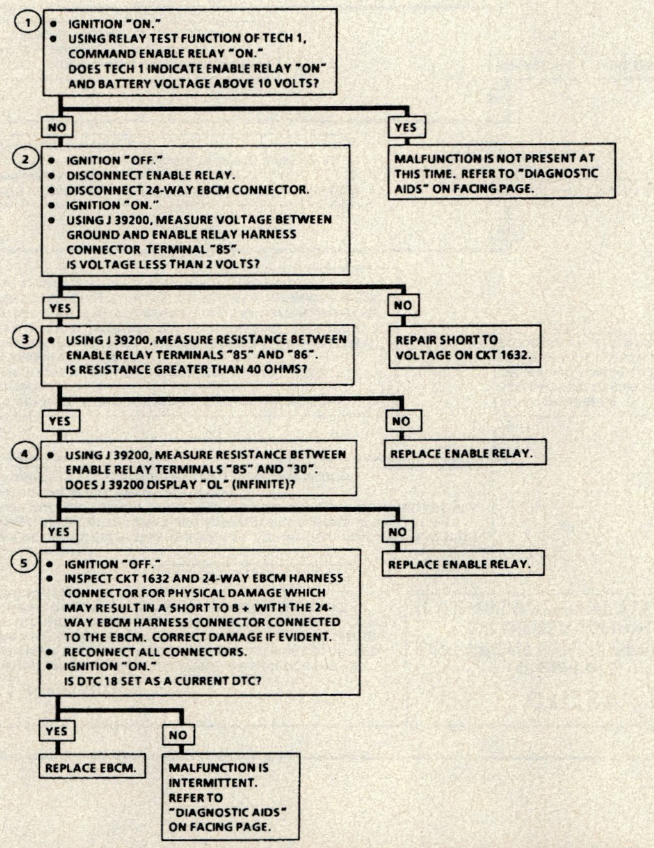

THIS CHART ASSUMES THAT A CURRENT DTC IS STORED INDICATING THAT THIS MALFUNCTION IS PRESENT.

DIAGNOSTIC CHART DTC 21 — 1994–95 ACHIEVA, GRAND AM AND SKYLARK WITHOUT VES

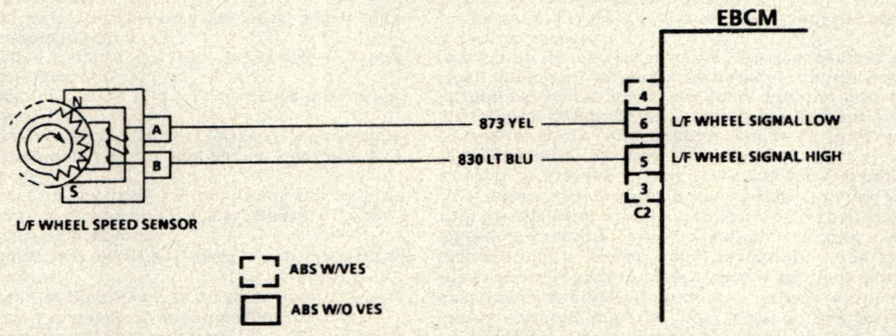

DTC 21
(Page 1 of 2)
LEFT FRONT WHEEL SPEED = 0 OR UNREASONABLE
(FOR ABS WITHOUT VES ONLY)

Circuit Description:
As a toothed ring passes by the wheel speed sensor, changes in the electromagnetic field cause the wheel speed sensor to produce a sinusoidal (AC) voltage signal whose frequency is proportional to wheel speed. The magnitude of this signal is directly related to wheel speed and the proximity of the wheel speed sensor to the toothed ring, often referred to as the air gap.

DTC Will Set When: DTC 21 can be set when the vehicle is not in an ABS stop. If the left front wheel speed is less than 1/2 of the vehicle speed and the vehicle speed is greater than 8 km/h (5 mph), a malfunction exists.

Action Taken: A malfunction DTC is stored, ABS is disabled and the ABS warning lamp is turned "ON."

DTC Chart Test Description: Number(s) below refer to circled number(s) on the diagnostic chart.

1. Identifies if a wheel speed sensor or circuitry is damaged and visibly apparent.
2. Checks wheel speed sensor for proper resistance values.
3. Ensures the wheel speed sensor and toothed ring generate the proper voltage.
4. Ensures the wheel speed sensor is not shorted to ground.

Diagnostic Aids: An "Intermittent" malfunction is most likely caused by a poor connection, rubbed through wire insulation, or a wire that is broken inside the insulation.

The frequency of the malfunction can be checked by using the enhanced diagnostic function of the Tech 1.

If the customer's comments reflect that the ABS amber warning lamp is "ON" only during moist environmental changes, (rain, snow, vehicle wash) all wheel speed sensor circuitry should be thoroughly inspected for signs of water intrusion. Use the following procedure. Spray down the suspected area with 5% salt water solution (two teaspoons of salt to 12 oz. of water). Drive vehicle above 24 km/h (15 mph) for at least 30 seconds. If DTC returns immediately, replace suspected harness.

Any circuitry, that is suspected as causing the intermittent complaint, should be thoroughly checked for backed out terminals, improper mating, broken locks, improperly formed or damaged terminals, poor terminal to wiring connections or physical damage to the wiring harness.

Resistance of the wheel speed sensor will increase with an increase in sensor temperature.

When replacing a wheel speed sensor, inspect the sensor terminals and harness connector for corrosion and/or water intrusion. If evidence of corrosion or water intrusion exists, replace wheel speed sensor harness. Refer to "On-Vehicle Service" in this section.

If wheel speed sensor harness shows evidence of water intrusion, inspect wheel speed sensor for water intrusion. If water intrusion is evident, replace wheel speed sensor.

DIAGNOSTIC CHART DTC 21 — 1994–95 ACHIEVA, GRAND AM AND SKYLARK WITHOUT VES

DTC 21
(Page 1 of 2)
LEFT FRONT WHEEL SPEED = 0 OR UNREASONABLE
(FOR ABS WITHOUT VES ONLY)

IMPORTANT:
- WHEEL SPEED SENSOR INTERMITTENT MALFUNCTIONS MAY BE DIFFICULT TO LOCATE. CARE SHOULD BE TAKEN NOT TO DISTURB ANY ELECTRICAL CONNECTIONS PRIOR TO AN INDICATED STEP OF THIS CHART. THIS WILL ENSURE THAT AN INTERMITTENT CONNECTION WILL NOT BE CORRECTED BEFORE THE SOURCE OF THE MALFUNCTION IS FOUND.

(1)
- IGNITION "OFF."
- INSPECT WHEEL SPEED SENSOR WIRING AND CONNECTORS FOR DAMAGE. INSPECT WHEEL SPEED SENSOR AND TOOTHED RING FOR LOOSENESS OR DAMAGE. PAY PARTICULAR ATTENTION TO THE TOOTHED RING. ANY SIGNIFICANT DAMAGE WILL AFFECT THE WHEEL SPEED INPUT SIGNAL.

IS DAMAGE OR SENSOR/TOOTHED RING LOOSENESS EVIDENT?

- YES → REPAIR DAMAGE OR SENSOR/TOOTHED RING LOOSENESS. DRIVE VEHICLE ABOVE 24 km/h (15 MPH) FOR AT LEAST 30 SECONDS TO ENSURE DTC 21 DOES NOT SET.
- NO → step 2

(2)
- DISCONNECT CONNECTOR FROM LEFT FRONT WHEEL SPEED SENSOR.
- USING J 39200, MEASURE RESISTANCE BETWEEN SENSOR TERMINALS "A" AND "B".

IS RESISTANCE BETWEEN 1628 AND 1889 OHMS WHEN SENSOR IS 25°C (77°F)? *[IF TEMPERATURE OF THE SENSOR IS SUSPECTED TO BE OTHER THAN 25°C (77°F), REFER TO TABLE BELOW FOR APPLICABLE SENSOR RESISTANCE VALUES; THE VALUES IN THE TABLE ARE FOR TEMPERATURE OF THE SENSOR NOT AIR TEMPERATURE.]*

- NO → REPLACE WHEEL SPEED SENSOR.
- YES → step 3

(3)
- WITH J 39200 STILL CONNECTED, SELECT AC VOLTAGE SCALE.
- SPIN WHEEL BY HAND WHILE MONITORING VOLTAGE.

IS VOLTAGE AT LEAST 100 mV (VOLTAGE WILL INCREASE AS WHEEL SPEED INCREASES)?

- NO → REPLACE WHEEL SPEED SENSOR.
- YES → step 4

(4)
- USING J 39200, MEASURE RESISTANCE BETWEEN SENSOR TERMINAL "A" AND GROUND.

DOES J 39200 DISPLAY "OL" (INFINITE)?

- NO → REPLACE WHEEL SPEED SENSOR.
- YES → GO TO PAGE 2 OF 2 OF THIS DTC CHART.

FRONT WHEEL SPEED SENSOR
TEMPERATURE OF SENSOR VS. SENSOR RESISTANCE (APPROXIMATE)

TEMP. (°C)	TEMP. (°F)	RESISTANCE (OHMS)
-34 TO 4	-30 TO 40	1258 TO 1739
5 TO 43	41 TO 110	1503 TO 2023
44 TO 93	111 TO 200	1747 TO 2384

THIS CHART ASSUMES THAT A CURRENT DTC IS STORED INDICATING THAT THIS MALFUNCTION IS PRESENT.

DIAGNOSTIC CHART DTC 21 — 1994–95 ACHIEVA, GRAND AM AND SKYLARK WITHOUT VES

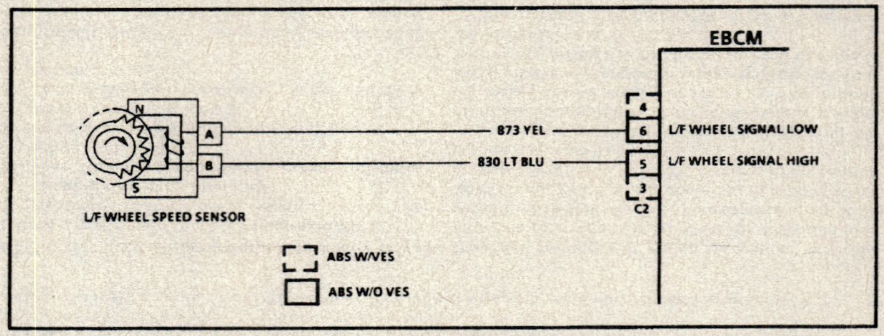

DTC 21
(Page 2 of 2)
LEFT FRONT WHEEL SPEED = 0 OR UNREASONABLE
(FOR ABS WITHOUT VES ONLY)

Circuit Description:
As a toothed ring passes by the wheel speed sensor, changes in the electromagnetic field cause the wheel speed sensor to produce a sinusoidal (AC) voltage signal whose frequency is proportional to wheel speed. The magnitude of this signal is directly related to wheel speed and the proximity of the wheel speed sensor to the toothed ring often referred to as the air gap.

DTC Will Set When: DTC 21 can be set when the vehicle is not in an ABS stop. If the left front wheel speed is less than 1/2 of the vehicle speed and the vehicle speed is greater than 8 km/h (5 mph), a malfunction exists.

Action Taken: A malfunction DTC is stored, ABS is disabled and the ABS warning lamp is turned "ON."

DTC Chart Test Description: Number(s) below refer to circled number(s) on the diagnostic chart.

5. Checks for proper voltages at speed sensor harness connector.
6. Ensures that the wheel speed sensor circuitry is not shorted to ground.
7. Checks for an open in the wheel signal low circuit.
8. Checks for an open in the wheel signal high circuit.
9. Ensures that DTC 21 was not set due to a poor connection at the EBCM or physical damage of the circuitry.
10. Checks for a short to B+ in either the high or low wheel signal circuits.
11. Ensures short to B+ was not due to physical damage of the circuitry.
12. Tests for a short between the wheel signal high and low circuits.
13. Ensures malfunction was not due to a poor connection at the wheel speed sensor or physical damage of the circuitry.

Diagnostic Aids: An "Intermittent" malfunction is most likely caused by a poor connection, rubbed through wire insulation, or a wire that is broken inside the insulation.

The frequency of the malfunction can be checked by using the enhanced diagnostic function of the Tech 1.

If the customer's comments reflect that the ABS amber warning lamp is "ON" only during moist environmental changes, (rain, snow, vehicle wash) all wheel speed sensor circuitry should be thoroughly inspected for signs of water intrusion. Use the following procedure. Spray down the suspected area with 5% salt water solution (two teaspoons of salt to 12 oz. of water). Drive vehicle above 24 km/h (15 mph) for at least 30 seconds. If DTC returns immediately, replace suspected harness.

Any circuitry, that is suspected as causing the intermittent complaint, should be thoroughly checked for backed out terminals, improper mating, broken locks, improperly formed or damaged terminals, poor terminal to wiring connections or physical damage to the wiring harness.

Resistance of the wheel speed sensor will increase with an increase in sensor temperature.

When replacing a wheel speed sensor, inspect the sensor terminals and harness connector for corrosion and/or water intrusion. If evidence of corrosion or water intrusion exists, replace wheel speed sensor harness.

If wheel speed sensor harness shows evidence of water intrusion, inspect wheel speed sensor for water intrusion. If water intrusion is evident, replace wheel speed sensor.

DIAGNOSTIC CHART DTC 21 — 1994–95 ACHIEVA, GRAND AM AND SKYLARK WITHOUT VES

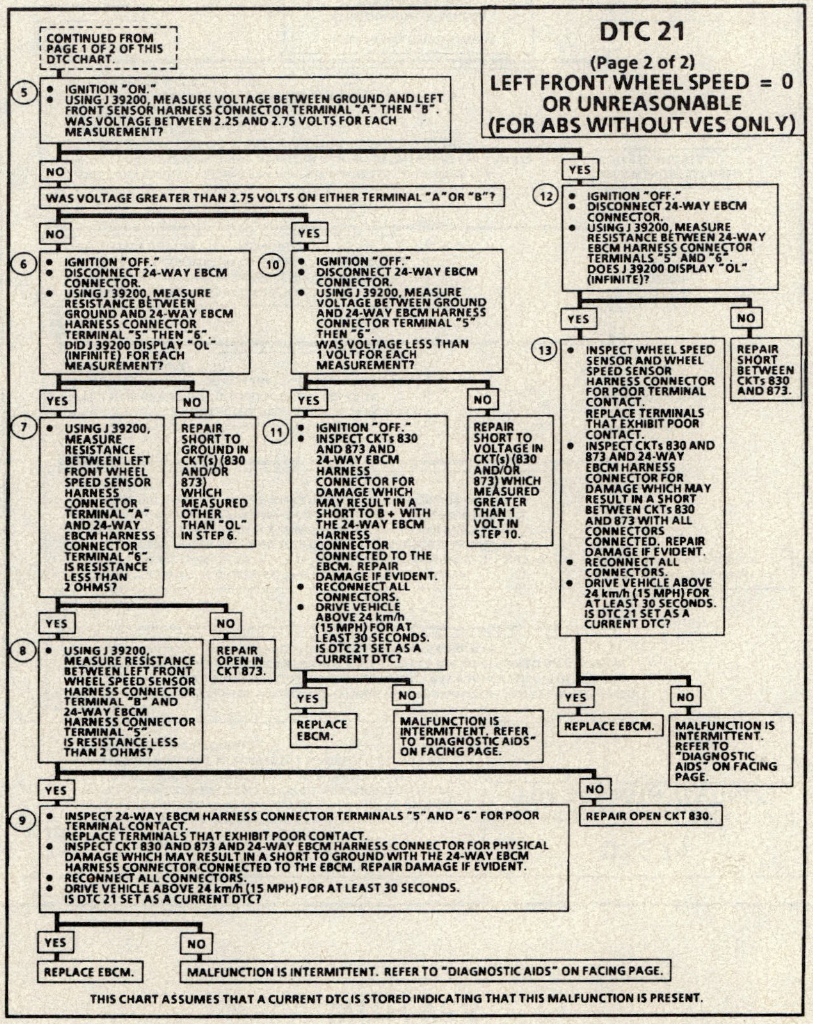

DIAGNOSTIC CHART DTC 21 — 1994–95 ACHIEVA, GRAND AM AND SKYLARK WITH VES

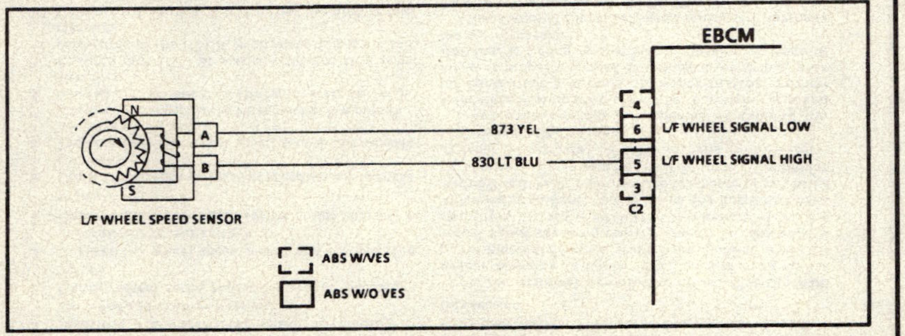

DTC 21

LEFT FRONT WHEEL SPEED = 0
(FOR ABS WITH VES ONLY)

Circuit Description:
As a toothed ring passes by the wheel speed sensor, changes in the electromagnetic field cause the wheel speed sensor to produce a sinusoidal (AC) voltage signal whose frequency is proportional to wheel speed. The magnitude of this signal is directly related to wheel speed and the proximity of the wheel speed sensor to the toothed ring, often referred to as the air gap.

DTC Will Set When: DTC 21 can be set when the vehicle is not in an ABS stop. A malfunction exists if the left front wheel speed = 0 and the vehicle speed is greater than 8 km/h (5 mph).

Action Taken: A malfunction DTC is stored, ABS/VES is disabled and the ABS warning lamp is turned "ON."

DTC Chart Test Description: Number(s) below refer to circled number(s) on the diagnostic chart.

1. Identifies if a wheel speed sensor or circuitry is damaged and visibly apparent.
2. Tests wheel speed sensor for proper resistance values.
3. Ensures the wheel speed sensor and toothed ring generate the proper voltage.
4. Ensures wheel speed sensor high and low circuits are not shorted together.
5. Ensures malfunction is not due to physical damage of the wheel speed sensor circuitry.

Diagnostic Aids: An "Intermittent" malfunction is most likely caused by a poor connection, rubbed through wire insulation, or a wire that is broken inside the insulation.

The frequency of the malfunction can be checked by using the enhanced diagnostic function of the Tech 1, as described in "Enhanced Diagnostics," found in this section.

If the customer's comments reflect that the ABS amber warning lamp is "ON" only during moist environmental changes, (rain, snow, vehicle wash) all wheel speed sensor circuitry should be thoroughly inspected for signs of water intrusion. Use the following procedure. Spray down the suspected area with 5% salt water solution (two teaspoons of salt to 12 oz. of water). Drive vehicle above 24 km/h (15 mph) for at least 30 seconds. If DTC returns immediately, replace suspected harness.

Any circuitry, that is suspected as causing the intermittent complaint, should be thoroughly checked for backed out terminals, improper mating, broken locks, improperly formed or damaged terminals, poor terminal to wiring connections or physical damage to the wiring harness.

Resistance of the wheel speed sensor will increase with an increase in sensor temperature.

When replacing a wheel speed sensor, inspect the sensor terminals and harness connector for corrosion and/or water intrusion. If evidence of corrosion or water intrusion exists, replace wheel speed sensor harness. Refer to "On-Vehicle Service" in this section.

If wheel speed sensor harness shows evidence of water intrusion, inspect wheel speed sensor for water intrusion. If water intrusion is evident, replace wheel speed sensor.

DIAGNOSTIC CHART DTC 21 — 1994–95 ACHIEVA, GRAND AM AND SKYLARK WITH VES

DTC 21

LEFT FRONT WHEEL SPEED = 0 (FOR ABS WITH VES ONLY)

Important

- WHEEL SPEED SENSOR INTERMITTENT MALFUNCTIONS MAY BE DIFFICULT TO LOCATE. CARE SHOULD BE TAKEN NOT TO DISTURB ANY ELECTRICAL CONNECTIONS PRIOR TO AN INDICATED STEP OF THIS CHART. THIS WILL ENSURE THAT AN INTERMITTENT CONNECTION WILL NOT BE CORRECTED BEFORE THE SOURCE OF THE MALFUNCTION IS FOUND.

(1)
- IGNITION "OFF."
- INSPECT WHEEL SPEED SENSOR WIRING AND CONNECTORS FOR DAMAGE. INSPECT WHEEL SPEED SENSOR AND TOOTHED RING FOR LOOSENESS OR DAMAGE. PAY PARTICULAR ATTENTION TO THE TOOTHED RING. ANY SIGNIFICANT DAMAGE WILL AFFECT THE WHEEL SPEED INPUT SIGNAL. IS PHYSICAL DAMAGE OR SENSOR/TOOTHED RING LOOSENESS EVIDENT?

YES: CORRECT PHYSICAL DAMAGE; CLEAR DTCs AND RECHECK.

NO:

(2)
- DISCONNECT CONNECTOR FROM LEFT FRONT WHEEL SPEED SENSOR.
- USING J 39200, MEASURE RESISTANCE BETWEEN SENSOR TERMINALS "A" AND "B". IS RESISTANCE BETWEEN 1628 AND 1889 OHMS WHEN SENSOR IS 25°C (77°F)? *[IF TEMPERATURE OF THE SENSOR IS SUSPECTED TO BE OTHER THAN 25°C (77°F), REFER TO THE TABLE BELOW FOR THE APPLICABLE SENSOR RESISTANCE VALUES; THE VALUES IN THE TABLE ARE FOR TEMPERATURE OF THE SENSOR NOT AIR TEMPERATURE.]*

NO: REPLACE WHEEL SPEED SENSOR.

YES:

(3)
- WITH J 39200 STILL CONNECTED, SELECT AC VOLTAGE SCALE.
- SPIN WHEEL BY HAND WHILE MONITORING VOLTAGE. IS VOLTAGE AT LEAST 100 mV (VOLTAGE WILL INCREASE AS WHEEL SPEED INCREASES)?

NO: REPLACE WHEEL SPEED SENSOR.

YES:

(4)
- DISCONNECT 24-WAY EBCM CONNECTOR.
- USING J 39200, MEASURE RESISTANCE BETWEEN 24-WAY EBCM HARNESS CONNECTOR TERMINALS "3" AND "4". DOES J 39200 DISPLAY "OL" (INFINITE)?

NO: REPAIR SHORT BETWEEN CKTs 830 AND 873.

YES:

(5)
- INSPECT CKT 830 AND 873 AND 24-WAY EBCM HARNESS CONNECTOR FOR PHYSICAL DAMAGE WHICH MAY RESULT IN A SHORT BETWEEN CKT 830 AND 873 WITH THE 24-WAY EBCM HARNESS CONNECTOR CONNECTED TO THE EBCM. CORRECT DAMAGE IF EVIDENT.
- RECONNECT ALL CONNECTORS.
- DRIVE VEHICLE ABOVE 24 km/h (15 MPH) FOR AT LEAST 30 SECONDS. IS DTC 21 SET AS A CURRENT DTC?

YES: REPLACE EBCM.

NO: MALFUNCTION IS INTERMITTENT. REFER TO "DIAGNOSTIC AIDS" ON FACING PAGE.

FRONT WHEEL SPEED SENSOR
TEMPERATURE OF SENSOR VS. SENSOR RESISTANCE (APPROXIMATE)

TEMP. (°C)	TEMP. (°F)	RESISTANCE (OHMS)
-34 TO 4	-30 TO 40	1258 TO 1739
5 TO 43	41 TO 110	1503 TO 2023
44 TO 93	111 TO 200	1747 TO 2384

THIS CHART ASSUMES THAT A CURRENT DTC IS STORED INDICATING THAT THIS MALFUNCTION IS PRESENT.

DIAGNOSTIC CHART DTC 22 — 1994–95 ACHIEVA, GRAND AM AND SKYLARK WITHOUT VES

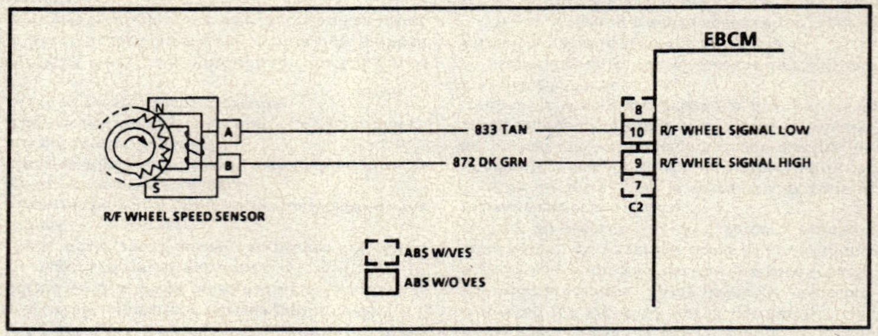

DTC 22

(Page 1 of 2)
RIGHT FRONT WHEEL SPEED = 0 OR UNREASONABLE (FOR ABS WITHOUT VES ONLY)

Circuit Description:

As a toothed ring passes by the wheel speed sensor, changes in the electromagnetic field cause the wheel speed sensor to produce a sinusoidal (AC) voltage signal whose frequency is proportional to wheel speed. The magnitude of this signal is directly related to wheel speed and the proximity of the wheel speed sensor to the toothed ring, often referred to as the air gap.

DTC Will Set When: DTC 22 can be set when the vehicle is not in an ABS stop. If the right front wheel speed is less than 1/2 of the vehicle speed and the vehicle speed is greater than 8 km/h (5 mph), a malfunction exists.

Action Taken: A malfunction DTC is stored, ABS is disabled and the ABS warning lamp is turned "ON."

DTC Chart Test Description: Number(s) below refer to circled number(s) on the diagnostic chart.

1. Identifies if a wheel speed sensor or circuitry is damaged and visibly apparent.
2. Checks wheel speed sensor for proper resistance values.
3. Ensures the wheel speed sensor and toothed ring generate the proper voltage.
4. Ensures the wheel speed sensor is not shorted to ground.
5. Checks for proper voltages at speed sensor harness connector.
6. Ensures that the wheel speed sensor circuitry is not shorted to ground.
7. Checks for an open in the wheel signal low circuit.
8. Checks for an open in the wheel signal high circuit.
9. Ensures that DTC 21 was not set due to a poor connection at the EBCM or physical damage of the circuitry.
10. Checks for a short to B+ in either the high or low wheel signal circuits.
11. Ensures short to B+ was not due to physical damage of the circuitry.
12. Tests for a short between the wheel signal high and low circuits.
13. Ensures malfunction was not due to a poor connection at the wheel speed sensor or physical damage of the circuitry.

Diagnostic Aids: An "Intermittent" malfunction is most likely caused by a poor connection, rubbed through wire insulation, or a wire that is broken inside the insulation.

The frequency of the malfunction can be checked by using the enhanced diagnostic function of the Tech 1, as described in "Enhanced Diagnostics," found in this section.

If the customer's comments reflect that the ABS amber warning lamp is "ON" only during moist environmental changes, (rain, snow, vehicle wash) all wheel speed sensor circuitry should be thoroughly inspected for signs of water intrusion. Use the following procedure. Spray down the suspected area with 5% salt water solution (two teaspoons of salt to 12 oz. of water). Drive vehicle above 24 km/h (15 mph) for at least 30 seconds. If DTC returns immediately, replace suspected harness.

Any circuitry, that is suspected as causing the intermittent complaint, should be thoroughly checked for backed out terminals, improper mating, broken locks, improperly formed or damaged terminals, poor terminal to wiring connections or physical damage to the wiring harness.

Resistance of the wheel speed sensor will increase with an increase in sensor temperature.

When replacing a wheel speed sensor, inspect the sensor terminals and harness connector for corrosion and/or water intrusion. If evidence of corrosion or water intrusion exists, replace wheel speed sensor harness. Refer to "On-Vehicle Service" in this section.

If wheel speed sensor harness shows evidence of water intrusion, inspect wheel speed sensor for water intrusion. If water intrusion is evident, replace wheel speed sensor.

DIAGNOSTIC CHART DTC 22 — 1994–95 ACHIEVA, GRAND AM AND SKYLARK WITHOUT VES

DTC 22
(Page 1 of 2)
RIGHT FRONT WHEEL SPEED = 0 OR UNREASONABLE
(FOR ABS WITHOUT VES ONLY)

IMPORTANT:
- WHEEL SPEED SENSOR INTERMITTENT MALFUNCTIONS MAY BE DIFFICULT TO LOCATE. CARE SHOULD BE TAKEN NOT TO DISTURB ANY ELECTRICAL CONNECTIONS PRIOR TO AN INDICATED STEP OF THIS CHART. THIS WILL ENSURE THAT AN INTERMITTENT CONNECTION WILL NOT BE CORRECTED BEFORE THE SOURCE OF THE MALFUNCTION IS FOUND.

(1)
- IGNITION "OFF."
- INSPECT WHEEL SPEED SENSOR WIRING AND CONNECTORS FOR DAMAGE. INSPECT WHEEL SPEED SENSOR AND TOOTHED RING FOR LOOSENESS OR DAMAGE. PAY PARTICULAR ATTENTION TO THE TOOTHED RING. ANY SIGNIFICANT DAMAGE WILL AFFECT THE WHEEL SPEED INPUT SIGNAL. IS DAMAGE OR SENSOR/TOOTHED RING LOOSENESS EVIDENT?

YES: REPAIR DAMAGE OR SENSOR/TOOTHED RING LOOSENESS. DRIVE VEHICLE ABOVE 24 km/h (15 MPH) FOR AT LEAST 30 SECONDS TO ENSURE DTC 22 DOES NOT SET.

NO:

(2)
- DISCONNECT CONNECTOR FROM RIGHT FRONT WHEEL SPEED SENSOR.
- USING J 39200, MEASURE RESISTANCE BETWEEN SENSOR TERMINALS "A" AND "B".
  IS RESISTANCE BETWEEN 1628 AND 1889 OHMS WHEN SENSOR IS 25°C (77°F)? *[IF TEMPERATURE OF THE SENSOR IS SUSPECTED TO BE OTHER THAN 25°C (77°F), REFER TO TABLE BELOW FOR APPLICABLE SENSOR RESISTANCE VALUES; THE VALUES IN THE TABLE ARE FOR TEMPERATURE OF THE SENSOR NOT AIR TEMPERATURE.]*

NO: REPLACE WHEEL SPEED SENSOR.

YES:

(3)
- WITH J 39200 STILL CONNECTED, SELECT AC VOLTAGE SCALE.
- SPIN WHEEL BY HAND WHILE MONITORING VOLTAGE.
  IS VOLTAGE AT LEAST 100 mV (VOLTAGE WILL INCREASE AS WHEEL SPEED INCREASES)?

NO: REPLACE WHEEL SPEED SENSOR.

YES:

(4)
- USING J 39200, MEASURE RESISTANCE BETWEEN SENSOR TERMINAL "A" AND GROUND.
  DOES J 39200 DISPLAY "OL" (INFINITE)?

NO: REPLACE WHEEL SPEED SENSOR.

YES: GO TO PAGE 2 OF 2 OF THIS DTC CHART.

FRONT WHEEL SPEED SENSOR
TEMPERATURE OF SENSOR VS. SENSOR RESISTANCE (APPROXIMATE)

TEMP. (°C)	TEMP. (°F)	RESISTANCE (OHMS)
-34 TO 4	-30 TO 40	1258 TO 1739
5 TO 43	41 TO 110	1503 TO 2023
44 TO 93	111 TO 200	1747 TO 2384

THIS CHART ASSUMES THAT A CURRENT DTC IS STORED INDICATING THAT THIS MALFUNCTION IS PRESENT.

DIAGNOSTIC CHART DTC 22 — 1994–95 ACHIEVA, GRAND AM AND SKYLARK WITHOUT VES

DTC 22
(Page 2 of 2)
RIGHT FRONT WHEEL SPEED = 0 OR UNREASONABLE
(FOR ABS WITHOUT VES ONLY)

CONTINUED FROM PAGE 1 OF 2 OF THIS DTC CHART.

(5)
- IGNITION "ON."
- USING J 39200, MEASURE VOLTAGE BETWEEN GROUND AND RIGHT FRONT SENSOR HARNESS CONNECTOR TERMINAL "A" THEN "B".
  WAS VOLTAGE BETWEEN 2.25 AND 2.75 VOLTS FOR EACH MEASUREMENT?

NO: WAS VOLTAGE GREATER THAN 2.75 VOLTS ON EITHER TERMINAL "A" OR "B"?

NO:

(6)
- IGNITION "OFF."
- DISCONNECT 24-WAY EBCM CONNECTOR.
- USING J 39200, MEASURE RESISTANCE BETWEEN GROUND AND 24-WAY EBCM HARNESS CONNECTOR TERMINAL "9" THEN "10".
  DID J 39200 DISPLAY "OL" (INFINITE) FOR EACH MEASUREMENT?

NO: REPAIR SHORT TO GROUND IN CKT(S) (833 AND/OR 872) WHICH MEASURED OTHER THAN "OL" IN STEP 6.

YES:

(7)
- USING J 39200, MEASURE RESISTANCE BETWEEN RIGHT FRONT WHEEL SPEED SENSOR HARNESS CONNECTOR TERMINAL "A" AND 24-WAY EBCM HARNESS CONNECTOR TERMINAL "10".
  IS RESISTANCE LESS THAN 2 OHMS?

NO: REPAIR OPEN IN CKT 833.

YES:

(8)
- USING J 39200, MEASURE RESISTANCE BETWEEN RIGHT FRONT WHEEL SPEED SENSOR HARNESS CONNECTOR TERMINAL "B" AND 24-WAY EBCM HARNESS CONNECTOR TERMINAL "9".
  IS RESISTANCE LESS THAN 2 OHMS?

NO: REPAIR OPEN IN CKT 872.

YES:

(9)
- INSPECT 24-WAY EBCM HARNESS CONNECTOR TERMINALS "9" AND "10" FOR POOR TERMINAL CONTACT. REPLACE TERMINALS THAT EXHIBIT POOR CONTACT.
- INSPECT CKT 833 AND 872 AND 24-WAY EBCM HARNESS CONNECTOR FOR PHYSICAL DAMAGE WHICH MAY RESULT IN A SHORT TO GROUND WITH THE 24-WAY EBCM HARNESS CONNECTOR CONNECTED TO THE EBCM. REPAIR DAMAGE IF EVIDENT.
- DRIVE VEHICLE ABOVE 24 km/h (15 MPH) FOR AT LEAST 30 SECONDS.
  IS DTC 22 SET AS A CURRENT DTC?

YES: REPLACE EBCM.

NO: MALFUNCTION IS INTERMITTENT. REFER TO "DIAGNOSTIC AIDS" ON FACING PAGE.

(Step 5 answer NO, voltage greater than 2.75 volts) YES:

(10)
- IGNITION "OFF."
- DISCONNECT 24-WAY EBCM CONNECTOR.
- IGNITION "ON."
- USING J 39200, MEASURE VOLTAGE BETWEEN GROUND AND 24-WAY EBCM HARNESS CONNECTOR TERMINAL "9" THEN "10".
  WAS VOLTAGE LESS THAN 1 VOLT FOR EACH MEASUREMENT?

NO: REPAIR SHORT TO VOLTAGE IN CKT(S) (833 AND/OR 872) WHICH MEASURED GREATER THAN 1 VOLT IN STEP 10.

YES:

(11)
- IGNITION "OFF."
- INSPECT CKTS 833 AND 872 AND 24-WAY EBCM HARNESS CONNECTOR FOR DAMAGE WHICH MAY RESULT IN A SHORT TO B+ WITH THE 24-WAY EBCM HARNESS CONNECTOR CONNECTED TO THE EBCM. REPAIR DAMAGE IF EVIDENT.
- RECONNECT ALL CONNECTORS.
- DRIVE VEHICLE ABOVE 24 km/h (15 MPH) FOR AT LEAST 30 SECONDS.
  IS DTC 22 SET AS A CURRENT DTC?

YES: REPLACE EBCM.

NO: MALFUNCTION IS INTERMITTENT. REFER TO "DIAGNOSTIC AIDS" ON FACING PAGE.

(Step 5) YES:

(12)
- IGNITION "OFF."
- DISCONNECT 24-WAY EBCM CONNECTOR.
- USING J 39200, MEASURE RESISTANCE BETWEEN 24-WAY EBCM HARNESS CONNECTOR TERMINALS "9" AND "10".
  DOES J 39200 DISPLAY "OL" (INFINITE)?

NO: REPAIR SHORT TO BETWEEN CKTS 833 AND 872.

YES:

(13)
- INSPECT WHEEL SPEED SENSOR AND WHEEL SPEED SENSOR HARNESS CONNECTOR FOR POOR TERMINAL CONTACT. REPLACE TERMINALS THAT EXHIBIT POOR CONTACT. INSPECT CKTs 833 AND 872 AND 24-WAY EBCM HARNESS CONNECTOR FOR DAMAGE WHICH MAY RESULT IN A SHORT BETWEEN CKTs 833 AND 872 WITH ALL CONNECTORS CONNECTED. REPAIR DAMAGE IF EVIDENT.
- RECONNECT ALL CONNECTORS.
- DRIVE VEHICLE ABOVE 24 km/h (15 MPH) FOR AT LEAST 30 SECONDS.
  IS DTC 22 SET AS A CURRENT DTC?

YES: REPLACE EBCM.

NO: MALFUNCTION IS INTERMITTENT. REFER TO "DIAGNOSTIC AIDS" ON FACING PAGE.

DIAGNOSTIC CHART DTC 22 — 1994–95 ACHIEVA, GRAND AM AND SKYLARK WITH VES

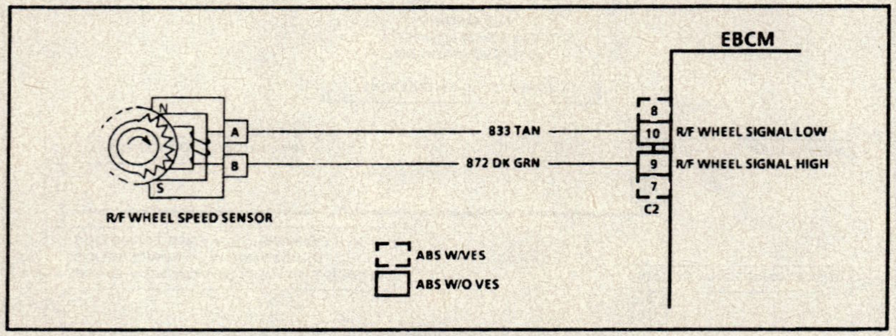

DTC 22

RIGHT FRONT WHEEL SPEED = 0 (FOR ABS WITH VES ONLY)

Circuit Description:

As a toothed ring passes by the wheel speed sensor, changes in the electromagnetic field cause the wheel speed sensor to produce a sinusoidal (AC) voltage signal whose frequency is proportional to wheel speed. The magnitude of this signal is directly related to wheel speed and the proximity of the wheel speed sensor to the toothed ring, often referred to as the air gap.

DTC Will Set When: DTC 22 can be set when the vehicle is not in an ABS stop. A malfunction exists if the right front wheel speed = 0 and the vehicle speed is greater than 8 km/h (5 mph).

Action Taken: A malfunction DTC is stored, ABS/VES is disabled and the ABS warning lamp is turned "ON."

DTC Chart Test Description: Number(s) below refer to circled number(s) on the diagnostic chart.

1. Identifies if a wheel speed sensor or circuitry is damaged and visibly apparent.
2. Tests wheel speed sensor for proper resistance values.
3. Ensures wheel speed sensor and toothed ring generate the proper voltage.
4. Ensures wheel speed sensor high and low circuits are not shorted together.
5. Ensures malfunction is not due to physical damage of the wheel speed sensor circuitry.

Diagnostic Aids: An "Intermittent" malfunction is most likely caused by a poor connection, rubbed through wire insulation, or a wire that is broken inside the insulation.

The frequency of the malfunction can be checked by using the enhanced diagnostic function of the Tech 1, as described in "Enhanced Diagnostics," found in this section.

If the customer's comments reflect that the ABS amber warning lamp is "ON" only during moist environmental changes, (rain, snow, vehicle wash) all wheel speed sensor circuitry should be thoroughly inspected for signs of water intrusion. Use the following procedure. Spray down the suspected area with 5% salt water solution (two teaspoons of salt to 12 oz. of water). Drive vehicle above 24 km/h (15 mph) for at least 30 seconds. If DTC returns immediately, replace suspected harness.

Any circuitry, that is suspected as causing the intermittent complaint, should be thoroughly checked for backed out terminals, improper mating, broken locks, improperly formed or damaged terminals, poor terminal to wiring connections or physical damage to the wiring harness.

Resistance of the wheel speed sensor will increase with an increase in sensor temperature.

When replacing a wheel speed sensor, inspect the sensor terminals and harness connector for corrosion and/or water intrusion. If evidence of corrosion or water intrusion exists, replace wheel speed sensor harness. Refer to "On-Vehicle Service" in this section.

If wheel speed sensor harness shows evidence of water intrusion, inspect wheel speed sensor for water intrusion. If water intrusion is evident, replace wheel speed sensor.

DIAGNOSTIC CHART DTC 22 — 1994–95 ACHIEVA, GRAND AM AND SKYLARK WITH VES

DTC 22

RIGHT FRONT WHEEL SPEED = 0 (FOR ABS WITH VES ONLY)

Important

- WHEEL SPEED SENSOR INTERMITTENT MALFUNCTIONS MAY BE DIFFICULT TO LOCATE. CARE SHOULD BE TAKEN NOT TO DISTURB ANY ELECTRICAL CONNECTIONS PRIOR TO AN INDICATED STEP OF THIS CHART. THIS WILL ENSURE THAT AN INTERMITTENT CONNECTION WILL NOT BE CORRECTED BEFORE THE SOURCE OF THE MALFUNCTION IS FOUND.

(1)
- IGNITION "OFF."
- INSPECT WHEEL SPEED SENSOR WIRING AND CONNECTORS FOR DAMAGE. INSPECT WHEEL SPEED SENSOR AND TOOTHED RING FOR LOOSENESS OR DAMAGE. PAY PARTICULAR ATTENTION TO THE TOOTHED RING. ANY SIGNIFICANT DAMAGE WILL AFFECT THE WHEEL SPEED INPUT SIGNAL.

IS PHYSICAL DAMAGE OR SENSOR/TOOTHED RING LOOSENESS EVIDENT?

YES: CORRECT PHYSICAL DAMAGE; CLEAR DTCs AND RECHECK.

NO: (2)
- DISCONNECT CONNECTOR FROM RIGHT FRONT WHEEL SPEED SENSOR.
- USING J 39200, MEASURE RESISTANCE BETWEEN SENSOR TERMINALS "A" AND "B".

IS RESISTANCE BETWEEN 1628 AND 1889 OHMS WHEN SENSOR IS 25°C (77°F)? *[IF TEMPERATURE OF THE SENSOR IS SUSPECTED TO BE OTHER THAN 25°C (77°F), REFER TO THE TABLE BELOW FOR THE APPLICABLE SENSOR RESISTANCE VALUES; THE VALUES IN THE TABLE ARE FOR TEMPERATURE OF THE SENSOR NOT AIR TEMPERATURE.]*

NO: REPLACE WHEEL SPEED SENSOR.

YES: (3)
- WITH J 39200 STILL CONNECTED, SELECT AC VOLTAGE SCALE.
- SPIN WHEEL BY HAND WHILE MONITORING VOLTAGE.

IS VOLTAGE AT LEAST 100 mV (VOLTAGE WILL INCREASE AS WHEEL SPEED INCREASES)?

NO: REPLACE WHEEL SPEED SENSOR.

YES: (4)
- DISCONNECT 24-WAY EBCM CONNECTOR.
- USING J 39200, MEASURE RESISTANCE BETWEEN 24-WAY EBCM HARNESS CONNECTOR TERMINALS "7" AND "8".

DOES J 39200 DISPLAY "OL" (INFINITE)?

NO: REPAIR SHORT BETWEEN CKTs 833 AND 872.

YES: (5)
- INSPECT CKT 833 AND 872 AND 24-WAY EBCM HARNESS CONNECTOR FOR PHYSICAL DAMAGE WHICH MAY RESULT IN A SHORT BETWEEN CKT 833 AND 872 WITH THE 24-WAY EBCM HARNESS CONNECTOR CONNECTED TO THE EBCM. CORRECT DAMAGE IF EVIDENT.
- RECONNECT ALL CONNECTORS.
- DRIVE VEHICLE ABOVE 24 km/h (15 MPH) FOR AT LEAST 30 SECONDS.

IS DTC 22 SET AS A CURRENT DTC?

YES: REPLACE EBCM.

NO: MALFUNCTION IS INTERMITTENT. REFER TO "DIAGNOSTIC AIDS" ON FACING PAGE.

FRONT WHEEL SPEED SENSOR

TEMPERATURE OF SENSOR VS. SENSOR RESISTANCE (APPROXIMATE)

TEMP. (°C)	TEMP. (°F)	RESISTANCE (OHMS)
-34 TO 4	-30 TO 40	1258 TO 1739
5 TO 43	41 TO 110	1503 TO 2023
44 TO 93	111 TO 200	1747 TO 2384

THIS CHART ASSUMES THAT A CURRENT DTC IS STORED INDICATING THAT THIS MALFUNCTION IS PRESENT.

DIAGNOSTIC CHART DTC 23 — 1994–95 ACHIEVA, GRAND AM AND SKYLARK WITHOUT VES

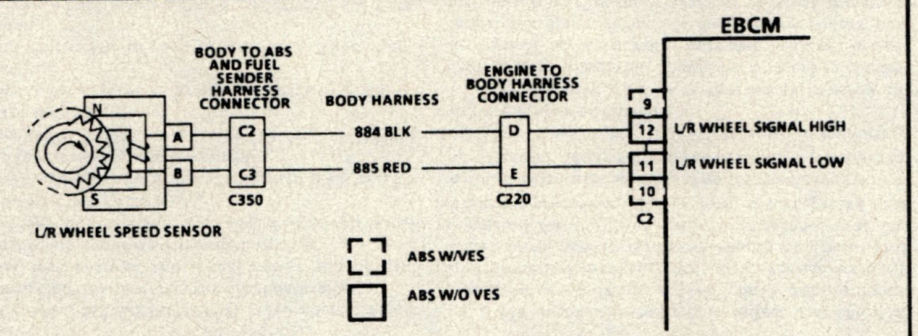

DTC 23
(Page 1 of 2)
LEFT REAR WHEEL SPEED = 0 OR UNREASONABLE (FOR ABS WITHOUT VES ONLY)

Circuit Description:
As a toothed ring passes by the wheel speed sensor, changes in the electromagnetic field cause the wheel speed sensor to produce a sinusoidal (AC) voltage signal whose frequency is proportional to wheel speed. The magnitude of this signal is directly related to wheel speed and the proximity of the wheel speed sensor to the toothed ring, often referred to as the air gap.

DTC Will Set When: DTC 23 can be set when the vehicle is not in an ABS stop. If the left rear wheel speed is less than 1/2 of the vehicle speed and the vehicle speed is greater than 8 km/h (5 mph), a malfunction exists.

Action Taken: A malfunction DTC is stored, ABS is disabled and the ABS warning lamp is turned "ON."

DTC Chart Test Description: Number(s) below refer to circled number(s) on the diagnostic chart.

1. This step will identify circuitry that is damaged and visibly apparent.
2. Checks for proper voltages at the rear sensor harness connector.
3. Checks for a short between the wheel signal high and low circuits.
4. This test checks wheel speed sensor for proper resistance values.
5. This step ensures the wheel speed sensor is not shorted to ground.

Diagnostic Aids: An "Intermittent" malfunction is most likely caused by a poor connection, rubbed through wire insulation, or a wire that is broken inside the insulation.

The frequency of the malfunction can be checked by using the enhanced diagnostic function of the Tech 1, as described in "Enhanced Diagnostics," found in this section.

If the customer's comments reflect that the ABS amber warning lamp is "ON" only during moist environmental changes, (rain, snow, vehicle wash) all wheel speed sensor circuitry should be thoroughly inspected for signs of water intrusion. Use the following procedure. Spray down the suspected area with 5% salt water solution (two teaspoons of salt to 12 oz. of water). Drive vehicle above 24 km/h (15 mph) for at least 30 seconds. If DTC returns immediately, replace suspected harness.

Any circuitry, that is suspected as causing the intermittent complaint, should be thoroughly checked for backed out terminals, improper mating, broken locks, improperly formed or damaged terminals, poor terminal to wiring connections or physical damage to the wiring harness.

Resistance of the wheel speed sensor will increase with an increase in sensor temperature.

When replacing a wheel speed sensor, inspect the sensor terminals and harness connector for corrosion and/or water intrusion. If evidence or corrosion or water intrusion exists, replace wheel speed sensor harness. Refer to "On-Vehicle Service" in this section.

If wheel speed sensor harness shows evidence of water intrusion, inspect wheel speed sensor for water intrusion. If water intrusion is evident, replace wheel speed sensor.

DIAGNOSTIC CHART DTC 23 — 1994–95 ACHIEVA, GRAND AM AND SKYLARK WITHOUT VES

DTC 23
(Page 1 of 4)
LEFT REAR WHEEL SPEED = 0 OR UNREASONABLE (FOR ABS WITHOUT VES ONLY)

Important
- WHEEL SPEED SENSOR INTERMITTENT MALFUNCTIONS MAY BE DIFFICULT TO LOCATE. CARE SHOULD BE TAKEN NOT TO DISTURB ANY ELECTRICAL CONNECTIONS PRIOR TO AN INDICATED STEP OF THIS CHART. THIS WILL ENSURE THAT AN INTERMITTENT CONNECTION WILL NOT BE CORRECTED BEFORE THE SOURCE OF THE MALFUNCTION IS FOUND.

(1)
- IGNITION "OFF."
- INSPECT WHEEL SPEED SENSOR WIRING AND CONNECTORS FOR DAMAGE. IS DAMAGE EVIDENT?

YES: REPAIR DAMAGE. DRIVE VEHICLE ABOVE 24 km/h (15 MPH) FOR AT LEAST 30 SECONDS TO ENSURE DTC 23 DOES NOT SET.

NO:

(2)
- REMOVE LEFT REAR INTEGRAL BEARING AND SPEED SENSOR ASSEMBLY (TUBULAR AXLE ONLY).
- DISCONNECT CONNECTOR FROM LEFT REAR WHEEL SPEED SENSOR.
- IGNITION "ON."
- USING J 39200, MEASURE VOLTAGE BETWEEN GROUND AND LEFT REAR SENSOR HARNESS CONNECTOR TERMINAL "A" THEN "B". WAS VOLTAGE BETWEEN 2.25 AND 2.75 VOLTS FOR EACH MEASUREMENT?

NO: REPAIR SHORT BETWEEN CKTs 884 AND 885.

YES:

(3)
- IGNITION "OFF."
- DISCONNECT 24-WAY EBCM CONNECTOR.
- USING J 39200, MEASURE RESISTANCE BETWEEN 24-WAY EBCM HARNESS CONNECTOR TERMINALS "11" AND "12." DOES J 39200 DISPLAY "OL" (INFINITE)?

NO: REPLACE WHEEL SPEED SENSOR.

YES:

(4)
- USING J 39200, MEASURE THE RESISTANCE BETWEEN LEFT REAR WHEEL SPEED SENSOR TERMINALS "A" AND "B". IS RESISTANCE BETWEEN 2100 AND 2400 OHMS WHEN SENSOR IS 25°C (77°F)? (IF TEMPERATURE OF THE SENSOR IS SUSPECTED TO BE OTHER THAN 25°C (77°F). REFER TO THE TABLE BELOW FOR APPLICABLE SENSOR RESISTANCE VALUES; THE VALUES IN THE TABLE ARE FOR TEMPERATURE OF THE SENSOR NOT AIR TEMPERATURE).

NO: REPAIR SHORT BETWEEN CKTs 830 AND 873.

YES:

(5)
- USING J 39200, MEASURE RESISTANCE BETWEEN SENSOR TERMINAL "A" AND INTEGRAL BEARING AND SPEED SENSOR ASSEMBLY MOUNTING FLANGE. DOES J 39200 DISPLAY "OL" (INFINITE)?

FRONT WHEEL SPEED SENSOR
TEMPERATURE OF SENSOR VS. SENSOR RESISTANCE (APPROXIMATE)

TEMP °C	TEMP °F	RESISTANCE (OHMS)
-34 TO 4	-30 TO 40	1258 TO 1739
5 TO 43	41 TO 110	1503 TO 2023
44 TO 93	111 TO 200	1747 TO 2384

THIS CHART ASSUMES THAT A CURRENT DTC IS STORED INDICATING THAT THIS MALFUNCTION IS PRESENT.

DIAGNOSTIC CHART DTC 23 — 1994–95 ACHIEVA, GRAND AM AND SKYLARK WITHOUT VES

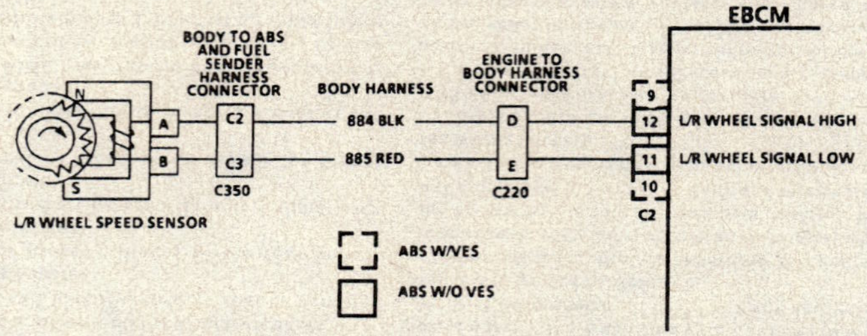

DTC 23
(Page 2 of 2)
LEFT REAR WHEEL SPEED = 0 OR UNREASONABLE (FOR ABS WITHOUT VES ONLY)

Circuit Description:

As a toothed ring passes by the wheel speed sensor, changes in the electromagnetic field cause the wheel speed sensor to produce a sinusoidal (AC) voltage signal whose frequency is proportional to wheel speed. The magnitude of this signal is directly related to wheel speed and the proximity of the wheel speed sensor to the toothed ring, often referred to as the air gap.

DTC Will Set When: DTC 23 can be set when the vehicle is not in an ABS stop. If the left rear wheel speed is less than 1/2 of the vehicle speed and the vehicle speed is greater than 8 km/h (5 mph), a malfunction exists.

Action Taken: A malfunction DTC is stored, ABS is disabled and the ABS warning lamp is turned "ON."

DTC Chart Test Description: Number(s) below refer to circled number(s) on the diagnostic chart.

6. This test ensures the wheel speed sensor and sensor ring generate the proper voltage.
7. Ensures malfunction was not due to physical damage of the circuitry.
8. This step checks for proper voltages at body to ABS and fuel sender harness connector.
9. Checks for proper voltages at the engine to body harness connector.
10. Checks for a short to ground in the wheel signal circuit.
11. This checks for an open in the wheel signal high circuit.
12. Checks wheel speed circuits for short to B+ at engine harness.
13. Checks for an open in the wheel signal low circuit.
14. Ensures malfunction was not due to physical damage of the circuitry.

Diagnostic Aids: An "Intermittent" malfunction is most likely caused by a poor connection, rubbed through wire insulation, or a wire that is broken inside the insulation.

The frequency of the malfunction can be checked by using the enhanced diagnostic function of the Tech 1, as described in "Enhanced Diagnostics," found in this section.

If the customer's comments reflect that the ABS amber warning lamp is "ON" only during moist environmental changes, (rain, snow, vehicle wash) all wheel speed sensor circuitry should be thoroughly inspected for signs of water intrusion. Use the following procedure. Spray down the suspected area with 5% salt water solution (two teaspoons of salt to 12 oz. of water). Drive vehicle above 24 km/h (15 mph) for at least 30 seconds. If DTC returns immediately, replace suspected harness.

Any circuitry, that is suspected as causing the intermittent complaint, should be thoroughly checked for backed out terminals, improper mating, broken locks, improperly formed or damaged terminals, poor terminal to wiring connections or physical damage to the wiring harness.

Resistance of the wheel speed sensor will increase with an increase in sensor temperature.

When replacing a wheel speed sensor, inspect the sensor terminals and harness connector for corrosion and/or water intrusion. If evidence or corrosion or water intrusion exists, replace wheel speed sensor harness. Refer to "On-Vehicle Service" in this section.

If wheel speed sensor harness shows evidence of water intrusion, inspect wheel speed sensor for water intrusion. If water intrusion is evident, replace wheel speed sensor.

DIAGNOSTIC CHART DTC 23 — 1994–95 ACHIEVA, GRAND AM AND SKYLARK WITHOUT VES

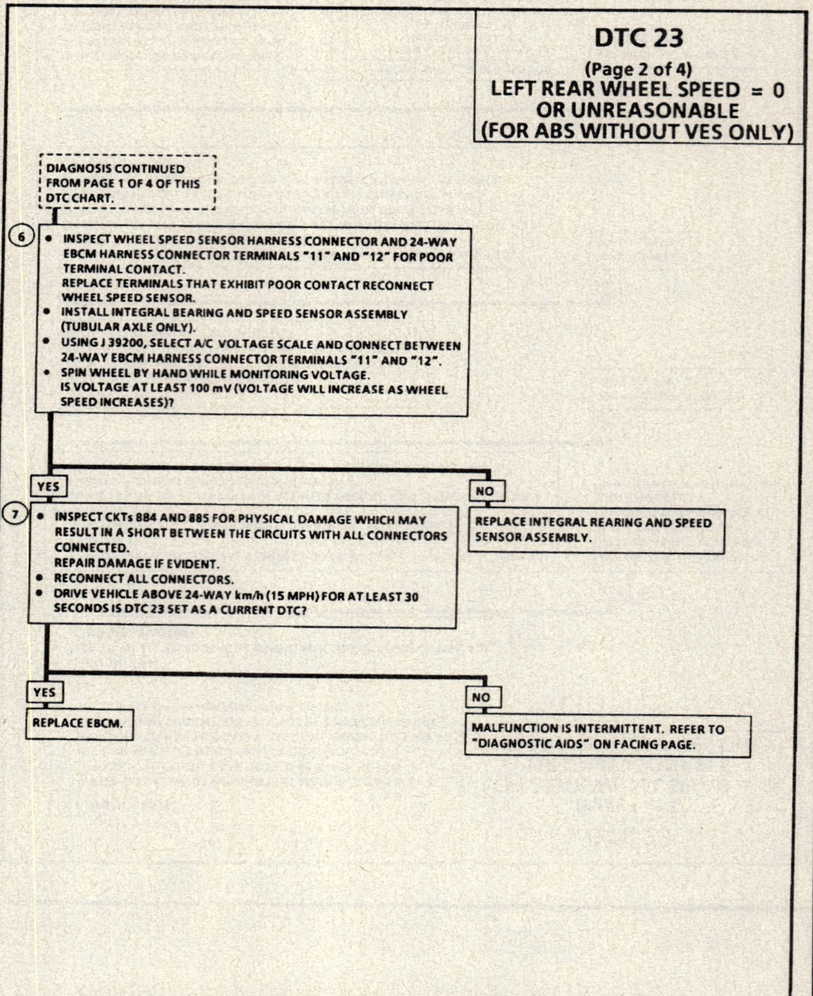

THIS CHART ASSUMES THAT A CURRENT DTC IS STORED INDICATING THAT THIS MALFUNCTION IS PRESENT.

DIAGNOSTIC CHART DTC 23 — 1994–95 ACHIEVA, GRAND AM AND SKYLARK WITHOUT VES

DTC 23
(Page 3 of 4)
LEFT REAR WHEEL SPEED = 0 OR UNREASONABLE
(FOR ABS WITHOUT VES ONLY)

CONTINUED FROM PAGE 1 OF 4 OF THIS DTC CHART.

8
- IGNITION "OFF."
- DISCONNECT REAR ABS HARNESS.
- IGNITION "ON."
- USING J 39200 MEASURE VOLTAGE BETWEEN GROUND AND BODY HARNESS CONNECTOR TERMINAL "C2" THE "C3". WAS VOLTAGE BETWEEN 2.25 AND 2.75 VOLTS FOR EACH MEASUREMENT?

YES: INSPECT BODY TO REAR ABS HARNESS CONNECTOR FOR POOR TERMINAL CONTACT. REPLACE TERMINALS THAT EXHIBIT POOR CONTACT. IF POOR TERMINAL CONTACT IS NOT EVIDENT, REPLACE REAR ABS HARNESS.

NO:

9
- IGNITION "OFF."
- DISCONNECT BODY HARNESS FROM ENGINE HARNESS.
- IGNITION "ON."
- USING J 39200, MEASURE VOLTAGE BETWEEN GROUND AND ENGINE HARNESS CONNECTOR TERMINAL "D" THEN "E". WAS VOLTAGE BETWEEN 2.25 AND 2.75 VOLTS FOR EACH MEASUREMENT?

YES: WAS VOLTAGE GREATER THAN 2.75 FOR ONE OR BOTH OF THE MEASUREMENTS IN STEP 8?

- NO: REPAIR OPEN OR SHORT TO GROUND IN BODY HARNESS CKT(s) (884 AND/OR 885) WHICH MEASURED LESS THAN 2.25 VOLTS IN STEP 8.
- YES: REPAIR SHORT TO VOLTAGE IN BODY HARNESS CKT(s) (884 AND/OR 885) WHICH MEASURED GREATER THAN 2.75 VOLTS IN STEP 8.

NO: IS VOLTAGE GREATER THAN 2.75 VOLTS ON EITHER TERMINAL "D" OR "E"?

YES:

12
- IGNITION "OFF."
- DISCONNECT 24-WAY EBCM CONNECTOR.
- IGNITION "ON."
- USING J 39200 MEASURE VOLTAGE BETWEEN GROUND AND 24-WAY EBCM HARNESS CONNECTOR TERMINAL "11" THEN "12". WAS VOLTAGE LESS THAN 1 VOLT FOR EACH MEASUREMENT?

- YES: CONTINUE DIAGNOSIS AT STEP 13 ON PAGE 4 OF 4 OF THIS DTC CHART.
- NO: REPAIR SHORT TO VOLTAGE IN ENGINE HARNESS CKT(s) (884 AND/OR 885 WHICH MEASURED GREATER THAN 1 VOLT.

NO:

10
- IGNITION "OFF."
- DISCONNECT 24-WAY EBCM CONNECTOR.
- USING J 39200, MEASURE RESISTANCE BETWEEN GROUND AND 24-WAY EBCM HARNESS CONNECTOR TERMINAL "11" THEN "12". DID J 39200 DISPLAY "OL" (INFINITE) FOR EACH MEASUREMENT?

NO: REPAIR SHORT TO GROUND IN ENGINE HARNESS CKT(s) (884 AND/OR 885) WHICH MEASURED OTHER THAN "OL" (INFINITE).

YES:

11
- USING J 39200 MEASURE RESISTANCE BETWEEN ENGINE HARNESS CONNECTOR TERMINAL "D" AND 24-WAY EBCM HARNESS CONNECTOR TERMINAL "12". IS RESISTANCE LESS THAN 2 OHMS?

- YES: CONTINUE DIAGNOSIS ON NEXT PAGE.
- NO: REPAIR OPEN IN ENGINE HARNESS CKT 884.

THIS CHART ASSUMES THAT A CURRENT DTC IS STORED INDICATING THAT THIS MALFUNCTION IS PRESENT.

DIAGNOSTIC CHART DTC 23 — 1994–95 ACHIEVA, GRAND AM AND SKYLARK WITHOUT VES

DTC 23
(Page 4 of 4)
LEFT REAR WHEEL SPEED = 0 OR UNREASONABLE
(FOR ABS WITHOUT VES ONLY)

CONTINUED FROM PAGE 3 OF 4 OF THIS DTC CHART.

13
- USING J 39200, MEASURE RESISTANCE BETWEEN ENGINE HARNESS CONNECTOR TERMINAL "E" AND 24-WAY EBCM HARNESS CONNECTOR TERMINAL "11". IS RESISTANCE LESS THAN 2 OHMS?

NO: REPAIR OPEN IN ENGINE HARNESS CKT 885.

YES:

14
- IGNITION "OFF."
- INSPECT ALL WHEEL SPEED SENSOR HARNESS CONNECTORS FOR POOR TERMINAL CONTACT OR CORROSION. REPLACE TERMINALS THAT EXHIBIT POOR CONTACT OR CORROSION.
- INSPECT ALL WHEEL SPEED SENSOR CONNECTORS AND HARNESSES FOR DAMAGE WHICH MAY RESULT IN AN OPEN OR SHORT TO GROUND/BATTERY WHEN ALL COMPONENTS ARE CONNECTED. REPAIR DAMAGE IF EVIDENT.
- RECONNECT ALL CONNECTORS.
- DRIVE VEHICLE ABOVE 24 km/h (15 MPH) FOR AT LEAST 30 SECONDS. IS DTC 23 SET AS A CURRENT DTC?

- YES: REPLACE EBCM.
- NO: MALFUNCTION IS INTERMITTENT. REFER TO "DIAGNOSTIC AIDS" ON FACING PAGE.

THIS CHART ASSUMES THAT A CURRENT DTC IS STORED INDICATING THAT THIS MALFUNCTION IS PRESENT.

DIAGNOSTIC CHART DTC 23 — 1994–95 ACHIEVA, GRAND AM AND SKYLARK WITH VES

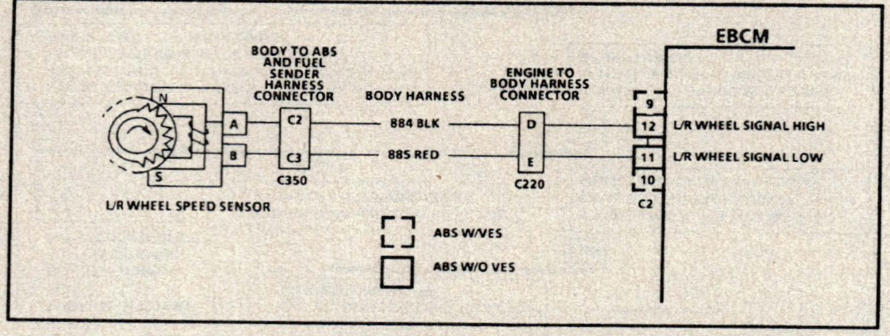

DTC 23

LEFT REAR WHEEL SPEED = 0 (FOR ABS WITH VES ONLY)

Circuit Description:

As a toothed ring passes by the wheel speed sensor, changes in the electromagnetic field cause the wheel speed sensor to produce a sinusoidal (AC) voltage signal whose frequency is proportional to wheel speed. The magnitude of this signal is directly related to wheel speed and the proximity of the wheel speed sensor to the toothed ring, often referred to as the air gap.

DTC Will Set When: DTC 23 can be set when the vehicle is not in an ABS stop. A malfunction exists if the left rear wheel speed = 0 and the vehicle speed is greater than 8 km/h (5 mph).

Action Taken: A malfunction DTC is stored, ABS/VES is disabled and the ABS warning lamp is turned "ON."

DTC Chart Test Description: Number(s) below refer to circled number(s) on the diagnostic chart.

1. Identifies if wheel speed sensor circuitry is damaged and visibly apparent.
2. Tests wheel speed sensor and rear ABS harness for proper resistance values.
3. Checks for a short between circuits in the rear ABS harness.
4. Tests wheel speed sensor for proper resistance values.
5. Ensures wheel speed sensor and toothed ring generate the proper voltage.
6. Ensures wheel speed sensor high and low circuits are not shorted together.
7. Ensures malfunction is not due to physical damage of the wheel speed sensor circuitry.

Diagnostic Aids: An "Intermittent" malfunction is most likely caused by a poor connection, rubbed through wire insulation, or a wire that is broken inside the insulation.

The frequency of the malfunction can be checked by using the enhanced diagnostic function of the Tech 1.

If the customer's comments reflect that the ABS amber warning lamp is "ON" only during moist environmental changes, (rain, snow, vehicle wash) all wheel speed sensor circuitry should be thoroughly inspected for signs of water intrusion. Use the following procedure. Spray down the suspected area with 5% salt water solution (two teaspoons of salt to 12 oz. of water). Drive vehicle above 24 km/h (15 mph) for at least 30 seconds. If DTC returns immediately, replace suspected harness.

Any circuitry, that is suspected as causing the intermittent complaint, should be thoroughly checked for backed out terminals, improper mating, broken locks, improperly formed or damaged terminals, poor terminal to wiring connections or physical damage to the wiring harness.

Resistance of the wheel speed sensor will increase with an increase in sensor temperature.

When replacing a wheel speed sensor, inspect the sensor terminals and harness connector for corrosion and/or water intrusion. If evidence or corrosion or water intrusion exists, replace wheel speed sensor harness.

If wheel speed sensor harness shows evidence of water intrusion, inspect wheel speed sensor for water intrusion. If water intrusion is evident, replace wheel speed sensor.

DIAGNOSTIC CHART DTC 23 — 1994–95 ACHIEVA, GRAND AM AND SKYLARK WITH VES

DTC 23

(Page 1 of 2)

LEFT REAR WHEEL SPEED = 0 (FOR ABS WITH VES ONLY)

Important

- WHEEL SPEED SENSOR INTERMITTENT MALFUNCTIONS MAY BE DIFFICULT TO LOCATE. CARE SHOULD BE TAKEN NOT TO DISTURB ANY ELECTRICAL CONNECTIONS PRIOR TO AN INDICATED STEP OF THIS CHART. THIS WILL ENSURE THAT AN INTERMITTENT CONNECTION WILL NOT BE CORRECTED BEFORE THE SOURCE OF THE MALFUNCTION IS FOUND.

(1)
- IGNITION "OFF."
- INSPECT WHEEL SPEED SENSOR WIRING AND CONNECTORS FOR DAMAGE. IS DAMAGE EVIDENT?

YES: REPAIR DAMAGE. DRIVE VEHICLE ABOVE 24 km/h (15 MPH) FOR AT LEAST 30 SECONDS TO ENSURE DTC 23 DOES NOT SET.

NO:

(2)
- DISCONNECT REAR ABS HARNESS FROM BODY TO ABS AND FUEL SENDER CONNECTOR.
- CONNECT J 39200, MEASURE RESISTANCE BETWEEN REAR ABS HARNESS CONNECTOR TERMINALS "C2" AND "C3".

IS RESISTANCE BETWEEN 2100 AND 2400 OHMS WHEN SENSOR IS 25°C (77°F)? *[IF TEMPERATURE OF THE SENSOR IS SUSPECTED TO BE OTHER THAN 25°C (77°F), REFER TO THE TABLE BELOW FOR THE APPLICABLE SENSOR RESISTANCE VALUES; THE VALUES IN THE TABLE ARE FOR TEMPERATURE OF THE SENSOR, NOT AIR TEMPERATURE.]*

YES: GO TO PAGE 2 OF 2 OF THIS CHART.

NO:

(3)
- REMOVE LEFT REAR INTEGRAL BEARING AND SPEED SENSOR ASSEMBLY (TUBULAR AXLE ONLY).
- DISCONNECT CONNECTOR FROM LEFT REAR WHEEL SPEED SENSOR.
- USING J 39200, MEASURE RESISTANCE BETWEEN REAR ABS HARNESS CONNECTOR TERMINALS "C2" AND "C3".

DOES J 39200 DISPLAY "OL" (INFINITE)?

NO: REPLACE REAR ABS HARNESS.

YES:

(4)
- USING J 39200, MEASURE RESISTANCE BETWEEN WHEEL SPEED SENSOR TERMINALS "A" AND "B."

IS RESISTANCE BETWEEN 2100 AND 2400 OHMS WHEN SENSOR IS 25°C (77°F)? *[IF TEMPERATURE OF THE SENSOR IS SUSPECTED TO BE OTHER THAN 25°C (77°F), REFER TO THE TABLE BELOW FOR THE APPLICABLE SENSOR RESISTANCE VALUES; THE VALUES IN THE TABLE ARE FOR TEMPERATURE OF THE SENSOR, NOT AIR TEMPERATURE.]*

YES: MALFUNCTION WAS DUE TO AN INTERMITTENT SHORT BETWEEN CKTs 884 AND 885. REPLACE REAR ABS HARNESS.

NO: REPLACE INTEGRAL BEARING AND SPEED SENSOR ASSEMBLY.

REAR WHEEL SPEED SENSOR

TEMPERATURE OF SENSOR VS. SENSOR RESISTANCE (APPROXIMATE)

TEMP. (°C)	TEMP. (°F)	RESISTANCE (OHMS)
-34 TO 4	-30 TO 40	1623 TO 2210
5 TO 43	41 TO 110	1939 TO 2570
44 TO 93	111 TO 200	2254 TO 3029

THIS CHART ASSUMES THAT A CURRENT DTC IS STORED INDICATING THAT THIS MALFUNCTION IS PRESENT.

DIAGNOSTIC CHART DTC 23 — 1994–95 ACHIEVA, GRAND AM AND SKYLARK WITH VES

DTC 23

(Page 2 of 2)
LEFT REAR WHEEL SPEED = 0
(FOR ABS WITH VES ONLY)

CONTINUED FROM PAGE 1 OF 2 OF THIS CHART.

(5)
- WITH J 39200 STILL CONNECTED, SELECT AC VOLTAGE SCALE.
- SPIN WHEEL BY HAND WHILE MONITORING VOLTAGE.

IS VOLTAGE AT LEAST 100 mV (VOLTAGE WILL INCREASE AS WHEEL SPEED INCREASES)?

NO → REPLACE INTEGRAL BEARING AND SPEED SENSOR ASSEMBLY.

YES →

(6)
- DISCONNECT 24-WAY EBCM CONNECTOR.
- USING J 39200, MEASURE RESISTANCE BETWEEN 24-WAY EBCM HARNESS CONNECTOR TERMINALS "9" AND "10".

DOES J 39200 DISPLAY "OL" (INFINITE)?

NO → REPAIR SHORT BETWEEN CKTs 884 AND 885.

YES →

(7)
- INSPECT CKTs 884 AND 885 AND ALL CONNECTORS FOR DAMAGE WHICH MAY RESULT IN A SHORT BETWEEN CKTs 884 AND 885 WITH ALL CONNECTORS CONNECTED. REPAIR DAMAGE IF EVIDENT.
- DRIVE VEHICLE ABOVE 24 km/h (15 MPH) FOR AT LEAST 30 SECONDS.

IS DTC 23 SET AS A CURRENT DTC?

YES → REPLACE EBCM.

NO → MALFUNCTION IS INTERMITTENT. REFER TO "DIAGNOSTIC AIDS" ON FACING PAGE.

THIS CHART ASSUMES THAT A CURRENT DTC IS STORED INDICATING THAT THIS MALFUNCTION IS PRESENT.

DIAGNOSTIC CHART DTC 24 — 1994–95 ACHIEVA, GRAND AM AND SKYLARK WITHOUT VES

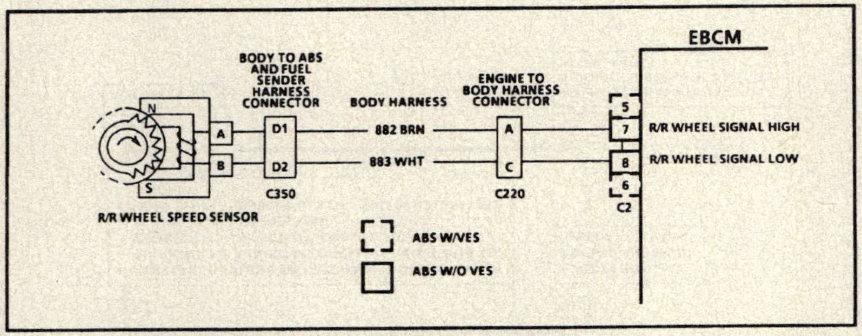

DTC 24

RIGHT REAR WHEEL SPEED = 0
OR UNREASONABLE
(FOR ABS WITHOUT VES ONLY)

Circuit Description:
As a toothed ring passes by the wheel speed sensor, changes in the electromagnetic field cause the wheel speed sensor to produce a sinusoidal (AC) voltage signal whose frequency is proportional to wheel speed. The magnitude of this signal is directly related to wheel speed and the proximity of the wheel speed sensor to the toothed ring, often referred to as the air gap.

DTC Will Set When: DTC 24 can be set when the vehicle is not in an ABS stop. If the right rear wheel speed is less than 1/2 of the vehicle speed and the vehicle speed is greater than 8 km/h (5 mph), a malfunction exists.

Action Taken: A malfunction DTC is stored, ABS is disabled and the ABS warning lamp is turned "ON."

DTC Chart Test Description: Number(s) below refer to circled number(s) on the diagnostic chart.

1. This step will identify circuitry that is damaged and visibly apparent.
2. Checks for proper voltages at the rear sensor harness connector.
3. Checks for a short between the wheel signal high and low circuits.
4. This test checks wheel speed sensor for proper resistance values.
5. This step ensures the wheel speed sensor is not shorted to ground.
6. This test ensures the wheel speed sensor and sensor ring generate the proper voltage.
7. Ensures malfunction was not due to physical damage of the circuitry.

Diagnostic Aids: An "Intermittent" malfunction is most likely caused by a poor connection, rubbed through wire insulation, or a wire that is broken inside the insulation.

The frequency of the malfunction can be checked by using the enhanced diagnostic function of the Tech 1.

If the customer's comments reflect that the ABS amber warning lamp is "ON" only during moist environmental changes, (rain, snow, vehicle wash) all wheel speed sensor circuitry should be thoroughly inspected for water intrusion. Use the following procedure. Spray down the suspected area with 5% salt water solution (two teaspoons of salt to 12 oz. of water). Drive vehicle above 24 km/h (15 mph) for at least 30 seconds. If DTC returns immediately, replace suspected harness.

Any circuitry, that is suspected as causing the intermittent complaint, should be thoroughly checked for backed out terminals, improper mating, broken locks, improperly formed or damaged terminals, poor terminal to wiring connections or physical damage to the wiring harness.

Resistance of the wheel speed sensor will increase with an increase in sensor temperature.

When replacing a wheel speed sensor, inspect the sensor terminals and harness connector for corrosion and/or water intrusion. If evidence of corrosion or water intrusion exists, replace wheel speed sensor harness.

If wheel speed sensor harness shows evidence of water intrusion, inspect wheel speed sensor for water intrusion. If water intrusion is evident, replace wheel speed sensor.

DIAGNOSTIC CHART 24 — 1994–95 ACHIEVA, GRAND AM AND SKYLARK WITHOUT VES

DTC 24
(Page 1 of 4)
RIGHT REAR WHEEL SPEED = 0 OR UNREASONABLE (FOR ABS WITHOUT VES ONLY)

Important
- WHEEL SPEED SENSOR INTERMITTENT MALFUNCTIONS MAY BE DIFFICULT TO LOCATE. CARE SHOULD BE TAKEN NOT TO DISTURB ANY ELECTRICAL CONNECTIONS PRIOR TO AN INDICATED STEP OF THIS CHART. THIS WILL ENSURE THAT AN INTERMITTENT CONNECTION WILL NOT BE CORRECTED BEFORE THE SOURCE OF THE MALFUNCTION IS FOUND.

1.
- IGNITION "OFF."
- INSPECT WHEEL SPEED SENSOR WIRING AND CONNECTORS FOR DAMAGE.

IS DAMAGE EVIDENT?

YES: REPAIR DAMAGE. DRIVE VEHICLE ABOVE 24 km/h (15MPH) FOR AT LEAST 30 SECONDS TO ENSURE DTC 24 DOES NOT SET.

NO:

2.
- REMOVE RIGHT REAR INTEGRAL BEARING AND SPEED SENSOR ASSEMBLY (TUBULAR AXLE ONLY).
- DISCONNECT CONNECTOR FROM RIGHT REAR WHEEL SPEED SENSOR.
- IGNITION "ON."
- USING J 39200, MEASURE VOLTAGE BETWEEN GROUND AND RIGHT REAR SENSOR HARNESS CONNECTOR TERMINAL "A" THEN "B". WAS VOLTAGE BETWEEN 2.25 AND 2.75 VOLTS FOR EACH MEASUREMENT?

YES: GO TO PAGE 3 OF 4 OF THIS CHART.

NO:

3.
- IGNITION "OFF."
- DISCONNECT 24-WAY EBCM CONNECTOR.
- USING J39200, MEASURE RESISTANCE BETWEEN 24-WAY EBCM HARNESS CONNECTOR TERMINALS "8" AND "7." DOES J 39200 DISPLAY "OL" (INFINITE).

NO: REPAIR SHORT BETWEEN CKTS 882 AND 883.

YES:

4.
- USING J 39200, MEASURE RESISTANCE BETWEEN RIGHT REAR WHEEL SPEED SENSOR TERMINALS "A" AND "B." IS RESISTANCE BETWEEN 2100 AND 2400 OHMS WHEN SENSOR IS 25°C (77°F)? (IF TEMPERATURE OF THE SENSOR IS SUSPECTED TO BE OTHER THAN 25°C (77°F). REFER TO THE TABLE BELOW FOR APPLICABLE SENSOR RESISTANCE VALUES; THE VALUES IN THE TABLE ARE FOR TEMPERATURE OF THE SENSOR NOT AIR TEMPERATURE).

NO: REPLACE INTEGRAL BEARING AND SPEED SENSOR ASSEMBLY.

YES: CONTINUE DIAGNOSIS ON NEXT PAGE.

REAR WHEEL SPEED SENSOR
TEMPERATURE OF SENSOR VS. SENSOR RESISTANCE (APPROXIMATE)

TEMP. (°C)	TEMP. (°F)	RESISTANCE (OHMS)
-34 TO 4	-30 TO 40	1623 TO 2210
5 TO 43	41 TO 110	1939 TO 2570
44 TO 93	111 TO 200	2254 TO 3029

THIS CHART ASSUMES THAT A CURRENT DTC IS STORED INDICATING THAT THIS MALFUNCTION IS PRESENT.

DIAGNOSTIC CHART 24 — 1994–95 ACHIEVA, GRAND AM AND SKYLARK WITHOUT VES

DTC 24
(Page 2 of 4)
RIGHT REAR WHEEL SPEED = 0 OR UNREASONABLE (FOR ABS WITHOUT VES ONLY)

CONTINUED FROM PAGE 1 OF 4 OF THIS DTC CHART.

5.
- USING J 39200, MEASURE RESISTANCE BETWEEN SENSOR TERMINAL "A" AND INTEGRAL BEARING AND SPEED SENSOR ASSEMBLY MOUNTING FLANGE. DOES J 39200 DISPLAY "OL" (INFINITE)?

NO: REPLACE INTEGRAL BEARING AND SPEED SENSOR ASSEMBLY.

YES:

6.
- INSPECT WHEEL SPEED SENSOR HARNESS CONNECTOR AND 24-WAY EBCM HARNESS CONNECTOR TERMINALS "8" AND "7" FOR POOR TERMINAL CONTACT. REPLACE TERMINALS THAT EXHIBIT POOR CONTACT.
- RECONNECT WHEEL SPEED SENSOR.
- INSTALL INTEGRAL BEARING AND SPEED SENSOR ASSEMBLY (TUBULAR AXLE ONLY).
- USING J 39200, SELECT A/C VOLTAGE SCALE AND CONNECT CONNECTOR TERMINALS "8" AND "7". SPIN WHEEL BY HAND WHILE MONITORING VOLTAGE. IS VOLTAGE AT LEAST 100 mV (VOLTAGE WILL INCREASE AS WHEEL SPEED INCREASES)?

NO: REPLACE INTEGRAL BEARING AND SPEED SENSOR ASSEMBLY.

YES:

7.
- INSPECT CKTs 882 AND 883 FOR PHYSICAL DAMAGE WHICH MAY RESULT IN A SHORT BETWEEN THE CIRCUITS WITH ALL CONNECTORS CONNECTED REPAIR DAMAGE IF EVIDENT.
- RECONNECT ALL CONNECTORS.
- DRIVE VEHICLE ABOVE 24 km/h (15 MPH) FOR AT LEAST 30 SECONDS. IS DTC 24 SET AS A CURRENT DTC?

YES: REPLACE EBCM.

NO: MALFUNCTION IS INTERMITTENT. REFER TO "DIAGNOSTIC AIDS" ON FACING PAGE.

THIS CHART ASSUMES THAT A CURRENT DTC IS STORED INDICATING THAT THIS MALFUNCTION IS PRESENT.

DIAGNOSTIC CHART 24 — 1994–95 ACHIEVA, GRAND AM AND SKYLARK WITHOUT VES

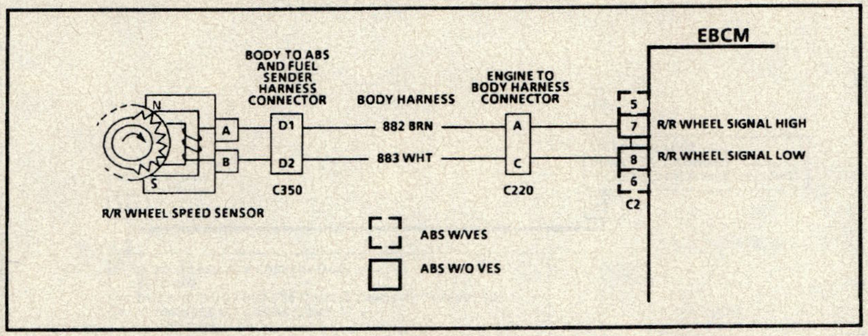

DTC 24

RIGHT REAR WHEEL SPEED = 0 OR UNREASONABLE (FOR ABS WITHOUT VES ONLY)

Circuit Description:

As a toothed ring passes by the wheel speed sensor, changes in the electromagnetic field cause the wheel speed sensor to produce a sinusoidal (AC) voltage signal whose frequency is proportional to wheel speed. The magnitude of this signal is directly related to wheel speed and the proximity of the wheel speed sensor to the toothed ring, often referred to as the air gap.

DTC Will Set When: DTC 24 can be set when the vehicle is not in an ABS stop. If the right rear wheel speed is less than 1/2 of the vehicle speed and the vehicle speed is greater than 8 km/h (5 mph), a malfunction exists.

Action Taken: A malfunction DTC is stored, ABS is disabled and the ABS warning lamp is turned "ON."

DTC Chart Test Description: Number(s) below refer to circled number(s) on the diagnostic chart.

8. This step checks for proper voltages at body to ABS and fuel sender harness connector.
9. Checks for proper voltages at the engine to body harness connector.
10. Checks for a short to ground in the wheel signal circuits.
11. This checks for an open in the wheel signal high circuit.
12. Checks wheel speed circuits for short to B+ at engine harness.
13. Checks for an open in the wheel signal low circuit.
14. Ensures malfunction was not due to physical damage of the circuitry.

Diagnostic Aids: An "Intermittent" malfunction is most likely caused by a poor connection, rubbed through wire insulation, or a wire that is broken inside the insulation.

The frequency of the malfunction can be checked by using the enhanced diagnostic function of the Tech 1.

If the customer's comments reflect that the ABS amber warning lamp is "ON" only during moist environmental changes, (rain, snow, vehicle wash) all wheel speed sensor circuitry should be thoroughly inspected for water intrusion. Use the following procedure. Spray down the suspected area with 5% salt water solution (two teaspoons of salt to 12 oz. of water). Drive vehicle above 24 km/h (15 mph) for at least 30 seconds. If DTC returns immediately, replace suspected harness.

Any circuitry, that is suspected as causing the intermittent complaint, should be thoroughly checked for backed out terminals, improper mating, broken locks, improperly formed or damaged terminals, poor terminal to wiring connections or physical damage to the wiring harness.

Resistance of the wheel speed sensor will increase with an increase in sensor temperature.

When replacing a wheel speed sensor, inspect the sensor terminals and harness connector for corrosion and/or water intrusion. If evidence of corrosion or water intrusion exists, replace wheel speed sensor harness.

If wheel speed sensor harness shows evidence of water intrusion, inspect wheel speed sensor for water intrusion. If water intrusion is evident, replace wheel speed sensor.

DIAGNOSTIC CHART 24 — 1994–95 ACHIEVA, GRAND AM AND SKYLARK WITHOUT VES

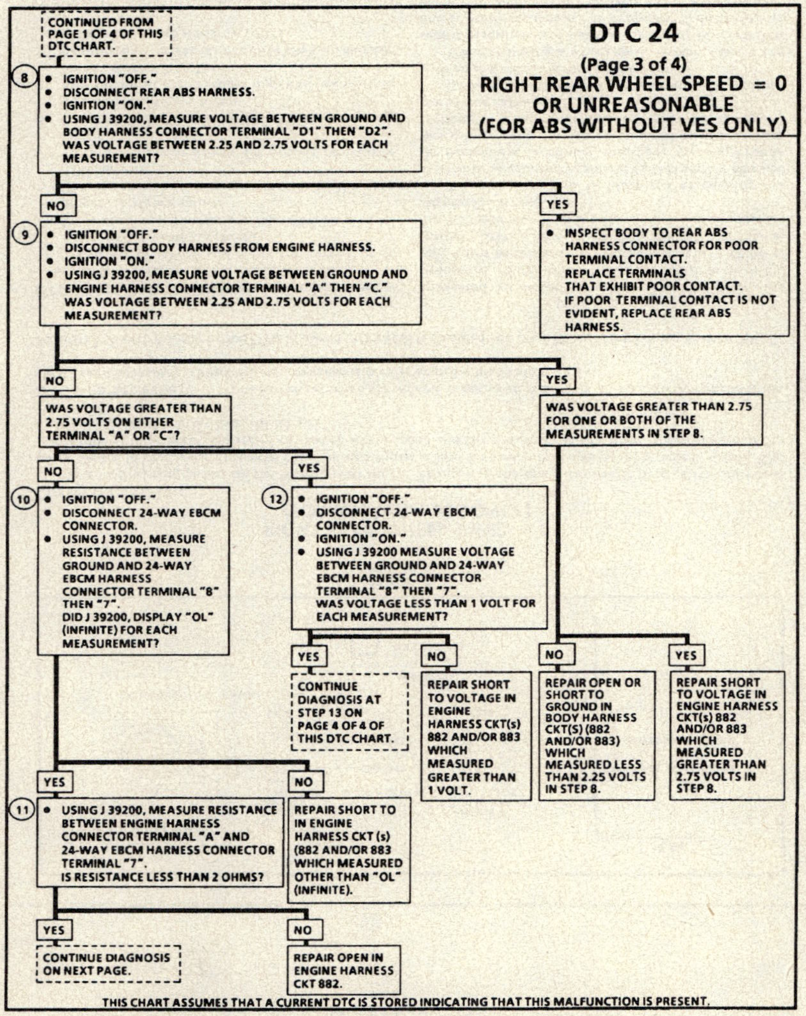

DIAGNOSTIC CHART 24 — 1994–95 ACHIEVA, GRAND AM AND SKYLARK WITHOUT VES

DTC 24
(Page 4 of 4)
RIGHT REAR WHEEL SPEED = 0 OR UNREASONABLE (FOR ABS WITHOUT VES ONLY)

CONTINUED FROM PAGE 3 OF 4 OF THIS DTC CHART.

YES

(13)
- USING J 39200, MEASURE RESISTANCE BETWEEN TERMINAL "C" AND 24-WAY EBCM HARNESS CONNECTOR TERMINAL "8". IS RESISTANCE LESS THAN 2 OHMS?

NO → REPAIR OPEN IN ENGINE HARNESS CKT 883.

YES

(14)
- IGNITION "OFF."
- INSPECT ALL WHEEL SPEED SENSOR HARNESS CONNECTORS FOR POOR TERMINAL CONTACT OR CORROSION. REPLACE TERMINALS THAT EXHIBIT POOR CONTACT OR CORROSION.
- INSPECT ALL WHEEL SPEED SENSOR CONNECTORS AND HARNESSES FOR DAMAGE WHICH MAY RESULT IN AN OPEN OR SHORT TO GROUND/BATTERY WHEN ALL COMPONENTS ARE CONNECTED. REPAIR DAMAGE IF EVIDENT.
- RECONNECT ALL CONNECTORS.
- DRIVE VEHICLE ABOVE 24 km/h (15 MPH) FOR AT LEAST 30 SECONDS. IS DTC 24 SET AS A CURRENT DTC?

YES → REPLACE EBCM.

NO → MALFUNCTION IS INTERMITTENT. REFER TO "DIAGNOSTIC AIDS" ON FACING PAGE.

THIS CHART ASSUMES THAT A CURRENT DTC IS STORED INDICATING THAT THIS MALFUNCTION IS PRESENT.

DIAGNOSTIC CHART 24 — 1994–95 ACHIEVA, GRAND AM AND SKYLARK WITH VES

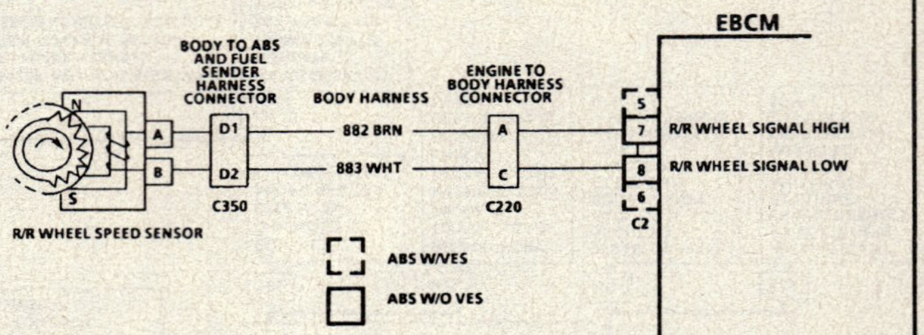

DTC 24

RIGHT REAR WHEEL SPEED = 0 (FOR ABS WITH VES ONLY)

Circuit Description:
As a toothed ring passes by the wheel speed sensor, changes in the electromagnetic field cause the wheel speed sensor to produce a sinusoidal (AC) voltage signal whose frequency is proportional to wheel speed. The magnitude of this signal is directly related to wheel speed and the proximity of the wheel speed sensor to the toothed ring, often referred to as the air gap.

DTC Will Set When: DTC 24 can be set when the vehicle is not in an ABS stop. A malfunction exists if the right rear wheel speed = 0 and the vehicle speed is greater than 8 km/h (5 mph).

Action Taken: A malfunction DTC is stored, ABS/VES is disabled and the ABS warning lamp is turned "ON."

DTC Chart Test Description: Number(s) below refer to circled number(s) on the diagnostic chart.
1. Identifies if wheel speed sensor circuitry is damaged and visibly apparent.
2. Tests wheel speed sensor and rear ABS harness for proper resistance values.
3. Checks for a short between circuits in the rear ABS harness.
4. Tests wheel speed sensor for proper resistance values.
5. Ensures wheel speed sensor and toothed ring generate the proper voltage.
6. Ensures wheel speed sensor high and low circuits are not shorted together.
7. Ensures malfunction is not due to physical damage of the wheel speed sensor circuitry.

Diagnostic Aids: An "Intermittent" malfunction is most likely caused by a poor connection, rubbed through wire insulation, or a wire that is broken inside the insulation.

The frequency of the malfunction can be checked by using the enhanced diagnostic function of the Tech 1, as described in "Enhanced Diagnostics," found in this section.

If the customer's comments reflect that the ABS amber warning lamp is "ON" only during moist environmental changes, (rain, snow, vehicle wash) all wheel speed sensor circuitry should be thoroughly inspected for water intrusion. Use the following procedure. Spray down the suspected area with 5% salt water solution (two teaspoons of salt to 12 oz. of water). Drive vehicle above 24 km/h (15 mph) for at least 30 seconds. If DTC returns immediately, replace suspected harness.

Any circuitry, that is suspected as causing the intermittent complaint, should be thoroughly checked for backed out terminals, improper mating, broken locks, improperly formed or damaged terminals, poor terminal to wiring connections or physical damage to the wiring harness.

Resistance of the wheel speed sensor will increase with an increase in sensor temperature.

When replacing a wheel speed sensor, inspect the sensor terminals and harness connector for corrosion and/or water intrusion. If evidence of corrosion or water intrusion exists, replace wheel speed sensor harness. Refer to "On-Vehicle Service" in this section.

If wheel speed sensor harness shows evidence of water intrusion, inspect wheel speed sensor for water intrusion. If water intrusion is evident, replace wheel speed sensor.

DIAGNOSTIC CHART 24 — 1994–95 ACHIEVA, GRAND AM AND SKYLARK WITH VES

DTC 24
(Page 1 of 2)
RIGHT REAR WHEEL SPEED = 0
(FOR ABS WITH VES ONLY)

Important

- WHEEL SPEED SENSOR INTERMITTENT MALFUNCTIONS MAY BE DIFFICULT TO LOCATE. CARE SHOULD BE TAKEN NOT TO DISTURB ANY ELECTRICAL CONNECTIONS PRIOR TO AN INDICATED STEP OF THIS CHART. THIS WILL ENSURE THAT AN INTERMITTENT CONNECTION WILL NOT BE CORRECTED BEFORE THE SOURCE OF THE MALFUNCTION IS FOUND.

(1)
- IGNITION "OFF."
- INSPECT WHEEL SPEED SENSOR WIRING AND CONNECTORS FOR DAMAGE.

IS DAMAGE EVIDENT?

- YES: REPAIR DAMAGE. DRIVE VEHICLE ABOVE 24 km/h (15MPH) FOR AT LEAST 30 SECONDS TO ENSURE DTC 24 DOES NOT SET.
- NO: go to step 2.

(2)
- DISCONNECT REAR ABS HARNESS FROM BODY TO ABS AND FUEL SENDER CONNECTOR.
- USING J 39200, MEASURE RESISTANCE BETWEEN REAR ABS HARNESS CONNECTOR TERMINALS "D1" AND "D2."

IS RESISTANCE BETWEEN 2100 AND 2400 OHMS WHEN SENSOR IS 25°C (77°F)? *[IF TEMPERATURE OF THE SENSOR IS SUSPECTED TO BE OTHER THAN 25°C (77°F), REFER TO THE TABLE BELOW FOR THE APPLICABLE SENSOR RESISTANCE VALUES; THE VALUES IN THE TABLE ARE FOR TEMPERATURE OF THE SENSOR, NOT AIR TEMPERATURE.]*

- YES: GO TO PAGE 2 OF 2 OF THIS CHART.
- NO: go to step 3.

(3)
- REMOVE RIGHT REAR INTEGRAL BEARING AND SPEED SENSOR ASSEMBLY (TUBULAR AXLE ONLY).
- DISCONNECT CONNECTOR FROM RIGHT REAR WHEEL SPEED SENSOR.
- USING J 39200, MEASURE RESISTANCE BETWEEN REAR ABS HARNESS CONNECTOR TERMINALS "D1" AND "D2."

DOES J 39200 DISPLAY "OL" (INFINITE)?

- NO: REPLACE REAR ABS HARNESS.
- YES: go to step 4.

(4)
- USING J 39200, MEASURE RESISTANCE BETWEEN WHEEL SPEED SENSOR TERMINALS "A" AND "B."

IS RESISTANCE BETWEEN 2100 AND 2400 OHMS WHEN SENSOR IS 25°C (77°F)? *[IF TEMPERATURE OF THE SENSOR IS SUSPECTED TO BE OTHER THAN 25°C (77°F), REFER TO THE TABLE BELOW FOR THE APPLICABLE SENSOR RESISTANCE VALUES; THE VALUES IN THE TABLE ARE FOR TEMPERATURE OF THE SENSOR, NOT AIR TEMPERATURE.]*

- YES: MALFUNCTION WAS DUE TO AN INTERMITTENT SHORT BETWEEN CKTs 884 AND 885. REPLACE REAR ABS HARNESS.
- NO: REPLACE INTEGRAL BEARING AND SPEED SENSOR ASSEMBLY.

REAR WHEEL SPEED SENSOR

TEMPERATURE OF SENSOR VS. SENSOR RESISTANCE (APPROXIMATE)

TEMP. (°C)	TEMP. (°F)	RESISTANCE (OHMS)
-34 TO 4	-30 TO 40	1623 TO 2210
5 TO 43	41 TO 110	1939 TO 2570
44 TO 93	111 TO 200	2254 TO 3029

THIS CHART ASSUMES THAT A CURRENT DTC IS STORED INDICATING THAT THIS MALFUNCTION IS PRESENT.

DIAGNOSTIC CHART 24 — 1994–95 ACHIEVA, GRAND AM AND SKYLARK WITH VES

DTC 24
(Page 2 of 2)
RIGHT REAR WHEEL SPEED = 0
(FOR ABS WITH VES ONLY)

CONTINUED FROM PAGE 1 OF 2 OF THIS CHART.

(5)
- WITH J 39200 STILL CONNECTED, SELECT AC VOLTAGE SCALE.
- SPIN WHEEL BY HAND WHILE MONITORING VOLTAGE.

IS VOLTAGE AT LEAST 100 mV (VOLTAGE WILL INCREASE AS WHEEL SPEED INCREASES)?

- NO: REPLACE INTEGRAL BEARING AND SPEED SENSOR ASSEMBLY.
- YES: go to step 6.

(6)
- DISCONNECT 24-WAY EBCM CONNECTOR.
- USING J 39200, MEASURE RESISTANCE BETWEEN 24-WAY EBCM HARNESS CONNECTOR TERMINALS "5" AND "6".

DOES J 39200 DISPLAY "OL" (INFINITE)?

- NO: REPAIR SHORT BETWEEN CKTs 882 AND 883.
- YES: go to step 7.

(7)
- INSPECT CKTs 882 AND 883 AND ALL CONNECTORS FOR DAMAGE WHICH MAY RESULT IN A SHORT BETWEEN CKTs 882 AND 883 WITH ALL CONNECTORS CONNECTED. REPAIR DAMAGE IF EVIDENT.
- DRIVE VEHICLE ABOVE 24 km/h (15 MPH) FOR AT LEAST 30 SECONDS.

IS DTC 24 SET AS A CURRENT DTC?

- NO: REPLACE EBCM.
- YES: MALFUNCTION IS INTERMITTENT. REFER TO "DIAGNOSTIC AIDS" ON FACING PAGE.

THIS CHART ASSUMES THAT A CURRENT DTC IS STORED INDICATING THAT THIS MALFUNCTION IS PRESENT.

DIAGNOSTIC CHART 25 — 1994–95 ACHIEVA, GRAND AM AND SKYLARK

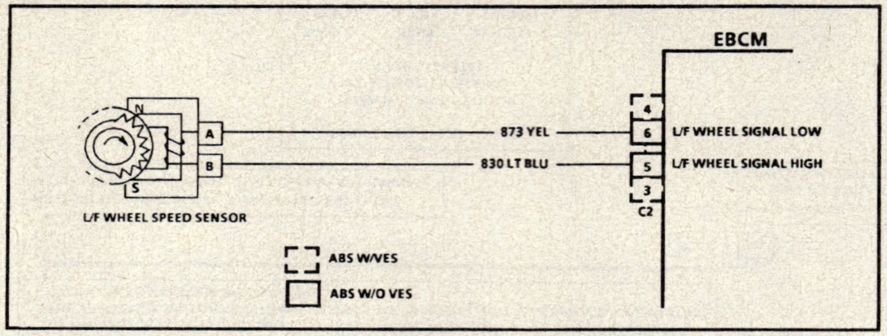

DTC 25

LEFT FRONT EXCESSIVE WHEEL SPEED VARIATION

Circuit Description:

As a toothed ring passes by the wheel speed sensor, changes in the electromagnetic field cause the wheel speed sensor to produce a sinusoidal (AC) voltage signal whose frequency is proportional to wheel speed. The magnitude of this signal is directly related to wheel speed and the proximity of the wheel speed sensor to the toothed ring, often referred to as the air gap.

DTC Will Set When: DTC 25 can be set when the brake is "OFF." The purpose of this test is to detect a situation in which the left front wheel acceleration or deceleration is beyond specified limits.

Action Taken: A malfunction DTC is stored, ABS/VES is disabled and the ABS warning lamp is turned "ON."

DTC Chart Test Description: Number(s) below refer to circled number(s) on the diagnostic chart.

1. This test checks wheel speed sensor for proper resistance values.
2. Checks for a short to ground in the wheel speed sensor circuitry.
3. Checks for a short to voltage in the wheel speed sensor circuitry.
4. Ensures malfunction was not due to physical damage of the circuitry.
5. Checks for a short to voltage in the wheel signal low circuit.
6. Checks for a short to voltage in the wheel signal high circuit.
7. Checks for a short between the wheel signal high and low circuits.
8. Checks for high resistance in the wheel signal low circuit.
9. Checks for high resistance in the wheel signal high circuit.
10. Checks for proper resistance of the wheel speed sensor.
11. Checks for a short to ground in either of the wheel signal high or low circuits.
12. Checks for a short to ground in the wheel speed sensor.

Diagnostic Aids: An "Intermittent" malfunction is most likely caused by a poor connection, rubbed through wire insulation, or a wire that is broken inside the insulation.

The frequency of the malfunction can be checked by using the enhanced diagnostic function of the Tech 1.

If the customer's comments reflect that the ABS amber warning lamp is "ON" only during moist environmental changes, (rain, snow, vehicle wash) all wheel speed sensor circuitry should be thoroughly inspected for water intrusion. Use the following procedure. Spray down the suspected area with 5% salt water solution (two teaspoons of salt to 12 oz. of water). Drive vehicle above 24 km/h (15 mph) for at least 30 seconds. If DTC returns immediately, replace suspected harness.

Any circuitry, that is suspected as causing the intermittent complaint, should be thoroughly checked for backed out terminals, improper mating, broken locks, improperly formed or damaged terminals, poor terminal to wiring connections or physical damage to the wiring harness.

A careful visual inspection of the toothed ring, CV joint, bearing and wheel speed sensor for physical damage should also be performed. If DTC sets at the same mph every time and wheel speed variation is noted above this mph on the Tech 1, the speed ring is most likely damaged. This can occur at any speed.

Resistance of the wheel speed sensor will increase with an increase in sensor temperature.

When replacing a wheel speed sensor, inspect the sensor terminals and harness connector for corrosion and/or water intrusion. If evidence of corrosion or water intrusion exists, replace wheel speed sensor harness.

If wheel speed sensor harness shows evidence of water intrusion, inspect wheel speed sensor for water intrusion. If water intrusion is evident, replace wheel speed sensor.

DIAGNOSTIC CHART 25 — 1994–95 ACHIEVA, GRAND AM AND SKYLARK

DTC 25
(Page 1 of 3)
LEFT FRONT EXCESSIVE WHEEL SPEED VARIATION

IMPORTANT:

- WHEEL SPEED SENSOR INTERMITTENT MALFUNCTIONS MAY BE DIFFICULT TO LOCATE. CARE SHOULD BE TAKEN NOT TO DISTURB ANY ELECTRICAL CONNECTIONS PRIOR TO AN INDICATED STEP OF THIS CHART. THIS WILL ENSURE THAT AN INTERMITTENT CONNECTION WILL NOT BE CORRECTED BEFORE THE SOURCE OF THE MALFUNCTION IS FOUND.

(1)
- IGNITION "OFF."
- USING A 5% SALT WATER SOLUTION (TWO TEASPOONS OF SALT PER 12 OZ. WATER), THOROUGHLY SPRAY THE LEFT FRONT WHEEL SPEED SENSOR HARNESS.
- DISCONNECT 24-WAY EBCM CONNECTOR.
- USING J 39200, MEASURE RESISTANCE BETWEEN 24-WAY EBCM HARNESS CONNECTOR TERMINALS "5" (ABS W/O VES) OR "3" (ABS W/VES) AND "6" (ABS W/O VES) OR "4" (ABS W/VES). IS RESISTANCE BETWEEN 1628 AND 1889 OHMS WHEN SENSOR IS 25°C (77°F)? *(IF TEMPERATURE OF THE SENSOR IS SUSPECTED TO BE OTHER THAN 25°C (77°F), REFER TO TABLE ON PAGE 2 OF 2 OF THIS DTC CHART FOR APPLICABLE SENSOR RESISTANCE VALUES; THE VALUES IN THE TABLE ARE FOR TEMPERATURE OF THE SENSOR NOT AIR TEMPERATURE.)*

YES → (2) NO → GO TO PAGE 2 OF 3 OF THIS DTC CHART.

(2)
- USING J 39200, MEASURE THE RESISTANCE BETWEEN GROUND AND EBCM HARNESS CONNECTOR TERMINAL "5" (ABS W/O VES) OR "3" (ABS W/VES). DOES J 39200 DISPLAY "OL" (INFINITE)?

YES → (3) NO → GO TO PAGE 3 OF 3 OF THIS DTC CHART.

(3)
- IGNITION "ON."
- USING J 39200, MEASURE THE VOLTAGE BETWEEN GROUND AND 24-WAY EBCM HARNESS CONNECTOR TERMINAL "6" (ABS W/O VES) OR "4" (ABS W/VES). IS VOLTAGE LESS THAN 1 VOLT?

YES → (4) NO → (5)

(4)
- IGNITION "OFF."
- INSPECT 24-WAY EBCM HARNESS CONNECTOR TERMINALS "5" (ABS W/O VES) OR "3" (ABS W/VES) AND "6" (ABS W/O VES) OR "4" (ABS W/VES) FOR POOR TERMINAL CONTACT OR CORROSION. INSPECT CKT(s) 830 AND 873 FOR DAMAGE WHICH MAY RESULT IN AN OPEN CIRCUIT, SHORT TO GROUND OR SHORT TO VOLTAGE. REPAIR DAMAGE IF EVIDENT. REPLACE TERMINALS IF POOR CONTACT OR CORROSION EXISTS.
- RECONNECT ALL CONNECTORS.
- IGNITION "ON."
IS DTC 25 SET AS A CURRENT DTC?

YES → REPLACE EBCM.

NO → MALFUNCTION IS INTERMITTENT. REFER TO "DIAGNOSTIC AIDS" ON FACING PAGE.

(5)
- IGNITION "OFF."
- DISCONNECT LEFT FRONT WHEEL SPEED SENSOR CONNECTOR.
- IGNITION "ON."
- USING J 39200, MEASURE THE VOLTAGE BETWEEN GROUND AND 24-WAY EBCM HARNESS CONNECTOR TERMINAL "6" (ABS W/O VES) OR "4" (ABS W/VES). IS VOLTAGE LESS THAN 1 VOLT?

YES → (6) NO → REPAIR SHORT TO VOLTAGE IN CKT 873.

(6)
- USING J 39200, MEASURE THE VOLTAGE BETWEEN GROUND AND 24-WAY EBCM HARNESS CONNECTOR TERMINAL "5" (ABS W/O VES) OR "3" (ABS W/VES). IS VOLTAGE LESS THAN 1 VOLT?

YES → MALFUNCTION IS INTERMITTENT. INSPECT ALL CONNECTORS AND HARNESSES FOR DAMAGE WHICH MAY RESULT IN A SHORT TO VOLTAGE WHEN ALL COMPONENTS ARE CONNECTED. REPAIR ALL DAMAGE IF FOUND. REFER TO "DIAGNOSTIC AIDS" ON FACING PAGE FOR MORE INFORMATION.

NO → REPAIR SHORT TO VOLTAGE IN CKT 830.

THIS CHART ASSUMES THAT A CURRENT DTC IS STORED INDICATING THAT THIS MALFUNCTION IS PRESENT.

DIAGNOSTIC CHART 25 — 1994–95 ACHIEVA, GRAND AM AND SKYLARK

DTC 25
(Page 2 of 3)
LEFT FRONT EXCESSIVE WHEEL SPEED VARIATION

CONTINUED FROM PAGE 1 OF 3 OF THIS DTC CHART.

(7)
- DISCONNECT LEFT FRONT WHEEL SPEED SENSOR CONNECTOR.
- USING J 39200, MEASURE THE RESISTANCE BETWEEN 24-WAY EBCM HARNESS CONNECTOR TERMINALS "5" (ABS W/O VES) OR "3" (ABS W/VES) AND "6" (ABS W/O VES) OR "4" (ABS W/VES). DOES J 39200 DISPLAY "OL" (INFINITE)?

NO: REPAIR SHORT BETWEEN CKTs 830 AND 873.

YES:

(8)
- USING J 39200, MEASURE THE RESISTANCE BETWEEN 24-WAY EBCM HARNESS CONNECTOR TERMINAL "6" (ABS W/O VES) OR "4" (ABS W/VES) AND LEFT FRONT WHEEL SPEED SENSOR HARNESS CONNECTOR TERMINAL "A." IS RESISTANCE LESS THAN 2 OHMS?

NO: REPAIR HIGH RESISTANCE IN CKT 873.

YES:

(9)
- USING J 39200, MEASURE THE RESISTANCE BETWEEN 24-WAY EBCM HARNESS CONNECTOR TERMINAL "5" (ABS W/O VES) OR "3" (ABS W/VES) AND LEFT FRONT WHEEL SPEED SENSOR HARNESS CONNECTOR TERMINAL "B." IS RESISTANCE LESS THAN 2 OHMS?

NO: REPAIR HIGH RESISTANCE IN CKT 830.

YES:

(10)
- USING J 39200, MEASURE THE RESISTANCE BETWEEN LEFT FRONT WHEEL SPEED SENSOR TERMINALS "A" AND "B." IS RESISTANCE BETWEEN 1628 AND 1889 OHMS WHEN SENSOR IS 25°C (77°F)? *(IF TEMPERATURE OF THE SENSOR IS SUSPECTED TO BE OTHER THAN 25°C (77°F), REFER TO THE TABLE BELOW FOR APPLICABLE SENSOR RESISTANCE VALUES; THE VALUES IN THE TABLE ARE FOR TEMPERATURE OF THE SENSOR NOT AIR TEMPERATURE).*

NO: REPLACE WHEEL SPEED SENSOR.

YES: MALFUNCTION IS INTERMITTENT. INSPECT ALL CONNECTORS AND HARNESSES FOR DAMAGES WHICH MAY RESULT IN HIGH RESISTANCE WHEN ALL COMPONENTS ARE CONNECTED. REFER TO "DIAGNOSTIC AIDS". REPAIR ALL DAMAGE IF FOUND.

FRONT WHEEL SPEED SENSOR

TEMPERATURE OF SENSOR VS. SENSOR RESISTANCE (APPROXIMATE)

TEMP. (°C)	TEMP. (°F)	RESISTANCE (OHMS)
-34 TO 4	-30 TO 40	1258 TO 1739
5 TO 43	41 TO 110	1503 TO 2023
44 TO 93	111 TO 200	1747 TO 2384

THIS CHART ASSUMES THAT A CURRENT DTC IS STORED INDICATING THAT THIS MALFUNCTION IS PRESENT.

DIAGNOSTIC CHART 25 — 1994–95 ACHIEVA, GRAND AM AND SKYLARK

DTC 25
(Page 3 of 3)
LEFT FRONT EXCESSIVE WHEEL SPEED VARIATION

CONTINUED FROM PAGE 1 OF 3 OF THIS DTC CHART.

(11)
- THOROUGHLY RESPRAY THE LEFT FRONT WHEEL SPEED SENSOR HARNESS WITH THE SALT WATER SOLUTION. IMPORTANT: DO NOT ALLOW WATER TO ENTER THE WHEEL SPEED SENSOR CONNECTOR.
- DISCONNECT LEFT FRONT WHEEL SPEED SENSOR CONNECTOR.
- USING J 39200, MEASURE THE RESISTANCE BETWEEN GROUND AND LEFT FRONT WHEEL SPEED SENSOR HARNESS CONNECTOR TERMINAL "A" THEN "B." DID J 39200 DISPLAY "OL" (INFINITE) FOR EACH MEASUREMENT?

NO: REPAIR SHORT TO GROUND IN CKT(s) (830 AND/OR 873) WHICH MEASURED OTHER THAN "OL" IN STEP 11.

YES:

(12)
- USING J 39200, MEASURE THE RESISTANCE BETWEEN GROUND AND LEFT FRONT WHEEL SPEED SENSOR TERMINAL "A." DOES J 39200 DISPLAY "OL" (INFINITE)?

NO: REPLACE WHEEL SPEED SENSOR.

YES: MALFUNCTION IS INTERMITTENT. INSPECT ALL CONNECTORS AND HARNESSES FOR DAMAGES WHICH MAY RESULT IN A SHORT TO GROUND WHEN ALL COMPONENTS ARE CONNECTED. SMALL NICKS OR CUTS IN THE HARNESS CAN CAUSE A MALFUNCTION. REPAIR ANY DAMAGE IF FOUND. JUMPER HARNESS MUST BE REPLACED IF DAMAGED.

THIS CHART ASSUMES THAT A CURRENT DTC IS STORED INDICATING THAT THIS MALFUNCTION IS PRESENT.

DIAGNOSTIC CHART 26 — 1994–95 ACHIEVA, GRAND AM AND SKYLARK

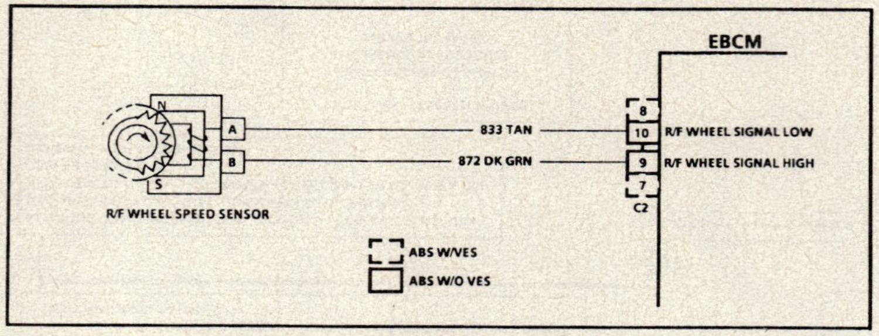

DTC 26

RIGHT FRONT EXCESSIVE WHEEL SPEED VARIATION

Circuit Description:

As a toothed ring passes by the wheel speed sensor, changes in the electromagnetic field cause the wheel speed sensor to produce a sinusoidal (AC) voltage signal whose frequency is proportional to wheel speed. The magnitude of this signal is directly related to wheel speed and the proximity of the wheel speed sensor to the toothed ring, often referred to as the air gap.

DTC Will Set When: DTC 26 can be set when the brake is "OFF." The purpose of this test is to detect a situation in which the right front wheel acceleration or deceleration is beyond specified limits.

Action Taken: A malfunction DTC is stored, ABS/VES is disabled and the ABS warning lamp is turned "ON."

DTC Chart Test Description: Number(s) below refer to circled number(s) on the diagnostic chart.

1. This test checks wheel speed sensor for proper resistance values.
2. Checks for a short to ground in the wheel speed sensor circuitry.
3. Checks for a short to voltage in the wheel speed sensor circuitry.
4. Ensures malfunction was not due to physical damage of the circuitry.
5. Checks for a short to voltage in the wheel signal low circuit.
6. Checks for a short to voltage in the wheel signal high circuit.
7. Checks for a short between the wheel signal high and low circuits.
8. Checks for high resistance in the wheel signal low circuit.
9. Checks for high resistance in the wheel signal high circuit.
10. Checks for proper resistance of the wheel speed sensor.
11. Checks for a short to ground in either of the wheel signal high or low circuits.
12. Checks for a short to ground in the wheel speed sensor.

Diagnostic Aids: An "Intermittent" malfunction is most likely caused by a poor connection, rubbed through wire insulation, or a wire that is broken inside the insulation.

The frequency of the malfunction can be checked by using the enhanced diagnostic function of the Tech 1.

If the customer's comments reflect that the ABS amber warning lamp is "ON" only during moist environmental changes, (rain, snow, vehicle wash) all wheel speed sensor circuitry should be thoroughly inspected for water intrusion. Use the following procedure. Spray down the suspected area with 5% salt water solution (two teaspoons of salt to 12 oz. of water). Drive vehicle above 24 km/h (15 mph) for at least 30 seconds. If DTC returns immediately, replace suspected harness.

Any circuitry, that is suspected as causing the intermittent complaint, should be thoroughly checked for backed out terminals, improper mating, broken locks, improperly formed or damaged terminals, poor terminal to wiring connections or physical damage to the wiring harness.

A careful visual inspection of the toothed ring, CV joint, bearing and wheel speed sensor for physical damage should also be performed. If DTC sets at the same mph every time and wheel speed variation is noted above this mph on the Tech 1, the speed ring is most likely damaged. This can occur at any speed.

Resistance of the wheel speed sensor will increase with an increase in sensor temperature.

When replacing a wheel speed sensor, inspect the sensor terminals and harness connector for corrosion and/or water intrusion. If evidence of corrosion or water intrusion exists, replace wheel speed sensor harness.

If wheel speed sensor harness shows evidence of water intrusion, inspect wheel speed sensor for water intrusion. If water intrusion is evident, replace wheel speed sensor.

DIAGNOSTIC CHART 26 — 1994–95 ACHIEVA, GRAND AM AND SKYLARK

DTC 26
(Page 1 of 3)
RIGHT FRONT EXCESSIVE WHEEL SPEED VARIATION

IMPORTANT:

- WHEEL SPEED SENSOR INTERMITTENT MALFUNCTIONS MAY BE DIFFICULT TO LOCATE. CARE SHOULD BE TAKEN NOT TO DISTURB ANY ELECTRICAL CONNECTIONS PRIOR TO AN INDICATED STEP OF THIS CHART. THIS WILL ENSURE THAT AN INTERMITTENT CONNECTION WILL NOT BE CORRECTED BEFORE THE SOURCE OF THE MALFUNCTION IS FOUND.

(1)
- IGNITION "OFF."
- USING A 5% SALT WATER SOLUTION (TWO TEASPOONS OF SALT PER 12 OZ. WATER), THOROUGHLY SPRAY THE RIGHT FRONT WHEEL SPEED SENSOR HARNESS.
- DISCONNECT 24-WAY EBCM CONNECTOR.
- USING J 39200, MEASURE RESISTANCE BETWEEN 24-WAY EBCM HARNESS CONNECTOR TERMINALS "9" (ABS W/O VES) OR "7" (ABS W/VES) AND "10" (ABS W/O VES) OR "8" (ABS W/VES). IS RESISTANCE BETWEEN 1628 AND 1889 OHMS WHEN THE SENSOR IS 25°C (77°F)? *[IF TEMPERATURE OF THE SENSOR IS SUSPECTED TO BE OTHER THAN 25°C (77°F), REFER TO TABLE ON PAGE 2 OF 2 OF THIS DTC CHART FOR APPLICABLE SENSOR RESISTANCE VALUES; THE VALUES IN THE TABLE ARE FOR TEMPERATURE OF THE SENSOR NOT AIR TEMPERATURE.]*

YES → (2)

NO → GO TO PAGE 2 OF 3 OF THIS DTC CHART.

(2)
- USING J 39200, MEASURE THE RESISTANCE BETWEEN GROUND AND EBCM HARNESS CONNECTOR TERMINAL "9" (ABS W/O VES) OR "7" (ABS W/VES). DOES J 39200 DISPLAY "OL" (INFINITE)?

YES → (3)

NO → GO TO PAGE 3 OF 3 OF THIS DTC CHART.

(3)
- IGNITION "ON."
- USING J 39200, MEASURE THE VOLTAGE BETWEEN GROUND AND 24-WAY EBCM HARNESS CONNECTOR TERMINAL "10" (ABS W/O VES) OR "8" (ABS W/VES). IS VOLTAGE LESS THAN 1 VOLT?

YES → (4)

NO → (5)

(4)
- IGNITION "OFF."
- INSPECT 24-WAY EBCM HARNESS CONNECTOR TERMINALS "9" (ABS W/O VES) OR "7" (ABS W/VES) AND "10" (ABS W/O VES) OR "8" (ABS W/VES) FOR POOR TERMINAL CONTACT OR CORROSION. INSPECT CKT(s) 833 AND 872 FOR DAMAGE WHICH MAY RESULT IN AN OPEN CIRCUIT, SHORT TO GROUND OR SHORT TO VOLTAGE. REPAIR DAMAGE IF EVIDENT. REPLACE TERMINALS IF POOR CONTACT OR CORROSION EXISTS.
- RECONNECT ALL CONNECTORS.
- IGNITION "ON."

IS DTC 26 SET AS A CURRENT DTC?

YES → REPLACE EBCM.

NO → MALFUNCTION IS INTERMITTENT. REFER TO "DIAGNOSTIC AIDS" ON FACING PAGE.

(5)
- IGNITION "OFF."
- DISCONNECT RIGHT FRONT WHEEL SPEED SENSOR CONNECTOR.
- IGNITION "ON."
- USING J 39200, MEASURE THE VOLTAGE BETWEEN GROUND AND 24-WAY EBCM HARNESS CONNECTOR TERMINAL "10" (ABS W/O VES) OR "8" (ABS W/VES).

IS VOLTAGE LESS THAN 1 VOLT?

YES → (6)

NO → REPAIR SHORT TO VOLTAGE IN CKT 833.

(6)
- USING J 39200, MEASURE THE VOLTAGE BETWEEN GROUND AND 24-WAY EBCM HARNESS CONNECTOR TERMINAL "9" (ABS W/O VES) OR "7" (ABS W/VES).

IS VOLTAGE LESS THAN 1 VOLT?

YES → MALFUNCTION IS INTERMITTENT. INSPECT ALL CONNECTORS AND HARNESSES FOR DAMAGE WHICH MAY RESULT IN A SHORT TO VOLTAGE WHEN ALL COMPONENTS ARE CONNECTED. REPAIR ALL DAMAGE IF FOUND. REFER TO "DIAGNOSTIC AIDS" ON FACING PAGE FOR MORE INFORMATION.

NO → REPAIR SHORT TO VOLTAGE IN CKT 872.

THIS CHART ASSUMES THAT A CURRENT DTC IS STORED INDICATING THAT THIS MALFUNCTION IS PRESENT.

DIAGNOSTIC CHART 26 — 1994–95 ACHIEVA, GRAND AM AND SKYLARK

DTC 26
(Page 2 of 3)
RIGHT FRONT EXCESSIVE WHEEL SPEED VARIATION

CONTINUED FROM PAGE 1 OF 3 OF THIS DTC CHART.

(7)
- DISCONNECT RIGHT FRONT WHEEL SPEED SENSOR CONNECTOR.
- USING J 39200, MEASURE THE RESISTANCE BETWEEN 24-WAY EBCM HARNESS CONNECTOR TERMINALS "9" (ABS W/O VES) OR "7" (ABS W/VES) AND "10" (ABS W/O VES) OR "8" (ABS W/VES). DOES J 39200 DISPLAY "OL" (INFINITE)?

NO: REPAIR SHORT BETWEEN CKTs 872 AND 833.

YES:

(8)
- USING J 39200, MEASURE THE RESISTANCE BETWEEN 24-WAY EBCM HARNESS CONNECTOR TERMINAL "10" (ABS W/O VES) OR "8" (ABS W/VES) AND RIGHT FRONT WHEEL SPEED SENSOR HARNESS CONNECTOR TERMINAL "A." IS RESISTANCE LESS THAN 2 OHMS?

NO: REPAIR HIGH RESISTANCE IN CKT 833.

YES:

(9)
- USING J 39200, MEASURE THE RESISTANCE BETWEEN 24-WAY EBCM HARNESS CONNECTOR TERMINAL "9" (ABS W/O VES) OR "7" (ABS W/VES) AND RIGHT FRONT WHEEL SPEED SENSOR HARNESS CONNECTOR TERMINAL "B." IS RESISTANCE LESS THAN 2 OHMS?

NO: REPAIR HIGH RESISTANCE IN CKT 872.

YES:

(10)
- USING J 39200, MEASURE THE RESISTANCE BETWEEN RIGHT FRONT WHEEL SPEED SENSOR TERMINALS "A" AND "B." IS RESISTANCE BETWEEN 1628 AND 1889 OHMS WHEN SENSOR IS 25°C (77°F)? *(IF TEMPERATURE OF THE SENSOR IS SUSPECTED TO BE OTHER THAN 25°C (77°F), REFER TO THE TABLE BELOW FOR APPLICABLE SENSOR RESISTANCE VALUES; THE VALUES IN THE TABLE ARE FOR TEMPERATURE OF THE SENSOR NOT AIR TEMPERATURE).*

NO: REPLACE WHEEL SPEED SENSOR.

YES: MALFUNCTION IS INTERMITTENT. INSPECT ALL CONNECTORS AND HARNESSES FOR DAMAGES WHICH MAY RESULT IN HIGH RESISTANCE WHEN ALL COMPONENTS ARE CONNECTED. REFER TO "DIAGNOSTIC AIDS" ON FACING PAGE FOR MORE INFORMATION. REPAIR ALL DAMAGE IF FOUND.

FRONT WHEEL SPEED SENSOR

TEMPERATURE OF SENSOR VS. SENSOR RESISTANCE (APPROXIMATE)

TEMP. (°C)	TEMP. (°F)	RESISTANCE (OHMS)
-34 TO 4	-30 TO 40	1258 TO 1739
5 TO 43	41 TO 110	1503 TO 2023
44 TO 93	111 TO 200	1747 TO 2384

THIS CHART ASSUMES THAT A CURRENT DTC IS STORED INDICATING THAT THIS MALFUNCTION IS PRESENT.

DIAGNOSTIC CHART 26 — 1994–95 ACHIEVA, GRAND AM AND SKYLARK

DTC 26
(Page 3 of 3)
RIGHT FRONT EXCESSIVE WHEEL SPEED VARIATION

CONTINUED FROM PAGE 1 OF 3 OF THIS DTC CHART.

(11)
- THOROUGHLY RESPRAY THE RIGHT FRONT WHEEL SPEED SENSOR HARNESS WITH THE SALT WATER SOLUTION. IMPORTANT: DO NOT ALLOW WATER TO ENTER THE WHEEL SPEED SENSOR CONNECTOR.
- DISCONNECT RIGHT FRONT WHEEL SPEED SENSOR CONNECTOR.
- USING J 39200, MEASURE THE RESISTANCE BETWEEN GROUND AND RIGHT FRONT WHEEL SPEED SENSOR HARNESS CONNECTOR TERMINAL "A" THEN "B." DID J 39200 DISPLAY "OL" (INFINITE) FOR EACH MEASUREMENT?

NO: REPAIR SHORT TO GROUND IN CKT(s) (833 AND/OR 872) WHICH MEASURED OTHER THAN "OL" IN STEP 11.

YES:

(12)
- USING J 39200, MEASURE THE RESISTANCE BETWEEN GROUND AND RIGHT FRONT WHEEL SPEED SENSOR TERMINAL "A." DOES J 39200 DISPLAY "OL" (INFINITE)?

NO: REPLACE WHEEL SPEED SENSOR.

YES: MALFUNCTION IS INTERMITTENT. INSPECT ALL CONNECTORS AND HARNESSES FOR DAMAGES WHICH MAY RESULT IN A SHORT TO GROUND WHEN ALL COMPONENTS ARE CONNECTED. SMALL NICKS OR CUTS IN THE HARNESS CAN CAUSE A MALFUNCTION. REPAIR ANY DAMAGE IF FOUND. JUMPER HARNESS MUST BE REPLACED IF DAMAGED. REFER TO "DIAGNOSTIC AIDS" ON FACING PAGE FOR MORE INFORMATION.

THIS CHART ASSUMES THAT A CURRENT DTC IS STORED INDICATING THAT THIS MALFUNCTION IS PRESENT.

DIAGNOSTIC CHART 27 — 1994–95 ACHIEVA, GRAND AM AND SKYLARK

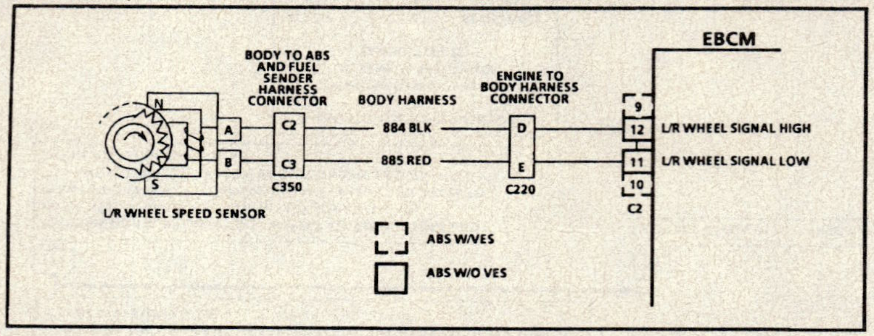

DTC 27

LEFT REAR EXCESSIVE WHEEL SPEED VARIATION

Circuit Description:

As a toothed ring passes by the wheel speed sensor, changes in the electromagnetic field cause the wheel speed sensor to produce a sinusoidal (AC) voltage signal whose frequency is proportional to wheel speed. The magnitude of this signal is directly related to wheel speed and the proximity of the wheel speed sensor to the toothed ring, often referred to as the air gap.

DTC Will Set When: DTC 27 can be set when the brake is "OFF." The purpose of this test is to detect a situation in which the left rear wheel acceleration or deceleration is beyond specified limits.

Action Taken: A malfunction DTC is stored, ABS/VES is disabled and the ABS warning lamp is turned "ON."

DTC Chart Test Description: Number(s) below refer to circled number(s) on the diagnostic chart.

1. This test checks wheel speed sensor for proper resistance values.
2. Checks for a short to ground in the wheel speed sensor circuitry.
3. Checks for a short to voltage in the wheel speed sensor circuitry.
4. Ensures malfunction was not due to physical damage of the circuitry.
5. Checks for a short to voltage in the wheel signal low circuit.
6. Checks for a short to voltage in the wheel signal high circuit.
7. Checks for a short between the wheel signal high and low circuits.
8. Checks for high resistance in the wheel signal low circuit.
9. Checks for high resistance in the wheel signal high circuit.
10. Checks for proper resistance of the wheel speed sensor.
11. Checks for a short to ground in either of the wheel signal high or low circuits.
12. Checks for a short to ground in the wheel speed sensor.

Diagnostic Aids: An "Intermittent" malfunction is most likely caused by a poor connection, rubbed through wire insulation, or a wire that is broken inside the insulation.

The frequency of the malfunction can be checked by using the enhanced diagnostic function of the Tech 1.

If the customer's comments reflect that the ABS amber warning lamp is "ON" only during moist environmental changes, (rain, snow, vehicle wash) all wheel speed sensor circuitry should be thoroughly inspected for water intrusion. Use the following procedure. Spray down the suspected area with 5% salt water solution (two teaspoons of salt to 12 oz. of water). Drive vehicle above 24 km/h (15 mph) for at least 30 seconds. If DTC returns immediately, replace suspected harness.

Any circuitry, that is suspected as causing the intermittent complaint, should be thoroughly checked for backed out terminals, improper mating, broken locks, improperly formed or damaged terminals, poor terminal to wiring connections or physical damage to the wiring harness.

A careful visual inspection of the toothed ring, CV joint, bearing and wheel speed sensor for physical damage should also be performed. If DTC sets at the same mph every time and wheel speed variation is noted above this mph on the Tech 1, the speed ring is most likely damaged. This can occur at any speed.

Resistance of the wheel speed sensor will increase with an increase in sensor temperature.

When replacing a wheel speed sensor, inspect the sensor terminals and harness connector for corrosion and/or water intrusion. If evidence of corrosion or water intrusion exists, replace wheel speed sensor harness.

If wheel speed sensor harness shows evidence of water intrusion, inspect wheel speed sensor for water intrusion. If water intrusion is evident, replace wheel speed sensor.

DIAGNOSTIC CHART 27 — 1994–95 ACHIEVA, GRAND AM AND SKYLARK

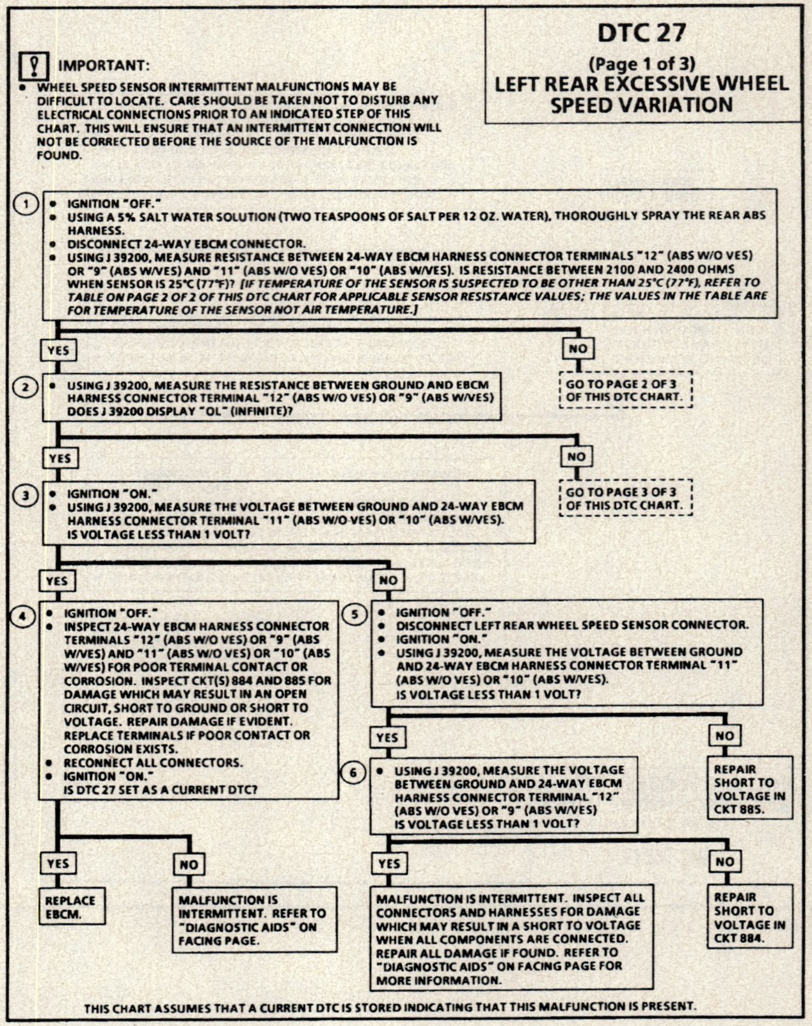

DIAGNOSTIC CHART 27 — 1994–95 ACHIEVA, GRAND AM AND SKYLARK

DTC 27 (Page 2 of 3) LEFT REAR EXCESSIVE WHEEL SPEED VARIATION

CONTINUED FROM PAGE 1 OF 3 OF THIS DTC CHART.

(7)
- DISCONNECT LEFT REAR WHEEL SPEED SENSOR CONNECTOR.
- USING J 39200, MEASURE THE RESISTANCE BETWEEN 24-WAY EBCM HARNESS CONNECTOR TERMINALS "11" (ABS W/O VES) OR "10" (ABS W/VES) AND "12" (ABS W/O VES) OR "9" (ABS W/VES). DOES J 39200 DISPLAY "OL" (INFINITE)?

NO: REPAIR SHORT BETWEEN CKTs 884 AND 885.

YES:

(8)
- USING J 39200, MEASURE THE RESISTANCE BETWEEN 24-WAY EBCM HARNESS CONNECTOR TERMINAL "12" (ABS W/O VES) OR "9" (ABS W/VES) AND LEFT REAR WHEEL SPEED SENSOR HARNESS CONNECTOR TERMINAL "A." IS RESISTANCE LESS THAN 2 OHMS?

NO: REPAIR HIGH RESISTANCE IN CKT 884.

YES:

(9)
- USING J 39200, MEASURE THE RESISTANCE BETWEEN 24-WAY EBCM HARNESS CONNECTOR TERMINAL "11" (ABS W/O VES) OR "10" (ABS W/VES) AND LEFT REAR WHEEL SPEED SENSOR HARNESS CONNECTOR TERMINAL "B." IS RESISTANCE LESS THAN 2 OHMS?

NO: REPAIR HIGH RESISTANCE IN CKT 885.

YES:

(10)
- USING J 39200, MEASURE THE RESISTANCE BETWEEN LEFT REAR WHEEL SPEED SENSOR TERMINALS "A" AND "B." IS RESISTANCE BETWEEN 2100 AND 2400 OHMS WHEN SENSOR IS 25°C (77°F)? *(IF TEMPERATURE OF THE SENSOR IS SUSPECTED TO BE OTHER THAN 25°C (77°F), REFER TO THE TABLE BELOW FOR APPLICABLE SENSOR RESISTANCE VALUES; THE VALUES IN THE TABLE ARE FOR TEMPERATURE OF THE SENSOR NOT AIR TEMPERATURE).*

NO: REPLACE INTEGRAL BEARING AND SPEED SENSOR ASSEMBLY.

YES: MALFUNCTION IS INTERMITTENT. INSPECT ALL CONNECTORS AND HARNESSES FOR DAMAGES WHICH MAY RESULT IN HIGH RESISTANCE WHEN ALL COMPONENTS ARE CONNECTED. REFER TO "DIAGNOSTIC AIDS" ON FACING PAGE FOR MORE INFORMATION. REPAIR ALL DAMAGE IF FOUND.

REAR WHEEL SPEED SENSOR

TEMPERATURE OF SENSOR VS. SENSOR RESISTANCE (APPROXIMATE)

TEMP. (°C)	TEMP. (°F)	RESISTANCE (OHMS)
-34 TO 4	-30 TO 40	1623 TO 2210
5 TO 43	41 TO 110	1939 TO 2570
44 TO 93	111 TO 200	2254 TO 3029

THIS CHART ASSUMES THAT A CURRENT DTC IS STORED INDICATING THAT THIS MALFUNCTION IS PRESENT.

DIAGNOSTIC CHART 27 — 1994–95 ACHIEVA, GRAND AM AND SKYLARK

DTC 27 (Page 3 of 3) LEFT REAR EXCESSIVE WHEEL SPEED VARIATION

CONTINUED FROM PAGE 1 OF 3 OF THIS DTC CHART.

(11)
- DISCONNECT REAR ABS HARNESS FROM BODY TO ABS AND FUEL SENDER HARNESS CONNECTOR.
- USING J 39200, MEASURE THE RESISTANCE BETWEEN GROUND AND 24-WAY EBCM HARNESS CONNECTOR TERMINAL "12" (ABS W/O VES) OR "9" (ABS W/VES). DOES J 39200 DISPLAY "OL" (INFINITE)?

NO: REPAIR SHORT TO GROUND IN CKT 884.

YES:

(12)
- USING J 39200, MEASURE THE RESISTANCE BETWEEN GROUND AND 24-WAY EBCM HARNESS CONNECTOR TERMINAL "11" (ABS W/O VES) OR "10" (ABS W/VES). DOES J 39200 DISPLAY "OL" (INFINITE)?

NO: REPAIR SHORT TO GROUND IN CKT 885.

YES:

(13)
- THOROUGHLY RESPRAY THE REAR ABS HARNESS WITH THE SALT WATER SOLUTION. IMPORTANT: DO NOT ALLOW WATER TO ENTER THE WHEEL SPEED SENSOR CONNECTOR.
- DISCONNECT LEFT REAR WHEEL SPEED SENSOR CONNECTOR.
- USING J 39200, MEASURE THE RESISTANCE BETWEEN GROUND AND LEFT REAR WHEEL SPEED SENSOR HARNESS CONNECTOR TERMINAL "A" THEN "B." DID J 39200 DISPLAY "OL" (INFINITE) FOR EACH MEASUREMENT?

NO: REPLACE REAR ABS HARNESS.

YES:

(14)
- USING J 39200, MEASURE THE RESISTANCE BETWEEN SENSOR TERMINAL "A" AND INTEGRAL BEARING AND SPEED SENSOR ASSEMBLY MOUNTING FLANGE. DOES J 39200 DISPLAY "OL" (INFINITE)?

NO: REPLACE INTEGRAL BEARING AND SPEED SENSOR ASSEMBLY.

YES: MALFUNCTION IS INTERMITTENT. INSPECT ALL CONNECTORS AND HARNESSES FOR DAMAGES WHICH MAY RESULT IN A SHORT TO GROUND WHEN ALL COMPONENTS ARE CONNECTED. SMALL NICKS OR CUTS IN THE HARNESS CAN CAUSE A MALFUNCTION. REPAIR ANY DAMAGE IF FOUND. JUMPER HARNESS MUST BE REPLACED IF DAMAGED. REFER TO "DIAGNOSTIC AIDS" ON FACING PAGE FOR MORE INFORMATION.

THIS CHART ASSUMES THAT A CURRENT DTC IS STORED INDICATING THAT THIS MALFUNCTION IS PRESENT.

DIAGNOSTIC CHART 28 — 1994–95 ACHIEVA, GRAND AM AND SKYLARK

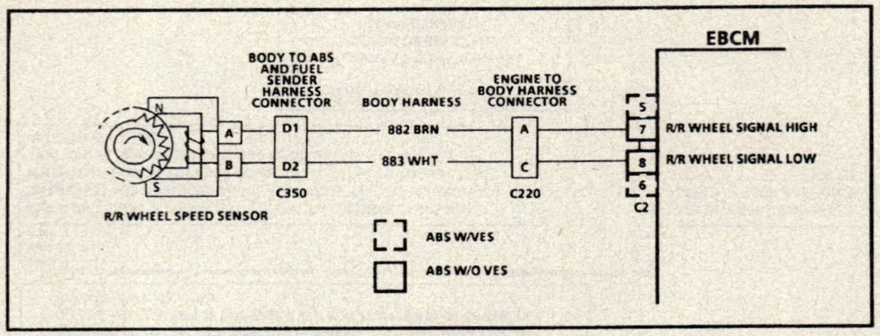

DTC 28

RIGHT REAR EXCESSIVE WHEEL SPEED VARIATION

Circuit Description:
As a toothed ring passes by the wheel speed sensor, changes in the electromagnetic field cause the wheel speed sensor to produce a sinusoidal (AC) voltage signal whose frequency is proportional to wheel speed. The magnitude of this signal is directly related to wheel speed and the proximity of the wheel speed sensor to the toothed ring, often referred to as the air gap.

DTC Will Set When: DTC 28 can be set when the brake is "OFF." The purpose of this test is to detect a situation in which the right rear wheel acceleration or deceleration is beyond specified limits.

Action Taken: A malfunction DTC is stored, ABS/VES is disabled and the ABS warning lamp is turned "ON."

DTC Chart Test Description: Number(s) below refer to circled number(s) on the diagnostic chart.

1. This test checks wheel speed sensor for proper resistance values.
2. Checks for a short to ground in the wheel speed sensor circuitry.
3. Checks for a short to voltage in the wheel speed sensor circuitry.
4. Ensures malfunction was not due to physical damage of the circuitry.
5. Checks for a short to voltage in the wheel signal low circuit.
6. Checks for a short to voltage in the wheel signal high circuit.
7. Checks for a short between the wheel signal high and low circuits.
8. Checks for high resistance in the wheel signal low circuit.
9. Checks for high resistance in the wheel signal high circuit.
10. Checks for proper resistance of the wheel speed sensor.
11. Checks for a short to ground in either of the wheel signal high or low circuits.
12. Checks for a short to ground in the wheel speed sensor.

Diagnostic Aids: An "Intermittent" malfunction is most likely caused by a poor connection, rubbed through wire insulation, or a wire that is broken inside the insulation.

The frequency of the malfunction can be checked by using the enhanced diagnostic function of the Tech 1.

If the customer's comments reflect that the ABS amber warning lamp is "ON" only during moist environmental changes, (rain, snow, vehicle wash) all wheel speed sensor circuitry should be thoroughly inspected for water intrusion. Use the following procedure. Spray down the suspected area with 5% salt water solution (two teaspoons of salt to 12 oz. of water). Drive vehicle above 24 km/h (15 mph) for at least 30 seconds. If DTC returns immediately, replace suspected harness.

Any circuitry, that is suspected as causing the intermittent complaint, should be thoroughly checked for backed out terminals, improper mating, broken locks, improperly formed or damaged terminals, poor terminal to wiring connections or physical damage to the wiring harness.

A careful visual inspection of the toothed ring, CV joint, bearing and wheel speed sensor for physical damage should also be performed. If DTC sets at the same mph every time and wheel speed variation is noted above this mph on the Tech 1, the speed ring is most likely damaged. This can occur at any speed.

Resistance of the wheel speed sensor will increase with an increase in sensor temperature.

When replacing a wheel speed sensor, inspect the sensor terminals and harness connector for corrosion and/or water intrusion. If evidence of corrosion or water intrusion exists, replace wheel speed sensor harness.

If wheel speed sensor harness shows evidence of water intrusion, inspect wheel speed sensor for water intrusion. If water intrusion is evident, replace wheel speed sensor.

DIAGNOSTIC CHART 28 — 1994–95 ACHIEVA, GRAND AM AND SKYLARK

DTC 28
(Page 1 of 3)
RIGHT REAR EXCESSIVE WHEEL SPEED VARIATION

IMPORTANT:
- WHEEL SPEED SENSOR INTERMITTENT MALFUNCTIONS MAY BE DIFFICULT TO LOCATE. CARE SHOULD BE TAKEN NOT TO DISTURB ANY ELECTRICAL CONNECTIONS PRIOR TO AN INDICATED STEP OF THIS CHART. THIS WILL ENSURE THAT AN INTERMITTENT CONNECTION WILL NOT BE CORRECTED BEFORE THE SOURCE OF THE MALFUNCTION IS FOUND.

(1)
- IGNITION "OFF."
- USING A 5% SALT WATER SOLUTION (TWO TEASPOONS OF SALT PER 12 OZ. WATER), THOROUGHLY SPRAY THE REAR ABS HARNESS.
- DISCONNECT 24-WAY EBCM CONNECTOR.
- USING J 39200, MEASURE RESISTANCE BETWEEN 24-WAY EBCM HARNESS CONNECTOR TERMINALS "7" (ABS W/O VES) OR "5" (ABS W/VES) AND "8" (ABS W/O VES) OR "6" (ABS W/VES). IS RESISTANCE BETWEEN 2100 AND 2400 OHMS WHEN SENSOR IS 25°C (77°F)? *[IF TEMPERATURE OF THE SENSOR IS SUSPECTED TO BE OTHER THAN 25°C (77°F), REFER TO TABLE ON PAGE 2 OF 2 OF THIS DTC CHART FOR APPLICABLE SENSOR RESISTANCE VALUES; THE VALUES IN THE TABLE ARE FOR TEMPERATURE OF THE SENSOR NOT AIR TEMPERATURE.]*

YES → (2)
NO → GO TO PAGE 2 OF 3 OF THIS DTC CHART.

(2)
- USING J 39200, MEASURE THE RESISTANCE BETWEEN GROUND AND EBCM HARNESS CONNECTOR TERMINAL "7" (ABS W/O VES) OR "5" (ABS W/VES). DOES J 39200 DISPLAY "OL" (INFINITE)?

YES → (3)
NO → GO TO PAGE 3 OF 3 OF THIS DTC CHART.

(3)
- IGNITION "ON."
- USING J 39200, MEASURE THE VOLTAGE BETWEEN GROUND AND 24-WAY EBCM HARNESS CONNECTOR TERMINAL "8" (ABS W/O VES) OR "6" (ABS W/VES). IS VOLTAGE LESS THAN 1 VOLT?

YES → (4)
NO → (5)

(4)
- IGNITION "OFF."
- INSPECT 24-WAY EBCM HARNESS CONNECTOR TERMINALS "7" (ABS W/O VES) OR "5" (ABS W/VES) AND "8" (ABS W/O VES) OR "6" (ABS W/VES) FOR POOR TERMINAL CONTACT OR CORROSION. INSPECT CKT(s) 882 AND 883 FOR DAMAGE WHICH MAY RESULT IN AN OPEN CIRCUIT, SHORT TO GROUND OR SHORT TO VOLTAGE. REPAIR DAMAGE IF EVIDENT. REPLACE TERMINALS IF POOR CONTACT OR CORROSION EXISTS.
- RECONNECT ALL CONNECTORS.
- IGNITION "ON."
IS DTC 35 SET AS A CURRENT DTC?

YES → REPLACE EBCM.
NO → MALFUNCTION IS INTERMITTENT. REFER TO "DIAGNOSTIC AIDS" ON FACING PAGE.

(5)
- IGNITION "OFF."
- DISCONNECT RIGHT REAR WHEEL SPEED SENSOR CONNECTOR.
- IGNITION "ON."
- USING J 39200, MEASURE THE VOLTAGE BETWEEN GROUND AND 24-WAY EBCM HARNESS CONNECTOR TERMINAL "8" (ABS W/O VES) OR "6" (ABS W/VES). IS VOLTAGE LESS THAN 1 VOLT?

YES → (6)
NO → REPAIR SHORT TO VOLTAGE IN CKT 883.

(6)
- USING J 39200, MEASURE THE VOLTAGE BETWEEN GROUND AND 24-WAY EBCM HARNESS CONNECTOR TERMINAL "7" (ABS W/O VES) OR "5" (ABS W/VES). IS VOLTAGE LESS THAN 1 VOLT?

YES → MALFUNCTION IS INTERMITTENT. INSPECT ALL CONNECTORS AND HARNESSES FOR DAMAGE WHICH MAY RESULT IN A SHORT TO VOLTAGE WHEN ALL COMPONENTS ARE CONNECTED. REPAIR ALL DAMAGE IF FOUND. REFER TO "DIAGNOSTIC AIDS" ON FACING PAGE FOR MORE INFORMATION.
NO → REPAIR SHORT TO VOLTAGE IN CKT 882.

THIS CHART ASSUMES THAT A CURRENT DTC IS STORED INDICATING THAT THIS MALFUNCTION IS PRESENT.

DIAGNOSTIC CHART 28 — 1994–95 ACHIEVA, GRAND AM AND SKYLARK

DTC 28 (Page 2 of 3) RIGHT REAR EXCESSIVE WHEEL SPEED VARIATION

CONTINUED FROM PAGE 1 OF 3 OF THIS DTC CHART.

(7)
- DISCONNECT RIGHT REAR WHEEL SPEED SENSOR CONNECTOR.
- USING J 39200, MEASURE THE RESISTANCE BETWEEN 24-WAY EBCM HARNESS CONNECTOR TERMINALS "7" (ABS W/O VES) OR "5" (ABS W/VES) AND "8" (ABS W/O VES) OR "6" (ABS W/VES). DOES J 39200 DISPLAY "OL" (INFINITE)?

NO: REPAIR SHORT BETWEEN CKTs 882 AND 883.

YES:

(8)
- USING J 39200, MEASURE THE RESISTANCE BETWEEN 24-WAY EBCM HARNESS CONNECTOR TERMINAL "7" (ABS W/O VES) OR "5" (ABS W/VES) AND RIGHT REAR WHEEL SPEED SENSOR HARNESS CONNECTOR TERMINAL "A." IS RESISTANCE LESS THAN 2 OHMS?

NO: REPAIR HIGH RESISTANCE IN CKT 882.

YES:

(9)
- USING J 39200, MEASURE THE RESISTANCE BETWEEN 24-WAY EBCM HARNESS CONNECTOR TERMINAL "8" (ABS W/O VES) OR "6" (ABS W/VES) AND RIGHT REAR WHEEL SPEED SENSOR HARNESS CONNECTOR TERMINAL "B." IS RESISTANCE LESS THAN 2 OHMS?

NO: REPAIR HIGH RESISTANCE IN CKT 883.

YES:

(10)
- USING J 39200, MEASURE THE RESISTANCE BETWEEN RIGHT REAR WHEEL SPEED SENSOR TERMINALS "A" AND "B." IS RESISTANCE BETWEEN 2100 AND 2400 OHMS WHEN SENSOR IS 25°C (77°F)? *(IF TEMPERATURE OF THE SENSOR IS SUSPECTED TO BE OTHER THAN 25°C (77°F), REFER TO THE TABLE BELOW FOR APPLICABLE SENSOR RESISTANCE VALUES; THE VALUES IN THE TABLE ARE FOR TEMPERATURE OF THE SENSOR NOT AIR TEMPERATURE).*

NO: REPLACE INTEGRAL BEARING AND SPEED SENSOR ASSEMBLY.

YES: MALFUNCTION IS INTERMITTENT. INSPECT ALL CONNECTORS AND HARNESSES FOR DAMAGES WHICH MAY RESULT IN HIGH RESISTANCE WHEN ALL COMPONENTS ARE CONNECTED. REFER TO "DIAGNOSTIC AIDS" ON FACING PAGE FOR MORE INFORMATION. REPAIR ALL DAMAGE IF FOUND.

REAR WHEEL SPEED SENSOR

TEMPERATURE OF SENSOR VS. SENSOR RESISTANCE (APPROXIMATE)

TEMP. (°C)	TEMP. (°F)	RESISTANCE (OHMS)
-34 TO 4	-30 TO 40	1623 TO 2210
5 TO 43	41 TO 110	1939 TO 2570
44 TO 93	111 TO 200	2254 TO 3029

THIS CHART ASSUMES THAT A CURRENT DTC IS STORED INDICATING THAT THIS MALFUNCTION IS PRESENT.

DIAGNOSTIC CHART 28 — 1994–95 ACHIEVA, GRAND AM AND SKYLARK

DTC 28 (Page 3 of 3) RIGHT REAR EXCESSIVE WHEEL SPEED VARIATION

CONTINUED FROM PAGE 1 OF 3 OF THIS DTC CHART.

(11)
- DISCONNECT REAR ABS HARNESS FROM BODY TO ABS AND FUEL SENDER HARNESS CONNECTOR.
- USING J 39200, MEASURE THE RESISTANCE BETWEEN GROUND AND 24-WAY EBCM HARNESS CONNECTOR TERMINAL "7" (ABS W/O VES) OR "5" (ABS W/VES). DOES J 39200 DISPLAY "OL" (INFINITE)?

NO: REPAIR SHORT TO GROUND IN CKT 882.

YES:

(12)
- USING J 39200, MEASURE THE RESISTANCE BETWEEN GROUND AND 24-WAY EBCM HARNESS CONNECTOR TERMINAL "8" (ABS W/O VES) OR "6" (ABS W/VES). DOES J 39200 DISPLAY "OL" (INFINITE)?

NO: REPAIR SHORT TO GROUND IN CKT 883.

YES:

(13)
- THOROUGHLY RESPRAY THE REAR ABS HARNESS WITH THE SALT WATER SOLUTION. IMPORTANT: DO NOT ALLOW WATER TO ENTER THE WHEEL SPEED SENSOR CONNECTOR.
- DISCONNECT RIGHT REAR WHEEL SPEED SENSOR CONNECTOR.
- USING J 39200, MEASURE THE RESISTANCE BETWEEN GROUND AND RIGHT REAR WHEEL SPEED SENSOR HARNESS CONNECTOR TERMINAL "A" THEN "B." DID J 39200 DISPLAY "OL" (INFINITE) FOR EACH MEASUREMENT?

NO: REPLACE REAR ABS HARNESS.

YES:

(14)
- USING J 39200, MEASURE THE RESISTANCE BETWEEN SENSOR TERMINAL "A" AND INTEGRAL BEARING AND SPEED SENSOR ASSEMBLY MOUNTING FLANGE. DOES J 39200 DISPLAY "OL" (INFINITE)?

NO: REPLACE INTEGRAL BEARING AND SPEED SENSOR ASSEMBLY.

YES: MALFUNCTION IS INTERMITTENT. INSPECT ALL CONNECTORS AND HARNESSES FOR DAMAGES WHICH MAY RESULT IN A SHORT TO GROUND WHEN ALL COMPONENTS ARE CONNECTED. SMALL NICKS OR CUTS IN THE HARNESS CAN CAUSE A MALFUNCTION. REPAIR ANY DAMAGE IF FOUND. JUMPER HARNESS MUST BE REPLACED IF DAMAGED. REFER TO "DIAGNOSTIC AIDS" ON FACING PAGE FOR MORE INFORMATION.

THIS CHART ASSUMES THAT A CURRENT DTC IS STORED INDICATING THAT THIS MALFUNCTION IS PRESENT.

DIAGNOSTIC CHART 32 — 1994–95 ACHIEVA, GRAND AM AND SKYLARK WITH VES

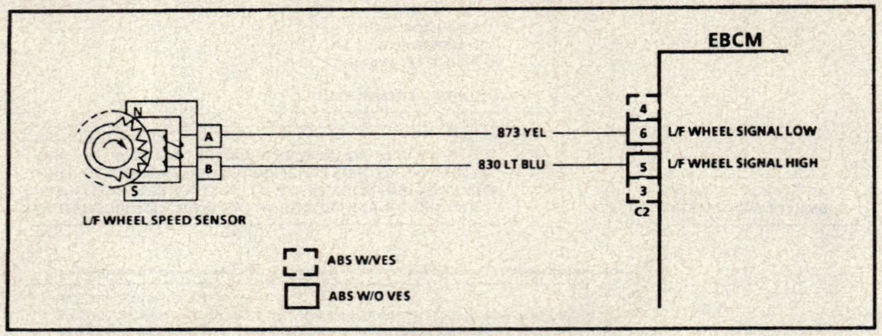

DTC 32

LEFT FRONT WHEEL SPEED SENSOR CIRCUIT OPEN OR SHORTED TO GROUND/BATTERY (FOR ABS WITH VES ONLY)

Circuit Description:
As a toothed ring passes by the wheel speed sensor, changes in the electromagnetic field cause the wheel speed sensor to produce a sinusoidal (AC) voltage signal whose frequency is proportional to wheel speed. The magnitude of this signal is directly related to wheel speed and the proximity of the wheel speed sensor to the toothed ring, often referred to as the air gap.

DTC Will Set When: DTC 32 can be set anytime after the ABS enable relay has been commanded "ON." The relay is commanded "ON" during the three second bulb check.

Action Taken: A malfunction DTC is stored, ABS/VES is disabled and the ABS warning lamp is turned "ON."

DTC Chart Test Description: Number(s) below refer to circled number(s) on the diagnostic chart.

1. Checks wheel speed sensor circuitry for proper resistance values.
2. Checks for a short to ground in the wheel speed sensor circuitry.
3. Checks for a short to voltage in the wheel speed sensor circuitry.
4. Ensures malfunction was not due to physical damage of the circuitry.
5. Checks for a short to voltage in the wheel signal low circuit.
6. Checks for a short to voltage in the wheel signal high circuit.
7. Checks for high resistance in the wheel signal low circuit.
8. Checks for high resistance in the wheel signal high circuit.
9. Checks for proper resistance of the wheel speed sensor.
10. Checks for a short to ground in either of the wheel signal high or low circuits.
11. Checks for a short to ground in the wheel speed sensor.

Diagnostic Aids: An "Intermittent" malfunction is most likely caused by a poor connection, rubbed through wire insulation, or a wire that is broken inside the insulation.

The frequency of the malfunction can be checked by using the enhanced diagnostic function of the Tech 1.

If the customer's comments reflect that the ABS amber warning lamp is "ON" only during moist environmental changes, (rain, snow, vehicle wash) all wheel speed sensor circuitry should be thoroughly inspected for signs of water intrusion. Use the following procedure. Spray down the suspected area with 5% salt water solution (two teaspoons of salt to 12 oz. of water). Drive vehicle above 24 km/h (15 mph) for at least 30 seconds. If DTC returns immediately, replace suspected harness.

Any circuitry, that is suspected as causing the intermittent complaint, should be thoroughly checked for backed out terminals, improper mating, broken locks, improperly formed or damaged terminals, poor terminal to wiring connections or physical damage to the wiring harness.

Resistance of the wheel speed sensor will increase with an increase in sensor temperature.

When replacing a wheel speed sensor, inspect the sensor terminals and harness connector for corrosion and/or water intrusion. If evidence of corrosion or water intrusion exists, replace wheel speed sensor harness.

If wheel speed sensor harness shows evidence of water intrusion, inspect wheel speed sensor for water intrusion. If water intrusion is evident, replace wheel speed sensor.

DIAGNOSTIC CHART 32 — 1994–95 ACHIEVA, GRAND AM AND SKYLARK WITH VES

DTC 32
(Page 1 of 3)
LEFT FRONT WHEEL SPEED SENSOR CIRCUIT OPEN OR SHORTED TO GROUND/BATTERY (FOR ABS WITH VES ONLY)

IMPORTANT:
- WHEEL SPEED SENSOR INTERMITTENT MALFUNCTIONS MAY BE DIFFICULT TO LOCATE. CARE SHOULD BE TAKEN NOT TO DISTURB ANY ELECTRICAL CONNECTIONS PRIOR TO AN INDICATED STEP OF THIS CHART. THIS WILL ENSURE THAT AN INTERMITTENT CONNECTION WILL NOT BE CORRECTED BEFORE THE SOURCE OF THE MALFUNCTION IS FOUND.

(1)
- IGNITION "OFF."
- USING A 5% SALT WATER SOLUTION (TWO TEASPOONS OF SALT PER 12 OZ. WATER), THOROUGHLY SPRAY THE LEFT FRONT WHEEL SPEED SENSOR HARNESS.
- DISCONNECT 24-WAY EBCM CONNECTOR.
- USING J 39200, MEASURE RESISTANCE BETWEEN 24-WAY EBCM HARNESS CONNECTOR TERMINALS "3" AND "4." IS RESISTANCE BETWEEN 1628 AND 1889 OHMS WHEN SENSOR IS 25°C (77°F)? *(IF TEMPERATURE OF THE SENSOR IS SUSPECTED TO BE OTHER THAN 25°C (77°F), REFER TO TABLE ON PAGE 2 OF 2 OF THIS DTC CHART FOR APPLICABLE SENSOR RESISTANCE VALUES; THE VALUES IN THE TABLE ARE FOR TEMPERATURE OF THE SENSOR NOT AIR TEMPERATURE.)*

YES → (2) NO → GO TO PAGE 2 OF 3 OF THIS DTC CHART.

(2)
- USING J 39200, MEASURE THE RESISTANCE BETWEEN GROUND AND EBCM HARNESS CONNECTOR TERMINAL "3." DOES J 39200 DISPLAY "OL" (INFINITE)?

YES → (3) NO → GO TO PAGE 3 OF 3 OF THIS DTC CHART.

(3)
- IGNITION "ON."
- USING J 39200, MEASURE THE VOLTAGE BETWEEN GROUND AND 24-WAY EBCM HARNESS CONNECTOR TERMINAL "4." IS VOLTAGE LESS THAN 1 VOLT?

YES → (4) NO → (5)

(4)
- IGNITION "OFF."
- INSPECT 24-WAY EBCM HARNESS CONNECTOR TERMINALS "3" AND "4" FOR POOR TERMINAL CONTACT OR CORROSION. INSPECT CKT(s) 830 AND 873 FOR DAMAGE WHICH MAY RESULT IN AN OPEN CIRCUIT, SHORT TO GROUND OR SHORT TO VOLTAGE. REPAIR DAMAGE IF EVIDENT. REPLACE TERMINALS IF POOR CONTACT OR CORROSION EXISTS.
- RECONNECT ALL CONNECTORS.
- IGNITION "ON."
- IS DTC 32 SET AS A CURRENT DTC?

YES → REPLACE EBCM.
NO → MALFUNCTION IS INTERMITTENT. REFER TO "DIAGNOSTIC AIDS" ON FACING PAGE.

(5)
- IGNITION "OFF."
- DISCONNECT LEFT FRONT WHEEL SPEED SENSOR CONNECTOR.
- IGNITION "ON."
- USING J 39200, MEASURE THE VOLTAGE BETWEEN GROUND AND 24-WAY EBCM HARNESS CONNECTOR TERMINAL "4." IS VOLTAGE LESS THAN 1 VOLT?

YES → (6) NO → REPAIR SHORT TO VOLTAGE IN CKT 873.

(6)
- USING J 39200, MEASURE THE VOLTAGE BETWEEN GROUND AND 24-WAY EBCM HARNESS CONNECTOR TERMINAL "3." IS VOLTAGE LESS THAN 1 VOLT?

YES → MALFUNCTION IS INTERMITTENT. INSPECT ALL CONNECTORS AND HARNESSES FOR DAMAGE WHICH MAY RESULT IN A SHORT TO VOLTAGE WHEN ALL COMPONENTS ARE CONNECTED. REPAIR ALL DAMAGE IF FOUND. REFER TO "DIAGNOSTIC AIDS" ON FACING PAGE FOR MORE INFORMATION.
NO → REPAIR SHORT TO VOLTAGE IN CKT 830.

THIS CHART ASSUMES THAT A CURRENT DTC IS STORED INDICATING THAT THIS MALFUNCTION IS PRESENT.

DIAGNOSTIC CHART 32 — 1994–95 ACHIEVA, GRAND AM AND SKYLARK WITH VES

DTC 32
(Page 2 of 3)
LEFT FRONT WHEEL SPEED SENSOR CIRCUIT OPEN OR SHORTED TO GROUND/BATTERY (FOR ABS WITH VES ONLY)

CONTINUED FROM PAGE 1 OF 3 OF THIS DTC CHART.

(7)
- DISCONNECT LEFT FRONT WHEEL SPEED SENSOR CONNECTOR.
- USING J 39200, MEASURE THE RESISTANCE BETWEEN 24-WAY EBCM HARNESS CONNECTOR TERMINAL "4" AND LEFT FRONT WHEEL SPEED SENSOR HARNESS CONNECTOR TERMINAL "A." IS RESISTANCE LESS THAN 2 OHMS?

NO: REPAIR HIGH RESISTANCE IN CKT 873.

YES:

(8)
- USING J 39200, MEASURE THE RESISTANCE BETWEEN 24-WAY EBCM HARNESS CONNECTOR TERMINAL "3" AND LEFT FRONT WHEEL SPEED SENSOR HARNESS CONNECTOR TERMINAL "B." IS RESISTANCE LESS THAN 2 OHMS?

NO: REPAIR HIGH RESISTANCE IN CKT 830.

YES:

(9)
- USING J 39200, MEASURE THE RESISTANCE BETWEEN LEFT FRONT WHEEL SPEED SENSOR TERMINALS "A" AND "B." IS RESISTANCE BETWEEN 1628 AND 1889 OHMS WHEN SENSOR IS 25°C (77°F)? *(IF TEMPERATURE OF THE SENSOR IS SUSPECTED TO BE OTHER THAN 25°C (77°F), REFER TO THE TABLE BELOW FOR APPLICABLE SENSOR RESISTANCE VALUES; THE VALUES IN THE TABLE ARE FOR TEMPERATURE OF THE SENSOR NOT AIR TEMPERATURE).*

NO: REPLACE WHEEL SPEED SENSOR.

YES: MALFUNCTION IS INTERMITTENT. INSPECT ALL CONNECTORS AND HARNESSES FOR DAMAGES WHICH MAY RESULT IN HIGH RESISTANCE WHEN ALL COMPONENTS ARE CONNECTED. REFER TO "DIAGNOSTIC AIDS" ON FACING PAGE FOR MORE INFORMATION. REPAIR ALL DAMAGE IF FOUND.

FRONT WHEEL SPEED SENSOR

TEMPERATURE OF SENSOR VS. SENSOR RESISTANCE (APPROXIMATE)

TEMP. (°C)	TEMP. (°F)	RESISTANCE (OHMS)
-34 TO 4	-30 TO 40	1258 TO 1739
5 TO 43	41 TO 110	1503 TO 2023
44 TO 93	111 TO 200	1747 TO 2384

THIS CHART ASSUMES THAT A CURRENT DTC IS STORED INDICATING THAT THIS MALFUNCTION IS PRESENT.

DIAGNOSTIC CHART 32 — 1994–95 ACHIEVA, GRAND AM AND SKYLARK WITH VES

DTC 32
(Page 3 of 3)
LEFT FRONT WHEEL SPEED SENSOR CIRCUIT OPEN OR SHORTED TO GROUND/BATTERY (FOR ABS WITH VES ONLY)

CONTINUED FROM PAGE 1 OF 3 OF THIS DTC CHART.

(10)
- THOROUGHLY RESPRAY THE LEFT FRONT WHEEL SPEED SENSOR HARNESS WITH THE SALT WATER SOLUTION. IMPORTANT: DO NOT ALLOW WATER TO ENTER THE WHEEL SPEED SENSOR CONNECTOR.
- DISCONNECT LEFT FRONT WHEEL SPEED SENSOR CONNECTOR.
- USING J 39200, MEASURE THE RESISTANCE BETWEEN GROUND AND LEFT FRONT WHEEL SPEED SENSOR HARNESS CONNECTOR TERMINAL "A" THEN "B."

DID J 39200 DISPLAY "OL" (INFINITE) FOR EACH MEASUREMENT?

NO: REPAIR SHORT TO GROUND IN CKT(s) (830 AND/OR 873) WHICH MEASURED OTHER THAN "OL" IN STEP 10.

YES:

(11)
- USING J 39200, MEASURE THE RESISTANCE BETWEEN GROUND AND LEFT FRONT WHEEL SPEED SENSOR TERMINAL "A." DOES J 39200 DISPLAY "OL" (INFINITE)?

NO: REPLACE WHEEL SPEED SENSOR.

YES: MALFUNCTION IS INTERMITTENT. INSPECT ALL CONNECTORS AND HARNESSES FOR DAMAGES WHICH MAY RESULT IN A SHORT TO GROUND WHEN ALL COMPONENTS ARE CONNECTED. SMALL NICKS OR CUTS IN THE HARNESS CAN CAUSE A MALFUNCTION. REPAIR ANY DAMAGE IF FOUND. JUMPER HARNESS MUST BE REPLACED IF DAMAGED. REFER TO "DIAGNOSTIC AIDS" ON FACING PAGE FOR MORE INFORMATION.

THIS CHART ASSUMES THAT A CURRENT DTC IS STORED INDICATING THAT THIS MALFUNCTION IS PRESENT.

DIAGNOSTIC CHART 33 — 1994–95 ACHIEVA, GRAND AM AND SKYLARK WITH VES

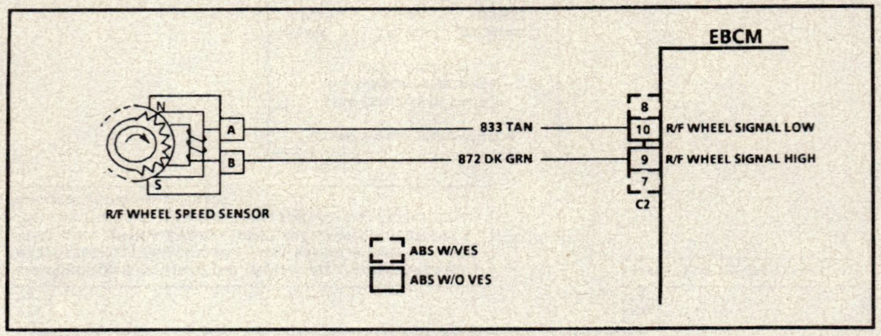

DTC 33

RIGHT FRONT WHEEL SPEED SENSOR CIRCUIT OPEN OR SHORTED TO GROUND/BATTERY (FOR ABS WITH VES ONLY)

Circuit Description:
As a toothed ring passes by the wheel speed sensor, changes in the electromagnetic field cause the wheel speed sensor to produce a sinusoidal (AC) voltage signal whose frequency is proportional to wheel speed. The magnitude of this signal is directly related to wheel speed and the proximity of the wheel speed sensor to the toothed ring, often referred to as the air gap.

DTC Will Set When: DTC 33 can be set anytime after the ABS enable relay has been commanded "ON." The relay is commanded "ON" during the three second bulb check.

Action Taken: A malfunction DTC is stored, ABS/VES is disabled and the ABS warning lamp is turned "ON."

DTC Chart Test Description: Number(s) below refer to circled number(s) on the diagnostic chart.

1. Checks wheel speed sensor circuitry for proper resistance values.
2. Checks for a short to ground in the wheel speed sensor circuitry.
3. Checks for a short to voltage in the wheel speed sensor circuitry.
4. Ensures malfunction was not due to physical damage of the circuitry.
5. Checks for a short to voltage in the wheel signal low circuit.
6. Checks for a short to voltage in the wheel signal high circuit.
7. Checks for high resistance in the wheel signal low circuit.
8. Checks for high resistance in the wheel signal high circuit.
9. Checks for proper resistance of the wheel speed sensor.
10. Checks for a short to ground in either of the wheel signal high or low circuits.
11. Checks for a short to ground in the wheel speed sensor.

Diagnostic Aids: An "Intermittent" malfunction is most likely caused by a poor connection, rubbed through wire insulation, or a wire that is broken inside the insulation.

The frequency of the malfunction can be checked by using the enhanced diagnostic function of the Tech 1, as described in "Enhanced Diagnostics," found in this section.

If the customer's comments reflect that the ABS amber warning lamp is "ON" only during moist environmental changes, (rain, snow, vehicle wash) all wheel speed sensor circuitry should be thoroughly inspected for signs of water intrusion. Use the following procedure. Spray down the suspected area with 5% salt water solution (two teaspoons of salt to 12 oz. of water). Drive vehicle above 24 km/h (15 mph) for at least 30 seconds. If DTC returns immediately, replace suspected harness.

Any circuitry, that is suspected as causing the intermittent complaint, should be thoroughly checked for backed out terminals, improper mating, broken locks, improperly formed or damaged terminals, poor terminal to wiring connections or physical damage to the wiring harness.

Resistance of the wheel speed sensor will increase with an increase in sensor temperature.

When replacing a wheel speed sensor, inspect the sensor terminals and harness connector for corrosion and/or water intrusion. If evidence of corrosion or water intrusion exists, replace wheel speed sensor harness. Refer to "On-Vehicle Service" in this section.

If wheel speed sensor harness shows evidence of water intrusion, inspect wheel speed sensor for water intrusion. If water intrusion is evident, replace wheel speed sensor.

DIAGNOSTIC CHART 33 — 1994–95 ACHIEVA, GRAND AM AND SKYLARK WITH VES

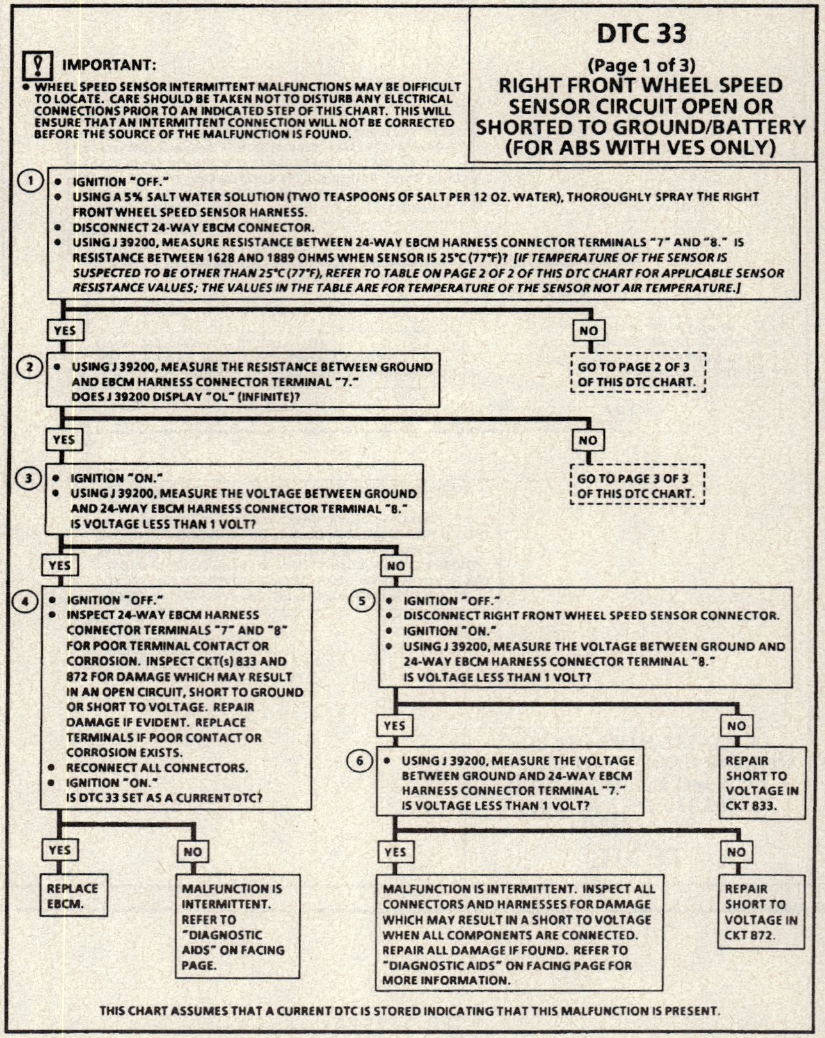

DIAGNOSTIC CHART 33 — 1994–95 ACHIEVA, GRAND AM AND SKYLARK WITH VES

DTC 33
(Page 2 of 3)
RIGHT FRONT WHEEL SPEED SENSOR CIRCUIT OPEN OR SHORTED TO GROUND/BATTERY (FOR ABS WITH VES ONLY)

CONTINUED FROM PAGE 1 OF 3 OF THIS DTC CHART.

(7)
- DISCONNECT RIGHT FRONT WHEEL SPEED SENSOR CONNECTOR.
- USING J 39200, MEASURE THE RESISTANCE BETWEEN 24-WAY EBCM HARNESS CONNECTOR TERMINAL "8" AND RIGHT FRONT WHEEL SPEED SENSOR HARNESS CONNECTOR TERMINAL "A." IS RESISTANCE LESS THAN 2 OHMS?

NO: REPAIR HIGH RESISTANCE IN CKT 833.

YES:

(8)
- USING J 39200, MEASURE THE RESISTANCE BETWEEN 24-WAY EBCM HARNESS CONNECTOR TERMINAL "7" AND RIGHT FRONT WHEEL SPEED SENSOR HARNESS CONNECTOR TERMINAL "B." IS RESISTANCE LESS THAN 2 OHMS?

NO: REPAIR HIGH RESISTANCE IN CKT 872.

YES:

(9)
- USING J 39200, MEASURE THE RESISTANCE BETWEEN RIGHT FRONT WHEEL SPEED SENSOR TERMINALS "A" AND "B." IS RESISTANCE BETWEEN 1628 AND 1889 OHMS WHEN SENSOR IS 25°C (77°F)? *(IF TEMPERATURE OF THE SENSOR IS SUSPECTED TO BE OTHER THAN 25°C (77°F), REFER TO THE TABLE BELOW FOR APPLICABLE SENSOR RESISTANCE VALUES; THE VALUES IN THE TABLE ARE FOR TEMPERATURE OF THE SENSOR NOT AIR TEMPERATURE).*

NO: REPLACE WHEEL SPEED SENSOR.

YES: MALFUNCTION IS INTERMITTENT. INSPECT ALL CONNECTORS AND HARNESSES FOR DAMAGES WHICH MAY RESULT IN HIGH RESISTANCE WHEN ALL COMPONENTS ARE CONNECTED. REFER TO "DIAGNOSTIC AIDS" ON FACING PAGE FOR MORE INFORMATION. REPAIR ALL DAMAGE IF FOUND.

FRONT WHEEL SPEED SENSOR

TEMPERATURE OF SENSOR VS. SENSOR RESISTANCE (APPROXIMATE)

TEMP. (°C)	TEMP. (°F)	RESISTANCE (OHMS)
-34 TO 4	-30 TO 40	1258 TO 1739
5 TO 43	41 TO 110	1503 TO 2023
44 TO 93	111 TO 200	1747 TO 2384

THIS CHART ASSUMES THAT A CURRENT DTC IS STORED INDICATING THAT THIS MALFUNCTION IS PRESENT.

DIAGNOSTIC CHART 33 — 1994–95 ACHIEVA, GRAND AM AND SKYLARK WITH VES

DTC 33
(Page 3 of 3)
RIGHT FRONT WHEEL SPEED SENSOR CIRCUIT OPEN OR SHORTED TO GROUND/BATTERY (FOR ABS WITH VES ONLY)

CONTINUED FROM PAGE 1 OF 3 OF THIS DTC CHART.

(10)
- THOROUGHLY RESPRAY THE RIGHT FRONT WHEEL SPEED SENSOR HARNESS WITH THE SALT WATER SOLUTION. IMPORTANT: DO NOT ALLOW WATER TO ENTER THE WHEEL SPEED SENSOR CONNECTOR.
- DISCONNECT RIGHT FRONT WHEEL SPEED SENSOR CONNECTOR.
- USING J 39200, MEASURE THE RESISTANCE BETWEEN GROUND AND RIGHT FRONT WHEEL SPEED SENSOR HARNESS CONNECTOR TERMINAL "A" THEN "B." DID J 39200 DISPLAY "OL" (INFINITE) FOR EACH MEASUREMENT?

NO: REPAIR SHORT TO GROUND IN CKT(s) (833 AND/OR 872) WHICH MEASURED OTHER THAN "OL" IN STEP 10.

YES:

(11)
- USING J 39200, MEASURE THE RESISTANCE BETWEEN GROUND AND RIGHT FRONT WHEEL SPEED SENSOR TERMINAL "A." DOES J 39200 DISPLAY "OL" (INFINITE)?

NO: REPLACE WHEEL SPEED SENSOR.

YES: MALFUNCTION IS INTERMITTENT. INSPECT ALL CONNECTORS AND HARNESSES FOR DAMAGES WHICH MAY RESULT IN A SHORT TO GROUND WHEN ALL COMPONENTS ARE CONNECTED. SMALL NICKS OR CUTS IN THE HARNESS CAN CAUSE A MALFUNCTION. REPAIR ANY DAMAGE IF FOUND. JUMPER HARNESS MUST BE REPLACED IF DAMAGED. REFER TO "DIAGNOSTIC AIDS" ON FACING PAGE FOR MORE INFORMATION.

THIS CHART ASSUMES THAT A CURRENT DTC IS STORED INDICATING THAT THIS MALFUNCTION IS PRESENT.

DIAGNOSTIC CHART 34 — 1994–95 ACHIEVA, GRAND AM AND SKYLARK WITH VES

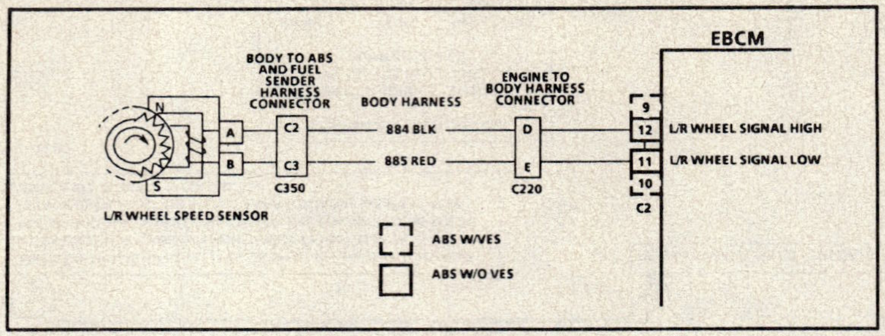

DTC 34

LEFT REAR WHEEL SPEED SENSOR CIRCUIT OPEN OR SHORTED TO GROUND/BATTERY (FOR ABS WITH VES ONLY)

Circuit Description:
As a toothed ring passes by the wheel speed sensor, changes in the electromagnetic field cause the wheel speed sensor to produce a sinusoidal (AC) voltage signal whose frequency is proportional to wheel speed. The magnitude of this signal is directly related to wheel speed and the proximity of the wheel speed sensor to the toothed ring, often referred to as the air gap.

DTC Will Set When: DTC 34 can be set anytime after the ABS enable relay has been commanded "ON." The relay is commanded "ON" during the three second bulb check.

Action Taken: A malfunction DTC is stored, ABS/VES is disabled and the ABS warning lamp is turned "ON."

DTC Chart Test Description: Number(s) below refer to circled number(s) on the diagnostic chart.

1. Checks wheel speed sensor circuitry for proper resistance values.
2. Checks for a short to ground in the wheel speed sensor circuitry.
3. Checks for a short to voltage in the wheel speed sensor circuitry.
4. Ensures malfunction was not due to physical damage of the circuitry.
5. Checks for a short to voltage in the wheel signal low circuit.
6. Checks for a short to voltage in the wheel signal high circuit.
7. Checks for high resistance in the wheel signal low circuit.
8. Checks for high resistance in the wheel signal high circuit.
9. Checks for proper resistance of the wheel speed sensor.
10. Checks for a short to ground in the wheel signal high circuit.
11. Checks for a short to ground in the wheel signal low circuit.
12. Checks for a short to ground in the rear ABS harness.
13. Checks for a short to ground in the wheel speed sensor.

Diagnostic Aids: An "Intermittent" malfunction is most likely caused by a poor connection, rubbed through wire insulation, or a wire that is broken inside the insulation.

The frequency of the malfunction can be checked by using the enhanced diagnostic function of the Tech 1, as described in "Enhanced Diagnostics," found in this section.

If the customer's comments reflect that the ABS amber warning lamp is "ON" only during moist environmental changes, (rain, snow, vehicle wash) all wheel speed sensor circuitry should be thoroughly inspected for signs of water intrusion. Use the following procedure. Spray down the suspected area with 5% salt water solution (two teaspoons of salt to 12 oz. of water). Drive vehicle above 24 km/h (15 mph) for at least 30 seconds. If DTC returns immediately, replace suspected harness.

Any circuitry, that is suspected as causing the intermittent complaint, should be thoroughly checked for backed out terminals, improper mating, broken locks, improperly formed or damaged terminals, poor terminal to wiring connections or physical damage to the wiring harness.

Resistance of the wheel speed sensor will increase with an increase in sensor temperature.

When replacing a wheel speed sensor, inspect the sensor terminals and harness connector for corrosion and/or water intrusion. If evidence of corrosion or water intrusion exists, replace wheel speed sensor harness. Refer to "On-Vehicle Service" in this section.

If wheel speed sensor harness shows evidence of water intrusion, inspect wheel speed sensor for water intrusion. If water intrusion is evident, replace wheel speed sensor.

DIAGNOSTIC CHART 34 — 1994–95 ACHIEVA, GRAND AM AND SKYLARK WITH VES

DTC 34
(Page 1 of 3)
LEFT REAR WHEEL SPEED SENSOR CIRCUIT OPEN OR SHORTED TO GROUND/BATTERY (FOR ABS WITH VES ONLY)

IMPORTANT:
- WHEEL SPEED SENSOR INTERMITTENT MALFUNCTIONS MAY BE DIFFICULT TO LOCATE. CARE SHOULD BE TAKEN NOT TO DISTURB ANY ELECTRICAL CONNECTIONS PRIOR TO AN INDICATED STEP OF THIS CHART. THIS WILL ENSURE THAT AN INTERMITTENT CONNECTION WILL NOT BE CORRECTED BEFORE THE SOURCE OF THE MALFUNCTION IS FOUND.

(1)
- IGNITION "OFF."
- USING A 5% SALT WATER SOLUTION (TWO TEASPOONS OF SALT PER 12 OZ. WATER), THOROUGHLY SPRAY THE REAR ABS HARNESS.
- DISCONNECT 24-WAY EBCM CONNECTOR.
- USING J 39200, MEASURE RESISTANCE BETWEEN 24-WAY EBCM HARNESS CONNECTOR TERMINALS "9" AND "10." IS RESISTANCE BETWEEN 2100 AND 2400 OHMS WHEN SENSOR IS 25°C (77°F)? *(IF TEMPERATURE OF THE SENSOR IS SUSPECTED TO BE OTHER THAN 25°C (77°F), REFER TO TABLE ON PAGE 2 OF 2 OF THIS DTC CHART FOR APPLICABLE SENSOR RESISTANCE VALUES; THE VALUES IN THE TABLE ARE FOR TEMPERATURE OF THE SENSOR NOT AIR TEMPERATURE.)*

YES → (2); NO → GO TO PAGE 2 OF 3 OF THIS DTC CHART.

(2)
- USING J 39200, MEASURE THE RESISTANCE BETWEEN GROUND AND EBCM HARNESS CONNECTOR TERMINAL "9." DOES J 39200 DISPLAY "OL" (INFINITE)?

YES → (3); NO → GO TO PAGE 3 OF 3 OF THIS DTC CHART.

(3)
- IGNITION "ON."
- USING J 39200, MEASURE THE VOLTAGE BETWEEN GROUND AND 24-WAY EBCM HARNESS CONNECTOR TERMINAL "10." IS VOLTAGE LESS THAN 1 VOLT?

YES → (4); NO → (5)

(4)
- IGNITION "OFF."
- INSPECT 24-WAY EBCM HARNESS CONNECTOR TERMINALS "9" AND "10" FOR POOR TERMINAL CONTACT OR CORROSION. INSPECT CKT(s) 884 AND 885 FOR DAMAGE WHICH MAY RESULT IN AN OPEN CIRCUIT, SHORT TO GROUND OR SHORT TO VOLTAGE. REPAIR DAMAGE IF EVIDENT. REPLACE TERMINALS IF POOR CONTACT OR CORROSION EXISTS.
- RECONNECT ALL CONNECTORS.
- IGNITION "ON."
IS DTC 34 SET AS A CURRENT DTC?

YES → REPLACE EBCM.
NO → MALFUNCTION IS INTERMITTENT. REFER TO "DIAGNOSTIC AIDS" ON FACING PAGE.

(5)
- IGNITION "OFF."
- DISCONNECT LEFT REAR WHEEL SPEED SENSOR CONNECTOR.
- IGNITION "ON."
- USING J 39200, MEASURE THE VOLTAGE BETWEEN GROUND AND 24-WAY EBCM HARNESS CONNECTOR TERMINAL "10." IS VOLTAGE LESS THAN 1 VOLT?

YES → (6); NO → REPAIR SHORT TO VOLTAGE IN CKT 885.

(6)
- USING J 39200, MEASURE THE VOLTAGE BETWEEN GROUND AND 24-WAY EBCM HARNESS CONNECTOR TERMINAL "9." IS VOLTAGE LESS THAN 1 VOLT?

YES → MALFUNCTION IS INTERMITTENT. INSPECT ALL CONNECTORS AND HARNESSES FOR DAMAGE WHICH MAY RESULT IN A SHORT TO VOLTAGE WHEN ALL COMPONENTS ARE CONNECTED. REPAIR ALL DAMAGE IF FOUND. REFER TO "DIAGNOSTIC AIDS" ON FACING PAGE FOR MORE INFORMATION.
NO → REPAIR SHORT TO VOLTAGE IN CKT 884.

THIS CHART ASSUMES THAT A CURRENT DTC IS STORED INDICATING THAT THIS MALFUNCTION IS PRESENT.

DIAGNOSTIC CHART 34 — 1994–95 ACHIEVA, GRAND AM AND SKYLARK WITH VES

DTC 34
(Page 2 of 3)
LEFT REAR WHEEL SPEED SENSOR CIRCUIT OPEN OR SHORTED TO GROUND/BATTERY (FOR ABS WITH VES ONLY)

CONTINUED FROM PAGE 1 OF 3 OF THIS DTC CHART.

(7)
- DISCONNECT LEFT REAR WHEEL SPEED SENSOR CONNECTOR.
- USING J 39200, MEASURE THE RESISTANCE BETWEEN 24-WAY EBCM HARNESS CONNECTOR TERMINAL "9" AND LEFT REAR WHEEL SPEED SENSOR HARNESS CONNECTOR TERMINAL "A." IS RESISTANCE LESS THAN 2 OHMS?

NO: REPAIR HIGH RESISTANCE IN CKT 884.

YES:

(8)
- USING J 39200, MEASURE THE RESISTANCE BETWEEN 24-WAY EBCM HARNESS CONNECTOR TERMINAL "10" AND LEFT REAR WHEEL SPEED SENSOR HARNESS CONNECTOR TERMINAL "B." IS RESISTANCE LESS THAN 2 OHMS?

NO: REPAIR HIGH RESISTANCE IN CKT 885.

YES:

(9)
- USING J 39200, MEASURE THE RESISTANCE BETWEEN LEFT REAR WHEEL SPEED SENSOR TERMINALS "A" AND "B." IS RESISTANCE BETWEEN 2100 AND 2400 OHMS WHEN SENSOR IS 25°C (77°F)? *(IF TEMPERATURE OF THE SENSOR IS SUSPECTED TO BE OTHER THAN 25°C (77°F), REFER TO THE TABLE BELOW FOR APPLICABLE SENSOR RESISTANCE VALUES; THE VALUES IN THE TABLE ARE FOR TEMPERATURE OF THE SENSOR NOT AIR TEMPERATURE).*

NO: REPLACE INTEGRAL BEARING AND SPEED SENSOR ASSEMBLY.

YES: MALFUNCTION IS INTERMITTENT. INSPECT ALL CONNECTORS AND HARNESSES FOR DAMAGES WHICH MAY RESULT IN HIGH RESISTANCE WHEN ALL COMPONENTS ARE CONNECTED. REFER TO "DIAGNOSTIC AIDS" ON FACING PAGE FOR MORE INFORMATION. REPAIR ALL DAMAGE IF FOUND.

REAR WHEEL SPEED SENSOR
TEMPERATURE OF SENSOR VS. SENSOR RESISTANCE (APPROXIMATE)

TEMP. (°C)	TEMP. (°F)	RESISTANCE (OHMS)
-34 TO 4	-30 TO 40	1623 TO 2210
5 TO 43	41 TO 110	1939 TO 2570
44 TO 93	111 TO 200	2254 TO 3029

THIS CHART ASSUMES THAT A CURRENT DTC IS STORED INDICATING THAT THIS MALFUNCTION IS PRESENT.

DIAGNOSTIC CHART 34 — 1994–95 ACHIEVA, GRAND AM AND SKYLARK WITH VES

DTC 34
(Page 3 of 3)
LEFT REAR WHEEL SPEED SENSOR CIRCUIT OPEN OR SHORTED TO GROUND/BATTERY (FOR ABS WITH VES ONLY)

CONTINUED FROM PAGE 1 OF 3 OF THIS DTC CHART.

(10)
- DISCONNECT REAR ABS HARNESS FROM BODY TO ABS AND FUEL SENDER HARNESS CONNECTOR.
- USING J 39200, MEASURE THE RESISTANCE BETWEEN GROUND AND 24-WAY EBCM HARNESS CONNECTOR TERMINAL "9." DOES J 39200 DISPLAY "OL" (INFINITE)?

NO: REPAIR SHORT TO GROUND IN CKT 884.

YES:

(11)
- USING J 39200, MEASURE THE RESISTANCE BETWEEN GROUND AND 24-WAY EBCM HARNESS CONNECTOR TERMINAL "10." DOES J 39200 DISPLAY "OL" (INFINITE)?

NO: REPAIR SHORT TO GROUND IN CKT 885.

YES:

(12)
- THOROUGHLY RESPRAY THE REAR ABS HARNESS WITH THE SALT WATER SOLUTION. IMPORTANT: DO NOT ALLOW WATER TO ENTER THE WHEEL SPEED SENSOR CONNECTOR.
- DISCONNECT LEFT REAR WHEEL SPEED SENSOR CONNECTOR.
- USING J 39200, MEASURE THE RESISTANCE BETWEEN GROUND AND LEFT REAR WHEEL SPEED SENSOR HARNESS CONNECTOR TERMINAL "A" THEN "B." DID J 39200 DISPLAY "OL" (INFINITE) FOR EACH MEASUREMENT?

NO: REPLACE REAR ABS HARNESS.

YES:

(13)
- USING J 39200, MEASURE THE RESISTANCE BETWEEN SENSOR TERMINAL "A" AND INTEGRAL BEARING AND SPEED SENSOR ASSEMBLY MOUNTING FLANGE. DOES J 39200 DISPLAY "OL" (INFINITE)?

NO: REPLACE INTEGRAL BEARING AND SPEED SENSOR ASSEMBLY.

YES: MALFUNCTION IS INTERMITTENT. INSPECT ALL CONNECTORS AND HARNESSES FOR DAMAGES WHICH MAY RESULT IN A SHORT TO GROUND WHEN ALL COMPONENTS ARE CONNECTED. SMALL NICKS OR CUTS IN THE HARNESS CAN CAUSE A MALFUNCTION. REPAIR ANY DAMAGE IF FOUND. JUMPER HARNESS MUST BE REPLACED IF DAMAGED. REFER TO "DIAGNOSTIC AIDS" ON FACING PAGE FOR MORE INFORMATION.

THIS CHART ASSUMES THAT A CURRENT DTC IS STORED INDICATING THAT THIS MALFUNCTION IS PRESENT.

DIAGNOSTIC CHART 35 — 1994–95 ACHIEVA, GRAND AM AND SKYLARK WITH VES

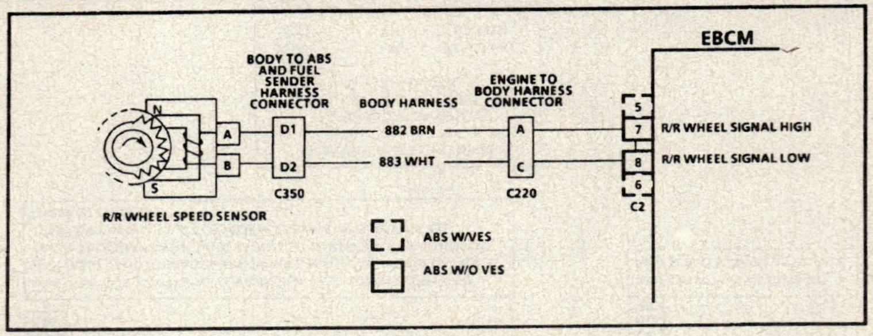

DTC 35

RIGHT REAR WHEEL SPEED SENSOR CIRCUIT OPEN OR SHORTED TO GROUND/BATTERY (FOR ABS WITH VES ONLY)

Circuit Description:
As a toothed ring passes by the wheel speed sensor, changes in the electromagnetic field cause the wheel speed sensor to produce a sinusoidal (AC) voltage signal whose frequency is proportional to wheel speed. The magnitude of this signal is directly related to wheel speed and the proximity of the wheel speed sensor to the toothed ring, often referred to as the air gap.

DTC Will Set When: DTC 35 can be set anytime after the ABS enable relay has been commanded "ON." The relay is commanded "ON" during the three second bulb check.

Action Taken: A malfunction DTC is stored, ABS/VES is disabled and the ABS warning lamp is turned "ON."

DTC Chart Test Description: Number(s) below refer to circled number(s) on the diagnostic chart.

1. Checks wheel speed sensor circuitry for proper resistance values.
2. Checks for a short to ground in the wheel speed sensor circuitry.
3. Checks for a short to voltage in the wheel speed sensor circuitry.
4. Ensures malfunction was not due to physical damage of the circuitry.
5. Checks for a short to voltage in the wheel signal low circuit.
6. Checks for a short to voltage in the wheel signal high circuit.
7. Checks for high resistance in the wheel signal low circuit.
8. Checks for high resistance in the wheel signal high circuit.
9. Checks for proper resistance of the wheel speed sensor.
10. Checks for a short to ground in the wheel signal high circuit.
11. Checks for a short to ground in the wheel signal low circuit.
12. Checks for a short to ground in the rear ABS harness.
13. Checks for a short to ground in the wheel speed sensor.

Diagnostic Aids: An "Intermittent" malfunction is most likely caused by a poor connection, rubbed through wire insulation, or a wire that is broken inside the insulation.

The frequency of the malfunction can be checked by using the enhanced diagnostic function of the Tech 1, as described in "Enhanced Diagnostics," found in this section.

If the customer's comments reflect that the ABS amber warning lamp is "ON" only during moist environmental changes, (rain, snow, vehicle wash) all wheel speed sensor circuitry should be thoroughly inspected for signs of water intrusion. Use the following procedure. Spray down the suspected area with 5% salt water solution (two teaspoons of salt to 12 oz. of water). Drive vehicle above 24 km/h (15 mph) for at least 30 seconds. If DTC returns immediately, replace suspected harness.

Any circuitry, that is suspected as causing the intermittent complaint, should be thoroughly checked for backed out terminals, improper mating, broken locks, improperly formed or damaged terminals, poor terminal to wiring connections or physical damage to the wiring harness.

Resistance of the wheel speed sensor will increase with an increase in sensor temperature.

When replacing a wheel speed sensor, inspect the sensor terminals and harness connector for corrosion and/or water intrusion. If evidence of corrosion or water intrusion exists, replace wheel speed sensor harness. Refer to "On-Vehicle Service" in this section.

If wheel speed sensor harness shows evidence of water intrusion, inspect wheel speed sensor for water intrusion. If water intrusion is evident, replace wheel speed sensor.

DIAGNOSTIC CHART 35 — 1994–95 ACHIEVA, GRAND AM AND SKYLARK WITH VES

DTC 35

(Page 1 of 3)

RIGHT REAR WHEEL SPEED SENSOR CIRCUIT OPEN OR SHORTED TO GROUND/BATTERY (FOR ABS WITH VES ONLY)

IMPORTANT:

- WHEEL SPEED SENSOR INTERMITTENT MALFUNCTIONS MAY BE DIFFICULT TO LOCATE. CARE SHOULD BE TAKEN NOT TO DISTURB ANY ELECTRICAL CONNECTIONS PRIOR TO AN INDICATED STEP OF THIS CHART. THIS WILL ENSURE THAT AN INTERMITTENT CONNECTION WILL NOT BE CORRECTED BEFORE THE SOURCE OF THE MALFUNCTION IS FOUND.

(1)
- IGNITION "OFF."
- USING A 5% SALT WATER SOLUTION (TWO TEASPOONS OF SALT PER 12 OZ. WATER), THOROUGHLY SPRAY THE REAR ABS HARNESS.
- DISCONNECT 24-WAY EBCM CONNECTOR.
- USING J 39200, MEASURE RESISTANCE BETWEEN 24-WAY EBCM HARNESS CONNECTOR TERMINALS "5" AND "6." IS RESISTANCE BETWEEN 2100 AND 2400 OHMS WHEN SENSOR IS 25°C (77°F)? *[IF TEMPERATURE OF THE SENSOR IS SUSPECTED TO BE OTHER THAN 25°C (77°F), REFER TO TABLE ON PAGE 2 OF 2 OF THIS DTC CHART FOR APPLICABLE SENSOR RESISTANCE VALUES; THE VALUES IN THE TABLE ARE FOR TEMPERATURE OF THE SENSOR NOT AIR TEMPERATURE.]*

NO: GO TO PAGE 2 OF 3 OF THIS DTC CHART.

YES:

(2)
- USING J 39200, MEASURE THE RESISTANCE BETWEEN GROUND AND EBCM HARNESS CONNECTOR TERMINAL "5." DOES J 39200 DISPLAY "OL" (INFINITE)?

NO: GO TO PAGE 3 OF 3 OF THIS DTC CHART.

YES:

(3)
- IGNITION "ON."
- USING J 39200, MEASURE THE VOLTAGE BETWEEN GROUND AND 24-WAY EBCM HARNESS CONNECTOR TERMINAL "6." IS VOLTAGE LESS THAN 1 VOLT?

YES:

(4)
- IGNITION "OFF."
- INSPECT 24-WAY EBCM HARNESS CONNECTOR TERMINALS "5" AND "6" FOR POOR TERMINAL CONTACT OR CORROSION. INSPECT CKT(s) 882 AND 883 FOR DAMAGE WHICH MAY RESULT IN AN OPEN CIRCUIT, SHORT TO GROUND OR SHORT TO VOLTAGE. REPAIR DAMAGE IF EVIDENT. REPLACE TERMINALS IF POOR CONTACT OR CORROSION EXISTS.
- RECONNECT ALL CONNECTORS.
- IGNITION "ON."

IS DTC 35 SET AS A CURRENT DTC?

YES: REPLACE EBCM.

NO: MALFUNCTION IS INTERMITTENT. REFER TO "DIAGNOSTIC AIDS" ON FACING PAGE.

NO:

(5)
- IGNITION "OFF."
- DISCONNECT RIGHT REAR WHEEL SPEED SENSOR CONNECTOR.
- IGNITION "ON."
- USING J 39200, MEASURE THE VOLTAGE BETWEEN GROUND AND 24-WAY EBCM HARNESS CONNECTOR TERMINAL "6." IS VOLTAGE LESS THAN 1 VOLT?

NO: REPAIR SHORT TO VOLTAGE IN CKT 883.

YES:

(6)
- USING J 39200, MEASURE THE VOLTAGE BETWEEN GROUND AND 24-WAY EBCM HARNESS CONNECTOR TERMINAL "5." IS VOLTAGE LESS THAN 1 VOLT?

YES: MALFUNCTION IS INTERMITTENT. INSPECT ALL CONNECTORS AND HARNESSES FOR DAMAGE WHICH MAY RESULT IN A SHORT TO VOLTAGE WHEN ALL COMPONENTS ARE CONNECTED. REPAIR ALL DAMAGE IF FOUND. REFER TO "DIAGNOSTIC AIDS" ON FACING PAGE FOR MORE INFORMATION.

NO: REPAIR SHORT TO VOLTAGE IN CKT 882.

THIS CHART ASSUMES THAT A CURRENT DTC IS STORED INDICATING THAT THIS MALFUNCTION IS PRESENT.

DIAGNOSTIC CHART 35 — 1994–95 ACHIEVA, GRAND AM AND SKYLARK WITH VES

DTC 35
(Page 2 of 3)
RIGHT REAR WHEEL SPEED SENSOR CIRCUIT OPEN OR SHORTED TO GROUND/BATTERY (FOR ABS WITH VES ONLY)

CONTINUED FROM PAGE 1 OF 3 OF THIS DTC CHART.

(7)
- DISCONNECT RIGHT REAR WHEEL SPEED SENSOR CONNECTOR.
- USING J 39200, MEASURE THE RESISTANCE BETWEEN 24-WAY EBCM HARNESS CONNECTOR TERMINAL "5" AND RIGHT REAR WHEEL SPEED SENSOR HARNESS CONNECTOR TERMINAL "A." IS RESISTANCE LESS THAN 2 OHMS?

NO: REPAIR HIGH RESISTANCE IN CKT 882.

YES:

(8)
- USING J 39200, MEASURE THE RESISTANCE BETWEEN 24-WAY EBCM HARNESS CONNECTOR TERMINAL "6" AND RIGHT REAR WHEEL SPEED SENSOR HARNESS CONNECTOR TERMINAL "B." IS RESISTANCE LESS THAN 2 OHMS?

NO: REPAIR HIGH RESISTANCE IN CKT 883.

YES:

(9)
- USING J 39200, MEASURE THE RESISTANCE BETWEEN RIGHT REAR WHEEL SPEED SENSOR TERMINALS "A" AND "B." IS RESISTANCE BETWEEN 2100 AND 2400 OHMS WHEN SENSOR IS 25°C (77°F)? *(IF TEMPERATURE OF THE SENSOR IS SUSPECTED TO BE OTHER THAN 25°C (77°F), REFER TO THE TABLE BELOW FOR APPLICABLE SENSOR RESISTANCE VALUES; THE VALUES IN THE TABLE ARE FOR TEMPERATURE OF THE SENSOR NOT AIR TEMPERATURE).*

NO: REPLACE INTEGRAL BEARING AND SPEED SENSOR ASSEMBLY.

YES: MALFUNCTION IS INTERMITTENT. INSPECT ALL CONNECTORS AND HARNESSES FOR DAMAGES WHICH MAY RESULT IN HIGH RESISTANCE WHEN ALL COMPONENTS ARE CONNECTED. REFER TO "DIAGNOSTIC AIDS" ON FACING PAGE FOR MORE INFORMATION. REPAIR ALL DAMAGE IF FOUND.

REAR WHEEL SPEED SENSOR
TEMPERATURE OF SENSOR VS. SENSOR RESISTANCE (APPROXIMATE)

TEMP. (°C)	TEMP. (°F)	RESISTANCE (OHMS)
-34 TO 4	-30 TO 40	1623 TO 2210
5 TO 43	41 TO 110	1939 TO 2570
44 TO 93	111 TO 200	2254 TO 3029

THIS CHART ASSUMES THAT A CURRENT DTC IS STORED INDICATING THAT THIS MALFUNCTION IS PRESENT.

DIAGNOSTIC CHART 35 — 1994–95 ACHIEVA, GRAND AM AND SKYLARK WITH VES

DTC 35
(Page 3 of 3)
RIGHT REAR WHEEL SPEED SENSOR CIRCUIT OPEN OR SHORTED TO GROUND/BATTERY (FOR ABS WITH VES ONLY)

CONTINUED FROM PAGE 1 OF 3 OF THIS DTC CHART.

(10)
- DISCONNECT REAR ABS HARNESS FROM BODY TO ABS AND FUEL SENDER HARNESS CONNECTOR.
- USING J 39200, MEASURE THE RESISTANCE BETWEEN GROUND AND 24-WAY EBCM HARNESS CONNECTOR TERMINAL "5." DOES J 39200 DISPLAY "OL" (INFINITE)?

NO: REPAIR SHORT TO GROUND IN CKT 882.

YES:

(11)
- USING J 39200, MEASURE THE RESISTANCE BETWEEN GROUND AND 24-WAY EBCM HARNESS CONNECTOR TERMINAL "6." DOES J 39200 DISPLAY "OL" (INFINITE)?

NO: REPAIR SHORT TO GROUND IN CKT 883.

YES:

(12)
- THOROUGHLY RESPRAY THE REAR ABS HARNESS WITH THE SALT WATER SOLUTION. IMPORTANT: DO NOT ALLOW WATER TO ENTER THE WHEEL SPEED SENSOR CONNECTOR.
- DISCONNECT RIGHT REAR WHEEL SPEED SENSOR CONNECTOR.
- USING J 39200, MEASURE THE RESISTANCE BETWEEN GROUND AND RIGHT REAR WHEEL SPEED SENSOR HARNESS CONNECTOR TERMINAL "A" THEN "B." DID J 39200 DISPLAY "OL" (INFINITE) FOR EACH MEASUREMENT?

NO: REPLACE REAR ABS HARNESS.

YES:

(13)
- USING J 39200, MEASURE THE RESISTANCE BETWEEN SENSOR TERMINAL "A" AND INTEGRAL BEARING AND SPEED SENSOR ASSEMBLY MOUNTING FLANGE. DOES J 39200 DISPLAY "OL" (INFINITE)?

NO: REPLACE INTEGRAL BEARING AND SPEED SENSOR ASSEMBLY.

YES: MALFUNCTION IS INTERMITTENT. INSPECT ALL CONNECTORS AND HARNESSES FOR DAMAGES WHICH MAY RESULT IN A SHORT TO GROUND WHEN ALL COMPONENTS ARE CONNECTED. SMALL NICKS OR CUTS IN THE HARNESS CAN CAUSE A MALFUNCTION. REPAIR ANY DAMAGE IF FOUND. JUMPER HARNESS MUST BE REPLACED IF DAMAGED. REFER TO "DIAGNOSTIC AIDS" ON FACING PAGE FOR MORE INFORMATION.

THIS CHART ASSUMES THAT A CURRENT DTC IS STORED INDICATING THAT THIS MALFUNCTION IS PRESENT.

DIAGNOSTIC CHART 36 — 1994–95 ACHIEVA, GRAND AM AND SKYLARK

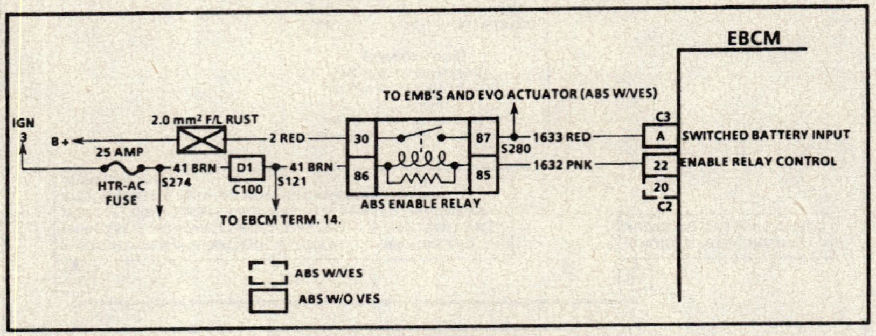

DTC 36

LOW SYSTEM VOLTAGE

Circuit Description:
This DTC is used to monitor the voltage level available to the ABS EBCM. If the voltage drops below 11 volts, full performance of the ABS system cannot be guaranteed. During ABS operation, there are several current requirements that will cause battery voltage to drop. Because of this, voltage is monitored prior to ABS operation to indicate good charging system condition and also during ABS operation when voltage may drop significantly.

DTC Will Set When: DTC 36 can only be set if the vehicle's speed is greater than 5 km/h (3 mph). If both the switched battery and switched ignition voltage is less than 11.2 volts non-ABS or less than 9.6 volts ABS, a malfunction exists.

Action Taken: A malfunction DTC is stored, ABS/VES is disabled and the ABS warning lamp is turned "ON."

DTC Chart Test Description: Number(s) below refer to circled number(s) on the diagnostic chart.

1. Tests to see if malfunction is present.
2. Checks charging system integrity.
3. Checks for excessive resistance in the switched battery input circuit.
4. Tests for high resistance in the enable relay control circuit.
5. Ensures low voltage condition was not due to a poor connection.

Diagnostic Aids: An "Intermittent" malfunction is most likely caused by a poor connection, rubbed through wire insulation, or a wire that is broken inside the insulation.

The frequency of the malfunction can be checked by using the enhanced diagnostic function of the Tech 1, as described in "Enhanced Diagnostics," found in this section.

Any circuitry, that is suspected as causing the intermittent complaint, should be thoroughly checked for backed out terminals, improper mating, broken locks, improperly formed or damaged terminals, poor terminal to wiring connections or physical damage to the wiring harness.

While performing a voltage load test, if it is noted that only ignition voltage drops below acceptable voltage levels, CKT 41 should be checked for high resistance or an open condition.

DIAGNOSTIC CHART 36 — 1994–95 ACHIEVA, GRAND AM AND SKYLARK

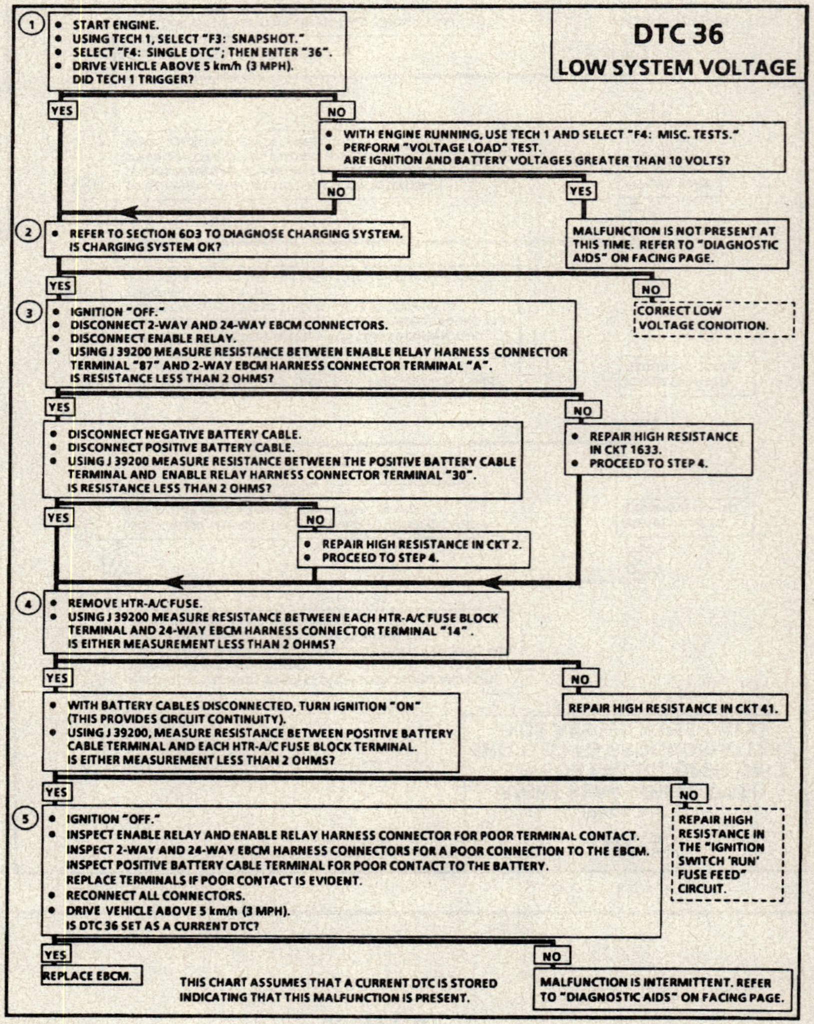

DIAGNOSTIC CHART 37 — 1994–95 ACHIEVA, GRAND AM AND SKYLARK

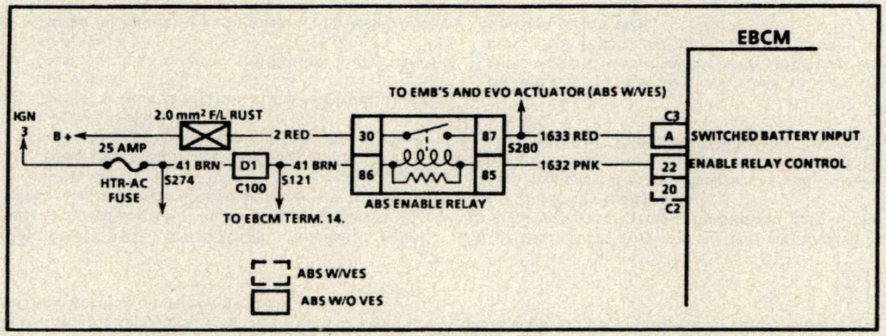

DTC 37
HIGH SYSTEM VOLTAGE

Circuit Description:
This DTC is designed to detect high vehicle voltage levels prior to any required motor movement (initialization or ABS). If excessive voltage exists, demagnetization of the motor magnets may occur, which would eventually affect or eliminate ABS performance.

DTC Will Set When: DTC 37 can only be set if the vehicle's speed is greater than 5 km/h (3 mph). If the switched battery voltage is greater than 17 volts, a malfunction exists.

Action Taken: A malfunction DTC is stored, ABS/VES is disabled and the ABS warning lamp is turned "ON."

DTC Chart Test Description: Number(s) below refer to circled number(s) on the diagnostic chart.
1. Checks the voltage level being received by the EBCM.
2. Isolates the malfunction to either the EBCM or the charging system.

Diagnostic Aids: An "Intermittent" malfunction is most likely caused by a poor connection, rubbed through wire insulation, or a wire that is broken inside the insulation.

The frequency of the malfunction can be checked by using the enhanced diagnostic function of the Tech 1, as described in "Enhanced Diagnostics," found in this section.

Any circuitry, that is suspected as causing the intermittent malfunction, should be thoroughly checked for backed out terminals, improper mating, broken locks, improperly formed or damaged terminals, poor terminal to wiring connections or physical damage to the wiring harness.

DIAGNOSTIC CHART 37 — 1994–95 ACHIEVA, GRAND AM AND SKYLARK

DTC 37
HIGH SYSTEM VOLTAGE

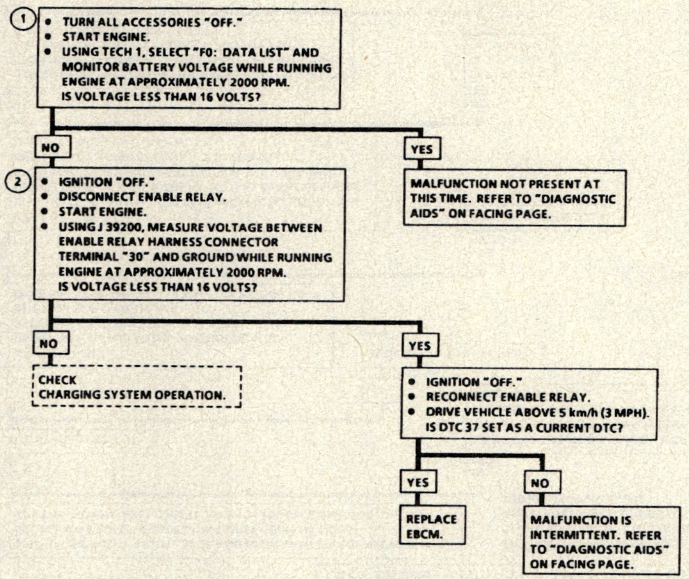

THIS CHART ASSUMES THAT A CURRENT DTC IS STORED INDICATING THAT THIS MALFUNCTION IS PRESENT.

DIAGNOSTIC CHART 38 — 1994–95 ACHIEVA, GRAND AM AND SKYLARK

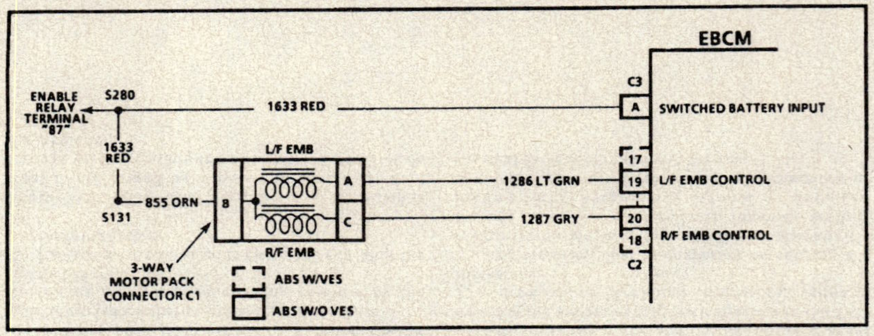

DTC 38

LEFT FRONT EMB WILL NOT HOLD MOTOR

Circuit Description:

Current is supplied to the EMBs through terminal "B" of the three-way motor pack connector. When a release of the EMB is commanded, the EBCM grounds the EMB input and allows the EMB to be energized to release the motor.

DTC Will Set When: DTC 38 can be set during initialization. The left front motor is commanded in the forward direction while the EMB is released (energized). A malfunction exists if the feedback current is less than the command current and the left front channel was expected to be in the home position. DTC 86 is always set with DTC 38.

Action Taken: If an EMB cannot hold a motor in the home position, the motor may be backdriven when the brake pedal is applied causing the brake pedal to drop. A malfunction DTC is stored, ABS/VES is disabled and the ABS warning lamp is turned "ON."

DTC Chart Test Description: Number(s) below refer to circled number(s) on the diagnostic chart.

1. Ensures motor circuits are in the proper connector cavity.
2. Checks the ability of the EMB to hold the motor. If the brake pedal moves during this test, it would indicate the EMB is slipping.
3. Checks for high resistance in the left front motor high circuit.
4. Checks for high resistance in the left front motor low circuit.
5. Checks for excessive resistance in the left front motor.
6. Ensures malfunction was not due to a poor connection.

Diagnostic Aids: An "Intermittent" malfunction in this DTC may result from a mechanical part of the system that sticks, binds, or slips.

The frequency of the malfunction can be checked by using the enhanced diagnostic function of the Tech 1, as described in "Enhanced Diagnostics," found in this section.

NOTICE: This test may fail if low voltage (10 volts or less) was present when initialization occurred. Ensure battery/charging system is functioning properly.

DIAGNOSTIC CHART 38 — 1994–95 ACHIEVA, GRAND AM AND SKYLARK

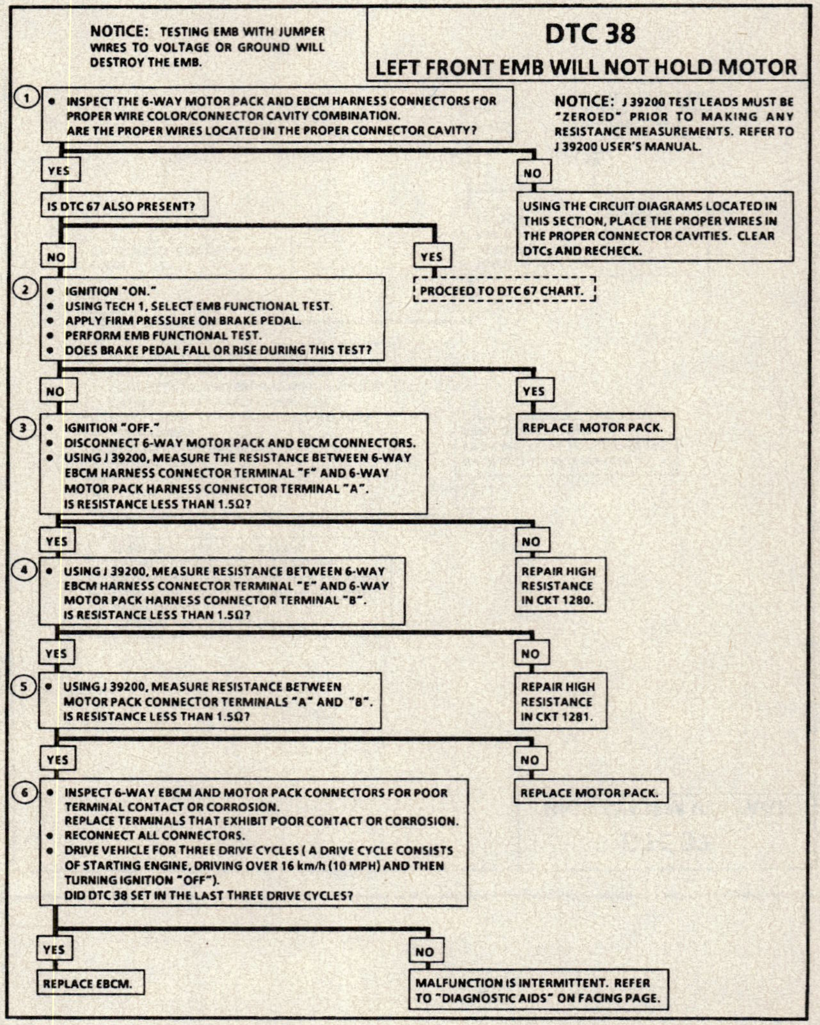

DIAGNOSTIC CHART 41 — 1994–95 ACHIEVA, GRAND AM AND SKYLARK

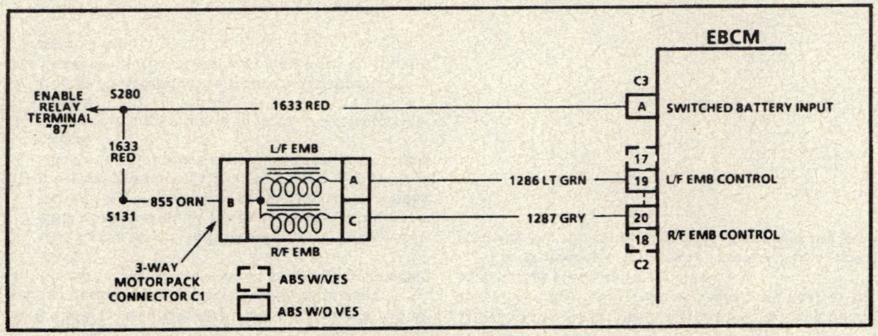

DTC 41

RIGHT FRONT EMB WILL NOT HOLD MOTOR

Circuit Description:
Current is supplied to the EMBs through terminal "B" of the three-way motor pack connector. When a release of the EMB is commanded, the EBCM grounds the EMB input and allows the EMB to be energized to release the motor.

DTC Will Set When: DTC 41 can be set during initialization. The right front motor is commanded in the forward direction while the EMB is released (energized). A malfunction exists if the feedback current is less than the command current and the right front channel was expected to be in the home position. DTC 86 is always set with DTC 41.

Action Taken: If an EMB cannot hold a motor in the home position, the motor may be backdriven when the brake pedal is applied causing the brake pedal to fall. A malfunction DTC is stored, ABS/VES is disabled and the ABS warning lamp is turned "ON."

DTC Chart Test Description: Number(s) below refer to circled number(s) on the diagnostic chart.

1. Ensures motor circuits are in the proper connector cavity.
2. Checks the ability of the EMB to hold the motor. If the brake pedal moves during this test, it would indicate the EMB is slipping.
3. Checks for high resistance in the right front motor high circuit.
4. Checks for high resistance in the right front motor low circuit.
5. Checks for excessive resistance in the right front motor.
6. Ensures malfunction was not due to a poor connection.

Diagnostic Aids: An "Intermittent" malfunction in this DTC may result from a mechanical part of the system that sticks, binds, or slips.

The frequency of the malfunction can be checked by using the enhanced diagnostic function of the Tech 1, as described in "Enhanced Diagnostics," found in this section.

NOTICE: This DTC may fail if low voltage (10 volts or less) was present when initialization occurred. Ensure battery/charging system is functioning properly.

DIAGNOSTIC CHART 41 — 1994–95 ACHIEVA, GRAND AM AND SKYLARK

DTC 41

RIGHT FRONT EMB WILL NOT HOLD MOTOR

NOTICE: TESTING EMB WITH JUMPER WIRES TO VOLTAGE OR GROUND WILL DESTROY THE EMB.

NOTICE: J 39200 TEST LEADS MUST BE "ZEROED" PRIOR TO MAKING ANY RESISTANCE MEASUREMENTS. REFER TO J 39200 USER'S MANUAL.

(1)
- INSPECT THE 6-WAY MOTOR PACK AND EBCM HARNESS CONNECTORS FOR PROPER WIRE COLOR/CONNECTOR CAVITY COMBINATION.
ARE THE PROPER WIRES LOCATED IN THE PROPER CONNECTOR CAVITY?

NO → USING THE CIRCUIT DIAGRAMS LOCATED IN THIS SECTION, PLACE THE PROPER WIRES IN THE PROPER CONNECTOR CAVITIES. CLEAR DTCs AND RECHECK.

YES → IS DTC 71 ALSO PRESENT?

YES → PROCEED TO DTC 71 CHART.

NO ↓

(2)
- IGNITION "ON."
- USING TECH 1, SELECT EMB FUNCTIONAL TEST.
- APPLY FIRM PRESSURE ON BRAKE PEDAL.
- PERFORM EMB FUNCTIONAL TEST.
DOES BRAKE PEDAL FALL OR RISE DURING THIS TEST?

YES → REPLACE MOTOR PACK.

NO ↓

(3)
- IGNITION "OFF."
- DISCONNECT 6-WAY MOTOR PACK AND EBCM CONNECTORS.
- USING J 39200, MEASURE THE RESISTANCE BETWEEN 6-WAY EBCM HARNESS CONNECTOR TERMINAL "H" AND 6-WAY MOTOR PACK HARNESS CONNECTOR TERMINAL "E".
IS RESISTANCE LESS THAN 1.5Ω?

NO → REPAIR HIGH RESISTANCE IN CKT 1282.

YES ↓

(4)
- USING J 39200, MEASURE RESISTANCE BETWEEN 6-WAY EBCM HARNESS CONNECTOR TERMINAL "G" AND 6-WAY MOTOR PACK HARNESS CONNECTOR TERMINAL "F".
IS RESISTANCE LESS THAN 1.5Ω?

NO → REPAIR HIGH RESISTANCE IN CKT 1283.

YES ↓

(5)
- USING J 39200, MEASURE RESISTANCE BETWEEN MOTOR PACK CONNECTOR TERMINALS "E" AND "F".
IS RESISTANCE LESS THAN 1.5Ω?

NO → REPLACE MOTOR PACK.

YES ↓

(6)
- INSPECT 6-WAY EBCM AND MOTOR PACK CONNECTORS FOR POOR TERMINAL CONTACT OR CORROSION.
REPLACE
TERMINALS THAT EXHIBIT POOR CONTACT OR CORROSION.
- RECONNECT ALL CONNECTORS.
- DRIVE VEHICLE FOR THREE DRIVE CYCLES (A DRIVE CYCLE CONSISTS OF STARTING ENGINE, DRIVING OVER 16 km/h (10 MPH) AND THEN TURNING IGNITION "OFF").
DID DTC 41 SET IN THE LAST THREE DRIVE CYCLES?

YES → REPLACE EBCM.

NO → MALFUNCTION IS INTERMITTENT. REFER TO "DIAGNOSTIC AIDS" ON FACING PAGE.

DIAGNOSTIC CHART 42 — 1994–95 ACHIEVA, GRAND AM AND SKYLARK

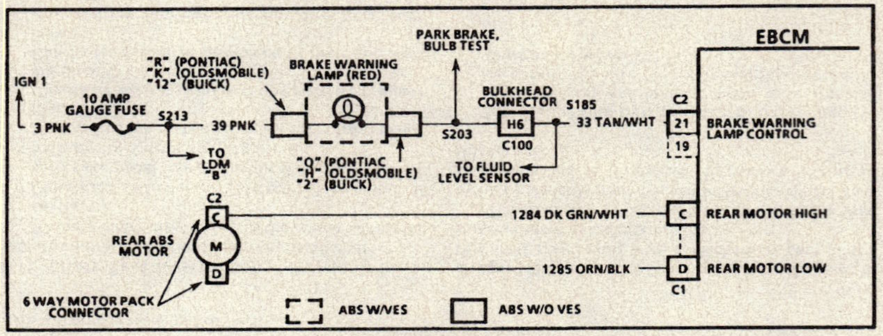

DTC 42
REAR ESB WILL NOT HOLD MOTOR

Circuit Description:
This DTC is designed to detect a slipping rear ESB. During initialization and braking, the rear motor is re-homed. If the ESB slips, the motor/piston will move. During the next key "ON" initialization, a re-home of the motor verifies the motor/piston remained at the home position. If motor movement is detected, the ESB must be slipping.

DTC Will Set When: DTC 42 can be set during initialization. DTC 86 is always set with DTC 42.

Action Taken: If an ESB cannot hold a motor in the home position, the motor may be backdriven when the brake pedal is applied causing the brake pedal to drop. A malfunction DTC is stored, ABS/VES is disabled and the ABS warning lamp is turned "ON."

DTC Chart Test Description: Number(s) below refer to circled number(s) on the diagnostic chart.
1. Ensures motor circuits are in the proper connector cavity.
2. Checks the rear axle ESB. A broken or defective ESB would result in the rear piston being back driven by hydraulic pressure, and wheel movement would result.
3. Checks for high resistance in the rear motor high circuit.
4. Checks for high resistance in the rear motor low circuit.
5. Checks for excessive resistance in the rear motor.
6. Ensures malfunction was not due to a poor connection.

Diagnostic Aids: An "Intermittent" malfunction in this DTC may result from a mechanical part of the system that sticks, binds, or slips.
The frequency of the malfunction can be checked by using the enhanced diagnostic function of the Tech 1.

DIAGNOSTIC CHART 42 — 1994–95 ACHIEVA, GRAND AM AND SKYLARK

DTC 42
REAR ESB WILL NOT HOLD MOTOR

NOTICE: J 39200 TEST LEADS MUST BE "ZEROED" PRIOR TO MAKING ANY RESISTANCE MEASUREMENTS. REFER TO J 39200 USER'S MANUAL.

(1)
- INSPECT THE 6-WAY EBCM AND MOTOR PACK CONNECTORS FOR PROPER WIRE COLOR/CONNECTOR CAVITY COMBINATION.
- ARE THE PROPER WIRES LOCATED IN THE PROPER CONNECTOR CAVITIES?

YES → go to (2)
NO → USING THE CIRCUIT DIAGRAMS LOCATED IN THE FRONT OF THIS SECTION, PLACE THE PROPER WIRES INTO THE PROPER CONNECTOR CAVITIES. CLEAR DTCs AND RECHECK SYSTEM.

(2)
- RAISE THE VEHICLE SUCH THAT THE REAR WHEELS ARE APPROXIMATELY SIX INCHES OFF THE FLOOR.
- IGNITION "ON."
- APPLY FIRM PRESSURE ON BRAKE PEDAL AND RELEASE.
- REAPPLY FIRM PRESSURE ON BRAKE PEDAL AND HAVE AN ASSISTANT TRY TO SPIN THE REAR WHEEL BY HAND.
- COULD ASSISTANT SPIN THE REAR WHEELS?

NO → go to (3)
YES → REPLACE MOTOR PACK.

(3)
- IGNITION "OFF."
- DISCONNECT 6-WAY MOTOR PACK AND EBCM CONNECTORS.
- USING J 39200, MEASURE THE RESISTANCE BETWEEN 6-WAY EBCM HARNESS CONNECTOR TERMINAL "C" AND 6-WAY MOTOR PACK HARNESS CONNECTOR TERMINAL "C".
- IS RESISTANCE LESS THAN 1.5Ω?

YES → go to (4)
NO → REPAIR HIGH RESISTANCE IN CKT 1284.

(4)
- USING J 39200, MEASURE RESISTANCE BETWEEN 6-WAY EBCM HARNESS CONNECTOR TERMINAL "D" AND 6-WAY MOTOR PACK HARNESS CONNECTOR TERMINAL "D".
- IS RESISTANCE LESS THAN 1.5Ω?

YES → go to (5)
NO → REPAIR HIGH RESISTANCE IN CKT 1285.

(5)
- USING J 39200, MEASURE RESISTANCE BETWEEN MOTOR PACK CONNECTOR TERMINALS "C" AND "D".
- IS RESISTANCE LESS THAN 1.5Ω?

YES → go to (6)
NO → REPLACE MOTOR PACK.

(6)
- INSPECT 6-WAY EBCM AND MOTOR PACK CONNECTORS FOR POOR TERMINAL CONTACT OR CORROSION.
- REPLACE TERMINALS THAT EXHIBIT POOR CONTACT OR CORROSION.
- RECONNECT ALL CONNECTORS.
- DRIVE VEHICLE FOR THREE DRIVE CYCLES (A DRIVE CYCLE CONSISTS OF STARTING ENGINE, DRIVING OVER 16 km/h (10 MPH) AND THEN TURNING IGNITION "OFF").
- DID DTC 42 SET IN THE LAST THREE DRIVE CYCLES?

YES → REPLACE EBCM.
NO → MALFUNCTION IS INTERMITTENT. REFER TO "DIAGNOSTIC AIDS" ON FACING PAGE.

DIAGNOSTIC CHART 43 — 1994–95 ACHIEVA, GRAND AM AND SKYLARK WITH VES

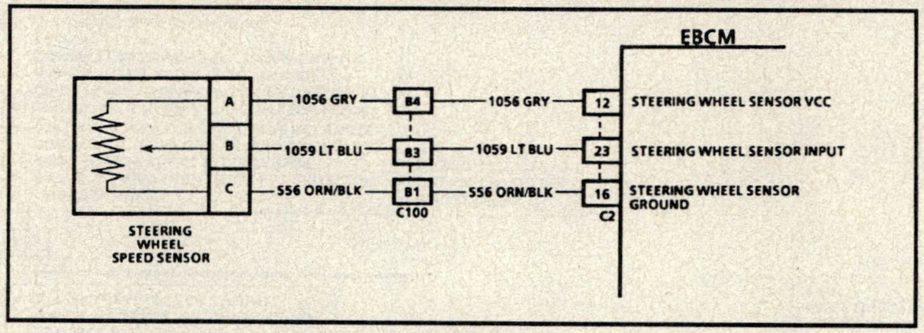

DTC 43

VES STEERING WHEEL SPEED SENSOR CIRCUIT MALFUNCTION (FOR ABS WITH VES ONLY)

Circuit Description:
A 5 volt reference is supplied to the steering wheel speed sensor through EBCM terminal "12" to sensor terminal "A". Steering wheel position input to the EBCM is provided through steering wheel speed sensor terminal "B" to EBCM terminal "23". Steering wheel speed sensor ground is provided through EBCM terminal "16".

DTC Will Set When: DTC 43 can be set anytime. A malfunction exists if the steering wheel speed sensor voltage is: less than .4 volt, greater than 4.67 volts or changes by more than 2 volts in 8 msec.

Action Taken: A malfunction DTC is stored, VES is disabled and power steering returns to full assist.

DTC Chart Test Description: Number(s) below refer to circled number(s) on the diagnostic chart.

1. Checks steering wheel speed sensor input to the EBCM.
2. Checks for a short between the steering wheel sensor input and ground circuits.
3. Checks for a short between the steering wheel sensor VCC and ground circuits.
4. Checks for a short to ground in the steering wheel sensor input circuit.
5. Checks for a short to ground in the steering wheel sensor VCC circuit.
6. Checks for high resistance in the steering wheel sensor VCC circuit.
7. Checks for high resistance in the steering wheel sensor input circuit.
8. Checks for poor terminal contact.
9. Checks for proper voltage output of the EBCM.
10. Ensures malfunction was not due to physical damage of the steering wheel speed sensor circuits.
11. Checks for proper voltage with the steering wheel speed sensor disconnected.
12. Checks for high resistance in the steering wheel sensor ground.
13. Isolates high resistance to the EBCM or the ground circuit.
14. Ensures open ground was not due to a short to battery.
15. Ensures malfunction was not due to circuit damage or poor terminal contact.
16. Checks for a short to B+ in the steering wheel sensor input circuit.
17. Ensures malfunction was not due to physical damage of the circuitry.
18. Checks for proper voltage with the steering wheel speed sensor connected.
19. Checks for proper voltage with the steering wheel speed sensor disconnected.
20. Checks for a short to voltage in the steering wheel sensor VCC circuit.
21. Ensures malfunction was not due to physical damage of the circuitry.
22. Checks for a short between the steering wheel sensor VCC and input circuits.
23. Checks for proper resistance of the steering wheel speed sensor.
24. Ensures malfunction was not due to physical damage of the circuits.
25. Checks for an open resistor in the steering wheel speed sensor.

Diagnostic Aids: An "Intermittent" malfunction is most likely caused by a poor connection, rubbed through wire insulation, or a wire that is broken inside the insulation.

The frequency of the malfunction can be checked by using the enhanced diagnostic function of the Tech 1, as described in "Enhanced Diagnostics," found in this section.

Any circuitry, that is suspected as causing the intermittent malfunction, should be thoroughly checked for backed out terminals, improper mating, broken locks, improperly formed or damaged terminals, poor terminal to wiring connections or physical damage to the wiring harness.

DIAGNOSTIC CHART 43 — 1994–95 ACHIEVA, GRAND AM AND SKYLARK WITH VES

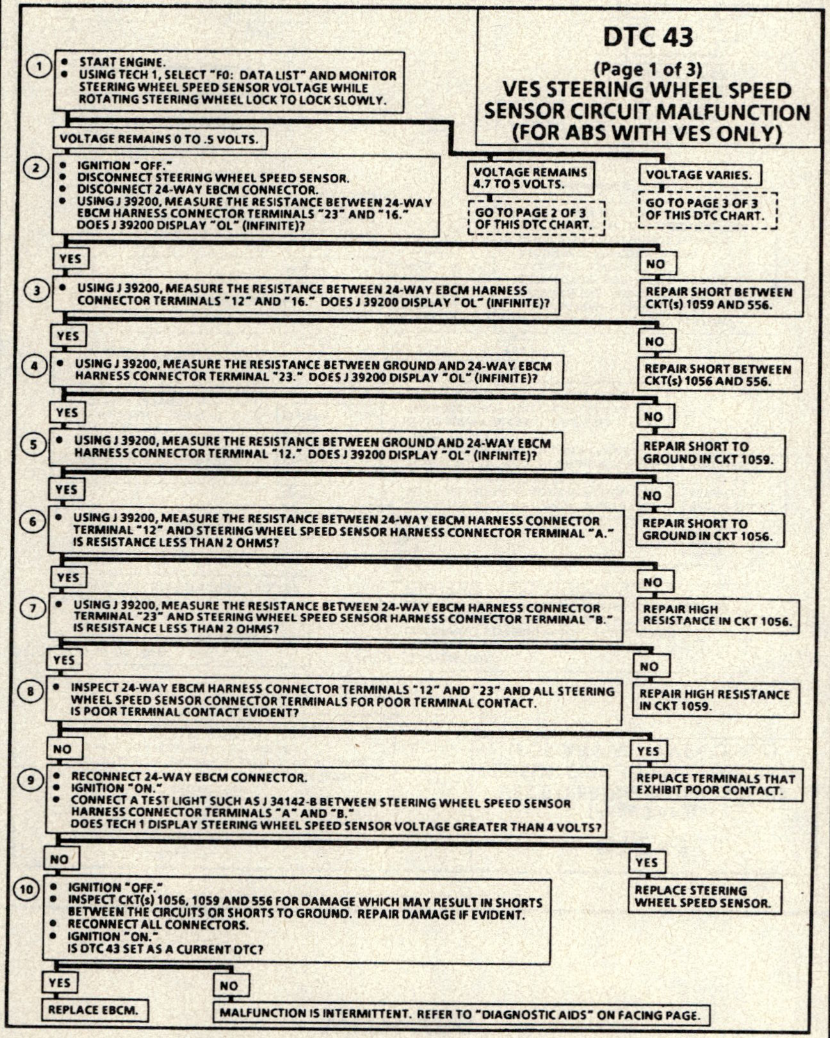

DIAGNOSTIC CHART 43 — 1994–95 ACHIEVA, GRAND AM AND SKYLARK WITH VES

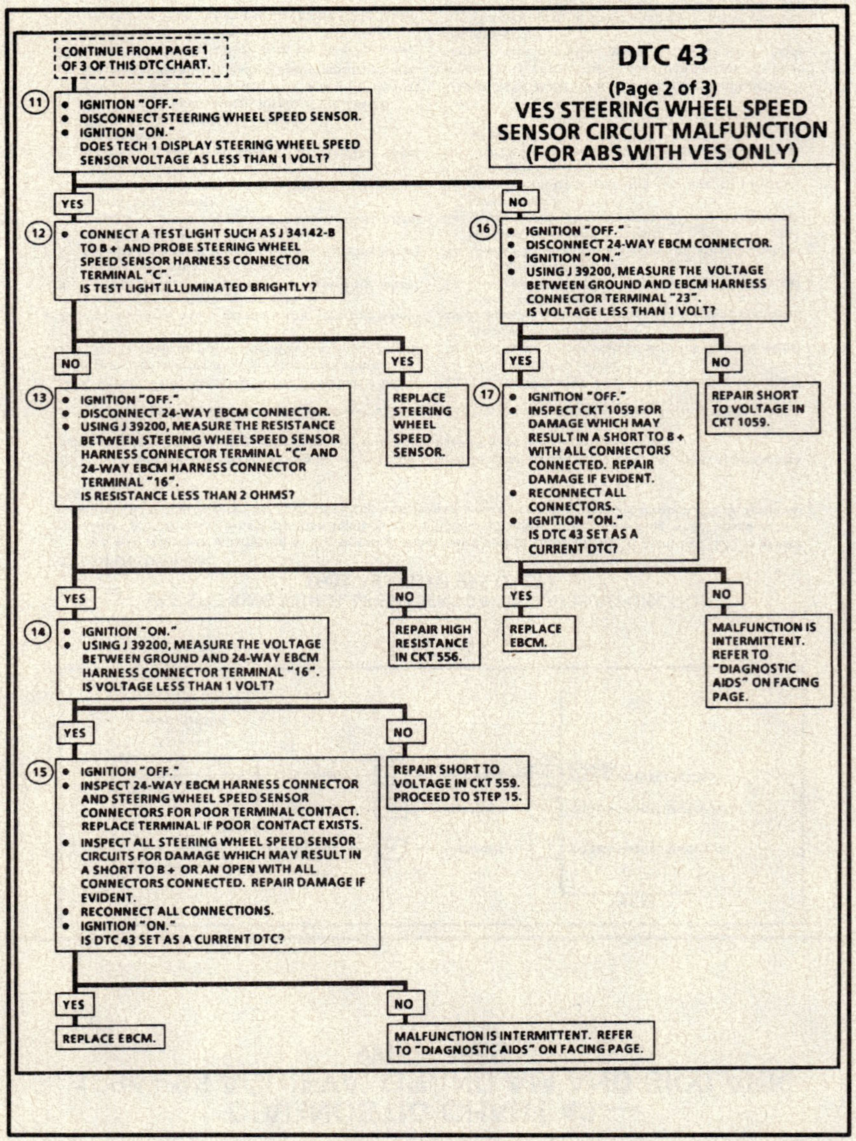

DIAGNOSTIC CHART 43 — 1994–95 ACHIEVA, GRAND AM AND SKYLARK WITH VES

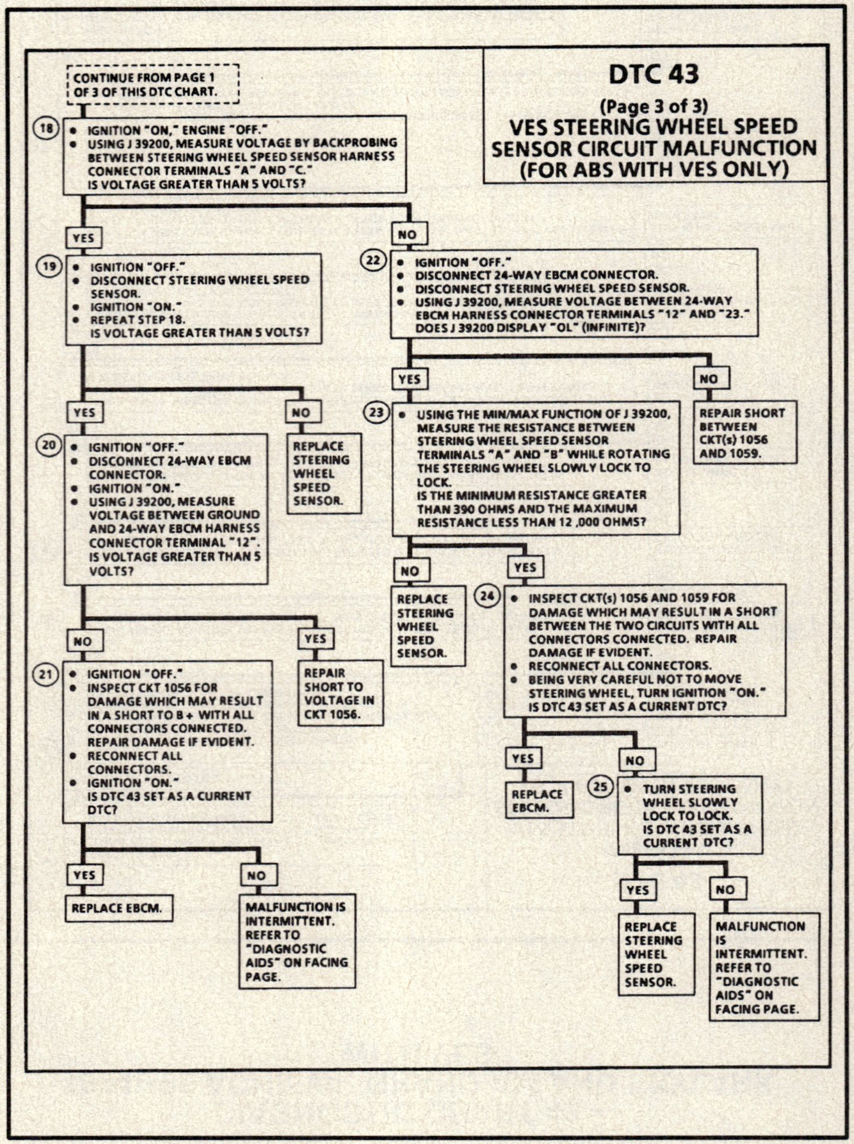

DIAGNOSTIC CHART DTC 44 — 1994–95 ACHIEVA, GRAND AM AND SKYLARK

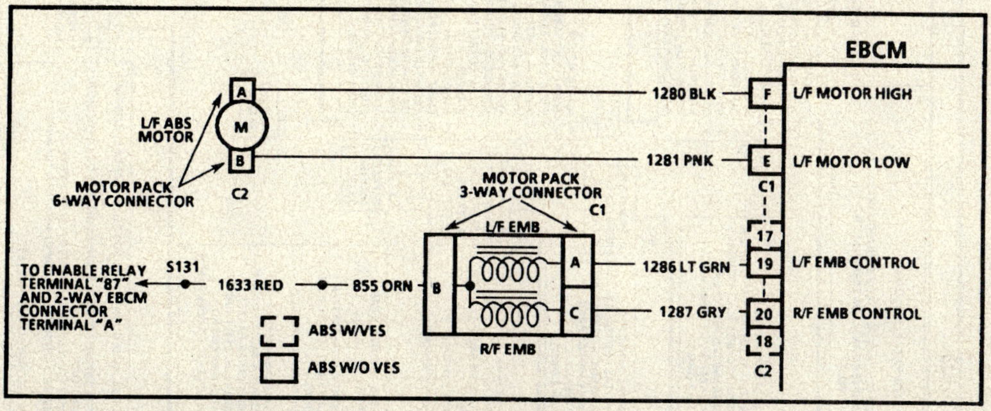

DTC 44

(Page 1 of 2)

LEFT FRONT ABS CHANNEL WILL NOT MOVE

Circuit Description:

Current is supplied to the EMB through terminal "B" of the three-way motor pack connector. When a release of the EMB is commanded, the EBCM grounds the EMB input and allows the EMB to be energized to release the motor. This DTC is designed to detect non-actuating EMB, stuck motor or a seized hydraulic modulator.

DTC Will Set When: DTC 44 can be set at initialization. The EBCM commands the left front motor in the reverse and forward direction to ensure the motor moves properly. If the EBCM detects a condition in which it cannot move the motor in either direction, a malfunction exists.

Action Taken: This malfunction indicates the channel cannot be moved properly. A malfunction DTC is stored, ABS/VES is disabled and the ABS warning lamp is turned "ON."

DTC Chart Test Description: Number(s) below refer to circled number(s) on the diagnostic chart.

1. Ensures motor circuits are in the proper connector cavity.
2. Checks for motor movement during release.
3. Compares feedback current to command current during apply.
4. Checks for a short between motor high and low circuits.
5. Checks for a shorted motor.
6. Ensures malfunction was not due to physical damage of the circuitry.
7. Checks if motor can apply properly under load.
8. Checks for a short to B+ in the left front EMB control circuit.
9. Checks for high resistance in the left front EBM control circuit.
10. Checks for high resistance in the EMB power feed circuit.
11. Checks for high resistance in the left front EMB.
12. Checks for proper operation of the hydraulic modulator.
13. Ensures malfunction was not due to poor terminal contact.
14. Isolates malfunction to either the left front EMB circuitry or the EBCM.
15. Ensures malfunction was not due to physical damage of the EMB circuitry.

Diagnostic Aids: An "Intermittent" malfunction in this DTC may result from a mechanical part of the system that sticks, binds, or slips.

The frequency of the malfunction can be checked by using the enhanced diagnostic function of the Tech 1, as described in "Enhanced Diagnostics," found in this section. DTC 44 may set after modulator disassembly if the modulator pistons are positioned at the bottom of their bore.

Any circuitry, that is suspected as causing the intermittent complaint, should be thoroughly checked for backed out terminals, improper mating, broken locks, improperly formed or damaged terminals, poor terminal to wiring connections or physical damage to the wiring harness.

NOTICE: This DTC may fail if low voltage (10 volts or less) was present when initialization occurred. Ensure battery/charging system is functioning properly.

DIAGNOSTIC CHART DTC 44 — 1994–95 ACHIEVA, GRAND AM AND SKYLARK, CONTINUED

DTC 44 (Page 1 of 2) LEFT FRONT ABS CHANNEL WILL NOT MOVE

NOTICE: J 39200 TEST LEADS MUST BE "ZEROED" PRIOR TO MAKING ANY RESISTANCE MEASUREMENTS. REFER TO J 39200 USER'S MANUAL.

1.
- VISUALLY INSPECT THE 6-WAY EBCM AND MOTOR PACK CONNECTORS FOR PROPER WIRE COLOR/CONNECTOR CAVITY COMBINATION.

ARE THE PROPER WIRES LOCATED IN THE PROPER CONNECTOR CAVITIES?

- YES: go to 2.
- NO: USING THE CIRCUIT DIAGRAMS LOCATED IN THE FRONT OF THIS SECTION, PLACE THE PROPER WIRES IN THE PROPER CONNECTOR CAVITIES. CLEAR DTCs AND RECHECK SYSTEM.

2.
- IGNITION "ON."
- USING TECH 1 MISC. TESTS, SELECT MANUAL CONTROL FUNCTION.
- SELECT LEFT FRONT MOTOR APPLY FUNCTION AND APPLY THE MOTOR.
- APPLY FIRM PRESSURE ON BRAKE PEDAL.
- SELECT LEFT FRONT MOTOR RELEASE FUNCTION AND RELEASE MOTOR.

DID BRAKE PEDAL FALL?

- NO: go to 3.
- YES: go to 7.

7.
- FIRM PRESSURE STILL ON BRAKE PEDAL.
- SELECT LEFT FRONT MOTOR APPLY FUNCTION AND APPLY THE MOTOR.

DID BRAKE PEDAL RISE?

- NO: go to 3.
- YES: MALFUNCTION IS INTERMITTENT. REFER TO "DIAGNOSTIC AIDS" ON FACING PAGE.

3.
- REMOVE FOOT FROM BRAKE PEDAL.
- SELECT LEFT FRONT MOTOR APPLY FUNCTION AND APPLY THE MOTOR WHILE CAREFULLY OBSERVING THE COMMANDED CURRENT AND THE FEEDBACK CURRENT ON THE TECH 1.

WAS THE FEEDBACK CURRENT HIGHER THAN THE COMMANDED CURRENT?

- YES: go to 4.
- NO: GO TO PAGE 2 OF 2 OF THIS DTC CHART.

4.
- IGNITION "OFF."
- DISCONNECT 6-WAY EBCM CONNECTOR AND THE 6-WAY MOTOR PACK CONNECTOR.
- USING J 39200, MEASURE RESISTANCE BETWEEN TERMINAL "E" AND TERMINAL "F" OF THE 6-WAY EBCM HARNESS CONNECTOR.

DOES J 39200 DISPLAY RESISTANCE AS "OL" (INFINITE)?

- YES: go to 5.
- NO: REPAIR SHORT BETWEEN CKT 1280 AND CKT 1281.

5.
- USING J 39200, MEASURE RESISTANCE BETWEEN MOTOR PACK TERMINALS "A" AND "B".

IS RESISTANCE GREATER THAN .4 OHMS?

- YES: go to 6.
- NO: REPLACE MOTOR PACK.

6.
- INSPECT CKTs 1280 AND 1281 FOR DAMAGE WHICH MAY RESULT IN A SHORT BETWEEN THE TWO CIRCUITS. REPAIR DAMAGE IF EVIDENT.
- RECONNECT ALL CONNECTORS.
- DRIVE VEHICLE FOR THREE DRIVE CYCLES (A DRIVE CYCLE CONSISTS OF STARTING ENGINE, DRIVING OVER 16 KM/H (10 MPH) AND THEN TURNING IGNITION "OFF").

DID DTC 44 SET IN THE LAST THREE DRIVE CYCLES?

- YES: REPLACE EBCM.
- NO: MALFUNCTION IS INTERMITTENT. REFER TO "DIAGNOSTIC AIDS" ON FACING PAGE.

* AFTER DIAGNOSIS IS COMPLETE, CLEAR DTCs AND TEST DRIVE VEHICLE FOR THREE (3) DRIVE CYCLES TO VERIFY DTC DOES NOT RESET. A DRIVE CYCLE CONSISTS OF STARTING THE VEHICLE, DRIVING OVER 10 MPH AND THEN KEYING DOWN.

DIAGNOSTIC CHART DTC 44 — 1994–95 ACHIEVA, GRAND AM AND SKYLARK, CONTINUED

DTC 44 (Page 2 of 2) LEFT FRONT ABS CHANNEL WILL NOT MOVE

8.
- IGNITION "OFF."
- DISCONNECT 3-WAY MOTOR PACK CONNECTOR.
- IGNITION "ON."
- USING J 39200, MEASURE VOLTAGE BETWEEN GROUND AND 3-WAY MOTOR PACK HARNESS CONNECTOR TERMINAL "A."

IS VOLTAGE LESS THAN 1 VOLT?

- YES: go to 9.
- NO: go to 14.

NOTICE: J 39200 TEST LEADS MUST BE "ZEROED" PRIOR TO MAKING ANY RESISTANCE MEASUREMENTS. REFER TO J 39200 USER'S MANUAL.

14.
- IGNITION "OFF."
- DISCONNECT 24-WAY EBCM CONNECTOR.
- IGNITION "ON."
- USING J 39200, MEASURE VOLTAGE BETWEEN GROUND AND 3-WAY MOTOR PACK HARNESS CONNECTOR TERMINAL "A."

IS VOLTAGE LESS THAN 1 VOLT?

- YES: go to 15.
- NO: REPAIR SHORT TO VOLTAGE IN CKT 1286.

15.
- IGNITION "OFF."
- INSPECT CKT 1286 AND 24-WAY EBCM HARNESS CONNECTOR FOR PHYSICAL DAMAGE WHICH MAY RESULT IN A SHORT TO B+ WITH THE 24-WAY EBCM HARNESS CONNECTOR CONNECTED TO THE EBCM. CORRECT DAMAGE IF EVIDENT.
- RECONNECT ALL CONNECTORS.
- DRIVE VEHICLE FOR THREE DRIVE CYCLES (A DRIVE CYCLE CONSISTS OF STARTING ENGINE, DRIVING OVER 16 KM/H (10 MPH) AND THEN TURNING IGNITION "OFF").

DID DTC 44 SET IN THE LAST THREE DRIVE CYCLES?

- YES: REPLACE EBCM.
- NO: MALFUNCTION IS INTERMITTENT. REFER TO "DIAGNOSTIC AIDS" ON FACING PAGE.

9.
- IGNITION "OFF."
- DISCONNECT 24-WAY EBCM CONNECTOR.
- USING J 39200, MEASURE THE RESISTANCE BETWEEN 3-WAY MOTOR PACK HARNESS CONNECTOR TERMINAL "A" AND 24-WAY EBCM HARNESS CONNECTOR TERMINAL "19" (ABS W/O VES) OR "17" (ABS W/VES).

IS RESISTANCE LESS THAN 2 OHMS?

- YES: go to 10.
- NO: REPAIR HIGH RESISTANCE IN CKT 1286.

10.
- DISCONNECT 2-WAY EBCM CONNECTOR.
- USING J 39200, MEASURE THE RESISTANCE BETWEEN 2-WAY EBCM HARNESS CONNECTOR TERMINAL "A" AND 3-WAY MOTOR PACK HARNESS CONNECTOR TERMINAL "B."

IS RESISTANCE LESS THAN 2 OHMS?

- YES: go to 11.
- NO: REPAIR HIGH RESISTANCE IN CKT 855 OR CKT 1633.

11.
- USING J 39200, MEASURE THE RESISTANCE BETWEEN MOTOR PACK TERMINALS "A" AND "B."

IS RESISTANCE LESS THAN 2 OHMS?

- YES: go to 12.
- NO: REPLACE MOTOR PACK.

12.
- USING TECH 1 MISC. TESTS, PERFORM GEAR TENSION RELIEF FUNCTION.
- IGNITION "OFF."
- REMOVE HYDRAULIC MODULATOR/MASTER CYLINDER ASSEMBLY (REFER TO "ON-VEHICLE SERVICE" IN THIS SECTION).
- SEPARATE MOTOR PACK FROM HYDRAULIC MODULATOR.
- GRASP THE GEAR ON THE HYDRAULIC MODULATOR (THE HYDRAULIC MODULATOR IS THE UNIT WITH THE THREE LARGE GEARS) FURTHEST FROM THE BRAKE BOOSTER AND ATTEMPT TO MOVE THE GEAR BY HAND.

CAN GEAR BE ROTATED AT LEAST 8.5 FULL TURNS LOCK TO LOCK?

- YES: go to 13.
- NO: REPLACE HYDRAULIC MODULATOR

13.
- INSPECT MOTOR PACK AND EBCM CONNECTORS FOR POOR TERMINAL CONTACT OR CORROSION.
  REPLACE TERMINALS THAT EXHIBIT POOR TERMINAL CONTACT OR CORROSION.
- RECONNECT ALL CONNECTORS.
- DRIVE VEHICLE FOR THREE DRIVE CYCLES (A DRIVE CYCLE CONSISTS OF STARTING ENGINE, DRIVING OVER 16 KM/H (10 MPH) AND THEN TURNING IGNITION "OFF").

DID DTC 44 SET IN THE LAST THREE DRIVE CYCLES?

- YES: REPLACE MOTOR PACK.
- NO: MALFUNCTION IS INTERMITTENT. REFER TO "DIAGNOSTIC AIDS" ON FACING PAGE.

*AFTER DIAGNOSIS IS COMPLETE, CLEAR DTCs AND TEST DRIVE VEHICLE FOR THREE (3) DRIVE CYCLES TO VERIFY DTC DOES NOT RESET. A DRIVE CYCLE CONSISTS OF STARTING THE VEHICLE, DRIVING OVER 10 MPH AND THEN KEYING DOWN.

DIAGNOSTIC CHART DTC 45 — 1994–95 ACHIEVA, GRAND AM AND SKYLARK

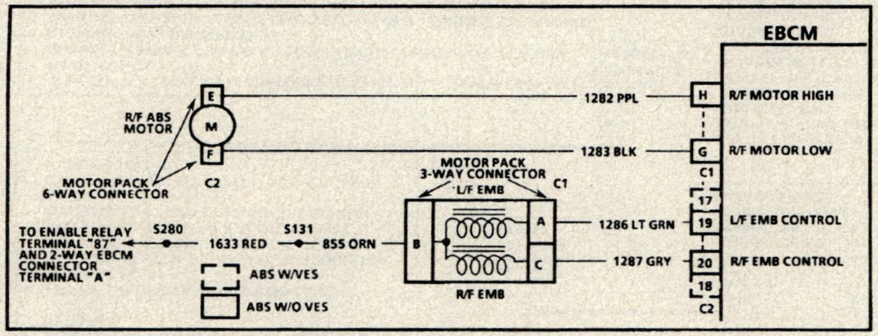

DTC 45

RIGHT FRONT ABS CHANNEL WILL NOT MOVE

Circuit Description:
Current is supplied to the EMB through terminal "B" of the three-way motor pack connector. When a release of the EMB is commanded, the EBCM grounds the EMB input and allows the EMB to be energized to release the motor. This DTC is designed to detect non-actuating EMB, stuck motor or a seized hydraulic modulator.

DTC Will Set When: DTC 45 can be set during initialization. The EBCM commands the right front motor in the reverse and forward direction to ensure the motor moves properly. If the EBCM detects a condition in which it cannot move the motor in either direction, a malfunction exists.

Action Taken: This malfunction indicates the channel cannot be moved properly. A malfunction DTC is stored, ABS/VES is disabled and the ABS warning lamp is turned "ON."

DTC Chart Test Description: Number(s) below refer to circled number(s) on the diagnostic chart.

1. Ensures motor circuits are in the proper connector cavity.
2. Checks for motor movement during release.
3. Compares feedback current to command current during apply.
4. Checks for a short between motor high and low circuits.
5. Checks for a shorted motor.
6. Ensures malfunction was not due to physical damage of the circuitry.
7. Checks if motor can apply properly under load.
8. Checks for a short to B+ in the right front EMB control circuit.
9. Checks for high resistance in the right front EBM control circuit.
10. Checks for high resistance in the EMB power feed circuit.
11. Checks for high resistance in the right front EMB.
12. Checks for proper operation of the hydraulic modulator.
13. Ensures malfunction was not due to poor terminal contact.
14. Isolates malfunction to either the right front EMB circuitry or the EBCM.
15. Ensures malfunction was not due to physical damage of the EMB circuitry.

Diagnostic Aids: An "Intermittent" malfunction in this DTC may result from a mechanical part of the system that sticks, binds, or slips.

The frequency of the malfunction can be checked by using the enhanced diagnostic function of the Tech 1, as described in "Enhanced Diagnostics," found in this section. DTC 45 may set after modulator disassembly if the modulator pistons are positioned at the bottom of their bore.

Any circuitry, that is suspected as causing the intermittent complaint, should be thoroughly checked for backed out terminals, improper mating, broken locks, improperly formed or damaged terminals, poor terminal to wiring connections or physical damage to the wiring harness.

NOTICE: This DTC may fail if low voltage (10 volts or less) was present when initialization occurred. Ensure battery/charging system is functioning properly.

DIAGNOSTIC CHART DTC 45 — 1994–95 ACHIEVA, GRAND AM AND SKYLARK, CONTINUED

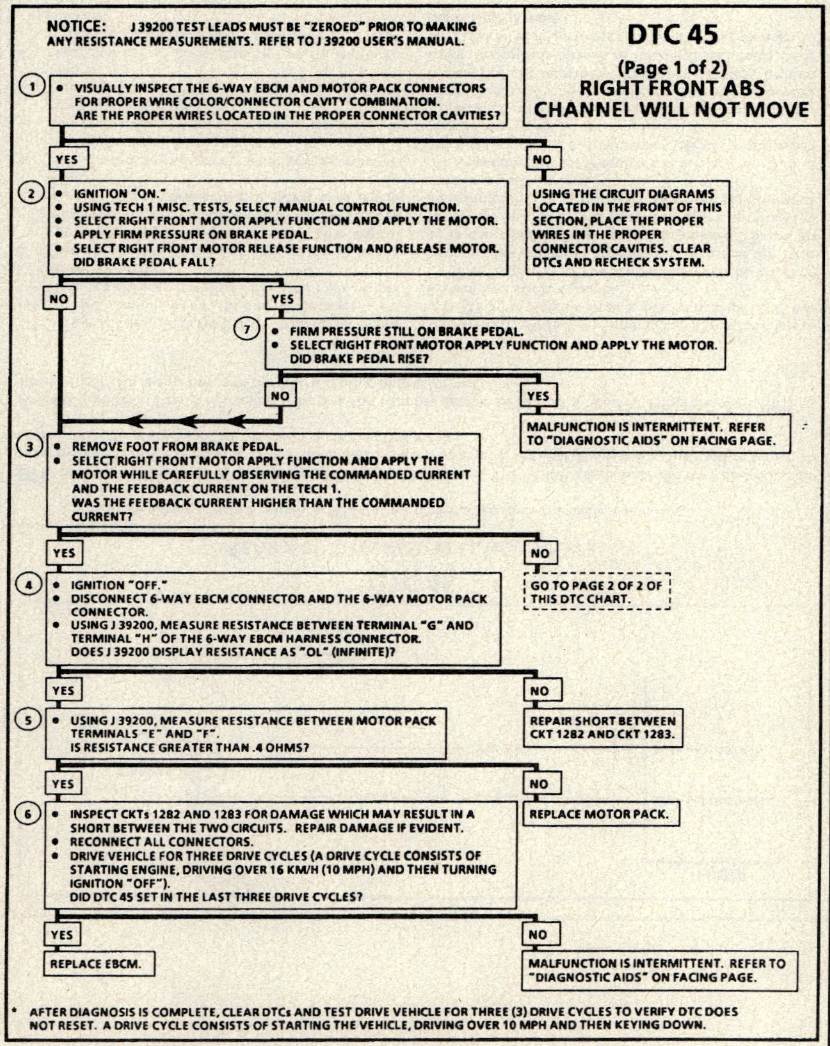

DIAGNOSTIC CHART DTC 45 — 1994–95 ACHIEVA, GRAND AM AND SKYLARK, CONTINUED

DTC 45
(Page 2 of 2)
RIGHT FRONT ABS CHANNEL WILL NOT MOVE

(8)
- IGNITION "OFF."
- DISCONNECT 3-WAY MOTOR PACK CONNECTOR.
- IGNITION "ON."
- USING J 39200, MEASURE VOLTAGE BETWEEN GROUND AND 3-WAY MOTOR PACK HARNESS CONNECTOR TERMINAL "C."

IS VOLTAGE LESS THAN 1 VOLT?

YES → (9); NO → (14)

NOTICE: J 39200 TEST LEADS MUST BE "ZEROED" PRIOR TO MAKING ANY RESISTANCE MEASUREMENTS. REFER TO J 39200 USER'S MANUAL.

(9)
- IGNITION "OFF."
- DISCONNECT 24-WAY EBCM CONNECTOR.
- USING J 39200, MEASURE THE RESISTANCE BETWEEN 3-WAY MOTOR PACK HARNESS CONNECTOR TERMINAL "C" AND 24-WAY EBCM HARNESS CONNECTOR TERMINAL "20" (ABS W/O VES) OR "18" (ABS W/VES).

IS RESISTANCE LESS THAN 2 OHMS?

YES → (10); NO → REPAIR HIGH RESISTANCE IN CKT 1287.

(10)
- DISCONNECT 2-WAY EBCM CONNECTOR.
- USING J 39200, MEASURE THE RESISTANCE BETWEEN 2-WAY EBCM HARNESS CONNECTOR TERMINAL "A" AND 3-WAY MOTOR PACK HARNESS CONNECTOR TERMINAL "B."

IS RESISTANCE LESS THAN 2 OHMS?

YES → (11); NO → REPAIR HIGH RESISTANCE IN CKT 855 OR CKT 1633.

(11)
- USING J 39200, MEASURE THE RESISTANCE BETWEEN MOTOR PACK TERMINALS "C" AND "B."

IS RESISTANCE LESS THAN 2 OHMS?

YES → (12); NO → REPLACE MOTOR PACK.

(12)
- USING TECH 1 MISC. TESTS, PERFORM GEAR TENSION RELIEF FUNCTION.
- IGNITION "OFF."
- REMOVE HYDRAULIC MODULATOR/MASTER CYLINDER ASSEMBLY
- SEPARATE MOTOR PACK FROM HYDRAULIC MODULATOR.
- GRASP THE GEAR ON THE HYDRAULIC MODULATOR (THE HYDRAULIC MODULATOR IS THE UNIT WITH THE THREE LARGE GEARS) CLOSEST TO THE BRAKE BOOSTER AND ATTEMPT TO MOVE THE GEAR BY HAND.

CAN GEAR BE ROTATED AT LEAST 8.5 FULL TURNS LOCK TO LOCK?

YES → (13); NO → REPLACE HYDRAULIC MODULATOR.

(13)
- INSPECT MOTOR PACK AND EBCM CONNECTORS FOR POOR TERMINAL CONTACT OR CORROSION. REPLACE TERMINALS THAT EXHIBIT POOR TERMINAL CONTACT OR CORROSION.
- RECONNECT ALL CONNECTORS.
- DRIVE VEHICLE FOR THREE DRIVE CYCLES (A DRIVE CYCLE CONSISTS OF STARTING ENGINE, DRIVING OVER 16 KM/H (10 MPH) AND THEN TURNING IGNITION "OFF").

DID DTC 45 SET IN THE LAST THREE DRIVE CYCLES?

YES → REPLACE MOTOR PACK.; NO → MALFUNCTION IS INTERMITTENT. REFER TO "DIAGNOSTIC AIDS" ON FACING PAGE.

(14)
- IGNITION "OFF."
- DISCONNECT 24-WAY EBCM CONNECTOR.
- IGNITION "ON."
- USING J 39200, MEASURE VOLTAGE BETWEEN GROUND AND 3-WAY MOTOR PACK HARNESS CONNECTOR TERMINAL "C."

IS VOLTAGE LESS THAN 1 VOLT?

YES → (15); NO → REPAIR SHORT TO VOLTAGE IN CKT 1287.

(15)
- IGNITION "OFF."
- INSPECT CKT 1287 AND 24-WAY EBCM HARNESS CONNECTOR FOR PHYSICAL DAMAGE WHICH MAY RESULT IN A SHORT TO B+ WITH THE 24-WAY EBCM HARNESS CONNECTOR CONNECTED TO THE EBCM. CORRECT DAMAGE IF EVIDENT.
- RECONNECT ALL CONNECTORS.
- DRIVE VEHICLE FOR THREE DRIVE CYCLES (A DRIVE CYCLE CONSISTS OF STARTING ENGINE, DRIVING OVER 16 KM/H (10 MPH) AND THEN TURNING IGNITION "OFF").

DID DTC 45 SET IN THE LAST THREE DRIVE CYCLES?

YES → REPLACE EBCM.; NO → MALFUNCTION IS INTERMITTENT. REFER TO "DIAGNOSTIC AIDS" ON FACING PAGE.

*AFTER DIAGNOSIS IS COMPLETE, CLEAR DTCs AND TEST DRIVE VEHICLE FOR THREE (3) DRIVE CYCLES TO VERIFY DTC DOES NOT RESET. A DRIVE CYCLE CONSISTS OF STARTING THE VEHICLE, DRIVING OVER 10 MPH AND THEN KEYING DOWN.

DIAGNOSTIC CHART DTC 46 — 1994–95 ACHIEVA, GRAND AM AND SKYLARK

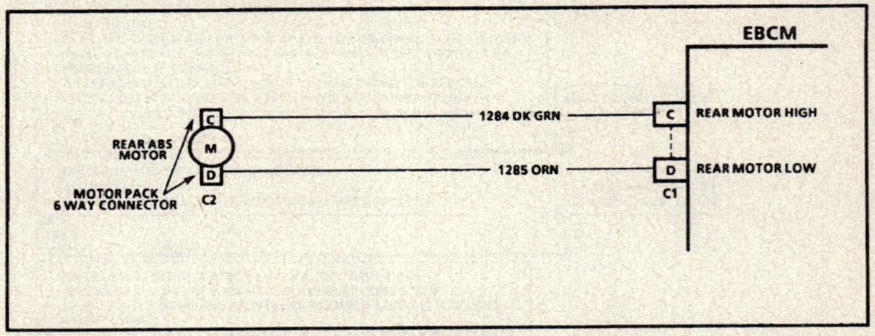

DTC 46
REAR ABS CHANNEL WILL NOT MOVE

Circuit Description:
This DTC is designed to detect bound-up ESBs, stuck motor or a seized hydraulic modulator.

DTC Will Set When: DTC 46 can be set during initialization. The EBCM commands the rear motor in the reverse and forward direction to ensure the motor moves properly. If the EBCM detects a condition in which it cannot move the motor in either direction, a malfunction exists.

Action Taken: This malfunction indicates the channel cannot be moved properly. A malfunction DTC is stored, ABS/VES is disabled, and the ABS warning lamp is turned "ON."

DTC Chart Test Description: Number(s) below refer to circled number(s) on the diagnostic chart.

1. This step ensures motor circuits are in the proper connector cavity.
2. This checks the ability of the motor to move applying hydraulic pressure to the wheel cylinders. A frozen piston or gear would result in no apply of rear brakes and a wheel that freely spins.
3. This procedure compares feedback current to command current during apply.
4. This step checks to see if hydraulic modulator is functioning properly.
5. This step checks if motor can apply properly under load.
6. This step serves to isolate the malfunction to either the EBCM or a short between CKTs 1284 and 1285.
7. Checks for a shorted motor.
8. Ensures malfunction was not due to physical damage of the circuitry.

Diagnostic Aids: An "Intermittent" malfunction in this DTC may result from a mechanical part of the system that sticks or binds.

The frequency of the malfunction can be checked by using the enhanced diagnostic function of the Tech 1, as described in "Enhanced Diagnostics," found in this section. DTC 46 may set after modulator disassembly if the modulator pistons are positioned at the bottom of their bore.

Depending on the frequency of the malfunction, a physical inspection of the mechanical parts suspected may be necessary.

Any circuitry, that is suspected as causing the intermittent complaint, should be thoroughly checked for backed out terminals, improper mating, broken locks, improperly formed or damaged terminals, poor terminal to wiring connections or physical damage to the wiring harness.

DIAGNOSTIC CHART DTC 46 — 1994–95 ACHIEVA, GRAND AM AND SKYLARK, CONTINUED

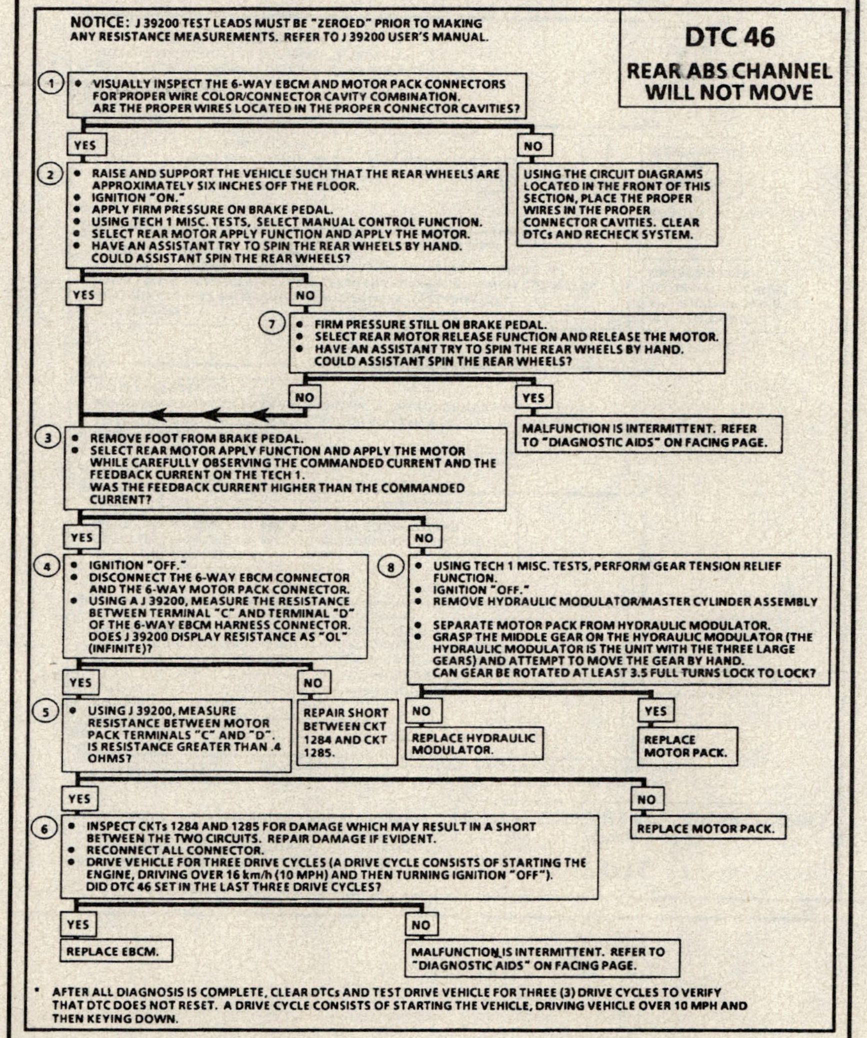

DIAGNOSTIC CHART DTC 47 — 1994–95 ACHIEVA, GRAND AM AND SKYLARK

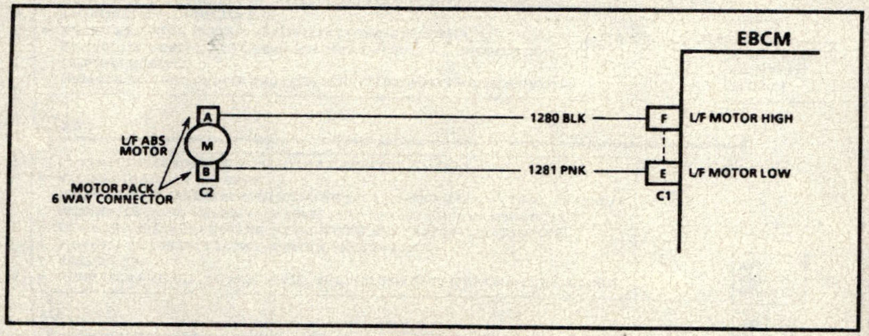

DTC 47

LEFT FRONT ABS MOTOR FREE SPINS

Circuit Description:
This DTC is designed to detect a stripped nut or gear assembly during initialization. During the homing sequence, the piston should reach the top of the bore resulting in a stalled motor. If this does not occur, the motor must be spinning with little or no resistance, indicating a nut/screw or gear malfunction.

DTC Will Set When: DTC 47 can only be set during initialization. If the feedback current is less than the command current for a specified period to time, a malfunction exists.

Action Taken: A malfunction DTC is stored, ABS/VES is disabled, and the ABS warning lamp is turned "ON."

DTC Chart Test Description: Number(s) below refer to circled number(s) on the diagnostic chart.

1. This step compares feedback current to command current during apply.
2. This step compares feedback current to command current during release.
3. Steps 1, 2 and 3 verify malfunction exists. This step also moves the piston to the top of the bore.
4. Checks for excessive resistance in the left front motor high circuit.
5. Checks for excessive resistance in the left front motor low circuit.
6. Checks for excessive resistance in the left front motor.
7. Checks for a stripped motor pack gear.
8. Checks for a stripped hydraulic modulator gear.
9. Checks for proper operation of motor pack.
10. Checks for proper operation of the hydraulic modulator.
11. Ensures malfunction was not due to poor terminal contact.

Diagnostic Aids: An "Intermittent" malfunction in this DTC may result from a mechanical part of the system that sticks, binds, or slips.

The frequency of the malfunction can be checked by using the enhanced diagnostic function of the Tech 1, as described in "Enhanced Diagnostics," found in this section. If DTC only occurs once and DTC 56 also occurs, see DTC 56. If intermittent and enhanced diagnostics shows this DTC occurs during ABS, see DTC 56.

Depending on the frequency of the malfunction, a physical inspection of the mechanical parts suspected may be necessary.

DIAGNOSTIC CHART DTC 47 — 1994–95 ACHIEVA, GRAND AM AND SKYLARK, CONTINUED

DTC 47 (Page 1 of 2) LEFT FRONT ABS MOTOR FREE SPINS

AFTER DIAGNOSIS IS COMPLETE, CLEAR DTCs AND TEST DRIVE VEHICLE FOR THREE(3) DRIVE CYCLES TO VERIFY THAT DTC DOES NOT RESET. A DRIVE CYCLE CONSISTS OF STARTING THE ENGINE, DRIVING THE VEHICLE ABOVE 16 km/h (10 MPH) AND THEN KEYING DOWN.

NOTICE: J 39200 TEST LEADS MUST BE "ZEROED" PRIOR TO MAKING ANY RESISTANCE MEASUREMENTS. REFER TO J 39200 USER'S MANUAL.

IS DTC 56 ALSO PRESENT?
- YES: PROCEED TO DTC 56 CHART FIRST.
- NO: go to step 1.

(1)
- IGNITION "ON," ENGINE "OFF."
- PUMP BRAKE PEDAL TO DEPLETE VACUUM POWER BOOSTER.
- USING TECH 1 MISC. TESTS, SELECT MANUAL CONTROL FUNCTION.
- SELECT LEFT FRONT MOTOR APPLY FUNCTION AND APPLY THE MOTOR WHILE CAREFULLY OBSERVING THE COMMAND CURRENT AND FEEDBACK CURRENT ON THE TECH 1.

WERE BOTH COMMAND AND FEEDBACK CURRENT 10 AMPS?
- YES: go to step 2.
- NO: GO TO STEP 4.

(2)
- APPLY FIRM PRESSURE ON BRAKE PEDAL.
- SELECT LEFT FRONT MOTOR RELEASE FUNCTION AND RELEASE MOTOR.

DID THE BRAKE PEDAL FALL AND WERE BOTH COMMAND AND FEEDBACK CURRENT 6 AMPS?
- YES: go to step 3.
- NO: GO TO STEP 4.

(3)
- FIRM PRESSURE STILL ON BRAKE PEDAL.
- SELECT LEFT FRONT MOTOR APPLY FUNCTION AND APPLY THE MOTOR (THIS MOVES THE HYDRAULIC MODULATOR PISTON TO THE TOP OF ITS BORE).

DID BRAKE PEDAL RISE?
- YES: MALFUNCTION IS INTERMITTENT. REFER TO "DIAGNOSTIC AIDS" ON FACING PAGE.
- NO: go to step 4.

(4)
- IGNITION "OFF."
- DISCONNECT 6-WAY EBCM AND MOTOR PACK HARNESS CONNECTORS.
- USING J 39200, MEASURE RESISTANCE BETWEEN 6-WAY EBCM HARNESS CONNECTOR TERMINAL "F" AND 6-WAY MOTOR PACK HARNESS CONNECTOR TERMINAL "A".

IS RESISTANCE LESS THAN 1.5 OHMS?
- YES: go to step 5.
- NO: REPAIR HIGH RESISTANCE IN CKT 1280.

(5)
- USING J 39200, MEASURE RESISTANCE BETWEEN 6-WAY EBCM HARNESS CONNECTOR TERMINAL "E" AND 6-WAY MOTOR PACK HARNESS CONNECTOR TERMINAL "B".

IS RESISTANCE LESS THAN 1.5 OHMS?
- YES: go to step 6.
- NO: REPAIR HIGH RESISTANCE IN CKT 1281.

(6)
- USING J 39200, MEASURE RESISTANCE BETWEEN 6-WAY MOTOR PACK CONNECTOR TERMINALS "A" AND "B".

IS RESISTANCE LESS THAN 1.5 OHMS?
- YES: GO TO PAGE 2 OF 2 OF THIS CHART.
- NO: REPLACE MOTOR PACK.

DIAGNOSTIC CHART DTC 47 — 1994–95 ACHIEVA, GRAND AM AND SKYLARK, CONTINUED

DTC 47 (Page 2 of 2) LEFT FRONT ABS MOTOR FREE SPINS

CONTINUED FROM PAGE 1 OF 2 OF THIS CHART.

(7)
- REMOVE HYDRAULIC MODULATOR/MASTER CYLINDER ASSEMBLY. (REFER TO "ON-VEHICLE SERVICE" IN THIS SECTION.)
- REMOVE GEAR COVER.
- CHECK FOR STRIPPED GEAR ON MOTOR PACK; THE MOTOR PACK IS THE UNIT WITH THE THREE SMALL GEARS. THE LEFT FRONT GEAR IS FURTHEST FROM THE BRAKE BOOSTER WHEN THE ASSEMBLY IS INSTALLED IN THE VEHICLE.

IS THE MOTOR PACK GEAR STRIPPED?
- YES: REPLACE MOTOR PACK.
- NO: go to step 8.

(8)
- CHECK FOR A STRIPPED GEAR ON THE HYDRAULIC MODULATOR; THE HYDRAULIC MODULATOR IS THE UNIT WITH THE THREE LARGE GEARS. THE LEFT FRONT GEAR IS FURTHEST FROM THE BRAKE BOOSTER WHEN THE ASSEMBLY IS INSTALLED IN THE VEHICLE.

IS THE HYDRAULIC MODULATOR GEAR STRIPPED?
- YES: REPLACE HYDRAULIC MODULATOR GEAR.
- NO: go to step 9.

(9)
- SECURELY POSITION HYDRAULIC MODULATOR/MASTER CYLINDER ASSEMBLY IN VEHICLE ALLOWING LEFT FRONT GEAR SET OBSERVATION. BE CAREFUL NOT TO DAMAGE GEARS OR ALLOW THE HYDRAULIC MODULATOR/MASTER CYLINDER ASSEMBLY TO MOVE DURING TESTING.
- RECONNECT ELECTRICAL CONNECTORS.
- IGNITION "ON."
- USING TECH 1 MISC. TESTS, SELECT MANUAL CONTROL FUNCTION.
- SELECT LEFT FRONT MOTOR APPLY AND APPLY THE MOTOR WHILE OBSERVING THE GEAR SET FURTHEST FORWARD FROM THE BOOSTER WHEN THE ASSEMBLY IS INSTALLED IN THE VEHICLE.
- SELECT LEFT FRONT MOTOR RELEASE FUNCTION AND RELEASE THE MOTOR WHILE STILL OBSERVING THE GEAR SET.
- SELECT LEFT FRONT MOTOR APPLY FUNCTION AND APPLY THE MOTOR.

DID GEAR SET MOVE IN BOTH DIRECTIONS FOR AT LEAST ONE REVOLUTION?
- YES: go to step 10.
- NO: REPLACE MOTOR PACK.

(10)
- USING TECH 1, SELECT "F4: MISC. TESTS" AND PERFORM GEAR TENSION RELIEF FUNCTION.
- IGNITION "OFF."
- SEPARATE MOTOR PACK FROM HYDRAULIC MODULATOR.
- GRASP THE LEFT FRONT GEAR ON THE HYDRAULIC MODULATOR; THE HYDRAULIC MODULATOR IS THE UNIT WITH THE THREE LARGE GEARS. THE LEFT FRONT GEAR IS FURTHEST FROM THE BRAKE BOOSTER WHEN THE ASSEMBLY IS INSTALLED IN THE VEHICLE.
- ATTEMPT TO ROTATE THE GEAR.

CAN THE GEAR BE ROTATED MORE THAN 9 TURNS LOCK TO LOCK?
- YES: REPLACE HYDRAULIC MODULATOR.
- NO: go to step 11.

(11)
- INSPECT 6-WAY MOTOR PACK AND EBCM CONNECTORS FOR CORROSION OR POOR TERMINAL CONTACT. REPLACE TERMINALS THAT EXHIBIT POOR TERMINAL CONTACT OR CORROSION.
- INSTALL HYDRAULIC MODULATOR/MASTER CYLINDER ASSEMBLY.
- RECONNECT ALL CONNECTORS.
- DRIVE VEHICLE FOR THREE DRIVE CYCLES (A DRIVE CYCLE CONSISTS OF STARTING ENGINE, DRIVING OVER 16 km/h (10 MPH) AND THEN TURNING IGNITION "OFF").

DID DTC 47 SET IN THE LAST THREE DRIVE CYCLES?
- YES: REPLACE EBCM.
- NO: MALFUNCTION WAS DUE TO A POOR CONNECTION.

DIAGNOSTIC CHART DTC 48 — 1994–95 ACHIEVA, GRAND AM AND SKYLARK

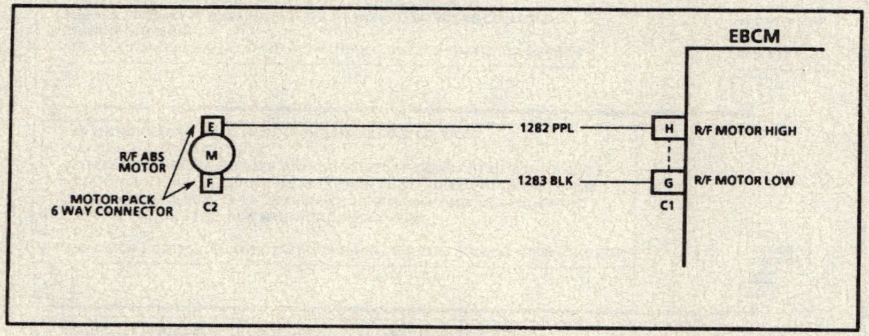

DTC 48

RIGHT FRONT ABS MOTOR FREE SPINS

Circuit Description:
This DTC is designed to detect a stripped nut or gear assembly during initialization. During the homing sequence, the piston should reach the top of the bore resulting in a stalled motor. If this does not occur, the motor must be spinning with little or no resistance, indicating a nut/screw or gear malfunction.

DTC Will Set When: DTC 48 can only be set during initialization. If the feedback current is less than the command current for a specified period of time, a malfunction exists.

Action Taken: A malfunction DTC is stored, ABS/VES is disabled, and the ABS warning lamp is turned "ON."

DTC Chart Test Description: Number(s) below refer to circled number(s) on the diagnostic chart.

1. This step compares feedback current to command current during apply.
2. This step compares feedback current to command current during release.
3. Steps 1, 2 and 3 verify malfunction exists. This step also moves the piston to the top of the bore.
4. Checks for excessive resistance in the right front motor high circuit.
5. Checks for excessive resistance in the right front motor low circuit.
6. Checks for excessive resistance in the right front motor.
7. Checks for a stripped motor pack gear.
8. Checks for a stripped hydraulic modulator gear.
9. Checks for proper operation of motor pack.
10. Checks for proper operation of the hydraulic modulator.
11. Ensures malfunction was not due to poor terminal contact.

Diagnostic Aids: An "Intermittent" malfunction in this DTC may result from a mechanical part of the system that sticks, binds, or slips.

The frequency of the malfunction can be checked by using the enhanced diagnostic function of the Tech 1, as described in "Enhanced Diagnostics," found in this section. If DTC only occurs once and DTC 61 also occurs, see DTC 61. If intermittent and enhanced diagnostics shows this DTC occurs during ABS, see DTC 61.

Depending on the frequency of the malfunction, a physical inspection of the mechanical parts suspected may be necessary.

DIAGNOSTIC CHART DTC 48 — 1994–95 ACHIEVA, GRAND AM AND SKYLARK, CONTINUED

AFTER DIAGNOSIS IS COMPLETE, CLEAR DTCs AND TEST DRIVE VEHICLE FOR THREE(3) DRIVE CYCLES TO VERIFY THAT DTC DOES NOT RESET. A DRIVE CYCLE CONSISTS OF STARTING THE ENGINE, DRIVING THE VEHICLE ABOVE 16 km/h (10 MPH) AND THEN KEYING DOWN.

DTC 48
(Page 1 of 2)
RIGHT FRONT ABS MOTOR FREE SPINS

NOTICE: J 39200 TEST LEADS MUST BE "ZEROED" PRIOR TO MAKING ANY RESISTANCE MEASUREMENTS. REFER TO J 39200 USER'S MANUAL.

IS DTC 61 ALSO PRESENT?
- YES: PROCEED TO DTC 61 CHART FIRST.
- NO: go to step 1.

(1)
- IGNITION "ON," ENGINE "OFF."
- PUMP BRAKE PEDAL TO DEPLETE VACUUM POWER BOOSTER.
- USING TECH 1 MISC. TESTS, SELECT MANUAL CONTROL FUNCTION.
- SELECT RIGHT FRONT MOTOR APPLY FUNCTION AND APPLY THE MOTOR WHILE CAREFULLY OBSERVING THE COMMAND CURRENT AND FEEDBACK CURRENT ON THE TECH 1. WERE BOTH COMMAND AND FEEDBACK CURRENT 10 AMPS?

YES: go to step 2. NO: GO TO STEP 4.

(2)
- APPLY FIRM PRESSURE ON BRAKE PEDAL.
- SELECT RIGHT FRONT MOTOR RELEASE FUNCTION AND RELEASE MOTOR. DID THE BRAKE PEDAL FALL AND WERE BOTH COMMAND AND FEEDBACK CURRENT 6 AMPS?

YES: go to step 3. NO: GO TO STEP 4.

(3)
- FIRM PRESSURE STILL ON BRAKE PEDAL.
- SELECT RIGHT FRONT MOTOR APPLY FUNCTION AND APPLY THE MOTOR (THIS MOVES THE HYDRAULIC MODULATOR PISTON TO THE TOP OF ITS BORE). DID BRAKE PEDAL RISE?

YES: MALFUNCTION IS INTERMITTENT. REFER TO "DIAGNOSTIC AIDS" ON FACING PAGE.
NO: go to step 4.

(4)
- IGNITION "OFF."
- DISCONNECT 6-WAY EBCM AND MOTOR PACK HARNESS CONNECTORS.
- USING J 39200, MEASURE RESISTANCE BETWEEN 6-WAY EBCM HARNESS CONNECTOR TERMINAL "H" AND 6-WAY MOTOR PACK HARNESS CONNECTOR TERMINAL "E". IS RESISTANCE LESS THAN 1.5 OHMS?

YES: go to step 5. NO: REPAIR HIGH RESISTANCE IN CKT 1282.

(5)
- USING J 39200, MEASURE RESISTANCE BETWEEN 6-WAY EBCM HARNESS CONNECTOR TERMINAL "G" AND 6-WAY MOTOR PACK HARNESS CONNECTOR TERMINAL "F". IS RESISTANCE LESS THAN 1.5 OHMS?

YES: go to step 6. NO: REPAIR HIGH RESISTANCE IN CKT 1283.

(6)
- USING J 39200, MEASURE RESISTANCE BETWEEN 6-WAY MOTOR PACK CONNECTOR TERMINALS "E" AND "F". IS RESISTANCE LESS THAN 1.5 OHMS?

YES: GO TO PAGE 2 OF 2 OF THIS CHART.
NO: REPLACE MOTOR PACK.

DIAGNOSTIC CHART DTC 48 — 1994–95 ACHIEVA, GRAND AM AND SKYLARK, CONTINUED

DTC 48
(Page 2 of 2)
RIGHT FRONT ABS MOTOR FREE SPINS

CONTINUED FROM PAGE 1 OF 2 OF THIS CHART.

(7)
- REMOVE HYDRAULIC MODULATOR/MASTER CYLINDER ASSEMBLY. (REFER TO "ON-VEHICLE SERVICE" IN THIS SECTION.)
- REMOVE GEAR COVER.
- CHECK FOR STRIPPED GEAR ON MOTOR PACK; THE MOTOR PACK IS THE UNIT WITH THE THREE SMALL GEARS. THE RIGHT FRONT GEAR IS CLOSEST TO THE BRAKE BOOSTER WHEN THE ASSEMBLY IS INSTALLED IN THE VEHICLE.

IS THE MOTOR PACK GEAR STRIPPED?

YES: REPLACE MOTOR PACK.

NO: (8)
- CHECK FOR A STRIPPED GEAR ON THE HYDRAULIC MODULATOR; THE HYDRAULIC MODULATOR IS THE UNIT WITH THE THREE LARGE GEARS. THE RIGHT FRONT GEAR IS CLOSEST TO THE BRAKE BOOSTER WHEN THE ASSEMBLY IS INSTALLED IN THE VEHICLE.

IS THE HYDRAULIC MODULATOR GEAR STRIPPED?

YES: REPLACE HYDRAULIC MODULATOR GEAR.

NO: (9)
- SECURELY POSITION HYDRAULIC MODULATOR/MASTER CYLINDER ASSEMBLY IN VEHICLE ALLOWING RIGHT FRONT GEAR SET OBSERVATION. BE CAREFUL NOT TO DAMAGE GEARS OR ALLOW THE HYDRAULIC MODULATOR/MASTER CYLINDER ASSEMBLY TO MOVE DURING TESTING.
- RECONNECT ELECTRICAL CONNECTORS.
- IGNITION "ON."
- USING TECH 1 MISC. TESTS, SELECT MANUAL CONTROL FUNCTION.
- SELECT RIGHT FRONT MOTOR APPLY AND APPLY THE MOTOR WHILE OBSERVING THE GEAR SET CLOSEST TO THE BOOSTER WHEN THE ASSEMBLY IS INSTALLED IN THE VEHICLE.
- SELECT RIGHT FRONT MOTOR RELEASE FUNCTION AND RELEASE THE MOTOR WHILE STILL OBSERVING THE GEAR SET.
- SELECT RIGHT FRONT MOTOR APPLY FUNCTION AND APPLY THE MOTOR.

DID GEAR SET MOVE IN BOTH DIRECTIONS FOR AT LEAST ONE REVOLUTION?

NO: REPLACE MOTOR PACK.

YES: (10)
- USING TECH 1, SELECT "F4: MISC. TESTS" AND PERFORM GEAR TENSION RELIEF FUNCTION.
- IGNITION "OFF."
- SEPARATE MOTOR PACK FROM HYDRAULIC MODULATOR.
- GRASP THE RIGHT FRONT GEAR ON THE HYDRAULIC MODULATOR; THE HYDRAULIC MODULATOR IS THE UNIT WITH THE THREE LARGE GEARS. THE RIGHT FRONT GEAR IS CLOSEST TO THE BRAKE BOOSTER WHEN THE ASSEMBLY IS INSTALLED IN THE VEHICLE.
- ATTEMPT TO ROTATE THE GEAR.

CAN THE GEAR BE ROTATED MORE THAN 9 TURNS LOCK TO LOCK?

YES: REPLACE HYDRAULIC MODULATOR.

NO: (11)
- INSPECT 6-WAY MOTOR PACK AND EBCM CONNECTORS FOR CORROSION OR POOR TERMINAL CONTACT.
  REPLACE TERMINALS THAT EXHIBIT POOR TERMINAL CONTACT OR CORROSION.
- INSTALL HYDRAULIC MODULATOR/MASTER CYLINDER ASSEMBLY.
- RECONNECT ALL CONNECTORS.
- DRIVE VEHICLE FOR THREE DRIVE CYCLES (A DRIVE CYCLE CONSISTS OF STARTING ENGINE, DRIVING OVER 16 km/h (10 MPH) AND THEN TURNING IGNITION "OFF").

DID DTC 48 SET IN THE LAST THREE DRIVE CYCLES?

YES: REPLACE EBCM.

NO: MALFUNCTION WAS DUE TO A POOR CONNECTION.

DIAGNOSTIC CHART DTC 51 — 1994–95 ACHIEVA, GRAND AM AND SKYLARK

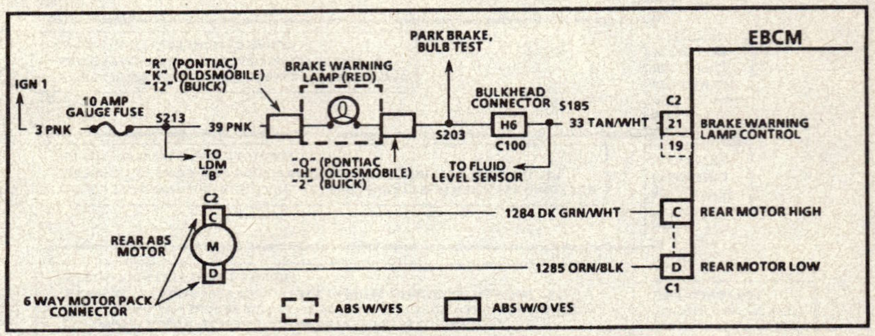

DTC 51

REAR ABS MOTOR FREE SPINS

Circuit Description:
This DTC is designed to detect a stripped nut or gear assembly during initialization. During the homing sequence, the piston should reach the top of the bore resulting in a stalled motor. If this does not occur, the motor must be spinning with little or no resistance, indicating a nut/screw or gear malfunction.

DTC Will Set When: DTC 51 can be set during initialization. If the feedback current is less than the command current for a specified period of time, a malfunction exists.

Action Taken: A malfunction DTC is stored, ABS/VES is disabled, and the ABS warning lamp is turned "ON."

DTC Chart Test Description: Number(s) below refer to circled number(s) on the diagnostic chart.

1. This step compares feedback current to command current during apply.
2. This step compares feedback current to command current during release.
3. Steps 1, 2 and 3 verify malfunction exists. This step also moves the piston to the top of the bore.
4. Checks for excessive resistance in the rear motor high circuit.
5. Checks for excessive resistance in the rear motor low circuit.
6. Checks for excessive resistance in the rear motor.
7. Checks for a stripped motor pack gear.
8. Checks for a stripped hydraulic modulator gear.
9. Checks for proper operation of motor pack.
10. Checks for proper operation of the hydraulic modulator.
11. Ensures malfunction was not due to poor terminal contact.

Diagnostic Aids: An "Intermittent" malfunction in this DTC may result from a mechanical part of the system that sticks, binds, or slips.

The frequency of the malfunction can be checked by using the enhanced diagnostic function of the Tech 1, as described in "Enhanced Diagnostics," found in this section. If DTC only occurs once and DTC 64 also occurs, see DTC 64. If intermittent and enhanced diagnostics show this DTC occurs during ABS, see DTC 64.

Depending on the frequency of the malfunction, a physical inspection of the mechanical parts suspected may be necessary.

DIAGNOSTIC CHART DTC 51 — 1994–95 ACHIEVA, GRAND AM AND SKYLARK, CONTINUED

DTC 51
(Page 1 of 2)
REAR ABS MOTOR FREE SPINS

AFTER DIAGNOSIS IS COMPLETE, CLEAR DTCs AND TEST DRIVE VEHICLE FOR THREE(3) DRIVE CYCLES TO VERIFY THAT DTC DOES NOT RESET. A DRIVE CYCLE CONSISTS OF STARTING THE ENGINE, DRIVING THE VEHICLE ABOVE 16 km/h (10 MPH) AND THEN KEYING DOWN.

NOTICE: J 39200 TEST LEADS MUST BE "ZEROED" PRIOR TO MAKING ANY RESISTANCE MEASUREMENTS. REFER TO J 39200 USER'S MANUAL.

IS DTC 64 ALSO PRESENT?
- YES: PROCEED TO DTC 64 CHART FIRST.
- NO: go to step 1.

1.
- IGNITION "ON," ENGINE "OFF."
- PUMP BRAKE PEDAL TO DEPLETE VACUUM POWER BOOSTER.
- USING TECH 1 MISC. TESTS, SELECT MANUAL CONTROL FUNCTION.
- SELECT REAR MOTOR APPLY FUNCTION AND APPLY THE MOTOR WHILE CAREFULLY OBSERVING THE COMMAND CURRENT AND FEEDBACK CURRENT ON THE TECH 1.

WERE BOTH COMMAND AND FEEDBACK CURRENT 10 AMPS?
- YES: go to step 2.
- NO: GO TO STEP 4.

2.
- APPLY FIRM PRESSURE ON BRAKE PEDAL.
- SELECT REAR MOTOR RELEASE FUNCTION AND RELEASE MOTOR.

DID THE BRAKE PEDAL FALL AND WERE BOTH COMMAND AND FEEDBACK CURRENT 6 AMPS?
- YES: go to step 3.
- NO: GO TO STEP 4.

3.
- FIRM PRESSURE STILL ON BRAKE PEDAL.
- SELECT REAR MOTOR APPLY FUNCTION AND APPLY THE MOTOR (THIS MOVES THE HYDRAULIC MODULATOR PISTON TO THE TOP OF ITS BORE).

DID BRAKE PEDAL RISE?
- YES: MALFUNCTION IS INTERMITTENT. REFER TO "DIAGNOSTIC AIDS" ON FACING PAGE.
- NO: go to step 4.

4.
- IGNITION "OFF."
- DISCONNECT 6-WAY EBCM AND MOTOR PACK HARNESS CONNECTORS.
- USING J 39200, MEASURE RESISTANCE BETWEEN 6-WAY EBCM HARNESS CONNECTOR TERMINAL "C" AND 6-WAY MOTOR PACK HARNESS CONNECTOR TERMINAL "C".

IS RESISTANCE LESS THAN 1.5 OHMS?
- YES: go to step 5.
- NO: REPAIR HIGH RESISTANCE IN CKT 1284.

5.
- USING J 39200, MEASURE RESISTANCE BETWEEN 6-WAY EBCM HARNESS CONNECTOR TERMINAL "D" AND 6-WAY MOTOR PACK HARNESS CONNECTOR TERMINAL "D".

IS RESISTANCE LESS THAN 1.5 OHMS?
- YES: go to step 6.
- NO: REPAIR HIGH RESISTANCE IN CKT 1285.

6.
- USING J 39200, MEASURE RESISTANCE BETWEEN 6-WAY MOTOR PACK CONNECTOR TERMINALS "C" AND "D".

IS RESISTANCE LESS THAN 1.5 OHMS?
- YES: GO TO PAGE 2 OF 2 OF THIS CHART.
- NO: REPLACE MOTOR PACK.

DIAGNOSTIC CHART DTC 51 — 1994–95 ACHIEVA, GRAND AM AND SKYLARK, CONTINUED

DTC 51
(Page 2 of 2)
REAR ABS MOTOR FREE SPINS

CONTINUED FROM PAGE 1 OF 2 OF THIS CHART.

7.
- REMOVE HYDRAULIC MODULATOR/MASTER CYLINDER ASSEMBLY. (REFER TO "ON-VEHICLE SERVICE" IN THIS SECTION.)
- REMOVE GEAR COVER.
- CHECK FOR STRIPPED GEAR ON MOTOR PACK; THE MOTOR PACK IS THE UNIT WITH THE THREE SMALL GEARS. THE REAR CHANNEL GEAR IS THE MIDDLE GEAR OF THE THREE.

IS THE MOTOR PACK GEAR STRIPPED?
- YES: REPLACE MOTOR PACK.
- NO: go to step 8.

8.
- CHECK FOR A STRIPPED GEAR ON THE HYDRAULIC MODULATOR; THE HYDRAULIC MODULATOR IS THE UNIT WITH THE THREE LARGE GEARS. THE REAR CHANNEL GEAR IS THE MIDDLE GEAR OF THE THREE.

IS THE HYDRAULIC MODULATOR GEAR STRIPPED?
- YES: REPLACE HYDRAULIC MODULATOR GEAR.
- NO: go to step 9.

9.
- SECURELY POSITION HYDRAULIC MODULATOR/MASTER CYLINDER ASSEMBLY IN VEHICLE ALLOWING REAR CHANNEL GEAR SET OBSERVATION. BE CAREFUL NOT TO DAMAGE GEARS OR ALLOW THE HYDRAULIC MODULATOR/MASTER CYLINDER ASSEMBLY TO MOVE DURING TESTING.
- RECONNECT ELECTRICAL CONNECTORS.
- IGNITION "ON."
- USING TECH 1 MISC. TESTS, SELECT MANUAL CONTROL FUNCTION.
- SELECT REAR MOTOR APPLY AND APPLY THE MOTOR WHILE OBSERVING THE MIDDLE GEAR SET.
- SELECT REAR MOTOR RELEASE FUNCTION AND RELEASE THE MOTOR WHILE STILL OBSERVING THE GEAR SET.
- SELECT REAR MOTOR APPLY FUNCTION AND APPLY THE MOTOR.

DID GEAR SET MOVE IN BOTH DIRECTIONS FOR AT LEAST ONE REVOLUTION?
- YES: go to step 10.
- NO: REPLACE MOTOR PACK.

10.
- USING TECH 1, SELECT "F4: MISC. TESTS" AND PERFORM GEAR TENSION RELIEF FUNCTION.
- IGNITION "OFF."
- SEPARATE MOTOR PACK FROM HYDRAULIC MODULATOR.
- GRASP THE REAR CHANNEL GEAR ON THE HYDRAULIC MODULATOR; THE HYDRAULIC MODULATOR IS THE UNIT WITH THE THREE LARGE GEARS. THE REAR CHANNEL GEAR IS THE MIDDLE GEAR OF THE THREE.
- ATTEMPT TO ROTATE THE GEAR.

CAN THE GEAR BE ROTATED MORE THAN 4 TURNS LOCK TO LOCK?
- YES: REPLACE HYDRAULIC MODULATOR.
- NO: go to step 11.

11.
- INSPECT 6-WAY MOTOR PACK AND EBCM CONNECTORS FOR CORROSION OR POOR TERMINAL CONTACT.

REPLACE TERMINALS THAT EXHIBIT POOR TERMINAL CONTACT OR CORROSION.
- INSTALL HYDRAULIC MODULATOR/MASTER CYLINDER ASSEMBLY.
- RECONNECT ALL CONNECTORS.
- DRIVE VEHICLE FOR THREE DRIVE CYCLES (A DRIVE CYCLE CONSISTS OF STARTING ENGINE, DRIVING OVER 16 km/h (10 MPH) AND THEN TURNING IGNITION "OFF").

DID DTC 51 SET IN THE LAST THREE DRIVE CYCLES?
- YES: REPLACE EBCM.
- NO: MALFUNCTION WAS DUE TO A POOR CONNECTION.

DIAGNOSTIC CHART DTC 52 — 1994–95 ACHIEVA, GRAND AM AND SKYLARK

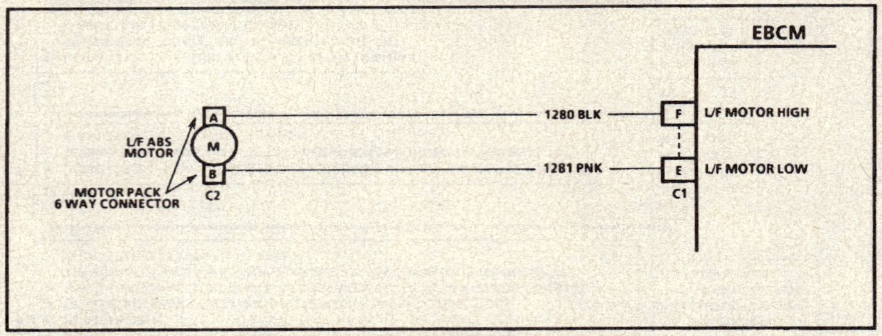

DTC 52

LEFT FRONT ABS CHANNEL IN RELEASE TOO LONG

Circuit Description:

This DTC will diagnose a motor that is energized longer than expected. This could occur if a wheel speed sensor is faulty, the motor does not turn, the left front solenoid mechanically fails open, the brakes are dragging, or the motor wires are crossed.

DTC Will Set When: DTC 52 can be set only during an ABS stop. If the EBCM commands the left front channel in release for three seconds, a malfunction exists.

Action Taken: A malfunction DTC is stored, ABS/VES is disabled, and the ABS warning lamp is turned "ON."

DTC Chart Test Description: Number(s) below refer to circled number(s) on the diagnostic chart.

1. This step identifies a malfunction in a wheel speed sensor that may cause the system to be in release too long.
2. This step identifies a motor or modulator malfunction.
3. This step checks for a solenoid that may have mechanically failed open.
4. Checks for a solenoid that may have failed closed.
5. Isolates malfunction to either the solenoid or the hydraulic modulator.
6. This step checks for proper operation of hydraulic modulator.

Diagnostic Aids: An "Intermittent" malfunction is most likely caused by a mechanical part of the system that sticks or binds.

The frequency of the malfunction can be checked by using the enhanced diagnostic function of the "Tech 1, as described in "Enhanced Diagnostics," found in this section. DTC 52 may occur if on ice and steering wheel is turned to lock during braking. Using Tech 1 perform hydraulic test to ensure total brake system is functional.

Any circuitry, that is suspected as causing the intermittent complaint, should be thoroughly checked for backed out terminals, improper mating, broken locks, improperly formed or damaged terminals, poor terminal to wiring connections or physical damage to the wiring harness.

DIAGNOSTIC CHART DTC 52 — 1994–95 ACHIEVA, GRAND AM AND SKYLARK, CONTINUED

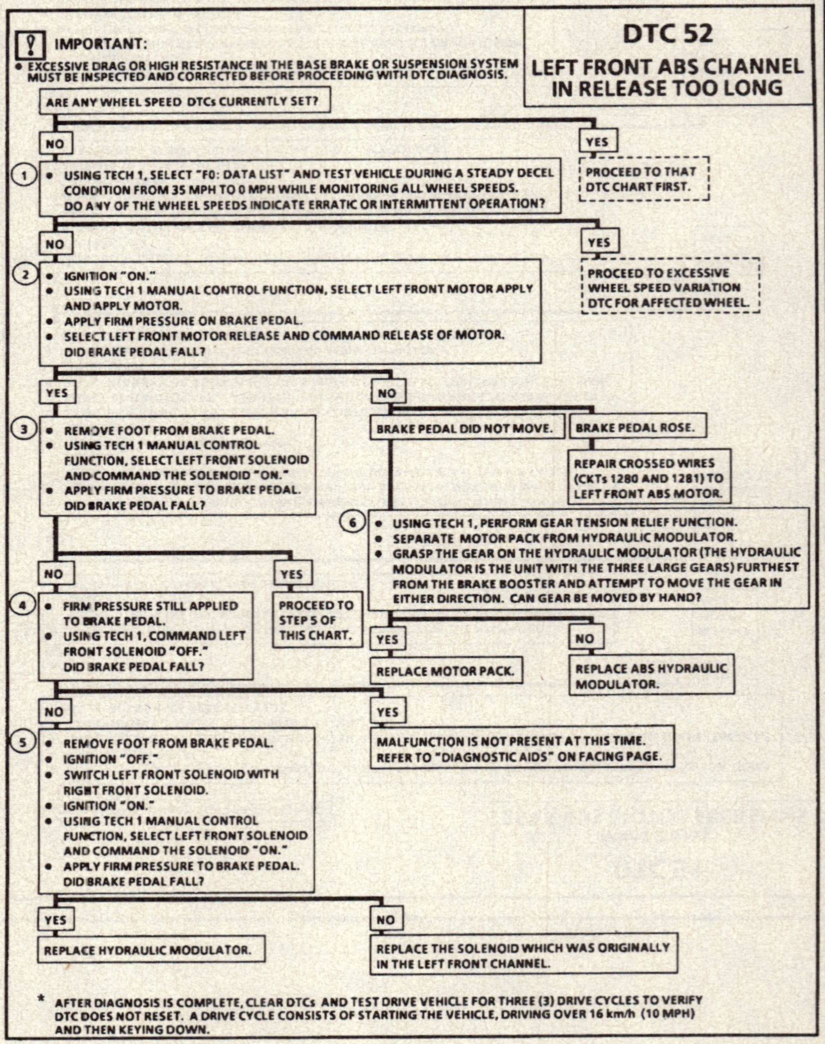

DIAGNOSTIC CHART DTC 53 — 1994–95 ACHIEVA, GRAND AM AND SKYLARK

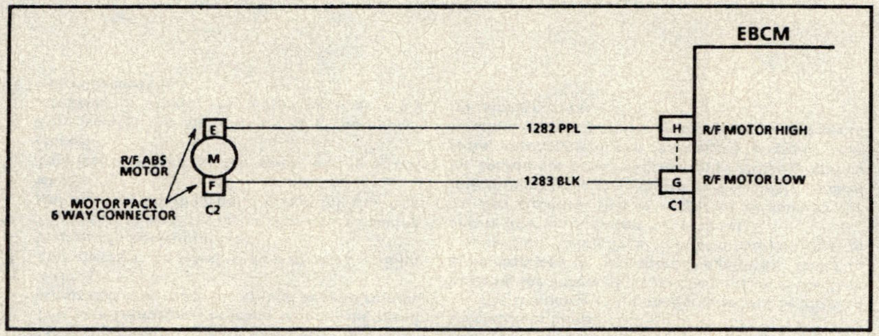

DTC 53

RIGHT FRONT ABS CHANNEL IN RELEASE TOO LONG

Circuit Description:

This DTC will diagnose a motor that is energized longer than expected. This could occur if a wheel speed sensor is faulty, the motor does not turn, the right front solenoid mechanically fails open, the brakes are dragging, or the motor wires are crossed.

DTC Will Set When: DTC 53 can be set only during an ABS stop. If the EBCM commands the right front channel in release for three seconds, a malfunction exists.

Action Taken: A malfunction DTC is stored, ABS/VES is disabled, and the ABS warning lamp is turned "ON."

DTC Chart Test Description: Number(s) below refer to circled number(s) on the diagnostic chart.

1. This step identifies a malfunction in a wheel speed sensor that may cause the system to be in release too long.
2. This step identifies a motor or modulator malfunction.
3. This step checks for a solenoid that may have mechanically failed open.
4. Checks for a solenoid that may have failed closed.
5. Isolates malfunction to either the solenoid or the hydraulic modulator.
6. This step checks for proper operation of hydraulic modulator.

Diagnostic Aids: An "Intermittent" malfunction is most likely caused by a mechanical part of the system that sticks or binds.

The frequency of the malfunction can be checked by using the enhanced diagnostic function of the Tech 1, as described in "Enhanced Diagnostics," found in this section. DTC 53 may occur on ice if steering wheel is turned to lock during braking. Using Tech 1 perform hydraulic test to ensure total brake system is functional.

Any circuitry, that is suspected as causing the intermittent complaint, should be thoroughly checked for backed out terminals, improper mating, broken locks, improperly formed or damaged terminals, poor terminal to wiring connections or physical damage to the wiring harness.

DIAGNOSTIC CHART DTC 53 — 1994–95 ACHIEVA, GRAND AM AND SKYLARK, CONTINUED

DTC 53

RIGHT FRONT ABS CHANNEL IN RELEASE TOO LONG

Important

- EXCESSIVE DRAG OR HIGH RESISTANCE IN THE BASE BRAKE OR SUSPENSION SYSTEM MUST BE INSPECTED AND CORRECTED BEFORE PROCEEDING WITH DTC DIAGNOSIS.

ARE ANY WHEEL SPEED DTCs CURRENTLY SET?

- YES → PROCEED TO THAT DTC CHART FIRST.
- NO → (1)

(1)
- USING TECH 1, SELECT "F0: DATA LIST" AND TEST VEHICLE DURING A STEADY DECEL CONDITION FROM 35 MPH TO 0 MPH WHILE MONITORING ALL WHEEL SPEEDS.

DO ANY OF THE WHEEL SPEEDS INDICATE ERRATIC OR INTERMITTENT OPERATION?

- YES → PROCEED TO EXCESSIVE WHEEL SPEED VARIATION DTC FOR AFFECTED WHEEL.
- NO → (2)

(2)
- IGNITION "ON."
- USING TECH 1 MANUAL CONTROL FUNCTION, SELECT RIGHT FRONT MOTOR APPLY AND APPLY MOTOR.
- APPLY FIRM PRESSURE ON BRAKE PEDAL.
- SELECT RIGHT FRONT MOTOR RELEASE AND COMMAND RELEASE OF MOTOR.

DID BRAKE PEDAL FALL?

- YES → (3)
- NO →
  - BRAKE PEDAL DID NOT MOVE. → (6)
  - BRAKE PEDAL ROSE → REPAIR CROSSED WIRES (CKTs 1282 AND 1283) TO RIGHT FRONT ABS MOTOR.

(3)
- REMOVE FOOT FROM BRAKE PEDAL.
- USING TECH 1 MANUAL CONTROL FUNCTION, SELECT RIGHT FRONT SOLENOID AND COMMAND THE SOLENOID "ON."
- APPLY FIRM PRESSURE TO BRAKE PEDAL.

DID BRAKE PEDAL FALL?

- YES → PROCEED TO STEP 5 OF THIS CHART.
- NO → (4)

(4)
- FIRM PRESSURE STILL APPLIED TO BRAKE PEDAL.
- USING TECH 1, COMMAND RIGHT FRONT SOLENOID "OFF."

DID BRAKE PEDAL FALL?

- YES → MALFUNCTION IS NOT PRESENT AT THIS TIME. REFER TO "DIAGNOSTIC AIDS" ON FACING PAGE.
- NO → (5)

(5)
- REMOVE FOOT FROM BRAKE PEDAL.
- IGNITION "OFF."
- SWITCH RIGHT FRONT SOLENOID WITH LEFT FRONT SOLENOID.
- IGNITION "ON."
- USING TECH 1 MANUAL CONTROL FUNCTION, SELECT RIGHT FRONT SOLENOID AND COMMAND THE SOLENOID "ON."
- APPLY FIRM PRESSURE TO BRAKE PEDAL.

DID BRAKE PEDAL FALL?

- YES → REPLACE HYDRAULIC MODULATOR.
- NO → REPLACE THE SOLENOID WHICH WAS ORIGINALLY IN THE RIGHT FRONT CHANNEL.

(6)
- USING TECH 1, PERFORM GEAR TENSION RELIEF FUNCTION.
- SEPARATE MOTOR PACK FROM HYDRAULIC MODULATOR. (THE HYDRAULIC MODULATOR IS THE UNIT WITH THE THREE LARGE GEARS.)
- GRASP THE GEAR ON THE HYDRAULIC MODULATOR CLOSEST TO THE BRAKE BOOSTER AND ATTEMPT TO MOVE THE GEAR IN EITHER DIRECTION.

CAN GEAR BE MOVED BY HAND?

- YES → REPLACE MOTOR PACK.
- NO → REPLACE ABS HYDRAULIC MODULATOR.

- AFTER DIAGNOSIS IS COMPLETE, CLEAR DTCs AND TEST DRIVE VEHICLE FOR THREE (3) DRIVE CYCLES TO VERIFY DTC DOES NOT RESET. A DRIVE CYCLE CONSISTS OF STARTING THE VEHICLE, DRIVING OVER 16km/h (10 MPH) AND THEN KEYING DOWN.

DIAGNOSTIC CHART DTC 54 — 1994–95 ACHIEVA, GRAND AM AND SKYLARK

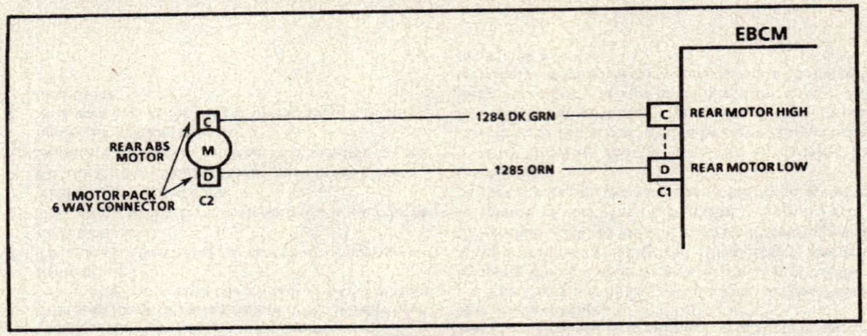

DTC 54

REAR ABS CHANNEL IN RELEASE TOO LONG

Circuit Description:
This DTC will diagnose a motor that is energized longer than expected. This could occur if a wheel speed sensor is faulty, the motor does not turn, the brakes are dragging, or the motor wires are crossed.

DTC Will Set When: DTC 54 can be set only during an ABS stop. If the EBCM commands the rear axle channel in release for three seconds, a malfunction exists.

Action Taken: A malfunction DTC is stored, ABS/VES is disabled, and the ABS warning lamp is turned "ON."

DTC Chart Test Description: Number(s) below refer to circled number(s) on the diagnostic chart.

1. This step identifies a problem in a wheel speed sensor that may cause the system to be in release too long.
2. This checks for a wheel that may stick or bind because of a mechanical fault.
3. This checks to see if the motor is capable of moving and applying the hydraulic piston for the rear wheels.
4. This step ensures that the motor wiring is not crossed.
5. This isolates the malfunction of a "no-apply" situation to either the motor pack or ABS hydraulic modulator.

Diagnostic Aids: An "Intermittent" malfunction is most likely caused by a mechanical part of the system that sticks or binds.

The frequency of the malfunction can be checked by using the enhanced diagnostic function of the Tech 1, as described in "Enhanced Diagnostics," found in this section. Using Tech 1 perform hydraulic test to ensure total brake system is functional.

Any circuitry, that is suspected as causing the intermittent complaint, should be thoroughly checked for backed out terminals, improper mating, broken locks, improperly formed or damaged terminals, poor terminal to wiring connections or physical damage to the wiring harness.

DIAGNOSTIC CHART DTC 54 — 1994–95 ACHIEVA, GRAND AM AND SKYLARK, CONTINUED

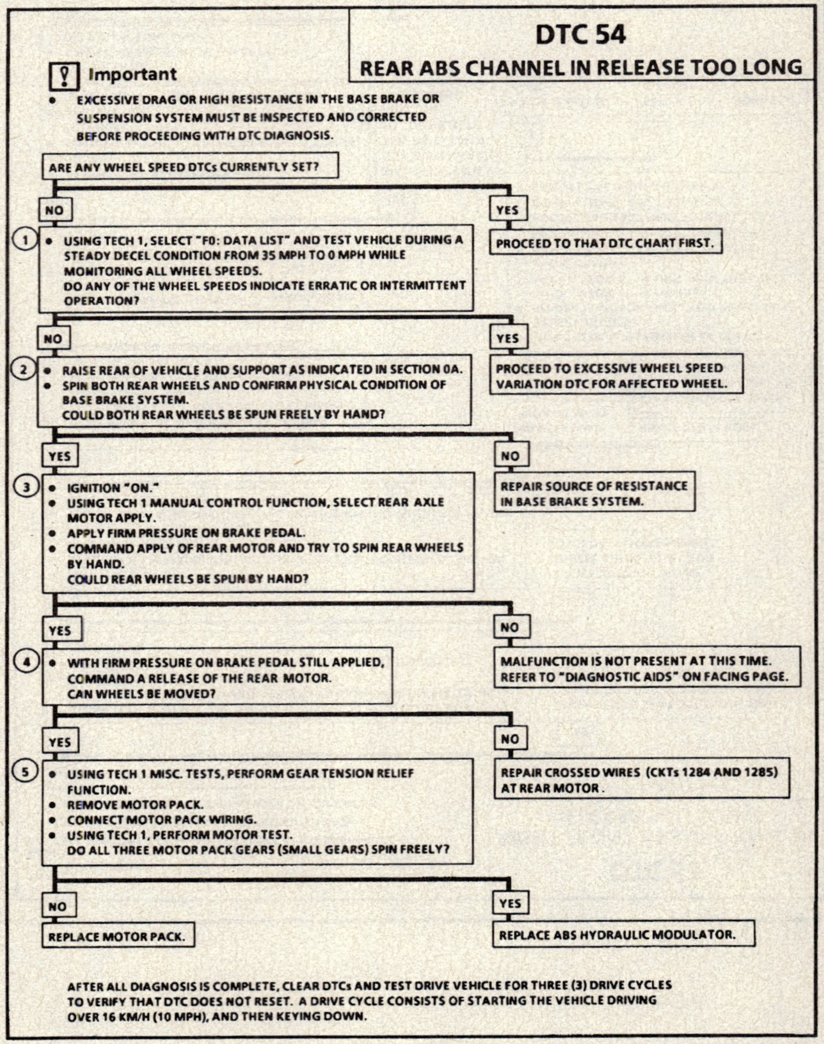

DIAGNOSTIC CHART DTC 55 — 1994–95 ACHIEVA, GRAND AM AND SKYLARK

D.M. A.B.S. VI
MODULE ASSEMBLY -
ELECTRONIC BRAKE CONTROL
(TYPICAL)

DTC 55
EBCM MALFUNCTION

Circuit Description:
This DTC identifies a malfunction detected by the MDI custom IC. It also ensures the cause of the malfunction is not a result of a malfunction with the enable relay under a DTC 55 failure.

DTC Will Set When: DTC 55 is set when an internal EBCM malfunction exists.

Action Taken: A malfunction DTC is stored, ABS/VES is disabled and the ABS warning lamp is turned "ON."

DTC Chart Test Description: Number(s) below refer to circled number(s) on the diagnostic chart.
1. Checks to see if there are any other malfunctions present.
2. This checks to see if the malfunction is still present.

Diagnostic Aids: An "Intermittent" malfunction is most likely caused by a poor connection, rubbed through wire insulation, or a wire that is broken inside the insulation.

The frequency of the malfunction can be checked by using the enhanced diagnostic function of the Tech 1, as described in "Enhanced Diagnostics," found in this section.

Any circuitry, that is suspected as causing the intermittent complaint, should be thoroughly checked for backed out terminals, improper mating, broken locks, improperly formed or damaged terminals, poor terminal to wiring connections or physical damage to the wiring harness.

DIAGNOSTIC CHART DTC 55 — 1994–95 ACHIEVA, GRAND AM AND SKYLARK, CONTINUED

DTC 55
EBCM MALFUNCTION

(1)
- IGNITION "ON," ENGINE "OFF."
- USING TECH 1, SCAN DTCs.

ARE THERE ANY OTHER DTCs SET?

- YES → PROCEED TO THAT DTC FIRST.
- NO →

(2)
- CLEAR DTCs.

DOES DTC 55 RE-SET?

- YES → REPLACE EBCM.
- NO → MALFUNCTION IS NOT PRESENT AT THIS TIME. REFER TO "DIAGNOSTIC AIDS" ON FACING PAGE.

* AFTER DIAGNOSIS IS COMPLETE, CLEAR DTCs AND TEST DRIVE VEHICLE FOR THREE (3) DRIVE CYCLES TO VERIFY DTC DOES NOT RESET. A DRIVE CYCLE CONSISTS OF STARTING THE VEHICLE, DRIVING OVER 16 km/h (10 MPH) AND THEN KEYING DOWN.

THIS CHART ASSUMES THAT A CURRENT DTC IS STORED INDICATING THAT THIS MALFUNCTION IS PRESENT.

DIAGNOSTIC CHART DTC 56 — 1994–95 ACHIEVA, GRAND AM AND SKYLARK

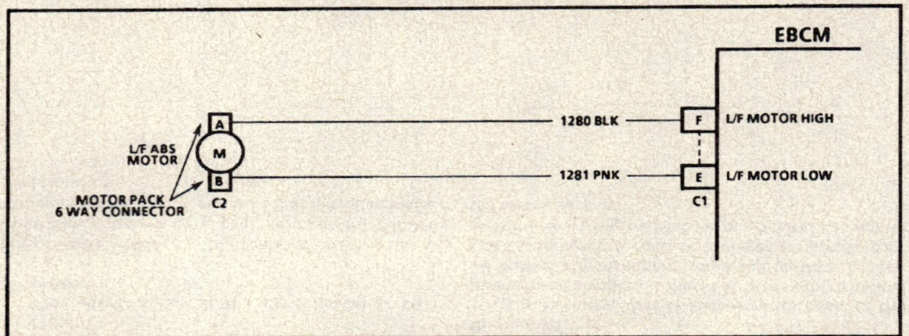

DTC 56

LEFT FRONT ABS MOTOR CIRCUIT OPEN

Circuit Description:
This DTC identifies a motor that cannot be energized due to an open in its circuitry. This malfunction will not allow proper front ABS operation.

DTC Will Set When: DTC 56 can be set only when the motor is commanded "OFF." If the EBCM detects an out of range voltage in either of the left front motor circuits indicating an open circuit, a malfunction exists.

Action Taken: An open motor will not activate when requested. A malfunction DTC is stored, ABS/VES is disabled and the ABS warning lamp is turned "ON."

DTC Chart Test Description: Number(s) below refer to circled number(s) on the diagnostic chart.
1. This step checks for proper resistance of the motor.
2. This step checks for an open in the motor "high" circuitry.
3. This step checks for an open in the motor "low" circuitry.
4. Ensures malfunction was not due to poor terminal contact.

Diagnostic Aids: Using Tech 1, select manual control function, and exercise motor movement of affected channel in both directions while applying light pressure on the brake pedal.

If erratic or "jumpy" brake pedal movement is detected while performing an "apply" or "release" function of the motor, an intermittent malfunction may be indicated.

An "Intermittent" malfunction is most likely caused by a poor connection, rubbed through wire insulation, or a wire that is broken inside the insulation.

If the malfunction is not current, wiggle the wires of the affected channel and check if the DTC resets. This will help to pinpoint an intermittent malfunction in the motor circuitry or connections.

The frequency of the malfunction can be checked by using the enhanced diagnostic function of the Tech 1, as described in "Enhanced Diagnostics," found in this section.

Any circuitry, that is suspected as causing the intermittent complaint, should be thoroughly checked for backed out terminals, improper mating, broken locks, improperly formed or damaged terminals, poor terminal to wiring connections or physical damage to the wiring harness.

DIAGNOSTIC CHART DTC 56 — 1994–95 ACHIEVA, GRAND AM AND SKYLARK, CONTINUED

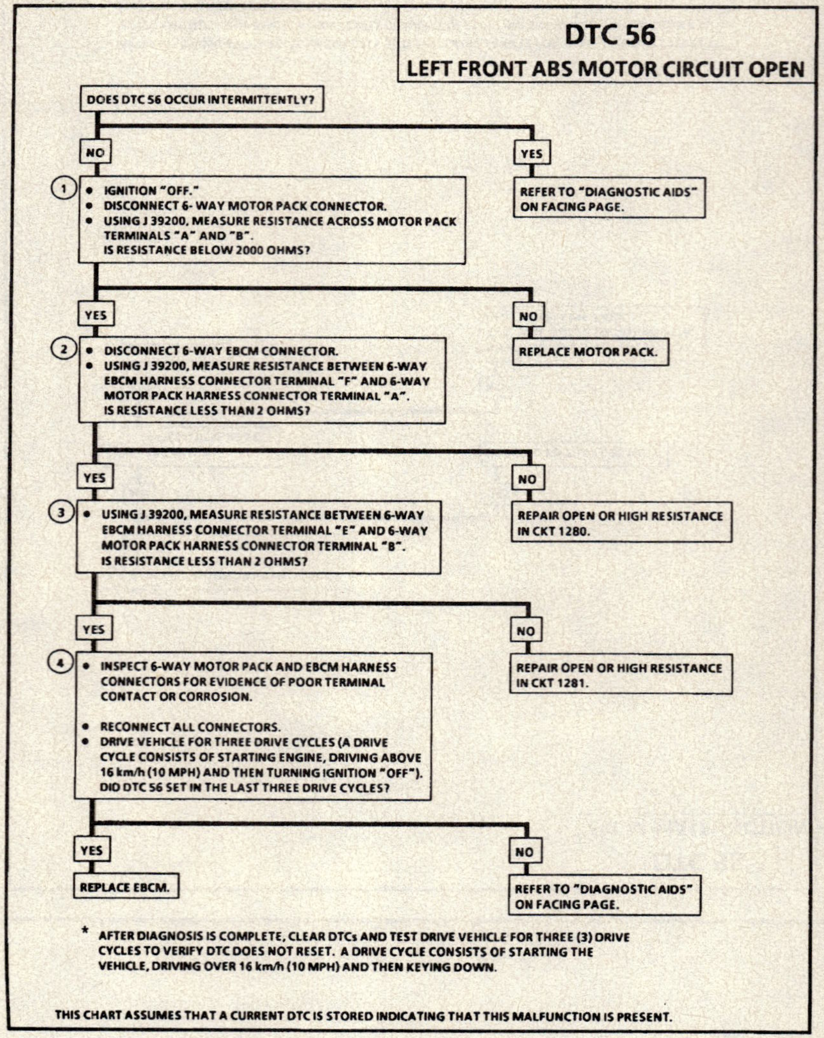

DIAGNOSTIC CHART DTC 57 — 1994–95 ACHIEVA, GRAND AM AND SKYLARK

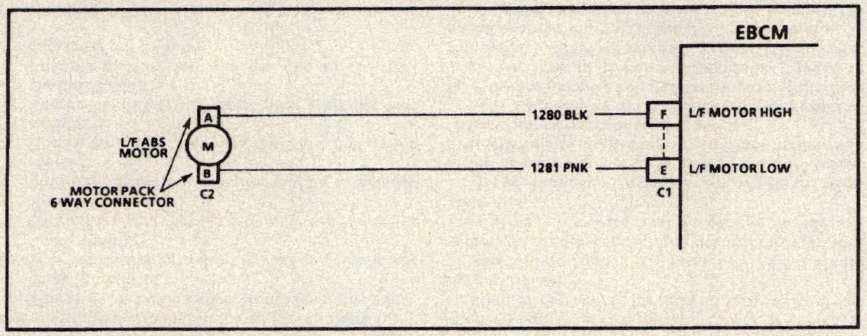

DTC 57

LEFT FRONT ABS MOTOR CIRCUIT SHORTED TO GROUND

Circuit Description:

This DTC identifies a motor circuit that is shorted to ground. This malfunction will not allow the motor to be controlled at the commanded current rate or will cause the driver circuit to allow current directly to ground.

DTC Will Set When: DTC 57 can be set anytime. If the EBCM detects an out of range voltage in either of the left front motor circuits indicating a circuit shorted to ground, a malfunction exists.

Action Taken: A malfunction DTC is stored, ABS/VES is disabled and the ABS warning lamp is turned "ON."

DTC Chart Test Description: Number(s) below refer to circled number(s) on the diagnostic chart.

1. This step checks for a short to ground in the motor "high" circuitry.
2. This step checks for a short to ground in the motor "low" circuitry.
3. This step checks for a motor that is internally shorted to ground.
4. Ensures malfunction was not due to physical damage of the circuitry.

Diagnostic Aids: Using Tech 1, select manual control function, and exercise motor movement of affected channel in both directions while applying light pressure on the brake pedal.

If erratic or "jumpy" brake pedal movement is detected while performing an "apply" or "release" function of the motor, an intermittent malfunction may be indicated.

An "Intermittent" malfunction is most likely caused by a poor connection, rubbed through wire insulation, or a wire that is broken inside the insulation.

If the malfunction is not current, wiggle the wires of the affected channel and check if the DTC resets. This will help to pinpoint an intermittent malfunction in the motor circuitry or connections.

The frequency of the malfunction can be checked by using the enhanced diagnostic function of the Tech 1, as described in "Enhanced Diagnostics," found in this section.

Any circuitry, that is suspected as causing the intermittent complaint, should be thoroughly checked for backed out terminals, improper mating, broken locks, improperly formed or damaged terminals, poor terminal to wiring connections or physical damage to the wiring harness.

DIAGNOSTIC CHART DTC 57 — 1994–95 ACHIEVA, GRAND AM AND SKYLARK, CONTINUED

DTC 57

LEFT FRONT ABS MOTOR CIRCUIT SHORTED TO GROUND

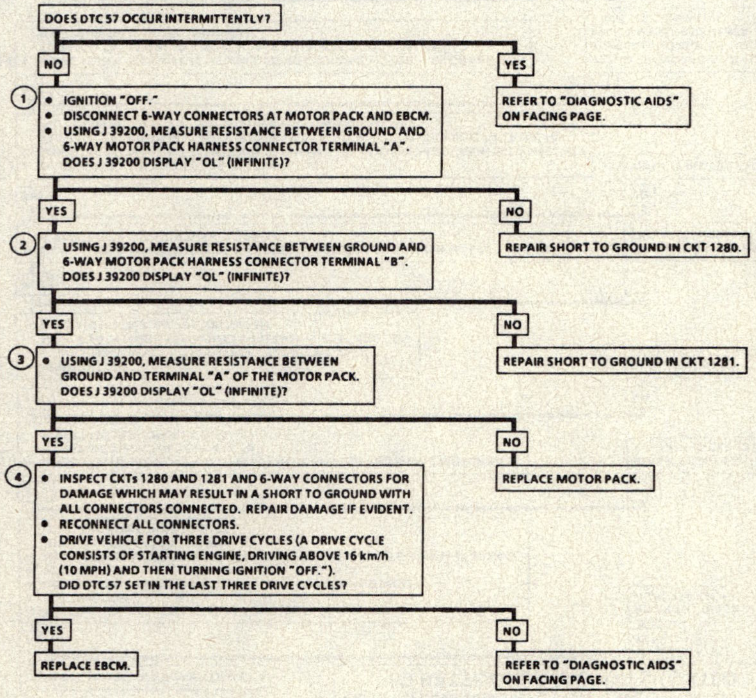

* AFTER DIAGNOSIS IS COMPLETE, CLEAR DTCs AND TEST DRIVE VEHICLE FOR THREE (3) DRIVE CYCLES TO VERIFY DTC DOES NOT RESET. A DRIVE CYCLE CONSISTS OF STARTING THE VEHICLE, DRIVING OVER 16 km/h (10 MPH) AND THEN KEYING DOWN.

THIS CHART ASSUMES THAT A CURRENT DTC IS STORED INDICATING THAT THIS MALFUNCTION IS PRESENT.

DIAGNOSTIC CHART DTC 58 — 1994–95 ACHIEVA, GRAND AM AND SKYLARK

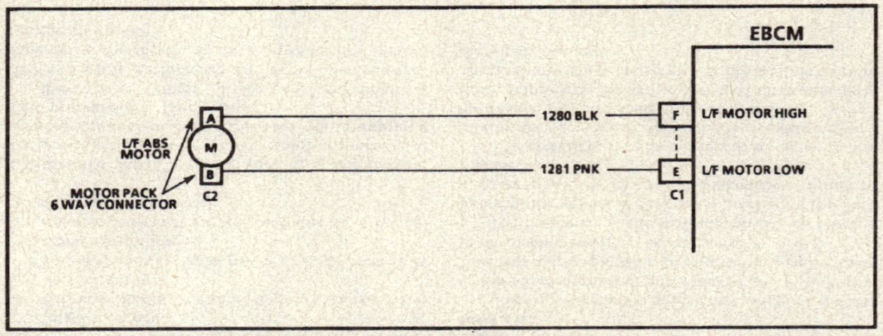

DTC 58

LEFT FRONT ABS MOTOR CIRCUIT SHORTED TO BATTERY OR MOTOR SHORTED

Circuit Description:
This DTC identifies a motor circuit that is shorted to battery or a motor that has low or no resistance. This malfunction will not allow the motor to be controlled at the commanded current rate or will cause the motor to turn in the opposite direction, or not at all.

DTC Will Set When: DTC 58 can be set anytime. If the EBCM detects an out of range voltage in either of the left front motor circuits indicating a circuit shorted to battery or a motor shorted, a malfunction exists.

Action Taken: A malfunction DTC is stored, ABS/VES is disabled and the ABS warning lamp is turned "ON."

DTC Chart Test Description: Number(s) below refer to circled number(s) on the diagnostic chart.

1. This step checks for a short to voltage in the motor "high" circuitry.
2. This step checks for a short to voltage in the motor "low" circuitry.
3. Checks for a short between the high and low motor circuits.
4. This step checks for a motor that is internally shorted.
5. Checks for a short between the EMB and motor circuits.
6. Checks for an internal motor pack short between the EMB and motor.
7. Ensures malfunction was not due to physical damage of the circuitry.

Diagnostic Aids: Using Tech 1, select manual control function, and exercise motor movement of affected channel in both directions while applying light pressure on the brake pedal.

If erratic or "jumpy" brake pedal movement is detected while performing an "apply" or "release" function of the motor, an intermittent malfunction may be indicated.

An "Intermittent" malfunction is most likely caused by a poor connection, rubbed through wire insulation, or a wire that is broken inside the insulation.

If the malfunction is not current, wiggle the wires of the affected channel and check if the DTC resets. This will help to pinpoint an intermittent malfunction in the motor circuitry or connections.

The frequency of the malfunction can be checked by using the enhanced diagnostic function of the Tech 1, as described in "Enhanced Diagnostics," found in this section. If DTC 38 sets as a current DTC and 58 is an infrequent history DTC, see DTC 38 diagnostics.

Any circuitry, that is suspected as causing the intermittent complaint, should be thoroughly checked for backed out terminals, improper mating, broken locks, improperly formed or damaged terminals, poor terminal to wiring connections or physical damage to the wiring harness.

DIAGNOSTIC CHART DTC 58 — 1994–95 ACHIEVA, GRAND AM AND SKYLARK, CONTINUED

DTC 58

LEFT FRONT ABS MOTOR CIRCUIT SHORTED TO BATTERY OR MOTOR SHORTED

NOTICE: J 39200 TEST LEADS MUST BE "ZEROED" PRIOR TO MAKING ANY RESISTANCE MEASUREMENTS. REFER TO J 39200 USER'S MANUAL.

DOES DTC 58 OCCUR INTERMITTENTLY?

- YES: REFER TO "DIAGNOSTIC AIDS" ON FACING PAGE.
- NO:

(1) NOTICE: THIS STEP MAY CAUSE ADDITIONAL DTCs TO SET.
- IGNITION "OFF."
- DISCONNECT 6-WAY CONNECTORS AT MOTOR PACK AND EBCM.
- IGNITION "ON."
- USING J 39200, MEASURE VOLTAGE BETWEEN GROUND AND 6-WAY MOTOR PACK HARNESS CONNECTOR TERMINAL "A".
IS VOLTAGE LESS THAN 1 VOLT?

- NO: REPAIR SHORT TO VOLTAGE IN CKT 1280.
- YES:

(2)
- USING J 39200, MEASURE VOLTAGE BETWEEN GROUND AND 6-WAY MOTOR PACK HARNESS CONNECTOR TERMINAL "B".
IS VOLTAGE LESS THAN 1 VOLT?

- NO: REPAIR SHORT TO VOLTAGE IN CKT 1281.
- YES:

(3)
- IGNITION "OFF."
- USING J 39200, MEASURE RESISTANCE BETWEEN 6-WAY MOTOR PACK HARNESS CONNECTOR TERMINALS "A" AND "B".
DOES J 39200 DISPLAY "OL" (INFINITE)?

- NO: REPAIR SHORT BETWEEN CKTs 1280 AND 1281.
- YES:

(4)
- USING J 39200, MEASURE RESISTANCE BETWEEN MOTOR PACK TERMINALS "A" AND "B".
IS RESISTANCE GREATER THAN .4 OHMS?

- NO: REPLACE MOTOR PACK.
- YES:

(5)
- DISCONNECT 3-WAY MOTOR PACK CONNECTOR.
- USING J 39200, MEASURE RESISTANCE BETWEEN 3-WAY MOTOR PACK HARNESS CONNECTOR TERMINAL "B" AND 6-WAY MOTOR PACK HARNESS CONNECTOR TERMINAL "A" THEN "B".
DID J 39200 DISPLAY "OL" (INFINITE) FOR EACH MEASUREMENT?

- NO: REPAIR SHORT BETWEEN CKT 855/1633 AND MOTOR CIRCUIT WHICH MEASURED OTHER THAN "OL" (INFINITE).
- YES:

(6)
- USING J 39200, MEASURE RESISTANCE BETWEEN 3-WAY MOTOR PACK CONNECTOR TERMINAL "B" AND 6-WAY MOTOR PACK CONNECTOR TERMINAL "A".
DID J 39200 DISPLAY "OL" (INFINITE)?

- NO: REPLACE MOTOR PACK.
- YES:

(7)
- INSPECT CKTs 1280 AND 1281 AND 6-WAY CONNECTORS FOR DAMAGE WHICH MAY RESULT IN A SHORT TO B+ WITH ALL CONNECTORS CONNECTED.
- REPAIR DAMAGE IF EVIDENT.
- RECONNECT ALL CONNECTORS.
- DRIVE VEHICLE FOR THREE DRIVE CYCLES (A DRIVE CYCLE CONSISTS OF STARTING ENGINE, DRIVE ABOVE 16 km/h (10 MPH) AND THEN TURNING IGNITION "OFF").
DID DTC 58 SET IN THE LAST THREE DRIVE CYCLES?

- YES: REPLACE EBCM.
- NO: REFER TO "DIAGNOSTIC AIDS" ON FACING PAGE.

* AFTER DIAGNOSIS IS COMPLETE, CLEAR DTCs AND TEST DRIVE VEHICLE FOR THREE (3) DRIVE CYCLES TO VERIFY DTC DOES NOT RESET. A DRIVE CYCLE CONSISTS OF STARTING THE VEHICLE, DRIVING OVER 16 km/h (10 MPH) AND THEN KEYING DOWN.

THIS CHART ASSUMES THAT A CURRENT DTC IS STORED INDICATING THAT THIS MALFUNCTION IS PRESENT.

DIAGNOSTIC CHART DTC 61 — 1994–95 ACHIEVA, GRAND AM AND SKYLARK

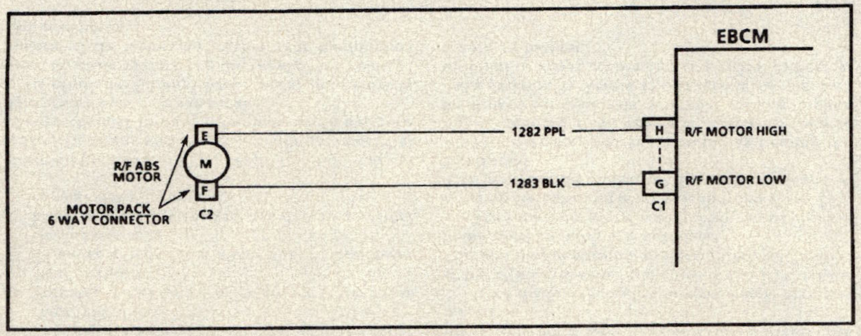

DTC 61

RIGHT FRONT ABS MOTOR CIRCUIT OPEN

Circuit Description:
This DTC identifies a motor that cannot be energized due to an open in its circuitry. This malfunction will not allow proper front ABS operation.

DTC Will Set When: DTC 61 can be set only when the motor is commanded "OFF." If the EBCM detects an out of range voltage in either of the right front motor circuits indicating an open circuit, a malfunction exists.

Action Taken: An open motor will not activate when requested. A malfunction DTC is stored, ABS/VES is disabled and the ABS warning lamp is turned "ON."

DTC Chart Test Description: Number(s) below refer to circled number(s) on the diagnostic chart.

1. This step checks for proper resistance of the motor.
2. This step checks for an open in the motor "high" circuitry.
3. This step checks for an open in the motor "low" circuitry.
4. Ensures malfunction was not due to poor terminal contact.

Diagnostic Aids: Using Tech 1, select manual control function, and exercise motor movement of affected channel in both directions while applying light pressure on the brake pedal.

If erratic or "jumpy" brake pedal movement is detected while performing an "apply" or "release" function of the motor, an intermittent malfunction may be indicated.

An "Intermittent" malfunction is most likely caused by a poor connection, rubbed through wire insulation, or a wire that is broken inside the insulation.

If the malfunction is not current, wiggle the wires of the affected channel and check if the DTC resets. This will help to pinpoint an intermittent malfunction in the motor circuitry or connections.

The frequency of the malfunction can be checked by using the enhanced diagnostic function of the Tech 1, as described in "Enhanced Diagnostics," found in this section.

Any circuitry, that is suspected as causing the intermittent complaint, should be thoroughly checked for backed out terminals, improper mating, broken locks, improperly formed or damaged terminals, poor terminal to wiring connections or physical damage to the wiring harness.

DIAGNOSTIC CHART DTC 61 — 1994–95 ACHIEVA, GRAND AM AND SKYLARK, CONTINUED

DTC 61

RIGHT FRONT ABS MOTOR CIRCUIT OPEN

DOES DTC 61 OCCUR INTERMITTENTLY?

- YES → REFER TO "DIAGNOSTIC AIDS" ON FACING PAGE.
- NO →

(1)
- IGNITION "OFF."
- DISCONNECT 6-WAY MOTOR PACK CONNECTOR.
- USING J 39200, MEASURE RESISTANCE ACROSS MOTOR PACK TERMINALS "E" AND "F".

IS RESISTANCE BELOW 2000 OHMS?

- NO → REPLACE MOTOR PACK.
- YES →

(2)
- DISCONNECT 6-WAY EBCM CONNECTOR.
- USING J 39200, MEASURE RESISTANCE BETWEEN 6-WAY EBCM HARNESS CONNECTOR TERMINAL "H" AND 6-WAY MOTOR PACK HARNESS CONNECTOR TERMINAL "E".

IS RESISTANCE LESS THAN 2 OHMS?

- NO → REPAIR OPEN OR HIGH RESISTANCE IN CKT 1282.
- YES →

(3)
- USING J 39200, MEASURE RESISTANCE BETWEEN 6-WAY EBCM HARNESS CONNECTOR TERMINAL "G" AND 6-WAY MOTOR PACK HARNESS CONNECTOR TERMINAL "F".

IS RESISTANCE LESS THAN 2 OHMS?

- NO → REPAIR OPEN OR HIGH RESISTANCE IN CKT 1283.
- YES →

(4)
- INSPECT 6-WAY MOTOR PACK AND EBCM HARNESS CONNECTORS FOR EVIDENCE OF POOR TERMINAL CONTACT OR CORROSION.
- RECONNECT ALL CONNECTORS.
- DRIVE VEHICLE FOR THREE DRIVE CYCLES (A DRIVE CYCLE CONSISTS OF STARTING ENGINE, DRIVING ABOVE 16 km/h (10 MPH) AND THEN TURNING IGNITION "OFF").

DID DTC 61 SET IN THE LAST THREE DRIVE CYCLES?

- YES → REPLACE EBCM.
- NO → REFER TO "DIAGNOSTIC AIDS" ON FACING PAGE.

* AFTER DIAGNOSIS IS COMPLETE, CLEAR DTCs AND TEST DRIVE VEHICLE FOR THREE (3) DRIVE CYCLES TO VERIFY DTC DOES NOT RESET. A DRIVE CYCLE CONSISTS OF STARTING THE VEHICLE, DRIVING OVER 16 km/h (10 MPH) AND THEN KEYING DOWN.

THIS CHART ASSUMES THAT A CURRENT DTC IS STORED INDICATING THAT THIS MALFUNCTION IS PRESENT.

DIAGNOSTIC CHART DTC 62 — 1994–95 ACHIEVA, GRAND AM AND SKYLARK

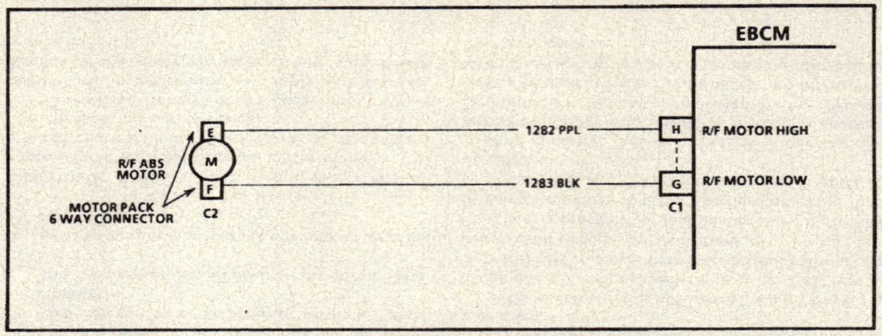

DTC 62

RIGHT FRONT ABS MOTOR CIRCUIT SHORTED TO GROUND

Circuit Description:
This DTC identifies a motor circuit that is shorted to ground. This malfunction will not allow the motor to be controlled at the commanded current rate or will cause the motor driver circuit to allow current directly to ground.

DTC Will Set When: DTC 62 can be set anytime. If the EBCM detects an out of range voltage in either of the right front motor circuits indicating a circuit shorted to ground, a malfunction exists.

Action Taken: A malfunction DTC is stored, ABS/VES is disabled and the ABS warning lamp is turned "ON."

DTC Chart Test Description: Number(s) below refer to circled number(s) on the diagnostic chart.

1. This step checks for a short to ground in the motor "high" circuitry.
2. This step checks for a short to ground in the motor "low" circuitry.
3. This step checks for a motor that is internally shorted to ground.
4. Ensures malfunction was not due to physical damage of the circuitry.

Diagnostic Aids: Using Tech 1, select manual control function, and exercise motor movement of affected channel in both directions while applying light pressure on the brake pedal.

If erratic or "jumpy" brake pedal movement is detected while performing an "apply" or "release" function of the motor, an intermittent malfunction may be indicated.

An "Intermittent" malfunction is most likely caused by a poor connection, rubbed through wire insulation, or a wire that is broken inside the insulation.

If the malfunction is not current, wiggle the wires of the affected channel and check if the DTC resets. This will help to pinpoint an intermittent malfunction in the motor circuitry or connections.

The frequency of the malfunction can be checked by using the enhanced diagnostic function of the Tech 1, as described in "Enhanced Diagnostics," found in this section.

Any circuitry, that is suspected as causing the intermittent complaint, should be thoroughly checked for backed out terminals, improper mating, broken locks, improperly formed or damaged terminals, poor terminal to wiring connections or physical damage to the wiring harness.

DIAGNOSTIC CHART DTC 62 — 1994–95 ACHIEVA, GRAND AM AND SKYLARK, CONTINUED

DTC 62

RIGHT FRONT ABS MOTOR CIRCUIT SHORTED TO GROUND

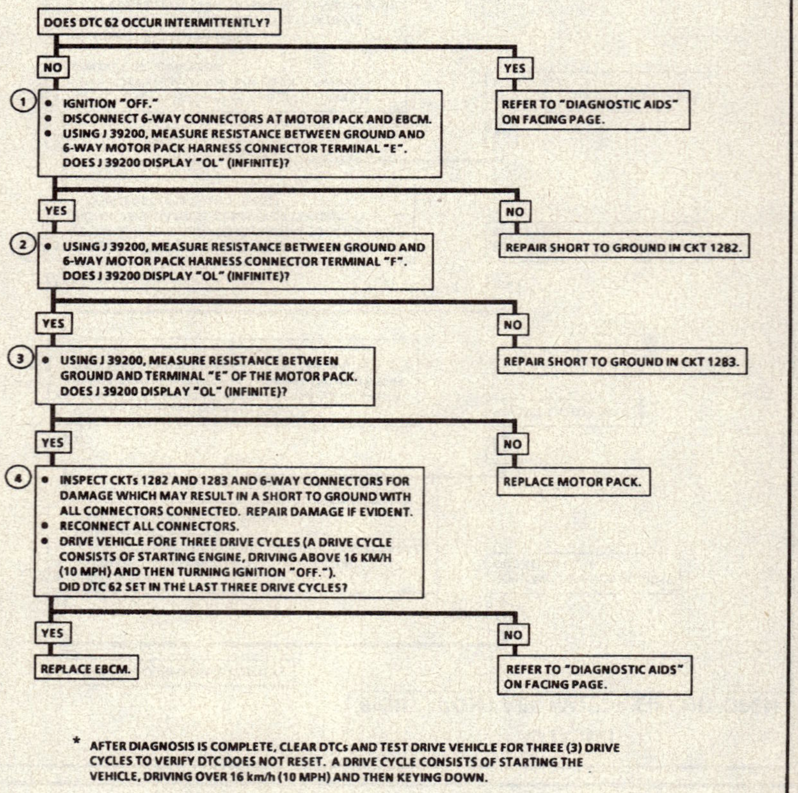

THIS CHART ASSUMES THAT A CURRENT DTC IS STORED INDICATING THAT THIS MALFUNCTION IS PRESENT.

DIAGNOSTIC CHART DTC 63 — 1994–95 ACHIEVA, GRAND AM AND SKYLARK

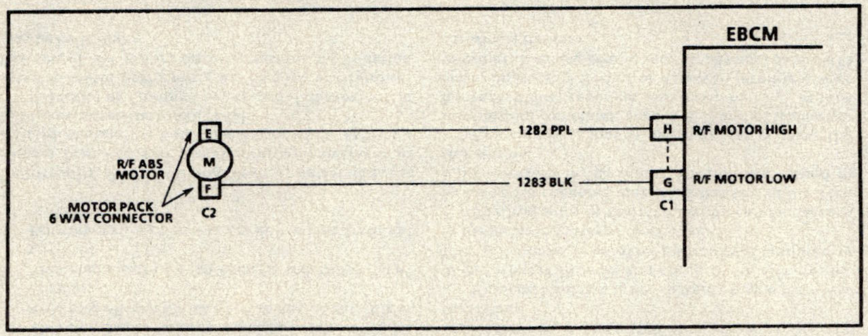

DTC 63

RIGHT FRONT ABS MOTOR CIRCUIT SHORTED TO BATTERY OR MOTOR SHORTED

Circuit Description:
This DTC identifies a motor circuit that is shorted to battery or a motor that has low or no resistance. This malfunction will not allow the motor to be controlled at the commanded current rate or will cause the motor to turn in the opposite direction or not at all.

DTC Will Set When: DTC 63 can be set anytime. If the EBCM detects an out of range voltage in either of the right front motor circuits indicating a circuit shorted to battery or a motor shorted, a malfunction exists.

Action Taken: A malfunction DTC is stored, ABS/VES is disabled and the ABS warning lamp is turned "ON."

DTC Chart Test Description: Number(s) below refer to circled number(s) on the diagnostic chart.

1. This step checks for a short to voltage in the motor "high" circuitry.
2. This step checks for a short to voltage in the motor "low" circuitry.
3. Checks for a short between the high and low motor circuits.
4. This step checks for a motor that is internally shorted.
5. Checks for a short between the EMB and motor circuits.
6. Checks for an internal motor pack short between the EMB and motor.
7. Ensures malfunction was not due to physical damage of the circuitry.

Diagnostic Aids: Using Tech 1, select manual control function, and exercise motor movement of affected channel in both directions while applying light pressure on the brake pedal.

If erratic or "jumpy" brake pedal movement is detected while performing an "apply" or "release" function of the motor, an intermittent malfunction may be indicated.

An "Intermittent" malfunction is most likely caused by a poor connection, rubbed through wire insulation, or a wire that is broken inside the insulation.

If the malfunction is not current, wiggle the wires of the affected channel and check if the DTC resets. This will help to pinpoint an intermittent malfunction in the motor circuitry or connections.

The frequency of the malfunction can be checked by using the enhanced diagnostic function of the Tech 1, as described in "Enhanced Diagnostics," found in this section. If DTC 41 sets as a current DTC and 63 is infrequent history DTC, see DTC 41 for diagnosis.

Any circuitry, that is suspected as causing the intermittent complaint, should be thoroughly checked for backed out terminals, improper mating, broken locks, improperly formed or damaged terminals, poor terminal to wiring connections or physical damage to the wiring harness.

DIAGNOSTIC CHART DTC 63 — 1994–95 ACHIEVA, GRAND AM AND SKYLARK, CONTINUED

DTC 63
RIGHT FRONT ABS MOTOR CIRCUIT SHORTED TO BATTERY OR MOTOR SHORTED

NOTICE: J 39200 TEST LEADS MUST BE "ZEROED" PRIOR TO MAKING ANY RESISTANCE MEASUREMENTS. REFER TO J 39200 USER'S MANUAL.

DOES DTC 63 OCCUR INTERMITTENTLY?
- YES: REFER TO "DIAGNOSTIC AIDS" ON FACING PAGE.
- NO: go to step 1.

1. NOTICE: THIS STEP MAY CAUSE ADDITIONAL DTCs TO SET.
   - IGNITION "OFF."
   - DISCONNECT 6-WAY CONNECTORS AT MOTOR PACK AND EBCM.
   - IGNITION "ON."
   - USING J 39200, MEASURE VOLTAGE BETWEEN GROUND AND 6-WAY MOTOR PACK HARNESS CONNECTOR TERMINAL "E".

   IS VOLTAGE LESS THAN 1 VOLT?
   - NO: REPAIR SHORT TO VOLTAGE IN CKT 1282.
   - YES: go to step 2.

2. - USING J 39200, MEASURE VOLTAGE BETWEEN GROUND AND 6-WAY MOTOR PACK HARNESS CONNECTOR TERMINAL "F".

   IS VOLTAGE LESS THAN 1 VOLT?
   - NO: REPAIR SHORT TO VOLTAGE IN CKT 1283.
   - YES: go to step 3.

3. - IGNITION "OFF."
   - USING J 39200, MEASURE RESISTANCE BETWEEN 6-WAY MOTOR PACK HARNESS CONNECTOR TERMINALS "E" AND "F".

   DOES J 39200 DISPLAY "OL" (INFINITE)?
   - NO: REPAIR SHORT BETWEEN CKTs 1282 AND 1283.
   - YES: go to step 4.

4. - USING J 39200, MEASURE RESISTANCE BETWEEN MOTOR PACK TERMINALS "E" AND "F".

   IS RESISTANCE GREATER THAN .4 OHMS?
   - NO: REPLACE MOTOR PACK.
   - YES: go to step 5.

5. - DISCONNECT 3-WAY MOTOR PACK CONNECTOR.
   - USING J 39200, MEASURE RESISTANCE BETWEEN 3-WAY MOTOR PACK HARNESS CONNECTOR TERMINAL "B" AND 6-WAY MOTOR PACK HARNESS CONNECTOR TERMINAL "E" THEN "F".

   DID J 39200 DISPLAY "OL" (INFINITE) FOR EACH MEASUREMENT?
   - NO: REPAIR SHORT BETWEEN CKT 855/1633 AND MOTOR CIRCUIT WHICH MEASURED OTHER THAN "OL" (INFINITE).
   - YES: go to step 6.

6. - USING J 39200, MEASURE RESISTANCE BETWEEN 3-WAY MOTOR PACK CONNECTOR TERMINAL "B" AND 6-WAY MOTOR PACK CONNECTOR TERMINAL "E".

   DID J 39200 DISPLAY "OL" (INFINITE)?
   - NO: REPLACE MOTOR PACK.
   - YES: go to step 7.

7. - INSPECT CKTs 1282 AND 1283 AND 6-WAY CONNECTORS FOR DAMAGE WHICH MAY RESULT IN A SHORT TO B+ WITH ALL CONNECTORS CONNECTED.
   - REPAIR DAMAGE IF EVIDENT.
   - RECONNECT ALL CONNECTORS.
   - DRIVE VEHICLE FOR THREE DRIVE CYCLES (A DRIVE CYCLE CONSISTS OF STARTING ENGINE, DRIVE ABOVE 16 km/h (10 MPH) AND THEN TURNING IGNITION "OFF").

   DID DTC 63 SET IN THE LAST THREE DRIVE CYCLES?
   - YES: REPLACE EBCM.
   - NO: REFER TO "DIAGNOSTIC AIDS" ON FACING PAGE.

* AFTER DIAGNOSIS IS COMPLETE, CLEAR DTCs AND TEST DRIVE VEHICLE FOR THREE (3) DRIVE CYCLES TO VERIFY DTC DOES NOT RESET. A DRIVE CYCLE CONSISTS OF STARTING THE VEHICLE, DRIVING OVER 16 km/h (10 MPH) AND THEN KEYING DOWN.

THIS CHART ASSUMES THAT A CURRENT DTC IS STORED INDICATING THAT THIS MALFUNCTION IS PRESENT.

DIAGNOSTIC CHART DTC 64 — 1994–95 ACHIEVA, GRAND AM AND SKYLARK

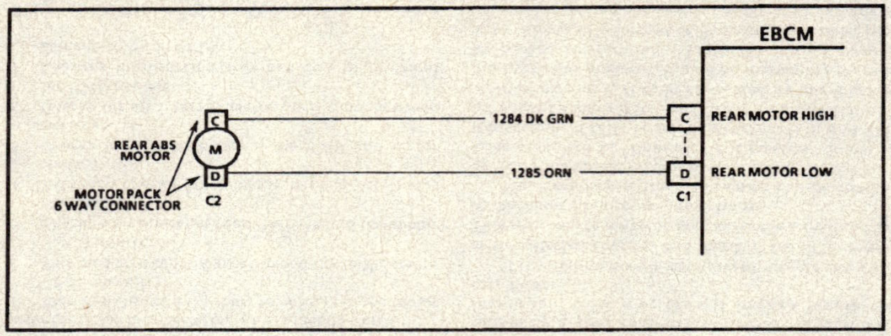

DTC 64
REAR ABS MOTOR CIRCUIT OPEN

Circuit Description:
This DTC identifies a motor that cannot be energized due to an open in its circuitry. This malfunction will not allow proper rear ABS operation.

DTC Will Set When: DTC 64 can be set only when the motor is commanded "OFF." If the EBCM detects an out of range voltage in either of the rear axle motor circuits indicating an open circuit, a malfunction exists.

Action Taken: An open motor will not activate when requested. A malfunction DTC is stored, ABS/VES is disabled and the ABS warning lamp is turned "ON."

DTC Chart Test Description: Number(s) below refer to circled number(s) on the diagnostic chart.

1. This step checks for proper resistance of the motor.
2. This step checks for an open in the motor "high" circuitry.
3. This step checks for an open in the motor "low" circuitry.
4. Ensures malfunction was not due to poor terminal contact.

Diagnostic Aids: Using Tech 1, select manual control function, and exercise motor movement of affected channel in both directions while applying light pressure on the brake pedal.

If erratic or "jumpy" brake pedal movement is detected while performing an "apply" or "release" function of the motor, an intermittent malfunction may be indicated.

An "Intermittent" malfunction is most likely caused by a poor connection, rubbed through wire insulation, or a wire that is broken inside the insulation.

If the malfunction is not current, wiggle the wires of the affected channel and check if the DTC resets. This will help to pinpoint an intermittent malfunction in the motor circuitry or connections.

The frequency of the malfunction can be checked by using the enhanced diagnostic function of the Tech 1, as described in "Enhanced Diagnostics," found in this section.

Any circuitry, that is suspected as causing the intermittent complaint, should be thoroughly checked for backed out terminals, improper mating, broken locks, improperly formed or damaged terminals, poor terminal to wiring connections or physical damage to the wiring harness.

DIAGNOSTIC CHART DTC 64 — 1994–95 ACHIEVA, GRAND AM AND SKYLARK, CONTINUED

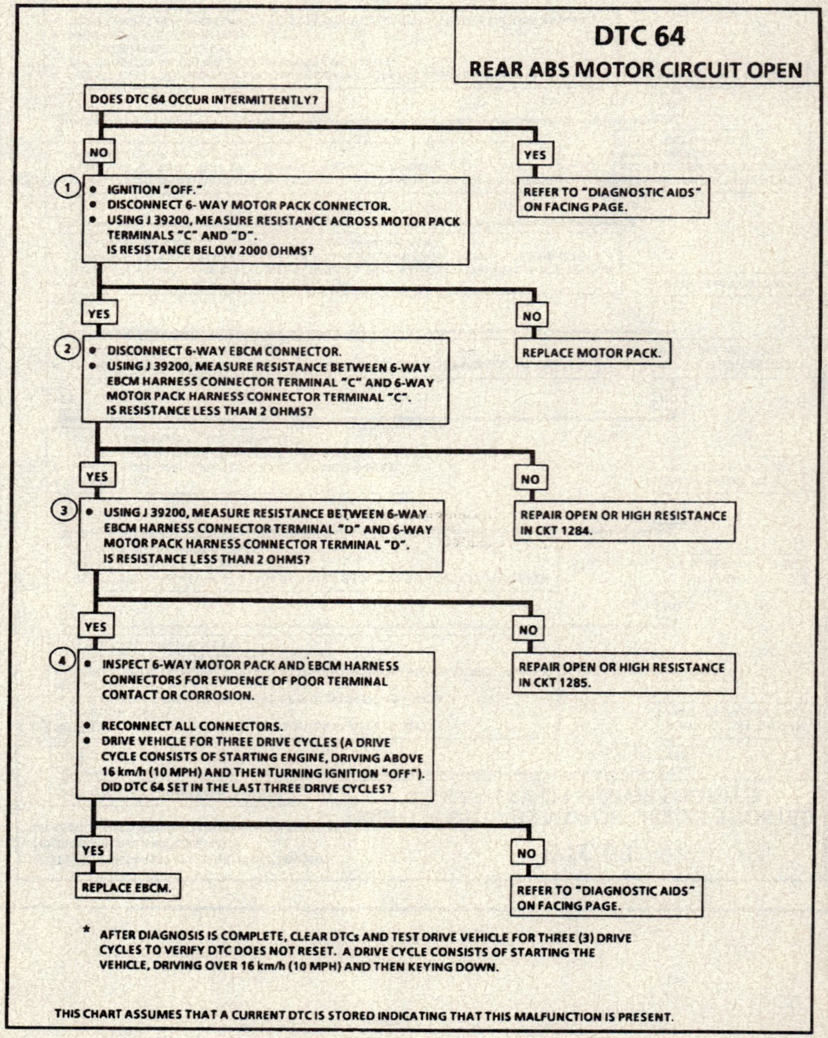

DIAGNOSTIC CHART DTC 65 — 1994–95 ACHIEVA, GRAND AM AND SKYLARK

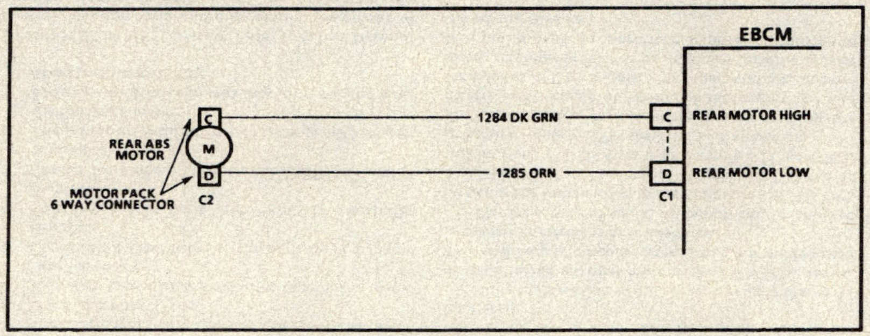

DTC 65

REAR ABS MOTOR CIRCUIT SHORTED TO GROUND

Circuit Description:
This DTC identifies a motor circuit that is shorted to ground. This malfunction will not allow the motor to be controlled at the commanded current rate or will cause the driver circuit to allow current directly to ground.

DTC Will Set When: DTC 65 can be set anytime. If the EBCM detects an out of range voltage in either of the rear axle motor circuits indicating a circuit shorted to ground, a malfunction exists.

Action Taken: A malfunction DTC is stored, ABS/VES is disabled and the ABS warning lamp is turned "ON."

DTC Chart Test Description: Number(s) below refer to circled number(s) on the diagnostic chart.

1. This step checks for a short to ground in the motor "high" circuitry.
2. This step checks for a short to ground in the motor "low" circuitry.
3. This step checks for a motor that is internally shorted to ground.
4. Ensures malfunction was not due to physical damage of the circuitry.

Diagnostic Aids: Using Tech 1, select manual control function, and exercise motor movement of affected channel in both directions while applying light pressure on the brake pedal.

If erratic or "jumpy" brake pedal movement is detected while performing an "apply" or "release" function of the motor, an intermittent malfunction may be indicated.

An "Intermittent" malfunction is most likely caused by a poor connection, rubbed through wire insulation, or a wire that is broken inside the insulation.

If the malfunction is not current, wiggle the wires of the affected channel and check if the DTC resets. This will help to pinpoint an intermittent malfunction in the motor circuitry or connections.

The frequency of the problem can be checked by using the enhanced diagnostic function of the Tech 1, as described in "Enhanced Diagnostics," found in this section.

Any circuitry, that is suspected as causing the intermittent complaint, should be thoroughly checked for backed out terminals, improper mating, broken locks, improperly formed or damaged terminals, poor terminal to wiring connections or physical damage to the wiring harness.

DIAGNOSTIC CHART DTC 65 — 1994–95 ACHIEVA, GRAND AM AND SKYLARK, CONTINUED

DTC 65

REAR ABS MOTOR CIRCUIT SHORTED TO GROUND

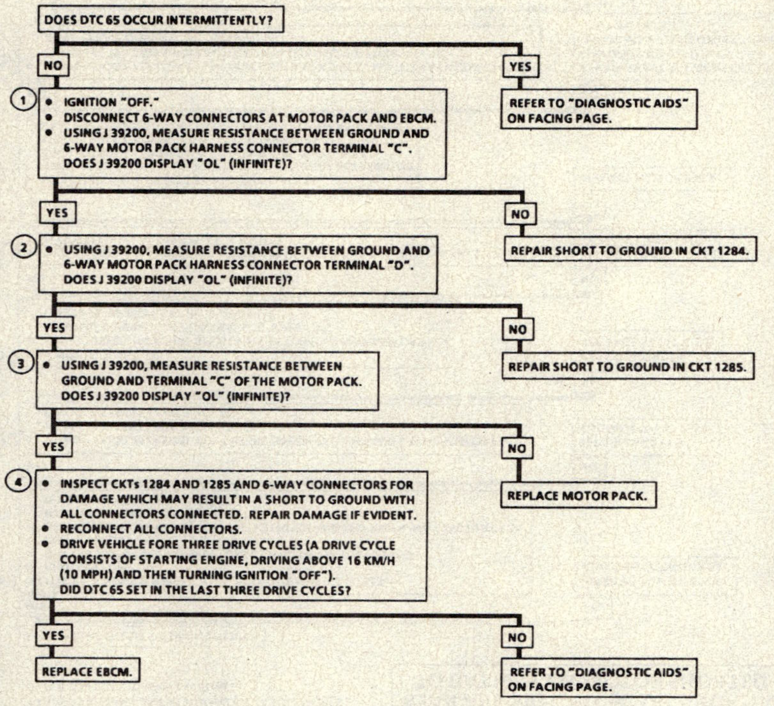

* AFTER DIAGNOSIS IS COMPLETE, CLEAR DTCs AND TEST DRIVE VEHICLE FOR THREE (3) DRIVE CYCLES TO VERIFY DTC DOES NOT RESET. A DRIVE CYCLE CONSISTS OF STARTING THE VEHICLE, DRIVING OVER 16 km/h (10 MPH) AND THEN KEYING DOWN.

THIS CHART ASSUMES THAT A CURRENT DTC IS STORED INDICATING THAT THIS MALFUNCTION IS PRESENT.

DIAGNOSTIC CHART DTC 66 — 1994–95 ACHIEVA, GRAND AM AND SKYLARK

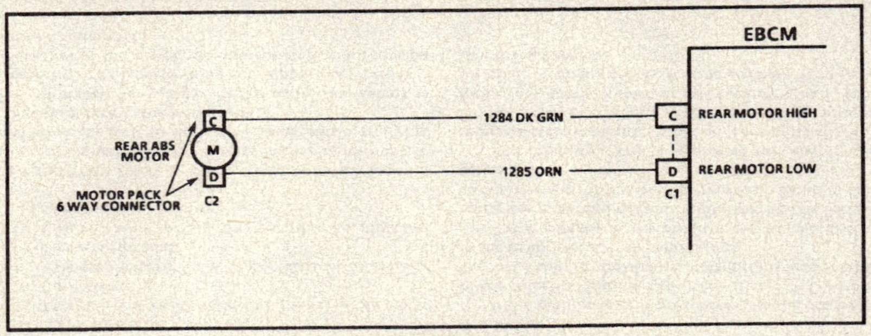

DTC 66
REAR ABS MOTOR CIRCUIT SHORTED TO BATTERY OR MOTOR SHORTED

Circuit Description:
This DTC identifies a motor circuit that is shorted to battery or a motor that has low or zero resistance. This malfunction will not allow the motor to be controlled at the commanded current rate or will cause the motor to turn in the opposite direction or not turn at all.

DTC Will Set When: DTC 66 can be set anytime. If the EBCM detects an out of range voltage in either of the rear axle motor circuits indicating a circuit shorted to battery or a motor shorted, a malfunction exists.

Action Taken: A malfunction DTC is stored, ABS/VES is disabled and the ABS warning lamp is turned "ON."

DTC Chart Test Description: Number(s) below refer to circled number(s) on the diagnostic chart.
1. This step checks for a short to voltage in the motor "high" circuitry.
2. This step checks for a short to voltage in the motor "low" circuitry.
3. Checks for a short between the high and low motor circuits.
4. This step checks for a motor that is internally shorted.
5. Checks for a short between the EMB and motor circuits.
6. Checks for an internal motor pack short between the EMB and motor.
7. Ensures malfunction was not due to physical damage of the circuitry.

Diagnostic Aids: Using Tech 1, select manual control function, and exercise motor movement of affected channel in both directions while applying light pressure on the brake pedal.

If erratic or "jumpy" brake pedal movement is detected while performing an "apply" or "release" function of the motor, an intermittent malfunction may be indicated.

An "Intermittent" malfunction is most likely caused by a poor connection, rubbed through wire insulation, or a wire that is broken inside the insulation.

If the malfunction is not current, wiggle the wires of the affected channel and check if the DTC resets. This will help to pinpoint an intermittent malfunction in the motor circuitry or connections.

The frequency of the malfunction can be checked by using the enhanced diagnostic function of the Tech 1, as described in "Enhanced Diagnostics," found in this section. If DTC 42 sets as current DTC and 66 is infrequent history DTC, see DTC 42 diagnostics.

Any circuitry, that is suspected as causing the intermittent complaint, should be thoroughly checked for backed out terminals, improper mating, broken locks, improperly formed or damaged terminals, poor terminal to wiring connections or physical damage to the wiring harness.

DIAGNOSTIC CHART DTC 66 — 1994–95 ACHIEVA, GRAND AM AND SKYLARK, CONTINUED

DTC 66
REAR ABS MOTOR CIRCUIT SHORTED TO BATTERY OR MOTOR SHORTED

NOTICE: J 39200 TEST LEADS MUST BE "ZEROED" PRIOR TO MAKING ANY RESISTANCE MEASUREMENTS. REFER TO J 39200 USER'S MANUAL.

DOES DTC 66 OCCUR INTERMITTENTLY?
- YES: REFER TO "DIAGNOSTIC AIDS" ON FACING PAGE.
- NO: go to step 1.

(1) NOTICE: THIS STEP MAY CAUSE ADDITIONAL DTCs TO SET.
- IGNITION "OFF."
- DISCONNECT 6-WAY CONNECTORS AT MOTOR PACK AND EBCM.
- IGNITION "ON."
- USING J 39200, MEASURE VOLTAGE BETWEEN GROUND AND 6-WAY MOTOR PACK HARNESS CONNECTOR TERMINAL "C".

IS VOLTAGE LESS THAN 1 VOLT?
- NO: REPAIR SHORT TO VOLTAGE IN CKT 1284.
- YES: go to step 2.

(2)
- USING J 39200, MEASURE VOLTAGE BETWEEN GROUND AND 6-WAY MOTOR PACK HARNESS CONNECTOR TERMINAL "D".

IS VOLTAGE LESS THAN 1 VOLT?
- NO: REPAIR SHORT TO VOLTAGE IN CKT 1285.
- YES: go to step 3.

(3)
- IGNITION "OFF."
- USING J 39200, MEASURE RESISTANCE BETWEEN 6-WAY MOTOR PACK HARNESS CONNECTOR TERMINALS "C" AND "D".

DOES J 39200 DISPLAY "OL" (INFINITE)?
- NO: REPAIR SHORT BETWEEN CKTs 1284 AND 1285.
- YES: go to step 4.

(4)
- USING J 39200, MEASURE RESISTANCE BETWEEN MOTOR PACK TERMINALS "C" AND "D".

IS RESISTANCE GREATER THAN .4 OHMS?
- NO: REPLACE MOTOR PACK.
- YES: go to step 5.

(5)
- DISCONNECT 3-WAY MOTOR PACK CONNECTOR.
- USING J 39200, MEASURE RESISTANCE BETWEEN 3-WAY MOTOR PACK HARNESS CONNECTOR TERMINAL "B" AND 6-WAY MOTOR PACK HARNESS CONNECTOR TERMINAL "C" THEN "D".

DID J 39200 DISPLAY "OL" (INFINITE) FOR EACH MEASUREMENT?
- NO: REPAIR SHORT BETWEEN CIRCUIT 855/1633 AND MOTOR CIRCUIT WHICH MEASURED OTHER THEN "OL" (INFINITE).
- YES: go to step 6.

(6)
- USING J 39200, MEASURE RESISTANCE BETWEEN 3-WAY MOTOR PACK CONNECTOR TERMINAL "B" AND 6-WAY MOTOR PACK CONNECTOR TERMINAL "C".

DOES J 39200 DISPLAY "OL" (INFINITE)?
- NO: REPLACE MOTOR PACK.
- YES: go to step 7.

(7)
- INSPECT CKTs 1284 AND 1285 AND 6-WAY CONNECTORS FOR DAMAGE WHICH MAY RESULT IN A SHORT TO B + WITH ALL CONNECTORS CONNECTED.
- REPAIR DAMAGE IF EVIDENT.
- RECONNECT ALL CONNECTORS.
- DRIVE VEHICLE FOR THREE DRIVE CYCLES (A DRIVE CYCLE CONSISTS OF STARTING ENGINE, DRIVE ABOVE 16 km/h (10 MPH) AND THEN TURNING IGNITION "OFF").

DID DTC 66 SET IN THE LAST THREE DRIVE CYCLES?
- YES: REPLACE EBCM.
- NO: REFER TO "DIAGNOSTIC AIDS" ON FACING PAGE.

* AFTER DIAGNOSIS IS COMPLETE, CLEAR DTCs AND TEST DRIVE VEHICLE FOR THREE (3) DRIVE CYCLES TO VERIFY DTC DOES NOT RESET. A DRIVE CYCLE CONSISTS OF STARTING THE VEHICLE, DRIVING OVER 16 km/h (10 MPH) AND THEN KEYING DOWN

THIS CHART ASSUMES THAT A CURRENT DTC IS STORED INDICATING THAT THIS MALFUNCTION IS PRESENT.

DIAGNOSTIC CHART DTC 67 — 1994–95 ACHIEVA, GRAND AM AND SKYLARK

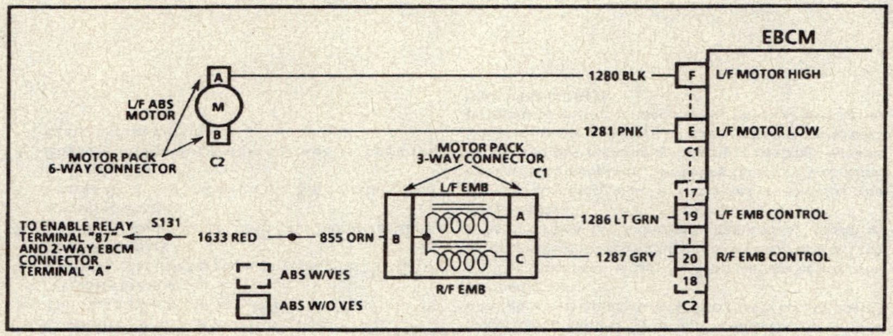

DTC 67

LEFT FRONT EMB CIRCUIT OPEN OR SHORTED TO GROUND

Circuit Description:

This DTC is designed to detect shorts to ground and opens in the EMB release circuit.

A short to ground between the EMB and the ABS EBCM will cause the EMB to be energized indefinitely at an uncontrolled current, causing the EMB to be destroyed.

DTC Will Set When: DTC 67 can be set only when the left front EMB is commanded "OFF" (not released). If the EBCM detects the left front EMB control circuit is constantly at ground, a malfunction exists.

Action Taken: A malfunction DTC is stored, ABS/VES is disabled and the ABS warning lamp is turned "ON.

DTC Chart Test Description: Number(s) below refer to circled number(s) on the diagnostic chart.

1. This step checks for a short to ground in the EMB control circuitry.
2. This step looks for an open in the EMB control circuitry.
3. Checks for an open in the EMB battery feed circuit.
4. This step ensures the EMB is not internally shorted to ground.
5. Ensures EMB is not internally open.
6. Checks for short to ground in the EMB battery feed circuit.
7. Ensures malfunction was not due to physical damage of the circuitry.

Diagnostic Aids: An "Intermittent" malfunction is most likely caused by a poor connection, rubbed through wire insulation, or a wire that is broken inside the insulation.

The frequency of the malfunction can be checked by using the enhanced diagnostic function of the Tech 1, as described in "Enhanced Diagnostics," found in this section.

Any circuitry, that is suspected as causing the intermittent complaint, should be thoroughly checked for backed out terminals, improper mating, broken locks, improperly formed or damaged terminals, poor terminal to wiring connections or physical damage to the wiring harness.

DIAGNOSTIC CHART DTC 67 — 1994–95 ACHIEVA, GRAND AM AND SKYLARK, CONTINUED

DTC 67

LEFT FRONT EMB CIRCUIT OPEN OR SHORTED TO GROUND

(1)
- IGNITION "OFF."
- DISCONNECT 24-WAY AND 2-WAY EBCM HARNESS CONNECTORS.
- DISCONNECT 3-WAY MOTOR PACK HARNESS CONNECTOR.
- USING J 39200, MEASURE RESISTANCE BETWEEN GROUND AND 24-WAY EBCM HARNESS CONNECTOR TERMINAL "19" (ABS W/O VES) OR "17" (ABS W/VES).

DOES J 39200 DISPLAY "OL" (INFINITE)?

- YES: go to step 2
- NO: REPAIR SHORT TO GROUND IN CKT 1286.

(2)
- USING J 39200, MEASURE RESISTANCE BETWEEN 3-WAY MOTOR PACK HARNESS CONNECTOR TERMINAL "A" AND 24-WAY EBCM HARNESS CONNECTOR TERMINAL "19" (ABS W/O VES) OR "17" (ABS W/VES).

IS RESISTANCE LESS THAN 2 OHMS?

- YES: go to step 3
- NO: REPAIR SHORT TO GROUND IN CKT 1286.

(3)
- USING J 39200, MEASURE RESISTANCE BETWEEN 2-WAY EBCM HARNESS CONNECTOR TERMINAL "A" AND 3-WAY MOTOR PACK HARNESS CONNECTOR TERMINAL "B".

IS RESISTANCE LESS THAN 2 OHMS?

- YES: go to step 4
- NO: REPAIR OPEN OR HIGH RESISTANCE IN CKT 1286.

(4)
- USING J 39200, MEASURE RESISTANCE BETWEEN MOTOR PACK TERMINAL "A" AND GROUND.

DOES J 39200 DISPLAY "OL" (INFINITE)?

- YES: go to step 5
- NO: REPAIR OPEN OR HIGH RESISTANCE IN CKT 1633/855.

(5)
- USING J 39200, MEASURE RESISTANCE BETWEEN MOTOR PACK TERMINALS "A" AND "B".

IS RESISTANCE LESS THAN 2 OHMS?

- YES: go to step 6
- NO: REPLACE MOTOR PACK.

(6)
- USING J 39200, MEASURE RESISTANCE BETWEEN GROUND AND 2-WAY EBCM HARNESS CONNECTOR TERMINAL "A".

DOES J 39200 DISPLAY "OL" (INFINITE)?

- YES: go to step 7
- NO: REPLACE MOTOR PACK.

(7)
- INSPECT 2-WAY AND 24-WAY EBCM AND 3-WAY MOTOR PACK CONNECTIONS FOR POOR TERMINAL CONTACT. REPLACE TERMINALS THAT EXHIBIT POOR CONTACT.
- INSPECT CKT 855/1633 OR 1294, CKT 1286, 3-WAY MOTOR PACK AND EBCM HARNESS CONNECTORS FOR PHYSICAL DAMAGE WHICH MAY RESULT IN A SHORT TO GROUND WITH ALL CONNECTORS CONNECTED. REPAIR DAMAGE IF EVIDENT.
- RECONNECT ALL CONNECTORS.
- TEST DRIVE VEHICLE FOR THREE DRIVE CYCLES. A DRIVE CYCLE CONSISTS OF STARTING THE ENGINE, DRIVING OVER 16 km/h (10 MPH) AND THEN KEYING DOWN.

DID DTC 67 SET IN THE LAST THREE DRIVE CYCLES?

- YES: REPLACE EBCM.
- NO: REPAIR SHORT TO GROUND IN CKT 855/1633 OR 1294 (ABS W/VES).
- NO: MALFUNCTION IS INTERMITTENT. REFER TO "DIAGNOSTIC AIDS" ON FACING PAGE.

* AFTER DIAGNOSIS IS COMPLETE, CLEAR DTCs AND TEST DRIVE VEHICLE FOR THREE (3) DRIVE CYCLES TO VERIFY DTC DOES NOT RESET. A DRIVE CYCLE CONSISTS OF STARTING THE VEHICLE, DRIVING OVER 16km/h (10 MPH) AND THEN KEYING DOWN.

THIS CHART ASSUMES THAT A CURRENT DTC IS STORED INDICATING THAT THIS MALFUNCTION IS PRESENT.

DIAGNOSTIC CHART DTC 68 — 1994–95 ACHIEVA, GRAND AM AND SKYLARK

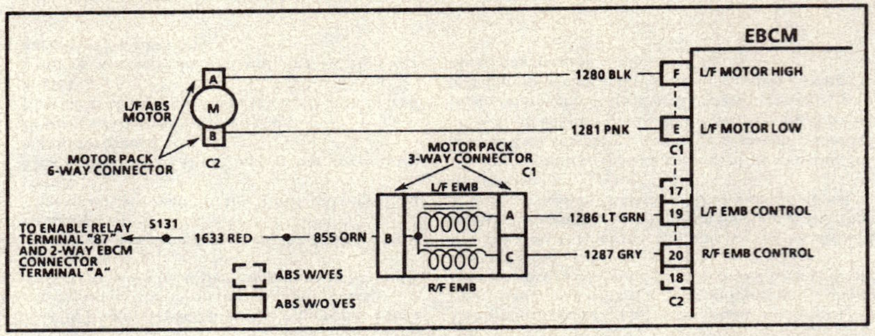

DTC 68

LEFT FRONT EMB CIRCUIT SHORTED TO BATTERY OR DRIVER OPEN

Circuit Description:
This DTC is designed to detect shorts to battery, shorts in the EMB release circuit, or an open EMB driver. These malfunctions will not allow the EMB to be energized (released) when needed.

DTC Will Set When: DTC 68 can be set only when the left front EMB is commanded "ON" (released). If an excessive voltage drop is across the motor driver for the left front EMB, a malfunction exists.

Action Taken: A malfunction DTC is stored, ABS/VES is disabled and the ABS warning lamp is turned "ON."

DTC Chart Test Description: Number(s) below refer to circled number(s) on the diagnostic chart.
1. This step checks to see if the malfunction is currently present.
2. This step identifies a short to voltage in the EMB control circuitry.
3. This step checks for an EMB that has lower than normal resistance.
4. Checks for a short between the EMB battery feed and control circuits.
5. Ensures malfunction was not due to physical damage of the circuitry.

Diagnostic Aids: An "Intermittent" malfunction is most likely caused by a poor connection, rubbed through wire insulation, or a wire that is broken inside the insulation.

The frequency of the malfunction can be checked by using the enhanced diagnostic function of the Tech 1, as described in "Enhanced Diagnostics," found in this section.

Any circuitry, that is suspected as causing the intermittent complaint, should be thoroughly checked for backed out terminals, improper mating, broken locks, improperly formed or damaged terminals, poor terminal to wiring connections or physical damage to the wiring harness.

DIAGNOSTIC CHART DTC 68 — 1994–95 ACHIEVA, GRAND AM AND SKYLARK, CONTINUED

*NOTICE: J 39200 TEST LEADS MUST BE "ZEROED" PRIOR TO MAKING ANY RESISTANCE MEASUREMENTS. REFER TO J 39200 USER'S MANUAL.

DTC 68

LEFT FRONT EMB CIRCUIT SHORTED TO BATTERY OR DRIVER OPEN

(1)
- IGNITION "ON."
- USING TECH 1 MISC. TESTS, SELECT MANUAL CONTROL FUNCTION.
- SELECT LEFT FRONT MOTOR APPLY FUNCTION AND APPLY MOTOR.
- APPLY FIRM PRESSURE ON BRAKE PEDAL.
- SELECT LEFT FRONT EMB RELEASE FUNCTION AND RELEASE EMB. DID BRAKE PEDAL FALL WITH INCREASING PEDAL FORCE?

NO → go to step 2.
YES → MALFUNCTION IS NOT PRESENT AT THIS TIME. REFER TO "DIAGNOSTIC AIDS" ON FACING PAGE.

(2) NOTICE: THIS STEP MAY CAUSE ADDITIONAL DTCs TO SET.
- IGNITION "OFF."
- DISCONNECT 3-WAY MOTOR PACK CONNECTOR.
- DISCONNECT 24-WAY EBCM CONNECTOR.
- IGNITION "ON."
- USING J 39200, MEASURE VOLTAGE BETWEEN GROUND AND 3-WAY MOTOR PACK HARNESS CONNECTOR TERMINAL "A". IS VOLTAGE LESS THAN 1 VOLT?

YES → go to step 3.
NO → REPAIR SHORT TO VOLTAGE IN CKT 1286.

(3)
- IGNITION "OFF."
- USING J 39200, MEASURE RESISTANCE BETWEEN MOTOR PACK TERMINALS "A" AND "B". IS RESISTANCE GREATER THAN .4 OHMS?

YES → go to step 4.
NO → REPLACE MOTOR PACK.

(4)
- USING J 39200, MEASURE THE RESISTANCE BETWEEN 3-WAY MOTOR PACK HARNESS CONNECTOR TERMINALS "B" AND "A". DOES J 39200 DISPLAY "OL" (INFINITE)?

YES → go to step 5.
NO → REPAIR SHORT BETWEEN CKTs 1286 AND 855/1633.

(5)
- INSPECT CKT 1286, 3-WAY MOTOR PACK AND 24-WAY EBCM CONNECTORS FOR DAMAGE WHICH MAY RESULT IN A SHORT TO B+ WITH ALL CONNECTORS CONNECTED. REPAIR DAMAGE IF EVIDENT.
- RECONNECT ALL CONNECTORS.
- DRIVE VEHICLE FOR THREE DRIVE CYCLES (A DRIVE CYCLE CONSISTS OF STARTING ENGINE, DRIVING ABOVE 16 km/h (10 MPH) AND THEN TURNING IGNITION "OFF").
- IGNITION "ON."
- USING TECH 1, SELECT "F1: DTC HISTORY." DID DTC 68 SET IN LAST THREE DRIVE CYCLES?

YES → REPLACE EBCM.
NO → MALFUNCTION IS INTERMITTENT. REFER TO "DIAGNOSTIC AIDS" ON FACING PAGE.

* AFTER DIAGNOSIS IS COMPLETE, CLEAR DTCs AND TEST DRIVE VEHICLE FOR THREE (3) DRIVE CYCLES TO VERIFY DTC DOES NOT RESET. A DRIVE CYCLE CONSISTS OF STARTING THE VEHICLE, DRIVING OVER 16 km/h (10 MPH) AND THEN KEYING DOWN.

THIS CHART ASSUMES THAT A CURRENT DTC IS STORED INDICATING THAT THIS MALFUNCTION IS PRESENT.

DIAGNOSTIC CHART DTC 71 — 1994–95 ACHIEVA, GRAND AM AND SKYLARK

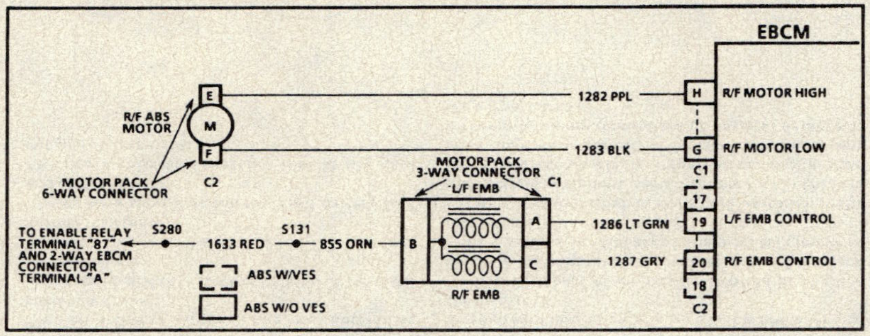

DTC 71

RIGHT FRONT EMB CIRCUIT OPEN OR SHORTED TO GROUND

Circuit Description:

This DTC is designed to detect shorts to ground and opens in the EMB release circuit.

A short to ground between the EMB and the ABS EBCM will cause the EMB to be energized indefinitely at an uncontrolled current causing the EMB to be destroyed.

DTC Will Set When: DTC 71 can be set only when the right front EMB is commanded "OFF" (not released). If the EBCM detects the right front EMB control circuit is constantly at ground, a malfunction exists.

Action Taken: A malfunction DTC is stored, ABS/VES is disabled and the ABS warning lamp is turned "ON."

DTC Chart Test Description: Number(s) below refer to circled number(s) on the diagnostic chart.

1. This step checks for a short to ground in the EMB control circuitry.
2. This step looks for an open in the EMB control circuitry.
3. Checks for an open in the EMB battery feed circuit.
4. This step ensures the EMB is not internally shorted to ground.
5. Ensures EMB is not internally open.
6. Checks for short to ground in the EMB battery feed circuit.
7. Ensures malfunction was not due to physical damage of the circuitry.

Diagnostic Aids: An "Intermittent" malfunction is most likely caused by a poor connection, rubbed through wire insulation, or a wire that is broken inside the insulation.

The frequency of the malfunction can be checked by using the enhanced diagnostic function of the Tech 1, as described in "Enhanced Diagnostics," found in this section.

Any circuitry, that is suspected as causing the intermittent complaint, should be thoroughly checked for backed out terminals, improper mating, broken locks, improperly formed or damaged terminals, poor terminal to wiring connections or physical damage to the wiring harness.

DIAGNOSTIC CHART DTC 71 — 1994–95 ACHIEVA, GRAND AM AND SKYLARK, CONTINUED

DTC 71

RIGHT FRONT EMB CIRCUIT OPEN OR SHORTED TO GROUND

(1)
- IGNITION "OFF."
- DISCONNECT 24-WAY AND 2-WAY EBCM HARNESS CONNECTORS.
- DISCONNECT 3-WAY MOTOR PACK HARNESS CONNECTOR.
- USING J 39200, MEASURE RESISTANCE BETWEEN GROUND AND 24-WAY EBCM HARNESS CONNECTOR TERMINAL "20" (ABS W/O VES) OR "18" (ABS W/VES). DOES J 39200 DISPLAY "OL" (INFINITE)?

NO → REPAIR SHORT TO GROUND IN CKT 1287.

YES →

(2)
- USING J 39200, MEASURE RESISTANCE BETWEEN 3-WAY MOTOR PACK HARNESS CONNECTOR TERMINAL "A" AND 24-WAY EBCM HARNESS CONNECTOR TERMINAL "20" (ABS W/O VES) OR "18" (ABS W/VES). IS RESISTANCE LESS THAN 2 OHMS?

NO → REPAIR OPEN OR HIGH RESISTANCE IN CKT 1287.

YES →

(3)
- USING J 39200, MEASURE RESISTANCE BETWEEN 2-WAY EBCM HARNESS CONNECTOR TERMINAL "A" AND 3-WAY MOTOR PACK HARNESS CONNECTOR TERMINAL "B". IS RESISTANCE LESS THAN 2 OHMS?

NO → REPAIR OPEN OR HIGH RESISTANCE IN CKT 1633/855.

YES →

(4)
- USING J 39200, MEASURE RESISTANCE BETWEEN MOTOR PACK TERMINAL "C" AND GROUND. DOES J 39200 DISPLAY "OL" (INFINITE)?

NO → REPLACE MOTOR PACK.

YES →

(5)
- USING J 39200, MEASURE RESISTANCE BETWEEN MOTOR PACK TERMINALS "C" AND "B". IS RESISTANCE LESS THAN 2 OHMS?

NO → REPLACE MOTOR PACK.

YES →

(6)
- USING J 39200, MEASURE RESISTANCE BETWEEN GROUND AND 2-WAY EBCM HARNESS CONNECTOR TERMINAL "A". DOES J 39200 DISPLAY "OL" (INFINITE)?

NO → REPAIR SHORT TO GROUND IN CKT 855/1633 OR 1294 (ABS W/VES).

YES →

(7)
- INSPECT 2-WAY AND 24-WAY EBCM AND 3-WAY MOTOR PACK CONNECTIONS FOR POOR TERMINAL CONTACT. REPLACE TERMINALS THAT EXHIBIT POOR CONTACT.
- INSPECT CKT 855/1633 OR 1294, CKT 1287, 3-WAY MOTOR PACK AND EBCM HARNESS CONNECTORS FOR PHYSICAL DAMAGE WHICH MAY RESULT IN A SHORT TO GROUND WITH ALL CONNECTORS CONNECTED. REPAIR DAMAGE IF EVIDENT.
- RECONNECT ALL CONNECTORS.
- TEST DRIVE VEHICLE FOR THREE DRIVE CYCLES. A DRIVE CYCLE CONSISTS OF STARTING THE ENGINE, DRIVING OVER 16 km/h (10 MPH) AND THEN KEYING DOWN. DID DTC 71 SET THE LAST THREE DRIVE CYCLES?

YES → REPLACE EBCM.

NO → MALFUNCTION IS INTERMITTENT. REFER TO "DIAGNOSTIC AIDS" ON FACING PAGE.

- AFTER DIAGNOSIS IS COMPLETE, CLEAR DTCs AND TEST DRIVE VEHICLE FOR THREE (3) DRIVE CYCLES TO VERIFY DTC DOES NOT RESET. A DRIVE CYCLE CONSISTS OF STARTING THE VEHICLE, DRIVING OVER 16km/h (10 MPH) AND THEN KEYING DOWN.

THIS CHART ASSUMES THAT A CURRENT DTC IS STORED INDICATING THAT THIS MALFUNCTION IS PRESENT.

DIAGNOSTIC CHART DTC 72 — 1994–95 ACHIEVA, GRAND AM AND SKYLARK

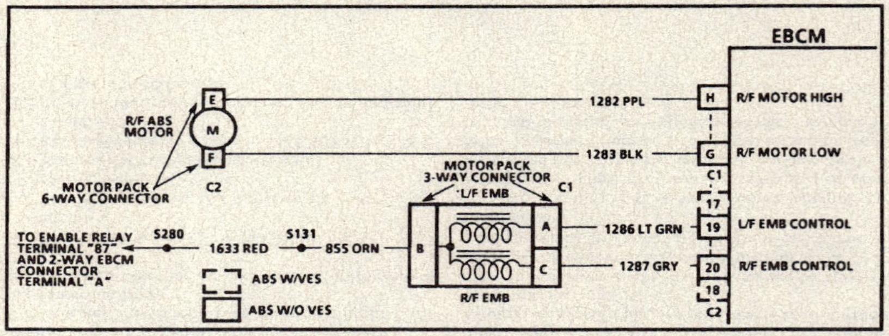

DTC 72

RIGHT FRONT EMB CIRCUIT SHORTED TO BATTERY OR DRIVER OPEN

Circuit Description:

This DTC is designed to detect shorts to battery, shorts in the EMB release circuit, or an open EMB driver. These malfunctions will not allow the EMB to be energized (released) when needed.

DTC Will Set When: DTC 72 can be set only when the right front EMB is commanded "ON" (released). If an excessive voltage drop is across the motor driver for the right front EMB, a malfunction exists.

Action Taken: A malfunction DTC is stored, ABS/VES is disabled and the ABS warning lamp is turned "ON."

DTC Chart Test Description: Number(s) below refer to circled number(s) on the diagnostic chart.

1. This step checks to see if the malfunction is currently present.
2. This step identifies a short to voltage in the EMB control circuitry.
3. This step checks for an EMB that has lower than normal resistance.
4. Checks for a short between the EMB battery feed and control circuits.
5. Ensures malfunction was not due to physical damage of the circuitry.

Diagnostic Aids: An "Intermittent" malfunction is most likely caused by a poor connection, rubbed through wire insulation, or a wire that is broken inside the insulation.

The frequency of the malfunction can be checked by using the enhanced diagnostic function of the Tech 1, as described in "Enhanced Diagnostics," found in this section.

Any circuitry, that is suspected as causing the intermittent complaint, should be thoroughly checked for backed out terminals, improper mating, broken locks, improperly formed or damaged terminals, poor terminal to wiring connections or physical damage to the wiring harness.

DIAGNOSTIC CHART DTC 72 — 1994–95 ACHIEVA, GRAND AM AND SKYLARK, CONTINUED

DTC 72

RIGHT FRONT EMB CIRCUIT SHORTED TO BATTERY OR DRIVER OPEN

*NOTICE: J 39200 TEST LEADS MUST BE "ZEROED" PRIOR TO MAKING ANY RESISTANCE MEASUREMENTS. REFER TO J 39200 USER'S MANUAL.

(1)
- IGNITION "ON."
- USING TECH 1 MISC. TESTS, SELECT MANUAL CONTROL FUNCTION.
- SELECT RIGHT FRONT MOTOR APPLY FUNCTION AND APPLY MOTOR.
- APPLY FIRM PRESSURE ON BRAKE PEDAL.
- SELECT RIGHT FRONT EMB RELEASE FUNCTION AND RELEASE EMB.

DID BRAKE PEDAL FALL WITH INCREASING PEDAL FORCE?

- YES: MALFUNCTION IS NOT PRESENT AT THIS TIME. REFER TO "DIAGNOSTIC AIDS" ON FACING PAGE.
- NO: go to step 2.

(2) NOTICE: THIS STEP MAY CAUSE ADDITIONAL DTCs TO SET.
- IGNITION "OFF."
- DISCONNECT 3-WAY MOTOR PACK CONNECTOR.
- DISCONNECT 24-WAY EBCM CONNECTOR.
- IGNITION "ON."
- USING J 39200, MEASURE VOLTAGE BETWEEN GROUND AND 3-WAY MOTOR PACK HARNESS CONNECTOR TERMINAL "C".

IS VOLTAGE LESS THAN 1 VOLT?

- NO: REPAIR SHORT TO VOLTAGE IN CKT 1287.
- YES: go to step 3.

(3)
- IGNITION "OFF."
- USING J 39200, MEASURE RESISTANCE BETWEEN MOTOR PACK TERMINALS "C" AND "B".

IS RESISTANCE GREATER THAN .4 OHMS?

- NO: REPLACE MOTOR PACK.
- YES: go to step 4.

(4)
- USING J 39200, MEASURE RESISTANCE BETWEEN 3-WAY MOTOR PACK HARNESS CONNECTOR TERMINALS "B" AND "C".

DOES J 39200 DISPLAY "OL" (INFINITE)?

- NO: REPAIR SHORT BETWEEN CKT(s) 1287 AND 855/1633.
- YES: go to step 5.

(5)
- INSPECT CKT 1287, 3-WAY MOTOR PACK AND 24-WAY EBCM CONNECTORS FOR DAMAGE WHICH MAY RESULT IN A SHORT TO B+ WITH ALL CONNECTORS CONNECTED. REPAIR DAMAGE IF EVIDENT.
- RECONNECT ALL CONNECTORS.
- DRIVE VEHICLE FOR THREE DRIVE CYCLES (A DRIVE CYCLE CONSISTS OF STARTING ENGINE, DRIVING ABOVE 16 KM/H (10 MPH) AND THEN TURNING IGNITION "OFF").
- IGNITION "ON."
- USING TECH 1, SELECT "F1: DTC HISTORY."

DID DTC 72 SET IN LAST THREE DRIVE CYCLES?

- YES: REPLACE EBCM.
- NO: MALFUNCTION IS INTERMITTENT. REFER TO "DIAGNOSTIC AIDS" ON FACING PAGE.

* AFTER DIAGNOSIS IS COMPLETE, CLEAR DTCs AND TEST DRIVE VEHICLE FOR THREE (3) DRIVE CYCLES TO VERIFY DTC DOES NOT RESET. A DRIVE CYCLE CONSISTS OF STARTING THE VEHICLE, DRIVING OVER 16 km/h (10 MPH) AND THEN KEYING DOWN.

THIS CHART ASSUMES THAT A CURRENT DTC IS STORED INDICATING THAT THIS MALFUNCTION IS PRESENT.

DIAGNOSTIC CHART DTC 73 — 1994–95 ACHIEVA, GRAND AM AND SKYLARK

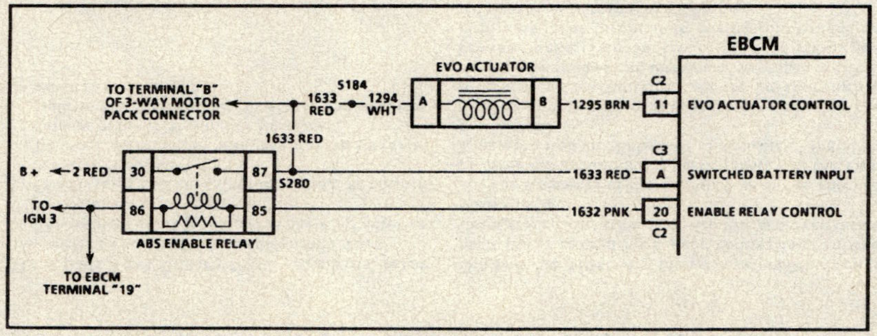

DTC 73

EVO ACTUATOR CIRCUIT OPEN OR SHORTED TO GROUND (FOR ABS WITH VES ONLY)

Circuit Description:
Battery voltage is supplied to EVO actuator terminal "A" when the EBCM commands the ABS enable relay "ON." Ground for the EVO actuator is provided through EBCM terminal "11" to actuator terminal "B". The EBCM controls the amount of current supplied to the EVO actuator based on input from the wheel speed sensors and the steering wheel speed sensor.

DTC Will Set When: DTC 73 can be set before ABS initialization or above 16 km/h (10 mph). Before initialization, if the EBCM detects the EVO actuator control circuit is at ground, a malfunction exists. Above 16 km/h (10 mph), if the EVO actuator sense current is less than half the command current, a malfunction exists.

Action Taken: A malfunction DTC is stored, ABS/VES is disabled, and the ABS warning lamp is turned "ON" and power steering returns to full assist.

DTC Will Clear When: DTC 73 setting conditions no longer exist and the ignition key is turned "OFF."

DTC Chart Test Description: Number(s) below refer to circled number(s) on the diagnostic chart.

1. Checks for proper operation of the EVO actuator.
2. Checks for a high resistance in the EVO actuator control circuit.
3. Checks for a short to ground in the EVO actuator control circuit.
4. Checks for high resistance in the EVO actuator power feed circuit.
5. Checks for a short to ground in the EVO actuator control circuit.
6. Checks for a short to ground in the EVO actuator.
7. Checks for an open in the EVO actuator.
8. Ensures open or short to ground is not due to physical damage of the circuitry.

Diagnostic Aids: An "Intermittent" malfunction is most likely caused by a poor connection, rubbed through wire insulation, or a wire that is broken inside the insulation.

The frequency of the malfunction can be checked by using the enhanced diagnostic function of the Tech 1, as described in "Enhanced Diagnostics," found in this section.

Any circuitry, that is suspected as causing the intermittent malfunction, should be thoroughly checked for backed out terminals, improper mating, broken locks, improperly formed or damaged terminals, poor terminal to wiring connections or physical damage to the wiring harness.

DIAGNOSTIC CHART DTC 73 — 1994–95 ACHIEVA, GRAND AM AND SKYLARK, CONTINUED

DTC 73

EVO ACTUATOR CIRCUIT OPEN OR SHORTED TO GROUND (FOR ABS WITH VES ONLY)

(1)
- IGNITION "ON."
- USING TECH 1 MISC. TESTS, SELECT VES MANUAL CONTROL FUNCTION AND COMMAND EVO ACTUATOR "ON."

IS FEEDBACK CURRENT GREATER THAN 100 mA?

YES: MALFUNCTION IS INTERMITTENT. REFER TO "DIAGNOSTIC AIDS" ON FACING PAGE.

NO: (2)
- IGNITION "OFF."
- DISCONNECT 24-WAY EBCM CONNECTOR.
- DISCONNECT 2-WAY EVO ACTUATOR CONNECTOR.
- USING J 39200, MEASURE RESISTANCE BETWEEN 24-WAY EBCM HARNESS CONNECTOR TERMINAL "11" AND 2-WAY EVO ACTUATOR HARNESS CONNECTOR TERMINAL "B".

IS RESISTANCE LESS THAN 2 OHMS?

NO: REPAIR OPEN OR HIGH RESISTANCE IN CKT 1295.

YES: (3)
- USING J 39200, MEASURE RESISTANCE BETWEEN GROUND AND 24-WAY EBCM HARNESS CONNECTOR TERMINAL "11".

DOES J 39200 DISPLAY "OL" (INFINITE)?

NO: REPAIR SHORT TO GROUND IN CKT 1295.

YES: (4)
- DISCONNECT 2-WAY EBCM CONNECTOR.
- USING J 39200, MEASURE RESISTANCE BETWEEN 2-WAY EBCM HARNESS CONNECTOR TERMINAL "A" AND 2-WAY EVO ACTUATOR HARNESS CONNECTOR TERMINAL "A".

IS RESISTANCE LESS THAN 2 OHMS?

NO: REPAIR OPEN OR HIGH RESISTANCE IN CKT 1294/1633.

YES: (5)
- USING J 39200, MEASURE RESISTANCE BETWEEN GROUND AND 2-WAY EBCM HARNESS CONNECTOR TERMINAL "A".

DOES J 39200 DISPLAY "OL" (INFINITE)?

NO: REPAIR SHORT TO GROUND IN CKT 1633/1294 OR 855.

YES: (6)
- RECONNECT EVO ACTUATOR CONNECTOR.
- USING J 39200, MEASURE RESISTANCE BETWEEN GROUND AND 24-WAY EBCM HARNESS CONNECTOR TERMINAL "11".

DOES J 39200 DISPLAY "OL" (INFINITE)?

NO: REPLACE EVO ACTUATOR.

YES: (7)
- USING J 39200, MEASURE RESISTANCE BETWEEN 24-WAY EBCM HARNESS CONNECTOR TERMINAL "11" AND 2-WAY EBCM HARNESS CONNECTER TERMINAL "A".

IS RESISTANCE BETWEEN 7 AND 19 OHMS?

NO: REPLACE EVO ACTUATOR.

YES: (8)
- INSPECT 2-WAY AND 24-WAY EBCM HARNESS CONNECTORS AND 2-WAY EVO ACTUATOR CONNECTORS FOR POOR TERMINAL CONTACT OR CORROSION. REPLACE TERMINALS THAT EXHIBIT POOR CONTACT OR CORROSION.
- INSPECT CKT 1295 FOR PHYSICAL DAMAGE WHICH MAY RESULT IN AN OPEN OR SHORT TO GROUND WITH THE 24-WAY EBCM HARNESS CONNECTOR OR 2-WAY EVO ACTUATOR CONNECTOR CONNECTED. REPAIR DAMAGE IF EVIDENT.
- RECONNECT ALL CONNECTORS.
- IGNITION "ON."

IS DTC 73 SET AS A CURRENT DTC?

YES: REPLACE EBCM.

NO: MALFUNCTION IS INTERMITTENT. REFER TO "DIAGNOSTIC AIDS" ON FACING PAGE.

DIAGNOSTIC CHART DTC 74 — 1994–95 ACHIEVA, GRAND AM AND SKYLARK

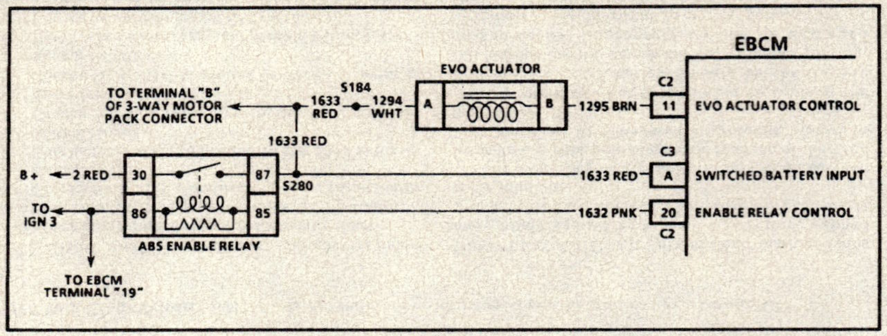

DTC 74

EVO ACTUATOR CIRCUIT SHORTED TO BATTERY OR SOLENOID SHORTED (FOR ABS WITH VES ONLY)

Circuit Description:

Battery voltage is supplied to EVO actuator terminal "A" when the EBCM commands the ABS enable relay "ON." Ground for the EVO actuator is provided through EBCM terminal "11" to actuator terminal "B". The EBCM controls the amount of current supplied to the EVO actuator based on input from the wheel speed sensors and the steering wheel speed sensor.

DTC Will Set When: DTC 74 can be set anytime after ABS initialization. A malfunction exists if an excessive voltage drop is across the EVO actuator driver.

Action Taken: A malfunction DTC is stored, VES is disabled and power steering returns to full assist.

DTC Will Clear When: DTC 74 setting conditions no longer exist and the ignition key is turned "OFF."

DTC Chart Test Description: Number(s) below refer to circled number(s) on the diagnostic chart.

1. Checks for a short to voltage in the EVO actuator control circuit.
2. Checks for a shorted actuator solenoid or shorted actuator circuits.
3. Ensures malfunction was not due to physical damage of the EVO actuator circuitry.
4. Isolates the malfunction to either the actuator solenoid or its circuitry.

Diagnostic Aids: An "Intermittent" malfunction is most likely caused by a poor connection, rubbed through wire insulation, or a wire that is broken inside the insulation.

The frequency of the malfunction can be checked by using the enhanced diagnostic function of the Tech 1, as described in "Enhanced Diagnostics," found in this section.

Any circuitry, that is suspected as causing the intermittent malfunction, should be thoroughly checked for backed out terminals, improper mating, broken locks, improperly formed or damaged terminals, poor terminal to wiring connections or physical damage to the wiring harness.

DIAGNOSTIC CHART DTC 74 — 1994–95 ACHIEVA, GRAND AM AND SKYLARK, CONTINUED

DTC 74

EVO ACTUATOR CIRCUIT SHORTED TO BATTERY OR SOLENOID SHORTED (FOR ABS WITH VES ONLY)

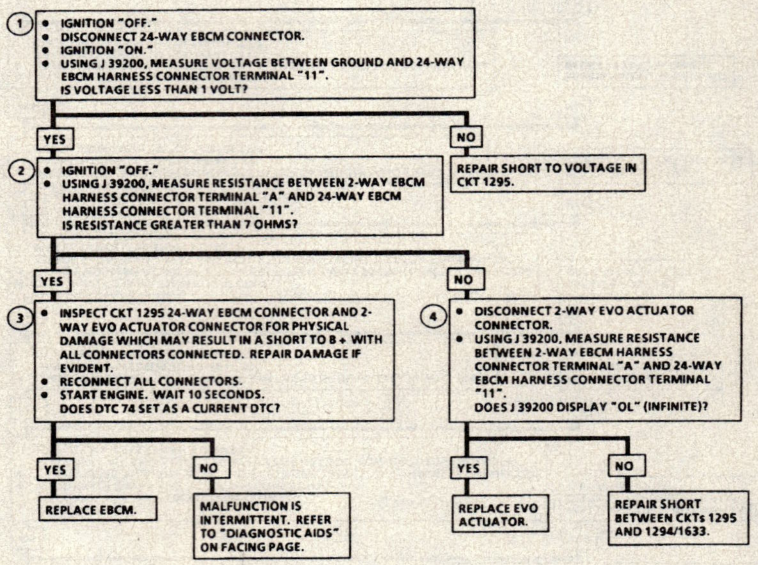

DIAGNOSTIC CHART DTC 76 — 1994–95 ACHIEVA, GRAND AM AND SKYLARK

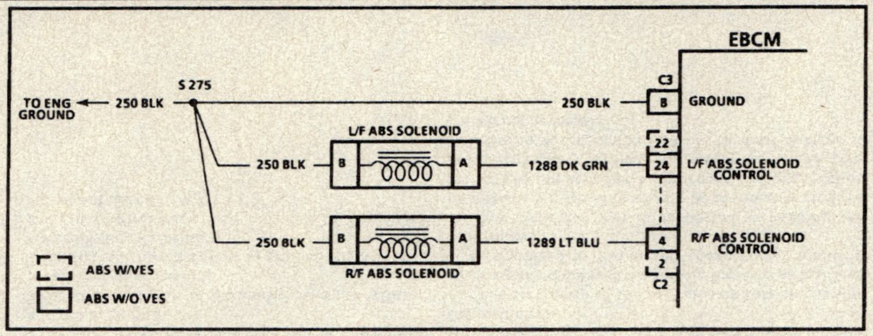

DTC 76

LEFT FRONT SOLENOID CIRCUIT OPEN OR SHORTED TO BATTERY

Circuit Description:

This DTC identifies a solenoid that cannot be energized due to an open in its circuitry, or a solenoid that is always energized due to a short to battery in its circuitry between the driver and the solenoid. An open will not allow proper ABS operation, but the short to battery simply turns "ON" the solenoid. A path for base brakes is still allowed once the motor re-homes and the check valve is lifted off its seat during key "ON" initialization.

DTC Will Set When: DTC 76 can be set only when the solenoid is commanded "OFF." If the EBCM detects an excessive voltage in the left front solenoid control circuit, a malfunction exists.

Action Taken: A malfunction DTC is stored, ABS/VES is disabled and the ABS warning lamp is turned "ON."

DTC Chart Test Description: Number(s) below refer to circled number(s) on the diagnostic chart.

1. This checks for a short to voltage in the control circuitry of the solenoid.
2. This checks for an open in the control circuit of the solenoid.
3. This step identifies an open in the solenoid or ground circuit.
4. Checks for proper resistance values of the solenoid.
5. Ensures malfunction was not due to physical damage of the circuitry.

Diagnostic Aids: An "Intermittent" malfunction is most likely caused by a poor connection, rubbed through wire insulation, or a wire that is broken inside the insulation.

The frequency of the malfunction can be checked by using the enhanced diagnostic function of the Tech 1, as described in "Enhanced Diagnostics," found in this section.

Any circuitry, that is suspected as causing the intermittent complaint, should be thoroughly checked for backed out terminals, improper mating, broken locks, improperly formed or damaged terminals, poor terminal to wiring connections or physical damage to the wiring harness.

DIAGNOSTIC CHART DTC 76 — 1994–95 ACHIEVA, GRAND AM AND SKYLARK, CONTINUED

DTC 76

LEFT FRONT SOLENOID CIRCUIT OPEN OR SHORTED TO BATTERY

*NOTICE: J 39200 TEST LEADS MUST BE "ZEROED" PRIOR TO MAKING ANY RESISTANCE MEASUREMENTS. REFER TO J 39200 USER'S MANUAL.

(1)
- IGNITION "OFF."
- DISCONNECT 24-WAY EBCM CONNECTOR.
- IGNITION "ON."
- USING J 39200, MEASURE VOLTAGE BETWEEN GROUND AND 24-WAY EBCM HARNESS CONNECTOR TERMINAL "24" (ABS W/O VES) OR "22" (ABS W/VES). IS VOLTAGE LESS THAN 1 VOLT?

YES → go to (2)
NO → REPAIR SHORT TO VOLTAGE IN CKT 1288.

(2)
- IGNITION "OFF."
- DISCONNECT LEFT FRONT SOLENOID CONNECTOR.
- USING J 39200, MEASURE RESISTANCE BETWEEN LEFT FRONT SOLENOID HARNESS CONNECTOR TERMINAL "A" AND 24-WAY EBCM HARNESS CONNECTOR TERMINAL "24" (ABS W/O VES) OR "22" (ABS W/ VES). IS RESISTANCE LESS THAN 2 OHMS?

YES → go to (3)
NO → REPAIR OPEN OR HIGH RESISTANCE IN CKT 1288.

(3)
- USING J 39200, MEASURE RESISTANCE BETWEEN GROUND AND LEFT FRONT SOLENOID HARNESS CONNECTOR TERMINAL "B". IS RESISTANCE LESS THAN 2 OHMS?

YES → go to (4)
NO → REPAIR OPEN OR HIGH RESISTANCE IN CKT 250.

(4)
- USING J 39200, MEASURE RESISTANCE BETWEEN LEFT FRONT SOLENOID TERMINALS "A" AND "B". IS RESISTANCE BETWEEN 2.5 AND 5 OHMS?

YES → go to (5)
NO → REPLACE LEFT FRONT SOLENOID.

(5)
- INSPECT LEFT FRONT SOLENOID AND 24-WAY EBCM HARNESS CONNECTORS FOR POOR TERMINAL CONTACT OR CORROSION. REPLACE TERMINALS THAT EXHIBIT POOR CONTACT OR CORROSION.
- INSPECT CKT 1288 FOR DAMAGE WHICH MAY RESULT IN A SHORT TO B + WITH ALL CONNECTORS CONNECTED. REPAIR DAMAGE IF EVIDENT.
- RECONNECT ALL CONNECTORS.
- IGNITION "ON." IS DTC 76 SET AS A CURRENT DTC?

YES → REPLACE EBCM.
NO → REFER TO "DIAGNOSTIC AIDS" ON FACING PAGE.

THIS CHART ASSUMES THAT A CURRENT DTC IS STORED INDICATING THAT THIS MALFUNCTION IS PRESENT.

DIAGNOSTIC CHART DTC 77 — 1994–95 ACHIEVA, GRAND AM AND SKYLARK

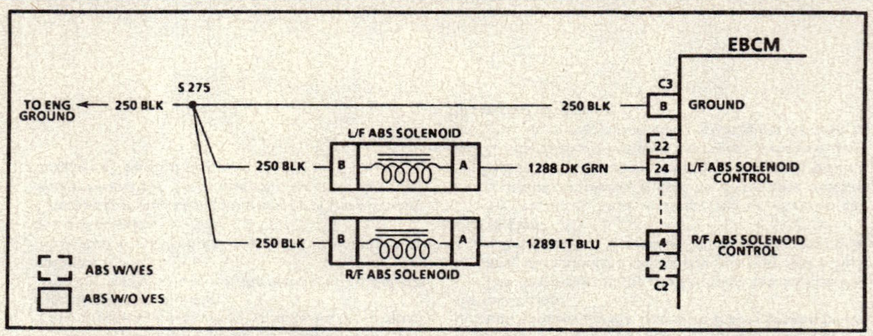

DTC 77

LEFT FRONT SOLENOID CIRCUIT SHORTED TO GROUND OR DRIVER OPEN

Circuit Description:

This DTC identifies a solenoid that cannot be energized due to an open in its driver circuitry, or a short to ground between the solenoid driver and the solenoid. These malfunctions can affect ABS operation since the flow of brake fluid to the wheel cylinder cannot be stopped, making ABS operation for that channel impossible.

DTC Will Set When: DTC 77 can be set only when the solenoid is commanded "ON." If the EBCM detects the left front solenoid control circuit voltage is out of specification, a malfunction exists.

Action Taken: A malfunction DTC is stored, ABS/VES is disabled and the ABS warning lamp is turned "ON."

DTC Chart Test Description: Number(s) below refer to circled number(s) on the diagnostic chart.

1. This step checks for a solenoid that is internally shorted to ground.
2. This step checks for a solenoid that is not within proper resistance values.
3. This step indicates if a short to ground exists in the solenoid control circuitry.
4. Ensures malfunction was not due to physical damage of the circuitry.

Diagnostic Aids: An "Intermittent" malfunction is most likely caused by a poor connection, rubbed through wire insulation, or a wire that is broken inside the insulation.

The frequency of the malfunction can be checked by using the enhanced diagnostic function of the Tech 1, as described in "Enhanced Diagnostics," found in this section.

Any circuitry, that is suspected as causing the intermittent complaint, should be thoroughly checked for backed out terminals, improper mating, broken locks, improperly formed or damaged terminals, poor terminal to wiring connections or physical damage to the wiring harness.

DIAGNOSTIC CHART DTC 77 — 1994–95 ACHIEVA, GRAND AM AND SKYLARK, CONTINUED

DTC 77

LEFT FRONT SOLENOID CIRCUIT SHORTED TO GROUND OR DRIVER OPEN

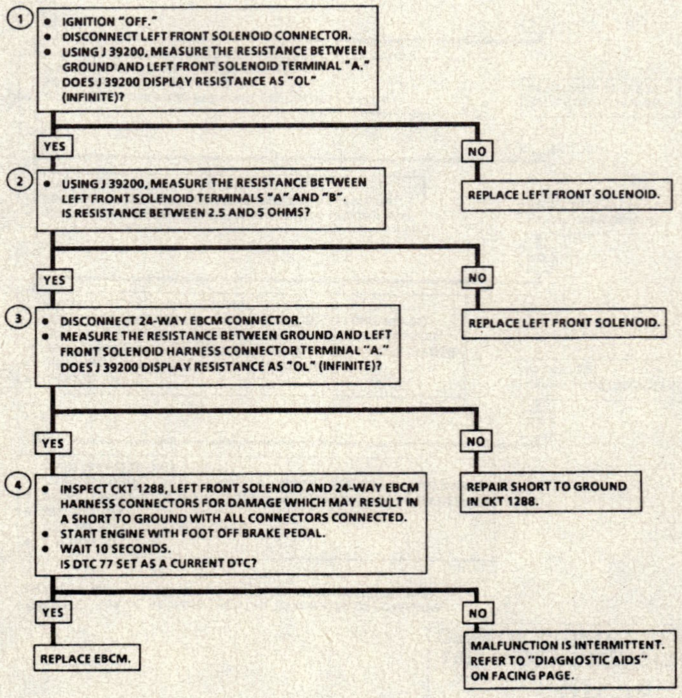

THIS CHART ASSUMES THAT A CURRENT DTC IS STORED INDICATING THAT THIS MALFUNCTION IS PRESENT.

DIAGNOSTIC CHART DTC 78 — 1994–95 ACHIEVA, GRAND AM AND SKYLARK

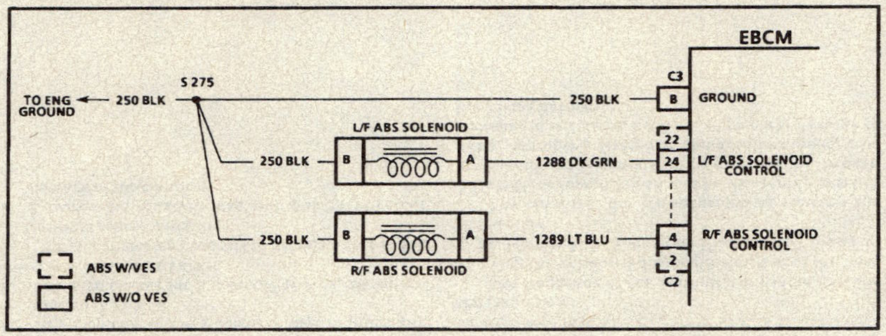

DTC 78

RIGHT FRONT SOLENOID CIRCUIT OPEN OR SHORTED TO BATTERY

Circuit Description:

This DTC identifies a solenoid that cannot be energized due to an open in its circuitry, or a solenoid that is always energized due to a short to battery in its circuitry between the driver and the solenoid. An open will not allow proper ABS operation, but the short to battery simply turns "ON" the solenoid. A path for base brakes is still allowed once the motor re-homes and the check valve is lifted off its seat during key "ON" initialization.

DTC Will Set When: DTC 78 can be set only when the solenoid is commanded "OFF." If the EBCM detects an excessive voltage in the right front solenoid control circuit, a malfunction exists.

Action Taken: A malfunction DTC is stored, ABS/VES is disabled and the ABS warning lamp is turned "ON."

DTC Chart Test Description: Number(s) below refer to circled number(s) on the diagnostic chart.

1. This checks for a short to voltage in the control circuitry of the solenoid.
2. This checks for an open in the control circuit of the solenoid.
3. This step identifies an open in the solenoid or ground circuit.
4. Checks for proper resistance values of the solenoid.
5. Ensures malfunction was not due to physical damage of the circuitry.

Diagnostic Aids: An "Intermittent" malfunction is most likely caused by a poor connection, rubbed through wire insulation, or a wire that is broken inside the insulation.

The frequency of the malfunction can be checked by using the enhanced diagnostic function of the Tech 1, as described in "Enhanced Diagnostics," found in this section.

Any circuitry, that is suspected as causing the intermittent complaint, should be thoroughly checked for backed out terminals, improper mating, broken locks, improperly formed or damaged terminals, poor terminal to wiring connections or physical damage to the wiring harness.

DIAGNOSTIC CHART DTC 78 — 1994–95 ACHIEVA, GRAND AM AND SKYLARK, CONTINUED

DTC 78

RIGHT FRONT SOLENOID CIRCUIT OPEN OR SHORTED TO BATTERY

*NOTICE: J39200 TEST LEADS MUST BE "ZEROED" PRIOR TO MAKING ANY RESISTANCE MEASUREMENTS. REFER TO J39200 USER'S MANUAL.

(1)
- IGNITION "OFF."
- DISCONNECT 24-WAY EBCM CONNECTOR.
- IGNITION "ON."
- MEASURE VOLTAGE BETWEEN GROUND AND 24-WAY EBCM HARNESS CONNECTOR TERMINAL "4" (ABS W/O VES) OR "2" (ABS W/VES). IS VOLTAGE LESS THAN 1 VOLT?

NO: REPAIR SHORT TO VOLTAGE IN CKT 1289.

YES:

(2)
- IGNITION "OFF."
- DISCONNECT RIGHT FRONT SOLENOID CONNECTOR.
- USING J 39200, MEASURE RESISTANCE BETWEEN RIGHT FRONT SOLENOID HARNESS CONNECTOR TERMINAL "A" AND 24-WAY EBCM HARNESS CONNECTOR TERMINAL "4" (ABS W/O VES) OR "2" (ABS W/ VES). IS RESISTANCE LESS THAN 2 OHMS?

NO: REPAIR OPEN OR HIGH RESISTANCE IN CKT 1289.

YES:

(3)
- USING J39200, MEASURE RESISTANCE BETWEEN GROUND AND RIGHT FRONT SOLENOID HARNESS CONNECTOR TERMINAL "B". IS RESISTANCE LESS THAN 2 OHMS?

NO: REPAIR OPEN IN CKT 250.

YES:

(4)
- USING J39200, MEASURE RESISTANCE BETWEEN RIGHT FRONT SOLENOID TERMINALS "A" AND "B". IS RESISTANCE BETWEEN 2.5 AND 5 OHMS?

NO: REPLACE RIGHT FRONT SOLENOID.

YES:

(5)
- INSPECT RIGHT FRONT SOLENOID AND 24-WAY EBCM HARNESS CONNECTORS FOR POOR TERMINAL CONTACT OR CORROSION. REPLACE TERMINALS THAT EXHIBIT POOR CONTACT OR CORROSION.
- INSPECT CKT 1289 FOR DAMAGE WHICH MAY RESULT IN A SHORT TO B+ WITH ALL CONNECTORS CONNECTED. REPAIR DAMAGE IF EVIDENT.
- RECONNECT ALL CONNECTORS.
- IGNITION "ON."
  IS DTC 78 SET AS A CURRENT DTC?

YES: REPLACE EBCM.

NO: MALFUNCTION IS INTERMITTENT, REFER TO "DIAGNOSTIC AIDS" ON FACING PAGE.

THIS CHART ASSUMES THAT A CURRENT DTC IS STORED INDICATING THAT THIS MALFUNCTION IS PRESENT.

DIAGNOSTIC CHART DTC 81 — 1994–95 ACHIEVA, GRAND AM AND SKYLARK

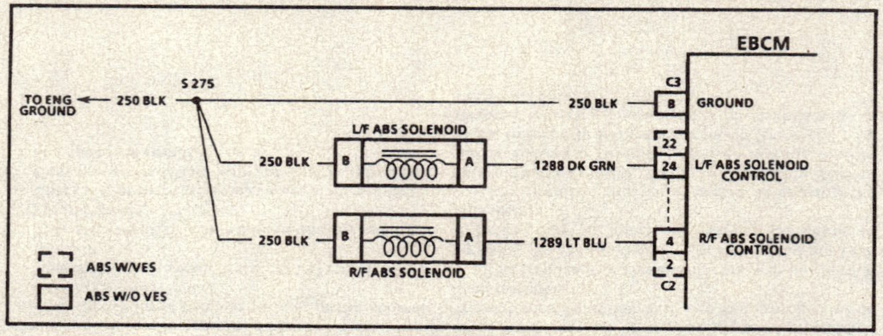

DTC 81

RIGHT FRONT SOLENOID CIRCUIT SHORTED TO GROUND OR DRIVER OPEN

Circuit Description:

This DTC identifies a solenoid that cannot be energized due to an open in its driver circuitry, or a short to ground between the solenoid driver and the solenoid. These malfunctions can affect ABS operation since the flow of brake fluid to the wheel cylinder cannot be stopped, making ABS operation for that channel impossible.

DTC Will Set When: DTC 81 can be set only when the solenoid is commanded "ON." If the EBCM detects the right front solenoid control circuit voltage is out of specification, a malfunction exists.

Action Taken: A malfunction DTC is stored, ABS/VES is disabled and the ABS warning lamp is turned "ON."

DTC Chart Test Description: Number(s) below refer to circled number(s) on the diagnostic chart.

1. This step checks for a solenoid that is internally shorted to ground.
2. This step checks for a solenoid that is not within proper resistance values.
3. This step indicates if a short to ground exists in the solenoid control circuitry.
4. Ensures malfunction was not due to physical damage of the circuitry.

Diagnostic Aids: An "Intermittent" malfunction is most likely caused by a poor connection, rubbed through wire insulation, or a wire that is broken inside the insulation.

The frequency of the malfunction can be checked by using the enhanced diagnostic function of the Tech 1, as described in "Enhanced Diagnostics," found in this section.

Any circuitry, that is suspected as causing the intermittent complaint, should be thoroughly checked for backed out terminals, improper mating, broken locks, improperly formed or damaged terminals, poor terminal to wiring connections or physical damage to the wiring harness.

DIAGNOSTIC CHART DTC 81 — 1994–95 ACHIEVA, GRAND AM AND SKYLARK, CONTINUED

DTC 81

RIGHT FRONT SOLENOID CIRCUIT SHORTED TO GROUND OR DRIVER OPEN

(1)
- IGNITION "OFF."
- DISCONNECT RIGHT FRONT SOLENOID CONNECTOR.
- USING J 39200, MEASURE THE RESISTANCE BETWEEN GROUND AND RIGHT FRONT SOLENOID TERMINAL "A". DOES J 39200 DISPLAY RESISTANCE AS "OL" (INFINITE)?

NO → REPLACE RIGHT FRONT SOLENOID.

YES ↓

(2)
- USING J 39200, MEASURE RESISTANCE BETWEEN RIGHT FRONT SOLENOID TERMINALS "A" AND "B". IS RESISTANCE BETWEEN 2.5 AND 5 OHMS?

NO → REPLACE RIGHT FRONT SOLENOID.

YES ↓

(3)
- DISCONNECT 24-WAY EBCM CONNECTOR.
- MEASURE THE RESISTANCE BETWEEN GROUND AND RIGHT FRONT SOLENOID HARNESS CONNECTOR TERMINAL "A". DOES J 39200 DISPLAY RESISTANCE AS "OL" (INFINITE)?

NO → REPAIR SHORT TO GROUND IN CKT 1289.

YES ↓

(4)
- INSPECT CKT 1289, RIGHT FRONT SOLENOID AND 24-WAY EBCM HARNESS CONNECTORS FOR DAMAGE WHICH MAY RESULT IN A SHORT TO GROUND WITH ALL CONNECTORS CONNECTED.
- START ENGINE WITH FOOT OFF BRAKE PEDAL.
- WAIT 10 SECONDS. IS DTC 81 SET AS A CURRENT DTC?

YES → REPLACE EBCM.

NO → MALFUNCTION IS INTERMITTENT. REFER TO "DIAGNOSTIC AIDS" ON FACING PAGE.

THIS CHART ASSUMES THAT A CURRENT DTC IS STORED INDICATING THAT THIS MALFUNCTION IS PRESENT.

DIAGNOSTIC CHART DTC 82 — 1994–95 ACHIEVA, GRAND AM AND SKYLARK

DTC 82

CALIBRATION MALFUNCTION

Circuit Description:

This DTC allows the EBCM to check for a calibration failure by comparing the calibration value to a known value stored in the EEPROM.

This test is also used as a security measure to prevent improper use of calibrations or changes to these calibrations that may alter the designed function of ABS.

DTC Will Set When: DTC 82 can be set at key-up. If the program identifier is incorrect or the memory checksum is incorrect, a malfunction exists.

Action Taken: A malfunction DTC is stored, ABS/VES is disabled and the ABS warning lamp is turned "ON."

DTC Chart Test Description: Number(s) below refer to circled number(s) on the diagnostic chart.

1. This step checks to see if the malfunction is present during diagnosis. If present, the EBCM is not functioning correctly and must be replaced.

Important:

- At the time of printing, 1994 vehicles with ABS VI were not being field reprogrammed to correct this failure. A service bulletin will be issued when EEPROM reprogramming is authorized.

Diagnostic Aids: An intermittent DTC 82 is most likely caused by a bad cell in the EEPROM that is sensitive to temperature changes. If DTC 82 failed more than once, but is intermittent, replace EBCM.

The frequency of the malfunction can be checked by using the enhanced diagnostic function of the Tech 1, as described in "Enhanced Diagnostics," found in this section.

DIAGNOSTIC CHART DTC 82 — 1994–95 ACHIEVA, GRAND AM AND SKYLARK, CONTINUED

DTC 82
CALIBRATION MALFUNCTION

(1) • IGNITION "ON."
IS DTC 82 SET AS A CURRENT DTC?

- YES → REPLACE EBCM.
- NO → MALFUNCTION IS NOT PRESENT AT THIS TIME. REFER TO "DIAGNOSTIC AIDS" ON FACING PAGE.

THIS CHART ASSUMES THAT A CURRENT DTC IS STORED INDICATING THAT THIS MALFUNCTION IS PRESENT.

DIAGNOSTIC CHART DTC 86 — 1994–95 ACHIEVA, GRAND AM AND SKYLARK

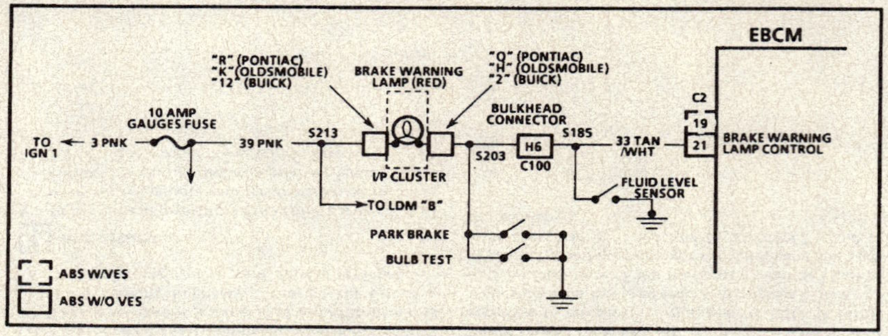

DTC 86

EBCM TURNED "ON" THE RED "BRAKE" WARNING LAMP

Circuit Description:

This DTC is provided as an information only test, and reflects the status of the command issued by the ABS EBCM to illuminate the red "BRAKE" warning lamp. If another diagnostic DTC issues a command to illuminate the red "BRAKE" warning lamp, DTC 86 will be stored in EEPROM as a history DTC at the conclusion of the ignition cycle.

DTC Chart Test Description: Number(s) below refer to circled number(s) on the diagnostic chart.

1. This step serves to identify if a DTC other than DTC 86 commanded the red "BRAKE" warning lamp "ON."

Diagnostic Aids: Any ABS mechanical DTC that issues a command to illuminate the red "BRAKE" warning lamp will also result in DTC 86 being stored in EEPROM during shutdown. These DTCs are: 38, 41, 42, 45, 46, 47, 48 and 51. If the motors are not in their home position, certain electrical DTCs will also command the red "BRAKE" warning lamp "ON." These DTCs are: 14, 18, 55, 56, 57, 58, 61, 62, 63, 64, 65, 66, 67, 68, 71 and 72.

If any of these DTCs are indicated along with DTC 86, they must be corrected prior to addressing a DTC 86 malfunction.

DIAGNOSTIC CHART DTC 86 — 1994–95 ACHIEVA, GRAND AM AND SKYLARK, CONTINUED

DTC 86

EBCM TURNED "ON" THE RED "BRAKE" WARNING LAMP

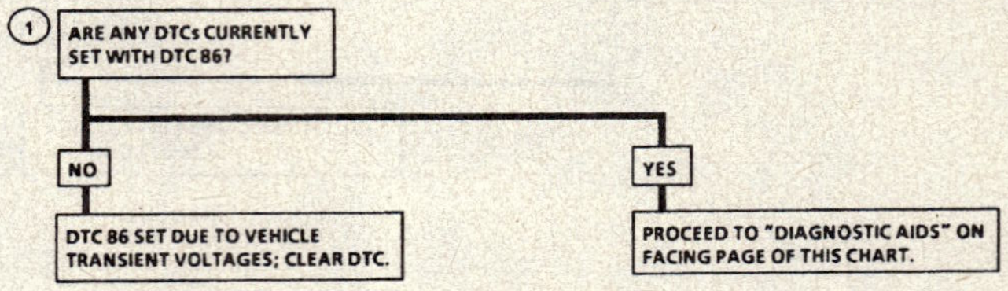

DIAGNOSTIC CHART DTC 87 — 1994–95 ACHIEVA, GRAND AM AND SKYLARK

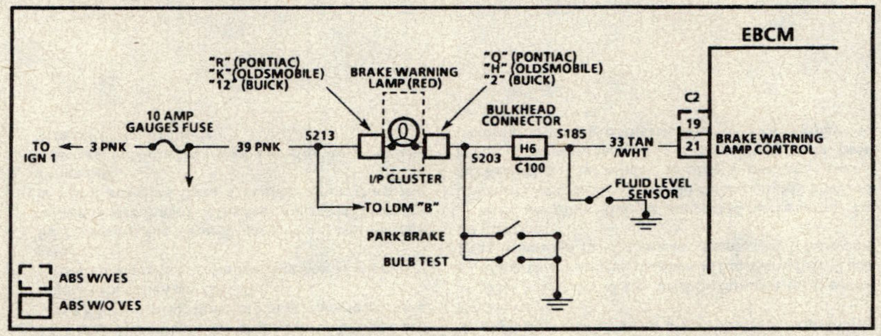

DTC 87

RED "BRAKE" WARNING LAMP CIRCUIT OPEN

Circuit Description:

This DTC is used to verify the EBCM has continuity to the brake warning lamp in case the EBCM must turn it "ON." This will only occur if an ABS malfunction is detected that may degrade base brake operation.

Because ABS is not the only device controlling the brake warning lamp (parking brake and low fluid sensor may also turn "ON" the lamp) a short to ground in this circuit can not necessarily be detected. Excessive resistance in the fluid level switch or parking brake circuitry may set this DTC.

DTC Will Set When: DTC 87 can be set after initialization is completed. If the red "BRAKE" warning lamp circuit voltage is out of specification indicating an open circuit, a malfunction exists.

Action Taken: A malfunction DTC is stored; however, ABS is not disabled.

DTC Chart Test Description: Number(s) below refer to circled number(s) on the diagnostic chart.

1. This step identifies if the red "BRAKE" warning lamp can be turned "ON."
2. This step tests for high resistance in park brake circuitry.
3. Ensures malfunction was not due to poor terminal contact.
4. This step identifies if high resistance is present in the fluid level switch circuitry.
5. Checks integrity of red "BRAKE" warning bulb.
6. Checks for high resistance in the brake warning lamp control circuit.

Diagnostic Aids: An "Intermittent" malfunction is most likely caused by a poor connection, rubbed through wire insulation, or a wire that is broken inside the insulation.

The frequency of the malfunction can be checked by using the enhanced diagnostic function of the Tech 1, as described in "Enhanced Diagnostics," found in this section. If the Tech 1 is not available, the parking brake handle may be lifted to verify proper lamp operation and continuity of CKTs 3 and 39.

Any circuitry, that is suspected as causing the intermittent complaint, should be thoroughly checked for backed out terminals, improper mating, broken locks, improperly formed or damaged terminals, poor terminal to wiring connections or physical damage to the wiring harness.

DIAGNOSTIC CHART DTC 87 — 1994–95 ACHIEVA, GRAND AM AND SKYLARK, CONTINUED

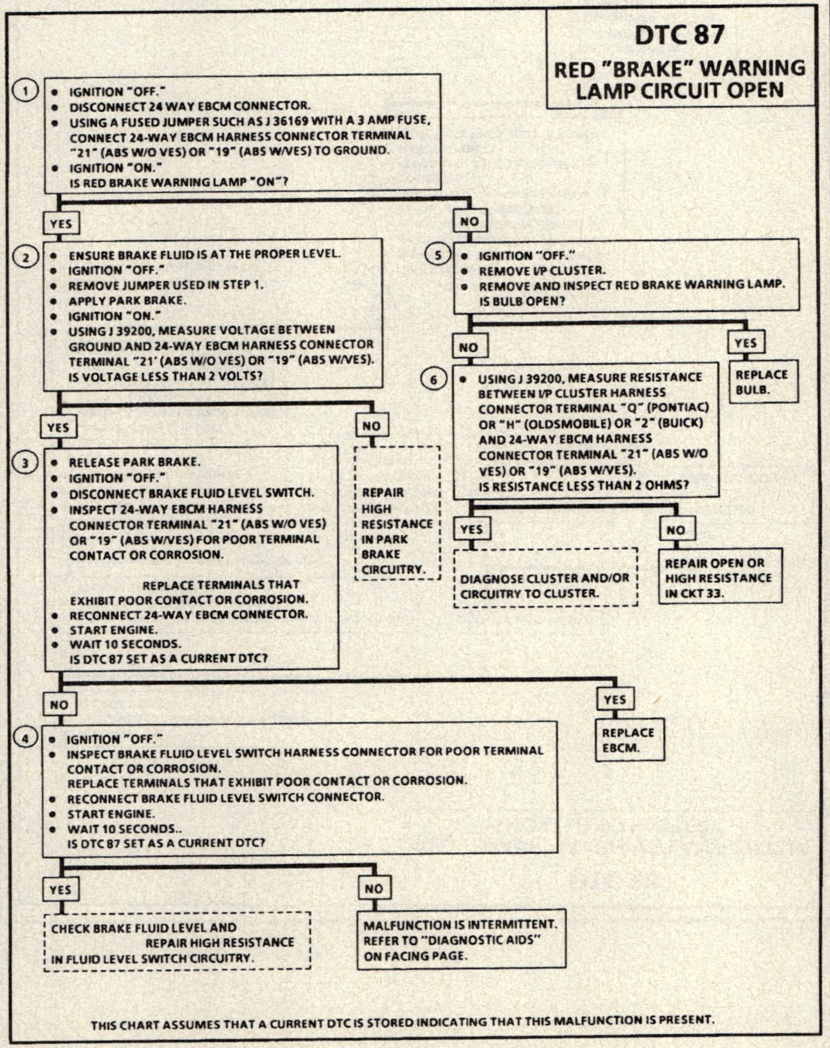

DIAGNOSTIC CHART DTC 88 — 1994–95 ACHIEVA, GRAND AM AND SKYLARK

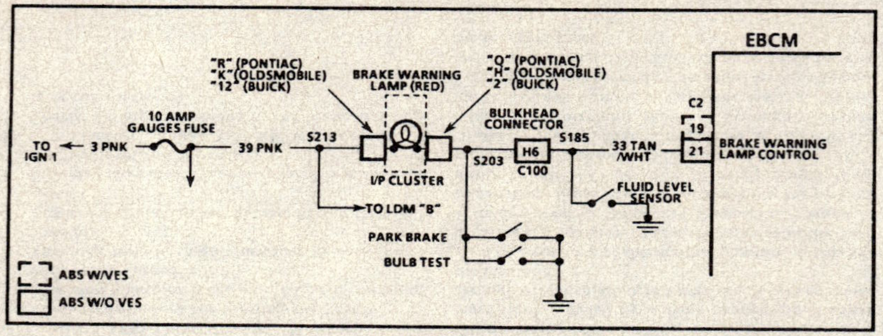

DTC 88

RED "BRAKE" WARNING LAMP CIRCUIT SHORTED TO BATTERY

Circuit Description:

This DTC identifies a short to battery between the EBCM and the red "BRAKE" warning lamp, or an open driver that does not allow the red "BRAKE" warning lamp to be illuminated by the EBCM. This will only occur if an ABS malfunction is detected that may degrade base brake operation.

DTC Will Set When: DTC 88 can be set only at the beginning of the three second bulb check. If the EBCM detects battery voltage on the red "BRAKE" warning lamp control circuit, a malfunction exists.

Action Taken: A malfunction DTC is stored. ABS/VES is not disabled; however, the ABS warning lamp will flash to indicate a malfunction exists.

DTC Chart Test Description: Number(s) below refer to circled number(s) on the diagnostic chart.

1. This step identifies if the ground circuit to the red "BRAKE" warning lamp is being completed by a source other than the EBCM.
2. This step identifies if the malfunction is currently present.
3. By removing the 10 amp gauge fuse, the voltage source is eliminated. This test indicates if voltage is being supplied from a source other than the gauge fuse.
4. Ensures malfunction was not due to physical damage of the circuitry.

Diagnostic Aids: An "Intermittent" malfunction is most likely caused by a poor connection, rubbed through wire insulation, or a wire that is broken inside the insulation.

The frequency of the malfunction can be checked by using the enhanced diagnostic function of the Tech 1, as described in "Enhanced Diagnostics," found in this section.

Any circuitry, that is suspected as causing the intermittent complaint, should be thoroughly checked for backed out terminals, improper mating, broken locks, improperly formed or damaged terminals, poor terminal to wiring connections or physical damage to the wiring harness.

DIAGNOSTIC CHART DTC 88 — 1994–95 ACHIEVA, GRAND AM AND SKYLARK, CONTINUED

DTC 88

RED "BRAKE" WARNING LAMP CIRCUIT SHORTED TO BATTERY

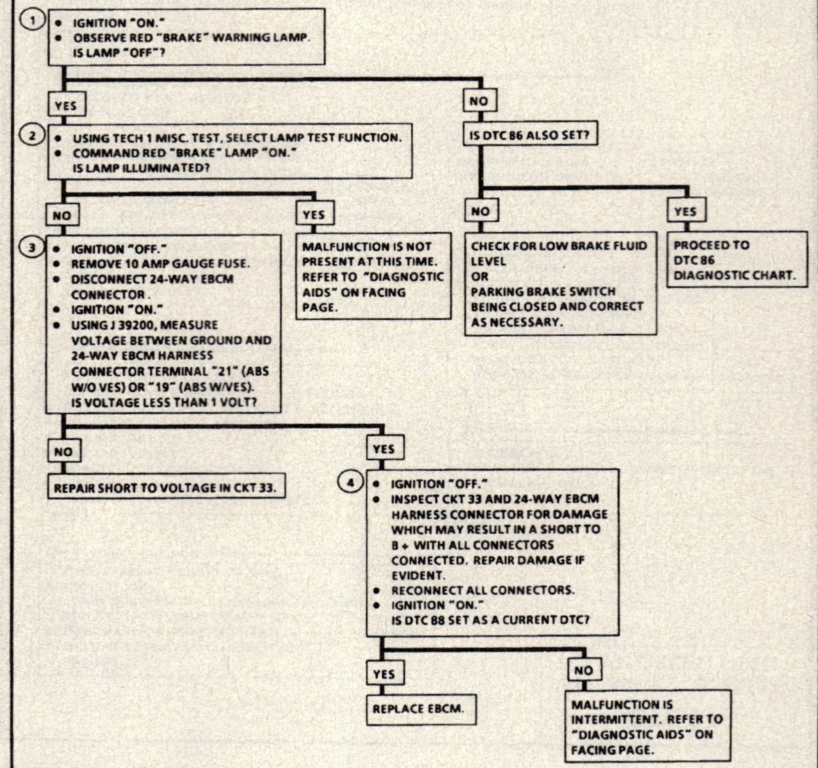

THIS CHART ASSUMES THAT A CURRENT DTC IS STORED INDICATING THAT THIS MALFUNCTION IS PRESENT.

DIAGNOSTIC CHART DTC 91 — 1994–95 ACHIEVA, GRAND AM AND SKYLARK

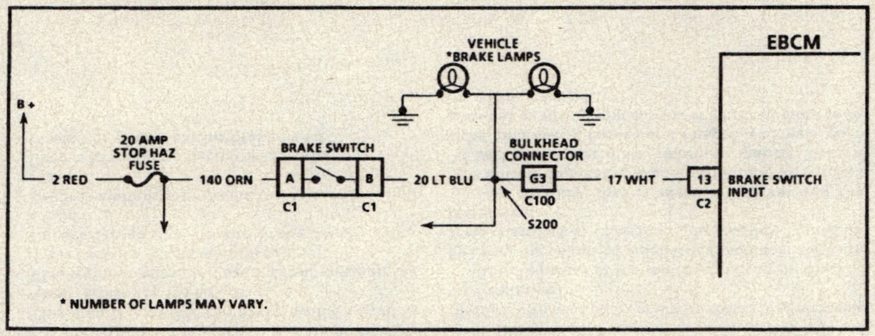

DTC 91

OPEN BRAKE SWITCH CONTACTS DURING DECELERATION

Circuit Description:

This DTC is used to detect an open brake switch in the non-ABS mode. The EBCM looks for decel rates that would indicate braking action and verifies this assumption by requiring several repeats of this detection method. In each case, ABS will not be available since no brake switch input is seen.

DTC Will Set When: DTC 91 can be set if three deceleration cycles occur with the brake switch "OFF."

Action Taken: A malfunction DTC is stored, ABS/VES is disabled and the ABS warning lamp is turned "ON."

DTC Chart Test Description: Number(s) below refer to circled number(s) on the diagnostic chart.

1. This step indicates if the brake switch signal is being received by the EBCM.
2. This step indicates if an open circuit exists in the brake switch or brake lamp circuitry.
3. Checks for an open in the brake switch input circuit.
4. Ensures malfunction was not due to poor terminal contact.
5. Checks for voltage entering brake switch.
6. Checks for voltage exiting brake switch.

Diagnostic Aids: An "Intermittent" malfunction is most likely caused by a poor connection, rubbed through wire insulation, or a wire that is broken inside the insulation.

The frequency of the malfunction can be checked by using the enhanced diagnostic function of the Tech 1, as described in "Enhanced Diagnostics," found in this section.

Any circuitry, that is suspected as causing the intermittent complaint, should be thoroughly checked for backed out terminals, improper mating, broken locks, improperly formed or damaged terminals, poor terminal to wiring connections or physical damage to the wiring harness.

DIAGNOSTIC CHART DTC 91 — 1994–95 ACHIEVA, GRAND AM AND SKYLARK, CONTINUED

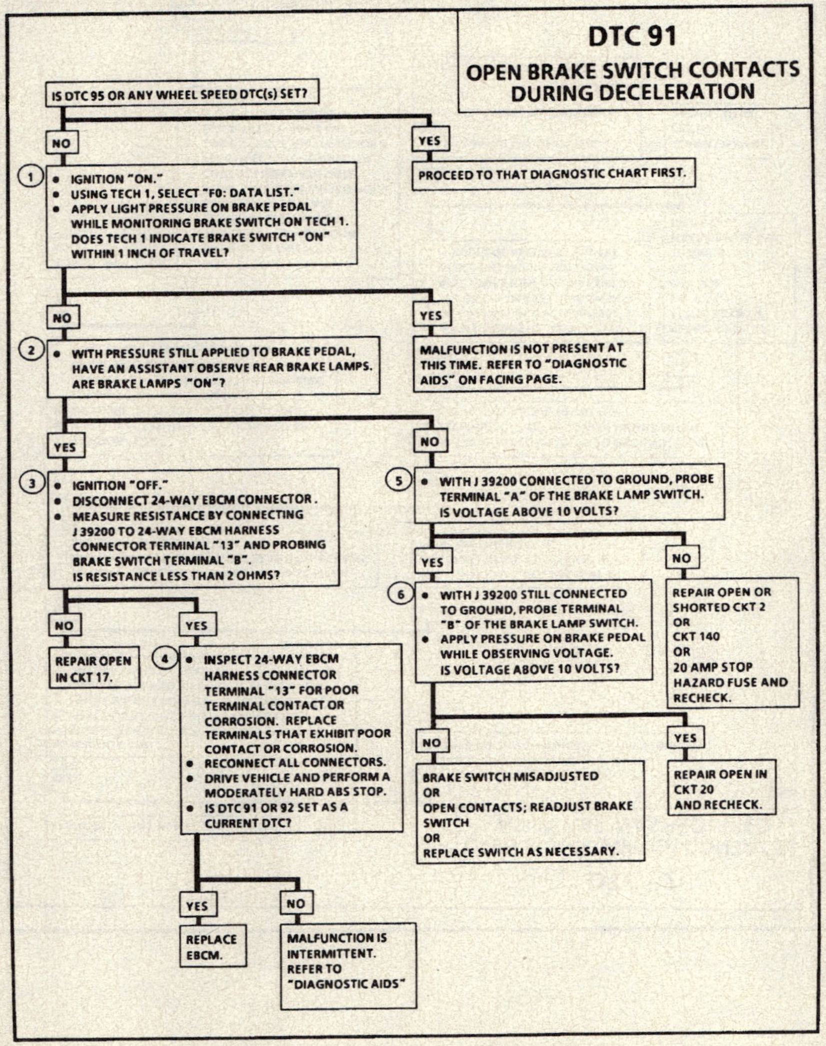

DIAGNOSTIC CHART DTC 92 — 1994–95 ACHIEVA, GRAND AM AND SKYLARK

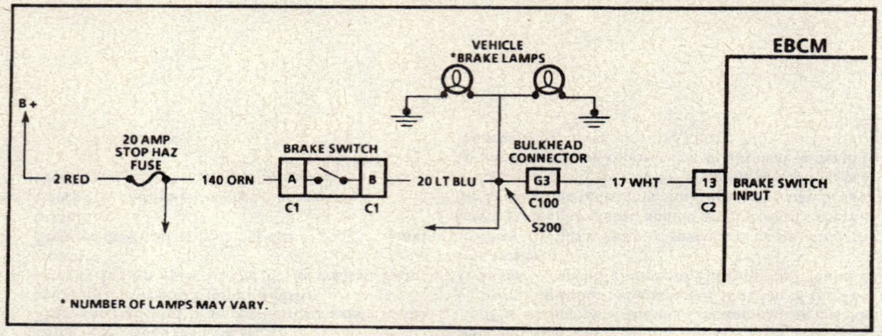

DTC 92

OPEN BRAKE SWITCH CONTACTS WHEN ABS WAS REQUIRED

Circuit Description:
This DTC is run to determine the proper operation of the brake switch. This is important because ABS is activated when the brake switch is "ON." If the brake switch is "OFF," ABS will never be activated. Since this malfunction is difficult to detect under normal braking conditions, this malfunction is only detected when ABS is required.

DTC Will Set When: DTC 92 can be set if the vehicle's speed is greater than 8 km/h (5 mph). If the brake was not "ON" and a release was required on two channels for .7 second, a malfunction exists.

Action Taken: A malfunction DTC is stored, ABS/VES is disabled and the ABS warning lamp is turned "ON."

DTC Chart Test Description: Number(s) below refer to circled number(s) on the diagnostic chart.

1. This step indicates if the brake switch signal is being received by the EBCM.
2. This step indicates if an open circuit exists in the brake switch or brake lamp circuitry.
3. Checks for an open in the brake switch input circuit.
4. Ensures malfunction was not due to poor terminal contact.
5. Checks for voltage entering brake switch.
6. Checks for voltage exiting brake switch.

Diagnostic Aids: An "Intermittent" malfunction is most likely caused by a poor connection, rubbed through wire insulation, or a wire that is broken inside the insulation.

The frequency of the malfunction can be checked by using the enhanced diagnostic function of the Tech 1, as described in "Enhanced Diagnostics," found in this section.

Any circuitry, that is suspected as causing the intermittent complaint, should be thoroughly checked for backed out terminals, improper mating, broken locks, improperly formed or damaged terminals, poor terminal to wiring connections or physical damage to the wiring harness.

DIAGNOSTIC CHART DTC 92 — 1994–95 ACHIEVA, GRAND AM AND SKYLARK, CONTINUED

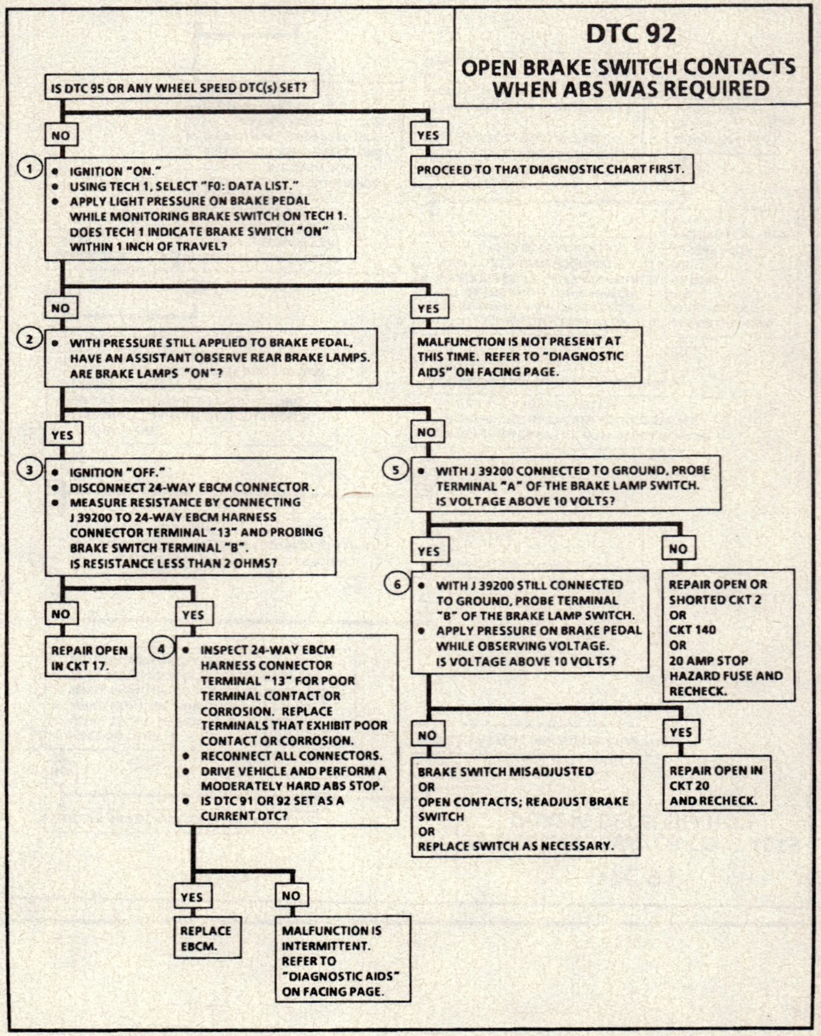

DIAGNOSTIC CHART DTC 93 — 1994–95 ACHIEVA, GRAND AM AND SKYLARK

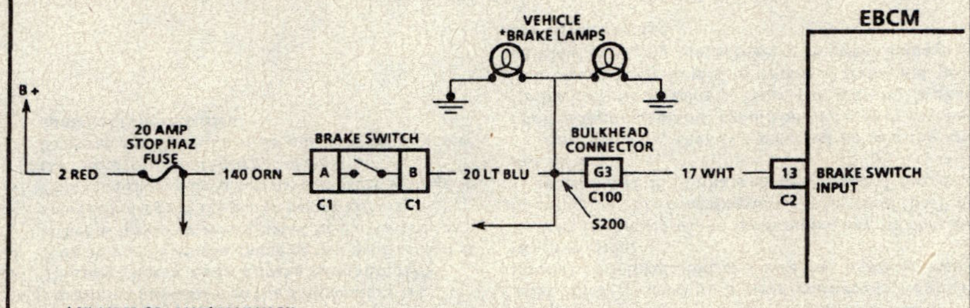

* NUMBER OF LAMPS MAY VARY.

DTC 93

DTCs 91 OR 92 SET IN CURRENT OR PREVIOUS IGNITION CYCLE

Circuit Description:

This DTC is the second portion of DTCs 91 and 92. If DTCs 91 or 92 failed during the last ignition cycle, DTC 93 becomes a current failure during the next ignition cycle, keeping ABS disabled until a brake switch "ON" state is seen. When a change is seen during an ignition cycle in which DTC 93 is a current failure, DTC 91 or 92 will clear itself at the end of the current ignition cycle, and ABS will enable itself at the start of the next ignition cycle. DTC 93 alone indicates DTCs 91 or 92 failed previously, but is intermittent, or has been corrected.

DTC Chart Test Description: Number(s) below refer to circled number(s) on the diagnostic chart.

1. This step indicates which DTC (either 91 or 92) caused DTC 93 to set.

Diagnostic Aids: An "Intermittent" malfunction is most likely caused by a poor connection, rubbed through wire insulation, or a wire that is broken inside the insulation.

The frequency of the malfunction can be checked by using the enhanced diagnostic function of the Tech 1, as described in "Enhanced Diagnostics," found in this section.

Any circuitry, that is suspected as causing the intermittent complaint, should be thoroughly checked for backed out terminals, improper mating, broken locks, improperly formed or damaged terminals, poor terminal to wiring connections or physical damage to the wiring harness.

Also, verify proper brake switch operation using the data list of the Tech 1. As the brake is applied, the data list should display the brake switch "ON" within 1 inch of travel.

DIAGNOSTIC CHART DTC 93 — 1994–95 ACHIEVA, GRAND AM AND SKYLARK, CONTINUED

DTC 93

DTCs 91 OR 92 SET IN CURRENT OR PREVIOUS IGNITION CYCLE

(1)
- IGNITION "ON."
- USING TECH 1, READ DTCs.
  IS DTC 91 OR 92 SET AS A HISTORY OR CURRENT DTC?

YES → PROCEED TO DTC CHART THAT SET.

NO → USING TECH 1, SELECT "F1: DTC HISTORY," VERIFY MALFUNCTION FREQUENCY WAS LOW AND CLEAR DTCs.

THIS CHART ASSUMES THAT A CURRENT DTC IS STORED INDICATING THAT THIS MALFUNCTION IS PRESENT.

DIAGNOSTIC CHART DTC 94 — 1994–95 ACHIEVA, GRAND AM AND SKYLARK

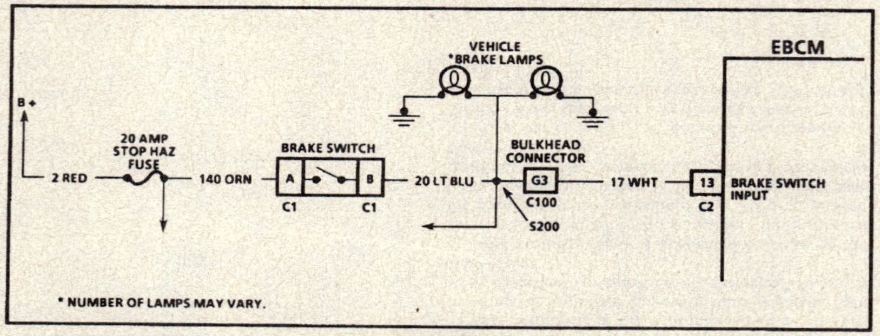

DTC 94

BRAKE SWITCH CONTACTS ALWAYS CLOSED

Circuit Description:

This DTC is run to determine the proper operation of the brake switch. This is important because ABS is activated when the brake switch is "ON" and turned "OFF" when the brake is "OFF." If the brake switch is always "ON," ABS operation will always be requested resulting in potential modulator cycling on rough roads. Additionally, this malfunction will most likely result in a dead battery (due to the brake lamps staying "ON") if the driver is not informed of this malfunction.

DTC Will Set When: DTC 94 can be set if the vehicle speed reaches at least 40 km/h (25 mph). If the brake was never "OFF" during two consecutive drive cycles, a malfunction exists.

Action Taken: A malfunction DTC is stored; however, ABS/VES is not disabled.

DTC Chart Test Description: Number(s) below refer to circled number(s) on the diagnostic chart.

1. This step checks to see if brake lamps are "ON."
2. This step isolates the cause of the malfunction to either a faulty or misadjusted brake switch, or a short to voltage in the brake switch circuitry.
3. This step checks for a resistive short to voltage in the brake switch circuitry.
4. Ensures malfunction was not due to physical damage of the circuitry.

Diagnostic Aids: An "Intermittent" malfunction is most likely caused by a poor connection, rubbed through wire insulation, or a wire that is broken inside the insulation.

The frequency of the malfunction can be checked by using the enhanced diagnostic function of the Tech 1, as described in "Enhanced Diagnostics," found in this section.

Any circuitry, that is suspected as causing the intermittent complaint, should be thoroughly checked for backed out terminals, improper mating, broken locks, improperly formed or damaged terminals, poor terminal to wiring connections or physical damage to the wiring harness.

DIAGNOSTIC CHART DTC 94 — 1994–95 ACHIEVA, GRAND AM AND SKYLARK, CONTINUED

DTC 94

BRAKE SWITCH CONTACTS ALWAYS CLOSED

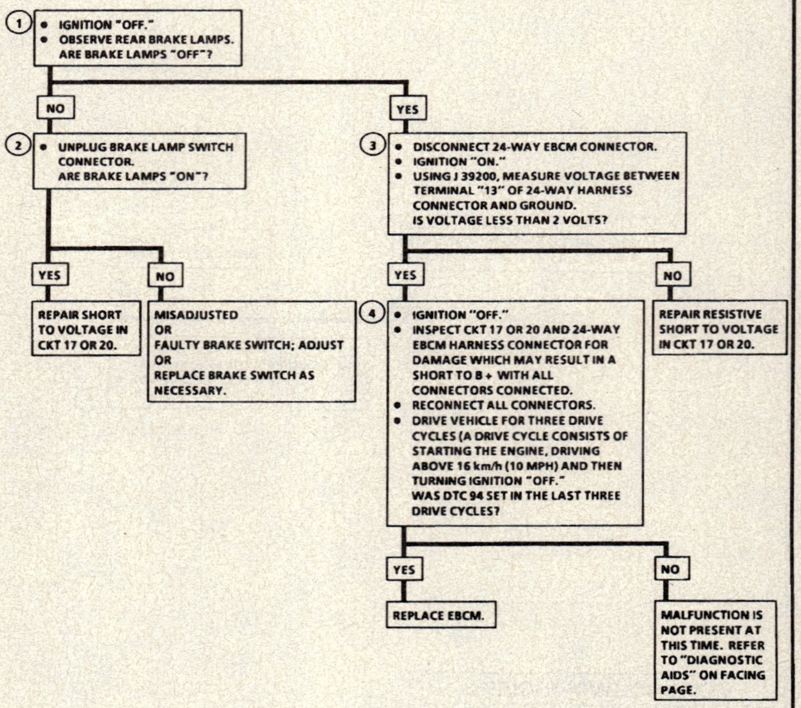

THIS CHART ASSUMES THAT A CURRENT DTC IS STORED INDICATING THAT THIS MALFUNCTION IS PRESENT.

DIAGNOSTIC CHART DTC 95 — 1994–95 ACHIEVA, GRAND AM AND SKYLARK

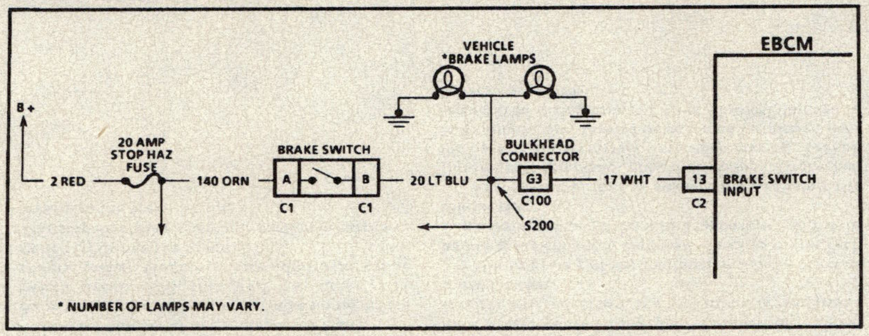

DTC 95

BRAKE SWITCH CIRCUIT OPEN

Circuit Description:
This DTC is used to identify open brake switch circuitry that prevents the brake switch input to the EBCM from changing states when the brake is applied. This DTC is used in conjunction with DTCs 91 and 92 to determine the cause of an open brake switch malfunction.

DTC Will Set When: DTC 95 can be set after initialization is completed. If the brake switch input voltage is out of specification for one second indicating an open circuit, a malfunction exists.

Action Taken: A malfunction DTC is stored, ABS/VES is disabled and the ABS warning lamp is turned "ON."

DTC Chart Test Description: Number(s) below refer to circled number(s) on the diagnostic chart.

1. This step is used to confirm that an open in the brake switch circuitry currently exists.
2. This step indicates if the brake switch signal is being received by the EBCM.
3. This step indicates if an open circuit exists in the brake switch or brake lamp circuitry.
4. This step isolates the open circuit to either the brake switch input circuitry, or the EBCM.
5. Ensures malfunction was not due to poor terminal contact.

Diagnostic Aids: An "Intermittent" malfunction is most likely caused by a poor connection, rubbed through wire insulation, or a wire that is broken inside the insulation.

The frequency of the malfunction can be checked by using the enhanced diagnostic function of the Tech 1, as described in "Enhanced Diagnostics," found in this section.

Any circuitry, that is suspected as causing the intermittent complaint, should be thoroughly checked for backed out terminals, improper mating, broken locks, improperly formed or damaged terminals, poor terminal to wiring connections or physical damage to the wiring harness.

DIAGNOSTIC CHART DTC 95 — 1994–95 ACHIEVA, GRAND AM AND SKYLARK, CONTINUED

DTC 95

BRAKE SWITCH CIRCUIT OPEN

IS DTC 96 CURRENTLY SET?

- YES: PROCEED TO DTC 96 DIAGNOSTIC CHART.
- NO:

(1)
- IGNITION "ON."
- USING TECH 1, SELECT "F0: DATA LIST" AND MONITOR BRAKE SWITCH. DOES BRAKE SWITCH INDICATE "CIRCUIT OPEN"?

  - NO: MALFUNCTION IS NOT PRESENT AT THIS TIME. REFER TO "DIAGNOSTIC AIDS" ON FACING PAGE.
  - YES:

(2)
- APPLY LIGHT PRESSURE ON BRAKE PEDAL WHILE MONITORING BRAKE SWITCH ON TECH 1. DOES TECH 1 INDICATE BRAKE SWITCH "ON" WITHIN 1 INCH OF TRAVEL?

  - YES: PROCEED TO DTC 96 DIAGNOSTIC CHART.
  - NO:

(3)
- APPLY FIRM PRESSURE TO BRAKE PEDAL AND OBSERVE CHMSL AND REAR BRAKE LAMPS. ARE ALL BRAKE LAMPS "ON"?

  - NO: REPAIR OPEN BRAKE LAMP CIRCUIT.
  - YES:

(4)
- IGNITION "OFF."
- DISCONNECT 24-WAY EBCM CONNECTOR.
- MEASURE RESISTANCE BY CONNECTING J 39200 TO 24-WAY EBCM HARNESS CONNECTOR TERMINAL "13" AND PROBING BRAKE SWITCH TERMINAL "B". IS RESISTANCE LESS THAN 2 OHMS?

  - NO: REPAIR OPEN IN CKT 17 OR 20.
  - YES:

(5)
- INSPECT 24-WAY EBCM HARNESS CONNECTOR TERMINAL "13" FOR POOR TERMINAL CONTACT OR CORROSION. REPLACE TERMINALS THAT EXHIBIT POOR CONTACT OR CORROSION.
- RECONNECT ALL CONNECTORS.
- WITH FOOT OFF OF BRAKE PEDAL, START ENGINE.
- WAIT 10 SECONDS. IS DTC 95 SET AS A CURRENT DTC?

  - YES: REPLACE EBCM.
  - NO: MALFUNCTION IS INTERMITTENT. REFER TO "DIAGNOSTIC AIDS" ON FACING PAGE.

THIS CHART ASSUMES THAT A CURRENT DTC IS STORED INDICATING THAT THIS MALFUNCTION IS PRESENT.

DIAGNOSTIC CHART DTC 96 — 1994–95 ACHIEVA, GRAND AM AND SKYLARK

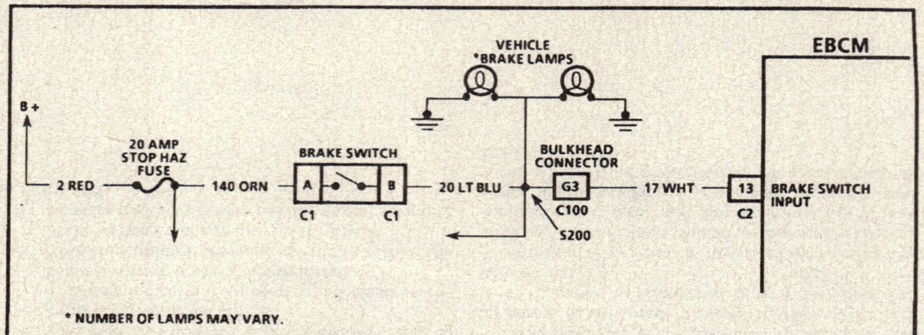

DTC 96

BRAKE LAMP CIRCUIT OPEN

Circuit Description:

This DTC is designed to isolate the cause of a DTC 95 failure and indicate to the driver ABS is still available. If DTC 95 fails with DTC 96, the brake lamps are open, the brake lamp grounds are open, or the Center High Mounted Stop Lamp (CHMSL) circuit is open during 4-way flasher use. The presence of battery voltage at the brake switch input indicates a valid brake switch input is still available.

DTC Will Set When: DTC 96 can be set only after DTC 95 has been set. If the EBCM detects battery voltage on the brake switch input circuit for one second, a malfunction exists.

Action Taken: A malfunction DTC is stored, ABS is disabled and the ABS warning lamp is turned "ON."

DTC Chart Test Description: Number(s) below refer to circled number(s) on the diagnostic chart.

1. As a result of a failure of an additional brake switch circuit DTC, this DTC may be set. To ensure proper diagnosis, any additional brake switch DTCs must be repaired first.
2. This step identifies if the malfunction is currently present in the brake circuit.

Diagnostic Aids: An "Intermittent" malfunction is most likely caused by a poor connection, rubbed through wire insulation, or a wire that is broken inside the insulation.

The frequency of the malfunction can be checked by using the enhanced diagnostic function of the Tech 1, as described in "Enhanced Diagnostics," found in this section.

Any circuitry, that is suspected as causing the intermittent complaint, should be thoroughly checked for backed out terminals, improper mating, broken locks, improperly formed or damaged terminals, poor terminal to wiring connections or physical damage to the wiring harness.

DIAGNOSTIC CHART DTC 96 — 1994–95 ACHIEVA, GRAND AM AND SKYLARK, CONTINUED

DTC 96

BRAKE LAMP CIRCUIT OPEN

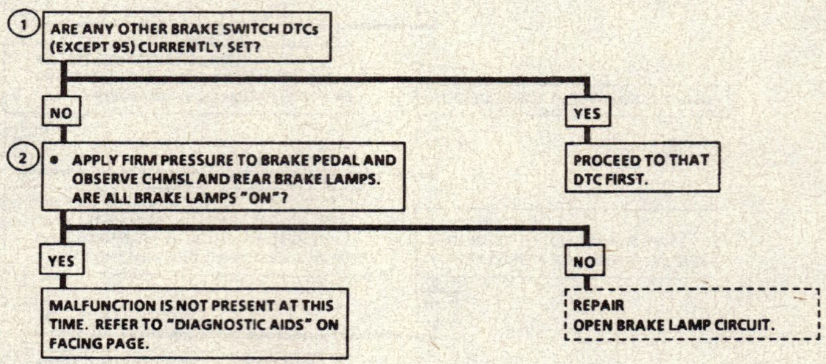

THIS CHART ASSUMES THAT A CURRENT DTC IS STORED INDICATING THAT THIS MALFUNCTION IS PRESENT.

DIAGNOSTIC CIRCUIT CHECK — 1993 CUTLASS SUPREME, GRAND PRIX, LUMINA AND REGAL

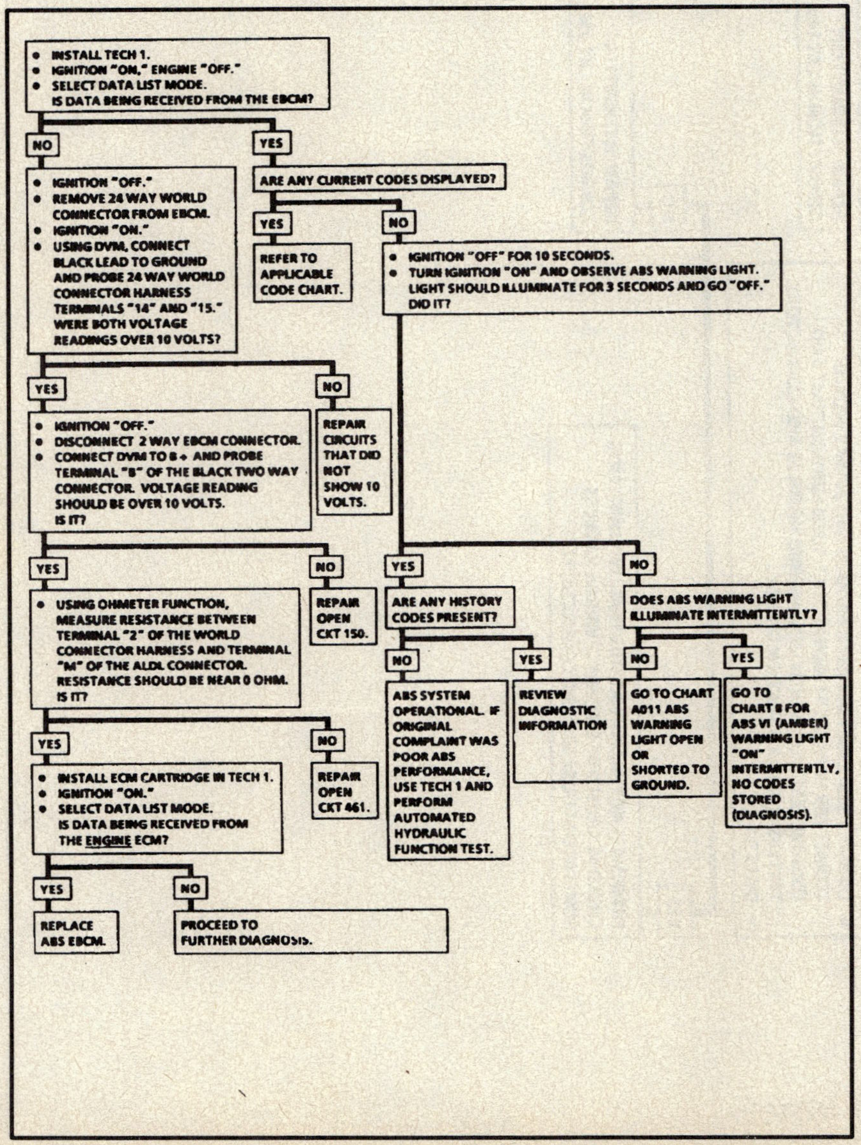

DIAGNOSTIC CHART A — 1993 CUTLASS SUPREME, GRAND PRIX, LUMINA AND REGAL

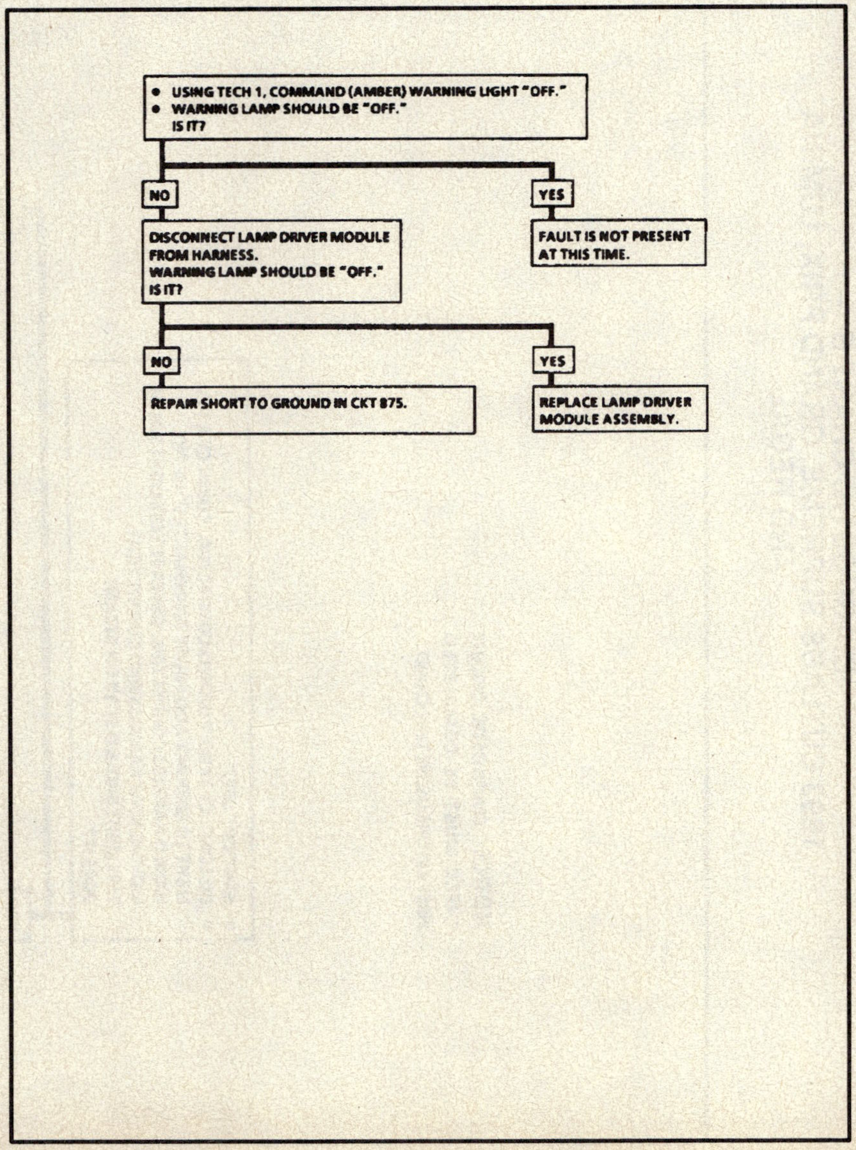

DIAGNOSTIC CHART B — 1993 CUTLASS SUPREME, GRAND PRIX, LUMINA AND REGAL

NOTICE: DIAGNOSTIC CIRCUIT CHECK MUST BE COMPLETED FIRST BEFORE USING THIS CHART.

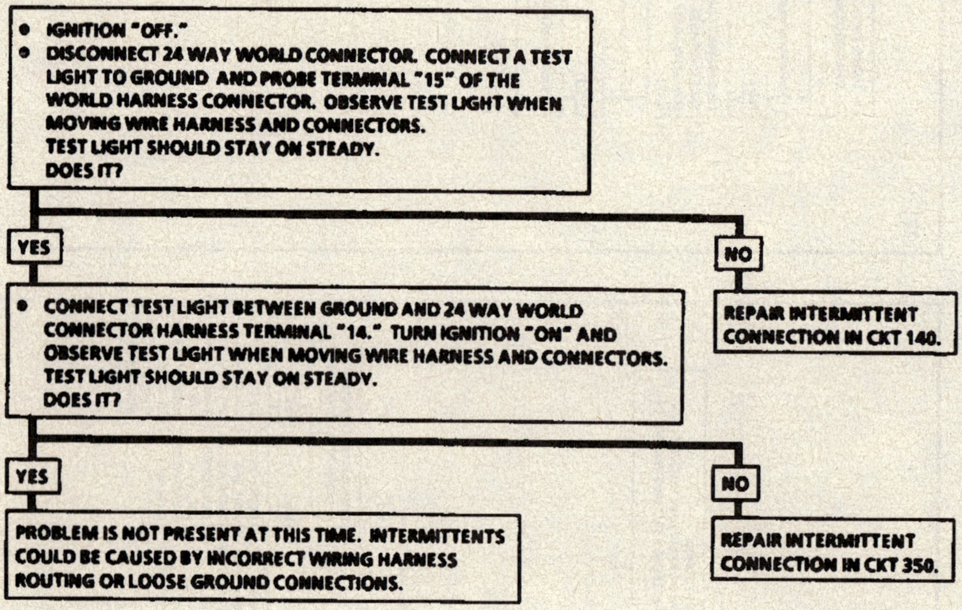

DIAGNOSTIC CHART DTC A011 — 1993 CUTLASS SUPREME, GRAND PRIX, LUMINA AND REGAL

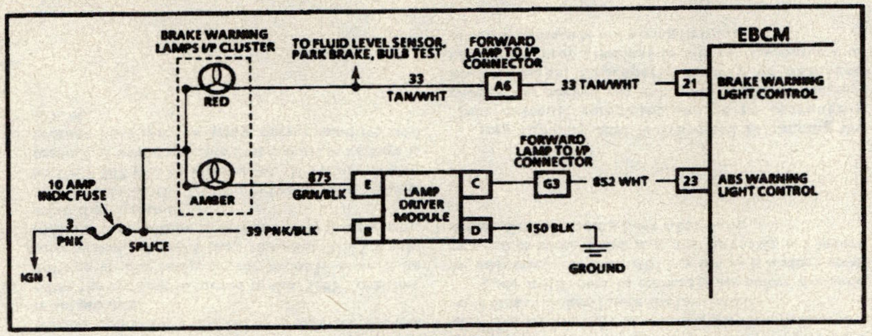

CODE A011

ABS WARNING LIGHT CIRCUIT OPEN OR SHORTED TO GROUND

Circuit Description:

This code checks the state of the ABS warning light to identify a situation in which the driver could not be warned of a system fault by the ABS warning light, or the warning light is always "ON." The Lamp Driver Module (LDM) turns the ABS warning light "ON," unless the EBCM provides ground to turn it "OFF." Because of the circuitry in the LDM, only external faults can be detected. As a result, the ABS warning light itself is not diagnosable, only the control line to the EBCM can be diagnosed. In the event of an open CKT 852, the ABS warning light will be on at all times, due to the loss of ground at the LDM input. If the control line is shorted to ground, the ABS warning light is kept "OFF," due to the LDM input being grounded.

Test Description: Number(s) below refer to circled number(s) on the diagnostic chart.

1. Checks to see if ABS light functions properly. Normal operation would indicate that the problem is not present.
2. This step determines if any other codes are present or if A018 is set falsely.
3. This step insures that the circuitry from the EBCM to the lamp driver module is complete.
4. This step insures that the ignition circuit to the lamp driver module is functional. An open indicator fuse or open CKT 3, 39, or 875 would result in inoperative indicator lights.
5. Isolates the cause of the ABS lamp remaining "OFF," by disconnecting the 24 way world connector, CKT 852 should be open and cause the ABS lamp to be "ON."

Diagnostic Aids:

An "Intermittent" problem may be caused by a poor connection, rubbed through wire insulation, or a wire that is broken inside the insulation.

The light test function of the "Scan" tool may be used to command the light "ON" while looking for an intermittent problem in the warning light circuitry.

Any circuitry that is suspected as causing the intermittent complaint should be thoroughly checked for backed out terminals, improper mating, broken locks, improperly formed or damaged terminals, poor terminal to wiring connections or physical damage to the wiring harness.

DIAGNOSTIC CHART DTC A011 — 1993 CUTLASS SUPREME, GRAND PRIX, LUMINA AND REGAL, CONTINUED

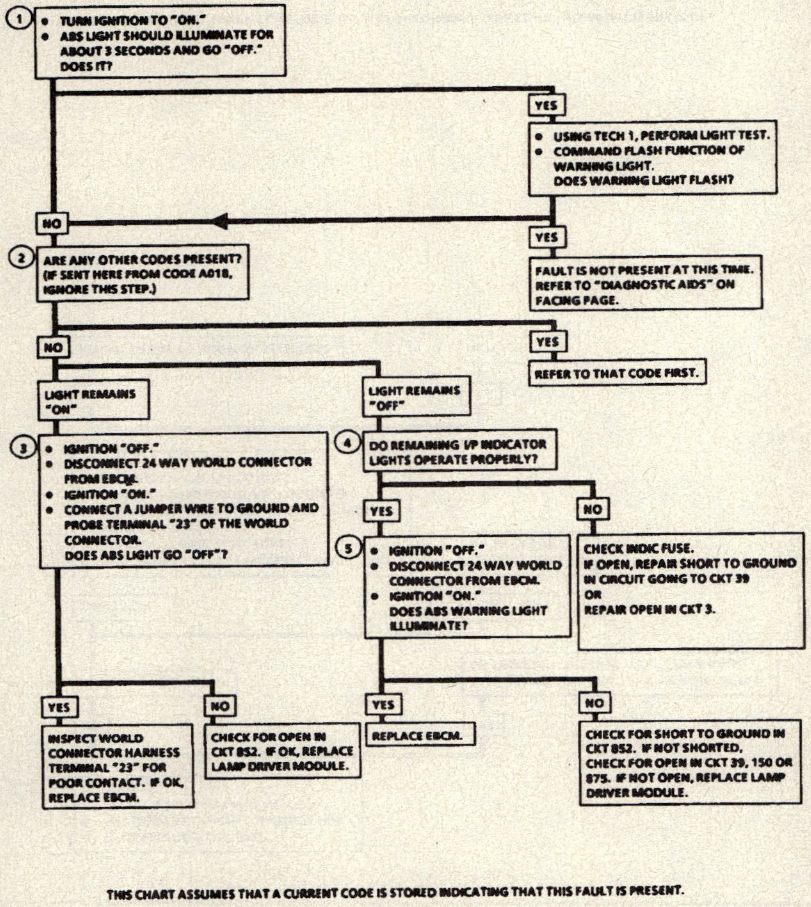

DIAGNOSTIC CHART DTC A013 — 1993 CUTLASS SUPREME, GRAND PRIX, LUMINA AND REGAL

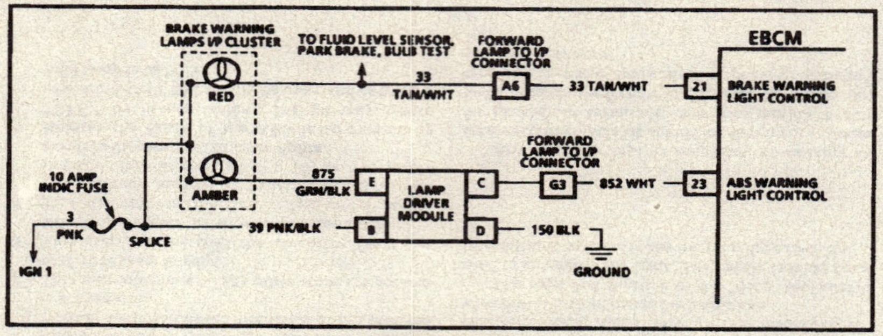

CODE A013

ABS WARNING LIGHT CIRCUIT SHORTED TO BATTERY

Circuit Description:

This code checks the state of the ABS warning light to identify a situation in which the driver could not be warned of a system fault by the ABS warning light. The Lamp Driver Module (LDM) turns the ABS warning light "ON," unless the EBCM provides ground to turn it "OFF." Because of the circuitry internal to the LDM, only external faults can be detected. As a result, the ABS warning light itself is not diagnosable, only the control line to the EBCM can be diagnosed.

If the ABS warning light control line is shorted to battery or the EBCM control circuit is open, the Electronic Brake Control Module (EBCM) is not able to turn "OFF" the ABS warning light.

Test Description: Number(s) below refer to circled number(s) on the diagnostic chart.

1. Checks to see if ABS light functions properly. Normal operation would indicate that the problem is not present.
2. After the ignition is turned to the "OFF" position, voltage to the lamp driver module should be interrupted. A light that remains "ON" would indicate that voltage to this circuit is from a point other than ignition.
3. After removing the 10 amp indicator fuse, voltage on CKT 852 will be eliminated. If voltage is still present, it would indicate that a short to voltage is present between the lamp driver module and EBCM.

Diagnostic Aids:

An "Intermittent" problem may be caused by a poor connection, rubbed through wire insulation or a wire that is broken inside the insulation.

Code A013 must be currently set before it should be addressed. If Code A013 is set as a history code, proceed to those codes that are currently set before attempting to repair a Code A013 fault.

Any circuitry that is suspected as causing the "Intermittent" complaint should be thoroughly checked for backed out terminals, improper mating, broken locks, improperly formed or damaged terminals, poor terminal to wiring connections or physical damage to the wiring harness.

DIAGNOSTIC CHART DTC A013 — 1993 CUTLASS SUPREME, GRAND PRIX, LUMINA AND REGAL, CONTINUED

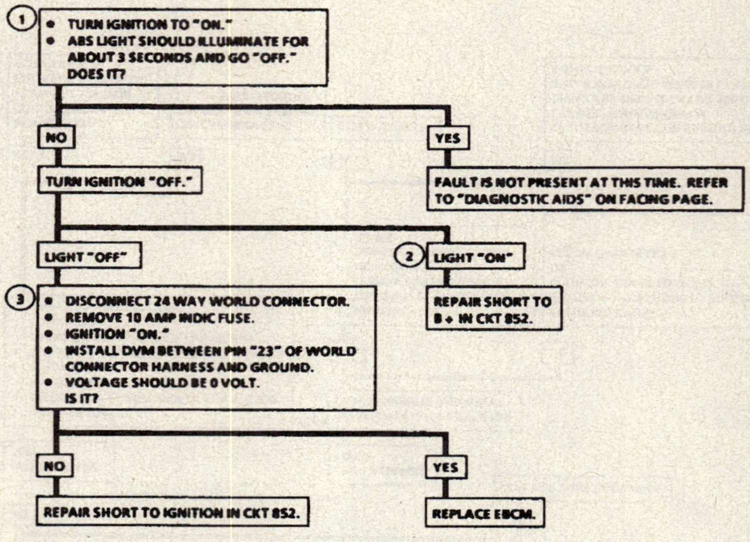

THIS CHART ASSUMES THAT A CURRENT CODE IS STORED INDICATING THAT THIS FAULT IS PRESENT.

DIAGNOSTIC CHART DTC A014 — 1993 CUTLASS SUPREME, GRAND PRIX, LUMINA AND REGAL

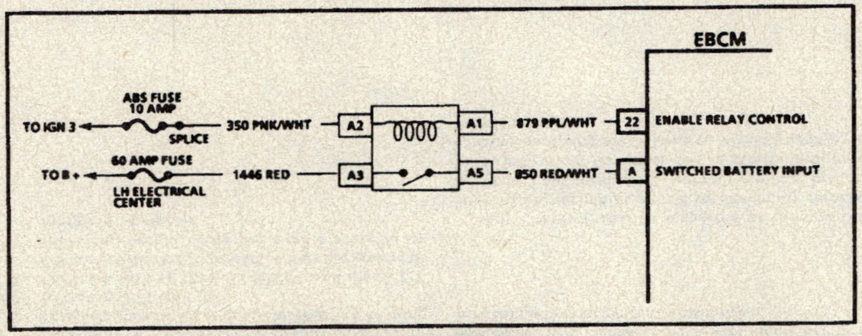

CODE A014

ENABLE RELAY CONTACTS OR FUSE OPEN

Circuit Description:

Ignition voltage is supplied through terminal "A2" of the ABS enable relay. The EBCM then is able to energize the pull-in coil by completing the ground circuit at Pin "22" of the EBCM. The magnetic field created closes the enable relay contacts and allows battery voltage and current to be supplied to the EBCM, which supplies power to the motors and solenoids.

This code monitors the availability of current/voltage to the motors and solenoids. This fault indicates voltage is not available and would therefore not allow ABS operation if required.

Test Description: Number(s) below refer to circled number(s) on the diagnostic chart.

1. This step would indicate that the EBCM is sensing battery voltage present through the relay contacts at terminal "A". This shows that the relay and circuitry is functional.
2. This insures that battery voltage is available to terminals "A2" and "A3" of the relay.
3. This step checks to make sure that the pull-in coil of the enable relay has continuity.
4. This checks the ability of the EBCM to energize the enable relay. An open circuit in the EBCM or CKT 879 would not allow the enable relay to be energized.
5. This step checks for an open CKT 850 between the enable relay and EBCM.

Diagnostic Aids:

An "Intermittent" problem may be caused by a poor connection, rubbed through wire insulation, or a wire that is broken inside the insulation.

If the frequency of the failure was low, repair any other codes first and then clear codes. After clearing codes, turn the ignition "ON" for 5 seconds, and then "OFF" for 5 seconds. Repeat this procedure 10 times. If the ABS light went "OFF" each time after 3 seconds, and when reading codes, Code A014 still exists, the Code A014 is being caused by vehicle transient voltage, and should be ignored.

Any circuitry that is suspected as causing the "Intermittent" complaint should be thoroughly checked for backed out terminals, improper mating, broken locks, improperly formed or damaged terminals, poor terminal to wiring connections or physical damage to the wiring harness.

VIBRATION, TEMPERATURE EFFECTS:

Check for vibration effects by performing the relay test function of the "Scan" tool. With the relay test commanded "ON," lightly tap the top and sides of the relay while monitoring relay voltage. If the relay voltage changes significantly, replace the relay.

If Code A014 only fails when the vehicle is initially started in cold ambient conditions (temperature less than 32°F - 0°C), replace the relay.

DIAGNOSTIC CHART DTC A014 — 1993 CUTLASS SUPREME, GRAND PRIX, LUMINA AND REGAL, CONTINUED

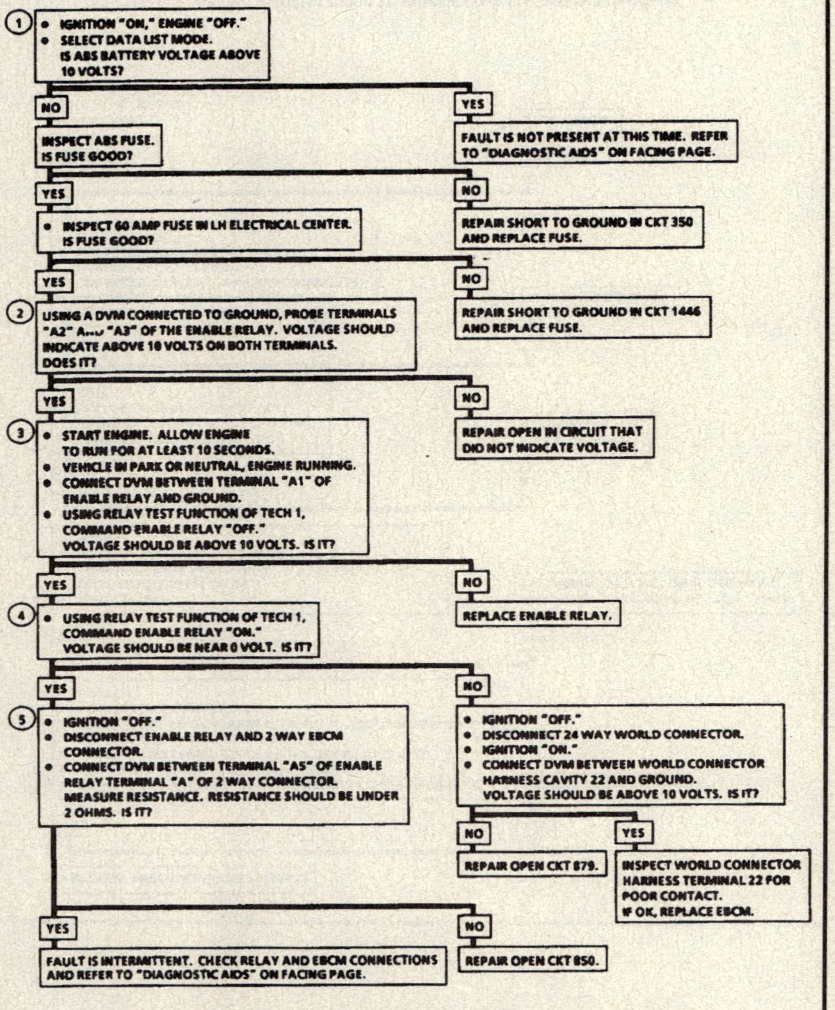

THIS CHART ASSUMES THAT A CURRENT CODE IS STORED INDICATING THAT THIS FAULT IS PRESENT.

DIAGNOSTIC CHART DTC A015 — 1993 CUTLASS SUPREME, GRAND PRIX, LUMINA AND REGAL

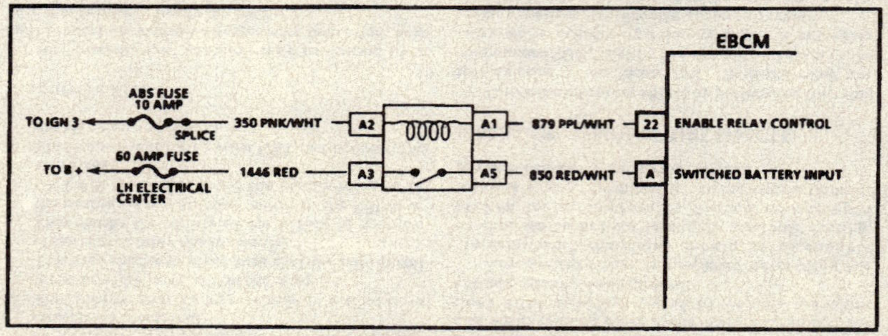

CODE A015

ENABLE RELAY CONTACTS SHORTED TO BATTERY

Circuit Description:

Ignition voltage is supplied through terminal "A2" of the ABS enable relay. The EBCM then is able to energize the pull-in coil by completing the ground circuit at Pin "22" of the EBCM. The magnetic field created closes the enable relay contacts and allows battery voltage and current to be supplied to the EBCM, which supplies power to the motors and solenoids.

This code determines if the enable relay is energized when it should not be. This fault would not allow the enable relay to remove power to the ABS system. If a second fault were to occur that requires the enable relay to be turned "OFF," that fault cannot be removed if the relay cannot be controlled.

Test Description: Number(s) below refer to circled number(s) on the diagnostic chart.

1. This step indicates that the EBCM is capable of controlling the enable relay as commanded.
2. This step checks for a possible short to battery voltage in CKT 850.
3. This step insures that the EBCM and also CKT 879 are not shorted to ground. A short to ground in either case would cause the enable relay to be energized constantly.

Diagnostic Aids:

An "Intermittent" problem may be caused by a poor connection, rubbed through wire insulation, or a wire that is broken inside the insulation.

Any circuitry, that is suspected as causing the intermittent complaint, should be thoroughly checked for backed out terminals, improper mating, broken locks, improperly formed or damaged terminals, poor terminal to wiring connections or physical damage to the wiring harness.

DIAGNOSTIC CHART DTC A015 — 1993 CUTLASS SUPREME, GRAND PRIX, LUMINA AND REGAL, CONTINUED

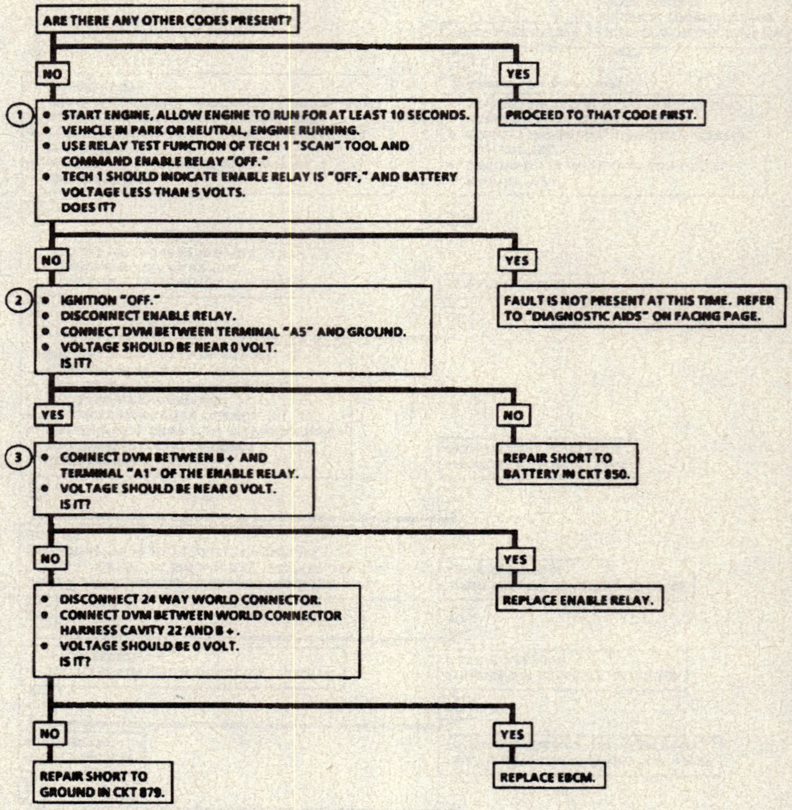

THIS CHART ASSUMES THAT A CURRENT CODE IS STORED INDICATING THAT THIS FAULT IS PRESENT.

DIAGNOSTIC CHART DTC A016 — 1993 CUTLASS SUPREME, GRAND PRIX, LUMINA AND REGAL

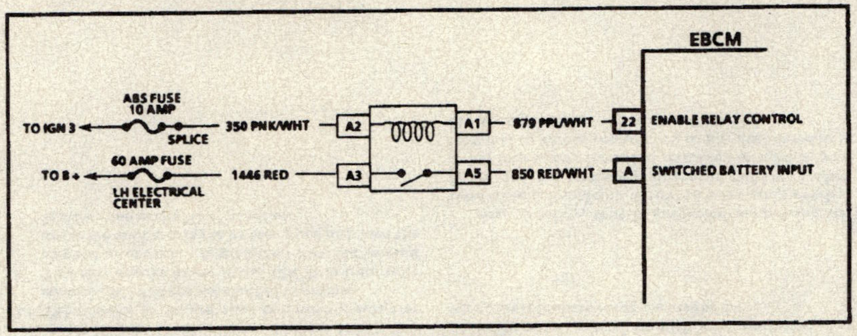

CODE A016

ENABLE RELAY COIL CIRCUIT OPEN

Circuit Description:

Ignition voltage is supplied through terminal "A2" of the ABS enable relay. The EBCM then is able to energize the pull-in coil by completing the ground circuit at Pin "22" of the EBCM. The magnetic field created closes the enable relay contacts and allows battery voltage and current to be supplied to the EBCM, which supplies power to the motors and solenoids.

This code detects opens in the enable relay coil circuit. An open in this circuit will not allow the enable relay to be energized, thus preventing voltage/current to the motors and solenoids.

Test Description: Number(s) below refer to circled number(s) on the diagnostic chart.

1. This step indicates the EBCM is capable of controlling the enable relay as commanded.
2. This step checks to insure that voltage is available to the pull-in coil of the enable relay.
3. This step insures that there is continuity through the pull-in coil of the enable relay.

Diagnostic Aids:

An "Intermittent" problem may be caused by a poor connection, rubbed through wire insulation, or a wire that is broken inside the insulation

If the frequency of the problem is high, but is currently intermittent, check for high coil resistance by measuring between relay terminals 5 and 2 using a DVM. If resistance shows greater than 100 ohms, replace the relay.

Any circuitry, that is suspected as causing the intermittent complaint, should be thoroughly checked for backed out terminals, improper mating, broken locks, improperly formed or damaged terminals, poor terminal to wiring connections or physical damage to the wiring harness.

DIAGNOSTIC CHART DTC A016 — 1993 CUTLASS SUPREME, GRAND PRIX, LUMINA AND REGAL, CONTINUED

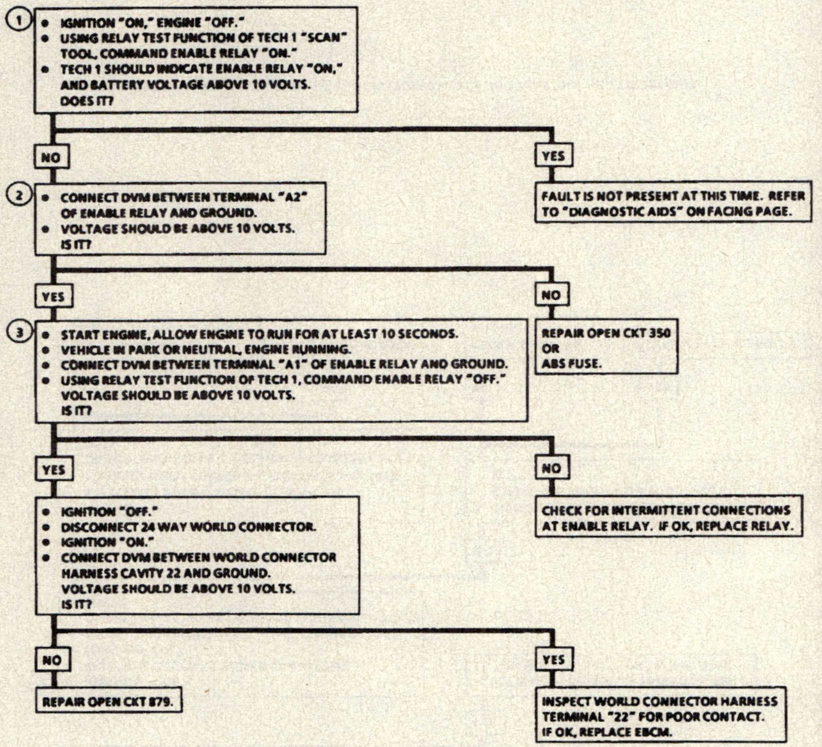

THIS CHART ASSUMES THAT A CURRENT CODE IS STORED INDICATING THAT THIS FAULT IS PRESENT.

DIAGNOSTIC CHART DTC A017 — 1993 CUTLASS SUPREME, GRAND PRIX, LUMINA AND REGAL

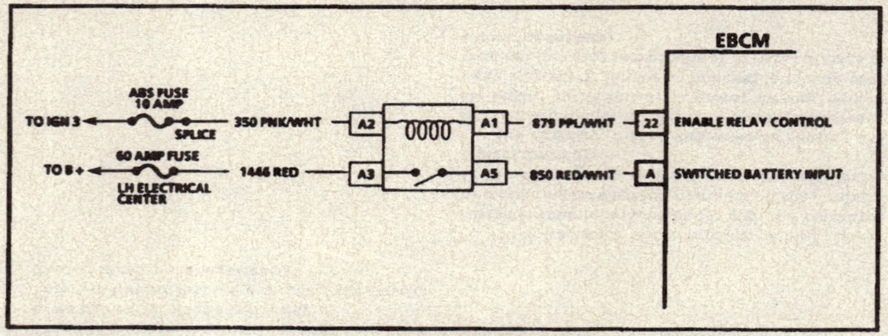

CODE A017

ENABLE RELAY COIL CIRCUIT SHORTED TO GROUND

Circuit Description:

Ignition voltage is supplied through terminal "A2" of the ABS enable relay. The EBCM then is able to energize the pull-in coil by completing the ground circuit at Pin "22" of the EBCM. The magnetic field created closes the enable relay contacts and allows battery voltage and current to be supplied to the EBCM, which supplies voltage to the motors and solenoids.

This code determines if the enable relay is energized when it should not be. This fault would not allow the enable relay to remove power to the ABS system. If a second fault were to occur that requires the enable relay to be turned "OFF," that fault cannot be removed if the relay cannot be controlled.

Test Description: Number(s) below refer to circled number(s) on the diagnostic chart.

1. This step indicates that the EBCM is capable of controlling the enable relay as commanded.
2. This checks to insure that the enable relay or control CKT 879 are not shorted to ground.
3. This step checks to see if the EBCM is internally shorted to ground. A grounded Pin "22" would cause the enable relay to be energized any time the ignition was in the "ON" position.

Diagnostic Aids:

An "Intermittent" problem may be caused by a poor connection, rubbed through wire insulation, or a wire that is broken inside the insulation

Any circuitry, that is suspected as causing the intermittent complaint, should be thoroughly checked for backed out terminals, improper mating, broken locks, improperly formed or damaged terminals, poor terminal to wiring connections or physical damage to the wiring harness.

DIAGNOSTIC CHART DTC A017 — 1993 CUTLASS SUPREME, GRAND PRIX, LUMINA AND REGAL, CONTINUED

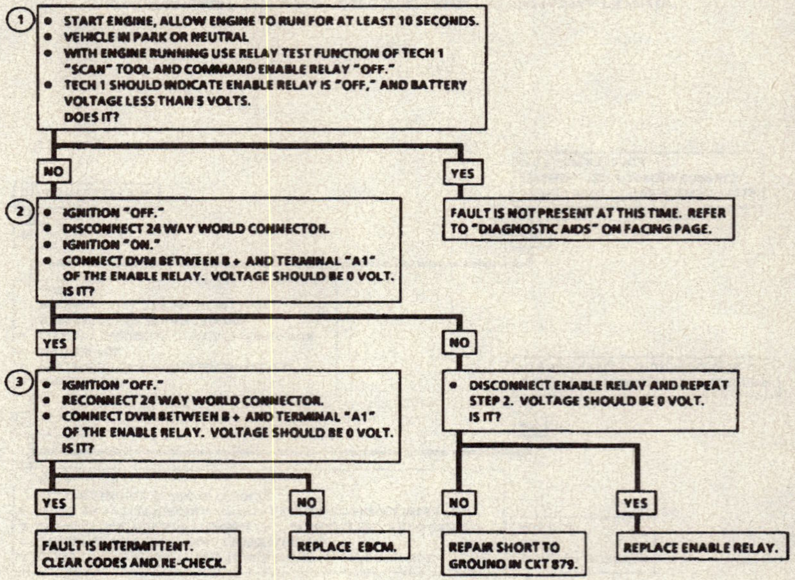

THIS CHART ASSUMES THAT A CURRENT CODE IS STORED INDICATING THAT THIS FAULT IS PRESENT.

DIAGNOSTIC CHART DTC A018 — 1993 CUTLASS SUPREME, GRAND PRIX, LUMINA AND REGAL

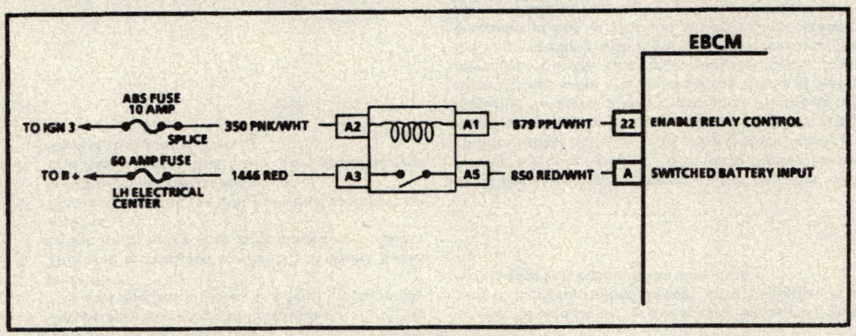

CODE A018

ENABLE RELAY COIL CIRCUIT SHORTED TO BATTERY

Circuit Description:

Ignition voltage is supplied through terminal "A2" of the ABS enable relay. The EBCM then is able to energise the pull-in coil by completing the ground circuit at Pin "22" of the EBCM. The magnetic field created closes the enable relay contacts and allows battery voltage and current to be supplied to the EBCM, which supplies voltage to the motors and solenoids.

This code monitors the availability of current/voltage to the motors, solenoids. A fault of the type specified above would not allow voltage to be available to the motors and solenoids and would, therefore, not allow ABS operation if required.

Test Description: Number(s) below refer to circled number(s) on the diagnostic chart.

1. This step indicates that the EBCM is capable of controlling the enable relay as commanded.
2. With the enable relay removed, voltage should not be available at cavity "A1". Any voltage at this point would indicate that CKT 879 was shorted to a voltage source.
3. This step checks for an open Quad-Driver in the EBCM.

Diagnostic Aids:

An "Intermittent" problem may be caused by a poor connection, rubbed through wire insulation, or a wire that is broken inside the insulation.

Any circuitry, that is suspected as causing the intermittent complaint, should be thoroughly checked for backed out terminals, improper mating, broken locks, improperly formed or damaged terminals, poor terminal to wiring connections or physical damage to the wiring harness.

DIAGNOSTIC CHART DTC A018 — 1993 CUTLASS SUPREME, GRAND PRIX, LUMINA AND REGAL, CONTINUED

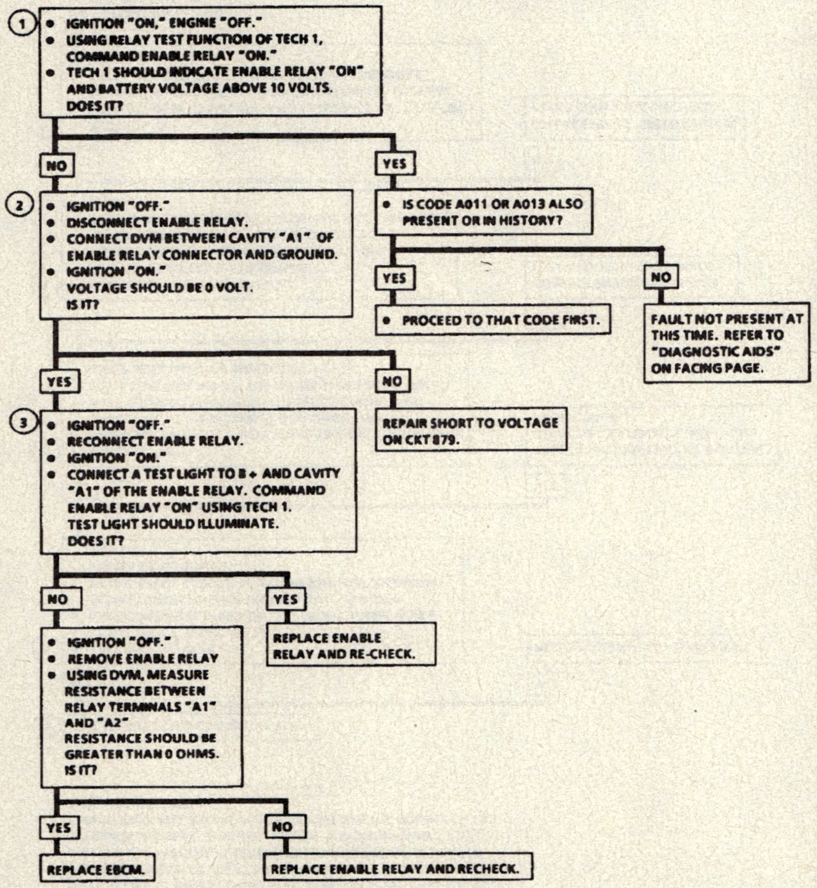

THIS CHART ASSUMES THAT A CURRENT CODE IS STORED INDICATING THAT THIS FAULT IS PRESENT.

DIAGNOSTIC CHART DTC A021 — 1993 CUTLASS SUPREME, GRAND PRIX, LUMINA AND REGAL

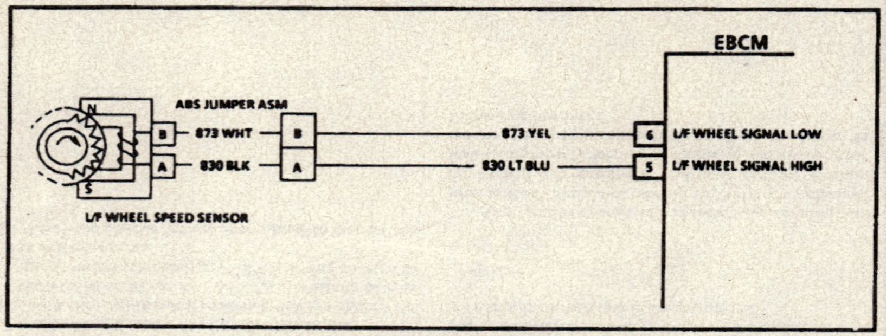

CODE A021

LEFT FRONT WHEEL SPEED = 0

Circuit Description:

As a toothed ring passes by the wheel speed sensor, changes in the electromagnetic field cause the wheel speed sensor to produce a sinusoidal (AC) voltage signal whose frequency is proportional to wheel speed. The magnitude of this signal is directly related to wheel speed and the proximity of the wheel speed sensor to the toothed ring, often referred to as the air gap.

The code detects opens or a low output condition that cause a wheel speed of 0 mph to be calculated by the ABS EBCM. The conditions present to fail this test are: (1) Three (3) wheels are rotating at least 5 mph, and (2) The fourth wheel is 0 mph.

Test Description: Number(s) below refer to circled number(s) on the diagnostic chart.

1. This determines if code A032 is present.
2. This test verifies whether the fault is currently present.
3. This step is designed to identify a wheel speed sensor or circuitry that is damaged and visibly apparent.
4. This test checks to see that the wheel speed sensor is within proper resistance values.
5. This test insures that the wheel speed sensor circuitry has good continuity.

Diagnostic Aids:

An "Intermittent" problem may be caused by a poor connection, rubbed through wire insulation, or a wire that is broken inside the insulation.

If the customers comments reflect that the ABS Amber warning lamp is "ON" only during moist environmental changes, (rain, snow, vehicle wash) all wheel speed sensor circuitry should be thoroughly inspected for signs of water intrusion. If code is not current clear codes and simulate the effects of water intrusion. Use the following procedure. Spray down the suspected area with a 5% salt water solution (two teaspoons of salt to 12 oz. of water). Start vehicle, allow vehicle to run for 10 seconds. If code returns immediately, replace suspected harness.

Any circuitry, that is suspected as causing the intermittent complaint, should be thoroughly checked for backed out terminals, improper mating, broken locks, improperly formed or damaged terminals, poor terminal to wiring connections or physical damage to the wiring harness.

DIAGNOSTIC CHART DTC A021 — 1993 CUTLASS SUPREME, GRAND PRIX, LUMINA AND REGAL, CONTINUED

IMPORTANT: WHEEL SPEED SENSOR INTERMITTENT PROBLEMS MAY BE DIFFICULT TO LOCATE. CARE SHOULD BE TAKEN NOT TO DISTURB ANY ELECTRICAL CONNECTIONS PRIOR TO AN INDICATED STEP OF THIS CHART. THIS WILL INSURE THAT AN INTERMITTENT CONNECTION WILL NOT BE CORRECTED BEFORE THE SOURCE OF THE PROBLEM IS FOUND.

(1) IS CODE A032 ALSO CURRENTLY SET?

- YES → PROCEED WITH THAT CODE FIRST.
- NO ↓

(2)
- TEST DRIVE VEHICLE.
- USING TECH 1, SELECT DATA LIST.
- MONITOR WHEEL SPEED ON LEFT FRONT WHEEL WHILE SLOWLY DECELERATING FROM 35 MPH TO 0 MPH.
- WHEEL SPEED SHOULD BE WITHIN 5 MPH OF REMAINING WHEEL SPEED INPUTS.

DOES IT?

- YES → FAULT IS NOT PRESENT AT THIS TIME. REFER TO "DIAGNOSTIC AIDS" ON FACING PAGE.
- NO ↓

(3)
- IGNITION "OFF."
- PHYSICALLY INSPECT WHEEL SPEED SENSOR, SENSOR RING, WIRING AND CONNECTORS FOR DAMAGE. PAY PARTICULAR ATTENTION TO THE SENSOR RING. ANY DEVIATION WILL AFFECT THE WHEEL SPEED INPUT SIGNAL.

IS PHYSICAL DAMAGE INDICATED?

- YES → CORRECT PHYSICAL DAMAGE, CLEAR CODES AND RECHECK.
- NO ↓

(4)
- DISCONNECT CONNECTOR FROM LEFT FRONT WHEEL SPEED SENSOR.
- CONNECT DVM BETWEEN TERMINALS "A" AND "B" OF SENSOR, AND MEASURE RESISTANCE. RESISTANCE SHOULD BE BETWEEN 1650 AND 1800 OHMS.

IS IT?

- NO → REPLACE WHEEL SPEED SENSOR, CLEAR CODES AND RECHECK.
- YES ↓

(5)
- IGNITION "ON."
- CONNECT DVM TO GROUND AND TERMINALS "A" THEN "B" OF LEFT FRONT SENSOR CONNECTOR HARNESS. VOLTAGE SHOULD BE 2.25 TO 2.75 VOLTS ON BOTH TERMINALS.

IS IT?

GO TO FOLLOWING PAGE OF THIS CODE CHART

THIS CHART ASSUMES THAT A CURRENT CODE IS STORED INDICATING THAT THIS FAULT IS PRESENT.

DIAGNOSTIC CHART DTC A021 — 1993 CUTLASS SUPREME, GRAND PRIX, LUMINA AND REGAL, CONTINUED

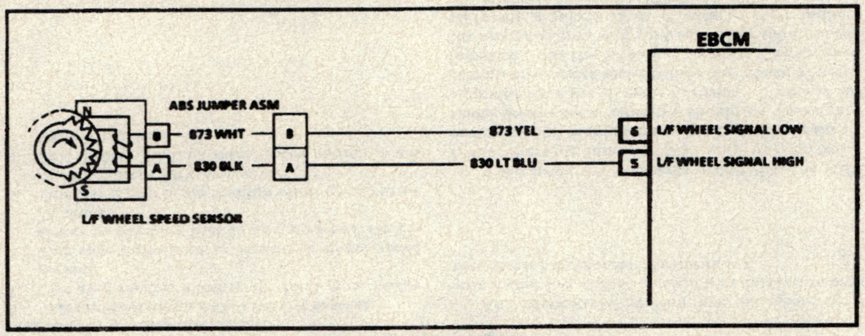

CODE A021

LEFT FRONT WHEEL SPEED = 0

Circuit Description:

As a toothed ring passes by the wheel speed sensor, changes in the electromagnetic field cause the wheel speed sensor to produce a sinusoidal (AC) voltage signal whose frequency is proportional to wheel speed. The magnitude of this signal is directly related to wheel speed and the proximity of the wheel speed sensor to the toothed ring often referred to as the air gap.

This code detects opens or a low output condition that cause a wheel speed of 0 mph to be calculated by the ABS EBCM. The conditions present to fail this test are: (1) Three (3) wheels are rotating at least 5 mph, and (2) The fourth wheel is 0 mph.

Test Description: Number(s) below refer to circled number(s) on the diagnostic chart.

6. This step insures that wheel speed sensor circuitry is not internally shorted
7. This checks for an open in the wheel signal high circuit.
8. This checks for an open in the wheel signal low circuit.
9. This step insures that Code A021 was not set due to a poor connection between the world connector and EBCM.

Diagnostic Aids:

An "Intermittent" problem may be caused by a poor connection, rubbed through wire insulation, or a wire that is broken inside the insulation.

If the customers comments reflect that the ABS Amber warning lamp is "ON" only during moist environmental changes, (rain, snow, vehicle wash) all wheel speed sensor circuitry should be thoroughly inspected for signs of water intrusion. If code is not current clear codes and simulate the effects of water intrusion. Use the following procedure. Spray down the suspected area with a 5% salt water solution (two teaspoons of salt to 12 oz. of water). Start vehicle, allow vehicle to run for 10 seconds. If code returns immediately, replace suspected harness.

Any circuitry, that is suspected as causing the intermittent complaint, should be thoroughly checked for backed out terminals, improper mating, broken locks, improperly formed or damaged terminals, poor terminal to wiring connections or physical damage to the wiring harness.

DIAGNOSTIC CHART DTC A021 — 1993 CUTLASS SUPREME, GRAND PRIX, LUMINA AND REGAL, CONTINUED

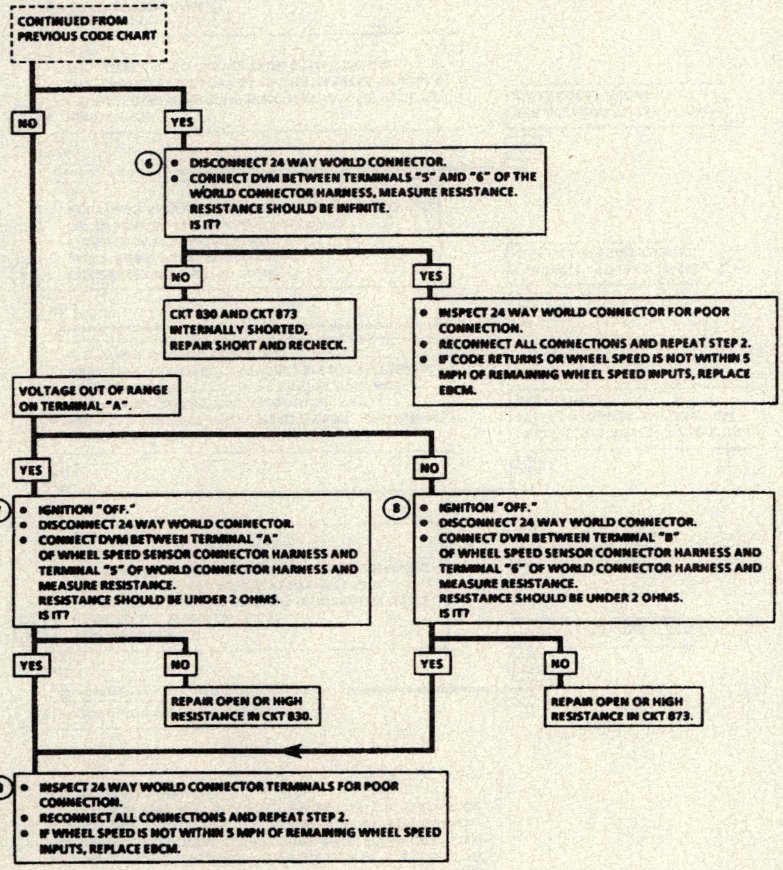

THIS CHART ASSUMES THAT A CURRENT CODE IS STORED INDICATING THAT THIS FAULT IS PRESENT.

DIAGNOSTIC CHART DTC A022 — 1993 CUTLASS SUPREME, GRAND PRIX, LUMINA AND REGAL

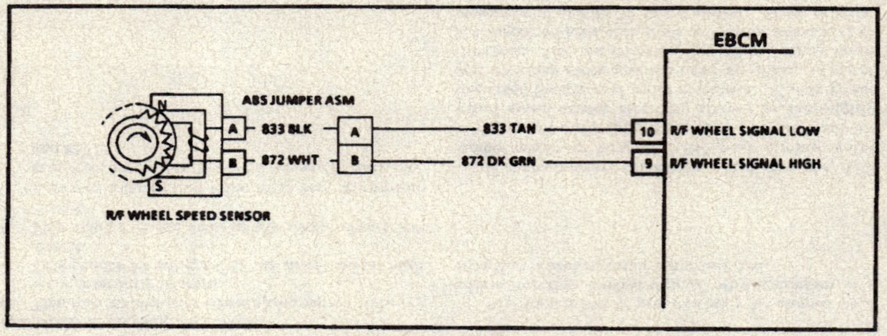

CODE A022

RIGHT FRONT WHEEL SPEED = 0

Circuit Description:

As a toothed ring passes by the wheel speed sensor, changes in the electromagnetic field cause the wheel speed sensor to produce a sinusoidal (AC) voltage signal whose frequency is proportional to wheel speed. The magnitude of this signal is directly related to wheel speed and the proximity of the wheel speed sensor to the toothed ring, often referred to as the air gap.

This code detects opens or a low output condition that cause a wheel speed of 0 mph to be calculated by the ABS EBCM. The conditions present to fail this test are: (1) Three (3) wheels are rotating at least 5 mph, and (2) The fourth wheel is 0 mph.

Test Description: Number(s) below refer to circled number(s) on the diagnostic chart.

1. This test determines if Code A033 is present.
2. This step verifies whether the fault is currently present.
3. This step is designed to identify a wheel speed sensor or circuitry that is damaged and visibly apparent.
4. This test checks to see that the wheel speed sensor is within proper resistance values.
5. This insures that the wheel speed sensor is not shorted to ground.

Diagnostic Aids:

An "Intermittent" problem may be caused by a poor connection, rubbed through wire insulation, or a wire that is broken inside the insulation.

If the customers comments reflect that the ABS Amber warning lamp is "ON" only during moist environmental changes, (rain, snow, vehicle wash) all wheel speed sensor circuitry should be thoroughly inspected for signs of water intrusion. If code is not current clear codes and simulate the effects of water intrusion. Use the following procedure. Spray down the suspected area with a 5% salt water solution (two teaspoons of salt to 12 oz. of water). Start vehicle, allow vehicle to run for 10 seconds. If code returns immediately, replace suspected harness.

Any circuitry, that is suspected as causing the intermittent complaint, should be thoroughly checked for backed out terminals, improper mating, broken locks, improperly formed or damaged terminals, poor terminal to wiring connections or physical damage to the wiring harness.

DIAGNOSTIC CHART DTC A022 — 1993 CUTLASS SUPREME, GRAND PRIX, LUMINA AND REGAL, CONTINUED

IMPORTANT: WHEEL SPEED SENSOR INTERMITTENT PROBLEMS MAY BE DIFFICULT TO LOCATE. CARE SHOULD BE TAKEN NOT TO DISTURB ANY ELECTRICAL CONNECTIONS PRIOR TO AN INDICATED STEP OF THIS CHART. THIS WILL INSURE THAT AN INTERMITTENT CONNECTION WILL NOT BE CORRECTED BEFORE THE SOURCE OF THE PROBLEM IS FOUND.

(1) IS CODE A033 ALSO CURRENTLY SET?

- YES → PROCEED WITH THAT CODE FIRST.
- NO ↓

(2)
- TEST DRIVE VEHICLE.
- USING TECH 1, SELECT DATA LIST.
- MONITOR WHEEL SPEED ON RIGHT FRONT WHEEL WHILE SLOWLY DECELERATING FROM 35 MPH TO 0 MPH.
- WHEEL SPEED SHOULD BE WITHIN 5 MPH OF REMAINING WHEEL SPEED INPUTS.

IS IT?

- YES → FAULT IS NOT PRESENT AT THIS TIME. REFER TO "DIAGNOSTIC AIDS" ON FACING PAGE.
- NO ↓

(3)
- IGNITION "OFF."
- PHYSICALLY INSPECT WHEEL SPEED SENSOR, SENSOR RING, WIRING AND CONNECTORS FOR DAMAGE. PAY PARTICULAR ATTENTION TO THE SENSOR RING. ANY DEVIATION WILL AFFECT THE WHEEL SPEED INPUT SIGNAL.

IS PHYSICAL DAMAGE INDICATED?

- YES → CORRECT PHYSICAL DAMAGE, CLEAR CODES AND RECHECK.
- NO ↓

(4)
- DISCONNECT CONNECTOR FROM RIGHT FRONT WHEEL SPEED SENSOR.
- CONNECT DVM BETWEEN TERMINALS "A" AND "B" OF SENSOR, AND MEASURE RESISTANCE. RESISTANCE SHOULD BE BETWEEN 1650 AND 1800 OHMS.

IS IT?

- NO → REPLACE WHEEL SPEED SENSOR, CLEAR CODES AND RECHECK.
- YES ↓

(5)
- IGNITION "ON."
- CONNECT DVM TO GROUND AND TERMINALS "A" THEN "B" OF RIGHT FRONT SENSOR CONNECTOR HARNESS. VOLTAGE SHOULD BE 2.25 TO 2.75 VOLTS ON BOTH TERMINALS.

IS IT?

GO TO FOLLOWING PAGE OF THIS CODE CHART

THIS CHART ASSUMES THAT A CURRENT CODE IS STORED INDICATING THAT THIS FAULT IS PRESENT.

DIAGNOSTIC CHART DTC A022 — 1993 CUTLASS SUPREME, GRAND PRIX, LUMINA AND REGAL, CONTINUED

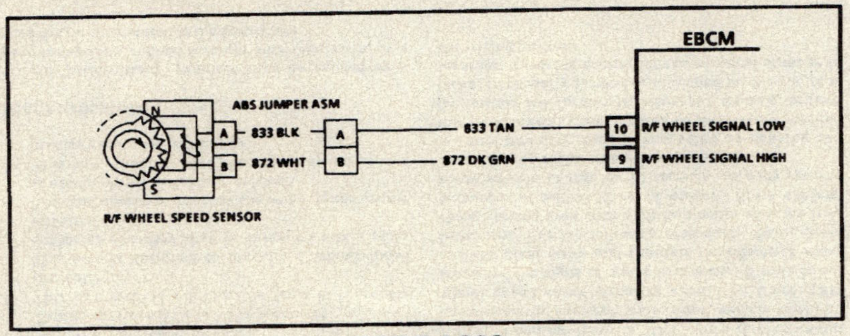

CODE A022

RIGHT FRONT WHEEL SPEED = 0

Circuit Description:

As a toothed ring passes by the wheel speed sensor, changes in the electromagnetic field cause the wheel speed sensor to produce a sinusoidal (AC) voltage signal whose frequency is proportional to wheel speed. The magnitude of this signal is directly related to wheel speed and the proximity of the wheel speed sensor to the toothed ring, often referred to as the air gap.

This code detects opens or a low output condition that cause a wheel speed of 0 mph to be calculated by the ABS EBCM. The conditions present to fail this test are: (1) Three (3) wheels are rotating at least 5 mph, and (2) The fourth wheel is 0 mph.

Test Description: Number(s) below refer to circled number(s) on the diagnostic chart.

6. This test insures that wheel speed sensor circuitry is not internally shorted.
7. This checks for an open in the wheel signal low circuit.
8. This checks for an open in the wheel signal high circuit.
9. This step insures that Code A022 was not set due to a poor connection between the world connector and EBCM.

Diagnostic Aids:

An "Intermittent" problem may be caused by a poor connection, rubbed through wire insulation, or a wire that is broken inside the insulation.

If the customers comments reflect that the ABS Amber warning lamp is "ON" only during moist environmental changes, (rain, snow, vehicle wash) all wheel speed sensor circuitry should be thoroughly inspected for signs of water intrusion. If code is not current clear codes and simulate the effects of water intrusion. Use the following procedure. Spray down the suspected area with a 5% salt water solution (two teaspoons of salt to 12 oz. of water). Start vehicle, allow vehicle to run for 10 seconds. If code returns immediately, replace suspected harness.

Any circuitry, that is suspected as causing the intermittent complaint, should be thoroughly checked for backed out terminals, improper mating, broken locks, improperly formed or damaged terminals, poor terminal to wiring connections or physical damage to the wiring harness.

DIAGNOSTIC CHART DTC A022 — 1993 CUTLASS SUPREME, GRAND PRIX, LUMINA AND REGAL, CONTINUED

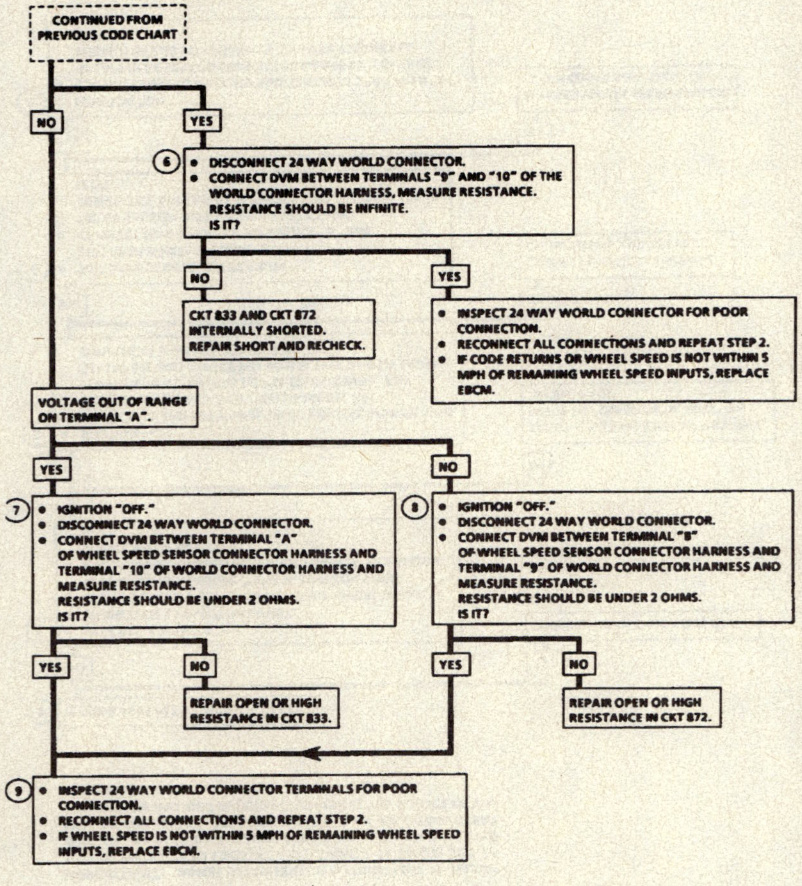

THIS CHART ASSUMES THAT A CURRENT CODE IS STORED INDICATING THAT THIS FAULT IS PRESENT.

DIAGNOSTIC CHART DTC A023 — 1993 CUTLASS SUPREME, GRAND PRIX, LUMINA AND REGAL

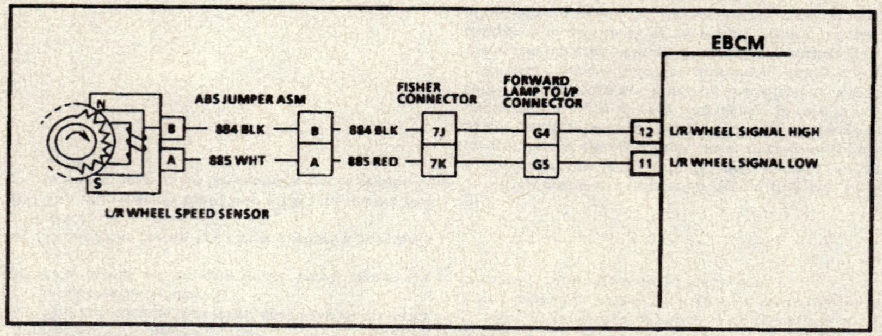

CODE A023

LEFT REAR WHEEL SPEED = 0

Circuit Description:

As a toothed ring passes by the wheel speed sensor, changes in the electromagnetic field cause the wheel speed sensor to produce a sinusoidal (AC) voltage signal whose frequency is proportional to wheel speed. The magnitude of this signal is directly related to wheel speed and the proximity of the wheel speed sensor to the toothed ring, often referred to as the air gap.

This code detects opens or a low output condition that cause a wheel speed of 0 mph to be calculated by the ABS EBCM. The conditions present to fail this test are: (1) Three (3) wheels are rotating at least 5 mph, (2) The fourth wheel is 0 mph.

Test Description: Number(s) below refer to circled number(s) on the diagnostic chart.

1. This determines if Code A034 is present.
2. This test verifies whether the fault is currently present.
3. This step is designed to identify a wheel speed sensor or circuitry that is damaged and visibly apparent.
4. This test checks to see that the wheel speed sensor is within proper resistance values.
5. This test insures that the wheel speed sensor circuitry has good continuity.

Diagnostic Aids:

An "Intermittent" problem may be caused by a poor connection, rubbed through wire insulation, or a wire that is broken inside the insulation.

If the customers comments reflect that the ABS Amber warning lamp is "ON" only during moist environmental changes, (rain, snow, vehicle wash) all wheel speed senor circuitry should be thoroughly inspected for signs of water intrusion. If code is not current, clear codes and simulate the effects of water intrusion. Use the following procedure. Spray down the suspected area with a 5% salt water solution (two teaspoons of salt to 12 oz. of water). Start vehicle, allow vehicle to run for 10 seconds. If code returns immediately, replace suspected harness.

Any circuitry, that is suspected as causing the intermittent complaint, should be thoroughly checked for backed out terminals, improper mating, broken locks, improperly formed or damaged terminals, poor terminal to wiring connections or physical damage to the wiring harness.

DIAGNOSTIC CHART DTC A023 — 1993 CUTLASS SUPREME, GRAND PRIX, LUMINA AND REGAL, CONTINUED

IMPORTANT: WHEEL SPEED SENSOR INTERMITTENT PROBLEMS MAY BE DIFFICULT TO LOCATE. CARE SHOULD BE TAKEN NOT TO DISTURB ANY ELECTRICAL CONNECTIONS PRIOR TO AN INDICATED STEP OF THIS CHART. THIS WILL INSURE THAT AN INTERMITTENT CONNECTION WILL NOT BE CORRECTED BEFORE THE SOURCE OF THE PROBLEM IS FOUND.

(1) IS CODE A034 ALSO CURRENTLY SET?

- YES: PROCEED WITH THAT CODE FIRST.
- NO:

(2)
- TEST DRIVE VEHICLE.
- USING TECH 1, SELECT DATA LIST.
- MONITOR WHEEL SPEED ON LEFT REAR WHEEL WHILE SLOWLY DECELERATING FROM 35 MPH TO 0 MPH.
- WHEEL SPEED SHOULD BE WITHIN 5 MPH OF REMAINING WHEEL SPEED INPUTS,

IS IT?

- YES: FAULT IS NOT PRESENT AT THIS TIME. REFER TO "DIAGNOSTIC AIDS" ON FACING PAGE.
- NO:

(3)
- IGNITION "OFF."
- PHYSICALLY INSPECT WHEEL SPEED SENSOR, SENSOR RING, WIRING AND CONNECTORS FOR DAMAGE. PAY PARTICULAR ATTENTION TO THE SENSOR RING. ANY DEVIATION WILL AFFECT THE WHEEL SPEED INPUT SIGNAL.

IS PHYSICAL DAMAGE INDICATED?

- YES: CORRECT PHYSICAL DAMAGE, CLEAR CODES AND RECHECK.
- NO:

(4)
- DISCONNECT CONNECTOR FROM LEFT REAR WHEEL SPEED SENSOR.
- CONNECT DVM BETWEEN TERMINALS "A" AND "B" OF SENSOR, AND MEASURE RESISTANCE. RESISTANCE SHOULD BE BETWEEN 2100 AND 2400 OHMS.

IS IT?

- NO: REPLACE WHEEL SPEED SENSOR, CLEAR CODES AND RECHECK.
- YES:

(5)
- IGNITION "ON."
- CONNECT DVM TO GROUND AND TERMINALS "A" THEN "B" OF LEFT REAR SENSOR CONNECTOR HARNESS. VOLTAGE SHOULD BE 2.25 TO 2.75 VOLTS ON BOTH TERMINALS.

IS IT?

GO TO FOLLOWING PAGE OF THIS CODE CHART

THIS CHART ASSUMES THAT A CURRENT CODE IS STORED INDICATING THAT THIS FAULT IS PRESENT.

DIAGNOSTIC CHART DTC A023 — 1993 CUTLASS SUPREME, GRAND PRIX, LUMINA AND REGAL, CONTINUED

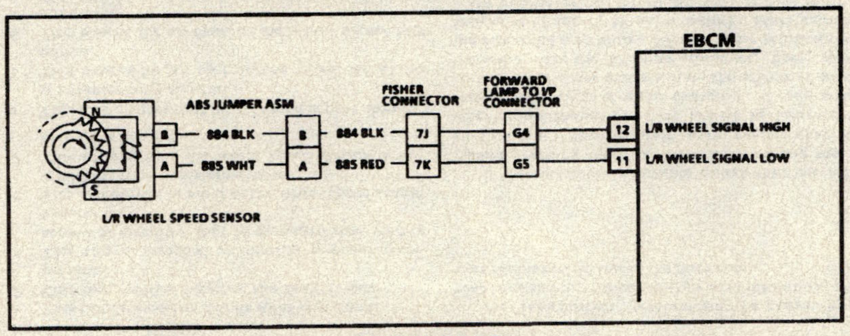

CODE A023

LEFT REAR WHEEL SPEED = 0

Circuit Description:

As a toothed ring passes by the wheel speed sensor, changes in the electromagnetic field cause the wheel speed sensor to produce a sinusoidal (AC) voltage signal whose frequency is proportional to wheel speed. The magnitude of this signal is directly related to wheel speed and the proximity of the wheel speed sensor to the toothed ring, often referred to as the air gap.

This code detects opens or a low output condition that cause a wheel speed of 0 mph to be calculated by the ABS EBCM. The conditions present to fail this test are: (1) Three (3) wheels are rotating at least 5 mph, and (2) The fourth wheel is 0 mph.

Test Description: Number(s) below refer to circled number(s) on the diagnostic chart.

6. This test insures that wheel speed sensor circuitry is not internally shorted.
7. This checks for an open in the wheel signal low circuit.
8. This checks for an open in the wheel signal high circuit.
9. This step insures that Code A023 was not set due to a poor connection between the world connector and EBCM.

Diagnostic Aids:

An "Intermittent" problem may be caused by a poor connection, rubbed through wire insulation, or a wire that is broken inside the insulation.

If the customers comments reflect that the ABS Amber warning lamp is "ON" only during moist environmental changes, (rain, snow, vehicle wash) all wheel speed sensor circuitry should be thoroughly inspected for signs of water intrusion. If code is not current clear codes and simulate the effects of water intrusion. Use the following procedure. Spray down the suspected area with a 5% salt water solution (two teaspoons of salt to 12 oz. of water). Start vehicle, allow vehicle to run for 10 seconds. If code returns immediately, replace suspected harness.

Any circuitry, that is suspected as causing the intermittent complaint, should be thoroughly checked for backed out terminals, improper mating, broken locks, improperly formed or damaged terminals, poor terminal to wiring connections or physical damage to the wiring harness.

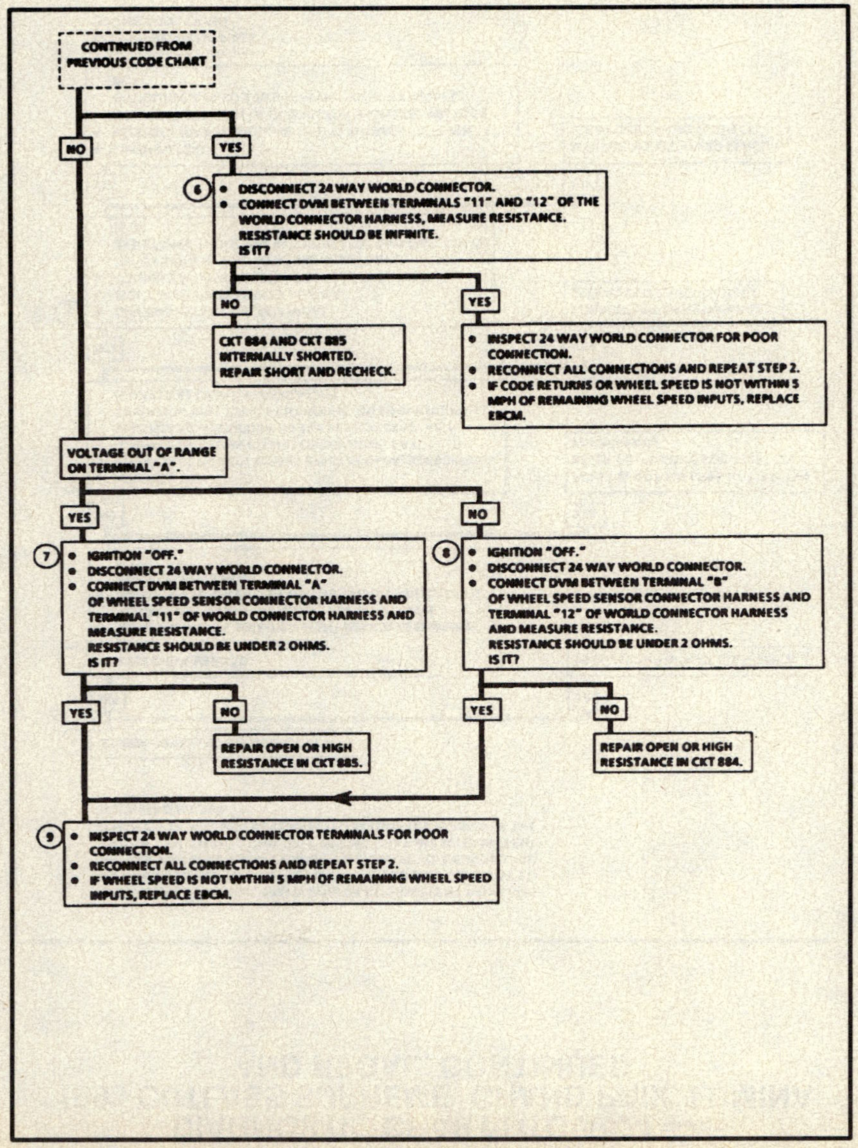

DIAGNOSTIC CHART DTC A024 — 1993 CUTLASS SUPREME, GRAND PRIX, LUMINA AND REGAL

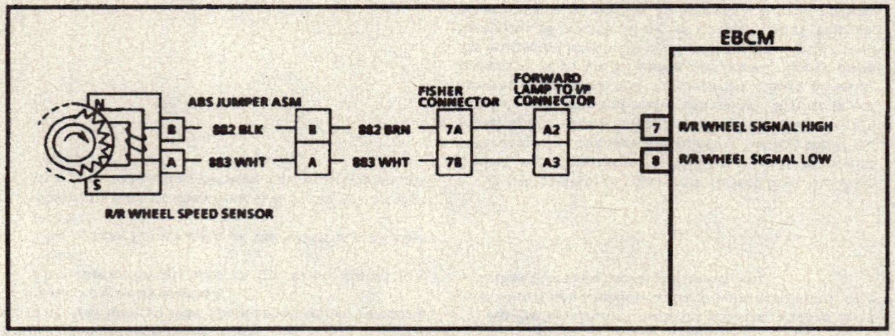

CODE A024

RIGHT REAR WHEEL SPEED = 0

Circuit Description:

As a toothed ring passes by the wheel speed sensor, changes in the electromagnetic field cause the wheel speed sensor to produce a sinusoidal (AC) voltage signal whose frequency is proportional to wheel speed. The magnitude of this signal is directly related to wheel speed and the proximity of the wheel speed sensor to the toothed ring, often referred to as the air gap.

This code detects opens or a low output condition that cause a wheel speed of 0 mph to be calculated by the ABS EBCM. The conditions present to fail this test are: (1) Three (3) wheels are rotating at least 5 mph, and (2) The fourth wheel is 0 mph.

Test Description: Number(s) below refer to circled number(s) on the diagnostic chart.

1. This test determines if Code A035 is present.
2. This test verifies whether the fault is currently present.
3. This step is designed to identify a wheel speed sensor or circuitry that is damaged and visibly apparent.
4. This test checks to see that the wheel speed sensor is within proper resistance values.
5. This insures that the wheel speed sensor is not shorted to ground.
6. This test insures that wheel speed sensor circuitry is not internally shorted.
7. This checks for an open in the wheel signal low circuit.
8. This checks for an open in the wheel signal high circuit.
9. This step insures that Code A024 was not set due to a poor connection between the world connector and EBCM.

Diagnostic Aids:

An "Intermittent" problem may be caused by a poor connection, rubbed through wire insulation, or a wire that is broken inside the insulation.

If the customers comments reflect that the ABS Amber warning lamp is "ON" only during moist environmental changes, (rain, snow, vehicle wash) all wheel speed sensor circuitry should be thoroughly inspected for signs of water intrusion. If code is not current clear codes and simulate the effects of water intrusion. Use the following procedure. Spray down the suspected area with a 5% salt water solution (two teaspoons of salt to 12 oz. of water). Start vehicle, allow vehicle to run for 10 seconds. If code returns immediately, replace suspected harness.

Any circuitry, that is suspected as causing the intermittent complaint, should be thoroughly checked for backed out terminals, improper mating, broken locks, improperly formed or damaged terminals, poor terminal to wiring connections or physical damage to the wiring harness.

DIAGNOSTIC CHART DTC A024 — 1993 CUTLASS SUPREME, GRAND PRIX, LUMINA AND REGAL, CONTINUED

IMPORTANT: WHEEL SPEED SENSOR INTERMITTENT PROBLEMS MAY BE DIFFICULT TO LOCATE. CARE SHOULD BE TAKEN NOT TO DISTURB ANY ELECTRICAL CONNECTIONS PRIOR TO AN INDICATED STEP OF THIS CHART. THIS WILL INSURE THAT AN INTERMITTENT CONNECTION WILL NOT BE CORRECTED BEFORE THE SOURCE OF THE PROBLEM IS FOUND.

(1) IS CODE A035 ALSO CURRENTLY SET?

- YES: PROCEED WITH THAT CODE FIRST.
- NO:

(2)
- TEST DRIVE VEHICLE.
- USING TECH 1, SELECT DATA LIST.
- MONITOR WHEEL SPEED ON RIGHT REAR WHEEL WHILE SLOWLY DECELERATING FROM 35 MPH TO 0 MPH.
- WHEEL SPEED SHOULD BE WITHIN 5 MPH OF REMAINING WHEEL SPEED INPUTS,

IS IT?

- YES: FAULT IS NOT PRESENT AT THIS TIME. REFER TO "DIAGNOSTIC AIDS" ON FACING PAGE.
- NO:

(3)
- IGNITION "OFF."
- PHYSICALLY INSPECT WHEEL SPEED SENSOR, SENSOR RING, WIRING AND CONNECTORS FOR DAMAGE. PAY PARTICULAR ATTENTION TO THE SENSOR RING. ANY DEVIATION WILL AFFECT THE WHEEL SPEED INPUT SIGNAL.

IS PHYSICAL DAMAGE INDICATED?

- YES: CORRECT PHYSICAL DAMAGE, CLEAR CODES AND RECHECK.
- NO:

(4)
- DISCONNECT CONNECTOR FROM RIGHT REAR WHEEL SPEED SENSOR.
- CONNECT DVM BETWEEN TERMINALS "A" AND "B" OF SENSOR, AND MEASURE RESISTANCE. RESISTANCE SHOULD BE BETWEEN 2100 AND 2400 OHMS.

IS IT?

- NO: REPLACE WHEEL SPEED SENSOR, CLEAR CODES AND RECHECK.
- YES:

(5)
- IGNITION "ON."
- CONNECT DVM TO GROUND AND TERMINALS "A" THEN "B" OF RIGHT REAR SENSOR CONNECTOR HARNESS. VOLTAGE SHOULD BE 2.25 TO 2.75 VOLTS ON BOTH TERMINALS.

IS IT?

GO TO FOLLOWING PAGE OF THIS CODE CHART

THIS CHART ASSUMES THAT A CURRENT CODE IS STORED INDICATING THAT THIS FAULT IS PRESENT.

DIAGNOSTIC CHART DTC A024 — 1993 CUTLASS SUPREME, GRAND PRIX, LUMINA AND REGAL, CONTINUED

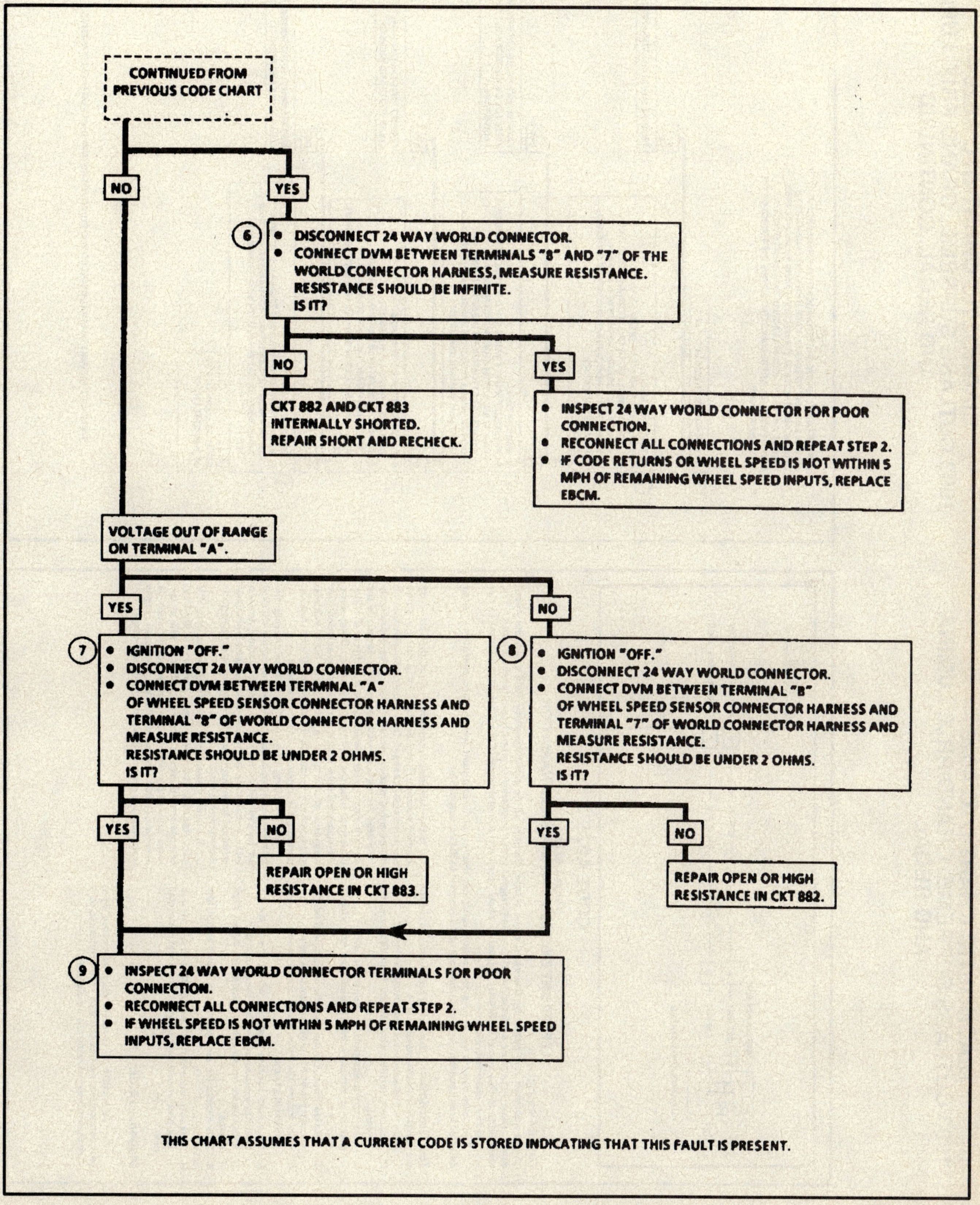

DIAGNOSTIC CHART DTC A025 — 1993 CUTLASS SUPREME, GRAND PRIX, LUMINA AND REGAL

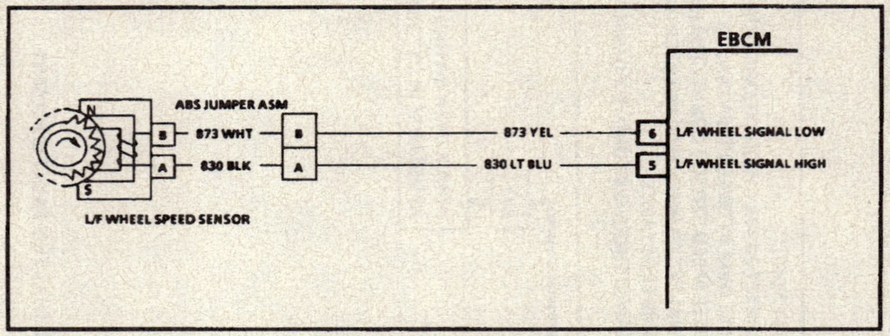

CODE A025

(Page 1 of 2)

LEFT FRONT EXCESSIVE WHEEL SPEED VARIATION

Circuit Description:

As a toothed ring passes by the wheel speed sensor, changes in the electromagnetic field cause the wheel speed sensor to produce a sinusoidal (AC) voltage signal whose frequency is proportional to wheel speed. The magnitude of this signal is directly related to wheel speed and the proximity of the wheel speed sensor to the toothed ring, often referred to as the air gap.

This code detects opens or a low output condition that cause intermittent wheel speed operation. Also detected is any sudden change in wheel speed determined to the unreasonable.

Test Description: Number(s) below refer to circled number(s) on the diagnostic chart.

1. This test determines if Code A032 is also set.
2. This test verifies whether the fault is currently present.
3. This step is designed to identify a wheel speed sensor or circuitry that is damaged and visibly apparent.
4. This test checks to see that the wheel speed sensor circuitry is within proper resistance value.
5. This insure that the wheel speed sensor is not shorted to ground.
6. This step insures that the wheel speed sensor circuitry resistance is within the proper value.
7. This checks for an open in the wheel signal high circuit.
8. This checks for an open in the wheel signal low circuit.
9. This step insures that Code A025 was not set due to a poor connection between the world connector and EBCM.

Diagnostic Aids:

An "Intermittent" problem may be caused by a poor connection, rubbed through wire insulation, or a wire that is broken inside the insulation.

If the customers comments reflect that the ABS Amber warning lamp is "ON" only during moist environmental changes, (rain, snow, vehicle wash) all wheel speed sensor circuitry should be thoroughly inspected for signs of water intrusion. If code is not current clear codes and simulate the effects of water intrusion. Use the following procedure. Spray down the suspected area with a 5% salt water solution (two teaspoons of salt to 12 oz. of water). Start vehicle, allow vehicle to run for 10 seconds. If code returns immediately, replace suspected harness.

Any circuitry, that is suspected as causing the intermittent complaint, should be thoroughly checked for backed out terminals, improper mating, broken locks, improperly formed or damaged terminals, poor terminal to wiring connections or physical damage to the wiring harness.

A careful visual inspection of the toothed ring, CV joint, bearing and wheel speed sensor for physical damage should also be performed. If code sets at the same mph every time and wheel speed variation is noted above this mph on the Tech 1, the speed ring is most likely damaged. This can occur at any speed.

DIAGNOSTIC CHART DTC A025 — 1993 CUTLASS SUPREME, GRAND PRIX, LUMINA AND REGAL, CONTINUED

IMPORTANT: WHEEL SPEED SENSOR INTERMITTENT PROBLEMS MAY BE DIFFICULT TO LOCATE. CARE SHOULD BE TAKEN NOT TO DISTURB ANY ELECTRICAL CONNECTIONS PRIOR TO AN INDICATED STEP OF THIS CHART. THIS WILL INSURE THAT AN INTERMITTENT CONNECTION WILL NOT BE CORRECTED BEFORE THE SOURCE OF THE PROBLEM IS FOUND.

(1) IS CODE A032 ALSO CURRENTLY SET?

- YES: PROCEED WITH THAT CODE FIRST.
- NO: go to (2)

(2)
- TEST DRIVE VEHICLE.
- USING TECH 1, SELECT DATA LIST.
- MONITOR WHEEL SPEED ON LEFT FRONT WHEEL WHILE SLOWLY DECELERATING FROM 35 MPH TO 0 MPH.
- WHEEL SPEED SHOULD BE WITHIN 5 MPH OF REMAINING WHEEL SPEED INPUTS.

DOES IT?

- YES: FAULT IS NOT PRESENT AT THIS TIME. REFER TO "DIAGNOSTIC AIDS" ON FACING PAGE.
- NO: go to (3)

(3)
- IGNITION "OFF."
- PHYSICALLY INSPECT WHEEL SPEED SENSOR, SENSOR RING, WIRING AND CONNECTORS FOR DAMAGE. PAY PARTICULAR ATTENTION TO THE SENSOR RING. ANY DEVIATION WILL AFFECT THE WHEEL SPEED INPUT SIGNAL.

IS PHYSICAL DAMAGE INDICATED?

- YES: CORRECT PHYSICAL DAMAGE, CLEAR CODES AND RECHECK.
- NO: go to (4)

(4)
- DISCONNECT CONNECTOR FROM LEFT FRONT WHEEL SPEED SENSOR.
- CONNECT DVM BETWEEN TERMINALS "A" AND "B" OF SENSOR, AND MEASURE RESISTANCE.

RESISTANCE SHOULD BE BETWEEN 1650 AND 1800 OHMS. IS IT?

- NO: REPLACE WHEEL SPEED SENSOR, CLEAR CODES AND RECHECK.
- YES: go to (5)

(5)
- IGNITION "ON."
- CONNECT DVM TO GROUND AND TERMINALS "A" THEN "B" OF LEFT FRONT SENSOR CONNECTOR HARNESS. VOLTAGE SHOULD BE 2.25 TO 2.75 VOLTS ON BOTH TERMINALS.

IS IT?

GO TO FOLLOWING PAGE OF THIS CODE CHART

THIS CHART ASSUMES THAT A CURRENT CODE IS STORED INDICATING THAT THIS FAULT IS PRESENT.

DIAGNOSTIC CHART DTC A025 — 1993 CUTLASS SUPREME, GRAND PRIX, LUMINA AND REGAL, CONTINUED

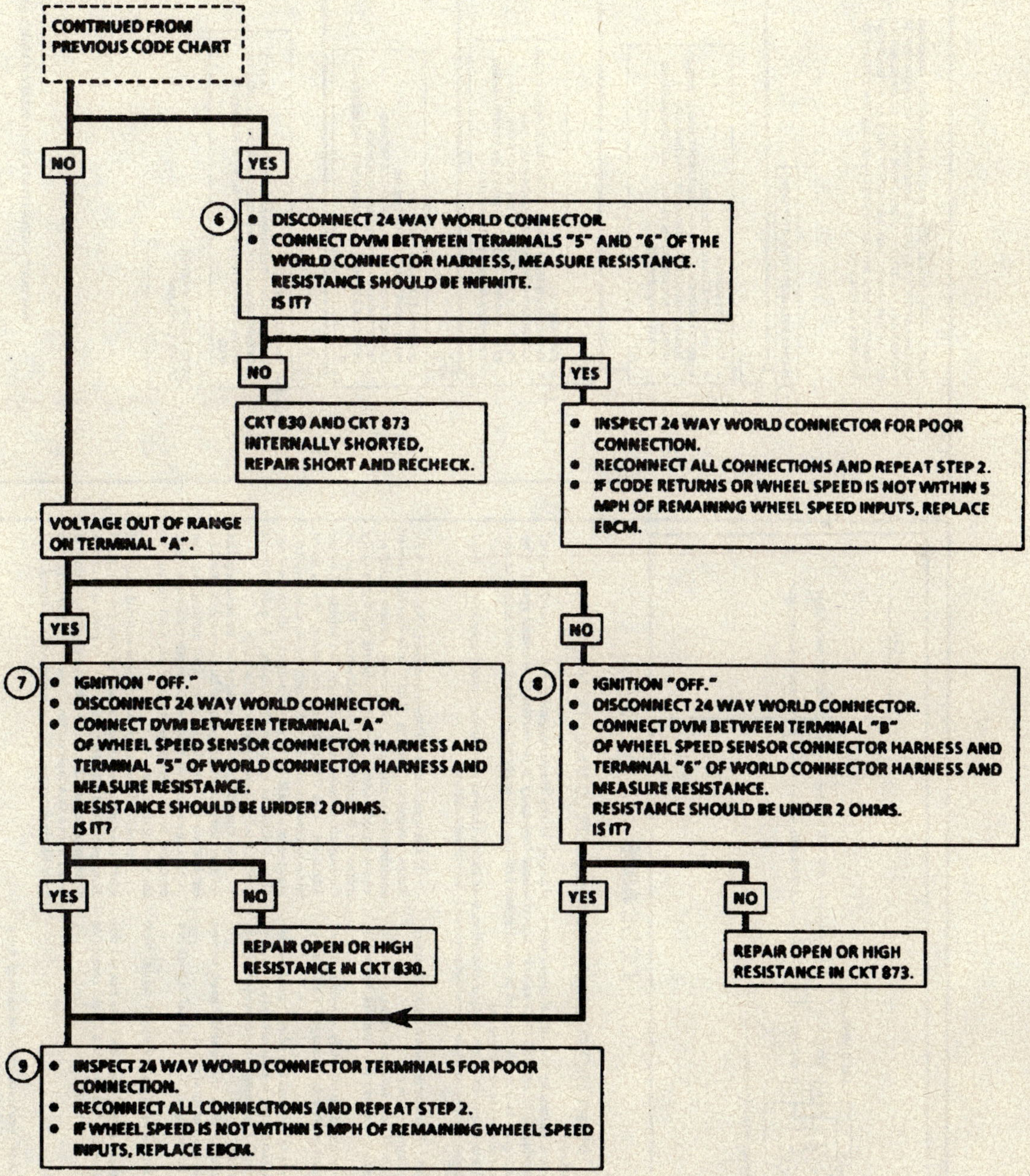

THIS CHART ASSUMES THAT A CURRENT CODE IS STORED INDICATING THAT THIS FAULT IS PRESENT.

DIAGNOSTIC CHART DTC A026 — 1993 CUTLASS SUPREME, GRAND PRIX, LUMINA AND REGAL

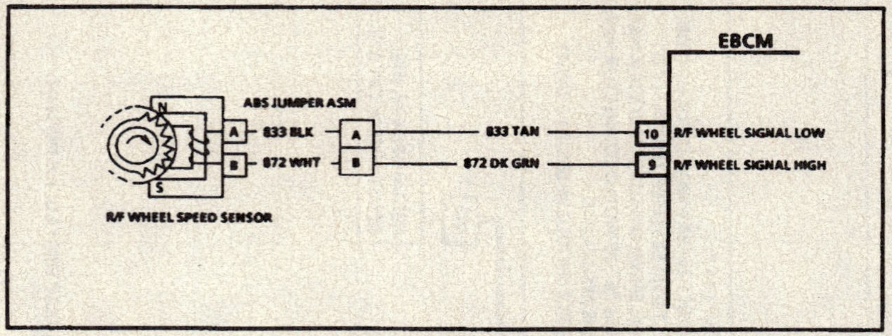

CODE A026

RIGHT FRONT EXCESSIVE WHEEL SPEED VARIATION

Circuit Description:

As a toothed ring passes by the wheel speed sensor, changes in the electromagnetic field cause the wheel speed sensor to produce a sinusoidal (AC) voltage signal whose frequency is proportional to wheel speed. The magnitude of this signal is directly related to wheel speed and the proximity of the wheel speed sensor to the toothed ring, often referred to as the air gap.

This code detects opens or a low output condition that cause intermittent wheel speed operation. Also detected is any sudden change in wheel speed determined to be unreasonable.

Test Description: Number(s) below refer to circled number(s) on the diagnostic chart.

1. This determines if Code A033 is present.
2. This test verifies whether the fault is currently present.
3. This step is designed to identify a wheel speed sensor or circuitry that is damaged and visibly apparent.
4. This test checks to see that the wheel speed sensor circuitry is within proper resistance value.
5. This insures that the wheel speed sensor is not shorted to ground.
6. This step insures that the wheel speed sensor circuitry is not internally shorted.
7. This checks for an open in the wheel signal low circuit.
8. This checks for an open in the wheel signal high circuit.
9. This step insures that Code A026 was not set due to a poor connection between the world connector and EBCM.

Diagnostic Aids:

An "Intermittent" problem may be caused by a poor connection, rubbed through wire insulation, or a wire that is broken inside the insulation.

If the customers comments reflect that the ABS Amber warning lamp is "ON" only during moist environmental changes, (rain, snow, vehicle wash) all wheel speed sensor circuitry should be thoroughly inspected for signs of water intrusion. If code is not current clear codes and simulate the effects of water intrusion. Use the following procedure. Spray down the suspected area with a 5% salt water solution (two teaspoons of salt to 12 oz. of water). Start vehicle, allow vehicle to run for 10 seconds. If code returns immediately, replace suspected harness.

Any circuitry, that is suspected as causing the intermittent complaint, should be thoroughly checked for backed out terminals, improper mating, broken locks, improperly formed or damaged terminals, poor terminal to wiring connections or physical damage to the wiring harness.

A careful visual inspection of the toothed ring, CV joint, bearing and wheel speed sensor for physical damage should also be performed. If code sets at the same mph every time and wheel speed variation is noted above this mph on the Tech 1, the speed ring is most likely damaged. This can occur at any speed.

DIAGNOSTIC CHART DTC A026 — 1993 CUTLASS SUPREME, GRAND PRIX, LUMINA AND REGAL, CONTINUED

IMPORTANT: WHEEL SPEED SENSOR INTERMITTENT PROBLEMS MAY BE DIFFICULT TO LOCATE. CARE SHOULD BE TAKEN NOT TO DISTURB ANY ELECTRICAL CONNECTIONS PRIOR TO AN INDICATED STEP OF THIS CHART. THIS WILL INSURE THAT AN INTERMITTENT CONNECTION WILL NOT BE CORRECTED BEFORE THE SOURCE OF THE PROBLEM IS FOUND.

(1) IS CODE A033 ALSO CURRENTLY SET?
- YES: PROCEED WITH THAT CODE FIRST.
- NO:

(2)
- TEST DRIVE VEHICLE.
- USING TECH 1, SELECT DATA LIST.
- MONITOR WHEEL SPEED ON RIGHT FRONT WHEEL WHILE SLOWLY DECELERATING FROM 35 MPH TO 0 MPH.
- WHEEL SPEED SHOULD BE WITHIN 5 MPH OF REMAINING WHEEL SPEED INPUTS.

IS IT?
- YES: FAULT IS NOT PRESENT AT THIS TIME. REFER TO "DIAGNOSTIC AIDS" ON FACING PAGE.
- NO:

(3)
- IGNITION "OFF."
- PHYSICALLY INSPECT WHEEL SPEED SENSOR, SENSOR RING, WIRING AND CONNECTORS FOR DAMAGE. PAY PARTICULAR ATTENTION TO THE SENSOR RING. ANY DEVIATION WILL AFFECT THE WHEEL SPEED INPUT SIGNAL.

IS PHYSICAL DAMAGE INDICATED?
- YES: CORRECT PHYSICAL DAMAGE, CLEAR CODES AND RECHECK.
- NO:

(4)
- DISCONNECT CONNECTOR FROM RIGHT FRONT WHEEL SPEED SENSOR.
- CONNECT DVM BETWEEN TERMINALS "A" AND "B" OF SENSOR, AND MEASURE RESISTANCE. RESISTANCE SHOULD BE BETWEEN 1650 AND 1800 OHMS.

IS IT?
- NO: REPLACE WHEEL SPEED SENSOR, CLEAR CODES AND RECHECK.
- YES:

(5)
- IGNITION "ON."
- CONNECT DVM TO GROUND AND TERMINALS "A" THEN "B" OF RIGHT FRONT SENSOR CONNECTOR HARNESS. VOLTAGE SHOULD BE 2.25 TO 2.75 VOLTS ON BOTH TERMINALS.

IS IT?

GO TO FOLLOWING PAGE OF THIS CODE CHART

THIS CHART ASSUMES THAT A CURRENT CODE IS STORED INDICATING THAT THIS FAULT IS PRESENT.

DIAGNOSTIC CHART DTC A026 — 1993 CUTLASS SUPREME, GRAND PRIX, LUMINA AND REGAL, CONTINUED

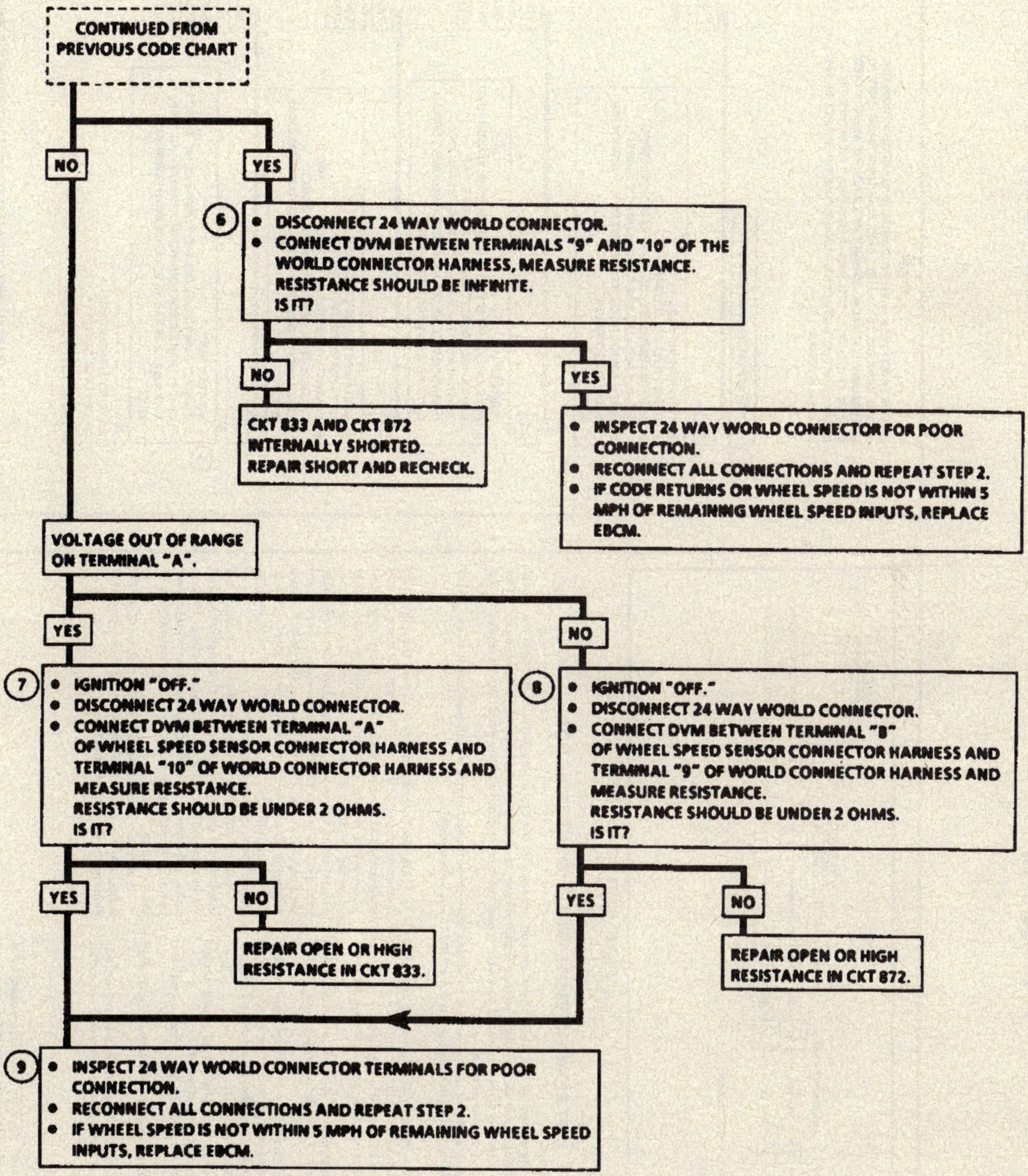

THIS CHART ASSUMES THAT A CURRENT CODE IS STORED INDICATING THAT THIS FAULT IS PRESENT.

DIAGNOSTIC CHART DTC A027 — 1993 CUTLASS SUPREME, GRAND PRIX, LUMINA AND REGAL

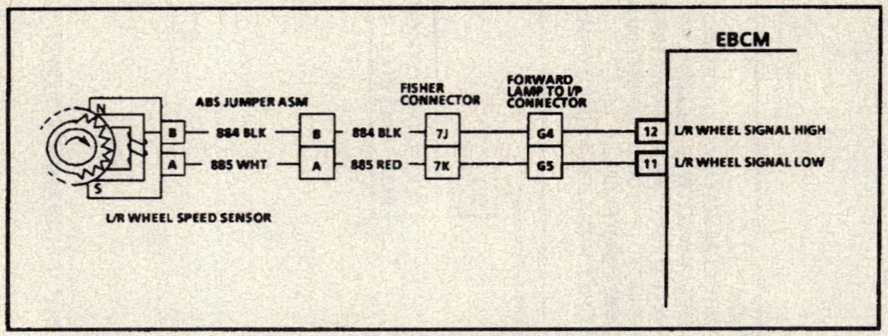

CODE A027

LEFT REAR EXCESSIVE WHEEL SPEED VARIATION

Circuit Description:

As a toothed ring passes by the wheel speed sensor, changes in the electromagnetic field cause the wheel speed sensor to produce a sinusoidal (AC) voltage signal whose frequency is proportional to wheel speed. The magnitude of this signal is directly related to wheel speed and the proximity of the wheel speed sensor to the toothed ring, often referred to as the air gap.

This code detects opens or a low output condition that cause intermittent wheel speed operation. Also detected is any sudden change in wheel speed determined to be unreasonable.

Test Description: Number(s) below refer to circled number(s) on the diagnostic chart.

1. This determines if Code A034 is present.
2. This test verifies whether the fault is currently present.
3. This step is designed to identify a wheel speed sensor or circuitry that is damaged and visibly apparent.
4. This test checks to see that the wheel speed sensor circuitry is within proper resistance value.
5. This test insures that the wheel speed sensor circuitry has good continuity.
6. This step insures that the wheel speed sensor circuitry is not internally shorted.
7. This checks for an open in the wheel signal low circuit.
8. This checks for an open in the wheel signal high circuit.
9. This step insures that Code A027 was not set due to a poor connection between the world connector and EBCM.

Diagnostic Aids:

An "Intermittent" problem may be caused by a poor connection, rubbed through wire insulation, or a wire that is broken inside the insulation.

If the customers comments reflect that the ABS Amber warning lamp is "ON" only during moist environmental changes, (rain, snow, vehicle wash) all wheel speed sensor circuitry should be thoroughly inspected for signs of water intrusion. If code is not current clear codes and simulate the effects of water intrusion. Use the following procedure. Spray down the suspected area with a 5% salt water solution (two teaspoons of salt to 12 oz. of water). Start vehicle, allow vehicle to run for 10 seconds. If code returns immediately, replace suspected harness.

Any circuitry, that is suspected as causing the intermittent complaint, should be thoroughly checked for backed out terminals, improper mating, broken locks, improperly formed or damaged terminals, poor terminal to wiring connections or physical damage to the wiring harness.

A careful visual inspection of the toothed ring, CV joint, bearing and wheel speed sensor for physical damage should also be performed.

DIAGNOSTIC CHART DTC A027 — 1993 CUTLASS SUPREME, GRAND PRIX, LUMINA AND REGAL, CONTINUED

IMPORTANT: WHEEL SPEED SENSOR INTERMITTENT PROBLEMS MAY BE DIFFICULT TO LOCATE. CARE SHOULD BE TAKEN NOT TO DISTURB ANY ELECTRICAL CONNECTIONS PRIOR TO AN INDICATED STEP OF THIS CHART. THIS WILL INSURE THAT AN INTERMITTENT CONNECTION WILL NOT BE CORRECTED BEFORE THE SOURCE OF THE PROBLEM IS FOUND.

(1) IS CODE A034 ALSO CURRENTLY SET?
- YES: PROCEED WITH THAT CODE FIRST.
- NO: go to step 2.

(2)
- TEST DRIVE VEHICLE.
- USING TECH 1, SELECT DATA LIST.
- MONITOR WHEEL SPEED ON LEFT REAR WHEEL WHILE SLOWLY DECELERATING FROM 35 MPH TO 0 MPH.
- WHEEL SPEED SHOULD BE WITHIN 5 MPH OF REMAINING WHEEL SPEED INPUTS,

IS IT?
- YES: FAULT IS NOT PRESENT AT THIS TIME. REFER TO "DIAGNOSTIC AIDS" ON FACING PAGE.
- NO: go to step 3.

(3)
- IGNITION "OFF."
- PHYSICALLY INSPECT WHEEL SPEED SENSOR, SENSOR RING, WIRING AND CONNECTORS FOR DAMAGE. PAY PARTICULAR ATTENTION TO THE SENSOR RING. ANY DEVIATION WILL AFFECT THE WHEEL SPEED INPUT SIGNAL.

IS PHYSICAL DAMAGE INDICATED?
- YES: CORRECT PHYSICAL DAMAGE, CLEAR CODES AND RECHECK.
- NO: go to step 4.

(4)
- DISCONNECT CONNECTOR FROM LEFT REAR WHEEL SPEED SENSOR.
- CONNECT DVM BETWEEN TERMINALS "A" AND "B" OF SENSOR, AND MEASURE RESISTANCE. RESISTANCE SHOULD BE BETWEEN 2100 AND 2400 OHMS.

IS IT?
- NO: REPLACE WHEEL SPEED SENSOR, CLEAR CODES AND RECHECK.
- YES: go to step 5.

(5)
- IGNITION "ON."
- CONNECT DVM TO GROUND AND TERMINALS "A" THEN "B" OF LEFT REAR SENSOR CONNECTOR HARNESS. VOLTAGE SHOULD BE 2.25 TO 2.75 VOLTS ON BOTH TERMINALS.

IS IT?

GO TO FOLLOWING PAGE OF THIS CODE CHART

THIS CHART ASSUMES THAT A CURRENT CODE IS STORED INDICATING THAT THIS FAULT IS PRESENT.

DIAGNOSTIC CHART DTC A027 — 1993 CUTLASS SUPREME, GRAND PRIX, LUMINA AND REGAL, CONTINUED

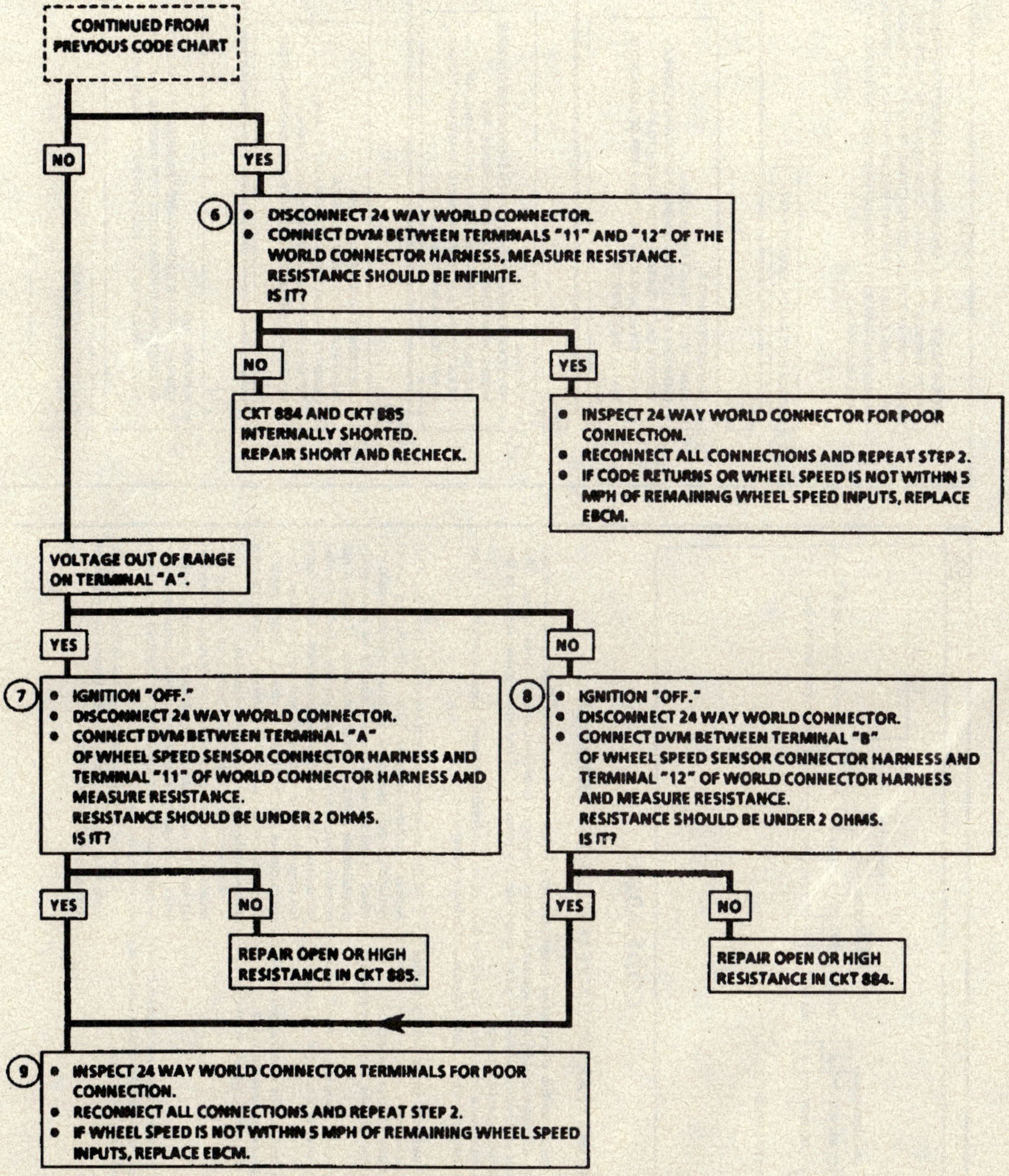

THIS CHART ASSUMES THAT A CURRENT CODE IS STORED INDICATING THAT THIS FAULT IS PRESENT.

DIAGNOSTIC CHART DTC A028 — 1993 CUTLASS SUPREME, GRAND PRIX, LUMINA AND REGAL

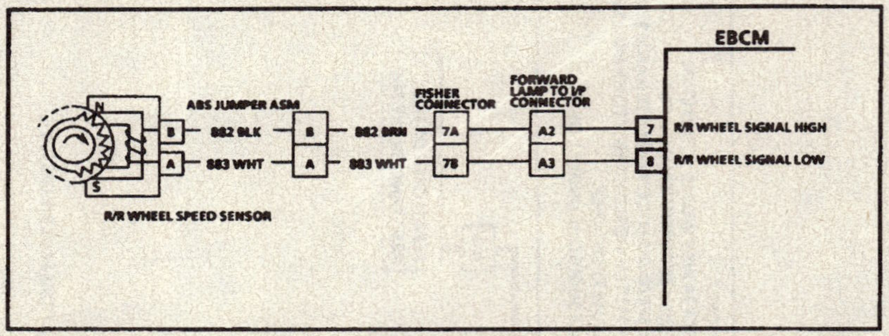

CODE A028

RIGHT REAR EXCESSIVE WHEEL SPEED VARIATION

Circuit Description:

As a toothed ring passes by the wheel speed sensor, changes in the electromagnetic field cause the wheel speed sensor to produce a sinusoidal (AC) voltage signal whose frequency is proportional to wheel speed. The magnitude of this signal is directly related to wheel speed and the proximity of the wheel speed sensor to the toothed ring, often referred to as the air gap.

This code detects opens or a low output condition that cause intermittent wheel speed operation. Also detected is any sudden change in wheel speed determined to be unreasonable.

Test Description: Number(s) below refer to circled number(s) on the diagnostic chart.

1. This determines if Code A035 is present.
2. This test verifies whether the fault is currently present.
3. This step is designed to identify a wheel speed sensor or circuitry that is damaged and visibly apparent.
4. This test checks to see that the wheel speed sensor circuitry is within proper resistance value.
5. This insures that the wheel speed sensor is not shorted to ground.

Diagnostic Aids:

An "Intermittent" problem may be caused by a poor connection, rubbed through wire insulation, or a wire that is broken inside the insulation.

If the customers comments reflect that the ABS Amber warning lamp is "ON" only during moist environmental changes, (rain, snow, vehicle wash) all wheel speed sensor circuitry should be thoroughly inspected for signs of water intrusion. If code is not current clear codes and simulate the effects of water intrusion. Use the following procedure. Spray down the suspected area with a 5% salt water solution (two teaspoons of salt to 12 oz. of water). Start vehicle, allow vehicle to run for 10 seconds. If code returns immediately, replace suspected harness.

Any circuitry, that is suspected as causing the intermittent complaint, should be thoroughly checked for backed out terminals, improper mating, broken locks, improperly formed or damaged terminals, poor terminal to wiring connections or physical damage to the wiring harness.

A careful visual inspection of the toothed ring, CV joint, bearing and wheel speed sensor for physical damage should also be performed.

DIAGNOSTIC CHART DTC A028 — 1993 CUTLASS SUPREME, GRAND PRIX, LUMINA AND REGAL, CONTINUED

IMPORTANT: WHEEL SPEED SENSOR INTERMITTENT PROBLEMS MAY BE DIFFICULT TO LOCATE. CARE SHOULD BE TAKEN NOT TO DISTURB ANY ELECTRICAL CONNECTIONS PRIOR TO AN INDICATED STEP OF THIS CHART. THIS WILL INSURE THAT AN INTERMITTENT CONNECTION WILL NOT BE CORRECTED BEFORE THE SOURCE OF THE PROBLEM IS FOUND.

(1) IS CODE A035 ALSO CURRENTLY SET?

- YES: PROCEED WITH THAT CODE FIRST.
- NO:

(2)
- TEST DRIVE VEHICLE.
- USING TECH 1, SELECT DATA LIST.
- MONITOR WHEEL SPEED ON RIGHT REAR WHEEL WHILE SLOWLY DECELERATING FROM 35 MPH TO 0 MPH.
- WHEEL SPEED SHOULD BE WITHIN 5 MPH OF REMAINING WHEEL SPEED INPUTS,

IS IT?

- YES: FAULT IS NOT PRESENT AT THIS TIME. REFER TO "DIAGNOSTIC AIDS" ON FACING PAGE.
- NO:

(3)
- IGNITION "OFF."
- PHYSICALLY INSPECT WHEEL SPEED SENSOR, SENSOR RING, WIRING AND CONNECTORS FOR DAMAGE. PAY PARTICULAR ATTENTION TO THE SENSOR RING. ANY DEVIATION WILL AFFECT THE WHEEL SPEED INPUT SIGNAL.

IS PHYSICAL DAMAGE INDICATED?

- YES: CORRECT PHYSICAL DAMAGE, CLEAR CODES AND RECHECK.
- NO:

(4)
- DISCONNECT CONNECTOR FROM RIGHT REAR WHEEL SPEED SENSOR.
- CONNECT DVM BETWEEN TERMINALS "A" AND "B" OF SENSOR, AND MEASURE RESISTANCE. RESISTANCE SHOULD BE BETWEEN 2100 AND 2400 OHMS.

IS IT?

- NO: REPLACE WHEEL SPEED SENSOR, CLEAR CODES AND RECHECK.
- YES:

(5)
- IGNITION "ON."
- CONNECT DVM TO GROUND AND TERMINALS "A" THEN "B" OF RIGHT REAR SENSOR CONNECTOR HARNESS. VOLTAGE SHOULD BE 2.25 TO 2.75 VOLTS ON BOTH TERMINALS.

IS IT?

GO TO FOLLOWING PAGE OF THIS CODE CHART

DIAGNOSTIC CHART DTC A028 — 1993 CUTLASS SUPREME, GRAND PRIX, LUMINA AND REGAL, CONTINUED

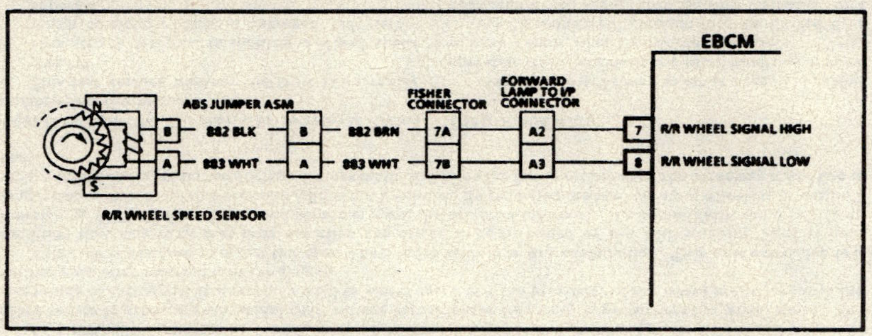

CODE A028

RIGHT REAR EXCESSIVE WHEEL SPEED VARIATION

Circuit Description:

As a toothed ring passes by the wheel speed sensor, changes in the electromagnetic field cause the wheel speed sensor to produce a sinusoidal (AC) voltage signal whose frequency is proportional to wheel speed. The magnitude of this signal is directly related to wheel speed and the proximity of the wheel speed sensor to the toothed ring, often referred to as the air gap.

This code detects opens or a low output condition that cause intermittent wheel speed operation. Also detected is any sudden change in wheel speed determined to be unreasonable.

Test Description: Number(s) below refer to circled number(s) on the diagnostic chart.

6. This step insures that the wheel speed sensor circuitry is not internally shorted.
7. This checks for an open in the wheel signal low circuit.
8. This checks for an open in the wheel signal high circuit.
9. This step insures that Code A028 was not set due to a poor connection between the world connector and EBCM.

Diagnostic Aids:

An "Intermittent" problem may be caused by a poor connection, rubbed through wire insulation, or a wire that is broken inside the insulation.

If the customers comments reflect that the ABS Amber warning lamp is "ON" only during moist environmental changes, (rain, snow, vehicle wash) all wheel speed sensor circuitry should be thoroughly inspected for signs of water intrusion. If code is not current clear codes and simulate the effects of water intrusion. Use the following procedure. Spray down the suspected area with a 5% salt water solution (two teaspoons of salt to 12 oz. of water). Start vehicle, allow vehicle to run for 10 seconds. If code returns immediately, replace suspected harness.

Any circuitry, that is suspected as causing the intermittent complaint, should be thoroughly checked for backed out terminals, improper mating, broken locks, improperly formed or damaged terminals, poor terminal to wiring connections or physical damage to the wiring harness.

A careful visual inspection of the toothed ring, CV joint, bearing and wheel speed sensor for physical damage should also be performed.

DIAGNOSTIC CHART DTC A028 — 1993 CUTLASS SUPREME, GRAND PRIX, LUMINA AND REGAL, CONTINUED

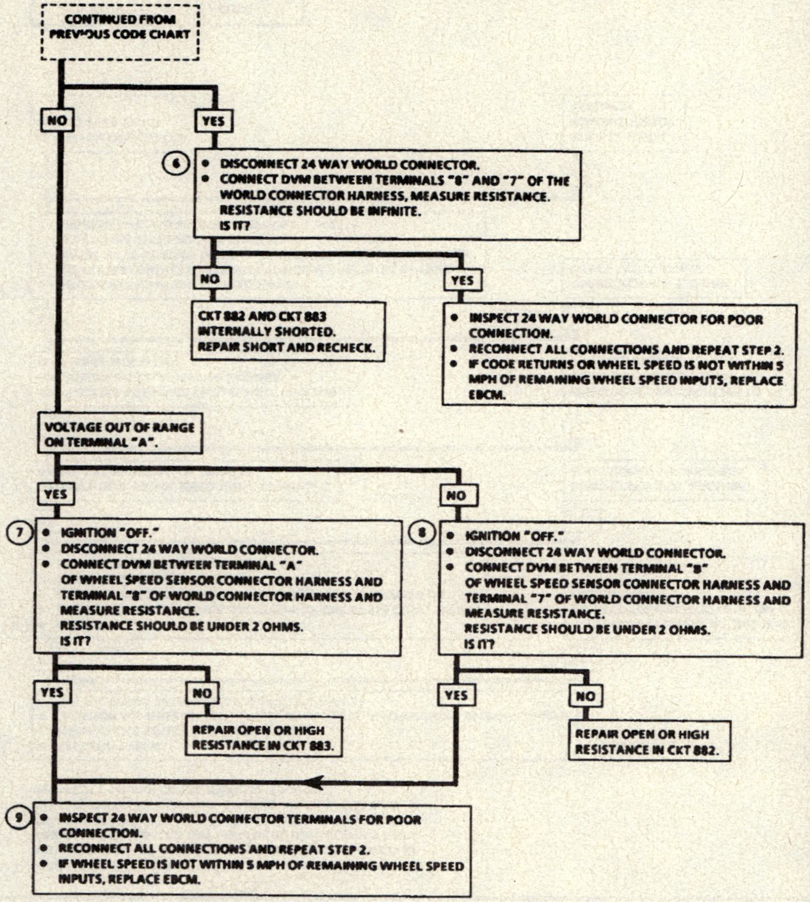

THIS CHART ASSUMES THAT A CURRENT CODE IS STORED INDICATING THAT THIS FAULT IS PRESENT.

DIAGNOSTIC CHART DTC A031 — 1993 CUTLASS SUPREME, GRAND PRIX, LUMINA AND REGAL

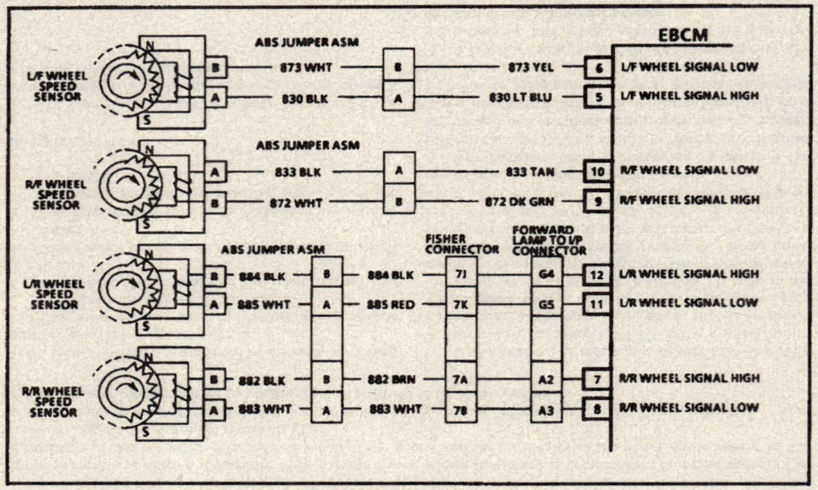

CODE A031

TWO WHEEL SPEEDS = 0

Circuit Description:

As a toothed ring passes by the wheel speed sensor, changes in the electromagnetic field cause the wheel speed sensor to produce a sinusoidal (AC) voltage signal whose frequency is proportional to wheel speed. The magnitude of this signal is directly related to wheel speed and the proximity of the wheel speed sensor to the toothed ring, often referred to as the air gap.

This code is designed to detect two open wheel speed sensors in any combination. This code exists due to a condition that may exist that does not allow the driver to be informed by the ABS warning light of non-functioning ABS. If one wheel speed sensor is open, the driver is informed. If a second fails, the driver is no longer informed without this code and ABS is not available. By requiring the two operating wheels to be within 7 mph, a false code is prevented during spin-ups while still allowing for detection when the stowaway spare tire is used.

Test Description: Number(s) below refer to circled number(s) on the diagnostic chart.

1. This test verifies whether the fault is currently present.
2. This step is designed to identify a wheel speed sensor or circuitry that is damaged and visibly apparent.
3. This step identifies if the cause of the problem is common to the rear wheel sensor circuitry.
4. This test checks to see if connectors at pass thru are connected properly.
5. This test checks to see that the wheel speed sensor is within proper resistance values.

Diagnostic Aids:

A false Code A031 may be set as a result of vehicle operation while on a hoist or a lifted condition or stuck in mud or snow, when the vehicle is in gear.

An "Intermittent" problem may be caused by a poor connection, rubbed through wire insulation, or a wire that is broken inside the insulation.

DIAGNOSTIC CHART DTC A031 — 1993 CUTLASS SUPREME, GRAND PRIX, LUMINA AND REGAL, CONTINUED

IMPORTANT: WHEEL SPEED SENSOR INTERMITTENT PROBLEMS MAY BE DIFFICULT TO LOCATE. CARE SHOULD BE TAKEN NOT TO DISTURB ANY ELECTRICAL CONNECTIONS PRIOR TO AN INDICATED STEP OF THIS CHART. THIS WILL INSURE THAT AN INTERMITTENT CONNECTION WILL NOT BE CORRECTED BEFORE THE SOURCE OF THE PROBLEM IS FOUND.

(1)
- TEST DRIVE VEHICLE.
- USING TECH 1, SELECT DATA LIST.
- MONITOR ALL WHEEL SPEEDS WHILE SLOWLY DECELERATING FROM 35 MPH TO 0 MPH.

ARE ALL WHEEL SPEEDS APPROXIMATELY EQUAL?

- YES: FAULT IS NOT PRESENT AT THIS TIME. REFER TO "DIAGNOSTIC AIDS" ON FACING PAGE.
- NO: go to 2.

(2)
- IGNITION "OFF."
- PHYSICALLY INSPECT WHEEL SPEED SENSOR CIRCUITS IN QUESTION FOR DAMAGE. ALSO, CHECK WIRE CONNECTIONS FOR CORROSION OR POOR MATING OF CONNECTORS.

IS PHYSICAL DAMAGE INDICATED?

- YES: CORRECT PHYSICAL DAMAGE, CLEAR CODES AND RE-CHECK.
- NO: go to 3.

(3) DID BOTH REAR WHEEL SPEED SENSORS INDICATE 0 MPH, OR ERRATIC WHEEL SPEEDS?

- NO: USING CHART BELOW, REFER TO SPEED SENSOR CODE CHART FOR AFFECTED WHEELS.
- YES: go to 4.

(4) INSPECT 2 REAR ABS PASS THRU CONNECTORS FOR CORROSION, POOR MATING OR DAMAGE. IS DAMAGE INDICATED?

- YES: CORRECT DAMAGE, CLEAR CODES AND RE-CHECK.
- NO: go to 5.

(5)
- DISCONNECT REAR WHEEL SPEED SENSORS.
- USING DVM CONNECTED BETWEEN TERMINALS "A" & "B" OF THE REAR WHEEL SPEED SENSORS MEASURE RESISTANCE. RESISTANCE SHOULD BE BETWEEN 2100 AND 2400 OHMS FOR REAR SENSORS.

ARE RESISTANCE VALUES IN ACCEPTABLE RANGES?

- YES: GO TO FOLLOWING PAGE OF THIS CODE CHART
- NO: REPLACE WHEEL BEARING/SENSOR ASSEMBLY.

SENSOR	CODE
L/F SENSOR	A021
R/F SENSOR	A022
L/R SENSOR	A023
R/R SENSOR	A024

DIAGNOSTIC CHART DTC A031 — 1993 CUTLASS SUPREME, GRAND PRIX, LUMINA AND REGAL, CONTINUED

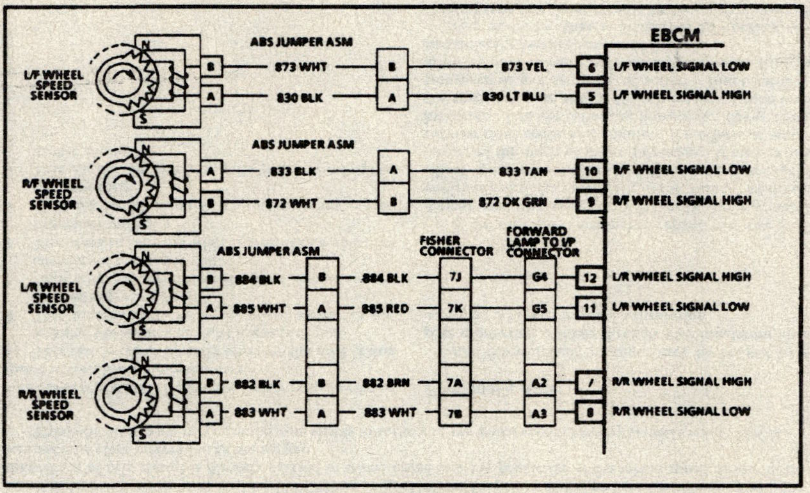

CODE A031

TWO WHEEL SPEEDS = 0

Circuit Description:

As a toothed ring passes by the wheel speed sen.or, changes in the electromagnetic field cause the wheel speed sensor to produce a sinusoidal (AC) voltage signal whose frequency is proportional to wheel speed. The magnitude of this signal is directly related to wheel speed and the proximity of the wheel speed sensor to the toothed ring, often referred to as the air gap.

This code is designed to detect two open wheel speed sensors in any combination. This code exists due to a condition that may exist that does not allow the driver to be informed by the ABS warning light of non-functioning ABS. If one wheel speed sensor is open, the driver is informed. If a second fails, the driver is no longer informed without this code and ABS is not available. By requiring the two operating wheels to be within 7 mph, a false code is prevented during spin-ups while still allowing for detection when the stowaway spare tire is used.

Test Description: Number(s) below refer to circled number(s) on the diagnostic chart.

6. This step checks for an open in the wheel speed sensor circuitry of one side of the sensor.
7. This step checks for an open on the other side of the sensor in question.
8. This insures that the circuitry is not shorted to voltage.

Diagnostic Aids:

A false Code A031 may be set as a result of vehicle operation while on a hoist or a lifted condition, or stuck in mud or snow when the vehicle is in gear.

An "Intermittent" problem may be caused by a poor connection, rubbed through wire insulation, or a wire that is broken inside the insulation.

DIAGNOSTIC CHART DTC A031 — 1993 CUTLASS SUPREME, GRAND PRIX, LUMINA AND REGAL, CONTINUED

CONTINUED FROM CHART CODE A031

(6)
- CONNECT DVM TO TERMINAL "A" OF WHEEL SPEED SENSOR HARNESS CONNECTOR IN QUESTION.
- DISCONNECT 24 WAY WORLD CONNECTOR.
- USING THE TABLE SUPPLIED AT THE BOTTOM OF THIS PAGE, CONNECT DVM TO THE APPROPRIATE TERMINAL OF THE WORLD CONNECTOR AND MEASURE RESISTANCE.

RESISTANCE SHOULD BE LESS THAN 2 OHMS.
IS IT?

NO: REPAIR CIRCUIT THAT SHOWED HIGH RESISTANCE OR OPEN.

YES:

(7)
- USING DVM, PROBE CONNECTOR TERMINAL "B" OF WHEEL SPEED SENSOR IN QUESTION.
- USING THE TABLE SUPPLIED, CONNECT DVM TO THE APPROPRIATE TERMINAL OF THE WORLD CONNECTOR AND MEASURE RESISTANCE.

RESISTANCE SHOULD BE LESS THAN 2 OHMS.
IS IT?

NO: REPAIR CIRCUIT THAT SHOWED HIGH RESISTANCE OR OPEN.

YES:

(8)
- DVM STILL CONNECTED.
- SELECT THE 20 VOLT SCALE OF THE DVM.
- IGNITION "ON."
- REPEAT STEP 8.

VOLTAGE SHOULD BE 0.
IS IT?

NO: REPAIR SHORT TO VOLTAGE IN CIRCUIT THAT INDICATED VOLTAGE ABOVE 0 VOLTS.

YES:
- INSPECT WORLD CONNECTOR TERMINALS FOR POOR CONNECTION.
- RE-CONNECT ALL CONNECTIONS AND REPEAT STEP 1.
- IF WHEEL SPEED DOES NOT MATCH REMAINING WHEEL SPEED INPUTS, REPLACE EBCM.

SENSOR TERMINAL	WORLD CONNECTOR	CIRCUIT NUMBER
L/R WHEEL B	PIN 12	884
L/R WHEEL A	PIN 11	885
R/R WHEEL B	PIN 7	882
R/R WHEEL A	PIN 8	883

THIS CHART ASSUMES THAT A CURRENT CODE IS STORED INDICATING THAT THIS FAULT IS PRESENT.

DIAGNOSTIC CHART DTC A032 — 1993 CUTLASS SUPREME, GRAND PRIX, LUMINA AND REGAL

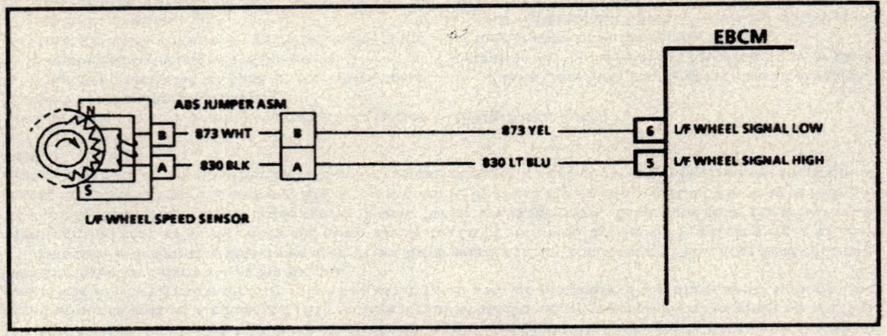

CODE A032

LEFT FRONT WHEEL SENSOR SHORTED TO BATTERY OR GROUND

Circuit Description:

As a toothed ring passes by the wheel speed sensor, changes in the electromagnetic field cause the wheel speed sensor to produce a sinusoidal (AC) voltage signal whose frequency is proportional to wheel speed. The magnitude of this signal is directly related to wheel speed and the proximity of the wheel speed sensor to the toothed ring often referred to as the air gap.

This code detects shorts to ground, and shorts to battery in the wheel speed circuitry sensed by ABS EBCM.

Test Description: Number(s) below refer to circled number(s) on the diagnostic chart.

1. This test helps to identify whether the code is due to a short to battery or short to ground.
2. This insures that the wheel speed sensor is not shorted to ground.
3. This step checks to see if wheel speed sensor circuitry is shorted to voltage.
4. This insures that the wheel speed sensor circuitry has good continuity.
5. This checks for short to ground in wheel speed sensor wiring.
6. This checks for short to voltage in wheel speed sensor wiring.

Diagnostic Aids:

An "Intermittent" problem may be caused by a poor connection, rubbed through wire insulation, or a wire that is broken inside the insulation.

If the customers comments reflect that the ABS Amber warning lamp is "ON" only during moist environmental changes, (rain, snow, vehicle wash) all wheel speed sensor circuitry should be thoroughly inspected for signs of water intrusion. If code is not current clear codes and simulate the effects of water intrusion. Use the following procedure. Spray down the suspected area with a 5% salt water solution (two teaspoons of salt to 12 oz. of water). Start vehicle, allow vehicle to run for 10 seconds. If code returns immediately, replace suspected harness.

Any circuitry, that is suspected as causing the intermittent complaint, should be thoroughly checked for backed out terminals, improper mating, broken locks, improperly formed or damaged terminals, poor terminal to wiring connections or physical damage to the wiring harness.

DIAGNOSTIC CHART DTC A032 — 1993 CUTLASS SUPREME, GRAND PRIX, LUMINA AND REGAL, CONTINUED

IMPORTANT: WHEEL SPEED SENSOR INTERMITTENT PROBLEMS MAY BE DIFFICULT TO LOCATE. CARE SHOULD BE TAKEN NOT TO DISTURB ANY ELECTRICAL CONNECTIONS PRIOR TO AN INDICATED STEP OF THIS CHART. THIS WILL INSURE THAT AN INTERMITTENT CONNECTION WILL NOT BE CORRECTED BEFORE THE SOURCE OF THE PROBLEM IS FOUND.

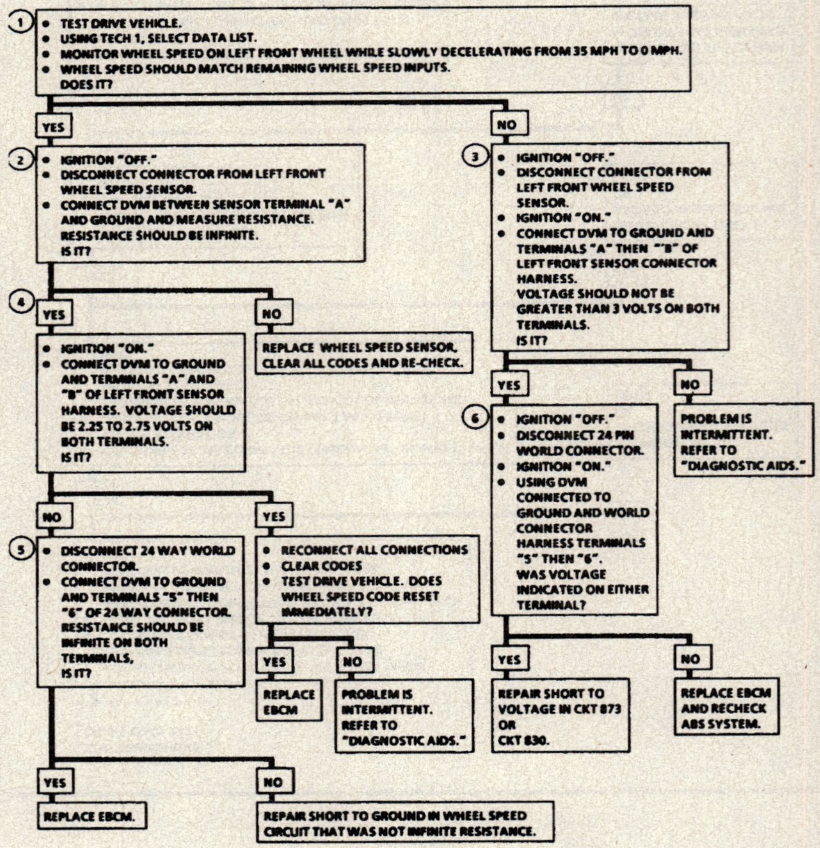

DIAGNOSTIC CHART DTC A033 — 1993 CUTLASS SUPREME, GRAND PRIX, LUMINA AND REGAL

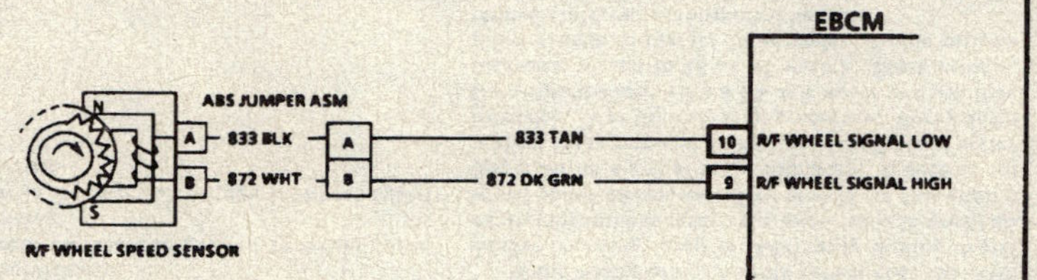

CODE A033

RIGHT FRONT WHEEL SENSOR SHORTED TO BATTERY OR GROUND

Circuit Description:

As a toothed ring passes by the wheel speed sensor, changes in the electromagnetic field cause the wheel speed sensor to produce a sinusoidal (AC) voltage signal whose frequency is proportional to wheel speed. The magnitude of this signal is directly related to wheel speed and the proximity of the wheel speed sensor to the toothed ring, often referred to as the air gap.

This code detects shorts to ground, and shorts to battery in the wheel speed circuitry sensed by the ABS EBCM.

Test Description: Number(s) below refer to circled number(s) on the diagnostic chart.

1. This test helps identify whether the code is due to a short to battery or short to ground.
2. This insures that the wheel speed sensor is not shorted to ground.
3. This step checks to see if wheel speed sensor circuitry is shorted to voltage.
4. This insures that the wheel speed sensor circuitry has good continuity.
5. This checks for short to ground in wheel speed sensor wiring.
6. This checks for short to voltage in wheel speed sensor wiring.

Diagnostic Aids:

An "Intermittent" problem may be caused by a poor connection, rubbed through wire insulation, or a wire that is broken inside the insulation.

If the customers comments reflect that the ABS Amber warning lamp is "ON" only during moist environmental changes, (rain, snow, vehicle wash) all wheel speed sensor circuitry should be thoroughly inspected for signs of water intrusion. If code is not current clear codes and simulate the effects of water intrusion. Use the following procedure. Spray down the suspected area with a 5% salt water solution (two teaspoons of salt to 12 oz. of water). Start vehicle, allow vehicle to run for 10 seconds. If code returns immediately, replace suspected harness.

Any circuitry, that is suspected as causing the intermittent complaint, should be thoroughly checked for backed out terminals, improper mating, broken locks, improperly formed or damaged terminals, poor terminal to wiring connections or physical damage to the wiring harness.

DIAGNOSTIC CHART DTC A033 — 1993 CUTLASS SUPREME, GRAND PRIX, LUMINA AND REGAL, CONTINUED

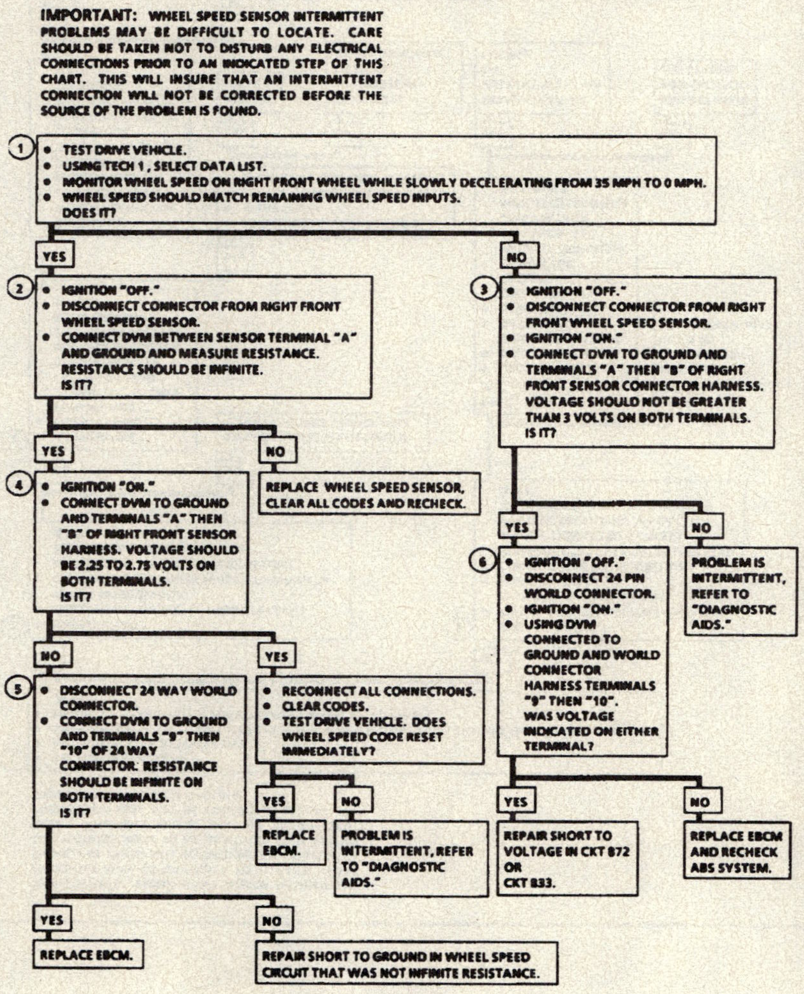

DIAGNOSTIC CHART DTC A034 — 1993 CUTLASS SUPREME, GRAND PRIX, LUMINA AND REGAL

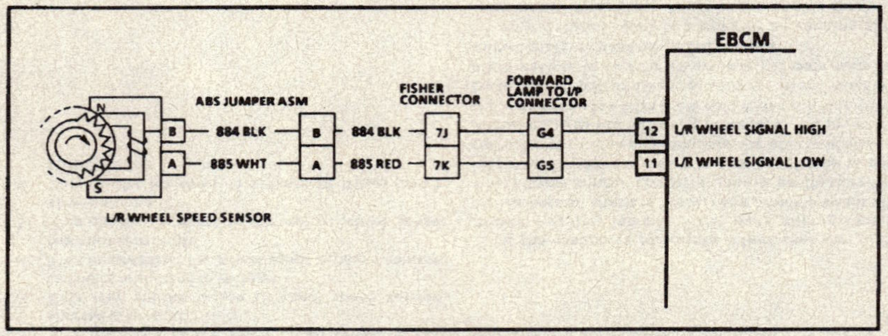

CODE A034

LEFT REAR WHEEL SENSOR SHORTED TO BATTERY OR GROUND

Circuit Description:

As a toothed ring passes by the wheel speed sensor, changes in the electromagnetic field cause the wheel speed sensor to produce a sinusoidal (AC) voltage signal whose frequency is proportional to wheel speed. The magnitude of this signal is directly related to wheel speed and the proximity of the wheel speed sensor to the toothed ring, often referred to as the air gap.

This code detects opens, shorts to ground, and shorts to battery in the wheel speed circuitry sensed by the ABS EBCM.

Test Description: Number(s) below refer to circled number(s) on the diagnostic chart.

1. This test helps to identify whether the code is due to a short to battery or short to ground.
2. This insures that the wheel speed sensor is not shorted to ground.
3. This step checks to see if wheel speed sensor circuitry is shorted to voltage.
4. This insures that the wheel speed sensor circuitry has good continuity.
5. This checks for short to ground in wheel speed sensor wiring.
6. This checks for short to voltage in wheel speed sensor wiring.

Diagnostic Aids:

An "Intermittent" problem may be caused by a poor connection, rubbed through wire insulation, or a wire that is broken inside the insulation.

If the customers comments reflect that the ABS Amber warning lamp is "ON" only during moist environmental changes, (rain, snow, vehicle wash) all wheel speed sensor circuitry should be thoroughly inspected for signs of water intrusion. If code is not current clear codes and simulate the effects of water intrusion. Use the following procedure. Spray down the suspected area with a 5% salt water solution (two teaspoons of salt to 12 oz. of water). Start vehicle, allow vehicle to run for 10 seconds. If code returns immediately, replace suspected harness.

Any circuitry, that is suspected as causing the intermittent complaint, should be thoroughly checked for backed out terminals, improper mating, broken locks, improperly formed or damaged terminals, poor terminal to wiring connections or physical damage to the wiring harness.

DIAGNOSTIC CHART DTC A034 — 1993 CUTLASS SUPREME, GRAND PRIX, LUMINA AND REGAL, CONTINUED

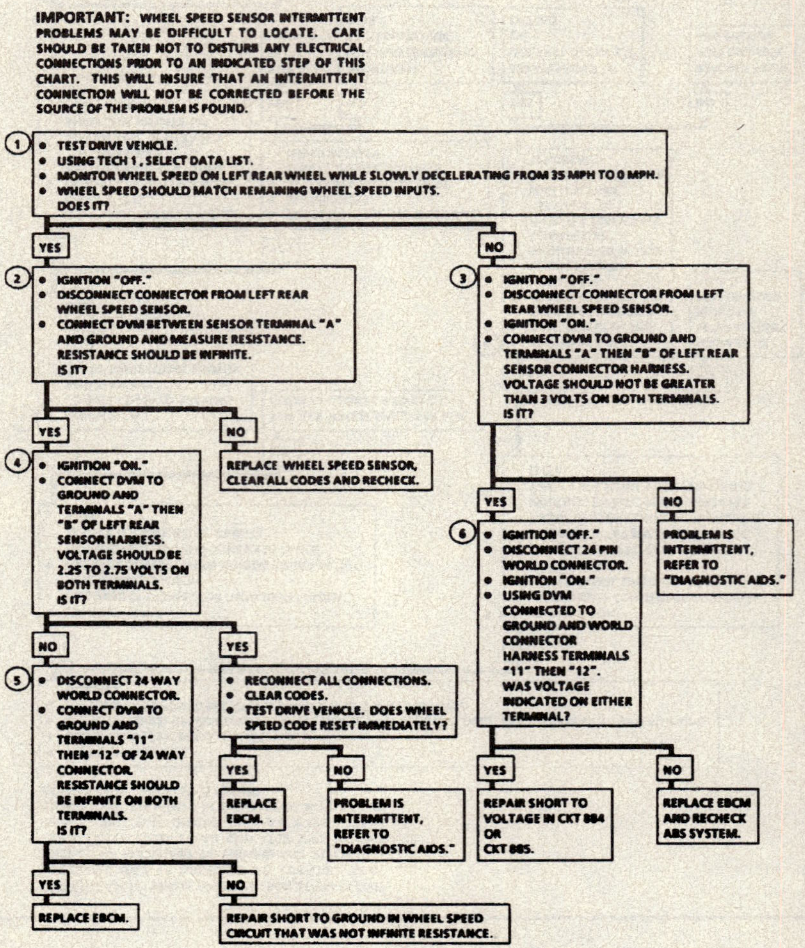

DIAGNOSTIC CHART DTC A035 — 1993 CUTLASS SUPREME, GRAND PRIX, LUMINA AND REGAL

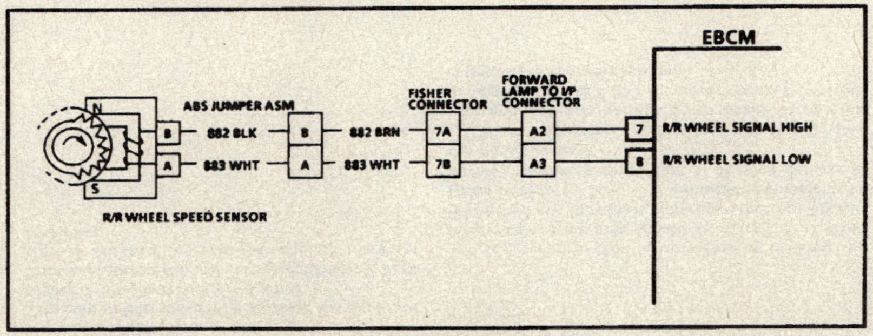

CODE A035

RIGHT REAR WHEEL SENSOR SHORTED TO BATTERY OR GROUND

Circuit Description:

As a toothed ring passes by the wheel speed sensor, changes in the electromagnetic field cause the wheel speed sensor to produce a sinusoidal (AC) voltage signal whose frequency is proportional to wheel speed. The magnitude of this signal is directly related to wheel speed and the proximity of the wheel speed sensor to the toothed ring, often referred to as the air gap.

This code detects opens, shorts to ground, and shorts to battery in the wheel speed circuitry sensed by the ABS EBCM.

Test Description: Number(s) below refer to circled number(s) on the diagnostic chart.

1. This helps to identify whether the code is due to a short to battery or short to ground.
2. This insures that the wheel speed sensor is not shorted to ground.
3. This step checks to see if wheel speed sensor circuitry is shorted to voltage.
4. This insures that the wheel speed sensor circuitry has good continuity.
5. This checks for short to ground in wheel speed sensor wiring.
6. This checks for short to voltage in wheel speed sensor wiring.

Diagnostic Aids:

An "Intermittent" problem may be caused by a poor connection, rubbed through wire insulation, or a wire that is broken inside the insulation.

If the customers comments reflect that the ABS Amber warning lamp is "ON" only during moist environmental changes, (rain, snow, vehicle wash) all wheel speed sensor circuitry should be thoroughly inspected for signs of water intrusion. If code is not current clear codes and simulate the effects of water intrusion. Use the following procedure. Spray down the suspected area with a 5% salt water solution (two teaspoons of salt to 12 oz. of water). Start vehicle, allow vehicle to run for 10 seconds. If code returns immediately, replace suspected harness.

Any circuitry, that is suspected as causing the intermittent complaint, should be thoroughly checked for backed out terminals, improper mating, broken locks, improperly formed or damaged terminals, poor terminal to wiring connections or physical damage to the wiring harness.

DIAGNOSTIC CHART DTC A035 — 1993 CUTLASS SUPREME, GRAND PRIX, LUMINA AND REGAL, CONTINUED

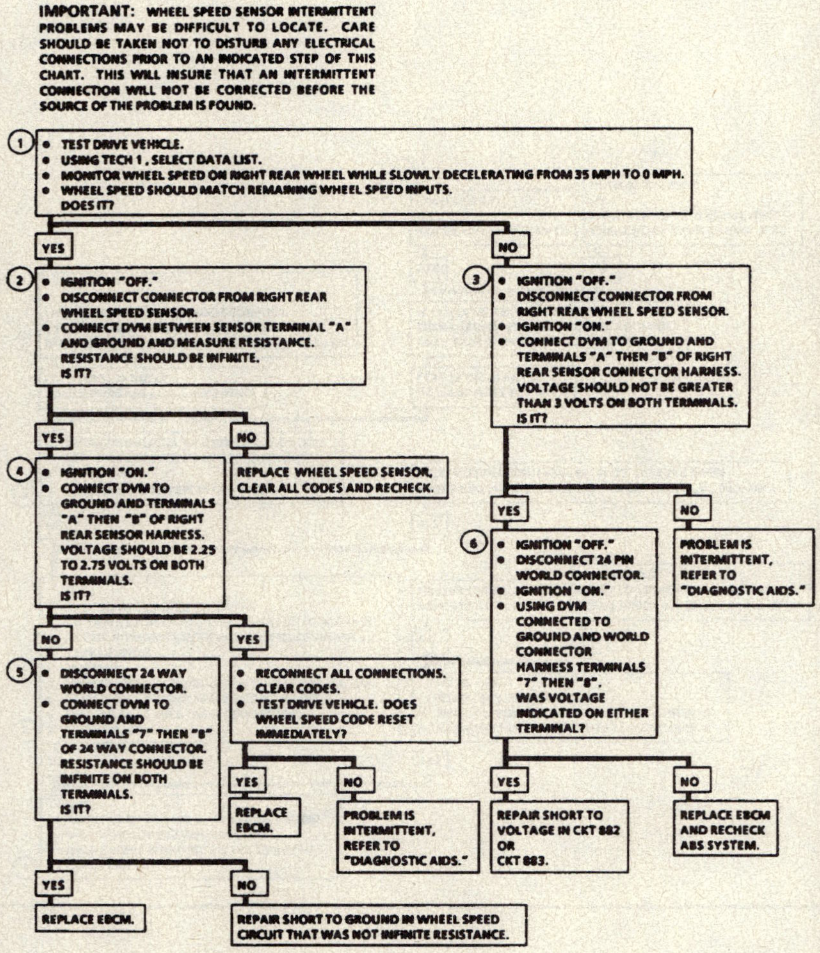

DIAGNOSTIC CHART DTC A036 — 1993 CUTLASS SUPREME, GRAND PRIX, LUMINA AND REGAL

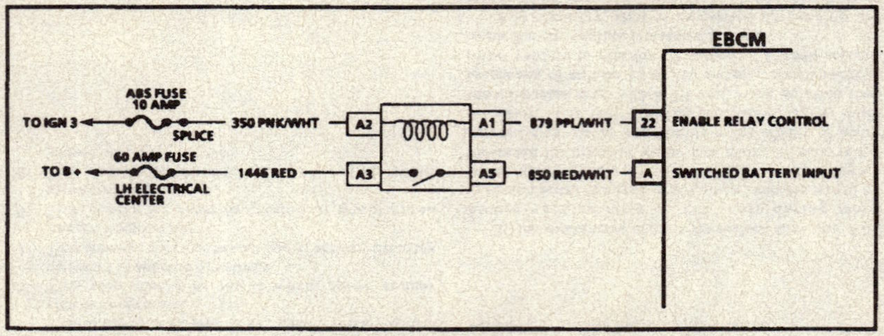

CODE A036

LOW SYSTEM VOLTAGE

Circuit Description:

This code is used to monitor the voltage level available to the EBCM controller. If the voltage drops below 11 volts, full performance of the ABS system cannot be guaranteed. During ABS operation, there are several current requirements that will cause battery voltage to drop. Because of this, voltage is monitored prior to ABS to indicate good charging system condition and also during ABS when voltage may drop significantly. If voltage falls below a level that will provide adequate ABS performance, Code A036 will result.

Test Description: Number(s) below refer to circled number(s) on the diagnostic chart.

1. This test checks to see what voltage is readily available at terminal "A" of the EBCM. If voltage is representative of good charging system condition, a fault is not present.
2. This step isolates the EBCM to check the condition of the circuitry and charging system.
3. This step isolates the low voltage condition to high circuit resistance, or improper charging system operation.

Diagnostic Aids:

An "Intermittent" problem may be caused by a poor connection, rubbed through wire insulation, or a wire that is broken inside the insulation.

Any circuitry, that is suspected as causing the intermittent complaint, should be thoroughly checked for backed out terminals, improper mating, broken locks, improperly formed or damaged terminals, poor terminal to wiring connections or physical damage to the wiring harness.

While performing a voltage load test, if it is noted that only ignition voltage drops below acceptable voltage levels, CKT 350 should be checked for high resistance or an open condition.

DIAGNOSTIC CHART DTC A036 — 1993 CUTLASS SUPREME, GRAND PRIX, LUMINA AND REGAL, CONTINUED

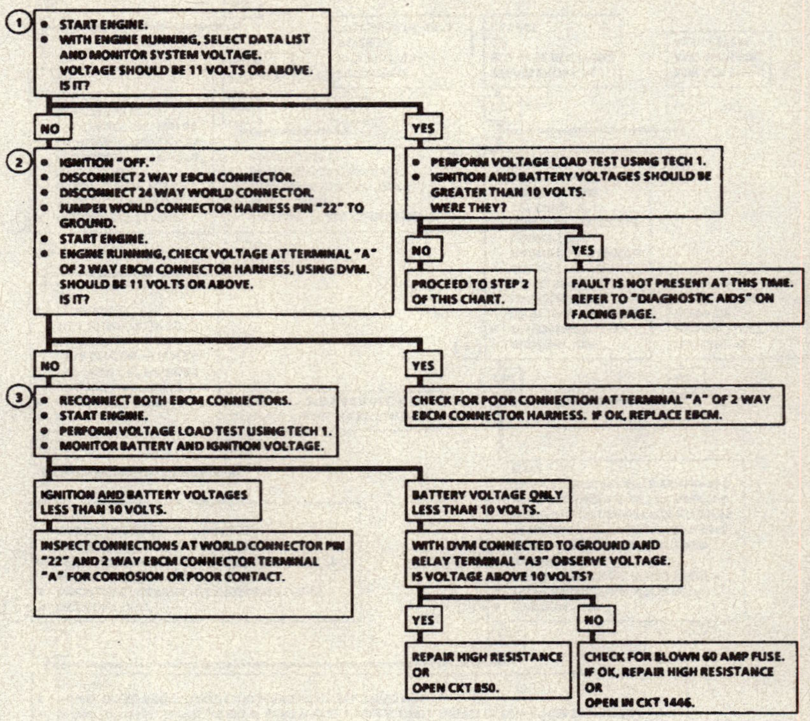

DIAGNOSTIC CHART DTC A037 — 1993 CUTLASS SUPREME, GRAND PRIX, LUMINA AND REGAL

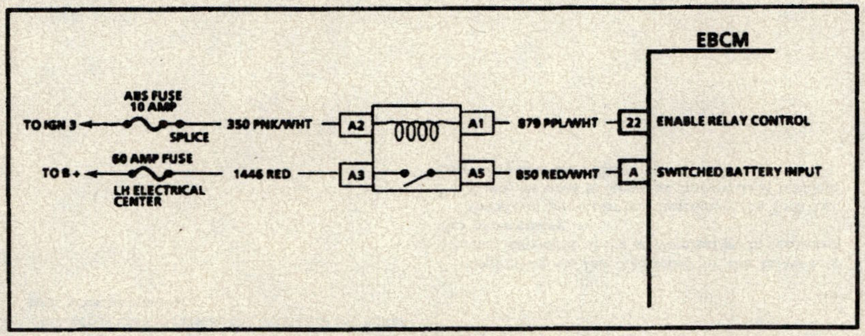

CODE A037
HIGH SYSTEM VOLTAGE

Circuit Description:

This code is designed to detect high vehicle voltage levels prior to any required motor movement (initialization or ABS). If excessive voltage exists, demagnetization of the motor magnets may occur, which would eventually affect or eliminate ABS performance.

Test Description: Number(s) below refer to circled number(s) on the diagnostic chart.

1. This step checks the voltage level being received by the EBCM. If high voltage is present, a charging system problem is indicated.

Diagnostic Aids:

An "Intermittent" problem may be caused by a poor connection, rubbed through wire insulation, or a wire that is broken inside the insulation.

Any circuitry, that is suspected as causing the intermittent complaint, should be thoroughly checked for backed out terminals, improper mating, broken locks, improperly formed or damaged terminals, poor terminal to wiring connections or physical damage to the wiring harness.

DIAGNOSTIC CHART DTC A037 — 1993 CUTLASS SUPREME, GRAND PRIX, LUMINA AND REGAL, CONTINUED

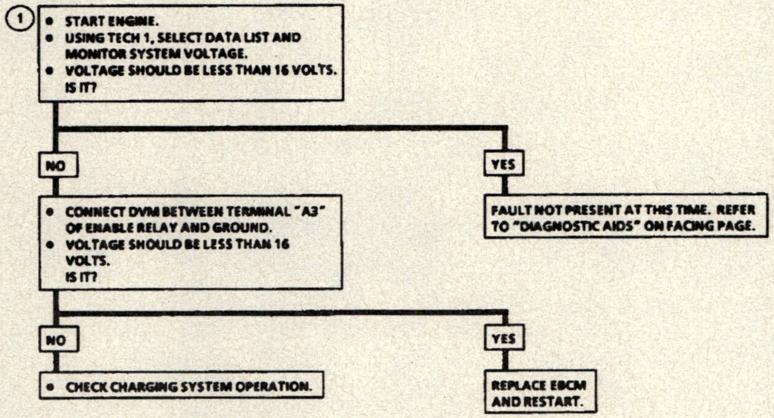

THIS CHART ASSUMES THAT A CURRENT CODE IS STORED INDICATING THAT THIS FAULT IS PRESENT.

DIAGNOSTIC CHART DTC A038 — 1993 CUTLASS SUPREME, GRAND PRIX, LUMINA AND REGAL

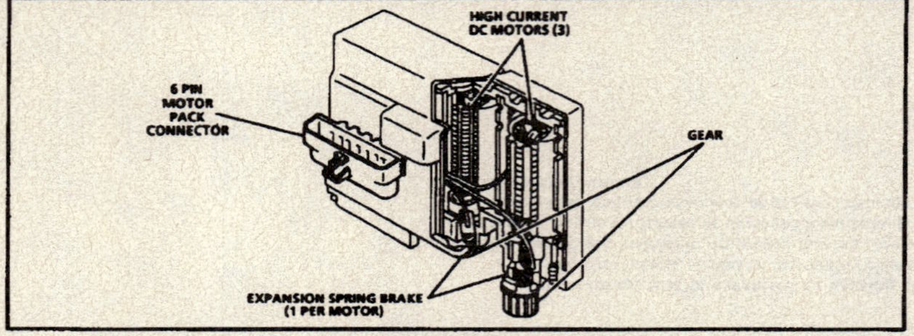

CODE A038

LEFT FRONT ESB WILL NOT HOLD MOTOR

Circuit Description:

This code is designed to detect a slipping left front ESB. During initialization and braking, the left front motor is re-homed. If the ESB slips, the motor/piston will move. During the next key on initialization, a re-home of the motor verifies the motor/piston remained at the home position. If motor movement is detected, the ESB must be slipping.

Test Description: Number(s) below refer to circled number(s) on the diagnostic chart.

1. This step checks the left front ESB. A broken or defective ESB would result in the left front piston being back driven by hydraulic pressure, pedal movement would result.
2. This step is used to release the tension of the motor pack prior to removal.

Diagnostic Aids:

An "Intermittent" fault in this code may result from a mechanical part of the system that sticks, binds, or slips.

Depending on the frequency of the failure, a physical inspection of the mechanical parts suspected may be necessary.

The static modulator test function of the Tech 1 or T-100 may be used to locate an intermittent problem associated with the ESB.

DIAGNOSTIC CHART DTC A038 — 1993 CUTLASS SUPREME, GRAND PRIX, LUMINA AND REGAL, CONTINUED

(1)
- IGNITION "ON," ENGINE "OFF."
- USING TECH 1, SELECT MANUAL CONTROL FUNCTION.
- SELECT MOTOR APPLY FUNCTION OF THE LEFT FRONT CHANNEL, AND APPLY MOTOR.
- WAIT 5 SECONDS.
- APPLY FIRM PRESSURE ON BRAKE PEDAL. PEDAL SHOULD NOT FALL.

DID IT?

YES → (2) USING TECH 1, PERFORM GEAR TENSION RELIEF FUNCTION. → REPLACE MOTOR PACK ASSEMBLY.

NO → FAULT IS NOT PRESENT AT THIS TIME. REFER TO "DIAGNOSTIC AIDS" ON FACING PAGE.

DIAGNOSTIC CHART DTC A041 — 1993 CUTLASS SUPREME, GRAND PRIX, LUMINA AND REGAL

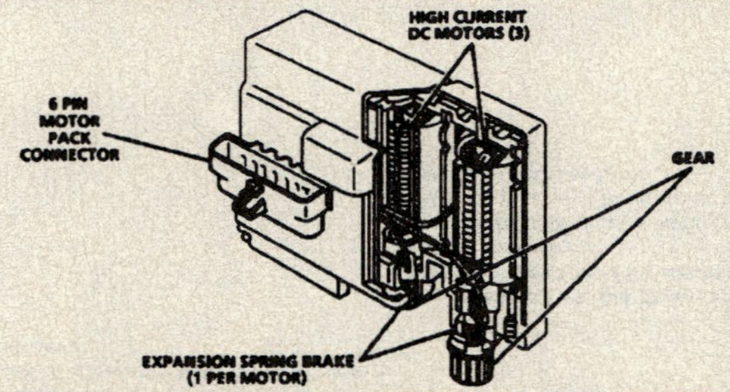

CODE A041

RIGHT FRONT ESB WILL NOT HOLD MOTOR

Circuit Description:

This code is designed to detect a slipping right front ESB. During initialization and braking, the right front motor is re-homed. If the ESB slips, the motor/piston will move. During the next key on initialization, a re-home of the motor verifies the motor/piston remained at the home position. If motor movement is detected, the ESB must be slipping.

Test Description: Number(s) below refer to circled number(s) on the diagnostic chart.

1. This step checks the right front ESB. A broken or defective ESB would result in the right front piston being back driven by hydraulic pressure, and wheel movement would result.
2. This step is used to release the tension of the motor pack prior to removal.

Diagnostic Aids:

An "Intermittent" fault in this code may result from a mechanical part of the system that sticks, binds, or slips.

Depending on the frequency of the failure, a physical inspection of the mechanical parts suspected may be necessary.

The static modulator test function of the Tech 1 or T-100 may be used to locate an intermittent problem associated with the ESB.

DIAGNOSTIC CHART DTC A041 — 1993 CUTLASS SUPREME, GRAND PRIX, LUMINA AND REGAL, CONTINUED

(1)
- IGNITION "ON," ENGINE "OFF."
- USING TECH 1, SELECT MANUAL CONTROL FUNCTION.
- SELECT MOTOR APPLY FUNCTION OF THE RIGHT FRONT CHANNEL, AND APPLY MOTOR.
- WAIT 5 SECONDS.
- APPLY FIRM PRESSURE ON BRAKE PEDAL. PEDAL SHOULD NOT FALL. DID IT?

YES → (2) USING TECH 1, PERFORM GEAR TENSION RELIEF FUNCTION. → REPLACE MOTOR PACK ASSEMBLY.

NO → FAULT IS NOT PRESENT AT THIS TIME. REFER TO "DIAGNOSTIC AIDS" ON FACING PAGE.

DIAGNOSTIC CHART DTC A042 — 1993 CUTLASS SUPREME, GRAND PRIX, LUMINA AND REGAL

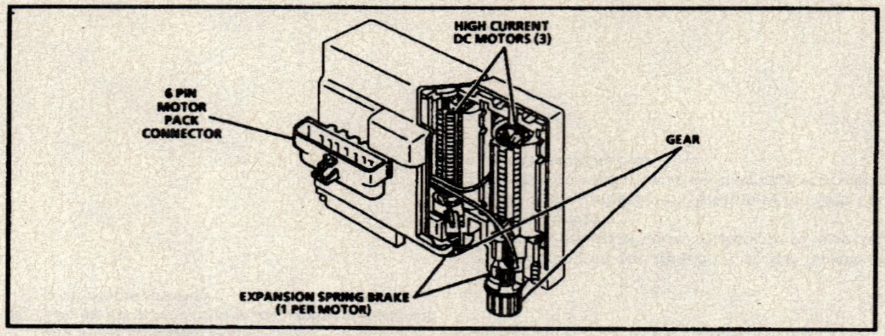

CODE A042

REAR AXLE ESB WILL NOT HOLD MOTOR

Circuit Description:

This code is designed to detect a slipping rear axle ESB. During initialization and braking, the rear motor is re-homed. If the ESB slips, the motor/piston will move. During the next key on initialization, a re-home of the motor verifies the motor/piston remained at the home position. If motor movement is detected, the ESB must be slipping.

Test Description: Number(s) below refer to circled number(s) on the diagnostic chart.

1. This step checks the rear axle ESB. A broken or defective ESB would result in the rear piston being back driven by hydraulic pressure, and wheel movement would result.
2. This step is used to release the tension of the motor pack prior to removal.

Diagnostic Aids:

An "Intermittent" fault in this code may result from a mechanical part of the system that sticks, binds, or slips.

Depending on the frequency of the failure, a physical inspection of the mechanical parts suspected may be necessary.

The static modulator test function of the Tech 1 or T-100 may be used to locate an intermittent problem associated with the ESB.

DIAGNOSTIC CHART DTC A042 — 1993 CUTLASS SUPREME, GRAND PRIX, LUMINA AND REGAL, CONTINUED

(1)
- IGNITION "ON," ENGINE "OFF."
- APPLY FIRM PRESSURE ON BRAKE PEDAL AND RELEASE.
- RE-APPLY FIRM PRESSURE ON BRAKE PEDAL AND TRY TO SPIN REAR WHEELS BY HAND, WHEELS SHOULD NOT BE ABLE TO BE MOVED.

DID WHEELS REMAIN STATIONARY?

YES → FAULT IS NOT PRESENT AT THIS TIME. REFER TO "DIAGNOSTIC AIDS" ON FACING PAGE.

NO → (2) USING TECH 1, PERFORM GEAR TENSION RELIEF FUNCTION.
→ REPLACE MOTOR PACK ASSEMBLY AND RE-CHECK ABS SYSTEM.

DIAGNOSTIC CHART DTC A044 — 1993 CUTLASS SUPREME, GRAND PRIX, LUMINA AND REGAL

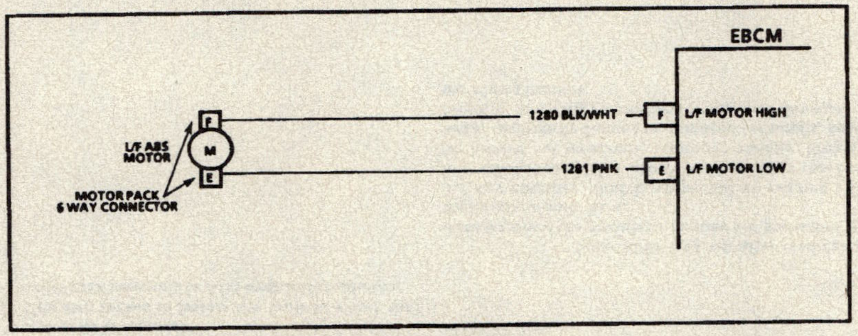

CODE A044

LEFT FRONT CHANNEL WILL NOT MOVE

Circuit Description:
This code is designed to detect bound-up ESB, a stuck motor, or a seized hydraulic modulator. When the release is commanded during initialization, the ESB should release the motor, resulting in sensed current being less than commanded current (motor is spinning freely). If the motor is not moving, sensed current will be equal to stall current.

Test Description: Number(s) below refer to circled number(s) on the diagnostic chart.

1. This step checks for motor movement.
2. This step is used to release the tension of the motor pack prior to removal.
3. This step serves to isolate the fault to either the motor pack assembly or ABS hydraulic modulator.

Diagnostic Aids:

An "Intermittent" fault in this code may result from a mechanical part of the system that sticks, binds, or slips.

Code A044 may set after modulator disassembly if the modulator pistons are positioned at the bottom of their bore.

Any circuitry, that is suspected as causing the intermittent complaint, should be thoroughly checked for backed out terminals, improper mating, broken locks, improperly formed or damaged terminals, poor terminal to wiring connections or physical damage to the wiring harness.

DIAGNOSTIC CHART DTC A044 — 1993 CUTLASS SUPREME, GRAND PRIX, LUMINA AND REGAL, CONTINUED

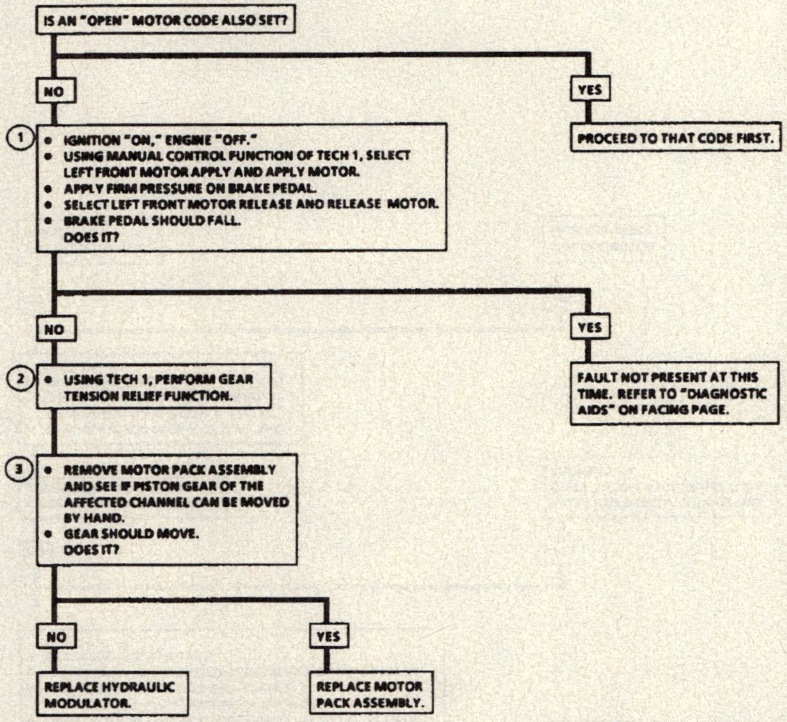

DIAGNOSTIC CHART DTC A046 — 1993 CUTLASS SUPREME, GRAND PRIX, LUMINA AND REGAL

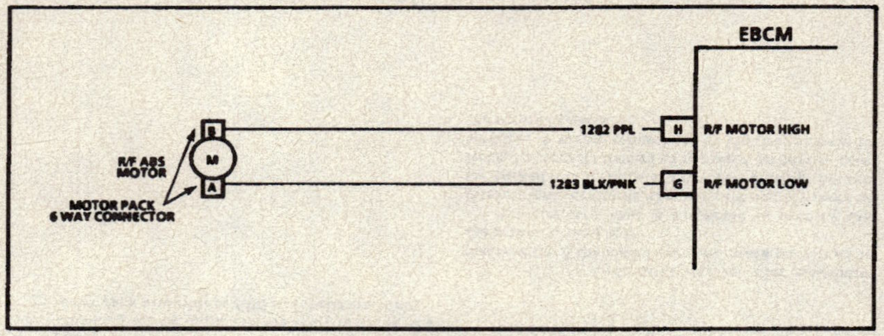

CODE A045

RIGHT FRONT CHANNEL WILL NOT MOVE

Circuit Description:

This code is designed to detect bound-up ESB, a stuck motor, or a seized hydraulic modulator. When the release is commanded during initialization, the ESB should release the motor, resulting in sensed current being less than commanded current (motor is spinning freely). If the motor is not moving, sensed current will be equal to stall current.

Test Description: Number(s) below refer to circled number(s) on the diagnostic chart.

1. This step checks for motor movement.
2. This step is used to release the tension of the motor pack prior to removal.
3. This step serves to isolate the fault to either the motor pack assembly or ABS hydraulic modulator.

Diagnostic Aids:

An "Intermittent" fault in this code may result from a mechanical part of the system that sticks, binds, or slips.

Code A045 may set after modulator disassembly if the modulator pistons are positioned at the bottom of their bore.

Any circuitry, that is suspected as causing the intermittent complaint, should be thoroughly checked for backed out terminals, improper mating, broken locks, improperly formed or damaged terminals, poor terminal to wiring connections or physical damage to the wiring harness.

DIAGNOSTIC CHART DTC A046 — 1993 CUTLASS SUPREME, GRAND PRIX, LUMINA AND REGAL, CONTINUED

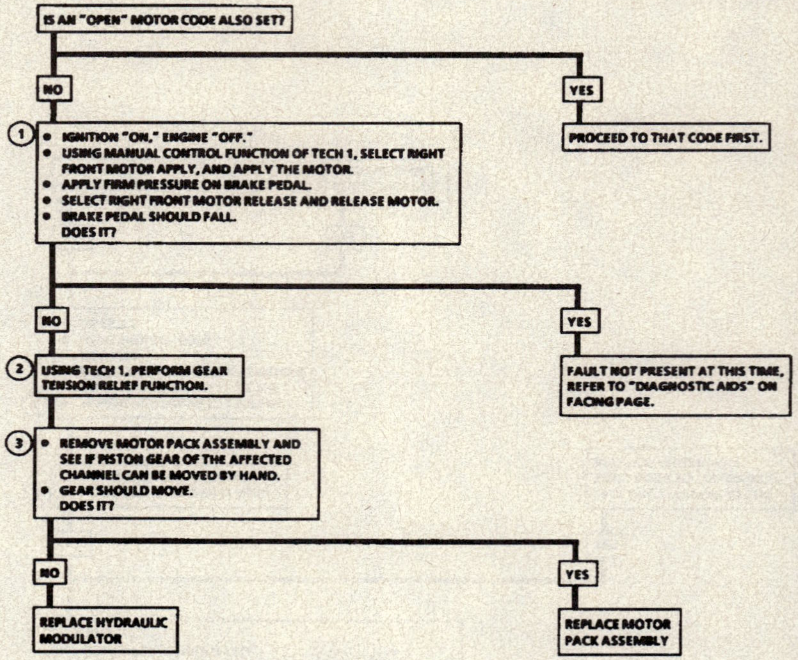

DIAGNOSTIC CHART DTC A046 — 1993 CUTLASS SUPREME, GRAND PRIX, LUMINA AND REGAL, CONTINUED

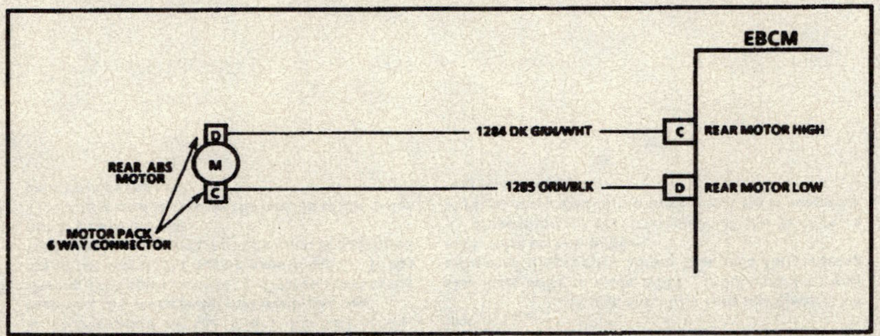

CODE A046

REAR AXLE CHANNEL WILL NOT MOVE

Circuit Description:

This code is designed to detect bound-up ESB, a stuck motor, or a seized hydraulic modulator. When the release is commanded during initialization, the ESB should release the motor, resulting in sensed current being less than commanded current (motor is spinning freely). If the motor is not moving, sensed current will be equal to stall current.

Test Description: Number(s) below refer to circled number(s) on the diagnostic chart.

1. This checks the ability of the motor to move, applying hydraulic pressure to the wheel cylinders. A frozen piston or gear would result in no apply of rear brakes and a wheel that freely spins.
2. This step checks to see that the motor can also move to the release position. If a motor were not able to release hydraulic pressure, the wheel would not spin.
3. This step is used to release the tension of the motor pack prior to removal.
4. Isolation of the faulty component is identified in this step.

Diagnostic Aids:

An "Intermittent" fault in this code may result from a mechanical part of the system that sticks or binds.

Code A046 may set after modulator disassembly if the modulator pistons are positioned at the bottom of their bore.

Depending on the frequency of the failure, a physical inspection of the mechanical parts suspected may be necessary.

Any circuitry, that is suspected as causing the intermittent complaint, should be thoroughly checked for backed out terminals, improper mating, broken locks, improperly formed or damaged terminals, poor terminal to wiring connections or physical damage to the wiring harness.

DIAGNOSTIC CHART DTC A046 — 1993 CUTLASS SUPREME, GRAND PRIX, LUMINA AND REGAL, CONTINUED

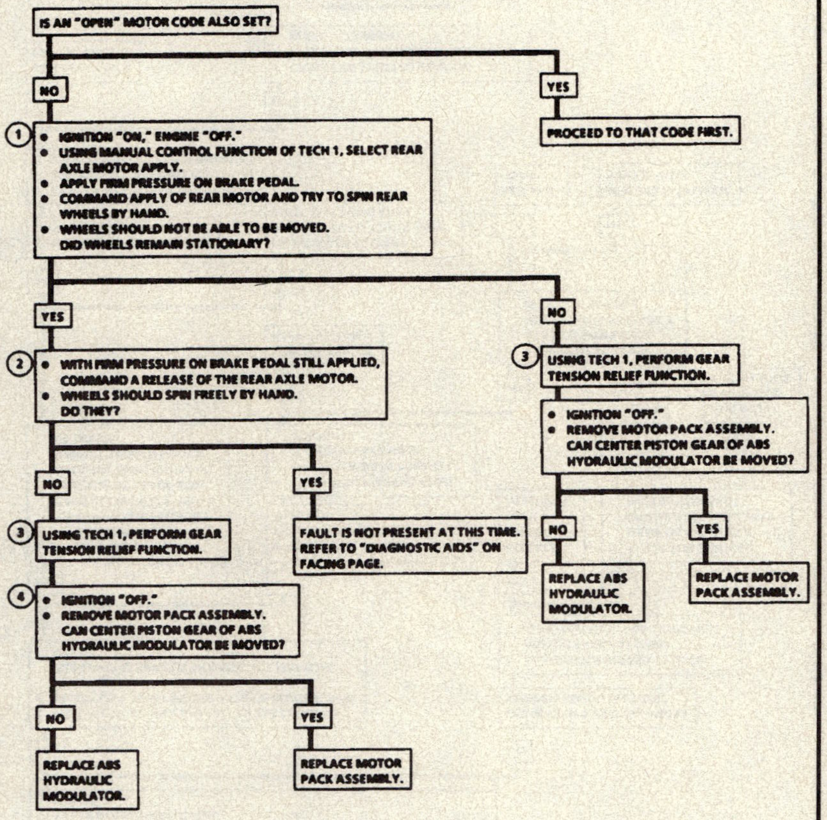

DIAGNOSTIC CHART DTC A047 — 1993 CUTLASS SUPREME, GRAND PRIX, LUMINA AND REGAL

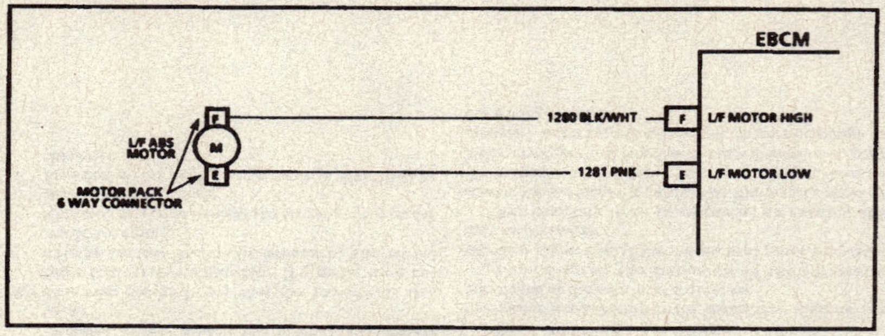

CODE A047

LEFT FRONT MOTOR FREE SPINS

Circuit Description:

This code is designed to detect a stripped nut or gear assembly during initialization. During the homing sequence, the piston should reach the top of the bore resulting in a stalled motor. If this does not occur, the motor must be spinning with little or no resistance, indicating a nut/screw or gear failure.

Test Description: Number(s) below refer to circled number(s) on the diagnostic chart.

1. This step positions the motor at the bottom of its travel.
2. This checks to see if the nut or gear is stripped. If either were stripped, the hydraulic pressure applied from the master cylinder would not be able to be overcome by the hydraulic ball screw, resulting in a brake pedal that would not rise.
3. During a full apply, once the ball screw reaches the top of its travel, the motor should stall. If the motor continues to run, an indication of a stripped nut or gear is present.
4. This step is used to release the tension of the motor pack prior to removal.

Diagnostic Aids:

An "Intermittent" fault in this code may result from a mechanical part of the system that sticks, binds, or slips.

The frequency of the problem can be checked by using the enhanced diagnostic function of the "Scan" tool

If code only fails once and Code A056 also fails, refer to Code A056. If intermittent and enhanced diagnostics shows this code fails during ABS, refer to Code A056.

Depending on the frequency of the failure, a physical inspection of the mechanical parts suspected may be necessary.

DIAGNOSTIC CHART DTC A047 — 1993 CUTLASS SUPREME, GRAND PRIX, LUMINA AND REGAL, CONTINUED

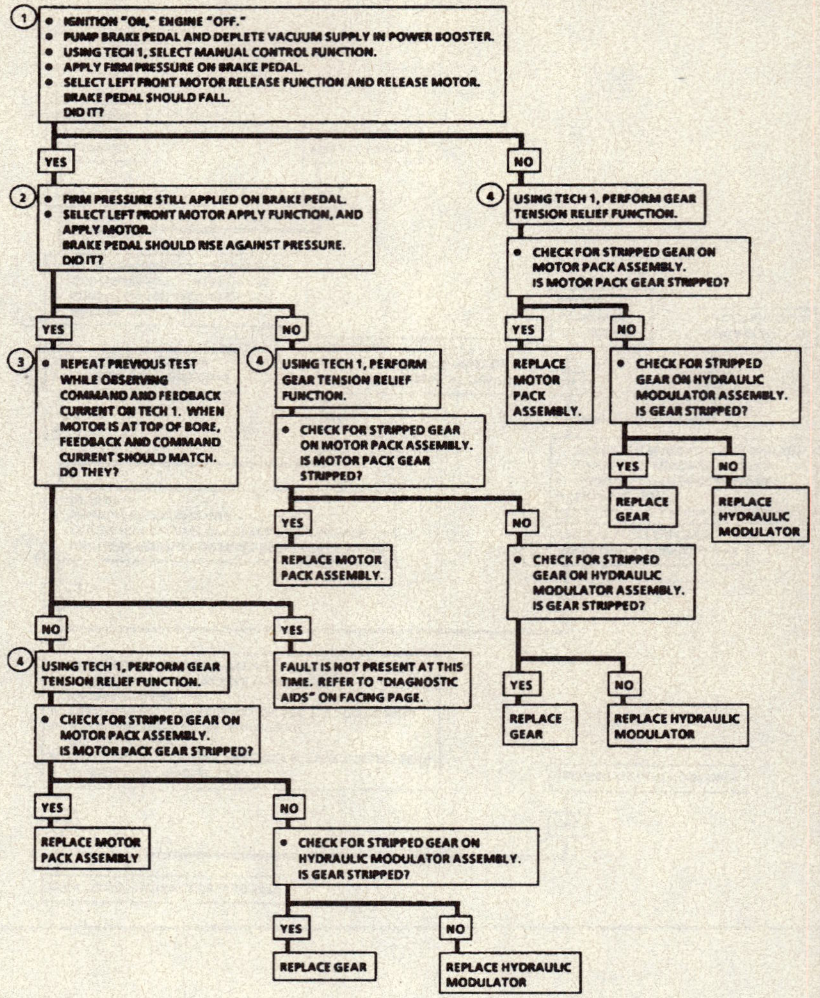

DIAGNOSTIC CHART DTC A048 — 1993 CUTLASS SUPREME, GRAND PRIX, LUMINA AND REGAL

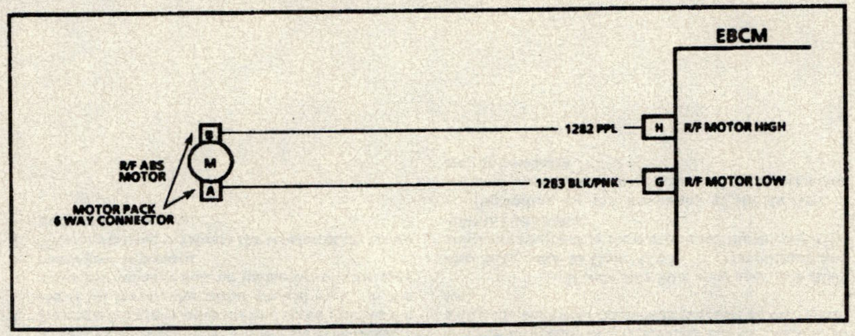

CODE A048

RIGHT FRONT MOTOR FREE SPINS

Circuit Description:
This code is designed to detect a stripped nut or gear assembly during initialization. During the homing sequence, the piston should reach the top of the bore resulting in a stalled motor. If this does not occur, the motor must be spinning with little or no resistance, indicating a nut/screw or gear failure.

Test Description: Number(s) below refer to circled number(s) on the diagnostic chart.

1. This step positions the motor at the bottom of its travel.
2. This checks to see if the nut or gear is stripped. If either were stripped, the hydraulic pressure applied from the master cylinder would not be able to be overcome by the hydraulic ball screw, resulting in a brake pedal that would not rise.
3. During a full apply, once the ball screw reaches the top of its travel, the motor should stall. If the motor continues to run, an indication of a stripped nut or gear is present.
4. This step is used to release the tension of the motor pack prior to removal.

Diagnostic Aids:

An "Intermittent" fault in this code may result from a mechanical part of the system that sticks, binds, or slips.

The frequency of the problem can be checked by using the enhanced diagnostic function of the "Scan" tool

If code only fails once and Code A061 also fails, refer to Code A061. If intermittent and enhanced diagnostics shows this code fails during ABS, refer to Code 61.

Depending on the frequency of the failure, a physical inspection of the mechanical parts suspected may be necessary.

DIAGNOSTIC CHART DTC A048 — 1993 CUTLASS SUPREME, GRAND PRIX, LUMINA AND REGAL, CONTINUED

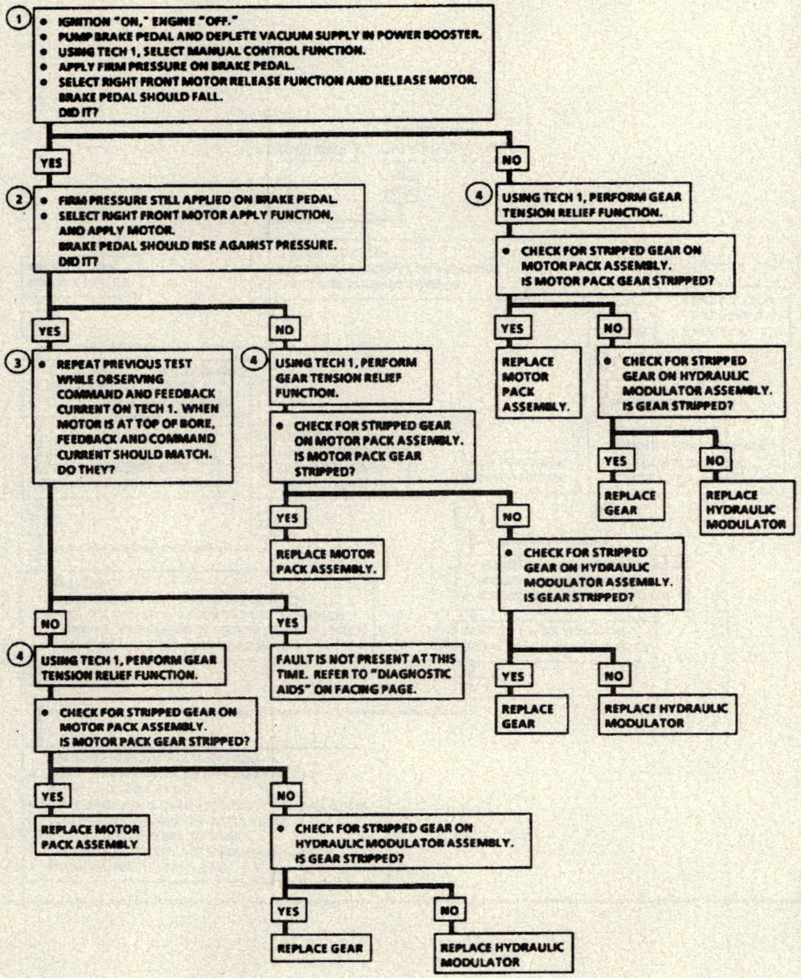

DIAGNOSTIC CHART DTC A051 — 1993 CUTLASS SUPREME, GRAND PRIX, LUMINA AND REGAL

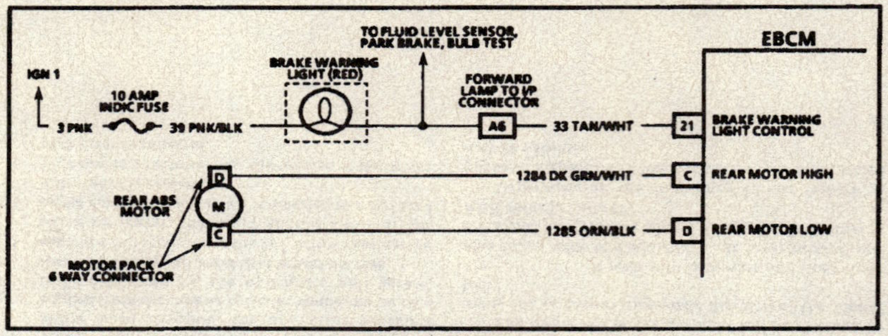

CODE A051

REAR AXLE MOTOR FREE SPINS

Circuit Description:

This code is designed to detect a stripped nut or gear assembly during initialization. During the homing sequence, the piston should reach the top of the bore resulting in a stalled motor. If this does not occur, the motor must be spinning with little or no resistance, indicating a nut/screw or gear failure.

Test Description: Number(s) below refer to circled number(s) on the diagnostic chart.

1. This step checks for a slipping nut or gear. If either were to slip, it would result in a lack of hydraulic pressure being applied at the rear wheels, causing the wheels to be able to be moved.
2. During a full apply, once the ball screw reaches the top of its travel, the motor should stall. If the motor continues to run, an indication of a stripped nut or gear is present.
3. This step is used to release the tension of the motor pack prior to removal.

Diagnostic Aids:

An "Intermittent" fault in this code may result from a mechanical part of the system that sticks, binds, or slips.

The frequency of the problem can be checked by using the enhanced diagnostic function of the "Scan" tool

If code only fails once and Code A064 also fails, refer to Code A064. If intermittent and enhanced diagnostics show this code fails during ABS, refer to Code A064.

Depending on the frequency of the failure, a physical inspection of the mechanical parts suspected may be necessary.

DIAGNOSTIC CHART DTC A051 — 1993 CUTLASS SUPREME, GRAND PRIX, LUMINA AND REGAL, CONTINUED

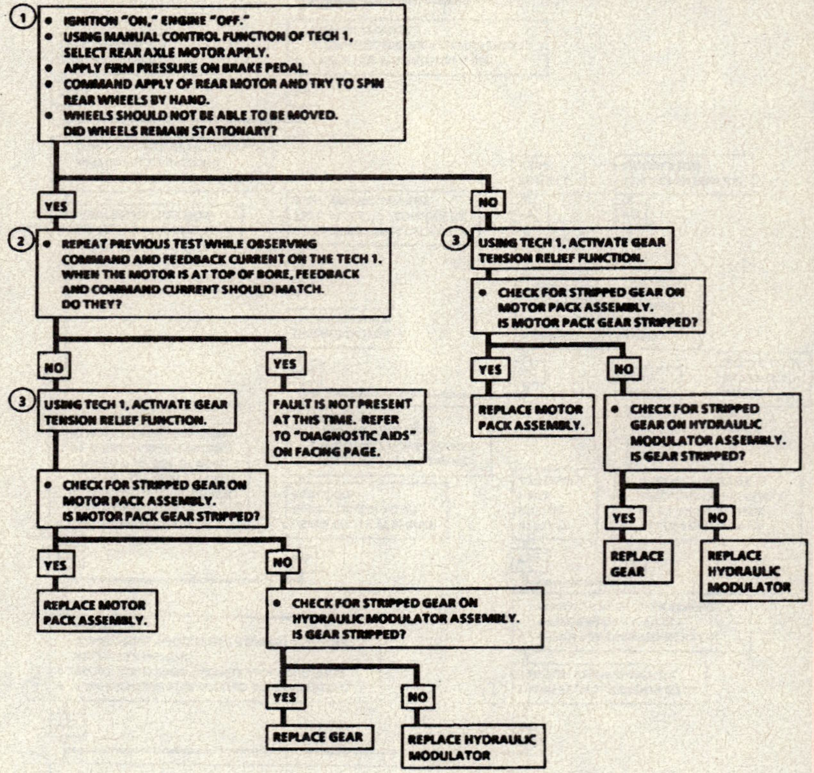

DIAGNOSTIC CHART DTC A052 — 1993 CUTLASS SUPREME, GRAND PRIX, LUMINA AND REGAL

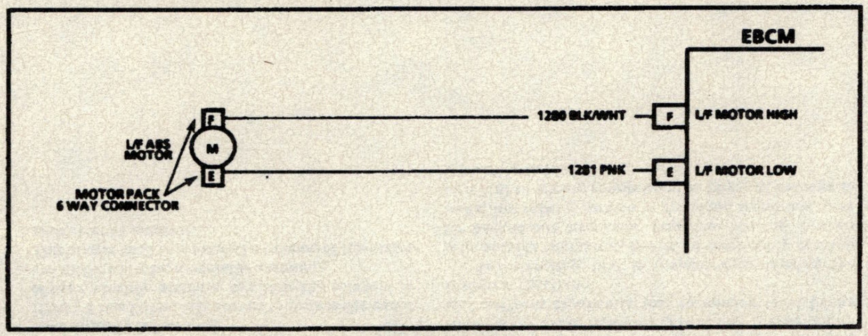

CODE A052

LEFT FRONT CHANNEL IN RELEASE TOO LONG

Circuit Description:

This code will diagnose a motor that is energized longer than expected. This could occur if a wheel speed sensor is faulty, the motor does not turn, the left front solenoid mechanically fails open, or the motor wires are crossed.

Test Description: Number(s) below refer to circled number(s) on the diagnostic chart.

1. This step identifies a problem in a wheel speed sensor that may cause the system to be in release too long.
2. This step identifies a motor fault as being failed or wired incorrectly.
3. This step checks for a solenoid that may have mechanically failed open.
4. If Step 3 has failed, this serves to isolate the cause of the hydraulic problem, to either the solenoid or the ABS hydraulic modulator assembly.
5. This step is used to release the tension of the motor pack.

Diagnostic Aids:

An "Intermittent" problem may be caused by a mechanical part of the system that sticks or binds.

The frequency of the problem can be checked by using the enhanced diagnostic function of the "Scan tool,

Code A052 may fail if on ice and steering wheel is turned to lock during braking. Using "Scan" tool perform Hydraulic test to ensure total brake system is functional.

Any circuitry, that is suspected as causing the intermittent complaint, should be thoroughly checked for backed out terminals, improper mating, broken locks, improperly formed or damaged terminals, poor terminal to wiring connections or physical damage to the wiring harness.

DIAGNOSTIC CHART DTC A052 — 1993 CUTLASS SUPREME, GRAND PRIX, LUMINA AND REGAL, CONTINUED

IMPORTANT: EXCESSIVE DRAG OR HIGH RESISTANCE IN THE BASE BRAKE OR SUSPENSION SYSTEM MUST BE INSPECTED AND CORRECTED BEFORE PROCEEDING WITH CODE DIAGNOSIS.

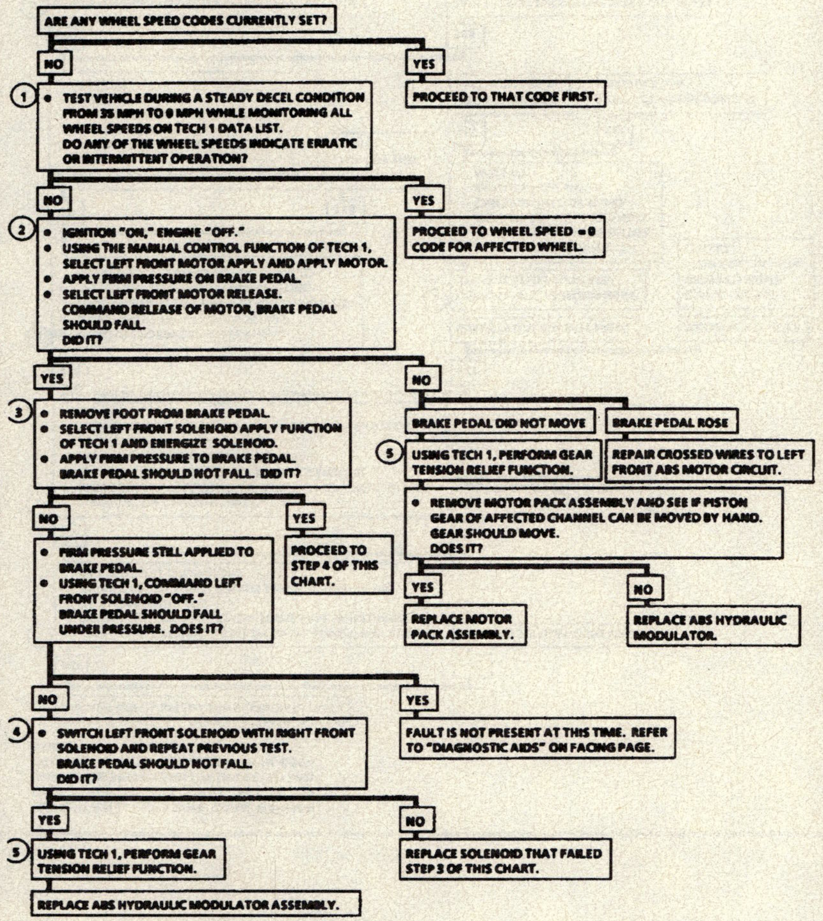

DIAGNOSTIC CHART DTC A053 — 1993 CUTLASS SUPREME, GRAND PRIX, LUMINA AND REGAL

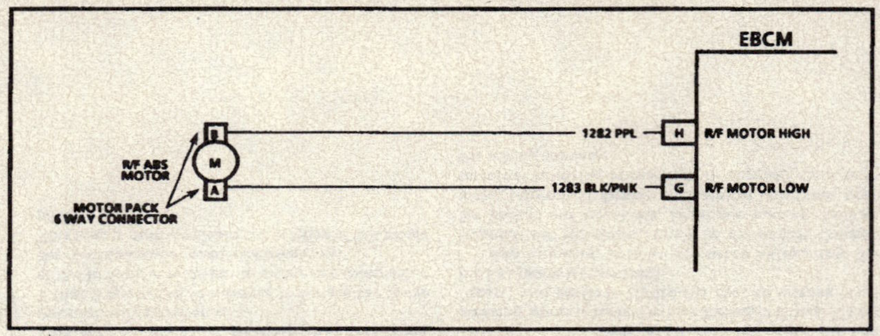

CODE A053

RIGHT FRONT CHANNEL IN RELEASE TOO LONG

Circuit Description:

This code will diagnose a motor that is energized longer than expected. This could occur if a wheel speed sensor is faulty, the motor does not turn, the right front solenoid mechanically fails open, or the motor wires are crossed.

Test Description: Number(s) below refer to circled number(s) on the diagnostic chart.

1. This step identifies a problem in a wheel speed sensor that may cause the system to be in release too long.
2. This step identifies a motor fault as being failed or wired incorrectly.
3. This step checks for a solenoid that may have mechanically failed open.
4. If Step 3 has failed, this serves to isolate the cause of the hydraulic problem, to either the solenoid or the ABS hydraulic modulator assembly.
5. This step is used to release the tension of the motor pack prior to removal.

Diagnostic Aids:

An "Intermittent" problem may be caused by a mechanical part of the system that sticks or binds.

The frequency of the problem can be checked by using the enhanced diagnostic function of the "Scan" tool,

Code A053 may fail on ice if steering wheel is turned to lock during braking. Using "Scan" tool perform Hydraulic test to ensure total brake system is functional.

Any circuitry, that is suspected as causing the intermittent complaint, should be thoroughly checked for backed out terminals, improper mating, broken locks, improperly formed or damaged terminals, poor terminal to wiring connections or physical damage to the wiring harness.

DIAGNOSTIC CHART DTC A053 — 1993 CUTLASS SUPREME, GRAND PRIX, LUMINA AND REGAL, CONTINUED

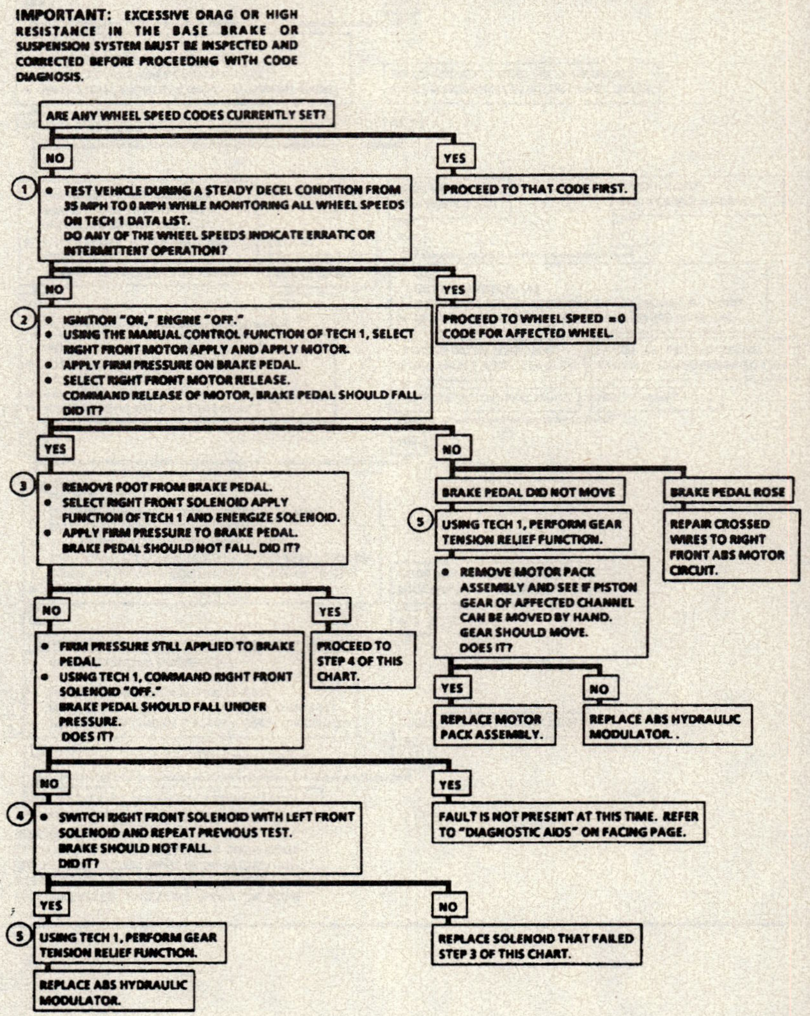

DIAGNOSTIC CHART DTC A054 — 1993 CUTLASS SUPREME, GRAND PRIX, LUMINA AND REGAL

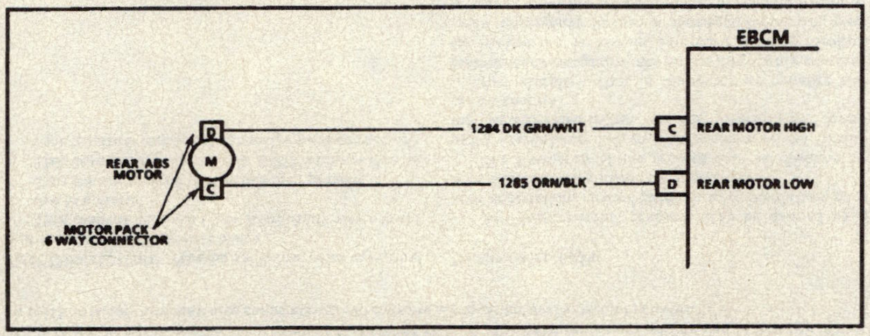

CODE A054

REAR AXLE CHANNEL IN RELEASE TOO LONG

Circuit Description:

This code will diagnose a motor that is energized longer than expected. This could occur if a wheel speed sensor is faulty, the motor does not turn, or the motor wires are crossed.

Test Description: Number(s) below refer to circled number(s) on the diagnostic chart.

1. This step identifies a problem in a wheel speed sensor that may cause the system to be in release too long.
2. This checks for a wheel that may stick or bind because of a mechanical fault.
3. This checks to see if the motor is capable of moving and applying the hydraulic piston for the rear wheels.
4. This step insures that the motor wiring is not crossed.
5. This step is used to release the tension of the motor pack prior to removal.
6. This isolates the fault of a "no-apply" situation to either the motor pack or ABS hydraulic modulator assembly.

Diagnostic Aids:

An "Intermittent" problem may be caused by a mechanical part of the system that sticks or binds.

The frequency of the problem can be checked by using the enhanced diagnostic function of the "Scan" tool,

Using "Scan" tool perform hydraulic test to ensure total brake system is functional.

Any circuitry, that is suspected as causing the intermittent complaint, should be thoroughly checked for backed out terminals, improper mating, broken locks, improperly formed or damaged terminals, poor terminal to wiring connections or physical damage to the wiring harness.

DIAGNOSTIC CHART DTC A054 — 1993 CUTLASS SUPREME, GRAND PRIX, LUMINA AND REGAL, CONTINUED

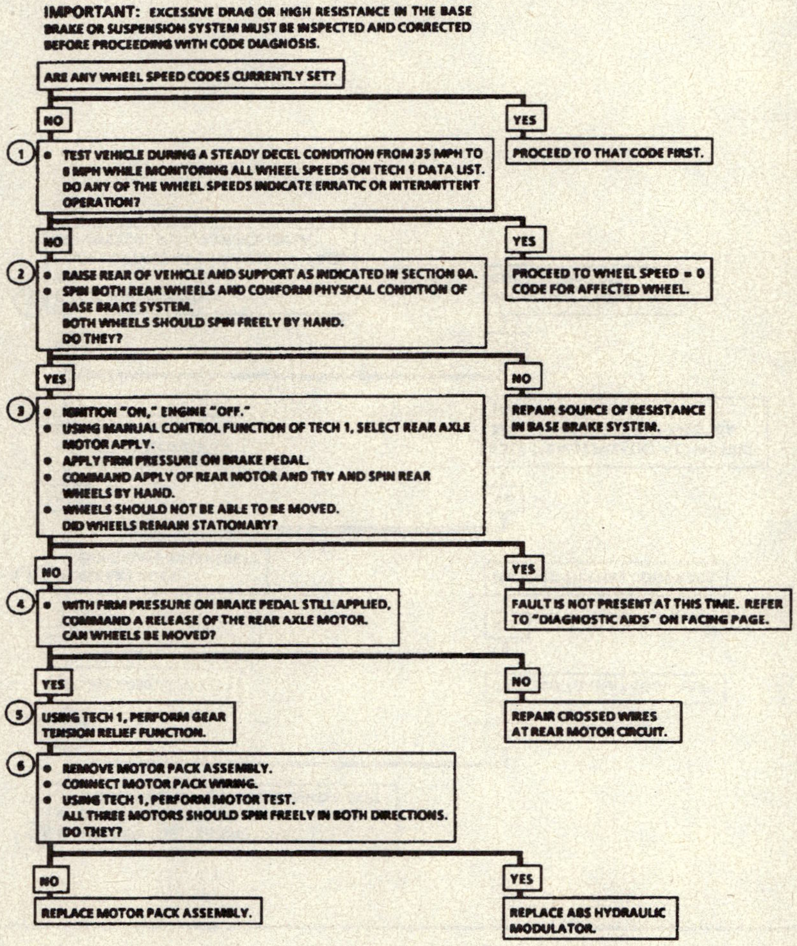

DIAGNOSTIC CHART DTC A055 — 1993 CUTLASS SUPREME, GRAND PRIX, LUMINA AND REGAL

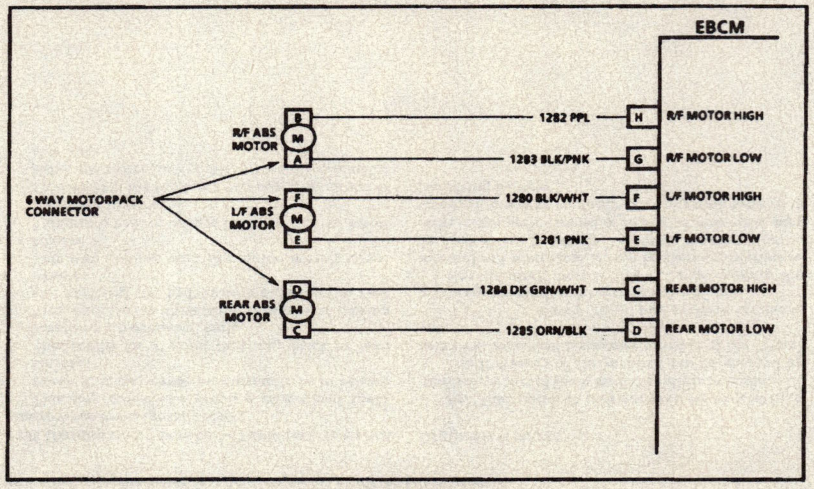

CODE A055

MOTOR DRIVER FAULT DETECTED

Circuit Description:

This code identifies a fault detected by the MDI custom IC. It also ensures the cause of the fault is not a result of a problem with the enable relay under a Code A055 failure. The MDI fault tests will be run to identify which circuit has failed, and may also cause additional codes to set to pinpoint the problem circuit.

Test Description: Number(s) below refer to circled number(s) on the diagnostic chart.

1. This insures that the Code A055 fault was not set due to a motor.
2. This checks to see if the fault is still present.
3. This step insures that Code A055 was not due to poor terminal contact at the motor pack connector.

Diagnostic Aids:

An "Intermittent" problem may be caused by a poor connection, rubbed through wire insulation, or a wire that is broken inside the insulation.

The frequency of the problem can be checked by using the enhanced diagnostic function of the "Scan" tool, as described in the "General Information," found in this section.

Any circuitry, that is suspected as causing the intermittent complaint, should be thoroughly checked for backed out terminals, improper mating, broken locks, improperly formed or damaged terminals, poor terminal to wiring connections or physical damage to the wiring harness.

DIAGNOSTIC CHART DTC A055 — 1993 CUTLASS SUPREME, GRAND PRIX, LUMINA AND REGAL, CONTINUED

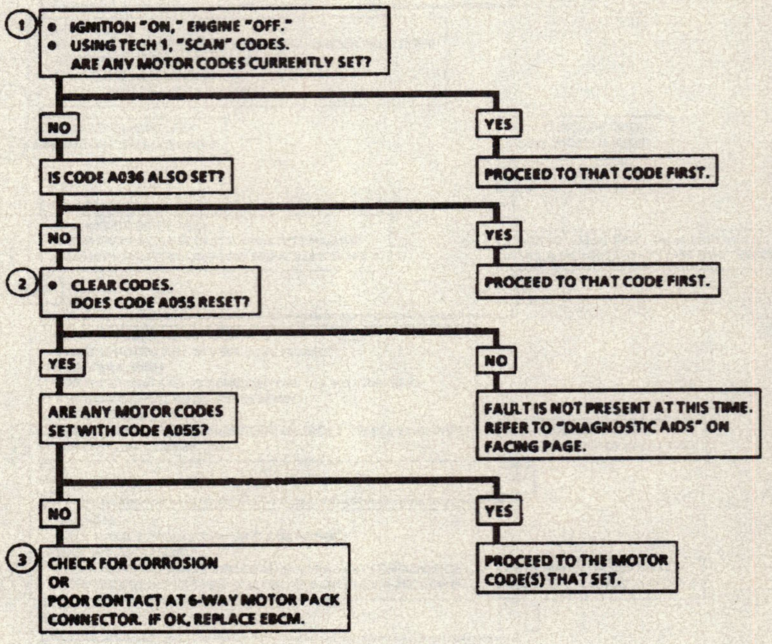

DIAGNOSTIC CHART DTC A056 — 1993 CUTLASS SUPREME, GRAND PRIX, LUMINA AND REGAL

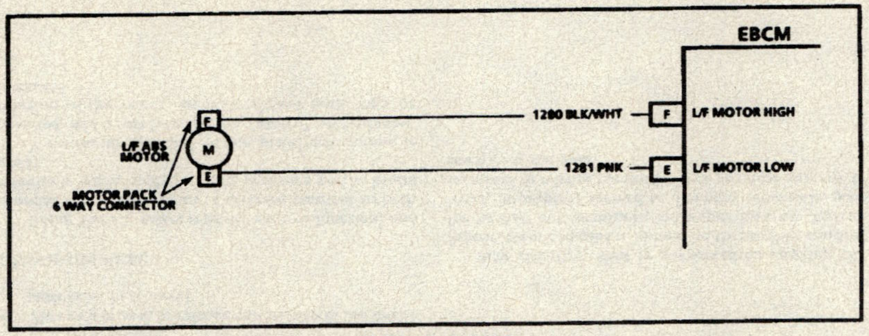

CODE A056

LEFT FRONT MOTOR CIRCUIT OPEN

Circuit Description:

This code identifies a motor that cannot be energized due to an open in its circuitry. This fault will not allow proper front ABS operation.

Test Description: Number(s) below refer to circled number(s) on the diagnostic chart.

1. This step checks for proper resistance of the motor.
2. This step is used to release the tension of the motor pack prior to removal.
3. This step checks for an open in the motor "high" circuitry.
4. This step checks for an open in the motor "low" circuitry.

An "Intermittent" problem may be caused by a poor connection, rubbed through wire insulation, or a wire that is broken inside the insulation.

If the fault is not current, wiggle the wires of the affected channel and check if the code resets. This will help to pinpoint an intermittent problem in the motor circuitry or connections.

Diagnostic Aids:

Using Tech 1, select manual control function, and exercise motor movement of affected channel in both directions while applying light pressure on the brake pedal.

If erratic or "jumpy" brake pedal movement is detected while performing an "apply" or "release" function of the motor, an intermittent fault may be indicated.

Any circuitry, that is suspected as causing the intermittent complaint, should be thoroughly checked for backed out terminals, improper mating, broken locks, improperly formed or damaged terminals, poor terminal to wiring connections or physical damage to the wiring harness.

DIAGNOSTIC CHART DTC A056 — 1993 CUTLASS SUPREME, GRAND PRIX, LUMINA AND REGAL, CONTINUED

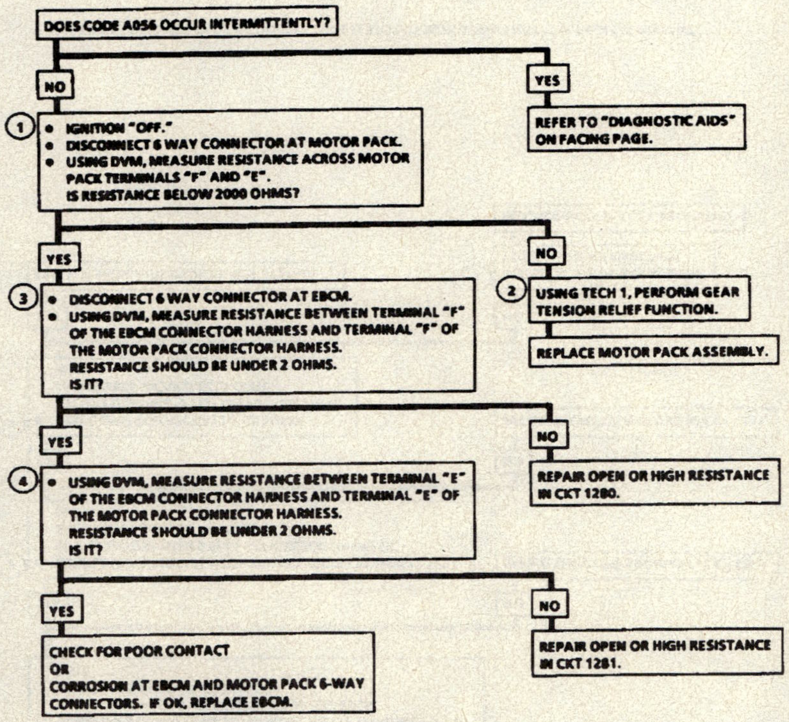

DIAGNOSTIC CHART DTC A057 — 1993 CUTLASS SUPREME, GRAND PRIX, LUMINA AND REGAL

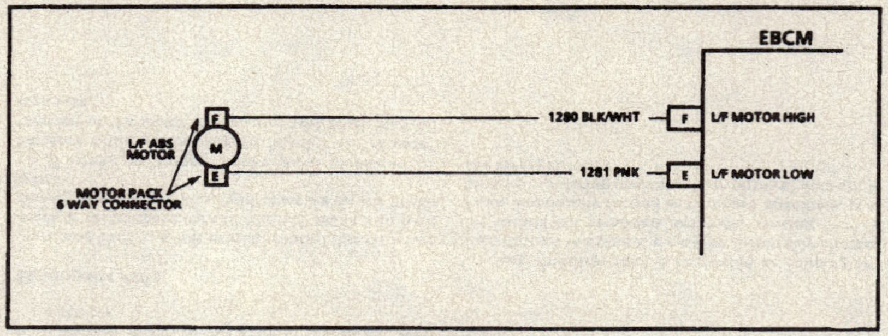

CODE A057

LEFT FRONT MOTOR CIRCUIT SHORTED TO GROUND

Circuit Description:

This code identifies a motor circuit that is shorted to ground. This fault will not allow the motor to be controlled at the commanded current rate or will cause the driver circuit to allow current directly to ground.

Test Description: Number(s) below refer to circled number(s) on the diagnostic chart.

1. This step checks for a short to ground in the motor "high" circuitry.
2. This step checks for a short to ground in the motor "low" circuitry.
3. This step checks for a motor that is internally shorted to ground.
4. This step is used to release the tension of the motor pack prior to removal.

An "Intermittent" problem may be caused by a poor connection, rubbed through wire insulation, or a wire that is broken inside the insulation.

If the fault is not current, wiggle the wires of the affected channel and check if the code resets. This will help to pinpoint an intermittent problem in the motor circuitry or connections.

Diagnostic Aids:

Using Tech 1, select manual control function, and exercise motor movement of affected channel in both directions while applying light pressure on the brake pedal.

If erratic or "jumpy" brake pedal movement is detected while performing an "apply" or "release" function of the motor, an intermittent fault may be indicated.

Any circuitry, that is suspected as causing the intermittent complaint, should be thoroughly checked for backed out terminals, improper mating, broken locks, improperly formed or damaged terminals, poor terminal to wiring connections or physical damage to the wiring harness.

DIAGNOSTIC CHART DTC A057 — 1993 CUTLASS SUPREME, GRAND PRIX, LUMINA AND REGAL, CONTINUED

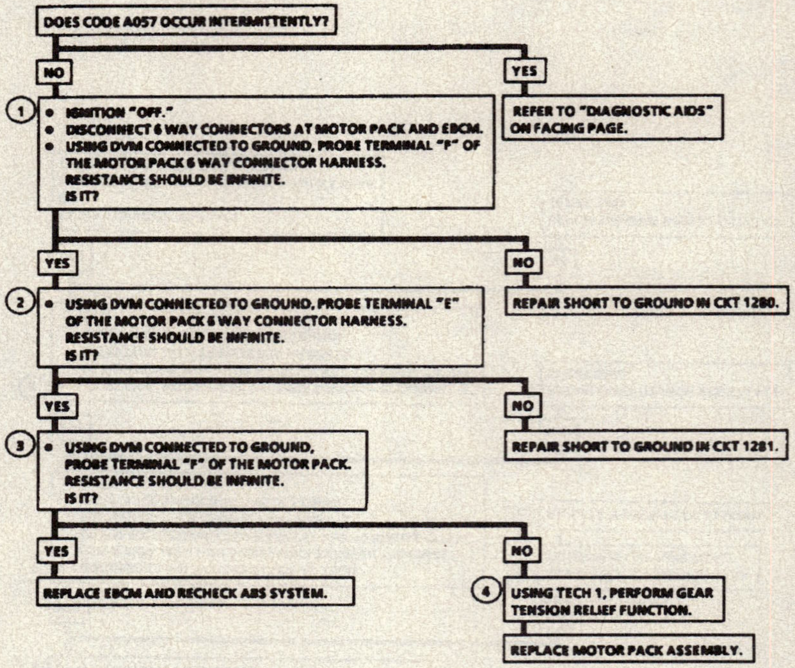

THIS CHART ASSUMES THAT A CURRENT CODE IS STORED INDICATING THAT THIS FAULT IS PRESENT.

DIAGNOSTIC CHART DTC A058 — 1993 CUTLASS SUPREME, GRAND PRIX, LUMINA AND REGAL

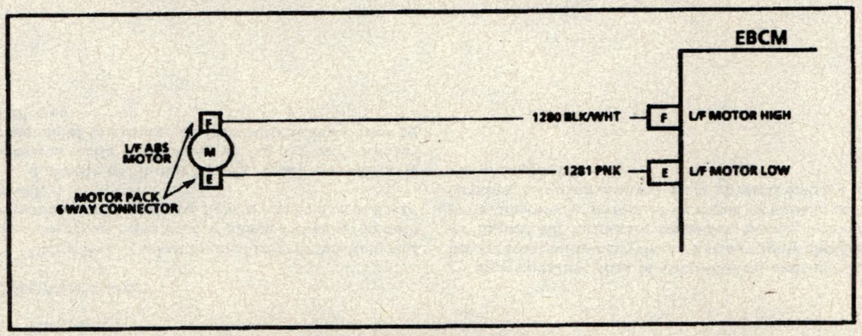

CODE A058

LEFT FRONT MOTOR CIRCUIT SHORTED TO BATTERY OR MOTOR SHORTED

Circuit Description:

This code identifies a motor circuit that is shorted to battery or a motor that has low or no resistance. This fault will not allow the motor to be controlled at the commanded current rate or will cause the motor to turn in the opposite direction, or not at all.

Test Description: Number(s) below refer to circled number(s) on the diagnostic chart.

1. This step checks for a short to voltage in the motor "high" circuitry.
2. This step checks for a short to voltage in the motor "low" circuitry.
3. This step checks for a motor that is internally shorted.
4. This step is used to release the tension of the motor pack prior to removal.

Diagnostic Aids:

Using Tech 1, select manual control function, and exercise motor movement of affected channel in both directions while applying light pressure on the brake pedal.

If erratic or "jumpy" brake pedal movement is detected while performing an "apply" or "release" function of the motor, an intermittent fault may be indicated.

An "Intermittent" problem may be caused by a poor connection, rubbed through wire insulation, or a wire that is broken inside the insulation.

If the fault is not current, wiggle the wires of the affected channel and check if the code resets. This will help to pinpoint an intermittent problem in the motor circuitry or connections.

Any circuitry, that is suspected as causing the intermittent complaint, should be thoroughly checked for backed out terminals, improper mating, broken locks, improperly formed or damaged terminals, poor terminal to wiring connections or physical damage to the wiring harness.

DIAGNOSTIC CHART DTC A058 — 1993 CUTLASS SUPREME, GRAND PRIX, LUMINA AND REGAL, CONTINUED

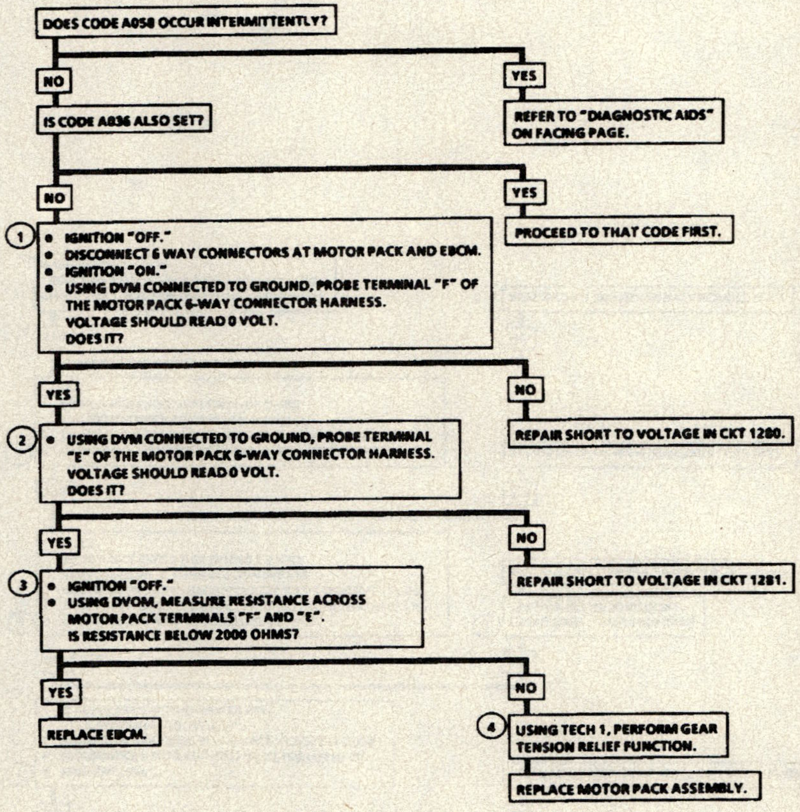

DIAGNOSTIC CHART DTC A061 — 1993 CUTLASS SUPREME, GRAND PRIX, LUMINA AND REGAL

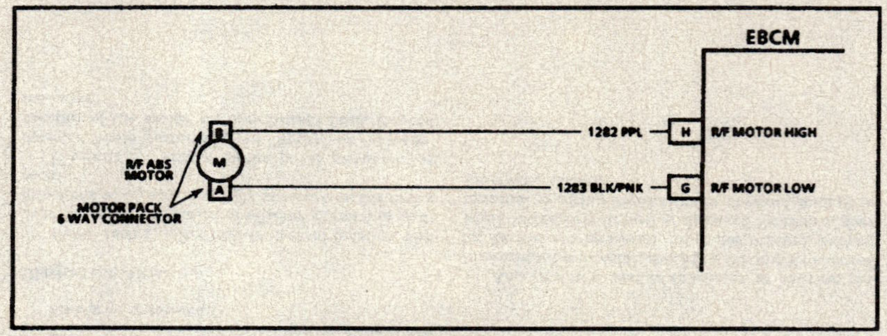

CODE A061

RIGHT FRONT MOTOR CIRCUIT OPEN

Circuit Description:

This code identifies a motor that cannot be energized due to an open in its circuitry. This fault will not allow proper front ABS operation.

Test Description: Number(s) below refer to circled number(s) on the diagnostic chart.

1. This step checks for proper resistance of the motor.
2. This step is used to release the tension of the motor pack prior to removal.
3. This step checks for an open in the motor "high" circuitry.
4. This step checks for an open in the motor "low" circuitry.

An "Intermittent" problem may be caused by a poor connection, rubbed through wire insulation, or a wire that is broken inside the insulation.

If the fault is not current, wiggle the wires of the affected channel and check if the code resets. This will help to pinpoint an intermittent problem in the motor circuitry or connections.

Diagnostic Aids:

Using Tech 1, select manual control function, and exercise motor movement of affected channel in both directions while applying light pressure on the brake pedal.

If erratic or "jumpy" brake pedal movement is detected while performing an "apply" or "release" function of the motor, an intermittent fault may be indicated.

Any circuitry, that is suspected as causing the intermittent complaint, should be thoroughly checked for backed out terminals, improper mating, broken locks, improperly formed or damaged terminals, poor terminal to wiring connections or physical damage to the wiring harness.

DIAGNOSTIC CHART DTC A061 — 1993 CUTLASS SUPREME, GRAND PRIX, LUMINA AND REGAL, CONTINUED

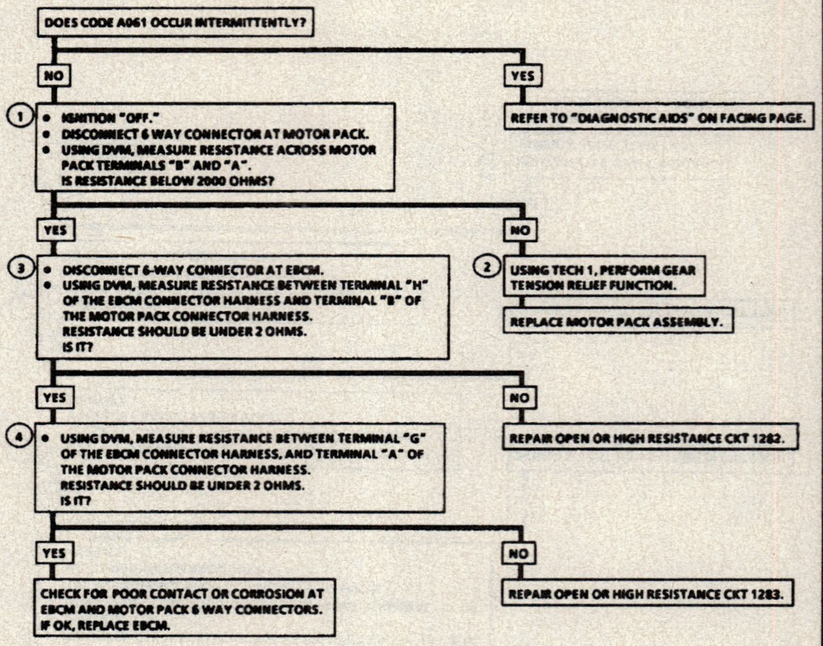

DIAGNOSTIC CHART DTC A062 — 1993 CUTLASS SUPREME, GRAND PRIX, LUMINA AND REGAL

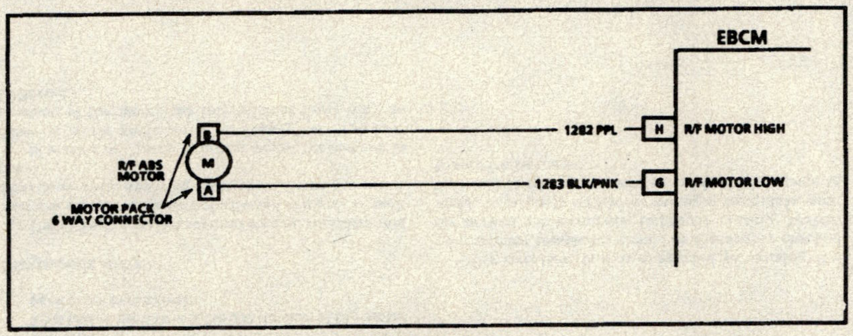

CODE A062

RIGHT FRONT MOTOR CIRCUIT SHORTED TO GROUND

Circuit Description:

This code identifies a motor circuit that is shorted to ground. This fault will not allow the motor to be controlled at the commanded current rate or will cause the motor driver circuit to allow current directly to ground.

Test Description: Number(s) below refer to circled number(s) on the diagnostic chart.

1. This step checks for a short to ground in the motor "high" circuitry.
2. This step checks for a short to ground in the motor "low" circuitry.
3. This step checks for a motor that is internally shorted to ground.
4. This step is used to release the tension of the motor pack prior to removal.

Diagnostic Aids:

Using Tech 1, select manual control function, and exercise motor movement of affected channel in both directions while applying light pressure on the brake pedal.

If erratic or "jumpy" brake pedal movement is detected while performing an "apply" or "release" function of the motor, an intermittent fault may be indicated.

An "Intermittent" problem may be caused by a poor connection, rubbed through wire insulation, or a wire that is broken inside the insulation.

If the fault is not current, wiggle the wires of the affected channel and check if the code resets. This will help to pinpoint an intermittent problem in the motor circuitry or connections.

Any circuitry, that is suspected as causing the intermittent complaint, should be thoroughly checked for backed out terminals, improper mating, broken locks, improperly formed or damaged terminals, poor terminal to wiring connections or physical damage to the wiring harness.

DIAGNOSTIC CHART DTC A062 — 1993 CUTLASS SUPREME, GRAND PRIX, LUMINA AND REGAL, CONTINUED

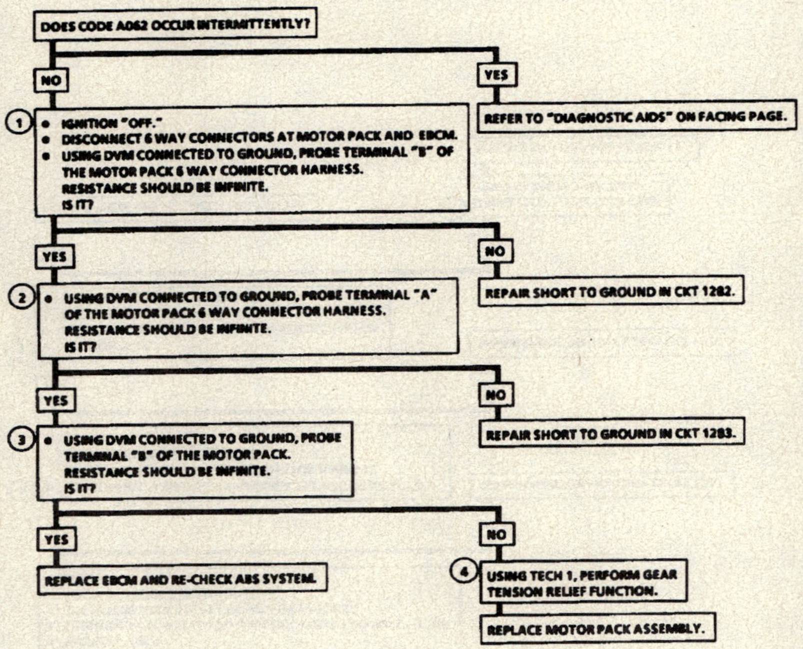

THIS CHART ASSUMES THAT A CURRENT CODE IS STORED INDICATING THAT THIS FAULT IS PRESENT.

DIAGNOSTIC CHART DTC A063 — 1993 CUTLASS SUPREME, GRAND PRIX, LUMINA AND REGAL

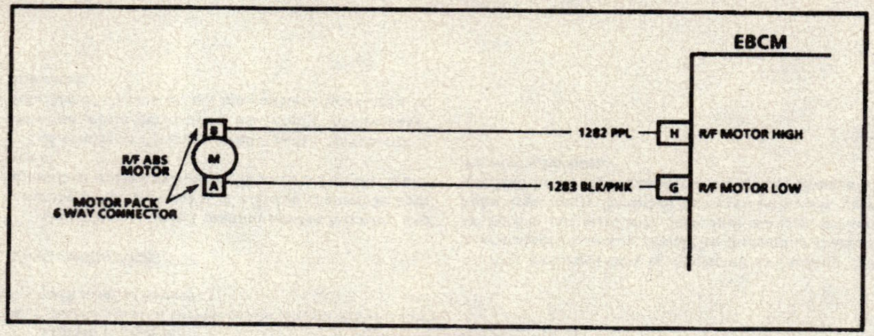

CODE A063

RIGHT FRONT MOTOR CIRCUIT SHORTED TO BATTERY OR MOTOR SHORTED

Circuit Description:

This code identifies a motor circuit that is shorted to battery or a motor that has low or no resistance. This fault will not allow the motor to be controlled at the commanded current rate or will cause the motor to turn in the opposite direction or not at all.

Test Description: Number(s) below refer to circled number(s) on the diagnostic chart.

1. This step checks for a short to voltage in the motor "high" circuitry.
2. This step checks for a short to voltage in the motor "low" circuitry.
3. This step checks for a motor that is internally shorted.
4. This test is used to release the tension of the motor pack prior to removal.

An "Intermittent" problem may be caused by a poor connection, rubbed through wire insulation, or a wire that is broken inside the insulation.

If the fault is not current, wiggle the wires of the affected channel and check if the code resets. This will help to pinpoint an intermittent problem in the motor circuitry or connections.

Diagnostic Aids:

Using Tech 1, select manual control function, and exercise motor movement of affected channel in both directions while applying light pressure on the brake pedal.

If erratic or "jumpy" brake pedal movement is detected while performing an "apply" or "release" function of the motor, an intermittent fault may be indicated.

Any circuitry, that is suspected as causing the intermittent complaint, should be thoroughly checked for backed out terminals, improper mating, broken locks, improperly formed or damaged terminals, poor terminal to wiring connections or physical damage to the wiring harness.

DIAGNOSTIC CHART DTC A063 — 1993 CUTLASS SUPREME, GRAND PRIX, LUMINA AND REGAL, CONTINUED

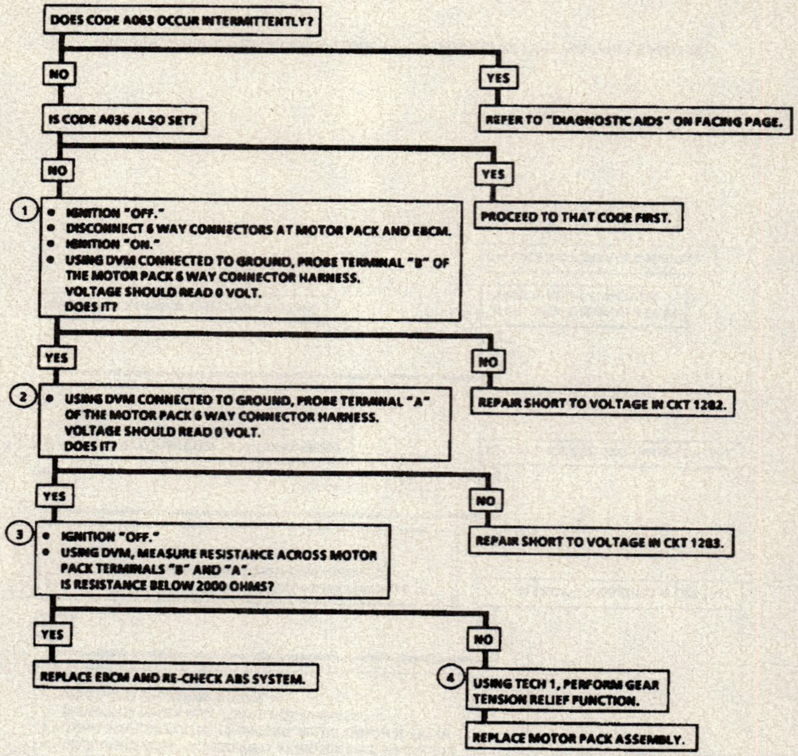

THIS CHART ASSUMES THAT A CURRENT CODE IS STORED INDICATING THAT THIS FAULT IS PRESENT.

DIAGNOSTIC CHART DTC A064 — 1993 CUTLASS SUPREME, GRAND PRIX, LUMINA AND REGAL

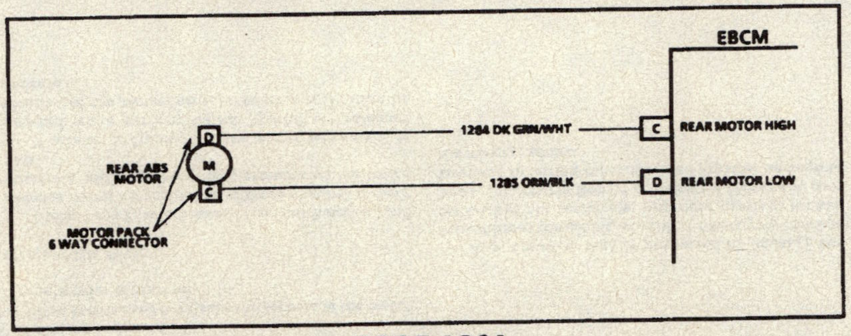

CODE A064

REAR AXLE MOTOR CIRCUIT OPEN

Circuit Description:

This code identifies a motor that cannot be energized due to an open in its circuitry. This fault will not allow proper rear ABS operation.

Test Description: Number(s) below refer to circled number(s) on the diagnostic chart.

1. This checks for an open circuit in the motor.
2. This step is used to release the tension of the motor pack prior to removal.
3. This step checks for an open in the motor "high" circuitry.
4. This step checks for an open in the motor "low" circuitry.

An "Intermittent" problem may be caused by a poor connection, rubbed through wire insulation, or a wire that is broken inside the insulation.

If the fault is not current, wiggle the wires of the affected channel and check if the code resets. This will help to pinpoint an intermittent problem in the motor circuitry or connections.

Diagnostic Aids:

Using Tech 1, select manual control function, and exercise motor movement of affected channel in both directions while applying light pressure on the brake pedal.

If erratic or "jumpy" brake pedal movement is detected while performing an "apply" or "release" function of the motor, an intermittent fault may be indicated.

Any circuitry, that is suspected as causing the intermittent complaint, should be thoroughly checked for backed out terminals, improper mating, broken locks, improperly formed or damaged terminals, poor terminal to wiring connections or physical damage to the wiring harness.

DIAGNOSTIC CHART DTC A064 — 1993 CUTLASS SUPREME, GRAND PRIX, LUMINA AND REGAL, CONTINUED

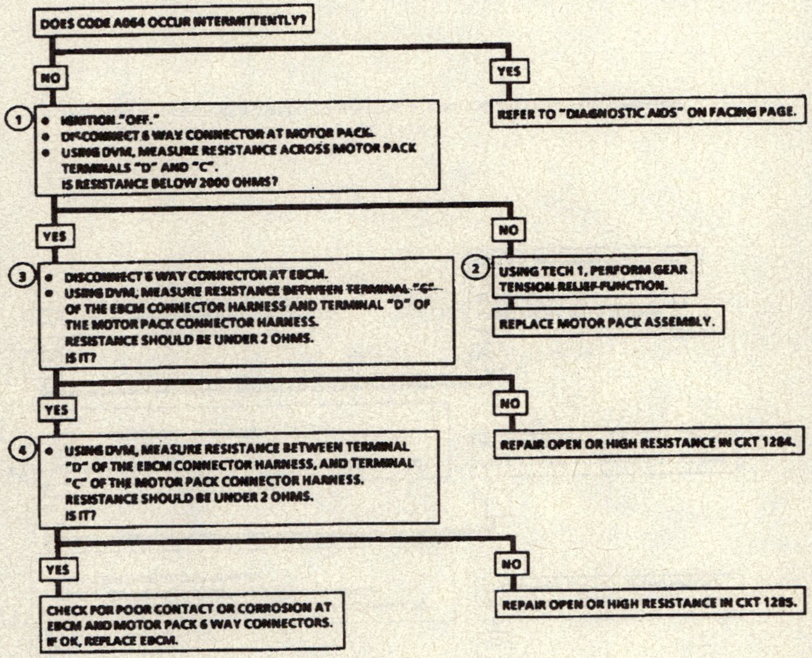

THIS CHART ASSUMES THAT A CURRENT CODE IS STORED INDICATING THAT THIS FAULT IS PRESENT.

DIAGNOSTIC CHART DTC A065 — 1993 CUTLASS SUPREME, GRAND PRIX, LUMINA AND REGAL

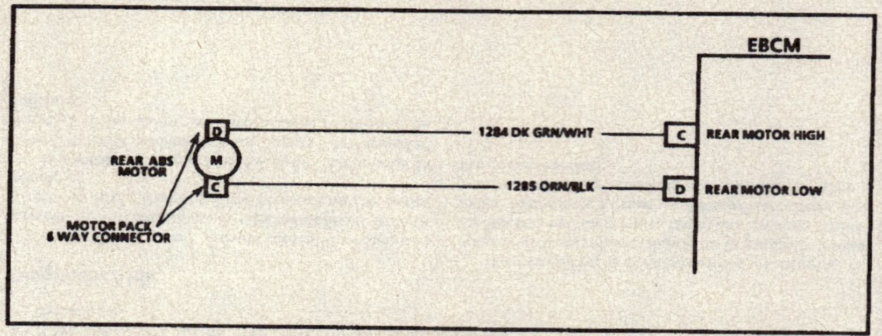

CODE A065

REAR AXLE MOTOR CIRCUIT SHORTED TO GROUND

Circuit Description:

This code identifies a motor circuit that is shorted to ground. This fault will not allow the motor to be controlled at the commanded current rate or will cause the driver circuit to allow current directly to ground.

Test Description: Number(s) below refer to circled number(s) on the diagnostic chart.

1. This step checks for a short to ground in the motor "high" circuitry.
2. This step checks for a short to ground in the motor "low" circuitry.
3. This step checks for a motor that is internally shorted to ground.
4. This step is used to release the tension of the motor pack prior to removal.

An "Intermittent" problem may be caused by a poor connection, rubbed through wire insulation, or a wire that is broken inside the insulation.

If the fault is not current, wiggle the wires of the affected channel and check if the code resets. This will help to pinpoint an intermittent problem in the motor circuitry or connections.

Diagnostic Aids:

Using Tech 1, select manual control function, and exercise motor movement of affected channel in both directions while applying light pressure on the brake pedal.

If erratic or "jumpy" brake pedal movement is detected while performing an "apply" or "release" function of the motor, an intermittent fault may be indicated.

Any circuitry, that is suspected as causing the intermittent complaint, should be thoroughly checked for backed out terminals, improper mating, broken locks, improperly formed or damaged terminals, poor terminal to wiring connections or physical damage to the wiring harness.

DIAGNOSTIC CHART DTC A065 — 1993 CUTLASS SUPREME, GRAND PRIX, LUMINA AND REGAL, CONTINUED

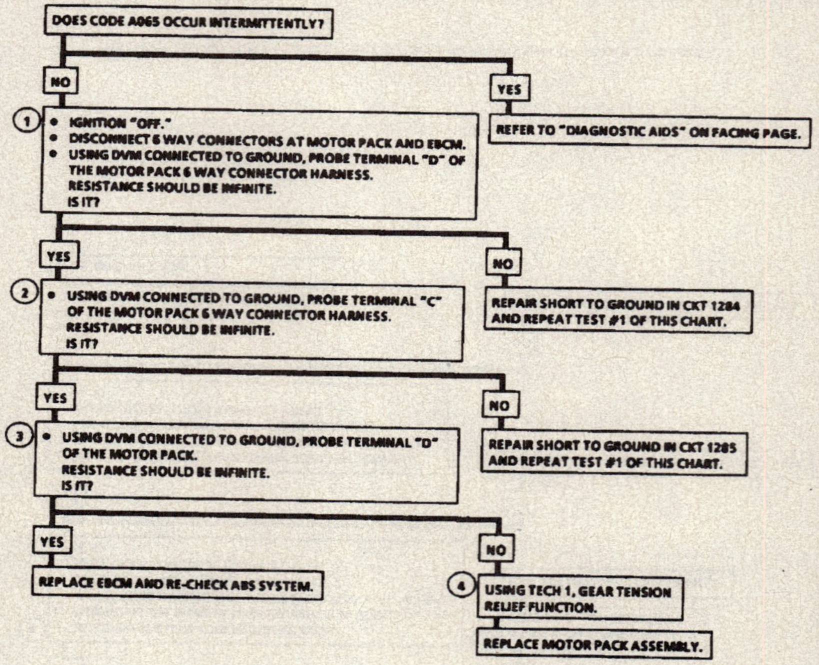

THIS CHART ASSUMES THAT A CURRENT CODE IS STORED INDICATING THAT THIS FAULT IS PRESENT.

DIAGNOSTIC CHART DTC A066 — 1993 CUTLASS SUPREME, GRAND PRIX, LUMINA AND REGAL

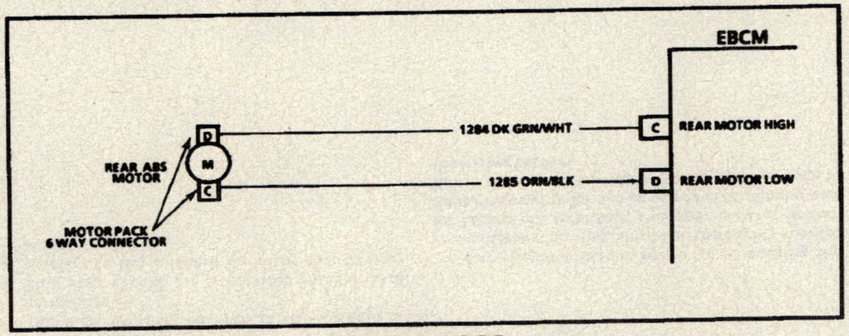

CODE A066

REAR AXLE MOTOR CIRCUIT SHORTED TO BATTERY OR MOTOR SHORTED

Circuit Description:

This code identifies a motor circuit that is shorted to battery or a motor that has low or zero resistance. This fault will not allow the motor to be controlled at the commanded current rate or will cause the motor to turn in the opposite direction or not turn at all.

Test Description: Number(s) below refer to circled number(s) on the diagnostic chart.

1. This step checks for a short to voltage in the motor "high" circuitry.
2. This step checks for a short to voltage in the motor "low" circuitry.
3. This step checks for a motor that is internally shorted.
4. This step is used to release the tension of the motor pack prior to removal.

Diagnostic Aids:

Using Tech 1, select manual control function, and exercise motor movement of affected channel in both directions while applying light pressure on the brake pedal.

If erratic or "jumpy" brake pedal movement is detected while performing an "apply" or "release" function of the motor, an intermittent fault may be indicated.

An "Intermittent" problem may be caused by a poor connection, rubbed through wire insulation, or a wire that is broken inside the insulation.

If the fault is not current, wiggle the wires of the affected channel and check if the code resets. This will help to pinpoint an intermittent problem in the motor circuitry or connections.

Any circuitry, that is suspected as causing the intermittent complaint, should be thoroughly checked for backed out terminals, improper mating, broken locks, improperly formed or damaged terminals, poor terminal to wiring connections or physical damage to the wiring harness.

DIAGNOSTIC CHART DTC A066 — 1993 CUTLASS SUPREME, GRAND PRIX, LUMINA AND REGAL, CONTINUED

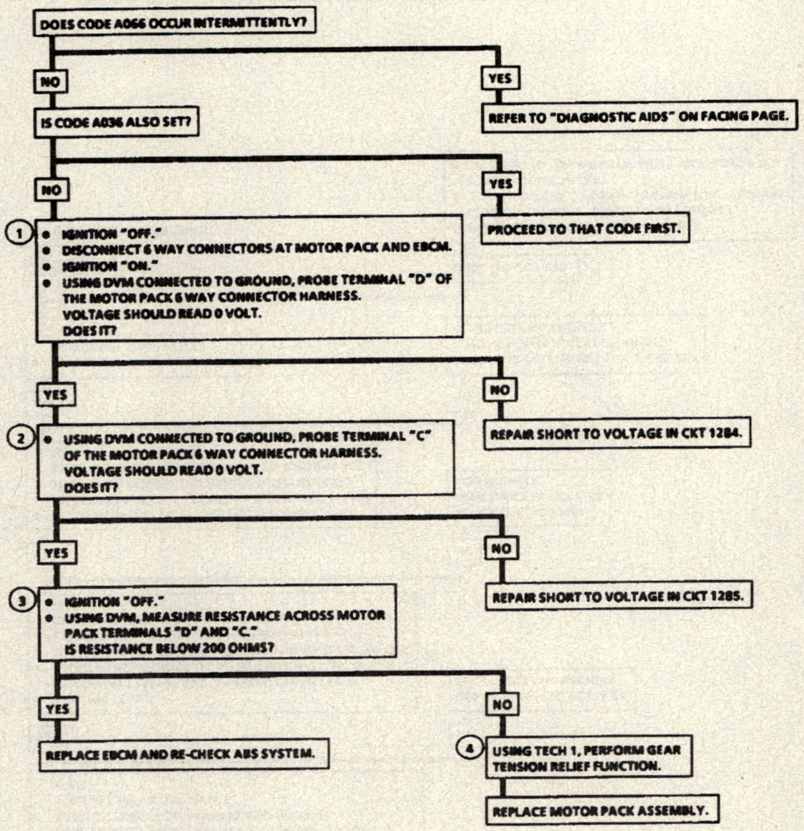

THIS CHART ASSUMES THAT A CURRENT CODE IS STORED INDICATING THAT THIS FAULT IS PRESENT.

DIAGNOSTIC CHART DTC A076 — 1993 CUTLASS SUPREME, GRAND PRIX, LUMINA AND REGAL

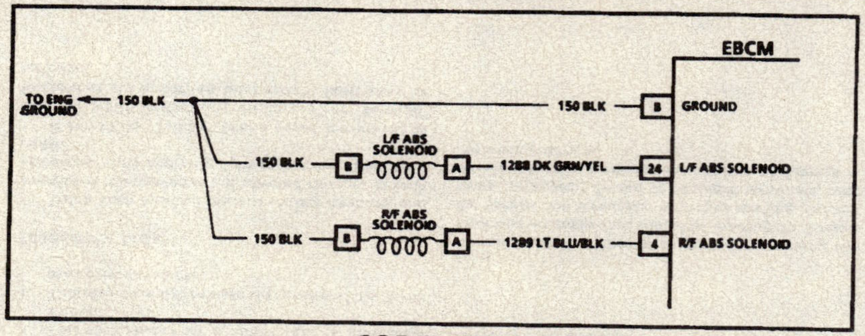

CODE A076

LEFT FRONT SOLENOID CIRCUIT SHORTED TO BATTERY OR OPEN

Circuit Description:

This code identifies a solenoid that cannot be energized due to an open in its circuitry, or a solenoid that is always energized due to a short to battery in its circuitry between the driver and the solenoid. An open will not allow proper ABS operation, but the short to battery simply turns "ON" the solenoid. A path for base brakes is still allowed once the motor re-homes and the check ball is lifted off its seat during key on initialization.

Test Description: Number(s) below refer to circled number(s) on the diagnostic chart.

1. This checks for a short to voltage in the control circuitry of the solenoid.
2. This checks for an open in the control circuit of the solenoid.
3. This step identifies an open in the solenoid or ground circuit.
4. This step checks for a possible intermittent problem in the solenoid circuitry due to poor terminal contact.

Diagnostic Aids:

An "Intermittent" problem may be caused by a poor connection, rubbed through wire insulation, or a wire that is broken inside the insulation.

Any circuitry, that is suspected as causing the intermittent complaint, should be thoroughly checked for backed out terminals, improper mating, broken locks, improperly formed or damaged terminals, poor terminal to wiring connections or physical damage to the wiring harness.

DIAGNOSTIC CHART DTC A076 — 1993 CUTLASS SUPREME, GRAND PRIX, LUMINA AND REGAL, CONTINUED

(1)
- IGNITION "OFF."
- DISCONNECT 24 WAY WORLD CONNECTOR FROM EBCM.
- IGNITION "ON," ENGINE "OFF."
- USING DVM, CONNECT BETWEEN TERMINAL "24" OF THE WORLD CONNECTOR HARNESS AND GROUND. VOLTAGE SHOULD BE 0 VOLT. IS IT?

NO: REPAIR SHORT TO VOLTAGE IN CKT 1288 AND RECHECK.

YES:

(2)
- IGNITION "OFF."
- DISCONNECT LEFT FRONT SOLENOID CONNECTOR.
- MEASURE RESISTANCE BETWEEN TERMINAL "24" OF THE WORLD CONNECTOR HARNESS AND TERMINAL "A" OF THE LEFT FRONT SOLENOID CONNECTOR HARNESS. RESISTANCE SHOULD BE 2 OHMS OR LESS. IS IT?

NO: REPAIR OPEN OR HIGH RESISTANCE IN CKT 1288 AND RECHECK.

YES:

(3)
- RECONNECT LEFT FRONT SOLENOID CONNECTOR.
- MEASURE RESISTANCE BETWEEN TERMINAL "24" OF THE WORLD CONNECTOR HARNESS AND GROUND. RESISTANCE SHOULD BE BETWEEN 2.5 AND 5 OHMS. IS IT?

NO: CHECK FOR OPEN IN CKT 150. IF CKT 150 IS NOT OPEN, REPLACE LEFT FRONT SOLENOID AND RECHECK.

YES:

(4)
- RECONNECT WORLD CONNECTOR AND VERIFY CODE RESETS.

CODE RESET: REPLACE EBCM AND RECHECK.

CODE DID NOT SET: CHECK SOLENOID CONNECTOR HARNESS TERMINALS "A" AND "B" FOR POOR CONTACT, ALSO CHECK WORLD CONNECTOR HARNESS TERMINAL "24" FOR POOR CONTACT. IF OK, REFER TO "DIAGNOSTIC AIDS" ON FACING PAGE.

THIS CHART ASSUMES THAT A CURRENT CODE IS STORED INDICATING THAT THIS FAULT IS PRESENT.

DIAGNOSTIC CHART DTC A077 — 1993 CUTLASS SUPREME, GRAND PRIX, LUMINA AND REGAL

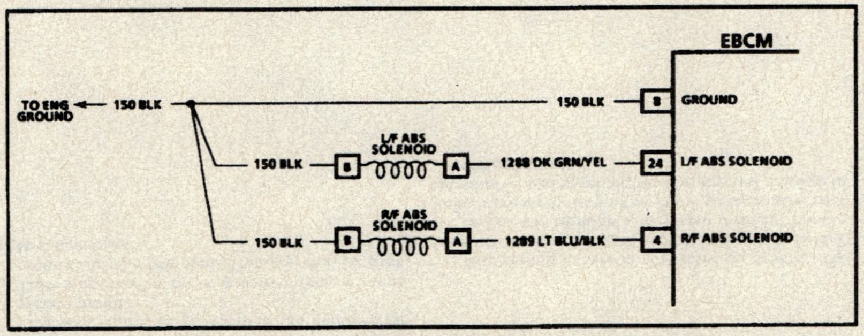

CODE A077

LEFT FRONT SOLENOID CIRCUIT SHORTED TO GROUND OR DRIVER OPEN

Circuit Description:

This code identifies a solenoid that cannot be energized due to an open in its driver circuitry, or a short to ground between the solenoid driver and the solenoid. These faults can affect ABS operation since the flow of brake fluid to the wheel cylinder cannot be stopped, making ABS operation for that channel impossible.

Test Description: Number(s) below refer to circled number(s) on the diagnostic chart.

1. This step checks for a solenoid that is internally shorted to ground.
2. This step checks for a solenoid that is not within proper resistance values.
3. This step indicates if a short to ground exists in the solenoid circuitry.
4. This checks for an open control circuit to the EBCM.

Diagnostic Aids:

An "Intermittent" problem may be caused by a poor connection, rubbed through wire insulation, or a wire that is broken inside the insulation.

Any circuitry, that is suspected as causing the intermittent complaint, should be thoroughly checked for backed out terminals, improper mating, broken locks, improperly formed or damaged terminals, poor terminal to wiring connections or physical damage to the wiring harness.

DIAGNOSTIC CHART DTC A077 — 1993 CUTLASS SUPREME, GRAND PRIX, LUMINA AND REGAL, CONTINUED

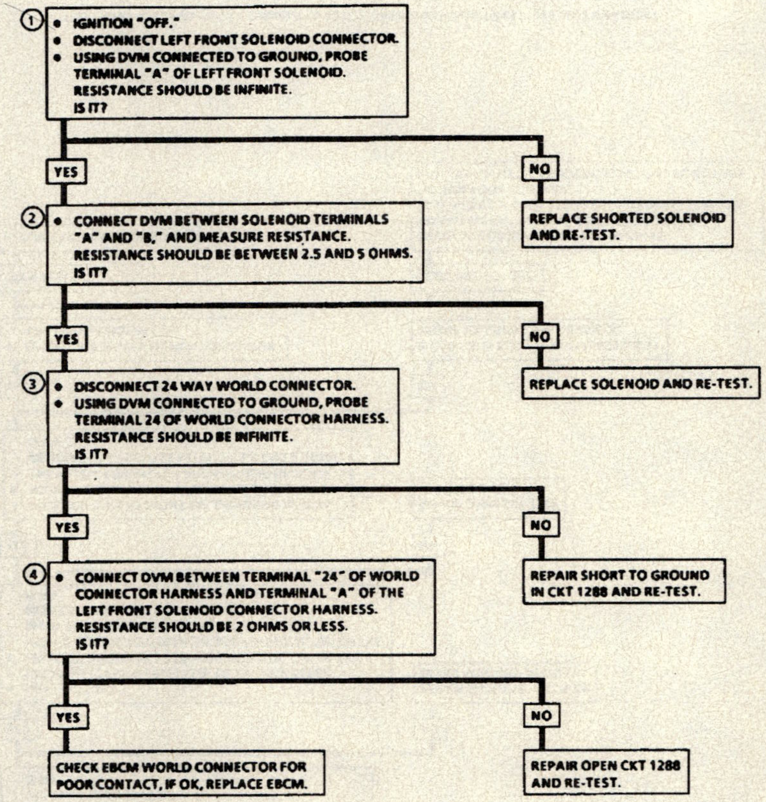

THIS CHART ASSUMES THAT A CURRENT CODE IS STORED INDICATING THAT THIS FAULT IS PRESENT.

DIAGNOSTIC CHART DTC AO78 — 1993 CUTLASS SUPREME, GRAND PRIX, LUMINA AND REGAL

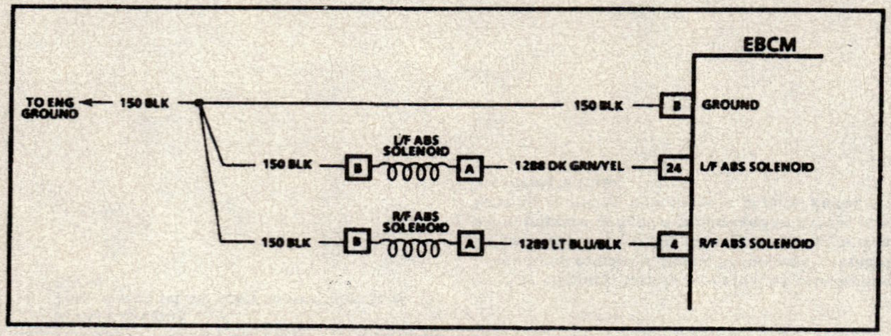

CODE A078

RIGHT FRONT SOLENOID CIRCUIT SHORTED TO BATTERY OR OPEN

Circuit Description:

This code identifies a solenoid that cannot be energized due to an open in its circuitry, or a solenoid that is always energized due to a short to battery in its circuitry between the driver and the solenoid. An open will not allow proper ABS operation, but the short to battery simply turns "ON" the solenoid. A path for base brakes is still allowed once the motor re-homes and the check ball is lifted off its seat during key on initialization.

Test Description: Number(s) below refer to circled number(s) on the diagnostic chart.

1. This checks for a short to voltage in the control circuitry of the solenoid.
2. This checks for an open in the control circuit of the solenoid.
3. This step identifies an open in the solenoid or ground circuit.
4. This step checks for a possible intermittent problem in the solenoid circuitry due to poor terminal contact.

Diagnostic Aids:

An "Intermittent" problem may be caused by a poor connection, rubbed through wire insulation, or a wire that is broken inside the insulation.

Any circuitry, that is suspected as causing the intermittent complaint, should be thoroughly checked for backed out terminals, improper mating, broken locks, improperly formed or damaged terminals, poor terminal to wiring connections or physical damage to the wiring harness.

DIAGNOSTIC CHART DTC AO78 — 1993 CUTLASS SUPREME, GRAND PRIX, LUMINA AND REGAL, CONTINUED

(1)
- IGNITION "OFF."
- DISCONNECT 24 WAY WORLD CONNECTOR FROM EBCM.
- IGNITION "ON," ENGINE "OFF."
- USING DVM, CONNECT BETWEEN TERMINAL "4" OF WORLD CONNECTOR AND GROUND. VOLTAGE SHOULD BE 0 VOLTS. IS IT?

NO → REPAIR SHORT TO VOLTAGE IN CKT 1289 AND RECHECK.

YES →

(2)
- IGNITION "OFF."
- DISCONNECT RIGHT FRONT SOLENOID CONNECTOR.
- MEASURE RESISTANCE BETWEEN TERMINAL "4" OF THE WORLD CONNECTOR HARNESS AND TERMINAL "A" OF THE RIGHT FRONT SOLENOID CONNECTOR HARNESS. RESISTANCE SHOULD BE 2 OHMS OR LESS. IS IT?

NO → REPAIR OPEN CKT 1289 AND RECHECK.

YES →

(3)
- RECONNECT RIGHT FRONT SOLENOID CONNECTOR.
- MEASURE RESISTANCE BETWEEN TERMINAL "4" OF THE WORLD CONNECTOR HARNESS AND GROUND. RESISTANCE SHOULD BE BETWEEN 2.5 AND 5 OHMS. IS IT?

NO → IF CKT 150 IS NOT OPEN, REPLACE RIGHT FRONT SOLENOID, AND RECHECK.

YES →

(4)
- RECONNECT 24 WAY WORLD CONNECTOR AND VERIFY CODE RESETS.

CODE RESET. → REPLACE EBCM AND RECHECK.

CODE DID NOT SET. → CHECK SOLENOID CONNECTOR HARNESS TERMINALS "A" AND "B" FOR POOR CONTACT. ALSO CHECK WORLD CONNECTOR HARNESS TERMINAL "4" FOR POOR CONTACT. IF OK, REFER TO "DIAGNOSTIC AIDS" ON FACING PAGE.

THIS CHART ASSUMES THAT A CURRENT CODE IS STORED INDICATING THAT THIS FAULT IS PRESENT.

DIAGNOSTIC CHART DTC A081 — 1993 CUTLASS SUPREME, GRAND PRIX, LUMINA AND REGAL

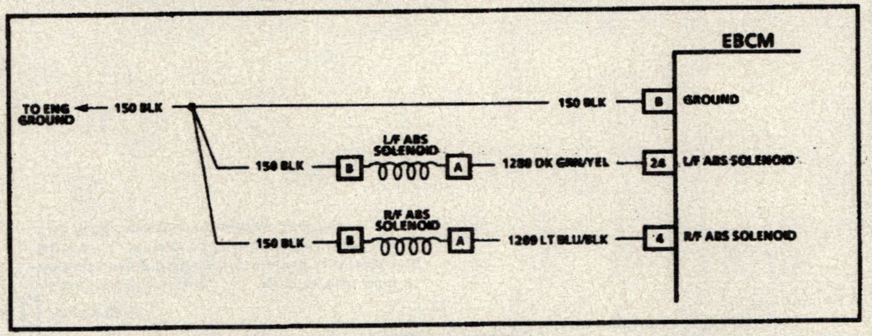

CODE A081

RIGHT FRONT SOLENOID CIRCUIT SHORTED TO GROUND OR DRIVER OPEN

Circuit Description:

This code identifies a solenoid that cannot be energized due to an open in its driver circuitry, or a short to ground between the solenoid driver and the solenoid. These faults can affect ABS operation since the flow of brake fluid to the wheel cylinder cannot be stopped, making ABS operation for that channel impossible.

Test Description: Number(s) below refer to circled number(s) on the diagnostic chart.

1. This step checks for a solenoid that is internally shorted to ground.
2. This step checks for a solenoid that is not within proper resistance values.
3. This step indicates if a short to ground exists in the solenoid circuitry.
4. This checks for an open control circuit to the EBCM.

Diagnostic Aids:

An "Intermittent" problem may be caused by a poor connection, rubbed through wire insulation, or a wire that is broken inside the insulation.

Any circuitry, that is suspected as causing the intermittent complaint, should be thoroughly checked for backed out terminals, improper mating, broken locks, improperly formed or damaged terminals, poor terminal to wiring connections or physical damage to the wiring harness.

DIAGNOSTIC CHART DTC A081 — 1993 CUTLASS SUPREME, GRAND PRIX, LUMINA AND REGAL, CONTINUED

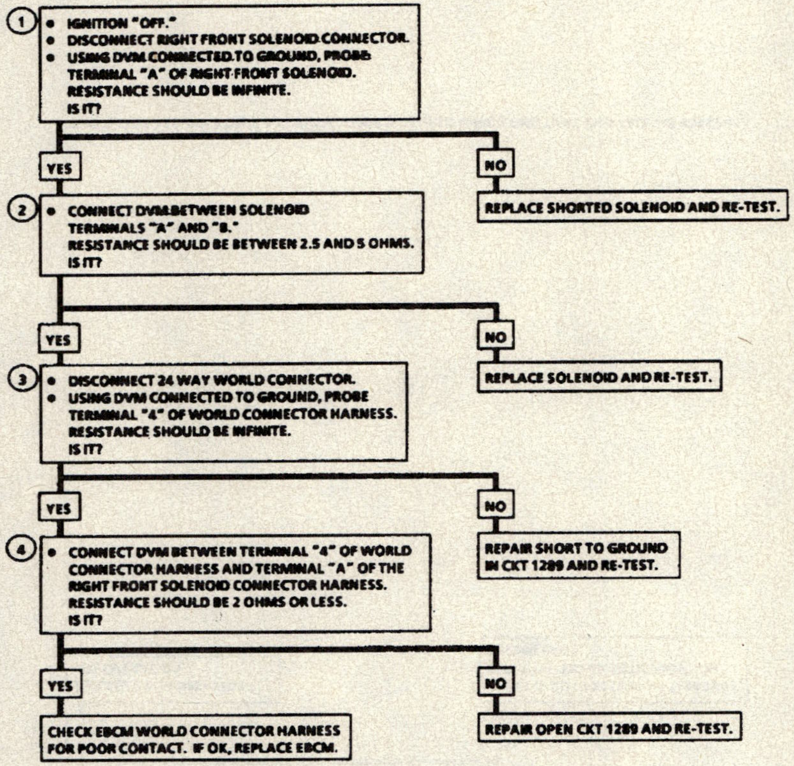

THIS CHART ASSUMES THAT A CURRENT CODE IS STORED INDICATING THAT THIS FAULT IS PRESENT.

DIAGNOSTIC CHART DTC A082 — 1993 CUTLASS SUPREME, GRAND PRIX, LUMINA AND REGAL

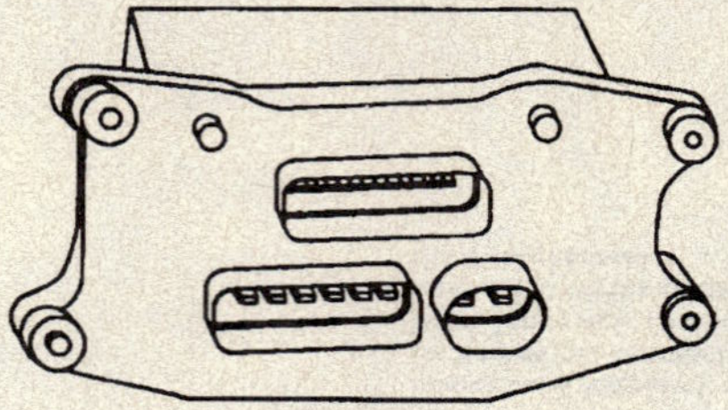

D.M. ABS. VI
ELECTRONIC CONTROL UNIT
(TYPICAL)

CODE A082

CALIBRATION MEMORY FAILURE

Circuit Description:

This code allows the EBCM to check for a calibration failure by comparing the calibration value to a known value stored in the EEPROM.

This code is also used as a security measure to prevent improper use of calibrations or changes to these calibrations that may alter the designed function of ABS.

Test Description: Number(s) below refer to circled number(s) on the diagnostic chart.

1. This step checks to see if the fault is present during diagnosis. If present, the EBCM is not functioning correctly and must be replaced.

Important

- At the time of printing, vehicles with ABS-VI were not being field reprogrammed to correct this failure. A service bulletin will be issued when EEPROM reprogramming is authorized.

Diagnostic Aids:

An intermittent Code A082 may be caused by a bad cell in the EEPROM that is sensitive to temperature changes. If Code A082 failed more than once, but is intermittent, replace EBCM.

DIAGNOSTIC CHART DTC A082 — 1993 CUTLASS SUPREME, GRAND PRIX, LUMINA AND REGAL, CONTINUED

(1)
- IGNITION "ON."
- USING TECH 1, CLEAR CODES.

DOES CODE A082 RESET?

YES → REPLACE EBCM AND VERIFY ABS OPERATION.

NO → FAULT IS NOT PRESENT AT THIS TIME. REFER TO "DIAGNOSTIC AIDS" ON FACING PAGE.

THIS CHART ASSUMES THAT A CURRENT CODE IS STORED INDICATING THAT THIS FAULT IS PRESENT.

DIAGNOSTIC CHART DTC A086 — 1993 CUTLASS SUPREME, GRAND PRIX, LUMINA AND REGAL

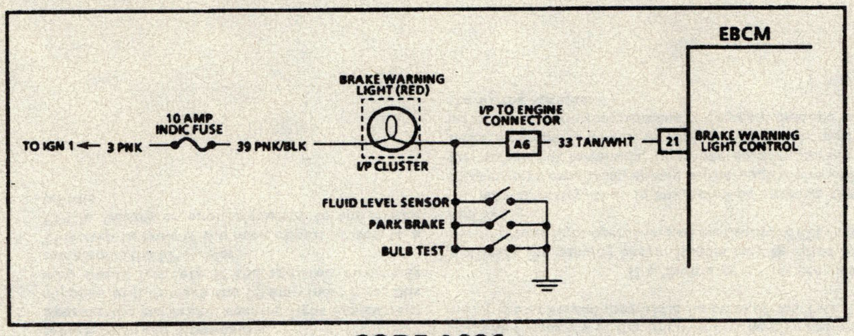

CODE A086

RED BRAKE WARNING LIGHT ACTIVATED BY ABS

Circuit Description:

This code is provided as an information only code, and reflects the status of the command issued by the ABS EBCM to illuminate the red brake warning light. If another diagnostic code issues a command to illuminate the red brake warning light, Code A086 will be stored in EEPROM as a history code at the conclusion of the ignition cycle.

Test Description: Number(s) below refer to circled number(s) on the diagnostic chart.

1. This step serves to identify if a code other than Code A086 commanded the red brake warning light "ON."

Diagnostic Aids:

Any ABS mechanical code that issues a command to illuminate the red brake telltale will also result in Code A086 being stored in EEPROM during shut down. These Codes are: A038, A041, A042, A046, and A051. If the motors are not in their home position, certain electrical codes will also command the red brake telltale "ON." These Codes are: A014, A016, A018, A055, A056, A057, A058, A061, A062, A063, A064, A065, and A066.

If any of these codes are indicated along with Code A086, they must be corrected prior to addressing a Code A086 fault.

DIAGNOSTIC CHART DTC A086 — 1993 CUTLASS SUPREME, GRAND PRIX, LUMINA AND REGAL, CONTINUED

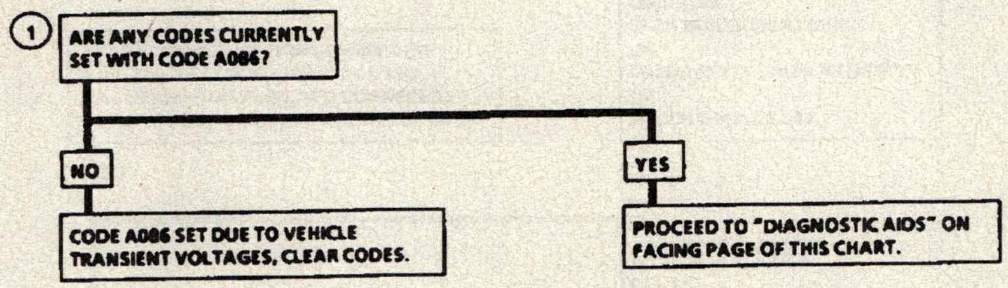

DIAGNOSTIC CHART DTC A087 — 1993 CUTLASS SUPREME, GRAND PRIX, LUMINA AND REGAL

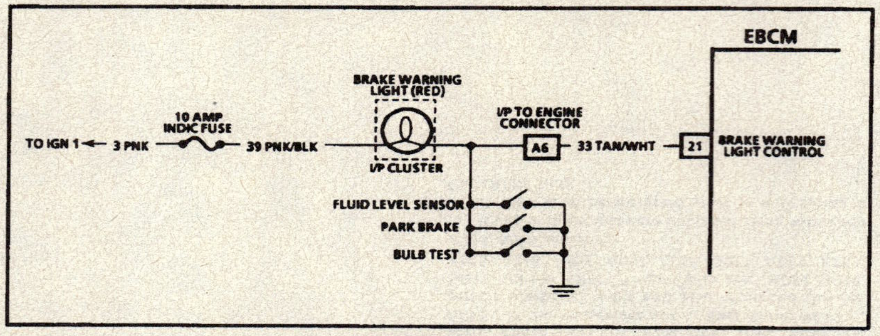

CODE A087

RED BRAKE WARNING LIGHT CIRCUIT OPEN

Circuit Description:

This code is used to verify the EBCM has continuity to the red brake warning light in case the EBCM must turn it "ON." This will only occur if an ABS fault is detected that may degrade base brake operation.

Because ABS is not the only device controlling the red brake warning light (parking brake and low fluid sensor may also turn "ON" the light) a short to ground in this circuit cannot be detected.

Test Description: Number(s) below refer to circled number(s) on the diagnostic chart.

1. This step identifies if the fault is currently present.
2. This step indicates if the EBCM and circuitry has the ability to complete the ground to the brake telltale lamp and illuminate it.
3. Because the red brake warning light has current supplied to it through the 10 amp indic fuse, this step would indicate if the ignition circuit is complete to the I/P cluster.
4. This step isolates if the open circuit is due to a EBCM failure or open circuitry to the brake telltale.

Diagnostic Aids:

An "Intermittent" problem may be caused by a poor connection, rubbed through wire insulation, or a wire that is broken inside the insulation.

The frequency of the problem can be checked by using the enhanced diagnostic function of the Tech 1

If a Tech 1 or T-100 are not available the parking brake handle may be lifted to verify proper light operation and continuity of CKTs 3 and 39.

Any circuitry, that is suspected as causing the intermittent complaint, should be thoroughly checked for backed out terminals, improper mating, broken locks, improperly formed or damaged terminals, poor terminal to wiring connections or physical damage to the wiring harness.

DIAGNOSTIC CHART DTC A087 — 1993 CUTLASS SUPREME, GRAND PRIX, LUMINA AND REGAL, CONTINUED

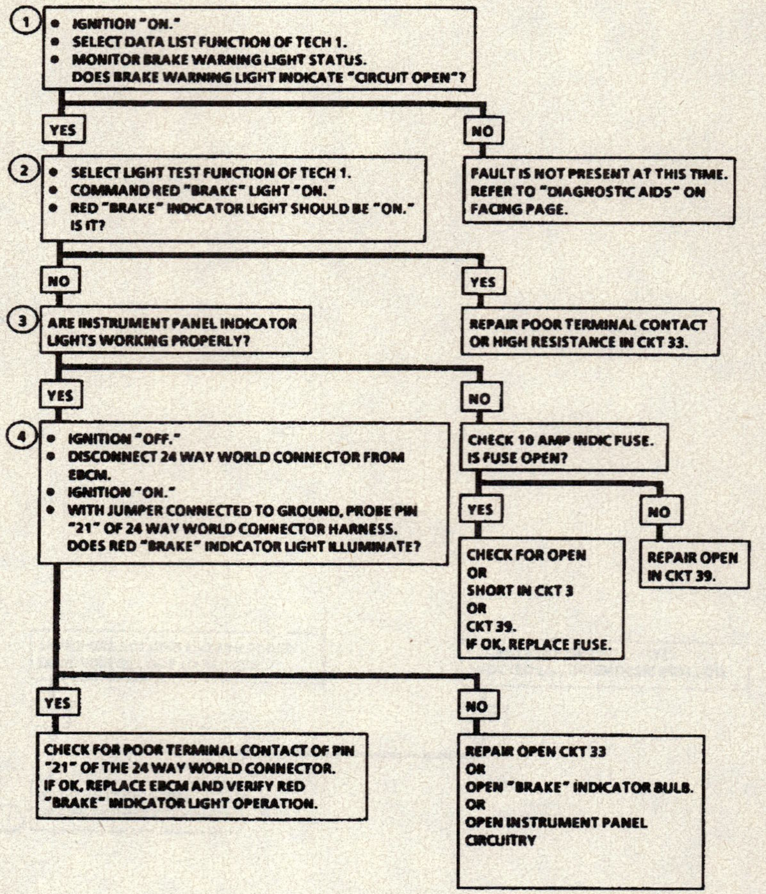

THIS CHART ASSUMES THAT A CURRENT CODE IS STORED INDICATING THAT THIS FAULT IS PRESENT.

DIAGNOSTIC CHART DTC A088 — 1993 CUTLASS SUPREME, GRAND PRIX, LUMINA AND REGAL

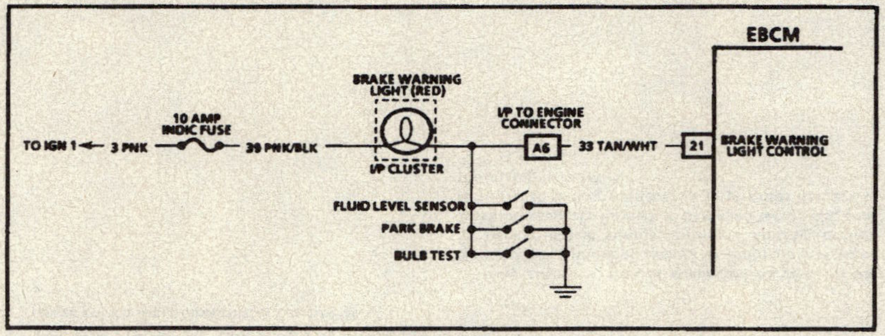

CODE A088

RED BRAKE WARNING LIGHT CIRCUIT SHORTED TO BATTERY

Circuit Description:

This code identifies a short to battery between the EBCM and the red brake warning light, or an open driver that does not allow the red brake warning light to be illuminated by the EBCM. This will only occur if an ABS fault is detected that may degrade base brake operation.

Test Description: Number(s) below refer to circled number(s) on the diagnostic chart.

1. This step identifies if the ground circuit to the red brake warning light is being completed by a source other than the EBCM.
2. This step identifies if the fault is currently present.
3. By removing the 10 amp indic fuse, the voltage source is eliminated. This test indicates if voltage is being supplied from a source other than the gauge fuse.

Diagnostic Aids:

An "Intermittent" problem may be caused by a poor connection, rubbed through wire insulation, or a wire that is broken inside the insulation.

Any circuitry, that is suspected as causing the intermittent complaint, should be thoroughly checked for backed out terminals, improper mating, broken locks, improperly formed or damaged terminals, poor terminal to wiring connections or physical damage to the wiring harness.

DIAGNOSTIC CHART DTC A088 — 1993 CUTLASS SUPREME, GRAND PRIX, LUMINA AND REGAL, CONTINUED

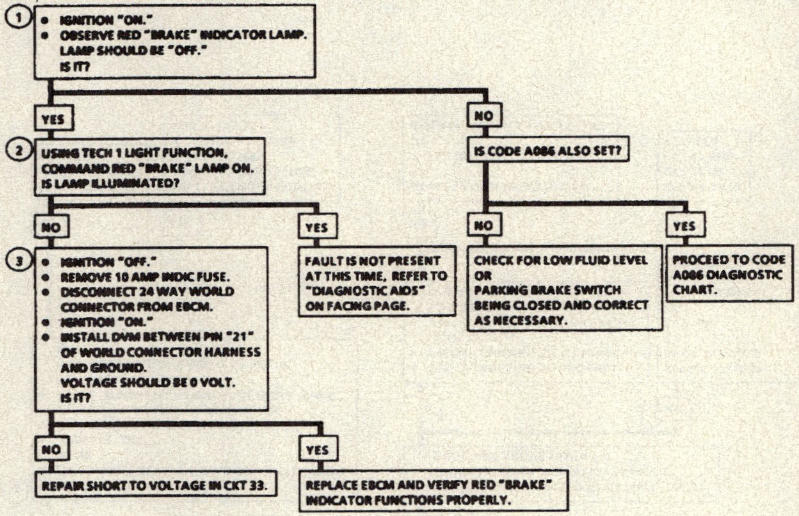

THIS CHART ASSUMES THAT A CURRENT CODE IS STORED INDICATING THAT THIS FAULT IS PRESENT.

DIAGNOSTIC CHART DTC A091 — 1993 CUTLASS SUPREME, GRAND PRIX, LUMINA AND REGAL

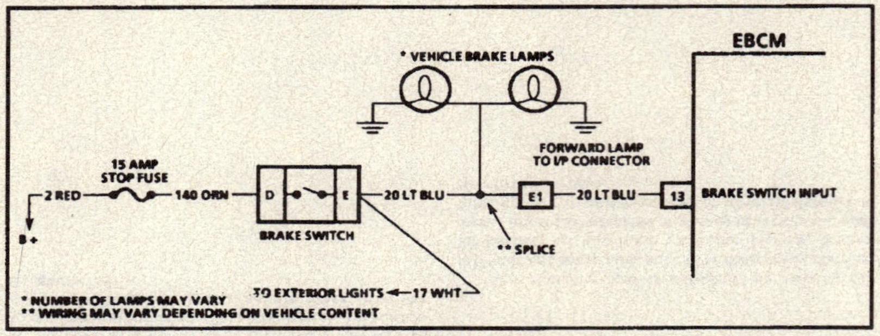

CODE A091

OPEN BRAKE SWITCH CONTACTS DURING DECELERATION

Circuit Description:

This code is used to detect an open brake switch in the non-ABS mode. The EBCM looks for decel rates that would indicate braking action and verifies this assumption by requiring several repeats of this detection method. In each case, ABS will not be available since no brake switch is seen.

Test Description: Number(s) below refer to circled number(s) on the diagnostic chart.

1. This step indicates if the brake switch signal is being received by the EBCM.
2. This step indicates if an open circuit exists in the brake switch or brake lamp circuitry.
3. This step isolates the open circuit to either the brake switch input circuitry, or the EBCM.

Diagnostic Aids:

An "Intermittent" problem may be caused by a poor connection, rubbed through wire insulation, or a wire that is broken inside the insulation.

Any circuitry, that is suspected as causing the intermittent complaint, should be thoroughly checked for backed out terminals, improper mating, broken locks, improperly formed or damaged terminals, poor terminal to wiring connections or physical damage to the wiring harness.

DIAGNOSTIC CHART DTC A091 — 1993 CUTLASS SUPREME, GRAND PRIX, LUMINA AND REGAL, CONTINUED

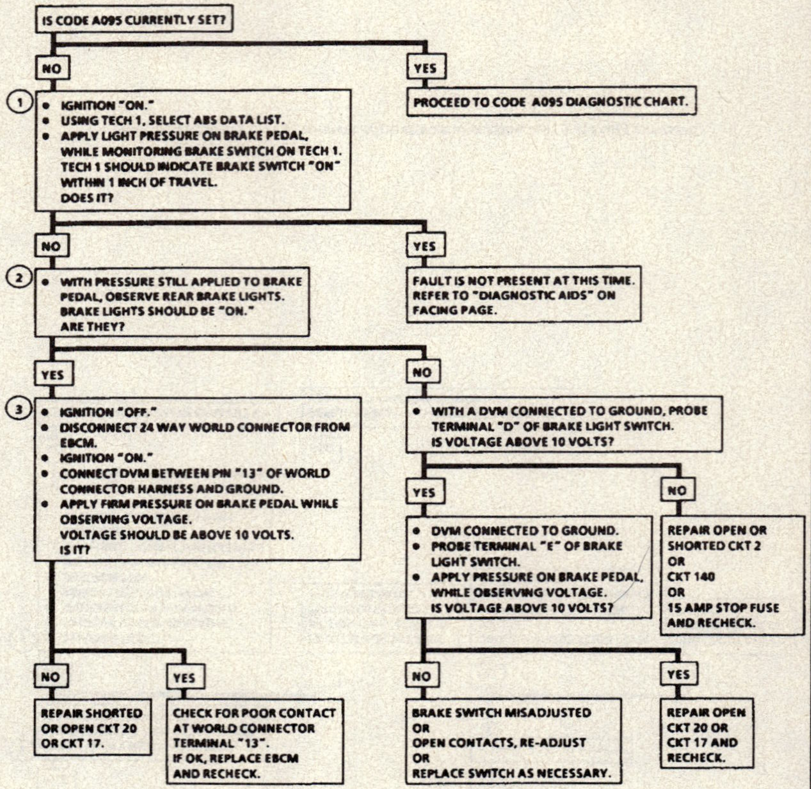

DIAGNOSTIC CHART DTC A092 — 1993 CUTLASS SUPREME, GRAND PRIX, LUMINA AND REGAL

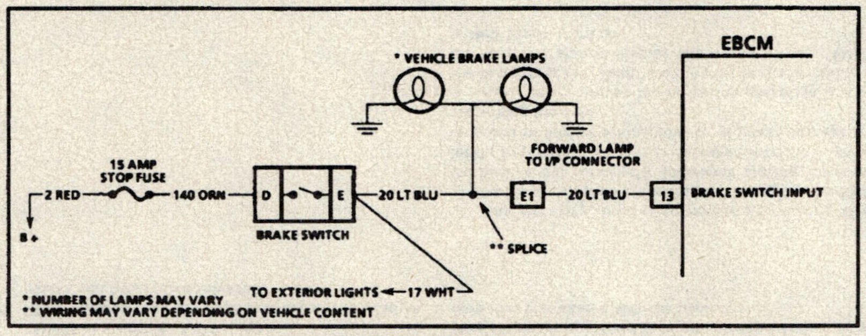

CODE A092

OPEN BRAKE SWITCH CONTACTS WHEN ABS WAS REQUIRED

Circuit Description:

This code is run to determine the proper operation of the brake switch. This is important because ABS is activated when the brake switch is "ON." If the brake switch is "OFF," ABS will never be activated. Since this fault is difficult to detect under normal braking conditions, this fault is only detected when ABS is required.

Test Description: Number(s) below refer to circled number(s) on the diagnostic chart.

1. This step indicates if the brake switch signal is being received by the EBCM.
2. This step indicates if an open circuit exists in the brake switch or brake lamp circuitry.
3. This step isolates the open circuit to either the brake switch input circuitry, or the EBCM.

Diagnostic Aids:

An "Intermittent" problem may be caused by a poor connection, rubbed through wire insulation, or a wire that is broken inside the insulation.

Any circuitry, that is suspected as causing the intermittent complaint, should be thoroughly checked for backed out terminals, improper mating, broken locks, improperly formed or damaged terminals, poor terminal to wiring connections or physical damage to the wiring harness.

DIAGNOSTIC CHART DTC A092 — 1993 CUTLASS SUPREME, GRAND PRIX, LUMINA AND REGAL, CONTINUED

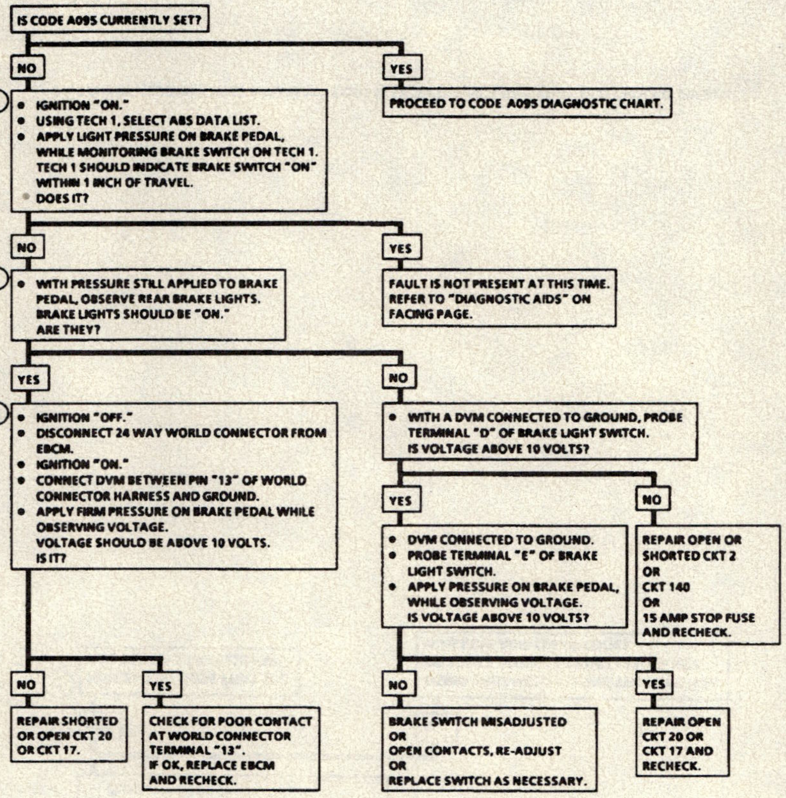

DIAGNOSTIC CHART DTC A093 — 1993 CUTLASS SUPREME, GRAND PRIX, LUMINA AND REGAL

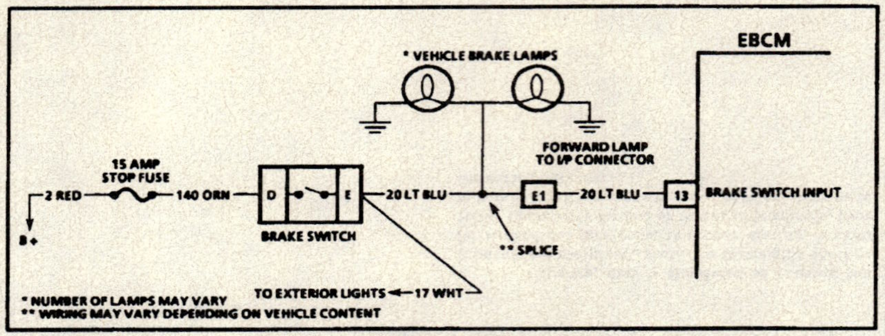

CODE A093

CODE A091 OR A092 SET IN CURRENT OR PREVIOUS IGNITION CYCLE

Circuit Description:

This code is the second portion of Codes A091 and A092. If Code A091 or A092 failed during the last ignition cycle, Code A093 becomes a current failure during the next ignition cycle, keeping ABS disabled until a brake switch "ON" state is seen. When a change is seen during an ignition cycle in which Code A093 is a current failure, Code A091 or A092 will clear itself at the end of the current ignition cycle, and ABS will enable itself at the start of the next ignition cycle. Code A093 alone indicates A091 or A092 failed previously, but is intermittent, or has been corrected.

Test Description: Number(s) below refer to circled number(s) on the diagnostic chart.

1. This step indicates which Code (either A091 or A092) caused Code A093 to set.
2. This insures that the code that set is repaired so that Code A093 can be cleared.

Diagnostic Aids:

An "Intermittent" problem may be caused by a poor connection, rubbed through wire insulation, or a wire that is broken inside the insulation.

Any circuitry, that is suspected as causing the intermittent complaint, should be thoroughly checked for backed out terminals, improper mating, broken locks, improperly formed or damaged terminals, poor terminal to wiring connections or physical damage to the wiring harness.

Also, verify proper brake switch operation using the data list of the "Scan" tool. As the brake is applied, the data list should display the brake switch "ON" within 1 inch of travel.

DIAGNOSTIC CHART DTC A093 — 1993 CUTLASS SUPREME, GRAND PRIX, LUMINA AND REGAL, CONTINUED

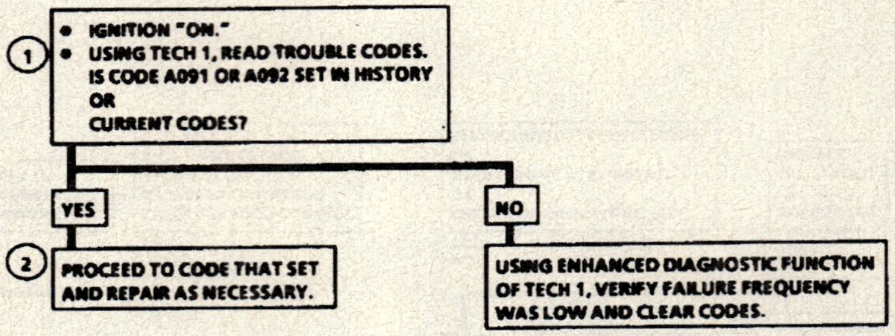

THIS CHART ASSUMES THAT A CURRENT CODE IS STORED INDICATING THAT THIS FAULT IS PRESENT.

DIAGNOSTIC CHART DTC A094 — 1993 CUTLASS SUPREME, GRAND PRIX, LUMINA AND REGAL

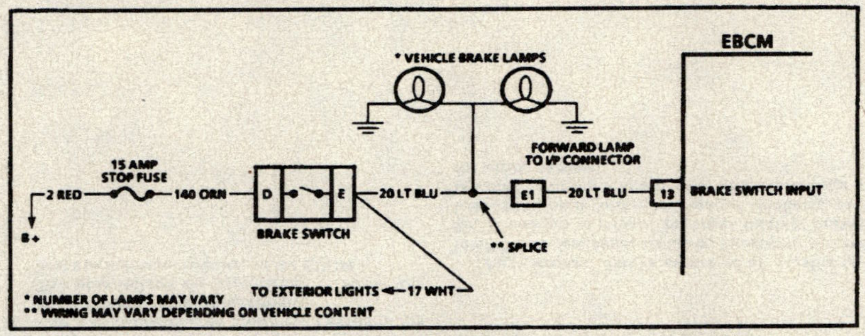

CODE A094

BRAKE SWITCH CONTACTS ALWAYS CLOSED

Circuit Description:

This code is run to determine the proper operation of the brake switch. This is important because ABS is activated when the brake switch is "ON" and turned "OFF" when the brake is "OFF." If the brake switch is always "ON," ABS operation will always be requested resulting in potential modulator cycling on rough roads. Additionally, this failure will most likely result in a dead battery (due to the brake lights staying "ON") if the driver is not informed of this fault.

Test Description: Number(s) below refer to circled number(s) on the diagnostic chart.

1. This checks if the fault is currently present.
2. This step isolates the cause of the fault to either a faulty or misadjusted brake switch, or a short to voltage in the brake switch circuitry.

Diagnostic Aids:

An "Intermittent" problem may be caused by a poor connection, rubbed through wire insulation, or a wire that is broken inside the insulation.

Any circuitry, that is suspected as causing the intermittent complaint, should be thoroughly checked for backed out terminals, improper mating, broken locks, improperly formed or damaged terminals, poor terminal to wiring connections or physical damage to the wiring harness.

DIAGNOSTIC CHART DTC A094 — 1993 CUTLASS SUPREME, GRAND PRIX, LUMINA AND REGAL, CONTINUED

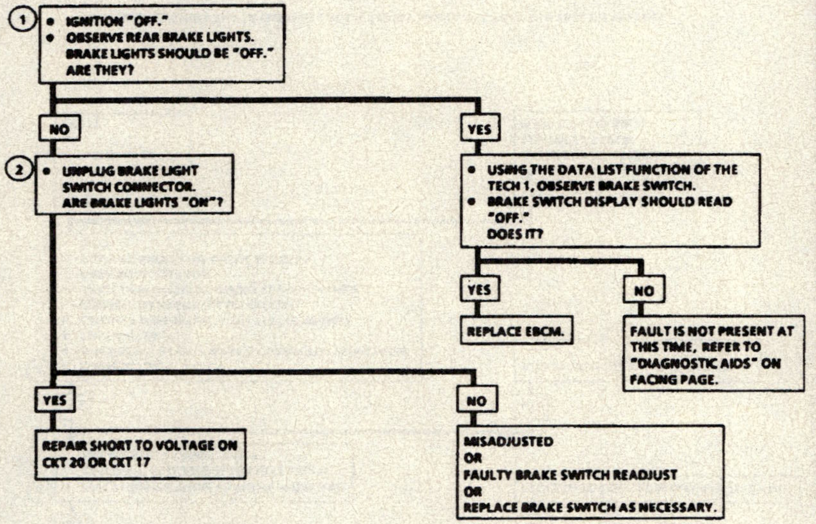

THIS CHART ASSUMES THAT A CURRENT CODE IS STORED INDICATING THAT THIS FAULT IS PRESENT.

DIAGNOSTIC CHART DTC A095 — 1993 CUTLASS SUPREME, GRAND PRIX, LUMINA AND REGAL

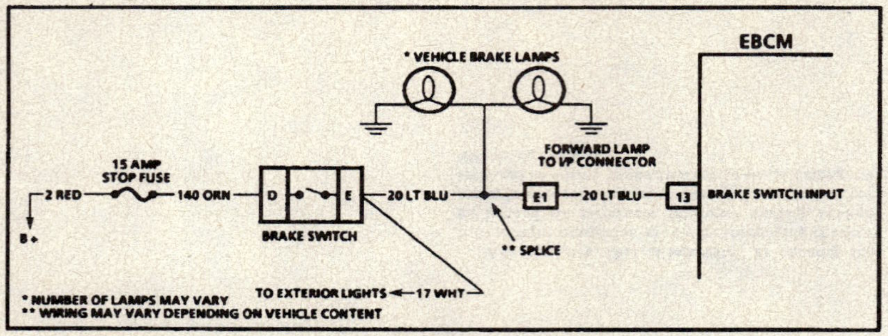

CODE A095

BRAKE SWITCH CIRCUIT OPEN

Circuit Description:

This code is used to identify open brake switch circuitry that prevents the brake switch input to the EBCM from changing states when the brake is applied. This code is used in conjunction with Codes A091 and A092 to determine the cause of an open brake switch fault.

Test Description: Number(s) below refer to circled number(s) on the diagnostic chart.

1. This step is used to confirm that an open in the brake switch circuitry currently exists.
2. This step indicates if the brake switch signal is being received by the EBCM.
3. This step indicates if an open circuit exists in the brake switch or brake lamp circuitry.
4. This step isolates the open circuit to either the brake switch input circuitry, or the EBCM.

Diagnostic Aids:

An "Intermittent" problem may be caused by a poor connection, rubbed through wire insulation, or a wire that is broken inside the insulation.

Any circuitry, that is suspected as causing the intermittent complaint, should be thoroughly checked for backed out terminals, improper mating, broken locks, improperly formed or damaged terminals, poor terminal to wiring connections or physical damage to the wiring harness.

DIAGNOSTIC CHART DTC A095 — 1993 CUTLASS SUPREME, GRAND PRIX, LUMINA AND REGAL, CONTINUED

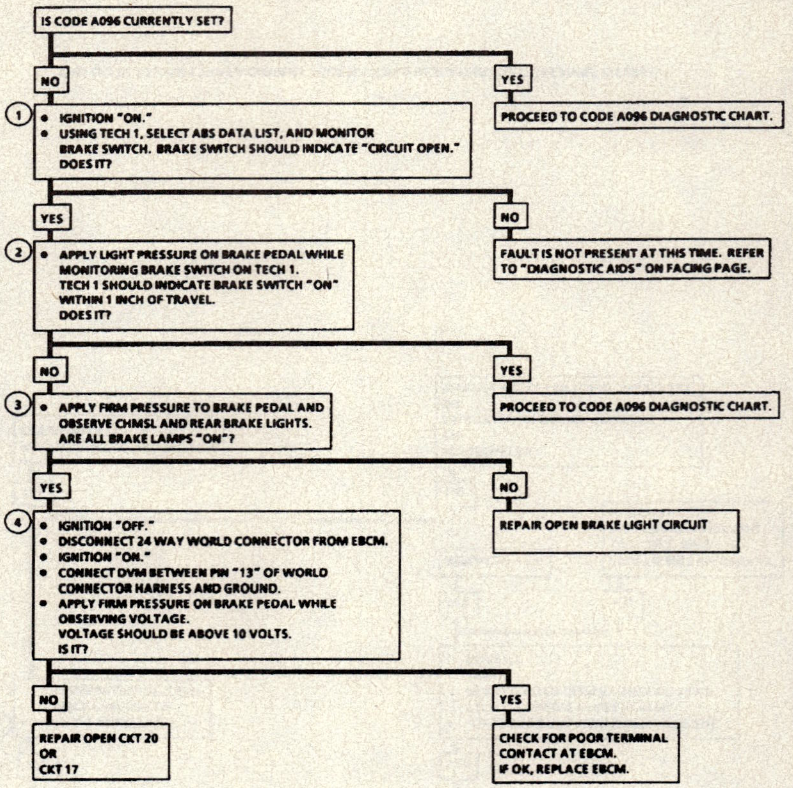

THIS CHART ASSUMES THAT A CURRENT CODE IS STORED INDICATING THAT THIS FAULT IS PRESENT.

DIAGNOSTIC CHART DTC A096 — 1993 CUTLASS SUPREME, GRAND PRIX, LUMINA AND REGAL

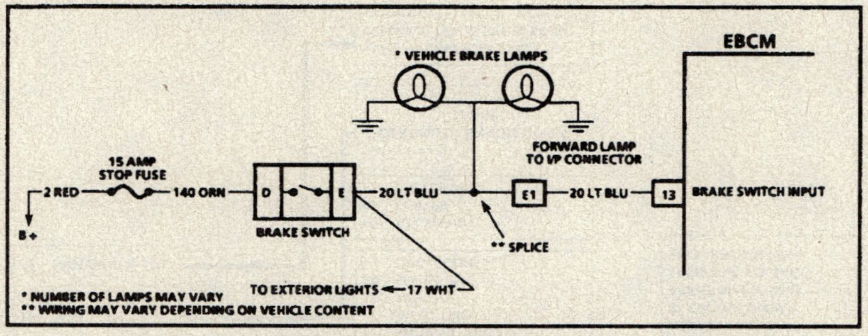

CODE A096

BRAKE LIGHTS CIRCUIT OPEN

Circuit Description:

This code is designed to isolate the cause of a Code A095 failure and indicate to the driver ABS is still available. If Code A095 fails with Code A096, the brake lights are open, the brake light grounds are open, or the Center High Mounted Stop Lamp (CHMSL) circuit is open during 4 way flasher use. The presence of battery voltage at the brake switch input indicates a valid brake switch input is still available.

Test Description: Number(s) below refer to circled number(s) on the diagnostic chart.

1. As a result of a failure of an additional brake switch circuit code, this code may be set. To insure proper diagnosis, any additional brake switch codes must be repaired first.
2. This step identifies if the fault is currently present in the brake circuit.

Diagnostic Aids:

An "Intermittent" problem may be caused by a poor connection, rubbed through wire insulation, or a wire that is broken inside the insulation.

Any circuitry, that is suspected as causing the intermittent complaint, should be thoroughly checked for backed out terminals, improper mating, broken locks, improperly formed or damaged terminals, poor terminal to wiring connections or physical damage to the wiring harness.

DIAGNOSTIC CHART DTC A096 — 1993 CUTLASS SUPREME, GRAND PRIX, LUMINA AND REGAL, CONTINUED

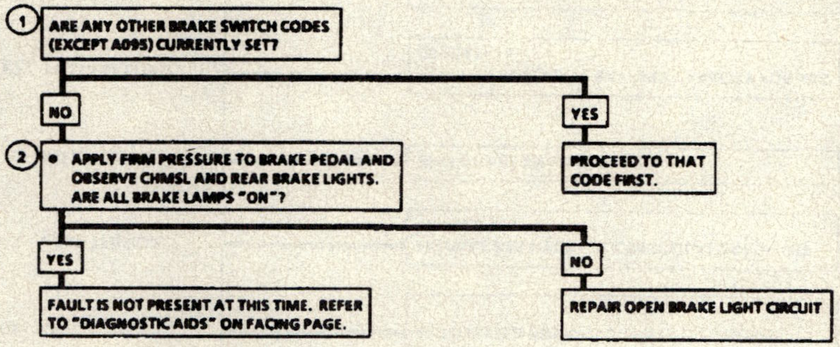

THIS CHART ASSUMES THAT A CURRENT CODE IS STORED INDICATING THAT THIS FAULT IS PRESENT.

DIAGNOSTIC PROCESS CHART — 1994 CUTLASS SUPREME AND REGAL

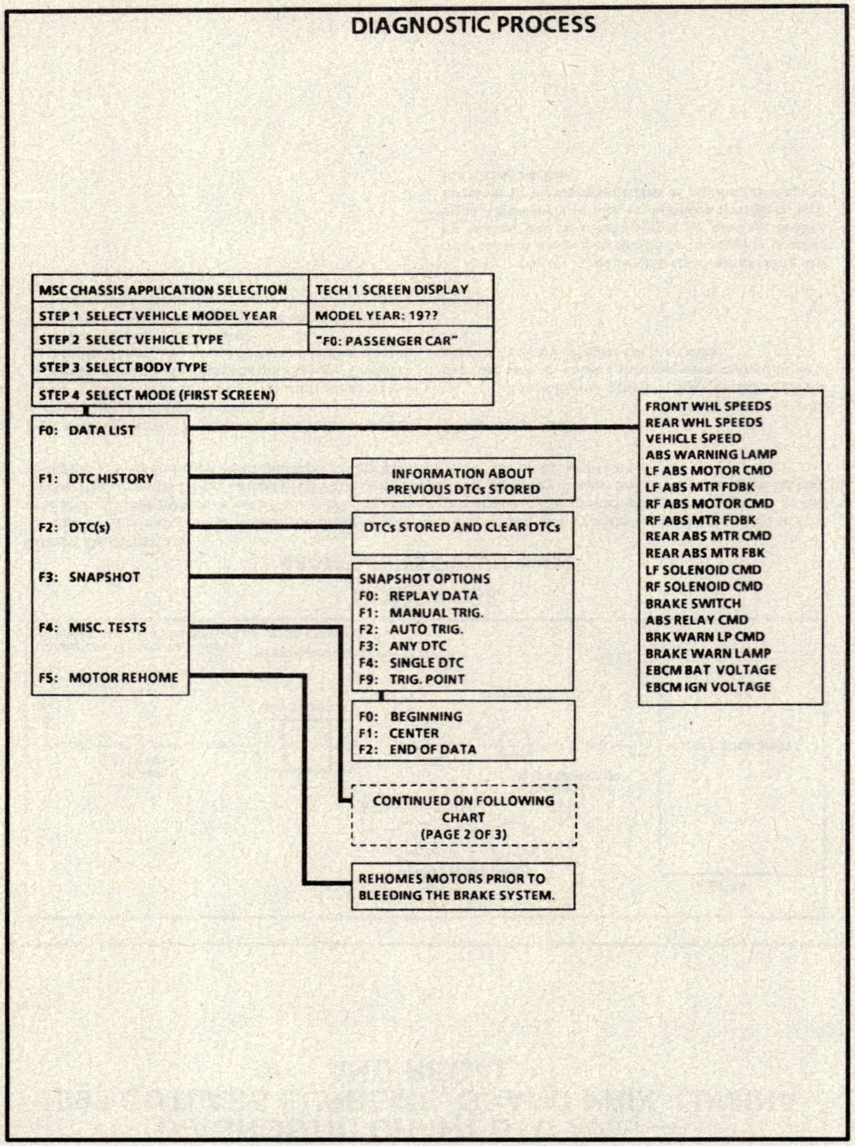

DIAGNOSTIC PROCESS CHART — 1994 CUTLASS SUPREME AND REGAL, CONTINUED

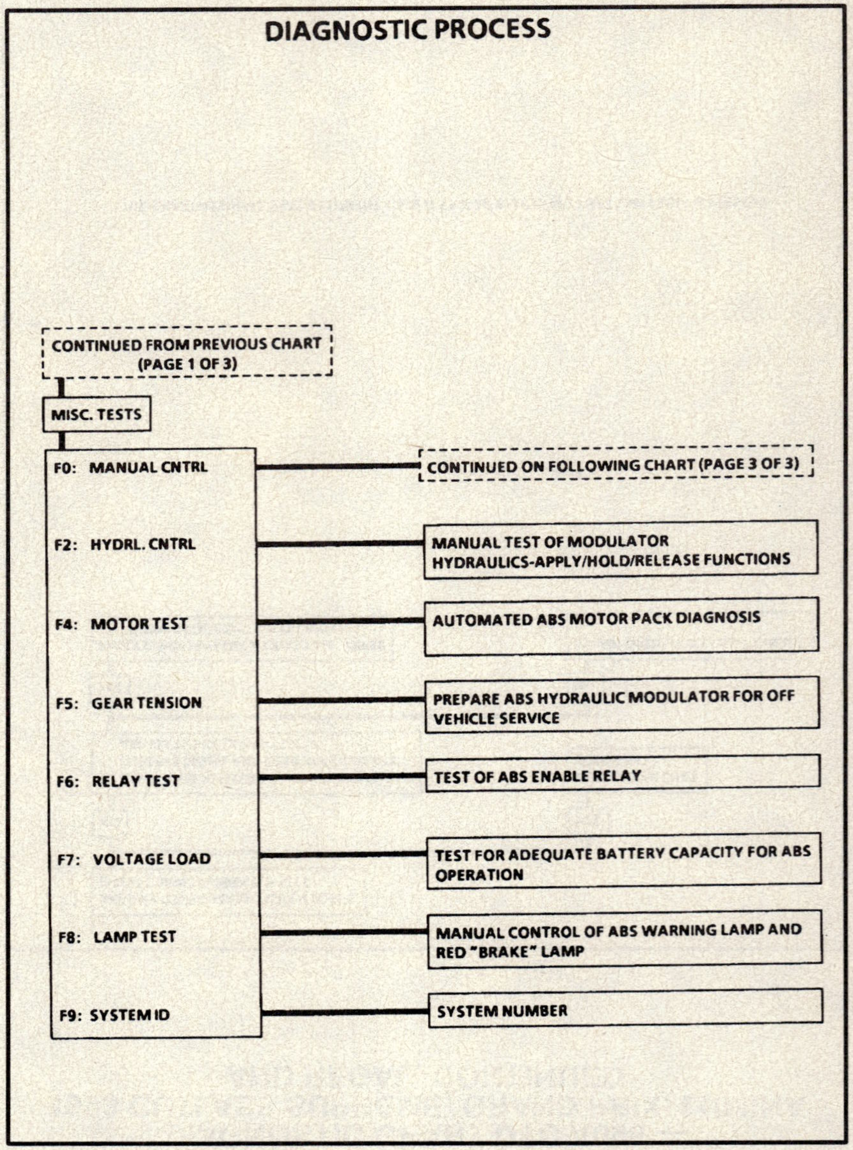

DIAGNOSTIC PROCESS CHART — 1994 CUTLASS SUPREME AND REGAL, CONTINUED

DIAGNOSTIC PROCESS

CONTINUED FROM PREVIOUS CHART (PAGE 2 OF 3)

MANUAL CONTROL OF SOLENOIDS AND MOTORS.

↑ = TURN ON
↓ = TURN OFF
YES TO CONTINUE

SELECT FUNCTION
F0: LF APPLY
F1: LF RELEASE
F2: LF SOLENOID
F4: RF APPLY
F5: RF RELEASE
F6: RF SOLENOID
F8: REAR APPLY
F9: REAR RELEASE

MOTOR APPLY
COMMAND: XX AMPS
FEEDBACK: XX AMPS
ABS RELAY: ON/OFF

MOTOR RELEASE
COMMAND: XX AMPS
FEEDBACK: XX AMPS
ABS RELAY: ON/OFF

SOLENOID
COMMAND: ON/OFF
ABS RELAY: ON/OFF

DIAGNOSTIC CHART SYSTEM CHECK — 1994 CUTLASS SUPREME AND REGAL

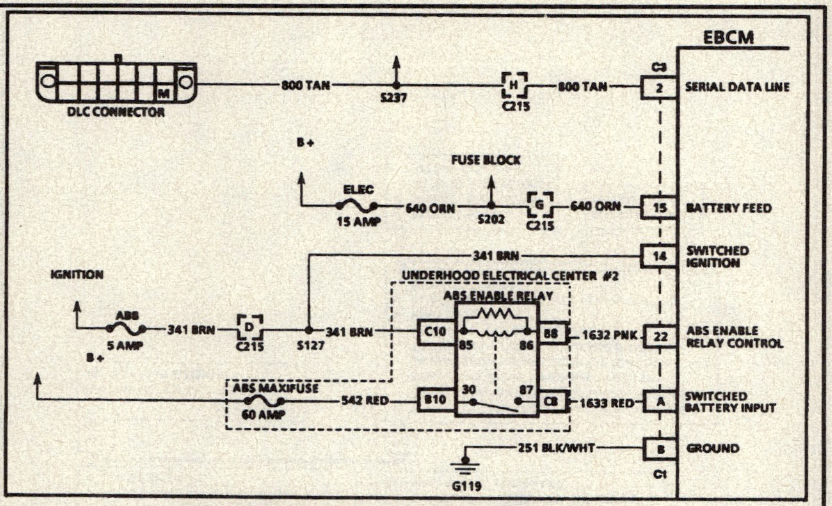

DIAGNOSTIC SYSTEM CHECK

System Description:

The diagnostic system check is an organized approach to identifying a problem created by an Antilock Brake System (ABS) malfunction. It must be the starting point for any ABS complaint diagnosis because it directs the service technician to the next logical step in diagnosing the complaint.

Serial data is transmitted by the EBCM through terminal "2". The EBCM is supplied switched ignition voltage through terminal "14", battery voltage through terminal "15" and switched battery voltage through terminal "A". The EBCM ground is provided through terminal "B".

Chart Test Description: Number(s) below refer to circled number(s) on the diagnostic chart.

1. Indicates if EBCM is transmitting data.
2. Checks for a serial data line malfunction.
3. Checks for proper Tech 1 connection to DLC.
4. Checks for open ELEC 15 amp fuse.
5. Checks for open ABS 5 amp fuse.
6. Checks for high resistance in the EBCM ground circuit.
7. Checks for current DTCs.
8. Checks for proper ABS warning lamp operation.
9. Checks for history DTCs.
10. Checks for possible short to ground in the battery feed circuitry.
11. Checks for possible short to ground in the EBCM.
12. Ensures short to ground is not due to physical damage of the circuitry.
13. Checks for possible short to ground in the switched ignition circuitry.
14. Isolates the short to ground to either the circuitry or the EBCM.
15. Ensures short to ground is not due to physical damage of the circuitry.
16. Checks for an open in CKT 640.
17. Checks for high resistance in battery feed circuitry.
18. Checks for high resistance in the switched ignition circuitry.
19. Ensures malfunction is not due to poor terminal contact.
20. Checks for high resistance in the switched ignition circuit between the fuse block and the EBCM.

Diagnostic Aids: Excessive resistance in the ground or power supply circuits will not allow communication with the EBCM. If communication with the EBCM is not possible, ensure the ABS ground connection is good and that there is no excessive resistance in any of the power supply circuits.

DIAGNOSTIC CHART SYSTEM CHECK — 1994 CUTLASS SUPREME AND REGAL, CONTINUED

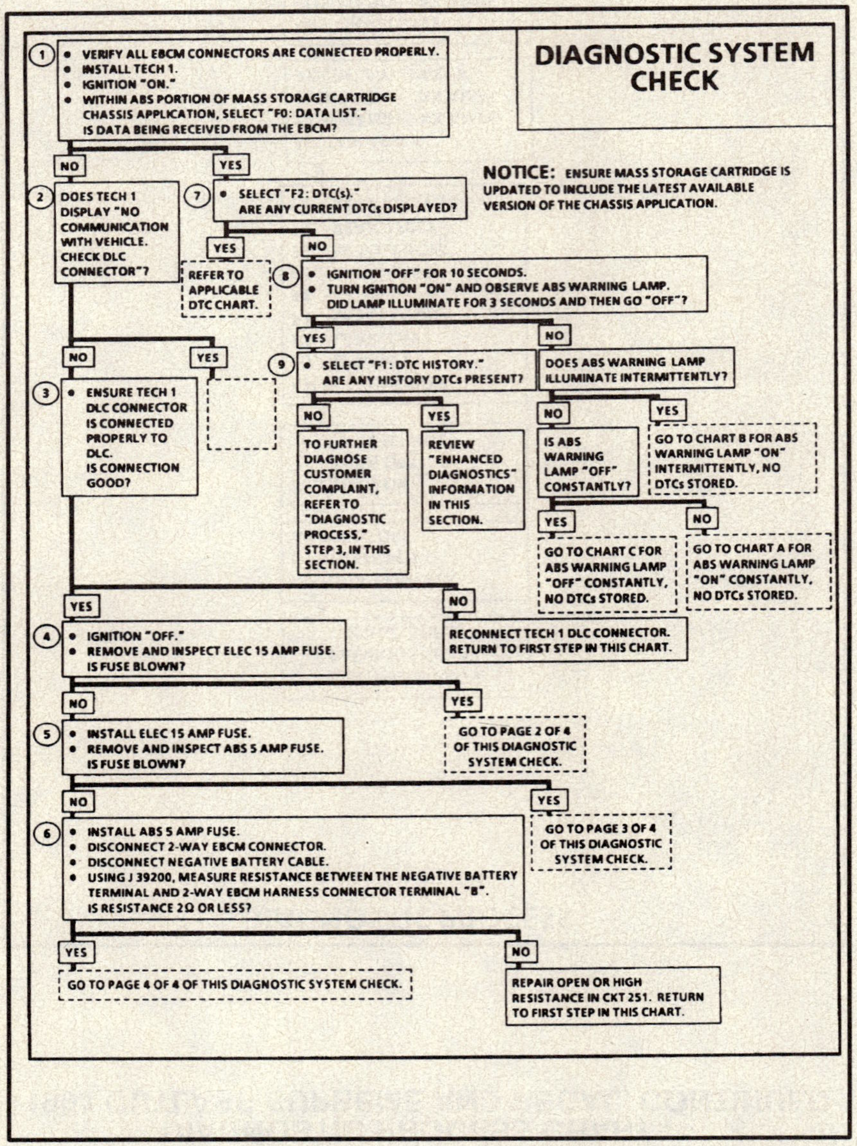

DIAGNOSTIC CHART SYSTEM CHECK — 1994 CUTLASS SUPREME AND REGAL, CONTINUED

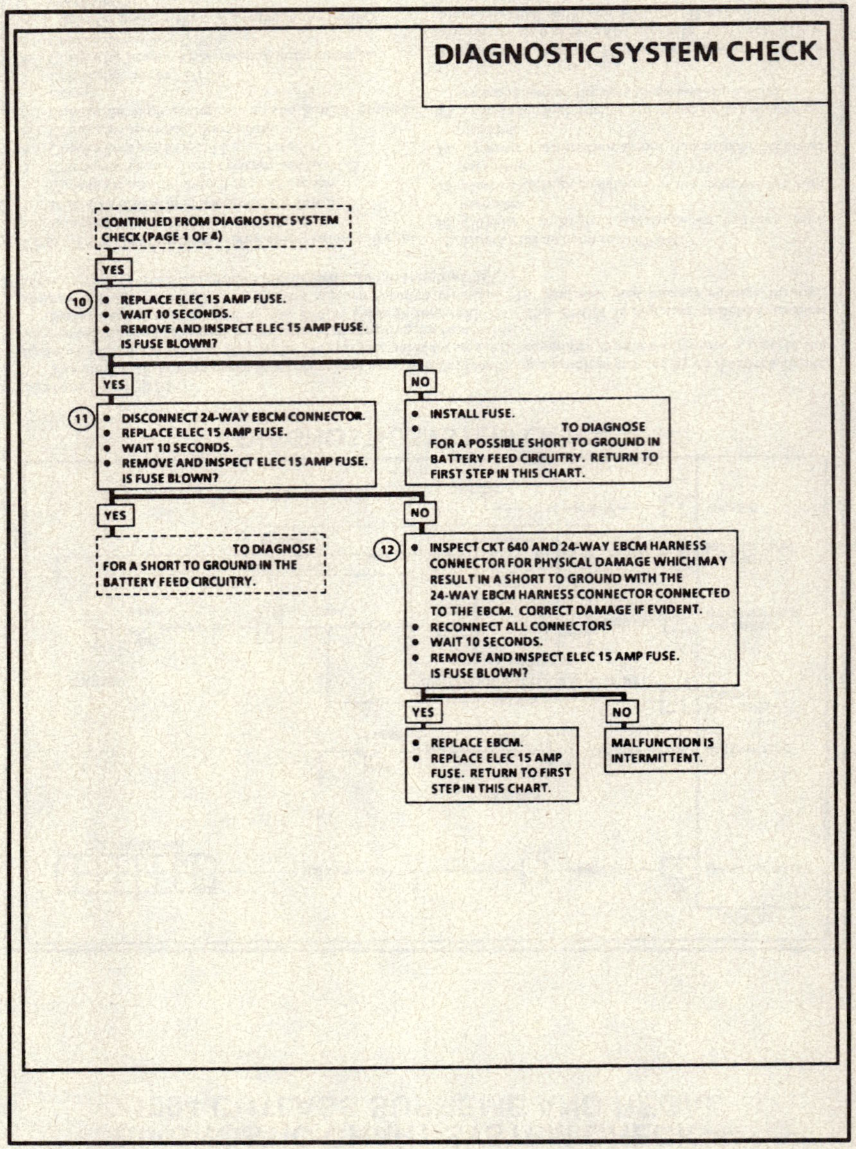

DIAGNOSTIC CHART SYSTEM CHECK — 1994 CUTLASS SUPREME AND REGAL, CONTINUED

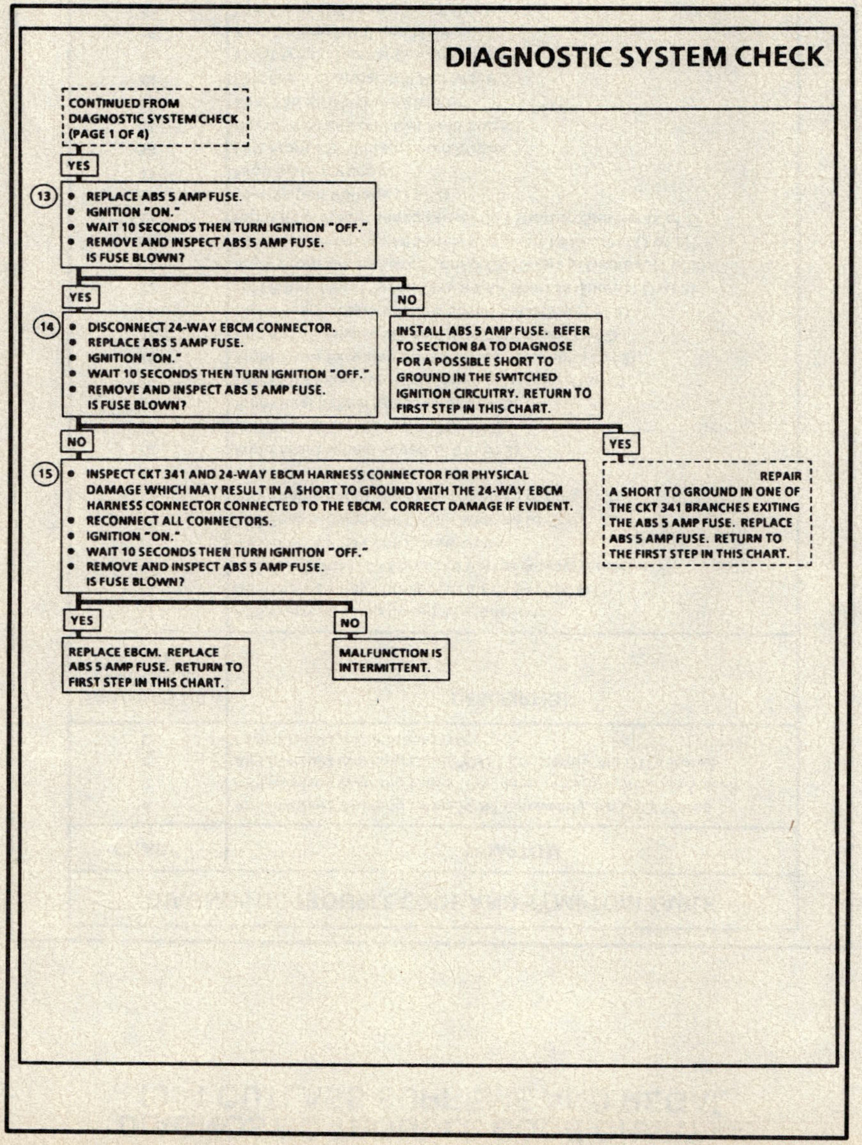

DIAGNOSTIC CHART SYSTEM CHECK — 1994 CUTLASS SUPREME AND REGAL, CONTINUED

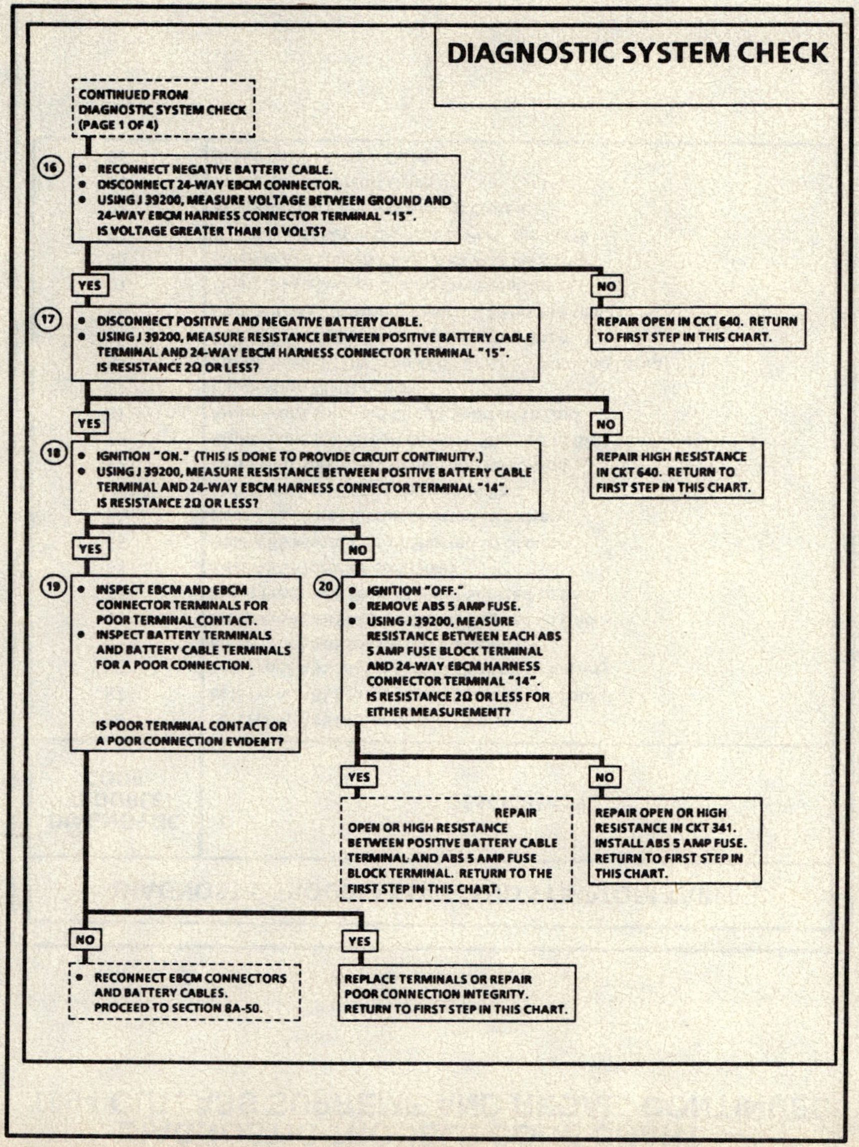

DIAGNOSTIC TROUBLE CODE CHART — 1994 CUTLASS SUPREME AND REGAL

DIAGNOSTIC TROUBLE CODE AND SYMPTOM TABLE

CHART	SYMPTOM
A	ABS (Amber) Warning Lamp "ON" Constantly, No DTCs Stored
B	ABS (Amber) Warning Lamp "ON" Intermittently, No DTCs Stored
C	ABS (Amber) Warning Lamp "OFF" Constantly, No DTCs Stored
D	Tech 1 Displays Undefined DTCs

DIAGNOSTIC TROUBLE CODE	DESCRIPTION
11	ABS Warning Lamp Circuit Malfunction
14	ABS Enable Relay Contact Circuit Open (1 of 3)
15	ABS Rly. Contact CKT Short to Batt or Always Closed
16	ABS Enable Relay Coil Circuit Open
17	ABS Enable Relay Coil Circuit Shorted to Gnd
18	ABS Enable Relay Coil Circuit Shorted to Batt
21	Left Front Wheel Speed = 0 (1 of 2)
22	Right Front Wheel Speed = 0 (1 of 2)
23	Left Rear Wheel Speed = 0 (1 of 2)
24	Right Rear Wheel Speed = 0 (1 of 2)
25	Left Front Excessive Wheel Speed Variation (1 of 2)
26	Right Front Excessive Wheel Speed Variation (1 of 2)
27	Left Rear Excessive Wheel Speed Variation (1 of 2)
28	Right Rear Excessive Wheel Speed Variation (1 of 2)
32	Left Front Wheel Speed Sensor CKT Short to Gnd/Bat (1 of 2)
33	Right Front Wheel Speed Sensor CKT Short to Gnd/Bat (1 of 2)
34	Left Rear Wheel Speed Sensor CKT Short to Gnd/Bat (1 of 2)
35	Right Rear Wheel Speed Sensor CKT Short to Gnd/Bat (1 of 2)
36	Low System Voltage (1 of 2)
37	High System Voltage
38	Left Front ESB Will Not Hold Motor
41	Right Front ESB Will Not Hold Motor
42	Rear ESB Will Not Hold Motor
44	Left Front Channel Will Not Move
45	Right Front Channel Will Not Move
46	Rear Channel Will Not Move
47	Left Front ABS Motor Free Spins (1 of 2)
48	Right Front ABS Motor Free Spins (1 of 2)
51	Rear ABS Motor Free Spins (1 of 2)
52	Left Front Channel in Release Too Long
53	Right Front Channel In Release Too Long
54	Rear Channel in Release Too Long
55	EBCM Malfunction

DIAGNOSTIC TROUBLE CODE CHART — 1994 CUTLASS SUPREME AND REGAL, CONTINUED

DIAGNOSTIC TROUBLE CODE AND SYMPTOM TABLE

DIAGNOSTIC TROUBLE CODE	DESCRIPTION
56	Left Front ABS Motor Circuit Open
57	Left Front ABS Motor Circuit Shorted to Ground
58	Left Front ABS Motor Circuit Shorted to Battery
61	Right Front ABS Motor Circuit Open
62	Right Front ABS Motor Circuit Shorted to Gnd
63	Right Front ABS Motor Circuit Shorted to Batt
64	Rear ABS Motor Circuit Open
65	Rear ABS Motor Circuit Shorted to Ground
66	Rear ABS Motor Circuit Shorted to Battery
76	Left Front Sol. Circuit Shorted to Batt
77	Left Front Solenoid Circuit Shorted to Gnd
78	Right Front Sol. Circuit Open or Shorted to Batt
81	Right Front Solenoid Circuit Shorted to Gnd
82	Calibration Malfunction
86	EBCM Turned "ON" the Red "BRAKE" Warning Lamp
87	Red "BRAKE" Warning Lamp Circuit Open
88	Red "BRAKE" Warning Lamp CKT Shorted to Batt
91	Open Brake Switch During Deceleration
92	Open Brake Switch When ABS Was Required
93	DTC 91 or 92 Set in Current or Prev. Ign Cycle
94	Brake Switch Contacts Always Closed
95	Brake Switch Circuit Open
96	Brake Lamps Circuit Open

ELECTRICAL CONNECTOR TERMINAL IDENTIFICATION — 1994 CUTLASS SUPREME AND REGAL

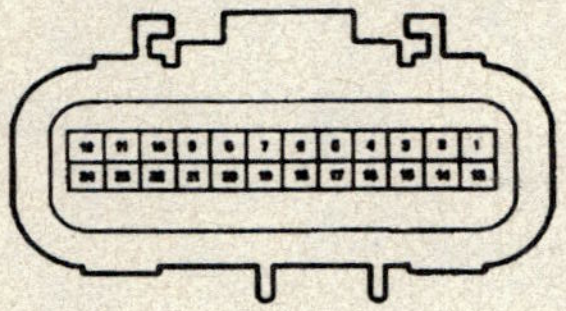

24-WAY F MICRO-PACK 100 SERIES
BLK

ELECTRONIC BRAKE CONTROL MODULE (EBCM) CONNECTOR C3

CAVITY	CIRCUIT NO.	COLOR	CIRCUIT
1	PLUGGED		NOT USED
2	461	ORN	SERIAL DATA LINE
3	PLUGGED		NOT USED
4	1289	LT BLU/BLK	RF ABS SOLENOID
5	830	LT BLU	LF WHEEL SIGNAL HIGH
6	873	YEL	LF WHEEL SIGNAL LOW
7	882	BRN	RR WHEEL SIGNAL HIGH
8	883	WHT	RR WHEEL SIGNAL LOW
9	872	DK GRN	RF WHEEL SIGNAL HIGH
10	833	TAN	RF WHEEL SIGNAL LOW
11	885	RED	LR WHEEL SIGNAL LOW
12	884	BLK	LR WHEEL SIGNAL HIGH
13	20	LT BLU	BRAKE SWITCH INPUT
14	350	PNK/WHT	SWITCHED IGNITION
15	140	ORN	B + FEED
16	PLUGGED		NOT USED
17	PLUGGED		NOT USED
18	VENT TUBE	BLK	VENT TUBE
19	PLUGGED		NOT USED
20	PLUGGED		NOT USED
21	33	TAN/WHT	BRAKE WARNING LAMP CONTROL
22	879	PPL/WHT	ENABLE RELAY CONTROL
23	852	WHT	ABS INDICATOR LAMP CONTROL
24	1288	DK GRN/YEL	LF ABS SOLENOID

ELECTRICAL CONNECTOR TERMINAL IDENTIFICATION — 1994 CUTLASS SUPREME AND REGAL, CONTINUED

6 WAY EBCM CONNECTOR

PIN	CIRCUIT NO.	COLOR	CIRCUIT
C	1284	DK GRN	REAR MOTOR HIGH
D	1285	ORN	REAR MOTOR LOW
E	1281	PNK	LF MOTOR LOW
F	1280	BLK	LF MOTOR HIGH
G	1283	BLK	RF MOTOR LOW
H	1282	PPL	RF MOTOR HIGH

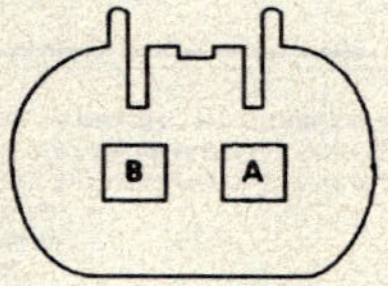

2 WAY EBCM CONNECTOR

PIN	CIRCUIT NO.	COLOR	CIRCUIT
A	1633	RED	SWITCHED BATTERY INPUT
B	251	BLK/WHT	GROUND

DIAGNOSTIC CHART A — 1994 CUTLASS SUPREME AND REGAL

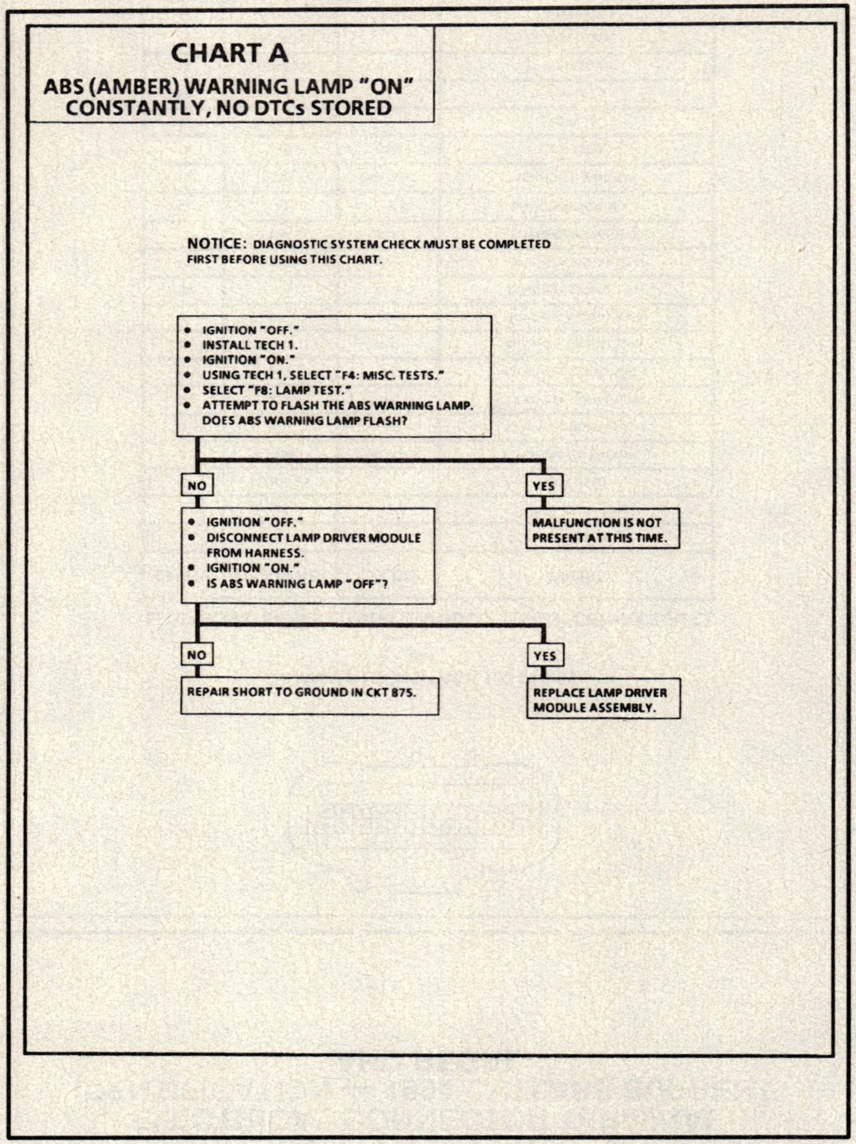

DIAGNOSTIC CHART B — 1994 CUTLASS SUPREME AND REGAL

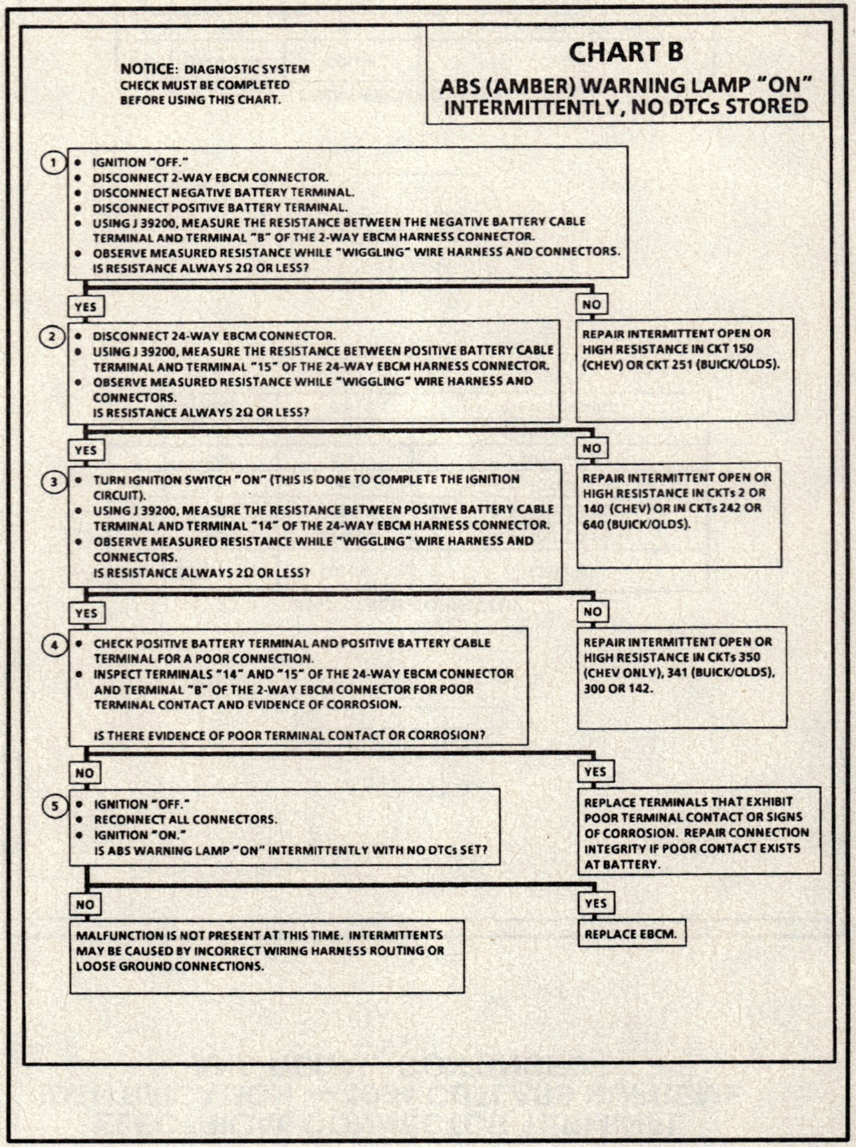

DIAGNOSTIC CHART C — 1994 CUTLASS SUPREME AND REGAL

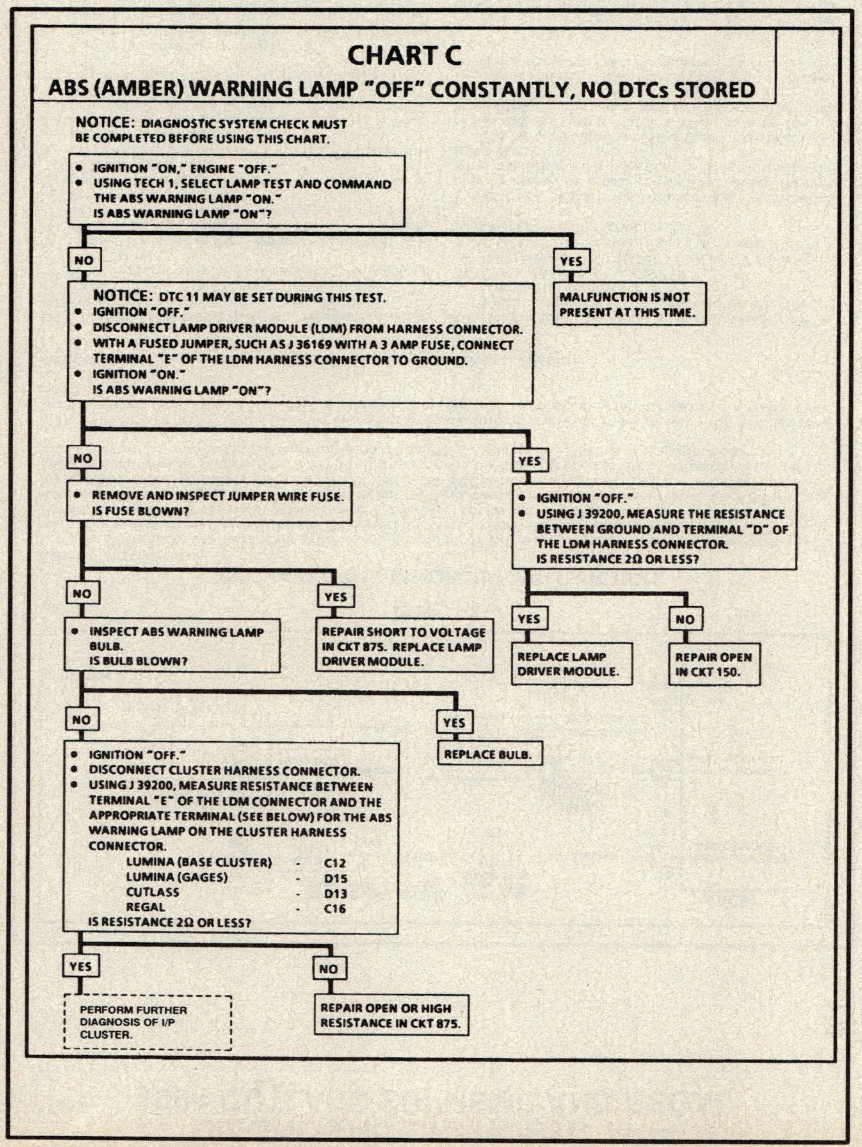

DIAGNOSTIC CHART D — 1994 CUTLASS SUPREME AND REGAL

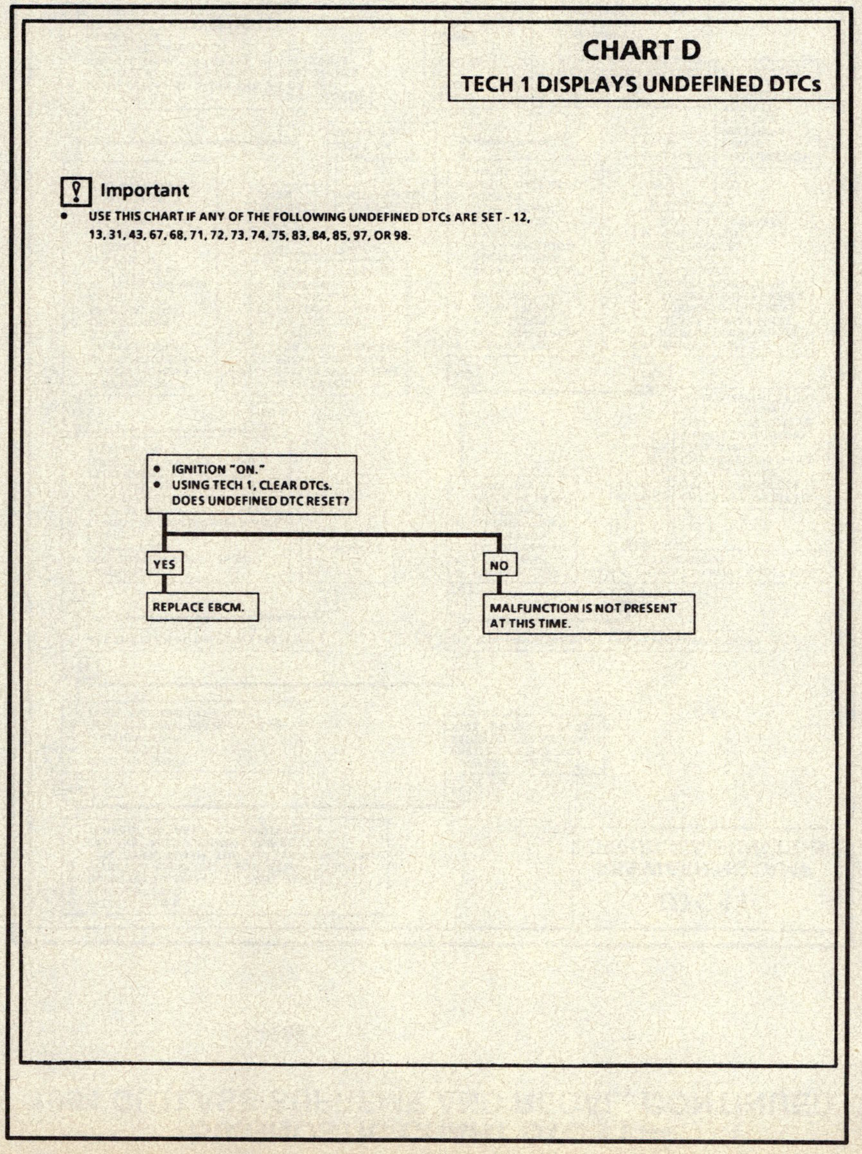

DIAGNOSTIC CHART DTC 11 — 1994 CUTLASS SUPREME AND REGAL

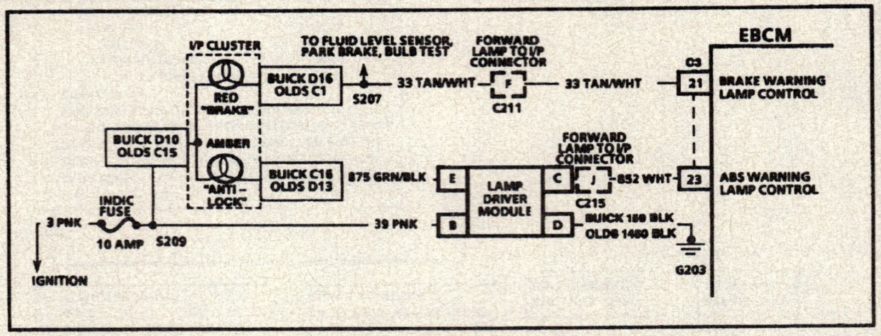

DTC 11

ABS WARNING LAMP CIRCUIT MALFUNCTION

Circuit Description:
This DTC checks the state of the ABS warning lamp to identify a situation in which the driver could not be warned of a system malfunction by the ABS warning lamp, or the warning lamp is always "ON." The Lamp Driver Module (LDM) turns the ABS warning lamp "ON," unless the EBCM provides ground to turn it "OFF." Because of the circuitry in the LDM, only external malfunctions can be detected. As a result, the ABS warning lamp itself is not diagnosable, only the control line to the EBCM can be diagnosed. In the event of an open CKT 852, the ABS warning lamp will be "ON" at all times, due to the loss of ground at the LDM input. If the control line is shorted to ground, the ABS warning lamp is kept "OFF," due to the LDM input being grounded.

DTC Will Set When: DTC 11 can be set only during the three second bulb check or when the ABS warning lamp is commanded "ON." If the EBCM cannot control the ABS warning lamp for two seconds, a malfunction exists.

Action Taken: A malfunction DTC is stored; however, ABS is not disabled.

DTC Chart Test Description: Number(s) below refer to circled number(s) on the diagnostic chart.

1. Checks to see if ABS lamp functions properly. Normal operation would indicate that the problem is not present.
2. Isolates the cause of the lamp remaining "ON" to a short to voltage, a short to the EBCM, or an open in CKT 39 which supplies voltage to the lamp driver module.
3. Determines if any other DTCs are present or if DTC 18 is set falsely.
4. Ensures that the circuitry from the EBCM to the lamp driver module is complete.
5. Determines whether the lamp remains "ON" due to a lamp driver malfunction or an open in CKT 852.
6. Checks to see if there is a short to voltage in CKT 852.
7. Checks for a malfunctioning EBCM.
8. Ensures that the ignition circuit to the lamp driver module is functional. An open indicator fuse or open CKT 3, 39, or 875 would result in inoperative warning lamp.
9. Isolates the cause of the ABS lamp remaining "OFF," by disconnecting the 24-way EBCM connector, CKT 852 should be open and cause the ABS lamp to be "ON."
10. Isolates the cause of the lamp remaining "OFF" to a short to ground in CKT 852 or a malfunctioning lamp driver module.
11. Checks for an open in CKT 39.
12. Determines if the "INDIC" fuse or CKT 3 is open, causing the I/P warning lamps to remain "OFF."
13. Checks for malfunctioning EBCM.

Diagnostic Aids: An "Intermittent" malfunction may be caused by a poor connection, rubbed through wire insulation, or a wire that is broken inside the insulation.

The lamp test function of the Tech 1 may be used to command the lamp "ON" while looking for an intermittent malfunction in the ABS warning lamp circuitry.

The frequency of the malfunction can be checked by using the enhanced diagnostic function of the Tech 1,

Any circuitry that is suspected as causing the intermittent complaint should be thoroughly checked for backed out terminals, improper mating, broken locks, improperly formed or damaged terminals, poor terminal to wiring connections or physical damage to the wiring harness.

DIAGNOSTIC CHART DTC 11 — 1994 CUTLASS SUPREME AND REGAL, CONTINUED

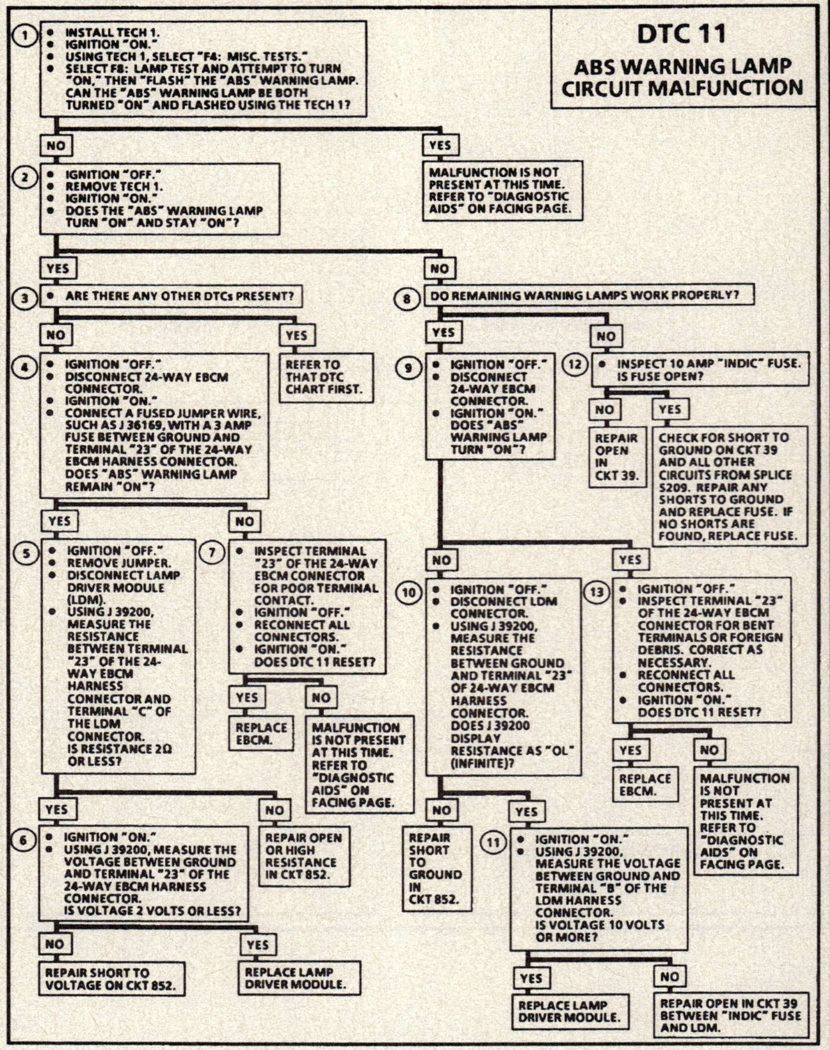

DIAGNOSTIC CHART DTC 14 — 1994 CUTLASS SUPREME AND REGAL

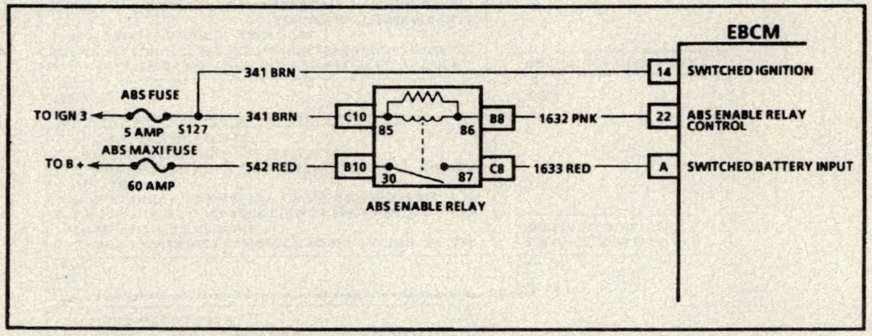

DTC 14

(Page 1 of 3)

ABS ENABLE RELAY CONTACT CIRCUIT OPEN

Circuit Description:

Ignition voltage is supplied through terminal "C10" of the ABS enable relay. The EBCM then is able to energize the pull-in coil by completing the ground circuit at terminal "22" of the EBCM. The magnetic field created closes the ABS enable relay contacts and allows battery voltage and current to be supplied to the EBCM, which supplies power to the motors and solenoids.

DTC Will Set When: DTC 14 can be set anytime after the EBCM commands the ABS enable relay "ON" (the relay is first commanded "ON" during the three second bulb check). This test monitors the availability of current/voltage to the motors and solenoids. This malfunction indicates voltage is not available and would therefore not allow ABS operation if required.

Action Taken: A malfunction DTC is stored, ABS is disabled and the ABS warning lamp is turned "ON."

DTC Chart Test Description: Number(s) below refer to circled number(s) on the diagnostic chart.

1. Checks to see if DTC 16 is also set as current or history. If DTC 16 is also set, proceed to that chart first.
2. Verifies that the malfunction is currently present.
3. Verifies that the relay can be turned "ON" and provide voltage to the EBCM.
4. Checks for high resistance in the switched battery circuit from the relay to the EBCM.
5. Checks for high resistance in the switched battery circuit from the battery to the relay.
6. Verifies the condition of the ABS enable relay.
7. Checks for poor terminal contact or corrosion at the connectors.
8. Determines if the EBCM is the cause of the malfunction.
9. Checks for high resistance in the entire relay coil circuit.
10. Verifies proper battery condition.
11. Verifies proper voltage is available at the ABS enable relay.
12. Checks for an open in the switched battery circuit from the relay to the EBCM.
13. Verifies the condition of the ABS enable relay.
14. Checks for proper resistance of the relay coil.
15. Checks for high resistance in the ABS enable relay control circuit between the relay and the EBCM.
16. Checks for high resistance in the ABS enable relay control circuit between the battery and the relay.

Diagnostic Aids: An "Intermittent" problem may be caused by a poor connection, rubbed through wire insulation, or a wire that is broken inside the insulation.

The frequency of the problem can be checked by using the enhanced diagnostic function of the Tech 1,

Any circuitry that is suspected as causing the "Intermittent" complaint should be thoroughly checked for backed out terminals, improper mating, broken locks, improperly formed or damaged terminals, poor terminal to wiring connections or physical damage to the wiring harness.

VIBRATION, TEMPERATURE EFFECTS:

Check for vibration effects by performing the relay test function of the Tech 1. With the relay test commanded "ON," lightly tap the top and sides of the relay while monitoring relay voltage. If the relay voltage changes significantly, replace the relay.

If DTC 14 only sets when the vehicle is initially started in cold ambient conditions (temperature less than 32°F-0°C), replace the relay.

DIAGNOSTIC CHART DTC 14 — 1994 CUTLASS SUPREME AND REGAL, CONTINUED

DTC 14

(Page 1 of 3)

ABS ENABLE RELAY CONTACT CIRCUIT OPEN

(1) IS DTC 16 ALSO PRESENT OR IN HISTORY?

- YES: PROCEED TO DTC 16 CHART FIRST.
- NO:

(2)
- IGNITION "ON," ENGINE "OFF."
- USING TECH 1, SELECT DATA LIST AND OBSERVE ABS BATTERY VOLTAGE .

DOES TECH 1 INDICATE THAT ABS BATTERY VOLTAGE IS GREATER THAN 10 VOLTS?

- YES: MALFUNCTION IS NOT PRESENT AT THIS TIME. REFER TO "DIAGNOSTIC AIDS" ON FACING PAGE.
- NO:

(3)
- IGNITION "OFF."
- DISCONNECT 24-WAY AND 2-WAY EBCM CONNECTORS.
- CONNECT A FUSED JUMPER, SUCH AS J 36169 WITH A 3 AMP FUSE BETWEEN GROUND AND TERMINAL "22" OF THE 24-WAY EBCM HARNESS CONNECTOR.
- IGNITION "ON."
- USING J 39200, MEASURE THE VOLTAGE BETWEEN GROUND AND TERMINAL "A" OF THE 2-WAY EBCM HARNESS CONNECTOR.

IS VOLTAGE 10 VOLTS OR GREATER?

- NO: DIAGNOSIS CONTINUED ON NEXT PAGE.
- YES:

(4)
- IGNITION "OFF."
- DISCONNECT JUMPER USED IN STEP 3.
- DISCONNECT ABS ENABLE RELAY.
- USING J 39200, MEASURE RESISTANCE BETWEEN TERMINAL "C8" OF THE ABS ENABLE RELAY HARNESS CONNECTOR AND TERMINAL "A" OF THE 2-WAY EBCM HARNESS CONNECTOR.

IS RESISTANCE 2Ω OR LESS?

- NO: REPAIR HIGH RESISTANCE IN CKT 1633.
- YES:

(5)
- DISCONNECT NEGATIVE BATTERY CABLE.
- DISCONNECT POSITIVE BATTERY CABLE.
- USING J 39200, MEASURE THE RESISTANCE BETWEEN POSITIVE BATTERY CABLE TERMINAL AND TERMINAL "B10" OF THE ABS ENABLE RELAY HARNESS CONNECTOR.

IS RESISTANCE 2Ω OR LESS?

- NO: REPAIR HIGH RESISTANCE IN CKT 542.
- YES:

(6)
- USING A JUMPER WIRE, CONNECT TERMINAL "86" ON THE ABS ENABLE RELAY TO GROUND.
- USING A FUSED JUMPER WIRE, SUCH AS J 36169, WITH A 3 AMP FUSE, CONNECT TERMINAL "85" ON THE ABS ENABLE RELAY TO VOLTAGE.
- USING J 39200, MEASURE THE RESISTANCE BETWEEN TERMINALS "30" AND "87" ON THE ABS ENABLE RELAY.

IS RESISTANCE 2Ω OR LESS?

- NO: REPLACE ABS ENABLE RELAY.
- YES:

(7)
- INSPECT ABS ENABLE RELAY AND ABS ENABLE RELAY HARNESS CONNECTOR FOR A POOR CONNECTION OR POOR TERMINAL CONTACT.
- INSPECT EBCM AND EBCM CONNECTOR TERMINALS "22" AND "A" FOR POOR TERMINAL CONTACT.
- INSPECT POSITIVE BATTERY TERMINAL AND POSITIVE BATTERY CABLE TERMINAL FOR A POOR CONNECTION.

IS POOR TERMINAL CONTACT OR A POOR CONNECTION EVIDENT?

- YES: REPLACE TERMINALS THAT EXHIBIT POOR TERMINAL CONTACT OR EVIDENCE OF CORROSION.
- NO:

(8)
- RECONNECT ALL CONNECTORS.
- IGNITION "ON."

DOES DTC 14 RESET?

- YES: REPLACE EBCM.
- NO: MALFUNCTION IS INTERMITTENT. REFER TO "DIAGNOSTIC AIDS" ON FACING PAGE.

DIAGNOSTIC CHART DTC 14 — 1994 CUTLASS SUPREME AND REGAL, CONTINUED

DTC 14
(Page 2 of 3)
ABS ENABLE RELAY CONTACT CIRCUIT OPEN

DIAGNOSIS CONTINUED FROM PREVIOUS PAGE.

9.
- IGNITION "OFF."
- REMOVE JUMPER WIRE INSTALLED IN STEP #3, IF NOT ALREADY REMOVED.
- DISCONNECT NEGATIVE BATTERY CABLE.
- DISCONNECT POSITIVE BATTERY CABLE.
- WITH BATTERY STILL DISCONNECTED, TURN IGNITION SWITCH TO "RUN." THIS COMPLETES THE IGNITION CIRCUIT TO THE ABS ENABLE RELAY.
- USING J 39200, MEASURE THE RESISTANCE BETWEEN TERMINAL "22" OF THE 24-WAY EBCM HARNESS CONNECTOR AND THE POSITIVE BATTERY CABLE TERMINAL. IS RESISTANCE 95Ω OR LESS WHEN MEASURED AT APPROXIMATELY 20°C (68°F)?

YES → Step 10. NO → DIAGNOSIS CONTINUED ON NEXT PAGE.

10.
- USING J 39200, MEASURE THE VOLTAGE BETWEEN THE NEGATIVE AND POSITIVE BATTERY TERMINALS. IS VOLTAGE 10 VOLTS OR GREATER?

YES → Step 11. NO → CORRECT LOW VOLTAGE CONDITION.

11.
- IGNITION "OFF."
- RECONNECT POSITIVE BATTERY CABLE.
- RECONNECT NEGATIVE BATTERY CABLE.
- DISCONNECT ABS ENABLE RELAY.
- MEASURE THE VOLTAGE BETWEEN TERMINAL "B10" OF THE ABS ENABLE RELAY HARNESS CONNECTOR AND GROUND. IS VOLTAGE 10 VOLTS OR GREATER?

YES → Step 12. NO → REPAIR OPEN OR HIGH RESISTANCE IN CKT 542.

12.
- USING J 39200, MEASURE THE RESISTANCE BETWEEN TERMINAL "C8" OF THE ABS ENABLE RELAY HARNESS CONNECTOR AND TERMINAL "A" OF THE 2-WAY EBCM HARNESS CONNECTOR. IS RESISTANCE 2Ω OR LESS?

YES → Step 13. NO → REPAIR OPEN OR HIGH RESISTANCE IN CKT 1633.

13.
- USING A JUMPER WIRE, CONNECT TERMINAL "86" ON THE ABS ENABLE RELAY TO GROUND.
- USING A FUSED JUMPER WIRE, SUCH AS J 36169, WITH A 3 AMP FUSE, CONNECT TERMINAL "85" ON THE ABS ENABLE RELAY TO VOLTAGE.
- USING J 39200, MEASURE THE RESISTANCE BETWEEN TERMINALS "30" AND "87" ON THE ABS ENABLE RELAY. IS RESISTANCE 2Ω OR LESS?

YES → MALFUNCTION WAS MOST LIKELY DUE TO A POOR CONNECTION. INSPECT ALL CONNECTORS AND TERMINALS FOR POOR TERMINAL CONTACT AND EVIDENCE OF CORROSION. REPLACE ALL TERMINALS THAT EXHIBIT EVIDENCE OF POOR TERMINAL CONTACT OR CORROSION.

NO → REPLACE ABS ENABLE RELAY.

DIAGNOSTIC CHART DTC 14 — 1994 CUTLASS SUPREME AND REGAL, CONTINUED

DTC 14
(Page 3 of 3)
ABS ENABLE RELAY CONTACT CIRCUIT OPEN

DIAGNOSIS CONTINUED FROM PREVIOUS PAGE.

14.
- DISCONNECT ABS ENABLE RELAY.
- USING J 39200, MEASURE THE RESISTANCE BETWEEN TERMINALS "85" AND "86" ON THE ABS ENABLE RELAY. IS RESISTANCE 95Ω OR LESS WHEN MEASURED AT APPROXIMATELY 20°C (68°F)?

YES → Step 15. NO → REPLACE ABS ENABLE RELAY.

15.
- MEASURE THE RESISTANCE BETWEEN TERMINAL "B8" OF THE ABS ENABLE RELAY HARNESS CONNECTOR AND TERMINAL "22" OF THE 24-WAY EBCM HARNESS CONNECTOR. IS RESISTANCE 2Ω OR LESS?

YES → Step 16. NO → REPAIR OPEN OR HIGH RESISTANCE IN CKT 1632.

16.
- MEASURE THE RESISTANCE BETWEEN TERMINAL "C10" OF THE ABS ENABLE RELAY HARNESS CONNECTOR AND THE POSITIVE BATTERY CABLE TERMINAL. IS RESISTANCE 2Ω OR LESS?

YES → MALFUNCTION WAS MOST LIKELY DUE TO A POOR CONNECTION. INSPECT ALL CONNECTORS AND TERMINALS FOR POOR TERMINAL CONTACT AND EVIDENCE OF CORROSION. REPLACE ALL TERMINALS THAT EXHIBIT EVIDENCE OF POOR TERMINAL CONTACT OR CORROSION.

NO → REPAIR OPEN OR HIGH RESISTANCE IN CKT 341, CKT 300, IGNITION SWITCH, OR BATTERY FEED CIRCUIT TO IGNITION SWITCH.

DIAGNOSTIC CHART DTC 15 — 1994 CUTLASS SUPREME AND REGAL

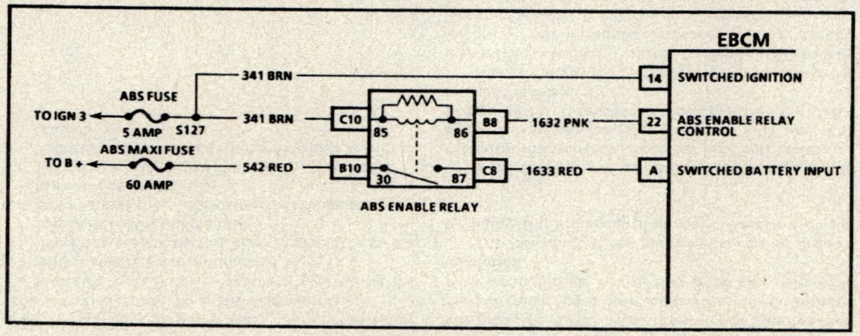

DTC 15

ABS RLY. CONTACT CKT SHORT TO BATT OR ALWAYS CLOSED

Circuit Description:

Ignition voltage is supplied through terminal "C10" of the ABS enable relay. The EBCM then is able to energize the pull-in coil by completing the ground circuit at terminal "22" of the EBCM. The magnetic field created closes the ABS enable relay contacts and allows battery voltage and current to be supplied to the EBCM, which supplies power to the motors and solenoids.

DTC Will Set When: DTC 15 can be set only before the EBCM commands the relay "ON." This test determines if the ABS enable relay is energized when it should not be. This malfunction would not allow the ABS enable relay to remove power to the ABS system. If a second malfunction were to occur that requires the ABS enable relay to be turned "OFF," that malfunction cannot be removed if the relay cannot be controlled. The malfunction must be present for three consecutive drive cycles before the DTC is set.

Action Taken: A malfunction DTC is stored. ABS is not disabled.

DTC Chart Test Description: Number(s) below refer to circled number(s) on the diagnostic chart.

1. Checks to see if other DTCs are set.
2. Indicates that the EBCM is capable of controlling the enable relay as commanded.
3. Checks for a possible short to battery voltage in CKT 1633.
4. Determines whether the short to voltage is in CKT 1633 or due to a malfunctioning relay.
5. Checks for a possible intermittent in wiring or connectors.

Diagnostic Aids: An "Intermittent" malfunction may be caused by a poor connection, rubbed through wire insulation, or a wire that is broken inside the insulation.

The frequency of the malfunction can be checked by using the enhanced diagnostic function of the Tech 1.

Any circuitry, that is suspected as causing the intermittent complaint, should be thoroughly checked for backed out terminals, improper mating, broken locks, improperly formed or damaged terminals, poor terminal to wiring connections or physical damage to the wiring harness.

DIAGNOSTIC CHART DTC 15 — 1994 CUTLASS SUPREME AND REGAL, CONTINUED

DTC 15

ABS RLY. CONTACT CKT SHORT TO BATT OR ALWAYS CLOSED

(1) ARE ANY OTHER DTCs PRESENT?

- YES: PROCEED TO THAT DTC CHART FIRST.
- NO:

(2)
- IGNITION "ON," ENGINE "OFF."
- USING THE RELAY TEST FUNCTION OF THE TECH 1, COMMAND ABS ENABLE RELAY "OFF."

DOES TECH 1 INDICATE THAT ABS ENABLE RELAY IS "OFF" AND BATTERY VOLTAGE IS LESS THAN 5 VOLTS?

- YES: MALFUNCTION IS NOT PRESENT AT THIS TIME. REFER TO "DIAGNOSTIC AIDS" ON FACING PAGE.
- NO:

(3)
- IGNITION "OFF."
- DISCONNECT 24-WAY AND 2-WAY EBCM HARNESS CONNECTORS.
- IGNITION "ON."
- USING J 39200, MEASURE THE VOLTAGE BETWEEN GROUND AND TERMINAL "A" OF THE EBCM 2-WAY CONNECTOR.

IS VOLTAGE 2 VOLTS OR LESS?

- NO:

(4)
- IGNITION "OFF."
- DISCONNECT ABS ENABLE RELAY.
- IGNITION "ON."
- MEASURE THE VOLTAGE BETWEEN GROUND AND TERMINAL "A" OF THE 2-WAY EBCM HARNESS CONNECTOR.

IS VOLTAGE 2 VOLTS OR LESS?

- YES: REPLACE ABS ENABLE RELAY.
- NO: REPAIR SHORT TO VOLTAGE IN CKT 850 (CHEV) OR 1633 (BUICK/OLDS).

- YES (from 3):

(5)
- IGNITION "OFF."
- INSPECT CKT 850 (CHEV) OR 1633 (BUICK/OLDS) AND 2-WAY EBCM HARNESS CONNECTOR FOR DAMAGE WHICH MAY RESULT IN A SHORT TO VOLTAGE WHEN ALL CONNECTORS ARE CONNECTED. REPAIR DAMAGE IF EVIDENT.
- RECONNECT 2-WAY AND 24-WAY EBCM CONNECTORS.
- IGNITION "ON."
- USING RELAY TEST FUNCTION OF TECH 1, COMMAND ABS ENABLE RELAY "OFF."

DOES TECH 1 INDICATE THAT ABS ENABLE RELAY IS "OFF" AND BATTERY VOLTAGE IS LESS THAN 5 VOLTS?

- NO: REPLACE EBCM.
- YES: MALFUNCTION IS INTERMITTENT. REFER TO "DIAGNOSTIC AIDS" ON FACING PAGE.

AFTER ALL DIAGNOSIS IS COMPLETE, CLEAR DTCs, AND TEST DRIVE VEHICLE FOR 3 DRIVE CYCLES TO VERIFY THAT DTC DOES NOT RESET. A DRIVE CYCLE CONSISTS OF STARTING THE VEHICLE, DRIVING ABOVE 16 km/h (10 MPH), AND THEN KEYING DOWN.

DIAGNOSTIC CHART DTC 16 — 1994 CUTLASS SUPREME AND REGAL

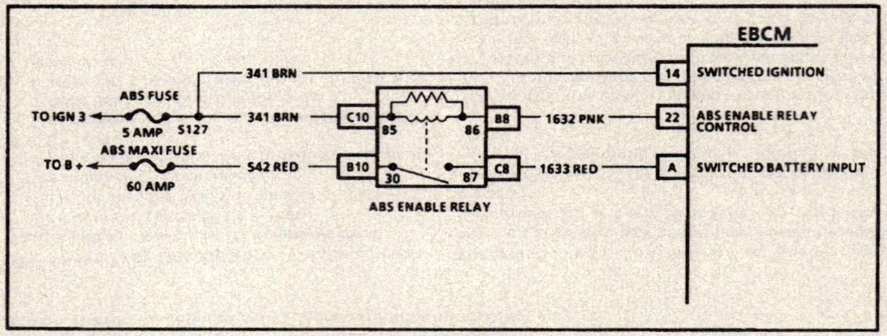

DTC 16

ABS ENABLE RELAY COIL CIRCUIT OPEN

Circuit Description:

Ignition voltage is supplied through terminal "C10" of the ABS enable relay. The EBCM then is able to energize the pull-in coil by completing the ground circuit at terminal "22" of the EBCM. The magnetic field created closes the ABS enable relay contacts and allows battery voltage and current to be supplied to the EBCM, which supplies power to the motors and solenoids.

DTC Will Set When: DTC 16 can be set only after DTC 14 has been set. This test detects an open in the ABS enable relay coil circuit. An open in this circuit will not allow the ABS enable relay to be energized thus preventing voltage/current to the motors and solenoids. If this malfunction is present and the ignition is turned "OFF" before 5 km/h (3 mph) is reached, DTC 14 is set alone.

Action Taken: A malfunction DTC is stored, ABS is disabled and the ABS warning lamp is turned "ON."

DTC Chart Test Description: Number(s) below refer to circled number(s) on the diagnostic chart.

1. Indicates the EBCM is capable of controlling the ABS enable relay as commanded.
2. Checks to ensure that voltage is available to the pull-in coil of the enable relay.
3. Ensures that there is continuity through the pull-in coil of the enable relay.
4. Verifies the integrity of the relay control circuit.
5. Ensures malfunction was not due to poor terminal contact.

Diagnostic Aids: An "Intermittent" malfunction may be caused by a poor connection, rubbed through wire insulation, or a wire that is broken inside the insulation.

The frequency of the malfunction can be checked by using the enhanced diagnostic function of the Tech 1.

If the frequency of the malfunction is high, but is currently intermittent, check for high coil resistance by measuring between relay terminals "C10" and "B8" using J 39200. If resistance shows greater than 100Ω, replace the relay.

Any circuitry, that is suspected as causing the intermittent complaint, should be thoroughly checked for backed out terminals, improper mating, broken locks, improperly formed or damaged terminals, poor terminal to wiring connections or physical damage to the wiring harness.

DIAGNOSTIC CHART DTC 16 — 1994 CUTLASS SUPREME AND REGAL, CONTINUED

DTC 16

ABS ENABLE RELAY COIL CIRCUIT OPEN

(1)
- IGNITION "ON," ENGINE "OFF."
- USING THE RELAY TEST FUNCTION OF THE TECH 1, COMMAND THE ABS ENABLE RELAY "ON."

DOES TECH 1 INDICATE ABS ENABLE RELAY "ON," AND BATTERY VOLTAGE ABOVE 10 VOLTS?

- YES: MALFUNCTION IS NOT PRESENT AT THIS TIME. REFER TO "DIAGNOSTIC AIDS" ON FACING PAGE.
- NO: go to (2)

(2)
- USING J 39200, MEASURE THE VOLTAGE BETWEEN GROUND AND TERMINAL "C10" OF THE ABS ENABLE RELAY CONNECTOR.

IS VOLTAGE 10 VOLTS OR GREATER?

- NO: REPAIR OPEN OR HIGH RESISTANCE IN CKT 341 OR ABS FUSE.
- YES: go to (3)

(3)
- USING THE RELAY TEST FUNCTION OF THE TECH 1, COMMAND THE ABS ENABLE RELAY "OFF."
- MEASURE THE VOLTAGE BETWEEN GROUND AND TERMINAL "B8" OF THE ABS ENABLE RELAY CONNECTOR.

IS VOLTAGE 10 VOLTS OR GREATER?

- NO: CHECK FOR INTERMITTENT CONNECTIONS AT ABS ENABLE RELAY. IF OK, REPLACE ABS RELAY.
- YES: go to (4)

(4)
- IGNITION "OFF."
- DISCONNECT 24-WAY EBCM CONNECTOR.
- IGNITION "ON."
- MEASURE THE VOLTAGE BETWEEN GROUND AND TERMINAL "22" OF THE 24-WAY EBCM HARNESS CONNECTOR.

IS VOLTAGE 10 VOLTS OR GREATER?

- NO: REPAIR OPEN OR HIGH RESISTANCE IN CKT 1632.
- YES: go to (5)

(5)
- IGNITION "OFF."
- INSPECT 24-WAY EBCM CONNECTOR TERMINAL "22" FOR POOR TERMINAL CONTACT OR EVIDENCE OF CORROSION.

REPLACE ALL TERMINALS THAT EXHIBIT SIGNS OF POOR TERMINAL CONTACT OR CORROSION.

- RECONNECT ALL CONNECTORS.
- IGNITION "ON."

DOES DTC 16 SET AS A CURRENT DTC?

- YES: REPLACE EBCM.
- NO: MALFUNCTION WAS MOST LIKELY CAUSED BY A POOR CONNECTION. REFER TO "DIAGNOSTIC AIDS" ON FACING PAGE.

AFTER ALL DIAGNOSIS IS COMPLETE, CLEAR DTCs AND TEST DRIVE VEHICLE FOR THREE (3) DRIVE CYCLES TO VERIFY THAT DTC DOES NOT RESET. A DRIVE CYCLE CONSISTS OF STARTING THE VEHICLE, DRIVING OVER 16 km/h (10 MPH), AND THEN KEYING DOWN.

DIAGNOSTIC CHART DTC 17 — 1994 CUTLASS SUPREME AND REGAL

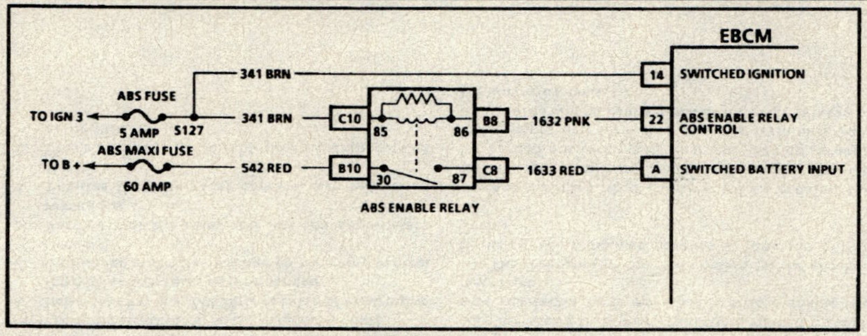

DTC 17

ABS ENABLE RELAY COIL CIRCUIT SHORTED TO GND

Circuit Description:
Ignition voltage is supplied through terminal "C10" of the ABS enable relay. The EBCM then is able to energize the pull-in coil by completing the ground circuit at terminal "22" of the EBCM. The magnetic field created closes the ABS enable relay contacts and allows battery voltage and current to be supplied to the EBCM, which supplies voltage to the motors and solenoids.

DTC Will Set When: DTC 17 can be set before the EBCM commands the ABS enable relay "ON." This test determines if the ABS enable relay is energized when it should not be. This malfunction would not allow the ABS enable relay to remove power to the ABS system. If a second malfunction were to occur that requires the ABS enable relay to be turned "OFF," that malfunction cannot be removed if the relay cannot be controlled.

Action Taken: A malfunction DTC is stored. ABS is not disabled.

DTC Chart Test Description: Number(s) below refer to circled number(s) on the diagnostic chart.

1. Indicates that the EBCM is capable of controlling the ABS enable relay as commanded.
2. Checks to ensure that the ABS enable relay or control CKT 1632 are not shorted to ground.
3. Checks to see if the EBCM is internally shorted to ground. A grounded terminal "22" would cause the ABS enable relay to be energized any time the ignition was in the "ON" position.
4. Checks for a short to ground in control CKT 1632 or a defective relay.

Diagnostic Aids: An "Intermittent" malfunction may be caused by a poor connection, rubbed through wire insulation, or a wire that is broken inside the insulation.

The frequency of the malfunction can be checked by using the enhanced diagnostic function of the Tech 1,

Any circuitry, that is suspected as causing the intermittent complaint, should be thoroughly checked for backed out terminals, improper mating, broken locks, improperly formed or damaged terminals, poor terminal to wiring connections or physical damage to the wiring harness.

DIAGNOSTIC CHART DTC 17 — 1994 CUTLASS SUPREME AND REGAL, CONTINUED

DTC 17

ABS ENABLE RELAY COIL CIRCUIT SHORTED TO GND

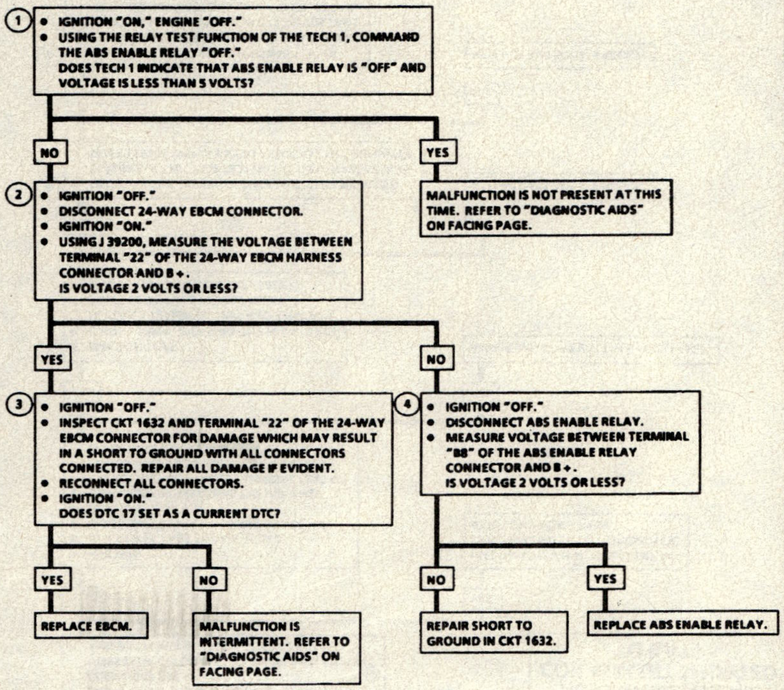

AFTER ALL DIAGNOSIS IS COMPLETE, CLEAR DTCs AND TEST DRIVE VEHICLE FOR THREE (3) DRIVE CYCLES TO VERIFY THAT DTC DOES NOT RESET. A DRIVE CYCLE CONSISTS OF STARTING THE VEHICLE, DRIVING OVER 16 km/h (10 MPH), AND THEN KEYING DOWN.

DIAGNOSTIC CHART DTC 18 — 1994 CUTLASS SUPREME AND REGAL

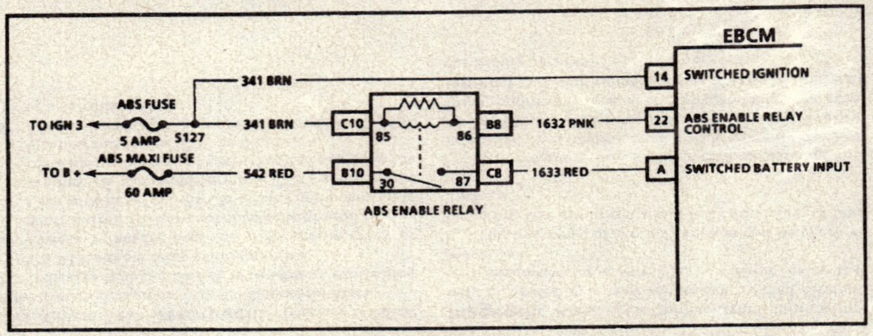

DTC 18

ABS ENABLE RELAY COIL CIRCUIT SHORTED TO BATT

Circuit Description:

Ignition voltage is supplied through terminal "C10" of the ABS enable relay. The EBCM then is able to energize the pull-in coil by completing the ground circuit at terminal "22" of the EBCM. The magnetic field created closes the ABS enable relay contacts and allows battery voltage and current to be supplied to the EBCM, which supplies voltage to the motors and solenoids.

DTC Will Set When: DTC 18 can be set after the EBCM commands the enable relay "ON." This test monitors the availability of current/voltage to the motors and solenoids. A malfunction of this type will not allow voltage to be available to the motors and solenoids; therefore, ABS operation would not be allowed if required.

Action Taken: A malfunction DTC is stored, ABS is disabled and the ABS warning lamp is turned "ON." The red BRAKE Warning Lamp is turned "ON" if the rear piston in the ABS Motor Pack is not in the home position.

DTC Chart Test Description: Number(s) below refer to circled number(s) on the diagnostic chart.

1. Indicates that the EBCM is capable of controlling the ABS enable relay as commanded.
2. Checks for a short to voltage on the relay control circuit.
3. Verifies that the relay coil has the appropriate resistance.
4. Verifies that the relay contacts are not failed closed.
5. This step checks for possible poor terminal contact at the EBCM.

Diagnostic Aids: An "Intermittent" malfunction may be caused by a poor connection, rubbed through wire insulation, or a wire that is broken inside the insulation.

The frequency of the malfunction can be checked by using the enhanced diagnostic function of the Tech 1.

Any circuitry, that is suspected as causing the intermittent complaint, should be thoroughly checked for backed out terminals, improper mating, broken locks, improperly formed or damaged terminals, poor terminal to wiring connections or physical damage to the wiring harness.

DIAGNOSTIC CHART DTC 18 — 1994 CUTLASS SUPREME AND REGAL, CONTINUED

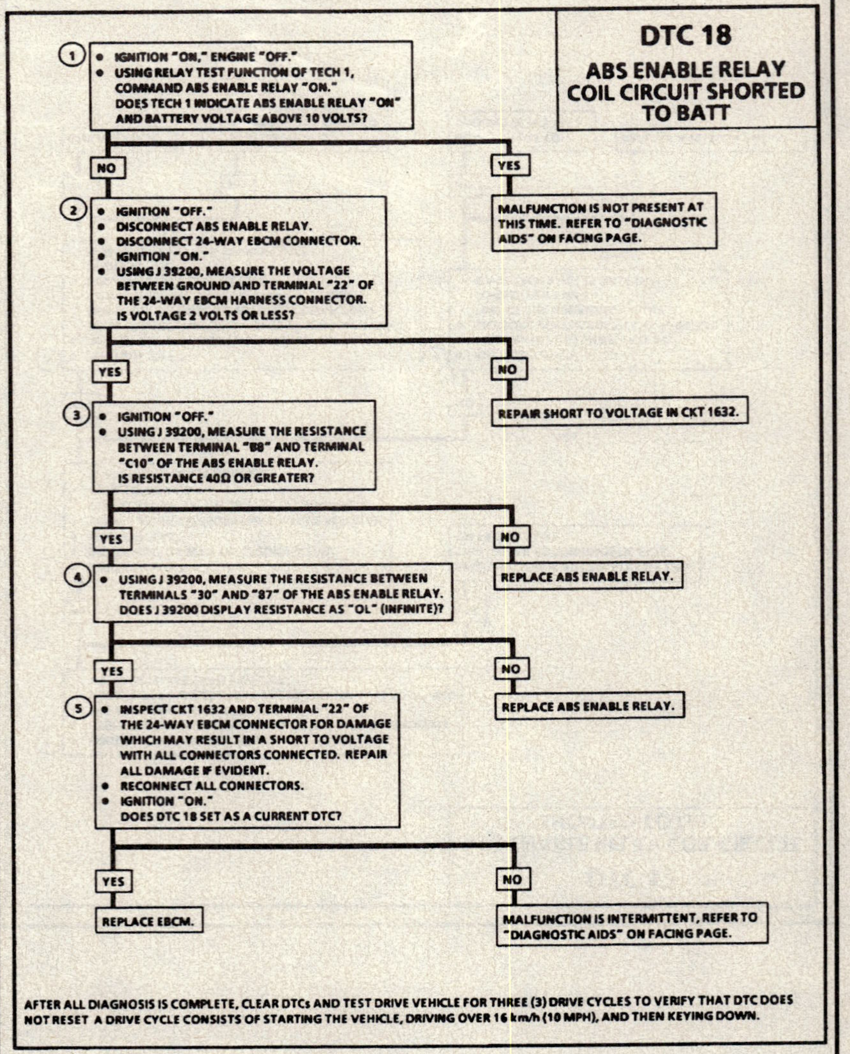

DIAGNOSTIC CHART DTC 21 — 1994 CUTLASS SUPREME AND REGAL

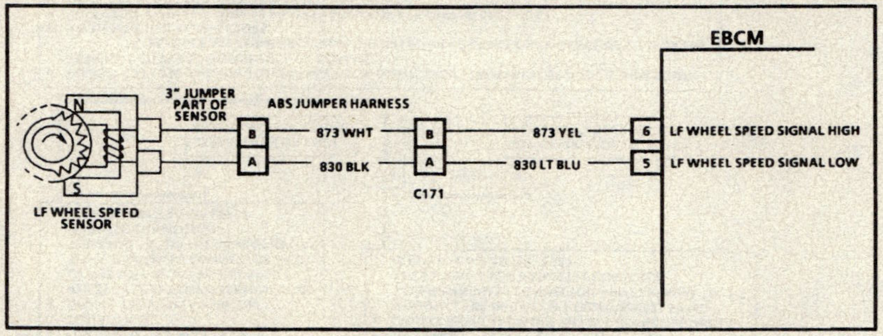

DTC 21

LEFT FRONT WHEEL SPEED = 0

Circuit Description:
As a toothed ring passes by the wheel speed sensor, changes in the electromagnetic field cause the wheel speed sensor to produce a sinusoidal (AC) voltage signal whose frequency is proportional to wheel speed. The magnitude of this signal is directly related to wheel speed and the proximity of the wheel speed sensor to the toothed ring, often referred to as the air gap.

DTC Will Set When: DTC 21 can be set when the vehicle is not in an ABS stop. If the left front wheel speed = 0 and the vehicle's reference speed is greater than 8 km/h (5 mph), a malfunction exists.

Action Taken: A malfunction DTC is stored, ABS is disabled and the ABS warning lamp is turned "ON."

DTC Chart Test Description: Number(s) below refer to circled number(s) on the diagnostic chart.

1. Determines if DTC 32 is present.
2. Designed to identify a wheel speed sensor or circuitry that is damaged and visibly apparent.
3. Checks to see that the wheel speed sensor is within proper resistance values.
4. Verifies that the wheel speed sensor can generate an appropriate output signal.
5. Checks for proper output voltage from the EBCM.
6. Checks for an open in the wheel speed sensor low circuit.
7. Ensures malfunction was not due to physical damage of the circuitry.
8. Ensures that wheel speed sensor circuitry is not shorted.
9. Ensures malfunction was not due to physical damage of the circuitry.
10. Checks for an open in the wheel speed sensor high circuit.

Diagnostic Aids: An "Intermittent" malfunction may be caused by a poor connection, rubbed through wire insulation, or a wire that is broken inside the insulation.

The frequency of the malfunction can be checked by using the enhanced diagnostic function of the Tech 1.

If the customer's comments reflect that the ABS warning lamp is "ON" only during moist environmental changes (rain, snow, vehicle wash), all wheel speed sensor circuitry should be thoroughly inspected for signs of water intrusion. If DTC is not current, clear DTCs and simulate the effects of water intrusion. Use the following procedure. Spray down the suspected area with a 5% salt water solution (two teaspoons of salt to 12 oz. of water). Test drive vehicle over various road surfaces (bumps, turns, etc.) above 24 km/h (15 mph) for at least 30 seconds. If DTC returns, replace suspected harness.

Any circuitry, that is suspected as causing the intermittent complaint, should be thoroughly checked for backed out terminals, improper mating, broken locks, improperly formed or damaged terminals, poor terminal to wiring connections or physical damage to the wiring harness.

Resistance of the wheel speed sensor will increase with an increase in sensor temperature.

When replacing a wheel speed sensor, inspect the sensor terminals and harness connector for corrosion and/or water intrusion. If evidence of corrosion or water intrusion exists, replace wheel speed sensor jumper harness. Likewise, if replacing a wheel speed sensor jumper harness, inspect sensor terminals. If evidence of corrosion or water intrusion exists, replace wheel speed sensor. Refer to "On-Vehicle Service" in this section.

DIAGNOSTIC CHART DTC 21 — 1994 CUTLASS SUPREME AND REGAL, CONTINUED

DTC 21
(Page 1 of 2)
LEFT FRONT WHEEL SPEED = 0

Important
- WHEEL SPEED SENSOR INTERMITTENT PROBLEMS MAY BE DIFFICULT TO LOCATE. CARE SHOULD BE TAKEN NOT TO DISTURB ANY ELECTRICAL CONNECTIONS PRIOR TO AN INDICATED STEP OF THIS CHART. THIS WILL ENSURE THAT AN INTERMITTENT CONNECTION WILL NOT BE CORRECTED BEFORE THE SOURCE OF THE PROBLEM IS FOUND.

(1) IS DTC 32 ALSO CURRENTLY SET?
- YES: PROCEED WITH THAT DTC CHART FIRST.
- NO: go to (2).

(2)
- IGNITION "OFF."
- PHYSICALLY INSPECT THE WHEEL SPEED SENSOR WIRING AND CONNECTORS FOR DAMAGE. INSPECT THE WHEEL SPEED SENSOR AND TOOTHED RING FOR LOOSENESS OR DAMAGE. PAY PARTICULAR ATTENTION TO THE TOOTHED RING, ANY DEVIATION WILL AFFECT THE WHEEL SPEED SENSOR OUTPUT SIGNAL.

IS PHYSICAL DAMAGE INDICATED?
- YES: REPAIR DAMAGE OR SENSOR/TOOTHED RING LOOSENESS. DRIVE VEHICLE ABOVE 24 km/h (15 MPH) TO ENSURE DTC DOES NOT SET.
- NO: go to (3).

(3)
- DISCONNECT LEFT FRONT WHEEL SPEED SENSOR DIRECTLY AT SENSOR.
- USING J 39200, MEASURE THE RESISTANCE BETWEEN SENSOR TERMINALS "A" AND "B".

IS RESISTANCE BETWEEN 1661 AND 1929Ω WHEN SENSOR IS 20°C (68°F)? *(IF TEMPERATURE OF THE SENSOR IS SUSPECTED TO BE OTHER THAN 20°C (68°F), REFER TO THE TABLE BELOW FOR APPLICABLE SENSOR RESISTANCE VALUES; THE VALUES IN THE TABLE ARE FOR TEMPERATURE OF THE SENSOR, NOT AIR TEMPERATURE.)*
- NO: REPLACE WHEEL SPEED SENSOR.
- YES: go to (4).

(4)
- WITH J 39200 STILL CONNECTED, SELECT A/C VOLTAGE SCALE.
- SPIN WHEEL BY HAND WHILE OBSERVING VOLTAGE.

IS VOLTAGE AT LEAST 100 mV (VOLTAGE WILL INCREASE AS WHEEL SPEED INCREASES)?
- YES: DIAGNOSIS CONTINUED ON PAGE 2 OF 2 OF THIS DTC CHART.
- NO: VERIFY THAT TOOTHED RING IS INTACT. IF OK, REPLACE WHEEL SPEED SENSOR.

FRONT WHEEL SPEED SENSOR

TEMPERATURE OF SENSOR VS. SENSOR RESISTANCE (APPROXIMATE)

TEMP. (°C)	TEMP. (°F)	RESISTANCE (OHMS)
-34 TO 4	-30 TO 40	1309 TO 1808
5 TO 43	41 TO 110	1557 TO 2103
44 TO 93	111 TO 200	1811 TO 2482

DIAGNOSTIC CHART DTC 21 — 1994 CUTLASS SUPREME AND REGAL, CONTINUED

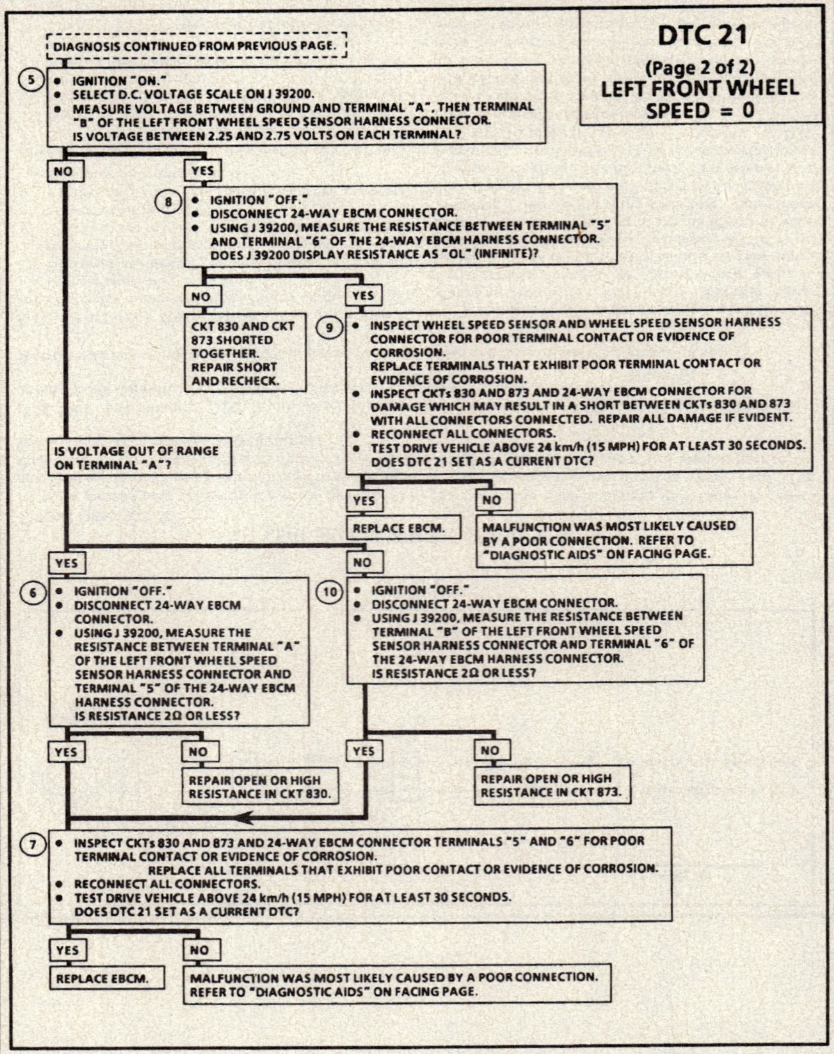

DIAGNOSTIC CHART DTC 22 — 1994 CUTLASS SUPREME AND REGAL

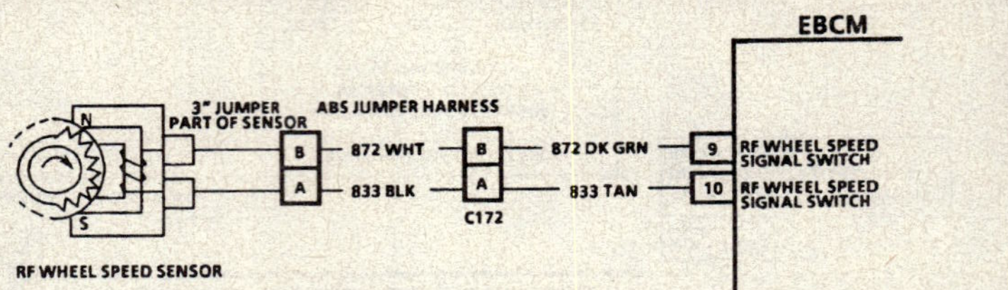

DTC 22

RIGHT FRONT WHEEL SPEED = 0

Circuit Description:

As a toothed ring passes by the wheel speed sensor, changes in the electromagnetic field cause the wheel speed sensor to produce a sinusoidal (AC) voltage signal whose frequency is proportional to wheel speed. The magnitude of this signal is directly related to wheel speed and the proximity of the wheel speed sensor to the toothed ring, often referred to as the air gap.

DTC Will Set When: DTC 22 can be set when the vehicle is not in an ABS stop. If the right front wheel speed = 0 and the vehicle's reference speed is greater than 8 km/h (5 mph), a malfunction exists.

Action Taken: A malfunction DTC is stored, ABS is disabled and the ABS warning lamp is turned "ON."

DTC Chart Test Description: Number(s) below refer to circled number(s) on the diagnostic chart.

1. Determines if DTC 33 is present.
2. Designed to identify a wheel speed sensor or circuitry that is damaged and visibly apparent.
3. Checks to see that the wheel speed sensor is within proper resistance values.
4. Verifies that the wheel speed sensor can generate an appropriate output signal.
5. Checks for proper output voltage from the EBCM.
6. Checks for an open in the wheel speed sensor low circuit.
7. Ensures malfunction was not due to physical damage of the circuitry.
8. Ensures that wheel speed sensor circuitry is not shorted.
9. Ensures malfunction was not due to physical damage of the circuitry.
10. Checks for an open in the wheel speed sensor high circuit.

Diagnostic Aids: An "Intermittent" malfunction may be caused by a poor connection, rubbed through wire insulation, or a wire that is broken inside the insulation.

The frequency of the malfunction can be checked by using the enhanced diagnostic function of the Tech 1.

If the customer's comments reflect that the ABS warning lamp is "ON" only during moist environmental changes (rain, snow, vehicle wash), all wheel speed sensor circuitry should be thoroughly inspected for signs of water intrusion. If DTC is not current, clear DTCs and simulate the effects of water intrusion. Use the following procedure. Spray down the suspected area with a 5% salt water solution (two teaspoons of salt to 12 oz. of water). Test drive vehicle over various road surfaces (bumps, turns, etc.) above 24 km/h (15 mph) for at least 30 seconds. If DTC returns, replace suspected harness.

Any circuitry, that is suspected as causing the intermittent complaint, should be thoroughly checked for backed out terminals, improper mating, broken locks, improperly formed or damaged terminals, poor terminal to wiring connections or physical damage to the wiring harness.

Resistance of the wheel speed sensor will increase with an increase in sensor temperature.

When replacing a wheel speed sensor, inspect the sensor terminals and harness connector for corrosion and/or water intrusion. If evidence of corrosion or water intrusion exists, replace wheel speed sensor jumper harness. Likewise, if replacing a wheel speed sensor jumper harness, inspect sensor terminals. If evidence of corrosion or water intrusion exists, replace wheel speed sensor. Refer to "On-Vehicle Service" in this section.

DIAGNOSTIC CHART DTC 22 — 1994 CUTLASS SUPREME AND REGAL, CONTINUED

DTC 22
(Page 1 of 2)
RIGHT FRONT WHEEL SPEED = 0

Important

- WHEEL SPEED SENSOR INTERMITTENT PROBLEMS MAY BE DIFFICULT TO LOCATE. CARE SHOULD BE TAKEN NOT TO DISTURB ANY ELECTRICAL CONNECTIONS PRIOR TO AN INDICATED STEP OF THIS CHART. THIS WILL ENSURE THAT AN INTERMITTENT CONNECTION WILL NOT BE CORRECTED BEFORE THE SOURCE OF THE PROBLEM IS FOUND.

(1) IS DTC 33 ALSO CURRENTLY SET?
- YES: PROCEED WITH THAT DTC CHART FIRST.
- NO: go to step 2.

(2)
- IGNITION "OFF."
- PHYSICALLY INSPECT THE WHEEL SPEED SENSOR WIRING AND CONNECTORS FOR DAMAGE. INSPECT THE WHEEL SPEED SENSOR AND TOOTHED RING FOR LOOSENESS OR DAMAGE. PAY PARTICULAR ATTENTION TO THE TOOTHED RING, ANY DEVIATION WILL AFFECT THE WHEEL SPEED SENSOR OUTPUT SIGNAL.

IS PHYSICAL DAMAGE INDICATED?
- YES: REPAIR DAMAGE OR SENSOR/TOOTHED RING LOOSENESS. DRIVE VEHICLE ABOVE 24 km/h (15 MPH) TO ENSURE DTC DOES NOT SET.
- NO: go to step 3.

(3)
- DISCONNECT RIGHT FRONT WHEEL SPEED SENSOR DIRECTLY AT SENSOR.
- USING J 39200, MEASURE THE RESISTANCE BETWEEN SENSOR TERMINALS "A" AND "B".

IS RESISTANCE BETWEEN 1661 AND 1929Ω WHEN SENSOR IS 20°C (68°F)? *(IF TEMPERATURE OF THE SENSOR IS SUSPECTED TO BE OTHER THAN 20°C (68°F), REFER TO THE TABLE BELOW FOR APPLICABLE SENSOR RESISTANCE VALUES; THE VALUES IN THE TABLE ARE FOR TEMPERATURE OF THE SENSOR, NOT AIR TEMPERATURE.)*
- NO: REPLACE WHEEL SPEED SENSOR.
- YES: go to step 4.

(4)
- WITH J 39200 STILL CONNECTED, SELECT A/C VOLTAGE SCALE.
- SPIN WHEEL BY HAND WHILE OBSERVING VOLTAGE.

IS VOLTAGE AT LEAST 100 mV (VOLTAGE WILL INCREASE AS WHEEL SPEED INCREASES)?
- YES: DIAGNOSIS CONTINUED ON PAGE 2 OF 2 OF THIS DTC CHART.
- NO: VERIFY THAT TOOTHED RING IS INTACT. IF OK, REPLACE WHEEL SPEED SENSOR.

FRONT WHEEL SPEED SENSOR

TEMPERATURE OF SENSOR VS. SENSOR RESISTANCE (APPROXIMATE)

TEMP. (°C)	TEMP. (°F)	RESISTANCE (OHMS)
-34 TO 4	-30 TO 40	1309 TO 1808
5 TO 43	41 TO 110	1557 TO 2103
44 TO 93	111 TO 200	1811 TO 2482

DIAGNOSTIC CHART DTC 22 — 1994 CUTLASS SUPREME AND REGAL, CONTINUED

DTC 22
(Page 2 of 2)
RIGHT FRONT WHEEL SPEED = 0

DIAGNOSIS CONTINUED FROM PREVIOUS PAGE.

(5)
- IGNITION "ON."
- SELECT D.C. VOLTAGE SCALE ON J 39200.
- MEASURE VOLTAGE BETWEEN GROUND AND TERMINAL "A", THEN TERMINAL "B" OF THE RIGHT FRONT WHEEL SPEED SENSOR HARNESS CONNECTOR.

IS VOLTAGE BETWEEN 2.25 AND 2.75 VOLTS ON EACH TERMINAL?
- NO: IS VOLTAGE OUT OF RANGE ON TERMINAL "A"?
  - YES: go to step 6.
  - NO: go to step 10.
- YES: go to step 8.

(8)
- IGNITION "OFF."
- DISCONNECT 24-WAY EBCM CONNECTOR.
- USING J 39200, MEASURE THE RESISTANCE BETWEEN TERMINAL "9" AND TERMINAL "10" OF THE 24-WAY EBCM HARNESS CONNECTOR.

DOES J 39200 DISPLAY RESISTANCE AS "OL" (INFINITE)?
- NO: CKT 833 AND CKT 872 SHORTED TOGETHER. REPAIR SHORT AND RECHECK.
- YES: go to step 9.

(9)
- INSPECT WHEEL SPEED SENSOR AND WHEEL SPEED SENSOR HARNESS CONNECTOR FOR POOR TERMINAL CONTACT OR EVIDENCE OF CORROSION. REPLACE ALL TERMINALS THAT EXHIBIT POOR TERMINAL CONTACT OR EVIDENCE OF CORROSION.
- INSPECT CKTs 833 AND 872 AND 24-WAY EBCM CONNECTOR FOR DAMAGE WHICH MAY RESULT IN A SHORT BETWEEN CKTs 833 AND 872 WITH ALL CONNECTORS CONNECTED. REPAIR ALL DAMAGE IF EVIDENT.
- RECONNECT ALL CONNECTORS.
- TEST DRIVE VEHICLE ABOVE 24 km/h (15 MPH) FOR AT LEAST 30 SECONDS.

DOES DTC 22 SET AS A CURRENT DTC?
- YES: REPLACE EBCM.
- NO: MALFUNCTION WAS MOST LIKELY CAUSED BY A POOR CONNECTION. REFER TO "DIAGNOSTIC AIDS" ON FACING PAGE.

(6)
- IGNITION "OFF."
- DISCONNECT 24-WAY EBCM CONNECTOR.
- USING J 39200, MEASURE THE RESISTANCE BETWEEN TERMINAL "A" OF THE RIGHT FRONT WHEEL SPEED SENSOR HARNESS CONNECTOR AND TERMINAL "10" OF THE 24-WAY EBCM HARNESS CONNECTOR.

IS RESISTANCE 2Ω OR LESS?
- YES: go to step 7.
- NO: REPAIR OPEN OR HIGH RESISTANCE IN CKT 833.

(10)
- IGNITION "OFF."
- DISCONNECT 24-WAY EBCM CONNECTOR.
- USING J 39200, MEASURE THE RESISTANCE BETWEEN TERMINAL "B" OF THE RIGHT FRONT WHEEL SPEED SENSOR HARNESS CONNECTOR AND TERMINAL "9" OF THE 24-WAY EBCM HARNESS CONNECTOR.

IS RESISTANCE 2Ω OR LESS?
- YES: go to step 7.
- NO: REPAIR OPEN OR HIGH RESISTANCE IN CKT 872.

(7)
- INSPECT CKTs 833 AND 872 AND 24-WAY EBCM CONNECTOR TERMINALS "9" AND "10" FOR POOR TERMINAL CONTACT OR EVIDENCE OF CORROSION. REPLACE ALL TERMINALS THAT EXHIBIT POOR TERMINAL CONTACT OR EVIDENCE OF CORROSION.
- RECONNECT ALL CONNECTORS.
- TEST DRIVE VEHICLE ABOVE 24 km/h (15 MPH) FOR AT LEAST 30 SECONDS.

DOES DTC 22 SET AS A CURRENT DTC?
- YES: REPLACE EBCM.
- NO: MALFUNCTION WAS MOST LIKELY CAUSED BY A POOR CONNECTION. REFER TO "DIAGNOSTIC AIDS" ON FACING PAGE.

DIAGNOSTIC CHART DTC 23 — 1994 CUTLASS SUPREME AND REGAL

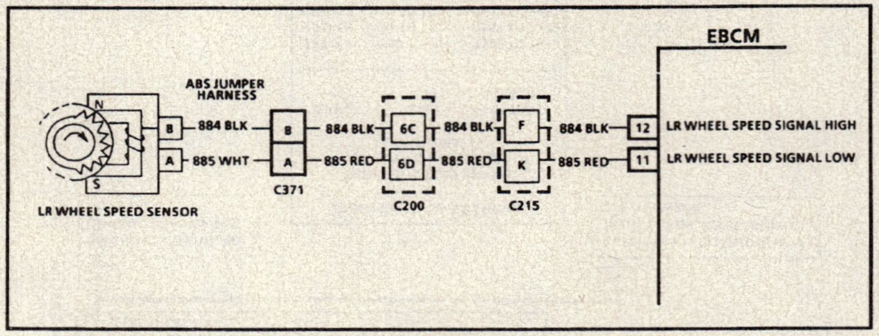

DTC 23

LEFT REAR WHEEL SPEED = 0

Circuit Description:

As a toothed ring passes by the wheel speed sensor, changes in the electromagnetic field cause the wheel speed sensor to produce a sinusoidal (AC) voltage signal whose frequency is proportional to wheel speed. The magnitude of this signal is directly related to wheel speed and the proximity of the wheel speed sensor to the toothed ring, often referred to as the air gap.

DTC Will Set When: DTC 23 can be set when the vehicle is not in an ABS stop. If the left rear wheel speed = 0 and the vehicle's reference speed is greater than 8 km/h (5 mph), a malfunction exists.

Action Taken: A malfunction DTC is stored, ABS is disabled and the ABS warning lamp is turned "ON."

DTC Chart Test Description: Number(s) below refer to circled number(s) on the diagnostic chart.

1. Determines if DTC 34 is present.
2. This step is designed to identify a wheel speed sensor or circuitry that is damaged and visibly apparent.
3. Checks to see that the wheel speed sensor is within proper resistance values.
4. Verifies that the wheel speed sensor can generate an appropriate output signal.
5. Checks for proper output voltage from the EBCM.
6. Checks for an open in the wheel speed sensor low circuit.
7. Ensures malfunction was not due to physical damage of the circuitry.
8. Ensures that wheel speed sensor circuitry is not shorted.
9. Ensures malfunction was not due to physical damage of the circuitry.
10. Checks for an incorrect or malfunctioning hub and bearing assembly.
11. Checks for an open in the wheel speed sensor high circuit.

Diagnostic Aids: An "Intermittent" malfunction may be caused by a poor connection, rubbed through wire insulation, or a wire that is broken inside the insulation.

The frequency of the malfunction can be checked by using the enhanced diagnostic function of the Tech 1, as described in "Tech 1 Diagnostics" found in this section.

If the customer's comments reflect that the ABS warning lamp is "ON" only during moist environmental changes (rain, snow, vehicle wash), all wheel speed sensor circuitry should be thoroughly inspected for signs of water intrusion. If DTC is not current, clear DTCs and simulate the effects of water intrusion. Use the following procedure. Spray down the suspected area with a 5% salt water solution (two teaspoons of salt to 12 oz. of water). Test drive vehicle over various road surfaces (bumps, turns, etc.) above 24 km/h (15 mph) for at least 30 seconds. If DTC returns, replace suspected harness.

Any circuitry, that is suspected as causing the intermittent complaint, should be thoroughly checked for backed out terminals, improper mating, broken locks, improperly formed or damaged terminals, poor terminal to wiring connections or physical damage to the wiring harness.

Resistance of the wheel speed sensor will increase with an increase in sensor temperature.

When replacing a wheel speed sensor, inspect the sensor terminals and harness connector for corrosion and/or water intrusion. If evidence of corrosion or water intrusion exists, replace wheel speed sensor jumper harness. Likewise, if replacing a wheel speed sensor jumper harness, inspect sensor terminals. If evidence of corrosion or water intrusion exists, replace wheel speed sensor. Refer to "On-Vehicle Service" in this section.

DIAGNOSTIC CHART DTC 23 — 1994 CUTLASS SUPREME AND REGAL, CONTINUED

DTC 23
(Page 1 of 2)
LEFT REAR WHEEL SPEED = 0

Important

- WHEEL SPEED SENSOR INTERMITTENT PROBLEMS MAY BE DIFFICULT TO LOCATE. CARE SHOULD BE TAKEN NOT TO DISTURB ANY ELECTRICAL CONNECTIONS PRIOR TO AN INDICATED STEP OF THIS CHART. THIS WILL ENSURE THAT AN INTERMITTENT CONNECTION WILL NOT BE CORRECTED BEFORE THE SOURCE OF THE PROBLEM IS FOUND.

(1) IS DTC 34 ALSO CURRENTLY SET?
- YES: PROCEED WITH THAT DTC CHART FIRST.
- NO:

(2)
- IGNITION "OFF."
- PHYSICALLY INSPECT THE WHEEL SPEED SENSOR, WHEEL SPEED SENSOR JUMPER HARNESS, AND CONNECTORS FOR DAMAGE.

IS PHYSICAL DAMAGE INDICATED?
- YES: REPLACE WHEEL SPEED SENSOR AND/OR WHEEL SPEED SENSOR JUMPER HARNESS IF DAMAGED. DRIVE VEHICLE ABOVE 24 km/h (15 MPH) TO ENSURE DTC DOES NOT SET.
- NO:

(3)
- DISCONNECT LEFT REAR WHEEL SPEED SENSOR DIRECTLY AT SENSOR.
- USING J 39200, MEASURE THE RESISTANCE BETWEEN SENSOR TERMINALS "A" AND "B".

IS RESISTANCE BETWEEN 2059 AND 2353Ω WHEN SENSOR IS 20°C (68°F)? *(IF TEMPERATURE OF THE SENSOR IS SUSPECTED TO BE OTHER THAN 20°C (68°F), REFER TO THE TABLE BELOW FOR APPLICABLE SENSOR RESISTANCE VALUES; THE VALUES IN THE TABLE ARE FOR TEMPERATURE OF THE SENSOR, NOT AIR TEMPERATURE.)*
- NO: REPLACE WHEEL SPEED SENSOR.
- YES:

(4)
- WITH J 39200 STILL CONNECTED, SELECT A/C VOLTAGE SCALE.
- SPIN WHEEL BY HAND WHILE OBSERVING VOLTAGE.

IS VOLTAGE AT LEAST 100 mV (VOLTAGE WILL INCREASE AS WHEEL SPEED INCREASES)?
- NO: REPLACE WHEEL SPEED SENSOR.
- YES: DIAGNOSIS CONTINUED ON PAGE 2 OF 2 OF THIS DTC CHART.

REAR WHEEL SPEED SENSOR

TEMPERATURE OF SENSOR VS. SENSOR RESISTANCE (APPROXIMATE)

TEMP. (°C)	TEMP. (°F)	RESISTANCE (OHMS)
-34 TO 4	-30 TO 40	1622 TO 2205
5 TO 43	41 TO 110	1930 TO 2566
44 TO 93	111 TO 200	2245 TO 3028

DIAGNOSTIC CHART DTC 23 — 1994 CUTLASS SUPREME AND REGAL, CONTINUED

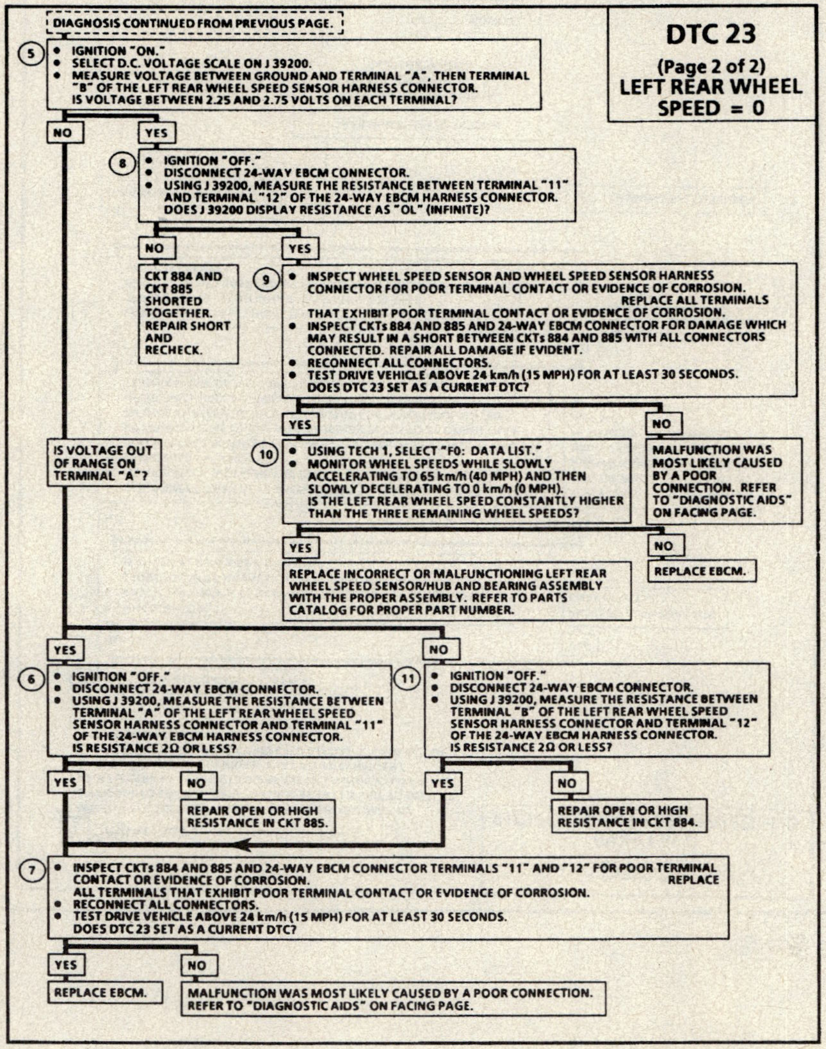

DIAGNOSTIC CHART DTC 24 — 1994 CUTLASS SUPREME AND REGAL

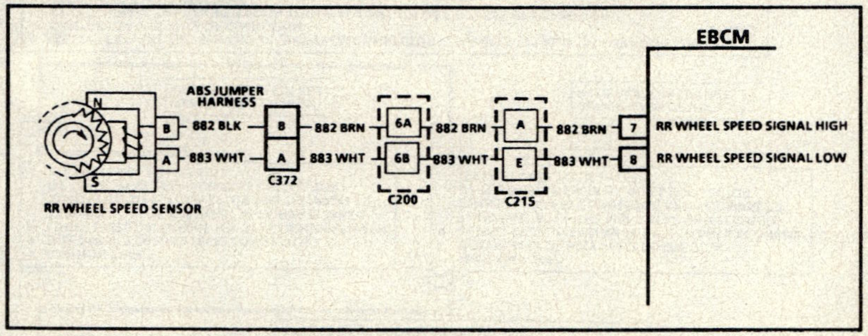

DTC 24

RIGHT REAR WHEEL SPEED = 0

Circuit Description:
As a toothed ring passes by the wheel speed sensor, changes in the electromagnetic field cause the wheel speed sensor to produce a sinusoidal (AC) voltage signal whose frequency is proportional to wheel speed. The magnitude of this signal is directly related to wheel speed and the proximity of the wheel speed sensor to the toothed ring, often referred to as the air gap.

DTC Will Set When: DTC 24 can be set when the vehicle is not in an ABS stop. If the right rear wheel speed = 0 and the vehicle's reference speed is greater than 8 km/h (5 mph), a malfunction exists.

Action Taken: A malfunction DTC is stored, ABS is disabled and the ABS warning lamp is turned "ON."

DTC Chart Test Description: Number(s) below refer to circled number(s) on the diagnostic chart.

1. Determines if DTC 35 is present.
2. Designed to identify a wheel speed sensor or circuitry that is damaged and visibly apparent.
3. Checks to see that the wheel speed sensor is within proper resistance values.
4. Verifies that the wheel speed sensor can generate an appropriate output signal.
5. Checks for proper output voltage from the EBCM.
6. Checks for an open in the wheel speed sensor low circuit.
7. Ensures malfunction was not due to physical damage of the circuitry.
8. Ensures that wheel speed sensor circuitry is not shorted.
9. Ensures malfunction was not due to physical damage of the circuitry.
10. Checks for an incorrect or malfunctioning hub and bearing assembly.
11. Checks for an open in the wheel speed sensor high circuit.

Diagnostic Aids: An "Intermittent" malfunction may be caused by a poor connection, rubbed through wire insulation, or a wire that is broken inside the insulation.

The frequency of the malfunction can be checked by using the enhanced diagnostic function of the Tech 1, as described in "Tech 1 Diagnostics" found in this section.

If the customer's comments reflect that the ABS warning lamp is "ON" only during moist environmental changes (rain, snow, vehicle wash), all wheel speed sensor circuitry should be thoroughly inspected for signs of water intrusion. If DTC is not current, clear DTCs and simulate the effects of water intrusion. Use the following procedure. Spray down the suspected area with a 5% salt water solution (two teaspoons of salt to 12 oz. of water). Test drive vehicle over various road surfaces (bumps, turns, etc.) above 24 km/h (15 mph) for at least 30 seconds. If DTC returns, replace suspected harness.

Any circuitry, that is suspected as causing the intermittent complaint, should be thoroughly checked for backed out terminals, improper mating, broken locks, improperly formed or damaged terminals, poor terminal to wiring connections or physical damage to the wiring harness.

Resistance of the wheel speed sensor will increase with an increase in sensor temperature.

When replacing a wheel speed sensor, inspect the sensor terminals and harness connector for corrosion and/or water intrusion. If evidence of corrosion or water intrusion exists, replace wheel speed sensor jumper harness. Likewise, if replacing a wheel speed sensor jumper harness, inspect sensor terminals. If evidence of corrosion or water intrusion exists, replace wheel speed sensor. Refer to "On-Vehicle Service" in this section.

DIAGNOSTIC CHART DTC 24 — 1994 CUTLASS SUPREME AND REGAL, CONTINUED

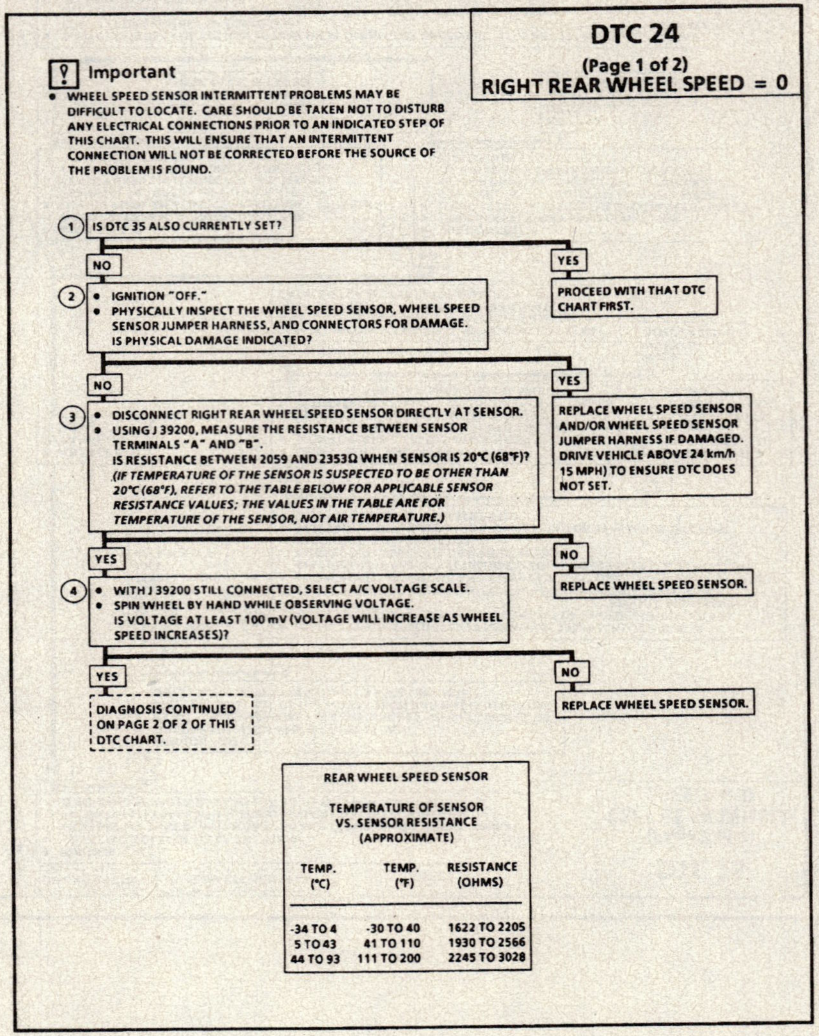

REAR WHEEL SPEED SENSOR

TEMPERATURE OF SENSOR VS. SENSOR RESISTANCE (APPROXIMATE)

TEMP. (°C)	TEMP. (°F)	RESISTANCE (OHMS)
-34 TO 4	-30 TO 40	1622 TO 2205
5 TO 43	41 TO 110	1930 TO 2566
44 TO 93	111 TO 200	2245 TO 3028

DIAGNOSTIC CHART DTC 24 — 1994 CUTLASS SUPREME AND REGAL, CONTINUED

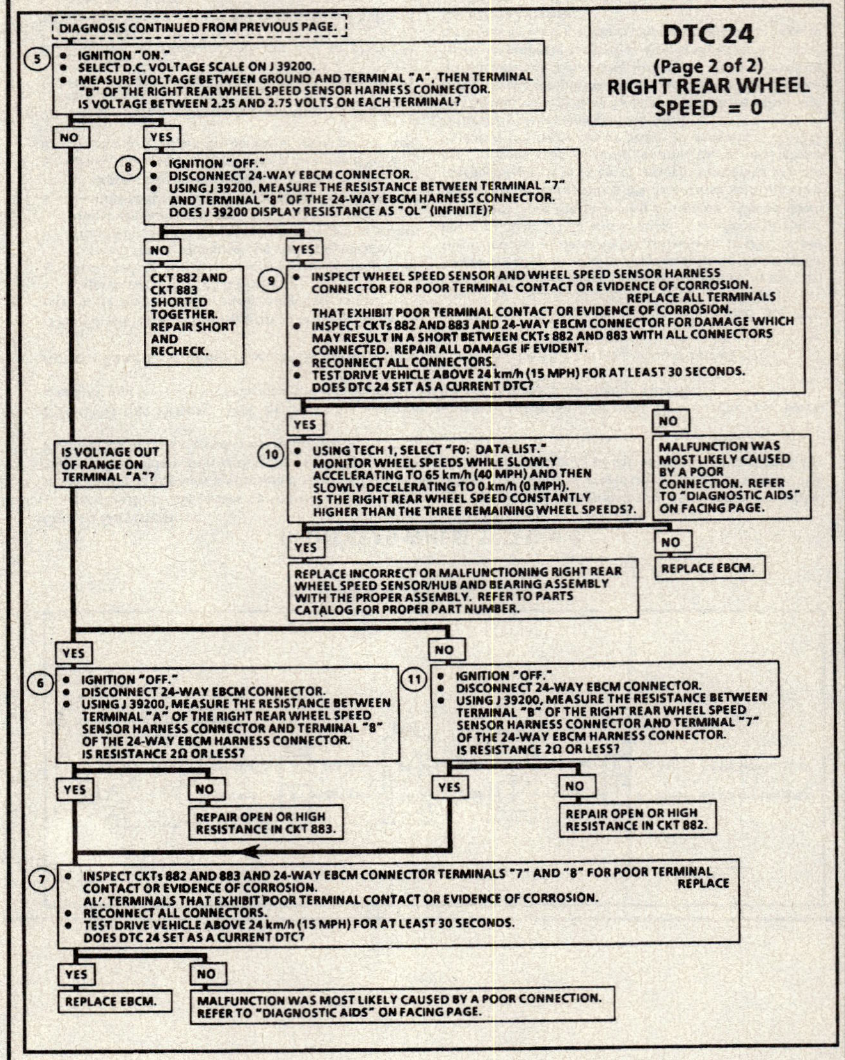

DIAGNOSTIC CHART DTC 25 — 1994 CUTLASS SUPREME AND REGAL

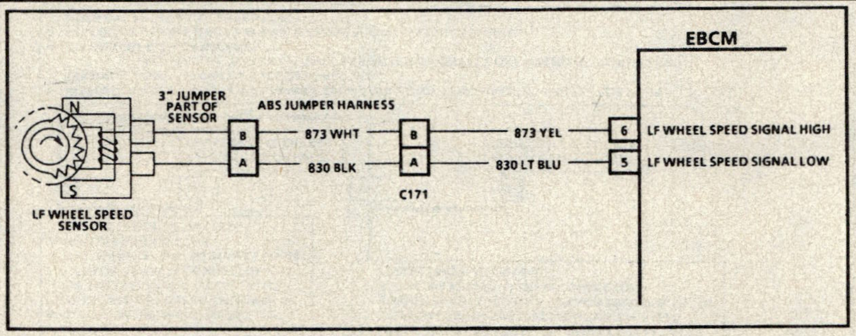

DTC 25

LEFT FRONT EXCESSIVE WHEEL SPEED VARIATION

Circuit Description:

As a toothed ring passes by the wheel speed sensor, changes in the electromagnetic field cause the wheel speed sensor to produce a sinusoidal (AC) voltage signal whose frequency is proportional to wheel speed. The magnitude of this signal is directly related to wheel speed and the proximity of the wheel speed sensor to the toothed ring, often referred to as the air gap.

DTC Will Set When: DTC 25 can be set when the brake is "OFF." The purpose of this test is to detect a situation in which the left front wheel acceleration or deceleration is beyond specified limits.

Action Taken: A malfunction DTC is stored, ABS is disabled and the ABS warning lamp is turned "ON."

DTC Chart Test Description: Number(s) below refer to circled number(s) on the diagnostic chart.

1. Determines if DTC 32 is present.
2. Designed to identify a wheel speed sensor or circuitry that is damaged and visibly apparent.
3. Checks to see that the wheel speed sensor is within proper resistance values.
4. Verifies that the wheel speed sensor can generate an appropriate output signal.
5. Checks for proper output voltage from the EBCM.
6. Checks for an open in the wheel speed sensor low circuit.
7. Ensures malfunction was not due to physical damage of the circuitry.
8. Ensures that wheel speed sensor circuitry is not shorted.
9. Ensures malfunction was not due to physical damage of the circuitry.
10. Checks for an open in the wheel speed sensor high circuit.

Diagnostic Aids: An "Intermittent" malfunction may be caused by a poor connection, rubbed through wire insulation, or a wire that is broken inside the insulation.

The frequency of the malfunction can be checked by using the enhanced diagnostic function of the Tech 1, as described in "Tech 1 Diagnostics" found in this section.

If the customer's comments reflect that the ABS warning lamp is "ON" only during moist environmental changes (rain, snow, vehicle wash), all wheel speed sensor circuitry should be thoroughly inspected for signs of water intrusion. If DTC is not current, clear DTCs and simulate the effects of water intrusion. Use the following procedure. Spray down the suspected area with a 5% salt water solution (two teaspoons of salt to 12 oz. of water). Test drive vehicle over various road surfaces (bumps, turns, etc.) above 24 km/h (15 mph) for at least 30 seconds. If DTC returns, replace suspected harness.

Any circuitry, that is suspected as causing the intermittent complaint, should be thoroughly checked for backed out terminals, improper mating, broken locks, improperly formed or damaged terminals, poor terminal to wiring connections or physical damage to the wiring harness.

Resistance of the wheel speed sensor will increase with an increase in sensor temperature.

When replacing a wheel speed sensor, inspect the sensor terminals and harness connector for corrosion and/or water intrusion. If evidence of corrosion or water intrusion exists, replace wheel speed sensor jumper harness. Likewise, if replacing a wheel speed sensor jumper harness, inspect sensor terminals. If evidence of corrosion or water intrusion exists, replace wheel speed sensor. Refer to "On-Vehicle Service" in this section.

DIAGNOSTIC CHART DTC 25 — 1994 CUTLASS SUPREME AND REGAL, CONTINUED

DTC 25
(Page 1 of 2)
LEFT FRONT EXCESSIVE WHEEL SPEED VARIATION

Important

- WHEEL SPEED SENSOR INTERMITTENT PROBLEMS MAY BE DIFFICULT TO LOCATE. CARE SHOULD BE TAKEN NOT TO DISTURB ANY ELECTRICAL CONNECTIONS PRIOR TO AN INDICATED STEP OF THIS CHART. THIS WILL ENSURE THAT AN INTERMITTENT CONNECTION WILL NOT BE CORRECTED BEFORE THE SOURCE OF THE PROBLEM IS FOUND.

(1) IS DTC 32 ALSO CURRENTLY SET?

- YES: PROCEED WITH THAT DTC CHART FIRST.
- NO: go to step 2.

(2)
- IGNITION "OFF."
- PHYSICALLY INSPECT THE WHEEL SPEED SENSOR WIRING AND CONNECTORS FOR DAMAGE. INSPECT THE WHEEL SPEED SENSOR AND TOOTHED RING FOR LOOSENESS OR DAMAGE. PAY PARTICULAR ATTENTION TO THE TOOTHED RING, ANY DEVIATION WILL AFFECT THE WHEEL SPEED SENSOR OUTPUT SIGNAL.

IS PHYSICAL DAMAGE INDICATED?

- YES: REPAIR DAMAGE OR SENSOR/TOOTHED RING LOOSENESS. DRIVE VEHICLE ABOVE 24 km/h (15 MPH) TO ENSURE DTC DOES NOT SET.
- NO: go to step 3.

(3)
- DISCONNECT LEFT FRONT WHEEL SPEED SENSOR DIRECTLY AT SENSOR.
- USING J 39200, MEASURE THE RESISTANCE BETWEEN SENSOR TERMINALS "A" AND "B".

IS RESISTANCE BETWEEN 1661 AND 1929Ω WHEN SENSOR IS 20°C (68°F)?

(IF TEMPERATURE OF THE SENSOR IS SUSPECTED TO BE OTHER THAN 20°C (68°F), REFER TO THE TABLE BELOW FOR APPLICABLE SENSOR RESISTANCE VALUES; THE VALUES IN THE TABLE ARE FOR TEMPERATURE OF THE SENSOR, NOT AIR TEMPERATURE.)

- NO: REPLACE WHEEL SPEED SENSOR.
- YES: go to step 4.

(4)
- WITH J 39200 STILL CONNECTED, SELECT A/C VOLTAGE SCALE.
- SPIN WHEEL BY HAND WHILE OBSERVING VOLTAGE.

IS VOLTAGE AT LEAST 100 mV (VOLTAGE WILL INCREASE AS WHEEL SPEED INCREASES)?

- YES: DIAGNOSIS CONTINUED ON PAGE 2 OF 2 OF THIS DTC CHART.
- NO: VERIFY THAT TOOTHED RING IS INTACT. IF OK, REPLACE WHEEL SPEED SENSOR.

FRONT WHEEL SPEED SENSOR

TEMPERATURE OF SENSOR VS. SENSOR RESISTANCE (APPROXIMATE)

TEMP. (°C)	TEMP. (°F)	RESISTANCE (OHMS)
-34 TO 4	-30 TO 40	1309 TO 1808
5 TO 43	41 TO 110	1557 TO 2103
44 TO 93	111 TO 200	1811 TO 2482

DIAGNOSTIC CHART DTC 25 — 1994 CUTLASS SUPREME AND REGAL, CONTINUED

DTC 25
(Page 2 of 2)
LEFT FRONT EXCESSIVE WHEEL SPEED VARIATION

DIAGNOSIS CONTINUED FROM PREVIOUS PAGE.

(5)
- IGNITION "ON."
- SELECT D.C. VOLTAGE SCALE ON J 39200.
- MEASURE VOLTAGE BETWEEN GROUND AND TERMINAL "A", THEN TERMINAL "B" OF THE LEFT FRONT WHEEL SPEED SENSOR HARNESS CONNECTOR.

IS VOLTAGE BETWEEN 2.25 AND 2.75 VOLTS ON EACH TERMINAL?

NO → IS VOLTAGE OUT OF RANGE ON TERMINAL "A"?

YES → (8)
- IGNITION "OFF."
- DISCONNECT 24-WAY EBCM CONNECTOR.
- USING J 39200, MEASURE THE RESISTANCE BETWEEN TERMINAL "5" AND TERMINAL "6" OF THE 24-WAY EBCM HARNESS CONNECTOR.

DOES J 39200 DISPLAY RESISTANCE AS "OL" (INFINITE)?

NO → CKT 830 AND CKT 873 SHORTED TOGETHER. REPAIR SHORT AND RECHECK.

YES → (9)
- INSPECT WHEEL SPEED SENSOR AND WHEEL SPEED SENSOR HARNESS CONNECTOR FOR POOR TERMINAL CONTACT OR EVIDENCE OF CORROSION. REPLACE TERMINALS THAT EXHIBIT POOR TERMINAL CONTACT OR EVIDENCE OF CORROSION.
- INSPECT CKTs 830 AND 873 AND 24-WAY EBCM CONNECTOR FOR DAMAGE WHICH MAY RESULT IN A SHORT BETWEEN CKTs 830 AND 873 WITH ALL CONNECTORS CONNECTED. REPAIR ALL DAMAGE IF EVIDENT.
- RECONNECT ALL CONNECTORS.
- TEST DRIVE VEHICLE ABOVE 24 km/h (15 MPH) FOR AT LEAST 30 SECONDS.

DOES DTC 25 SET AS A CURRENT DTC?

YES → REPLACE EBCM.

NO → MALFUNCTION WAS MOST LIKELY CAUSED BY A POOR CONNECTION. REFER TO "DIAGNOSTIC AIDS" ON FACING PAGE.

IS VOLTAGE OUT OF RANGE ON TERMINAL "A"?

YES → (6)
- IGNITION "OFF."
- DISCONNECT 24-WAY EBCM CONNECTOR.
- USING J 39200, MEASURE THE RESISTANCE BETWEEN TERMINAL "A" OF THE LEFT FRONT WHEEL SPEED SENSOR HARNESS CONNECTOR AND TERMINAL "5" OF THE 24-WAY EBCM HARNESS CONNECTOR.

IS RESISTANCE 2Ω OR LESS?

YES → go to (7)

NO → REPAIR OPEN OR HIGH RESISTANCE IN CKT 830.

NO → (10)
- IGNITION "OFF."
- DISCONNECT 24-WAY EBCM CONNECTOR.
- USING J 39200, MEASURE THE RESISTANCE BETWEEN TERMINAL "B" OF THE LEFT FRONT WHEEL SPEED SENSOR HARNESS CONNECTOR AND TERMINAL "6" OF THE 24-WAY EBCM HARNESS CONNECTOR.

IS RESISTANCE 2Ω OR LESS?

YES → go to (7)

NO → REPAIR OPEN OR HIGH RESISTANCE IN CKT 873.

(7)
- INSPECT CKTs 830 AND 873 AND 24-WAY EBCM CONNECTOR TERMINALS "5" AND "6" FOR POOR TERMINAL CONTACT OR EVIDENCE OF CORROSION. REPLACE ALL TERMINALS THAT EXHIBIT POOR CONTACT OR EVIDENCE OF CORROSION.
- RECONNECT ALL CONNECTORS.
- TEST DRIVE VEHICLE ABOVE 24 km/h (15 MPH) FOR AT LEAST 30 SECONDS.

DOES DTC 25 SET AS A CURRENT DTC?

YES → REPLACE EBCM.

NO → MALFUNCTION WAS MOST LIKELY CAUSED BY A POOR CONNECTION. REFER TO "DIAGNOSTIC AIDS" ON FACING PAGE.

DIAGNOSTIC CHART DTC 26 — 1994 CUTLASS SUPREME AND REGAL

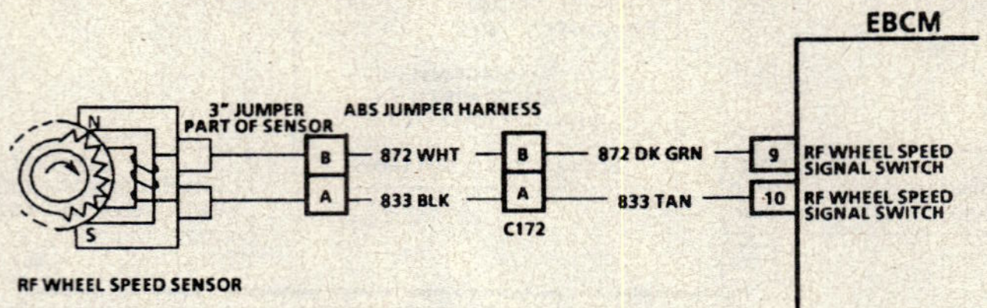

DTC 26

RIGHT FRONT EXCESSIVE WHEEL SPEED VARIATION

Circuit Description:

As a toothed ring passes by the wheel speed sensor, changes in the electromagnetic field cause the wheel speed sensor to produce a sinusoidal (AC) voltage signal whose frequency is proportional to wheel speed. The magnitude of this signal is directly related to wheel speed and the proximity of the wheel speed sensor to the toothed ring, often referred to as the air gap.

DTC Will Set When: DTC 26 can be set when the brake is "OFF." The purpose of this test is to detect a situation in which the right front wheel acceleration or deceleration is beyond specified limits.

Action Taken: A malfunction DTC is stored, ABS is disabled and the ABS warning lamp is turned "ON."

DTC Chart Test Description: Number(s) below refer to circled number(s) on the diagnostic chart.

1. Determines if DTC 33 is present.
2. Designed to identify a wheel speed sensor or circuitry that is damaged and visibly apparent.
3. Checks to see that the wheel speed sensor is within proper resistance values.
4. Verifies that the wheel speed sensor can generate an appropriate output signal.
5. Checks for proper output voltage from the EBCM.
6. Checks for an open in the wheel speed sensor low circuit.
7. Ensures malfunction was not due to physical damage of the circuitry.
8. Ensures that wheel speed sensor circuitry is not shorted.
9. Ensures malfunction was not due to physical damage of the circuitry.
10. Checks for an open in the wheel speed sensor high circuit.

Diagnostic Aids: An "Intermittent" malfunction may be caused by a poor connection, rubbed through wire insulation, or a wire that is broken inside the insulation.

The frequency of the malfunction can be checked by using the enhanced diagnostic function of the Tech 1, as described in "Tech 1 Diagnostics" found in this section.

If the customer's comments reflect that the ABS warning lamp is "ON" only during moist environmental changes (rain, snow, vehicle wash), all wheel speed sensor circuitry should be thoroughly inspected for signs of water intrusion. If DTC is not current, clear DTCs and simulate the effects of water intrusion. Use the following procedure. Spray down the suspected area with a 5% salt water solution (two teaspoons of salt to 12 oz. of water). Test drive vehicle over various road surfaces (bumps, turns, etc.) above 24 km/h (15 mph) for at least 30 seconds. If DTC returns, replace suspected harness.

Any circuitry, that is suspected as causing the intermittent complaint, should be thoroughly checked for backed out terminals, improper mating, broken locks, improperly formed or damaged terminals, poor terminal to wiring connections or physical damage to the wiring harness.

Resistance of the wheel speed sensor will increase with an increase in sensor temperature.

When replacing a wheel speed sensor, inspect the sensor terminals and harness connector for corrosion and/or water intrusion. If evidence of corrosion or water intrusion exists, replace wheel speed sensor jumper harness. Likewise, if replacing a wheel speed sensor jumper harness, inspect sensor terminals. If evidence of corrosion or water intrusion exists, replace wheel speed sensor. Refer to "On-Vehicle Service" in this section.

DIAGNOSTIC CHART DTC 26 — 1994 CUTLASS SUPREME AND REGAL, CONTINUED

DTC 26 (Page 1 of 2) RIGHT FRONT EXCESSIVE WHEEL SPEED VARIATION

Important

- WHEEL SPEED SENSOR INTERMITTENT PROBLEMS MAY BE DIFFICULT TO LOCATE. CARE SHOULD BE TAKEN NOT TO DISTURB ANY ELECTRICAL CONNECTIONS PRIOR TO AN INDICATED STEP OF THIS CHART. THIS WILL ENSURE THAT AN INTERMITTENT CONNECTION WILL NOT BE CORRECTED BEFORE THE SOURCE OF THE PROBLEM IS FOUND.

1. IS DTC 33 ALSO CURRENTLY SET?
   - YES: PROCEED WITH THAT DTC CHART FIRST.
   - NO: Go to step 2.

2. - IGNITION "OFF."
   - PHYSICALLY INSPECT THE WHEEL SPEED SENSOR WIRING AND CONNECTORS FOR DAMAGE. INSPECT THE WHEEL SPEED SENSOR AND TOOTHED RING FOR LOOSENESS OR DAMAGE. PAY PARTICULAR ATTENTION TO THE TOOTHED RING, ANY DEVIATION WILL AFFECT THE WHEEL SPEED SENSOR OUTPUT SIGNAL.

   IS PHYSICAL DAMAGE INDICATED?
   - YES: REPAIR DAMAGE OR SENSOR/TOOTHED RING LOOSENESS. DRIVE VEHICLE ABOVE 24 km/h (15 MPH) TO ENSURE DTC DOES NOT SET.
   - NO: Go to step 3.

3. - DISCONNECT RIGHT FRONT WHEEL SPEED SENSOR DIRECTLY AT SENSOR.
   - USING J 39200, MEASURE THE RESISTANCE BETWEEN SENSOR TERMINALS "A" AND "B"

   IS RESISTANCE BETWEEN 1661 AND 1929Ω WHEN SENSOR IS 20°C (68°F)?

   (IF TEMPERATURE OF THE SENSOR IS SUSPECTED TO BE OTHER THAN 20°C (68°F), REFER TO THE TABLE BELOW FOR APPLICABLE SENSOR RESISTANCE VALUES; THE VALUES IN THE TABLE ARE FOR TEMPERATURE OF THE SENSOR, NOT AIR TEMPERATURE.)
   - NO: REPLACE WHEEL SPEED SENSOR.
   - YES: Go to step 4.

4. - WITH J 39200 STILL CONNECTED, SELECT A/C VOLTAGE SCALE.
   - SPIN WHEEL BY HAND WHILE OBSERVING VOLTAGE.

   IS VOLTAGE AT LEAST 100 mV (VOLTAGE WILL INCREASE AS WHEEL SPEED INCREASES)?
   - NO: VERIFY THAT TOOTHED RING IS INTACT. IF OK, REPLACE WHEEL SPEED SENSOR.
   - YES: DIAGNOSIS CONTINUED ON PAGE 2 OF 2 OF THIS DTC CHART.

FRONT WHEEL SPEED SENSOR

TEMPERATURE OF SENSOR VS. SENSOR RESISTANCE (APPROXIMATE)

TEMP. (°C)	TEMP. (°F)	RESISTANCE (OHMS)
-34 TO 4	-30 TO 40	1309 TO 1808
5 TO 43	41 TO 110	1557 TO 2103
44 TO 93	111 TO 200	1811 TO 2482

DIAGNOSTIC CHART DTC 26 — 1994 CUTLASS SUPREME AND REGAL, CONTINUED

DTC 26 (Page 2 of 2) RIGHT FRONT EXCESSIVE WHEEL SPEED VARIATION

DIAGNOSIS CONTINUED FROM PREVIOUS PAGE.

5. - IGNITION "ON."
   - SELECT D.C. VOLTAGE SCALE ON J 39200.
   - MEASURE VOLTAGE BETWEEN GROUND AND TERMINAL "A", THEN TERMINAL "B" OF THE RIGHT FRONT WHEEL SPEED SENSOR HARNESS CONNECTOR.

   IS VOLTAGE BETWEEN 2.25 AND 2.75 VOLTS ON EACH TERMINAL?
   - NO: IS VOLTAGE OUT OF RANGE ON TERMINAL "A"?
     - YES: Go to step 6.
     - NO: Go to step 10.
   - YES: Go to step 8.

6. - IGNITION "OFF."
   - DISCONNECT 24-WAY EBCM CONNECTOR.
   - USING J 39200, MEASURE THE RESISTANCE BETWEEN TERMINAL "A" OF THE RIGHT FRONT WHEEL SPEED SENSOR HARNESS CONNECTOR AND TERMINAL "10" OF THE 24-WAY EBCM HARNESS CONNECTOR.

   IS RESISTANCE 2Ω OR LESS?
   - YES: Go to step 7.
   - NO: REPAIR OPEN OR HIGH RESISTANCE IN CKT 833.

7. - INSPECT CKTs 833 AND 872 AND 24-WAY EBCM CONNECTOR TERMINALS "9" AND "10" FOR POOR TERMINAL CONTACT OR EVIDENCE OF CORROSION. REPLACE ALL TERMINALS THAT EXHIBIT POOR CONTACT OR EVIDENCE OF CORROSION.
   - RECONNECT ALL CONNECTORS.
   - TEST DRIVE VEHICLE ABOVE 24 km/h (15 MPH) FOR AT LEAST 30 SECONDS.

   DOES DTC 26 SET AS A CURRENT DTC?
   - YES: REPLACE EBCM.
   - NO: MALFUNCTION WAS MOST LIKELY CAUSED BY A POOR CONNECTION. REFER TO "DIAGNOSTIC AIDS" ON FACING PAGE.

8. - IGNITION "OFF."
   - DISCONNECT 24-WAY EBCM CONNECTOR.
   - USING J 39200, MEASURE THE RESISTANCE BETWEEN TERMINAL "9" AND TERMINAL "10" OF THE 24-WAY EBCM HARNESS CONNECTOR.

   DOES J 39200 DISPLAY RESISTANCE AS "OL" (INFINITE)?
   - NO: CKT 833 AND CKT 872 SHORTED TOGETHER. REPAIR SHORT AND RECHECK.
   - YES: Go to step 9.

9. - INSPECT WHEEL SPEED SENSOR AND WHEEL SPEED SENSOR HARNESS CONNECTOR FOR POOR TERMINAL CONTACT OR EVIDENCE OF CORROSION. REPLACE TERMINALS THAT EXHIBIT POOR TERMINAL CONTACT OR EVIDENCE OF CORROSION.
   - INSPECT CKTs 833 AND 872 AND 24-WAY EBCM CONNECTOR FOR DAMAGE WHICH MAY RESULT IN A SHORT BETWEEN CKTs 833 AND 872 WITH ALL CONNECTORS CONNECTED. REPAIR ALL DAMAGE IF EVIDENT.
   - RECONNECT ALL CONNECTORS.
   - TEST DRIVE VEHICLE ABOVE 24 km/h (15 MPH) FOR AT LEAST 30 SECONDS.

   DOES DTC 26 SET AS A CURRENT DTC?
   - YES: REPLACE EBCM.
   - NO: MALFUNCTION WAS MOST LIKELY CAUSED BY A POOR CONNECTION. REFER TO "DIAGNOSTIC AIDS" ON FACING PAGE.

10. - IGNITION "OFF."
    - DISCONNECT 24-WAY EBCM CONNECTOR.
    - USING J 39200, MEASURE THE RESISTANCE BETWEEN TERMINAL "B" OF THE RIGHT FRONT WHEEL SPEED SENSOR HARNESS CONNECTOR AND TERMINAL "9" OF THE 24-WAY EBCM HARNESS CONNECTOR.

    IS RESISTANCE 2Ω OR LESS?
    - YES: Go to step 7.
    - NO: REPAIR OPEN OR HIGH RESISTANCE IN CKT 872.

DIAGNOSTIC CHART DTC 27 — 1994 CUTLASS SUPREME AND REGAL

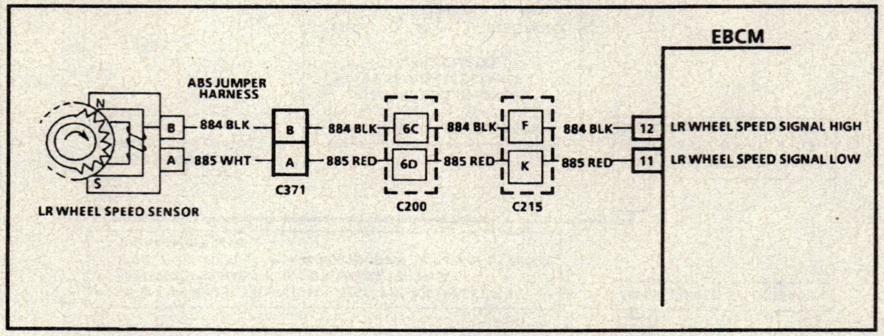

DTC 27

LEFT REAR EXCESSIVE WHEEL SPEED VARIATION

Circuit Description:

As a toothed ring passes by the wheel speed sensor, changes in the electromagnetic field cause the wheel speed sensor to produce a sinusoidal (AC) voltage signal whose frequency is proportional to wheel speed. The magnitude of this signal is directly related to wheel speed and the proximity of the wheel speed sensor to the toothed ring, often referred to as the air gap.

DTC Will Set When: DTC 27 can be set when the brake is "OFF." The purpose of this test is to detect a situation in which the left rear wheel acceleration or deceleration is beyond specified limits.

Action Taken: A malfunction DTC is stored, ABS is di

DTC Chart Test Description: Number(s) below refer to circled number(s) on the diagnostic chart.

1. Determines if DTC 34 is present.
2. Designed to identify a wheel speed sensor or circuitry that is damaged and visibly apparent.
3. Checks to see that the wheel speed sensor is within proper resistance values.
4. Verifies that the wheel speed sensor can generate an appropriate output signal.
5. Checks for proper output voltage from the EBCM.
6. Checks for an open in the wheel speed sensor low circuit.
7. Ensures malfunction was not due to physical damage of the circuitry.
8. Ensures that wheel speed sensor circuitry is not shorted.
9. Ensures malfunction was not due to physical damage of the circuitry.
10. Checks for an open in the wheel speed sensor high circuit.

Diagnostic Aids: An "Intermittent" malfunction may be caused by a poor connection, rubbed through wire insulation, or a wire that is broken inside the insulation.

The frequency of the malfunction can be checked by using the enhanced diagnostic function of the Tech 1, as described in "Tech 1 Diagnostics" found in this section.

If the customer's comments reflect that the ABS warning lamp is "ON" only during moist environmental changes (rain, snow, vehicle wash), all wheel speed sensor circuitry should be thoroughly inspected for signs of water intrusion. If DTC is not current, clear DTCs and simulate the effects of water intrusion. Use the following procedure. Spray down the suspected area with a 5% salt water solution (two teaspoons of salt to 12 oz. of water). Test drive vehicle over various road surfaces (bumps, turns, etc.) above 24 km/h (15 mph) for at least 30 seconds. If DTC returns, replace suspected harness.

Any circuitry, that is suspected as causing the intermittent complaint, should be thoroughly checked for backed out terminals, improper mating, broken locks, improperly formed or damaged terminals, poor terminal to wiring connections or physical damage to the wiring harness.

Resistance of the wheel speed sensor will increase with an increase in sensor temperature.

When replacing a wheel speed sensor, inspect the sensor terminals and harness connector for corrosion and/or water intrusion. If evidence of corrosion or water intrusion exists, replace wheel speed sensor jumper harness. Likewise, if replacing a wheel speed sensor jumper harness, inspect sensor terminals. If evidence of corrosion or water intrusion exists, replace wheel speed sensor. Refer to "On-Vehicle Service" in this section.

DIAGNOSTIC CHART DTC 27 — 1994 CUTLASS SUPREME AND REGAL, CONTINUED

DTC 27
(Page 1 of 2)
LEFT REAR EXCESSIVE WHEEL SPEED VARIATION

Important
- WHEEL SPEED SENSOR INTERMITTENT PROBLEMS MAY BE DIFFICULT TO LOCATE. CARE SHOULD BE TAKEN NOT TO DISTURB ANY ELECTRICAL CONNECTIONS PRIOR TO AN INDICATED STEP OF THIS CHART. THIS WILL ENSURE THAT AN INTERMITTENT CONNECTION WILL NOT BE CORRECTED BEFORE THE SOURCE OF THE PROBLEM IS FOUND.

(1) IS DTC 34 ALSO CURRENTLY SET?
- YES: PROCEED WITH THAT DTC CHART FIRST.
- NO: go to 2.

(2)
- IGNITION "OFF."
- PHYSICALLY INSPECT THE WHEEL SPEED SENSOR, WHEEL SPEED SENSOR JUMPER HARNESS, AND CONNECTORS FOR DAMAGE.

IS PHYSICAL DAMAGE INDICATED?
- YES: REPLACE WHEEL SPEED SENSOR AND/OR WHEEL SPEED SENSOR JUMPER HARNESS IF DAMAGED. DRIVE VEHICLE ABOVE 24 km/h (15 MPH) TO ENSURE DTC DOES NOT SET.
- NO: go to 3.

(3)
- DISCONNECT LEFT REAR WHEEL SPEED SENSOR DIRECTLY AT SENSOR.
- USING J 39200, MEASURE THE RESISTANCE BETWEEN SENSOR TERMINALS "A" AND "B".

IS RESISTANCE BETWEEN 2059 AND 2353Ω WHEN SENSOR IS 20°C (68°F)?
(IF TEMPERATURE OF THE SENSOR IS SUSPECTED TO BE OTHER THAN 20°C (68°F), REFER TO THE TABLE BELOW FOR APPLICABLE SENSOR RESISTANCE VALUES; THE VALUES IN THE TABLE ARE FOR TEMPERATURE OF THE SENSOR, NOT AIR TEMPERATURE.)
- NO: REPLACE WHEEL SPEED SENSOR.
- YES: go to 4.

(4)
- WITH J 39200 STILL CONNECTED, SELECT A/C VOLTAGE SCALE.
- SPIN WHEEL BY HAND WHILE OBSERVING VOLTAGE.

IS VOLTAGE AT LEAST 100 mV (VOLTAGE WILL INCREASE AS WHEEL SPEED INCREASES)?
- NO: REPLACE WHEEL SPEED SENSOR.
- YES: DIAGNOSIS CONTINUED ON PAGE 2 OF 2 OF THIS DTC CHART.

REAR WHEEL SPEED SENSOR
TEMPERATURE OF SENSOR VS. SENSOR RESISTANCE (APPROXIMATE)

TEMP. (°C)	TEMP. (°F)	RESISTANCE (OHMS)
-34 TO 4	-30 TO 40	1622 TO 2205
5 TO 43	41 TO 110	1930 TO 2566
44 TO 93	111 TO 200	2245 TO 3028

DIAGNOSTIC CHART DTC 27 — 1994 CUTLASS SUPREME AND REGAL, CONTINUED

DTC 27
(Page 2 of 2)
LEFT REAR EXCESSIVE WHEEL SPEED VARIATION

DIAGNOSIS CONTINUED FROM PREVIOUS PAGE.

(5)
- IGNITION "ON."
- SELECT D.C. VOLTAGE SCALE ON J 39200.
- MEASURE VOLTAGE BETWEEN GROUND AND TERMINAL "A", THEN TERMINAL "B" OF THE LEFT REAR WHEEL SPEED SENSOR HARNESS CONNECTOR. IS VOLTAGE BETWEEN 2.25 AND 2.75 VOLTS ON EACH TERMINAL?

NO → IS VOLTAGE OUT OF RANGE ON TERMINAL "A"?

YES →

(8)
- IGNITION "OFF."
- DISCONNECT 24-WAY EBCM CONNECTOR.
- USING J 39200, MEASURE THE RESISTANCE BETWEEN TERMINAL "11" AND TERMINAL "12" OF THE 24-WAY EBCM HARNESS CONNECTOR. DOES J 39200 DISPLAY RESISTANCE AS "OL" (INFINITE)?

NO → CKT 884 AND CKT 885 SHORTED TOGETHER. REPAIR SHORT AND RECHECK.

YES →

(9)
- INSPECT WHEEL SPEED SENSOR AND WHEEL SPEED SENSOR HARNESS CONNECTOR FOR POOR TERMINAL CONTACT OR EVIDENCE OF CORROSION. REPLACE TERMINALS THAT EXHIBIT POOR TERMINAL CONTACT OR EVIDENCE OF CORROSION.
- INSPECT CKTs 884 AND 885 AND 24-WAY EBCM CONNECTOR FOR DAMAGE WHICH MAY RESULT IN A SHORT BETWEEN CKTs 884 AND 885 WITH ALL CONNECTORS CONNECTED. REPAIR ALL DAMAGE IF EVIDENT.
- RECONNECT ALL CONNECTORS.
- TEST DRIVE VEHICLE ABOVE 24 km/h (15 MPH) FOR AT LEAST 30 SECONDS. DOES DTC 27 SET AS A CURRENT DTC?

YES → REPLACE EBCM.

NO → MALFUNCTION WAS MOST LIKELY CAUSED BY A POOR CONNECTION. REFER TO "DIAGNOSTIC AIDS" ON FACING PAGE.

IS VOLTAGE OUT OF RANGE ON TERMINAL "A"?

YES →

(6)
- IGNITION "OFF."
- DISCONNECT 24-WAY EBCM CONNECTOR.
- USING J 39200, MEASURE THE RESISTANCE BETWEEN TERMINAL "A" OF THE LEFT REAR WHEEL SPEED SENSOR HARNESS CONNECTOR AND TERMINAL "11" OF THE 24-WAY EBCM HARNESS CONNECTOR. IS RESISTANCE 2Ω OR LESS?

YES → (7)

NO → REPAIR OPEN OR HIGH RESISTANCE IN CKT 885.

NO →

(10)
- IGNITION "OFF."
- DISCONNECT 24-WAY EBCM CONNECTOR.
- USING J 39200, MEASURE THE RESISTANCE BETWEEN TERMINAL "B" OF THE LEFT REAR WHEEL SPEED SENSOR HARNESS CONNECTOR AND TERMINAL "12" OF THE 24-WAY EBCM HARNESS CONNECTOR. IS RESISTANCE 2Ω OR LESS?

YES → (7)

NO → REPAIR OPEN OR HIGH RESISTANCE IN CKT 884.

(7)
- INSPECT CKTs 884 AND 885 AND 24-WAY EBCM CONNECTOR TERMINALS "11" AND "12" FOR POOR TERMINAL CONTACT OR EVIDENCE OF CORROSION. REPLACE ALL TERMINALS THAT EXHIBIT POOR CONTACT OR EVIDENCE OF CORROSION.
- RECONNECT ALL CONNECTORS.
- TEST DRIVE VEHICLE ABOVE 24 km/h (15 MPH) FOR AT LEAST 30 SECONDS. DOES DTC 27 SET AS A CURRENT DTC?

YES → REPLACE EBCM.

NO → MALFUNCTION WAS MOST LIKELY CAUSED BY A POOR CONNECTION. REFER TO "DIAGNOSTIC AIDS" ON FACING PAGE.

DIAGNOSTIC CHART DTC 28 — 1994 CUTLASS SUPREME AND REGAL

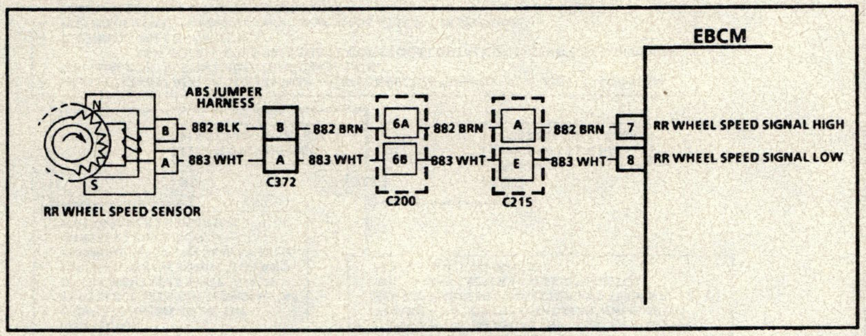

DTC 28

RIGHT REAR EXCESSIVE WHEEL SPEED VARIATION

Circuit Description:

As a toothed ring passes by the wheel speed sensor, changes in the electromagnetic field cause the wheel speed sensor to produce a sinusoidal (AC) voltage signal whose frequency is proportional to wheel speed. The magnitude of this signal is directly related to wheel speed and the proximity of the wheel speed sensor to the toothed ring, often referred to as the air gap.

DTC Will Set When: DTC 28 can be set when the brake is "OFF." The purpose of this test is to detect a situation in which the right rear wheel acceleration or deceleration is beyond specified limits.

Action Taken: A malfunction DTC is stored, ABS is disabled and the ABS warning lamp is turned "ON."

DTC Chart Test Description: Number(s) below refer to circled number(s) on the diagnostic chart.

1. Determines if DTC 35 is present.
2. Designed to identify a wheel speed sensor or circuitry that is damaged and visibly apparent.
3. Checks to see that the wheel speed sensor is within proper resistance values.
4. Verifies that the wheel speed sensor can generate an appropriate output signal.
5. Checks for proper output voltage from the EBCM.
6. Checks for an open in the wheel speed sensor low circuit.
7. Ensures malfunction was not due to physical damage of the circuitry.
8. Ensures that wheel speed sensor circuitry is not shorted.
9. Ensures malfunction was not due to physical damage of the circuitry.
10. Checks for an open in the wheel speed sensor high circuit.

Diagnostic Aids: An "Intermittent" malfunction may be caused by a poor connection, rubbed through wire insulation, or a wire that is broken inside the insulation.

The frequency of the malfunction can be checked by using the enhanced diagnostic function of the Tech 1, as described in "Tech 1 Diagnostics" found in this section.

If the customer's comments reflect that the ABS warning lamp is "ON" only during moist environmental changes (rain, snow, vehicle wash), all wheel speed sensor circuitry should be thoroughly inspected for signs of water intrusion. If DTC is not current, clear DTCs and simulate the effects of water intrusion. Use the following procedure. Spray down the suspected area with a 5% salt water solution (two teaspoons of salt to 12 oz. of water). Test drive vehicle over various road surfaces (bumps, turns, etc.) above 24 km/h (15 mph) for at least 30 seconds. If DTC returns, replace suspected harness.

Any circuitry, that is suspected as causing the intermittent complaint, should be thoroughly checked for backed out terminals, improper mating, broken locks, improperly formed or damaged terminals, poor terminal to wiring connections or physical damage to the wiring harness.

Resistance of the wheel speed sensor will increase with an increase in sensor temperature.

When replacing a wheel speed sensor, inspect the sensor terminals and harness connector for corrosion and/or water intrusion. If evidence of corrosion or water intrusion exists, replace wheel speed sensor jumper harness. Likewise, if replacing a wheel speed sensor jumper harness, inspect sensor terminals. If evidence of corrosion or water intrusion exists, replace wheel speed sensor. Refer to "On-Vehicle Service" in this section.

DIAGNOSTIC CHART DTC 28 — 1994 CUTLASS SUPREME AND REGAL, CONTINUED

DTC 28 (Page 1 of 2) RIGHT REAR EXCESSIVE WHEEL SPEED VARIATION

Important
- WHEEL SPEED SENSOR INTERMITTENT PROBLEMS MAY BE DIFFICULT TO LOCATE. CARE SHOULD BE TAKEN NOT TO DISTURB ANY ELECTRICAL CONNECTIONS PRIOR TO AN INDICATED STEP OF THIS CHART. THIS WILL ENSURE THAT AN INTERMITTENT CONNECTION WILL NOT BE CORRECTED BEFORE THE SOURCE OF THE PROBLEM IS FOUND.

1. IS DTC 35 ALSO CURRENTLY SET?
   - YES: PROCEED WITH THAT DTC CHART FIRST.
   - NO: Go to step 2.

2. - IGNITION "OFF."
   - PHYSICALLY INSPECT THE WHEEL SPEED SENSOR, WHEEL SPEED SENSOR JUMPER HARNESS, AND CONNECTORS FOR DAMAGE.
   - IS PHYSICAL DAMAGE INDICATED?
   - YES: REPLACE WHEEL SPEED SENSOR AND/OR WHEEL SPEED SENSOR JUMPER HARNESS IF DAMAGED. DRIVE VEHICLE ABOVE 24 km/h (15 MPH) TO ENSURE DTC DOES NOT SET.
   - NO: Go to step 3.

3. - DISCONNECT RIGHT REAR WHEEL SPEED SENSOR DIRECTLY AT SENSOR.
   - USING J 39200, MEASURE THE RESISTANCE BETWEEN SENSOR TERMINALS "A" AND "B".
   - IS RESISTANCE BETWEEN 2059 AND 2353Ω WHEN SENSOR IS 20°C (68°F)? *(IF TEMPERATURE OF THE SENSOR IS SUSPECTED TO BE OTHER THAN 20°C (68°F), REFER TO THE TABLE BELOW FOR APPLICABLE SENSOR RESISTANCE VALUES; THE VALUES IN THE TABLE ARE FOR TEMPERATURE OF THE SENSOR, NOT AIR TEMPERATURE.)*
   - NO: REPLACE WHEEL SPEED SENSOR.
   - YES: Go to step 4.

4. - WITH J 39200 STILL CONNECTED, SELECT A/C VOLTAGE SCALE.
   - SPIN WHEEL BY HAND WHILE OBSERVING VOLTAGE.
   - IS VOLTAGE AT LEAST 100 mV (VOLTAGE WILL INCREASE AS WHEEL SPEED INCREASES)?
   - NO: REPLACE WHEEL SPEED SENSOR.
   - YES: DIAGNOSIS CONTINUED ON PAGE 2 OF 2 OF THIS DTC CHART.

REAR WHEEL SPEED SENSOR
TEMPERATURE OF SENSOR VS. SENSOR RESISTANCE (APPROXIMATE)

TEMP. (°C)	TEMP. (°F)	RESISTANCE (OHMS)
-34 TO 4	-30 TO 40	1622 TO 2205
5 TO 43	41 TO 110	1930 TO 2566
44 TO 93	111 TO 200	2245 TO 3028

DIAGNOSTIC CHART DTC 28 — 1994 CUTLASS SUPREME AND REGAL, CONTINUED

DTC 28 (Page 2 of 2) RIGHT REAR EXCESSIVE WHEEL SPEED VARIATION

DIAGNOSIS CONTINUED FROM PREVIOUS PAGE.

5. - IGNITION "ON."
   - SELECT D.C. VOLTAGE SCALE ON J 39200.
   - MEASURE VOLTAGE BETWEEN GROUND AND TERMINAL "A", THEN TERMINAL "B" OF THE RIGHT REAR WHEEL SPEED SENSOR HARNESS CONNECTOR.
   - IS VOLTAGE BETWEEN 2.25 AND 2.75 VOLTS ON EACH TERMINAL?
   - YES: Go to step 8.
   - NO: IS VOLTAGE OUT OF RANGE ON TERMINAL "A"?
     - YES: Go to step 6.
     - NO: Go to step 10.

8. - IGNITION "OFF."
   - DISCONNECT 24-WAY EBCM CONNECTOR.
   - USING J 39200, MEASURE THE RESISTANCE BETWEEN TERMINAL "7" AND TERMINAL "8" OF THE 24-WAY EBCM HARNESS CONNECTOR.
   - DOES J 39200 DISPLAY RESISTANCE AS "OL" (INFINITE)?
   - NO: CKT 882 AND CKT 883 SHORTED TOGETHER. REPAIR SHORT AND RECHECK.
   - YES: Go to step 9.

9. - INSPECT WHEEL SPEED SENSOR AND WHEEL SPEED SENSOR HARNESS CONNECTOR FOR POOR TERMINAL CONTACT OR EVIDENCE OF CORROSION.
   - REPLACE TERMINALS THAT EXHIBIT POOR TERMINAL CONTACT OR EVIDENCE OF CORROSION.
   - INSPECT CKTs 882 AND 883 AND 24-WAY EBCM CONNECTOR FOR DAMAGE WHICH MAY RESULT IN A SHORT BETWEEN CKTs 882 AND 883 WITH ALL CONNECTORS CONNECTED. REPAIR ALL DAMAGE IF EVIDENT.
   - RECONNECT ALL CONNECTORS.
   - TEST DRIVE VEHICLE ABOVE 24 km/h (15 MPH) FOR AT LEAST 30 SECONDS.
   - DOES DTC 28 SET AS A CURRENT DTC?
   - YES: REPLACE EBCM.
   - NO: MALFUNCTION WAS MOST LIKELY CAUSED BY A POOR CONNECTION. REFER TO "DIAGNOSTIC AIDS" ON FACING PAGE.

6. - IGNITION "OFF."
   - DISCONNECT 24-WAY EBCM CONNECTOR.
   - USING J 39200, MEASURE THE RESISTANCE BETWEEN TERMINAL "A" OF THE RIGHT REAR WHEEL SPEED SENSOR HARNESS CONNECTOR AND TERMINAL "8" OF THE 24-WAY EBCM HARNESS CONNECTOR.
   - IS RESISTANCE 2Ω OR LESS?
   - NO: REPAIR OPEN OR HIGH RESISTANCE IN CKT 883.
   - YES: Go to step 7.

10. - IGNITION "OFF."
    - DISCONNECT 24-WAY EBCM CONNECTOR.
    - USING J 39200, MEASURE THE RESISTANCE BETWEEN TERMINAL "B" OF THE RIGHT REAR WHEEL SPEED SENSOR HARNESS CONNECTOR AND TERMINAL "7" OF THE 24-WAY EBCM HARNESS CONNECTOR.
    - IS RESISTANCE 2Ω OR LESS?
    - NO: REPAIR OPEN OR HIGH RESISTANCE IN CKT 882.
    - YES: Go to step 7.

7. - INSPECT CKTs 882 AND 883 AND 24-WAY EBCM CONNECTOR TERMINALS "7" AND "8" FOR POOR TERMINAL CONTACT OR EVIDENCE OF CORROSION. REPLACE ALL TERMINALS THAT EXHIBIT POOR CONTACT OR EVIDENCE OF CORROSION.
   - RECONNECT ALL CONNECTORS.
   - TEST DRIVE VEHICLE ABOVE 24 km/h (15 MPH) FOR AT LEAST 30 SECONDS.
   - DOES DTC 28 SET AS A CURRENT DTC?
   - YES: REPLACE EBCM.
   - NO: MALFUNCTION WAS MOST LIKELY CAUSED BY A POOR CONNECTION. REFER TO "DIAGNOSTIC AIDS" ON FACING PAGE.

DIAGNOSTIC CHART DTC 32 — 1994 CUTLASS SUPREME AND REGAL

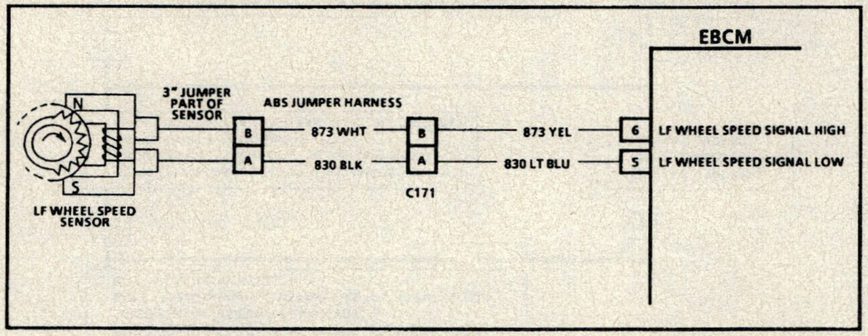

DTC 32

LEFT FRONT WHEEL SPEED SENSOR CKT SHORT TO GND/BAT

Circuit Description:

As a toothed ring passes by the wheel speed sensor, changes in the electromagnetic field cause the wheel speed sensor to produce a sinusoidal (AC) voltage signal whose frequency is proportional to wheel speed. The magnitude of this signal is directly related to wheel speed and the proximity of the wheel speed sensor to the toothed ring often referred to as the air gap.

DTC Will Set When: DTC 32 can be set anytime after initialization. If either of the left front wheel speed circuits are shorted to voltage or ground, a malfunction exists.

Action Taken: A malfunction DTC is stored, ABS is disabled and the ABS warning lamp is turned "ON."

DTC Chart Test Description: Number(s) below refer to circled number(s) on the diagnostic chart.

1. Checks for high resistance in wheel speed sensor wiring circuitry.
2. Checks for short to voltage in wheel speed sensor wiring circuitry.
3. Checks for short to voltage in CKT 830.
4. Checks for short to voltage in CKT 873.
5. Checks for short to ground or voltage in CKTs 830 or 873.
6. Checks for short to ground in CKT 830.
7. Checks for short to ground in CKT 873.
8. Checks for high resistance in wheel speed sensor jumper harness.
9. Checks for high resistance in wheel speed sensor circuitry.

Diagnostic Aids: An "Intermittent" malfunction may be caused by a poor connection, rubbed through wire insulation, or a wire that is broken inside the insulation.

The frequency of the malfunction can be checked by using the enhanced diagnostic function of the Tech 1, as described in "Tech 1 Diagnostics" found in this section.

If the customer's comments reflect that the ABS warning lamp is "ON" only during moist environmental changes (rain, snow, vehicle wash), all wheel speed sensor circuitry should be thoroughly inspected for signs of water intrusion. If DTC is not current, clear DTCs and simulate the effects of water intrusion. Use the following procedure. Spray down the suspected area with a 5% salt water solution (two teaspoons of salt to 12 oz. of water). Test drive vehicle over various road surfaces (bumps, turns, etc.) above 24 km/h (15 mph) for at least 30 seconds. If DTC returns, replace suspected harness.

Any circuitry, that is suspected as causing the intermittent complaint, should be thoroughly checked for backed out terminals, improper mating, broken locks, improperly formed or damaged terminals, poor terminal to wiring connections or physical damage to the wiring harness.

Resistance of the wheel speed sensor will increase with an increase in sensor temperature.

When replacing a wheel speed sensor, inspect the sensor terminals and harness connector for corrosion and/or water intrusion. If evidence of corrosion or water intrusion exists, replace wheel speed sensor jumper harness. Likewise, if replacing a wheel speed sensor jumper harness, inspect sensor terminals. If evidence of corrosion or water intrusion exists, replace wheel speed sensor. Refer to "On-Vehicle Service" in this section.

DIAGNOSTIC CHART DTC 32 — 1994 CUTLASS SUPREME AND REGAL, CONTINUED

DTC 32
(Page 1 of 2)
LEFT FRONT WHEEL SPEED SENSOR CKT SHORT TO GND/BAT

Important

- WHEEL SPEED SENSOR INTERMITTENT MALFUNCTIONS MAY BE DIFFICULT TO LOCATE. CARE SHOULD BE TAKEN NOT TO DISTURB ANY ELECTRICAL CONNECTIONS PRIOR TO AN INDICATED STEP OF THIS CHART. THIS WILL ENSURE THAT AN INTERMITTENT CONNECTION WILL NOT BE CORRECTED BEFORE THE SOURCE OF THE MALFUNCTION IS FOUND.

(1)
- IGNITION "OFF."
- USING A 5% SALT WATER SOLUTION (TWO TEASPOONS OF SALT TO 12 OZ. WATER), THOROUGHLY SPRAY THE LEFT FRONT WHEEL SPEED SENSOR JUMPER HARNESS AND SURROUNDING AREA.
- DISCONNECT 24-WAY EBCM HARNESS.
- USING J 39200, MEASURE THE RESISTANCE BETWEEN GROUND AND TERMINAL "5" OF THE 24-WAY EBCM HARNESS CONNECTOR. DOES J 39200 DISPLAY RESISTANCE AS "OL" (INFINITE)?

YES → (2)

NO → DIAGNOSIS CONTINUED ON PAGE 2 OF 2 OF THIS DTC CHART.

(2)
- IGNITION "ON."
- USING J 39200, MEASURE THE VOLTAGE BETWEEN GROUND AND TERMINAL "5" OF THE 24-WAY EBCM HARNESS CONNECTOR. IS VOLTAGE 1 VOLT OR LESS?

NO → (3)

YES → (5)

(3)
- IGNITION "OFF."
- DISCONNECT LEFT FRONT WHEEL SPEED SENSOR DIRECTLY AT SENSOR.
- IGNITION "ON."
- USING J 39200, MEASURE THE VOLTAGE BETWEEN GROUND AND TERMINAL "5" OF THE 24-WAY EBCM HARNESS CONNECTOR. IS VOLTAGE 1 VOLT OR LESS?

YES → (4)

NO → REPAIR SHORT TO VOLTAGE IN CKT 830.

(4)
- USING J 39200, MEASURE THE VOLTAGE BETWEEN GROUND AND TERMINAL "6" OF THE 24-WAY EBCM HARNESS CONNECTOR. IS VOLTAGE 1 VOLT OR LESS?

YES → MALFUNCTION IS INTERMITTENT. INSPECT ALL CONNECTORS AND HARNESSES FOR DAMAGE WHICH MAY RESULT IN A SHORT TO VOLTAGE WHEN ALL COMPONENTS ARE CONNECTED. REPAIR ALL DAMAGE IF FOUND. REFER TO "DIAGNOSTIC AIDS" ON FACING PAGE FOR MORE INFORMATION.

NO → REPAIR SHORT TO VOLTAGE IN CKT 873.

(5)
- IGNITION "OFF."
- INSPECT TERMINALS "5" AND "6" OF THE 24-WAY EBCM CONNECTOR AND CKTs 830 AND 873 FOR DAMAGE WHICH MAY RESULT IN A SHORT TO VOLTAGE OR GROUND WHEN ALL COMPONENTS ARE CONNECTED. PAY PARTICULAR ATTENTION TO THE JUMPER HARNESS; SMALL NICKS OR CUTS CAN CAUSE A MALFUNCTION. REPAIR ANY DAMAGE IF FOUND. JUMPER HARNESSES MUST BE REPLACED IF DAMAGED.
- RECONNECT ALL CONNECTORS.
- IGNITION "ON."
- DOES DTC 32 SET AS A CURRENT DTC?

YES → REPLACE EBCM.

NO → MALFUNCTION IS INTERMITTENT. REFER TO "DIAGNOSTIC AIDS" ON FACING PAGE.

DIAGNOSTIC CHART DTC 32 — 1994 CUTLASS SUPREME AND REGAL, CONTINUED

DTC 32
(Page 2 of 2)
LEFT FRONT WHEEL SPEED SENSOR CKT SHORT TO GND/BAT

Important
- WHEEL SPEED SENSOR INTERMITTENT MALFUNCTIONS MAY BE DIFFICULT TO LOCATE. CARE SHOULD BE TAKEN NOT TO DISTURB ANY ELECTRICAL CONNECTIONS PRIOR TO AN INDICATED STEP OF THIS CHART. THIS WILL ENSURE THAT AN INTERMITTENT CONNECTION WILL NOT BE CORRECTED BEFORE THE SOURCE OF THE MALFUNCTION IS FOUND.

DIAGNOSIS CONTINUED FROM PAGE 1 OF 2 OF THIS DTC CHART.

(6)
- DISCONNECT CONNECTOR C171 (MOUNTED ON LH SHOCK TOWER).
- USING J 39200, MEASURE THE RESISTANCE BETWEEN GROUND AND TERMINAL "5", OF THE 24-WAY EBCM HARNESS CONNECTOR.
DOES J 39200 DISPLAY RESISTANCE AS "OL" (INFINITE)?

NO: REPAIR SHORT TO GROUND IN CKT 830.

YES:

(7)
- USING J 39200, MEASURE THE RESISTANCE BETWEEN GROUND AND TERMINAL "6" OF THE 24-WAY EBCM HARNESS CONNECTOR.
DOES J 39200 DISPLAY RESISTANCE AS "OL" (INFINITE)?

NO: REPAIR SHORT TO GROUND IN CKT 873.

YES:

(8)
- THOROUGHLY RESPRAY THE LEFT FRONT WHEEL SPEED SENSOR JUMPER HARNESS WITH THE SALT WATER SOLUTION.
IMPORTANT: DO NOT ALLOW WATER TO ENTER ANY WHEEL SPEED SENSOR CONNECTOR.
- DISCONNECT LEFT FRONT WHEEL SPEED SENSOR DIRECTLY AT SENSOR.
- USING J 39200, MEASURE THE RESISTANCE BETWEEN GROUND AND TERMINAL "A", THEN "B" OF THE LEFT FRONT WHEEL SPEED SENSOR HARNESS CONNECTOR.
DOES J 39200 DISPLAY RESISTANCE AS "OL" (INFINITE) FOR BOTH MEASUREMENTS?

NO: REPLACE LEFT FRONT WHEEL SPEED SENSOR JUMPER HARNESS.

YES:

(9)
- USING J 39200, MEASURE THE RESISTANCE BETWEEN GROUND AND TERMINAL "A" OF THE LEFT FRONT WHEEL SPEED SENSOR CONNECTOR.
DOES J 39200 DISPLAY RESISTANCE AS "OL" (INFINITE)?

NO: REPLACE WHEEL SPEED SENSOR.

YES: MALFUNCTION IS INTERMITTENT. INSPECT ALL CONNECTORS AND HARNESSES FOR DAMAGE WHICH MAY RESULT IN A SHORT TO GROUND WHEN ALL COMPONENTS ARE CONNECTED. PAY PARTICULAR ATTENTION TO THE JUMPER HARNESS. SMALL NICKS OR CUTS CAN CAUSE A MALFUNCTION. REPAIR ANY DAMAGE IF FOUND. JUMPER HARNESS MUST BE REPLACED IF DAMAGED. REFER TO "DIAGNOSTIC AIDS" ON FACING PAGE FOR MORE INFORMATION.

DIAGNOSTIC CHART DTC 33 — 1994 CUTLASS SUPREME AND REGAL

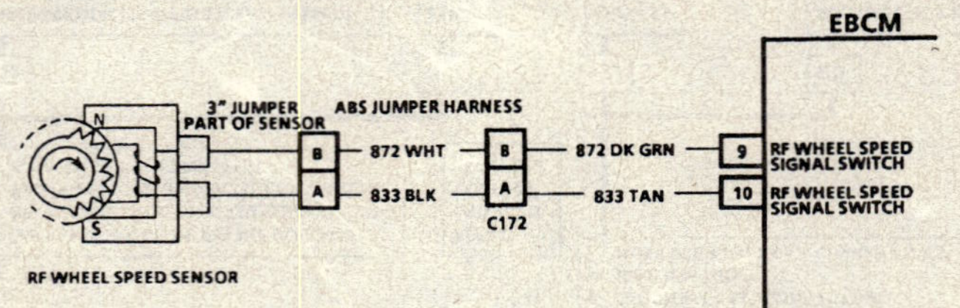

DTC 33

RIGHT FRONT WHEEL SPEED SENSOR CKT SHORT TO GND/BAT

Circuit Description:
As a toothed ring passes by the wheel speed sensor, changes in the electromagnetic field cause the wheel speed sensor to produce a sinusoidal (AC) voltage signal whose frequency is proportional to wheel speed. The magnitude of this signal is directly related to wheel speed and the proximity of the wheel speed sensor to the toothed ring, often referred to as the air gap.

DTC Will Set When: DTC 33 can be set anytime after initialization. If either of the right front wheel speed circuits are shorted to voltage or ground, a malfunction exists.

Action Taken: A malfunction DTC is stored, ABS is disabled and the ABS warning lamp is turned "ON."

DTC Chart Test Description: Number(s) below refer to circled number(s) on the diagnostic chart.

1. Checks for high resistance in wheel speed sensor circuitry.
2. Checks for short to voltage in wheel speed sensor wiring circuitry.
3. Checks for short to voltage in CKT 872.
4. Checks for short to voltage in CKT 833.
5. Checks for short to ground or voltage in CKTs 872 or 833.
6. Checks for short to ground in CKT 872.
7. Checks for short to ground in CKT 833.
8. Checks for high resistance in wheel speed sensor jumper harness.
9. Checks for high resistance in wheel speed sensor circuitry.

Diagnostic Aids: An "Intermittent" malfunction may be caused by a poor connection, rubbed through wire insulation, or a wire that is broken inside the insulation.

The frequency of the malfunction can be checked by using the enhanced diagnostic function of the Tech 1, as described in "Tech 1 Diagnostics" found in this section.

If the customer's comments reflect that the ABS warning lamp is "ON" only during moist environmental changes (rain, snow, vehicle wash), all wheel speed sensor circuitry should be thoroughly inspected for signs of water intrusion. If DTC is not current, clear DTCs and simulate the effects of water intrusion. Use the following procedure. Spray down the suspected area with a 5% salt water solution (two teaspoons of salt to 12 oz. of water). Test drive vehicle over various road surfaces (bumps, turns, etc.) above 24 km/h (15 mph) for at least 30 seconds. If DTC returns, replace suspected harness.

Any circuitry, that is suspected as causing the intermittent complaint, should be thoroughly checked for backed out terminals, improper mating, broken locks, improperly formed or damaged terminals, poor terminal to wiring connections or physical damage to the wiring harness.

Resistance of the wheel speed sensor will increase with an increase in sensor temperature.

When replacing a wheel speed sensor, inspect the sensor terminals and harness connector for corrosion and/or water intrusion. If evidence of corrosion or water intrusion exists, replace wheel speed sensor jumper harness. Likewise, if replacing a wheel speed sensor jumper harness, inspect sensor terminals. If evidence of corrosion or water intrusion exists, replace wheel speed sensor. Refer to "On-Vehicle Service" in this section.

DIAGNOSTIC CHART DTC 33 — 1994 CUTLASS SUPREME AND REGAL, CONTINUED

DTC 33 (Page 1 of 2) RIGHT FRONT WHEEL SPEED SENSOR CKT SHORT TO GND/BAT

Important

- WHEEL SPEED SENSOR INTERMITTENT MALFUNCTIONS MAY BE DIFFICULT TO LOCATE. CARE SHOULD BE TAKEN NOT TO DISTURB ANY ELECTRICAL CONNECTIONS PRIOR TO AN INDICATED STEP OF THIS CHART. THIS WILL ENSURE THAT AN INTERMITTENT CONNECTION WILL NOT BE CORRECTED BEFORE THE SOURCE OF THE MALFUNCTION IS FOUND.

1
- IGNITION "OFF."
- USING A 5% SALT WATER SOLUTION (TWO TEASPOONS OF SALT TO 12 OZ. WATER), THOROUGHLY SPRAY THE RIGHT FRONT WHEEL SPEED SENSOR JUMPER HARNESS AND SURROUNDING AREA.
- DISCONNECT 24-WAY EBCM HARNESS.
- USING J 39200, MEASURE THE RESISTANCE BETWEEN GROUND AND TERMINAL "9" OF THE 24-WAY EBCM HARNESS CONNECTOR.

DOES J 39200 DISPLAY RESISTANCE AS "OL" (INFINITE)?

- YES → 2
- NO → DIAGNOSIS CONTINUED ON PAGE 2 OF 2 OF THIS DTC CHART.

2
- IGNITION "ON."
- USING J 39200, MEASURE THE VOLTAGE BETWEEN GROUND AND TERMINAL "9" OF THE 24-WAY EBCM HARNESS CONNECTOR.

IS VOLTAGE 1 VOLT OR LESS?

- NO → 3
- YES → 5

3
- IGNITION "OFF."
- DISCONNECT RIGHT FRONT WHEEL SPEED SENSOR DIRECTLY AT SENSOR.
- IGNITION "ON."
- USING J 39200, MEASURE THE VOLTAGE BETWEEN GROUND AND TERMINAL "9" OF THE 24-WAY EBCM HARNESS CONNECTOR.

IS VOLTAGE 1 VOLT OR LESS?

- YES → 4
- NO → REPAIR SHORT TO VOLTAGE IN CKT 872.

4
- USING J 39200, MEASURE THE VOLTAGE BETWEEN GROUND AND TERMINAL "10" OF THE 24-WAY EBCM HARNESS CONNECTOR.

IS VOLTAGE 1 VOLT OR LESS?

- YES → MALFUNCTION IS INTERMITTENT. INSPECT ALL CONNECTORS AND HARNESSES FOR DAMAGE WHICH MAY RESULT IN A SHORT TO VOLTAGE WHEN ALL COMPONENTS ARE CONNECTED. REPAIR ALL DAMAGE IF FOUND. REFER TO "DIAGNOSTIC AIDS" ON FACING PAGE FOR MORE INFORMATION.
- NO → REPAIR SHORT TO VOLTAGE IN CKT 833.

5
- IGNITION "OFF."
- INSPECT TERMINALS "9" AND "10" OF THE 24-WAY EBCM CONNECTOR AND CKTs 872 AND 833 FOR DAMAGE WHICH MAY RESULT IN A SHORT TO VOLTAGE OR GROUND WHEN ALL COMPONENTS ARE CONNECTED. PAY PARTICULAR ATTENTION TO THE JUMPER HARNESS; SMALL NICKS OR CUTS CAN CAUSE A MALFUNCTION. REPAIR ANY DAMAGE IF FOUND. JUMPER HARNESSES MUST BE REPLACED IF DAMAGED.
- RECONNECT ALL CONNECTORS.
- IGNITION "ON."

DOES DTC 33 SET AS A CURRENT DTC?

- YES → REPLACE EBCM.
- NO → MALFUNCTION IS INTERMITTENT. REFER TO "DIAGNOSTIC AIDS" ON FACING PAGE.

DIAGNOSTIC CHART DTC 33 — 1994 CUTLASS SUPREME AND REGAL, CONTINUED

DTC 33 (Page 2 of 2) RIGHT FRONT WHEEL SPEED SENSOR CKT SHORT TO GND/BAT

Important

- WHEEL SPEED SENSOR INTERMITTENT MALFUNCTIONS MAY BE DIFFICULT TO LOCATE. CARE SHOULD BE TAKEN NOT TO DISTURB ANY ELECTRICAL CONNECTIONS PRIOR TO AN INDICATED STEP OF THIS CHART. THIS WILL ENSURE THAT AN INTERMITTENT CONNECTION WILL NOT BE CORRECTED BEFORE THE SOURCE OF THE MALFUNCTION IS FOUND.

DIAGNOSIS CONTINUED FROM PAGE 1 OF 2 OF THIS DTC CHART.

6
- DISCONNECT CONNECTOR C172 (MOUNTED ON RH SHOCK TOWER).
- USING J 39200, MEASURE THE RESISTANCE BETWEEN GROUND AND TERMINAL "9", OF THE 24-WAY EBCM HARNESS CONNECTOR.

DOES J 39200 DISPLAY RESISTANCE AS "OL" (INFINITE)?

- YES → 7
- NO → REPAIR SHORT TO GROUND IN CKT 872.

7
- USING J 39200, MEASURE THE RESISTANCE BETWEEN GROUND AND TERMINAL "10" OF THE 24-WAY EBCM HARNESS CONNECTOR.

DOES J 39200 DISPLAY RESISTANCE AS "OL" (INFINITE)?

- YES → 8
- NO → REPAIR SHORT TO GROUND IN CKT 833.

8
- THOROUGHLY RESPRAY THE RIGHT FRONT WHEEL SPEED SENSOR JUMPER HARNESS WITH THE SALT WATER SOLUTION. IMPORTANT: DO NOT ALLOW WATER TO ENTER ANY WHEEL SPEED SENSOR CONNECTOR.
- DISCONNECT RIGHT FRONT WHEEL SPEED SENSOR DIRECTLY AT SENSOR.
- USING J 39200, MEASURE THE RESISTANCE BETWEEN GROUND AND TERMINAL "A", THEN "B" OF THE RIGHT FRONT WHEEL SPEED SENSOR HARNESS CONNECTOR.

DOES J 39200 DISPLAY RESISTANCE AS "OL" (INFINITE) FOR BOTH MEASUREMENTS?

- YES → 9
- NO → REPLACE RIGHT FRONT WHEEL SPEED SENSOR JUMPER HARNESS.

9
- USING J 39200, MEASURE THE RESISTANCE BETWEEN GROUND AND TERMINAL "A" OF THE RIGHT FRONT WHEEL SPEED SENSOR CONNECTOR.

DOES J 39200 DISPLAY RESISTANCE AS "OL" (INFINITE)?

- YES → MALFUNCTION IS INTERMITTENT. INSPECT ALL CONNECTORS AND HARNESSES FOR DAMAGE WHICH MAY RESULT IN A SHORT TO GROUND WHEN ALL COMPONENTS ARE CONNECTED. PAY PARTICULAR ATTENTION TO THE JUMPER HARNESS, SMALL NICKS OR CUTS CAN CAUSE A MALFUNCTION. REPAIR ANY DAMAGE IF FOUND. JUMPER HARNESS MUST BE REPLACED IF DAMAGED. REFER TO "DIAGNOSTIC AIDS" ON FACING PAGE FOR MORE INFORMATION.
- NO → REPLACE WHEEL SPEED SENSOR.

DIAGNOSTIC CHART DTC 34 — 1994 CUTLASS SUPREME AND REGAL

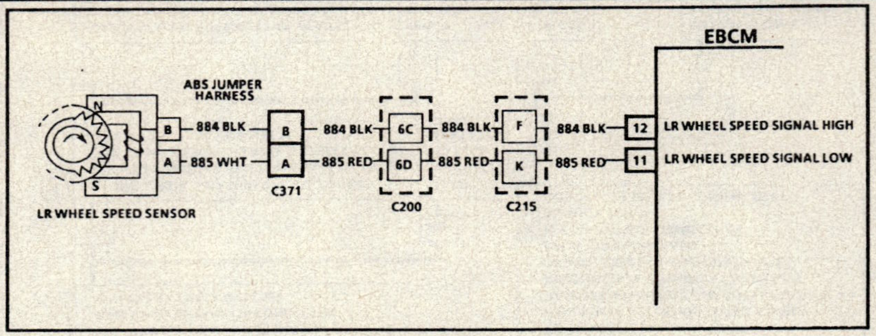

DTC 34

(Page 1 of 2)

LEFT REAR WHEEL SPEED SENSOR CKT SHORT TO GND/BAT

Circuit Description:

As a toothed ring passes by the wheel speed sensor, changes in the electromagnetic field cause the wheel speed sensor to produce a sinusoidal (AC) voltage signal whose frequency is proportional to wheel speed. The magnitude of this signal is directly related to wheel speed and the proximity of the wheel speed sensor to the toothed ring, often referred to as the air gap.

DTC Will Set When: DTC 34 can be set anytime after initialization. If either of the left rear wheel speed circuits are shorted to voltage or ground, a malfunction exists.

Action Taken: A malfunction DTC is stored, ABS is disabled and the ABS warning lamp is turned "ON."

DTC Chart Test Description: Number(s) below refer to circled number(s) on the diagnostic chart.

1. Checks for high resistance in wheel speed sensor wiring circuitry.
2. Checks for short to voltage in wheel speed sensor wiring circuitry.
3. Checks for short to voltage in CKT 885.
4. Checks for short to voltage in CKT 884.
5. Checks for short to ground or voltage in CKTs 885 or 884.
6. Checks for short to ground in CKT 885.
7. Checks for short to ground in CKT 884.
8. Checks for high resistance in wheel speed sensor jumper harness.
9. Checks for high resistance in wheel speed sensor circuitry.

Diagnostic Aids: An "Intermittent" malfunction may be caused by a poor connection, rubbed through wire insulation, or a wire that is broken inside the insulation.

The frequency of the malfunction can be checked by using the enhanced diagnostic function of the Tech 1, as described in "Tech 1 Diagnostics" found in tnis section.

If the customer's comments reflect that the ABS warning lamp is "ON" only during moist environmental changes (rain, snow, vehicle wash), all wheel speed sensor circuitry should be thoroughly inspected for signs of water intrusion. If DTC is not current, clear DTCs and simulate the effects of water intrusion. Use the following procedure. Spray down the suspected area with a 5% salt water solution (two teaspoons of salt to 12 oz. of water). Test drive vehicle over various road surfaces (bumps, turns, etc.) above 24 km/h (15 mph) for at least 30 seconds. If DTC returns, replace suspected harness.

Any circuitry, that is suspected as causing the intermittent complaint, should be thoroughly checked for backed out terminals, improper mating, broken locks, improperly formed or damaged terminals, poor terminal to wiring connections or physical damage to the wiring harness.

Resistance of the wheel speed sensor will increase with an increase in sensor temperature.

When replacing a wheel speed sensor, inspect the sensor terminals and harness connector for corrosion and/or water intrusion. If evidence of corrosion or water intrusion exists, replace wheel speed sensor jumper harness. Likewise, if replacing a wheel speed sensor jumper harness, inspect sensor terminals. If evidence of corrosion or water intrusion exists, replace wheel speed sensor. Refer to "On-Vehicle Service" in this section.

DIAGNOSTIC CHART DTC 34 — 1994 CUTLASS SUPREME AND REGAL, CONTINUED

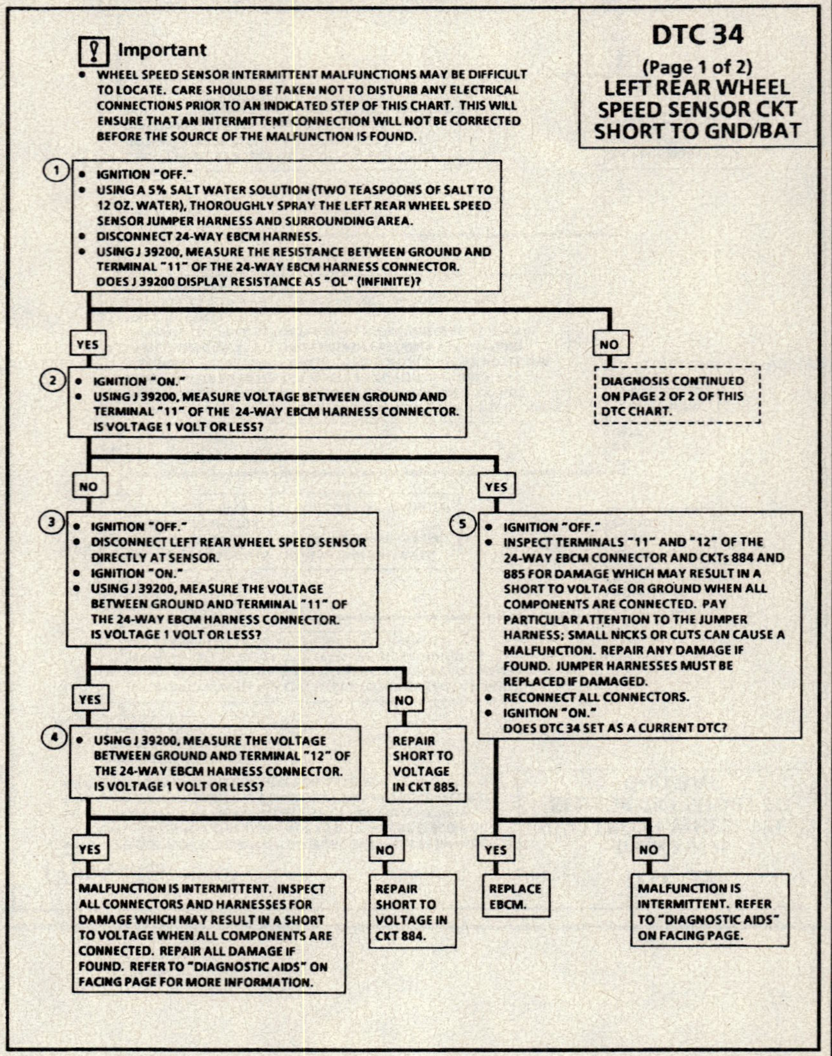

DIAGNOSTIC CHART DTC 34 — 1994 CUTLASS SUPREME AND REGAL, CONTINUED

DTC 34

(Page 2 of 2)

LEFT REAR WHEEL SPEED SENSOR CKT SHORT TO GND/BAT

Important

- WHEEL SPEED SENSOR INTERMITTENT MALFUNCTIONS MAY BE DIFFICULT TO LOCATE. CARE SHOULD BE TAKEN NOT TO DISTURB ANY ELECTRICAL CONNECTIONS PRIOR TO AN INDICATED STEP OF THIS CHART. THIS WILL ENSURE THAT AN INTERMITTENT CONNECTION WILL NOT BE CORRECTED BEFORE THE SOURCE OF THE MALFUNCTION IS FOUND.

DIAGNOSIS CONTINUED FROM PAGE 1 OF 2 OF THIS DTC CHART.

(6)
- DISCONNECT CONNECTOR C371 (LOCATED BEHIND LH SIDE OF REAR SEAT).
- USING J 39200, MEASURE THE RESISTANCE BETWEEN GROUND AND TERMINAL "11", OF THE 24-WAY EBCM HARNESS CONNECTOR.

DOES J 39200 DISPLAY RESISTANCE AS "OL" (INFINITE)?

- NO: REPAIR SHORT TO GROUND IN CKT 885.
- YES: go to 7.

(7)
- USING J 39200, MEASURE THE RESISTANCE BETWEEN GROUND AND TERMINAL "12" OF THE 24-WAY EBCM HARNESS CONNECTOR.

DOES J 39200 DISPLAY RESISTANCE AS "OL" (INFINITE)?

- NO: REPAIR SHORT TO GROUND IN CKT 884.
- YES: go to 8.

(8)
- THOROUGHLY RESPRAY THE LEFT REAR WHEEL SPEED SENSOR JUMPER HARNESS WITH THE SALT WATER SOLUTION. IMPORTANT: DO NOT ALLOW WATER TO ENTER ANY WHEEL SPEED SENSOR CONNECTOR.
- DISCONNECT LEFT REAR WHEEL SPEED SENSOR DIRECTLY AT SENSOR.
- USING J 39200, MEASURE THE RESISTANCE BETWEEN GROUND AND TERMINAL "A", THEN "B" OF THE LEFT REAR WHEEL SPEED SENSOR HARNESS CONNECTOR.

DOES J 39200 DISPLAY RESISTANCE AS "OL" (INFINITE) FOR BOTH MEASUREMENTS?

- NO: REPLACE LEFT REAR WHEEL SPEED SENSOR JUMPER HARNESS.
- YES: go to 9.

(9)
- USING J 39200, MEASURE THE RESISTANCE BETWEEN GROUND AND TERMINAL "A" OF THE LEFT REAR WHEEL SPEED SENSOR CONNECTOR.

DOES J 39200 DISPLAY RESISTANCE AS "OL" (INFINITE)?

- NO: REPLACE WHEEL SPEED SENSOR.
- YES: MALFUNCTION IS INTERMITTENT. INSPECT ALL CONNECTORS AND HARNESSES FOR DAMAGE WHICH MAY RESULT IN A SHORT TO GROUND WHEN ALL COMPONENTS ARE CONNECTED. PAY PARTICULAR ATTENTION TO THE JUMPER HARNESS, SMALL NICKS OR CUTS CAN CAUSE A MALFUNCTION. REPAIR ANY DAMAGE IF FOUND. JUMPER HARNESS MUST BE REPLACED IF DAMAGED. REFER TO "DIAGNOSTIC AIDS" ON FACING PAGE FOR MORE INFORMATION.

DIAGNOSTIC CHART DTC 35 — 1994 CUTLASS SUPREME AND REGAL

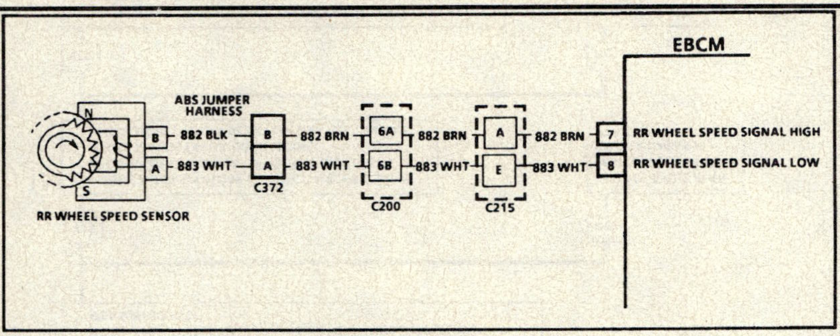

DTC 35

RIGHT REAR WHEEL SPEED SENSOR CKT SHORT TO GND/BAT

Circuit Description:

As a toothed ring passes by the wheel speed sensor, changes in the electromagnetic field cause the wheel speed sensor to produce a sinusoidal (AC) voltage signal whose frequency is proportional to wheel speed. The magnitude of this signal is directly related to wheel speed and the proximity of the wheel speed sensor to the toothed ring, often referred to as the air gap.

DTC Will Set When: DTC 35 can be set anytime after initialization. If either of the right rear wheel speed circuits are shorted to voltage or ground, a malfunction exists.

Action Taken: A malfunction DTC is stored, ABS is disabled and the ABS warning lamp is turned "ON."

DTC Chart Test Description: Number(s) below refer to circled number(s) on the diagnostic chart.

1. Checks for high resistance in wheel speed sensor wiring.
2. Checks for short to voltage in wheel speed sensor wiring.
3. Checks for short to voltage in CKT 882.
4. Checks for short to voltage in CKT 883.
5. Checks for short to ground or voltage in CKTs 882 or 883.
6. Checks for short to ground in CKT 882.
7. Checks for short to ground in CKT 883.
8. Checks for short to ground in wheel speed sensor jumper harness.
9. Checks for short to ground in wheel speed sensor circuitry.

Diagnostic Aids: An "Intermittent" malfunction may be caused by a poor connection, rubbed through wire insulation, or a wire that is broken inside the insulation.

The frequency of the malfunction can be checked by using the enhanced diagnostic function of the Tech 1, as described in "Tech 1 Diagnostics" found in this section.

If the customer's comments reflect that the ABS warning lamp is "ON" only during moist environmental changes (rain, snow, vehicle wash), all wheel speed sensor circuitry should be thoroughly inspected for signs of water intrusion. If DTC is not current, clear DTCs and simulate the effects of water intrusion. Use the following procedure. Spray down the suspected area with a 5% salt water solution (two teaspoons of salt to 12 oz. of water). Test drive vehicle over various road surfaces (bumps, turns, etc.) above 24 km/h (15 mph) for at least 30 seconds. If DTC returns, replace suspected harness.

Any circuitry, that is suspected as causing the intermittent complaint, should be thoroughly checked for backed out terminals, improper mating, broken locks, improperly formed or damaged terminals, poor terminal to wiring connections or physical damage to the wiring harness.

Resistance of the wheel speed sensor will increase with an increase in sensor temperature.

When replacing a wheel speed sensor, inspect the sensor terminals and harness connector for corrosion and/or water intrusion. If evidence of corrosion or water intrusion exists, replace wheel speed sensor jumper harness. Likewise, if replacing a wheel speed sensor jumper harness, inspect sensor terminals. If evidence of corrosion or water intrusion exists, replace wheel speed sensor. Refer to "On-Vehicle Service" in this section.

DIAGNOSTIC CHART DTC 35 — 1994 CUTLASS SUPREME AND REGAL, CONTINUED

DTC 35 (Page 1 of 2) RIGHT REAR WHEEL SPEED SENSOR CKT SHORT TO GND/BAT

Important

WHEEL SPEED SENSOR INTERMITTENT MALFUNCTIONS MAY BE DIFFICULT TO LOCATE. CARE SHOULD BE TAKEN NOT TO DISTURB ANY ELECTRICAL CONNECTIONS PRIOR TO AN INDICATED STEP OF THIS CHART. THIS WILL ENSURE THAT AN INTERMITTENT CONNECTION WILL NOT BE CORRECTED BEFORE THE SOURCE OF THE MALFUNCTION IS FOUND.

1.
- IGNITION "OFF."
- USING A 5% SALT WATER SOLUTION (TWO TEASPOONS OF SALT TO 12 OZ. WATER), THOROUGHLY SPRAY THE RIGHT REAR WHEEL SPEED SENSOR JUMPER HARNESS AND SURROUNDING AREA.
- DISCONNECT 24-WAY EBCM HARNESS.
- USING J 39200, MEASURE THE RESISTANCE BETWEEN GROUND AND TERMINAL "7" OF THE 24-WAY EBCM HARNESS CONNECTOR.

DOES J 39200 DISPLAY RESISTANCE AS "OL" (INFINITE)?

- YES: Go to step 2.
- NO: DIAGNOSIS CONTINUED ON PAGE 2 OF 2 OF THIS DTC CHART.

2.
- IGNITION "ON."
- USING J 39200, MEASURE THE VOLTAGE BETWEEN GROUND AND TERMINAL "7" OF THE 24-WAY EBCM HARNESS CONNECTOR.

IS VOLTAGE 1 VOLT OR LESS?

- NO: Go to step 3.
- YES: Go to step 5.

3.
- IGNITION "OFF."
- DISCONNECT RIGHT REAR WHEEL SPEED SENSOR DIRECTLY AT SENSOR.
- IGNITION "ON."
- USING J 39200, MEASURE THE VOLTAGE BETWEEN GROUND AND TERMINAL "7" OF THE 24-WAY EBCM HARNESS CONNECTOR.

IS VOLTAGE 1 VOLT OR LESS?

- YES: Go to step 4.
- NO: REPAIR SHORT TO VOLTAGE IN CKT 882.

4.
- USING J 39200, MEASURE THE VOLTAGE BETWEEN GROUND AND TERMINAL "8" OF THE 24-WAY EBCM HARNESS CONNECTOR.

IS VOLTAGE 1 VOLT OR LESS?

- YES: MALFUNCTION IS INTERMITTENT. INSPECT ALL CONNECTORS AND HARNESSES FOR DAMAGE WHICH MAY RESULT IN A SHORT TO VOLTAGE WHEN ALL COMPONENTS ARE CONNECTED. REPAIR ALL DAMAGE IF FOUND. REFER TO "DIAGNOSTIC AIDS" ON FACING PAGE FOR MORE INFORMATION.
- NO: REPAIR SHORT TO VOLTAGE IN CKT 883.

5.
- IGNITION "OFF."
- INSPECT TERMINALS "7" AND "8" OF THE 24-WAY EBCM CONNECTOR AND CKTs 882 AND 883 FOR DAMAGE WHICH MAY RESULT IN A SHORT TO VOLTAGE OR GROUND WHEN ALL COMPONENTS ARE CONNECTED. PAY PARTICULAR ATTENTION TO THE JUMPER HARNESS; SMALL NICKS OR CUTS CAN CAUSE A MALFUNCTION. REPAIR ANY DAMAGE IF FOUND. JUMPER HARNESSES MUST BE REPLACED IF DAMAGED.
- RECONNECT ALL CONNECTORS.
- IGNITION "ON."

DOES DTC 35 SET AS A CURRENT DTC?

- YES: REPLACE EBCM.
- NO: MALFUNCTION IS INTERMITTENT. REFER TO "DIAGNOSTIC AIDS" ON FACING PAGE.

DIAGNOSTIC CHART DTC 35 — 1994 CUTLASS SUPREME AND REGAL, CONTINUED

DTC 35 (Page 2 of 2) RIGHT REAR WHEEL SPEED SENSOR CKT SHORT TO GND/BAT

Important

- WHEEL SPEED SENSOR INTERMITTENT MALFUNCTIONS MAY BE DIFFICULT TO LOCATE. CARE SHOULD BE TAKEN NOT TO DISTURB ANY ELECTRICAL CONNECTIONS PRIOR TO AN INDICATED STEP OF THIS CHART. THIS WILL ENSURE THAT AN INTERMITTENT CONNECTION WILL NOT BE CORRECTED BEFORE THE SOURCE OF THE MALFUNCTION IS FOUND.

DIAGNOSIS CONTINUED FROM PAGE 1 OF 2 OF THIS DTC CHART.

6.
- DISCONNECT CONNECTOR C372 (LOCATED BEHIND RH SIDE OF REAR SEAT).
- USING J 39200, MEASURE THE RESISTANCE BETWEEN GROUND AND TERMINAL "7", OF THE 24-WAY EBCM HARNESS CONNECTOR.

DOES J 39200 DISPLAY RESISTANCE AS "OL" (INFINITE)?

- YES: Go to step 7.
- NO: REPAIR SHORT TO GROUND IN CKT 882.

7.
- USING J 39200, MEASURE THE RESISTANCE BETWEEN GROUND AND TERMINAL "8" OF THE 24-WAY EBCM HARNESS CONNECTOR.

DOES J 39200 DISPLAY RESISTANCE AS "OL" (INFINITE)?

- YES: Go to step 8.
- NO: REPAIR SHORT TO GROUND IN CKT 883.

8.
- THOROUGHLY RESPRAY THE RIGHT REAR WHEEL SPEED SENSOR JUMPER HARNESS WITH THE SALT WATER SOLUTION. IMPORTANT: DO NOT ALLOW WATER TO ENTER ANY WHEEL SPEED SENSOR CONNECTOR.
- DISCONNECT RIGHT REAR WHEEL SPEED SENSOR DIRECTLY AT SENSOR.
- USING J 39200, MEASURE THE RESISTANCE BETWEEN GROUND AND TERMINAL "A", THEN "B" OF THE RIGHT REAR WHEEL SPEED SENSOR HARNESS CONNECTOR.

DOES J 39200 DISPLAY RESISTANCE AS "OL" (INFINITE) FOR BOTH MEASUREMENTS?

- YES: Go to step 9.
- NO: REPLACE RIGHT REAR WHEEL SPEED SENSOR JUMPER HARNESS.

9.
- USING J 39200, MEASURE THE RESISTANCE BETWEEN GROUND AND TERMINAL "A" OF THE RIGHT REAR WHEEL SPEED SENSOR CONNECTOR.

DOES J 39200 DISPLAY RESISTANCE AS "OL" (INFINITE)?

- YES: MALFUNCTION IS INTERMITTENT. INSPECT ALL CONNECTORS AND HARNESSES FOR DAMAGE WHICH MAY RESULT IN A SHORT TO GROUND WHEN ALL COMPONENTS ARE CONNECTED. PAY PARTICULAR ATTENTION TO THE JUMPER HARNESS, SMALL NICKS OR CUTS CAN CAUSE A MALFUNCTION. REPAIR ANY DAMAGE IF FOUND. JUMPER HARNESS MUST BE REPLACED IF DAMAGED. REFER TO "DIAGNOSTIC AIDS" ON FACING PAGE FOR MORE INFORMATION.
- NO: REPLACE WHEEL SPEED SENSOR.

DIAGNOSTIC CHART DTC 36 — 1994 CUTLASS SUPREME AND REGAL

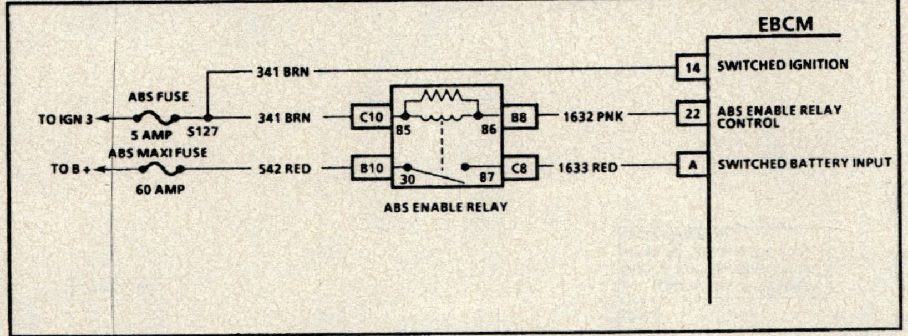

DTC 36

LOW SYSTEM VOLTAGE

Circuit Description:
This DTC is used to monitor the voltage level available to the EBCM controller. If the voltage drops below 11 volts, full performance of the ABS system cannot be guaranteed. During ABS operation, there are several current requirements that will cause battery voltage to drop. Because of this, voltage is monitored prior to ABS to indicate good charging system condition and also during ABS when voltage may drop significantly.

DTC Will Set When: DTC 36 can only be set if the vehicle's speed is greater than 5 km/h (3 mph). If the switched battery voltage is less than 11.2 volts non-ABS or less than 9.6 volts ABS, a malfunction exists.

Action Taken: A malfunction DTC is stored, ABS is disabled and the ABS warning lamp is turned "ON."

DTC Chart Test Description: Number(s) below refer to circled number(s) on the diagnostic chart.
1. Checks to see what voltage is readily available at terminal "A" of the EBCM. If voltage is representative of good charging system condition, a malfunction is not present.
2. Isolates the condition to high circuit resistance, or improper charging system operation.
3. Isolates which circuit is the source of the high resistance condition.
4. Checks for high resistance in CKTs 542 or 2.
5. Checks for high resistance in CKT 341.
6. Checks for high resistance in the ignition fuse feed circuit.
7. Checks for malfunctioning EBCM.

Diagnostic Aids: An "Intermittent" malfunction may be caused by a poor connection, rubbed through wire insulation, or a wire that is broken inside the insulation.

The frequency of the malfunction can be checked by using the enhanced diagnostic function of the Tech 1, as described in "Tech 1 Diagnostics" found in this section.

Any circuitry, that is suspected as causing the intermittent complaint, should be thoroughly checked for backed out terminals, improper mating, broken locks, improperly formed or damaged terminals, poor terminal to wiring connections or physical damage to the wiring harness.

While performing a voltage load test, if it is noted that only ignition voltage drops below acceptable voltage levels, CKT 341 should be checked for high resistance or an open condition.

DIAGNOSTIC CHART DTC 36 — 1994 CUTLASS SUPREME AND REGAL, CONTINUED

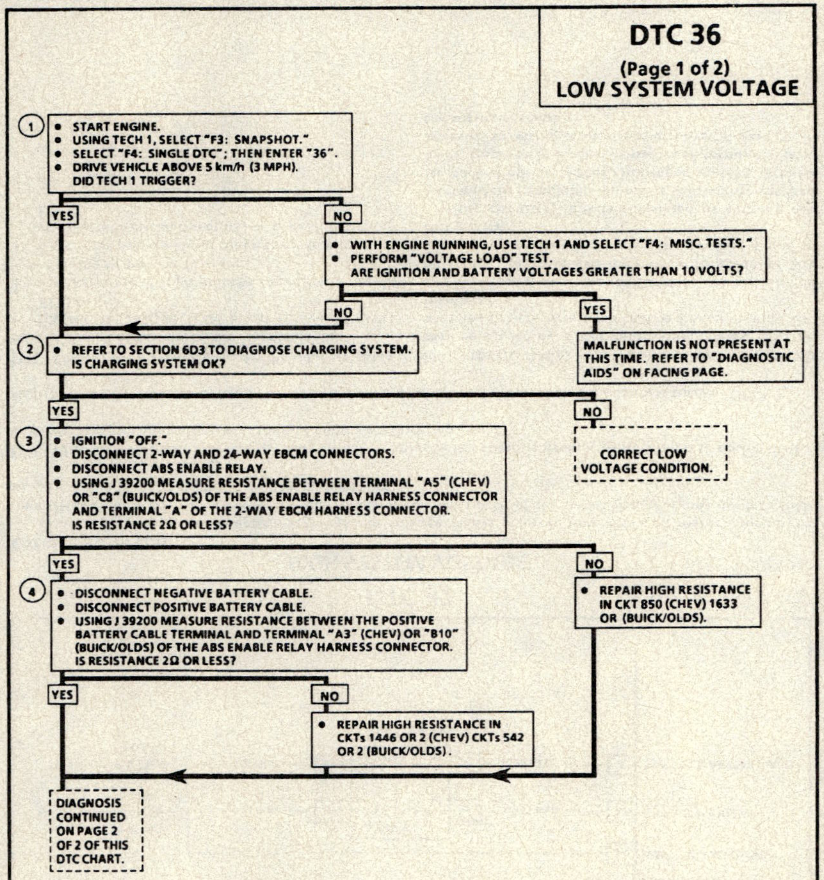

DIAGNOSTIC CHART DTC 36 — 1994 CUTLASS SUPREME AND REGAL, CONTINUED

DTC 36
(Page 2 of 2)
LOW SYSTEM VOLTAGE

DIAGNOSIS CONTINUED FROM PAGE 1 OF 2 OF THIS DTC CHART.

(5)
- REMOVE ABS FUSE.
- USING J 39200 MEASURE RESISTANCE BETWEEN EACH ABS FUSE BLOCK TERMINAL AND TERMINAL "14" OF THE 24-WAY EBCM HARNESS CONNECTOR. IS EITHER MEASUREMENT 2Ω OR LESS?

NO → REPAIR HIGH RESISTANCE IN CKT 350 (CHEV) OR 341 (BUICK/OLDS).

YES →

(6)
- WITH BATTERY CABLES DISCONNECTED, TURN IGNITION "ON" (THIS PROVIDES IGNITION CIRCUIT CONTINUITY).
- USING J 39200, MEASURE RESISTANCE BETWEEN POSITIVE BATTERY CABLE TERMINAL AND EACH ABS FUSE BLOCK TERMINAL. IS EITHER MEASUREMENT 2Ω OR LESS ?

NO → REPAIR HIGH RESISTANCE IN THE "IGNITION SWITCH 'RUN' FUSE FEED" CIRCUIT.

YES →

(7)
- IGNITION "OFF."
- INSPECT ABS ENABLE RELAY AND ABS ENABLE RELAY HARNESS CONNECTOR FOR POOR TERMINAL CONTACT. INSPECT 2-WAY AND 24-WAY EBCM HARNESS CONNECTORS FOR A POOR CONNECTION TO THE EBCM. INSPECT POSITIVE BATTERY CABLE TERMINAL FOR POOR CONTACT TO THE BATTERY.

REPLACE TERMINALS IF POOR CONTACT IS EVIDENT.
- RECONNECT ALL CONNECTORS.
- DRIVE VEHICLE ABOVE 5 km/h (3 MPH). DOES DTC 36 SET AS A CURRENT DTC?

YES → REPLACE EBCM.

NO → MALFUNCTION IS MOST LIKELY DUE TO A POOR CONNECTION. REFER TO "DIAGNOSTIC AIDS" ON FACING PAGE.

DIAGNOSTIC CHART DTC 37 — 1994 CUTLASS SUPREME AND REGAL

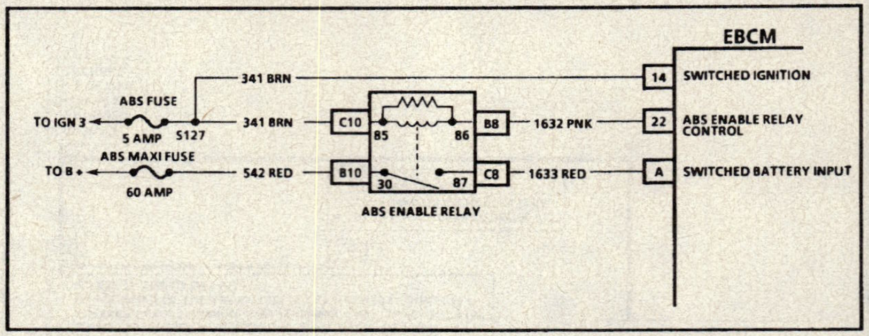

DTC 37
HIGH SYSTEM VOLTAGE

Circuit Description:
This DTC is designed to detect high vehicle voltage levels prior to any required motor movement (initialization or ABS). If excessive voltage exists, demagnetization of the motor magnets may occur, which would eventually affect or eliminate ABS performance.

DTC Will Set When: DTC 37 can only be set if the vehicle's speed is greater than 5 km/h (3 mph). If the switched battery voltage is greater than 17 volts, a malfunction exists.

Action Taken: A malfunction DTC is stored, ABS is disabled and the ABS warning lamp is turned "ON."

DTC Chart Test Description: Number(s) below refer to circled number(s) on the diagnostic chart.
1. Checks the voltage level being received by the EBCM.
2. Checks to see if high voltage condition is due to charging system malfunction.
3. Checks for poor terminal contact which may cause an intermittent malfunction, or a malfunctioning EBCM.

Diagnostic Aids: An "Intermittent" malfunction may be caused by a poor connection, rubbed through wire insulation, or a wire that is broken inside the insulation.

The frequency of the malfunction can be checked by using the enhanced diagnostic function of the Tech 1, as described in "Tech 1 Diagnostics" found in this section.

Any circuitry, that is suspected as causing the intermittent complaint, should be thoroughly checked for backed out terminals, improper mating, broken locks, improperly formed or damaged terminals, poor terminal to wiring connections or physical damage to the wiring harness.

DIAGNOSTIC CHART DTC 37 — 1994 CUTLASS SUPREME AND REGAL, CONTINUED

DTC 37
HIGH SYSTEM VOLTAGE

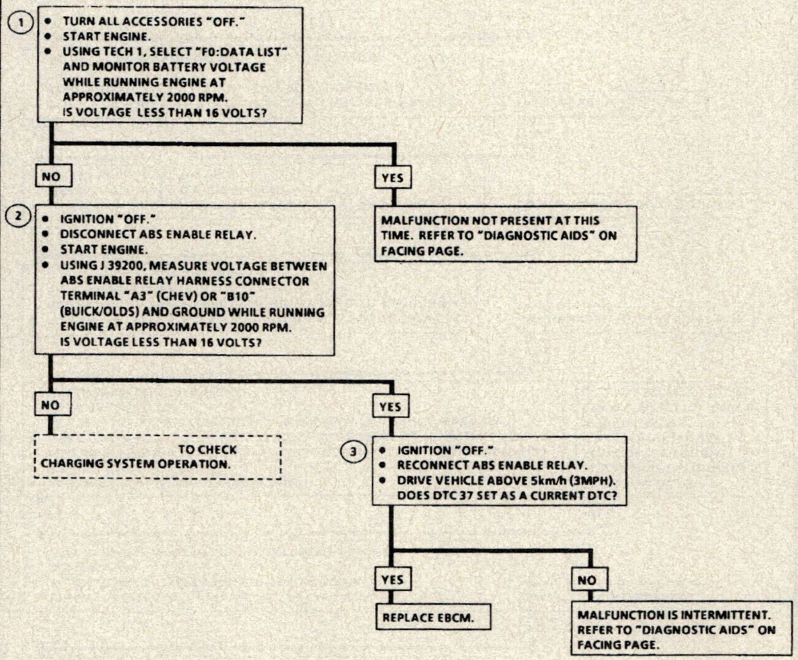

DIAGNOSTIC CHART DTC 38 — 1994 CUTLASS SUPREME AND REGAL

DTC 38
LEFT FRONT ESB WILL NOT HOLD MOTOR

Circuit Description:
This DTC is designed to detect a slipping left front ESB. During initialization and braking, the left front motor is re-homed. If the ESB slips, the motor/piston will move. During the next key "ON" initialization, a re-home of the motor verifies the motor/piston remained at the home position. If motor movement is detected, the ESB must be slipping.

DTC Will Set When: DTC 38 can be set during initialization. If the EBCM detects that the ESB could not hold the piston in the home position, a malfunction exists. DTC 86 is always set with DTC 38.

Action Taken: If an ESB cannot hold a piston in the home position, the piston may be backdriven when the brake pedal is applied, causing the brake pedal to drop. A malfunction DTC is stored, ABS is disabled, and the ABS warning lamp and the red "BRAKE" warning lamp are turned "ON."

DTC Chart Test Description: Number(s) below refer to circled number(s) on the diagnostic chart.

1. Checks the left front ESB. A broken or defective ESB would result in the left front piston being backdriven by hydraulic pressure and pedal movement would result.
2. Ensures that the proper motor circuits are located in the proper connector cavities.
3. Checks for high resistance in the motor "HIGH" circuitry.
4. Checks for high resistance in the motor "LOW" circuitry.
5. Checks for proper resistance of the motor windings.
6. Checks for poor terminal contact or corrosion in the connectors.
7. Checks for a malfunctioning EBCM.

Diagnostic Aids: An "Intermittent" malfunction in this DTC may result from a mechanical part of the system that sticks, binds, or slips.

The frequency of the malfunction can be checked by using the enhanced diagnostic function of the Tech 1, as described in "Tech 1 Diagnostics" found in this section.

The static modulator test function of the Tech 1 or T-100 may be used to locate an intermittent malfunction associated with the ESB.

DIAGNOSTIC CHART DTC 38 — 1994 CUTLASS SUPREME AND REGAL, CONTINUED

DTC 38
LEFT FRONT ESB WILL NOT HOLD MOTOR

1.
- IGNITION "ON," ENGINE "OFF."
- PUMP BRAKE PEDAL UNTIL FIRM TO DEPLETE VACUUM RESERVE.
- USING TECH 1, SELECT MANUAL CONTROL FUNCTION.
- SELECT ABS MOTOR APPLY FUNCTION OF THE LEFT FRONT ABS CHANNEL AND APPLY THE MOTOR.
- WAIT 5 SECONDS.
- APPLY FIRM PRESSURE ON BRAKE PEDAL.

DID BRAKE PEDAL FALL?

- YES: REPLACE ABS MOTOR PACK.
- NO: go to step 2.

2.
- VISUALLY INSPECT THE 6-WAY ABS MOTOR PACK AND 6-WAY EBCM CONNECTORS FOR PROPER WIRE COLOR/CONNECTOR CAVITY COMBINATION.

ARE THE PROPER WIRES LOCATED IN THE PROPER CONNECTOR CAVITIES?

- NO: USING THE CIRCUIT DIAGRAMS LOCATED IN THE FRONT OF THIS SECTION, PLACE THE PROPER WIRES INTO THE PROPER CONNECTOR CAVITIES. CLEAR DTCs AND RECHECK SYSTEM.
- YES: go to step 3.

3.
- EXIT TECH 1 MANUAL CONTROL FUNCTION.
- IGNITION "OFF."
- DISCONNECT 6-WAY ABS MOTOR PACK CONNECTOR AND 6-WAY EBCM CONNECTOR.
- USING J 39200, MEASURE THE RESISTANCE BETWEEN TERMINAL "F" OF THE 6-WAY ABS MOTOR PACK HARNESS CONNECTOR AND TERMINAL "F" OF THE 6-WAY EBCM HARNESS CONNECTOR.

IS RESISTANCE LESS THAN 1.5Ω?

- NO: REPAIR HIGH RESISTANCE IN CKT 1280.
- YES: go to step 4.

4.
- MEASURE THE RESISTANCE BETWEEN TERMINAL "E" OF THE 6-WAY ABS MOTOR PACK HARNESS CONNECTOR AND TERMINAL "E" OF THE 6-WAY EBCM HARNESS CONNECTOR.

IS RESISTANCE LESS THAN 1.5Ω?

- NO: REPAIR HIGH RESISTANCE IN CKT 1281.
- YES: go to step 5.

5.
- MEASURE THE RESISTANCE BETWEEN TERMINAL "E" AND TERMINAL "F" OF THE ABS MOTOR PACK CONNECTOR.

IS RESISTANCE LESS THAN 1.5Ω?

- NO: REPLACE ABS MOTOR PACK.
- YES: go to step 6.

6.
- INSPECT THE 6-WAY ABS MOTOR PACK CONNECTOR AND 6-WAY EBCM CONNECTOR FOR POOR TERMINAL CONTACT OR CORROSION.

DO ANY TERMINALS EXHIBIT POOR CONTACT OR EVIDENCE OF CORROSION?

- YES: REPLACE TERMINALS THAT EXHIBIT POOR TERMINAL CONTACT OR EVIDENCE OF CORROSION.
- NO: go to step 7.

7.
- IGNITION "OFF."
- RECONNECT ALL CONNECTORS.
- START ENGINE WITH FOOT OFF BRAKE.
- ALLOW ENGINE TO RUN FOR AT LEAST 10 SECONDS.
- REPEAT ABOVE IGNITION CYCLE SEQUENCE TWO MORE TIMES.

DID DTC 38 SET IN THE LAST THREE IGNITION CYCLES?

- YES: REPLACE EBCM.
- NO: MALFUNCTION IS INTERMITTENT. REFER TO "DIAGNOSTIC AIDS" ON FACING PAGE.

AFTER DIAGNOSIS IS COMPLETE, CLEAR DTCs AND TEST DRIVE VEHICLE FOR THREE (3) DRIVE CYCLES TO VERIFY THAT DTC DOES NOT RESET. A DRIVE CYCLE CONSISTS OF STARTING THE VEHICLE, DRIVING OVER 16 km/h (10 MPH) AND THEN KEYING DOWN.

DIAGNOSTIC CHART DTC 41 — 1994 CUTLASS SUPREME AND REGAL

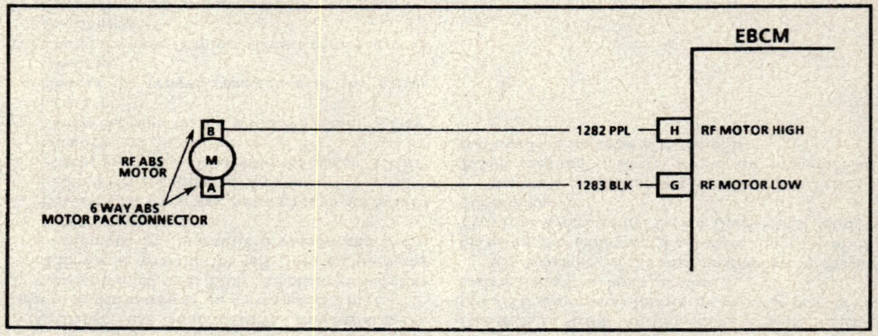

DTC 41
RIGHT FRONT ESB WILL NOT HOLD MOTOR

Circuit Description:
This DTC is designed to detect a slipping right front ESB. During initialization and braking, the right front motor is re-homed. If the ESB slips, the motor/piston will move. During the next key "ON" initialization, a re-home of the motor verifies the motor/piston remained at the home position. If motor movement is detected, the ESB must be slipping.

DTC Will Set When: DTC 41 can be set during initialization. If the EBCM detects that the ESB could not hold the piston in the home position, a malfunction exists. DTC 86 is always set with DTC 41.

Action Taken: If an ESB cannot hold a piston in the home position, the piston may be backdriven when the brake pedal is applied, causing the brake pedal to drop. A malfunction DTC is stored, ABS is disabled, and the ABS warning lamp and the red "BRAKE" warning lamp are turned "ON."

DTC Chart Test Description: Number(s) below refer to circled number(s) on the diagnostic chart.

1. Checks the right front ESB. A broken or defective ESB would result in the right front piston being backdriven by hydraulic pressure, and pedal movement would result.
2. Ensures that the proper motor circuits are located in the proper connector cavities.
3. Checks for high resistance in the motor "HIGH" circuitry.
4. Checks for high resistance in the motor "LOW" circuitry.
5. Checks for proper resistance of the motor windings.
6. Checks for poor terminal contact or corrosion in the connectors.
7. Checks for a malfunctioning EBCM.

Diagnostic Aids: An "Intermittent" malfunction in this DTC may result from a mechanical part of the system that sticks, binds, or slips.

The frequency of the malfunction can be checked by using the enhanced diagnostic function of the Tech 1, as described in "Tech 1 Diagnostics" found in this section.

The static modulator test function of the Tech 1 or T-100 may be used to locate an intermittent malfunction associated with the ESB.

DIAGNOSTIC CHART DTC 41 — 1994 CUTLASS SUPREME AND REGAL, CONTINUED

DTC 41
RIGHT FRONT ESB WILL NOT HOLD MOTOR

(1)
- IGNITION "ON," ENGINE "OFF."
- PUMP BRAKE PEDAL UNTIL FIRM TO DEPLETE VACUUM RESERVE.
- USING TECH 1 MISC TESTS, SELECT MANUAL CONTROL FUNCTION.
- SELECT ABS MOTOR APPLY FUNCTION OF THE RIGHT FRONT ABS CHANNEL AND APPLY THE MOTOR.
- WAIT 5 SECONDS.
- APPLY FIRM PRESSURE ON BRAKE PEDAL.

DID BRAKE PEDAL FALL?

- YES: REPLACE ABS MOTOR PACK.
- NO: go to step 2.

(2)
- VISUALLY INSPECT THE 6-WAY ABS MOTOR PACK AND 6-WAY EBCM CONNECTORS FOR PROPER WIRE COLOR/CONNECTOR CAVITY COMBINATION.

ARE THE PROPER WIRES LOCATED IN THE PROPER CONNECTOR CAVITIES?

- NO: USING THE CIRCUIT DIAGRAMS LOCATED IN THE FRONT OF THIS SECTION, PLACE THE PROPER WIRES INTO THE PROPER CONNECTOR CAVITIES. CLEAR DTCs AND RECHECK SYSTEM.
- YES: go to step 3.

(3)
- EXIT TECH 1 MANUAL CONTROL FUNCTION.
- IGNITION "OFF."
- DISCONNECT 6-WAY ABS MOTOR PACK CONNECTOR AND 6-WAY EBCM CONNECTOR.
- USING J 39200, MEASURE THE RESISTANCE BETWEEN TERMINAL "B" OF THE 6-WAY ABS MOTOR PACK HARNESS CONNECTOR AND TERMINAL "H" OF THE 6-WAY EBCM HARNESS CONNECTOR.

IS RESISTANCE LESS THAN 1.5Ω?

- NO: REPAIR HIGH RESISTANCE IN CKT 1282.
- YES: go to step 4.

(4)
- MEASURE THE RESISTANCE BETWEEN TERMINAL "A" OF THE 6-WAY ABS MOTOR PACK HARNESS CONNECTOR AND TERMINAL "G" OF THE 6-WAY EBCM HARNESS CONNECTOR.

IS RESISTANCE LESS THAN 1.5Ω?

- NO: REPAIR HIGH RESISTANCE IN CKT 1283.
- YES: go to step 5.

(5)
- MEASURE THE RESISTANCE BETWEEN TERMINAL "A" AND TERMINAL "B" OF THE ABS MOTOR PACK CONNECTOR.

IS RESISTANCE LESS THAN 1.5Ω?

- NO: REPLACE ABS MOTOR PACK.
- YES: go to step 6.

(6)
- INSPECT THE 6-WAY ABS MOTOR PACK CONNECTOR AND 6-WAY EBCM CONNECTOR FOR POOR TERMINAL CONTACT OR CORROSION.

DO ANY TERMINALS EXHIBIT POOR CONTACT OR EVIDENCE OF CORROSION?

- YES: REPLACE TERMINALS THAT EXHIBIT POOR TERMINAL CONTACT OR EVIDENCE OF CORROSION.
- NO: go to step 7.

(7)
- IGNITION "OFF."
- RECONNECT ALL CONNECTORS.
- START ENGINE WITH FOOT OFF BRAKE.
- ALLOW ENGINE TO RUN FOR AT LEAST 10 SECONDS.
- REPEAT ABOVE IGNITION CYCLE TWO MORE TIMES.

DID DTC 41 SET IN THE LAST THREE IGNITION CYCLES?

- YES: REPLACE EBCM.
- NO: MALFUNCTION IS INTERMITTENT. REFER TO "DIAGNOSTIC AIDS" ON FACING PAGE.

AFTER DIAGNOSIS IS COMPLETE, CLEAR DTCs AND TEST DRIVE VEHICLE FOR THREE (3) DRIVE CYCLES TO VERIFY THAT DTC DOES NOT RESET. A DRIVE CYCLE CONSISTS OF STARTING THE VEHICLE, DRIVING OVER 16 KM/H (10 MPH) AND THEN KEYING DOWN.

DIAGNOSTIC CHART DTC 42 — 1994 CUTLASS SUPREME AND REGAL

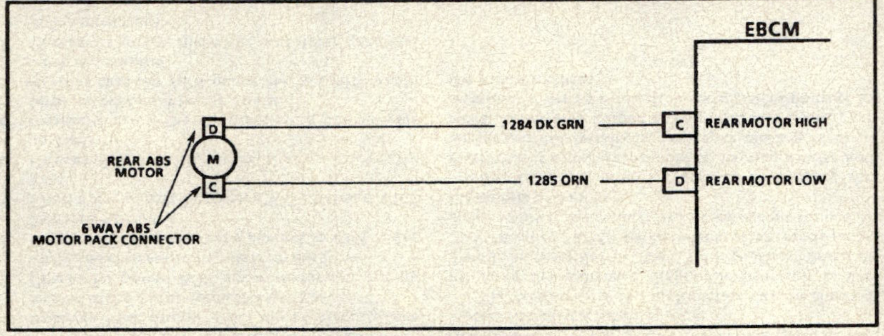

DTC 42
REAR ESB WILL NOT HOLD MOTOR

Circuit Description:
This DTC is designed to detect a slipping rear axle ESB. During initialization and braking, the rear motor is re-homed. If the ESB slips, the motor/piston will move. During the next key "ON" initialization, a re-home of the motor verifies the motor/piston remained at the home position. If motor movement is detected, the ESB must be slipping.

DTC Will Set When: DTC 42 can be set during initialization. If the EBCM detects that the ESB could not hold the piston in the home position, a malfunction exists. DTC 86 is always set with DTC 42.

Action Taken: If an ESB cannot hold a piston in the home position, the piston may be backdriven when the brake pedal is applied, causing the brake pedal to drop. A malfunction DTC is stored, ABS is disabled, and the ABS warning lamp and the red "BRAKE" warning lamp are turned "ON."

DTC Chart Test Description: Number(s) below refer to circled number(s) on the diagnostic chart.

1. Checks the rear axle ESB. A broken or defective ESB would result in the rear axle pistons being backdriven by hydraulic pressure and wheel movement would result.
2. Ensures that the proper motor circuits are located in the proper connector cavities.
3. Checks for high resistance in the motor "HIGH" circuitry.
4. Checks for high resistance in the motor "LOW" circuitry.
5. Checks for proper resistance of the motor windings.
6. Checks for poor terminal contact or corrosion in the connectors.
7. Checks for a malfunctioning EBCM.

Diagnostic Aids: An "Intermittent" malfunction in this DTC may result from a mechanical part of the system that sticks, binds, or slips.

The frequency of the malfunction can be checked by using the enhanced diagnostic function of the Tech 1, as described in "Tech 1 Diagnostics" found in this section.

The static modulator test function of the Tech 1 or T-100 may be used to locate an intermittent malfunction associated with the ESB.

DIAGNOSTIC CHART DTC 42 — 1994 CUTLASS SUPREME AND REGAL, CONTINUED

DTC 42
REAR ESB WILL NOT HOLD MOTOR

1.
   - RAISE THE VEHICLE SUCH THAT THE REAR WHEELS ARE APPROXIMATELY SIX INCHES OFF THE FLOOR.
   - IGNITION "ON," ENGINE "OFF."
   - PUMP BRAKE PEDAL UNTIL FIRM TO DEPLETE VACUUM RESERVE.
   - APPLY FIRM PRESSURE ON BRAKE PEDAL AND RELEASE.
   - REAPPLY VERY FIRM PRESSURE ON BRAKE PEDAL AND HAVE AN ASSISTANT TRY TO SPIN THE REAR WHEELS BY HAND.

   COULD ASSISTANT SPIN THE REAR WHEELS?
   - YES: REPLACE ABS MOTOR PACK.
   - NO: go to 2.

2.
   - VISUALLY INSPECT THE 6-WAY ABS MOTOR PACK AND 6-WAY EBCM CONNECTORS FOR PROPER WIRE COLOR/CONNECTOR CAVITY COMBINATION.

   ARE THE PROPER WIRES LOCATED IN THE PROPER CONNECTOR CAVITIES?
   - NO: USING THE CIRCUIT DIAGRAMS LOCATED IN THE FRONT OF THIS SECTION, PLACE THE PROPER WIRES INTO THE PROPER CONNECTOR CAVITIES. CLEAR DTCs AND RECHECK SYSTEM.
   - YES: go to 3.

3.
   - IGNITION "OFF."
   - DISCONNECT 6-WAY ABS MOTOR PACK CONNECTOR AND 6-WAY EBCM CONNECTOR.
   - USING J 39200, MEASURE THE RESISTANCE BETWEEN TERMINAL "D" OF THE 6-WAY ABS MOTOR PACK HARNESS CONNECTOR AND TERMINAL "C" OF THE 6-WAY EBCM HARNESS CONNECTOR.

   IS RESISTANCE LESS THAN 1.5Ω?
   - NO: REPAIR HIGH RESISTANCE IN CKT 1284.
   - YES: go to 4.

4.
   - MEASURE THE RESISTANCE BETWEEN TERMINAL "C" OF THE 6-WAY ABS MOTOR PACK HARNESS CONNECTOR AND TERMINAL "D" OF THE 6-WAY EBCM HARNESS CONNECTOR.

   IS RESISTANCE LESS THAN 1.5Ω?
   - NO: REPAIR HIGH RESISTANCE IN CKT 1285.
   - YES: go to 5.

5.
   - MEASURE THE RESISTANCE BETWEEN TERMINAL "C" AND TERMINAL "D" OF THE ABS MOTOR PACK CONNECTOR.

   IS RESISTANCE LESS THAN 1.5Ω?
   - NO: REPLACE ABS MOTOR PACK.
   - YES: go to 6.

6.
   - INSPECT THE 6-WAY ABS MOTOR PACK CONNECTOR AND 6-WAY EBCM CONNECTOR FOR POOR TERMINAL CONTACT OR CORROSION.

   DO ANY TERMINALS EXHIBIT POOR CONTACT OR EVIDENCE OF CORROSION?
   - YES: REPLACE TERMINALS THAT EXHIBIT POOR TERMINAL CONTACT OR EVIDENCE OF CORROSION.
   - NO: go to 7.

7.
   - IGNITION "OFF."
   - RECONNECT ALL CONNECTORS.
   - START ENGINE WITH FOOT OFF BRAKE.
   - ALLOW ENGINE TO RUN FOR AT LEAST 10 SECONDS.
   - REPEAT ABOVE IGNITION CYCLE SEQUENCE TWO MORE TIMES.

   DID DTC 42 SET IN THIS LAST THREE IGNITION CYCLES?
   - YES: REPLACE EBCM.
   - NO: MALFUNCTION IS INTERMITTENT. REFER TO "DIAGNOSTIC AIDS" ON FACING PAGE.

AFTER DIAGNOSIS IS COMPLETE, CLEAR DTCs AND TEST DRIVE VEHICLE FOR THREE (3) DRIVE CYCLES TO VERIFY DTC DOES NOT RESET. A DRIVE CYCLE CONSISTS OF STARTING THE VEHICLE, DRIVING OVER 16 km/h (10 MPH) AND THEN KEYING DOWN.

DIAGNOSTIC CHART DTC 44 — 1994 CUTLASS SUPREME AND REGAL

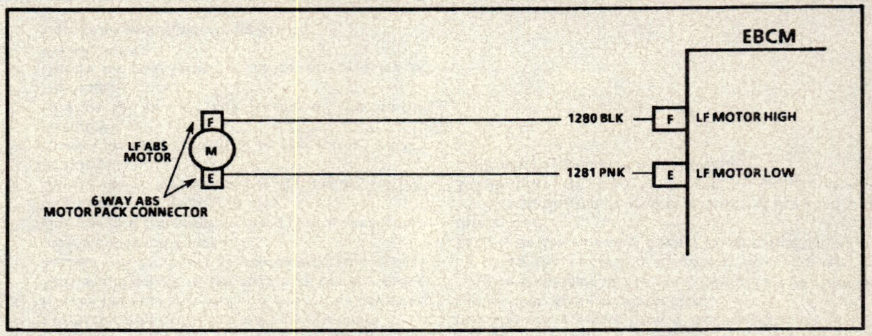

DTC 44
LEFT FRONT CHANNEL WILL NOT MOVE

Circuit Description:
This DTC is designed to detect bound-up ESB, a stuck ABS motor, or a seized hydraulic modulator. When the release is commanded during initialization, the ESB should release the ABS motor, resulting in sensed current being less than commanded current (motor is spinning freely). If the ABS motor is not moving, sensed current will be equal to stall current.

DTC Will Set When: DTC 44 can be set during normal operation. If the EBCM detects a condition in which it cannot move the ABS motor in either direction, a malfunction exists.

Action Taken: This malfunction indicates the channel cannot be moved properly. A malfunction DTC is stored, ABS is disabled and the ABS warning lamp is turned "ON."

DTC Chart Test Description: Number(s) below refer to circled number(s) on the diagnostic chart.

1. Ensures that proper ABS motor circuits are located in the proper connector cavities.
2. Checks for proper ABS motor movement during apply and release commands from the Tech 1.
3. Compares EBCM command current to ABS motor feedback current.
4. Checks for a short between CKT 1280 and CKT 1281.
5. Checks for an ABS motor that is internally shorted.
6. Checks to see if the malfunction is due to poor terminal contact or faulty EBCM.
7. Verifies that the ABS motor can actually apply when commanded.
8. Checks for proper hydraulic modulator gear and piston movement.

Diagnostic Aids: An "Intermittent" malfunction in this DTC may result from a mechanical part of the system that sticks, binds, or slips.

The frequency of the malfunction can be checked by using the enhanced diagnostic function of the Tech 1, as described in "Tech 1 Diagnostics" found in this section. DTC 44 may set after modulator disassembly if the modulator pistons are positioned at the bottom of their bore.

Any circuitry, that is suspected as causing the intermittent complaint, should be thoroughly checked for backed out terminals, improper mating, broken locks, improperly formed or damaged terminals, poor terminal to wiring connections or physical damage to the wiring harness.

DIAGNOSTIC CHART DTC 44 — 1994 CUTLASS SUPREME AND REGAL, CONTINUED

DTC 44
LEFT FRONT CHANNEL WILL NOT MOVE

(1)
- VISUALLY INSPECT THE 6-WAY ABS MOTOR PACK AND THE 6-WAY EBCM CONNECTORS FOR PROPER WIRE COLOR/CONNECTOR CAVITY COMBINATION.

ARE THE PROPER WIRES LOCATED IN THE PROPER CONNECTOR CAVITIES?

NO → USING THE CIRCUIT DIAGRAMS LOCATED IN THE FRONT OF THIS SECTION, PLACE THE PROPER WIRES IN THE PROPER CONNECTOR CAVITIES. CLEAR DTCs AND RECHECK SYSTEM.

YES →

(2)
- IGNITION "ON," ENGINE "OFF."
- PUMP BRAKE PEDAL UNTIL FIRM TO DEPLETE VACUUM RESERVE.
- USING TECH 1 SELECT MANUAL CONTROL FUNCTION.
- SELECT LEFT FRONT ABS MOTOR APPLY FUNCTION AND APPLY THE MOTOR.
- APPLY FIRM PRESSURE ON BRAKE PEDAL.
- SELECT LEFT FRONT ABS MOTOR RELEASE FUNCTION AND RELEASE MOTOR.

DID BRAKE PEDAL FALL?

NO → go to step 3.

YES →

(7)
- FIRM PRESSURE STILL "ON" BRAKE PEDAL.
- SELECT LEFT FRONT ABS MOTOR APPLY FUNCTION AND APPLY THE MOTOR.

DID BRAKE PEDAL RISE?

YES → MALFUNCTION IS INTERMITTENT. REFER TO "DIAGNOSTIC AIDS" ON FACING PAGE.

NO → go to step 3.

(3)
- REMOVE FOOT FROM BRAKE PEDAL.
- SELECT LEFT FRONT ABS MOTOR APPLY FUNCTION AND APPLY THE MOTOR WHILE CAREFULLY OBSERVING THE COMMANDED CURRENT AND THE FEEDBACK CURRENT ON THE TECH 1.

WAS THE FEEDBACK CURRENT HIGHER THAN THE COMMANDED CURRENT?

NO →

(8)
- USING TECH 1 PERFORM GEAR TENSION RELIEF FUNCTION.
- IGNITION "OFF."
- REMOVE HYDRAULIC MODULATOR/MASTER CYLINDER ASSEMBLY FROM VEHICLE. REFER TO "ON VEHICLE SERVICE" IN THIS SECTION.
- SEPARATE ABS MOTOR PACK FROM HYDRAULIC MODULATOR.
- GRASP THE GEAR ON THE HYDRAULIC MODULATOR (THE HYDRAULIC MODULATOR IS THE UNIT WITH THE THREE LARGE GEARS) FURTHEST FORWARD WHEN THE UNIT IS INSTALLED IN THE VEHICLE AND ATTEMPT TO MOVE THE GEAR.

CAN THE GEAR BE ROTATED AT LEAST 10 FULL TURNS LOCK TO LOCK?

NO → REPLACE HYDRAULIC MODULATOR.

YES → REPLACE ABS MOTOR PACK.

YES →

(4)
- EXIT TECH 1 MANUAL CONTROL FUNCTION.
- IGNITION "OFF."
- DISCONNECT THE 6-WAY EBCM CONNECTOR AND THE 6-WAY ABS MOTOR PACK CONNECTOR.
- USING J 39200, MEASURE THE RESISTANCE BETWEEN TERMINAL "E" AND TERMINAL "F" OF THE 6-WAY EBCM HARNESS CONNECTOR.

DOES J 39200 DISPLAY RESISTANCE AS "OL" (INFINITE)?

NO → REPAIR SHORT BETWEEN CKT 1280 AND CKT 1281.

YES →

(5)
- MEASURE THE RESISTANCE BETWEEN TERMINAL "E" AND TERMINAL "F" OF THE ABS MOTOR PACK CONNECTOR.

IS RESISTANCE GREATER THAN 0.4Ω?

NO → REPLACE ABS MOTOR PACK.

YES →

(6)
- IGNITION "OFF."
- RECONNECT ALL CONNECTORS.
- START ENGINE WITH FOOT OFF BRAKE.
- ALLOW ENGINE TO RUN FOR AT LEAST 10 SECONDS.
- REPEAT ABOVE IGNITION CYCLE SEQUENCE TWO MORE TIMES.

DID DTC 44 SET IN THE LAST THREE IGNITION CYCLES?

YES → REPLACE EBCM.

NO → MALFUNCTION IS INTERMITTENT. INSPECT ALL CONNECTORS FOR CONTAMINATION AND POOR TERMINAL CONTACT. REPLACE ANY TERMINALS THAT EXHIBIT POOR TERMINAL CONTACT OR EVIDENCE OF CORROSION.

AFTER DIAGNOSIS IS COMPLETE, CLEAR DTCs AND TEST DRIVE VEHICLE FOR THREE (3) DRIVE CYCLES TO VERIFY THAT DTC DOES NOT RESET. A DRIVE CYCLE CONSISTS OF STARTING THE VEHICLE, DRIVING OVER 16 KM/H (10 MPH), AND THEN KEYING DOWN.

DIAGNOSTIC CHART DTC 45 — 1994 CUTLASS SUPREME AND REGAL

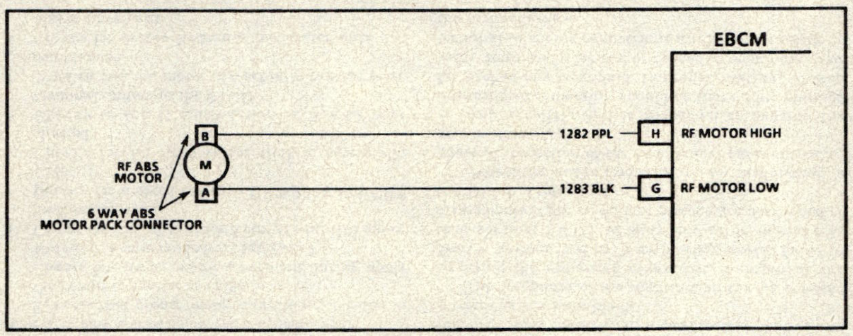

DTC 45
RIGHT FRONT CHANNEL WILL NOT MOVE

Circuit Description:

This DTC is designed to detect bound-up ESB, a stuck ABS motor, or a seized hydraulic modulator. When the release is commanded during initialization, the ESB should release the ABS motor, resulting in sensed current being less than commanded current (motor is spinning freely). If the ABS motor is not moving, sensed current will be equal to stall current.

DTC Will Set When: DTC 45 can be set during normal operation. If the EBCM detects a condition in which it cannot move the motor in either direction, a malfunction exists.

Action Taken: This malfunction indicates the channel cannot be moved properly. A malfunction DTC is stored, ABS is disabled and the ABS warning lamp is turned "ON."

DTC Chart Test Description: Number(s) below refer to circled number(s) on the diagnostic chart.

1. Ensures that proper motor circuits are located in the proper connector cavities.
2. Checks for proper ABS motor movement during apply and release commands from the Tech 1.
3. Compares EBCM command current to ABS motor feedback current.
4. Checks for a short between CKT 1282 and CKT 1283.
5. Checks for an ABS motor that is internally shorted.
6. Checks to see if malfunction is due to poor terminal contact or the EBCM.
7. Verifies that the motor can actually apply when commanded.
8. Checks for proper hydraulic modulator gear and piston movement.

Diagnostic Aids: An "Intermittent" malfunction in this DTC may result from a mechanical part of the system that sticks, binds, or slips.

The frequency of the malfunction can be checked by using the enhanced diagnostic function of the Tech 1, as described in "Tech 1 Diagnostics" found in this section. DTC 45 may set after modulator disassembly if the modulator pistons are positioned at the bottom of their bore.

Any circuitry, that is suspected as causing the intermittent complaint, should be thoroughly checked for backed out terminals, improper mating, broken locks, improperly formed or damaged terminals, poor terminal to wiring connections or physical damage to the wiring harness.

DIAGNOSTIC CHART DTC 45 — 1994 CUTLASS SUPREME AND REGAL, CONTINUED

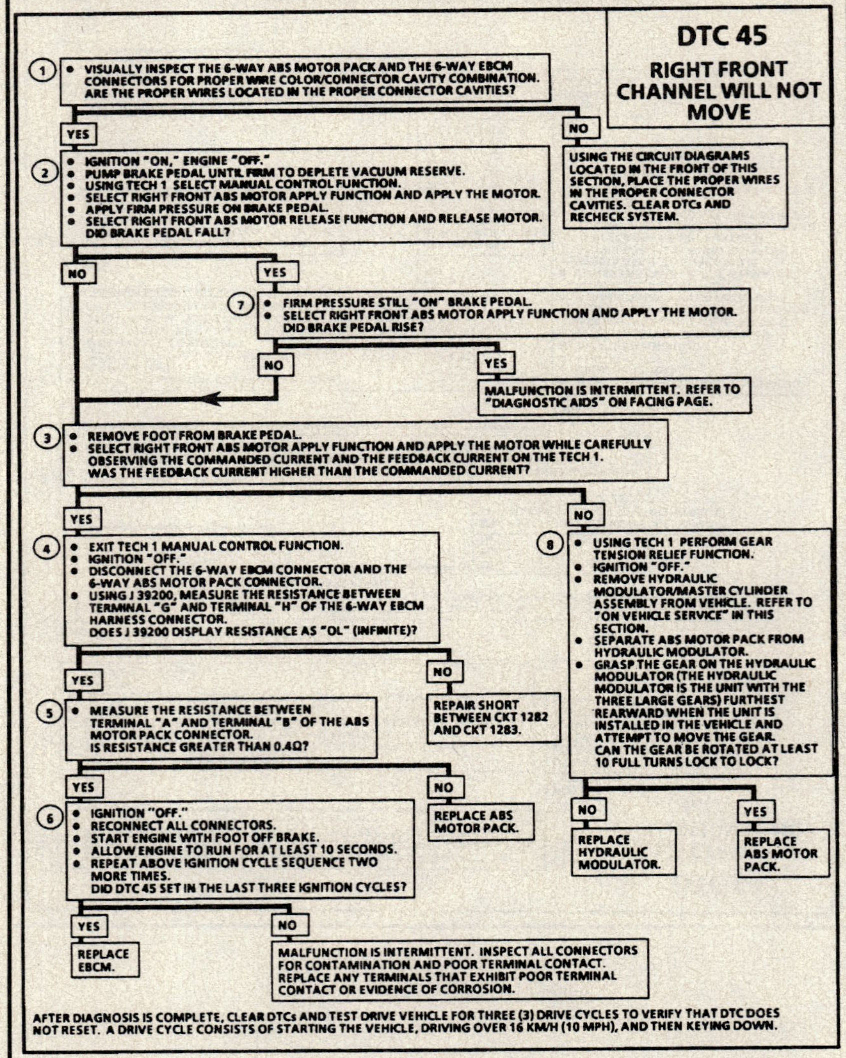

DIAGNOSTIC CHART DTC 46 — 1994 CUTLASS SUPREME AND REGAL

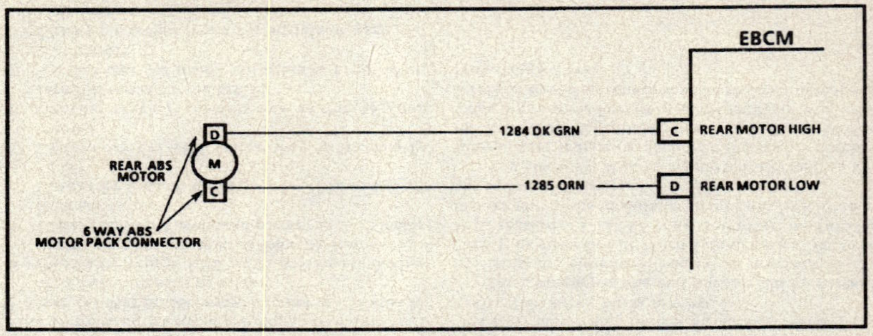

DTC 46

REAR CHANNEL WILL NOT MOVE

Circuit Description:

This DTC is designed to detect bound-up ESB, a stuck ABS motor, or a seized hydraulic modulator. When the release is commanded during initialization, the ESB should release the ABS motor, resulting in sensed current being less than commanded current (motor is spinning freely). If the ABS motor is not moving, sensed current will be equal to stall current.

DTC Will Set When: DTC 46 can be set during normal operation. If the EBCM detects a condition in which it cannot move the ABS motor in either direction, a malfunction exists.

Action Taken: This malfunction indicates the channel cannot be moved properly. A malfunction DTC is stored, ABS is disabled and the ABS warning lamp is turned "ON."

DTC Chart Test Description: Number(s) below refer to circled number(s) on the diagnostic chart.

1. Ensures that proper motor circuits are located in the proper connector cavities.
2. Checks for proper motor movement during apply and release commands from the Tech 1.
3. Compares EBCM command current to ABS motor feedback current.
4. Checks for a short between CKT 1284 and CKT 1285.
5. Checks for an ABS motor that is internally shorted.
6. Checks to see if malfunction is due to poor terminal contact or the EBCM.
7. Verifies that the motor can actually release when commanded.
8. Checks for proper hydraulic modulator gear and piston movement.

Diagnostic Aids: An "Intermittent" malfunction in this DTC may result from a mechanical part of the system that sticks or binds.

The frequency of the malfunction can be checked by using the enhanced diagnostic function of the Tech 1, as described in "Tech 1 Diagnostics" found in this section. DTC 46 may set after modulator disassembly if the modulator pistons are positioned at the bottom of their bore.

Depending on the frequency of the malfunction, a physical inspection of the mechanical parts suspected may be necessary.

Any circuitry, that is suspected as causing the intermittent complaint, should be thoroughly checked for backed out terminals, improper mating, broken locks, improperly formed or damaged terminals, poor terminal to wiring connections or physical damage to the wiring harness.

DIAGNOSTIC CHART DTC 46 — 1994 CUTLASS SUPREME AND REGAL, CONTINUED

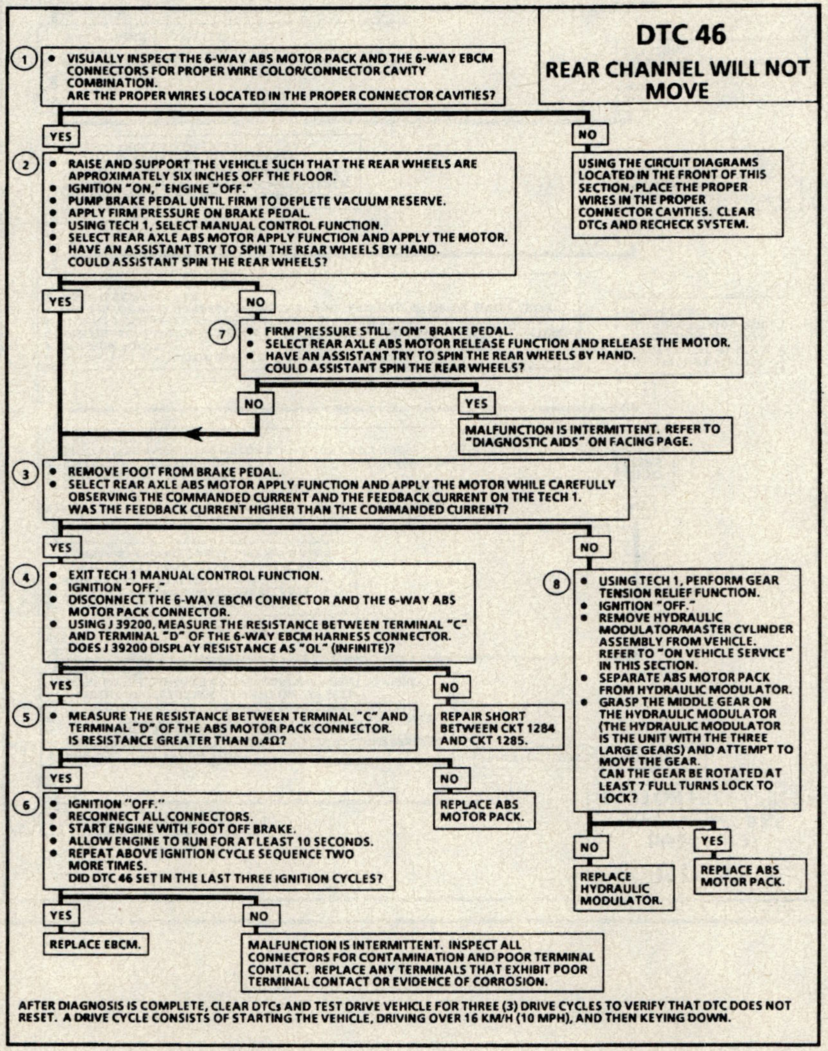

DIAGNOSTIC CHART DTC 47 — 1994 CUTLASS SUPREME AND REGAL

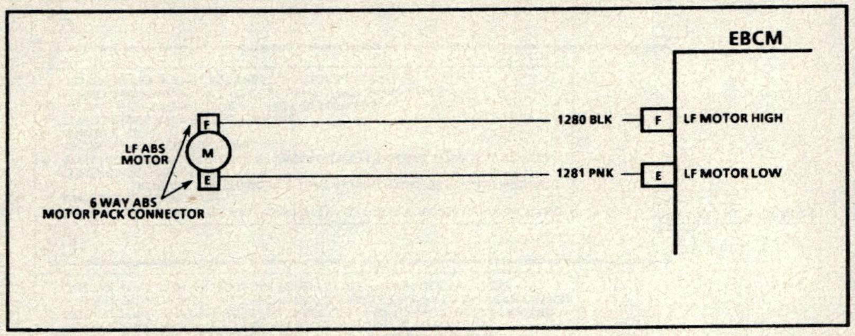

DTC 47

LEFT FRONT ABS MOTOR FREE SPINS

Circuit Description:

This DTC is designed to detect a stripped nut or gear assembly during initialization. During the homing sequence, the piston should reach the top of the bore resulting in a stalled ABS motor. If this does not occur, the ABS motor must be spinning with little or no resistance, indicating a nut/screw or gear malfunction.

DTC Will Set When: DTC 47 can only be set during initialization. If the feedback current is less than the command current for a specified period of time, a malfunction exists.

Action Taken: A malfunction DTC is stored, ABS is disabled, and the ABS warning lamp is turned "ON."

DTC Chart Test Description: Number(s) below refer to circled number(s) on the diagnostic chart.

1. Checks to see if the corresponding open motor DTC is also set.
2. Verifies that the ABS motor was actually applied as command by observing feedback current.
3. Verifies that the ABS motor can release.
4. Verifies that the ABS motor can be applied by observing pedal movement.
5. Checks for an open in the ABS motor "HIGH" circuitry.
6. Checks for an open in the ABS motor "LOW" circuitry.
7. Checks for an ABS motor that is internally shorted.
8. Checks for a stripped gear on the ABS motor pack (unit with three small gears).
9. Checks for a stripped gear on the hydraulic modulator (unit with three large gears).
10. Checks to see if the ABS motor pack has an internal malfunction.
11. Checks for a malfunctioning hydraulic modulator.
12. Ensures malfunction was not due to poor terminal contact.

Diagnostic Aids: An "Intermittent" malfunction in this DTC may result from a mechanical part of the system that sticks, binds, or slips.

The frequency of the malfunction can be checked by using the enhanced diagnostic function of the Tech 1, as described in "Tech 1 Diagnostics" found in this section. If the DTC only occurs once and DTC 56 also occurs, refer to DTC 56. If intermittent and enhanced diagnostics shows this DTC occurs during ABS, refer to DTC 56.

Depending on the frequency of the malfunction, a physical inspection of the mechanical parts suspected may be necessary.

DIAGNOSTIC CHART DTC 47 — 1994 CUTLASS SUPREME AND REGAL, CONTINUED

DTC 47 (Page 1 of 2) LEFT FRONT ABS MOTOR FREE SPINS

1. IS DTC 56 ALSO CURRENTLY SET?
   - YES: PROCEED TO DTC 56 CHART FIRST.
   - NO: Go to 2.

2.
   - IGNITION "ON," ENGINE "OFF."
   - PUMP BRAKE PEDAL UNTIL FIRM TO DEPLETE VACUUM RESERVE.
   - USING TECH 1, SELECT MANUAL CONTROL FUNCTION.
   - SELECT LEFT FRONT ABS MOTOR APPLY FUNCTION AND APPLY THE MOTOR WHILE CAREFULLY OBSERVING THE COMMAND CURRENT AND FEEDBACK CURRENT ON THE TECH 1.

   WERE BOTH COMMAND AND FEEDBACK CURRENT 10 AMPS?
   - YES: Go to 3.
   - NO: GO TO STEP 5.

3.
   - APPLY FIRM PRESSURE ON BRAKE PEDAL.
   - SELECT LEFT FRONT ABS MOTOR RELEASE FUNCTION AND RELEASE THE MOTOR.

   DID THE BRAKE PEDAL FALL AND WERE BOTH COMMAND AND FEEDBACK CURRENT 6 AMPS?
   - YES: Go to 4.
   - NO: GO TO STEP 5.

4.
   - FIRM PRESSURE STILL ON BRAKE PEDAL.
   - SELECT LEFT FRONT ABS MOTOR APPLY FUNCTION AND APPLY THE MOTOR (THIS MOVES THE PISTON TO THE TOP OF ITS BORE).

   DID BRAKE PEDAL RISE?
   - YES: MALFUNCTION IS MOST LIKELY CAUSED BY POOR CONNECTION. REFER TO "DIAGNOSTIC AIDS" ON FACING PAGE.
   - NO: Go to 5.

5.
   - USING TECH 1, PERFORM GEAR TENSION RELIEF FUNCTION.
   - IGNITION "OFF."
   - DISCONNECT 6-WAY ABS MOTOR PACK CONNECTOR AND 6-WAY EBCM CONNECTOR.
   - USING J 39200, MEASURE THE RESISTANCE BETWEEN TERMINAL "F" OF THE 6-WAY ABS MOTOR PACK HARNESS CONNECTOR AND TERMINAL "F" OF THE 6-WAY EBCM HARNESS CONNECTOR.

   IS RESISTANCE 1.5Ω OR LESS?
   - YES: Go to 6.
   - NO: REPAIR HIGH RESISTANCE IN CKT 1280.

6.
   - MEASURE THE RESISTANCE BETWEEN TERMINAL "E" OF THE 6-WAY ABS MOTOR PACK HARNESS CONNECTOR AND TERMINAL "E" OF THE 6-WAY EBCM HARNESS CONNECTOR.

   IS RESISTANCE 1.5Ω OR LESS?
   - YES: Go to 7.
   - NO: REPAIR HIGH RESISTANCE IN CKT 1281.

7.
   - MEASURE THE RESISTANCE BETWEEN TERMINAL "E" AND TERMINAL "F" OF THE ABS MOTOR PACK CONNECTOR.

   IS RESISTANCE 1.5Ω OR LESS?
   - YES: DIAGNOSIS CONTINUED ON NEXT PAGE.
   - NO: REPLACE ABS MOTOR PACK.

AFTER DIAGNOSIS IS COMPLETE, CLEAR DTCs AND TEST DRIVE VEHICLE FOR THREE (3) DRIVE CYCLES TO VERIFY THAT DTC DOES NOT RESET. A DRIVE CYCLE CONSISTS OF STARTING THE VEHICLE, DRIVING OVER 16 km/h (10 MPH), AND THEN KEYING DOWN.

DIAGNOSTIC CHART DTC 47 — 1994 CUTLASS SUPREME AND REGAL, CONTINUED

DTC 47 (Page 2 of 2) LEFT FRONT ABS MOTOR FREE SPINS

DIAGNOSIS CONTINUED FROM PREVIOUS PAGE.

8.
   - REMOVE HYDRAULIC MODULATOR/MASTER CYLINDER ASSEMBLY FROM VEHICLE (REFER TO "ON-VEHICLE SERVICE" IN THIS SECTION).
   - REMOVE GEAR COVER.
   - CHECK FOR STRIPPED GEAR ON ABS MOTOR PACK; ABS MOTOR PACK IS THE UNIT WITH THE THREE SMALL GEARS. THE LEFT FRONT GEAR IS THE GEAR FURTHEST FORWARD WHEN THE ASSEMBLY IS INSTALLED IN THE VEHICLE.

   IS THE ABS MOTOR PACK GEAR STRIPPED?
   - YES: REPLACE ABS MOTOR PACK.
   - NO: Go to 9.

9.
   - CHECK FOR A STRIPPED GEAR ON HYDRAULIC MODULATOR; MODULATOR IS THE UNIT WITH THE THREE LARGE GEARS. THE LEFT FRONT GEAR IS THE GEAR FURTHEST FORWARD WHEN THE UNIT IS INSTALLED IN THE VEHICLE.

   IS THE MODULATOR GEAR STRIPPED?
   - YES: REPLACE HYDRAULIC MODULATOR GEAR.
   - NO: Go to 10.

10.
    - RECONNECT ELECTRICAL CONNECTORS
    - SECURELY POSITION HYDRAULIC MODULATOR/MASTER CYLINDER ASSEMBLY IN VEHICLE WITH GEAR COVER REMOVED SUCH THAT THE GEAR SET CAN BE OBSERVED. BE CAREFUL NOT TO DAMAGE GEAR SET OR ALLOW THE HYDRAULIC MODULATOR TO MOVE WHILE TESTING.
    - IGNITION "ON."
    - USING TECH 1 SELECT MANUAL CONTROL FUNCTION.
    - SELECT LEFT FRONT ABS MOTOR APPLY AND APPLY THE MOTOR WHILE OBSERVING THE GEAR SET FURTHEST FORWARD WHEN THE UNIT IS INSTALLED IN THE VEHICLE.
    - SELECT LEFT FRONT ABS MOTOR RELEASE FUNCTION AND RELEASE THE MOTOR WHILE STILL OBSERVING THE GEAR SET.
    - SELECT LEFT FRONT ABS MOTOR APPLY FUNCTION AND APPLY THE MOTOR.

    DID GEAR SET MOVE IN BOTH DIRECTIONS FOR AT LEAST ONE REVOLUTION?
    - YES: Go to 11.
    - NO: REPLACE ABS MOTOR PACK.

11.
    - USING TECH 1 PERFORM GEAR TENSION RELIEF FUNCTION.
    - IGNITION "OFF."
    - SEPARATE THE ABS MOTOR PACK FROM THE HYDRAULIC MODULATOR.
    - GRASP THE GEAR ON THE HYDRAULIC MODULATOR (THE HYDRAULIC MODULATOR IS THE UNIT WITH THE THREE LARGE GEARS) FURTHEST FORWARD WHEN THE UNIT IS INSTALLED IN THE VEHICLE AND ATTEMPT TO MOVE THE GEAR.

    CAN THE GEAR BE ROTATED MORE THAN 13 FULL TURNS LOCK TO LOCK?
    - YES: REPLACE THE HYDRAULIC MODULATOR.
    - NO: Go to 12.

12.
    - INSPECT ALL CONNECTORS AND TERMINALS FOR POOR TERMINAL CONTACT AND EVIDENCE OF CORROSION.
      REPLACE ALL TERMINALS THAT EXHIBIT SIGNS OF POOR TERMINAL CONTACT OR CORROSION.
    - REINSTALL HYDRAULIC MODULATOR/MASTER CYLINDER ASSEMBLY IN VEHICLE.
    - RECONNECT ALL CONNECTORS.
    - IGNITION "OFF."
    - START ENGINE WITH FOOT OFF BRAKE.
    - ALLOW ENGINE TO RUN FOR AT LEAST 10 SECONDS.
    - REPEAT ABOVE IGNITION CYCLE SEQUENCE TWO MORE TIMES.

    DID DTC 47 SET IN THE LAST THREE IGNITION CYCLES?
    - YES: REPLACE EBCM.
    - NO: MALFUNCTION IS MOST LIKELY CAUSED BY A POOR CONNECTION. REFER TO "DIAGNOSTIC AIDS" ON FACING PAGE.

AFTER DIAGNOSIS IS COMPLETE, CLEAR DTCs AND TEST DRIVE VEHICLE FOR THREE (3) DRIVE CYCLES TO VERIFY THAT DTC DOES NOT RESET. A DRIVE CYCLE CONSISTS OF STARTING THE VEHICLE, DRIVING OVER 16 km/h (10 MPH), AND THEN KEYING DOWN.

DIAGNOSTIC CHART DTC 48 — 1994 CUTLASS SUPREME AND REGAL

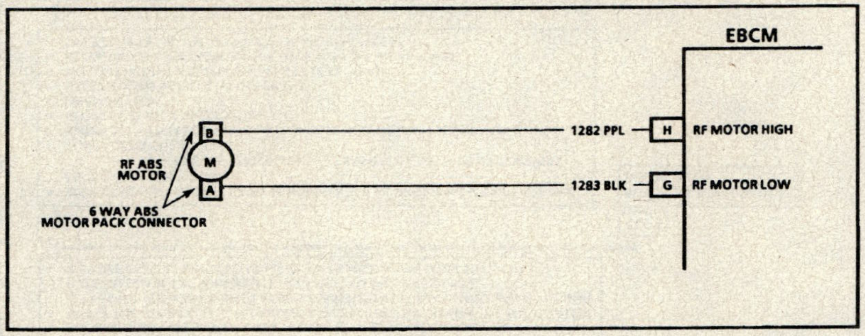

DTC 48

RIGHT FRONT ABS MOTOR FREE SPINS

Circuit Description:
This DTC is designed to detect a stripped nut or gear assembly during initialization. During the homing sequence, the piston should reach the top of the bore resulting in a stalled ABS motor. If this does not occur, the ABS motor must be spinning with little or no resistance, indicating a nut/screw or gear malfunction.

DTC Will Set When: DTC 48 can only be set during initialization. If the feedback current is less than the command current for a specified period of time, a malfunction exists.

Action Taken: A malfunction DTC is stored, ABS is disabled, and the ABS warning lamp is turned "ON."

DTC Chart Test Description: Number(s) below refer to circled number(s) on the diagnostic chart.

1. Checks to see if the corresponding open ABS motor DTC is also set.
2. Verifies that the ABS motor was actually applied as commanded by observing feedback current.
3. Verifies that the ABS motor can release.
4. Verifies that the ABS motor can be applied by observing pedal movement.
5. Checks for high resistance in the ABS motor "HIGH" circuitry.
6. Checks for high resistance in the ABS motor "LOW" circuitry.
7. Checks for proper resistance of the ABS motor windings.
8. Checks for a stripped gear on the ABS motor pack (unit with three small gears).
9. Checks for a stripped gear on the hydraulic modulator (unit with three large gears).
10. Verifies that the ABS motor can be applied under a load situation.
11. Verifies that the hydraulic modulator is functioning properly.
12. Ensures malfunction was not due to poor terminal contact.

Diagnostic Aids: An "Intermittent" malfunction in this DTC may result from a mechanical part of the system that sticks, binds, or slips.

The frequency of the malfunction can be checked by using the enhanced diagnostic function of the Tech 1, as described in "Tech 1 Diagnostics" found in this section. If the DTC only occurs once and DTC 61 also occurs, refer to DTC 61. If intermittent and enhanced diagnostics shows this DTC occurs during ABS, refer to DTC 61.

Depending on the frequency of the malfunction, a physical inspection of the mechanical parts suspected may be necessary.

DIAGNOSTIC CHART DTC 48 — 1994 CUTLASS SUPREME AND REGAL, CONTINUED

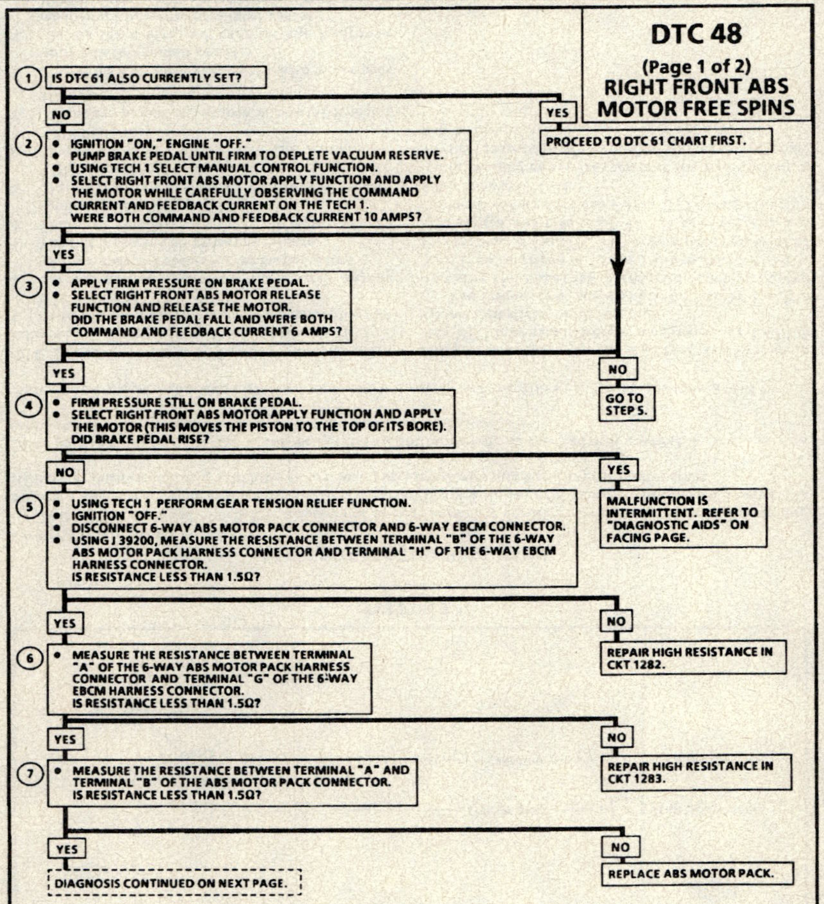

AFTER DIAGNOSIS IS COMPLETE, CLEAR DTCs AND TEST DRIVE VEHICLE FOR THREE (3) DRIVE CYCLES TO VERIFY THAT DTC DOES NOT RESET. A DRIVE CYCLE CONSISTS OF STARTING THE VEHICLE, DRIVING OVER 16 km/h (10 MPH), AND THEN KEYING DOWN.

DIAGNOSTIC CHART DTC 48 — 1994 CUTLASS SUPREME AND REGAL, CONTINUED

DTC 48 (Page 2 of 2) RIGHT FRONT ABS MOTOR FREE SPINS

DIAGNOSIS CONTINUED FROM PREVIOUS PAGE.

(8)
- REMOVE HYDRAULIC MODULATOR/MASTER CYLINDER ASSEMBLY FROM VEHICLE.
- REMOVE GEAR COVER.
- CHECK FOR STRIPPED GEAR ON ABS MOTOR PACK; ABS MOTOR PACK IS THE UNIT WITH THE THREE SMALL GEARS. THE RIGHT FRONT GEAR IS THE GEAR FURTHEST REARWARD WHEN THE ASSEMBLY IS INSTALLED IN THE VEHICLE. IS THE ABS MOTOR PACK GEAR STRIPPED?

YES: REPLACE ABS MOTOR PACK.

NO:

(9)
- CHECK FOR A STRIPPED GEAR ON HYDRAULIC MODULATOR; MODULATOR IS THE UNIT WITH THE THREE LARGE GEARS. THE RIGHT FRONT GEAR IS THE GEAR FURTHEST REARWARD WHEN THE UNIT IS INSTALLED IN THE VEHICLE. IS THE MODULATOR GEAR STRIPPED?

YES: REPLACE HYDRAULIC MODULATOR GEAR.

NO:

(10)
- RECONNECT ELECTRICAL CONNECTORS.
- SECURELY POSITION HYDRAULIC MODULATOR/MASTER CYLINDER ASSEMBLY IN VEHICLE WITH GEAR COVER REMOVED SUCH THAT THE GEAR SET CAN BE OBSERVED. BE CAREFUL NOT TO DAMAGE GEAR SET OR ALLOW THE HYDRAULIC MODULATOR TO MOVE WHILE TESTING.
- IGNITION "ON."
- USING TECH 1, SELECT MANUAL CONTROL FUNCTION.
- SELECT RIGHT FRONT ABS MOTOR APPLY AND APPLY THE MOTOR WHILE OBSERVING THE GEAR SET FURTHEST REARWARD WHEN THE UNIT IS INSTALLED IN THE VEHICLE.
- SELECT RIGHT FRONT ABS MOTOR RELEASE FUNCTION AND RELEASE THE MOTOR WHILE STILL OBSERVING THE GEAR SET.
- SELECT RIGHT FRONT ABS MOTOR APPLY FUNCTION AND APPLY THE MOTOR. DID GEAR SET MOVE IN BOTH DIRECTIONS FOR AT LEAST ONE REVOLUTION?

NO: REPLACE ABS MOTOR PACK.

YES:

(11)
- USING TECH 1, PERFORM GEAR TENSION RELIEF FUNCTION.
- IGNITION "OFF."
- SEPARATE THE ABS MOTOR PACK FROM THE HYDRAULIC MODULATOR.
- GRASP THE GEAR ON THE HYDRAULIC MODULATOR (THE HYDRAULIC MODULATOR IS THE UNIT WITH THE THREE LARGE GEARS) FURTHEST REARWARD WHEN THE UNIT IS INSTALLED IN THE VEHICLE AND ATTEMPT TO MOVE THE GEAR. CAN THE GEAR BE ROTATED MORE THAN 13 FULL TURNS LOCK TO LOCK?

YES: REPLACE THE HYDRAULIC MODULATOR.

NO:

(12)
- INSPECT ALL CONNECTORS AND TERMINALS FOR POOR TERMINAL CONTACT AND EVIDENCE OF CORROSION. REPLACE ALL TERMINALS THAT EXHIBIT SIGNS OF POOR TERMINAL CONTACT OR CORROSION.
- REINSTALL HYDRAULIC MODULATOR/MASTER CYLINDER ASSEMBLY IN VEHICLE.
- RECONNECT ALL CONNECTORS.
- IGNITION "OFF."
- START ENGINE WITH FOOT OFF BRAKE.
- ALLOW ENGINE TO RUN FOR AT LEAST 10 SECONDS.
- REPEAT ABOVE IGNITION CYCLE SEQUENCE TWO MORE TIMES. DID DTC 48 SET IN THE LAST THREE IGNITION CYCLES?

YES: REPLACE EBCM.

NO: MALFUNCTION WAS MOST LIKELY CAUSED BY A POOR CONNECTION. REFER TO "DIAGNOSTIC AIDS" ON FACING PAGE.

AFTER DIAGNOSIS IS COMPLETE, CLEAR DTCs AND TEST DRIVE VEHICLE FOR THREE (3) DRIVE CYCLES TO VERIFY THAT DTC DOES NOT RESET. A DRIVE CYCLE CONSISTS OF STARTING THE VEHICLE, DRIVING OVER 16 km/h (10 MPH), AND THEN KEYING DOWN.

DIAGNOSTIC CHART DTC 51 — 1994 CUTLASS SUPREME AND REGAL

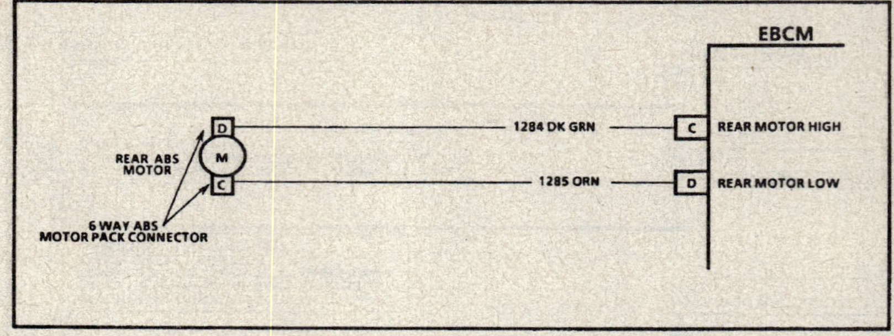

DTC 51

REAR ABS MOTOR FREE SPINS

Circuit Description:
This DTC is designed to detect a stripped nut or gear assembly during initialization. During the homing sequence, the piston should reach the top of the bore resulting in a stalled ABS motor. If this does not occur, the ABS motor must be spinning with little or no resistance, indicating a nut/screw or gear malfunction.

DTC Will Set When: DTC 51 can be set during initialization. If the feedback current is less than the command current for a specified period of time, a malfunction exists.

Action Taken: A malfunction DTC is stored, ABS is disabled, and the ABS warning lamp is turned "ON."

DTC Chart Test Description: Number(s) below refer to circled number(s) on the diagnostic chart.

1. Checks to see if the corresponding open ABS motor DTC is also set.
2. Verifies that the ABS motor was actually applied as commanded by observing feedback current.
3. Verifies that the ABS motor can release.
4. This step verifies that the ABS motor can be applied by observing pedal movement.
5. Checks for high resistance in the ABS motor "HIGH" circuitry.
6. Checks for high resistance in the ABS motor "LOW" circuitry.
7. Checks for proper resistance of the ABS motor windings.
8. Checks for a stripped gear on the ABS motor pack (unit with three small gears).
9. Checks for a stripped gear on the hydraulic modulator (unit with three large gears).
10. Verifies that the ABS motor can be applied under a load situation.
11. Verifies that the hydraulic modulator is functioning properly.
12. Ensures malfunction was not due to poor terminal contact.

Diagnostic Aids: An "Intermittent" malfunction in this DTC may result from a mechanical part of the system that sticks, binds, or slips.

The frequency of the malfunction can be checked by using the enhanced diagnostic function of the Tech 1, as described in "Tech 1 Diagnostics" found in this section. If the DTC only occurs once and DTC 64 also occurs, refer to DTC 64. If intermittent and enhanced diagnostics show this DTC occurs during ABS, refer to DTC 64.

Depending on the frequency of the malfunction, a physical inspection of the mechanical parts suspected may be necessary.

DIAGNOSTIC CHART DTC 51 — 1994 CUTLASS SUPREME AND REGAL, CONTINUED

DTC 51
(Page 1 of 2)
REAR ABS MOTOR FREE SPINS

1. IS DTC 64 ALSO CURRENTLY SET?
   - YES: PROCEED TO DTC 64 CHART FIRST.
   - NO: Go to step 2.
2. • IGNITION "ON," ENGINE "OFF."
   • PUMP BRAKE PEDAL UNTIL FIRM TO DEPLETE VACUUM RESERVE.
   • USING TECH 1, SELECT MANUAL CONTROL FUNCTION.
   • SELECT REAR AXLE ABS MOTOR APPLY FUNCTION AND APPLY THE MOTOR WHILE CAREFULLY OBSERVING THE COMMAND CURRENT AND FEEDBACK CURRENT ON THE TECH 1.
   WERE BOTH COMMAND AND FEEDBACK CURRENT 10 AMPS?
   - YES: Go to step 3.
   - NO: GO TO STEP 5
3. • APPLY FIRM PRESSURE ON BRAKE PEDAL.
   • SELECT REAR AXLE ABS MOTOR RELEASE FUNCTION AND RELEASE THE MOTOR.
   DID THE BRAKE PEDAL FALL AND WERE BOTH COMMAND AND FEEDBACK CURRENT 6 AMPS?
   - YES: Go to step 4.
   - NO: GO TO STEP 5
4. • FIRM PRESSURE STILL ON BRAKE PEDAL.
   • SELECT REAR AXLE ABS MOTOR APPLY FUNCTION AND APPLY THE MOTOR (THIS MOVES THE PISTON TO THE TOP OF ITS BORE).
   DID BRAKE PEDAL RISE?
   - YES: MALFUNCTION IS INTERMITTENT. REFER TO "DIAGNOSTIC AIDS" ON FACING PAGE.
   - NO: Go to step 5.
5. • USING TECH 1, PERFORM GEAR TENSION RELIEF FUNCTION.
   • IGNITION "OFF."
   • DISCONNECT 6-WAY ABS MOTOR PACK CONNECTOR AND 6-WAY EBCM CONNECTOR.
   • USING J 39200, MEASURE THE RESISTANCE BETWEEN TERMINAL "D" OF THE 6-WAY ABS MOTOR PACK HARNESS CONNECTOR AND TERMINAL "C" OF THE 6-WAY EBCM HARNESS CONNECTOR.
   IS RESISTANCE LESS THAN 1.5Ω?
   - YES: Go to step 6.
   - NO: REPAIR HIGH RESISTANCE IN CKT 1284.
6. • MEASURE THE RESISTANCE BETWEEN TERMINAL "C" OF THE 6-WAY ABS MOTOR PACK HARNESS CONNECTOR AND TERMINAL "D" OF THE 6-WAY EBCM HARNESS CONNECTOR.
   IS RESISTANCE LESS THAN 1.5Ω?
   - YES: Go to step 7.
   - NO: REPAIR HIGH RESISTANCE IN CKT 1285.
7. • MEASURE THE RESISTANCE BETWEEN TERMINAL "C" AND TERMINAL "D" OF THE ABS MOTOR PACK CONNECTOR.
   IS RESISTANCE LESS THAN 1.5Ω?
   - YES: DIAGNOSIS CONTINUED ON NEXT PAGE.
   - NO: REPLACE ABS MOTOR PACK.

AFTER DIAGNOSIS IS COMPLETE, CLEAR DTCs AND TEST DRIVE VEHICLE FOR THREE (3) DRIVE CYCLES TO VERIFY THAT DTC DOES NOT RESET. A DRIVE CYCLE CONSISTS OF STARTING THE VEHICLE, DRIVING OVER 16 km/h (10 MPH), AND THEN KEYING DOWN.

DIAGNOSTIC CHART DTC 51 — 1994 CUTLASS SUPREME AND REGAL, CONTINUED

DTC 51
(Page 2 of 2)
REAR ABS MOTOR FREE SPINS

DIAGNOSIS CONTINUED FROM PREVIOUS PAGE.

8. • REMOVE HYDRAULIC MODULATOR/MASTER CYLINDER ASSEMBLY FROM VEHICLE (REFER TO "ON-VEHICLE SERVICE" IN THIS SECTION).
   • REMOVE GEAR COVER.
   • CHECK FOR STRIPPED GEAR ON ABS MOTOR PACK; ABS MOTOR PACK IS THE UNIT WITH THE THREE SMALL GEARS. THE REAR AXLE GEAR IS THE MIDDLE GEAR OF THE THREE.
   IS THE ABS MOTOR PACK GEAR STRIPPED?
   - YES: REPLACE ABS MOTOR PACK.
   - NO: Go to step 9.
9. • CHECK FOR A STRIPPED GEAR ON HYDRAULIC MODULATOR; MODULATOR IS THE UNIT WITH THE THREE LARGE GEARS. THE REAR AXLE GEAR IS THE MIDDLE GEAR OF THE THREE.
   IS THE MODULATOR GEAR STRIPPED?
   - YES: REPLACE HYDRAULIC MODULATOR GEAR.
   - NO: Go to step 10.
10. • RECONNECT ELECTRICAL CONNECTORS
    • SECURELY POSITION HYDRAULIC MODULATOR/MASTER CYLINDER ASSEMBLY IN VEHICLE WITH GEAR COVER REMOVED SUCH THAT THE GEAR SET CAN BE OBSERVED. BE CAREFUL NOT TO DAMAGE GEAR SET OR ALLOW THE HYDRAULIC MODULATOR TO MOVE WHILE TESTING.
    • IGNITION "ON."
    • USING TECH 1, SELECT MANUAL CONTROL FUNCTION.
    • SELECT REAR AXLE ABS MOTOR APPLY AND APPLY THE MOTOR WHILE OBSERVING THE MIDDLE GEAR SET.
    • SELECT REAR AXLE ABS MOTOR RELEASE FUNCTION AND RELEASE THE MOTOR WHILE STILL OBSERVING THE GEAR SET.
    • SELECT REAR AXLE ABS MOTOR APPLY FUNCTION AND APPLY THE MOTOR.
    DID GEAR SET MOVE IN BOTH DIRECTIONS FOR AT LEAST ONE REVOLUTION?
    - YES: Go to step 11.
    - NO: REPLACE ABS MOTOR PACK.
11. • USING TECH 1, PERFORM GEAR TENSION RELIEF FUNCTION.
    • IGNITION "OFF."
    • SEPARATE THE ABS MOTOR PACK FROM THE HYDRAULIC MODULATOR.
    • GRASP THE MIDDLE GEAR ON THE HYDRAULIC MODULATOR (THE HYDRAULIC MODULATOR IS THE UNIT WITH THE THREE LARGE GEARS) AND ATTEMPT TO MOVE THE GEAR.
    CAN THE GEAR BE ROTATED MORE THAN 8 FULL TURNS LOCK TO LOCK?
    - YES: REPLACE THE HYDRAULIC MODULATOR.
    - NO: Go to step 12.
12. • INSPECT ALL CONNECTORS AND TERMINALS FOR POOR TERMINAL CONTACT AND EVIDENCE OF CORROSION. REFER TO SECTION 8A-4 "CHECKING TERMINAL CONTACT." REPLACE ALL TERMINALS THAT EXHIBIT SIGNS OF POOR TERMINAL CONTACT OR CORROSION.
    • REINSTALL HYDRAULIC MODULATOR/MASTER CYLINDER ASSEMBLY IN VEHICLE.
    • RECONNECT ALL CONNECTORS.
    • IGNITION "OFF."
    • START ENGINE WITH FOOT OFF BRAKE.
    • ALLOW ENGINE TO RUN FOR AT LEAST 10 SECONDS.
    • REPEAT ABOVE IGNITION CYCLE SEQUENCE TWO MORE TIMES.
    DID DTC 51 SET IN THE LAST THREE IGNITION CYCLES?
    - YES: REPLACE EBCM.
    - NO: MALFUNCTION WAS MOST LIKELY CAUSED BY A POOR CONNECTION. REFER TO "DIAGNOSTIC AIDS" ON FACING PAGE.

AFTER DIAGNOSIS IS COMPLETE, CLEAR DTCs AND TEST DRIVE VEHICLE FOR THREE (3) DRIVE CYCLES TO VERIFY THAT DTC DOES NOT RESET. A DRIVE CYCLE CONSISTS OF STARTING THE VEHICLE, DRIVING OVER 16 km/h (10 MPH), AND THEN KEYING DOWN.

DIAGNOSTIC CHART DTC 52 — 1994 CUTLASS SUPREME AND REGAL

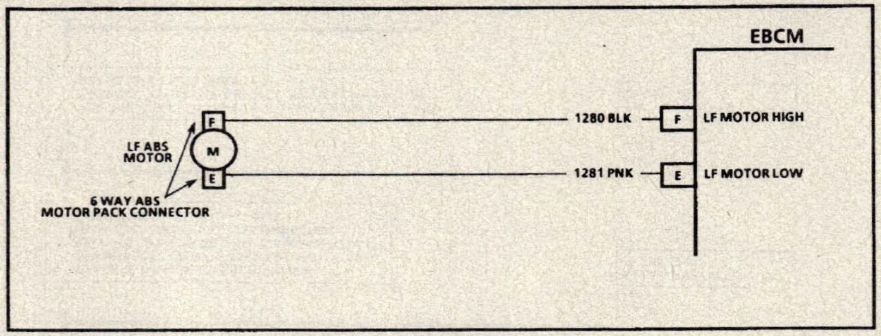

DTC 52

LEFT FRONT CHANNEL IN RELEASE TOO LONG

Circuit Description:

This DTC will diagnose an ABS motor that is energized longer than expected. This could occur if a wheel speed sensor is malfunctioning, the ABS motor does not turn, the left front solenoid mechanically fails open, or the ABS motor wires are crossed.

DTC Will Set When: DTC 52 can be set only during an ABS stop. If the EBCM commands the left front ABS channel in release for three seconds, a malfunction exists.

Action Taken: A malfunction DTC is stored, ABS is disabled and the ABS warning lamp is turned "ON."

DTC Chart Test Description: Number(s) below refer to circled number(s) on the diagnostic chart.

1. Identifies a problem in a wheel speed sensor that may cause the system to be in release too long.
2. Identifies an ABS motor as being failed or wired incorrectly.
3. Checks for a solenoid that may have mechanically failed open.
4. Verifies proper solenoid operation.
5. This serves to isolate the cause of the hydraulic problem, to either the solenoid or the ABS hydraulic modulator assembly.
6. Determines whether a malfunctioning ABS motor pack or ABS hydraulic modulator is the reason for DTC 52 being set.

Diagnostic Aids: An "Intermittent" malfunction may be caused by a mechanical part of the system that sticks or binds.

The frequency of the malfunction can be checked by using the enhanced diagnostic function of the Tech 1, as described in "Tech 1 Diagnostics" found in this section. DTC 52 may occur if on ice and steering wheel is turned to lock during braking. Using the Tech 1, perform hydraulic test to ensure total brake system is functional.

Any circuitry, that is suspected as causing the intermittent complaint, should be thoroughly checked for backed out terminals, improper mating, broken locks, improperly formed or damaged terminals, poor terminal to wiring connections or physical damage to the wiring harness.

DIAGNOSTIC CHART DTC 52 — 1994 CUTLASS SUPREME AND REGAL, CONTINUED

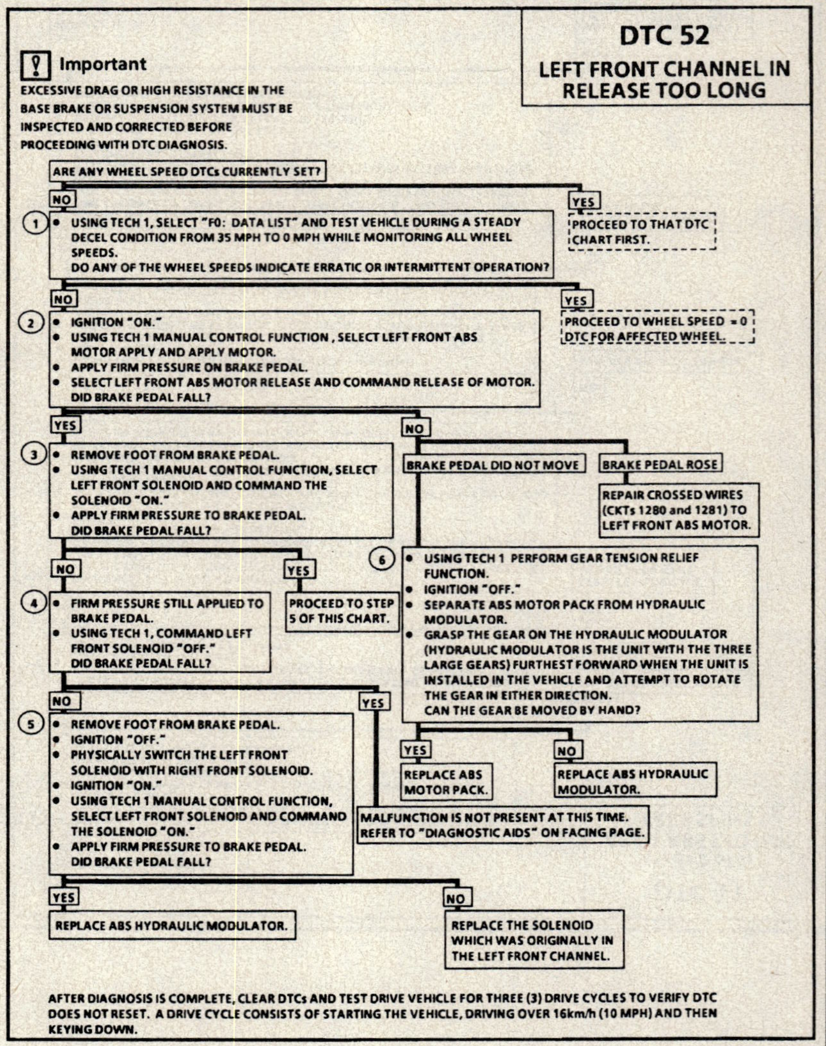

DIAGNOSTIC CHART DTC 53 — 1994 CUTLASS SUPREME AND REGAL

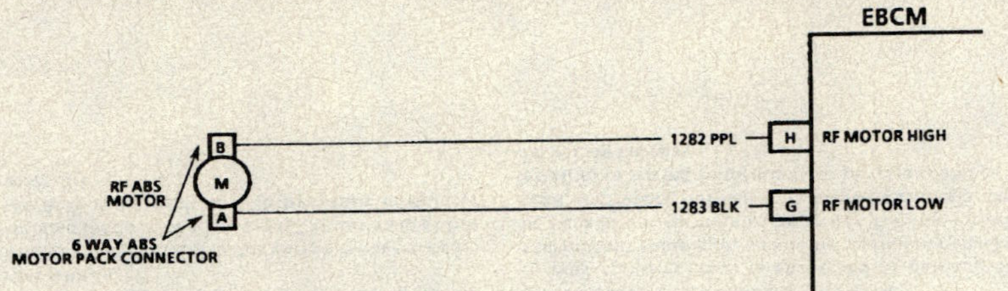

DTC 53

RIGHT FRONT CHANNEL IN RELEASE TOO LONG

Circuit Description:
This DTC will diagnose an ABS motor that is energized longer than expected. This could occur if a wheel speed sensor is malfunctioning, the ABS motor does not turn, the right front solenoid mechanically fails open, or the ABS motor wires are crossed.

DTC Will Set When: DTC 53 can be set only during an ABS stop. If the EBCM commands the right front channel in release for three seconds, a malfunction exists.

Action Taken: A malfunction DTC is stored, ABS is disabled and the ABS warning lamp is turned "ON."

DTC Chart Test Description: Number(s) below refer to circled number(s) on the diagnostic chart.

1. Identifies a problem in a wheel speed sensor that may cause the system to be in release too long.
2. Identifies an ABS motor as being failed or wired incorrectly.
3. Checks for a solenoid that may have mechanically failed open.
4. Checks for a solenoid that may have mechanically failed closed.
5. This serves to isolate the cause of the hydraulic problem, to either the solenoid or the ABS hydraulic modulator assembly.
6. Determines whether a malfunction ABS motor pack or ABS hydraulic modulator is the reason for DTC 53 being set.

Diagnostic Aids: An "Intermittent" malfunction may be caused by a mechanical part of the system that sticks or binds.

The frequency of the malfunction can be checked by using the enhanced diagnostic function of the Tech 1, as described in "Tech 1 Diagnostics" found in this section. DTC 53 may occur on ice if steering wheel is turned to lock during braking. Using the Tech 1 perform hydraulic test to ensure total brake system is functional.

Any circuitry, that is suspected as causing the intermittent complaint, should be thoroughly checked for backed out terminals, improper mating, broken locks, improperly formed or damaged terminals, poor terminal to wiring connections or physical damage to the wiring harness.

DIAGNOSTIC CHART DTC 53 — 1994 CUTLASS SUPREME AND REGAL, CONTINUED

DTC 53

RIGHT FRONT CHANNEL IN RELEASE TOO LONG

Important
EXCESSIVE DRAG OR HIGH RESISTANCE IN THE BASE BRAKE OR SUSPENSION SYSTEM MUST BE INSPECTED AND CORRECTED BEFORE PROCEEDING WITH DTC DIAGNOSIS.

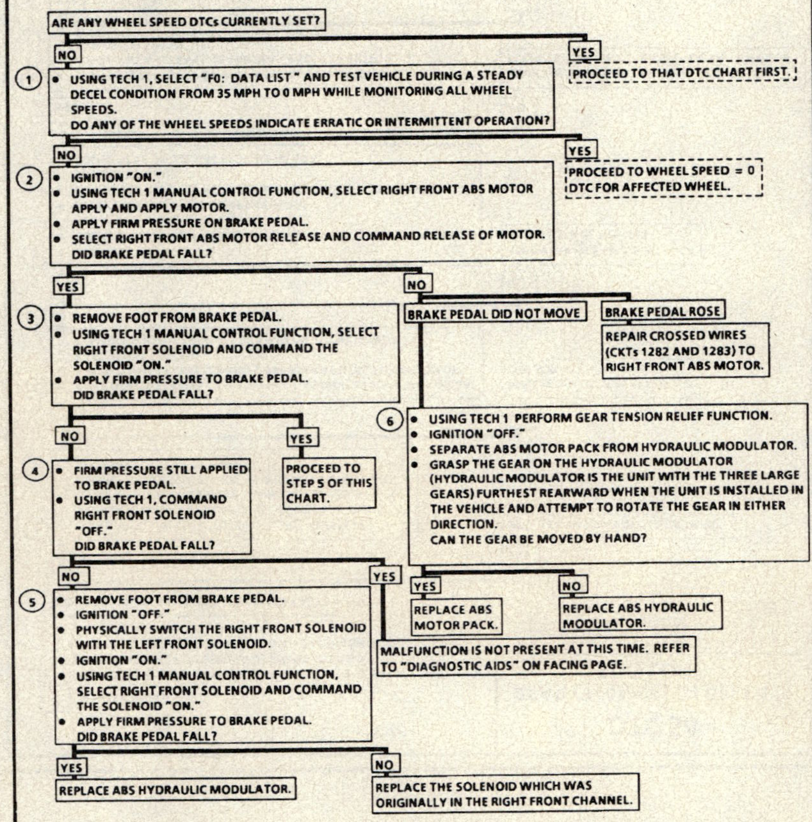

AFTER DIAGNOSIS IS COMPLETE, CLEAR DTCs AND TEST DRIVE VEHICLE FOR THREE (3) DRIVE CYCLES TO VERIFY DTC DOES NOT RESET. A DRIVE CYCLE CONSISTS OF STARTING THE VEHICLE, DRIVING OVER 16km/h (10 MPH) AND THEN KEYING DOWN.

DIAGNOSTIC CHART DTC 54 — 1994 CUTLASS SUPREME AND REGAL

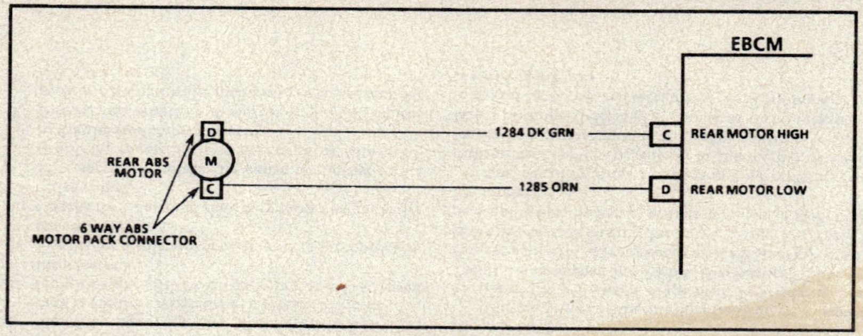

DTC 54
REAR CHANNEL IN RELEASE TOO LONG

Circuit Description:
This DTC will diagnose an ABS motor that is energized longer than expected. This could occur if a wheel speed sensor is malfunctioning, the ABS motor does not turn, or the ABS motor wires are crossed.

DTC Will Set When: DTC 54 can be set only during an ABS stop. If the EBCM commands the rear axle channel in release for three seconds, a malfunction exists.

Action Taken: A malfunction DTC is stored, ABS is disabled and the ABS warning lamp is turned "ON."

DTC Chart Test Description: Number(s) below refer to circled number(s) on the diagnostic chart.

1. Identifies a problem in a wheel speed sensor that may cause the system to be in release too long.
2. Checks for a wheel that may stick or bind because of a mechanical fault.
3. Checks to see if the ABS motor is capable of moving and applying the hydraulic piston for the rear wheels.
4. Ensures that the ABS motor wiring is not crossed.
5. Isolates the fault of a "no-apply" situation to either the ABS motor pack or ABS hydraulic modulator assembly.

Diagnostic Aids: An "Intermittent" malfunction may be caused by a mechanical part of the system that sticks or binds.

The frequency of the malfunction can be checked by using the enhanced diagnostic function of the Tech 1, as described in "Tech 1 Diagnostics" found in this section. Using the Tech 1, perform hydraulic test to ensure total brake system is functional.

Any circuitry, that is suspected as causing the intermittent complaint, should be thoroughly checked for backed out terminals, improper mating, broken locks, improperly formed or damaged terminals, poor terminal to wiring connections or physical damage to the wiring harness.

DIAGNOSTIC CHART DTC 54 — 1994 CUTLASS SUPREME AND REGAL, CONTINUED

DTC 54
REAR CHANNEL IN RELEASE TOO LONG

Important
- EXCESSIVE DRAG OR HIGH RESISTANCE IN THE BASE BRAKE OR SUSPENSION SYSTEM MUST BE INSPECTED AND CORRECTED BEFORE PROCEEDING WITH DTC DIAGNOSIS.

ARE ANY WHEEL SPEED DTCs CURRENTLY SET?

- YES: PROCEED TO THAT DTC CHART FIRST.
- NO:

(1)
- USING TECH 1, SELECT "F0: DATA LIST" AND TEST VEHICLE DURING A STEADY DECEL CONDITION FROM 35 MPH TO 0 MPH WHILE MONITORING ALL WHEEL SPEEDS. DO ANY OF THE WHEEL SPEEDS INDICATE ERRATIC OR INTERMITTENT OPERATION?

- YES: PROCEED TO WHEEL SPEED = 0 DTC FOR AFFECTED WHEEL.
- NO:

(2)
- RAISE REAR OF VEHICLE AND SUPPORT AS INDICATED IN SECTION 0A.
- SPIN BOTH REAR WHEELS AND CONFIRM PHYSICAL CONDITION OF BASE BRAKE SYSTEM.

COULD BOTH REAR WHEELS BE SPUN FREELY BY HAND?

- NO: REPAIR SOURCE OF RESISTANCE IN BASE BRAKE SYSTEM.
- YES:

(3)
- IGNITION "ON."
- USING MANUAL CONTROL FUNCTION OF TECH 1, SELECT REAR AXLE ABS MOTOR APPLY.
- APPLY FIRM PRESSURE ON BRAKE PEDAL.
- COMMAND APPLY OF REAR ABS MOTOR AND TRY TO SPIN REAR WHEELS BY HAND.

COULD REAR WHEELS BE SPUN BY HAND?

- NO: MALFUNCTION IS NOT PRESENT AT THIS TIME. REFER TO "DIAGNOSTIC AIDS" ON FACING PAGE.
- YES:

(4)
- WITH FIRM PRESSURE ON BRAKE PEDAL STILL APPLIED, COMMAND A RELEASE OF THE REAR AXLE ABS MOTOR.

CAN WHEELS BE MOVED?

- NO: REPAIR CROSSED WIRES AT REAR ABS MOTOR (CKTs 1284 AND 1285).
- YES:

(5)
- USING TECH 1, PERFORM GEAR TENSION RELIEF FUNCTION.
- SEPARATE ABS MOTOR PACK FROM HYDRAULIC MODULATOR.
- CONNECT ABS MOTOR PACK WIRING.
- IGNITION "ON."
- USING TECH 1, PERFORM MOTOR TEST.

DO ALL THREE ABS MOTOR PACK GEARS (SMALL GEARS) SPIN FREELY?

- NO: REPLACE ABS MOTOR PACK.
- YES: REPLACE ABS HYDRAULIC MODULATOR.

AFTER ALL DIAGNOSIS IS COMPLETE, CLEAR DTCs AND TEST DRIVE VEHICLE FOR THREE (3) DRIVE CYCLES TO VERIFY THAT DTC DOES NOT RESET. A DRIVE CYCLE CONSISTS OF STARTING THE VEHICLE DRIVING OVER 16 KM/H (10 MPH), AND THEN KEYING DOWN.

DIAGNOSTIC CHART DTC 55 — 1994 CUTLASS SUPREME AND REGAL

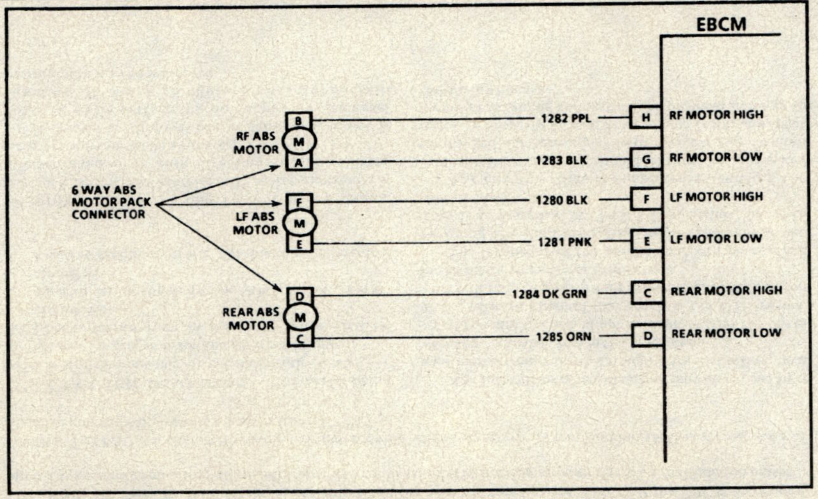

DTC 55

EBCM MALFUNCTION

Circuit Description:

This DTC identifies a malfunction detected by the MDI custom IC. It also ensures the cause of the malfunction is not a result of a problem with the ABS enable relay under a DTC 55 malfunction.

Action Taken: A malfunction DTC is stored, ABS is disabled and the ABS warning lamp is turned "ON."

DTC Chart Test Description: Number(s) below refer to circled number(s) on the diagnostic chart.

1. Checks to see if the malfunction is still present.
2. Checks to see if the malfunction is intermittent.

Diagnostic Aids: An "Intermittent" malfunction may be caused by a poor connection, rubbed through wire insulation, or a wire that is broken inside the insulation.

The frequency of the malfunction can be checked by using the enhanced diagnostic function of the Tech 1, as described in "Tech 1 Diagnostics" found in this section.

Any circuitry, that is suspected as causing the intermittent complaint, should be thoroughly checked for backed out terminals, improper mating, broken locks, improperly formed or damaged terminals, poor terminal to wiring connections or physical damage to the wiring harness.

DIAGNOSTIC CHART DTC 55 — 1994 CUTLASS SUPREME AND REGAL, CONTINUED

DTC 55

EBCM MALFUNCTION

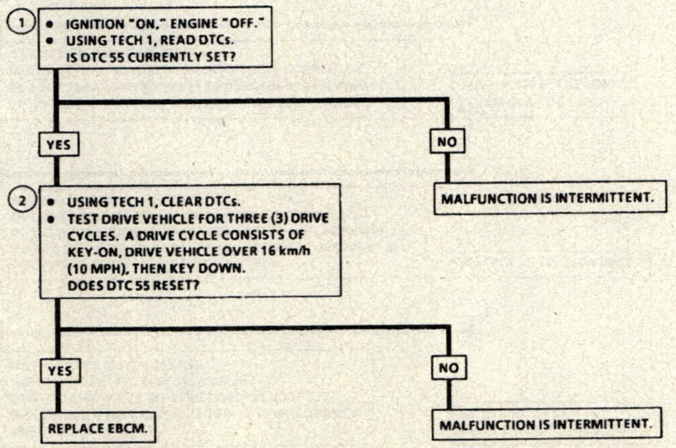

DIAGNOSTIC CHART DTC 56 — 1994 CUTLASS SUPREME AND REGAL

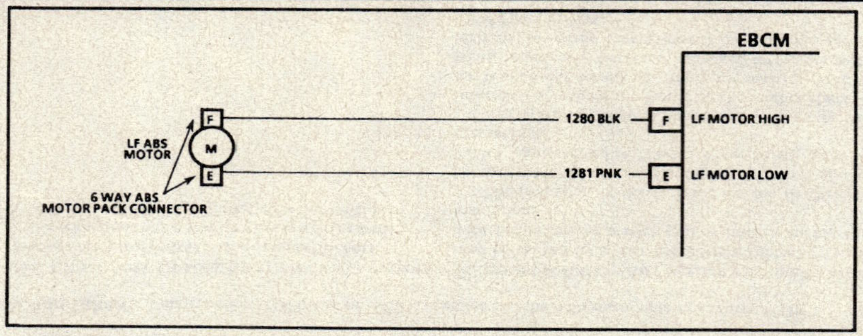

DTC 56

LEFT FRONT ABS MOTOR CIRCUIT OPEN

Circuit Description:
This DTC identifies a motor that cannot be energized due to an open in its circuitry.

DTC Will Set When: DTC 56 can be set only when the motor is commanded "OFF." If the EBCM detects an out of range voltage on either of the left front ABS motor circuits indicating an open circuit, a malfunction exists.

Action Taken: An open ABS motor will not activate when requested. A malfunction DTC is stored, ABS is disabled and the ABS warning lamp is turned "ON."

DTC Chart Test Description: Number(s) below refer to circled number(s) on the diagnostic chart.

1. Checks for proper resistance of the ABS motor.
2. Checks for an open in the ABS motor "HIGH" circuitry.
3. Checks for an open in the ABS motor "LOW" circuitry.
4. Ensures malfunction was not due to poor terminal contact.

Diagnostic Aids: Using Tech 1, select manual control function, and exercise ABS motor movement of affected channel in both directions while applying light pressure on the brake pedal.

If erratic or "jumpy" brake pedal movement is detected while performing an "apply" or "release" function of the ABS motor, an intermittent malfunction may be indicated.

An "Intermittent" malfunction may be caused by a poor connection, rubbed through wire insulation, or a wire that is broken inside the insulation.

If the malfunction is not current, wiggle the wires of the affected channel and check if the DTC resets. This will help to pinpoint an intermittent malfunction in the motor circuitry or connections.

The frequency of the malfunction can be checked by using the enhanced diagnostic function of the Tech 1, as described in "Tech 1 Diagnostics" found in this section.

Any circuitry, that is suspected as causing the intermittent complaint, should be thoroughly checked for backed out terminals, improper mating, broken locks, improperly formed or damaged terminals, poor terminal to wiring connections or physical damage to the wiring harness.

DIAGNOSTIC CHART DTC 56 — 1994 CUTLASS SUPREME AND REGAL, CONTINUED

DTC 56

LEFT FRONT ABS MOTOR CIRCUIT OPEN

DOES DTC 56 OCCUR INTERMITTENTLY?

- YES: REFER TO "DIAGNOSTIC AIDS" ON FACING PAGE.
- NO:

(1)
- IGNITION "OFF."
- DISCONNECT 6-WAY CONNECTOR AT ABS MOTOR PACK.
- USING J 39200, MEASURE RESISTANCE ACROSS ABS MOTOR PACK TERMINALS "F" AND "E".

IS RESISTANCE BELOW 2000Ω?

- NO: REPLACE ABS MOTOR PACK.
- YES:

(2)
- DISCONNECT 6-WAY CONNECTOR AT EBCM.
- USING J 39200, MEASURE RESISTANCE BETWEEN TERMINAL "F" OF THE 6-WAY EBCM HARNESS CONNECTOR AND TERMINAL "F" OF THE 6-WAY ABS MOTOR PACK HARNESS CONNECTOR.

IS RESISTANCE 2Ω OR LESS?

- NO: REPAIR OPEN OR HIGH RESISTANCE IN CKT 1280.
- YES:

(3)
- USING J 39200, MEASURE RESISTANCE BETWEEN TERMINAL "E" OF THE 6-WAY EBCM HARNESS CONNECTOR AND TERMINAL "E" OF THE 6-WAY ABS MOTOR PACK HARNESS CONNECTOR.

IS RESISTANCE 2Ω OR LESS?

- NO: REPAIR OPEN OR HIGH RESISTANCE IN CKT 1281.
- YES:

(4)
- INSPECT THE 6-WAY ABS MOTOR PACK AND 6-WAY EBCM CONNECTORS FOR POOR TERMINAL CONTACT AND EVIDENCE OF CORROSION.

REPLACE ALL TERMINALS THAT EXHIBIT SIGNS OF POOR TERMINAL CONTACT OR CORROSION.

- RECONNECT ALL CONNECTORS.
- IGNITION "OFF."
- START ENGINE AND TEST DRIVE VEHICLE, ACHIEVING AT LEAST 16 km/h (10 MPH).
- REPEAT ABOVE DRIVE CYCLE SEQUENCE TWO MORE TIMES.

DID DTC 56 SET IN THE LAST THREE DRIVE CYCLES?

- YES: REPLACE EBCM.
- NO: MALFUNCTION WAS MOST LIKELY CAUSED BY A POOR CONNECTION. REFER TO "DIAGNOSTIC AIDS" ON FACING PAGE.

AFTER ALL DIAGNOSIS IS COMPLETE, CLEAR DTCs AND TEST DRIVE VEHICLE FOR THREE (3) DRIVE CYCLES TO VERIFY THAT DTC DOES NOT RESET. A DRIVE CYCLE CONSISTS OF STARTING THE VEHICLE, DRIVING OVER 16 KM/H (10 MPH), AND THEN KEYING DOWN.

DIAGNOSTIC CHART DTC 57 — 1994 CUTLASS SUPREME AND REGAL

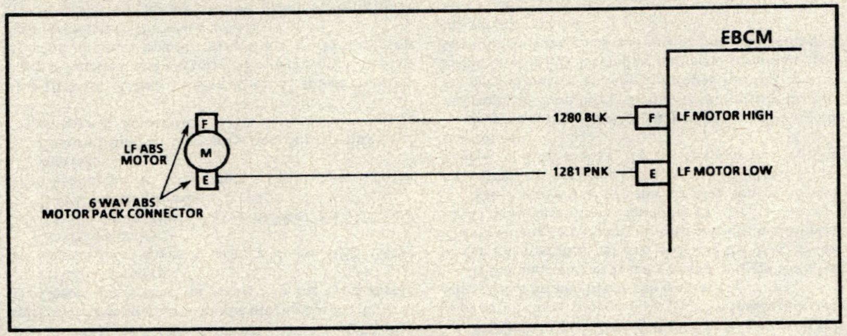

DTC 57

LEFT FRONT ABS MOTOR CIRCUIT SHORTED TO GROUND

Circuit Description:
This DTC identifies an ABS motor circuit that is shorted to ground. This malfunction will not allow the ABS motor to be controlled at the commanded current rate or will cause the driver circuit to allow current directly to ground.

DTC Will Set When: DTC 57 can be set anytime. If the EBCM detects an out of range voltage on either of the left front ABS motor circuits indicating a circuit shorted to ground, a malfunction exists.

Action Taken: A malfunction DTC is stored, ABS is disabled and the ABS warning lamp is turned "ON."

DTC Chart Test Description: Number(s) below refer to circled number(s) on the diagnostic chart.

1. Checks for a short to ground in the ABS motor "HIGH" circuitry.
2. Checks for a short to ground in the ABS motor "LOW" circuitry.
3. Checks for an ABS motor that is internally shorted to ground.
4. Ensures malfunction was not due to physical damage of the circuitry.

Diagnostic Aids: Using Tech 1, select manual control function, and exercise ABS motor movement of affected channel in both directions while applying light pressure on the brake pedal.

If erratic or "jumpy" brake pedal movement is detected while performing an "apply" or "release" function of the ABS motor, an intermittent malfunction may be indicated.

An "Intermittent" malfunction may be caused by a poor connection, rubbed through wire insulation, or a wire that is broken inside the insulation.

If the malfunction is not current, wiggle the wires of the affected channel and check if the DTC resets. This will help to pinpoint an intermittent malfunction in the motor circuitry or connections.

The frequency of the malfunction can be checked by using the enhanced diagnostic function of the Tech 1, as described in "Tech 1 Diagnostics" found in this section.

Any circuitry, that is suspected as causing the intermittent complaint, should be thoroughly checked for backed out terminals, improper mating, broken locks, improperly formed or damaged terminals, poor terminal to wiring connections or physical damage to the wiring harness.

DIAGNOSTIC CHART DTC 57 — 1994 CUTLASS SUPREME AND REGAL, CONTINUED

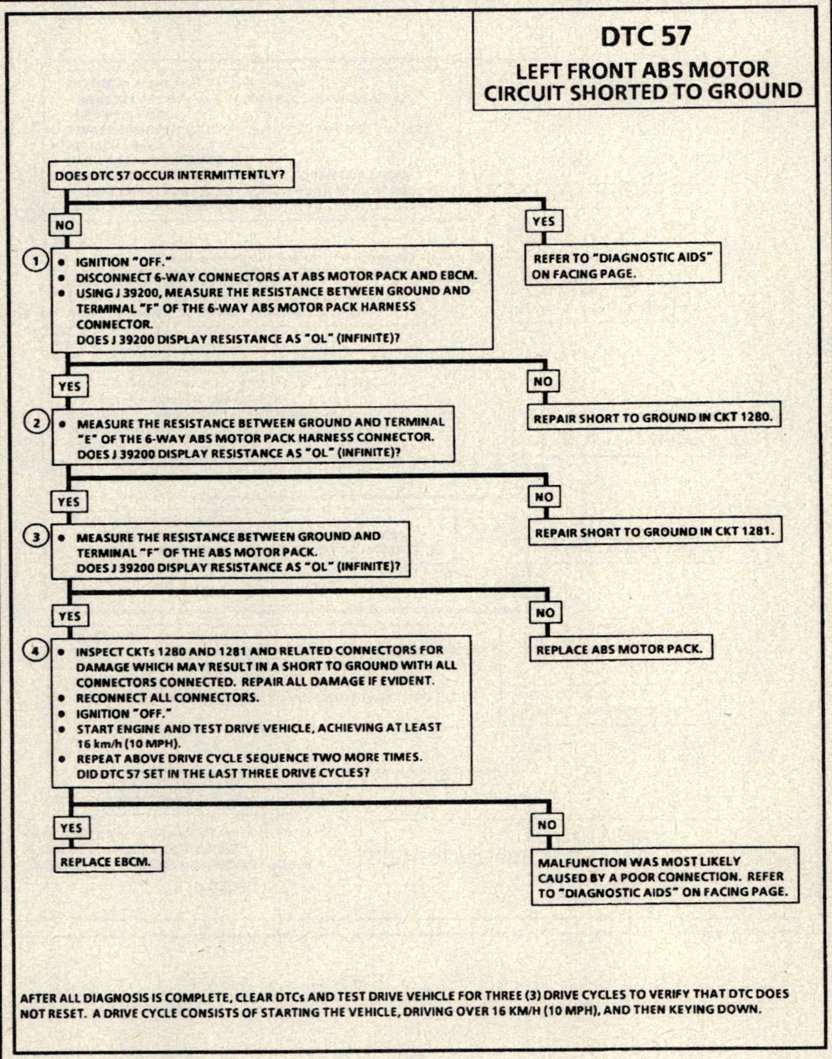

DIAGNOSTIC CHART DTC 58 — 1994 CUTLASS SUPREME AND REGAL

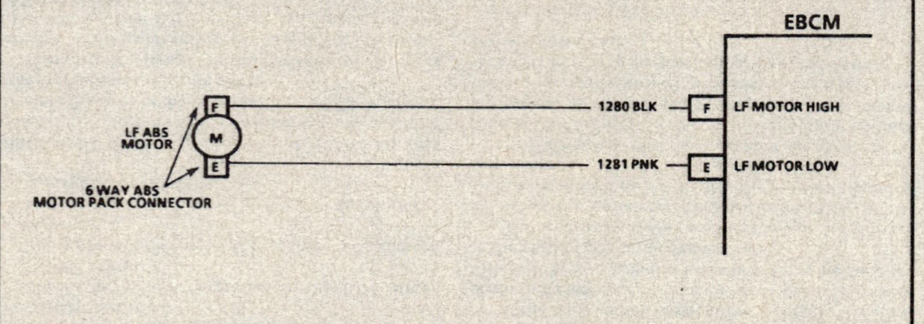

DTC 58

LEFT FRONT ABS MOTOR CIRCUIT SHORTED TO BATTERY

Circuit Description:

This DTC identifies an ABS motor circuit that is shorted to battery or an ABS motor that has low or no resistance. This malfunction will not allow the ABS motor to be controlled at the commanded current rate or will cause the ABS motor to turn in the opposite direction, or not at all.

DTC Will Set When: DTC 58 can be set only when the ABS motor is commanded "OFF." If the EBCM detects an out of range voltage on either of the left front ABS motor circuits indicating a circuit shorted to battery or an ABS motor shorted, a malfunction exists.

Action Taken: A malfunction DTC is stored, ABS is disabled and the ABS warning lamp is turned "ON."

DTC Chart Test Description: Number(s) below refer to circled number(s) on the diagnostic chart.

1. Checks for a short to voltage in the ABS motor "HIGH" circuitry.
2. Checks for a short to voltage in the ABS motor "LOW" circuitry.
3. Checks for a short between the "HIGH" and "LOW" motor circuits.
4. Checks for an ABS motor that is internally shorted.
5. Ensures malfunction was not due to physical damage of the circuitry.

Diagnostic Aids: Using Tech 1, select manual control function, and exercise ABS motor movement of affected channel in both directions while applying light pressure on the brake pedal.

If erratic or "jumpy" brake pedal movement is detected while performing an "apply" or "release" function of the ABS motor, an intermittent malfunction may be indicated.

An "Intermittent" malfunction may be caused by a poor connection, rubbed through wire insulation, or a wire that is broken inside the insulation.

If the malfunction is not current, wiggle the wires of the affected channel and check if the DTC resets. This will help to pinpoint an intermittent malfunction in the motor circuitry or connections.

The frequency of the malfunction can be checked by using the enhanced diagnostic function of the Tech 1, as described in "Tech 1 Diagnostics" found in this section.

Any circuitry, that is suspected as causing the intermittent complaint, should be thoroughly checked for backed out terminals, improper mating, broken locks, improperly formed or damaged terminals, poor terminal to wiring connections or physical damage to the wiring harness.

DIAGNOSTIC CHART DTC 58 — 1994 CUTLASS SUPREME AND REGAL, CONTINUED

DTC 58

LEFT FRONT ABS MOTOR CIRCUIT SHORTED TO BATTERY

NOTICE: J 39200 TEST LEADS MUST BE "ZEROED" PRIOR TO MAKING ANY RESISTANCE MEASUREMENTS. REFER TO J 39200 USER'S MANUAL.

DOES DTC 58 OCCUR INTERMITTENTLY?

- YES: REFER TO "DIAGNOSTIC AIDS" ON FACING PAGE.
- NO:

(1)
- IGNITION "OFF."
- DISCONNECT 6-WAY CONNECTORS AT ABS MOTOR PACK AND EBCM.
- IGNITION "ON."
- USING J 39200, MEASURE THE VOLTAGE BETWEEN GROUND AND TERMINAL "F" OF THE 6-WAY ABS MOTOR PACK HARNESS CONNECTOR. IS VOLTAGE 1 VOLT OR LESS?

- NO: REPAIR SHORT TO VOLTAGE IN CKT 1280.
- YES:

(2)
- MEASURE THE VOLTAGE BETWEEN GROUND AND TERMINAL "E" OF THE 6-WAY ABS MOTOR PACK HARNESS CONNECTOR. IS VOLTAGE 1 VOLT OR LESS?

- NO: REPAIR SHORT TO VOLTAGE IN CKT 1281.
- YES:

(3)
- IGNITION "OFF."
- USING J 39200, MEASURE THE RESISTANCE BETWEEN TERMINAL "E" AND TERMINAL "F" OF THE 6-WAY ABS MOTOR PACK HARNESS CONNECTOR. DOES J 39200 DISPLAY RESISTANCE AS "OL" (INFINITE)?

- NO: REPAIR SHORT BETWEEN CKTs 1280 AND 1281.
- YES:

(4)
- USING J 39200, MEASURE THE RESISTANCE BETWEEN TERMINAL "E" AND TERMINAL "F" OF THE ABS MOTOR PACK. IS RESISTANCE GREATER THAN 0.4Ω?

- NO: REPLACE ABS MOTOR PACK.
- YES:

(5)
- INSPECT CKTs 1280 AND 1281 AND RELATED CONNECTORS FOR DAMAGE WHICH MAY RESULT IN A SHORT TO GROUND WITH ALL CONNECTORS CONNECTED. REPAIR ALL DAMAGE IF EVIDENT.
- RECONNECT ALL CONNECTORS.
- IGNITION "OFF."
- START ENGINE AND TEST DRIVE VEHICLE, ACHIEVING AT LEAST 16 km/h (10 MPH).
- REPEAT ABOVE DRIVE CYCLE SEQUENCE TWO MORE TIMES. DID DTC 58 SET IN THE LAST THREE DRIVE CYCLES?

- YES: REPLACE EBCM.
- NO: MALFUNCTION WAS MOST LIKELY CAUSED BY A POOR CONNECTION. REFER TO "DIAGNOSTIC AIDS" ON FACING PAGE.

AFTER ALL DIAGNOSIS IS COMPLETE, CLEAR DTCs AND TEST DRIVE VEHICLE FOR THREE (3) DRIVE CYCLES TO VERIFY THAT DTC DOES NOT RESET. A DRIVE CYCLE CONSISTS OF STARTING THE VEHICLE, DRIVING OVER 16 km/h (10 MPH), AND THEN KEYING DOWN.

DIAGNOSTIC CHART DTC 61 — 1994 CUTLASS SUPREME AND REGAL

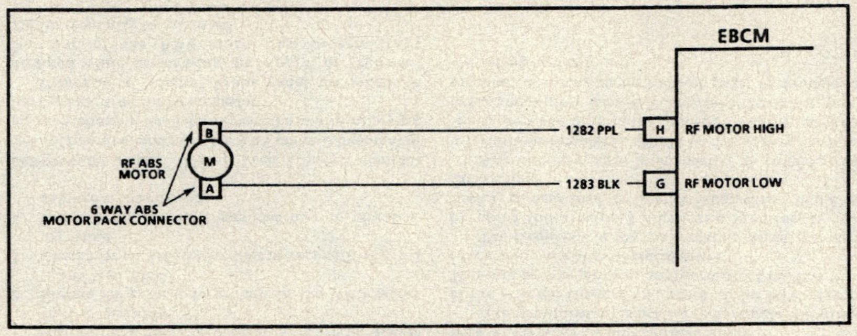

DTC 61

RIGHT FRONT ABS MOTOR CIRCUIT OPEN

Circuit Description:
This DTC identifies an ABS motor that cannot be energized due to an open in its circuitry. This malfunction will not allow proper front ABS operation.

DTC Will Set When: DTC 61 can be set only when the ABS motor is commanded "OFF." If the EBCM detects an out of range voltage on either of the right front ABS motor circuits indicating an open circuit, a malfunction exists.

Action Taken: An open ABS motor will not activate when requested. A malfunction DTC is stored, ABS is disabled and the ABS warning lamp is turned "ON."

DTC Chart Test Description: Number(s) below refer to circled number(s) on the diagnostic chart.

1. Checks for proper resistance of the ABS motor.
2. Checks for an open in the ABS motor "HIGH" circuitry.
3. Checks for an open in the ABS motor "LOW" circuitry.
4. Ensures malfunction was not due to physical damage of the circuitry.

Diagnostic Aids: Using Tech 1, select manual control function, and exercise ABS motor movement of affected channel in both directions while applying light pressure on the brake pedal.

If erratic or "jumpy" brake pedal movement is detected while performing an "apply" or "release" function of the ABS motor, an intermittent malfunction may be indicated.

An "Intermittent" malfunction may be caused by a poor connection, rubbed through wire insulation, or a wire that is broken inside the insulation.

If the malfunction is not current, wiggle the wires of the affected channel and check if the DTC resets. This will help to pinpoint an intermittent malfunction in the motor circuitry or connections.

The frequency of the malfunction can be checked by using the enhanced diagnostic function of the Tech 1, as described in "Tech 1 Diagnostics" found in this section.

Any circuitry, that is suspected as causing the intermittent complaint, should be thoroughly checked for backed out terminals, improper mating, broken locks, improperly formed or damaged terminals, poor terminal to wiring connections or physical damage to the wiring harness.

DIAGNOSTIC CHART DTC 61 — 1994 CUTLASS SUPREME AND REGAL, CONTINUED

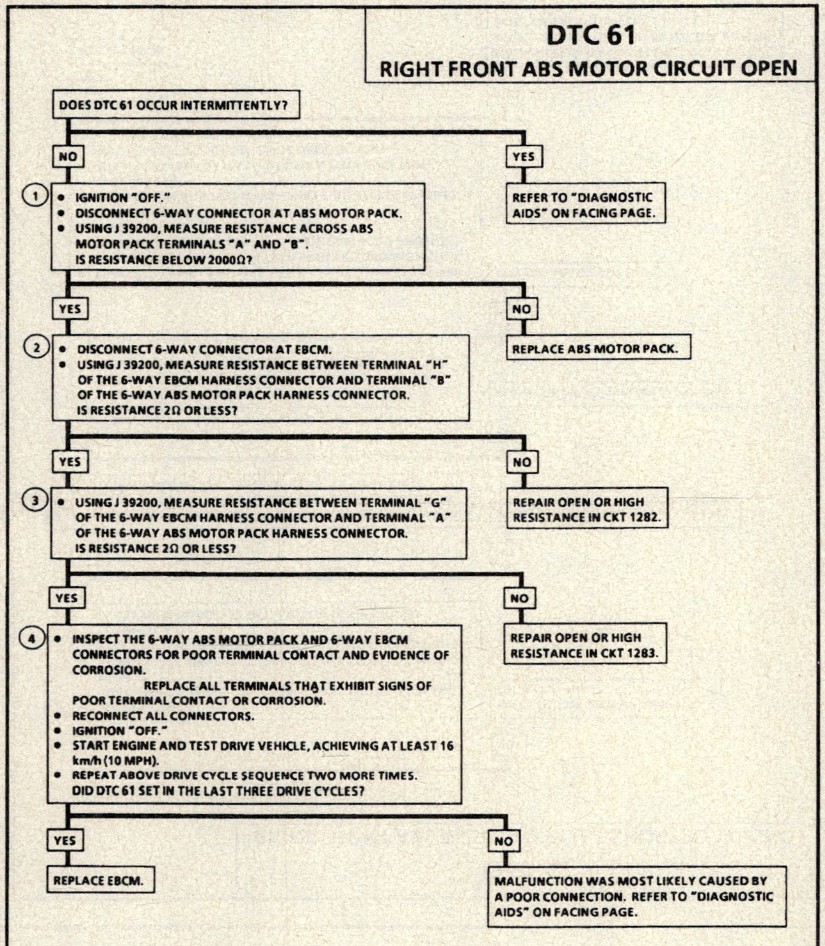

AFTER ALL DIAGNOSIS IS COMPLETE, CLEAR DTCs AND TEST DRIVE VEHICLE FOR THREE (3) DRIVE CYCLES TO VERIFY THAT DTC DOES NOT RESET. A DRIVE CYCLE CONSISTS OF STARTING THE VEHICLE, DRIVING OVER 16 km/h (10 MPH), AND THEN KEYING DOWN.

DIAGNOSTIC CHART DTC 62 — 1994 CUTLASS SUPREME AND REGAL

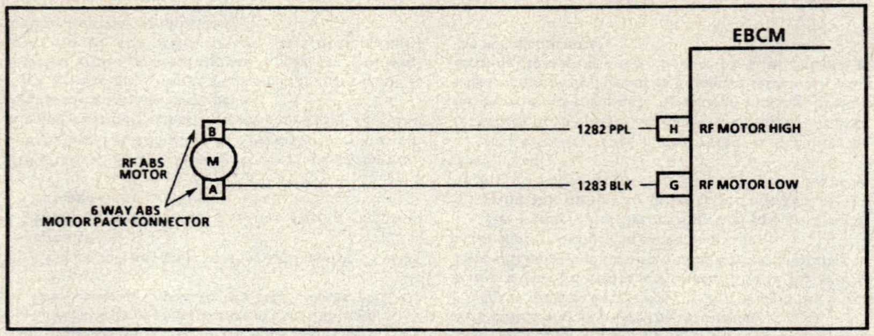

DTC 62

RIGHT FRONT ABS MOTOR CIRCUIT SHORTED TO GND

Circuit Description:

This DTC identifies an ABS motor circuit that is shorted to ground. This malfunction will not allow the ABS motor to be controlled at the commanded current rate or will cause the ABS motor driver circuit to allow current directly to ground.

DTC Will Set When: DTC 62 can be set anytime. If the EBCM detects an out of range voltage on either of the right front ABS motor circuits indicating a circuit shorted to ground, a malfunction exists.

Action Taken: A malfunction DTC is stored, ABS is disabled and the ABS warning lamp is turned "ON."

DTC Chart Test Description: Number(s) below refer to circled number(s) on the diagnostic chart.

1. Checks for a short to ground in the ABS motor "HIGH" circuitry.
2. Checks for a short to ground in the ABS motor "LOW" circuitry.
3. Checks for an ABS motor that is internally shorted to ground.
4. Ensures malfunction was not due to physical damage of the circuitry.

Diagnostic Aids: Using Tech 1, select manual control function, and exercise ABS motor movement of affected channel in both directions while applying light pressure on the brake pedal.

If erratic or "jumpy" brake pedal movement is detected while performing an "apply" or "release" function of the ABS motor, an intermittent malfunction may be indicated.

An "Intermittent" malfunction may be caused by a poor connection, rubbed through wire insulation, or a wire that is broken inside the insulation.

If the malfunction is not current, wiggle the wires of the affected channel and check if the DTC resets. This will help to pinpoint an intermittent malfunction in the motor circuitry or connections.

The frequency of the malfunction can be checked by using the enhanced diagnostic function of the Tech 1, as described in "Tech 1 Diagnostics" found in this section.

Any circuitry, that is suspected as causing the intermittent complaint, should be thoroughly checked for backed out terminals, improper mating, broken locks, improperly formed or damaged terminals, poor terminal to wiring connections or physical damage to the wiring harness.

DIAGNOSTIC CHART DTC 62 — 1994 CUTLASS SUPREME AND REGAL, CONTINUED

DTC 62

RIGHT FRONT ABS MOTOR CIRCUIT SHORTED TO GND

DOES DTC 62 OCCUR INTERMITTENTLY?

- YES: REFER TO "DIAGNOSTIC AIDS" ON FACING PAGE.
- NO:

(1)
- IGNITION "OFF."
- DISCONNECT 6-WAY CONNECTORS AT ABS MOTOR PACK AND EBCM.
- USING J 39200, MEASURE THE RESISTANCE BETWEEN GROUND AND TERMINAL "B" OF THE 6-WAY ABS MOTOR PACK HARNESS CONNECTOR.

DOES J 39200 DISPLAY RESISTANCE AS "OL" (INFINITE)?

- NO: REPAIR SHORT TO GROUND IN CKT 1282.
- YES:

(2)
- MEASURE THE RESISTANCE BETWEEN GROUND AND TERMINAL "A" OF THE 6-WAY ABS MOTOR PACK HARNESS CONNECTOR.

DOES J 39200 DISPLAY RESISTANCE AS "OL" (INFINITE)?

- NO: REPAIR SHORT TO GROUND IN CKT 1283.
- YES:

(3)
- MEASURE THE RESISTANCE BETWEEN GROUND AND TERMINAL "B" OF THE ABS MOTOR PACK.

DOES J 39200 DISPLAY RESISTANCE AS "OL" (INFINITE)?

- NO: REPLACE ABS MOTOR PACK.
- YES:

(4)
- INSPECT CKTs 1282 AND 1283 AND RELATED CONNECTORS FOR DAMAGE WHICH MAY RESULT IN A SHORT TO GROUND WITH ALL CONNECTORS CONNECTED. REPAIR ALL DAMAGE IF EVIDENT.
- RECONNECT ALL CONNECTORS.
- IGNITION "OFF."
- START ENGINE AND TEST DRIVE VEHICLE, ACHIEVING AT LEAST 16 km/h (10 MPH).
- REPEAT ABOVE DRIVE CYCLE SEQUENCE TWO MORE TIMES.

DID DTC 62 SET IN THE LAST THREE DRIVE CYCLES?

- YES: REPLACE EBCM.
- NO: MALFUNCTION WAS MOST LIKELY CAUSED BY A POOR CONNECTION. REFER TO "DIAGNOSTIC AIDS" ON FACING PAGE.

AFTER ALL DIAGNOSIS IS COMPLETE, CLEAR DTCs AND TEST DRIVE VEHICLE FOR THREE (3) DRIVE CYCLES TO VERIFY THAT DTC DOES NOT RESET. A DRIVE CYCLE CONSISTS OF STARTING THE VEHICLE, DRIVING OVER 16 KM/H (10 MPH), AND THEN KEYING DOWN.

DIAGNOSTIC CHART DTC 63 — 1994 CUTLASS SUPREME AND REGAL

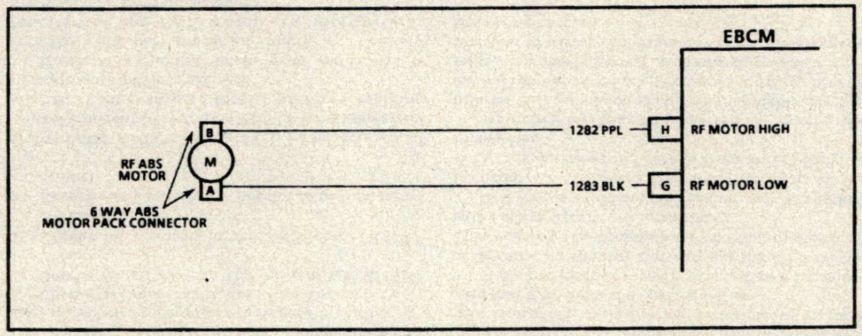

DTC 63

RIGHT FRONT ABS MOTOR CIRCUIT SHORTED TO BATT

Circuit Description:

This DTC identifies an ABS motor circuit that is shorted to battery or an ABS motor that has low or no resistance. This malfunction will not allow the ABS motor to be controlled at the commanded current rate or will cause the ABS motor to turn in the opposite direction or not at all.

DTC Will Set When: DTC 63 can be set only when the ABS motor is commanded "OFF." If the EBCM detects an out of range voltage on either of the right front ABS motor circuits indicating a circuit shorted to battery or an ABS motor shorted, a malfunction exists.

Action Taken: A malfunction DTC is stored, ABS is disabled and the ABS warning lamp is turned "ON."

DTC Chart Test Description: Number(s) below refer to circled number(s) on the diagnostic chart.

1. Checks for a short to voltage in the ABS motor "HIGH" circuitry.
2. Checks for a short to voltage in the ABS motor "LOW" circuitry.
3. Checks for a short circuit between motor "HIGH" and motor "LOW."
4. Checks for proper resistance of the ABS motor.
5. Ensures malfunction was not due to physical damage of the circuitry.

Diagnostic Aids: Using Tech 1, select manual control function, and exercise ABS motor movement of affected channel in both directions while applying light pressure on the brake pedal.

If erratic or "jumpy" brake pedal movement is detected while performing an "apply" or "release" function of the ABS motor, an intermittent malfunction may be indicated.

An "Intermittent" malfunction may be caused by a poor connection, rubbed through wire insulation, or a wire that is broken inside the insulation.

If the malfunction is not current, wiggle the wires of the affected channel and check if the DTC resets. This will help to pinpoint an intermittent malfunction in the motor circuitry or connections.

The frequency of the malfunction can be checked by using the enhanced diagnostic function of the Tech 1, as described in "Tech 1 Diagnostics" found in this section.

Any circuitry, that is suspected as causing the intermittent complaint, should be thoroughly checked for backed out terminals, improper mating, broken locks, improperly formed or damaged terminals, poor terminal to wiring connections or physical damage to the wiring harness.

DIAGNOSTIC CHART DTC 63 — 1994 CUTLASS SUPREME AND REGAL, CONTINUED

DTC 63
RIGHT FRONT ABS MOTOR CIRCUIT SHORTED TO BATT

NOTICE: J 39200 TEST LEADS MUST BE "ZEROED" PRIOR TO MAKING ANY RESISTANCE MEASUREMENTS. REFER TO J 39200 USER'S MANUAL.

DOES DTC 63 OCCUR INTERMITTENTLY?

- YES → REFER TO "DIAGNOSTIC AIDS" ON FACING PAGE.
- NO ↓

(1)
- IGNITION "OFF."
- DISCONNECT 6-WAY CONNECTORS AT ABS MOTOR PACK AND EBCM.
- IGNITION "ON."
- USING J 39200, MEASURE THE VOLTAGE BETWEEN GROUND AND TERMINAL "B" OF THE 6-WAY ABS MOTOR PACK HARNESS CONNECTOR. IS VOLTAGE 1 VOLT OR LESS?

- NO → REPAIR SHORT TO VOLTAGE IN CKT 1282.
- YES ↓

(2)
- MEASURE THE VOLTAGE BETWEEN GROUND AND TERMINAL "A" OF THE 6-WAY ABS MOTOR PACK HARNESS CONNECTOR. IS VOLTAGE 1 VOLT OR LESS?

- NO → REPAIR SHORT TO VOLTAGE IN CKT 1283.
- YES ↓

(3)
- IGNITION "OFF."
- USING J 39200, MEASURE THE RESISTANCE BETWEEN TERMINAL "A" AND TERMINAL "B" OF THE 6-WAY ABS MOTOR PACK HARNESS CONNECTOR. DOES J 39200 DISPLAY RESISTANCE AS "OL" (INFINITE)?

- NO → REPAIR SHORT BETWEEN CKTs 1282 AND 1283.
- YES ↓

(4)
- USING J 39200, MEASURE THE RESISTANCE BETWEEN TERMINAL "A" AND TERMINAL "B" OF THE ABS MOTOR PACK. IS RESISTANCE GREATER THAN 0.4Ω?

- NO → REPLACE ABS MOTOR PACK.
- YES ↓

(5)
- INSPECT CKTs 1282 AND 1283 AND RELATED CONNECTORS FOR DAMAGE WHICH MAY RESULT IN A SHORT TO GROUND WITH ALL CONNECTORS CONNECTED. REPAIR ALL DAMAGE IF EVIDENT.
- RECONNECT ALL CONNECTORS.
- IGNITION "OFF."
- START ENGINE AND TEST DRIVE VEHICLE, ACHIEVING AT LEAST 16 km/h (10 MPH).
- REPEAT ABOVE DRIVE CYCLE SEQUENCE TWO MORE TIMES. DID DTC 63 SET IN THE LAST THREE DRIVE CYCLES?

- YES → REPLACE EBCM.
- NO → MALFUNCTION WAS MOST LIKELY CAUSED BY A POOR CONNECTION. REFER TO "DIAGNOSTIC AIDS" ON FACING PAGE.

AFTER ALL DIAGNOSIS IS COMPLETE, CLEAR DTCs AND TEST DRIVE VEHICLE FOR THREE (3) DRIVE CYCLES TO VERIFY THAT DTC DOES NOT RESET. A DRIVE CYCLE CONSISTS OF STARTING THE VEHICLE, DRIVING OVER 16 km/h (10 MPH), AND THEN KEYING DOWN.

DIAGNOSTIC CHART DTC 64 — 1994 CUTLASS SUPREME AND REGAL

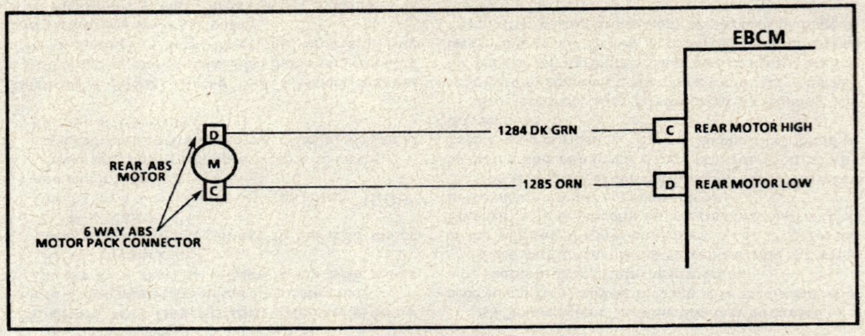

DTC 64

REAR ABS MOTOR CIRCUIT OPEN

Circuit Description:
This DTC identifies an ABS motor that cannot be energized due to an open in its circuitry. This malfunction will not allow proper rear ABS operation.

DTC Will Set When: DTC 64 can be set only when the ABS motor is commanded "OFF." If the EBCM detects an out of range voltage on either of the rear ABS motor circuits indicating an open circuit, a malfunction exists.

Action Taken: An open motor will not activate when requested. A malfunction DTC is stored, ABS is disabled and the ABS warning lamp is turned "ON."

DTC Chart Test Description: Number(s) below refer to circled number(s) on the diagnostic chart.

1. Checks for proper resistance of the ABS motor.
2. Checks for an open in the ABS motor "HIGH" circuitry.
3. Checks for an open in the ABS motor "LOW" circuitry.
4. Ensures malfunction was not due to poor terminal contact.

Diagnostic Aids: Using Tech 1, select manual control function, and exercise ABS motor movement of affected channel in both directions while applying light pressure on the brake pedal.

If erratic or "jumpy" brake pedal movement is detected while performing an "apply" or "release" function of the ABS motor, an intermittent malfunction may be indicated.

An "Intermittent" malfunction may be caused by a poor connection, rubbed through wire insulation, or a wire that is broken inside the insulation.

If the malfunction is not current, wiggle the wires of the affected channel and check if the DTC resets. This will help to pinpoint an intermittent malfunction in the motor circuitry or connections.

The frequency of the malfunction can be checked by using the enhanced diagnostic function of the Tech 1, as described in "Tech 1 Diagnostics" found in this section.

Any circuitry, that is suspected as causing the intermittent complaint, should be thoroughly checked for backed out terminals, improper mating, broken locks, improperly formed or damaged terminals, poor terminal to wiring connections or physical damage to the wiring harness.

DIAGNOSTIC CHART DTC 64 — 1994 CUTLASS SUPREME AND REGAL, CONTINUED

DTC 64

REAR ABS MOTOR CIRCUIT OPEN

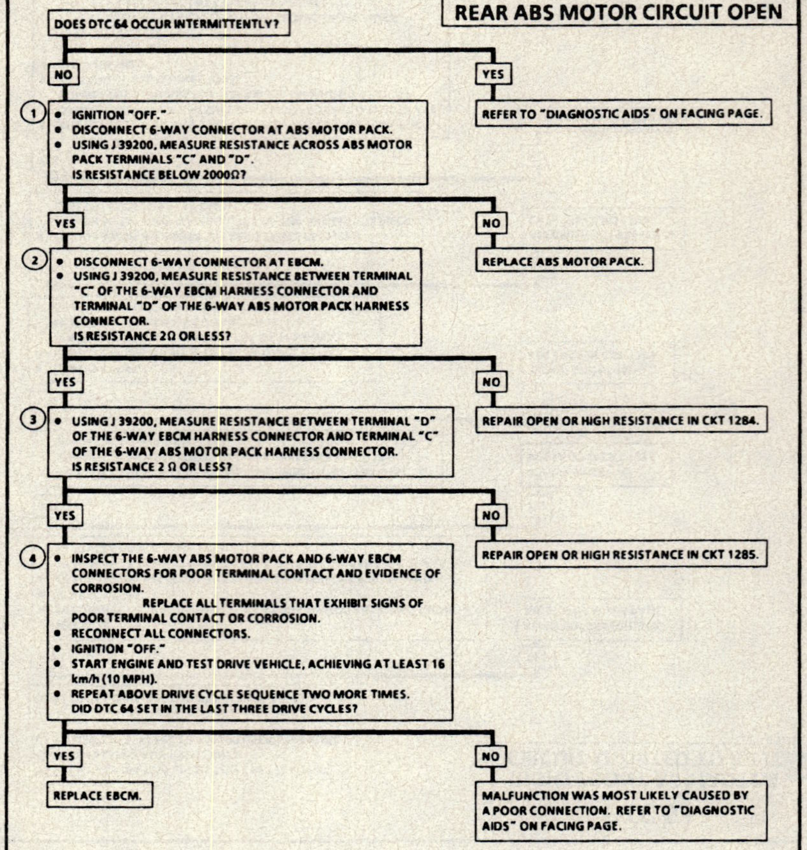

AFTER ALL DIAGNOSIS IS COMPLETE, CLEAR DTCs AND TEST DRIVE VEHICLE FOR THREE (3) DRIVE CYCLES TO VERIFY THAT DTC DOES NOT RESET. A DRIVE CYCLE CONSISTS OF STARTING THE VEHICLE, DRIVING OVER 16 km/h (10 MPH), AND THEN KEYING DOWN.

DIAGNOSTIC CHART DTC 65 — 1994 CUTLASS SUPREME AND REGAL

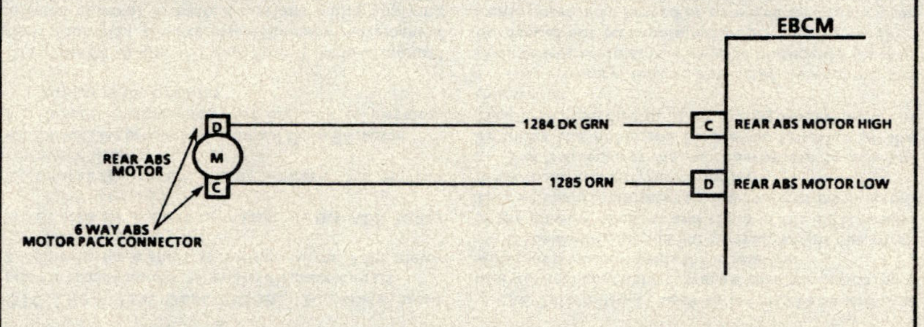

DTC 65

REAR ABS MOTOR CIRCUIT SHORTED TO GROUND

Circuit Description:
This DTC identifies an ABS motor circuit that is shorted to ground. This malfunction will not allow the ABS motor to be controlled at the commanded current rate or will cause the driver circuit to allow current directly to ground.

DTC Will Set When: DTC 65 can be set anytime. If the EBCM detects an out of range voltage on either of the rear axle ABS motor circuits indicating a circuit shorted to ground, a malfunction exists.

Action Taken: A malfunction DTC is stored, ABS is disabled and the ABS warning lamp is turned "ON." The red BRAKE Warning Lamp is turned "ON" if the rear piston in the ABS Motor Pack is not in the home position.

DTC Chart Test Description: Number(s) below refer to circled number(s) on the diagnostic chart.

1. Checks for a short to ground in the ABS motor "HIGH" circuitry.
2. Checks for a short to ground in the ABS motor "LOW" circuitry.
3. Checks for an ABS motor that is internally shorted to ground.
4. Ensures malfunction was not due to physical damage.

Diagnostic Aids: Using Tech 1, select manual control function, and exercise ABS motor movement of affected channel in both directions while applying light pressure on the brake pedal.

If erratic or "jumpy" brake pedal movement is detected while performing an "apply" or "release" function of the ABS motor, an intermittent malfunction may be indicated.

An "Intermittent" malfunction may be caused by a poor connection, rubbed through wire insulation, or a wire that is broken inside the insulation.

If the malfunction is not current, wiggle the wires of the affected channel and check if the DTC resets. This will help to pinpoint an intermittent malfunction in the motor circuitry or connections.

The frequency of the malfunction can be checked by using the enhanced diagnostic function of the Tech 1, as described in "Tech 1 Diagnostics" found in this section.

Any circuitry, that is suspected as causing the intermittent complaint, should be thoroughly checked for backed out terminals, improper mating, broken locks, improperly formed or damaged terminals, poor terminal to wiring connections or physical damage to the wiring harness.

DIAGNOSTIC CHART DTC 65 — 1994 CUTLASS SUPREME AND REGAL, CONTINUED

DTC 65

REAR ABS MOTOR CIRCUIT SHORTED TO GROUND

DOES DTC 65 OCCUR INTERMITTENTLY?

- YES: REFER TO "DIAGNOSTIC AIDS" ON FACING PAGE.
- NO:

(1)
- IGNITION "OFF."
- DISCONNECT 6-WAY CONNECTORS AT ABS MOTOR PACK AND EBCM.
- USING A J 39200, MEASURE THE RESISTANCE BETWEEN GROUND AND TERMINAL "D" OF THE 6-WAY ABS MOTOR PACK HARNESS CONNECTOR.

DOES J 39200 DISPLAY RESISTANCE AS "OL" (INFINITE)?

- NO: REPAIR SHORT TO GROUND IN CKT 1284.
- YES:

(2)
- MEASURE THE RESISTANCE BETWEEN GROUND AND TERMINAL "C" OF THE 6-WAY ABS MOTOR PACK HARNESS CONNECTOR.

DOES J 39200 DISPLAY RESISTANCE AS "OL" (INFINITE)?

- NO: REPAIR SHORT TO GROUND IN CKT 1285.
- YES:

(3)
- MEASURE THE RESISTANCE BETWEEN GROUND AND TERMINAL "D" OF THE ABS MOTOR PACK.

DOES J 39200 DISPLAY RESISTANCE AS "OL" (INFINITE)?

- NO: REPLACE ABS MOTOR PACK.
- YES:

(4)
- INSPECT CKTs 1284 AND 1285 AND RELATED CONNECTORS FOR DAMAGE WHICH MAY RESULT IN A SHORT TO GROUND WITH ALL CONNECTORS CONNECTED. REPAIR ALL DAMAGE IF EVIDENT.
- RECONNECT ALL CONNECTORS.
- IGNITION "OFF."
- START ENGINE AND TEST DRIVE VEHICLE, ACHIEVING AT LEAST 16 km/h (10 MPH).
- REPEAT ABOVE DRIVE CYCLE SEQUENCE TWO MORE TIMES.

DID DTC 65 SET IN THE LAST THREE DRIVE CYCLES?

- YES: REPLACE EBCM.
- NO: MALFUNCTION WAS MOST LIKELY CAUSED BY A POOR CONNECTION. REFER TO "DIAGNOSTIC AIDS" ON FACING PAGE.

AFTER ALL DIAGNOSIS IS COMPLETE, CLEAR DTCs AND TEST DRIVE VEHICLE FOR THREE (3) DRIVE CYCLES TO VERIFY THAT DTC DOES NOT RESET. A DRIVE CYCLE CONSISTS OF STARTING THE VEHICLE, DRIVING OVER 16 km/h (10 MPH), AND THEN KEYING DOWN.

DIAGNOSTIC CHART DTC 66 — 1994 CUTLASS SUPREME AND REGAL

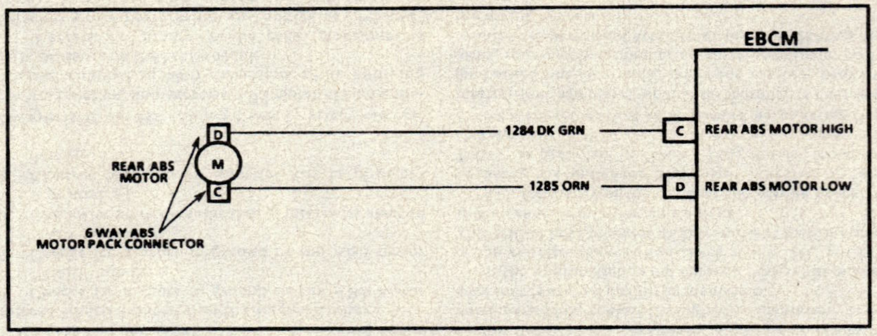

DTC 66

REAR ABS MOTOR CIRCUIT SHORTED TO BATTERY

Circuit Description:
This DTC identifies an ABS motor circuit that is shorted to battery or an ABS motor that has low or zero resistance. This malfunction will not allow the ABS motor to be controlled at the commanded current rate or will cause the ABS motor to turn in the opposite direction or not turn at all.

DTC Will Set When: DTC 66 can be set only when the ABS motor is commanded "OFF." If the EBCM detects an out of range voltage on either of the rear ABS motor circuits indicating a circuit shorted to battery or an ABS motor shorted, a malfunction exists.

Action Taken: A malfunction DTC is stored, ABS is disabled and the ABS warning lamp is turned "ON." The red BRAKE Warning Lamp is turned "ON" if the rear piston in the ABS Motor Pack is not in the home position.

DTC Chart Test Description: Number(s) below refer to circled number(s) on the diagnostic chart.

1. Checks for a short to voltage in the ABS motor "HIGH" circuitry.
2. Checks for a short to voltage in the ABS motor "LOW" circuitry.
3. Checks for a short circuit between "HIGH" and motor "LOW."
4. Checks for proper resistance of the ABS motor.
5. Ensures malfunction was not due to physical damage of the circuitry.

Diagnostic Aids: Using Tech 1, select manual control function, and exercise ABS motor movement of affected channel in both directions while applying light pressure on the brake pedal.

If erratic or "jumpy" brake pedal movement is detected while performing an "apply" or "release" function of the ABS motor, an intermittent malfunction may be indicated.

An "Intermittent" malfunction may be caused by a poor connection, rubbed through wire insulation, or a wire that is broken inside the insulation.

If the malfunction is not current, wiggle the wires of the affected channel and check if the DTC resets. This will help to pinpoint an intermittent malfunction in the motor circuitry or connections.

The frequency of the malfunction can be checked by using the enhanced diagnostic function of the Tech 1, as described in "Tech 1 Diagnostics" found in this section.

Any circuitry, that is suspected as causing the intermittent complaint, should be thoroughly checked for backed out terminals, improper mating, broken locks, improperly formed or damaged terminals, poor terminal to wiring connections or physical damage to the wiring harness.

DIAGNOSTIC CHART DTC 66 — 1994 CUTLASS SUPREME AND REGAL, CONTINUED

DTC 66

REAR ABS MOTOR CIRCUIT SHORTED TO BATTERY

NOTICE: J 39200 TEST LEADS MUST BE "ZEROED" PRIOR TO MAKING ANY RESISTANCE MEASUREMENTS. REFER TO J 39200 USER'S MANUAL.

DOES DTC 66 OCCUR INTERMITTENTLY?

- YES: REFER TO "DIAGNOSTIC AIDS" ON FACING PAGE.
- NO:

(1)
- IGNITION "OFF."
- DISCONNECT 6-WAY CONNECTORS AT ABS MOTOR PACK AND EBCM.
- IGNITION "ON."
- USING J 39200, MEASURE THE VOLTAGE BETWEEN GROUND AND TERMINAL "D" OF THE 6-WAY ABS MOTOR PACK HARNESS CONNECTOR.

IS VOLTAGE 1 VOLT OR LESS?

- NO: REPAIR SHORT TO VOLTAGE IN CKT 1284.
- YES:

(2)
- MEASURE THE VOLTAGE BETWEEN GROUND AND TERMINAL "C" OF THE 6-WAY ABS MOTOR PACK HARNESS CONNECTOR.

IS VOLTAGE 1 VOLT OR LESS?

- NO: REPAIR SHORT TO VOLTAGE IN CKT 1285.
- YES:

(3)
- IGNITION "OFF."
- USING J 39200, MEASURE THE VOLTAGE BETWEEN TERMINAL "C" AND TERMINAL "D" OF THE 6-WAY ABS MOTOR PACK HARNESS CONNECTOR.

DOES J 39200 DISPLAY RESISTANCE AS "OL" (INFINITE)?

- NO: REPAIR SHORT BETWEEN CKTs 1284 AND 1285.
- YES:

(4)
- USING J 39200, MEASURE THE RESISTANCE BETWEEN TERMINAL "C" AND TERMINAL "D" OF THE ABS MOTOR PACK.

IS RESISTANCE GREATER THAN 0.4Ω?

- NO: REPLACE ABS MOTOR PACK.
- YES:

(5)
- INSPECT CKTs 1284 AND 1285 ABS RELATED CONNECTORS FOR DAMAGE WHICH MAY RESULT IN A SHORT TO VOLTAGE WITH ALL CONNECTORS CONNECTED. REPAIR ALL DAMAGE IF EVIDENT.
- RECONNECT ALL CONNECTORS.
- IGNITION "OFF."
- START ENGINE AND TEST DRIVE VEHICLE, ACHIEVING AT LEAST 16 km/h (10 MPH).
- REPEAT ABOVE DRIVE CYCLE SEQUENCE TWO MORE TIMES.

DID DTC 66 SET IN THE LAST THREE DRIVE CYCLES?

- YES: REPLACE EBCM.
- NO: MALFUNCTION WAS MOSTLY CAUSED BY A POOR CONNECTION. REFER TO "DIAGNOSTIC AIDS" ON FACING PAGE.

AFTER ALL DIAGNOSIS IS COMPLETE, CLEAR DTCs AND TEST DRIVE VEHICLE FOR THREE (3) DRIVE CYCLES TO VERIFY THAT DTC DOES NOT RESET. A DRIVE CYCLE CONSISTS OF STARTING THE VEHICLE, DRIVING OVER 16 km/h (10 MPH), AND THEN KEYING DOWN.

DIAGNOSTIC CHART DTC 76 — 1994 CUTLASS SUPREME AND REGAL

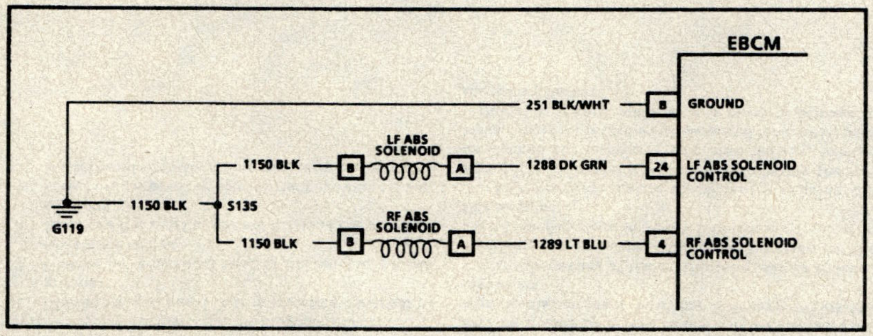

DTC 76

LEFT FRONT SOL. CIRCUIT SHORTED TO BATT

Circuit Description:

This DTC identifies a solenoid that cannot be energized due to an open in its circuitry, or a solenoid that is always energized due to a short to battery in its circuitry between the driver and the solenoid. An open will not allow proper ABS operation, but the short to battery simply turns "ON" the solenoid. A path for base brakes is still allowed once the motor re-homes and the check ball is lifted off its seat during key "ON" initialization.

DTC Will Set When: DTC 76 can be set only when the solenoid is commanded "OFF." If the EBCM detects an excessive voltage in the left front solenoid control circuit, a malfunction exists.

Action Taken: A malfunction DTC is stored, ABS is disabled and the ABS warning lamp is turned "ON."

DTC Chart Test Description: Number(s) below refer to circled number(s) on the diagnostic chart.

1. Checks for a short to voltage in the control circuitry of the solenoid.
2. Checks for an open in the control circuit of the solenoid.
3. Checks the solenoid coil for proper resistance.
4. Checks for an open in the solenoid ground circuit.
5. Checks for a possible intermittent malfunction in the solenoid circuitry due to poor terminal contact.

Diagnostic Aids: An "Intermittent" malfunction may be caused by a poor connection, rubbed through wire insulation, or a wire that is broken inside the insulation.

The frequency of the malfunction can be checked by using the enhanced diagnostic function of the Tech 1, as described in "Tech 1 Diagnostics" found in this section.

Any circuitry, that is suspected as causing the intermittent complaint, should be thoroughly checked for backed out terminals, improper mating, broken locks, improperly formed or damaged terminals, poor terminal to wiring connections or physical damage to the wiring harness.

DIAGNOSTIC CHART DTC 76 — 1994 CUTLASS SUPREME AND REGAL, CONTINUED

DTC 76

LEFT FRONT SOL. CIRCUIT SHORTED TO BATT

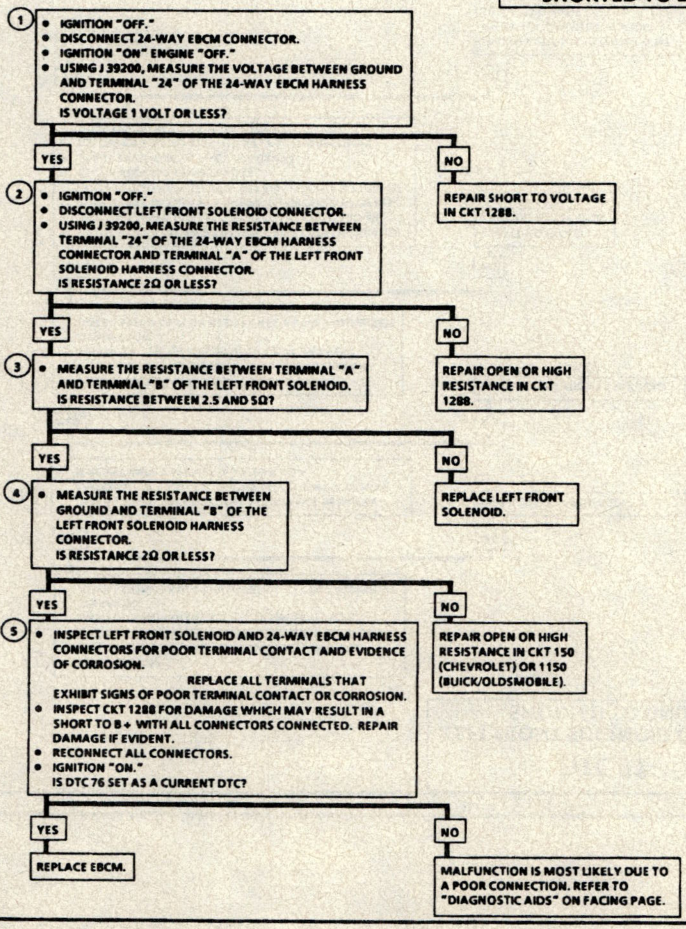

DIAGNOSTIC CHART DTC 77 — 1994 CUTLASS SUPREME AND REGAL

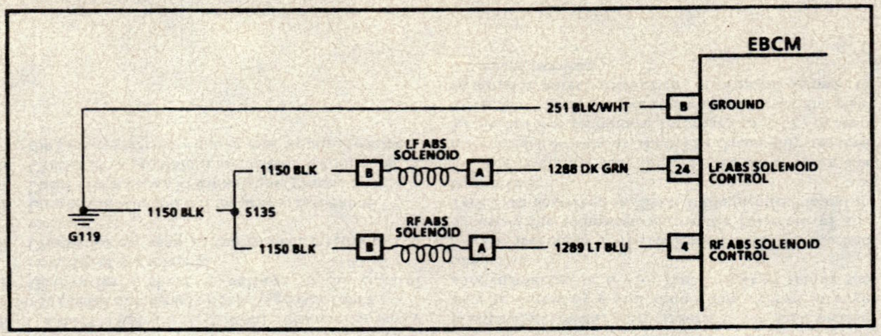

DTC 77

LEFT FRONT SOLENOID CIRCUIT SHORTED TO GND

Circuit Description:
This DTC identifies a solenoid that cannot be energized due to an open in its driver circuitry, or a short to ground between the solenoid driver and the solenoid. These malfunctions can affect ABS operation since the flow of brake fluid to the wheel cylinder cannot be stopped, making ABS operation for that channel impossible.

DTC Will Set When: DTC 77 can be set only when the solenoid is commanded "ON." If the EBCM detects the left front solenoid control circuit voltage is out of specification, a malfunction exists.

Action Taken: A malfunction DTC is stored, ABS is disabled and the ABS warning lamp is turned "ON."

DTC Chart Test Description: Number(s) below refer to circled number(s) on the diagnostic chart.

1. Checks for a solenoid that is internally shorted to ground.
2. Checks for a solenoid that is not within proper resistance values.
3. Indicates if a short to ground exists in the solenoid circuitry.
4. Checks for a possible intermittent malfunction in the solenoid circuitry due to poor terminal contact.

Diagnostic Aids: An "Intermittent" malfunction may be caused by a poor connection, rubbed through wire insulation, or a wire that is broken inside the insulation.

The frequency of the malfunction can be checked by using the enhanced diagnostic function of the Tech 1, as described in "Tech 1 Diagnostics" found in this section.

Any circuitry, that is suspected as causing the intermittent complaint, should be thoroughly checked for backed out terminals, improper mating, broken locks, improperly formed or damaged terminals, poor terminal to wiring connections or physical damage to the wiring harness.

DIAGNOSTIC CHART DTC 77 — 1994 CUTLASS SUPREME AND REGAL, CONTINUED

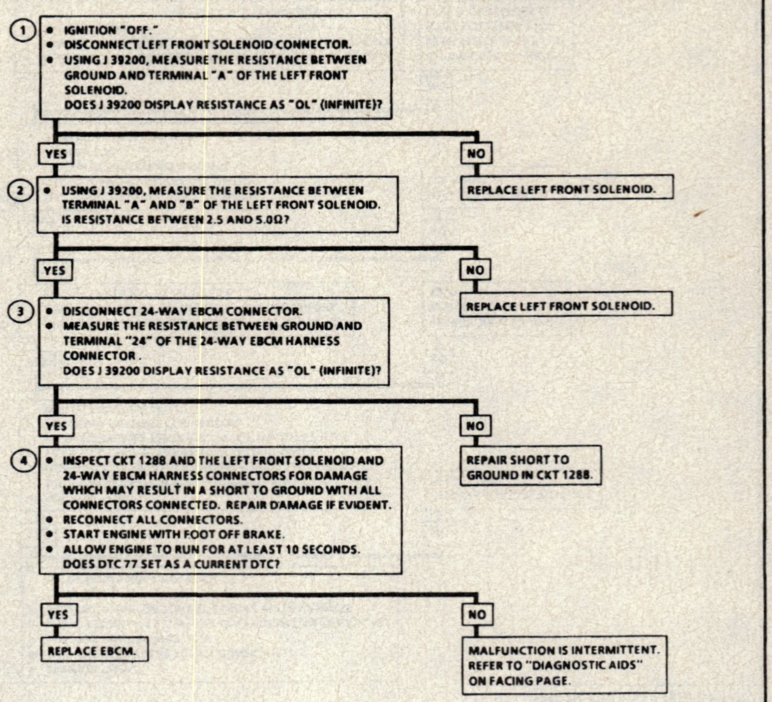

DIAGNOSTIC CHART DTC 78 — 1994 CUTLASS SUPREME AND REGAL

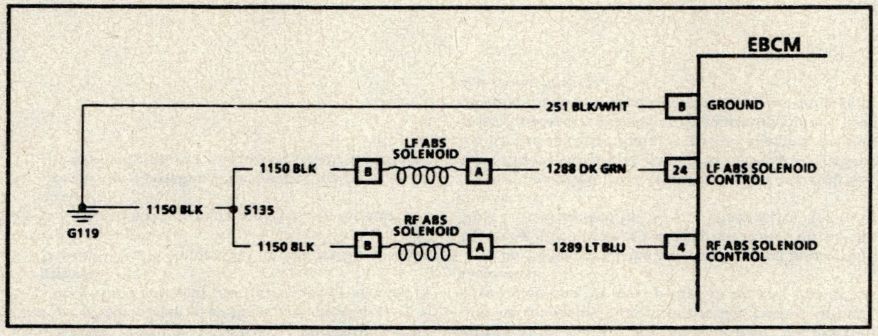

DTC 78

RIGHT FRONT SOL. CIRCUIT OPEN OR SHORTED TO BATT

Circuit Description:

This DTC identifies a solenoid that cannot be energized due to an open in its circuitry, or a solenoid that is always energized due to a short to battery in its circuitry between the driver and the solenoid. An open will not allow proper ABS operation, but the short to battery simply turns "ON" the solenoid. A path for base brakes is still allowed once the motor re-homes and the check ball is lifted off its seat during key "ON" initialization.

DTC Will Set When: DTC 78 can be set only when the solenoid is commanded "OFF." If the EBCM detects an excessive voltage in the right front solenoid control circuit, a malfunction exists.

Action Taken: A malfunction DTC is stored, ABS is disabled and the ABS warning lamp is turned "ON."

DTC Chart Test Description: Number(s) below refer to circled number(s) on the diagnostic chart.

1. Checks for a short to voltage in the control circuitry of the solenoid.
2. Checks for an open in the control circuit of the solenoid.
3. Checks the solenoid coil for proper resistance.
4. Checks for an open in the solenoid ground circuit.
5. Checks for a possible intermittent malfunction in the solenoid circuitry due to poor terminal contact.

Diagnostic Aids: An "Intermittent" malfunction may be caused by a poor connection, rubbed through wire insulation, or a wire that is broken inside the insulation.

The frequency of the malfunction can be checked by using the enhanced diagnostic function of the Tech 1, as described in "Tech 1 Diagnostics" found in this section.

Any circuitry, that is suspected as causing the intermittent complaint, should be thoroughly checked for backed out terminals, improper mating, broken locks, improperly formed or damaged terminals, poor terminal to wiring connections or physical damage to the wiring harness.

DIAGNOSTIC CHART DTC 78 — 1994 CUTLASS SUPREME AND REGAL, CONTINUED

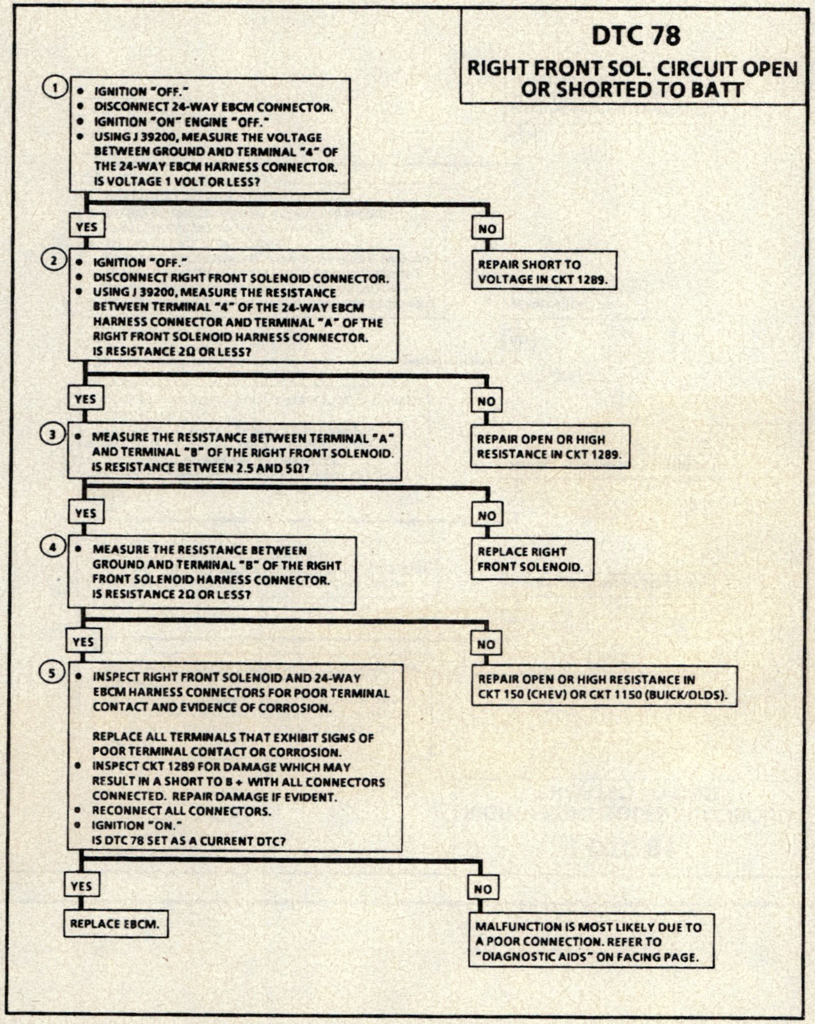

DIAGNOSTIC CHART DTC 81 — 1994 CUTLASS SUPREME AND REGAL

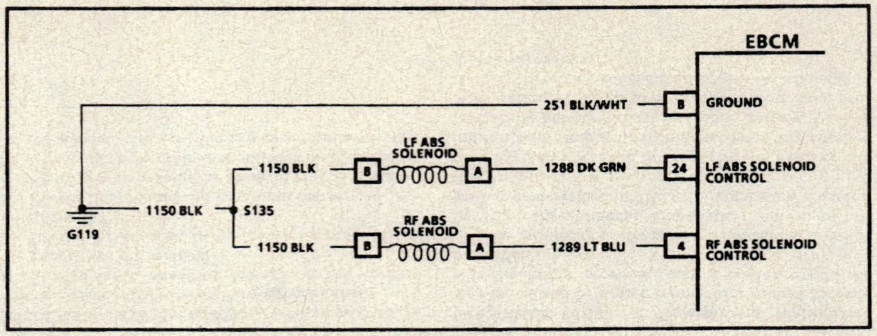

DTC 81

RIGHT FRONT SOLENOID CIRCUIT SHORTED TO GND

Circuit Description:

This DTC identifies a solenoid that cannot be energized due to an open in its driver circuitry, or a short to ground between the solenoid driver and the solenoid. These malfunctions can affect ABS operation since the flow of brake fluid to the wheel cylinder cannot be stopped, making ABS operation for that channel impossible.

DTC Will Set When: DTC 81 can be set only when the solenoid is commanded "ON." If the EBCM detects the right front solenoid control circuit voltage is out of specification, a malfunction exists.

Action Taken: A malfunction DTC is stored, ABS is disabled and the ABS warning lamp is turned "ON."

DTC Chart Test Description: Number(s) below refer to circled number(s) on the diagnostic chart.

1. Checks for a solenoid that is internally shorted to ground.
2. Checks for a solenoid that is not within proper resistance values.
3. Indicates if a short to ground exists in the solenoid circuitry.
4. Checks for a possible intermittent malfunction in the solenoid circuitry due to poor terminal contact.

Diagnostic Aids: An "Intermittent" malfunction may be caused by a poor connection, rubbed through wire insulation, or a wire that is broken inside the insulation.

The frequency of the malfunction can be checked by using the enhanced diagnostic function of the Tech 1, as described in "Tech 1 Diagnostics" found in this section.

Any circuitry, that is suspected as causing the intermittent complaint, should be thoroughly checked for backed out terminals, improper mating, broken locks, improperly formed or damaged terminals, poor terminal to wiring connections or physical damage to the wiring harness.

DIAGNOSTIC CHART DTC 81 — 1994 CUTLASS SUPREME AND REGAL, CONTINUED

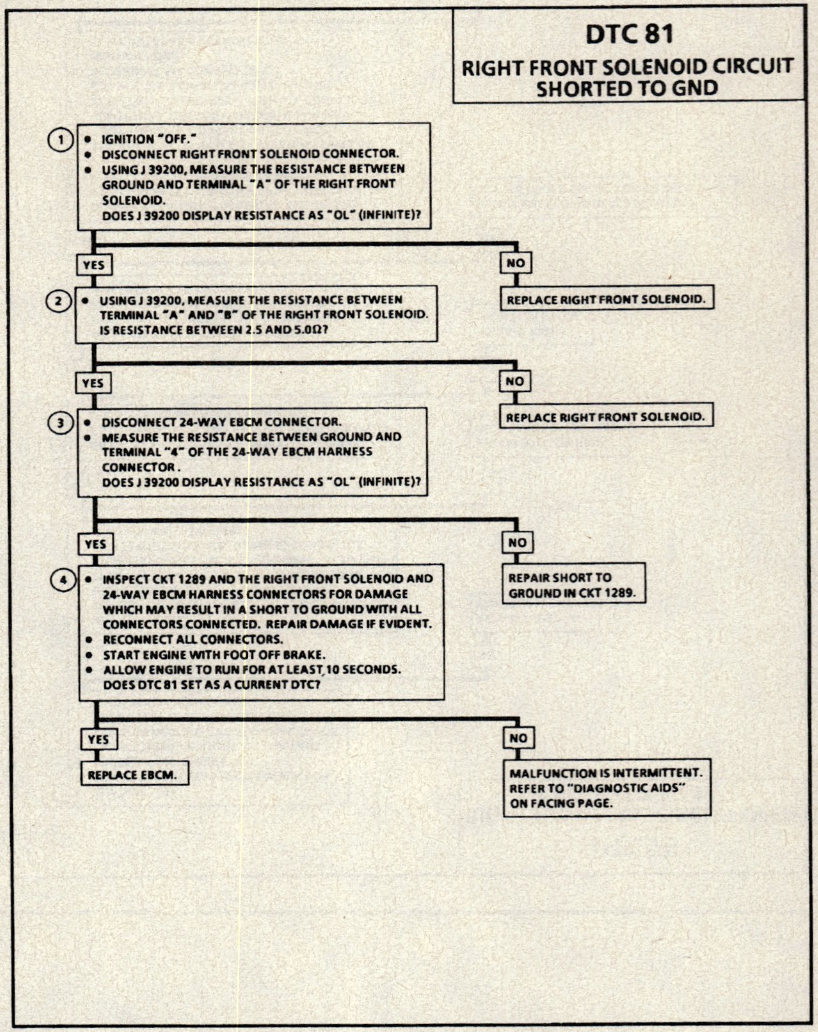

DIAGNOSTIC CHART DTC 82 — 1994 CUTLASS SUPREME AND REGAL

DTC 82
CALIBRATION MALFUNCTION

Circuit Description:
This DTC allows the EBCM to check for a calibration malfunction by comparing the calibration value to a known value stored in the EEPROM.
This DTC is also used as a security measure to prevent improper use of calibrations or changes to these calibrations that may alter the designed function of ABS.

DTC Will Set When: DTC 82 can be set at key-up. If the program identifier is incorrect or the memory checksum is incorrect, a malfunction exists.

Action Taken: A malfunction DTC is stored, ABS is disabled and the ABS warning lamp is turned "ON."

DTC Chart Test Description: Number(s) below refer to circled number(s) on the diagnostic chart.

1. Checks to see if the malfunction is present during diagnosis. If present, the EBCM is not functioning correctly and must be replaced.

Important

- At the time of printing, 1994 vehicles with ABS VI were not being field reprogrammed to correct this failure. A service bulletin will be issued when EEPROM reprogramming is authorized.

Diagnostic Aids: An intermittent DTC 82 may be caused by a bad cell in the EEPROM that is sensitive to temperature changes. If DTC 82 failed more than once, but is intermittent, replace EBCM.
The frequency of the malfunction can be checked by using the enhanced diagnostic function of the Tech 1, as described in "Tech 1 Diagnostics" found in this section.

DIAGNOSTIC CHART DTC 82 — 1994 CUTLASS SUPREME AND REGAL, CONTINUED

DTC 82
CALIBRATION MALFUNCTION

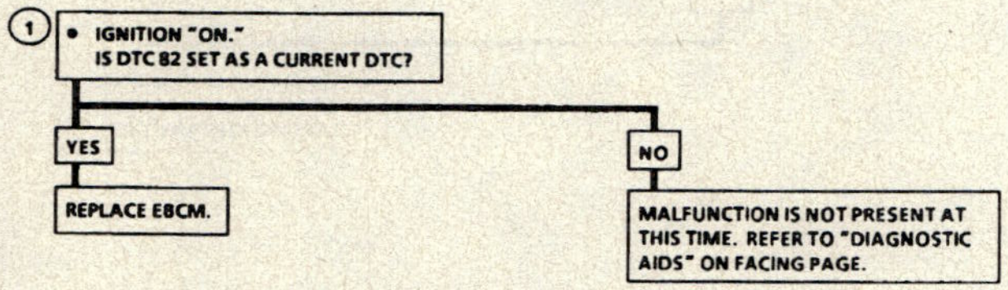

DIAGNOSTIC CHART DTC 86 — 1994 CUTLASS SUPREME AND REGAL

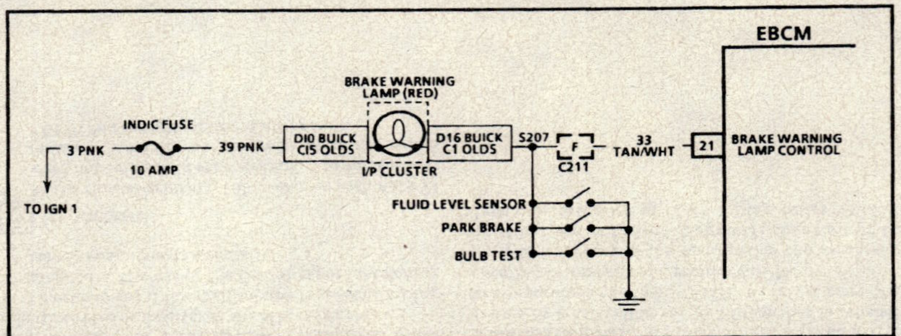

DTC 86

EBCM TURNED "ON" THE RED "BRAKE" WARNING LAMP

Circuit Description:

This DTC is provided as an information only DTC, and reflects the status of the command issued by the ABS EBCM to illuminate the red brake warning lamp. If another DTC issues a command to illuminate the red brake warning lamp, DTC 86 will be stored in EEPROM as a history DTC at the conclusion of the ignition cycle.

DTC Chart Test Description: Number(s) below refer to circled number(s) on the diagnostic chart.

1. This step serves to identify if a DTC other than DTC 86 commanded the red brake warning lamp "ON."

Diagnostic Aids: Any ABS mechanical DTC that issues a command to illuminate the red "BRAKE" telltale will also result in DTC 86 being stored in EEPROM during shutdown. These DTCs are: 38, 41, 42, 46, and 51. If the motors are not in their home position, certain electrical DTCs will also command the red "BRAKE" telltale "ON." These DTCs are: 14, 16, 18, 55, 56, 57, 58, 61, 62, 63, 64, 65, and 66.

If any of these DTCs are indicated along with DTC 86, they must be corrected prior to addressing a DTC 86 malfunction.

DIAGNOSTIC CHART DTC 86 — 1994 CUTLASS SUPREME AND REGAL, CONTINUED

DTC 86

EBCM TURNED "ON" THE RED "BRAKE" WARNING LAMP

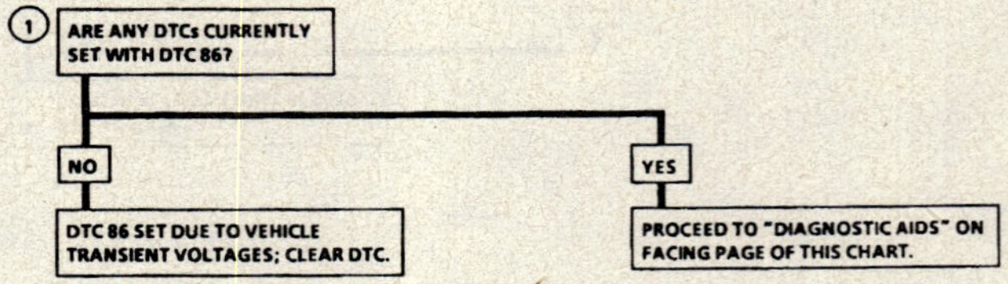

DIAGNOSTIC CHART DTC 87 — 1994 CUTLASS SUPREME AND REGAL

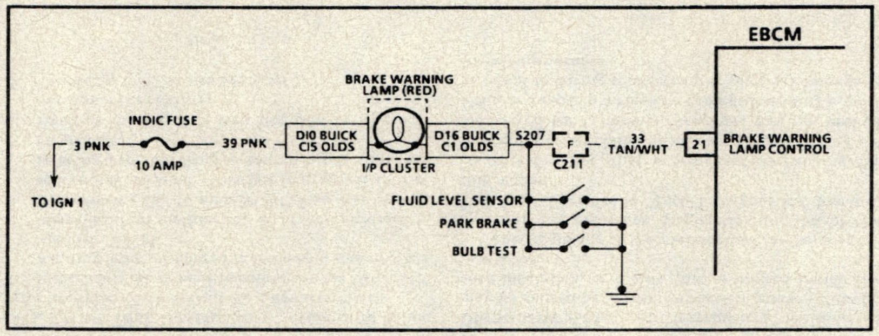

DTC 87

RED "BRAKE" WARNING LAMP CIRCUIT OPEN

Circuit Description:

This DTC is used to verify the EBCM has continuity to the red "BRAKE" warning lamp in case the EBCM must turn it "ON." This will only occur if an ABS malfunction is detected that may degrade base brake operation.

Because ABS is not the only device controlling the red "BRAKE" warning lamp (parking brake and low fluid sensor may also turn "ON" the lamp) a short to ground in this circuit cannot be detected.

DTC Will Set When: DTC 87 can be set after initialization is completed. If the red "BRAKE" warning lamp circuit voltage is out of specification indicating an open circuit, a malfunction exists.

Action Taken: A malfunction DTC is stored, ABS is not disabled.

DTC Chart Test Description: Number(s) below refer to circled number(s) on the diagnostic chart.

1. Identifies if the ground circuit to the red "BRAKE" warning lamp is completed by a source other than the EBCM.
2. Checks for high resistance in park brake circuitry.
3. Checks for a malfunctioning EBCM.
4. Checks for high resistance in fluid level switch circuitry.
5. Checks for a malfunctioning warning lamp bulb.
6. Checks for open or high resistance in CKT 33.

Diagnostic Aids: An "Intermittent" malfunction may be caused by a poor connection, rubbed through wire insulation, or a wire that is broken inside the insulation.

The frequency of the malfunction can be checked by using the enhanced diagnostic function of the Tech 1, as described in "Tech 1 Diagnostics" found in this section. If a Tech 1 or T-100 are not available, the parking brake handle may be lifted to verify proper lamp operation and continuity of CKTs 3 and 39.

Any circuitry, that is suspected as causing the intermittent complaint, should be thoroughly checked for backed out terminals, improper mating, broken locks, improperly formed or damaged terminals, poor terminal to wiring connections or physical damage to the wiring harness.

DIAGNOSTIC CHART DTC 87 — 1994 CUTLASS SUPREME AND REGAL, CONTINUED

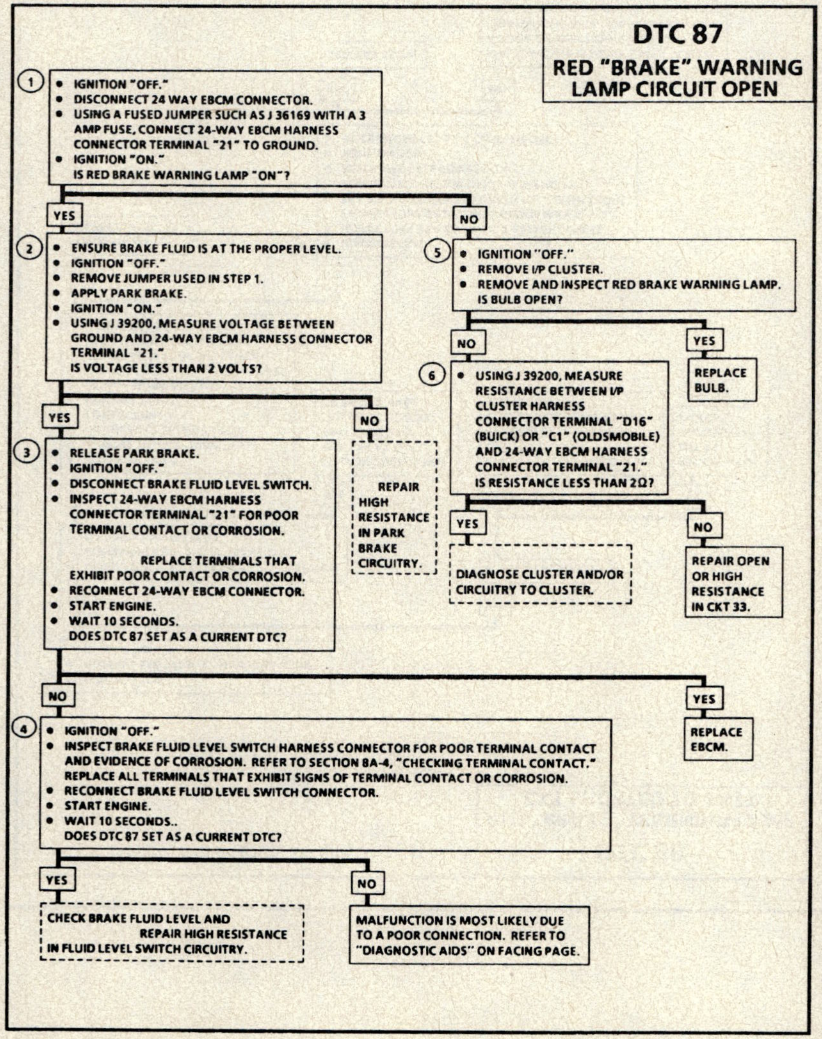

DIAGNOSTIC CHART DTC 88 — 1994 CUTLASS SUPREME AND REGAL

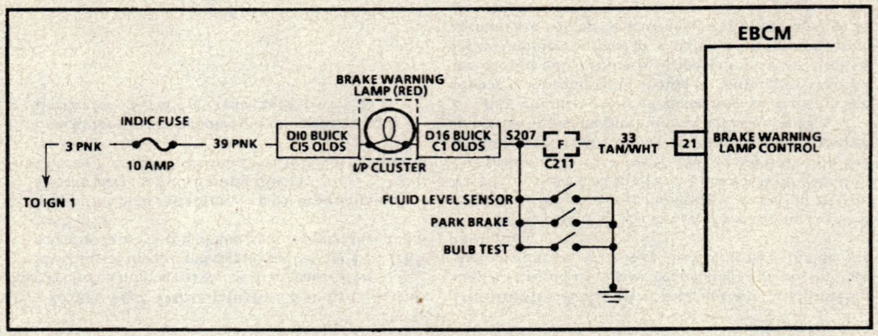

DTC 88

RED "BRAKE" WARNING LAMP CKT SHORTED TO BATT

Circuit Description:

This DTC identifies a short to battery between the EBCM and the red "BRAKE" warning lamp, or an open driver that does not allow the red "BRAKE" warning lamp to be illuminated by the EBCM. This will only occur if an ABS fault is detected that may degrade base brake operation.

DTC Will Set When: DTC 88 can be set only at the beginning of the three second bulb check. If the EBCM detects battery voltage on the red "BRAKE" warning lamp control circuit, a malfunction exists.

Action Taken: A malfunction DTC is stored. ABS is not disabled.

DTC Chart Test Description: Number(s) below refer to circled number(s) on the diagnostic chart.

1. Identifies if the ground circuit to the red "BRAKE" warning lamp is being completed by a source other than the EBCM.
2. Identifies if the malfunction is currently present.
3. By removing the 10 amp INDIC fuse, the voltage source is eliminated. This test indicates if voltage is being supplied from a source other than the gauge fuse.
4. Ensures malfunction was not due to physical damage of the circuitry.
5. Checks for DTC 86 also being set.

Diagnostic Aids: An "Intermittent" malfunction may be caused by a poor connection, rubbed through wire insulation, or a wire that is broken inside the insulation.

The frequency of the malfunction can be checked by using the enhanced diagnostic function of the Tech 1, as described in "Tech 1 Diagnostics" found in this section.

Any circuitry, that is suspected as causing the intermittent complaint, should be thoroughly checked for backed out terminals, improper mating, broken locks, improperly formed or damaged terminals, poor terminal to wiring connections or physical damage to the wiring harness.

DIAGNOSTIC CHART DTC 88 — 1994 CUTLASS SUPREME AND REGAL, CONTINUED

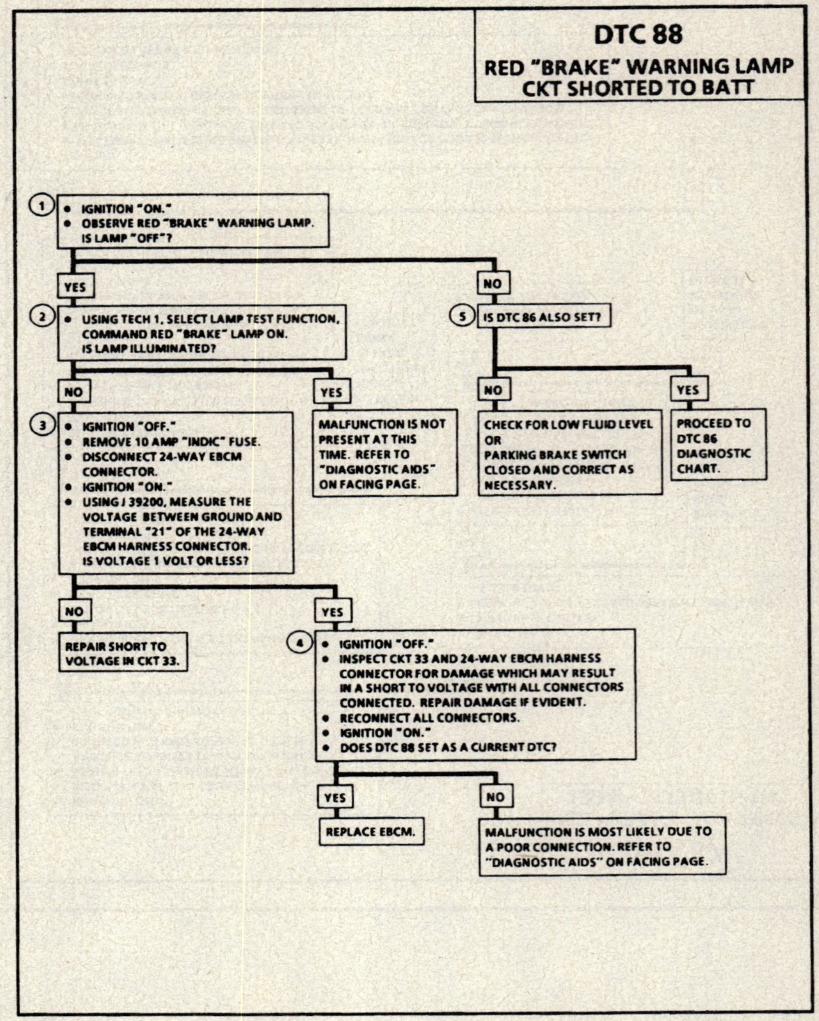

DIAGNOSTIC CHART DTC 93 — 1994 CUTLASS SUPREME AND REGAL

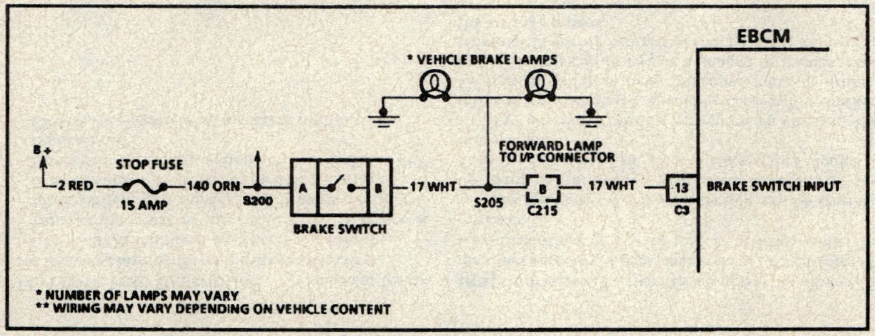

DTC 93

DTC 91 OR 92 SET IN CURRENT OR PREV. IGN CYCLE

Circuit Description:

This DTC is the second portion of DTCs 91 and 92. If DTCs 91 or 92 failed during the last ignition cycle, DTC 93 becomes a current failure during the next ignition cycle, keeping ABS disabled until a brake switch "ON" state is seen. When a change is seen during an ignition cycle in which DTC 93 is a current malfunction, DTC 91 or 92 will clear itself at the end of the current ignition cycle, and ABS will enable itself at the start of the next ignition cycle. DTC 93 alone indicates 91 or 92 failed previously, but is intermittent, or has been corrected.

DTC Chart Test Description: Number(s) below refer to circled number(s) on the diagnostic chart.

1. Indicates which DTC (either 91 or 92) caused DTC 93 to set.

Diagnostic Aids: An "Intermittent" malfunction may be caused by a poor connection, rubbed through wire insulation, or a wire that is broken inside the insulation.

The frequency of the malfunction can be checked by using the enhanced diagnostic function of the Tech 1, as described in "Tech 1 Diagnostics" found in this section.

Any circuitry, that is suspected as causing the intermittent complaint, should be thoroughly checked for backed out terminals, improper mating, broken locks, improperly formed or damaged terminals, poor terminal to wiring connections or physical damage to the wiring harness.

Also, verify proper brake switch operation using the data list of the Tech 1. As the brake is applied, the data list should display the brake switch "ON" within 1 inch of travel.

DIAGNOSTIC CHART DTC 93 — 1994 CUTLASS SUPREME AND REGAL, CONTINUED

DTC 93

DTC 91 OR 92 SET IN CURRENT OR PREV. IGN CYCLE

(1)
- IGNITION "ON."
- USING TECH 1, READ DTCs. IS DTC 91 OR 92 SET AS A HISTORY OR CURRENT DTC?

YES: PROCEED TO DTC CHART THAT SET.

NO: USING TECH 1, SELECT "F1: DTC HISTORY." VERIFY MALFUNCTION FREQUENCY WAS LOW AND CLEAR DTCs.

DIAGNOSTIC CHART DTC 94 — 1994 CUTLASS SUPREME AND REGAL

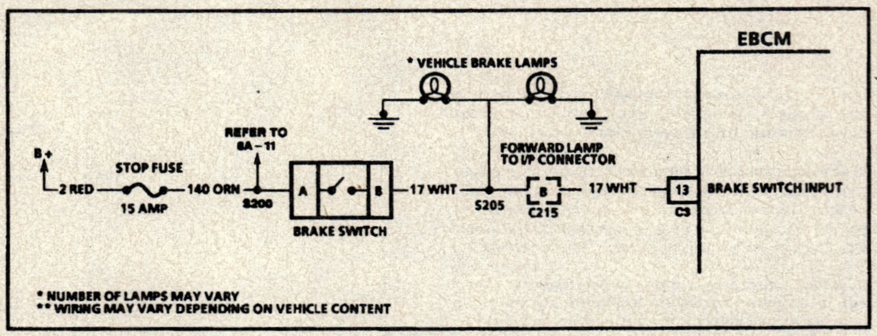

DTC 94

BRAKE SWITCH CONTACTS ALWAYS CLOSED

Circuit Description:

This DTC is run to determine the proper operation of the brake switch. This is important because ABS is activated when the brake switch is "ON" and turned "OFF" when the brake is "OFF." If the brake switch is always "ON," ABS operation will always be requested resulting in potential modulator cycling on rough roads. Additionally, this malfunction will most likely result in a dead battery (due to the brake lights staying "ON") if the driver is not informed of this malfunction.

DTC Will Set When: DTC 94 can be set when the vehicle reaches at least 40 km/h (25 mph). If the brake was never "OFF" during two consecutive drive cycles, a malfunction exists.

Action Taken: A malfunction DTC is stored. ABS is not disabled.

DTC Chart Test Description: Number(s) below refer to circled number(s) on the diagnostic chart.

1. Checks if the malfunction is currently present.
2. Isolates the cause of the malfunction to either a malfunctioning or misadjusted brake switch, or a short to voltage in the brake switch circuitry.
3. Checks for unwanted voltage on the brake switch input circuit.
4. Checks for a possible intermittent malfunction.

Diagnostic Aids: An "Intermittent" malfunction may be caused by a poor connection, rubbed through wire insulation, or a wire that is broken inside the insulation.

The frequency of the malfunction can be checked by using the enhanced diagnostic function of the Tech 1, as described in "Tech 1 Diagnostics" found in this section.

Any circuitry, that is suspected as causing the intermittent complaint, should be thoroughly checked for backed out terminals, improper mating, broken locks, improperly formed or damaged terminals, poor terminal to wiring connections or physical damage to the wiring harness.

DIAGNOSTIC CHART DTC 94 — 1994 CUTLASS SUPREME AND REGAL, CONTINUED

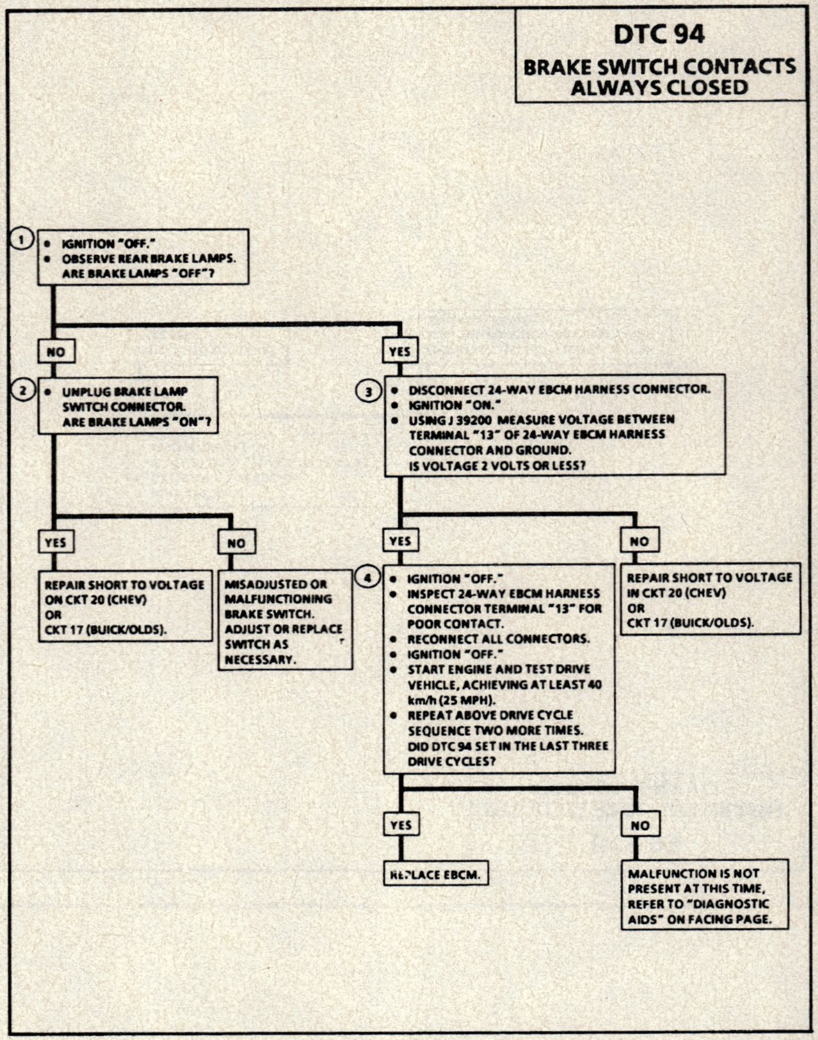

DIAGNOSTIC CHART DTC 95 — 1994 CUTLASS SUPREME AND REGAL

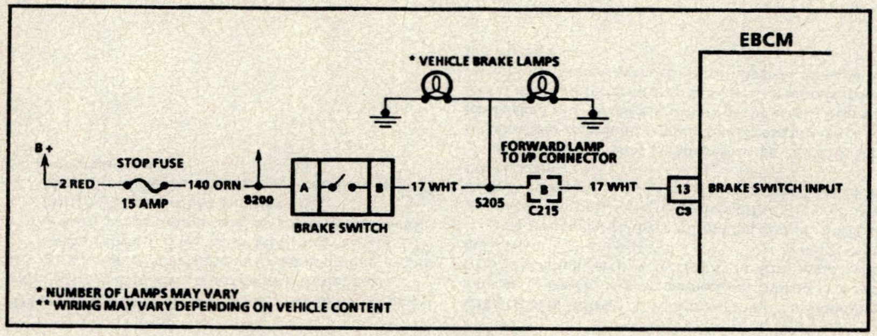

DTC 95
BRAKE SWITCH CIRCUIT OPEN

Circuit Description:
This DTC is used to identify open brake switch circuitry that prevents the brake switch input to the EBCM from changing states when the brake is applied. This DTC is used in conjunction with DTCs 91 and 92 to determine the cause of an open brake switch malfunction.

DTC Will Set When: DTC 95 can be set after initialization is completed. If the brake switch input voltage is out of specification for one second indicating an open circuit, a malfunction exists.

Action Taken: A malfunction DTC is stored, ABS is disabled and the ABS warning lamp is turned "ON."

DTC Chart Test Description: Number(s) below refer to circled number(s) on the diagnostic chart.

1. This step is used to confirm that an open in the brake switch circuitry currently exists.
2. Indicates if the brake switch signal is being received by the EBCM.
3. Indicates if an open circuit exists in the brake switch or brake lamp circuitry.
4. Isolates the open circuit to either the brake switch input circuitry, or the EBCM.

Diagnostic Aids: An "Intermittent" malfunction may be caused by a poor connection, rubbed through wire insulation, or a wire that is broken inside the insulation.

The frequency of the malfunction can be checked by using the enhanced diagnostic function of the Tech 1, as described in "Tech 1 Diagnostics" found in this section.

Any circuitry, that is suspected as causing the intermittent complaint, should be thoroughly checked for backed out terminals, improper mating, broken locks, improperly formed or damaged terminals, poor terminal to wiring connections or physical damage to the wiring harness.

DIAGNOSTIC CHART DTC 95 — 1994 CUTLASS SUPREME AND REGAL, CONTINUED

DTC 95
BRAKE SWITCH CIRCUIT OPEN

IS DTC 96 CURRENTLY SET?

- YES → PROCEED TO DTC 96 DIAGNOSTIC CHART.
- NO →

(1)
- IGNITION "ON."
- USING TECH 1, SELECT "FO" ABS DATA LIST, AND MONITOR BRAKE SWITCH.

DOES BRAKE SWITCH INDICATE "CIRCUIT OPEN"?

- NO → MALFUNCTION IS NOT PRESENT AT THIS TIME. REFER TO "DIAGNOSTIC AIDS" ON FACING PAGE.
- YES →

(2)
- APPLY LIGHT PRESSURE ON BRAKE PEDAL WHILE MONITORING BRAKE SWITCH STATUS ON TECH 1.

DOES TECH 1 INDICATE BRAKE SWITCH "ON" WITHIN 1 INCH OF TRAVEL?

- YES → PROCEED TO DTC 96 DIAGNOSTIC CHART.
- NO →

(3)
- APPLY FIRM PRESSURE TO BRAKE PEDAL AND OBSERVE CHMSL AND REAR BRAKE LAMPS.

ARE ALL BRAKE LAMPS "ON"?

- NO → REPAIR OPEN BRAKE LAMP CIRCUIT AS DIRECTED IN SECTION 8A.
- YES →

(4)
- IGNITION "OFF."
- DISCONNECT 24-WAY EBCM CONNECTOR.
- IGNITION "ON."
- USING J 39200, MEASURE THE VOLTAGE BETWEEN GROUND AND TERMINAL "13" OF THE 24-WAY EBCM HARNESS CONNECTOR WHILE APPLYING FIRM PRESSURE ON BRAKE PEDAL.

IS VOLTAGE 10 VOLTS OR GREATER?

- NO → REPAIR OPEN IN CKT 20 (CHEV) OR CKT 17 (BUICK/OLDS).
- YES → CHECK FOR POOR TERMINAL CONTACT AT EBCM. IF OK, REPLACE EBCM.

DIAGNOSTIC CHART DTC 96 — 1994 CUTLASS SUPREME AND REGAL

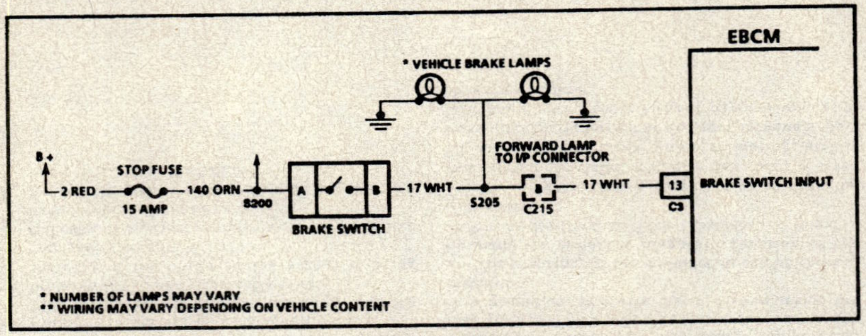

DTC 96

BRAKE LAMPS CIRCUIT OPEN

Circuit Description:
This DTC is designed to isolate the cause of a DTC 95 malfunction and indicate to the driver ABS is still available. If DTC 95 fails with DTC 96, the brake lamps are open, the brake lamp grounds are open, or the Center High Mounted Stop Lamp (CHMSL) circuit is open during 4-way flasher use. The presence of battery voltage at the brake switch input indicates a valid brake switch input is still available.

DTC Will Set When: DTC 96 can be set only after DTC 95 has been set. If the EBCM detects battery voltage on the brake switch input circuit for 0.5 second, a malfunction exists.

Action Taken: A malfunction DTC is stored. ABS is not disabled.

DTC Chart Test Description: Number(s) below refer to circled number(s) on the diagnostic chart.

1. As a result of a malfunction of an additional brake switch circuit DTC, this DTC may be set. To ensure proper diagnosis, any additional brake switch DTCs must be repaired first.
2. Identifies if the malfunction is currently present in the brake circuit.

Diagnostic Aids: An "Intermittent" malfunction may be caused by a poor connection, rubbed through wire insulation, or a wire that is broken inside the insulation.

The frequency of the malfunction can be checked by using the enhanced diagnostic function of the Tech 1, as described in "Tech 1 Diagnostics" found in this section.

Any circuitry, that is suspected as causing the intermittent complaint, should be thoroughly checked for backed out terminals, improper mating, broken locks, improperly formed or damaged terminals, poor terminal to wiring connections or physical damage to the wiring harness.

DIAGNOSTIC CHART DTC 96 — 1994 CUTLASS SUPREME AND REGAL, CONTINUED

DTC 96

BRAKE LAMPS CIRCUIT OPEN

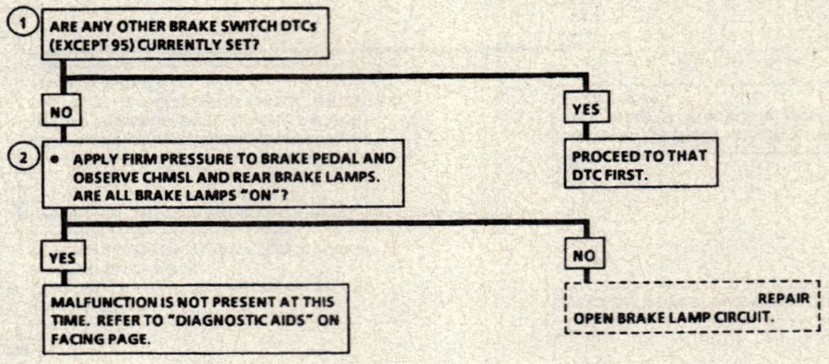

DIAGNOSTIC PROCESS CHART — 1994 GRAND PRIX

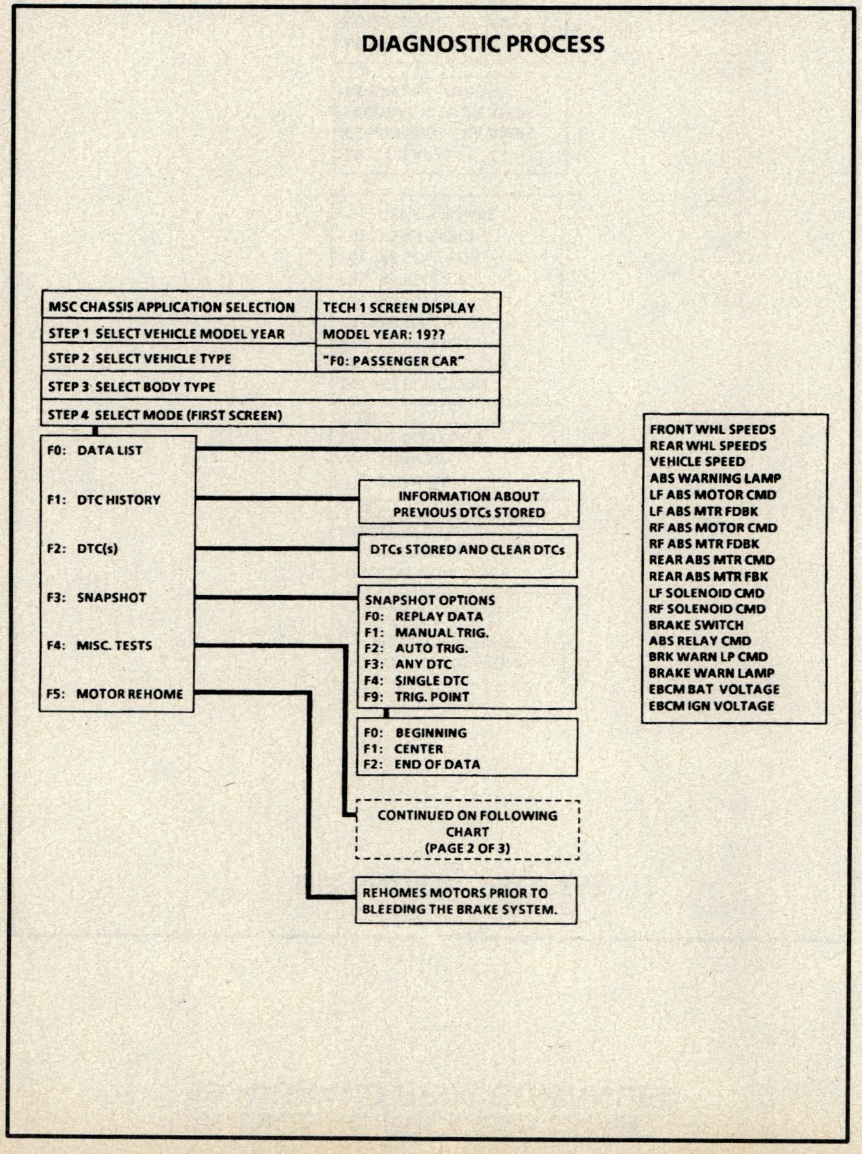

DIAGNOSTIC PROCESS CHART — 1994 GRAND PRIX, CONTINUED

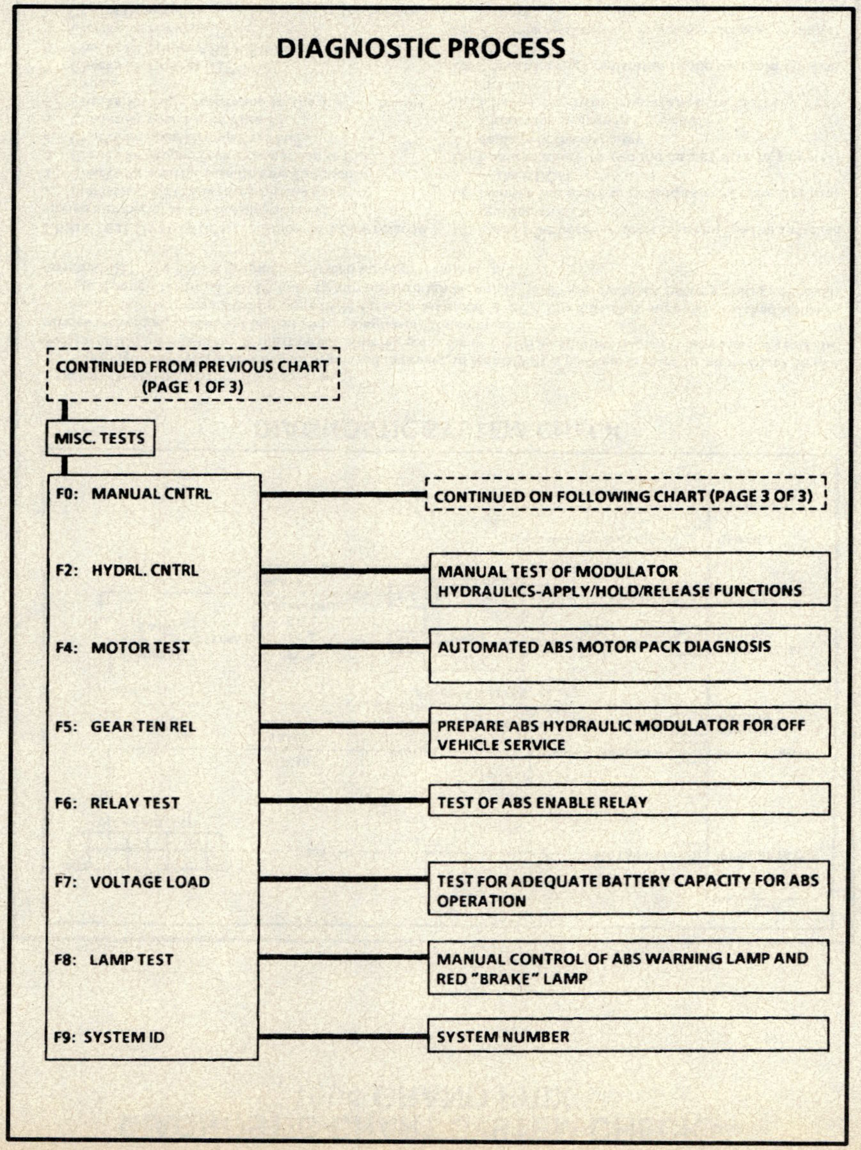

DIAGNOSTIC PROCESS CHART — 1994 GRAND PRIX, CONTINUED

DIAGNOSTIC PROCESS

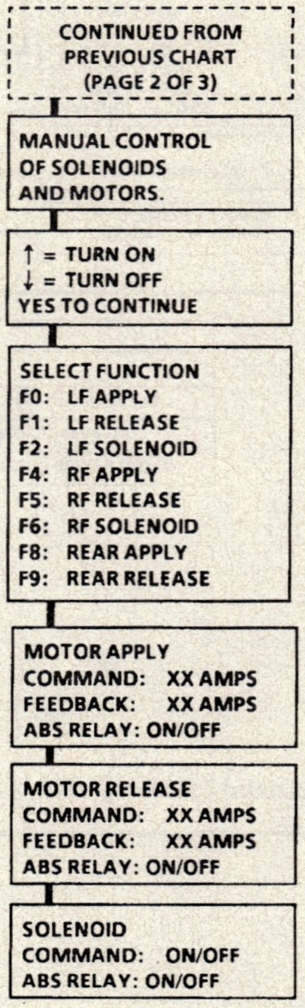

DIAGNOSTIC CHART SYSTEM CHECK — 1994 GRAND PRIX

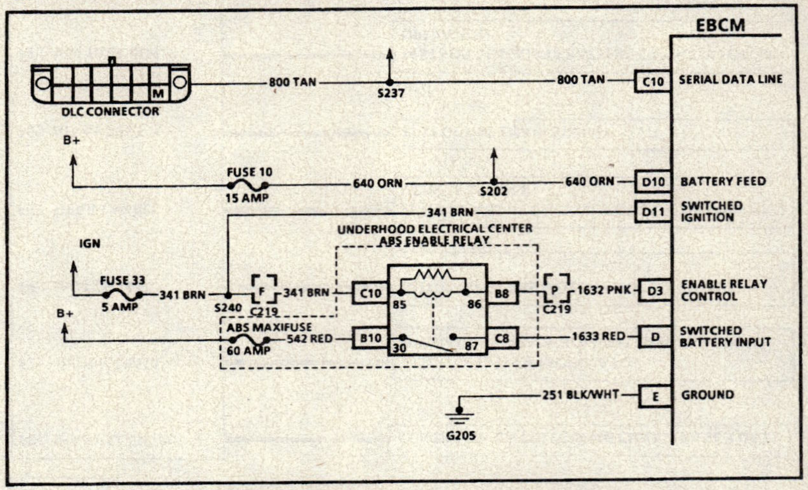

DIAGNOSTIC SYSTEM CHECK

System Description:

The diagnostic system check is an organized approach to identifying a problem created by an Antilock Brake System (ABS) malfunction. It must be the starting point for any ABS complaint diagnosis because it directs the service technician to the next logical step in diagnosing the complaint.

Serial data is transmitted by the EBCM through terminal "C10". The EBCM is supplied switched ignition voltage through terminal "D11", battery voltage through terminal "D10" and switched battery voltage through terminal "D". The EBCM ground is provided through terminal "E".

Chart Test Description: Number(s) below refer to circled number(s) on the diagnostic chart.

1. Indicates if EBCM is transmitting data.
2. Checks for a serial data line malfunction.
3. Checks for proper Tech 1 connection to DLC.
4. Checks for open 15 amp Fuse 10.
5. Checks for open 5 amp Fuse 33.
6. Checks for high resistance in the EBCM ground circuit.
7. Checks for current DTCs.
8. Checks for proper ABS warning lamp operation.
9. Checks for history DTCs.
10. Checks for possible short to ground in the battery feed circuitry.
11. Checks for possible short to ground in the EBCM.
12. Ensures short to ground is not due to physical damage of the circuitry.
13. Checks for possible short to ground in the switched ignition circuitry.
14. Isolates the short to ground to either the circuitry or the EBCM.
15. Ensures short to ground is not due to physical damage of the circuitry.
16. Checks for an open in CKT 640.
17. Checks for high resistance in battery feed circuitry.
18. Checks for high resistance in the switched ignition circuitry.
19. Ensures malfunction is not due to poor terminal contact.
20. Checks for high resistance in the switched ignition circuit between the fuse block and the EBCM.

Diagnostic Aids: Excessive resistance in the ground or power supply circuits will not allow communication with the EBCM. If communication with the EBCM is not possible, ensure the ABS ground connection is good and that there is no excessive resistance in any of the power supply circuits.

DIAGNOSTIC CHART SYSTEM CHECK — 1994 GRAND PRIX, CONTINUED

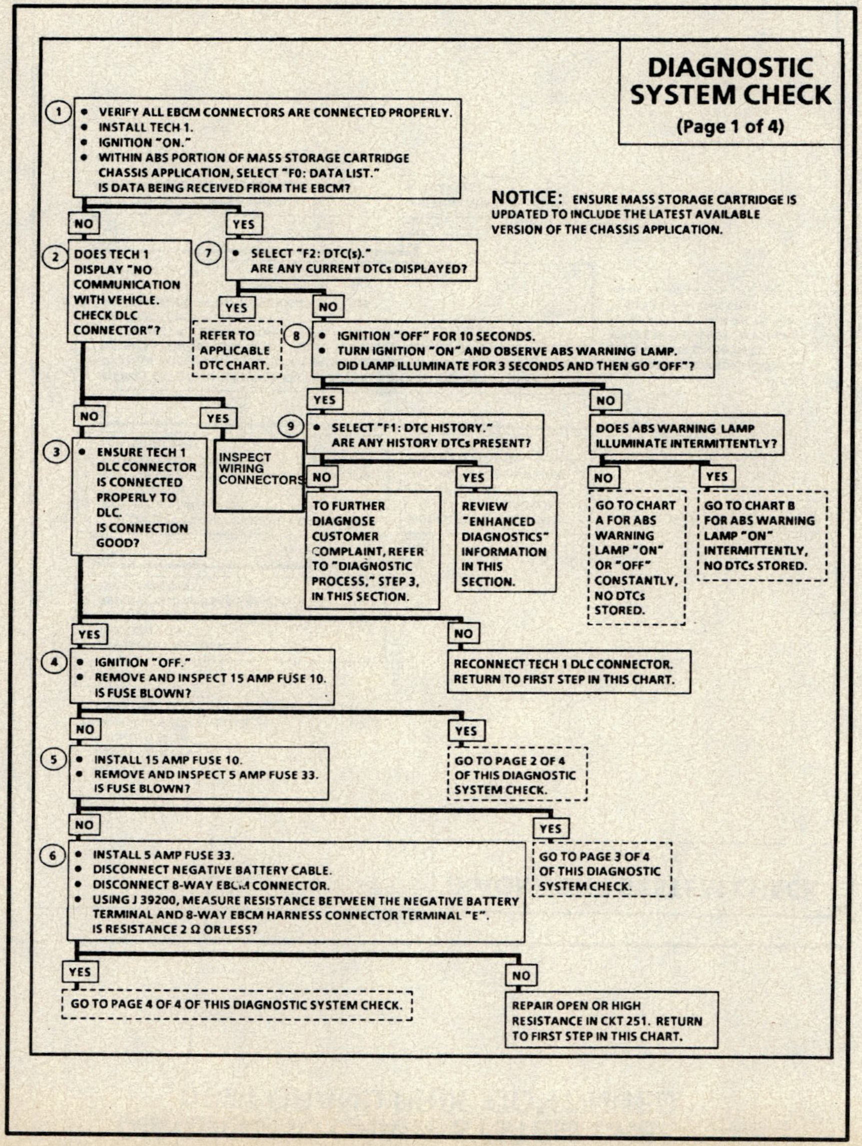

DIAGNOSTIC CHART SYSTEM CHECK — 1994 GRAND PRIX, CONTINUED

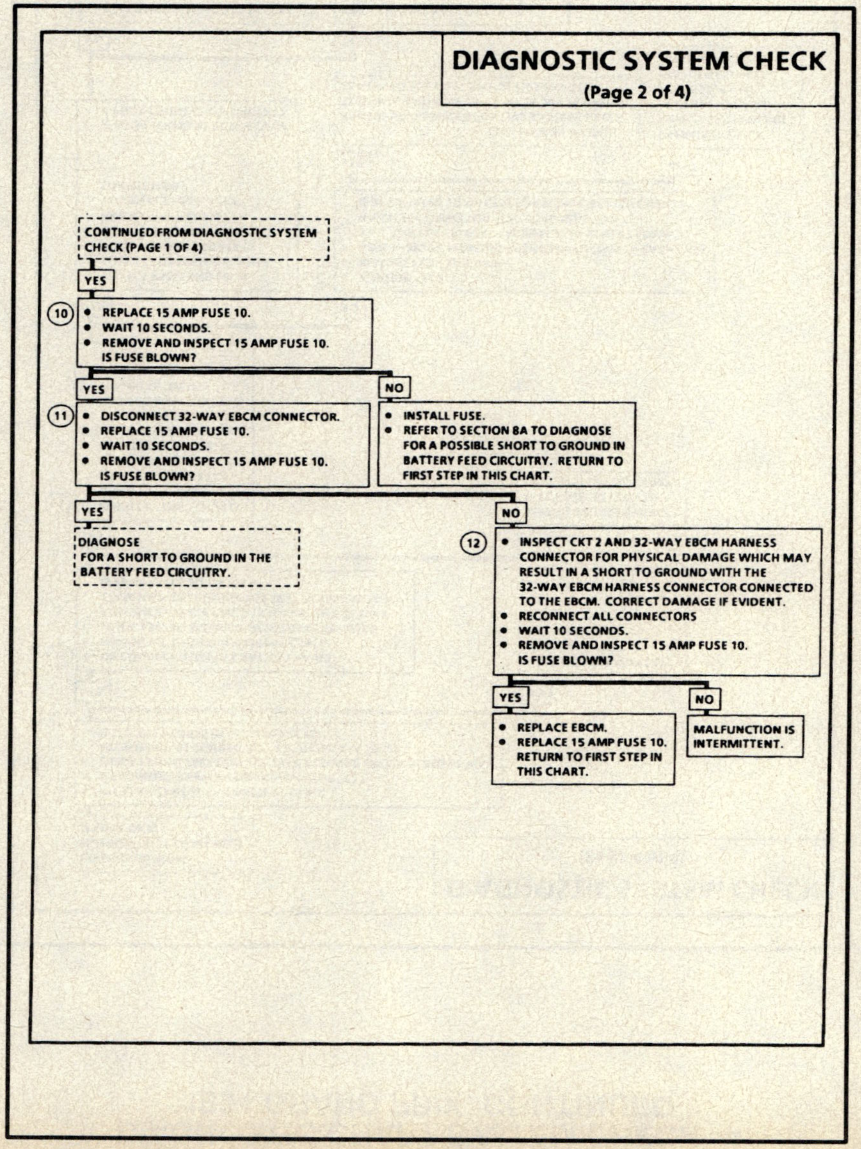

DIAGNOSTIC CHART SYSTEM CHECK — 1994 GRAND PRIX, CONTINUED

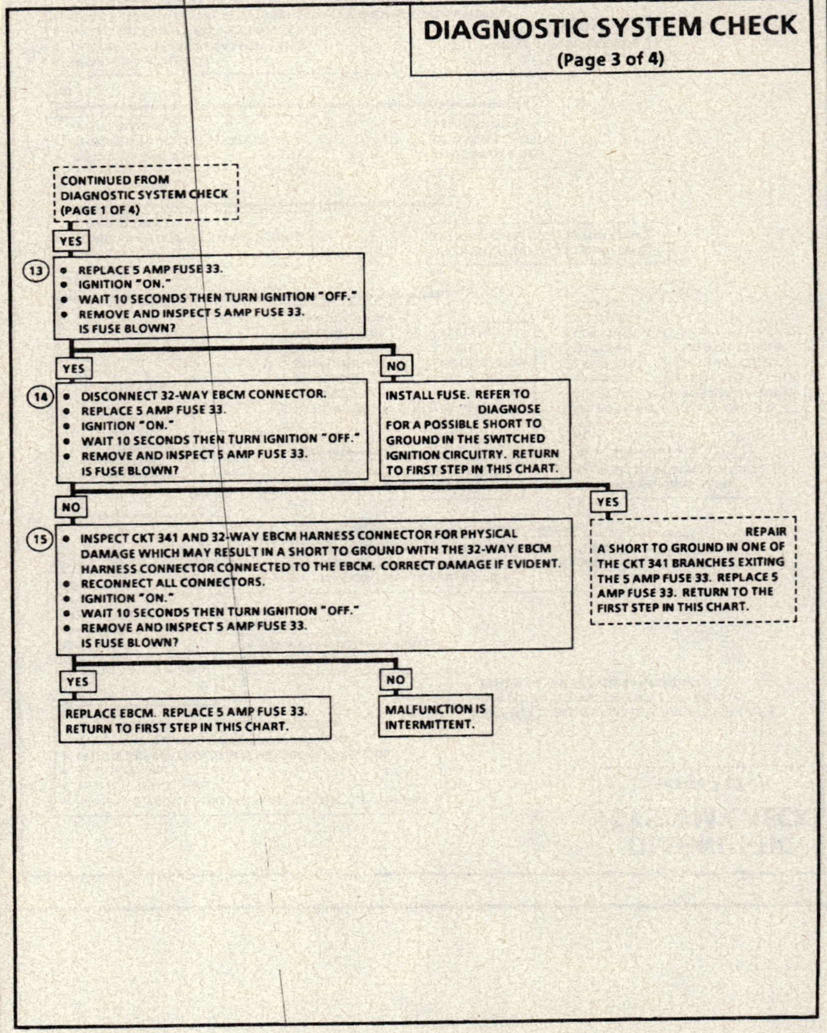

DIAGNOSTIC CHART SYSTEM CHECK — 1994 GRAND PRIX, CONTINUED

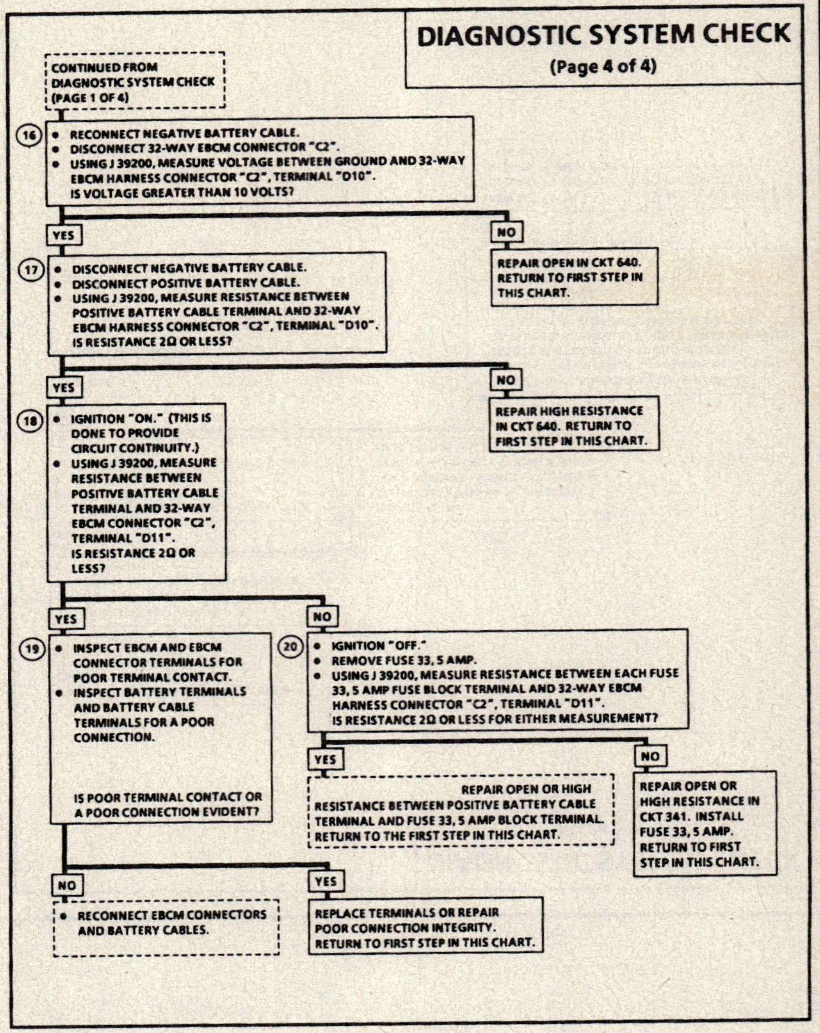

DIAGNOSTIC TROUBLE CODE CHART — 1994 GRAND PRIX

DIAGNOSTIC TROUBLE CODE AND SYMPTOM TABLE

CHART	SYMPTOM
A	ABS (Amber) Warning Lamp "ON" Constantly, No DTCs Stored
B	ABS (Amber) Warning Lamp "ON" Intermittently, No DTCs Stored
C	Tech 1 Displays Undefined Codes

DIAGNOSTIC TROUBLE CODE	DESCRIPTION
11	ABS Warning Lamp Circuit Malfunction
14	ABS Enable Relay Contact Circuit Open (1 of 3)
15	ABS Rly Contact Ckt Short to Batt or Always Closed
16	ABS Enable Relay Coil Circuit Open
17	ABS Enable Relay Coil Circuit Shorted to Gnd
18	ABS Enable Relay Coil Circuit Shorted to Batt
21	Left Front Wheel Speed = 0
22	Right Front Wheel Speed = 0
23	Left Rear Wheel Speed = 0
24	Right Rear Wheel Speed = 0
25	Left Front Excessive Wheel Speed Variation
26	Right Front Excessive Wheel Speed Variation
27	Left Rear Excessive Wheel Speed Variation
28	Right Rear Excessive Wheel Speed Variation
32	LF Wheel Speed Snr Ckt Open or Short to Gnd/Batt (1 of 3)
33	RF Wheel Speed Snr Ckt Open or Short to Gnd/Batt (1 of 3)
34	LR Wheel Speed Snr Ckt Open or Short to Gnd/Batt (1 of 3)
35	RR Wheel Speed Snr Ckt Open or Short to Gnd/Batt (1 of 3)
36	Low System Voltage (1 of 2)
37	High System Voltage
38	Left Front ESB Will Not Hold Motor
41	Right Front ESB Will Not Hold Motor
42	Rear ESB Will Not Hold Motor
44	Left Front Channel Will Not Move
45	Right Front Channel Will Not Move
46	Rear Channel Will Not Move
47	Left Front ABS Motor Free Spins (1 of 2)
48	Right Front ABS Motor Free Spins (1 of 2)
51	Rear ABS Motor Free Spins (1 of 2)
52	Left Front Channel in Release Too Long
53	Right Front Channel In Release Too Long
54	Rear Channel in Release Too Long
55	EBCM Malfunction

DIAGNOSTIC TROUBLE CODE CHART — 1994 GRAND PRIX

DIAGNOSTIC TROUBLE CODE AND SYMPTOM TABLE

DIAGNOSTIC TROUBLE CODE	DESCRIPTION
56	Left Front ABS Motor Circuit Open
57	Left Front ABS Motor Circuit Shorted to Ground
58	Left Front ABS Motor Circuit Shorted to Battery
61	Right Front ABS Motor Circuit Open
62	Right Front ABS Motor Circuit Shorted to Gnd
63	Right Front ABS Motor Circuit Shorted to Batt
64	Rear ABS Motor Circuit Open
65	Rear ABS Motor Circuit Shorted to Ground
66	Rear ABS Motor Circuit Shorted to Battery
76	Left Front Sol Circuit Open or Shorted to Batt
77	Left Front Solenoid Circuit Shorted to Gnd
78	Right Front Sol Circuit Open or Shorted to Batt
81	Right Front Solenoid Circuit Shorted to Gnd
82	Calibration Malfunction
86	EBCM Turned "ON" the Red "BRAKE" Warning Lamp
87	Red "BRAKE" Warning Lamp Circuit Open
88	Red "BRAKE" Warning Lamp Ckt Shorted to Batt
91	Open Brake Switch During Deceleration
92	Open Brake Switch When ABS Was Required
93	DTC 91 or 92 Set in Current or Prev. Ign Cycle
94	Brake Switch Contacts Always Closed
95	Brake Switch Circuit Open
96	Brake Lamps Circuit Open

WIRING SCHEMATIC — 1994 GRAND PRIX, CONTINUED

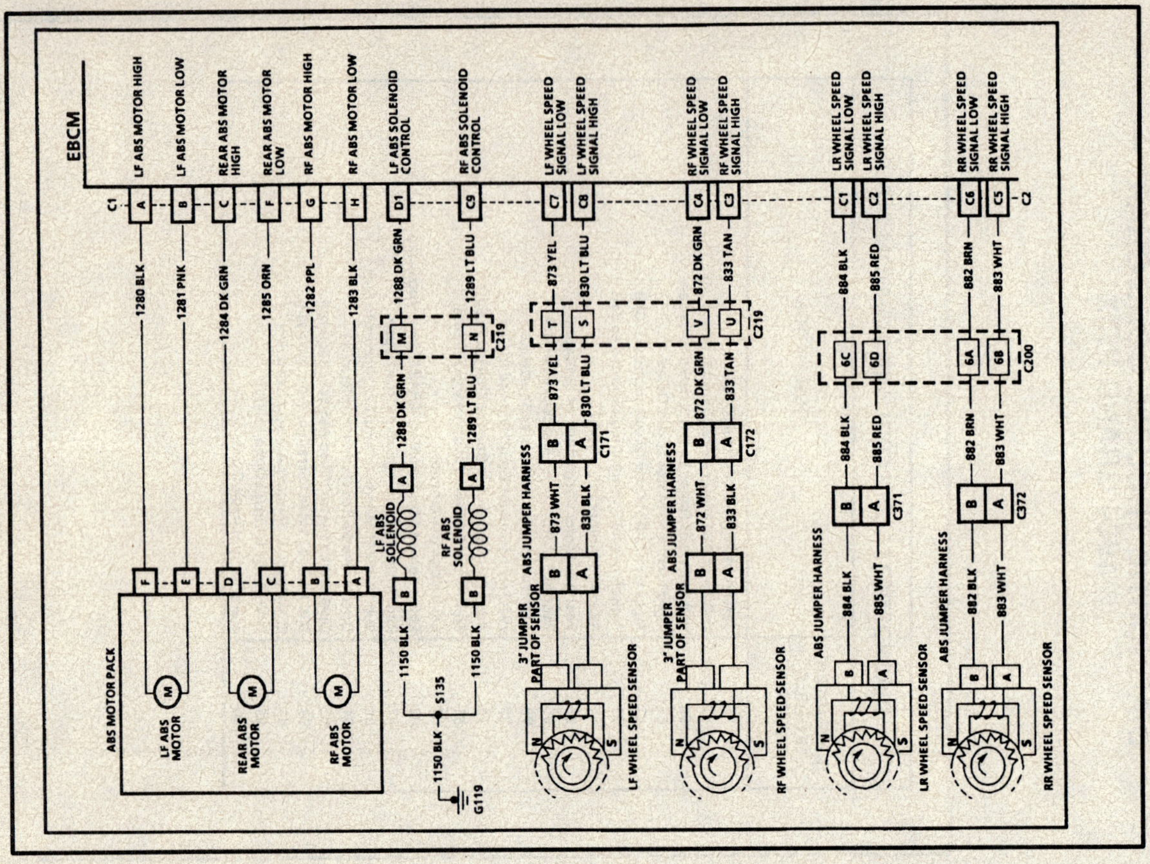

WIRING SCHEMATIC — 1994 GRAND PRIX

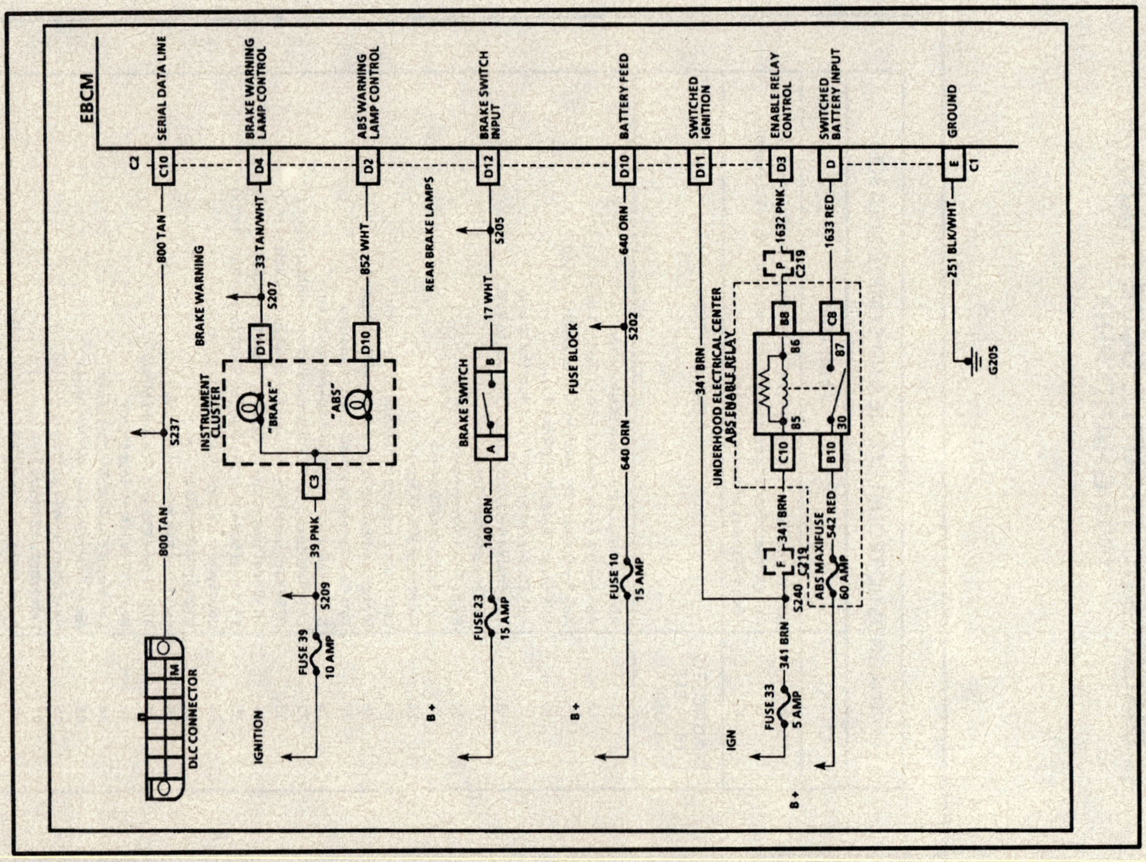

ELECTRICAL CONNECTOR TERMINAL IDENTIFICATION — 1994 GRAND PRIX

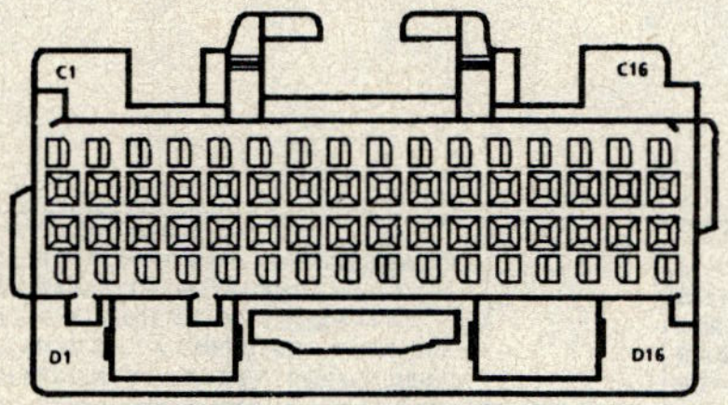

EBCM 32-WAY CONNECTOR

CONNECTOR TERMINAL	CIRCUIT NO.	COLOR	CIRCUIT DESCRIPTION
C1	884	BLK	LR WHEEL SIGNAL LOW
C2	885	RED	LR WHEEL SIGNAL HIGH
C3	833	TAN	RF WHEEL SIGNAL HIGH
C4	872	DK GRN	RF WHEEL SIGNAL LOW
C5	883	WHT	RR WHEEL SIGNAL HIGH
C6	882	BRN	RR WHEEL SIGNAL LOW
C7	873	YEL	LF WHEEL SIGNAL LOW
C8	830	LT BLU	LF WHEEL SIGNAL HIGH
C9	1289	LT BLU	RF ABS SOLENOID CONTROL
C10	800	TAN	SERIAL DATA LINE
C11	PLUGGED		NOT USED
C12	PLUGGED		NOT USED
C13	PLUGGED		NOT USED
C14	PLUGGED		NOT USED
C15	PLUGGED		NOT USED
C16	PLUGGED		NOT USED
D1	1288	DK GRN	LF ABS SOLENOID CONTROL
D2	852	WHT	ABS WARNING LAMP CONTROL
D3	1632	PNK	ABS ENABLE RELAY CONTROL
D4	33	TAN/WHT	BRAKE WARNING LAMP CONTROL
D5	PLUGGED		NOT USED
D6	PLUGGED		NOT USED
D7	PLUGGED		NOT USED
D8	PLUGGED		NOT USED
D9	PLUGGED		NOT USED
D10	640	ORN	BATTERY FEED
D11	341	BRN	SWITCHED IGNITION
D12	17	WHT	BRAKE SWITCH INPUT
D13	PLUGGED		NOT USED
D14	PLUGGED		NOT USED
D15	PLUGGED		NOT USED
D16	PLUGGED		NOT USED

ELECTRICAL CONNECTOR TERMINAL IDENTIFICATION — 1994 GRAND PRIX, CONTINUED

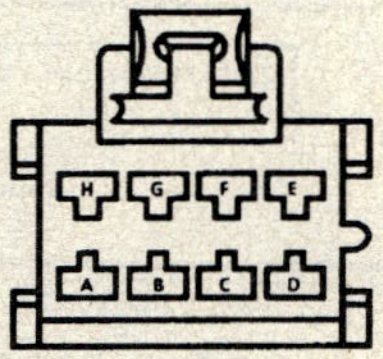

EBCM 8-WAY CONNECTOR

PIN	CIRCUIT NO.	COLOR	CIRCUIT
A	1280	BLK	LF ABS MOTOR HIGH
B	1281	PNK	LF ABS MOTOR LOW
C	1284	DK GRN	REAR ABS MOTOR HIGH
D	1633	RED	SWITCHED BATTERY INPUT
E	251	BLK/WHT	GROUND
F	1285	ORN	REAR ABS MOTOR LOW
G	1282	PPL	RF ABS MOTOR HIGH
H	1283	BLK	RF ABS MOTOR LOW

DIAGNOSTIC CHART A — 1994 GRAND PRIX

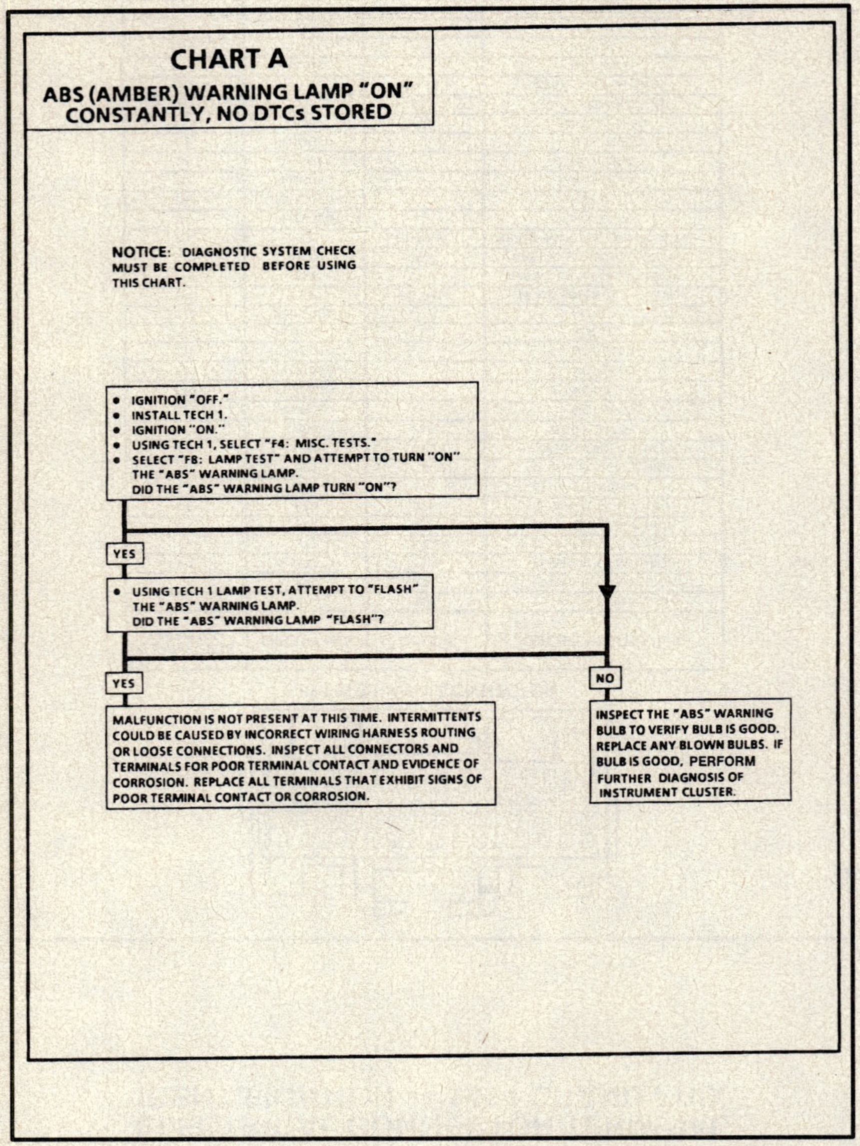

DIAGNOSTIC CHART B — 1994 GRAND PRIX

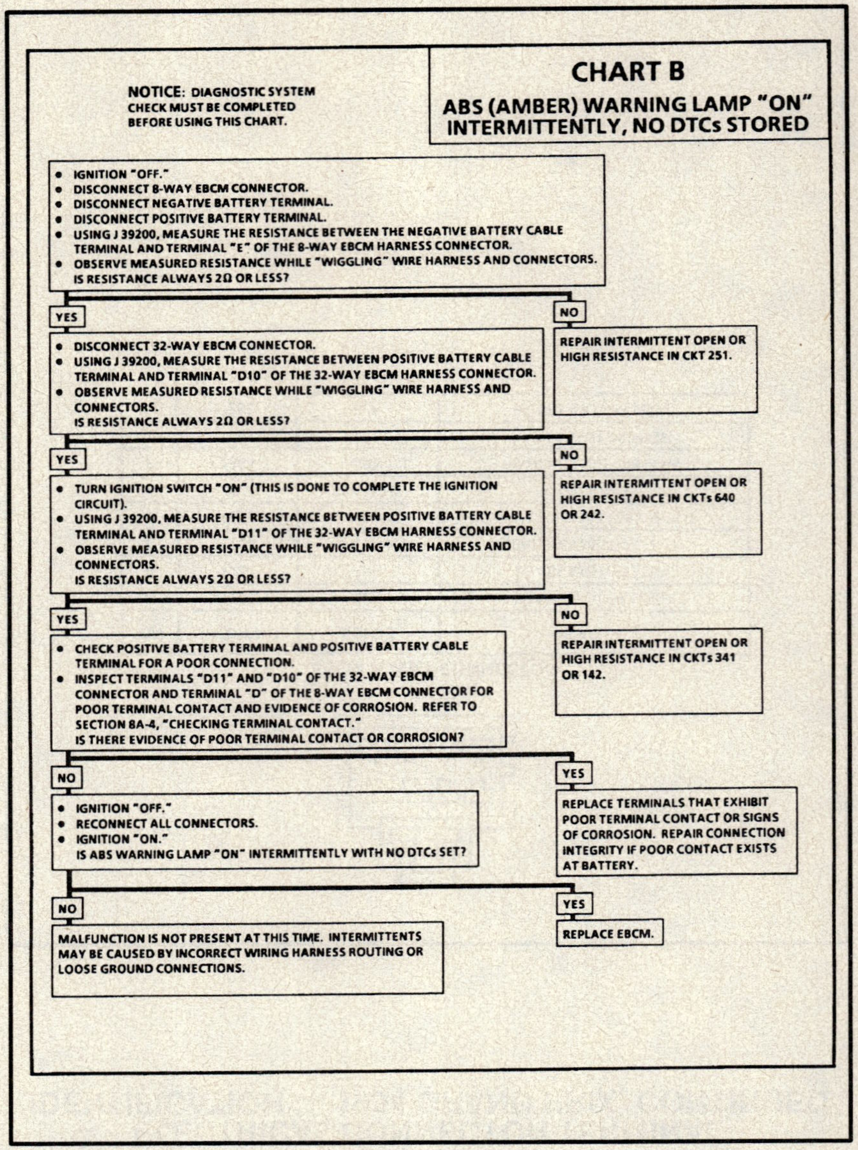

DIAGNOSTIC CHART C — 1994 GRAND PRIX

CHART C
TECH 1 DISPLAYS UNDEFINED CODES

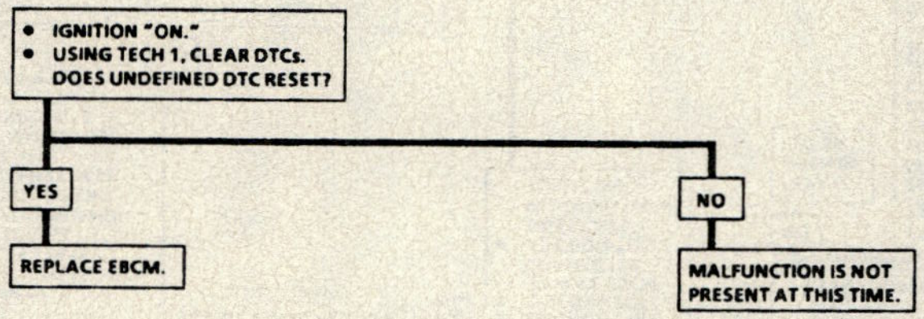

DIAGNOSTIC CHART DTC 11 — 1994 GRAND PRIX

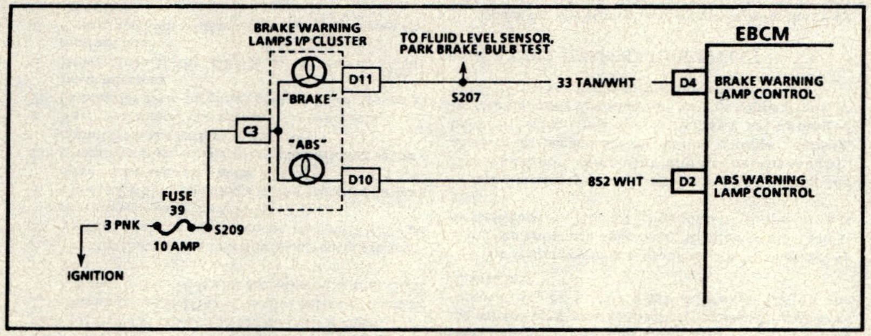

DTC 11
ABS WARNING LAMP CIRCUIT MALFUNCTION

Circuit Description:
This DTC checks the state of both the ABS warning lamp "BRAKE" lamp to identify a situation in which the driver could not be alerted to a system malfunction or of possible slippery driving conditions. Due to an integral lamp driver module within the instrument panel cluster, the EBCM must provide a ground to turn the ABS warning lamp "OFF." The "BRAKE" lamp circuit is such that the EBCM provides a ground to turn "ON" the "BRAKE" lamp.

DTC Will Set When: DTC 11 can only be set during the three second bulb check or when either of the lamps are commanded "ON." If the EBCM cannot control either the ABS warning lamp or the "BRAKE" lamp for two seconds, a malfunction exists.

Action Taken: A malfunction DTC is stored; however, ABS is not disabled.

DTC Chart Test Description: Number(s) below refer to circled number(s) on the diagnostic chart.
1. Determines if the ABS warning lamp can be properly controlled.
2. Determines if the lamp is "ON" constantly.
3. Checks for other DTCs which could turn "ON" the ABS warning lamp.
4. Determines if the ABS warning lamp can be turned "OFF" manually.
5. Checks the integrity of the ABS warning lamp circuitry.
6. Checks for a short to voltage on the ABS warning lamp circuitry.
7. Determines if a malfunctioning EBCM is the cause of DTC 11.
8. Verifies that the other I/P indicator lamps function properly.
9. Determines if the ABS warning lamp can be turned "ON" manually.
10. Checks for a short to ground in the ABS warning lamp circuitry.
11. Verifies condition of the 10 amp Fuse 39.
12. Determines if DTC 11 was set due to an EBCM malfunction.

Diagnostic Aids: An "Intermittent" malfunction may be caused by a poor connection, rubbed through wire insulation, or a wire that is broken inside the insulation.

The lamp test function of the Tech 1 may be used to command the lamp "ON" while looking for an intermittent malfunction in the ABS warning lamp circuitry.

The frequency of the malfunction can be checked by using the enhanced diagnostic function of the Tech 1, as described in "Tech 1 Diagnostics" found in this section.

Any circuitry that is suspected as causing the intermittent complaint should be thoroughly checked for backed out terminals, improper mating, broken locks, improperly formed or damaged terminals, poor terminal to wiring connections or physical damage to the wiring harness.

DIAGNOSTIC CHART DTC 11 — 1994 GRAND PRIX, CONTINUED

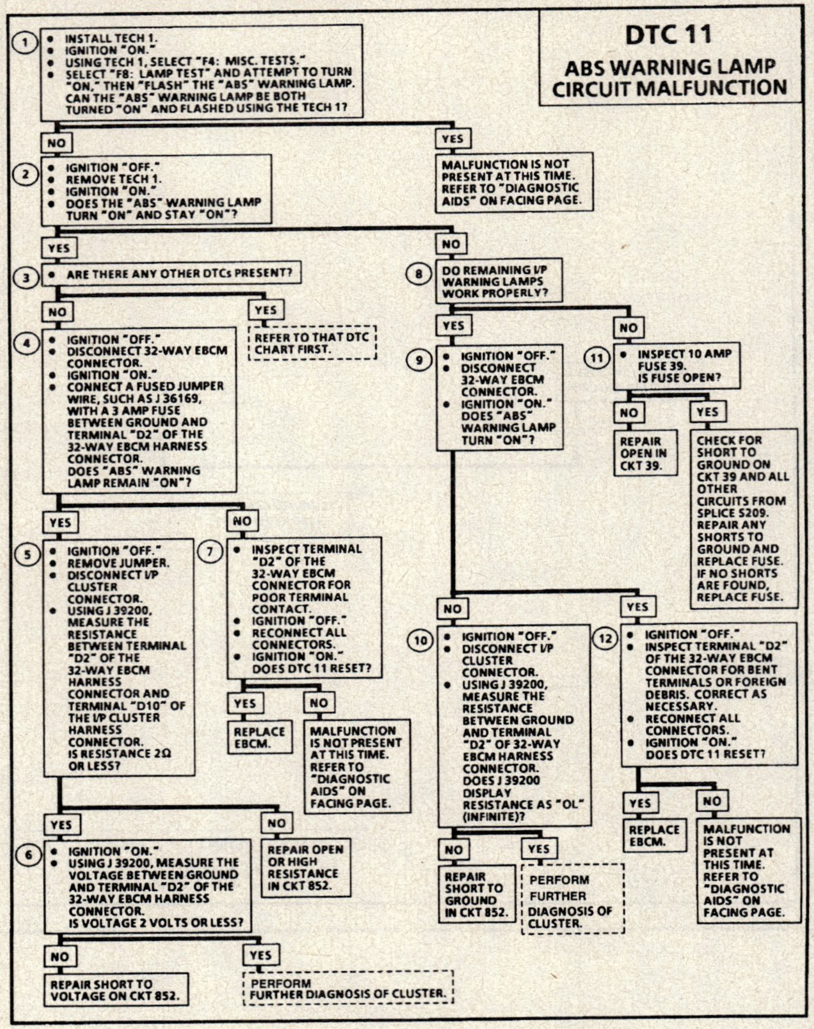

DIAGNOSTIC CHART DTC 14 — 1994 GRAND PRIX

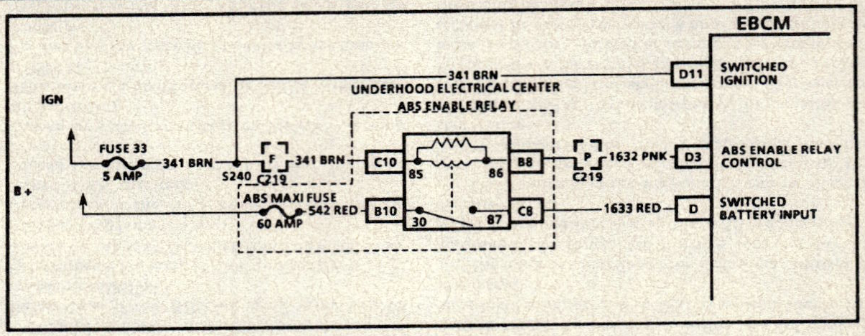

DTC 14

ABS ENABLE RELAY CONTACT CIRCUIT OPEN

Circuit Description:

Ignition voltage is supplied through terminal "C10" of the ABS enable relay. The EBCM then is able to energize the pull-in coil by completing the ground circuit at terminal "D3" of the EBCM. The magnetic field created closes the ABS enable relay contacts and allows battery voltage and current to be supplied to the EBCM, which supplies power to the motors and solenoids.

DTC Will Set When: DTC 14 can be set anytime after the EBCM commands the ABS enable relay "ON" (the relay is first commanded "ON" during the three second bulb check). This test monitors the availability of current/voltage to the motors and solenoids. This malfunction indicates voltage is not available and would therefore not allow ABS operation if required.

Action Taken: A malfunction DTC is stored, ABS is disabled and the ABS warning lamp is turned "ON."

DTC Chart Test Description: Number(s) below refer to circled number(s) on the diagnostic chart.

1. Checks to see if DTC 16 is also set as current or history. If DTC 16 is also set, proceed to that chart first.
2. Verifies that the malfunction is currently present.
3. Verifies that the relay can be turned "ON" and provide voltage to the EBCM.
4. Checks for high resistance in the switched battery circuit from the relay to the EBCM.
5. Checks for high resistance in the switched battery circuit from the battery to the relay.
6. Verifies the condition of the ABS enable relay.
7. Checks for poor terminal contact or corrosion at the connectors.
8. Determines if the EBCM is the cause of the malfunction.
9. Checks for high resistance in the entire relay coil circuit.
10. Verifies proper battery condition.
11. Verifies proper voltage is available at the ABS enable relay.
12. Checks for an open in the switched battery circuit from the relay to the EBCM.
13. Verifies the condition of the ABS enable relay.
14. Checks for proper resistance of the relay coil.
15. Checks for high resistance in the ABS enable relay control circuit between the relay and the EBCM.
16. Checks for high resistance in the ABS enable relay control circuit between the battery and the relay.

Diagnostic Aids: An "Intermittent" problem may be caused by a poor connection, rubbed through wire insulation, or a wire that is broken inside the insulation.

The frequency of the problem can be checked by using the enhanced diagnostic function of the Tech 1, as described in "Tech 1 Diagnostics" found in this section.

Any circuitry that is suspected as causing the "Intermittent" complaint should be thoroughly checked for backed out terminals, improper mating, broken locks, improperly formed or damaged terminals, poor terminal to wiring connections or physical damage to the wiring harness.

VIBRATION, TEMPERATURE EFFECTS:

Check for vibration effects by performing the relay test function of the Tech 1. With the relay test commanded "ON," lightly tap the top and sides of the relay while monitoring relay voltage. If the relay voltage changes significantly, replace the relay.

If DTC 14 only sets when the vehicle is initially started in cold ambient conditions (temperature less than 32°F-0°C), replace the relay.

DIAGNOSTIC CHART DTC 14 — 1994 GRAND PRIX, CONTINUED

DTC 14 (Page 1 of 3) ABS ENABLE RELAY CONTACT CIRCUIT OPEN

1. IS DTC 16 ALSO PRESENT OR IN HISTORY?
   - YES: PROCEED TO DTC 16 CHART FIRST.
   - NO: go to 2.

2.
   - IGNITION "ON," ENGINE "OFF."
   - USING TECH 1, SELECT DATA LIST AND OBSERVE EBCM BATTERY VOLTAGE.

   DOES TECH 1 INDICATE THAT EBCM BATTERY VOLTAGE IS GREATER THAN 10 VOLTS?
   - YES: MALFUNCTION IS NOT PRESENT AT THIS TIME. REFER TO "DIAGNOSTIC AIDS" ON FACING PAGE.
   - NO: go to 3.

3.
   - IGNITION "OFF."
   - DISCONNECT 32-WAY AND 8-WAY EBCM CONNECTORS.
   - CONNECT A FUSED JUMPER, SUCH AS J 36169 WITH A 3 AMP FUSE BETWEEN GROUND AND TERMINAL "D3" OF THE 32-WAY EBCM HARNESS CONNECTOR.
   - IGNITION "ON."
   - USING J 39200, MEASURE THE VOLTAGE BETWEEN GROUND AND TERMINAL "D" OF THE 8-WAY EBCM HARNESS CONNECTOR.

   IS VOLTAGE 10 VOLTS OR GREATER?
   - NO: DIAGNOSIS CONTINUED ON NEXT PAGE.
   - YES: go to 4.

4.
   - IGNITION "OFF."
   - DISCONNECT JUMPER USED IN STEP 3.
   - DISCONNECT ABS ENABLE RELAY.
   - USING J 39200, MEASURE RESISTANCE BETWEEN TERMINAL "C8" OF THE ABS ENABLE RELAY HARNESS CONNECTOR AND TERMINAL "D" OF THE 8-WAY EBCM HARNESS CONNECTOR.

   IS RESISTANCE 2Ω OR LESS?
   - NO: REPAIR HIGH RESISTANCE IN CKT 1633.
   - YES: go to 5.

5.
   - DISCONNECT NEGATIVE BATTERY CABLE.
   - DISCONNECT POSITIVE BATTERY CABLE.
   - USING J 39200, MEASURE THE RESISTANCE BETWEEN POSITIVE BATTERY CABLE TERMINAL AND TERMINAL "B10" OF THE ABS ENABLE RELAY HARNESS CONNECTOR.

   IS RESISTANCE 2Ω OR LESS?
   - NO: REPAIR HIGH RESISTANCE IN CKT 542.
   - YES: go to 6.

6.
   - USING A JUMPER WIRE, CONNECT TERMINAL "86" ON THE ABS ENABLE RELAY TO GROUND.
   - USING A FUSED JUMPER WIRE, SUCH AS J 36169, WITH A 3 AMP FUSE, CONNECT TERMINAL "85" ON THE ABS ENABLE RELAY TO B+.
   - USING J 39200, MEASURE THE RESISTANCE BETWEEN TERMINALS "30" AND "87" ON THE ABS ENABLE RELAY.

   IS RESISTANCE 2Ω OR LESS?
   - NO: REPLACE ABS ENABLE RELAY.
   - YES: go to 7.

7.
   - INSPECT ABS ENABLE RELAY AND ABS ENABLE RELAY HARNESS CONNECTOR FOR A POOR CONNECTION OR POOR TERMINAL CONTACT.
   - INSPECT EBCM AND EBCM CONNECTOR TERMINALS "D3" AND "D" FOR POOR TERMINAL CONTACT.
   - INSPECT POSITIVE BATTERY TERMINAL AND POSITIVE BATTERY CABLE TERMINAL FOR A POOR CONNECTION.

   IS POOR TERMINAL CONTACT OR A POOR CONNECTION EVIDENT?
   - YES: REPLACE TERMINALS THAT EXHIBIT POOR TERMINAL CONTACT OR EVIDENCE OF CORROSION.
   - NO: go to 8.

8.
   - RECONNECT ALL CONNECTORS.
   - IGNITION "ON."

   DOES DTC 14 RESET?
   - YES: REPLACE EBCM.
   - NO: MALFUNCTION IS INTERMITTENT. REFER TO "DIAGNOSTIC AIDS" ON FACING PAGE.

DIAGNOSTIC CHART DTC 14 — 1994 GRAND PRIX, CONTINUED

DTC 14 (Page 2 of 3) ABS ENABLE RELAY CONTACT CIRCUIT OPEN

DIAGNOSIS CONTINUED FROM PREVIOUS PAGE.

9.
   - IGNITION "OFF."
   - REMOVE JUMPER WIRE INSTALLED IN STEP #3, IF NOT ALREADY REMOVED.
   - DISCONNECT NEGATIVE BATTERY CABLE.
   - DISCONNECT POSITIVE BATTERY CABLE.
   - WITH BATTERY STILL DISCONNECTED, TURN IGNITION SWITCH TO "RUN." THIS COMPLETES THE IGNITION CIRCUIT TO THE ABS ENABLE RELAY.
   - USING J 39200, MEASURE THE RESISTANCE BETWEEN TERMINAL "D3" OF THE 32-WAY EBCM HARNESS CONNECTOR AND THE POSITIVE BATTERY CABLE TERMINAL.

   IS RESISTANCE 95Ω OR LESS WHEN MEASURED AT APPROXIMATELY 20°C (68°F)?
   - NO: DIAGNOSIS CONTINUED ON NEXT PAGE.
   - YES: go to 10.

10.
    - USING J 39200, MEASURE THE VOLTAGE BETWEEN THE NEGATIVE AND POSITIVE BATTERY TERMINALS.

    IS VOLTAGE 10 VOLTS OR GREATER?
    - NO: CORRECT LOW VOLTAGE CONDITION.
    - YES: go to 11.

11.
    - IGNITION "OFF."
    - RECONNECT POSITIVE BATTERY CABLE.
    - RECONNECT NEGATIVE BATTERY CABLE.
    - DISCONNECT ABS ENABLE RELAY.
    - MEASURE THE VOLTAGE BETWEEN TERMINAL "B10" OF THE ABS ENABLE RELAY HARNESS CONNECTOR AND GROUND.

    IS VOLTAGE 10 VOLTS OR GREATER?
    - NO: REPAIR OPEN OR HIGH RESISTANCE IN CKT 542.
    - YES: go to 12.

12.
    - USING J 39200, MEASURE THE RESISTANCE BETWEEN TERMINAL "C8" OF THE ABS ENABLE RELAY HARNESS CONNECTOR AND TERMINAL "D" OF THE 8-WAY EBCM HARNESS CONNECTOR.

    IS RESISTANCE 2Ω OR LESS?
    - NO: REPAIR OPEN OR HIGH RESISTANCE IN CKT 1633.
    - YES: go to 13.

13.
    - USING A JUMPER WIRE, CONNECT TERMINAL "86" ON THE ABS ENABLE RELAY TO GROUND.
    - USING A FUSED JUMPER WIRE, SUCH AS J 36169, WITH A 3 AMP FUSE, CONNECT TERMINAL "85" ON THE ABS ENABLE RELAY TO B+.
    - USING J 39200, MEASURE THE RESISTANCE BETWEEN TERMINALS "30" AND "87" ON THE ABS ENABLE RELAY.

    IS RESISTANCE 2Ω OR LESS?
    - NO: REPLACE ABS ENABLE RELAY.
    - YES: MALFUNCTION WAS MOST LIKELY DUE TO A POOR CONNECTION. INSPECT ALL CONNECTORS AND TERMINALS FOR POOR TERMINAL CONTACT AND EVIDENCE OF CORROSION. REPLACE ALL TERMINALS THAT EXHIBIT EVIDENCE OF POOR TERMINAL CONTACT OR CORROSION.

DIAGNOSTIC CHART DTC 14 — 1994 GRAND PRIX, CONTINUED

DTC 14
(Page 3 of 3)
ABS ENABLE RELAY CONTACT CIRCUIT OPEN

DIAGNOSIS CONTINUED FROM PREVIOUS PAGE.

(14)
- DISCONNECT ABS ENABLE RELAY.
- USING J 39200, MEASURE THE RESISTANCE BETWEEN TERMINALS "85" AND "86" ON THE ABS ENABLE RELAY. IS RESISTANCE 95Ω OR LESS WHEN MEASURED AT APPROXIMATELY 20°C (68°F)?

NO: REPLACE ABS ENABLE RELAY.

YES:

(15)
- MEASURE THE RESISTANCE BETWEEN TERMINAL "B8" OF THE ABS ENABLE RELAY HARNESS CONNECTOR AND TERMINAL "D3" OF THE 32-WAY EBCM HARNESS CONNECTOR. IS RESISTANCE 2Ω OR LESS?

NO: REPAIR OPEN OR HIGH RESISTANCE IN CKT 1632.

YES:

(16)
- MEASURE THE RESISTANCE BETWEEN TERMINAL "C10" OF THE ABS ENABLE RELAY HARNESS CONNECTOR AND THE POSITIVE BATTERY CABLE TERMINAL. IS RESISTANCE 2Ω OR LESS?

NO: REPAIR OPEN OR HIGH RESISTANCE IN CKT 341, CKT 300, IGNITION SWITCH, OR BATTERY FEED CIRCUIT TO IGNITION SWITCH.

YES: MALFUNCTION WAS MOST LIKELY DUE TO A POOR CONNECTION. INSPECT ALL CONNECTORS AND TERMINALS FOR POOR TERMINAL CONTACT AND EVIDENCE OF CORROSION. REPLACE ALL TERMINALS THAT EXHIBIT EVIDENCE OF POOR TERMINAL CONTACT OR CORROSION.

DIAGNOSTIC CHART DTC 15 — 1994 GRAND PRIX

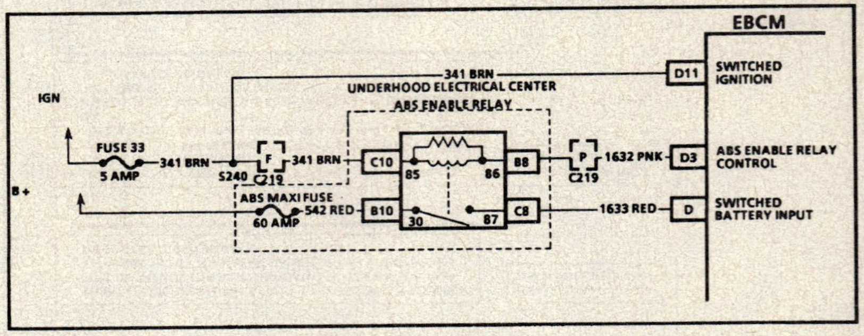

DTC 15

ABS RLY CONTACT CKT SHORT TO BATT OR ALWAYS CLOSED

Circuit Description:

Ignition voltage is supplied through terminal "C10" of the ABS enable relay. The EBCM then is able to energize the pull-in coil by completing the ground circuit at terminal "D3" of the EBCM. The magnetic field created closes the ABS enable relay contacts and allows battery voltage and current to be supplied to the EBCM, which supplies power to the motors and solenoids.

DTC Will Set When: DTC 15 can be set only before the EBCM commands the relay "ON." This test determines if the ABS enable relay is energized when it should not be. This malfunction would not allow the ABS enable relay to remove power to the ABS system. If a second malfunction were to occur that requires the ABS enable relay to be turned "OFF," that malfunction cannot be removed if the relay cannot be controlled. The malfunction must be present for three consecutive drive cycles before the DTC is set.

Action Taken: A malfunction DTC is stored. ABS is not disabled.

DTC Chart Test Description: Number(s) below refer to circled number(s) on the diagnostic chart.

1. Checks to see if other DTCs are set.
2. Indicates that the EBCM is capable of controlling the ABS enable relay as commanded.
3. Checks for a possible short to battery voltage in CKT 1633.
4. Determines whether the short to voltage is in CKT 1633 or due to a malfunctioning relay.
5. Checks for a possible intermittent in wiring or connectors.

Diagnostic Aids: An "Intermittent" malfunction may be caused by a poor connection, rubbed through wire insulation, or a wire that is broken inside the insulation.

The frequency of the malfunction can be checked by using the enhanced diagnostic function of the Tech 1, as described in "Tech 1 Diagnostics" found in this section.

Any circuitry, that is suspected as causing the intermittent complaint, should be thoroughly checked for backed out terminals, improper mating, broken locks, improperly formed or damaged terminals, poor terminal to wiring connections or physical damage to the wiring harness.

DIAGNOSTIC CHART DTC 15 — 1994 GRAND PRIX, CONTINUED

DTC 15
ABS RLY CONTACT SHORT TO BATT OR ALWAYS CLOSED

1. ARE ANY OTHER DTCs PRESENT?
   - YES: PROCEED TO THAT DTC CHART FIRST.
   - NO: go to 2.
2. • IGNITION "ON," ENGINE "OFF."
   • USING THE RELAY TEST FUNCTION OF THE TECH 1, COMMAND ABS ENABLE RELAY "OFF."
   DOES TECH 1 INDICATE THAT ABS ENABLE RELAY IS "OFF," AND BATTERY VOLTAGE IS LESS THAN 5 VOLTS?
   - YES: MALFUNCTION IS NOT PRESENT AT THIS TIME. REFER TO "DIAGNOSTIC AIDS" ON FACING PAGE.
   - NO: go to 3.
3. • IGNITION "OFF."
   • DISCONNECT 32-WAY AND 8-WAY EBCM HARNESS CONNECTORS.
   • IGNITION "ON."
   • USING J 39200, MEASURE THE VOLTAGE BETWEEN GROUND AND TERMINAL "D" OF THE EBCM 8-WAY CONNECTOR.
   IS VOLTAGE 2 VOLTS OR LESS?
   - NO: go to 4.
   - YES: go to 5.
4. • IGNITION "OFF."
   • DISCONNECT ABS ENABLE RELAY.
   • IGNITION "ON."
   • MEASURE THE VOLTAGE BETWEEN GROUND AND TERMINAL "D" OF THE 8-WAY EBCM HARNESS CONNECTOR.
   IS VOLTAGE 2 VOLTS OR LESS?
   - YES: REPLACE ABS ENABLE RELAY.
   - NO: REPAIR SHORT TO VOLTAGE IN CKT 1633.
5. • IGNITION "OFF."
   • INSPECT CKT 1633 AND 8-WAY EBCM HARNESS CONNECTOR FOR DAMAGE WHICH MAY RESULT IN A SHORT TO VOLTAGE WHEN ALL CONNECTORS ARE CONNECTED. REPAIR DAMAGE IF EVIDENT.
   • RECONNECT 8-WAY AND 32-WAY EBCM CONNECTORS.
   • IGNITION "ON."
   • USING RELAY TEST FUNCTION OF TECH 1, COMMAND ABS ENABLE RELAY "OFF."
   DOES TECH 1 INDICATE THAT ABS ENABLE RELAY IS "OFF," AND BATTERY VOLTAGE IS LESS THAN 5 VOLTS?
   - NO: REPLACE EBCM.
   - YES: MALFUNCTION MOST LIKELY CAUSED BY A POOR CONNECTION. REFER TO "DIAGNOSTIC AIDS" ON FACING PAGE.

AFTER ALL DIAGNOSIS IS COMPLETE, CLEAR DTCs, AND TEST DRIVE VEHICLE FOR 3 DRIVE CYCLES TO VERIFY THAT DTC DOES NOT RESET. A DRIVE CYCLE CONSISTS OF STARTING THE VEHICLE, DRIVING ABOVE 16 km/h (10 MPH), AND THEN KEYING DOWN.

DIAGNOSTIC CHART DTC 16 — 1994 GRAND PRIX

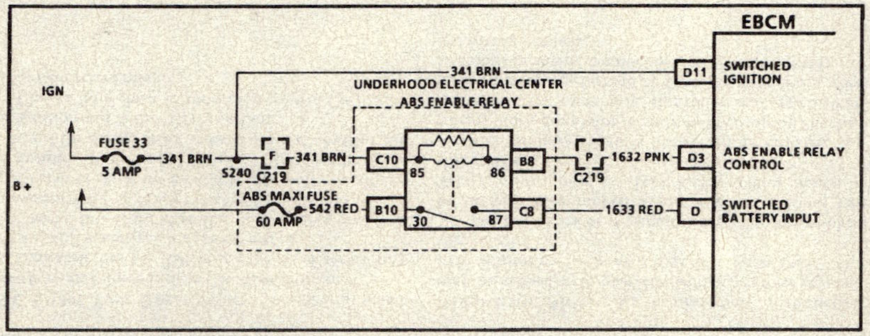

DTC 16
ABS ENABLE RELAY COIL CIRCUIT OPEN

Circuit Description:
Ignition voltage is supplied through terminal "C10" of the ABS enable relay. The EBCM then is able to energize the pull-in coil by completing the ground circuit at terminal "D3" of the EBCM. The magnetic field created closes the ABS enable relay contacts and allows battery voltage and current to be supplied to the EBCM, which supplies power to the motors and solenoids.

DTC Will Set When: DTC 16 can be set only after DTC 14 has been set. This test detects an open in the ABS enable relay coil circuit. An open in this circuit will not allow the ABS enable relay to be energized thus preventing voltage/current to the motors and solenoids. If this malfunction is present and the ignition is turned "OFF" before 5 km/h (3 mph) is reached, DTC 14 is set alone.

Action Taken: A malfunction DTC is stored, ABS is disabled and the ABS warning lamp is turned "ON."

DTC Chart Test Description: Number(s) below refer to circled number(s) on the diagnostic chart.
1. Indicates the EBCM is capable of controlling the ABS enable relay as commanded.
2. Checks to ensure that voltage is available to the pull-in coil of the enable relay.
3. Ensures that there is continuity through the pull-in coil of the enable relay.
4. Verifies the integrity of the relay control circuit.
5. Ensures malfunction was not due to poor terminal contact.

Diagnostic Aids: An "Intermittent" malfunction may be caused by a poor connection, rubbed through wire insulation, or a wire that is broken inside the insulation.

The frequency of the malfunction can be checked by using the enhanced diagnostic function of the Tech 1, as described in "Tech 1 Diagnostics" found in this section.

If the frequency of the malfunction is high, but is currently intermittent, check for high coil resistance by measuring between relay terminals "C10" and "B8" using J 39200. If resistance shows greater than 100Ω, replace the relay.

Any circuitry, that is suspected as causing the intermittent complaint, should be thoroughly checked for backed out terminals, improper mating, broken locks, improperly formed or damaged terminals, poor terminal to wiring connections or physical damage to the wiring harness.

DIAGNOSTIC CHART DTC 16 — 1994 GRAND PRIX, CONTINUED

DTC 16
ABS ENABLE RELAY COIL CIRCUIT OPEN

(1)
- IGNITION "ON," ENGINE "OFF."
- USING THE RELAY TEST FUNCTION OF THE TECH 1, COMMAND THE ABS ENABLE RELAY "ON." DOES TECH 1 INDICATE ABS ENABLE RELAY "ON," AND BATTERY VOLTAGE ABOVE 10 VOLTS?

YES: MALFUNCTION IS NOT PRESENT AT THIS TIME. REFER TO "DIAGNOSTIC AIDS" ON FACING PAGE.

NO:

(2)
- USING J 39200, MEASURE THE VOLTAGE BETWEEN GROUND AND TERMINAL "C10" OF THE ABS ENABLE RELAY CONNECTOR. IS VOLTAGE 10 VOLTS OR GREATER?

NO: REPAIR OPEN OR HIGH RESISTANCE IN CKT 341 OR FUSE 33.

YES:

(3)
- USING THE RELAY TEST FUNCTION OF THE TECH 1, COMMAND THE ABS ENABLE RELAY "OFF."
- MEASURE THE VOLTAGE BETWEEN GROUND AND TERMINAL "B8" OF THE ABS ENABLE RELAY CONNECTOR. IS VOLTAGE 10 VOLTS OR GREATER?

NO: CHECK FOR INTERMITTENT CONNECTIONS AT ABS ENABLE RELAY. IF OK, REPLACE ABS RELAY.

YES:

(4)
- IGNITION "OFF."
- DISCONNECT 32-WAY EBCM CONNECTOR.
- IGNITION "ON."
- MEASURE THE VOLTAGE BETWEEN GROUND AND TERMINAL "D3" OF THE 32-WAY EBCM HARNESS CONNECTOR. IS VOLTAGE 10 VOLTS OR GREATER?

NO: REPAIR OPEN OR HIGH RESISTANCE IN CKT 1632.

YES:

(5)
- IGNITION "OFF."
- INSPECT 32-WAY EBCM CONNECTOR TERMINAL "D3" FOR POOR TERMINAL CONTACT OR EVIDENCE OF CORROSION. REPLACE ALL TERMINALS THAT EXHIBIT SIGNS OF POOR TERMINAL CONTACT OR CORROSION.
- RECONNECT ALL CONNECTORS.
- IGNITION "ON." DOES DTC 16 SET AS A CURRENT DTC?

YES: REPLACE EBCM.

NO: MALFUNCTION WAS MOST LIKELY CAUSED BY A POOR CONNECTION. REFER TO "DIAGNOSTIC AIDS" ON FACING PAGE.

AFTER ALL DIAGNOSIS IS COMPLETE, CLEAR DTCs AND TEST DRIVE VEHICLE FOR THREE (3) DRIVE CYCLES TO VERIFY THAT DTC DOES NOT RESET. A DRIVE CYCLE CONSISTS OF STARTING THE VEHICLE, DRIVING OVER 16 km/h (10 MPH), AND THEN KEYING DOWN.

DIAGNOSTIC CHART DTC 17 — 1994 GRAND PRIX

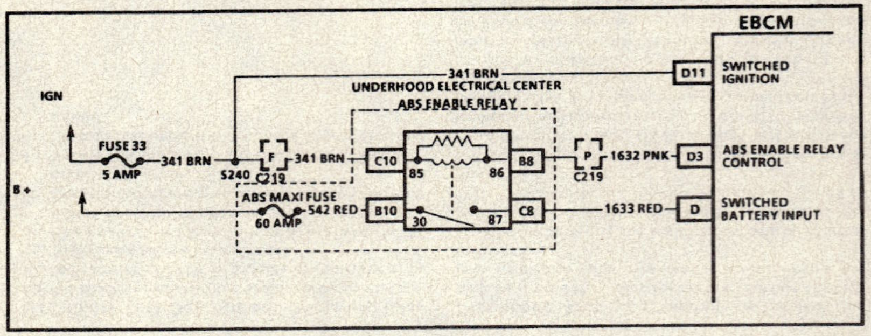

DTC 17
ABS ENABLE RELAY COIL CIRCUIT SHORTED TO GND

Circuit Description:
Ignition voltage is supplied through terminal "C10" of the ABS enable relay. The EBCM then is able to energize the pull-in coil by completing the ground circuit at terminal "D3" of the EBCM. The magnetic field created closes the ABS enable relay contacts and allows battery voltage and current to be supplied to the EBCM, which supplies voltage to the motors and solenoids.

DTC Will Set When: DTC 17 can be set before the EBCM commands the ABS enable relay "ON." This test determines if the ABS enable relay is energized when it should not be. This malfunction would not allow the ABS enable relay to remove power to the ABS system. If a second malfunction were to occur that requires the ABS enable relay to be turned "OFF," that malfunction cannot be removed if the relay cannot be controlled.

Action Taken: A malfunction DTC is stored. ABS is not disabled.

DTC Chart Test Description: Number(s) below refer to circled number(s) on the diagnostic chart.

1. Indicates that the EBCM is capable of controlling the ABS enable relay as commanded.
2. Checks to ensure that the ABS enable relay or control CKT 1632 are not shorted to ground.
3. Checks to see if the EBCM is internally shorted to ground. A grounded terminal "D3" would cause the ABS enable relay to be energized anytime the ignition was in the "ON" position.
4. Checks for a short to ground in control CKT 1632 or a defective relay.

Diagnostic Aids: An "Intermittent" malfunction may be caused by a poor connection, rubbed through wire insulation, or a wire that is broken inside the insulation.

The frequency of the malfunction can be checked by using the enhanced diagnostic function of the Tech 1, as described in "Tech 1 Diagnostics" found in this section.

Any circuitry, that is suspected as causing the intermittent complaint, should be thoroughly checked for backed out terminals, improper mating, broken locks, improperly formed or damaged terminals, poor terminal to wiring connections or physical damage to the wiring harness.

DIAGNOSTIC CHART DTC 17 — 1994 GRAND PRIX, CONTINUED

DTC 17
ABS ENABLE RELAY COIL CIRCUIT SHORTED TO GND

(1)
- IGNITION "ON," ENGINE "OFF."
- USING THE RELAY TEST FUNCTION OF THE TECH 1, COMMAND THE ABS ENABLE RELAY "OFF."

DOES TECH 1 INDICATE THAT ABS ENABLE RELAY IS "OFF" AND B+ VOLTAGE IS LESS THAN 5 VOLTS?

- YES: MALFUNCTION IS NOT PRESENT AT THIS TIME. REFER TO "DIAGNOSTIC AIDS" ON FACING PAGE.
- NO: go to step 2.

(2)
- IGNITION "OFF."
- DISCONNECT 32-WAY EBCM CONNECTOR.
- IGNITION "ON."
- USING J 39200, MEASURE THE VOLTAGE BETWEEN TERMINAL "D3" OF THE 32-WAY EBCM HARNESS CONNECTOR AND B+.

IS VOLTAGE 2 VOLTS OR LESS?

- YES: go to step 3.
- NO: go to step 4.

(3)
- IGNITION "OFF."
- INSPECT CKT 1632 AND TERMINAL "D3" OF THE 32-WAY EBCM CONNECTOR FOR DAMAGE WHICH MAY RESULT IN A SHORT TO GROUND WITH ALL CONNECTORS CONNECTED. REPAIR ALL DAMAGE IF EVIDENT.
- RECONNECT ALL CONNECTORS.
- IGNITION "ON."

DOES DTC 17 SET AS A CURRENT DTC?

- YES: REPLACE EBCM.
- NO: MALFUNCTION WAS MOST LIKELY CAUSED BY A POOR CONNECTION. REFER TO "DIAGNOSTIC AIDS" ON FACING PAGE.

(4)
- IGNITION "OFF."
- DISCONNECT ABS ENABLE RELAY.
- MEASURE VOLTAGE BETWEEN TERMINAL "B8" OF THE ABS ENABLE RELAY CONNECTOR AND B+.

IS VOLTAGE 2 VOLTS OR LESS?

- NO: REPAIR SHORT TO GROUND IN CKT 1632.
- YES: REPLACE ABS ENABLE RELAY.

AFTER ALL DIAGNOSIS IS COMPLETE, CLEAR DTCs AND TEST DRIVE VEHICLE FOR THREE (3) DRIVE CYCLES TO VERIFY THAT DTC DOES NOT RESET. A DRIVE CYCLE CONSISTS OF STARTING THE VEHICLE, DRIVING OVER 16 km/h (10 MPH), AND THEN KEYING DOWN.

DIAGNOSTIC CHART DTC 18 — 1994 GRAND PRIX

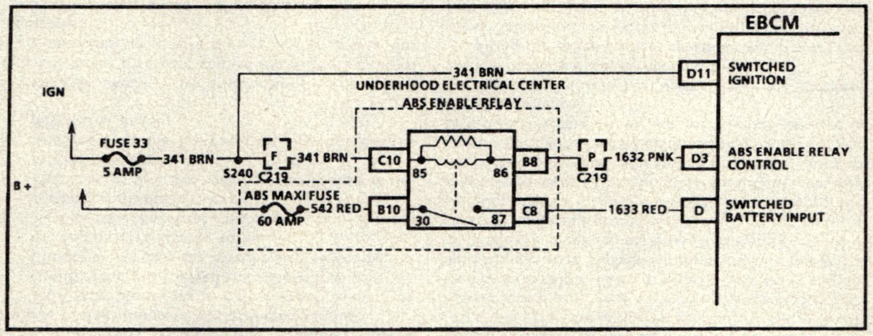

DTC 18
ABS ENABLE RELAY COIL CIRCUIT SHORTED TO BATT

Circuit Description:

Ignition voltage is supplied through terminal "C10" of the ABS enable relay. The EBCM then is able to energize the pull-in coil by completing the ground circuit at terminal "D3" of the EBCM. The magnetic field created closes the ABS enable relay contacts and allows battery voltage and current to be supplied to the EBCM, which supplies voltage to the motors and solenoids.

DTC Will Set When: DTC 18 can be set after the EBCM commands the enable relay "ON." This test monitors the availability of current/voltage to the motors and solenoids. A malfunction of this type will not allow voltage to be available to the motors and solenoids; therefore, ABS operation would not be allowed if required.

Action Taken: A malfunction DTC is stored, ABS is disabled and the ABS warning lamp is turned "ON."

DTC Chart Test Description: Number(s) below refer to circled number(s) on the diagnostic chart.

1. Indicates that the EBCM is capable of controlling the ABS enable relay as commanded.
2. Checks for a short to voltage on the relay control circuit.
3. Verifies that the relay coil has the appropriate resistance.
4. Verifies that the relay contacts are not failed closed.
5. Checks for possible poor terminal contact at the EBCM.

Diagnostic Aids: An "Intermittent" malfunction may be caused by a poor connection, rubbed through wire insulation, or a wire that is broken inside the insulation.

The frequency of the malfunction can be checked by using the enhanced diagnostic function of the Tech 1, as described in "Tech 1 Diagnostics" found in this section.

Any circuitry, that is suspected as causing the intermittent complaint, should be thoroughly checked for backed out terminals, improper mating, broken locks, improperly formed or damaged terminals, poor terminal to wiring connections or physical damage to the wiring harness.

DIAGNOSTIC CHART DTC 18 — 1994 GRAND PRIX, CONTINUED

DTC 18
ABS ENABLE RELAY COIL CIRCUIT SHORTED TO BATT

1
- IGNITION "ON," ENGINE "OFF."
- USING RELAY TEST FUNCTION OF TECH 1, COMMAND ABS ENABLE RELAY "ON."
- DOES TECH 1 INDICATE ABS ENABLE RELAY "ON" AND BATTERY VOLTAGE ABOVE 10 VOLTS?

YES: MALFUNCTION IS NOT PRESENT AT THIS TIME. REFER TO "DIAGNOSTIC AIDS" ON FACING PAGE.

NO:

2
- IGNITION "OFF."
- DISCONNECT ABS ENABLE RELAY.
- DISCONNECT 32-WAY EBCM CONNECTOR.
- IGNITION "ON."
- USING J 39200, MEASURE THE VOLTAGE BETWEEN GROUND AND TERMINAL "D3" OF THE 32-WAY EBCM HARNESS CONNECTOR.

IS VOLTAGE 2 VOLTS OR LESS?

NO: REPAIR SHORT TO VOLTAGE IN CKT 1632.

YES:

3
- IGNITION "OFF."
- USING J 39200, MEASURE THE RESISTANCE BETWEEN TERMINAL "85" AND TERMINAL "86" OF THE ABS ENABLE RELAY.

IS RESISTANCE 40Ω OR GREATER?

NO: REPLACE ABS ENABLE RELAY.

YES:

4
- USING J 39200, MEASURE THE RESISTANCE BETWEEN TERMINALS "30" AND "87" OF THE ABS ENABLE RELAY. DOES J 39200 DISPLAY RESISTANCE AS "OL" (INFINITE)?

NO: REPLACE ABS ENABLE RELAY.

YES:

5
- INSPECT CKT 1632 AND TERMINAL "D3" OF THE 32-WAY EBCM CONNECTOR FOR DAMAGE WHICH MAY RESULT IN A SHORT TO VOLTAGE WITH ALL CONNECTORS CONNECTED. REPAIR ALL DAMAGE IF EVIDENT.
- RECONNECT ALL CONNECTORS.
- IGNITION "ON."

DOES DTC 18 SET AS A CURRENT DTC?

YES: REPLACE EBCM.

NO: MALFUNCTION IS MOST LIKELY DUE TO A POOR CONNECTION. REFER TO "DIAGNOSTIC AIDS" ON FACING PAGE.

AFTER ALL DIAGNOSIS IS COMPLETE, CLEAR DTCs AND TEST DRIVE VEHICLE FOR THREE (3) DRIVE CYCLES TO VERIFY THAT DTC DOES NOT RESET. A DRIVE CYCLE CONSISTS OF STARTING THE VEHICLE, DRIVING OVER 16 km/h (10 MPH), AND THEN KEYING DOWN.

DIAGNOSTIC CHART DTC 21 — 1994 GRAND PRIX

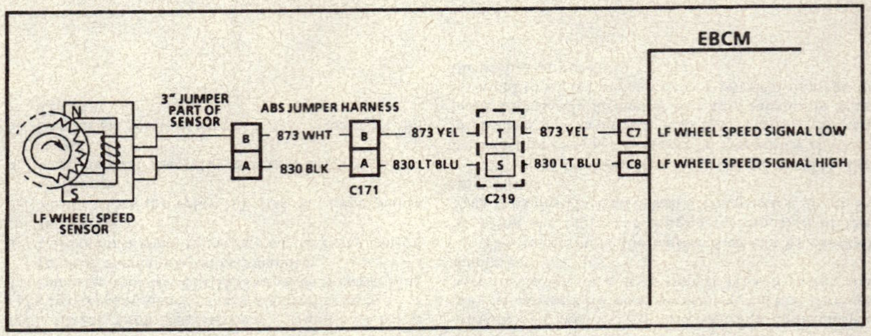

DTC 21
LEFT FRONT WHEEL SPEED = 0

Circuit Description:
As a toothed ring passes by the wheel speed sensor, changes in the electromagnetic field cause the wheel speed sensor to produce a sinusoidal (AC) voltage signal whose frequency is proportional to wheel speed. The magnitude of this signal is directly related to wheel speed and the proximity of the wheel speed sensor to the toothed ring, often referred to as the air gap.

DTC Will Set When: DTC 21 can be set when the vehicle is not in an ABS stop. If the left front wheel speed = 0 and the vehicle's reference speed is greater than 8 km/h (5 mph), a malfunction exists.

Action Taken: A malfunction DTC is stored, ABS is disabled and the ABS warning lamp is turned "ON."

DTC Chart Test Description: Number(s) below refer to circled number(s) on the diagnostic chart.

1. This step identifies a wheel speed sensor or circuitry that is damaged and visibly apparent.
2. Checks to see that the wheel speed sensor coil is within proper resistance values.
3. Verifies that the wheel speed sensor can generate an appropriate output signal.
4. This ensures that the wheel speed circuitry is not shorted together.
5. Ensures that the malfunction was not due to physical damage.

Diagnostic Aids: An "Intermittent" malfunction may be caused by a poor connection, rubbed through wire insulation, or a wire that is broken inside the insulation.

The frequency of the malfunction can be checked by using the enhanced diagnostic function of the Tech 1, as described in "Tech 1 Diagnostics" found in this section.

If the customer's comments reflect that the ABS warning lamp is "ON" only during moist environmental changes (rain, snow, vehicle wash), all wheel speed sensor circuitry should be thoroughly inspected for signs of water intrusion. If DTC is not current, clear DTCs and simulate the effects of water intrusion. Use the following procedure. Spray down the suspected area with a 5% salt water solution (two teaspoons of salt to 12 oz. of water). Test drive vehicle over various road surfaces (bumps, turns, etc.) above 24 km/h (15 mph) for at least 30 seconds. If DTC returns, replace suspected harness.

Any circuitry, that is suspected as causing the intermittent complaint, should be thoroughly checked for backed out terminals, improper mating, broken locks, improperly formed or damaged terminals, poor terminal to wiring connections or physical damage to the wiring harness.

Resistance of the wheel speed sensor will increase with an increase in sensor temperature.

When replacing a wheel speed sensor, inspect the sensor terminals and harness connector for corrosion and/or water intrusion. If evidence of corrosion or water intrusion exists, replace wheel speed sensor jumper harness. Likewise, if replacing a wheel speed sensor jumper harness, inspect sensor terminals. If evidence of corrosion or water intrusion exists, replace wheel speed sensor. Refer to "On-Vehicle Service" in this section.

DIAGNOSTIC CHART DTC 21 — 1994 GRAND PRIX, CONTINUED

DTC 21
LEFT FRONT WHEEL SPEED = 0

Important
- WHEEL SPEED SENSOR INTERMITTENT MALFUNCTIONS MAY BE DIFFICULT TO LOCATE. CARE SHOULD BE TAKEN NOT TO DISTURB ANY ELECTRICAL CONNECTIONS PRIOR TO AN INDICATED STEP OF THIS CHART. THIS WILL ENSURE THAT AN INTERMITTENT CONNECTION WILL NOT BE CORRECTED BEFORE THE SOURCE OF THE MALFUNCTION IS FOUND.

(1)
- IGNITION "OFF."
- PHYSICALLY INSPECT THE LEFT FRONT WHEEL SPEED SENSOR, LEFT FRONT WHEEL SPEED SENSOR JUMPER HARNESS, AND CONNECTORS FOR DAMAGE. INSPECT THE WHEEL SPEED SENSOR AND TOOTHED RING FOR LOOSENESS OR DAMAGE. PAY PARTICULAR ATTENTION TO THE TOOTHED RING, ANY DEVIATION WILL AFFECT THE WHEEL SPEED SENSOR OUTPUT SIGNAL.
  IS PHYSICAL DAMAGE INDICATED?

YES: REPLACE WHEEL SPEED SENSOR AND/OR JUMPER HARNESS IF DAMAGED. REPAIR ANY SENSOR OR TOOTHED RING LOOSENESS. DRIVE VEHICLE ABOVE 24 km/h (15 MPH) TO ENSURE DTC 21 DOES NOT SET.

NO:

(2)
- DISCONNECT LEFT FRONT WHEEL SPEED SENSOR DIRECTLY AT SENSOR.
- USING J 39200, MEASURE THE RESISTANCE BETWEEN SENSOR TERMINALS "A" AND "B."
  IS RESISTANCE BETWEEN 1661 AND 1929Ω WHEN SENSOR IS 20°C (68°F)?
  (IF TEMPERATURE OF THE SENSOR IS SUSPECTED TO BE OTHER THAN 20°C (68°F), REFER TO THE TABLE BELOW FOR APPLICABLE SENSOR RESISTANCE VALUES; THE VALUES IN THE TABLE ARE FOR TEMPERATURE OF THE SENSOR, NOT AIR TEMPERATURE.)

NO: REPLACE WHEEL SPEED SENSOR.

YES:

(3)
- WITH J 39200 STILL CONNECTED, SELECT A/C VOLTAGE SCALE.
- SPIN WHEEL BY HAND WHILE OBSERVING VOLTAGE.
  IS VOLTAGE AT LEAST 100 mV (VOLTAGE WILL INCREASE AS WHEEL SPEED INCREASES)?

NO: VERIFY THAT TOOTHED RING IS INTACT. IF OK, REPLACE WHEEL SPEED SENSOR

YES:

(4)
- IGNITION "OFF."
- DISCONNECT 32-WAY EBCM CONNECTOR.
- USING J 39200, MEASURE THE RESISTANCE BETWEEN TERMINAL "C7" AND "C8" OF THE 32-WAY EBCM HARNESS CONNECTOR.
  DOES J 39200 DISPLAY RESISTANCE AS "OL" (INFINITE)?

NO: REPAIR SHORT BETWEEN CKTs 830 AND 873.

YES:

(5)
- INSPECT CKTs 830 AND 873 AND 32-WAY EBCM CONNECTOR FOR DAMAGE WHICH MAY RESULT IN A SHORT BETWEEN CKTs 830 AND 873 WITH ALL CONNECTORS CONNECTED. REPAIR ALL DAMAGE IF EVIDENT.
- RECONNECT ALL CONNECTORS.
- TEST DRIVE VEHICLE ABOVE 24 km/h (15 MPH) FOR AT LEAST 30 SECONDS.
  DOES DTC 21 SET AS A CURRENT DTC?

YES: REPLACE EBCM.

NO: MALFUNCTION IS INTERMITTENT. REFER TO "DIAGNOSTIC AIDS" ON FACING PAGE.

FRONT WHEEL SPEED SENSOR
TEMPERATURE OF SENSOR VS. SENSOR RESISTANCE (APPROXIMATE)

TEMP. (°C)	TEMP. (°F)	RESISTANCE (OHMS)
-34 TO 4	-30 TO 40	1309 TO 1808
5 TO 43	41 TO 110	1557 TO 2103
44 TO 93	111 TO 200	1811 TO 2482

DIAGNOSTIC CHART DTC 22 — 1994 GRAND PRIX

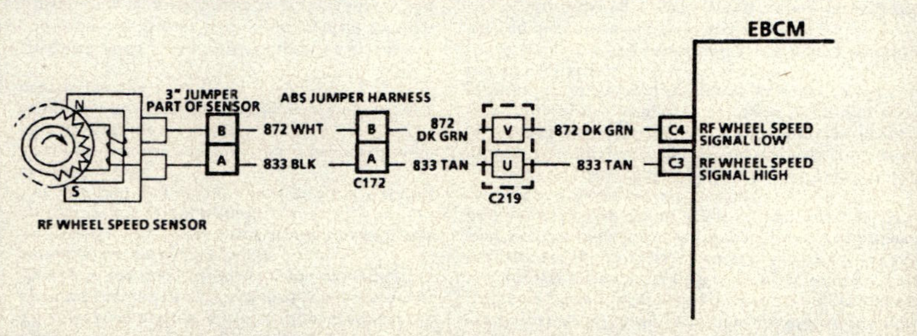

DTC 22
RIGHT FRONT WHEEL SPEED = 0

Circuit Description:
As a toothed ring passes by the wheel speed sensor, changes in the electromagnetic field cause the wheel speed sensor to produce a sinusoidal (AC) voltage signal whose frequency is proportional to wheel speed. The magnitude of this signal is directly related to wheel speed and the proximity of the wheel speed sensor to the toothed ring, often referred to as the air gap.

DTC Will Set When: DTC 22 can be set when the vehicle is not in an ABS stop. If the right front wheel speed = 0 and the vehicle's reference speed is greater than 8 km/h (5 mph), a malfunction exists.

Action Taken: A malfunction DTC is stored, ABS is disabled and the ABS warning lamp is turned "ON."

DTC Chart Test Description: Number(s) below refer to circled number(s) on the diagnostic chart.
1. This step identifies a wheel speed sensor or circuitry that is damaged and visibly apparent.
2. Checks to see that the wheel speed sensor coil is within proper resistance values.
3. Verifies that the wheel speed sensor can generate an appropriate output signal.
4. This ensures that the wheel speed circuitry is not shorted together.
5. Ensures that the malfunction was not due to physical damage.

Diagnostic Aids: An "Intermittent" malfunction may be caused by a poor connection, rubbed through wire insulation, or a wire that is broken inside the insulation.

The frequency of the malfunction can be checked by using the enhanced diagnostic function of the Tech 1, as described in "Tech 1 Diagnostics" found in this section.

If the customer's comments reflect that the ABS warning lamp is "ON" only during moist environmental changes (rain, snow, vehicle wash), all wheel speed sensor circuitry should be thoroughly inspected for signs of water intrusion. If DTC is not current, clear DTCs and simulate the effects of water intrusion. Use the following procedure. Spray down the suspected area with a 5% salt water solution (two teaspoons of salt to 12 oz. of water). Test drive vehicle over various road surfaces (bumps, turns, etc.) above 24 km/h (15 mph) for at least 30 seconds. If DTC returns, replace suspected harness.

Any circuitry, that is suspected as causing the intermittent complaint, should be thoroughly checked for backed out terminals, improper mating, broken locks, improperly formed or damaged terminals, poor terminal to wiring connections or physical damage to the wiring harness.

Resistance of the wheel speed sensor will increase with an increase in sensor temperature.

When replacing a wheel speed sensor, inspect the sensor terminals and harness connector for corrosion and/or water intrusion. If evidence of corrosion or water intrusion exists, replace wheel speed sensor jumper harness. Likewise, if replacing a wheel speed sensor jumper harness, inspect sensor terminals. If evidence of corrosion or water intrusion exists, replace wheel speed sensor. Refer to "On-Vehicle Service" in this section.

DIAGNOSTIC CHART DTC 22 — 1994 GRAND PRIX, CONTINUED

DTC 22
RIGHT FRONT WHEEL SPEED = 0

Important
- WHEEL SPEED SENSOR INTERMITTENT MALFUNCTIONS MAY BE DIFFICULT TO LOCATE. CARE SHOULD BE TAKEN NOT TO DISTURB ANY ELECTRICAL CONNECTIONS PRIOR TO AN INDICATED STEP OF THIS CHART. THIS WILL ENSURE THAT AN INTERMITTENT CONNECTION WILL NOT BE CORRECTED BEFORE THE SOURCE OF THE MALFUNCTION IS FOUND.

(1)
- IGNITION "OFF."
- PHYSICALLY INSPECT THE RIGHT FRONT WHEEL SPEED SENSOR, RIGHT FRONT WHEEL SPEED SENSOR JUMPER HARNESS, AND CONNECTORS FOR DAMAGE. INSPECT THE WHEEL SPEED SENSOR AND TOOTHED RING FOR LOOSENESS OR DAMAGE. PAY PARTICULAR ATTENTION TO THE TOOTHED RING, ANY DEVIATION WILL AFFECT THE WHEEL SPEED SENSOR OUTPUT SIGNAL. IS PHYSICAL DAMAGE INDICATED?

YES: REPLACE WHEEL SPEED SENSOR AND/OR JUMPER HARNESS IF DAMAGED. REPAIR ANY SENSOR OR TOOTHED RING LOOSENESS. DRIVE VEHICLE ABOVE 24 km/h (15 MPH) TO ENSURE DTC 22 DOES NOT SET.

NO:

(2)
- DISCONNECT RIGHT FRONT WHEEL SPEED SENSOR DIRECTLY AT SENSOR.
- USING J 39200, MEASURE THE RESISTANCE BETWEEN SENSOR TERMINALS "A" AND "B."
IS RESISTANCE BETWEEN 1661 AND 1929Ω WHEN SENSOR IS 20°C (68°F)?
(IF TEMPERATURE OF THE SENSOR IS SUSPECTED TO BE OTHER THAN 20°C (68°F), REFER TO THE TABLE BELOW FOR APPLICABLE SENSOR RESISTANCE VALUES; THE VALUES IN THE TABLE ARE FOR TEMPERATURE OF THE SENSOR, NOT AIR TEMPERATURE.)

NO: REPLACE WHEEL SPEED SENSOR.

YES:

(3)
- WITH J 39200 STILL CONNECTED, SELECT A/C VOLTAGE SCALE.
- SPIN WHEEL BY HAND WHILE OBSERVING VOLTAGE.
IS VOLTAGE AT LEAST 100 mV (VOLTAGE WILL INCREASE AS WHEEL SPEED INCREASES)?

NO: VERIFY THAT TOOTHED RING IS INTACT. IF OK, REPLACE WHEEL SPEED SENSOR

YES:

(4)
- IGNITION "OFF."
- DISCONNECT 32-WAY EBCM CONNECTOR.
- USING J 39200, MEASURE THE RESISTANCE BETWEEN TERMINAL "C3" AND "C4" OF THE 32-WAY EBCM HARNESS CONNECTOR. DOES J 39200 DISPLAY RESISTANCE AS "OL" (INFINITE)?

NO: REPAIR SHORT BETWEEN CKTs 833 AND 872.

YES:

(5)
- INSPECT CKTs 833 AND 872 AND 32-WAY EBCM CONNECTOR FOR DAMAGE WHICH MAY RESULT IN A SHORT BETWEEN CKTs 833 AND 872 WITH ALL CONNECTORS CONNECTED. REPAIR ALL DAMAGE IF EVIDENT.
- RECONNECT ALL CONNECTORS.
- TEST DRIVE VEHICLE ABOVE 24 km/h (15 MPH) FOR AT LEAST 30 SECONDS. DOES DTC 22 SET AS A CURRENT DTC?

YES: REPLACE EBCM.

NO: MALFUNCTION IS INTERMITTENT. REFER TO "DIAGNOSTIC AIDS" ON FACING PAGE.

FRONT WHEEL SPEED SENSOR
TEMPERATURE OF SENSOR VS. SENSOR RESISTANCE (APPROXIMATE)

TEMP. (°C)	TEMP. (°F)	RESISTANCE (OHMS)
-34 TO 4	-30 TO 40	1309 TO 1808
5 TO 43	41 TO 110	1557 TO 2103
44 TO 93	111 TO 200	1811 TO 2482

DIAGNOSTIC CHART DTC 23 — 1994 GRAND PRIX

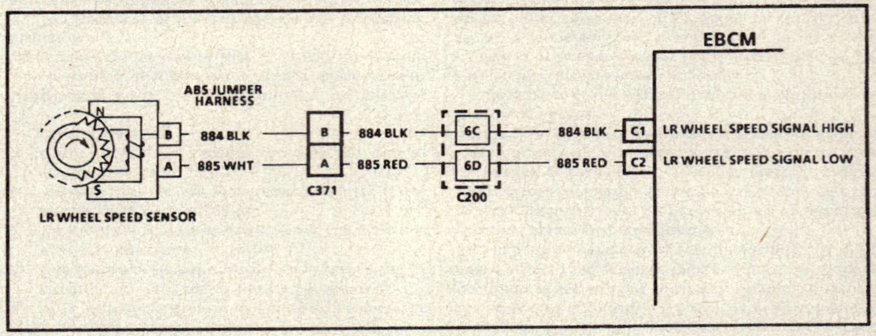

DTC 23
LEFT REAR WHEEL SPEED = 0

Circuit Description:
As a toothed ring passes by the wheel speed sensor, changes in the electromagnetic field cause the wheel speed sensor to produce a sinusoidal (AC) voltage signal whose frequency is proportional to wheel speed. The magnitude of this signal is directly related to wheel speed and the proximity of the wheel speed sensor to the toothed ring, often referred to as the air gap.

DTC Will Set When: DTC 23 can be set when the vehicle is not in an ABS stop. If the left rear wheel speed = 0 and the vehicle's reference speed is greater than 8 km/h (5 mph), a malfunction exists.

Action Taken: A malfunction DTC is stored, ABS is disabled and the ABS warning lamp is turned "ON."

DTC Chart Test Description: Number(s) below refer to circled number(s) on the diagnostic chart.
1. This step identifies a wheel speed sensor or circuitry that is damaged and visibly apparent.
2. Checks to see that the wheel speed sensor coil is within proper resistance values.
3. Verifies that the wheel speed sensor can generate an appropriate output signal.
4. Verifies that the wheel speed circuitry is not shorted together.
5. Ensures that the malfunction was not due to physical damage.
6. Checks for an incorrect or malfunctioning hub and bearing assembly.

Diagnostic Aids: An "Intermittent" malfunction may be caused by a poor connection, rubbed through wire insulation, or a wire that is broken inside the insulation.

The frequency of the malfunction can be checked by using the enhanced diagnostic function of the Tech 1, as described in "Tech 1 Diagnostics" found in this section.

If the customer's comments reflect that the ABS warning lamp is "ON" only during moist environmental changes (rain, snow, vehicle wash), all wheel speed sensor circuitry should be thoroughly inspected for signs of water intrusion. If DTC is not current, clear DTCs and simulate the effects of water intrusion. Use the following procedure. Spray down the suspected area with a 5% salt water solution (two teaspoons of salt to 12 oz. of water). Test drive vehicle over various road surfaces (bumps, turns, etc.) above 24 km/h (15 mph) for at least 30 seconds. If DTC returns, replace suspected harness.

Any circuitry, that is suspected as causing the intermittent complaint, should be thoroughly checked for backed out terminals, improper mating, broken locks, improperly formed or damaged terminals, poor terminal to wiring connections or physical damage to the wiring harness.

Resistance of the wheel speed sensor will increase with an increase in sensor temperature.

When replacing a wheel speed sensor, inspect the sensor terminals and harness connector for corrosion and/or water intrusion. If evidence of corrosion or water intrusion exists, replace wheel speed sensor jumper harness. Likewise, if replacing a wheel speed sensor jumper harness, inspect sensor terminals. If evidence of corrosion or water intrusion exists, replace wheel speed sensor. Refer to "On-Vehicle Service" in this section.

DIAGNOSTIC CHART DTC 23 — 1994 GRAND PRIX, CONTINUED

DTC 23
LEFT REAR WHEEL SPEED = 0

Important
- WHEEL SPEED SENSOR INTERMITTENT MALFUNCTIONS MAY BE DIFFICULT TO LOCATE. CARE SHOULD BE TAKEN NOT TO DISTURB ANY ELECTRICAL CONNECTIONS PRIOR TO AN INDICATED STEP OF THIS CHART. THIS WILL ENSURE THAT AN INTERMITTENT CONNECTION WILL NOT BE CORRECTED BEFORE THE SOURCE OF THE MALFUNCTION IS FOUND.

(1)
- IGNITION "OFF."
- PHYSICALLY INSPECT THE LEFT REAR WHEEL SPEED SENSOR, WHEEL SPEED SENSOR JUMPER HARNESS, AND CONNECTORS FOR DAMAGE. IS PHYSICAL DAMAGE INDICATED?

NO → go to step 2. YES → REPLACE WHEEL SPEED SENSOR AND/OR WHEEL SPEED SENSOR JUMPER HARNESS IF DAMAGED. DRIVE VEHICLE ABOVE 24 km/h (15 MPH) TO ENSURE DTC 23 DOES NOT SET.

(2)
- DISCONNECT LEFT REAR WHEEL SPEED SENSOR DIRECTLY AT SENSOR.
- USING J 39200, MEASURE THE RESISTANCE BETWEEN SENSOR TERMINALS "A" AND "B." IS RESISTANCE BETWEEN 2059 AND 2353Ω WHEN SENSOR IS 20°C (68°F)? *(IF TEMPERATURE OF THE SENSOR IS SUSPECTED TO BE OTHER THAN 20°C (68°F), REFER TO THE TABLE BELOW FOR APPLICABLE SENSOR RESISTANCE VALUES; THE VALUES IN THE TABLE ARE FOR TEMPERATURE OF THE SENSOR, NOT AIR TEMPERATURE.)*

YES → go to step 3. NO → REPLACE WHEEL SPEED SENSOR.

(3)
- WITH J 39200 STILL CONNECTED, SELECT A/C VOLTAGE SCALE.
- SPIN WHEEL BY HAND WHILE OBSERVING VOLTAGE. IS VOLTAGE AT LEAST 100 mV (VOLTAGE WILL INCREASE AS WHEEL SPEED INCREASES)?

YES → go to step 4. NO → REPLACE WHEEL SPEED SENSOR.

(4)
- IGNITION "OFF."
- DISCONNECT 32-WAY EBCM CONNECTOR.
- USING J 39200, MEASURE THE RESISTANCE BETWEEN TERMINAL "C1" AND "C2" OF THE 32-WAY EBCM HARNESS CONNECTOR. DOES J 39200 DISPLAY RESISTANCE AS "OL" (INFINITE)?

YES → go to step 5. NO → REPAIR SHORT BETWEEN CKTs 884 AND 885.

(5)
- INSPECT CKTs 884 AND 885 AND 32-WAY EBCM CONNECTOR FOR DAMAGE WHICH MAY RESULT IN A SHORT BETWEEN CKTs 884 AND 885 WITH ALL CONNECTORS CONNECTED. REPAIR ALL DAMAGE IF EVIDENT.
- RECONNECT ALL CONNECTORS.
- TEST DRIVE VEHICLE ABOVE 24 km/h (15 MPH) FOR AT LEAST 30 SECONDS. DOES DTC 23 SET AS A CURRENT DTC?

YES → go to step 6. NO → MALFUNCTION WAS MOST LIKELY CAUSED BY A POOR CONNECTION. REFER TO "DIAGNOSTIC AIDS" ON FACING PAGE.

(6)
- USING TECH 1, SELECT "F0: DATA LIST."
- MONITOR WHEEL SPEEDS WHILE SLOWLY ACCELERATING TO 65 km/h (40 MPH) AND THEN SLOWLY DECELERATING TO 0 km/h (0 MPH). IS THE LEFT REAR WHEEL SPEED CONSTANTLY HIGHER THAN THE THREE REMAINING WHEEL SPEEDS?

YES → REPLACE INCORRECT OR MALFUNCTIONING LEFT WHEEL SPEED SENSOR/HUB AND BEARING ASSEMBLY WITH THE PROPER ASSEMBLY. REFER TO PARTS CATALOG FOR PROPER PART NUMBER. NO → REPLACE EBCM.

REAR WHEEL SPEED SENSOR
TEMPERATURE OF SENSOR VS. SENSOR RESISTANCE (APPROXIMATE)

TEMP. (°C)	TEMP. (°F)	RESISTANCE (OHMS)
-34 TO 4	-30 TO 40	1622 TO 2205
5 TO 43	41 TO 110	1930 TO 2566
44 TO 93	111 TO 200	2245 TO 3028

DIAGNOSTIC CHART DTC 24 — 1994 GRAND PRIX

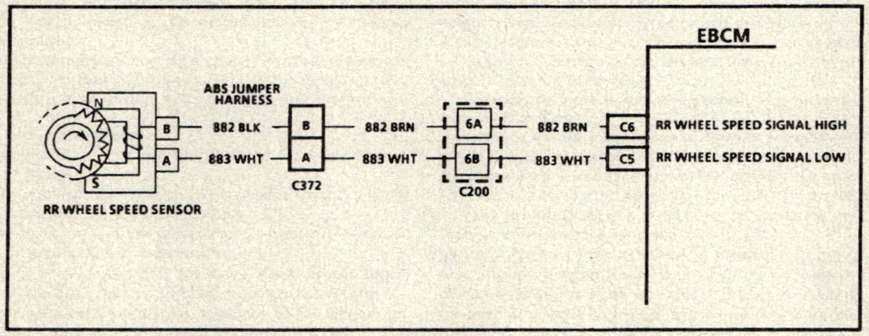

DTC 24
RIGHT REAR WHEEL SPEED = 0

Circuit Description:
As a toothed ring passes by the wheel speed sensor, changes in the electromagnetic field cause the wheel speed sensor to produce a sinusoidal (AC) voltage signal whose frequency is proportional to wheel speed. The magnitude of this signal is directly related to wheel speed and the proximity of the wheel speed sensor to the toothed ring, often referred to as the air gap.

DTC Will Set When: DTC 24 can be set when the vehicle is not in an ABS stop. If the right rear wheel speed = 0 and the vehicle's reference speed is greater than 8 km/h (5 mph), a malfunction exists.

Action Taken: A malfunction DTC is stored, ABS is disabled and the ABS warning lamp is turned "ON."

DTC Chart Test Description: Number(s) below refer to circled number(s) on the diagnostic chart.
1. This step identifies a wheel speed sensor or circuitry that is damaged and visibly apparent.
2. Checks to see that the wheel speed sensor coil is within proper resistance values.
3. Verifies that the wheel speed sensor can generate an appropriate output signal.
4. Verifies that the wheel speed circuitry is not shorted together.
5. Ensures that the malfunction was not due to physical damage.
6. Checks for an incorrect or malfunctioning hub and bearing assembly.

Diagnostic Aids: An "Intermittent" malfunction may be caused by a poor connection, rubbed through wire insulation, or a wire that is broken inside the insulation.

The frequency of the malfunction can be checked by using the enhanced diagnostic function of the Tech 1, as described in "Tech 1 Diagnostics" found in this section.

If the customer's comments reflect that the ABS warning lamp is "ON" only during moist environmental changes (rain, snow, vehicle wash), all wheel speed sensor circuitry should be thoroughly inspected for signs of water intrusion. If DTC is not current, clear DTCs and simulate the effects of water intrusion. Use the following procedure. Spray down the suspected area with a 5% salt water solution (two teaspoons of salt to 12 oz. of water). Test drive vehicle over various road surfaces (bumps, turns, etc.) above 24 km/h (15 mph) for at least 30 seconds. If DTC returns, replace suspected harness.

Any circuitry, that is suspected as causing the intermittent complaint, should be thoroughly checked for backed out terminals, improper mating, broken locks, improperly formed or damaged terminals, poor terminal to wiring connections or physical damage to the wiring harness.

Resistance of the wheel speed sensor will increase with an increase in sensor temperature.

When replacing a wheel speed sensor, inspect the sensor terminals and harness connector for corrosion and/or water intrusion. If evidence of corrosion or water intrusion exists, replace wheel speed sensor jumper harness. Likewise, if replacing a wheel speed sensor jumper harness, inspect sensor terminals. If evidence of corrosion or water intrusion exists, replace wheel speed sensor. Refer to "On-Vehicle Service" in this section.

DIAGNOSTIC CHART DTC 24 — 1994 GRAND PRIX, CONTINUED

DTC 24
RIGHT REAR WHEEL SPEED = 0

Important

- WHEEL SPEED SENSOR INTERMITTENT MALFUNCTIONS MAY BE DIFFICULT TO LOCATE. CARE SHOULD BE TAKEN NOT TO DISTURB ANY ELECTRICAL CONNECTIONS PRIOR TO AN INDICATED STEP OF THIS CHART. THIS WILL ENSURE THAT AN INTERMITTENT CONNECTION WILL NOT BE CORRECTED BEFORE THE SOURCE OF THE MALFUNCTION IS FOUND.

(1)
- IGNITION "OFF."
- PHYSICALLY INSPECT THE RIGHT REAR WHEEL SPEED SENSOR, WHEEL SPEED SENSOR JUMPER HARNESS, AND CONNECTORS FOR DAMAGE. IS PHYSICAL DAMAGE INDICATED?

YES: REPLACE WHEEL SPEED SENSOR AND/OR WHEEL SPEED SENSOR JUMPER HARNESS IF DAMAGED. DRIVE VEHICLE ABOVE 24 km/h (15 MPH) TO ENSURE DTC 24 DOES NOT SET.

NO:

(2)
- DISCONNECT RIGHT REAR WHEEL SPEED SENSOR DIRECTLY AT SENSOR.
- USING J 39200, MEASURE THE RESISTANCE BETWEEN SENSOR TERMINALS "A" AND "B." IS RESISTANCE BETWEEN 2059 AND 2353Ω WHEN SENSOR IS 20°C (68°F)? *(IF TEMPERATURE OF THE SENSOR IS SUSPECTED TO BE OTHER THAN 20°C (68°F), REFER TO THE TABLE BELOW FOR APPLICABLE SENSOR RESISTANCE VALUES; THE VALUES IN THE TABLE ARE FOR TEMPERATURE OF THE SENSOR, NOT AIR TEMPERATURE.)*

NO: REPLACE WHEEL SPEED SENSOR.

YES:

(3)
- WITH J 39200 STILL CONNECTED, SELECT A/C VOLTAGE SCALE.
- SPIN WHEEL BY HAND WHILE OBSERVING VOLTAGE. IS VOLTAGE AT LEAST 100 mV (VOLTAGE WILL INCREASE AS WHEEL SPEED INCREASES)?

NO: REPLACE WHEEL SPEED SENSOR.

YES:

(4)
- IGNITION "OFF."
- DISCONNECT 32-WAY EBCM CONNECTOR.
- USING J 39200, MEASURE THE RESISTANCE BETWEEN TERMINAL "C5" AND "C6" OF THE 32-WAY EBCM HARNESS CONNECTOR. DOES J 39200 DISPLAY RESISTANCE AS "OL" (INFINITE)?

NO: REPAIR SHORT BETWEEN CKTs 883 AND 883.

YES:

(5)
- INSPECT CKTs 882 AND 883 AND 32-WAY EBCM CONNECTOR FOR DAMAGE WHICH MAY RESULT IN A SHORT BETWEEN CKTs 882 AND 883 WITH ALL CONNECTORS CONNECTED. REPAIR ALL DAMAGE IF EVIDENT.
- RECONNECT ALL CONNECTORS.
- TEST DRIVE VEHICLE ABOVE 24 km/h (15 MPH) FOR AT LEAST 30 SECONDS. DOES DTC 24 SET AS A CURRENT DTC?

NO: MALFUNCTION WAS MOST LIKELY CAUSED BY A POOR CONNECTION. REFER TO "DIAGNOSTIC AIDS" ON FACING PAGE.

YES:

(6)
- USING TECH 1, SELECT "F0: DATA LIST."
- MONITOR WHEEL SPEEDS WHILE SLOWLY ACCELERATING TO 65 km/h (40 MPH) AND THEN SLOWLY DECELERATING TO 0 km/h (0 MPH). IS THE RIGHT REAR WHEEL SPEED CONSTANTLY HIGHER THAN THE THREE REMAINING WHEEL SPEEDS?

NO: REPLACE EBCM.

YES: REPLACE INCORRECT OR MALFUNCTIONING RIGHT WHEEL SPEED SENSOR/HUB AND BEARING ASSEMBLY WITH THE PROPER ASSEMBLY. REFER TO PARTS CATALOG FOR PROPER PART NUMBER.

REAR WHEEL SPEED SENSOR
TEMPERATURE OF SENSOR VS. SENSOR RESISTANCE (APPROXIMATE)

TEMP. (°C)	TEMP. (°F)	RESISTANCE (OHMS)
-34 TO 4	-30 TO 40	1622 TO 2205
5 TO 43	41 TO 110	1930 TO 2566
44 TO 93	111 TO 200	2245 TO 3028

DIAGNOSTIC CHART DTC 25 — 1994 GRAND PRIX

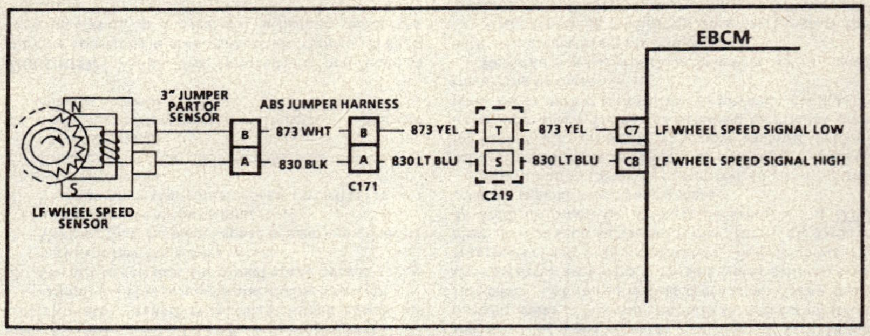

DTC 25

LEFT FRONT EXCESSIVE WHEEL SPEED VARIATION

Circuit Description:
As a toothed ring passes by the wheel speed sensor, changes in the electromagnetic field cause the wheel speed sensor to produce a sinusoidal (AC) voltage signal whose frequency is proportional to wheel speed. The magnitude of this signal is directly related to wheel speed and the proximity of the wheel speed sensor to the toothed ring, often referred to as the air gap.

DTC Will Set When: DTC 25 can be set when the brake is "OFF." The purpose of this test is to detect a situation in which the left front wheel acceleration or deceleration is beyond specified limits.

Action Taken: A malfunction DTC is stored, ABS is disabled and the ABS warning lamp is turned "ON."

DTC Chart Test Description: Number(s) below refer to circled number(s) on the diagnostic chart.

1. This step identifies a wheel speed sensor or circuitry that is damaged and visibly apparent.
2. Checks to see that the wheel speed sensor coil is within proper resistance values.
3. Verifies that the wheel speed sensor can generate an appropriate output signal.
4. This ensures that the wheel speed circuitry is not shorted together.
5. Ensures that the malfunction was not due to physical damage.

Diagnostic Aids: An "Intermittent" malfunction may be caused by a poor connection, rubbed through wire insulation, or a wire that is broken inside the insulation.

The frequency of the malfunction can be checked by using the enhanced diagnostic function of the Tech 1, as described in "Tech 1 Diagnostics" found in this section.

If the customer's comments reflect that the ABS warning lamp is "ON" only during moist environmental changes (rain, snow, vehicle wash), all wheel speed sensor circuitry should be thoroughly inspected for signs of water intrusion. If DTC is not current, clear DTCs and simulate the effects of water intrusion. Use the following procedure. Spray down the suspected area with a 5% salt water solution (two teaspoons of salt to 12 oz. of water). Test drive vehicle over various road surfaces (bumps, turns, etc.) above 24 km/h (15 mph) for at least 30 seconds. If DTC returns, replace suspected harness.

Any circuitry, that is suspected as causing the intermittent complaint, should be thoroughly checked for backed out terminals, improper mating, broken locks, improperly formed or damaged terminals, poor terminal to wiring connections or physical damage to the wiring harness.

Resistance of the wheel speed sensor will increase with an increase in sensor temperature.

When replacing a wheel speed sensor, inspect the sensor terminals and harness connector for corrosion and/or water intrusion. If evidence of corrosion or water intrusion exists, replace wheel speed sensor jumper harness. Likewise, if replacing a wheel speed sensor jumper harness, inspect sensor terminals. If evidence of corrosion or water intrusion exists, replace wheel speed sensor. Refer to "On-Vehicle Service" in this section.

DIAGNOSTIC CHART DTC 25 — 1994 GRAND PRIX, CONTINUED

DTC 25
LEFT FRONT EXCESSIVE WHEEL SPEED VARIATION

Important

- WHEEL SPEED SENSOR INTERMITTENT MALFUNCTIONS MAY BE DIFFICULT TO LOCATE. CARE SHOULD BE TAKEN NOT TO DISTURB ANY ELECTRICAL CONNECTIONS PRIOR TO AN INDICATED STEP OF THIS CHART. THIS WILL ENSURE THAT AN INTERMITTENT CONNECTION WILL NOT BE CORRECTED BEFORE THE SOURCE OF THE MALFUNCTION IS FOUND.

(1)
- IGNITION "OFF."
- PHYSICALLY INSPECT THE LEFT FRONT WHEEL SPEED SENSOR, LEFT FRONT WHEEL SPEED SENSOR JUMPER HARNESS, AND CONNECTORS FOR DAMAGE. INSPECT THE WHEEL SPEED SENSOR AND TOOTHED RING FOR LOOSENESS OR DAMAGE. PAY PARTICULAR ATTENTION TO THE TOOTHED RING, ANY DEVIATION WILL AFFECT THE WHEEL SPEED SENSOR OUTPUT SIGNAL. IS PHYSICAL DAMAGE INDICATED?

YES: REPLACE WHEEL SPEED SENSOR AND/OR JUMPER HARNESS IF DAMAGED. REPAIR ANY SENSOR OR TOOTHED RING LOOSENESS. DRIVE VEHICLE ABOVE 24 km/h (15 MPH) TO ENSURE DTC 25 DOES NOT SET.

NO:

(2)
- DISCONNECT LEFT FRONT WHEEL SPEED SENSOR DIRECTLY AT SENSOR.
- USING J 39200, MEASURE THE RESISTANCE BETWEEN SENSOR TERMINALS "A" AND "B." IS RESISTANCE BETWEEN 1661 AND 1929Ω WHEN SENSOR IS 20°C (68°F)? *(IF TEMPERATURE OF THE SENSOR IS SUSPECTED TO BE OTHER THAN 20°C (68°F), REFER TO THE TABLE BELOW FOR APPLICABLE SENSOR RESISTANCE VALUES; THE VALUES IN THE TABLE ARE FOR TEMPERATURE OF THE SENSOR, NOT AIR TEMPERATURE.)*

NO: REPLACE WHEEL SPEED SENSOR.

YES:

(3)
- WITH J 39200 STILL CONNECTED, SELECT A/C VOLTAGE SCALE.
- SPIN WHEEL BY HAND WHILE OBSERVING VOLTAGE. IS VOLTAGE AT LEAST 100 mV (VOLTAGE WILL INCREASE AS WHEEL SPEED INCREASES)?

NO: VERIFY THAT TOOTHED RING IS INTACT. IF OK, REPLACE WHEEL SPEED SENSOR

YES:

(4)
- IGNITION "OFF."
- DISCONNECT 32-WAY EBCM CONNECTOR.
- USING J 39200, MEASURE THE RESISTANCE BETWEEN TERMINAL "C7" AND "C8" OF THE 32-WAY EBCM HARNESS CONNECTOR. DOES J 39200 DISPLAY RESISTANCE AS "OL" (INFINITE)?

NO: REPAIR SHORT BETWEEN CKTs 830 AND 873.

YES:

(5)
- INSPECT CKTs 830 AND 873 AND 32-WAY EBCM CONNECTOR FOR DAMAGE WHICH MAY RESULT IN A SHORT BETWEEN CKTs 830 AND 873 WITH ALL CONNECTORS CONNECTED. REPAIR ALL DAMAGE IF EVIDENT.
- RECONNECT ALL CONNECTORS.
- TEST DRIVE VEHICLE ABOVE 24 km/h (15 MPH) FOR AT LEAST 30 SECONDS. DOES DTC 25 SET AS A CURRENT DTC?

YES: REPLACE EBCM.

NO: MALFUNCTION IS INTERMITTENT. REFER TO "DIAGNOSTIC AIDS" ON FACING PAGE.

FRONT WHEEL SPEED SENSOR
TEMPERATURE OF SENSOR VS. SENSOR RESISTANCE (APPROXIMATE)

TEMP. (°C)	TEMP. (°F)	RESISTANCE (OHMS)
-34 TO 4	-30 TO 40	1309 TO 1808
5 TO 43	41 TO 110	1557 TO 2103
44 TO 93	111 TO 200	1811 TO 2482

DIAGNOSTIC CHART DTC 26 — 1994 GRAND PRIX

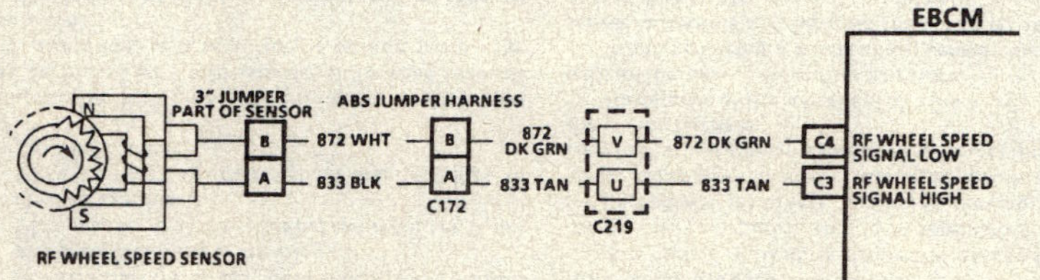

DTC 26

RIGHT FRONT EXCESSIVE WHEEL SPEED VARIATION

Circuit Description:
As a toothed ring passes by the wheel speed sensor, changes in the electromagnetic field cause the wheel speed sensor to produce a sinusoidal (AC) voltage signal whose frequency is proportional to wheel speed. The magnitude of this signal is directly related to wheel speed and the proximity of the wheel speed sensor to the toothed ring, often referred to as the air gap.

DTC Will Set When: DTC 26 can be set when the brake is "OFF." The purpose of this test is to detect a situation in which the right front wheel acceleration or deceleration is beyond specified limits.

Action Taken: A malfunction DTC is stored, ABS is disabled and the ABS warning lamp is turned "ON."

DTC Chart Test Description: Number(s) below refer to circled number(s) on the diagnostic chart.

1. This step identifies a wheel speed sensor or circuitry that is damaged and visibly apparent.
2. Checks to see that the wheel speed sensor-coil is within proper resistance values.
3. Verifies that the wheel speed sensor can generate an appropriate output signal.
4. This ensures that the wheel speed circuitry is not shorted together.
5. Ensures that the malfunction was not due to physical damage.

Diagnostic Aids: An "Intermittent" malfunction may be caused by a poor connection, rubbed through wire insulation, or a wire that is broken inside the insulation.

The frequency of the malfunction can be checked by using the enhanced diagnostic function of the Tech 1, as described in "Tech 1 Diagnostics" found in this section.

If the customer's comments reflect that the ABS warning lamp is "ON" only during moist environmental changes (rain, snow, vehicle wash), all wheel speed sensor circuitry should be thoroughly inspected for signs of water intrusion. If DTC is not current, clear DTCs and simulate the effects of water intrusion. Use the following procedure. Spray down the suspected area with a 5% salt water solution (two teaspoons of salt to 12 oz. of water). Test drive vehicle over various road surfaces (bumps, turns, etc.) above 24 km/h (15 mph) for at least 30 seconds. If DTC returns, replace suspected harness.

Any circuitry, that is suspected as causing the intermittent complaint, should be thoroughly checked for backed out terminals, improper mating, broken locks, improperly formed or damaged terminals, poor terminal to wiring connections or physical damage to the wiring harness.

Resistance of the wheel speed sensor will increase with an increase in sensor temperature.

When replacing a wheel speed sensor, inspect the sensor terminals and harness connector for corrosion and/or water intrusion. If evidence of corrosion or water intrusion exists, replace wheel speed sensor jumper harness. Likewise, if replacing a wheel speed sensor jumper harness, inspect sensor terminals. If evidence of corrosion or water intrusion exists, replace wheel speed sensor. Refer to "On-Vehicle Service" in this section.

DIAGNOSTIC CHART DTC 26 — 1994 GRAND PRIX, CONTINUED

DTC 26
RIGHT FRONT EXCESSIVE WHEEL SPEED VARIATION

Important

- WHEEL SPEED SENSOR INTERMITTENT MALFUNCTIONS MAY BE DIFFICULT TO LOCATE. CARE SHOULD BE TAKEN NOT TO DISTURB ANY ELECTRICAL CONNECTIONS PRIOR TO AN INDICATED STEP OF THIS CHART. THIS WILL ENSURE THAT AN INTERMITTENT CONNECTION WILL NOT BE CORRECTED BEFORE THE SOURCE OF THE MALFUNCTION IS FOUND.

(1)
- IGNITION "OFF."
- PHYSICALLY INSPECT THE RIGHT FRONT WHEEL SPEED SENSOR, RIGHT FRONT WHEEL SPEED SENSOR JUMPER HARNESS, AND CONNECTORS FOR DAMAGE. INSPECT THE WHEEL SPEED SENSOR AND TOOTHED RING FOR LOOSENESS OR DAMAGE. PAY PARTICULAR ATTENTION TO THE TOOTHED RING, ANY DEVIATION WILL AFFECT THE WHEEL SPEED SENSOR OUTPUT SIGNAL.

IS PHYSICAL DAMAGE INDICATED?

YES: REPLACE WHEEL SPEED SENSOR AND/OR JUMPER HARNESS IF DAMAGED. REPAIR ANY SENSOR OR TOOTHED RING LOOSENESS. DRIVE VEHICLE ABOVE 24 km/h (15 MPH) TO ENSURE DTC 26 DOES NOT SET.

NO:

(2)
- DISCONNECT RIGHT FRONT WHEEL SPEED SENSOR DIRECTLY AT SENSOR.
- USING J 39200, MEASURE THE RESISTANCE BETWEEN SENSOR TERMINALS "A" AND "B."

IS RESISTANCE BETWEEN 1661 AND 1929Ω WHEN SENSOR IS 20°C (68°F)? *(IF TEMPERATURE OF THE SENSOR IS SUSPECTED TO BE OTHER THAN 20°C (68°F), REFER TO THE TABLE BELOW FOR APPLICABLE SENSOR RESISTANCE VALUES; THE VALUES IN THE TABLE ARE FOR TEMPERATURE OF THE SENSOR, NOT AIR TEMPERATURE.)*

NO: REPLACE WHEEL SPEED SENSOR.

YES:

(3)
- WITH J 39200 STILL CONNECTED, SELECT A/C VOLTAGE SCALE.
- SPIN WHEEL BY HAND WHILE OBSERVING VOLTAGE.

IS VOLTAGE AT LEAST 100 mV (VOLTAGE WILL INCREASE AS WHEEL SPEED INCREASES)?

NO: VERIFY THAT TOOTHED RING IS INTACT. IF OK, REPLACE WHEEL SPEED SENSOR

YES:

(4)
- IGNITION "OFF."
- DISCONNECT 32-WAY EBCM CONNECTOR.
- USING J 39200, MEASURE THE RESISTANCE BETWEEN TERMINAL "C3" AND "C4" OF THE 32-WAY EBCM HARNESS CONNECTOR.

DOES J 39200 DISPLAY RESISTANCE AS "OL" (INFINITE)?

NO: REPAIR SHORT BETWEEN CKTs 833 AND 872.

YES:

(5)
- INSPECT CKTs 833 AND 872 AND 32-WAY EBCM CONNECTOR FOR DAMAGE WHICH MAY RESULT IN A SHORT BETWEEN CKTs 833 AND 872 WITH ALL CONNECTORS CONNECTED. REPAIR ALL DAMAGE IF EVIDENT.
- RECONNECT ALL CONNECTORS.
- TEST DRIVE VEHICLE ABOVE 24 km/h (15 MPH) FOR AT LEAST 30 SECONDS.

DOES DTC 26 SET AS A CURRENT DTC?

YES: REPLACE EBCM.

NO: MALFUNCTION IS INTERMITTENT. REFER TO "DIAGNOSTIC AIDS" ON FACING PAGE.

FRONT WHEEL SPEED SENSOR
TEMPERATURE OF SENSOR VS. SENSOR RESISTANCE (APPROXIMATE)

TEMP. (°C)	TEMP. (°F)	RESISTANCE (OHMS)
-34 TO 4	-30 TO 40	1309 TO 1808
5 TO 43	41 TO 110	1557 TO 2103
44 TO 93	111 TO 200	1811 TO 2482

DIAGNOSTIC CHART DTC 27 — 1994 GRAND PRIX

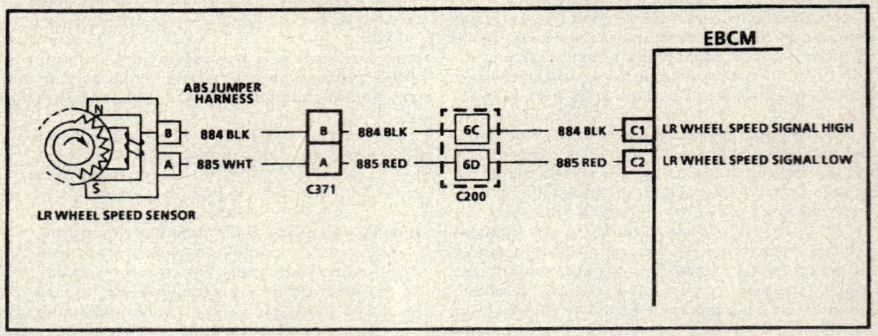

DTC 27

LEFT REAR EXCESSIVE WHEEL SPEED VARIATION

Circuit Description:

As a toothed ring passes by the wheel speed sensor, changes in the electromagnetic field cause the wheel speed sensor to produce a sinusoidal (AC) voltage signal whose frequency is proportional to wheel speed. The magnitude of this signal is directly related to wheel speed and the proximity of the wheel speed sensor to the toothed ring, often referred to as the air gap.

DTC Will Set When: DTC 27 can be set when the brake is "OFF." The purpose of this test is to detect a situation in which the left rear wheel acceleration or deceleration is beyond specified limits.

Action Taken: A malfunction DTC is stored, ABS is disabled and the ABS warning lamp is turned "ON."

DTC Chart Test Description: Number(s) below refer to circled number(s) on the diagnostic chart.

1. This step identifies a wheel speed sensor or circuitry that is damaged and visibly apparent.
2. Checks to see that the wheel speed sensor coil is within proper resistance values.
3. Verifies that the wheel speed sensor can generate an appropriate output signal.
4. This ensures that the wheel speed circuitry is not shorted together.
5. Ensures that the malfunction was not due to physical damage.

Diagnostic Aids: An "Intermittent" malfunction may be caused by a poor connection, rubbed through wire insulation, or a wire that is broken inside the insulation.

The frequency of the malfunction can be checked by using the enhanced diagnostic function of the Tech 1, as described in "Tech 1 Diagnostics" found in this section.

If the customer's comments reflect that the ABS warning lamp is "ON" only during moist environmental changes (rain, snow, vehicle wash), all wheel speed sensor circuitry should be thoroughly inspected for signs of water intrusion. If DTC is not current, clear DTCs and simulate the effects of water intrusion. Use the following procedure. Spray down the suspected area with a 5% salt water solution (two teaspoons of salt to 12 oz. of water). Test drive vehicle over various road surfaces (bumps, turns, etc.) above 24 km/h (15 mph) for at least 30 seconds. If DTC returns, replace suspected harness.

Any circuitry, that is suspected as causing the intermittent complaint, should be thoroughly checked for backed out terminals, improper mating, broken locks, improperly formed or damaged terminals, poor terminal to wiring connections or physical damage to the wiring harness.

Resistance of the wheel speed sensor will increase with an increase in sensor temperature.

When replacing a wheel speed sensor, inspect the sensor terminals and harness connector for corrosion and/or water intrusion. If evidence of corrosion or water intrusion exists, replace wheel speed sensor jumper harness. Likewise, if replacing a wheel speed sensor jumper harness, inspect sensor terminals. If evidence of corrosion or water intrusion exists, replace wheel speed sensor. Refer to "On-Vehicle Service" in this section.

DIAGNOSTIC CHART DTC 27 — 1994 GRAND PRIX, CONTINUED

DTC 27
LEFT REAR EXCESSIVE WHEEL SPEED VARIATION

Important

- WHEEL SPEED SENSOR INTERMITTENT MALFUNCTIONS MAY BE DIFFICULT TO LOCATE. CARE SHOULD BE TAKEN NOT TO DISTURB ANY ELECTRICAL CONNECTIONS PRIOR TO AN INDICATED STEP OF THIS CHART. THIS WILL ENSURE THAT AN INTERMITTENT CONNECTION WILL NOT BE CORRECTED BEFORE THE SOURCE OF THE MALFUNCTION IS FOUND.

(1)
- IGNITION "OFF."
- PHYSICALLY INSPECT THE LEFT REAR WHEEL SPEED SENSOR, WHEEL SPEED SENSOR JUMPER HARNESS, AND CONNECTORS FOR DAMAGE.

IS PHYSICAL DAMAGE INDICATED?

YES → REPLACE WHEEL SPEED SENSOR AND/OR WHEEL SPEED SENSOR JUMPER HARNESS IF DAMAGED. DRIVE VEHICLE ABOVE 24 km/h (15 MPH) TO ENSURE DTC 27 DOES NOT SET.

NO →

(2)
- DISCONNECT LEFT REAR WHEEL SPEED SENSOR DIRECTLY AT SENSOR.
- USING J 39200, MEASURE THE RESISTANCE BETWEEN SENSOR TERMINALS "A" AND "B."

IS RESISTANCE BETWEEN 2059 AND 2353Ω WHEN SENSOR IS 20°C (68°F)? *(IF TEMPERATURE OF THE SENSOR IS SUSPECTED TO BE OTHER THAN 20°C (68°F), REFER TO THE TABLE BELOW FOR APPLICABLE SENSOR RESISTANCE VALUES; THE VALUES IN THE TABLE ARE FOR TEMPERATURE OF THE SENSOR, NOT AIR TEMPERATURE.)*

NO → REPLACE WHEEL SPEED SENSOR.

YES →

(3)
- WITH J 39200 STILL CONNECTED, SELECT A/C VOLTAGE SCALE.
- SPIN WHEEL BY HAND WHILE OBSERVING VOLTAGE.

IS VOLTAGE AT LEAST 100 mV (VOLTAGE WILL INCREASE AS WHEEL SPEED INCREASES)?

NO → REPLACE WHEEL SPEED SENSOR.

YES →

(4)
- IGNITION "OFF."
- DISCONNECT 32-WAY EBCM CONNECTOR.
- USING J 39200, MEASURE THE RESISTANCE BETWEEN TERMINAL "C1" AND "C2" OF THE 32-WAY EBCM HARNESS CONNECTOR.

DOES J 39200 DISPLAY RESISTANCE AS "OL" (INFINITE)?

NO → REPAIR SHORT BETWEEN CKTs 884 AND 885.

YES →

(5)
- INSPECT CKTs 884 AND 885 AND 32-WAY EBCM CONNECTOR FOR DAMAGE WHICH MAY RESULT IN A SHORT BETWEEN CKTs 884 AND 885 WITH ALL CONNECTORS CONNECTED. REPAIR ALL DAMAGE IF EVIDENT.
- RECONNECT ALL CONNECTORS.
- TEST DRIVE VEHICLE ABOVE 24 km/h (15 MPH) FOR AT LEAST 30 SECONDS.

DOES DTC 27 SET AS A CURRENT DTC?

YES → REPLACE EBCM.

NO → MALFUNCTION IS INTERMITTENT. REFER TO "DIAGNOSTIC AIDS" ON FACING PAGE.

REAR WHEEL SPEED SENSOR
TEMPERATURE OF SENSOR VS. SENSOR RESISTANCE (APPROXIMATE)

TEMP. (°C)	TEMP. (°F)	RESISTANCE (OHMS)
-34 TO 4	-30 TO 40	1622 TO 2205
5 TO 43	41 TO 110	1930 TO 2566
44 TO 93	111 TO 200	2245 TO 3028

DIAGNOSTIC CHART DTC 28 — 1994 GRAND PRIX

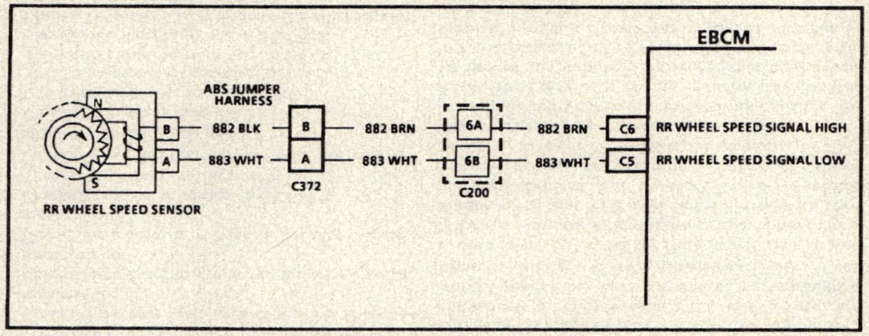

DTC 28

RIGHT REAR EXCESSIVE WHEEL SPEED VARIATION

Circuit Description:

As a toothed ring passes by the wheel speed sensor, changes in the electromagnetic field cause the wheel speed sensor to produce a sinusoidal (AC) voltage signal whose frequency is proportional to wheel speed. The magnitude of this signal is directly related to wheel speed and the proximity of the wheel speed sensor to the toothed ring, often referred to as the air gap.

DTC Will Set When: DTC 28 can be set when the brake is "OFF." The purpose of this test is to detect a situation in which the right rear wheel acceleration or deceleration is beyond specified limits.

Action Taken: A malfunction DTC is stored, ABS is disabled and the ABS warning lamp is turned "ON."

DTC Chart Test Description: Number(s) below refer to circled number(s) on the diagnostic chart.

1. This step identifies a wheel speed sensor or circuitry that is damaged and visibly apparent.
2. Checks to see that the wheel speed sensor coil is within proper resistance values.
3. Verifies that the wheel speed sensor can generate an appropriate output signal.
4. This ensures that the wheel speed circuitry is not shorted together.
5. Ensures that the malfunction was not due to physical damage.

Diagnostic Aids: An "Intermittent" malfunction may be caused by a poor connection, rubbed through wire insulation, or a wire that is broken inside the insulation.

The frequency of the malfunction can be checked by using the enhanced diagnostic function of the Tech 1, as described in "Tech 1 Diagnostics" found in this section.

If the customer's comments reflect that the ABS warning lamp is "ON" only during moist environmental changes (rain, snow, vehicle wash), all wheel speed sensor circuitry should be thoroughly inspected for signs of water intrusion. If DTC is not current, clear DTCs and simulate the effects of water intrusion. Use the following procedure. Spray down the suspected area with a 5% salt water solution (two teaspoons of salt to 12 oz. of water). Test drive vehicle over various road surfaces (bumps, turns, etc.) above 24 km/h (15 mph) for at least 30 seconds. If DTC returns, replace suspected harness.

Any circuitry, that is suspected as causing the intermittent complaint, should be thoroughly checked for backed out terminals, improper mating, broken locks, improperly formed or damaged terminals, poor terminal to wiring connections or physical damage to the wiring harness.

Resistance of the wheel speed sensor will increase with an increase in sensor temperature.

When replacing a wheel speed sensor, inspect the sensor terminals and harness connector for corrosion and/or water intrusion. If evidence of corrosion or water intrusion exists, replace wheel speed sensor jumper harness. Likewise, if replacing a wheel speed sensor jumper harness, inspect sensor terminals. If evidence of corrosion or water intrusion exists, replace wheel speed sensor. Refer to "On-Vehicle Service" in this section.

DIAGNOSTIC CHART DTC 28 — 1994 GRAND PRIX, CONTINUED

DTC 28
RIGHT REAR EXCESSIVE WHEEL SPEED VARIATION

Important
- WHEEL SPEED SENSOR INTERMITTENT MALFUNCTIONS MAY BE DIFFICULT TO LOCATE. CARE SHOULD BE TAKEN NOT TO DISTURB ANY ELECTRICAL CONNECTIONS PRIOR TO AN INDICATED STEP OF THIS CHART. THIS WILL ENSURE THAT AN INTERMITTENT CONNECTION WILL NOT BE CORRECTED BEFORE THE SOURCE OF THE MALFUNCTION IS FOUND.

(1)
- IGNITION "OFF."
- PHYSICALLY INSPECT THE RIGHT REAR WHEEL SPEED SENSOR, WHEEL SPEED SENSOR JUMPER HARNESS, AND CONNECTORS FOR DAMAGE. IS PHYSICAL DAMAGE INDICATED?

YES: REPLACE WHEEL SPEED SENSOR AND/OR WHEEL SPEED SENSOR JUMPER HARNESS IF DAMAGED. DRIVE VEHICLE ABOVE 24 km/h (15 MPH) TO ENSURE DTC 28 DOES NOT SET.

NO:

(2)
- DISCONNECT RIGHT REAR WHEEL SPEED SENSOR DIRECTLY AT SENSOR.
- USING J 39200, MEASURE THE RESISTANCE BETWEEN SENSOR TERMINALS "A" AND "B."

IS RESISTANCE BETWEEN 2059 AND 2353Ω WHEN SENSOR IS 20°C (68°F)? *(IF TEMPERATURE OF THE SENSOR IS SUSPECTED TO BE OTHER THAN 20°C (68°F), REFER TO THE TABLE BELOW FOR APPLICABLE SENSOR RESISTANCE VALUES; THE VALUES IN THE TABLE ARE FOR TEMPERATURE OF THE SENSOR, NOT AIR TEMPERATURE.)*

NO: REPLACE WHEEL SPEED SENSOR.

YES:

(3)
- WITH J 39200 STILL CONNECTED, SELECT A/C VOLTAGE SCALE.
- SPIN WHEEL BY HAND WHILE OBSERVING VOLTAGE.

IS VOLTAGE AT LEAST 100 mV (VOLTAGE WILL INCREASE AS WHEEL SPEED INCREASES)?

NO: REPLACE WHEEL SPEED SENSOR.

YES:

(4)
- IGNITION "OFF."
- DISCONNECT 32-WAY EBCM CONNECTOR.
- USING J 39200, MEASURE THE RESISTANCE BETWEEN TERMINAL "C5" AND "C6" OF THE 32-WAY EBCM HARNESS CONNECTOR. DOES J 39200 DISPLAY RESISTANCE AS "OL" (INFINITE)?

NO: REPAIR SHORT BETWEEN CKTs 882 AND 883.

YES:

(5)
- INSPECT CKTs 882 AND 883 AND 32-WAY EBCM CONNECTOR FOR DAMAGE WHICH MAY RESULT IN A SHORT BETWEEN CKTs 882 AND 883 WITH ALL CONNECTORS CONNECTED. REPAIR ALL DAMAGE IF EVIDENT.
- RECONNECT ALL CONNECTORS.
- TEST DRIVE VEHICLE ABOVE 24 km/h (15 MPH) FOR AT LEAST 30 SECONDS. DOES DTC 28 SET AS A CURRENT DTC?

YES: REPLACE EBCM.

NO: MALFUNCTION IS INTERMITTENT. REFER TO "DIAGNOSTIC AIDS" ON FACING PAGE.

REAR WHEEL SPEED SENSOR
TEMPERATURE OF SENSOR VS. SENSOR RESISTANCE (APPROXIMATE)

TEMP. (°C)	TEMP. (°F)	RESISTANCE (OHMS)
-34 TO 4	-30 TO 40	1622 TO 2205
5 TO 43	41 TO 110	1930 TO 2566
44 TO 93	111 TO 200	2245 TO 3028

DIAGNOSTIC CHART DTC 32 — 1994 GRAND PRIX

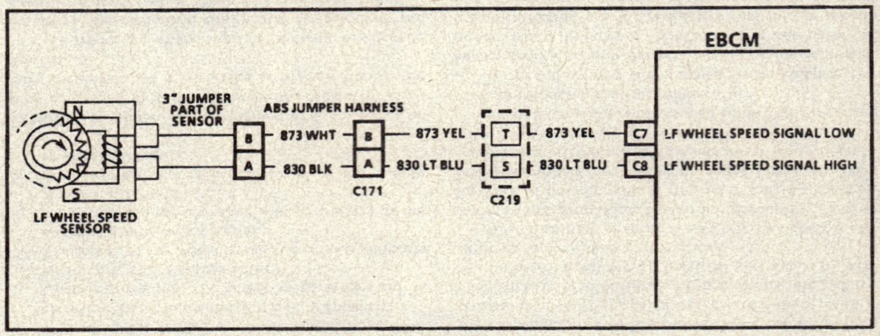

DTC 32

LF WHEEL SPEED SNR CKT OPEN OR SHORT TO GND/BATT

Circuit Description:
As a toothed ring passes by the wheel speed sensor, changes in the electromagnetic field cause the wheel speed sensor to produce a sinusoidal (AC) voltage signal whose frequency is proportional to wheel speed. The magnitude of this signal is directly related to wheel speed and the proximity of the wheel speed sensor to the toothed ring often referred to as the air gap.

DTC Will Set When: DTC 32 can be set anytime after initialization. If either of the left front wheel speed circuits are open or shorted to voltage or ground, a malfunction exists.

Action Taken: A malfunction DTC is stored, ABS is disabled and the ABS warning lamp is turned "ON."

DTC Chart Test Description: Number(s) below refer to circled number(s) on the diagnostic chart.

1. Checks for proper resistance of the wheel speed sensor coil.
2. Checks for a short to ground in the wheel speed sensor wiring.
3. Checks for a short to voltage in the wheel speed sensor wiring.
4. Checks for a short to voltage in CKT 873.
5. Checks for a short to voltage in CKT 830.
6. Ensures that the malfunction was not due to physical damage of the circuitry.
7. Checks for high resistance in CKT 873.
8. Checks for high resistance in CKT 830.
9. Verifies proper wheel speed sensor resistance.
10. Checks for a short to ground in CKT 873.
11. Checks for a short to ground in CKT 830.
12. Checks for a short to ground in the wheel speed sensor jumper harness.
13. Verifies that wheel speed sensor is not internally shorted to ground.

Diagnostic Aids: An "Intermittent" malfunction may be caused by a poor connection, rubbed through wire insulation, or a wire that is broken inside the insulation.

The frequency of the malfunction can be checked by using the enhanced diagnostic function of the Tech 1, as described in "Tech 1 Diagnostics" found in this section.

If the customer's comments reflect that the ABS warning lamp is "ON" only during moist environmental changes (rain, snow, vehicle wash), all wheel speed sensor circuitry should be thoroughly inspected for signs of water intrusion. If DTC is not current, clear DTCs and simulate the effects of water intrusion. Use the following procedure. Spray down the suspected area with a 5% salt water solution (two teaspoons of salt to 12 oz. of water). Test drive vehicle over various road surfaces (bumps, turns, etc.) above 24 km/h (15 mph) for at least 30 seconds. If DTC returns, replace suspected harness.

Any circuitry, that is suspected as causing the intermittent complaint, should be thoroughly checked for backed out terminals, improper mating, broken locks, improperly formed or damaged terminals, poor terminal to wiring connections or physical damage to the wiring harness.

Resistance of the wheel speed sensor will increase with an increase in sensor temperature.

When replacing a wheel speed sensor, inspect the sensor terminals and harness connector for corrosion and/or water intrusion. If evidence of corrosion or water intrusion exists, replace wheel speed sensor jumper harness. Likewise, if replacing a wheel speed sensor jumper harness, inspect sensor terminals. If evidence of corrosion or water intrusion exists, replace wheel speed sensor. Refer to "On-Vehicle Service" in this section.

DIAGNOSTIC CHART DTC 32 — 1994 GRAND PRIX, CONTINUED

DTC 32
(Page 1 of 3)
LF WHEEL SPEED SNR CKT OPEN OR SHORT TO GND/BATT

Important

- WHEEL SPEED SENSOR INTERMITTENT MALFUNCTIONS MAY BE DIFFICULT TO LOCATE. CARE SHOULD BE TAKEN NOT TO DISTURB ANY ELECTRICAL CONNECTIONS PRIOR TO AN INDICATED STEP OF THIS CHART. THIS WILL ENSURE THAT AN INTERMITTENT CONNECTION WILL NOT BE CORRECTED BEFORE THE SOURCE OF THE MALFUNCTION IS FOUND.

1.
- IGNITION "OFF."
- USING A 5% SALT WATER SOLUTION (TWO TEASPOONS OF SALT TO 12 OZ. WATER), THOROUGHLY SPRAY THE LEFT FRONT WHEEL SPEED SENSOR JUMPER HARNESS.
- DISCONNECT 32-WAY EBCM HARNESS CONNECTOR.
- USING J 39200, MEASURE THE RESISTANCE BETWEEN TERMINALS "C7" AND "C8" OF THE 32-WAY EBCM HARNESS CONNECTOR.

IS RESISTANCE BETWEEN 1661 AND 1929Ω WHEN SENSOR IS 20°C (68°F)? *(IF TEMPERATURE OF SENSOR IS SUSPECTED TO OTHER THAN 20°C (68°F), REFER TO THE TABLE ON PAGE 2 OF 2 OF THIS DTC CHART FOR APPLICABLE SENSOR RESISTANCE VALUES; THE VALUES IN THE TABLE ARE FOR TEMPERATURE OF THE SENSOR, NOT AIR TEMPERATURE.)*

- YES → step 2
- NO → DIAGNOSIS CONTINUED ON PAGE 2 OF 3 OF THIS DTC CHART.

2.
- MEASURE THE RESISTANCE BETWEEN GROUND AND TERMINAL "C7" OF THE 32-WAY EBCM HARNESS CONNECTOR.

DOES J 39200 DISPLAY RESISTANCE AS "OL" (INFINITE)?

- YES → step 3
- NO → DIAGNOSIS CONTINUED ON PAGE 3 OF 3 OF THIS DTC CHART.

3.
- IGNITION "ON."
- USING J 39200, MEASURE THE VOLTAGE BETWEEN GROUND AND TERMINAL "C7" OF THE 32-WAY EBCM HARNESS CONNECTOR.

IS VOLTAGE 1 VOLT OR LESS?

- NO → step 4
- YES → step 6

4.
- IGNITION "OFF."
- DISCONNECT LEFT FRONT WHEEL SPEED SENSOR DIRECTLY AT SENSOR.
- IGNITION "ON."
- USING J 39200, MEASURE THE VOLTAGE BETWEEN GROUND AND TERMINAL "C7" OF THE 32-WAY EBCM HARNESS CONNECTOR.

IS VOLTAGE 1 VOLT OR LESS?

- YES → step 5
- NO → REPAIR SHORT TO VOLTAGE IN CKT 873.

5.
- USING J 39200, MEASURE THE VOLTAGE BETWEEN GROUND AND TERMINAL "C8" OF THE 32-WAY EBCM HARNESS CONNECTOR.

IS VOLTAGE 1 VOLT OR LESS?

- YES → MALFUNCTION IS INTERMITTENT. INSPECT ALL CONNECTORS AND HARNESSES FOR DAMAGE WHICH MAY RESULT IN A SHORT TO VOLTAGE WHEN ALL COMPONENTS ARE CONNECTED. REPAIR ALL DAMAGE IF FOUND. REFER TO "DIAGNOSTIC AIDS" ON FACING PAGE FOR MORE INFORMATION.
- NO → REPAIR SHORT TO VOLTAGE IN CKT 830.

6.
- IGNITION "OFF."
- INSPECT TERMINALS "C7" AND "C8" OF THE 32-WAY EBCM HARNESS CONNECTOR FOR POOR TERMINAL CONTACT AND CONTAMINATION. ALSO INSPECT CKTs 830 AND 873 FOR DAMAGE WHICH MAY RESULT IN AN OPEN CIRCUIT, SHORT TO GROUND, OR SHORT TO VOLTAGE. REPAIR ALL DAMAGE IF EVIDENT.
- RECONNECT ALL CONNECTORS.
- IGNITION "ON."

DOES DTC 32 SET AS A CURRENT DTC?

- YES → REPLACE EBCM.
- NO → MALFUNCTION IS INTERMITTENT. REFER TO "DIAGNOSTIC AIDS" ON FACING PAGE.

DIAGNOSTIC CHART DTC 32 — 1994 GRAND PRIX, CONTINUED

DTC 32
(Page 2 of 3)
LF WHEEL SPEED SNR CKT OPEN OR SHORT TO GND/BATT

DIAGNOSIS CONTINUED FROM PAGE 1 OF 3 OF THIS DTC CHART.

7.
- DISCONNECT LEFT FRONT WHEEL SPEED SENSOR DIRECTLY AT SENSOR.
- USING J 39200, MEASURE THE RESISTANCE BETWEEN TERMINAL "C7" OF THE 32-WAY EBCM HARNESS CONNECTOR AND TERMINAL "B" OF THE LEFT FRONT WHEEL SPEED SENSOR HARNESS CONNECTOR.

IS RESISTANCE 2Ω OR LESS?

- YES → step 8
- NO → REPAIR HIGH RESISTANCE IN CKT 873 (WHEEL SPEED SENSOR JUMPER MUST BE REPLACED IF DAMAGED).

8.
- USING J 39200, MEASURE THE RESISTANCE BETWEEN TERMINAL "C8" OF THE 32-WAY EBCM HARNESS CONNECTOR AND TERMINAL "A" OF THE LEFT FRONT WHEEL SPEED SENSOR HARNESS CONNECTOR.

IS RESISTANCE 2Ω OR LESS?

- YES → step 9
- NO → REPAIR HIGH RESISTANCE IN CKT 830 (WHEEL SPEED SENSOR JUMPER MUST BE REPLACED IF DAMAGED).

9.
- USING J 39200, MEASURE THE RESISTANCE BETWEEN TERMINAL "A" AND TERMINAL "B" OF THE LEFT FRONT WHEEL SPEED SENSOR CONNECTOR.

IS RESISTANCE BETWEEN 1661 AND 1929Ω WHEN SENSOR IS 20°C (68°F)? *(IF TEMPERATURE OF SENSOR IS SUSPECTED TO OTHER THAN 20°C (68°F), REFER TO THE TABLE BELOW FOR APPLICABLE SENSOR RESISTANCE VALUES; THE VALUES IN THE TABLE ARE FOR TEMPERATURE OF THE SENSOR, NOT AIR TEMPERATURE.)*

- YES → MALFUNCTION IS INTERMITTENT. INSPECT ALL CONNECTORS AND HARNESSES FOR DAMAGE WHICH MAY RESULT IN HIGH RESISTANCE WHEN ALL COMPONENTS ARE CONNECTED. REFER TO "DIAGNOSTIC AIDS" ON FACING PAGE FOR MORE INFORMATION. REPAIR ALL DAMAGE IF FOUND.
- NO → REPLACE WHEEL SPEED SENSOR.

FRONT WHEEL SPEED SENSOR

TEMPERATURE OF SENSOR VS. SENSOR RESISTANCE (APPROXIMATE)

TEMP. (°C)	TEMP. (°F)	RESISTANCE (OHMS)
-34 TO 4	-30 TO 40	1309 TO 1808
5 TO 43	41 TO 110	1557 TO 2103
44 TO 93	111 TO 200	1811 TO 2482

DIAGNOSTIC CHART DTC 33 — 1994 GRAND PRIX

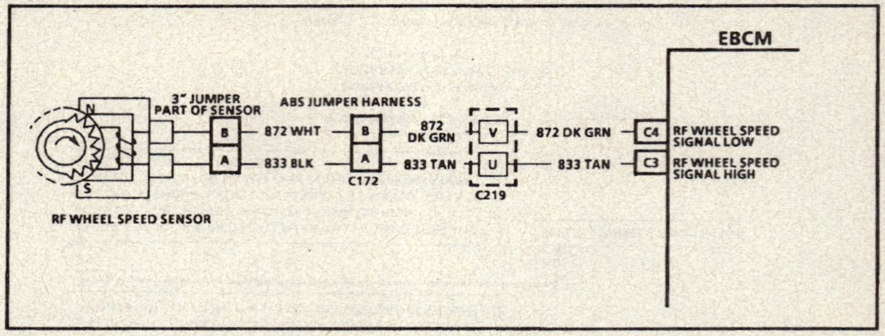

DTC 33

RF WHEEL SPEED SNR CKT OPEN OR SHORT TO GND/BATT

Circuit Description:

As a toothed ring passes by the wheel speed sensor, changes in the electromagnetic field cause the wheel speed sensor to produce a sinusoidal (AC) voltage signal whose frequency is proportional to wheel speed. The magnitude of this signal is directly related to wheel speed and the proximity of the wheel speed sensor to the toothed ring, often referred to as the air gap.

DTC Will Set When: DTC 33 can be set anytime after initialization. If either of the right front wheel speed circuits are open or shorted to voltage or ground, a malfunction exists.

Action Taken: A malfunction DTC is stored, ABS is disabled and the ABS warning lamp is turned "ON."

DTC Chart Test Description: Number(s) below refer to circled number(s) on the diagnostic chart.

1. Checks for proper resistance of the wheel speed sensor coil.
2. Checks for a short to ground in the wheel speed sensor wiring.
3. Checks for a short to voltage in the wheel speed sensor wiring.
4. Checks for a short to voltage in CKT 833.
5. Checks for short to voltage in CKT 872.
6. Ensures that the malfunction was not due to physical damage of the circuitry.
7. Checks for high resistance in CKT 833.
8. Checks for high resistance in CKT 872.
9. Verifies proper wheel speed sensor resistance.
10. Checks for a short to ground in CKT 833.
11. Checks for a short to ground in CKT 872.
12. Checks for a short to ground in the wheel speed sensor jumper harness.
13. Verifies that wheel speed sensor is not internally shorted to ground.

Diagnostic Aids: An "Intermittent" malfunction may be caused by a poor connection, rubbed through wire insulation, or a wire that is broken inside the insulation.

The frequency of the malfunction can be checked by using the enhanced diagnostic function of the Tech 1, as described in "Tech 1 Diagnostics" found in this section.

If the customer's comments reflect that the ABS warning lamp is "ON" only during moist environmental changes (rain, snow, vehicle wash), all wheel speed sensor circuitry should be thoroughly inspected for signs of water intrusion. If DTC is not current, clear DTCs and simulate the effects of water intrusion. Use the following procedure. Spray down the suspected area with a 5% salt water solution (two teaspoons of salt to 12 oz. of water). Test drive vehicle over various road surfaces (bumps, turns, etc.) above 24 km/h (15 mph) for at least 30 seconds. If DTC returns, replace suspected harness.

Any circuitry, that is suspected as causing the intermittent complaint, should be thoroughly checked for backed out terminals, improper mating, broken locks, improperly formed or damaged terminals, poor terminal to wiring connections or physical damage to the wiring harness.

Resistance of the wheel speed sensor will increase with an increase in sensor temperature.

When replacing a wheel speed sensor, inspect the sensor terminals and harness connector for corrosion and/or water intrusion. If evidence of corrosion or water intrusion exists, replace wheel speed sensor jumper harness. Likewise, if replacing a wheel speed sensor jumper harness, inspect sensor terminals. If evidence of corrosion or water intrusion exists, replace wheel speed sensor. Refer to "On-Vehicle Service" in this section.

DIAGNOSTIC CHART DTC 33 — 1994 GRAND PRIX, CONTINUED

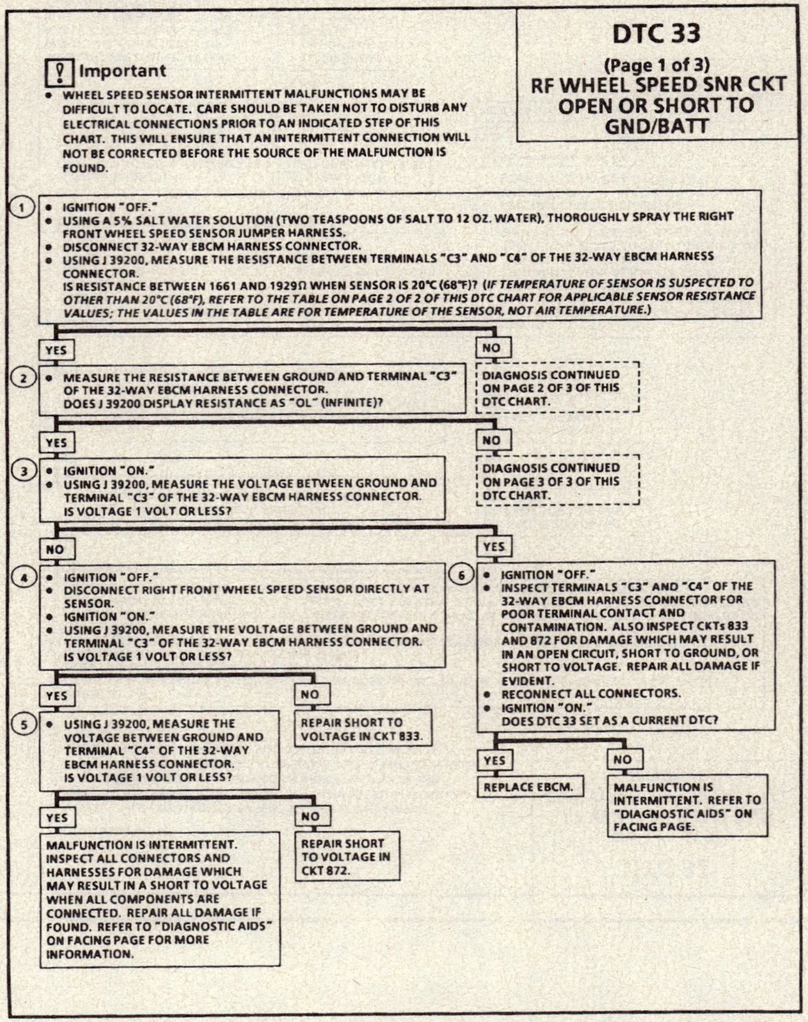

DIAGNOSTIC CHART DTC 33 — 1994 GRAND PRIX, CONTINUED

DTC 33

(Page 2 of 3)
RF WHEEL SPEED SNR CKT OPEN OR SHORT TO GND/BATT

DIAGNOSIS CONTINUED FROM PAGE 1 OF 3 OF THIS DTC CHART.

(7)
- DISCONNECT RIGHT FRONT WHEEL SPEED SENSOR DIRECTLY AT SENSOR.
- USING J 39200, MEASURE THE RESISTANCE BETWEEN TERMINAL "C3" OF THE 32-WAY EBCM HARNESS CONNECTOR AND TERMINAL "A" OF THE RIGHT FRONT WHEEL SPEED SENSOR HARNESS CONNECTOR.

IS RESISTANCE 2Ω OR LESS?

- YES: go to step 8.
- NO: REPAIR HIGH RESISTANCE IN CKT 833 (WHEEL SPEED SENSOR JUMPER MUST BE REPLACED IF DAMAGED).

(8)
- USING J 39200, MEASURE THE RESISTANCE BETWEEN TERMINAL "C4" OF THE 32-WAY EBCM HARNESS CONNECTOR AND TERMINAL "B" OF THE RIGHT FRONT WHEEL SPEED SENSOR HARNESS CONNECTOR.

IS RESISTANCE 2Ω OR LESS?

- YES: go to step 9.
- NO: REPAIR HIGH RESISTANCE IN CKT 872 (WHEEL SPEED SENSOR JUMPER MUST BE REPLACED IF DAMAGED).

(9)
- USING J 39200, MEASURE THE RESISTANCE BETWEEN TERMINAL "A" AND TERMINAL "B" OF THE RIGHT FRONT WHEEL SPEED SENSOR CONNECTOR.

IS RESISTANCE BETWEEN 1661 AND 1929Ω WHEN SENSOR IS 20°C (68°F)? *(IF TEMPERATURE OF SENSOR IS SUSPECTED TO OTHER THAN 20°C (68°F), REFER TO THE TABLE BELOW FOR APPLICABLE SENSOR RESISTANCE VALUES; THE VALUES IN THE TABLE ARE FOR TEMPERATURE OF THE SENSOR, NOT AIR TEMPERATURE.)*

- YES: MALFUNCTION IS INTERMITTENT. INSPECT ALL CONNECTORS AND HARNESSES FOR DAMAGE WHICH MAY RESULT IN HIGH RESISTANCE WHEN ALL COMPONENTS ARE CONNECTED. REFER TO "DIAGNOSTIC AIDS" ON FACING PAGE FOR MORE INFORMATION. REPAIR ALL DAMAGE IF FOUND.
- NO: REPLACE WHEEL SPEED SENSOR.

FRONT WHEEL SPEED SENSOR

TEMPERATURE OF SENSOR VS. SENSOR RESISTANCE (APPROXIMATE)

TEMP. (°C)	TEMP. (°F)	RESISTANCE (OHMS)
-34 TO 4	-30 TO 40	1309 TO 1808
5 TO 43	41 TO 110	1557 TO 2103
44 TO 93	111 TO 200	1811 TO 2482

DIAGNOSTIC CHART DTC 33 — 1994 GRAND PRIX, CONTINUED

DTC 33

(Page 3 of 3)
RF WHEEL SPEED SNR CKT OPEN OR SHORT TO GND/BATT

DIAGNOSIS CONTINUED FROM PAGE 1 OF 3 OF THIS DTC CHART.

(10)
- DISCONNECT CONNECTOR C172 (LOCATED RH REAR OF ENGINE COMPARTMENT).
- USING J 39200, MEASURE THE RESISTANCE BETWEEN GROUND AND TERMINAL "C3" OF THE 32-WAY EBCM HARNESS CONNECTOR.

DOES J 39200 DISPLAY RESISTANCE AS "OL" (INFINITE)?

- YES: go to step 11.
- NO: REPAIR SHORT TO GROUND IN CKT 833.

(11)
- USING J 39200, MEASURE THE RESISTANCE BETWEEN GROUND AND TERMINAL "C4" OF THE 32-WAY EBCM HARNESS CONNECTOR.

DOES J 39200 DISPLAY RESISTANCE AS "OL" (INFINITE)?

- YES: go to step 12.
- NO: REPAIR SHORT TO GROUND IN CKT 872.

(12)
- THOROUGHLY RESPRAY THE RIGHT FRONT WHEEL SPEED SENSOR JUMPER HARNESS WITH THE SALT WATER SOLUTION.

IMPORTANT: DO NOT ALLOW WATER TO ENTER ANY WHEEL SPEED SENSOR CONNECTOR.

- DISCONNECT RIGHT FRONT WHEEL SPEED SENSOR DIRECTLY AT SENSOR.
- USING J 39200, MEASURE THE RESISTANCE BETWEEN GROUND AND TERMINAL "A," THEN "B" OF THE RIGHT FRONT WHEEL SPEED SENSOR HARNESS CONNECTOR.

DOES J 39200 DISPLAY RESISTANCE AS "OL" (INFINITE) FOR BOTH MEASUREMENTS?

- YES: go to step 13.
- NO: REPLACE RIGHT FRONT WHEEL SPEED SENSOR JUMPER HARNESS.

(13)
- USING J 39200, MEASURE THE RESISTANCE BETWEEN GROUND AND TERMINAL "A" OF THE RIGHT FRONT WHEEL SPEED SENSOR CONNECTOR.

DOES J 39200 DISPLAY RESISTANCE AS "OL" (INFINITE)?

- YES: MALFUNCTION IS INTERMITTENT. INSPECT ALL CONNECTORS AND HARNESSES FOR DAMAGE WHICH MAY RESULT IN A SHORT TO GROUND WHEN ALL COMPONENTS ARE CONNECTED. PAY PARTICULAR ATTENTION TO THE JUMPER HARNESS, SMALL NICKS OR CUTS CAN CAUSE A MALFUNCTION. REPAIR ANY DAMAGE IF FOUND. JUMPER HARNESS MUST BE REPLACED IF DAMAGED. REFER TO "DIAGNOSTIC AIDS" ON FACING PAGE FOR MORE INFORMATION.
- NO: REPLACE WHEEL SPEED SENSOR.

DIAGNOSTIC CHART DTC 34 — 1994 GRAND PRIX

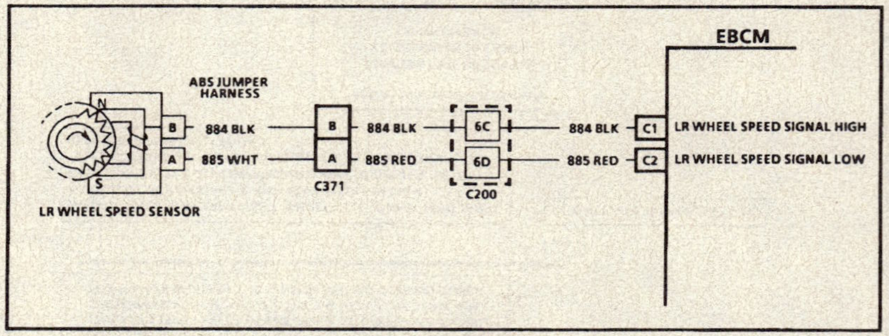

DTC 34

LR WHEEL SPEED SNR CKT OPEN OR SHORT TO GND/BATT

Circuit Description:
As a toothed ring passes by the wheel speed sensor, changes in the electromagnetic field cause the wheel speed sensor to produce a sinusoidal (AC) voltage signal whose frequency is proportional to wheel speed. The magnitude of this signal is directly related to wheel speed and the proximity of the wheel speed sensor to the toothed ring, often referred to as the air gap.

DTC Will Set When: DTC 34 can be set anytime after initialization. If either of the left rear wheel speed circuits are open or shorted to voltage or ground, a malfunction exists.

Action Taken: A malfunction DTC is stored, ABS is disabled and the ABS warning lamp is turned "ON."

DTC Chart Test Description: Number(s) below refer to circled number(s) on the diagnostic chart.

1. Checks for proper resistance of the wheel speed sensor coil.
2. Checks for a short to ground in wheel speed sensor wiring.
3. Checks for a short to voltage in the wheel speed sensor wiring.
4. Checks for a short to voltage in CKT 884.
5. Checks for a short to voltage in CKT 885.
6. Ensures that the malfunction was not due to physical damage of the circuitry.
7. Checks for high resistance in CKT 884.
8. Checks for high resistance in CKT 885.
9. Verifies proper wheel speed sensor resistance.
10. Checks for a short to ground in CKT 884.
11. Checks for a short to ground in CKT 885.
12. Checks for a short to ground in the wheel speed sensor jumper harness.
13. Verifies that wheel speed sensor is not internally shorted to ground.

Diagnostic Aids: An "Intermittent" malfunction may be caused by a poor connection, rubbed through wire insulation, or a wire that is broken inside the insulation.

The frequency of the malfunction can be checked by using the enhanced diagnostic function of the Tech 1, as described in "Tech 1 Diagnostics" found in this section.

If the customer's comments reflect that the ABS warning lamp is "ON" only during moist environmental changes (rain, snow, vehicle wash), all wheel speed sensor circuitry should be thoroughly inspected for signs of water intrusion. If DTC is not current, clear DTCs and simulate the effects of water intrusion. Use the following procedure. Spray down the suspected area with a 5% salt water solution (two teaspoons of salt to 12 oz. of water). Test drive vehicle over various road surfaces (bumps, turns, etc.) above 24 km/h (15 mph) for at least 30 seconds. If DTC returns, replace suspected harness.

Any circuitry, that is suspected as causing the intermittent complaint, should be thoroughly checked for backed out terminals, improper mating, broken locks, improperly formed or damaged terminals, poor terminal to wiring connections or physical damage to the wiring harness.

Resistance of the wheel speed sensor will increase with an increase in sensor temperature.

When replacing a wheel speed sensor, inspect the sensor terminals and harness connector for corrosion and/or water intrusion. If evidence of corrosion or water intrusion exists, replace wheel speed sensor jumper harness. Likewise, if replacing a wheel speed sensor jumper harness, inspect sensor terminals. If evidence of corrosion or water intrusion exists, replace wheel speed sensor. Refer to "On-Vehicle Service" in this section.

DIAGNOSTIC CHART DTC 34 — 1994 GRAND PRIX, CONTINUED

DTC 34
(Page 1 of 3)
LR WHEEL SPEED SNR CKT OPEN OR SHORT TO GND/BATT

Important
- WHEEL SPEED SENSOR INTERMITTENT MALFUNCTIONS MAY BE DIFFICULT TO LOCATE. CARE SHOULD BE TAKEN NOT TO DISTURB ANY ELECTRICAL CONNECTIONS PRIOR TO AN INDICATED STEP OF THIS CHART. THIS WILL ENSURE THAT AN INTERMITTENT CONNECTION WILL NOT BE CORRECTED BEFORE THE SOURCE OF THE MALFUNCTION IS FOUND.

(1)
- IGNITION "OFF."
- USING A 5% SALT WATER SOLUTION (TWO TEASPOONS OF SALT TO 12 OZ. WATER), THOROUGHLY SPRAY THE LEFT REAR WHEEL SPEED SENSOR JUMPER HARNESS.
- DISCONNECT 32-WAY EBCM HARNESS CONNECTOR.
- USING J 39200, MEASURE THE RESISTANCE BETWEEN TERMINALS "C1" AND "C2" OF THE 32-WAY EBCM HARNESS CONNECTOR.

IS RESISTANCE BETWEEN 2059 AND 2353Ω WHEN SENSOR IS 20°C (68°F)? *(IF TEMPERATURE OF SENSOR IS SUSPECTED TO OTHER THAN 20°C (68°F), REFER TO THE TABLE ON PAGE 2 OF 2 OF THIS DTC CHART FOR APPLICABLE SENSOR RESISTANCE VALUES; THE VALUES IN THE TABLE ARE FOR TEMPERATURE OF THE SENSOR, NOT AIR TEMPERATURE.)*

YES → (2); NO → DIAGNOSIS CONTINUED ON PAGE 2 OF 3 OF THIS DTC CHART.

(2)
- MEASURE THE RESISTANCE BETWEEN GROUND AND TERMINAL "C1" OF THE 32-WAY EBCM HARNESS CONNECTOR.

DOES J 39200 DISPLAY RESISTANCE AS "OL" (INFINITE)?

YES → (3); NO → DIAGNOSIS CONTINUED ON PAGE 3 OF 3 OF THIS DTC CHART.

(3)
- IGNITION "ON."
- USING J 39200, MEASURE THE VOLTAGE BETWEEN GROUND AND TERMINAL "C1" OF THE 32-WAY EBCM HARNESS CONNECTOR.

IS VOLTAGE 1 VOLT OR LESS?

NO → (4); YES → (6)

(4)
- IGNITION "OFF."
- DISCONNECT LEFT REAR WHEEL SPEED SENSOR DIRECTLY AT SENSOR.
- IGNITION "ON."
- USING J 39200, MEASURE THE VOLTAGE BETWEEN GROUND AND TERMINAL "C1" OF THE 32-WAY EBCM HARNESS CONNECTOR.

IS VOLTAGE 1 VOLT OR LESS?

YES → (5); NO → REPAIR SHORT TO VOLTAGE IN CKT 884.

(5)
- USING J 39200, MEASURE THE VOLTAGE BETWEEN GROUND AND TERMINAL "C2" OF THE 32-WAY EBCM HARNESS CONNECTOR.

IS VOLTAGE 1 VOLT OR LESS?

YES → MALFUNCTION IS INTERMITTENT. INSPECT ALL CONNECTORS AND HARNESSES FOR DAMAGE WHICH MAY RESULT IN A SHORT TO VOLTAGE WHEN ALL COMPONENTS ARE CONNECTED. REPAIR ALL DAMAGE IF FOUND. REFER TO "DIAGNOSTIC AIDS" ON FACING PAGE FOR MORE INFORMATION.

NO → REPAIR SHORT TO VOLTAGE IN CKT 885.

(6)
- IGNITION "OFF."
- INSPECT TERMINALS "C1" AND "C2" OF THE 32-WAY EBCM HARNESS CONNECTOR FOR POOR TERMINAL CONTACT AND CONTAMINATION. ALSO INSPECT CKTs 884 AND 885 FOR DAMAGE WHICH MAY RESULT IN AN OPEN CIRCUIT, SHORT TO GROUND, OR SHORT TO VOLTAGE. REPAIR ALL DAMAGE IF EVIDENT.
- RECONNECT ALL CONNECTORS.
- IGNITION "ON."

DOES DTC 34 SET AS A CURRENT DTC?

YES → REPLACE EBCM.

NO → MALFUNCTION IS INTERMITTENT. REFER TO "DIAGNOSTIC AIDS" ON FACING PAGE.

DIAGNOSTIC CHART DTC 34 — 1994 GRAND PRIX, CONTINUED

DTC 34
(Page 2 of 3)
LR WHEEL SPEED SNR CKT OPEN OR SHORT TO GND/BATT

DIAGNOSIS CONTINUED FROM PAGE 1 OF 3 OF THIS DTC CHART.

(7)
- DISCONNECT LEFT REAR WHEEL SPEED SENSOR DIRECTLY AT SENSOR.
- USING J 39200, MEASURE THE RESISTANCE BETWEEN TERMINAL "C1" OF THE 32-WAY EBCM HARNESS CONNECTOR AND TERMINAL "B" OF THE LEFT REAR WHEEL SPEED SENSOR HARNESS CONNECTOR.
IS RESISTANCE 2Ω OR LESS?

NO: REPAIR HIGH RESISTANCE IN CKT 884 (WHEEL SPEED SENSOR JUMPER MUST BE REPLACED IF DAMAGED).

YES:

(8)
- USING J 39200, MEASURE THE RESISTANCE BETWEEN TERMINAL "C2" OF THE 32-WAY EBCM HARNESS CONNECTOR AND TERMINAL "A" OF THE LEFT REAR WHEEL SPEED SENSOR HARNESS CONNECTOR.
IS RESISTANCE 2Ω OR LESS?

NO: REPAIR HIGH RESISTANCE IN CKT 885 (WHEEL SPEED SENSOR JUMPER MUST BE REPLACED IF DAMAGED).

YES:

(9)
- USING J 39200, MEASURE THE RESISTANCE BETWEEN TERMINAL "A" AND TERMINAL "B" OF THE LEFT REAR WHEEL SPEED SENSOR CONNECTOR.
IS RESISTANCE BETWEEN 2059 AND 2353Ω WHEN SENSOR IS 20°C (68°F)? *(IF TEMPERATURE OF SENSOR IS SUSPECTED TO OTHER THAN 20°C (68°F), REFER TO THE TABLE BELOW FOR APPLICABLE SENSOR RESISTANCE VALUES; THE VALUES IN THE TABLE ARE FOR TEMPERATURE OF THE SENSOR, NOT AIR TEMPERATURE.)*

NO: REPLACE WHEEL SPEED SENSOR.

YES: MALFUNCTION IS INTERMITTENT. INSPECT ALL CONNECTORS AND HARNESSES FOR DAMAGE WHICH MAY RESULT IN HIGH RESISTANCE WHEN ALL COMPONENTS ARE CONNECTED. REFER TO "DIAGNOSTIC AIDS" ON FACING PAGE FOR MORE INFORMATION. REPAIR ALL DAMAGE IF FOUND.

REAR WHEEL SPEED SENSOR

TEMPERATURE OF SENSOR VS. SENSOR RESISTANCE (APPROXIMATE)

TEMP. (°C)	TEMP. (°F)	RESISTANCE (OHMS)
-34 TO 4	-30 TO 40	1622 TO 2205
5 TO 43	41 TO 110	1930 TO 2566
44 TO 93	111 TO 200	2245 TO 3028

DIAGNOSTIC CHART DTC 34 — 1994 GRAND PRIX, CONTINUED

DTC 34
(Page 3 of 3)
LR WHEEL SPEED SNR CKT OPEN OR SHORT TO GND/BATT

DIAGNOSIS CONTINUED FROM PAGE 1 OF 3 OF THIS DTC CHART.

(10)
- DISCONNECT CONNECTOR C371 (LOCATED BEHIND LH SIDE OF REAR SEAT).
- USING J 39200, MEASURE THE RESISTANCE BETWEEN GROUND AND TERMINAL "C1" OF THE 32-WAY EBCM HARNESS CONNECTOR.
DOES J 39200 DISPLAY RESISTANCE AS "OL" (INFINITE)?

NO: REPAIR SHORT TO GROUND IN CKT 884.

YES:

(11)
- USING J 39200, MEASURE THE RESISTANCE BETWEEN GROUND AND TERMINAL "C2" OF THE 32-WAY EBCM HARNESS CONNECTOR.
DOES J 39200 DISPLAY RESISTANCE AS "OL" (INFINITE)?

NO: REPAIR SHORT TO GROUND IN CKT 885.

YES:

(12)
- THOROUGHLY RESPRAY THE LEFT REAR WHEEL SPEED SENSOR JUMPER HARNESS WITH THE SALT WATER SOLUTION.

IMPORTANT: DO NOT ALLOW WATER TO ENTER ANY WHEEL SPEED SENSOR CONNECTOR.

- DISCONNECT LEFT REAR WHEEL SPEED SENSOR DIRECTLY AT SENSOR.
- USING J 39200, MEASURE THE RESISTANCE BETWEEN GROUND AND TERMINAL "A," THEN "B" OF THE LEFT REAR WHEEL SPEED SENSOR HARNESS CONNECTOR.
DOES J 39200 DISPLAY RESISTANCE AS "OL" (INFINITE) FOR BOTH MEASUREMENTS?

NO: REPLACE LEFT REAR WHEEL SPEED SENSOR JUMPER HARNESS.

YES:

(13)
- USING J 39200, MEASURE THE RESISTANCE BETWEEN GROUND AND TERMINAL "A" OF THE LEFT REAR WHEEL SPEED SENSOR CONNECTOR.
DOES J 39200 DISPLAY RESISTANCE AS "OL" (INFINITE)?

NO: REPLACE WHEEL SPEED SENSOR.

YES: MALFUNCTION IS INTERMITTENT. INSPECT ALL CONNECTORS AND HARNESSES FOR DAMAGE WHICH MAY RESULT IN A SHORT TO GROUND WHEN ALL COMPONENTS ARE CONNECTED. PAY PARTICULAR ATTENTION TO THE JUMPER HARNESS, SMALL NICKS OR CUTS CAN CAUSE A MALFUNCTION. REPAIR ANY DAMAGE IF FOUND. JUMPER HARNESS MUST BE REPLACED IF DAMAGED. REFER TO "DIAGNOSTIC AIDS" ON FACING PAGE FOR MORE INFORMATION.

DIAGNOSTIC CHART DTC 35 — 1994 GRAND PRIX

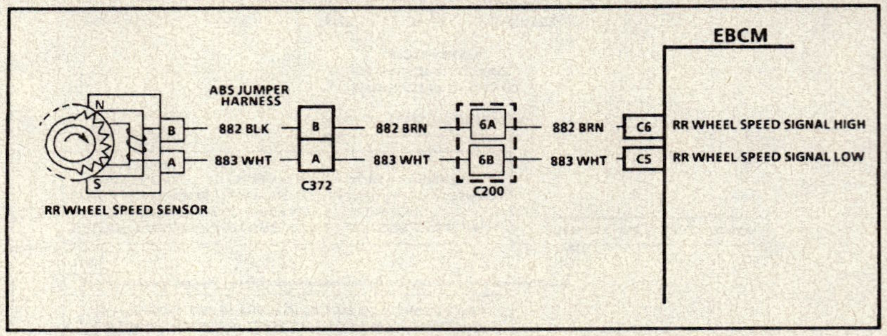

DTC 35

RR WHEEL SPEED SNR CKT OPEN OR SHORT TO GND/BATT

Circuit Description:

As a toothed ring passes by the wheel speed sensor, changes in the electromagnetic field cause the wheel speed sensor to produce a sinusoidal (AC) voltage signal whose frequency is proportional to wheel speed. The magnitude of this signal is directly related to wheel speed and the proximity of the wheel speed sensor to the toothed ring, often referred to as the air gap.

DTC Will Set When: DTC 35 can be set anytime after initialization. If either of the right rear wheel speed circuits are open or shorted to voltage or ground, a malfunction exists.

Action Taken: A malfunction DTC is stored, ABS is disabled and the ABS warning lamp is turned "ON."

DTC Chart Test Description: Number(s) below refer to circled number(s) on the diagnostic chart.

1. Checks for proper resistance of the wheel speed sensor coil.
2. Checks for a short to ground in the wheel speed sensor wiring.
3. Checks for a short to voltage in the wheel speed sensor wiring.
4. Checks for a short to voltage in CKT 883.
5. Checks for a short to voltage in CKT 882.
6. Ensures that the malfunction was not due to physical damage of the circuitry.
7. Checks for high resistance in CKT 883.
8. Checks for high resistance in CKT 882.
9. Verifies proper wheel speed sensor resistance.
10. Checks for a short to ground in CKT 883.
11. Checks for a short to ground in CKT 882.
12. Checks for a short to ground in the wheel speed sensor jumper harness.
13. Verifies that wheel speed sensor is not internally shorted to ground.

Diagnostic Aids: An "Intermittent" malfunction may be caused by a poor connection, rubbed through wire insulation, or a wire that is broken inside the insulation.

The frequency of the malfunction can be checked by using the enhanced diagnostic function of the Tech 1, as described in "Tech 1 Diagnostics" found in this section.

If the customer's comments reflect that the ABS warning lamp is "ON" only during moist environmental changes (rain, snow, vehicle wash), all wheel speed sensor circuitry should be thoroughly inspected for signs of water intrusion. If DTC is not current, clear DTCs and simulate the effects of water intrusion. Use the following procedure. Spray down the suspected area with a 5% salt water solution (two teaspoons of salt to 12 oz. of water). Test drive vehicle over various road surfaces (bumps, turns, etc.) above 24 km/h (15 mph) for at least 30 seconds. If DTC returns, replace suspected harness.

Any circuitry, that is suspected as causing the intermittent complaint, should be thoroughly checked for backed out terminals, improper mating, broken locks, improperly formed or damaged terminals, poor terminal to wiring connections or physical damage to the wiring harness.

Resistance of the wheel speed sensor will increase with an increase in sensor temperature.

When replacing a wheel speed sensor, inspect the sensor terminals and harness connector for corrosion and/or water intrusion. If evidence of corrosion or water intrusion exists, replace wheel speed sensor jumper harness. Likewise, if replacing a wheel speed sensor jumper harness, inspect sensor terminals. If evidence of corrosion or water intrusion exists, replace wheel speed sensor. Refer to "On-Vehicle Service" in this section.

DIAGNOSTIC CHART DTC 35 — 1994 GRAND PRIX, CONTINUED

DTC 35
(Page 1 of 3)
RR WHEEL SPEED SNR CKT OPEN OR SHORT TO GND/BATT

Important

- WHEEL SPEED SENSOR INTERMITTENT MALFUNCTIONS MAY BE DIFFICULT TO LOCATE. CARE SHOULD BE TAKEN NOT TO DISTURB ANY ELECTRICAL CONNECTIONS PRIOR TO AN INDICATED STEP OF THIS CHART. THIS WILL ENSURE THAT AN INTERMITTENT CONNECTION WILL NOT BE CORRECTED BEFORE THE SOURCE OF THE MALFUNCTION IS FOUND.

(1)
- IGNITION "OFF."
- USING A 5% SALT WATER SOLUTION (TWO TEASPOONS OF SALT TO 12 OZ. WATER), THOROUGHLY SPRAY THE RIGHT REAR WHEEL SPEED SENSOR JUMPER HARNESS.
- DISCONNECT 32-WAY EBCM HARNESS CONNECTOR.
- USING J 39200, MEASURE THE RESISTANCE BETWEEN TERMINALS "C5" AND "C6" OF THE 32-WAY EBCM HARNESS CONNECTOR.

IS RESISTANCE BETWEEN 2059 AND 2353Ω WHEN SENSOR IS 20°C (68°F)? *(IF TEMPERATURE OF SENSOR IS SUSPECTED TO OTHER THAN 20°C (68°F), REFER TO THE TABLE ON PAGE 2 OF 2 OF THIS DTC CHART FOR APPLICABLE SENSOR RESISTANCE VALUES; THE VALUES IN THE TABLE ARE FOR TEMPERATURE OF THE SENSOR, NOT AIR TEMPERATURE.)*

- NO: DIAGNOSIS CONTINUED ON PAGE 2 OF 3 OF THIS DTC CHART.
- YES:

(2)
- MEASURE THE RESISTANCE BETWEEN GROUND AND TERMINAL "C5" OF THE 32-WAY EBCM HARNESS CONNECTOR.

DOES J 39200 DISPLAY RESISTANCE AS "OL" (INFINITE)?

- NO: DIAGNOSIS CONTINUED ON PAGE 3 OF 3 OF THIS DTC CHART.
- YES:

(3)
- IGNITION "ON."
- USING J 39200, MEASURE THE VOLTAGE BETWEEN GROUND AND TERMINAL "C5" OF THE 32-WAY EBCM HARNESS CONNECTOR.

IS VOLTAGE 1 VOLT OR LESS?

- NO:

(4)
- IGNITION "OFF."
- DISCONNECT RIGHT REAR WHEEL SPEED SENSOR DIRECTLY AT SENSOR.
- IGNITION "ON."
- USING J 39200, MEASURE THE VOLTAGE BETWEEN GROUND AND TERMINAL "C5" OF THE 32-WAY EBCM HARNESS CONNECTOR.

IS VOLTAGE 1 VOLT OR LESS?

- NO: REPAIR SHORT TO VOLTAGE IN CKT 883.
- YES:

(5)
- USING J 39200, MEASURE THE VOLTAGE BETWEEN GROUND AND TERMINAL "C6" OF THE 32-WAY EBCM HARNESS CONNECTOR.

IS VOLTAGE 1 VOLT OR LESS?

- NO: REPAIR SHORT TO VOLTAGE IN CKT 882.
- YES: MALFUNCTION IS INTERMITTENT. INSPECT ALL CONNECTORS AND HARNESSES FOR DAMAGE WHICH MAY RESULT IN A SHORT TO VOLTAGE WHEN ALL COMPONENTS ARE CONNECTED. REPAIR ALL DAMAGE IF FOUND. REFER TO "DIAGNOSTIC AIDS" ON FACING PAGE FOR MORE INFORMATION.

(3) YES:

(6)
- IGNITION "OFF."
- INSPECT TERMINALS "C5" AND "C6" OF THE 32-WAY EBCM HARNESS CONNECTOR FOR POOR TERMINAL CONTACT AND CONTAMINATION. ALSO INSPECT CKTs 882 AND 883 FOR DAMAGE WHICH MAY RESULT IN AN OPEN CIRCUIT, SHORT TO GROUND, OR SHORT TO VOLTAGE. REPAIR ALL DAMAGE IF EVIDENT.
- RECONNECT ALL CONNECTORS.
- IGNITION "ON."

DOES DTC 35 SET AS A CURRENT DTC?

- YES: REPLACE EBCM.
- NO: MALFUNCTION IS INTERMITTENT. REFER TO "DIAGNOSTIC AIDS" ON FACING PAGE.

DIAGNOSTIC CHART DTC 35 — 1994 GRAND PRIX, CONTINUED

DTC 35
(Page 2 of 3)
RR WHEEL SPEED SNR CKT OPEN OR SHORT TO GND/BATT

DIAGNOSIS CONTINUED FROM PAGE 1 OF 3 OF THIS DTC CHART.

(7)
- DISCONNECT RIGHT REAR WHEEL SPEED SENSOR DIRECTLY AT SENSOR.
- USING J 39200, MEASURE THE RESISTANCE BETWEEN TERMINAL "C5" OF THE 32-WAY EBCM HARNESS CONNECTOR AND TERMINAL "A" OF THE RIGHT REAR WHEEL SPEED SENSOR HARNESS CONNECTOR.

IS RESISTANCE 2Ω OR LESS?

- NO: REPAIR HIGH RESISTANCE IN CKT 883 (WHEEL SPEED SENSOR JUMPER MUST BE REPLACED IF DAMAGED).
- YES:

(8)
- USING J 39200, MEASURE THE RESISTANCE BETWEEN TERMINAL "C6" OF THE 32-WAY EBCM HARNESS CONNECTOR AND TERMINAL "B" OF THE RIGHT REAR WHEEL SPEED SENSOR HARNESS CONNECTOR.

IS RESISTANCE 2Ω OR LESS?

- NO: REPAIR HIGH RESISTANCE IN CKT 882 (WHEEL SPEED SENSOR JUMPER MUST BE REPLACED IF DAMAGED).
- YES:

(9)
- USING J 39200, MEASURE THE RESISTANCE BETWEEN TERMINAL "A" AND TERMINAL "B" OF THE RIGHT REAR WHEEL SPEED SENSOR CONNECTOR.

IS RESISTANCE BETWEEN 2059 AND 2353Ω WHEN SENSOR IS 20°C (68°F)? *(IF TEMPERATURE OF SENSOR IS SUSPECTED TO OTHER THAN 20°C (68°F), REFER TO THE TABLE BELOW FOR APPLICABLE SENSOR RESISTANCE VALUES; THE VALUES IN THE TABLE ARE FOR TEMPERATURE OF THE SENSOR, NOT AIR TEMPERATURE.)*

- NO: REPLACE WHEEL SPEED SENSOR.
- YES: MALFUNCTION IS INTERMITTENT. INSPECT ALL CONNECTORS AND HARNESSES FOR DAMAGE WHICH MAY RESULT IN HIGH RESISTANCE WHEN ALL COMPONENTS ARE CONNECTED. REFER TO "DIAGNOSTIC AIDS" ON FACING PAGE FOR MORE INFORMATION. REPAIR ALL DAMAGE IF FOUND.

REAR WHEEL SPEED SENSOR

TEMPERATURE OF SENSOR VS. SENSOR RESISTANCE (APPROXIMATE)

TEMP. (°C)	TEMP. (°F)	RESISTANCE (OHMS)
-34 TO 4	-30 TO 40	1622 TO 2205
5 TO 43	41 TO 110	1930 TO 2566
44 TO 93	111 TO 200	2245 TO 3028

DIAGNOSTIC CHART DTC 35 — 1994 GRAND PRIX, CONTINUED

DTC 35
(Page 3 of 3)
RR WHEEL SPEED SNR CKT OPEN OR SHORT TO GND/BATT

DIAGNOSIS CONTINUED FROM PAGE 1 OF 3 OF THIS DTC CHART.

(10)
- DISCONNECT CONNECTOR C372 (LOCATED BEHIND RH SIDE OF REAR SEAT).
- USING J 39200, MEASURE THE RESISTANCE BETWEEN GROUND AND TERMINAL "C5" OF THE 32-WAY EBCM HARNESS CONNECTOR.

DOES J 39200 DISPLAY RESISTANCE AS "OL" (INFINITE)?

- NO: REPAIR SHORT TO GROUND IN CKT 883.
- YES:

(11)
- USING J 39200, MEASURE THE RESISTANCE BETWEEN GROUND AND TERMINAL "C6" OF THE 32-WAY EBCM HARNESS CONNECTOR.

DOES J 39200 DISPLAY RESISTANCE AS "OL" (INFINITE)?

- NO: REPAIR SHORT TO GROUND IN CKT 882.
- YES:

(12)
- THOROUGHLY RESPRAY THE RIGHT REAR WHEEL SPEED SENSOR JUMPER HARNESS WITH THE SALT WATER SOLUTION.

IMPORTANT: DO NOT ALLOW WATER TO ENTER ANY WHEEL SPEED SENSOR CONNECTOR.

- DISCONNECT RIGHT REAR WHEEL SPEED SENSOR DIRECTLY AT SENSOR.
- USING J 39200, MEASURE THE RESISTANCE BETWEEN GROUND AND TERMINAL "A," THEN "B" OF THE RIGHT REAR WHEEL SPEED SENSOR HARNESS CONNECTOR.

DOES J 39200 DISPLAY RESISTANCE AS "OL" (INFINITE) FOR BOTH MEASUREMENTS?

- NO: REPLACE RIGHT REAR WHEEL SPEED SENSOR JUMPER HARNESS.
- YES:

(13)
- USING J 39200, MEASURE THE RESISTANCE BETWEEN GROUND AND TERMINAL "A" OF THE RIGHT REAR WHEEL SPEED SENSOR CONNECTOR.

DOES J 39200 DISPLAY RESISTANCE AS "OL" (INFINITE)?

- NO: REPLACE WHEEL SPEED SENSOR.
- YES: MALFUNCTION IS INTERMITTENT. INSPECT ALL CONNECTORS AND HARNESSES FOR DAMAGE WHICH MAY RESULT IN A SHORT TO GROUND WHEN ALL COMPONENTS ARE CONNECTED. PAY PARTICULAR ATTENTION TO THE JUMPER HARNESS, SMALL NICKS OR CUTS CAN CAUSE A MALFUNCTION. REPAIR ANY DAMAGE IF FOUND. JUMPER HARNESS MUST BE REPLACED IF DAMAGED. REFER TO "DIAGNOSTIC AIDS" ON FACING PAGE FOR MORE INFORMATION.

DIAGNOSTIC CHART DTC 36 — 1994 GRAND PRIX

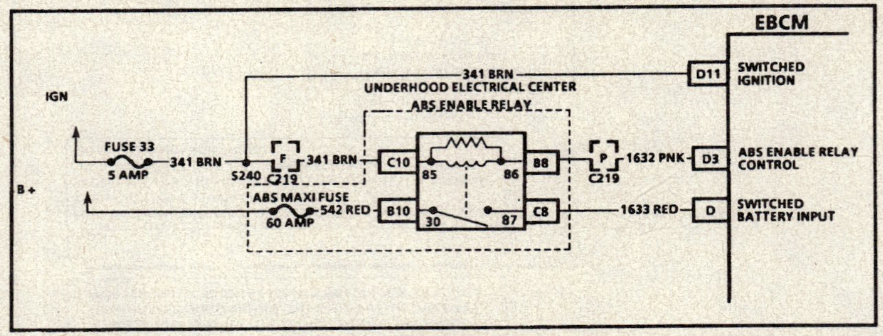

DTC 36

LOW SYSTEM VOLTAGE

Circuit Description:

This DTC is used to monitor the voltage level available to the EBCM controller. If the voltage drops below 11 volts, full performance of the ABS system cannot be guaranteed. During ABS operation, there are several current requirements that will cause battery voltage to drop. Because of this, voltage is monitored prior to ABS to indicate good charging system condition and also during ABS when voltage may drop significantly.

DTC Will Set When: DTC 36 can only be set if the vehicle's speed is greater than 5 km/h (3 mph). If the switched battery voltage is less than 11.2 volts non-ABS or less than 9.6 volts ABS, a malfunction exists.

Action Taken: A malfunction DTC is stored, ABS is disabled and the ABS warning lamp is turned "ON."

DTC Chart Test Description: Number(s) below refer to circled number(s) on the diagnostic chart.

1. Checks to see what voltage is readily available at terminal "D" of the EBCM. If voltage is representative of good charging system condition, a malfunction is not present.
2. Isolates the condition to high circuit resistance, or improper charging system operation.
3. Isolates which circuit is the source of the high resistance condition.
4. Checks for high resistance in CKTs 542 or 2.
5. Checks for high resistance in CKT 341.
6. Checks for high resistance in the ignition fuse feed circuit.
7. Checks for possible intermittent malfunction or a malfunctioning EBCM.

Diagnostic Aids: An "Intermittent" malfunction may be caused by a poor connection, rubbed through wire insulation, or a wire that is broken inside the insulation.

The frequency of the malfunction can be checked by using the enhanced diagnostic function of the Tech 1, as described in "Tech 1 Diagnostics" found in this section.

Any circuitry, that is suspected as causing the intermittent complaint, should be thoroughly checked for backed out terminals, improper mating, broken locks, improperly formed or damaged terminals, poor terminal to wiring connections or physical damage to the wiring harness.

While performing a voltage load test, if it is noted that only ignition voltage drops below acceptable voltage levels, CKT 341 should be checked for high resistance or an open condition.

DIAGNOSTIC CHART DTC 36 — 1994 GRAND PRIX, CONTINUED

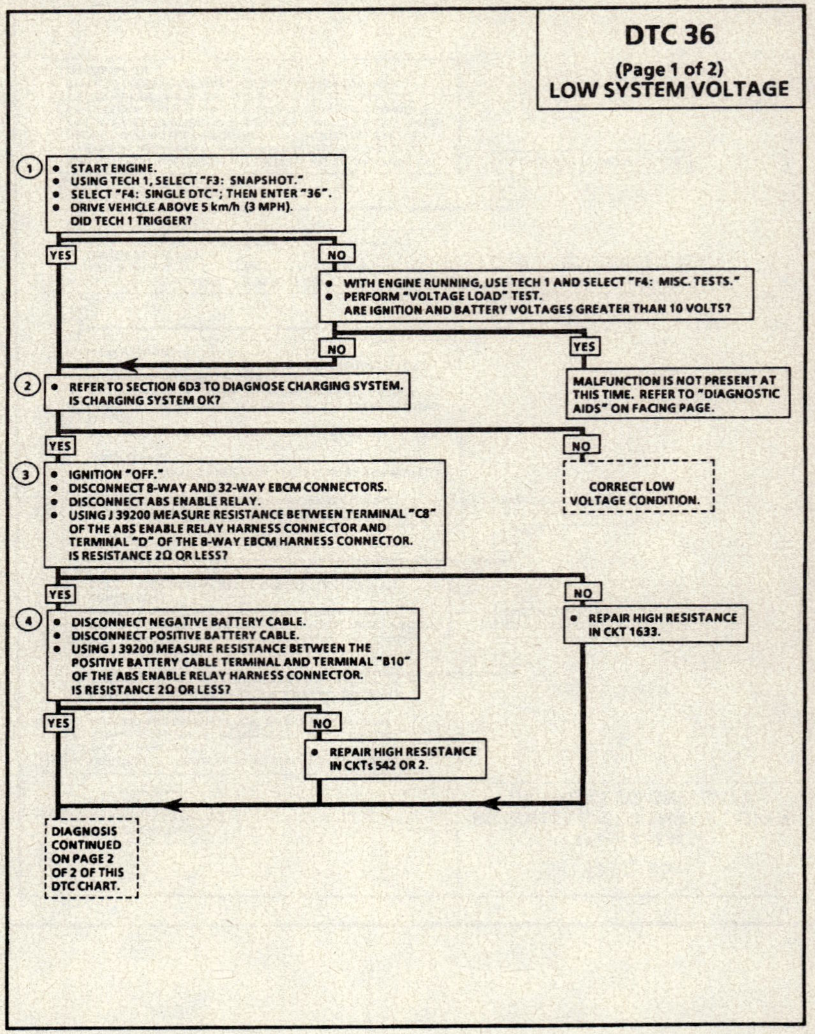

DIAGNOSTIC CHART DTC 36 — 1994 GRAND PRIX, CONTINUED

DTC 36
(Page 2 of 2)
LOW SYSTEM VOLTAGE

DIAGNOSIS CONTINUED FROM PAGE 1 OF 2 OF THIS DTC CHART.

(5)
- REMOVE FUSE 33.
- USING J 39200 MEASURE RESISTANCE BETWEEN EACH FUSE 33 FUSE BLOCK TERMINAL AND TERMINAL "D11" OF THE 32-WAY EBCM HARNESS CONNECTOR. IS EITHER MEASUREMENT 2Ω OR LESS?

NO → REPAIR HIGH RESISTANCE IN CKT 341.

YES ↓

(6)
- WITH BATTERY CABLES DISCONNECTED, TURN IGNITION "ON" (THIS PROVIDES IGNITION CIRCUIT CONTINUITY).
- USING J 39200, MEASURE RESISTANCE BETWEEN POSITIVE BATTERY CABLE TERMINAL AND EACH FUSE 33 FUSE BLOCK TERMINAL. IS EITHER MEASUREMENT 2Ω OR LESS?

NO → REPAIR HIGH RESISTANCE IN THE "IGNITION SWITCH 'RUN' FUSE FEED" CIRCUIT.

YES ↓

(7)
- IGNITION "OFF."
- INSPECT ABS ENABLE RELAY AND ABS ENABLE RELAY HARNESS CONNECTOR FOR POOR TERMINAL CONTACT. INSPECT 8-WAY AND 32-WAY EBCM HARNESS CONNECTORS FOR A POOR CONNECTION TO THE EBCM. INSPECT POSITIVE BATTERY CABLE TERMINAL FOR POOR CONTACT TO THE BATTERY.

REPLACE TERMINALS IF POOR CONTACT IS EVIDENT.
- RECONNECT ALL CONNECTORS.
- DRIVE VEHICLE ABOVE 5 km/h (3 MPH). DOES DTC 36 SET AS A CURRENT DTC?

YES → REPLACE EBCM.

NO → MALFUNCTION IS MOST LIKELY DUE TO A POOR CONNECTION. REFER TO "DIAGNOSTIC AIDS" ON FACING PAGE.

DIAGNOSTIC CHART DTC 37 — 1994 GRAND PRIX

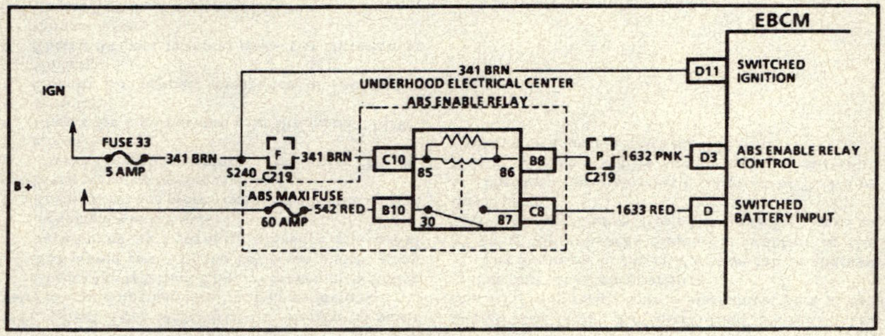

DTC 37
HIGH SYSTEM VOLTAGE

Circuit Description:
This DTC is designed to detect high vehicle voltage levels prior to any required motor movement (initialization or ABS). If excessive voltage exists, demagnetization of the motor magnets may occur, which would eventually affect or eliminate ABS performance.

DTC Will Set When: DTC 37 can only be set if the vehicle's speed is greater than 5 km/h (3 mph). If the switched battery voltage is greater than 17 volts, a malfunction exists.

Action Taken: A malfunction DTC is stored, ABS is disabled and the ABS warning lamp is turned "ON."

DTC Chart Test Description: Number(s) below refer to circled number(s) on the diagnostic chart.
1. Checks the voltage level being received by the EBCM.
2. Checks to see if high voltage condition is due to charging system malfunction.
3. Checks for poor terminal contact which may cause an intermittent malfunction, or a malfunctioning EBCM.

Diagnostic Aids: An "Intermittent" malfunction may be caused by a poor connection, rubbed through wire insulation, or a wire that is broken inside the insulation.

The frequency of the malfunction can be checked by using the enhanced diagnostic function of the Tech 1, as described in "Tech 1 Diagnostics" found in this section.

Any circuitry, that is suspected as causing the intermittent complaint, should be thoroughly checked

DIAGNOSTIC CHART DTC 37 — 1994 GRAND PRIX, CONTINUED

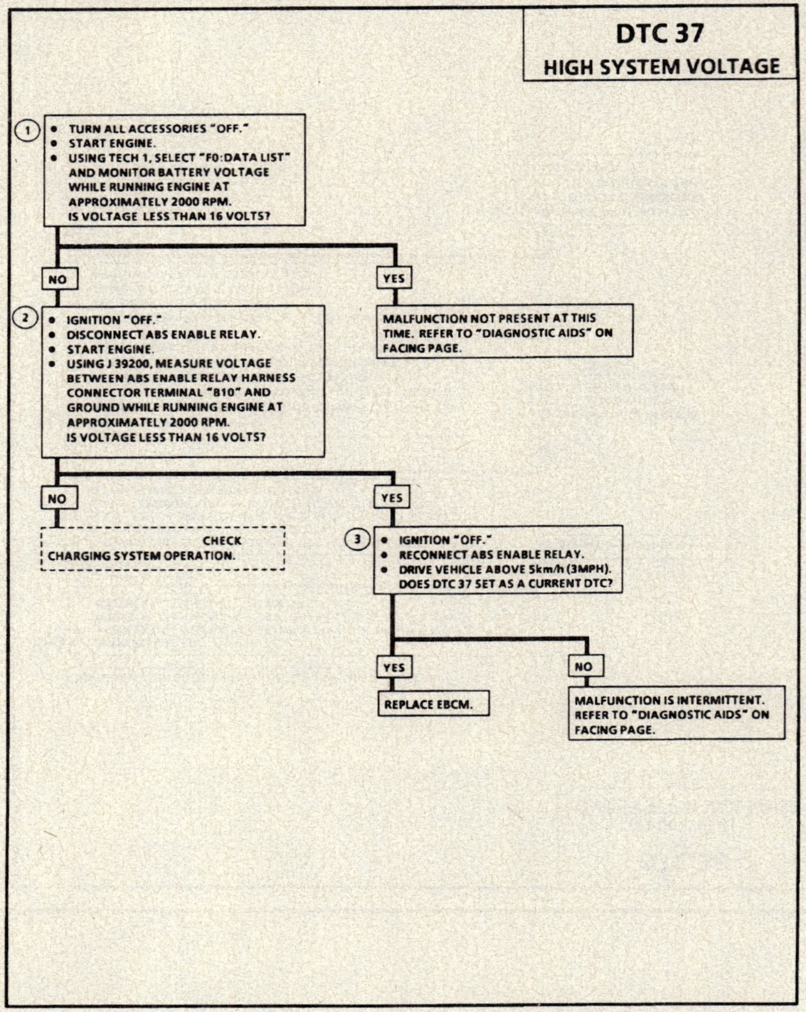

DIAGNOSTIC CHART DTC 38 — 1994 GRAND PRIX

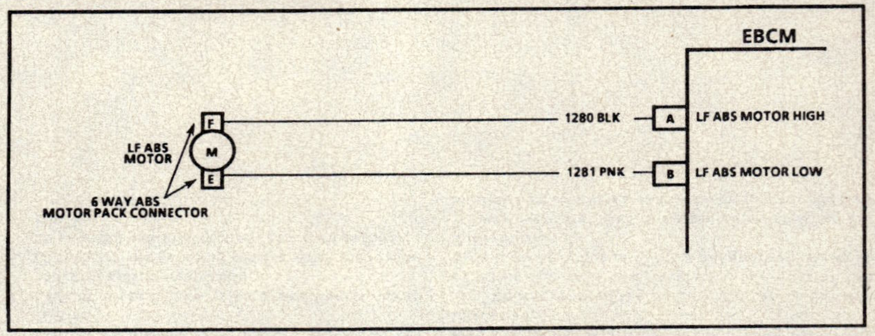

DTC 38
LEFT FRONT ESB WILL NOT HOLD MOTOR

Circuit Description:
This DTC is designed to detect a slipping left front ESB. During initialization and braking, the left front motor is rehomed. If the ESB slips, the motor/piston will move. During the next key "ON" initialization, a rehome of the motor verifies the motor/piston remained at the home position. If motor movement is detected, the ESB must be slipping.

DTC Will Set When: DTC 38 can be set during initialization. If the EBCM detects that the ESB could not hold the piston in the home position, a malfunction exists. DTC 86 is always set with DTC 38.

Action Taken: If an ESB cannot hold a piston in the home position, the piston may be backdriven when the brake pedal is applied, causing the brake pedal to drop. A malfunction DTC is stored, ABS is disabled, and the ABS warning lamp and the red "BRAKE" warning lamp are turned "ON."

DTC Chart Test Description: Number(s) below refer to circled number(s) on the diagnostic chart.

1. Checks the left front ESB. A broken or defective ESB would result in the left front piston being backdriven by hydraulic pressure and pedal movement would result.
2. Ensures that the proper motor circuits are located in the proper connector cavities.
3. Checks for high resistance in the motor "HIGH" circuitry.
4. Checks for high resistance in the motor "LOW" circuitry.
5. Checks for proper resistance of the motor windings.
6. Checks for poor terminal contact or corrosion in the connectors.
7. Checks for a malfunctioning EBCM.

Diagnostic Aids: An "Intermittent" malfunction in this DTC may result from a mechanical part of the system that sticks, binds, or slips.

The frequency of the malfunction can be checked by using the enhanced diagnostic function of the Tech 1, as described in "Tech 1 Diagnostics" found in this section.

The static modulator test function of the Tech 1 or T-100 may be used to locate an intermittent malfunction associated with the ESB.

DIAGNOSTIC CHART DTC 38 — 1994 GRAND PRIX, CONTINUED

DTC 38
LEFT FRONT ESB WILL NOT HOLD MOTOR

(1)
- IGNITION "ON," ENGINE "OFF."
- PUMP BRAKE PEDAL UNTIL FIRM TO DEPLETE VACUUM RESERVE.
- USING TECH 1 MISC TESTS, SELECT MANUAL CONTROL FUNCTION.
- SELECT ABS MOTOR APPLY FUNCTION OF THE LEFT FRONT ABS CHANNEL AND APPLY THE MOTOR.
- WAIT 5 SECONDS.
- APPLY FIRM PRESSURE ON BRAKE PEDAL.

DID BRAKE PEDAL FALL?

YES → REPLACE ABS MOTOR PACK.

NO → (2)
- VISUALLY INSPECT THE 6-WAY ABS MOTOR PACK AND 8-WAY EBCM CONNECTORS FOR PROPER WIRE COLOR/CONNECTOR CAVITY COMBINATION.

ARE THE PROPER WIRES LOCATED IN THE PROPER CONNECTOR CAVITIES?

NO → USING THE CIRCUIT DIAGRAMS LOCATED IN THE FRONT OF THIS SECTION, PLACE THE PROPER WIRES INTO THE PROPER CONNECTOR CAVITIES. CLEAR DTCs AND RECHECK SYSTEM.

YES → (3)
- EXIT TECH 1 MANUAL CONTROL FUNCTION.
- IGNITION "OFF."
- DISCONNECT 6-WAY ABS MOTOR PACK CONNECTOR AND 8-WAY EBCM CONNECTOR.
- USING J 39200, MEASURE THE RESISTANCE BETWEEN TERMINAL "F" OF THE 6-WAY ABS MOTOR PACK HARNESS CONNECTOR AND TERMINAL "A" OF THE 8-WAY EBCM HARNESS CONNECTOR.

IS RESISTANCE 1.5Ω OR LESS?

NO → REPAIR HIGH RESISTANCE IN CKT 1280.

YES → (4)
- MEASURE THE RESISTANCE BETWEEN TERMINAL "E" OF THE 6-WAY ABS MOTOR PACK HARNESS CONNECTOR AND TERMINAL "B" OF THE 8-WAY EBCM HARNESS CONNECTOR.

IS RESISTANCE 1.5Ω OR LESS?

NO → REPAIR HIGH RESISTANCE IN CKT 1281.

YES → (5)
- MEASURE THE RESISTANCE BETWEEN TERMINAL "E" AND TERMINAL "F" OF THE ABS MOTOR PACK CONNECTOR.

IS RESISTANCE 1.5Ω OR LESS?

NO → REPLACE ABS MOTOR PACK.

YES → (6)
- INSPECT THE 6-WAY ABS MOTOR PACK CONNECTOR AND 8-WAY EBCM CONNECTOR FOR POOR TERMINAL CONTACT OR CORROSION.

DO ANY TERMINALS EXHIBIT POOR CONTACT OR EVIDENCE OF CORROSION?

YES → REPLACE TERMINALS THAT EXHIBIT POOR TERMINAL CONTACT OR EVIDENCE OF CORROSION.

NO → (7)
- IGNITION "OFF."
- RECONNECT ALL CONNECTORS.
- START ENGINE WITH FOOT OFF BRAKE.
- ALLOW ENGINE TO RUN FOR AT LEAST 10 SECONDS.
- REPEAT ABOVE IGNITION CYCLE SEQUENCE TWO MORE TIMES.

DID DTC 38 SET IN THE LAST THREE IGNITION CYCLES?

YES → REPLACE EBCM.

NO → MALFUNCTION IS INTERMITENT. REFER TO "DIAGNOSTIC AIDS" ON FACING PAGE.

AFTER DIAGNOSIS IS COMPLETE, CLEAR DTCs AND TEST DRIVE VEHICLE FOR THREE (3) DRIVE CYCLES TO VERIFY THAT DTC DOES NOT RESET. A DRIVE CYCLE CONSISTS OF STARTING THE VEHICLE, DRIVING OVER 16 km/h (10 MPH) AND THEN KEYING DOWN.

DIAGNOSTIC CHART DTC 41 — 1994 GRAND PRIX

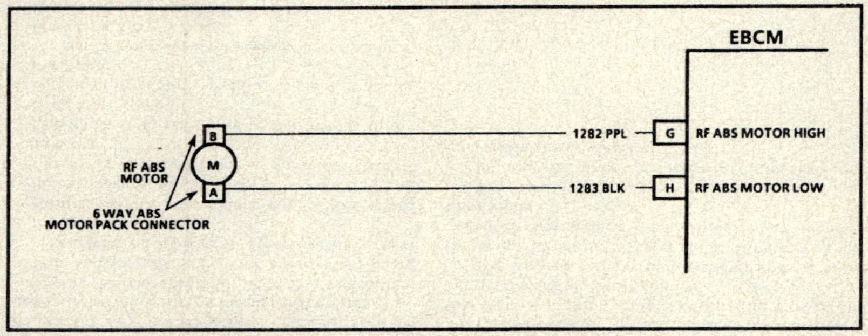

DTC 41
RIGHT FRONT ESB WILL NOT HOLD MOTOR

Circuit Description:
This DTC is designed to detect a slipping right front ESB. During initialization and braking, the right front motor is rehomed. If the ESB slips, the motor/piston will move. During the next key "ON" initialization, a rehome of the motor verifies the motor/piston remained at the home position. If motor movement is detected, the ESB must be slipping.

DTC Will Set When: DTC 41 can be set during initialization. If the EBCM detects that the ESB could not hold the piston in the home position, a malfunction exists. DTC 86 is always set with DTC 41.

Action Taken: If an ESB cannot hold a piston in the home position, the piston may be backdriven when the brake pedal is applied, causing the brake pedal to drop. A malfunction DTC is stored, ABS is disabled, and the ABS warning lamp and the red "BRAKE" warning lamp are turned "ON."

DTC Chart Test Description: Number(s) below refer to circled number(s) on the diagnostic chart.

1. Checks the right front ESB. A broken or defective ESB would result in the right front piston being backdriven by hydraulic pressure, and pedal movement would result.
2. Ensures that the proper motor circuits are located in the proper connector cavities.
3. Checks for high resistance in the motor "HIGH" circuitry.
4. Checks for high resistance in the motor "LOW" circuitry.
5. Checks for proper resistance of the motor windings.
6. Checks for poor terminal contact or corrosion in the connectors.
7. Checks for a malfunctioning EBCM.

Diagnostic Aids: An "Intermittent" malfunction in this DTC may result from a mechanical part of the system that sticks, binds, or slips.

The frequency of the malfunction can be checked by using the enhanced diagnostic function of the Tech 1, as described in "Tech 1 Diagnostics" found in this section.

The static modulator test function of the Tech 1 or T-100 may be used to locate an intermittent malfunction associated with the ESB.

DIAGNOSTIC CHART DTC 41 — 1994 GRAND PRIX, CONTINUED

DTC 41
RIGHT FRONT ESB WILL NOT HOLD MOTOR

(1)
- IGNITION "ON," ENGINE "OFF."
- PUMP BRAKE PEDAL UNTIL FIRM TO DEPLETE VACUUM RESERVE.
- USING TECH 1 MISC. TESTS, SELECT MANUAL CONTROL FUNCTION.
- SELECT ABS MOTOR APPLY FUNCTION OF THE RIGHT FRONT CHANNEL AND APPLY THE MOTOR.
- WAIT 5 SECONDS.
- APPLY FIRM PRESSURE ON BRAKE PEDAL.

DID BRAKE PEDAL FALL?

- YES: REPLACE ABS MOTOR PACK.
- NO: go to (2)

(2)
- VISUALLY INSPECT THE 6-WAY ABS MOTOR PACK AND 8-WAY EBCM CONNECTORS FOR PROPER WIRE COLOR/CONNECTOR CAVITY COMBINATION.

ARE THE PROPER WIRES LOCATED IN THE PROPER CONNECTOR CAVITIES?

- NO: USING THE CIRCUIT DIAGRAMS LOCATED IN THE FRONT OF THIS SECTION, PLACE THE PROPER WIRES INTO THE PROPER CONNECTOR CAVITIES. CLEAR DTCs AND RECHECK SYSTEM.
- YES: go to (3)

(3)
- EXIT TECH 1 MANUAL CONTROL FUNCTION.
- IGNITION "OFF."
- DISCONNECT 6-WAY ABS MOTOR PACK CONNECTOR AND 8-WAY EBCM CONNECTOR.
- USING J 39200, MEASURE THE RESISTANCE BETWEEN TERMINAL "B" OF THE 6-WAY ABS MOTOR PACK HARNESS CONNECTOR AND TERMINAL "G" OF THE 8-WAY EBCM HARNESS CONNECTOR.

IS RESISTANCE 1.5Ω OR LESS?

- NO: REPAIR HIGH RESISTANCE IN CKT 1282.
- YES: go to (4)

(4)
- MEASURE THE RESISTANCE BETWEEN TERMINAL "A" OF THE 6-WAY ABS MOTOR PACK HARNESS CONNECTOR AND TERMINAL "H" OF THE 8-WAY EBCM HARNESS CONNECTOR.

IS RESISTANCE 1.5Ω OR LESS?

- NO: REPAIR HIGH RESISTANCE IN CKT 1283.
- YES: go to (5)

(5)
- MEASURE THE RESISTANCE BETWEEN TERMINAL "A" AND TERMINAL "B" OF THE ABS MOTOR PACK CONNECTOR.

IS RESISTANCE 1.5Ω OR LESS?

- NO: REPLACE ABS MOTOR PACK.
- YES: go to (6)

(6)
- INSPECT THE 6-WAY MOTOR PACK CONNECTOR AND 8-WAY EBCM CONNECTOR FOR POOR TERMINAL CONTACT OR CORROSION.

DO ANY TERMINALS EXHIBIT POOR CONTACT OR EVIDENCE OF CORROSION?

- YES: REPLACE TERMINALS THAT EXHIBIT POOR TERMINAL CONTACT OR EVIDENCE OF CORROSION.
- NO: go to (7)

(7)
- IGNITION "OFF."
- RECONNECT ALL CONNECTORS.
- START ENGINE WITH FOOT OFF BRAKE.
- ALLOW ENGINE TO RUN FOR AT LEAST 10 SECONDS.
- REPEAT ABOVE IGNITION CYCLE TWO MORE TIMES

DID DTC 41 SET IN THE LAST THREE IGNITION CYCLES?

- YES: REPLACE EBCM.
- NO: MALFUNCTION IS INTERMITTENT. REFER TO "DIAGNOSTIC AIDS" ON FACING PAGE.

AFTER DIAGNOSIS IS COMPLETE, CLEAR DTCs AND TEST DRIVE VEHICLE FOR THREE (3) DRIVE CYCLES TO VERIFY THAT DTC DOES NOT RESET. A DRIVE CYCLE CONSISTS OF STARTING THE VEHICLE, DRIVING OVER 16 KM/H (10 MPH) AND THEN KEYING DOWN.

DIAGNOSTIC CHART DTC 42 — 1994 GRAND PRIX

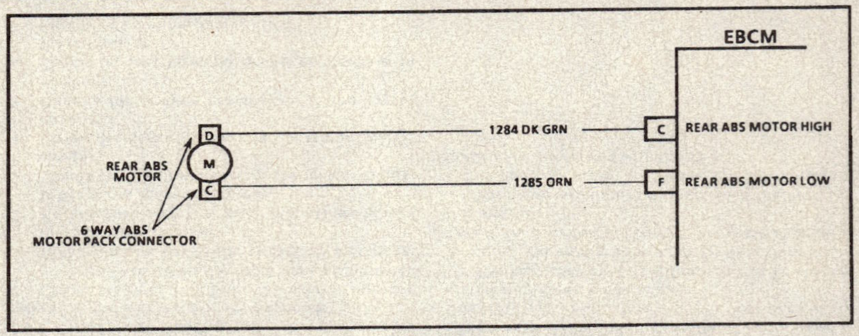

DTC 42
REAR ESB WILL NOT HOLD MOTOR

Circuit Description:

This DTC is designed to detect a slipping rear axle ESB. During initialization and braking, the rear motor is rehomed. If the ESB slips, the motor/piston will move. During the next key "ON" initialization, a rehome of the motor verifies the motor/piston remained at the home position. If motor movement is detected, the ESB must be slipping.

DTC Will Set When: DTC 42 can be set during initialization. If the EBCM detects that the ESB could not hold the piston in the home position, a malfunction exists. DTC 86 is always set with DTC 42.

Action Taken: If an ESB cannot hold a piston in the home position, the piston may be backdriven when the brake pedal is applied, causing the brake pedal to drop. A malfunction DTC is stored, ABS is disabled, and the ABS warning lamp and the red "BRAKE" warning lamp are turned "ON."

DTC Chart Test Description: Number(s) below refer to circled number(s) on the diagnostic chart.

1. Checks the rear axle ESB. A broken or defective ESB would result in the rear axle pistons being backdriven by hydraulic pressure and wheel movement would result.
2. Ensures that the proper motor circuits are located in the proper connector cavities.
3. Checks for high resistance in the motor "HIGH" circuitry.
4. Checks for high resistance in the motor "LOW" circuitry.
5. Checks for proper resistance of the motor windings.
6. Checks for poor terminal contact or corrosion in the connectors.
7. Checks for a malfunctioning EBCM.

Diagnostic Aids: An "Intermittent" malfunction in this DTC may result from a mechanical part of the system that sticks, binds, or slips.

The frequency of the malfunction can be checked by using the enhanced diagnostic function of the Tech 1, as described in "Tech 1 Diagnostics" found in this section.

The static modulator test function of the Tech 1 or T-100 may be used to locate an intermittent malfunction associated with the ESB.

DIAGNOSTIC CHART DTC 42 — 1994 GRAND PRIX, CONTINUED

DTC 42
REAR ESB WILL NOT HOLD MOTOR

(1)
- RAISE THE VEHICLE SUCH THAT THE REAR WHEELS ARE APPROXIMATELY SIX INCHES OFF THE FLOOR.
- IGNITION "ON," ENGINE "OFF."
- PUMP BRAKE PEDAL UNTIL FIRM TO DEPLETE VACUUM RESERVE.
- APPLY FIRM PRESSURE ON BRAKE PEDAL AND RELEASE.
- REAPPLY VERY FIRM PRESSURE ON BRAKE PEDAL AND HAVE AN ASSISTANT TRY TO SPIN THE REAR WHEELS BY HAND.

COULD ASSISTANT SPIN THE REAR WHEELS?

- YES: REPLACE ABS MOTOR PACK.
- NO: go to (2)

(2)
- VISUALLY INSPECT THE 6-WAY ABS MOTOR PACK AND 8-WAY EBCM CONNECTORS FOR PROPER WIRE COLOR/CONNECTOR CAVITY COMBINATION.

ARE THE PROPER WIRES LOCATED IN THE PROPER CONNECTOR CAVITIES?

- NO: USING THE CIRCUIT DIAGRAMS LOCATED IN THE FRONT OF THIS SECTION, PLACE THE PROPER WIRES INTO THE PROPER CONNECTOR CAVITIES. CLEAR DTCs AND RECHECK SYSTEM.
- YES: go to (3)

(3)
- IGNITION "OFF."
- DISCONNECT 6-WAY MOTOR PACK CONNECTOR AND 8-WAY EBCM CONNECTOR.
- USING J 39200, MEASURE THE RESISTANCE BETWEEN TERMINAL "D" OF THE 6-WAY ABS MOTOR PACK HARNESS CONNECTOR AND TERMINAL "C" OF THE 8-WAY EBCM HARNESS CONNECTOR.

IS RESISTANCE 1.5Ω OR LESS?

- NO: REPAIR HIGH RESISTANCE IN CKT 1284.
- YES: go to (4)

(4)
- MEASURE THE RESISTANCE BETWEEN TERMINAL "C" OF THE 6-WAY ABS MOTOR PACK HARNESS CONNECTOR AND TERMINAL "F" OF THE 8-WAY EBCM HARNESS CONNECTOR.

IS RESISTANCE 1.5Ω OR LESS?

- NO: REPAIR HIGH RESISTANCE IN CKT 1285.
- YES: go to (5)

(5)
- MEASURE THE RESISTANCE BETWEEN TERMINAL "C" AND TERMINAL "D" OF THE ABS MOTOR PACK CONNECTOR.

IS RESISTANCE 1.5Ω OR LESS?

- NO: REPLACE ABS MOTOR PACK.
- YES: go to (6)

(6)
- INSPECT THE 6-WAY ABS MOTOR PACK CONNECTOR AND 8-WAY EBCM CONNECTOR FOR POOR TERMINAL CONTACT OR CORROSION.

DO ANY TERMINALS EXHIBIT POOR CONTACT OR EVIDENCE OF CORROSION?

- YES: REPLACE TERMINALS THAT EXHIBIT POOR TERMINAL CONTACT OR EVIDENCE OF CORROSION.
- NO: go to (7)

(7)
- IGNITION "OFF."
- RECONNECT ALL CONNECTORS.
- START ENGINE WITH FOOT OFF BRAKE.
- ALLOW ENGINE TO RUN FOR AT LEAST 10 SECONDS.
- REPEAT ABOVE IGNITION CYCLE SEQUENCE TWO MORE TIMES.

DID DTC 42 SET IN THE LAST THREE IGNITION CYCLES?

- YES: REPLACE EBCM.
- NO: MALFUNCTION IS INTERMITTENT. REFER TO "DIAGNOSTIC AIDS" ON FACING PAGE.

AFTER DIAGNOSIS IS COMPLETE, CLEAR DTCs AND TEST DRIVE VEHICLE FOR THREE (3) DRIVE CYCLES TO VERIFY DTC DOES NOT RESET. A DRIVE CYCLE CONSISTS OF STARTING THE VEHICLE, DRIVING OVER 16 km/h (10 MPH) AND THEN KEYING DOWN.

DIAGNOSTIC CHART DTC 44 — 1994 GRAND PRIX

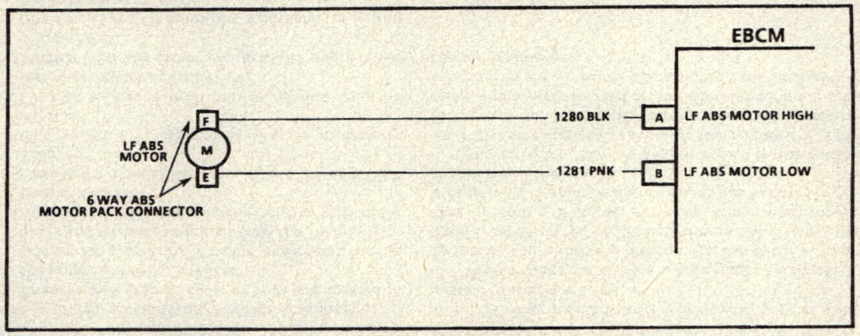

DTC 44
LEFT FRONT CHANNEL WILL NOT MOVE

Circuit Description:

This DTC is designed to detect bound-up ESB, a stuck ABS motor, or a seized hydraulic modulator. When the release is commanded during initialization, the ESB should release the ABS motor, resulting in sensed current being less than commanded current (motor is spinning freely). If the ABS motor is not moving, sensed current will be equal to stall current.

DTC Will Set When: DTC 44 can be set during normal operation. If the EBCM detects a condition in which it cannot move the ABS motor in either direction, a malfunction exists.

Action Taken: This malfunction indicates the channel cannot be moved properly. A malfunction DTC is stored, ABS is disabled and the ABS warning lamp is turned "ON."

DTC Chart Test Description: Number(s) below refer to circled number(s) on the diagnostic chart.

1. Ensures that proper ABS motor circuits are located in the proper connector cavities.
2. Checks for proper ABS motor movement during apply and release commands from the Tech 1.
3. Compares EBCM command current to ABS motor feedback current.
4. Checks for a short between CKT 1280 and CKT 1281.
5. Checks for an ABS motor that is internally shorted.
6. Checks to see if the malfunction is due to poor terminal contact or a faulty EBCM.
7. Verifies that the ABS motor can actually apply when commanded.
8. Checks for proper hydraulic modulator gear and piston movement.

Diagnostic Aids: An "Intermittent" malfunction in this DTC may result from a mechanical part of the system that sticks, binds, or slips.

The frequency of the malfunction can be checked by using the enhanced diagnostic function of the Tech 1, as described in "Tech 1 Diagnostics" found in this section. DTC 44 may set after modulator disassembly if the modulator pistons are positioned at the bottom of their bore.

Any circuitry, that is suspected as causing the intermittent complaint, should be thoroughly checked for backed out terminals, improper mating, broken locks, improperly formed or damaged terminals, poor terminal to wiring connections or physical damage to the wiring harness.

DIAGNOSTIC CHART DTC 44 — 1994 GRAND PRIX, CONTINUED

DTC 44
LEFT FRONT CHANNEL WILL NOT MOVE

(1)
- VISUALLY INSPECT THE 6-WAY ABS MOTOR PACK AND THE 8-WAY EBCM CONNECTORS FOR PROPER WIRE COLOR/CONNECTOR CAVITY COMBINATION.
ARE THE PROPER WIRES LOCATED IN THE PROPER CONNECTOR CAVITIES?

YES → go to (2)

NO → USING THE CIRCUIT DIAGRAMS LOCATED IN THE FRONT OF THIS SECTION, PLACE THE PROPER WIRES IN THE PROPER CONNECTOR CAVITIES. CLEAR DTCs AND RECHECK SYSTEM.

(2)
- IGNITION "ON," ENGINE "OFF."
- PUMP BRAKE PEDAL UNTIL FIRM TO DEPLETE VACUUM RESERVE.
- USING TECH 1 MISC.TESTS, SELECT MANUAL CONTROL FUNCTION.
- SELECT LEFT FRONT ABS MOTOR APPLY FUNCTION AND APPLY THE ABS MOTOR.
- APPLY FIRM PRESSURE ON BRAKE PEDAL.
- SELECT LEFT FRONT ABS MOTOR RELEASE FUNCTION AND RELEASE MOTOR.
DID BRAKE PEDAL FALL?

NO → go to (3)

YES → go to (7)

(7)
- FIRM PRESSURE STILL "ON" BRAKE PEDAL.
- SELECT LEFT FRONT ABS MOTOR APPLY FUNCTION AND APPLY THE MOTOR.
DID BRAKE PEDAL RISE?

NO → go to (3)

YES → MALFUNCTION IS INTERMITTENT. REFER TO "DIAGNOSTIC AIDS" ON FACING PAGE.

(3)
- REMOVE FOOT FROM BRAKE PEDAL.
- SELECT LEFT FRONT ABS MOTOR APPLY FUNCTION AND APPLY THE MOTOR WHILE CAREFULLY OBSERVING THE COMMANDED CURRENT AND THE FEEDBACK CURRENT ON THE TECH 1.
WAS THE FEEDBACK CURRENT HIGHER THAN THE COMMANDED CURRENT?

YES → go to (4)

NO → go to (8)

(4)
- EXIT TECH 1 MANUAL CONTROL FUNCTION.
- IGNITION "OFF."
- DISCONNECT THE 8-WAY EBCM CONNECTOR AND THE 6-WAY MOTOR PACK CONNECTOR.
- USING J 39200, MEASURE THE RESISTANCE BETWEEN TERMINAL "A" AND TERMINAL "B" OF THE 8-WAY EBCM HARNESS CONNECTOR.
DOES J 39200 DISPLAY RESISTANCE AS "OL" (INFINITE)?

YES → go to (5)

NO → REPAIR SHORT BETWEEN CKT 1280 AND CKT 1281.

(5)
- MEASURE THE RESISTANCE BETWEEN TERMINAL "E" AND TERMINAL "F" OF THE ABS MOTOR PACK CONNECTOR.
IS RESISTANCE 0.4Ω OR GREATER?

YES → go to (6)

NO → REPLACE ABS MOTOR PACK.

(6)
- IGNITION "OFF."
- RECONNECT ALL CONNECTORS.
- START ENGINE WITH FOOT OFF BRAKE.
- ALLOW ENGINE TO RUN FOR AT LEAST 10 SECONDS.
- REPEAT ABOVE IGNITION CYCLE SEQUENCE TWO MORE TIMES.
DID DTC 44 SET IN THE LAST THREE IGNITION CYCLES?

YES → REPLACE EBCM.

NO → MALFUNCTION IS INTERMITTENT. INSPECT ALL CONNECTORS FOR CONTAMINATION AND POOR TERMINAL CONTACT. REPLACE ANY TERMINALS THAT EXHIBIT POOR TERMINAL CONTACT OR EVIDENCE OF CORROSION.

(8)
- USING TECH 1 MISC. TESTS, PERFORM GEAR TENSION RELIEF FUNCTION.
- IGNITION "OFF."
- REMOVE HYDRAULIC MODULATOR/MASTER CYLINDER ASSEMBLY FROM VEHICLE. REFER TO "ON VEHICLE SERVICE" IN THIS SECTION.
- SEPARATE ABS MOTOR PACK FROM HYDRAULIC MODULATOR.
- GRASP THE GEAR ON THE HYDRAULIC MODULATOR (THE HYDRAULIC MODULATOR IS THE UNIT WITH THE THREE LARGE GEARS) FURTHEST FORWARD WHEN THE UNIT IS INSTALLED IN THE VEHICLE AND ATTEMPT TO MOVE THE GEAR.
CAN THE GEAR BE ROTATED AT LEAST 10 FULL TURNS LOCK TO LOCK?

NO → REPLACE HYDRAULIC MODULATOR.

YES → REPLACE ABS MOTOR PACK.

AFTER DIAGNOSIS IS COMPLETE, CLEAR DTCs AND TEST DRIVE VEHICLE FOR THREE (3) DRIVE CYCLES TO VERIFY THAT DTC DOES NOT RESET. A DRIVE CYCLE CONSISTS OF STARTING THE VEHICLE, DRIVING OVER 16 KM/H (10 MPH), AND THEN KEYING DOWN.

DIAGNOSTIC CHART DTC 45 — 1994 GRAND PRIX

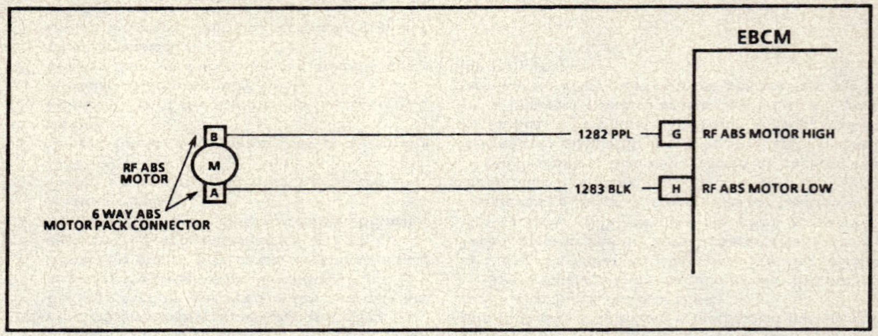

DTC 45
RIGHT FRONT CHANNEL WILL NOT MOVE

Circuit Description:
This DTC is designed to detect bound-up ESB, a stuck ABS motor, or a seized hydraulic modulator. When the release is commanded during initialization, the ESB should release the ABS motor, resulting in sensed current being less than commanded current (motor is spinning freely). If the ABS motor is not moving, sensed current will be equal to stall current.

DTC Will Set When: DTC 45 can be set during normal operation. If the EBCM detects a condition in which it cannot move the motor in either direction, a malfunction exists.

Action Taken: This malfunction indicates the channel cannot be moved properly. A malfunction DTC is stored, ABS is disabled and the ABS warning lamp is turned "ON."

DTC Chart Test Description: Number(s) below refer to circled number(s) on the diagnostic chart.
1. Ensures that proper motor circuits are located in the proper connector cavities.
2. Checks for proper ABS motor movement during apply and release commands from the Tech 1.
3. Compares EBCM command current to ABS motor feedback current.
4. Checks for a short between CKT 1282 and CKT 1283.
5. Checks for an ABS motor that is internally shorted.
6. Checks to see if malfunction is due to poor terminal contact or the EBCM.
7. Verifies that the motor can actually apply when commanded.
8. Checks for proper hydraulic modulator gear and piston movement.

Diagnostic Aids: An "Intermittent" malfunction in this DTC may result from a mechanical part of the system that sticks, binds, or slips.

The frequency of the malfunction can be checked by using the enhanced diagnostic function of the Tech 1, as described in "Tech 1 Diagnostics" found in this section. DTC 45 may set after modulator disassembly if the modulator pistons are positioned at the bottom of their bore.

Any circuitry, that is suspected as causing the intermittent complaint, should be thoroughly checked for backed out terminals, improper mating, broken locks, improperly formed or damaged terminals, poor terminal to wiring connections or physical damage to the wiring harness.

DIAGNOSTIC CHART DTC 45 — 1994 GRAND PRIX, CONTINUED

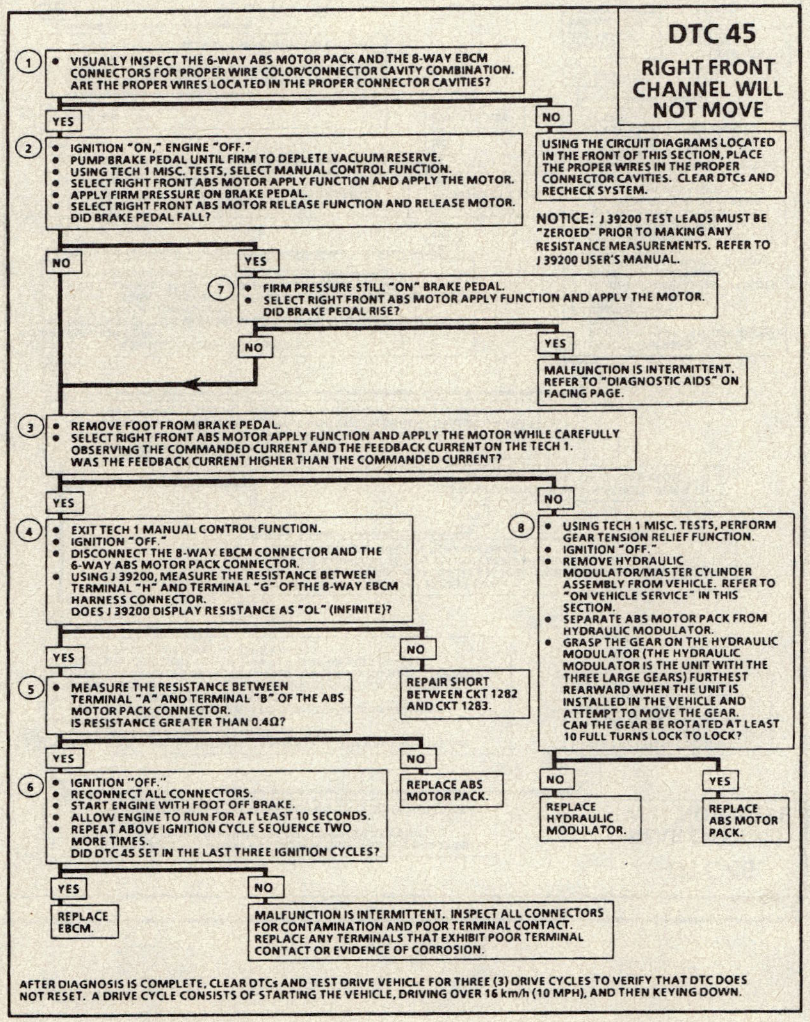

DIAGNOSTIC CHART DTC 46 — 1994 GRAND PRIX

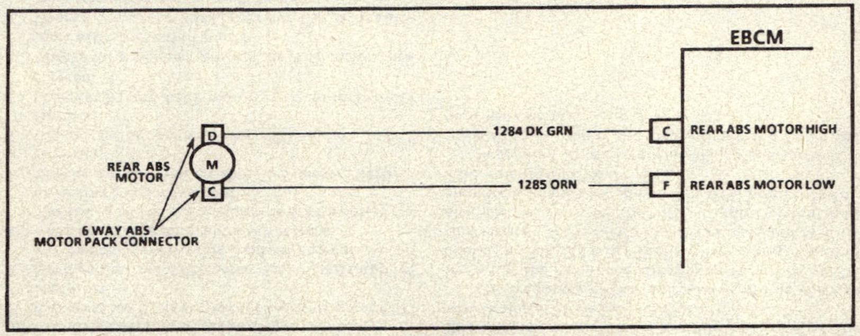

DTC 46

REAR CHANNEL WILL NOT MOVE

Circuit Description:
This DTC is designed to detect bound-up ESB, a stuck ABS motor, or a seized hydraulic modulator. When the release is commanded during initialization, the ESB should release the ABS motor, resulting in sensed current being less than commanded current (motor is spinning freely). If the ABS motor is not moving, sensed current will be equal to stall current.

DTC Will Set When: DTC 46 can be set during normal operation. If the EBCM detects a condition in which it cannot move the ABS motor in either direction, a malfunction exists.

Action Taken: This malfunction indicates the channel cannot be moved properly. A malfunction DTC is stored, ABS is disabled and the ABS warning lamp is turned "ON."

DTC Chart Test Description: Number(s) below refer to circled number(s) on the diagnostic chart.

1. Ensures that proper motor circuits are located in the proper connector cavities.
2. Checks for proper motor movement during apply and release commands from the Tech 1.
3. Compares EBCM command current to ABS motor feedback current.
4. Checks for a short between CKT 1284 and CKT 1285.
5. Checks for an ABS motor that is internally shorted.
6. Checks to see if malfunction is due to poor terminal contact or the EBCM.
7. Verifies that the motor can actually release when commanded.
8. Checks for proper hydraulic modulator gear and piston movement.

Diagnostic Aids: An "Intermittent" malfunction in this DTC may result from a mechanical part of the system that sticks or binds.

The frequency of the malfunction can be checked by using the enhanced diagnostic function of the Tech 1, as described in "Tech 1 Diagnostics" found in this section. DTC 46 may set after modulator disassembly if the modulator pistons are positioned at the bottom of their bore.

Depending on the frequency of the malfunction, a physical inspection of the mechanical parts suspected may be necessary.

Any circuitry, that is suspected as causing the intermittent complaint, should be thoroughly checked for backed out terminals, improper mating, broken locks, improperly formed or damaged terminals, poor terminal to wiring connections or physical damage to the wiring harness.

DIAGNOSTIC CHART DTC 46 — 1994 GRAND PRIX, CONTINUED

DTC 46
REAR CHANNEL WILL NOT MOVE

(1)
- VISUALLY INSPECT THE 6-WAY ABS MOTOR PACK AND THE 8-WAY EBCM CONNECTORS FOR PROPER WIRE COLOR/CONNECTOR CAVITY COMBINATION.

ARE THE PROPER WIRES LOCATED IN THE PROPER CONNECTOR CAVITIES?

NO: USING THE CIRCUIT DIAGRAMS LOCATED IN THE FRONT OF THIS SECTION, PLACE THE PROPER WIRES IN THE PROPER CONNECTOR CAVITIES. CLEAR DTCs AND RECHECK SYSTEM.

YES: (2)
- RAISE AND SUPPORT THE VEHICLE SUCH THAT THE REAR WHEELS ARE APPROXIMATELY SIX INCHES OFF THE FLOOR.
- IGNITION "ON," ENGINE "OFF."
- PUMP BRAKE PEDAL UNTIL FIRM TO DEPLETE VACUUM RESERVE.
- APPLY FIRM PRESSURE ON BRAKE PEDAL.
- USING TECH 1 MISC. TESTS, SELECT MANUAL CONTROL FUNCTION.
- SELECT REAR AXLE ABS MOTOR APPLY FUNCTION AND APPLY THE MOTOR.
- HAVE AN ASSISTANT TRY TO SPIN THE REAR WHEELS BY HAND.

COULD ASSISTANT SPIN THE REAR WHEELS?

NO: (7)
- FIRM PRESSURE STILL "ON" BRAKE PEDAL.
- SELECT REAR AXLE ABS MOTOR RELEASE FUNCTION AND RELEASE THE MOTOR.
- HAVE AN ASSISTANT TRY TO SPIN THE REAR WHEELS BY HAND.

COULD ASSISTANT SPIN THE REAR WHEELS?

YES: MALFUNCTION IS INTERMITTENT. REFER TO "DIAGNOSTIC AIDS" ON FACING PAGE.

NO: (go to step 3)

YES: (3)
- REMOVE FOOT FROM BRAKE PEDAL.
- SELECT REAR AXLE ABS MOTOR APPLY FUNCTION AND APPLY THE MOTOR WHILE CAREFULLY OBSERVING THE COMMANDED CURRENT AND THE FEEDBACK CURRENT ON THE TECH 1.

WAS THE FEEDBACK CURRENT HIGHER THAN THE COMMANDED CURRENT?

YES: (4)
- EXIT TECH 1 MANUAL CONTROL FUNCTION.
- IGNITION "OFF."
- DISCONNECT THE 8-WAY EBCM CONNECTOR AND THE 6-WAY ABS MOTOR PACK CONNECTOR.
- USING J 39200, MEASURE THE RESISTANCE BETWEEN TERMINAL "C" AND TERMINAL "F" OF THE 8-WAY EBCM HARNESS CONNECTOR.

DOES J 39200 DISPLAY RESISTANCE AS "OL" (INFINITE)?

NO: REPAIR SHORT BETWEEN CKT 1284 AND CKT 1285.

YES: (5)
- MEASURE THE RESISTANCE BETWEEN TERMINAL "C" AND TERMINAL "D" OF THE ABS MOTOR PACK CONNECTOR.

IS RESISTANCE 0.4Ω OR GREATER?

NO: REPLACE ABS MOTOR PACK.

YES: (6)
- IGNITION "OFF."
- RECONNECT ALL CONNECTORS.
- START ENGINE WITH FOOT OFF BRAKE.
- ALLOW ENGINE TO RUN FOR AT LEAST 10 SECONDS.
- REPEAT ABOVE IGNITION CYCLE SEQUENCE TWO MORE TIMES.

DID DTC 46 SET IN THE LAST THREE IGNITION CYCLES?

YES: REPLACE EBCM.

NO: MALFUNCTION IS INTERMITTENT. INSPECT ALL CONNECTORS FOR CONTAMINATION AND POOR TERMINAL CONTACT. REPLACE ANY TERMINALS THAT EXHIBIT POOR TERMINAL CONTACT OR EVIDENCE OF CORROSION.

NO (from step 3): (8)
- USING TECH 1 MISC. TESTS, PERFORM GEAR TENSION RELIEF FUNCTION.
- IGNITION "OFF."
- REMOVE HYDRAULIC MODULATOR/MASTER CYLINDER ASSEMBLY FROM VEHICLE. REFER TO "ON VEHICLE SERVICE" IN THIS SECTION.
- SEPARATE ABS MOTOR PACK FROM HYDRAULIC MODULATOR.
- GRASP THE MIDDLE GEAR ON THE HYDRAULIC MODULATOR (THE HYDRAULIC MODULATOR IS THE UNIT WITH THE THREE LARGE GEARS) AND ATTEMPT TO MOVE THE GEAR.

CAN THE GEAR BE ROTATED AT LEAST 7 FULL TURNS LOCK TO LOCK?

NO: REPLACE HYDRAULIC MODULATOR.

YES: REPLACE ABS MOTOR PACK.

AFTER DIAGNOSIS IS COMPLETE, CLEAR DTCs AND TEST DRIVE VEHICLE FOR THREE (3) DRIVE CYCLES TO VERIFY THAT DTC DOES NOT RESET. A DRIVE CYCLE CONSISTS OF STARTING THE VEHICLE, DRIVING OVER 16 KM/H (10 MPH), AND THEN KEYING DOWN.

DIAGNOSTIC CHART DTC 47 — 1994 GRAND PRIX

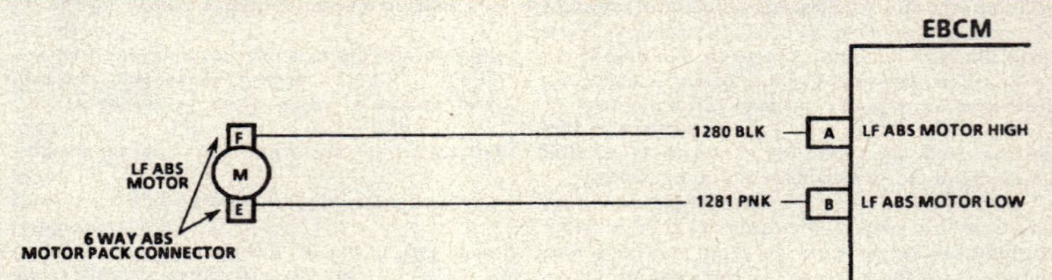

DTC 47

LEFT FRONT ABS MOTOR FREE SPINS

Circuit Description:

This DTC is designed to detect a stripped nut or gear assembly during initialization. During the homing sequence, the piston should reach the top of the bore resulting in a stalled ABS motor. If this does not occur, the ABS motor must be spinning with little or no resistance, indicating a nut/screw or gear malfunction.

DTC Will Set When: DTC 47 can only be set during initialization. If the feedback current is less than the command current for a specified period of time, a malfunction exists.

Action Taken: A malfunction DTC is stored, ABS is disabled, and the ABS warning lamp is turned "ON."

DTC Chart Test Description: Number(s) below refer to circled number(s) on the diagnostic chart.

1. Checks to see if the corresponding open motor DTC is also set.
2. Verifies that the ABS motor was actually applied as command by observing feedback current.
3. Verifies that the ABS motor can release.
4. Verifies that the ABS motor can be applied by observing pedal movement.
5. Checks for an open in the ABS motor "HIGH" circuitry.
6. Checks for an open in the ABS motor "LOW" circuitry.
7. Checks for an ABS motor that is internally shorted.
8. Checks for a stripped gear on the ABS motor pack (unit with three small gears).
9. Checks for a stripped gear on the hydraulic modulator (unit with three large gears).
10. Checks to see if the ABS motor pack has an internal malfunction.
11. Checks for a malfunctioning hydraulic modulator.
12. Ensures malfunction was not due to poor terminal contact.

Diagnostic Aids: An "Intermittent" malfunction in this DTC may result from a mechanical part of the system that sticks, binds, or slips.

The frequency of the malfunction can be checked by using the enhanced diagnostic function of the Tech 1, as described in "Tech 1 Diagnostics" found in this section. If the DTC only occurs once and DTC 56 also occurs, refer to DTC 56. If intermittent and enhanced diagnostics shows this DTC occurs during ABS, refer to DTC 56.

Depending on the frequency of the malfunction, a physical inspection of the mechanical parts suspected may be necessary.

DIAGNOSTIC CHART DTC 47 — 1994 GRAND PRIX, CONTINUED

DTC 47 (Page 1 of 2) LEFT FRONT ABS MOTOR FREE SPINS

1. IS DTC 56 ALSO CURRENTLY SET?
   - YES: PROCEED TO DTC 56 CHART FIRST.
   - NO: go to step 2.

2.
   - IGNITION "ON," ENGINE "OFF."
   - PUMP BRAKE PEDAL UNTIL FIRM TO DEPLETE VACUUM RESERVE.
   - USING TECH 1 MISC. TESTS, SELECT MANUAL CONTROL FUNCTION.
   - SELECT LEFT FRONT ABS MOTOR APPLY FUNCTION AND APPLY THE MOTOR WHILE CAREFULLY OBSERVING THE COMMAND CURRENT AND FEEDBACK CURRENT ON THE TECH 1.

   WERE BOTH COMMAND AND FEEDBACK CURRENT 10 AMPS?
   - YES: go to step 3.
   - NO: GO TO STEP 5.

3.
   - APPLY FIRM PRESSURE ON BRAKE PEDAL.
   - SELECT LEFT FRONT ABS MOTOR RELEASE FUNCTION AND RELEASE THE MOTOR.

   DID THE BRAKE PEDAL FALL AND WERE BOTH COMMAND AND FEEDBACK CURRENT 6 AMPS?
   - YES: go to step 4.
   - NO: GO TO STEP 5.

4.
   - FIRM PRESSURE STILL ON BRAKE PEDAL.
   - SELECT LEFT FRONT ABS MOTOR APPLY FUNCTION AND APPLY THE MOTOR (THIS MOVES THE PISTON TO THE TOP OF ITS BORE).

   DID BRAKE PEDAL RISE?
   - YES: MALFUNCTION IS MOST LIKELY CAUSED BY POOR CONNECTION. REFER TO "DIAGNOSTIC AIDS" ON FACING PAGE.
   - NO: go to step 5.

5.
   - USING TECH 1 MISC. TESTS, PERFORM GEAR TENSION RELIEF FUNCTION.
   - IGNITION "OFF."
   - DISCONNECT 6-WAY ABS MOTOR PACK CONNECTOR AND 8-WAY EBCM CONNECTOR.
   - USING J 39200, MEASURE THE RESISTANCE BETWEEN TERMINAL "F" OF THE 6-WAY ABS MOTOR PACK HARNESS CONNECTOR AND TERMINAL "A" OF THE 8-WAY EBCM HARNESS CONNECTOR.

   IS RESISTANCE 1.5Ω OR LESS?
   - YES: go to step 6.
   - NO: REPAIR HIGH RESISTANCE IN CKT 1280.

6.
   - MEASURE THE RESISTANCE BETWEEN TERMINAL "E" OF THE 6-WAY ABS MOTOR PACK HARNESS CONNECTOR AND TERMINAL "B" OF THE 8-WAY EBCM HARNESS CONNECTOR.

   IS RESISTANCE 1.5Ω OR LESS?
   - YES: go to step 7.
   - NO: REPAIR HIGH RESISTANCE IN CKT 1281.

7.
   - MEASURE THE RESISTANCE BETWEEN TERMINAL "E" AND TERMINAL "F" OF THE ABS MOTOR PACK CONNECTOR.

   IS RESISTANCE 1.5Ω OR LESS?
   - YES: DIAGNOSIS CONTINUED ON NEXT PAGE.
   - NO: REPLACE ABS MOTOR PACK.

AFTER DIAGNOSIS IS COMPLETE, CLEAR DTCs AND TEST DRIVE VEHICLE FOR THREE (3) DRIVE CYCLES TO VERIFY THAT DTC DOES NOT RESET. A DRIVE CYCLE CONSISTS OF STARTING THE VEHICLE, DRIVING OVER 16 km/h (10 MPH), AND THEN KEYING DOWN.

DIAGNOSTIC CHART DTC 47 — 1994 GRAND PRIX, CONTINUED

DTC 47 (Page 2 of 2) LEFT FRONT ABS MOTOR FREE SPINS

DIAGNOSIS CONTINUED FROM PREVIOUS PAGE.

8.
   - REMOVE HYDRAULIC MODULATOR/MASTER CYLINDER ASSEMBLY FROM VEHICLE.
   - REMOVE GEAR COVER.
   - CHECK FOR STRIPPED GEAR ON ABS MOTOR PACK; ABS MOTOR PACK IS THE UNIT WITH THE THREE SMALL GEARS. THE LEFT FRONT GEAR IS THE GEAR FURTHEST FORWARD WHEN THE ASSEMBLY IS INSTALLED IN THE VEHICLE.

   IS THE ABS MOTOR PACK GEAR STRIPPED?
   - YES: REPLACE ABS MOTOR PACK.
   - NO: go to step 9.

9.
   - CHECK FOR A STRIPPED GEAR ON HYDRAULIC MODULATOR; MODULATOR IS THE UNIT WITH THE THREE LARGE GEARS. THE LEFT FRONT GEAR IS THE GEAR FURTHEST FORWARD WHEN THE UNIT IS INSTALLED IN THE VEHICLE.

   IS THE MODULATOR GEAR STRIPPED?
   - YES: REPLACE HYDRAULIC MODULATOR GEAR.
   - NO: go to step 10.

10.
    - RECONNECT ELECTRICAL CONNECTORS
    - SECURELY POSITION HYDRAULIC MODULATOR/MASTER CYLINDER ASSEMBLY IN VEHICLE WITH GEAR COVER REMOVED SUCH THAT THE GEAR SET CAN BE OBSERVED. BE CAREFUL NOT TO DAMAGE GEAR SET OR ALLOW THE HYDRAULIC MODULATOR TO MOVE WHILE TESTING.
    - IGNITION "ON."
    - USING TECH 1 SELECT MANUAL CONTROL FUNCTION.
    - SELECT LEFT FRONT ABS MOTOR APPLY AND APPLY THE MOTOR WHILE OBSERVING THE GEAR SET FURTHEST FORWARD WHEN THE UNIT IS INSTALLED IN THE VEHICLE.
    - SELECT LEFT FRONT ABS MOTOR RELEASE FUNCTION AND RELEASE THE MOTOR WHILE STILL OBSERVING THE GEAR SET.
    - SELECT LEFT FRONT ABS MOTOR APPLY FUNCTION AND APPLY THE MOTOR.

    DID GEAR SET MOVE IN BOTH DIRECTIONS FOR AT LEAST ONE REVOLUTION?
    - YES: go to step 11.
    - NO: REPLACE ABS MOTOR PACK.

11.
    - USING TECH 1 PERFORM GEAR TENSION RELIEF FUNCTION.
    - IGNITION "OFF."
    - SEPARATE THE ABS MOTOR PACK FROM THE HYDRAULIC MODULATOR.
    - GRASP THE GEAR ON THE HYDRAULIC MODULATOR (THE HYDRAULIC MODULATOR IS THE UNIT WITH THE THREE LARGE GEARS) FURTHEST FORWARD WHEN THE UNIT IS INSTALLED IN THE VEHICLE AND ATTEMPT TO MOVE THE GEAR.

    CAN THE GEAR BE ROTATED MORE THAN 13 FULL TURNS LOCK TO LOCK?
    - YES: REPLACE THE HYDRAULIC MODULATOR.
    - NO: go to step 12.

12.
    - INSPECT ALL CONNECTORS AND TERMINALS FOR POOR TERMINAL CONTACT AND EVIDENCE OF CORROSION.
      REPLACE ALL TERMINALS THAT EXHIBIT SIGNS OF POOR TERMINAL CONTACT OR CORROSION.
    - REINSTALL HYDRAULIC MODULATOR/MASTER CYLINDER ASSEMBLY IN VEHICLE.
    - RECONNECT ALL CONNECTORS.
    - IGNITION "OFF."
    - START ENGINE WITH FOOT OFF BRAKE.
    - ALLOW ENGINE TO RUN FOR AT LEAST 10 SECONDS.
    - REPEAT ABOVE IGNITION CYCLE SEQUENCE TWO MORE TIMES.

    DID DTC 47 SET IN THE LAST THREE IGNITION CYCLES?
    - YES: REPLACE EBCM.
    - NO: MALFUNCTION IS MOST LIKELY CAUSED BY A POOR CONNECTION. REFER TO "DIAGNOSTIC AIDS" ON FACING PAGE.

AFTER DIAGNOSIS IS COMPLETE, CLEAR DTCs AND TEST DRIVE VEHICLE FOR THREE (3) DRIVE CYCLES TO VERIFY THAT DTC DOES NOT RESET. A DRIVE CYCLE CONSISTS OF STARTING THE VEHICLE, DRIVING OVER 16 km/h (10 MPH), AND THEN KEYING DOWN.

DIAGNOSTIC CHART DTC 48 — 1994 GRAND PRIX

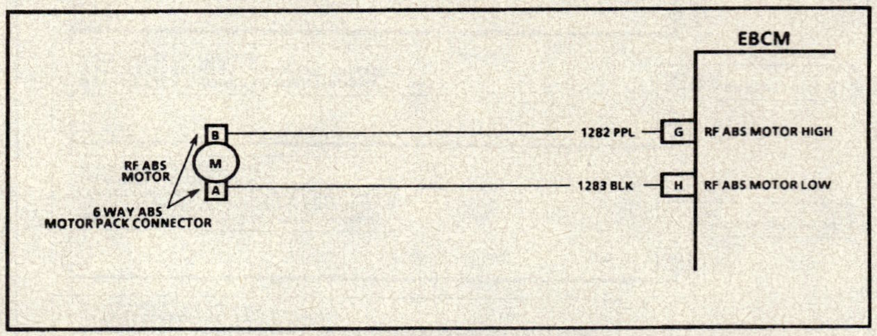

DTC 48

RIGHT FRONT ABS MOTOR FREE SPINS

Circuit Description:
This DTC is designed to detect a stripped nut or gear assembly during initialization. During the homing sequence, the piston should reach the top of the bore resulting in a stalled ABS motor. If this does not occur, the ABS motor must be spinning with little or no resistance, indicating a nut/screw or gear malfunction.

DTC Will Set When: DTC 48 can only be set during initialization. If the feedback current is less than the command current for a specified period of time, a malfunction exists.

Action Taken: A malfunction DTC is stored, ABS is disabled, and the ABS warning lamp is turned "ON."

DTC Chart Test Description: Number(s) below refer to circled number(s) on the diagnostic chart.

1. Checks to see if the corresponding open ABS motor DTC is also set.
2. Verifies that the ABS motor was actually applied as commanded by observing feedback current.
3. Verifies that the ABS motor can release.
4. Verifies that the ABS motor can be applied by observing pedal movement.
5. Checks for high resistance in the ABS motor "HIGH" circuitry.
6. Checks for high resistance in the ABS motor "LOW" circuitry.
7. Checks for proper resistance of the ABS motor windings.
8. Checks for a stripped gear on the ABS motor pack (unit with three small gears).
9. Checks for a stripped gear on the hydraulic modulator (unit with three large gears).
10. Verifies that the ABS motor can be applied under a load situation.
11. Verifies that the hydraulic modulator is functioning properly.
12. Ensures malfunction was not due to poor terminal contact.

Diagnostic Aids: An "Intermittent" malfunction in this DTC may result from a mechanical part of the system that sticks, binds, or slips.

The frequency of the malfunction can be checked by using the enhanced diagnostic function of the Tech 1, as described in "Tech 1 Diagnostics" found in this section. If the DTC only occurs once and DTC 61 also occurs, refer to DTC 61. If intermittent and enhanced diagnostics shows this DTC occurs during ABS, refer to DTC 61.

Depending on the frequency of the malfunction, a physical inspection of the mechanical parts suspected may be necessary.

DIAGNOSTIC CHART DTC 48 — 1994 GRAND PRIX, CONTINUED

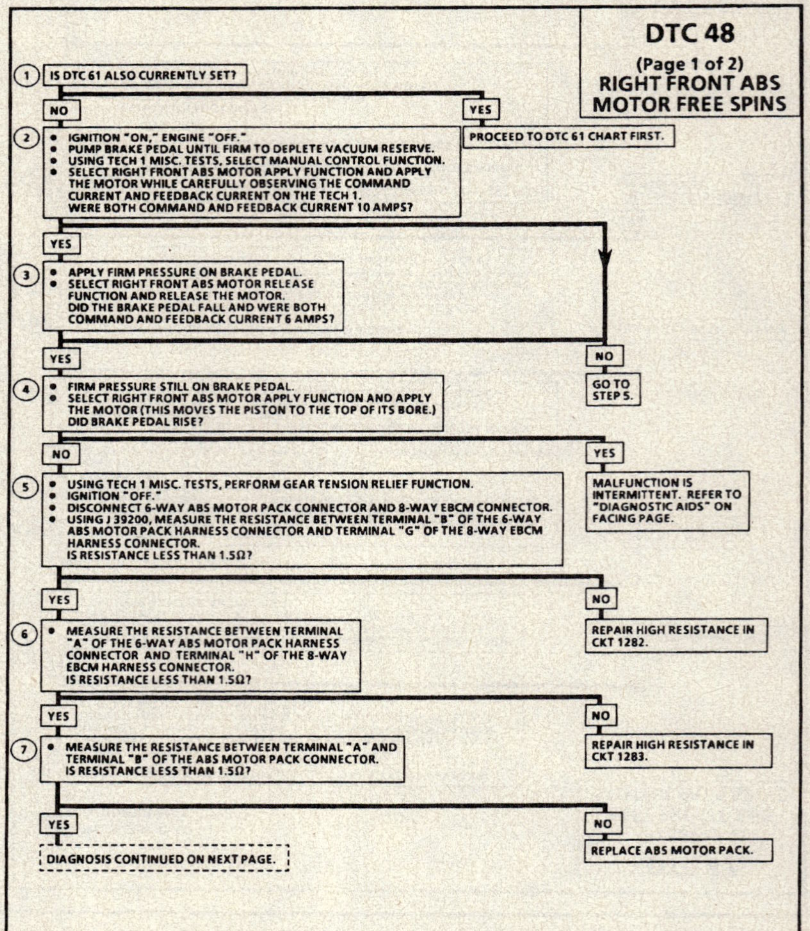

AFTER DIAGNOSIS IS COMPLETE, CLEAR DTCs AND TEST DRIVE VEHICLE FOR THREE (3) DRIVE CYCLES TO VERIFY THAT DTC DOES NOT RESET. A DRIVE CYCLE CONSISTS OF STARTING THE VEHICLE, DRIVING OVER 16 km/h (10 MPH), AND THEN KEYING DOWN.

DIAGNOSTIC CHART DTC 48 — 1994 GRAND PRIX, CONTINUED

DTC 48
(Page 2 of 2)
RIGHT FRONT ABS MOTOR FREE SPINS

DIAGNOSIS CONTINUED FROM PREVIOUS PAGE.

(8)
- REMOVE HYDRAULIC MODULATOR/MASTER CYLINDER ASSEMBLY FROM VEHICLE.
- REMOVE GEAR COVER.
- CHECK FOR STRIPPED GEAR ON ABS MOTOR PACK; ABS MOTOR PACK IS THE UNIT WITH THE THREE SMALL GEARS. THE RIGHT FRONT GEAR IS THE GEAR FURTHEST REARWARD WHEN THE ASSEMBLY IS INSTALLED IN THE VEHICLE. IS THE ABS MOTOR PACK GEAR STRIPPED?

YES: REPLACE ABS MOTOR PACK.

NO:

(9)
- CHECK FOR A STRIPPED GEAR ON HYDRAULIC MODULATOR; MODULATOR IS THE UNIT WITH THE THREE LARGE GEARS. THE RIGHT FRONT GEAR IS THE GEAR FURTHEST REARWARD WHEN THE UNIT IS INSTALLED IN THE VEHICLE. IS THE MODULATOR GEAR STRIPPED?

YES: REPLACE HYDRAULIC MODULATOR GEAR.

NO:

(10)
- RECONNECT ELECTRICAL CONNECTORS.
- SECURELY POSITION HYDRAULIC MODULATOR/MASTER CYLINDER ASSEMBLY IN VEHICLE WITH GEAR COVER REMOVED SUCH THAT THE GEAR SET CAN BE OBSERVED. BE CAREFUL NOT TO DAMAGE GEAR SET OR ALLOW THE HYDRAULIC MODULATOR TO MOVE WHILE TESTING.
- IGNITION "ON."
- USING TECH 1, SELECT MANUAL CONTROL FUNCTION.
- SELECT RIGHT FRONT ABS MOTOR APPLY AND APPLY THE MOTOR WHILE OBSERVING THE GEAR SET FURTHEST REARWARD WHEN THE UNIT IS INSTALLED IN THE VEHICLE.
- SELECT RIGHT FRONT ABS MOTOR RELEASE FUNCTION AND RELEASE THE MOTOR WHILE STILL OBSERVING THE GEAR SET.
- SELECT RIGHT FRONT ABS MOTOR APPLY FUNCTION AND APPLY THE MOTOR. DID GEAR SET MOVE IN BOTH DIRECTIONS FOR AT LEAST ONE REVOLUTION?

NO: REPLACE ABS MOTOR PACK.

YES:

(11)
- USING TECH 1, PERFORM GEAR TENSION RELIEF FUNCTION.
- IGNITION "OFF."
- SEPARATE THE ABS MOTOR PACK FROM THE HYDRAULIC MODULATOR.
- GRASP THE GEAR ON THE HYDRAULIC MODULATOR (THE HYDRAULIC MODULATOR IS THE UNIT WITH THE THREE LARGE GEARS) FURTHEST REARWARD WHEN THE UNIT IS INSTALLED IN THE VEHICLE AND ATTEMPT TO MOVE THE GEAR. CAN THE GEAR BE ROTATED MORE THAN ±3 FULL TURNS LOCK TO LOCK?

YES: REPLACE THE HYDRAULIC MODULATOR.

NO:

(12)
- INSPECT ALL CONNECTORS AND TERMINALS FOR POOR TERMINAL CONTACT AND EVIDENCE OF CORROSION.
  REPLACE ALL TERMINALS THAT EXHIBIT SIGNS OF POOR TERMINAL CONTACT OR CORROSION.
- REINSTALL HYDRAULIC MODULATOR/MASTER CYLINDER ASSEMBLY IN VEHICLE.
- RECONNECT ALL CONNECTORS.
- IGNITION "OFF."
- START ENGINE WITH FOOT OFF BRAKE.
- ALLOW ENGINE TO RUN FOR AT LEAST 10 SECONDS.
- REPEAT ABOVE IGNITION CYCLE SEQUENCE TWO MORE TIMES. DID DTC 48 SET IN THE LAST THREE IGNITION CYCLES?

YES: REPLACE EBCM.

NO: MALFUNCTION WAS MOST LIKELY CAUSED BY A POOR CONNECTION. REFER TO "DIAGNOSTIC AIDS" ON FACING PAGE.

AFTER DIAGNOSIS IS COMPLETE, CLEAR DTCs AND TEST DRIVE VEHICLE FOR THREE (3) DRIVE CYCLES TO VERIFY THAT DTC DOES NOT RESET. A DRIVE CYCLE CONSISTS OF STARTING THE VEHICLE, DRIVING OVER 16 km/h (10 MPH), AND THEN KEYING DOWN.

DIAGNOSTIC CHART DTC 51 — 1994 GRAND PRIX

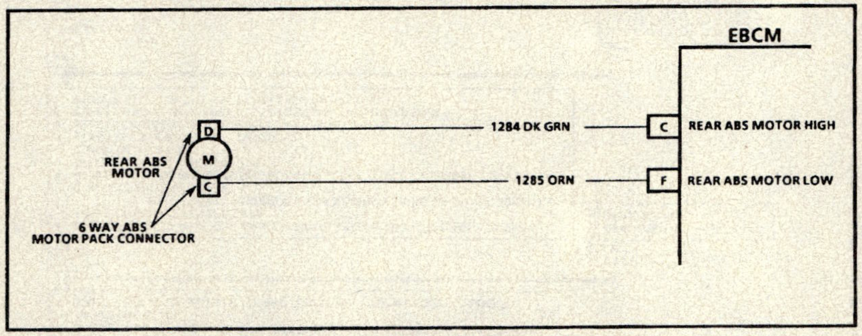

DTC 51

REAR ABS MOTOR FREE SPINS

Circuit Description:
This DTC is designed to detect a stripped nut or gear assembly during initialization. During the homing sequence, the piston should reach the top of the bore resulting in a stalled ABS motor. If this does not occur, the ABS motor must be spinning with little or no resistance, indicating a nut/screw or gear malfunction.

DTC Will Set When: DTC 51 can be set during initialization. If the feedback current is less than the command current for a specified period of time, a malfunction exists.

Action Taken: A malfunction DTC is stored, ABS is disabled, and the ABS warning lamp is turned "ON."

DTC Chart Test Description: Number(s) below refer to circled number(s) on the diagnostic chart.

1. Checks to see if the corresponding open ABS motor DTC is also set.
2. Verifies that the ABS motor was actually applied as commanded by observing feedback current.
3. Verifies that the ABS motor can release.
4. Verifies that the ABS motor can be applied by observing pedal movement.
5. Checks for high resistance in the ABS motor "HIGH" circuitry.
6. Checks for high resistance in the ABS motor "LOW" circuitry.
7. Checks for proper resistance of the ABS motor windings.
8. Checks for a stripped gear on the ABS motor pack (unit with three small gears).
9. Checks for a stripped gear on the hydraulic modulator (unit with three large gears).
10. Verifies that the ABS motor can be applied under a load situation.
11. Verifies that the hydraulic modulator is functioning properly.
12. Ensures malfunction was not due to poor terminal contact.

Diagnostic Aids: An "Intermittent" malfunction in this DTC may result from a mechanical part of the system that sticks, binds, or slips.

The frequency of the malfunction can be checked by using the enhanced diagnostic function of the Tech 1, as described in "Tech 1 Diagnostics" found in this section. If the DTC only occurs once and DTC 64 also occurs, refer to DTC 64. If intermittent and enhanced diagnostics show this DTC occurs during ABS, refer to DTC 64.

Depending on the frequency of the malfunction, a physical inspection of the mechanical parts suspected may be necessary.

DIAGNOSTIC CHART DTC 51 — 1994 GRAND PRIX, CONTINUED

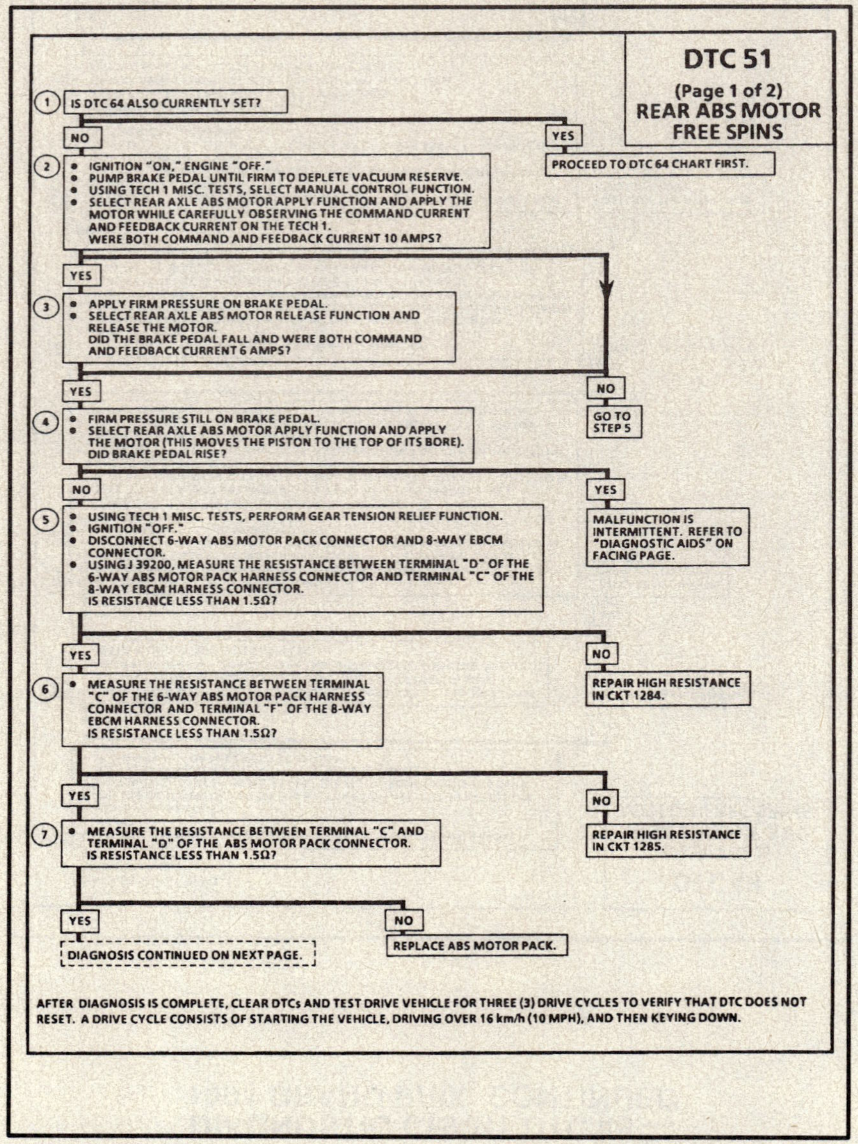

DIAGNOSTIC CHART DTC 51 — 1994 GRAND PRIX, CONTINUED

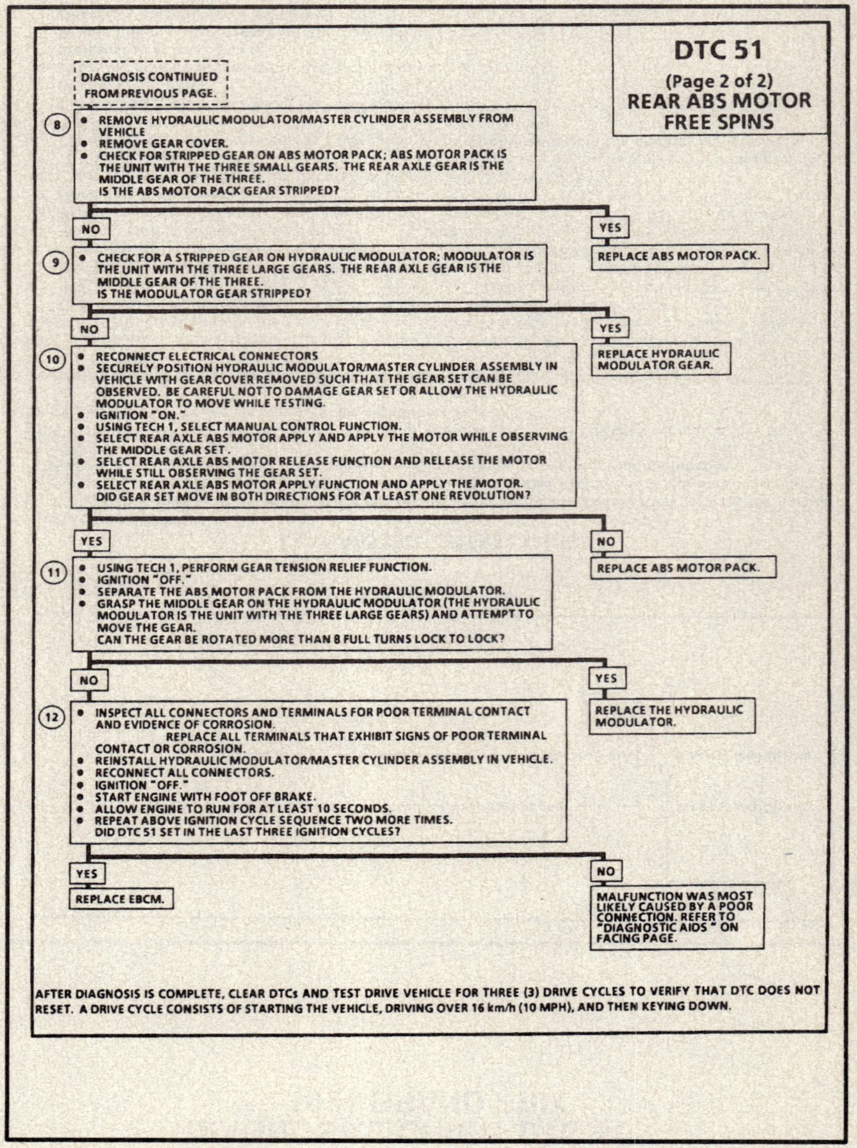

DIAGNOSTIC CHART DTC 52 — 1994 GRAND PRIX

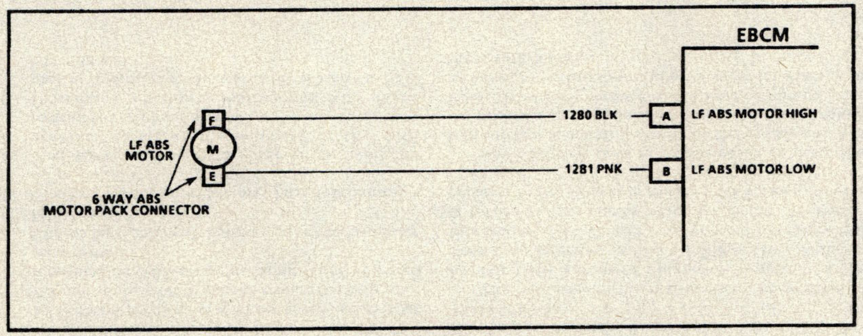

DTC 52

LEFT FRONT CHANNEL IN RELEASE TOO LONG

Circuit Description:

This DTC will diagnose an ABS motor that is energized longer than expected. This could occur if a wheel speed sensor is malfunctioning, the ABS motor does not turn, the left front solenoid mechanically fails open, or the ABS motor wires are crossed.

DTC Will Set When: DTC 52 can be set only during an ABS stop. If the EBCM commands the left front ABS channel in release for three seconds, a malfunction exists.

Action Taken: A malfunction DTC is stored, ABS is disabled and the ABS warning lamp is turned "ON."

DTC Chart Test Description: Number(s) below refer to circled number(s) on the diagnostic chart.

1. Identifies a problem in a wheel speed sensor that may cause the system to be in release too long.
2. Identifies an ABS motor as being failed or wired incorrectly.
3. Checks for a solenoid that may have mechanically failed open.
4. Verifies proper solenoid operation.
5. This serves to isolate the cause of the hydraulic problem, to either the solenoid or the ABS hydraulic modulator assembly.
6. Determines whether a malfunctioning ABS motor pack or hydraulic modulator is the reason for DTC 52 being set.

Diagnostic Aids: An "Intermittent" malfunction may be caused by a mechanical part of the system that sticks or binds.

The frequency of the malfunction can be checked by using the enhanced diagnostic function of the Tech 1, as described in "Tech 1 Diagnostics" found in this section. DTC 52 may occur if on ice and steering wheel is turned to lock during braking. Using the Tech 1, perform hydraulic test to ensure total brake system is functional.

Any circuitry, that is suspected as causing the intermittent complaint, should be thoroughly checked for backed out terminals, improper mating, broken locks, improperly formed or damaged terminals, poor terminal to wiring connections or physical damage to the wiring harness.

DIAGNOSTIC CHART DTC 52 — 1994 GRAND PRIX, CONTINUED

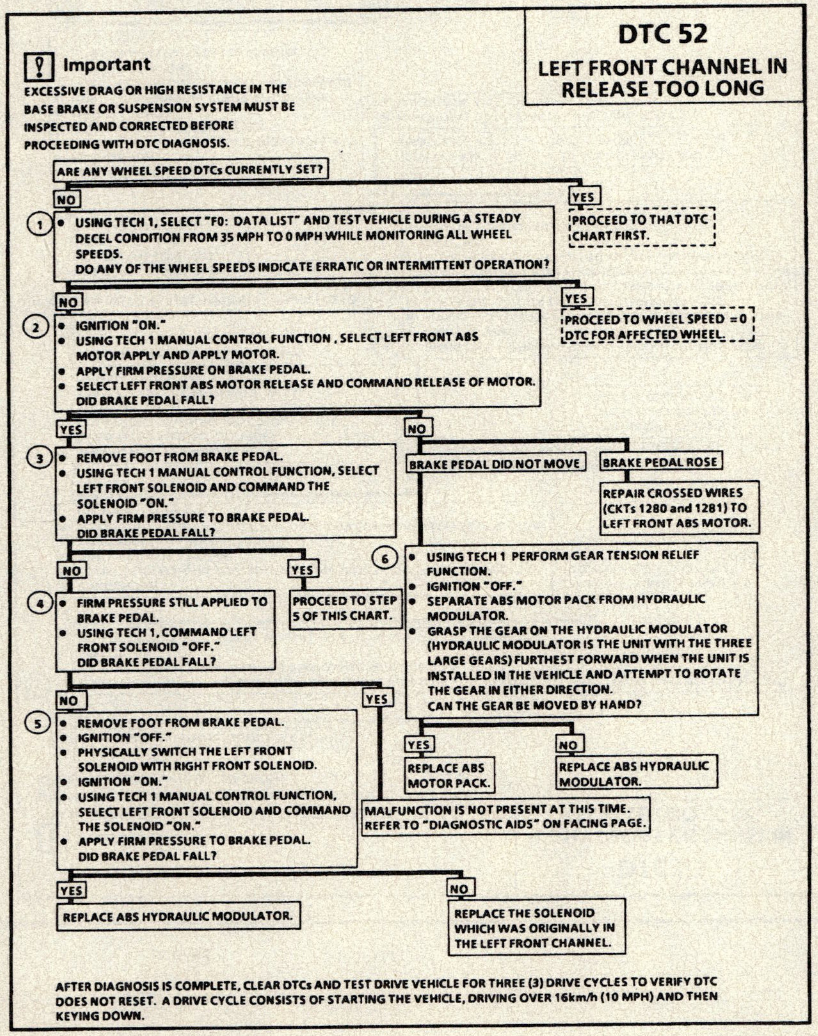

DIAGNOSTIC CHART DTC 53 — 1994 GRAND PRIX

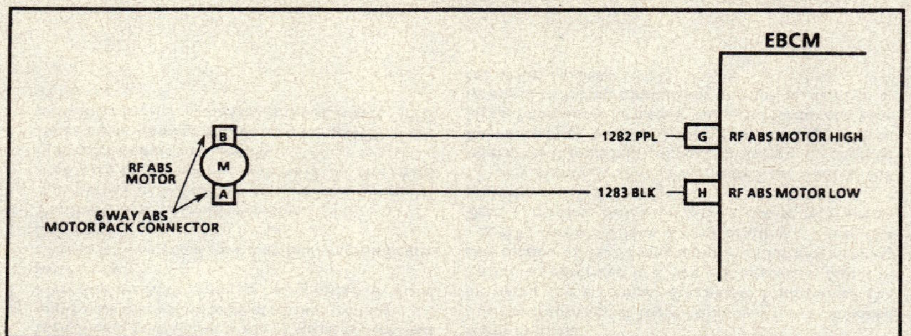

DTC 53

RIGHT FRONT CHANNEL IN RELEASE TOO LONG

Circuit Description:

This DTC will diagnose an ABS motor that is energized longer than expected. This could occur if a wheel speed sensor is malfunctioning, the ABS motor does not turn, the right front solenoid mechanically fails open, or the ABS motor wires are crossed.

DTC Will Set When: DTC 53 can be set only during an ABS stop. If the EBCM commands the right front channel in release for three seconds, a malfunction exists.

Action Taken: A malfunction DTC is stored, ABS is disabled and the ABS warning lamp is turned "ON."

DTC Chart Test Description: Number(s) below refer to circled number(s) on the diagnostic chart.

1. Identifies a problem in a wheel speed sensor that may cause the system to be in release too long.
2. Identifies an ABS motor as being failed or wired incorrectly.
3. Checks for a solenoid that may have mechanically failed open.
4. Checks for a solenoid that may have mechanically failed closed.
5. This serves to isolate the cause of the hydraulic problem, to either the solenoid or the ABS hydraulic modulator assembly.
6. Determines whether a malfunction ABS motor pack or hydraulic modulator is the reason for DTC 53 being set.

Diagnostic Aids: An "Intermittent" malfunction may be caused by a mechanical part of the system that sticks or binds.

The frequency of the malfunction can be checked by using the enhanced diagnostic function of the Tech 1, as described in "Tech 1 Diagnostics" found in this section. DTC 53 may occur on ice if steering wheel is turned to lock during braking. Using the Tech 1 perform hydraulic test to ensure total brake system is functional.

Any circuitry, that is suspected as causing the intermittent complaint, should be thoroughly checked for backed out terminals, improper mating, broken locks, improperly formed or damaged terminals, poor terminal to wiring connections or physical damage to the wiring harness.

DIAGNOSTIC CHART DTC 53 — 1994 GRAND PRIX, CONTINUED

DTC 53

RIGHT FRONT CHANNEL IN RELEASE TOO LONG

Important

EXCESSIVE DRAG OR HIGH RESISTANCE IN THE BASE BRAKE OR SUSPENSION SYSTEM MUST BE INSPECTED AND CORRECTED BEFORE PROCEEDING WITH DTC DIAGNOSIS.

ARE ANY WHEEL SPEED DTCs CURRENTLY SET?

- YES: PROCEED TO THAT DTC CHART FIRST.
- NO: go to step 1.

1.
   - USING TECH 1, SELECT "F0: DATA LIST" AND TEST VEHICLE DURING A STEADY DECEL CONDITION FROM 35 MPH TO 0 MPH WHILE MONITORING ALL WHEEL SPEEDS.
   - DO ANY OF THE WHEEL SPEEDS INDICATE ERRATIC OR INTERMITTENT OPERATION?
   - YES: PROCEED TO WHEEL SPEED = 0 DTC FOR AFFECTED WHEEL.
   - NO: go to step 2.

2.
   - IGNITION "ON."
   - USING TECH 1 MANUAL CONTROL FUNCTION, SELECT RIGHT FRONT ABS MOTOR APPLY AND APPLY MOTOR.
   - APPLY FIRM PRESSURE ON BRAKE PEDAL.
   - SELECT RIGHT FRONT ABS MOTOR RELEASE AND COMMAND RELEASE OF MOTOR.
   - DID BRAKE PEDAL FALL?
   - YES: go to step 3.
   - NO:
     - BRAKE PEDAL DID NOT MOVE: go to step 6.
     - BRAKE PEDAL ROSE: REPAIR CROSSED WIRES (CKTs 1282 AND 1283) TO RIGHT FRONT ABS MOTOR.

3.
   - REMOVE FOOT FROM BRAKE PEDAL.
   - USING TECH 1 MANUAL CONTROL FUNCTION, SELECT RIGHT FRONT SOLENOID AND COMMAND THE SOLENOID "ON."
   - APPLY FIRM PRESSURE TO BRAKE PEDAL.
   - DID BRAKE PEDAL FALL?
   - YES: PROCEED TO STEP 5 OF THIS CHART.
   - NO: go to step 4.

4.
   - FIRM PRESSURE STILL APPLIED TO BRAKE PEDAL.
   - USING TECH 1, COMMAND RIGHT FRONT SOLENOID "OFF."
   - DID BRAKE PEDAL FALL?
   - YES: MALFUNCTION IS NOT PRESENT AT THIS TIME. REFER TO "DIAGNOSTIC AIDS" ON FACING PAGE.
   - NO: go to step 5.

5.
   - REMOVE FOOT FROM BRAKE PEDAL.
   - IGNITION "OFF."
   - PHYSICALLY SWITCH THE RIGHT FRONT SOLENOID WITH THE LEFT FRONT SOLENOID.
   - IGNITION "ON."
   - USING TECH 1 MANUAL CONTROL FUNCTION, SELECT RIGHT FRONT SOLENOID AND COMMAND THE SOLENOID "ON."
   - APPLY FIRM PRESSURE TO BRAKE PEDAL.
   - DID BRAKE PEDAL FALL?
   - YES: REPLACE ABS HYDRAULIC MODULATOR.
   - NO: REPLACE THE SOLENOID WHICH WAS ORIGINALLY IN THE RIGHT FRONT CHANNEL.

6.
   - USING TECH 1 PERFORM GEAR TENSION RELIEF FUNCTION.
   - IGNITION "OFF."
   - SEPARATE ABS MOTOR PACK FROM HYDRAULIC MODULATOR.
   - GRASP THE GEAR ON THE HYDRAULIC MODULATOR (HYDRAULIC MODULATOR IS THE UNIT WITH THE THREE LARGE GEARS) FURTHEST REARWARD WHEN THE UNIT IS INSTALLED IN THE VEHICLE AND ATTEMPT TO ROTATE THE GEAR IN EITHER DIRECTION.
   - CAN THE GEAR BE MOVED BY HAND?
   - YES: REPLACE ABS MOTOR PACK.
   - NO: REPLACE ABS HYDRAULIC MODULATOR.

AFTER DIAGNOSIS IS COMPLETE, CLEAR DTCs AND TEST DRIVE VEHICLE FOR THREE (3) DRIVE CYCLES TO VERIFY DTC DOES NOT RESET. A DRIVE CYCLE CONSISTS OF STARTING THE VEHICLE, DRIVING OVER 16km/h (10 MPH) AND THEN KEYING DOWN.

DIAGNOSTIC CHART DTC 54 — 1994 GRAND PRIX

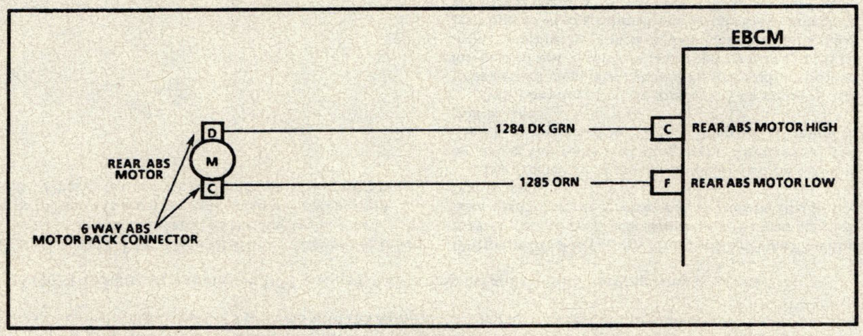

DTC 54

REAR CHANNEL IN RELEASE TOO LONG

Circuit Description:

This DTC will diagnose an ABS motor that is energized longer than expected. This could occur if a wheel speed sensor is malfunctioning, the ABS motor does not turn, or the ABS motor wires are crossed.

DTC Will Set When: DTC 54 can be set only during an ABS stop. If the EBCM commands the rear channel in release for three seconds, a malfunction exists.

Action Taken: A malfunction DTC is stored, ABS is disabled and the ABS warning lamp is turned "ON."

DTC Chart Test Description: Number(s) below refer to circled number(s) on the diagnostic chart.

1. Identifies a problem in a wheel speed sensor that may cause the system to be in release too long.
2. Checks for a wheel that may stick or bind because of a mechanical fault.
3. Checks to see if the ABS motor is capable of moving and applying the hydraulic piston for the rear wheels.
4. Ensures that the ABS motor wiring is not crossed.
5. Isolates the fault of a "no-apply" situation to either the ABS motor pack or ABS hydraulic modulator assembly.

Diagnostic Aids: An "Intermittent" malfunction may be caused by a mechanical part of the system that sticks or binds.

The frequency of the malfunction can be checked by using the enhanced diagnostic function of the Tech 1, as described in "Tech 1 Diagnostics" found in this section. Using the Tech 1, perform hydraulic test to ensure total brake system is functional.

Any circuitry, that is suspected as causing the intermittent complaint, should be thoroughly checked for backed out terminals, improper mating, broken locks, improperly formed or damaged terminals, poor terminal to wiring connections or physical damage to the wiring harness.

DIAGNOSTIC CHART DTC 54 — 1994 GRAND PRIX, CONTINUED

DTC 54

REAR CHANNEL IN RELEASE TOO LONG

Important

- EXCESSIVE DRAG OR HIGH RESISTANCE IN THE BASE BRAKE OR SUSPENSION SYSTEM MUST BE INSPECTED AND CORRECTED BEFORE PROCEEDING WITH DTC DIAGNOSIS.

ARE ANY WHEEL SPEED DTCs CURRENTLY SET?

- YES: PROCEED TO THAT DTC CHART FIRST.
- NO:

(1)
- USING TECH 1, SELECT "F0: DATA LIST" AND TEST VEHICLE DURING A STEADY DECEL CONDITION FROM 35 MPH TO 0 MPH WHILE MONITORING ALL WHEEL SPEEDS. DO ANY OF THE WHEEL SPEEDS INDICATE ERRATIC OR INTERMITTENT OPERATION?

- YES: PROCEED TO WHEEL SPEED = 0 DTC FOR AFFECTED WHEEL.
- NO:

(2)
- RAISE REAR OF VEHICLE AND SUPPORT AS INDICATED IN SECTION 0A.
- SPIN BOTH REAR WHEELS AND CONFIRM PHYSICAL CONDITION OF BASE BRAKE SYSTEM.

COULD BOTH REAR WHEELS BE SPUN FREELY BY HAND?

- NO: REPAIR SOURCE OF RESISTANCE IN BASE BRAKE SYSTEM.
- YES:

(3)
- IGNITION "ON."
- USING MANUAL CONTROL FUNCTION OF TECH 1, SELECT REAR AXLE ABS MOTOR APPLY.
- APPLY FIRM PRESSURE ON BRAKE PEDAL.
- COMMAND APPLY OF REAR ABS MOTOR AND TRY TO SPIN REAR WHEELS BY HAND.

COULD REAR WHEELS BE SPUN BY HAND?

- NO: MALFUNCTION IS NOT PRESENT AT THIS TIME. REFER TO "DIAGNOSTIC AIDS" ON FACING PAGE.
- YES:

(4)
- WITH FIRM PRESSURE ON BRAKE PEDAL STILL APPLIED, COMMAND A RELEASE OF THE REAR AXLE ABS MOTOR.

CAN WHEELS BE MOVED?

- NO: REPAIR CROSSED WIRES AT REAR ABS MOTOR (CKTs 1284 AND 1285).
- YES:

(5)
- USING TECH 1, PERFORM GEAR TENSION RELIEF FUNCTION.
- SEPARATE ABS MOTOR PACK FROM HYDRAULIC MODULATOR.
- CONNECT ABS MOTOR PACK WIRING.
- IGNITION "ON."
- USING TECH 1, PERFORM MOTOR TEST.

DO ALL THREE ABS MOTOR PACK GEARS (SMALL GEARS) SPIN FREELY?

- NO: REPLACE ABS MOTOR PACK.
- YES: REPLACE ABS HYDRAULIC MODULATOR.

AFTER ALL DIAGNOSIS IS COMPLETE, CLEAR DTCs AND TEST DRIVE VEHICLE FOR THREE (3) DRIVE CYCLES TO VERIFY THAT DTC DOES NOT RESET. A DRIVE CYCLE CONSISTS OF STARTING THE VEHICLE DRIVING OVER 16 KM/H (10 MPH), AND THEN KEYING DOWN.

DIAGNOSTIC CHART DTC 55 — 1994 GRAND PRIX

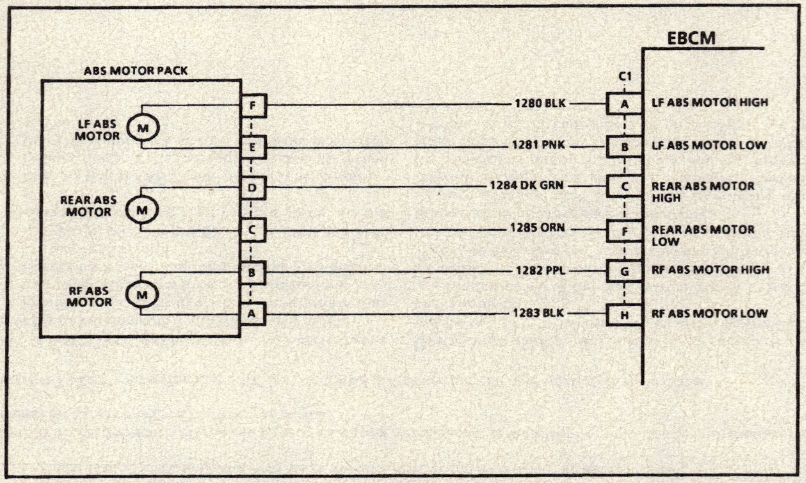

DTC 55
EBCM MALFUNCTION

Circuit Description:
This DTC identifies a malfunction detected by the MDI custom IC. It also ensures the cause of the malfunction is not a result of a problem with the ABS enable relay under a DTC 55 malfunction.

Action Taken: A malfunction DTC is stored, ABS is disabled and the ABS warning lamp is turned "ON."

DTC Chart Test Description: Number(s) below refer to circled number(s) on the diagnostic chart.
1. Checks to see if the malfunction is still present.
2. Checks to see if the malfunction is intermittent.

Diagnostic Aids: An "Intermittent" malfunction may be caused by a poor connection, rubbed through wire insulation, or a wire that is broken inside the insulation.

The frequency of the malfunction can be checked by using the enhanced diagnostic function of the Tech 1, as described in "Tech 1 Diagnostics" found in this section.

Any circuitry, that is suspected as causing the intermittent complaint, should be thoroughly checked for backed out terminals, improper mating, broken locks, improperly formed or damaged terminals, poor terminal to wiring connections or physical damage to the wiring harness.

DIAGNOSTIC CHART DTC 55 — 1994 GRAND PRIX, CONTINUED

DTC 55
EBCM MALFUNCTION

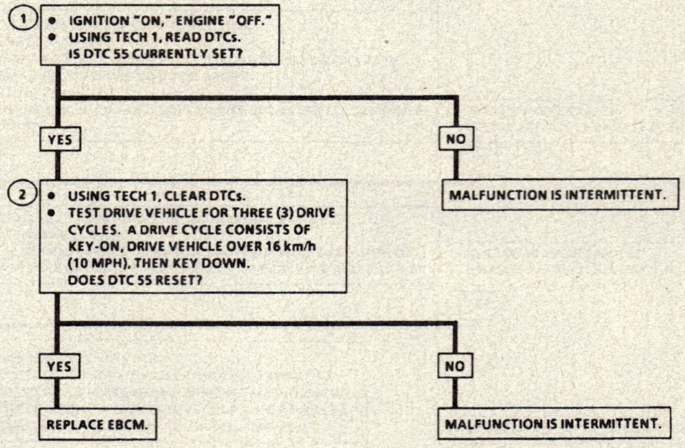

DIAGNOSTIC CHART DTC 56 — 1994 GRAND PRIX

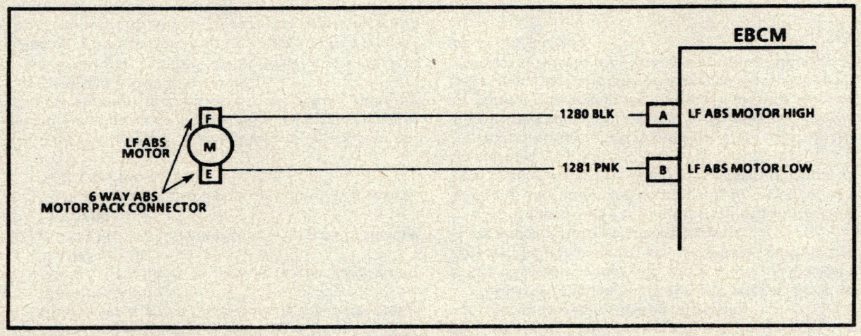

DTC 56

LEFT FRONT ABS MOTOR CIRCUIT OPEN

Circuit Description:
This DTC identifies a motor that cannot be energized due to an open in its circuitry.

DTC Will Set When: DTC 56 can be set only when the motor is commanded "OFF." If the EBCM detects an out of range voltage on either of the left front ABS motor circuits indicating an open circuit, a malfunction exists.

Action Taken: An open ABS motor will not activate when requested. A malfunction DTC is stored, ABS is disabled and the ABS warning lamp is turned "ON."

DTC Chart Test Description: Number(s) below refer to circled number(s) on the diagnostic chart.

1. Checks for proper resistance of the ABS motor.
2. Checks for an open in the ABS motor "HIGH" circuitry.
3. Checks for an open in the ABS motor "LOW" circuitry.
4. Ensures malfunction was not due to poor terminal contact.

Diagnostic Aids: Using Tech 1, select manual control function, and exercise ABS motor movement of affected channel in both directions while applying light pressure on the brake pedal.

If erratic or "jumpy" brake pedal movement is detected while performing an "apply" or "release" function of the ABS motor, an intermittent malfunction may be indicated.

An "Intermittent" malfunction may be caused by a poor connection, rubbed through wire insulation, or a wire that is broken inside the insulation.

If the malfunction is not current, wiggle the wires of the affected channel and check if the DTC resets. This will help to pinpoint an intermittent malfunction in the motor circuitry or connections.

The frequency of the malfunction can be checked by using the enhanced diagnostic function of the Tech 1, as described in "Tech 1 Diagnostics" found in this section.

Any circuitry, that is suspected as causing the intermittent complaint, should be thoroughly checked for backed out terminals, improper mating, broken locks, improperly formed or damaged terminals, poor terminal to wiring connections or physical damage to the wiring harness.

DIAGNOSTIC CHART DTC 56 — 1994 GRAND PRIX, CONTINUED

DTC 56

LEFT FRONT ABS MOTOR CIRCUIT OPEN

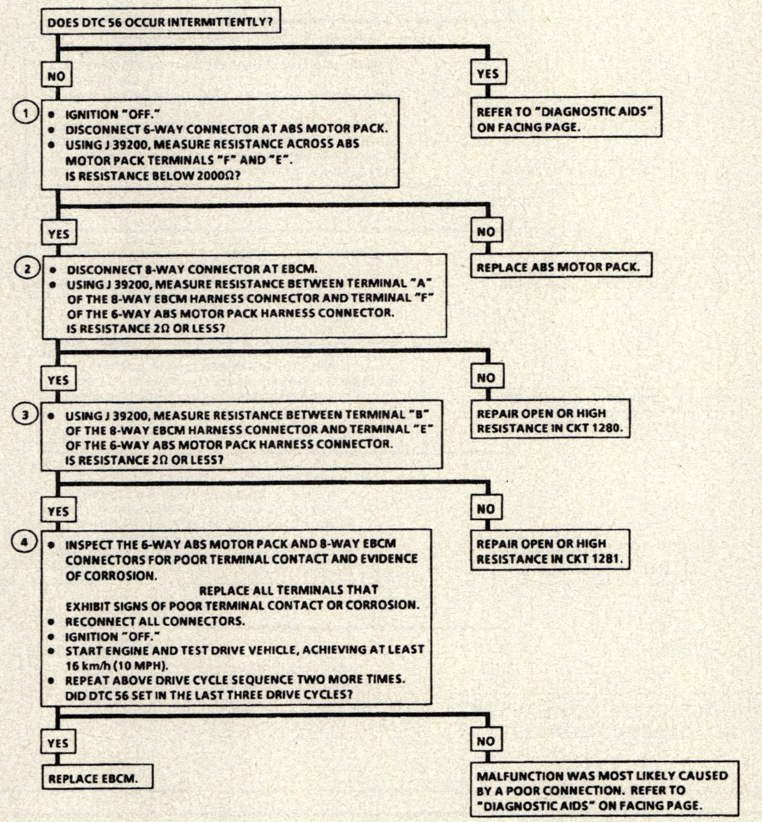

AFTER ALL DIAGNOSIS IS COMPLETE, CLEAR DTCs AND TEST DRIVE VEHICLE FOR THREE (3) DRIVE CYCLES TO VERIFY THAT DTC DOES NOT RESET. A DRIVE CYCLE CONSISTS OF STARTING THE VEHICLE, DRIVING OVER 16 KM/H (10 MPH), AND THEN KEYING DOWN.

DIAGNOSTIC CHART DTC 57 — 1994 GRAND PRIX

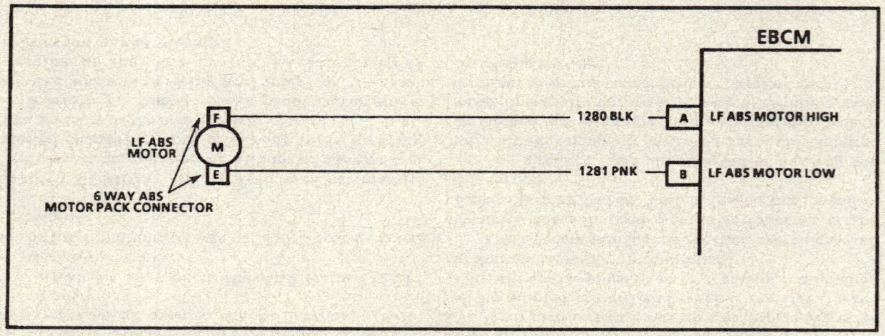

DTC 57

LEFT FRONT ABS MOTOR CIRCUIT SHORTED TO GROUND

Circuit Description:

This DTC identifies an ABS motor circuit that is shorted to ground. This malfunction will not allow the ABS motor to be controlled at the commanded current rate or will cause the driver circuit to allow current directly to ground.

DTC Will Set When: DTC 57 can be set anytime. If the EBCM detects an out of range voltage on either of the left front ABS motor circuits indicating a circuit shorted to ground, a malfunction exists.

Action Taken: A malfunction DTC is stored, ABS is disabled and the ABS warning lamp is turned "ON."

DTC Chart Test Description: Number(s) below refer to circled number(s) on the diagnostic chart.

1. Checks for a short to ground in the ABS motor "HIGH" circuitry.
2. Checks for a short to ground in the ABS motor "LOW" circuitry.
3. Checks for an ABS motor that is internally shorted to ground.
4. Ensures malfunction was not due to physical damage of the circuitry.

Diagnostic Aids: Using Tech 1, select manual control function, and exercise ABS motor movement of affected channel in both directions while applying light pressure on the brake pedal.

If erratic or "jumpy" brake pedal movement is detected while performing an "apply" or "release" function of the ABS motor, an intermittent malfunction may be indicated.

An "Intermittent" malfunction may be caused by a poor connection, rubbed through wire insulation, or a wire that is broken inside the insulation.

If the malfunction is not current, wiggle the wires of the affected channel and check if the DTC resets. This will help to pinpoint an intermittent malfunction in the motor circuitry or connections.

The frequency of the malfunction can be checked by using the enhanced diagnostic function of the Tech 1, as described in "Tech 1 Diagnostics" found in this section.

Any circuitry, that is suspected as causing the intermittent complaint, should be thoroughly checked for backed out terminals, improper mating, broken locks, improperly formed or damaged terminals, poor terminal to wiring connections or physical damage to the wiring harness.

DIAGNOSTIC CHART DTC 57 — 1994 GRAND PRIX, CONTINUED

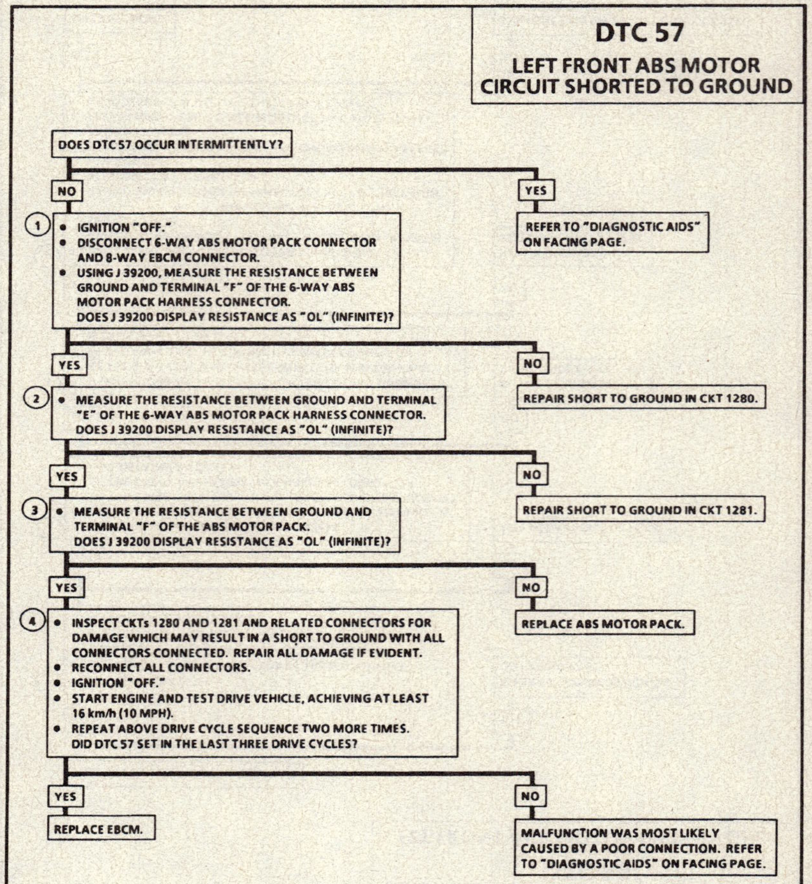

AFTER ALL DIAGNOSIS IS COMPLETE, CLEAR DTCs AND TEST DRIVE VEHICLE FOR THREE (3) DRIVE CYCLES TO VERIFY THAT DTC DOES NOT RESET. A DRIVE CYCLE CONSISTS OF STARTING THE VEHICLE, DRIVING OVER 16 KM/H (10 MPH), AND THEN KEYING DOWN.

DIAGNOSTIC CHART DTC 58 — 1994 GRAND PRIX

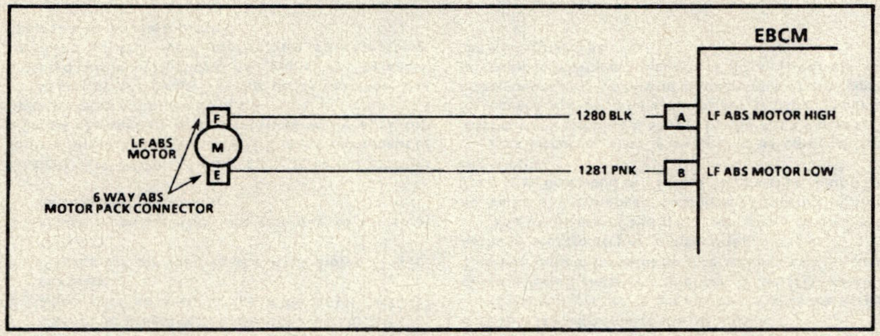

DTC 58

LEFT FRONT ABS MOTOR CIRCUIT SHORTED TO BATTERY

Circuit Description:
This DTC identifies an ABS motor circuit that is shorted to battery or an ABS motor that has low or no resistance. This malfunction will not allow the ABS motor to be controlled at the commanded current rate or will cause the ABS motor to turn in the opposite direction, or not at all.

DTC Will Set When: DTC 58 can be set only when the ABS motor is commanded "OFF." If the EBCM detects an out of range voltage on either of the left front ABS motor circuits indicating a circuit shorted to battery or an ABS motor shorted, a malfunction exists.

Action Taken: A malfunction DTC is stored, ABS is disabled and the ABS warning lamp is turned "ON."

DTC Chart Test Description: Number(s) below refer to circled number(s) on the diagnostic chart.

1. Checks for a short to voltage in the ABS motor "HIGH" circuitry.
2. Checks for a short to voltage in the ABS motor "LOW" circuitry.
3. Checks for a short between the "HIGH" and "LOW" ABS motor circuits.
4. Checks for an ABS motor that is internally shorted.
5. Ensures malfunction was not due to physical damage of the circuitry.

Diagnostic Aids: Using Tech 1, select manual control function, and exercise ABS motor movement of affected channel in both directions while applying light pressure on the brake pedal.

If erratic or "jumpy" brake pedal movement is detected while performing an "apply" or "release" function of the ABS motor, an intermittent malfunction may be indicated.

An "Intermittent" malfunction may be caused by a poor connection, rubbed through wire insulation, or a wire that is broken inside the insulation.

If the malfunction is not current, wiggle the wires of the affected channel and check if the DTC resets. This will help to pinpoint an intermittent malfunction in the motor circuitry or connections.

The frequency of the malfunction can be checked by using the enhanced diagnostic function of the Tech 1, as described in "Tech 1 Diagnostics" found in this section.

Any circuitry, that is suspected as causing the intermittent complaint, should be thoroughly checked for backed out terminals, improper mating, broken locks, improperly formed or damaged terminals, poor terminal to wiring connections or physical damage to the wiring harness.

DIAGNOSTIC CHART DTC 58 — 1994 GRAND PRIX, CONTINUED

DTC 58
LEFT FRONT ABS MOTOR CIRCUIT SHORTED TO BATTERY

NOTICE: J 39200 TEST LEADS MUST BE "ZEROED" PRIOR TO MAKING ANY RESISTANCE MEASUREMENTS. REFER TO J 39200 USER'S MANUAL.

DOES DTC 58 OCCUR INTERMITTENTLY?

- YES → REFER TO "DIAGNOSTIC AIDS" ON FACING PAGE.
- NO → ①

①
- IGNITION "OFF."
- DISCONNECT 6-WAY ABS MOTOR PACK CONNECTOR AND 8-WAY EBCM CONNECTOR.
- IGNITION "ON."
- USING J 39200, MEASURE THE VOLTAGE BETWEEN GROUND AND TERMINAL "F" OF THE 6-WAY ABS MOTOR PACK HARNESS CONNECTOR.

IS VOLTAGE 1 VOLT OR LESS?

- YES → ②
- NO → REPAIR SHORT TO VOLTAGE IN CKT 1280.

②
- MEASURE THE VOLTAGE BETWEEN GROUND AND TERMINAL "E" OF THE 6-WAY ABS MOTOR PACK HARNESS CONNECTOR.

IS VOLTAGE 1 VOLT OR LESS?

- YES → ③
- NO → REPAIR SHORT TO VOLTAGE IN CKT 1281.

③
- IGNITION "OFF."
- USING J 39200, MEASURE THE RESISTANCE BETWEEN TERMINAL "E" AND TERMINAL "F" OF THE 6-WAY ABS MOTOR PACK HARNESS CONNECTOR.

DOES J 39200 DISPLAY RESISTANCE AS "OL" (INFINITE)?

- YES → ④
- NO → REPAIR SHORT BETWEEN CKTs 1280 AND 1281.

④
- USING J 39200, MEASURE THE RESISTANCE BETWEEN TERMINAL "E" AND TERMINAL "F" OF THE ABS MOTOR PACK.

IS RESISTANCE GREATER THAN 0.4Ω?

- YES → ⑤
- NO → REPLACE ABS MOTOR PACK.

⑤
- INSPECT CKTs 1280 AND 1281 AND RELATED CONNECTORS FOR DAMAGE WHICH MAY RESULT IN A SHORT TO GROUND WITH ALL CONNECTORS CONNECTED. REPAIR ALL DAMAGE IF EVIDENT.
- RECONNECT ALL CONNECTORS.
- IGNITION "OFF."
- START ENGINE AND TEST DRIVE VEHICLE, ACHIEVING AT LEAST 16 km/h (10 MPH).
- REPEAT ABOVE DRIVE CYCLE SEQUENCE TWO MORE TIMES.

DID DTC 58 SET IN THE LAST THREE DRIVE CYCLES?

- YES → REPLACE EBCM.
- NO → MALFUNCTION WAS MOST LIKELY CAUSED BY A POOR CONNECTION. REFER TO "DIAGNOSTIC AIDS" ON FACING PAGE.

AFTER ALL DIAGNOSIS IS COMPLETE, CLEAR DTCs AND TEST DRIVE VEHICLE FOR THREE (3) DRIVE CYCLES TO VERIFY THAT DTC DOES NOT RESET. A DRIVE CYCLE CONSISTS OF STARTING THE VEHICLE, DRIVING OVER 16 km/h (10 MPH), AND THEN KEYING DOWN.

DIAGNOSTIC CHART DTC 61 — 1994 GRAND PRIX

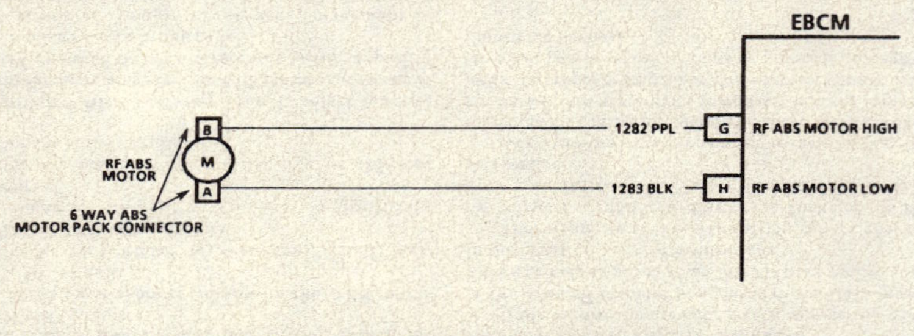

DTC 61
RIGHT FRONT ABS MOTOR CIRCUIT OPEN

Circuit Description:
This DTC identifies an ABS motor that cannot be energized due to an open in its circuitry. This malfunction will not allow proper front ABS operation.

DTC Will Set When: DTC 61 can be set only when the ABS motor is commanded "OFF." If the EBCM detects an out of range voltage on either of the right front ABS motor circuits indicating an open circuit, a malfunction exists.

Action Taken: An open ABS motor will not activate when requested. A malfunction DTC is stored, ABS is disabled and the ABS warning lamp is turned "ON."

DTC Chart Test Description: Number(s) below refer to circled number(s) on the diagnostic chart.
1. Checks for proper resistance of the ABS motor.
2. Checks for an open in the ABS motor "HIGH" circuitry.
3. Checks for an open in the ABS motor "LOW" circuitry.
4. Ensures malfunction was not due to physical damage of the circuitry.

Diagnostic Aids: Using Tech 1, select manual control function, and exercise ABS motor movement of affected channel in both directions while applying light pressure on the brake pedal.

If erratic or "jumpy" brake pedal movement is detected while performing an "apply" or "release" function of the ABS motor, an intermittent malfunction may be indicated.

An "Intermittent" malfunction may be caused by a poor connection, rubbed through wire insulation, or a wire that is broken inside the insulation.

If the malfunction is not current, wiggle the wires of the affected channel and check if the DTC resets. This will help to pinpoint an intermittent malfunction in the motor circuitry or connections.

The frequency of the malfunction can be checked by using the enhanced diagnostic function of the Tech 1, as described in "Tech 1 Diagnostics" found in this section.

Any circuitry, that is suspected as causing the intermittent complaint, should be thoroughly checked for backed out terminals, improper mating, broken locks, improperly formed or damaged terminals, poor terminal to wiring connections or physical damage to the wiring harness.

DIAGNOSTIC CHART DTC 61 — 1994 GRAND PRIX, CONTINUED

DTC 61
RIGHT FRONT ABS MOTOR CIRCUIT OPEN

DOES DTC 61 OCCUR INTERMITTENTLY?

- YES → REFER TO "DIAGNOSTIC AIDS" ON FACING PAGE.
- NO →

(1)
- IGNITION "OFF."
- DISCONNECT 6-WAY CONNECTOR AT ABS MOTOR PACK.
- USING J 39200, MEASURE RESISTANCE ACROSS ABS MOTOR PACK TERMINALS "A" AND "B".

IS RESISTANCE BELOW 2000Ω?

- NO → REPLACE ABS MOTOR PACK.
- YES →

(2)
- DISCONNECT 8-WAY CONNECTOR AT EBCM.
- USING J 39200, MEASURE RESISTANCE BETWEEN TERMINAL "G" OF THE 8-WAY EBCM HARNESS CONNECTOR AND TERMINAL "B" OF THE 6-WAY ABS MOTOR PACK HARNESS CONNECTOR.

IS RESISTANCE 2Ω OR LESS?

- NO → REPAIR OPEN OR HIGH RESISTANCE IN CKT 1282.
- YES →

(3)
- USING J 39200, MEASURE RESISTANCE BETWEEN TERMINAL "H" OF THE 8-WAY EBCM HARNESS CONNECTOR AND TERMINAL "A" OF THE 6-WAY ABS MOTOR PACK HARNESS CONNECTOR.

IS RESISTANCE 2Ω OR LESS?

- NO → REPAIR OPEN OR HIGH RESISTANCE IN CKT 1283.
- YES →

(4)
- INSPECT THE 6-WAY ABS MOTOR PACK AND 8-WAY EBCM CONNECTORS FOR POOR TERMINAL CONTACT AND EVIDENCE OF CORROSION.

REPLACE ALL TERMINALS THAT EXHIBIT SIGNS OF POOR TERMINAL CONTACT OR CORROSION.
- RECONNECT ALL CONNECTORS.
- IGNITION "OFF."
- START ENGINE AND TEST DRIVE VEHICLE, ACHIEVING AT LEAST 16 km/h (10 MPH).
- REPEAT ABOVE DRIVE CYCLE SEQUENCE TWO MORE TIMES.

DID DTC 61 SET IN THE LAST THREE DRIVE CYCLES?

- YES → REPLACE EBCM.
- NO → MALFUNCTION WAS MOST LIKELY CAUSED BY A POOR CONNECTION. REFER TO "DIAGNOSTIC AIDS" ON FACING PAGE.

AFTER ALL DIAGNOSIS IS COMPLETE, CLEAR DTCs AND TEST DRIVE VEHICLE FOR THREE (3) DRIVE CYCLES TO VERIFY THAT DTC DOES NOT RESET. A DRIVE CYCLE CONSISTS OF STARTING THE VEHICLE, DRIVING OVER 16 km/h (10 MPH), AND THEN KEYING DOWN.

DIAGNOSTIC CHART DTC 62 — 1994 GRAND PRIX

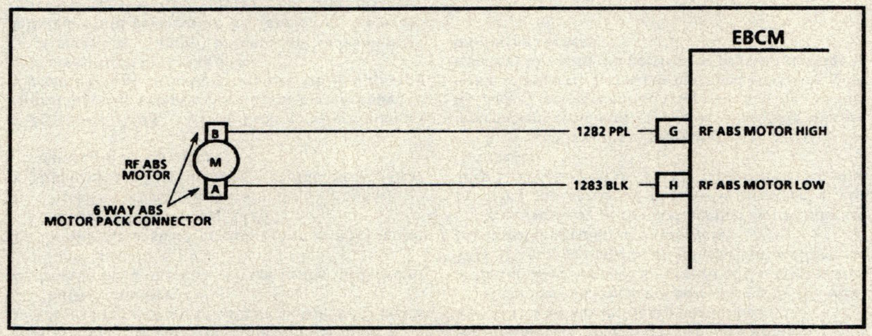

DTC 62

RIGHT FRONT ABS MOTOR CIRCUIT SHORTED TO GND

Circuit Description:
This DTC identifies an ABS motor circuit that is shorted to ground. This malfunction will not allow the ABS motor to be controlled at the commanded current rate or will cause the ABS motor driver circuit to allow current directly to ground.

DTC Will Set When: DTC 62 can be set anytime. If the EBCM detects an out of range voltage on either of the right front ABS motor circuits indicating a circuit shorted to ground, a malfunction exists.

Action Taken: A malfunction DTC is stored, ABS is disabled and the ABS warning lamp is turned "ON."

DTC Chart Test Description: Number(s) below refer to circled number(s) on the diagnostic chart.

1. Checks for a short to ground in the ABS motor "HIGH" circuitry.
2. Checks for a short to ground in the ABS motor "LOW" circuitry.
3. Checks for an ABS motor that is internally shorted to ground.
4. Ensures malfunction was not due to physical damage of the circuitry.

Diagnostic Aids: Using Tech 1, select manual control function, and exercise ABS motor movement of affected channel in both directions while applying light pressure on the brake pedal.

If erratic or "jumpy" brake pedal movement is detected while performing an "apply" or "release" function of the ABS motor, an intermittent malfunction may be indicated.

An "Intermittent" malfunction may be caused by a poor connection, rubbed through wire insulation, or a wire that is broken inside the insulation.

If the malfunction is not current, wiggle the wires of the affected channel and check if the DTC resets. This will help to pinpoint an intermittent malfunction in the motor circuitry or connections.

The frequency of the malfunction can be checked by using the enhanced diagnostic function of the Tech 1, as described in "Tech 1 Diagnostics" found in this section.

Any circuitry, that is suspected as causing the intermittent complaint, should be thoroughly checked for backed out terminals, improper mating, broken locks, improperly formed or damaged terminals, poor terminal to wiring connections or physical damage to the wiring harness.

DIAGNOSTIC CHART DTC 62 — 1994 GRAND PRIX, CONTINUED

DTC 62

RIGHT FRONT ABS MOTOR CIRCUIT SHORTED TO GND

DOES DTC 62 OCCUR INTERMITTENTLY?

- YES → REFER TO "DIAGNOSTIC AIDS" ON FACING PAGE.
- NO →

(1)
- IGNITION "OFF."
- DISCONNECT 6-WAY ABS MOTOR PACK CONNECTOR AND 8-WAY EBCM CONNECTOR.
- USING J 39200, MEASURE THE RESISTANCE BETWEEN GROUND AND TERMINAL "B" OF THE 6-WAY ABS MOTOR PACK HARNESS CONNECTOR.

DOES J 39200 DISPLAY RESISTANCE AS "OL" (INFINITE)?

- NO → REPAIR SHORT TO GROUND IN CKT 1282.
- YES →

(2)
- MEASURE THE RESISTANCE BETWEEN GROUND AND TERMINAL "A" OF THE 6-WAY ABS MOTOR PACK HARNESS CONNECTOR.

DOES J 39200 DISPLAY RESISTANCE AS "OL" (INFINITE)?

- NO → REPAIR SHORT TO GROUND IN CKT 1283.
- YES →

(3)
- MEASURE THE RESISTANCE BETWEEN GROUND AND TERMINAL "B" OF THE ABS MOTOR PACK.

DOES J 39200 DISPLAY RESISTANCE AS "OL" (INFINITE)?

- NO → REPLACE ABS MOTOR PACK.
- YES →

(4)
- INSPECT CKTs 1282 AND 1283 AND RELATED CONNECTORS FOR DAMAGE WHICH MAY RESULT IN A SHORT TO GROUND WITH ALL CONNECTORS CONNECTED. REPAIR ALL DAMAGE IF EVIDENT.
- RECONNECT ALL CONNECTORS.
- IGNITION "OFF."
- START ENGINE AND TEST DRIVE VEHICLE, ACHIEVING AT LEAST 16 km/h (10 MPH).
- REPEAT ABOVE DRIVE CYCLE SEQUENCE TWO MORE TIMES.

DID DTC 62 SET IN THE LAST THREE DRIVE CYCLES?

- YES → REPLACE EBCM.
- NO → MALFUNCTION WAS MOST LIKELY CAUSED BY A POOR CONNECTION. REFER TO "DIAGNOSTIC AIDS" ON FACING PAGE.

AFTER ALL DIAGNOSIS IS COMPLETE, CLEAR DTCs AND TEST DRIVE VEHICLE FOR THREE (3) DRIVE CYCLES TO VERIFY THAT DTC DOES NOT RESET. A DRIVE CYCLE CONSISTS OF STARTING THE VEHICLE, DRIVING OVER 16 KM/H (10 MPH), AND THEN KEYING DOWN.

DIAGNOSTIC CHART DTC 63 — 1994 GRAND PRIX

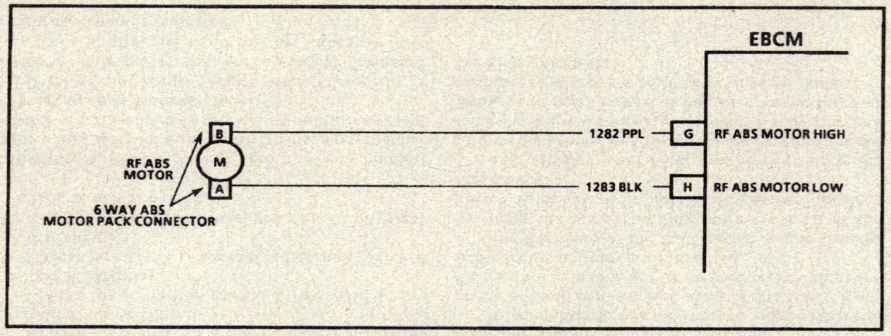

DTC 63

RIGHT FRONT ABS MOTOR CIRCUIT SHORTED TO BATT

Circuit Description:

This DTC identifies an ABS motor circuit that is shorted to battery or an ABS motor that has low or no resistance. This malfunction will not allow the ABS motor to be controlled at the commanded current rate or will cause the ABS motor to turn in the opposite direction or not at all.

DTC Will Set When: DTC 63 can be set only when the ABS motor is commanded "OFF." If the EBCM detects an out of range voltage on either of the right front ABS motor circuits indicating a circuit shorted to battery or an ABS motor shorted, a malfunction exists.

Action Taken: A malfunction DTC is stored, ABS is disabled and the ABS warning lamp is turned "ON."

DTC Chart Test Description: Number(s) below refer to circled number(s) on the diagnostic chart.

1. Checks for a short to voltage in the ABS motor "HIGH" circuitry.
2. Checks for a short to voltage in the ABS motor "LOW" circuitry.
3. Checks for a short circuit between ABS motor "HIGH" and ABS motor "LOW."
4. Checks for proper resistance of the ABS motor.
5. Ensures malfunction was not due to physical damage of the circuitry.

Diagnostic Aids: Using Tech 1, select manual control function, and exercise ABS motor movement of affected channel in both directions while applying light pressure on the brake pedal.

If erratic or "jumpy" brake pedal movement is detected while performing an "apply" or "release" function of the ABS motor, an intermittent malfunction may be indicated.

An "Intermittent" malfunction may be caused by a poor connection, rubbed through wire insulation, or a wire that is broken inside the insulation.

If the malfunction is not current, wiggle the wires of the affected channel and check if the DTC resets. This will help to pinpoint an intermittent malfunction in the motor circuitry or connections.

The frequency of the malfunction can be checked by using the enhanced diagnostic function of the Tech 1, as described in "Tech 1 Diagnostics" found in this section.

Any circuitry, that is suspected as causing the intermittent complaint, should be thoroughly checked for backed out terminals, improper mating, broken locks, improperly formed or damaged terminals, poor terminal to wiring connections or physical damage to the wiring harness.

DIAGNOSTIC CHART DTC 63 — 1994 GRAND PRIX, CONTINUED

DTC 63

RIGHT FRONT ABS MOTOR CIRCUIT SHORTED TO BATT

NOTICE: J 39200 TEST LEADS MUST BE "ZEROED" PRIOR TO MAKING ANY RESISTANCE MEASUREMENTS. REFER TO J 39200 USER'S MANUAL.

DOES DTC 63 OCCUR INTERMITTENTLY?

- YES: REFER TO "DIAGNOSTIC AIDS" ON FACING PAGE.
- NO:

(1)
- IGNITION "OFF."
- DISCONNECT 6-WAY ABS MOTOR PACK CONNECTOR AND 8-WAY EBCM CONNECTOR.
- IGNITION "ON."
- USING J 39200, MEASURE THE VOLTAGE BETWEEN GROUND AND TERMINAL "B" OF THE 6-WAY ABS MOTOR PACK HARNESS CONNECTOR. IS VOLTAGE 1 VOLT OR LESS?

- NO: REPAIR SHORT TO VOLTAGE IN CKT 1282.
- YES:

(2)
- MEASURE THE VOLTAGE BETWEEN GROUND AND TERMINAL "A" OF THE 6-WAY ABS MOTOR PACK HARNESS CONNECTOR. IS VOLTAGE 1 VOLT OR LESS?

- NO: REPAIR SHORT TO VOLTAGE IN CKT 1283.
- YES:

(3)
- IGNITION "OFF."
- USING J 39200, MEASURE THE RESISTANCE BETWEEN TERMINAL "A" AND TERMINAL "B" OF THE 6-WAY ABS MOTOR PACK HARNESS CONNECTOR. DOES J 39200 DISPLAY RESISTANCE AS "OL" (INFINITE)?

- NO: REPAIR SHORT BETWEEN CKTs 1282 AND 1283.
- YES:

(4)
- USING J 39200, MEASURE THE RESISTANCE BETWEEN TERMINAL "A" AND TERMINAL "B" OF THE ABS MOTOR PACK. IS RESISTANCE GREATER THAN 0.4Ω?

- NO: REPLACE ABS MOTOR PACK.
- YES:

(5)
- INSPECT CKTs 1282 AND 1283 AND RELATED CONNECTORS FOR DAMAGE WHICH MAY RESULT IN A SHORT TO GROUND WITH ALL CONNECTORS CONNECTED. REPAIR ALL DAMAGE IF EVIDENT.
- RECONNECT ALL CONNECTORS.
- IGNITION "OFF."
- START ENGINE AND TEST DRIVE VEHICLE, ACHIEVING AT LEAST 16 km/h (10 MPH).
- REPEAT ABOVE DRIVE CYCLE SEQUENCE TWO MORE TIMES. DID DTC 63 SET IN THE LAST THREE DRIVE CYCLES?

- YES: REPLACE EBCM.
- NO: MALFUNCTION WAS MOST LIKELY CAUSED BY A POOR CONNECTION. REFER TO "DIAGNOSTIC AIDS" ON FACING PAGE.

AFTER ALL DIAGNOSIS IS COMPLETE, CLEAR DTCs AND TEST DRIVE VEHICLE FOR THREE (3) DRIVE CYCLES TO VERIFY THAT DTC DOES NOT RESET. A DRIVE CYCLE CONSISTS OF STARTING THE VEHICLE, DRIVING OVER 16 km/h (10 MPH), AND THEN KEYING DOWN

DIAGNOSTIC CHART DTC 64 — 1994 GRAND PRIX

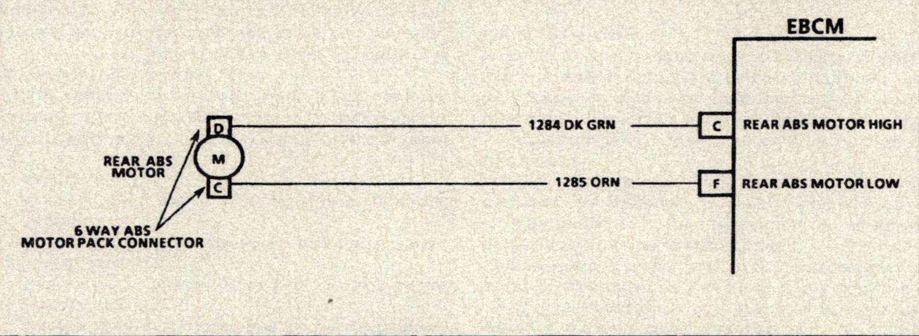

DTC 64
REAR ABS MOTOR CIRCUIT OPEN

Circuit Description:
This DTC identifies an ABS motor that cannot be energized due to an open in its circuitry. This malfunction will not allow proper rear ABS operation.

DTC Will Set When: DTC 64 can be set only when the ABS motor is commanded "OFF." If the EBCM detects an out of range voltage on either of the rear ABS motor circuits indicating an open circuit, a malfunction exists.

Action Taken: An open motor will not activate when requested. A malfunction DTC is stored, ABS is disabled and the ABS warning lamp is turned "ON."

DTC Chart Test Description: Number(s) below refer to circled number(s) on the diagnostic chart.

1. Checks for proper resistance of the ABS motor.
2. Checks for an open in the ABS motor "HIGH" circuitry.
3. Checks for an open in the ABS motor "LOW" circuitry.
4. Ensures malfunction was not due to poor terminal contact.

Diagnostic Aids: Using Tech 1, select manual control function, and exercise ABS motor movement of affected channel in both directions while applying light pressure on the brake pedal.

If erratic or "jumpy" brake pedal movement is detected while performing an "apply" or "release" function of the ABS motor, an intermittent malfunction may be indicated.

An "Intermittent" malfunction may be caused by a poor connection, rubbed through wire insulation, or a wire that is broken inside the insulation.

If the malfunction is not current, wiggle the wires of the affected channel and check if the DTC resets. This will help to pinpoint an intermittent malfunction in the motor circuitry or connections.

The frequency of the malfunction can be checked by using the enhanced diagnostic function of the Tech 1, as described in "Tech 1 Diagnostics" found in this section.

Any circuitry, that is suspected as causing the intermittent complaint, should be thoroughly checked for backed out terminals, improper mating, broken locks, improperly formed or damaged terminals, poor terminal to wiring connections or physical damage to the wiring harness.

DIAGNOSTIC CHART DTC 64 — 1994 GRAND PRIX, CONTINUED

DTC 64
REAR ABS MOTOR CIRCUIT OPEN

DOES DTC 64 OCCUR INTERMITTENTLY?

- YES: REFER TO "DIAGNOSTIC AIDS" ON FACING PAGE.
- NO:

(1)
- IGNITION "OFF."
- DISCONNECT 6-WAY ABS MOTOR PACK CONNECTOR.
- USING J 39200, MEASURE RESISTANCE ACROSS ABS MOTOR PACK TERMINALS "C" AND "D".

IS RESISTANCE BELOW 2000Ω?

- NO: REPLACE ABS MOTOR PACK.
- YES:

(2)
- DISCONNECT 8-WAY CONNECTOR AT EBCM.
- USING J 39200, MEASURE RESISTANCE BETWEEN TERMINAL "C" OF THE 8-WAY EBCM HARNESS CONNECTOR AND TERMINAL "D" OF THE 6-WAY ABS MOTOR PACK HARNESS CONNECTOR.

IS RESISTANCE 2Ω OR LESS?

- NO: REPAIR OPEN OR HIGH RESISTANCE IN CKT 1284.
- YES:

(3)
- USING J 39200, MEASURE RESISTANCE BETWEEN TERMINAL "F" OF THE 8-WAY EBCM HARNESS CONNECTOR AND TERMINAL "C" OF THE 6-WAY ABS MOTOR PACK HARNESS CONNECTOR.

IS RESISTANCE 2Ω OR LESS?

- NO: REPAIR OPEN OR HIGH RESISTANCE IN CKT 1285.
- YES:

(4)
- INSPECT THE 6-WAY ABS MOTOR PACK AND 8-WAY EBCM CONNECTORS FOR POOR TERMINAL CONTACT AND EVIDENCE OF CORROSION.

  REPLACE ALL TERMINALS THAT EXHIBIT SIGNS OF POOR TERMINAL CONTACT OR CORROSION.
- RECONNECT ALL CONNECTORS.
- IGNITION "OFF."
- START ENGINE AND TEST DRIVE VEHICLE, ACHIEVING AT LEAST 16 km/h (10 MPH).
- REPEAT ABOVE DRIVE CYCLE SEQUENCE TWO MORE TIMES.

DID DTC 64 SET IN THE LAST THREE DRIVE CYCLES?

- YES: REPLACE EBCM.
- NO: MALFUNCTION WAS MOST LIKELY CAUSED BY A POOR CONNECTION. REFER TO "DIAGNOSTIC AIDS" ON FACING PAGE.

AFTER ALL DIAGNOSIS IS COMPLETE, CLEAR DTCs AND TEST DRIVE VEHICLE FOR THREE (3) DRIVE CYCLES TO VERIFY THAT DTC DOES NOT RESET. A DRIVE CYCLE CONSISTS OF STARTING THE VEHICLE, DRIVING OVER 16 km/h (10 MPH), AND THEN KEYING DOWN.

DIAGNOSTIC CHART DTC 65 — 1994 GRAND PRIX

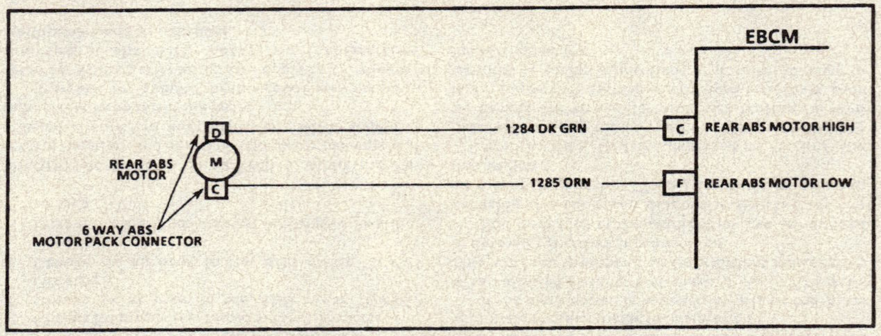

DTC 65

REAR ABS MOTOR CIRCUIT SHORTED TO GROUND

Circuit Description:

This DTC identifies an ABS motor circuit that is shorted to ground. This malfunction will not allow the ABS motor to be controlled at the commanded current rate or will cause the driver circuit to allow current directly to ground.

DTC Will Set When: DTC 65 can be set anytime. If the EBCM detects an out of range voltage on either of the rear ABS motor circuits indicating a circuit shorted to ground, a malfunction exists.

Action Taken: A malfunction DTC is stored, ABS is disabled and the ABS warning lamp is turned "ON."

DTC Chart Test Description: Number(s) below refer to circled number(s) on the diagnostic chart.

1. Checks for a short to ground in the ABS motor "HIGH" circuitry.
2. Checks for a short to ground in the ABS motor "LOW" circuitry.
3. Checks for an ABS motor that is internally shorted to ground.
4. Ensures malfunction was not due to physical damage.

Diagnostic Aids: Using Tech 1, select manual control function, and exercise ABS motor movement of affected channel in both directions while applying light pressure on the brake pedal.

If erratic or "jumpy" brake pedal movement is detected while performing an "apply" or "release" function of the ABS motor, an intermittent malfunction may be indicated.

An "Intermittent" malfunction may be caused by a poor connection, rubbed through wire insulation, or a wire that is broken inside the insulation.

If the malfunction is not current, wiggle the wires of the affected channel and check if the DTC resets. This will help to pinpoint an intermittent malfunction in the motor circuitry or connections.

The frequency of the malfunction can be checked by using the enhanced diagnostic function of the Tech 1, as described in "Tech 1 Diagnostics" found in this section.

Any circuitry, that is suspected as causing the intermittent complaint, should be thoroughly checked for backed out terminals, improper mating, broken locks, improperly formed or damaged terminals, poor terminal to wiring connections or physical damage to the wiring harness.

DIAGNOSTIC CHART DTC 65 — 1994 GRAND PRIX, CONTINUED

DTC 65

REAR ABS MOTOR CIRCUIT SHORTED TO GROUND

DOES DTC 65 OCCUR INTERMITTENTLY?

- YES → REFER TO "DIAGNOSTIC AIDS" ON FACING PAGE.
- NO →

(1)
- IGNITION "OFF."
- DISCONNECT 6-WAY ABS MOTOR PACK CONNECTOR AND 8-WAY EBCM CONNECTOR.
- USING A J 39200, MEASURE THE RESISTANCE BETWEEN GROUND AND TERMINAL "D" OF THE 6-WAY ABS MOTOR PACK HARNESS CONNECTOR.

DOES J 39200 DISPLAY RESISTANCE AS "OL" (INFINITE)?

- NO → REPAIR SHORT TO GROUND IN CKT 1284.
- YES →

(2)
- MEASURE THE RESISTANCE BETWEEN GROUND AND TERMINAL "C" OF THE 6-WAY ABS MOTOR PACK HARNESS CONNECTOR.

DOES J 39200 DISPLAY RESISTANCE AS "OL" (INFINITE)?

- NO → REPAIR SHORT TO GROUND IN CKT 1285.
- YES →

(3)
- MEASURE THE RESISTANCE BETWEEN GROUND AND TERMINAL "D" OF THE ABS MOTOR PACK.

DOES J 39200 DISPLAY RESISTANCE AS "OL" (INFINITE)?

- NO → REPLACE ABS MOTOR PACK.
- YES →

(4)
- INSPECT CKTs 1284 AND 1285 AND RELATED CONNECTORS FOR DAMAGE WHICH MAY RESULT IN A SHORT TO GROUND WITH ALL CONNECTORS CONNECTED. REPAIR ALL DAMAGE IF EVIDENT.
- RECONNECT ALL CONNECTORS.
- IGNITION "OFF."
- START ENGINE AND TEST DRIVE VEHICLE, ACHIEVING AT LEAST 16 km/h (10 MPH).
- REPEAT ABOVE DRIVE CYCLE SEQUENCE TWO MORE TIMES.

DID DTC 65 SET IN THE LAST THREE DRIVE CYCLES?

- YES → REPLACE EBCM.
- NO → MALFUNCTION WAS MOST LIKELY CAUSED BY A POOR CONNECTION. REFER TO "DIAGNOSTIC AIDS" ON FACING PAGE.

AFTER ALL DIAGNOSIS IS COMPLETE, CLEAR DTCs AND TEST DRIVE VEHICLE FOR THREE (3) DRIVE CYCLES TO VERIFY THAT DTC DOES NOT RESET. A DRIVE CYCLE CONSISTS OF STARTING THE VEHICLE, DRIVING OVER 16 km/h (10 MPH), AND THEN KEYING DOWN.

DIAGNOSTIC CHART DTC 66 — 1994 GRAND PRIX

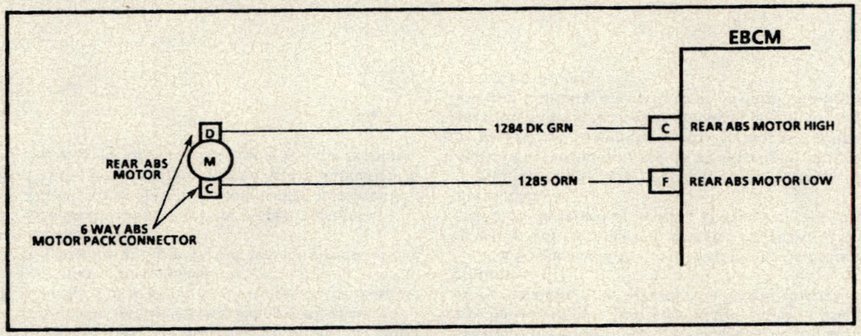

DTC 66

REAR ABS MOTOR CIRCUIT SHORTED TO BATTERY

Circuit Description:

This DTC identifies an ABS motor circuit that is shorted to battery or an ABS motor that has low or zero resistance. This malfunction will not allow the ABS motor to be controlled at the commanded current rate or will cause the ABS motor to turn in the opposite direction or not turn at all.

DTC Will Set When: DTC 66 can be set only when the ABS motor is commanded "OFF." If the EBCM detects an out of range voltage on either of the rear ABS motor circuits indicating a circuit shorted to battery or an ABS motor shorted, a malfunction exists.

Action Taken: A malfunction DTC is stored, ABS is disabled and the ABS warning lamp is turned "ON."

DTC Chart Test Description: Number(s) below refer to circled number(s) on the diagnostic chart.

1. Checks for a short to voltage in the ABS motor "HIGH" circuitry.
2. Checks for a short to voltage in the ABS motor "LOW" circuitry.
3. Checks for a short circuit between ABS motor "HIGH" and ABS motor "LOW."
4. Checks for proper resistance of the ABS motor.
5. Ensures malfunction was not due to physical damage of the circuitry.

Diagnostic Aids: Using Tech 1, select manual control function, and exercise ABS motor movement of affected channel in both directions while applying light pressure on the brake pedal.

If erratic or "jumpy" brake pedal movement is detected while performing an "apply" or "release" function of the ABS motor, an intermittent malfunction may be indicated.

An "Intermittent" malfunction may be caused by a poor connection, rubbed through wire insulation, or a wire that is broken inside the insulation.

If the malfunction is not current, wiggle the wires of the affected channel and check if the DTC resets. This will help to pinpoint an intermittent malfunction in the motor circuitry or connections.

The frequency of the malfunction can be checked by using the enhanced diagnostic function of the Tech 1, as described in "Tech 1 Diagnostics" found in this section.

Any circuitry, that is suspected as causing the intermittent complaint, should be thoroughly checked for backed out terminals, improper mating, broken locks, improperly formed or damaged terminals, poor terminal to wiring connections or physical damage to the wiring harness.

DIAGNOSTIC CHART DTC 66 — 1994 GRAND PRIX, CONTINUED

DTC 66
REAR ABS MOTOR CIRCUIT SHORTED TO BATTERY

NOTICE: J 39200 TEST LEADS MUST BE "ZEROED" PRIOR TO MAKING ANY RESISTANCE MEASUREMENTS. REFER TO J 39200 USER'S MANUAL.

DOES DTC 66 OCCUR INTERMITTENTLY?
- YES → REFER TO "DIAGNOSTIC AIDS" ON FACING PAGE.
- NO ↓

(1)
- IGNITION "OFF."
- DISCONNECT 6-WAY ABS MOTOR PACK CONNECTOR AND 8-WAY EBCM CONNECTORS.
- IGNITION "ON."
- USING J 39200, MEASURE THE VOLTAGE BETWEEN GROUND AND TERMINAL "D" OF THE 6-WAY ABS MOTOR PACK HARNESS CONNECTOR.

IS VOLTAGE 1 VOLT OR LESS?
- NO → REPAIR SHORT TO VOLTAGE IN CKT 1284.
- YES ↓

(2)
- MEASURE THE VOLTAGE BETWEEN GROUND AND TERMINAL "C" OF THE 6-WAY ABS MOTOR PACK HARNESS CONNECTOR.

IS VOLTAGE 1 VOLT OR LESS?
- NO → REPAIR SHORT TO VOLTAGE IN CKT 1285.
- YES ↓

(3)
- IGNITION "OFF."
- USING J 39200, MEASURE THE VOLTAGE BETWEEN TERMINAL "C" AND TERMINAL "D" OF THE 6-WAY ABS MOTOR PACK HARNESS CONNECTOR.

DOES J 39200 DISPLAY RESISTANCE AS "OL" (INFINITE)?
- NO → REPAIR SHORT BETWEEN CKTs 1284 AND 1285.
- YES ↓

(4)
- USING J 39200, MEASURE THE RESISTANCE BETWEEN TERMINAL "C" AND TERMINAL "D" OF THE ABS MOTOR PACK.

IS RESISTANCE GREATER THAN 0.4Ω?
- NO → REPLACE ABS MOTOR PACK.
- YES ↓

(5)
- INSPECT CKTs 1284 AND 1285 RELATED CONNECTORS FOR DAMAGE WHICH MAY RESULT IN A SHORT TO VOLTAGE WITH ALL CONNECTORS CONNECTED. REPAIR ALL DAMAGE IF EVIDENT.
- RECONNECT ALL CONNECTORS.
- IGNITION "OFF."
- START ENGINE AND TEST DRIVE VEHICLE, ACHIEVING AT LEAST 16 km/h (10 MPH).
- REPEAT ABOVE DRIVE CYCLE SEQUENCE TWO MORE TIMES.

DID DTC 66 SET IN THE LAST THREE DRIVE CYCLES.
- YES → REPLACE EBCM.
- NO → MALFUNCTION WAS MOSTLY CAUSED BY A POOR CONNECTION. REFER TO "DIAGNOSTIC AIDS" ON FACING PAGE.

AFTER ALL DIAGNOSIS IS COMPLETE, CLEAR DTCs AND TEST DRIVE VEHICLE FOR THREE (3) DRIVE CYCLES TO VERIFY THAT DTC DOES NOT RESET. A DRIVE CYCLE CONSISTS OF STARTING THE VEHICLE, DRIVING OVER 16 km/h (10 MPH), AND THEN KEYING DOWN.

DIAGNOSTIC CHART DTC 76 — 1994 GRAND PRIX

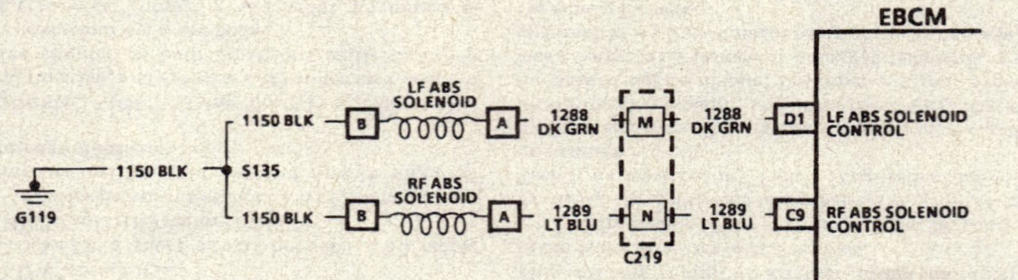

DTC 76

LEFT FRONT SOL CIRCUIT OPEN OR SHORTED TO BATT

Circuit Description:

This DTC identifies a solenoid that cannot be energized due to an open in its circuitry, or a solenoid that is always energized due to a short to battery in its circuitry between the driver and the solenoid. An open will not allow proper ABS operation, but the short to battery simply turns "ON" the solenoid. A path for base brakes is still allowed once the motor rehomes and the check ball is lifted off its seat during key "ON" initialization.

DTC Will Set When: DTC 76 can be set only when the solenoid is commanded "OFF." If the EBCM detects an excessive voltage in the left front solenoid control circuit, a malfunction exists.

Action Taken: A malfunction DTC is stored, ABS is disabled and the ABS warning lamp is turned "ON."

DTC Chart Test Description: Number(s) below refer to circled number(s) on the diagnostic chart.

1. Checks for a short to voltage in the control circuitry of the solenoid.
2. Checks for an open in the control circuit of the solenoid.
3. Checks the solenoid coil for proper resistance.
4. Checks for an open in the solenoid ground circuit.
5. Checks for a possible intermittent malfunction in the solenoid circuitry due to poor terminal contact.

Diagnostic Aids: An "Intermittent" malfunction may be caused by a poor connection, rubbed through wire insulation, or a wire that is broken inside the insulation.

The frequency of the malfunction can be checked by using the enhanced diagnostic function of the Tech 1, as described in "Tech 1 Diagnostics" found in this section.

Any circuitry, that is suspected as causing the intermittent complaint, should be thoroughly checked for backed out terminals, improper mating, broken locks, improperly formed or damaged terminals, poor terminal to wiring connections or physical damage to the wiring harness.

DIAGNOSTIC CHART DTC 76 — 1994 GRAND PRIX, CONTINUED

DTC 76

LEFT FRONT SOL CIRCUIT OPEN OR SHORTED TO BATT

(1)
- IGNITION "OFF."
- DISCONNECT 32-WAY EBCM CONNECTOR.
- IGNITION "ON," ENGINE "OFF."
- USING J 39200, MEASURE THE VOLTAGE BETWEEN GROUND AND TERMINAL "D1" OF THE 32-WAY EBCM HARNESS CONNECTOR.

IS VOLTAGE 1 VOLT OR LESS?

- YES: go to 2
- NO: REPAIR SHORT TO VOLTAGE IN CKT 1288.

(2)
- IGNITION "OFF."
- DISCONNECT LEFT FRONT SOLENOID CONNECTOR.
- USING J 39200, MEASURE THE RESISTANCE BETWEEN TERMINAL "D1" OF THE 32-WAY EBCM HARNESS CONNECTOR AND TERMINAL "A" OF THE LEFT FRONT SOLENOID HARNESS CONNECTOR.

IS RESISTANCE 2Ω OR LESS?

- YES: go to 3
- NO: REPAIR OPEN OR HIGH RESISTANCE IN CKT 1288.

(3)
- MEASURE THE RESISTANCE BETWEEN TERMINAL "A" AND TERMINAL "B" OF THE LEFT FRONT SOLENOID.

IS RESISTANCE BETWEEN 2.5 AND 5Ω?

- YES: go to 4
- NO: REPLACE LEFT FRONT SOLENOID.

(4)
- MEASURE THE RESISTANCE BETWEEN GROUND AND TERMINAL "B" OF THE LEFT FRONT SOLENOID HARNESS CONNECTOR.

IS RESISTANCE 2Ω OR LESS?

- YES: go to 5
- NO: REPAIR OPEN OR HIGH RESISTANCE IN CKT 1150.

(5)
- INSPECT LEFT FRONT SOLENOID AND 32-WAY EBCM HARNESS CONNECTORS FOR POOR TERMINAL CONTACT AND EVIDENCE OF CORROSION.

REPLACE ALL TERMINALS THAT EXHIBIT SIGNS OF POOR TERMINAL CONTACT OR CORROSION.
- INSPECT CKT 1288 FOR DAMAGE WHICH MAY RESULT IN A SHORT TO VOLTAGE WITH ALL CONNECTORS CONNECTED. REPAIR DAMAGE IF EVIDENT.
- RECONNECT ALL CONNECTORS.
- IGNITION "ON."

DOES DTC 76 SET AS A CURRENT DTC?

- YES: REPLACE EBCM.
- NO: MALFUNCTION IS MOST LIKELY DUE TO A POOR CONNECTION. REFER TO "DIAGNOSTIC AIDS" ON FACING PAGE.

DIAGNOSTIC CHART DTC 77 — 1994 GRAND PRIX

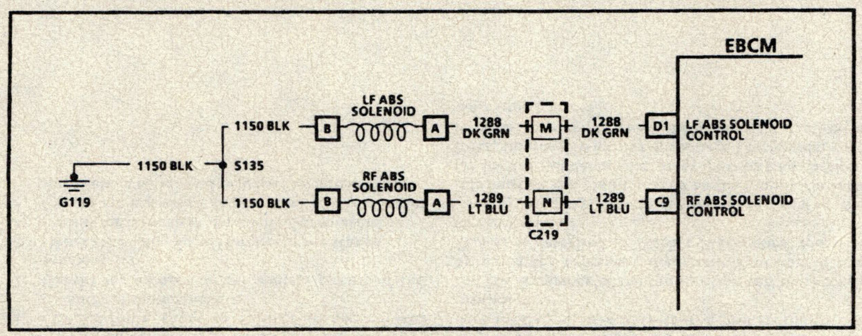

DTC 77

LEFT FRONT SOLENOID CIRCUIT SHORTED TO GND

Circuit Description:

This DTC identifies a solenoid that cannot be energized due to an open in its driver circuitry, or a short to ground between the solenoid driver and the solenoid. These malfunctions can affect ABS operation since the flow of brake fluid to the wheel cylinder cannot be stopped, making ABS operation for that channel impossible.

DTC Will Set When: DTC 77 can be set only when the solenoid is commanded "ON." If the EBCM detects the left front solenoid control circuit voltage is out of specification, a malfunction exists.

Action Taken: A malfunction DTC is stored, ABS is disabled and the ABS warning lamp is turned "ON."

DTC Chart Test Description: Number(s) below refer to circled number(s) on the diagnostic chart.

1. Checks for a solenoid that is internally shorted to ground.
2. Checks for a solenoid that is not within proper resistance values.
3. Indicates if a short to ground exists in the solenoid circuitry.
4. Checks for a possible intermittent malfunction in the solenoid circuitry.

Diagnostic Aids: An "Intermittent" malfunction may be caused by a poor connection, rubbed through wire insulation, or a wire that is broken inside the insulation.

The frequency of the malfunction can be checked by using the enhanced diagnostic function of the Tech 1, as described in "Tech 1 Diagnostics" found in this section.

Any circuitry, that is suspected as causing the intermittent complaint, should be thoroughly checked for backed out terminals, improper mating, broken locks, improperly formed or damaged terminals, poor terminal to wiring connections or physical damage to the wiring harness.

DIAGNOSTIC CHART DTC 77 — 1994 GRAND PRIX, CONTINUED

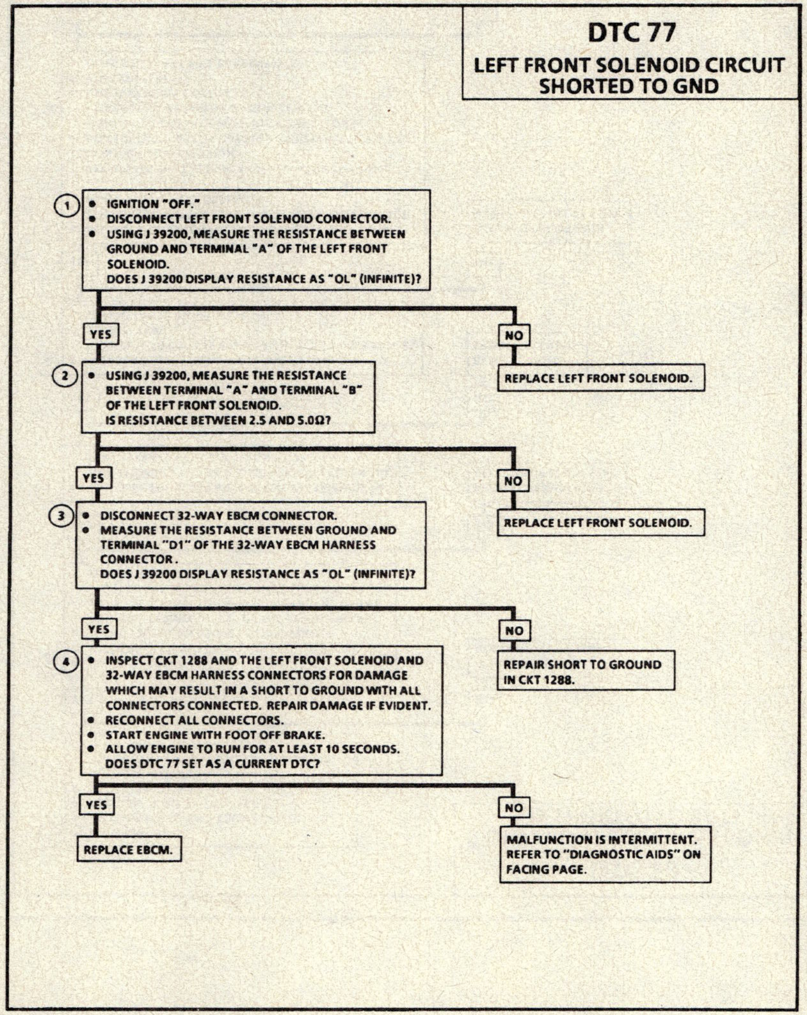

DIAGNOSTIC CHART DTC 78 — 1994 GRAND PRIX

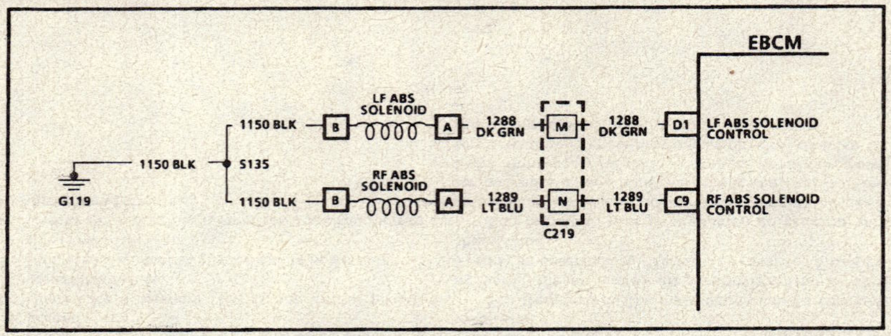

DTC 78

RIGHT FRONT SOL CIRCUIT OPEN OR SHORTED TO BATT

Circuit Description:
This DTC identifies a solenoid that cannot be energized due to an open in its circuitry, or a solenoid that is always energized due to a short to battery in its circuitry between the driver and the solenoid. An open will not allow proper ABS operation, but the short to battery simply turns "ON" the solenoid. A path for base brakes is still allowed once the motor rehomes and the check ball is lifted off its seat during key "ON" initialization.

DTC Will Set When: DTC 78 can be set only when the solenoid is commanded "OFF." If the EBCM detects an excessive voltage in the right front solenoid control circuit, a malfunction exists.

Action Taken: A malfunction DTC is stored, ABS is disabled and the ABS warning lamp is turned "ON."

DTC Chart Test Description: Number(s) below refer to circled number(s) on the diagnostic chart.
1. Checks for a short to voltage in the control circuitry of the solenoid.
2. Checks for an open in the control circuit of the solenoid.
3. Checks the solenoid coil for proper resistance.
4. Checks for an open in the solenoid ground circuit.
5. Checks for a possible intermittent malfunction in the solenoid circuitry due to poor terminal contact.

Diagnostic Aids: An "Intermittent" malfunction may be caused by a poor connection, rubbed through wire insulation, or a wire that is broken inside the insulation.

The frequency of the malfunction can be checked by using the enhanced diagnostic function of the Tech 1, as described in "Tech 1 Diagnostics" found in this section.

Any circuitry, that is suspected as causing the intermittent complaint, should be thoroughly checked for backed out terminals, improper mating, broken locks, improperly formed or damaged terminals, poor terminal to wiring connections or physical damage to the wiring harness.

DIAGNOSTIC CHART DTC 78 — 1994 GRAND PRIX, CONTINUED

DTC 78

RIGHT FRONT SOL CIRCUIT OPEN OR SHORTED TO BATT

(1)
- IGNITION "OFF."
- DISCONNECT 32-WAY EBCM CONNECTOR.
- IGNITION "ON," ENGINE "OFF."
- USING J 39200, MEASURE THE VOLTAGE BETWEEN GROUND AND TERMINAL "C9" OF THE 32-WAY EBCM HARNESS CONNECTOR.

IS VOLTAGE 1 VOLT OR LESS?

YES → go to (2)

NO → REPAIR SHORT TO VOLTAGE IN CKT 1289.

(2)
- IGNITION "OFF."
- DISCONNECT RIGHT FRONT SOLENOID CONNECTOR.
- USING J 39200, MEASURE THE RESISTANCE BETWEEN TERMINAL "C9" OF THE 32-WAY EBCM HARNESS CONNECTOR AND TERMINAL "A" OF THE RIGHT FRONT SOLENOID HARNESS CONNECTOR.

IS RESISTANCE 2Ω OR LESS?

YES → go to (3)

NO → REPAIR OPEN OR HIGH RESISTANCE IN CKT 1289.

(3)
- MEASURE THE RESISTANCE BETWEEN TERMINAL "A" AND TERMINAL "B" OF THE RIGHT FRONT SOLENOID.

IS RESISTANCE BETWEEN 2.5 AND 5Ω?

YES → go to (4)

NO → REPLACE RIGHT FRONT SOLENOID.

(4)
- MEASURE THE RESISTANCE BETWEEN GROUND AND TERMINAL "B" OF THE RIGHT FRONT SOLENOID HARNESS CONNECTOR.

IS RESISTANCE 2Ω OR LESS?

YES → go to (5)

NO → REPAIR OPEN OR HIGH RESISTANCE IN CKT 1150.

(5)
- INSPECT RIGHT FRONT SOLENOID AND 32-WAY EBCM HARNESS CONNECTORS FOR POOR TERMINAL CONTACT AND EVIDENCE OF CORROSION. REPLACE ALL TERMINALS THAT EXHIBIT SIGNS OF POOR TERMINAL CONTACT OR CORROSION.
- INSPECT CKT 1289 FOR DAMAGE WHICH MAY RESULT IN A SHORT TO VOLTAGE WITH ALL CONNECTORS CONNECTED. REPAIR DAMAGE IF EVIDENT.
- RECONNECT ALL CONNECTORS.
- IGNITION "ON."

DOES DTC 78 SET AS A CURRENT DTC?

YES → REPLACE EBCM.

NO → MALFUNCTION IS MOST LIKELY DUE TO A POOR CONNECTION. REFER TO "DIAGNOSTIC AIDS" ON FACING PAGE.

DIAGNOSTIC CHART DTC 81 — 1994 GRAND PRIX

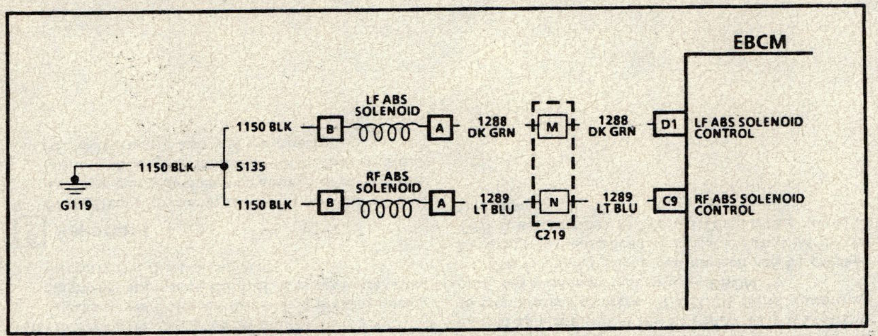

DTC 81

RIGHT FRONT SOLENOID CIRCUIT SHORTED TO GND

Circuit Description:

This DTC identifies a solenoid that cannot be energized due to an open in its driver circuitry, or a short to ground between the solenoid driver and the solenoid. These malfunctions can affect ABS operation since the flow of brake fluid to the wheel cylinder cannot be stopped, making ABS operation for that channel impossible.

DTC Will Set When: DTC 81 can be set only when the solenoid is commanded "ON." If the EBCM detects the right front solenoid control circuit voltage is out of specification, a malfunction exists.

Action Taken: A malfunction DTC is stored, ABS is disabled and the ABS warning lamp is turned "ON."

DTC Chart Test Description: Number(s) below refer to circled number(s) on the diagnostic chart.

1. Checks for a solenoid that is internally shorted to ground.
2. Checks for a solenoid that is not within proper resistance values.
3. Indicates if a short to ground exists in the solenoid circuitry.
4. Checks for a possible intermittent malfunction in the solenoid circuitry.

Diagnostic Aids: An "Intermittent" malfunction may be caused by a poor connection, rubbed through wire insulation, or a wire that is broken inside the insulation.

The frequency of the malfunction can be checked by using the enhanced diagnostic function of the Tech 1, as described in "Tech 1 Diagnostics" found in this section.

Any circuitry, that is suspected as causing the intermittent complaint, should be thoroughly checked for backed out terminals, improper mating, broken locks, improperly formed or damaged terminals, poor terminal to wiring connections or physical damage to the wiring harness.

DIAGNOSTIC CHART DTC 81 — 1994 GRAND PRIX, CONTINUED

DTC 81

RIGHT FRONT SOLENOID CIRCUIT SHORTED TO GND

(1)
- IGNITION "OFF."
- DISCONNECT RIGHT FRONT SOLENOID CONNECTOR.
- USING J 39200, MEASURE THE RESISTANCE BETWEEN GROUND AND TERMINAL "A" OF THE RIGHT FRONT SOLENOID.

DOES T 39200 DISPLAY RESISTANCE AS "OL" (INFINITE)?

NO: REPLACE RIGHT FRONT SOLENOID.

YES:

(2)
- USING J 39200, MEASURE THE RESISTANCE BETWEEN TERMINAL "A" AND TERMINAL "B" OF THE RIGHT FRONT SOLENOID.

IS RESISTANCE BETWEEN 2.5 AND 5.0Ω?

NO: REPLACE RIGHT FRONT SOLENOID.

YES:

(3)
- DISCONNECT 32-WAY EBCM CONNECTOR.
- MEASURE THE RESISTANCE BETWEEN GROUND AND TERMINAL "C9" OF THE 32-WAY EBCM HARNESS CONNECTOR.

DOES J 39200 DISPLAY RESISTANCE AS "OL" (INFINITE)?

NO: REPAIR SHORT TO GROUND IN CKT 1289.

YES:

(4)
- INSPECT CKT 1289 AND THE RIGHT FRONT SOLENOID AND 32-WAY EBCM HARNESS CONNECTORS FOR DAMAGE WHICH MAY RESULT IN A SHORT TO GROUND WITH ALL CONNECTORS CONNECTED. REPAIR DAMAGE IF EVIDENT.
- RECONNECT ALL CONNECTORS.
- START ENGINE WITH FOOT OFF BRAKE.
- ALLOW ENGINE TO RUN FOR AT LEAST 10 SECONDS.

DOES DTC 81 SET AS A CURRENT DTC?

YES: REPLACE EBCM.

NO: MALFUNCTION IS INTERMITTENT. REFER TO "DIAGNOSTIC AIDS" ON FACING PAGE.

DIAGNOSTIC CHART DTC 82 — 1994 GRAND PRIX

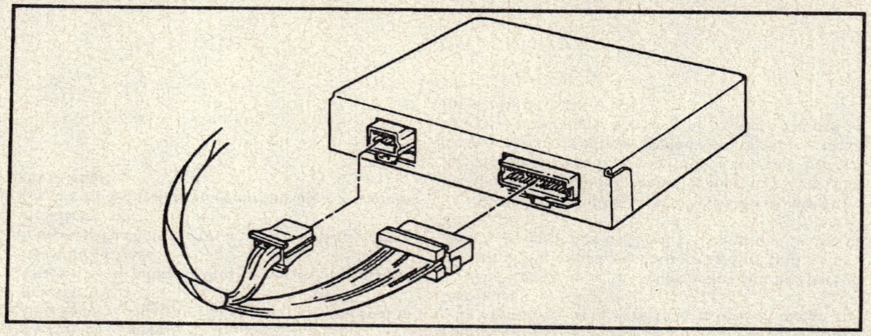

DTC 82
CALIBRATION MALFUNCTION

Circuit Description:
This DTC allows the EBCM to check for a calibration malfunction by comparing the calibration value to a known value stored in the EEPROM.
This DTC is also used as a security measure to prevent improper use of calibrations or changes to these calibrations that may alter the designed function of ABS.

DTC Will Set When: DTC 82 can be set at key-up. If the program identifier is incorrect or the memory checksum is incorrect, a malfunction exists.

Action Taken: A malfunction DTC is stored, ABS is disabled and the ABS warning lamp is turned "ON."

DTC Chart Test Description: Number(s) below refer to circled number(s) on the diagnostic chart.
1. Checks to see if the malfunction is present during diagnosis. If present, the EBCM is not functioning correctly and must be replaced.

Important
- At the time of printing, 1994 vehicles with ABS VI were not being field reprogrammed to correct this failure. A service bulletin will be issued when EEPROM reprogramming is authorized.

Diagnostic Aids: An intermittent DTC 82 may be caused by a bad cell in the EEPROM that is sensitive to temperature changes. If DTC 82 failed more than once, but is intermittent, replace EBCM.
The frequency of the malfunction can be checked by using the enhanced diagnostic function of the Tech 1, as described in "Tech 1 Diagnostics" found in this section.

DIAGNOSTIC CHART DTC 82 — 1994 GRAND PRIX, CONTINUED

DTC 82
CALIBRATION MALFUNCTION

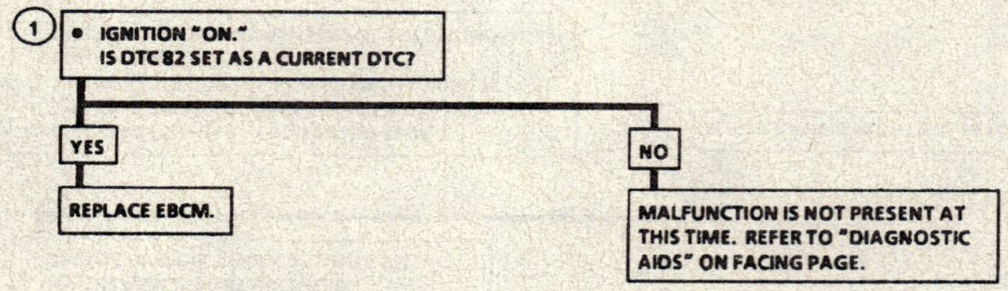

DIAGNOSTIC CHART DTC 86 — 1994 GRAND PRIX

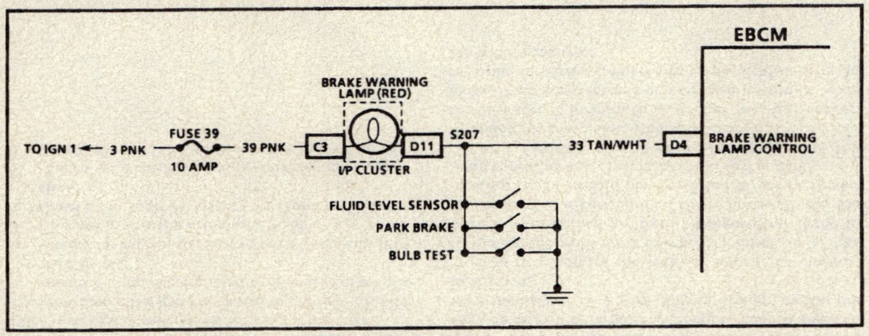

DTC 86

EBCM TURNED "ON" THE RED "BRAKE" WARNING LAMP

Circuit Description:

This DTC is provided as an information only DTC, and reflects the status of the command issued by the ABS EBCM to illuminate the red "BRAKE" warning lamp. If another DTC issues a command to illuminate the red "BRAKE" warning lamp, DTC 86 will be stored in EEPROM as a history DTC at the conclusion of the ignition cycle.

DTC Chart Test Description: Number(s) below refer to circled number(s) on the diagnostic chart.

1. This step serves to identify if a DTC other than DTC 86 commanded the red "BRAKE" warning lamp "ON."

Diagnostic Aids: Any ABS mechanical DTC that issues a command to illuminate the red "BRAKE" telltale will also result in DTC 86 being stored in EEPROM during shutdown. These DTCs are: 38, 41, 42, 46, and 51. If the motors are not in their home position, certain electrical DTCs will also command the red "BRAKE" telltale "ON." These DTCs are: 14, 16, 18, 55, 56, 57, 58, 61, 62, 63, 64, 65, and 66.

If any of these DTCs are indicated along with DTC 86, they must be corrected prior to addressing a DTC 86 malfunction.

DIAGNOSTIC CHART DTC 86 — 1994 GRAND PRIX, CONTINUED

DTC 86

EBCM TURNED "ON" THE RED "BRAKE" WARNING LAMP

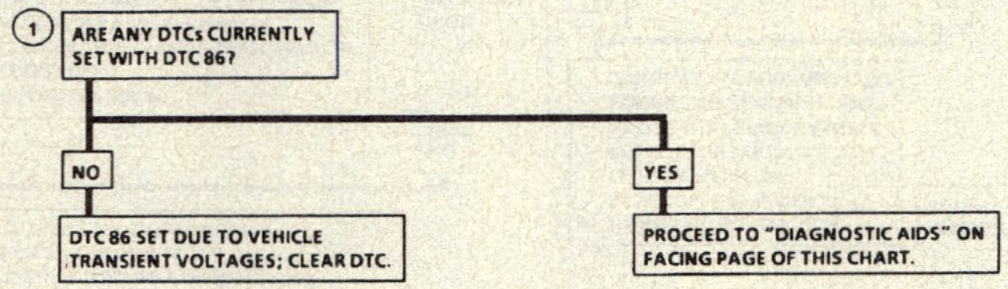

DIAGNOSTIC CHART DTC 87 — 1994 GRAND PRIX

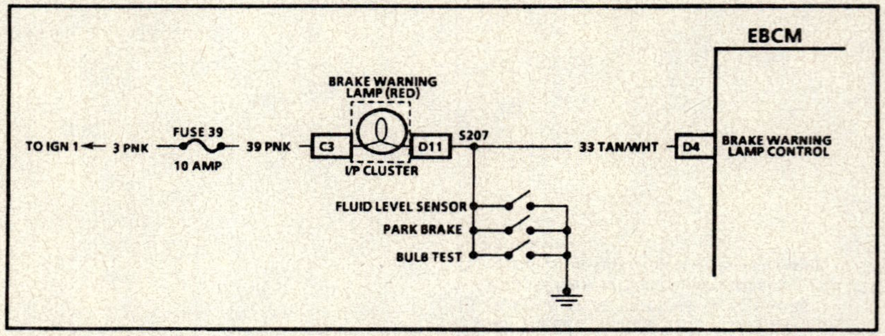

DTC 87

RED "BRAKE" WARNING LAMP CIRCUIT OPEN

Circuit Description:

This DTC is used to verify the EBCM has continuity to the red "BRAKE" warning lamp in case the EBCM must turn it "ON." This will only occur if an ABS malfunction is detected that may degrade base brake operation.

Because ABS is not the only device controlling the red "BRAKE" warning lamp (parking brake and low fluid sensor may also turn "ON" the lamp) a short to ground in this circuit cannot be detected.

DTC Will Set When: DTC 87 can be set after initialization is completed. If the red "BRAKE" warning lamp circuit voltage is out of specification indicating an open circuit, a malfunction exists.

Action Taken: A malfunction DTC is stored, ABS is not disabled.

DTC Chart Test Description: Number(s) below refer to circled number(s) on the diagnostic chart.

1. Identifies if the ground circuit to the red "BRAKE" warning lamp is completed by a source other than the EBCM.
2. Checks for high resistance in park brake circuitry.
3. Checks for a malfunctioning EBCM.
4. Checks for high resistance in fluid level switch circuitry.
5. Checks for a malfunctioning warning lamp bulb.
6. Checks for open or high resistance in CKT 33.

Diagnostic Aids: An "Intermittent" malfunction may be caused by a poor connection, rubbed through wire insulation, or a wire that is broken inside the insulation.

The frequency of the malfunction can be checked by using the enhanced diagnostic function of the Tech 1, as described in "Tech 1 Diagnostics" found in this section. If a Tech 1 or T-100 are not available, the parking brake handle may be lifted to verify proper lamp operation and continuity of CKTs 3 and 39.

Any circuitry, that is suspected as causing the intermittent complaint, should be thoroughly checked for backed out terminals, improper mating, broken locks, improperly formed or damaged terminals, poor terminal to wiring connections or physical damage to the wiring harness.

DIAGNOSTIC CHART DTC 87 — 1994 GRAND PRIX, CONTINUED

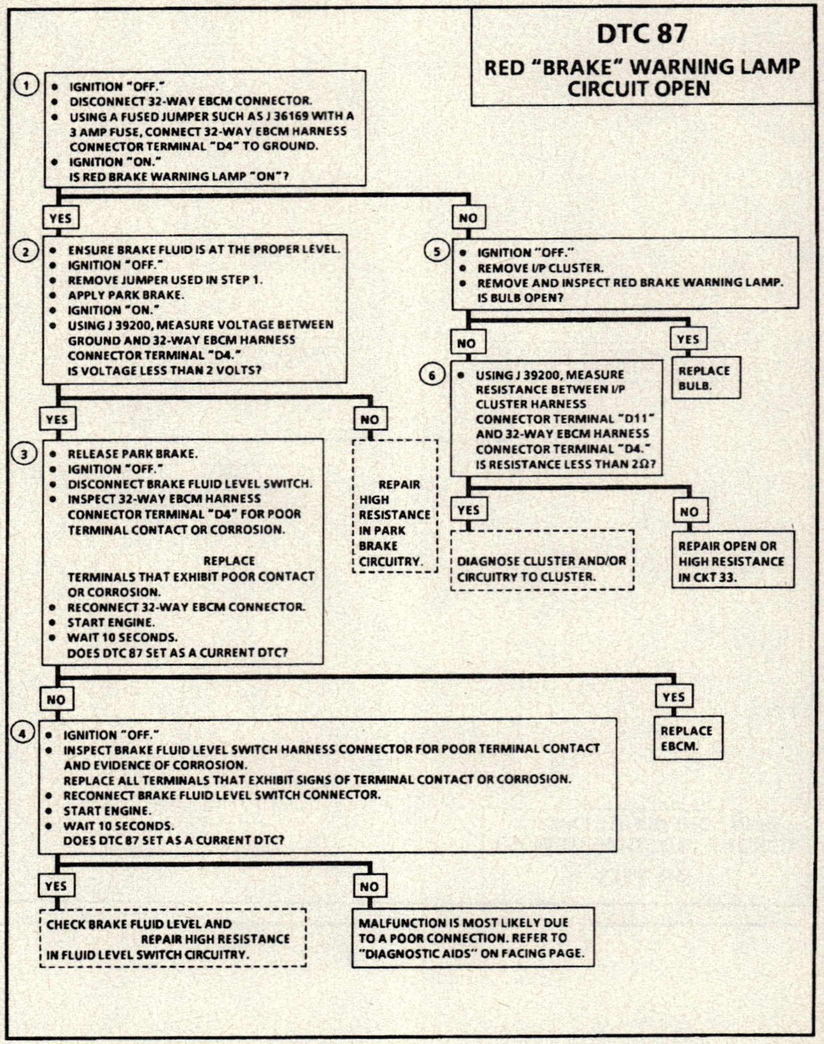

DIAGNOSTIC CHART DTC 88 — 1994 GRAND PRIX

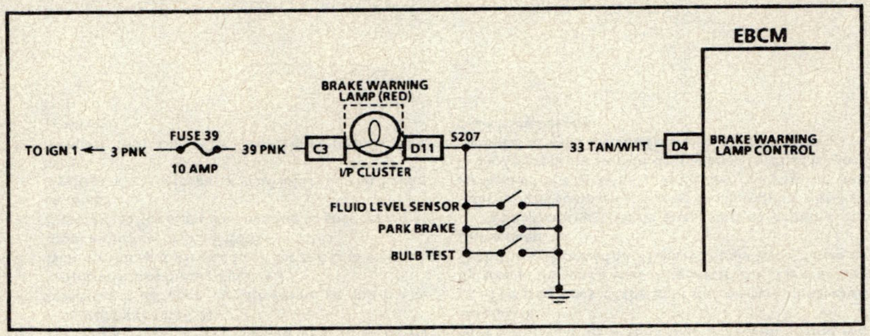

DTC 88

RED "BRAKE" WARNING LAMP CKT SHORTED TO BATT

Circuit Description:

This DTC identifies a short to battery between the EBCM and the red "BRAKE" warning lamp, or an open driver that does not allow the red "BRAKE" warning lamp to be illuminated by the EBCM. This will only occur if an ABS fault is detected that may degrade base brake operation.

DTC Will Set When: DTC 88 can be set only at the beginning of the three second bulb check. If the EBCM detects battery voltage on the red "BRAKE" warning lamp control circuit, a malfunction exists.

Action Taken: A malfunction DTC is stored. ABS is not disabled.

DTC Chart Test Description: Number(s) below refer to circled number(s) on the diagnostic chart.

1. Identifies if the ground circuit to the red "BRAKE" warning lamp is being completed by a source other than the EBCM.
2. Identifies if the malfunction is currently present.
3. By removing Fuse 39, the voltage source is eliminated. This test indicates if voltage is being supplied from a source other than the gauge fuse.
4. Ensures malfunction was not due to physical damage of the circuitry.
5. Checks for DTC 86 also being set.

Diagnostic Aids: An "Intermittent" malfunction may be caused by a poor connection, rubbed through wire insulation, or a wire that is broken inside the insulation.

The frequency of the malfunction can be checked by using the enhanced diagnostic function of the Tech 1, as described in "Tech 1 Diagnostics" found in this section.

Any circuitry, that is suspected as causing the intermittent complaint, should be thoroughly checked for backed out terminals, improper mating, broken locks, improperly formed or damaged terminals, poor terminal to wiring connections or physical damage to the wiring harness.

DIAGNOSTIC CHART DTC 88 — 1994 GRAND PRIX, CONTINUED

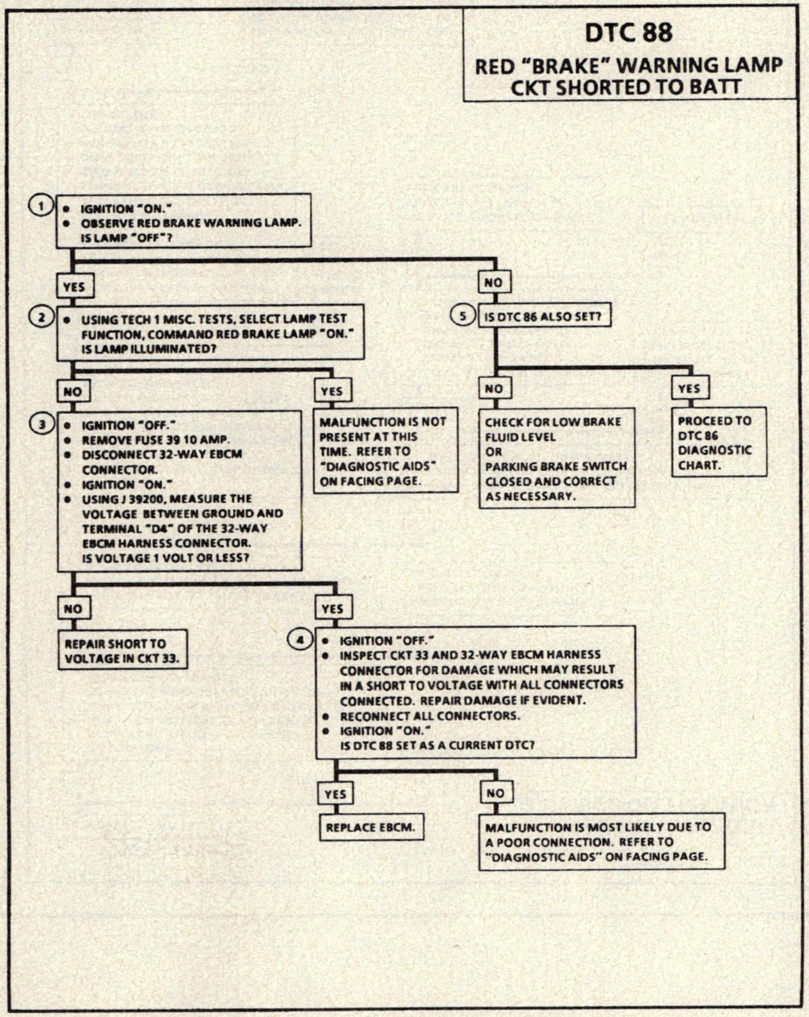

DIAGNOSTIC CHART DTC 91 — 1994 GRAND PRIX

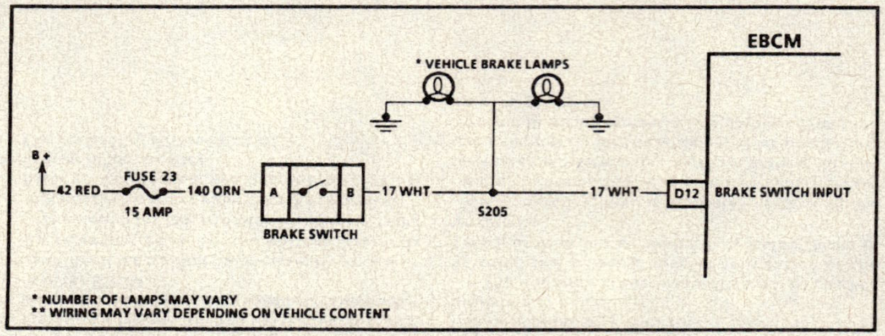

DTC 91

OPEN BRAKE SWITCH DURING DECELERATION

Circuit Description:
This DTC is used to detect an open brake switch in the non-ABS mode. The EBCM looks for deceleration rates that would indicate braking action and verifies this assumption by requiring several repeats of this detection method. In each case, ABS will not be available since no brake switch is seen.

DTC Will Set When: DTC 91 can be set if three deceleration cycles occur with the brake switch "OFF."

Action Taken: A malfunction DTC is stored, ABS is disabled and the ABS warning lamp is turned "ON."

DTC Chart Test Description: Number(s) below refer to circled number(s) on the diagnostic chart.

1. Indicates if the brake switch signal is being received by the EBCM.
2. Indicates if an open circuit exists in the brake switch or brake lamp circuitry.
3. Isolates the open circuit to either the brake switch input circuitry, or the EBCM.
4. Ensures malfunction was not due to poor terminal contact.
5. Verifies that voltage is available at the brake switch.
6. Verifies that the brake switch is functioning properly.

Diagnostic Aids: An "Intermittent" malfunction may be caused by a poor connection, rubbed through wire insulation, or a wire that is broken inside the insulation.

The frequency of the malfunction can be checked by using the enhanced diagnostic function of the Tech 1, as described in "Tech 1 Diagnostics" found in this section.

Any circuitry, that is suspected as causing the intermittent complaint, should be thoroughly checked for backed out terminals, improper mating, broken locks, improperly formed or damaged terminals, poor terminal to wiring connections or physical damage to the wiring harness.

DIAGNOSTIC CHART DTC 91 — 1994 GRAND PRIX, CONTINUED

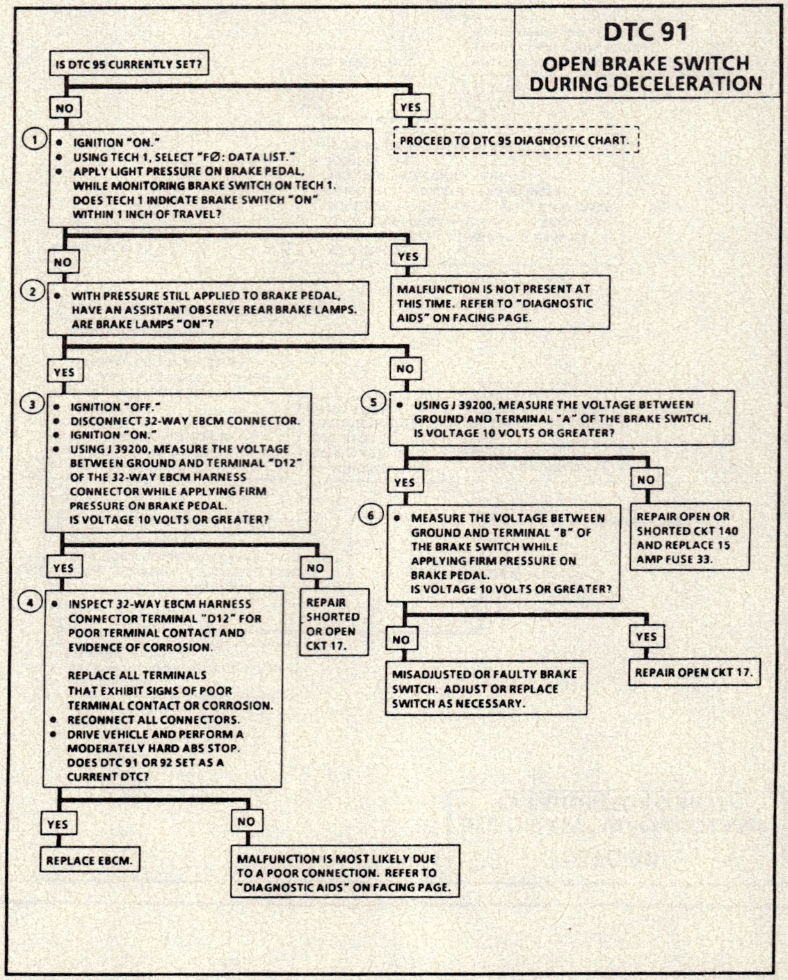

DIAGNOSTIC CHART DTC 92 — 1994 GRAND PRIX

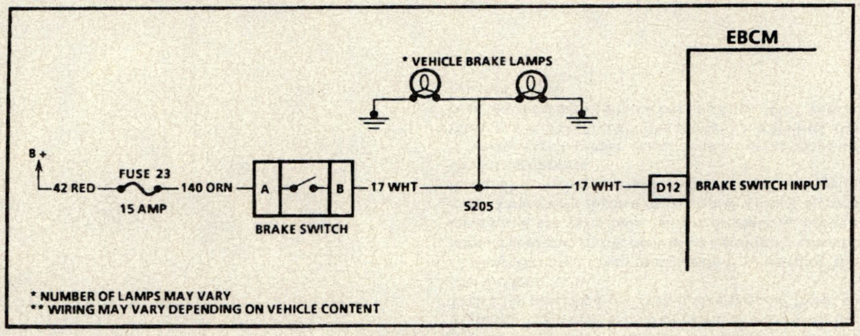

DTC 92

OPEN BRAKE SWITCH WHEN ABS WAS REQUIRED

Circuit Description:

This DTC is run to determine the proper operation of the brake switch. This is important because ABS is activated when the brake switch is "ON." If the brake switch is "OFF," ABS will never be activated. Since this malfunction is difficult to detect under normal braking conditions, this malfunction is only detected when ABS is required.

DTC Will Set When: DTC 92 can be set if the vehicle's speed is greater than 8 km/h (5 mph). If the brake was not "ON" and a release was required on two channels for 0.5 second, a malfunction exists.

Action Taken: A malfunction DTC is stored, ABS is disabled and the ABS warning lamp is turned "ON."

DTC Chart Test Description: Number(s) below refer to circled number(s) on the diagnostic chart.

1. Indicates if the brake switch signal is being received by the EBCM.
2. Indicates if an open circuit exists in the brake switch or brake lamp circuitry.
3. Isolates the open circuit to either the brake switch input circuitry, or the EBCM.
4. Ensures malfunction was not due to poor terminal contact.
5. Verifies the voltage is available at the brake switch.
6. Verifies that the brake switch is functioning properly.

Diagnostic Aids: An "Intermittent" malfunction may be caused by a poor connection, rubbed through wire insulation, or a wire that is broken inside the insulation.

The frequency of the malfunction can be checked by using the enhanced diagnostic function of the Tech 1, as described in "Tech 1 Diagnostics" found in this section.

Any circuitry, that is suspected as causing the intermittent complaint, should be thoroughly checked for backed out terminals, improper mating, broken locks, improperly formed or damaged terminals, poor terminal to wiring connections or physical damage to the wiring harness.

DIAGNOSTIC CHART DTC 91 — 1994 GRAND PRIX, CONTINUED

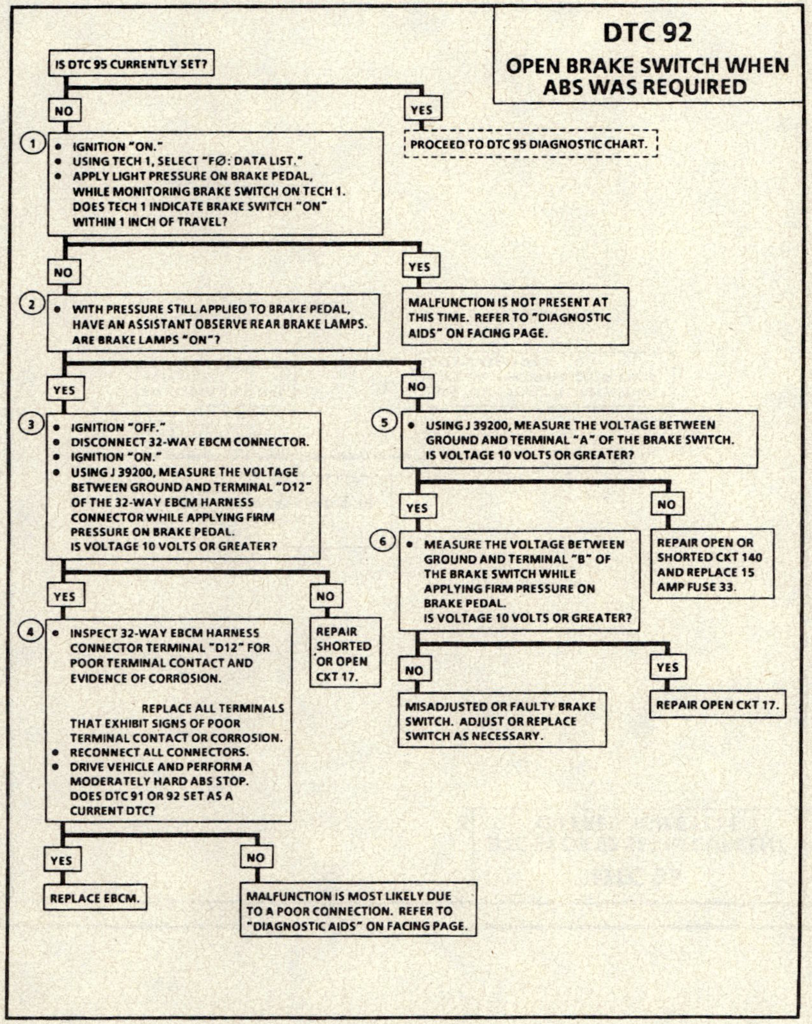

DIAGNOSTIC CHART DTC 93 — 1994 GRAND PRIX

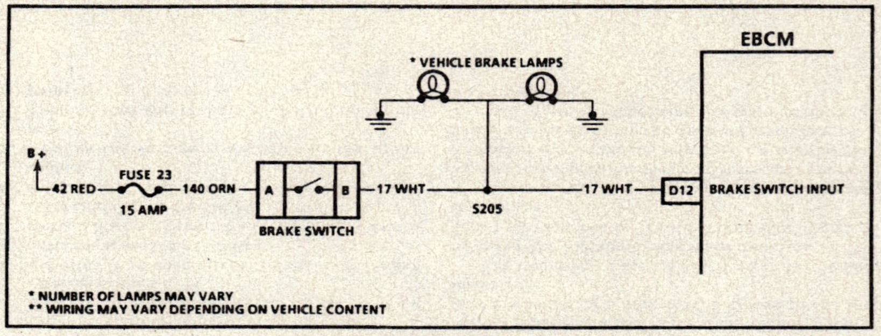

DTC 93

DTC 91 OR 92 SET IN CURRENT OR PREV. IGN CYCLE

Circuit Description:

This DTC is the second portion of DTCs 91 and 92. If DTCs 91 or 92 failed during the last ignition cycle, DTC 93 becomes a current failure during the next ignition cycle, keeping ABS disabled until a brake switch "ON" state is seen. When a change is seen during an ignition cycle in which DTC 93 is a current malfunction, DTC 91 or 92 will clear itself at the end of the current ignition cycle, and ABS will enable itself at the start of the next ignition cycle. DTC 93 alone indicates 91 or 92 failed previously, but is intermittent, or has been corrected.

DTC Chart Test Description: Number(s) below refer to circled number(s) on the diagnostic chart.

1. Indicates which DTC (either 91 or 92) caused DTC 93 to set.

Diagnostic Aids: An "Intermittent" malfunction may be caused by a poor connection, rubbed through wire insulation, or a wire that is broken inside the insulation.

The frequency of the malfunction can be checked by using the enhanced diagnostic function of the Tech 1, as described in "Tech 1 Diagnostics" found in this section.

Any circuitry, that is suspected as causing the intermittent complaint, should be thoroughly checked for backed out terminals, improper mating, broken locks, improperly formed or damaged terminals, poor terminal to wiring connections or physical damage to the wiring harness.

Also, verify proper brake switch operation using the data list of the Tech 1. As the brake is applied, the data list should display the brake switch "ON" within 1 inch of travel.

DIAGNOSTIC CHART DTC 93 — 1994 GRAND PRIX, CONTINUED

DTC 93

DTC 91 OR 92 SET IN CURRENT OR PREV. IGN CYCLE

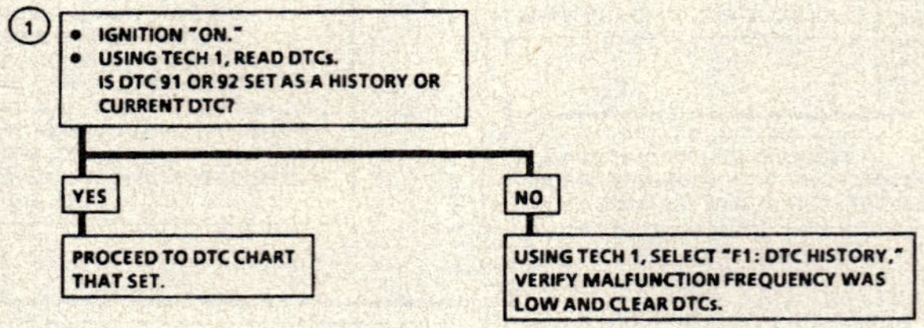

DIAGNOSTIC CHART DTC 94 — 1994 GRAND PRIX

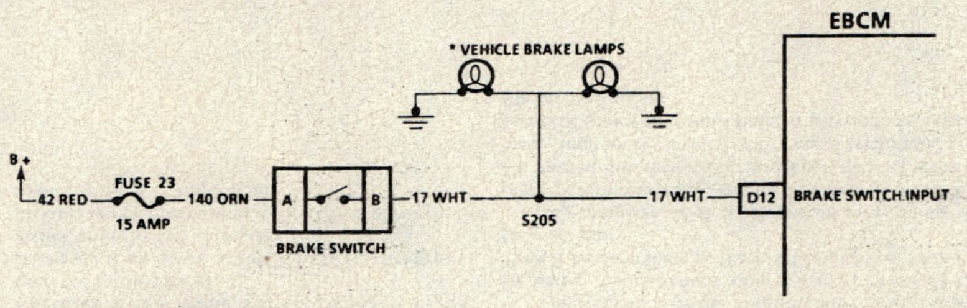

* NUMBER OF LAMPS MAY VARY
** WIRING MAY VARY DEPENDING ON VEHICLE CONTENT

DTC 94

BRAKE SWITCH CONTACTS ALWAYS CLOSED

Circuit Description:
This DTC is run to determine the proper operation of the brake switch. This is important because ABS is activated when the brake switch is "ON" and turned "OFF" when the brake is "OFF." If the brake switch is always "ON," ABS operation will always be requested resulting in potential modulator cycling on rough roads. Additionally, this malfunction will most likely result in a dead battery (due to the brake lights staying "ON") if the driver is not informed of this malfunction.

DTC Will Set When: DTC 94 can be set when the vehicle reaches at least 40 km/h (25 mph). If the brake was never "OFF" during two consecutive drive cycles, a malfunction exists.

Action Taken: A malfunction DTC is stored. ABS is not disabled.

DTC Chart Test Description: Number(s) below refer to circled number(s) on the diagnostic chart.

1. Checks if the malfunction is currently present.
2. Isolates the cause of the malfunction to either a malfunctioning or misadjusted brake switch, or a short to voltage in the brake switch circuitry.
3. Checks for unwanted voltage on the brake switch input circuit.
4. Checks for a possible intermittent malfunction.

Diagnostic Aids: An "Intermittent" malfunction may be caused by a poor connection, rubbed through wire insulation, or a wire that is broken inside the insulation.

The frequency of the malfunction can be checked by using the enhanced diagnostic function of the Tech 1, as described in "Tech 1 Diagnostics" found in this section.

Any circuitry, that is suspected as causing the intermittent complaint, should be thoroughly checked for backed out terminals, improper mating, broken locks, improperly formed or damaged terminals, poor terminal to wiring connections or physical damage to the wiring harness.

DIAGNOSTIC CHART DTC 94 — 1994 GRAND PRIX, CONTINUED

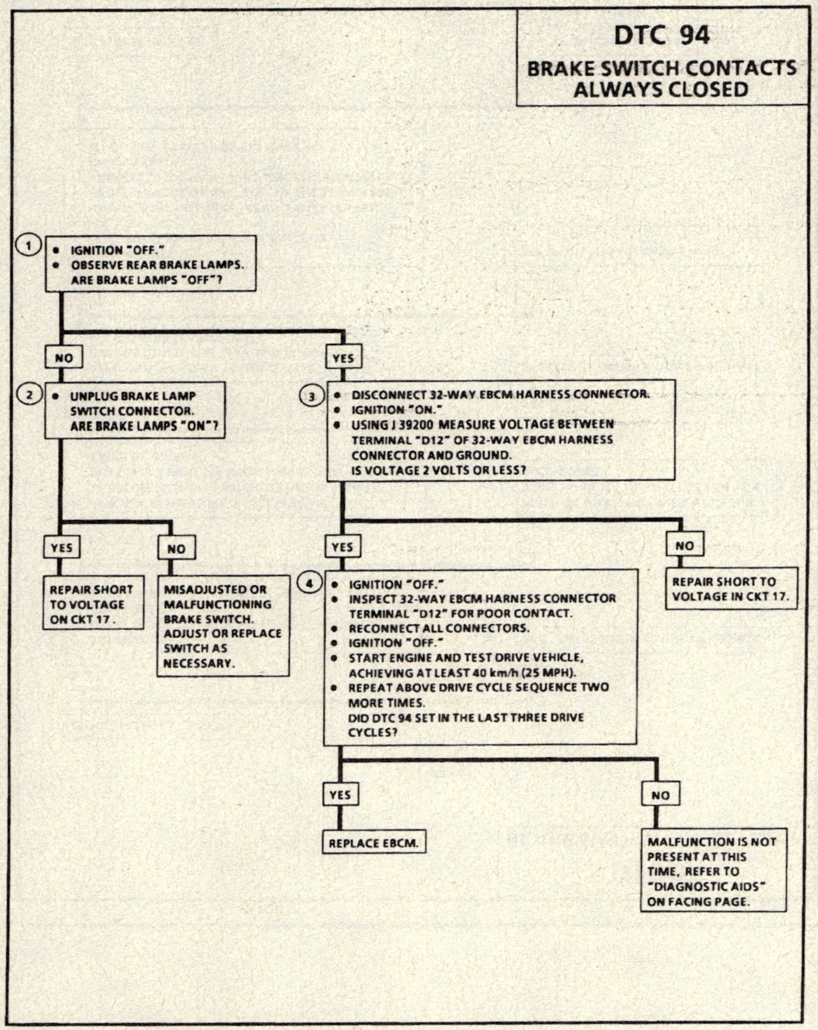

DIAGNOSTIC CHART DTC 95 — 1994 GRAND PRIX

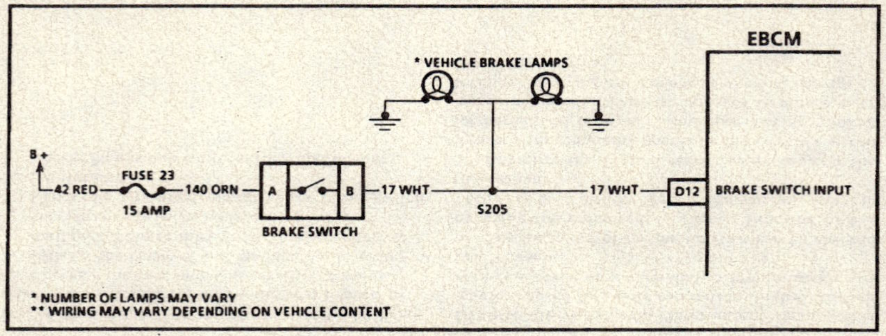

DTC 95
BRAKE SWITCH CIRCUIT OPEN

Circuit Description:
This DTC is used to identify open brake switch circuitry that prevents the brake switch input to the EBCM from changing states when the brake is applied. This DTC is used in conjunction with DTCs 91 and 92 to determine the cause of an open brake switch malfunction.

DTC Will Set When: DTC 95 can be set after initialization is completed. If the brake switch input voltage is out of specification for one second indicating an open circuit, a malfunction exists.

Action Taken: A malfunction DTC is stored, ABS is disabled and the ABS warning lamp is turned "ON."

DTC Chart Test Description: Number(s) below refer to circled number(s) on the diagnostic chart.

1. This step is used to confirm that an open in the brake switch circuitry currently exists.
2. Indicates if the brake switch signal is being received by the EBCM.
3. Indicates if an open circuit exists in the brake switch or brake lamp circuitry.
4. Isolates the open circuit to either the brake switch input circuitry, or the EBCM.

Diagnostic Aids: An "Intermittent" malfunction may be caused by a poor connection, rubbed through wire insulation, or a wire that is broken inside the insulation.

The frequency of the malfunction can be checked by using the enhanced diagnostic function of the Tech 1, as described in "Tech 1 Diagnostics" found in this section.

Any circuitry, that is suspected as causing the intermittent complaint, should be thoroughly checked for backed out terminals, improper mating, broken locks, improperly formed or damaged terminals, poor terminal to wiring connections or physical damage to the wiring harness.

DIAGNOSTIC CHART DTC 95 — 1994 GRAND PRIX, CONTINUED

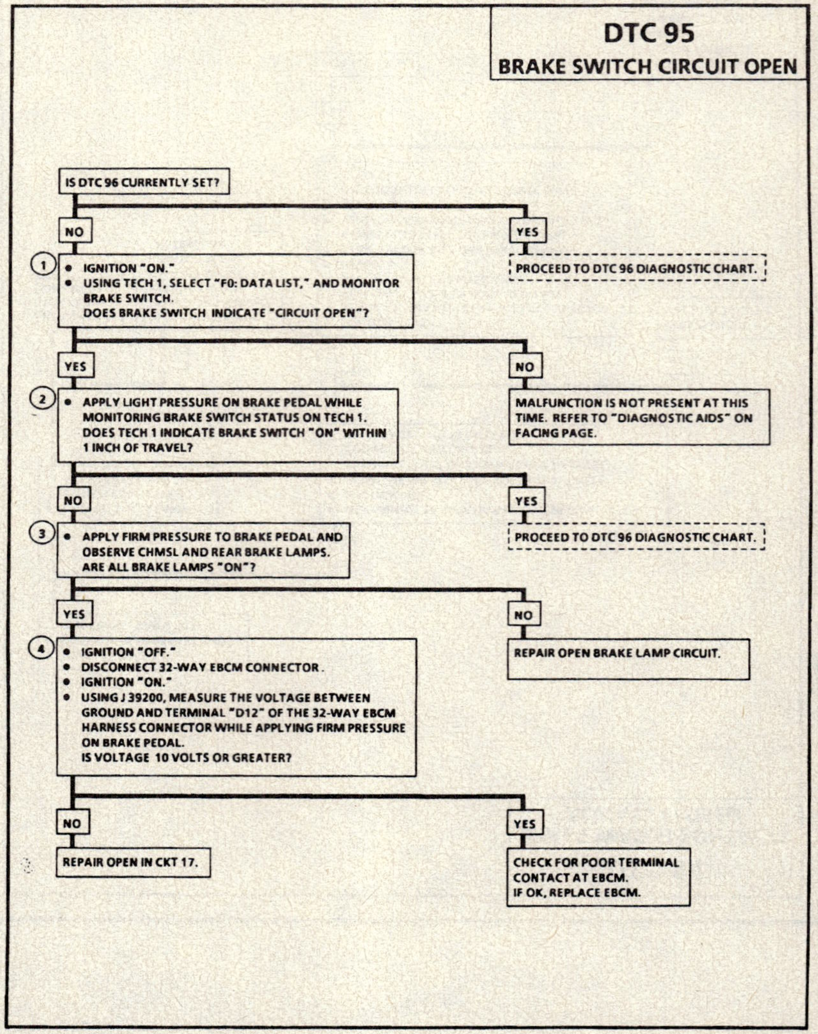

DIAGNOSTIC CHART DTC 96 — 1994 GRAND PRIX

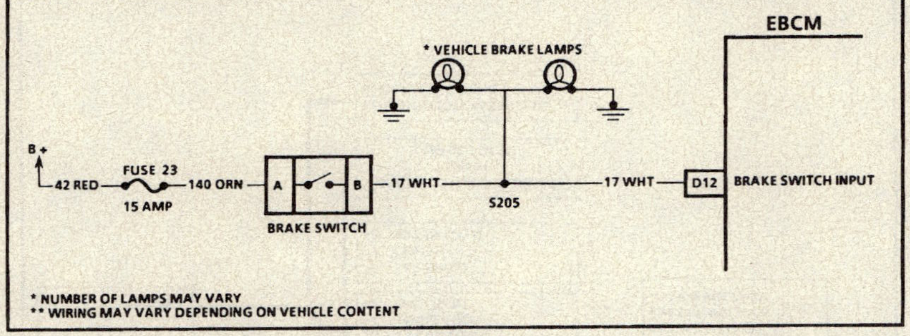

DTC 96

BRAKE LAMPS CIRCUIT OPEN

Circuit Description:
This DTC is designed to isolate the cause of a DTC 95 malfunction and indicate to the driver ABS is still available. If DTC 95 occurs with DTC 96, the brake lamps are open, the brake lamp grounds are open, or the Center High Mounted Stop Lamp (CHMSL) circuit is open during 4-way flasher use. The presence of battery voltage at the brake switch input indicates a valid brake switch input is still available.

DTC Will Set When: DTC 96 can be set only after DTC 95 has been set. If the EBCM detects battery voltage on the brake switch input circuit for 0.5 second, a malfunction exists.

Action Taken: A malfunction DTC is stored. ABS is not disabled.

DTC Chart Test Description: Number(s) below refer to circled number(s) on the diagnostic chart.

1. As a result of a malfunction of an additional brake switch circuit DTC, this DTC may be set. To ensure proper diagnosis, any additional brake switch DTCs must be repaired first.
2. Identifies if the malfunction is currently present in the brake circuit.

Diagnostic Aids: An "Intermittent" malfunction may be caused by a poor connection, rubbed through wire insulation, or a wire that is broken inside the insulation.

The frequency of the malfunction can be checked by using the enhanced diagnostic function of the Tech 1, as described in "Tech 1 Diagnostics" found in this section.

Any circuitry, that is suspected as causing the intermittent complaint, should be thoroughly checked for backed out terminals, improper mating, broken locks, improperly formed or damaged terminals, poor terminal to wiring connections or physical damage to the wiring harness.

DIAGNOSTIC CHART DTC 96 — 1994 GRAND PRIX, CONTINUED

DTC 96

BRAKE LAMPS CIRCUIT OPEN

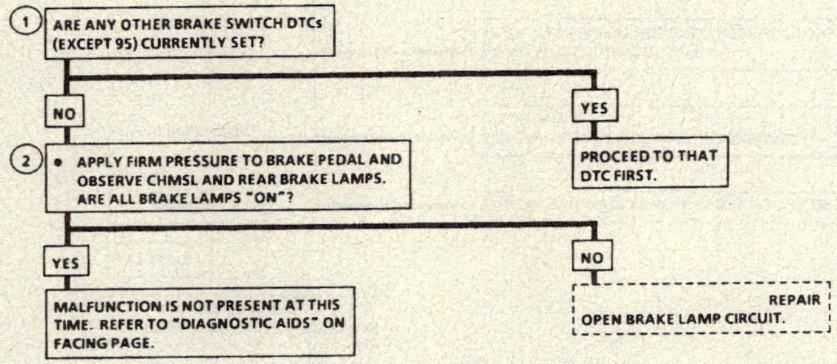

DIAGNOSTIC PROCESS CHART — 1994 LUMINA

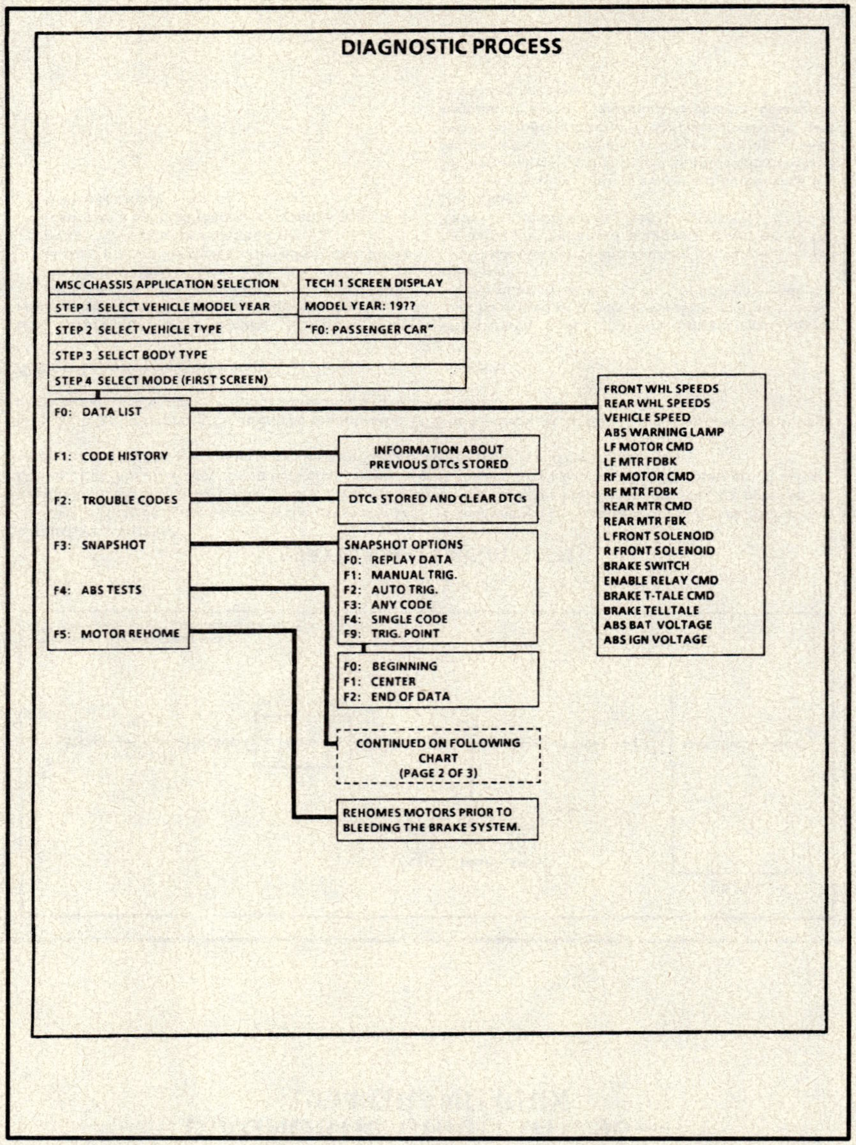

DIAGNOSTIC PROCESS CHART — 1994 LUMINA, CONTINUED

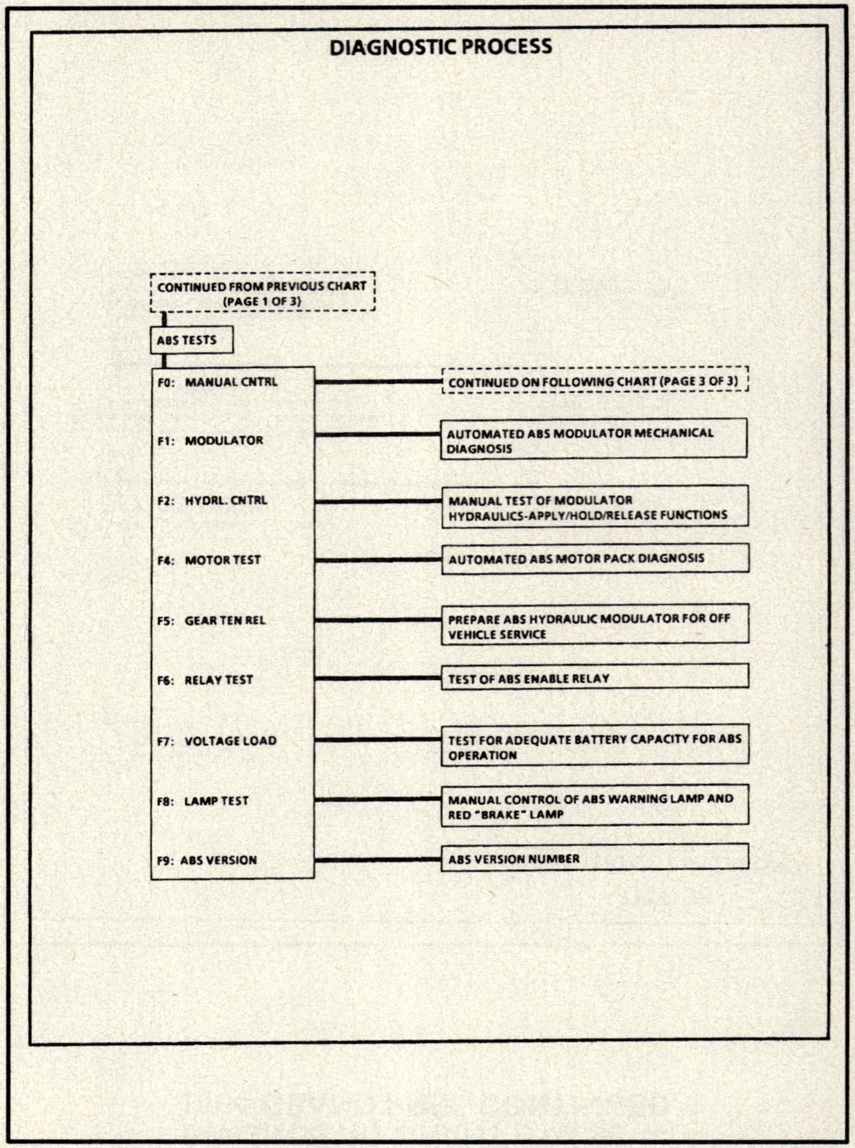

DIAGNOSTIC PROCESS CHART — 1994 LUMINA, CONTINUED

DIAGNOSTIC PROCESS

CONTINUED FROM PREVIOUS CHART (PAGE 2 OF 3)

MANUAL CONTROL OF ENABLE RELAY SOLENOIDS AND MOTORS.

↑ = TURN ON
↓ = TURN OFF
ENTER = RLY ON/OFF
YES TO CONTINUE

SELECT FUNCTION
F0: LF APPLY
F1: LF RELEASE
F2: LF SOLENOID
F4: RF APPLY
F5: RF RELEASE
F6: RF SOLENOID
F8: REAR APPLY
F9: REAR RELEASE

MOTOR APPLY
COMMAND: XX AMPS
FEEDBACK: XX AMPS
ENAB RELAY: ON/OFF

MOTOR RELEASE
COMMAND: XX AMPS
FEEDBACK: XX AMPS
ENAB RELAY: ON/OFF

SOLENOID
COMMAND: ON/OFF
ENAB RELAY: ON/OFF

DIAGNOSTIC CHART SYSTEM CHECK — 1994 LUMINA

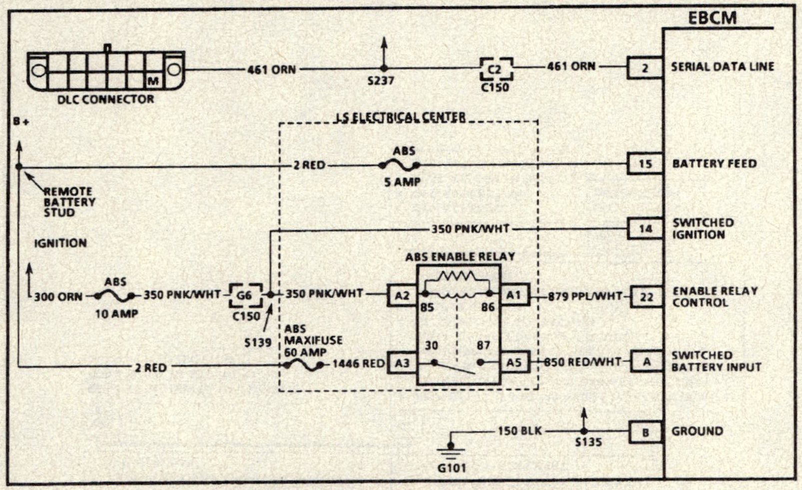

DIAGNOSTIC SYSTEM CHECK

System Description:
The diagnostic system check is an organized approach to identifying a problem created by an Antilock Brake System (ABS) malfunction. It must be the starting point for any ABS complaint diagnosis because it directs the service technician to the next logical step in diagnosing the complaint.

Serial data is transmitted by the EBCM through terminal "2". The EBCM is supplied switched ignition voltage through terminal "14", battery voltage through terminal "15" and switched battery voltage through terminal "A". The EBCM ground is provided through terminal "B".

Chart Test Description: Number(s) below refer to circled number(s) on the diagnostic chart.

1. Indicates if EBCM is transmitting data.
2. Checks for a serial data line malfunction.
3. Checks for proper Tech 1 connection to DLC.
4. Checks for open ABS 5 amp fuse.
5. Checks for open ABS 10 amp fuse.
6. Checks for high resistance in the EBCM ground circuit.
7. Checks for current DTCs.
8. Checks for proper ABS warning lamp operation.
9. Checks for history DTCs.
10. Checks for possible short to ground in the battery feed circuitry.
11. Checks for possible short to ground in the EBCM.
12. Ensures short to ground is not due to physical damage of the circuitry.
13. Checks for possible short to ground in the switched ignition circuitry.
14. Isolates the short to ground to either the circuitry or the EBCM.
15. Ensures short to ground is not due to physical damage of the circuitry.
16. Checks for an open in CKT 2 or 140.
17. Checks for high resistance in battery feed circuitry.
18. Checks for high resistance in the switched ignition circuitry.
19. Ensures malfunction is not due to poor terminal contact.
20. Checks for high resistance in the switched ignition circuit between the fuse block and the EBCM.

Diagnostic Aids: Excessive resistance in the ground or power supply circuits will not allow communication with the EBCM. If communication with the EBCM is not possible, ensure the ABS ground connection is good and that there is no excessive resistance in any of the power supply circuits.

DIAGNOSTIC CHART SYSTEM CHECK — 1994 LUMINA, CONTINUED

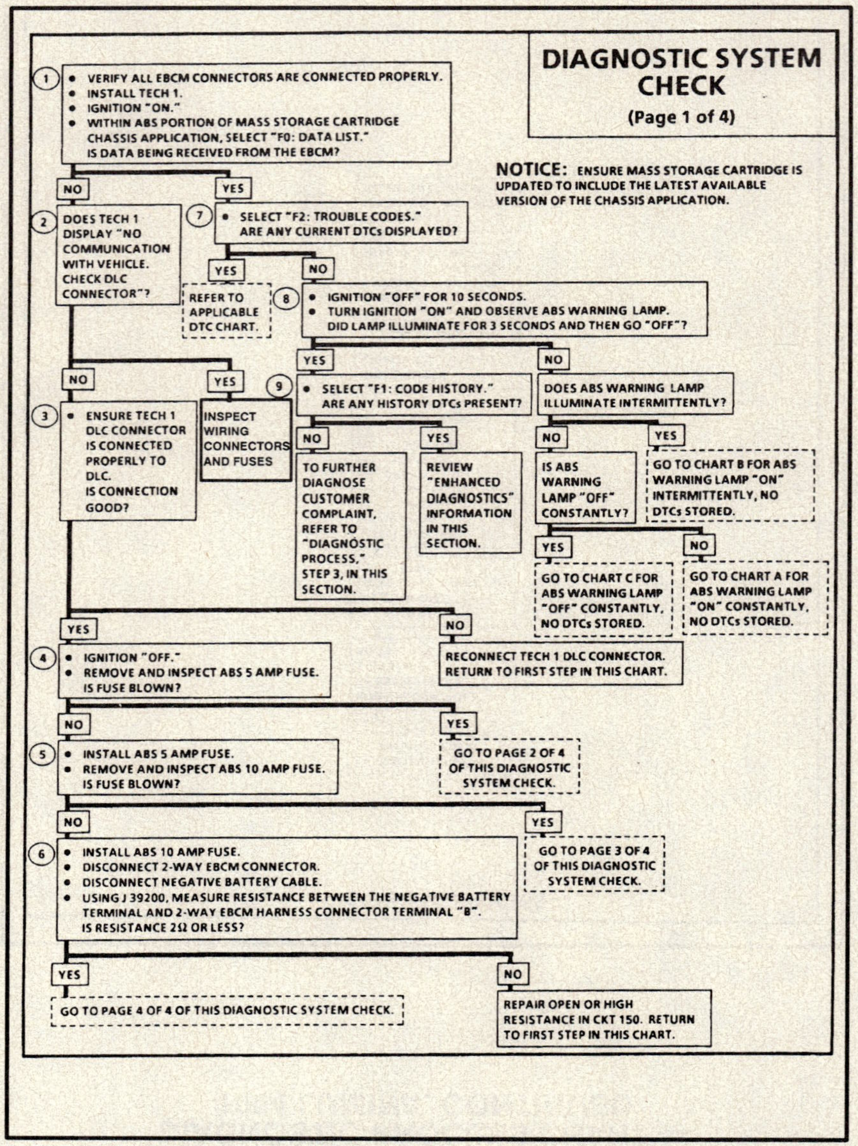

DIAGNOSTIC CHART SYSTEM CHECK — 1994 LUMINA, CONTINUED

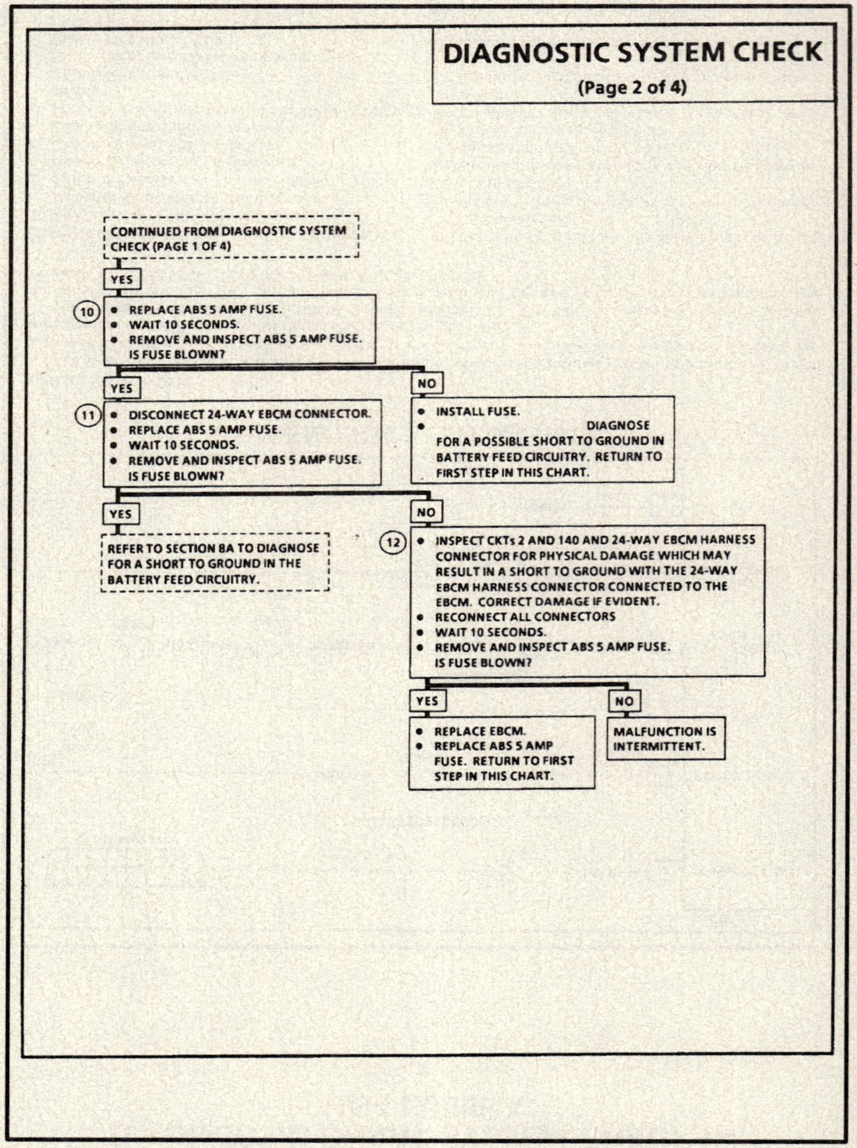

DIAGNOSTIC CHART SYSTEM CHECK — 1994 LUMINA, CONTINUED

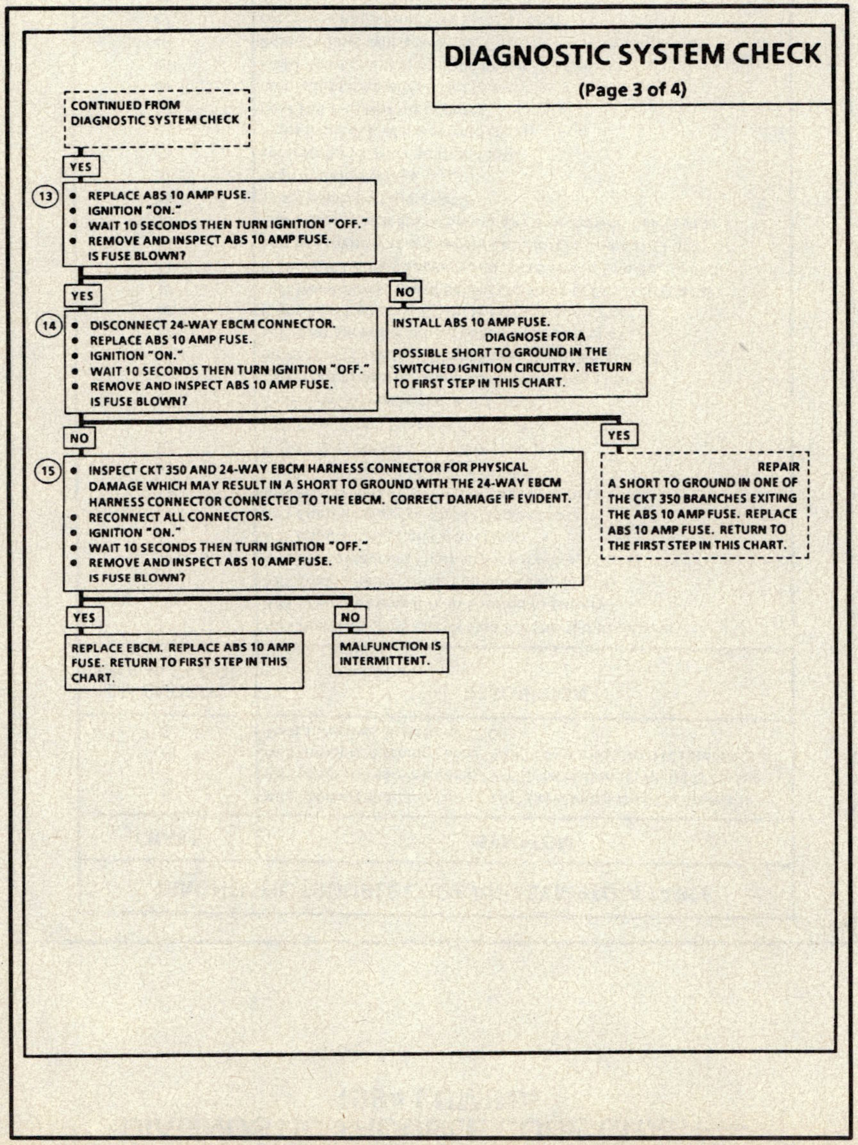

DIAGNOSTIC CHART SYSTEM CHECK — 1994 LUMINA, CONTINUED

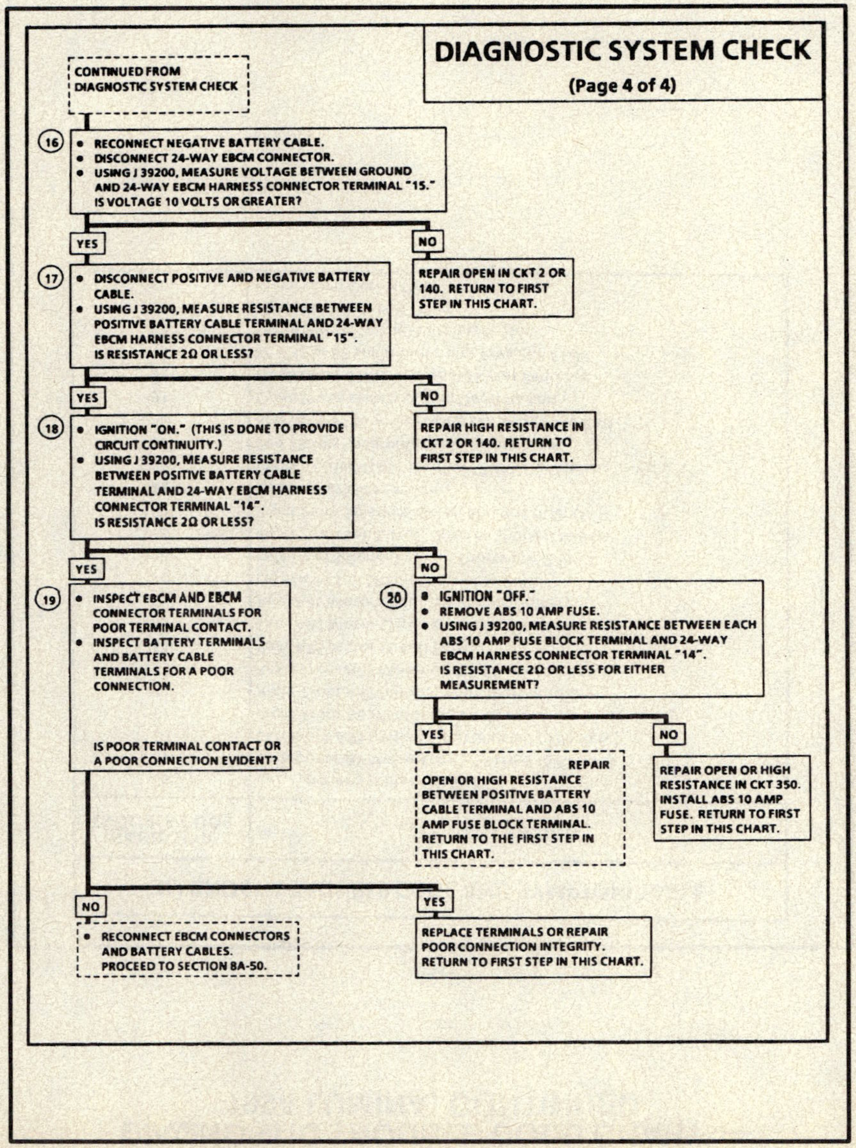

DIAGNOSTIC TROUBLE CODE CHART — 1994 LUMINA

DIAGNOSTIC TROUBLE CODE AND SYMPTOM TABLE

CHART	SYMPTOM
A	ABS (Amber) Warning Lamp "ON" Constantly, No DTCs Stored
B	ABS (Amber) Warning Lamp "ON" Intermittently, No DTCs Stored
C	ABS (Amber) Warning Lamp "OFF" Constantly, No DTCs Stored
D	Tech 1 Displays Undefined DTCs

DIAGNOSTIC TROUBLE CODE	DESCRIPTION
11	ABS Warning Lamp Circuit Open or Shorted to Ground
13	ABS Warning Lamp Circuit Shorted to Battery
14	ABS Enable Relay Contact Circuit Open (1 of 3)
15	ABS Rly Contact Ckt Short to Batt or Always Closed
16	ABS Enable Relay Coil Circuit Open
17	ABS Enable Relay Coil Circuit Shorted to Gnd
18	ABS Enable Relay Coil Circuit Shorted to Batt
21	Left Front Wheel Speed = 0 (1 of 2)
22	Right Front Wheel Speed = 0 (1 of 2)
23	Left Rear Wheel Speed = 0 (1 of 2)
24	Right Rear Wheel Speed = 0 (1 of 2)
25	Left Front Excessive Wheel Speed Variation (1 of 2)
26	Right Front Excessive Wheel Speed Variation (1 of 2)
27	Left Rear Excessive Wheel Speed Variation (1 of 2)
28	Right Rear Excessive Wheel Speed Variation (1 of 2)
32	Left Front Wheel Speed Sensor Ckt Short to Gnd/Bat (1 of 2)
33	Right Front Wheel Speed Sensor Ckt Short to Gnd/Bat (1 of 2)
34	Left Rear Wheel Speed Sensor Ckt Short to Gnd/Bat (1 of 2)
35	Right Rear Wheel Speed Sensor Ckt Short to Gnd/Bat (1 of 2)
36	Low System Voltage (1 of 2)
37	High System Voltage
38	Left Front ESB Will Not Hold Motor
41	Right Front ESB Will Not Hold Motor
42	Rear ESB Will Not Hold Motor
44	Left Front Channel Will Not Move
45	Right Front Channel Will Not Move
46	Rear Channel Will Not Move
47	Left Front ABS Motor Free Spins (1 of 2)
48	Right Front ABS Motor Free Spins (1 of 2)
51	Rear ABS Motor Free Spins (1 of 2)
52	Left Front Channel in Release Too Long
53	Right Front Channel In Release Too Long
54	Rear Channel in Release Too Long
55	EBCM Malfunction

DIAGNOSTIC TROUBLE CODE CHART — 1994 LUMINA, CONTINUED

DIAGNOSTIC TROUBLE CODE AND SYMPTOM TABLE

DIAGNOSTIC TROUBLE CODE	DESCRIPTION
56	Left Front ABS Motor Circuit Open
57	Left Front ABS Motor Circuit Shorted to Ground
58	Left Front ABS Motor Circuit Shorted to Battery
61	Right Front ABS Motor Circuit Open
62	Right Front ABS Motor Circuit Shorted to Gnd
63	Right Front ABS Motor Circuit Shorted to Batt
64	Rear ABS Motor Circuit Open
65	Rear ABS Motor Circuit Shorted to Ground
66	Rear ABS Motor Circuit Shorted to Battery
76	Left Front Sol. Circuit Shorted to Batt
77	Left Front Solenoid Circuit Shorted to Gnd
78	Right Front Sol. Circuit Open or Shorted to Batt
81	Right Front Solenoid Circuit Shorted to Gnd
82	Calibration Malfunction
86	EBCM Turned "ON" the Red "BRAKE" Warning Lamp
87	Red "Brake" Warning Lamp Circuit Open
88	Red "Brake" Warning Lamp Ckt Shorted to Batt
91	Open Brake Switch During Deceleration
92	Open Brake Switch When ABS Was Required
93	DTC 91 or 92 Set in Current or Prev. Ign Cycle
94	Brake Switch Contacts Always Closed
95	Brake Switch Circuit Open
96	Brake Lamp Circuit Open

ELECTRICAL CONNECTOR TERMINAL IDENTIFICATION — 1994 LUMINA

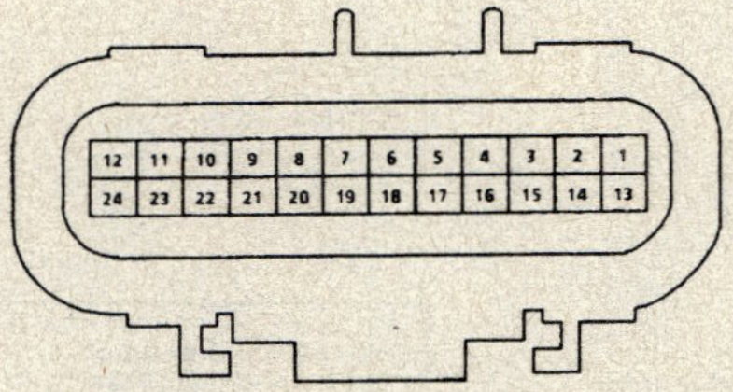

EBCM 24 PIN WORLD CONNECTOR

PIN	CIRCUIT NO.	COLOR	CIRCUIT
1	PLUGGED		NOT USED
2	461	ORN	SERIAL DATA LINE
3	PLUGGED		NOT USED
4	1289	LT BLU/BLK	R/F ABS SOLENOID
5	830	LT BLU	L/F WHEEL SIGNAL HIGH
6	873	YEL	L/F WHEEL SIGNAL LOW
7	882	BRN	R/R WHEEL SIGNAL HIGH
8	883	WHT	R/R WHEEL SIGNAL LOW
9	872	DK GRN	R/F WHEEL SIGNAL HIGH
10	833	TAN	R/F WHEEL SIGNAL LOW
11	885	RED	L/R WHEEL SIGNAL LOW
12	884	BLK	L/R WHEEL SIGNAL HIGH
13	20	LT BLU	BRAKE SWITCH INPUT
14	350	PNK/WHT	SWITCHED IGNITION
15	140	ORN	B + FEED
16	PLUGGED		NOT USED
17	PLUGGED		NOT USED
18	VENT TUBE	BLK	VENT TUBE
19	PLUGGED		NOT USED
20	PLUGGED		NOT USED
21	33	TAN/WHT	BRAKE WARNING LAMP CONTROL
22	879	PPL/WHT	ENABLE RELAY CONTROL
23	852	WHT	ABS INDICATOR LAMP CONTROL
24	1288	DK GRN/YEL	L/F ABS SOLENOID

ELECTRICAL CONNECTOR TERMINAL IDENTIFICATION — 1994 LUMINA

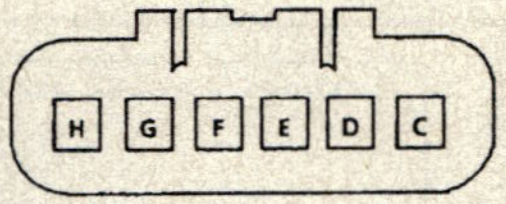

6 WAY EBCM CONNECTOR

PIN	CIRCUIT NO.	COLOR	CIRCUIT
C	1284	DK GRN/WHT	REAR MOTOR HIGH
D	1285	ORN/BLK	REAR MOTOR LOW
E	1281	PNK	LF MOTOR LOW
F	1280	BLK/WHT	LF MOTOR HIGH
G	1283	BLK/PNK	RF MOTOR LOW
H	1282	PPL	RF MOTOR HIGH

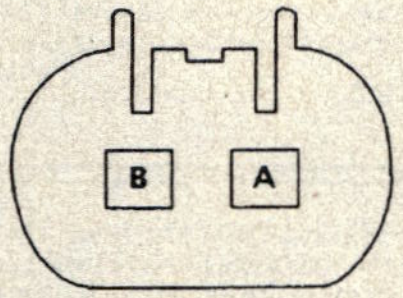

2 WAY EBCM CONNECTOR

PIN	CIRCUIT NO.	COLOR	CIRCUIT
A	850	RED/WHT	SWITCHED BATTERY INPUT
B	150	BLK	GROUND

DIAGNOSTIC CHART A — 1994 LUMINA

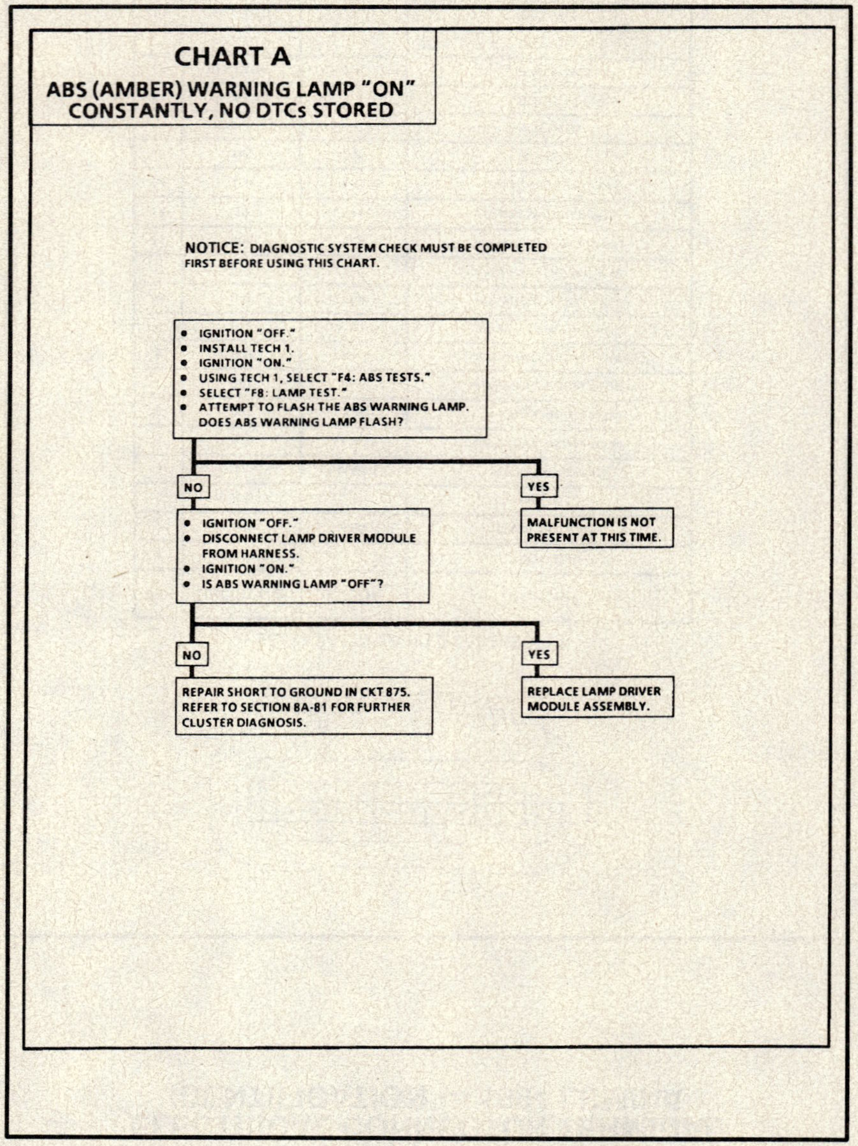

DIAGNOSTIC CHART B — 1994 LUMINA

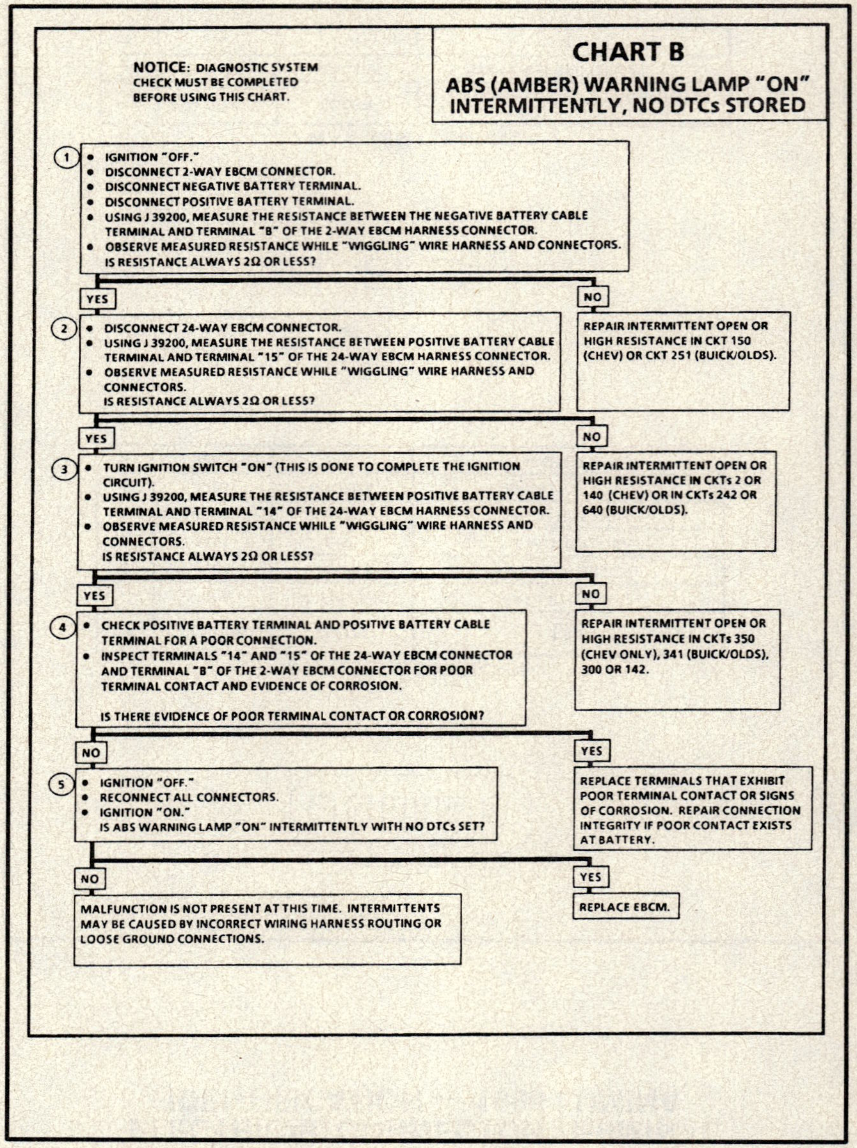

DIAGNOSTIC CHART C — 1994 LUMINA

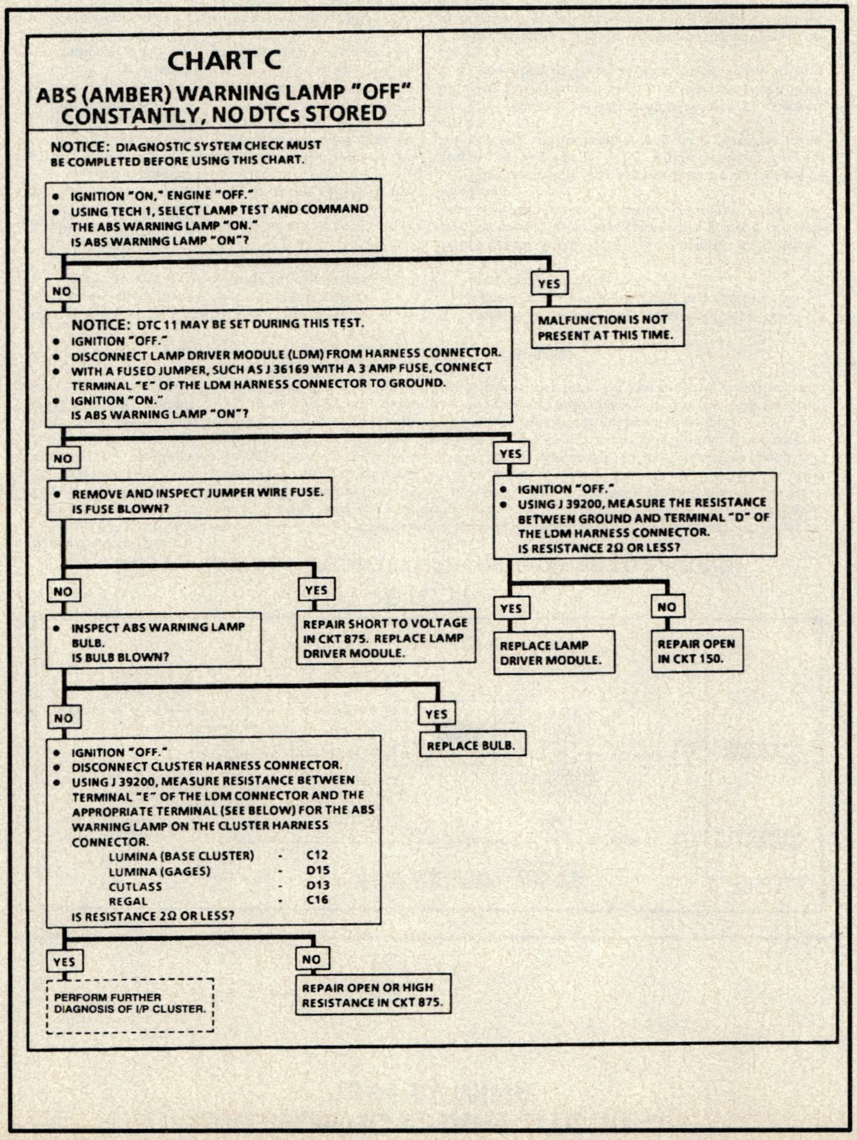

DIAGNOSTIC CHART D — 1994 LUMINA

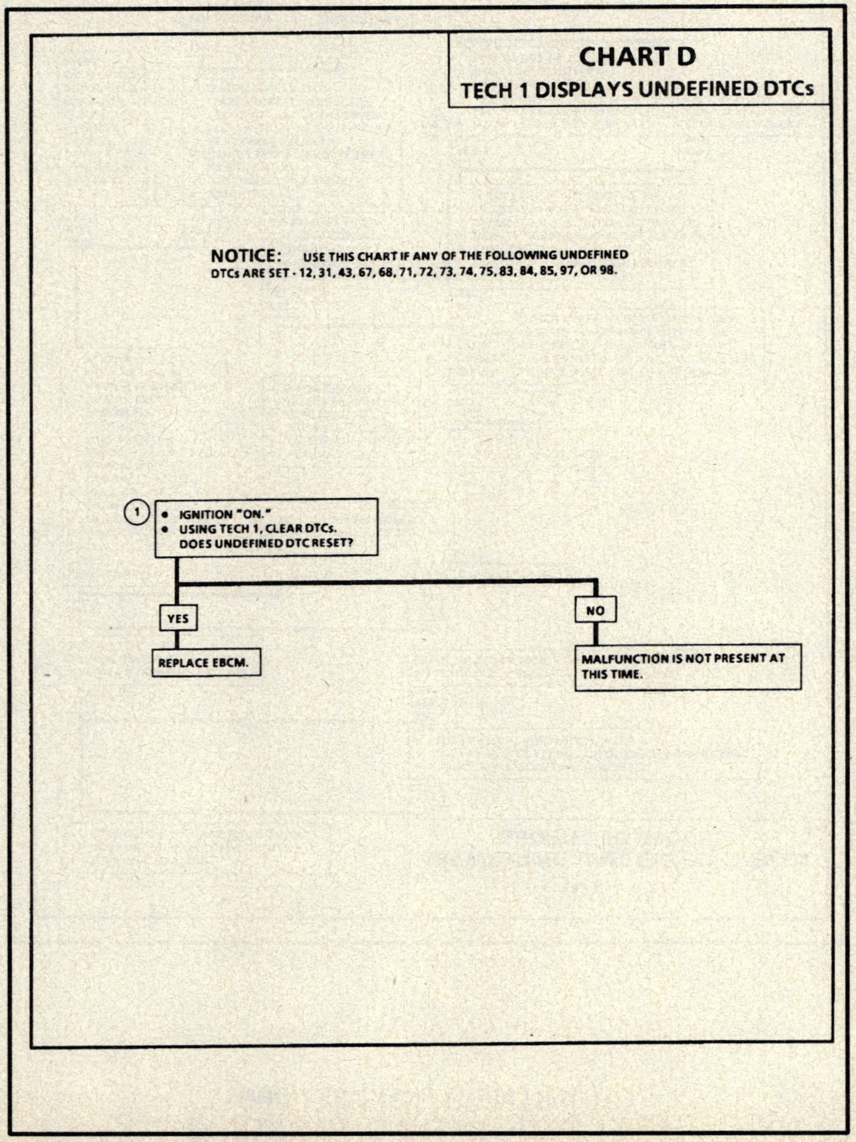

DIAGNOSTIC CHART DTC 11 — 1994 LUMINA

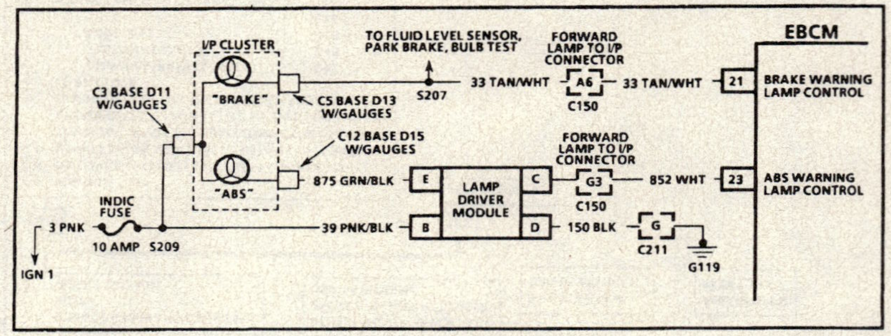

DTC 11

ABS WARNING LAMP CIRCUIT OPEN OR SHORTED TO GROUND

Circuit Description:

This DTC checks the state of the ABS warning lamp to identify a situation in which the driver could not be warned of a system malfunction by the ABS warning lamp, or the warning lamp is always "ON." The Lamp Driver Module (LDM) turns the ABS warning lamp "ON," unless the EBCM provides ground to turn it "OFF." Because of the circuitry in the LDM, only external malfunctions can be detected. As a result, the ABS warning lamp itself is not diagnosable, only the control line to the EBCM can be diagnosed. In the event of an open CKT 852, the ABS warning lamp will be "ON" at all times, due to the loss of ground at the LDM input. If the control line is shorted to ground, the ABS warning lamp is kept "OFF," due to the LDM input being grounded.

DTC Will Set When: DTC 11 can be set only during the three second bulb check or when the ABS warning lamp is commanded "ON." If the EBCM cannot control the ABS warning lamp for two seconds, a malfunction exists.

Action Taken: A malfunction DTC is stored; however, ABS is not disabled.

DTC Chart Test Description: Number(s) below refer to circled number(s) on the diagnostic chart.

1. Checks to see if ABS lamp functions properly. Normal operation would indicate that the problem is not present.
2. Determines if any other DTCs are present or if DTC 18 is set falsely.
3. Ensures that the circuitry from the EBCM to the lamp driver module is complete.
4. Ensures that the ignition circuit to the lamp driver module is functional. An open indicator fuse or open CKT 3, 39 or 875 would result in inoperative indicator lamp.
5. Isolates the cause of the ABS lamp remaining "OFF," by disconnecting the 24-way EBCM connector, CKT 852 should be open and cause the ABS lamp to be "ON."
6. Determines whether the lamp remains "ON" due to a lamp driver malfunction or an open in CKT 852.
7. Determines if the "INDIC" fuse or CKT 3 is open, causing the I/P indicator lamps to remain "OFF."
8. Checks for an open in CKT 39 which supplies voltage to the lamp driver module.
9. Isolates the cause of the lamp remaining "OFF" to a short to ground in CKT 852 or a malfunctioning lamp driver module.

Diagnostic Aids: An "Intermittent" malfunction may be caused by a poor connection, rubbed through wire insulation, or a wire that is broken inside the insulation.

The lamp test function of the Tech 1 may be used to command the lamp "ON" while looking for an intermittent malfunction in the ABS warning lamp circuitry.

The frequency of the malfunction can be checked by using the enhanced diagnostic function of the Tech 1, as described in "Tech 1 Diagnostics" found in this section.

Any circuitry that is suspected as causing the intermittent complaint should be thoroughly checked for backed out terminals, improper mating, broken locks, improperly formed or damaged terminals, poor terminal to wiring connections or physical damage to the wiring harness.

DIAGNOSTIC CHART DTC 11 — 1994 LUMINA, CONTINUED

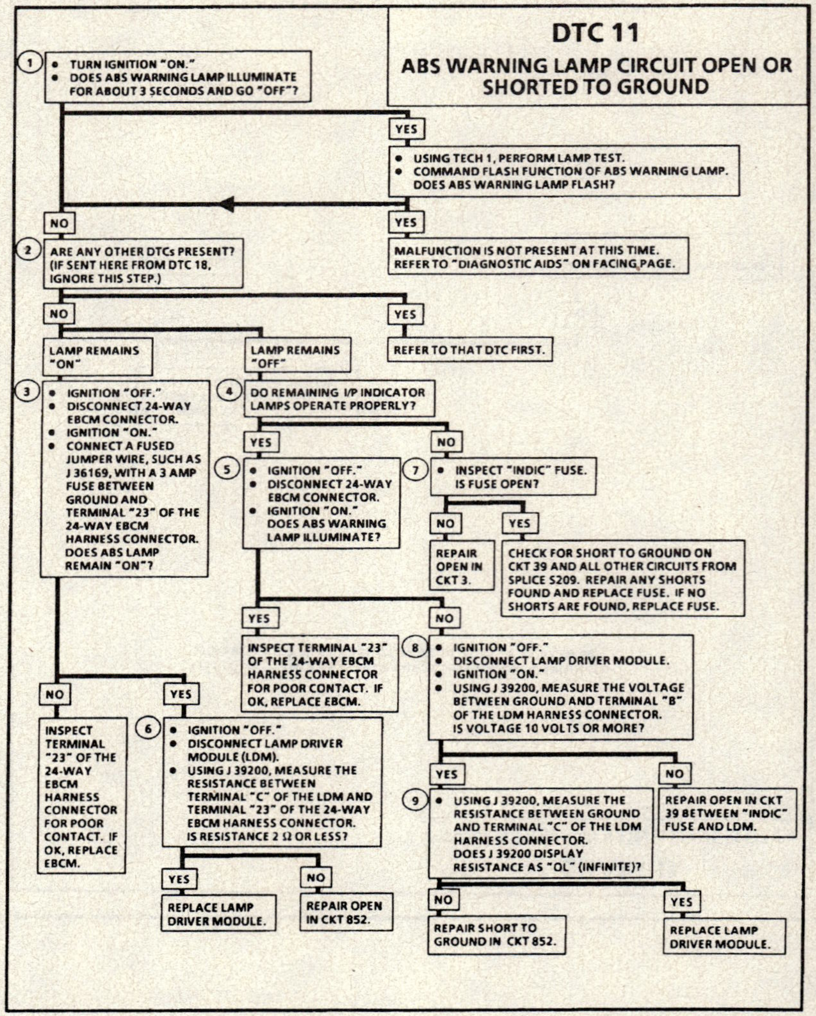

DIAGNOSTIC CHART DTC 13 — 1994 LUMINA

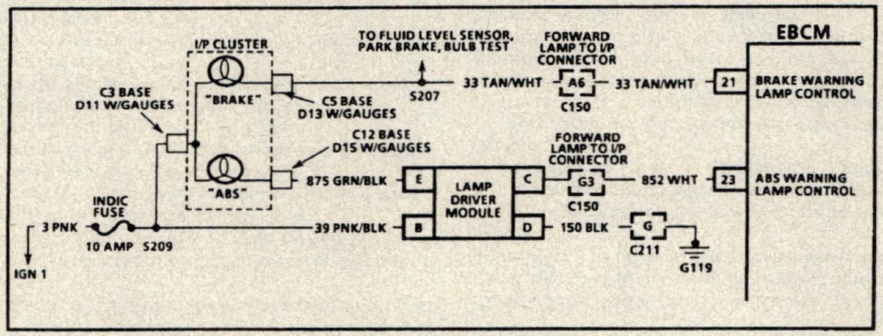

DTC 13

ABS WARNING LAMP CIRCUIT SHORTED TO BATTERY

Circuit Description:

This DTC checks the state of the ABS warning lamp to identify a situation in which the driver could not be warned of a system malfunction by the ABS warning lamp. The Lamp Driver Module (LDM) turns the ABS warning lamp "ON," unless the EBCM provides ground to turn it "OFF." Because of the circuitry internal to the LDM, only external malfunctions can be detected. As a result, the ABS warning lamp itself is not diagnosable, only the control line to the EBCM can be diagnosed.

If the ABS warning lamp control line is shorted to battery or the EBCM control circuit is open, the Electronic Brake Control Module (EBCM) is not able to turn "OFF" the ABS warning lamp.

DTC Will Set When: DTC 13 can be set anytime the EBCM commands the ABS indicator lamp "OFF." If the EBCM cannot control the ABS warning lamp for two seconds, a malfunction exists.

Action Taken: A malfunction DTC is stored; however, ABS is not disabled.

DTC Chart Test Description: Number(s) below refer to circled number(s) on the diagnostic chart.

1. Checks to see if ABS lamp functions properly. Normal operation would indicate that the problem is not present.
2. After removing the 10 amp indicator fuse, voltage on CKT 852 will be eliminated. If voltage is still present, it would indicate that a short to voltage is present between the lamp driver module and EBCM.

Diagnostic Aids: An "Intermittent" malfunction may be caused by a poor connection, rubbed through wire insulation or a wire that is broken inside the insulation.

DTC 13 must be currently set before it should be addressed. If DTC 13 is set as a history DTC, proceed to those DTCs that are currently set before attempting to repair a DTC 13 malfunction.

The frequency of the malfunction can be checked by using the enhanced diagnostic function of the Tech 1, as described in "Tech 1 Diagnostics" found in this section.

Any circuitry that is suspected as causing the "Intermittent" complaint should be thoroughly checked for backed out terminals, improper mating, broken locks, improperly formed or damaged terminals, poor terminal to wiring connections or physical damage to the wiring harness.

DIAGNOSTIC CHART DTC 13 — 1994 LUMINA, CONTINUED

DTC 13

ABS WARNING LAMP CIRCUIT SHORTED TO BATTERY

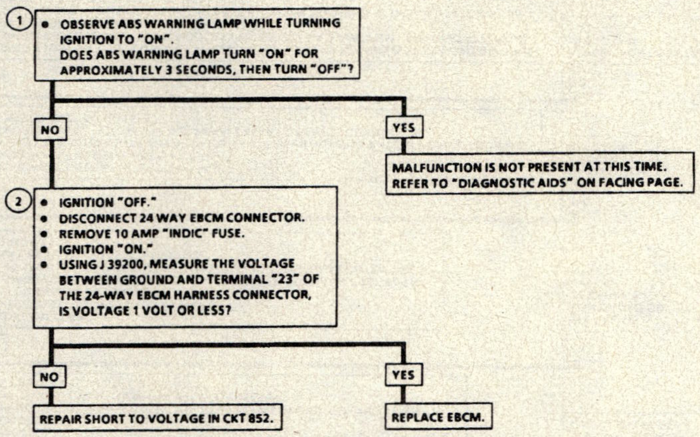

DIAGNOSTIC CHART DTC 14 — 1994 LUMINA

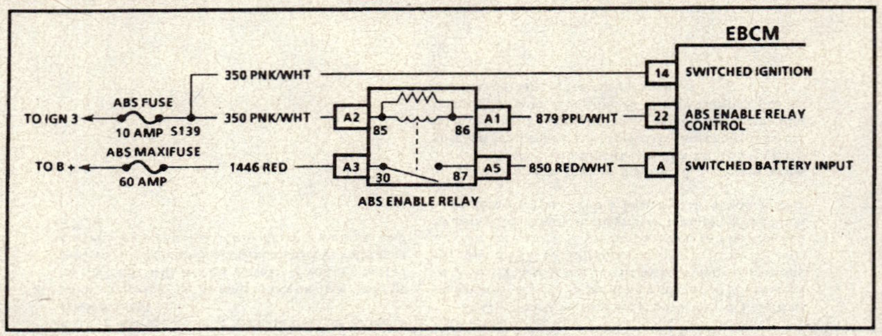

DTC 14

ABS ENABLE RELAY CONTACT CIRCUIT OPEN

Circuit Description:

Ignition voltage is supplied through terminal "A2" of the ABS enable relay. The EBCM then is able to energize the pull-in coil by completing the ground circuit at terminal "22" of the EBCM. The magnetic field created closes the ABS enable relay contacts and allows battery voltage and current to be supplied to the EBCM, which supplies power to the motors and solenoids.

DTC Will Set When: DTC 14 can be set anytime after the EBCM commands the ABS enable relay "ON" (the relay is first commanded "ON" during the three second bulb check). This test monitors the availability of current/voltage to the motors and solenoids. This malfunction indicates voltage is not available and would therefore not allow ABS operation if required.

Action Taken: A malfunction DTC is stored, ABS is disabled and the ABS warning lamp is turned "ON."

DTC Chart Test Description: Number(s) below refer to circled number(s) on the diagnostic chart.

1. Checks to see if DTC 16 is also set as current or history. If DTC 16 is also set, proceed to that chart first.
2. Verifies that the malfunction is currently present.
3. Verifies that the relay can be turned "ON" and provide voltage to the EBCM.
4. Checks for high resistance in the switched battery circuit from the relay to the EBCM.
5. Checks for high resistance in the switched battery circuit from the battery to the relay.
6. Verifies the condition of the ABS enable relay.
7. Checks for poor terminal contact or corrosion at the connectors.
8. Determines if the EBCM is the cause of the malfunction.
9. Checks for high resistance in the entire relay coil circuit.
10. Verifies proper battery condition.
11. Verifies proper voltage is available at the ABS enable relay.
12. Checks for an open in the switched battery circuit from the relay to the EBCM.
13. Verifies the condition of the ABS enable relay.
14. Checks for proper resistance of the relay coil.
15. Checks for high resistance in the ABS enable relay control circuit between the relay and the EBCM.
16. Checks for high resistance in the ABS enable relay control circuit between the battery and the relay.

Diagnostic Aids: An "Intermittent" problem may be caused by a poor connection, rubbed through wire insulation, or a wire that is broken inside the insulation.

The frequency of the problem can be checked by using the enhanced diagnostic function of the Tech 1, as described in "Tech 1 Diagnostics" found in this section.

Any circuitry that is suspected as causing the "Intermittent" complaint should be thoroughly checked for backed out terminals, improper mating, broken locks, improperly formed or damaged terminals, poor terminal to wiring connections or physical damage to the wiring harness.

VIBRATION, TEMPERATURE EFFECTS:

Check for vibration effects by performing the relay test function of the Tech 1. With the relay test commanded "ON," lightly tap the top and sides of the relay while monitoring relay voltage. If the relay voltage changes significantly, replace the relay.

If DTC 14 only sets when the vehicle is initially started in cold ambient conditions (temperature less than 32°F-0°C), replace the relay.

DIAGNOSTIC CHART DTC 14 — 1994 LUMINA, CONTINUED

DTC 14
(Page 1 of 3)
ABS ENABLE RELAY CONTACT CIRCUIT OPEN

1. IS DTC 16 ALSO PRESENT OR IN HISTORY?
   - YES: PROCEED TO DTC 16 CHART FIRST.
   - NO: go to 2.
2. - IGNITION "ON," ENGINE "OFF."
   - USING TECH 1, SELECT DATA LIST AND OBSERVE ABS BATTERY VOLTAGE.

   DOES TECH 1 INDICATE THAT ABS BATTERY VOLTAGE IS GREATER THAN 10 VOLTS?
   - YES: MALFUNCTION IS NOT PRESENT AT THIS TIME. REFER TO "DIAGNOSTIC AIDS" ON FACING PAGE.
   - NO: go to 3.
3. - IGNITION "OFF."
   - DISCONNECT 24-WAY AND 2-WAY EBCM CONNECTORS.
   - CONNECT A FUSED JUMPER, SUCH AS J 36169 WITH A 3 AMP FUSE BETWEEN GROUND AND TERMINAL "22" OF THE 24-WAY EBCM HARNESS CONNECTOR.
   - IGNITION "ON."
   - USING J 39200, MEASURE THE VOLTAGE BETWEEN GROUND AND TERMINAL "A" OF THE 2-WAY EBCM HARNESS CONNECTOR.

   IS VOLTAGE 10 VOLTS OR GREATER?
   - NO: DIAGNOSIS CONTINUED ON NEXT PAGE.
   - YES: go to 4.
4. - IGNITION "OFF."
   - DISCONNECT JUMPER USED IN STEP 3.
   - DISCONNECT ABS ENABLE RELAY.
   - USING J 39200, MEASURE RESISTANCE BETWEEN TERMINAL "A5" OF THE ABS ENABLE RELAY HARNESS CONNECTOR AND TERMINAL "A" OF THE 2-WAY EBCM HARNESS CONNECTOR.

   IS RESISTANCE 2Ω OR LESS?
   - NO: REPAIR HIGH RESISTANCE IN CKT 850.
   - YES: go to 5.
5. - DISCONNECT NEGATIVE BATTERY CABLE.
   - DISCONNECT POSITIVE BATTERY CABLE.
   - USING J 39200, MEASURE THE RESISTANCE BETWEEN POSITIVE BATTERY CABLE TERMINAL AND TERMINAL "A3" OF THE ABS ENABLE RELAY HARNESS CONNECTOR.

   IS RESISTANCE 2Ω OR LESS?
   - NO: REPAIR HIGH RESISTANCE IN CKT 1446.
   - YES: go to 6.
6. - USING A JUMPER WIRE, CONNECT TERMINAL "86" ON THE ABS ENABLE RELAY TO GROUND.
   - USING A FUSED JUMPER WIRE, SUCH AS J 36169, WITH A 3 AMP FUSE, CONNECT TERMINAL "85" ON THE ABS ENABLE RELAY TO VOLTAGE.
   - USING J 39200, MEASURE THE RESISTANCE BETWEEN TERMINALS "30" AND "87" ON THE ABS ENABLE RELAY.

   IS RESISTANCE 2Ω OR LESS?
   - NO: REPLACE ABS ENABLE RELAY.
   - YES: go to 7.
7. - INSPECT ABS ENABLE RELAY AND ABS ENABLE RELAY HARNESS CONNECTOR FOR A POOR CONNECTION OR POOR TERMINAL CONTACT.
   - INSPECT EBCM AND EBCM CONNECTOR TERMINALS "22" AND "A" FOR POOR TERMINAL CONTACT.
   - INSPECT POSITIVE BATTERY TERMINAL AND POSITIVE BATTERY CABLE TERMINAL FOR A POOR CONNECTION.

   IS POOR TERMINAL CONTACT OR A POOR CONNECTION EVIDENT?
   - YES: REPLACE TERMINALS THAT EXHIBIT POOR TERMINAL CONTACT OR EVIDENCE OF CORROSION.
   - NO: go to 8.
8. - RECONNECT ALL CONNECTORS.
   - IGNITION "ON."

   DOES DTC 14 RESET?
   - YES: REPLACE EBCM.
   - NO: MALFUNCTION IS INTERMITTENT. REFER TO "DIAGNOSTIC AIDS" ON FACING PAGE.

DIAGNOSTIC CHART DTC 14 — 1994 LUMINA, CONTINUED

DTC 14
(Page 2 of 3)
ABS ENABLE RELAY CONTACT CIRCUIT OPEN

DIAGNOSIS CONTINUED FROM PREVIOUS PAGE.

(9)
- IGNITION "OFF."
- REMOVE JUMPER WIRE INSTALLED IN STEP #3, IF NOT ALREADY REMOVED.
- DISCONNECT NEGATIVE BATTERY CABLE.
- DISCONNECT POSITIVE BATTERY CABLE.
- WITH BATTERY STILL DISCONNECTED, TURN IGNITION SWITCH TO "RUN." THIS COMPLETES THE IGNITION CIRCUIT TO THE ABS ENABLE RELAY.
- USING J 39200, MEASURE THE RESISTANCE BETWEEN TERMINAL "22" OF THE 24-WAY EBCM HARNESS CONNECTOR AND THE POSITIVE BATTERY CABLE TERMINAL. IS RESISTANCE 95Ω OR LESS WHEN MEASURED AT APPROXIMATELY 20°C (68°F)?

NO: DIAGNOSIS CONTINUED ON NEXT PAGE.

YES:

(10)
- USING J 39200, MEASURE THE VOLTAGE BETWEEN THE NEGATIVE AND POSITIVE BATTERY TERMINALS. IS VOLTAGE 10 VOLTS OR GREATER?

NO: CORRECT LOW VOLTAGE CONDITION.

YES:

(11)
- IGNITION "OFF."
- RECONNECT POSITIVE BATTERY CABLE.
- RECONNECT NEGATIVE BATTERY CABLE.
- DISCONNECT ABS ENABLE RELAY.
- MEASURE THE VOLTAGE BETWEEN TERMINAL "A3" OF THE ABS ENABLE RELAY HARNESS CONNECTOR AND GROUND. IS VOLTAGE 10 VOLTS OR GREATER?

NO: REPAIR OPEN OR HIGH RESISTANCE IN CKT 1446.

YES:

(12)
- USING J 39200, MEASURE THE RESISTANCE BETWEEN TERMINAL "A5" OF THE ABS ENABLE RELAY HARNESS CONNECTOR AND TERMINAL "A" OF THE 2-WAY EBCM HARNESS CONNECTOR. IS RESISTANCE 2Ω OR LESS?

NO: REPAIR OPEN OR HIGH RESISTANCE IN CKT 850.

YES:

(13)
- USING A JUMPER WIRE, CONNECT TERMINAL "86" ON THE ABS ENABLE RELAY TO GROUND.
- USING A FUSED JUMPER WIRE, SUCH AS J 36169, WITH A 3 AMP FUSE, CONNECT TERMINAL "85" ON THE ABS ENABLE RELAY TO VOLTAGE.
- USING J 39200, MEASURE THE RESISTANCE BETWEEN TERMINALS "30" AND "87" ON THE ABS ENABLE RELAY. IS RESISTANCE 2Ω OR LESS?

NO: REPLACE ABS ENABLE RELAY.

YES: MALFUNCTION WAS MOST LIKELY DUE TO A POOR CONNECTION. INSPECT ALL CONNECTORS AND TERMINALS FOR POOR TERMINAL CONTACT AND EVIDENCE OF CORROSION. REPLACE ALL TERMINALS THAT EXHIBIT EVIDENCE OF POOR TERMINAL CONTACT OR CORROSION.

DIAGNOSTIC CHART DTC 14 — 1994 LUMINA, CONTINUED

DTC 14
(Page 3 of 3)
ABS ENABLE RELAY CONTACT CIRCUIT OPEN

DIAGNOSIS CONTINUED FROM PREVIOUS PAGE.

(14)
- DISCONNECT ABS ENABLE RELAY.
- USING J 39200, MEASURE THE RESISTANCE BETWEEN TERMINALS "85" AND "86" ON THE ABS ENABLE RELAY. IS RESISTANCE 95Ω OR LESS WHEN MEASURED AT APPROXIMATELY 20°C (68°F)?

NO: REPLACE ABS ENABLE RELAY.

YES:

(15)
- MEASURE THE RESISTANCE BETWEEN TERMINAL "A1" OF THE ABS ENABLE RELAY HARNESS CONNECTOR AND TERMINAL "22" OF THE 24-WAY EBCM HARNESS CONNECTOR. IS RESISTANCE 2Ω OR LESS?

NO: REPAIR OPEN OR HIGH RESISTANCE IN CKT 879.

YES:

(16)
- MEASURE THE RESISTANCE BETWEEN TERMINAL "A2" OF THE ABS ENABLE RELAY HARNESS CONNECTOR AND THE POSITIVE BATTERY CABLE TERMINAL. IS RESISTANCE 2Ω OR LESS?

NO: REPAIR OPEN OR HIGH RESISTANCE IN CKT 350, CKT 300, IGNITION SWITCH, OR BATTERY FEED CIRCUIT TO IGNITION SWITCH.

YES: MALFUNCTION WAS MOST LIKELY DUE TO A POOR CONNECTION. INSPECT ALL CONNECTORS AND TERMINALS FOR POOR TERMINAL CONTACT AND EVIDENCE OF CORROSION. REPLACE ALL TERMINALS THAT EXHIBIT EVIDENCE OF POOR TERMINAL CONTACT OR CORROSION.

DIAGNOSTIC CHART DTC 15 — 1994 LUMINA

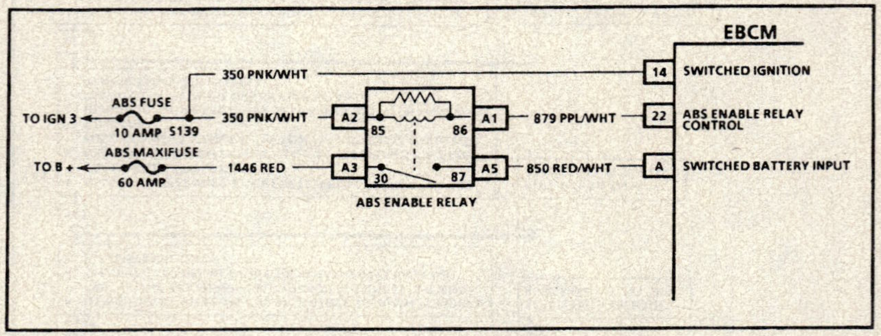

DTC 15

ABS RLY CONTACT CKT SHORT TO BATT OR ALWAYS CLOSED

Circuit Description:

Ignition voltage is supplied through terminal "A2" of the ABS enable relay. The EBCM then is able to energize the pull-in coil by completing the ground circuit at terminal "22" of the EBCM. The magnetic field created closes the ABS enable relay contacts and allows battery voltage and current to be supplied to the EBCM, which supplies power to the motors and solenoids.

DTC Will Set When: DTC 15 can be set only before the EBCM commands the relay "ON." This test determines if the ABS enable relay is energized when it should not be. This malfunction would not allow the ABS enable relay to remove power to the ABS system. If a second malfunction were to occur that requires the ABS enable relay to be turned "OFF," that malfunction cannot be removed if the relay cannot be controlled. The malfunction must be present for three consecutive drive cycles before the DTC is set.

Action Taken: A malfunction DTC is stored. ABS is not disabled; however, the ABS warning lamp will flash to indicate a malfunction exists.

DTC Chart Test Description: Number(s) below refer to circled number(s) on the diagnostic chart.

1. Checks to see if other DTCs are set.
2. Indicates that the EBCM is capable of controlling the ABS enable relay as commanded.
3. Checks for a possible short to battery voltage in CKT 850.
4. Determines whether the short to voltage is in CKT 850 due to a malfunctioning relay.
5. Checks for a possible intermittent in wiring or connectors.

Diagnostic Aids: An "Intermittent" malfunction may be caused by a poor connection, rubbed through wire insulation, or a wire that is broken inside the insulation.

The frequency of the malfunction can be checked by using the enhanced diagnostic function of the Tech 1, as described in "Tech 1 Diagnostics" found in this section.

Any circuitry, that is suspected as causing the intermittent complaint, should be thoroughly checked for backed out terminals, improper mating, broken locks, improperly formed or damaged terminals, poor terminal to wiring connections or physical damage to the wiring harness.

DIAGNOSTIC CHART DTC 15 — 1994 LUMINA, CONTINUED

DTC 15

ABS RLY CONTACT CKT SHORT TO BATT OR ALWAYS CLOSED

(1) ARE ANY OTHER DTCs PRESENT?

- YES → PROCEED TO THAT DTC CHART FIRST.
- NO →

(2)
- IGNITION "ON," ENGINE "OFF."
- USING THE RELAY TEST FUNCTION OF THE TECH 1, COMMAND ABS ENABLE RELAY "OFF."

DOES TECH 1 INDICATE THAT ABS ENABLE RELAY IS "OFF" AND BATTERY VOLTAGE IS LESS THAN 5 VOLTS?

- YES → MALFUNCTION IS NOT PRESENT AT THIS TIME. REFER TO "DIAGNOSTIC AIDS" ON FACING PAGE.
- NO →

(3)
- IGNITION "OFF."
- DISCONNECT 24-WAY AND 2-WAY EBCM HARNESS CONNECTORS.
- IGNITION "ON."
- USING J 39200, MEASURE THE VOLTAGE BETWEEN GROUND AND TERMINAL "A" OF THE EBCM 2-WAY CONNECTOR.

IS VOLTAGE 2 VOLTS OR LESS?

- YES → (5)
- NO → (4)

(4)
- IGNITION "OFF."
- DISCONNECT ABS ENABLE RELAY.
- IGNITION "ON."
- MEASURE THE VOLTAGE BETWEEN GROUND AND TERMINAL "A" OF THE 2-WAY EBCM HARNESS CONNECTOR.

IS VOLTAGE 2 VOLTS OR LESS?

- YES → REPLACE ABS ENABLE RELAY.
- NO → REPAIR SHORT TO VOLTAGE IN CKT 850 (CHEV) OR 1633 (BUICK/OLDS).

(5)
- IGNITION "OFF."
- INSPECT CKT 850 (CHEV) OR 1633 (BUICK/OLDS) AND 2-WAY EBCM HARNESS CONNECTOR FOR DAMAGE WHICH MAY RESULT IN A SHORT TO VOLTAGE WHEN ALL CONNECTORS ARE CONNECTED. REPAIR DAMAGE IF EVIDENT.
- RECONNECT 2-WAY AND 24-WAY EBCM CONNECTORS.
- IGNITION "ON."
- USING RELAY TEST FUNCTION OF TECH 1, COMMAND ABS ENABLE RELAY "OFF."

DOES TECH 1 INDICATE THAT ABS ENABLE RELAY IS "OFF" AND BATTERY VOLTAGE IS LESS THAN 5 VOLTS?

- NO → REPLACE EBCM.
- YES → MALFUNCTION IS INTERMITTENT. REFER TO "DIAGNOSTIC AIDS" ON FACING PAGE.

AFTER ALL DIAGNOSIS IS COMPLETE, CLEAR DTCs, AND TEST DRIVE VEHICLE FOR 3 DRIVE CYCLES TO VERIFY THAT DTC DOES NOT RESET. A DRIVE CYCLE CONSISTS OF STARTING THE VEHICLE, DRIVING ABOVE 16 km/h (10 MPH), AND THEN KEYING DOWN.

DIAGNOSTIC CHART DTC 16 — 1994 LUMINA

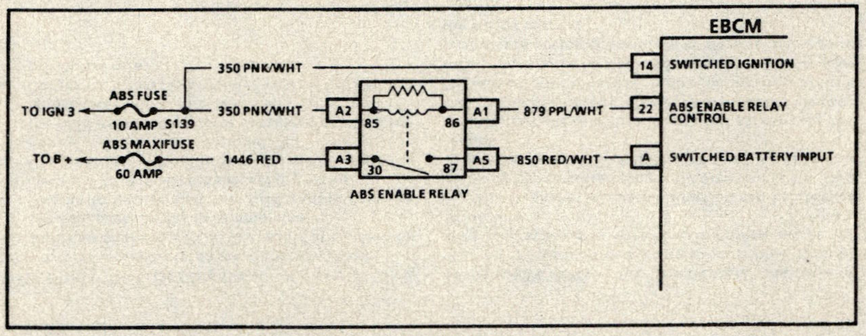

DTC 16

ABS ENABLE RELAY COIL CIRCUIT OPEN

Circuit Description:

Ignition voltage is supplied through terminal "A2" of the ABS enable relay. The EBCM then is able to energize the pull-in coil by completing the ground circuit at terminal "22" of the EBCM. The magnetic field created closes the ABS enable relay contacts and allows battery voltage and current to be supplied to the EBCM, which supplies power to the motors and solenoids.

DTC Will Set When: DTC 16 can be set only after DTC 14 has been set. This test detects an open in the ABS enable relay coil circuit. An open in this circuit will not allow the ABS enable relay to be energized thus preventing voltage/current to the motors and solenoids. If this malfunction is present and the ignition is turned "OFF" before 5 km/h (3 mph) is reached, DTC 14 is set alone.

Action Taken: A malfunction DTC is stored, ABS is disabled and the ABS warning lamp is turned "ON."

DTC Chart Test Description: Number(s) below refer to circled number(s) on the diagnostic chart.

1. Indicates the EBCM is capable of controlling the ABS enable relay as commanded.
2. Checks to ensure that voltage is available to the pull-in coil of the enable relay.
3. Ensures that there is continuity through the pull-in coil of the enable relay.
4. Verifies the integrity of the relay control circuit.
5. Ensures malfunction was not due to poor terminal contact.

Diagnostic Aids: An "Intermittent" malfunction may be caused by a poor connection, rubbed through wire insulation, or a wire that is broken inside the insulation.

The frequency of the malfunction can be checked by using the enhanced diagnostic function of the Tech 1, as described in "Tech 1 Diagnostics" found in this section.

If the frequency of the malfunction is high, but is currently intermittent, check for high coil resistance by measuring between relay terminals "A1" and "A2" using a J 39200. If resistance shows greater than 100Ω, replace the relay.

Any circuitry, that is suspected as causing the intermittent complaint, should be thoroughly checked for backed out terminals, improper mating, broken locks, improperly formed or damaged terminals, poor terminal to wiring connections or physical damage to the wiring harness.

DIAGNOSTIC CHART DTC 16 — 1994 LUMINA, CONTINUED

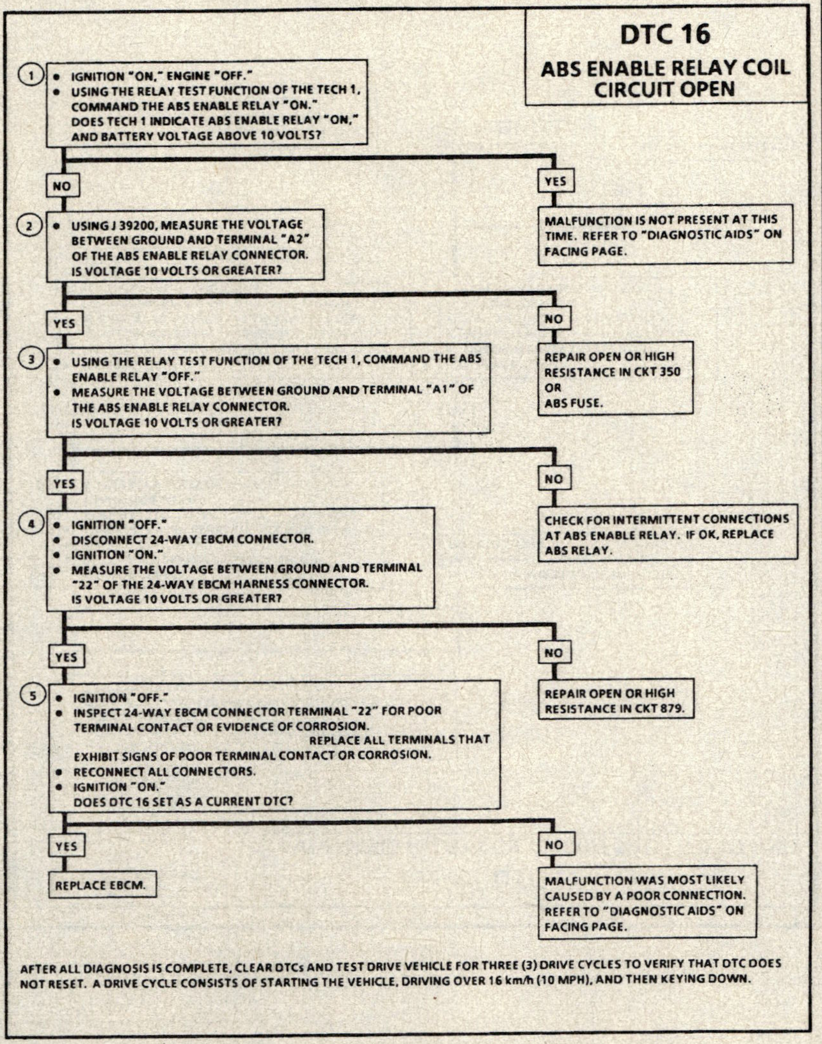

DIAGNOSTIC CHART DTC 17 — 1994 LUMINA

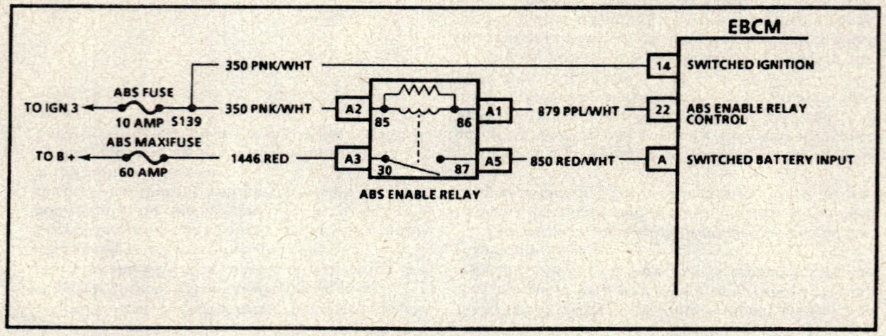

DTC 17

ABS ENABLE RELAY COIL CIRCUIT SHORTED TO GND

Circuit Description:

Ignition voltage is supplied through terminal "A2" of the ABS enable relay. The EBCM then is able to energize the pull-in coil by completing the ground circuit at terminal "22" of the EBCM. The magnetic field created closes the ABS enable relay contacts and allows battery voltage and current to be supplied to the EBCM, which supplies voltage to the motors and solenoids.

DTC Will Set When: DTC 17 can be set before the EBCM commands the ABS enable relay "ON." This test determines if the ABS enable relay is energized when it should not be. This malfunction would not allow the ABS enable relay to remove power to the ABS system. If a second malfunction were to occur that requires the ABS enable relay to be turned "OFF," that malfunction cannot be removed if the relay cannot be controlled.

Action Taken: A malfunction DTC is stored. ABS is not disabled; however the ABS warning lamp is flashed to indicate a malfunction exists.

DTC Chart Test Description: Number(s) below refer to circled number(s) on the diagnostic chart.

1. Indicates that the EBCM is capable of controlling the ABS enable relay as commanded.
2. Checks to ensure that the ABS enable relay or control CKT 879 are not shorted to ground.
3. Checks to see if the EBCM is internally shorted to ground. A grounded terminal "22" would cause the ABS enable relay to be energized anytime the ignition was in the "ON" position.
4. Checks for a short to ground in control CKT 879 or a defective relay.

Diagnostic Aids: An "Intermittent" malfunction may be caused by a poor connection, rubbed through wire insulation, or a wire that is broken inside the insulation.

The frequency of the malfunction can be checked by using the enhanced diagnostic function of the Tech 1, as described in "Tech 1 Diagnostics" found in this section.

Any circuitry, that is suspected as causing the intermittent complaint, should be thoroughly checked for backed out terminals, improper mating, broken locks, improperly formed or damaged terminals, poor terminal to wiring connections or physical damage to the wiring harness.

DIAGNOSTIC CHART DTC 17 — 1994 LUMINA, CONTINUED

DTC 17

ABS ENABLE RELAY COIL CIRCUIT SHORTED TO GND

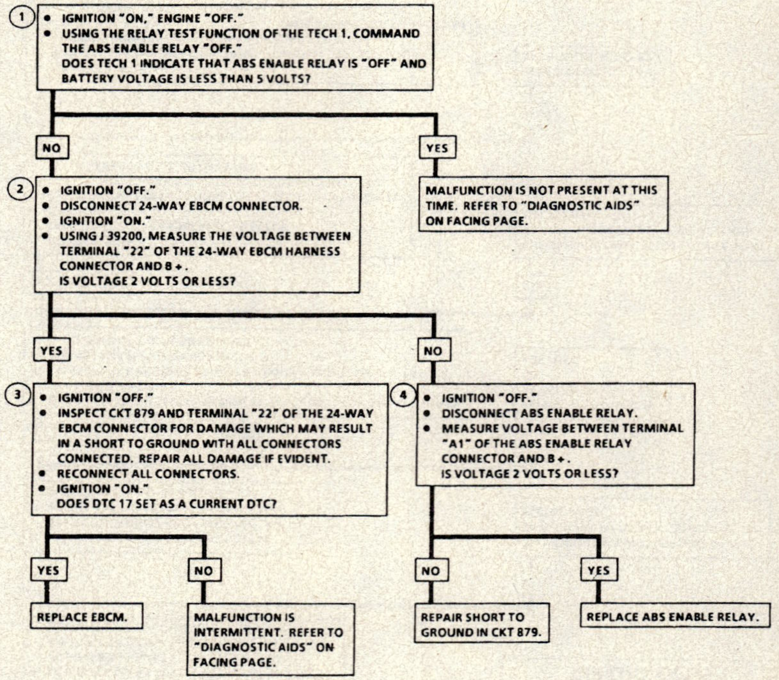

AFTER ALL DIAGNOSIS IS COMPLETE, CLEAR DTCs AND TEST DRIVE VEHICLE FOR THREE (3) DRIVE CYCLES TO VERIFY THAT DTC DOES NOT RESET. A DRIVE CYCLE CONSISTS OF STARTING THE VEHICLE, DRIVING OVER 16 km/h (10 MPH), AND THEN KEYING DOWN.

DIAGNOSTIC CHART DTC 18 — 1994 LUMINA

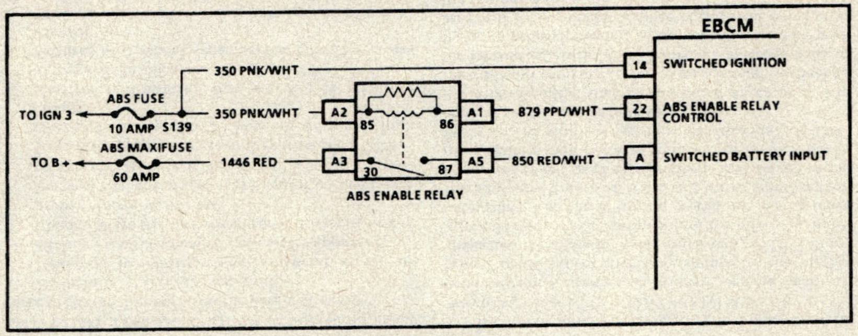

DTC 18

ABS ENABLE RELAY COIL CIRCUIT SHORTED TO BATT

Circuit Description:

Ignition voltage is supplied through terminal "A2" of the ABS enable relay. The EBCM then is able to energize the pull-in coil by completing the ground circuit at terminal "22" of the EBCM. The magnetic field created closes the ABS enable relay contacts and allows battery voltage and current to be supplied to the EBCM, which supplies voltage to the motors and solenoids.

DTC Will Set When: DTC 18 can be set after the EBCM commands the enable relay "ON." This test monitors the availability of current/voltage to the motors and solenoids. A malfunction of this type will not allow voltage to be available to the motors and solenoids; therefore, ABS operation would not be allowed if required.

Action Taken: A malfunction DTC is stored, ABS is disabled and the ABS warning lamp is turned "ON."

DTC Chart Test Description: Number(s) below refer to circled number(s) on the diagnostic chart.

1. Indicates that the EBCM is capable of controlling the ABS enable relay as commanded.
2. Checks for a short to voltage on the relay control circuit.
3. Verifies that the relay coil has the appropriate resistance.
4. Verifies that the relay contacts are not failed closed.
5. Checks for possible poor terminal contact at the EBCM.

Diagnostic Aids: An "Intermittent" malfunction may be caused by a poor connection, rubbed through wire insulation, or a wire that is broken inside the insulation.

The frequency of the malfunction can be checked by using the enhanced diagnostic function of the Tech 1, as described in "Tech 1 Diagnostics" found in this section.

Any circuitry, that is suspected as causing the intermittent complaint, should be thoroughly checked for backed out terminals, improper mating, broken locks, improperly formed or damaged terminals, poor terminal to wiring connections or physical damage to the wiring harness.

DIAGNOSTIC CHART DTC 18 — 1994 LUMINA, CONTINUED

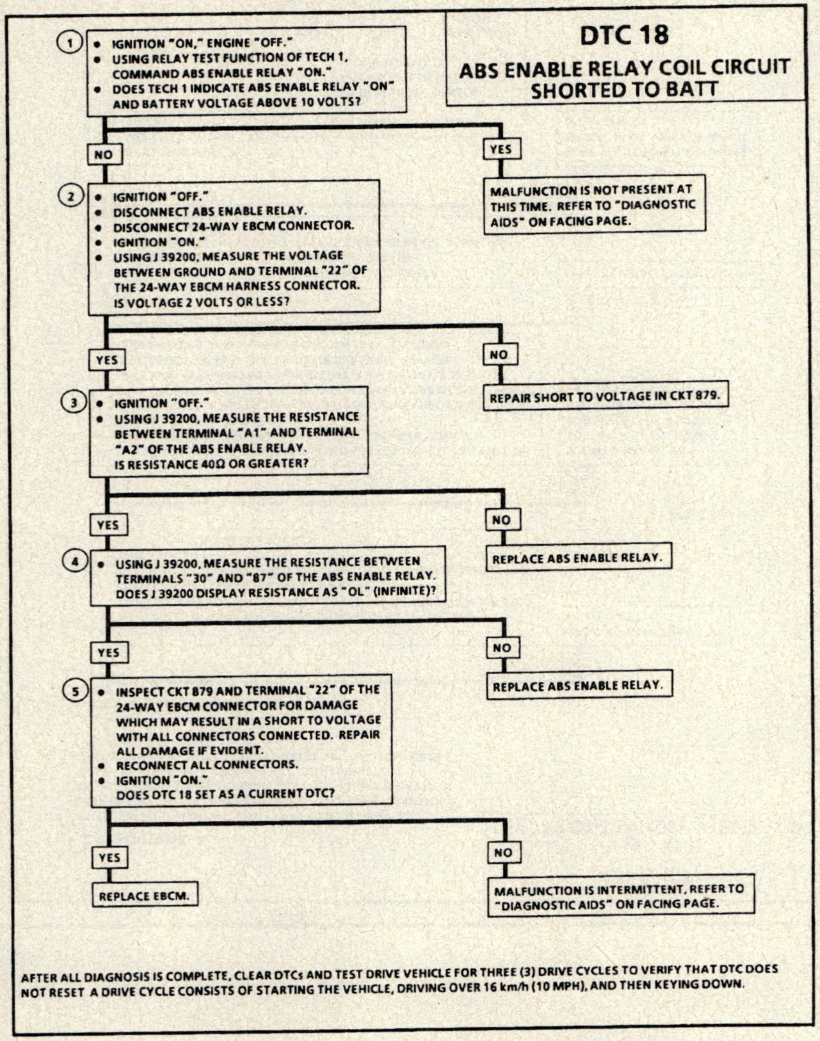

DIAGNOSTIC CHART DTC 21 — 1994 LUMINA

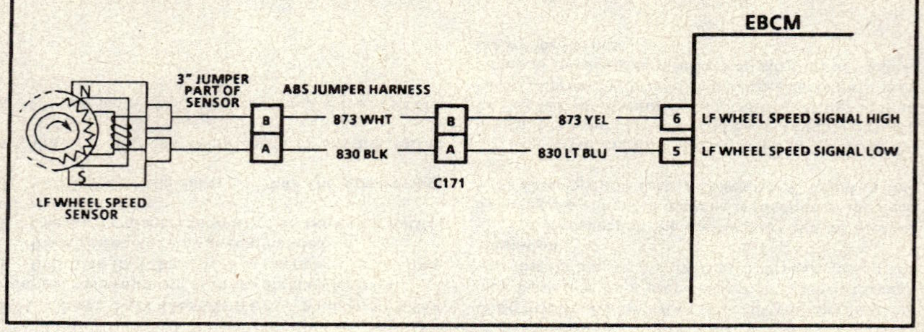

DTC 21

LEFT FRONT WHEEL SPEED = 0

Circuit Description:
As a toothed ring passes by the wheel speed sensor, changes in the electromagnetic field cause the wheel speed sensor to produce a sinusoidal (AC) voltage signal whose frequency is proportional to wheel speed. The magnitude of this signal is directly related to wheel speed and the proximity of the wheel speed sensor to the toothed ring, often referred to as the air gap.

DTC Will Set When: DTC 21 can be set when the vehicle is not in an ABS stop. If the left front wheel speed = 0 and the vehicle's reference speed is greater than 8 km/h (5 mph), a malfunction exists.

Action Taken: A malfunction DTC is stored, ABS is disabled and the ABS warning lamp is turned "ON."

DTC Chart Test Description: Number(s) below refer to circled number(s) on the diagnostic chart.

1. Determines if DTC 32 is present.
2. Designed to identify a wheel speed sensor or circuitry that is damaged and visibly apparent.
3. Checks to see that the wheel speed sensor is within proper resistance values.
4. Verifies that the wheel speed sensor can generate an appropriate output signal.
5. Checks for proper output voltage from the EBCM.
6. Checks for an open in the wheel speed sensor high circuit.
7. Ensures malfunction was not due to physical damage of the circuitry.
8. Ensures that wheel speed sensor circuitry is not shorted.
9. Ensures malfunction was not due to physical damage of the circuitry.
10. Checks for an open in the wheel speed sensor low circuit.

Diagnostic Aids: An "Intermittent" malfunction may be caused by a poor connection, rubbed through wire insulation, or a wire that is broken inside the insulation.

The frequency of the malfunction can be checked by using the enhanced diagnostic function of the Tech 1, as described in "Tech 1 Diagnostics" found in this section.

If the customer's comments reflect that the ABS warning lamp is "ON" only during moist environmental changes (rain, snow, vehicle wash), all wheel speed sensor circuitry should be thoroughly inspected for signs of water intrusion. If DTC is not current, clear DTCs and simulate the effects of water intrusion. Use the following procedure. Spray down the suspected area with a 5% salt water solution (two teaspoons of salt to 12 oz. of water). Test drive vehicle over various road surfaces (bumps, turns, etc.) above 24 km/h (15 mph) for at least 30 seconds. If DTC returns replace suspected harness.

Any circuitry, that is suspected as causing the intermittent complaint, should be thoroughly checked for backed out terminals, improper mating, broken locks, improperly formed or damaged terminals, poor terminal to wiring connections or physical damage to the wiring harness.

Resistance of the wheel speed sensor will increase with an increase in sensor temperature.

When replacing a wheel speed sensor, inspect the sensor terminals and harness connector for corrosion and/or water intrusion. If evidence of corrosion or water intrusion exists, replace wheel speed sensor jumper harness. Likewise, if replacing a wheel speed sensor jumper harness, inspect sensor terminals. If evidence of corrosion or water intrusion exists, replace wheel speed sensor. Refer to "On-Vehicle Service" in this section.

DIAGNOSTIC CHART DTC 21 — 1994 LUMINA, CONTINUED

DTC 21
(Page 1 of 2)
LEFT FRONT WHEEL SPEED = 0

Important
- WHEEL SPEED SENSOR INTERMITTENT PROBLEMS MAY BE DIFFICULT TO LOCATE. CARE SHOULD BE TAKEN NOT TO DISTURB ANY ELECTRICAL CONNECTIONS PRIOR TO AN INDICATED STEP OF THIS CHART. THIS WILL ENSURE THAT AN INTERMITTENT CONNECTION WILL NOT BE CORRECTED BEFORE THE SOURCE OF THE PROBLEM IS FOUND.

(1) IS DTC 32 ALSO CURRENTLY SET?
- YES: PROCEED WITH THAT DTC CHART FIRST.
- NO: go to 2.

(2)
- IGNITION "OFF."
- PHYSICALLY INSPECT THE WHEEL SPEED SENSOR WIRING AND CONNECTORS FOR DAMAGE. INSPECT THE WHEEL SPEED SENSOR AND TOOTHED RING FOR LOOSENESS OR DAMAGE. PAY PARTICULAR ATTENTION TO THE TOOTHED RING, ANY DEVIATION WILL AFFECT THE WHEEL SPEED SENSOR OUTPUT SIGNAL.

IS PHYSICAL DAMAGE INDICATED?
- YES: REPAIR DAMAGE OR SENSOR/TOOTHED RING LOOSENESS. DRIVE VEHICLE ABOVE 24 km/h (15 MPH) TO ENSURE DTC DOES NOT SET.
- NO: go to 3.

(3)
- DISCONNECT LEFT FRONT WHEEL SPEED SENSOR DIRECTLY AT SENSOR.
- USING J 39200, MEASURE THE RESISTANCE BETWEEN SENSOR TERMINALS "A" AND "B".

IS RESISTANCE BETWEEN 1661 AND 1929Ω WHEN SENSOR IS 20°C (68°F)?
(IF TEMPERATURE OF THE SENSOR IS SUSPECTED TO BE OTHER THAN 20°C (68°F), REFER TO THE TABLE BELOW FOR APPLICABLE SENSOR RESISTANCE VALUES; THE VALUES IN THE TABLE ARE FOR TEMPERATURE OF THE SENSOR, NOT AIR TEMPERATURE.)
- NO: REPLACE WHEEL SPEED SENSOR.
- YES: go to 4.

(4)
- WITH J 39200 STILL CONNECTED, SELECT A/C VOLTAGE SCALE.
- SPIN WHEEL BY HAND WHILE OBSERVING VOLTAGE.

IS VOLTAGE AT LEAST 100 mV (VOLTAGE WILL INCREASE AS WHEEL SPEED INCREASES)?
- YES: DIAGNOSIS CONTINUED ON PAGE 2 OF 2 OF THIS DTC CHART.
- NO: VERIFY THAT TOOTHED RING IS INTACT. IF OK, REPLACE WHEEL SPEED SENSOR.

FRONT WHEEL SPEED SENSOR
TEMPERATURE OF SENSOR VS. SENSOR RESISTANCE (APPROXIMATE)

TEMP. (°C)	TEMP. (°F)	RESISTANCE (OHMS)
-34 TO 4	-30 TO 40	1309 TO 1808
5 TO 43	41 TO 110	1557 TO 2103
44 TO 93	111 TO 200	1811 TO 2482

DIAGNOSTIC CHART DTC 21 — 1994 LUMINA, CONTINUED

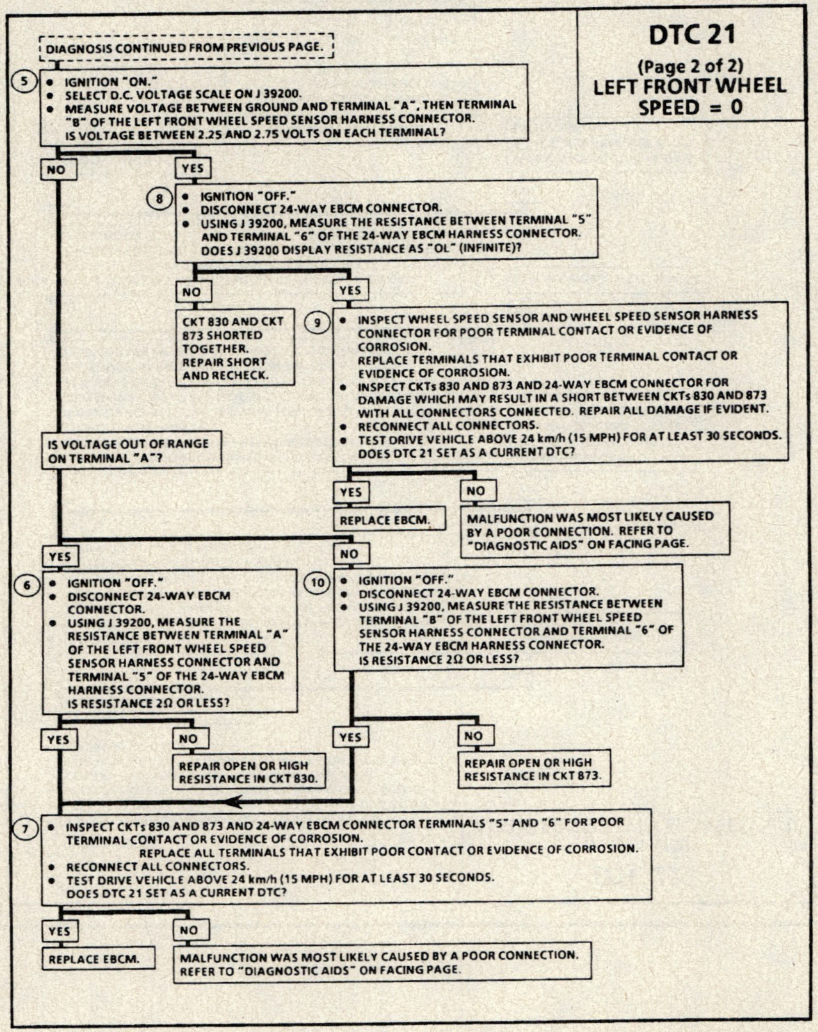

DIAGNOSTIC CHART DTC 22 — 1994 LUMINA

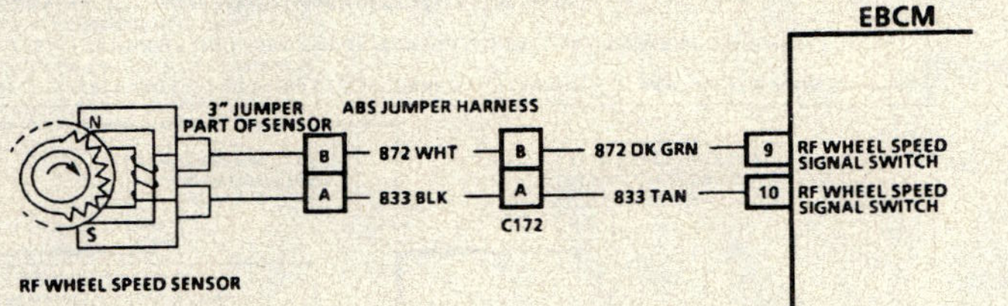

DTC 22

RIGHT FRONT WHEEL SPEED = 0

Circuit Description:
As a toothed ring passes by the wheel speed sensor, changes in the electromagnetic field cause the wheel speed sensor to produce a sinusoidal (AC) voltage signal whose frequency is proportional to wheel speed. The magnitude of this signal is directly related to wheel speed and the proximity of the wheel speed sensor to the toothed ring, often referred to as the air gap.

DTC Will Set When: DTC 22 can be set when the vehicle is not in an ABS stop. If the right front wheel speed = 0 and the vehicle's reference speed is greater than 8 km/h (5 mph), a malfunction exists.

Action Taken: A malfunction DTC is stored, ABS is disabled and the ABS warning lamp is turned "ON."

DTC Chart Test Description: Number(s) below refer to circled number(s) on the diagnostic chart.

1. Determines if DTC 33 is present.
2. Designed to identify a wheel speed sensor or circuitry that is damaged and visibly apparent.
3. Checks to see that the wheel speed sensor is within proper resistance values.
4. Verifies that the wheel speed sensor can generate an appropriate output signal.
5. Checks for proper output voltage from the EBCM.
6. Checks for an open in the wheel speed sensor high circuit.
7. Ensures malfunction was not due to physical damage of the circuitry.
8. Ensures that wheel speed sensor circuitry is not shorted.
9. Ensures malfunction was not due to physical damage of the circuitry.
10. Checks for an open in the wheel speed sensor low circuit.

Diagnostic Aids: An "Intermittent" malfunction may be caused by a poor connection, rubbed through wire insulation, or a wire that is broken inside the insulation.

The frequency of the malfunction can be checked by using the enhanced diagnostic function of the Tech 1, as described in "Tech 1 Diagnostics" found in this section.

If the customer's comments reflect that the ABS warning lamp is "ON" only during moist environmental changes (rain, snow, vehicle wash), all wheel speed sensor circuitry should be thoroughly inspected for signs of water intrusion. If DTC is not current, clear DTCs and simulate the effects of water intrusion. Use the following procedure. Spray down the suspected area with a 5% salt water solution (two teaspoons of salt to 12 oz. of water). Test drive vehicle over various road surfaces (bumps, turns, etc.) above 24 km/h (15 mph) for at least 30 seconds. If DTC returns replace suspected harness.

Any circuitry, that is suspected as causing the intermittent complaint, should be thoroughly checked for backed out terminals, improper mating, broken locks, improperly formed or damaged terminals, poor terminal to wiring connections or physical damage to the wiring harness.

Resistance of the wheel speed sensor will increase with an increase in sensor temperature.

When replacing a wheel speed sensor, inspect the sensor terminals and harness connector for corrosion and/or water intrusion. If evidence of corrosion or water intrusion exists, replace wheel speed sensor jumper harness. Likewise, if replacing a wheel speed sensor jumper harness, inspect sensor terminals. If evidence of corrosion or water intrusion exists, replace wheel speed sensor. Refer to "On-Vehicle Service" in this section.

DIAGNOSTIC CHART DTC 22 — 1994 LUMINA, CONTINUED

DTC 22
(Page 1 of 2)
RIGHT FRONT WHEEL SPEED = 0

Important

- WHEEL SPEED SENSOR INTERMITTENT PROBLEMS MAY BE DIFFICULT TO LOCATE. CARE SHOULD BE TAKEN NOT TO DISTURB ANY ELECTRICAL CONNECTIONS PRIOR TO AN INDICATED STEP OF THIS CHART. THIS WILL ENSURE THAT AN INTERMITTENT CONNECTION WILL NOT BE CORRECTED BEFORE THE SOURCE OF THE PROBLEM IS FOUND.

(1) IS DTC 33 ALSO CURRENTLY SET?

- YES: PROCEED WITH THAT DTC CHART FIRST.
- NO: go to 2.

(2)
- IGNITION "OFF."
- PHYSICALLY INSPECT THE WHEEL SPEED SENSOR WIRING AND CONNECTORS FOR DAMAGE. INSPECT THE WHEEL SPEED SENSOR AND TOOTHED RING FOR LOOSENESS OR DAMAGE. PAY PARTICULAR ATTENTION TO THE TOOTHED RING, ANY DEVIATION WILL AFFECT THE WHEEL SPEED SENSOR OUTPUT SIGNAL.

IS PHYSICAL DAMAGE INDICATED?

- YES: REPAIR DAMAGE OR SENSOR/TOOTHED RING LOOSENESS. DRIVE VEHICLE ABOVE 24 km/h (15 MPH) TO ENSURE DTC DOES NOT SET.
- NO: go to 3.

(3)
- DISCONNECT RIGHT FRONT WHEEL SPEED SENSOR DIRECTLY AT SENSOR.
- USING J 39200, MEASURE THE RESISTANCE BETWEEN SENSOR TERMINALS "A" AND "B".

IS RESISTANCE BETWEEN 1661 AND 1929Ω WHEN SENSOR IS 20°C (68°F)? *(IF TEMPERATURE OF THE SENSOR IS SUSPECTED TO BE OTHER THAN 20°C (68°F), REFER TO THE TABLE BELOW FOR APPLICABLE SENSOR RESISTANCE VALUES; THE VALUES IN THE TABLE ARE FOR TEMPERATURE OF THE SENSOR, NOT AIR TEMPERATURE.)*

- NO: REPLACE WHEEL SPEED SENSOR.
- YES: go to 4.

(4)
- WITH J 39200 STILL CONNECTED, SELECT A/C VOLTAGE SCALE.
- SPIN WHEEL BY HAND WHILE OBSERVING VOLTAGE.

IS VOLTAGE AT LEAST 100 mV (VOLTAGE WILL INCREASE AS WHEEL SPEED INCREASES)?

- YES: DIAGNOSIS CONTINUED ON PAGE 2 OF 2 OF THIS DTC CHART.
- NO: VERIFY THAT TOOTHED RING IS INTACT. IF OK, REPLACE WHEEL SPEED SENSOR.

FRONT WHEEL SPEED SENSOR

TEMPERATURE OF SENSOR VS. SENSOR RESISTANCE (APPROXIMATE)

TEMP. (°C)	TEMP. (°F)	RESISTANCE (OHMS)
-34 TO 4	-30 TO 40	1309 TO 1808
5 TO 43	41 TO 110	1557 TO 2103
44 TO 93	111 TO 200	1811 TO 2482

DIAGNOSTIC CHART DTC 22 — 1994 LUMINA, CONTINUED

DTC 22
(Page 2 of 2)
RIGHT FRONT WHEEL SPEED = 0

DIAGNOSIS CONTINUED FROM PREVIOUS PAGE.

(5)
- IGNITION "ON."
- SELECT D.C. VOLTAGE SCALE ON J 39200.
- MEASURE VOLTAGE BETWEEN GROUND AND TERMINAL "A", THEN TERMINAL "B" OF THE RIGHT FRONT WHEEL SPEED SENSOR HARNESS CONNECTOR.

IS VOLTAGE BETWEEN 2.25 AND 2.75 VOLTS ON EACH TERMINAL?

- YES: go to 8.
- NO: IS VOLTAGE OUT OF RANGE ON TERMINAL "A"?
  - YES: go to 6.
  - NO: go to 10.

(8)
- IGNITION "OFF."
- DISCONNECT 24-WAY EBCM CONNECTOR.
- USING J 39200, MEASURE THE RESISTANCE BETWEEN TERMINAL "9" AND TERMINAL "10" OF THE 24-WAY EBCM HARNESS CONNECTOR.

DOES J 39200 DISPLAY RESISTANCE AS "OL" (INFINITE)?

- NO: CKT 833 AND CKT 872 SHORTED TOGETHER. REPAIR SHORT AND RECHECK.
- YES: go to 9.

(9)
- INSPECT WHEEL SPEED SENSOR AND WHEEL SPEED SENSOR HARNESS CONNECTOR FOR POOR TERMINAL CONTACT OR EVIDENCE OF CORROSION.

  REPLACE ALL TERMINALS THAT EXHIBIT POOR TERMINAL CONTACT OR EVIDENCE OF CORROSION.
- INSPECT CKTs 833 AND 872 AND 24-WAY EBCM CONNECTOR FOR DAMAGE WHICH MAY RESULT IN A SHORT BETWEEN CKTs 833 AND 872 WITH ALL CONNECTORS CONNECTED. REPAIR ALL DAMAGE IF EVIDENT.
- RECONNECT ALL CONNECTORS.
- TEST DRIVE VEHICLE ABOVE 24 km/h (15 MPH) FOR AT LEAST 30 SECONDS.

DOES DTC 22 SET AS A CURRENT DTC?

- YES: REPLACE EBCM.
- NO: MALFUNCTION WAS MOST LIKELY CAUSED BY A POOR CONNECTION. REFER TO "DIAGNOSTIC AIDS" ON FACING PAGE.

(6)
- IGNITION "OFF."
- DISCONNECT 24-WAY EBCM CONNECTOR.
- USING J 39200, MEASURE THE RESISTANCE BETWEEN TERMINAL "A" OF THE RIGHT FRONT WHEEL SPEED SENSOR HARNESS CONNECTOR AND TERMINAL "10" OF THE 24-WAY EBCM HARNESS CONNECTOR.

IS RESISTANCE 2Ω OR LESS?

- YES: go to 7.
- NO: REPAIR OPEN OR HIGH RESISTANCE IN CKT 833.

(10)
- IGNITION "OFF."
- DISCONNECT 24-WAY EBCM CONNECTOR.
- USING J 39200, MEASURE THE RESISTANCE BETWEEN TERMINAL "B" OF THE RIGHT FRONT WHEEL SPEED SENSOR HARNESS CONNECTOR AND TERMINAL "9" OF THE 24-WAY EBCM HARNESS CONNECTOR.

IS RESISTANCE 2Ω OR LESS?

- YES: go to 7.
- NO: REPAIR OPEN OR HIGH RESISTANCE IN CKT 872.

(7)
- INSPECT CKTs 833 AND 872 AND 24-WAY EBCM CONNECTOR TERMINALS "9" AND "10" FOR POOR TERMINAL CONTACT OR EVIDENCE OF CORROSION.

  REPLACE ALL TERMINALS THAT EXHIBIT POOR TERMINAL CONTACT OR EVIDENCE OF CORROSION.
- RECONNECT ALL CONNECTORS.
- TEST DRIVE VEHICLE ABOVE 24 km/h (15 MPH) FOR AT LEAST 30 SECONDS.

DOES DTC 22 SET AS A CURRENT DTC?

- YES: REPLACE EBCM.
- NO: MALFUNCTION WAS MOST LIKELY CAUSED BY A POOR CONNECTION. REFER TO "DIAGNOSTIC AIDS" ON FACING PAGE.

DIAGNOSTIC CHART DTC 23 — 1994 LUMINA

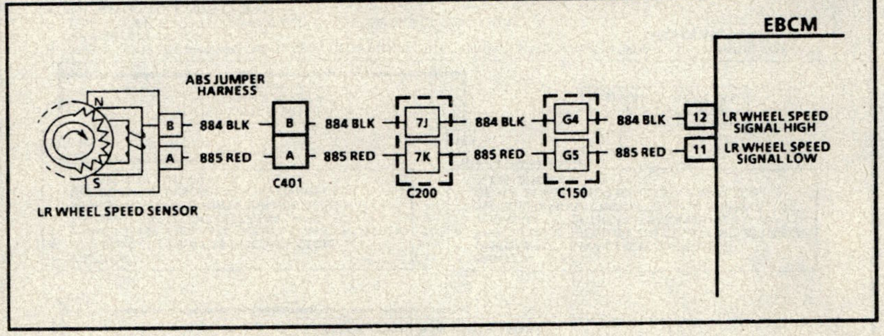

DTC 23

LEFT REAR WHEEL SPEED = 0

Circuit Description:

As a toothed ring passes by the wheel speed sensor, changes in the electromagnetic field cause the wheel speed sensor to produce a sinusoidal (AC) voltage signal whose frequency is proportional to wheel speed. The magnitude of this signal is directly related to wheel speed and the proximity of the wheel speed sensor to the toothed ring, often referred to as the air gap.

DTC Will Set When: DTC 23 can be set when the vehicle is not in an ABS stop. If the left rear wheel speed = 0 and the vehicle's reference speed is greater than 8 km/h (5 mph), a malfunction exists.

Action Taken: A malfunction DTC is stored, ABS is disabled and the ABS warning lamp is turned "ON."

DTC Chart Test Description: Number(s) below refer to circled number(s) on the diagnostic chart.

1. Determines if DTC 34 is present.
2. Designed to identify a wheel speed sensor or circuitry that is damaged and visibly apparent.
3. Checks to see that the wheel speed sensor is within proper resistance values.
4. Verifies that the wheel speed sensor can generate an appropriate output signal.
5. Checks for proper output voltage from the EBCM.
6. Checks for an open in the wheel speed sensor high circuit.
7. Ensures malfunction was not due to physical damage of the circuitry.
8. Ensures that wheel speed sensor circuitry is not shorted.
9. Ensures malfunction was not due to physical damage of the circuitry.
10. Checks for an open in the wheel speed sensor low circuit.

Diagnostic Aids: An "Intermittent" malfunction may be caused by a poor connection, rubbed through wire insulation, or a wire that is broken inside the insulation.

The frequency of the malfunction can be checked by using the enhanced diagnostic function of the Tech 1, as described in "Tech 1 Diagnostics" found in this section.

If the customer's comments reflect that the ABS warning lamp is "ON" only during moist environmental changes (rain, snow, vehicle wash), all wheel speed sensor circuitry should be thoroughly inspected for signs of water intrusion. If DTC is not current, clear DTCs and simulate the effects of water intrusion. Use the following procedure. Spray down the suspected area with a 5% salt water solution (two teaspoons of salt to 12 oz. of water). Test drive vehicle over various road surfaces (bumps, turns, etc.) above 24 km/h (15 mph) for at least 30 seconds. If DTC returns replace suspected harness.

Any circuitry, that is suspected as causing the intermittent complaint, should be thoroughly checked for backed out terminals, improper mating, broken locks, improperly formed or damaged terminals, poor terminal to wiring connections or physical damage to the wiring harness.

Resistance of the wheel speed sensor will increase with an increase in sensor temperature.

When replacing a wheel speed sensor, inspect the sensor terminals and harness connector for corrosion and/or water intrusion. If evidence of corrosion or water intrusion exists, replace wheel speed sensor jumper harness. Likewise, if replacing a wheel speed sensor jumper harness, inspect sensor terminals. If evidence of corrosion or water intrusion exists, replace wheel speed sensor. Refer to "On-Vehicle Service" in this section.

DIAGNOSTIC CHART DTC 23 — 1994 LUMINA, CONTINUED

DTC 23
(Page 1 of 2)
LEFT REAR WHEEL SPEED = 0

Important
- WHEEL SPEED SENSOR INTERMITTENT PROBLEMS MAY BE DIFFICULT TO LOCATE. CARE SHOULD BE TAKEN NOT TO DISTURB ANY ELECTRICAL CONNECTIONS PRIOR TO AN INDICATED STEP OF THIS CHART. THIS WILL ENSURE THAT AN INTERMITTENT CONNECTION WILL NOT BE CORRECTED BEFORE THE SOURCE OF THE PROBLEM IS FOUND.

(1) IS DTC 34 ALSO CURRENTLY SET?
- YES: PROCEED WITH THAT DTC CHART FIRST.
- NO: go to (2)

(2)
- IGNITION "OFF."
- PHYSICALLY INSPECT THE WHEEL SPEED SENSOR, WHEEL SPEED SENSOR JUMPER HARNESS, AND CONNECTORS FOR DAMAGE.

IS PHYSICAL DAMAGE INDICATED?
- YES: REPLACE WHEEL SPEED SENSOR AND/OR WHEEL SPEED SENSOR JUMPER HARNESS IF DAMAGED. DRIVE VEHICLE ABOVE 24 km/h 15 MPH) TO ENSURE DTC DOES NOT SET.
- NO: go to (3)

(3)
- DISCONNECT LEFT REAR WHEEL SPEED SENSOR DIRECTLY AT SENSOR.
- USING J 39200, MEASURE THE RESISTANCE BETWEEN SENSOR TERMINALS "A" AND "B".

IS RESISTANCE BETWEEN 2059 AND 2353Ω WHEN SENSOR IS 20°C (68°F)? *(IF TEMPERATURE OF THE SENSOR IS SUSPECTED TO BE OTHER THAN 20°C (68°F), REFER TO THE TABLE BELOW FOR APPLICABLE SENSOR RESISTANCE VALUES; THE VALUES IN THE TABLE ARE FOR TEMPERATURE OF THE SENSOR, NOT AIR TEMPERATURE.)*
- NO: REPLACE WHEEL SPEED SENSOR.
- YES: go to (4)

(4)
- WITH J 39200 STILL CONNECTED, SELECT A/C VOLTAGE SCALE.
- SPIN WHEEL BY HAND WHILE OBSERVING VOLTAGE.

IS VOLTAGE AT LEAST 100 mV (VOLTAGE WILL INCREASE AS WHEEL SPEED INCREASES)?
- NO: REPLACE WHEEL SPEED SENSOR.
- YES: DIAGNOSIS CONTINUED ON PAGE 2 OF 2 OF THIS DTC CHART.

REAR WHEEL SPEED SENSOR

TEMPERATURE OF SENSOR VS. SENSOR RESISTANCE (APPROXIMATE)

TEMP. (°C)	TEMP. (°F)	RESISTANCE (OHMS)
-34 TO 4	-30 TO 40	1622 TO 2205
5 TO 43	41 TO 110	1930 TO 2566
44 TO 93	111 TO 200	2245 TO 3028

DIAGNOSTIC CHART DTC 23 — 1994 LUMINA, CONTINUED

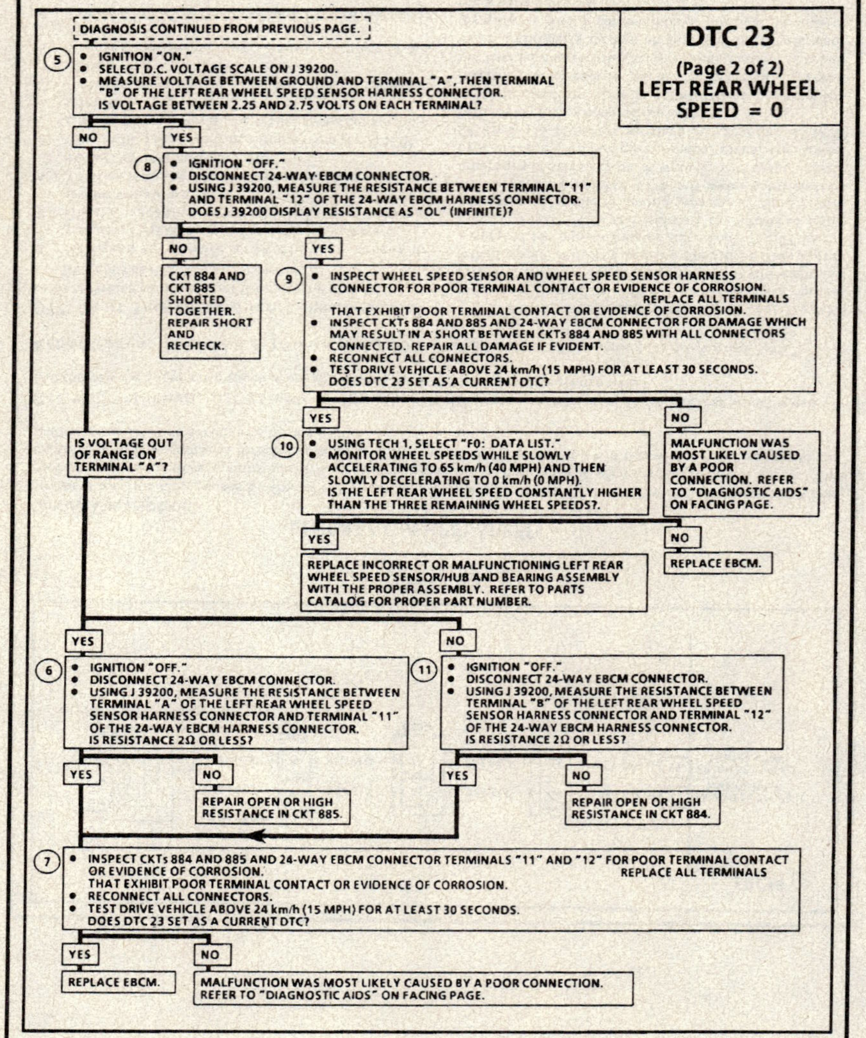

DIAGNOSTIC CHART DTC 24 — 1994 LUMINA

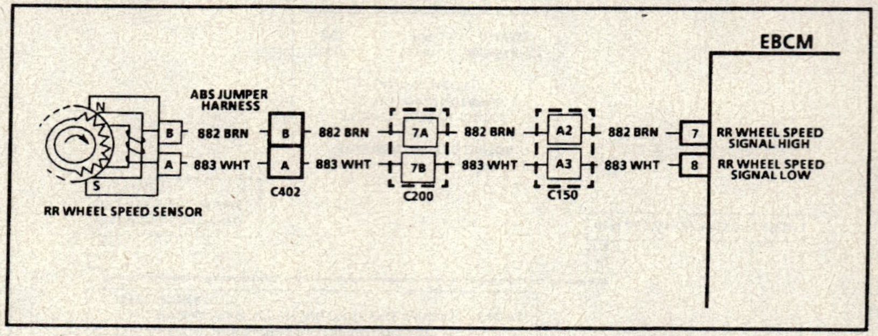

DTC 24

RIGHT REAR WHEEL SPEED = 0

Circuit Description:

As a toothed ring passes by the wheel speed sensor, changes in the electromagnetic field cause the wheel speed sensor to produce a sinusoidal (AC) voltage signal whose frequency is proportional to wheel speed. The magnitude of this signal is directly related to wheel speed and the proximity of the wheel speed sensor to the toothed ring, often referred to as the air gap.

DTC Will Set When: DTC 24 can be set when the vehicle is not in an ABS stop. If the right rear wheel speed = 0 and the vehicle's reference speed is greater than 8 km/h (5 mph), a malfunction exists.

Action Taken: A malfunction DTC is stored, ABS is disabled and the ABS warning lamp is turned "ON."

DTC Chart Test Description: Number(s) below refer to circled number(s) on the diagnostic chart.

1. Determines if DTC 35 is present.
2. Designed to identify a wheel speed sensor or circuitry that is damaged and visibly apparent.
3. Checks to see that the wheel speed sensor is within proper resistance values.
4. Verifies that the wheel speed sensor can generate an appropriate output signal.
5. Checks for proper output voltage from the EBCM.
6. Checks for an open in the wheel speed sensor high circuit.
7. Ensures malfunction was not due to physical damage of the circuitry.
8. Ensures that wheel speed sensor circuitry is not shorted.
9. Ensures malfunction was not due to physical damage of the circuitry.
10. Checks for an open in the wheel speed sensor low circuit.

Diagnostic Aids: An "Intermittent" malfunction may be caused by a poor connection, rubbed through wire insulation, or a wire that is broken inside the insulation.

The frequency of the malfunction can be checked by using the enhanced diagnostic function of the Tech 1, as described in "Tech 1 Diagnostics" found in this section.

If the customer's comments reflect that the ABS warning lamp is "ON" only during moist environmental changes (rain, snow, vehicle wash), all wheel speed sensor circuitry should be thoroughly inspected for signs of water intrusion. If DTC is not current, clear DTCs and simulate the effects of water intrusion. Use the following procedure. Spray down the suspected area with a 5% salt water solution (two teaspoons of salt to 12 oz. of water). Test drive vehicle over various road surfaces (bumps, turns, etc.) above 24 km/h (15 mph) for at least 30 seconds. If DTC returns replace suspected harness.

Any circuitry, that is suspected as causing the intermittent complaint, should be thoroughly checked for backed out terminals, improper mating, broken locks, improperly formed or damaged terminals, poor terminal to wiring connections or physical damage to the wiring harness.

Resistance of the wheel speed sensor will increase with an increase in sensor temperature.

When replacing a wheel speed sensor, inspect the sensor terminals and harness connector for corrosion and/or water intrusion. If evidence of corrosion or water intrusion exists, replace wheel speed sensor jumper harness. Likewise, if replacing a wheel speed sensor jumper harness, inspect sensor terminals. If evidence of corrosion or water intrusion exists, replace wheel speed sensor. Refer to "On-Vehicle Service" in this section.

DIAGNOSTIC CHART DTC 24 — 1994 LUMINA, CONTINUED

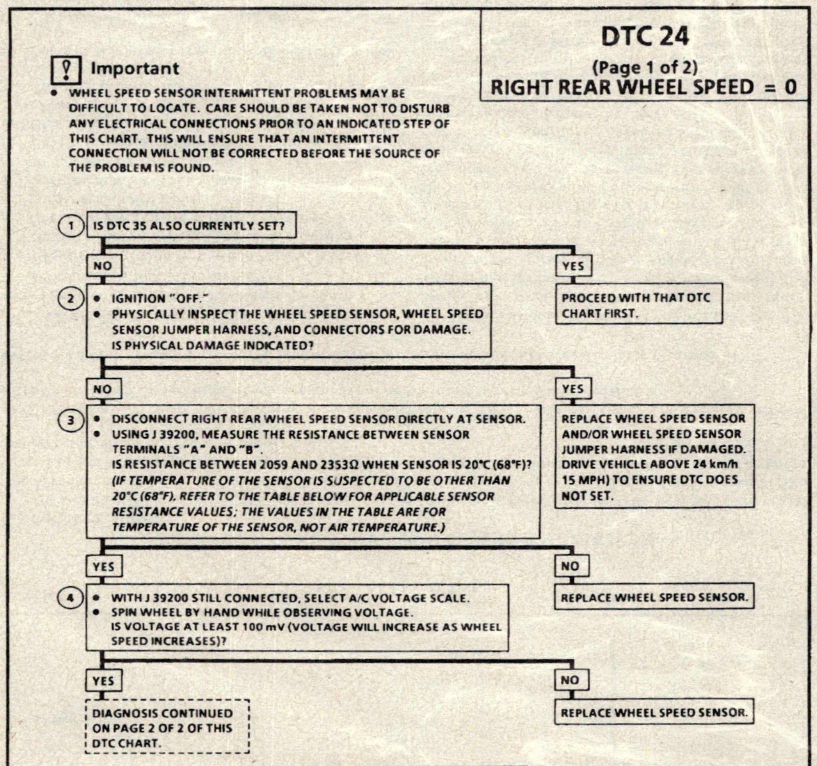

REAR WHEEL SPEED SENSOR
TEMPERATURE OF SENSOR VS. SENSOR RESISTANCE (APPROXIMATE)

TEMP. (°C)	TEMP. (°F)	RESISTANCE (OHMS)
-34 TO 4	-30 TO 40	1622 TO 2205
5 TO 43	41 TO 110	1930 TO 2566
44 TO 93	111 TO 200	2245 TO 3028

DIAGNOSTIC CHART DTC 24 — 1994 LUMINA, CONTINUED

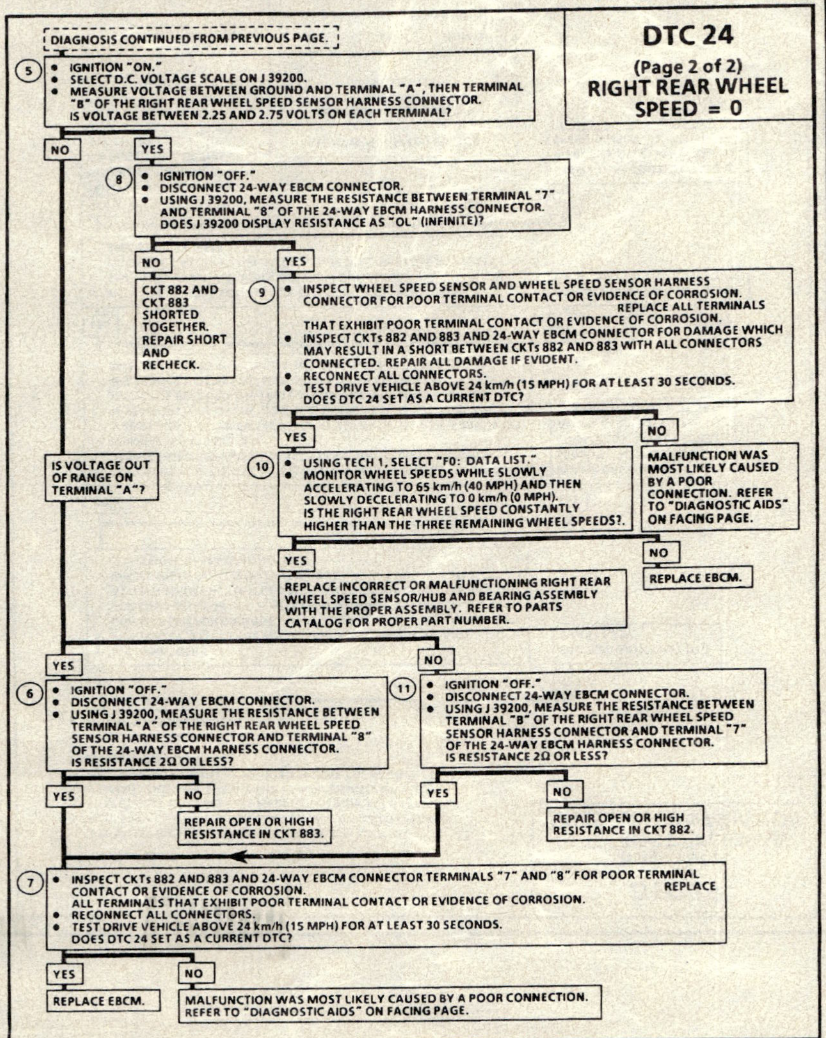

DIAGNOSTIC CHART DTC 25 — 1994 LUMINA

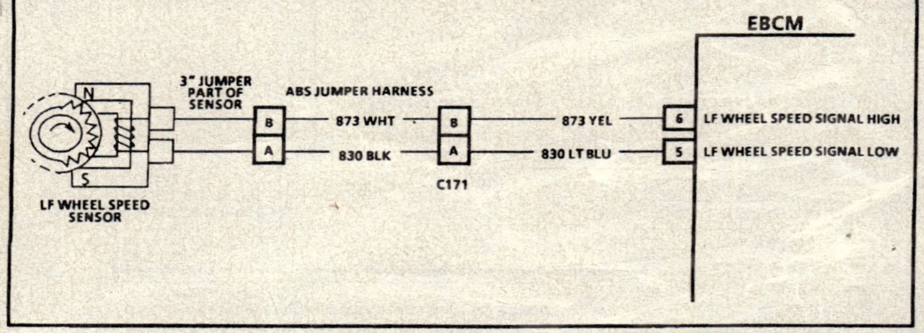

DTC 25

LEFT FRONT EXCESSIVE WHEEL SPEED VARIATION

Circuit Description:

As a toothed ring passes by the wheel speed sensor, changes in the electromagnetic field cause the wheel speed sensor to produce a sinusoidal (AC) voltage signal whose frequency is proportional to wheel speed. The magnitude of this signal is directly related to wheel speed and the proximity of the wheel speed sensor to the toothed ring, often referred to as the air gap.

DTC Will Set When: DTC 25 can be set when the brake is "OFF." The purpose of this test is to detect a situation in which the left front wheel acceleration or deceleration is beyond specified limits.

Action Taken: A malfunction DTC is stored, ABS is disabled and the ABS warning lamp is turned "ON."

DTC Chart Test Description: Number(s) below refer to circled number(s) on the diagnostic chart.

1. Determines if DTC 32 is present.
2. Designed to identify a wheel speed sensor or circuitry that is damaged and visibly apparent.
3. Checks to see that the wheel speed sensor is within proper resistance values.
4. Verifies that the wheel speed sensor can generate an appropriate output signal.
5. Checks for proper output voltage from the EBCM.
6. Checks for an open in the wheel speed sensor high circuit.
7. Ensures malfunction was not due to physical damage of the circuitry.
8. Ensures that wheel speed sensor circuitry is not shorted.
9. Ensures malfunction was not due to physical damage of the circuitry.
10. Checks for an open in the wheel speed sensor low circuit.

Diagnostic Aids: An "Intermittent" malfunction may be caused by a poor connection, rubbed through wire insulation, or a wire that is broken inside the insulation.

The frequency of the malfunction can be checked by using the enhanced diagnostic function of the Tech 1, as described in "Tech 1 Diagnostics" found in this section.

If the customer's comments reflect that the ABS warning lamp is "ON" only during moist environmental changes (rain, snow, vehicle wash), all wheel speed sensor circuitry should be thoroughly inspected for signs of water intrusion. If DTC is not current, clear DTCs and simulate the effects of water intrusion. Use the following procedure. Spray down the suspected area with a 5% salt water solution (two teaspoons of salt to 12 oz. of water). Test drive vehicle over various road surfaces (bumps, turns, etc.) above 24 km/h (15 mph) for at least 30 seconds. If DTC returns replace suspected harness.

Any circuitry, that is suspected as causing the intermittent complaint, should be thoroughly checked for backed out terminals, improper mating, broken locks, improperly formed or damaged terminals, poor terminal to wiring connections or physical damage to the wiring harness.

Resistance of the wheel speed sensor will increase with an increase in sensor temperature.

When replacing a wheel speed sensor, inspect the sensor terminals and harness connector for corrosion and/or water intrusion. If evidence of corrosion or water intrusion exists, replace wheel speed sensor jumper harness. Likewise, if replacing a wheel speed sensor jumper harness, inspect sensor terminals. If evidence of corrosion or water intrusion exists, replace wheel speed sensor. Refer to "On-Vehicle Service" in this section.

DIAGNOSTIC CHART DTC 25 — 1994 LUMINA, CONTINUED

DTC 25

(Page 1 of 2)

LEFT FRONT EXCESSIVE WHEEL SPEED VARIATION

Important

- WHEEL SPEED SENSOR INTERMITTENT PROBLEMS MAY BE DIFFICULT TO LOCATE. CARE SHOULD BE TAKEN NOT TO DISTURB ANY ELECTRICAL CONNECTIONS PRIOR TO AN INDICATED STEP OF THIS CHART. THIS WILL ENSURE THAT AN INTERMITTENT CONNECTION WILL NOT BE CORRECTED BEFORE THE SOURCE OF THE PROBLEM IS FOUND.

(1) IS DTC 32 ALSO CURRENTLY SET?

- YES: PROCEED WITH THAT DTC CHART FIRST.
- NO: go to (2).

(2)
- IGNITION "OFF."
- PHYSICALLY INSPECT THE WHEEL SPEED SENSOR WIRING AND CONNECTORS FOR DAMAGE. INSPECT THE WHEEL SPEED SENSOR AND TOOTHED RING FOR LOOSENESS OR DAMAGE. PAY PARTICULAR ATTENTION TO THE TOOTHED RING, ANY DEVIATION WILL AFFECT THE WHEEL SPEED SENSOR OUTPUT SIGNAL.

IS PHYSICAL DAMAGE INDICATED?

- YES: REPAIR DAMAGE OR SENSOR/TOOTHED RING LOOSENESS. DRIVE VEHICLE ABOVE 24 km/h (15 MPH) TO ENSURE DTC DOES NOT SET.
- NO: go to (3).

(3)
- DISCONNECT LEFT FRONT WHEEL SPEED SENSOR DIRECTLY AT SENSOR.
- USING J 39200, MEASURE THE RESISTANCE BETWEEN SENSOR TERMINALS "A" AND "B".

IS RESISTANCE BETWEEN 1661 AND 1929Ω WHEN SENSOR IS 20°C (68°F)? *(IF TEMPERATURE OF THE SENSOR IS SUSPECTED TO BE OTHER THAN 20°C (68°F), REFER TO THE TABLE BELOW FOR APPLICABLE SENSOR RESISTANCE VALUES; THE VALUES IN THE TABLE ARE FOR TEMPERATURE OF THE SENSOR, NOT AIR TEMPERATURE.)*

- NO: REPLACE WHEEL SPEED SENSOR.
- YES: go to (4).

(4)
- WITH J 39200 STILL CONNECTED, SELECT A/C VOLTAGE SCALE.
- SPIN WHEEL BY HAND WHILE OBSERVING VOLTAGE.

IS VOLTAGE AT LEAST 100 mV (VOLTAGE WILL INCREASE AS WHEEL SPEED INCREASES)?

- NO: VERIFY THAT TOOTHED RING IS INTACT. IF OK, REPLACE WHEEL SPEED SENSOR.
- YES: DIAGNOSIS CONTINUED ON PAGE 2 OF 2 OF THIS DTC CHART.

FRONT WHEEL SPEED SENSOR

TEMPERATURE OF SENSOR VS. SENSOR RESISTANCE (APPROXIMATE)

TEMP. (°C)	TEMP. (°F)	RESISTANCE (OHMS)
-34 TO 4	-30 TO 40	1309 TO 1808
5 TO 43	41 TO 110	1557 TO 2103
44 TO 93	111 TO 200	1811 TO 2482

DIAGNOSTIC CHART DTC 25 — 1994 LUMINA, CONTINUED

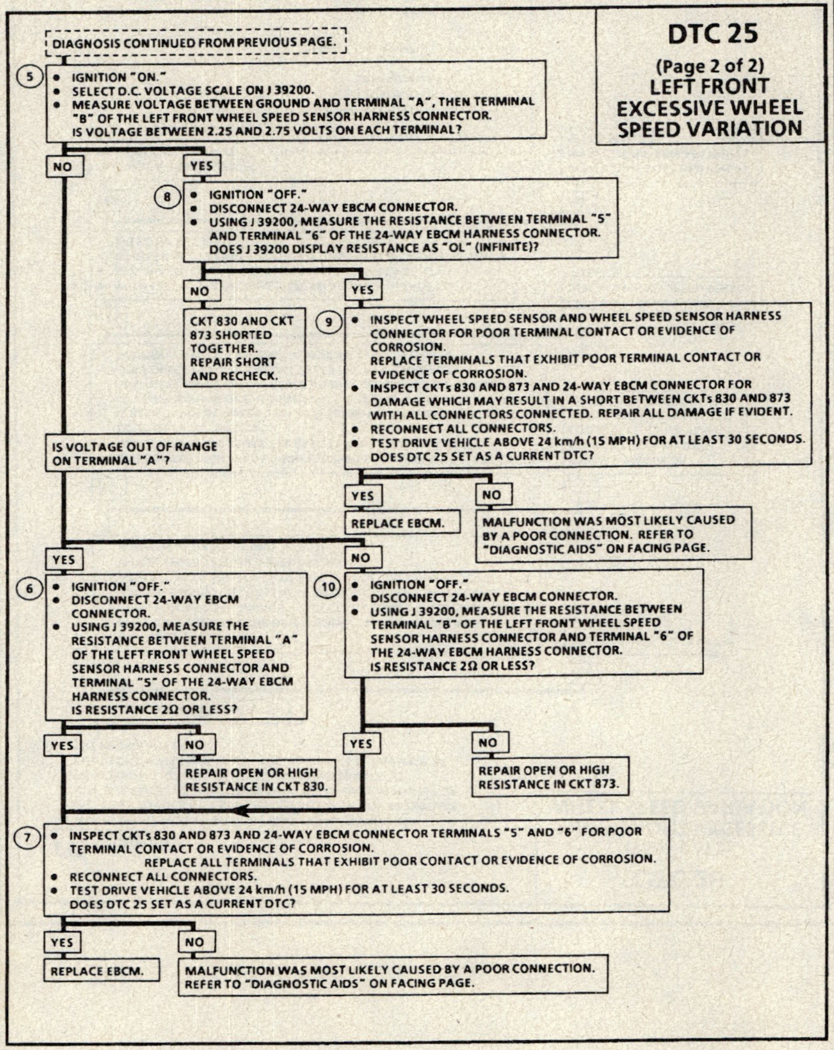

DIAGNOSTIC CHART DTC 26 — 1994 LUMINA

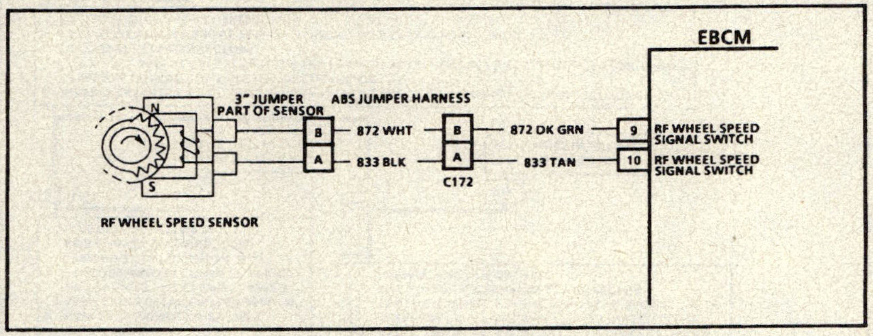

DTC 26

RIGHT FRONT EXCESSIVE WHEEL SPEED VARIATION

Circuit Description:

As a toothed ring passes by the wheel speed sensor, changes in the electromagnetic field cause the wheel speed sensor to produce a sinusoidal (AC) voltage signal whose frequency is proportional to wheel speed. The magnitude of this signal is directly related to wheel speed and the proximity of the wheel speed sensor to the toothed ring, often referred to as the air gap.

DTC Will Set When: DTC 26 can be set when the brake is "OFF." The purpose of this test is to detect a situation in which the right front wheel acceleration or deceleration is beyond specified limits.

Action Taken: A malfunction DTC is stored, ABS is disabled and the ABS warning lamp is turned "ON."

DTC Chart Test Description: Number(s) below refer to circled number(s) on the diagnostic chart.

1. Determines if DTC 33 is present.
2. Designed to identify a wheel speed sensor or circuitry that is damaged and visibly apparent.
3. Checks to see that the wheel speed sensor is within proper resistance values.
4. Verifies that the wheel speed sensor can generate an appropriate output signal.
5. Checks for proper output voltage from the EBCM.
6. Checks for an open in the wheel speed sensor high circuit.
7. Ensures malfunction was not due to physical damage of the circuitry.
8. Ensures that wheel speed sensor circuitry is not shorted.
9. Ensures malfunction was not due to physical damage of the circuitry.
10. Checks for an open in the wheel speed sensor low circuit.

Diagnostic Aids: An "Intermittent" malfunction may be caused by a poor connection, rubbed through wire insulation, or a wire that is broken inside the insulation.

The frequency of the malfunction can be checked by using the enhanced diagnostic function of the Tech 1, as described in "Tech 1 Diagnostics" found in this section.

If the customer's comments reflect that the ABS warning lamp is "ON" only during moist environmental changes (rain, snow, vehicle wash), all wheel speed sensor circuitry should be thoroughly inspected for signs of water intrusion. If DTC is not current, clear DTCs and simulate the effects of water intrusion. Use the following procedure. Spray down the suspected area with a 5% salt water solution (two teaspoons of salt to 12 oz. of water). Test drive vehicle over various road surfaces (bumps, turns, etc.) above 24 km/h (15 mph) for at least 30 seconds. If DTC returns replace suspected harness.

Any circuitry, that is suspected as causing the intermittent complaint, should be thoroughly checked for backed out terminals, improper mating, broken locks, improperly formed or damaged terminals, poor terminal to wiring connections or physical damage to the wiring harness.

Resistance of the wheel speed sensor will increase with an increase in sensor temperature.

When replacing a wheel speed sensor, inspect the sensor terminals and harness connector for corrosion and/or water intrusion. If evidence of corrosion or water intrusion exists, replace wheel speed sensor jumper harness. Likewise, if replacing a wheel speed sensor jumper harness, inspect sensor terminals. If evidence of corrosion or water intrusion exists, replace wheel speed sensor. Refer to "On-Vehicle Service" in this section.

DIAGNOSTIC CHART DTC 26 — 1994 LUMINA, CONTINUED

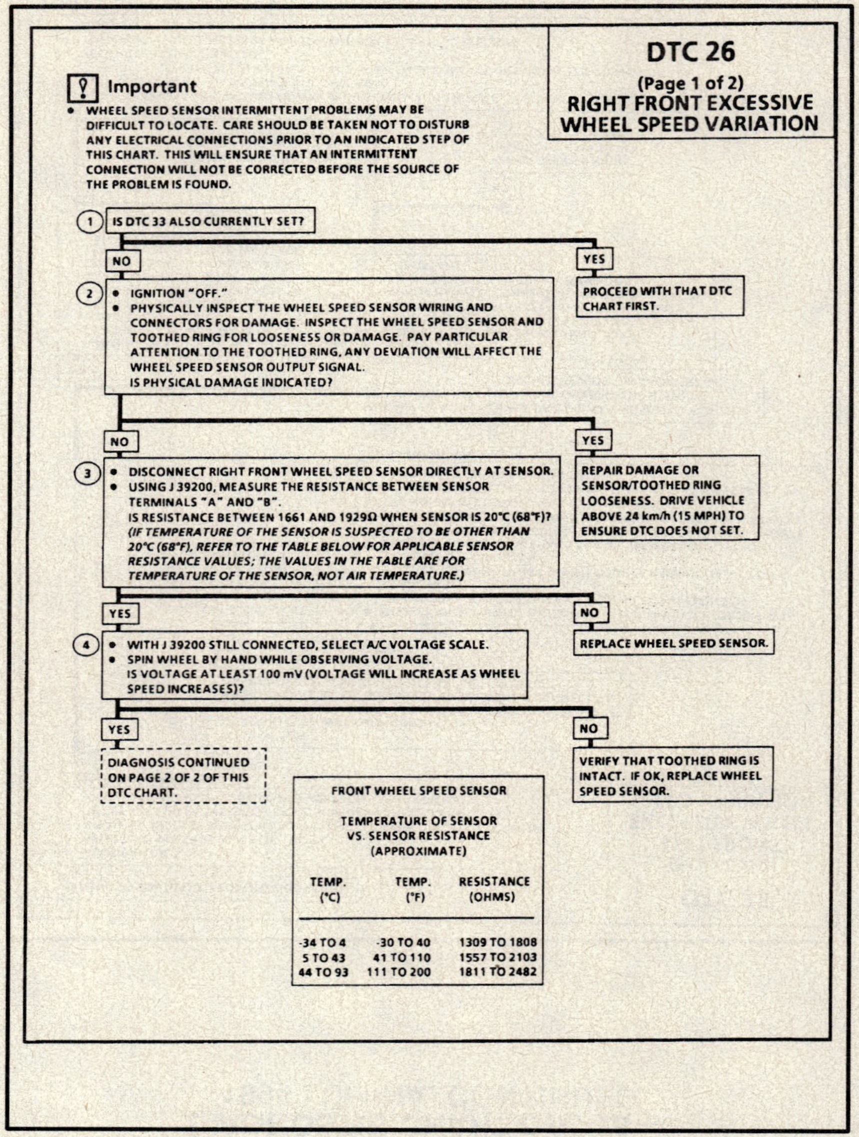

FRONT WHEEL SPEED SENSOR

TEMPERATURE OF SENSOR VS. SENSOR RESISTANCE (APPROXIMATE)

TEMP. (°C)	TEMP. (°F)	RESISTANCE (OHMS)
-34 TO 4	-30 TO 40	1309 TO 1808
5 TO 43	41 TO 110	1557 TO 2103
44 TO 93	111 TO 200	1811 TO 2482

DIAGNOSTIC CHART DTC 26 — 1994 LUMINA, CONTINUED

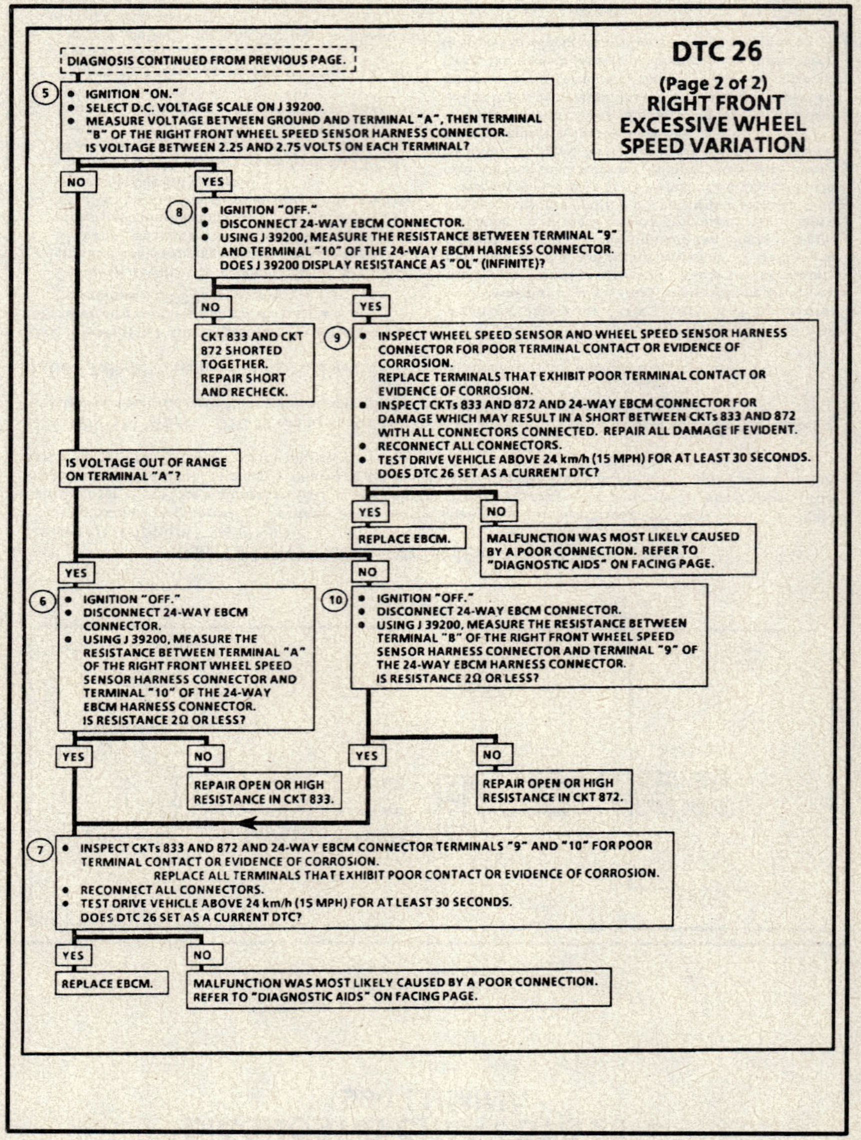

DIAGNOSTIC CHART DTC 27 — 1994 LUMINA

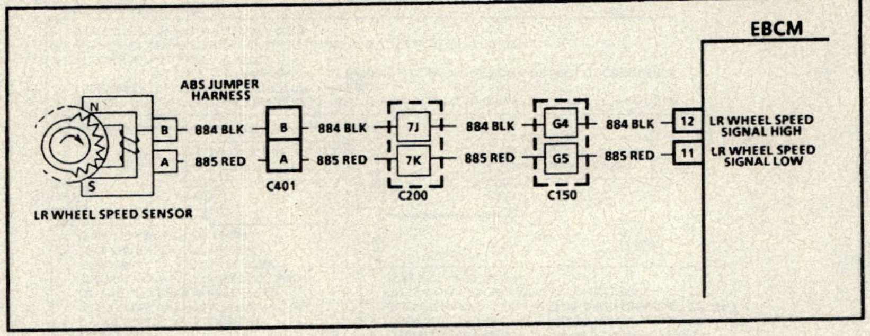

DTC 27

LEFT REAR EXCESSIVE WHEEL SPEED VARIATION

Circuit Description:
As a toothed ring passes by the wheel speed sensor, changes in the electromagnetic field cause the wheel speed sensor to produce a sinusoidal (AC) voltage signal whose frequency is proportional to wheel speed. The magnitude of this signal is directly related to wheel speed and the proximity of the wheel speed sensor to the toothed ring, often referred to as the air gap.

DTC Will Set When: DTC 27 can be set when the brake is "OFF." The purpose of this test is to detect a situation in which the left rear wheel acceleration or deceleration is beyond specified limits.

Action Taken: A malfunction DTC is stored, ABS is disabled and the ABS warning lamp is turned "ON."

DTC Chart Test Description: Number(s) below refer to circled number(s) on the diagnostic chart.

1. Determines if DTC 34 is present.
2. Designed to identify a wheel speed sensor or circuitry that is damaged and visibly apparent.
3. Checks to see that the wheel speed sensor is within proper resistance values.
4. Verifies that the wheel speed sensor can generate an appropriate output signal.
5. Checks for proper output voltage from the EBCM.
6. Checks for an open in the wheel speed sensor high circuit.
7. Ensures malfunction was not due to physical damage of the circuitry.
8. Ensures that wheel speed sensor circuitry is not shorted.
9. Ensures malfunction was not due to physical damage of the circuitry.
10. Checks for an open in the wheel speed sensor low circuit.

Diagnostic Aids: An "Intermittent" malfunction may be caused by a poor connection, rubbed through wire insulation, or a wire that is broken inside the insulation.

The frequency of the malfunction can be checked by using the enhanced diagnostic function of the Tech 1, as described in "Tech 1 Diagnostics" found in this section.

If the customer's comments reflect that the ABS warning lamp is "ON" only during moist environmental changes (rain, snow, vehicle wash), all wheel speed sensor circuitry should be thoroughly inspected for signs of water intrusion. If DTC is not current, clear DTCs and simulate the effects of water intrusion. Use the following procedure. Spray down the suspected area with a 5% salt water solution (two teaspoons of salt to 12 oz. of water). Test drive vehicle over various road surfaces (bumps, turns, etc.) above 24 km/h (15 mph) for at least 30 seconds. If DTC returns replace suspected harness.

Any circuitry, that is suspected as causing the intermittent complaint, should be thoroughly checked for backed out terminals, improper mating, broken locks, improperly formed or damaged terminals, poor terminal to wiring connections or physical damage to the wiring harness.

Resistance of the wheel speed sensor will increase with an increase in sensor temperature.

When replacing a wheel speed sensor, inspect the sensor terminals and harness connector for corrosion and/or water intrusion. If evidence of corrosion or water intrusion exists, replace wheel speed sensor jumper harness. Likewise, if replacing a wheel speed sensor jumper harness, inspect sensor terminals. If evidence of corrosion or water intrusion exists, replace wheel speed sensor. Refer to "On-Vehicle Service" in this section.

DIAGNOSTIC CHART DTC 27 — 1994 LUMINA, CONTINUED

DTC 27
(Page 1 of 2)
LEFT REAR EXCESSIVE WHEEL SPEED VARIATION

Important
- WHEEL SPEED SENSOR INTERMITTENT PROBLEMS MAY BE DIFFICULT TO LOCATE. CARE SHOULD BE TAKEN NOT TO DISTURB ANY ELECTRICAL CONNECTIONS PRIOR TO AN INDICATED STEP OF THIS CHART. THIS WILL ENSURE THAT AN INTERMITTENT CONNECTION WILL NOT BE CORRECTED BEFORE THE SOURCE OF THE PROBLEM IS FOUND.

(1) IS DTC 34 ALSO CURRENTLY SET?
- YES: PROCEED WITH THAT DTC CHART FIRST.
- NO:

(2)
- IGNITION "OFF."
- PHYSICALLY INSPECT THE WHEEL SPEED SENSOR, WHEEL SPEED SENSOR JUMPER HARNESS, AND CONNECTORS FOR DAMAGE.

IS PHYSICAL DAMAGE INDICATED?
- YES: REPLACE WHEEL SPEED SENSOR AND/OR WHEEL SPEED SENSOR JUMPER HARNESS IF DAMAGED. DRIVE VEHICLE ABOVE 24 km/h (15 MPH) TO ENSURE DTC DOES NOT SET.
- NO:

(3)
- DISCONNECT LEFT REAR WHEEL SPEED SENSOR DIRECTLY AT SENSOR.
- USING J 39200, MEASURE THE RESISTANCE BETWEEN SENSOR TERMINALS "A" AND "B".

IS RESISTANCE BETWEEN 2059 AND 2353Ω WHEN SENSOR IS 20°C (68°F)? *(IF TEMPERATURE OF THE SENSOR IS SUSPECTED TO BE OTHER THAN 20°C (68°F), REFER TO THE TABLE BELOW FOR APPLICABLE SENSOR RESISTANCE VALUES; THE VALUES IN THE TABLE ARE FOR TEMPERATURE OF THE SENSOR, NOT AIR TEMPERATURE.)*
- NO: REPLACE WHEEL SPEED SENSOR.
- YES:

(4)
- WITH J 39200 STILL CONNECTED, SELECT A/C VOLTAGE SCALE.
- SPIN WHEEL BY HAND WHILE OBSERVING VOLTAGE.

IS VOLTAGE AT LEAST 100 mV (VOLTAGE WILL INCREASE AS WHEEL SPEED INCREASES)?
- NO: REPLACE WHEEL SPEED SENSOR.
- YES: DIAGNOSIS CONTINUED ON PAGE 2 OF 2 OF THIS DTC CHART.

REAR WHEEL SPEED SENSOR

TEMPERATURE OF SENSOR VS. SENSOR RESISTANCE (APPROXIMATE)

TEMP. (°C)	TEMP. (°F)	RESISTANCE (OHMS)
-34 TO 4	-30 TO 40	1622 TO 2205
5 TO 43	41 TO 110	1930 TO 2566
44 TO 93	111 TO 200	2245 TO 3028

DIAGNOSTIC CHART DTC 27 — 1994 LUMINA, CONTINUED

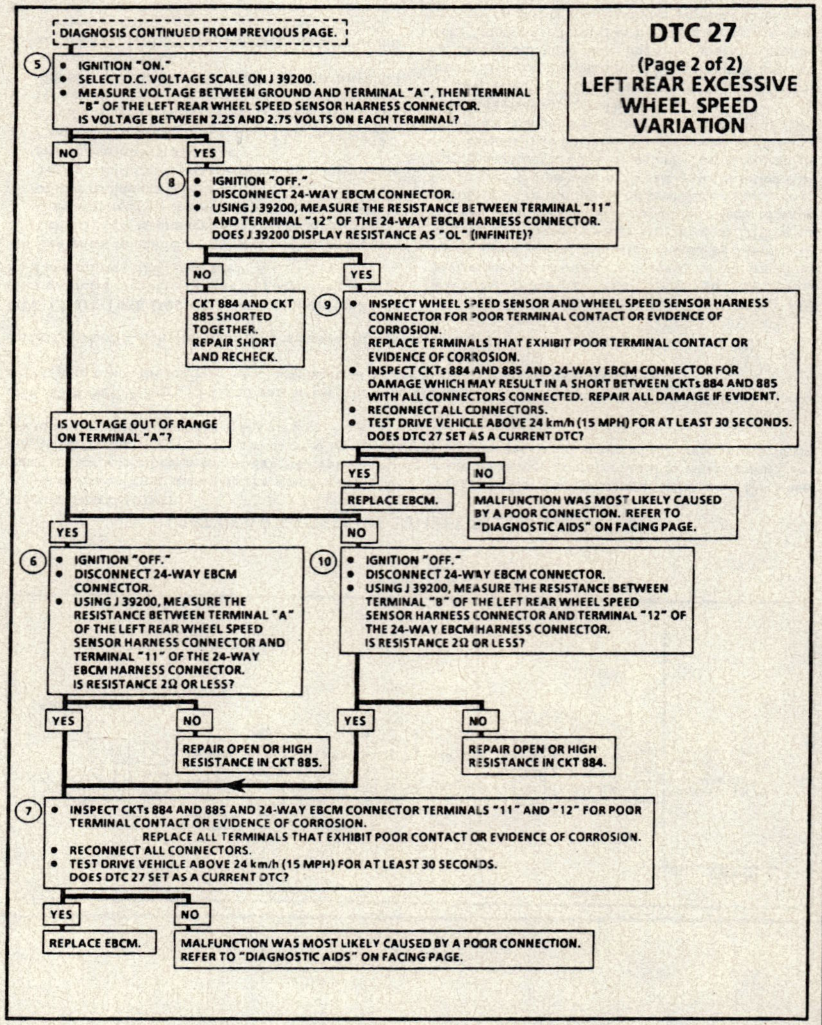

DIAGNOSTIC CHART DTC 28 — 1994 LUMINA

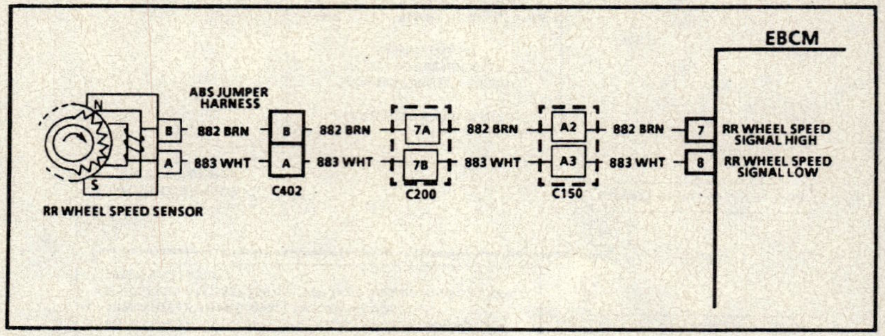

DTC 28

RIGHT REAR EXCESSIVE WHEEL SPEED VARIATION

Circuit Description:

As a toothed ring passes by the wheel speed sensor, changes in the electromagnetic field cause the wheel speed sensor to produce a sinusoidal (AC) voltage signal whose frequency is proportional to wheel speed. The magnitude of this signal is directly related to wheel speed and the proximity of the wheel speed sensor to the toothed ring, often referred to as the air gap.

DTC Will Set When: DTC 28 can be set when the brake is "OFF." The purpose of this test is to detect a situation in which the right rear wheel acceleration or deceleration is beyond specified limits.

Action Taken: A malfunction DTC is stored, ABS is disabled and the ABS warning lamp is turned "ON."

DTC Chart Test Description: Number(s) below refer to circled number(s) on the diagnostic chart.

1. Determines if DTC 35 is present.
2. Designed to identify a wheel speed sensor or circuitry that is damaged and visibly apparent.
3. Checks to see that the wheel speed sensor is within proper resistance values.
4. Verifies that the wheel speed sensor can generate an appropriate output signal.
5. Checks for proper output voltage from the EBCM.
6. Checks for an open in the wheel speed sensor high circuit.
7. Ensures malfunction was not due to physical damage of the circuitry.
8. Ensures that wheel speed sensor circuitry is not shorted.
9. Ensures malfunction was not due to physical damage of the circuitry.
10. Checks for an open in the wheel speed sensor low circuit.

Diagnostic Aids: An "Intermittent" malfunction may be caused by a poor connection, rubbed through wire insulation, or a wire that is broken inside the insulation.

The frequency of the malfunction can be checked by using the enhanced diagnostic function of the Tech 1, as described in "Tech 1 Diagnostics" found in this section.

If the customer's comments reflect that the ABS warning lamp is "ON" only during moist environmental changes (rain, snow, vehicle wash), all wheel speed sensor circuitry should be thoroughly inspected for signs of water intrusion. If DTC is not current, clear DTCs and simulate the effects of water intrusion. Use the following procedure. Spray down the suspected area with a 5% salt water solution (two teaspoons of salt to 12 oz. of water). Test drive vehicle over various road surfaces (bumps, turns, etc.) above 24 km/h (15 mph) for at least 30 seconds. If DTC returns replace suspected harness.

Any circuitry, that is suspected as causing the intermittent complaint, should be thoroughly checked for backed out terminals, improper mating, broken locks, improperly formed or damaged terminals, poor terminal to wiring connections or physical damage to the wiring harness.

Resistance of the wheel speed sensor will increase with an increase in sensor temperature.

When replacing a wheel speed sensor, inspect the sensor terminals and harness connector for corrosion and/or water intrusion. If evidence of corrosion or water intrusion exists, replace wheel speed sensor jumper harness. Likewise, if replacing a wheel speed sensor jumper harness, inspect sensor terminals. If evidence of corrosion or water intrusion exists, replace wheel speed sensor. Refer to "On-Vehicle Service" in this section.

DIAGNOSTIC CHART DTC 28 — 1994 LUMINA, CONTINUED

DTC 28
(Page 1 of 2)
RIGHT REAR EXCESSIVE WHEEL SPEED VARIATION

Important
- WHEEL SPEED SENSOR INTERMITTENT PROBLEMS MAY BE DIFFICULT TO LOCATE. CARE SHOULD BE TAKEN NOT TO DISTURB ANY ELECTRICAL CONNECTIONS PRIOR TO AN INDICATED STEP OF THIS CHART. THIS WILL ENSURE THAT AN INTERMITTENT CONNECTION WILL NOT BE CORRECTED BEFORE THE SOURCE OF THE PROBLEM IS FOUND.

1. IS DTC 35 ALSO CURRENTLY SET?
   - YES: PROCEED WITH THAT DTC CHART FIRST.
   - NO: Go to step 2.
2. - IGNITION "OFF."
   - PHYSICALLY INSPECT THE WHEEL SPEED SENSOR, WHEEL SPEED SENSOR JUMPER HARNESS, AND CONNECTORS FOR DAMAGE.
   IS PHYSICAL DAMAGE INDICATED?
   - YES: REPLACE WHEEL SPEED SENSOR AND/OR WHEEL SPEED SENSOR JUMPER HARNESS IF DAMAGED. DRIVE VEHICLE ABOVE 24 km/h (15 MPH) TO ENSURE DTC DOES NOT SET.
   - NO: Go to step 3.
3. - DISCONNECT RIGHT REAR WHEEL SPEED SENSOR DIRECTLY AT SENSOR.
   - USING J 39200, MEASURE THE RESISTANCE BETWEEN SENSOR TERMINALS "A" AND "B".
   IS RESISTANCE BETWEEN 2059 AND 2353Ω WHEN SENSOR IS 20°C (68°F)? *(IF TEMPERATURE OF THE SENSOR IS SUSPECTED TO BE OTHER THAN 20°C (68°F), REFER TO THE TABLE BELOW FOR APPLICABLE SENSOR RESISTANCE VALUES; THE VALUES IN THE TABLE ARE FOR TEMPERATURE OF THE SENSOR, NOT AIR TEMPERATURE.)*
   - NO: REPLACE WHEEL SPEED SENSOR.
   - YES: Go to step 4.
4. - WITH J 39200 STILL CONNECTED, SELECT A/C VOLTAGE SCALE.
   - SPIN WHEEL BY HAND WHILE OBSERVING VOLTAGE.
   IS VOLTAGE AT LEAST 100 mV (VOLTAGE WILL INCREASE AS WHEEL SPEED INCREASES)?
   - NO: REPLACE WHEEL SPEED SENSOR.
   - YES: DIAGNOSIS CONTINUED ON PAGE 2 OF 2 OF THIS DTC CHART.

REAR WHEEL SPEED SENSOR
TEMPERATURE OF SENSOR VS. SENSOR RESISTANCE (APPROXIMATE)

TEMP. (°C)	TEMP. (°F)	RESISTANCE (OHMS)
-34 TO 4	-30 TO 40	1622 TO 2205
5 TO 43	41 TO 110	1930 TO 2566
44 TO 93	111 TO 200	2245 TO 3028

DIAGNOSTIC CHART DTC 28 — 1994 LUMINA, CONTINUED

DTC 28
(Page 2 of 2)
RIGHT REAR EXCESSIVE WHEEL SPEED VARIATION

DIAGNOSIS CONTINUED FROM PREVIOUS PAGE.

5. - IGNITION "ON."
   - SELECT D.C. VOLTAGE SCALE ON J 39200.
   - MEASURE VOLTAGE BETWEEN GROUND AND TERMINAL "A", THEN TERMINAL "B" OF THE RIGHT REAR WHEEL SPEED SENSOR HARNESS CONNECTOR.
   IS VOLTAGE BETWEEN 2.25 AND 2.75 VOLTS ON EACH TERMINAL?
   - YES: Go to step 8.
   - NO: IS VOLTAGE OUT OF RANGE ON TERMINAL "A"?
     - YES: Go to step 6.
     - NO: Go to step 10.
8. - IGNITION "OFF."
   - DISCONNECT 24-WAY EBCM CONNECTOR.
   - USING J 39200, MEASURE THE RESISTANCE BETWEEN TERMINAL "7" AND TERMINAL "8" OF THE 24-WAY EBCM HARNESS CONNECTOR.
   DOES J 39200 DISPLAY RESISTANCE AS "OL" (INFINITE)?
   - NO: CKT 882 AND CKT 883 SHORTED TOGETHER. REPAIR SHORT AND RECHECK.
   - YES: Go to step 9.
9. - INSPECT WHEEL SPEED SENSOR AND WHEEL SPEED SENSOR HARNESS CONNECTOR FOR POOR TERMINAL CONTACT OR EVIDENCE OF CORROSION.
   REPLACE TERMINALS THAT EXHIBIT POOR TERMINAL CONTACT OR EVIDENCE OF CORROSION.
   - INSPECT CKTs 882 AND 883 AND 24-WAY EBCM CONNECTOR FOR DAMAGE WHICH MAY RESULT IN A SHORT BETWEEN CKTs 882 AND 883 WITH ALL CONNECTORS CONNECTED. REPAIR ALL DAMAGE IF EVIDENT.
   - RECONNECT ALL CONNECTORS.
   - TEST DRIVE VEHICLE ABOVE 24 km/h (15 MPH) FOR AT LEAST 30 SECONDS.
   DOES DTC 28 SET AS A CURRENT DTC?
   - YES: REPLACE EBCM.
   - NO: MALFUNCTION WAS MOST LIKELY CAUSED BY A POOR CONNECTION. REFER TO "DIAGNOSTIC AIDS" ON FACING PAGE.
6. - IGNITION "OFF."
   - DISCONNECT 24-WAY EBCM CONNECTOR.
   - USING J 39200, MEASURE THE RESISTANCE BETWEEN TERMINAL "A" OF THE RIGHT REAR WHEEL SPEED SENSOR HARNESS CONNECTOR AND TERMINAL "8" OF THE 24-WAY EBCM HARNESS CONNECTOR.
   IS RESISTANCE 2Ω OR LESS?
   - YES: Go to step 7.
   - NO: REPAIR OPEN OR HIGH RESISTANCE IN CKT 883.
10. - IGNITION "OFF."
    - DISCONNECT 24-WAY EBCM CONNECTOR.
    - USING J 39200, MEASURE THE RESISTANCE BETWEEN TERMINAL "B" OF THE RIGHT REAR WHEEL SPEED SENSOR HARNESS CONNECTOR AND TERMINAL "7" OF THE 24-WAY EBCM HARNESS CONNECTOR.
    IS RESISTANCE 2Ω OR LESS?
    - YES: Go to step 7.
    - NO: REPAIR OPEN OR HIGH RESISTANCE IN CKT 882.
7. - INSPECT CKTs 882 AND 883 AND 24-WAY EBCM CONNECTOR TERMINALS "7" AND "8" FOR POOR TERMINAL CONTACT OR EVIDENCE OF CORROSION.
   REPLACE ALL TERMINALS THAT EXHIBIT POOR CONTACT OR EVIDENCE OF CORROSION.
   - RECONNECT ALL CONNECTORS.
   - TEST DRIVE VEHICLE ABOVE 24 km/h (15 MPH) FOR AT LEAST 30 SECONDS.
   DOES DTC 28 SET AS A CURRENT DTC?
   - YES: REPLACE EBCM.
   - NO: MALFUNCTION WAS MOST LIKELY CAUSED BY A POOR CONNECTION. REFER TO "DIAGNOSTIC AIDS" ON FACING PAGE.

DIAGNOSTIC CHART DTC 32 — 1994 LUMINA

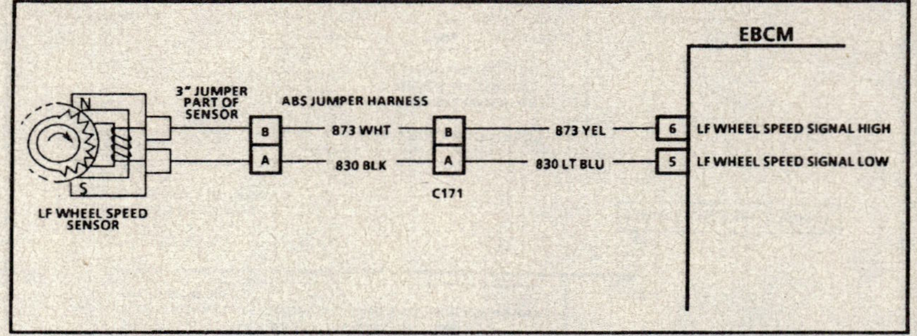

DTC 32

LEFT FRONT WHEEL SPEED SENSOR CKT SHORT TO GND/BAT

Circuit Description:

As a toothed ring passes by the wheel speed sensor, changes in the electromagnetic field cause the wheel speed sensor to produce a sinusoidal (AC) voltage signal whose frequency is proportional to wheel speed. The magnitude of this signal is directly related to wheel speed and the proximity of the wheel speed sensor to the toothed ring often referred to as the air gap.

DTC Will Set When: DTC 32 can be set anytime after initialization. If either of the left front wheel speed circuits are shorted to voltage or ground, a malfunction exists.

Action Taken: A malfunction DTC is stored, ABS is disabled and the ABS warning lamp is turned "ON."

DTC Chart Test Description: Number(s) below refer to circled number(s) on the diagnostic chart.

1. Checks for short to ground in wheel speed sensor wiring circuitry.
2. Checks for short to voltage in wheel speed sensor wiring circuitry.
3. Checks for short to voltage in CKT 830.
4. Checks for short to voltage in CKT 873.
5. Checks for short to ground in CKT 830.
6. Checks for short to ground in CKT 830.
7. Checks for short to ground in CKT 873.
8. Checks for short to ground in wheel speed sensor jumper harness.
9. Checks for proper resistance in wheel speed sensor.

Diagnostic Aids: An "Intermittent" malfunction may be caused by a poor connection, rubbed through wire insulation, or a wire that is broken inside the insulation.

The frequency of the malfunction can be checked by using the enhanced diagnostic function of the Tech 1, as described in "Tech 1 Diagnostics" found in this section.

If the customer's comments reflect that the ABS warning lamp is "ON" only during moist environmental changes (rain, snow, vehicle wash), all wheel speed sensor circuitry should be thoroughly inspected for signs of water intrusion. If DTC is not current, clear DTCs and simulate the effects of water intrusion. Use the following procedure Spray down the suspected area with a 5% salt water solution (two teaspoons of salt to 12 oz. of water). Test drive vehicle over various road surfaces (bumps, turns, etc.) above 24 km/h (15 mph) for at least 30 seconds. If DTC returns replace suspected harness.

Any circuitry, that is suspected as causing the intermittent complaint, should be thoroughly checked for backed out terminals, improper mating, broken locks, improperly formed or damaged terminals, poor terminal to wiring connections or physical damage to the wiring harness.

Resistance of the wheel speed sensor will increase with an increase in sensor temperature.

When replacing a wheel speed sensor, inspect the sensor terminals and harness connector for corrosion and/or water intrusion. If evidence of corrosion or water intrusion exists, replace wheel speed sensor jumper harness. Likewise, if replacing a wheel speed sensor jumper harness, inspect sensor terminals. If evidence of corrosion or water intrusion exists, replace wheel speed sensor. Refer to "On-Vehicle Service" in this section.

DIAGNOSTIC CHART DTC 32 — 1994 LUMINA, CONTINUED

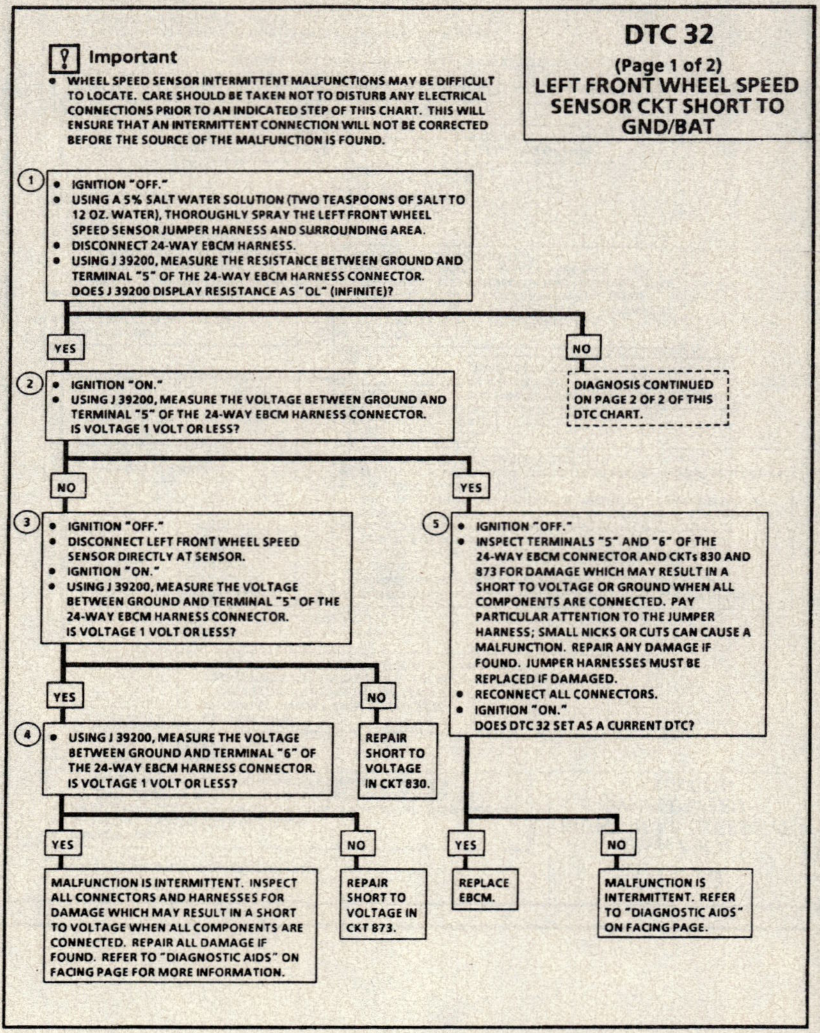

DIAGNOSTIC CHART DTC 32 — 1994 LUMINA, CONTINUED

DTC 32
(Page 2 of 2)
LEFT FRONT WHEEL SPEED SENSOR CKT SHORT TO GND/BAT

Important
- WHEEL SPEED SENSOR INTERMITTENT MALFUNCTIONS MAY BE DIFFICULT TO LOCATE. CARE SHOULD BE TAKEN NOT TO DISTURB ANY ELECTRICAL CONNECTIONS PRIOR TO AN INDICATED STEP OF THIS CHART. THIS WILL ENSURE THAT AN INTERMITTENT CONNECTION WILL NOT BE CORRECTED BEFORE THE SOURCE OF THE MALFUNCTION IS FOUND.

DIAGNOSIS CONTINUED FROM PAGE 1 OF 2 OF THIS DTC CHART.

(6)
- DISCONNECT CONNECTOR C171 (MOUNTED ON LH SHOCK TOWER).
- USING J 39200, MEASURE THE RESISTANCE BETWEEN GROUND AND TERMINAL "5", OF THE 24-WAY EBCM HARNESS CONNECTOR.

DOES J 39200 DISPLAY RESISTANCE AS "OL" (INFINITE)?

- NO: REPAIR SHORT TO GROUND IN CKT 830.
- YES:

(7)
- USING J 39200, MEASURE THE RESISTANCE BETWEEN GROUND AND TERMINAL "6" OF THE 24-WAY EBCM HARNESS CONNECTOR.

DOES J 39200 DISPLAY RESISTANCE AS "OL" (INFINITE)?

- NO: REPAIR SHORT TO GROUND IN CKT 873.
- YES:

(8)
- THOROUGHLY RESPRAY THE LEFT FRONT WHEEL SPEED SENSOR JUMPER HARNESS WITH THE SALT WATER SOLUTION. IMPORTANT: DO NOT ALLOW WATER TO ENTER ANY WHEEL SPEED SENSOR CONNECTOR.
- DISCONNECT LEFT FRONT WHEEL SPEED SENSOR DIRECTLY AT SENSOR.
- USING J 39200, MEASURE THE RESISTANCE BETWEEN GROUND AND TERMINAL "A", THEN "B" OF THE LEFT FRONT WHEEL SPEED SENSOR HARNESS CONNECTOR.

DOES J 39200 DISPLAY RESISTANCE AS "OL" (INFINITE) FOR BOTH MEASUREMENTS?

- NO: REPLACE LEFT FRONT WHEEL SPEED SENSOR JUMPER HARNESS.
- YES:

(9)
- USING J 39200, MEASURE THE RESISTANCE BETWEEN GROUND AND TERMINAL "A" OF THE LEFT FRONT WHEEL SPEED SENSOR CONNECTOR.

DOES J 39200 DISPLAY RESISTANCE AS "OL" (INFINITE)?

- NO: REPLACE WHEEL SPEED SENSOR.
- YES: MALFUNCTION IS INTERMITTENT. INSPECT ALL CONNECTORS AND HARNESSES FOR DAMAGE WHICH MAY RESULT IN A SHORT TO GROUND WHEN ALL COMPONENTS ARE CONNECTED. PAY PARTICULAR ATTENTION TO THE JUMPER HARNESS, SMALL NICKS OR CUTS CAN CAUSE A MALFUNCTION. REPAIR ANY DAMAGE IF FOUND. JUMPER HARNESS MUST BE REPLACED IF DAMAGED. REFER TO "DIAGNOSTIC AIDS" ON FACING PAGE FOR MORE INFORMATION.

DIAGNOSTIC CHART DTC 33 — 1994 LUMINA

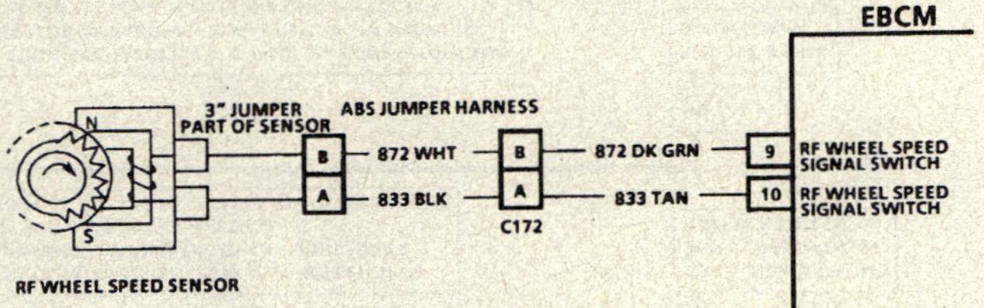

DTC 33

RIGHT FRONT WHEEL SPEED SENSOR CKT SHORT TO GND/BAT

Circuit Description:
As a toothed ring passes by the wheel speed sensor, changes in the electromagnetic field cause the wheel speed sensor to produce a sinusoidal (AC) voltage signal whose frequency is proportional to wheel speed. The magnitude of this signal is directly related to wheel speed and the proximity of the wheel speed sensor to the toothed ring, often referred to as the air gap.

DTC Will Set When: DTC 33 can be set anytime after initialization. If either of the right front wheel speed circuits are shorted to voltage or ground, a malfunction exists.

Action Taken: A malfunction DTC is stored, ABS is disabled and the ABS warning lamp is turned "ON."

DTC Chart Test Description: Number(s) below refer to circled number(s) on the diagnostic chart.

1. Checks for short to ground in wheel speed sensor wiring circuitry.
2. Checks for short to voltage in wheel speed sensor wiring circuitry.
3. Checks for short to voltage in CKT 872.
4. Checks for short to voltage in CKT 833.
5. Checks for short to ground or voltage in CKTs 872 or 833.
6. Checks for short to ground in CKT 872.
7. Checks for short to ground in CKT 833.
8. Checks for short to ground in wheel speed sensor jumper harness.
9. Checks for proper resistance in wheel speed sensor circuitry.

Diagnostic Aids: An "Intermittent" malfunction may be caused by a poor connection, rubbed through wire insulation, or a wire that is broken inside the insulation.

The frequency of the malfunction can be checked by using the enhanced diagnostic function of the Tech 1, as described in "Tech 1 Diagnostics" found in this section.

If the customer's comments reflect that the ABS warning lamp is "ON" only during moist environmental changes (rain, snow, vehicle wash), all wheel speed sensor circuitry should be thoroughly inspected for signs of water intrusion. If DTC is not current, clear DTCs and simulate the effects of water intrusion. Use the following procedure. Spray down the suspected area with a 5% salt water solution (two teaspoons of salt to 12 oz. of water). Test drive vehicle over various road surfaces (bumps, turns, etc.) above 24 km/h (15 mph) for at least 30 seconds. If DTC returns replace suspected harness.

Any circuitry, that is suspected as causing the intermittent complaint, should be thoroughly checked for backed out terminals, improper mating, broken locks, improperly formed or damaged terminals, poor terminal to wiring connections or physical damage to the wiring harness.

Resistance of the wheel speed sensor will increase with an increase in sensor temperature.

When replacing a wheel speed sensor, inspect the sensor terminals and harness connector for corrosion and/or water intrusion. If evidence of corrosion or water intrusion exists, replace wheel speed sensor jumper harness. Likewise, if replacing a wheel speed sensor jumper harness, inspect sensor terminals. If evidence of corrosion or water intrusion exists, replace wheel speed sensor. Refer to "On-Vehicle Service" in this section.

DIAGNOSTIC CHART DTC 33 — 1994 LUMINA, CONTINUED

DTC 33
(Page 1 of 2)
RIGHT FRONT WHEEL SPEED SENSOR CKT SHORT TO GND/BAT

Important
- WHEEL SPEED SENSOR INTERMITTENT MALFUNCTIONS MAY BE DIFFICULT TO LOCATE. CARE SHOULD BE TAKEN NOT TO DISTURB ANY ELECTRICAL CONNECTIONS PRIOR TO AN INDICATED STEP OF THIS CHART. THIS WILL ENSURE THAT AN INTERMITTENT CONNECTION WILL NOT BE CORRECTED BEFORE THE SOURCE OF THE MALFUNCTION IS FOUND.

(1)
- IGNITION "OFF."
- USING A 5% SALT WATER SOLUTION (TWO TEASPOONS OF SALT TO 12 OZ. WATER), THOROUGHLY SPRAY THE RIGHT FRONT WHEEL SPEED SENSOR JUMPER HARNESS AND SURROUNDING AREA.
- DISCONNECT 24-WAY EBCM HARNESS.
- USING J 39200, MEASURE THE RESISTANCE BETWEEN GROUND AND TERMINAL "9" OF THE 24-WAY EBCM HARNESS CONNECTOR.

DOES J 39200 DISPLAY RESISTANCE AS "OL" (INFINITE)?

- YES → go to (2)
- NO → DIAGNOSIS CONTINUED ON PAGE 2 OF 2 OF THIS DTC CHART.

(2)
- IGNITION "ON."
- USING J 39200, MEASURE THE VOLTAGE BETWEEN GROUND AND TERMINAL "9" OF THE 24-WAY EBCM HARNESS CONNECTOR.

IS VOLTAGE 1 VOLT OR LESS?

- NO → go to (3)
- YES → go to (5)

(3)
- IGNITION "OFF."
- DISCONNECT RIGHT FRONT WHEEL SPEED SENSOR DIRECTLY AT SENSOR.
- IGNITION "ON."
- USING J 39200, MEASURE THE VOLTAGE BETWEEN GROUND AND TERMINAL "9" OF THE 24-WAY EBCM HARNESS CONNECTOR.

IS VOLTAGE 1 VOLT OR LESS?

- YES → go to (4)
- NO → REPAIR SHORT TO VOLTAGE IN CKT 872.

(4)
- USING J 39200, MEASURE THE VOLTAGE BETWEEN GROUND AND TERMINAL "10" OF THE 24-WAY EBCM HARNESS CONNECTOR.

IS VOLTAGE 1 VOLT OR LESS?

- YES → MALFUNCTION IS INTERMITTENT. INSPECT ALL CONNECTORS AND HARNESSES FOR DAMAGE WHICH MAY RESULT IN A SHORT TO VOLTAGE WHEN ALL COMPONENTS ARE CONNECTED. REPAIR ALL DAMAGE IF FOUND. REFER TO "DIAGNOSTIC AIDS" ON FACING PAGE FOR MORE INFORMATION.
- NO → REPAIR SHORT TO VOLTAGE IN CKT 833.

(5)
- IGNITION "OFF."
- INSPECT TERMINALS "9" AND "10" OF THE 24-WAY EBCM CONNECTOR AND CKTs 872 AND 833 FOR DAMAGE WHICH MAY RESULT IN A SHORT TO VOLTAGE OR GROUND WHEN ALL COMPONENTS ARE CONNECTED. PAY PARTICULAR ATTENTION TO THE JUMPER HARNESS; SMALL NICKS OR CUTS CAN CAUSE A MALFUNCTION. REPAIR ANY DAMAGE IF FOUND. JUMPER HARNESSES MUST BE REPLACED IF DAMAGED.
- RECONNECT ALL CONNECTORS.
- IGNITION "ON."

DOES DTC 33 SET AS A CURRENT DTC?

- YES → REPLACE EBCM.
- NO → MALFUNCTION IS INTERMITTENT. REFER TO "DIAGNOSTIC AIDS" ON FACING PAGE.

DIAGNOSTIC CHART DTC 33 — 1994 LUMINA, CONTINUED

DTC 33
(Page 2 of 2)
RIGHT FRONT WHEEL SPEED SENSOR CKT SHORT TO GND/BAT

Important
- WHEEL SPEED SENSOR INTERMITTENT MALFUNCTIONS MAY BE DIFFICULT TO LOCATE. CARE SHOULD BE TAKEN NOT TO DISTURB ANY ELECTRICAL CONNECTIONS PRIOR TO AN INDICATED STEP OF THIS CHART. THIS WILL ENSURE THAT AN INTERMITTENT CONNECTION WILL NOT BE CORRECTED BEFORE THE SOURCE OF THE MALFUNCTION IS FOUND.

DIAGNOSIS CONTINUED FROM PAGE 1 OF 2 OF THIS DTC CHART.

(6)
- DISCONNECT CONNECTOR C172 (MOUNTED ON RH SHOCK TOWER).
- USING J 39200, MEASURE THE RESISTANCE BETWEEN GROUND AND TERMINAL "9", OF THE 24-WAY EBCM HARNESS CONNECTOR.

DOES J 39200 DISPLAY RESISTANCE AS "OL" (INFINITE)?

- YES → go to (7)
- NO → REPAIR SHORT TO GROUND IN CKT 872.

(7)
- USING J 39200, MEASURE THE RESISTANCE BETWEEN GROUND AND TERMINAL "10" OF THE 24-WAY EBCM HARNESS CONNECTOR.

DOES J 39200 DISPLAY RESISTANCE AS "OL" (INFINITE)?

- YES → go to (8)
- NO → REPAIR SHORT TO GROUND IN CKT 833.

(8)
- THOROUGHLY RESPRAY THE RIGHT FRONT WHEEL SPEED SENSOR JUMPER HARNESS WITH THE SALT WATER SOLUTION. IMPORTANT: DO NOT ALLOW WATER TO ENTER ANY WHEEL SPEED SENSOR CONNECTOR.
- DISCONNECT RIGHT FRONT WHEEL SPEED SENSOR DIRECTLY AT SENSOR.
- USING J 39200, MEASURE THE RESISTANCE BETWEEN GROUND AND TERMINAL "A", THEN "B" OF THE RIGHT FRONT WHEEL SPEED SENSOR HARNESS CONNECTOR.

DOES J 39200 DISPLAY RESISTANCE AS "OL" (INFINITE) FOR BOTH MEASUREMENTS?

- YES → go to (9)
- NO → REPLACE RIGHT FRONT WHEEL SPEED SENSOR JUMPER HARNESS.

(9)
- USING J 39200, MEASURE THE RESISTANCE BETWEEN GROUND AND TERMINAL "A" OF THE RIGHT FRONT WHEEL SPEED SENSOR CONNECTOR.

DOES J 39200 DISPLAY RESISTANCE AS "OL" (INFINITE)?

- YES → MALFUNCTION IS INTERMITTENT. INSPECT ALL CONNECTORS AND HARNESSES FOR DAMAGE WHICH MAY RESULT IN A SHORT TO GROUND WHEN ALL COMPONENTS ARE CONNECTED. PAY PARTICULAR ATTENTION TO THE JUMPER HARNESS, SMALL NICKS OR CUTS CAN CAUSE A MALFUNCTION. REPAIR ANY DAMAGE IF FOUND. JUMPER HARNESS MUST BE REPLACED IF DAMAGED. REFER TO "DIAGNOSTIC AIDS" ON FACING PAGE FOR MORE INFORMATION.
- NO → REPLACE WHEEL SPEED SENSOR.

DIAGNOSTIC CHART DTC 34 — 1994 LUMINA

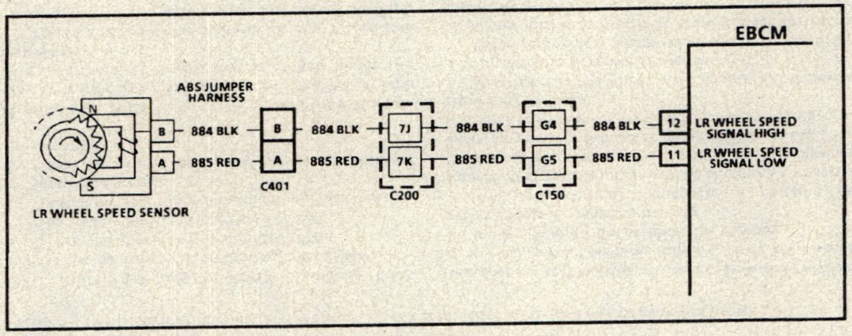

DTC 34

(Page 1 of 2)

LEFT REAR WHEEL SPEED SENSOR CKT SHORT TO GND/BAT

Circuit Description:

As a toothed ring passes by the wheel speed sensor, changes in the electromagnetic field cause the wheel speed sensor to produce a sinusoidal (AC) voltage signal whose frequency is proportional to wheel speed. The magnitude of this signal is directly related to wheel speed and the proximity of the wheel speed sensor to the toothed ring, often referred to as the air gap.

DTC Will Set When: DTC 34 can be set anytime after initialization. If either of the left rear wheel speed circuits are shorted to voltage or ground, a malfunction exists.

Action Taken: A malfunction DTC is stored, ABS is disabled and the ABS warning lamp is turned "ON."

DTC Chart Test Description: Number(s) below refer to circled number(s) on the diagnostic chart.

1. Checks for short to ground in wheel speed sensor wiring circuitry.
2. Checks for short to voltage in wheel speed sensor wiring circuitry.
3. Checks for short to voltage in CKT 885.
4. Checks for short to voltage in CKT 884.
5. Checks for short to ground or voltage in CKTs 885 or 884.

Diagnostic Aids: An "Intermittent" malfunction may be caused by a poor connection, rubbed through wire insulation, or a wire that is broken inside the insulation.

The frequency of the malfunction can be checked by using the enhanced diagnostic function of the Tech 1, as described in "Tech 1 Diagnostics" found in this section.

If the customer's comments reflect that the ABS warning lamp is "ON" only during moist environmental changes (rain, snow, vehicle wash), all wheel speed sensor circuitry should be thoroughly inspected for signs of water intrusion. If DTC is not current, clear DTCs and simulate the effects of water intrusion. Use the following procedure; spray down the suspected area with a 5% salt water solution (two teaspoons of salt to 12 oz. of water). Test drive vehicle over various road surfaces (bumps, turns, etc.) above 24 km/h (15 mph) for at least 30 seconds. If DTC returns replace suspected harness.

Any circuitry, that is suspected as causing the intermittent complaint, should be thoroughly checked for backed out terminals, improper mating, broken locks, improperly formed or damaged terminals, poor terminal to wiring connections or physical damage to the wiring harness.

Resistance of the wheel speed sensor will increase with an increase in sensor temperature.

When replacing a wheel speed sensor, inspect the sensor terminals and harness connector for corrosion and/or water intrusion. If evidence of corrosion or water intrusion exists, replace wheel speed sensor jumper harness. Likewise, if replacing a wheel speed sensor jumper harness, inspect sensor terminals. If evidence of corrosion or water intrusion exists, replace wheel speed sensor. Refer to "On-Vehicle Service" in this section.

DIAGNOSTIC CHART DTC 34 — 1994 LUMINA

DTC 34

(Page 1 of 2)

LEFT REAR WHEEL SPEED SENSOR CKT SHORT TO GND/BAT

Important

- WHEEL SPEED SENSOR INTERMITTENT MALFUNCTIONS MAY BE DIFFICULT TO LOCATE. CARE SHOULD BE TAKEN NOT TO DISTURB ANY ELECTRICAL CONNECTIONS PRIOR TO AN INDICATED STEP OF THIS CHART. THIS WILL ENSURE THAT AN INTERMITTENT CONNECTION WILL NOT BE CORRECTED BEFORE THE SOURCE OF THE MALFUNCTION IS FOUND.

(1)
- IGNITION "OFF."
- USING A 5% SALT WATER SOLUTION (TWO TEASPOONS OF SALT TO 12 OZ. WATER), THOROUGHLY SPRAY THE LEFT REAR WHEEL SPEED SENSOR JUMPER HARNESS AND SURROUNDING AREA.
- DISCONNECT 24-WAY EBCM HARNESS.
- USING J 39200, MEASURE THE RESISTANCE BETWEEN GROUND AND TERMINAL "11" OF THE 24-WAY EBCM HARNESS CONNECTOR.
- DOES J 39200 DISPLAY RESISTANCE AS "OL" (INFINITE)?

YES → (2)
NO → DIAGNOSIS CONTINUED ON PAGE 2 OF 2 OF THIS DTC CHART.

(2)
- IGNITION "ON."
- USING J 39200, MEASURE VOLTAGE BETWEEN GROUND AND TERMINAL "11" OF THE 24-WAY EBCM HARNESS CONNECTOR.
- IS VOLTAGE 1 VOLT OR LESS?

NO → (3)
YES → (5)

(3)
- IGNITION "OFF."
- DISCONNECT LEFT REAR WHEEL SPEED SENSOR DIRECTLY AT SENSOR.
- IGNITION "ON."
- USING J 39200, MEASURE THE VOLTAGE BETWEEN GROUND AND TERMINAL "11" OF THE 24-WAY EBCM HARNESS CONNECTOR.
- IS VOLTAGE 1 VOLT OR LESS?

YES → (4)
NO → REPAIR SHORT TO VOLTAGE IN CKT 885.

(4)
- USING J 39200, MEASURE THE VOLTAGE BETWEEN GROUND AND TERMINAL "12" OF THE 24-WAY EBCM HARNESS CONNECTOR.
- IS VOLTAGE 1 VOLT OR LESS?

YES → MALFUNCTION IS INTERMITTENT. INSPECT ALL CONNECTORS AND HARNESSES FOR DAMAGE WHICH MAY RESULT IN A SHORT TO VOLTAGE WHEN ALL COMPONENTS ARE CONNECTED. REPAIR ALL DAMAGE IF FOUND. REFER TO "DIAGNOSTIC AIDS" ON FACING PAGE FOR MORE INFORMATION.
NO → REPAIR SHORT TO VOLTAGE IN CKT 884.

(5)
- IGNITION "OFF."
- INSPECT TERMINALS "11" AND "12" OF THE 24-WAY EBCM CONNECTOR AND CKTs 884 AND 885 FOR DAMAGE WHICH MAY RESULT IN A SHORT TO VOLTAGE OR GROUND WHEN ALL COMPONENTS ARE CONNECTED. PAY PARTICULAR ATTENTION TO THE JUMPER HARNESS; SMALL NICKS OR CUTS CAN CAUSE A MALFUNCTION. REPAIR ANY DAMAGE IF FOUND. JUMPER HARNESSES MUST BE REPLACED IF DAMAGED.
- RECONNECT ALL CONNECTORS.
- IGNITION "ON."
- DOES DTC 34 SET AS A CURRENT DTC?

YES → REPLACE EBCM.
NO → MALFUNCTION IS INTERMITTENT. REFER TO "DIAGNOSTIC AIDS" ON FACING PAGE.

DIAGNOSTIC CHART DTC 34 — 1994 LUMINA

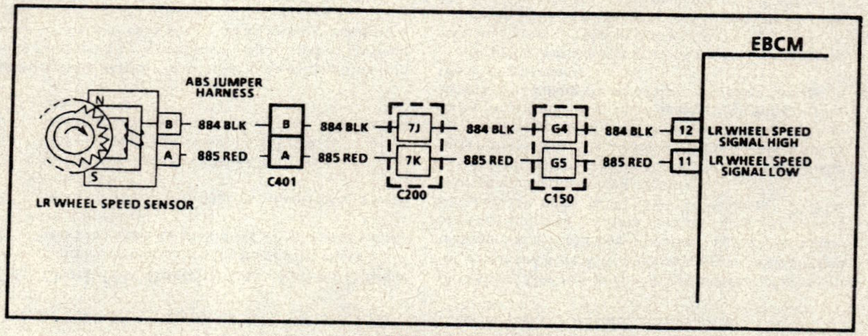

DTC 34
(Page 2 of 2)
LEFT REAR WHEEL SPEED SENSOR CKT SHORT TO GND/BAT

Circuit Description:
As a toothed ring passes by the wheel speed sensor, changes in the electromagnetic field cause the wheel speed sensor to produce a sinusoidal (AC) voltage signal whose frequency is proportional to wheel speed. The magnitude of this signal is directly related to wheel speed and the proximity of the wheel speed sensor to the toothed ring, often referred to as the air gap.

DTC Will Set When: DTC 34 can be set anytime after initialization. If either of the left rear wheel speed circuits are shorted to voltage or ground, a malfunction exists.

Action Taken: A malfunction DTC is stored, ABS is disabled and the ABS warning lamp is turned "ON."

DTC Chart Test Description: Number(s) below refer to circled number(s) on the diagnostic chart.
6. Checks for short to ground in CKT 885.
7. Checks for short to ground in CKT 884.
8. Checks for short to ground in wheel speed sensor jumper harness.
9. Checks for proper resistance in wheel speed sensor circuitry.

Diagnostic Aids: An "Intermittent" malfunction may be caused by a poor connection, rubbed through wire insulation, or a wire that is broken inside the insulation.

The frequency of the malfunction can be checked by using the enhanced diagnostic function of the Tech 1, as described in "Tech 1 Diagnostics" found in this section.

If the customer's comments reflect that the ABS warning lamp is "ON" only during moist environmental changes (rain, snow, vehicle wash), all wheel speed sensor circuitry should be thoroughly inspected for signs of water intrusion. If DTC is not current, clear DTCs and simulate the effects of water intrusion. Use the following procedure. Spray down the suspected area with a 5% salt water solution (two teaspoons of salt to 12 oz. of water). Test drive vehicle over various road surfaces (bumps, turns, etc.) above 24 km/h (15 mph) for at least 30 seconds. If DTC returns replace suspected harness.

Any circuitry, that is suspected as causing the intermittent complaint, should be thoroughly checked for backed out terminals, improper mating, broken locks, improperly formed or damaged terminals, poor terminal to wiring connections or physical damage to the wiring harness.

Resistance of the wheel speed sensor will increase with an increase in sensor temperature.

When replacing a wheel speed sensor, inspect the sensor terminals and harness connector for corrosion and/or water intrusion. If evidence of corrosion or water intrusion exists, replace wheel speed sensor jumper harness. Likewise, if replacing a wheel speed sensor jumper harness, inspect sensor terminals. If evidence of corrosion or water intrusion exists, replace wheel speed sensor. Refer to "On-Vehicle Service" in this section.

DIAGNOSTIC CHART DTC 34 — 1994 LUMINA

DTC 34
(Page 2 of 2)
LEFT REAR WHEEL SPEED SENSOR CKT SHORT TO GND/BAT

Important
- WHEEL SPEED SENSOR INTERMITTENT MALFUNCTIONS MAY BE DIFFICULT TO LOCATE. CARE SHOULD BE TAKEN NOT TO DISTURB ANY ELECTRICAL CONNECTIONS PRIOR TO AN INDICATED STEP OF THIS CHART. THIS WILL ENSURE THAT AN INTERMITTENT CONNECTION WILL NOT BE CORRECTED BEFORE THE SOURCE OF THE MALFUNCTION IS FOUND.

DIAGNOSIS CONTINUED FROM PAGE 1 OF 2 OF THIS DTC CHART.

(6)
- DISCONNECT CONNECTOR C371 (LOCATED BEHIND LH SIDE OF REAR SEAT).
- USING J 39200, MEASURE THE RESISTANCE BETWEEN GROUND AND TERMINAL "11", OF THE 24-WAY EBCM HARNESS CONNECTOR.

DOES J 39200 DISPLAY RESISTANCE AS "OL" (INFINITE)?

- NO → REPAIR SHORT TO GROUND IN CKT 885.
- YES ↓

(7)
- USING J 39200, MEASURE THE RESISTANCE BETWEEN GROUND AND TERMINAL "12" OF THE 24-WAY EBCM HARNESS CONNECTOR.

DOES J 39200 DISPLAY RESISTANCE AS "OL" (INFINITE)?

- NO → REPAIR SHORT TO GROUND IN CKT 884.
- YES ↓

(8)
- THOROUGHLY RESPRAY THE LEFT REAR WHEEL SPEED SENSOR JUMPER HARNESS WITH THE SALT WATER SOLUTION. IMPORTANT: DO NOT ALLOW WATER TO ENTER ANY WHEEL SPEED SENSOR CONNECTOR.
- DISCONNECT LEFT REAR WHEEL SPEED SENSOR DIRECTLY AT SENSOR.
- USING J 39200, MEASURE THE RESISTANCE BETWEEN GROUND AND TERMINAL "A", THEN "B" OF THE LEFT REAR WHEEL SPEED SENSOR HARNESS CONNECTOR.

DOES J 39200 DISPLAY RESISTANCE AS "OL" (INFINITE) FOR BOTH MEASUREMENTS?

- NO → REPLACE LEFT REAR WHEEL SPEED SENSOR JUMPER HARNESS.
- YES ↓

(9)
- USING J 39200, MEASURE THE RESISTANCE BETWEEN GROUND AND TERMINAL "A" OF THE LEFT REAR WHEEL SPEED SENSOR CONNECTOR.

DOES J 39200 DISPLAY RESISTANCE AS "OL" (INFINITE)?

- NO → REPLACE WHEEL SPEED SENSOR.
- YES → MALFUNCTION IS INTERMITTENT. INSPECT ALL CONNECTORS AND HARNESSES FOR DAMAGE WHICH MAY RESULT IN A SHORT TO GROUND WHEN ALL COMPONENTS ARE CONNECTED. PAY PARTICULAR ATTENTION TO THE JUMPER HARNESS. SMALL NICKS OR CUTS CAN CAUSE A MALFUNCTION. REPAIR ANY DAMAGE IF FOUND. JUMPER HARNESS MUST BE REPLACED IF DAMAGED. REFER TO "DIAGNOSTIC AIDS" ON FACING PAGE FOR MORE INFORMATION.

DIAGNOSTIC CHART DTC 35 — 1994 LUMINA

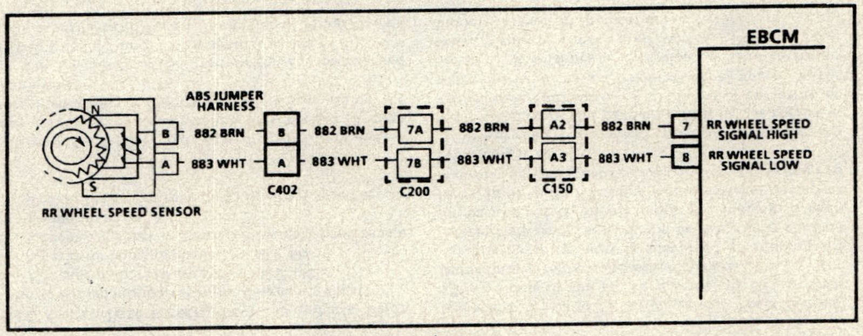

DTC 35
(Page 1 of 2)
RIGHT REAR WHEEL SPEED SENSOR CKT SHORT TO GND/BAT

Circuit Description:

As a toothed ring passes by the wheel speed sensor, changes in the electromagnetic field cause the wheel speed sensor to produce a sinusoidal (AC) voltage signal whose frequency is proportional to wheel speed. The magnitude of this signal is directly related to wheel speed and the proximity of the wheel speed sensor to the toothed ring, often referred to as the air gap.

DTC Will Set When: DTC 35 can be set anytime after initialization. If either of the right rear wheel speed circuits are shorted to voltage or ground, a malfunction exists.

Action Taken: A malfunction DTC is stored, ABS is disabled and the ABS warning lamp is turned "ON."

DTC Chart Test Description: Number(s) below refer to circled number(s) on the diagnostic chart.
1. Checks for short to ground in wheel speed sensor wiring circuitry.
2. Checks for short to voltage in wheel speed sensor wiring circuitry.
3. Checks for short to voltage in CKT 882.
4. Checks for short to voltage in CKT 883.
5. Checks for short to ground or voltage in CKTs 882 or 883.

Diagnostic Aids: An "Intermittent" malfunction may be caused by a poor connection, rubbed through wire insulation, or a wire that is broken inside the insulation.

The frequency of the malfunction can be checked by using the enhanced diagnostic function of the Tech 1, as described in "Tech 1 Diagnostics" found in this section.

If the customer's comments reflect that the ABS warning lamp is "ON" only during moist environmental changes (rain, snow, vehicle wash), all wheel speed sensor circuitry should be thoroughly inspected for signs of water intrusion. If DTC is not current, clear DTCs and simulate the effects of water intrusion. Use the following procedure; spray down the suspected area with a 5% salt water solution (two teaspoons of salt to 12 oz. of water). Test drive vehicle over various road surfaces (bumps, turns, etc.) above 24 km/h (15 mph) for at least 30 seconds. If DTC returns replace suspected harness.

Any circuitry, that is suspected as causing the intermittent complaint, should be thoroughly checked for backed out terminals, improper mating, broken locks, improperly formed or damaged terminals, poor terminal to wiring connections or physical damage to the wiring harness.

Resistance of the wheel speed sensor will increase with an increase in sensor temperature.

When replacing a wheel speed sensor, inspect the sensor terminals and harness connector for corrosion and/or water intrusion. If evidence of corrosion or water intrusion exists, replace wheel speed sensor jumper harness. Likewise, if replacing a wheel speed sensor jumper harness, inspect sensor terminals. If evidence of corrosion or water intrusion exists, replace wheel speed sensor. Refer to "On-Vehicle Service" in this section.

DIAGNOSTIC CHART DTC 35 — 1994 LUMINA

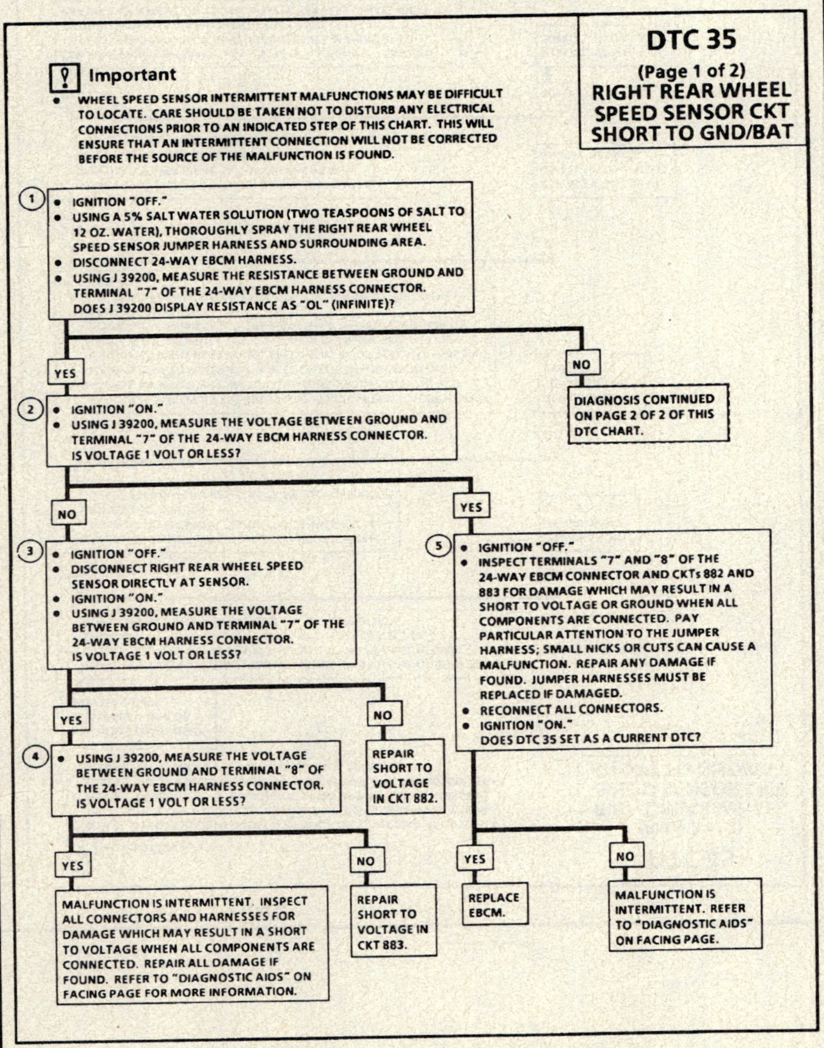

DIAGNOSTIC CHART DTC 35 — 1994 LUMINA

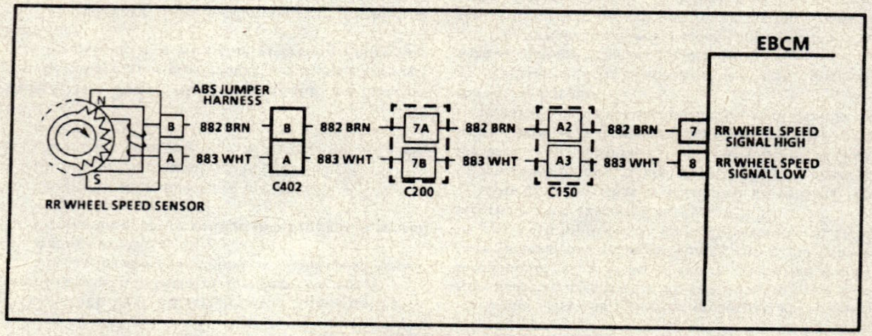

DTC 35

(Page 2 of 2)

RIGHT REAR WHEEL SPEED SENSOR CKT SHORT TO GND/BAT

Circuit Description:

As a toothed ring passes by the wheel speed sensor, changes in the electromagnetic field cause the wheel speed sensor to produce a sinusoidal (AC) voltage signal whose frequency is proportional to wheel speed. The magnitude of this signal is directly related to wheel speed and the proximity of the wheel speed sensor to the toothed ring, often referred to as the air gap.

DTC Will Set When: DTC 35 can be set anytime after initialization. If either of the right rear wheel speed circuits are shorted to voltage or ground, a malfunction exists.

Action Taken: A malfunction DTC is stored, ABS is disabled and the ABS warning lamp is turned "ON."

DTC Chart Test Description: Number(s) below refer to circled number(s) on the diagnostic chart.

6. Checks for short to ground in CKT 882.
7. Checks for short to ground in CKT 883.
8. Checks for short to ground in wheel speed sensor jumper harness.
9. Checks for proper resistance in wheel speed sensor circuitry.

Diagnostic Aids: An "Intermittent" malfunction may be caused by a poor connection, rubbed through wire insulation, or a wire that is broken inside the insulation.

The frequency of the malfunction can be checked by using the enhanced diagnostic function of the Tech 1, as described in "Tech 1 Diagnostics" found in this section.

If the customer's comments reflect that the ABS warning lamp is "ON" only during moist environmental changes (rain, snow, vehicle wash), all wheel speed sensor circuitry should be thoroughly inspected for signs of water intrusion. If DTC is not current, clear DTCs and simulate the effects of water intrusion. Use the following procedure. Spray down the suspected area with a 5% salt water solution (two teaspoons of salt to 12 oz. of water). Start vehicle, allow vehicle to run for 10 seconds. If DTC returns immediately, replace suspected harness.

Any circuitry, that is suspected as causing the intermittent complaint, should be thoroughly checked for backed out terminals, improper mating, broken locks, improperly formed or damaged terminals, poor terminal to wiring connections or physical damage to the wiring harness.

When replacing a wheel speed sensor, inspect the sensor terminals and harness connector for corrosion and/or water intrusion. If evidence of corrosion or water intrusion exists, replace wheel speed sensor harness. If evidence of corrosion or water intrusion exists, replace wheel speed sensor. Refer to "On-Vehicle Service" in this section.

DIAGNOSTIC CHART DTC 35 — 1994 LUMINA

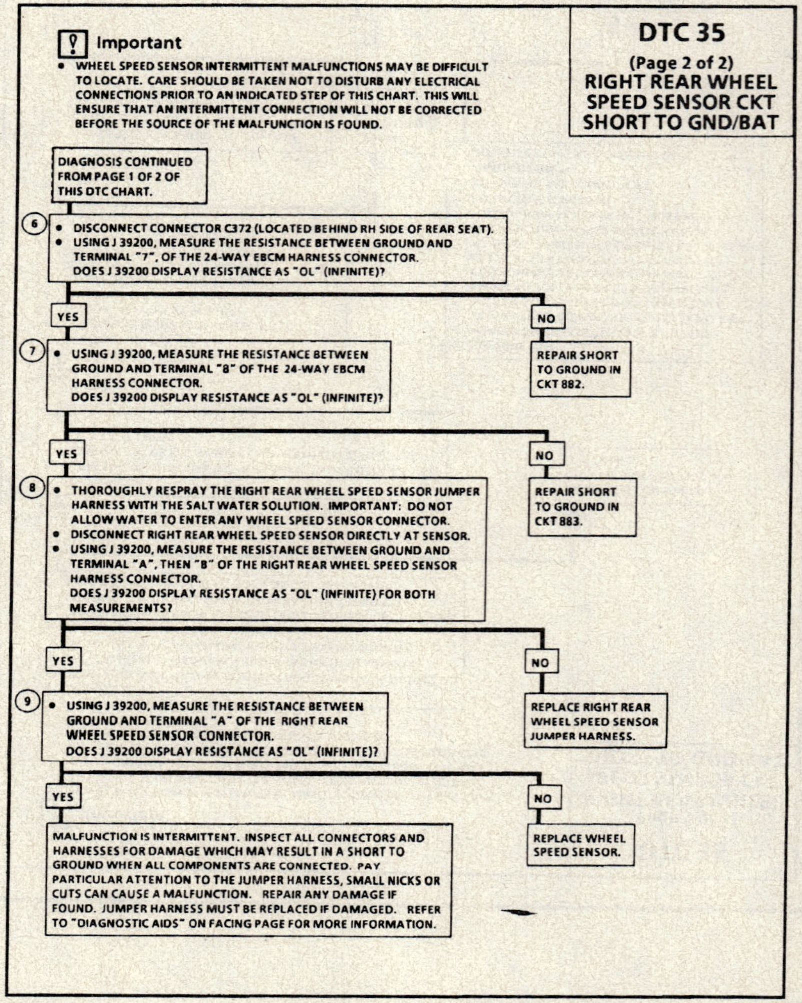

DIAGNOSTIC CHART DTC 36 — 1994 LUMINA

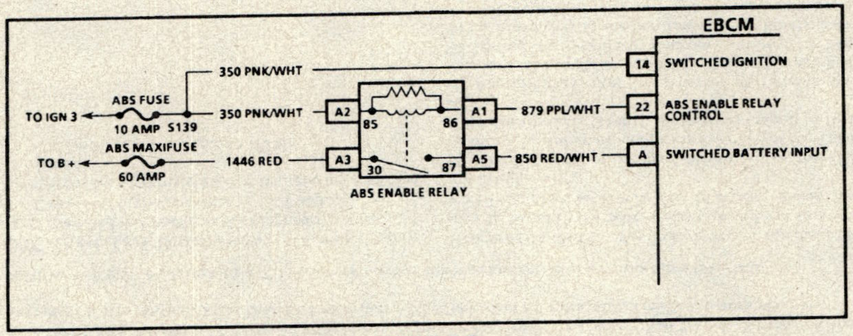

DTC 36

(Page 1 of 2)
LOW SYSTEM VOLTAGE

Circuit Description:
This DTC is used to monitor the voltage level available to the EBCM controller. If the voltage drops below 11 volts, full performance of the ABS system cannot be guaranteed. During ABS operation, there are several current requirements that will cause battery voltage to drop. Because of this, voltage is monitored prior to ABS to indicate good charging system condition and also during ABS when voltage may drop significantly.

DTC Will Set When: DTC 36 can only be set if the vehicle's speed is greater than 5 km/h (3 mph). If the switched battery voltage is less than 11.2 volts non-ABS or less than 9.6 volts ABS, a malfunction exists.

Action Taken: A malfunction DTC is stored, ABS is disabled and the ABS warning lamp is turned "ON."

DTC Chart Test Description: Number(s) below refer to circled number(s) on the diagnostic chart.

1. Checks to see what voltage is readily available at terminal "A" of the EBCM. If voltage is representative of good charging system condition, a malfunction is not present.
2. Isolates the condition to high circuit resistance, or improper charging system operation.
3. Isolates which circuit is the source of the high resistance condition.
4. Checks for high resistance in CKTs 1446 or 2.

Diagnostic Aids: An "Intermittent" malfunction may be caused by a poor connection, rubbed through wire insulation, or a wire that is broken inside the insulation.

The frequency of the malfunction can be checked by using the enhanced diagnostic function of the Tech 1, as described in "Tech 1 Diagnostics" found in this section.

Any circuitry, that is suspected as causing the intermittent complaint, should be thoroughly checked for backed out terminals, improper mating, broken locks, improperly formed or damaged terminals, poor terminal to wiring connections or physical damage to the wiring harness.

While performing a voltage load test, if it is noted that only ignition voltage drops below acceptable voltage levels, CKT 350 should be checked for high resistance or an open condition.

DIAGNOSTIC CHART DTC 36 — 1994 LUMINA

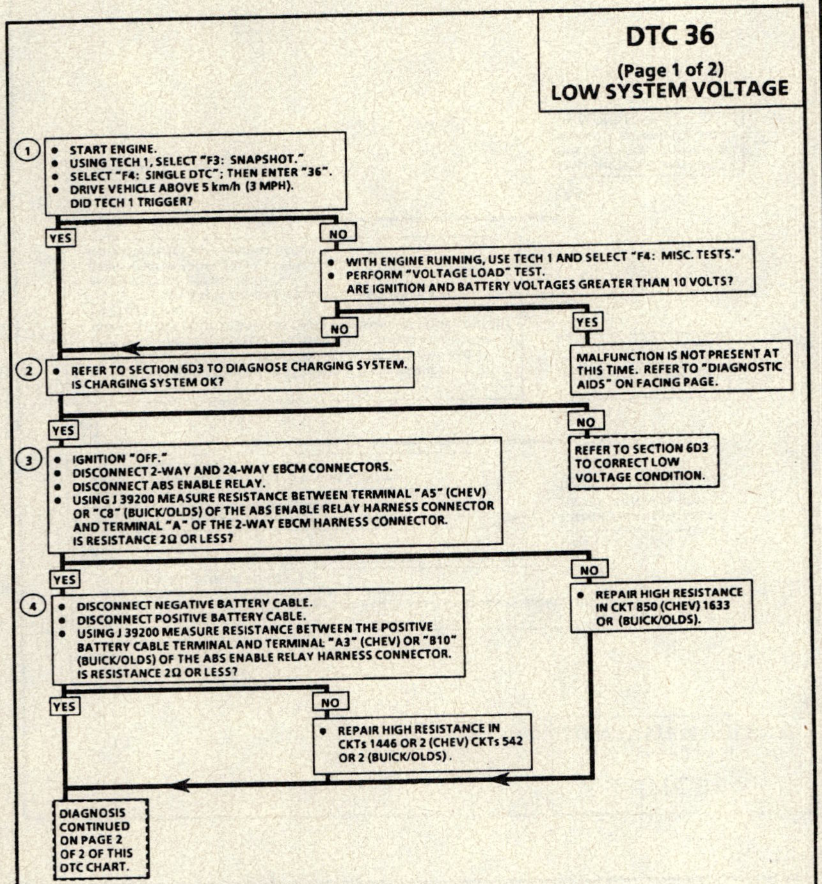

DIAGNOSTIC CHART DTC 36 — 1994 LUMINA

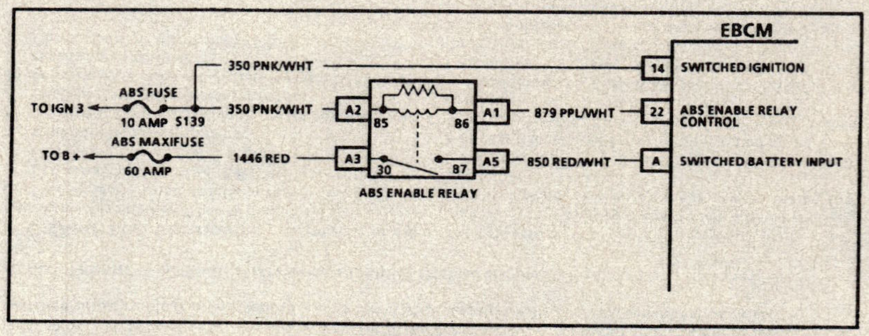

DTC 36
(Page 2 of 2)
LOW SYSTEM VOLTAGE

Circuit Description:
This DTC is used to monitor the voltage level available to the EBCM controller. If the voltage drops below 11 volts, full performance of the ABS system cannot be guaranteed. During ABS operation, there are several current requirements that will cause battery voltage to drop. Because of this, voltage is monitored prior to ABS to indicate good charging system condition and also during ABS when voltage may drop significantly.

DTC Will Set When: DTC 36 can only be set if the vehicle's speed is greater than 5 km/h (3 mph). If the switched battery voltage is less than 11.2 volts non-ABS or less than 9.6 volts ABS, a malfunction exists.

Action Taken: A malfunction DTC is stored, ABS is disabled and the ABS warning lamp is turned "ON."

DTC Chart Test Description: Number(s) below refer to circled number(s) on the diagnostic chart.

5. Checks for high resistance in CKT 350.
6. Checks for high resistance in the ignition feed fuse circuit.
7. Checks for malfunctioning EBCM.

Diagnostic Aids: An "Intermittent" malfunction may be caused by a poor connection, rubbed through wire insulation, or a wire that is broken inside the insulation.

The frequency of the malfunction can be checked by using the enhanced diagnostic function of the Tech 1, as described in "Tech 1 Diagnostics" found in this section.

Any circuitry, that is suspected as causing the intermittent complaint, should be thoroughly checked for backed out terminals, improper mating, broken locks, improperly formed or damaged terminals, poor terminal to wiring connections or physical damage to the wiring harness.

While performing a voltage load test, if it is noted that only ignition voltage drops below acceptable voltage levels, CKT 350 should be checked for high resistance or an open condition.

DIAGNOSTIC CHART DTC 36 — 1994 LUMINA

DTC 36
(Page 2 of 2)
LOW SYSTEM VOLTAGE

DIAGNOSIS CONTINUED FROM PAGE 1 OF 2 OF THIS DTC

(5)
- REMOVE ABS FUSE.
- USING J 39200 MEASURE RESISTANCE BETWEEN EACH ABS FUSE BLOCK TERMINAL AND TERMINAL "14" OF THE 24-WAY EBCM HARNESS CONNECTOR.
  IS EITHER MEASUREMENT 2Ω OR LESS?

YES → (6)
NO → REPAIR HIGH RESISTANCE IN CKT 350 (CHEV) OR 341 (BUICK/OLDS).

(6)
- WITH BATTERY CABLES DISCONNECTED, TURN IGNITION "ON" (THIS PROVIDES IGNITION CIRCUIT CONTINUITY).
- USING J 39200, MEASURE RESISTANCE BETWEEN POSITIVE BATTERY CABLE TERMINAL AND EACH ABS FUSE BLOCK TERMINAL.
  IS EITHER MEASUREMENT 2Ω OR LESS?

YES → (7)
NO → REPAIR HIGH RESISTANCE IN THE "IGNITION SWITCH 'RUN' FUSE FEED" CIRCUIT.

(7)
- IGNITION "OFF."
- INSPECT ABS ENABLE RELAY AND ABS ENABLE RELAY HARNESS CONNECTOR FOR POOR TERMINAL CONTACT. INSPECT 2-WAY AND 24-WAY EBCM HARNESS CONNECTORS FOR A POOR CONNECTION TO THE EBCM. INSPECT POSITIVE BATTERY CABLE TERMINAL FOR POOR CONTACT TO THE BATTERY.
  REPLACE TERMINALS IF POOR CONTACT IS EVIDENT.
- RECONNECT ALL CONNECTORS.
- DRIVE VEHICLE ABOVE 5 km/h (3 MPH).
  DOES DTC 36 SET AS A CURRENT DTC?

YES → REPLACE EBCM.
NO → MALFUNCTION IS MOST LIKELY DUE TO A POOR CONNECTION. REFER TO "DIAGNOSTIC AIDS" ON FACING PAGE.

DIAGNOSTIC CHART DTC 37 — 1994 LUMINA

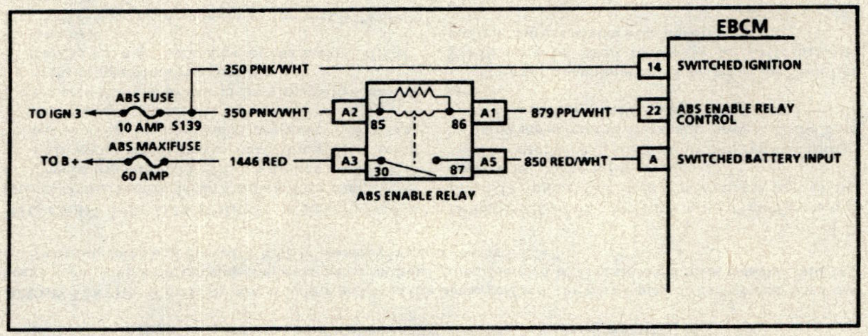

DTC 37
HIGH SYSTEM VOLTAGE

Circuit Description:
This DTC is designed to detect high vehicle voltage levels prior to any required motor movement (initialization or ABS). If excessive voltage exists, demagnetization of the motor magnets may occur, which would eventually affect or eliminate ABS performance.

DTC Will Set When: DTC 37 can only be set if the vehicle's speed is greater than 5 km/h (3 mph). If the switched battery voltage is greater than 17 volts, a malfunction exists.

Action Taken: A malfunction DTC is stored, ABS is disabled and the ABS warning lamp is turned "ON."

DTC Chart Test Description: Number(s) below refer to circled number(s) on the diagnostic chart.
1. Checks the voltage level being received by the EBCM.
2. Checks to see if high voltage condition is due to charging system malfunction.
3. Checks for poor terminal contact which may cause an intermittent malfunction or a malfunctioning EBCM.

Diagnostic Aids: An "Intermittent" malfunction may be caused by a poor connection, rubbed through wire insulation, or a wire that is broken inside the insulation.

The frequency of the malfunction can be checked by using the enhanced diagnostic function of the Tech 1, as described in "Tech 1 Diagnostics" found in this section.

Any circuitry, that is suspected as causing the intermittent complaint, should be thoroughly checked for backed out terminals, improper mating, broken locks, improperly formed or damaged terminals, poor terminal to wiring connections or physical damage to the wiring harness.

DIAGNOSTIC CHART DTC 37 — 1994 LUMINA

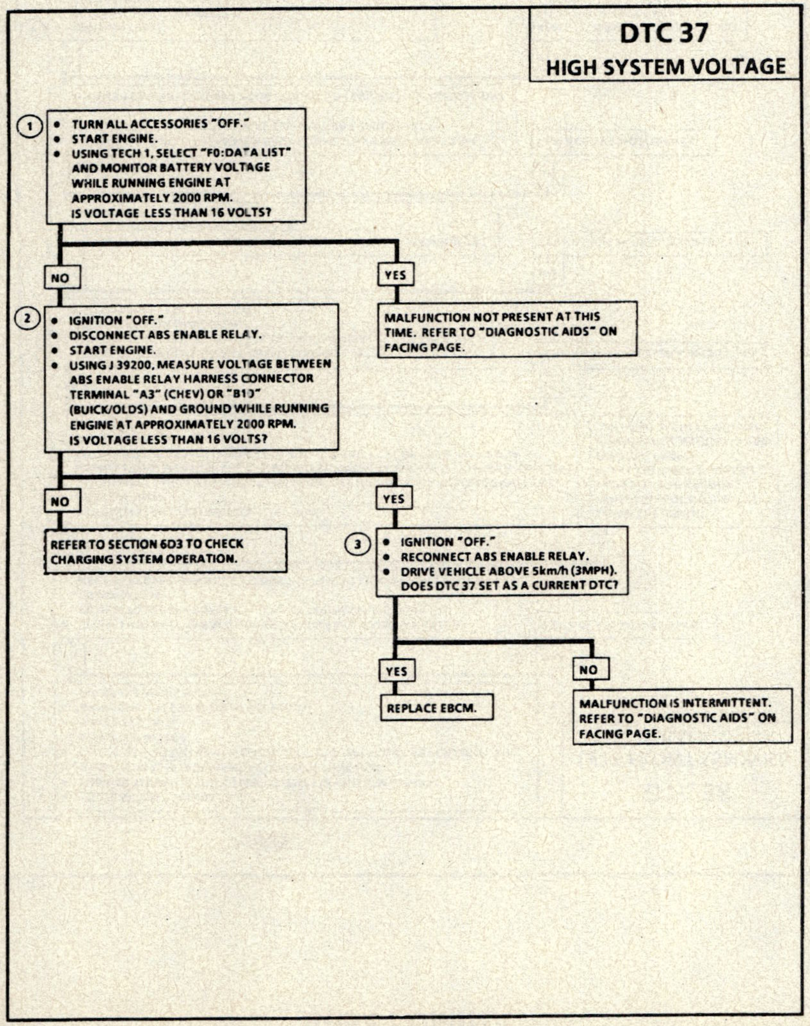

DIAGNOSTIC CHART DTC 38 — 1994 LUMINA

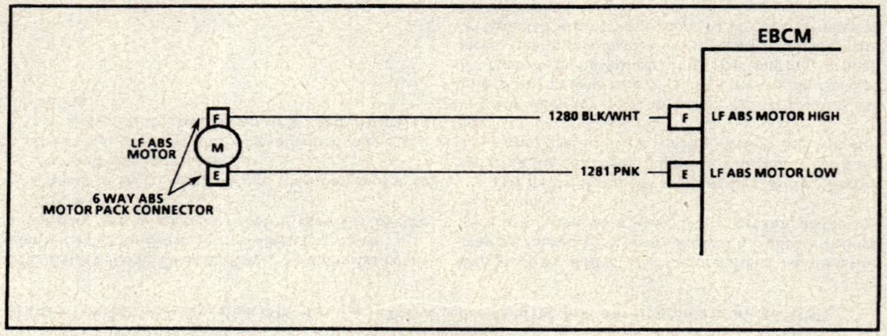

DTC 38

LEFT FRONT ESB WILL NOT HOLD MOTOR

Circuit Description:
This DTC is designed to detect a slipping left front ESB. During initialization and braking, the left front motor is rehomed. If the ESB slips, the motor/piston will move. During the next key "ON" initialization, a rehome of the motor verifies the motor/piston remained at the home position. If motor movement is detected, the ESB must be slipping.

DTC Will Set When: DTC 38 can be set during initialization. If the EBCM detects that the ESB could not hold the piston in the home position, a malfunction exists. DTC 86 is always set with DTC 38.

Action Taken: If an ESB cannot hold a piston in the home position, the piston may be backdriven when the brake pedal is applied, causing the brake pedal to drop. A malfunction DTC is stored, ABS is disabled, and the ABS warning lamp and the red "BRAKE" warning lamp are turned "ON."

DTC Chart Test Description: Number(s) below refer to circled number(s) on the diagnostic chart.

1. Checks the left front ESB. A broken or defective ESB would result in the left front piston being backdriven by hydraulic pressure and pedal movement would result.
2. Ensures that the proper motor circuits are located in the proper connector cavities.
3. Checks for high resistance in the motor "HIGH" circuitry.
4. Checks for high resistance in the motor "LOW" circuitry.
5. Checks for proper resistance of the motor windings.
6. Checks for poor terminal contact or corrosion in the connectors.
7. Checks for a malfunctioning EBCM.

Diagnostic Aids: An "Intermittent" malfunction in this DTC may result from a mechanical part of the system that sticks, binds, or slips.

The frequency of the malfunction can be checked by using the enhanced diagnostic function of the Tech 1, as described in "Tech 1 Diagnostics" found in this section.

The static modulator test function of the Tech 1 or T-100 may be used to locate an intermittent malfunction associated with the ESB.

DIAGNOSTIC CHART DTC 38 — 1994 LUMINA

DTC 38
LEFT FRONT ESB WILL NOT HOLD MOTOR

(1)
- IGNITION "ON," ENGINE "OFF."
- PUMP BRAKE PEDAL UNTIL FIRM TO DEPLETE VACUUM RESERVE.
- USING TECH 1, SELECT MANUAL CONTROL FUNCTION.
- SELECT ABS MOTOR APPLY FUNCTION OF THE LEFT FRONT ABS CHANNEL AND APPLY THE MOTOR.
- WAIT 5 SECONDS.
- APPLY FIRM PRESSURE ON BRAKE PEDAL.

DID BRAKE PEDAL FALL?

- YES: REPLACE ABS MOTOR PACK.
- NO: go to (2)

(2)
- VISUALLY INSPECT THE 6-WAY ABS MOTOR PACK AND 6-WAY EBCM CONNECTORS FOR PROPER WIRE COLOR/CONNECTOR CAVITY COMBINATION.

ARE THE PROPER WIRES LOCATED IN THE PROPER CONNECTOR CAVITIES?

- NO: USING THE CIRCUIT DIAGRAMS LOCATED IN THE FRONT OF THIS SECTION, PLACE THE PROPER WIRES INTO THE PROPER CONNECTOR CAVITIES. CLEAR DTCs AND RECHECK SYSTEM.
- YES: go to (3)

(3)
- EXIT TECH 1 MANUAL CONTROL FUNCTION.
- IGNITION "OFF."
- DISCONNECT 6-WAY ABS MOTOR PACK CONNECTOR AND 6-WAY EBCM CONNECTOR.
- USING J 39200, MEASURE THE RESISTANCE BETWEEN TERMINAL "F" OF THE 6-WAY ABS MOTOR PACK HARNESS CONNECTOR AND TERMINAL "F" OF THE 6-WAY EBCM HARNESS CONNECTOR.

IS RESISTANCE LESS THAN 1.5Ω?

- NO: REPAIR HIGH RESISTANCE IN CKT 1280.
- YES: go to (4)

(4)
- MEASURE THE RESISTANCE BETWEEN TERMINAL "E" OF THE 6-WAY ABS MOTOR PACK HARNESS CONNECTOR AND TERMINAL "E" OF THE 6-WAY EBCM HARNESS CONNECTOR.

IS RESISTANCE LESS THAN 1.5Ω?

- NO: REPAIR HIGH RESISTANCE IN CKT 1281.
- YES: go to (5)

(5)
- MEASURE THE RESISTANCE BETWEEN TERMINAL "E" AND TERMINAL "F" OF THE ABS MOTOR PACK CONNECTOR.

IS RESISTANCE LESS THAN 1.5Ω?

- NO: REPLACE ABS MOTOR PACK.
- YES: go to (6)

(6)
- INSPECT THE 6-WAY ABS MOTOR PACK CONNECTOR AND 6-WAY EBCM CONNECTOR FOR POOR TERMINAL CONTACT OR CORROSION.

DO ANY TERMINALS EXHIBIT POOR CONTACT OR EVIDENCE OF CORROSION?

- YES: REPLACE TERMINALS THAT EXHIBIT POOR TERMINAL CONTACT OR EVIDENCE OF CORROSION.
- NO: go to (7)

(7)
- IGNITION "OFF."
- RECONNECT ALL CONNECTORS.
- START ENGINE WITH FOOT OFF BRAKE.
- ALLOW ENGINE TO RUN FOR AT LEAST 10 SECONDS.
- REPEAT ABOVE IGNITION CYCLE SEQUENCE TWO MORE TIMES.

DID DTC 38 SET IN THE LAST THREE IGNITION CYCLES?

- YES: REPLACE EBCM.
- NO: MALFUNCTION IS INTERMITTENT. REFER TO "DIAGNOSTIC AIDS" ON FACING PAGE.

AFTER DIAGNOSIS IS COMPLETE, CLEAR DTCs AND TEST DRIVE VEHICLE FOR THREE (3) DRIVE CYCLES TO VERIFY THAT DTC DOES NOT RESET. A DRIVE CYCLE CONSISTS OF STARTING THE VEHICLE, DRIVING OVER 16 km/h (10 MPH) AND THEN KEYING DOWN.

DIAGNOSTIC CHART DTC 41 — 1994 LUMINA

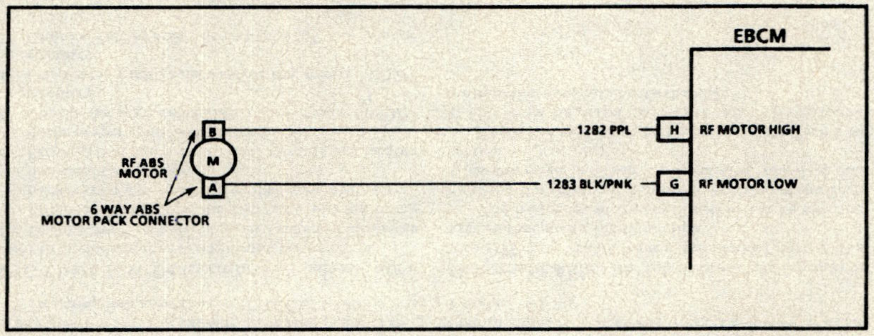

DTC 41

RIGHT FRONT ESB WILL NOT HOLD MOTOR

Circuit Description:
This DTC is designed to detect a slipping right front ESB. During initialization and braking, the right front motor is rehomed. If the ESB slips, the motor/piston will move. During the next key "ON" initialization, a rehome of the motor verifies the motor/piston remained at the home position. If motor movement is detected, the ESB must be slipping.

DTC Will Set When: DTC 41 can be set during initialization. If the EBCM detects that the ESB could not hold the piston in the home position, a malfunction exists. DTC 86 is always set with DTC 41.

Action Taken: If an ESB cannot hold a piston in the home position, the piston may be backdriven when the brake pedal is applied, causing the brake pedal to drop. A malfunction DTC is stored, ABS is disabled, and the ABS warning lamp and the red "BRAKE" warning lamp are turned "ON."

DTC Chart Test Description: Number(s) below refer to circled number(s) on the diagnostic chart.

1. Checks the right front ESB. A broken or defective ESB would result in the right front piston being backdriven by hydraulic pressure, and pedal movement would result.
2. Ensures that the proper motor circuits are located in the proper connector cavities.
3. Checks for high resistance in the motor "HIGH" circuitry.
4. Checks for high resistance in the motor "LOW" circuitry.
5. Checks for proper resistance of the motor windings.
6. Checks for poor terminal contact or corrosion in the connectors.
7. Checks for a malfunctioning EBCM.

Diagnostic Aids: An "Intermittent" malfunction in this DTC may result from a mechanical part of the system that sticks, binds, or slips.

The frequency of the malfunction can be checked by using the enhanced diagnostic function of the Tech 1, as described in "Tech 1 Diagnostics" found in this section.

The static modulator test function of the Tech 1 or T-100 may be used to locate an intermittent malfunction associated with the ESB.

DIAGNOSTIC CHART DTC 41 — 1994 LUMINA

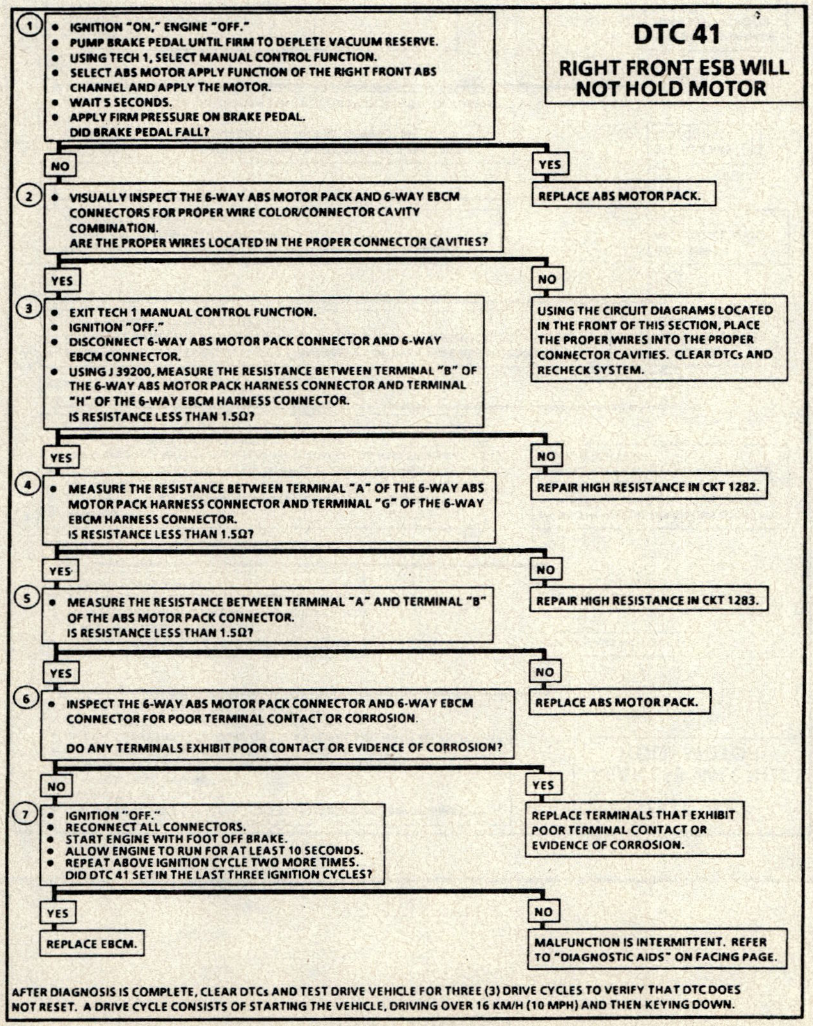

DIAGNOSTIC CHART DTC 42 — 1994 LUMINA

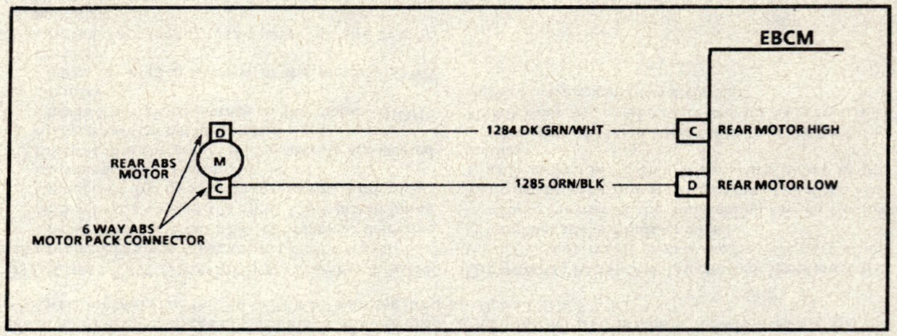

DTC 42
REAR ESB WILL NOT HOLD MOTOR

Circuit Description:
This DTC is designed to detect a slipping rear axle ESB. During initialization and braking, the rear motor is rehomed. If the ESB slips, the motor/piston will move. During the next key "ON" initialization, a rehome of the motor verifies the motor/piston remained at the home position. If motor movement is detected, the ESB must be slipping.

DTC Will Set When: DTC 42 can be set during initialization. If the EBCM detects that the ESB could not hold the piston in the home position, a malfunction exists. DTC 86 is always set with DTC 42.

Action Taken: If an ESB cannot hold a piston in the home position, the piston may be backdriven when the brake pedal is applied, causing the brake pedal to drop. A malfunction DTC is stored, ABS is disabled, and the ABS warning lamp and the red "BRAKE" warning lamp are turned "ON."

DTC Chart Test Description: Number(s) below refer to circled number(s) on the diagnostic chart.
1. Checks the rear axle ESB. A broken or defective ESB would result in the rear axle pistons being backdriven by hydraulic pressure and wheel movement would result.
2. Ensures that the proper motor circuits are located in the proper connector cavities.
3. Checks for high resistance in the motor "HIGH" circuitry.
4. Checks for high resistance in the motor "LOW" circuitry.
5. Checks for proper resistance of the motor windings.
6. Checks for poor terminal contact or corrosion in the connectors.
7. Checks for a malfunctioning EBCM.

Diagnostic Aids: An "Intermittent" malfunction in this DTC may result from a mechanical part of the system that sticks, binds, or slips.

The frequency of the malfunction can be checked by using the enhanced diagnostic function of the Tech 1, as described in "Tech 1 Diagnostics" found in this section.

The static modulator test function of the Tech 1 or T-100 may be used to locate an intermittent malfunction associated with the ESB.

DIAGNOSTIC CHART DTC 42 — 1994 LUMINA

DTC 42
REAR ESB WILL NOT HOLD MOTOR

(1)
- RAISE THE VEHICLE SUCH THAT THE REAR WHEELS ARE APPROXIMATELY SIX INCHES OFF THE FLOOR.
- IGNITION "ON," ENGINE "OFF."
- PUMP BRAKE PEDAL UNTIL FIRM TO DEPLETE VACUUM RESERVE.
- APPLY FIRM PRESSURE ON BRAKE PEDAL AND RELEASE.
- REAPPLY VERY FIRM PRESSURE ON BRAKE PEDAL AND HAVE AN ASSISTANT TRY TO SPIN THE REAR WHEELS BY HAND.

COULD ASSISTANT SPIN THE REAR WHEELS?

YES: REPLACE ABS MOTOR PACK.

NO:

(2)
- VISUALLY INSPECT THE 6-WAY ABS MOTOR PACK AND 6-WAY EBCM CONNECTORS FOR PROPER WIRE COLOR/CONNECTOR CAVITY COMBINATION.

ARE THE PROPER WIRES LOCATED IN THE PROPER CONNECTOR CAVITIES?

NO: USING THE CIRCUIT DIAGRAMS LOCATED IN THE FRONT OF THIS SECTION, PLACE THE PROPER WIRES INTO THE PROPER CONNECTOR CAVITIES. CLEAR DTCs AND RECHECK SYSTEM.

YES:

(3)
- IGNITION "OFF."
- DISCONNECT 6-WAY ABS MOTOR PACK CONNECTOR AND 6-WAY EBCM CONNECTOR.
- USING J 39200, MEASURE THE RESISTANCE BETWEEN TERMINAL "D" OF THE 6-WAY ABS MOTOR PACK HARNESS CONNECTOR AND TERMINAL "C" OF THE 6-WAY EBCM HARNESS CONNECTOR.

IS RESISTANCE LESS THAN 1.5Ω?

NO: REPAIR HIGH RESISTANCE IN CKT 1284.

YES:

(4)
- MEASURE THE RESISTANCE BETWEEN TERMINAL "C" OF THE 6-WAY ABS MOTOR PACK HARNESS CONNECTOR AND TERMINAL "D" OF THE 6-WAY EBCM HARNESS CONNECTOR.

IS RESISTANCE LESS THAN 1.5Ω?

NO: REPAIR HIGH RESISTANCE IN CKT 1285.

YES:

(5)
- MEASURE THE RESISTANCE BETWEEN TERMINAL "C" AND TERMINAL "D" OF THE ABS MOTOR PACK CONNECTOR.

IS RESISTANCE LESS THAN 1.5Ω?

NO: REPLACE ABS MOTOR PACK.

YES:

(6)
- INSPECT THE 6-WAY ABS MOTOR PACK CONNECTOR AND 6-WAY EBCM CONNECTOR FOR POOR TERMINAL CONTACT OR CORROSION.

DO ANY TERMINALS EXHIBIT POOR CONTACT OR EVIDENCE OF CORROSION?

YES: REPLACE TERMINALS THAT EXHIBIT POOR TERMINAL CONTACT OR EVIDENCE OF CORROSION.

NO:

(7)
- IGNITION "OFF."
- RECONNECT ALL CONNECTORS.
- START ENGINE WITH FOOT OFF BRAKE.
- ALLOW ENGINE TO RUN FOR AT LEAST 10 SECONDS.
- REPEAT ABOVE IGNITION CYCLE SEQUENCE TWO MORE TIMES.

DID DTC 42 SET IN THIS LAST THREE IGNITION CYCLES?

YES: REPLACE EBCM.

NO: MALFUNCTION IS INTERMITTENT. REFER TO "DIAGNOSTIC AIDS" ON FACING PAGE.

AFTER DIAGNOSIS IS COMPLETE, CLEAR DTCs AND TEST DRIVE VEHICLE FOR THREE (3) DRIVE CYCLES TO VERIFY DTC DOES NOT RESET. A DRIVE CYCLE CONSISTS OF STARTING THE VEHICLE, DRIVING OVER 16 km/h (10 MPH) AND THEN KEYING DOWN.

DIAGNOSTIC CHART DTC 44 — 1994 LUMINA

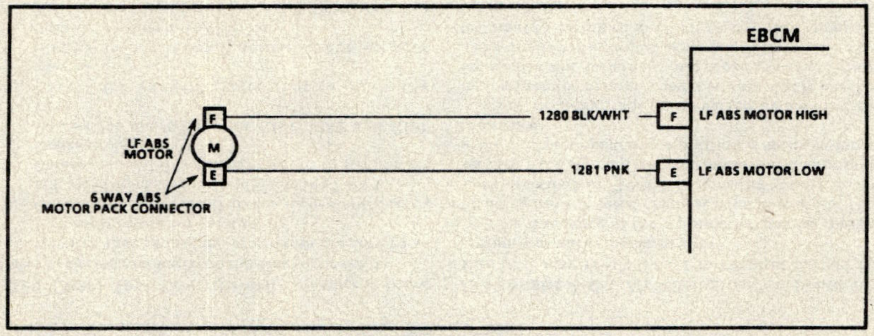

DTC 44

LEFT FRONT CHANNEL WILL NOT MOVE

Circuit Description:

This DTC is designed to detect bound-up ESB, a stuck ABS motor, or a seized hydraulic modulator. When the release is commanded during initialization, the ESB should release the ABS motor, resulting in sensed current being less than commanded current (motor is spinning freely). If the ABS motor is not moving, sensed current will be equal to stall current.

DTC Will Set When: DTC 44 can be set during normal operation. If the EBCM detects a condition in which it cannot move the ABS motor in either direction, a malfunction exists.

Action Taken: This malfunction indicates the channel cannot be moved properly. A malfunction DTC is stored, ABS is disabled and the ABS warning lamp is turned "ON."

DTC Chart Test Description: Number(s) below refer to circled number(s) on the diagnostic chart.

1. Ensures that proper motor circuits are located in the proper connector cavities.
2. Checks for proper ABS motor movement during apply and release commands from the Tech 1.
3. Compares EBCM command current to ABS motor feedback current.
4. Checks for short between CKTs 1280 and 1281.
5. Checks for an ABS motor that is internally shorted.
6. Determines if the malfunction is caused by a defective EBCM or by a short circuit.
7. Verifies that the ABS motor can actually apply when commanded.
8. Checks for proper hydraulic modulator gear and piston movement.

Diagnostic Aids: An "Intermittent" malfunction in this DTC may result from a mechanical part of the system that sticks, binds, or slips.

The frequency of the malfunction can be checked by using the enhanced diagnostic function of the Tech 1, as described in "Tech 1 Diagnostics" found in this section. DTC 44 may set after modulator disassembly if the modulator pistons are positioned at the bottom of their bore.

Any circuitry, that is suspected as causing the intermittent complaint, should be thoroughly checked for backed out terminals, improper mating, broken locks, improperly formed or damaged terminals, poor terminal to wiring connections or physical damage to the wiring harness.

DIAGNOSTIC CHART DTC 44 — 1994 LUMINA

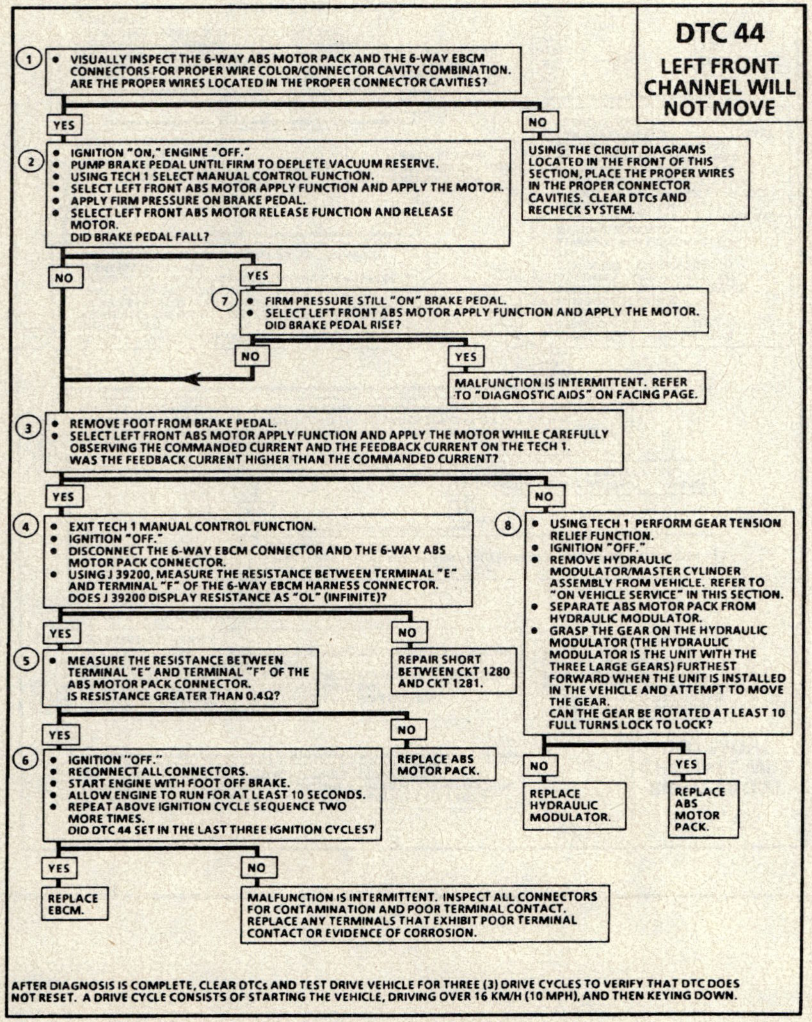

DIAGNOSTIC CHART DTC 45 — 1994 LUMINA

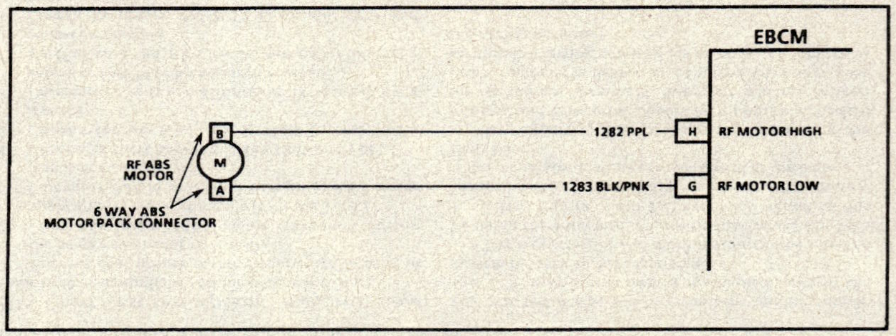

DTC 45

RIGHT FRONT CHANNEL WILL NOT MOVE

Circuit Description:

This DTC is designed to detect bound-up ESB, a stuck ABS motor, or a seized hydraulic modulator. When the release is commanded during initialization, the ESB should release the ABS motor, resulting in sensed current being less than commanded current (motor is spinning freely). If the ABS motor is not moving, sensed current will be equal to stall current.

DTC Will Set When: DTC 45 can be set during normal operation. If the EBCM detects a condition in which it cannot move the ABS motor in either direction, a malfunction exists.

Action Taken: This malfunction indicates the channel cannot be moved properly. A malfunction DTC is stored, ABS is disabled and the ABS warning lamp is turned "ON."

DTC Chart Test Description: Number(s) below refer to circled number(s) on the diagnostic chart.

1. Ensures that proper motor circuits are located in the proper connector cavities.
2. Checks for proper ABS motor movement during apply and release commands from the Tech 1.
3. Compares EBCM command current to ABS motor feedback current.
4. Checks for short between CKT 1282 and CKT 1283.
5. Checks for an ABS motor that is internally shorted.
6. Checks to see if malfunction is due to poor terminal contact or the EBCM.
7. Verifies that the motor can actually apply when commanded.
8. Checks for proper hydraulic modulator gear and piston movement.

Diagnostic Aids: An "Intermittent" malfunction in this DTC may result from a mechanical part of the system that sticks, binds, or slips.

The frequency of the malfunction can be checked by using the enhanced diagnostic function of the Tech 1, as described in "Tech 1 Diagnostics" found in this section. DTC 45 may set after modulator disassembly if the modulator pistons are positioned at the bottom of their bore.

Any circuitry, that is suspected as causing the intermittent complaint, should be thoroughly checked for backed out terminals, improper mating, broken locks, improperly formed or damaged terminals, poor terminal to wiring connections or physical damage to the wiring harness.

DIAGNOSTIC CHART DTC 45 — 1994 LUMINA

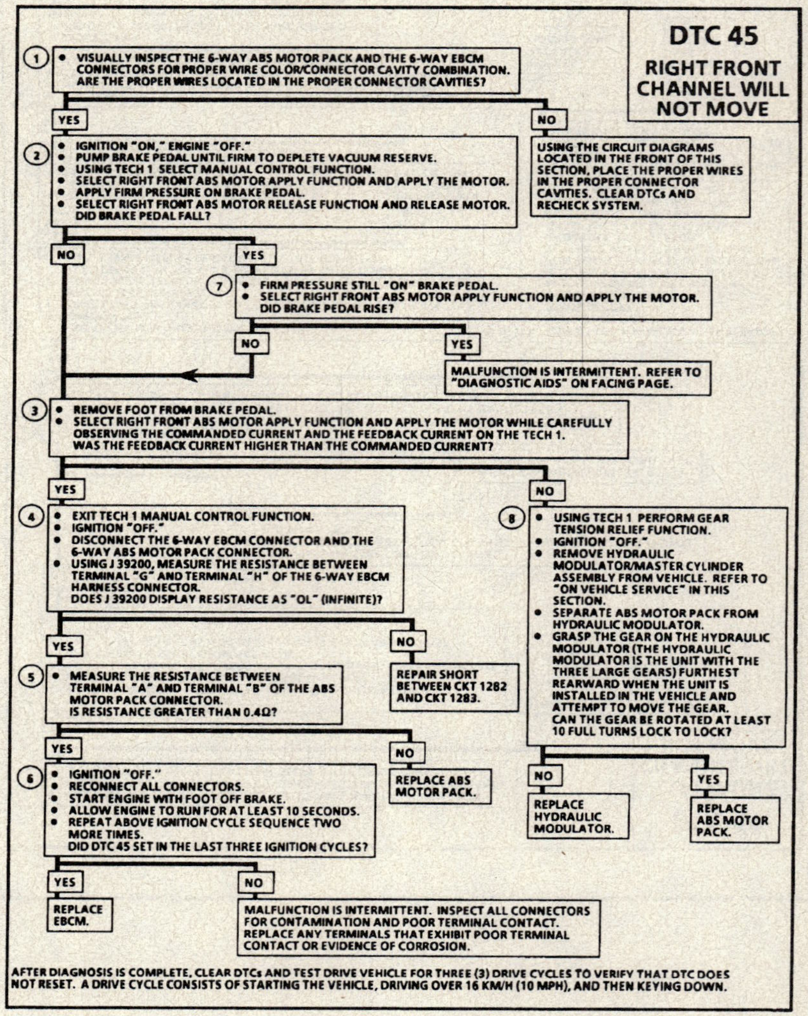

DIAGNOSTIC CHART DTC 46 — 1994 LUMINA

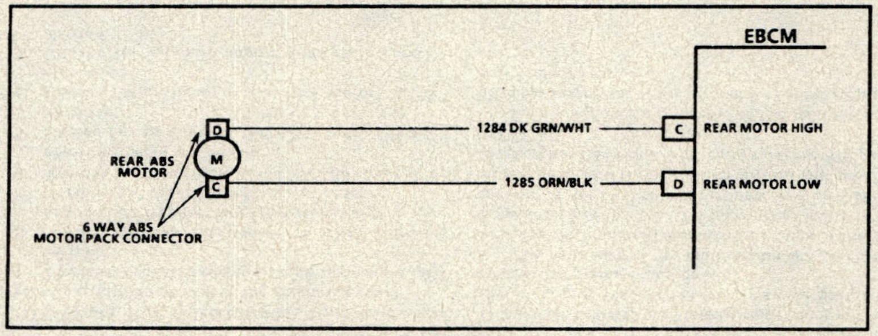

DTC 46

REAR CHANNEL WILL NOT MOVE

Circuit Description:

This DTC is designed to detect bound-up ESB, a stuck ABS motor, or a seized hydraulic modulator. When the release is commanded during initialization, the ESB should release the ABS motor, resulting in sensed current being less than commanded current (motor is spinning freely). If the ABS motor is not moving, sensed current will be equal to stall current.

DTC Will Set When: DTC 46 can be set during normal operation. If the EBCM detects a condition in which it cannot move the ABS motor in either direction, a malfunction exists.

Action Taken: This malfunction indicates the channel cannot be moved properly. A malfunction DTC is stored, ABS is disabled and the ABS warning lamp is turned "ON."

DTC Chart Test Description: Number(s) below refer to circled number(s) on the diagnostic chart.

1. Ensures that proper motor circuits are located in the proper connector cavities.
2. Checks for proper motor movement during apply and release commands from the Tech 1.
3. Compares EBCM command current to ABS motor feedback current.
4. Checks for short between CKT 1284 and CKT 1285.
5. Checks for an ABS motor that is internally shorted.
6. Checks to see if malfunction is due to poor terminal contact or the EBCM.
7. Verifies that the motor can actually release when commanded.
8. Checks for proper hydraulic modulator gear and piston movement.

Diagnostic Aids: An "Intermittent" malfunction in this DTC may result from a mechanical part of the system that sticks or binds.

The frequency of the malfunction can be checked by using the enhanced diagnostic function of the Tech 1, as described in "Tech 1 Diagnostics" found in this section. DTC 46 may set after modulator disassembly if the modulator pistons are positioned at the bottom of their bore.

Depending on the frequency of the malfunction, a physical inspection of the mechanical parts suspected may be necessary.

Any circuitry, that is suspected as causing the intermittent complaint, should be thoroughly checked for backed out terminals, improper mating, broken locks, improperly formed or damaged terminals, poor terminal to wiring connections or physical damage to the wiring harness.

DIAGNOSTIC CHART DTC 46 — 1994 LUMINA

DTC 46
REAR CHANNEL WILL NOT MOVE

(1)
- VISUALLY INSPECT THE 6-WAY ABS MOTOR PACK AND THE 6-WAY EBCM CONNECTORS FOR PROPER WIRE COLOR/CONNECTOR CAVITY COMBINATION.

ARE THE PROPER WIRES LOCATED IN THE PROPER CONNECTOR CAVITIES?

- NO: USING THE CIRCUIT DIAGRAMS LOCATED IN THE FRONT OF THIS SECTION, PLACE THE PROPER WIRES IN THE PROPER CONNECTOR CAVITIES. CLEAR DTCs AND RECHECK SYSTEM.
- YES:

(2)
- RAISE AND SUPPORT THE VEHICLE SUCH THAT THE REAR WHEELS ARE APPROXIMATELY SIX INCHES OFF THE FLOOR.
- IGNITION "ON," ENGINE "OFF."
- PUMP BRAKE PEDAL UNTIL FIRM TO DEPLETE VACUUM RESERVE.
- APPLY FIRM PRESSURE ON BRAKE PEDAL.
- USING TECH 1, SELECT MANUAL CONTROL FUNCTION.
- SELECT REAR AXLE ABS MOTOR APPLY FUNCTION AND APPLY THE MOTOR.
- HAVE AN ASSISTANT TRY TO SPIN THE REAR WHEELS BY HAND.

COULD ASSISTANT SPIN THE REAR WHEELS?

- YES: go to (3)
- NO:

(7)
- FIRM PRESSURE STILL "ON" BRAKE PEDAL.
- SELECT REAR AXLE ABS MOTOR RELEASE FUNCTION AND RELEASE THE MOTOR.
- HAVE AN ASSISTANT TRY TO SPIN THE REAR WHEELS BY HAND.

COULD ASSISTANT SPIN THE REAR WHEELS?

- YES: MALFUNCTION IS INTERMITTENT. REFER TO "DIAGNOSTIC AIDS" ON FACING PAGE.
- NO: go to (3)

(3)
- REMOVE FOOT FROM BRAKE PEDAL.
- SELECT REAR AXLE ABS MOTOR APPLY FUNCTION AND APPLY THE MOTOR WHILE CAREFULLY OBSERVING THE COMMANDED CURRENT AND THE FEEDBACK CURRENT ON THE TECH 1.

WAS THE FEEDBACK CURRENT HIGHER THAN THE COMMANDED CURRENT?

- YES: go to (4)
- NO: go to (8)

(4)
- EXIT TECH 1 MANUAL CONTROL FUNCTION.
- IGNITION "OFF."
- DISCONNECT THE 6-WAY EBCM CONNECTOR AND THE 6-WAY ABS MOTOR PACK CONNECTOR.
- USING J 39200, MEASURE THE RESISTANCE BETWEEN TERMINAL "C" AND TERMINAL "D" OF THE 6-WAY EBCM HARNESS CONNECTOR.

DOES J 39200 DISPLAY RESISTANCE AS "OL" (INFINITE)?

- NO: REPAIR SHORT BETWEEN CKT 1284 AND CKT 1285.
- YES:

(5)
- MEASURE THE RESISTANCE BETWEEN TERMINAL "C" AND TERMINAL "D" OF THE ABS MOTOR PACK CONNECTOR.

IS RESISTANCE GREATER THAN 0.4Ω?

- NO: REPLACE ABS MOTOR PACK.
- YES:

(6)
- IGNITION "OFF."
- RECONNECT ALL CONNECTORS.
- START ENGINE WITH FOOT OFF BRAKE.
- ALLOW ENGINE TO RUN FOR AT LEAST 10 SECONDS.
- REPEAT ABOVE IGNITION CYCLE SEQUENCE TWO MORE TIMES.

DID DTC 46 SET IN THE LAST THREE IGNITION CYCLES?

- YES: REPLACE EBCM.
- NO: MALFUNCTION IS INTERMITTENT. INSPECT ALL CONNECTORS FOR CONTAMINATION AND POOR TERMINAL CONTACT. REPLACE ANY TERMINALS THAT EXHIBIT POOR TERMINAL CONTACT OR EVIDENCE OF CORROSION.

(8)
- USING TECH 1, PERFORM GEAR TENSION RELIEF FUNCTION.
- IGNITION "OFF."
- REMOVE HYDRAULIC MODULATOR/MASTER CYLINDER ASSEMBLY FROM VEHICLE. REFER TO "ON VEHICLE SERVICE" IN THIS SECTION.
- SEPARATE ABS MOTOR PACK FROM HYDRAULIC MODULATOR.
- GRASP THE MIDDLE GEAR ON THE HYDRAULIC MODULATOR (THE HYDRAULIC MODULATOR IS THE UNIT WITH THE THREE LARGE GEARS) AND ATTEMPT TO MOVE THE GEAR.

CAN THE GEAR BE ROTATED AT LEAST 7 FULL TURNS LOCK TO LOCK?

- NO: REPLACE HYDRAULIC MODULATOR.
- YES: REPLACE ABS MOTOR PACK.

AFTER DIAGNOSIS IS COMPLETE, CLEAR DTCs AND TEST DRIVE VEHICLE FOR THREE (3) DRIVE CYCLES TO VERIFY THAT DTC DOES NOT RESET. A DRIVE CYCLE CONSISTS OF STARTING THE VEHICLE, DRIVING OVER 16 KM/H (10 MPH), AND THEN KEYING DOWN.

DIAGNOSTIC CHART DTC 47 — 1994 LUMINA

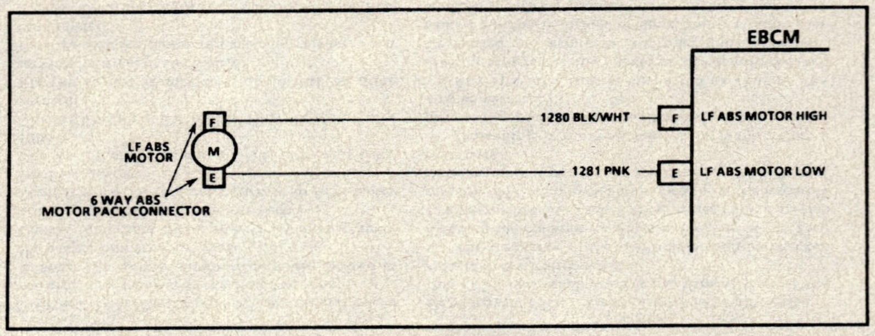

DTC 47

(Page 1 of 2)

LEFT FRONT ABS MOTOR FREE SPINS

Circuit Description:

This DTC is designed to detect a stripped nut or gear assembly during initialization. During the homing sequence, the piston should reach the top of the bore resulting in a stalled ABS motor. If this does not occur, the ABS motor must be spinning with little or no resistance, indicating a nut/screw or gear malfunction.

DTC Will Set When: DTC 47 can only be set during initialization. If the feedback current is less than the command current for a specified period of time, a malfunction exists.

Action Taken: A malfunction DTC is stored, ABS is disabled, and the ABS warning lamp is turned "ON."

DTC Chart Test Description: Number(s) below refer to circled number(s) on the diagnostic chart.

1. Checks to see if the corresponding open motor DTC 56 is also set.
2. Verifies that the ABS motor was actually applied as command by observing feedback current.
3. Verifies that the ABS motor can release.
4. Verifies that the ABS motor can be applied by observing pedal movement.
5. Checks for an open in the ABS motor "HIGH" circuitry.
6. Checks for an open in the ABS motor "LOW" circuitry.
7. Checks for an ABS motor that is internally shorted.

Diagnostic Aids: An "Intermittent" malfunction in this DTC may result from a mechanical part of the system that sticks, binds, or slips.

The frequency of the malfunction can be checked by using the enhanced diagnostic function of the Tech 1, as described in "Tech 1 Diagnostics" found in this section. If the DTC only occurs once and DTC 56 also occurs, refer to DTC 56. If intermittent and enhanced diagnostics shows this DTC occurs during ABS, refer to DTC 56.

Depending on the frequency of the malfunction, a physical inspection of the mechanical parts suspected may be necessary.

DIAGNOSTIC CHART DTC 47 — 1994 LUMINA

DTC 47

(Page 1 of 2)

LEFT FRONT ABS MOTOR FREE SPINS

(1) IS DTC 56 ALSO CURRENTLY SET?

- YES: PROCEED TO DTC 56 CHART FIRST.
- NO:

(2)
- IGNITION "ON," ENGINE "OFF."
- PUMP BRAKE PEDAL UNTIL FIRM TO DEPLETE VACUUM RESERVE.
- USING TECH 1, SELECT MANUAL CONTROL FUNCTION.
- SELECT LEFT FRONT ABS MOTOR APPLY FUNCTION AND APPLY THE MOTOR WHILE CAREFULLY OBSERVING THE COMMAND CURRENT AND FEEDBACK CURRENT ON THE TECH 1.

WERE BOTH COMMAND AND FEEDBACK CURRENT 10 AMPS?

- NO: GO TO STEP 5.
- YES:

(3)
- APPLY FIRM PRESSURE ON BRAKE PEDAL.
- SELECT LEFT FRONT ABS MOTOR RELEASE FUNCTION AND RELEASE THE MOTOR.

DID THE BRAKE PEDAL FALL AND WERE BOTH COMMAND AND FEEDBACK CURRENT 6 AMPS?

- NO: GO TO STEP 5.
- YES:

(4)
- FIRM PRESSURE STILL ON BRAKE PEDAL.
- SELECT LEFT FRONT ABS MOTOR APPLY FUNCTION AND APPLY THE MOTOR (THIS MOVES THE PISTON TO THE TOP OF ITS BORE).

DID BRAKE PEDAL RISE?

- YES: MALFUNCTION IS MOST LIKELY CAUSED BY POOR CONNECTION. REFER TO "DIAGNOSTIC AIDS" ON FACING PAGE.
- NO:

(5)
- USING TECH 1, PERFORM GEAR TENSION RELIEF FUNCTION.
- IGNITION "OFF."
- DISCONNECT 6-WAY ABS MOTOR PACK CONNECTOR AND 6-WAY EBCM CONNECTOR.
- USING J 39200, MEASURE THE RESISTANCE BETWEEN TERMINAL "F" OF THE 6-WAY ABS MOTOR PACK HARNESS CONNECTOR AND TERMINAL "F" OF THE 6-WAY EBCM HARNESS CONNECTOR.

IS RESISTANCE 1.5Ω OR LESS?

- NO: REPAIR HIGH RESISTANCE IN CKT 1280.
- YES:

(6)
- MEASURE THE RESISTANCE BETWEEN TERMINAL "E" OF THE 6-WAY ABS MOTOR PACK HARNESS CONNECTOR AND TERMINAL "E" OF THE 6-WAY EBCM HARNESS CONNECTOR.

IS RESISTANCE 1.5Ω OR LESS?

- NO: REPAIR HIGH RESISTANCE IN CKT 1281.
- YES:

(7)
- MEASURE THE RESISTANCE BETWEEN TERMINAL "E" AND TERMINAL "F" OF THE ABS MOTOR PACK CONNECTOR.

IS RESISTANCE 1.5Ω OR LESS?

- NO: REPLACE ABS MOTOR PACK.
- YES: DIAGNOSIS CONTINUED ON NEXT PAGE.

AFTER DIAGNOSIS IS COMPLETE, CLEAR DTCs AND TEST DRIVE VEHICLE FOR THREE (3) DRIVE CYCLES TO VERIFY THAT DTC DOES NOT RESET. A DRIVE CYCLE CONSISTS OF STARTING THE VEHICLE, DRIVING OVER 16 km/h (10 MPH), AND THEN KEYING DOWN.

DIAGNOSTIC CHART DTC 47 — 1994 LUMINA

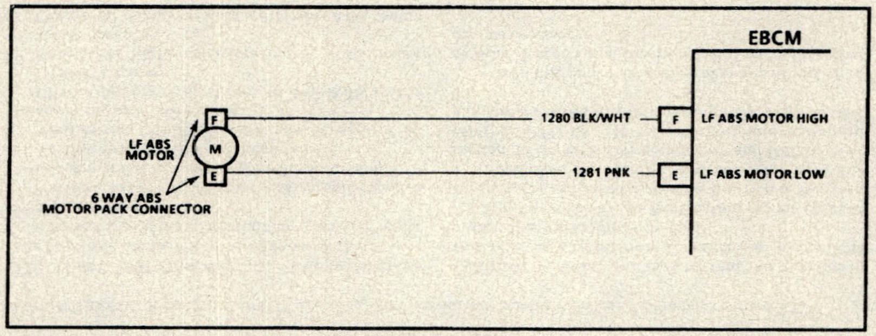

DTC 47
(Page 2 of 2)
LEFT FRONT ABS MOTOR FREE SPINS

Circuit Description:
This DTC is designed to detect a stripped nut or gear assembly during initialization. During the homing sequence, the piston should reach the top of the bore resulting in a stalled ABS motor. If this does not occur, the ABS motor must be spinning with little or no resistance, indicating a nut/screw or gear malfunction.

DTC Will Set When: DTC 47 can only be set during initialization. If the feedback current is less than the command current for a specified period of time, a malfunction exists.

Action Taken: A malfunction DTC is stored, ABS is disabled, and the ABS warning lamp is turned "ON."

DTC Chart Test Description: Number(s) below refer to circled number(s) on the diagnostic chart.

8. Checks for a stripped gear on the ABS motor pack (unit with three small gears).
9. Checks for a stripped gear on the hydraulic modulator (unit with three large gears).
10. Checks to see if the ABS motor pack has an internal malfunction.
11. Checks for a malfunctioning hydraulic modulator.
12. Ensures malfunction was not due to poor terminal contact.

Diagnostic Aids: An "Intermittent" malfunction in this DTC may result from a mechanical part of the system that sticks, binds, or slips.

The frequency of the malfunction can be checked by using the enhanced diagnostic function of the Tech 1, as described in "Tech 1 Diagnostics" found in this section. If the DTC only occurs once and DTC 56 also occurs, refer to DTC 56. If intermittent and enhanced diagnostics shows this DTC occurs during ABS, refer to DTC 56.

Depending on the frequency of the malfunction, a physical inspection of the mechanical parts suspected may be necessary.

DIAGNOSTIC CHART DTC 47 — 1994 LUMINA

DTC 47
(Page 2 of 2)
LEFT FRONT ABS MOTOR FREE SPINS

DIAGNOSIS CONTINUED FROM PREVIOUS PAGE.

(8)
- REMOVE HYDRAULIC MODULATOR/MASTER CYLINDER ASSEMBLY FROM VEHICLE (REFER TO "ON-VEHICLE SERVICE" IN THIS SECTION).
- REMOVE GEAR COVER.
- CHECK FOR STRIPPED GEAR ON ABS MOTOR PACK; ABS MOTOR PACK IS THE UNIT WITH THE THREE SMALL GEARS. THE LEFT FRONT GEAR IS THE GEAR FURTHEST FORWARD WHEN THE ASSEMBLY IS INSTALLED IN THE VEHICLE. IS THE ABS MOTOR PACK GEAR STRIPPED?

YES: REPLACE ABS MOTOR PACK.

NO:

(9)
- CHECK FOR A STRIPPED GEAR ON HYDRAULIC MODULATOR; MODULATOR IS THE UNIT WITH THE THREE LARGE GEARS. THE LEFT FRONT GEAR IS THE GEAR FURTHEST FORWARD WHEN THE UNIT IS INSTALLED IN THE VEHICLE. IS THE MODULATOR GEAR STRIPPED?

YES: REPLACE HYDRAULIC MODULATOR GEAR.

NO:

(10)
- RECONNECT ELECTRICAL CONNECTORS
- SECURELY POSITION HYDRAULIC MODULATOR/MASTER CYLINDER ASSEMBLY IN VEHICLE WITH GEAR COVER REMOVED SUCH THAT THE GEAR SET CAN BE OBSERVED. BE CAREFUL NOT TO DAMAGE GEAR SET OR ALLOW THE HYDRAULIC MODULATOR TO MOVE WHILE TESTING.
- IGNITION "ON."
- USING TECH 1 SELECT MANUAL CONTROL FUNCTION.
- SELECT LEFT FRONT ABS MOTOR APPLY AND APPLY THE MOTOR WHILE OBSERVING THE GEAR SET FURTHEST FORWARD WHEN THE UNIT IS INSTALLED IN THE VEHICLE.
- SELECT LEFT FRONT ABS MOTOR RELEASE FUNCTION AND RELEASE THE MOTOR WHILE STILL OBSERVING THE GEAR SET.
- SELECT LEFT FRONT ABS MOTOR APPLY FUNCTION AND APPLY THE MOTOR. DID GEAR SET MOVE IN BOTH DIRECTIONS FOR AT LEAST ONE REVOLUTION?

NO: REPLACE ABS MOTOR PACK.

YES:

(11)
- USING TECH 1 PERFORM GEAR TENSION RELIEF FUNCTION.
- IGNITION "OFF."
- SEPARATE THE ABS MOTOR PACK FROM THE HYDRAULIC MODULATOR.
- GRASP THE GEAR ON THE HYDRAULIC MODULATOR (THE HYDRAULIC MODULATOR IS THE UNIT WITH THE THREE LARGE GEARS) FURTHEST FORWARD WHEN THE UNIT IS INSTALLED IN THE VEHICLE AND ATTEMPT TO MOVE THE GEAR. CAN THE GEAR BE ROTATED MORE THAN 13 FULL TURNS LOCK TO LOCK?

YES: REPLACE THE HYDRAULIC MODULATOR.

NO:

(12)
- INSPECT ALL CONNECTORS AND TERMINALS FOR POOR TERMINAL CONTACT AND EVIDENCE OF CORROSION.
  REPLACE ALL TERMINALS THAT EXHIBIT SIGNS OF POOR TERMINAL CONTACT OR CORROSION.
- REINSTALL HYDRAULIC MODULATOR/MASTER CYLINDER ASSEMBLY IN VEHICLE.
- RECONNECT ALL CONNECTORS.
- IGNITION "OFF."
- START ENGINE WITH FOOT OFF BRAKE.
- ALLOW ENGINE TO RUN FOR AT LEAST 10 SECONDS.
- REPEAT ABOVE IGNITION CYCLE SEQUENCE TWO MORE TIMES. DID DTC 47 SET IN THE LAST THREE IGNITION CYCLES?

YES: REPLACE EBCM.

NO: MALFUNCTION IS MOST LIKELY CAUSED BY A POOR CONNECTION. REFER TO "DIAGNOSTIC AIDS" ON FACING PAGE.

AFTER DIAGNOSIS IS COMPLETE, CLEAR DTCs AND TEST DRIVE VEHICLE FOR THREE (3) DRIVE CYCLES TO VERIFY THAT DTC DOES NOT RESET. A DRIVE CYCLE CONSISTS OF STARTING THE VEHICLE, DRIVING OVER 16 km/h (10 MPH), AND THEN KEYING DOWN.

DIAGNOSTIC CHART DTC 48 — 1994 LUMINA

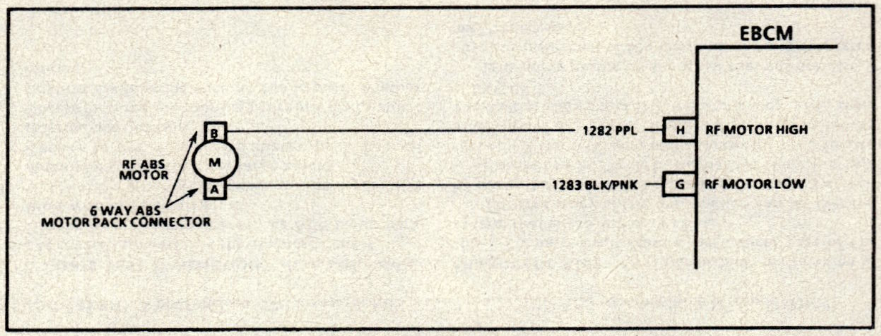

DTC 48
(Page 1 of 2)
RIGHT FRONT ABS MOTOR FREE SPINS

Circuit Description:
This DTC is designed to detect a stripped nut or gear assembly during initialization. During the homing sequence, the piston should reach the top of the bore resulting in a stalled ABS motor. If this does not occur, the ABS motor must be spinning with little or no resistance, indicating a nut/screw or gear malfunction.

DTC Will Set When: DTC 48 can only be set during initialization. If the feedback current is less than the command current for a specified period of time, a malfunction exists.

Action Taken: A malfunction DTC is stored, ABS is disabled, and the ABS warning lamp is turned "ON."

DTC Chart Test Description: Number(s) below refer to circled number(s) on the diagnostic chart.

1. Checks to see if the corresponding open ABS motor DTC is also set.
2. Verifies that the ABS motor was actually applied as commanded by observing feedback current.
3. Verifies that the ABS motor can release.
4. Verifies that the ABS motor can be applied by observing pedal movement.
5. Checks for high resistance in the ABS motor "HIGH" circuitry.
6. Checks for high resistance in the ABS motor "LOW" circuitry.
7. Checks for proper resistance of the ABS motor windings.

Diagnostic Aids: An "Intermittent" malfunction in this DTC may result from a mechanical part of the system that sticks, binds, or slips.

The frequency of the malfunction can be checked by using the enhanced diagnostic function of the Tech 1, as described in "Tech 1 Diagnostics" found in this section. If the DTC only occurs once and DTC 61 also, refer to DTC 61. If intermittent and enhanced diagnostics shows this DTC during ABS, refer to DTC 61.

Depending on the frequency of the malfunction, a physical inspection of the mechanical parts suspected may be necessary.

DIAGNOSTIC CHART DTC 48 — 1994 LUMINA

DTC 48
(Page 1 of 2)
RIGHT FRONT ABS MOTOR FREE SPINS

(1) IS DTC 61 ALSO CURRENTLY SET?
- YES: PROCEED TO DTC 61 CHART FIRST.
- NO: go to (2)

(2)
- IGNITION "ON," ENGINE "OFF."
- PUMP BRAKE PEDAL UNTIL FIRM TO DEPLETE VACUUM RESERVE.
- USING TECH 1 SELECT MANUAL CONTROL FUNCTION.
- SELECT RIGHT FRONT ABS MOTOR APPLY FUNCTION AND APPLY THE MOTOR WHILE CAREFULLY OBSERVING THE COMMAND CURRENT AND FEEDBACK CURRENT ON THE TECH 1. WERE BOTH COMMAND AND FEEDBACK CURRENT 10 AMPS?

YES: go to (3) · NO: GO TO STEP 5.

(3)
- APPLY FIRM PRESSURE ON BRAKE PEDAL.
- SELECT RIGHT FRONT ABS MOTOR RELEASE FUNCTION AND RELEASE THE MOTOR. DID THE BRAKE PEDAL FALL AND WERE BOTH COMMAND AND FEEDBACK CURRENT 6 AMPS?

YES: go to (4) · NO: GO TO STEP 5.

(4)
- FIRM PRESSURE STILL ON BRAKE PEDAL.
- SELECT RIGHT FRONT ABS MOTOR APPLY FUNCTION AND APPLY THE MOTOR (THIS MOVES THE PISTON TO THE TOP OF ITS BORE). DID BRAKE PEDAL RISE?

- YES: MALFUNCTION IS INTERMITTENT. REFER TO "DIAGNOSTIC AIDS" ON FACING PAGE.
- NO: go to (5)

(5)
- USING TECH 1 PERFORM GEAR TENSION RELIEF FUNCTION.
- IGNITION "OFF."
- DISCONNECT 6-WAY ABS MOTOR PACK CONNECTOR AND 6-WAY EBCM CONNECTOR.
- USING J 39200, MEASURE THE RESISTANCE BETWEEN TERMINAL "B" OF THE 6-WAY ABS MOTOR PACK HARNESS CONNECTOR AND TERMINAL "H" OF THE 6-WAY EBCM HARNESS CONNECTOR. IS RESISTANCE LESS THAN 1.5Ω?

- YES: go to (6)
- NO: REPAIR HIGH RESISTANCE IN CKT 1282.

(6)
- MEASURE THE RESISTANCE BETWEEN TERMINAL "A" OF THE 6-WAY ABS MOTOR PACK HARNESS CONNECTOR AND TERMINAL "G" OF THE 6-WAY EBCM HARNESS CONNECTOR. IS RESISTANCE LESS THAN 1.5Ω?

- YES: go to (7)
- NO: REPAIR HIGH RESISTANCE IN CKT 1283.

(7)
- MEASURE THE RESISTANCE BETWEEN TERMINAL "A" AND TERMINAL "B" OF THE ABS MOTOR PACK CONNECTOR. IS RESISTANCE LESS THAN 1.5Ω?

- YES: DIAGNOSIS CONTINUED ON NEXT PAGE.
- NO: REPLACE ABS MOTOR PACK.

AFTER DIAGNOSIS IS COMPLETE, CLEAR DTCs AND TEST DRIVE VEHICLE FOR THREE (3) DRIVE CYCLES TO VERIFY THAT DTC DOES NOT RESET. A DRIVE CYCLE CONSISTS OF STARTING THE VEHICLE, DRIVING OVER 16 km/h (10 MPH), AND THEN KEYING DOWN.

DIAGNOSTIC CHART DTC 48 — 1994 LUMINA

DTC 48
(Page 2 of 2)
RIGHT FRONT ABS MOTOR FREE SPINS

Circuit Description:
This DTC is designed to detect a stripped nut or gear assembly during initialization. During the homing sequence, the piston should reach the top of the bore resulting in a stalled ABS motor. If this does not occur, the ABS motor must be spinning with little or no resistance, indicating a nut/screw or gear malfunction.

DTC Will Set When: DTC 48 can only be set during initialization. If the feedback current is less than the command current for a specified period of time, a malfunction exists.

Action Taken: A malfunction DTC is stored, ABS is disabled, and the ABS warning lamp is turned "ON."

DTC Chart Test Description: Number(s) below refer to circled number(s) on the diagnostic chart.

8. Checks for a stripped gear on the ABS motor pack (unit with three small gears).
9. Checks for a stripped gear on the hydraulic modulator (unit with three large gears).
10. Verifies that the ABS motor can be applied under a load situation.
11. Verifies that the hydraulic modulator is functioning properly.
12. Ensures malfunction was not due to poor terminal contact.

Diagnostic Aids: An "Intermittent" malfunction in this DTC may result from a mechanical part of the system that sticks, binds, or slips.

The frequency of the malfunction can be checked by using the enhanced diagnostic function of the Tech 1, as described in "Tech 1 Diagnostics" found in this section. If the DTC only occurs once and DTC 61 also occurs, refer to DTC 61. If intermittent and enhanced diagnostics shows this DTC occurs during ABS, refer to DTC 61.

Depending on the frequency of the malfunction, a physical inspection of the mechanical parts suspected may be necessary.

DIAGNOSTIC CHART DTC 48 — 1994 LUMINA

DTC 48
(Page 2 of 2)
RIGHT FRONT ABS MOTOR FREE SPINS

DIAGNOSIS CONTINUED FROM PREVIOUS PAGE.

(8)
- REMOVE HYDRAULIC MODULATOR/MASTER CYLINDER ASSEMBLY FROM VEHICLE.
- REMOVE GEAR COVER.
- CHECK FOR STRIPPED GEAR ON ABS MOTOR PACK; ABS MOTOR PACK IS THE UNIT WITH THE THREE SMALL GEARS. THE RIGHT FRONT GEAR IS THE GEAR FURTHEST REARWARD WHEN THE ASSEMBLY IS INSTALLED IN THE VEHICLE. IS THE ABS MOTOR PACK GEAR STRIPPED?

YES: REPLACE ABS MOTOR PACK.

NO:

(9)
- CHECK FOR A STRIPPED GEAR ON HYDRAULIC MODULATOR; MODULATOR IS THE UNIT WITH THE THREE LARGE GEARS. THE RIGHT FRONT GEAR IS THE GEAR FURTHEST REARWARD WHEN THE UNIT IS INSTALLED IN THE VEHICLE. IS THE MODULATOR GEAR STRIPPED?

YES: REPLACE HYDRAULIC MODULATOR GEAR.

NO:

(10)
- RECONNECT ELECTRICAL CONNECTORS.
- SECURELY POSITION HYDRAULIC MODULATOR/MASTER CYLINDER ASSEMBLY IN VEHICLE WITH GEAR COVER REMOVED SUCH THAT THE GEAR SET CAN BE OBSERVED. BE CAREFUL NOT TO DAMAGE GEAR SET OR ALLOW THE HYDRAULIC MODULATOR TO MOVE WHILE TESTING.
- IGNITION "ON."
- USING TECH 1, SELECT MANUAL CONTROL FUNCTION.
- SELECT RIGHT FRONT ABS MOTOR APPLY AND APPLY THE MOTOR WHILE OBSERVING THE GEAR SET FURTHEST REARWARD WHEN THE UNIT IS INSTALLED IN THE VEHICLE.
- SELECT RIGHT FRONT ABS MOTOR RELEASE FUNCTION AND RELEASE THE MOTOR WHILE STILL OBSERVING THE GEAR SET.
- SELECT RIGHT FRONT ABS MOTOR APPLY FUNCTION AND APPLY THE MOTOR. DID GEAR SET MOVE IN BOTH DIRECTIONS FOR AT LEAST ONE REVOLUTION?

NO: REPLACE ABS MOTOR PACK.

YES:

(11)
- USING TECH 1, PERFORM GEAR TENSION RELIEF FUNCTION.
- IGNITION "OFF."
- SEPARATE THE ABS MOTOR PACK FROM THE HYDRAULIC MODULATOR.
- GRASP THE GEAR ON THE HYDRAULIC MODULATOR (THE HYDRAULIC MODULATOR IS THE UNIT WITH THE THREE LARGE GEARS) FURTHEST REARWARD WHEN THE UNIT IS INSTALLED IN THE VEHICLE AND ATTEMPT TO MOVE THE GEAR. CAN THE GEAR BE ROTATED MORE THAN 13 FULL TURNS LOCK TO LOCK?

YES: REPLACE THE HYDRAULIC MODULATOR.

NO:

(12)
- INSPECT ALL CONNECTORS AND TERMINALS FOR POOR TERMINAL CONTACT AND EVIDENCE OF CORROSION.
  REPLACE ALL TERMINALS THAT EXHIBIT SIGNS OF POOR TERMINAL CONTACT OR CORROSION.
- REINSTALL HYDRAULIC MODULATOR/MASTER CYLINDER ASSEMBLY IN VEHICLE.
- RECONNECT ALL CONNECTORS.
- IGNITION "OFF."
- START ENGINE WITH FOOT OFF BRAKE.
- ALLOW ENGINE TO RUN FOR AT LEAST 10 SECONDS.
- REPEAT ABOVE IGNITION CYCLE SEQUENCE TWO MORE TIMES. DID DTC 48 SET IN THE LAST THREE IGNITION CYCLES?

YES: REPLACE EBCM.

NO: MALFUNCTION WAS MOST LIKELY CAUSED BY A POOR CONNECTION. REFER TO "DIAGNOSTIC AIDS" ON FACING PAGE.

AFTER DIAGNOSIS IS COMPLETE, CLEAR DTCs AND TEST DRIVE VEHICLE FOR THREE (3) DRIVE CYCLES TO VERIFY THAT DTC DOES NOT RESET. A DRIVE CYCLE CONSISTS OF STARTING THE VEHICLE, DRIVING OVER 16 km/h (10 MPH), AND THEN KEYING DOWN.

DIAGNOSTIC CHART DTC 51 — 1994 LUMINA

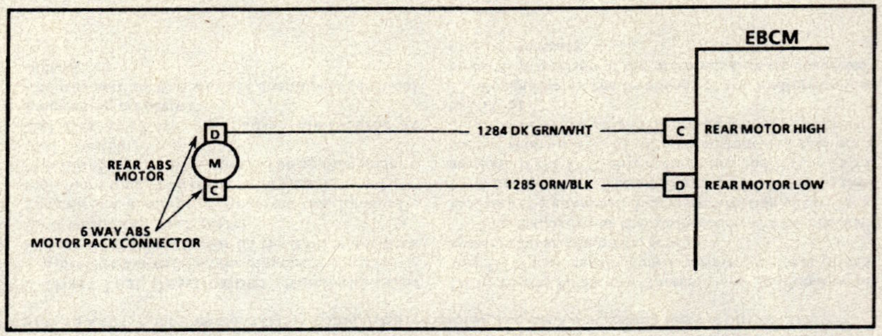

DTC 51
(Page 1 of 2)
REAR ABS MOTOR FREE SPINS

Circuit Description:

This DTC is designed to detect a stripped nut or gear assembly during initialization. During the homing sequence, the piston should reach the top of the bore resulting in a stalled ABS motor. If this does not occur, the ABS motor must be spinning with little or no resistance, indicating a nut/screw or gear malfunction.

DTC Will Set When: DTC 51 can be set during initialization. If the feedback current is less than the command current for a specified period of time, a malfunction exists.

Action Taken: A malfunction DTC is stored, ABS is disabled, and the ABS warning lamp is turned "ON."

DTC Chart Test Description: Number(s) below refer to circled number(s) on the diagnostic chart.

1. Checks to see if the corresponding open ABS motor DTC 64 is also set.
2. Verifies that the ABS motor was actually applied as commanded by observing feedback current.
3. Verifies that the ABS motor can release.
4. Verifies that the ABS motor can be applied by observing pedal movement.
5. Checks for high resistance in the ABS motor "HIGH" circuitry.
6. Checks for high resistance in the ABS motor "LOW" circuitry.
7. Checks for proper resistance of the ABS motor windings.

Diagnostic Aids: An "Intermittent" malfunction in this DTC may result from a mechanical part of the system that sticks, binds, or slips.

The frequency of the malfunction can be checked by using the enhanced diagnostic function of the Tech 1, as described in "Tech 1 Diagnostics" found in this section. If the DTC only occurs once and DTC 64 also occurs, refer to DTC 64. If intermittent and enhanced diagnostics show this DTC occurs during ABS, refer to DTC 64.

Depending on the frequency of the malfunction, a physical inspection of the mechanical parts suspected may be necessary.

DIAGNOSTIC CHART DTC 51 — 1994 LUMINA

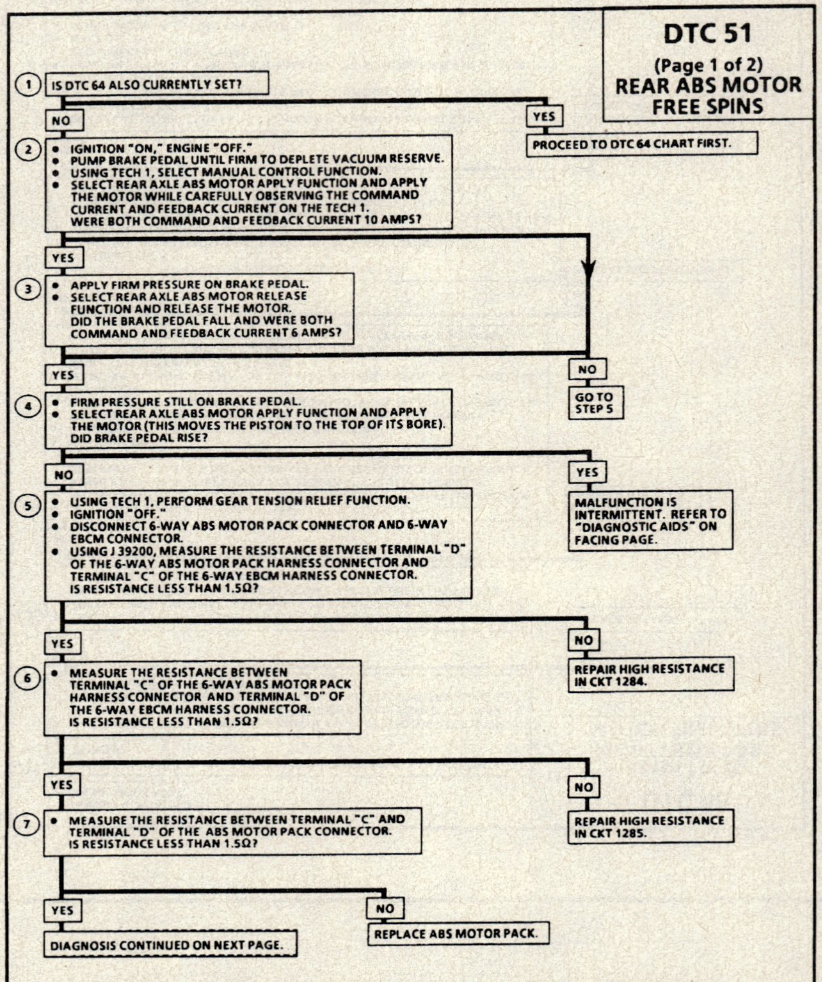

AFTER DIAGNOSIS IS COMPLETE, CLEAR DTCs AND TEST DRIVE VEHICLE FOR THREE (3) DRIVE CYCLES TO VERIFY THAT DTC DOES NOT RESET. A DRIVE CYCLE CONSISTS OF STARTING THE VEHICLE, DRIVING OVER 16 km/h (10 MPH), AND THEN KEYING DOWN.

DIAGNOSTIC CHART DTC 51 — 1994 LUMINA

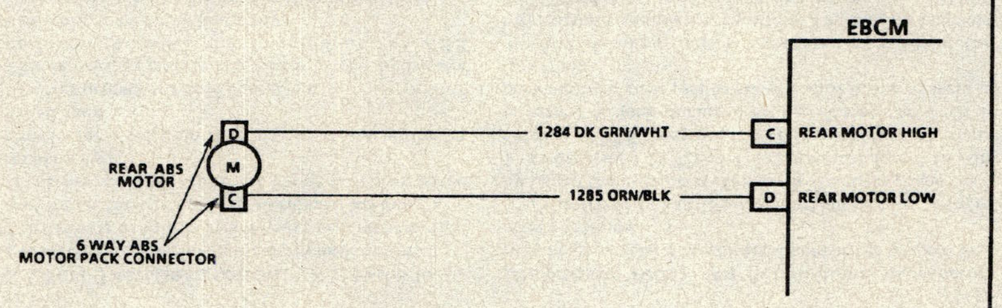

DTC 51
(Page 2 of 2)
REAR ABS MOTOR FREE SPINS

Circuit Description:
This DTC is designed to detect a stripped nut or gear assembly during initialization. During the homing sequence, the piston should reach the top of the bore resulting in a stalled motor. If this does not occur, the ABS motor must be spinning with little or no resistance, indicating a nut/screw or gear malfunction.

DTC Will Set When: DTC 51 can be set during initialization. If the feedback current is less than the command current for a specified period of time, a malfunction exists.

Action Taken: A malfunction DTC is stored, ABS is disabled, and the ABS warning lamp is turned "ON."

DTC Chart Test Description: Number(s) below refer to circled number(s) on the diagnostic chart.
8. Checks for a stripped gear on the ABS motor pack (unit with three small gears).
9. Checks for a stripped gear on the hydraulic modulator (unit with three large gears).
10. Verifies that the ABS motor can be applied under a load situation.
11. Verifies that the hydraulic modulator is functioning properly.
12. Ensures malfunction was not due to poor terminal contact.

Diagnostic Aids: An "Intermittent" malfunction in this DTC may result from a mechanical part of the system that sticks, binds, or slips.

The frequency of the malfunction can be checked by using the enhanced diagnostic function of the Tech 1, as described in "Tech 1 Diagnostics" found in this section. If the DTC only occurs once and DTC 64 also occurs, refer to DTC 64. If intermittent and enhanced diagnostics show this DTC occurs during ABS, refer to DTC 64.

Depending on the frequency of the malfunction, a physical inspection of the mechanical parts suspected may be necessary.

DIAGNOSTIC CHART DTC 51 — 1994 LUMINA

DTC 51
(Page 2 of 2)
REAR ABS MOTOR FREE SPINS

DIAGNOSIS CONTINUED FROM PREVIOUS PAGE.

(8)
- REMOVE HYDRAULIC MODULATOR/MASTER CYLINDER ASSEMBLY FROM VEHICLE (REFER TO "ON-VEHICLE SERVICE" IN THIS SECTION).
- REMOVE GEAR COVER.
- CHECK FOR STRIPPED GEAR ON ABS MOTOR PACK; ABS MOTOR PACK IS THE UNIT WITH THE THREE SMALL GEARS. THE REAR AXLE GEAR IS THE MIDDLE GEAR OF THE THREE.
IS THE ABS MOTOR PACK GEAR STRIPPED?

YES: REPLACE ABS MOTOR PACK.
NO: go to (9)

(9)
- CHECK FOR A STRIPPED GEAR ON HYDRAULIC MODULATOR; MODULATOR IS THE UNIT WITH THE THREE LARGE GEARS. THE REAR AXLE GEAR IS THE MIDDLE GEAR OF THE THREE.
IS THE MODULATOR GEAR STRIPPED?

YES: REPLACE HYDRAULIC MODULATOR GEAR.
NO: go to (10)

(10)
- RECONNECT ELECTRICAL CONNECTORS
- SECURELY POSITION HYDRAULIC MODULATOR/MASTER CYLINDER ASSEMBLY IN VEHICLE WITH GEAR COVER REMOVED SUCH THAT THE GEAR SET CAN BE OBSERVED. BE CAREFUL NOT TO DAMAGE GEAR SET OR ALLOW THE HYDRAULIC MODULATOR TO MOVE WHILE TESTING.
- IGNITION "ON."
- USING TECH 1, SELECT MANUAL CONTROL FUNCTION.
- SELECT REAR AXLE ABS MOTOR APPLY AND APPLY THE MOTOR WHILE OBSERVING THE MIDDLE GEAR SET.
- SELECT REAR AXLE ABS MOTOR RELEASE FUNCTION AND RELEASE THE MOTOR WHILE STILL OBSERVING THE GEAR SET.
- SELECT REAR AXLE ABS MOTOR APPLY FUNCTION AND APPLY THE MOTOR.
DID GEAR SET MOVE IN BOTH DIRECTIONS FOR AT LEAST ONE REVOLUTION?

YES: go to (11)
NO: REPLACE ABS MOTOR PACK.

(11)
- USING TECH 1, PERFORM GEAR TENSION RELIEF FUNCTION.
- IGNITION "OFF."
- SEPARATE THE ABS MOTOR PACK FROM THE HYDRAULIC MODULATOR.
- GRASP THE MIDDLE GEAR ON THE HYDRAULIC MODULATOR (THE HYDRAULIC MODULATOR IS THE UNIT WITH THE THREE LARGE GEARS) AND ATTEMPT TO MOVE THE GEAR.
CAN THE GEAR BE ROTATED MORE THAN 8 FULL TURNS LOCK TO LOCK?

YES: REPLACE THE HYDRAULIC MODULATOR.
NO: go to (12)

(12)
- INSPECT ALL CONNECTORS AND TERMINALS FOR POOR TERMINAL CONTACT AND EVIDENCE OF CORROSION.
REPLACE ALL TERMINALS THAT EXHIBIT SIGNS OF POOR TERMINAL CONTACT OR CORROSION.
- REINSTALL HYDRAULIC MODULATOR/MASTER CYLINDER ASSEMBLY IN VEHICLE.
- RECONNECT ALL CONNECTORS.
- IGNITION "OFF."
- START ENGINE WITH FOOT OFF BRAKE.
- ALLOW ENGINE TO RUN FOR AT LEAST 10 SECONDS.
- REPEAT ABOVE IGNITION CYCLE SEQUENCE TWO MORE TIMES.
DID DTC 51 SET IN THE LAST THREE IGNITION CYCLES?

YES: REPLACE EBCM.
NO: MALFUNCTION WAS MOST LIKELY CAUSED BY A POOR CONNECTION. REFER TO "DIAGNOSTIC AIDS" ON FACING PAGE.

AFTER DIAGNOSIS IS COMPLETE, CLEAR DTCs AND TEST DRIVE VEHICLE FOR THREE (3) DRIVE CYCLES TO VERIFY THAT DTC DOES NOT RESET. A DRIVE CYCLE CONSISTS OF STARTING THE VEHICLE, DRIVING OVER 16 km/h (10 MPH), AND THEN KEYING DOWN.

DIAGNOSTIC CHART DTC 52 — 1994 LUMINA

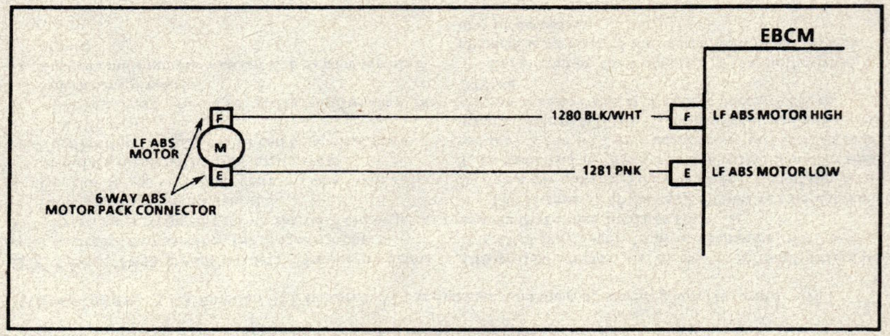

DTC 52

LEFT FRONT CHANNEL IN RELEASE TOO LONG

Circuit Description:

This DTC will diagnose an ABS motor that is energized longer than expected. This could occur if a wheel speed sensor is malfunctioning, the ABS motor does not turn, the left front solenoid mechanically fails open, or the ABS motor wires are crossed.

DTC Will Set When: DTC 52 can be set only during an ABS stop. If the EBCM commands the left front ABS channel in release for three seconds, a malfunction exists.

Action Taken: A malfunction DTC is stored, ABS is disabled and the ABS warning lamp is turned "ON."

DTC Chart Test Description: Number(s) below refer to circled number(s) on the diagnostic chart.

1. Identifies a problem in a wheel speed sensor that may cause the system to be in release too long.
2. Identifies an ABS motor as being failed or wired incorrectly.
3. Checks for a solenoid that may have mechanically failed open.
4. Verifies proper solenoid operation.
5. This serves to isolate the cause of the hydraulic problem, to either the solenoid or the ABS hydraulic modulator assembly.
6. Determines whether a malfunctioning ABS motor pack or hydraulic modulator is the reason for DTC 52 being set.

Diagnostic Aids: An "Intermittent" malfunction may be caused by a mechanical part of the system that sticks or binds.

The frequency of the malfunction can be checked by using the enhanced diagnostic function of the Tech 1, as described in "Tech 1 Diagnostics" found in this section. DTC 52 may occur if on ice and steering wheel is turned to lock during braking. Using the Tech 1, perform hydraulic test to ensure total brake system is functional.

Any circuitry, that is suspected as causing the intermittent complaint, should be thoroughly checked for backed out terminals, improper mating, broken locks, improperly formed or damaged terminals, poor terminal to wiring connections or physical damage to the wiring harness.

DIAGNOSTIC CHART DTC 52 — 1994 LUMINA

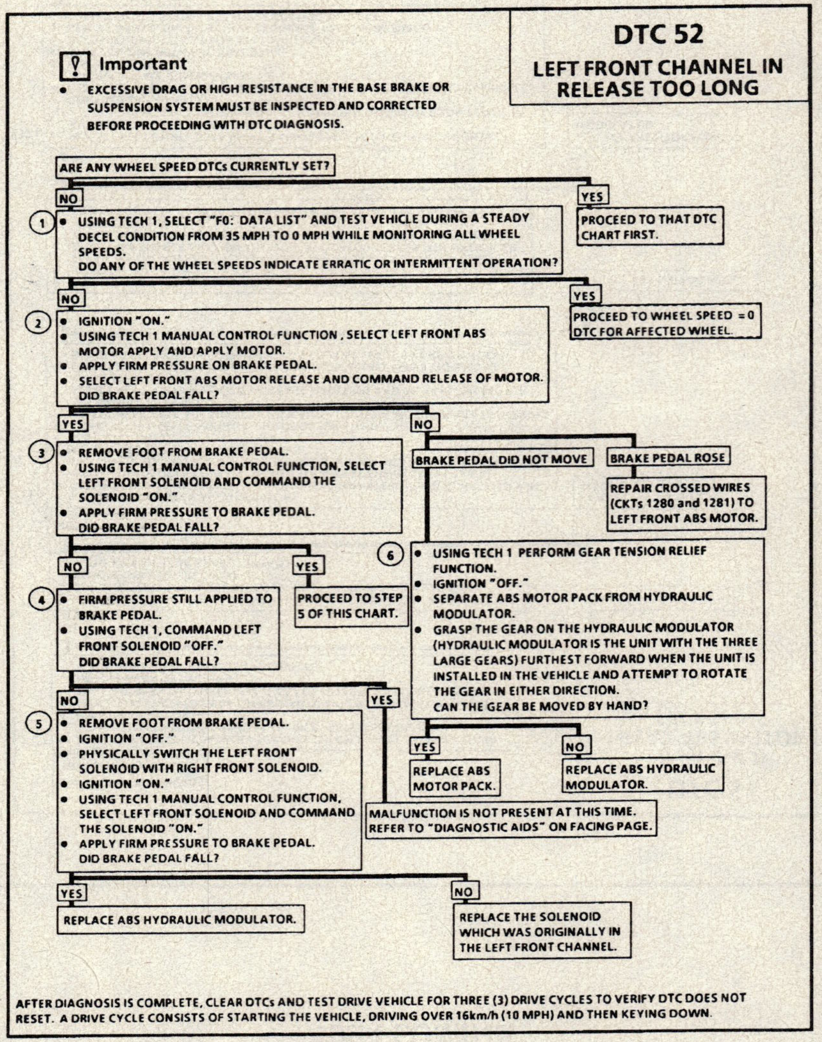

DIAGNOSTIC CHART DTC 53 — 1994 LUMINA

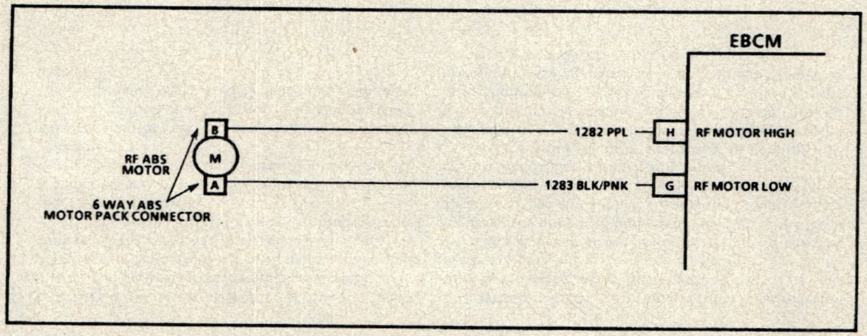

DTC 53

RIGHT FRONT CHANNEL IN RELEASE TOO LONG

Circuit Description:

This DTC will diagnose an ABS motor that is energized longer than expected. This could occur if a wheel speed sensor is malfunctioning, the ABS motor does not turn, the right front solenoid mechanically fails open, or the ABS motor wires are crossed.

DTC Will Set When: DTC 53 can be set only during an ABS stop. If the EBCM commands the right front channel in release for three seconds, a malfunction exists.

Action Taken: A malfunction DTC is stored, ABS is disabled and the ABS warning lamp is turned "ON."

DTC Chart Test Description: Number(s) below refer to circled number(s) on the diagnostic chart.

1. Identifies a problem in a wheel speed sensor that may cause the system to be in release too long.
2. Identifies an ABS motor as being failed or wired incorrectly.
3. Checks for a solenoid that may have mechanically failed open.
4. Checks for a solenoid that may have mechanically failed closed.
5. This serves to isolate the cause of the hydraulic problem, to either the solenoid or the ABS hydraulic modulator assembly.
6. Determines whether a malfunction ABS motor pack or hydraulic modulator is the reason for DTC 53 being set.

Diagnostic Aids: An "Intermittent" malfunction may be caused by a mechanical part of the system that sticks or binds.

The frequency of the malfunction can be checked by using the enhanced diagnostic function of the Tech 1, as described in "Tech 1 Diagnostics" found in this section. DTC 53 may occur on ice if steering wheel is turned to lock during braking. Using the Tech 1 perform hydraulic test to ensure total brake system is functional.

Any circuitry, that is suspected as causing the intermittent complaint, should be thoroughly checked for backed out terminals, improper mating, broken locks, improperly formed or damaged terminals, poor terminal to wiring connections or physical damage to the wiring harness.

DIAGNOSTIC CHART DTC 53 — 1994 LUMINA

DTC 53

RIGHT FRONT CHANNEL IN RELEASE TOO LONG

Important

- EXCESSIVE DRAG OR HIGH RESISTANCE IN THE BASE BRAKE OR SUSPENSION SYSTEM MUST BE INSPECTED AND CORRECTED BEFORE PROCEEDING WITH DTC DIAGNOSIS.

ARE ANY WHEEL SPEED DTCs CURRENTLY SET?

- YES: PROCEED TO THAT DTC CHART FIRST.
- NO: go to step 1.

(1)
- USING TECH 1, SELECT "F0: DATA LIST" AND TEST VEHICLE DURING A STEADY DECEL CONDITION FROM 35 MPH TO 0 MPH WHILE MONITORING ALL WHEEL SPEEDS.

DO ANY OF THE WHEEL SPEEDS INDICATE ERRATIC OR INTERMITTENT OPERATION?

- YES: PROCEED TO WHEEL SPEED = 0 DTC FOR AFFECTED WHEEL.
- NO: go to step 2.

(2)
- IGNITION "ON."
- USING TECH 1 MANUAL CONTROL FUNCTION, SELECT RIGHT FRONT ABS MOTOR APPLY AND APPLY MOTOR.
- APPLY FIRM PRESSURE ON BRAKE PEDAL.
- SELECT RIGHT FRONT ABS MOTOR RELEASE AND COMMAND RELEASE OF MOTOR.

DID BRAKE PEDAL FALL?

- YES: go to step 3.
- NO:
  - BRAKE PEDAL DID NOT MOVE: go to step 6.
  - BRAKE PEDAL ROSE: REPAIR CROSSED WIRES (CKTs 1282 AND 1283) TO RIGHT FRONT ABS MOTOR.

(3)
- REMOVE FOOT FROM BRAKE PEDAL.
- USING TECH 1 MANUAL CONTROL FUNCTION, SELECT RIGHT FRONT SOLENOID AND COMMAND THE SOLENOID "ON."
- APPLY FIRM PRESSURE TO BRAKE PEDAL.

DID BRAKE PEDAL FALL?

- YES: PROCEED TO STEP 5 OF THIS CHART.
- NO: go to step 4.

(4)
- FIRM PRESSURE STILL APPLIED TO BRAKE PEDAL.
- USING TECH 1, COMMAND RIGHT FRONT SOLENOID "OFF."

DID BRAKE PEDAL FALL?

- YES: MALFUNCTION IS NOT PRESENT AT THIS TIME. REFER TO "DIAGNOSTIC AIDS" ON FACING PAGE.
- NO: go to step 5.

(5)
- REMOVE FOOT FROM BRAKE PEDAL.
- IGNITION "OFF."
- PHYSICALLY SWITCH THE RIGHT FRONT SOLENOID WITH THE LEFT FRONT SOLENOID.
- IGNITION "ON."
- USING TECH 1 MANUAL CONTROL FUNCTION, SELECT RIGHT FRONT SOLENOID AND COMMAND THE SOLENOID "ON."
- APPLY FIRM PRESSURE TO BRAKE PEDAL.

DID BRAKE PEDAL FALL?

- YES: REPLACE ABS HYDRAULIC MODULATOR.
- NO: REPLACE THE SOLENOID WHICH WAS ORIGINALLY IN THE RIGHT FRONT CHANNEL.

(6)
- USING TECH 1 PERFORM GEAR TENSION RELIEF FUNCTION.
- IGNITION "OFF."
- SEPARATE ABS MOTOR PACK FROM HYDRAULIC MODULATOR.
- GRASP THE GEAR ON THE HYDRAULIC MODULATOR (HYDRAULIC MODULATOR IS THE UNIT WITH THE THREE LARGE GEARS) FURTHEST REARWARD WHEN THE UNIT IS INSTALLED IN THE VEHICLE AND ATTEMPT TO ROTATE THE GEAR IN EITHER DIRECTION.

CAN THE GEAR BE MOVED BY HAND?

- YES: REPLACE ABS MOTOR PACK.
- NO: REPLACE ABS HYDRAULIC MODULATOR.

AFTER DIAGNOSIS IS COMPLETE, CLEAR DTCs AND TEST DRIVE VEHICLE FOR THREE (3) DRIVE CYCLES TO VERIFY DTC DOES NOT RESET. A DRIVE CYCLE CONSISTS OF STARTING THE VEHICLE, DRIVING OVER 16km/h (10 MPH) AND THEN KEYING DOWN.

DIAGNOSTIC CHART DTC 54 — 1994 LUMINA

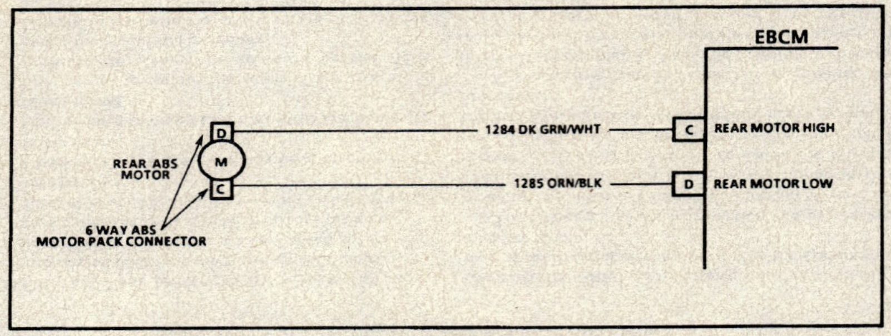

DTC 54
REAR CHANNEL IN RELEASE TOO LONG

Circuit Description:
This DTC will diagnose an ABS motor that is energized longer than expected. This could occur if a wheel speed sensor is malfunctioning, the ABS motor does not turn, or the ABS motor wires are crossed.

DTC Will Set When: DTC 54 can be set only during an ABS stop. If the EBCM commands the rear ABS channel in release for three seconds, a malfunction exists.

Action Taken: A malfunction DTC is stored, ABS is disabled and the ABS warning lamp is turned "ON."

DTC Chart Test Description: Number(s) below refer to circled number(s) on the diagnostic chart.
1. Identifies a problem in a wheel speed sensor that may cause the system to be in release too long.
2. Checks for a wheel that may stick or bind because of a mechanical fault.
3. Checks to see if the ABS motor is capable of moving and applying the hydraulic piston for the rear wheels.
4. Ensures that the ABS motor wiring is not crossed.
5. Isolates the fault of a "no-apply" situation to either the ABS motor pack or ABS hydraulic modulator assembly.

Diagnostic Aids: An "Intermittent" malfunction may be caused by a mechanical part of the system that sticks or binds.

The frequency of the malfunction can be checked by using the enhanced diagnostic function of the Tech 1, as described in "Tech 1 Diagnostics" found in this section. Using the Tech 1, perform hydraulic test to ensure total brake system is functional.

Any circuitry, that is suspected as causing the intermittent complaint, should be thoroughly checked for backed out terminals, improper mating, broken locks, improperly formed or damaged terminals, poor terminal to wiring connections or physical damage to the wiring harness.

DIAGNOSTIC CHART DTC 54 — 1994 LUMINA

DTC 54
REAR CHANNEL IN RELEASE TOO LONG

Important
- EXCESSIVE DRAG OR HIGH RESISTANCE IN THE BASE BRAKE OR SUSPENSION SYSTEM MUST BE INSPECTED AND CORRECTED BEFORE PROCEEDING WITH DTC DIAGNOSIS.

ARE ANY WHEEL SPEED DTCs CURRENTLY SET?
- YES → PROCEED TO THAT DTC CHART FIRST.
- NO ↓

(1)
- USING TECH 1, SELECT "F0: DATA LIST" AND TEST VEHICLE DURING A STEADY DECEL CONDITION FROM 35 MPH TO 0 MPH WHILE MONITORING ALL WHEEL SPEEDS. DO ANY OF THE WHEEL SPEEDS INDICATE ERRATIC OR INTERMITTENT OPERATION?
  - YES → PROCEED TO WHEEL SPEED = 0 DTC FOR AFFECTED WHEEL.
  - NO ↓

(2)
- RAISE REAR OF VEHICLE AND SUPPORT AS INDICATED IN SECTION 0A.
- SPIN BOTH REAR WHEELS AND CONFIRM PHYSICAL CONDITION OF BASE BRAKE SYSTEM.
  COULD BOTH REAR WHEELS BE SPUN FREELY BY HAND?
  - NO → REPAIR SOURCE OF RESISTANCE IN BASE BRAKE SYSTEM.
  - YES ↓

(3)
- IGNITION "ON."
- USING MANUAL CONTROL FUNCTION OF TECH 1, SELECT REAR AXLE ABS MOTOR APPLY.
- APPLY FIRM PRESSURE ON BRAKE PEDAL.
- COMMAND APPLY OF REAR ABS MOTOR AND TRY TO SPIN REAR WHEELS BY HAND.
  COULD REAR WHEELS BE SPUN BY HAND?
  - NO → MALFUNCTION IS NOT PRESENT AT THIS TIME. REFER TO "DIAGNOSTIC AIDS" ON FACING PAGE.
  - YES ↓

(4)
- WITH FIRM PRESSURE ON BRAKE PEDAL STILL APPLIED, COMMAND A RELEASE OF THE REAR AXLE ABS MOTOR. CAN WHEELS BE MOVED?
  - NO → REPAIR CROSSED WIRES AT REAR ABS MOTOR (CKTs 1284 AND 1285).
  - YES ↓

(5)
- USING TECH 1, PERFORM GEAR TENSION RELIEF FUNCTION.
- SEPARATE ABS MOTOR PACK FROM HYDRAULIC MODULATOR.
- CONNECT ABS MOTOR PACK WIRING.
- IGNITION "ON."
- USING TECH 1, PERFORM MOTOR TEST.
  DO ALL THREE ABS MOTOR PACK GEARS (SMALL GEARS) SPIN FREELY?
  - NO → REPLACE ABS MOTOR PACK.
  - YES → REPLACE ABS HYDRAULIC MODULATOR.

AFTER ALL DIAGNOSIS IS COMPLETE, CLEAR DTCs AND TEST DRIVE VEHICLE FOR THREE (3) DRIVE CYCLES TO VERIFY THAT DTC DOES NOT RESET. A DRIVE CYCLE CONSISTS OF STARTING THE VEHICLE DRIVING OVER 16 KM/H (10 MPH), AND THEN KEYING DOWN.

DIAGNOSTIC CHART DTC 55 — 1994 LUMINA

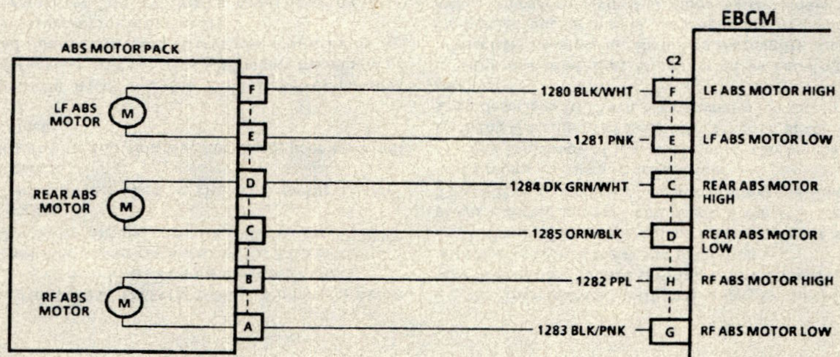

DTC 55
EBCM MALFUNCTION

Circuit Description:
This DTC identifies a malfunction detected by the MDI custom IC. It also ensures the cause of the malfunction is not a result of a problem with the ABS enable relay under a DTC 55 malfunction.

Action Taken: A malfunction DTC is stored, ABS is disabled and the ABS warning lamp is turned "ON."

DTC Chart Test Description: Number(s) below refer to circled number(s) on the diagnostic chart.
1. Checks to see if the malfunction is still present.
2. Checks to see if the malfunction is intermittent.

Diagnostic Aids: An "Intermittent" malfunction may be caused by a poor connection, rubbed through wire insulation, or a wire that is broken inside the insulation.

The frequency of the malfunction can be checked by using the enhanced diagnostic function of the Tech 1, as described in "Tech 1 Diagnostics" found in this section.

Any circuitry, that is suspected as causing the intermittent complaint, should be thoroughly checked for backed out terminals, improper mating, broken locks, improperly formed or damaged terminals, poor terminal to wiring connections or physical damage to the wiring harness.

DIAGNOSTIC CHART DTC 55 — 1994 LUMINA

DTC 55
EBCM MALFUNCTION

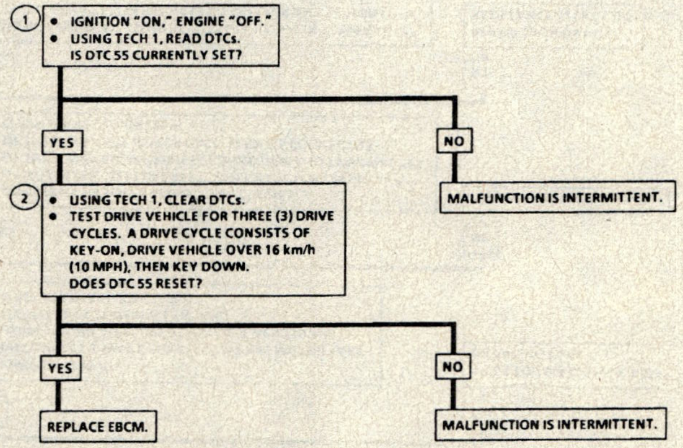

DIAGNOSTIC CHART DTC 56 — 1994 LUMINA

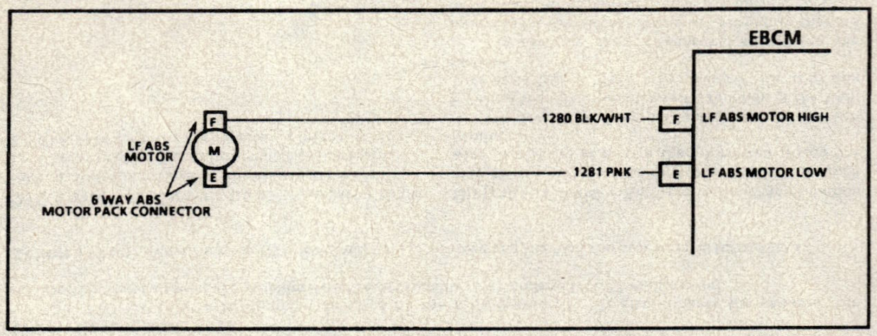

DTC 56

LEFT FRONT ABS MOTOR CIRCUIT OPEN

Circuit Description:
This DTC identifies an ABS motor that cannot be energized due to an open in its circuitry.

DTC Will Set When: DTC 56 can be set only when the motor is commanded "OFF." If the EBCM detects an out of range voltage on either of the left front ABS motor circuits indicating an open circuit, a malfunction exists.

Action Taken: An open ABS motor will not activate when requested. A malfunction DTC is stored, ABS is disabled and the ABS warning lamp is turned "ON."

DTC Chart Test Description: Number(s) below refer to circled number(s) on the diagnostic chart.

1. Checks for proper resistance of the ABS motor.
2. Checks for an open in the ABS motor "HIGH" circuitry.
3. Checks for an open in the ABS motor "LOW" circuitry.
4. Ensures malfunction was not due to poor terminal contact.

Diagnostic Aids: Using Tech 1, select manual control function, and exercise ABS motor movement of affected channel in both directions while applying light pressure on the brake pedal.

If erratic or "jumpy" brake pedal movement is detected while performing an "apply" or "release" function of the ABS motor, an intermittent malfunction may be indicated.

An "Intermittent" malfunction may be caused by a poor connection, rubbed through wire insulation, or a wire that is broken inside the insulation.

If the malfunction is not current, wiggle the wires of the affected channel and check if the DTC resets. This will help to pinpoint an intermittent malfunction in the motor circuitry or connections.

The frequency of the malfunction can be checked by using the enhanced diagnostic function of the Tech 1, as described in "Tech 1 Diagnostics" found in this section.

Any circuitry, that is suspected as causing the intermittent complaint, should be thoroughly checked for backed out terminals, improper mating, broken locks, improperly formed or damaged terminals, poor terminal to wiring connections or physical damage to the wiring harness.

DIAGNOSTIC CHART DTC 56 — 1994 LUMINA

DTC 56

LEFT FRONT ABS MOTOR CIRCUIT OPEN

DOES DTC 56 OCCUR INTERMITTENTLY?

- YES → REFER TO "DIAGNOSTIC AIDS" ON FACING PAGE.
- NO →

(1)
- IGNITION "OFF."
- DISCONNECT 6-WAY CONNECTOR AT ABS MOTOR PACK.
- USING J 39200, MEASURE RESISTANCE ACROSS ABS MOTOR PACK TERMINALS "F" AND "E".

IS RESISTANCE BELOW 2000Ω?

- NO → REPLACE ABS MOTOR PACK.
- YES →

(2)
- DISCONNECT 6-WAY CONNECTOR AT EBCM.
- USING J 39200, MEASURE RESISTANCE BETWEEN TERMINAL "F" OF THE 6-WAY EBCM HARNESS CONNECTOR AND TERMINAL "F" OF THE 6-WAY ABS MOTOR PACK HARNESS CONNECTOR.

IS RESISTANCE 2Ω OR LESS?

- NO → REPAIR OPEN OR HIGH RESISTANCE IN CKT 1280.
- YES →

(3)
- USING J 39200, MEASURE RESISTANCE BETWEEN TERMINAL "E" OF THE 6-WAY EBCM HARNESS CONNECTOR AND TERMINAL "E" OF THE 6-WAY ABS MOTOR PACK HARNESS CONNECTOR.

IS RESISTANCE 2Ω OR LESS?

- NO → REPAIR OPEN OR HIGH RESISTANCE IN CKT 1281.
- YES →

(4)
- INSPECT THE 6-WAY ABS MOTOR PACK AND 6-WAY EBCM CONNECTORS FOR POOR TERMINAL CONTACT AND EVIDENCE OF CORROSION.

REPLACE ALL TERMINALS THAT EXHIBIT SIGNS OF POOR TERMINAL CONTACT OR CORROSION.
- RECONNECT ALL CONNECTORS.
- IGNITION "OFF."
- START ENGINE AND TEST DRIVE VEHICLE, ACHIEVING AT LEAST 16 km/h (10 MPH).
- REPEAT ABOVE DRIVE CYCLE SEQUENCE TWO MORE TIMES.

DID DTC 56 SET IN THE LAST THREE DRIVE CYCLES?

- YES → REPLACE EBCM.
- NO → MALFUNCTION WAS MOST LIKELY CAUSED BY A POOR CONNECTION. REFER TO "DIAGNOSTIC AIDS" ON FACING PAGE.

AFTER ALL DIAGNOSIS IS COMPLETE, CLEAR DTCs AND TEST DRIVE VEHICLE FOR THREE (3) DRIVE CYCLES TO VERIFY THAT DTC DOES NOT RESET. A DRIVE CYCLE CONSISTS OF STARTING THE VEHICLE, DRIVING OVER 16 KM/H (10 MPH), AND THEN KEYING DOWN.

DIAGNOSTIC CHART DTC 57 — 1994 LUMINA

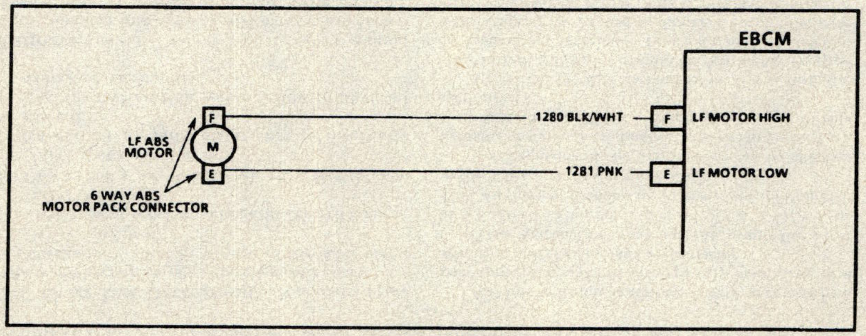

DTC 57
LEFT FRONT ABS MOTOR CIRCUIT SHORTED TO GROUND

Circuit Description:
This DTC identifies an ABS motor circuit that is shorted to ground. This malfunction will not allow the ABS motor to be controlled at the commanded current rate or will cause the driver circuit to allow current directly to ground.

DTC Will Set When: DTC 57 can be set anytime. If the EBCM detects an out of range voltage on either of the left front ABS motor circuits indicating a circuit shorted to ground, a malfunction exists.

Action Taken: A malfunction DTC is stored, ABS is disabled and the ABS warning lamp is turned "ON."

DTC Chart Test Description: Number(s) below refer to circled number(s) on the diagnostic chart.

1. Checks for a short to ground in the ABS motor "HIGH" circuitry.
2. Checks for a short to ground in the ABS motor "LOW" circuitry.
3. Checks for an ABS motor that is internally shorted to ground.
4. Ensures malfunction was not due to physical damage of the circuitry.

Diagnostic Aids: Using Tech 1, select manual control function, and exercise ABS motor movement of affected channel in both directions while applying light pressure on the brake pedal.

If erratic or "jumpy" brake pedal movement is detected while performing an "apply" or "release" function of the ABS motor, an intermittent malfunction may be indicated.

An "Intermittent" malfunction may be caused by a poor connection, rubbed through wire insulation, or a wire that is broken inside the insulation.

If the malfunction is not current, wiggle the wires of the affected channel and check if the DTC resets. This will help to pinpoint an intermittent malfunction in the motor circuitry or connections.

The frequency of the malfunction can be checked by using the enhanced diagnostic function of the Tech 1, as described in "Tech 1 Diagnostics" found in this section.

Any circuitry, that is suspected as causing the intermittent complaint, should be thoroughly checked for backed out terminals, improper mating, broken locks, improperly formed or damaged terminals, poor terminal to wiring connections or physical damage to the wiring harness.

DIAGNOSTIC CHART DTC 57 — 1994 LUMINA

DTC 57
LEFT FRONT ABS MOTOR CIRCUIT SHORTED TO GROUND

DOES DTC 57 OCCUR INTERMITTENTLY?

- YES → REFER TO "DIAGNOSTIC AIDS" ON FACING PAGE.
- NO →

(1)
- IGNITION "OFF."
- DISCONNECT 6-WAY CONNECTORS AT ABS MOTOR PACK AND EBCM.
- USING J 39200, MEASURE THE RESISTANCE BETWEEN GROUND AND TERMINAL "F" OF THE 6-WAY ABS MOTOR PACK HARNESS CONNECTOR.

DOES J 39200 DISPLAY RESISTANCE AS "OL" (INFINITE)?

- NO → REPAIR SHORT TO GROUND IN CKT 1280.
- YES →

(2)
- MEASURE THE RESISTANCE BETWEEN GROUND AND TERMINAL "E" OF THE 6-WAY ABS MOTOR PACK HARNESS CONNECTOR.

DOES J 39200 DISPLAY RESISTANCE AS "OL" (INFINITE)?

- NO → REPAIR SHORT TO GROUND IN CKT 1281.
- YES →

(3)
- MEASURE THE RESISTANCE BETWEEN GROUND AND TERMINAL "F" OF THE ABS MOTOR PACK.

DOES J 39200 DISPLAY RESISTANCE AS "OL" (INFINITE)?

- NO → REPLACE ABS MOTOR PACK.
- YES →

(4)
- INSPECT CKTs 1280 AND 1281 AND RELATED CONNECTORS FOR DAMAGE WHICH MAY RESULT IN A SHORT TO GROUND WITH ALL CONNECTORS CONNECTED. REPAIR ALL DAMAGE IF EVIDENT.
- RECONNECT ALL CONNECTORS.
- IGNITION "OFF."
- START ENGINE AND TEST DRIVE VEHICLE, ACHIEVING AT LEAST 16 km/h (10 MPH).
- REPEAT ABOVE DRIVE CYCLE SEQUENCE TWO MORE TIMES.

DID DTC 57 SET IN THE LAST THREE DRIVE CYCLES?

- YES → REPLACE EBCM.
- NO → MALFUNCTION WAS MOST LIKELY CAUSED BY A POOR CONNECTION. REFER TO "DIAGNOSTIC AIDS" ON FACING PAGE.

AFTER ALL DIAGNOSIS IS COMPLETE, CLEAR DTCs AND TEST DRIVE VEHICLE FOR THREE (3) DRIVE CYCLES TO VERIFY THAT DTC DOES NOT RESET. A DRIVE CYCLE CONSISTS OF STARTING THE VEHICLE, DRIVING OVER 16 KM/H (10 MPH), AND THEN KEYING DOWN.

DIAGNOSTIC CHART DTC 58 — 1994 LUMINA

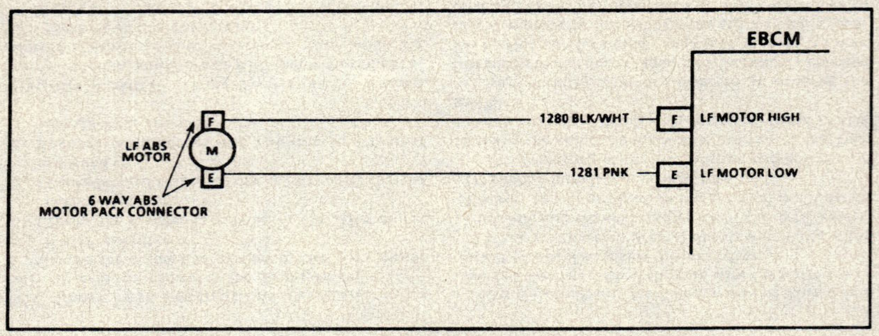

DTC 58

LEFT FRONT ABS MOTOR CIRCUIT SHORTED TO BATTERY

Circuit Description:
This DTC identifies an ABS motor circuit that is shorted to battery or an ABS motor that has low or no resistance. This malfunction will not allow the ABS motor to be controlled at the commanded current rate or will cause the ABS motor to turn in the opposite direction, or not at all.

DTC Will Set When: DTC 58 can be set only when the ABS motor is commanded "OFF." If the EBCM detects an out of range voltage on either of the left front ABS motor circuits indicating a circuit shorted to battery or an ABS motor shorted, a malfunction exists.

Action Taken: A malfunction DTC is stored, ABS is disabled and the ABS warning lamp is turned "ON."

DTC Chart Test Description: Number(s) below refer to circled number(s) on the diagnostic chart.

1. Checks for a short to voltage in the ABS motor "HIGH" circuitry.
2. Checks for a short to voltage in the ABS motor "LOW" circuitry.
3. Checks for a short between the "HIGH" and "LOW" motor circuits.
4. Checks for an ABS motor that is internally shorted.
5. Ensures malfunction was not due to physical damage of the circuitry.

Diagnostic Aids: Using Tech 1, select manual control function, and exercise ABS motor movement of affected channel in both directions while applying light pressure on the brake pedal.

If erratic or "jumpy" brake pedal movement is detected while performing an "apply" or "release" function of the ABS motor, an intermittent malfunction may be indicated.

An "Intermittent" malfunction may be caused by a poor connection, rubbed through wire insulation, or a wire that is broken inside the insulation.

If the malfunction is not current, wiggle the wires of the affected channel and check if the DTC resets. This will help to pinpoint an intermittent malfunction in the motor circuitry or connections.

The frequency of the malfunction can be checked by using the enhanced diagnostic function of the Tech 1, as described in "Tech 1 Diagnostics" found in this section.

Any circuitry, that is suspected as causing the intermittent complaint, should be thoroughly checked for backed out terminals, improper mating, broken locks, improperly formed or damaged terminals, poor terminal to wiring connections or physical damage to the wiring harness.

DIAGNOSTIC CHART DTC 58 — 1994 LUMINA

DTC 58

LEFT FRONT ABS MOTOR CIRCUIT SHORTED TO BATTERY

NOTICE: J 39200 TEST LEADS MUST BE "ZEROED" PRIOR TO MAKING ANY RESISTANCE MEASUREMENTS. REFER TO J 39200 USER'S MANUAL.

DOES DTC 58 OCCUR INTERMITTENTLY?

- YES: REFER TO "DIAGNOSTIC AIDS" ON FACING PAGE.
- NO:

(1)
- IGNITION "OFF."
- DISCONNECT 6-WAY CONNECTORS AT ABS MOTOR PACK AND EBCM.
- IGNITION "ON."
- USING J 39200, MEASURE THE VOLTAGE BETWEEN GROUND AND TERMINAL "F" OF THE 6-WAY ABS MOTOR PACK HARNESS CONNECTOR. IS VOLTAGE 1 VOLT OR LESS?

- NO: REPAIR SHORT TO VOLTAGE IN CKT 1280.
- YES:

(2)
- MEASURE THE VOLTAGE BETWEEN GROUND AND TERMINAL "E" OF THE 6-WAY ABS MOTOR PACK HARNESS CONNECTOR. IS VOLTAGE 1 VOLT OR LESS?

- NO: REPAIR SHORT TO VOLTAGE IN CKT 1281.
- YES:

(3)
- IGNITION "OFF."
- USING J 39200, MEASURE THE RESISTANCE BETWEEN TERMINAL "E" AND TERMINAL "F" OF THE 6-WAY ABS MOTOR PACK HARNESS CONNECTOR. DOES J 39200 DISPLAY RESISTANCE AS "OL" (INFINITE)?

- NO: REPAIR SHORT BETWEEN CKTs 1280 AND 1281.
- YES:

(4)
- USING J 39200, MEASURE THE RESISTANCE BETWEEN TERMINAL "E" AND TERMINAL "F" OF THE ABS MOTOR PACK. IS RESISTANCE GREATER THAN 0.4Ω?

- NO: REPLACE ABS MOTOR PACK.
- YES:

(5)
- INSPECT CKTs 1280 AND 1281 AND RELATED CONNECTORS FOR DAMAGE WHICH MAY RESULT IN A SHORT TO GROUND WITH ALL CONNECTORS CONNECTED. REPAIR ALL DAMAGE IF EVIDENT.
- RECONNECT ALL CONNECTORS.
- IGNITION "OFF."
- START ENGINE AND TEST DRIVE VEHICLE, ACHIEVING AT LEAST 16 km/h (10 MPH).
- REPEAT ABOVE DRIVE CYCLE SEQUENCE TWO MORE TIMES. DID DTC 58 SET IN THE LAST THREE DRIVE CYCLES?

- YES: REPLACE EBCM.
- NO: MALFUNCTION WAS MOST LIKELY CAUSED BY A POOR CONNECTION. REFER TO "DIAGNOSTIC AIDS" ON FACING PAGE.

AFTER ALL DIAGNOSIS IS COMPLETE, CLEAR DTCs AND TEST DRIVE VEHICLE FOR THREE (3) DRIVE CYCLES TO VERIFY THAT DTC DOES NOT RESET. A DRIVE CYCLE CONSISTS OF STARTING THE VEHICLE, DRIVING OVER 16 km/h (10 MPH), AND THEN KEYING DOWN.

DIAGNOSTIC CHART DTC 61 — 1994 LUMINA

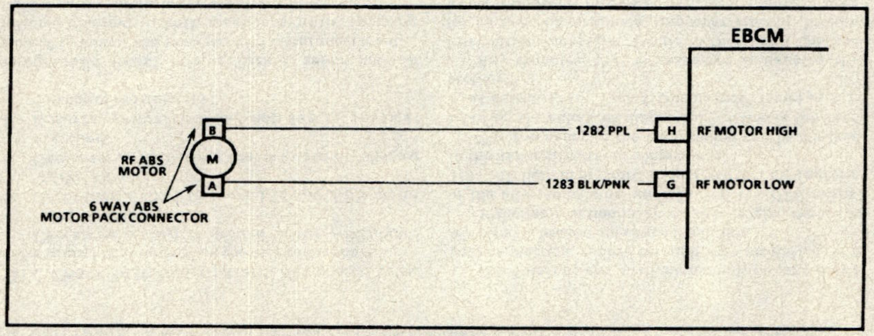

DTC 61
RIGHT FRONT ABS MOTOR CIRCUIT OPEN

Circuit Description:
This DTC identifies an ABS motor that cannot be energized due to an open in its circuitry. This malfunction will not allow proper front ABS operation.

DTC Will Set When: DTC 61 can be set only when the ABS motor is commanded "OFF." If the EBCM detects an out of range voltage on either of the right front ABS motor circuits indicating an open circuit, a malfunction exists.

Action Taken: An open ABS motor will not activate when requested. A malfunction DTC is stored, ABS is disabled and the ABS warning lamp is turned "ON."

DTC Chart Test Description: Number(s) below refer to circled number(s) on the diagnostic chart.

1. Checks for proper resistance of the ABS motor.
2. Checks for an open in the ABS motor "HIGH" circuitry.
3. Checks for an open in the ABS motor "LOW" circuitry.
4. Ensures malfunction was not due to physical damage of the circuitry.

Diagnostic Aids: Using Tech 1, select manual control function, and exercise ABS motor movement of affected channel in both directions while applying light pressure on the brake pedal.

If erratic or "jumpy" brake pedal movement is detected while performing an "apply" or "release" function of the ABS motor, an intermittent malfunction may be indicated.

An "Intermittent" malfunction may be caused by a poor connection, rubbed through wire insulation, or a wire that is broken inside the insulation.

If the malfunction is not current, wiggle the wires of the affected channel and check if the DTC resets. This will help to pinpoint an intermittent malfunction in the motor circuitry or connections.

The frequency of the malfunction can be checked by using the enhanced diagnostic function of the Tech 1, as described in "Tech 1 Diagnostics" found in this section.

Any circuitry, that is suspected as causing the intermittent complaint, should be thoroughly checked for backed out terminals, improper mating, broken locks, improperly formed or damaged terminals, poor terminal to wiring connections or physical damage to the wiring harness.

DIAGNOSTIC CHART DTC 61 — 1994 LUMINA

DTC 61
RIGHT FRONT ABS MOTOR CIRCUIT OPEN

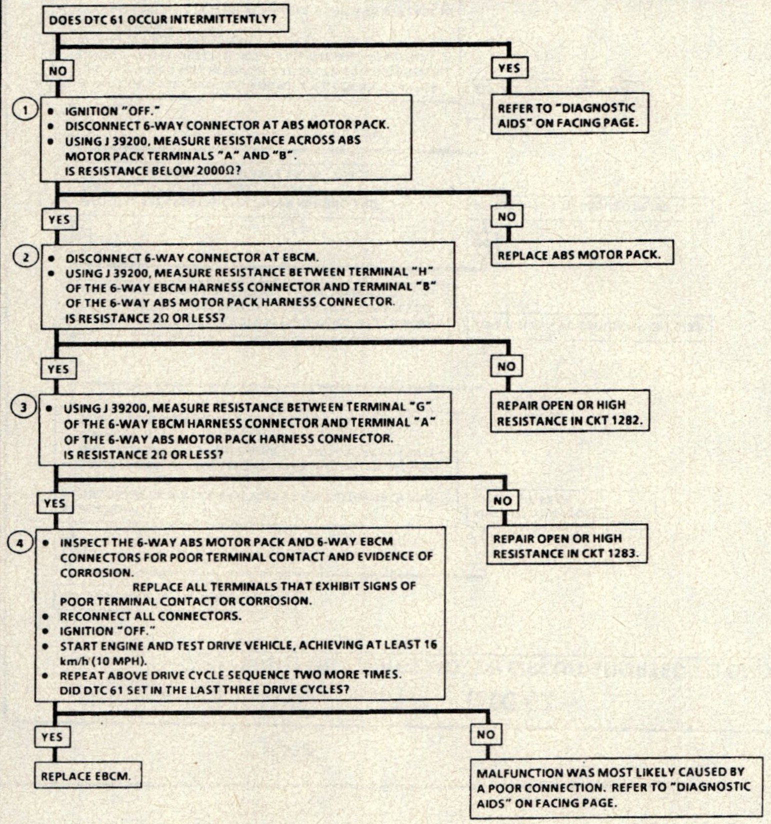

AFTER ALL DIAGNOSIS IS COMPLETE, CLEAR DTCs AND TEST DRIVE VEHICLE FOR THREE (3) DRIVE CYCLES TO VERIFY THAT DTC DOES NOT RESET. A DRIVE CYCLE CONSISTS OF STARTING THE VEHICLE, DRIVING OVER 16 km/h (10 MPH), AND THEN KEYING DOWN.

DIAGNOSTIC CHART DTC 62 — 1994 LUMINA

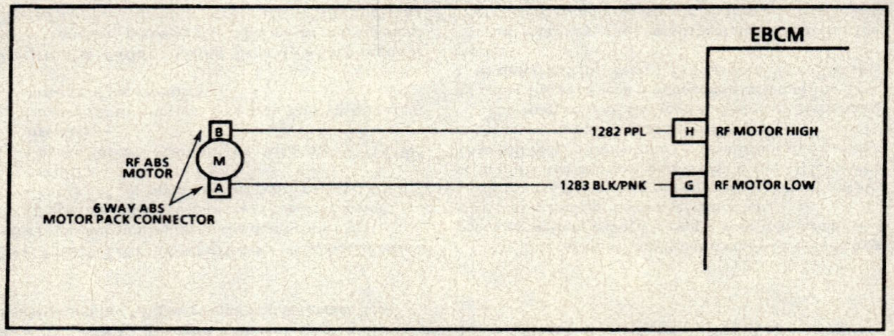

DTC 62
RIGHT FRONT ABS MOTOR CIRCUIT SHORTED TO GND

Circuit Description:
This DTC identifies an ABS motor circuit that is shorted to ground. This malfunction will not allow the ABS motor to be controlled at the commanded current rate or will cause the ABS motor driver circuit to allow current directly to ground.

DTC Will Set When: DTC 62 can be set anytime. If the EBCM detects an out of range voltage on either of the right front ABS motor circuits indicating a circuit shorted to ground, a malfunction exists.

Action Taken: A malfunction DTC is stored, ABS is disabled and the ABS warning lamp is turned "ON."

DTC Chart Test Description: Number(s) below refer to circled number(s) on the diagnostic chart.

1. Checks for a short to ground in the ABS motor "HIGH" circuitry.
2. Checks for a short to ground in the ABS motor "LOW" circuitry.
3. Checks for an ABS motor that is internally shorted to ground.
4. Ensures malfunction was not due to physical damage of the circuitry.

Diagnostic Aids: Using Tech 1, select manual control function, and exercise ABS motor movement of affected channel in both directions while applying light pressure on the brake pedal.

If erratic or "jumpy" brake pedal movement is detected while performing an "apply" or "release" function of the ABS motor, an intermittent malfunction may be indicated.

An "Intermittent" malfunction may be caused by a poor connection, rubbed through wire insulation, or a wire that is broken inside the insulation.

If the malfunction is not current, wiggle the wires of the affected channel and check if the DTC resets. This will help to pinpoint an intermittent malfunction in the motor circuitry or connections.

The frequency of the malfunction can be checked by using the enhanced diagnostic function of the Tech 1, as described in "Tech 1 Diagnostics" found in this section.

Any circuitry, that is suspected as causing the intermittent complaint, should be thoroughly checked for backed out terminals, improper mating, broken locks, improperly formed or damaged terminals, poor terminal to wiring connections or physical damage to the wiring harness.

DIAGNOSTIC CHART DTC 62 — 1994 LUMINA

DTC 62
RIGHT FRONT ABS MOTOR CIRCUIT SHORTED TO GND

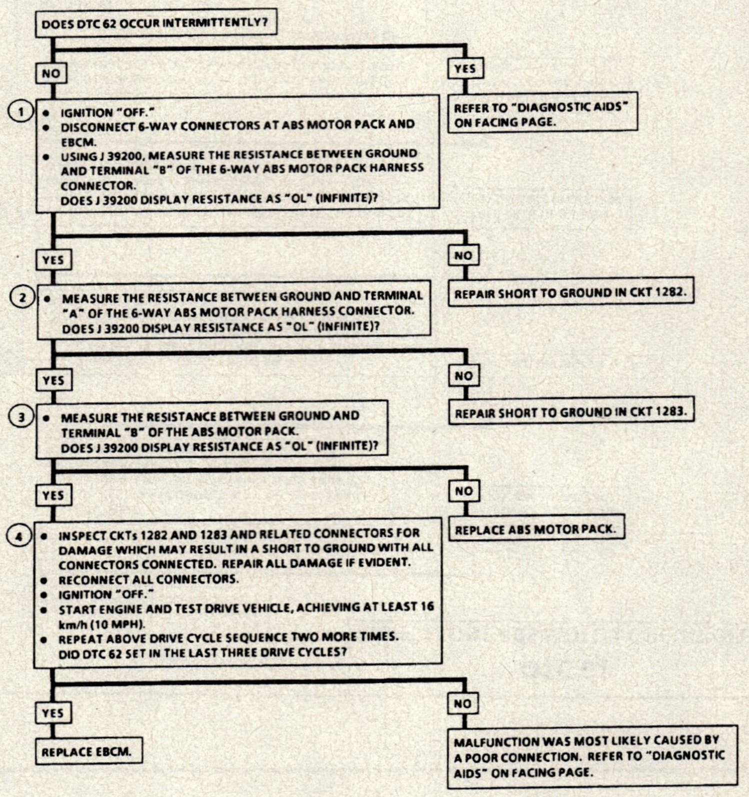

AFTER ALL DIAGNOSIS IS COMPLETE, CLEAR DTCs AND TEST DRIVE VEHICLE FOR THREE (3) DRIVE CYCLES TO VERIFY THAT DTC DOES NOT RESET. A DRIVE CYCLE CONSISTS OF STARTING THE VEHICLE, DRIVING OVER 16 KM/H (10 MPH), AND THEN KEYING DOWN.

DIAGNOSTIC CHART DTC 63 — 1994 LUMINA

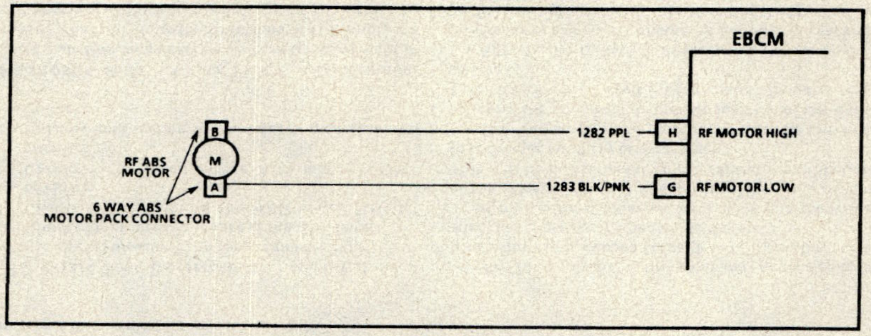

DTC 63

RIGHT FRONT ABS MOTOR CIRCUIT SHORTED TO BATT

Circuit Description:
This DTC identifies an ABS motor circuit that is shorted to battery or an ABS motor that has low or no resistance. This malfunction will not allow the ABS motor to be controlled at the commanded current rate or will cause the ABS motor to turn in the opposite direction or not at all.

DTC Will Set When: DTC 63 can be set only when the ABS motor is commanded "OFF." If the EBCM detects an out of range voltage on either of the right front ABS motor circuits indicating a circuit shorted to battery or an ABS motor shorted, a malfunction exists.

Action Taken: A malfunction DTC is stored, ABS is disabled and the ABS warning lamp is turned "ON."

DTC Chart Test Description: Number(s) below refer to circled number(s) on the diagnostic chart.

1. Checks for a short to voltage in the ABS motor "HIGH" circuitry.
2. Checks for a short to voltage in the ABS motor "LOW" circuitry.
3. Checks for a short circuit between ABS motor "HIGH" and motor "LOW."
4. Checks for proper resistance of the ABS motor.
5. Ensures malfunction was not due to physical damage of the circuitry.

Diagnostic Aids: Using Tech 1, select manual control function, and exercise ABS motor movement of affected channel in both directions while applying light pressure on the brake pedal.

If erratic or "jumpy" brake pedal movement is detected while performing an "apply" or "release" function of the ABS motor, an intermittent malfunction may be indicated.

An "Intermittent" malfunction may be caused by a poor connection, rubbed through wire insulation, or a wire that is broken inside the insulation.

If the malfunction is not current, wiggle the wires of the affected channel and check if the DTC resets. This will help to pinpoint an intermittent malfunction in the motor circuitry or connections.

The frequency of the malfunction can be checked by using the enhanced diagnostic function of the Tech 1, as described in "Tech 1 Diagnostics" found in this section.

Any circuitry, that is suspected as causing the intermittent complaint, should be thoroughly checked for backed out terminals, improper mating, broken locks, improperly formed or damaged terminals, poor terminal to wiring connections or physical damage to the wiring harness.

DIAGNOSTIC CHART DTC 63 — 1994 LUMINA

DTC 63
RIGHT FRONT ABS MOTOR CIRCUIT SHORTED TO BATT

NOTICE: J 39200 TEST LEADS MUST BE "ZEROED" PRIOR TO MAKING ANY RESISTANCE MEASUREMENTS. REFER TO J 39200 USER'S MANUAL.

DOES DTC 63 OCCUR INTERMITTENTLY?

- YES → REFER TO "DIAGNOSTIC AIDS" ON FACING PAGE.
- NO →

(1)
- IGNITION "OFF."
- DISCONNECT 6-WAY CONNECTORS AT ABS MOTOR PACK AND EBCM.
- IGNITION "ON."
- USING J 39200, MEASURE THE VOLTAGE BETWEEN GROUND AND TERMINAL "B" OF THE 6-WAY ABS MOTOR PACK HARNESS CONNECTOR. IS VOLTAGE 1 VOLT OR LESS?

- NO → REPAIR SHORT TO VOLTAGE IN CKT 1282.
- YES →

(2)
- MEASURE THE VOLTAGE BETWEEN GROUND AND TERMINAL "A" OF THE 6-WAY ABS MOTOR PACK HARNESS CONNECTOR. IS VOLTAGE 1 VOLT OR LESS?

- NO → REPAIR SHORT TO VOLTAGE IN CKT 1283.
- YES →

(3)
- IGNITION "OFF."
- USING J 39200, MEASURE THE RESISTANCE BETWEEN TERMINAL "A" AND TERMINAL "B" OF THE 6-WAY ABS MOTOR PACK HARNESS CONNECTOR. DOES J 39200 DISPLAY RESISTANCE AS "OL" (INFINITE)?

- NO → REPAIR SHORT BETWEEN CKTs 1282 AND 1283.
- YES →

(4)
- USING J 39200, MEASURE THE RESISTANCE BETWEEN TERMINAL "A" AND TERMINAL "B" OF THE ABS MOTOR PACK. IS RESISTANCE GREATER THAN 0.4Ω?

- NO → REPLACE ABS MOTOR PACK.
- YES →

(5)
- INSPECT CKTs 1282 AND 1283 AND RELATED CONNECTORS FOR DAMAGE WHICH MAY RESULT IN A SHORT TO GROUND WITH ALL CONNECTORS CONNECTED. REPAIR ALL DAMAGE IF EVIDENT.
- RECONNECT ALL CONNECTORS.
- IGNITION "OFF."
- START ENGINE AND TEST DRIVE VEHICLE, ACHIEVING AT LEAST 16 km/h (10 MPH).
- REPEAT ABOVE DRIVE CYCLE SEQUENCE TWO MORE TIMES. DID DTC 63 SET IN THE LAST THREE DRIVE CYCLES?

- YES → REPLACE EBCM.
- NO → MALFUNCTION WAS MOST LIKELY CAUSED BY A POOR CONNECTION. REFER TO "DIAGNOSTIC AIDS" ON FACING PAGE.

AFTER ALL DIAGNOSIS IS COMPLETE, CLEAR DTCs AND TEST DRIVE VEHICLE FOR THREE (3) DRIVE CYCLES TO VERIFY THAT DTC DOES NOT RESET. A DRIVE CYCLE CONSISTS OF STARTING THE VEHICLE, DRIVING OVER 16 km/h (10 MPH), AND THEN KEYING DOWN.

DIAGNOSTIC CHART DTC 64 — 1994 LUMINA

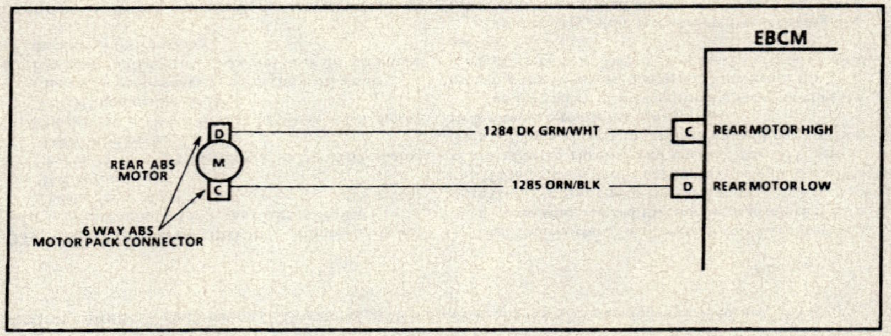

DTC 64
REAR ABS MOTOR CIRCUIT OPEN

Circuit Description:
This DTC identifies an ABS motor that cannot be energized due to an open in its circuitry. This malfunction will not allow proper rear ABS operation.

DTC Will Set When: DTC 64 can be set only when the ABS motor is commanded "OFF." If the EBCM detects an out of range voltage on either of the rear ABS motor circuits indicating an open circuit, a malfunction exists.

Action Taken: An open motor will not activate when requested. A malfunction DTC is stored, ABS is disabled and the ABS warning lamp is turned "ON."

DTC Chart Test Description: Number(s) below refer to circled number(s) on the diagnostic chart.

1. Checks for proper resistance of the ABS motor.
2. Checks for an open in the ABS motor "HIGH" circuitry.
3. Checks for an open in the ABS motor "LOW" circuitry.
4. Ensures malfunction was not due to poor terminal contact.

Diagnostic Aids: Using Tech 1, select manual control function, and exercise ABS motor movement of affected channel in both directions while applying light pressure on the brake pedal.

If erratic or "jumpy" brake pedal movement is detected while performing an "apply" or "release" function of the ABS motor, an intermittent malfunction may be indicated.

An "Intermittent" malfunction may be caused by a poor connection, rubbed through wire insulation, or a wire that is broken inside the insulation.

If the malfunction is not current, wiggle the wires of the affected channel and check if the DTC resets. This will help to pinpoint an intermittent malfunction in the motor circuitry or connections.

The frequency of the malfunction can be checked by using the enhanced diagnostic function of the Tech 1, as described in "Tech 1 Diagnostics" found in this section.

Any circuitry, that is suspected as causing the intermittent complaint, should be thoroughly checked for backed out terminals, improper mating, broken locks, improperly formed or damaged terminals, poor terminal to wiring connections or physical damage to the wiring harness.

DIAGNOSTIC CHART DTC 64 — 1994 LUMINA

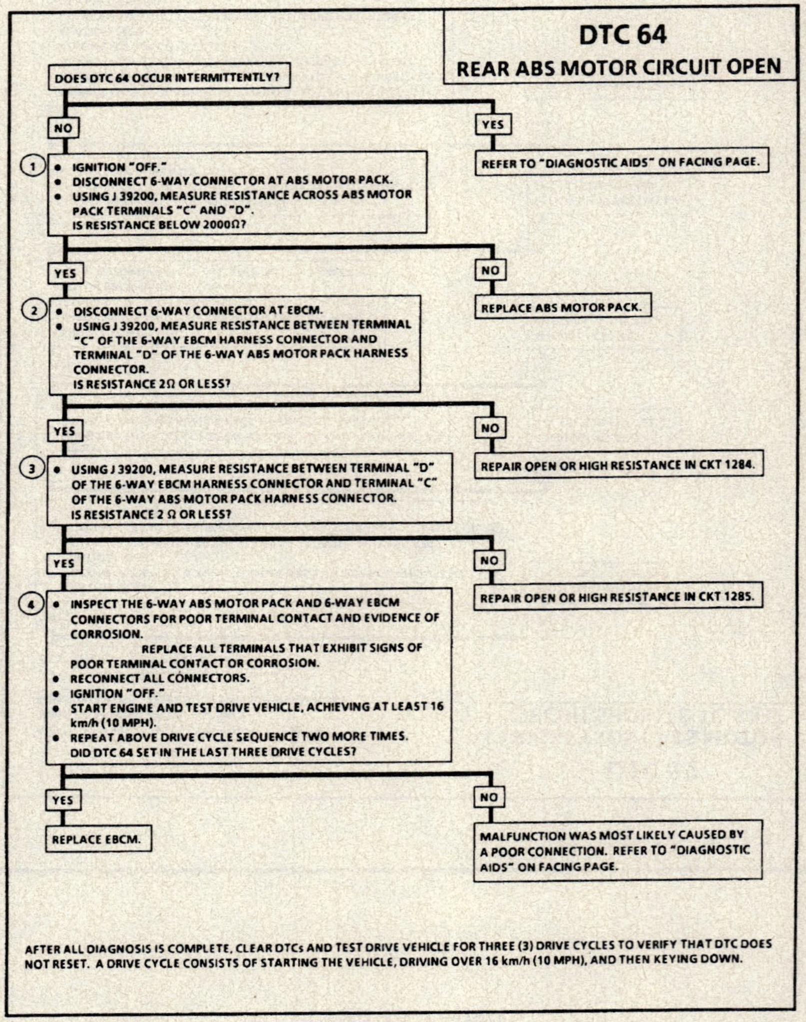

DIAGNOSTIC CHART DTC 65 — 1994 LUMINA

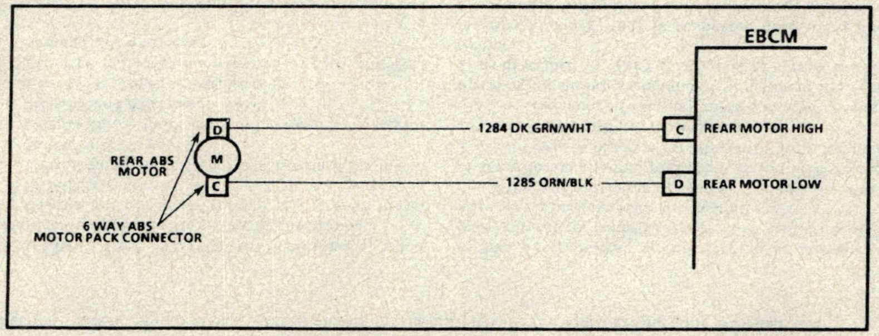

DTC 65

REAR ABS MOTOR CIRCUIT SHORTED TO GROUND

Circuit Description:
This DTC identifies an ABS motor circuit that is shorted to ground. This malfunction will not allow the ABS motor to be controlled at the commanded current rate or will cause the driver circuit to allow current directly to ground.

DTC Will Set When: DTC 65 can be set anytime. If the EBCM detects an out of range voltage on either of the rear ABS motor circuits indicating a circuit shorted to ground, a malfunction exists.

Action Taken: A malfunction DTC is stored, ABS is disabled and the ABS warning lamp is turned "ON."

DTC Chart Test Description: Number(s) below refer to circled number(s) on the diagnostic chart.

1. Checks for a short to ground in the ABS motor "HIGH" circuitry.
2. Checks for a short to ground in the ABS motor "LOW" circuitry.
3. Checks for an ABS motor that is internally shorted to ground.
4. Ensures malfunction was not due to physical damage of the circuitry.

Diagnostic Aids: Using Tech 1, select manual control function, and exercise ABS motor movement of affected channel in both directions while applying light pressure on the brake pedal.

If erratic or "jumpy" brake pedal movement is detected while performing an "apply" or "release" function of the ABS motor, an intermittent malfunction may be indicated.

An "Intermittent" malfunction may be caused by a poor connection, rubbed through wire insulation, or a wire that is broken inside the insulation.

If the malfunction is not current, wiggle the wires of the affected channel and check if the DTC resets. This will help to pinpoint an intermittent malfunction in the motor circuitry or connections.

The frequency of the malfunction can be checked by using the enhanced diagnostic function of the Tech 1, as described in "Tech 1 Diagnostics" found in this section.

Any circuitry, that is suspected as causing the intermittent complaint, should be thoroughly checked for backed out terminals, improper mating, broken locks, improperly formed or damaged terminals, poor terminal to wiring connections or physical damage to the wiring harness.

DIAGNOSTIC CHART DTC 65 — 1994 LUMINA

DTC 65

REAR ABS MOTOR CIRCUIT SHORTED TO GROUND

DOES DTC 65 OCCUR INTERMITTENTLY?

- YES → REFER TO "DIAGNOSTIC AIDS" ON FACING PAGE.
- NO →

(1)
- IGNITION "OFF."
- DISCONNECT 6-WAY CONNECTORS AT ABS MOTOR PACK AND EBCM.
- USING A J 39200, MEASURE THE RESISTANCE BETWEEN GROUND AND TERMINAL "D" OF THE 6-WAY ABS MOTOR PACK HARNESS CONNECTOR.

DOES J 39200 DISPLAY RESISTANCE AS "OL" (INFINITE)?

- NO → REPAIR SHORT TO GROUND IN CKT 1284.
- YES →

(2)
- MEASURE THE RESISTANCE BETWEEN GROUND AND TERMINAL "C" OF THE 6-WAY ABS MOTOR PACK HARNESS CONNECTOR.

DOES J 39200 DISPLAY RESISTANCE AS "OL" (INFINITE)?

- NO → REPAIR SHORT TO GROUND IN CKT 1285.
- YES →

(3)
- MEASURE THE RESISTANCE BETWEEN GROUND AND TERMINAL "D" OF THE ABS MOTOR PACK.

DOES J 39200 DISPLAY RESISTANCE AS "OL" (INFINITE)?

- NO → REPLACE ABS MOTOR PACK.
- YES →

(4)
- INSPECT CKTs 1284 AND 1285 AND RELATED CONNECTORS FOR DAMAGE WHICH MAY RESULT IN A SHORT TO GROUND WITH ALL CONNECTORS CONNECTED. REPAIR ALL DAMAGE IF EVIDENT.
- RECONNECT ALL CONNECTORS.
- IGNITION "OFF."
- START ENGINE AND TEST DRIVE VEHICLE, ACHIEVING AT LEAST 16 km/h (10 MPH).
- REPEAT ABOVE DRIVE CYCLE SEQUENCE TWO MORE TIMES.

DID DTC 65 SET IN THE LAST THREE DRIVE CYCLES?

- YES → REPLACE EBCM.
- NO → MALFUNCTION WAS MOST LIKELY CAUSED BY A POOR CONNECTION. REFER TO "DIAGNOSTIC AIDS" ON FACING PAGE.

AFTER ALL DIAGNOSIS IS COMPLETE, CLEAR DTCs AND TEST DRIVE VEHICLE FOR THREE (3) DRIVE CYCLES TO VERIFY THAT DTC DOES NOT RESET. A DRIVE CYCLE CONSISTS OF STARTING THE VEHICLE, DRIVING OVER 16 km/h (10 MPH), AND THEN KEYING DOWN.

DIAGNOSTIC CHART DTC 66 — 1994 LUMINA

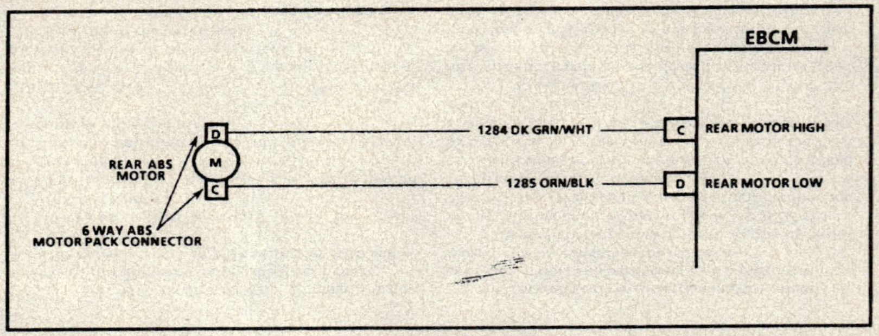

DTC 66

REAR ABS MOTOR CIRCUIT SHORTED TO BATTERY

Circuit Description:

This DTC identifies an ABS motor circuit that is shorted to battery or an ABS motor that has low or zero resistance. This malfunction will not allow the ABS motor to be controlled at the commanded current rate or will cause the ABS motor to turn in the opposite direction or not turn at all.

DTC Will Set When: DTC 66 can be set only when the ABS motor is commanded "OFF." If the EBCM detects an out of range voltage on either of the rear ABS motor circuits indicating a circuit shorted to battery or an ABS motor shorted, a malfunction exists.

Action Taken: A malfunction DTC is stored, ABS is disabled and the ABS warning lamp is turned "ON."

DTC Chart Test Description: Number(s) below refer to circled number(s) on the diagnostic chart.

1. Checks for a short to voltage in the ABS motor "HIGH" circuitry.
2. Checks for a short to voltage in the ABS motor "LOW" circuitry.
3. Checks for a short circuit between ABS motor "HIGH" and ABS motor "LOW."
4. Checks for proper resistance of the ABS motor.
5. Ensures malfunction was not due to physical damage of the circuitry.

Diagnostic Aids: Using Tech 1, select manual control function, and exercise ABS motor movement of affected channel in both directions while applying light pressure on the brake pedal.

If erratic or "jumpy" brake pedal movement is detected while performing an "apply" or "release" function of the ABS motor, an intermittent malfunction may be indicated.

An "Intermittent" malfunction may be caused by a poor connection, rubbed through wire insulation, or a wire that is broken inside the insulation.

If the malfunction is not current, wiggle the wires of the affected channel and check if the DTC resets. This will help to pinpoint an intermittent malfunction in the motor circuitry or connections.

The frequency of the malfunction can be checked by using the enhanced diagnostic function of the Tech 1, as described in "Tech 1 Diagnostics" found in this section.

Any circuitry, that is suspected as causing the intermittent complaint, should be thoroughly checked for backed out terminals, improper mating, broken locks, improperly formed or damaged terminals, poor terminal to wiring connections or physical damage to the wiring harness.

DIAGNOSTIC CHART DTC 66 — 1994 LUMINA

DTC 66

REAR ABS MOTOR CIRCUIT SHORTED TO BATTERY

NOTICE: J 39200 TEST LEADS MUST BE "ZEROED" PRIOR TO MAKING ANY RESISTANCE MEASUREMENTS. REFER TO J 39200 USER'S MANUAL.

DOES DTC 66 OCCUR INTERMITTENTLY?

- YES: REFER TO "DIAGNOSTIC AIDS" ON FACING PAGE.
- NO:

(1)
- IGNITION "OFF."
- DISCONNECT 6-WAY CONNECTORS AT ABS MOTOR PACK AND EBCM.
- IGNITION "ON."
- USING J 39200, MEASURE THE VOLTAGE BETWEEN GROUND AND TERMINAL "D" OF THE 6-WAY ABS MOTOR PACK HARNESS CONNECTOR.

IS VOLTAGE 1 VOLT OR LESS?

- NO: REPAIR SHORT TO VOLTAGE IN CKT 1284.
- YES:

(2)
- MEASURE THE VOLTAGE BETWEEN GROUND AND TERMINAL "C" OF THE 6-WAY ABS MOTOR PACK HARNESS CONNECTOR.

- NO: REPAIR SHORT TO VOLTAGE IN CKT 1285.
- YES:

(3)
- IGNITION "OFF."
- USING J 39200, MEASURE THE VOLTAGE BETWEEN TERMINAL "C" AND TERMINAL "D" OF THE 6-WAY ABS MOTOR PACK HARNESS CONNECTOR.

DOES J 39200 DISPLAY RESISTANCE AS "OL"

- NO: REPAIR SHORT BETWEEN CKTs 1284 AND 1285.
- YES:

(4)
- USING J 39200, MEASURE THE RESISTANCE BETWEEN TERMINAL "C" AND TERMINAL "D" OF THE ABS MOTOR PACK.

IS RESISTANCE GREATER THAN 0.4Ω?

- NO: REPLACE ABS MOTOR PACK.
- YES:

(5)
- INSPECT CKTs 1284 AND 1285 ABS RELATED CONNECTORS FOR DAMAGE WHICH MAY RESULT IN A SHORT TO VOLTAGE WITH ALL CONNECTORS CONNECTED. REPAIR ALL DAMAGE IF EVIDENT.
- RECONNECT ALL CONNECTORS.
- IGNITION "OFF."
- START ENGINE AND TEST DRIVE VEHICLE, ACHIEVING AT LEAST 16 km/h (10 MPH).
- REPEAT ABOVE DRIVE CYCLE SEQUENCE TWO MORE TIMES.

DID DTC 66 SET IN THE LAST THREE DRIVE CYCLES.

- YES: REPLACE EBCM.
- NO: MALFUNCTION WAS MOSTLY CAUSED BY A POOR CONNECTION. REFER TO "DIAGNOSTIC AIDS" ON FACING PAGE.

AFTER ALL DIAGNOSIS IS COMPLETE, CLEAR DTCs AND TEST DRIVE VEHICLE FOR THREE (3) DRIVE CYCLES TO VERIFY THAT DTC DOES NOT RESET. A DRIVE CYCLE CONSISTS OF STARTING THE VEHICLE, DRIVING OVER 16 km/h (10 MPH), AND THEN KEYING DOWN.

DIAGNOSTIC CHART DTC 76 — 1994 LUMINA

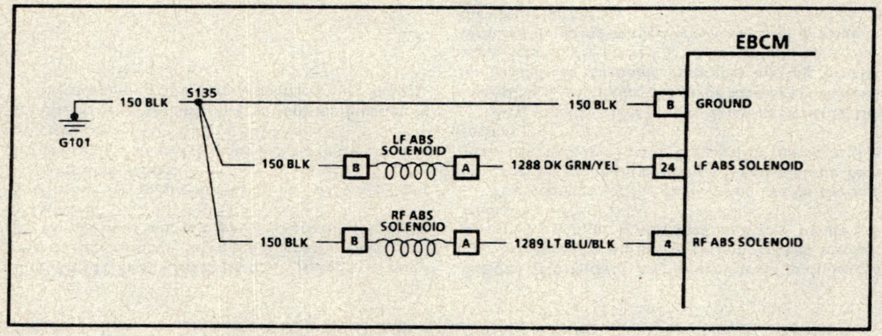

DTC 76

LEFT FRONT SOL. CIRCUIT SHORTED TO BATT

Circuit Description:

This DTC identifies a solenoid that cannot be energized due to an open in its circuitry, or a solenoid that is always energized due to a short to battery in its circuitry between the driver and the solenoid. An open will not allow proper ABS operation, but the short to battery simply turns "ON" the solenoid. A path for base brakes is still allowed once the motor rehomes and the check ball is lifted off its seat during key "ON" initialization.

DTC Will Set When: DTC 76 can be set only when the solenoid is commanded "OFF." If the EBCM detects an excessive voltage in the left front solenoid control circuit, a malfunction exists.

Action Taken: A malfunction DTC is stored, ABS is disabled and the ABS warning lamp is turned "ON."

DTC Chart Test Description: Number(s) below refer to circled number(s) on the diagnostic chart.

1. Checks for a short to voltage in the control circuitry of the solenoid.
2. Checks for an open in the control circuit of the solenoid.
3. Checks the solenoid coil for proper resistance.
4. Checks for an open in the solenoid ground circuit.
5. Checks for a possible intermittent malfunction in the solenoid circuitry due to poor terminal contact.

Diagnostic Aids: An "Intermittent" malfunction may be caused by a poor connection, rubbed through wire insulation, or a wire that is broken inside the insulation.

The frequency of the malfunction can be checked by using the enhanced diagnostic function of the Tech 1, as described in "Tech 1 Diagnostics" found in this section.

Any circuitry, that is suspected as causing the intermittent complaint, should be thoroughly checked for backed out terminals, improper mating, broken locks, improperly formed or damaged terminals, poor terminal to wiring connections or physical damage to the wiring harness.

DIAGNOSTIC CHART DTC 76 — 1994 LUMINA

DTC 76

LEFT FRONT SOL. CIRCUIT SHORTED TO BATT

(1)
- IGNITION "OFF."
- DISCONNECT 24-WAY EBCM CONNECTOR.
- IGNITION "ON" ENGINE "OFF."
- USING J 39200, MEASURE THE VOLTAGE BETWEEN GROUND AND TERMINAL "24" OF THE 24-WAY EBCM HARNESS CONNECTOR.

IS VOLTAGE 1 VOLT OR LESS?

NO → REPAIR SHORT TO VOLTAGE IN CKT 1288.

YES ↓

(2)
- IGNITION "OFF."
- DISCONNECT LEFT FRONT SOLENOID CONNECTOR.
- USING J 39200, MEASURE THE RESISTANCE BETWEEN TERMINAL "24" OF THE 24-WAY EBCM HARNESS CONNECTOR AND TERMINAL "A" OF THE LEFT FRONT SOLENOID HARNESS CONNECTOR.

IS RESISTANCE 2Ω OR LESS?

NO → REPAIR OPEN OR HIGH RESISTANCE IN CKT 1288.

YES ↓

(3)
- MEASURE THE RESISTANCE BETWEEN TERMINAL "A" AND TERMINAL "B" OF THE LEFT FRONT SOLENOID.

IS RESISTANCE BETWEEN 2.5 AND 5Ω?

NO → REPLACE LEFT FRONT SOLENOID.

YES ↓

(4)
- MEASURE THE RESISTANCE BETWEEN GROUND AND TERMINAL "B" OF THE LEFT FRONT SOLENOID HARNESS CONNECTOR.

IS RESISTANCE 2Ω OR LESS?

NO → REPAIR OPEN OR HIGH RESISTANCE IN CKT 150 (CHEVROLET) OR 1150 (BUICK/OLDSMOBILE).

YES ↓

(5)
- INSPECT LEFT FRONT SOLENOID AND 24-WAY EBCM HARNESS CONNECTORS FOR POOR TERMINAL CONTACT AND EVIDENCE OF CORROSION.
  REPLACE ALL TERMINALS THAT EXHIBIT SIGNS OF POOR TERMINAL CONTACT OR CORROSION.
- INSPECT CKT 1288 FOR DAMAGE WHICH MAY RESULT IN A SHORT TO B+ WITH ALL CONNECTORS CONNECTED. REPAIR DAMAGE IF EVIDENT.
- RECONNECT ALL CONNECTORS.
- IGNITION "ON."

IS DTC 76 SET AS A CURRENT DTC?

YES → REPLACE EBCM.

NO → MALFUNCTION IS MOST LIKELY DUE TO A POOR CONNECTION. REFER TO "DIAGNOSTIC AIDS" ON FACING PAGE.

DIAGNOSTIC CHART DTC 77 — 1994 LUMINA

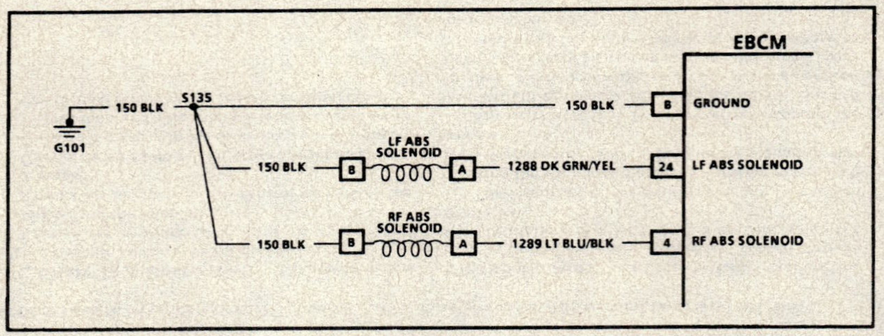

DTC 77

LEFT FRONT SOLENOID CIRCUIT SHORTED TO GND

Circuit Description:

This DTC identifies a solenoid that cannot be energized due to an open in its driver circuitry, or a short to ground between the solenoid driver and the solenoid. These malfunctions can affect ABS operation since the flow of brake fluid to the wheel cylinder cannot be stopped, making ABS operation for that channel impossible.

DTC Will Set When: DTC 77 can be set only when the solenoid is commanded "ON." If the EBCM detects the left front solenoid control circuit voltage is out of specification, a malfunction exists.

Action Taken: A malfunction DTC is stored, ABS is disabled and the ABS warning lamp is turned "ON."

DTC Chart Test Description: Number(s) below refer to circled number(s) on the diagnostic chart.

1. Checks for a solenoid that is internally shorted to ground.
2. Checks for solenoid that is not within proper resistance values.
3. Indicates if a short to ground exists in the solenoid circuitry.
4. Checks for a possible intermittent malfunction in the solenoid circuitry due to poor terminal contact.

Diagnostic Aids: An "Intermittent" malfunction may be caused by a poor connection, rubbed through wire insulation, or a wire that is broken inside the insulation.

The frequency of the malfunction can be checked by using the enhanced diagnostic function of the Tech 1, as described in "Tech 1 Diagnostics" found in this section.

Any circuitry, that is suspected as causing the intermittent complaint, should be thoroughly checked for backed out terminals, improper mating, broken locks, improperly formed or damaged terminals, poor terminal to wiring connections or physical damage to the wiring harness.

DIAGNOSTIC CHART DTC 77 — 1994 LUMINA

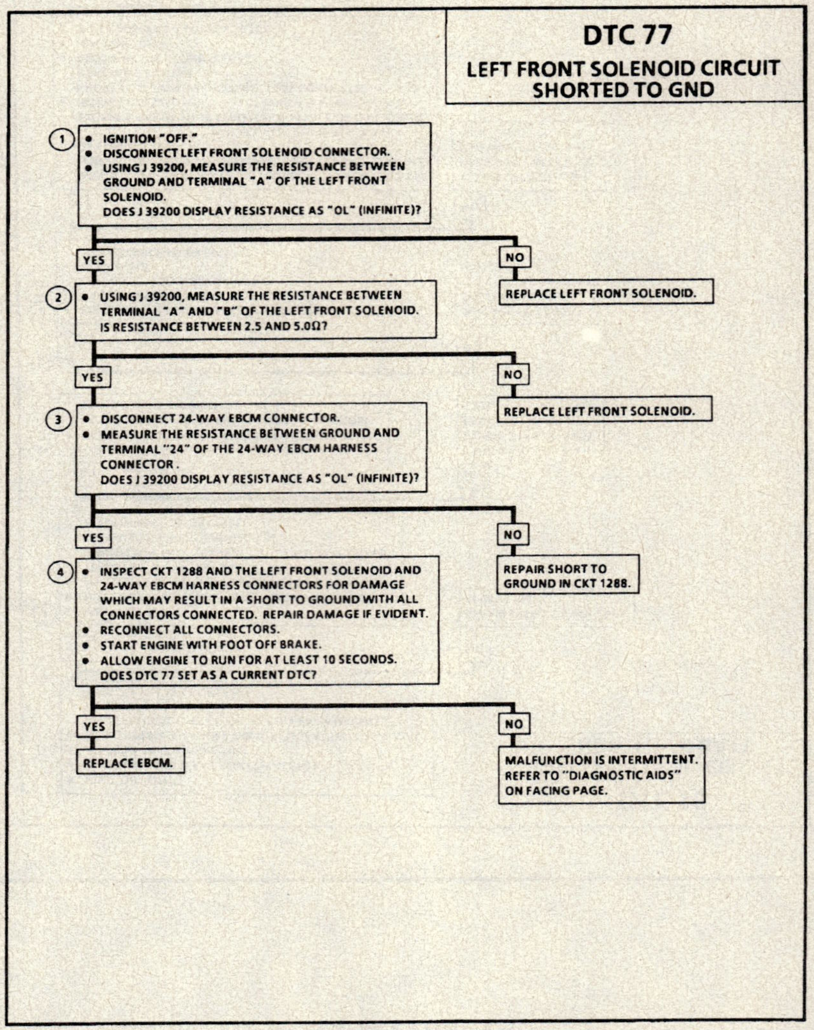

DIAGNOSTIC CHART DTC 78 — 1994 LUMINA

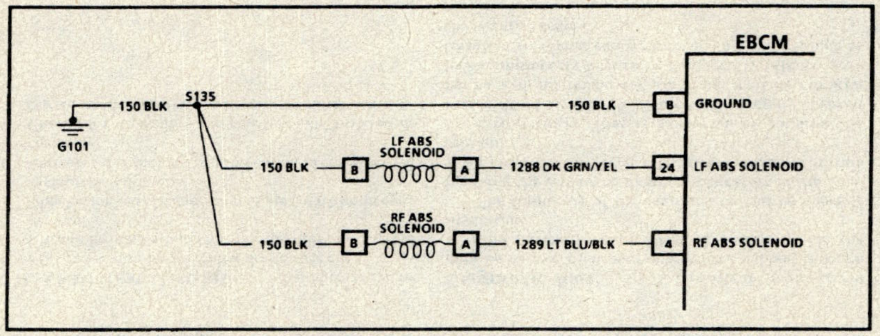

DTC 78

RIGHT FRONT SOL. CIRCUIT OPEN OR SHORTED TO BATT

Circuit Description:
This DTC identifies a solenoid that cannot be energized due to an open in its circuitry, or a solenoid that is always energized due to a short to battery in its circuitry between the driver and the solenoid. An open will not allow proper ABS operation, but the short to battery simply turns "ON" the solenoid. A path for base brakes is still allowed once the motor rehomes and the check ball is lifted off its seat during key "ON" initialization.

DTC Will Set When: DTC 78 can be set only when the solenoid is commanded "OFF." If the EBCM detects an excessive voltage in the right front solenoid control circuit, a malfunction exists.

Action Taken: A malfunction DTC is stored, ABS is disabled and the ABS warning lamp is turned "ON."

DTC Chart Test Description: Number(s) below refer to circled number(s) on the diagnostic chart.

1. Checks for a short to voltage in the control circuitry of the solenoid.
2. Checks for an open in the control circuit of the solenoid.
3. Checks the solenoid coil for proper resistance.
4. Checks for an open in the solenoid ground circuit.
5. Checks for a possible intermittent malfunction in the solenoid circuitry due to poor terminal contact.

Diagnostic Aids: An "Intermittent" malfunction may be caused by a poor connection, rubbed through wire insulation, or a wire that is broken inside the insulation.

The frequency of the malfunction can be checked by using the enhanced diagnostic function of the Tech 1, as described in "Tech 1 Diagnostics" found in this section.

Any circuitry, that is suspected as causing the intermittent complaint, should be thoroughly checked for backed out terminals, improper mating, broken locks, improperly formed or damaged terminals, poor terminal to wiring connections or physical damage to the wiring harness.

DIAGNOSTIC CHART DTC 78 — 1994 LUMINA

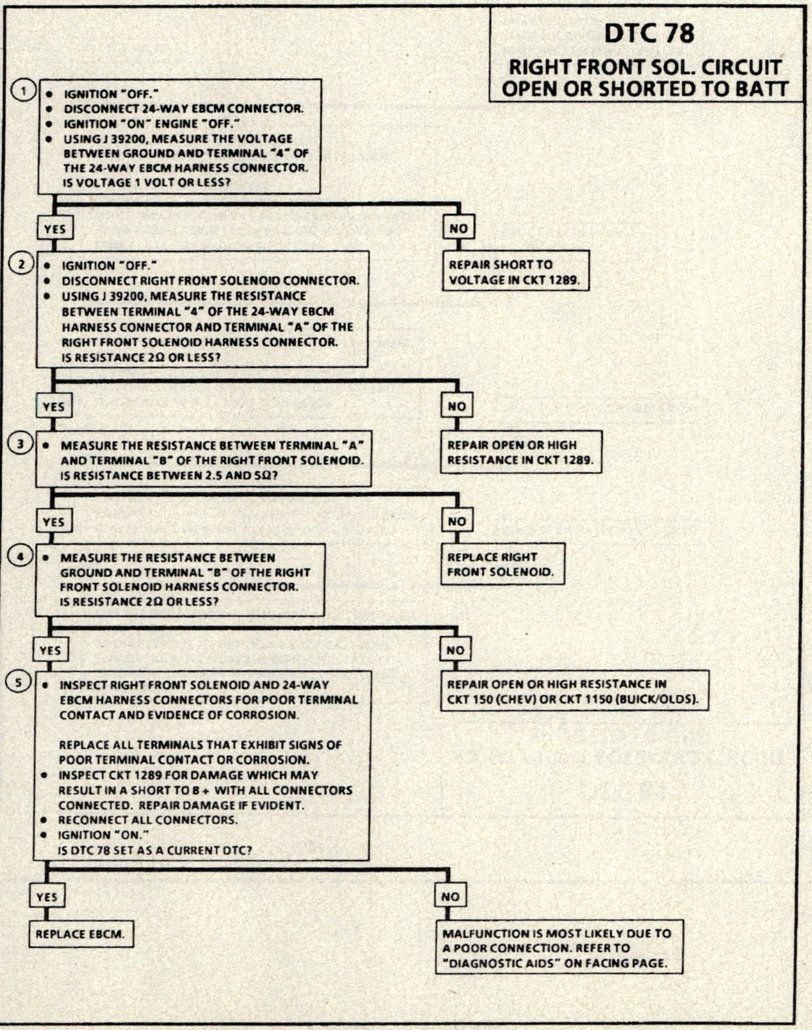

DIAGNOSTIC CHART DTC 81 — 1994 LUMINA

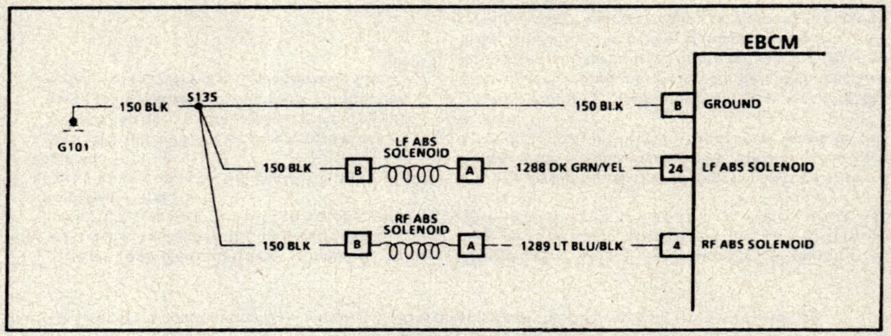

DTC 81

RIGHT FRONT SOLENOID CIRCUIT SHORTED TO GND

Circuit Description:

This DTC identifies a solenoid that cannot be energized due to an open in its driver circuitry, or a short to ground between the solenoid driver and the solenoid. These malfunctions can affect ABS operation since the flow of brake fluid to the wheel cylinder cannot be stopped, making ABS operation for that channel impossible.

DTC Will Set When: DTC 81 can be set only when the solenoid is commanded "ON." If the EBCM detects the right front solenoid control circuit voltage is out of specification, a malfunction exists.

Action Taken: A malfunction DTC is stored, ABS is disabled and the ABS warning lamp is turned "ON."

DTC Chart Test Description: Number(s) below refer to circled number(s) on the diagnostic chart.

1. Checks for a solenoid that is internally shorted to ground.
2. Checks for a solenoid that is not within proper resistance values.
3. Indicates if a short to ground exists in the solenoid circuitry.
4. Checks for a possible intermittent malfunction in the solenoid circuitry due to poor terminal contact.

Diagnostic Aids: An "Intermittent" malfunction may be caused by a poor connection, rubbed through wire insulation, or a wire that is broken inside the insulation.

The frequency of the malfunction can be checked by using the enhanced diagnostic function of the Tech 1, as described in "Tech 1 Diagnostics" found in this section.

Any circuitry, that is suspected as causing the intermittent complaint, should be thoroughly checked for backed out terminals, improper mating, broken locks, improperly formed or damaged terminals, poor terminal to wiring connections or physical damage to the wiring harness.

DIAGNOSTIC CHART DTC 81 — 1994 LUMINA

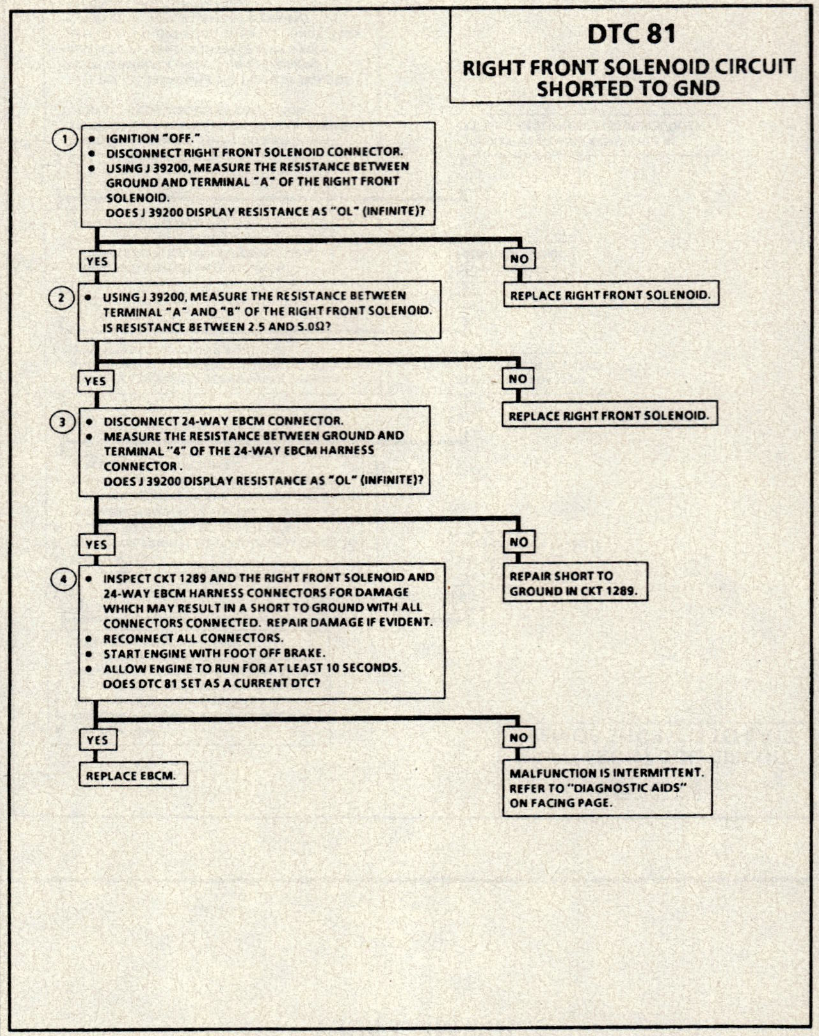

DIAGNOSTIC CHART DTC 82 — 1994 LUMINA

DTC 82
CALIBRATION MALFUNCTION

Circuit Description:

This DTC allows the EBCM to check for a calibration malfunction by comparing the calibration value to a known value stored in the EEPROM.

This DTC is also used as a security measure to prevent improper use of calibrations or changes to these calibrations that may alter the designed function of ABS.

DTC Will Set When: DTC 82 can be set at key-up. If the program identifier is incorrect or the memory checksum is incorrect, a malfunction exists.

Action Taken: A malfunction DTC is stored, ABS is disabled and the ABS warning lamp is turned "ON."

DTC Chart Test Description: Number(s) below refer to circled number(s) on the diagnostic chart.

1. Checks to see if the malfunction is present during diagnosis. If present, the EBCM is not functioning correctly and must be replaced.

Important

- At the time of printing, 1994 vehicles with ABS VI were not being field reprogrammed to correct this failure. A service bulletin will be issued when EEPROM reprogramming is authorized.

Diagnostic Aids: An intermittent DTC 82 may be caused by a bad cell in the EEPROM that is sensitive to temperature changes. If DTC 82 occurs more than once, but is intermittent, replace EBCM.

The frequency of the malfunction can be checked by using the enhanced diagnostic function of the Tech 1, as described in "Tech 1 Diagnostics" found in this section.

DIAGNOSTIC CHART DTC 82 — 1994 LUMINA

DTC 82
CALIBRATION MALFUNCTION

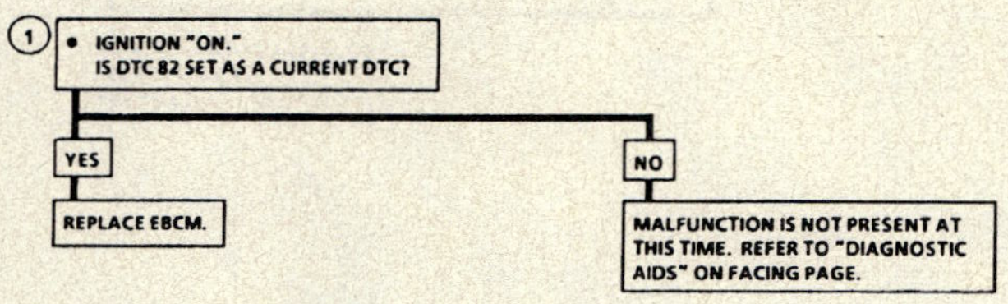

DIAGNOSTIC CHART DTC 86 — 1994 LUMINA

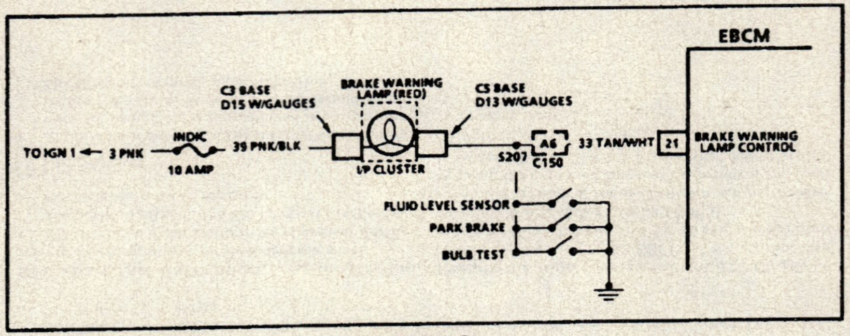

DTC 86

EBCM TURNED "ON" THE RED "BRAKE" WARNING LAMP

Circuit Description:

This DTC is provided as an information only DTC, and reflects the status of the command issued by the ABS EBCM to illuminate the red "BRAKE" warning lamp. If another DTC issues a command to illuminate the red "BRAKE" warning lamp, DTC 86 will be stored in EEPROM as a history DTC at the conclusion of the ignition cycle.

DTC Chart Test Description: Number(s) below refer to circled number(s) on the diagnostic chart.

1. This step serves to identify if a DTC other than DTC 86 commanded the red "BRAKE" warning lamp "ON."

Diagnostic Aids: Any ABS mechanical DTC that issues a command to illuminate the red "BRAKE" telltale will also result in DTC 86 being stored in EEPROM during shutdown. These DTCs are: 38, 41, 42, 46, and 51. If the motors are not in their home position, certain electrical DTCs will also command the red "BRAKE" telltale "ON." These DTCs are: 14, 16, 18, 55, 56, 57, 58, 61, 62, 63, 64, 65, and 66.

If any of these DTCs are indicated along with DTC 86, they must be corrected prior to addressing a DTC 86 malfunction.

DIAGNOSTIC CHART DTC 86 — 1994 LUMINA

DTC 86

EBCM TURNED "ON" THE RED "BRAKE" WARNING LAMP

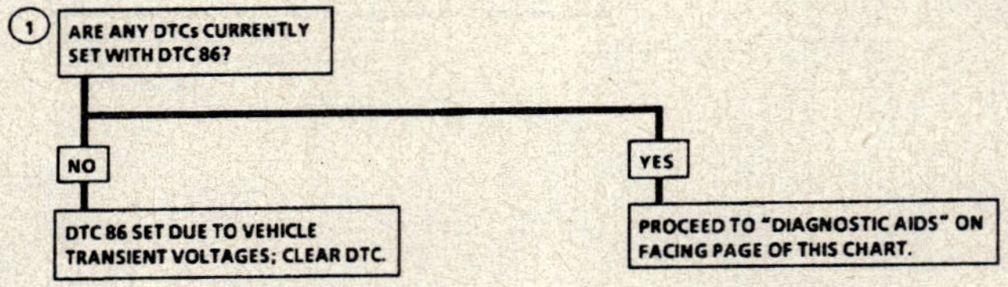

DIAGNOSTIC CHART DTC 87 — 1994 LUMINA

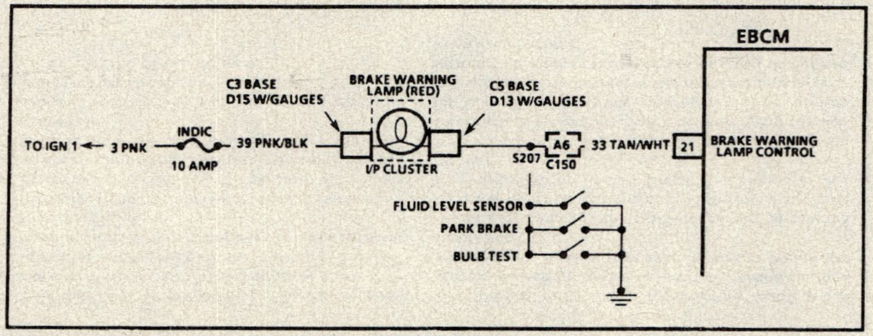

DTC 87

RED "BRAKE" WARNING LAMP CIRCUIT OPEN

Circuit Description:

This DTC is used to verify the EBCM has continuity to the red "BRAKE" warning lamp in case the EBCM must turn it "ON." This will only occur if an ABS malfunction is detected that may degrade base brake operation.

Because ABS is not the only device controlling the red "BRAKE" warning lamp (parking brake and low fluid sensor may also turn "ON" the lamp) a short to ground in this circuit cannot be detected.

DTC Will Set When: DTC 87 can be set after initialization is completed. If the red "BRAKE" warning lamp circuit voltage is out of specification indicating an open circuit, a malfunction exists.

Action Taken: A malfunction DTC is stored, ABS is not disabled; however, the ABS warning lamp will flash to indicate a malfunction exists.

DTC Chart Test Description: Number(s) below refer to circled number(s) on the diagnostic chart.

1. Identifies if the ground circuit to the red "BRAKE" warning lamp is completed by a source other than the EBCM.
2. Checks for high resistance in park brake circuitry.
3. Checks for a malfunctioning EBCM.
4. Checks for high resistance in fluid level switch circuitry.
5. Checks for a malfunctioning warning lamp bulb.
6. Checks for open or high resistance in CKT 33.

Diagnostic Aids: An "Intermittent" malfunction may be caused by a poor connection, rubbed through wire insulation, or a wire that is broken inside the insulation.

The frequency of the malfunction can be checked by using the enhanced diagnostic function of the Tech 1, as described in "Tech 1 Diagnostics" found in this section. If a Tech 1 or T-100 are not available, the parking brake handle may be lifted to verify proper lamp operation and continuity of CKTs 3 and 39.

Any circuitry, that is suspected as causing the intermittent complaint, should be thoroughly checked for backed out terminals, improper mating, broken locks, improperly formed or damaged terminals, poor terminal to wiring connections or physical damage to the wiring harness.

DIAGNOSTIC CHART DTC 87 — 1994 LUMINA

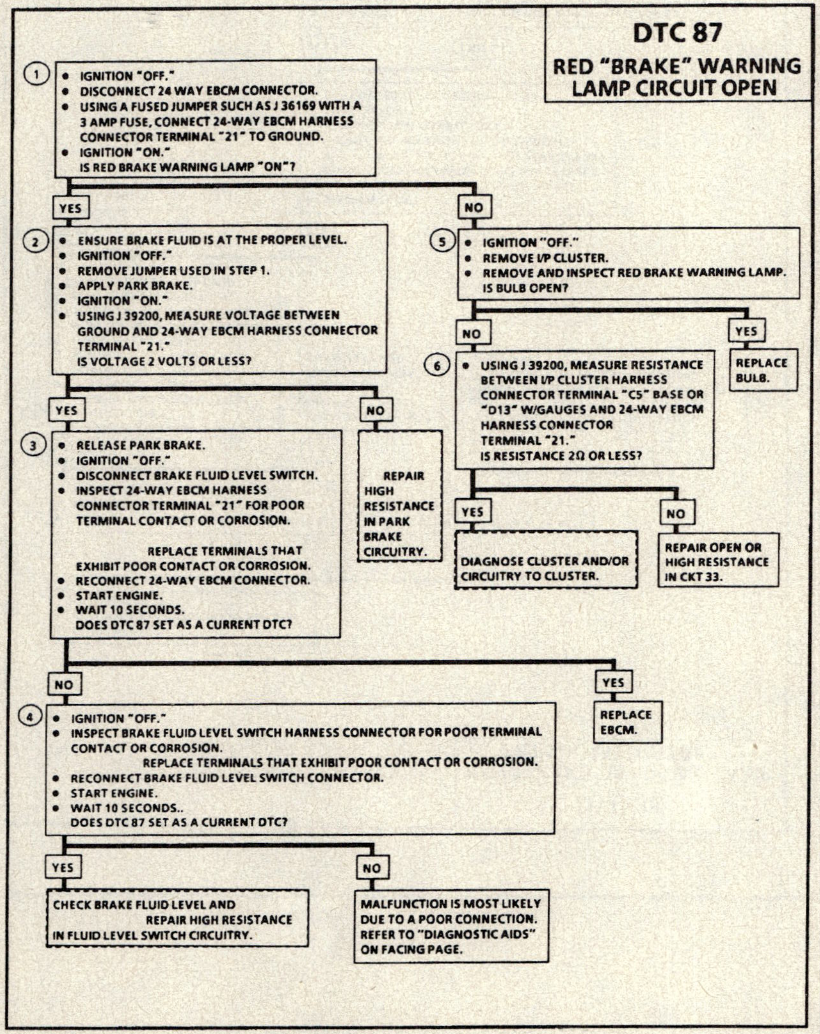

DIAGNOSTIC CHART DTC 88 — 1994 LUMINA

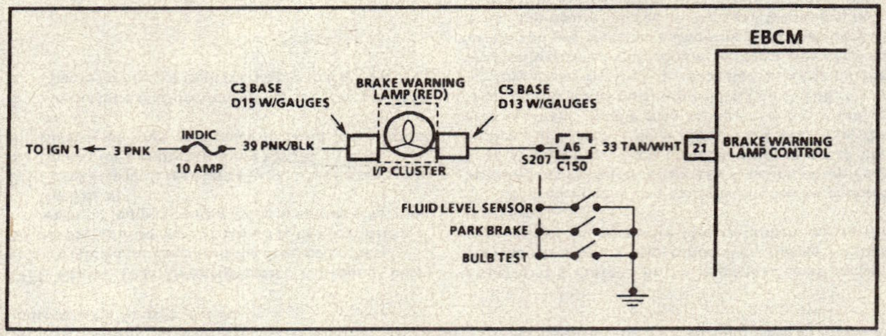

DTC 88

RED "BRAKE" WARNING LAMP CKT SHORTED TO BATT

Circuit Description:

This DTC identifies a short to battery between the EBCM and the red "BRAKE" warning lamp, or an open driver that does not allow the red "BRAKE" warning lamp to be illuminated by the EBCM. This will only occur if an ABS fault is detected that may degrade base brake operation.

DTC Will Set When: DTC 88 can be set only at the beginning of the three second bulb check. If the EBCM detects battery voltage on the red "BRAKE" warning lamp control circuit, a malfunction exists.

Action Taken: A malfunction DTC is stored. ABS is not disabled; however, the ABS warning lamp will flash to indicate a malfunction exists.

DTC Chart Test Description: Number(s) below refer to circled number(s) on the diagnostic chart.

1. Identifies if the ground circuit to the red "BRAKE" warning lamp is being completed by a source other than the EBCM.
2. Identifies if the malfunction is currently present.
3. By removing the 10 amp INDIC fuse, the voltage source is eliminated. This test indicates if voltage is being supplied from a source other than the gauge fuse.
4. Ensures malfunction was not due to physical damage of the circuitry.
5. Checks for DTC 86 also being set.

Diagnostic Aids: An "Intermittent" malfunction may be caused by a poor connection, rubbed through wire insulation, or a wire that is broken inside the insulation.

The frequency of the malfunction can be checked by using the enhanced diagnostic function of the Tech 1, as described in "Tech 1 Diagnostics" found in this section.

Any circuitry, that is suspected as causing the intermittent complaint, should be thoroughly checked for backed out terminals, improper mating, broken locks, improperly formed or damaged terminals, poor terminal to wiring connections or physical damage to the wiring harness.

DIAGNOSTIC CHART DTC 88 — 1994 LUMINA

DTC 88

RED "BRAKE" WARNING LAMP CKT SHORTED TO BATT

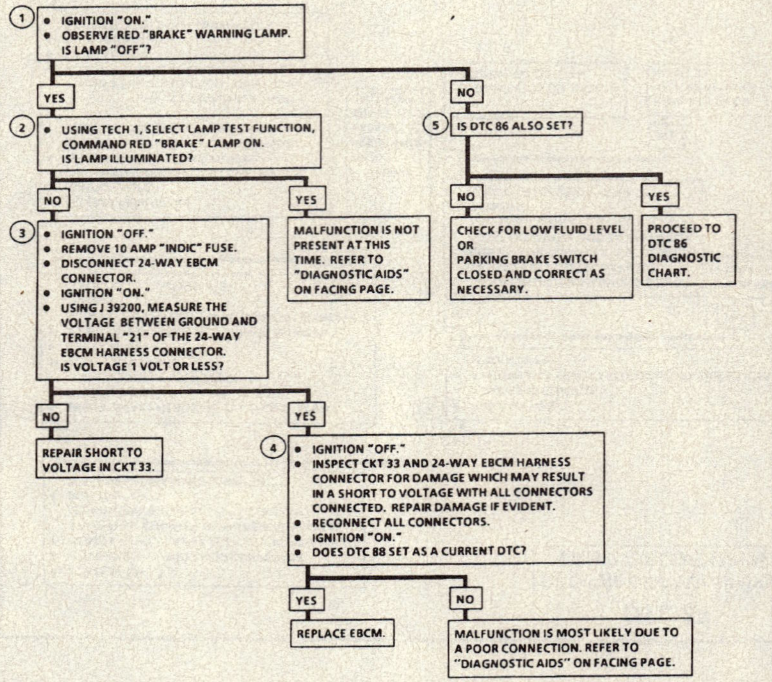

DIAGNOSTIC CHART DTC 91 — 1994 LUMINA

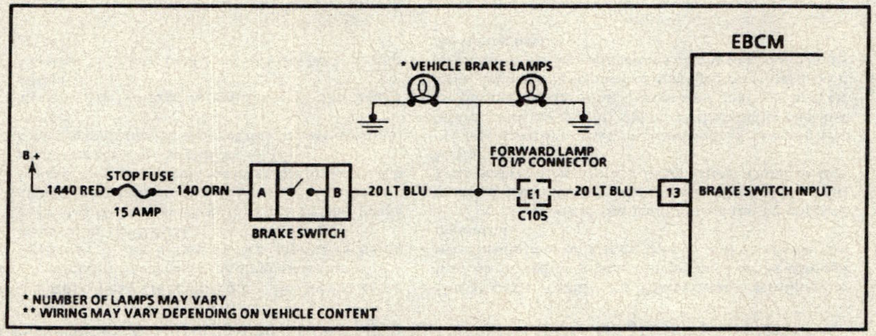

DTC 91

OPEN BRAKE SWITCH DURING DECELERATION

Circuit Description:

This DTC is used to detect an open brake switch in the non-ABS mode. The EBCM looks for deceleration rates that would indicate braking action and verifies this assumption by requiring several repeats of this detection method. In each case, ABS will not be available since no brake switch is seen.

DTC Will Set When: DTC 91 can be set if three deceleration cycles occur with the brake switch "OFF."

Action Taken: A malfunction DTC is stored, ABS is disabled and the ABS warning lamp is turned "ON."

DTC Chart Test Description: Number(s) below refer to circled number(s) on the diagnostic chart.

1. Indicates if the brake switch signal is being received by the EBCM.
2. Indicates if an open circuit exists in the brake switch or brake lamp circuitry.
3. Isolates the open circuit to either the brake switch input circuitry, or the EBCM.
4. Ensures malfunction was not due to poor terminal contact.
5. Verifies that voltage is available at the brake switch.
6. Verifies that the brake switch is functioning properly.

Diagnostic Aids: An "Intermittent" malfunction may be caused by a poor connection, rubbed through wire insulation, or a wire that is broken inside the insulation.

The frequency of the malfunction can be checked by using the enhanced diagnostic function of the Tech 1, as described in "Tech 1 Diagnostics" found in this section.

Any circuitry, that is suspected as causing the intermittent complaint, should be thoroughly checked for backed out terminals, improper mating, broken locks, improperly formed or damaged terminals, poor terminal to wiring connections or physical damage to the wiring harness.

DIAGNOSTIC CHART DTC 91 — 1994 LUMINA

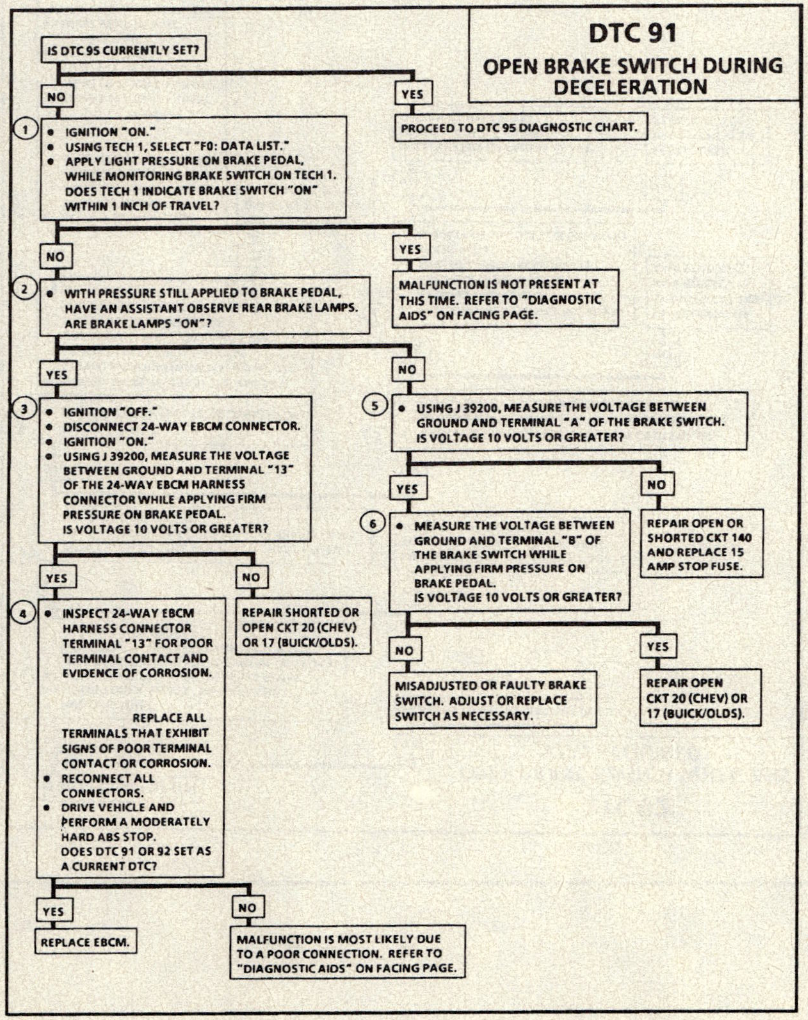

DIAGNOSTIC CHART DTC 92 — 1994 LUMINA

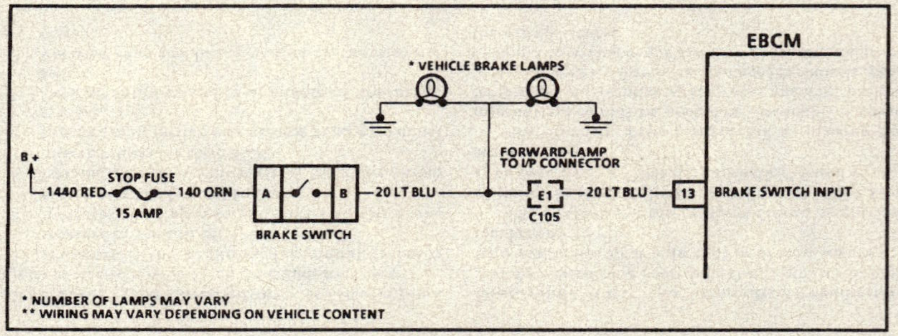

DTC 92

OPEN BRAKE SWITCH WHEN ABS WAS REQUIRED

Circuit Description:

This DTC is run to determine the proper operation of the brake switch. This is important because ABS is activated when the brake switch is "ON." If the brake switch is "OFF," ABS will never be activated. Since this malfunction is difficult to detect under normal braking conditions, this malfunction is only detected when ABS is required.

DTC Will Set When: DTC 92 can be set if the vehicle's speed is greater than 8 km/h (5 mph). If the brake was not "ON" and a release was required on two channels for 0.5 second, a malfunction exists.

Action Taken: A malfunction DTC is stored, ABS is disabled and the ABS warning lamp is turned "ON."

DTC Chart Test Description: Number(s) below refer to circled number(s) on the diagnostic chart.

1. Indicates if the brake switch signal is being received by the EBCM.
2. Indicates if an open circuit exists in the brake switch or brake lamp circuitry.
3. Isolates the open circuit to either the brake switch input circuitry, or the EBCM.
4. Ensures malfunction was not due to poor terminal contact.
5. Verifies the voltage is available at the brake switch.
6. Verifies that the brake switch is functioning properly.

Diagnostic Aids: An "Intermittent" malfunction may be caused by a poor connection, rubbed through wire insulation, or a wire that is broken inside the insulation.

The frequency of the malfunction can be checked by using the enhanced diagnostic function of the Tech 1, as described in "Tech 1 Diagnostics" found in this section.

Any circuitry, that is suspected as causing the intermittent complaint, should be thoroughly checked for backed out terminals, improper mating, broken locks, improperly formed or damaged terminals, poor terminal to wiring connections or physical damage to the wiring harness.

DIAGNOSTIC CHART DTC 92 — 1994 LUMINA

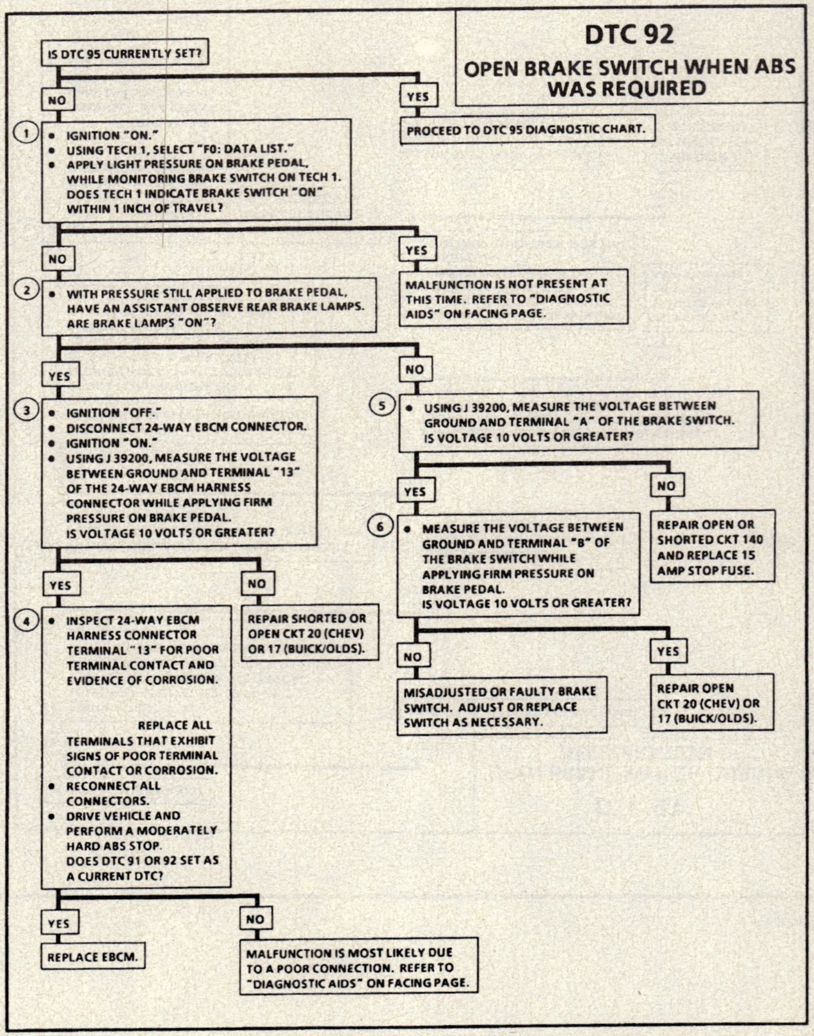

DIAGNOSTIC CHART DTC 93 — 1994 LUMINA

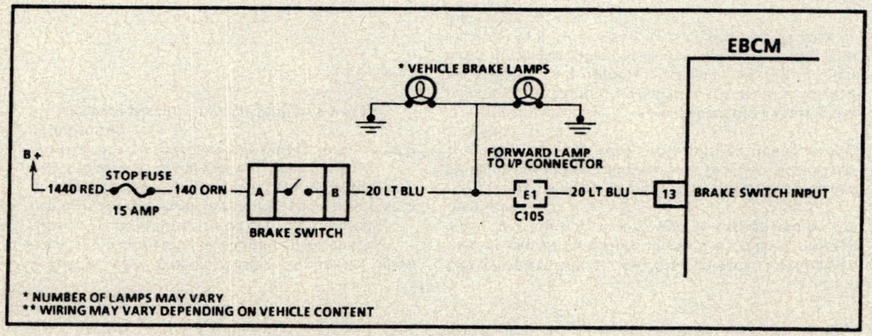

DTC 93

DTC 91 OR 92 SET IN CURRENT OR PREV. IGN CYCLE

Circuit Description:

This DTC is the second portion of DTCs 91 and 92. If DTCs 91 or 92 failed during the last ignition cycle, DTC 93 becomes a current failure during the next ignition cycle, keeping ABS disabled until a brake switch "ON" state is seen. When a change is seen during an ignition cycle in which DTC 93 is a current malfunction, DTC 91 or 92 will clear itself at the end of the current ignition cycle, and ABS will enable itself at the start of the next ignition cycle. DTC 93 alone indicates 91 or 92 failed previously, but is intermittent, or has been corrected.

DTC Chart Test Description: Number(s) below refer to circled number(s) on the diagnostic chart.

1. Indicates which DTC (either 91 or 92) caused DTC 93 to set.

Diagnostic Aids: An "Intermittent" malfunction may be caused by a poor connection, rubbed through wire insulation, or a wire that is broken inside the insulation.

The frequency of the malfunction can be checked by using the enhanced diagnostic function of the Tech 1, as described in "Tech 1 Diagnostics" found in this section.

Any circuitry, that is suspected as causing the intermittent complaint, should be thoroughly checked for backed out terminals, improper mating, broken locks, improperly formed or damaged terminals, poor terminal to wiring connections or physical damage to the wiring harness.

Also, verify proper brake switch operation using the data list of the Tech 1. As the brake is applied, the data list should display the brake switch "ON" within 1 inch of travel.

DIAGNOSTIC CHART DTC 93 — 1994 LUMINA

DTC 93

DTC 91 OR 92 SET IN CURRENT OR PREV. IGN CYCLE

(1)
- IGNITION "ON."
- USING TECH 1, READ DTCs.

IS DTC 91 OR 92 SET AS A HISTORY OR CURRENT DTC?

YES: PROCEED TO DTC CHART THAT SET.

NO: USING TECH 1, SELECT "F1: DTC HISTORY." VERIFY MALFUNCTION FREQUENCY WAS LOW AND CLEAR DTCs.

DIAGNOSTIC CHART DTC 94 — 1994 LUMINA

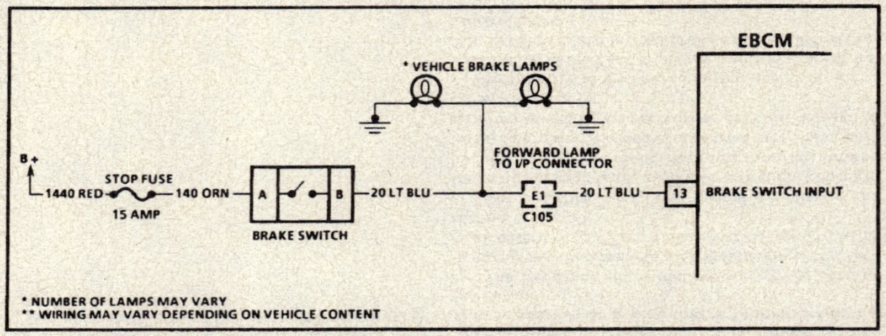

DTC 94

BRAKE SWITCH CONTACTS ALWAYS CLOSED

Circuit Description:

This DTC is run to determine the proper operation of the brake switch. This is important because ABS is activated when the brake switch is "ON" and turned "OFF" when the brake is "OFF." If the brake switch is always "ON," ABS operation will always be requested resulting in potential modulator cycling on rough roads. Additionally, this malfunction will most likely result in a dead battery (due to the brake lights staying "ON") if the driver is not informed of this malfunction.

DTC Will Set When: DTC 94 can be set when the vehicle reaches at least 40 km/h (25 mph). If the brake was never "OFF" during two consecutive drive cycles, a malfunction exists.

Action Taken: A malfunction DTC is stored. ABS is not disabled; however, the ABS warning will flash to indicate a malfunction exists.

DTC Chart Test Description: Number(s) below refer to circled number(s) on the diagnostic chart.

1. Checks if the malfunction is currently present.
2. Isolates the cause of the malfunction to either a malfunctioning or misadjusted brake switch, or a short to voltage in the brake switch circuitry.
3. Checks for unwanted voltage on the brake switch input circuit.
4. Checks for a possible intermittent malfunction.

Diagnostic Aids: An "Intermittent" malfunction may be caused by a poor connection, rubbed through wire insulation, or a wire that is broken inside the insulation.

The frequency of the malfunction can be checked by using the enhanced diagnostic function of the Tech 1, as described in "Tech 1 Diagnostics" found in this section.

Any circuitry, that is suspected as causing the intermittent complaint, should be thoroughly checked for backed out terminals, improper mating, broken locks, improperly formed or damaged terminals, poor terminal to wiring connections or physical damage to the wiring harness.

DIAGNOSTIC CHART DTC 94 — 1994 LUMINA

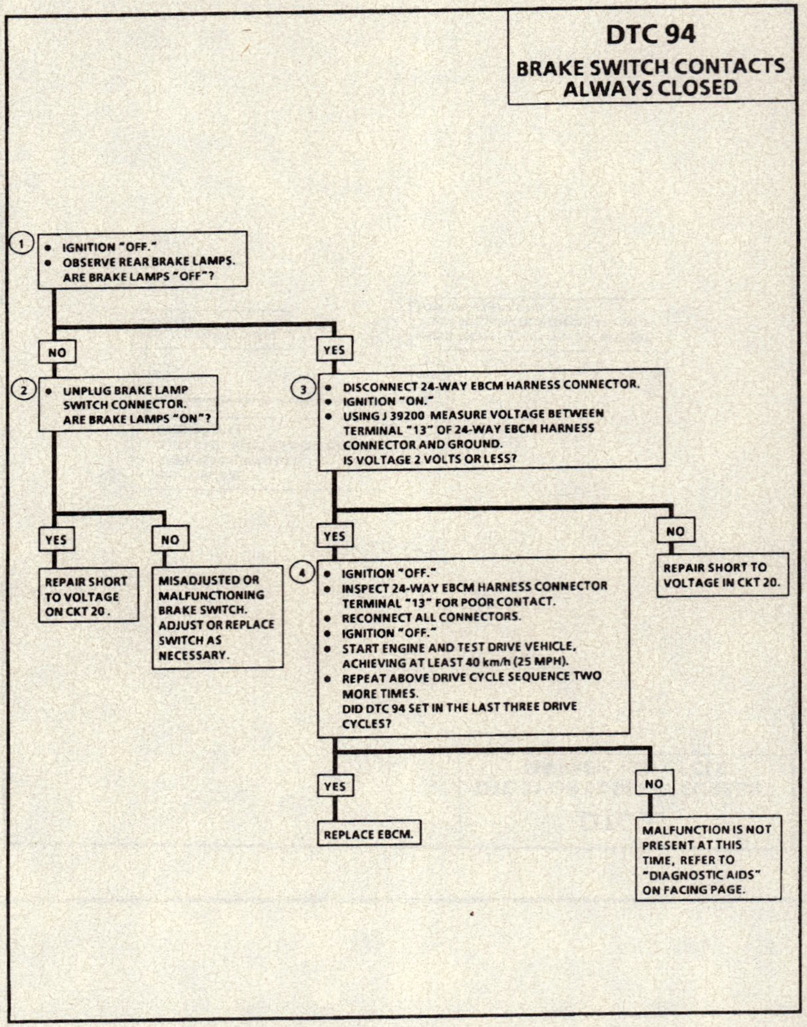

DIAGNOSTIC CHART DTC 95 — 1994 LUMINA

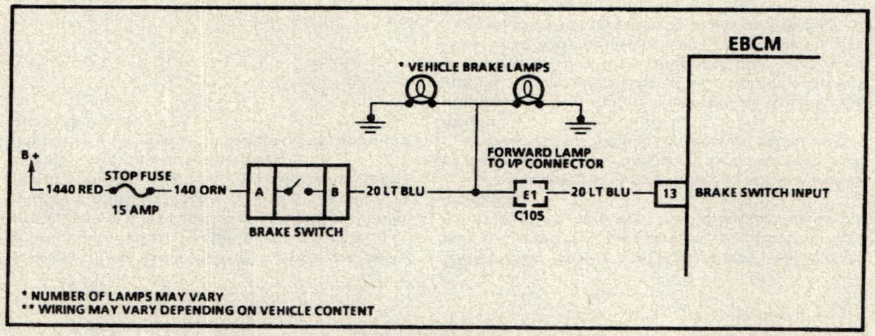

DTC 95

BRAKE SWITCH CIRCUIT OPEN

Circuit Description:

This DTC is used to identify open brake switch circuitry that prevents the brake switch input to the EBCM from changing states when the brake is applied. This DTC is used in conjunction with DTCs 91 and 92 to determine the cause of an open brake switch malfunction.

DTC Will Set When: DTC 95 can be set after initialization is completed. If the brake switch input voltage is out of specification for one second indicating an open circuit, a malfunction exists.

Action Taken: A malfunction DTC is stored, ABS is disabled and the ABS warning lamp is turned "ON."

DTC Chart Test Description: Number(s) below refer to circled number(s) on the diagnostic chart.

1. This step is used to confirm that an open in the brake switch circuitry currently exists.
2. Indicates if the brake switch signal is being received by the EBCM.
3. Indicates if an open circuit exists in the brake switch or brake lamp circuitry.
4. Isolates the open circuit to either the brake switch input circuitry, or the EBCM.

Diagnostic Aids: An "Intermittent" malfunction may be caused by a poor connection, rubbed through wire insulation, or a wire that is broken inside the insulation.

The frequency of the malfunction can be checked by using the enhanced diagnostic function of the Tech 1, as described in "Tech 1 Diagnostics" found in this section.

Any circuitry, that is suspected as causing the intermittent complaint, should be thoroughly checked for backed out terminals, improper mating, broken locks, improperly formed or damaged terminals, poor terminal to wiring connections or physical damage to the wiring harness.

DIAGNOSTIC CHART DTC 95 — 1994 LUMINA

DTC 95

BRAKE SWITCH CIRCUIT OPEN

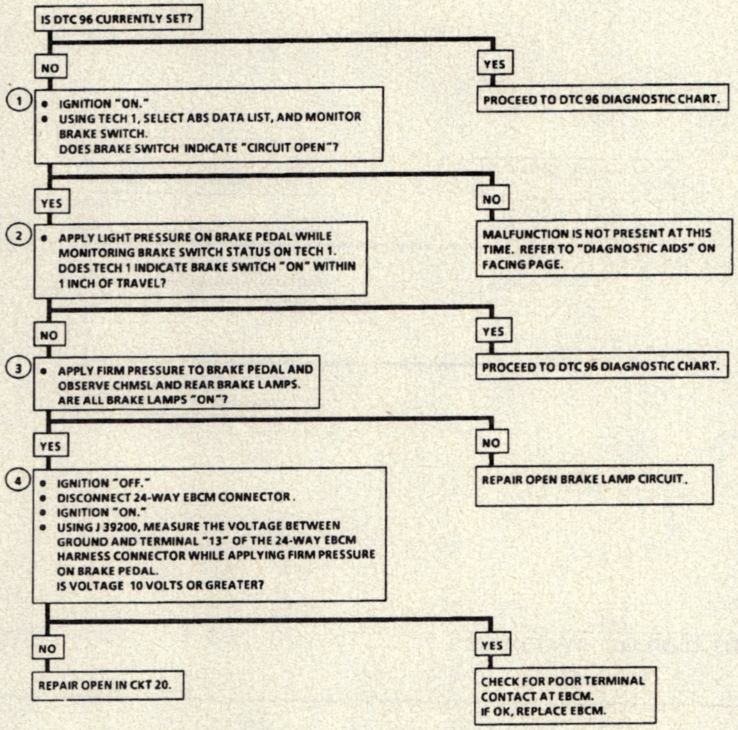

DIAGNOSTIC CHART DTC 96 — 1994 LUMINA

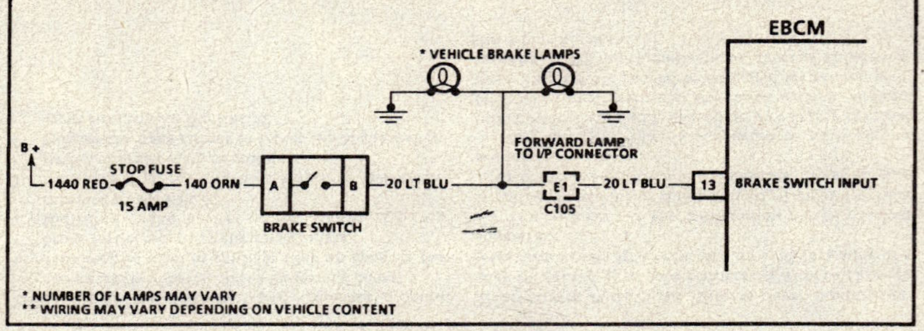

DTC 96
BRAKE LAMP CIRCUIT OPEN

Circuit Description:
This DTC is designed to isolate the cause of a DTC 95 malfunction and indicate to the driver ABS is still available. If DTC 95 fails with DTC 96, the brake lamps are open, the brake lamp grounds are open, or the Center High Mounted Stop Lamp (CHMSL) circuit is open during 4-way flasher use. The presence of battery voltage at the brake switch input indicates a valid brake switch input is still available.

DTC Will Set When: DTC 96 can be set only after DTC 95 has been set. If the EBCM detects battery voltage on the brake switch input circuit for 0.5 second, a malfunction exists.

Action Taken: A malfunction DTC is stored. ABS is not disabled; however, the ABS warning lamp will flash indicating a malfunction exists.

DTC Chart Test Description: Number(s) below refer to circled number(s) on the diagnostic chart.
1. As a result of a malfunction of an additional brake switch circuit DTC, this DTC may be set. To ensure proper diagnosis, any additional brake switch DTCs must be repaired first.
2. Identifies if the malfunction is currently present in the brake circuit.

Diagnostic Aids: An "Intermittent" malfunction may be caused by a poor connection, rubbed through wire insulation, or a wire that is broken inside the insulation.

The frequency of the malfunction can be checked by using the enhanced diagnostic function of the Tech 1, as described in "Tech 1 Diagnostics" found in this section.

Any circuitry, that is suspected as causing the intermittent complaint, should be thoroughly checked for backed out terminals, improper mating, broken locks, improperly formed or damaged terminals, poor terminal to wiring connections or physical damage to the wiring harness.

DIAGNOSTIC CHART DTC 96 — 1994 LUMINA

DTC 96
BRAKE LAMP CIRCUIT OPEN

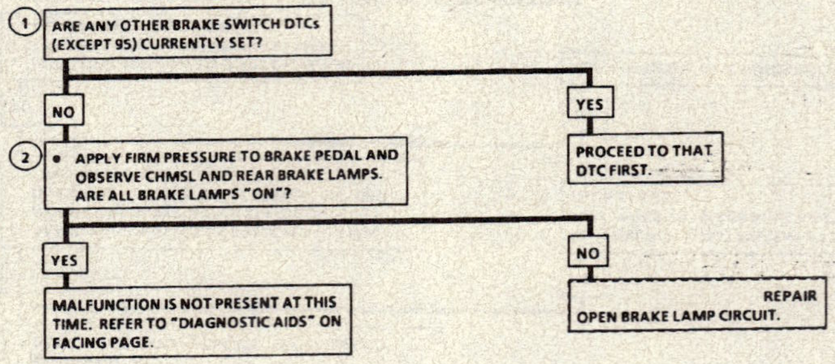

DIAGNOSTIC PROCESS CHART — 1995 CUTLASS SUPREME, GRAND PRIX, LUMINA, MONTE CARLO AND REGAL

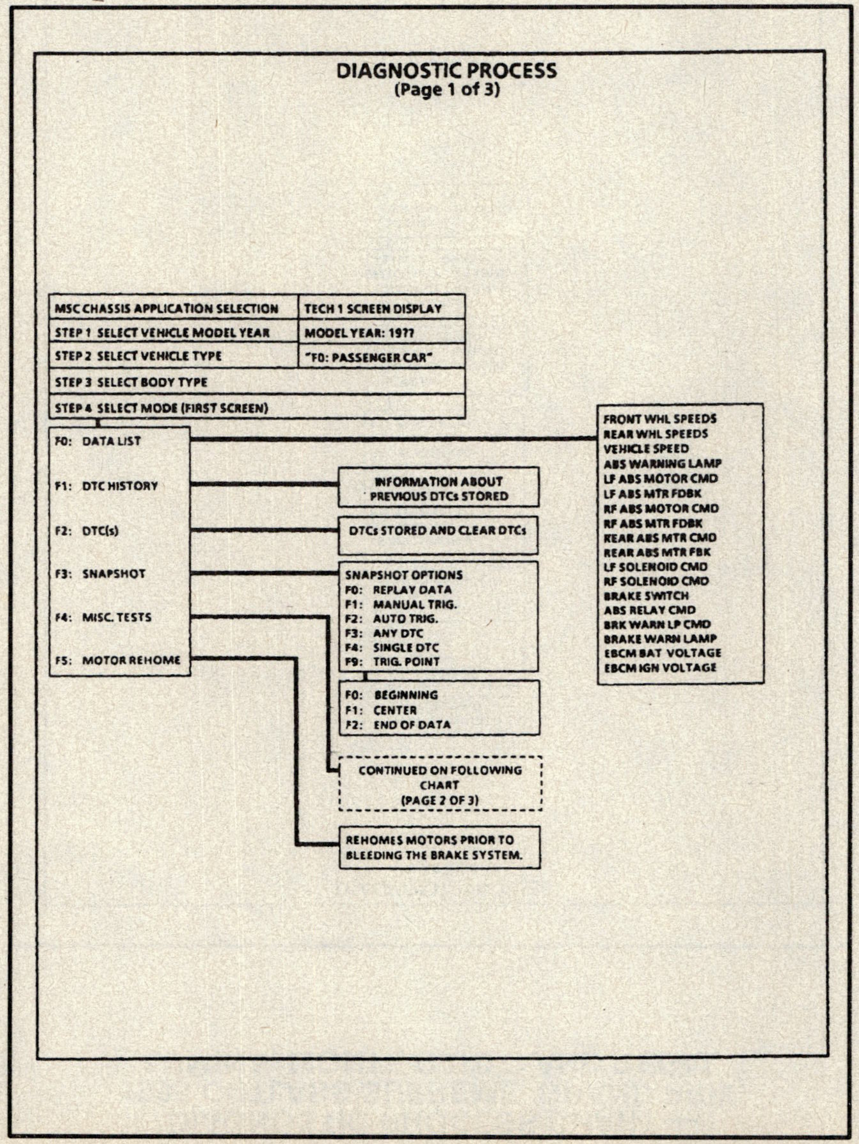

DIAGNOSTIC PROCESS CHART — 1995 CUTLASS SUPREME, GRAND PRIX, LUMINA, MONTE CARLO AND REGAL

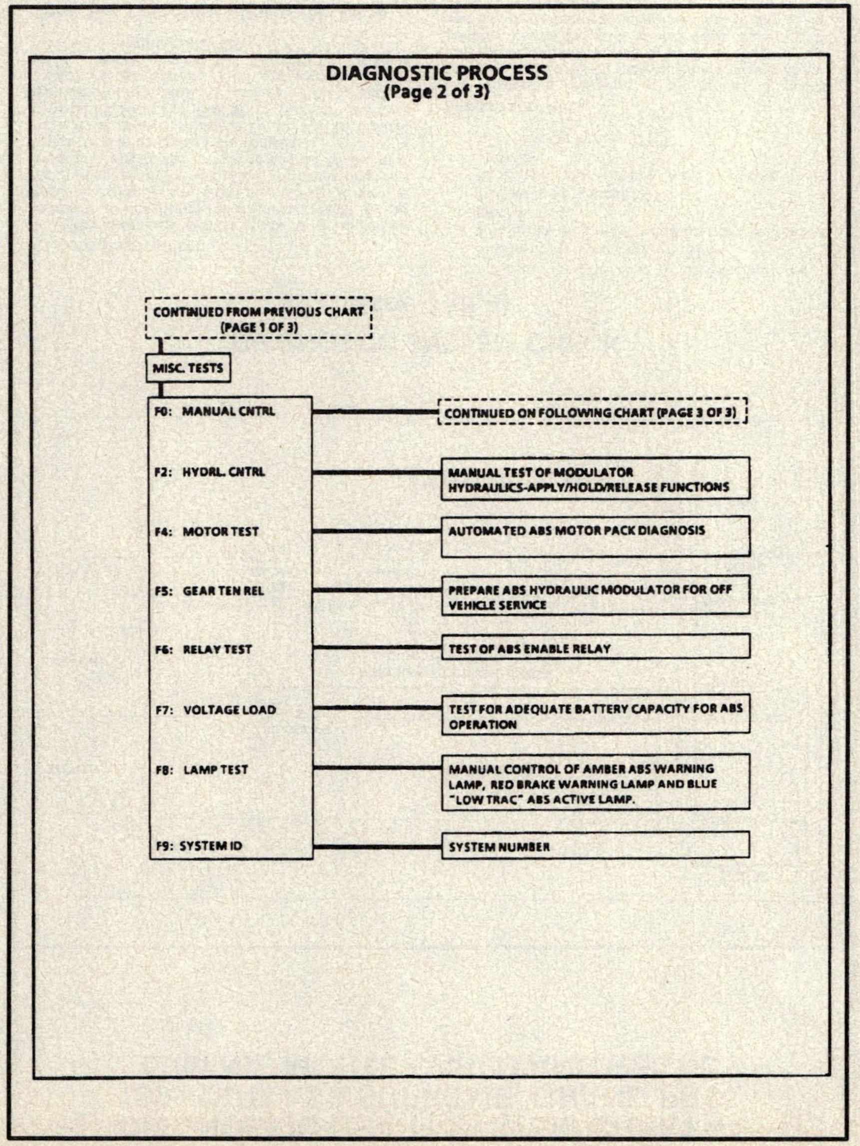

DIAGNOSTIC CHART SYSTEM CHART — 1995 CUTLASS SUPREME, GRAND PRIX, LUMINA, MONTE CARLO AND REGAL

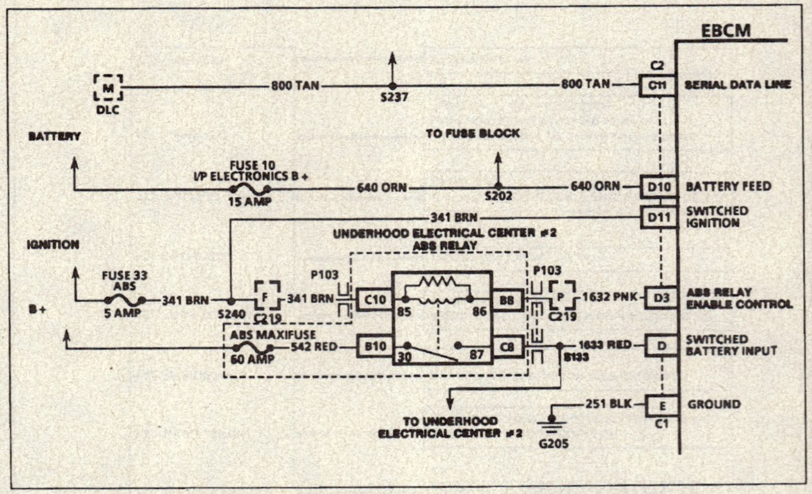

DIAGNOSTIC SYSTEM CHECK
(Page 1 of 4)

System Description:

The diagnostic system check is an organized approach to identifying a problem created by an Antilock Brake System (ABS) malfunction. It must be the starting point for any ABS complaint diagnosis because it directs the service technician to the next logical step in diagnosing the complaint.

Serial data is exchanged by the EBCM through terminal "C10". The EBCM is supplied switched ignition voltage through terminal "D11", battery voltage through terminal "D10" and switched battery voltage through terminal "D". The EBCM ground is provided through terminal "E".

Chart Test Description: Number(s) below refer to circled number(s) on the diagnostic chart.

1. Indicates if EBCM is transmitting data.
2. Checks for a serial data line malfunction.
3. Checks for proper Tech 1 connection to DLC.
4. Checks for open Fuse 10, I/P electronics B+.
5. Checks for open Fuse 33, ABS.
6. Checks for high resistance in the EBCM ground circuit.
7. Checks for current DTCs.
8. Checks for proper ABS warning lamp operation.
9. Checks for history DTCs.

Diagnostic Aids:

Excessive resistance in the ground or power supply circuits will not allow communication with the EBCM. If communication with the EBCM is not possible, ensure the ABS ground connection is good and that there is no excessive resistance in any of the power supply circuits.

DIAGNOSTIC PROCESS CHART — 1995 CUTLASS SUPREME, GRAND PRIX, LUMINA, MONTE CARLO AND REGAL

DIAGNOSTIC PROCESS
(Page 3 of 3)

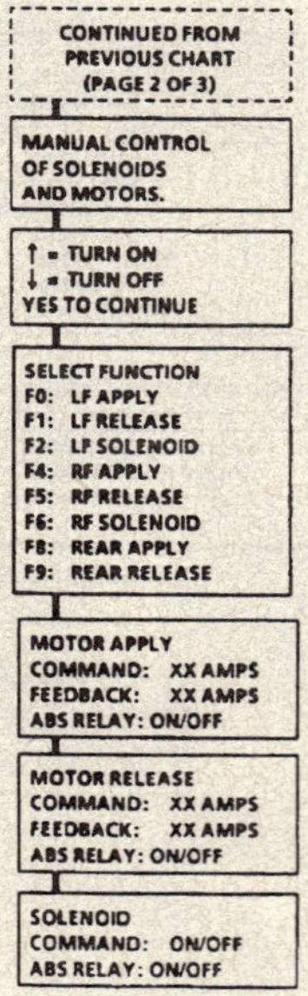

DIAGNOSTIC CHART SYSTEM CHART — 1995 CUTLASS SUPREME, GRAND PRIX, LUMINA, MONTE CARLO AND REGAL

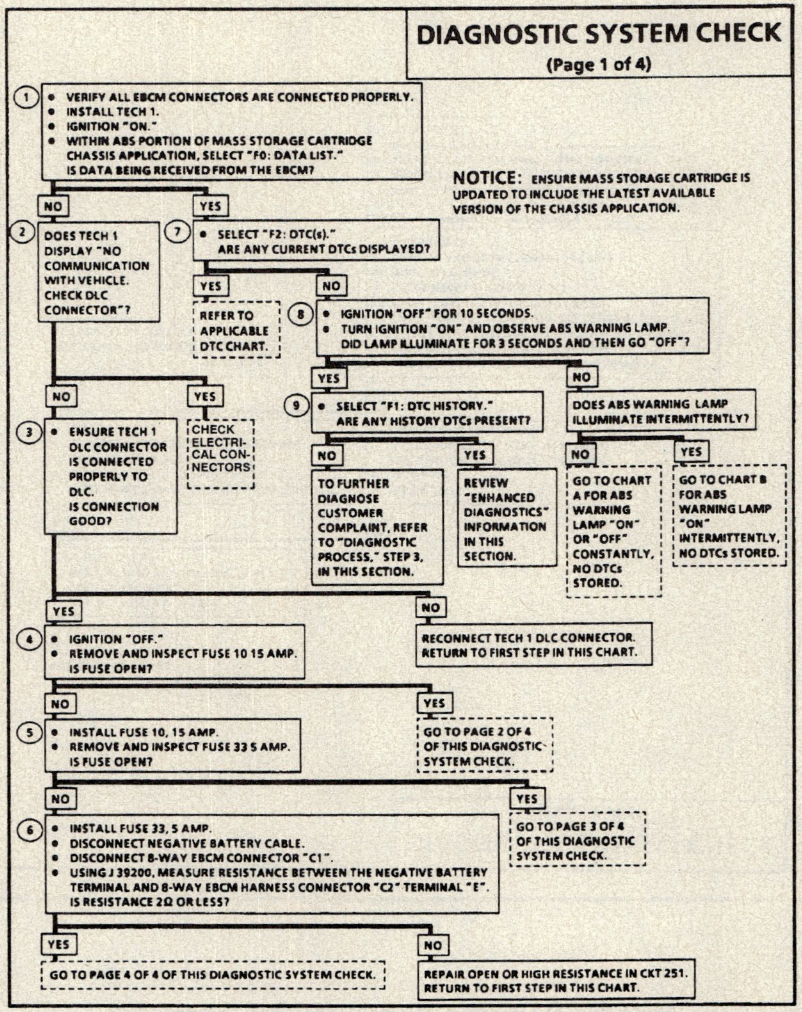

DIAGNOSTIC CHART SYSTEM CHART — 1995 CUTLASS SUPREME, GRAND PRIX, LUMINA, MONTE CARLO AND REGAL

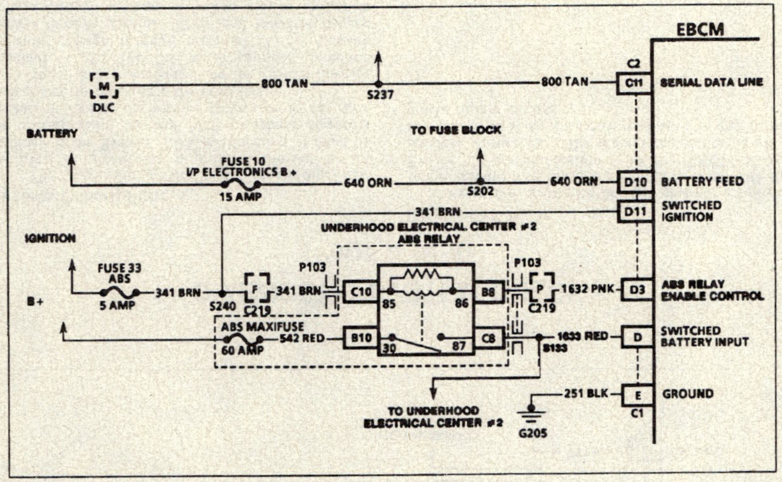

DIAGNOSTIC SYSTEM CHECK

System Description:

The diagnostic system check is an organized approach to identifying a problem created by an Antilock Brake System (ABS) malfunction. It must be the starting point for any ABS complaint diagnosis because it directs the service technician to the next logical step in diagnosing the complaint.

Serial data is exchanged by the EBCM through terminal "C10". The EBCM is supplied switched ignition voltage through terminal "D11", battery voltage through terminal "D10" and switched battery voltage through terminal "D". The EBCM ground is provided through terminal "E".

Chart Test Description: Number(s) below refer to circled number(s) on the diagnostic chart.

10. Checks for possible short to ground in the battery feed circuitry.
11. Checks for possible short to ground in the EBCM.
12. Ensures short to ground is not due to physical damage of the circuitry.

Diagnostic Aids:

Excessive resistance in the ground or power supply circuits will not allow communication with the EBCM. If communication with the EBCM is not possible, ensure the ABS ground connection is good and that there is no excessive resistance in any of the power supply circuits.

DIAGNOSTIC CHART SYSTEM CHART — 1995 CUTLASS SUPREME, GRAND PRIX, LUMINA, MONTE CARLO AND REGAL

DIAGNOSTIC SYSTEM CHECK
(Page 2 of 4)

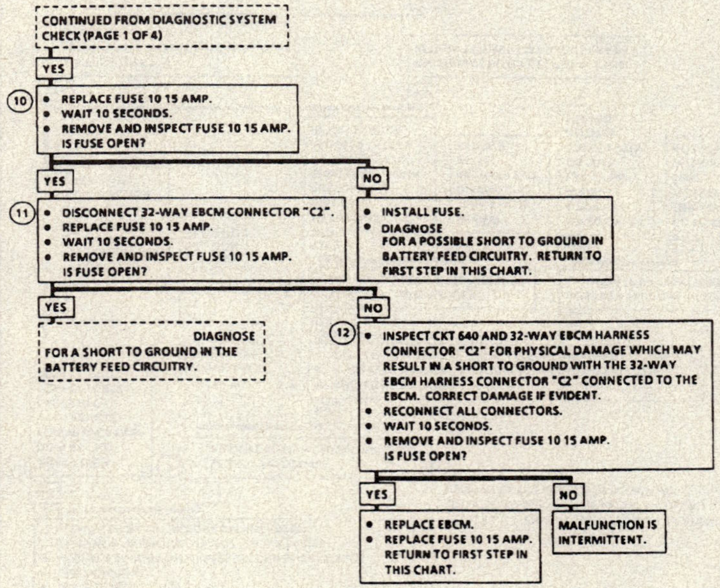

DIAGNOSTIC CHART SYSTEM CHART — 1995 CUTLASS SUPREME, GRAND PRIX, LUMINA, MONTE CARLO AND REGAL

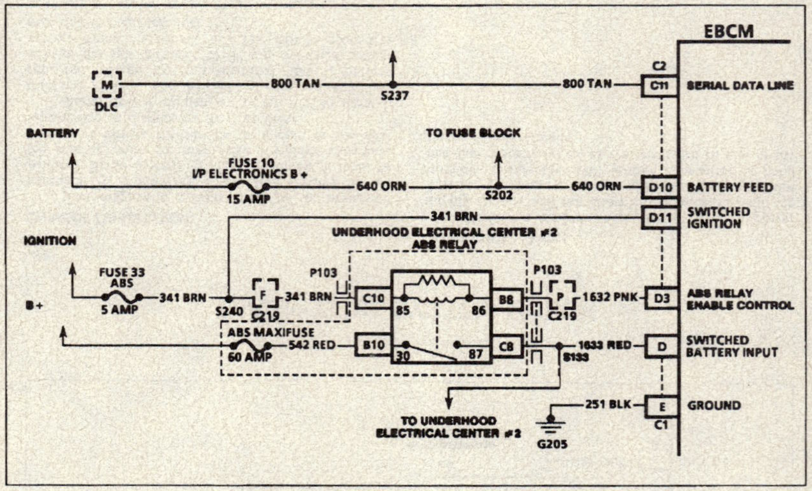

DIAGNOSTIC SYSTEM CHECK
(Page 3 of 4)

System Description:

The diagnostic system check is an organized approach to identifying a problem created by an Antilock Brake System (ABS) malfunction. It must be the starting point for any ABS complaint diagnosis because it directs the service technician to the next logical step in diagnosing the complaint.

Serial data is exchanged by the EBCM through terminal "C10". The EBCM is supplied switched ignition voltage through terminal "D11", battery voltage through terminal "D10" and switched battery voltage through terminal "D". The EBCM ground is provided through terminal "E".

Chart Test Description: Number(s) below refer to circled number(s) on the diagnostic chart.

13. Checks for possible short to ground in the switched ignition circuitry.

14. Isolates the short to ground to either the circuitry or the EBCM.

15. Ensures short to ground is not due to physical damage of the circuitry.

Diagnostic Aids:

Excessive resistance in the ground or power supply circuits will not allow communication with the EBCM. If communication with the EBCM is not possible, ensure the ABS ground connection is good and that there is no excessive resistance in any of the power supply circuits.

DIAGNOSTIC CHART SYSTEM CHART — 1995 CUTLASS SUPREME, GRAND PRIX, LUMINA, MONTE CARLO AND REGAL

DIAGNOSTIC SYSTEM CHECK
(Page 3 of 4)

CONTINUED FROM DIAGNOSTIC SYSTEM CHECK (PAGE 1 OF 4)

YES

(13)
- REPLACE FUSE 33, 5 AMP.
- IGNITION "ON."
- WAIT 10 SECONDS THEN TURN IGNITION "OFF."
- REMOVE AND INSPECT FUSE 33, 5 AMP. IS FUSE OPEN?

NO: INSTALL FUSE. REFER TO SECTION 8A TO DIAGNOSE FOR A POSSIBLE SHORT TO GROUND IN THE SWITCHED IGNITION CIRCUITRY. RETURN TO FIRST STEP IN THIS CHART.

YES:

(14)
- DISCONNECT 32-WAY EBCM CONNECTOR "C2".
- REPLACE FUSE 33, 5 AMP.
- IGNITION "ON."
- WAIT 10 SECONDS THEN TURN IGNITION "OFF."
- REMOVE AND INSPECT FUSE 33, 5 AMP. IS FUSE OPEN?

YES: REPAIR A SHORT TO GROUND IN ONE OF THE CKT 341 BRANCHES EXITING THE FUSE 33, 5 AMP. REPLACE FUSE 33, 5 AMP. RETURN TO THE FIRST STEP IN THIS CHART.

NO:

(15)
- INSPECT CKT 341 AND 32-WAY EBCM HARNESS CONNECTOR "C2" FOR PHYSICAL DAMAGE WHICH MAY RESULT IN A SHORT TO GROUND WITH THE 32-WAY EBCM HARNESS CONNECTOR "C2" CONNECTED TO THE EBCM. CORRECT DAMAGE IF EVIDENT.
- RECONNECT ALL CONNECTORS.
- IGNITION "ON."
- WAIT 10 SECONDS THEN TURN IGNITION "OFF."
- REMOVE AND INSPECT FUSE 33, 5 AMP. IS FUSE OPEN?

YES: REPLACE EBCM. REPLACE FUSE 33, 5 AMP. RETURN TO FIRST STEP IN THIS CHART.

NO: MALFUNCTION IS INTERMITTENT.

DIAGNOSTIC CHART SYSTEM CHART — 1995 CUTLASS SUPREME, GRAND PRIX, LUMINA, MONTE CARLO AND REGAL

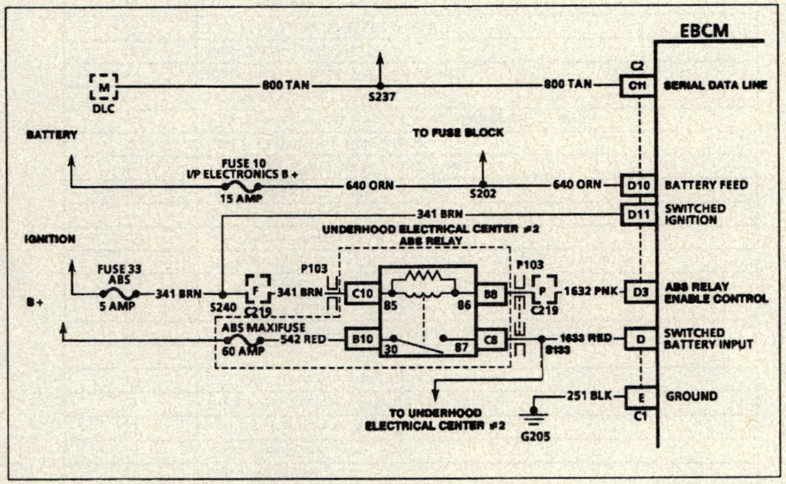

DIAGNOSTIC SYSTEM CHECK
(Page 4 of 4)

System Description:

The diagnostic system check is an organized approach to identifying a problem created by an Antilock Brake System (ABS) malfunction. It must be the starting point for any ABS complaint diagnosis because it directs the service technician to the next logical step in diagnosing the complaint.

Serial data is exchanged by the EBCM through terminal "C10". The EBCM is supplied switched ignition voltage through terminal "D11", battery voltage through terminal "D10" and switched battery voltage through terminal "D". The EBCM ground is provided through terminal "E".

Chart Test Description: Number(s) below refer to circled number(s) on the diagnostic chart.

16. Checks for an open in CKT 640.
17. Checks for high resistance in battery feed circuitry.
18. Checks for high resistance in the switched ignition circuitry.
19. Ensures malfunction is not due to poor terminal contact.
20. Checks for high resistance in the switched ignition circuit between the fuse block and the EBCM.

Diagnostic Aids:

Excessive resistance in the ground or power supply circuits will not allow communication with the EBCM. If communication with the EBCM is not possible, ensure the ABS ground connection is good and that there is no excessive resistance in any of the power supply circuits.

DIAGNOSTIC CHART SYSTEM CHART — 1995 CUTLASS SUPREME, GRAND PRIX, LUMINA, MONTE CARLO AND REGAL

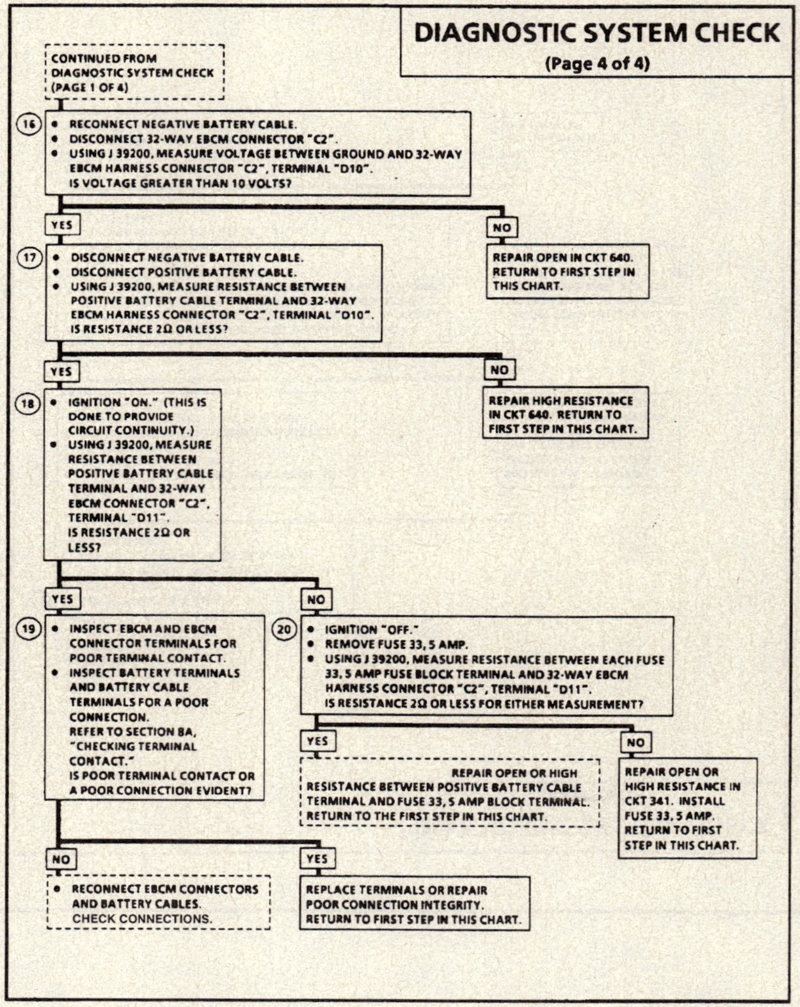

DIAGNOSTIC CHART SYSTEM CHART — 1995 CUTLASS SUPREME, GRAND PRIX, LUMINA, MONTE CARLO AND REGAL

DIAGNOSTIC TROUBLE CODE AND SYMPTOM TABLE	
CHART	**SYMPTOM**
A	ABS (Amber) Warning Lamp "ON" Constantly, No DTCs Stored
B	ABS (Amber) Warning Lamp "ON" Intermittently, No DTCs Stored
C	ABS (Amber) Warning Lamp "OFF" Constantly, No DTCs Stored
D	LOW TRAC Indicator Lamp "ON" with Ignition "OFF," No DTCs Stored
E	Tech 1 Displays Undefined DTCs
DTC	**DESCRIPTION**
11	ABS Warning or Low Trac Lamp Circuit Malfunction (1 of 2)
14	ABS Enable Relay Contact Circuit Open (1 of 3)
15	ABS Rly Contact CKT Short to Batt or Always Closed
16	ABS Enable Relay Coil Circuit Open
17	ABS Enable Relay Coil Circuit Shorted to Gnd
18	ABS Enable Relay Coil Circuit Shorted to Batt
21	Left Front Wheel Speed = 0
22	Right Front Wheel Speed = 0
23	Left Rear Wheel Speed = 0
24	Right Rear Wheel Speed = 0
25	Left Front Excessive Wheel Speed Variation
26	Right Front Excessive Wheel Speed Variation
27	Left Rear Excessive Wheel Speed Variation
28	Right Rear Excessive Wheel Speed Variation
32	LF Wheel Speed Snr CKT Open or Short to Gnd/Batt (1 of 3)
33	RF Wheel Speed Snr CKT Open or Short to Gnd/Batt (1 of 3)
34	LR Wheel Speed Snr CKT Open or Short to Gnd/Batt (1 of 3)
35	RR Wheel Speed Snr CKT Open or Short to Gnd/Batt (1 of 3)
36	Low System Voltage (1 of 2)
37	High System Voltage
38	Left Front ESB Will Not Hold Motor
41	Right Front ESB Will Not Hold Motor
42	Rear ESB Will Not Hold Motor
43	VES Steering Wheel Sensor Circuit Malfunction (1 of 3)
44	Left Front Channel Will Not Move
45	Right Front Channel Will Not Move
46	Rear Channel Will Not Move
47	Left Front Motor Free Spins (1 of 2)
48	Right Front Motor Free Spins (1 of 2)
51	Rear ABS Motor Free Spins (1 of 2)
52	Left Front Channel in Release Too Long
53	Right Front Channel in Release Too Long
54	Rear Channel in Release Too Long
55	EBCM Malfunction
56	Left Front ABS Motor Circuit Open
57	Left Front ABS Motor Circuit Shorted to Ground
58	Left Front ABS Motor Circuit Shorted to Battery
61	Right Front ABS Motor Circuit Open

DIAGNOSTIC CHART SYSTEM CHART — 1995 CUTLASS SUPREME, GRAND PRIX, LUMINA, MONTE CARLO AND REGAL

DIAGNOSTIC TROUBLE CODE AND SYMPTOM TABLE	
62	Right Front ABS Motor Circuit Shorted to Gnd
63	Right Front ABS Motor Circuit Shorted to Batt
64	Rear ABS Motor Circuit Open
65	Rear ABS Motor Circuit Shorted to Ground
66	Rear ABS Motor Circuit Shorted to Battery
73	EVO Actuator Circuit Open or Shorted to Ground
74	EVO Actuator Circuit Shorted to Battery
76	Left Front Sol Circuit Open or Shorted to Batt
77	Left Front Solenoid Circuit Shorted to Gnd
78	Right Front Sol Circuit Open or Shorted to Batt
81	Right Front Solenoid Circuit Shorted to Gnd
82	Calibration Malfunction
86	EBCM Turned "ON" the Red "BRAKE" Warning Lamp
87	Red "BRAKE" Warning Lamp Circuit Open
88	Red "BRAKE" Warning Lamp Ckt Shorted to Batt
91	Open Brake Switch During Deceleration
92	Open Brake Switch When ABS Was Required
93	DTC 91 or 92 Set in Current or Prev. Ign Cycle
94	Brake Switch Contacts Always Closed
95	Brake Switch Circuit Open
96	Brake Lamps Circuit Open

ELECTRICAL CONNECTOR TERMINAL IDENTIFICATION — 1995 CUTLASS SUPREME, GRAND PRIX, LUMINA, MONTE CARLO AND REGAL

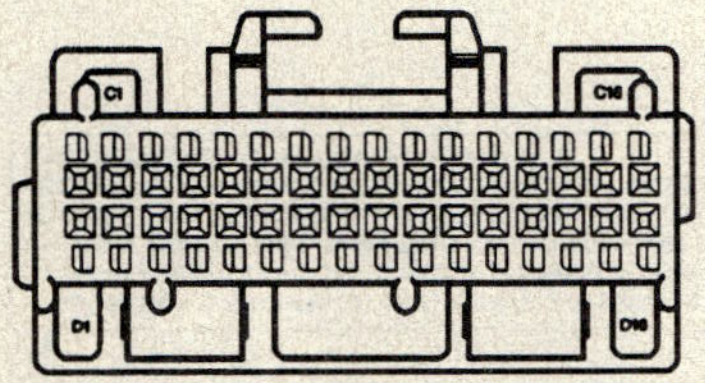

32-WAY F MICRO - PACK 100 SERIES
WHT
ELECTRONIC BRAKE CONTROL MODULE
(EBCM) CONNECTOR "C2"

TERMINAL	CIRCUIT NO.	COLOR	CIRCUIT DESCRIPTION
C1	884	BLK	LR WHEEL SIGNAL HIGH
C2	885	RED	LR WHEEL SIGNAL LOW
C3	833	TAN	RF WHEEL SIGNAL LOW
C4	872	DK GRN	RF WHEEL SIGNAL HIGH
C5	883	WHT	RR WHEEL SIGNAL LOW
C6	882	BRN	RR WHEEL SIGNAL HIGH
C7	873	YEL	LF WHEEL SIGNAL LOW
C8	830	LT BLU	LF WHEEL SIGNAL HIGH
C9	1289	LT BLU	RF ABS SOLENOID CONTROL
C10	800	TAN	SERIAL DATA LINE
C11	PLUGGED		NOT USED
C12	PLUGGED		NOT USED
C13	PLUGGED		NOT USED
C14	PLUGGED		NOT USED
C15	PLUGGED		NOT USED
C16	PLUGGED		NOT USED
D1	1228	DK GRN	LF ABS SOLENOID CONTROL
D2	852	WHT	ABS WARNING LAMP CONTROL
D3	1632	PNK	ABS ENABLE RELAY CONTROL
D4	33	TAN/WHT	BRAKE WARNING LAMP CONTROL
D5	1656	LT GRN	ABS ACTIVE INDICATOR LAMP CONTROL
D6	PLUGGED		NOT USED
D7	PLUGGED		NOT USED
D8	PLUGGED		NOT USED
D9	PLUGGED		NOT USED
D10	640	ORN	BATTERY FEED
D11	341	BRN	SWITCHED IGNITION
D12	17	WHT	BRAKE SWITCH INPUT
D13	PLUGGED		NOT USED
D14	PLUGGED		NOT USED
D15	PLUGGED		NOT USED
D16	PLUGGED		NOT USED

ELECTRICAL CONNECTOR TERMINAL IDENTIFICATION — 1995 CUTLASS SUPREME, GRAND PRIX, LUMINA, MONTE CARLO AND REGAL

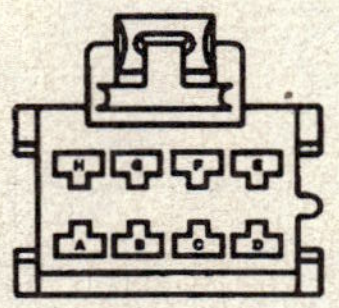

8-WAY F METRI-PACK 280 SERIES
BLK
ELECTRONIC BRAKE CONTROL MODULE
(EBCM) CONNECTOR "C1"

TERMINAL	CIRCUIT NO.	COLOR	CIRCUIT
A	1280	BLK	LF ABS MOTOR HIGH
B	1281	PNK	LF ABS MOTOR LOW
C	1284	DK GRN	REAR ABS MOTOR HIGH
D	1633	RED	SWITCHED BATTERY INPUT
E	251	BLK	GROUND
F	1285	ORN	REAR ABS MOTOR LOW
G	1282	PPL	RF ABS MOTOR HIGH
H	1283	BLK	RF ABS MOTOR LOW

DIAGNOSTIC CHART A — 1995 CUTLASS SUPREME, GRAND PRIX, LUMINA, MONTE CARLO AND REGAL

CHART A
ABS (AMBER) WARNING LAMP "ON" CONSTANTLY, NO DTCs STORED

NOTICE: DIAGNOSTIC SYSTEM CHECK MUST BE COMPLETED BEFORE USING THIS CHART.

- IGNITION "OFF."
- INSTALL TECH 1.
- IGNITION "ON."
- USING TECH 1, SELECT "F4: MISC. TESTS."
- SELECT "F8: LAMP TEST" AND ATTEMPT TO TURN "ON" THE ABS WARNING LAMP.

DID THE ABS WARNING LAMP TURN "ON"?

YES:

- USING TECH 1 LAMP TEST, ATTEMPT TO "FLASH" THE ABS WARNING LAMP.

DID THE ABS WARNING LAMP "FLASH"?

YES: MALFUNCTION IS NOT PRESENT AT THIS TIME. INTERMITTENTS COULD BE CAUSED BY INCORRECT WIRING HARNESS ROUTING OR LOOSE CONNECTIONS. INSPECT ALL CONNECTORS AND TERMINALS FOR POOR TERMINAL CONTACT AND EVIDENCE OF CORROSION. REPLACE ALL TERMINALS THAT EXHIBIT SIGNS OF POOR TERMINAL CONTACT OR CORROSION.

NO: INSPECT THE ABS WARNING LAMP BULB TO VERIFY BULB IS NOT OPEN. REPLACE ANY OPEN BULBS. IF BULB IS GOOD, PERFORM FURTHER DIAGNOSIS OF INSTRUMENT CLUSTER.

DIAGNOSTIC CHART B — 1995 CUTLASS SUPREME, GRAND PRIX, LUMINA, MONTE CARLO AND REGAL

CHART B
ABS (AMBER) WARNING LAMP "ON" INTERMITTENTLY, NO DTCs STORED

NOTICE: DIAGNOSTIC SYSTEM CHECK MUST BE COMPLETED BEFORE USING THIS CHART.

- IGNITION "OFF."
- DISCONNECT 8-WAY EBCM CONNECTOR.
- DISCONNECT NEGATIVE BATTERY TERMINAL.
- DISCONNECT POSITIVE BATTERY TERMINAL.
- USING J 39200, MEASURE THE RESISTANCE BETWEEN THE NEGATIVE BATTERY CABLE TERMINAL AND TERMINAL "E" OF THE 8-WAY EBCM HARNESS CONNECTOR.
- OBSERVE MEASURED RESISTANCE WHILE "WIGGLING" WIRE HARNESS AND CONNECTORS.

IS RESISTANCE ALWAYS 2Ω OR LESS?

NO: REPAIR INTERMITTENT OPEN OR HIGH RESISTANCE IN CKT 251.

YES:

- DISCONNECT 32-WAY EBCM CONNECTOR.
- USING J 39200, MEASURE THE RESISTANCE BETWEEN POSITIVE BATTERY CABLE TERMINAL AND TERMINAL "D10" OF THE 32-WAY EBCM HARNESS CONNECTOR.
- OBSERVE MEASURED RESISTANCE WHILE "WIGGLING" WIRE HARNESS AND CONNECTORS.

IS RESISTANCE ALWAYS 2Ω OR LESS?

NO: REPAIR INTERMITTENT OPEN OR HIGH RESISTANCE IN CKTs 640 OR 242.

YES:

- TURN IGNITION SWITCH "ON" (THIS IS DONE TO COMPLETE THE IGNITION CIRCUIT).
- USING J 39200, MEASURE THE RESISTANCE BETWEEN POSITIVE BATTERY CABLE TERMINAL AND TERMINAL "D11" OF THE 32-WAY EBCM HARNESS CONNECTOR.
- OBSERVE MEASURED RESISTANCE WHILE "WIGGLING" WIRE HARNESS AND CONNECTORS.

IS RESISTANCE ALWAYS 2Ω OR LESS?

NO: REPAIR INTERMITTENT OPEN OR HIGH RESISTANCE IN CKTs 341, 300 OR 142.

YES:

- CHECK POSITIVE BATTERY TERMINAL AND POSITIVE BATTERY CABLE TERMINAL FOR A POOR CONNECTION.
- INSPECT TERMINALS "D11" AND "D10" OF THE 32-WAY EBCM CONNECTOR "C2" AND TERMINAL "D" OF THE 8-WAY EBCM CONNECTOR FOR POOR TERMINAL CONTACT AND EVIDENCE OF CORROSION.

IS THERE EVIDENCE OF POOR TERMINAL CONTACT OR CORROSION?

YES: REPLACE TERMINALS THAT EXHIBIT POOR TERMINAL CONTACT OR SIGNS OF CORROSION. REPAIR CONNECTION INTEGRITY IF POOR CONTACT EXISTS AT BATTERY.

NO:

- IGNITION "OFF."
- RECONNECT ALL CONNECTORS.
- IGNITION "ON."

IS ABS WARNING LAMP "ON" INTERMITTENTLY WITH NO DTCs SET?

YES: REPLACE EBCM.

NO: MALFUNCTION IS NOT PRESENT AT THIS TIME. INTERMITTENTS MAY BE CAUSED BY INCORRECT WIRING HARNESS ROUTING OR LOOSE GROUND CONNECTIONS.

DIAGNOSTIC CHART C — 1995 CUTLASS SUPREME, GRAND PRIX, LUMINA, MONTE CARLO AND REGAL

CHART C
ABS (AMBER) WARNING LAMP "OFF" CONSTANTLY, NO DTCs STORED

NOTICE: DIAGNOSTIC SYSTEM CHECK MUST BE COMPLETED BEFORE USING THIS CHART.

- IGNITION "OFF."
- INSTALL TECH 1.
- IGNITION "ON."
- USING TECH 1, SELECT "F4: MISC. TESTS."
- SELECT "F8: LAMP TEST" AND ATTEMPT TO TURN "ON" THE ABS (AMBER) WARNING LAMP.

DID THE ABS (AMBER) WARNING LAMP TURN "ON"?

YES:

- USING TECH 1 LAMP TEST, ATTEMPT TO "FLASH" THE ABS (AMBER) WARNING LAMP.

DID THE ABS (AMBER) WARNING LAMP "FLASH"?

YES: MALFUNCTION IS NOT PRESENT AT THIS TIME. INTERMITTENTS COULD BE CAUSED BY INCORRECT WIRING HARNESS ROUTING OR LOOSE CONNECTIONS. INSPECT ALL CONNECTORS AND TERMINALS FOR POOR TERMINAL CONTACT AND EVIDENCE OF CORROSION. REPLACE ALL TERMINALS THAT EXHIBIT SIGNS OF POOR TERMINAL CONTACT OR CORROSION.

NO (lamp did not turn "ON", or did not "FLASH"): INSPECT THE ABS (AMBER) WARNING BULB TO VERIFY BULB IS GOOD. REPLACE ANY BLOWN BULBS. IF BULB IS GOOD, REFER TO SECTION 8A-81 FOR FURTHER DIAGNOSIS OF INSTRUMENT CLUSTER.

DIAGNOSTIC CHART D — 1995 CUTLASS SUPREME, GRAND PRIX, LUMINA, MONTE CARLO AND REGAL

CHART D
"LOW TRAC" INDICATOR LAMP "ON" WITH IGNITION "OFF," NO DTCs STORED

NOTICE: DIAGNOSTIC SYSTEM CHECK MUST BE COMPLETED BEFORE USING THIS CHART.

- IGNITION "OFF."
- OBSERVE THE "LOW TRAC" WARNING LAMP. IS THE "LOW TRAC" WARNING LAMP "ON"?

NO: MALFUNCTION IS NOT PRESENT AT THIS TIME. INTERMITTENTS COULD BE CAUSED BY INCORRECT WIRING HARNESS ROUTING OR LOOSE CONNECTIONS. INSPECT ALL CONNECTORS AND TERMINALS FOR POOR TERMINAL CONTACT AND EVIDENCE OF CORROSION. REPLACE ALL TERMINALS THAT EXHIBIT SIGNS OF POOR TERMINAL CONTACT OR CORROSION.

YES: REPAIR SHORT TO VOLTAGE ON CKT 1656.

DIAGNOSTIC CHART E — 1995 CUTLASS SUPREME, GRAND PRIX, LUMINA, MONTE CARLO AND REGAL

CHART E
TECH 1 DISPLAYS UNDEFINED DTCs

Important

- USE THIS CHART IF ANY OF THE FOLLOWING UNDEFINED DTCs ARE SET - 12, 13, 31, 43, 67, 68, 71, 72, 73, 75, 83, 84, 85, 97, OR 98.

- IGNITION "ON."
- USING TECH 1, CLEAR DTCs. DOES UNDEFINED DTC RESET?

YES: REPLACE EBCM.

NO: MALFUNCTION IS NOT PRESENT AT THIS TIME.

DIAGNOSTIC CHART DTC 11 — 1995 CUTLASS SUPREME, GRAND PRIX, LUMINA, MONTE CARLO AND REGAL

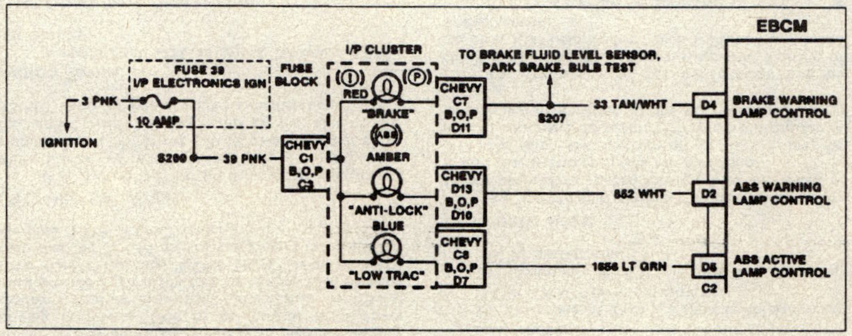

DTC 11

WARNING OR LOW TRAC CKT MALFUNCTION

Circuit Description:

This DTC checks the state of both the amber ABS warning lamp and the blue "LOW TRAC" lamp to identify a situation in which the driver could not be alerted to a system malfunction or of possible slippery driving conditions. Due to an integral lamp driver module within the instrument panel cluster, the EBCM must provide a ground to turn the amber ABS warning lamp "OFF." The blue "LOW TRAC" lamp circuit is such that the EBCM provides a ground to turn "ON" the blue "LOW TRAC" lamp.

DTC Will Set When:

DTC 11 can only be set during the three second bulb check or when either of the lamps are commanded "ON." If the EBCM cannot control either the amber ABS warning lamp or the blue "LOW TRAC" lamp for two seconds, a malfunction exists.

Action Taken:

A malfunction DTC is stored; however, ABS is not disabled.

DTC Chart Test Description:

Number(s) below refer to circled number(s) on the diagnostic chart.

1. Determines if the ABS warning lamp can be properly controlled.
2. Determines if the lamp is "ON" constantly.
3. Checks for other DTCs which could turn "ON" the ABS warning lamp.
4. Determines if the ABS warning lamp can be turned "OFF" manually.
5. Checks the integrity of the ABS warning lamp circuitry.
6. Checks for a short to voltage on the ABS warning lamp circuitry.
7. Determines if a malfunctioning EBCM is the cause of DTC 11.
8. Verifies that the other I/P indicator lamps function properly.
9. Determines if the ABS warning lamp can be turned "ON" manually.
10. Checks for a short to ground in the ABS warning lamp circuitry.
11. Verifies condition of the Fuse 39, I/P Electronics IGN.
12. Determines if DTC 11 was set due to an EBCM malfunction.

Diagnostic Aids:

An "Intermittent" malfunction may be caused by a poor connection, rubbed through wire insulation, or a wire that is broken inside the insulation.

The lamp test function of the Tech 1 may be used to command the lamp "ON" while looking for an intermittent malfunction in the ABS warning lamp circuitry.

The frequency of the malfunction can be checked by using the enhanced diagnostic function of the Tech 1, as described in "Tech 1 Diagnostics" found in this section.

Any circuitry that is suspected of causing the intermittent complaint should be thoroughly checked for backed out terminals, improper mating, broken locks, improperly formed or damaged terminals, poor terminal to wiring connections or physical damage to the wiring harness.

DIAGNOSTIC CHART DTC 11 — 1995 CUTLASS SUPREME, GRAND PRIX, LUMINA, MONTE CARLO AND REGAL

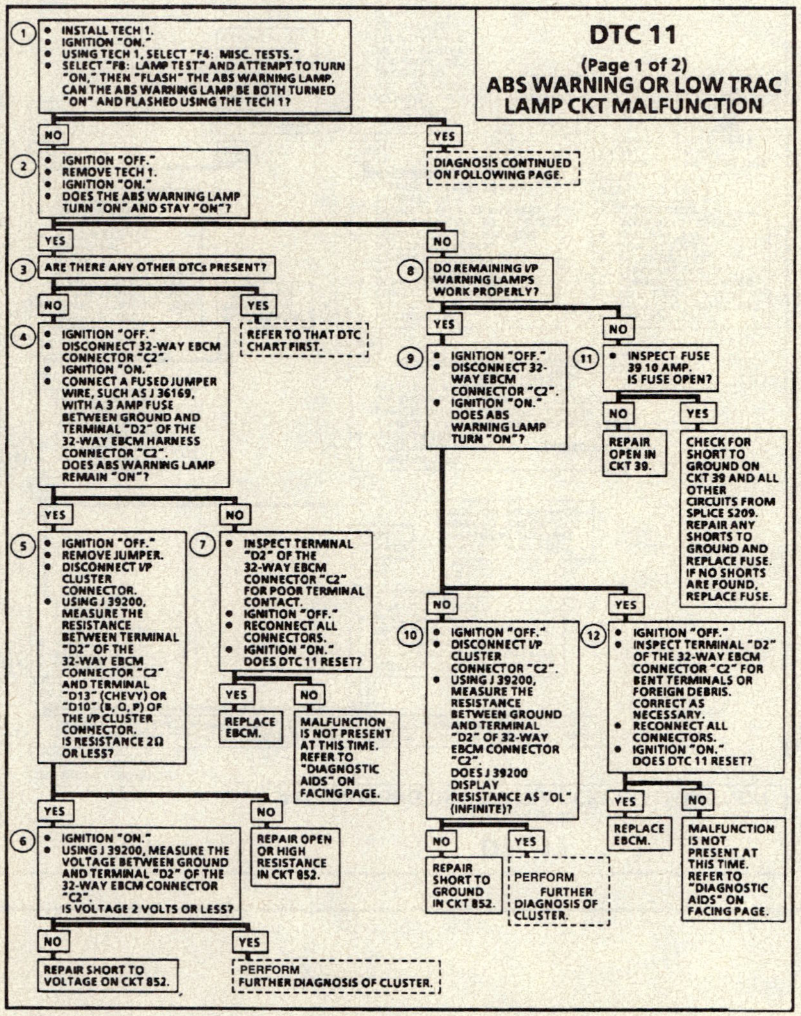

DIAGNOSTIC CHART DTC 11 — 1995 CUTLASS SUPREME, GRAND PRIX, LUMINA, MONTE CARLO AND REGAL

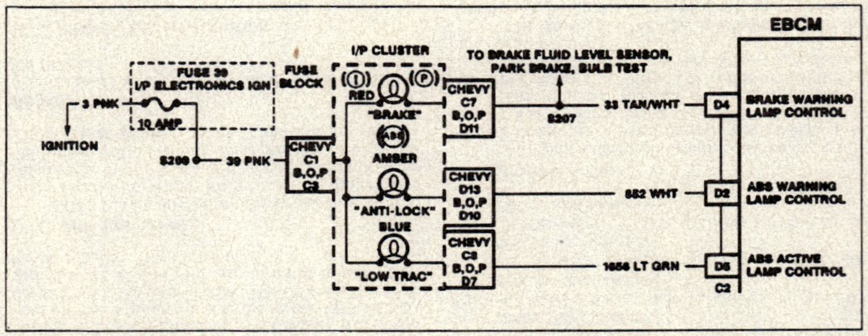

DTC 11
(Page 2 of 2)
WARNING OR LOW TRAC LAMP CKT MALFUNCTION

Circuit Description:

This DTC checks the state of both the amber ABS warning lamp and the blue "LOW TRAC" lamp to identify a situation in which the driver could not be alerted to a system malfunction or of possible slippery driving conditions. Due to an integral lamp driver module within the instrument panel cluster, the EBCM must provide a ground to turn the amber ABS warning lamp "OFF." The blue "LOW TRAC" lamp circuit is such that the EBCM provides a ground to turn "ON" the blue "LOW TRAC" lamp.

DTC Will Set When:

DTC 11 can only be set during the three second bulb check or when either of the lamps are commanded "ON." If the EBCM cannot control either the amber ABS warning lamp or the blue "LOW TRAC" lamp for two seconds, a malfunction exists.

Action Taken:

A malfunction DTC is stored; however, ABS is not disabled.

DTC Chart Test Description:

Number(s) below refer to circled number(s) on the diagnostic chart.

13. Determines if the "LOW TRAC" lamp can be properly controlled.
14. Determines if the "LOW TRAC" lamp is "ON" constantly.
15. Checks for a short to ground in the "LOW TRAC" lamp circuitry.
16. Determines if DTC 11 was set due to a EBCM malfunction.
17. Determines if DTC 11 was set due to a EBCM malfunction.
18. Determines if the "LOW TRAC" lamp can be controlled manually.
19. Verifies the integrity of the "LOW TRAC" lamp circuitry.
20. Determines if there is a short to voltage in the "LOW TRAC" lamp circuitry.
21. Determines if DTC 11 was set due to an EBCM malfunction.

Diagnostic Aids:

An "Intermittent" malfunction may be caused by a poor connection, rubbed through wire insulation, or a wire that is broken inside the insulation.

The lamp test function of the Tech 1 may be used to command the lamp "ON" while looking for an intermittent malfunction in the "LOW TRAC" lamp circuitry.

The frequency of the malfunction can be checked by using the enhanced diagnostic function of the Tech 1, as described in "Tech 1 Diagnostics" found in this section.

Any circuitry that is suspected of causing the intermittent complaint should be thoroughly checked for backed out terminals, improper mating, broken locks, improperly formed or damaged terminals, poor terminal to wiring connections or physical damage to the wiring harness.

DIAGNOSTIC CHART DTC 11 — 1995 CUTLASS SUPREME, GRAND PRIX, LUMINA, MONTE CARLO AND REGAL

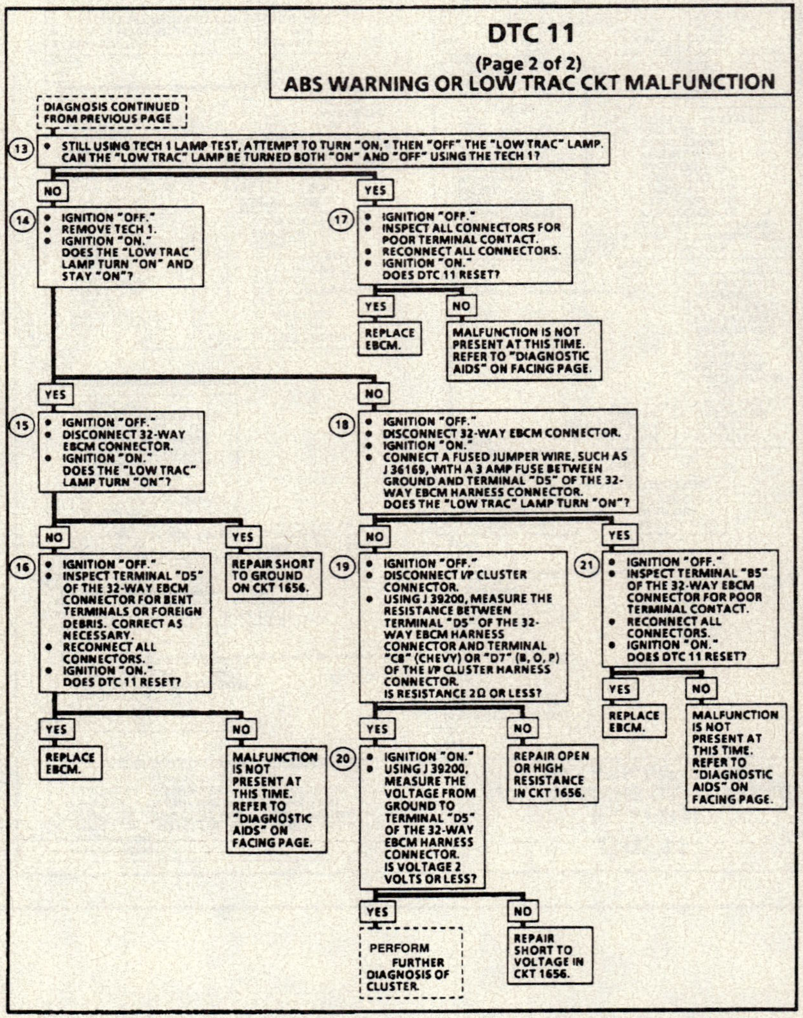

DIAGNOSTIC CHART DTC 14 — 1995 CUTLASS SUPREME, GRAND PRIX, LUMINA, MONTE CARLO AND REGAL

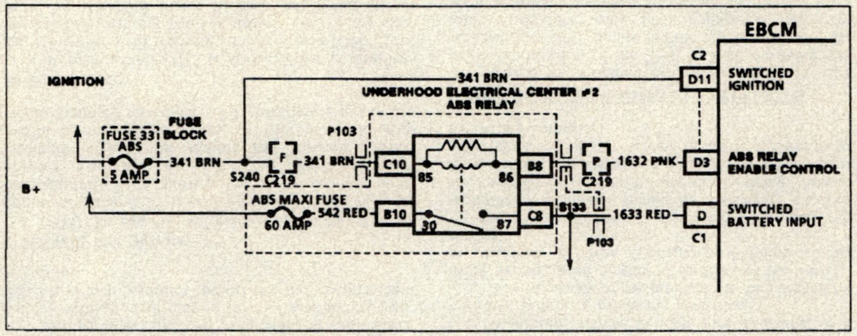

DTC 14

ABS ENABLE RELAY CONTACT CIRCUIT OPEN

Circuit Description:

Ignition voltage is supplied through terminal "C10" of the ABS relay. The EBCM then is able to energize the pull-in coil by completing the ground circuit at terminal "D3" of the EBCM. The magnetic field created closes the ABS relay contacts and allows battery voltage and current to be supplied to the EBCM, which supplies power to the motors and solenoids.

DTC Will Set When:

DTC 14 can be set anytime after the EBCM commands the ABS relay "ON" (the relay is first commanded "ON" during the three second bulb check). This test monitors the availability of current/voltage to the motors and solenoids. This malfunction indicates voltage is not available and would therefore not allow ABS operation if required.

Action Taken:

A malfunction DTC is stored, ABS is disabled and the amber ABS warning lamp is turned "ON." The red BRAKE warning lamp is turned "ON" if the rear piston in the ABS motor pack is not in the home position.

DTC Chart Test Description:

Number(s) below refer to circled number(s) on the diagnostic chart.

1. Checks to see if DTC 16 is also set as current or history. If DTC 16 is also set, proceed to that chart first.
2. Verifies that the malfunction is currently present.
3. Verifies that the relay can be turned "ON" and provide voltage to the EBCM.
4. Checks for high resistance in the switched battery circuit from the relay to the EBCM.
5. Checks for high resistance in the switched battery circuit from the battery to the relay.
6. Verifies the condition of the ABS relay.
7. Checks for poor terminal contact or corrosion at the connectors.
8. Determines if the EBCM is the cause of the malfunction.

Diagnostic Aids:

An "Intermittent" problem may be caused by a poor connection, rubbed through wire insulation, or a wire that is broken inside the insulation.

The frequency of the problem can be checked by using the enhanced diagnostic function of the Tech 1, as described in "Tech 1 Diagnostics" found in this section.

Any circuitry that is suspected of causing the intermittent complaint should be thoroughly checked for backed out terminals, improper mating, broken locks, improperly formed or damaged terminals, poor terminal to wiring connections or physical damage to the wiring harness.

VIBRATION, TEMPERATURE EFFECTS:

Check for vibration effects by performing the relay test function of the Tech 1. With the relay test commanded "ON," lightly tap the top and sides of the relay while monitoring relay voltage. If the relay voltage changes significantly, replace the relay.

If DTC 14 only sets when the vehicle is initially started in cold ambient conditions (temperature less than 32°F-0°C), replace the relay.

DIAGNOSTIC CHART DTC 14 — 1995 CUTLASS SUPREME, GRAND PRIX, LUMINA, MONTE CARLO AND REGAL

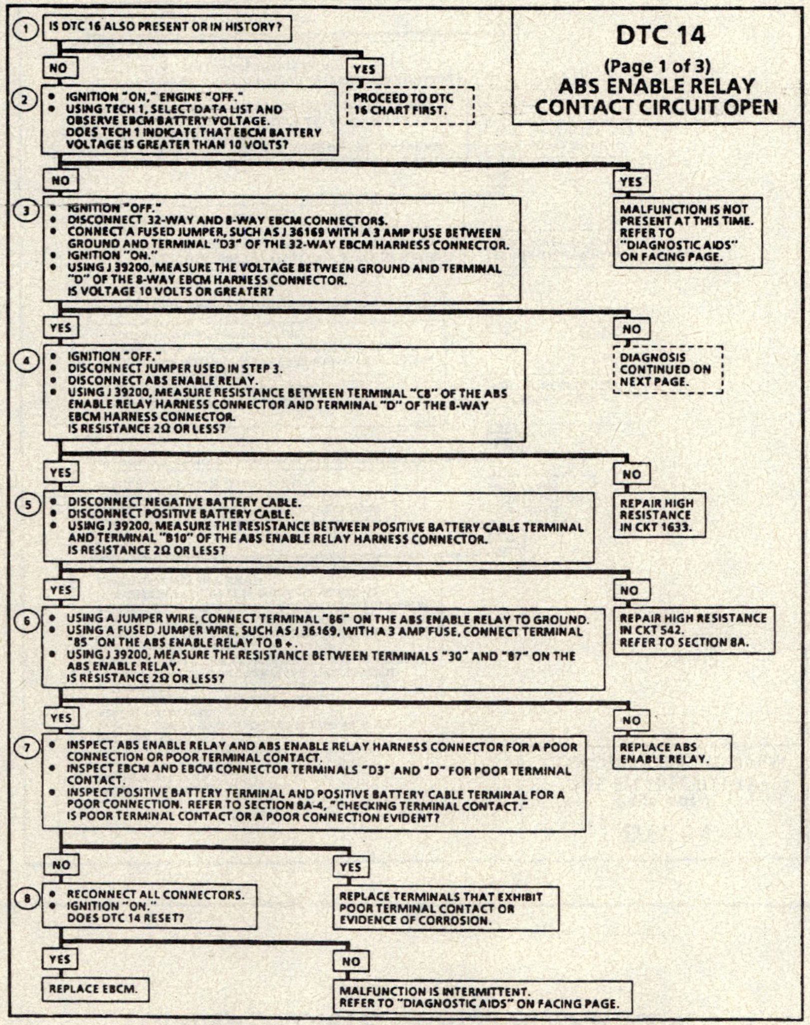

DIAGNOSTIC CHART DTC 14 — 1995 CUTLASS SUPREME, GRAND PRIX, LUMINA, MONTE CARLO AND REGAL

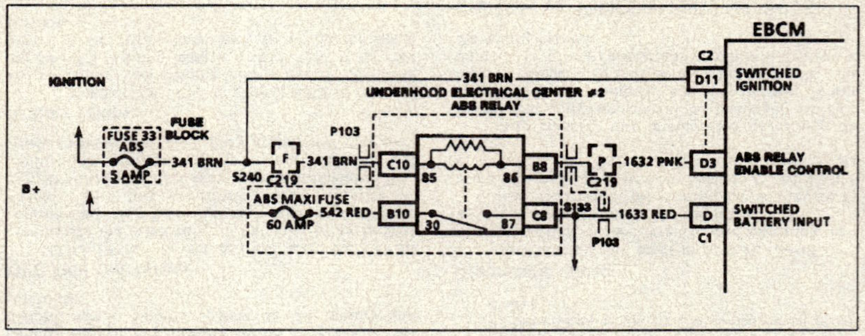

DTC 14

ABS ENABLE RELAY CONTACT CIRCUIT OPEN

Circuit Description:

Ignition voltage is supplied through terminal "C10" of the ABS relay. The EBCM then is able to energize the pull-in coil by completing the ground circuit at terminal "D3" of the EBCM. The magnetic field created closes the ABS relay contacts and allows battery voltage and current to be supplied to the EBCM, which supplies power to the motors and solenoids.

DTC Will Set When:

DTC 14 can be set anytime after the EBCM commands the ABS relay "ON" (the relay is first commanded "ON" during the three second bulb check). This test monitors the availability of current/voltage to the motors and solenoids. This malfunction indicates voltage is not available and would therefore not allow ABS operation if required.

Action Taken:

A malfunction DTC is stored, ABS is disabled and the amber ABS warning lamp is turned "ON." The red BRAKE warning lamp is turned "ON" if the rear piston in the ABS motor pack is not in the home position.

DTC Chart Test Description:

Number(s) below refer to circled number(s) on the diagnostic chart.

9 Checks for high resistance in the entire relay coil circuit.

10 Verifies proper battery condition.

11. Verifies proper voltage is available at the ABS relay.

12. Checks for an open in the switched battery circuit from the relay to the EBCM.

13. Verifies the condition of the ABS relay.

Diagnostic Aids:

An "Intermittent" problem may be caused by a poor connection, rubbed through wire insulation, or a wire that is broken inside the insulation.

The frequency of the problem can be checked by using the enhanced diagnostic function of the Tech 1, as described in "Tech 1 Diagnostics" found in this section.

Any circuitry that is suspected of causing the intermittent complaint should be thoroughly checked for backed out terminals, improper mating, broken locks, improperly formed or damaged terminals, poor terminal to wiring connections or physical damage to the wiring harness.

VIBRATION, TEMPERATURE EFFECTS:

Check for vibration effects by performing the relay test function of the Tech 1. With the relay test commanded "ON," lightly tap the top and sides of the relay while monitoring relay voltage. If the relay voltage changes significantly, replace the relay.

If DTC 14 only sets when the vehicle is initially started in cold ambient conditions (temperature less than 32°F-0°C), replace the relay.

DIAGNOSTIC CHART DTC 14 — 1995 CUTLASS SUPREME, GRAND PRIX, LUMINA, MONTE CARLO AND REGAL

DTC 14
(Page 2 of 3)
ABS ENABLE RELAY CONTACT CIRCUIT OPEN

DIAGNOSIS CONTINUED FROM PREVIOUS PAGE.

(9)
- IGNITION "OFF."
- REMOVE JUMPER WIRE INSTALLED IN STEP #3, IF NOT ALREADY REMOVED.
- DISCONNECT NEGATIVE BATTERY CABLE.
- DISCONNECT POSITIVE BATTERY CABLE.
- WITH BATTERY STILL DISCONNECTED, TURN IGNITION SWITCH TO "ON." THIS COMPLETES THE IGNITION CIRCUIT TO THE ABS ENABLE RELAY.
- USING J 39200, MEASURE THE RESISTANCE BETWEEN TERMINAL "D3" OF THE 32-WAY EBCM HARNESS CONNECTOR AND THE POSITIVE BATTERY CABLE TERMINAL. IS RESISTANCE 95Ω OR LESS WHEN MEASURED AT APPROXIMATELY 20°C (68°F)?

YES → (10) | NO → DIAGNOSIS CONTINUED ON NEXT PAGE.

(10)
- USING J 39200, MEASURE THE VOLTAGE BETWEEN THE NEGATIVE AND POSITIVE BATTERY TERMINALS. IS VOLTAGE 10 VOLTS OR GREATER?

YES → (11) | NO → CORRECT LOW VOLTAGE CONDITION.

(11)
- IGNITION "OFF."
- RECONNECT POSITIVE BATTERY CABLE.
- RECONNECT NEGATIVE BATTERY CABLE.
- DISCONNECT ABS ENABLE RELAY.
- MEASURE THE VOLTAGE BETWEEN TERMINAL "B10" OF THE ABS ENABLE RELAY HARNESS CONNECTOR AND GROUND. IS VOLTAGE 10 VOLTS OR GREATER?

YES → (12) | NO → REPAIR OPEN OR HIGH RESISTANCE IN CKT 542.

(12)
- USING J 39200, MEASURE THE RESISTANCE BETWEEN TERMINAL "C8" OF THE ABS ENABLE RELAY HARNESS CONNECTOR AND TERMINAL "D" OF THE 8-WAY EBCM HARNESS CONNECTOR. IS RESISTANCE 2Ω OR LESS?

YES → (13) | NO → REPAIR OPEN OR HIGH RESISTANCE IN CKT 1633.

(13)
- USING A JUMPER WIRE, CONNECT TERMINAL "86" ON THE ABS ENABLE RELAY TO GROUND.
- USING A FUSED JUMPER WIRE, SUCH AS J 36169, WITH A 3 AMP FUSE, CONNECT TERMINAL "85" ON THE ABS ENABLE RELAY TO B+.
- USING J 39200, MEASURE THE RESISTANCE BETWEEN TERMINALS "30" AND "87" ON THE ABS ENABLE RELAY. IS RESISTANCE 2Ω OR LESS?

YES → MALFUNCTION WAS MOST LIKELY DUE TO A POOR CONNECTION. INSPECT ALL CONNECTORS AND TERMINALS FOR POOR TERMINAL CONTACT AND EVIDENCE OF CORROSION. REPLACE ALL TERMINALS THAT EXHIBIT EVIDENCE OF POOR TERMINAL CONTACT OR CORROSION.

NO → REPLACE ABS ENABLE RELAY.

DIAGNOSTIC CHART DTC 14 — 1995 CUTLASS SUPREME, GRAND PRIX, LUMINA, MONTE CARLO AND REGAL

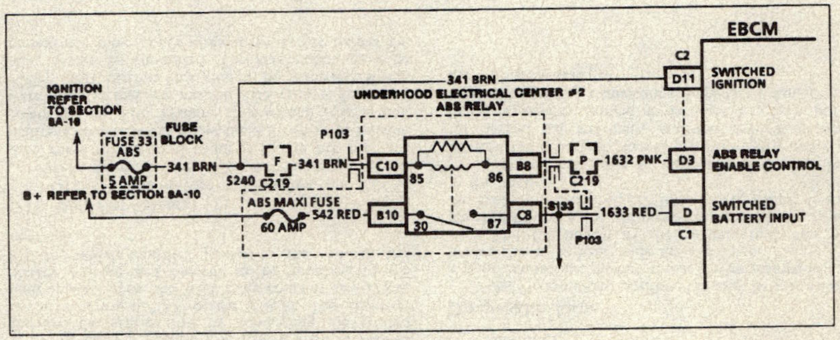

DTC 14
(Page 3 of 3)
ABS ENABLE RELAY CONTACT CIRCUIT OPEN

Circuit Description:

Ignition voltage is supplied through terminal "C10" of the ABS relay. The EBCM then is able to energize the pull-in coil by completing the ground circuit at terminal "D3" of the EBCM. The magnetic field created closes the ABS relay contacts and allows battery voltage and current to be supplied to the EBCM, which supplies power to the motors and solenoids.

DTC Will Set When:

DTC 14 can be set anytime after the EBCM commands the ABS enable relay "ON" (the relay is first commanded "ON" during the three second bulb check). This test monitors the availability of current/voltage to the motors and solenoids. This malfunction indicates voltage is not available and would therefore not allow ABS operation if required.

Action Taken:

A malfunction DTC is stored, ABS is disabled and the amber ABS warning lamp is turned "ON." The red BRAKE warning lamp is turned "ON" if the rear piston in the ABS motor pack is not in the home position.

DTC Chart Test Description:

Number(s) below refer to circled number(s) on the diagnostic chart.

14. Checks for proper resistance of the relay coil.
15. Checks for high resistance in the ABS relay control circuit between the relay and the EBCM.
16. Checks for high resistance in the ABS relay control circuit between the battery andthe relay.

Diagnostic Aids:

An "Intermittent" problem may be caused by a poor connection, rubbed through wire insulation, or a wire that is broken inside the insulation.

The frequency of the problem can be checked by using the enhanced diagnostic function of the Tech 1, as described in "Tech 1 Diagnostics" found in this section.

Any circuitry that is suspected of causing the intermittent complaint should be thoroughly checked for backed out terminals, improper mating, broken locks, improperly formed or damaged terminals, poor terminal to wiring connections or physical damage to the wiring harness.

VIBRATION, TEMPERATURE EFFECTS:

Check for vibration effects by performing the relay test function of the Tech 1. With the relay test commanded "ON," lightly tap the top and sides of the relay while monitoring relay voltage. If the relay voltage changes significantly, replace the relay.

If DTC 14 only sets when the vehicle is initially started in cold ambient conditions (temperature less than 32°F-0°C), replace the relay.

DIAGNOSTIC CHART DTC 14 — 1995 CUTLASS SUPREME, GRAND PRIX, LUMINA, MONTE CARLO AND REGAL

DTC 14
(Page 3 of 3)
ABS ENABLE RELAY CONTACT CIRCUIT OPEN

DIAGNOSIS CONTINUED FROM PREVIOUS PAGE.

(14)
- DISCONNECT ABS ENABLE RELAY.
- USING J 39200, MEASURE THE RESISTANCE BETWEEN TERMINALS "85" AND "86" ON THE ABS ENABLE RELAY. IS RESISTANCE 95Ω OR LESS WHEN MEASURED AT APPROXIMATELY 20°C (68°F)?

NO: REPLACE ABS ENABLE RELAY.

YES:

(15)
- MEASURE THE RESISTANCE BETWEEN TERMINAL "B8" OF THE ABS ENABLE RELAY HARNESS CONNECTOR AND TERMINAL "D3" OF THE 32-WAY EBCM HARNESS CONNECTOR. IS RESISTANCE 2Ω OR LESS?

NO: REPAIR OPEN OR HIGH RESISTANCE IN CKT 1632.

YES:

(16)
- MEASURE THE RESISTANCE BETWEEN TERMINAL "C10" OF THE ABS ENABLE RELAY HARNESS CONNECTOR AND THE POSITIVE BATTERY CABLE TERMINAL. IS RESISTANCE 2Ω OR LESS?

NO: REPAIR OPEN OR HIGH RESISTANCE IN CKT 341, CKT 300, IGNITION SWITCH, OR BATTERY FEED CKT 142 TO IGNITION SWITCH.

YES: MALFUNCTION WAS MOST LIKELY DUE TO A POOR CONNECTION. INSPECT ALL CONNECTORS AND TERMINALS FOR POOR TERMINAL CONTACT AND EVIDENCE OF CORROSION. REPLACE ALL TERMINALS THAT EXHIBIT EVIDENCE OF POOR TERMINAL CONTACT OR CORROSION.

DIAGNOSTIC CHART DTC 15 — 1995 CUTLASS SUPREME, GRAND PRIX, LUMINA, MONTE CARLO AND REGAL

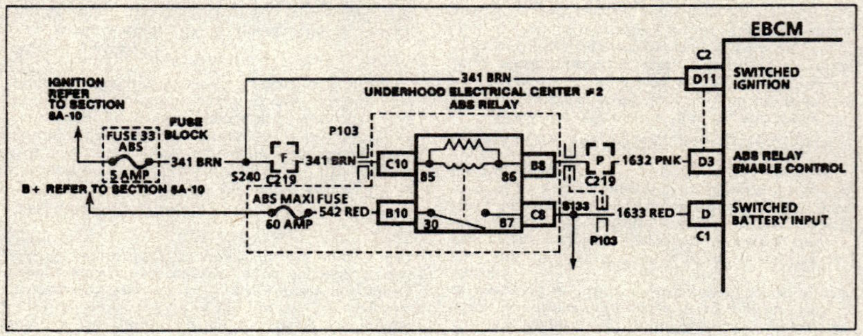

DTC 15

ABS RLY CONTACT CKT SHORT TO BATT OR ALWAYS CLOSED

Circuit Description:

Ignition voltage is supplied through terminal "C10" of the ABS relay. The EBCM then is able to energize the pull-in coil by completing the ground circuit at terminal "D3" of the EBCM. The magnetic field created closes the ABS relay contacts and allows battery voltage and current to be supplied to the EBCM, which supplies power to the motors and solenoids.

DTC Will Set When:

DTC 15 can be set only before the EBCM commands the relay "ON." This test determines if the ABS relay is energized when it should not be. This malfunction would not allow the ABS relay to remove power to the ABS system. If a second malfunction were to occur that requires the ABS relay to be turned "OFF," that malfunction cannot be removed if the relay cannot be controlled. The malfunction must be present for three consecutive drive cycles before the DTC is set.

Action Taken:

A malfunction DTC is stored. ABS is not disabled.

DTC Chart Test Description:

Number(s) below refer to circled number(s) on the diagnostic chart.

1. Checks to see if other DTCs are set.
2. Indicates that the EBCM is capable of controlling the ABS relay as commanded.
3. Checks for a possible short to battery voltage in CKT 1633.
4. Determines whether the short to voltage is in CKT 1633 or due to a malfunctioning relay.
5. Checks for a possible intermittent in wiring or connectors.

Diagnostic Aids:

An "Intermittent" malfunction may be caused by a poor connection, rubbed through wire insulation, or a wire that is broken inside the insulation.

The frequency of the malfunction can be checked by using the enhanced diagnostic function of the Tech 1, as described in "Tech 1 Diagnostics" found in this section.

Any circuitry that is suspected of causing the intermittent complaint should be thoroughly checked for backed out terminals, improper mating, broken locks, improperly formed or damaged terminals, poor terminal to wiring connections or physical damage to the wiring harness.

DIAGNOSTIC CHART DTC 15 — 1995 CUTLASS SUPREME, GRAND PRIX, LUMINA, MONTE CARLO AND REGAL

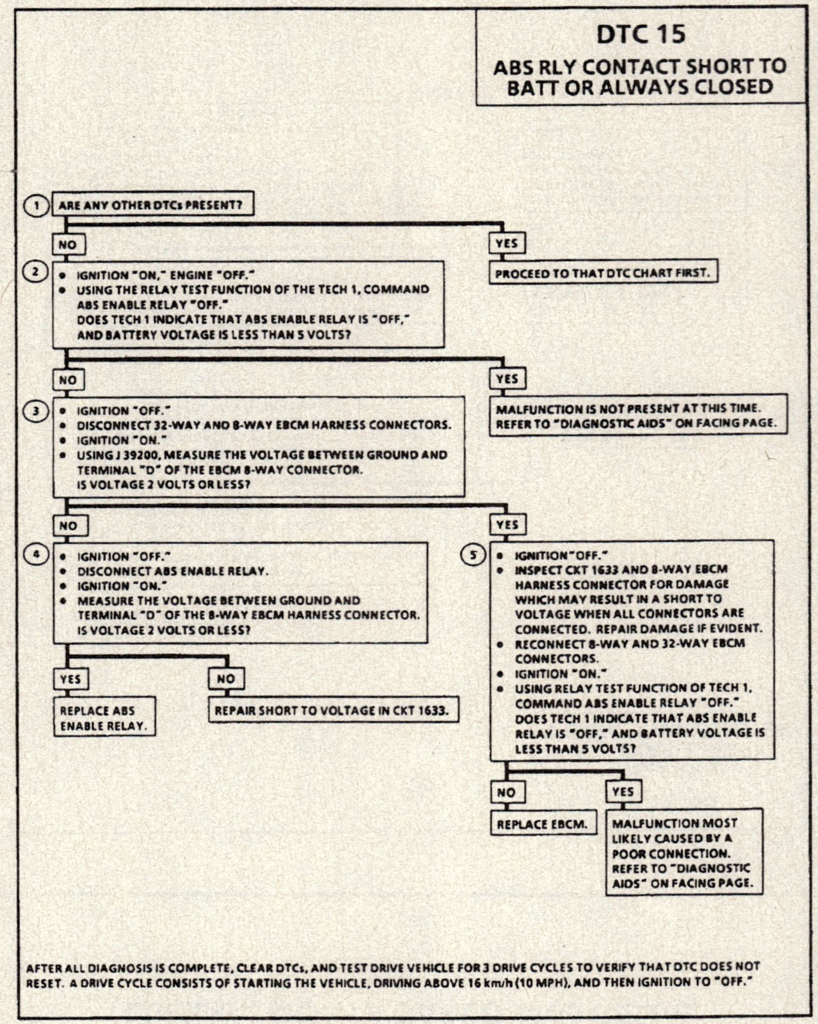

DIAGNOSTIC CHART DTC 16 — 1995 CUTLASS SUPREME, GRAND PRIX, LUMINA, MONTE CARLO AND REGAL

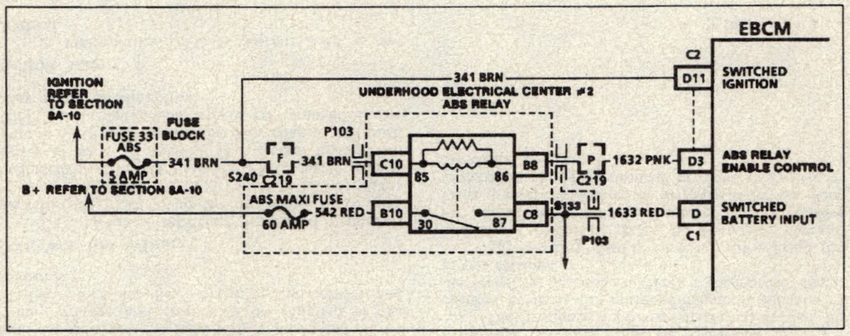

DTC 16
ABS ENABLE RELAY COIL CIRCUIT OPEN

Circuit Description:

Ignition voltage is supplied through terminal "C10" of the ABS relay. The EBCM then is able to energize the pull-in coil by completing the ground circuit at terminal "D3" of the EBCM. The magnetic field created closes the ABS relay contacts and allows battery voltage and current to be supplied to the EBCM, which supplies power to the motors and solenoids.

DTC Will Set When:

DTC 16 can be set only after DTC 14 has been set. This test detects an open in the ABS relay coil circuit. An open in this circuit will not allow the ABS relay to be energized thus preventing voltage/current to the motors and solenoids. If this malfunction is present and the ignition is turned "OFF" before 5 km/h (3 mph) is reached, DTC 14 is set alone.

Action Taken:

A malfunction DTC is stored, ABS is disabled and the amber ABS warning lamp is turned "ON." The red BRAKE warning lamp is turned "ON" if the rear piston in the ABS motor pack is not in the home position.

DTC Chart Test Description:

Number(s) below refer to circled number(s) on the diagnostic chart.

1. Indicates the EBCM is capable of controlling the ABS relay as commanded.
2. Checks to ensure that voltage is available to the pull-in coil of the ABS relay.
3. Ensures that there is continuity through the pull-in coil of the ABS relay.
4. Verifies the integrity of the ABS relay control circuit.
5. Ensures malfunction was not due to poor terminal contact.

Diagnostic Aids:

An "Intermittent" malfunction may be caused by a poor connection, rubbed through wire insulation, or a wire that is broken inside the insulation.

The frequency of the malfunction can be checked by using the enhanced diagnostic function of the Tech 1, as described in "Tech 1 Diagnostics" found in this section.

If the frequency of the malfunction is high, but is currently intermittent, check for high coil resistance by measuring between relay terminals "C10" and "B8" using J 39200. If resistance shows greater than 100 ohms, replace the relay.

Any circuitry that is suspected of causing the intermittent complaint should be thoroughly checked for backed out terminals, improper mating, broken locks, improperly formed or damaged terminals, poor terminal to wiring connections or physical damage to the wiring harness.

DIAGNOSTIC CHART DTC 16 — 1995 CUTLASS SUPREME, GRAND PRIX, LUMINA, MONTE CARLO AND REGAL

DTC 16
ABS ENABLE RELAY COIL CIRCUIT OPEN

(1)
- IGNITION "ON," ENGINE "OFF."
- USING THE RELAY TEST FUNCTION OF THE TECH 1, COMMAND THE ABS ENABLE RELAY "ON." DOES TECH 1 INDICATE ABS ENABLE RELAY "ON," AND IS VOLTAGE 10 VOLTS OR GREATER?

YES: MALFUNCTION IS NOT PRESENT AT THIS TIME. REFER TO "DIAGNOSTIC AIDS" ON FACING PAGE.

NO:

(2)
- USING J 39200, MEASURE THE VOLTAGE BETWEEN GROUND AND TERMINAL "C10" OF THE ABS ENABLE RELAY CONNECTOR. IS VOLTAGE 10 VOLTS OR GREATER?

NO: REPAIR OPEN OR HIGH RESISTANCE IN CKT 341 OR FUSE 33.

YES:

(3)
- USING THE RELAY TEST FUNCTION OF THE TECH 1, COMMAND THE ABS ENABLE RELAY "OFF."
- MEASURE THE VOLTAGE BETWEEN GROUND AND TERMINAL "B8" OF THE ABS ENABLE RELAY CONNECTOR. IS VOLTAGE 10 VOLTS OR GREATER?

NO: CHECK FOR INTERMITTENT CONNECTIONS AT ABS ENABLE RELAY. IF OK, REPLACE ABS RELAY.

YES:

(4)
- IGNITION "OFF."
- DISCONNECT 32-WAY EBCM CONNECTOR.
- IGNITION "ON."
- MEASURE THE VOLTAGE BETWEEN GROUND AND TERMINAL "D3" OF THE 32-WAY EBCM HARNESS CONNECTOR. IS VOLTAGE 10 VOLTS OR GREATER?

NO: REPAIR OPEN OR HIGH RESISTANCE IN CKT 1632.

YES:

(5)
- IGNITION "OFF."
- INSPECT 32-WAY EBCM CONNECTOR TERMINAL "D3" FOR POOR TERMINAL CONTACT OR EVIDENCE OF CORROSION. REPLACE ALL TERMINALS THAT EXHIBIT SIGNS OF POOR TERMINAL CONTACT OR CORROSION.
- RECONNECT ALL CONNECTORS.
- IGNITION "ON." DOES DTC 16 SET AS A CURRENT DTC?

YES: REPLACE EBCM.

NO: MALFUNCTION WAS MOST LIKELY CAUSED BY A POOR CONNECTION. REFER TO "DIAGNOSTIC AIDS" ON FACING PAGE.

AFTER ALL DIAGNOSIS IS COMPLETE, CLEAR DTCs AND TEST DRIVE VEHICLE FOR THREE (3) DRIVE CYCLES TO VERIFY THAT DTC DOES NOT RESET. A DRIVE CYCLE CONSISTS OF STARTING THE VEHICLE, DRIVING OVER 16 km/h (10 MPH), AND THEN IGNITION TO "OFF."

DIAGNOSTIC CHART DTC 17 — 1995 CUTLASS SUPREME, GRAND PRIX, LUMINA, MONTE CARLO AND REGAL

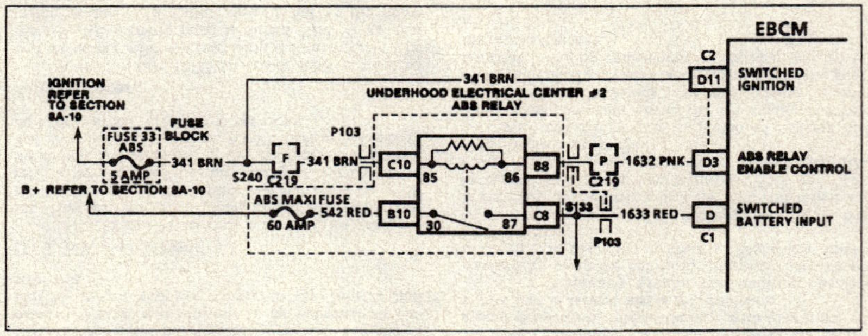

DTC 17
ABS ENABLE RELAY COIL CIRCUIT SHORTED TO GND

Circuit Description:

Ignition voltage is supplied through terminal "C10" of the ABS relay. The EBCM then is able to energize the pull-in coil by completing the ground circuit at terminal "D3" of the EBCM. The magnetic field created closes the ABS relay contacts and allows battery voltage and current to be supplied to the EBCM, which supplies voltage to the motors and solenoids.

DTC Will Set When:

DTC 17 can be set before the EBCM commands the ABS relay "ON." This test determines if the ABS relay is energized when it should not be. This malfunction would not allow the ABS relay to remove power to the ABS system. If a second malfunction were to occur that requires the ABS relay to be turned "OFF," that malfunction cannot be removed if the relay cannot be controlled.

Action Taken:

A malfunction DTC is stored. ABS is not disabled.

DTC Chart Test Description:

Number(s) below refer to circled number(s) on the diagnostic chart.

1. Indicates that the EBCM is capable of controlling the ABS relay as commanded.
2. Checks to ensure that the ABS relay or control CKT 1632 are not shorted to ground.
3. Checks to see if the EBCM is internally shorted to ground. A grounded terminal "D3" would cause the ABS relay to be energized anytime the ignition was in the "RUN," position.
4. Checks for a short to ground in control CKT 1632 or a defective ABS relay.

Diagnostic Aids:

An "Intermittent" malfunction may be caused by a poor connection, rubbed through wire insulation, or a wire that is broken inside the insulation.

The frequency of the malfunction can be checked by using the enhanced diagnostic function of the Tech 1, as described in "Tech 1 Diagnostics" found in this section.

Any circuitry that is suspected of causing the intermittent complaint should be thoroughly checked for backed out terminals, improper mating, broken locks, improperly formed or damaged terminals, poor terminal to wiring connections or physical damage to the wiring harness.

DIAGNOSTIC CHART DTC 17 — 1995 CUTLASS SUPREME, GRAND PRIX, LUMINA, MONTE CARLO AND REGAL

DTC 17
ABS ENABLE RELAY COIL CIRCUIT SHORTED TO GND

(1)
- IGNITION "ON," ENGINE "OFF."
- USING THE RELAY TEST FUNCTION OF THE TECH 1, COMMAND THE ABS ENABLE RELAY "OFF."

DOES TECH 1 INDICATE THAT ABS ENABLE RELAY IS "OFF" AND BATTERY VOLTAGE IS LESS THAN 5 VOLTS?

YES → MALFUNCTION IS NOT PRESENT AT THIS TIME. REFER TO "DIAGNOSTIC AIDS" ON FACING PAGE.

NO → (2)
- IGNITION "OFF."
- DISCONNECT 32-WAY EBCM CONNECTOR.
- IGNITION "ON."
- USING J 39200, MEASURE THE VOLTAGE BETWEEN TERMINAL "D3" OF THE 32-WAY EBCM HARNESS CONNECTOR AND B+.

IS VOLTAGE 2 VOLTS OR LESS?

YES → (3)
- IGNITION "OFF."
- INSPECT CKT 1632 AND TERMINAL "D3" OF THE 32-WAY EBCM CONNECTOR FOR DAMAGE WHICH MAY RESULT IN A SHORT TO GROUND WITH ALL CONNECTORS CONNECTED. REPAIR ALL DAMAGE IF EVIDENT.
- RECONNECT ALL CONNECTORS.
- IGNITION "ON."

DOES DTC 17 SET AS A CURRENT DTC?

- YES → REPLACE EBCM.
- NO → MALFUNCTION WAS MOST LIKELY CAUSED BY A POOR CONNECTION. REFER TO "DIAGNOSTIC AIDS" ON FACING PAGE.

NO → (4)
- IGNITION "OFF."
- DISCONNECT ABS ENABLE RELAY.
- MEASURE VOLTAGE BETWEEN TERMINAL "88" OF THE ABS ENABLE RELAY CONNECTOR AND B+.

IS VOLTAGE 2 VOLTS OR LESS?

- NO → REPAIR SHORT TO GROUND IN CKT 1632.
- YES → REPLACE ABS ENABLE RELAY.

AFTER ALL DIAGNOSIS IS COMPLETE, CLEAR DTCs AND TEST DRIVE VEHICLE FOR THREE (3) DRIVE CYCLES TO VERIFY THAT DTC DOES NOT RESET. A DRIVE CYCLE CONSISTS OF STARTING THE VEHICLE, DRIVING OVER 16 km/h (10 MPH), AND THEN IGNITION "OFF."

DIAGNOSTIC CHART DTC 18 — 1995 CUTLASS SUPREME, GRAND PRIX, LUMINA, MONTE CARLO AND REGAL

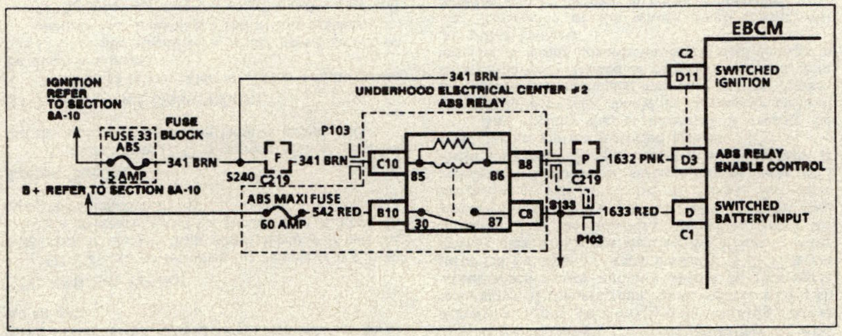

DTC 18

ABS ENABLE RELAY COIL CIRCUIT SHORTED TO BATT

Circuit Description:

Ignition voltage is supplied through terminal "C10" of the ABS relay. The EBCM then is able to energize the pull-in coil by completing the ground circuit at terminal "D3" of the EBCM. The magnetic field created closes the ABS relay contacts and allows battery voltage and current to be supplied to the EBCM, which supplies voltage to the motors and solenoids.

DTC Will Set When:

DTC 18 can be set after the EBCM commands the ABS relay "ON." This test monitors the availability of current/voltage to the motors and solenoids. A malfunction of this type will not allow voltage to be available to the motors and solenoids; therefore, ABS operation would not be allowed if required.

Action Taken:

A malfunction DTC is stored, ABS is disabled and the amber ABS warning lamp is turned "ON." The red BRAKE warning lamp is turned "ON" if the rear piston in the ABS motor pack is not in the home position.

DTC Chart Test Description:

Number(s) below refer to circled number(s) on the diagnostic chart.

1. Indicates that the EBCM is capable of controlling the ABS relay as commanded.
2. Checks for a short to voltage on the ABS relay control circuit.
3. Verifies that the ABS relay coil has the appropriate resistance.
4. Verifies that the relay contacts are not failed closed.
5. Checks for possible poor terminal contact at the EBCM.

Diagnostic Aids:

An "Intermittent" malfunction may be caused by a poor connection, rubbed through wire insulation, or a wire that is broken inside the insulation.

The frequency of the malfunction can be checked by using the enhanced diagnostic function of the Tech 1, as described in "Tech 1 Diagnostics" found in this section.

Any circuitry that is suspected of causing the intermittent complaint should be thoroughly checked for backed out terminals, improper mating, broken locks, improperly formed or damaged terminals, poor terminal to wiring connections or physical damage to the wiring harness.

DIAGNOSTIC CHART DTC 18 — 1995 CUTLASS SUPREME, GRAND PRIX, LUMINA, MONTE CARLO AND REGAL

DTC 18
ABS ENABLE RELAY COIL CIRCUIT SHORTED TO BATT

(1)
- IGNITION "ON," ENGINE "OFF."
- USING RELAY TEST FUNCTION OF TECH 1, COMMAND ABS ENABLE RELAY "ON."
- DOES TECH 1 INDICATE ABS ENABLE RELAY "ON" AND BATTERY VOLTAGE ABOVE 10 VOLTS?

YES: MALFUNCTION IS NOT PRESENT AT THIS TIME. REFER TO "DIAGNOSTIC AIDS" ON FACING PAGE.

NO: go to (2)

(2)
- IGNITION "OFF."
- DISCONNECT ABS ENABLE RELAY.
- DISCONNECT 32-WAY EBCM CONNECTOR.
- IGNITION "ON."
- USING J 39200, MEASURE THE VOLTAGE BETWEEN GROUND AND TERMINAL "D3" OF THE 32-WAY EBCM HARNESS CONNECTOR.
  IS VOLTAGE 2 VOLTS OR LESS?

NO: REPAIR SHORT TO VOLTAGE IN CKT 1632.

YES: go to (3)

(3)
- IGNITION "OFF."
- USING J 39200, MEASURE THE RESISTANCE BETWEEN TERMINAL "85" AND TERMINAL "86" OF THE ABS ENABLE RELAY.
  IS RESISTANCE 40Ω OR GREATER?

NO: REPLACE ABS ENABLE RELAY.

YES: go to (4)

(4)
- USING J 39200, MEASURE THE RESISTANCE BETWEEN TERMINALS "30" AND "87" OF THE ABS ENABLE RELAY.
  DOES J 39200 DISPLAY RESISTANCE AS "OL" (INFINITE)?

NO: REPLACE ABS ENABLE RELAY.

YES: go to (5)

(5)
- INSPECT CKT 1632 AND TERMINAL "D3" OF THE 32-WAY EBCM CONNECTOR FOR DAMAGE WHICH MAY RESULT IN A SHORT TO VOLTAGE WITH ALL CONNECTORS CONNECTED. REPAIR ALL DAMAGE IF EVIDENT.
- RECONNECT ALL CONNECTORS.
- IGNITION "ON."
  DOES DTC 18 SET AS A CURRENT DTC?

YES: REPLACE EBCM.

NO: MALFUNCTION IS MOST LIKELY DUE TO A POOR CONNECTION. REFER TO "DIAGNOSTIC AIDS" ON FACING PAGE.

AFTER ALL DIAGNOSIS IS COMPLETE, CLEAR DTCs AND TEST DRIVE VEHICLE FOR THREE (3) DRIVE CYCLES TO VERIFY THAT DTC DOES NOT RESET. A DRIVE CYCLE CONSISTS OF STARTING THE VEHICLE, DRIVING OVER 16 km/h (10 MPH), AND THEN IGNITION "OFF."

DIAGNOSTIC CHART DTC 21 — 1995 CUTLASS SUPREME, GRAND PRIX, LUMINA, MONTE CARLO AND REGAL

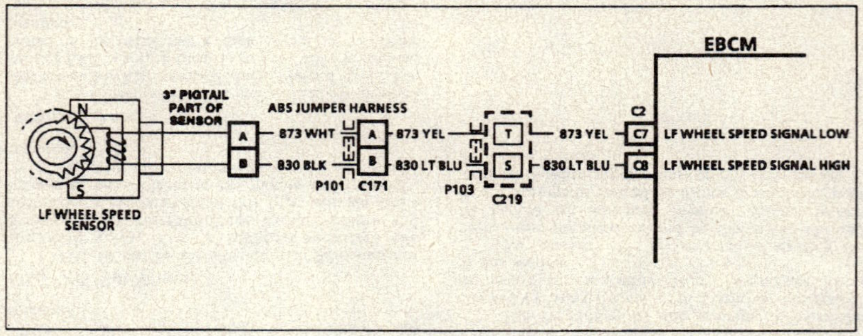

DTC 21
LEFT FRONT WHEEL SPEED = 0

Circuit Description:

As a toothed ring passes by the wheel speed sensor, changes in the electromagnetic field cause the wheel speed sensor to produce a sinusoidal (AC) voltage signal whose frequency is proportional to wheel speed. The magnitude of this signal is directly related to wheel speed and the proximity of the wheel speed sensor to the toothed ring, often referred to as the air gap.

DTC Will Set When:

DTC 21 can be set when the vehicle is not in an ABS stop. If the left front wheel speed = 0 and the vehicle's reference speed is greater than 8 km/h (5 mph), a malfunction exists.

Action Taken:

A malfunction DTC is stored, ABS is disabled and the amber ABS warning lamp is turned "ON."

DTC Chart Test Description:

Number(s) below refer to circled number(s) on the diagnostic chart.

- 1. This step identifies a wheel speed sensor or circuitry that is damaged and visibly apparent.
- 2. Checks to see that the wheel speed sensor coil is within proper resistance values.
- 3. Verifies that the wheel speed sensor can generate an appropriate output signal.
- 4. This ensures that the wheel speed circuitry is not shorted together.
- 5. Ensures that the malfunction was not due to physical damage.

Diagnostic Aids:

An "Intermittent" malfunction may be caused by a poor connection, rubbed through wire insulation, or a wire that is broken inside the insulation.

The frequency of the malfunction can be checked by using the enhanced diagnostic function of the Tech 1, as described in "Tech 1 Diagnostics" found in this section.

If the customer's comments reflect that the ABS warning lamp is "ON" only during moist environmental changes (rain, snow, vehicle wash), all wheel speed sensor circuitry should be thoroughly inspected for signs of water intrusion. If DTC is not current, clear DTCs and simulate the effects of water intrusion. Use the following procedure. Spray the suspected area with a 5% salt water solution (two teaspoons of salt to 12 oz. of water). Test drive vehicle over various road surfaces (bumps, turns, etc.) above 24 km/h (15 mph) for at least 30 seconds. If DTC returns, replace suspected harness.

Any circuitry that is suspected of causing the intermittent complaint should be thoroughly checked for backed out terminals, improper mating, broken locks, improperly formed or damaged terminals, poor terminal to wiring connections or physical damage to the wiring harness.

Resistance of the wheel speed sensor will increase with an increase in sensor temperature.

When replacing a wheel speed sensor, inspect the sensor terminals and harness connector for corrosion and/or water intrusion. If evidence of corrosion or water intrusion exists, replace wheel speed sensor jumper harness. Likewise, if replacing a wheel speed sensor jumper harness, inspect sensor terminals. If evidence of corrosion or water intrusion exists, replace wheel speed sensor. Refer to "On-Vehicle Service" in this section.

DIAGNOSTIC CHART DTC 21 — 1995 CUTLASS SUPREME, GRAND PRIX, LUMINA, MONTE CARLO AND REGAL

DTC 21
LEFT FRONT WHEEL SPEED = 0

Important

- WHEEL SPEED SENSOR INTERMITTENT MALFUNCTIONS MAY BE DIFFICULT TO LOCATE. CARE SHOULD BE TAKEN NOT TO DISTURB ANY ELECTRICAL CONNECTIONS PRIOR TO AN INDICATED STEP OF THIS CHART. THIS WILL ENSURE THAT AN INTERMITTENT CONNECTION WILL NOT BE CORRECTED BEFORE THE SOURCE OF THE MALFUNCTION IS FOUND.

(1)
- IGNITION "OFF."
- PHYSICALLY INSPECT THE LEFT FRONT WHEEL SPEED SENSOR, LEFT FRONT WHEEL SPEED SENSOR JUMPER HARNESS, AND CONNECTORS FOR DAMAGE. INSPECT THE WHEEL SPEED SENSOR AND TOOTHED RING FOR LOOSENESS OR DAMAGE. PAY PARTICULAR ATTENTION TO THE TOOTHED RING, ANY DEVIATION WILL AFFECT THE WHEEL SPEED SENSOR OUTPUT SIGNAL. IS PHYSICAL DAMAGE INDICATED?

YES: REPLACE WHEEL SPEED SENSOR AND/OR JUMPER HARNESS IF DAMAGED. REPAIR ANY SENSOR OR TOOTHED RING LOOSENESS. DRIVE VEHICLE ABOVE 24 km/h (15 MPH) TO ENSURE DTC 21 DOES NOT SET.

NO:

(2)
- DISCONNECT LEFT FRONT WHEEL SPEED SENSOR DIRECTLY AT SENSOR.
- USING J 39200, MEASURE THE RESISTANCE BETWEEN SENSOR TERMINALS "A" AND "B".
  IS RESISTANCE BETWEEN 1661 AND 1929Ω WHEN SENSOR IS 20°C (68°F)? *(IF TEMPERATURE OF THE SENSOR IS SUSPECTED TO BE OTHER THAN 20°C (68°F), REFER TO THE TABLE BELOW FOR APPLICABLE SENSOR RESISTANCE VALUES; THE VALUES IN THE TABLE ARE FOR TEMPERATURE OF THE SENSOR, NOT AIR TEMPERATURE.)*

NO: REPLACE WHEEL SPEED SENSOR.

YES:

(3)
- WITH J 39200 STILL CONNECTED, SELECT A/C VOLTAGE SCALE.
- SPIN WHEEL BY HAND WHILE OBSERVING VOLTAGE.
  IS VOLTAGE AT LEAST 100 mV (VOLTAGE WILL INCREASE AS WHEEL SPEED INCREASES)?

NO: VERIFY THAT TOOTHED RING IS INTACT. IF OK, REPLACE WHEEL SPEED SENSOR

YES:

(4)
- IGNITION "OFF."
- DISCONNECT 32-WAY EBCM CONNECTOR.
- USING J 39200, MEASURE THE RESISTANCE BETWEEN TERMINAL "C7" AND "C8" OF THE 32-WAY EBCM HARNESS CONNECTOR.
  DOES J 39200 DISPLAY RESISTANCE AS "OL" (INFINITE)?

NO: REPAIR SHORT BETWEEN CKTs 830 AND 873.

YES:

(5)
- INSPECT CKTs 830 AND 873 AND 32-WAY EBCM CONNECTOR FOR DAMAGE WHICH MAY RESULT IN A SHORT BETWEEN CKTs 830 AND 873 WITH ALL CONNECTORS CONNECTED. REPAIR ALL DAMAGE IF EVIDENT.
- RECONNECT ALL CONNECTORS.
- TEST DRIVE VEHICLE ABOVE 24 km/h (15 MPH) FOR AT LEAST 30 SECONDS.
  DOES DTC 21 SET AS A CURRENT DTC?

YES: REPLACE EBCM.

NO: MALFUNCTION IS INTERMITTENT. REFER TO "DIAGNOSTIC AIDS" ON FACING PAGE.

FRONT WHEEL SPEED SENSOR
TEMPERATURE OF SENSOR VS. SENSOR RESISTANCE (APPROXIMATE)

TEMP. (°C)	TEMP. (°F)	RESISTANCE (OHMS)
-34 TO 4	-30 TO 40	1258 TO 1808
5 TO 43	41 TO 110	1502 TO 2103
44 TO 93	111 TO 200	1747 TO 2482

DIAGNOSTIC CHART DTC 22 — 1995 CUTLASS SUPREME, GRAND PRIX, LUMINA, MONTE CARLO AND REGAL

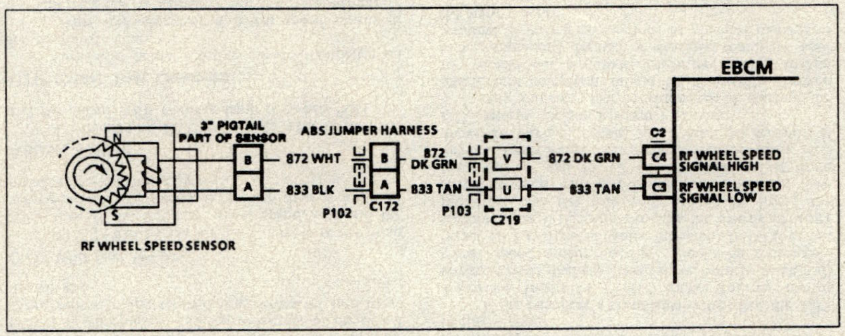

DTC 22
RIGHT FRONT WHEEL SPEED = 0

Circuit Description:

As a toothed ring passes by the wheel speed sensor, changes in the electromagnetic field cause the wheel speed sensor to produce a sinusoidal (AC) voltage signal whose frequency is proportional to wheel speed. The magnitude of this signal is directly related to wheel speed and the proximity of the wheel speed sensor to the toothed ring, often referred to as the air gap.

DTC Will Set When:

DTC 22 can be set when the vehicle is not in an ABS stop. If the right front wheel speed = 0 and the vehicle's reference speed is greater than 8 km/h (5 mph), a malfunction exists.

Action Taken:

A malfunction DTC is stored, ABS is disabled and the amber ABS warning lamp is turned "ON."

DTC Chart Test Description:

Number(s) below refer to circled number(s) on the diagnostic chart.

1. This step identifies a wheel speed sensor or circuitry that is damaged and visibly apparent.
2. Checks to see that the wheel speed sensor coil is within proper resistance values.
3. Verifies that the wheel speed sensor can generate an appropriate output signal.
4. This ensures that the wheel speed circuitry is not shorted together.
5. Ensures that the malfunction was not due to physical damage.

Diagnostic Aids:

An "Intermittent" malfunction may be caused by a poor connection, rubbed through wire insulation, or a wire that is broken inside the insulation.

The frequency of the malfunction can be checked by using the enhanced diagnostic function of the Tech 1, as described in "Tech 1 Diagnostics" found in this section.

If the customer's comments reflect that the ABS warning lamp is "ON" only during moist environmental changes (rain, snow, vehicle wash), all wheel speed sensor circuitry should be thoroughly inspected for signs of water intrusion. If DTC is not current, clear DTCs and simulate the effects of water intrusion. Use the following procedure. Spray the suspected area with a 5% salt water solution (two teaspoons of salt to 12 oz. of water). Test drive vehicle over various road surfaces (bumps, turns, etc.) above 24 km/h (15 mph) for at least 30 seconds. If DTC returns, replace suspected harness.

Any circuitry that is suspected of causing the intermittent complaint should be thoroughly checked for backed out terminals, improper mating, broken locks, improperly formed or damaged terminals, poor terminal to wiring connections or physical damage to the wiring harness.

Resistance of the wheel speed sensor will increase with an increase in sensor temperature.

When replacing a wheel speed sensor, inspect the sensor terminals and harness connector for corrosion and/or water intrusion. If evidence of corrosion or water intrusion exists, replace wheel speed sensor jumper harness. Likewise, if replacing a wheel speed sensor jumper harness, inspect sensor terminals. If evidence of corrosion or water intrusion exists, replace wheel speed sensor. Refer to "On-Vehicle Service" in this section.

DIAGNOSTIC CHART DTC 22 — 1995 CUTLASS SUPREME, GRAND PRIX, LUMINA, MONTE CARLO AND REGAL

DTC 22
RIGHT FRONT WHEEL SPEED = 0

Important

- WHEEL SPEED SENSOR INTERMITTENT MALFUNCTIONS MAY BE DIFFICULT TO LOCATE. CARE SHOULD BE TAKEN NOT TO DISTURB ANY ELECTRICAL CONNECTIONS PRIOR TO AN INDICATED STEP OF THIS CHART. THIS WILL ENSURE THAT AN INTERMITTENT CONNECTION WILL NOT BE CORRECTED BEFORE THE SOURCE OF THE MALFUNCTION IS FOUND.

(1)
- IGNITION "OFF."
- PHYSICALLY INSPECT THE RIGHT FRONT WHEEL SPEED SENSOR, RIGHT FRONT WHEEL SPEED SENSOR JUMPER HARNESS, AND CONNECTORS FOR DAMAGE. INSPECT THE WHEEL SPEED SENSOR AND TOOTHED RING FOR LOOSENESS OR DAMAGE. PAY PARTICULAR ATTENTION TO THE TOOTHED RING, ANY DEVIATION WILL AFFECT THE WHEEL SPEED SENSOR OUTPUT SIGNAL.

IS PHYSICAL DAMAGE INDICATED?

YES: REPLACE WHEEL SPEED SENSOR AND/OR JUMPER HARNESS IF DAMAGED. REPAIR ANY SENSOR OR TOOTHED RING LOOSENESS. DRIVE VEHICLE ABOVE 24 km/h (15 MPH) TO ENSURE DTC 22 DOES NOT SET.

NO:

(2)
- DISCONNECT RIGHT FRONT WHEEL SPEED SENSOR DIRECTLY AT SENSOR.
- USING J 39200, MEASURE THE RESISTANCE BETWEEN SENSOR TERMINALS "A" AND "B."

IS RESISTANCE BETWEEN 1661 AND 1929Ω WHEN SENSOR IS 20°C (68°F)? *(IF TEMPERATURE OF THE SENSOR IS SUSPECTED TO BE OTHER THAN 20°C (68°F), REFER TO THE TABLE BELOW FOR APPLICABLE SENSOR RESISTANCE VALUES; THE VALUES IN THE TABLE ARE FOR TEMPERATURE OF THE SENSOR, NOT AIR TEMPERATURE.)*

NO: REPLACE WHEEL SPEED SENSOR.

YES:

(3)
- WITH J 39200 STILL CONNECTED, SELECT A/C VOLTAGE SCALE.
- SPIN WHEEL BY HAND WHILE OBSERVING VOLTAGE.

IS VOLTAGE AT LEAST 100 mV (VOLTAGE WILL INCREASE AS WHEEL SPEED INCREASES)?

NO: VERIFY THAT TOOTHED RING IS INTACT. IF OK, REPLACE WHEEL SPEED SENSOR

YES:

(4)
- IGNITION "OFF."
- DISCONNECT 32-WAY EBCM CONNECTOR.
- USING J 39200, MEASURE THE RESISTANCE BETWEEN TERMINAL "C3" AND "C4" OF THE 32-WAY EBCM HARNESS CONNECTOR.

DOES J 39200 DISPLAY RESISTANCE AS "OL" (INFINITE)?

NO: REPAIR SHORT BETWEEN CKTs 833 AND 872.

YES:

(5)
- INSPECT CKTs 833 AND 872 AND 32-WAY EBCM CONNECTOR FOR DAMAGE WHICH MAY RESULT IN A SHORT BETWEEN CKTs 833 AND 872 WITH ALL CONNECTORS CONNECTED. REPAIR ALL DAMAGE IF EVIDENT.
- RECONNECT ALL CONNECTORS.
- TEST DRIVE VEHICLE ABOVE 24 km/h (15 MPH) FOR AT LEAST 30 SECONDS.

DOES DTC 22 SET AS A CURRENT DTC?

YES: REPLACE EBCM.

NO: MALFUNCTION IS INTERMITTENT. REFER TO "DIAGNOSTIC AIDS" ON FACING PAGE.

FRONT WHEEL SPEED SENSOR
TEMPERATURE OF SENSOR VS. SENSOR RESISTANCE (APPROXIMATE)

TEMP. (°C)	TEMP. (°F)	RESISTANCE (OHMS)
-34 TO 4	-30 TO 40	1258 TO 1808
5 TO 43	41 TO 110	1502 TO 2103
44 TO 93	111 TO 200	1747 TO 2482

DIAGNOSTIC CHART DTC 23 — 1995 CUTLASS SUPREME, GRAND PRIX, LUMINA, MONTE CARLO AND REGAL

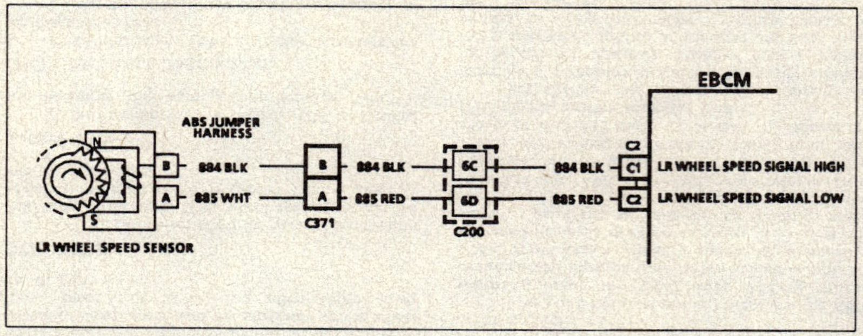

DTC 23
LEFT REAR WHEEL SPEED = 0

Circuit Description:

As a toothed ring passes by the wheel speed sensor, changes in the electromagnetic field cause the wheel speed sensor to produce a sinusoidal (AC) voltage signal whose frequency is proportional to wheel speed. The magnitude of this signal is directly related to wheel speed and the proximity of the wheel speed sensor to the toothed ring, often referred to as the air gap.

DTC Will Set When:

DTC 23 can be set when the vehicle is not in an ABS stop. If the left rear wheel speed = 0 and the vehicle's reference speed is greater than 8 km/h (5 mph), a malfunction exists.

Action Taken:

A malfunction DTC is stored, ABS is disabled and the amber ABS warning lamp is turned "ON."

DTC Chart Test Description:

Number(s) below refer to circled number(s) on the diagnostic chart.

1. This step identifies a wheel speed sensor or circuitry that is damaged and visibly apparent.
2. Checks to see that the wheel speed sensor coil is within proper resistance values.
3. Verifies that the wheel speed sensor can generate an appropriate output signal.
4. Verifies that the wheel speed circuitry is not shorted together.
5. Ensures that the malfunction was not due to physical damage.
6. Checks for an incorrect or malfunctioning hub and bearing assembly.

Diagnostic Aids:

An "Intermittent" malfunction may be caused by a poor connection, rubbed through wire insulation, or a wire that is broken inside the insulation.

The frequency of the malfunction can be checked by using the enhanced diagnostic function of the Tech 1, as described in "Tech 1 Diagnostics" found in this section.

If the customer's comments reflect that the ABS warning lamp is "ON" only during moist environmental changes (rain, snow, vehicle wash), all wheel speed sensor circuitry should be thoroughly inspected for signs of water intrusion. If DTC is not current, clear DTCs and simulate the effects of water intrusion. Use the following procedure. Spray the suspected area with a 5% salt water solution (two teaspoons of salt to 12 oz. of water). Test drive vehicle over various road surfaces (bumps, turns, etc.) above 24 km/h (15 mph) for at least 30 seconds. If DTC returns, replace suspected harness.

Any circuitry that is suspected of causing the intermittent complaint should be thoroughly checked for backed out terminals, improper mating, broken locks, improperly formed or damaged terminals, poor terminal to wiring connections or physical damage to the wiring harness.

Resistance of the wheel speed sensor will increase with an increase in sensor temperature.

When replacing a wheel speed sensor, inspect the sensor terminals and harness connector for corrosion and/or water intrusion. If evidence of corrosion or water intrusion exists, replace wheel speed sensor jumper harness. Likewise, if replacing a wheel speed sensor jumper harness, inspect sensor terminals. If evidence of corrosion or water intrusion exists, replace wheel speed sensor. Refer to "On-Vehicle Service" in this section.

DIAGNOSTIC CHART DTC 23 — 1995 CUTLASS SUPREME, GRAND PRIX, LUMINA, MONTE CARLO AND REGAL

DTC 23
LEFT REAR WHEEL SPEED = 0

Important

- WHEEL SPEED SENSOR INTERMITTENT MALFUNCTIONS MAY BE DIFFICULT TO LOCATE. CARE SHOULD BE TAKEN NOT TO DISTURB ANY ELECTRICAL CONNECTIONS PRIOR TO AN INDICATED STEP OF THIS CHART. THIS WILL ENSURE THAT AN INTERMITTENT CONNECTION WILL NOT BE CORRECTED BEFORE THE SOURCE OF THE MALFUNCTION IS FOUND.

(1)
- IGNITION "OFF."
- PHYSICALLY INSPECT THE LEFT REAR WHEEL SPEED SENSOR, WHEEL SPEED SENSOR JUMPER HARNESS, AND CONNECTORS FOR DAMAGE. IS PHYSICAL DAMAGE INDICATED?

YES: REPLACE WHEEL SPEED SENSOR AND/OR WHEEL SPEED SENSOR JUMPER HARNESS IF DAMAGED. DRIVE VEHICLE ABOVE 24 km/h (15 MPH) TO ENSURE DTC 23 DOES NOT SET.

NO:

(2)
- DISCONNECT LEFT REAR WHEEL SPEED SENSOR DIRECTLY AT SENSOR.
- USING J 39200, MEASURE THE RESISTANCE BETWEEN SENSOR TERMINALS "A" AND "B." IS RESISTANCE BETWEEN 2059 AND 2353Ω WHEN SENSOR IS 20°C (68°F)? *(IF TEMPERATURE OF THE SENSOR IS SUSPECTED TO BE OTHER THAN 20°C (68°F), REFER TO THE TABLE BELOW FOR APPLICABLE SENSOR RESISTANCE VALUES; THE VALUES IN THE TABLE ARE FOR TEMPERATURE OF THE SENSOR, NOT AIR TEMPERATURE.)*

NO: REPLACE WHEEL SPEED SENSOR.

YES:

(3)
- WITH J 39200 STILL CONNECTED, SELECT A/C VOLTAGE SCALE.
- SPIN WHEEL BY HAND WHILE OBSERVING VOLTAGE. IS VOLTAGE AT LEAST 100 mV (VOLTAGE WILL INCREASE AS WHEEL SPEED INCREASES)?

NO: REPLACE WHEEL SPEED SENSOR.

YES:

(4)
- IGNITION "OFF."
- DISCONNECT 32-WAY EBCM CONNECTOR.
- USING J 39200, MEASURE THE RESISTANCE BETWEEN TERMINAL "C1" AND "C2" OF THE 32-WAY EBCM HARNESS CONNECTOR. DOES J 39200 DISPLAY RESISTANCE AS "OL" (INFINITE)?

NO: REPAIR SHORT BETWEEN CKTs 884 AND 885.

YES:

(5)
- INSPECT CKTs 884 AND 885 AND 32-WAY EBCM CONNECTOR FOR DAMAGE WHICH MAY RESULT IN A SHORT BETWEEN CKTs 884 AND 885 WITH ALL CONNECTORS CONNECTED. REPAIR ALL DAMAGE IF EVIDENT.
- RECONNECT ALL CONNECTORS.
- TEST DRIVE VEHICLE ABOVE 24 km/h (15 MPH) FOR AT LEAST 30 SECONDS. DOES DTC 23 SET AS A CURRENT DTC?

NO: MALFUNCTION WAS MOST LIKELY CAUSED BY A POOR CONNECTION. REFER TO "DIAGNOSTIC AIDS" ON FACING PAGE.

YES:

(6)
- USING TECH 1, SELECT "F0: DATA LIST."
- MONITOR WHEEL SPEEDS WHILE SLOWLY ACCELERATING TO 65 km/h (40 MPH) AND THEN SLOWLY DECELERATING TO 0 km/h (0 MPH). IS THE LEFT REAR WHEEL SPEED CONSTANTLY HIGHER THAN THE THREE REMAINING WHEEL SPEEDS?

NO: REPLACE EBCM.

YES: REPLACE INCORRECT OR MALFUNCTIONING LEFT WHEEL SPEED SENSOR/HUB AND BEARING ASSEMBLY WITH THE PROPER ASSEMBLY. REFER TO PARTS CATALOG FOR PROPER PART NUMBER.

REAR WHEEL SPEED SENSOR
TEMPERATURE OF SENSOR VS. SENSOR RESISTANCE (APPROXIMATE)

TEMP. (°C)	TEMP. (°F)	RESISTANCE (OHMS)
-34 TO 4	-30 TO 40	1622 TO 2205
5 TO 43	41 TO 110	1930 TO 2566
44 TO 93	111 TO 200	2245 TO 3028

DIAGNOSTIC CHART DTC 24 — 1995 CUTLASS SUPREME, GRAND PRIX, LUMINA, MONTE CARLO AND REGAL

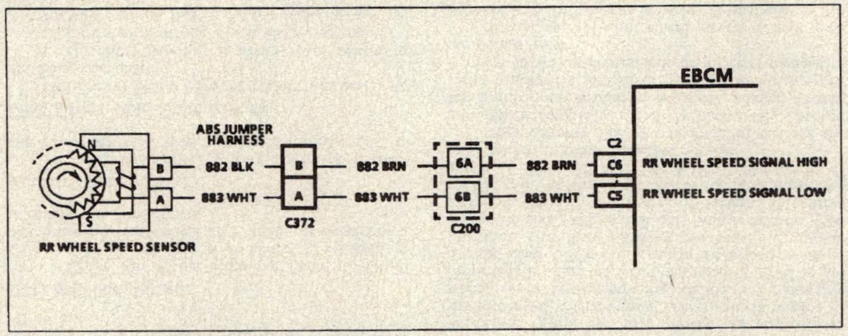

DTC 24

RIGHT REAR WHEEL SPEED = 0

Circuit Description:

As a toothed ring passes by the wheel speed sensor, changes in the electromagnetic field cause the wheel speed sensor to produce a sinusoidal (AC) voltage signal whose frequency is proportional to wheel speed. The magnitude of this signal is directly related to wheel speed and the proximity of the wheel speed sensor to the toothed ring, often referred to as the air gap.

DTC Will Set When:

DTC 24 can be set when the vehicle is not in an ABS stop. If the right rear wheel speed = 0 and the vehicle's reference speed is greater than 8 km/h (5 mph), a malfunction exists.

Action Taken:

A malfunction DTC is stored, ABS is disabled and the amber ABS warning lamp is turned "ON."

DTC Chart Test Description:

Number(s) below refer to circled number(s) on the diagnostic chart.

1. This step identifies a wheel speed sensor or circuitry that is damaged and visibly apparent.
2. Checks to see that the wheel speed sensor coil is within proper resistance values.
3. Verifies that the wheel speed sensor can generate an appropriate output signal.
4. Verifies that the wheel speed circuitry is not shorted together.
5. Ensures that the malfunction was not due to physical damage.
6. Checks for an incorrect or malfunctioning hub and bearing assembly.

Diagnostic Aids:

An "Intermittent" malfunction may be caused by a poor connection, rubbed through wire insulation, or a wire that is broken inside the insulation.

The frequency of the malfunction can be checked by using the enhanced diagnostic function of the Tech 1, as described in "Tech 1 Diagnostics" found in this section.

If the customer's comments reflect that the ABS warning lamp is "ON" only during moist environmental changes (rain, snow, vehicle wash), all wheel speed sensor circuitry should be thoroughly inspected for signs of water intrusion. If DTC is not current, clear DTCs and simulate the effects of water intrusion. Use the following procedure. Spray the suspected area with a 5% salt water solution (two teaspoons of salt to 12 oz. of water). Test drive vehicle over various road surfaces (bumps, turns, etc.) above 24 km/h (15 mph) for at least 30 seconds. If DTC returns, replace suspected harness.

Any circuitry that is suspected of causing the intermittent complaint should be thoroughly checked for backed out terminals, improper mating, broken locks, improperly formed or damaged terminals, poor terminal to wiring connections or physical damage to the wiring harness.

Resistance of the wheel speed sensor will increase with an increase in sensor temperature.

When replacing a wheel speed sensor, inspect the sensor terminals and harness connector for corrosion and/or water intrusion. If evidence of corrosion or water intrusion exists, replace wheel speed sensor jumper harness. Likewise, if replacing a wheel speed sensor jumper harness, inspect sensor terminals. If evidence of corrosion or water intrusion exists, replace wheel speed sensor. Refer to "On-Vehicle Service" in this section.

DIAGNOSTIC CHART DTC 24 — 1995 CUTLASS SUPREME, GRAND PRIX, LUMINA, MONTE CARLO AND REGAL

DTC 24
RIGHT REAR WHEEL SPEED = 0

Important

- WHEEL SPEED SENSOR INTERMITTENT MALFUNCTIONS MAY BE DIFFICULT TO LOCATE. CARE SHOULD BE TAKEN NOT TO DISTURB ANY ELECTRICAL CONNECTIONS PRIOR TO AN INDICATED STEP OF THIS CHART. THIS WILL ENSURE THAT AN INTERMITTENT CONNECTION WILL NOT BE CORRECTED BEFORE THE SOURCE OF THE MALFUNCTION IS FOUND.

(1)
- IGNITION "OFF."
- PHYSICALLY INSPECT THE RIGHT REAR WHEEL SPEED SENSOR, WHEEL SPEED SENSOR JUMPER HARNESS, AND CONNECTORS FOR DAMAGE.

IS PHYSICAL DAMAGE INDICATED?

YES: REPLACE WHEEL SPEED SENSOR AND/OR WHEEL SPEED SENSOR JUMPER HARNESS IF DAMAGED. DRIVE VEHICLE ABOVE 24 km/h (15 MPH) TO ENSURE DTC 24 DOES NOT SET.

NO:

(2)
- DISCONNECT RIGHT REAR WHEEL SPEED SENSOR DIRECTLY AT SENSOR.
- USING J 39200, MEASURE THE RESISTANCE BETWEEN SENSOR TERMINALS "A" AND "B". IS RESISTANCE BETWEEN 2059 AND 2353Ω WHEN SENSOR IS 20°C (68°F)?
(IF TEMPERATURE OF THE SENSOR IS SUSPECTED TO BE OTHER THAN 20°C (68°F), REFER TO THE TABLE BELOW FOR APPLICABLE SENSOR RESISTANCE VALUES; THE VALUES IN THE TABLE ARE FOR TEMPERATURE OF THE SENSOR, NOT AIR TEMPERATURE.)

NO: REPLACE WHEEL SPEED SENSOR.

YES:

(3)
- WITH J 39200 STILL CONNECTED, SELECT A/C VOLTAGE SCALE.
- SPIN WHEEL BY HAND WHILE OBSERVING VOLTAGE.

IS VOLTAGE AT LEAST 100 mV (VOLTAGE WILL INCREASE AS WHEEL SPEED INCREASES)?

NO: REPLACE WHEEL SPEED SENSOR.

YES:

(4)
- IGNITION "OFF."
- DISCONNECT 32-WAY EBCM CONNECTOR.
- USING J 39200, MEASURE THE RESISTANCE BETWEEN TERMINAL "C5" AND "C6" OF THE 32-WAY EBCM HARNESS CONNECTOR.

DOES J 39200 DISPLAY RESISTANCE AS "OL" (INFINITE)?

NO: REPAIR SHORT BETWEEN CKTs 882 AND 883.

YES:

(5)
- INSPECT CKTs 882 AND 883 AND 32-WAY EBCM CONNECTOR FOR DAMAGE WHICH MAY RESULT IN A SHORT BETWEEN CKTs 882 AND 883 WITH ALL CONNECTORS CONNECTED. REPAIR ALL DAMAGE IF EVIDENT.
- RECONNECT ALL CONNECTORS.
- TEST DRIVE VEHICLE ABOVE 24 km/h (15 MPH) FOR AT LEAST 30 SECONDS.

DOES DTC 24 SET AS A CURRENT DTC?

NO: MALFUNCTION WAS MOST LIKELY CAUSED BY A POOR CONNECTION. REFER TO "DIAGNOSTIC AIDS" ON FACING PAGE.

YES:

(5)
- USING TECH 1, SELECT "F0: DATA LIST."
- MONITOR WHEEL SPEEDS WHILE SLOWLY ACCELERATING TO 65 km/h (40 MPH) AND THEN SLOWLY DECELERATING TO 0 km/h (0 MPH).

IS THE RIGHT REAR WHEEL SPEED CONSTANTLY HIGHER THAN THE THREE REMAINING WHEEL SPEEDS?

NO: REPLACE EBCM.

YES: REPLACE INCORRECT OR MALFUNCTIONING RIGHT WHEEL SPEED SENSOR/HUB AND BEARING ASSEMBLY WITH THE PROPER ASSEMBLY. REFER TO PARTS CATALOG FOR PROPER PART NUMBER.

REAR WHEEL SPEED SENSOR
TEMPERATURE OF SENSOR VS. SENSOR RESISTANCE (APPROXIMATE)

TEMP. (°C)	TEMP. (°F)	RESISTANCE (OHMS)
-34 TO 4	-30 TO 40	1622 TO 2205
5 TO 43	41 TO 110	1930 TO 2566
44 TO 93	111 TO 200	2245 TO 3028

DIAGNOSTIC CHART DTC 25 — 1995 CUTLASS SUPREME, GRAND PRIX, LUMINA, MONTE CARLO AND REGAL

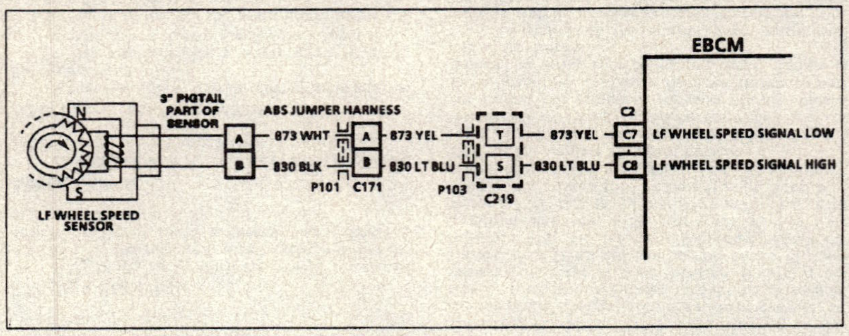

DTC 25
LEFT FRONT EXCESSIVE WHEEL SPEED VARIATION

Circuit Description:

As a toothed ring passes by the wheel speed sensor, changes in the electromagnetic field cause the wheel speed sensor to produce a sinusoidal (AC) voltage signal whose frequency is proportional to wheel speed. The magnitude of this signal is directly related to wheel speed and the proximity of the wheel speed sensor to the toothed ring, often referred to as the air gap.

DTC Will Set When:

DTC 25 can be set when the brake is "OFF." The purpose of this test is to detect a situation in which the left front wheel acceleration or deceleration is beyond specified limits.

Action Taken:

A malfunction DTC is stored, ABS is disabled and the amber ABS warning lamp is turned "ON."

DTC Chart Test Description:

Number(s) below refer to circled number(s) on the diagnostic chart.

1. This step identifies a wheel speed sensor or circuitry that is damaged and visibly apparent.
2. Checks to see that the wheel speed sensor coil is within proper resistance values.
3. Verifies that the wheel speed sensor can generate an appropriate output signal.
4. This ensures that the wheel speed circuitry is not shorted together.
5. Ensures that the malfunction was not due to physical damage.

Diagnostic Aids:

An "Intermittent" malfunction may be caused by a poor connection, rubbed through wire insulation, or a wire that is broken inside the insulation.

The frequency of the malfunction can be checked by using the enhanced diagnostic function of the Tech 1, as described in "Tech 1 Diagnostics" found in this section.

If the customer's comments reflect that the ABS warning lamp is "ON" only during moist environmental changes (rain, snow, vehicle wash), all wheel speed sensor circuitry should be thoroughly inspected for signs of water intrusion. If DTC is not current, clear DTCs and simulate the effects of water intrusion. Use the following procedure. Spray the suspected area with a 5% salt water solution (two teaspoons of salt to 12 oz. of water). Test drive vehicle over various road surfaces (bumps, turns, etc.) above 24 km/h (15 mph) for at least 30 seconds. If DTC returns, replace suspected harness.

Any circuitry that is suspected of causing the intermittent complaint should be thoroughly checked for backed out terminals, improper mating, broken locks, improperly formed or damaged terminals, poor terminal to wiring connections or physical damage to the wiring harness.

Resistance of the wheel speed sensor will increase with an increase in sensor temperature.

When replacing a wheel speed sensor, inspect the sensor terminals and harness connector for corrosion and/or water intrusion. If evidence of corrosion or water intrusion exists, replace wheel speed sensor jumper harness. Likewise, if replacing a wheel speed sensor jumper harness, inspect sensor terminals. If evidence of corrosion or water intrusion exists, replace wheel speed sensor. Refer to "On-Vehicle Service" in this section.

DIAGNOSTIC CHART DTC 25 — 1995 CUTLASS SUPREME, GRAND PRIX, LUMINA, MONTE CARLO AND REGAL

DTC 25 — LEFT FRONT EXCESSIVE WHEEL SPEED VARIATION

Important

- WHEEL SPEED SENSOR INTERMITTENT MALFUNCTIONS MAY BE DIFFICULT TO LOCATE. CARE SHOULD BE TAKEN NOT TO DISTURB ANY ELECTRICAL CONNECTIONS PRIOR TO AN INDICATED STEP OF THIS CHART. THIS WILL ENSURE THAT AN INTERMITTENT CONNECTION WILL NOT BE CORRECTED BEFORE THE SOURCE OF THE MALFUNCTION IS FOUND.

(1)
- IGNITION "OFF."
- PHYSICALLY INSPECT THE LEFT FRONT WHEEL SPEED SENSOR, LEFT FRONT WHEEL SPEED SENSOR JUMPER HARNESS, AND CONNECTORS FOR DAMAGE. INSPECT THE WHEEL SPEED SENSOR AND TOOTHED RING FOR LOOSENESS OR DAMAGE. PAY PARTICULAR ATTENTION TO THE TOOTHED RING, ANY DEVIATION WILL AFFECT THE WHEEL SPEED SENSOR OUTPUT SIGNAL.

IS PHYSICAL DAMAGE INDICATED?

YES: REPLACE WHEEL SPEED SENSOR AND/OR JUMPER HARNESS IF DAMAGED. REPAIR ANY SENSOR OR TOOTHED RING LOOSENESS. DRIVE VEHICLE ABOVE 24 km/h (15 MPH) TO ENSURE DTC 25 DOES NOT SET.

NO: go to (2)

(2)
- DISCONNECT LEFT FRONT WHEEL SPEED SENSOR DIRECTLY AT SENSOR.
- USING J 39200, MEASURE THE RESISTANCE BETWEEN SENSOR TERMINALS "A" AND "B".

IS RESISTANCE BETWEEN 1661 AND 1929Ω WHEN SENSOR IS 20°C (68°F)? *(IF TEMPERATURE OF THE SENSOR IS SUSPECTED TO BE OTHER THAN 20°C (68°F), REFER TO THE TABLE BELOW FOR APPLICABLE SENSOR RESISTANCE VALUES; THE VALUES IN THE TABLE ARE FOR TEMPERATURE OF THE SENSOR, NOT AIR TEMPERATURE.)*

NO: REPLACE WHEEL SPEED SENSOR.

YES: go to (3)

(3)
- WITH J 39200 STILL CONNECTED, SELECT A/C VOLTAGE SCALE.
- SPIN WHEEL BY HAND WHILE OBSERVING VOLTAGE.

IS VOLTAGE AT LEAST 100 mV (VOLTAGE WILL INCREASE AS WHEEL SPEED INCREASES)?

NO: VERIFY THAT TOOTHED RING IS INTACT. IF OK, REPLACE WHEEL SPEED SENSOR

YES: go to (4)

(4)
- IGNITION "OFF."
- DISCONNECT 32-WAY EBCM CONNECTOR.
- USING J 39200, MEASURE THE RESISTANCE BETWEEN TERMINAL "C7" AND "C8" OF THE 32-WAY EBCM HARNESS CONNECTOR.

DOES J 39200 DISPLAY RESISTANCE AS "OL" (INFINITE)?

NO: REPAIR SHORT BETWEEN CKTs 830 AND 873.

YES: go to (5)

(5)
- INSPECT CKTs 830 AND 873 AND 32-WAY EBCM CONNECTOR FOR DAMAGE WHICH MAY RESULT IN A SHORT BETWEEN CKTs 830 AND 873 WITH ALL CONNECTORS CONNECTED. REPAIR ALL DAMAGE IF EVIDENT.
- RECONNECT ALL CONNECTORS.
- TEST DRIVE VEHICLE ABOVE 24 km/h (15 MPH) FOR AT LEAST 30 SECONDS.

DOES DTC 25 SET AS A CURRENT DTC?

YES: REPLACE EBCM.

NO: MALFUNCTION IS INTERMITTENT. REFER TO "DIAGNOSTIC AIDS" ON FACING PAGE.

FRONT WHEEL SPEED SENSOR
TEMPERATURE OF SENSOR VS. SENSOR RESISTANCE (APPROXIMATE)

TEMP. (°C)	TEMP. (°F)	RESISTANCE (OHMS)
-34 TO 4	-30 TO 40	1258 TO 1808
5 TO 43	41 TO 110	1502 TO 2103
44 TO 93	111 TO 200	1747 TO 2482

DIAGNOSTIC CHART DTC 26 — 1995 CUTLASS SUPREME, GRAND PRIX, LUMINA, MONTE CARLO AND REGAL

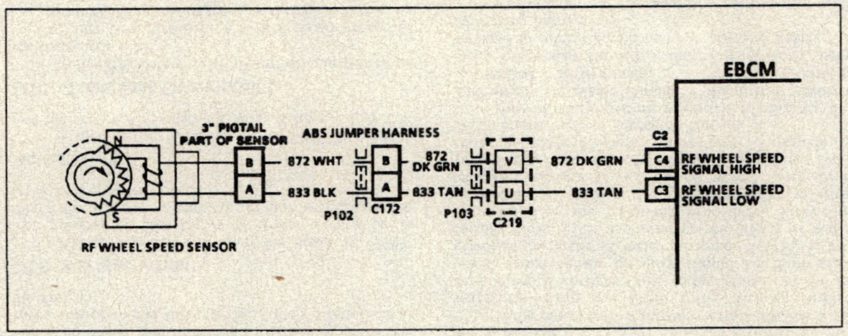

DTC 26
RIGHT FRONT EXCESSIVE WHEEL SPEED VARIATION

Circuit Description:

As a toothed ring passes by the wheel speed sensor, changes in the electromagnetic field cause the wheel speed sensor to produce a sinusoidal (AC) voltage signal whose frequency is proportional to wheel speed. The magnitude of this signal is directly related to wheel speed and the proximity of the wheel speed sensor to the toothed ring, often referred to as the air gap.

DTC Will Set When:

DTC 26 can be set when the brake is "OFF." The purpose of this test is to detect a situation in which the right front wheel acceleration or deceleration is beyond specified limits.

Action Taken:

A malfunction DTC is stored, ABS is disabled and the amber ABS warning lamp is turned "ON."

DTC Chart Test Description:

Number(s) below refer to circled number(s) on the diagnostic chart.

1. This step identifies a wheel speed sensor or circuitry that is damaged and visibly apparent.
2. Checks to see that the wheel speed sensor coil is within proper resistance values.
3. Verifies that the wheel speed sensor can generate an appropriate output signal.
4. This ensures that the wheel speed circuitry is not shorted together.
5. Ensures that the malfunction was not due to physical damage.

Diagnostic Aids:

An "Intermittent" malfunction may be caused by a poor connection, rubbed through wire insulation, or a wire that is broken inside the insulation.

The frequency of the malfunction can be checked by using the enhanced diagnostic function of the Tech 1, as described in "Tech 1 Diagnostics" found in this section.

If the customer's comments reflect that the ABS warning lamp is "ON" only during moist environmental changes (rain, snow, vehicle wash), all wheel speed sensor circuitry should be thoroughly inspected for signs of water intrusion. If DTC is not current, clear DTCs and simulate the effects of water intrusion. Use the following procedure. Spray the suspected area with a 5% salt water solution (two teaspoons of salt to 12 oz. of water). Test drive vehicle over various road surfaces (bumps, turns, etc.) above 24 km/h (15 mph) for at least 30 seconds. If DTC returns, replace suspected harness.

Any circuitry that is suspected of causing the intermittent complaint should be thoroughly checked for backed out terminals, improper mating, broken locks, improperly formed or damaged terminals, poor terminal to wiring connections or physical damage to the wiring harness.

Resistance of the wheel speed sensor will increase with an increase in sensor temperature.

When replacing a wheel speed sensor, inspect the sensor terminals and harness connector for corrosion and/or water intrusion. If evidence of corrosion or water intrusion exists, replace wheel speed sensor jumper harness. Likewise, if replacing a wheel speed sensor jumper harness, inspect sensor terminals. If evidence of corrosion or water intrusion exists, replace wheel speed sensor. Refer to "On-Vehicle Service" in this section.

DIAGNOSTIC CHART DTC 26 — 1995 CUTLASS SUPREME, GRAND PRIX, LUMINA, MONTE CARLO AND REGAL

DTC 26
RIGHT FRONT EXCESSIVE WHEEL SPEED VARIATION

Important
- WHEEL SPEED SENSOR INTERMITTENT MALFUNCTIONS MAY BE DIFFICULT TO LOCATE. CARE SHOULD BE TAKEN NOT TO DISTURB ANY ELECTRICAL CONNECTIONS PRIOR TO AN INDICATED STEP OF THIS CHART. THIS WILL ENSURE THAT AN INTERMITTENT CONNECTION WILL NOT BE CORRECTED BEFORE THE SOURCE OF THE MALFUNCTION IS FOUND.

(1)
- IGNITION "OFF."
- PHYSICALLY INSPECT THE RIGHT FRONT WHEEL SPEED SENSOR, RIGHT FRONT WHEEL SPEED SENSOR JUMPER HARNESS, AND CONNECTORS FOR DAMAGE. INSPECT THE WHEEL SPEED SENSOR AND TOOTHED RING FOR LOOSENESS OR DAMAGE. PAY PARTICULAR ATTENTION TO THE TOOTHED RING, ANY DEVIATION WILL AFFECT THE WHEEL SPEED SENSOR OUTPUT SIGNAL. IS PHYSICAL DAMAGE INDICATED?

YES: REPLACE WHEEL SPEED SENSOR AND/OR JUMPER HARNESS IF DAMAGED. REPAIR ANY SENSOR OR TOOTHED RING LOOSENESS. DRIVE VEHICLE ABOVE 24 km/h (15 MPH) TO ENSURE DTC 26 DOES NOT SET.

NO: go to (2)

(2)
- DISCONNECT RIGHT FRONT WHEEL SPEED SENSOR DIRECTLY AT SENSOR.
- USING J 39200, MEASURE THE RESISTANCE BETWEEN SENSOR TERMINALS "A" AND "B". IS RESISTANCE BETWEEN 1661 AND 1929Ω WHEN SENSOR IS 20°C (68°F)? *(IF TEMPERATURE OF THE SENSOR IS SUSPECTED TO BE OTHER THAN 20°C (68°F), REFER TO THE TABLE BELOW FOR APPLICABLE SENSOR RESISTANCE VALUES; THE VALUES IN THE TABLE ARE FOR TEMPERATURE OF THE SENSOR, NOT AIR TEMPERATURE.)*

NO: REPLACE WHEEL SPEED SENSOR.

YES: go to (3)

(3)
- WITH J 39200 STILL CONNECTED, SELECT A/C VOLTAGE SCALE.
- SPIN WHEEL BY HAND WHILE OBSERVING VOLTAGE. IS VOLTAGE AT LEAST 100 mV (VOLTAGE WILL INCREASE AS WHEEL SPEED INCREASES)?

NO: VERIFY THAT TOOTHED RING IS INTACT. IF OK, REPLACE WHEEL SPEED SENSOR

YES: go to (4)

(4)
- IGNITION "OFF."
- DISCONNECT 32-WAY EBCM CONNECTOR.
- USING J 39200, MEASURE THE RESISTANCE BETWEEN TERMINAL "C3" AND "C4" OF THE 32-WAY EBCM HARNESS CONNECTOR. DOES J 39200 DISPLAY RESISTANCE AS "OL" (INFINITE)?

NO: REPAIR SHORT BETWEEN CKTs 833 AND 872.

YES: go to (5)

(5)
- INSPECT CKTs 833 AND 872 AND 32-WAY EBCM CONNECTOR FOR DAMAGE WHICH MAY RESULT IN A SHORT BETWEEN CKTs 833 AND 872 WITH ALL CONNECTORS CONNECTED. REPAIR ALL DAMAGE IF EVIDENT.
- RECONNECT ALL CONNECTORS.
- TEST DRIVE VEHICLE ABOVE 24 km/h (15 MPH) FOR AT LEAST 30 SECONDS. DOES DTC 26 SET AS A CURRENT DTC?

YES: REPLACE EBCM.

NO: MALFUNCTION IS INTERMITTENT. REFER TO "DIAGNOSTIC AIDS" ON FACING PAGE.

FRONT WHEEL SPEED SENSOR
TEMPERATURE OF SENSOR VS. SENSOR RESISTANCE (APPROXIMATE)

TEMP. (°C)	TEMP. (°F)	RESISTANCE (OHMS)
-34 TO 4	-30 TO 40	1258 TO 1808
5 TO 43	41 TO 110	1502 TO 2103
44 TO 93	111 TO 200	1747 TO 2482

DIAGNOSTIC CHART DTC 27 — 1995 CUTLASS SUPREME, GRAND PRIX, LUMINA, MONTE CARLO AND REGAL

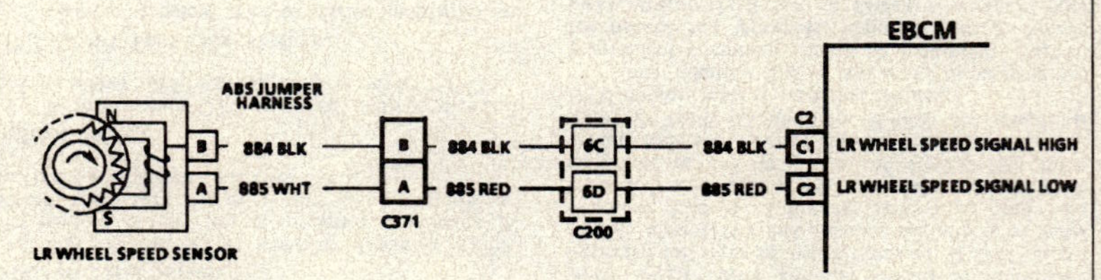

DTC 27
LEFT REAR EXCESSIVE WHEEL SPEED VARIATION

Circuit Description:

As a toothed ring passes by the wheel speed sensor, changes in the electromagnetic field cause the wheel speed sensor to produce a sinusoidal (AC) voltage signal whose frequency is proportional to wheel speed. The magnitude of this signal is directly related to wheel speed and the proximity of the wheel speed sensor to the toothed ring, often referred to as the air gap.

DTC Will Set When:

DTC 27 can be set when the brake is "OFF." The purpose of this test is to detect a situation in which the left rear wheel acceleration or deceleration is beyond specified limits.

Action Taken:

A malfunction DTC is stored, ABS is disabled and the amber ABS warning lamp is turned "ON."

DTC Chart Test Description:

Number(s) below refer to circled number(s) on the diagnostic chart.

1. This step identifies a wheel speed sensor or circuitry that is damaged and visibly apparent.
2. Checks to see that the wheel speed sensor coil is within proper resistance values.
3. Verifies that the wheel speed sensor can generate an appropriate output signal.
4. This ensures that the wheel speed circuitry is not shorted together.
5. Ensures that the malfunction was not due to physical damage.

Diagnostic Aids:

An "Intermittent" malfunction may be caused by a poor connection, rubbed through wire insulation, or a wire that is broken inside the insulation.

The frequency of the malfunction can be checked by using the enhanced diagnostic function of the Tech 1, as described in "Tech 1 Diagnostics" found in this section.

If the customer's comments reflect that the ABS warning lamp is "ON" only during moist environmental changes (rain, snow, vehicle wash), all wheel speed sensor circuitry should be thoroughly inspected for signs of water intrusion. If DTC is not current, clear DTCs and simulate the effects of water intrusion. Use the following procedure. Spray the suspected area with a 5% salt water solution (two teaspoons of salt to 12 oz. of water). Test drive vehicle over various road surfaces (bumps, turns, etc.) above 24 km/h (15 mph) for at least 30 seconds. If DTC returns, replace suspected harness.

Any circuitry that is suspected of causing the intermittent complaint should be thoroughly checked for backed out terminals, improper mating, broken locks, improperly formed or damaged terminals, poor terminal to wiring connections or physical damage to the wiring harness.

Resistance of the wheel speed sensor will increase with an increase in sensor temperature.

When replacing a wheel speed sensor, inspect the sensor terminals and harness connector for corrosion and/or water intrusion. If evidence of corrosion or water intrusion exists, replace wheel speed sensor jumper harness. Likewise, if replacing a wheel speed sensor jumper harness, inspect sensor terminals. If evidence of corrosion or water intrusion exists, replace wheel speed sensor. Refer to "On-Vehicle Service" in this section.

DIAGNOSTIC CHART DTC 27 — 1995 CUTLASS SUPREME, GRAND PRIX, LUMINA, MONTE CARLO AND REGAL

DTC 27
LEFT REAR EXCESSIVE WHEEL SPEED VARIATION

Important
- WHEEL SPEED SENSOR INTERMITTENT MALFUNCTIONS MAY BE DIFFICULT TO LOCATE. CARE SHOULD BE TAKEN NOT TO DISTURB ANY ELECTRICAL CONNECTIONS PRIOR TO AN INDICATED STEP OF THIS CHART. THIS WILL ENSURE THAT AN INTERMITTENT CONNECTION WILL NOT BE CORRECTED BEFORE THE SOURCE OF THE MALFUNCTION IS FOUND.

(1)
- IGNITION "OFF."
- PHYSICALLY INSPECT THE LEFT REAR WHEEL SPEED SENSOR, WHEEL SPEED SENSOR JUMPER HARNESS, AND CONNECTORS FOR DAMAGE. IS PHYSICAL DAMAGE INDICATED?

YES: REPLACE WHEEL SPEED SENSOR AND/OR WHEEL SPEED SENSOR JUMPER HARNESS IF DAMAGED. DRIVE VEHICLE ABOVE 24 km/h (15 MPH) TO ENSURE DTC 27 DOES NOT SET.

NO:

(2)
- DISCONNECT LEFT REAR WHEEL SPEED SENSOR DIRECTLY AT SENSOR.
- USING J 39200, MEASURE THE RESISTANCE BETWEEN SENSOR TERMINALS "A" AND "B."

IS RESISTANCE BETWEEN 2059 AND 2353Ω WHEN SENSOR IS 20°C (68°F)? *(IF TEMPERATURE OF THE SENSOR IS SUSPECTED TO BE OTHER THAN 20°C (68°F), REFER TO THE TABLE BELOW FOR APPLICABLE SENSOR RESISTANCE VALUES; THE VALUES IN THE TABLE ARE FOR TEMPERATURE OF THE SENSOR, NOT AIR TEMPERATURE.)*

NO: REPLACE WHEEL SPEED SENSOR.

YES:

(3)
- WITH J 39200 STILL CONNECTED, SELECT A/C VOLTAGE SCALE.
- SPIN WHEEL BY HAND WHILE OBSERVING VOLTAGE.

IS VOLTAGE AT LEAST 100 mV (VOLTAGE WILL INCREASE AS WHEEL SPEED INCREASES)?

NO: REPLACE WHEEL SPEED SENSOR.

YES:

(4)
- IGNITION "OFF."
- DISCONNECT 32-WAY EBCM CONNECTOR.
- USING J 39200, MEASURE THE RESISTANCE BETWEEN TERMINAL "C1" AND "C2" OF THE 32-WAY EBCM HARNESS CONNECTOR.

DOES J 39200 DISPLAY RESISTANCE AS "OL" (INFINITE)?

NO: REPAIR SHORT BETWEEN CKTs 884 AND 885.

YES:

(5)
- INSPECT CKTs 884 AND 885 AND 32-WAY EBCM CONNECTOR FOR DAMAGE WHICH MAY RESULT IN A SHORT BETWEEN CKTs 884 AND 885 WITH ALL CONNECTORS CONNECTED. REPAIR ALL DAMAGE IF EVIDENT.
- RECONNECT ALL CONNECTORS.
- TEST DRIVE VEHICLE ABOVE 24 km/h (15 MPH) FOR AT LEAST 30 SECONDS.

DOES DTC 27 SET AS A CURRENT DTC?

YES: REPLACE EBCM.

NO: MALFUNCTION IS INTERMITTENT. REFER TO "DIAGNOSTIC AIDS" ON FACING PAGE.

REAR WHEEL SPEED SENSOR
TEMPERATURE OF SENSOR VS. SENSOR RESISTANCE (APPROXIMATE)

TEMP. (°C)	TEMP. (°F)	RESISTANCE (OHMS)
-34 TO 4	-30 TO 40	1622 TO 2205
5 TO 43	41 TO 110	1930 TO 2566
44 TO 93	111 TO 200	2245 TO 3028

DIAGNOSTIC CHART DTC 28 — 1995 CUTLASS SUPREME, GRAND PRIX, LUMINA, MONTE CARLO AND REGAL

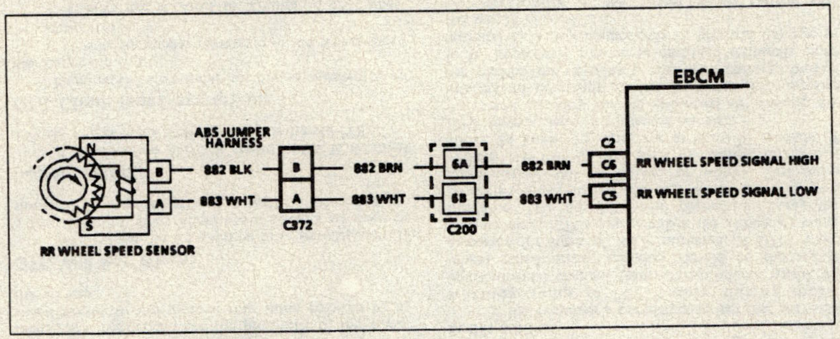

DTC 28

RIGHT REAR EXCESSIVE WHEEL SPEED VARIATION

Circuit Description:

As a toothed ring passes by the wheel speed sensor, changes in the electromagnetic field cause the wheel speed sensor to produce a sinusoidal (AC) voltage signal whose frequency is proportional to wheel speed. The magnitude of this signal is directly related to wheel speed and the proximity of the wheel speed sensor to the toothed ring, often referred to as the air gap.

DTC Will Set When:

DTC 28 can be set when the brake is "OFF." The purpose of this test is to detect a situation in which the right rear wheel acceleration or deceleration is beyond specified limits.

Action Taken:

A malfunction DTC is stored, ABS is disabled and the amber ABS warning lamp is turned "ON."

DTC Chart Test Description:

Number(s) below refer to circled number(s) on the diagnostic chart.

1. This step identifies a wheel speed sensor or circuitry that is damaged and visibly apparent.
2. Checks to see that the wheel speed sensor coil is within proper resistance values.
3. Verifies that the wheel speed sensor can generate an appropriate output signal.
4. This ensures that the wheel speed circuitry is not shorted together.
5. Ensures that the malfunction was not due to physical damage.

Diagnostic Aids:

An "Intermittent" malfunction may be caused by a poor connection, rubbed through wire insulation, or a wire that is broken inside the insulation.

The frequency of the malfunction can be checked by using the enhanced diagnostic function of the Tech 1, as described in "Tech 1 Diagnostics" found in this section.

If the customer's comments reflect that the ABS warning lamp is "ON" only during moist environmental changes (rain, snow, vehicle wash), all wheel speed sensor circuitry should be thoroughly inspected for signs of water intrusion. If DTC is not current, clear DTCs and simulate the effects of water intrusion. Use the following procedure. Spray the suspected area with a 5% salt water solution (two teaspoons of salt to 12 oz. of water). Test drive vehicle over various road surfaces (bumps, turns, etc.) above 24 km/h (15 mph) for at least 30 seconds. If DTC returns, replace suspected harness.

Any circuitry that is suspected of causing the intermittent complaint should be thoroughly checked for backed out terminals, improper mating, broken locks, improperly formed or damaged terminals, poor terminal to wiring connections or physical damage to the wiring harness.

Resistance of the wheel speed sensor will increase with an increase in sensor temperature.

When replacing a wheel speed sensor, inspect the sensor terminals and harness connector for corrosion and/or water intrusion. If evidence of corrosion or water intrusion exists, replace wheel speed sensor jumper harness. Likewise, if replacing a wheel speed sensor jumper harness, inspect sensor terminals. If evidence of corrosion or water intrusion exists, replace wheel speed sensor. Refer to "On-Vehicle Service" in this section.

DIAGNOSTIC CHART DTC 28 — 1995 CUTLASS SUPREME, GRAND PRIX, LUMINA, MONTE CARLO AND REGAL

DTC 28
RIGHT REAR EXCESSIVE WHEEL SPEED VARIATION

Important

- WHEEL SPEED SENSOR INTERMITTENT MALFUNCTIONS MAY BE DIFFICULT TO LOCATE. CARE SHOULD BE TAKEN NOT TO DISTURB ANY ELECTRICAL CONNECTIONS PRIOR TO AN INDICATED STEP OF THIS CHART. THIS WILL ENSURE THAT AN INTERMITTENT CONNECTION WILL NOT BE CORRECTED BEFORE THE SOURCE OF THE MALFUNCTION IS FOUND.

(1)
- IGNITION "OFF."
- PHYSICALLY INSPECT THE RIGHT REAR WHEEL SPEED SENSOR, WHEEL SPEED SENSOR JUMPER HARNESS, AND CONNECTORS FOR DAMAGE.

IS PHYSICAL DAMAGE INDICATED?

YES → REPLACE WHEEL SPEED SENSOR AND/OR WHEEL SPEED SENSOR JUMPER HARNESS IF DAMAGED. DRIVE VEHICLE ABOVE 24 km/h (15 MPH) TO ENSURE DTC 28 DOES NOT SET.

NO ↓

(2)
- DISCONNECT RIGHT REAR WHEEL SPEED SENSOR DIRECTLY AT SENSOR.
- USING J 39200, MEASURE THE RESISTANCE BETWEEN SENSOR TERMINALS "A" AND "B."

IS RESISTANCE BETWEEN 2059 AND 2353Ω WHEN SENSOR IS 20°C (68°F)? *(IF TEMPERATURE OF THE SENSOR IS SUSPECTED TO BE OTHER THAN 20°C (68°F), REFER TO THE TABLE BELOW FOR APPLICABLE SENSOR RESISTANCE VALUES; THE VALUES IN THE TABLE ARE FOR TEMPERATURE OF THE SENSOR, NOT AIR TEMPERATURE.)*

NO → REPLACE WHEEL SPEED SENSOR.

YES ↓

(3)
- WITH J 39200 STILL CONNECTED, SELECT A/C VOLTAGE SCALE.
- SPIN WHEEL BY HAND WHILE OBSERVING VOLTAGE.

IS VOLTAGE AT LEAST 100 mV (VOLTAGE WILL INCREASE AS WHEEL SPEED INCREASES)?

NO → REPLACE WHEEL SPEED SENSOR.

YES ↓

(4)
- IGNITION "OFF."
- DISCONNECT 32-WAY EBCM CONNECTOR.
- USING J 39200, MEASURE THE RESISTANCE BETWEEN TERMINAL "C5" AND "C6" OF THE 32-WAY EBCM HARNESS CONNECTOR.

DOES J 39200 DISPLAY RESISTANCE AS "OL" (INFINITE)?

NO → REPAIR SHORT BETWEEN CKTs 882 AND 883.

YES ↓

(5)
- INSPECT CKTs 882 AND 883 AND 32-WAY EBCM CONNECTOR FOR DAMAGE WHICH MAY RESULT IN A SHORT BETWEEN CKTs 882 AND 883 WITH ALL CONNECTORS CONNECTED. REPAIR ALL DAMAGE IF EVIDENT.
- RECONNECT ALL CONNECTORS.
- TEST DRIVE VEHICLE ABOVE 24 km/h (15 MPH) FOR AT LEAST 30 SECONDS.

DOES DTC 28 SET AS A CURRENT DTC?

YES → REPLACE EBCM.

NO → MALFUNCTION IS INTERMITTENT. REFER TO "DIAGNOSTIC AIDS" ON FACING PAGE.

REAR WHEEL SPEED SENSOR
TEMPERATURE OF SENSOR VS. SENSOR RESISTANCE (APPROXIMATE)

TEMP. (°C)	TEMP. (°F)	RESISTANCE (OHMS)
-34 TO 4	-30 TO 40	1622 TO 2205
5 TO 43	41 TO 110	1930 TO 2566
44 TO 93	111 TO 200	2245 TO 3028

DIAGNOSTIC CHART DTC 32 — 1995 CUTLASS SUPREME, GRAND PRIX, LUMINA, MONTE CARLO AND REGAL

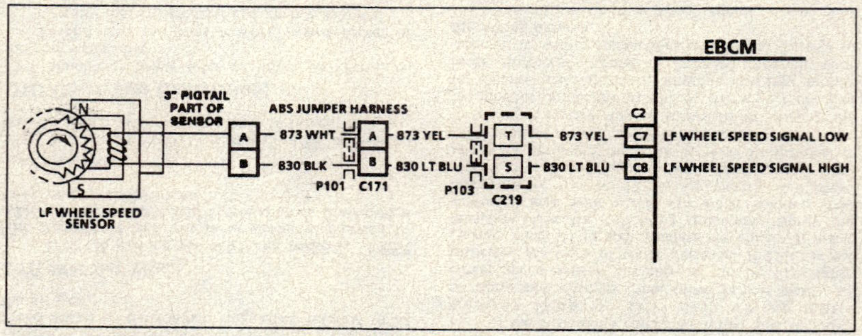

DTC 32
(Page 1 of 3)
LF WHEEL SPEED SNR CKT OPEN OR SHORT TO GND/BATT

Circuit Description:

As a toothed ring passes by the wheel speed sensor, changes in the electromagnetic field cause the wheel speed sensor to produce a sinusoidal (AC) voltage signal whose frequency is proportional to wheel speed. The magnitude of this signal is directly related to wheel speed and the proximity of the wheel speed sensor to the toothed ring often referred to as the air gap.

DTC Will Set When:

DTC 32 can be set anytime after initialization. If either of the left front wheel speed circuits are open or shorted to voltage or ground, a malfunction exists.

Action Taken:

A malfunction DTC is stored, ABS is disabled and the amber ABS warning lamp is turned "ON."

DTC Chart Test Description:

Number(s) below refer to circled number(s) on the diagnostic chart.

1. Checks for proper resistance of the wheel speed sensor coil.
2. Checks for a short to ground in the wheel speed sensor wiring.
3. Checks for a short to voltage in the wheel speed sensor wiring.
4. Checks for a short to voltage in CKT 873.
5. Checks for a short to voltage in CKT 830.
6. Ensures that the malfunction was not due to physical damage of the circuitry.

Diagnostic Aids:

An "Intermittent" malfunction may be caused by a poor connection, rubbed through wire insulation, or a wire that is broken inside the insulation.

The frequency of the malfunction can be checked by using the enhanced diagnostic function of the Tech 1, as described in "Tech 1 Diagnostics" found in this section.

If the customer's comments reflect that the ABS warning lamp is "ON" only during moist environmental changes (rain, snow, vehicle wash), all wheel speed sensor circuitry should be thoroughly inspected for signs of water intrusion. If DTC is not current, clear DTCs and simulate the effects of water intrusion. Use the following procedure. Spray the suspected area with a 5% salt water solution (two teaspoons of salt to 12 oz. of water). Test drive vehicle over various road surfaces (bumps, turns, etc.) above 24 km/h (15 mph) for at least 30 seconds. If DTC returns, replace suspected harness.

Any circuitry that is suspected of causing the intermittent complaint should be thoroughly checked for backed out terminals, improper mating, broken locks, improperly formed or damaged terminals, poor terminal to wiring connections or physical damage to the wiring harness.

Resistance of the wheel speed sensor will increase with an increase in sensor temperature.

When replacing a wheel speed sensor, inspect the sensor terminals and harness connector for corrosion and/or water intrusion. If evidence of corrosion or water intrusion exists, replace wheel speed sensor jumper harness. Likewise, if replacing a wheel speed sensor jumper harness, inspect sensor terminals. If evidence of corrosion or water intrusion exists, replace wheel speed sensor. Refer to "On-Vehicle Service" in this section.

DIAGNOSTIC CHART DTC 32 — 1995 CUTLASS SUPREME, GRAND PRIX, LUMINA, MONTE CARLO AND REGAL

DTC 32
(Page 1 of 3)
LF WHEEL SPEED SNR CKT OPEN OR SHORT TO GND/BATT

Important

- WHEEL SPEED SENSOR INTERMITTENT MALFUNCTIONS MAY BE DIFFICULT TO LOCATE. CARE SHOULD BE TAKEN NOT TO DISTURB ANY ELECTRICAL CONNECTIONS PRIOR TO AN INDICATED STEP OF THIS CHART. THIS WILL ENSURE THAT AN INTERMITTENT CONNECTION WILL NOT BE CORRECTED BEFORE THE SOURCE OF THE MALFUNCTION IS FOUND.

(1)
- IGNITION "OFF."
- USING A 5% SALT WATER SOLUTION (TWO TEASPOONS OF SALT TO 12 OZ. WATER), THOROUGHLY SPRAY THE LEFT FRONT WHEEL SPEED SENSOR JUMPER HARNESS.
- DISCONNECT 32-WAY EBCM HARNESS CONNECTOR.
- USING J 39200, MEASURE THE RESISTANCE BETWEEN TERMINALS "C7" AND "C8" OF THE 32-WAY EBCM HARNESS CONNECTOR.

IS RESISTANCE BETWEEN 1661 AND 1929Ω WHEN SENSOR IS 20°C (68°F)? *(IF TEMPERATURE OF SENSOR IS SUSPECTED TO OTHER THAN 20°C (68°F), REFER TO THE TABLE ON PAGE 2 OF 2 OF THIS DTC CHART FOR APPLICABLE SENSOR RESISTANCE VALUES; THE VALUES IN THE TABLE ARE FOR TEMPERATURE OF THE SENSOR, NOT AIR TEMPERATURE.)*

- YES → (2)
- NO → DIAGNOSIS CONTINUED ON PAGE 2 OF 3 OF THIS DTC CHART.

(2)
- MEASURE THE RESISTANCE BETWEEN GROUND AND TERMINAL "C7" OF THE 32-WAY EBCM HARNESS CONNECTOR.

DOES J 39200 DISPLAY RESISTANCE AS "OL" (INFINITE)?

- YES → (3)
- NO → DIAGNOSIS CONTINUED ON PAGE 3 OF 3 OF THIS DTC CHART.

(3)
- IGNITION "ON."
- USING J 39200, MEASURE THE VOLTAGE BETWEEN GROUND AND TERMINAL "C7" OF THE 32-WAY EBCM HARNESS CONNECTOR.

IS VOLTAGE 1 VOLT OR LESS?

- NO → (4)
- YES → (6)

(4)
- IGNITION "OFF."
- DISCONNECT LEFT FRONT WHEEL SPEED SENSOR DIRECTLY AT SENSOR.
- IGNITION "ON."
- USING J 39200, MEASURE THE VOLTAGE BETWEEN GROUND AND TERMINAL "C7" OF THE 32-WAY EBCM HARNESS CONNECTOR.

IS VOLTAGE 1 VOLT OR LESS?

- YES → (5)
- NO → REPAIR SHORT TO VOLTAGE IN CKT 873.

(5)
- USING J 39200, MEASURE THE VOLTAGE BETWEEN GROUND AND TERMINAL "C8" OF THE 32-WAY EBCM HARNESS CONNECTOR.

IS VOLTAGE 1 VOLT OR LESS?

- YES → MALFUNCTION IS INTERMITTENT. INSPECT ALL CONNECTORS AND HARNESSES FOR DAMAGE WHICH MAY RESULT IN A SHORT TO VOLTAGE WHEN ALL COMPONENTS ARE CONNECTED. REPAIR ALL DAMAGE IF FOUND. REFER TO "DIAGNOSTIC AIDS" ON FACING PAGE FOR MORE INFORMATION.
- NO → REPAIR SHORT TO VOLTAGE IN CKT 830.

(6)
- IGNITION "OFF."
- INSPECT TERMINALS "C7" AND "C8" OF THE 32-WAY EBCM HARNESS CONNECTOR FOR POOR TERMINAL CONTACT AND CONTAMINATION. ALSO INSPECT CKTs 830 AND 873 FOR DAMAGE WHICH MAY RESULT IN AN OPEN CIRCUIT, SHORT TO GROUND, OR SHORT TO VOLTAGE. REPAIR ALL DAMAGE IF EVIDENT.
- RECONNECT ALL CONNECTORS.
- IGNITION "ON."

DOES DTC 32 SET AS A CURRENT DTC?

- YES → REPLACE EBCM.
- NO → MALFUNCTION IS INTERMITTENT. REFER TO "DIAGNOSTIC AIDS" ON FACING PAGE.

DIAGNOSTIC CHART DTC 32 — 1995 CUTLASS SUPREME, GRAND PRIX, LUMINA, MONTE CARLO AND REGAL

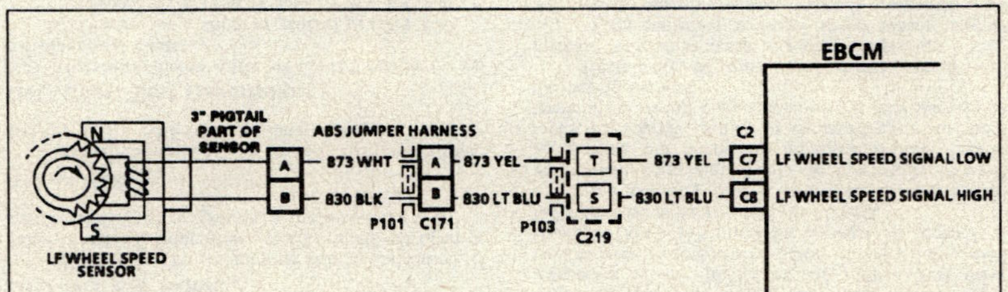

DTC 32 (Page 2 of 3) LF WHEEL SPEED SNR CKT OPEN OR SHORT TO GND/BATT

Circuit Description:

As a toothed ring passes by the wheel speed sensor, changes in the electromagnetic field cause the wheel speed sensor to produce a sinusoidal (AC) voltage signal whose frequency is proportional to wheel speed. The magnitude of this signal is directly related to wheel speed and the proximity of the wheel speed sensor to the toothed ring often referred to as the air gap.

DTC Will Set When:

DTC 32 can be set anytime after initialization. If either of the left front wheel speed circuits are open or shorted to voltage or ground, a malfunction exists.

Action Taken:

A malfunction DTC is stored, ABS is disabled and the amber ABS warning lamp is turned "ON."

DTC Chart Test Description:

Number(s) below refer to circled number(s) on the diagnostic chart.

7. Checks for high resistance in CKT 873.
8. Checks for high resistance in CKT 830.
9. Verifies proper wheel speed sensor resistance.

Diagnostic Aids:

An "Intermittent" malfunction may be caused by a poor connection, rubbed through wire insulation, or a wire that is broken inside the insulation.

The frequency of the malfunction can be checked by using the enhanced diagnostic function of the Tech 1, as described in "Tech 1 Diagnostics" found in this section.

If the customer's comments reflect that the ABS warning lamp is "ON" only during moist environmental changes (rain, snow, vehicle wash), all wheel speed sensor circuitry should be thoroughly inspected for signs of water intrusion. If DTC is not current, clear DTCs and simulate the effects of water intrusion. Use the following procedure. Spray the suspected area with a 5% salt water solution (two teaspoons of salt to 12 oz. of water). Test drive vehicle over various road surfaces (bumps, turns, etc.) above 24 km/h (15 mph) for at least 30 seconds. If DTC returns, replace suspected harness.

Any circuitry that is suspected of causing the intermittent complaint should be thoroughly checked for backed out terminals, improper mating, broken locks, improperly formed or damaged terminals, poor terminal to wiring connections or physical damage to the wiring harness.

Resistance of the wheel speed sensor will increase with an increase in sensor temperature.

When replacing a wheel speed sensor, inspect the sensor terminals and harness connector for corrosion and/or water intrusion. If evidence of corrosion or water intrusion exists, replace wheel speed sensor jumper harness. Likewise, if replacing a wheel speed sensor jumper harness, inspect sensor terminals. If evidence of corrosion or water intrusion exists, replace wheel speed sensor. Refer to "On-Vehicle Service" in this section.

DIAGNOSTIC CHART DTC 32 — 1995 CUTLASS SUPREME, GRAND PRIX, LUMINA, MONTE CARLO AND REGAL

DTC 32 (Page 2 of 3) LF WHEEL SPEED SNR CKT OPEN OR SHORT TO GND/BATT

DIAGNOSIS CONTINUED FROM PAGE 1 OF 3 OF THIS DTC CHART.

(7)
- DISCONNECT LEFT FRONT WHEEL SPEED SENSOR DIRECTLY AT SENSOR.
- USING J 39200, MEASURE THE RESISTANCE BETWEEN TERMINAL "C7" OF THE 32-WAY EBCM HARNESS CONNECTOR AND TERMINAL "B" OF THE LEFT FRONT WHEEL SPEED SENSOR HARNESS CONNECTOR. IS RESISTANCE 2Ω OR LESS?

NO: REPAIR HIGH RESISTANCE IN CKT 873 (WHEEL SPEED SENSOR JUMPER MUST BE REPLACED IF DAMAGED).

YES:

(8)
- USING J 39200, MEASURE THE RESISTANCE BETWEEN TERMINAL "C8" OF THE 32-WAY EBCM HARNESS CONNECTOR AND TERMINAL "A" OF THE LEFT FRONT WHEEL SPEED SENSOR HARNESS CONNECTOR. IS RESISTANCE 2Ω OR LESS?

NO: REPAIR HIGH RESISTANCE IN CKT 830 (WHEEL SPEED SENSOR JUMPER MUST BE REPLACED IF DAMAGED).

YES:

(9)
- USING J 39200, MEASURE THE RESISTANCE BETWEEN TERMINAL "A" AND TERMINAL "B" OF THE LEFT FRONT WHEEL SPEED SENSOR CONNECTOR. IS RESISTANCE BETWEEN 1661 AND 1929Ω WHEN SENSOR IS 20°C (68°F)? *(IF TEMPERATURE OF SENSOR IS SUSPECTED TO OTHER THAN 20°C (68°F), REFER TO THE TABLE BELOW FOR APPLICABLE SENSOR RESISTANCE VALUES; THE VALUES IN THE TABLE ARE FOR TEMPERATURE OF THE SENSOR, NOT AIR TEMPERATURE.)*

NO: REPLACE WHEEL SPEED SENSOR.

YES: MALFUNCTION IS INTERMITTENT. INSPECT ALL CONNECTORS AND HARNESSES FOR DAMAGE WHICH MAY RESULT IN HIGH RESISTANCE WHEN ALL COMPONENTS ARE CONNECTED. REFER TO "DIAGNOSTIC AIDS" ON FACING PAGE FOR MORE INFORMATION. REPAIR ALL DAMAGE IF FOUND.

FRONT WHEEL SPEED SENSOR

TEMPERATURE OF SENSOR VS. SENSOR RESISTANCE (APPROXIMATE)

TEMP. (°C)	TEMP. (°F)	RESISTANCE (OHMS)
-34 TO 4	-30 TO 40	1258 TO 1806
5 TO 43	41 TO 110	1502 TO 2103
44 TO 93	111 TO 200	1747 TO 2482

DIAGNOSTIC CHART DTC 32 — 1995 CUTLASS SUPREME, GRAND PRIX, LUMINA, MONTE CARLO AND REGAL

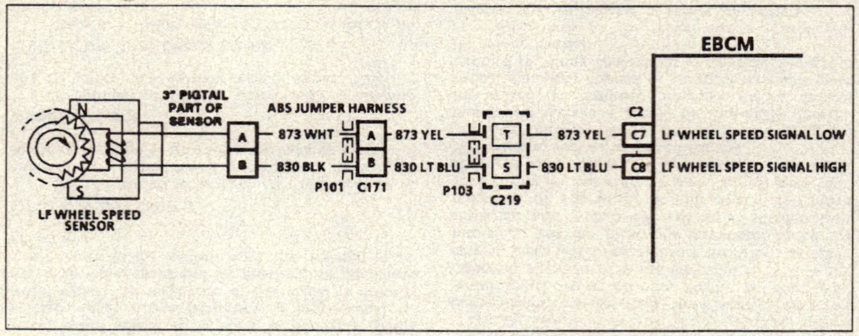

DTC 32
(Page 3 of 3)
LF WHEEL SPEED SNR CKT OPEN OR SHORT TO GND/BATT

Circuit Description:

As a toothed ring passes by the wheel speed sensor, changes in the electromagnetic field cause the wheel speed sensor to produce a sinusoidal (AC) voltage signal whose frequency is proportional to wheel speed. The magnitude of this signal is directly related to wheel speed and the proximity of the wheel speed sensor to the toothed ring often referred to as the air gap.

DTC Will Set When:

DTC 32 can be set anytime after initialization. If either of the left front wheel speed circuits are open or shorted to voltage or ground, a malfunction exists.

Action Taken:

A malfunction DTC is stored, ABS is disabled and the amber ABS warning lamp is turned "ON."

DTC Chart Test Description:

Number(s) below refer to circled number(s) on the diagnostic chart.

10. Checks for a short to ground in CKT 873.
11. Checks for a short to ground in CKT 830.
12. Checks for a short to ground in the wheel speed sensor jumper harness.
13. Verifies that wheel speed sensor is not internally shorted to ground.

Diagnostic Aids:

An "Intermittent" malfunction may be caused by a poor connection, rubbed through wire insulation, or a wire that is broken inside the insulation.

The frequency of the malfunction can be checked by using the enhanced diagnostic function of the Tech 1, as described in "Tech 1 Diagnostics" found in this section.

If the customer's comments reflect that the ABS warning lamp is "ON" only during moist environmental changes (rain, snow, vehicle wash), all wheel speed sensor circuitry should be thoroughly inspected for signs of water intrusion. If DTC is not current, clear DTCs and simulate the effects of water intrusion. Use the following procedure. Spray the suspected area with a 5% salt water solution (two teaspoons of salt to 12 oz. of water). Test drive vehicle over various road surfaces (bumps, turns, etc.) above 24 km/h (15 mph) for at least 30 seconds. If DTC returns, replace suspected harness.

Any circuitry that is suspected of causing the intermittent complaint should be thoroughly checked for backed out terminals, improper mating, broken locks, improperly formed or damaged terminals, poor terminal to wiring connections or physical damage to the wiring harness.

Resistance of the wheel speed sensor will increase with an increase in sensor temperature.

When replacing a wheel speed sensor, inspect the sensor terminals and harness connector for corrosion and/or water intrusion. If evidence of corrosion or water intrusion exists, replace wheel speed sensor jumper harness. Likewise, if replacing a wheel speed sensor jumper harness, inspect sensor terminals. If evidence of corrosion or water intrusion exists, replace wheel speed sensor. Refer to "On-Vehicle Service" in this section.

DIAGNOSTIC CHART DTC 32 — 1995 CUTLASS SUPREME, GRAND PRIX, LUMINA, MONTE CARLO AND REGAL

DTC 32
(Page 3 of 3)
LF WHEEL SPEED SNR CKT OPEN OR SHORT TO GND/BATT

DIAGNOSIS CONTINUED FROM PAGE 1 OF 3 OF THIS DTC CHART.

(10)
- DISCONNECT CONNECTOR C171 (LOCATED LH REAR OF ENGINE COMPARTMENT).
- USING J 39200, MEASURE THE RESISTANCE BETWEEN GROUND AND TERMINAL "C7" OF THE 32-WAY EBCM HARNESS CONNECTOR.

DOES J 39200 DISPLAY RESISTANCE AS "OL" (INFINITE)?

- NO: REPAIR SHORT TO GROUND IN CKT 873.
- YES:

(11)
- USING J 39200, MEASURE THE RESISTANCE BETWEEN GROUND AND TERMINAL "C8" OF THE 32-WAY EBCM HARNESS CONNECTOR.

DOES J 39200 DISPLAY RESISTANCE AS "OL" (INFINITE)?

- NO: REPAIR SHORT TO GROUND IN CKT 830.
- YES:

(12)
- THOROUGHLY RESPRAY THE LEFT FRONT WHEEL SPEED SENSOR JUMPER HARNESS WITH THE SALT WATER SOLUTION.

IMPORTANT: DO NOT ALLOW WATER TO ENTER ANY WHEEL SPEED SENSOR CONNECTOR.

- DISCONNECT LEFT FRONT WHEEL SPEED SENSOR DIRECTLY AT SENSOR.
- USING J 39200, MEASURE THE RESISTANCE BETWEEN GROUND AND TERMINAL "A," THEN "B" OF THE LEFT FRONT WHEEL SPEED SENSOR HARNESS CONNECTOR.

DOES J 39200 DISPLAY RESISTANCE AS "OL" (INFINITE) FOR BOTH MEASUREMENTS?

- NO: REPLACE LEFT FRONT WHEEL SPEED SENSOR JUMPER HARNESS.
- YES:

(13)
- USING J 39200, MEASURE THE RESISTANCE BETWEEN GROUND AND TERMINAL "A" OF THE LEFT FRONT WHEEL SPEED SENSOR CONNECTOR.

DOES J 39200 DISPLAY RESISTANCE AS "OL" (INFINITE)?

- NO: REPLACE WHEEL SPEED SENSOR.
- YES: MALFUNCTION IS INTERMITTENT. INSPECT ALL CONNECTORS AND HARNESSES FOR DAMAGE WHICH MAY RESULT IN A SHORT TO GROUND WHEN ALL COMPONENTS ARE CONNECTED. PAY PARTICULAR ATTENTION TO THE JUMPER HARNESS, SMALL NICKS OR CUTS CAN CAUSE A MALFUNCTION. REPAIR ANY DAMAGE IF FOUND. JUMPER HARNESS MUST BE REPLACED IF DAMAGED. REFER TO "DIAGNOSTIC AIDS" ON FACING PAGE FOR MORE INFORMATION.

DIAGNOSTIC CHART DTC 33 — 1995 CUTLASS SUPREME, GRAND PRIX, LUMINA, MONTE CARLO AND REGAL

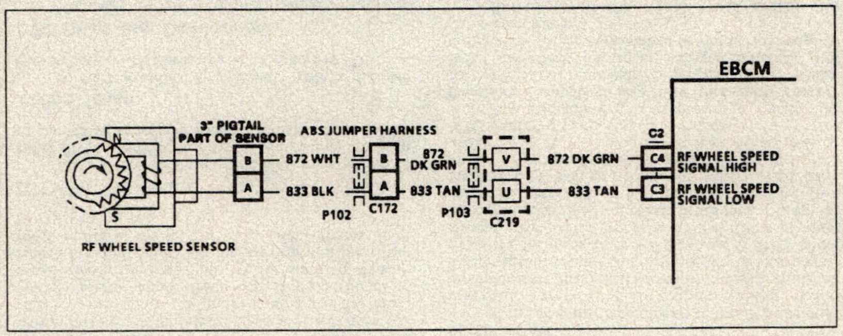

DTC 33
(Page 1 of 3)
RF WHEEL SPEED SNR CKT OPEN OR SHORT TO GND/BATT

Circuit Description:

As a toothed ring passes by the wheel speed sensor, changes in the electromagnetic field cause the wheel speed sensor to produce a sinusoidal (AC) voltage signal whose frequency is proportional to wheel speed. The magnitude of this signal is directly related to wheel speed and the proximity of the wheel speed sensor to the toothed ring, often referred to as the air gap.

DTC Will Set When:

DTC 33 can be set anytime after initialization. If either of the right front wheel speed circuits are open or shorted to voltage or ground, a malfunction exists.

Action Taken:

A malfunction DTC is stored, ABS is disabled and the amber ABS warning lamp is turned "ON."

DTC Chart Test Description:

Number(s) below refer to circled number(s) on the diagnostic chart.

1. Checks for proper resistance of the wheel speed sensor coil.
2. Checks for a short to ground in the wheel speed sensor wiring.
3. Checks for a short to voltage in the wheel speed sensor wiring.
4. Checks for a short to voltage in CKT 833.
5. Checks for short to voltage in CKT 872.
6. Ensures that the malfunction was not due to physical damage of the circuitry.

Diagnostic Aids:

An "Intermittent" malfunction may be caused by a poor connection, rubbed through wire insulation, or a wire that is broken inside the insulation.

The frequency of the malfunction can be checked by using the enhanced diagnostic function of the Tech 1, as described in "Tech 1 Diagnostics" found in this section.

If the customer's comments reflect that the ABS warning lamp is "ON" only during moist environmental changes (rain, snow, vehicle wash), all wheel speed sensor circuitry should be thoroughly inspected for signs of water intrusion. If DTC is not current, clear DTCs and simulate the effects of water intrusion. Use the following procedure. Spray the suspected area with a 5% salt water solution (two teaspoons of salt to 12 oz. of water). Test drive vehicle over various road surfaces (bumps, turns, etc.) above 24 km/h (15 mph) for at least 30 seconds. If DTC returns, replace suspected harness.

Any circuitry that is suspected of causing the intermittent complaint should be thoroughly checked for backed out terminals, improper mating, broken locks, improperly formed or damaged terminals, poor terminal to wiring connections or physical damage to the wiring harness.

Resistance of the wheel speed sensor will increase with an increase in sensor temperature.

When replacing a wheel speed sensor, inspect the sensor terminals and harness connector for corrosion and/or water intrusion. If evidence of corrosion or water intrusion exists, replace wheel speed sensor jumper harness. Likewise, if replacing a wheel speed sensor jumper harness, inspect sensor terminals. If evidence of corrosion or water intrusion exists, replace wheel speed sensor. Refer to "On-Vehicle Service" in this section.

DIAGNOSTIC CHART DTC 33 — 1995 CUTLASS SUPREME, GRAND PRIX, LUMINA, MONTE CARLO AND REGAL

DTC 33
(Page 1 of 3)
RF WHEEL SPEED SNR CKT OPEN OR SHORT TO GND/BATT

Important
- WHEEL SPEED SENSOR INTERMITTENT MALFUNCTIONS MAY BE DIFFICULT TO LOCATE. CARE SHOULD BE TAKEN NOT TO DISTURB ANY ELECTRICAL CONNECTIONS PRIOR TO AN INDICATED STEP OF THIS CHART. THIS WILL ENSURE THAT AN INTERMITTENT CONNECTION WILL NOT BE CORRECTED BEFORE THE SOURCE OF THE MALFUNCTION IS FOUND.

(1)
- IGNITION "OFF."
- USING A 5% SALT WATER SOLUTION (TWO TEASPOONS OF SALT TO 12 OZ. WATER), THOROUGHLY SPRAY THE RIGHT FRONT WHEEL SPEED SENSOR JUMPER HARNESS.
- DISCONNECT 32-WAY EBCM HARNESS CONNECTOR.
- USING J 39200, MEASURE THE RESISTANCE BETWEEN TERMINALS "C3" AND "C4" OF THE 32-WAY EBCM HARNESS CONNECTOR.

IS RESISTANCE BETWEEN 1661 AND 1929Ω WHEN SENSOR IS 20°C (68°F)? *(IF TEMPERATURE OF SENSOR IS SUSPECTED TO OTHER THAN 20°C (68°F), REFER TO THE TABLE ON PAGE 2 OF 2 OF THIS DTC CHART FOR APPLICABLE SENSOR RESISTANCE VALUES; THE VALUES IN THE TABLE ARE FOR TEMPERATURE OF THE SENSOR, NOT AIR TEMPERATURE.)*

- YES → (2)
- NO → DIAGNOSIS CONTINUED ON PAGE 2 OF 3 OF THIS DTC CHART.

(2)
- MEASURE THE RESISTANCE BETWEEN GROUND AND TERMINAL "C3" OF THE 32-WAY EBCM HARNESS CONNECTOR.

DOES J 39200 DISPLAY RESISTANCE AS "OL" (INFINITE)?

- YES → (3)
- NO → DIAGNOSIS CONTINUED ON PAGE 3 OF 3 OF THIS DTC CHART.

(3)
- IGNITION "ON."
- USING J 39200, MEASURE THE VOLTAGE BETWEEN GROUND AND TERMINAL "C3" OF THE 32-WAY EBCM HARNESS CONNECTOR.

IS VOLTAGE 1 VOLT OR LESS?

- NO → (4)
- YES → (6)

(4)
- IGNITION "OFF."
- DISCONNECT RIGHT FRONT WHEEL SPEED SENSOR DIRECTLY AT SENSOR.
- IGNITION "ON."
- USING J 39200, MEASURE THE VOLTAGE BETWEEN GROUND AND TERMINAL "C3" OF THE 32-WAY EBCM HARNESS CONNECTOR.

IS VOLTAGE 1 VOLT OR LESS?

- YES → (5)
- NO → REPAIR SHORT TO VOLTAGE IN CKT 833.

(5)
- USING J 39200, MEASURE THE VOLTAGE BETWEEN GROUND AND TERMINAL "C4" OF THE 32-WAY EBCM HARNESS CONNECTOR.

IS VOLTAGE 1 VOLT OR LESS?

- YES → MALFUNCTION IS INTERMITTENT. INSPECT ALL CONNECTORS AND HARNESSES FOR DAMAGE WHICH MAY RESULT IN A SHORT TO VOLTAGE WHEN ALL COMPONENTS ARE CONNECTED. REPAIR ALL DAMAGE IF FOUND. REFER TO "DIAGNOSTIC AIDS" ON FACING PAGE FOR MORE INFORMATION.
- NO → REPAIR SHORT TO VOLTAGE IN CKT 872.

(6)
- IGNITION "OFF."
- INSPECT TERMINALS "C3" AND "C4" OF THE 32-WAY EBCM HARNESS CONNECTOR FOR POOR TERMINAL CONTACT AND CONTAMINATION. ALSO INSPECT CKTs 833 AND 872 FOR DAMAGE WHICH MAY RESULT IN AN OPEN CIRCUIT, SHORT TO GROUND, OR SHORT TO VOLTAGE. REPAIR ALL DAMAGE IF EVIDENT.
- RECONNECT ALL CONNECTORS.
- IGNITION "ON."

DOES DTC 33 SET AS A CURRENT DTC?

- YES → REPLACE EBCM.
- NO → MALFUNCTION IS INTERMITTENT. REFER TO "DIAGNOSTIC AIDS" ON FACING PAGE.

DIAGNOSTIC CHART DTC 33 — 1995 CUTLASS SUPREME, GRAND PRIX, LUMINA, MONTE CARLO AND REGAL

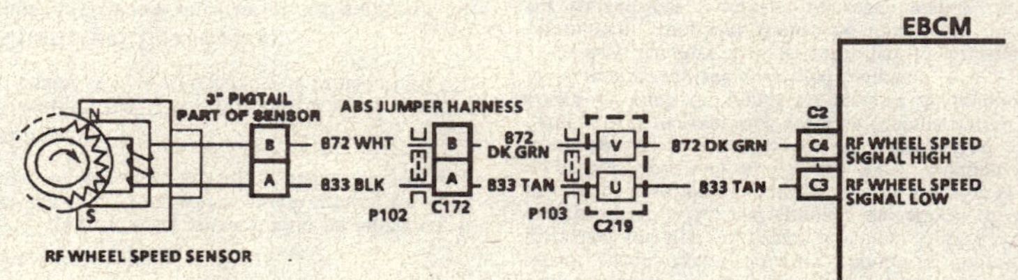

DTC 33
(Page 2 of 3)
RF WHEEL SPEED SNR CKT OPEN OR SHORT TO GND/BATT

Circuit Description:

As a toothed ring passes by the wheel speed sensor, changes in the electromagnetic field cause the wheel speed sensor to produce a sinusoidal (AC) voltage signal whose frequency is proportional to wheel speed. The magnitude of this signal is directly related to wheel speed and the proximity of the wheel speed sensor to the toothed ring, often referred to as the air gap.

DTC Will Set When:

DTC 33 can be set anytime after initialization. If either of the right front wheel speed circuits are open or shorted to voltage or ground, a malfunction exists.

Action Taken:

A malfunction DTC is stored, ABS is disabled and the amber ABS warning lamp is turned "ON."

DTC Chart Test Description:

Number(s) below refer to circled number(s) on the diagnostic chart.

7. Checks for high resistance in CKT 833.
8. Checks for high resistance in CKT 872.
9. Verifies proper wheel speed sensor resistance.

Diagnostic Aids:

An "Intermittent" malfunction may be caused by a poor connection, rubbed through wire insulation, or a wire that is broken inside the insulation.

The frequency of the malfunction can be checked by using the enhanced diagnostic function of the Tech 1, as described in "Tech 1 Diagnostics" found in this section.

If the customer's comments reflect that the ABS warning lamp is "ON" only during moist environmental changes (rain, snow, vehicle wash), all wheel speed sensor circuitry should be thoroughly inspected for signs of water intrusion. If DTC is not current, clear DTCs and simulate the effects of water intrusion. Use the following procedure. Spray the suspected area with a 5% salt water solution (two teaspoons of salt to 12 oz. of water). Test drive vehicle over various road surfaces (bumps, turns, etc.) above 24 km/h (15 mph) for at least 30 seconds. If DTC returns, replace suspected harness.

Any circuitry that is suspected of causing the intermittent complaint should be thoroughly checked for backed out terminals, improper mating, broken locks, improperly formed or damaged terminals, poor terminal to wiring connections or physical damage to the wiring harness.

Resistance of the wheel speed sensor will increase with an increase in sensor temperature.

When replacing a wheel speed sensor, inspect the sensor terminals and harness connector for corrosion and/or water intrusion. If evidence of corrosion or water intrusion exists, replace wheel speed sensor jumper harness. Likewise, if replacing a wheel speed sensor jumper harness, inspect sensor terminals. If evidence of corrosion or water intrusion exists, replace wheel speed sensor. Refer to "On-Vehicle Service" in this section.

DIAGNOSTIC CHART DTC 33 — 1995 CUTLASS SUPREME, GRAND PRIX, LUMINA, MONTE CARLO AND REGAL

DTC 33
(Page 2 of 3)
RF WHEEL SPEED SNR CKT OPEN OR SHORT TO GND/BATT

DIAGNOSIS CONTINUED FROM PAGE 1 OF 3 OF THIS DTC CHART.

(7)
- DISCONNECT RIGHT FRONT WHEEL SPEED SENSOR DIRECTLY AT SENSOR.
- USING J 39200, MEASURE THE RESISTANCE BETWEEN TERMINAL "C3" OF THE 32-WAY EBCM HARNESS CONNECTOR AND TERMINAL "A" OF THE RIGHT FRONT WHEEL SPEED SENSOR HARNESS CONNECTOR.

IS RESISTANCE 2Ω OR LESS?

NO: REPAIR HIGH RESISTANCE IN CKT 833 (WHEEL SPEED SENSOR JUMPER MUST BE REPLACED IF DAMAGED).

YES:

(8)
- USING J 39200, MEASURE THE RESISTANCE BETWEEN TERMINAL "C4" OF THE 32-WAY EBCM HARNESS CONNECTOR AND TERMINAL "B" OF THE RIGHT FRONT WHEEL SPEED SENSOR HARNESS CONNECTOR.

IS RESISTANCE 2Ω OR LESS?

NO: REPAIR HIGH RESISTANCE IN CKT 872 (WHEEL SPEED SENSOR JUMPER MUST BE REPLACED IF DAMAGED).

YES:

(9)
- USING J 39200, MEASURE THE RESISTANCE BETWEEN TERMINAL "A" AND TERMINAL "B" OF THE RIGHT FRONT WHEEL SPEED SENSOR CONNECTOR.

IS RESISTANCE BETWEEN 1661 AND 1929Ω WHEN SENSOR IS 20°C (68°F)? *(IF TEMPERATURE OF SENSOR IS SUSPECTED TO OTHER THAN 20°C (68°F), REFER TO THE TABLE BELOW FOR APPLICABLE SENSOR RESISTANCE VALUES; THE VALUES IN THE TABLE ARE FOR TEMPERATURE OF THE SENSOR, NOT AIR TEMPERATURE.)*

NO: REPLACE WHEEL SPEED SENSOR.

YES: MALFUNCTION IS INTERMITTENT. INSPECT ALL CONNECTORS AND HARNESSES FOR DAMAGE WHICH MAY RESULT IN HIGH RESISTANCE WHEN ALL COMPONENTS ARE CONNECTED. REFER TO "DIAGNOSTIC AIDS" ON FACING PAGE FOR MORE INFORMATION. REPAIR ALL DAMAGE IF FOUND.

FRONT WHEEL SPEED SENSOR

TEMPERATURE OF SENSOR VS. SENSOR RESISTANCE (APPROXIMATE)

TEMP. (°C)	TEMP. (°F)	RESISTANCE (OHMS)
-34 TO 4	-30 TO 40	1258 TO 1808
5 TO 43	41 TO 110	1502 TO 2103
44 TO 93	111 TO 200	1747 TO 2482

DIAGNOSTIC CHART DTC 33 — 1995 CUTLASS SUPREME, GRAND PRIX, LUMINA, MONTE CARLO AND REGAL

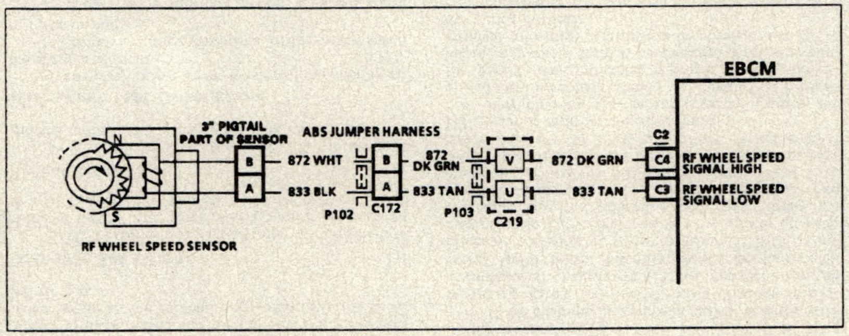

DTC 33
(Page 3 of 3)
RF WHEEL SPEED SNR CKT OPEN OR SHORT TO GND/BATT

Circuit Description:

As a toothed ring passes by the wheel speed sensor, changes in the electromagnetic field cause the wheel speed sensor to produce a sinusoidal (AC) voltage signal whose frequency is proportional to wheel speed. The magnitude of this signal is directly related to wheel speed and the proximity of the wheel speed sensor to the toothed ring, often referred to as the air gap.

DTC Will Set When:

DTC 33 can be set anytime after initialization. If either of the right front wheel speed circuits are open or shorted to voltage or ground, a malfunction exists.

Action Taken:

A malfunction DTC is stored, ABS is disabled and the amber ABS warning lamp is turned "ON."

DTC Chart Test Description:

Number(s) below refer to circled number(s) on the diagnostic chart.

10. Checks for a short to ground in CKT 833.
11. Checks for a short to ground in CKT 872.
12. Checks for a short to ground in the wheel speed sensor jumper harness.
13. Verifies that wheel speed sensor is not internally shorted to ground.

Diagnostic Aids:

An "Intermittent" malfunction may be caused by a poor connection, rubbed through wire insulation, or a wire that is broken inside the insulation.

The frequency of the malfunction can be checked by using the enhanced diagnostic function of the Tech 1, as described in "Tech 1 Diagnostics" found in this section.

If the customer's comments reflect that the ABS warning lamp is "ON" only during moist environmental changes (rain, snow, vehicle wash), all wheel speed sensor circuitry should be thoroughly inspected for signs of water intrusion. If DTC is not current, clear DTCs and simulate the effects of water intrusion. Use the following procedure. Spray the suspected area with a 5% salt water solution (two teaspoons of salt to 12 oz. of water). Test drive vehicle over various road surfaces (bumps, turns, etc.) above 24 km/h (15 mph) for at least 30 seconds. If DTC returns, replace suspected harness.

Any circuitry that is suspected of causing the intermittent complaint should be thoroughly checked for backed out terminals, improper mating, broken locks, improperly formed or damaged terminals, poor terminal to wiring connections or physical damage to the wiring harness.

Resistance of the wheel speed sensor will increase with an increase in sensor temperature.

When replacing a wheel speed sensor, inspect the sensor terminals and harness connector for corrosion and/or water intrusion. If evidence of corrosion or water intrusion exists, replace wheel speed sensor jumper harness. Likewise, if replacing a wheel speed sensor jumper harness, inspect sensor terminals. If evidence of corrosion or water intrusion exists, replace wheel speed sensor. Refer to "On-Vehicle Service" in this section.

DIAGNOSTIC CHART DTC 33 — 1995 CUTLASS SUPREME, GRAND PRIX, LUMINA, MONTE CARLO AND REGAL

DTC 33
(Page 3 of 3)
RF WHEEL SPEED SNR CKT OPEN OR SHORT TO GND/BATT

DIAGNOSIS CONTINUED FROM PAGE 1 OF 3 OF THIS DTC CHART.

(10)
- DISCONNECT CONNECTOR C172 (LOCATED RH REAR OF ENGINE COMPARTMENT).
- USING J 39200, MEASURE THE RESISTANCE BETWEEN GROUND AND TERMINAL "C3" OF THE 32-WAY EBCM HARNESS CONNECTOR.

DOES J 39200 DISPLAY RESISTANCE AS "OL" (INFINITE)?

NO → REPAIR SHORT TO GROUND IN CKT 833.

YES ↓

(11)
- USING J 39200, MEASURE THE RESISTANCE BETWEEN GROUND AND TERMINAL "C4" OF THE 32-WAY EBCM HARNESS CONNECTOR.

DOES J 39200 DISPLAY RESISTANCE AS "OL" (INFINITE)?

NO → REPAIR SHORT TO GROUND IN CKT 872.

YES ↓

(12)
- THOROUGHLY RESPRAY THE RIGHT FRONT WHEEL SPEED SENSOR JUMPER HARNESS WITH THE SALT WATER SOLUTION.

IMPORTANT: DO NOT ALLOW WATER TO ENTER ANY WHEEL SPEED SENSOR CONNECTOR.

- DISCONNECT RIGHT FRONT WHEEL SPEED SENSOR DIRECTLY AT SENSOR.
- USING J 39200, MEASURE THE RESISTANCE BETWEEN GROUND AND TERMINAL "A," THEN "B" OF THE RIGHT FRONT WHEEL SPEED SENSOR HARNESS CONNECTOR.

DOES J 39200 DISPLAY RESISTANCE AS "OL" (INFINITE) FOR BOTH MEASUREMENTS?

NO → REPLACE RIGHT FRONT WHEEL SPEED SENSOR JUMPER HARNESS.

YES ↓

(13)
- USING J 39200, MEASURE THE RESISTANCE BETWEEN GROUND AND TERMINAL "A" OF THE RIGHT FRONT WHEEL SPEED SENSOR CONNECTOR.

DOES J 39200 DISPLAY RESISTANCE AS "OL" (INFINITE)?

NO → REPLACE WHEEL SPEED SENSOR.

YES → MALFUNCTION IS INTERMITTENT. INSPECT ALL CONNECTORS AND HARNESSES FOR DAMAGE WHICH MAY RESULT IN A SHORT TO GROUND WHEN ALL COMPONENTS ARE CONNECTED. PAY PARTICULAR ATTENTION TO THE JUMPER HARNESS. SMALL NICKS OR CUTS CAN CAUSE A MALFUNCTION. REPAIR ANY DAMAGE IF FOUND. JUMPER HARNESS MUST BE REPLACED IF DAMAGED. REFER TO "DIAGNOSTIC AIDS" ON FACING PAGE FOR MORE INFORMATION.

DIAGNOSTIC CHART DTC 34 — 1995 CUTLASS SUPREME, GRAND PRIX, LUMINA, MONTE CARLO AND REGAL

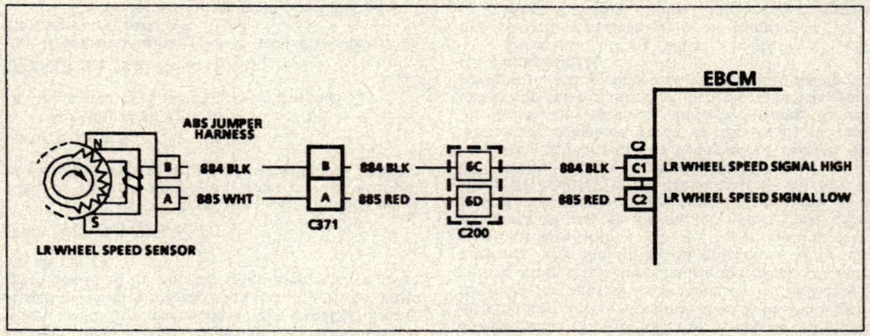

DTC 34
(Page 1 of 3)
LR WHEEL SPEED SNR CKT OPEN OR SHORT TO GND/BATT

Circuit Description:

As a toothed ring passes by the wheel speed sensor, changes in the electromagnetic field cause the wheel speed sensor to produce a sinusoidal (AC) voltage signal whose frequency is proportional to wheel speed. The magnitude of this signal is directly related to wheel speed and the proximity of the wheel speed sensor to the toothed ring, often referred to as the air gap.

DTC Will Set When:

DTC 34 can be set anytime after initialization. If either of the left rear wheel speed circuits are open or shorted to voltage or ground, a malfunction exists.

Action Taken:

A malfunction DTC is stored, ABS is disabled and the amber ABS warning lamp is turned "ON."

DTC Chart Test Description:

Number(s) below refer to circled number(s) on the diagnostic chart.

1. Checks for proper resistance of the wheel speed sensor coil.
2. Checks for a short to ground in wheel speed sensor wiring.
3. Checks for a short to voltage in the wheel speed sensor wiring.
4. Checks for a short to voltage in CKT 884.
5. Checks for a short to voltage in CKT 885.
6. Ensures that the malfunction was not due to physical damage of the circuitry.

Diagnostic Aids:

An "Intermittent" malfunction may be caused by a poor connection, rubbed through wire insulation, or a wire that is broken inside the insulation.

The frequency of the malfunction can be checked by using the enhanced diagnostic function of the Tech 1, as described in "Tech 1 Diagnostics" found in this section.

If the customer's comments reflect that the ABS warning lamp is "ON" only during moist environmental changes (rain, snow, vehicle wash), all wheel speed sensor circuitry should be thoroughly inspected for signs of water intrusion. If DTC is not current, clear DTCs and simulate the effects of water intrusion. Use the following procedure. Spray the suspected area with a 5% salt water solution (two teaspoons of salt to 12 oz. of water). Test drive vehicle over various road surfaces (bumps, turns, etc.) above 24 km/h (15 mph) for at least 30 seconds. If DTC returns, replace suspected harness.

Any circuitry that is suspected of causing the intermittent complaint should be thoroughly checked for backed out terminals, improper mating, broken locks, improperly formed or damaged terminals, poor terminal to wiring connections or physical damage to the wiring harness.

Resistance of the wheel speed sensor will increase with an increase in sensor temperature.

When replacing a wheel speed sensor, inspect the sensor terminals and harness connector for corrosion and/or water intrusion. If evidence of corrosion or water intrusion exists, replace wheel speed sensor jumper harness. Likewise, if replacing a wheel speed sensor jumper harness, inspect sensor terminals. If evidence of corrosion or water intrusion exists, replace wheel speed sensor. Refer to "On-Vehicle Service" in this section.

DIAGNOSTIC CHART DTC 34 — 1995 CUTLASS SUPREME, GRAND PRIX, LUMINA, MONTE CARLO AND REGAL

DTC 34
(Page 1 of 3)
LR WHEEL SPEED SNR CKT OPEN OR SHORT TO GND/BATT

Important

- WHEEL SPEED SENSOR INTERMITTENT MALFUNCTIONS MAY BE DIFFICULT TO LOCATE. CARE SHOULD BE TAKEN NOT TO DISTURB ANY ELECTRICAL CONNECTIONS PRIOR TO AN INDICATED STEP OF THIS CHART. THIS WILL ENSURE THAT AN INTERMITTENT CONNECTION WILL NOT BE CORRECTED BEFORE THE SOURCE OF THE MALFUNCTION IS FOUND.

(1)
- IGNITION "OFF."
- USING A 5% SALT WATER SOLUTION (TWO TEASPOONS OF SALT TO 12 OZ. WATER), THOROUGHLY SPRAY THE LEFT REAR WHEEL SPEED SENSOR JUMPER HARNESS.
- DISCONNECT 32-WAY EBCM HARNESS CONNECTOR.
- USING J 39200, MEASURE THE RESISTANCE BETWEEN TERMINALS "C1" AND "C2" OF THE 32-WAY EBCM HARNESS CONNECTOR.

IS RESISTANCE BETWEEN 2059 AND 2353Ω WHEN SENSOR IS 20°C (68°F)? *(IF TEMPERATURE OF SENSOR IS SUSPECTED TO OTHER THAN 20°C (68°F), REFER TO THE TABLE ON PAGE 2 OF 3 OF THIS DTC CHART FOR APPLICABLE SENSOR RESISTANCE VALUES; THE VALUES IN THE TABLE ARE FOR TEMPERATURE OF THE SENSOR, NOT AIR TEMPERATURE.)*

- NO → DIAGNOSIS CONTINUED ON PAGE 2 OF 3 OF THIS DTC CHART.
- YES →

(2)
- MEASURE THE RESISTANCE BETWEEN GROUND AND TERMINAL "C1" OF THE 32-WAY EBCM HARNESS CONNECTOR.

DOES J 39200 DISPLAY RESISTANCE AS "OL" (INFINITE)?

- NO → DIAGNOSIS CONTINUED ON PAGE 3 OF 3 OF THIS DTC CHART.
- YES →

(3)
- IGNITION "ON."
- USING J 39200, MEASURE THE VOLTAGE BETWEEN GROUND AND TERMINAL "C1" OF THE 32-WAY EBCM HARNESS CONNECTOR.

IS VOLTAGE 1 VOLT OR LESS?

- NO →

(4)
- IGNITION "OFF."
- DISCONNECT LEFT REAR WHEEL SPEED SENSOR DIRECTLY AT SENSOR.
- IGNITION "ON."
- USING J 39200, MEASURE THE VOLTAGE BETWEEN GROUND AND TERMINAL "C1" OF THE 32-WAY EBCM HARNESS CONNECTOR.

IS VOLTAGE 1 VOLT OR LESS?

- NO → REPAIR SHORT TO VOLTAGE IN CKT 884.
- YES →

(5)
- USING J 39200, MEASURE THE VOLTAGE BETWEEN GROUND AND TERMINAL "C2" OF THE 32-WAY EBCM HARNESS CONNECTOR.

IS VOLTAGE 1 VOLT OR LESS?

- YES → MALFUNCTION IS INTERMITTENT. INSPECT ALL CONNECTORS AND HARNESSES FOR DAMAGE WHICH MAY RESULT IN A SHORT TO VOLTAGE WHEN ALL COMPONENTS ARE CONNECTED. REPAIR ALL DAMAGE IF FOUND. REFER TO "DIAGNOSTIC AIDS" ON FACING PAGE FOR MORE INFORMATION.
- NO → REPAIR SHORT TO VOLTAGE IN CKT 885.

(3) YES →

(6)
- IGNITION "OFF."
- INSPECT TERMINALS "C1" AND "C2" OF THE 32-WAY EBCM HARNESS CONNECTOR FOR POOR TERMINAL CONTACT AND CONTAMINATION. ALSO INSPECT CKTs 884 AND 885 FOR DAMAGE WHICH MAY RESULT IN AN OPEN CIRCUIT, SHORT TO GROUND, OR SHORT TO VOLTAGE. REPAIR ALL DAMAGE IF EVIDENT.
- RECONNECT ALL CONNECTORS.
- IGNITION "ON."

DOES DTC 34 SET AS A CURRENT DTC?

- YES → REPLACE EBCM.
- NO → MALFUNCTION IS INTERMITTENT. REFER TO "DIAGNOSTIC AIDS" ON FACING PAGE.

DIAGNOSTIC CHART DTC 34 — 1995 CUTLASS SUPREME, GRAND PRIX, LUMINA, MONTE CARLO AND REGAL

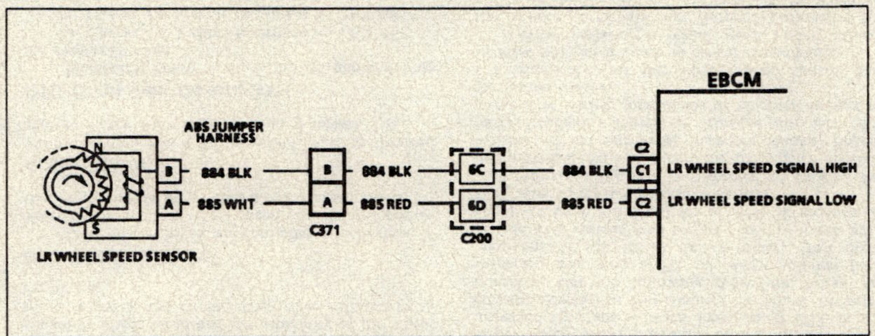

DTC 34
(Page 2 of 3)
LR WHEEL SPEED SNR CKT OPEN OR SHORT TO GND/BATT

Circuit Description:

As a toothed ring passes by the wheel speed sensor, changes in the electromagnetic field cause the wheel speed sensor to produce a sinusoidal (AC) voltage signal whose frequency is proportional to wheel speed. The magnitude of this signal is directly related to wheel speed and the proximity of the wheel speed sensor to the toothed ring, often referred to as the air gap.

DTC Will Set When:

DTC 34 can be set anytime after initialization. If either of the left rear wheel speed circuits are open or shorted to voltage or ground, a malfunction exists.

Action Taken:

A malfunction DTC is stored, ABS is disabled and the amber ABS warning lamp is turned "ON."

DTC Chart Test Description:

Number(s) below refer to circled number(s) on the diagnostic chart.

7. Checks for high resistance in CKT 884.
8. Checks for high resistance in CKT 885.
9. Verifies proper wheel speed sensor resistance.

Diagnostic Aids:

An "Intermittent" malfunction may be caused by a poor connection, rubbed through wire insulation, or a wire that is broken inside the insulation.

The frequency of the malfunction can be checked by using the enhanced diagnostic function of the Tech 1, as described in "Tech 1 Diagnostics" found in this section.

If the customer's comments reflect that the ABS warning lamp is "ON" only during moist environmental changes (rain, snow, vehicle wash), all wheel speed sensor circuitry should be thoroughly inspected for signs of water intrusion. If DTC is not current, clear DTCs and simulate the effects of water intrusion. Use the following procedure. Spray the suspected area with a 5% salt water solution (two teaspoons of salt to 12 oz. of water). Test drive vehicle over various road surfaces (bumps, turns, etc.) above 24 km/h (15 mph) for at least 30 seconds. If DTC returns, replace suspected harness.

Any circuitry that is suspected of causing the intermittent complaint should be thoroughly checked for backed out terminals, improper mating, broken locks, improperly formed or damaged terminals, poor terminal to wiring connections or physical damage to the wiring harness.

Resistance of the wheel speed sensor will increase with an increase in sensor temperature.

When replacing a wheel speed sensor, inspect the sensor terminals and harness connector for corrosion and/or water intrusion. If evidence of corrosion or water intrusion exists, replace wheel speed sensor jumper harness. Likewise, if replacing a wheel speed sensor jumper harness, inspect sensor terminals. If evidence of corrosion or water intrusion exists, replace wheel speed sensor. Refer to "On-Vehicle Service" in this section.

DIAGNOSTIC CHART DTC 34 — 1995 CUTLASS SUPREME, GRAND PRIX, LUMINA, MONTE CARLO AND REGAL

DTC 34
(Page 2 of 3)
LR WHEEL SPEED SNR CKT OPEN OR SHORT TO GND/BATT

DIAGNOSIS CONTINUED FROM PAGE 1 OF 3 OF THIS DTC CHART.

(7)
- DISCONNECT LEFT REAR WHEEL SPEED SENSOR DIRECTLY AT SENSOR.
- USING J 39200, MEASURE THE RESISTANCE BETWEEN TERMINAL "C1" OF THE 32-WAY EBCM HARNESS CONNECTOR AND TERMINAL "B" OF THE LEFT REAR WHEEL SPEED SENSOR HARNESS CONNECTOR. IS RESISTANCE 2Ω OR LESS?

NO: REPAIR HIGH RESISTANCE IN CKT 884 (WHEEL SPEED SENSOR JUMPER MUST BE REPLACED IF DAMAGED).

YES:

(8)
- USING J 39200, MEASURE THE RESISTANCE BETWEEN TERMINAL "C2" OF THE 32-WAY EBCM HARNESS CONNECTOR AND TERMINAL "A" OF THE LEFT REAR WHEEL SPEED SENSOR HARNESS CONNECTOR. IS RESISTANCE 2Ω OR LESS?

NO: REPAIR HIGH RESISTANCE IN CKT 885 (WHEEL SPEED SENSOR JUMPER MUST BE REPLACED IF DAMAGED).

YES:

(9)
- USING J 39200, MEASURE THE RESISTANCE BETWEEN TERMINAL "A" AND TERMINAL "B" OF THE LEFT REAR WHEEL SPEED SENSOR CONNECTOR. IS RESISTANCE BETWEEN 2059 AND 2353Ω WHEN SENSOR IS 20°C (68°F)? *(IF TEMPERATURE OF SENSOR IS SUSPECTED TO OTHER THAN 20°C (68°F), REFER TO THE TABLE BELOW FOR APPLICABLE SENSOR RESISTANCE VALUES; THE VALUES IN THE TABLE ARE FOR TEMPERATURE OF THE SENSOR, NOT AIR TEMPERATURE.)*

NO: REPLACE WHEEL SPEED SENSOR.

YES: MALFUNCTION IS INTERMITTENT. INSPECT ALL CONNECTORS AND HARNESSES FOR DAMAGE WHICH MAY RESULT IN HIGH RESISTANCE WHEN ALL COMPONENTS ARE CONNECTED. REFER TO "DIAGNOSTIC AIDS" ON FACING PAGE FOR MORE INFORMATION. REPAIR ALL DAMAGE IF FOUND.

REAR WHEEL SPEED SENSOR

TEMPERATURE OF SENSOR VS. SENSOR RESISTANCE (APPROXIMATE)

TEMP. (°C)	TEMP. (°F)	RESISTANCE (OHMS)
-34 TO 4	-30 TO 40	1622 TO 2205
5 TO 43	41 TO 110	1930 TO 2566
44 TO 93	111 TO 200	2245 TO 3028

DIAGNOSTIC CHART DTC 34 — 1995 CUTLASS SUPREME, GRAND PRIX, LUMINA, MONTE CARLO AND REGAL

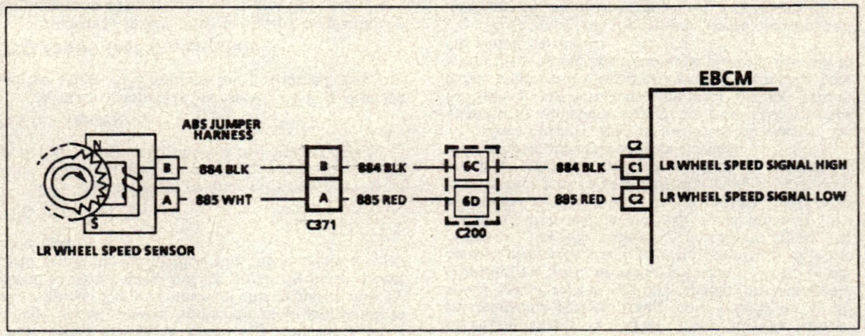

DTC 34

LR WHEEL SPEED SNR CKT OPEN OR SHORT TO GND/BATT

Circuit Description:

As a toothed ring passes by the wheel speed sensor, changes in the electromagnetic field cause the wheel speed sensor to produce a sinusoidal (AC) voltage signal whose frequency is proportional to wheel speed. The magnitude of this signal is directly related to wheel speed and the proximity of the wheel speed sensor to the toothed ring, often referred to as the air gap.

DTC Will Set When:

DTC 34 can be set anytime after initialization. If either of the left rear wheel speed circuits are open or shorted to voltage or ground, a malfunction exists.

Action Taken:

A malfunction DTC is stored, ABS is disabled and the amber ABS warning lamp is turned "ON."

DTC Chart Test Description:

Number(s) below refer to circled number(s) on the diagnostic chart.

10. Checks for a short to ground in CKT 884.
11. Checks for a short to ground in CKT 885.
12. Checks for a short to ground in the wheel speed sensor jumper harness.
13. Verifies that wheel speed sensor is not internally shorted to ground.

Diagnostic Aids:

An "Intermittent" malfunction may be caused by a poor connection, rubbed through wire insulation, or a wire that is broken inside the insulation.

The frequency of the malfunction can be checked by using the enhanced diagnostic function of the Tech 1, as described in "Tech 1 Diagnostics" found in this section.

If the customer's comments reflect that the ABS warning lamp is "ON" only during moist environmental changes (rain, snow, vehicle wash), all wheel speed sensor circuitry should be thoroughly inspected for signs of water intrusion. If DTC is not current, clear DTCs and simulate the effects of water intrusion. Use the following procedure. Spray the suspected area with a 5% salt water solution (two teaspoons of salt to 12 oz. of water). Test drive vehicle over various road surfaces (bumps, turns, etc.) above 24 km/h (15 mph) for at least 30 seconds. If DTC returns, replace suspected harness.

Any circuitry that is suspected of causing the intermittent complaint should be thoroughly checked for backed out terminals, improper mating, broken locks, improperly formed or damaged terminals, poor terminal to wiring connections or physical damage to the wiring harness.

Resistance of the wheel speed sensor will increase with an increase in sensor temperature.

When replacing a wheel speed sensor, inspect the sensor terminals and harness connector for corrosion and/or water intrusion. If evidence of corrosion or water intrusion exists, replace wheel speed sensor jumper harness. Likewise, if replacing a wheel speed sensor jumper harness, inspect sensor terminals. If evidence of corrosion or water intrusion exists, replace wheel speed sensor. Refer to "On-Vehicle Service" in this section.

DIAGNOSTIC CHART DTC 34 — 1995 CUTLASS SUPREME, GRAND PRIX, LUMINA, MONTE CARLO AND REGAL

DTC 34

(Page 3 of 3)

LR WHEEL SPEED SNR CKT OPEN OR SHORT TO GND/BATT

DIAGNOSIS CONTINUED FROM PAGE 1 OF 3 OF THIS DTC CHART.

(10)
- DISCONNECT CONNECTOR C371 (LOCATED BEHIND LH SIDE OF REAR SEAT).
- USING J 39200, MEASURE THE RESISTANCE BETWEEN GROUND AND TERMINAL "C1" OF THE 32-WAY EBCM HARNESS CONNECTOR.

DOES J 39200 DISPLAY RESISTANCE AS "OL" (INFINITE)?

NO → REPAIR SHORT TO GROUND IN CKT 884.

YES ↓

(11)
- USING J 39200, MEASURE THE RESISTANCE BETWEEN GROUND AND TERMINAL "C2" OF THE 32-WAY EBCM HARNESS CONNECTOR.

DOES J 39200 DISPLAY RESISTANCE AS "OL" (INFINITE)?

NO → REPAIR SHORT TO GROUND IN CKT 885.

YES ↓

(12)
- THOROUGHLY RESPRAY THE LEFT REAR WHEEL SPEED SENSOR JUMPER HARNESS WITH THE SALT WATER SOLUTION.

IMPORTANT: DO NOT ALLOW WATER TO ENTER ANY WHEEL SPEED SENSOR CONNECTOR.

- DISCONNECT LEFT REAR WHEEL SPEED SENSOR DIRECTLY AT SENSOR.
- USING J 39200, MEASURE THE RESISTANCE BETWEEN GROUND AND TERMINAL "A," THEN "B" OF THE LEFT REAR WHEEL SPEED SENSOR HARNESS CONNECTOR.

DOES J 39200 DISPLAY RESISTANCE AS "OL" (INFINITE) FOR BOTH MEASUREMENTS?

NO → REPLACE LEFT REAR WHEEL SPEED SENSOR JUMPER HARNESS.

YES ↓

(13)
- USING J 39200, MEASURE THE RESISTANCE BETWEEN GROUND AND TERMINAL "A" OF THE LEFT REAR WHEEL SPEED SENSOR CONNECTOR.

DOES J 39200 DISPLAY RESISTANCE AS "OL" (INFINITE)?

NO → REPLACE WHEEL SPEED SENSOR.

YES → MALFUNCTION IS INTERMITTENT. INSPECT ALL CONNECTORS AND HARNESSES FOR DAMAGE WHICH MAY RESULT IN A SHORT TO GROUND WHEN ALL COMPONENTS ARE CONNECTED. PAY PARTICULAR ATTENTION TO THE JUMPER HARNESS, SMALL NICKS OR CUTS CAN CAUSE A MALFUNCTION. REPAIR ANY DAMAGE IF FOUND. JUMPER HARNESS MUST BE REPLACED IF DAMAGED. REFER TO "DIAGNOSTIC AIDS" ON FACING PAGE FOR MORE INFORMATION.

DIAGNOSTIC CHART DTC 35 — 1995 CUTLASS SUPREME, GRAND PRIX, LUMINA, MONTE CARLO AND REGAL

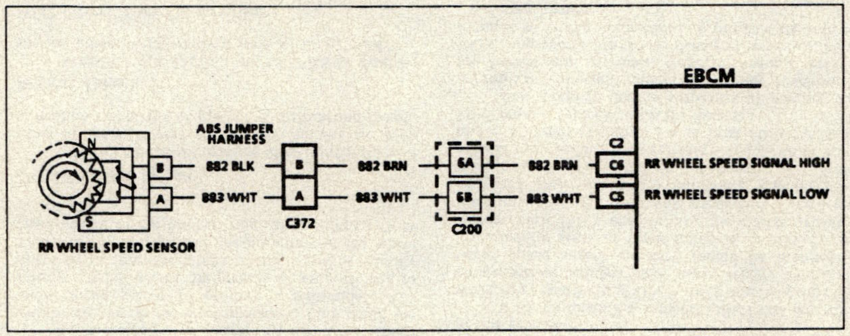

DTC 35

RR WHEEL SPEED SNR CKT OPEN OR SHORT TO GND/BATT

Circuit Description:

As a toothed ring passes by the wheel speed sensor, changes in the electromagnetic field cause the wheel speed sensor to produce a sinusoidal (AC) voltage signal whose frequency is proportional to wheel speed. The magnitude of this signal is directly related to wheel speed and the proximity of the wheel speed sensor to the toothed ring, often referred to as the air gap.

DTC Will Set When:

DTC 35 can be set anytime after initialization. If either of the right rear wheel speed circuits are open or shorted to voltage or ground, a malfunction exists.

Action Taken:

A malfunction DTC is stored, ABS is disabled and the amber ABS warning lamp is turned "ON."

DTC Chart Test Description:

Number(s) below refer to circled number(s) on the diagnostic chart.

1. Checks for proper resistance of the wheel speed sensor coil.
2. Checks for a short to ground in the wheel speed sensor wiring.
3. Checks for a short to voltage in the wheel speed sensor wiring.
4. Checks for a short to voltage in CKT 883.
5. Checks for a short to voltage in CKT 882.
6. Ensures that the malfunction was not due to physical damage of the circuitry.

Diagnostic Aids:

An "Intermittent" malfunction may be caused by a poor connection, rubbed through wire insulation, or a wire that is broken inside the insulation.

The frequency of the malfunction can be checked by using the enhanced diagnostic function of the Tech 1, as described in "Tech 1 Diagnostics" found in this section.

If the customer's comments reflect that the ABS warning lamp is "ON" only during moist environmental changes (rain, snow, vehicle wash), all wheel speed sensor circuitry should be thoroughly inspected for signs of water intrusion. If DTC is not current, clear DTCs and simulate the effects of water intrusion. Use the following procedure. Spray the suspected area with a 5% salt water solution (two teaspoons of salt to 12 oz. of water). Test drive vehicle over various road surfaces (bumps, turns, etc.) above 24 km/h (15 mph) for at least 30 seconds. If DTC returns, replace suspected harness.

Any circuitry that is suspected of causing the intermittent complaint should be thoroughly checked for backed out terminals, improper mating, broken locks, improperly formed or damaged terminals, poor terminal to wiring connections or physical damage to the wiring harness.

Resistance of the wheel speed sensor will increase with an increase in sensor temperature.

When replacing a wheel speed sensor, inspect the sensor terminals and harness connector for corrosion and/or water intrusion. If evidence of corrosion or water intrusion exists, replace wheel speed sensor jumper harness. Likewise, if replacing a wheel speed sensor jumper harness, inspect sensor terminals. If evidence of corrosion or water intrusion exists, replace wheel speed sensor. Refer to "On-Vehicle Service" in this section.

DIAGNOSTIC CHART DTC 35 — 1995 CUTLASS SUPREME, GRAND PRIX, LUMINA, MONTE CARLO AND REGAL

DTC 35
(Page 1 of 3)
RR WHEEL SPEED SNR CKT OPEN OR SHORT TO GND/BATT

Important

- WHEEL SPEED SENSOR INTERMITTENT MALFUNCTIONS MAY BE DIFFICULT TO LOCATE. CARE SHOULD BE TAKEN NOT TO DISTURB ANY ELECTRICAL CONNECTIONS PRIOR TO AN INDICATED STEP OF THIS CHART. THIS WILL ENSURE THAT AN INTERMITTENT CONNECTION WILL NOT BE CORRECTED BEFORE THE SOURCE OF THE MALFUNCTION IS FOUND.

(1)
- IGNITION "OFF."
- USING A 5% SALT WATER SOLUTION (TWO TEASPOONS OF SALT TO 12 OZ. WATER), THOROUGHLY SPRAY THE RIGHT REAR WHEEL SPEED SENSOR JUMPER HARNESS.
- DISCONNECT 32-WAY EBCM HARNESS CONNECTOR.
- USING J 39200, MEASURE THE RESISTANCE BETWEEN TERMINALS "C5" AND "C6" OF THE 32-WAY EBCM HARNESS CONNECTOR.

IS RESISTANCE BETWEEN 2059 AND 2353Ω WHEN SENSOR IS 20°C (68°F)? *(IF TEMPERATURE OF SENSOR IS SUSPECTED TO OTHER THAN 20°C (68°F), REFER TO THE TABLE ON PAGE 2 OF 2 OF THIS DTC CHART FOR APPLICABLE SENSOR RESISTANCE VALUES; THE VALUES IN THE TABLE ARE FOR TEMPERATURE OF THE SENSOR, NOT AIR TEMPERATURE.)*

- YES → (2)
- NO → DIAGNOSIS CONTINUED ON PAGE 2 OF 3 OF THIS DTC CHART.

(2)
- MEASURE THE RESISTANCE BETWEEN GROUND AND TERMINAL "C5" OF THE 32-WAY EBCM HARNESS CONNECTOR.

DOES J 39200 DISPLAY RESISTANCE AS "OL" (INFINITE)?

- YES → (3)
- NO → DIAGNOSIS CONTINUED ON PAGE 3 OF 3 OF THIS DTC CHART.

(3)
- IGNITION "ON."
- USING J 39200, MEASURE THE VOLTAGE BETWEEN GROUND AND TERMINAL "C5" OF THE 32-WAY EBCM HARNESS CONNECTOR.

IS VOLTAGE 1 VOLT OR LESS?

- NO → (4)
- YES → (6)

(4)
- IGNITION "OFF."
- DISCONNECT RIGHT REAR WHEEL SPEED SENSOR DIRECTLY AT SENSOR.
- IGNITION "ON."
- USING J 39200, MEASURE THE VOLTAGE BETWEEN GROUND AND TERMINAL "C5" OF THE 32-WAY EBCM HARNESS CONNECTOR.

IS VOLTAGE 1 VOLT OR LESS?

- YES → (5)
- NO → REPAIR SHORT TO VOLTAGE IN CKT 883.

(5)
- USING J 39200, MEASURE THE VOLTAGE BETWEEN GROUND AND TERMINAL "C6" OF THE 32-WAY EBCM HARNESS CONNECTOR.

IS VOLTAGE 1 VOLT OR LESS?

- YES → MALFUNCTION IS INTERMITTENT. INSPECT ALL CONNECTORS AND HARNESSES FOR DAMAGE WHICH MAY RESULT IN A SHORT TO VOLTAGE WHEN ALL COMPONENTS ARE CONNECTED. REPAIR ALL DAMAGE IF FOUND. REFER TO "DIAGNOSTIC AIDS" ON FACING PAGE FOR MORE INFORMATION.
- NO → REPAIR SHORT TO VOLTAGE IN CKT 882.

(6)
- IGNITION "OFF."
- INSPECT TERMINALS "C5" AND "C6" OF THE 32-WAY EBCM HARNESS CONNECTOR FOR POOR TERMINAL CONTACT AND CONTAMINATION. ALSO INSPECT CKTs 882 AND 883 FOR DAMAGE WHICH MAY RESULT IN AN OPEN CIRCUIT, SHORT TO GROUND, OR SHORT TO VOLTAGE. REPAIR ALL DAMAGE IF EVIDENT.
- RECONNECT ALL CONNECTORS.
- IGNITION "ON."

DOES DTC 35 SET AS A CURRENT DTC?

- YES → REPLACE EBCM.
- NO → MALFUNCTION IS INTERMITTENT. REFER TO "DIAGNOSTIC AIDS" ON FACING PAGE.

DIAGNOSTIC CHART DTC 35 — 1995 CUTLASS SUPREME, GRAND PRIX, LUMINA, MONTE CARLO AND REGAL

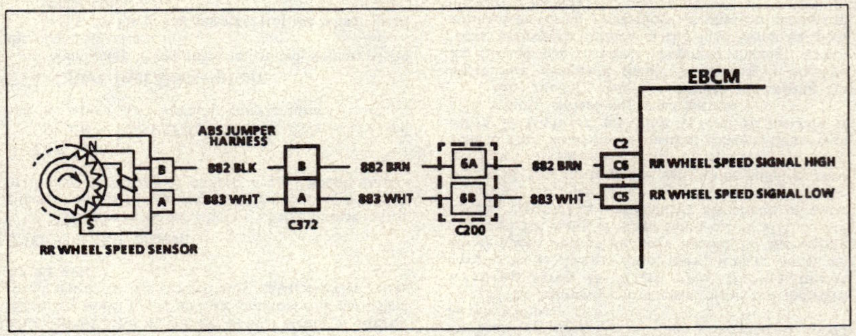

DTC 35

RR WHEEL SPEED SNR CKT OPEN OR SHORT TO GND/BATT

Circuit Description:

As a toothed ring passes by the wheel speed sensor, changes in the electromagnetic field cause the wheel speed sensor to produce a sinusoidal (AC) voltage signal whose frequency is proportional to wheel speed. The magnitude of this signal is directly related to wheel speed and the proximity of the wheel speed sensor to the toothed ring, often referred to as the air gap.

DTC Will Set When:

DTC 35 can be set anytime after initialization. If either of the right rear wheel speed circuits are open or shorted to voltage or ground, a malfunction exists.

Action Taken:

A malfunction DTC is stored, ABS is disabled and the amber ABS warning lamp is turned "ON."

DTC Chart Test Description:

Number(s) below refer to circled number(s) on the diagnostic chart.

7. Checks for high resistance in CKT 883.
8. Checks for high resistance in CKT 882.
9. Verifies proper wheel speed sensor resistance.

Diagnostic Aids:

An "Intermittent" malfunction may be caused by a poor connection, rubbed through wire insulation, or a wire that is broken inside the insulation.

The frequency of the malfunction can be checked by using the enhanced diagnostic function of the Tech 1, as described in "Tech 1 Diagnostics" found in this section.

If the customer's comments reflect that the ABS warning lamp is "ON" only during moist environmental changes (rain, snow, vehicle wash), all wheel speed sensor circuitry should be thoroughly inspected for signs of water intrusion. If DTC is not current, clear DTCs and simulate the effects of water intrusion. Use the following procedure. Spray the suspected area with a 5% salt water solution (two teaspoons of salt to 12 oz. of water). Test drive vehicle over various road surfaces (bumps, turns, etc.) above 24 km/h (15 mph) for at least 30 seconds. If DTC returns, replace suspected harness.

Any circuitry that is suspected of causing the intermittent complaint should be thoroughly checked for backed out terminals, improper mating, broken locks, improperly formed or damaged terminals, poor terminal to wiring connections or physical damage to the wiring harness.

Resistance of the wheel speed sensor will increase with an increase in sensor temperature.

When replacing a wheel speed sensor, inspect the sensor terminals and harness connector for corrosion and/or water intrusion. If evidence of corrosion or water intrusion exists, replace wheel speed sensor jumper harness. Likewise, if replacing a wheel speed sensor jumper harness, inspect sensor terminals. If evidence of corrosion or water intrusion exists, replace wheel speed sensor. Refer to "On-Vehicle Service" in this section.

DIAGNOSTIC CHART DTC 35 — 1995 CUTLASS SUPREME, GRAND PRIX, LUMINA, MONTE CARLO AND REGAL

DTC 35
(Page 2 of 3)
RR WHEEL SPEED SNR CKT OPEN OR SHORT TO GND/BATT

DIAGNOSIS CONTINUED FROM PAGE 1 OF 3 OF THIS DTC CHART.

(7)
- DISCONNECT RIGHT REAR WHEEL SPEED SENSOR DIRECTLY AT SENSOR.
- USING J 39200, MEASURE THE RESISTANCE BETWEEN TERMINAL "C5" OF THE 32-WAY EBCM HARNESS CONNECTOR AND TERMINAL "A" OF THE RIGHT REAR WHEEL SPEED SENSOR HARNESS CONNECTOR.
IS RESISTANCE 2Ω OR LESS?

NO: REPAIR HIGH RESISTANCE IN CKT 883 (WHEEL SPEED SENSOR JUMPER MUST BE REPLACED IF DAMAGED).

YES:

(8)
- USING J 39200, MEASURE THE RESISTANCE BETWEEN TERMINAL "C6" OF THE 32-WAY EBCM HARNESS CONNECTOR AND TERMINAL "B" OF THE RIGHT REAR WHEEL SPEED SENSOR HARNESS CONNECTOR.
IS RESISTANCE 2Ω OR LESS?

NO: REPAIR HIGH RESISTANCE IN CKT 882 (WHEEL SPEED SENSOR JUMPER MUST BE REPLACED IF DAMAGED).

YES:

(9)
- USING J 39200, MEASURE THE RESISTANCE BETWEEN TERMINAL "A" AND TERMINAL "B" OF THE RIGHT REAR WHEEL SPEED SENSOR CONNECTOR.
IS RESISTANCE BETWEEN 2059 AND 2353Ω WHEN SENSOR IS 20°C (68°F)? *(IF TEMPERATURE OF SENSOR IS SUSPECTED TO OTHER THAN 20°C (68°F), REFER TO THE TABLE BELOW FOR APPLICABLE SENSOR RESISTANCE VALUES; THE VALUES IN THE TABLE ARE FOR TEMPERATURE OF THE SENSOR, NOT AIR TEMPERATURE.)*

NO: REPLACE WHEEL SPEED SENSOR.

YES: MALFUNCTION IS INTERMITTENT. INSPECT ALL CONNECTORS AND HARNESSES FOR DAMAGE WHICH MAY RESULT IN HIGH RESISTANCE WHEN ALL COMPONENTS ARE CONNECTED. REFER TO "DIAGNOSTIC AIDS" ON FACING PAGE FOR MORE INFORMATION. REPAIR ALL DAMAGE IF FOUND.

REAR WHEEL SPEED SENSOR
TEMPERATURE OF SENSOR VS. SENSOR RESISTANCE (APPROXIMATE)

TEMP. (°C)	TEMP. (°F)	RESISTANCE (OHMS)
-34 TO 4	-30 TO 40	1622 TO 2205
5 TO 43	41 TO 110	1930 TO 2566
44 TO 93	111 TO 200	2245 TO 3028

DIAGNOSTIC CHART DTC 35 — 1995 CUTLASS SUPREME, GRAND PRIX, LUMINA, MONTE CARLO AND REGAL

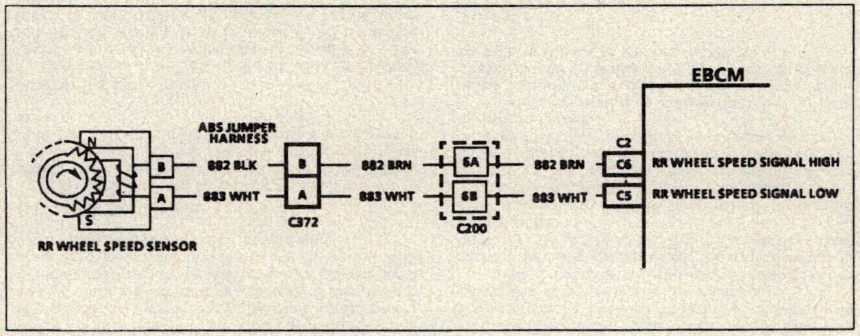

DTC 35

RR WHEEL SPEED SNR CKT OPEN OR SHORT TO GND/BATT

Circuit Description:

As a toothed ring passes by the wheel speed sensor, changes in the electromagnetic field cause the wheel speed sensor to produce a sinusoidal (AC) voltage signal whose frequency is proportional to wheel speed. The magnitude of this signal is directly related to wheel speed and the proximity of the wheel speed sensor to the toothed ring, often referred to as the air gap.

DTC Will Set When:

DTC 35 can be set anytime after initialization. If either of the right rear wheel speed circuits are open or shorted to voltage or ground, a malfunction exists.

Action Taken:

A malfunction DTC is stored, ABS is disabled and the amber ABS warning lamp is turned "ON."

DTC Chart Test Description:

Number(s) below refer to circled number(s) on the diagnostic chart.

10. Checks for a short to ground in CKT 883.
11. Checks for a short to ground in CKT 882.
12. Checks for a short to ground in the wheel speed sensor jumper harness.
13. Verifies that wheel speed sensor is not internally shorted to ground.

Diagnostic Aids:

An "Intermittent" malfunction may be caused by a poor connection, rubbed through wire insulation, or a wire that is broken inside the insulation.

The frequency of the malfunction can be checked by using the enhanced diagnostic function of the Tech 1, as described in "Tech 1 Diagnostics" found in this section.

If the customer's comments reflect that the ABS warning lamp is "ON" only during moist environmental changes (rain, snow, vehicle wash), all wheel speed sensor circuitry should be thoroughly inspected for signs of water intrusion. If DTC is not current, clear DTCs and simulate the effects of water intrusion. Use the following procedure. Spray the suspected area with a 5% salt water solution (two teaspoons of salt to 12 oz. of water). Test drive vehicle over various road surfaces (bumps, turns, etc.) above 24 km/h (15 mph) for at least 30 seconds. If DTC returns, replace suspected harness.

Any circuitry that is suspected of causing the intermittent complaint should be thoroughly checked for backed out terminals, improper mating, broken locks, improperly formed or damaged terminals, poor terminal to wiring connections or physical damage to the wiring harness.

Resistance of the wheel speed sensor will increase with an increase in sensor temperature.

When replacing a wheel speed sensor, inspect the sensor terminals and harness connector for corrosion and/or water intrusion. If evidence of corrosion or water intrusion exists, replace wheel speed sensor jumper harness. Likewise, if replacing a wheel speed sensor jumper harness, inspect sensor terminals. If evidence of corrosion or water intrusion exists, replace wheel speed sensor. Refer to "On-Vehicle Service" in this section.

DIAGNOSTIC CHART DTC 35 — 1995 CUTLASS SUPREME, GRAND PRIX, LUMINA, MONTE CARLO AND REGAL

DTC 35
(Page 3 of 3)
RR WHEEL SPEED SNR CKT OPEN OR SHORT TO GND/BATT

DIAGNOSIS CONTINUED FROM PAGE 1 OF 3 OF THIS DTC CHART.

(10)
- DISCONNECT CONNECTOR C372 (LOCATED BEHIND RH SIDE OF REAR SEAT).
- USING J 39200, MEASURE THE RESISTANCE BETWEEN GROUND AND TERMINAL "C5" OF THE 32-WAY EBCM HARNESS CONNECTOR.

DOES J 39200 DISPLAY RESISTANCE AS "OL" (INFINITE)?

- NO: REPAIR SHORT TO GROUND IN CKT 883.
- YES: go to 11.

(11)
- USING J 39200, MEASURE THE RESISTANCE BETWEEN GROUND AND TERMINAL "C6" OF THE 32-WAY EBCM HARNESS CONNECTOR.

DOES J 39200 DISPLAY RESISTANCE AS "OL" (INFINITE)?

- NO: REPAIR SHORT TO GROUND IN CKT 882.
- YES: go to 12.

(12)
- THOROUGHLY RESPRAY THE RIGHT REAR WHEEL SPEED SENSOR JUMPER HARNESS WITH THE SALT WATER SOLUTION.

IMPORTANT: DO NOT ALLOW WATER TO ENTER ANY WHEEL SPEED SENSOR CONNECTOR.

- DISCONNECT RIGHT REAR WHEEL SPEED SENSOR DIRECTLY AT SENSOR.
- USING J 39200, MEASURE THE RESISTANCE BETWEEN GROUND AND TERMINAL "A," THEN "B" OF THE RIGHT REAR WHEEL SPEED SENSOR HARNESS CONNECTOR.

DOES J 39200 DISPLAY RESISTANCE AS "OL" (INFINITE) FOR BOTH MEASUREMENTS?

- NO: REPLACE RIGHT REAR WHEEL SPEED SENSOR JUMPER HARNESS.
- YES: go to 13.

(13)
- USING J 39200, MEASURE THE RESISTANCE BETWEEN GROUND AND TERMINAL "A" OF THE RIGHT REAR WHEEL SPEED SENSOR CONNECTOR.

DOES J 39200 DISPLAY RESISTANCE AS "OL" (INFINITE)?

- NO: REPLACE WHEEL SPEED SENSOR.
- YES: MALFUNCTION IS INTERMITTENT. INSPECT ALL CONNECTORS AND HARNESSES FOR DAMAGE WHICH MAY RESULT IN A SHORT TO GROUND WHEN ALL COMPONENTS ARE CONNECTED. PAY PARTICULAR ATTENTION TO THE JUMPER HARNESS, SMALL NICKS OR CUTS CAN CAUSE A MALFUNCTION. REPAIR ANY DAMAGE IF FOUND. JUMPER HARNESS MUST BE REPLACED IF DAMAGED. REFER TO "DIAGNOSTIC AIDS" ON FACING PAGE FOR MORE INFORMATION.

DIAGNOSTIC CHART DTC 36 — 1995 CUTLASS SUPREME, GRAND PRIX, LUMINA, MONTE CARLO AND REGAL

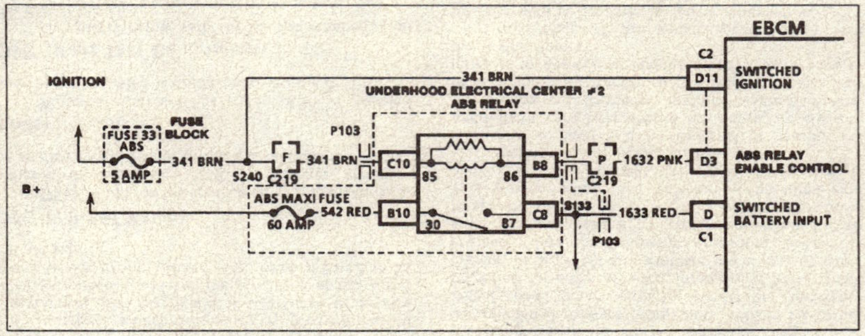

DTC 36

LOW SYSTEM VOLTAGE

Circuit Description:

This DTC is used to monitor the voltage level available to the EBCM controller. If the voltage drops below 11 volts, full performance of the ABS system cannot be guaranteed. During ABS operation, there are several current requirements that will cause battery voltage at the EBCM to drop. Because of this, voltage is monitored prior to an ABS event to indicate good charging system condition and also during an ABS event when voltage may drop significantly.

DTC Will Set When:

DTC 36 can only be set if the vehicle's speed is greater than 5 km/h (3 mph). If the switched battery voltage is less than 11.2 volts non-ABS event or less than 9.6 volts during a ABS event, a malfunction exists.

Action Taken:

A malfunction DTC is stored, ABS is disabled and the amber ABS warning lamp is turned "ON." The red BRAKE warning lamp is turned "ON" if the rear piston in the ABS motor pack is not in the home position.

DTC Chart Test Description:

Number(s) below refer to circled number(s) on the diagnostic chart.

1. Checks to see what voltage is readily available at connector C1, terminal "D" of the EBCM. If voltage is representative of good charging system condition, a malfunction is not present.
2. Isolates the condition to high circuit resistance, or improper charging system operation.
3. Isolates which circuit is the source of the high resistance condition.
4. Checks for high resistance in CKT 542.

Diagnostic Aids:

An "Intermittent" malfunction may be caused by a poor connection, rubbed through wire insulation, or a wire that is broken inside the insulation.

The frequency of the malfunction can be checked by using the enhanced diagnostic function of the Tech 1, as described in "Tech 1 Diagnostics" found in this section.

Any circuitry that is suspected of causing the intermittent complaint should be thoroughly checked for backed out terminals, improper mating, broken locks, improperly formed or damaged terminals, poor terminal to wiring connections or physical damage to the wiring harness.

While performing a voltage load test, if it is noted that only the ignition voltage drops below acceptable voltage levels, CKT 341 should be checked for high resistance or an open condition.

DIAGNOSTIC CHART DTC 36 — 1995 CUTLASS SUPREME, GRAND PRIX, LUMINA, MONTE CARLO AND REGAL

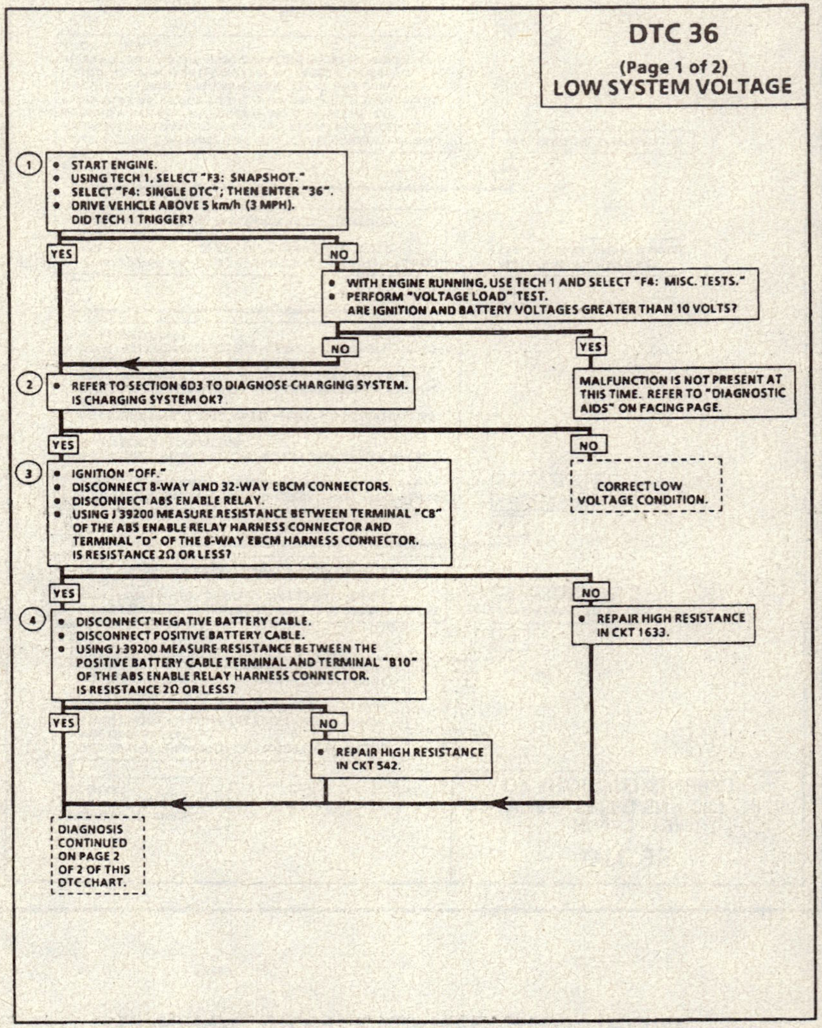

DIAGNOSTIC CHART DTC 38 — 1995 CUTLASS SUPREME, GRAND PRIX, LUMINA, MONTE CARLO AND REGAL

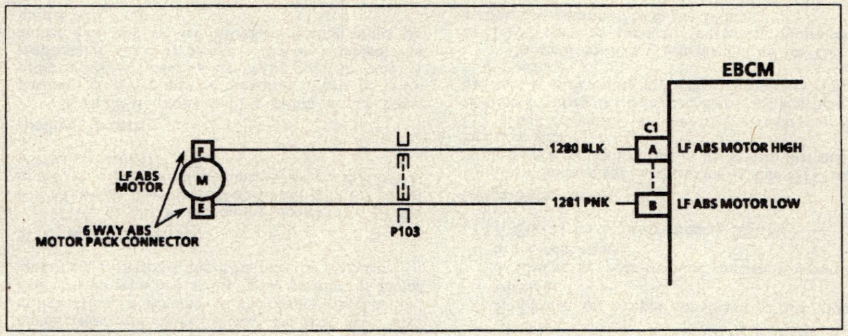

DTC 38
LEFT FRONT ESB WILL NOT HOLD MOTOR

Circuit Description:

This DTC is designed to detect a slipping left front ESB. During initialization and braking, the left front motor is rehomed. If the ESB slips, the motor/piston will move. During the next key "ON" initialization, a rehome of the motor verifies the motor/piston remained at the home position. If motor movement is detected, the ESB must be slipping.

DTC Will Set When:

DTC 38 can be set during initialization. If the EBCM detects that the ESB could not hold the piston in the home position, a malfunction exists. DTC 86 is always set with DTC 38.

Action Taken:

If an ESB cannot hold a piston in the home position, the piston may be backdriven when the brake pedal is applied, causing the brake pedal to drop. A malfunction DTC is stored, ABS is disabled, the amber ABS warning lamp and the red BRAKE warning lamp are turned "ON."

DTC Chart Test Description:

Number(s) below refer to circled number(s) on the diagnostic chart.

1. Checks the left front ESB. A broken or defective ESB would result in the left front piston being backdriven by hydraulic pressure and pedal movement would result.
2. Ensures that the proper motor circuits are located in the proper connector cavities.
3. Checks for high resistance in the motor "HIGH" circuitry.
4. Checks for high resistance in the motor "LOW" circuitry.
5. Checks for proper resistance of the motor windings.
6. Checks for poor terminal contact or corrosion in the connectors.
7. Checks for a malfunctioning EBCM.

Diagnostic Aids:

An "Intermittent" malfunction in this DTC may result from a mechanical part of the system that sticks, binds, or slips.

The frequency of the malfunction can be checked by using the enhanced diagnostic function of the Tech 1, as described in "Tech 1 Diagnostics" found in this section.

The static modulator test function of the Tech 1 or T-100 may be used to locate an intermittent malfunction associated with the ESB.

DIAGNOSTIC CHART DTC 38 — 1995 CUTLASS SUPREME, GRAND PRIX, LUMINA, MONTE CARLO AND REGAL

DTC 38
LEFT FRONT ESB WILL NOT HOLD MOTOR

(1)
- IGNITION "ON," ENGINE "OFF."
- PUMP BRAKE PEDAL UNTIL FIRM TO DEPLETE VACUUM RESERVE.
- USING TECH 1 MISC TESTS, SELECT MANUAL CONTROL FUNCTION.
- SELECT ABS MOTOR APPLY FUNCTION OF THE LEFT FRONT ABS CHANNEL AND APPLY THE MOTOR.
- WAIT 5 SECONDS.
- APPLY FIRM PRESSURE ON BRAKE PEDAL.

DID BRAKE PEDAL FALL?

- YES → REPLACE ABS MOTOR PACK.
- NO → (2)

(2)
- VISUALLY INSPECT THE 6-WAY ABS MOTOR PACK AND 8-WAY EBCM CONNECTORS FOR PROPER WIRE COLOR/CONNECTOR CAVITY COMBINATION.

ARE THE PROPER WIRES LOCATED IN THE PROPER CONNECTOR CAVITIES?

- NO → USING THE CIRCUIT DIAGRAMS LOCATED IN THE FRONT OF THIS SECTION, PLACE THE PROPER WIRES INTO THE PROPER CONNECTOR CAVITIES. CLEAR DTCs AND RECHECK SYSTEM.
- YES → (3)

(3)
- EXIT TECH 1 MANUAL CONTROL FUNCTION.
- IGNITION "OFF."
- DISCONNECT 6-WAY ABS MOTOR PACK CONNECTOR AND 8-WAY EBCM CONNECTOR.
- USING J 39200, MEASURE THE RESISTANCE BETWEEN TERMINAL "F" OF THE 6-WAY ABS MOTOR PACK HARNESS CONNECTOR AND TERMINAL "A" OF THE 8-WAY EBCM HARNESS CONNECTOR.

IS RESISTANCE 1.5Ω OR LESS?

- NO → REPAIR HIGH RESISTANCE IN CKT 1280.
- YES → (4)

(4)
- MEASURE THE RESISTANCE BETWEEN TERMINAL "E" OF THE 6-WAY ABS MOTOR PACK HARNESS CONNECTOR AND TERMINAL "B" OF THE 8-WAY EBCM HARNESS CONNECTOR.

IS RESISTANCE 1.5Ω OR LESS?

- NO → REPAIR HIGH RESISTANCE IN CKT 1281.
- YES → (5)

(5)
- MEASURE THE RESISTANCE BETWEEN TERMINAL "E" AND TERMINAL "F" OF THE ABS MOTOR PACK CONNECTOR.

IS RESISTANCE 1.5Ω OR LESS?

- NO → REPLACE ABS MOTOR PACK.
- YES → (6)

(6)
- INSPECT THE 6-WAY ABS MOTOR PACK CONNECTOR AND 8-WAY EBCM CONNECTOR FOR POOR TERMINAL CONTACT OR CORROSION.

DO ANY TERMINALS EXHIBIT POOR CONTACT OR EVIDENCE OF CORROSION?

- YES → REPLACE TERMINALS THAT EXHIBIT POOR TERMINAL CONTACT OR EVIDENCE OF CORROSION.
- NO → (7)

(7)
- IGNITION "OFF."
- RECONNECT ALL CONNECTORS.
- START ENGINE WITH FOOT OFF BRAKE.
- ALLOW ENGINE TO RUN FOR AT LEAST 10 SECONDS.
- REPEAT ABOVE IGNITION CYCLE SEQUENCE TWO MORE TIMES.

DID DTC 38 SET IN THE LAST THREE IGNITION CYCLES?

- YES → REPLACE EBCM.
- NO → MALFUNCTION IS INTERMITTENT. REFER TO "DIAGNOSTIC AIDS" ON FACING PAGE.

AFTER DIAGNOSIS IS COMPLETE, CLEAR DTCs AND TEST DRIVE VEHICLE FOR THREE (3) DRIVE CYCLES TO VERIFY THAT DTC DOES NOT RESET. A DRIVE CYCLE CONSISTS OF STARTING THE VEHICLE, DRIVING OVER 16 km/h (10 MPH) AND THEN IGNITION "OFF."

DIAGNOSTIC CHART DTC 41 — 1995 CUTLASS SUPREME, GRAND PRIX, LUMINA, MONTE CARLO AND REGAL

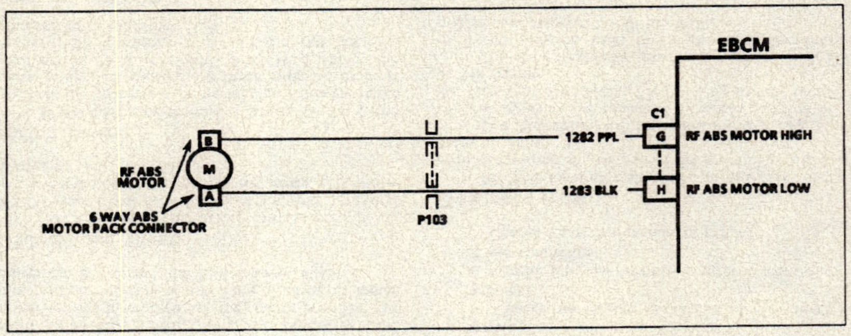

DTC 41
RIGHT FRONT ESB WILL NOT HOLD MOTOR

Circuit Description:

This DTC is designed to detect a slipping right front ESB. During initialization and braking, the right front motor is rehomed. If the ESB slips, the motor/piston will move. During the next key "ON" initialization, a rehome of the motor verifies the motor/piston remained at the home position. If motor movement is detected, the ESB must be slipping.

DTC Will Set When:

DTC 41 can be set during initialization. If the EBCM detects that the ESB could not hold the piston in the home position, a malfunction exists. DTC 86 is always set with DTC 41.

Action Taken:

If an ESB cannot hold a piston in the home position, the piston may be backdriven when the brake pedal is applied, causing the brake pedal to drop. A malfunction DTC is stored, ABS is disabled, the amber ABS and the red BRAKE warning lamps are turned "ON."

DTC Chart Test Description:

Number(s) below refer to circled number(s) on the diagnostic chart.

1. Checks the right front ESB. A broken or defective ESB would result in the right front piston being backdriven by hydraulic pressure, and pedal movement would result.
2. Ensures that the proper motor circuits are located in the proper connector cavities.
3. Checks for high resistance in the motor "HIGH" circuitry.
4. Checks for high resistance in the motor "LOW" circuitry.
5. Checks for proper resistance of the motor windings.
6. Checks for poor terminal contact or corrosion in the connectors.
7. Checks for a malfunctioning EBCM.

Diagnostic Aids:

An "Intermittent" malfunction in this DTC may result from a mechanical part of the system that sticks, binds, or slips.

The frequency of the malfunction can be checked by using the enhanced diagnostic function of the Tech 1, as described in "Tech 1 Diagnostics" found in this section.

The static modulator test function of the Tech 1 or T-100 may be used to locate an intermittent malfunction associated with the ESB.

DIAGNOSTIC CHART DTC 41 — 1995 CUTLASS SUPREME, GRAND PRIX, LUMINA, MONTE CARLO AND REGAL

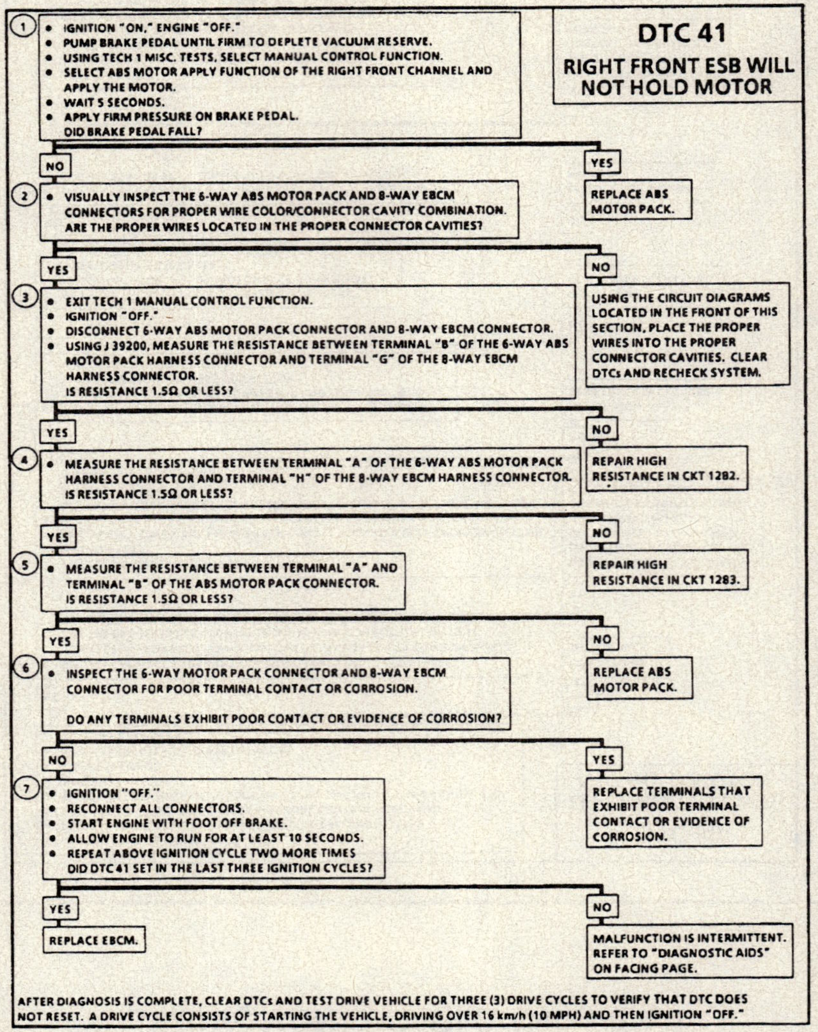

DIAGNOSTIC CHART DTC 42 — 1995 CUTLASS SUPREME, GRAND PRIX, LUMINA, MONTE CARLO AND REGAL

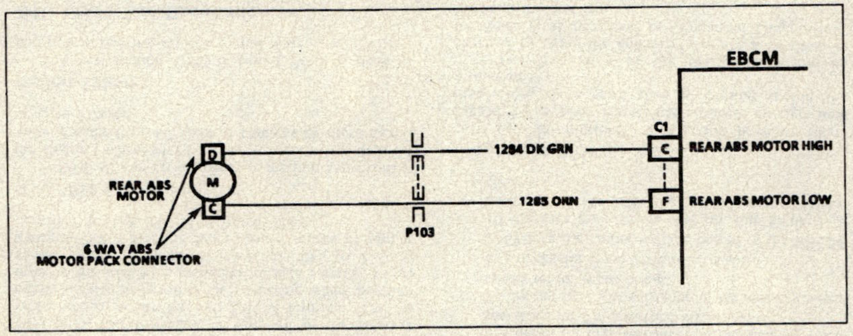

DTC 42
REAR ESB WILL NOT HOLD MOTOR

Circuit Description:

This DTC is designed to detect a slipping rear axle ESB. During initialization and braking, the rear motor is rehomed. If the ESB slips, the motor/piston will move. During the next key "ON" initialization, a rehome of the motor verifies the motor/piston remained at the home position. If motor movement is detected, the ESB must be slipping.

DTC Will Set When:

DTC 42 can be set during initialization. If the EBCM detects that the ESB could not hold the piston in the home position, a malfunction exists. DTC 86 is always set with DTC 42.

Action Taken:

If an ESB cannot hold a piston in the home position, the piston may be backdriven when the brake pedal is applied, causing the brake pedal to drop. A malfunction DTC is stored, ABS is disabled, the amber ABS and the red BRAKE warning lamps are turned "ON."

DTC Chart Test Description:

Number(s) below refer to circled number(s) on the diagnostic chart.

1. Checks the rear axle ESB. A broken or defective ESB would result in the rear axle pistons being backdriven by hydraulic pressure and wheel movement would result.
2. Ensures that the proper motor circuits are located in the proper connector cavities.
3. Checks for high resistance in the motor "HIGH" circuitry.
4. Checks for high resistance in the motor "LOW" circuitry.
5. Checks for proper resistance of the motor windings.
6. Checks for poor terminal contact or corrosion in the connectors.
7. Checks for a malfunctioning EBCM.

Diagnostic Aids:

An "Intermittent" malfunction in this DTC may result from a mechanical part of the system that sticks, binds, or slips.

The frequency of the malfunction can be checked by using the enhanced diagnostic function of the Tech 1, as described in "Tech 1 Diagnostics" found in this section.

The static modulator test function of the Tech 1 or T-100 may be used to locate an intermittent malfunction associated with the ESB.

DIAGNOSTIC CHART DTC 42 — 1995 CUTLASS SUPREME, GRAND PRIX, LUMINA, MONTE CARLO AND REGAL

DTC 42
REAR ESB WILL NOT HOLD MOTOR

(1)
- RAISE THE VEHICLE SUCH THAT THE REAR WHEELS ARE APPROXIMATELY SIX INCHES OFF THE FLOOR.
- IGNITION "ON," ENGINE "OFF."
- PUMP BRAKE PEDAL UNTIL FIRM TO DEPLETE VACUUM RESERVE.
- APPLY FIRM PRESSURE ON BRAKE PEDAL AND RELEASE.
- REAPPLY VERY FIRM PRESSURE ON BRAKE PEDAL AND HAVE AN ASSISTANT TRY TO SPIN THE REAR WHEELS BY HAND.

COULD ASSISTANT SPIN THE REAR WHEELS?

- YES: REPLACE ABS MOTOR PACK.
- NO: go to (2)

(2)
- VISUALLY INSPECT THE 6-WAY ABS MOTOR PACK AND 8-WAY EBCM CONNECTORS FOR PROPER WIRE COLOR/CONNECTOR CAVITY COMBINATION.

ARE THE PROPER WIRES LOCATED IN THE PROPER CONNECTOR CAVITIES?

- NO: USING THE CIRCUIT DIAGRAMS LOCATED IN THE FRONT OF THIS SECTION, PLACE THE PROPER WIRES INTO THE PROPER CONNECTOR CAVITIES. CLEAR DTCs AND RECHECK SYSTEM.
- YES: go to (3)

(3)
- IGNITION "OFF."
- DISCONNECT 6-WAY MOTOR PACK CONNECTOR AND 8-WAY EBCM CONNECTOR.
- USING J 39200, MEASURE THE RESISTANCE BETWEEN TERMINAL "D" OF THE 6-WAY ABS MOTOR PACK HARNESS CONNECTOR AND TERMINAL "C" OF THE 8-WAY EBCM HARNESS CONNECTOR.

IS RESISTANCE 1.5Ω OR LESS?

- NO: REPAIR HIGH RESISTANCE IN CKT 1284.
- YES: go to (4)

(4)
- MEASURE THE RESISTANCE BETWEEN TERMINAL "C" OF THE 6-WAY ABS MOTOR PACK HARNESS CONNECTOR AND TERMINAL "F" OF THE 8-WAY EBCM HARNESS CONNECTOR "C1."

IS RESISTANCE 1.5Ω OR LESS?

- NO: REPAIR HIGH RESISTANCE IN CKT 1285.
- YES: go to (5)

(5)
- MEASURE THE RESISTANCE BETWEEN TERMINAL "C" AND TERMINAL "D" OF THE ABS MOTOR PACK CONNECTOR.

IS RESISTANCE 1.5Ω OR LESS?

- NO: REPLACE ABS MOTOR PACK.
- YES: go to (6)

(6)
- INSPECT THE 6-WAY ABS MOTOR PACK CONNECTOR AND 8-WAY EBCM CONNECTOR FOR POOR TERMINAL CONTACT OR CORROSION.

DO ANY TERMINALS EXHIBIT POOR CONTACT OR EVIDENCE OF CORROSION?

- YES: REPLACE TERMINALS THAT EXHIBIT POOR TERMINAL CONTACT OR EVIDENCE OF CORROSION.
- NO: go to (7)

(7)
- IGNITION "OFF."
- RECONNECT ALL CONNECTORS.
- START ENGINE WITH FOOT OFF BRAKE.
- ALLOW ENGINE TO RUN FOR AT LEAST 10 SECONDS.
- REPEAT ABOVE IGNITION CYCLE SEQUENCE TWO MORE TIMES.

DID DTC 42 SET IN THE LAST THREE IGNITION CYCLES?

- YES: REPLACE EBCM.
- NO: MALFUNCTION IS INTERMITTENT. REFER TO "DIAGNOSTIC AIDS" ON FACING PAGE.

AFTER DIAGNOSIS IS COMPLETE, CLEAR DTCs AND TEST DRIVE VEHICLE FOR THREE (3) DRIVE CYCLES TO VERIFY DTC DOES NOT RESET. A DRIVE CYCLE CONSISTS OF STARTING THE VEHICLE, DRIVING OVER 16 km/h (10 MPH) AND THEN IGNITION "OFF."

DIAGNOSTIC CHART DTC 43 — 1995 CUTLASS SUPREME, GRAND PRIX, LUMINA, MONTE CARLO AND REGAL

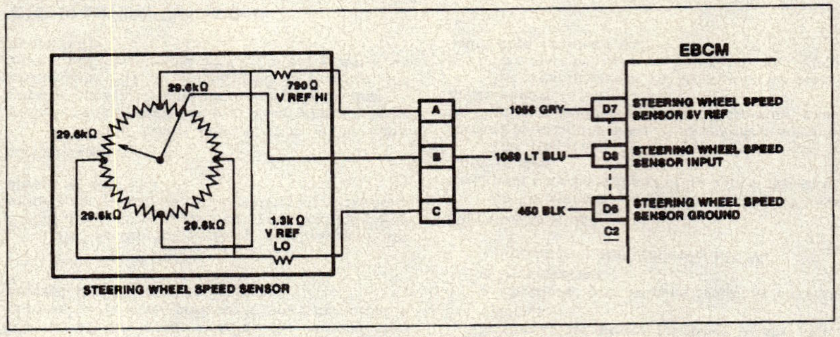

DTC 43

VES STEERING WHEEL SPEED SENSOR CIRCUIT MALFUNCTION (FOR ABS WITH VES ONLY)

Circuit Description:

A 5 volt reference is supplied to the steering wheel speed sensor through EBCM terminal "D7" to sensor connector terminal "A". Steering wheel position input to the EBCM is provided through steering wheel speed sensor connector terminal "B" to EBCM terminal "D8." Steering wheel speed sensor ground is provided through EBCM terminal "D6."

DTC Will Set When:

DTC 43 can set anytime. A malfunction exists if the steering wheel speed sensor voltage is less than 0.5 volts, greater than 4.7 volts or changes by more than 2 volts in 5 msec.

Action Taken:

A malfunction DTC is stored, VES is disabled and power steering returns to full assist.

DTC Chart Test Description:

Number(s) below refer to circled number(s) on the diagnostic chart.

1. Checks steering wheel speed sensor input to the EBCM.
2. Checks for a short between the steering wheel speed sensor input and ground circuits.
3. Checks for a short between the steering wheel speed sensor 5V reference signal and ground circuits.
4. Checks for a short to ground in the steering wheel speed sensor input circuit.
5. Checks for a short to ground in the steering wheel speed sensor 5V reference signal circuit.
6. Checks for high resistance in the steering wheel speed sensor 5V reference signal circuit.
7. Checks for high resistance in the steering wheel speed sensor input circuit.
8. Checks for poor terminal contact.
9. Checks for proper voltage output of the EBCM.
10. Ensures malfunction was not due to physical damage of the steering wheel speed sensor circuits.

Diagnostic Aids:

An "Intermittent" malfunction is most likely caused by a poor connection, rubbed through wire insulation, or a wire that is broken inside the insulation.

The frequency of the malfunction can be checked by using the enhanced diagnostic function of the Tech 1, as described in "Enhanced Diagnostics," found in this section.

Any circuitry that is suspected of causing the intermittent complaint should be thoroughly checked for backed out terminals, improper mating, broken locks, improperly formed or damaged terminals, poor terminal to wiring connections or physical damage to the wiring harness.

DIAGNOSTIC CHART DTC 43 — 1995 CUTLASS SUPREME, GRAND PRIX, LUMINA, MONTE CARLO AND REGAL

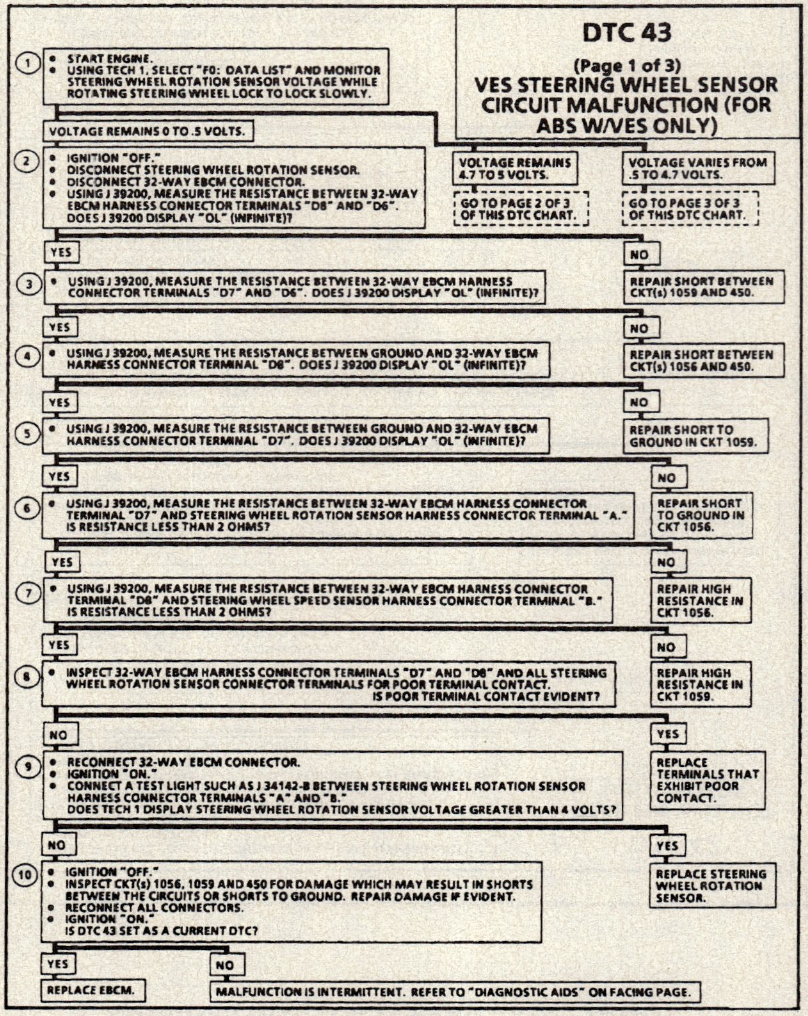

DIAGNOSTIC CHART DTC 43 — 1995 CUTLASS SUPREME, GRAND PRIX, LUMINA, MONTE CARLO AND REGAL

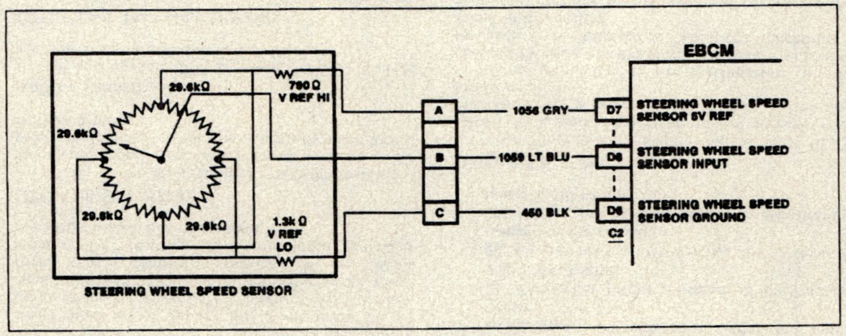

DTC 43

VES STEERING WHEEL SPEED SENSOR CIRCUIT MALFUNCTION (FOR ABS WITH VES ONLY)

Circuit Description:

A 5 volt reference is supplied to the steering wheel speed sensor through EBCM terminal "D7" to sensor connector terminal "A". Steering wheel position input to the EBCM is provided through steering wheel speed sensor connector terminal "B" to EBCM terminal "D8." Steering wheel speed sensor ground is provided through EBCM terminal "D6."

DTC Will Set When:

DTC 43 can set anytime. A malfunction exists if the steering wheel speed sensor voltage is less than 0.5 volts, greater than 4.7 volts or changes by more than 2 volts in 5 msec.

Action Taken:

A malfunction DTC is stored, VES is disabled and power steering returns to full assist.

DTC Chart Test Description:

Number(s) below refer to circled number(s) on the diagnostic chart.

11. Checks for proper voltage with the steering wheel speed sensor disconnected.
12. Checks for high resistance in the steering wheel speed sensor ground.
13. Isolates high resistance to the EBCM or the ground circuit.
14. Ensures open ground was not due to a short to battery.
15. Ensures malfunction was not due to circuit damage or poor terminal contact.
16. Checks for a short to voltage in the steering wheel speed sensor circuit.
17. Ensures malfunction was not due to physical damage of the circuitry.

Diagnostic Aids:

An "Intermittent" malfunction is most likely caused by a poor connection, rubbed through wire insulation, or a wire that is broken inside the insulation.

The frequency of the malfunction can be checked by using the enhanced diagnostic function of the Tech 1, as described in "Enhanced Diagnostics," found in this section.

Any circuitry that is suspected of causing the intermittent complaint should be thoroughly checked for backed out terminals, improper mating, broken locks, improperly formed or damaged terminals, poor terminal to wiring connections or physical damage to the wiring harness.

DIAGNOSTIC CHART DTC 43 — 1995 CUTLASS SUPREME, GRAND PRIX, LUMINA, MONTE CARLO AND REGAL

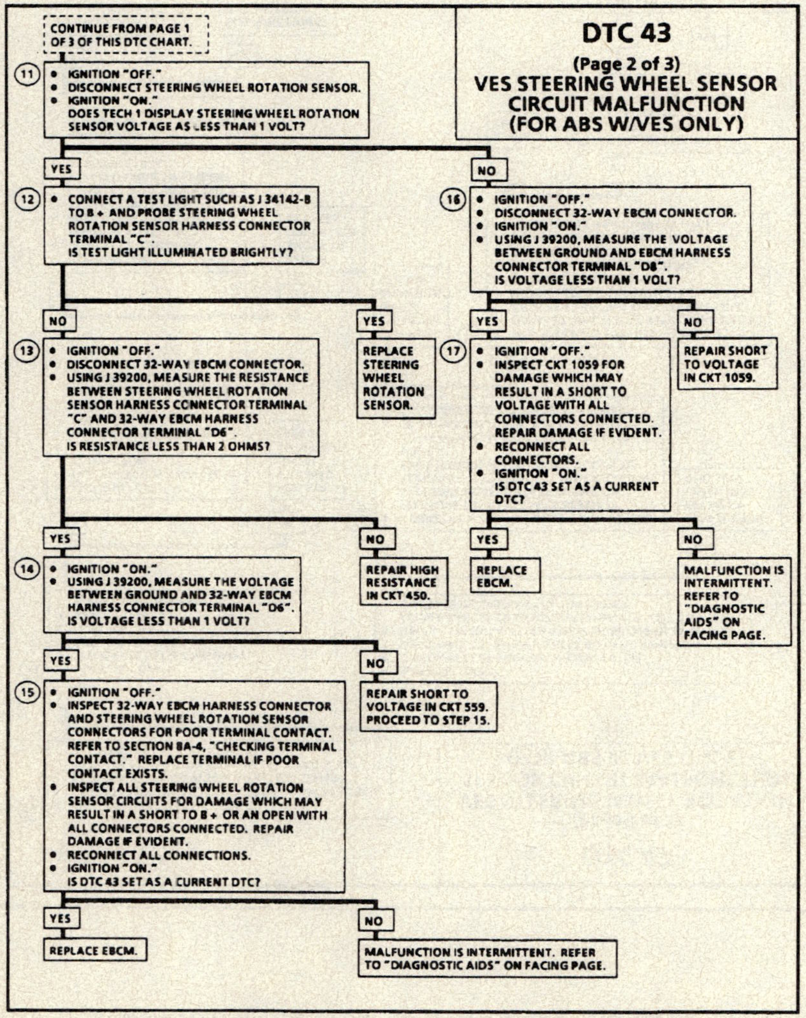

DIAGNOSTIC CHART DTC 43 — 1995 CUTLASS SUPREME, GRAND PRIX, LUMINA, MONTE CARLO AND REGAL

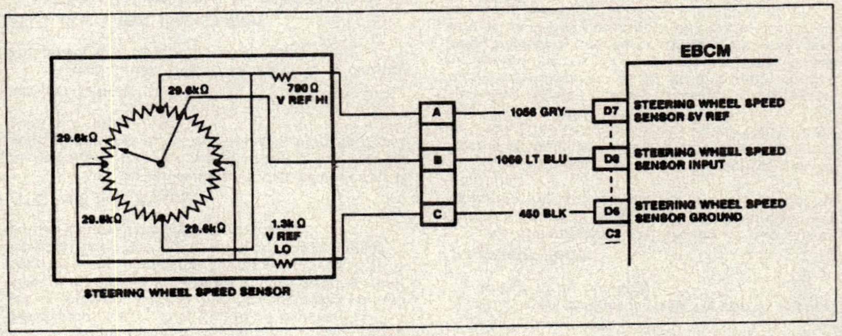

DTC 43

VES STEERING WHEEL SPEED SENSOR CIRCUIT MALFUNCTION (FOR ABS WITH VES ONLY)

Circuit Description:

A 5 volt reference is supplied to the steering wheel speed sensor through EBCM terminal "D7" to sensor connector terminal "A". Steering wheel position input to the EBCM is provided through steering wheel speed sensor connector terminal "B" to EBCM terminal "D8." Steering wheel speed sensor ground is provided through EBCM terminal "D6."

DTC Will Set When:

DTC 43 can set anytime. A malfunction exists if the steering wheel speed sensor voltage is less than 0.5 volts, greater than 4.7 volts or changes by more than 2 volts in 5 msec.

Action Taken:

A malfunction DTC is stored, VES is disabled and power steering returns to full assist.

DTC Chart Test Description:

Number(s) below refer to circled number(s) on the diagnostic chart.

18. Checks for proper voltage with the steering wheel speed sensor connected.
19. Checks for proper voltage with the steering wheel speed sensor disconnected.
20. Checks for a short between the steering wheel speed sensor 5V reference signal and ground circuits.
21. Ensures malfunction was not due to physical damage of the circuitry.
22. Checks for a short between the steering wheel speed sensor 5V reference signal and input circuits.
23. Checks for proper resistance of the steering wheel speed sensor.
24. Ensures malfunction was not due to physical damage of the circuits.
25. Checks for an open resistor in the steering wheel speed sensor.

Diagnostic Aids:

An "Intermittent" malfunction is most likely caused by a poor connection, rubbed through wire insulation, or a wire that is broken inside the insulation.

The frequency of the malfunction can be checked by using the enhanced diagnostic function of the Tech 1, as described in "Enhanced Diagnostics," found in this section.

Any circuitry that is suspected of causing the intermittent complaint should be thoroughly checked for backed out terminals, improper mating, broken locks, improperly formed or damaged terminals, poor terminal to wiring connections or physical damage to the wiring harness.

DIAGNOSTIC CHART DTC 43 — 1995 CUTLASS SUPREME, GRAND PRIX, LUMINA, MONTE CARLO AND REGAL

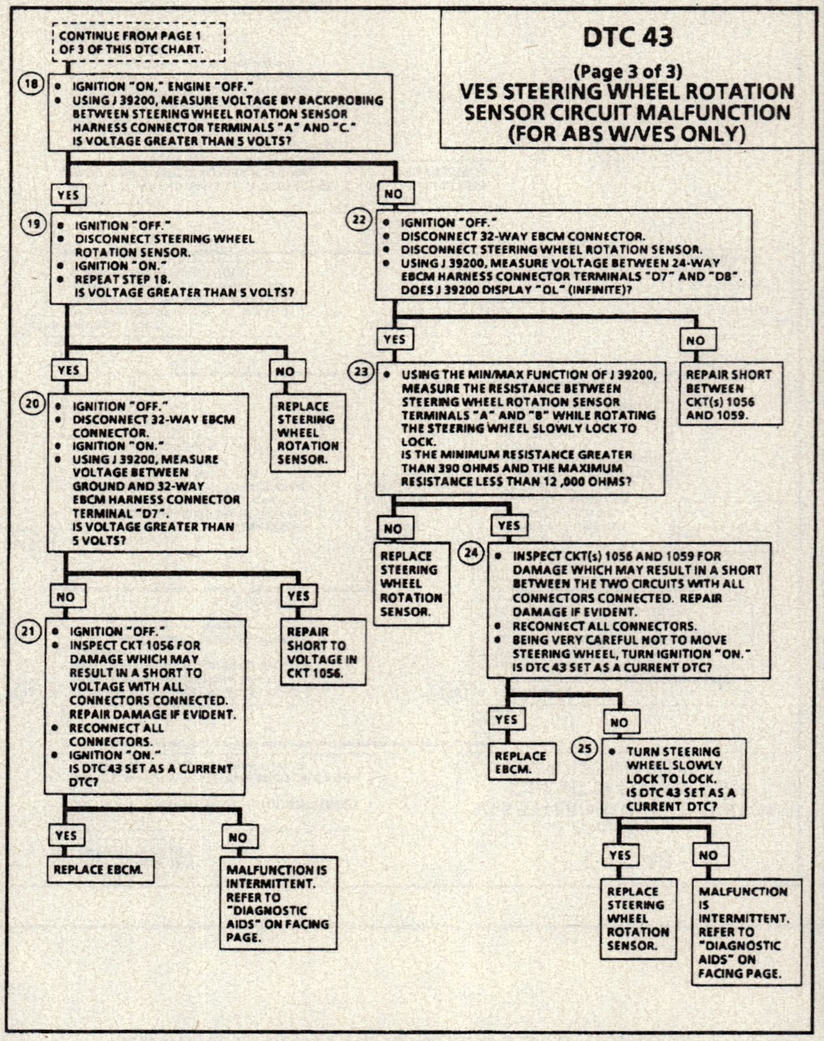

DIAGNOSTIC CHART DTC 44 — 1995 CUTLASS SUPREME, GRAND PRIX, LUMINA, MONTE CARLO AND REGAL

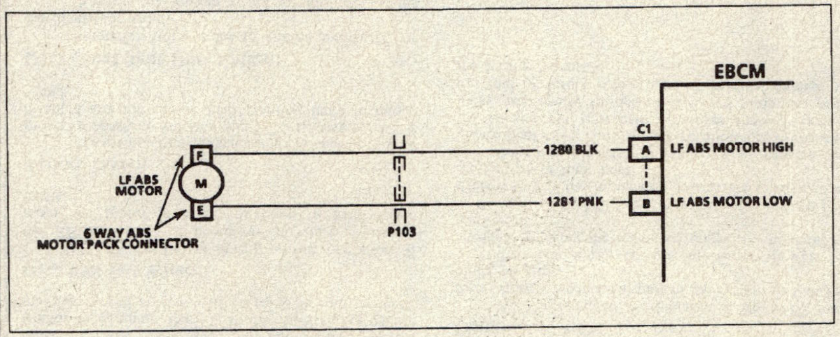

DTC 44
LEFT FRONT CHANNEL WILL NOT MOVE

Circuit Description:

This DTC is designed to detect bound-up ESB, a stuck ABS motor, or a seized hydraulic modulator. When the release is commanded during initialization, the ESB should release the ABS motor, resulting in sensed current being less than commanded current (motor is spinning freely). If the ABS motor is not moving, sensed current will be equal to stall current.

DTC Will Set When:

DTC 44 can be set during normal operation. If the EBCM detects a condition in which it cannot move the ABS motor in either direction, a malfunction exists.

Action Taken:

This malfunction indicates the channel cannot be moved properly. A malfunction DTC is stored, ABS is disabled and the amber ABS warning lamp is turned "ON."

DTC Chart Test Description:

Number(s) below refer to circled number(s) on the diagnostic chart.

1. Ensures that proper ABS motor circuits are located in the proper connector cavities.
2. Checks for proper ABS motor movement during apply and release commands from the Tech 1.
3. Compares EBCM command current to ABS motor feedback current.
4. Checks for a short between CKT 1280 and CKT 1281.
5. Checks for an ABS motor that is internally shorted.
6. Checks to see if the malfunction is due to poor terminal contact or a faulty EBCM.
7. Verifies that the ABS motor can actually apply when commanded.
8. Checks for proper hydraulic modulator gear and piston movement.

Diagnostic Aids:

An "Intermittent" malfunction in this DTC may result from a mechanical part of the system that sticks, binds, or slips.

The frequency of the malfunction can be checked by using the enhanced diagnostic function of the Tech 1.

DTC 44 may set after modulator disassembly if the modulator pistons are positioned at the bottom of their bore.

Any circuitry that is suspected of causing the intermittent complaint should be thoroughly checked for backed out terminals, improper mating, broken locks, improperly formed or damaged terminals, poor terminal to wiring connections or physical damage to the wiring harness.

DIAGNOSTIC CHART DTC 44 — 1995 CUTLASS SUPREME, GRAND PRIX, LUMINA, MONTE CARLO AND REGAL

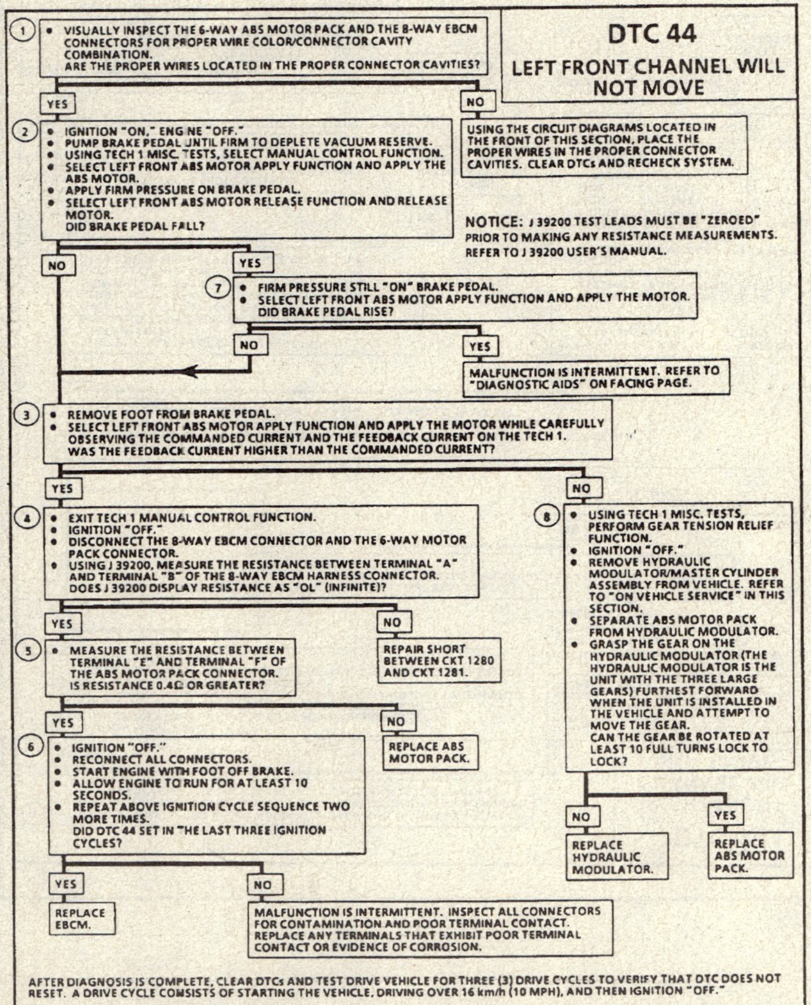

DIAGNOSTIC CHART DTC 45 — 1995 CUTLASS SUPREME, GRAND PRIX, LUMINA, MONTE CARLO AND REGAL

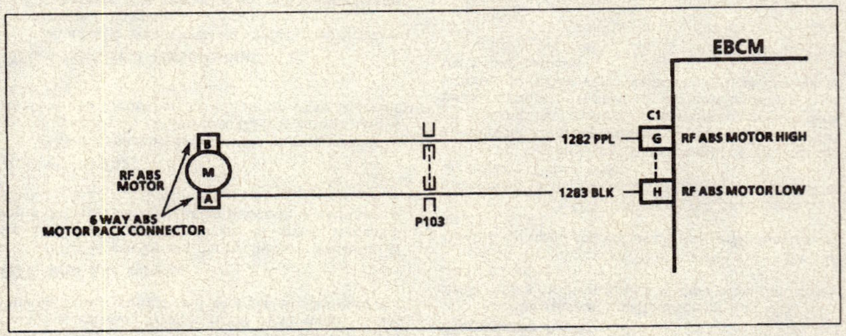

DTC 45
RIGHT FRONT CHANNEL WILL NOT MOVE

Circuit Description:

This DTC is designed to detect bound-up ESB, a stuck ABS motor, or a seized hydraulic modulator. When the release is commanded during initialization, the ESB should release the ABS motor, resulting in sensed current being less than commanded current (motor is spinning freely). If the ABS motor is not moving, sensed current will be equal to stall current.

DTC Will Set When:

DTC 45 can be set during normal operation. If the EBCM detects a condition in which it cannot move the motor in either direction, a malfunction exists.

Action Taken:

This malfunction indicates the channel cannot be moved properly. A malfunction DTC is stored, ABS is disabled and the amber ABS warning lamp is turned "ON."

DTC Chart Test Description:

Number(s) below refer to circled number(s) on the diagnostic chart.

1. Ensures that proper motor circuits are located in the proper connector cavities.
2. Checks for proper ABS motor movement during apply and release commands from the Tech 1.
3. Compares EBCM command current to ABS motor feedback current.
4. Checks for a short between CKT 1282 and CKT 1283.
5. Checks for an ABS motor that is internally shorted.
6. Checks to see if malfunction is due to poor terminal contact or the EBCM.
7. Verifies that the motor can actually apply when commanded.
8. Checks for proper hydraulic modulator gear and piston movement.

Diagnostic Aids:

An "Intermittent" malfunction in this DTC may result from a mechanical part of the system that sticks, binds, or slips.

The frequency of the malfunction can be checked by using the enhanced diagnostic function of the Tech 1.

DTC 45 may set after modulator disassembly if the modulator pistons are positioned at the bottom of their bore.

Any circuitry that is suspected of causing the intermittent complaint should be thoroughly checked for backed out terminals, improper mating, broken locks, improperly formed or damaged terminals, poor terminal to wiring connections or physical damage to the wiring harness.

DIAGNOSTIC CHART DTC 45 — 1995 CUTLASS SUPREME, GRAND PRIX, LUMINA, MONTE CARLO AND REGAL

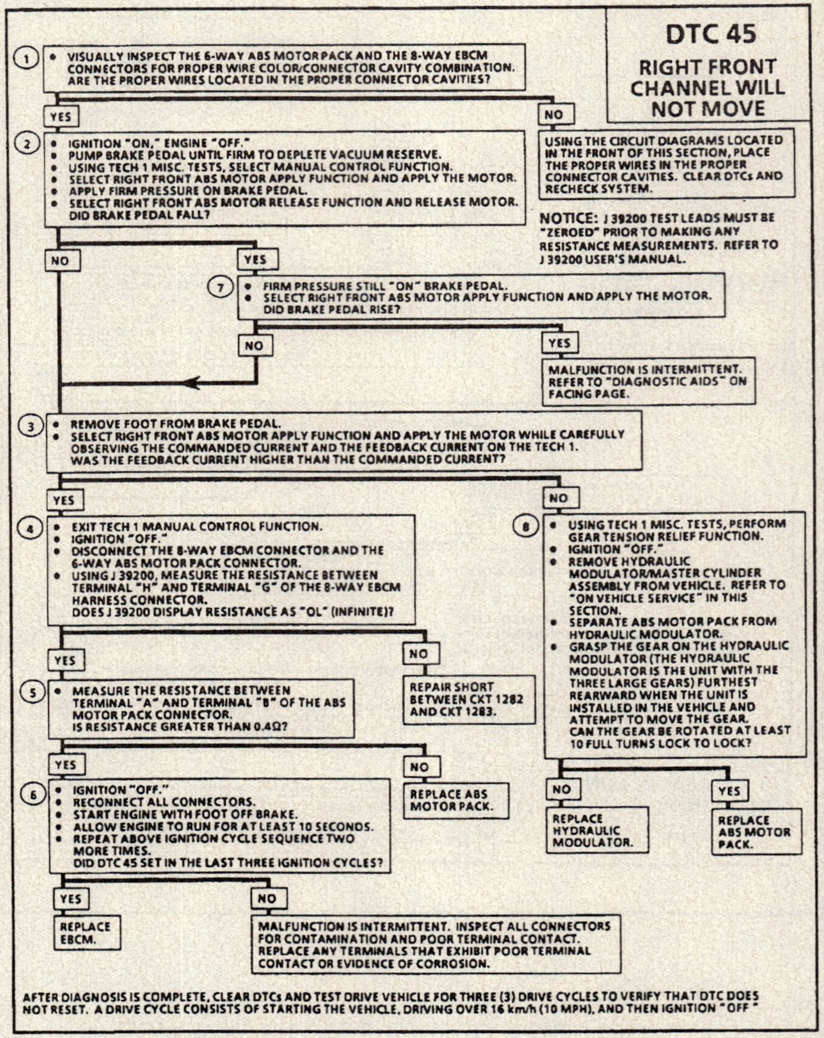

DIAGNOSTIC CHART DTC 46 — 1995 CUTLASS SUPREME, GRAND PRIX, LUMINA, MONTE CARLO AND REGAL

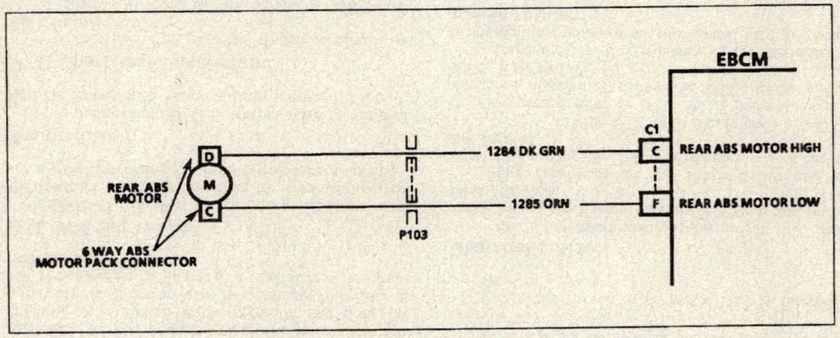

DTC 46
REAR CHANNEL WILL NOT MOVE

Circuit Description:

This DTC is designed to detect bound-up ESB, a stuck ABS motor, or a seized hydraulic modulator. When the release is commanded during initialization, the ESB should release the ABS motor, resulting in sensed current being less than commanded current (motor is spinning freely). If the ABS motor is not moving, sensed current will be equal to stall current.

DTC Will Set When:

DTC 46 can be set during normal operation. If the EBCM detects a condition in which it cannot move the ABS motor in either direction, a malfunction exists.

Action Taken:

This malfunction indicates the channel cannot be moved properly. A malfunction DTC is stored, ABS is disabled, the amber ABS warning lamp and the red BRAKE warning lamp are turned "ON."

DTC Chart Test Description:

Number(s) below refer to circled number(s) on the diagnostic chart.

1. Ensures that proper motor circuits are located in the proper connector cavities.
2. Checks for proper motor movement during apply and release commands from the Tech 1.
3. Compares EBCM command current to ABS motor feedback current.
4. Checks for a short between CKT 1284 and CKT 1285.
5. Checks for an ABS motor that is internally shorted.
6. Checks to see if malfunction is due to poor terminal contact or the EBCM.
7. Verifies that the motor can actually release when commanded.
8. Checks for proper hydraulic modulator gear and piston movement.

Diagnostic Aids:

An "Intermittent" malfunction in this DTC may result from a mechanical part of the system that sticks or binds.

The frequency of the malfunction can be checked by using the enhanced diagnostic function of the Tech 1.

DTC 46 may set after modulator disassembly if the modulator pistons are positioned at the bottom of their bore. Depending on the frequency of the malfunction, a physical inspection of the mechanical parts suspected may be necessary.

Any circuitry that is suspected of causing the intermittent complaint should be thoroughly checked for backed out terminals, improper mating, broken locks, improperly formed or damaged terminals, poor terminal to wiring connections or physical damage to the wiring harness.

DIAGNOSTIC CHART DTC 46 — 1995 CUTLASS SUPREME, GRAND PRIX, LUMINA, MONTE CARLO AND REGAL

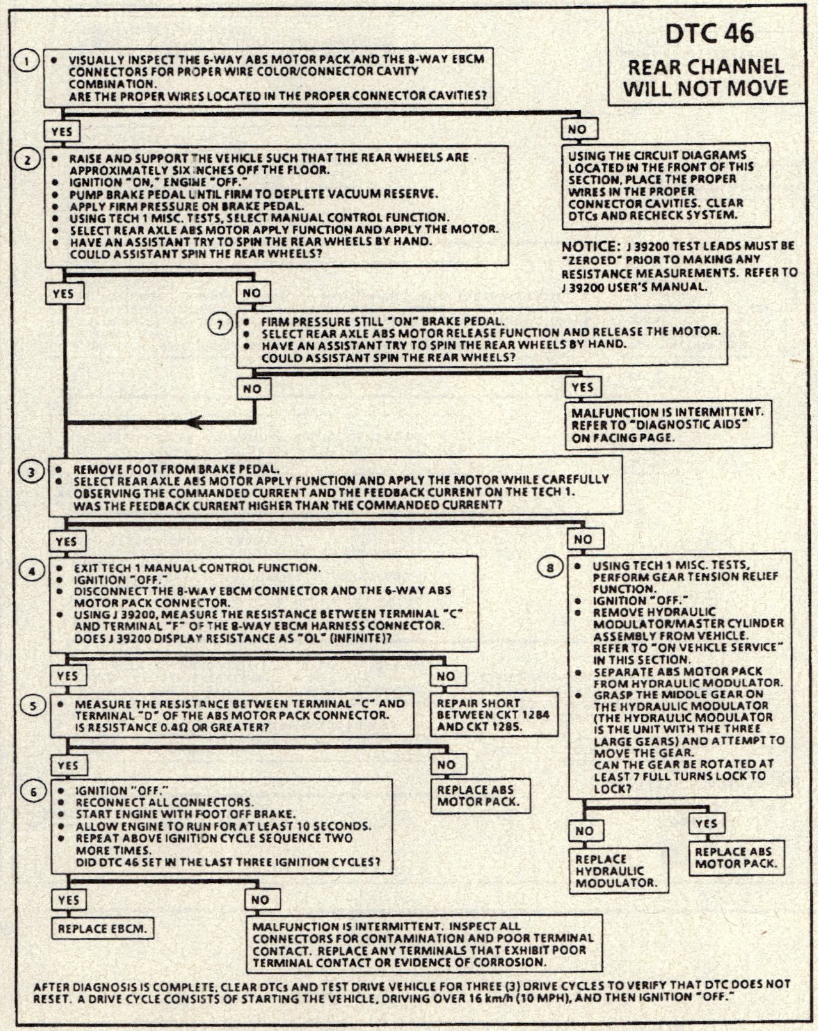

DIAGNOSTIC CHART DTC 47 — 1995 CUTLASS SUPREME, GRAND PRIX, LUMINA, MONTE CARLO AND REGAL

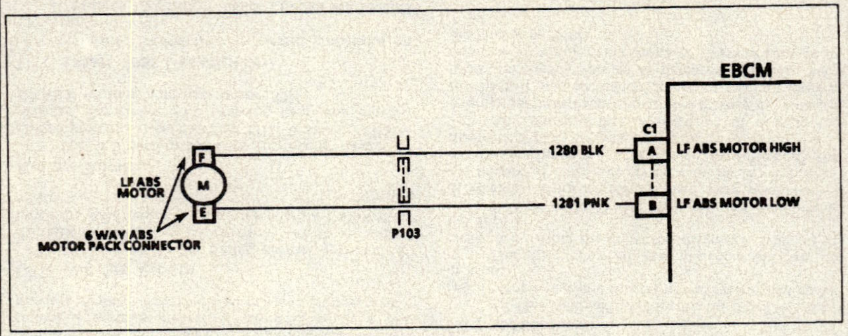

DTC 47

LEFT FRONT ABS MOTOR FREE SPINS

Circuit Description:

This DTC is designed to detect a stripped nut or gear assembly during initialization. During the homing sequence, the piston should reach the top of the bore resulting in a stalled ABS motor. If this does not occur, the ABS motor must be spinning with little or no resistance, indicating a nut/screw or gear malfunction.

DTC Will Set When:

DTC 47 can only be set during initialization. If the feedback current is less than the command current for a specified period of time, a malfunction exists.

Action Taken:

A malfunction DTC is stored, ABS is disabled, and the amber ABS warning lamp is turned "ON."

DTC Chart Test Description:

Number(s) below refer to circled number(s) on the diagnostic chart.

1. Checks to see if the corresponding open motor DTC is also set.
2. Verifies that the ABS motor was actually applied as command by observing feedback current.
3. Verifies that the ABS motor can release.
4. Verifies that the ABS motor can be applied by observing pedal movement.
5. Checks for an open in the ABS motor "HIGH" circuitry.
6. Checks for an open in the ABS motor "LOW" circuitry.
7. Checks for an ABS motor that is internally shorted.

Diagnostic Aids:

An "Intermittent" malfunction in this DTC may result from a mechanical part of the system that sticks, binds, or slips.

The frequency of the malfunction can be checked by using the enhanced diagnostic function of the Tech 1.

If the DTC only occurs once and DTC 56 also occurs, refer to DTC 56. If intermittent and enhanced diagnostics shows this DTC occurs during ABS, refer to DTC 56.

Depending on the frequency of the malfunction, a physical inspection of the mechanical parts suspected may be necessary.

DIAGNOSTIC CHART DTC 47 — 1995 CUTLASS SUPREME, GRAND PRIX, LUMINA, MONTE CARLO AND REGAL

DTC 47
(Page 1 of 2)
LEFT FRONT ABS MOTOR FREE SPINS

(1) IS DTC 56 ALSO CURRENTLY SET?

- YES: PROCEED TO DTC 56 CHART FIRST.
- NO: go to (2).

NOTICE: J 39200 TEST LEADS MUST BE "ZEROED" PRIOR TO MAKING ANY RESISTANCE MEASUREMENTS. REFER TO J 39200 USER'S MANUAL.

(2)
- IGNITION "ON," ENGINE "OFF."
- PUMP BRAKE PEDAL UNTIL FIRM TO DEPLETE VACUUM RESERVE.
- USING TECH 1 MISC. TESTS, SELECT MANUAL CONTROL FUNCTION.
- SELECT LEFT FRONT ABS MOTOR APPLY FUNCTION AND APPLY THE MOTOR WHILE CAREFULLY OBSERVING THE COMMAND CURRENT AND FEEDBACK CURRENT ON THE TECH 1.

WERE BOTH COMMAND AND FEEDBACK CURRENT 10 AMPS?

- YES: go to (3).
- NO: go to (5).

(3)
- APPLY FIRM PRESSURE ON BRAKE PEDAL.
- SELECT LEFT FRONT ABS MOTOR RELEASE FUNCTION AND RELEASE THE MOTOR.

DID THE BRAKE PEDAL FALL AND WERE BOTH COMMAND AND FEEDBACK CURRENT 6 AMPS?

- YES: go to (4).
- NO: GO TO STEP 5.

(4)
- FIRM PRESSURE STILL ON BRAKE PEDAL.
- SELECT LEFT FRONT ABS MOTOR APPLY FUNCTION AND APPLY THE MOTOR (THIS MOVES THE PISTON TO THE TOP OF ITS BORE).

DID BRAKE PEDAL RISE?

- YES: MALFUNCTION IS MOST LIKELY CAUSED BY POOR CONNECTION. REFER TO "DIAGNOSTIC AIDS" ON FACING PAGE.
- NO: go to (5).

(5)
- USING TECH 1 MISC. TESTS, PERFORM GEAR TENSION RELIEF FUNCTION.
- IGNITION "OFF."
- DISCONNECT 6-WAY ABS MOTOR PACK CONNECTOR AND 8-WAY EBCM CONNECTOR.
- USING J 39200, MEASURE THE RESISTANCE BETWEEN TERMINAL "F" OF THE 6-WAY ABS MOTOR PACK HARNESS CONNECTOR AND TERMINAL "A" OF THE 8-WAY EBCM HARNESS CONNECTOR.

IS RESISTANCE 1.5Ω OR LESS?

- YES: go to (6).
- NO: REPAIR HIGH RESISTANCE IN CKT 1280.

(6)
- MEASURE THE RESISTANCE BETWEEN TERMINAL "E" OF THE 6-WAY ABS MOTOR PACK HARNESS CONNECTOR AND TERMINAL "B" OF THE 8-WAY EBCM HARNESS CONNECTOR.

IS RESISTANCE 1.5Ω OR LESS?

- YES: go to (7).
- NO: REPAIR HIGH RESISTANCE IN CKT 1281.

(7)
- MEASURE THE RESISTANCE BETWEEN TERMINAL "E" AND TERMINAL "F" OF THE ABS MOTOR PACK CONNECTOR.

IS RESISTANCE 1.5Ω OR LESS?

- YES: DIAGNOSIS CONTINUED ON NEXT PAGE.
- NO: REPLACE ABS MOTOR PACK.

AFTER DIAGNOSIS IS COMPLETE, CLEAR DTCs AND TEST DRIVE VEHICLE FOR THREE (3) DRIVE CYCLES TO VERIFY THAT DTC DOES NOT RESET. A DRIVE CYCLE CONSISTS OF STARTING THE VEHICLE, DRIVING OVER 16 km/h (10 MPH), AND THEN IGNITION "OFF."

DIAGNOSTIC CHART DTC 47 — 1995 CUTLASS SUPREME, GRAND PRIX, LUMINA, MONTE CARLO AND REGAL

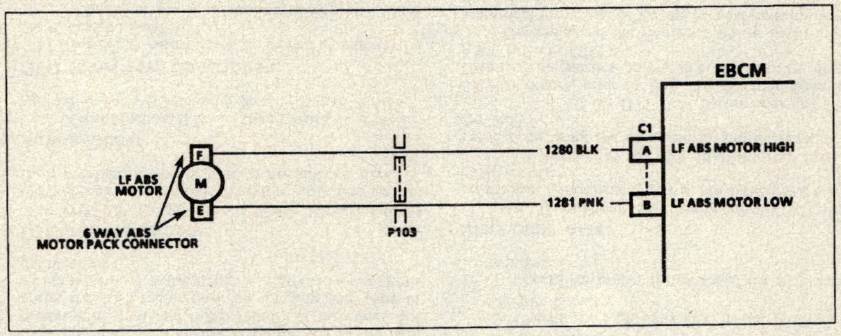

DTC 47

LEFT FRONT ABS MOTOR FREE SPINS

Circuit Description:

This DTC is designed to detect a stripped nut or gear assembly during initialization. During the homing sequence, the piston should reach the top of the bore resulting in a stalled ABS motor. If this does not occur, the ABS motor must be spinning with little or no resistance, indicating a nut/screw or gear malfunction.

DTC Will Set When:

DTC 47 can only be set during initialization. If the feedback current is less than the command current for a specified period of time, a malfunction exists.

Action Taken:

A malfunction DTC is stored, ABS is disabled, and the amber ABS warning lamp is turned "ON."

DTC Chart Test Description:

Number(s) below refer to circled number(s) on the diagnostic chart.

8. Checks for a stripped gear on the ABS motor pack (unit with three small gears).
9. Checks for a stripped gear on the hydraulic modulator (unit with three large gears).
10. Checks to see if the ABS motor pack has an internal malfunction.
11. Checks for a malfunctioning hydraulic modulator.
12. Ensures malfunction was not due to poor terminal contact.

Diagnostic Aids:

An "Intermittent" malfunction in this DTC may result from a mechanical part of the system that sticks, binds, or slips.

The frequency of the malfunction can be checked by using the enhanced diagnostic function of the Tech 1.

If the DTC only occurs once and DTC 56 also occurs, refer to DTC 56. If intermittent and enhanced diagnostics shows this DTC occurs during ABS, refer to DTC 56.

Depending on the frequency of the malfunction, a physical inspection of the mechanical parts suspected may be necessary.

DIAGNOSTIC CHART DTC 47 — 1995 CUTLASS SUPREME, GRAND PRIX, LUMINA, MONTE CARLO AND REGAL

DTC 47
(Page 2 of 2)
LEFT FRONT ABS MOTOR FREE SPINS

DIAGNOSIS CONTINUED FROM PREVIOUS PAGE.

(8)
- REMOVE HYDRAULIC MODULATOR/MASTER CYLINDER ASSEMBLY FROM VEHICLE
- REMOVE GEAR COVER.
- CHECK FOR STRIPPED GEAR ON ABS MOTOR PACK; ABS MOTOR PACK IS THE UNIT WITH THE THREE SMALL GEARS. THE LEFT FRONT GEAR IS THE GEAR FURTHEST FORWARD WHEN THE ASSEMBLY IS INSTALLED IN THE VEHICLE. IS THE ABS MOTOR PACK GEAR STRIPPED?

YES → REPLACE ABS MOTOR PACK.

NO ↓

(9)
- CHECK FOR A STRIPPED GEAR ON HYDRAULIC MODULATOR; MODULATOR IS THE UNIT WITH THE THREE LARGE GEARS. THE LEFT FRONT GEAR IS THE GEAR FURTHEST FORWARD WHEN THE UNIT IS INSTALLED IN THE VEHICLE. IS THE MODULATOR GEAR STRIPPED?

YES → REPLACE HYDRAULIC MODULATOR GEAR.

NO ↓

(10)
- RECONNECT ELECTRICAL CONNECTORS.
- SECURELY POSITION HYDRAULIC MODULATOR/MASTER CYLINDER ASSEMBLY IN VEHICLE WITH GEAR COVER REMOVED SUCH THAT THE GEAR SET CAN BE OBSERVED. BE CAREFUL NOT TO DAMAGE GEAR SET OR ALLOW THE HYDRAULIC MODULATOR TO MOVE WHILE TESTING.
- IGNITION "ON."
- USING TECH 1 SELECT MANUAL CONTROL FUNCTION.
- SELECT LEFT FRONT ABS MOTOR APPLY AND APPLY THE MOTOR WHILE OBSERVING THE GEAR SET FURTHEST FORWARD WHEN THE UNIT IS INSTALLED IN THE VEHICLE.
- SELECT LEFT FRONT ABS MOTOR RELEASE FUNCTION AND RELEASE THE MOTOR WHILE STILL OBSERVING THE GEAR SET.
- SELECT LEFT FRONT ABS MOTOR APPLY FUNCTION AND APPLY THE MOTOR. DID GEAR SET MOVE IN BOTH DIRECTIONS FOR AT LEAST ONE REVOLUTION?

NO → REPLACE ABS MOTOR PACK.

YES ↓

(11)
- USING TECH 1 PERFORM GEAR TENSION RELIEF FUNCTION.
- IGNITION "OFF."
- SEPARATE THE ABS MOTOR PACK FROM THE HYDRAULIC MODULATOR.
- GRASP THE GEAR ON THE HYDRAULIC MODULATOR (THE HYDRAULIC MODULATOR IS THE UNIT WITH THE THREE LARGE GEARS) FURTHEST FORWARD WHEN THE UNIT IS INSTALLED IN THE VEHICLE AND ATTEMPT TO MOVE THE GEAR. CAN THE GEAR BE ROTATED MORE THAN 13 FULL TURNS LOCK TO LOCK?

YES → REPLACE THE HYDRAULIC MODULATOR.

NO ↓

(12)
- INSPECT ALL CONNECTORS AND TERMINALS FOR POOR TERMINAL CONTACT AND EVIDENCE OF CORROSION.
  REPLACE ALL TERMINALS THAT EXHIBIT SIGNS OF POOR TERMINAL CONTACT OR CORROSION.
- REINSTALL HYDRAULIC MODULATOR/MASTER CYLINDER ASSEMBLY IN VEHICLE.
- RECONNECT ALL CONNECTORS.
- IGNITION "OFF."
- START ENGINE WITH FOOT OFF BRAKE.
- ALLOW ENGINE TO RUN FOR AT LEAST 10 SECONDS.
- REPEAT ABOVE IGNITION CYCLE SEQUENCE TWO MORE TIMES. DID DTC 47 SET IN THE LAST THREE IGNITION CYCLES?

YES → REPLACE EBCM.

NO → MALFUNCTION IS MOST LIKELY CAUSED BY A POOR CONNECTION. REFER TO "DIAGNOSTIC AIDS" ON FACING PAGE.

AFTER DIAGNOSIS IS COMPLETE, CLEAR DTCs AND TEST DRIVE VEHICLE FOR THREE (3) DRIVE CYCLES TO VERIFY THAT DTC DOES NOT RESET. A DRIVE CYCLE CONSISTS OF STARTING THE VEHICLE, DRIVING OVER 16 km/h (10 MPH), AND THEN IGNITION "OFF."

DIAGNOSTIC CHART DTC 48 — 1995 CUTLASS SUPREME, GRAND PRIX, LUMINA, MONTE CARLO AND REGAL

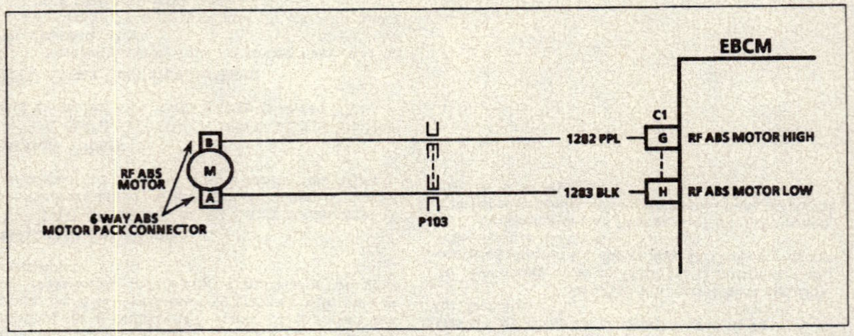

DTC 48

RIGHT FRONT ABS MOTOR FREE SPINS

Circuit Description:

This DTC is designed to detect a stripped nut or gear assembly during initialization. During the homing sequence, the piston should reach the top of the bore resulting in a stalled ABS motor. If this does not occur, the ABS motor must be spinning with little or no resistance, indicating a nut/screw or gear malfunction.

DTC Will Set When:

DTC 48 can only be set during initialization. If the feedback current is less than the command current for a specified period of time, a malfunction exists.

Action Taken:

A malfunction DTC is stored, ABS is disabled, and the amber ABS warning lamp is turned "ON."

DTC Chart Test Description:

Number(s) below refer to circled number(s) on the diagnostic chart.

1. Checks to see if the corresponding open ABS motor DTC is also set.
2. Verifies that the ABS motor was actually applied as commanded by observing feedback current.
3. Verifies that the ABS motor can release.
4. Verifies that the ABS motor can be applied by observing pedal movement.
5. Checks for high resistance in the ABS motor "HIGH" circuitry.
6. Checks for high resistance in the ABS motor "LOW" circuitry.
7. Checks for proper resistance of the ABS motor windings.

Diagnostic Aids:

An "Intermittent" malfunction in this DTC may result from a mechanical part of the system that sticks, binds, or slips.

The frequency of the malfunction can be checked by using the enhanced diagnostic function of the Tech 1.

If the DTC only occurs once and DTC 61 also occurs, refer to DTC 61. If intermittent and enhanced diagnostics shows this DTC occurs during ABS, refer to DTC 61.

Depending on the frequency of the malfunction, a physical inspection of the mechanical parts suspected may be necessary.

DIAGNOSTIC CHART DTC 48 — 1995 CUTLASS SUPREME, GRAND PRIX, LUMINA, MONTE CARLO AND REGAL

DTC 48
(Page 1 of 2)
RIGHT FRONT ABS MOTOR FREE SPINS

(1) IS DTC 61 ALSO CURRENTLY SET?

- YES → PROCEED TO DTC 61 CHART FIRST.
- NO → step 2

NOTICE: J 39200 TEST LEADS MUST BE "ZEROED" PRIOR TO MAKING ANY RESISTANCE MEASUREMENTS.

(2)
- IGNITION "ON," ENGINE "OFF."
- PUMP BRAKE PEDAL UNTIL FIRM TO DEPLETE VACUUM RESERVE.
- USING TECH 1 MISC. TESTS, SELECT MANUAL CONTROL FUNCTION.
- SELECT RIGHT FRONT ABS MOTOR APPLY FUNCTION AND APPLY THE MOTOR WHILE CAREFULLY OBSERVING THE COMMAND CURRENT AND FEEDBACK CURRENT ON THE TECH 1.

WERE BOTH COMMAND AND FEEDBACK CURRENT 10 AMPS?

- YES → step 3
- NO → GO TO STEP 5.

(3)
- APPLY FIRM PRESSURE ON BRAKE PEDAL.
- SELECT RIGHT FRONT ABS MOTOR RELEASE FUNCTION AND RELEASE THE MOTOR.

DID THE BRAKE PEDAL FALL AND WERE BOTH COMMAND AND FEEDBACK CURRENT 6 AMPS?

- YES → step 4
- NO → GO TO STEP 5.

(4)
- FIRM PRESSURE STILL ON BRAKE PEDAL.
- SELECT RIGHT FRONT ABS MOTOR APPLY FUNCTION AND APPLY THE MOTOR (THIS MOVES THE PISTON TO THE TOP OF ITS BORE.)

DID BRAKE PEDAL RISE?

- YES → MALFUNCTION IS INTERMITTENT. REFER TO "DIAGNOSTIC AIDS" ON FACING PAGE.
- NO → step 5

(5)
- USING TECH 1 MISC. TESTS, PERFORM GEAR TENSION RELIEF FUNCTION.
- IGNITION "OFF."
- DISCONNECT 6-WAY ABS MOTOR PACK CONNECTOR AND 8-WAY EBCM CONNECTOR.
- USING J 39200, MEASURE THE RESISTANCE BETWEEN TERMINAL "B" OF THE 6-WAY ABS MOTOR PACK HARNESS CONNECTOR AND TERMINAL "G" OF THE 8-WAY EBCM HARNESS CONNECTOR.

IS RESISTANCE LESS THAN 1.5Ω?

- YES → step 6
- NO → REPAIR HIGH RESISTANCE IN CKT 1282.

(6)
- MEASURE THE RESISTANCE BETWEEN TERMINAL "A" OF THE 6-WAY ABS MOTOR PACK HARNESS CONNECTOR AND TERMINAL "H" OF THE 8-WAY EBCM HARNESS CONNECTOR.

IS RESISTANCE LESS THAN 1.5Ω?

- YES → step 7
- NO → REPAIR HIGH RESISTANCE IN CKT 1283.

(7)
- MEASURE THE RESISTANCE BETWEEN TERMINAL "A" AND TERMINAL "B" OF THE ABS MOTOR PACK CONNECTOR.

IS RESISTANCE LESS THAN 1.5Ω?

- YES → DIAGNOSIS CONTINUED ON NEXT PAGE.
- NO → REPLACE ABS MOTOR PACK.

AFTER DIAGNOSIS IS COMPLETE, CLEAR DTCs AND TEST DRIVE VEHICLE FOR THREE (3) DRIVE CYCLES TO VERIFY THAT DTC DOES NOT RESET. A DRIVE CYCLE CONSISTS OF STARTING THE VEHICLE, DRIVING OVER 16 km/h (10 MPH), AND THEN IGNITION "OFF."

DIAGNOSTIC CHART DTC 48 — 1995 CUTLASS SUPREME, GRAND PRIX, LUMINA, MONTE CARLO AND REGAL

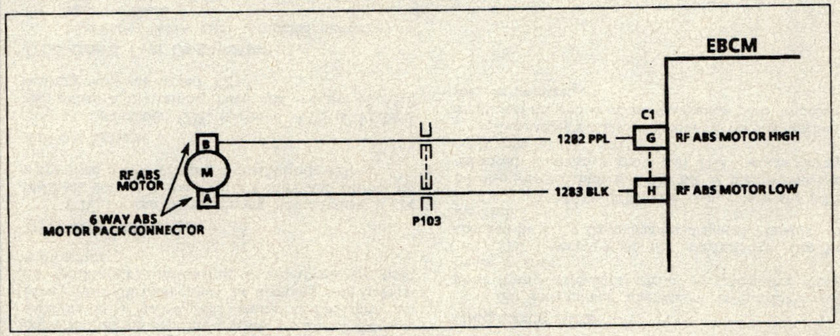

DTC 48

RIGHT FRONT ABS MOTOR FREE SPINS

Circuit Description:

This DTC is designed to detect a stripped nut or gear assembly during initialization. During the homing sequence, the piston should reach the top of the bore resulting in a stalled ABS motor. If this does not occur, the ABS motor must be spinning with little or no resistance, indicating a nut/screw or gear malfunction.

DTC Will Set When:

DTC 48 can only be set during initialization. If the feedback current is less than the command current for a specified period of time, a malfunction exists.

Action Taken:

A malfunction DTC is stored, ABS is disabled, and the amber ABS warning lamp is turned "ON."

DTC Chart Test Description:

Number(s) below refer to circled number(s) on the diagnostic chart.

8. Checks for a stripped gear on the ABS motor pack (unit with three small gears).
9. Checks for a stripped gear on the hydraulic modulator (unit with three large gears).
10. Verifies that the ABS motor can be applied under a load situation.
11. Verifies that the hydraulic modulator is functioning properly.
12. Ensures malfunction was not due to poor terminal contact.

Diagnostic Aids:

An "Intermittent" malfunction in this DTC may result from a mechanical part of the system that sticks, binds, or slips.

The frequency of the malfunction can be checked by using the enhanced diagnostic function of the Tech 1.

If the DTC only occurs once and DTC 61 also occurs, refer to DTC 61. If intermittent and enhanced diagnostics shows this DTC occurs during ABS, refer to DTC 61.

Depending on the frequency of the malfunction, a physical inspection of the mechanical parts suspected may be necessary.

DIAGNOSTIC CHART DTC 48 — 1995 CUTLASS SUPREME, GRAND PRIX, LUMINA, MONTE CARLO AND REGAL

DTC 48
(Page 2 of 2)
RIGHT FRONT ABS MOTOR FREE SPINS

DIAGNOSIS CONTINUED FROM PREVIOUS PAGE.

(8)
- REMOVE HYDRAULIC MODULATOR/MASTER CYLINDER ASSEMBLY FROM VEHICLE. (REFER TO "ON-VEHICLE SERVICE" IN THIS SECTION.)
- REMOVE GEAR COVER.
- CHECK FOR STRIPPED GEAR ON ABS MOTOR PACK; ABS MOTOR PACK IS THE UNIT WITH THE THREE SMALL GEARS. THE RIGHT FRONT GEAR IS THE GEAR FURTHEST REARWARD WHEN THE ASSEMBLY IS INSTALLED IN THE VEHICLE. IS THE ABS MOTOR PACK GEAR STRIPPED?

YES: REPLACE ABS MOTOR PACK.

NO:

(9)
- CHECK FOR A STRIPPED GEAR ON HYDRAULIC MODULATOR; MODULATOR IS THE UNIT WITH THE THREE LARGE GEARS. THE RIGHT FRONT GEAR IS THE GEAR FURTHEST REARWARD WHEN THE UNIT IS INSTALLED IN THE VEHICLE. IS THE MODULATOR GEAR STRIPPED?

YES: REPLACE HYDRAULIC MODULATOR GEAR.

NO:

(10)
- RECONNECT ELECTRICAL CONNECTORS.
- SECURELY POSITION HYDRAULIC MODULATOR/MASTER CYLINDER ASSEMBLY IN VEHICLE WITH GEAR COVER REMOVED SUCH THAT THE GEAR SET CAN BE OBSERVED. BE CAREFUL NOT TO DAMAGE GEAR SET OR ALLOW THE HYDRAULIC MODULATOR TO MOVE WHILE TESTING.
- IGNITION "ON."
- USING TECH 1, SELECT MANUAL CONTROL FUNCTION.
- SELECT RIGHT FRONT ABS MOTOR APPLY AND APPLY THE MOTOR WHILE OBSERVING THE GEAR SET FURTHEST REARWARD WHEN THE UNIT IS INSTALLED IN THE VEHICLE.
- SELECT RIGHT FRONT ABS MOTOR RELEASE FUNCTION AND RELEASE THE MOTOR WHILE STILL OBSERVING THE GEAR SET.
- SELECT RIGHT FRONT ABS MOTOR APPLY FUNCTION AND APPLY THE MOTOR. DID GEAR SET MOVE IN BOTH DIRECTIONS FOR AT LEAST ONE REVOLUTION?

NO: REPLACE ABS MOTOR PACK.

YES:

(11)
- USING TECH 1, PERFORM GEAR TENSION RELIEF FUNCTION.
- IGNITION "OFF."
- SEPARATE THE ABS MOTOR PACK FROM THE HYDRAULIC MODULATOR.
- GRASP THE GEAR ON THE HYDRAULIC MODULATOR (THE HYDRAULIC MODULATOR IS THE UNIT WITH THE THREE LARGE GEARS) FURTHEST REARWARD WHEN THE UNIT IS INSTALLED IN THE VEHICLE AND ATTEMPT TO MOVE THE GEAR. CAN THE GEAR BE ROTATED MORE THAN 13 FULL TURNS LOCK TO LOCK?

YES: REPLACE THE HYDRAULIC MODULATOR.

NO:

(12)
- INSPECT ALL CONNECTORS AND TERMINALS FOR POOR TERMINAL CONTACT AND EVIDENCE OF CORROSION.
  REPLACE ALL TERMINALS THAT EXHIBIT SIGNS OF POOR TERMINAL CONTACT OR CORROSION.
- REINSTALL HYDRAULIC MODULATOR/MASTER CYLINDER ASSEMBLY IN VEHICLE.
- RECONNECT ALL CONNECTORS.
- IGNITION "OFF."
- START ENGINE WITH FOOT OFF BRAKE.
- ALLOW ENGINE TO RUN FOR AT LEAST 10 SECONDS.
- REPEAT ABOVE IGNITION CYCLE SEQUENCE TWO MORE TIMES. DID DTC 48 SET IN THE LAST THREE IGNITION CYCLES?

YES: REPLACE EBCM.

NO: MALFUNCTION WAS MOST LIKELY CAUSED BY A POOR CONNECTION. REFER TO "DIAGNOSTIC AIDS" ON FACING PAGE.

AFTER DIAGNOSIS IS COMPLETE, CLEAR DTCs AND TEST DRIVE VEHICLE FOR THREE (3) DRIVE CYCLES TO VERIFY THAT DTC DOES NOT RESET. A DRIVE CYCLE CONSISTS OF STARTING THE VEHICLE, DRIVING OVER 16 km/h (10 MPH), AND THEN IGNITION "OFF."

DIAGNOSTIC CHART DTC 51 — 1995 CUTLASS SUPREME, GRAND PRIX, LUMINA, MONTE CARLO AND REGAL

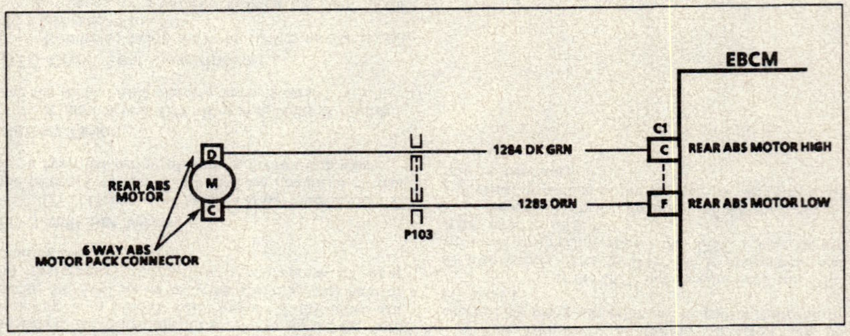

DTC 51

REAR ABS MOTOR FREE SPINS

Circuit Description:

This DTC is designed to detect a stripped nut or gear assembly during initialization. During the homing sequence, the piston should reach the top of the bore resulting in a stalled ABS motor. If this does not occur, the ABS motor must be spinning with little or no resistance, indicating a nut/screw or gear malfunction.

DTC Will Set When:

DTC 51 can be set during initialization. If the feedback current is less than the command current for a specified period of time, a malfunction exists.

Action Taken:

A malfunction DTC is stored, ABS is disabled, the amber ABS warning lamp and the red BRAKE warning lamp are turned "ON."

DTC Chart Test Description:

Number(s) below refer to circled number(s) on the diagnostic chart.

1. Checks to see if the corresponding open ABS motor DTC is also set.
2. Verifies that the ABS motor was actually applied as commanded by observing feedback current.
3. Verifies that the ABS motor can release.
4. Verifies that the ABS motor can be applied by observing pedal movement.
5. Checks for high resistance in the ABS motor "HIGH" circuitry.
6. Checks for high resistance in the ABS motor "LOW" circuitry.
7. Checks for proper resistance of the ABS motor windings.

Diagnostic Aids:

An "Intermittent" malfunction in this DTC may result from a mechanical part of the system that sticks, binds, or slips.

The frequency of the malfunction can be checked by using the enhanced diagnostic function of the Tech 1.

If the DTC only occurs once and DTC 64 also occurs, refer to DTC 64. If intermittent and enhanced diagnostics show this DTC occurs during ABS, refer to DTC 64.

Depending on the frequency of the malfunction, a physical inspection of the mechanical parts suspected may be necessary.

DIAGNOSTIC CHART DTC 51 — 1995 CUTLASS SUPREME, GRAND PRIX, LUMINA, MONTE CARLO AND REGAL

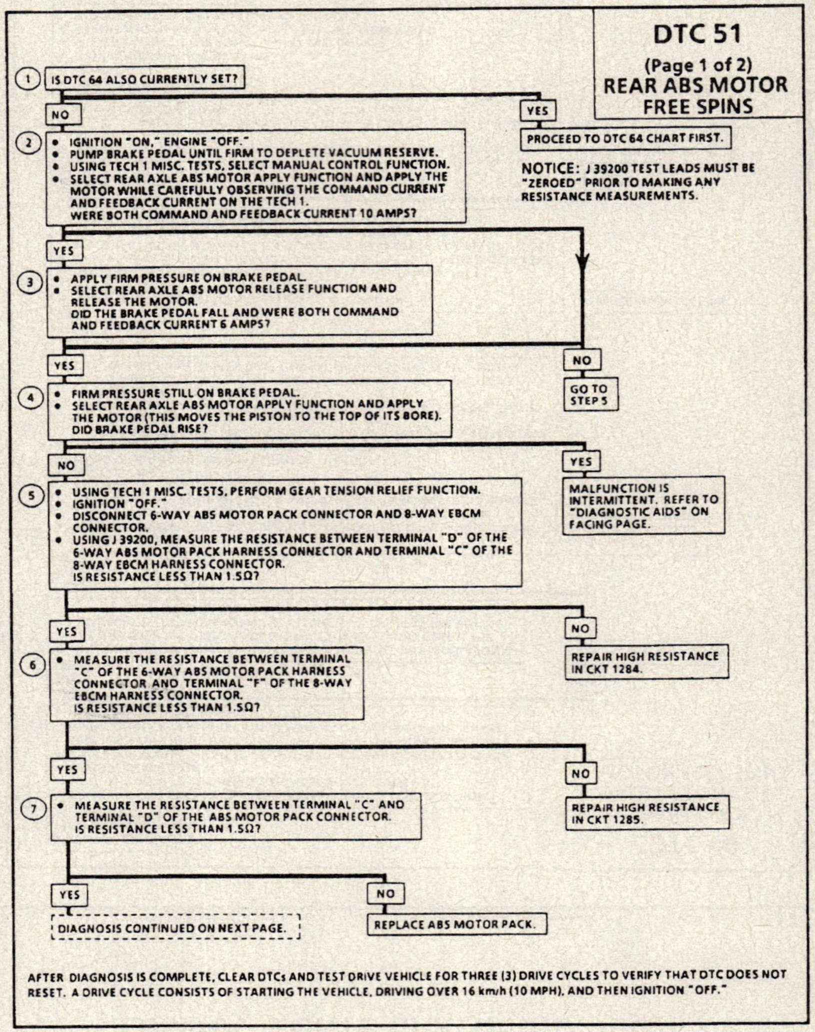

DIAGNOSTIC CHART DTC 51 — 1995 CUTLASS SUPREME, GRAND PRIX, LUMINA, MONTE CARLO AND REGAL

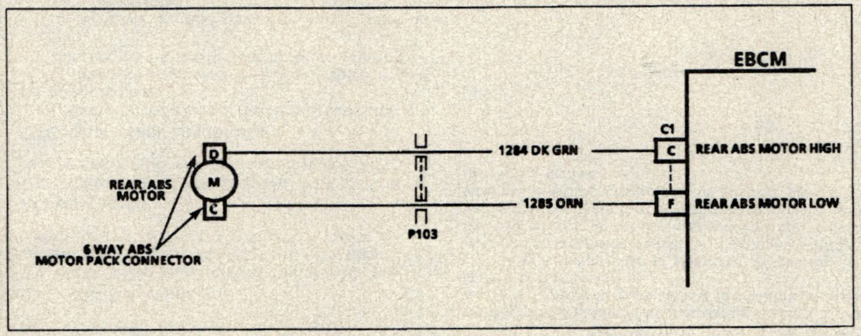

DTC 51

REAR ABS MOTOR FREE SPINS

Circuit Description:

This DTC is designed to detect a stripped nut or gear assembly during initialization. During the homing sequence, the piston should reach the top of the bore resulting in a stalled ABS motor. If this does not occur, the ABS motor must be spinning with little or no resistance, indicating a nut/screw or gear malfunction.

DTC Will Set When:

DTC 51 can be set during initialization. If the feedback current is less than the command current for a specified period of time, a malfunction exists.

Action Taken:

A malfunction DTC is stored, ABS is disabled, the amber ABS warning lamp and the red BRAKE warning lamp are turned "ON."

DTC Chart Test Description:

Number(s) below refer to circled number(s) on the diagnostic chart.

8. Checks for a stripped gear on the ABS motor pack (unit with three small gears).
9. Checks for a stripped gear on the hydraulic modulator (unit with three large gears).
10. Verifies that the ABS motor can be applied under a load situation.
11. Verifies that the hydraulic modulator is functioning properly.
12. Ensures malfunction was not due to poor terminal contact.

Diagnostic Aids:

An "Intermittent" malfunction in this DTC may result from a mechanical part of the system that sticks, binds, or slips.

The frequency of the malfunction can be checked by using the enhanced diagnostic function of the Tech 1.

If the DTC only occurs once and DTC 64 also occurs, refer to DTC 64. If intermittent and enhanced diagnostics show this DTC occurs during ABS, refer to DTC 64.

Depending on the frequency of the malfunction, a physical inspection of the mechanical parts suspected may be necessary.

DIAGNOSTIC CHART DTC 51 — 1995 CUTLASS SUPREME, GRAND PRIX, LUMINA, MONTE CARLO AND REGAL

DTC 51
(Page 2 of 2)
REAR ABS MOTOR FREE SPINS

DIAGNOSIS CONTINUED FROM PREVIOUS PAGE.

(8)
- REMOVE HYDRAULIC MODULATOR/MASTER CYLINDER ASSEMBLY FROM VEHICLE
- REMOVE GEAR COVER.
- CHECK FOR STRIPPED GEAR ON ABS MOTOR PACK; ABS MOTOR PACK IS THE UNIT WITH THE THREE SMALL GEARS. THE REAR AXLE GEAR IS THE MIDDLE GEAR OF THE THREE.

IS THE ABS MOTOR PACK GEAR STRIPPED?

YES: REPLACE ABS MOTOR PACK.

NO:

(9)
- CHECK FOR A STRIPPED GEAR ON HYDRAULIC MODULATOR; MODULATOR IS THE UNIT WITH THE THREE LARGE GEARS. THE REAR AXLE GEAR IS THE MIDDLE GEAR OF THE THREE.

IS THE MODULATOR GEAR STRIPPED?

YES: REPLACE HYDRAULIC MODULATOR GEAR.

NO:

(10)
- RECONNECT ELECTRICAL CONNECTORS.
- SECURELY POSITION HYDRAULIC MODULATOR/MASTER CYLINDER ASSEMBLY IN VEHICLE WITH GEAR COVER REMOVED SUCH THAT THE GEAR SET CAN BE OBSERVED. BE CAREFUL NOT TO DAMAGE GEAR SET OR ALLOW THE HYDRAULIC MODULATOR TO MOVE WHILE TESTING.
- IGNITION "ON."
- USING TECH 1, SELECT MANUAL CONTROL FUNCTION.
- SELECT REAR AXLE ABS MOTOR APPLY AND APPLY THE MOTOR WHILE OBSERVING THE MIDDLE GEAR SET.
- SELECT REAR AXLE ABS MOTOR RELEASE FUNCTION AND RELEASE THE MOTOR WHILE STILL OBSERVING THE GEAR SET.
- SELECT REAR AXLE ABS MOTOR APPLY FUNCTION AND APPLY THE MOTOR.

DID GEAR SET MOVE IN BOTH DIRECTIONS FOR AT LEAST ONE REVOLUTION?

NO: REPLACE ABS MOTOR PACK.

YES:

(11)
- USING TECH 1, PERFORM GEAR TENSION RELIEF FUNCTION.
- IGNITION "OFF."
- SEPARATE THE ABS MOTOR PACK FROM THE HYDRAULIC MODULATOR.
- GRASP THE MIDDLE GEAR ON THE HYDRAULIC MODULATOR (THE HYDRAULIC MODULATOR IS THE UNIT WITH THE THREE LARGE GEARS) AND ATTEMPT TO MOVE THE GEAR.

CAN THE GEAR BE ROTATED MORE THAN 8 FULL TURNS LOCK TO LOCK?

YES: REPLACE THE HYDRAULIC MODULATOR.

NO:

(12)
- INSPECT ALL CONNECTORS AND TERMINALS FOR POOR TERMINAL CONTACT AND EVIDENCE OF CORROSION.
  REPLACE ALL TERMINALS THAT EXHIBIT SIGNS OF POOR TERMINAL CONTACT OR CORROSION.
- REINSTALL HYDRAULIC MODULATOR/MASTER CYLINDER ASSEMBLY IN VEHICLE.
- RECONNECT ALL CONNECTORS.
- IGNITION "OFF."
- START ENGINE WITH FOOT OFF BRAKE.
- ALLOW ENGINE TO RUN FOR AT LEAST 10 SECONDS.
- REPEAT ABOVE IGNITION CYCLE SEQUENCE TWO MORE TIMES.

DID DTC 51 SET IN THE LAST THREE IGNITION CYCLES?

YES: REPLACE EBCM.

NO: MALFUNCTION WAS MOST LIKELY CAUSED BY A POOR CONNECTION. REFER TO "DIAGNOSTIC AIDS" ON FACING PAGE.

AFTER DIAGNOSIS IS COMPLETE, CLEAR DTCs AND TEST DRIVE VEHICLE FOR THREE (3) DRIVE CYCLES TO VERIFY THAT DTC DOES NOT RESET. A DRIVE CYCLE CONSISTS OF STARTING THE VEHICLE, DRIVING OVER 16 km/h (10 MPH), AND THEN IGNITION "OFF."

DIAGNOSTIC CHART DTC 52 — 1995 CUTLASS SUPREME, GRAND PRIX, LUMINA, MONTE CARLO AND REGAL

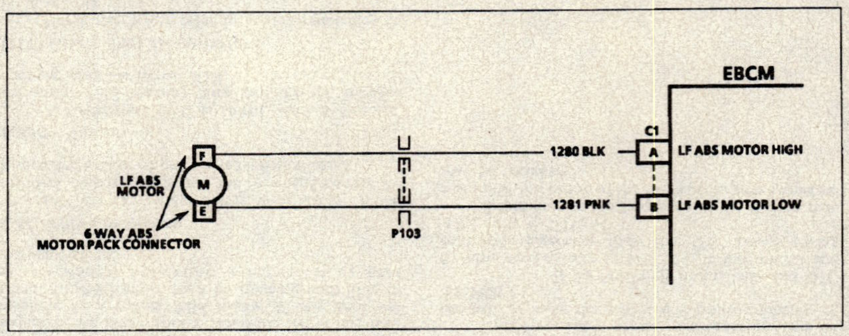

DTC 52
LEFT FRONT CHANNEL IN RELEASE TOO LONG

Circuit Description:

This DTC will diagnose an ABS motor that is energized longer than expected. This could occur if a wheel speed sensor is malfunctioning, the ABS motor does not turn, the left front solenoid mechanically fails open, or the ABS motor wires are crossed.

DTC Will Set When:

DTC 52 can be set only during an ABS stop. If the EBCM commands the left front ABS channel in release for three seconds, a malfunction exists.

Action Taken:

A malfunction DTC is stored, ABS is disabled and the amber ABS warning lamp is turned "ON."

DTC Chart Test Description:

Number(s) below refer to circled number(s) on the diagnostic chart.

1. Identifies a problem in a wheel speed sensor that may cause the system to be in release too long.
2. Identifies an ABS motor as being failed or wired incorrectly.
3. Checks for a solenoid that may have mechanically failed open.
4. Verifies proper solenoid operation.
5. This serves to isolate the cause of the hydraulic problem, to either the solenoid or the ABS hydraulic modulator assembly.
6. Determines whether a malfunctioning ABS motor pack or hydraulic modulator is the reason for DTC 52 being set.

Diagnostic Aids:

An "Intermittent" malfunction may be caused by a mechanical part of the system that sticks or binds.

The frequency of the malfunction can be checked by using the enhanced diagnostic function of the Tech 1.

DTC 52 may occur if on ice and steering wheel is turned to lock during braking. Using the Tech 1, perform hydraulic test to ensure total brake system is functional.

Any circuitry that is suspected of causing the intermittent complaint should be thoroughly checked for backed out terminals, improper mating, broken locks, improperly formed or damaged terminals, poor terminal to wiring connections or physical damage to the wiring harness.

DIAGNOSTIC CHART DTC 52 — 1995 CUTLASS SUPREME, GRAND PRIX, LUMINA, MONTE CARLO AND REGAL

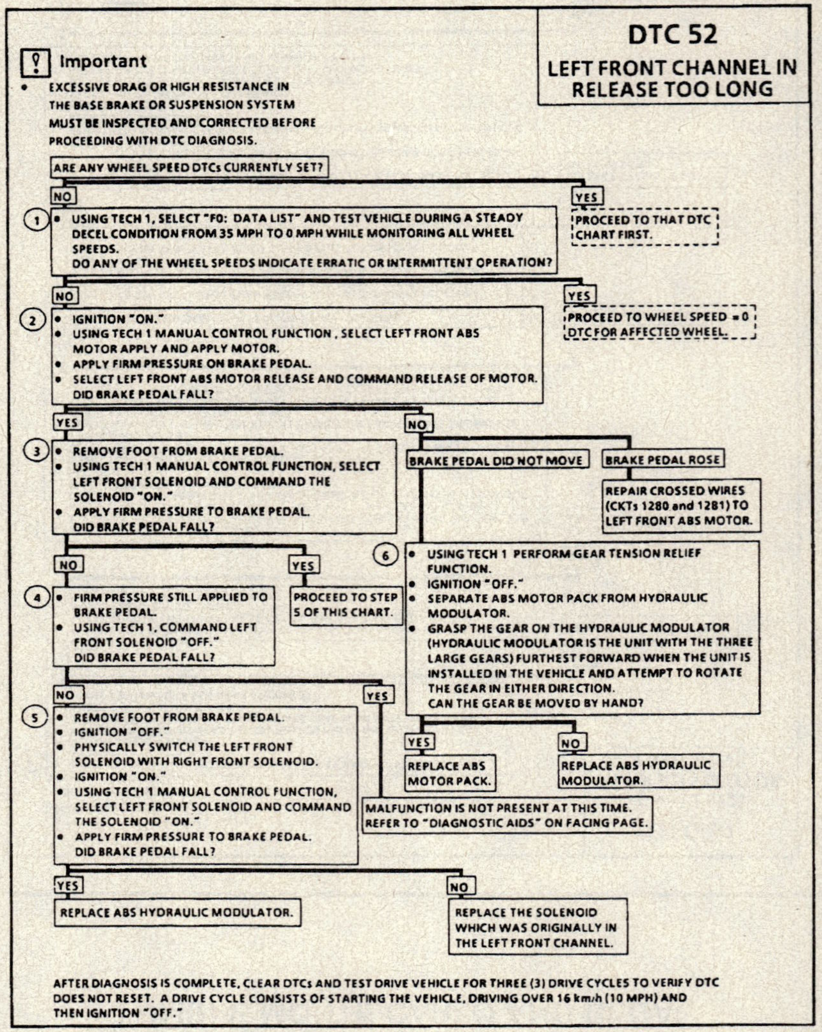

DIAGNOSTIC CHART DTC 53 — 1995 CUTLASS SUPREME, GRAND PRIX, LUMINA, MONTE CARLO AND REGAL

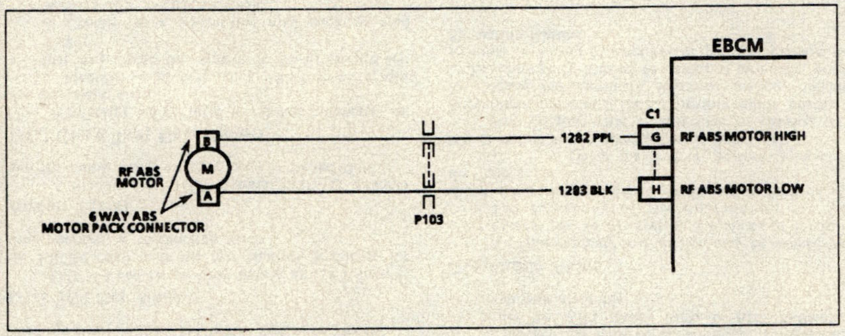

DTC 53

RIGHT FRONT CHANNEL IN RELEASE TOO LONG

Circuit Description:

This DTC will diagnose an ABS motor that is energized longer than expected. This could occur if a wheel speed sensor is malfunctioning, the ABS motor does not turn, the right front solenoid mechanically fails open, or the ABS motor wires are crossed.

DTC Will Set When:

DTC 53 can be set only during an ABS stop. If the EBCM commands the right front channel in release for three seconds, a malfunction exists.

Action Taken:

A malfunction DTC is stored, ABS is disabled and the amber ABS warning lamp is turned "ON."

DTC Chart Test Description:

Number(s) below refer to circled number(s) on the diagnostic chart.

1. Identifies a problem in a wheel speed sensor that may cause the system to be in release too long.
2. Identifies an ABS motor as being failed or wired incorrectly.
3. Checks for a solenoid that may have mechanically failed open.
4. Checks for a solenoid that may have mechanically failed closed.
5. This serves to isolate the cause of the hydraulic problem, to either the solenoid or the ABS hydraulic modulator assembly.
6. Determines whether a malfunction ABS motor pack or hydraulic modulator is the reason for DTC 53 being set.

Diagnostic Aids:

An "Intermittent" malfunction may be caused by a mechanical part of the system that sticks or binds.

The frequency of the malfunction can be checked by using the enhanced diagnostic function of the Tech 1.

DTC 53 may occur on ice if steering wheel is turned to lock during braking. Using the Tech 1 perform hydraulic test to ensure total brake system is functional.

Any circuitry that is suspected of causing the intermittent complaint should be thoroughly checked for backed out terminals, improper mating, broken locks, improperly formed or damaged terminals, poor terminal to wiring connections or physical damage to the wiring harness.

DIAGNOSTIC CHART DTC 53 — 1995 CUTLASS SUPREME, GRAND PRIX, LUMINA, MONTE CARLO AND REGAL

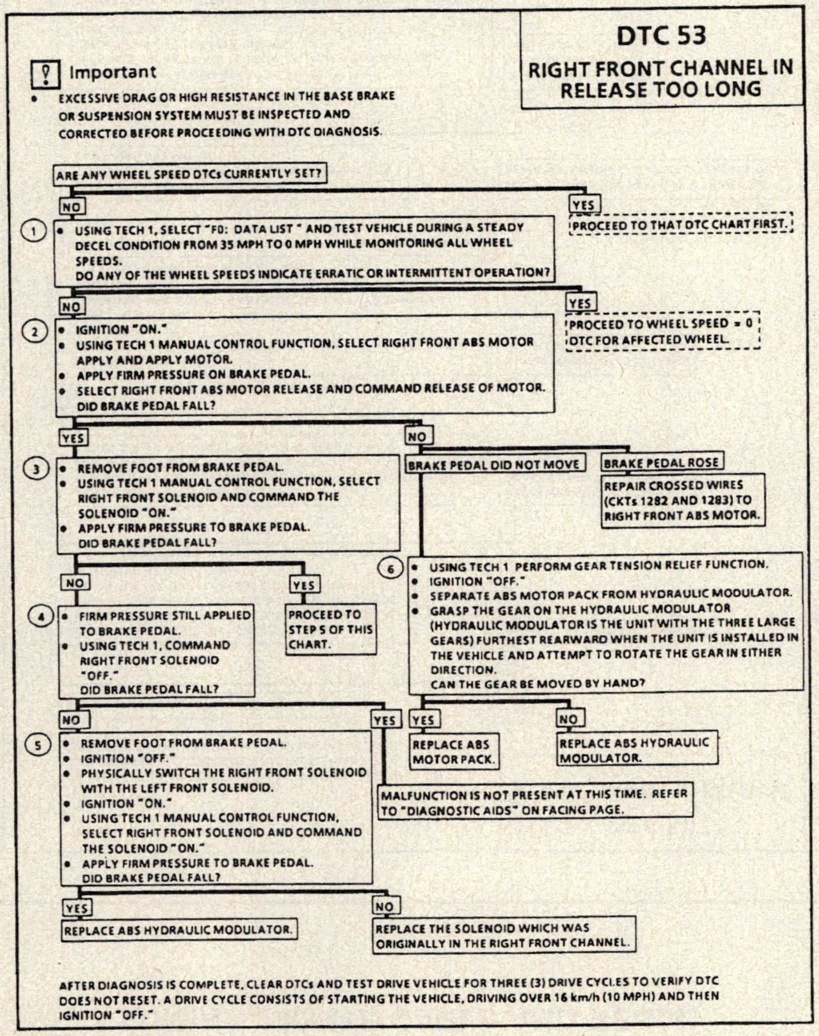

DIAGNOSTIC CHART DTC 54 — 1995 CUTLASS SUPREME, GRAND PRIX, LUMINA, MONTE CARLO AND REGAL

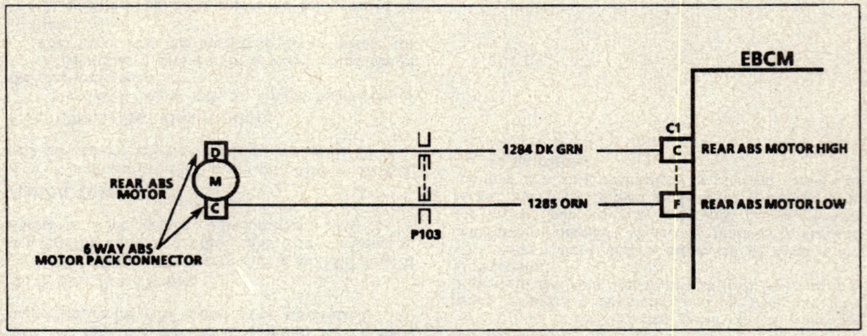

DTC 54
REAR CHANNEL IN RELEASE TOO LONG

Circuit Description:

This DTC will diagnose an ABS motor that is energized longer than expected. This could occur if a wheel speed sensor is malfunctioning, the ABS motor does not turn, or the ABS motor wires are crossed.

DTC Will Set When:

DTC 54 can be set only during an ABS stop. If the EBCM commands the rear channel in release for three seconds, a malfunction exists.

Action Taken:

A malfunction DTC is stored, ABS is disabled and the amber ABS warning lamp is turned "ON."

DTC Chart Test Description:

Number(s) below refer to circled number(s) on the diagnostic chart.

1. Identifies a problem in a wheel speed sensor that may cause the system to be in release too long.
2. Checks for a wheel that may stick or bind because of a mechanical fault.
3. Checks to see if the ABS motor is capable of moving and applying the hydraulic piston for the rear wheels.
4. Ensures that the ABS motor wiring is not crossed.
5. Isolates the fault of a "no-apply" situation to either the ABS motor pack or ABS hydraulic modulator assembly.

Diagnostic Aids:

An "Intermittent" malfunction may be caused by a mechanical part of the system that sticks or binds.

The frequency of the malfunction can be checked by using the enhanced diagnostic function of the Tech 1.

Using the Tech 1, perform hydraulic test to ensure total brake system is functional.

Any circuitry that is suspected of causing the intermittent complaint should be thoroughly checked for backed out terminals, improper mating, broken locks, improperly formed or damaged terminals, poor terminal to wiring connections or physical damage to the wiring harness.

DIAGNOSTIC CHART DTC 54 — 1995 CUTLASS SUPREME, GRAND PRIX, LUMINA, MONTE CARLO AND REGAL

DTC 54
REAR CHANNEL IN RELEASE TOO LONG

Important

- EXCESSIVE DRAG OR HIGH RESISTANCE IN THE BASE BRAKE OR SUSPENSION SYSTEM MUST BE INSPECTED AND CORRECTED BEFORE PROCEEDING WITH DTC DIAGNOSIS.

ARE ANY WHEEL SPEED DTCs CURRENTLY SET?

- YES: PROCEED TO THAT DTC CHART FIRST.
- NO: go to 1.

(1)
- USING TECH 1, SELECT "F0: DATA LIST" AND TEST VEHICLE DURING A STEADY DECEL CONDITION FROM 35 MPH TO 0 MPH WHILE MONITORING ALL WHEEL SPEEDS.

DO ANY OF THE WHEEL SPEEDS INDICATE ERRATIC OR INTERMITTENT OPERATION?

- YES: PROCEED TO WHEEL SPEED = 0 DTC FOR AFFECTED WHEEL.
- NO: go to 2.

(2)
- RAISE REAR OF VEHICLE AND SUPPORT
- SPIN BOTH REAR WHEELS AND CONFIRM PHYSICAL CONDITION OF BASE BRAKE SYSTEM.

COULD BOTH REAR WHEELS BE SPUN FREELY BY HAND?

- NO: REPAIR SOURCE OF RESISTANCE IN BASE BRAKE SYSTEM.
- YES: go to 3.

(3)
- IGNITION "ON."
- USING MANUAL CONTROL FUNCTION OF TECH 1, SELECT REAR AXLE ABS MOTOR APPLY.
- APPLY FIRM PRESSURE ON BRAKE PEDAL.
- COMMAND APPLY OF REAR ABS MOTOR AND TRY TO SPIN REAR WHEELS BY HAND.

COULD REAR WHEELS BE SPUN BY HAND?

- NO: MALFUNCTION IS NOT PRESENT AT THIS TIME. REFER TO "DIAGNOSTIC AIDS" ON FACING PAGE.
- YES: go to 4.

(4)
- WITH FIRM PRESSURE ON BRAKE PEDAL STILL APPLIED, COMMAND A RELEASE OF THE REAR AXLE ABS MOTOR.

CAN WHEELS BE MOVED?

- NO: REPAIR CROSSED WIRES AT REAR ABS MOTOR (CKTs 1284 AND 1285).
- YES: go to 5.

(5)
- USING TECH 1, PERFORM GEAR TENSION RELIEF FUNCTION.
- SEPARATE ABS MOTOR PACK FROM HYDRAULIC MODULATOR.
- CONNECT ABS MOTOR PACK WIRING.
- IGNITION "ON."
- USING TECH 1, PERFORM MOTOR TEST.

DO ALL THREE ABS MOTOR PACK GEARS (SMALL GEARS) SPIN FREELY?

- NO: REPLACE ABS MOTOR PACK.
- YES: REPLACE ABS HYDRAULIC MODULATOR.

AFTER ALL DIAGNOSIS IS COMPLETE, CLEAR DTCs AND TEST DRIVE VEHICLE FOR THREE (3) DRIVE CYCLES TO VERIFY THAT DTC DOES NOT RESET. A DRIVE CYCLE CONSISTS OF STARTING THE VEHICLE DRIVING OVER 16 KM/H (10 MPH), AND THEN IGNITION "OFF."

DIAGNOSTIC CHART DTC 55 — 1995 CUTLASS SUPREME, GRAND PRIX, LUMINA, MONTE CARLO AND REGAL

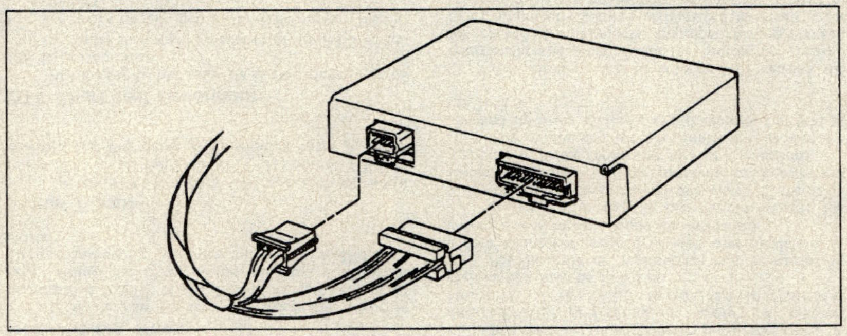

DTC 55
EBCM MALFUNCTION

Circuit Description:
This DTC identifies an internal malfunction detected by the EBCM ABS motor driver interface, (MDI) custom integrated circuit (IC).It also ensures the cause of the malfunction is not a result of a problem with the ABS enable relay under a DTC 55 malfunction.

Action Taken:
A malfunction DTC is stored, ABS is disabled and the amber ABS warning lamp is turned "ON." The red BRAKE warning lamp is turned "ON" if the rear piston in the ABS motor pack is not in the home position.

DTC Chart Test Description:
Number(s) below refer to circled number(s) on the diagnostic chart.
1. Checks to see if the malfunction is still present.
2. Checks to see if the malfunction is intermittent.

Diagnostic Aids:
An "Intermittent" malfunction may be caused by a poor connection, rubbed through wire insulation, or a wire that is broken inside the insulation.
The frequency of the malfunction can be checked by using the enhanced diagnostic function of the Tech 1.

Any circuitry that is suspected of causing the intermittent complaint should be thoroughly checked for backed out terminals, improper mating, broken locks, improperly formed or damaged terminals, poor terminal to wiring connections or physical damage to the wiring harness.

DIAGNOSTIC CHART DTC 55 — 1995 CUTLASS SUPREME, GRAND PRIX, LUMINA, MONTE CARLO AND REGAL

DTC 55
EBCM MALFUNCTION

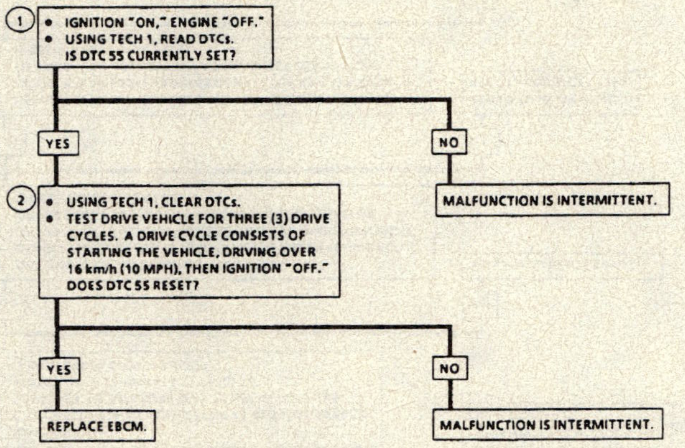

DIAGNOSTIC CHART DTC 56 — 1995 CUTLASS SUPREME, GRAND PRIX, LUMINA, MONTE CARLO AND REGAL

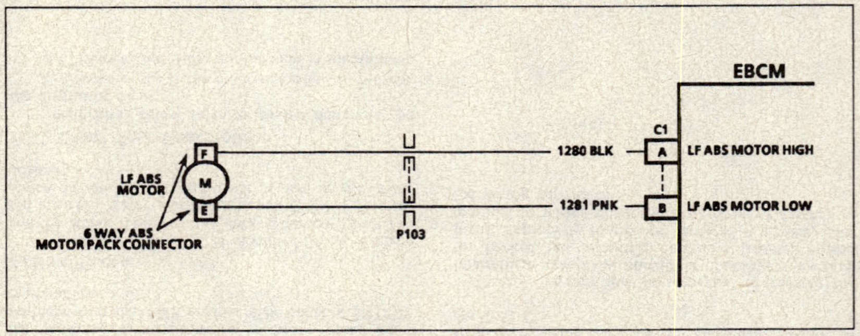

DTC 56
LEFT FRONT ABS MOTOR CIRCUIT OPEN

Circuit Description:

This DTC identifies a motor that cannot be energized due to an open in its circuitry.

DTC Will Set When:

DTC 56 can be set only when the motor is commanded "OFF." If the EBCM detects an out of range voltage on either of the left front ABS motor circuits indicating an open circuit, a malfunction exists.

Action Taken:

An open ABS motor will not activate when requested. A malfunction DTC is stored, ABS is disabled and the amber ABS warning lamp is turned "ON."

DTC Chart Test Description:

Number(s) below refer to circled number(s) on the diagnostic chart.

1. Checks for proper resistance of the ABS motor.
2. Checks for an open in the ABS motor "HIGH" circuitry.
3. Checks for an open in the ABS motor "LOW" circuitry.
4. Ensures malfunction was not due to poor terminal contact.

Diagnostic Aids:

Using Tech 1, select manual control function, and exercise ABS motor movement of affected channel in both directions while applying light pressure on the brake pedal.

If erratic or "jumpy" brake pedal movement is detected while performing an "apply" or "release" function of the ABS motor, an intermittent malfunction may be indicated.

An "Intermittent" malfunction may be caused by a poor connection, rubbed through wire insulation, or a wire that is broken inside the insulation.

If the malfunction is not current, wiggle the wires of the affected channel and check if the DTC resets. This will help to pinpoint an intermittent malfunction in the motor circuitry or connections.

The frequency of the malfunction can be checked by using the enhanced diagnostic function of the Tech 1.

Any circuitry that is suspected of causing the intermittent complaint should be thoroughly checked for backed out terminals, improper mating, broken locks, improperly formed or damaged terminals, poor terminal to wiring connections or physical damage to the wiring harness.

DIAGNOSTIC CHART DTC 56 — 1995 CUTLASS SUPREME, GRAND PRIX, LUMINA, MONTE CARLO AND REGAL

DTC 56
LEFT FRONT ABS MOTOR CIRCUIT OPEN

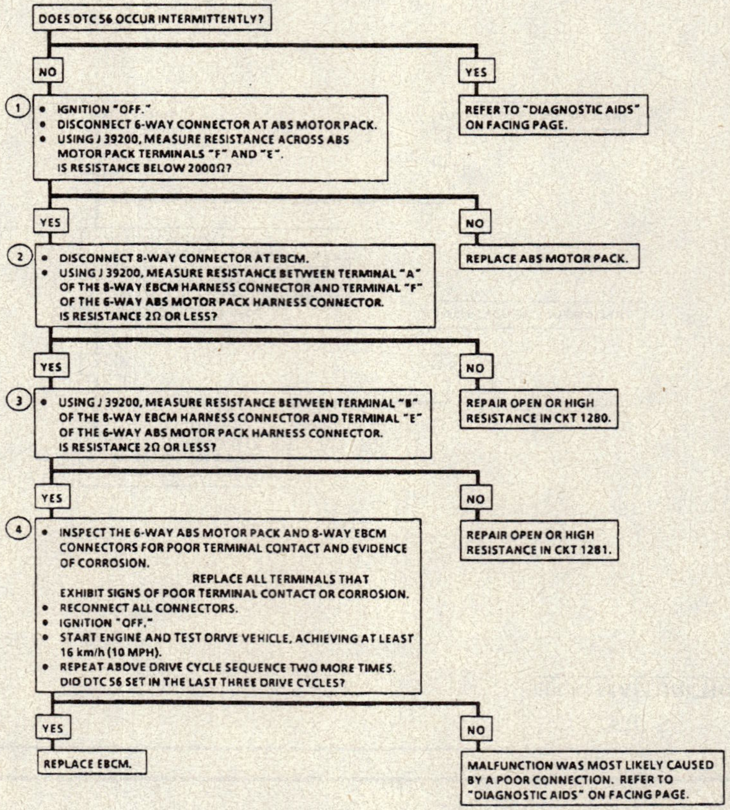

AFTER ALL DIAGNOSIS IS COMPLETE, CLEAR DTCs AND TEST DRIVE VEHICLE FOR THREE (3) DRIVE CYCLES TO VERIFY THAT DTC DOES NOT RESET. A DRIVE CYCLE CONSISTS OF STARTING THE VEHICLE, DRIVING OVER 16 km/h (10 MPH), AND THEN IGNITION "OFF."

DIAGNOSTIC CHART DTC 57 — 1995 CUTLASS SUPREME, GRAND PRIX, LUMINA, MONTE CARLO AND REGAL

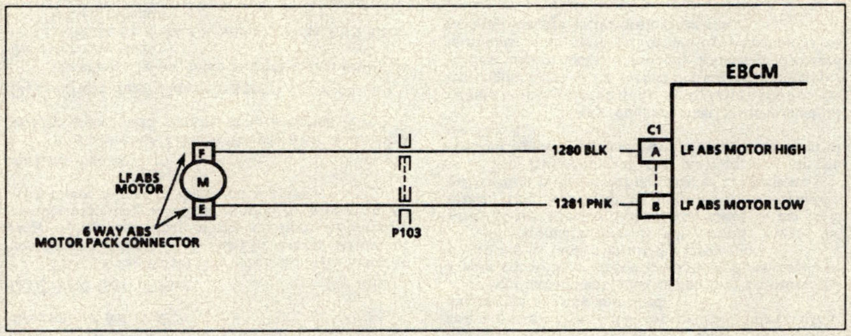

DTC 57
LEFT FRONT ABS MOTOR CIRCUIT SHORTED TO GROUND

Circuit Description:

This DTC identifies an ABS motor circuit that is shorted to ground. This malfunction will not allow the ABS motor to be controlled at the commanded current rate or will cause the driver circuit to allow current directly to ground.

DTC Will Set When:

DTC 57 can be set anytime. If the EBCM detects an out of range voltage on either of the left front ABS motor circuits indicating a circuit shorted to ground, a malfunction exists.

Action Taken:

A malfunction DTC is stored, ABS is disabled and the amber ABS warning lamp is turned "ON."

DTC Chart Test Description:

Number(s) below refer to circled number(s) on the diagnostic chart.

1. Checks for a short to ground in the ABS motor "HIGH" circuitry.
2. Checks for a short to ground in the ABS motor "LOW" circuitry.
3. Checks for an ABS motor that is internally shorted to ground.
4. Ensures malfunction was not due to physical damage of the circuitry.

Diagnostic Aids:

Using Tech 1, select manual control function, and exercise ABS motor movement of affected channel in both directions while applying light pressure on the brake pedal.

If erratic or "jumpy" brake pedal movement is detected while performing an "apply" or "release" function of the ABS motor, an intermittent malfunction may be indicated.

An "Intermittent" malfunction may be caused by a poor connection, rubbed through wire insulation, or a wire that is broken inside the insulation.

If the malfunction is not current, wiggle the wires of the affected channel and check if the DTC resets. This will help to pinpoint an intermittent malfunction in the motor circuitry or connections.

The frequency of the malfunction can be checked by using the enhanced diagnostic function of the Tech 1.

Any circuitry that is suspected of causing the intermittent complaint should be thoroughly checked for backed out terminals, improper mating, broken locks, improperly formed or damaged terminals, poor terminal to wiring connections or physical damage to the wiring harness.

DIAGNOSTIC CHART DTC 57 — 1995 CUTLASS SUPREME, GRAND PRIX, LUMINA, MONTE CARLO AND REGAL

DTC 57
LEFT FRONT ABS MOTOR CIRCUIT SHORTED TO GROUND

DOES DTC 57 OCCUR INTERMITTENTLY?

- YES → REFER TO "DIAGNOSTIC AIDS" ON FACING PAGE.
- NO →

(1)
- IGNITION "OFF."
- DISCONNECT 6-WAY ABS MOTOR PACK CONNECTOR AND 8-WAY EBCM CONNECTOR.
- USING J 39200, MEASURE THE RESISTANCE BETWEEN GROUND AND TERMINAL "F" OF THE 6-WAY ABS MOTOR PACK HARNESS CONNECTOR.

DOES J 39200 DISPLAY RESISTANCE AS "OL" (INFINITE)?

- NO → REPAIR SHORT TO GROUND IN CKT 1280.
- YES →

(2)
- MEASURE THE RESISTANCE BETWEEN GROUND AND TERMINAL "E" OF THE 6-WAY ABS MOTOR PACK HARNESS CONNECTOR.

DOES J 39200 DISPLAY RESISTANCE AS "OL" (INFINITE)?

- NO → REPAIR SHORT TO GROUND IN CKT 1281.
- YES →

(3)
- MEASURE THE RESISTANCE BETWEEN GROUND AND TERMINAL "F" OF THE ABS MOTOR PACK.

DOES J 39200 DISPLAY RESISTANCE AS "OL" (INFINITE)?

- NO → REPLACE ABS MOTOR PACK.
- YES →

(4)
- INSPECT CKTs 1280 AND 1281 AND RELATED CONNECTORS FOR DAMAGE WHICH MAY RESULT IN A SHORT TO GROUND WITH ALL CONNECTORS CONNECTED. REPAIR ALL DAMAGE IF EVIDENT.
- RECONNECT ALL CONNECTORS.
- IGNITION "OFF."
- START ENGINE AND TEST DRIVE VEHICLE, ACHIEVING AT LEAST 16 km/h (10 MPH).
- REPEAT ABOVE DRIVE CYCLE SEQUENCE TWO MORE TIMES.

DID DTC 57 SET IN THE LAST THREE DRIVE CYCLES?

- YES → REPLACE EBCM.
- NO → MALFUNCTION WAS MOST LIKELY CAUSED BY A POOR CONNECTION. REFER TO "DIAGNOSTIC AIDS" ON FACING PAGE.

AFTER ALL DIAGNOSIS IS COMPLETE, CLEAR DTCs AND TEST DRIVE VEHICLE FOR THREE (3) DRIVE CYCLES TO VERIFY THAT DTC DOES NOT RESET. A DRIVE CYCLE CONSISTS OF STARTING THE VEHICLE, DRIVING OVER 16 km/h (10 MPH), AND THEN IGNITION "OFF."

DIAGNOSTIC CHART DTC 58 — 1995 CUTLASS SUPREME, GRAND PRIX, LUMINA, MONTE CARLO AND REGAL

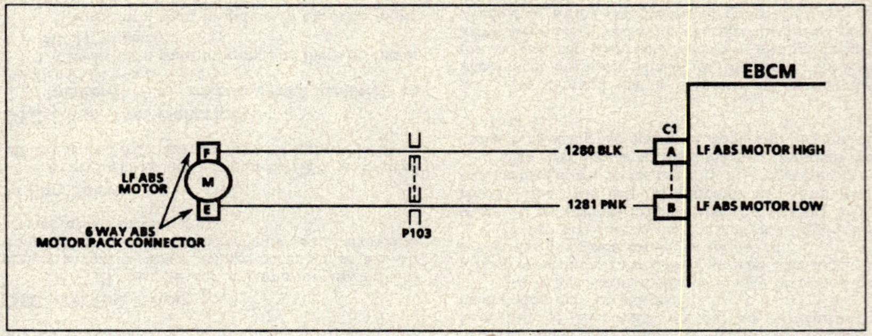

DTC 58
LEFT FRONT ABS MOTOR CIRCUIT SHORTED TO BATTERY

Circuit Description:

This DTC identifies an ABS motor circuit that is shorted to battery or an ABS motor that has low or no resistance. This malfunction will not allow the ABS motor to be controlled at the commanded current rate or will cause the ABS motor to turn in the opposite direction, or not at all.

DTC Will Set When:

DTC 58 can be set only when the ABS motor is commanded "OFF." If the EBCM detects an out of range voltage on either of the left front ABS motor circuits indicating a circuit shorted to battery or an ABS motor shorted, a malfunction exists.

Action Taken:

A malfunction DTC is stored, ABS is disabled and the amber ABS warning lamp is turned "ON."

DTC Chart Test Description:

Number(s) below refer to circled number(s) on the diagnostic chart.

1. Checks for a short to voltage in the ABS motor "HIGH" circuitry.
2. Checks for a short to voltage in the ABS motor "LOW" circuitry.
3. Checks for a short between the "HIGH" and "LOW" ABS motor circuits.
4. Checks for an ABS motor that is internally shorted.
5. Ensures malfunction was not due to physical damage of the circuitry.

Diagnostic Aids:

Using Tech 1, select manual control function, and exercise ABS motor movement of affected channel in both directions while applying light pressure on the brake pedal.

If erratic or "jumpy" brake pedal movement is detected while performing an "apply" or "release" function of the ABS motor, an intermittent malfunction may be indicated.

An "Intermittent" malfunction may be caused by a poor connection, rubbed through wire insulation, or a wire that is broken inside the insulation.

If the malfunction is not current, wiggle the wires of the affected channel and check if the DTC resets. This will help to pinpoint an intermittent malfunction in the motor circuitry or connections.

The frequency of the malfunction can be checked by using the enhanced diagnostic function of the Tech 1.

Any circuitry that is suspected of causing the intermittent complaint should be thoroughly checked for backed out terminals, improper mating, broken locks, improperly formed or damaged terminals, poor terminal to wiring connections or physical damage to the wiring harness.

DIAGNOSTIC CHART DTC 58 — 1995 CUTLASS SUPREME, GRAND PRIX, LUMINA, MONTE CARLO AND REGAL

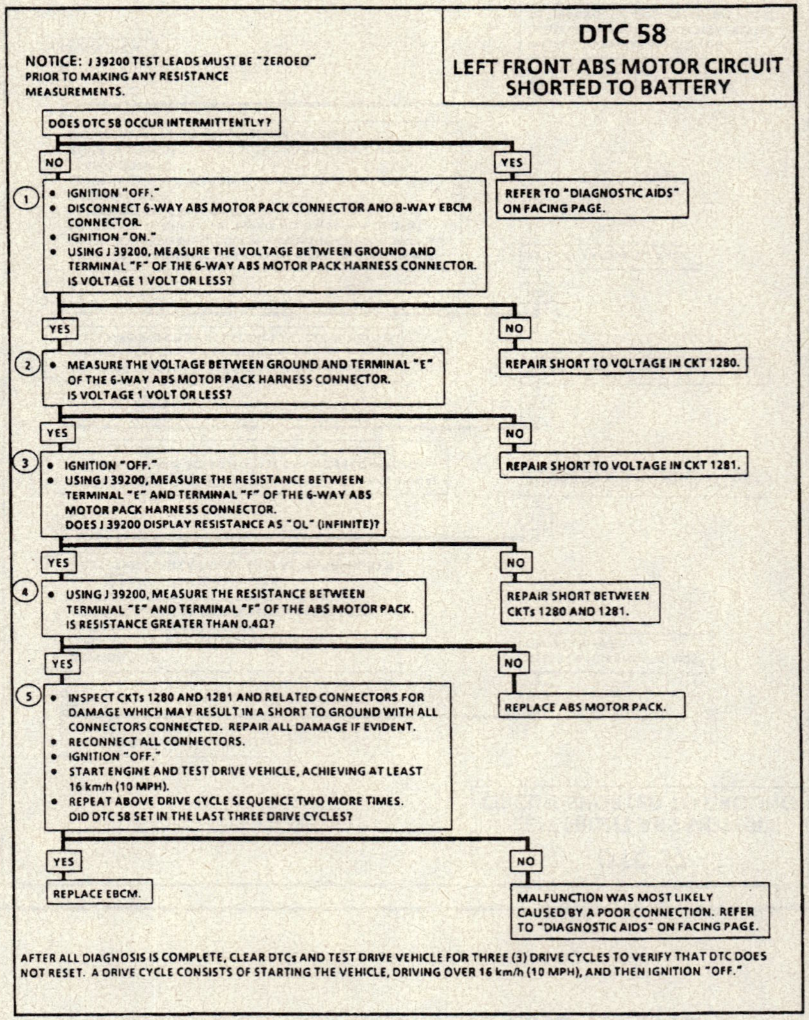

DIAGNOSTIC CHART DTC 61 — 1995 CUTLASS SUPREME, GRAND PRIX, LUMINA, MONTE CARLO AND REGAL

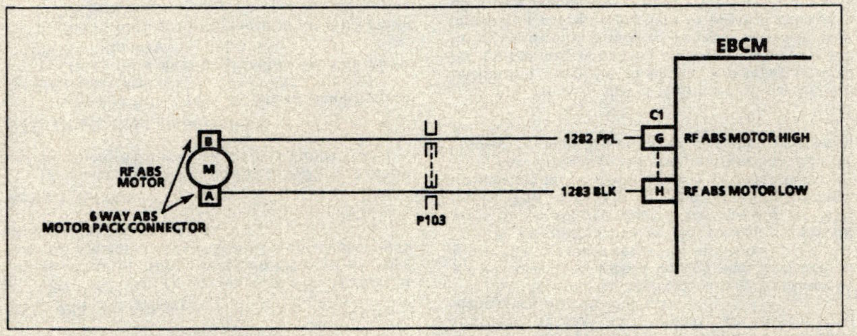

DTC 61
RIGHT FRONT ABS MOTOR CIRCUIT OPEN

Circuit Description:

This DTC identifies an ABS motor that cannot be energized due to an open in its circuitry. This malfunction will not allow proper front ABS operation.

DTC Will Set When:

DTC 61 can be set only when the ABS motor is commanded "OFF." If the EBCM detects an out of range voltage on either of the right front ABS motor circuits indicating an open circuit, a malfunction exists.

Action Taken:

An open ABS motor will not activate when requested. A malfunction DTC is stored, ABS is disabled and the amber ABS warning lamp is turned "ON."

DTC Chart Test Description:

Number(s) below refer to circled number(s) on the diagnostic chart.

1. Checks for proper resistance of the ABS motor.
2. Checks for an open in the ABS motor "HIGH" circuitry.
3. Checks for an open in the ABS motor "LOW" circuitry.
4. Ensures malfunction was not due to physical damage of the circuitry.

Diagnostic Aids:

Using Tech 1, select manual control function, and exercise ABS motor movement of affected channel in both directions while applying light pressure on the brake pedal.

If erratic or "jumpy" brake pedal movement is detected while performing an "apply" or "release" function of the ABS motor, an intermittent malfunction may be indicated.

An "Intermittent" malfunction may be caused by a poor connection, rubbed through wire insulation, or a wire that is broken inside the insulation.

If the malfunction is not current, wiggle the wires of the affected channel and check if the DTC resets. This will help to pinpoint an intermittent malfunction in the motor circuitry or connections.

The frequency of the malfunction can be checked by using the enhanced diagnostic function of the Tech 1.

Any circuitry that is suspected of causing the intermittent complaint should be thoroughly checked for backed out terminals, improper mating, broken locks, improperly formed or damaged terminals, poor terminal to wiring connections or physical damage to the wiring harness.

DIAGNOSTIC CHART DTC 61 — 1995 CUTLASS SUPREME, GRAND PRIX, LUMINA, MONTE CARLO AND REGAL

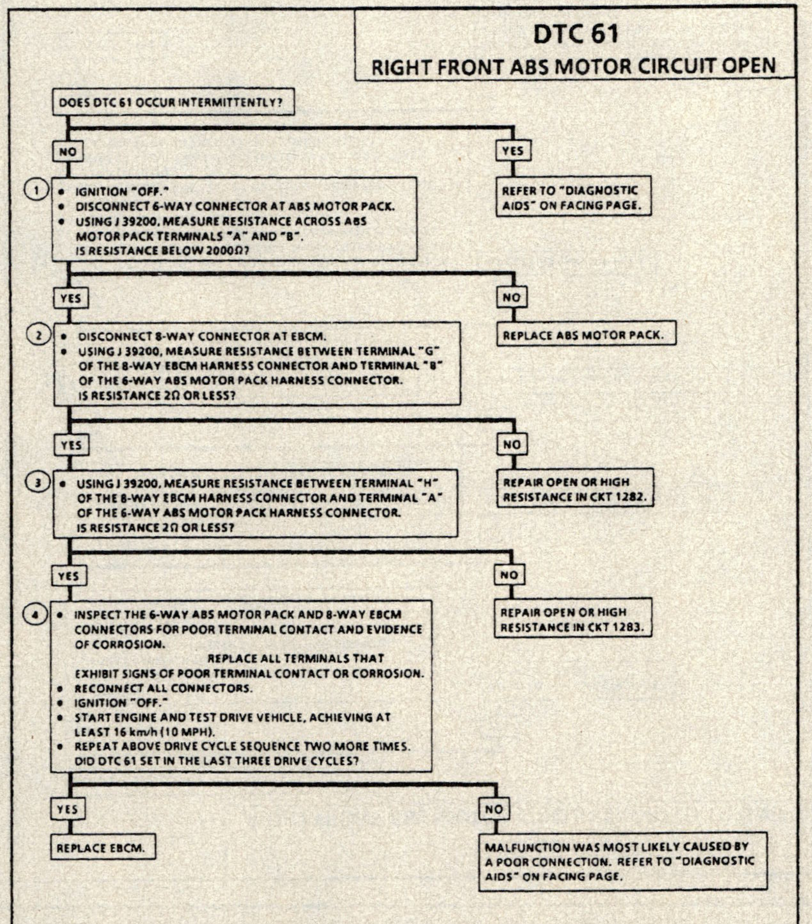

AFTER ALL DIAGNOSIS IS COMPLETE, CLEAR DTCs AND TEST DRIVE VEHICLE FOR THREE (3) DRIVE CYCLES TO VERIFY THAT DTC DOES NOT RESET. A DRIVE CYCLE CONSISTS OF STARTING THE VEHICLE, DRIVING OVER 16 km/h (10 MPH), AND THEN IGNITION "OFF."

DIAGNOSTIC CHART DTC 62 — 1995 CUTLASS SUPREME, GRAND PRIX, LUMINA, MONTE CARLO AND REGAL

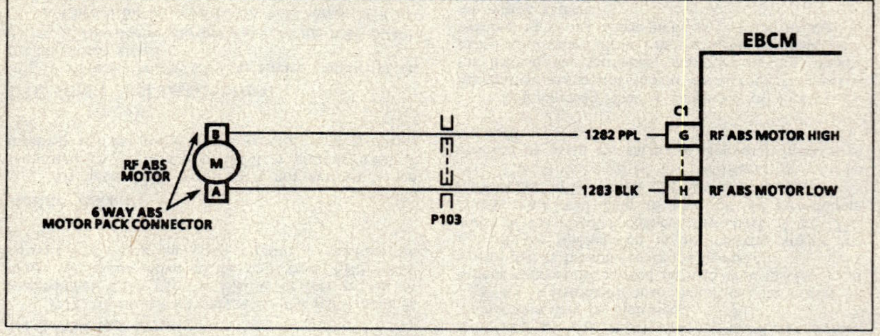

DTC 62
RIGHT FRONT ABS MOTOR CIRCUIT SHORTED TO GND

Circuit Description:

This DTC identifies an ABS motor circuit that is shorted to ground. This malfunction will not allow the ABS motor to be controlled at the commanded current rate or will cause the ABS motor driver circuit to allow current directly to ground.

DTC Will Set When:

DTC 62 can be set anytime. If the EBCM detects an out of range voltage on either of the right front ABS motor circuits indicating a circuit shorted to ground, a malfunction exists.

Action Taken:

A malfunction DTC is stored, ABS is disabled and the amber ABS warning lamp is turned "ON."

DTC Chart Test Description:

Number(s) below refer to circled number(s) on the diagnostic chart.

1. Checks for a short to ground in the ABS motor "HIGH" circuitry.
2. Checks for a short to ground in the ABS motor "LOW" circuitry.
3. Checks for an ABS motor that is internally shorted to ground.
4. Ensures malfunction was not due to physical damage of the circuitry.

Diagnostic Aids:

Using Tech 1, select manual control function, and exercise ABS motor movement of affected channel in both directions while applying light pressure on the brake pedal.

If erratic or "jumpy" brake pedal movement is detected while performing an "apply" or "release" function of the ABS motor, an intermittent malfunction may be indicated.

An "Intermittent" malfunction may be caused by a poor connection, rubbed through wire insulation, or a wire that is broken inside the insulation.

If the malfunction is not current, wiggle the wires of the affected channel and check if the DTC resets. This will help to pinpoint an intermittent malfunction in the motor circuitry or connections.

The frequency of the malfunction can be checked by using the enhanced diagnostic function of the Tech 1.

Any circuitry that is suspected of causing the intermittent complaint should be thoroughly checked for backed out terminals, improper mating, broken locks, improperly formed or damaged terminals, poor terminal to wiring connections or physical damage to the wiring harness.

DIAGNOSTIC CHART DTC 62 — 1995 CUTLASS SUPREME, GRAND PRIX, LUMINA, MONTE CARLO AND REGAL

DTC 62
RIGHT FRONT ABS MOTOR CIRCUIT SHORTED TO GND

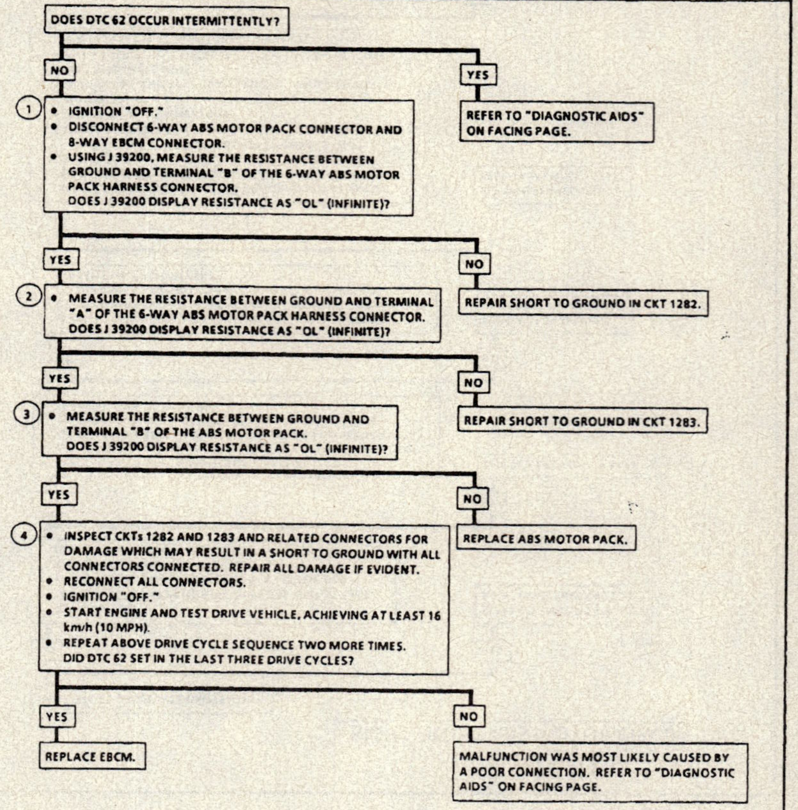

AFTER ALL DIAGNOSIS IS COMPLETE, CLEAR DTCs AND TEST DRIVE VEHICLE FOR THREE (3) DRIVE CYCLES TO VERIFY THAT DTC DOES NOT RESET. A DRIVE CYCLE CONSISTS OF STARTING THE VEHICLE, DRIVING OVER 16 km/h (10 MPH), AND THEN IGNITION "OFF."

DIAGNOSTIC CHART DTC 63 — 1995 CUTLASS SUPREME, GRAND PRIX, LUMINA, MONTE CARLO AND REGAL

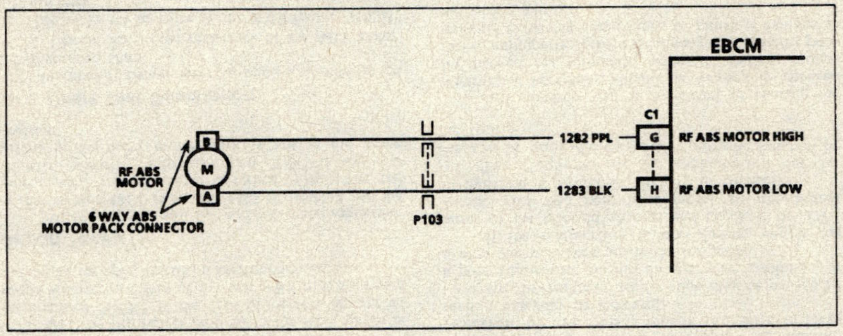

DTC 63
RIGHT FRONT ABS MOTOR CIRCUIT SHORTED TO BATT

Circuit Description:

This DTC identifies an ABS motor circuit that is shorted to battery or an ABS motor that has low or no resistance. This malfunction will not allow the ABS motor to be controlled at the commanded current rate or will cause the ABS motor to turn in the opposite direction or not at all.

DTC Will Set When:

DTC 63 can be set only when the ABS motor is commanded "OFF." If the EBCM detects an out of range voltage on either of the right front ABS motor circuits indicating a circuit shorted to battery or an ABS motor shorted, a malfunction exists.

Action Taken:

A malfunction DTC is stored, ABS is disabled and the amber ABS warning lamp is turned "ON."

DTC Chart Test Description:

Number(s) below refer to circled number(s) on the diagnostic chart.

1. Checks for a short to voltage in the ABS motor "HIGH" circuitry.
2. Checks for a short to voltage in the ABS motor "LOW" circuitry.
3. Checks for a short circuit between ABS motor "HIGH" and ABS motor "LOW."
4. Checks for proper resistance of the ABS motor.
5. Ensures malfunction was not due to physical damage of the circuitry.

Diagnostic Aids:

Using Tech 1, select manual control function, and exercise ABS motor movement of affected channel in both directions while applying light pressure on the brake pedal.

If erratic or "jumpy" brake pedal movement is detected while performing an "apply" or "release" function of the ABS motor, an intermittent malfunction may be indicated.

An "Intermittent" malfunction may be caused by a poor connection, rubbed through wire insulation, or a wire that is broken inside the insulation.

If the malfunction is not current, wiggle the wires of the affected channel and check if the DTC resets. This will help to pinpoint an intermittent malfunction in the motor circuitry or connections.

The frequency of the malfunction can be checked by using the enhanced diagnostic function of the Tech 1.

Any circuitry that is suspected of causing the intermittent complaint should be thoroughly checked for backed out terminals, improper mating, broken locks, improperly formed or damaged terminals, poor terminal to wiring connections or physical damage to the wiring harness.

DIAGNOSTIC CHART DTC 63 — 1995 CUTLASS SUPREME, GRAND PRIX, LUMINA, MONTE CARLO AND REGAL

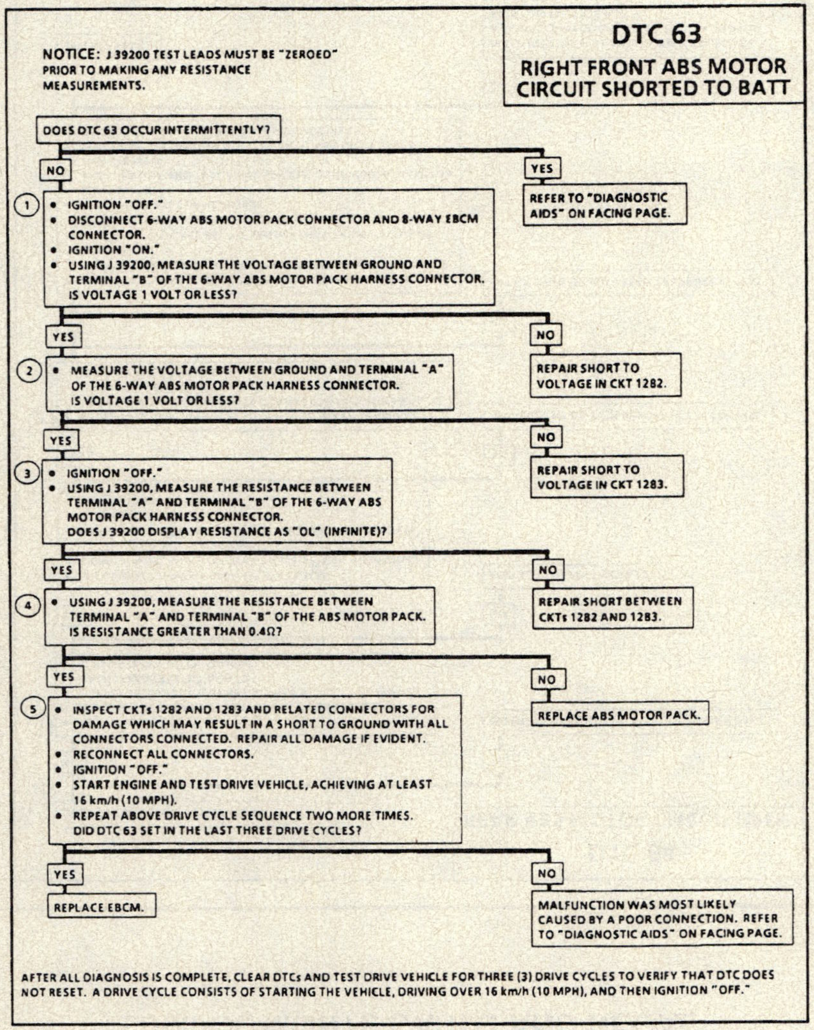

DIAGNOSTIC CHART DTC 64 — 1995 CUTLASS SUPREME, GRAND PRIX, LUMINA, MONTE CARLO AND REGAL

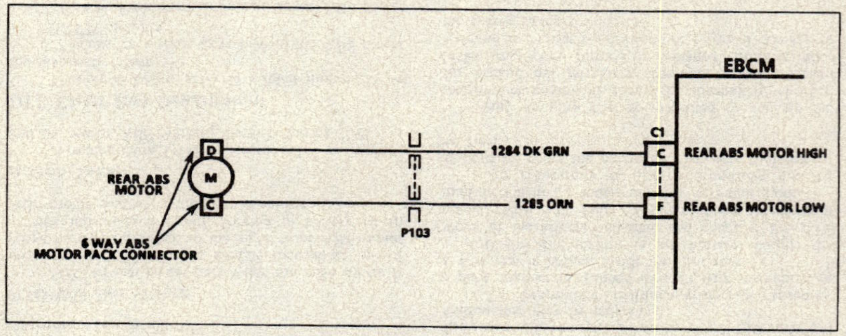

DTC 64
REAR ABS MOTOR CIRCUIT OPEN

Circuit Description:

This DTC identifies an ABS motor that cannot be energized due to an open in its circuitry. This malfunction will not allow proper rear ABS operation.

DTC Will Set When:

DTC 64 can be set only when the ABS motor is commanded "OFF." If the EBCM detects an out of range voltage on either of the rear ABS motor circuits indicating an open circuit, a malfunction exists.

Action Taken:

An open motor will not activate when requested. A malfunction DTC is stored, ABS is disabled and the amber ABS warning lamp is turned "ON." The red BRAKE warning lamp is turned "ON" if the rear piston in the ABS motor pack is not in the home position.

DTC Chart Test Description:

Number(s) below refer to circled number(s) on the diagnostic chart.

1. Checks for proper resistance of the ABS motor.
2. Checks for an open in the ABS motor "HIGH" circuitry.
3. Checks for an open in the ABS motor "LOW" circuitry.
4. Ensures malfunction was not due to poor terminal contact.

Diagnostic Aids:

Using Tech 1, select manual control function, and exercise ABS motor movement of affected channel in both directions while applying light pressure on the brake pedal.

If erratic or "jumpy" brake pedal movement is detected while performing an "apply" or "release" function of the ABS motor, an intermittent malfunction may be indicated.

An "Intermittent" malfunction may be caused by a poor connection, rubbed through wire insulation, or a wire that is broken inside the insulation.

If the malfunction is not current, wiggle the wires of the affected channel and check if the DTC resets. This will help to pinpoint an intermittent malfunction in the motor circuitry or connections.

The frequency of the malfunction can be checked by using the enhanced diagnostic function of the Tech 1.

Any circuitry that is suspected of causing the intermittent complaint should be thoroughly checked for backed out terminals, improper mating, broken locks, improperly formed or damaged terminals, poor terminal to wiring connections or physical damage to the wiring harness.

DIAGNOSTIC CHART DTC 64 — 1995 CUTLASS SUPREME, GRAND PRIX, LUMINA, MONTE CARLO AND REGAL

DTC 64
REAR ABS MOTOR CIRCUIT OPEN

DOES DTC 64 OCCUR INTERMITTENTLY?

- YES: REFER TO "DIAGNOSTIC AIDS" ON FACING PAGE.
- NO:

(1)
- IGNITION "OFF."
- DISCONNECT 6-WAY ABS MOTOR PACK CONNECTOR.
- USING J 39200, MEASURE RESISTANCE ACROSS ABS MOTOR PACK TERMINALS "C" AND "D".

IS RESISTANCE BELOW 2000Ω?

- NO: REPLACE ABS MOTOR PACK.
- YES:

(2)
- DISCONNECT 8-WAY CONNECTOR AT EBCM.
- USING J 39200, MEASURE RESISTANCE BETWEEN TERMINAL "C" OF THE 8-WAY EBCM HARNESS CONNECTOR AND TERMINAL "D" OF THE 6-WAY ABS MOTOR PACK HARNESS CONNECTOR.

IS RESISTANCE 2Ω OR LESS?

- NO: REPAIR OPEN OR HIGH RESISTANCE IN CKT 1284
- YES:

(3)
- USING J 39200, MEASURE RESISTANCE BETWEEN TERMINAL "F" OF THE 8-WAY EBCM HARNESS CONNECTOR AND TERMINAL "C" OF THE 6-WAY ABS MOTOR PACK HARNESS CONNECTOR.

IS RESISTANCE 2Ω OR LESS?

- NO: REPAIR OPEN OR HIGH RESISTANCE IN CKT 1285
- YES:

(4)
- INSPECT THE 6-WAY ABS MOTOR PACK AND 8-WAY EBCM CONNECTORS FOR POOR TERMINAL CONTACT AND EVIDENCE OF CORROSION.

REPLACE ALL TERMINALS THAT EXHIBIT SIGNS OF POOR TERMINAL CONTACT OR CORROSION.
- RECONNECT ALL CONNECTORS.
- IGNITION "OFF."
- START ENGINE AND TEST DRIVE VEHICLE, ACHIEVING AT LEAST 16 km/h (10 MPH).
- REPEAT ABOVE DRIVE CYCLE SEQUENCE TWO MORE TIMES.

DID DTC 64 SET IN THE LAST THREE DRIVE CYCLES?

- YES: REPLACE EBCM.
- NO: MALFUNCTION WAS MOST LIKELY CAUSED BY A POOR CONNECTION. REFER TO "DIAGNOSTIC AIDS" ON FACING PAGE

AFTER ALL DIAGNOSIS IS COMPLETE, CLEAR DTCs AND TEST DRIVE VEHICLE FOR THREE (3) DRIVE CYCLES TO VERIFY THAT DTC DOES NOT RESET. A DRIVE CYCLE CONSISTS OF STARTING THE VEHICLE, DRIVING OVER 16 km/h (10 MPH), AND THEN IGNITION "OFF."

DIAGNOSTIC CHART DTC 65 — 1995 CUTLASS SUPREME, GRAND PRIX, LUMINA, MONTE CARLO AND REGAL

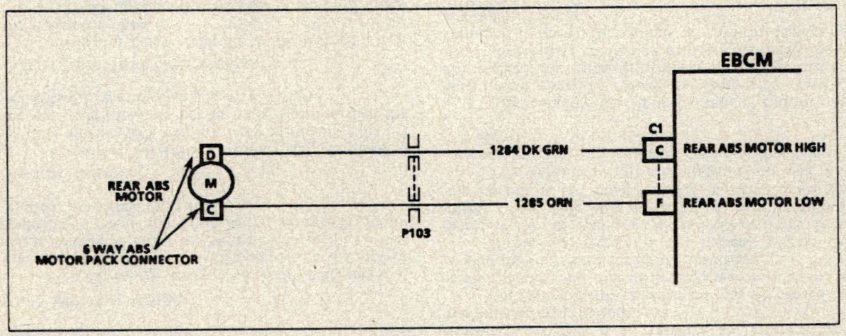

DTC 65
REAR ABS MOTOR CIRCUIT SHORTED TO GROUND

Circuit Description:

This DTC identifies an ABS motor circuit that is shorted to ground. This malfunction will not allow the ABS motor to be controlled at the commanded current rate or will cause the driver circuit to allow current directly to ground.

DTC Will Set When:

DTC 65 can be set anytime. If the EBCM detects an out of range voltage on either of the rear ABS motor circuits indicating a circuit shorted to ground, a malfunction exists.

Action Taken:

A malfunction DTC is stored, ABS is disabled, the amber ABS warning lamp and the red BRAKE warning lamp are turned "ON."

DTC Chart Test Description:

Number(s) below refer to circled number(s) on the diagnostic chart.

1. Checks for a short to ground in the ABS motor "HIGH" circuitry.
2. Checks for a short to ground in the ABS motor "LOW" circuitry.
3. Checks for an ABS motor that is internally shorted to ground.
4. Ensures malfunction was not due to physical damage.

Diagnostic Aids:

Using Tech 1, select manual control function, and exercise ABS motor movement of affected channel in both directions while applying light pressure on the brake pedal.

If erratic or "jumpy" brake pedal movement is detected while performing an "apply" or "release" function of the ABS motor, an intermittent malfunction may be indicated.

An "Intermittent" malfunction may be caused by a poor connection, rubbed through wire insulation, or a wire that is broken inside the insulation.

If the malfunction is not current, wiggle the wires of the affected channel and check if the DTC resets. This will help to pinpoint an intermittent malfunction in the motor circuitry or connections.

The frequency of the malfunction can be checked by using the enhanced diagnostic function of the Tech 1.

Any circuitry that is suspected of causing the intermittent complaint should be thoroughly checked for backed out terminals, improper mating, broken locks, improperly formed or damaged terminals, poor terminal to wiring connections or physical damage to the wiring harness.

DIAGNOSTIC CHART DTC 65 — 1995 CUTLASS SUPREME, GRAND PRIX, LUMINA, MONTE CARLO AND REGAL

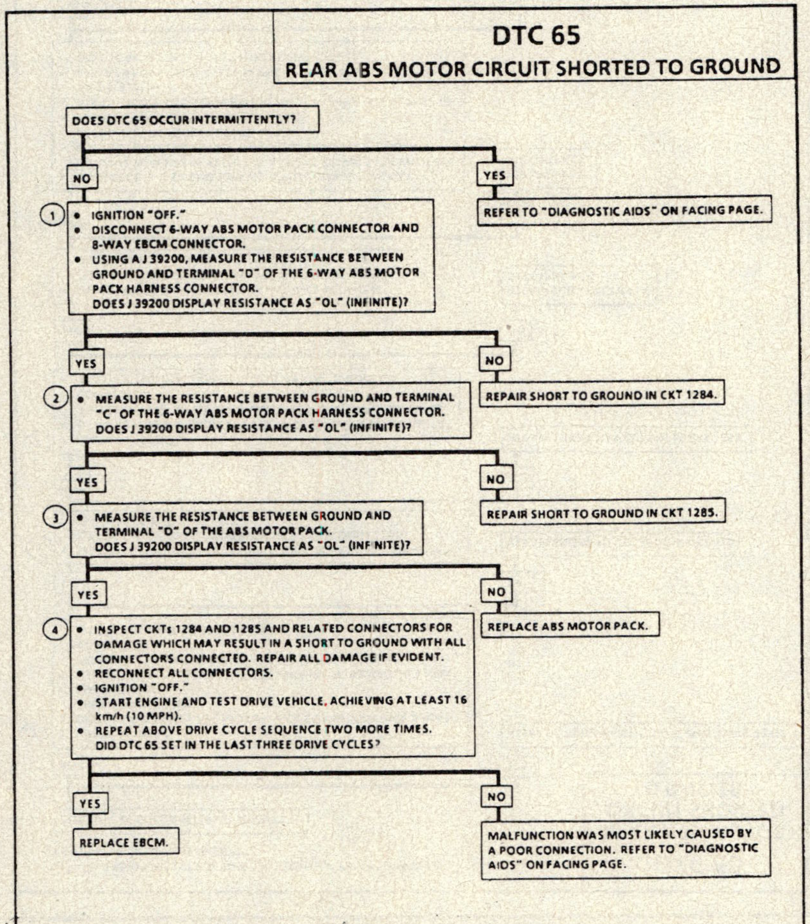

AFTER ALL DIAGNOSIS IS COMPLETE, CLEAR DTCs AND TEST DRIVE VEHICLE FOR THREE (3) DRIVE CYCLES TO VERIFY THAT DTC DOES NOT RESET. A DRIVE CYCLE CONSISTS OF STARTING THE VEHICLE, DRIVING OVER 16 km/h (10 MPH), AND THEN IGNITION "OFF."

DIAGNOSTIC CHART DTC 66 — 1995 CUTLASS SUPREME, GRAND PRIX, LUMINA, MONTE CARLO AND REGAL

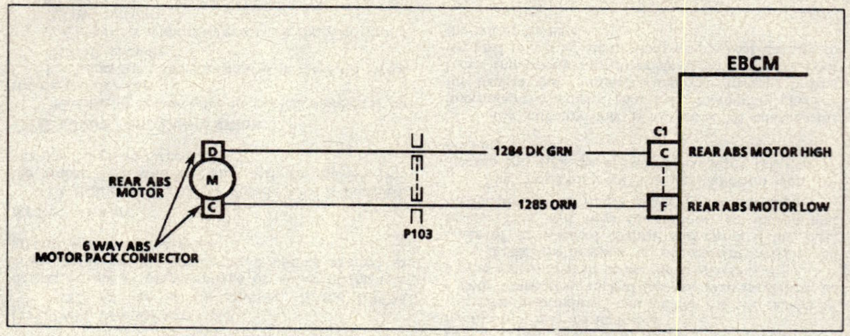

DTC 66
REAR ABS MOTOR CIRCUIT SHORTED TO BATTERY

Circuit Description:

This DTC identifies an ABS motor circuit that is shorted to battery or an ABS motor that has low or zero resistance. This malfunction will not allow the ABS motor to be controlled at the commanded current rate or will cause the ABS motor to turn in the opposite direction or not turn at all.

DTC Will Set When:

DTC 66 can be set only when the ABS motor is commanded "OFF." If the EBCM detects an out of range voltage on either of the rear ABS motor circuits indicating a circuit shorted to battery or an ABS motor shorted, a malfunction exists.

Action Taken:

A malfunction DTC is stored, ABS is disabled and the amber ABS warning lamp is turned "ON." If the rear axle motor is not in home position then the red BRAKE warning lamp is turned "ON."

DTC Chart Test Description:

Number(s) below refer to circled number(s) on the diagnostic chart.

1. Checks for a short to voltage in the ABS motor "HIGH" circuitry.
2. Checks for a short to voltage in the ABS motor "LOW" circuitry.
3. Checks for a short circuit between ABS motor "HIGH" and ABS motor "LOW."
4. Checks for proper resistance of the ABS motor.
5. Ensures malfunction was not due to physical damage of the circuitry.

Diagnostic Aids:

Using Tech 1, select manual control function, and exercise ABS motor movement of affected channel in both directions while applying light pressure on the brake pedal.

If erratic or "jumpy" brake pedal movement is detected while performing an "apply" or "release" function of the ABS motor, an intermittent malfunction may be indicated.

An "Intermittent" malfunction may be caused by a poor connection, rubbed through wire insulation, or a wire that is broken inside the insulation.

If the malfunction is not current, wiggle the wires of the affected channel and check if the DTC resets. This will help to pinpoint an intermittent malfunction in the motor circuitry or connections.

The frequency of the malfunction can be checked by using the enhanced diagnostic function of the Tech 1.

Any circuitry that is suspected of causing the intermittent complaint should be thoroughly checked for backed out terminals, improper mating, broken locks, improperly formed or damaged terminals, poor terminal to wiring connections or physical damage to the wiring harness.

DIAGNOSTIC CHART DTC 66 — 1995 CUTLASS SUPREME, GRAND PRIX, LUMINA, MONTE CARLO AND REGAL

DTC 66
REAR ABS MOTOR CIRCUIT SHORTED TO BATTERY

NOTICE: J 39200 TEST LEADS MUST BE "ZEROED" PRIOR TO MAKING ANY RESISTANCE MEASUREMENTS.

DOES DTC 66 OCCUR INTERMITTENTLY?

- YES: REFER TO "DIAGNOSTIC AIDS" ON FACING PAGE.
- NO:

(1)
- IGNITION "OFF."
- DISCONNECT 6-WAY ABS MOTOR PACK CONNECTOR AND 8-WAY EBCM CONNECTORS.
- IGNITION "ON."
- USING J 39200, MEASURE THE VOLTAGE BETWEEN GROUND AND TERMINAL "D" OF THE 6-WAY ABS MOTOR PACK HARNESS CONNECTOR.

IS VOLTAGE 1 VOLT OR LESS?

- NO: REPAIR SHORT TO VOLTAGE IN CKT 1284.
- YES:

(2)
- MEASURE THE VOLTAGE BETWEEN GROUND AND TERMINAL "C" OF THE 6-WAY ABS MOTOR PACK HARNESS CONNECTOR.

IS VOLTAGE 1 VOLT OR LESS?

- NO: REPAIR SHORT TO VOLTAGE IN CKT 1285.
- YES:

(3)
- IGNITION "OFF."
- USING J 39200, MEASURE THE VOLTAGE BETWEEN TERMINAL "C" AND TERMINAL "D" OF THE 6-WAY ABS MOTOR PACK HARNESS CONNECTOR.

DOES J 39200 DISPLAY RESISTANCE AS "OL" (INFINITE)?

- NO: REPAIR SHORT BETWEEN CKTs 1284 AND 1285.
- YES:

(4)
- USING J 39200, MEASURE THE RESISTANCE BETWEEN TERMINAL "C" AND TERMINAL "D" OF THE ABS MOTOR PACK.

IS RESISTANCE GREATER THAN 0.4Ω?

- NO: REPLACE ABS MOTOR PACK.
- YES:

(5)
- INSPECT CKTs 1284 AND 1285 RELATED CONNECTORS FOR DAMAGE WHICH MAY RESULT IN A SHORT TO VOLTAGE WITH ALL CONNECTORS CONNECTED. REPAIR ALL DAMAGE IF EVIDENT.
- RECONNECT ALL CONNECTORS.
- IGNITION "OFF."
- START ENGINE AND TEST DRIVE VEHICLE, ACHIEVING AT LEAST 16 km/h (10 MPH).
- REPEAT ABOVE DRIVE CYCLE SEQUENCE TWO MORE TIMES.

DID DTC 66 SET IN THE LAST THREE DRIVE CYCLES.

- YES: REPLACE EBCM.
- NO: MALFUNCTION WAS MOSTLY CAUSED BY A POOR CONNECTION. REFER TO "DIAGNOSTIC AIDS" ON FACING PAGE.

AFTER ALL DIAGNOSIS IS COMPLETE, CLEAR DTCs AND TEST DRIVE VEHICLE FOR THREE (3) DRIVE CYCLES TO VERIFY THAT DTC DOES NOT RESET. A DRIVE CYCLE CONSISTS OF STARTING THE VEHICLE, DRIVING OVER 16 km/h (10 MPH), AND THEN IGNITION "OFF."

DIAGNOSTIC CHART DTC 73 — 1995 CUTLASS SUPREME, GRAND PRIX, LUMINA, MONTE CARLO AND REGAL

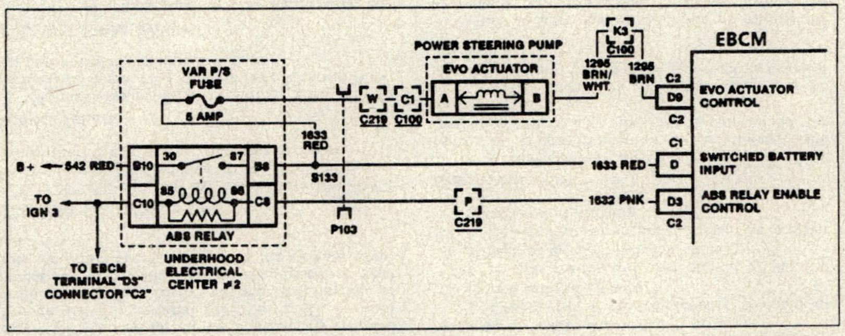

DTC 73

EVO ACTUATOR CIRCUIT OPEN OR SHORTED TO GROUND (FOR ABS WITH VES ONLY)

Circuit Description:

Battery voltage is supplied to EVO actuator terminal "A" when the EBCM commands the ABS relay "ON." Ground for the EVO actuator is provided through EBCM terminal "D9" to EVO actuator terminal "B." The EBCM controls the amount of current supplied to the EVO actuator based on input from the wheel speed sensors and the steering wheel speed sensor.

DTC Will Set When:

DTC 73 can be set before ABS initialization or above 16 km/h (10 mph). Before initialization, if the EBCM detects the EVO actuator control circuit is at ground, a malfunction exists. Above 16 km/h (10 mph), if the EVO actuator sense current is less than half the command current, a malfunction exists.

Action Taken:

A malfunction DTC is stored, ABS/VES is disabled, the amber ABS warning lamp is turned "ON" and power steering returns to full assist.

DTC Will Clear When:

DTC 73 setting conditions no longer exist and the ignition key is turned "OFF".

DTC Chart Test Description:

Number(s) below refer to circled number(s) on the diagnostic chart.

1. Checks for proper operation of the EVO actuator.
2. Checks for high resistance in the EVO actuator control circuit.
3. Checks for a short to ground in the EVO actuator control circuit.
4. Checks for high resistance in the EVO actuator control circuit.
5. Checks for a short to ground in the EVO actuator control circuit.
6. Checks for a short to ground in the EVO actuator.
7. Checks for an open in the EVO actuator.
8. Ensures open or short to ground is not due to physical damage of the circuitry.

Diagnostic Aids:

An "Intermittent" malfunction is most likely caused by a poor connection, rubbed through wire insulation, or a wire that is broken inside the insulation.

The frequency of the malfunction can be checked by using the enhanced diagnostic function of the Tech 1.

Any circuitry that is suspected of causing the intermittent complaint should be thoroughly checked for backed out terminals, improper mating, broken locks, improperly formed or damaged terminals, poor terminal to wiring connections or physical damage to the wiring harness.

DIAGNOSTIC CHART DTC 73 — 1995 CUTLASS SUPREME, GRAND PRIX, LUMINA, MONTE CARLO AND REGAL

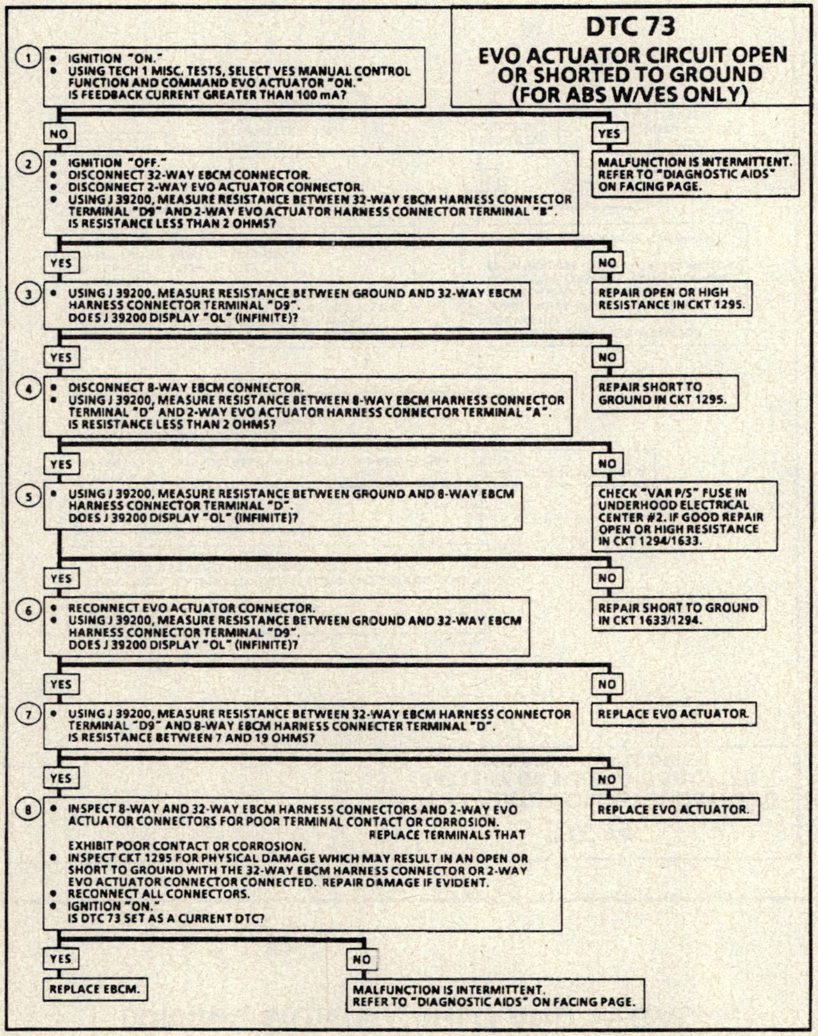

DIAGNOSTIC CHART DTC 74 — 1995 CUTLASS SUPREME, GRAND PRIX, LUMINA, MONTE CARLO AND REGAL

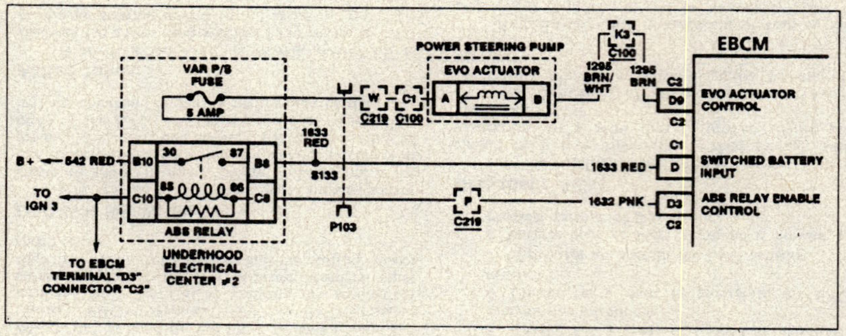

DTC 74
EVO ACTUATOR CIRCUIT SHORTED TO BATTERY OR SOLENOID SHORTED (FOR ABS WITH VES ONLY)

Circuit Description:

Battery voltage is supplied to EVO actuator terminal "A" when the EBCM commands the ABS relay "ON". Ground for the EVO actuator is provided through EBCM terminal "D9" to EVO actuator terminal "B". The EBCM controls the amount of current supplied to the EVO actuator based on input from the wheel speed sensors and the steering wheel speed sensor.

DTC Will Set When:

DTC 74 can be set anytime after ABS initialization. A malfunction exists if an excessive voltage drop is across the EVO actuator driver.

Action Taken:

A malfunction DTC is stored, ABS/VES is disabled, the amber ABS warning lamp is turned "ON" and power steering returns to full assist.

DTC Will Clear When:

DTC 74 setting conditions no longer exist and the ignition key is turned "OFF".

DTC Chart Test Description:

Number(s) below refer to circled number(s) on the diagnostic chart.

1. Checks for a short to voltage in the EVO actuator control circuit.
2. Checks fro a shorted actuator solenoid or shorted actuator circuits.
3. Ensures malfunction was not du to physical damage of the EVO actuator circuitry.
4. Isolates the malfunction to either the actuator solenoid or its circuitry.

Diagnostic Aids:

An "Intermittent" malfunction is most likely caused by a poor connection, rubbed through wire insulation, or a wire that is broken inside the insulation.

The frequency of the malfunction can be checked by using the enhanced diagnostic function of the Tech 1.

Any circuitry that is suspected of causing the intermittent complaint should be thoroughly checked for backed out terminals, improper mating, broken locks, improperly formed or damaged terminals, poor terminal to wiring connections or physical damage to the wiring harness.

DIAGNOSTIC CHART DTC 74 — 1995 CUTLASS SUPREME, GRAND PRIX, LUMINA, MONTE CARLO AND REGAL

DTC 74
EVO ACTUATOR CIRCUIT SHORTED TO BATTERY OR SOLENOID SHORTED (FOR ABS WITH VES ONLY)

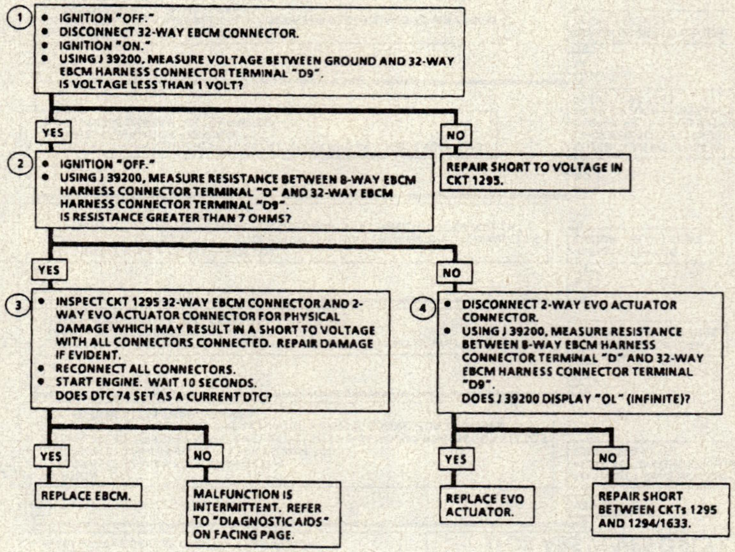

DIAGNOSTIC CHART DTC 76 — 1995 CUTLASS SUPREME, GRAND PRIX, LUMINA, MONTE CARLO AND REGAL

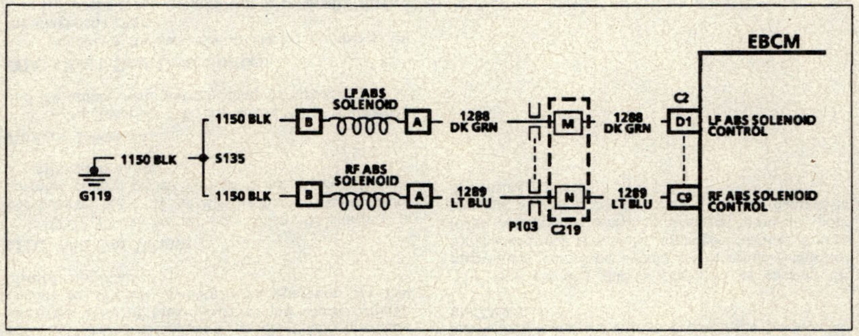

DTC 76
LEFT FRONT SOL CIRCUIT OPEN OR SHORTED TO BATT

Circuit Description:

This DTC identifies a solenoid that cannot be energized due to an open in its circuitry, or a solenoid that is always energized due to a short to battery in its circuitry between the driver and the solenoid. An open will not allow proper ABS operation, but the short to battery simply turns "ON" the solenoid. A path for base brakes is still allowed once the motor rehomes and the check ball is lifted off its seat during key "ON" initialization.

DTC Will Set When:

DTC 76 can be set only when the solenoid is commanded "OFF." If the EBCM detects an excessive voltage in the left front solenoid control circuit, a malfunction exists.

Action Taken:

A malfunction DTC is stored, ABS is disabled and the amber ABS warning lamp is turned "ON."

DTC Chart Test Description:

Number(s) below refer to circled number(s) on the diagnostic chart.

1. Checks for a short to voltage in the control circuitry of the solenoid.
2. Checks for an open in the control circuit of the solenoid.
3. Checks the solenoid coil for proper resistance.
4. Checks for an open in the solenoid ground circuit.
5. Checks for a possible intermittent malfunction in the solenoid circuitry due to poor terminal contact.

Diagnostic Aids:

An "Intermittent" malfunction may be caused by a poor connection, rubbed through wire insulation, or a wire that is broken inside the insulation.

The frequency of the malfunction can be checked by using the enhanced diagnostic function of the Tech 1.

Any circuitry that is suspected of causing the intermittent complaint should be thoroughly checked for backed out terminals, improper mating, broken locks, improperly formed or damaged terminals, poor terminal to wiring connections or physical damage to the wiring harness.

DIAGNOSTIC CHART DTC 76 — 1995 CUTLASS SUPREME, GRAND PRIX, LUMINA, MONTE CARLO AND REGAL

DTC 76
LEFT FRONT SOL CIRCUIT OPEN OR SHORTED TO BATT

(1)
- IGNITION "OFF."
- DISCONNECT 32-WAY EBCM CONNECTOR.
- IGNITION "ON," ENGINE "OFF."
- USING J 39200, MEASURE THE VOLTAGE BETWEEN GROUND AND TERMINAL "D1" OF THE 32-WAY EBCM HARNESS CONNECTOR.

IS VOLTAGE 1 VOLT OR LESS?

NO: REPAIR SHORT TO VOLTAGE IN CKT 1288.

YES:

(2)
- IGNITION "OFF."
- DISCONNECT LEFT FRONT SOLENOID CONNECTOR.
- USING J 39200, MEASURE THE RESISTANCE BETWEEN TERMINAL "D1" OF THE 32-WAY EBCM HARNESS CONNECTOR AND TERMINAL "A" OF THE LEFT FRONT SOLENOID HARNESS CONNECTOR.

IS RESISTANCE 2Ω OR LESS?

NO: REPAIR OPEN OR HIGH RESISTANCE IN CKT 1288.

YES:

(3)
- MEASURE THE RESISTANCE BETWEEN TERMINAL "A" AND TERMINAL "B" OF THE LEFT FRONT SOLENOID.

IS RESISTANCE BETWEEN 2.5 AND 5Ω?

NO: REPLACE LEFT FRONT SOLENOID.

YES:

(4)
- MEASURE THE RESISTANCE BETWEEN GROUND AND TERMINAL "B" OF THE LEFT FRONT SOLENOID HARNESS CONNECTOR.

IS RESISTANCE 2Ω OR LESS?

NO: REPAIR OPEN OR HIGH RESISTANCE IN CKT 1150.

YES:

(5)
- INSPECT LEFT FRONT SOLENOID AND 32-WAY EBCM HARNESS CONNECTORS FOR POOR TERMINAL CONTACT AND EVIDENCE OF CORROSION.

REPLACE ALL TERMINALS THAT EXHIBIT SIGNS OF POOR TERMINAL CONTACT OR CORROSION.
- INSPECT CKT 1288 FOR DAMAGE WHICH MAY RESULT IN A SHORT TO VOLTAGE WITH ALL CONNECTORS CONNECTED. REPAIR DAMAGE IF EVIDENT.
- RECONNECT ALL CONNECTORS.
- IGNITION "ON."

DOES DTC 76 SET AS A CURRENT DTC?

YES: REPLACE EBCM.

NO: MALFUNCTION IS MOST LIKELY DUE TO A POOR CONNECTION. REFER TO "DIAGNOSTIC AIDS" ON FACING PAGE.

DIAGNOSTIC CHART DTC 77 — 1995 CUTLASS SUPREME, GRAND PRIX, LUMINA, MONTE CARLO AND REGAL

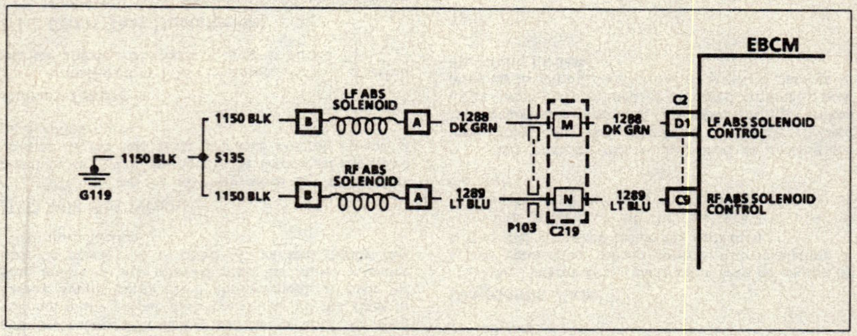

DTC 77

LEFT FRONT SOLENOID CIRCUIT SHORTED TO GND

Circuit Description:

This DTC identifies a solenoid that cannot be energized due to an open in its driver circuitry, or a short to ground between the solenoid driver and the solenoid. These malfunctions can affect ABS operation since the flow of brake fluid to the wheel cylinder cannot be stopped, making ABS operation for that channel impossible.

DTC Will Set When:

DTC 77 can be set only when the solenoid is commanded "ON." If the EBCM detects the left front solenoid control circuit voltage is out of specification, a malfunction exists.

Action Taken:

A malfunction DTC is stored, ABS is disabled and the amber ABS warning lamp is turned "ON."

DTC Chart Test Description:

Number(s) below refer to circled number(s) on the diagnostic chart.

1. Checks for a solenoid that is internally shorted to ground.
2. Checks for a solenoid that is not within proper resistance values.
3. Indicates if a short to ground exists in the solenoid circuitry.
4. Checks for a possible intermittent malfunction in the solenoid circuitry.

Diagnostic Aids:

An "Intermittent" malfunction may be caused by a poor connection, rubbed through wire insulation, or a wire that is broken inside the insulation.

The frequency of the malfunction can be checked by using the enhanced diagnostic function of the Tech 1.

Any circuitry that is suspected of causing the intermittent complaint should be thoroughly checked for backed out terminals, improper mating, broken locks, improperly formed or damaged terminals, poor terminal to wiring connections or physical damage to the wiring harness.

DIAGNOSTIC CHART DTC 77 — 1995 CUTLASS SUPREME, GRAND PRIX, LUMINA, MONTE CARLO AND REGAL

DTC 77

LEFT FRONT SOLENOID CIRCUIT SHORTED TO GND

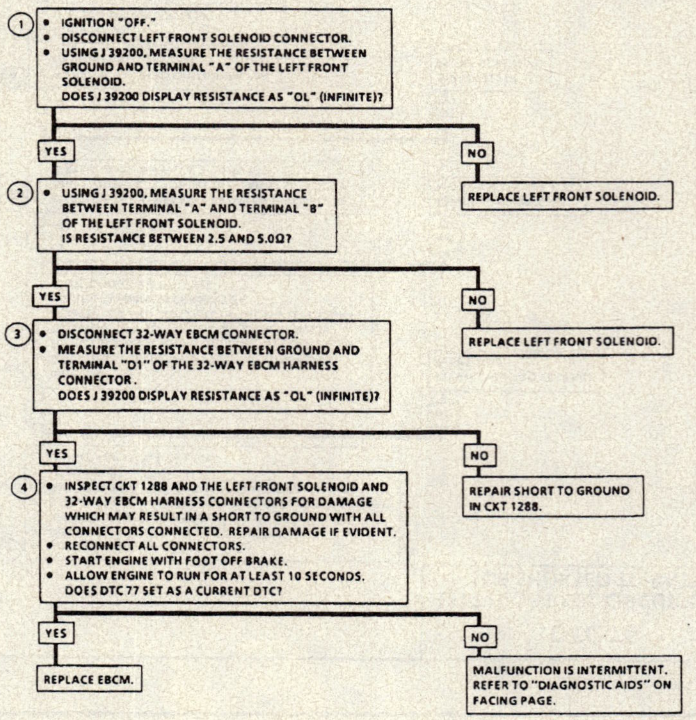

DIAGNOSTIC CHART DTC 78 — 1995 CUTLASS SUPREME, GRAND PRIX, LUMINA, MONTE CARLO AND REGAL

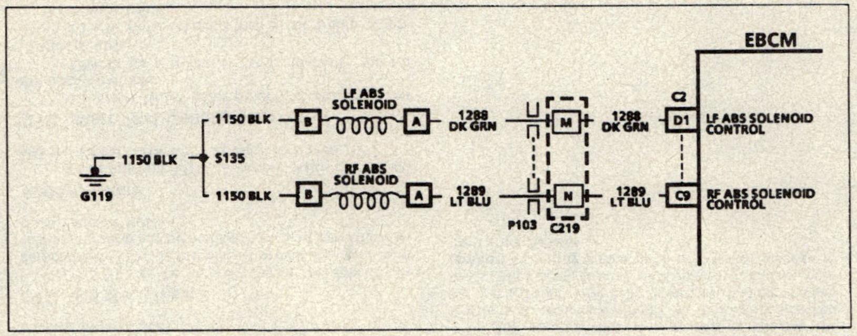

DTC 78

RIGHT FRONT SOL CIRCUIT OPEN OR SHORTED TO BATT

Circuit Description:

This DTC identifies a solenoid that cannot be energized due to an open in its circuitry, or a solenoid that is always energized due to a short to battery in its circuitry between the driver and the solenoid. An open will not allow proper ABS operation, but the short to battery simply turns "ON" the solenoid. A path for base brakes is still allowed once the motor rehomes and the check ball is lifted off its seat during key "ON" initialization.

DTC Will Set When:

DTC 78 can be set only when the solenoid is commanded "OFF." If the EBCM detects an excessive voltage in the right front solenoid control circuit, a malfunction exists.

Action Taken:

A malfunction DTC is stored, ABS is disabled and the amber ABS warning lamp is turned "ON."

DTC Chart Test Description:

Number(s) below refer to circled number(s) on the diagnostic chart.

1. Checks for a short to voltage in the control circuitry of the solenoid.
2. Checks for an open in the control circuit of the solenoid.
3. Checks the solenoid coil for proper resistance.
4. Checks for an open in the solenoid ground circuit.
5. Checks for a possible intermittent malfunction in the solenoid circuitry due to poor terminal contact.

Diagnostic Aids:

An "Intermittent" malfunction may be caused by a poor connection, rubbed through wire insulation, or a wire that is broken inside the insulation.

The frequency of the malfunction can be checked by using the enhanced diagnostic function of the Tech 1.

Any circuitry that is suspected of causing the intermittent complaint should be thoroughly checked for backed out terminals, improper mating, broken locks, improperly formed or damaged terminals, poor terminal to wiring connections or physical damage to the wiring harness.

DIAGNOSTIC CHART DTC 78 — 1995 CUTLASS SUPREME, GRAND PRIX, LUMINA, MONTE CARLO AND REGAL

DTC 78
RIGHT FRONT SOL CIRCUIT OPEN OR SHORTED TO BATT

(1)
- IGNITION "OFF."
- DISCONNECT 32-WAY EBCM CONNECTOR.
- IGNITION "ON," ENGINE "OFF."
- USING J 39200, MEASURE THE VOLTAGE BETWEEN GROUND AND TERMINAL "C9" OF THE 32-WAY EBCM HARNESS CONNECTOR. IS VOLTAGE 1 VOLT OR LESS?

NO → REPAIR SHORT TO VOLTAGE IN CKT 1289.

YES ↓

(2)
- IGNITION "OFF."
- DISCONNECT RIGHT FRONT SOLENOID CONNECTOR.
- USING J 39200, MEASURE THE RESISTANCE BETWEEN TERMINAL "C9" OF THE 32-WAY EBCM HARNESS CONNECTOR AND TERMINAL "A" OF THE RIGHT FRONT SOLENOID HARNESS CONNECTOR. IS RESISTANCE 2Ω OR LESS?

NO → REPAIR OPEN OR HIGH RESISTANCE IN CKT 1289.

YES ↓

(3)
- MEASURE THE RESISTANCE BETWEEN TERMINAL "A" AND TERMINAL "B" OF THE RIGHT FRONT SOLENOID. IS RESISTANCE BETWEEN 2.5 AND 5Ω?

NO → REPLACE RIGHT FRONT SOLENOID.

YES ↓

(4)
- MEASURE THE RESISTANCE BETWEEN GROUND AND TERMINAL "B" OF THE RIGHT FRONT SOLENOID HARNESS CONNECTOR. IS RESISTANCE 2Ω OR LESS?

NO → REPAIR OPEN OR HIGH RESISTANCE IN CKT 1150.

YES ↓

(5)
- INSPECT RIGHT FRONT SOLENOID AND 32-WAY EBCM HARNESS CONNECTORS FOR POOR TERMINAL CONTACT AND EVIDENCE OF CORROSION. REPLACE ALL TERMINALS THAT EXHIBIT SIGNS OF POOR TERMINAL CONTACT OR CORROSION.
- INSPECT CKT 1289 FOR DAMAGE WHICH MAY RESULT IN A SHORT TO VOLTAGE WITH ALL CONNECTORS CONNECTED. REPAIR DAMAGE IF EVIDENT.
- RECONNECT ALL CONNECTORS.
- IGNITION "ON."

DOES DTC 78 SET AS A CURRENT DTC?

YES → REPLACE EBCM.

NO → MALFUNCTION IS MOST LIKELY DUE TO A POOR CONNECTION. REFER TO "DIAGNOSTIC AIDS" ON FACING PAGE.

DIAGNOSTIC CHART DTC 81 — 1995 CUTLASS SUPREME, GRAND PRIX, LUMINA, MONTE CARLO AND REGAL

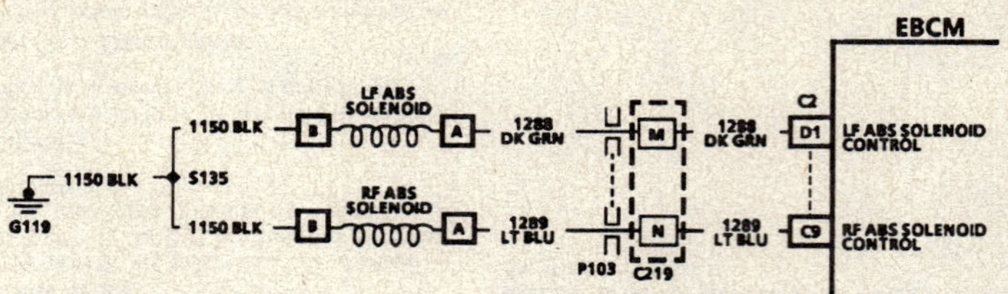

DTC 81
RIGHT FRONT SOLENOID CIRCUIT SHORTED TO GND

Circuit Description:

This DTC identifies a solenoid that cannot be energized due to an open in its driver circuitry, or a short to ground between the solenoid driver and the solenoid. These malfunctions can affect ABS operation since the flow of brake fluid to the wheel cylinder cannot be stopped, making ABS operation for that channel impossible.

DTC Will Set When:

DTC 81 can be set only when the solenoid is commanded "ON." If the EBCM detects the right front solenoid control circuit voltage is out of specification, a malfunction exists.

Action Taken:

A malfunction DTC is stored, ABS is disabled and the amber ABS warning lamp is turned "ON."

DTC Chart Test Description:

Number(s) below refer to circled number(s) on the diagnostic chart.

1. Checks for a solenoid that is internally shorted to ground.
2. Checks for a solenoid that is not within proper resistance values.
3. Indicates if a short to ground exists in the solenoid circuitry.
4. Checks for a possible intermittent malfunction in the solenoid circuitry.

Diagnostic Aids:

An "Intermittent" malfunction may be caused by a poor connection, rubbed through wire insulation, or a wire that is broken inside the insulation.

The frequency of the malfunction can be checked by using the enhanced diagnostic function of the Tech 1.

Any circuitry that is suspected of causing the intermittent complaint should be thoroughly checked for backed out terminals, improper mating, broken locks, improperly formed or damaged terminals, poor terminal to wiring connections or physical damage to the wiring harness.

DIAGNOSTIC CHART DTC 81 — 1995 CUTLASS SUPREME, GRAND PRIX, LUMINA, MONTE CARLO AND REGAL

DTC 81
RIGHT FRONT SOLENOID CIRCUIT SHORTED TO GND

(1)
- IGNITION "OFF."
- DISCONNECT RIGHT FRONT SOLENOID CONNECTOR.
- USING J 39200, MEASURE THE RESISTANCE BETWEEN GROUND AND TERMINAL "A" OF THE RIGHT FRONT SOLENOID.

DOES T 39200 DISPLAY RESISTANCE AS "OL" (INFINITE)?

NO: REPLACE RIGHT FRONT SOLENOID.

YES:

(2)
- USING J 39200, MEASURE THE RESISTANCE BETWEEN TERMINAL "A" AND TERMINAL "B" OF THE RIGHT FRONT SOLENOID.

IS RESISTANCE BETWEEN 2.5 AND 5.0Ω?

NO: REPLACE RIGHT FRONT SOLENOID.

YES:

(3)
- DISCONNECT 32-WAY EBCM CONNECTOR.
- MEASURE THE RESISTANCE BETWEEN GROUND AND TERMINAL "C9" OF THE 32-WAY EBCM HARNESS CONNECTOR.

DOES J 39200 DISPLAY RESISTANCE AS "OL" (INFINITE)?

NO: REPAIR SHORT TO GROUND IN CKT 1289.

YES:

(4)
- INSPECT CKT 1289 AND THE RIGHT FRONT SOLENOID AND 32-WAY EBCM HARNESS CONNECTORS FOR DAMAGE WHICH MAY RESULT IN A SHORT TO GROUND WITH ALL CONNECTORS CONNECTED. REPAIR DAMAGE IF EVIDENT.
- RECONNECT ALL CONNECTORS.
- START ENGINE WITH FOOT OFF BRAKE.
- ALLOW ENGINE TO RUN FOR AT LEAST 10 SECONDS.

DOES DTC 81 SET AS A CURRENT DTC?

YES: REPLACE EBCM.

NO: MALFUNCTION IS INTERMITTENT. REFER TO "DIAGNOSTIC AIDS" ON FACING PAGE.

DIAGNOSTIC CHART DTC 82 — 1995 CUTLASS SUPREME, GRAND PRIX, LUMINA, MONTE CARLO AND REGAL

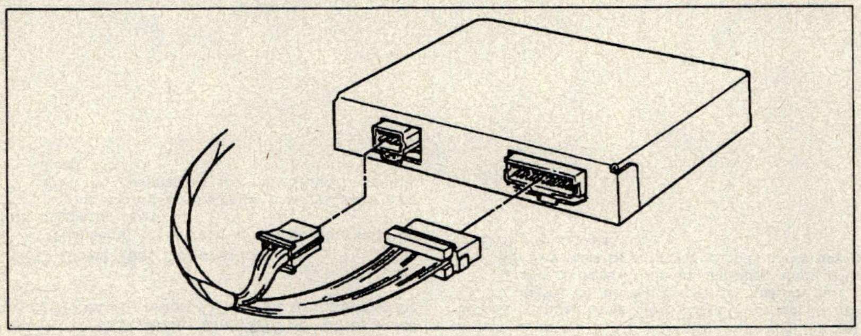

DTC 82
CALIBRATION MALFUNCTION

Circuit Description:

This DTC allows the EBCM to check for a calibration malfunction by comparing the calibration value to a known value stored in the EEPROM.

This DTC is also used as a security measure to prevent improper use of calibrations or changes to these calibrations that may alter the designed function of ABS.

DTC Will Set When:

DTC 82 can be set at key-up. If the program identifier is incorrect or the memory checksum is incorrect, a malfunction exists.

Action Taken:

A malfunction DTC is stored, ABS is disabled and the amber ABS warning lamp is turned "ON."

DTC Chart Test Description:

Number(s) below refer to circled number(s) on the diagnostic chart.

1. Checks to see if the malfunction is present during diagnosis. If present, the EBCM is not functioning correctly and must be replaced.

Important

- At the time of printing, 1995 vehicles with ABS VI were not being field reprogrammed to correct this failure. A service bulletin will be issued when EEPROM reprogramming is authorized.

Diagnostic Aids:

An intermittent DTC 82 may be caused by a bad cell in the EEPROM that is sensitive to temperature changes. If DTC 82 failed more than once, but is intermittent, replace EBCM.

The frequency of the malfunction can be checked by using the enhanced diagnostic function of the Tech 1.

DIAGNOSTIC CHART DTC 82 — 1995 CUTLASS SUPREME, GRAND PRIX, LUMINA, MONTE CARLO AND REGAL

DTC 82
CALIBRATION MALFUNCTION

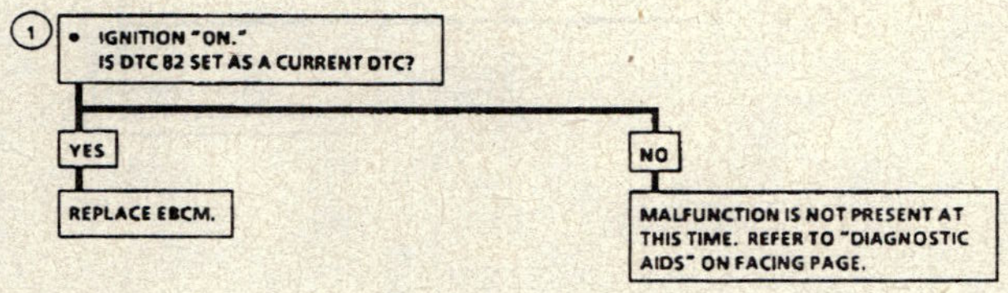

DIAGNOSTIC CHART DTC 86 — 1995 CUTLASS SUPREME, GRAND PRIX, LUMINA, MONTE CARLO AND REGAL

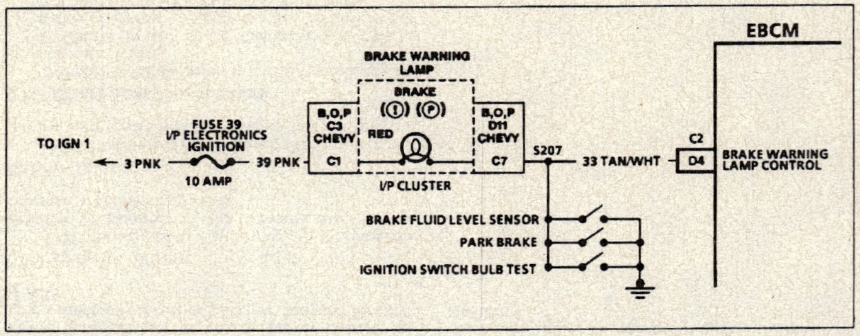

DTC 86
EBCM TURNED "ON" THE RED BRAKE WARNING LAMP

Circuit Description:

This DTC is provided as an information only DTC, and reflects the status of the command issued by the ABS EBCM to illuminate the red BRAKE warning lamp. If another DTC issues a command to illuminate the red BRAKE warning lamp, DTC 86 will be stored in EEPROM as a history DTC at the conclusion of the ignition cycle.

DTC Chart Test Description:

Number(s) below refer to circled number(s) on the diagnostic chart.

1. This step serves to identify if a DTC other than DTC 86 commanded the red BRAKE warning lamp "ON."

Diagnostic Aids:

Any ABS mechanical DTC that issues a command to illuminate the red BRAKE warning lamp will also result in DTC 86 being stored in EEPROM during shutdown. These DTCs are: 38, 41, 42, 46, and 51. If the motors are not in their home position, certain electrical DTCs will also command the red BRAKE warning lamp "ON." These DTCs are: 14, 16, 18, 55, 56, 57, 58, 61, 62, 63, 64, 65, and 66.

If any of these DTCs are indicated along with DTC 86, they must be corrected prior to addressing a DTC 86 malfunction.

DIAGNOSTIC CHART DTC 86 — 1995 CUTLASS SUPREME, GRAND PRIX, LUMINA, MONTE CARLO AND REGAL

DTC 86
EBCM TURNED "ON" THE RED BRAKE WARNING LAMP

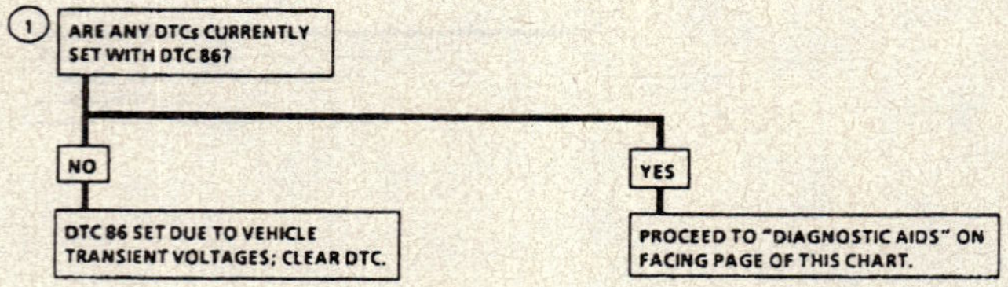

DIAGNOSTIC CHART DTC 87 — 1995 CUTLASS SUPREME, GRAND PRIX, LUMINA, MONTE CARLO AND REGAL

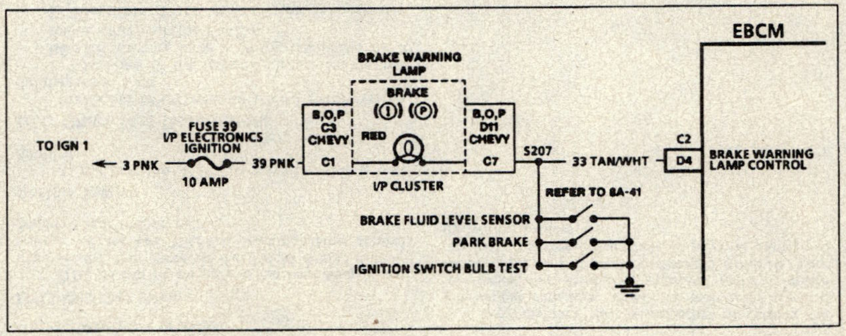

DTC 87
RED BRAKE WARNING LAMP CIRCUIT OPEN

Circuit Description:

This DTC is used to verify the EBCM has continuity to the red BRAKE warning lamp in case the EBCM must turn it "ON." This will only occur if an ABS malfunction is detected that may degrade base brake operation.

Because ABS is not the only device controlling the red BRAKE warning lamp (parking brake and brake fluid level sensor may also turn "ON" the lamp) a short to ground in this circuit cannot be detected.

DTC Will Set When:

DTC 87 can be set after initialization is completed. If the red BRAKE warning lamp circuit voltage is out of specification indicating an open circuit, a malfunction exists.

Action Taken:

A malfunction DTC is stored, ABS is not disabled.

DTC Chart Test Description:

Number(s) below refer to circled number(s) on the diagnostic chart.

1. Identifies if the ground circuit to the red BRAKE warning lamp is completed by a source other than the EBCM.
2. Checks for high resistance in park brake circuitry.
3. Checks for a malfunctioning EBCM.
4. Checks for high resistance in brake fluid level sensor circuitry.
5. Checks for a malfunctioning red BRAKE warning lamp bulb.
6. Checks for open or high resistance in CKT 33.

Diagnostic Aids:

An "Intermittent" malfunction may be caused by a poor connection, rubbed through wire insulation, or a wire that is broken inside the insulation.

The frequency of the malfunction can be checked by using the enhanced diagnostic function of the Tech 1.

If a Tech 1 or T-100 are not available, the parking brake may be applied to verify proper lamp operation and continuity of CKTs 3 and 39.

Any circuitry that is suspected of causing the intermittent complaint should be thoroughly checked for backed out terminals, improper mating, broken locks, improperly formed or damaged terminals, poor terminal to wiring connections or physical damage to the wiring harness.

DIAGNOSTIC CHART DTC 87 — 1995 CUTLASS SUPREME, GRAND PRIX, LUMINA, MONTE CARLO AND REGAL

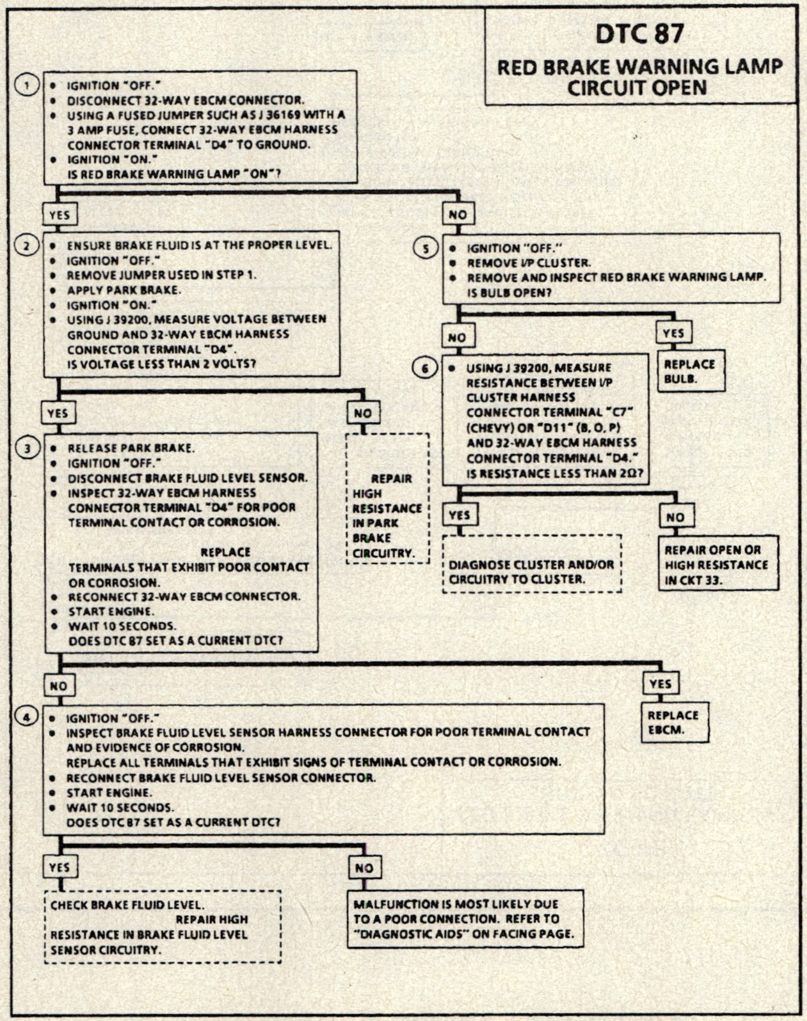

DIAGNOSTIC CHART DTC 88 — 1995 CUTLASS SUPREME, GRAND PRIX, LUMINA, MONTE CARLO AND REGAL

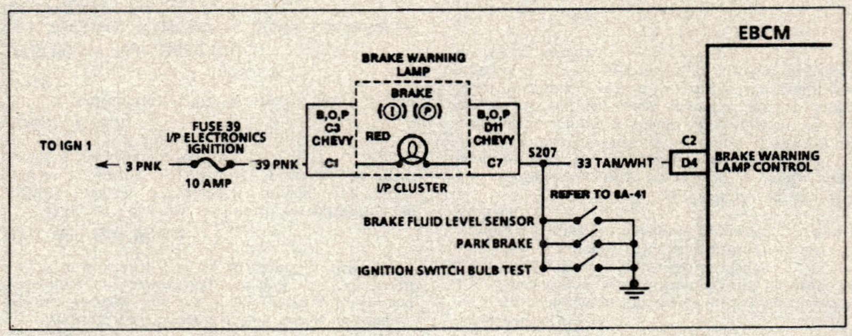

DTC 88
RED BRAKE WARNING LAMP CKT SHORTED TO BATT

Circuit Description:

This DTC identifies a short to battery between the EBCM and the red BRAKE warning lamp, or an open driver that does not allow the red BRAKE warning lamp to be illuminated by the EBCM. This will only occur if an ABS fault is detected that may degrade base brake operation.

DTC Will Set When:

DTC 88 can be set only at the beginning of the three second bulb check. If the EBCM detects battery voltage on the red BRAKE warning lamp control circuit, a malfunction exists.

Action Taken:

A malfunction DTC is stored. ABS is not disabled.

DTC Chart Test Description:

Number(s) below refer to circled number(s) on the diagnostic chart.

1. Identifies if the ground circuit to the red BRAKE warning lamp is being completed by a source other than the EBCM.
2. Identifies if the malfunction is currently present.
3. By removing fuse 39, the voltage source is eliminated. This test indicates if voltage is being supplied from a source other than fuse 39.
4. Ensures malfunction was not due to physical damage of the circuitry.
5. Checks for DTC 86 also being set.

Diagnostic Aids:

An "Intermittent" malfunction may be caused by a poor connection, rubbed through wire insulation, or a wire that is broken inside the insulation.

The frequency of the malfunction can be checked by using the enhanced diagnostic function of the Tech 1.

Any circuitry that is suspected of causing the intermittent complaint should be thoroughly checked for backed out terminals, improper mating, broken locks, improperly formed or damaged terminals, poor terminal to wiring connections or physical damage to the wiring harness.

DIAGNOSTIC CHART DTC 88 — 1995 CUTLASS SUPREME, GRAND PRIX, LUMINA, MONTE CARLO AND REGAL

DTC 88
RED BRAKE WARNING LAMP CKT SHORTED TO BATT

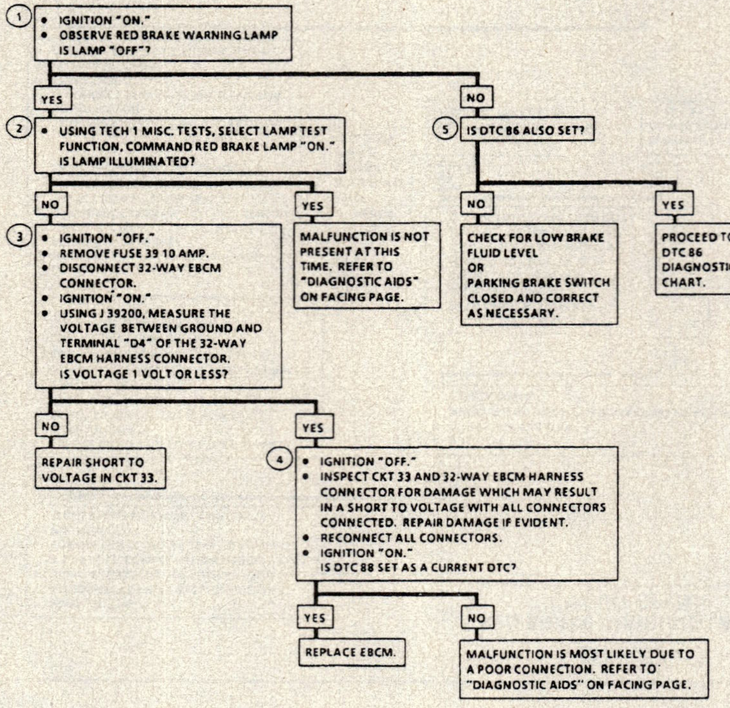

DIAGNOSTIC CHART DTC 91 — 1995 CUTLASS SUPREME, GRAND PRIX, LUMINA, MONTE CARLO AND REGAL

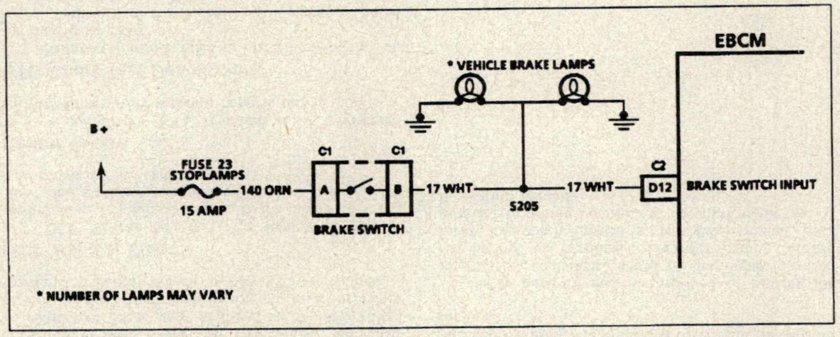

DTC 91
OPEN BRAKE SWITCH DURING DECELERATION

Circuit Description:

This DTC is used to detect an open brake switch in the non-ABS mode. The EBCM looks for deceleration rates that would indicate braking action and verifies this assumption by requiring several repeats of this detection method. In each case, ABS will not be available since no brake switch is seen.

DTC Will Set When:

DTC 91 can be set if three deceleration cycles within the specified rate occur during the current ignition cycle with the brake switch "OFF."

Action Taken:

A malfunction DTC is stored, ABS is disabled and the amber ABS warning lamp is turned "ON."

DTC Chart Test Description:

Number(s) below refer to circled number(s) on the diagnostic chart.

1. Indicates if the brake switch signal is being received by the EBCM.
2. Indicates if an open circuit exists in the brake switch or brake lamp circuitry.
3. Isolates the open circuit to either the brake switch input circuitry, or the EBCM.
4. Ensures malfunction was not due to poor terminal contact.
5. Verifies that voltage is available at the brake switch.
6. Verifies that the brake switch is functioning properly.

Diagnostic Aids:

An "Intermittent" malfunction may be caused by a poor connection, rubbed through wire insulation, or a wire that is broken inside the insulation.

The frequency of the malfunction can be checked by using the enhanced diagnostic function of the Tech 1.

Any circuitry that is suspected of causing the intermittent complaint should be thoroughly checked for backed out terminals, improper mating, broken locks, improperly formed or damaged terminals, poor terminal to wiring connections or physical damage to the wiring harness.

DIAGNOSTIC CHART DTC 91 — 1995 CUTLASS SUPREME, GRAND PRIX, LUMINA, MONTE CARLO AND REGAL

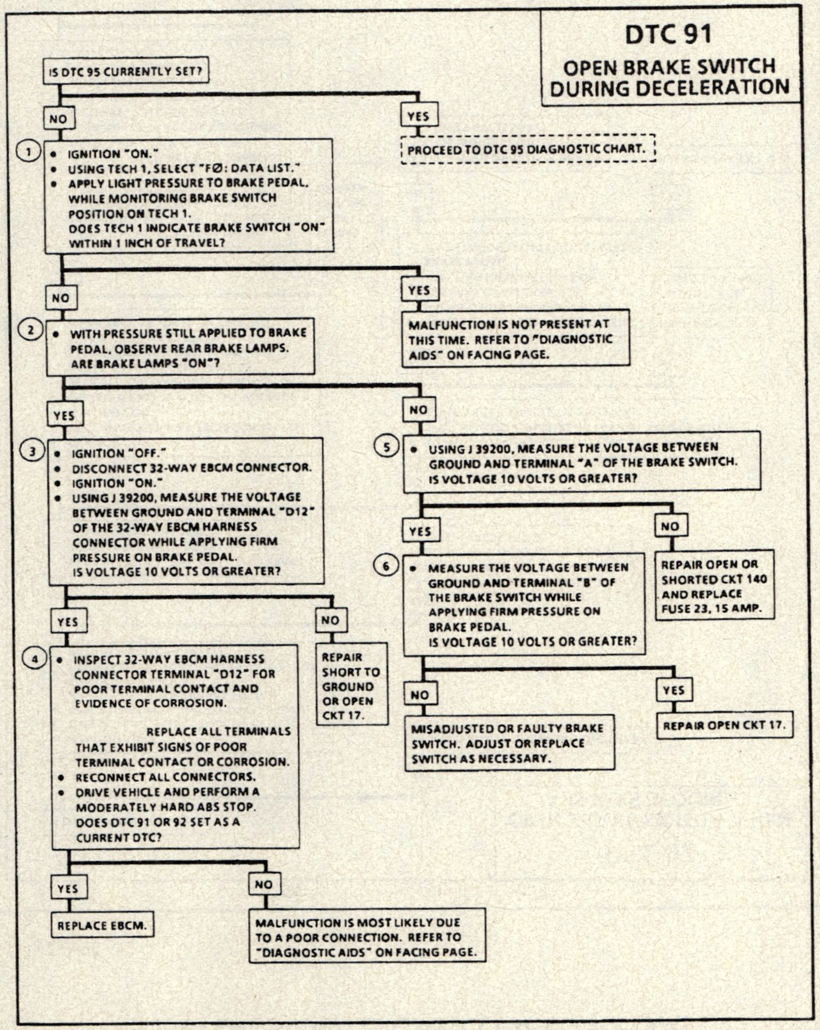

DIAGNOSTIC CHART DTC 92 — 1995 CUTLASS SUPREME, GRAND PRIX, LUMINA, MONTE CARLO AND REGAL

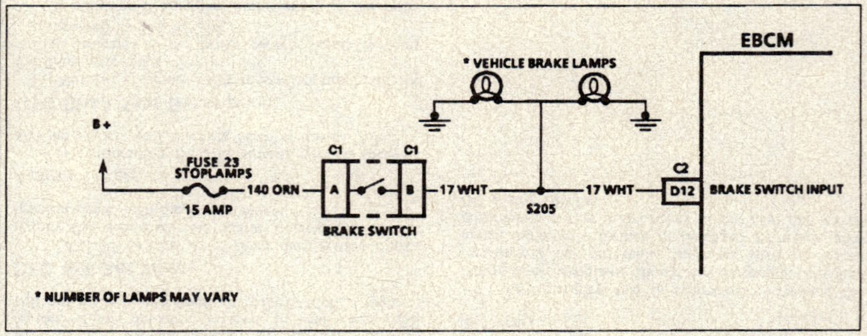

DTC 92
OPEN BRAKE SWITCH WHEN ABS WAS REQUIRED

Circuit Description:

This DTC is run to determine the proper operation of the brake switch. This is important because ABS is activated when the brake switch is "ON." If the brake switch is "OFF," ABS will never be activated. Since this malfunction is difficult to detect under normal braking conditions, this malfunction is only detected when ABS is required.

DTC Will Set When:

DTC 92 can be set if the vehicle's speed is greater than 8 km/h (5 mph). If the brake was not "ON" and a release was required on two channels for 0.5 second, a malfunction exists.

Action Taken:

A malfunction DTC is stored, ABS is disabled and the amber ABS warning lamp is turned "ON."

DTC Chart Test Description:

Number(s) below refer to circled number(s) on the diagnostic chart.

1. Indicates if the brake switch signal is being received by the EBCM.
2. Indicates if an open circuit exists in the brake switch or brake lamp circuitry.
3. Isolates the open circuit to either the brake switch input circuitry, or the EBCM.
4. Ensures malfunction was not due to poor terminal contact.
5. Verifies the voltage is available at the brake s witch.
6. Verifies that the brake switch is functioning properly.

Diagnostic Aids:

An "Intermittent" malfunction may be caused by a poor connection, rubbed through wire insulation, or a wire that is broken inside the insulation.

The frequency of the malfunction can be checked by using the enhanced diagnostic function of the Tech 1.

Any circuitry that is suspected of causing the intermittent complaint should be thoroughly checked for backed out terminals, improper mating, broken locks, improperly formed or damaged terminals, poor terminal to wiring connections or physical damage to the wiring harness.

DIAGNOSTIC CHART DTC 92 — 1995 CUTLASS SUPREME, GRAND PRIX, LUMINA, MONTE CARLO AND REGAL

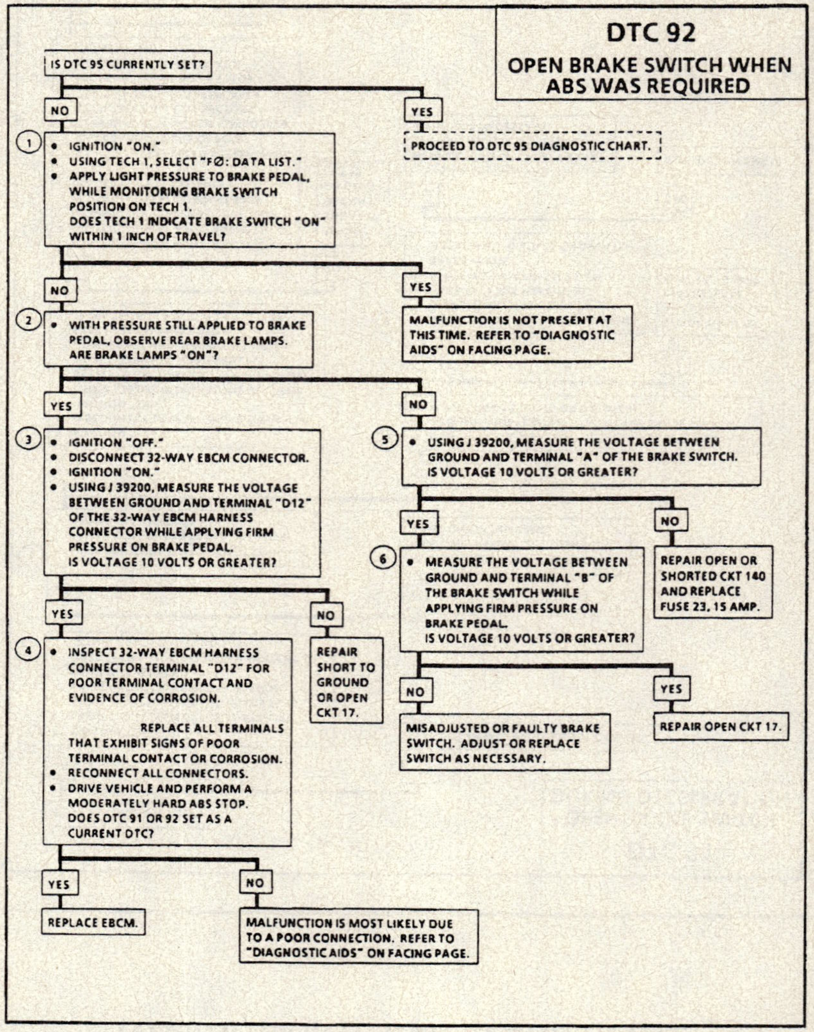

DIAGNOSTIC CHART DTC 93 — 1995 CUTLASS SUPREME, GRAND PRIX, LUMINA, MONTE CARLO AND REGAL

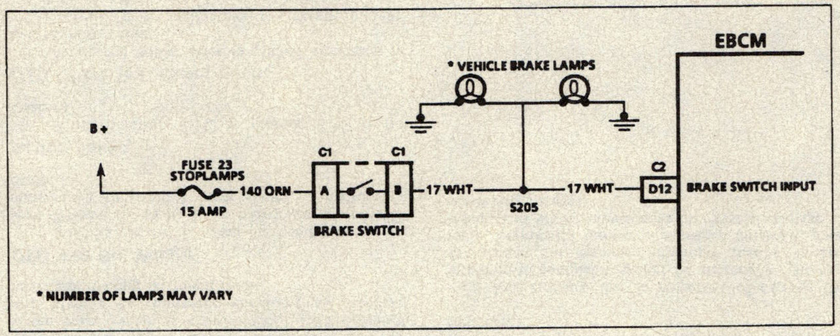

DTC 93

DTC 91 OR 92 SET IN CURRENT OR PREV. IGN CYCLE

Circuit Description:

This DTC is the second portion of DTCs 91 and 92. If DTCs 91 or 92 failed during the last ignition cycle, DTC 93 becomes a current failure during the next ignition cycle, keeping ABS disabled until a brake switch "ON" state is seen. When a change is seen during an ignition cycle in which DTC 93 is a current malfunction, DTC 91 or 92 will clear itself at the end of the current ignition cycle, and ABS will enable itself at the start of the next ignition cycle. DTC 93 alone indicates 91 or 92 failed previously, but is intermittent, or has been corrected.

DTC Chart Test Description:

Number(s) below refer to circled number(s) on the diagnostic chart.

1. Indicates which DTC (either 91 or 92) caused DTC 93 to set.

Diagnostic Aids:

An "Intermittent" malfunction may be caused by a poor connection, rubbed through wire insulation, or a wire that is broken inside the insulation.

The frequency of the malfunction can be checked by using the enhanced diagnostic function of the Tech 1.

Any circuitry that is suspected of causing the intermittent complaint should be thoroughly checked for backed out terminals, improper mating, broken locks, improperly formed or damaged terminals, poor terminal to wiring connections or physical damage to the wiring harness.

Also, verify proper brake switch operation using the data list of the Tech 1. As the brake is applied, the data list should display the brake switch "ON" within 1 inch of travel.

DIAGNOSTIC CHART DTC 93 — 1995 CUTLASS SUPREME, GRAND PRIX, LUMINA, MONTE CARLO AND REGAL

DTC 93
DTC 91 OR 92 SET IN CURRENT OR PREV. IGN CYCLE

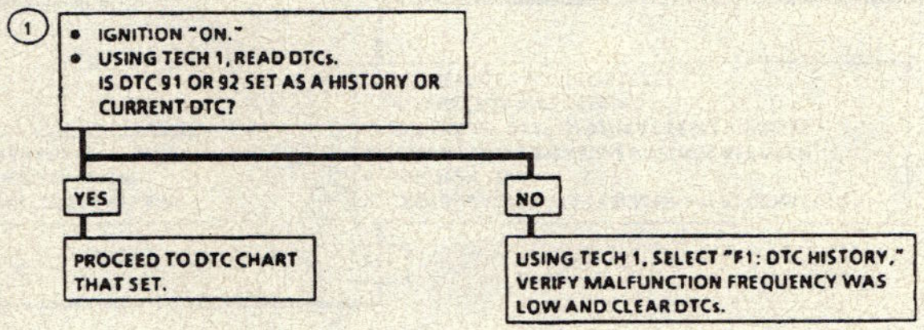

DIAGNOSTIC CHART DTC 94 — 1995 CUTLASS SUPREME, GRAND PRIX, LUMINA, MONTE CARLO AND REGAL

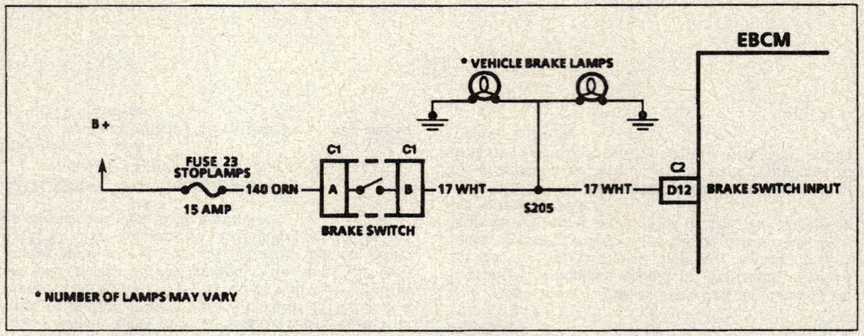

DTC 94
BRAKE SWITCH CONTACTS ALWAYS CLOSED

Circuit Description:

This DTC is run to determine the proper operation of the brake switch. This is important because ABS is activated when the brake switch is "ON" and deactivated when the brake switch is "OFF." If the brake switch is always "ON," ABS operation will always be requested resulting in potential modulator cycling on rough roads.

DTC Will Set When:

DTC 94 can be set when the vehicle reaches at least 40 km/h (25 mph). If the brake was never "OFF" during two consecutive drive cycles, a malfunction exists.

Action Taken:

A malfunction DTC is stored. ABS is not disabled.

DTC Chart Test Description:

Number(s) below refer to circled number(s) on the diagnostic chart.

1. Checks if the malfunction is currently present.
2. Isolates the cause of the malfunction to either a malfunctioning or misadjusted brake switch, or a short to voltage in the brake switch circuitry.
3. Checks for unwanted voltage on the brake s witch input circuit.
4. Checks for a possible intermittent malfunction.

Diagnostic Aids:

An "Intermittent" malfunction may be caused by a poor connection, rubbed through wire insulation, or a wire that is broken inside the insulation.

The frequency of the malfunction can be checked by using the enhanced diagnostic function of the Tech 1.

Any circuitry that is suspected of causing the intermittent complaint should be thoroughly checked for backed out terminals, improper mating, broken locks, improperly formed or damaged terminals, poor terminal to wiring connections or physical damage to the wiring harness.

DIAGNOSTIC CHART DTC 94 — 1995 CUTLASS SUPREME, GRAND PRIX, LUMINA, MONTE CARLO AND REGAL

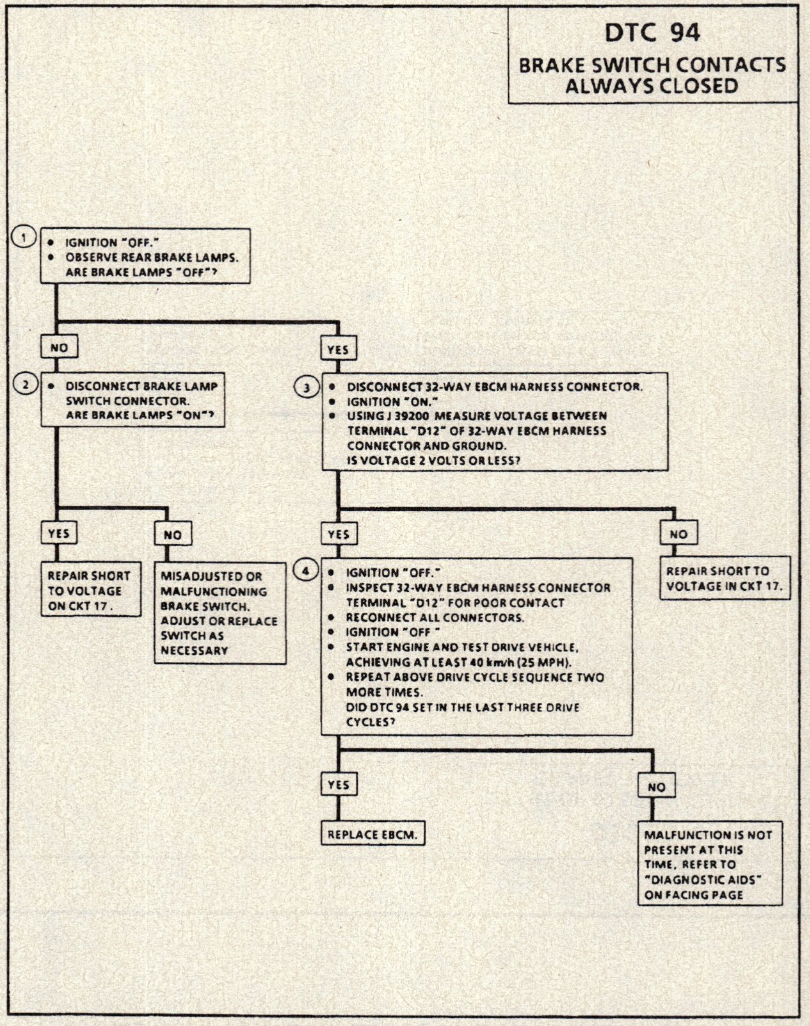

DIAGNOSTIC CHART DTC 95 — 1995 CUTLASS SUPREME, GRAND PRIX, LUMINA, MONTE CARLO AND REGAL

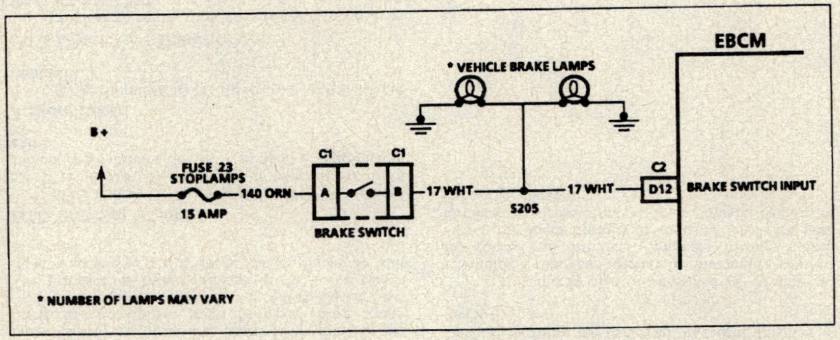

DTC 95
BRAKE SWITCH CIRCUIT OPEN

Circuit Description:

This DTC is used to identify open brake switch circuitry that prevents the brake switch input to the EBCM from changing states when the brake is applied. This DTC is used in conjunction with DTCs 91 and 92 to determine the cause of an open brake switch malfunction.

DTC Will Set When:

DTC 95 can be set after initialization is completed. If the brake switch input voltage is out of specification for one second, indicating an open circuit, a malfunction exists.

Action Taken:

A malfunction DTC is stored, ABS is disabled and the amber ABS warning lamp is turned "ON."

DTC Chart Test Description:

Number(s) below refer to circled number(s) on the diagnostic chart.

1. This step is used to confirm that an open in the brake switch circuitry currently exists.
2. Indicates if the brake switch signal is being received by the EBCM.
3. Indicates if an open circuit exists in the brake switch or brake lamp circuitry.
4. Isolates the open circuit to either the brake switch input circuitry, or the EBCM.

Diagnostic Aids:

An "Intermittent" malfunction may be caused by a poor connection, rubbed through wire insulation, or a wire that is broken inside the insulation.

The frequency of the malfunction can be checked by using the enhanced diagnostic function of the Tech 1.

Any circuitry that is suspected of causing the intermittent complaint should be thoroughly checked for backed out terminals, improper mating, broken locks, improperly formed or damaged terminals, poor terminal to wiring connections or physical damage to the wiring harness.

DIAGNOSTIC CHART DTC 95 — 1995 CUTLASS SUPREME, GRAND PRIX, LUMINA, MONTE CARLO AND REGAL

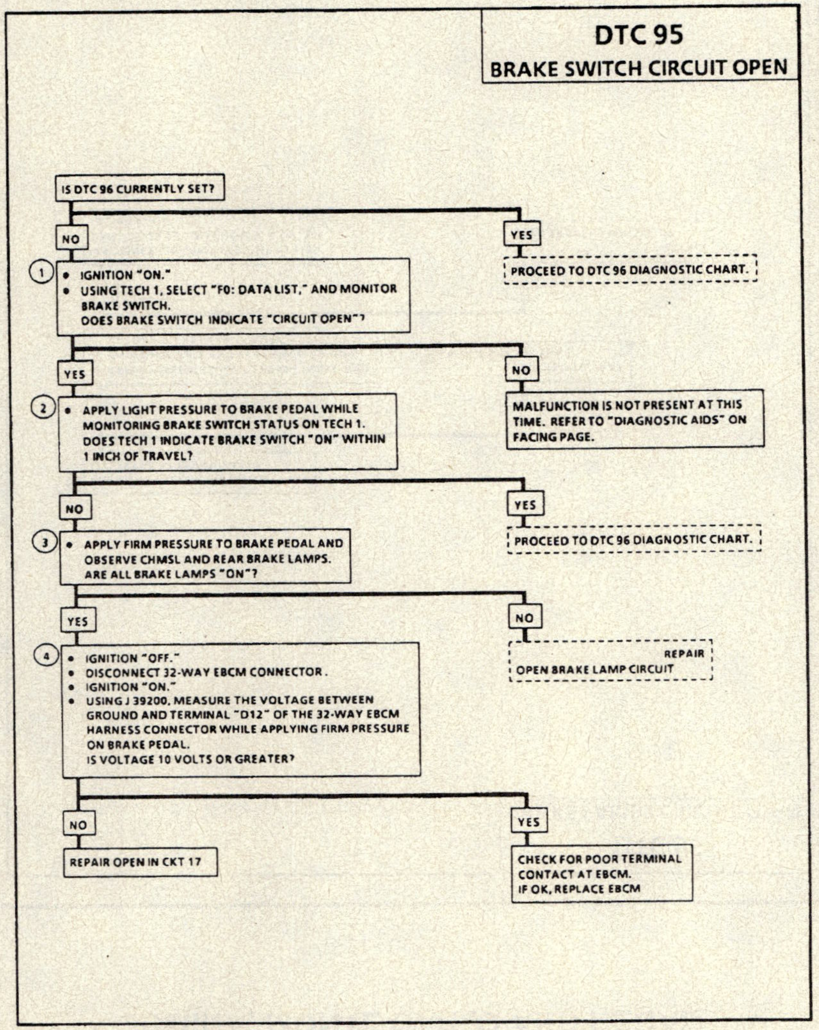

DIAGNOSTIC CHART DTC 96 — 1995 CUTLASS SUPREME, GRAND PRIX, LUMINA, MONTE CARLO AND REGAL

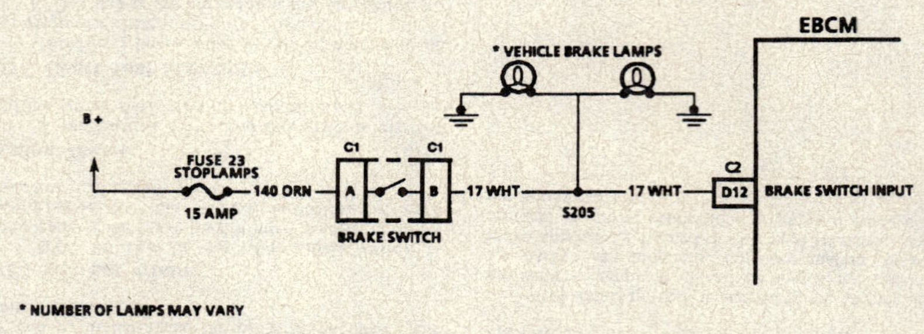

DTC 96
BRAKE LAMPS CIRCUIT OPEN

Circuit Description:

This DTC is designed to isolate the cause of a DTC 95 malfunction and indicate to the driver ABS is still available. If DTC 95 occurs with DTC 96, the brake lamps are open, the brake lamp grounds are open, or the Center High Mounted Stop Lamp (CHMSL) circuit is open during 4-way flasher use. The presence of battery voltage at the brake switch input indicates a valid brake switch input is still available.

DTC Will Set When:

DTC 96 can be set only after DTC 95 has been set. If the EBCM detects battery voltage on the brake switch input circuit for 0.5 second, a malfunction exists.

Action Taken:

A malfunction DTC is stored. ABS is not disabled.

DTC Chart Test Description:

Number(s) below refer to circled number(s) on the diagnostic chart.

1. As a result of a malfunction of an additional brake switch circuit DTC, this DTC may be set. To ensure proper diagnosis, any additional brake switch DTCs must be repaired first.
2. Identifies if the malfunction is currently present in the brake circuit.

Diagnostic Aids:

An "Intermittent" malfunction may be caused by a poor connection, rubbed through wire insulation, or a wire that is broken inside the insulation.

The frequency of the malfunction can be checked by using the enhanced diagnostic function of the Tech 1.

Any circuitry that is suspected of causing the intermittent complaint should be thoroughly checked for backed out terminals, improper mating, broken locks, improperly formed or damaged terminals, poor terminal to wiring connections or physical damage to the wiring harness.

DIAGNOSTIC CHART DTC 96 — 1995 CUTLASS SUPREME, GRAND PRIX, LUMINA, MONTE CARLO AND REGAL

DTC 96
BRAKE LAMPS CIRCUIT OPEN

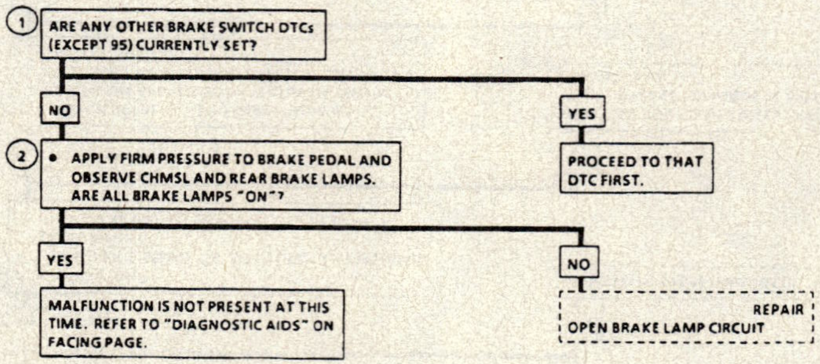

DIAGNOSTIC PROCESS CHART — 1993 SILHOUETTE, TRANS SPORT AND LUMINA APV

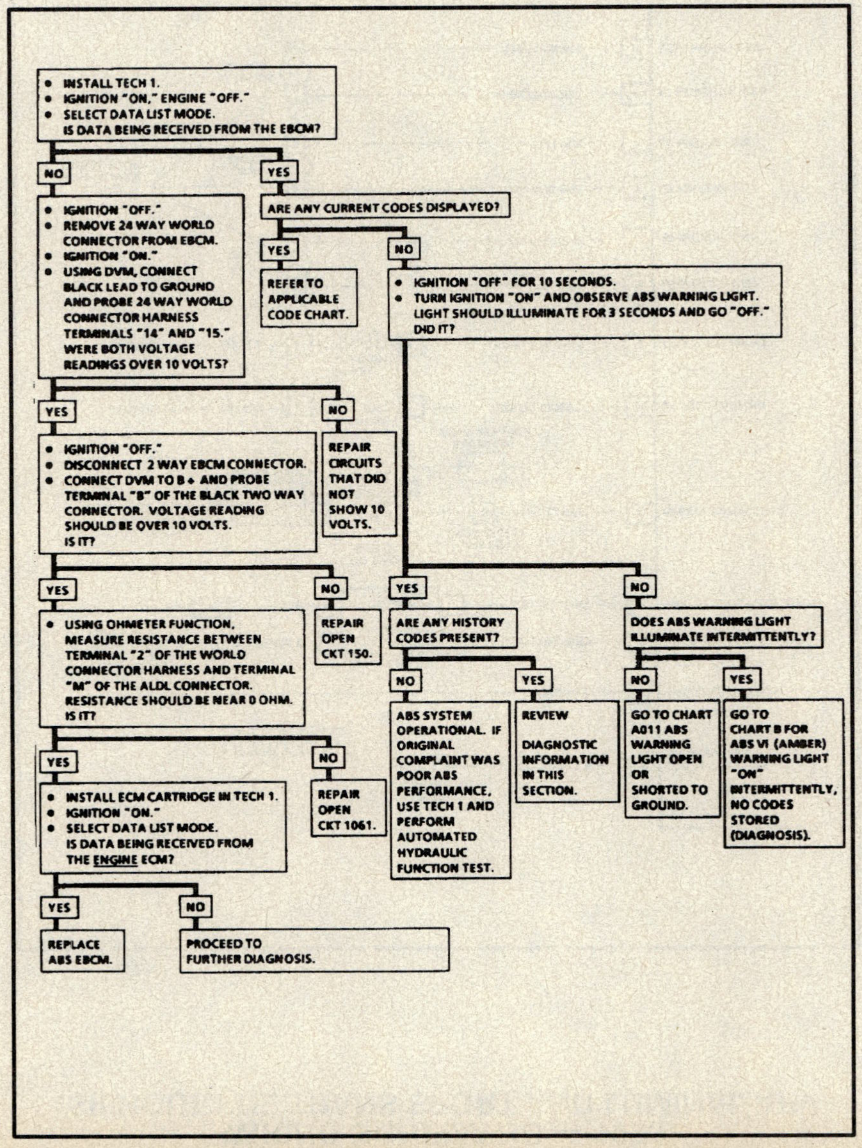

WIRING SCHEMATIC — 1993 SILHOUETTE, TRANS SPORT AND LUMINAL APV

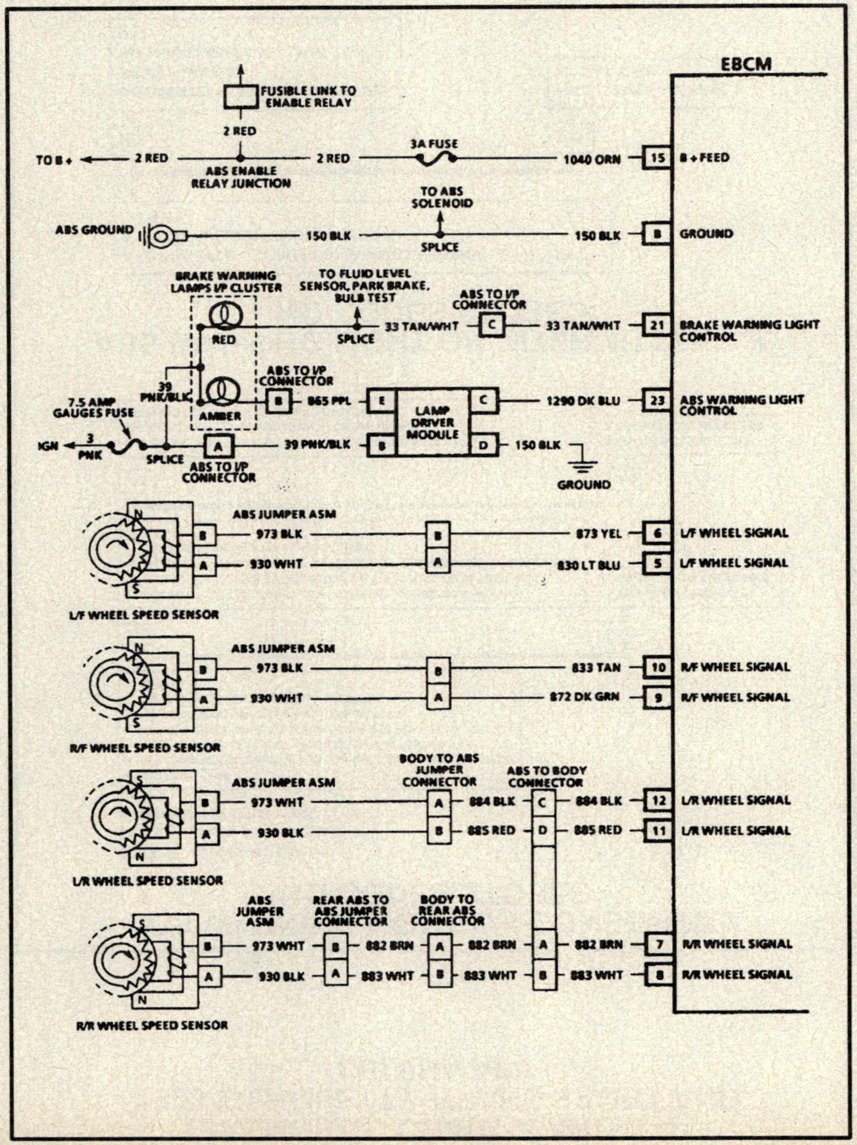

WIRING SCHEMATIC — 1993 SILHOUETTE, TRANS SPORT AND LUMINAL APV

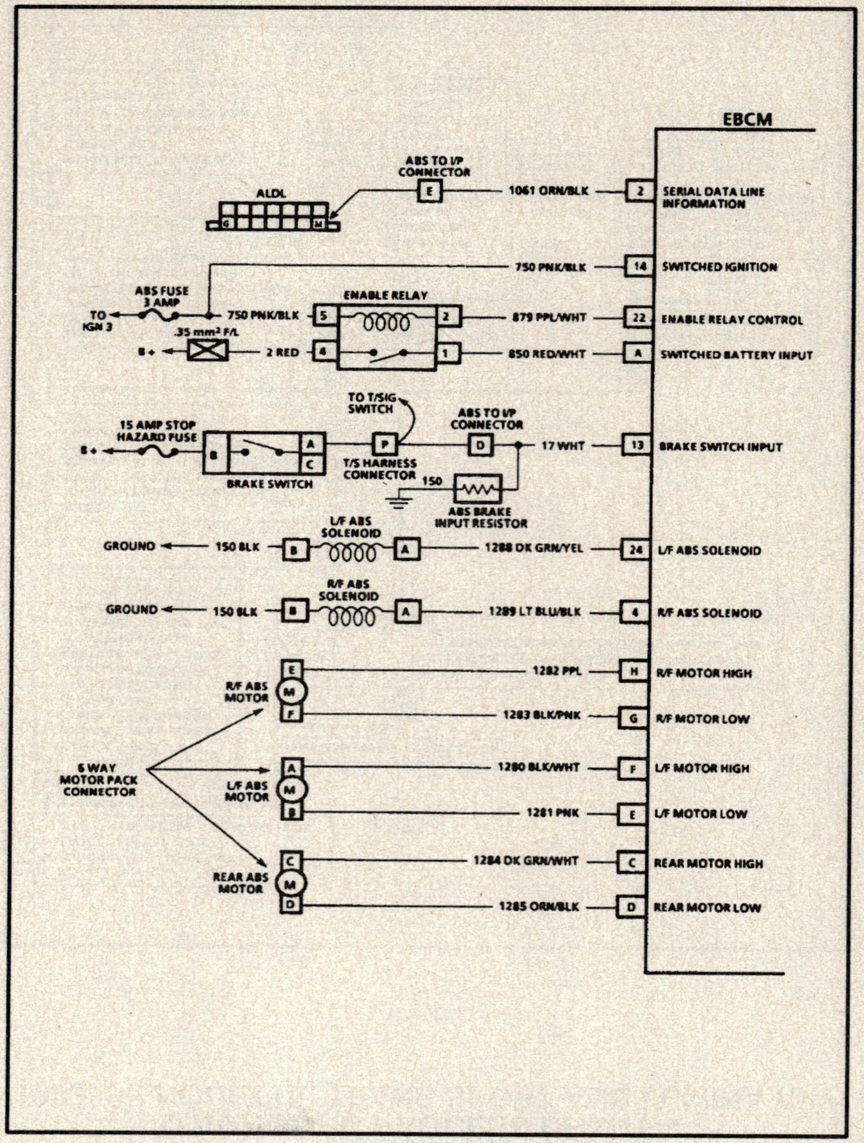

DIAGNOSTIC CHART A AND B — 1993 SILHOUETTE, TRANS SPORT AND LUMINA APV

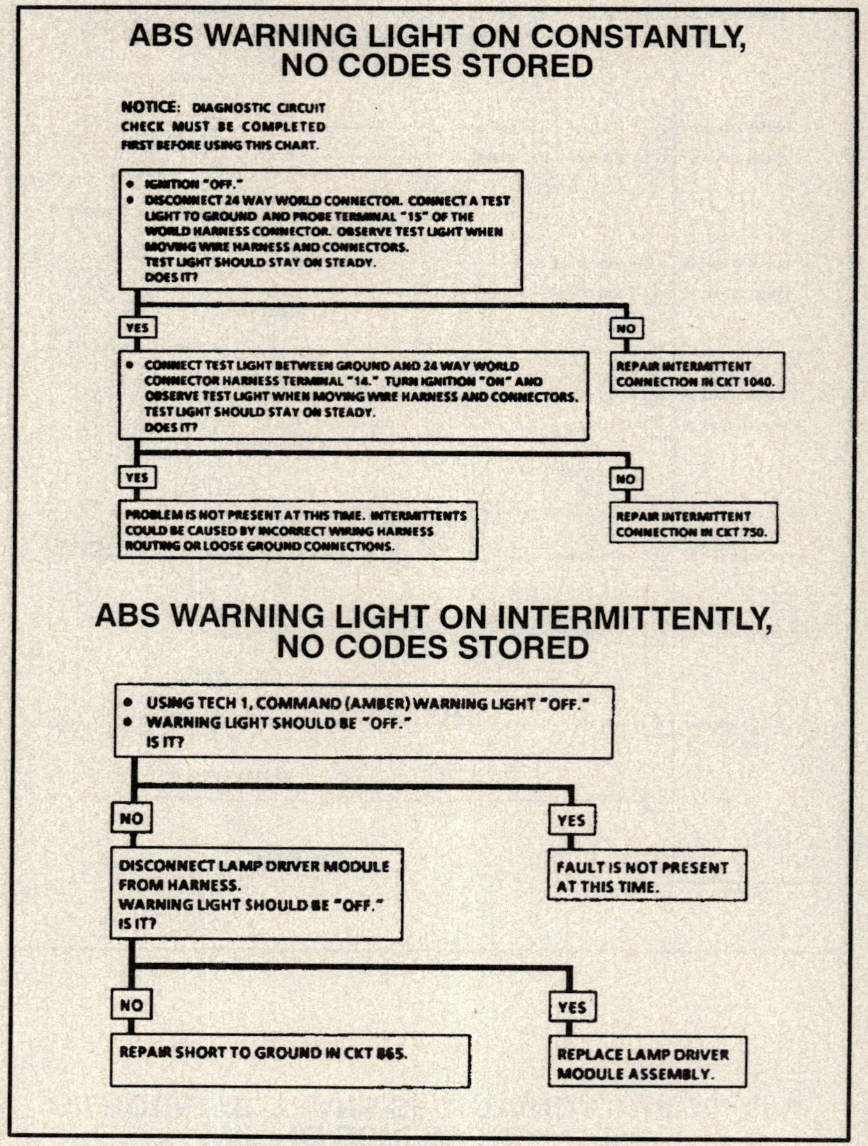

DIAGNOSTIC CHART DTC A011 — 1993 SILHOUETTE, TRANS SPORT AND LUMINA APV

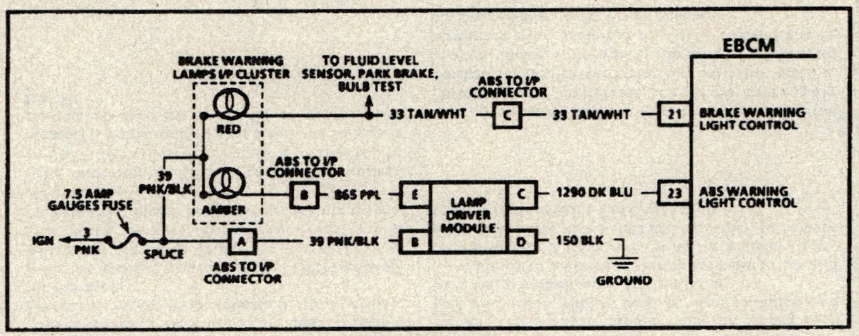

CODE A011

ABS WARNING LIGHT CIRCUIT OPEN OR SHORTED TO GROUND

Circuit Description:

This code checks the state of the ABS warning light to identify a situation in which the driver could not be warned of a system fault by the ABS warning light, or the warning light is always "ON." The Lamp Driver Module (LDM) turns the ABS warning light "ON," unless the EBCM provides ground to turn it "OFF." Because of the circuitry in the LDM, only external faults can be detected. As a result, the ABS warning light itself is not diagnosable, only the control line to the EBCM can be diagnosed. In the event of an open CKT 1290, the ABS warning light will be on at all times, due to the loss of ground at the LDM input. If the control line is shorted to ground, the ABS warning light is kept "OFF," due to the LDM input being grounded.

Test Description: Number(s) below refer to circled number(s) on the diagnostic chart.

1. Checks to see if ABS light functions properly. Normal operation would indicate that the problem is not present.
2. This step determines if any other codes are present or if A018 is set falsely.
3. This step insures that the circuitry from the EBCM to the lamp driver module is complete.
4. This step insures that the ignition circuit to the lamp driver module is functional. An open gauges fuse or open CKT 3, 39, or 865 would result in inoperative indicator lights.
5. Isolates the cause of the ABS lamp remaining "OFF," by disconnecting the 24 way world connector, CKT 1290 should be open and cause the ABS lamp to be "ON."

Diagnostic Aids:

An "Intermittent" problem may be caused by a poor connection, rubbed through wire insulation, or a wire that is broken inside the insulation.

The light test function of the "Scan" tool may be used to command the light "ON" while looking for an intermittent problem in the warning light circuitry.

Any circuitry that is suspected as causing the intermittent complaint should be thoroughly checked for backed out terminals, improper mating, broken locks, improperly formed or damaged terminals, poor terminal to wiring connections or physical damage to the wiring harness.

DIAGNOSTIC CHART DTC A011 — 1993 SILHOUETTE, TRANS SPORT AND LUMINA APV

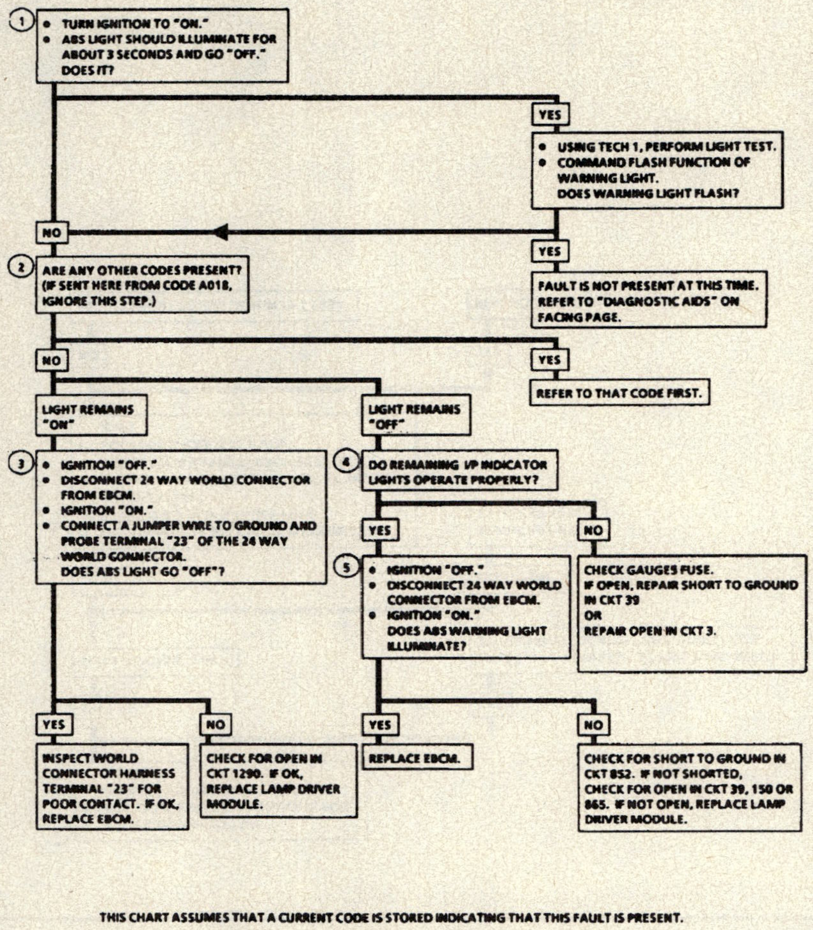

DIAGNOSTIC CHART DTC A013 — 1993 SILHOUETTE, TRANS SPORT AND LUMINA APV

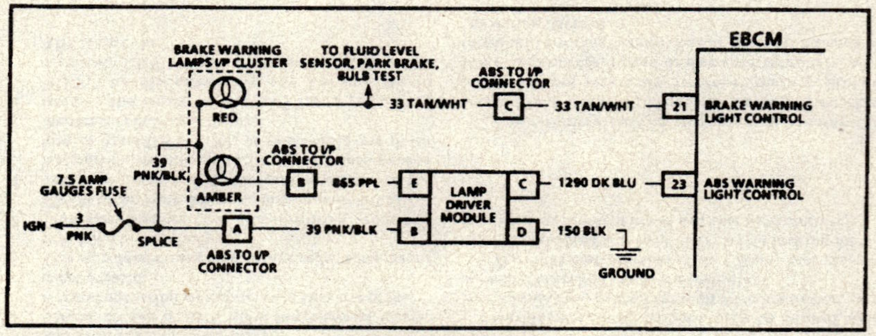

CODE A013

ABS WARNING LIGHT CIRCUIT SHORTED TO BATTERY

Circuit Description:

This code checks the state of the ABS warning light to identify a situation in which the driver could not be warned of a system fault by the ABS warning light. The Lamp Driver Module (LDM) turns the ABS warning light "ON," unless the EBCM provides ground to turn it "OFF." Because of the circuitry internal to the LDM, only external faults can be detected. As a result, the ABS warning light itself is not diagnosable, only the control line to the EBCM can be diagnosed.

If the ABS warning light control line is shorted to battery or the EBCM control circuit is open, the Electronic Brake Control Module (EBCM) is not able to turn "OFF" the ABS warning light.

Test Description: Number(s) below refer to circled number(s) on the diagnostic chart.

1. Checks to see if ABS light functions properly. Normal operation would indicate that the problem is not present.
2. After the ignition is turned to the "OFF" position, voltage to the lamp driver module should be interrupted. A light that remains "ON" would indicate that voltage to this circuit is from a point other than ignition.
3. After removing the 7.5 amp gauges fuse, voltage on CKT 1290 will be eliminated. If voltage is still present, it would indicate that a short to voltage is present between the lamp driver module and EBCM.

Diagnostic Aids:

An "Intermittent" problem may be caused by a poor connection, rubbed through wire insulation or a wire that is broken inside the insulation.

Code A013 must be currently set before it should be addressed. If Code A013 is set as a history code, proceed to those codes that are currently set before attempting to repair a Code A013 fault.

Any circuitry that is suspected as causing the "Intermittent" complaint should be thoroughly checked for backed out terminals, improper mating, broken locks, improperly formed or damaged terminals, poor terminal to wiring connections or physical damage to the wiring harness.

DIAGNOSTIC CHART DTC A013 — 1993 SILHOUETTE, TRANS SPORT AND LUMINA APV

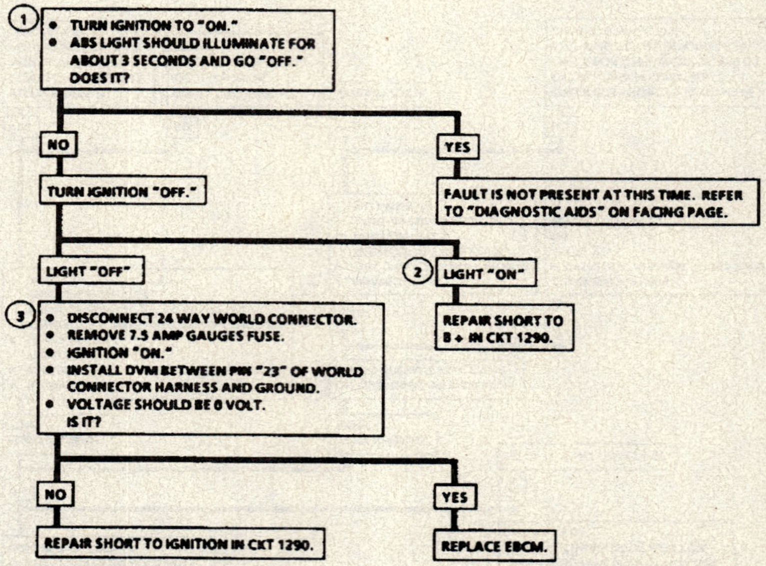

DIAGNOSTIC CHART DTC A014 — 1993 SILHOUETTE, TRANS SPORT AND LUMINA APV

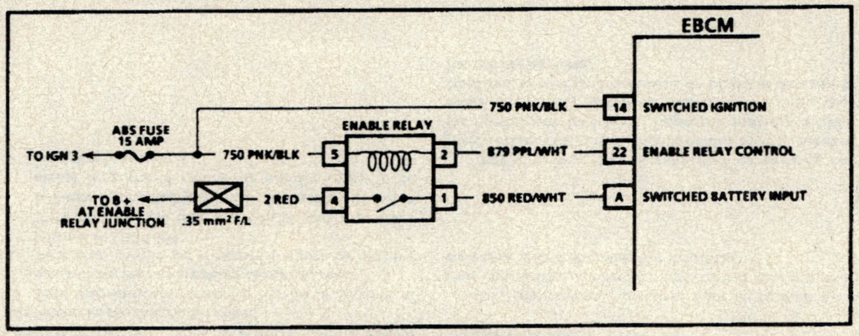

CODE A014

ENABLE RELAY CONTACTS OR FUSE OPEN

Circuit Description:

Ignition voltage is supplied through terminal "5" of the ABS enable relay. The EBCM then is able to energize the pull-in coil by completing the ground circuit at Pin "22" of the EBCM. The magnetic field created closes the enable relay contacts and allows battery voltage and current to be supplied to the EBCM, which supplies power to the motors and solenoids.

This code monitors the availability of current/voltage to the motors and solenoids. This fault indicates voltage is not available and would therefore not allow ABS operation if required.

Test Description: Number(s) below refer to circled number(s) on the diagnostic chart.

1. This step would indicate that the EBCM is sensing battery voltage present through the relay contacts at terminal "A". This shows that the relay and circuitry is functional.
2. This insures that battery voltage is available to terminals "5" and "4" of the relay.
3. This step checks to make sure that the pull-in coil of the enable relay has continuity.
4. This checks the ability of the EBCM to energize the enable relay. An open circuit in the EBCM or CKT 879 would not allow the enable relay to be energized.
5. This step checks for an open CKT 850 between the enable relay and EBCM.

Diagnostic Aids:

An "Intermittent" problem may be caused by a poor connection, rubbed through wire insulation, or a wire that is broken inside the insulation.

Any circuitry that is suspected as causing the "Intermittent" complaint should be thoroughly checked for backed out terminals, improper mating, broken locks, improperly formed or damaged terminals, poor terminal to wiring connections or physical damage to the wiring harness.

VIBRATION, TEMPERATURE EFFECTS:

Check for vibration effects by performing the relay test function of the "Scan" tool. With the relay test commanded "ON," lightly tap the top and sides of the relay while monitoring relay voltage. If the relay voltage changes significantly, replace the relay.

If Code A014 only fails when the vehicle is initially started in cold ambient conditions (temperature less than 32°F - 0°C), replace the relay.

DIAGNOSTIC CHART DTC A014 — 1993 SILHOUETTE, TRANS SPORT AND LUMINA APV

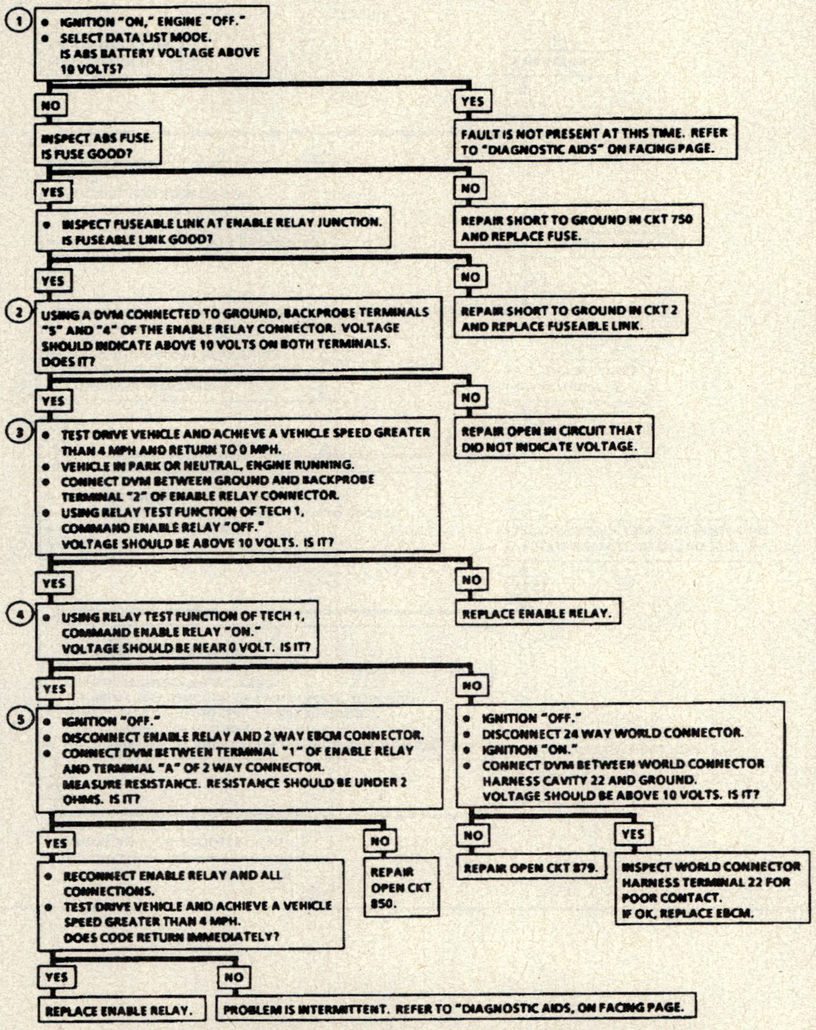

THIS CHART ASSUMES THAT A CURRENT CODE IS STORED INDICATING THAT THIS FAULT IS PRESENT.

DIAGNOSTIC CHART DTC A015 — 1993 SILHOUETTE, TRANS SPORT AND LUMINA APV

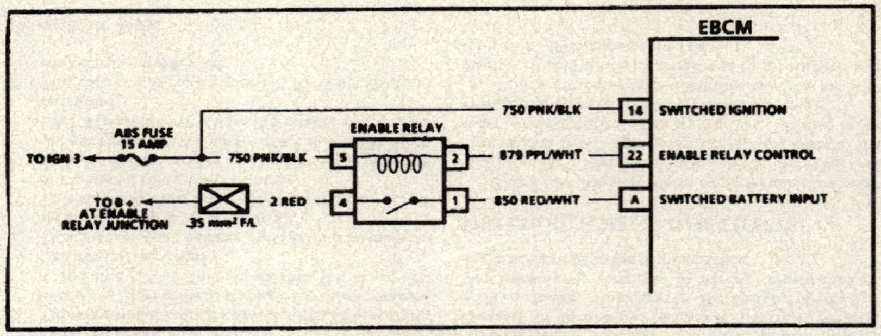

CODE A015

ENABLE RELAY CONTACTS SHORTED TO BATTERY

Circuit Description:

Ignition voltage is supplied through terminal "5" of the ABS enable relay. The EBCM then is able to energize the pull-in coil by completing the ground circuit at Pin "22" of the EBCM. The magnetic field created closes the enable relay contacts and allows battery voltage and current to be supplied to the EBCM, which supplies power to the motors and solenoids.

This code determines if the enable relay is energized when it should not be. This fault would not allow the enable relay to remove power to the ABS system. If a second fault were to occur that requires the enable relay to be turned "OFF," that fault cannot be removed if the relay cannot be controlled.

Test Description: Number(s) below refer to circled number(s) on the diagnostic chart.

1. This step indicates that the EBCM is capable of controlling the enable relay as commanded.
2. This step checks for a possible short to battery voltage in CKT 850.
3. This step insures that the EBCM and also CKT 879 are not shorted to ground. A short to ground in either case would cause the enable relay to be energized constantly.

Diagnostic Aids:

An "Intermittent" problem may be caused by a poor connection, rubbed through wire insulation, or a wire that is broken inside the insulation.

Any circuitry, that is suspected as causing the intermittent complaint, should be thoroughly checked for backed out terminals, improper mating, broken locks, improperly formed or damaged terminals, poor terminal to wiring connections or physical damage to the wiring harness.

DIAGNOSTIC CHART DTC A015 — 1993 SILHOUETTE, TRANS SPORT AND LUMINA APV

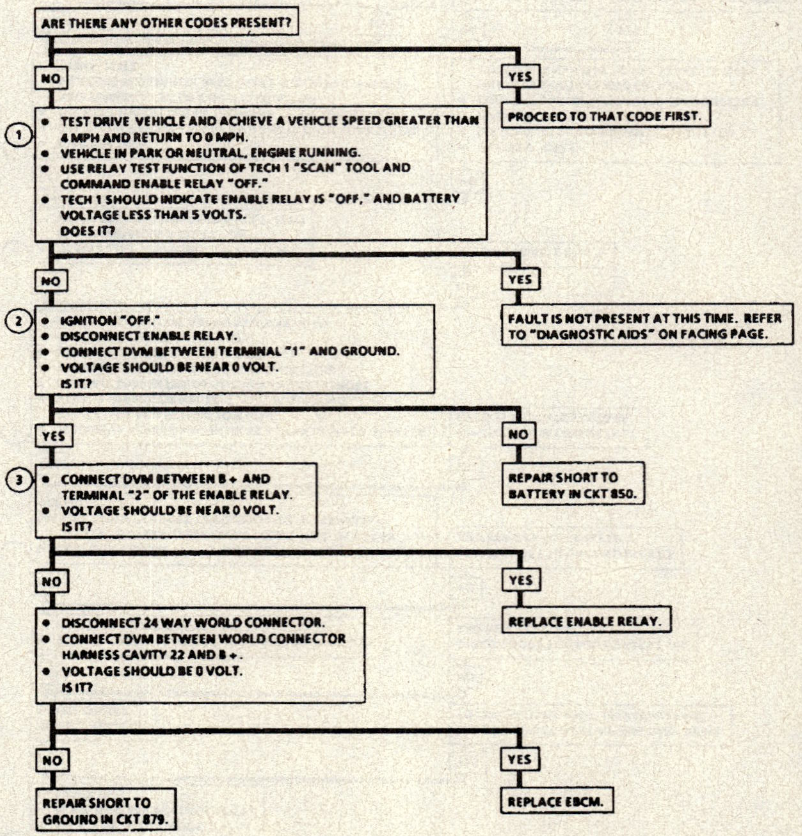

THIS CHART ASSUMES THAT A CURRENT CODE IS STORED INDICATING THAT THIS FAULT IS PRESENT.

DIAGNOSTIC CHART DTC A016 — 1993 SILHOUETTE, TRANS SPORT AND LUMINA APV

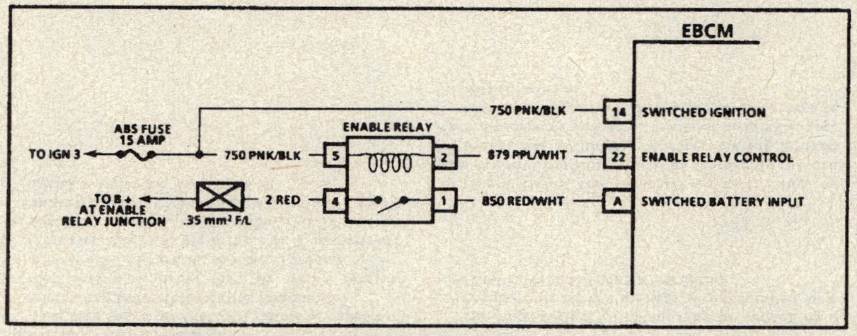

CODE A016

ENABLE RELAY COIL CIRCUIT OPEN

Circuit Description:

Ignition voltage is supplied through terminal "5" of the ABS enable relay. The EBCM then is able to energize the pull-in coil by completing the ground circuit at Pin "22" of the EBCM. The magnetic field created closes the enable relay contacts and allows battery voltage and current to be supplied to the EBCM, which supplies power to the motors and solenoids.

This code detects opens in the enable relay coil circuit. An open in this circuit will not allow the enable relay to be energized, thus preventing voltage/current to the motors and solenoids.

Test Description: Number(s) below refer to circled number(s) on the diagnostic chart.

1. This step indicates the EBCM is capable of controlling the enable relay as commanded.
2. This step checks to insure that voltage is available to the pull-in coil of the enable relay.
3. This step insures that there is continuity through the pull-in coil of the enable relay.

Diagnostic Aids:

An "Intermittent" problem may be caused by a poor connection, rubbed through wire insulation, or a wire that is broken inside the insulation.

If the frequency of the problem is high, but is currently intermittent, check for high coil resistance by measuring between relay terminals "5" and "2" using a DVM. If resistance shows greater than 100 ohms, replace the relay.

Any circuitry, that is suspected as causing the intermittent complaint, should be thoroughly checked for backed out terminals, improper mating, broken locks, improperly formed or damaged terminals, poor terminal to wiring connections or physical damage to the wiring harness.

DIAGNOSTIC CHART DTC A016 — 1993 SILHOUETTE, TRANS SPORT AND LUMINA APV

(1)
- IGNITION "ON," ENGINE "OFF."
- USING RELAY TEST FUNCTION OF TECH 1 "SCAN" TOOL, COMMAND ENABLE RELAY "ON."
- TECH 1 SHOULD INDICATE ENABLE RELAY "ON," AND BATTERY VOLTAGE ABOVE 10 VOLTS.

DOES IT?

- YES → FAULT IS NOT PRESENT AT THIS TIME. REFER TO "DIAGNOSTIC AIDS" ON FACING PAGE.
- NO ↓

(2)
- CONNECT DVM BETWEEN GROUND AND BACKPROBE TERMINAL "5" OF ENABLE RELAY CONNECTOR.
- VOLTAGE SHOULD BE ABOVE 10 VOLTS.

IS IT?

- NO → REPAIR OPEN CKT 750 OR ABS FUSE.
- YES ↓

(3)
- TEST DRIVE VEHICLE AND ACHIEVE A VEHICLE SPEED GREATER THAN 4 MPH AND RETURN TO 0 MPH.
- VEHICLE IN PARK OR NEUTRAL, ENGINE RUNNING.
- CONNECT DVM TO GROUND AND BACKPROBE TERMINAL "2" OF ENABLE RELAY CONNECTOR.
- USING RELAY TEST FUNCTION OF TECH 1, COMMAND ENABLE RELAY "OFF."

- NO → CHECK FOR INTERMITTENT CONNECTIONS AT ENABLE RELAY. IF OK, REPLACE RELAY.
- YES ↓

- IGNITION "OFF."
- DISCONNECT 24 WAY WORLD CONNECTOR.
- IGNITION "ON."
- CONNECT DVM BETWEEN WORLD CONNECTOR HARNESS CAVITY 22 AND GROUND. VOLTAGE SHOULD BE ABOVE 10 VOLTS.

IS IT?

- NO → REPAIR OPEN CKT 879.
- YES → INSPECT WORLD CONNECTOR HARNESS TERMINAL "22" FOR POOR CONTACT. IF OK, REPLACE EBCM.

THIS CHART ASSUMES THAT A CURRENT CODE IS STORED INDICATING THAT THIS FAULT IS PRESENT.

DIAGNOSTIC CHART DTC A017 — 1993 SILHOUETTE, TRANS SPORT AND LUMINA APV

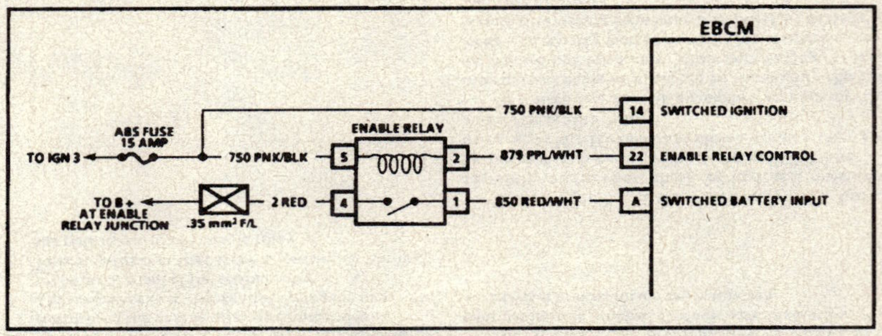

CODE A017

ENABLE RELAY COIL CIRCUIT SHORTED TO GROUND

Circuit Description:

Ignition voltage is supplied through terminal "5" of the ABS enable relay. The EBCM then is able to energize the pull-in coil by completing the ground circuit at Pin "22" of the EBCM. The magnetic field created closes the enable relay contacts and allows battery voltage and current to be supplied to the EBCM, which supplies voltage to the motors and solenoids.

This code determines if the enable relay is energized when it should not be. This fault would not allow the enable relay to remove power to the ABS system. If a second fault were to occur that requires the enable relay to be turned "OFF," that fault cannot be removed if the relay cannot be controlled.

Test Description: Number(s) below refer to circled number(s) on the diagnostic chart.

1. This step indicates that the EBCM is capable of controlling the enable relay as commanded.
2. This checks to insure that the enable relay or control CKT 879 are not shorted to ground.
3. This step checks to see if the EBCM is internally shorted to ground. A grounded Pin "22" would cause the enable relay to be energized any time the ignition was in the "ON" position.

Diagnostic Aids:

An "Intermittent" problem may be caused by a poor connection, rubbed through wire insulation, or a wire that is broken inside the insulation.

Any circuitry, that is suspected as causing the intermittent complaint, should be thoroughly checked for backed out terminals, improper mating, broken locks, improperly formed or damaged terminals, poor terminal to wiring connections or physical damage to the wiring harness.

DIAGNOSTIC CHART DTC A017 — 1993 SILHOUETTE, TRANS SPORT AND LUMINA APV

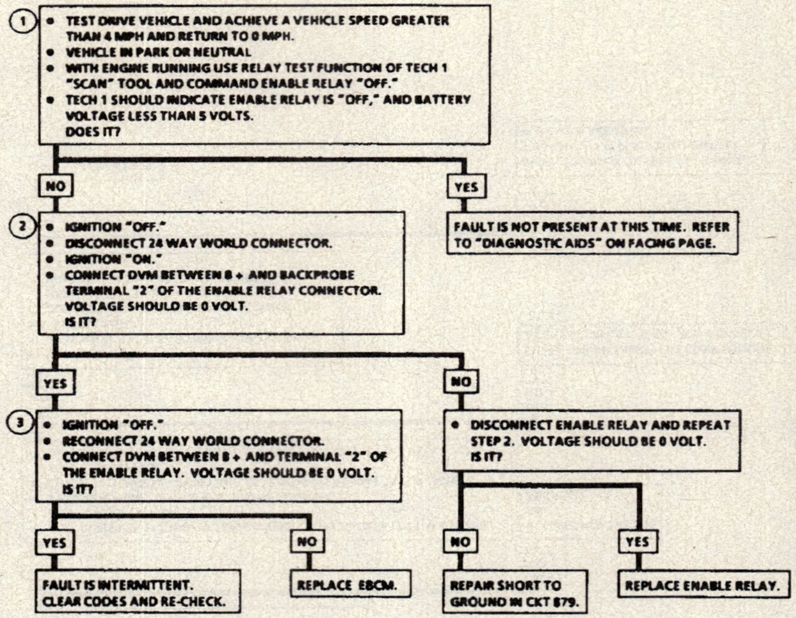

THIS CHART ASSUMES THAT A CURRENT CODE IS STORED INDICATING THAT THIS FAULT IS PRESENT.

DIAGNOSTIC CHART DTC A018 — 1993 SILHOUETTE, TRANS SPORT AND LUMINA APV

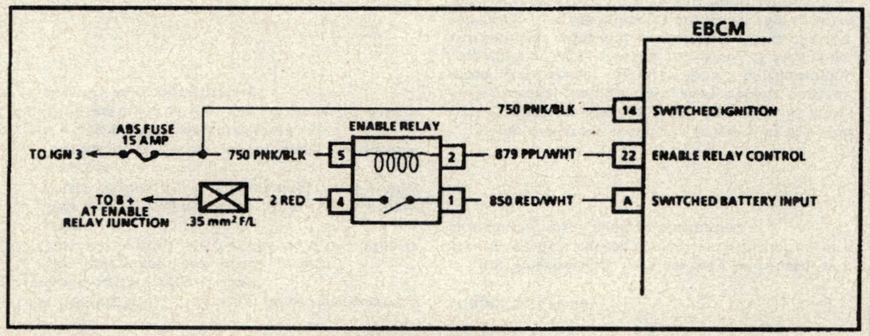

CODE A018

ENABLE RELAY COIL CIRCUIT SHORTED TO BATTERY

Circuit Description:

Ignition voltage is supplied through terminal "5" of the ABS enable relay. The EBCM then is able to energize the pull-in coil by completing the ground circuit at Pin "22" of the EBCM. The magnetic field created closes the enable relay contacts and allows battery voltage and current to be supplied to the EBCM, which supplies voltage to the motors and solenoids.

This code monitors the availability of current/voltage to the motors, solenoids. A fault of the type specified above would not allow voltage to be available to the motors and solenoids and would, therefore, not allow ABS operation if required.

Test Description: Number(s) below refer to circled number(s) on the diagnostic chart.

1. This step indicates that the EBCM is capable of controlling the enable relay as commanded.
2. With the enable relay removed, voltage should not be available at cavity "2". Any voltage at this point would indicate that CKT 879 was shorted to a voltage source.
3. This step checks for an open Quad-Driver in the EBCM.

Diagnostic Aids:

An "Intermittent" problem may be caused by a poor connection, rubbed through wire insulation, or a wire that is broken inside the insulation.

Any circuitry, that is suspected as causing the intermittent complaint, should be thoroughly checked for backed out terminals, improper mating, broken locks, improperly formed or damaged terminals, poor terminal to wiring connections or physical damage to the wiring harness.

DIAGNOSTIC CHART DTC A018 — 1993 SILHOUETTE, TRANS SPORT AND LUMINA APV

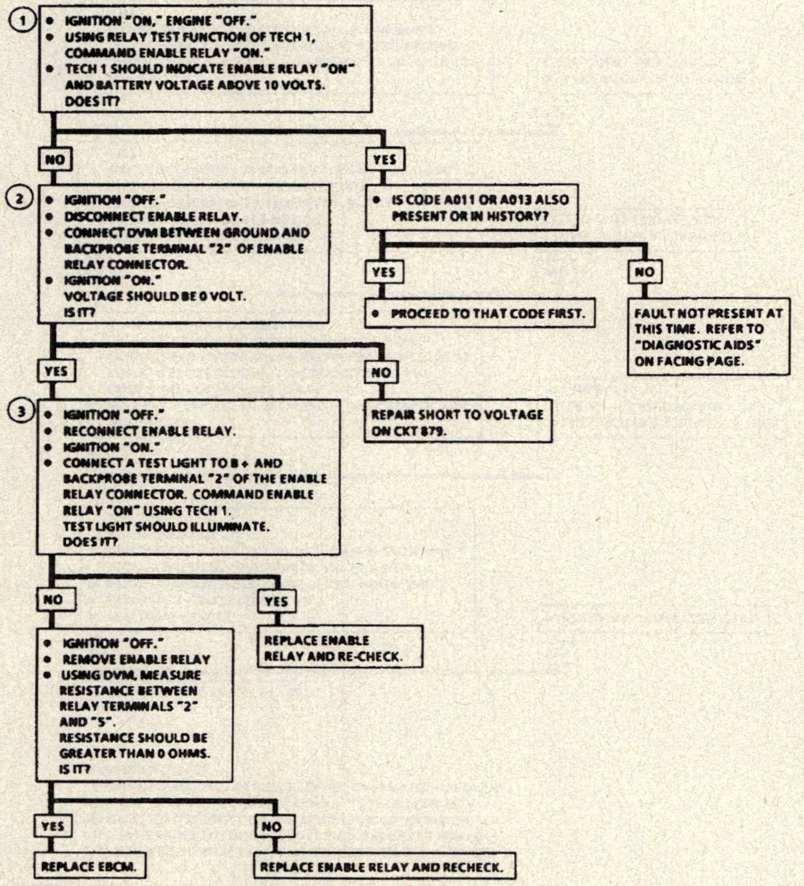

THIS CHART ASSUMES THAT A CURRENT CODE IS STORED INDICATING THAT THIS FAULT IS PRESENT.

DIAGNOSTIC CHART DTC A021 — 1993 SILHOUETTE, TRANS SPORT AND LUMINA APV

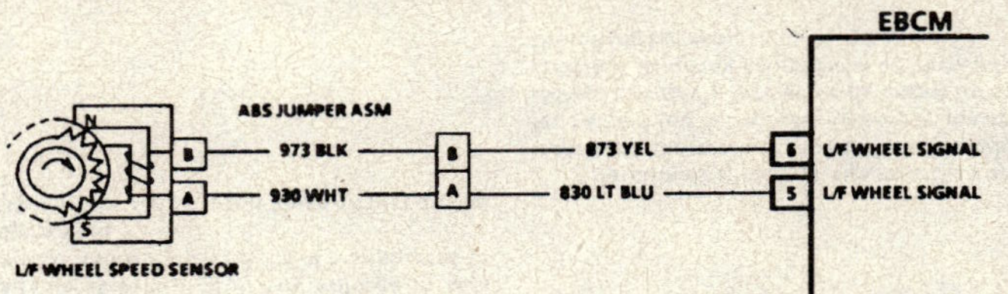

CODE A021

LEFT FRONT WHEEL SPEED = 0

Circuit Description:

As a toothed ring passes by the wheel speed sensor, changes in the electromagnetic field cause the wheel speed sensor to produce a sinusoidal (AC) voltage signal whose frequency is proportional to wheel speed. The magnitude of this signal is directly related to wheel speed and the proximity of the wheel speed sensor to the toothed ring, often referred to as the air gap.

The code detects opens or a low output condition that cause a wheel speed of 0 mph to be calculated by the ABS EBCM. The conditions present to fail this test are: (1) Three (3) wheels are rotating at least 5 mph, and (2) The fourth wheel is 0 mph.

Test Description: Number(s) below refer to circled number(s) on the diagnostic chart.

1. This determines if code A032 is present.
2. This test verifies whether the fault is currently present.
3. This step is designed to identify a wheel speed sensor or circuitry that is damaged and visibly apparent.
4. This test checks to see that the wheel speed sensor is within proper resistance values.
5. This test insures that the wheel speed sensor circuitry has good continuity.

Diagnostic Aids:

An "Intermittent" problem may be caused by a poor connection, rubbed through wire insulation, or a wire that is broken inside the insulation.

If the customers comments reflect that the ABS Amber warning lamp is "ON" only during moist environmental changes, (rain, snow, vehicle wash) all wheel speed sensor circuitry should be thoroughly inspected for signs of water intrusion. If code is not current clear codes and simulate the effects of water intrusion. Use the following procedure. Spray down the suspected area with a 5% salt water solution (two teaspoons of salt to 12 oz. of water). Test drive vehicle. Vehicle must achieve 4 mph. If code returns immediately, replace suspected harness.

Any circuitry, that is suspected as causing the intermittent complaint, should be thoroughly checked for backed out terminals, improper mating, broken locks, improperly formed or damaged terminals, poor terminal to wiring connections or physical damage to the wiring harness.

DIAGNOSTIC CHART DTC A021 — 1993 SILHOUETTE, TRANS SPORT AND LUMINA APV

IMPORTANT: WHEEL SPEED SENSOR INTERMITTENT PROBLEMS MAY BE DIFFICULT TO LOCATE. CARE SHOULD BE TAKEN NOT TO DISTURB ANY ELECTRICAL CONNECTIONS PRIOR TO AN INDICATED STEP OF THIS CHART. THIS WILL INSURE THAT AN INTERMITTENT CONNECTION WILL NOT BE CORRECTED BEFORE THE SOURCE OF THE PROBLEM IS FOUND.

(1) IS CODE A032 ALSO CURRENTLY SET?
- YES → PROCEED WITH THAT CODE FIRST.
- NO ↓

(2)
- TEST DRIVE VEHICLE.
- USING TECH 1, SELECT DATA LIST.
- MONITOR WHEEL SPEED ON LEFT FRONT WHEEL WHILE SLOWLY DECELERATING FROM 35 MPH TO 0 MPH.
- WHEEL SPEED SHOULD BE WITHIN 5 MPH OF REMAINING WHEEL SPEED INPUTS.

DOES IT?
- YES → FAULT IS NOT PRESENT AT THIS TIME. REFER TO "DIAGNOSTIC AIDS" ON FACING PAGE.
- NO ↓

(3)
- IGNITION "OFF."
- PHYSICALLY INSPECT WHEEL SPEED SENSOR, SENSOR RING, WIRING AND CONNECTORS FOR DAMAGE. PAY PARTICULAR ATTENTION TO THE SENSOR RING. ANY SIGNIFICANT DAMAGE (OTHER THAN NICKS FROM STONES, ETC.) WILL AFFECT THE WHEEL SPEED INPUT SIGNAL.

IS PHYSICAL DAMAGE INDICATED?
- YES → CORRECT PHYSICAL DAMAGE, CLEAR CODES AND RECHECK.
- NO ↓

(4)
- DISCONNECT CONNECTOR FROM LEFT FRONT WHEEL SPEED SENSOR.
- CONNECT DVM BETWEEN TERMINALS "A" AND "B" OF SENSOR, AND MEASURE RESISTANCE. RESISTANCE SHOULD BE BETWEEN 988 AND 1200 OHMS.

IS IT?
- NO → REPLACE WHEEL SPEED SENSOR, CLEAR CODES AND RECHECK.
- YES ↓

(5)
- IGNITION "ON."
- CONNECT DVM TO GROUND AND TERMINALS "A" THEN "B" OF LEFT FRONT SENSOR CONNECTOR HARNESS. VOLTAGE SHOULD BE 2.25 TO 2.75 VOLTS ON BOTH TERMINALS.

IS IT?

GO TO FOLLOWING PAGE OF THIS CODE CHART

THIS CHART ASSUMES THAT A CURRENT CODE IS STORED INDICATING THAT THIS FAULT IS PRESENT.

DIAGNOSTIC CHART DTC A021 — 1993 SILHOUETTE, TRANS SPORT AND LUMINA APV

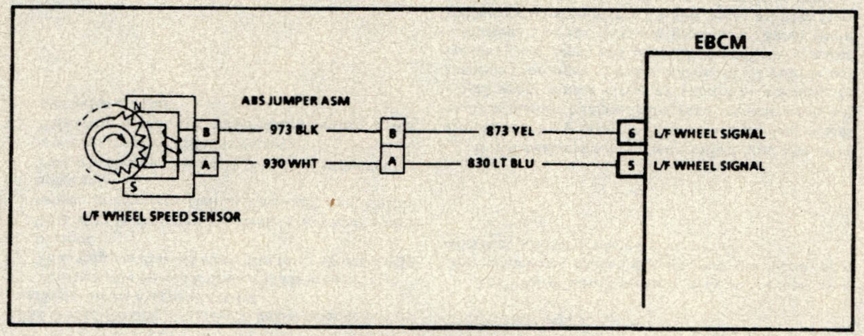

CODE A021

LEFT FRONT WHEEL SPEED = 0

Circuit Description:

As a toothed ring passes by the wheel speed sensor, changes in the electromagnetic field cause the wheel speed sensor to produce a sinusoidal (AC) voltage signal whose frequency is proportional to wheel speed. The magnitude of this signal is directly related to wheel speed and the proximity of the wheel speed sensor to the toothed ring often referred to as the air gap.

This code detects opens or a low output condition that cause a wheel speed of 0 mph to be calculated by the ABS EBCM. The conditions present to fail this test are: (1) Three (3) wheels are rotating at least 5 mph, and (2) The fourth wheel is 0 mph.

Test Description: Number(s) below refer to circled number(s) on the diagnostic chart.

6. This step insures that wheel speed sensor circuitry is not internally shorted
7. This checks for an open in the wheel signal high circuit.
8. This checks for an open in the wheel signal low circuit.
9. This step insures that Code A021 was not set due to a poor connection between the world connector and EBCM.

Diagnostic Aids:

An "Intermittent" problem may be caused by a poor connection, rubbed through wire insulation, or a wire that is broken inside the insulation.

If the customers comments reflect that the ABS Amber warning lamp is "ON" only during moist environmental changes, (rain, snow, vehicle wash) all wheel speed sensor circuitry should be thoroughly inspected for signs of water intrusion. If code is not current clear codes and simulate the effects of water intrusion. Use the following procedure. Spray down the suspected area with a 5% salt water solution (two teaspoons of salt to 12 oz. of water). Test drive vehicle. Vehicle must achieve 4 mph. If code returns immediately, replace suspected harness.

Any circuitry, that is suspected as causing the intermittent complaint, should be thoroughly checked for backed out terminals, improper mating, broken locks, improperly formed or damaged terminals, poor terminal to wiring connections or physical damage to the wiring harness.

DIAGNOSTIC CHART DTC A021 — 1993 SILHOUETTE, TRANS SPORT AND LUMINA APV

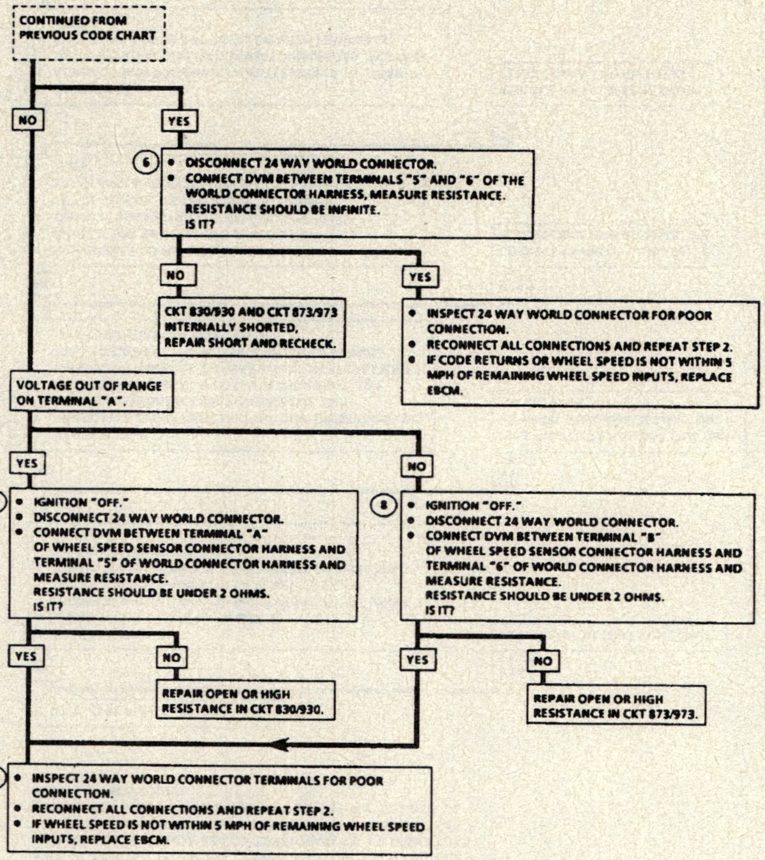

THIS CHART ASSUMES THAT A CURRENT CODE IS STORED INDICATING THAT THIS FAULT IS PRESENT.

DIAGNOSTIC CHART DTC A022 — 1993 SILHOUETTE, TRANS SPORT AND LUMINA APV

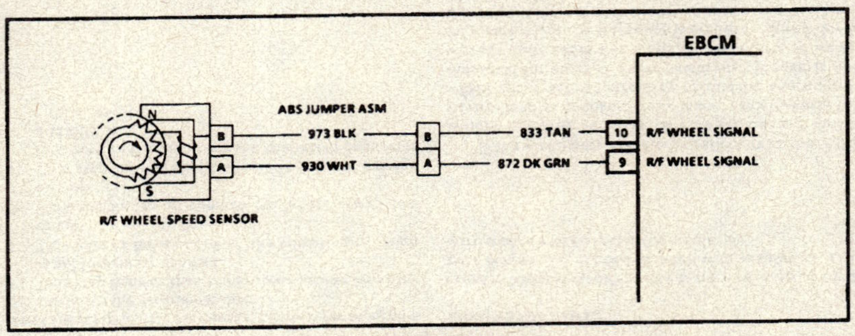

CODE A022

RIGHT FRONT WHEEL SPEED = 0

Circuit Description:

As a toothed ring passes by the wheel speed sensor, changes in the electromagnetic field cause the wheel speed sensor to produce a sinusoidal (AC) voltage signal whose frequency is proportional to wheel speed. The magnitude of this signal is directly related to wheel speed and the proximity of the wheel speed sensor to the toothed ring, often referred to as the air gap.

This code detects opens or a low output condition that cause a wheel speed of 0 mph to be calculated by the ABS EBCM. The conditions present to fail this test are: (1) Three (3) wheels are rotating at least 5 mph, and (2) The fourth wheel is 0 mph.

Test Description: Number(s) below refer to circled number(s) on the diagnostic chart.

1. This test determines if Code A033 is present.
2. This step verifies whether the fault is currently present.
3. This step is designed to identify a wheel speed sensor or circuitry that is damaged and visibly apparent.
4. This test checks to see that the wheel speed sensor is within proper resistance values.
5. This insures that the wheel speed sensor is not shorted to ground.

Diagnostic Aids:

An "Intermittent" problem may be caused by a poor connection, rubbed through wire insulation, or a wire that is broken inside the insulation.

If the customers comments reflect that the ABS Amber warning lamp is "ON" only during moist environmental changes, (rain, snow, vehicle wash) all wheel speed sensor circuitry should be thoroughly inspected for signs of water intrusion. If code is not current clear codes and simulate the effects of water intrusion. Use the following procedure. Spray down the suspected area with a 5% salt water solution (two teaspoons of salt to 12 oz. of water). Test drive vehicle. Vehicle must achieve 4 mph. If code returns immediately, replace suspected harness.

Any circuitry, that is suspected as causing the intermittent complaint, should be thoroughly checked for backed out terminals, improper mating, broken locks, improperly formed or damaged terminals, poor terminal to wiring connections or physical damage to the wiring harness.

DIAGNOSTIC CHART DTC A022 — 1993 SILHOUETTE, TRANS SPORT AND LUMINA APV

IMPORTANT: WHEEL SPEED SENSOR INTERMITTENT PROBLEMS MAY BE DIFFICULT TO LOCATE. CARE SHOULD BE TAKEN NOT TO DISTURB ANY ELECTRICAL CONNECTIONS PRIOR TO AN INDICATED STEP OF THIS CHART. THIS WILL INSURE THAT AN INTERMITTENT CONNECTION WILL NOT BE CORRECTED BEFORE THE SOURCE OF THE PROBLEM IS FOUND.

(1) IS CODE A033 ALSO CURRENTLY SET?

- YES: PROCEED WITH THAT CODE FIRST.
- NO:

(2)
- TEST DRIVE VEHICLE.
- USING TECH 1, SELECT DATA LIST.
- MONITOR WHEEL SPEED ON RIGHT FRONT WHEEL WHILE SLOWLY DECELERATING FROM 35 MPH TO 0 MPH.
- WHEEL SPEED SHOULD BE WITHIN 5 MPH OF REMAINING WHEEL SPEED INPUTS,

IS IT?

- YES: FAULT IS NOT PRESENT AT THIS TIME. REFER TO "DIAGNOSTIC AIDS" ON FACING PAGE.
- NO:

(3)
- IGNITION "OFF."
- PHYSICALLY INSPECT WHEEL SPEED SENSOR, SENSOR RING, WIRING AND CONNECTORS FOR DAMAGE. PAY PARTICULAR ATTENTION TO THE SENSOR RING. ANY SIGNIFICANT DAMAGE (OTHER THAN NICKS FROM STONES, ETC.) WILL AFFECT THE WHEEL SPEED INPUT SIGNAL.

IS PHYSICAL DAMAGE INDICATED?

- YES: CORRECT PHYSICAL DAMAGE, CLEAR CODES AND RECHECK.
- NO:

(4)
- DISCONNECT CONNECTOR FROM RIGHT FRONT WHEEL SPEED SENSOR.
- CONNECT DVM BETWEEN TERMINALS "A" AND "B" OF SENSOR, AND MEASURE RESISTANCE. RESISTANCE SHOULD BE BETWEEN 988 AND 1208 OHMS.

IS IT?

- NO: REPLACE WHEEL SPEED SENSOR, CLEAR CODES AND RECHECK.
- YES:

(5)
- IGNITION "ON."
- CONNECT DVM TO GROUND AND TERMINALS "A" THEN "B" OF RIGHT FRONT SENSOR CONNECTOR HARNESS. VOLTAGE SHOULD BE 2.25 TO 2.75 VOLTS ON BOTH TERMINALS.

IS IT?

GO TO FOLLOWING PAGE OF THIS CODE CHART

THIS CHART ASSUMES THAT A CURRENT CODE IS STORED INDICATING THAT THIS FAULT IS PRESENT.

DIAGNOSTIC CHART DTC A022 — 1993 SILHOUETTE, TRANS SPORT AND LUMINA APV

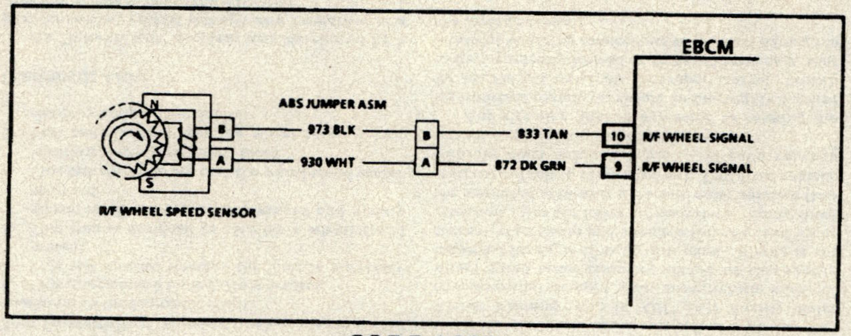

CODE A022

RIGHT FRONT WHEEL SPEED = 0

Circuit Description:

As a toothed ring passes by the wheel speed sensor, changes in the electromagnetic field cause the wheel speed sensor to produce a sinusoidal (AC) voltage signal whose frequency is proportional to wheel speed. The magnitude of this signal is directly related to wheel speed and the proximity of the wheel speed sensor to the toothed ring, often referred to as the air gap.

This code detects opens or a low output condition that cause a wheel speed of 0 mph to be calculated by the ABS EBCM. The conditions present to fail this test are: (1) Three (3) wheels are rotating at least 5 mph, and (2) The fourth wheel is 0 mph.

Test Description: Number(s) below refer to circled number(s) on the diagnostic chart.

6. This test insures that wheel speed sensor circuitry is not internally shorted.
7. This checks for an open in the wheel signal low circuit.
8. This checks for an open in the wheel signal high circuit.
9. This step insures that Code A022 was not set due to a poor connection between the world connector and EBCM.

Diagnostic Aids:

An "Intermittent" problem may be caused by a poor connection, rubbed through wire insulation, or a wire that is broken inside the insulation.

If the customers comments reflect that the ABS Amber warning lamp is "ON" only during moist environmental changes, (rain, snow, vehicle wash) all wheel speed sensor circuitry should be thoroughly inspected for signs of water intrusion. If code is not current clear codes and simulate the effects of water intrusion. Use the following procedure. Spray down the suspected area with a 5% salt water solution (two teaspoons of salt to 12 oz. of water). Test drive vehicle. Vehicle must achieve 4 mph. If code returns immediately, replace suspected harness.

Any circuitry, that is suspected as causing the intermittent complaint, should be thoroughly checked for backed out terminals, improper mating, broken locks, improperly formed or damaged terminals, poor terminal to wiring connections or physical damage to the wiring harness.

DIAGNOSTIC CHART DTC A022 — 1993 SILHOUETTE, TRANS SPORT AND LUMINA APV

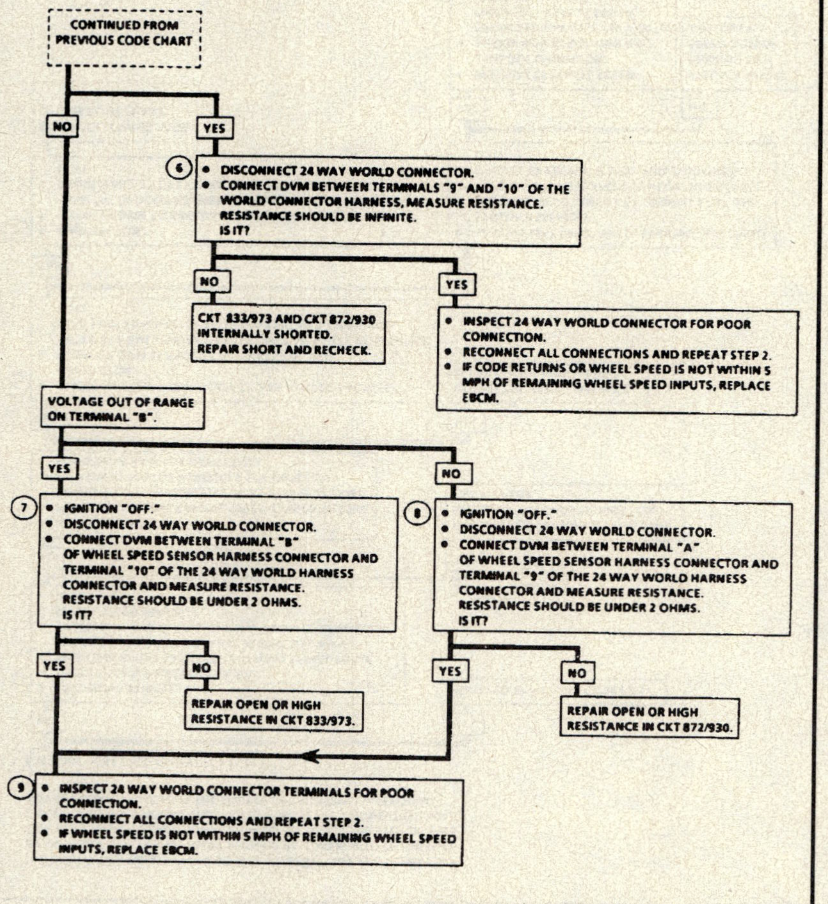

THIS CHART ASSUMES THAT A CURRENT CODE IS STORED INDICATING THAT THIS FAULT IS PRESENT.

DIAGNOSTIC CHART DTC A023 — 1993 SILHOUETTE, TRANS SPORT AND LUMINA APV

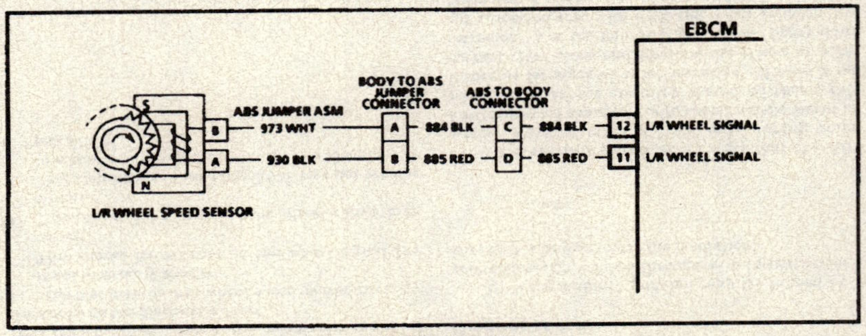

CODE A023

LEFT REAR WHEEL SPEED = 0

Circuit Description:

As a toothed ring passes by the wheel speed sensor, changes in the electromagnetic field cause the wheel speed sensor to produce a sinusoidal (AC) voltage signal whose frequency is proportional to wheel speed. The magnitude of this signal is directly related to wheel speed and the proximity of the wheel speed sensor to the toothed ring, often referred to as the air gap.

This code detects opens or a low output condition that cause a wheel speed of 0 mph to be calculated by the ABS EBCM. The conditions present to fail this test are: (1) Three (3) wheels are rotating at least 5 mph, (2) The fourth wheel is 0 mph.

Test Description: Number(s) below refer to circled number(s) on the diagnostic chart.

1. This determines if Code A034 is present.
2. This test verifies whether the fault is currently present.
3. This step is designed to identify a wheel speed sensor or circuitry that is damaged and visibly apparent.
4. This test checks to see that the wheel speed sensor is within proper resistance values.
5. This test insures that the wheel speed sensor circuitry has good continuity.

Diagnostic Aids:

An "Intermittent" problem may be caused by a poor connection, rubbed through wire insulation, or a wire that is broken inside the insulation.

If the customers comments reflect that the ABS Amber warning lamp is "ON" only during moist environmental changes, (rain, snow, vehicle wash) all wheel speed senor circuitry should be thoroughly inspected for signs of water intrusion. If code is not current, clear codes and simulate the effects of water intrusion. Use the following procedure. Spray down the suspected area with a 5% salt water solution (two teaspoons of salt to 12 oz. of water). Test drive vehicle. Vehicle must achieve 4 mph. If code returns immediately, replace suspected harness.

Any circuitry, that is suspected as causing the intermittent complaint, should be thoroughly checked for backed out terminals, improper mating, broken locks, improperly formed or damaged terminals, poor terminal to wiring connections or physical damage to the wiring harness.

DIAGNOSTIC CHART DTC A023 — 1993 SILHOUETTE, TRANS SPORT AND LUMINA APV

IMPORTANT: WHEEL SPEED SENSOR INTERMITTENT PROBLEMS MAY BE DIFFICULT TO LOCATE. CARE SHOULD BE TAKEN NOT TO DISTURB ANY ELECTRICAL CONNECTIONS PRIOR TO AN INDICATED STEP OF THIS CHART. THIS WILL INSURE THAT AN INTERMITTENT CONNECTION WILL NOT BE CORRECTED BEFORE THE SOURCE OF THE PROBLEM IS FOUND.

(1) IS CODE A034 ALSO CURRENTLY SET?
- YES: PROCEED WITH THAT CODE FIRST.
- NO:

(2)
- TEST DRIVE VEHICLE.
- USING TECH 1, SELECT DATA LIST.
- MONITOR WHEEL SPEED ON LEFT REAR WHEEL WHILE SLOWLY DECELERATING FROM 35 MPH TO 0 MPH.
- WHEEL SPEED SHOULD BE WITHIN 5 MPH OF REMAINING WHEEL SPEED INPUTS,

IS IT?
- YES: FAULT IS NOT PRESENT AT THIS TIME. REFER TO "DIAGNOSTIC AIDS" ON FACING PAGE.
- NO:

(3)
- IGNITION "OFF."
- PHYSICALLY INSPECT WHEEL SPEED SENSOR WIRING, CONNECTORS AND CONNECTOR AT WHEEL BEARING AND SPEED SENSOR ASSEMBLY FOR DAMAGE.

IS PHYSICAL DAMAGE INDICATED?
- YES: CORRECT PHYSICAL DAMAGE, CLEAR CODES AND RECHECK.
- NO:

(4)
- DISCONNECT LEFT REAR BODY TO ABS JUMPER HARNESS CONNECTOR.
- CONNECT DVM BETWEEN TERMINALS "A" AND "B" OF THE ABS JUMPER HARNESS AND MEASURE RESISTANCE. RESISTANCE SHOULD BE BETWEEN 2100 AND 2400 OHMS.

IS IT?
- YES:
  - (5)
    - IGNITION "ON."
    - CONNECT DVM TO GROUND AND TERMINALS "A" THEN "B" OF BODY HARNESS CONNECTOR. VOLTAGE SHOULD BE 2.25 TO 2.75 VOLTS ON BOTH TERMINALS.

    IS IT?
  - GO TO FOLLOWING PAGE OF THIS CODE CHART
- NO:
  - REMOVE LEFT REAR WHEEL BEARING AND SPEED SENSOR ASSEMBLY.
  - MEASURE RESISTANCE AT TERMINALS "A" AND "B" OF THE WHEEL SPEED SENSOR. RESISTANCE SHOULD BE BETWEEN 2100 AND 2400 OHMS.

  IS IT?
  - NO: REPLACE WHEEL BEARING AND SPEED SENSOR ASSEMBLY.
  - YES:
    - RECONNECT WHEEL SPEED SENSOR CONNECTOR.
    - RETAIN WHEEL BEARING AND SPEED SENSOR ASSEMBLY WITH ONE BOLT. REPEAT STEP 4.

    IS RESISTANCE OK?
    - YES: HAD POOR CONNECTION AT SPEED SENSOR CONNECTOR.
    - NO: REPLACE LEFT REAR ABS JUMPER HARNESS.

THIS CHART ASSUMES THAT A CURRENT CODE IS STORED INDICATING THAT THIS FAULT IS PRESENT.

DIAGNOSTIC CHART DTC A023 — 1993 SILHOUETTE, TRANS SPORT AND LUMINA APV

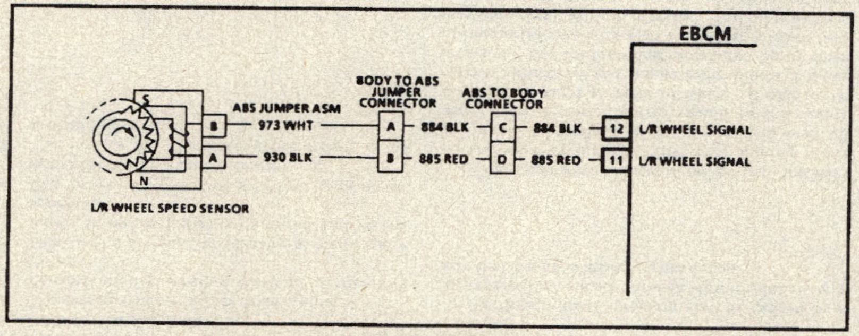

CODE A023

LEFT REAR WHEEL SPEED = 0

Circuit Description:

As a toothed ring passes by the wheel speed sensor, changes in the electromagnetic field cause the wheel speed sensor to produce a sinusoidal (AC) voltage signal whose frequency is proportional to wheel speed. The magnitude of this signal is directly related to wheel speed and the proximity of the wheel speed sensor to the toothed ring, often referred to as the air gap.

This code detects opens or a low output condition that cause a wheel speed of 0 mph to be calculated by the ABS EBCM. The conditions present to fail this test are: (1) Three (3) wheels are rotating at least 5 mph, and (2) The fourth wheel is 0 mph.

Test Description: Number(s) below refer to circled number(s) on the diagnostic chart.

6. This test insures that wheel speed sensor circuitry is not internally shorted.
7. This checks for an open in the wheel signal low circuit.
8. This checks for an open in the wheel signal high circuit.
9. This step insures that Code A023 was not set due to a poor connection between the world connector and EBCM.

Diagnostic Aids:

An "Intermittent" problem may be caused by a poor connection, rubbed through wire insulation, or a wire that is broken inside the insulation.

If the customers comments reflect that the ABS Amber warning lamp is "ON" only during moist environmental changes, (rain, snow, vehicle wash) all wheel speed sensor circuitry should be thoroughly inspected for signs of water intrusion. If code is not current clear codes and simulate the effects of water intrusion. Use the following procedure. Spray down the suspected area with a 5% salt water solution (two teaspoons of salt to 12 oz. of water). Test drive vehicle. Vehicle must achieve 4 mph. If code returns immediately, replace suspected harness.

Any circuitry, that is suspected as causing the intermittent complaint, should be thoroughly checked for backed out terminals, improper mating, broken locks, improperly formed or damaged terminals, poor terminal to wiring connections or physical damage to the wiring harness.

DIAGNOSTIC CHART DTC A023 — 1993 SILHOUETTE, TRANS SPORT AND LUMINA APV

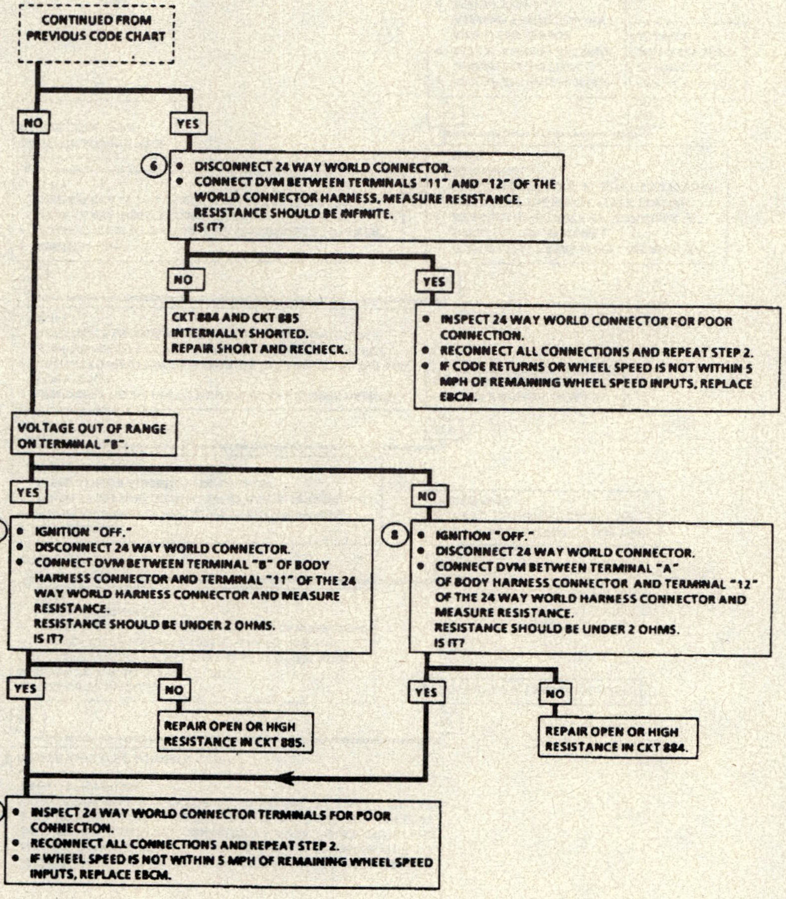

THIS CHART ASSUMES THAT A CURRENT CODE IS STORED INDICATING THAT THIS FAULT IS PRESENT.

DIAGNOSTIC CHART DTC A024 — 1993 SILHOUETTE, TRANS SPORT AND LUMINA APV

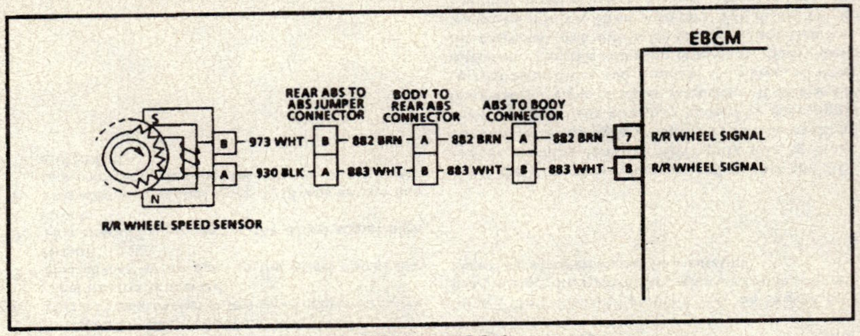

CODE A024

RIGHT REAR WHEEL SPEED = 0

Circuit Description:

As a toothed ring passes by the wheel speed sensor, changes in the electromagnetic field cause the wheel speed sensor to produce a sinusoidal (AC) voltage signal whose frequency is proportional to wheel speed. The magnitude of this signal is directly related to wheel speed and the proximity of the wheel speed sensor to the toothed ring, often referred to as the air gap.

This code detects opens or a low output condition that cause a wheel speed of 0 mph to be calculated by the ABS EBCM. The conditions present to fail this test are: (1) Three (3) wheels are rotating at least 5 mph, and (2) The fourth wheel is 0 mph.

Test Description: Number(s) below refer to circled number(s) on the diagnostic chart.

1. This test determines if Code A035 is present.
2. This test verifies whether the fault is currently present.
3. This step is designed to identify a wheel speed sensor or circuitry that is damaged and visibly apparent.
4. This test checks to see that the wheel speed sensor is within proper resistance values.
5. This insures that the wheel speed sensor is not shorted to ground.

Diagnostic Aids:

An "Intermittent" problem may be caused by a poor connection, rubbed through wire insulation, or a wire that is broken inside the insulation.

If the customers comments reflect that the ABS Amber warning lamp is "ON" only during moist environmental changes, (rain, snow, vehicle wash) all wheel speed sensor circuitry should be thoroughly inspected for signs of water intrusion. If code is not current clear codes and simulate the effects of water intrusion. Use the following procedure. Spray down the suspected area with a 5% salt water solution (two teaspoons of salt to 12 oz. of water). Test drive vehicle. Vehicle must achieve 4 mph. If code returns immediately, replace suspected harness.

Any circuitry, that is suspected as causing the intermittent complaint, should be thoroughly checked for backed out terminals, improper mating, broken locks, improperly formed or damaged terminals, poor terminal to wiring connections or physical damage to the wiring harness.

DIAGNOSTIC CHART DTC A024 — 1993 SILHOUETTE, TRANS SPORT AND LUMINA APV

IMPORTANT: WHEEL SPEED SENSOR INTERMITTENT PROBLEMS MAY BE DIFFICULT TO LOCATE. CARE SHOULD BE TAKEN NOT TO DISTURB ANY ELECTRICAL CONNECTIONS PRIOR TO AN INDICATED STEP OF THIS CHART. THIS WILL INSURE THAT AN INTERMITTENT CONNECTION WILL NOT BE CORRECTED BEFORE THE SOURCE OF THE PROBLEM IS FOUND.

(1) IS CODE A035 ALSO CURRENTLY SET?

- YES: PROCEED WITH THAT CODE FIRST.
- NO:

(2)
- TEST DRIVE VEHICLE.
- USING TECH 1, SELECT DATA LIST.
- MONITOR WHEEL SPEED ON RIGHT REAR WHEEL WHILE SLOWLY DECELERATING FROM 35 MPH TO 0 MPH.
- WHEEL SPEED SHOULD BE WITHIN 5 MPH OF REMAINING WHEEL SPEED INPUTS,

IS IT?

- YES: FAULT IS NOT PRESENT AT THIS TIME. REFER TO "DIAGNOSTIC AIDS" ON FACING PAGE.
- NO:

(3)
- IGNITION "OFF."
- PHYSICALLY INSPECT WHEEL SPEED SENSOR WIRING, CONNECTORS AND CONNECTOR AT WHEEL BEARING AND SPEED SENSOR ASSEMBLY FOR DAMAGE.

IS PHYSICAL DAMAGE INDICATED?

- YES: CORRECT PHYSICAL DAMAGE, CLEAR CODES AND RECHECK.
- NO:

(4)
- DISCONNECT REAR ABS TO RIGHT REAR ABS JUMPER HARNESS CONNECTOR.
- CONNECT DVM BETWEEN TERMINALS "A" AND "B" OF THE ABS JUMPER HARNESS CONNECTOR AND MEASURE RESISTANCE. RESISTANCE SHOULD BE BETWEEN 2100 AND 2400 OHMS.

IS IT?

- YES:

(5)
- IGNITION "ON."
- CONNECT DVM TO GROUND AND TERMINALS "A" THEN "B" OF REAR ABS HARNESS CONNECTOR. SHOULD BE 2.25 TO 2.75 VOLTS ON BOTH TERMINALS.

IS IT?

GO TO FOLLOWING PAGE OF THIS CODE CHART

- NO (from 4):
  - REMOVE RIGHT REAR WHEEL BEARING AND SPEED SENSOR ASSEMBLY.
  - MEASURE RESISTANCE AT TERMINALS "A" AND "B" OF THE WHEEL SPEED SENSOR. RESISTANCE SHOULD BE BETWEEN 2100 AND 2400 OHMS.

  IS IT?

  - NO: REPLACE WHEEL BEARING AND SPEED SENSOR ASSEMBLY.
  - YES:
    - RECONNECT WHEEL SPEED SENSOR CONNECTOR.
    - RETAIN A WHEEL BEARING AND SPEED SENSOR ASSEMBLY WITH ONE BOLT.
    - REPEAT STEP 4.

    IS RESISTANCE OK?

    - YES: POOR CONNECTION AT WHEEL SPEED SENSOR CONNECTOR. REPAIR AS NECESSARY.
    - NO: REPLACE RIGHT REAR ABS JUMPER HARNESS.

THIS CHART ASSUMES THAT A CURRENT CODE IS STORED INDICATING THAT THIS FAULT IS PRESENT.

DIAGNOSTIC CHART DTC A024 — 1993 SILHOUETTE, TRANS SPORT AND LUMINA APV

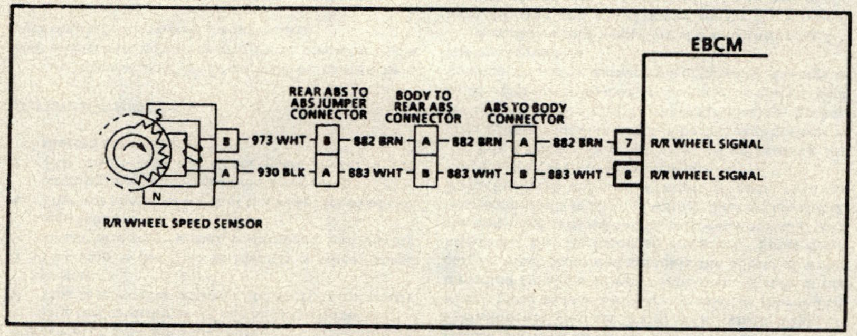

CODE A024

RIGHT REAR WHEEL SPEED = 0

Circuit Description:

As a toothed ring passes by the wheel speed sensor, changes in the electromagnetic field cause the wheel speed sensor to produce a sinusoidal (AC) voltage signal whose frequency is proportional to wheel speed. The magnitude of this signal is directly related to wheel speed and the proximity of the wheel speed sensor to the toothed ring, often referred to as the air gap.

This code detects opens or a low output condition that cause a wheel speed of 0 mph to be calculated by the ABS EBCM. The conditions present to fail this test are: (1) Three (3) wheels are rotating at least 5 mph, and (2) The fourth wheel is 0 mph.

Test Description: Number(s) below refer to circled number(s) on the diagnostic chart.

6. This test insures that wheel speed sensor circuitry is not internally shorted.
7. This checks for an open in the wheel signal low circuit.
8. This checks for an open in the wheel signal high circuit.
9. This step insures that Code A024 was not set due to a poor connection between the world connector and EBCM.

Diagnostic Aids:

An "Intermittent" problem may be caused by a poor connection, rubbed through wire insulation, or a wire that is broken inside the insulation.

If the customers comments reflect that the ABS Amber warning lamp is "ON" only during moist environmental changes, (rain, snow, vehicle wash) all wheel speed sensor circuitry should be thoroughly inspected for signs of water intrusion. If code is not current clear codes and simulate the effects of water intrusion. Use the following procedure. Spray down the suspected area with a 5% salt water solution (two teaspoons of salt to 12 oz. of water). Test drive vehicle. Vehicle must achieve 4 mph. If code returns immediately, replace suspected harness.

Any circuitry, that is suspected as causing the intermittent complaint, should be thoroughly checked for backed out terminals, improper mating, broken locks, improperly formed or damaged terminals, poor terminal to wiring connections or physical damage to the wiring harness.

DIAGNOSTIC CHART DTC A024 — 1993 SILHOUETTE, TRANS SPORT AND LUMINA APV

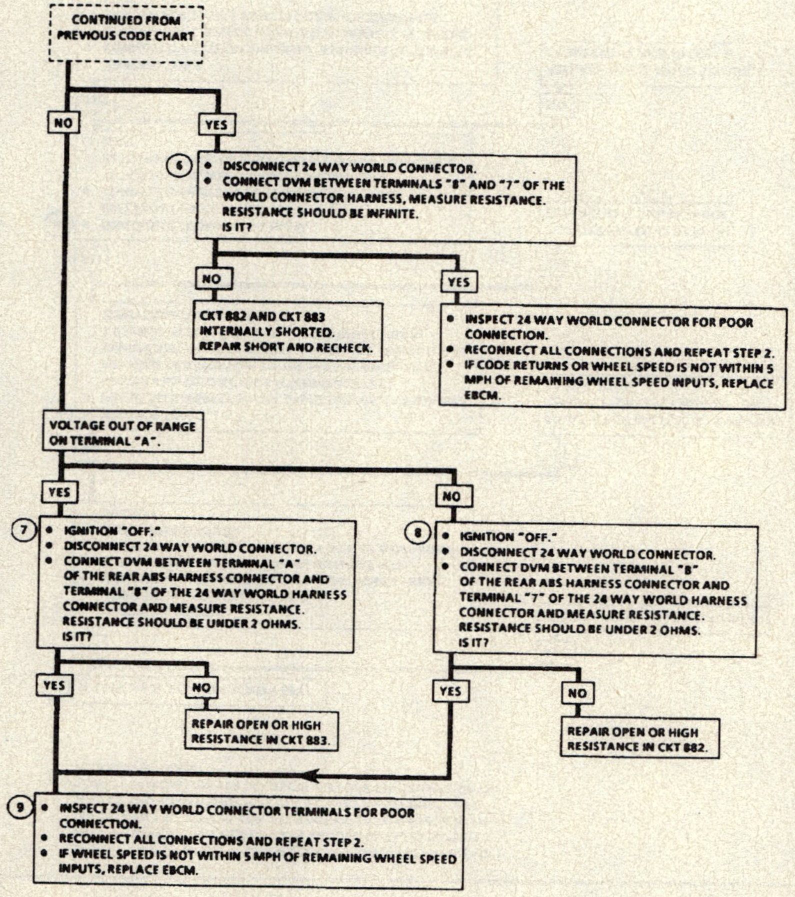

THIS CHART ASSUMES THAT A CURRENT CODE IS STORED INDICATING THAT THIS FAULT IS PRESENT.

DIAGNOSTIC CHART DTC A025 — 1993 SILHOUETTE, TRANS SPORT AND LUMINA APV

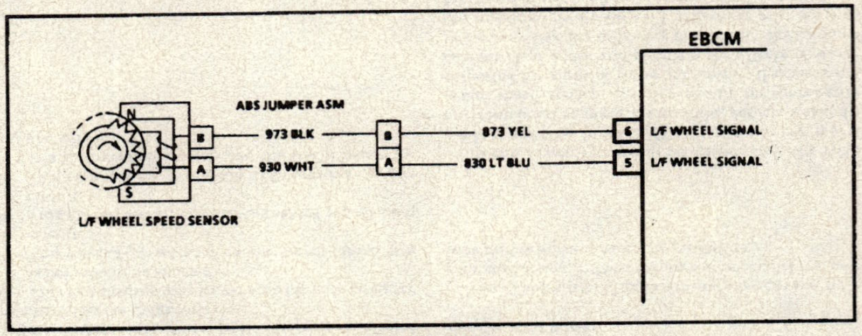

CODE A025

LEFT FRONT EXCESSIVE WHEEL SPEED VARIATION

Circuit Description:

As a toothed ring passes by the wheel speed sensor, changes in the electromagnetic field cause the wheel speed sensor to produce a sinusoidal (AC) voltage signal whose frequency is proportional to wheel speed. The magnitude of this signal is directly related to wheel speed and the proximity of the wheel speed sensor to the toothed ring, often referred to as the air gap.

This code detects opens or a low output condition that cause intermittent wheel speed operation. Also detected is any sudden change in wheel speed determined to the unreasonable.

Test Description: Number(s) below refer to circled number(s) on the diagnostic chart.

1. This test determines if Code A032 is also set.
2. This test verifies whether the fault is currently present.
3. This step is designed to identify a wheel speed sensor or circuitry that is damaged and visibly apparent.
4. This test checks to see that the wheel speed sensor circuitry is within proper resistance value.
5. This insure that the wheel speed sensor is not shorted to ground.

Diagnostic Aids:

An "Intermittent" problem may be caused by a poor connection, rubbed through wire insulation, or a wire that is broken inside the insulation.

If the customers comments reflect that the ABS Amber warning lamp is "ON" only during moist environmental changes, (rain, snow, vehicle wash) all wheel speed sensor circuitry should be thoroughly inspected for signs of water intrusion. If code is not current clear codes and simulate the effects of water intrusion. Use the following procedure. Spray down the suspected area with a 5% salt water solution (two teaspoons of salt to 12 oz. of water). Test drive vehicle. Vehicle must achieve 4 mph. If code returns immediately, replace suspected harness.

Any circuitry, that is suspected as causing the intermittent complaint, should be thoroughly checked for backed out terminals, improper mating, broken locks, improperly formed or damaged terminals, poor terminal to wiring connections or physical damage to the wiring harness.

A careful visual inspection of the toothed ring, CV joint, bearing and wheel speed sensor for physical damage should also be performed. If code sets at the same mph every time and wheel speed variation is noted above this mph on the Tech 1, the speed ring is most likely damaged. This can occur at any speed.

DIAGNOSTIC CHART DTC A025 — 1993 SILHOUETTE, TRANS SPORT AND LUMINA APV

IMPORTANT: WHEEL SPEED SENSOR INTERMITTENT PROBLEMS MAY BE DIFFICULT TO LOCATE. CARE SHOULD BE TAKEN NOT TO DISTURB ANY ELECTRICAL CONNECTIONS PRIOR TO AN INDICATED STEP OF THIS CHART. THIS WILL INSURE THAT AN INTERMITTENT CONNECTION WILL NOT BE CORRECTED BEFORE THE SOURCE OF THE PROBLEM IS FOUND.

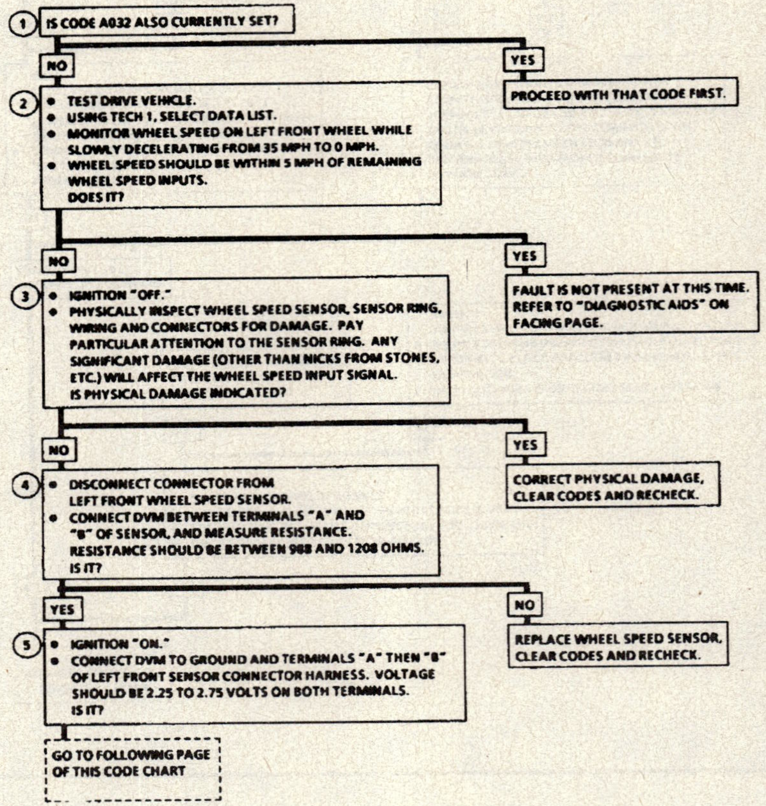

THIS CHART ASSUMES THAT A CURRENT CODE IS STORED INDICATING THAT THIS FAULT IS PRESENT.

DIAGNOSTIC CHART DTC A025 — 1993 SILHOUETTE, TRANS SPORT AND LUMINA APV

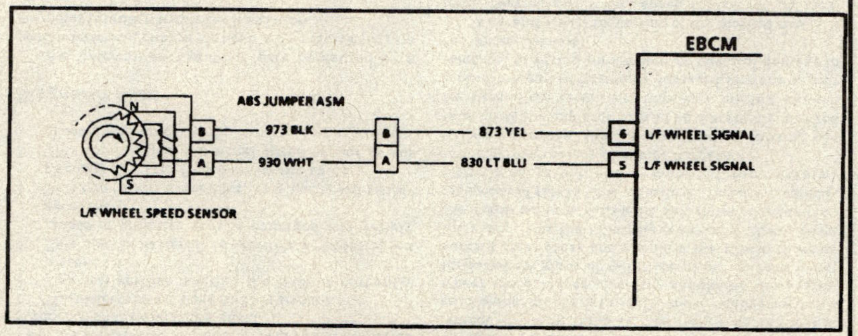

CODE A025

LEFT FRONT EXCESSIVE WHEEL SPEED VARIATION

Circuit Description:

As a toothed ring passes by the wheel speed sensor, changes in the electromagnetic field cause the wheel speed sensor to produce a sinusoidal (AC) voltage signal whose frequency is proportional to wheel speed. The magnitude of this signal is directly related to wheel speed and the proximity of the wheel speed sensor to the toothed ring, often referred to as the air gap.

This code detects opens or a low output condition that cause intermittent wheel speed operational. Also detected is any sudden change in wheel speed determined to be unreasonable.

Test Description: Number(s) below refer to circled number(s) on the diagnostic chart.

6. This step insures that the wheel speed sensor circuitry resistance is within the proper value.
7. This checks for an open in the wheel signal high circuit.
8. This checks for an open in the wheel signal low circuit.
9. This step insures that Code A025 was not set due to a poor connection between the world connector and EBCM.

Diagnostic Aids:

An "Intermittent" problem may be caused by a poor connection, rubbed through wire insulation, or a wire that is broken inside the insulation.

If the customers comments reflect that the ABS Amber warning lamp is "ON" only during moist environmental changes, (rain, snow, vehicle wash) all wheel speed sensor circuitry should be thoroughly inspected for signs of water intrusion. If code is not current clear codes and simulate the effects of water intrusion. Use the following procedure. Spray down the suspected area with a 5% salt water solution (two teaspoons of salt to 12 oz. of water). Test drive vehicle. Vehicle must achieve 4 mph. If code returns immediately, replace suspected harness.

Any circuitry, that is suspected as causing the intermittent complaint, should be thoroughly checked for backed out terminals, improper mating, broken locks, improperly formed or damaged terminals, poor terminal to wiring connections or physical damage to the wiring harness.

A careful visual inspection of the toothed ring, CV joint, bearing and wheel speed sensor for physical damage should also be performed. If code sets at the same mph every time and wheel speed variation is noted above this mph on the Tech 1, the speed ring is most likely damaged. This can occur at any speed.

DIAGNOSTIC CHART DTC A025 — 1993 SILHOUETTE, TRANS SPORT AND LUMINA APV

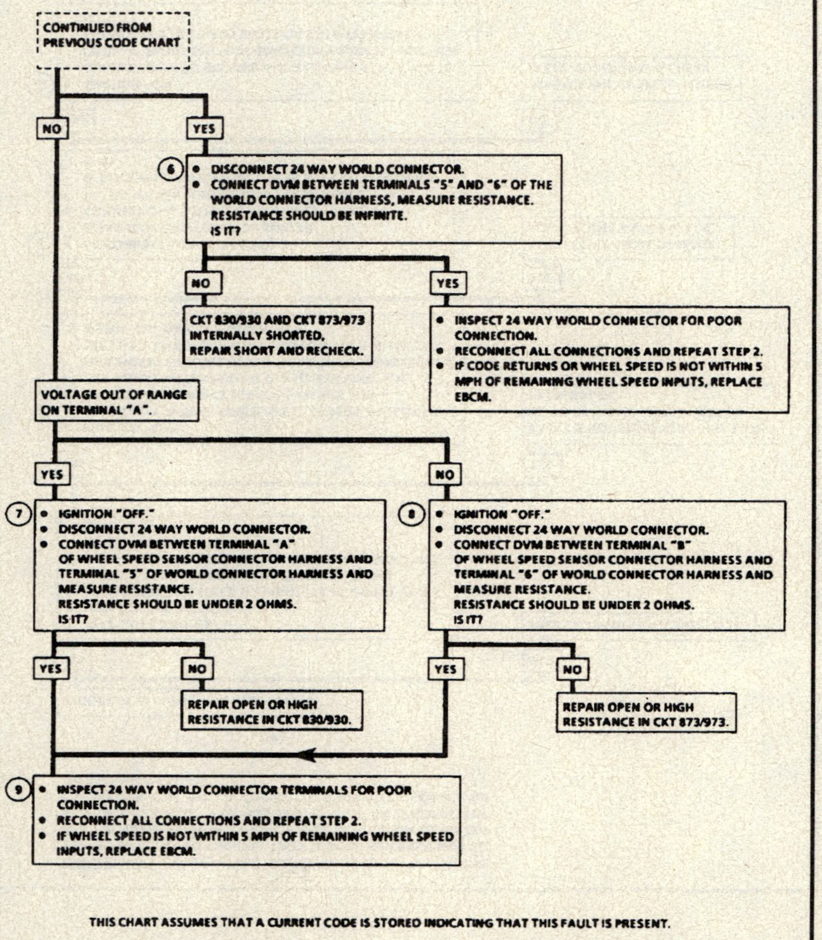

DIAGNOSTIC CHART DTC A026 — 1993 SILHOUETTE, TRANS SPORT AND LUMINA APV

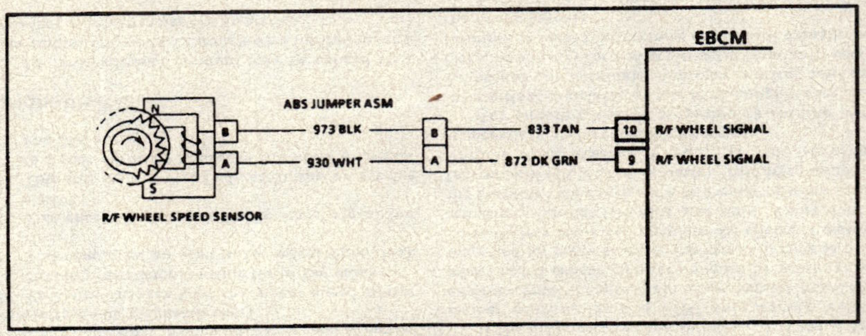

CODE A026

RIGHT FRONT EXCESSIVE WHEEL SPEED VARIATION

Circuit Description:

As a toothed ring passes by the wheel speed sensor, changes in the electromagnetic field cause the wheel speed sensor to produce a sinusoidal (AC) voltage signal whose frequency is proportional to wheel speed. The magnitude of this signal is directly related to wheel speed and the proximity of the wheel speed sensor to the toothed ring, often referred to as the air gap.

This code detects opens or a low output condition that cause intermittent wheel speed operation. Also detected is any sudden change in wheel speed determined to be unreasonable.

Test Description: Number(s) below refer to circled number(s) on the diagnostic chart.

1. This determines if Code A033 is present.
2. This test verifies whether the fault is currently present.
3. This step is designed to identify a wheel speed sensor or circuitry that is damaged and visibly apparent.
4. This test checks to see that the wheel speed sensor circuitry is within proper resistance value.
5. This insures that the wheel speed sensor is not shorted to ground.

Diagnostic Aids:

An "Intermittent" problem may be caused by a poor connection, rubbed through wire insulation, or a wire that is broken inside the insulation.

If the customers comments reflect that the ABS Amber warning lamp is "ON" only during moist environmental changes, (rain, snow, vehicle wash) all wheel speed sensor circuitry should be thoroughly inspected for signs of water intrusion. If code is not current clear codes and simulate the effects of water intrusion. Use the following procedure. Spray down the suspected area with a 5% salt water solution (two teaspoons of salt to 12 oz. of water). Test drive vehicle. Vehicle must achieve 4 mph. If code returns immediately, replace suspected harness.

Any circuitry, that is suspected as causing the intermittent complaint, should be thoroughly checked for backed out terminals, improper mating, broken locks, improperly formed or damaged terminals, poor terminal to wiring connections or physical damage to the wiring harness.

A careful visual inspection of the toothed ring, CV joint, bearing and wheel speed sensor for physical damage should also be performed. If code sets at the same mph every time and wheel speed variation is noted above this mph on the Tech 1, the speed ring is most likely damaged. This can occur at any speed.

DIAGNOSTIC CHART DTC A026 — 1993 SILHOUETTE, TRANS SPORT AND LUMINA APV

IMPORTANT: WHEEL SPEED SENSOR INTERMITTENT PROBLEMS MAY BE DIFFICULT TO LOCATE. CARE SHOULD BE TAKEN NOT TO DISTURB ANY ELECTRICAL CONNECTIONS PRIOR TO AN INDICATED STEP OF THIS CHART. THIS WILL INSURE THAT AN INTERMITTENT CONNECTION WILL NOT BE CORRECTED BEFORE THE SOURCE OF THE PROBLEM IS FOUND.

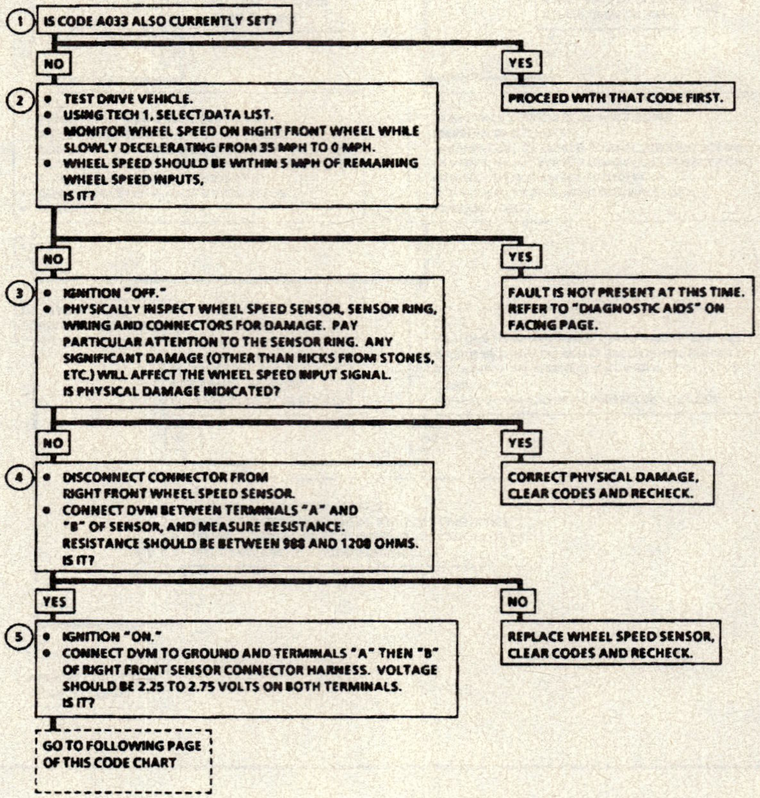

THIS CHART ASSUMES THAT A CURRENT CODE IS STORED INDICATING THAT THIS FAULT IS PRESENT.

DIAGNOSTIC CHART DTC A026 — 1993 SILHOUETTE, TRANS SPORT AND LUMINA APV

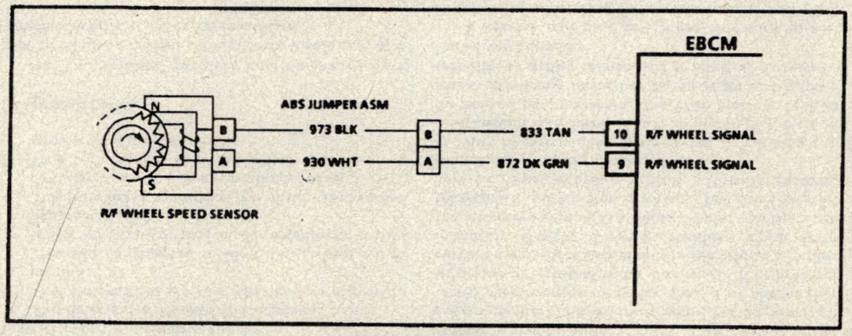

CODE A026

RIGHT FRONT EXCESSIVE WHEEL SPEED VARIATION

Circuit Description:

As a toothed ring passes by the wheel speed sensor, changes in the electromagnetic field cause the wheel speed sensor to produce a sinusoidal (AC) voltage signal whose frequency is proportional to wheel speed. The magnitude of this signal is directly related to wheel speed and the proximity of the wheel speed sensor to the toothed ring, often referred to as the air gap.

This code detects opens or a low output condition that cause intermittent wheel speed operation. Also detected is any sudden change in wheel speed determined to be unreasonable.

Test Description: Number(s) below refer to circled number(s) on the diagnostic chart.

6. This step insures that the wheel speed sensor circuitry is not internally shorted.
7. This checks for an open in the wheel signal low circuit.
8. This checks for an open in the wheel signal high circuit.
9. This step insures that Code A026 was not set due to a poor connection between the world connector and EBCM.

Diagnostic Aids:

An "Intermittent" problem may be caused by a poor connection, rubbed through wire insulation, or a wire that is broken inside the insulation.

If the customers comments reflect that the ABS Amber warning lamp is "ON" only during moist environmental changes, (rain, snow, vehicle wash) all wheel speed sensor circuitry should be thoroughly inspected for signs of water intrusion. If code is not current clear codes and simulate the effects of water intrusion. Use the following procedure. Spray down the suspected area with a 5% salt water solution (two teaspoons of salt to 12 oz. of water). Test drive vehicle. Vehicle must achieve 4 mph. If code returns immediately, replace suspected harness.

Any circuitry, that is suspected as causing the intermittent complaint, should be thoroughly checked for backed out terminals, improper mating, broken locks, improperly formed or damaged terminals, poor terminal to wiring connections or physical damage to the wiring harness.

A careful visual inspection of the toothed ring, CV joint, bearing and wheel speed sensor for physical damage should also be performed. If code sets at the same mph every time and wheel speed variation is noted above this mph on the Tech 1, the speed ring is most likely damaged. This can occur at any speed.

DIAGNOSTIC CHART DTC A026 — 1993 SILHOUETTE, TRANS SPORT AND LUMINA APV

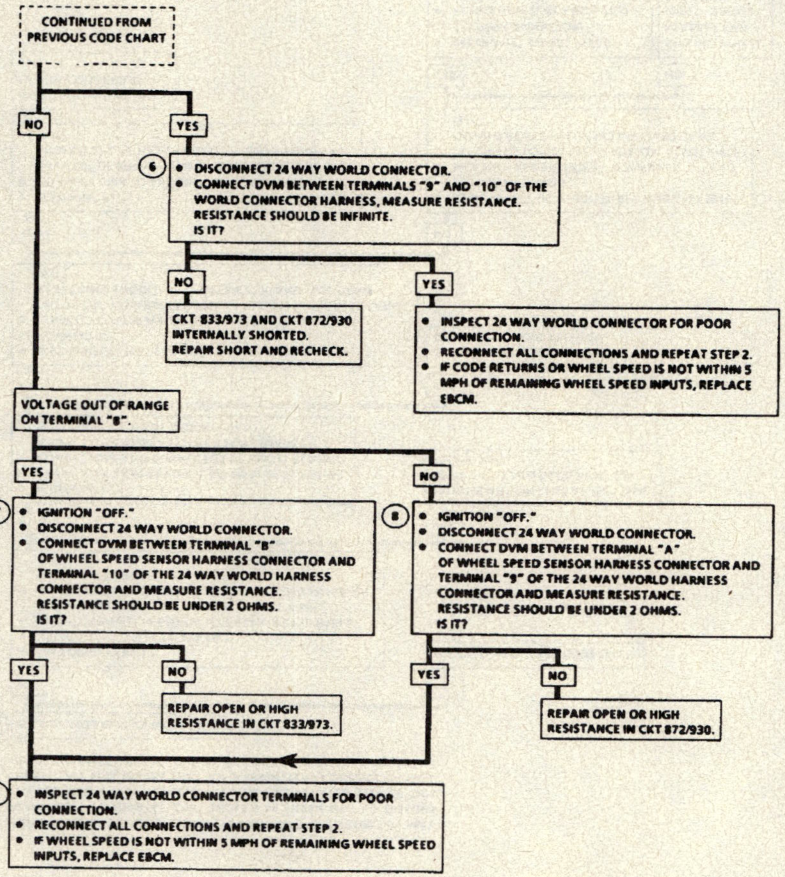

THIS CHART ASSUMES THAT A CURRENT CODE IS STORED INDICATING THAT THIS FAULT IS PRESENT.

DIAGNOSTIC CHART DTC A027 — 1993 SILHOUETTE, TRANS SPORT AND LUMINA APV

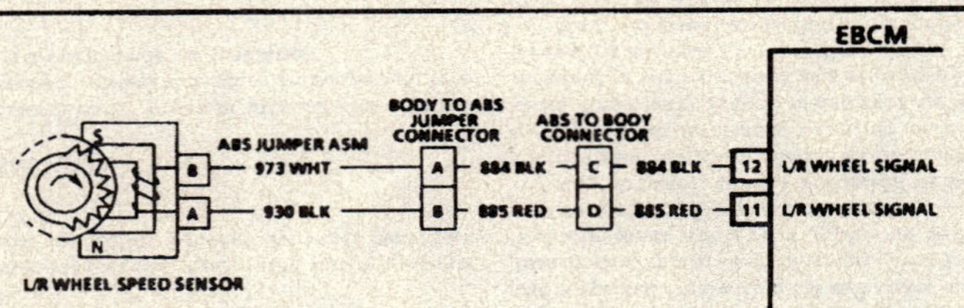

CODE A027

LEFT REAR EXCESSIVE WHEEL SPEED VARIATION

Circuit Description:

As a toothed ring passes by the wheel speed sensor, changes in the electromagnetic field cause the wheel speed sensor to produce a sinusoidal (AC) voltage signal whose frequency is proportional to wheel speed. The magnitude of this signal is directly related to wheel speed and the proximity of the wheel speed sensor to the toothed ring, often referred to as the air gap.

This code detects opens or a low output condition that cause intermittent wheel speed operation. Also detected is any sudden change in wheel speed determined to be unreasonable.

Test Description: Number(s) below refer to circled number(s) on the diagnostic chart.

1. This determines if Code A034 is present.
2. This test verifies whether the fault is currently present.
3. This step is designed to identify a wheel speed sensor or circuitry that is damaged and visibly apparent.
4. This test checks to see that the wheel speed sensor circuitry is within proper resistance value.
5. This test insures that the wheel speed sensor circuitry has good continuity.

Diagnostic Aids:

An "Intermittent" problem may be caused by a poor connection, rubbed through wire insulation, or a wire that is broken inside the insulation.

If the customers comments reflect that the ABS Amber warning lamp is "ON" only during moist environmental changes, (rain, snow, vehicle wash) all wheel speed sensor circuitry should be thoroughly inspected for signs of water intrusion. If code is not current clear codes and simulate the effects of water intrusion. Use the following procedure. Spray down the suspected area with a 5% salt water solution (two teaspoons of salt to 12 oz. of water). Test drive vehicle. Vehicle must achieve 4 mph. If code returns immediately, replace suspected harness.

Any circuitry, that is suspected as causing the intermittent complaint, should be thoroughly checked for backed out terminals, improper mating, broken locks, improperly formed or damaged terminals, poor terminal to wiring connections or physical damage to the wiring harness.

Excessive end play can cause this code to set. Check integral bearing for excessive axial end play,

DIAGNOSTIC CHART DTC A027 — 1993 SILHOUETTE, TRANS SPORT AND LUMINA APV

IMPORTANT: WHEEL SPEED SENSOR INTERMITTENT PROBLEMS MAY BE DIFFICULT TO LOCATE. CARE SHOULD BE TAKEN NOT TO DISTURB ANY ELECTRICAL CONNECTIONS PRIOR TO AN INDICATED STEP OF THIS CHART. THIS WILL INSURE THAT AN INTERMITTENT CONNECTION WILL NOT BE CORRECTED BEFORE THE SOURCE OF THE PROBLEM IS FOUND.

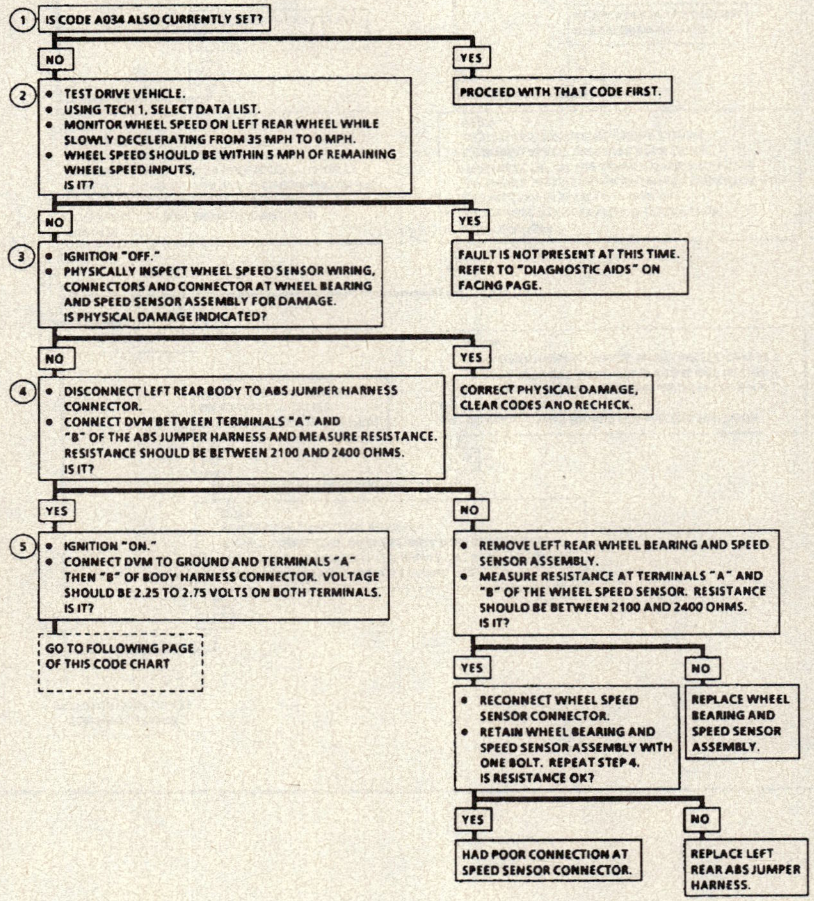

THIS CHART ASSUMES THAT A CURRENT CODE IS STORED INDICATING THAT THIS FAULT IS PRESENT.

DIAGNOSTIC CHART DTC A027 — 1993 SILHOUETTE, TRANS SPORT AND LUMINA APV

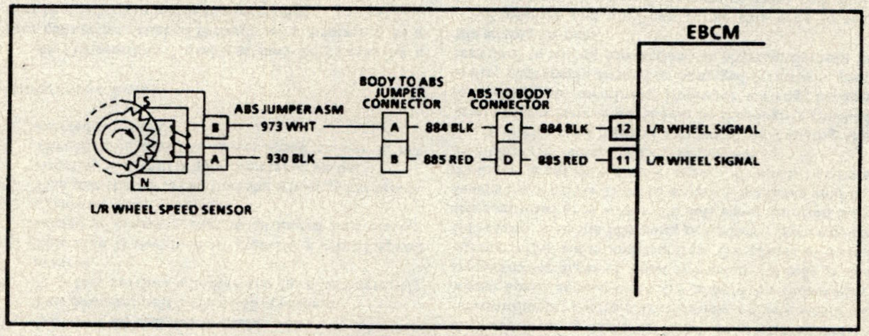

CODE A027

LEFT REAR EXCESSIVE WHEEL SPEED VARIATION

Circuit Description:

As a toothed ring passes by the wheel speed sensor, changes in the electromagnetic field cause the wheel speed sensor to produce a sinusoidal (AC) voltage signal whose frequency is proportional to wheel speed. The magnitude of this signal is directly related to wheel speed and the proximity of the wheel speed sensor to the toothed ring, often referred to as the air gap.

This code detects opens or a low output condition that cause intermittent wheel speed operation. Also detected is any sudden change in wheel speed determined to be unreasonable.

Test Description: Number(s) below refer to circled number(s) on the diagnostic chart.

6. This step insures that the wheel speed sensor circuitry is not internally shorted.
7. This checks for an open in the wheel signal low circuit.
8. This checks for an open in the wheel signal high circuit.
9. This step insures that Code A027 was not set due to a poor connection between the world connector and EBCM.

Diagnostic Aids:

An "Intermittent" problem may be caused by a poor connection, rubbed through wire insulation, or a wire that is broken inside the insulation.

If the customers comments reflect that the ABS Amber warning lamp is "ON" only during moist environmental changes, (rain, snow, vehicle wash) all wheel speed sensor circuitry should be thoroughly inspected for signs of water intrusion. If code is not current clear codes and simulate the effects of water intrusion. Use the following procedure. Spray down the suspected area with a 5% salt water solution (two teaspoons of salt to 12 oz. of water). Test drive vehicle. Vehicle must achieve 4 mph. If code returns immediately, replace suspected harness.

Any circuitry, that is suspected as causing the intermittent complaint, should be thoroughly checked for backed out terminals, improper mating, broken locks, improperly formed or damaged terminals, poor terminal to wiring connections or physical damage to the wiring harness.

Excessive end play can cause this code to set. Check integral bearing for excessive axial end play,

DIAGNOSTIC CHART DTC A027 — 1993 SILHOUETTE, TRANS SPORT AND LUMINA APV

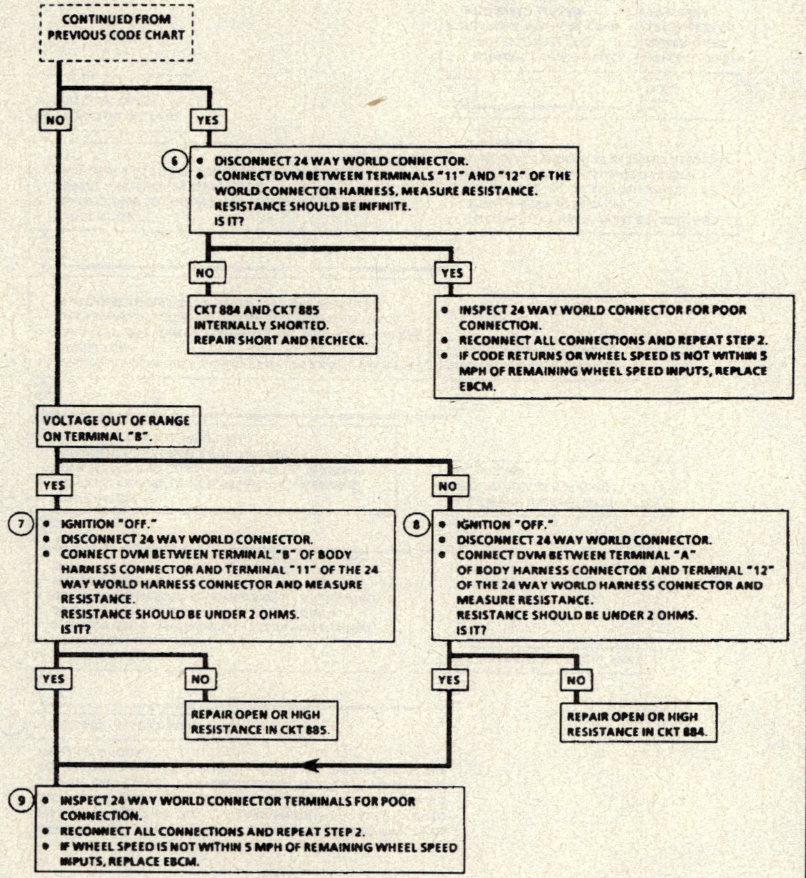

THIS CHART ASSUMES THAT A CURRENT CODE IS STORED INDICATING THAT THIS FAULT IS PRESENT.

DIAGNOSTIC CHART DTC A028 — 1993 SILHOUETTE, TRANS SPORT AND LUMINA APV

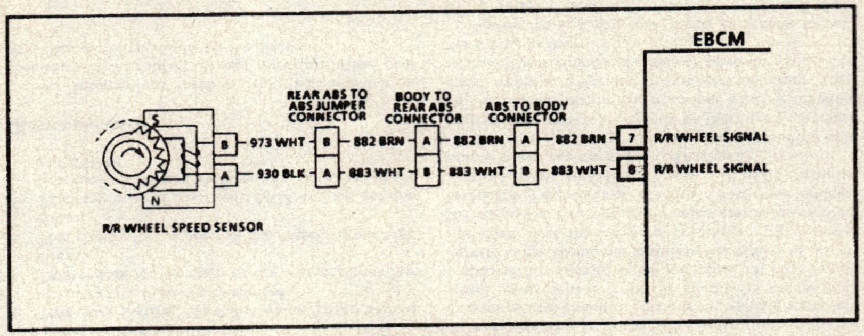

CODE A028

RIGHT REAR EXCESSIVE WHEEL SPEED VARIATION

Circuit Description:

As a toothed ring passes by the wheel speed sensor, changes in the electromagnetic field cause the wheel speed sensor to produce a sinusoidal (AC) voltage signal whose frequency is proportional to wheel speed. The magnitude of this signal is directly related to wheel speed and the proximity of the wheel speed sensor to the toothed ring, often referred to as the air gap.

This code detects opens or a low output condition that cause intermittent wheel speed operation. Also detected is any sudden change in wheel speed determined to be unreasonable.

Test Description: Number(s) below refer to circled number(s) on the diagnostic chart.

1. This determines if Code A035 is present.
2. This test verifies whether the fault is currently present.
3. This step is designed to identify a wheel speed sensor or circuitry that is damaged and visibly apparent.
4. This test checks to see that the wheel speed sensor circuitry is within proper resistance value.
5. This insures that the wheel speed sensor is not shorted to ground.

Diagnostic Aids:

An "Intermittent" problem may be caused by a poor connection, rubbed through wire insulation, or a wire that is broken inside the insulation.

If the customers comments reflect that the ABS Amber warning lamp is "ON" only during moist environmental changes, (rain, snow, vehicle wash) all wheel speed sensor circuitry should be thoroughly inspected for signs of water intrusion. If code is not current clear codes and simulate the effects of water intrusion. Use the following procedure. Spray down the suspected area with a 5% salt water solution (two teaspoons of salt to 12 oz. of water). Test drive vehicle. Vehicle must achieve 4 mph. If code returns immediately, replace suspected harness.

Any circuitry, that is suspected as causing the intermittent complaint, should be thoroughly checked for backed out terminals, improper mating, broken locks, improperly formed or damaged terminals, poor terminal to wiring connections or physical damage to the wiring harness.

Excessive end play can cause this code to set. Check integral bearing for excessive axial end play,

DIAGNOSTIC CHART DTC A028 — 1993 SILHOUETTE, TRANS SPORT AND LUMINA APV

IMPORTANT: WHEEL SPEED SENSOR INTERMITTENT PROBLEMS MAY BE DIFFICULT TO LOCATE. CARE SHOULD BE TAKEN NOT TO DISTURB ANY ELECTRICAL CONNECTIONS PRIOR TO AN INDICATED STEP OF THIS CHART. THIS WILL INSURE THAT AN INTERMITTENT CONNECTION WILL NOT BE CORRECTED BEFORE THE SOURCE OF THE PROBLEM IS FOUND.

(1) IS CODE A035 ALSO CURRENTLY SET?

- YES: PROCEED WITH THAT CODE FIRST.
- NO:

(2)
- TEST DRIVE VEHICLE.
- USING TECH 1, SELECT DATA LIST.
- MONITOR WHEEL SPEED ON RIGHT REAR WHEEL WHILE SLOWLY DECELERATING FROM 35 MPH TO 0 MPH.
- WHEEL SPEED SHOULD BE WITHIN 5 MPH OF REMAINING WHEEL SPEED INPUTS.

IS IT?

- YES: FAULT IS NOT PRESENT AT THIS TIME. REFER TO "DIAGNOSTIC AIDS" ON FACING PAGE.
- NO:

(3)
- IGNITION "OFF."
- PHYSICALLY INSPECT WHEEL SPEED SENSOR WIRING, CONNECTORS AND CONNECTOR AT WHEEL BEARING AND SPEED SENSOR ASSEMBLY FOR DAMAGE.

IS PHYSICAL DAMAGE INDICATED?

- YES: CORRECT PHYSICAL DAMAGE, CLEAR CODES AND RECHECK.
- NO:

(4)
- DISCONNECT REAR ABS TO RIGHT REAR ABS JUMPER HARNESS CONNECTOR.
- CONNECT DVM BETWEEN TERMINALS "A" AND "B" OF THE ABS JUMPER HARNESS CONNECTOR AND MEASURE RESISTANCE. RESISTANCE SHOULD BE BETWEEN 2100 AND 2400 OHMS.

IS IT?

- YES:

(5)
- IGNITION "ON."
- CONNECT DVM TO GROUND AND TERMINALS "A" THEN "B" OF REAR ABS HARNESS CONNECTOR. SHOULD BE 2.25 TO 2.75 VOLTS ON BOTH TERMINALS.

IS IT?

GO TO FOLLOWING PAGE OF THIS CODE CHART

- NO:
  - REMOVE RIGHT REAR WHEEL BEARING AND SPEED SENSOR ASSEMBLY.
  - MEASURE RESISTANCE AT TERMINALS "A" AND "B" OF THE WHEEL SPEED SENSOR. RESISTANCE SHOULD BE BETWEEN 2100 AND 2400 OHMS.

  IS IT?

  - NO: REPLACE WHEEL BEARING AND SPEED SENSOR ASSEMBLY.
  - YES:
    - RECONNECT WHEEL SPEED SENSOR CONNECTOR.
    - RETAIN A WHEEL BEARING AND SPEED SENSOR ASSEMBLY WITH ONE BOLT.
    - REPEAT STEP 4.

    IS RESISTANCE OK?

    - YES: HAD POOR CONNECTION AT WHEEL SPEED SENSOR CONNECTOR.
    - NO: REPLACE RIGHT REAR ABS JUMPER HARNESS.

DIAGNOSTIC CHART DTC A028 — 1993 SILHOUETTE, TRANS SPORT AND LUMINA APV

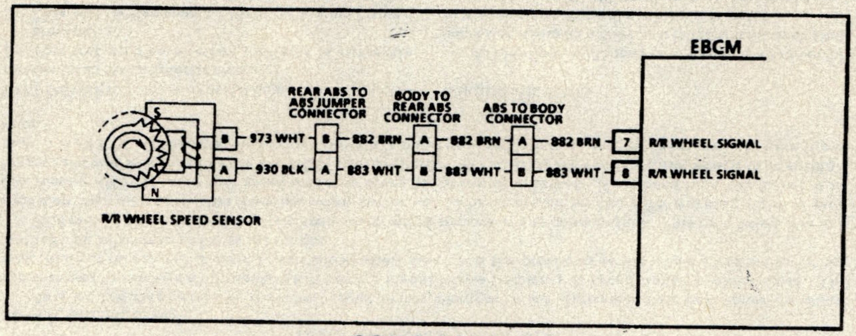

CODE A028

RIGHT REAR EXCESSIVE WHEEL SPEED VARIATION

Circuit Description:

As a toothed ring passes by the wheel speed sensor, changes in the electromagnetic field cause the wheel speed sensor to produce a sinusoidal (AC) voltage signal whose frequency is proportional to wheel speed. The magnitude of this signal is directly related to wheel speed and the proximity of the wheel speed sensor to the toothed ring, often referred to as the air gap.

This code detects opens or a low output condition that cause intermittent wheel speed operation. Also detected is any sudden change in wheel speed determined to be unreasonable.

Test Description: Number(s) below refer to circled number(s) on the diagnostic chart.

6. This step insures that the wheel speed sensor circuitry is not internally shorted.
7. This checks for an open in the wheel signal low circuit.
8. This checks for an open in the wheel signal high circuit.
9. This step insures that Code A028 was not set due to a poor connection between the world connector and EBCM.

Diagnostic Aids:

An "Intermittent" problem may be caused by a poor connection, rubbed through wire insulation, or a wire that is broken inside the insulation.

If the customers comments reflect that the ABS Amber warning lamp is "ON" only during moist environmental changes, (rain, snow, vehicle wash) all wheel speed sensor circuitry should be thoroughly inspected for signs of water intrusion. If code is not current clear codes and simulate the effects of water intrusion. Use the following procedure. Spray down the suspected area with a 5% salt water solution (two teaspoons of salt to 12 oz. of water). Test drive vehicle. Vehicle must achieve 4 mph. If code returns immediately, replace suspected harness.

Any circuitry, that is suspected as causing the intermittent complaint, should be thoroughly checked for backed out terminals, improper mating, broken locks, improperly formed or damaged terminals, poor terminal to wiring connections or physical damage to the wiring harness.

Excessive end play can cause this code to set. Check integral bearing for excessive axial end play.

DIAGNOSTIC CHART DTC A028 — 1993 SILHOUETTE, TRANS SPORT AND LUMINA APV

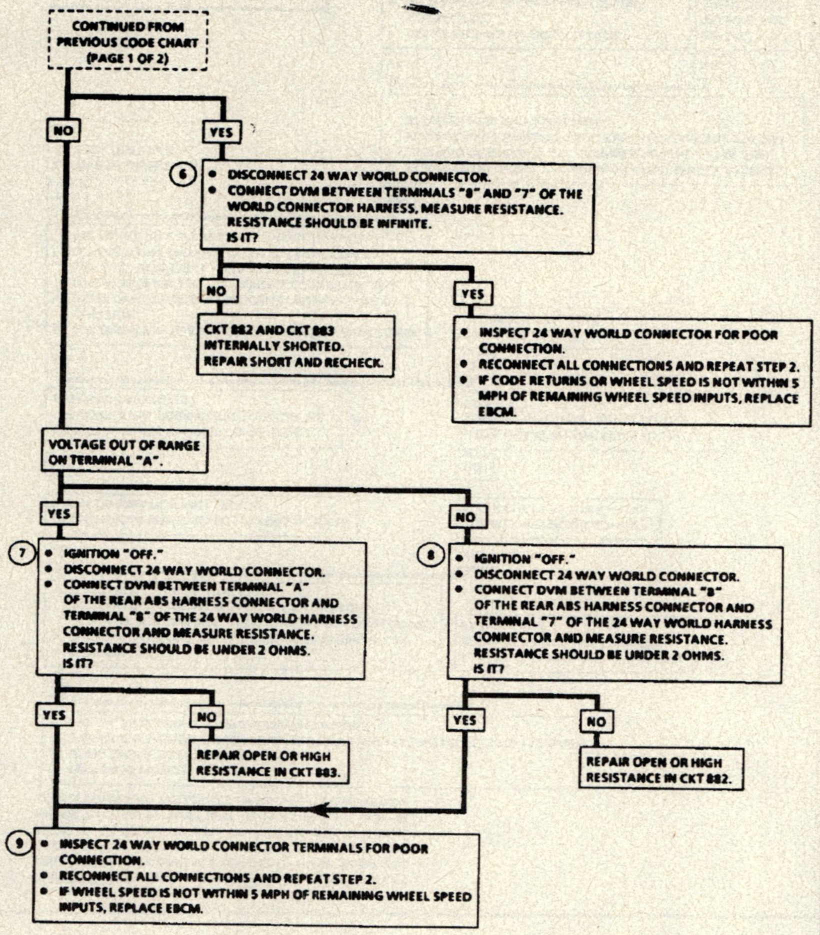

THIS CHART ASSUMES THAT A CURRENT CODE IS STORED INDICATING THAT THIS FAULT IS PRESENT.

DIAGNOSTIC CHART DTC A031 — 1993 SILHOUETTE, TRANS SPORT AND LUMINA APV

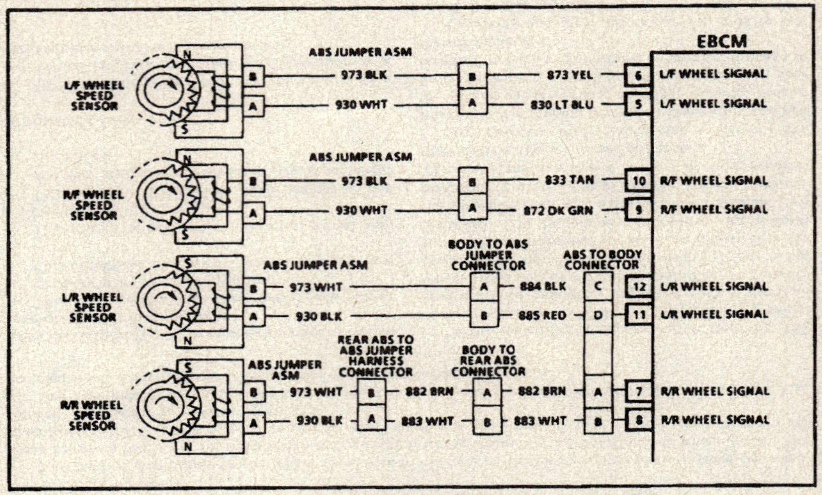

CODE A031

TWO WHEEL SPEEDS = 0

Circuit Description:

As a toothed ring passes by the wheel speed sensor, changes in the electromagnetic field cause the wheel speed sensor to produce a sinusoidal (AC) voltage signal whose frequency is proportional to wheel speed. The magnitude of this signal is directly related to wheel speed and the proximity of the wheel speed sensor to the toothed ring, often referred to as the air gap.

This code is designed to detect two open wheel speed sensors in any combination. This code exists due to a condition that may exist that does not allow the driver to be informed by the ABS warning light of non-functioning ABS. If one wheel speed sensor is open, the driver is informed. If a second fails, the driver is no longer informed without this code and ABS is not available. By requiring the two operating wheels to be within 7 mph, a false code is prevented during spin-ups while still allowing for detection when the stowaway spare tire is used.

Test Description: Number(s) below refer to circled number(s) on the diagnostic chart.

1. This test verifies whether the fault is currently present.
2. This step is designed to identify a wheel speed sensor or circuitry that is damaged and visibly apparent.
3. This step identifies if the cause of the problem is common to the rear wheel sensor circuitry.
4. This test checks to see if connectors at pass thru are connected properly.
5. This test checks to see that the wheel speed sensor is within proper resistance values.

Diagnostic Aids:

A false Code A031 may be set as a result of vehicle operation while on a hoist or a lifted condition or stuck in mud or snow, when the vehicle is in gear.

An "Intermittent" problem may be caused by a poor connection, rubbed through wire insulation, or a wire that is broken inside the insulation.

DIAGNOSTIC CHART DTC A031 — 1993 SILHOUETTE, TRANS SPORT AND LUMINA APV

IMPORTANT: WHEEL SPEED SENSOR INTERMITTENT PROBLEMS MAY BE DIFFICULT TO LOCATE. CARE SHOULD BE TAKEN NOT TO DISTURB ANY ELECTRICAL CONNECTIONS PRIOR TO AN INDICATED STEP OF THIS CHART. THIS WILL INSURE THAT AN INTERMITTENT CONNECTION WILL NOT BE CORRECTED BEFORE THE SOURCE OF THE PROBLEM IS FOUND.

(1)
- TEST DRIVE VEHICLE.
- USING TECH 1, SELECT DATA LIST.
- MONITOR ALL WHEEL SPEEDS WHILE SLOWLY DECELERATING FROM 35 MPH TO 0 MPH. ARE ALL WHEEL SPEEDS APPROXIMATELY EQUAL?

YES: FAULT IS NOT PRESENT AT THIS TIME. REFER TO "DIAGNOSTIC AIDS" ON FACING PAGE.

NO:

(2)
- IGNITION "OFF."
- PHYSICALLY INSPECT WHEEL SPEED SENSOR CIRCUITS IN QUESTION FOR DAMAGE. ALSO, CHECK WIRE CONNECTIONS FOR CORROSION OR POOR MATING OF CONNECTORS. IS PHYSICAL DAMAGE INDICATED?

YES: CORRECT PHYSICAL DAMAGE, CLEAR CODES AND RE-CHECK.

NO:

(3) DID BOTH REAR WHEEL SPEED SENSORS INDICATE 0 MPH, OR ERRATIC WHEEL SPEEDS?

NO: USING CHART BELOW, REFER TO SPEED SENSOR CODE CHART FOR AFFECTED WHEELS.

YES:

(4) INSPECT REAR ABS HARNESS CONNECTORS FOR CORROSION, POOR MATING OR DAMAGE. IS DAMAGE INDICATED?

YES: CORRECT DAMAGE, CLEAR CODES AND RE-CHECK.

NO:

(5)
- DISCONNECT REAR ABS JUMPER HARNESS CONNECTOR IN QUESTION.
- USING DVM CONNECTED BETWEEN TERMINALS "A" & "B" OF THE REAR ABS JUMPER HARNESS CONNECTOR AND MEASURE RESISTANCE. RESISTANCE SHOULD BE BETWEEN 2100 AND 2400 OHMS FOR REAR SENSORS. ARE RESISTANCE VALUES IN ACCEPTABLE RANGES?

YES: GO TO FOLLOWING PAGE OF THIS CODE CHART

NO:
- REMOVE REAR WHEEL BEARING AND SPEED SENSOR ASSEMBLY.
- MEASURE RESISTANCE AT TERMINALS "A" AND "B" OF THE WHEEL SPEED SENSOR IN QUESTION. RESISTANCE SHOULD BE BETWEEN 2100 AND 2400 OHMS. IS IT?

NO: REPLACE WHEEL BEARING AND SPEED SENSOR ASSEMBLY.

YES:
- RECONNECT WHEEL SPEED SENSOR CONNECTOR.
- RETAIN WHEEL BEARING AND SENSOR ASSEMBLY WITH ONE BOLT. REPEAT STEP 5. IS RESISTANCE OK?

YES: HAD POOR CONNECTION AT SPEED SENSOR CONNECTOR.

NO: REPLACE ABS JUMPER HARNESS.

SENSOR	CODE
L/F SENSOR	A021
R/F SENSOR	A022
L/R SENSOR	A023
R/R SENSOR	A024

DIAGNOSTIC CHART DTC A031 — 1993 SILHOUETTE, TRANS SPORT AND LUMINA APV

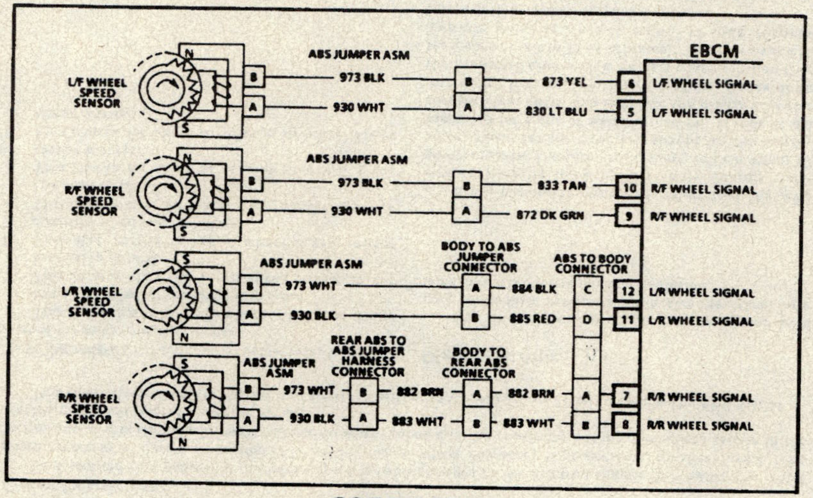

CODE A031

TWO WHEEL SPEEDS = 0

Circuit Description:

As a toothed ring passes by the wheel speed sensor, changes in the electromagnetic field cause the wheel speed sensor to produce a sinusoidal (AC) voltage signal whose frequency is proportional to wheel speed. The magnitude of this signal is directly related to wheel speed and the proximity of the wheel speed sensor to the toothed ring, often referred to as the air gap.

This code is designed to detect two open wheel speed sensors in any combination. This code exists due to a condition that may exist that does not allow the driver to be informed by the ABS warning light of non-functioning ABS. If one wheel speed sensor is open, the driver is informed. If a second fails, the driver is no longer informed without this code and ABS is not available. By requiring the two operating wheels to be within 7 mph, a false code is prevented during spin-ups while still allowing for detection when the stowaway spare tire is used.

Test Description: Number(s) below refer to circled number(s) on the diagnostic chart.

6. This step checks for an open in the wheel speed sensor circuitry of one side of the sensor.
7. This step checks for an open on the other side of the sensor in question.
8. This insures that the circuitry is not shorted to voltage.

Diagnostic Aids:

A false Code A031 may be set as a result of vehicle operation while on a hoist or a lifted condition, or stuck in mud or snow when the vehicle is in gear.

An "Intermittent" problem may be caused by a poor connection, rubbed through wire insulation, or a wire that is broken inside the insulation.

DIAGNOSTIC CHART DTC A031 — 1993 SILHOUETTE, TRANS SPORT AND LUMINA APV

CONTINUED FROM CHART CODE A031

6
- CONNECT DVM TO TERMINAL "A" OF LEFT BODY HARNESS OR REAR ABS HARNESS CONNECTOR IN QUESTION.
- DISCONNECT 24 WAY WORLD CONNECTOR.
- USING THE TABLE SUPPLIED AT THE BOTTOM OF THIS PAGE, CONNECT DVM TO THE APPROPRIATE TERMINAL OF THE WORLD CONNECTOR AND MEASURE RESISTANCE. RESISTANCE SHOULD BE LESS THAN 2 OHMS. IS IT?

NO → REPAIR CIRCUIT THAT SHOWED HIGH RESISTANCE OR OPEN.

YES ↓

7
- CONNECT DVM TO TERMINAL "B" OF LEFT BODY HARNESS OR REAR ABS HARNESS IN QUESTION.
- USING THE TABLE SUPPLIED, CONNECT DVM TO THE APPROPRIATE TERMINAL OF THE WORLD CONNECTOR AND MEASURE RESISTANCE. RESISTANCE SHOULD BE LESS THAN 2 OHMS. IS IT?

NO → REPAIR CIRCUIT THAT SHOWED HIGH RESISTANCE OR OPEN.

YES ↓

8
- DVM STILL CONNECTED.
- SELECT THE 20 VOLT SCALE OF THE DVM.
- IGNITION "ON."
- REPEAT STEP 7. VOLTAGE SHOULD BE 0. IS IT?

NO → REPAIR SHORT TO VOLTAGE IN CIRCUIT THAT INDICATED VOLTAGE ABOVE 0 VOLTS.

YES ↓

- INSPECT WORLD CONNECTOR TERMINALS FOR POOR CONNECTION.
- RE-CONNECT ALL CONNECTIONS AND REPEAT STEP 1.
- IF WHEEL SPEED DOES NOT MATCH REMAINING WHEEL SPEED INPUTS, REPLACE EBCM.

LEFT BODY HARNESS CONNECTOR TERMINAL	WORLD CONNECTOR	CIRCUIT NUMBER
L/R WHEEL B	PIN 12	884
L/R WHEEL A	PIN 11	885

REAR ABS HARNESS CONNECTOR TERMINAL	WORLD CONNECTOR	CIRCUIT NUMBER
R/R WHEEL B	PIN 7	882
R/R WHEEL A	PIN 8	883

THIS CHART ASSUMES THAT A CURRENT CODE IS STORED INDICATING THAT THIS FAULT IS PRESENT.

DIAGNOSTIC CHART DTC A032 — 1993 SILHOUETTE, TRANS SPORT AND LUMINA APV

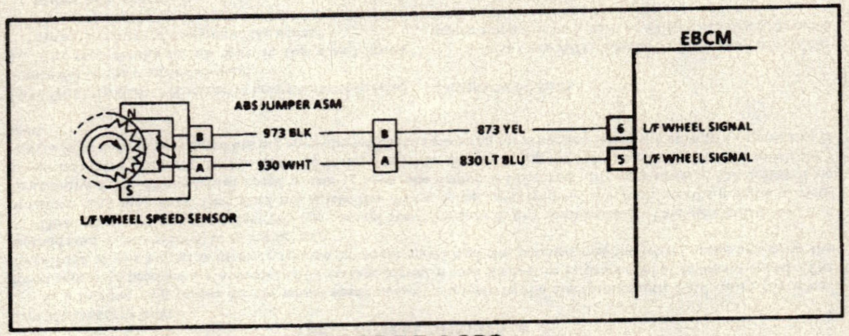

CODE A032

LEFT FRONT WHEEL SENSOR SHORTED TO BATTERY OR GROUND

Circuit Description:

As a toothed ring passes by the wheel speed sensor, changes in the electromagnetic field cause the wheel speed sensor to produce a sinusoidal (AC) voltage signal whose frequency is proportional to wheel speed. The magnitude of this signal is directly related to wheel speed and the proximity of the wheel speed sensor to the toothed ring often referred to as the air gap.

This code detects shorts to ground, and shorts to battery in the wheel speed circuitry sensed by ABS EBCM.

Test Description: Number(s) below refer to circled number(s) on the diagnostic chart.

1. This test helps to identify whether the code is due to a short to battery or short to ground.
2. This insures that the wheel speed sensor is not shorted to ground.
3. This step checks to see if wheel speed sensor circuitry is shorted to voltage.
4. This insures that the wheel speed sensor circuitry has good continuity.
5. This checks for short to ground in wheel speed sensor wiring.
6. This checks for short to voltage in wheel speed sensor wiring.

Diagnostic Aids:

An "Intermittent" problem may be caused by a poor connection, rubbed through wire insulation, or a wire that is broken inside the insulation.

If the customers comments reflect that the ABS Amber warning lamp is "ON" only during moist environmental changes, (rain, snow, vehicle wash) all wheel speed sensor circuitry should be thoroughly inspected for signs of water intrusion. If code is not current clear codes and simulate the effects of water intrusion. Use the following procedure. Spray down the suspected area with a 5% salt water solution (two teaspoons of salt to 12 oz. of water). Test drive vehicle. Vehicle must achieve 4 mph. If code returns immediately, replace suspected harness.

Any circuitry, that is suspected as causing the intermittent complaint, should be thoroughly checked for backed out terminals, improper mating, broken locks, improperly formed or damaged terminals, poor terminal to wiring connections or physical damage to the wiring harness.

DIAGNOSTIC CHART DTC A032 — 1993 SILHOUETTE, TRANS SPORT AND LUMINA APV

IMPORTANT: WHEEL SPEED SENSOR INTERMITTENT PROBLEMS MAY BE DIFFICULT TO LOCATE. CARE SHOULD BE TAKEN NOT TO DISTURB ANY ELECTRICAL CONNECTIONS PRIOR TO AN INDICATED STEP OF THIS CHART. THIS WILL INSURE THAT AN INTERMITTENT CONNECTION WILL NOT BE CORRECTED BEFORE THE SOURCE OF THE PROBLEM IS FOUND.

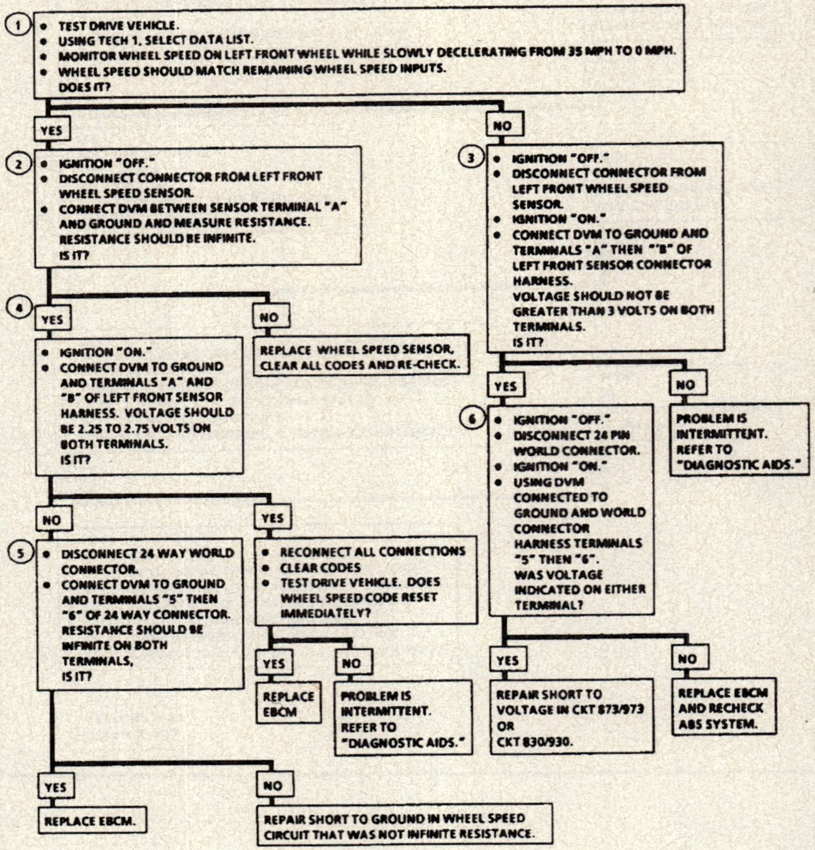

DIAGNOSTIC CHART DTC A033 — 1993 SILHOUETTE, TRANS SPORT AND LUMINA APV

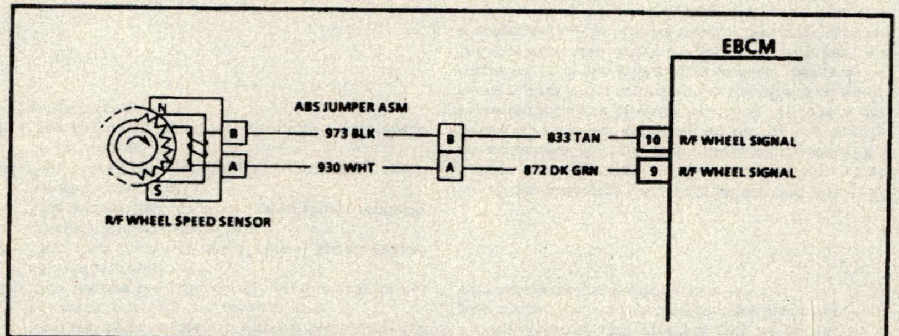

CODE A033

RIGHT FRONT WHEEL SENSOR SHORTED TO BATTERY OR GROUND

Circuit Description:

As a toothed ring passes by the wheel speed sensor, changes in the electromagnetic field cause the wheel speed sensor to produce a sinusoidal (AC) voltage signal whose frequency is proportional to wheel speed. The magnitude of this signal is directly related to wheel speed and the proximity of the wheel speed sensor to the toothed ring, often referred to as the air gap.

This code detects shorts to ground, and shorts to battery in the wheel speed circuitry sensed by the ABS EBCM.

Test Description: Number(s) below refer to circled number(s) on the diagnostic chart.

1. This test helps identify whether the code is due to a short to battery or short to ground.
2. This insures that the wheel speed sensor is not shorted to ground.
3. This step checks to see if wheel speed sensor circuitry is shorted to voltage.
4. This insures that the wheel speed sensor circuitry has good continuity.
5. This checks for short to ground in wheel speed sensor wiring.
6. This checks for short to voltage in wheel speed sensor wiring.

Diagnostic Aids:

An "Intermittent" problem may be caused by a poor connection, rubbed through wire insulation, or a wire that is broken inside the insulation.

If the customers comments reflect that the ABS Amber warning lamp is "ON" only during moist environmental changes, (rain, snow, vehicle wash) all wheel speed sensor circuitry should be thoroughly inspected for signs of water intrusion. If code is not current clear codes and simulate the effects of water intrusion. Use the following procedure. Spray down the suspected area with a 5% salt water solution (two teaspoons of salt to 12 oz. of water). Test drive vehicle. Vehicle must achieve 4 mph. If code returns immediately, replace suspected harness.

Any circuitry, that is suspected as causing the intermittent complaint, should be thoroughly checked for backed out terminals, improper mating, broken locks, improperly formed or damaged terminals, poor terminal to wiring connections or physical damage to the wiring harness.

DIAGNOSTIC CHART DTC A033 — 1993 SILHOUETTE, TRANS SPORT AND LUMINA APV

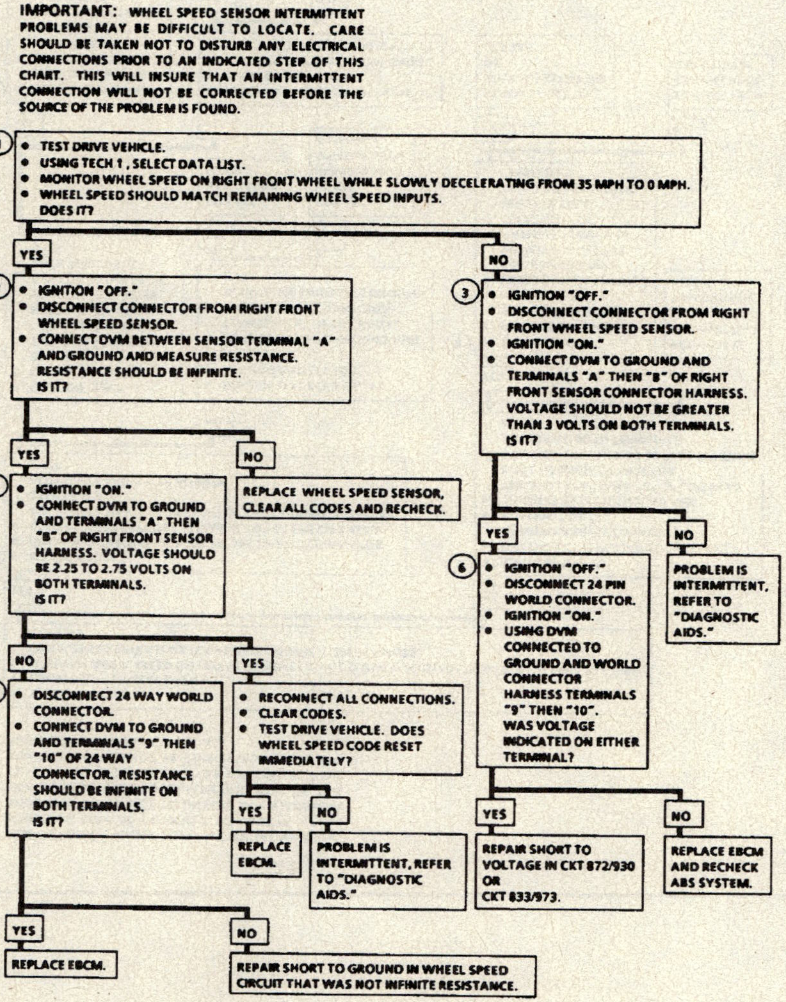

DIAGNOSTIC CHART DTC A034 — 1993 SILHOUETTE, TRANS SPORT AND LUMINA APV

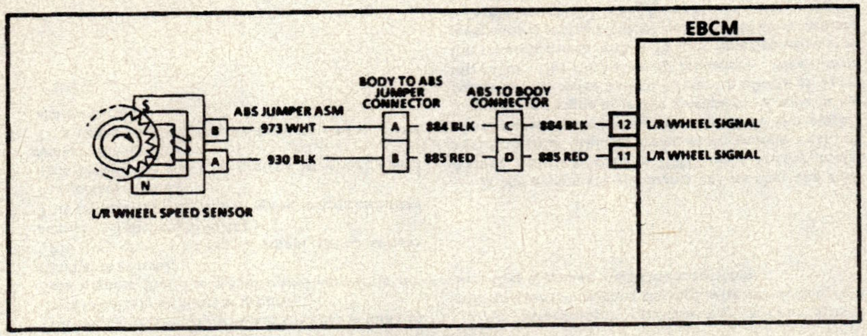

CODE A034

LEFT REAR WHEEL SENSOR SHORTED TO BATTERY OR GROUND

Circuit Description:

As a toothed ring passes by the wheel speed sensor, changes in the electromagnetic field cause the wheel speed sensor to produce a sinusoidal (AC) voltage signal whose frequency is proportional to wheel speed. The magnitude of this signal is directly related to wheel speed and the proximity of the wheel speed sensor to the toothed ring, often referred to as the air gap.

This code detects opens, shorts to ground, and shorts to battery in the wheel speed circuitry sensed by the ABS EBCM.

Test Description: Number(s) below refer to circled number(s) on the diagnostic chart.

1. This test helps to identify whether the code is due to a short to battery or short to ground.
2. This insures that the wheel speed sensor is not shorted to ground.
3. This step checks to see if wheel speed sensor circuitry is shorted to voltage.
4. This insures that the wheel speed sensor circuitry has good continuity.
5. This checks for short to ground in wheel speed sensor wiring.
6. This checks for short to voltage in wheel speed sensor wiring.

Diagnostic Aids:

An "Intermittent" problem may be caused by a poor connection, rubbed through wire insulation, or a wire that is broken inside the insulation.

If the customers comments reflect that the ABS Amber warning lamp is "ON" only during moist environmental changes, (rain, snow, vehicle wash) all wheel speed sensor circuitry should be thoroughly inspected for signs of water intrusion. If code is not current clear codes and simulate the effects of water intrusion. Use the following procedure. Spray down the suspected area with a 5% salt water solution (two teaspoons of salt to 12 oz. of water). Test drive vehicle. Vehicle must achieve 4 mph. If code returns immediately, replace suspected harness.

Any circuitry, that is suspected as causing the intermittent complaint, should be thoroughly checked for backed out terminals, improper mating, broken locks, improperly formed or damaged terminals, poor terminal to wiring connections or physical damage to the wiring harness.

DIAGNOSTIC CHART DTC A034 — 1993 SILHOUETTE, TRANS SPORT AND LUMINA APV

IMPORTANT: WHEEL SPEED SENSOR INTERMITTENT PROBLEMS MAY BE DIFFICULT TO LOCATE. CARE SHOULD BE TAKEN NOT TO DISTURB ANY ELECTRICAL CONNECTIONS PRIOR TO AN INDICATED STEP OF THIS CHART. THIS WILL INSURE THAT AN INTERMITTENT CONNECTION WILL NOT BE CORRECTED BEFORE THE SOURCE OF THE PROBLEM IS FOUND.

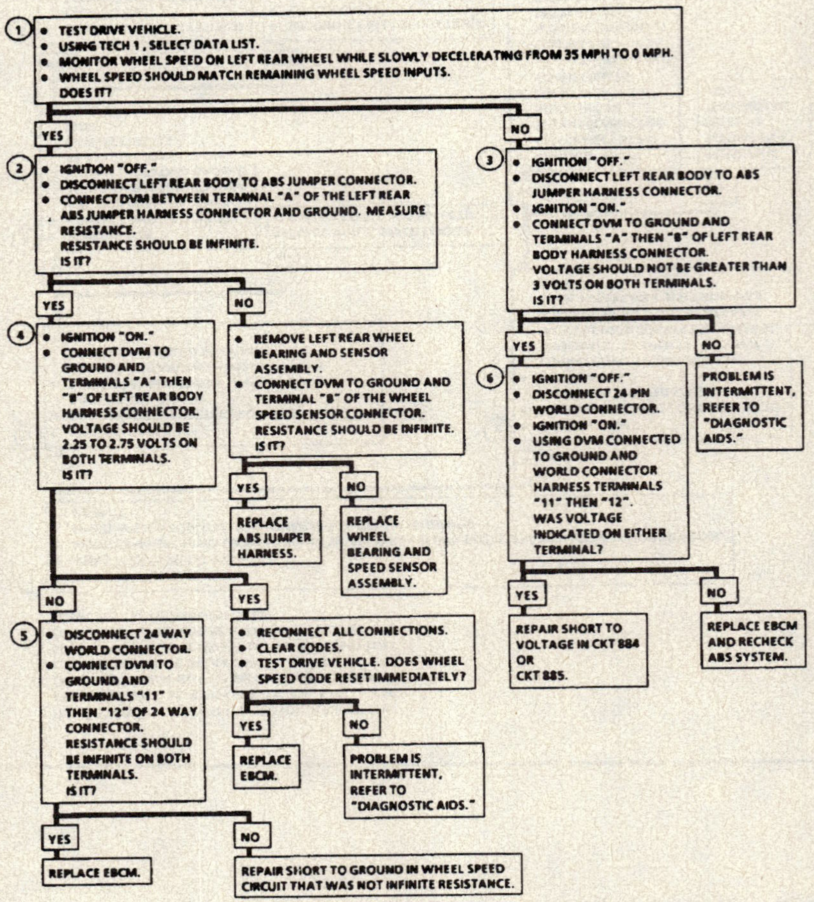

DIAGNOSTIC CHART DTC A035 — 1993 SILHOUETTE, TRANS SPORT AND LUMINA APV

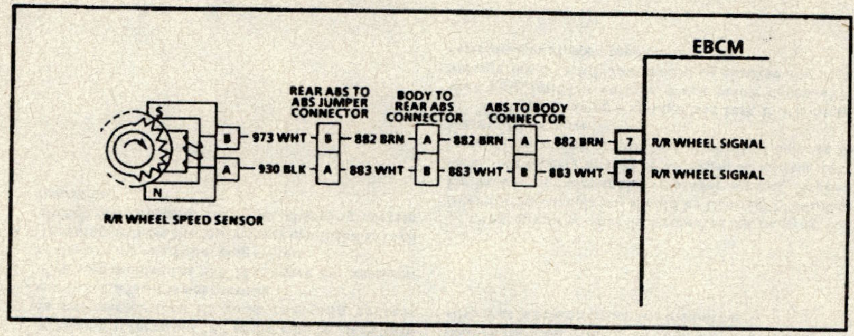

CODE A035

RIGHT REAR WHEEL SENSOR SHORTED TO BATTERY OR GROUND

Circuit Description:

As a toothed ring passes by the wheel speed sensor, changes in the electromagnetic field cause the wheel speed sensor to produce a sinusoidal (AC) voltage signal whose frequency is proportional to wheel speed. The magnitude of this signal is directly related to wheel speed and the proximity of the wheel speed sensor to the toothed ring, often referred to as the air gap.

This code detects opens, shorts to ground, and shorts to battery in the wheel speed circuitry sensed by the ABS EBCM.

Test Description: Number(s) below refer to circled number(s) on the diagnostic chart.

1. This helps to identify whether the code is due to a short to battery or short to ground.
2. This insures that the wheel speed sensor is not shorted to ground.
3. This step checks to see if wheel speed sensor circuitry is shorted to voltage.
4. This insures that the wheel speed sensor circuitry has good continuity.
5. This checks for short to ground in wheel speed sensor wiring.
6. This checks for short to voltage in wheel speed sensor wiring.

Diagnostic Aids:

An "Intermittent" problem may be caused by a poor connection, rubbed through wire insulation, or a wire that is broken inside the insulation.

If the customers comments reflect that the ABS Amber warning lamp is "ON" only during moist environmental changes, (rain, snow, vehicle wash) all wheel speed sensor circuitry should be thoroughly inspected for signs of water intrusion. If code is not current clear codes and simulate the effects of water intrusion. Use the following procedure. Spray down the suspected area with a 5% salt water solution (two teaspoons of salt to 12 oz. of water). Test drive vehicle. Vehicle must achieve 4 mph. If code returns immediately, replace suspected harness.

Any circuitry, that is suspected as causing the intermittent complaint, should be thoroughly checked for backed out terminals, improper mating, broken locks, improperly formed or damaged terminals, poor terminal to wiring connections or physical damage to the wiring harness.

DIAGNOSTIC CHART DTC A035 — 1993 SILHOUETTE, TRANS SPORT AND LUMINA APV

IMPORTANT: WHEEL SPEED SENSOR INTERMITTENT PROBLEMS MAY BE DIFFICULT TO LOCATE. CARE SHOULD BE TAKEN NOT TO DISTURB ANY ELECTRICAL CONNECTIONS PRIOR TO AN INDICATED STEP OF THIS CHART. THIS WILL INSURE THAT AN INTERMITTENT CONNECTION WILL NOT BE CORRECTED BEFORE THE SOURCE OF THE PROBLEM IS FOUND.

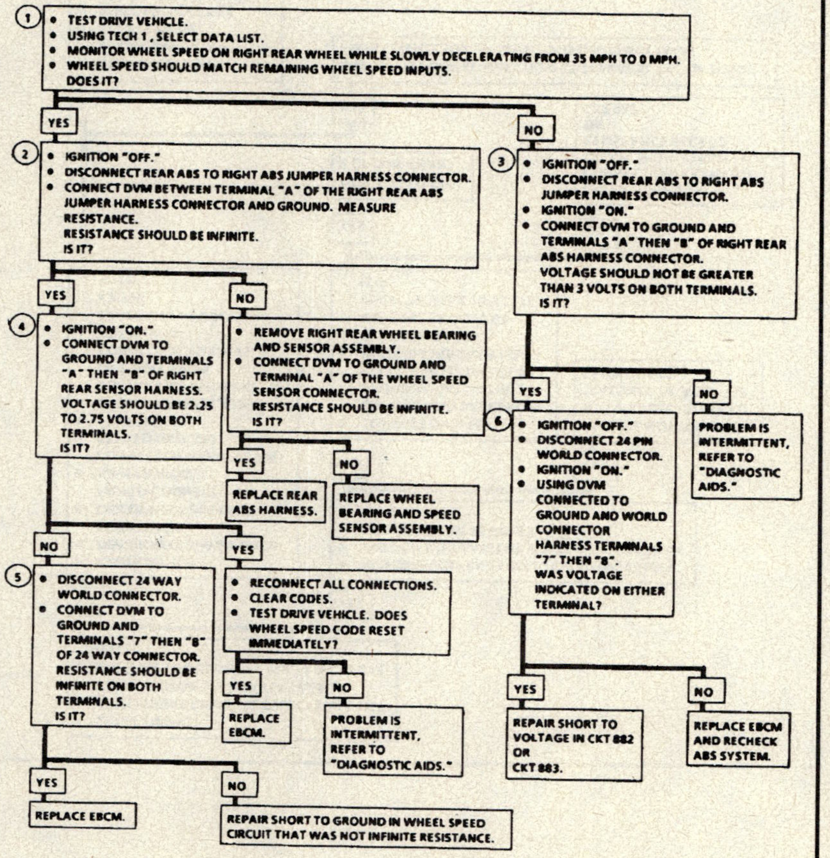

DIAGNOSTIC CHART DTC A036 — 1993 SILHOUETTE, TRANS SPORT AND LUMINA APV

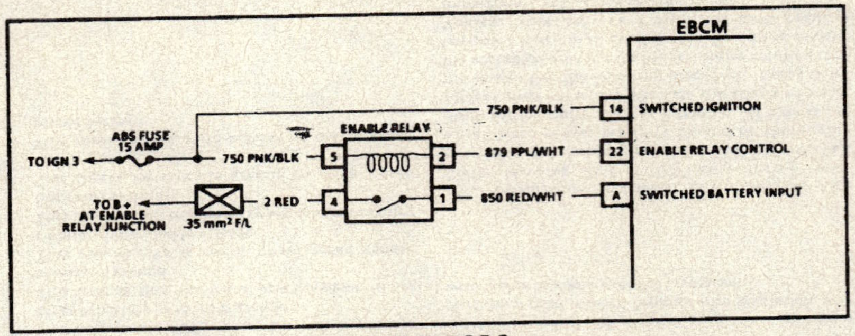

CODE A036

LOW SYSTEM VOLTAGE

Circuit Description:

This code is used to monitor the voltage level available to the EBCM controller. If the voltage drops below 11 volts, full performance of the ABS system cannot be guaranteed. During ABS operation, there are several current requirements that will cause battery voltage to drop. Because of this, voltage is monitored prior to ABS to indicate good charging system condition and also during ABS when voltage may drop significantly. If voltage falls below a level that will provide adequate ABS performance, Code A036 will result.

Test Description: Number(s) below refer to circled number(s) on the diagnostic chart.

1. This test checks to see what voltage is readily available at terminal "A" of the EBCM. If voltage is representative of good charging system condition, a fault is not present.
2. This step isolates the EBCM to check the condition of the circuitry and charging system.
3. This step isolates the low voltage condition to high circuit resistance, or improper charging system operation.

Diagnostic Aids:

An "Intermittent" problem may be caused by a poor connection, rubbed through wire insulation, or a wire that is broken inside the insulation.

Any circuitry, that is suspected as causing the intermittent complaint, should be thoroughly checked for backed out terminals, improper mating, broken locks, improperly formed or damaged terminals, poor terminal to wiring connections or physical damage to the wiring harness.

While performing a voltage load test, if it is noted that only ignition voltage drops below acceptable voltage levels, CKT 750 should be checked for high resistance or an open condition.

DIAGNOSTIC CHART DTC A036 — 1993 SILHOUETTE, TRANS SPORT AND LUMINA APV

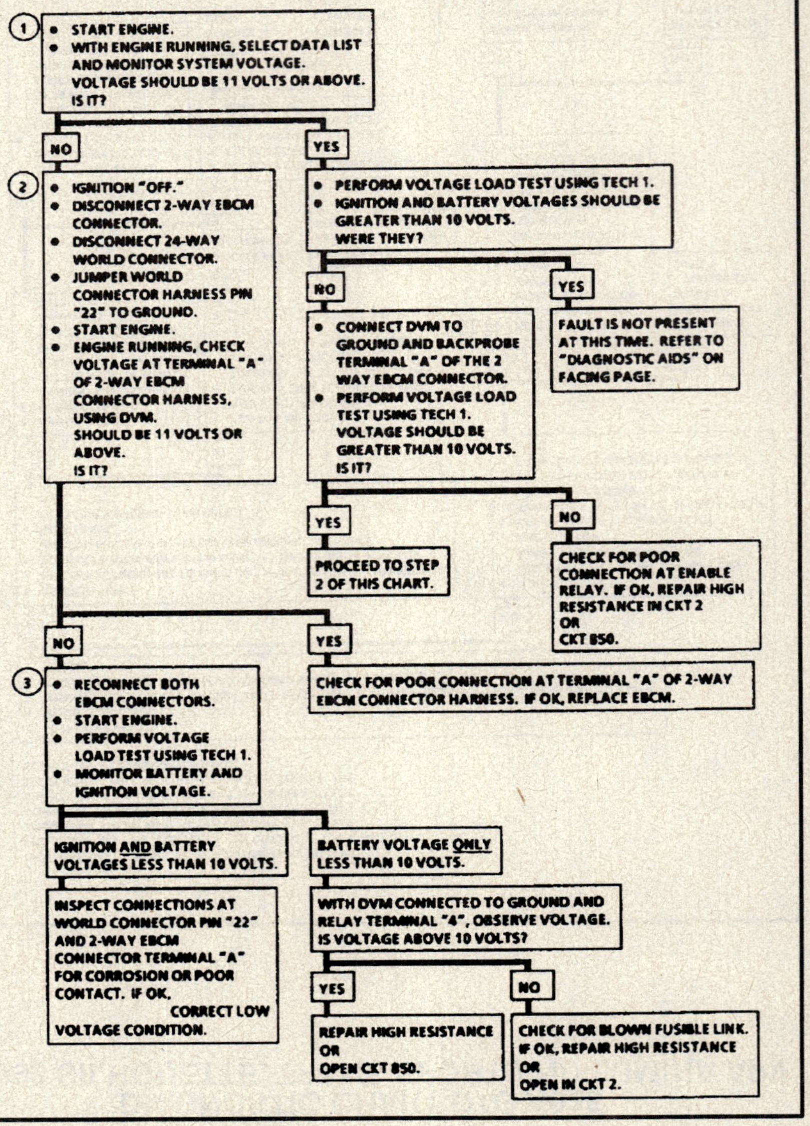

DIAGNOSTIC CHART DTC A037 — 1993 SILHOUETTE, TRANS SPORT AND LUMINA APV

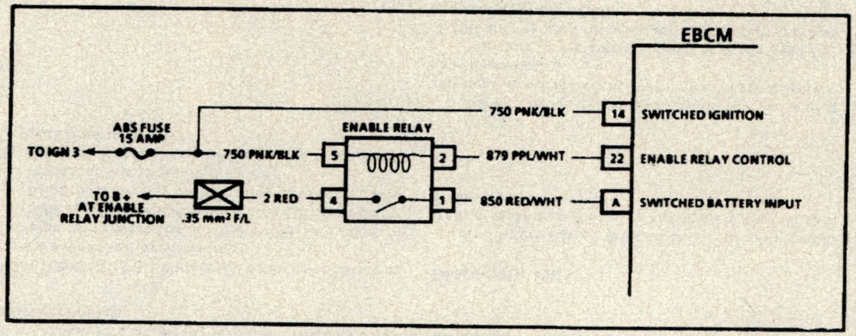

CODE A037

HIGH SYSTEM VOLTAGE

Circuit Description:

This code is designed to detect high vehicle voltage levels prior to any required motor movement (initialization or ABS). If excessive voltage exists, demagnetization of the motor magnets may occur, which would eventually affect or eliminate ABS performance.

Test Description: Number(s) below refer to circled number(s) on the diagnostic chart.

1. This step checks the voltage level being received by the EBCM. If high voltage is present, a charging system problem is indicated.

Diagnostic Aids:

An "Intermittent" problem may be caused by a poor connection, rubbed through wire insulation, or a wire that is broken inside the insulation.

Any circuitry, that is suspected as causing the intermittent complaint, should be thoroughly checked for backed out terminals, improper mating, broken locks, improperly formed or damaged terminals, poor terminal to wiring connections or physical damage to the wiring harness.

DIAGNOSTIC CHART DTC A037 — 1993 SILHOUETTE, TRANS SPORT AND LUMINA APV

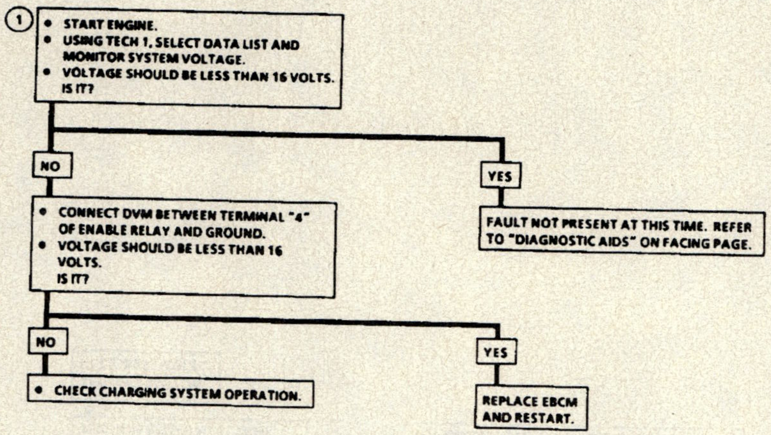

DIAGNOSTIC CHART DTC A038 — 1993 SILHOUETTE, TRANS SPORT AND LUMINA APV

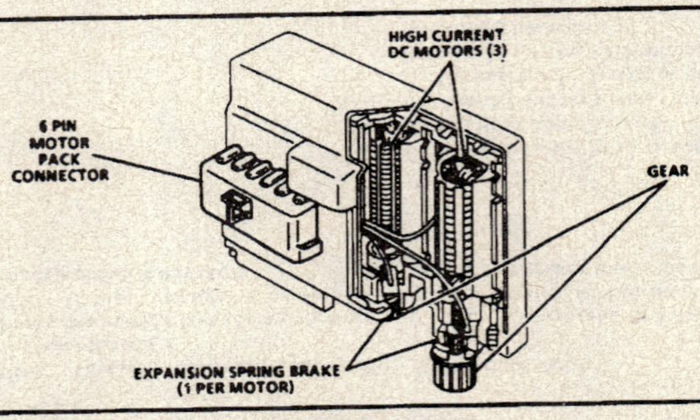

CODE A038

LEFT FRONT ESB WILL NOT HOLD MOTOR

Circuit Description:

This code is designed to detect a slipping left front ESB. During initialization and braking, the left front motor is re-homed. If the ESB slips, the motor/piston will move. During the next key on initialization, a re-home of the motor verifies the motor/piston remained at the home position. If motor movement is detected, the ESB must be slipping.

Test Description: Number(s) below refer to circled number(s) on the diagnostic chart.

1. This step checks the left front ESB. A broken or defective ESB would result in the left front piston being back driven by hydraulic pressure, pedal movement would result.
2. This step is used to release the tension of the motor pack prior to removal.

Diagnostic Aids:

An "Intermittent" fault in this code may result from a mechanical part of the system that sticks, binds, or slips.

Depending on the frequency of the failure, a physical inspection of the mechanical parts suspected may be necessary.

The static modulator test function of the Tech 1 or T-100 may be used to locate an intermittent problem associated with the ESB.

DIAGNOSTIC CHART DTC A038 — 1993 SILHOUETTE, TRANS SPORT AND LUMINA APV

(1)
- IGNITION "ON," ENGINE "OFF."
- USING TECH 1, SELECT MANUAL CONTROL FUNCTION.
- SELECT MOTOR APPLY FUNCTION OF THE LEFT FRONT CHANNEL, AND APPLY MOTOR.
- WAIT 5 SECONDS.
- APPLY FIRM PRESSURE ON BRAKE PEDAL. PEDAL SHOULD NOT FALL.

DID IT?

YES

(2) USING TECH 1, PERFORM GEAR TENSION RELIEF FUNCTION.

REPLACE MOTOR PACK ASSEMBLY.

NO

FAULT IS NOT PRESENT AT THIS TIME. REFER TO "DIAGNOSTIC AIDS" ON FACING PAGE.

DIAGNOSTIC CHART DTC A041 — 1993 SILHOUETTE, TRANS SPORT AND LUMINA APV

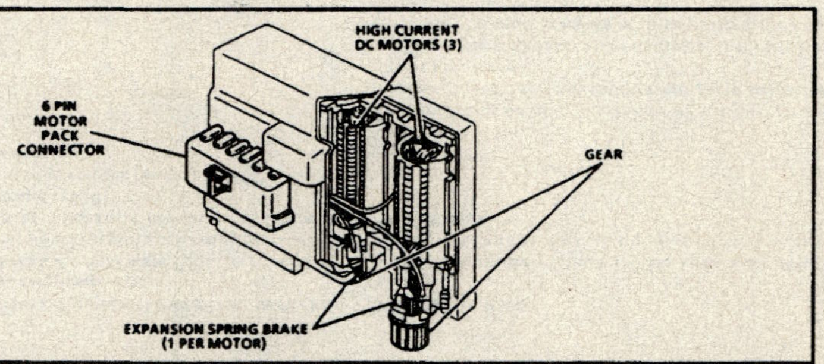

CODE A041

RIGHT FRONT ESB WILL NOT HOLD MOTOR

Circuit Description:

This code is designed to detect a slipping right front ESB. During initialization and braking, the right front motor is re-homed. If the ESB slips, the motor/piston will move. During the next key on initialization, a re-home of the motor verifies the motor/piston remained at the home position. If motor movement is detected, the ESB must be slipping.

Test Description: Number(s) below refer to circled number(s) on the diagnostic chart.

1. This step checks the right front ESB. A broken or defective ESB would result in the right front piston being back driven by hydraulic pressure, and wheel movement would result.
2. This step is used to release the tension of the motor pack prior to removal.

Diagnostic Aids:

An "Intermittent" fault in this code may result from a mechanical part of the system that sticks, binds, or slips.

Depending on the frequency of the failure, a physical inspection of the mechanical parts suspected may be necessary.

The static modulator test function of the Tech 1 or T-100 may be used to locate an intermittent problem associated with the ESB.

DIAGNOSTIC CHART DTC A041 — 1993 SILHOUETTE, TRANS SPORT AND LUMINA APV

(1)
- IGNITION "ON," ENGINE "OFF."
- USING TECH 1, SELECT MANUAL CONTROL FUNCTION.
- SELECT MOTOR APPLY FUNCTION OF THE RIGHT FRONT CHANNEL, AND APPLY MOTOR.
- WAIT 5 SECONDS.
- APPLY FIRM PRESSURE ON BRAKE PEDAL. PEDAL SHOULD NOT FALL. DID IT?

YES

(2) USING TECH 1, PERFORM GEAR TENSION RELIEF FUNCTION.

REPLACE MOTOR PACK ASSEMBLY.

NO

FAULT IS NOT PRESENT AT THIS TIME. REFER TO "DIAGNOSTIC AIDS" ON FACING PAGE.

DIAGNOSTIC CHART DTC A042 — 1993 SILHOUETTE, TRANS SPORT AND LUMINA APV

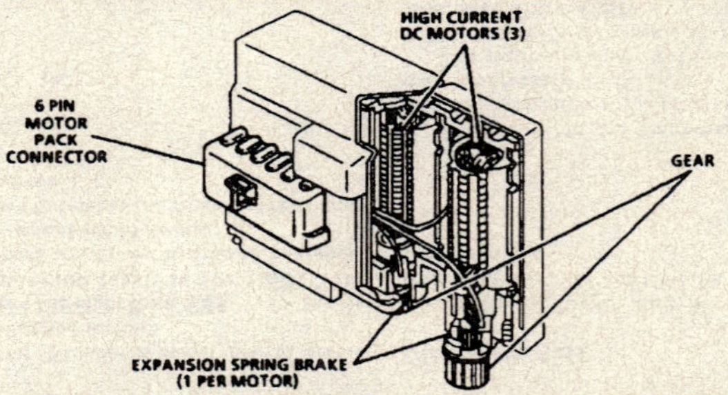

CODE A042

REAR AXLE ESB WILL NOT HOLD MOTOR

Circuit Description:

This code is designed to detect a slipping rear axle ESB. During initialization and braking, the rear motor is re-homed. If the ESB slips, the motor/piston will move. During the next key on initialization, a re-home of the motor verifies the motor/piston remained at the home position. If motor movement is detected, the ESB must be slipping.

Test Description: Number(s) below refer to circled number(s) on the diagnostic chart.

1. This step checks the rear axle ESB. A broken or defective ESB would result in the rear piston being back driven by hydraulic pressure, and wheel movement would result.
2. This step is used to release the tension of the motor pack prior to removal.

Diagnostic Aids:

An "Intermittent" fault in this code may result from a mechanical part of the system that sticks, binds, or slips.

Depending on the frequency of the failure, a physical inspection of the mechanical parts suspected may be necessary.

The static modulator test function of the Tech 1 or T-100 may be used to locate an intermittent problem associated with the ESB.

DIAGNOSTIC CHART DTC A042 — 1993 SILHOUETTE, TRANS SPORT AND LUMINA APV

(1)
- IGNITION "ON," ENGINE "OFF."
- APPLY FIRM PRESSURE ON BRAKE PEDAL AND RELEASE.
- RE-APPLY FIRM PRESSURE ON BRAKE PEDAL AND TRY TO SPIN REAR WHEELS BY HAND, WHEELS SHOULD NOT BE ABLE TO BE MOVED. DID WHEELS REMAIN STATIONARY?

YES → FAULT IS NOT PRESENT AT THIS TIME. REFER TO "DIAGNOSTIC AIDS" ON FACING PAGE.

NO → (2) USING TECH 1, PERFORM GEAR TENSION RELIEF FUNCTION. → REPLACE MOTOR PACK ASSEMBLY AND RE-CHECK ABS SYSTEM.

DIAGNOSTIC CHART DTC A044 — 1993 SILHOUETTE, TRANS SPORT AND LUMINA APV

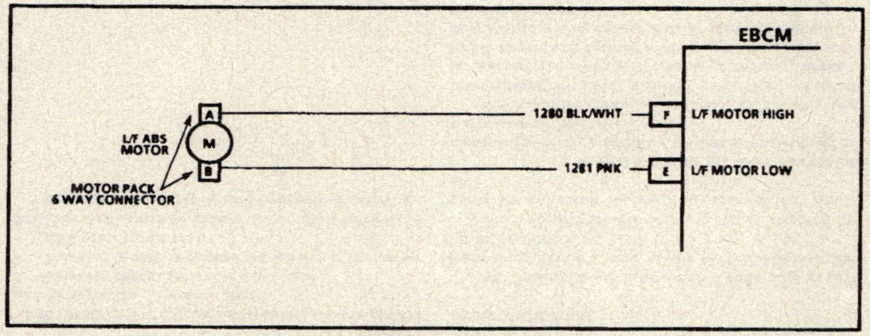

CODE A044

LEFT FRONT CHANNEL WILL NOT MOVE

Circuit Description:

This code is designed to detect bound-up ESB, a stuck motor, or a seized hydraulic modulator. When the release is commanded during initialization, the ESB should release the motor, resulting in sensed current being less than commanded current (motor is spinning freely). If the motor is not moving, sensed current will be equal to stall current.

Test Description: Number(s) below refer to circled number(s) on the diagnostic chart.

1. This step checks for motor movement.
2. This step is used to release the tension of the motor pack prior to removal.
3. This step serves to isolate the fault to either the motor pack assembly or ABS hydraulic modulator.

Diagnostic Aids:

An "Intermittent" fault in this code may result from a mechanical part of the system that sticks, binds, or slips.

The frequency of the problem can be checked by using the enhanced diagnostic function of the "Scan" tool

Code A044 may set after modulator disassembly if the modulator pistons are positioned at the bottom of their bore.

Any circuitry, that is suspected as causing the intermittent complaint, should be thoroughly checked for backed out terminals, improper mating, broken locks, improperly formed or damaged terminals, poor terminal to wiring connections or physical damage to the wiring harness.

DIAGNOSTIC CHART DTC A044 — 1993 SILHOUETTE, TRANS SPORT AND LUMINA APV

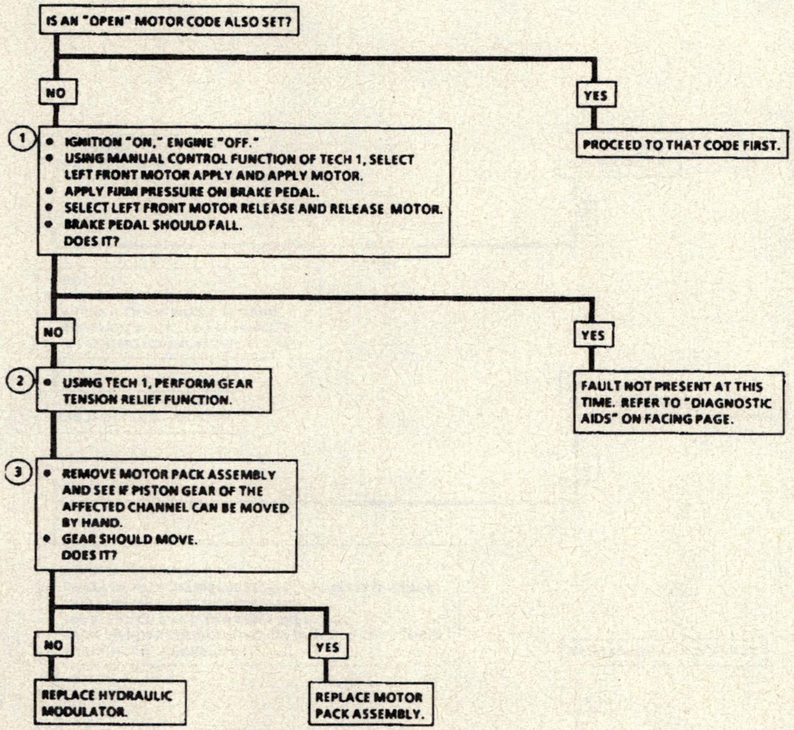

DIAGNOSTIC CHART DTC A045 — 1993 SILHOUETTE, TRANS SPORT AND LUMINA APV

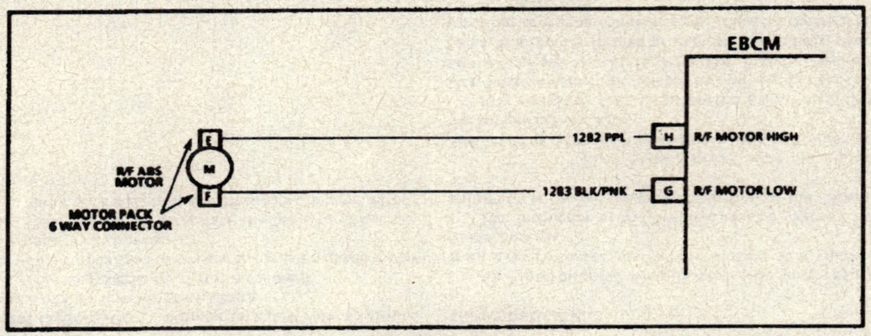

CODE A045

RIGHT FRONT CHANNEL WILL NOT MOVE

Circuit Description:

This code is designed to detect bound-up ESB, a stuck motor, or a seized hydraulic modulator. When the release is commanded during initialization, the ESB should release the motor, resulting in sensed current being less than commanded current (motor is spinning freely). If the motor is not moving, sensed current will be equal to stall current.

Test Description: Number(s) below refer to circled number(s) on the diagnostic chart.

1. This step checks for motor movement.
2. This step is used to release the tension of the motor pack prior to removal.
3. This step serves to isolate the fault to either the motor pack assembly or ABS hydraulic modulator.

Diagnostic Aids:

An "Intermittent" fault in this code may result from a mechanical part of the system that sticks, binds, or slips.

The frequency of the problem can be checked by using the enhanced diagnostic function of the "Scan" tool

Code A045 may set after modulator disassembly if the modulator pistons are positioned at the bottom of their bore.

Any circuitry, that is suspected as causing the intermittent complaint, should be thoroughly checked for backed out terminals, improper mating, broken locks, improperly formed or damaged terminals, poor terminal to wiring connections or physical damage to the wiring harness.

DIAGNOSTIC CHART DTC A045 — 1993 SILHOUETTE, TRANS SPORT AND LUMINA APV

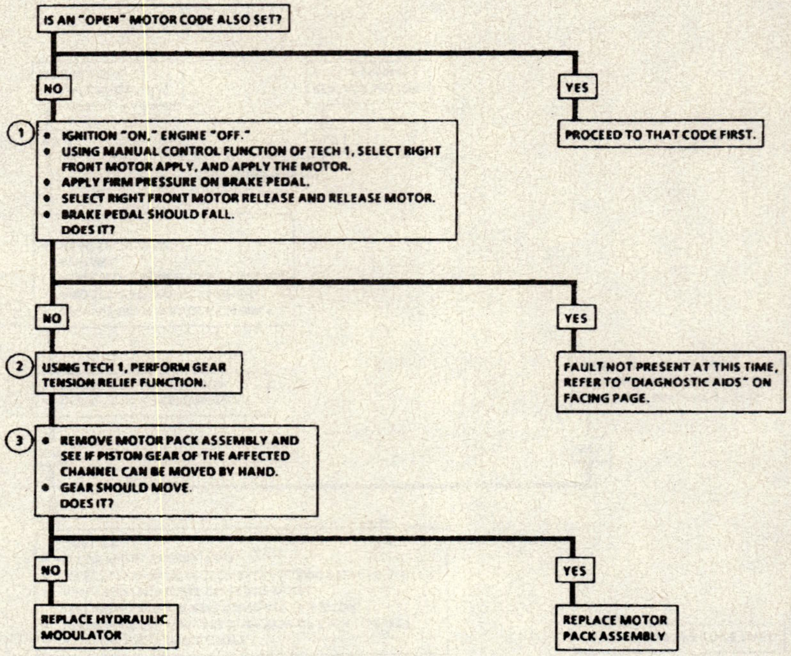

DIAGNOSTIC CHART DTC A046 — 1993 SILHOUETTE, TRANS SPORT AND LUMINA APV

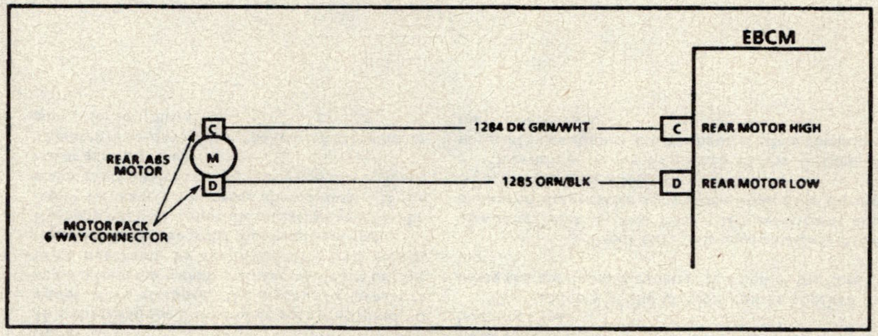

CODE A046

REAR AXLE CHANNEL WILL NOT MOVE

Circuit Description:

This code is designed to detect bound-up ESB, a stuck motor, or a seized hydraulic modulator. When the release is commanded during initialization, the ESB should release the motor, resulting in sensed current being less than commanded current (motor is spinning freely). If the motor is not moving, sensed current will be equal to stall current.

Test Description: Number(s) below refer to circled number(s) on the diagnostic chart.

1. This checks the ability of the motor to move, applying hydraulic pressure to the wheel cylinders. A frozen piston or gear would result in no apply of rear brakes and a wheel that freely spins.
2. This step checks to see that the motor can also move to the release position. If a motor were not able to release hydraulic pressure, the wheel would not spin.
3. This step is used to release the tension of the motor pack prior to removal.
4. Isolation of the faulty component is identified in this step.

Diagnostic Aids:

An "Intermittent" fault in this code may result from a mechanical part of the system that sticks or binds.

The frequency of the problem can be checked by using the enhanced diagnostic function of the Tech 1 "Scan" tool.

Code A046 may set after modulator disassembly if the modulator pistons are positioned at the bottom of their bore.

Depending on the frequency of the failure, a physical inspection of the mechanical parts suspected may be necessary.

Any circuitry, that is suspected as causing the intermittent complaint, should be thoroughly checked for backed out terminals, improper mating, broken locks, improperly formed or damaged terminals, poor terminal to wiring connections or physical damage to the wiring harness.

DIAGNOSTIC CHART DTC A046 — 1993 SILHOUETTE, TRANS SPORT AND LUMINA APV

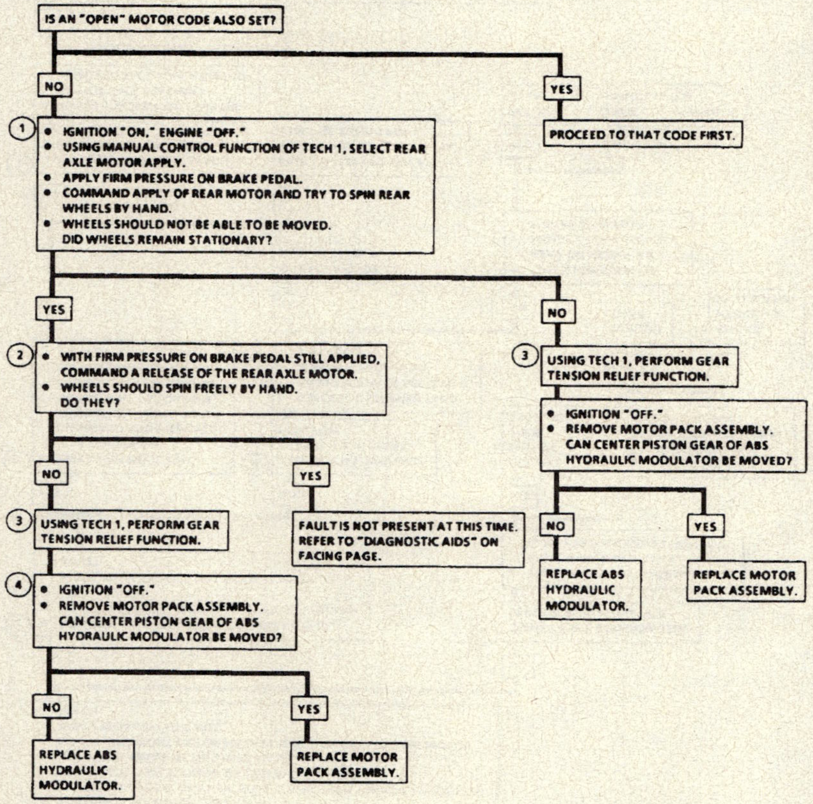

DIAGNOSTIC CHART DTC A047 — 1993 SILHOUETTE, TRANS SPORT AND LUMINA APV

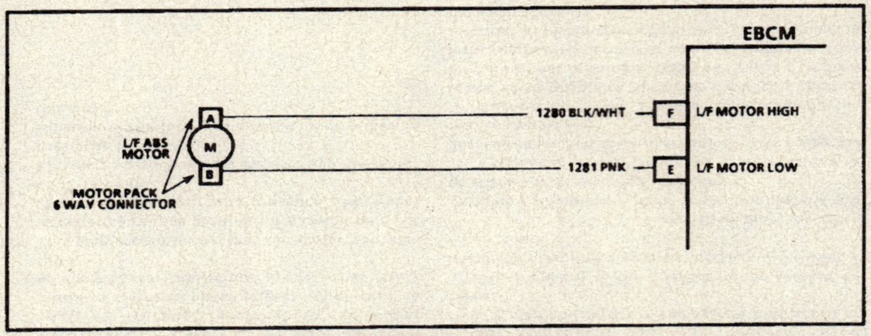

CODE A047

LEFT FRONT MOTOR FREE SPINS

Circuit Description:

This code is designed to detect a stripped nut or gear assembly during initialization. During the homing sequence, the piston should reach the top of the bore resulting in a stalled motor. If this does not occur, the motor must be spinning with little or no resistance, indicating a nut/screw or gear failure.

Test Description: Number(s) below refer to circled number(s) on the diagnostic chart.

1. This step positions the motor at the bottom of its travel.
2. This checks to see if the nut or gear is stripped. If either were stripped, the hydraulic pressure applied from the master cylinder would not be able to be overcome by the hydraulic ball screw, resulting in a brake pedal that would not rise.
3. During a full apply, once the ball screw reaches the top of its travel, the motor should stall. If the motor continues to run, an indication of a stripped nut or gear is present.
4. This step is used to release the tension of the motor pack prior to removal.

Diagnostic Aids:

An "Intermittent" fault in this code may result from a mechanical part of the system that sticks, binds, or slips.

The frequency of the problem can be checked by using the enhanced diagnostic function of the "Scan" tool

If code only fails once and Code A056 also fails, refer to Code A056. If intermittent and enhanced diagnostics shows this code fails during ABS, refer to Code A056.

Depending on the frequency of the failure, a physical inspection of the mechanical parts suspected may be necessary.

DIAGNOSTIC CHART DTC A047 — 1993 SILHOUETTE, TRANS SPORT AND LUMINA APV

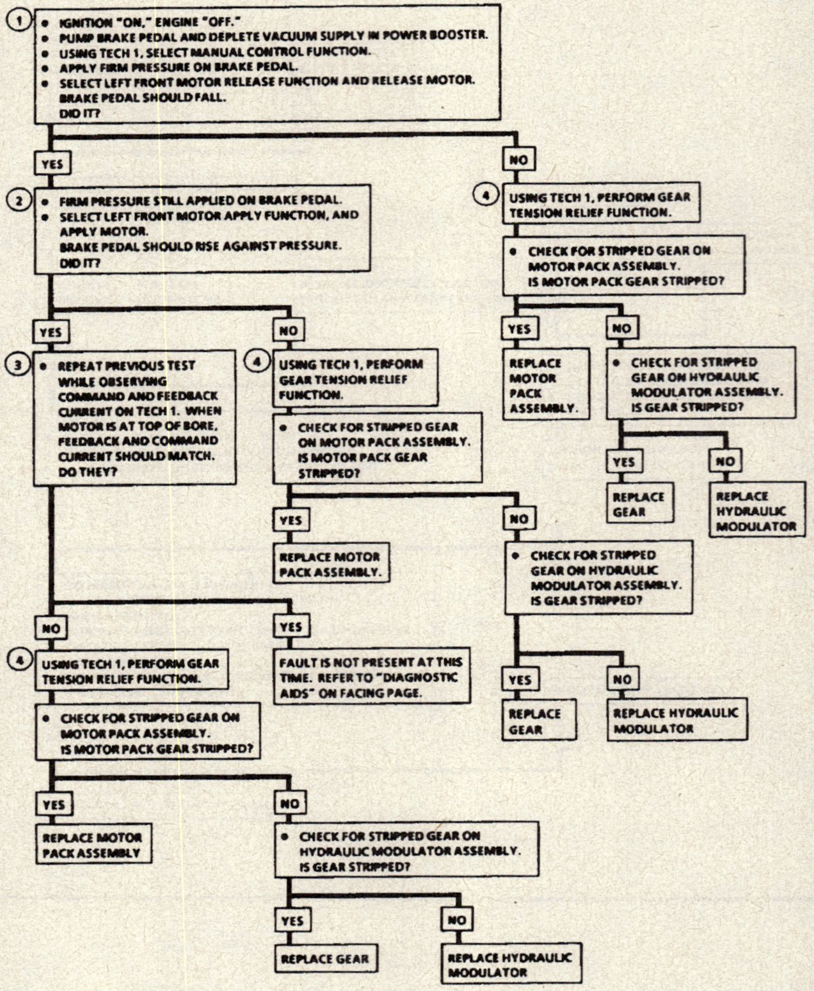

DIAGNOSTIC CHART DTC A048 — 1993 SILHOUETTE, TRANS SPORT AND LUMINA APV

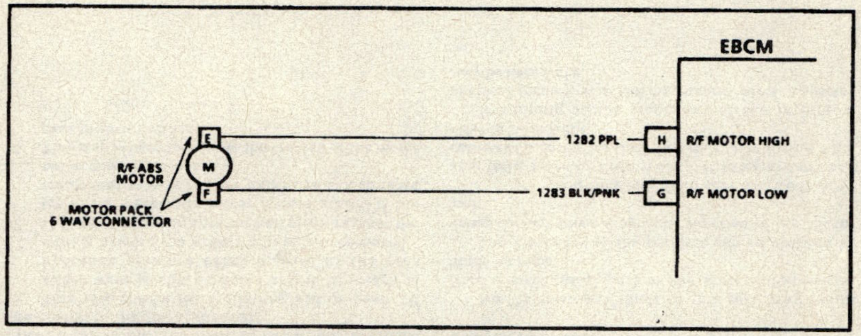

CODE A048

RIGHT FRONT MOTOR FREE SPINS

Circuit Description:

This code is designed to detect a stripped nut or gear assembly during initialization. During the homing sequence, the piston should reach the top of the bore resulting in a stalled motor. If this does not occur, the motor must be spinning with little or no resistance, indicating a nut/screw or gear failure.

Test Description: Number(s) below refer to circled number(s) on the diagnostic chart.

1. This step positions the motor at the bottom of its travel.
2. This checks to see if the nut or gear is stripped. If either were stripped, the hydraulic pressure applied from the master cylinder would not be able to be overcome by the hydraulic ball screw, resulting in a brake pedal that would not rise.
3. During a full apply, once the ball screw reaches the top of its travel, the motor should stall. If the motor continues to run, an indication of a stripped nut or gear is present.
4. This step is used to release the tension of the motor pack prior to removal.

Diagnostic Aids:

An "Intermittent" fault in this code may result from a mechanical part of the system that sticks, binds, or slips.

The frequency of the problem can be checked by using the enhanced diagnostic function of the "Scan" tool

If code only fails once and Code A061 also fails, refer to Code A061. If intermittent and enhanced diagnostics shows this code fails during ABS, refer to Code A061.

Depending on the frequency of the failure, a physical inspection of the mechanical parts suspected may be necessary.

DIAGNOSTIC CHART DTC A048 — 1993 SILHOUETTE, TRANS SPORT AND LUMINA APV

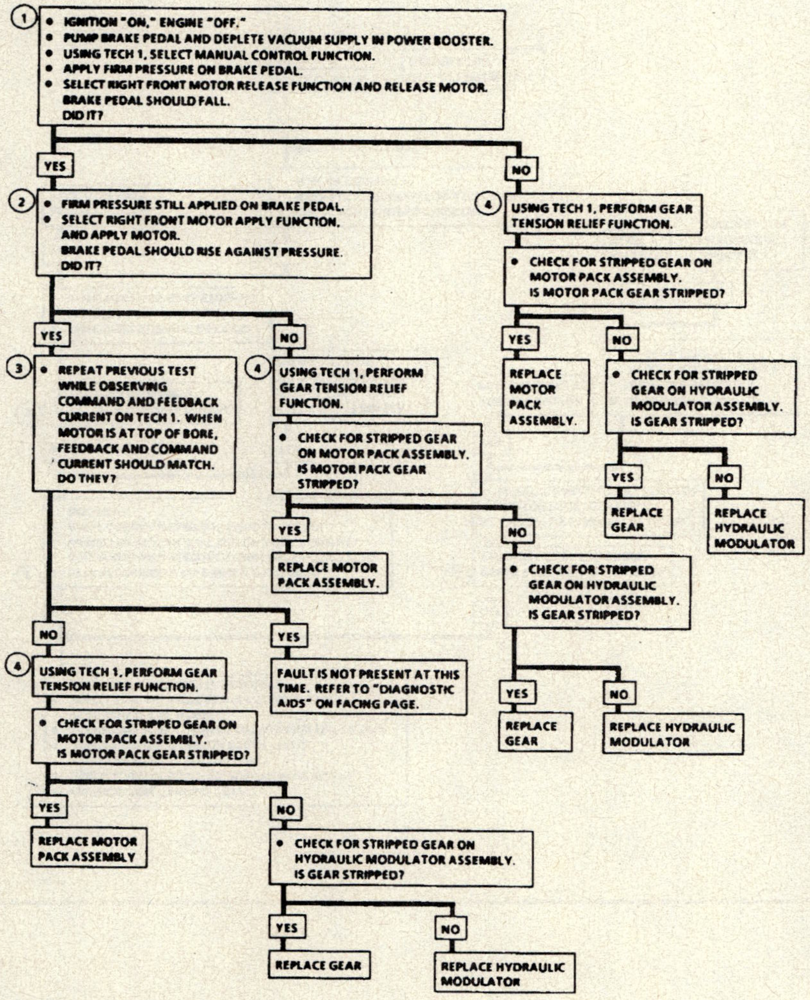

DIAGNOSTIC CHART DTC A051 — 1993 SILHOUETTE, TRANS SPORT AND LUMINA APV

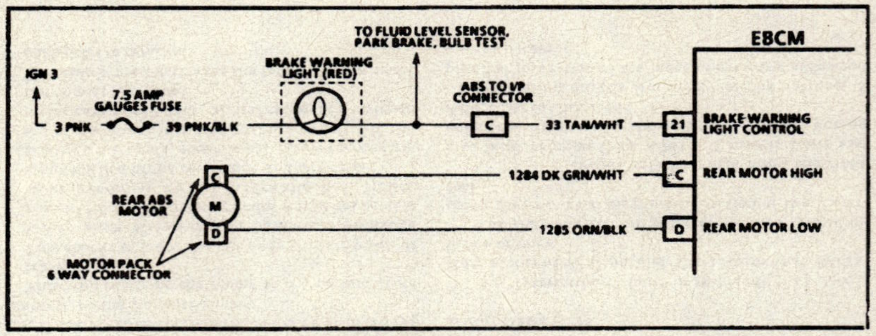

CODE A051

REAR AXLE MOTOR FREE SPINS

Circuit Description:

This code is designed to detect a stripped nut or gear assembly during initialization. During the homing sequence, the piston should reach the top of the bore resulting in a stalled motor. If this does not occur, the motor must be spinning with little or no resistance, indicating a nut/screw or gear failure.

Test Description: Number(s) below refer to circled number(s) on the diagnostic chart.

1. This step checks for a slipping nut or gear. If either were to slip, it would result in a lack of hydraulic pressure being applied at the rear wheels, causing the wheels to be able to be moved.
2. During a full apply, once the ball screw reaches the top of its travel, the motor should stall. If the motor continues to run, an indication of a stripped nut or gear is present.
3. This step is used to release the tension of the motor pack prior to removal.

Diagnostic Aids:

An "Intermittent" fault in this code may result from a mechanical part of the system that sticks, binds, or slips.

The frequency of the problem can be checked by using the enhanced diagnostic function of the "Scan" tool

If code only fails once and Code A064 also fails, refer to Code A064. If intermittent and enhanced diagnostics show this code fails during ABS, refer to Code A064.

Depending on the frequency of the failure, a physical inspection of the mechanical parts suspected may be necessary.

DIAGNOSTIC CHART DTC A051 — 1993 SILHOUETTE, TRANS SPORT AND LUMINA APV

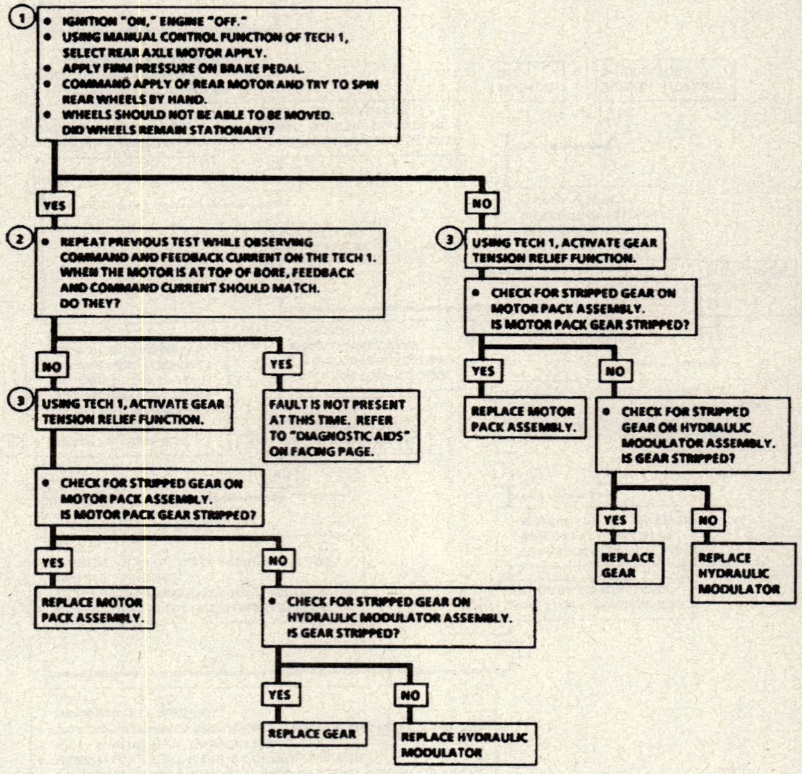

DIAGNOSTIC CHART DTC A052 — 1993 SILHOUETTE, TRANS SPORT AND LUMINA APV

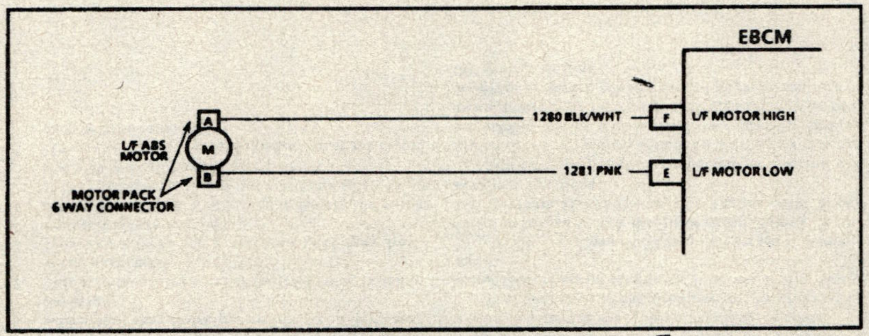

CODE A052

LEFT FRONT CHANNEL IN RELEASE TOO LONG

Circuit Description:

This code will diagnose a motor that is energized longer than expected. This could occur if a wheel speed sensor is faulty, the motor does not turn, the left front solenoid mechanically fails open, or the motor wires are crossed.

Test Description: Number(s) below refer to circled number(s) on the diagnostic chart.

1. This step identifies a problem in a wheel speed sensor that may cause the system to be in release too long.
2. This step identifies a motor fault as being failed or wired incorrectly.
3. This step checks for a solenoid that may have mechanically failed open.
4. If Step 3 has failed, this serves to isolate the cause of the hydraulic problem, to either the solenoid or the ABS hydraulic modulator assembly.
5. This step is used to release the tension of the motor pack.

Diagnostic Aids:

An "Intermittent" problem may be caused by a mechanical part of the system that sticks or binds.

The frequency of the problem can be checked by using the enhanced diagnostic function of the "Scan" tool,

Code A052 may fail if on ice and steering wheel is turned to lock during braking. Using "Scan" tool perform Hydraulic test to ensure total brake system is functional.

Any circuitry, that is suspected as causing the intermittent complaint, should be thoroughly checked for backed out terminals, improper mating, broken locks, improperly formed or damaged terminals, poor terminal to wiring connections or physical damage to the wiring harness.

DIAGNOSTIC CHART DTC A052 — 1993 SILHOUETTE, TRANS SPORT AND LUMINA APV

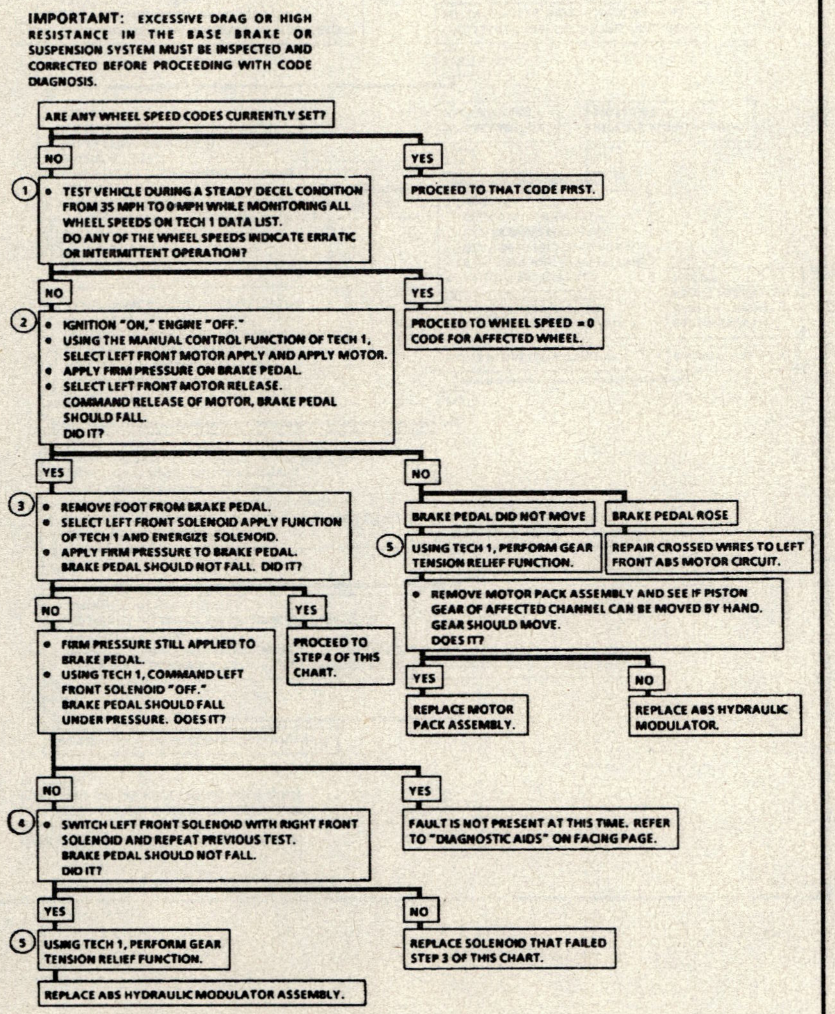

DIAGNOSTIC CHART DTC A053 — 1993 SILHOUETTE, TRANS SPORT AND LUMINA APV

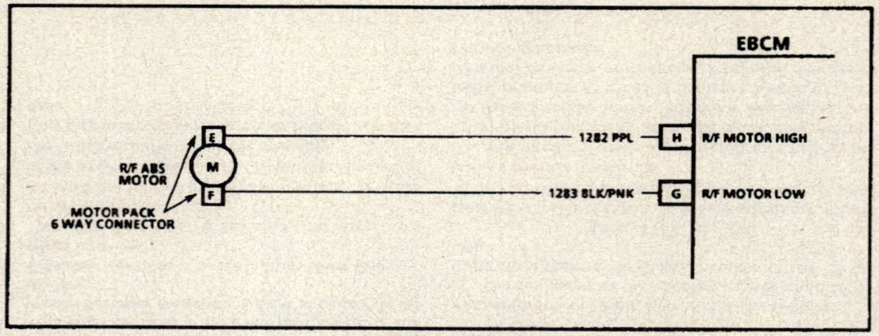

CODE A053

RIGHT FRONT CHANNEL IN RELEASE TOO LONG

Circuit Description:

This code will diagnose a motor that is energized longer than expected. This could occur if a wheel speed sensor is faulty, the motor does not turn, the right front solenoid mechanically fails open, or the motor wires are crossed.

Test Description: Number(s) below refer to circled number(s) on the diagnostic chart.

1. This step identifies a problem in a wheel speed sensor that may cause the system to be in release too long.
2. This step identifies a motor fault as being failed or wired incorrectly.
3. This step checks for a solenoid that may have mechanically failed open.
4. If Step 3 has failed, this serves to isolate the cause of the hydraulic problem, to either the solenoid or the ABS hydraulic modulator assembly.
5. This step is used to release the tension of the motor pack prior to removal.

Diagnostic Aids:

An "Intermittent" problem may be caused by a mechanical part of the system that sticks or binds.

The frequency of the problem can be checked by using the enhanced diagnostic function of the "Scan" tool,

Code A053 may fail on ice if steering wheel is turned to lock during braking. Using "Scan" tool perform Hydraulic test to ensure total brake system is functional.

Any circuitry, that is suspected as causing the intermittent complaint, should be thoroughly checked for backed out terminals, improper mating, broken locks, improperly formed or damaged terminals, poor terminal to wiring connections or physical damage to the wiring harness.

DIAGNOSTIC CHART DTC A053 — 1993 SILHOUETTE, TRANS SPORT AND LUMINA APV

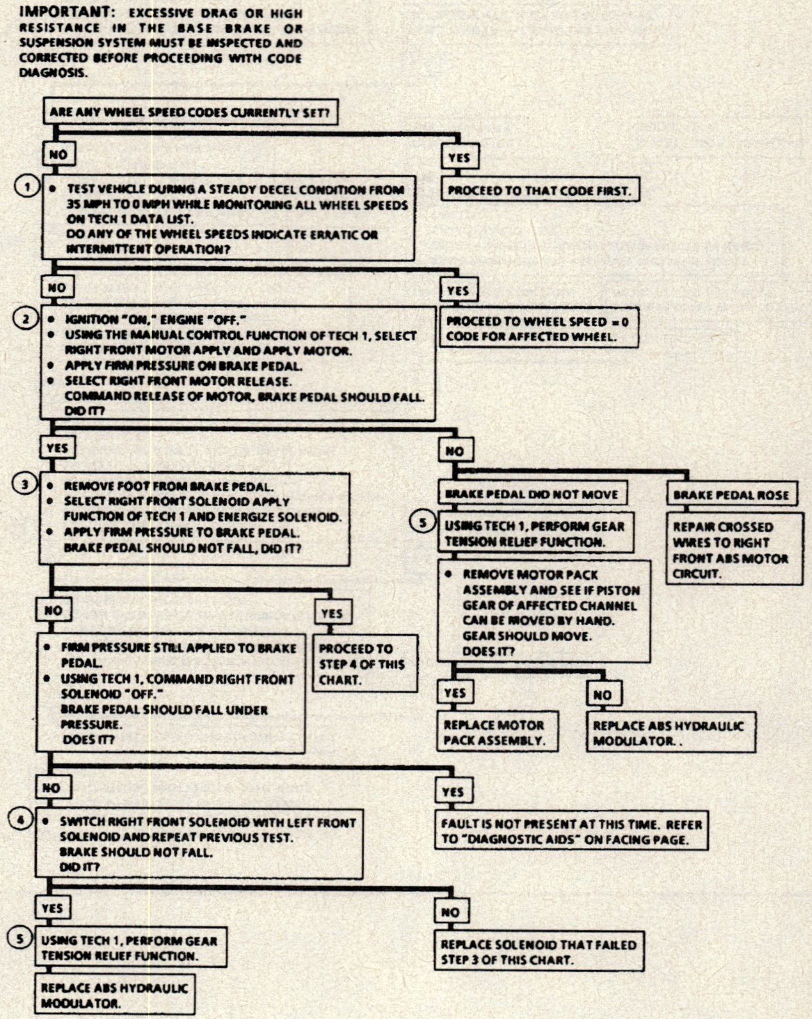

DIAGNOSTIC CHART DTC A054 — 1993 SILHOUETTE, TRANS SPORT AND LUMINA APV

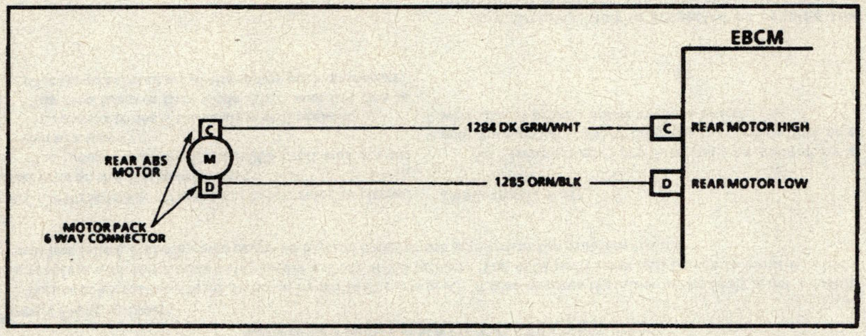

CODE A054

REAR AXLE CHANNEL IN RELEASE TOO LONG

Circuit Description:

This code will diagnose a motor that is energized longer than expected. This could occur if a wheel speed sensor is faulty, the motor does not turn, or the motor wires are crossed.

Test Description: Number(s) below refer to circled number(s) on the diagnostic chart.

1. This step identifies a problem in a wheel speed sensor that may cause the system to be in release too long.
2. This checks for a wheel that may stick or bind because of a mechanical fault.
3. This checks to see if the motor is capable of moving and applying the hydraulic piston for the rear wheels.
4. This step insures that the motor wiring is not crossed.
5. This step is used to release the tension of the motor pack prior to removal.
6. This isolates the fault of a "no-apply" situation to either the motor pack or ABS hydraulic modulator assembly.

Diagnostic Aids:

An "Intermittent" problem may be caused by a mechanical part of the system that sticks or binds.

Any circuitry, that is suspected as causing the intermittent complaint, should be thoroughly checked for backed out terminals, improper mating, broken locks, improperly formed or damaged terminals, poor terminal to wiring connections or physical damage to the wiring harness.

DIAGNOSTIC CHART DTC A054 — 1993 SILHOUETTE, TRANS SPORT AND LUMINA APV

IMPORTANT: EXCESSIVE DRAG OR HIGH RESISTANCE IN THE BASE BRAKE OR SUSPENSION SYSTEM MUST BE INSPECTED AND CORRECTED BEFORE PROCEEDING WITH CODE DIAGNOSIS.

ARE ANY WHEEL SPEED CODES CURRENTLY SET?

- YES: PROCEED TO THAT CODE FIRST.
- NO:

(1)
- TEST VEHICLE DURING A STEADY DECEL CONDITION FROM 35 MPH TO 0 MPH WHILE MONITORING ALL WHEEL SPEEDS ON TECH 1 DATA LIST. DO ANY OF THE WHEEL SPEEDS INDICATE ERRATIC OR INTERMITTENT OPERATION?

- YES: PROCEED TO WHEEL SPEED = 0 CODE FOR AFFECTED WHEEL.
- NO:

(2)
- RAISE REAR OF VEHICLE AND SUPPORT AS INDICATED IN SECTION 0A.
- SPIN BOTH REAR WHEELS AND CONFORM PHYSICAL CONDITION OF BASE BRAKE SYSTEM.
  BOTH WHEELS SHOULD SPIN FREELY BY HAND.
  DO THEY?

- NO: REPAIR SOURCE OF RESISTANCE IN BASE BRAKE SYSTEM.
- YES:

(3)
- IGNITION "ON," ENGINE "OFF."
- USING MANUAL CONTROL FUNCTION OF TECH 1, SELECT REAR AXLE MOTOR APPLY.
- APPLY FIRM PRESSURE ON BRAKE PEDAL.
- COMMAND APPLY OF REAR MOTOR AND TRY AND SPIN REAR WHEELS BY HAND.
- WHEELS SHOULD NOT BE ABLE TO BE MOVED.
  DID WHEELS REMAIN STATIONARY?

- YES: FAULT IS NOT PRESENT AT THIS TIME. REFER TO "DIAGNOSTIC AIDS" ON FACING PAGE.
- NO:

(4)
- WITH FIRM PRESSURE ON BRAKE PEDAL STILL APPLIED, COMMAND A RELEASE OF THE REAR AXLE MOTOR. CAN WHEELS BE MOVED?

- NO: REPAIR CROSSED WIRES AT REAR MOTOR CIRCUIT.
- YES:

(5) USING TECH 1, PERFORM GEAR TENSION RELIEF FUNCTION.

(6)
- REMOVE MOTOR PACK ASSEMBLY.
- CONNECT MOTOR PACK WIRING.
- USING TECH 1, PERFORM MOTOR TEST.
  ALL THREE MOTORS SHOULD SPIN FREELY IN BOTH DIRECTIONS.
  DO THEY?

- NO: REPLACE MOTOR PACK ASSEMBLY.
- YES: REPLACE ABS HYDRAULIC MODULATOR.

DIAGNOSTIC CHART DTC A055 — 1993 SILHOUETTE, TRANS SPORT AND LUMINA APV

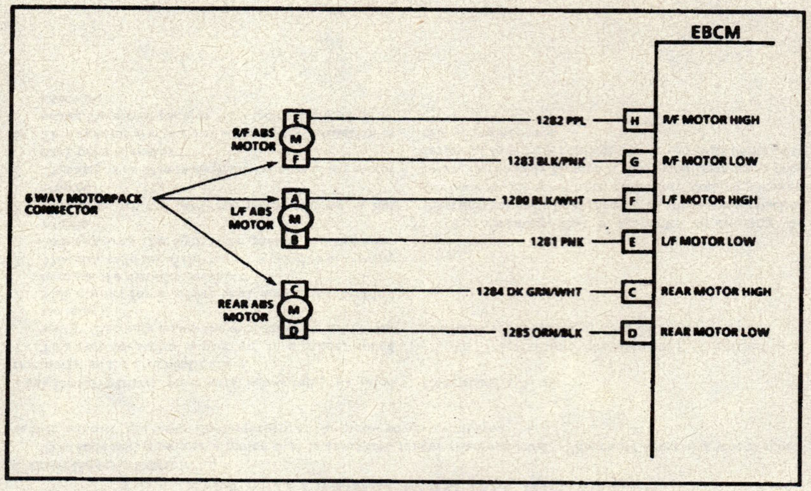

CODE A055

MOTOR DRIVER FAULT DETECTED

Circuit Description:

This code identifies a fault detected by the MDI custom IC. It also ensures the cause of the fault is not a result of a problem with the enable relay under a Code A055 failure. The MDI fault tests will be run to identify which circuit has failed, and may also cause additional codes to set to pinpoint the problem circuit.

Test Description: Number(s) below refer to circled number(s) on the diagnostic chart.

1. This insures that the Code A055 fault was not set due to a motor.
2. This checks to see if the fault is still present.
3. This step insures that Code A055 was not due to poor terminal contact at the motor pack connector.

Diagnostic Aids:

An "Intermittent" problem may be caused by a poor connection, rubbed through wire insulation, or a wire that is broken inside the insulation.

Any circuitry, that is suspected as causing the intermittent complaint, should be thoroughly checked for backed out terminals, improper mating, broken locks, improperly formed or damaged terminals, poor terminal to wiring connections or physical damage to the wiring harness.

DIAGNOSTIC CHART DTC A055 — 1993 SILHOUETTE, TRANS SPORT AND LUMINA APV

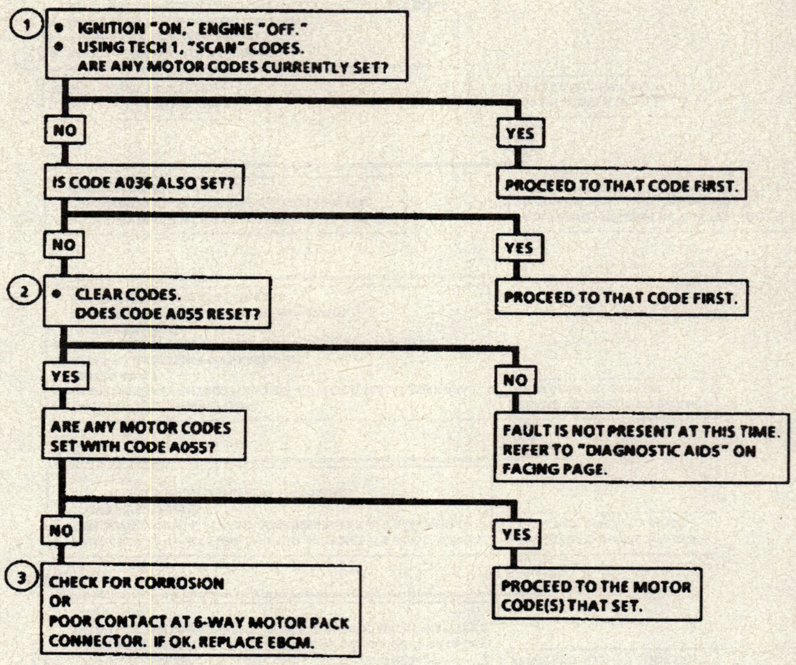

DIAGNOSTIC CHART DTC A056 — 1993 SILHOUETTE, TRANS SPORT AND LUMINA APV

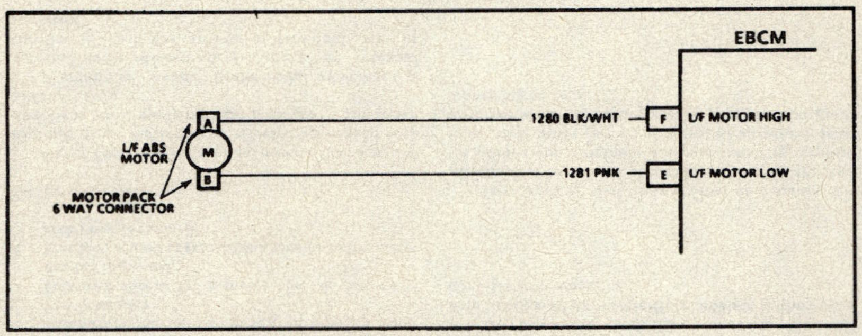

CODE A056

LEFT FRONT MOTOR CIRCUIT OPEN

Circuit Description:

This code identifies a motor that cannot be energized due to an open in its circuitry. This fault will not allow proper front ABS operation.

Test Description: Number(s) below refer to circled number(s) on the diagnostic chart.

1. This step checks for proper resistance of the motor.
2. This step is used to release the tension of the motor pack prior to removal.
3. This step checks for an open in the motor "high" circuitry.
4. This step checks for an open in the motor "low" circuitry.

An "Intermittent" problem may be caused by a poor connection, rubbed through wire insulation, or a wire that is broken inside the insulation.

If the fault is not current, wiggle the wires of the affected channel and check if the code resets. This will help to pinpoint an intermittent problem in the motor circuitry or connections.

Diagnostic Aids:

Using Tech 1, select manual control function, and exercise motor movement of affected channel in both directions while applying light pressure on the brake pedal.

If erratic or "jumpy" brake pedal movement is detected while performing an "apply" or "release" function of the motor, an intermittent fault may be indicated.

Any circuitry, that is suspected as causing the intermittent complaint, should be thoroughly checked for backed out terminals, improper mating, broken locks, improperly formed or damaged terminals, poor terminal to wiring connections or physical damage to the wiring harness.

DIAGNOSTIC CHART DTC A056 — 1993 SILHOUETTE, TRANS SPORT AND LUMINA APV

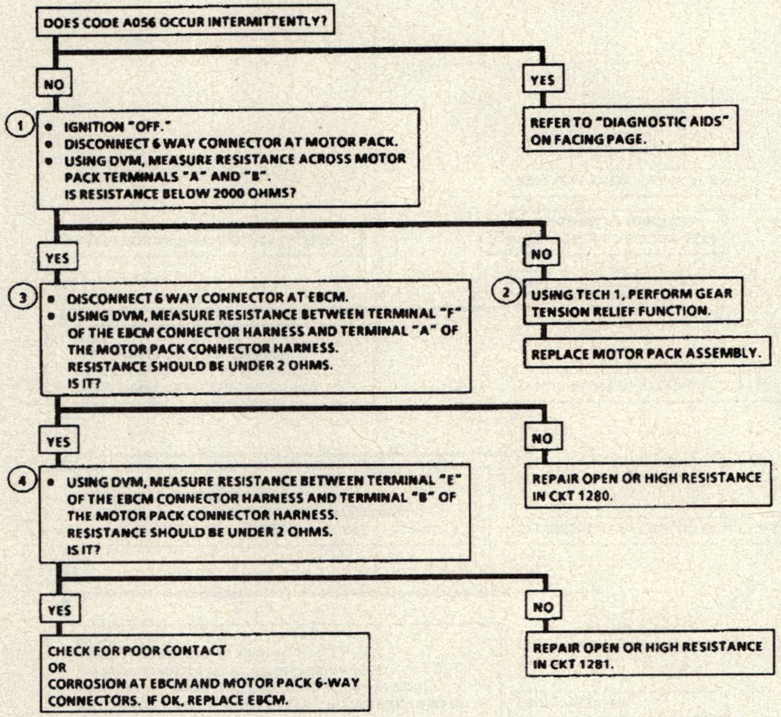

DIAGNOSTIC CHART DTC A057 — 1993 SILHOUETTE, TRANS SPORT AND LUMINA APV

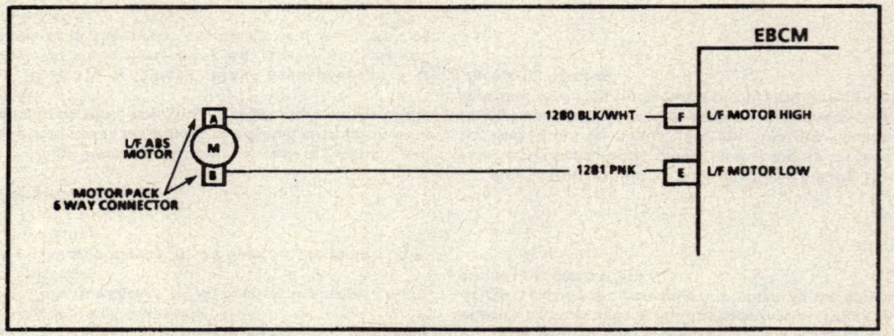

CODE A057

LEFT FRONT MOTOR CIRCUIT SHORTED TO GROUND

Circuit Description:

This code identifies a motor circuit that is shorted to ground. This fault will not allow the motor to be controlled at the commanded current rate or will cause the driver circuit to allow current directly to ground.

Test Description: Number(s) below refer to circled number(s) on the diagnostic chart.

1. This step checks for a short to ground in the motor "high" circuitry.
2. This step checks for a short to ground in the motor "low" circuitry.
3. This step checks for a motor that is internally shorted to ground.
4. This step is used to release the tension of the motor pack prior to removal.

An "Intermittent" problem may be caused by a poor connection, rubbed through wire insulation, or a wire that is broken inside the insulation.

If the fault is not current, wiggle the wires of the affected channel and check if the code resets. This will help to pinpoint an intermittent problem in the motor circuitry or connections.

Diagnostic Aids:

Using Tech 1, select manual control function, and exercise motor movement of affected channel in both directions while applying light pressure on the brake pedal.

If erratic or "jumpy" brake pedal movement is detected while performing an "apply" or "release" function of the motor, an intermittent fault may be indicated.

Any circuitry, that is suspected as causing the intermittent complaint, should be thoroughly checked for backed out terminals, improper mating, broken locks, improperly formed or damaged terminals, poor terminal to wiring connections or physical damage to the wiring harness.

DIAGNOSTIC CHART DTC A057 — 1993 SILHOUETTE, TRANS SPORT AND LUMINA APV

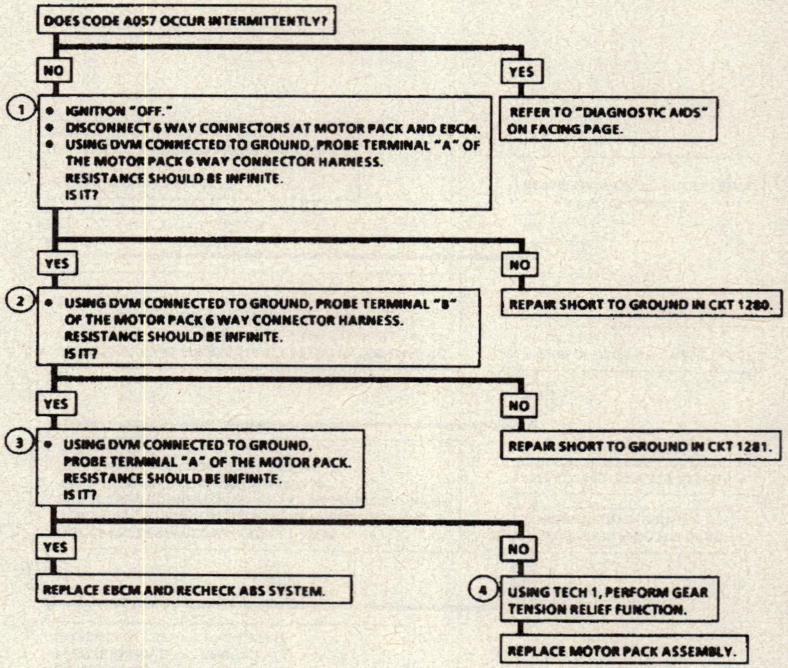

THIS CHART ASSUMES THAT A CURRENT CODE IS STORED INDICATING THAT THIS FAULT IS PRESENT.

DIAGNOSTIC CHART DTC A058 — 1993 SILHOUETTE, TRANS SPORT AND LUMINA APV

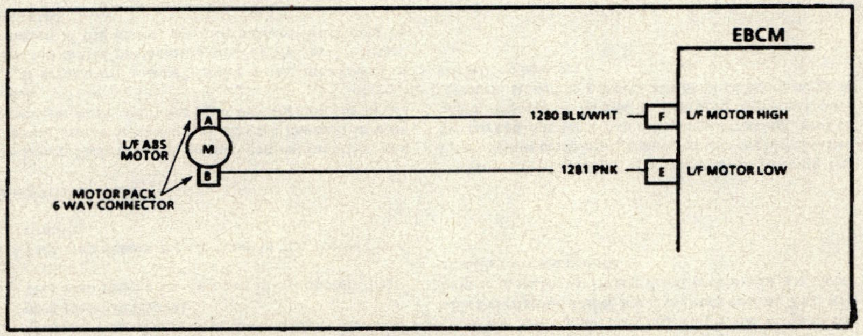

CODE A058

LEFT FRONT MOTOR CIRCUIT SHORTED TO BATTERY OR MOTOR SHORTED

Circuit Description:

This code identifies a motor circuit that is shorted to battery or a motor that has low or no resistance. This fault will not allow the motor to be controlled at the commanded current rate or will cause the motor to turn in the opposite direction, or not at all.

Test Description: Number(s) below refer to circled number(s) on the diagnostic chart.

1. This step checks for a short to voltage in the motor "high" circuitry.
2. This step checks for a short to voltage in the motor "low" circuitry.
3. This step checks for a motor that is internally shorted.
4. This step is used to release the tension of the motor pack prior to removal.

Diagnostic Aids:

Using Tech 1, select manual control function, and exercise motor movement of affected channel in both directions while applying light pressure on the brake pedal.

If erratic or "jumpy" brake pedal movement is detected while performing an "apply" or "release" function of the motor, an intermittent fault may be indicated.

An "Intermittent" problem may be caused by a poor connection, rubbed through wire insulation, or a wire that is broken inside the insulation.

If the fault is not current, wiggle the wires of the affected channel and check if the code resets. This will help to pinpoint an intermittent problem in the motor circuitry or connections.

Any circuitry, that is suspected as causing the intermittent complaint, should be thoroughly checked for backed out terminals, improper mating, broken locks, improperly formed or damaged terminals, poor terminal to wiring connections or physical damage to the wiring harness.

DIAGNOSTIC CHART DTC A058 — 1993 SILHOUETTE, TRANS SPORT AND LUMINA APV

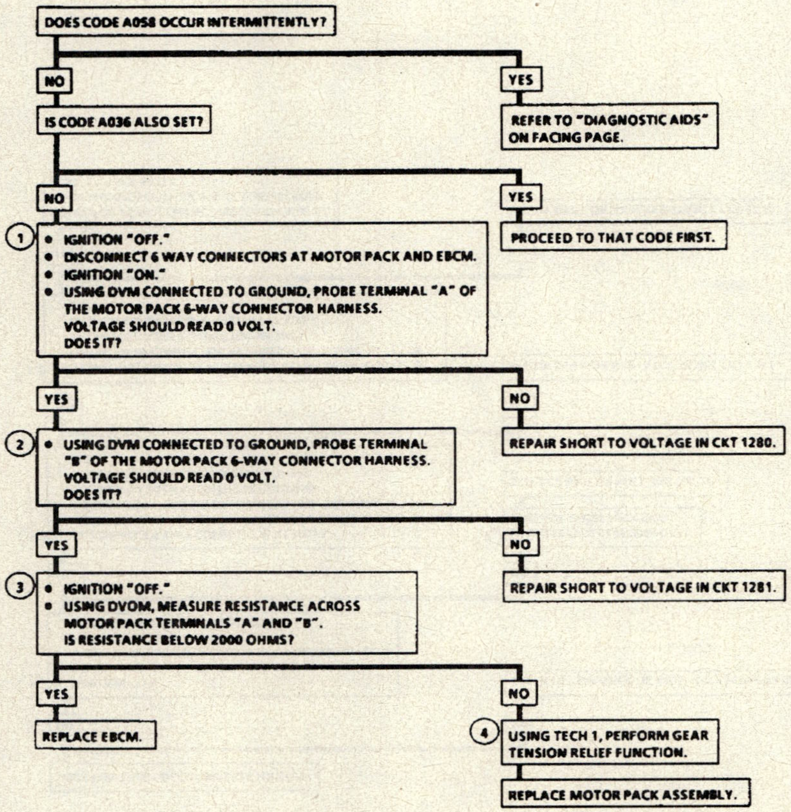

DIAGNOSTIC CHART DTC A061 — 1993 SILHOUETTE, TRANS SPORT AND LUMINA APV

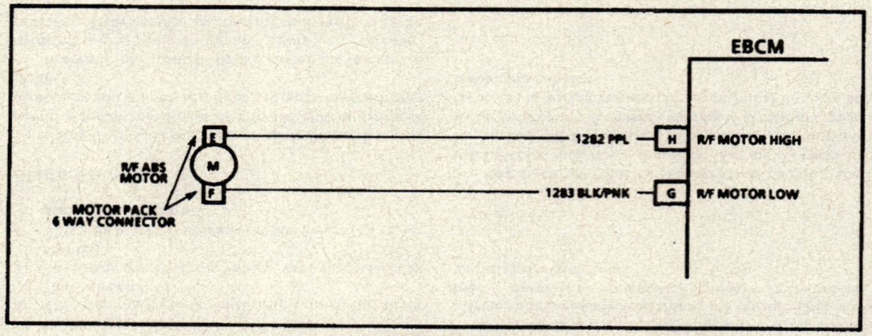

CODE A061

RIGHT FRONT MOTOR CIRCUIT OPEN

Circuit Description:

This code identifies a motor that cannot be energized due to an open in its circuitry. This fault will not allow proper front ABS operation.

Test Description: Number(s) below refer to circled number(s) on the diagnostic chart.

1. This step checks for proper resistance of the motor.
2. This step is used to release the tension of the motor pack prior to removal.
3. This step checks for an open in the motor "high" circuitry.
4. This step checks for an open in the motor "low" circuitry.

An "Intermittent" problem may be caused by a poor connection, rubbed through wire insulation, or a wire that is broken inside the insulation.

If the fault is not current, wiggle the wires of the affected channel and check if the code resets. This will help to pinpoint an intermittent problem in the motor circuitry or connections.

Diagnostic Aids:

Using Tech 1, select manual control function, and exercise motor movement of affected channel in both directions while applying light pressure on the brake pedal.

If erratic or "jumpy" brake pedal movement is detected while performing an "apply" or "release" function of the motor, an intermittent fault may be indicated.

Any circuitry, that is suspected as causing the intermittent complaint, should be thoroughly checked for backed out terminals, improper mating, broken locks, improperly formed or damaged terminals, poor terminal to wiring connections or physical damage to the wiring harness.

DIAGNOSTIC CHART DTC A061 — 1993 SILHOUETTE, TRANS SPORT AND LUMINA APV

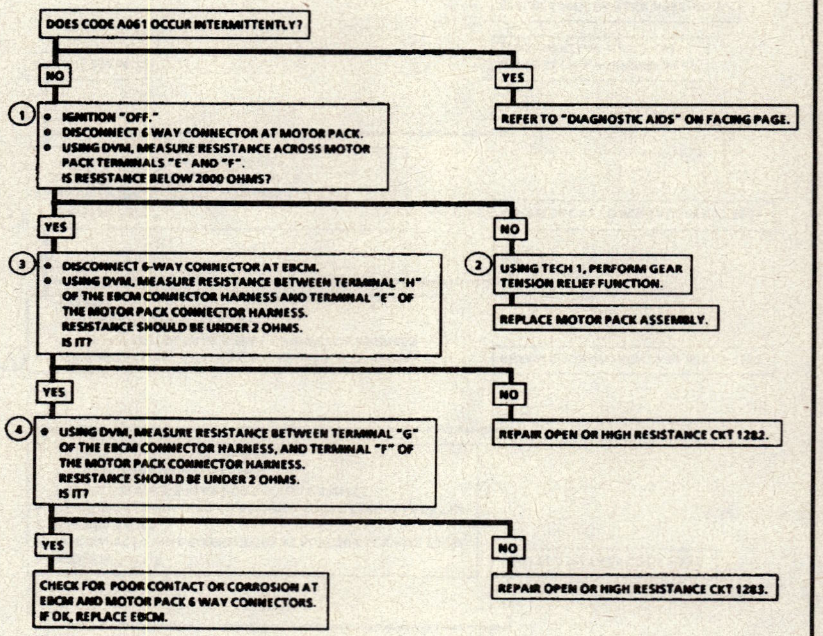

DIAGNOSTIC CHART DTC A062 — 1993 SILHOUETTE, TRANS SPORT AND LUMINA APV

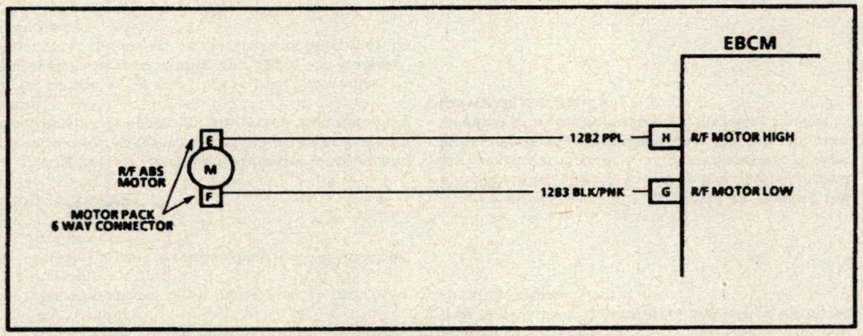

CODE A062

RIGHT FRONT MOTOR CIRCUIT SHORTED TO GROUND

Circuit Description:

This code identifies a motor circuit that is shorted to ground. This fault will not allow the motor to be controlled at the commanded current rate or will cause the motor driver circuit to allow current directly to ground.

Test Description: Number(s) below refer to circled number(s) on the diagnostic chart.

1. This step checks for a short to ground in the motor "high" circuitry.
2. This step checks for a short to ground in the motor "low" circuitry.
3. This step checks for a motor that is internally shorted to ground.
4. This step is used to release the tension of the motor pack prior to removal.

An "Intermittent" problem may be caused by a poor connection, rubbed through wire insulation, or a wire that is broken inside the insulation.

If the fault is not current, wiggle the wires of the affected channel and check if the code resets. This will help to pinpoint an intermittent problem in the motor circuitry or connections.

Diagnostic Aids:

Using Tech 1, select manual control function, and exercise motor movement of affected channel in both directions while applying light pressure on the brake pedal.

If erratic or "jumpy" brake pedal movement is detected while performing an "apply" or "release" function of the motor, an intermittent fault may be indicated.

Any circuitry, that is suspected as causing the intermittent complaint, should be thoroughly checked for backed out terminals, improper mating, broken locks, improperly formed or damaged terminals, poor terminal to wiring connections or physical damage to the wiring harness.

DIAGNOSTIC CHART DTC A062 — 1993 SILHOUETTE, TRANS SPORT AND LUMINA APV

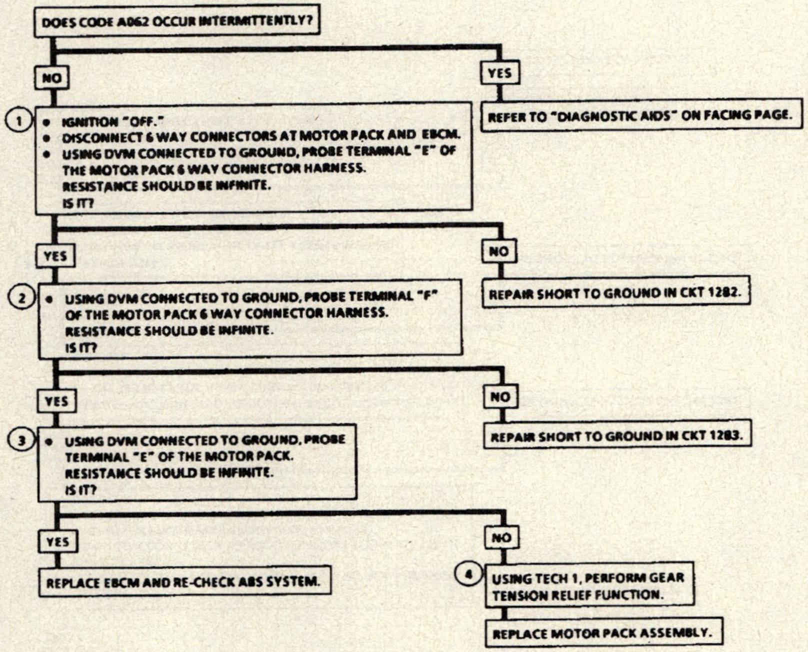

THIS CHART ASSUMES THAT A CURRENT CODE IS STORED INDICATING THAT THIS FAULT IS PRESENT.

DIAGNOSTIC CHART DTC A063 — 1993 SILHOUETTE, TRANS SPORT AND LUMINA APV

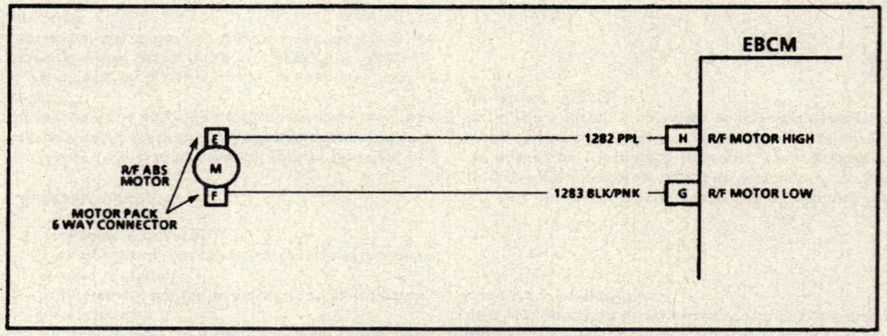

CODE A063

RIGHT FRONT MOTOR CIRCUIT SHORTED TO BATTERY OR MOTOR SHORTED

Circuit Description:

This code identifies a motor circuit that is shorted to battery or a motor that has low or no resistance. This fault will not allow the motor to be controlled at the commanded current rate or will cause the motor to turn in the opposite direction or not at all.

Test Description: Number(s) below refer to circled number(s) on the diagnostic chart.

1. This step checks for a short to voltage in the motor "high" circuitry.
2. This step checks for a short to voltage in the motor "low" circuitry.
3. This step checks for a motor that is internally shorted.
4. This test is used to release the tension of the motor pack prior to removal.

An "Intermittent" problem may be caused by a poor connection, rubbed through wire insulation, or a wire that is broken inside the insulation.

If the fault is not current, wiggle the wires of the affected channel and check if the code resets. This will help to pinpoint an intermittent problem in the motor circuitry or connections.

Diagnostic Aids:

Using Tech 1, select manual control function, and exercise motor movement of affected channel in both directions while applying light pressure on the brake pedal.

If erratic or "jumpy" brake pedal movement is detected while performing an "apply" or "release" function of the motor, an intermittent fault may be indicated.

Any circuitry, that is suspected as causing the intermittent complaint, should be thoroughly checked for backed out terminals, improper mating, broken locks, improperly formed or damaged terminals, poor terminal to wiring connections or physical damage to the wiring harness.

DIAGNOSTIC CHART DTC A063 — 1993 SILHOUETTE, TRANS SPORT AND LUMINA APV

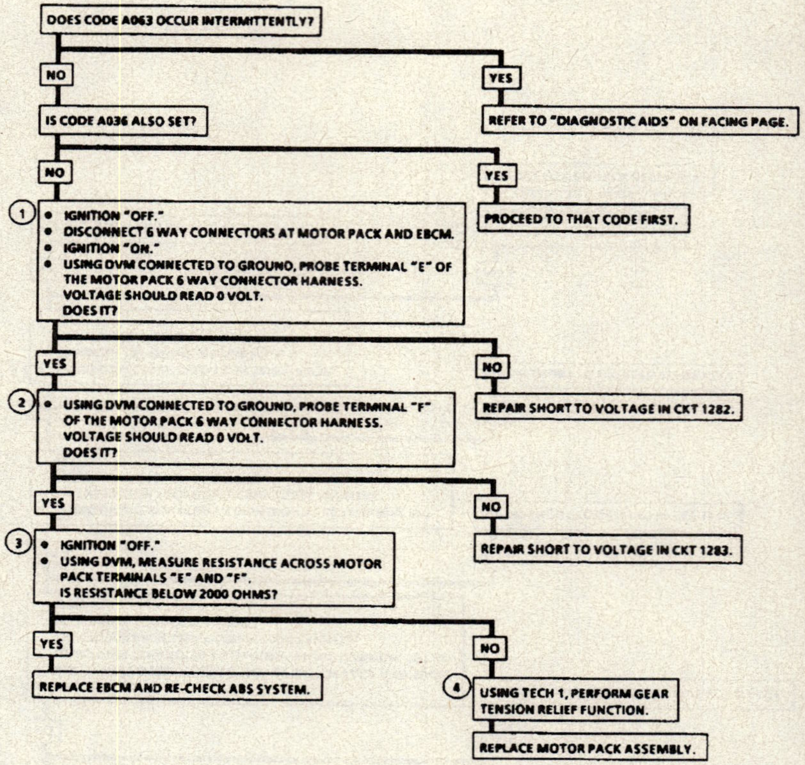

THIS CHART ASSUMES THAT A CURRENT CODE IS STORED INDICATING THAT THIS FAULT IS PRESENT.

DIAGNOSTIC CHART DTC A064 — 1993 SILHOUETTE, TRANS SPORT AND LUMINA APV

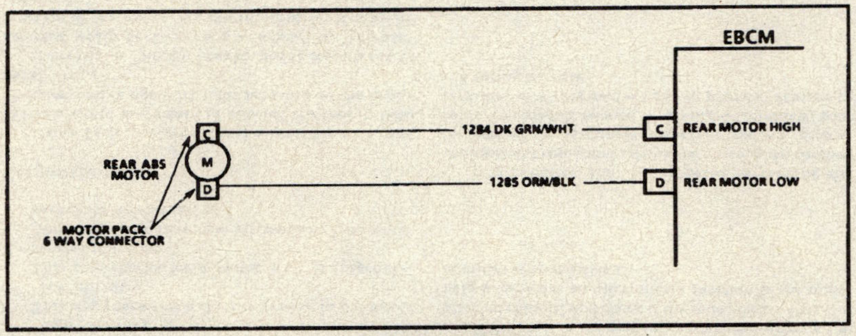

CODE A064

REAR AXLE MOTOR CIRCUIT OPEN

Circuit Description:

This code identifies a motor that cannot be energized due to an open in its circuitry. This fault will not allow proper rear ABS operation.

Test Description: Number(s) below refer to circled number(s) on the diagnostic chart.

1. This checks for an open circuit in the motor.
2. This step is used to release the tension of the motor pack prior to removal.
3. This step checks for an open in the motor "high" circuitry.
4. This step checks for an open in the motor "low" circuitry.

An "Intermittent" problem may be caused by a poor connection, rubbed through wire insulation, or a wire that is broken inside the insulation.

If the fault is not current, wiggle the wires of the affected channel and check if the code resets. This will help to pinpoint an intermittent problem in the motor circuitry or connections.

Diagnostic Aids:

Using Tech 1, select manual control function, and exercise motor movement of affected channel in both directions while applying light pressure on the brake pedal.

If erratic or "jumpy" brake pedal movement is detected while performing an "apply" or "release" function of the motor, an intermittent fault may be indicated.

Any circuitry, that is suspected as causing the intermittent complaint, should be thoroughly checked for backed out terminals, improper mating, broken locks, improperly formed or damaged terminals, poor terminal to wiring connections or physical damage to the wiring harness.

DIAGNOSTIC CHART DTC A064 — 1993 SILHOUETTE, TRANS SPORT AND LUMINA APV

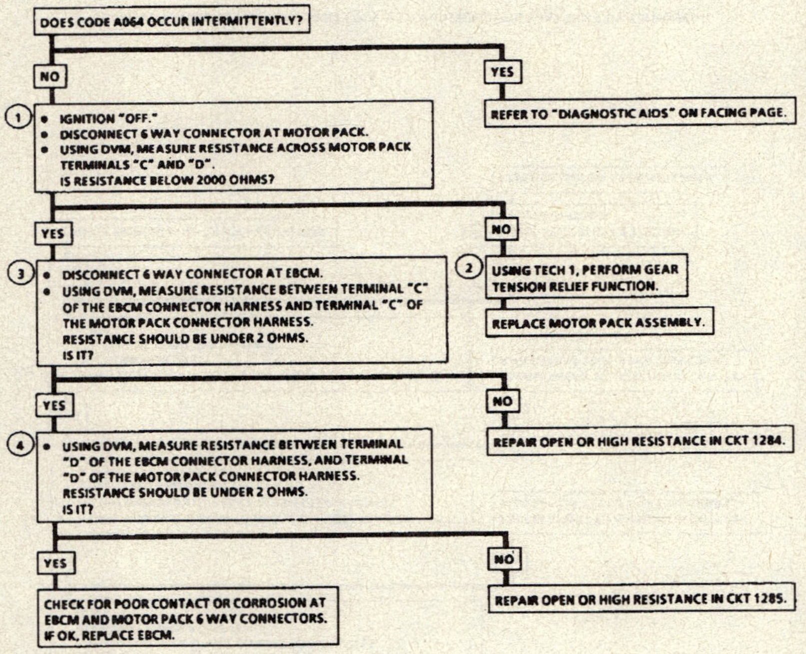

THIS CHART ASSUMES THAT A CURRENT CODE IS STORED INDICATING THAT THIS FAULT IS PRESENT.

DIAGNOSTIC CHART DTC A065 — 1993 SILHOUETTE, TRANS SPORT AND LUMINA APV

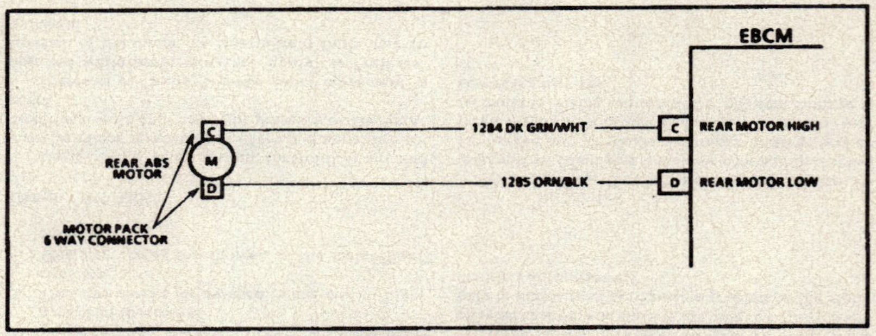

CODE A065

REAR AXLE MOTOR CIRCUIT SHORTED TO GROUND

Circuit Description:

This code identifies a motor circuit that is shorted to ground. This fault will not allow the motor to be controlled at the commanded current rate or will cause the driver circuit to allow current directly to ground.

Test Description: Number(s) below refer to circled number(s) on the diagnostic chart.

1. This step checks for a short to ground in the motor "high" circuitry.
2. This step checks for a short to ground in the motor "low" circuitry.
3. This step checks for a motor that is internally shorted to ground.
4. This step is used to release the tension of the motor pack prior to removal.

Diagnostic Aids:

Using Tech 1, select manual control function, and exercise motor movement of affected channel in both directions while applying light pressure on the brake pedal.

If erratic or "jumpy" brake pedal movement is detected while performing an "apply" or "release" function of the motor, an intermittent fault may be indicated.

An "Intermittent" problem may be caused by a poor connection, rubbed through wire insulation, or a wire that is broken inside the insulation.

If the fault is not current, wiggle the wires of the affected channel and check if the code resets. This will help to pinpoint an intermittent problem in the motor circuitry or connections.

Any circuitry, that is suspected as causing the intermittent complaint, should be thoroughly checked for backed out terminals, improper mating, broken locks, improperly formed or damaged terminals, poor terminal to wiring connections or physical damage to the wiring harness.

DIAGNOSTIC CHART DTC A065 — 1993 SILHOUETTE, TRANS SPORT AND LUMINA APV

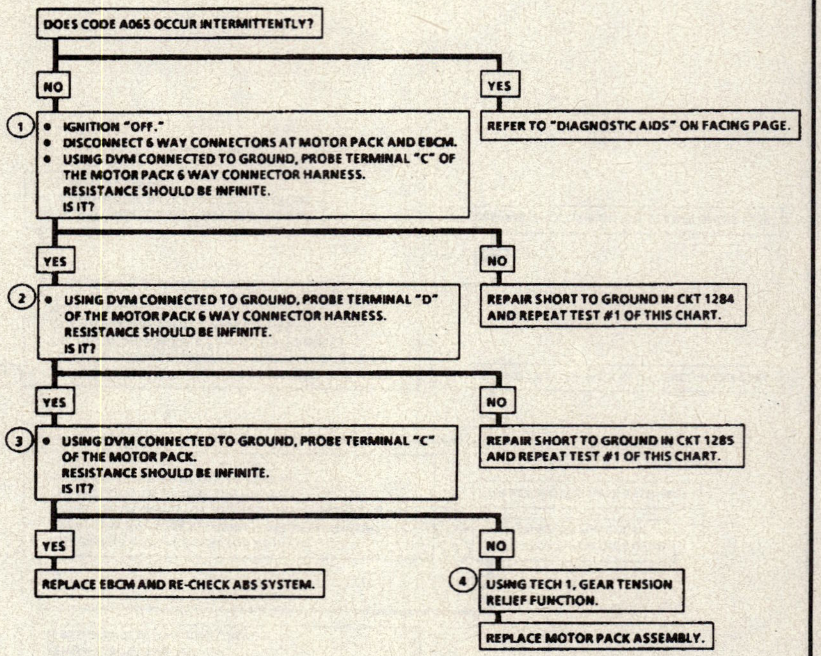

THIS CHART ASSUMES THAT A CURRENT CODE IS STORED INDICATING THAT THIS FAULT IS PRESENT.

DIAGNOSTIC CHART DTC A066 — 1993 SILHOUETTE, TRANS SPORT AND LUMINA APV

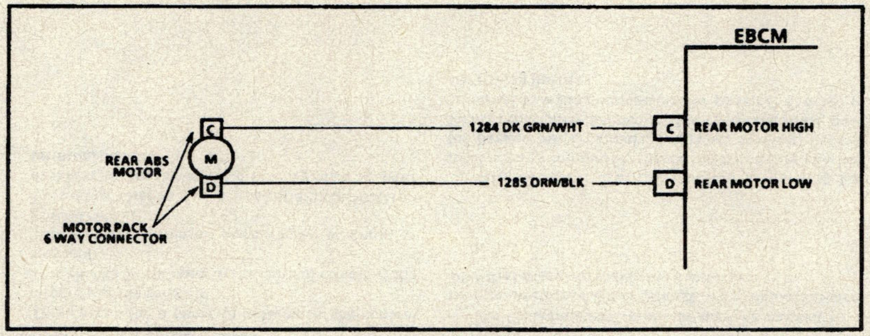

CODE A066

REAR AXLE MOTOR CIRCUIT SHORTED TO BATTERY OR MOTOR SHORTED

Circuit Description:

This code identifies a motor circuit that is shorted to battery or a motor that has low or zero resistance. This fault will not allow the motor to be controlled at the commanded current rate or will cause the motor to turn in the opposite direction or not turn at all.

Test Description: Number(s) below refer to circled number(s) on the diagnostic chart.

1. This step checks for a short to voltage in the motor "high" circuitry.
2. This step checks for a short to voltage in the motor "low" circuitry.
3. This step checks for a motor that is internally shorted.
4. This step is used to release the tension of the motor pack prior to removal.

An "Intermittent" problem may be caused by a poor connection, rubbed through wire insulation, or a wire that is broken inside the insulation.

If the fault is not current, wiggle the wires of the affected channel and check if the code resets. This will help to pinpoint an intermittent problem in the motor circuitry or connections.

Diagnostic Aids:

Using Tech 1, select manual control function, and exercise motor movement of affected channel in both directions while applying light pressure on the brake pedal.

If erratic or "jumpy" brake pedal movement is detected while performing an "apply" or "release" function of the motor, an intermittent fault may be indicated.

Any circuitry, that is suspected as causing the intermittent complaint, should be thoroughly checked for backed out terminals, improper mating, broken locks, improperly formed or damaged terminals, poor terminal to wiring connections or physical damage to the wiring harness.

DIAGNOSTIC CHART DTC A066 — 1993 SILHOUETTE, TRANS SPORT AND LUMINA APV

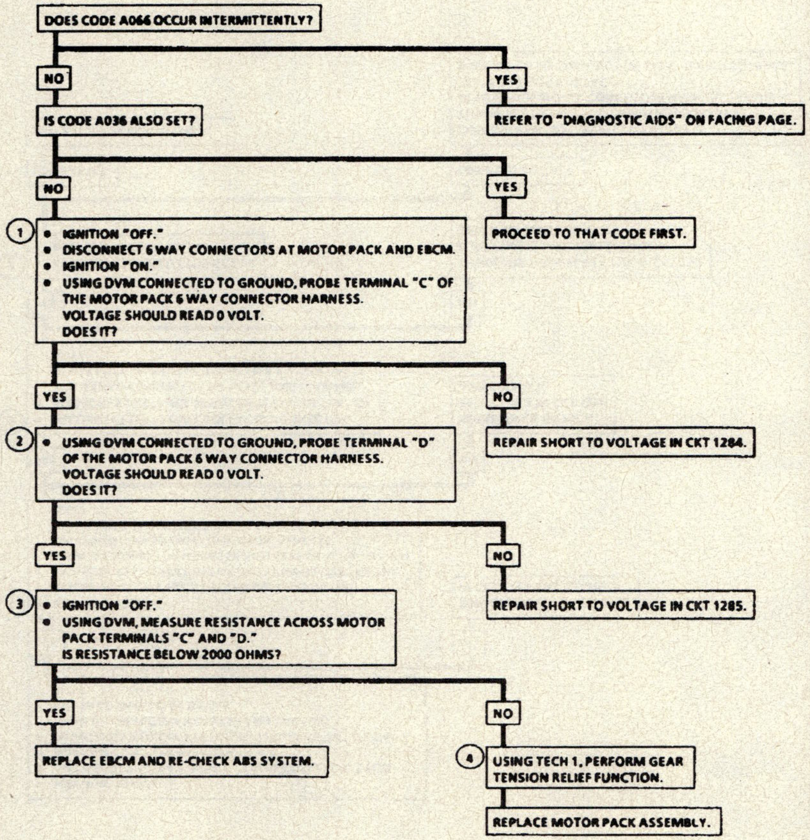

THIS CHART ASSUMES THAT A CURRENT CODE IS STORED INDICATING THAT THIS FAULT IS PRESENT.

DIAGNOSTIC CHART DTC A076 — 1993 SILHOUETTE, TRANS SPORT AND LUMINA APV

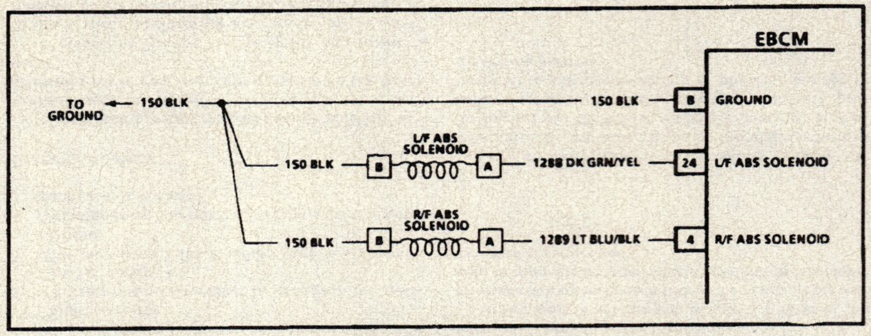

CODE A076

LEFT FRONT SOLENOID CIRCUIT SHORTED TO BATTERY OR OPEN

Circuit Description:

This code identifies a solenoid that cannot be energized due to an open in its circuitry, or a solenoid that is always energized due to a short to battery in its circuitry between the driver and the solenoid. An open will not allow proper ABS operation, but the short to battery simply turns "ON" the solenoid. A path for base brakes is still allowed once the motor re-homes and the check ball is lifted off its seat during key on initialization.

Test Description: Number(s) below refer to circled number(s) on the diagnostic chart.

1. This checks for a short to voltage in the control circuitry of the solenoid.
2. This checks for an open in the control circuit of the solenoid.
3. This step identifies an open in the solenoid or ground circuit.
4. This step checks for a possible intermittent problem in the solenoid circuitry due to poor terminal contact.

Diagnostic Aids:

An "Intermittent" problem may be caused by a poor connection, rubbed through wire insulation, or a wire that is broken inside the insulation.

Any circuitry, that is suspected as causing the intermittent complaint, should be thoroughly checked for backed out terminals, improper mating, broken locks, improperly formed or damaged terminals, poor terminal to wiring connections or physical damage to the wiring harness.

DIAGNOSTIC CHART DTC A076 — 1993 SILHOUETTE, TRANS SPORT AND LUMINA APV

(1)
- IGNITION "OFF."
- DISCONNECT 24 WAY WORLD CONNECTOR FROM EBCM.
- IGNITION "ON," ENGINE "OFF."
- USING DVM, CONNECT BETWEEN TERMINAL "24" OF THE WORLD CONNECTOR HARNESS AND GROUND. VOLTAGE SHOULD BE 0 VOLT. IS IT?

YES → (2) | NO → REPAIR SHORT TO VOLTAGE IN CKT 1288 AND RECHECK.

(2)
- IGNITION "OFF."
- DISCONNECT LEFT FRONT SOLENOID CONNECTOR.
- MEASURE RESISTANCE BETWEEN TERMINAL "24" OF THE WORLD CONNECTOR HARNESS AND TERMINAL "A" OF THE LEFT FRONT SOLENOID CONNECTOR HARNESS. RESISTANCE SHOULD BE 2 OHMS OR LESS. IS IT?

YES → (3) | NO → REPAIR OPEN OR HIGH RESISTANCE IN CKT 1288 AND RECHECK.

(3)
- RECONNECT LEFT FRONT SOLENOID CONNECTOR.
- MEASURE RESISTANCE BETWEEN TERMINAL "24" OF THE WORLD CONNECTOR HARNESS AND GROUND. RESISTANCE SHOULD BE BETWEEN 2.5 AND 5 OHMS. IS IT?

YES → (4) | NO → CHECK FOR OPEN IN CKT 150. IF CKT 150 IS NOT OPEN, REPLACE LEFT FRONT SOLENOID AND RECHECK.

(4)
- RECONNECT WORLD CONNECTOR AND VERIFY CODE RESETS.

CODE RESET → REPLACE EBCM AND RECHECK.

CODE DID NOT SET → CHECK SOLENOID CONNECTOR HARNESS TERMINALS "A" AND "B" FOR POOR CONTACT, ALSO CHECK WORLD CONNECTOR HARNESS TERMINAL "24" FOR POOR CONTACT. IF OK, REFER TO "DIAGNOSTIC AIDS" ON FACING PAGE.

THIS CHART ASSUMES THAT A CURRENT CODE IS STORED INDICATING THAT THIS FAULT IS PRESENT.

DIAGNOSTIC CHART DTC A077 — 1993 SILHOUETTE, TRANS SPORT AND LUMINA APV

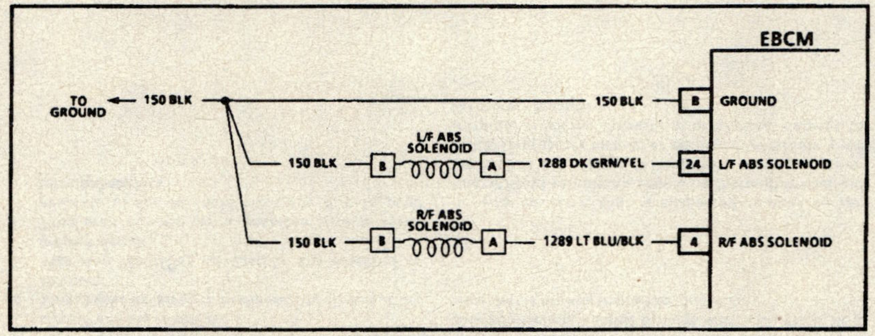

CODE A077

LEFT FRONT SOLENOID CIRCUIT SHORTED TO GROUND OR DRIVER OPEN

Circuit Description:

This code identifies a solenoid that cannot be energized due to an open in its driver circuitry, or a short to ground between the solenoid driver and the solenoid. These faults can affect ABS operation since the flow of brake fluid to the wheel cylinder cannot be stopped, making ABS operation for that channel impossible.

Test Description: Number(s) below refer to circled number(s) on the diagnostic chart.

1. This step checks for a solenoid that is internally shorted to ground.
2. This step checks for a solenoid that is not within proper resistance values.
3. This step indicates if a short to ground exists in the solenoid circuitry.
4. This checks for an open control circuit to the EBCM.

Diagnostic Aids:

An "Intermittent" problem may be caused by a poor connection, rubbed through wire insulation, or a wire that is broken inside the insulation.

Any circuitry, that is suspected as causing the intermittent complaint, should be thoroughly checked for backed out terminals, improper mating, broken locks, improperly formed or damaged terminals, poor terminal to wiring connections or physical damage to the wiring harness.

DIAGNOSTIC CHART DTC A077 — 1993 SILHOUETTE, TRANS SPORT AND LUMINA APV

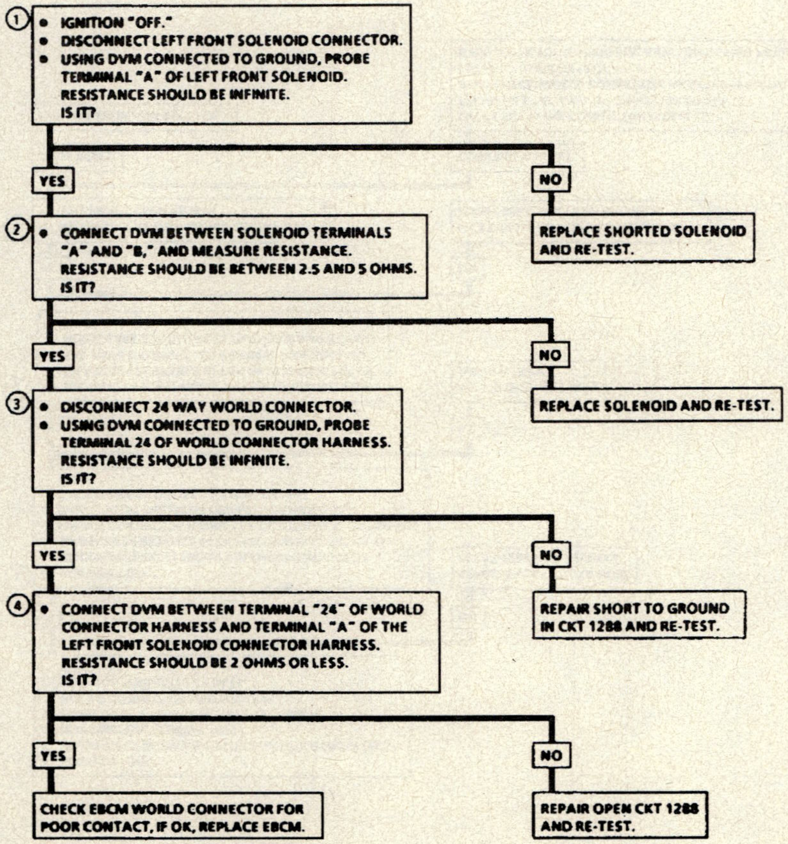

THIS CHART ASSUMES THAT A CURRENT CODE IS STORED INDICATING THAT THIS FAULT IS PRESENT.

DIAGNOSTIC CHART DTC A078 — 1993 SILHOUETTE, TRANS SPORT AND LUMINA APV

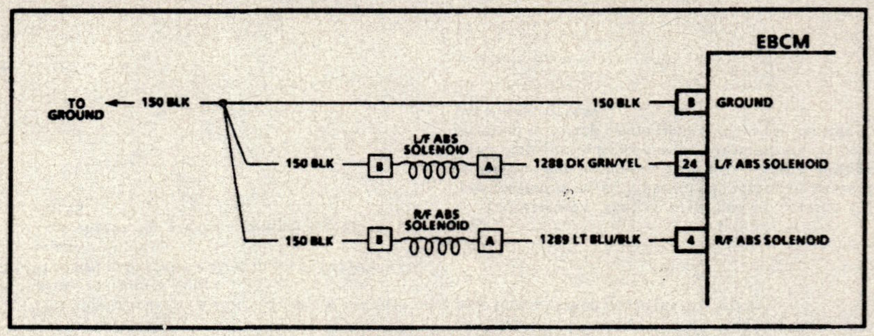

CODE A078

RIGHT FRONT SOLENOID CIRCUIT SHORTED TO BATTERY OR OPEN

Circuit Description:

This code identifies a solenoid that cannot be energized due to an open in its circuitry, or a solenoid that is always energized due to a short to battery in its circuitry between the driver and the solenoid. An open will not allow proper ABS operation, but the short to battery simply turns "ON" the solenoid. A path for base brakes is still allowed once the motor re-homes and the check ball is lifted off its seat during key on initialization.

Test Description: Number(s) below refer to circled number(s) on the diagnostic chart.

1. This checks for a short to voltage in the control circuitry of the solenoid.
2. This checks for an open in the control circuit of the solenoid.
3. This step identifies an open in the solenoid or ground circuit.
4. This step checks for a possible intermittent problem in the solenoid circuitry due to poor terminal contact.

Diagnostic Aids:

An "Intermittent" problem may be caused by a poor connection, rubbed through wire insulation, or a wire that is broken inside the insulation.

Any circuitry, that is suspected as causing the intermittent complaint, should be thoroughly checked for backed out terminals, improper mating, broken locks, improperly formed or damaged terminals, poor terminal to wiring connections or physical damage to the wiring harness.

DIAGNOSTIC CHART DTC A078 — 1993 SILHOUETTE, TRANS SPORT AND LUMINA APV

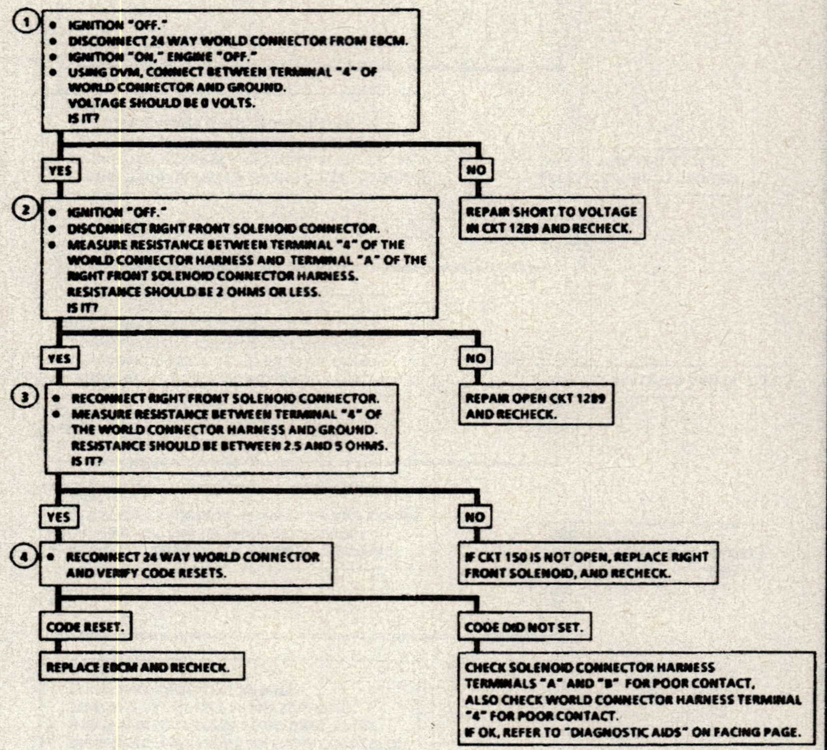

THIS CHART ASSUMES THAT A CURRENT CODE IS STORED INDICATING THAT THIS FAULT IS PRESENT.

DIAGNOSTIC CHART DTC A081 — 1993 SILHOUETTE, TRANS SPORT AND LUMINA APV

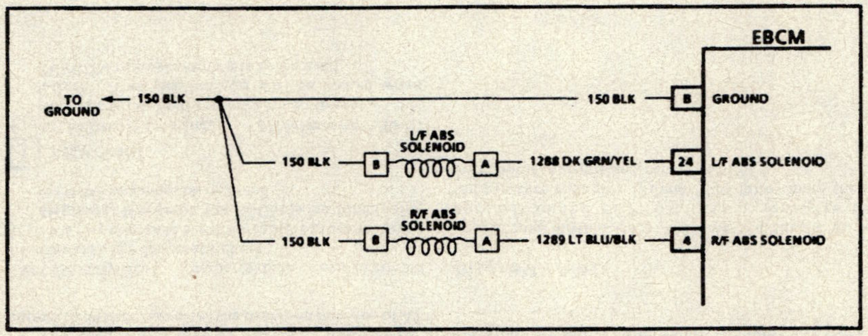

CODE A081

RIGHT FRONT SOLENOID CIRCUIT SHORTED TO GROUND OR DRIVER OPEN

Circuit Description:

This code identifies a solenoid that cannot be energized due to an open in its driver circuitry, or a short to ground between the solenoid driver and the solenoid. These faults can affect ABS operation since the flow of brake fluid to the wheel cylinder cannot be stopped, making ABS operation for that channel impossible.

Test Description: Number(s) below refer to circled number(s) on the diagnostic chart.

1. This step checks for a solenoid that is internally shorted to ground.
2. This step checks for a solenoid that is not within proper resistance values.
3. This step indicates if a short to ground exists in the solenoid circuitry.
4. This checks for an open control circuit to the EBCM.

Diagnostic Aids:

An "Intermittent" problem may be caused by a poor connection, rubbed through wire insulation, or a wire that is broken inside the insulation.

Any circuitry, that is suspected as causing the intermittent complaint, should be thoroughly checked for backed out terminals, improper mating, broken locks, improperly formed or damaged terminals, poor terminal to wiring connections or physical damage to the wiring harness.

DIAGNOSTIC CHART DTC A081 — 1993 SILHOUETTE, TRANS SPORT AND LUMINA APV

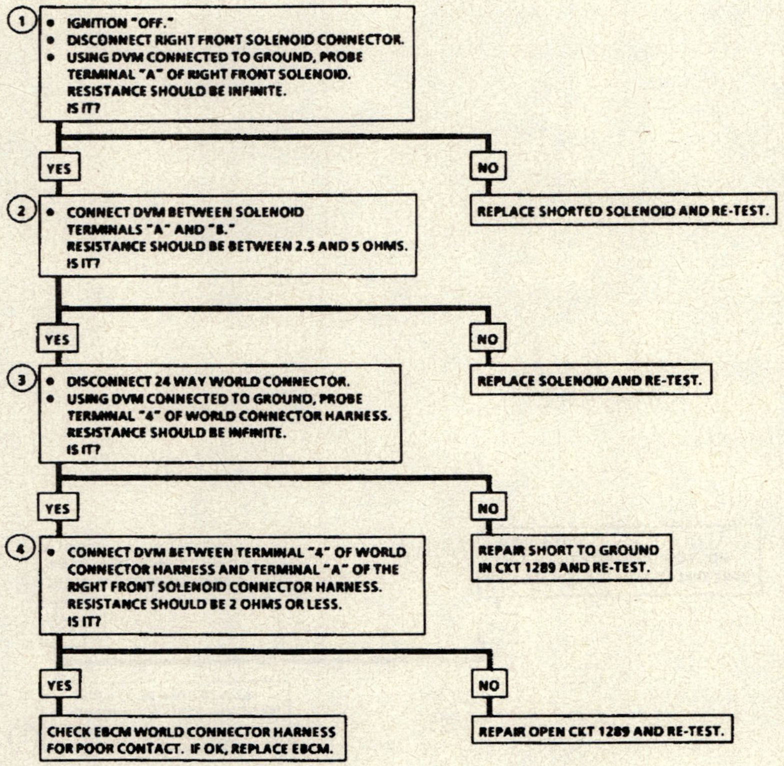

DIAGNOSTIC CHART DTC A082 — 1993 SILHOUETTE, TRANS SPORT AND LUMINA APV

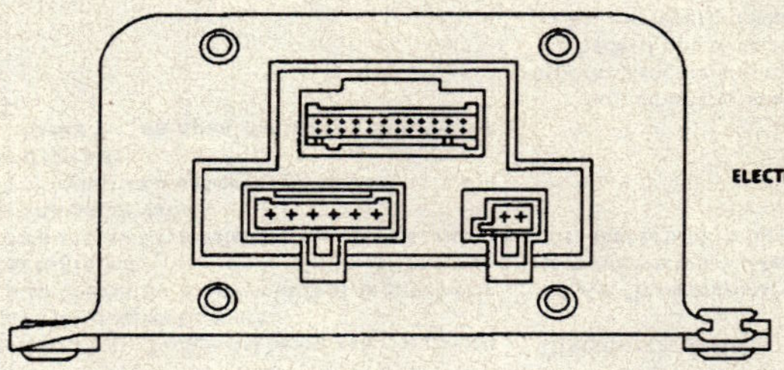

CODE A082

CALIBRATION MEMORY FAILURE

Circuit Description:

This code allows the EBCM to check for a calibration failure by comparing the calibration value to a known value stored in the EEPROM.

This code is also used as a security measure to prevent improper use of calibrations or changes to these calibrations that may alter the designed function of ABS.

Test Description: Number(s) below refer to circled number(s) on the diagnostic chart.

1. This step checks to see if the fault is present during diagnosis. If present, the EBCM is not functioning correctly and must be replaced.

[!] Important

- At the time of printing, vehicles with ABS-VI were not being field reprogrammed to correct this failure. A service bulletin will be issued when EEPROM reprogramming is authorized.

Diagnostic Aids:

An intermittent Code A082 may be caused by a bad cell in the EEPROM that is sensitive to temperature changes. If Code A082 failed more than once, but is intermittent, replace EBCM.

DIAGNOSTIC CHART DTC A082 — 1993 SILHOUETTE, TRANS SPORT AND LUMINA APV

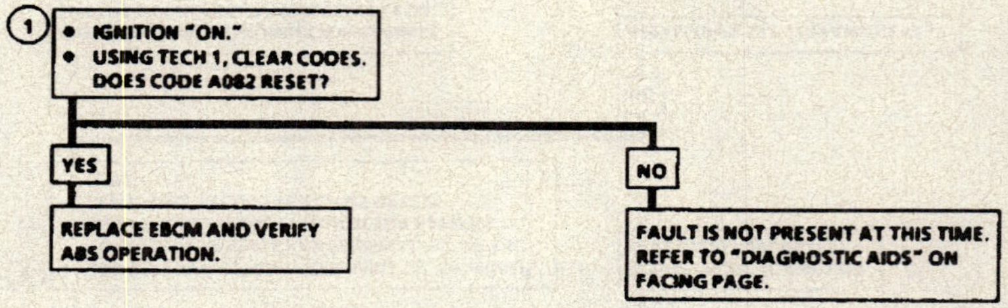

THIS CHART ASSUMES THAT A CURRENT CODE IS STORED INDICATING THAT THIS FAULT IS PRESENT.

DIAGNOSTIC CHART DTC A086 — 1993 SILHOUETTE, TRANS SPORT AND LUMINA APV

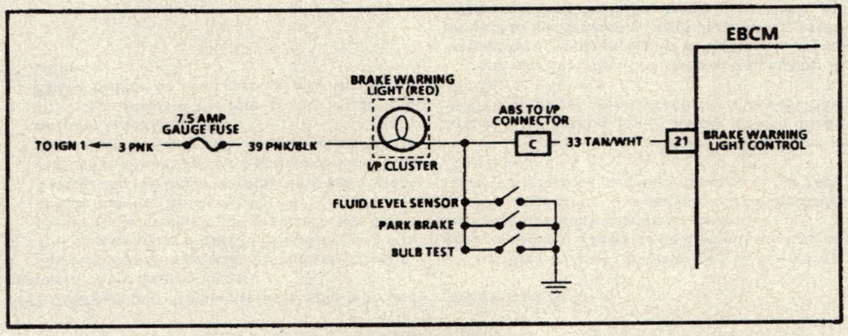

CODE A086

RED BRAKE WARNING LIGHT ACTIVATED BY ABS

Circuit Description:

This code is provided as an information only code, and reflects the status of the command issued by the ABS EBCM to illuminate the red brake warning light. If another diagnostic code issues a command to illuminate the red brake warning light, Code A086 will be stored in EEPROM as a history code at the conclusion of the ignition cycle.

Test Description: Number(s) below refer to circled number(s) on the diagnostic chart.

1. This step serves to identify if a code other than Code A086 commanded the red brake warning light "ON."

Diagnostic Aids:

Any ABS mechanical code that issues a command to illuminate the red brake warning light will also result in Code A086 being stored in EEPROM during shut down. These Codes are: A038, A041, A042, A046, and A051. If the motors are not in their home position, certain electrical codes will also command the red brake warning light "ON." These Codes are: A014, A016, A018, A055, A056, A057, A058, A061, A062, A063, A064, A065, and A066.

If any of these codes are indicated along with Code A086, they must be corrected prior to addressing a Code A086 fault.

DIAGNOSTIC CHART DTC A086 — 1993 SILHOUETTE, TRANS SPORT AND LUMINA APV

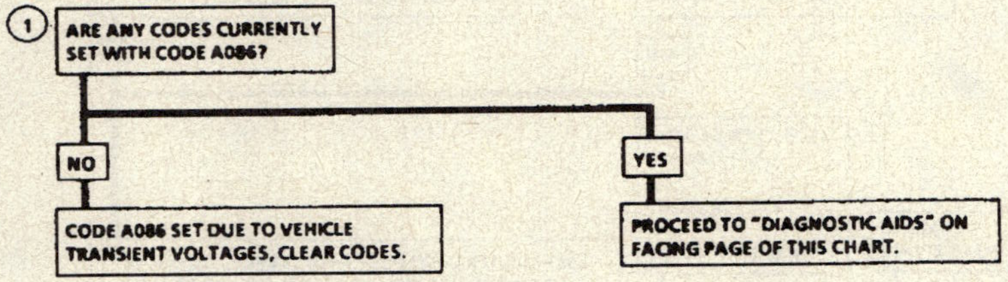

DIAGNOSTIC CHART DTC A087 — 1993 SILHOUETTE, TRANS SPORT AND LUMINA APV

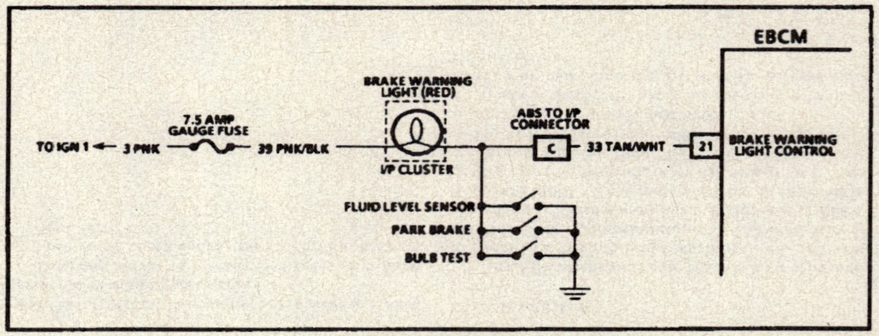

CODE A087

RED BRAKE WARNING LIGHT CIRCUIT OPEN

Circuit Description:

This code is used to verify the EBCM has continuity to the red brake warning light in case the EBCM must turn it "ON." This will only occur if an ABS fault is detected that may degrade base brake operation.

Because ABS is not the only device controlling the red brake warning light (parking brake and low fluid sensor may also turn "ON" the light) a short to ground in this circuit cannot be detected.

Test Description: Number(s) below refer to circled number(s) on the diagnostic chart.

1. This step identifies if the fault is currently present.
2. This step indicates if the EBCM and circuitry has the ability to complete the ground to the brake telltale lamp and illuminate it.
3. Because the red brake warning light has current supplied to it through the 7.5 amp gauges fuse, this step would indicate if the ignition circuit is complete to the I/P cluster.
4. This step isolates if the open circuit is due to a EBCM failure or open circuitry to the brake telltale.

Diagnostic Aids:

An "Intermittent" problem may be caused by a poor connection, rubbed through wire insulation, or a wire that is broken inside the insulation.

The frequency of the problem can be checked by using the enhanced diagnostic function of the Tech 1 "Scan" tool,

If a Tech 1 or T-100 are not available the parking brake handle may be lifted to verify proper light operation and continuity of CKTs 3 and 39.

Any circuitry, that is suspected as causing the intermittent complaint, should be thoroughly checked for backed out terminals, improper mating, broken locks, improperly formed or damaged terminals, poor terminal to wiring connections or physical damage to the wiring harness.

DIAGNOSTIC CHART DTC A087 — 1993 SILHOUETTE, TRANS SPORT AND LUMINA APV

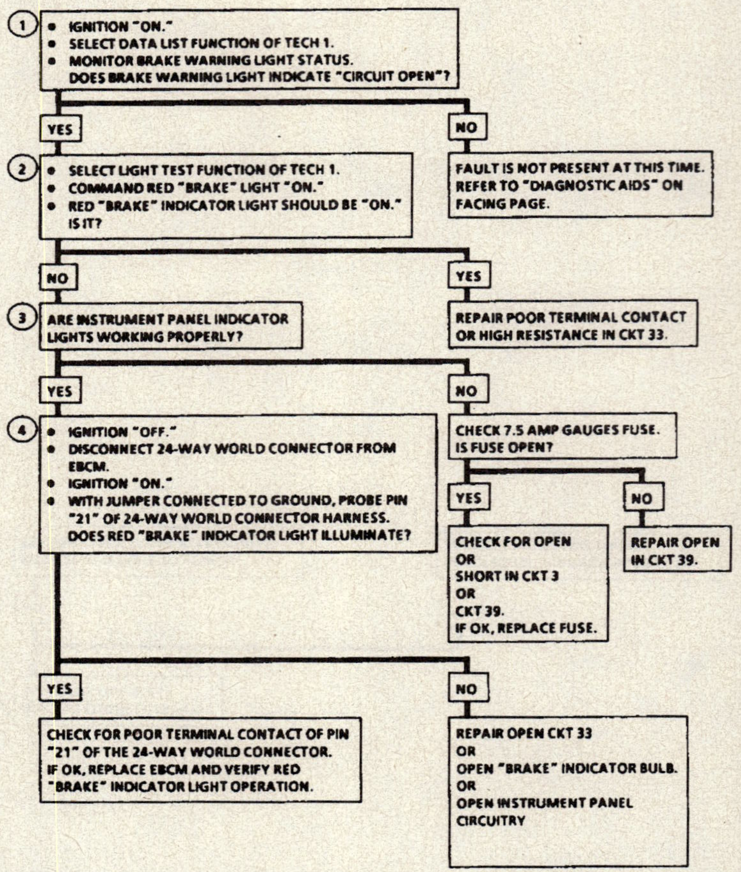

THIS CHART ASSUMES THAT A CURRENT CODE IS STORED INDICATING THAT THIS FAULT IS PRESENT.

DIAGNOSTIC CHART DTC A088 — 1993 SILHOUETTE, TRANS SPORT AND LUMINA APV

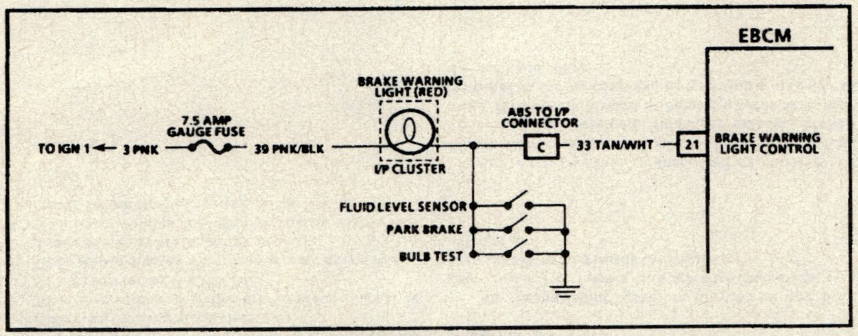

CODE A088

RED BRAKE WARNING LIGHT CIRCUIT SHORTED TO BATTERY

Circuit Description:

This code identifies a short to battery between the EBCM and the red brake warning light, or an open driver that does not allow the red brake warning light to be illuminated by the EBCM. This will only occur if an ABS fault is detected that may degrade base brake operation.

Test Description: Number(s) below refer to circled number(s) on the diagnostic chart.

1. This step identifies if the ground circuit to the red brake warning light is being completed by a source other than the EBCM.
2. This step identifies if the fault is currently present.
3. By removing the 10 amp indic fuse, the voltage source is eliminated. This test indicates if voltage is being supplied from a source other than the gauge fuse.

Diagnostic Aids:

An "Intermittent" problem may be caused by a poor connection, rubbed through wire insulation, or a wire that is broken inside the insulation.

Any circuitry, that is suspected as causing the intermittent complaint, should be thoroughly checked for backed out terminals, improper mating, broken locks, improperly formed or damaged terminals, poor terminal to wiring connections or physical damage to the wiring harness.

DIAGNOSTIC CHART DTC A088 — 1993 SILHOUETTE, TRANS SPORT AND LUMINA APV

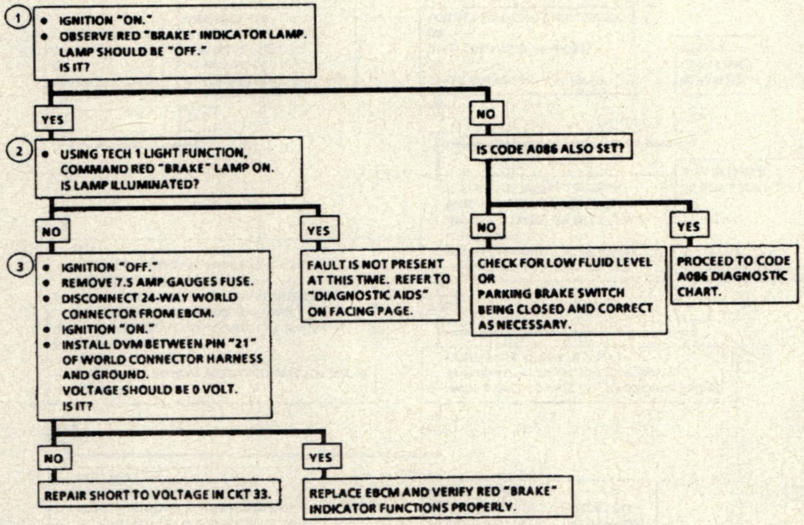

THIS CHART ASSUMES THAT A CURRENT CODE IS STORED INDICATING THAT THIS FAULT IS PRESENT.

DIAGNOSTIC CHART DTC A091 — 1993 SILHOUETTE, TRANS SPORT AND LUMINA APV

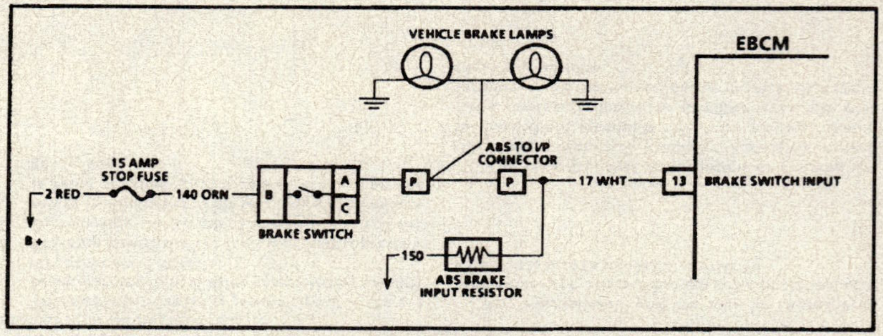

CODE A091

OPEN BRAKE SWITCH CONTACTS DURING DECELERATION

Circuit Description:

This code is used to detect an open brake switch in the non-ABS mode. The EBCM looks for decel rates that would indicate braking action and verifies this assumption by requiring several repeats of this detection method. In each case, ABS will not be available since no brake switch is seen.

Test Description: Number(s) below refer to circled number(s) on the diagnostic chart.

1. This step indicates if the brake switch signal is being received by the EBCM.
2. This step indicates if an open circuit exists in the brake switch or brake lamp circuitry.
3. This step isolates the open circuit to either the brake switch input circuitry, or the EBCM.

Diagnostic Aids:

An "Intermittent" problem may be caused by a poor connection, rubbed through wire insulation, or a wire that is broken inside the insulation.

Any circuitry, that is suspected as causing the intermittent complaint, should be thoroughly checked for backed out terminals, improper mating, broken locks, improperly formed or damaged terminals, poor terminal to wiring connections or physical damage to the wiring harness.

DIAGNOSTIC CHART DTC A091 — 1993 SILHOUETTE, TRANS SPORT AND LUMINA APV

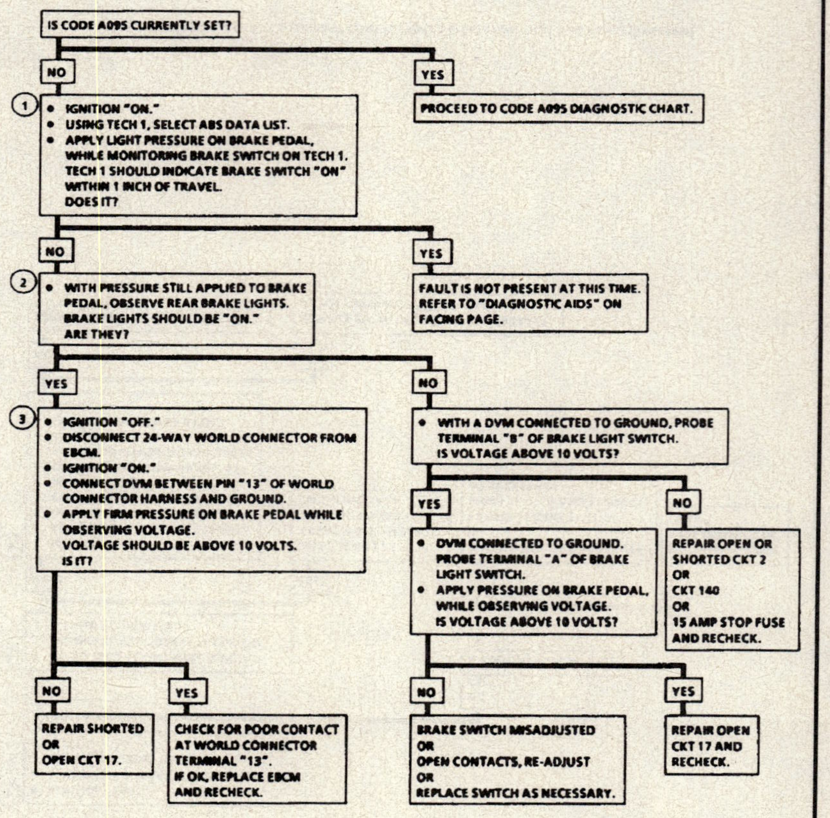

DIAGNOSTIC CHART DTC A092 — 1993 SILHOUETTE, TRANS SPORT AND LUMINA APV

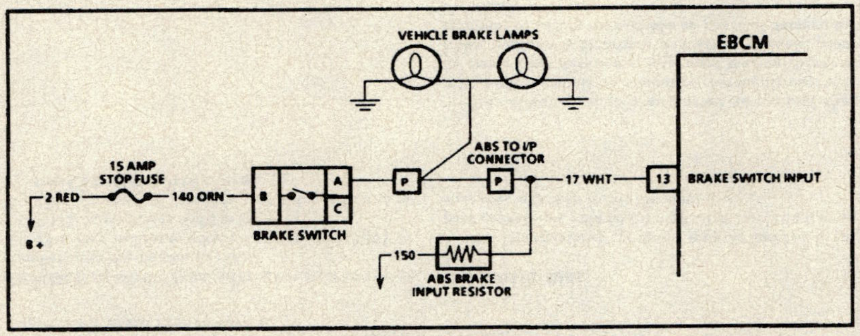

CODE A092

OPEN BRAKE SWITCH CONTACTS WHEN ABS WAS REQUIRED

Circuit Description:

This code is run to determine the proper operation of the brake switch. This is important because ABS is activated when the brake switch is "ON." If the brake switch is "OFF," ABS will never be activated. Since this fault is difficult to detect under normal braking conditions, this fault is only detected when ABS is required.

Test Description: Number(s) below refer to circled number(s) on the diagnostic chart.

1. This step indicates if the brake switch signal is being received by the EBCM.
2. This step indicates if an open circuit exists in the brake switch or brake lamp circuitry.
3. This step isolates the open circuit to either the brake switch input circuitry, or the EBCM.

Diagnostic Aids:

An "Intermittent" problem may be caused by a poor connection, rubbed through wire insulation, or a wire that is broken inside the insulation.

Any circuitry, that is suspected as causing the intermittent complaint, should be thoroughly checked for backed out terminals, improper mating, broken locks, improperly formed or damaged terminals, poor terminal to wiring connections or physical damage to the wiring harness.

DIAGNOSTIC CHART DTC A092 — 1993 SILHOUETTE, TRANS SPORT AND LUMINA APV

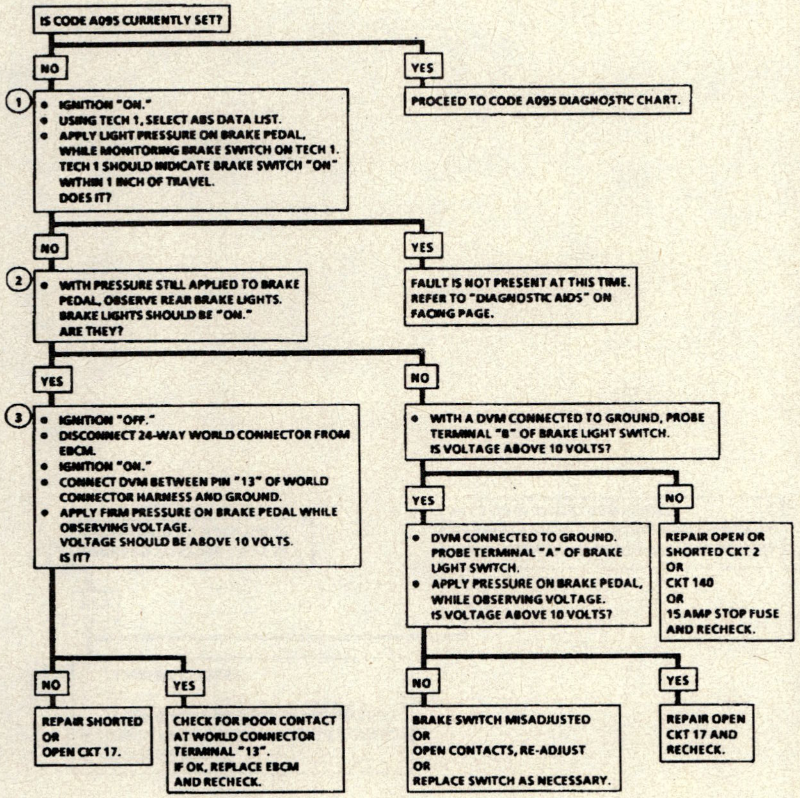

DIAGNOSTIC CHART DTC A093 — 1993 SILHOUETTE, TRANS SPORT AND LUMINA APV

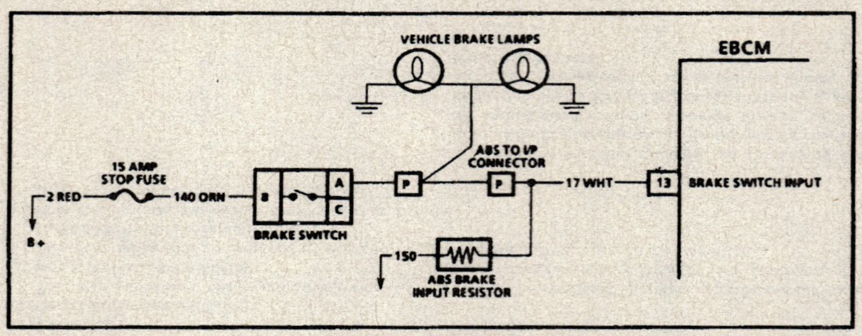

CODE A093

CODE A091 OR A092 SET IN CURRENT OR PREVIOUS IGNITION CYCLE

Circuit Description:

This code is the second portion of Codes A091 and A092. If Code A091 or A092 failed during the last ignition cycle, Code A093 becomes a current failure during the next ignition cycle, keeping ABS disabled until a brake switch "ON" state is seen. When a change is seen during an ignition cycle in which Code A093 is a current failure, Code A091 or A092 will clear itself at the end of the current ignition cycle, and ABS will enable itself at the start of the next ignition cycle. Code A093 alone indicates A091 or A092 failed previously, but is intermittent, or has been corrected.

Test Description: Number(s) below refer to circled number(s) on the diagnostic chart.

1. This step indicates which Code (either A091 or A092) caused Code A093 to set.
2. This insures that the code that set is repaired so that Code A093 can be cleared.

Diagnostic Aids:

An "Intermittent" problem may be caused by a poor connection, rubbed through wire insulation, or a wire that is broken inside the insulation.

Any circuitry, that is suspected as causing the intermittent complaint, should be thoroughly checked for backed out terminals, improper mating, broken locks, improperly formed or damaged terminals, poor terminal to wiring connections or physical damage to the wiring harness.

Also, verify proper brake switch operation using the data list of the "Scan" tool. As the brake is applied, the data list should display the brake switch "ON" within 1 inch of travel.

DIAGNOSTIC CHART DTC A093 — 1993 SILHOUETTE, TRANS SPORT AND LUMINA APV

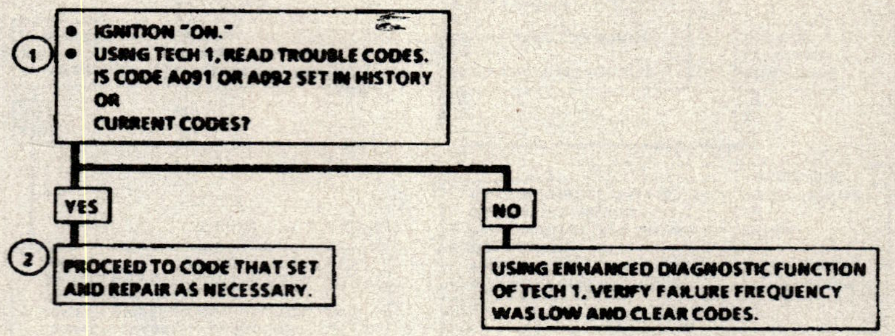

THIS CHART ASSUMES THAT A CURRENT CODE IS STORED INDICATING THAT THIS FAULT IS PRESENT.

DIAGNOSTIC CHART DTC A094 — 1993 SILHOUETTE, TRANS SPORT AND LUMINA APV

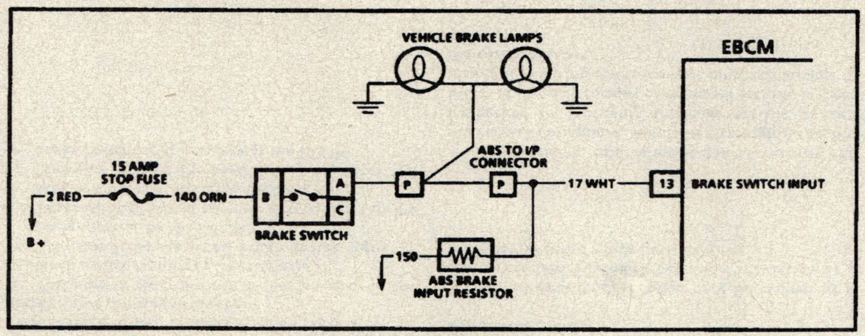

CODE A094

BRAKE SWITCH CONTACTS ALWAYS CLOSED

Circuit Description:

This code is run to determine the proper operation of the brake switch. This is important because ABS is activated when the brake switch is "ON" and turned "OFF" when the brake is "OFF." If the brake switch is always "ON," ABS operation will always be requested resulting in potential modulator cycling on rough roads. Additionally, this failure will most likely result in a dead battery (due to the brake lights staying "ON") if the driver is not informed of this fault.

Test Description: Number(s) below refer to circled number(s) on the diagnostic chart.

1. This checks if the fault is currently present.
2. This step isolates the cause of the fault to either a faulty or misadjusted brake switch, or a short to voltage in the brake switch circuitry.

Diagnostic Aids:

An "Intermittent" problem may be caused by a poor connection, rubbed through wire insulation, or a wire that is broken inside the insulation.

Any circuitry, that is suspected as causing the intermittent complaint, should be thoroughly checked for backed out terminals, improper mating, broken locks, improperly formed or damaged terminals, poor terminal to wiring connections or physical damage to the wiring harness.

DIAGNOSTIC CHART DTC A094 — 1993 SILHOUETTE, TRANS SPORT AND LUMINA APV

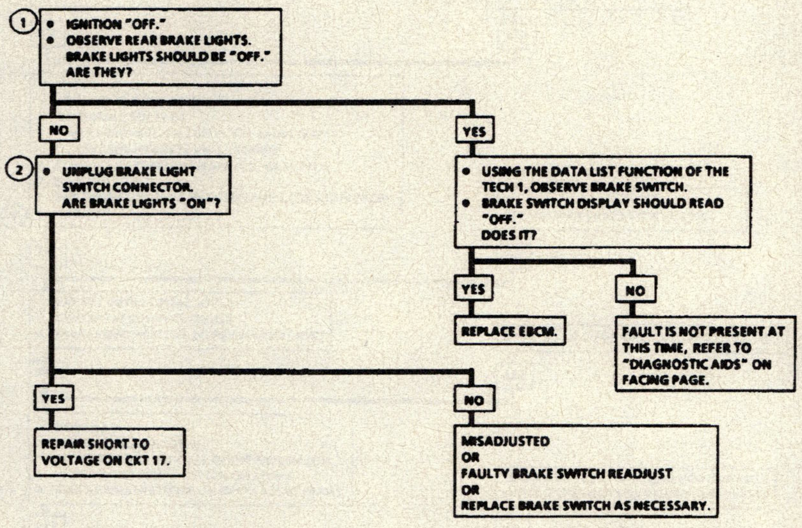

THIS CHART ASSUMES THAT A CURRENT CODE IS STORED INDICATING THAT THIS FAULT IS PRESENT.

DIAGNOSTIC CHART DTC A095 — 1993 SILHOUETTE, TRANS SPORT AND LUMINA APV

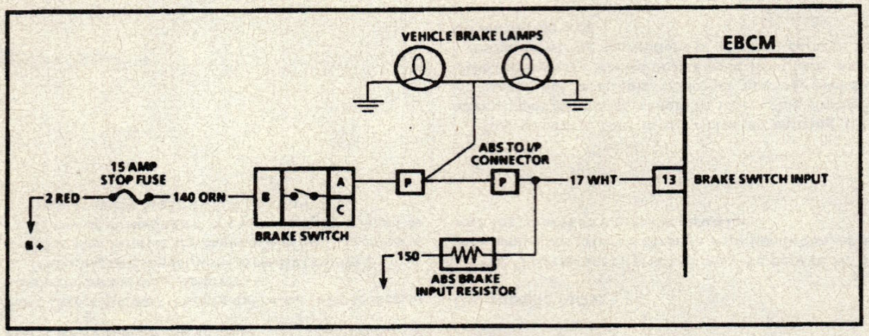

CODE A095

BRAKE SWITCH CIRCUIT OPEN

Circuit Description:

This code is used to identify open brake switch circuitry that prevents the brake switch input to the EBCM from changing states when the brake is applied. This code is used in conjunction with Codes A091 and A092 to determine the cause of an open brake switch fault.

Test Description: Number(s) below refer to circled number(s) on the diagnostic chart.

1. This step is used to confirm that an open in the brake switch circuitry currently exists.
2. This step indicates if the brake switch signal is being received by the EBCM.
3. This step indicates if an open circuit exists in the brake switch or brake lamp circuitry.
4. This step isolates the open circuit to either the brake switch input circuitry,or the EBCM.

Diagnostic Aids:

An "Intermittent" problem may be caused by a poor connection, rubbed through wire insulation, or a wire that is broken inside the insulation.

Any circuitry, that is suspected as causing the intermittent complaint, should be thoroughly checked for backed out terminals, improper mating, broken locks, improperly formed or damaged terminals, poor terminal to wiring connections or physical damage to the wiring harness.

DIAGNOSTIC CHART DTC A095 — 1993 SILHOUETTE, TRANS SPORT AND LUMINA APV

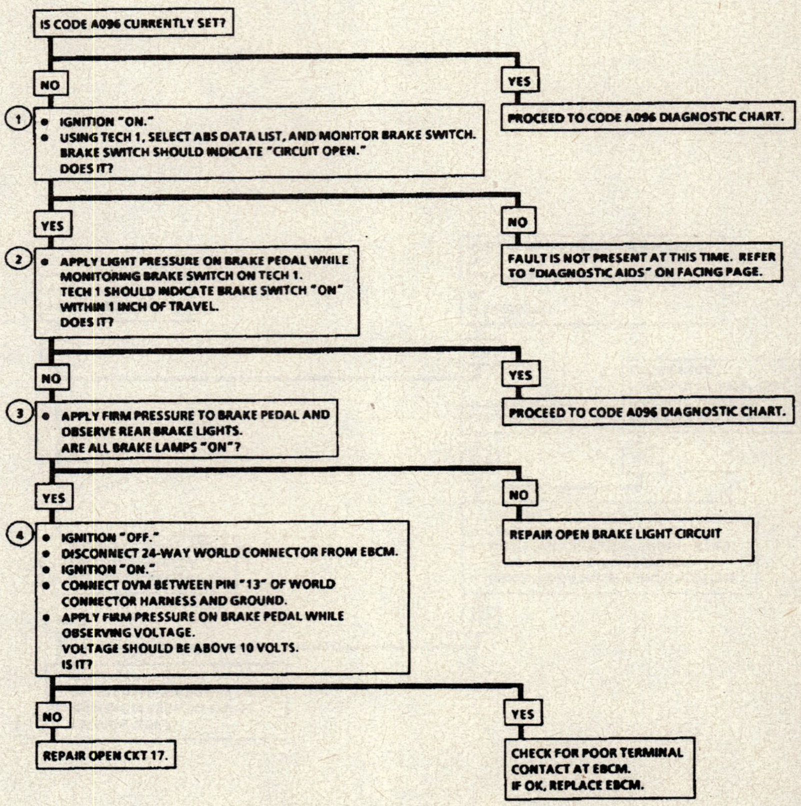

THIS CHART ASSUMES THAT A CURRENT CODE IS STORED INDICATING THAT THIS FAULT IS PRESENT.

DIAGNOSTIC CHART DTC A096 — 1993 SILHOUETTE, TRANS SPORT AND LUMINA APV

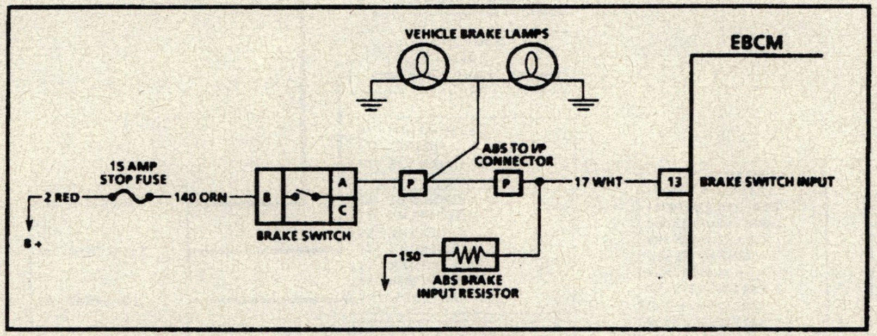

CODE A096

BRAKE LIGHTS CIRCUIT OPEN

Circuit Description:

This code is designed to isolate the cause of a Code A095 failure and indicate to the driver ABS is still available. If Code A095 fails with Code A096, the brake lights are open, the brake light grounds are open, or the ABS brake input resistor is open during 4 way flasher use. The presence of battery voltage at the brake switch input indicates a valid brake switch input is still available.

Test Description: Number(s) below refer to circled number(s) on the diagnostic chart.

1. As a result of a failure of an additional brake switch circuit code, this code may be set. To insure proper diagnosis, any additional brake switch codes must be repaired first.
2. This step identifies if the fault is currently present in the brake circuit.

Diagnostic Aids:

An "Intermittent" problem may be caused by a poor connection, rubbed through wire insulation, or a wire that is broken inside the insulation.

Any circuitry, that is suspected as causing the intermittent complaint, should be thoroughly checked for backed out terminals, improper mating, broken locks, improperly formed or damaged terminals, poor terminal to wiring connections or physical damage to the wiring harness.

DIAGNOSTIC CHART DTC A096 — 1993 SILHOUETTE, TRANS SPORT AND LUMINA APV

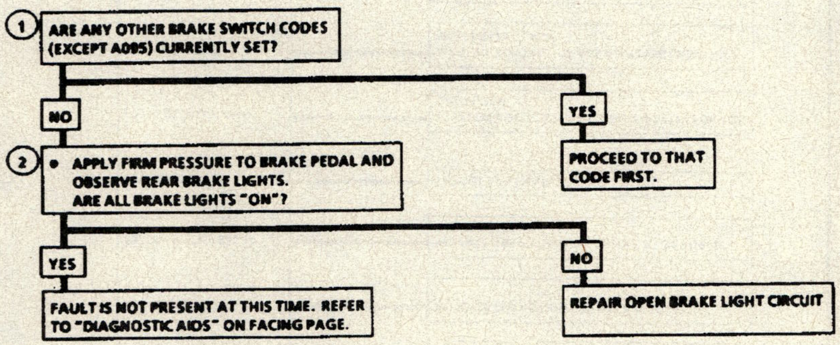

THIS CHART ASSUMES THAT A CURRENT CODE IS STORED INDICATING THAT THIS FAULT IS PRESENT.

DIAGNOSTIC PROCESS CHART — 1994 SILHOUETTE, TRANS SPORT AND LUMINA APV

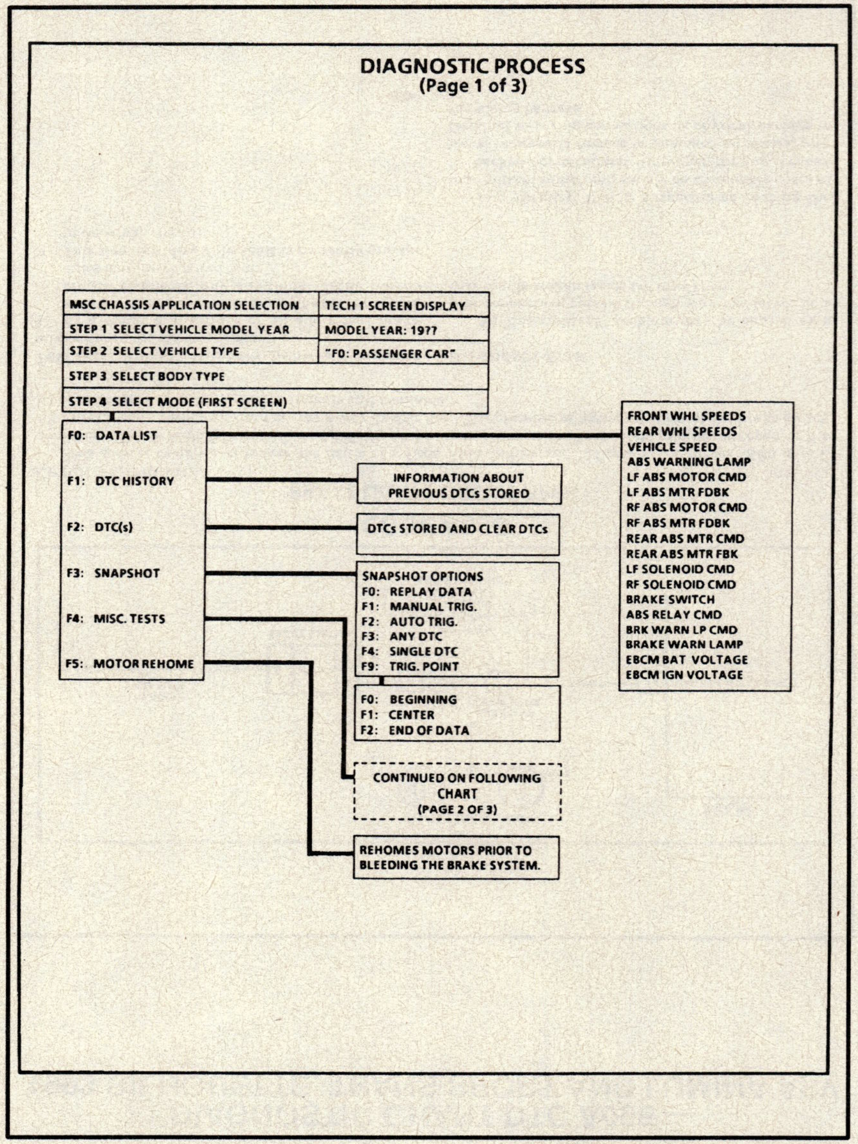

DIAGNOSTIC PROCESS CHART — 1994 SILHOUETTE, TRANS SPORT AND LUMINA APV

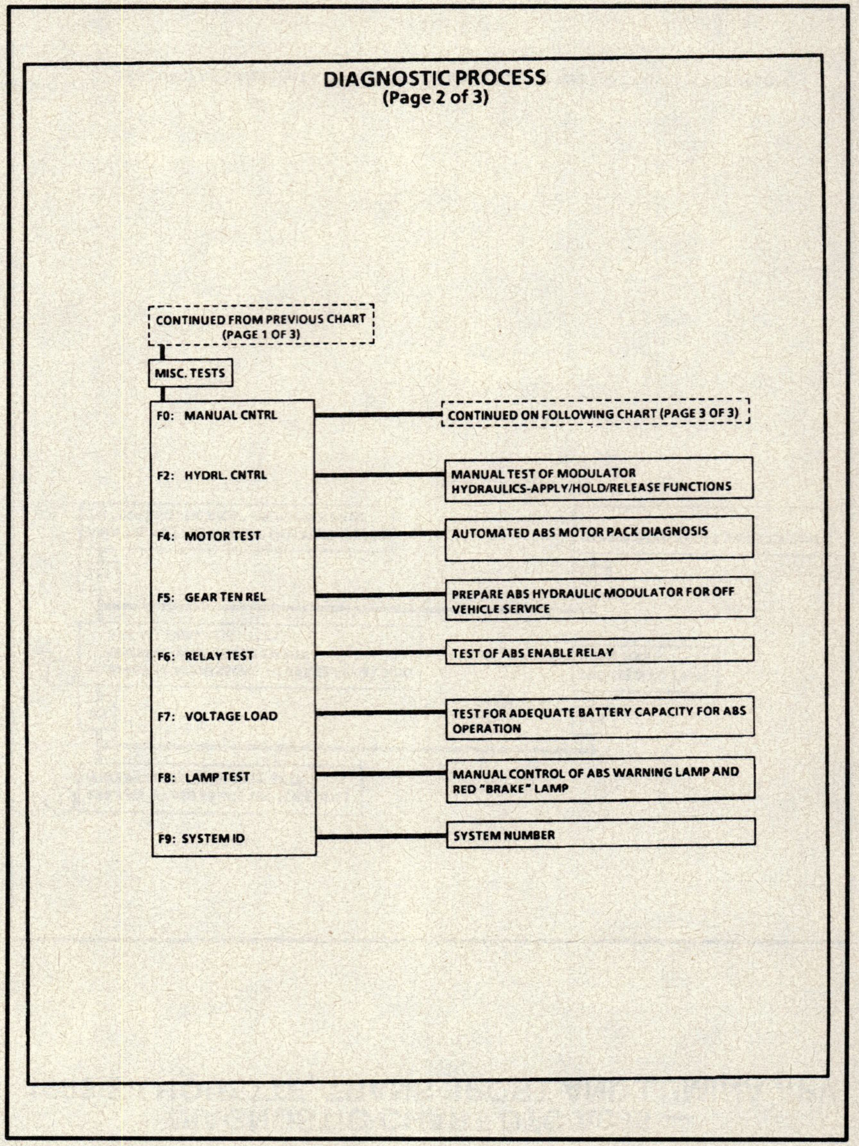

DIAGNOSTIC PROCESS CHART — 1994 SILHOUETTE, TRANS SPORT AND LUMINA APV

DIAGNOSTIC PROCESS
(Page 3 of 3)

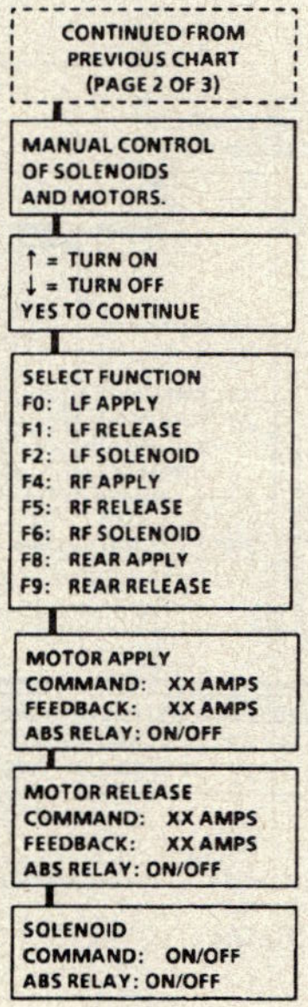

DIAGNOSTIC CHART SYSTEM CHECK — 1994 SILHOUETTE, TRANS SPORT AND LUMINA APV

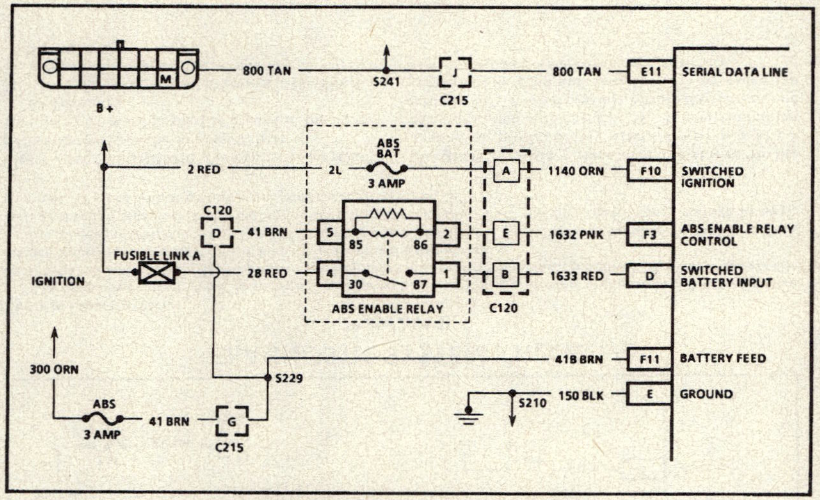

DIAGNOSTIC SYSTEM CHECK
(Page 1 of 4)

System Description:

The diagnostic system check is an organized approach to identifying a problem created by an Antilock Brake System (ABS) malfunction. It must be the starting point for any ABS complaint diagnosis because it directs the service technician to the next logical step in diagnosing the complaint.

Serial data is transmitted by the EBCM through terminal "E11". The EBCM is supplied switched ignition voltage through terminal "F11", battery voltage through terminal "F10" and switched battery voltage through terminal "D". The EBCM ground is provided through terminal "E".

Chart Test Description: Number(s) below refer to circled number(s) on the diagnostic chart.

1. Indicates if EBCM is transmitting data.
2. Checks for a serial data line malfunction.
3. Checks for proper Tech 1 connection to DLC.
4. Checks for open ABS BAT 3 amp fuse.
5. Checks for open ABS 3 amp fuse.
6. Checks for high resistance in the EBCM ground circuit.
7. Checks for current DTCs.
8. Checks for proper ABS warning lamp operation.
9. Checks for history DTCs.

Diagnostic Aids: Excessive resistance in the ground or power supply circuits will not allow communication with the EBCM. If communication with the EBCM is not possible, ensure the ABS ground connection is good and that there is no excessive resistance in any of the power supply circuits.

DIAGNOSTIC CHART SYSTEM CHECK — 1994 SILHOUETTE, TRANS SPORT AND LUMINA APV

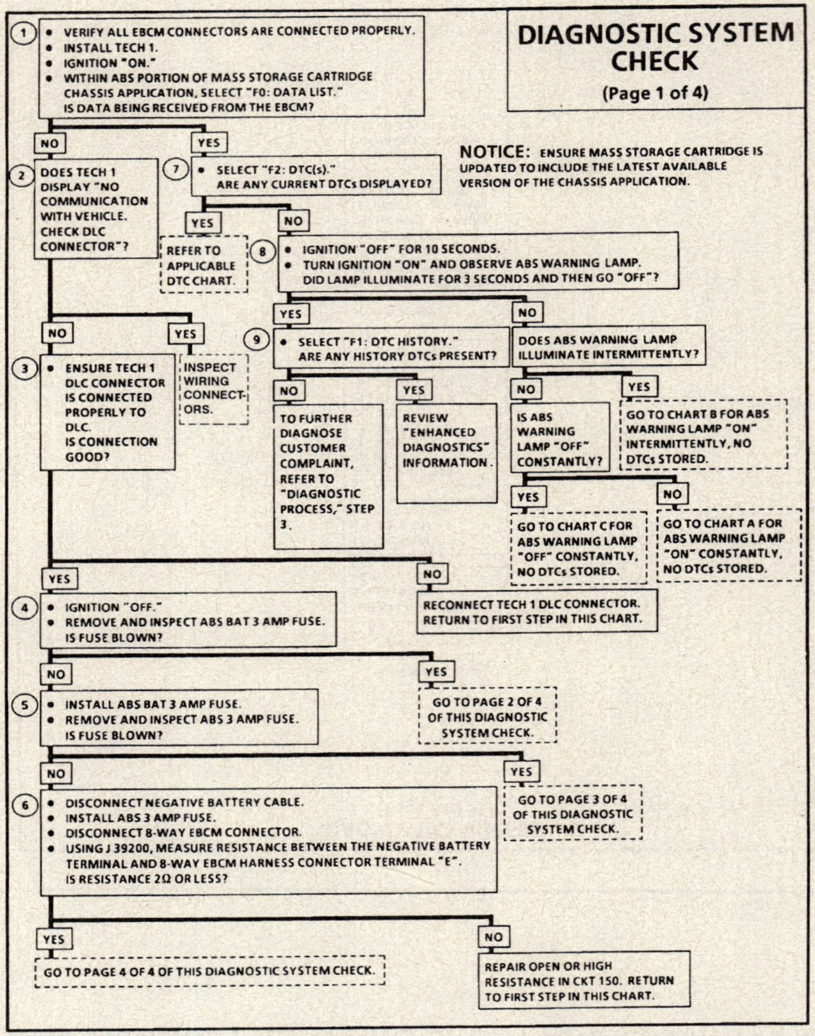

DIAGNOSTIC CHART SYSTEM CHECK — 1994 SILHOUETTE, TRANS SPORT AND LUMINA APV

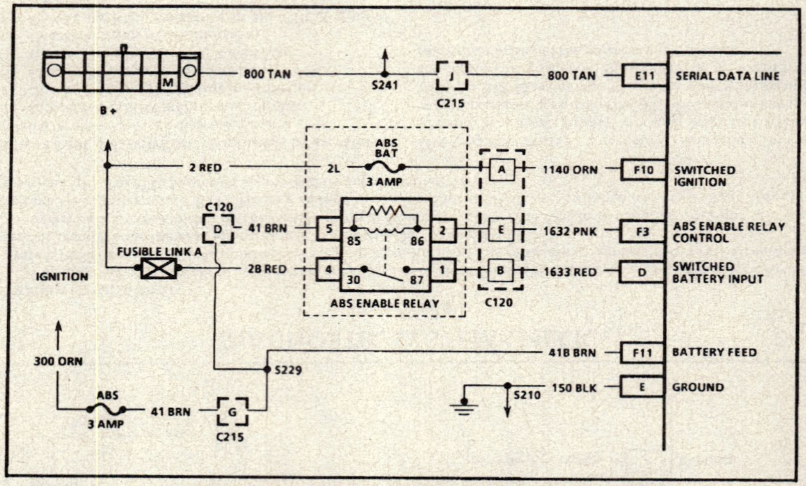

DIAGNOSTIC SYSTEM CHECK

(Page 2 of 4)

System Description:

The diagnostic system check is an organized approach to identifying a problem created by an Antilock Brake System (ABS) malfunction. It must be the starting point for any ABS complaint diagnosis because it directs the service technician to the next logical step in diagnosing the complaint.

Serial data is transmitted by the EBCM through terminal "E11". The EBCM is supplied switched ignition voltage through terminal "F11", battery voltage through terminal "F10" and switched battery voltage through terminal "D". The EBCM ground is provided through terminal "E".

Chart Test Description: Number(s) below refer to circled number(s) on the diagnostic chart.

10. Checks for possible short to ground in the battery feed circuitry.
11. Checks for possible short to ground in the EBCM.
12. Ensures short to ground is not due to physical damage of the circuitry.

Diagnostic Aids: Excessive resistance in the ground or power supply circuits will not allow communication with the EBCM. If communication with the EBCM is not possible, ensure the ABS ground connection is good and that there is no excessive resistance in any of the power supply circuits.

DIAGNOSTIC CHART SYSTEM CHECK — 1994 SILHOUETTE, TRANS SPORT AND LUMINA APV

DIAGNOSTIC SYSTEM CHECK

(Page 2 of 4)

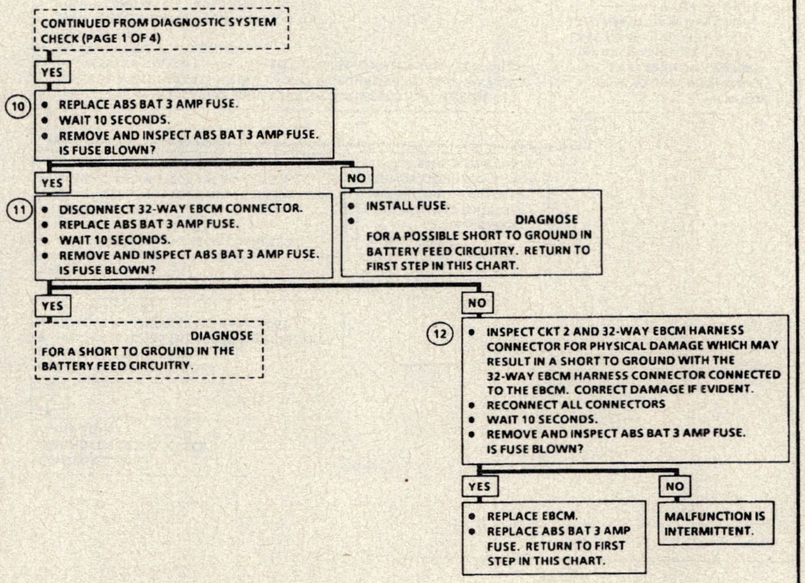

DIAGNOSTIC CHART SYSTEM CHECK — 1994 SILHOUETTE, TRANS SPORT AND LUMINA APV

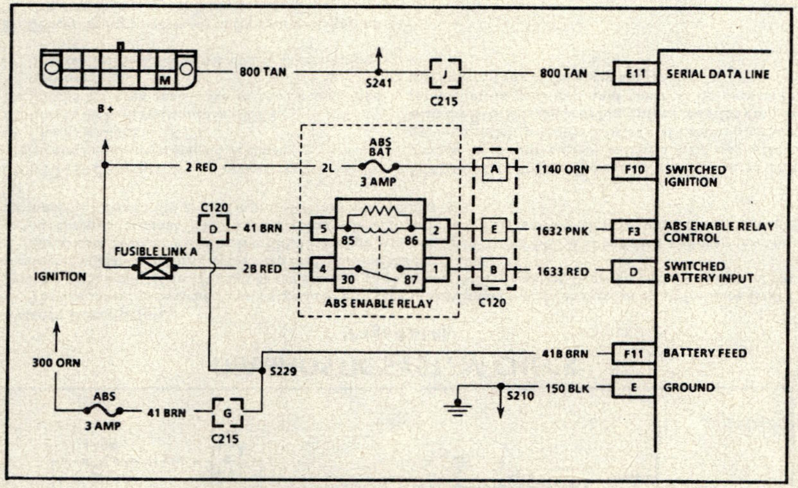

DIAGNOSTIC SYSTEM CHECK

(Page 3 of 4)

System Description:

The diagnostic system check is an organized approach to identifying a problem created by an Antilock Brake System (ABS) malfunction. It must be the starting point for any ABS complaint diagnosis because it directs the service technician to the next logical step in diagnosing the complaint.

Serial data is transmitted by the EBCM through terminal "E11". The EBCM is supplied switched ignition voltage through terminal "F11", battery voltage through terminal "F10" and switched battery voltage through terminal "D". The EBCM ground is provided through terminal "E".

Chart Test Description: Number(s) below refer to circled number(s) on the diagnostic chart.

13. Checks for possible short to ground in the switched ignition circuitry.
14. Isolates the short to ground to either the circuitry or the EBCM.
15. Ensures short to ground is not due to physical damage of the circuitry.

Diagnostic Aids: Excessive resistance in the ground or power supply circuits will not allow communication with the EBCM. If communication with the EBCM is not possible, ensure the ABS ground connection is good and that there is no excessive resistance in any of the power supply circuits.

DIAGNOSTIC CHART SYSTEM CHECK — 1994 SILHOUETTE, TRANS SPORT AND LUMINA APV

DIAGNOSTIC SYSTEM CHECK

(Page 3 of 4)

CONTINUED FROM DIAGNOSTIC SYSTEM CHECK (PAGE 1 OF 4)

YES

(13)
- REPLACE ABS 3 AMP FUSE.
- IGNITION "ON."
- WAIT 10 SECONDS THEN TURN IGNITION "OFF."
- REMOVE AND INSPECT ABS 3 AMP FUSE. IS FUSE BLOWN?

YES / NO

NO: INSTALL FUSE. DIAGNOSE FOR A POSSIBLE SHORT TO GROUND IN THE SWITCHED IGNITION CIRCUITRY. RETURN TO FIRST STEP IN THIS CHART.

YES:

(14)
- DISCONNECT 32-WAY EBCM CONNECTOR.
- REPLACE ABS 3 AMP FUSE.
- IGNITION "ON."
- WAIT 10 SECONDS THEN TURN IGNITION "OFF."
- REMOVE AND INSPECT ABS 3 AMP FUSE. IS FUSE BLOWN?

NO / YES

YES: REPAIR A SHORT TO GROUND IN ONE OF THE CKT 41 BRANCHES EXITING THE ABS 3 AMP FUSE. REPLACE ABS 3 AMP FUSE. RETURN TO THE FIRST STEP IN THIS CHART.

NO:

(15)
- INSPECT CKT 41 AND 32-WAY EBCM HARNESS CONNECTOR FOR PHYSICAL DAMAGE WHICH MAY RESULT IN A SHORT TO GROUND WITH THE 32-WAY EBCM HARNESS CONNECTOR CONNECTED TO THE EBCM. CORRECT DAMAGE IF EVIDENT.
- RECONNECT ALL CONNECTORS.
- IGNITION "ON."
- WAIT 10 SECONDS THEN TURN IGNITION "OFF."
- REMOVE AND INSPECT ABS 3 AMP FUSE. IS FUSE BLOWN?

YES: REPLACE EBCM. REPLACE ABS 3 AMP FUSE. RETURN TO FIRST STEP IN THIS CHART.

NO: MALFUNCTION IS INTERMITTENT.

DIAGNOSTIC CHART SYSTEM CHECK — 1994 SILHOUETTE, TRANS SPORT AND LUMINA APV

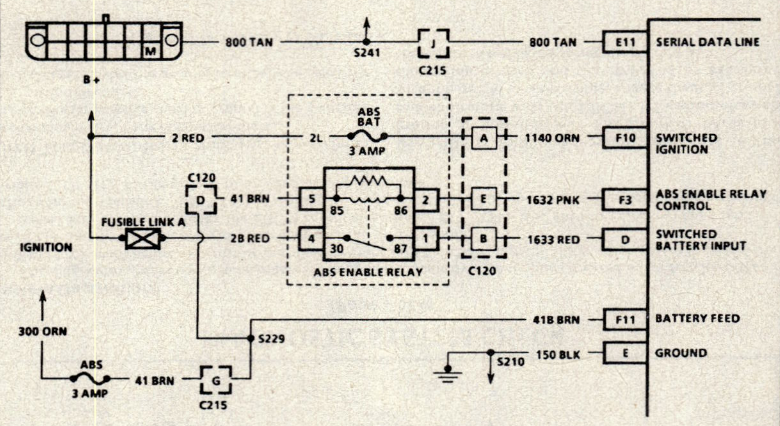

DIAGNOSTIC SYSTEM CHECK

(Page 4 of 4)

System Description:

The diagnostic system check is an organized approach to identifying a problem created by an Antilock Brake System (ABS) malfunction. It must be the starting point for any ABS complaint diagnosis because it directs the service technician to the next logical step in diagnosing the complaint.

Serial data is transmitted by the EBCM through terminal "E11". The EBCM is supplied switched ignition voltage through terminal "F11", battery voltage through terminal "F10" and switched battery voltage through terminal "D". The EBCM ground is provided through terminal "E".

Chart Test Description: Number(s) below refer to circled number(s) on the diagnostic chart.

16. Checks for open in CKT 2.
17. Checks for high resistance in CKT 2.
18. Checks for high resistance in the battery feed circuitry.
19. Ensures malfunction is not due to poor terminal contact.
20. Checks for high resistance in the switched ignition circuit between the fuse block and the EBCM.

Diagnostic Aids: Excessive resistance in the ground or power supply circuits will not allow communication with the EBCM. If communication with the EBCM is not possible, ensure the ABS ground connection is good and that there is no excessive resistance in any of the power supply circuits.

DIAGNOSTIC CHART SYSTEM CHECK — 1994 SILHOUETTE, TRANS SPORT AND LUMINA APV

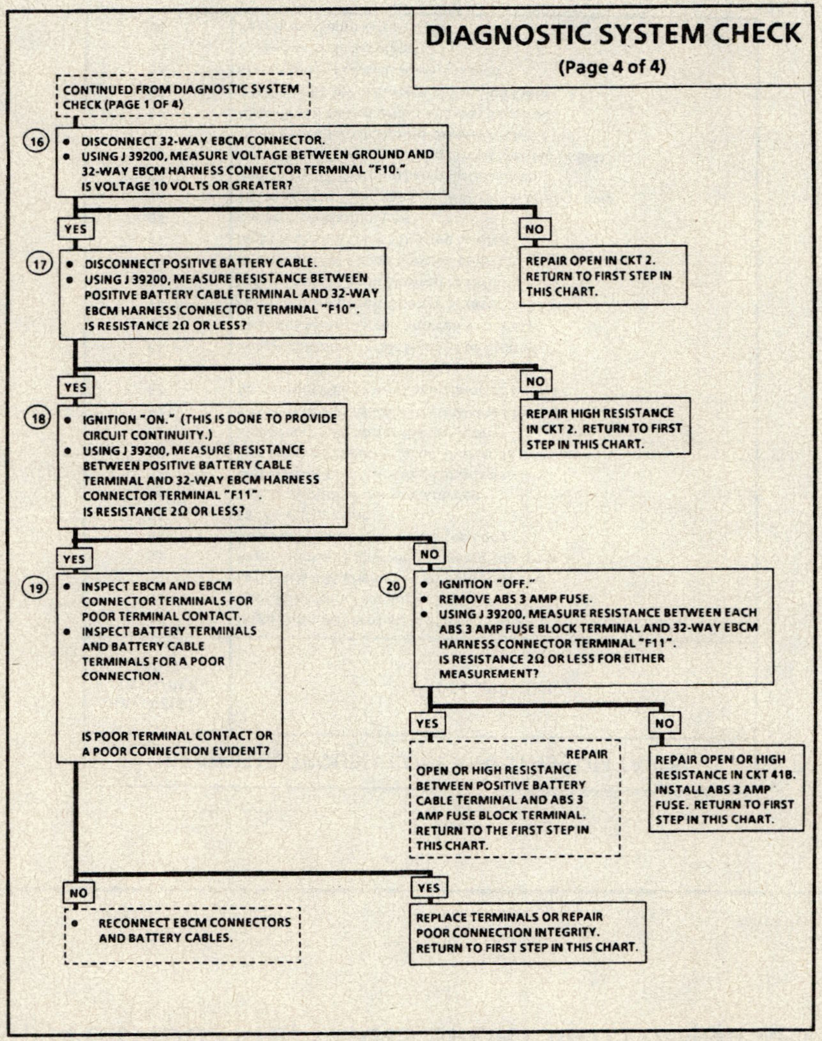

DIAGNOSTIC TROUBLE CODE CHART — 1994 SILHOUETTE, TRANS SPORT AND LUMINA APV

DIAGNOSTIC TROUBLE CODE AND SYMPTOM TABLE

CHART	SYMPTOM
A	ABS (Amber) Warning Lamp "ON" Constantly, No DTCs Stored (Chev Only)
A	ABS (Amber) Warning Lamp "ON" or "OFF" Constantly, No DTCs Stored (Pont/Olds Only)
B	ABS (Amber) Warning Lamp "ON" Intermittently, No DTCs Stored
C	ABS (Amber) Warning Lamp "OFF" Constantly, No DTCs Stored (Chev Only)
D	Tech 1 Displays Undefined DTCs

DIAGNOSTIC TROUBLE CODE	DESCRIPTION
11	ABS Warning Lamp Circuit Malfunction (Chev Only)
11	ABS Warning Lamp Circuit Malfunction (Pont/Olds Only)
14	ABS Enable Relay Contact Circuit Open (1 of 3)
15	ABS Enable RLY Contact CKT Short to Batt or Always Closed
16	ABS Enable Relay Coil Circuit Open
17	ABS Enable Relay Coil Circuit Shorted to GND
18	ABS Enable Relay Coil Circuit Shorted to Batt
21	Left Front Wheel Speed = 0
22	Right Front Wheel Speed = 0
23	Left Rear Wheel Speed = 0
24	Right Rear Wheel Speed = 0
25	Left Front Excessive Wheel Speed Variation
26	Right Front Excessive Wheel Speed Variation
27	Left Rear Excessive Wheel Speed Variation
28	Right Rear Excessive Wheel Speed Variation
32	LF Wheel Speed SNR CKT Open or Short to GND/BAT (1 of 3)
33	RF Wheel Speed SNR CKT Open or Short to GND/BAT (1 of 3)
34	LR Wheel Speed SNR CKT Open or Short to GND/BAT (1 of 3)
35	RR Wheel Speed SNR CKT Open or Short to GND/BAT (1 of 3)
36	Low System Voltage (1 of 2)
37	High System Voltage
38	Left Front ESB Will Not Hold Motor
41	Right Front ESB Will Not Hold Motor
42	Rear ESB Will Not Hold Motor
44	Left Front ABS Channel Will Not Move
45	Right Front ABS Channel Will Not Move
46	Rear ABS Channel Will Not Move
47	Left Front ABS Motor Free Spins (1 of 2)

DIAGNOSTIC TROUBLE CODE CHART — 1994 SILHOUETTE, TRANS SPORT AND LUMINA APV

DIAGNOSTIC TROUBLE CODE AND SYMPTOM TABLE

DIAGNOSTIC TROUBLE CODE	DESCRIPTION
48	Right Front ABS Motor Free Spins (1 of 2)
51	Rear ABS Motor Free Spins (1 of 2)
52	Left Front ABS Channel in Release Too Long
53	Right Front ABS Channel in Release Too Long
54	Rear ABS Channel in Release Too Long
55	EBCM Malfunction
56	Left Front ABS Motor Circuit Open
57	Left Front ABS Motor Circuit Shorted to GND
58	Left Front ABS Motor Circuit Shorted to BATT
61	Right Front ABS Motor Circuit Open
62	Right Front ABS Motor Circuit Shorted to GND
63	Right Front ABS Motor Circuit Shorted to BATT
64	Rear ABS Motor Circuit Open
65	Rear ABS Motor Circuit Shorted to Ground
66	Rear ABS Motor Circuit Shorted to Battery
76	Left Front Sol Circuit Open or Shorted to BATT
77	Left Front Sol Circuit Shorted to GND
78	Right Front Sol Circuit Shorted to BATT
81	Right Front Sol Circuit Shorted to GND
82	Calibration Malfunction
86	EBCM Turned "ON" the Red "BRAKE" Warning Lamp
87	Red "BRAKE" Warning Lamp Circuit Open
88	Red "BRAKE" Warning Lamp Circuit Shorted to Batt
91	Open Brake Switch During Deceleration
92	Open Brake Switch When ABS was Required
93	DTC 91 or 92 Set in Current or Prev Ign Cycle
94	Brake Switch Contacts Always Closed
95	Brake Switch Circuit Open
96	Rear Brake Lamp Circuit Open

DIAGNOSTIC CHART A — 1994 LUMINA APV

CHART A

ABS (AMBER) WARNING LAMP "ON" CONSTANTLY, NO DTCs STORED (CHEV ONLY)

NOTICE: DIAGNOSTIC SYSTEM CHECK MUST BE COMPLETED FIRST BEFORE USING THIS CHART.

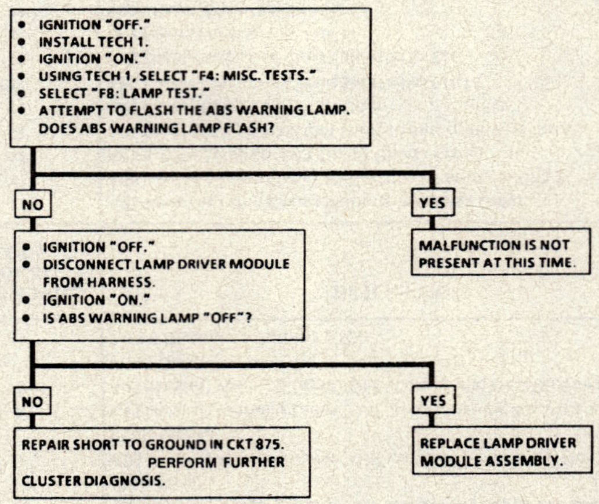

DIAGNOSTIC CHART A — 1994 SILHOUETTE AND TRANS SPORT

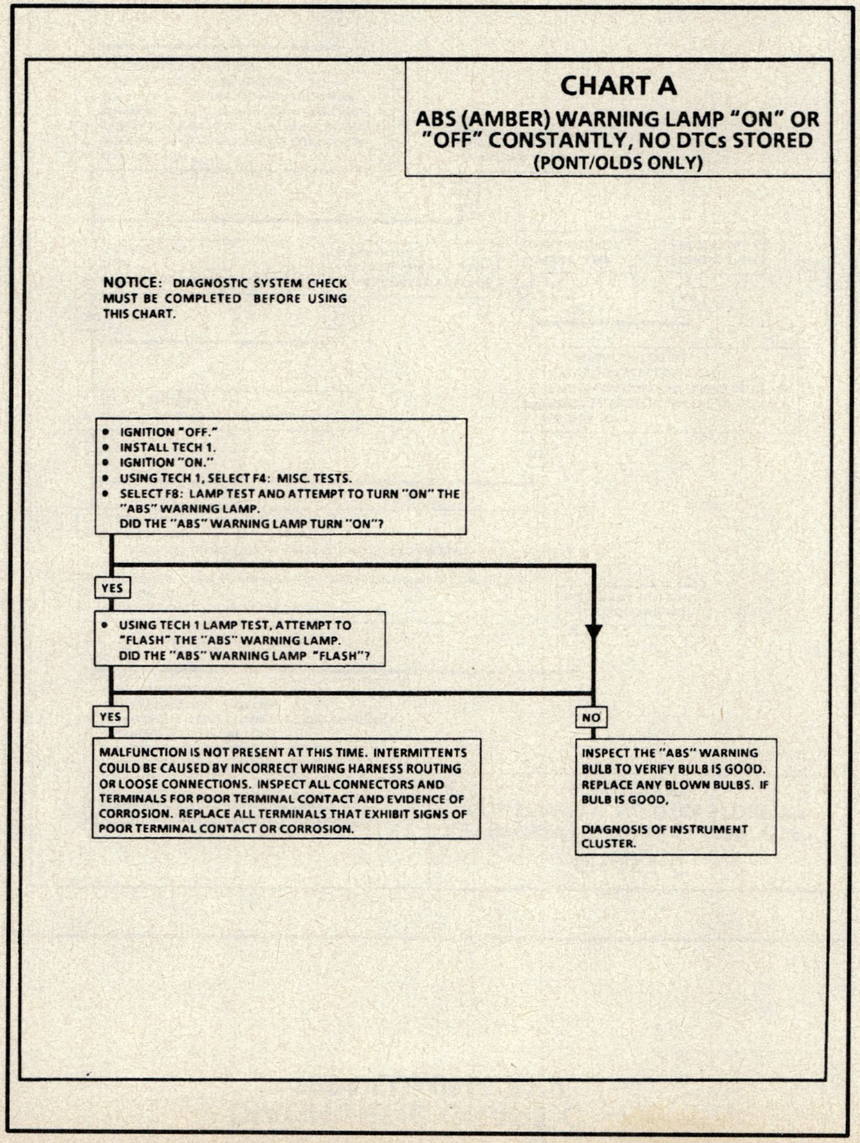

DIAGNOSTIC CHART B — 1994 SILHOUETTE, TRANS SPORT AND LUMINA APV

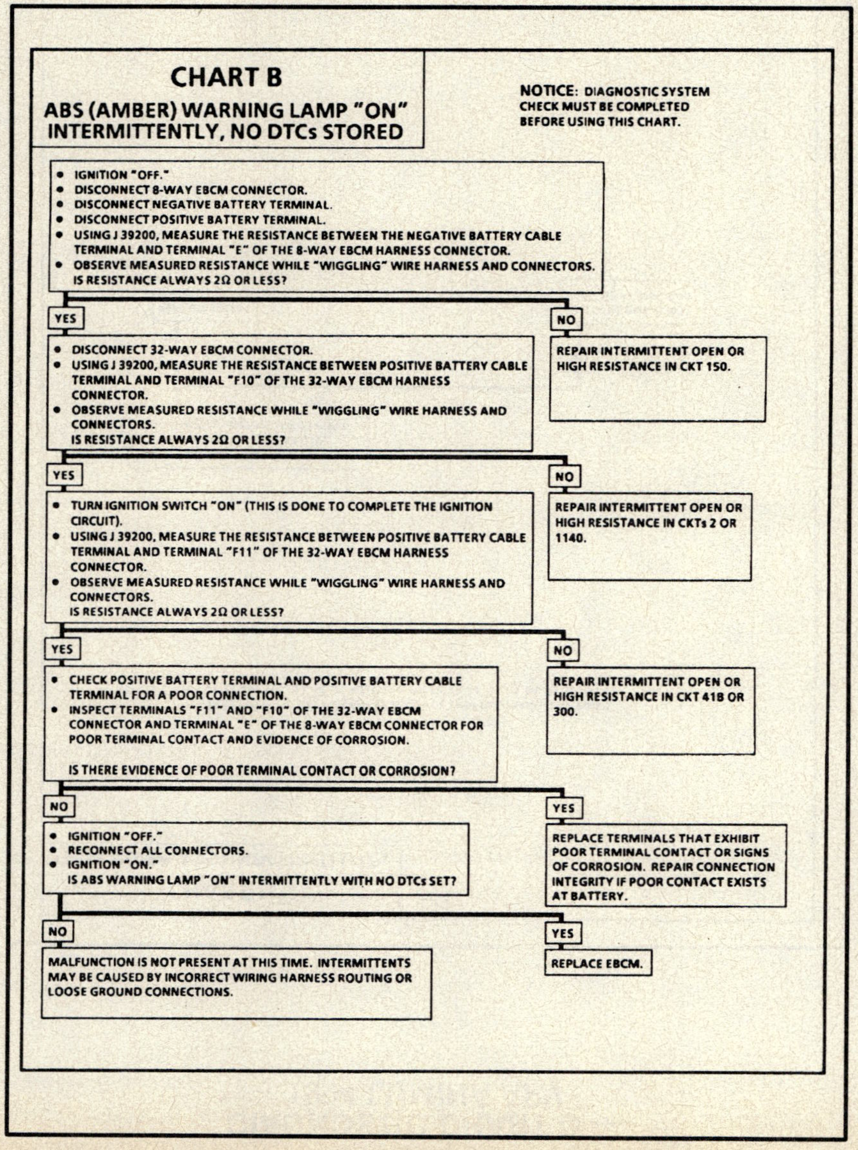

DIAGNOSTIC CHART C — 1994 LUMINA APV

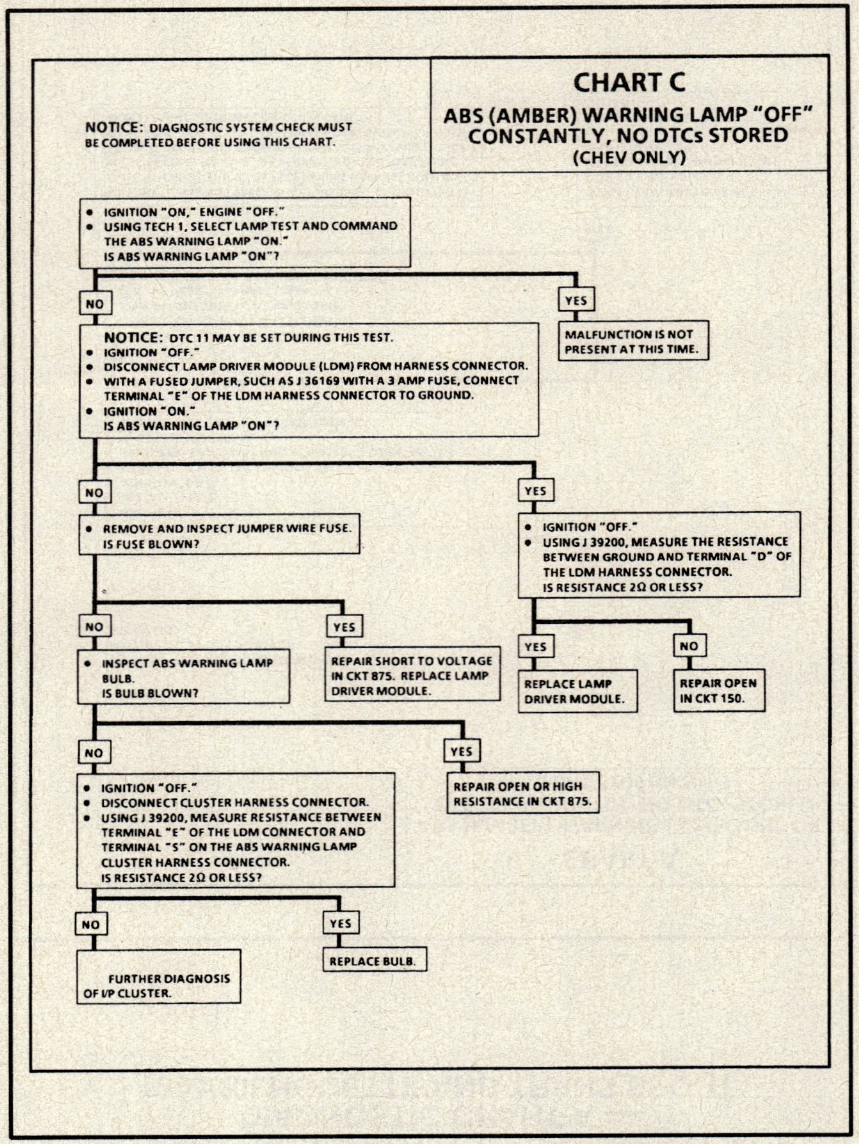

DIAGNOSTIC CHART D — 1994 LUMINA APV

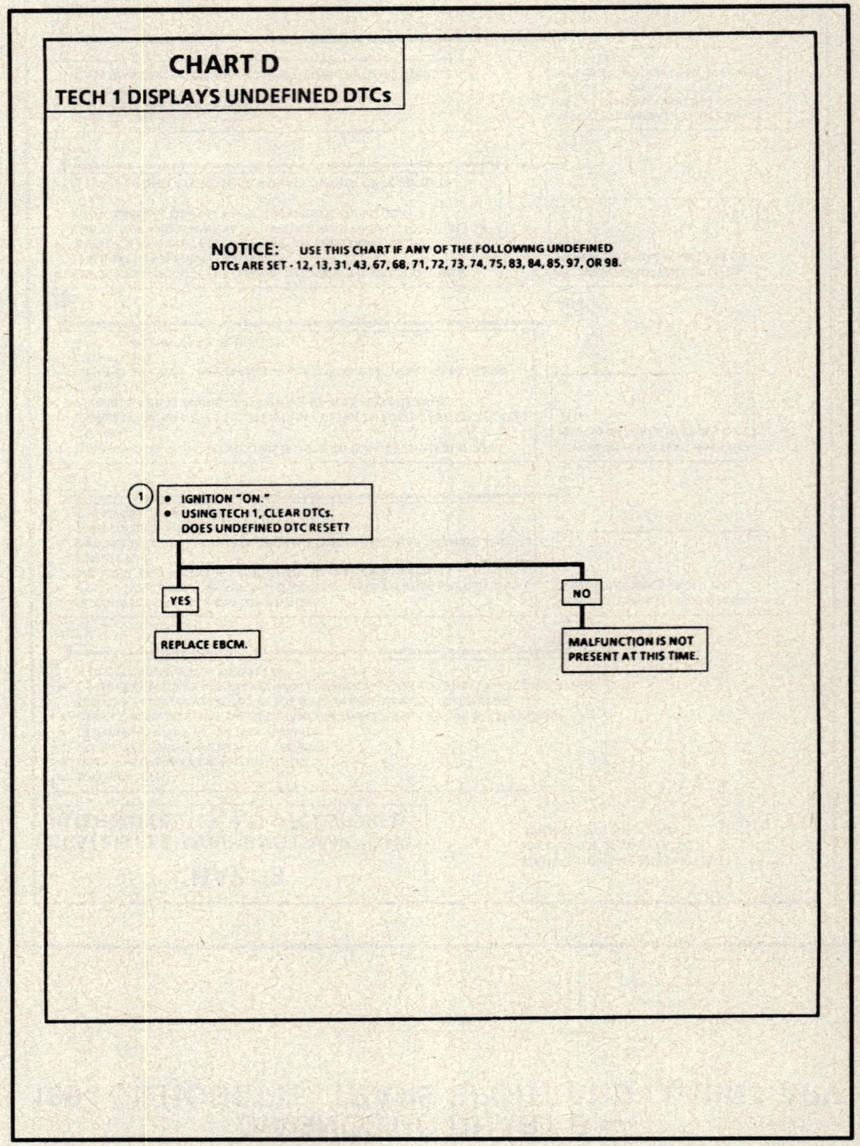

DIAGNOSTIC CHART DTC 11 — 1994 LUMINA APV

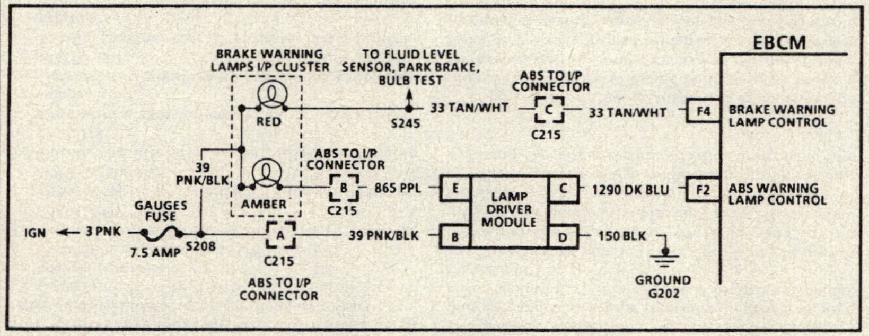

DTC 11

ABS WARNING LAMP CIRCUIT MALFUNCTION
(CHEV ONLY)

Circuit Description:

This DTC checks the state of the ABS warning lamp to identify a situation in which the driver could not be warned of a system malfunction by the ABS warning lamp, or the warning lamp is always "ON." The Lamp Driver Module (LDM) turns the ABS warning lamp "ON," unless the EBCM provides ground to turn it "OFF." Because of the circuitry in the LDM, only external malfunctions can be detected. As a result, the ABS warning lamp itself is not diagnosable, only the control line to the EBCM can be diagnosed. In the event of an open CKT 1290, the ABS warning lamp will be "ON" at all times, due to the loss of ground at the LDM input. If the control line is shorted to ground, the ABS warning lamp is kept "OFF," due to the LDM input being grounded.

DTC Will Set When: DTC 11 can be set only during the three second bulb check or when the ABS warning lamp is commanded "ON." If the EBCM cannot control the ABS warning lamp for two seconds, a malfunction exists.

Action Taken: A malfunction DTC is stored; however, ABS is not disabled.

DTC Chart Test Description: Number(s) below refer to circled number(s) on the diagnostic chart.

1. Checks to see if ABS lamp functions properly. Normal operation would indicate that the problem is not present.
2. Determines if any other DTCs are present or if DTC 18 is set falsely.
3. Ensures that the circuitry from the EBCM to the lamp driver module is complete.
4. Ensures that the ignition circuit to the lamp driver module is functional. An open indicator fuse or open CKT 3, 39, or 865 would result in inoperative warning lamp.
5. Isolates the cause of the ABS lamp remaining "OFF," by disconnecting the 32-way EBCM connector, CKT 1290 should be open and cause the ABS lamp to be "ON."
6. Determines whether the lamp remains "ON" due to a lamp driver malfunction or an open in CKT 1290.
7. Determines if the "GAUGES" fuse or CKT 3 is open, causing the I/P indicator lamps to remain "OFF."
8. Checks for an open in CKT 39 which supplies voltage to the lamp driver module.
9. Isolates the cause of the lamp remaining "OFF" to a short to ground in CKT 1290 or a malfunctioning lamp driver module.

Diagnostic Aids: An "Intermittent" malfunction may be caused by a poor connection, rubbed through wire insulation, or a wire that is broken inside the insulation.

The lamp test function of the Tech 1 may be used to command the lamp "ON" while looking for an intermittent malfunction in the ABS warning lamp circuitry.

The frequency of the malfunction can be checked by using the enhanced diagnostic function of the Tech 1.

Any circuitry that is suspected as causing the intermittent complaint should be thoroughly checked for backed out terminals, improper mating, broken locks, improperly formed or damaged terminals, poor terminal to wiring connections or physical damage to the wiring harness.

DIAGNOSTIC CHART DTC 11 — 1994 LUMINA APV

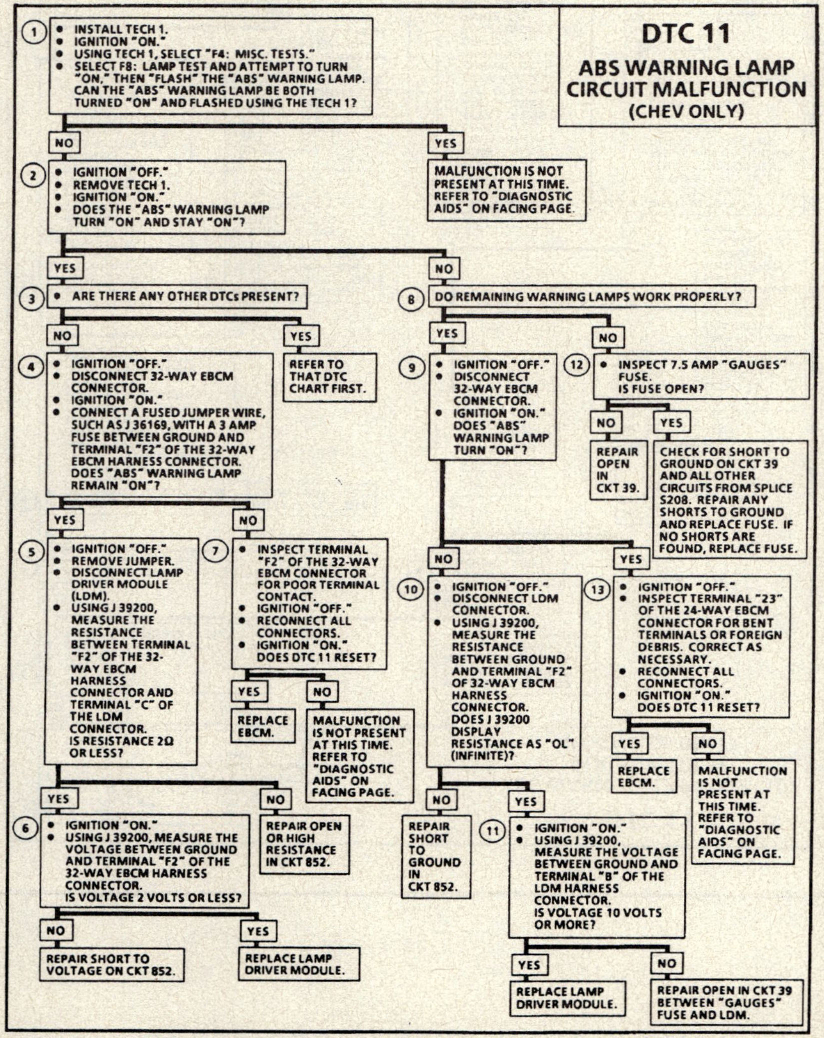

DIAGNOSTIC CHART DTC 11 — 1994 SILHOUETTE AND TRANS SPORT

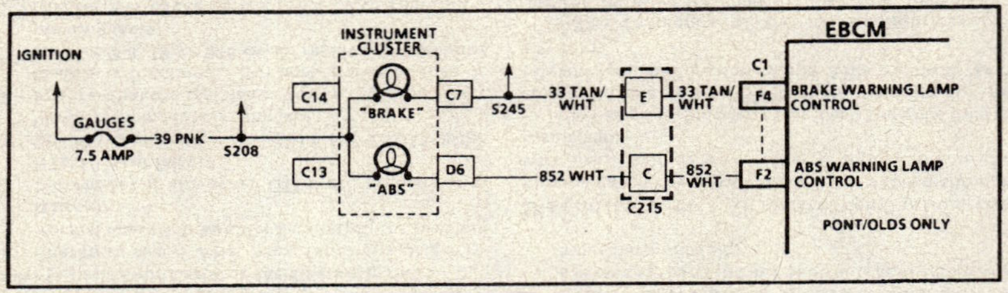

DTC 11

ABS WARNING LAMP CIRCUIT MALFUNCTION (PONT/OLDS ONLY)

Circuit Description:
This DTC checks the state of both the ABS warning lamp and the "BRAKE" warning lamp to identify a situation in which the driver could not be alerted to a system malfunction or of possible slippery driving conditions. Due to an integral lamp driver module within the instrument panel cluster, the EBCM must provide a ground to turn the ABS warning lamp "OFF." The "BRAKE" warning lamp circuit is such that the EBCM provides a ground to turn "ON" the "BRAKE" warning lamp.

DTC Will Set When: DTC 11 can be set only during the three second bulb check or when the ABS warning lamp is commanded "ON." If the EBCM cannot control the ABS warning lamp for two seconds, a malfunction exists.

Action Taken: A malfunction DTC is stored; however, ABS is not disabled.

DTC Chart Test Description: Number(s) below refer to circled number(s) on the diagnostic chart.

1. Determines if the "ABS INOP" lamp can be properly controlled.
2. Determines if the lamp is "ON" constantly.
3. Checks for other DTCs which could turn "ON" the "ABS INOP" lamp.
4. Determines if the "ABS INOP" lamp can be turned "OFF" manually.
5. Checks the integrity of the "ABS INOP" lamp circuitry.
6. Checks for a short to voltage on the "ABS INOP" lamp circuitry.
7. Determines if a malfunctioning EBCM is the cause of DTC A011.
8. Verifies that the other I/P indicator lamps function properly.
9. Determines if the "ABS INOP" lamp can be turned "ON" manually.
10. Checks for a short to ground in the "ABS INOP" lamp circuitry.
11. Verifies condition of the 10 amp "GAUGES" Fuse 9.
12. Determines if DTC A011 was set due to an EBCM malfunction.

Diagnostic Aids: An "Intermittent" malfunction may be caused by a poor connection, rubbed through wire insulation, or a wire that is broken inside the insulation.

The lamp test function of the Tech 1 may be used to command the lamp "ON" while looking for an intermittent malfunction in the "ABS INOP" lamp circuitry.

The frequency of the malfunction can be checked by using the enhanced diagnostic function of the Tech 1.

Any circuitry that is suspected as causing the intermittent complaint should be thoroughly checked for backed out terminals, improper mating, broken locks, improperly formed or damaged terminals, poor terminal to wiring connections or physical damage to the wiring harness.

DIAGNOSTIC CHART DTC 11 — 1994 SILHOUETTE AND TRANS SPORT

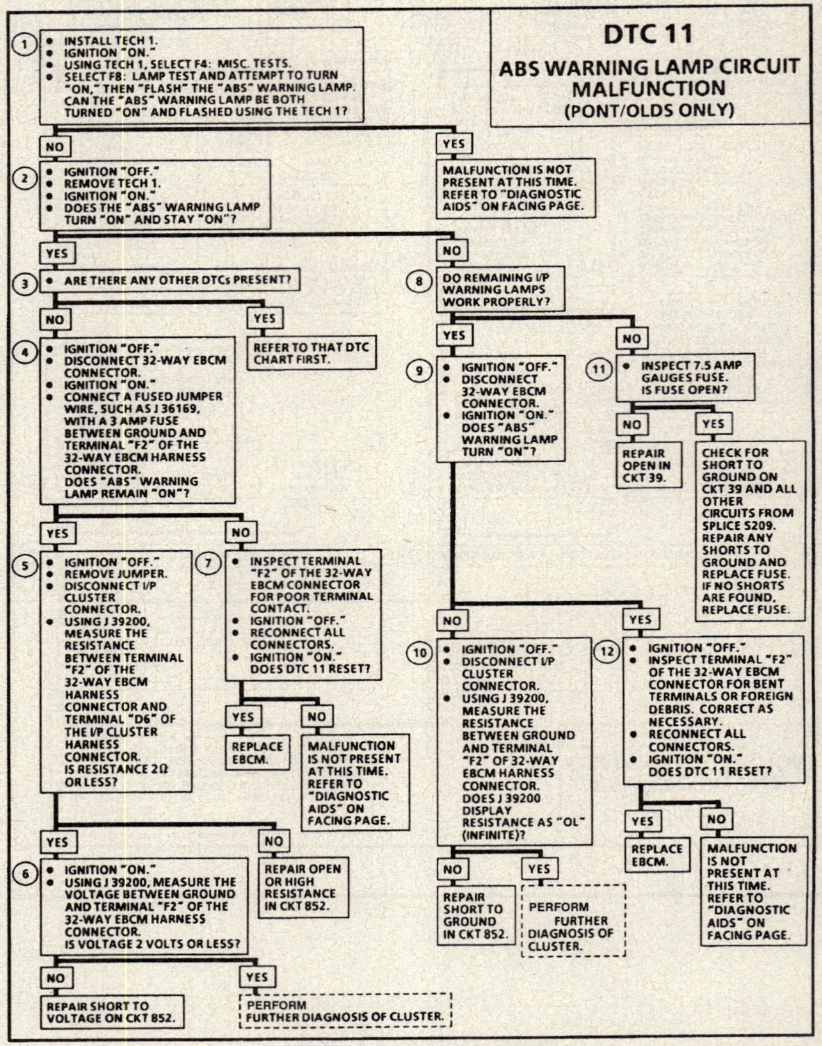

DIAGNOSTIC CHART DTC 14 — 1994 SILHOUETTE, TRANS SPORT AND LUMINA APV

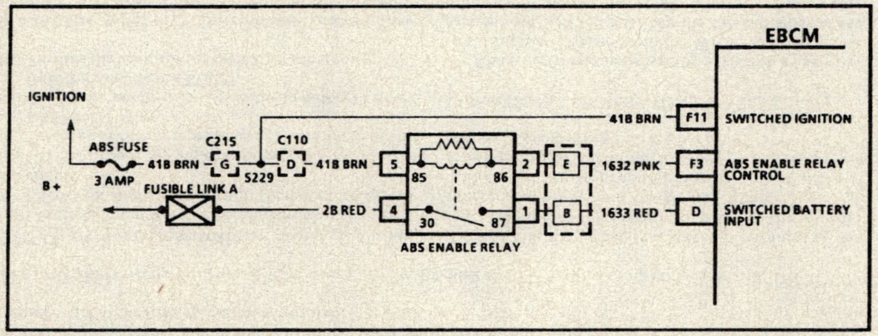

DTC 14
(Page 1 of 3)
ABS ENABLE RELAY CONTACT CIRCUIT OPEN

Circuit Description:

Ignition voltage is supplied through terminal "5" of the ABS enable relay. The EBCM then is able to energize the pull-in coil by completing the ground circuit at terminal "F3" of the EBCM. The magnetic field created closes the ABS enable relay contacts and allows battery voltage and current to be supplied to the EBCM, which supplies power to the motors and solenoids.

DTC Will Set When: DTC 14 can be set anytime after the EBCM commands the ABS enable relay "ON" (the relay is first commanded "ON" during the three second bulb check). This test monitors the availability of current/voltage to the motors and solenoids. This malfunction indicates voltage is not available and would therefore not allow ABS operation if required.

Action Taken: A malfunction DTC is stored, ABS is disabled and the ABS warning lamp is turned "ON."

DTC Chart Test Description: Number(s) below refer to circled number(s) on the diagnostic chart.

1. Checks to see if DTC 16 is also set as current or history. If DTC 16 is also set, proceed to that chart first.
2. Verifies that the malfunction is currently present.
3. Verifies that the relay can be turned "ON" and provide voltage to the EBCM.
4. Checks for high resistance in the switched battery circuit from the relay to the EBCM.
5. Checks for high resistance in the switched battery circuit from the battery to the relay.
6. Verifies the condition of the enable relay.
7. Checks for poor terminal contact or corrosion at the connectors.
8. Determines if the EBCM is the cause of the malfunction.

Diagnostic Aids: An "Intermittent" problem may be caused by a poor connection, rubbed through wire insulation, or a wire that is broken inside the insulation.

The frequency of the malfunction can be checked by using the enhanced diagnostic function of the Tech 1.

Any circuitry that is suspected as causing the "Intermittent" complaint should be thoroughly checked for backed out terminals, improper mating, broken locks, improperly formed or damaged terminals, poor terminal to wiring connections or physical damage to the wiring harness.

VIBRATION, TEMPERATURE EFFECTS:

Check for vibration effects by performing the relay test function of the Tech 1. With the relay test commanded "ON," lightly tap the top and sides of the relay while monitoring relay voltage. If the relay voltage changes significantly, replace the relay.

If DTC 14 only sets when the vehicle is initially started in cold ambient conditions (temperature less than 32°F-0°C), replace the relay.

DIAGNOSTIC CHART DTC 14 — 1994 SILHOUETTE, TRANS SPORT AND LUMINA APV

DTC 14
(Page 1 of 3)
ABS ENABLE RELAY CONTACT CIRCUIT OPEN

(1) IS DTC 16 ALSO PRESENT IN HISTORY?

- YES: PROCEED TO DTC 16 CHART FIRST.
- NO:

(2)
- IGNITION "ON," ENGINE "OFF."
- USING TECH 1, SELECT DATA LIST AND OBSERVE EBCM BATTERY VOLTAGE. DOES TECH 1 INDICATE THAT EBCM BATTERY VOLTAGE IS GREATER THAN 10 VOLTS?

- YES: MALFUNCTION IS NOT PRESENT AT THIS TIME. REFER TO "DIAGNOSTIC AIDS" ON FACING PAGE.
- NO:

(3)
- IGNITION "OFF."
- DISCONNECT 32-WAY AND 8-WAY EBCM CONNECTORS.
- CONNECT A FUSED JUMPER, SUCH AS J 36169 WITH A 3 AMP FUSE BETWEEN GROUND AND TERMINAL "F3" OF THE 32-WAY EBCM HARNESS CONNECTOR.
- IGNITION "ON."
- USING J 39200, MEASURE THE VOLTAGE BETWEEN GROUND AND TERMINAL "D" OF THE 8-WAY EBCM HARNESS CONNECTOR. IS VOLTAGE 10 VOLTS OR GREATER?

- NO: DIAGNOSIS CONTINUED ON NEXT PAGE.
- YES:

(4)
- IGNITION "OFF."
- DISCONNECT JUMPER USED IN STEP 3.
- DISCONNECT ABS ENABLE RELAY.
- USING J 39200, MEASURE RESISTANCE BETWEEN TERMINAL "1" OF THE ABS ENABLE RELAY HARNESS CONNECTOR AND TERMINAL "D" OF THE 8-WAY EBCM HARNESS CONNECTOR. IS RESISTANCE 2Ω OR LESS?

- NO: REPAIR HIGH RESISTANCE IN CKT 1633.
- YES:

(5)
- DISCONNECT NEGATIVE BATTERY CABLE.
- DISCONNECT POSITIVE BATTERY CABLE.
- USING J 39200, MEASURE THE RESISTANCE BETWEEN POSITIVE BATTERY CABLE TERMINAL AND TERMINAL "4" OF THE ABS ENABLE RELAY HARNESS CONNECTOR. IS RESISTANCE 2Ω OR LESS?

- NO: REPAIR HIGH RESISTANCE IN CKT 28.
- YES:

(6)
- USING A JUMPER WIRE, CONNECT TERMINAL "86" ON THE ABS ENABLE RELAY TO GROUND.
- USING A FUSED JUMPER WIRE, SUCH AS J 36169, WITH A 3 AMP FUSE, CONNECT TERMINAL "85" ON THE ABS ENABLE RELAY TO VOLTAGE.
- USING J 39200, MEASURE THE RESISTANCE BETWEEN TERMINALS "30" AND "87" ON THE ABS ENABLE RELAY. IS RESISTANCE 2Ω OR LESS?.

- NO: REPLACE ABS ENABLE RELAY.
- YES:

(7)
- INSPECT ABS ENABLE RELAY AND ABS ENABLE RELAY HARNESS CONNECTOR FOR A POOR CONNECTION OR POOR TERMINAL CONTACT.
- INSPECT EBCM AND EBCM CONNECTOR TERMINALS "F3" AND "D" FOR POOR TERMINAL CONTACT.
- INSPECT POSITIVE BATTERY TERMINAL AND POSITIVE BATTERY CABLE TERMINAL FOR A POOR CONNECTION. IS POOR TERMINAL CONTACT OR A POOR CONNECTION EVIDENT?

- YES: REPLACE TERMINALS THAT EXHIBIT POOR TERMINAL CONTACT OR EVIDENCE OF CORROSION.
- NO:

(8)
- RECONNECT ALL CONNECTORS.
- IGNITION "ON." DOES DTC 14 RESET?

- NO: MALFUNCTION IS INTERMITTENT. REFER TO "DIAGNOSTIC AIDS" ON FACING PAGE.
- YES: REPLACE EBCM.

DIAGNOSTIC CHART DTC 14 — 1994 SILHOUETTE, TRANS SPORT AND LUMINA APV

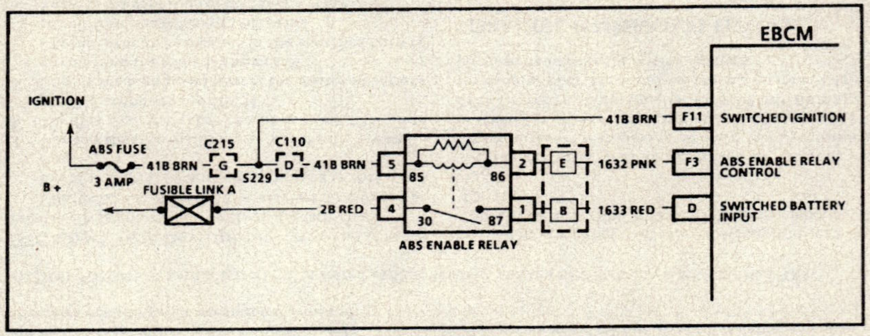

DTC 14

ABS ENABLE RELAY CONTACT CIRCUIT OPEN

Circuit Description:
Ignition voltage is supplied through terminal "5" of the ABS enable relay. The EBCM then is able to energize the pull-in coil by completing the ground circuit at terminal "F3" of the EBCM. The magnetic field created closes the ABS enable relay contacts and allows battery voltage and current to be supplied to the EBCM, which supplies power to the motors and solenoids.

DTC Will Set When: DTC 14 can be set anytime after the EBCM commands the ABS enable relay "ON" (the relay is first commanded "ON" during the three second bulb check). This test monitors the availability of current/voltage to the motors and solenoids. This malfunction indicates voltage is not available and would therefore not allow ABS operation if required.

Action Taken: A malfunction DTC is stored, ABS is disabled and the ABS warning lamp is turned "ON."

DTC Chart Test Description: Number(s) below refer to circled number(s) on the diagnostic chart.

9. Checks for high resistance in the entire relay coil circuit.
10. Verifies proper battery condition.
11. This step verifies proper voltage is available at the ABS enable relay.
12. Checks for an open in the switched battery circuit from the relay to the EBCM.
13. Verifies the condition of the ABS enable relay.

Diagnostic Aids: An "Intermittent" problem may be caused by a poor connection, rubbed through wire insulation, or a wire that is broken inside the insulation.

The frequency of the problem can be checked by using the enhanced diagnostic function of the Tech 1.

Any circuitry that is suspected as causing the "Intermittent" complaint should be thoroughly checked for backed out terminals, improper mating, broken locks, improperly formed or damaged terminals, poor terminal to wiring connections or physical damage to the wiring harness.

VIBRATION, TEMPERATURE EFFECTS:

Check for vibration effects by performing the relay test function of the Tech 1. With the relay test commanded "ON," lightly tap the top and sides of the relay while monitoring relay voltage. If the relay voltage changes significantly, replace the relay.

If DTC 14 only sets when the vehicle is initially started in cold ambient conditions (temperature less than 32°F-0°C), replace the relay.

DIAGNOSTIC CHART DTC 14 — 1994 SILHOUETTE, TRANS SPORT AND LUMINA APV

DTC 14
(Page 2 of 3)
ABS ENABLE RELAY CONTACT CIRCUIT OPEN

DIAGNOSIS CONTINUED FROM PREVIOUS PAGE.

(9)
- IGNITION "OFF."
- REMOVE JUMPER WIRE INSTALLED IN STEP #3 IF NOT ALREADY REMOVED.
- DISCONNECT NEGATIVE BATTERY CABLE.
- DISCONNECT POSITIVE BATTERY CABLE.
- WITH BATTERY STILL DISCONNECTED, TURN IGNITION SWITCH TO "RUN." THIS COMPLETES THE IGNITION CIRCUIT TO THE ABS ENABLE RELAY.
- USING J 39200, MEASURE THE RESISTANCE BETWEEN TERMINAL "F3" OF THE 32-WAY EBCM HARNESS CONNECTOR AND THE POSITIVE BATTERY CABLE TERMINAL.

IS RESISTANCE 95Ω OR LESS WHEN MEASURED AT APPROXIMATELY 20 °C (68°F)?

NO → DIAGNOSIS CONTINUED ON NEXT PAGE.

YES ↓

(10)
- USING J 39200, MEASURE THE VOLTAGE BETWEEN THE NEGATIVE AND POSITIVE BATTERY TERMINALS.

IS VOLTAGE 10 VOLTS OR GREATER?

NO → CORRECT LOW VOLTAGE CONDITION.

YES ↓

(11)
- IGNITION "OFF."
- RECONNECT POSITIVE BATTERY CABLE.
- RECONNECT NEGATIVE BATTERY CABLE.
- DISCONNECT ABS ENABLE RELAY.
- MEASURE THE VOLTAGE BETWEEN TERMINAL "4" OF THE ABS ENABLE RELAY HARNESS CONNECTOR AND GROUND.

IS VOLTAGE 10 VOLTS OR GREATER?

NO → REPAIR OPEN OR HIGH RESISTANCE IN CKT 2B.

YES ↓

(12)
- USING J 39200, MEASURE THE RESISTANCE BETWEEN TERMINAL "1" OF THE ABS ENABLE RELAY HARNESS CONNECTOR AND TERMINAL "D" OF THE 8-WAY EBCM HARNESS CONNECTOR.

IS RESISTANCE 2Ω OR LESS?

NO → REPAIR OPEN OR HIGH RESISTANCE IN CKT 1633.

YES ↓

(13)
- USING A JUMPER WIRE, CONNECT TERMINAL "86" ON THE ABS ENABLE RELAY TO GROUND.
- USING A FUSED JUMPER WIRE, SUCH AS J 36169, WITH A 3 AMP FUSE, CONNECT TERMINAL "85" ON THE ABS ENABLE RELAY TO B+.
- USING J 39200, MEASURE THE RESISTANCE BETWEEN TERMINALS "30" AND "87" ON THE ABS ENABLE RELAY.

IS RESISTANCE 2Ω OR LESS?

NO → REPLACE ABS ENABLE RELAY.

YES ↓

MALFUNCTION WAS MOST LIKELY DUE TO A POOR CONNECTION. INSPECT ALL CONNECTORS AND TERMINALS FOR POOR TERMINAL CONTACT AND EVIDENCE OF CORROSION. REPLACE ALL TERMINALS THAT EXHIBIT EVIDENCE OF POOR TERMINAL CONTACT OR CORROSION.

DIAGNOSTIC CHART DTC 14 — 1994 SILHOUETTE, TRANS SPORT AND LUMINA APV

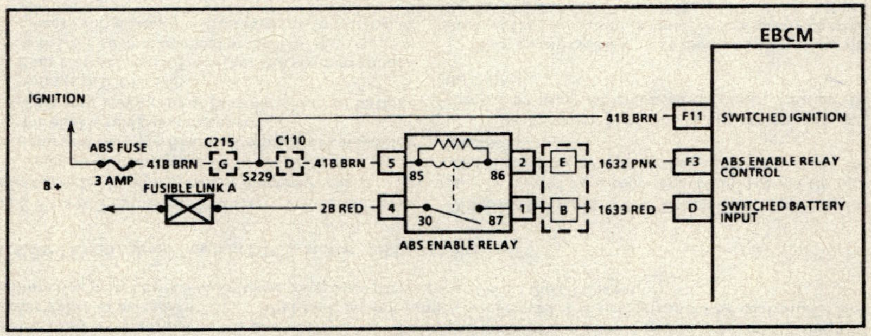

DTC 14

ABS ENABLE RELAY CONTACT CIRCUIT OPEN

Circuit Description:

Ignition voltage is supplied through terminal "5" of the ABS enable relay. The EBCM then is able to energize the pull-in coil by completing the ground circuit at terminal "F3" of the EBCM. The magnetic field created closes the ABS enable relay contacts and allows battery voltage and current to be supplied to the EBCM, which supplies power to the motors and solenoids.

DTC Will Set When: DTC 14 can be set anytime after the EBCM commands the ABS enable relay "ON" (the relay is first commanded "ON" during the three second bulb check). This test monitors the availability of current/voltage to the motors and solenoids. This malfunction indicates voltage is not available and would therefore not allow ABS operation if required.

Action Taken: A malfunction DTC is stored, ABS is disabled and the ABS warning lamp is turned "ON."

DTC Chart Test Description: Number(s) below refer to circled number(s) on the diagnostic chart.

14. Checks for proper resistance of the relay coil.
15. Checks for high resistance in the ABS enable relay control circuit between the relay and the EBCM.
16. Checks for high resistance in the ABS enable relay control circuit between the battery and the relay.

Diagnostic Aids: An "Intermittent" problem may be caused by a poor connection, rubbed through wire insulation, or a wire that is broken inside the insulation.

The frequency of the problem can be checked by using the enhanced diagnostic function of the Tech 1.

Any circuitry that is suspected as causing the "Intermittent" complaint should be thoroughly checked for backed out terminals, improper mating, broken locks, improperly formed or damaged terminals, poor terminal to wiring connections or physical damage to the wiring harness.

VIBRATION, TEMPERATURE EFFECTS:

Check for vibration effects by performing the relay test function of the Tech 1. With the relay test commanded "ON," lightly tap the top and sides of the relay while monitoring relay voltage. If the relay voltage changes significantly, replace the relay.

If DTC 14 only sets when the vehicle is initially started in cold ambient conditions (temperature less than 32°F-0°C), replace the relay.

DIAGNOSTIC CHART DTC 14 — 1994 SILHOUETTE, TRANS SPORT AND LUMINA APV

DTC 14
(Page 3 of 3)
ABS ENABLE RELAY CONTACT CIRCUIT OPEN

DIAGNOSIS CONTINUED FROM PREVIOUS PAGE.

(14)
- DISCONNECT ABS ENABLE RELAY.
- USING J 39200, MEASURE THE RESISTANCE BETWEEN TERMINALS "85" AND "86" ON THE ABS ENABLE RELAY.

IS RESISTANCE 95Ω OR LESS WHEN MEASURED AT APPROXIMATELY 20°C (68°F)?

- NO → REPLACE ABS ENABLE RELAY.
- YES ↓

(15)
- MEASURE THE RESISTANCE BETWEEN TERMINAL "2" OF THE ABS ENABLE RELAY HARNESS CONNECTOR AND TERMINAL "F3" OF THE 32-WAY EBCM HARNESS CONNECTOR.

IS RESISTANCE 2Ω OR LESS?

- NO → REPAIR OPEN OR HIGH RESISTANCE IN CKT 1632.
- YES ↓

(16)
- MEASURE THE RESISTANCE BETWEEN TERMINAL "5" OF THE ABS ENABLE RELAY HARNESS CONNECTOR AND THE POSITIVE BATTERY CABLE TERMINAL.

IS RESISTANCE 2Ω OR LESS?

- NO → REPAIR OPEN OR HIGH RESISTANCE IN CKT 418, CKT 300, IGNITION SWITCH, OR BATTERY FEED CIRCUIT TO IGNITION SWITCH.
- YES → MALFUNCTION WAS MOST LIKELY DUE TO A POOR CONNECTION. INSPECT ALL CONNECTORS AND TERMINALS FOR POOR TERMINAL CONTACT AND EVIDENCE OF CORROSION.
REPLACE ALL TERMINALS THAT EXHIBIT EVIDENCE OF POOR TERMINAL CONTACT OR CORROSION.

DIAGNOSTIC CHART DTC 15 — 1994 SILHOUETTE, TRANS SPORT AND LUMINA APV

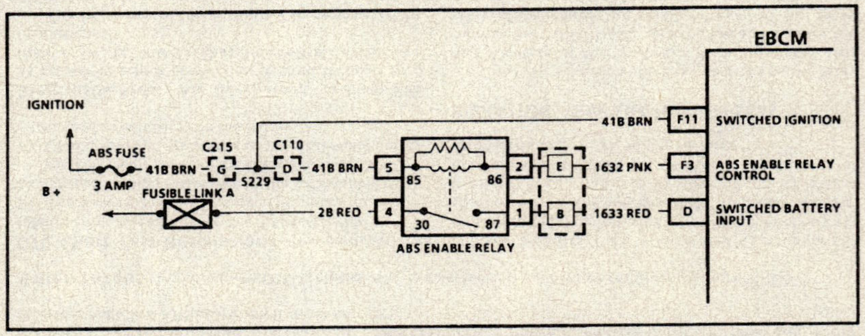

DTC 15

ABS ENABLE RLY CONTACT CKT SHORT TO BATT OR ALWAYS CLOSED

Circuit Description:

Ignition voltage is supplied through terminal "5" of the ABS enable relay. The EBCM then is able to energize the pull-in coil by completing the ground circuit at terminal "F3" of the EBCM. The magnetic field created closes the ABS enable relay contacts and allows battery voltage and current to be supplied to the EBCM, which supplies power to the motors and solenoids.

DTC Will Set When: DTC 15 can be set only before the EBCM commands the relay "ON." This test determines if the ABS enable relay is energized when it should not be. This malfunction would not allow the ABS enable relay to remove power to the ABS system. If a second malfunction were to occur that requires the ABS enable relay to be turned "OFF," that malfunction cannot be removed if the relay cannot be controlled. The malfunction must be present for three consecutive drive cycles before the DTC is set.

Action Taken: A malfunction DTC is stored. ABS is not disabled.

DTC Chart Test Description: Number(s) below refer to circled number(s) on the diagnostic chart.

1. Determines if any other DTCs are present.
2. Indicates that the EBCM is capable of controlling the enable relay as commanded.
3. This step checks for a possible short to battery voltage in CKT 1633.
4. This step determines whether the short to voltage is in CKT 1633 or internal to the EBCM.
5. Checks for a possible intermittent in wiring or connectors.

Diagnostic Aids: An "Intermittent" malfunction may be caused by a poor connection, rubbed through wire insulation, or a wire that is broken inside the insulation.

The frequency of the malfunction can be checked by using the enhanced diagnostic function of the Tech 1.

Any circuitry, that is suspected as causing the intermittent complaint, should be thoroughly checked for backed out terminals, improper mating, broken locks, improperly formed or damaged terminals, poor terminal to wiring connections or physical damage to the wiring harness.

DIAGNOSTIC CHART DTC 15 — 1994 SILHOUETTE, TRANS SPORT AND LUMINA APV

DTC 15

ABS ENABLE RLY CONTACT CKT SHORT TO BATT OR ALWAYS CLOSED

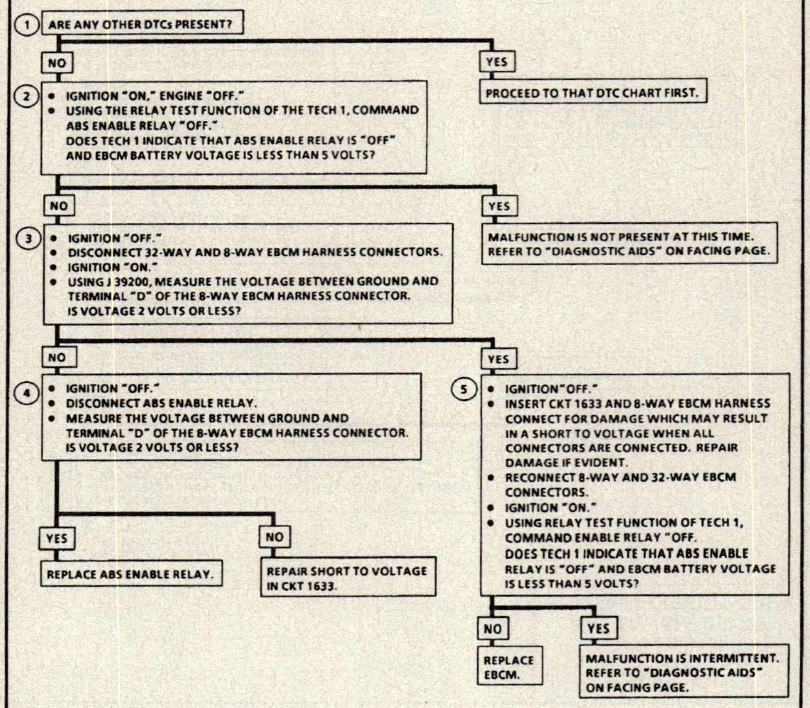

AFTER DIAGNOSIS IS COMPLETE, CLEAR DTCs, AND TEST DRIVE VEHICLE FOR 3 DRIVE CYCLES TO VERIFY THAT DTC DOES NOT RESET. A DRIVE CYCLE CONSISTS OF STARTING THE VEHICLE, DRIVING ABOVE 16 km/h (10 MPH), AND THEN KEYING DOWN.

DIAGNOSTIC CHART DTC 16 — 1994 SILHOUETTE, TRANS SPORT AND LUMINA APV

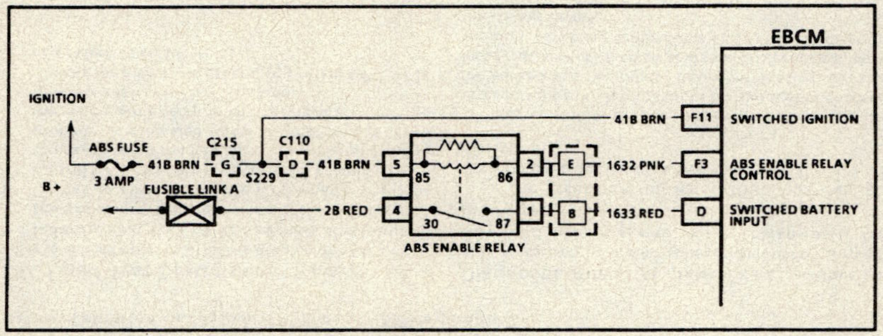

DTC 16

ABS ENABLE RELAY COIL CIRCUIT OPEN

Circuit Description:
Ignition voltage is supplied through terminal "5" of the ABS enable relay. The EBCM then is able to energize the pull-in coil by completing the ground circuit at terminal "F3" of the EBCM. The magnetic field created closes the ABS enable relay contacts and allows battery voltage and current to be supplied to the EBCM, which supplies power to the motors and solenoids.

DTC Will Set When: DTC 16 can be set only after DTC 14 has been set. This test detects an open in the ABS enable relay coil circuit. An open in this circuit will not allow the ABS enable relay to be energized thus preventing voltage/current to the motors and solenoids. If this malfunction is present and the ignition is turned "OFF" before 5 km/h (3 mph) is reached, DTC 14 is set alone.

Action Taken: A malfunction DTC is stored, ABS is disabled and the ABS warning lamp is turned "ON."

DTC Chart Test Description: Number(s) below refer to circled number(s) on the diagnostic chart.
1. Indicates the EBCM is capable of controlling the ABS enable relay as commanded.
2. Checks to ensure that voltage is available to the pull-in coil of the ABS enable relay.
3. Ensures that there is continuity through the pull-in coil of the ABS enable relay.
4. Verifies the integrity of the relay control circuit.
5. Ensures malfunction was not due to poor terminal contact.

Diagnostic Aids: An "Intermittent" malfunction may be caused by a poor connection, rubbed through wire insulation, or a wire that is broken inside the insulation.

The frequency of the malfunction can be checked by using the enhanced diagnostic function of the Tech 1.

If the frequency of the malfunction is high, but is currently intermittent, check for high coil resistance by measuring between relay terminals "2" and "5" using a J 39200. If resistance shows greater than 100Ω, replace the relay.

Any circuitry, that is suspected as causing the intermittent complaint, should be thoroughly checked for backed out terminals, improper mating, broken locks, improperly formed or damaged terminals, poor terminal to wiring connections or physical damage to the wiring harness.

DIAGNOSTIC CHART DTC 16 — 1994 SILHOUETTE, TRANS SPORT AND LUMINA APV

DTC 16
ABS ENABLE RELAY COIL CIRCUIT OPEN

(1)
- IGNITION "ON," ENGINE "OFF."
- USING THE RELAY TEST FUNCTION OF THE TECH 1, COMMAND THE ABS ENABLE RELAY "ON." DOES TECH 1 INDICATE ABS ENABLE RELAY "ON," AND EBCM BATTERY VOLTAGE ABOVE 10 VOLTS?

YES: MALFUNCTION IS NOT PRESENT AT THIS TIME. REFER TO "DIAGNOSTIC AIDS" ON FACING PAGE.

NO:

(2)
- USING J 39200, MEASURE THE VOLTAGE BETWEEN GROUND AND TERMINAL "5" OF THE ABS ENABLE RELAY CONNECTOR. IS VOLTAGE 10 VOLTS OR GREATER?

NO: REPAIR OPEN OR HIGH RESISTANCE IN CKT 418 OR ABS FUSE.

YES:

(3)
- USING THE RELAY TEST FUNCTION OF THE TECH 1, COMMAND THE ABS ENABLE RELAY "OFF."
- MEASURE THE VOLTAGE BETWEEN GROUND AND TERMINAL "2" OF THE ABS ENABLE RELAY CONNECTOR. IS VOLTAGE 10 VOLTS OR GREATER?

NO: CHECK FOR INTERMITTENT CONNECTIONS AT ABS ENABLE RELAY. IF OK, REPLACE ABS RELAY.

YES:

(4)
- IGNITION "OFF."
- DISCONNECT 32-WAY EBCM CONNECTOR.
- IGNITION "ON."
- MEASURE THE VOLTAGE BETWEEN GROUND AND TERMINAL "F3" OF THE 32-WAY EBCM HARNESS CONNECTOR. IS VOLTS 10 VOLTS OR GREATER?

NO: REPAIR OPEN OR HIGH RESISTANCE IN CKT 1632.

YES:

(5)
- IGNITION "OFF."
- INSPECT 32-WAY EBCM CONNECTOR TERMINAL "F3" FOR POOR TERMINAL CONTACT OR EVIDENCE OF CORROSION.

REPLACE ALL TERMINALS THAT EXHIBIT SIGNS OF POOR TERMINAL CONTACT OR CORROSION.
- IGNITION "ON."

DOES DTC 16 SET AS A CURRENT DTC?

YES: REPLACE EBCM.

NO: MALFUNCTION WAS MOST LIKELY CAUSED BY A POOR CONNECTION. REFER TO "DIAGNOSTIC AIDS" ON FACING PAGE.

AFTER ALL DIAGNOSIS IS COMPLETE, CLEAR DTCs AND TEST DRIVE VEHICLE FOR THREE (3) DRIVE CYCLES TO VERIFY THAT DTC DOES NOT RESET. A DRIVE CYCLE CONSISTS OF STARTING THE VEHICLE, DRIVING OVER 16 km/h (10 MPH), AND THEN KEYING DOWN.

DIAGNOSTIC CHART DTC 17 — 1994 SILHOUETTE, TRANS SPORT AND LUMINA APV

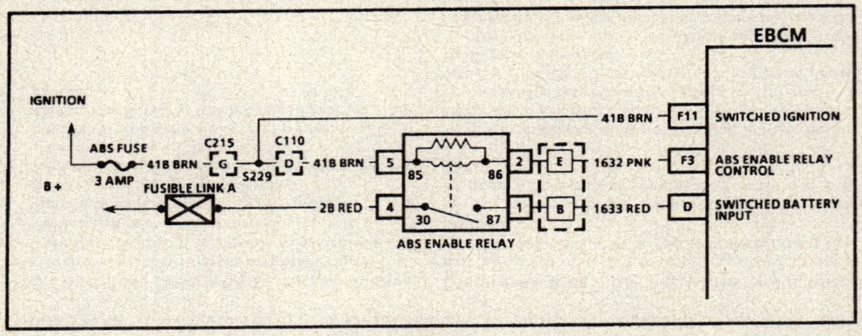

DTC 17

ABS ENABLE RELAY COIL CIRCUIT SHORTED TO GND

Circuit Description:

Ignition voltage is supplied through terminal "5" of the ABS enable relay. The EBCM then is able to energize the pull-in coil by completing the ground circuit at terminal "F3" of the EBCM. The magnetic field created closes the ABS enable relay contacts and allows battery voltage and current to be supplied to the EBCM, which supplies voltage to the motors and solenoids.

DTC Will Set When: DTC 17 can be set before the EBCM commands the ABS enable relay "ON." This test determines if the ABS enable relay is energized when it should not be. This malfunction would not allow the ABS enable relay to remove power to the ABS system. If a second malfunction were to occur that requires the ABS enable relay to be turned "OFF," that malfunction cannot be removed if the relay cannot be controlled.

Action Taken: A malfunction DTC is stored. ABS is not disabled.

DTC Chart Test Description: Number(s) below refer to circled number(s) on the diagnostic chart.

1. Indicates that the EBCM is capable of controlling the ABS enable relay as commanded.
2. Checks to ensure that the ABS enable relay or control CKT 1632 are not shorted to ground.
3. Checks to see if the EBCM is internally shorted to ground. A grounded terminal "F3" would cause the ABS enable relay to be energized anytime the ignition was in the "ON" position.
4. Checks for a short to ground in control CKT 1632 or a defective relay.

Diagnostic Aids: An "Intermittent" malfunction may be caused by a poor connection, rubbed through wire insulation, or a wire that is broken inside the insulation.

The frequency of the malfunction can be checked by using the enhanced diagnostic function of the Tech 1.

Any circuitry, that is suspected as causing the intermittent complaint, should be thoroughly checked for backed out terminals, improper mating, broken locks, improperly formed or damaged terminals, poor terminal to wiring connections or physical damage to the wiring harness.

DIAGNOSTIC CHART DTC 17 — 1994 SILHOUETTE, TRANS SPORT AND LUMINA APV

DTC 17

ABS ENABLE RELAY COIL CIRCUIT SHORTED TO GND

(1)
- IGNITION "ON," ENGINE "OFF."
- USING THE RELAY TEST FUNCTION OF THE TECH 1, COMMAND THE ABS ENABLE RELAY "OFF." DOES TECH 1 INDICATE THAT ABS ENABLE RELAY IS "OFF" AND EBCM BATTERY VOLTAGE IS 5 VOLTS OR LESS?

YES: MALFUNCTION IS NOT PRESENT AT THIS TIME. REFER TO "DIAGNOSTIC AIDS" ON FACING PAGE.

NO: (2)
- IGNITION "OFF."
- DISCONNECT 32-WAY EBCM CONNECTOR.
- IGNITION "ON."
- USING J 39200, MEASURE THE VOLTAGE BETWEEN TERMINAL "F3" OF THE 32-WAY EBCM HARNESS CONNECTOR AND B+. IS VOLTAGE 2 VOLTS OR LESS?

YES: (3)
- IGNITION "OFF."
- INSPECT CKT 1632 AND TERMINAL "F3" OF THE 32-WAY EBCM CONNECTOR FOR DAMAGE WHICH MAY RESULT IN A SHORT TO GROUND WITH ALL CONNECTORS CONNECTED. REPAIR ALL DAMAGE IF EVIDENT.
- RECONNECT ALL CONNECTORS.
- IGNITION "ON." DOES DTC 17 SET AS CURRENT DTC?

YES: REPLACE EBCM.

NO: MALFUNCTION WAS MOST LIKELY CAUSED BY A POOR CONNECTION. REFER TO "DIAGNOSTIC AIDS" ON FACING PAGE.

NO: (4)
- IGNITION "OFF."
- DISCONNECT ABS ENABLE RELAY.
- MEASURE VOLTAGE BETWEEN TERMINAL "2" OF THE ABS ENABLE RELAY CONNECTOR AND B+. IS VOLTAGE 2 VOLTS OR LESS?

NO: REPAIR SHORT TO GROUND IN CKT 1632.

YES: REPLACE ABS ENABLE RELAY.

AFTER ALL DIAGNOSIS IS COMPLETE, CLEAR DTCs AND TEST DRIVE VEHICLE FOR THREE (3) DRIVE CYCLES TO VERIFY THAT DTC DOES NOT RESET. A DRIVE CYCLE CONSISTS OF STARTING THE VEHICLE, DRIVING OVER 16 km/h (10 MPH), AND THEN KEYING DOWN.

DIAGNOSTIC CHART DTC 18 — 1994 SILHOUETTE, TRANS SPORT AND LUMINA APV

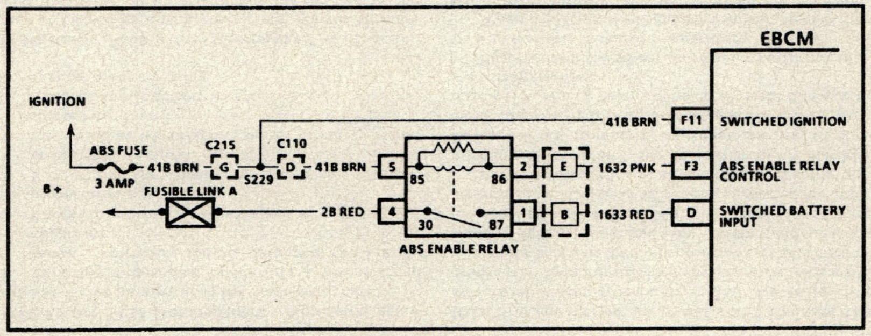

DTC 18

ABS ENABLE RELAY COIL CIRCUIT SHORTED TO BATT

Circuit Description:

Ignition voltage is supplied through terminal "5" of the ABS enable relay. The EBCM then is able to energize the pull-in coil by completing the ground circuit at terminal "F3" of the EBCM. The magnetic field created closes the ABS enable relay contacts and allows battery voltage and current to be supplied to the EBCM, which supplies voltage to the motors and solenoids.

DTC Will Set When: DTC 18 can be set after the EBCM commands the enable relay "ON." This test monitors the availability of current/voltage to the motors and solenoids. A malfunction of this type will not allow voltage to be available to the motors and solenoids; therefore, ABS operation would not be allowed if required.

Action Taken: A malfunction DTC is stored, ABS is disabled and the ABS warning lamp is turned "ON."

DTC Chart Test Description: Number(s) below refer to circled number(s) on the diagnostic chart.

1. Indicates that the EBCM is capable of controlling the ABS enable relay as commanded.
2. Checks for a short to voltage on the relay control circuit.
3. Verifies that the relay coil has the appropriate resistance.
4. Verifies that relay contacts have not failed closed.
5. Checks for possible poor terminal contact at the EBCM.

Diagnostic Aids: An "Intermittent" malfunction may be caused by a poor connection, rubbed through wire insulation, or a wire that is broken inside the insulation.

The frequency of the malfunction can be checked by using the enhanced diagnostic function of the Tech 1.

Any circuitry, that is suspected as causing the intermittent complaint, should be thoroughly checked for backed out terminals, improper mating, broken locks, improperly formed or damaged terminals, poor terminal to wiring connections or physical damage to the wiring harness.

DIAGNOSTIC CHART DTC 18 — 1994 SILHOUETTE, TRANS SPORT AND LUMINA APV

DTC 18

ABS ENABLE RELAY COIL CIRCUIT SHORTED TO BATT

(1)
- IGNITION "ON," ENGINE "OFF."
- USING RELAY TEST FUNCTION OF TECH 1, COMMAND ABS ENABLE RELAY "ON."
- DOES TECH 1 INDICATE ABS ENABLE RELAY "ON" AND BATTERY VOLTAGE ABOVE 10 VOLTS?

YES: MALFUNCTION IS NOT PRESENT AT THIS TIME. REFER TO "DIAGNOSTIC AIDS" ON FACING PAGE.

NO:

(2)
- IGNITION "OFF."
- DISCONNECT ABS ENABLE RELAY.
- DISCONNECT 32-WAY EBCM CONNECTOR.
- IGNITION "ON."
- USING J 39200, MEASURE THE VOLTAGE BETWEEN GROUND AND TERMINAL "F3" OF THE 32-WAY EBCM HARNESS CONNECTOR. IS VOLTAGE 2 VOLTS OR LESS?

NO: REPAIR SHORT TO VOLTAGE CKT 1632.

YES:

(3)
- IGNITION "ON."
- USING J 39200, MEASURE THE RESISTANCE BETWEEN TERMINAL "2" AND TERMINAL "5" OF THE ABS ENABLE RELAY. IS RESISTANCE 40Ω OR GREATER?

NO: REPLACE ABS ENABLE RELAY.

YES:

(4)
- USING J 39200, MEASURE THE RESISTANCE BETWEEN TERMINALS "30" AND "87" OF THE ABS ENABLE RELAY. DOES J 39200 DISPLAY RESISTANCE AS "OL" (INFINITE)?

NO: REPLACE ABS ENABLE RELAY.

YES:

(5)
- INSPECT CKT 1632 AND TERMINAL "F3" OF THE 32-WAY EBCM CONNECTOR FOR DAMAGE WHICH MAY RESULT IN A SHORT TO VOLTAGE WITH ALL CONNECTORS CONNECTED. REPAIR ALL DAMAGE IF EVIDENT.
- RECONNECT ALL CONNECTORS.
- IGNITION "ON."

DID DTC 18 SET AS A CURRENT DTC?

YES: REPLACE EBCM.

NO: MALFUNCTION WAS MOST LIKELY CAUSED BY A POOR CONNECTION. REFER TO "DIAGNOSTIC AIDS" ON FACING PAGE.

AFTER ALL DIAGNOSIS IS COMPLETE, CLEAR DTCs AND TEST DRIVE VEHICLE FOR THREE (3) DRIVE CYCLES TO VERIFY THAT DTC DOES NOT RESET. A DRIVE CYCLE CONSISTS OF STARTING THE VEHICLE, DRIVING OVER 16 km/h (10 MPH), AND THEN KEYING DOWN.

DIAGNOSTIC CHART DTC 21 — 1994 SILHOUETTE, TRANS SPORT AND LUMINA APV

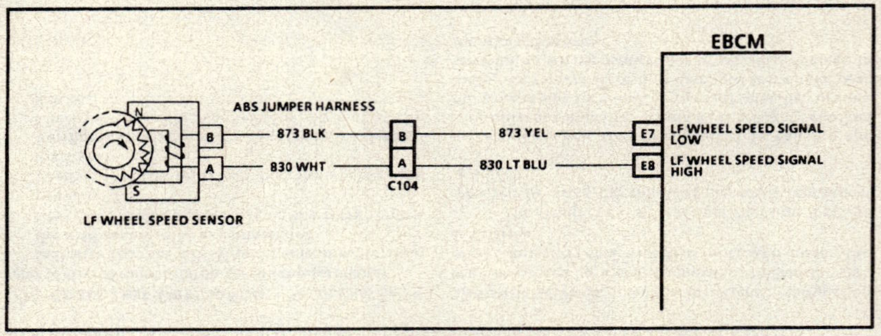

DTC 21

LEFT FRONT WHEEL SPEED = 0

Circuit Description:

As a toothed ring passes by the wheel speed sensor, changes in the electromagnetic field cause the wheel speed sensor to produce a sinusoidal (AC) voltage signal whose frequency is proportional to wheel speed. The magnitude of this signal is directly related to wheel speed and the proximity of the wheel speed sensor to the toothed ring, often referred to as the air gap.

DTC Will Set When: DTC 21 can be set when the vehicle is not in an ABS stop. If the left front wheel speed = 0 and the vehicle's reference speed is greater than 8 km/h (5 mph), a malfunction exists.

Action Taken: A malfunction DTC is stored, ABS is disabled and the ABS warning lamp is turned "ON."

DTC Chart Test Description: Number(s) below refer to circled number(s) on the diagnostic chart.

1. This step is designed to identify a wheel speed sensor or circuitry that is damaged and visibly apparent.
2. Checks to see that the wheel speed sensor is within proper resistance values.
3. Verifies that the wheel speed sensor can generate an appropriate output signal.
4. Ensures that wheel speed sensor circuitry is not shorted together.
5. Ensures malfunction was not due to physical damage of the circuitry.

Diagnostic Aids: An "Intermittent" malfunction may be caused by a poor connection, rubbed through wire insulation, or a wire that is broken inside the insulation.

The frequency of the malfunction can be checked by using the enhanced diagnostic function of the Tech 1.

If the customer's comments reflect that the ABS warning lamp is "ON" only during moist environmental changes (rain, snow, vehicle wash), all wheel speed sensor circuitry should be thoroughly inspected for signs of water intrusion.

If DTC is not current, clear DTCs and simulate the effects of water intrusion. Use the following procedure. Spray down the suspected area with a 5% salt water solution (two teaspoons of salt to 12 oz. of water). Test drive vehicle over various road surfaces (bumps, turns, etc.) above 24 km/h (15 mph) for at least 30 seconds. If DTC returns, replace suspected harness.

Any circuitry, that is suspected as causing the intermittent complaint, should be thoroughly checked for backed out terminals, improper mating, broken locks, improperly formed or damaged terminals, poor terminal to wiring connections or physical damage to the wiring harness.

Resistance of the wheel speed sensor will increase with an increase in sensor temperature.

When replacing a wheel speed sensor, inspect the sensor terminals and harness connector for corrosion and/or water intrusion. If evidence of corrosion or water intrusion exists, replace wheel speed sensor jumper harness. Likewise, if replacing a wheel speed sensor jumper harness, inspect sensor terminals. If evidence of corrosion or water intrusion exists, replace wheel speed sensor.

DIAGNOSTIC CHART DTC 21 — 1994 SILHOUETTE, TRANS SPORT AND LUMINA APV

DTC 21

LEFT FRONT WHEEL SPEED = 0

Important

- WHEEL SPEED SENSOR INTERMITTENT MALFUNCTIONS MAY BE DIFFICULT TO LOCATE. CARE SHOULD BE TAKEN NOT TO DISTURB ANY ELECTRICAL CONNECTIONS PRIOR TO AN INDICATED STEP OF THIS CHART. THIS WILL ENSURE THAT AN INTERMITTENT CONNECTION WILL NOT BE CORRECTED BEFORE THE SOURCE OF THE MALFUNCTION IS FOUND.

(1)
- IGNITION "OFF."
- PHYSICALLY INSPECT THE LEFT FRONT WHEEL SPEED SENSOR, LEFT FRONT WHEEL SPEED SENSOR JUMPER HARNESS, AND CONNECTORS FOR DAMAGE. IS PHYSICAL DAMAGE INDICATED?

YES: REPLACE WHEEL SPEED SENSOR AND/OR JUMPER HARNESS IF DAMAGED. DRIVE VEHICLE ABOVE 24 km/h (15 MPH) TO ENSURE DTC 21 DOES NOT SET.

NO:

(2)
- DISCONNECT LEFT FRONT WHEEL SPEED SENSOR DIRECTLY AT SENSOR.
- USING J 39200, MEASURE THE RESISTANCE BETWEEN SENSOR TERMINALS "A" AND "B." IS RESISTANCE BETWEEN 1020 AND 1137Ω WHEN SENSOR IS 20°C (68°F)? *(IF TEMPERATURE OF THE SENSOR IS SUSPECTED TO BE OTHER THAN 20°C (68°F), REFER TO THE TABLE BELOW FOR APPLICABLE SENSOR RESISTANCE VALUES; THE VALUES IN THE TABLE ARE FOR TEMPERATURE OF THE SENSOR, NOT AIR TEMPERATURE.)*

NO: REPLACE WHEEL SPEED SENSOR.

YES:

(3)
- WITH J 39200 STILL CONNECTED, SELECT A/C VOLTAGE SCALE.
- SPIN WHEEL BY HAND WHILE OBSERVING VOLTAGE. IS VOLTAGE AT LEAST 100 mV (VOLTAGE WILL INCREASE AS WHEEL SPEED INCREASES)?

NO: REPLACE WHEEL SPEED SENSOR.

YES:

(4)
- IGNITION "OFF."
- DISCONNECT 32-WAY EBCM CONNECTOR.
- USING J 39200, MEASURE THE RESISTANCE BETWEEN TERMINAL "E7" AND "E8" OF THE 32-WAY EBCM HARNESS CONNECTOR. DOES J 39200 DISPLAY RESISTANCE AS "OL" (INFINITE)?

NO: REPAIR SHORT BETWEEN CKTs 830 AND 873.

YES:

(5)
- INSPECT CKTs 830 AND 873 AND 32-WAY EBCM CONNECTOR FOR DAMAGE WHICH MAY RESULT IN A SHORT BETWEEN CKTs 830 AND 873 WITH ALL CONNECTORS CONNECTED. REPAIR ALL DAMAGE IF EVIDENT.
- RECONNECT ALL CONNECTORS.
- TEST DRIVE VEHICLE ABOVE 24 km/h (15 MPH) FOR AT LEAST 30 SECONDS. DOES DTC 21 SET AS A CURRENT DTC?

YES: REPLACE EBCM.

NO: MALFUNCTION IS INTERMITTENT. REFER TO "DIAGNOSTIC AIDS" ON FACING PAGE.

FRONT WHEEL SPEED SENSOR
TEMPERATURE OF SENSOR VS. SENSOR RESISTANCE (APPROXIMATE)

TEMP. (°C)	TEMP. (°F)	RESISTANCE (OHMS)
-34 TO 4	-30 TO 40	804 TO 1066
5 TO 43	41 TO 110	956 TO 1240
44 TO 93	111 TO 200	1112 TO 1463

DIAGNOSTIC CHART DTC 22 — 1994 SILHOUETTE, TRANS SPORT AND LUMINA APV

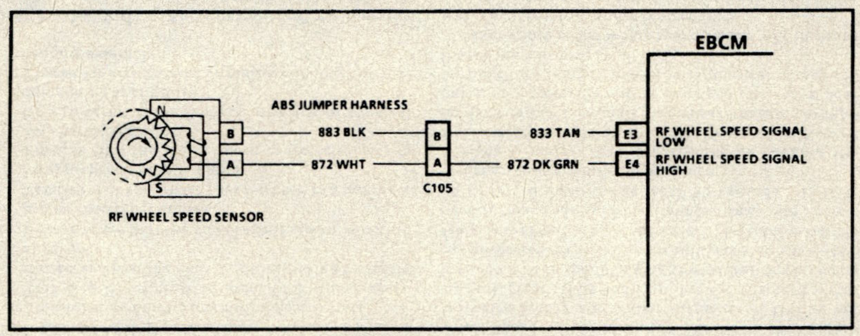

DTC 22

RIGHT FRONT WHEEL SPEED = 0

Circuit Description:

As a toothed ring passes by the wheel speed sensor, changes in the electromagnetic field cause the wheel speed sensor to produce a sinusoidal (AC) voltage signal whose frequency is proportional to wheel speed. The magnitude of this signal is directly related to wheel speed and the proximity of the wheel speed sensor to the toothed ring, often referred to as the air gap.

DTC Will Set When: DTC 22 can be set when the vehicle is not in an ABS stop. If the right front wheel speed = 0 and the vehicle's reference speed is greater than 8 km/h (5 mph), a malfunction exists.

Action Taken: A malfunction DTC is stored, ABS is disabled and the ABS warning lamp is turned "ON."

DTC Chart Test Description: Number(s) below refer to circled number(s) on the diagnostic chart.

1. This step is designed to identify a wheel speed sensor or circuitry that is damaged and visibly apparent.
2. This test checks to see that the wheel speed sensor is within proper resistance values.
3. Verifies that the wheel speed sensor can generate an appropriate output signal.
4. Ensures that wheel speed sensor circuitry is not shorted together.
5. Ensures malfunction was not due to physical damage of the circuitry.

Diagnostic Aids: An "Intermittent" malfunction may be caused by a poor connection, rubbed through wire insulation, or a wire that is broken inside the insulation.

The frequency of the malfunction can be checked by using the enhanced diagnostic function of the Tech 1.

If the customer's comments reflect that the ABS warning lamp is "ON" only during moist environmental changes (rain, snow, vehicle wash), all wheel speed sensor circuitry should be thoroughly inspected for signs of water intrusion. If DTC is not current, clear DTCs and simulate the effects of water intrusion. Use the following procedure. Spray down the suspected area with a 5% salt water solution (two teaspoons of salt to 12 oz. of water). Test drive vehicle over various road surfaces (bumps, turns, etc.) above 24 km/h (15 mph) for at least 30 seconds. If DTC returns, replace suspected harness.

Any circuitry, that is suspected as causing the intermittent complaint, should be thoroughly checked for backed out terminals, improper mating, broken locks, improperly formed or damaged terminals, poor terminal to wiring connections or physical damage to the wiring harness.

Resistance of the wheel speed sensor will increase with an increase in sensor temperature.

When replacing a wheel speed sensor, inspect the sensor terminals and harness connector for corrosion and/or water intrusion. If evidence of corrosion or water intrusion exists, replace wheel speed sensor jumper harness. Likewise, if replacing a wheel speed sensor jumper harness, inspect sensor terminals. If evidence of corrosion or water intrusion exists, replace wheel speed sensor.

DIAGNOSTIC CHART DTC 22 — 1994 SILHOUETTE, TRANS SPORT AND LUMINA APV

DTC 22

RIGHT FRONT WHEEL SPEED = 0

Important

- WHEEL SPEED SENSOR INTERMITTENT MALFUNCTIONS MAY BE DIFFICULT TO LOCATE. CARE SHOULD BE TAKEN NOT TO DISTURB ANY ELECTRICAL CONNECTIONS PRIOR TO AN INDICATED STEP OF THIS CHART. THIS WILL ENSURE THAT AN INTERMITTENT CONNECTION WILL NOT BE CORRECTED BEFORE THE SOURCE OF THE MALFUNCTION IS FOUND.

(1)
- IGNITION "OFF."
- PHYSICALLY INSPECT THE RIGHT FRONT WHEEL SPEED SENSOR, RIGHT FRONT WHEEL SPEED SENSOR JUMPER HARNESS, AND CONNECTORS FOR DAMAGE. IS PHYSICAL DAMAGE INDICATED?

YES: REPLACE WHEEL SPEED SENSOR AND/OR JUMPER HARNESS IF DAMAGED. DRIVE VEHICLE ABOVE 24 km/h (15 MPH) TO ENSURE DTC 22 DOES NOT SET.

NO:

(2)
- DISCONNECT RIGHT FRONT WHEEL SPEED SENSOR DIRECTLY AT SENSOR.
- USING J 39200, MEASURE THE RESISTANCE BETWEEN SENSOR TERMINALS "A" AND "B."

IS RESISTANCE BETWEEN 1020 AND 1137Ω WHEN SENSOR IS 20°C (68°F)? *(IF TEMPERATURE OF THE SENSOR IS SUSPECTED TO BE OTHER THAN 20°C (68°F), REFER TO THE TABLE BELOW FOR APPLICABLE SENSOR RESISTANCE VALUES; THE VALUES IN THE TABLE ARE FOR TEMPERATURE OF THE SENSOR, NOT AIR TEMPERATURE.)*

NO: REPLACE WHEEL SPEED SENSOR.

YES:

(3)
- WITH J 39200 STILL CONNECTED, SELECT A/C VOLTAGE SCALE.
- SPIN WHEEL BY HAND WHILE OBSERVING VOLTAGE.

IS VOLTAGE AT LEAST 100 mV (VOLTAGE WILL INCREASE AS WHEEL SPEED INCREASES)?

NO: REPLACE WHEEL SPEED SENSOR.

YES:

(4)
- IGNITION "OFF."
- DISCONNECT 32-WAY EBCM CONNECTOR.
- USING J 39200, MEASURE THE RESISTANCE BETWEEN TERMINAL "E3" AND "E4" OF THE 32-WAY EBCM HARNESS CONNECTOR.

DOES J 39200 DISPLAY RESISTANCE AS "OL" (INFINITE)?

NO: REPAIR SHORT BETWEEN CKTs 833 AND 872.

YES:

(5)
- INSPECT CKTs 833 AND 872 AND 32-WAY EBCM CONNECTOR FOR DAMAGE WHICH MAY RESULT IN A SHORT BETWEEN CKTs 833 AND 872 WITH ALL CONNECTORS CONNECTED. REPAIR ALL DAMAGE IF EVIDENT.
- RECONNECT ALL CONNECTORS.
- TEST DRIVE VEHICLE ABOVE 24 km/h (15 MPH) FOR AT LEAST 30 SECONDS.

DOES DTC 22 SET AS A CURRENT DTC?

YES: REPLACE EBCM.

NO: MALFUNCTION IS INTERMITTENT. REFER TO "DIAGNOSTIC AIDS" ON FACING PAGE.

FRONT WHEEL SPEED SENSOR
TEMPERATURE OF SENSOR VS. SENSOR RESISTANCE (APPROXIMATE)

TEMP. (°C)	TEMP. (°F)	RESISTANCE (OHMS)
-34 TO 4	-30 TO 40	804 TO 1066
5 TO 43	41 TO 110	956 TO 1240
44 TO 93	111 TO 200	1112 TO 1463

DIAGNOSTIC CHART DTC 23 — 1994 SILHOUETTE, TRANS SPORT AND LUMINA APV

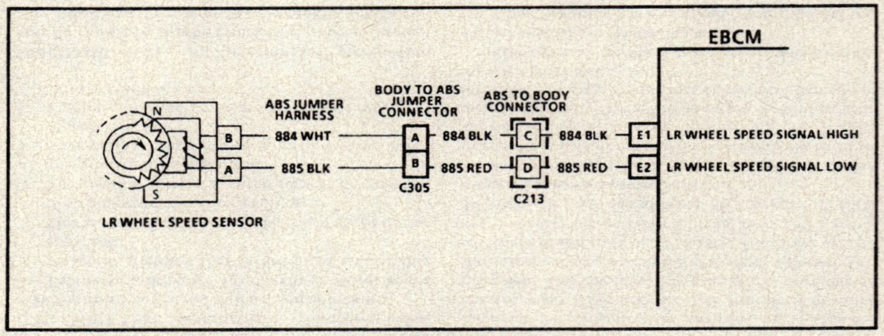

DTC 23

LEFT REAR WHEEL SPEED = 0

Circuit Description:

As a toothed ring passes by the wheel speed sensor, changes in the electromagnetic field cause the wheel speed sensor to produce a sinusoidal (AC) voltage signal whose frequency is proportional to wheel speed. The magnitude of this signal is directly related to wheel speed and the proximity of the wheel speed sensor to the toothed ring, often referred to as the air gap.

DTC Will Set When: DTC 23 can be set when the vehicle is not in an ABS stop. If the left rear wheel speed = 0 and the vehicle's reference speed is greater than 8 km/h (5 mph), a malfunction exists.

Action Taken: A malfunction DTC is stored, ABS is disabled and the ABS warning lamp is turned "ON."

DTC Chart Test Description: Number(s) below refer to circled number(s) on the diagnostic chart.

1. This step is designed to identify a wheel speed sensor or circuitry that is damaged and visibly apparent.
2. Checks to see that the wheel speed sensor is within proper resistance values.
3. Verifies that the wheel speed sensor can generate an appropriate output signal.
4. Ensures that the wheel speed sensor circuitry is not shorted together.
5. Checks to see if malfunction was due to physical damage of the circuitry.
6. Checks for an incorrect or malfunctioning hub and bearing assembly.

Diagnostic Aids: An "Intermittent" malfunction may be caused by a poor connection, rubbed through wire insulation, or a wire that is broken inside the insulation.

The frequency of the malfunction can be checked by using the enhanced diagnostic function of the Tech 1.

If the customer's comments reflect that the ABS warning lamp is "ON" only during moist environmental changes (rain, snow, vehicle wash), all wheel speed sensor circuitry should be thoroughly inspected for signs of water intrusion. If DTC is not current, clear DTCs and simulate the effects of water intrusion. Use the following procedure. Spray down the suspected area with a 5% salt water solution (two teaspoons of salt to 12 oz. of water). Test drive vehicle over various road surfaces (bumps, turns, etc.) above 24 km/h (15 mph) for at least 30 seconds. If DTC returns, replace suspected harness.

Any circuitry, that is suspected as causing the intermittent complaint, should be thoroughly checked for backed out terminals, improper mating, broken locks, improperly formed or damaged terminals, poor terminal to wiring connections or physical damage to the wiring harness.

Resistance of the wheel speed sensor will increase with an increase in sensor temperature.

When replacing a wheel speed sensor, inspect the sensor terminals and harness connector for corrosion and/or water intrusion. If evidence of corrosion or water intrusion exists, replace wheel speed sensor jumper harness. Likewise, if replacing a wheel speed sensor jumper harness, inspect sensor terminals. If evidence of corrosion or water intrusion exists, replace wheel speed sensor. Refer to "On-Vehicle Service" in this section.

DIAGNOSTIC CHART DTC 23 — 1994 SILHOUETTE, TRANS SPORT AND LUMINA APV

DTC 23

LEFT REAR WHEEL SPEED = 0

Important

- WHEEL SPEED SENSOR INTERMITTENT MALFUNCTIONS MAY BE DIFFICULT TO LOCATE. CARE SHOULD BE TAKEN NOT TO DISTURB ANY ELECTRICAL CONNECTIONS PRIOR TO AN INDICATED STEP OF THIS CHART. THIS WILL ENSURE THAT AN INTERMITTENT CONNECTION WILL NOT BE CORRECTED BEFORE THE SOURCE OF THE MALFUNCTION IS FOUND.

(1)
- IGNITION "OFF."
- PHYSICALLY INSPECT THE LEFT REAR WHEEL SPEED SENSOR, WHEEL SPEED SENSOR JUMPER HARNESS, AND CONNECTORS FOR DAMAGE.

IS PHYSICAL DAMAGE INDICATED?

YES: REPLACE WHEEL SPEED SENSOR AND/OR WHEEL SPEED SENSOR JUMPER HARNESS IF DAMAGED. DRIVE VEHICLE ABOVE 24 km/h (15 MPH) TO ENSURE DTC 23 DOES NOT SET.

NO:

(2)
- DISCONNECT LEFT REAR WHEEL SPEED SENSOR DIRECTLY AT SENSOR.
- USING J 39200, MEASURE THE RESISTANCE BETWEEN SENSOR TERMINALS "A" AND "B."

IS RESISTANCE BETWEEN 2059 AND 2353Ω WHEN SENSOR IS 20°C (68°F)?

(IF TEMPERATURE OF THE SENSOR IS SUSPECTED TO BE OTHER THAN 20°C (68°F), REFER TO THE TABLE BELOW FOR APPLICABLE SENSOR RESISTANCE VALUES; THE VALUES IN THE TABLE ARE FOR TEMPERATURE OF THE SENSOR, NOT AIR TEMPERATURE.)

NO: REPLACE WHEEL SPEED SENSOR.

YES:

(3)
- WITH J 39200 STILL CONNECTED, SELECT A/C VOLTAGE SCALE.
- SPIN WHEEL BY HAND WHILE OBSERVING VOLTAGE.

IS VOLTAGE AT LEAST 100 mV (VOLTAGE WILL INCREASE AS WHEEL SPEED INCREASES)?

NO: REPLACE WHEEL SPEED SENSOR.

YES:

(4)
- IGNITION "OFF."
- DISCONNECT 32-WAY EBCM CONNECTOR.
- USING J 39200, MEASURE THE RESISTANCE BETWEEN TERMINAL "E1" AND "E2" OF THE 32-WAY EBCM HARNESS CONNECTOR.

DOES J 39200 DISPLAY RESISTANCE AS "OL" (INFINITE)?

NO: REPAIR SHORT BETWEEN CKTs 884 AND 885.

YES:

(5)
- INSPECT CKTs 884 AND 885 AND 32-WAY EBCM CONNECTOR FOR DAMAGE WHICH MAY RESULT IN A SHORT BETWEEN CKTs 884 AND 885 WITH ALL CONNECTORS CONNECTED. REPAIR ALL DAMAGE IF EVIDENT.
- RECONNECT ALL CONNECTORS.
- TEST DRIVE VEHICLE ABOVE 24 km/h (15 MPH) FOR AT LEAST 30 SECONDS.

DOES DTC 23 SET AS A CURRENT DTC?

NO: MALFUNCTION WAS MOST LIKELY CAUSED BY A POOR CONNECTION. REFER TO "DIAGNOSTIC AIDS" ON FACING PAGE.

YES:

(6)
- USING TECH 1, SELECT "F0: DATA LIST."
- MONITOR WHEEL SPEEDS WHILE SLOWLY ACCELERATING TO 65 km/h (40 MPH) AND THEN SLOWLY DECELERATING TO 0 km/h (0 MPH).

IS THE LEFT REAR WHEEL SPEED CONSTANTLY HIGHER THAN THE THREE REMAINING WHEEL SPEEDS?

NO: REPLACE EBCM.

YES: REPLACE INCORRECT OR MALFUNCTIONING LEFT WHEEL SPEED SENSOR/HUB AND BEARING ASSEMBLY WITH THE PROPER ASSEMBLY. REFER TO PARTS CATALOG FOR PROPER PART NUMBER.

REAR WHEEL SPEED SENSOR
TEMPERATURE OF SENSOR VS. SENSOR RESISTANCE (APPROXIMATE)

TEMP. (°C)	TEMP. (°F)	RESISTANCE (OHMS)
-34 TO 4	-30 TO 40	1622 TO 2205
5 TO 43	41 TO 110	1930 TO 2566
44 TO 93	111 TO 200	2245 TO 3028

DIAGNOSTIC CHART DTC 24 — 1994 SILHOUETTE, TRANS SPORT AND LUMINA APV

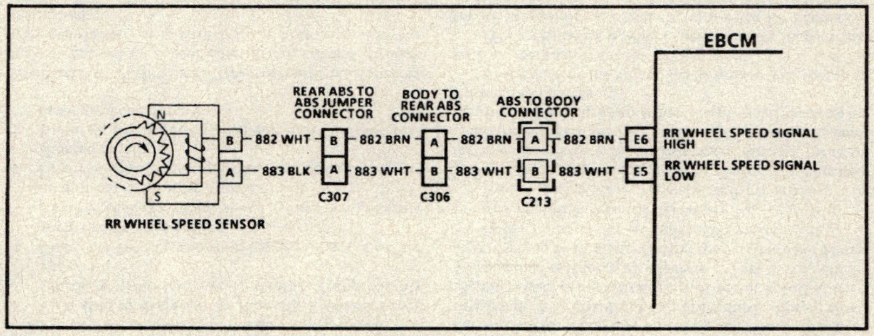

DTC 24

RIGHT REAR WHEEL SPEED = 0

Circuit Description:

As a toothed ring passes by the wheel speed sensor, changes in the electromagnetic field cause the wheel speed sensor to produce a sinusoidal (AC) voltage signal whose frequency is proportional to wheel speed. The magnitude of this signal is directly related to wheel speed and the proximity of the wheel speed sensor to the toothed ring, often referred to as the air gap.

DTC Will Set When: DTC 24 can be set when the vehicle is not in an ABS stop. If the right rear wheel speed = 0 and the vehicle's reference speed is greater than 8 km/h (5 mph), a malfunction exists.

Action Taken: A malfunction DTC is stored, ABS is disabled and the ABS warning lamp is turned "ON."

DTC Chart Test Description: Number(s) below refer to circled number(s) on the diagnostic chart.

1. This step is designed to identify a wheel speed sensor or circuitry that is damaged and visibly apparent.
2. Checks to see that the wheel speed sensor is within proper resistance values.
3. Verifies that the wheel speed sensor can generate an appropriate output signal.
4. Ensures that the wheel speed sensor circuitry is not shorted together.
5. Checks to see if malfunction was due to physical damage of the circuitry.
6. Checks for an incorrect or malfunctioning hub and bearing assembly.

Diagnostic Aids: An "Intermittent" malfunction may be caused by a poor connection, rubbed through wire insulation, or a wire that is broken inside the insulation.

The frequency of the malfunction can be checked by using the enhanced diagnostic function of the Tech 1.

If the customer's comments reflect that the ABS warning lamp is "ON" only during moist environmental changes (rain, snow, vehicle wash), all wheel speed sensor circuitry should be thoroughly inspected for signs of water intrusion. If DTC is not current, clear DTCs and simulate the effects of water intrusion. Use the following procedure. Spray down the suspected area with a 5% salt water solution (two teaspoons of salt to 12 oz. of water). Test drive vehicle over various road surfaces (bumps, turns, etc.) above 24 km/h (15 mph) for at least 30 seconds. If DTC returns, replace suspected harness.

Any circuitry, that is suspected as causing the intermittent complaint, should be thoroughly checked for backed out terminals, improper mating, broken locks, improperly formed or damaged terminals, poor terminal to wiring connections or physical damage to the wiring harness.

Resistance of the wheel speed sensor will increase with an increase in sensor temperature.

When replacing a wheel speed sensor, inspect the sensor terminals and narness connector for corrosion and/or water intrusion. If evidence of corrosion or water intrusion exists, replace wheel speed sensor jumper harness. Likewise, if replacing a wheel speed sensor jumper harness, inspect sensor terminals. If evidence of corrosion or water intrusion exists, replace wheel speed sensor. Refer to "On-Vehicle Service" in this section.

DIAGNOSTIC CHART DTC 24 — 1994 SILHOUETTE, TRANS SPORT AND LUMINA APV

DTC 24
RIGHT REAR WHEEL SPEED = 0

Important

- WHEEL SPEED SENSOR INTERMITTENT MALFUNCTIONS MAY BE DIFFICULT TO LOCATE. CARE SHOULD BE TAKEN NOT TO DISTURB ANY ELECTRICAL CONNECTIONS PRIOR TO AN INDICATED STEP OF THIS CHART. THIS WILL ENSURE THAT AN INTERMITTENT CONNECTION WILL NOT BE CORRECTED BEFORE THE SOURCE OF THE MALFUNCTION IS FOUND.

(1)
- IGNITION "OFF."
- PHYSICALLY INSPECT THE RIGHT REAR WHEEL SPEED SENSOR, WHEEL SPEED SENSOR JUMPER HARNESS, AND CONNECTORS FOR DAMAGE. IS PHYSICAL DAMAGE INDICATED?

YES: REPLACE WHEEL SPEED SENSOR AND/OR WHEEL SPEED SENSOR JUMPER HARNESS IF DAMAGED. DRIVE VEHICLE ABOVE 24 km/h (15 MPH) TO ENSURE DTC 24 DOES NOT SET.

NO:

(2)
- DISCONNECT RIGHT REAR WHEEL SPEED SENSOR DIRECTLY AT SENSOR.
- USING J 39200, MEASURE THE RESISTANCE BETWEEN SENSOR TERMINALS "A" AND "B." IS RESISTANCE BETWEEN 2059 AND 2353Ω WHEN SENSOR IS 20°C (68°F)?
(IF TEMPERATURE OF THE SENSOR IS SUSPECTED TO BE OTHER THAN 20°C (68°F), REFER TO THE TABLE BELOW FOR APPLICABLE SENSOR RESISTANCE VALUES; THE VALUES IN THE TABLE ARE FOR TEMPERATURE OF THE SENSOR, NOT AIR TEMPERATURE.)

NO: REPLACE WHEEL SPEED SENSOR.

YES:

(3)
- WITH J 39200 STILL CONNECTED, SELECT A/C VOLTAGE SCALE.
- SPIN WHEEL BY HAND WHILE OBSERVING VOLTAGE. IS VOLTAGE AT LEAST 100 mV (VOLTAGE WILL INCREASE AS WHEEL SPEED INCREASES)?

NO: REPLACE WHEEL SPEED SENSOR.

YES:

(4)
- IGNITION "OFF."
- DISCONNECT 32-WAY EBCM CONNECTOR.
- USING J 39200, MEASURE THE RESISTANCE BETWEEN TERMINAL "E5" AND "E6" OF THE 32-WAY EBCM HARNESS CONNECTOR. DOES J 39200 DISPLAY RESISTANCE AS "OL" (INFINITE)?

NO: REPAIR SHORT BETWEEN CKTs 883 AND 883.

YES:

(5)
- INSPECT CKTs 882 AND 883 AND 32-WAY EBCM CONNECTOR FOR DAMAGE WHICH MAY RESULT IN A SHORT BETWEEN CKTs 882 AND 883 WITH ALL CONNECTORS CONNECTED. REPAIR ALL DAMAGE IF EVIDENT.
- RECONNECT ALL CONNECTORS.
- TEST DRIVE VEHICLE ABOVE 24 km/h (15 MPH) FOR AT LEAST 30 SECONDS. DOES DTC 24 SET AS A CURRENT DTC?

NO: MALFUNCTION WAS MOST LIKELY CAUSED BY A POOR CONNECTION. REFER TO "DIAGNOSTIC AIDS" ON FACING PAGE.

YES:

(6)
- USING TECH 1, SELECT "F0: DATA LIST."
- MONITOR WHEEL SPEEDS WHILE SLOWLY ACCELERATING TO 65 km/h (40 MPH) AND THEN SLOWLY DECELERATING TO 0 km/h (0 MPH). IS THE RIGHT REAR WHEEL SPEED CONSTANTLY HIGHER THAN THE THREE REMAINING WHEEL SPEEDS?

NO: REPLACE EBCM.

YES: REPLACE INCORRECT OR MALFUNCTIONING RIGHT WHEEL SPEED SENSOR/HUB AND BEARING ASSEMBLY WITH THE PROPER ASSEMBLY. REFER TO PARTS CATALOG FOR PROPER PART NUMBER.

REAR WHEEL SPEED SENSOR
TEMPERATURE OF SENSOR VS. SENSOR RESISTANCE (APPROXIMATE)

TEMP. (°C)	TEMP. (°F)	RESISTANCE (OHMS)
-34 TO 4	-30 TO 40	1622 TO 2205
5 TO 43	41 TO 110	1930 TO 2566
44 TO 93	111 TO 200	2245 TO 3028

DIAGNOSTIC CHART DTC 25 — 1994 SILHOUETTE, TRANS SPORT AND LUMINA APV

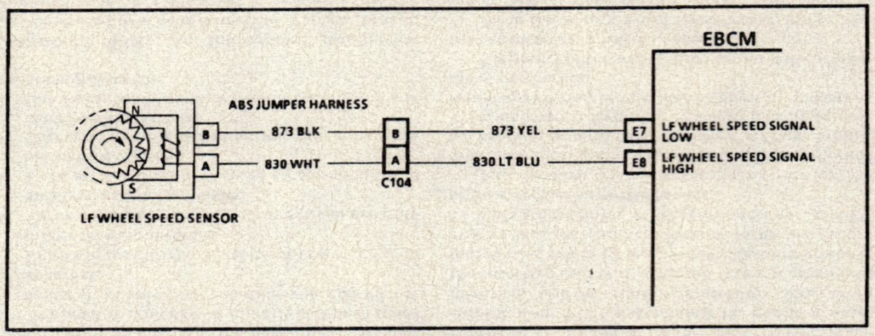

DTC 25

LEFT FRONT EXCESSIVE WHEEL SPEED VARIATION

Circuit Description:

As a toothed ring passes by the wheel speed sensor, changes in the electromagnetic field cause the wheel speed sensor to produce a sinusoidal (AC) voltage signal whose frequency is proportional to wheel speed. The magnitude of this signal is directly related to wheel speed and the proximity of the wheel speed sensor to the toothed ring, often referred to as the air gap.

DTC Will Set When: DTC 25 can be set when the brake is "OFF." The purpose of this test is to detect a situation in which the left front wheel acceleration or deceleration is beyond specified limits.

Action Taken: A malfunction DTC is stored, ABS is disabled and the ABS warning lamp is turned "ON."

DTC Chart Test Description: Number(s) below refer to circled number(s) on the diagnostic chart.

1. This step is designed to identify a wheel speed sensor or circuitry that is damaged and visibly apparent.
2. Checks to see that the wheel speed sensor is within proper resistance values.
3. Verifies that the wheel speed sensor can generate an appropriate output signal.
4. Ensures that wheel speed sensor circuitry is not shorted together.
5. Ensures malfunction was not due to physical damage of the circuitry.

Diagnostic Aids: An "Intermittent" malfunction may be caused by a poor connection, rubbed through wire insulation, or a wire that is broken inside the insulation.

The frequency of the malfunction can be checked by using the enhanced diagnostic function of the Tech 1.

If the customer's comments reflect that the ABS warning lamp is "ON" only during moist environmental changes (rain, snow, vehicle wash), all wheel speed sensor circuitry should be thoroughly inspected for signs of water intrusion. If DTC is not current, clear DTCs and simulate the effects of water intrusion. Use the following procedure. Spray down the suspected area with a 5% salt water solution (two teaspoons of salt to 12 oz. of water). Test drive vehicle over various road surfaces (bumps, turns, etc.) above 24 km/h (15 mph) for at least 30 seconds. If DTC returns, replace suspected harness.

Any circuitry, that is suspected as causing the intermittent complaint, should be thoroughly checked for backed out terminals, improper mating, broken locks, improperly formed or damaged terminals, poor terminal to wiring connections or physical damage to the wiring harness.

Resistance of the wheel speed sensor will increase with an increase in sensor temperature.

When replacing a wheel speed sensor, inspect the sensor terminals and harness connector for corrosion and/or water intrusion. If evidence of corrosion or water intrusion exists, replace wheel speed sensor jumper harness. Likewise, if replacing a wheel speed sensor jumper harness, inspect sensor terminals. If evidence of corrosion or water intrusion exists, replace wheel speed sensor.

DIAGNOSTIC CHART DTC 25 — 1994 SILHOUETTE, TRANS SPORT AND LUMINA APV

DTC 25

LEFT FRONT EXCESSIVE WHEEL SPEED VARIATION

Important

- WHEEL SPEED SENSOR INTERMITTENT MALFUNCTIONS MAY BE DIFFICULT TO LOCATE. CARE SHOULD BE TAKEN NOT TO DISTURB ANY ELECTRICAL CONNECTIONS PRIOR TO AN INDICATED STEP OF THIS CHART. THIS WILL ENSURE THAT AN INTERMITTENT CONNECTION WILL NOT BE CORRECTED BEFORE THE SOURCE OF THE MALFUNCTION IS FOUND.

(1)
- IGNITION "OFF."
- PHYSICALLY INSPECT THE LEFT FRONT WHEEL SPEED SENSOR, LEFT FRONT WHEEL SPEED SENSOR JUMPER HARNESS, AND CONNECTORS FOR DAMAGE. IS PHYSICAL DAMAGE INDICATED?

YES: REPLACE WHEEL SPEED SENSOR AND/OR JUMPER HARNESS IF DAMAGED. DRIVE VEHICLE ABOVE 24 km/h (15 MPH) TO ENSURE DTC 25 DOES NOT SET.

NO:

(2)
- DISCONNECT LEFT FRONT WHEEL SPEED SENSOR DIRECTLY AT SENSOR.
- USING J 39200, MEASURE THE RESISTANCE BETWEEN SENSOR TERMINALS "A" AND "B."

IS RESISTANCE BETWEEN 1020 AND 1137Ω WHEN SENSOR IS 20°C (68°F)? *(IF TEMPERATURE OF THE SENSOR IS SUSPECTED TO BE OTHER THAN 20°C (68°F), REFER TO THE TABLE BELOW FOR APPLICABLE SENSOR RESISTANCE VALUES; THE VALUES IN THE TABLE ARE FOR TEMPERATURE OF THE SENSOR, NOT AIR TEMPERATURE.)*

NO: REPLACE WHEEL SPEED SENSOR.

YES:

(3)
- WITH J 39200 STILL CONNECTED, SELECT A/C VOLTAGE SCALE.
- SPIN WHEEL BY HAND WHILE OBSERVING VOLTAGE.

IS VOLTAGE AT LEAST 100 mV (VOLTAGE WILL INCREASE AS WHEEL SPEED INCREASES)?

NO: REPLACE WHEEL SPEED SENSOR.

YES:

(4)
- IGNITION "OFF."
- DISCONNECT 32-WAY EBCM CONNECTOR.
- USING J 39200, MEASURE THE RESISTANCE BETWEEN TERMINAL "E7" AND "E8" OF THE 32-WAY EBCM HARNESS CONNECTOR.

DOES J 39200 DISPLAY RESISTANCE AS "OL" (INFINITE)?

NO: REPAIR SHORT BETWEEN CKTs 830 AND 873.

YES:

(5)
- INSPECT CKTs 830 AND 873 AND 32-WAY EBCM CONNECTOR FOR DAMAGE WHICH MAY RESULT IN A SHORT BETWEEN CKTs 830 AND 873 WITH ALL CONNECTORS CONNECTED. REPAIR ALL DAMAGE IF EVIDENT.
- RECONNECT ALL CONNECTORS.
- TEST DRIVE VEHICLE ABOVE 24 km/h (15 MPH) FOR AT LEAST 30 SECONDS. DOES DTC 25 SET AS A CURRENT DTC?

YES: REPLACE EBCM.

NO: MALFUNCTION IS INTERMITTENT. REFER TO "DIAGNOSTIC AIDS" ON FACING PAGE.

FRONT WHEEL SPEED SENSOR
TEMPERATURE OF SENSOR VS. SENSOR RESISTANCE (APPROXIMATE)

TEMP. (°C)	TEMP. (°F)	RESISTANCE (OHMS)
-34 TO 4	-30 TO 40	804 TO 1066
5 TO 43	41 TO 110	956 TO 1240
44 TO 93	111 TO 200	1112 TO 1463

DIAGNOSTIC CHART DTC 26 — 1994 SILHOUETTE, TRANS SPORT AND LUMINA APV

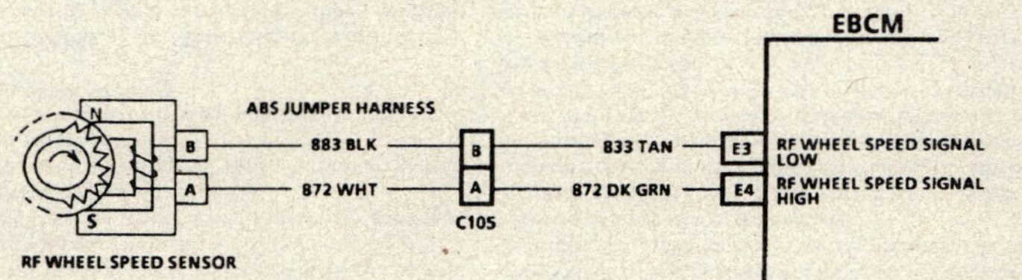

DTC 26

RIGHT FRONT EXCESSIVE WHEEL SPEED VARIATION

Circuit Description:

As a toothed ring passes by the wheel speed sensor, changes in the electromagnetic field cause the wheel speed sensor to produce a sinusoidal (AC) voltage signal whose frequency is proportional to wheel speed. The magnitude of this signal is directly related to wheel speed and the proximity of the wheel speed sensor to the toothed ring, often referred to as the air gap.

DTC Will Set When: DTC 26 can be set when the brake is "OFF." The purpose of this test is to detect a situation in which the right front wheel acceleration or deceleration is beyond specified limits.

Action Taken: A malfunction DTC is stored, ABS is disabled and the ABS warning lamp is turned "ON."

DTC Chart Test Description: Number(s) below refer to circled number(s) on the diagnostic chart.

1. This step is designed to identify a wheel speed sensor or circuitry that is damaged and visibly apparent.
2. Checks to see that the wheel speed sensor is within proper resistance values.
3. Verifies that the wheel speed sensor can generate an appropriate output signal.
4. Ensures that wheel speed sensor circuitry is not shorted together.
5. Ensures malfunction was not due to physical damage of the circuitry.

Diagnostic Aids: An "Intermittent" malfunction may be caused by a poor connection, rubbed through wire insulation, or a wire that is broken inside the insulation.

The frequency of the malfunction can be checked by using the enhanced diagnostic function of the Tech 1.

If the customer's comments reflect that the ABS warning lamp is "ON" only during moist environmental changes (rain, snow, vehicle wash), all wheel speed sensor circuitry should be thoroughly inspected for signs of water intrusion. If DTC is not current, clear DTCs and simulate the effects of water intrusion. Use the following procedure. Spray down the suspected area with a 5% salt water solution (two teaspoons of salt to 12 oz. of water). Test drive vehicle over various road surfaces (bumps, turns, etc.) above 24 km/h (15 mph) for at least 30 seconds. If DTC returns, replace suspected harness.

Any circuitry, that is suspected as causing the intermittent complaint, should be thoroughly checked for backed out terminals, improper mating, broken locks, improperly formed or damaged terminals, poor terminal to wiring connections or physical damage to the wiring harness.

Resistance of the wheel speed sensor will increase with an increase in sensor temperature.

When replacing a wheel speed sensor, inspect the sensor terminals and harness connector for corrosion and/or water intrusion. If evidence of corrosion or water intrusion exists, replace wheel speed sensor jumper harness. Likewise, if replacing a wheel speed sensor jumper harness, inspect sensor terminals. If evidence of corrosion or water intrusion exists, replace wheel speed sensor. Refer to "On-Vehicle Service" in this section.

DIAGNOSTIC CHART DTC 26 — 1994 SILHOUETTE, TRANS SPORT AND LUMINA APV

DTC 26
RIGHT FRONT EXCESSIVE WHEEL SPEED VARIATION

Important
- WHEEL SPEED SENSOR INTERMITTENT MALFUNCTIONS MAY BE DIFFICULT TO LOCATE. CARE SHOULD BE TAKEN NOT TO DISTURB ANY ELECTRICAL CONNECTIONS PRIOR TO AN INDICATED STEP OF THIS CHART. THIS WILL ENSURE THAT AN INTERMITTENT CONNECTION WILL NOT BE CORRECTED BEFORE THE SOURCE OF THE MALFUNCTION IS FOUND.

(1)
- IGNITION "OFF."
- PHYSICALLY INSPECT THE RIGHT FRONT WHEEL SPEED SENSOR, RIGHT FRONT WHEEL SPEED SENSOR JUMPER HARNESS, AND CONNECTORS FOR DAMAGE. IS PHYSICAL DAMAGE INDICATED?

YES: REPLACE WHEEL SPEED SENSOR AND/OR JUMPER HARNESS IF DAMAGED. DRIVE VEHICLE ABOVE 24 km/h (15 MPH) TO ENSURE DTC 26 DOES NOT SET.

NO:

(2)
- DISCONNECT RIGHT FRONT WHEEL SPEED SENSOR DIRECTLY AT SENSOR.
- USING J 39200, MEASURE THE RESISTANCE BETWEEN SENSOR TERMINALS "A" AND "B."
IS RESISTANCE BETWEEN 1020 AND 1137Ω WHEN SENSOR IS 20°C (68°F)?
(IF TEMPERATURE OF THE SENSOR IS SUSPECTED TO BE OTHER THAN 20°C (68°F), REFER TO THE TABLE BELOW FOR APPLICABLE SENSOR RESISTANCE VALUES; THE VALUES IN THE TABLE ARE FOR TEMPERATURE OF THE SENSOR, NOT AIR TEMPERATURE.)

NO: REPLACE WHEEL SPEED SENSOR.

YES:

(3)
- WITH J 39200 STILL CONNECTED, SELECT A/C VOLTAGE SCALE.
- SPIN WHEEL BY HAND WHILE OBSERVING VOLTAGE.
IS VOLTAGE AT LEAST 100 mV (VOLTAGE WILL INCREASE AS WHEEL SPEED INCREASES)?

NO: REPLACE WHEEL SPEED SENSOR.

YES:

(4)
- IGNITION "OFF."
- DISCONNECT 32-WAY EBCM CONNECTOR.
- USING J 39200, MEASURE THE RESISTANCE BETWEEN TERMINAL "E3" AND "E4" OF THE 32-WAY EBCM HARNESS CONNECTOR.
DOES J 39200 DISPLAY RESISTANCE AS "OL" (INFINITE)?

NO: REPAIR SHORT BETWEEN CKTs 833 AND 872.

YES:

(5)
- INSPECT CKTs 833 AND 872 AND 32-WAY EBCM CONNECTOR FOR DAMAGE WHICH MAY RESULT IN A SHORT BETWEEN CKTs 833 AND 872 WITH ALL CONNECTORS CONNECTED. REPAIR ALL DAMAGE IF EVIDENT.
- RECONNECT ALL CONNECTORS.
- TEST DRIVE VEHICLE ABOVE 24 km/h (15 MPH) FOR AT LEAST 30 SECONDS.
DOES DTC 26 SET AS A CURRENT DTC?

YES: REPLACE EBCM.

NO: MALFUNCTION IS INTERMITTENT. REFER TO "DIAGNOSTIC AIDS" ON FACING PAGE.

FRONT WHEEL SPEED SENSOR
TEMPERATURE OF SENSOR VS. SENSOR RESISTANCE (APPROXIMATE)

TEMP. (°C)	TEMP. (°F)	RESISTANCE (OHMS)
-34 TO 4	-30 TO 40	804 TO 1066
5 TO 43	41 TO 110	956 TO 1240
44 TO 93	111 TO 200	1112 TO 1463

DIAGNOSTIC CHART DTC 27 — 1994 SILHOUETTE, TRANS SPORT AND LUMINA APV

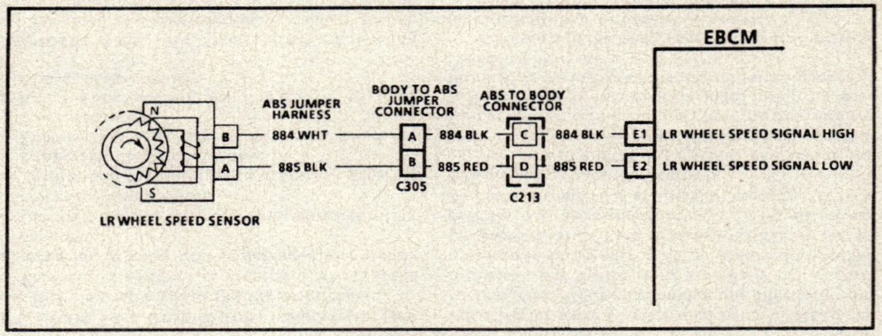

DTC 27

LEFT REAR EXCESSIVE WHEEL SPEED VARIATION

Circuit Description:

As a toothed ring passes by the wheel speed sensor, changes in the electromagnetic field cause the wheel speed sensor to produce a sinusoidal (AC) voltage signal whose frequency is proportional to wheel speed. The magnitude of this signal is directly related to wheel speed and the proximity of the wheel speed sensor to the toothed ring, often referred to as the air gap.

DTC Will Set When: DTC 27 can be set when the brake is "OFF." The purpose of this test is to detect a situation in which the left rear wheel acceleration or deceleration is beyond specified limits.

Action Taken: A malfunction DTC is stored, ABS is disabled and the ABS warning lamp is turned "ON."

DTC Chart Test Description: Number(s) below refer to circled number(s) on the diagnostic chart.

1. This step is designed to identify a wheel speed sensor or circuitry that is damaged and visibly apparent.
2. Checks to see that the wheel speed sensor is within proper resistance values.
3. Verifies that the wheel speed sensor can generate an appropriate output signal.
4. Ensures that wheel speed sensor circuitry is not shorted together.
5. Ensures malfunction was not due to physical damage of the circuitry.

Diagnostic Aids: An "Intermittent" malfunction may be caused by a poor connection, rubbed through wire insulation, or a wire that is broken inside the insulation.

The frequency of the malfunction can be checked by using the enhanced diagnostic function of the Tech 1.

If the customer's comments reflect that the ABS warning lamp is "ON" only during moist environmental changes (rain, snow, vehicle wash), all wheel speed sensor circuitry should be thoroughly inspected for signs of water intrusion. If DTC is not current, clear DTCs and simulate the effects of water intrusion. Use the following procedure. Spray down the suspected area with a 5% salt water solution (two teaspoons of salt to 12 oz. of water). Test drive vehicle over various road surfaces (bumps, turns, etc.) above 24 km/h (15 mph) for at least 30 seconds. If DTC returns, replace suspected harness.

Any circuitry, that is suspected as causing the intermittent complaint, should be thoroughly checked for backed out terminals, improper mating, broken locks, improperly formed or damaged terminals, poor terminal to wiring connections or physical damage to the wiring harness.

Resistance of the wheel speed sensor will increase with an increase in sensor temperature.

When replacing a wheel speed sensor, inspect the sensor terminals and harness connector for corrosion and/or water intrusion. If evidence of corrosion or water intrusion exists, replace wheel speed sensor jumper harness. Likewise, if replacing a wheel speed sensor jumper harness, inspect sensor terminals. If evidence of corrosion or water intrusion exists, replace wheel speed sensor.

DIAGNOSTIC CHART DTC 27 — 1994 SILHOUETTE, TRANS SPORT AND LUMINA APV

DTC 27
LEFT REAR EXCESSIVE WHEEL SPEED VARIATION

Important

- WHEEL SPEED SENSOR INTERMITTENT MALFUNCTIONS MAY BE DIFFICULT TO LOCATE. CARE SHOULD BE TAKEN NOT TO DISTURB ANY ELECTRICAL CONNECTIONS PRIOR TO AN INDICATED STEP OF THIS CHART. THIS WILL ENSURE THAT AN INTERMITTENT CONNECTION WILL NOT BE CORRECTED BEFORE THE SOURCE OF THE MALFUNCTION IS FOUND.

(1)
- IGNITION "OFF."
- PHYSICALLY INSPECT THE LEFT REAR WHEEL SPEED SENSOR, WHEEL SPEED SENSOR JUMPER HARNESS, AND CONNECTORS FOR DAMAGE. IS PHYSICAL DAMAGE INDICATED?

YES: REPLACE WHEEL SPEED SENSOR AND/OR WHEEL SPEED SENSOR JUMPER HARNESS IF DAMAGED. DRIVE VEHICLE ABOVE 24 km/h (15 MPH) TO ENSURE DTC 27 DOES NOT SET.

NO:

(2)
- DISCONNECT LEFT REAR WHEEL SPEED SENSOR DIRECTLY AT SENSOR.
- USING J 39200, MEASURE THE RESISTANCE BETWEEN SENSOR TERMINALS "A" AND "B."

IS RESISTANCE BETWEEN 2059 AND 2353Ω WHEN SENSOR IS 20°C (68°F)? *(IF TEMPERATURE OF THE SENSOR IS SUSPECTED TO BE OTHER THAN 20°C (68°F), REFER TO THE TABLE BELOW FOR APPLICABLE SENSOR RESISTANCE VALUES; THE VALUES IN THE TABLE ARE FOR TEMPERATURE OF THE SENSOR, NOT AIR TEMPERATURE.)*

NO: REPLACE WHEEL SPEED SENSOR.

YES:

(3)
- WITH J 39200 STILL CONNECTED, SELECT A/C VOLTAGE SCALE.
- SPIN WHEEL BY HAND WHILE OBSERVING VOLTAGE.

IS VOLTAGE AT LEAST 100 mV (VOLTAGE WILL INCREASE AS WHEEL SPEED INCREASES)?

NO: REPLACE WHEEL SPEED SENSOR.

YES:

(4)
- IGNITION "OFF."
- DISCONNECT 32-WAY EBCM CONNECTOR.
- USING J 39200, MEASURE THE RESISTANCE BETWEEN TERMINAL "E1" AND "E2" OF THE 32-WAY EBCM HARNESS CONNECTOR.

DOES J 39200 DISPLAY RESISTANCE AS "OL" (INFINITE)?

NO: REPAIR SHORT BETWEEN CKTs 884 AND 885.

YES:

(5)
- INSPECT CKTs 884 AND 885 AND 32-WAY EBCM CONNECTOR FOR DAMAGE WHICH MAY RESULT IN A SHORT BETWEEN CKTs 884 AND 885 WITH ALL CONNECTORS CONNECTED. REPAIR ALL DAMAGE IF EVIDENT.
- RECONNECT ALL CONNECTORS.
- TEST DRIVE VEHICLE ABOVE 24 km/h (15 MPH) FOR AT LEAST 30 SECONDS. DOES DTC 27 SET AS A CURRENT DTC?

YES: REPLACE EBCM.

NO: MALFUNCTION IS INTERMITTENT. REFER TO "DIAGNOSTIC AIDS" ON FACING PAGE.

REAR WHEEL SPEED SENSOR
TEMPERATURE OF SENSOR VS. SENSOR RESISTANCE (APPROXIMATE)

TEMP. (°C)	TEMP. (°F)	RESISTANCE (OHMS)
-34 TO 4	-30 TO 40	1622 TO 2205
5 TO 43	41 TO 110	1930 TO 2566
44 TO 93	111 TO 200	2245 TO 3028

DIAGNOSTIC CHART DTC 28 — 1994 SILHOUETTE, TRANS SPORT AND LUMINA APV

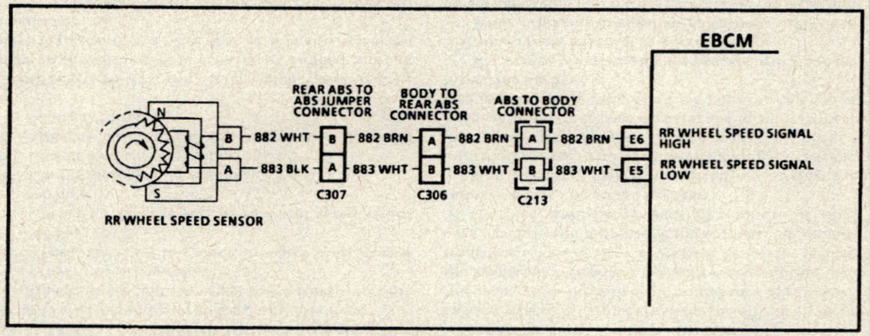

DTC 28

RIGHT REAR EXCESSIVE WHEEL SPEED VARIATION

Circuit Description:

As a toothed ring passes by the wheel speed sensor, changes in the electromagnetic field cause the wheel speed sensor to produce a sinusoidal (AC) voltage signal whose frequency is proportional to wheel speed. The magnitude of this signal is directly related to wheel speed and the proximity of the wheel speed sensor to the toothed ring, often referred to as the air gap.

DTC Will Set When: DTC 28 can be set when the brake is "OFF." The purpose of this test is to detect a situation in which the right rear wheel acceleration or deceleration is beyond specified limits.

Action Taken: A malfunction DTC is stored, ABS is disabled and the ABS warning lamp is turned "ON."

DTC Chart Test Description: Number(s) below refer to circled number(s) on the diagnostic chart.

1. This step is designed to identify a wheel speed sensor or circuitry that is damaged and visibly apparent.
2. Checks to see that the wheel speed sensor is within proper resistance values.
3. Verifies that the wheel speed sensor can generate an appropriate output signal.
4. Ensures that wheel speed sensor circuitry is not shorted together.
5. Ensures malfunction was not due to physical damage of the circuitry.

Diagnostic Aids: An "Intermittent" malfunction may be caused by a poor connection, rubbed through wire insulation, or a wire that is broken inside the insulation.

The frequency of the malfunction can be checked by using the enhanced diagnostic function of the Tech 1.

If the customer's comments reflect that the ABS warning lamp is "ON" only during moist environmental changes (rain, snow, vehicle wash), all wheel speed sensor circuitry should be thoroughly inspected for signs of water intrusion. If DTC is not current, clear DTCs and simulate the effects of water intrusion. Use the following procedure. Spray down the suspected area with a 5% salt water solution (two teaspoons of salt to 12 oz. of water). Test drive vehicle over various road surfaces (bumps, turns, etc.) above 24 km/h (15 mph) for at least 30 seconds. If DTC returns, replace suspected harness.

Any circuitry, that is suspected as causing the intermittent complaint, should be thoroughly checked for backed out terminals, improper mating, broken locks, improperly formed or damaged terminals, poor terminal to wiring connections or physical damage to the wiring harness.

Resistance of the wheel speed sensor will increase with an increase in sensor temperature.

When replacing a wheel speed sensor, inspect the sensor terminals and harness connector for corrosion and/or water intrusion. If evidence of corrosion or water intrusion exists, replace wheel speed sensor jumper harness. Likewise, if replacing a wheel speed sensor jumper harness, inspect sensor terminals. If evidence of corrosion or water intrusion exists, replace wheel speed sensor.

DIAGNOSTIC CHART DTC 28 — 1994 SILHOUETTE, TRANS SPORT AND LUMINA APV

DTC 28
RIGHT REAR EXCESSIVE WHEEL SPEED VARIATION

Important

- WHEEL SPEED SENSOR INTERMITTENT MALFUNCTIONS MAY BE DIFFICULT TO LOCATE. CARE SHOULD BE TAKEN NOT TO DISTURB ANY ELECTRICAL CONNECTIONS PRIOR TO AN INDICATED STEP OF THIS CHART. THIS WILL ENSURE THAT AN INTERMITTENT CONNECTION WILL NOT BE CORRECTED BEFORE THE SOURCE OF THE MALFUNCTION IS FOUND.

(1)
- IGNITION "OFF."
- PHYSICALLY INSPECT THE RIGHT REAR WHEEL SPEED SENSOR, WHEEL SPEED SENSOR JUMPER HARNESS, AND CONNECTORS FOR DAMAGE. IS PHYSICAL DAMAGE INDICATED?

YES: REPLACE WHEEL SPEED SENSOR AND/OR WHEEL SPEED SENSOR JUMPER HARNESS IF DAMAGED. DRIVE VEHICLE ABOVE 24 km/h (15 MPH) TO ENSURE DTC 28 DOES NOT SET.

NO:

(2)
- DISCONNECT RIGHT REAR WHEEL SPEED SENSOR DIRECTLY AT SENSOR.
- USING J 39200, MEASURE THE RESISTANCE BETWEEN SENSOR TERMINALS "A" AND "B." IS RESISTANCE BETWEEN 2059 AND 2353Ω WHEN SENSOR IS 20°C (68°F)? *(IF TEMPERATURE OF THE SENSOR IS SUSPECTED TO BE OTHER THAN 20°C (68°F), REFER TO THE TABLE BELOW FOR APPLICABLE SENSOR RESISTANCE VALUES; THE VALUES IN THE TABLE ARE FOR TEMPERATURE OF THE SENSOR, NOT AIR TEMPERATURE.)*

NO: REPLACE WHEEL SPEED SENSOR.

YES:

(3)
- WITH J 39200 STILL CONNECTED, SELECT A/C VOLTAGE SCALE.
- SPIN WHEEL BY HAND WHILE OBSERVING VOLTAGE. IS VOLTAGE AT LEAST 100 mV (VOLTAGE WILL INCREASE AS WHEEL SPEED INCREASES)?

NO: REPLACE WHEEL SPEED SENSOR.

YES:

(4)
- IGNITION "OFF."
- DISCONNECT 32-WAY EBCM CONNECTOR.
- USING J 39200, MEASURE THE RESISTANCE BETWEEN TERMINAL "E5" AND "E6" OF THE 32-WAY EBCM HARNESS CONNECTOR. DOES J 39200 DISPLAY RESISTANCE AS "OL" (INFINITE)?

NO: REPAIR SHORT BETWEEN CKTs 882 AND 883.

YES:

(5)
- INSPECT CKTs 882 AND 883 AND 32-WAY EBCM CONNECTOR FOR DAMAGE WHICH MAY RESULT IN A SHORT BETWEEN CKTs 882 AND 883 WITH ALL CONNECTORS CONNECTED. REPAIR ALL DAMAGE IF EVIDENT.
- RECONNECT ALL CONNECTORS.
- TEST DRIVE VEHICLE ABOVE 24 km/h (15 MPH) FOR AT LEAST 30 SECONDS. DOES DTC 28 SET AS A CURRENT DTC?

YES: REPLACE EBCM.

NO: MALFUNCTION IS INTERMITTENT. REFER TO "DIAGNOSTIC AIDS" ON FACING PAGE.

REAR WHEEL SPEED SENSOR
TEMPERATURE OF SENSOR VS. SENSOR RESISTANCE (APPROXIMATE)

TEMP. (°C)	TEMP. (°F)	RESISTANCE (OHMS)
-34 TO 4	-30 TO 40	1622 TO 2205
5 TO 43	41 TO 110	1930 TO 2566
44 TO 93	111 TO 200	2245 TO 3028

DIAGNOSTIC CHART DTC 32 — 1994 SILHOUETTE, TRANS SPORT AND LUMINA APV

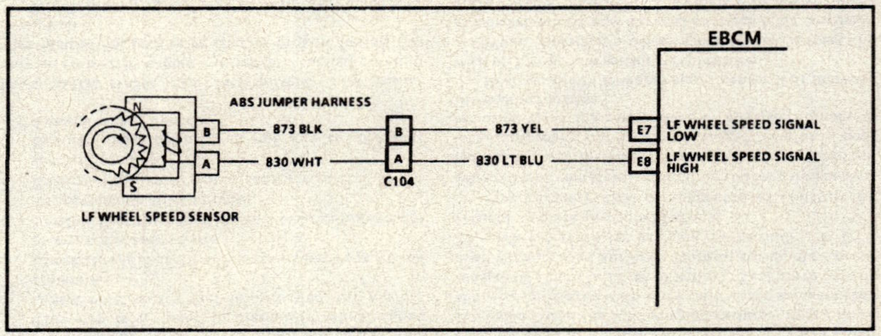

DTC 32

(Page 1 of 3)

LF WHEEL SPEED SNR CKT OPEN OR SHORT TO GND/BAT

Circuit Description:

As a toothed ring passes by the wheel speed sensor, changes in the electromagnetic field cause the wheel speed sensor to produce a sinusoidal (AC) voltage signal whose frequency is proportional to wheel speed. The magnitude of this signal is directly related to wheel speed and the proximity of the wheel speed sensor to the toothed ring often referred to as the air gap.

DTC Will Set When: DTC 32 can be set anytime after initialization. If either of the left front wheel speed circuits are shorted to voltage or ground, a malfunction exists.

Action Taken: A malfunction DTC is stored, ABS is disabled and the ABS warning lamp is turned "ON."

DTC Chart Test Description: Number(s) below refer to circled number(s) on the diagnostic chart.

1. Checks to see that the wheel speed sensor is within proper resistance values.
2. Checks for short to ground in wheel speed sensor wiring.
3. Checks for short to voltage in wheel speed sensor wiring.
4. Checks for a short to voltage in CKT 873.
5. Checks for a short to voltage in CKT 830.
6. Ensures malfunction was not due to physical damage of circuitry.

Diagnostic Aids: An "Intermittent" malfunction may be caused by a poor connection, rubbed through wire insulation, or a wire that is broken inside the insulation.

The frequency of the malfunction can be checked by using the enhanced diagnostic function of the Tech 1.

If the customer's comments reflect that the ABS warning lamp is "ON" only during moist environmental changes (rain, snow, vehicle wash), all wheel speed sensor circuitry should be thoroughly inspected for signs of water intrusion. If DTC is not current, clear DTCs and simulate the effects of water intrusion. Use the following procedure. Spray down the suspected area with a 5% salt water solution (two teaspoons of salt to 12 oz. of water). Test drive vehicle over various road surfaces (bumps, turns, etc.) above 24 km/h (15 mph) for at least 30 seconds. If DTC returns, replace suspected harness.

Any circuitry, that is suspected as causing the intermittent complaint, should be thoroughly checked for backed out terminals, improper mating, broken locks, improperly formed or damaged terminals, poor terminal to wiring connections or physical damage to the wiring harness.

Resistance of the wheel speed sensor will increase with an increase in sensor temperature.

When replacing a wheel speed sensor, inspect the sensor terminals and harness connector for corrosion and/or water intrusion. If evidence of corrosion or water intrusion exists, replace wheel speed sensor jumper harness. Likewise, if replacing a wheel speed sensor jumper harness, inspect sensor terminals. If evidence of corrosion or water intrusion exists, replace wheel speed sensor.

DIAGNOSTIC CHART DTC 32 — 1994 SILHOUETTE, TRANS SPORT AND LUMINA APV

DTC 32

(Page 1 of 3)

LF WHEEL SPEED SNR CKT OPEN OR SHORT TO GND/BAT

Important

- WHEEL SPEED SENSOR INTERMITTENT MALFUNCTIONS MAY BE DIFFICULT TO LOCATE. CARE SHOULD BE TAKEN NOT TO DISTURB ANY ELECTRICAL CONNECTIONS PRIOR TO AN INDICATED STEP OF THIS CHART. THIS WILL ENSURE THAT AN INTERMITTENT CONNECTION WILL NOT BE CORRECTED BEFORE THE SOURCE OF THE MALFUNCTION IS FOUND.

(1)
- IGNITION "OFF."
- USING A 5% SALT WATER SOLUTION (TWO TEASPOONS OF SALT TO 12 OZ. WATER), THOROUGHLY SPRAY THE LEFT FRONT WHEEL SPEED SENSOR JUMPER HARNESS.
- DISCONNECT 32-WAY EBCM HARNESS CONNECTOR.
- USING J 39200, MEASURE THE RESISTANCE BETWEEN TERMINALS "E7" AND "E8" OF THE 32-WAY EBCM HARNESS CONNECTOR.

IS RESISTANCE BETWEEN 1020 AND 1137Ω WHEN SENSOR IS 20°C (68°F)? *(IF TEMPERATURE OF SENSOR IS SUSPECTED TO OTHER THAN 20°C (68°F), REFER TO THE TABLE ON PAGE 2 OF 2 OF THIS DTC CHART FOR APPLICABLE SENSOR RESISTANCE VALUES; THE VALUES IN THE TABLE ARE FOR TEMPERATURE OF THE SENSOR, NOT AIR TEMPERATURE.)*

- YES → (2)
- NO → DIAGNOSIS CONTINUED ON PAGE 2 OF 3 OF THIS DTC CHART.

(2)
- MEASURE THE RESISTANCE BETWEEN GROUND AND TERMINAL "E7" OF THE 32-WAY EBCM HARNESS CONNECTOR.

DOES J 39200 DISPLAY RESISTANCE AS "OL" (INFINITE)?

- YES → (3)
- NO → DIAGNOSIS CONTINUED ON PAGE 3 OF 3 OF THIS DTC CHART.

(3)
- IGNITION "ON."
- USING J 39200, MEASURE THE VOLTAGE BETWEEN GROUND AND TERMINAL "E7" OF THE 32-WAY EBCM HARNESS CONNECTOR.

IS VOLTAGE 1 VOLT OR LESS?

- NO → (4)
- YES → (6)

(4)
- IGNITION "OFF."
- DISCONNECT LEFT FRONT WHEEL SPEED SENSOR DIRECTLY AT SENSOR.
- IGNITION "ON."
- USING J 39200, MEASURE THE VOLTAGE BETWEEN GROUND AND TERMINAL "E7" OF THE 32-WAY EBCM HARNESS CONNECTOR.

IS VOLTAGE 1 VOLT OR LESS?

- YES → (5)
- NO → REPAIR SHORT TO VOLTAGE IN CKT 873.

(5)
- USING J 39200, MEASURE THE VOLTAGE BETWEEN GROUND AND TERMINAL "E8" OF THE 32-WAY EBCM HARNESS CONNECTOR.

IS VOLTAGE 1 VOLT OR LESS?

- YES → MALFUNCTION IS INTERMITTENT. INSPECT ALL CONNECTORS AND HARNESSES FOR DAMAGE WHICH MAY RESULT IN A SHORT TO VOLTAGE WHEN ALL COMPONENTS ARE CONNECTED. REPAIR ALL DAMAGE IF FOUND. REFER TO "DIAGNOSTIC AIDS" ON FACING PAGE FOR MORE INFORMATION.
- NO → REPAIR SHORT TO VOLTAGE IN CKT 830.

(6)
- IGNITION "OFF."
- INSPECT TERMINALS "E7" AND "E8" OF THE 32-WAY EBCM HARNESS CONNECTOR FOR POOR TERMINAL CONTACT AND CONTAMINATION. ALSO INSPECT CKTs 830 AND 873 FOR DAMAGE WHICH MAY RESULT IN AN OPEN CIRCUIT, SHORT TO GROUND, OR SHORT TO VOLTAGE. REPAIR ALL DAMAGE IF EVIDENT.
- RECONNECT ALL CONNECTORS.
- IGNITION "ON."

DOES DTC 32 SET AS A CURRENT DTC?

- YES → REPLACE EBCM.
- NO → MALFUNCTION IS INTERMITTENT. REFER TO "DIAGNOSTIC AIDS" ON FACING PAGE.

DIAGNOSTIC CHART DTC 32 — 1994 SILHOUETTE, TRANS SPORT AND LUMINA APV

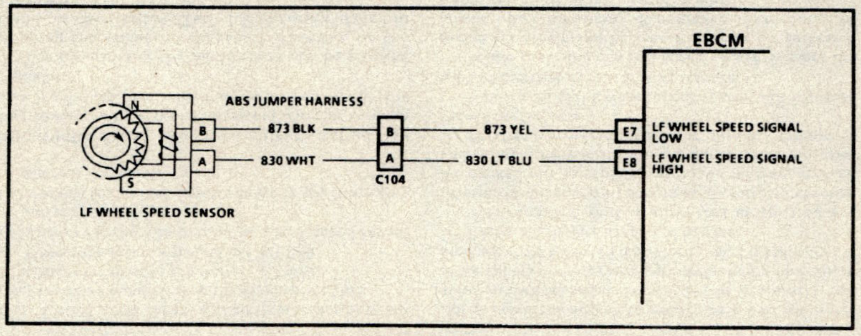

DTC 32
(Page 2 of 3)
LF WHEEL SPEED SNR CKT OPEN OR SHORT TO GND/BAT

Circuit Description:

As a toothed ring passes by the wheel speed sensor, changes in the electromagnetic field cause the wheel speed sensor to produce a sinusoidal (AC) voltage signal whose frequency is proportional to wheel speed. The magnitude of this signal is directly related to wheel speed and the proximity of the wheel speed sensor to the toothed ring often referred to as the air gap.

DTC Will Set When: DTC 32 can be set anytime after initialization. If either of the left front wheel speed circuits are shorted to voltage or ground, a malfunction exists.

Action Taken: A malfunction DTC is stored, ABS is disabled and the ABS warning lamp is turned "ON."

DTC Chart Test Description: Number(s) below refer to circled number(s) on the diagnostic chart.

7. Checks for high resistance in CKT 873.
8. Checks for high resistance in CKT 830.
9. Verifies proper wheel speed sensor resistance.

Diagnostic Aids: An "Intermittent" malfunction may be caused by a poor connection, rubbed through wire insulation, or a wire that is broken inside the insulation.

The frequency of the malfunction can be checked by using the enhanced diagnostic function of the Tech 1, as described in "Tech 1 Diagnostics" found in this section.

If the customer's comments reflect that the ABS warning lamp is "ON" only during moist environmental changes (rain, snow, vehicle wash), all wheel speed sensor circuitry should be thoroughly inspected for signs of water intrusion. If DTC is not current, clear DTCs and simulate the effects of water intrusion. Use the following procedure. Spray down the suspected area with a 5% salt water solution (two teaspoons of salt to 12 oz. of water). Test drive vehicle over various road surfaces (bumps, turns, etc.) above 24 km/h (15 mph) for at least 30 seconds. If DTC returns, replace suspected harness.

Any circuitry, that is suspected as causing the intermittent complaint, should be thoroughly checked for backed out terminals, improper mating, broken locks, improperly formed or damaged terminals, poor terminal to wiring connections or physical damage to the wiring harness.

Resistance of the wheel speed sensor will increase with an increase in sensor temperature.

When replacing a wheel speed sensor, inspect the sensor terminals and harness connector for corrosion and/or water intrusion. If evidence of corrosion or water intrusion exists, replace wheel speed sensor jumper harness. Likewise, if replacing a wheel speed sensor jumper harness, inspect sensor terminals. If evidence of corrosion or water intrusion exists, replace wheel speed sensor. Refer to "On-Vehicle Service" in this section.

DIAGNOSTIC CHART DTC 32 — 1994 SILHOUETTE, TRANS SPORT AND LUMINA APV

DTC 32
(Page 2 of 3)
LF WHEEL SPEED SNR CKT OPEN OR SHORT TO GND/BATT

DIAGNOSIS CONTINUED FROM PAGE 1 OF 3 OF THIS DTC CHART.

(7)
- DISCONNECT LEFT FRONT WHEEL SPEED SENSOR DIRECTLY AT SENSOR.
- USING J 39200, MEASURE THE RESISTANCE BETWEEN TERMINAL "E7" OF THE 32-WAY EBCM HARNESS CONNECTOR AND TERMINAL "B" OF THE LEFT FRONT WHEEL SPEED SENSOR HARNESS CONNECTOR. IS RESISTANCE 2Ω OR LESS?

NO: REPAIR HIGH RESISTANCE IN CKT 873 (WHEEL SPEED SENSOR JUMPER MUST BE REPLACED IF DAMAGED).

YES:

(8)
- USING J 39200, MEASURE THE RESISTANCE BETWEEN TERMINAL "E8" OF THE 32-WAY EBCM HARNESS CONNECTOR AND TERMINAL "A" OF THE LEFT FRONT WHEEL SPEED SENSOR HARNESS CONNECTOR. IS RESISTANCE 2Ω OR LESS?

NO: REPAIR HIGH RESISTANCE IN CKT 830 (WHEEL SPEED SENSOR JUMPER MUST BE REPLACED IF DAMAGED).

YES:

(9)
- USING J 39200, MEASURE THE RESISTANCE BETWEEN TERMINAL "A" AND TERMINAL "B" OF THE LEFT FRONT WHEEL SPEED SENSOR CONNECTOR. IS RESISTANCE BETWEEN 1020 AND 1137Ω WHEN SENSOR IS 20°C (68°F)? *(IF TEMPERATURE OF SENSOR IS SUSPECTED TO OTHER THAN 20°C (68°F), REFER TO THE TABLE BELOW FOR APPLICABLE SENSOR RESISTANCE VALUES; THE VALUES IN THE TABLE ARE FOR TEMPERATURE OF THE SENSOR, NOT AIR TEMPERATURE.)*

NO: REPLACE WHEEL SPEED SENSOR.

YES: MALFUNCTION IS INTERMITTENT. INSPECT ALL CONNECTORS AND HARNESSES FOR DAMAGE WHICH MAY RESULT IN HIGH RESISTANCE WHEN ALL COMPONENTS ARE CONNECTED. REFER TO "DIAGNOSTIC AIDS" ON FACING PAGE FOR MORE INFORMATION. REPAIR ALL DAMAGE IF FOUND.

FRONT WHEEL SPEED SENSOR

TEMPERATURE OF SENSOR VS. SENSOR RESISTANCE (APPROXIMATE)

TEMP. (°C)	TEMP. (°F)	RESISTANCE (OHMS)
-34 TO 4	-30 TO 40	804 TO 1066
5 TO 43	41 TO 110	956 TO 1240
44 TO 93	111 TO 200	1112 TO 1463

DIAGNOSTIC CHART DTC 32 — 1994 SILHOUETTE, TRANS SPORT AND LUMINA APV

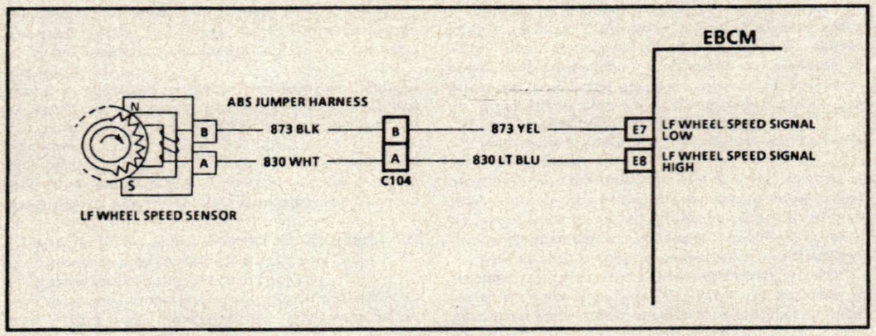

DTC 32
(Page 3 of 3)
LF WHEEL SPEED SNR CKT OPEN OR SHORT TO GND/BAT

Circuit Description:
As a toothed ring passes by the wheel speed sensor, changes in the electromagnetic field cause the wheel speed sensor to produce a sinusoidal (AC) voltage signal whose frequency is proportional to wheel speed. The magnitude of this signal is directly related to wheel speed and the proximity of the wheel speed sensor to the toothed ring often referred to as the air gap.

DTC Will Set When: DTC 32 can be set anytime after initialization. If either of the left front wheel speed circuits are shorted to voltage or ground, a malfunction exists.

Action Taken: A malfunction DTC is stored, ABS is disabled and the ABS warning lamp is turned "ON."

DTC Chart Test Description: Number(s) below refer to circled number(s) on the diagnostic chart.

10. Checks for short to ground in CKT 873.
11. Checks for short to ground in CKT 830.
12. Checks for short to ground in wheel speed sensor jumper harness.
13. Verifies that wheel speed sensor is not internally shorted to ground.

Diagnostic Aids: An "Intermittent" malfunction may be caused by a poor connection, rubbed through wire insulation, or a wire that is broken inside the insulation.

The frequency of the malfunction can be checked by using the enhanced diagnostic function of the Tech 1, as described in "Tech 1 Diagnostics" found in this section.

If the customer's comments reflect that the ABS warning lamp is "ON" only during moist environmental changes (rain, snow, vehicle wash), all wheel speed sensor circuitry should be thoroughly inspected for signs of water intrusion. If DTC is not current, clear DTCs and simulate the effects of water intrusion. Use the following procedure.

Spray down the suspected area with a 5% salt water solution (two teaspoons of salt to 12 oz. of water). Test drive vehicle over various road surfaces (bumps, turns, etc.) above 24 km/h (15 mph) for at least 30 seconds. If DTC returns, replace suspected harness.

Any circuitry, that is suspected as causing the intermittent complaint, should be thoroughly checked for backed out terminals, improper mating, broken locks, improperly formed or damaged terminals, poor terminal to wiring connections or physical damage to the wiring harness.

Resistance of the wheel speed sensor will increase with an increase in sensor temperature.

When replacing a wheel speed sensor, inspect the sensor terminals and harness connector for corrosion and/or water intrusion. If evidence of corrosion or water intrusion exists, replace wheel speed sensor jumper harness. Likewise, if replacing a wheel speed sensor jumper harness, inspect sensor terminals. If evidence of corrosion or water intrusion exists, replace wheel speed sensor. Refer to "On-Vehicle Service" in this section.

DIAGNOSTIC CHART DTC 32 — 1994 SILHOUETTE, TRANS SPORT AND LUMINA APV

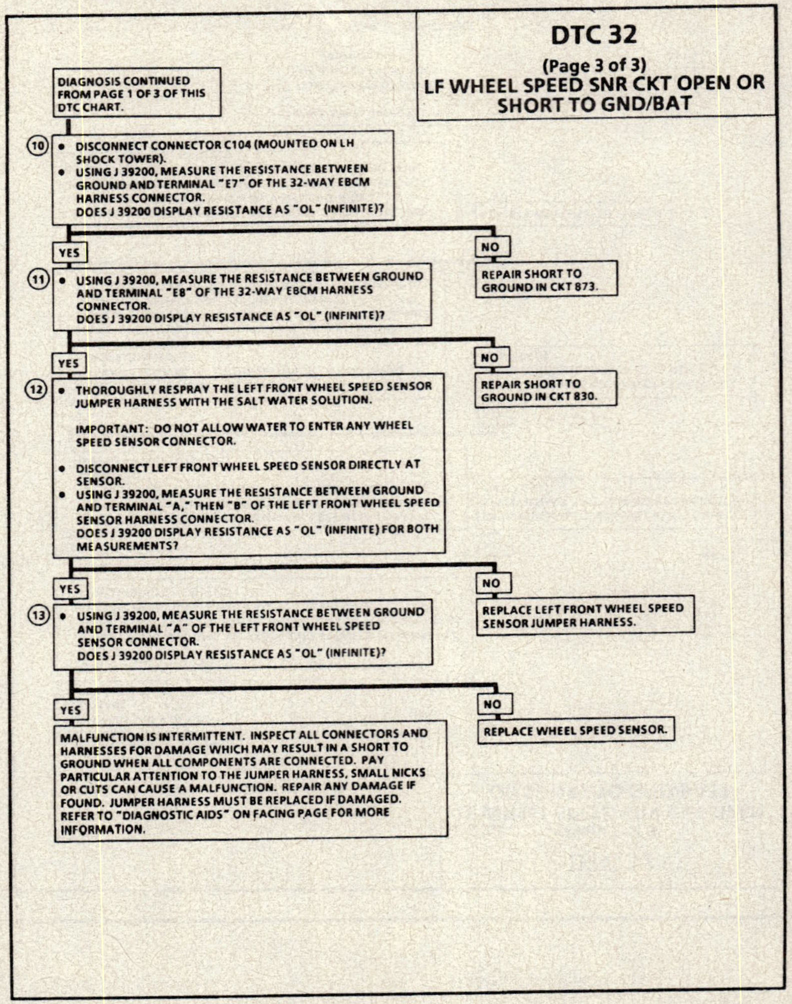

DIAGNOSTIC CHART DTC 33 — 1994 SILHOUETTE, TRANS SPORT AND LUMINA APV

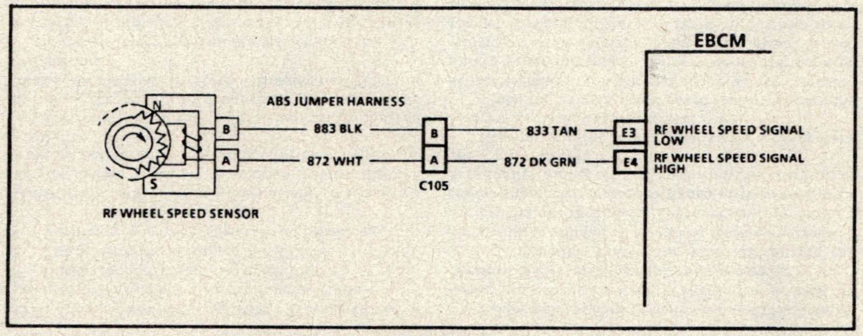

DTC 33

(Page 1 of 3)
RF WHEEL SPEED SNR CKT OPEN OR SHORT TO GND/BAT

Circuit Description:
As a toothed ring passes by the wheel speed sensor, changes in the electromagnetic field cause the wheel speed sensor to produce a sinusoidal (AC) voltage signal whose frequency is proportional to wheel speed. The magnitude of this signal is directly related to wheel speed and the proximity of the wheel speed sensor to the toothed ring, often referred to as the air gap.

DTC Will Set When: DTC 33 can be set anytime after initialization. If either of the right front wheel speed circuits are shorted to voltage or ground, a malfunction exists.

Action Taken: A malfunction DTC is stored, ABS is disabled and the ABS warning lamp is turned "ON."

DTC Chart Test Description: Number(s) below refer to circled number(s) on the diagnostic chart.

1. Checks to see that the wheel speed sensor is within proper resistance values.
2. Checks for short to ground in wheel speed sensor wiring.
3. Checks for short to voltage in wheel speed sensor wiring.
4. Checks for a short to voltage in CKT 833.
5. Checks for a short to voltage in CKT 872.
6. Ensures malfunction was not due to physical damage of circuitry.

Diagnostic Aids: An "Intermittent" malfunction may be caused by a poor connection, rubbed through wire insulation, or a wire that is broken inside the insulation.

The frequency of the malfunction can be checked by using the enhanced diagnostic function of the Tech 1, as described in "Tech 1 Diagnostics" found in this section.

If the customer's comments reflect that the ABS warning lamp is "ON" only during moist environmental changes (rain, snow, vehicle wash), all wheel speed sensor circuitry should be thoroughly inspected for signs of water intrusion. If DTC is not current, clear DTCs and simulate the effects of water intrusion. Use the following procedure. Spray down the suspected area with a 5% salt water solution (two teaspoons of salt to 12 oz. of water). Test drive vehicle over various road surfaces (bumps, turns, etc.) above 24 km/h (15 mph) for at least 30 seconds. If DTC returns, replace suspected harness.

Any circuitry, that is suspected as causing the intermittent complaint, should be thoroughly checked for backed out terminals, improper mating, broken locks, improperly formed or damaged terminals, poor terminal to wiring connections or physical damage to the wiring harness.

Resistance of the wheel speed sensor will increase with an increase in sensor temperature.

When replacing a wheel speed sensor, inspect the sensor terminals and harness connector for corrosion and/or water intrusion. If evidence of corrosion or water intrusion exists, replace wheel speed sensor jumper harness. Likewise, if replacing a wheel speed sensor jumper harness, inspect sensor terminals. If evidence of corrosion or water intrusion exists, replace wheel speed sensor. Refer to "On-Vehicle Service" in this section.

DIAGNOSTIC CHART DTC 33 — 1994 SILHOUETTE, TRANS SPORT AND LUMINA APV

DTC 33

(Page 1 of 3)
RF WHEEL SPEED SNR CKT OPEN OR SHORT TO GND/BAT

Important
- WHEEL SPEED SENSOR INTERMITTENT MALFUNCTIONS MAY BE DIFFICULT TO LOCATE. CARE SHOULD BE TAKEN NOT TO DISTURB ANY ELECTRICAL CONNECTIONS PRIOR TO AN INDICATED STEP OF THIS CHART. THIS WILL ENSURE THAT AN INTERMITTENT CONNECTION WILL NOT BE CORRECTED BEFORE THE SOURCE OF THE MALFUNCTION IS FOUND.

(1)
- IGNITION "OFF."
- USING A 5% SALT WATER SOLUTION (TWO TEASPOONS OF SALT TO 12 OZ. WATER), THOROUGHLY SPRAY THE RIGHT FRONT WHEEL SPEED SENSOR JUMPER HARNESS.
- DISCONNECT 32-WAY EBCM HARNESS CONNECTOR.
- USING J 39200, MEASURE THE RESISTANCE BETWEEN TERMINALS "E3" AND "E4" OF THE 32-WAY EBCM HARNESS CONNECTOR.

IS RESISTANCE BETWEEN 1020 AND 1137Ω WHEN SENSOR IS 20°C (68°F)? *(IF TEMPERATURE OF SENSOR IS SUSPECTED TO OTHER THAN 20°C (68°F), REFER TO THE TABLE ON PAGE 2 OF 2 OF THIS DTC CHART FOR APPLICABLE SENSOR RESISTANCE VALUES; THE VALUES IN THE TABLE ARE FOR TEMPERATURE OF THE SENSOR, NOT AIR TEMPERATURE.)*

- YES → (2)
- NO → DIAGNOSIS CONTINUED ON PAGE 2 OF 3 OF THIS DTC CHART.

(2)
- MEASURE THE RESISTANCE BETWEEN GROUND AND TERMINAL "E3" OF THE 32-WAY EBCM HARNESS CONNECTOR.

DOES J 39200 DISPLAY RESISTANCE AS "OL" (INFINITE)?

- YES → (3)
- NO → DIAGNOSIS CONTINUED ON PAGE 3 OF 3 OF THIS DTC CHART.

(3)
- IGNITION "ON."
- USING J 39200, MEASURE THE VOLTAGE BETWEEN GROUND AND TERMINAL "E3" OF THE 32-WAY EBCM HARNESS CONNECTOR.

IS VOLTAGE 1 VOLT OR LESS?

- NO → (4)
- YES → (6)

(4)
- IGNITION "OFF."
- DISCONNECT RIGHT FRONT WHEEL SPEED SENSOR DIRECTLY AT SENSOR.
- IGNITION "ON."
- USING J 39200, MEASURE THE VOLTAGE BETWEEN GROUND AND TERMINAL "E3" OF THE 32-WAY EBCM HARNESS CONNECTOR.

IS VOLTAGE 1 VOLT OR LESS?

- YES → (5)
- NO → REPAIR SHORT TO VOLTAGE IN CKT 833.

(5)
- USING J 39200, MEASURE THE VOLTAGE BETWEEN GROUND AND TERMINAL "E4" OF THE 32-WAY EBCM HARNESS CONNECTOR.

IS VOLTAGE 1 VOLT OR LESS?

- YES → MALFUNCTION IS INTERMITTENT. INSPECT ALL CONNECTORS AND HARNESSES FOR DAMAGE WHICH MAY RESULT IN A SHORT TO VOLTAGE WHEN ALL COMPONENTS ARE CONNECTED. REPAIR ALL DAMAGE IF FOUND. REFER TO "DIAGNOSTIC AIDS" ON FACING PAGE FOR MORE INFORMATION.
- NO → REPAIR SHORT TO VOLTAGE IN CKT 872.

(6)
- IGNITION "OFF."
- INSPECT TERMINALS "E3" AND "E4" OF THE 32-WAY EBCM HARNESS CONNECTOR FOR POOR TERMINAL CONTACT AND CONTAMINATION. ALSO INSPECT CKTs 833 AND 872 FOR DAMAGE WHICH MAY RESULT IN AN OPEN CIRCUIT, SHORT TO GROUND, OR SHORT TO VOLTAGE. REPAIR ALL DAMAGE IF EVIDENT.
- RECONNECT ALL CONNECTORS.
- IGNITION "ON."

DOES DTC 33 SET AS A CURRENT DTC?

- YES → REPLACE EBCM.
- NO → MALFUNCTION IS INTERMITTENT. REFER TO "DIAGNOSTIC AIDS" ON FACING PAGE.

DIAGNOSTIC CHART DTC 33 — 1994 SILHOUETTE, TRANS SPORT AND LUMINA APV

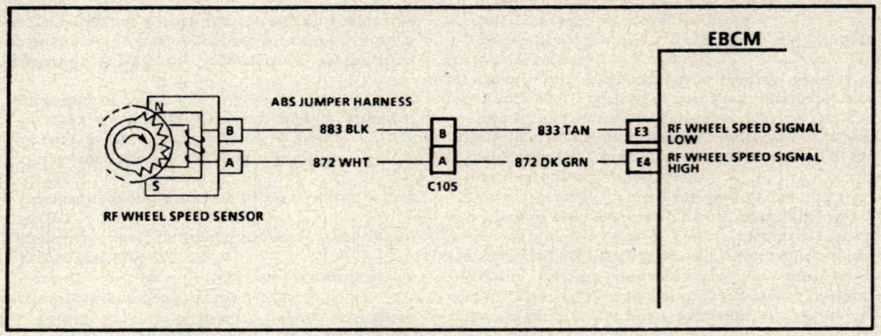

DTC 33

(Page 2 of 3)

RF WHEEL SPEED SNR CKT OPEN OR SHORT TO GND/BAT

Circuit Description:

As a toothed ring passes by the wheel speed sensor, changes in the electromagnetic field cause the wheel speed sensor to produce a sinusoidal (AC) voltage signal whose frequency is proportional to wheel speed. The magnitude of this signal is directly related to wheel speed and the proximity of the wheel speed sensor to the toothed ring, often referred to as the air gap.

DTC Will Set When: DTC 33 can be set anytime after initialization. If either of the right front wheel speed circuits are shorted to voltage or ground, a malfunction exists.

Action Taken: A malfunction DTC is stored, ABS is disabled and the ABS warning lamp is turned "ON."

DTC Chart Test Description: Number(s) below refer to circled number(s) on the diagnostic chart.

7. Checks for high resistance in CKT 833.
8. Checks for high resistance in CKT 872.
9. Verifies proper wheel speed sensor resistance.

Diagnostic Aids: An "Intermittent" malfunction may be caused by a poor connection, rubbed through wire insulation, or a wire that is broken inside the insulation.

The frequency of the malfunction can be checked by using the enhanced diagnostic function of the Tech 1, as described in "Tech 1 Diagnostics" found in this section.

If the customer's comments reflect that the ABS warning lamp is "ON" only during moist environmental changes (rain, snow, vehicle wash), all wheel speed sensor circuitry should be thoroughly inspected for signs of water intrusion. If DTC is not current, clear DTCs and simulate the effects of water intrusion. Use the following procedure. Spray down the suspected area with a 5% salt water solution (two teaspoons of salt to 12 oz. of water). Test drive vehicle over various road surfaces (bumps, turns, etc.) above 24 km/h (15 mph) for at least 30 seconds. If DTC returns, replace suspected harness.

Any circuitry, that is suspected as causing the intermittent complaint, should be thoroughly checked for backed out terminals, improper mating, broken locks, improperly formed or damaged terminals, poor terminal to wiring connections or physical damage to the wiring harness.

Resistance of the wheel speed sensor will increase with an increase in sensor temperature.

When replacing a wheel speed sensor, inspect the sensor terminals and harness connector for corrosion and/or water intrusion. If evidence of corrosion or water intrusion exists, replace wheel speed sensor jumper harness. Likewise, if replacing a wheel speed sensor jumper harness, inspect sensor terminals. If evidence of corrosion or water intrusion exists, replace wheel speed sensor. Refer to "On-Vehicle Service" in this section.

DIAGNOSTIC CHART DTC 33 — 1994 SILHOUETTE, TRANS SPORT AND LUMINA APV

DTC 33

(Page 2 of 3)

RF WHEEL SPEED SNR CKT SHORT TO GND/BAT

DIAGNOSIS CONTINUED FROM PAGE 1 OF 3 OF THIS DTC CHART.

(7)
- DISCONNECT RIGHT FRONT WHEEL SPEED SENSOR DIRECTLY AT SENSOR.
- USING J 39200, MEASURE THE RESISTANCE BETWEEN TERMINAL "E3" OF THE 32-WAY EBCM HARNESS CONNECTOR AND TERMINAL "B" OF THE RIGHT FRONT WHEEL SPEED SENSOR HARNESS CONNECTOR.

IS RESISTANCE 2Ω OR LESS?

NO: REPAIR HIGH RESISTANCE IN CKT 833 (WHEEL SPEED SENSOR JUMPER MUST BE REPLACED IF DAMAGED).

YES:

(8)
- USING J 39200, MEASURE THE RESISTANCE BETWEEN TERMINAL "E4" OF THE 32-WAY EBCM HARNESS CONNECTOR AND TERMINAL "A" OF THE RIGHT FRONT WHEEL SPEED SENSOR HARNESS CONNECTOR.

IS RESISTANCE 2Ω OR LESS?

NO: REPAIR HIGH RESISTANCE IN CKT 872 (WHEEL SPEED SENSOR JUMPER MUST BE REPLACED IF DAMAGED).

YES:

(9)
- USING J 39200, MEASURE THE RESISTANCE BETWEEN TERMINAL "A" AND TERMINAL "B" OF THE RIGHT FRONT WHEEL SPEED SENSOR CONNECTOR.

IS RESISTANCE BETWEEN 1020 AND 1137Ω WHEN SENSOR IS 20°C (68°F)? *(IF TEMPERATURE OF SENSOR IS SUSPECTED TO OTHER THAN 20°C (68°F), REFER TO THE TABLE BELOW FOR APPLICABLE SENSOR RESISTANCE VALUES; THE VALUES IN THE TABLE ARE FOR TEMPERATURE OF THE SENSOR, NOT AIR TEMPERATURE.)*

NO: REPLACE WHEEL SPEED SENSOR.

YES: MALFUNCTION IS INTERMITTENT. INSPECT ALL CONNECTORS AND HARNESSES FOR DAMAGE WHICH MAY RESULT IN HIGH RESISTANCE WHEN ALL COMPONENTS ARE CONNECTED. REFER TO "DIAGNOSTIC AIDS" ON FACING PAGE FOR MORE INFORMATION. REPAIR ALL DAMAGE IF FOUND.

FRONT WHEEL SPEED SENSOR

TEMPERATURE OF SENSOR VS. SENSOR RESISTANCE (APPROXIMATE)

TEMP. (°C)	TEMP. (°F)	RESISTANCE (OHMS)
-34 TO 4	-30 TO 40	804 TO 1066
5 TO 43	41 TO 110	956 TO 1240
44 TO 93	111 TO 200	1112 TO 1463

DIAGNOSTIC CHART DTC 33 — 1994 SILHOUETTE, TRANS SPORT AND LUMINA APV

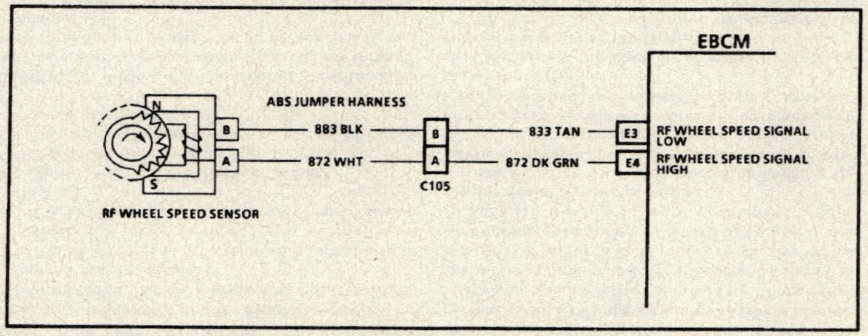

DTC 33
(Page 3 of 3)
RF WHEEL SPEED SNR CKT OPEN OR SHORT TO GND/BAT

Circuit Description:

As a toothed ring passes by the wheel speed sensor, changes in the electromagnetic field cause the wheel speed sensor to produce a sinusoidal (AC) voltage signal whose frequency is proportional to wheel speed. The magnitude of this signal is directly related to wheel speed and the proximity of the wheel speed sensor to the toothed ring, often referred to as the air gap.

DTC Will Set When: DTC 33 can be set anytime after initialization. If either of the right front wheel speed circuits are shorted to voltage or ground, a malfunction exists.

Action Taken: A malfunction DTC is stored, ABS is disabled and the ABS warning lamp is turned "ON."

DTC Chart Test Description: Number(s) below refer to circled number(s) on the diagnostic chart.

10. Checks for short to ground in CKT 833.
11. Checks for short to ground in CKT 872.
12. Checks for short to ground in wheel speed sensor jumper harness.
13. Verifies that wheel speed sensor is not internally shorted to ground.

Diagnostic Aids: An "Intermittent" malfunction may be caused by a poor connection, rubbed through wire insulation, or a wire that is broken inside the insulation.

The frequency of the malfunction can be checked by using the enhanced diagnostic function of the Tech 1, as described in "Tech 1 Diagnostics" found in this section.

If the customer's comments reflect that the ABS warning lamp is "ON" only during moist environmental changes (rain, snow, vehicle wash), all wheel speed sensor circuitry should be thoroughly inspected for signs of water intrusion. If DTC is not current, clear DTCs and simulate the effects of water intrusion. Use the following procedure.

Spray down the suspected area with a 5% salt water solution (two teaspoons of salt to 12 oz. of water). Test drive vehicle over various road surfaces (bumps, turns, etc.) above 24 km/h (15 mph) for at least 30 seconds. If DTC returns, replace suspected harness.

Any circuitry, that is suspected as causing the intermittent complaint, should be thoroughly checked for backed out terminals, improper mating, broken locks, improperly formed or damaged terminals, poor terminal to wiring connections or physical damage to the wiring harness.

Resistance of the wheel speed sensor will increase with an increase in sensor temperature.

When replacing a wheel speed sensor, inspect the sensor terminals and harness connector for corrosion and/or water intrusion. If evidence of corrosion or water intrusion exists, replace wheel speed sensor jumper harness. Likewise, if replacing a wheel speed sensor jumper harness, inspect sensor terminals. If evidence of corrosion or water intrusion exists, replace wheel speed sensor. Refer to "On-Vehicle Service" in this section.

DIAGNOSTIC CHART DTC 33 — 1994 SILHOUETTE, TRANS SPORT AND LUMINA APV

DTC 33
(Page 3 of 3)
RF WHEEL SPEED SNR CKT OPEN OR SHORT TO GND/BAT

DIAGNOSIS CONTINUED FROM PAGE 1 OF 3 OF THIS DTC CHART.

(10)
- DISCONNECT CONNECTOR C105 (MOUNTED ON RH SHOCK TOWER).
- USING J 39200, MEASURE THE RESISTANCE BETWEEN GROUND AND TERMINAL "E3" OF THE 32-WAY EBCM HARNESS CONNECTOR.

DOES J 39200 DISPLAY RESISTANCE AS "OL" (INFINITE)?

- YES → go to 11
- NO → REPAIR SHORT TO GROUND IN CKT 833.

(11)
- USING J 39200, MEASURE THE RESISTANCE BETWEEN GROUND AND TERMINAL "E4" OF THE 32-WAY EBCM HARNESS CONNECTOR.

DOES J 39200 DISPLAY RESISTANCE AS "OL" (INFINITE)?

- YES → go to 12
- NO → REPAIR SHORT TO GROUND IN CKT 872.

(12)
- THOROUGHLY RESPRAY THE RIGHT FRONT WHEEL SPEED SENSOR JUMPER HARNESS WITH THE SALT WATER SOLUTION.

IMPORTANT: DO NOT ALLOW WATER TO ENTER ANY WHEEL SPEED SENSOR CONNECTOR.

- DISCONNECT RIGHT FRONT WHEEL SPEED SENSOR DIRECTLY AT SENSOR.
- USING J 39200, MEASURE THE RESISTANCE BETWEEN GROUND AND TERMINAL "A," THEN "B" OF THE RIGHT FRONT WHEEL SPEED SENSOR HARNESS CONNECTOR.

DOES J 39200 DISPLAY RESISTANCE AS "OL" (INFINITE) FOR BOTH MEASUREMENTS?

- YES → go to 13
- NO → REPLACE RIGHT FRONT WHEEL SPEED SENSOR JUMPER HARNESS.

(13)
- USING J 39200, MEASURE THE RESISTANCE BETWEEN GROUND AND TERMINAL "A" OF THE RIGHT FRONT WHEEL SPEED SENSOR CONNECTOR.

DOES J 39200 DISPLAY RESISTANCE AS "OL" (INFINITE)?

- YES → MALFUNCTION IS INTERMITTENT. INSPECT ALL CONNECTORS AND HARNESSES FOR DAMAGE WHICH MAY RESULT IN A SHORT TO GROUND WHEN ALL COMPONENTS ARE CONNECTED. PAY PARTICULAR ATTENTION TO THE JUMPER HARNESS, SMALL NICKS OR CUTS CAN CAUSE A MALFUNCTION. REPAIR ANY DAMAGE IF FOUND. JUMPER HARNESS MUST BE REPLACED IF DAMAGED. REFER TO "DIAGNOSTIC AIDS" ON FACING PAGE FOR MORE INFORMATION.
- NO → REPLACE WHEEL SPEED SENSOR.

DIAGNOSTIC CHART DTC 34 — 1994 SILHOUETTE, TRANS SPORT AND LUMINA APV

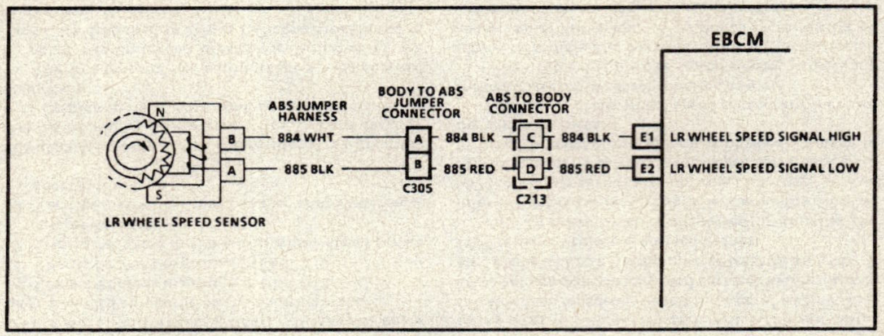

DTC 34

(Page 1 of 3)

LR WHEEL SPEED SNR CKT OPEN OR SHORT TO GND/BAT

Circuit Description:

As a toothed ring passes by the wheel speed sensor, changes in the electromagnetic field cause the wheel speed sensor to produce a sinusoidal (AC) voltage signal whose frequency is proportional to wheel speed. The magnitude of this signal is directly related to wheel speed and the proximity of the wheel speed sensor to the toothed ring, often referred to as the air gap.

DTC Will Set When: DTC 34 can be set anytime after initialization. If either of the left rear wheel speed circuits are shorted to voltage or ground, a malfunction exists.

Action Taken: A malfunction DTC is stored, ABS is disabled and the ABS warning lamp is turned "ON."

DTC Chart Test Description: Number(s) below refer to circled number(s) on the diagnostic chart.

1. Checks to see that the wheel speed sensor is within proper resistance values.
2. Checks for short to ground in wheel speed sensor wiring.
3. Checks for short to voltage in wheel speed sensor wiring.
4. Checks for a short to voltage in CKT 884.
5. Checks for a short to voltage in CKT 885.
6. Ensures malfunction was not due to physical damage of circuitry.

Diagnostic Aids: An "Intermittent" malfunction may be caused by a poor connection, rubbed through wire insulation, or a wire that is broken inside the insulation.

The frequency of the malfunction can be checked by using the enhanced diagnostic function of the Tech 1, as described in "Tech 1 Diagnostics" found in this section.

If the customer's comments reflect that the ABS warning lamp is "ON" only during moist environmental changes (rain, snow, vehicle wash), all wheel speed sensor circuitry should be thoroughly inspected for signs of water intrusion. If DTC is not current, clear DTCs and simulate the effects of water intrusion. Use the following procedure. Spray down the suspected area with a 5% salt water solution (two teaspoons of salt to 12 oz. of water). Test drive vehicle over various road surfaces (bumps, turns, etc.) above 24 km/h (15 mph) for at least 30 seconds. If DTC returns, replace suspected harness.

Any circuitry, that is suspected as causing the intermittent complaint, should be thoroughly checked for backed out terminals, improper mating, broken locks, improperly formed or damaged terminals, poor terminal to wiring connections or physical damage to the wiring harness.

Resistance of the wheel speed sensor will increase with an increase in sensor temperature.

When replacing a wheel speed sensor, inspect the sensor terminals and harness connector for corrosion and/or water intrusion. If evidence of corrosion or water intrusion exists, replace wheel speed sensor jumper harness. Likewise, if replacing a wheel speed sensor jumper harness, inspect sensor terminals. If evidence of corrosion or water intrusion exists, replace wheel speed sensor. Refer to "On-Vehicle Service" in this section.

DIAGNOSTIC CHART DTC 34 — 1994 SILHOUETTE, TRANS SPORT AND LUMINA APV

DTC 34

(Page 1 of 3)

LR WHEEL SPEED SNR CKT OPEN OR SHORT TO GND/BAT

Important

- WHEEL SPEED SENSOR INTERMITTENT MALFUNCTIONS MAY BE DIFFICULT TO LOCATE. CARE SHOULD BE TAKEN NOT TO DISTURB ANY ELECTRICAL CONNECTIONS PRIOR TO AN INDICATED STEP OF THIS CHART. THIS WILL ENSURE THAT AN INTERMITTENT CONNECTION WILL NOT BE CORRECTED BEFORE THE SOURCE OF THE MALFUNCTION IS FOUND.

(1)
- IGNITION "OFF."
- USING A 5% SALT WATER SOLUTION (TWO TEASPOONS OF SALT TO 12 OZ. WATER), THOROUGHLY SPRAY THE LEFT REAR WHEEL SPEED SENSOR JUMPER HARNESS.
- DISCONNECT 32-WAY EBCM HARNESS CONNECTOR.
- USING J 39200, MEASURE THE RESISTANCE BETWEEN TERMINALS "E1" AND "E2" OF THE 32-WAY EBCM HARNESS CONNECTOR.

IS RESISTANCE BETWEEN 2059 AND 2353Ω WHEN SENSOR IS 20°C (68°F)? *(IF TEMPERATURE OF SENSOR IS SUSPECTED TO OTHER THAN 20°C (68°F), REFER TO THE TABLE ON PAGE 2 OF 2 OF THIS DTC CHART FOR APPLICABLE SENSOR RESISTANCE VALUES; THE VALUES IN THE TABLE ARE FOR TEMPERATURE OF THE SENSOR, NOT AIR TEMPERATURE.)*

- NO → DIAGNOSIS CONTINUED ON PAGE 2 OF 3 OF THIS DTC CHART.
- YES → (2)

(2)
- MEASURE THE RESISTANCE BETWEEN GROUND AND TERMINAL "E1" OF THE 32-WAY EBCM HARNESS CONNECTOR.

DOES J 39200 DISPLAY RESISTANCE AS "OL" (INFINITE)?

- NO → DIAGNOSIS CONTINUED ON PAGE 3 OF 3 OF THIS DTC CHART.
- YES → (3)

(3)
- IGNITION "ON."
- USING J 39200, MEASURE THE VOLTAGE BETWEEN GROUND AND TERMINAL "E1" OF THE 32-WAY EBCM HARNESS CONNECTOR.

IS VOLTAGE 1 VOLT OR LESS?

- NO → (4)
- YES → (6)

(4)
- IGNITION "OFF."
- DISCONNECT LEFT REAR WHEEL SPEED SENSOR DIRECTLY AT SENSOR.
- IGNITION "ON."
- USING J 39200, MEASURE THE VOLTAGE BETWEEN GROUND AND TERMINAL "E1" OF THE 32-WAY EBCM HARNESS CONNECTOR.

IS VOLTAGE 1 VOLT OR LESS?

- YES → (5)
- NO → REPAIR SHORT TO VOLTAGE IN CKT 884.

(5)
- USING J 39200, MEASURE THE VOLTAGE BETWEEN GROUND AND TERMINAL "E2" OF THE 32-WAY EBCM HARNESS CONNECTOR.

IS VOLTAGE 1 VOLT OR LESS?

- YES → MALFUNCTION IS INTERMITTENT. INSPECT ALL CONNECTORS AND HARNESSES FOR DAMAGE WHICH MAY RESULT IN A SHORT TO VOLTAGE WHEN ALL COMPONENTS ARE CONNECTED. REPAIR ALL DAMAGE IF FOUND. REFER TO "DIAGNOSTIC AIDS" ON FACING PAGE FOR MORE INFORMATION.
- NO → REPAIR SHORT TO VOLTAGE IN CKT 885.

(6)
- IGNITION "OFF."
- INSPECT TERMINALS "E1" AND "E2" OF THE 32-WAY EBCM HARNESS CONNECTOR FOR POOR TERMINAL CONTACT AND CONTAMINATION. ALSO INSPECT CKTs 884 AND 885 FOR DAMAGE WHICH MAY RESULT IN AN OPEN CIRCUIT, SHORT TO GROUND, OR SHORT TO VOLTAGE. REPAIR ALL DAMAGE IF EVIDENT.
- RECONNECT ALL CONNECTORS.
- IGNITION "ON."

DOES DTC 34 SET AS A CURRENT DTC?

- YES → REPLACE EBCM.
- NO → MALFUNCTION IS INTERMITTENT. REFER TO "DIAGNOSTIC AIDS" ON FACING PAGE.

DIAGNOSTIC CHART DTC 34 — 1994 SILHOUETTE, TRANS SPORT AND LUMINA APV

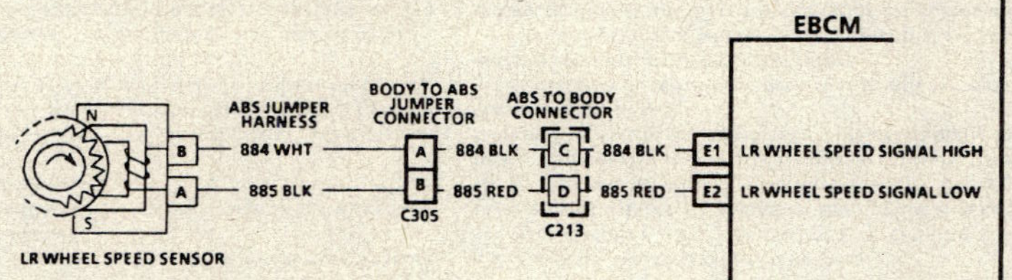

DTC 34
(Page 2 of 3)
LR WHEEL SPEED SNR CKT OPEN OR SHORT TO GND/BAT

Circuit Description:

As a toothed ring passes by the wheel speed sensor, changes in the electromagnetic field cause the wheel speed sensor to produce a sinusoidal (AC) voltage signal whose frequency is proportional to wheel speed. The magnitude of this signal is directly related to wheel speed and the proximity of the wheel speed sensor to the toothed ring, often referred to as the air gap.

DTC Will Set When: DTC 34 can be set anytime after initialization. If either of the left rear wheel speed circuits are shorted to voltage or ground, a malfunction exists.

Action Taken: A malfunction DTC is stored, ABS is disabled and the ABS warning lamp is turned "ON."

DTC Chart Test Description: Number(s) below refer to circled number(s) on the diagnostic chart.

7. Checks for high resistance in CKT 884.
8. Checks for high resistance in CKT 885.
9. Verifies proper wheel speed sensor resistance.

Diagnostic Aids: An "Intermittent" malfunction may be caused by a poor connection, rubbed through wire insulation, or a wire that is broken inside the insulation.

The frequency of the malfunction can be checked by using the enhanced diagnostic function of the Tech 1, as described in "Tech 1 Diagnostics" found in this section.

If the customer's comments reflect that the ABS warning lamp is "ON" only during moist environmental changes (rain, snow, vehicle wash), all wheel speed sensor circuitry should be thoroughly inspected for signs of water intrusion. If DTC is not current, clear DTCs and simulate the effects of water intrusion. Use the following procedure. Spray down the suspected area with a 5% salt water solution (two teaspoons of salt to 12 oz. of water). Test drive vehicle over various road surfaces (bumps, turns, etc.) above 24 km/h (15 mph) for at least 30 seconds. If DTC returns, replace suspected harness.

Any circuitry, that is suspected as causing the intermittent complaint, should be thoroughly checked for backed out terminals, improper mating, broken locks, improperly formed or damaged terminals, poor terminal to wiring connections or physical damage to the wiring harness.

Resistance of the wheel speed sensor will increase with an increase in sensor temperature.

When replacing a wheel speed sensor, inspect the sensor terminals and harness connector for corrosion and/or water intrusion. If evidence of corrosion or water intrusion exists, replace wheel speed sensor jumper harness. Likewise, if replacing a wheel speed sensor jumper harness, inspect sensor terminals. If evidence of corrosion or water intrusion exists, replace wheel speed sensor. Refer to "On-Vehicle Service" in this section.

DIAGNOSTIC CHART DTC 34 — 1994 SILHOUETTE, TRANS SPORT AND LUMINA APV

DTC 34
(Page 2 of 3)
LR WHEEL SPEED SNR CKT OPEN OR SHORT TO GND/BAT

DIAGNOSIS CONTINUED FROM PAGE 1 OF 3 OF THIS DTC CHART.

(7)
- DISCONNECT LEFT REAR WHEEL SPEED SENSOR DIRECTLY AT SENSOR.
- USING J 39200, MEASURE THE RESISTANCE BETWEEN TERMINAL "E1" OF THE 32-WAY EBCM HARNESS CONNECTOR AND TERMINAL "B" OF THE LEFT REAR WHEEL SPEED SENSOR HARNESS CONNECTOR.

IS RESISTANCE 2Ω OR LESS?

NO: REPAIR HIGH RESISTANCE IN CKT 884 (WHEEL SPEED SENSOR JUMPER MUST BE REPLACED IF DAMAGED).

YES:

(8)
- USING J 39200, MEASURE THE RESISTANCE BETWEEN TERMINAL "E2" OF THE 32-WAY EBCM HARNESS CONNECTOR AND TERMINAL "A" OF THE LEFT REAR WHEEL SPEED SENSOR HARNESS CONNECTOR.

IS RESISTANCE 2Ω OR LESS?

NO: REPAIR HIGH RESISTANCE IN CKT 885 (WHEEL SPEED SENSOR JUMPER MUST BE REPLACED IF DAMAGED).

YES:

(9)
- USING J 39200, MEASURE THE RESISTANCE BETWEEN TERMINAL "A" AND TERMINAL "B" OF THE LEFT REAR WHEEL SPEED SENSOR CONNECTOR.

IS RESISTANCE BETWEEN 2059 AND 2353Ω WHEN SENSOR IS 20°C (68°F)? *(IF TEMPERATURE OF SENSOR IS SUSPECTED TO OTHER THAN 20°C (68°F), REFER TO THE TABLE BELOW FOR APPLICABLE SENSOR RESISTANCE VALUES; THE VALUES IN THE TABLE ARE FOR TEMPERATURE OF THE SENSOR, NOT AIR TEMPERATURE.)*

NO: REPLACE WHEEL SPEED SENSOR.

YES: MALFUNCTION IS INTERMITTENT. INSPECT ALL CONNECTORS AND HARNESSES FOR DAMAGE WHICH MAY RESULT IN HIGH RESISTANCE WHEN ALL COMPONENTS ARE CONNECTED. REFER TO "DIAGNOSTIC AIDS" ON FACING PAGE FOR MORE INFORMATION. REPAIR ALL DAMAGE IF FOUND.

REAR WHEEL SPEED SENSOR

TEMPERATURE OF SENSOR VS. SENSOR RESISTANCE (APPROXIMATE)

TEMP. (°C)	TEMP. (°F)	RESISTANCE (OHMS)
-34 TO 4	-30 TO 40	1622 TO 2205
5 TO 43	41 TO 110	1930 TO 2566
44 TO 93	111 TO 200	2245 TO 3028

DIAGNOSTIC CHART DTC 34 — 1994 SILHOUETTE, TRANS SPORT AND LUMINA APV

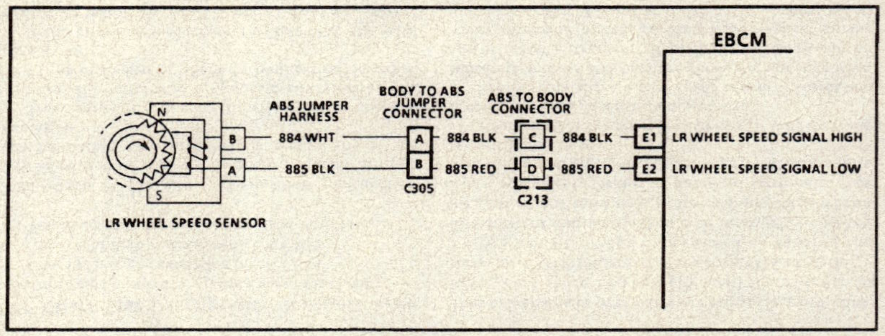

DTC 34
(Page 3 of 3)
LR WHEEL SPEED SNR CKT OPEN OR SHORT TO GND/BAT

Circuit Description:

As a toothed ring passes by the wheel speed sensor, changes in the electromagnetic field cause the wheel speed sensor to produce a sinusoidal (AC) voltage signal whose frequency is proportional to wheel speed. The magnitude of this signal is directly related to wheel speed and the proximity of the wheel speed sensor to the toothed ring, often referred to as the air gap.

DTC Will Set When: DTC 34 can be set anytime after initialization. If either of the left rear wheel speed circuits are shorted to voltage or ground, a malfunction exists.

Action Taken: A malfunction DTC is stored, ABS is disabled and the ABS warning lamp is turned "ON."

DTC Chart Test Description: Number(s) below refer to circled number(s) on the diagnostic chart.

10. Checks for short to ground in CKT 884.
11. Checks for short to ground in CKT 885.
12. Checks for short to ground in wheel speed sensor jumper harness.
13. Verifies that wheel speed sensor is not internally shorted to ground.

Diagnostic Aids: An "Intermittent" malfunction may be caused by a poor connection, rubbed through wire insulation, or a wire that is broken inside the insulation.

The frequency of the malfunction can be checked by using the enhanced diagnostic function of the Tech 1, as described in "Tech 1 Diagnostics" found in this section.

If the customer's comments reflect that the ABS warning lamp is "ON" only during moist environmental changes (rain, snow, vehicle wash), all wheel speed sensor circuitry should be thoroughly inspected for signs of water intrusion. If DTC is not current, clear DTCs and simulate the effects of water intrusion. Use the following procedure.

Spray down the suspected area with a 5% salt water solution (two teaspoons of salt to 12 oz. of water). Test drive vehicle over various road surfaces (bumps, turns, etc.) above 24 km/h (15 mph) for at least 30 seconds. If DTC returns, replace suspected harness.

Any circuitry, that is suspected as causing the intermittent complaint, should be thoroughly checked for backed out terminals, improper mating, broken locks, improperly formed or damaged terminals, poor terminal to wiring connections or physical damage to the wiring harness.

Resistance of the wheel speed sensor will increase with an increase in sensor temperature.

When replacing a wheel speed sensor, inspect the sensor terminals and harness connector for corrosion and/or water intrusion. If evidence of corrosion or water intrusion exists, replace wheel speed sensor jumper harness. Likewise, if replacing a wheel speed sensor jumper harness, inspect sensor terminals. If evidence of corrosion or water intrusion exists, replace wheel speed sensor. Refer to "On-Vehicle Service" in this section.

DIAGNOSTIC CHART DTC 34 — 1994 SILHOUETTE, TRANS SPORT AND LUMINA APV

DTC 34
(Page 3 of 3)
LR WHEEL SPEED SNR CKT OPEN OR SHORT TO GND/BAT

DIAGNOSIS CONTINUED FROM PAGE 1 OF 3 OF THIS DTC CHART.

(10)
- DISCONNECT CONNECTOR C305 (LH REAR UNDER BODY ABOVE FRONT SPRING HANGER).
- USING J 39200, MEASURE THE RESISTANCE BETWEEN GROUND AND TERMINAL "E1" OF THE 32-WAY EBCM HARNESS CONNECTOR.

DOES J 39200 DISPLAY RESISTANCE AS "OL" (INFINITE)?

NO → REPAIR SHORT TO GROUND IN CKT 884.

YES ↓

(11)
- USING J 39200, MEASURE THE RESISTANCE BETWEEN GROUND AND TERMINAL "E2" OF THE 32-WAY EBCM HARNESS CONNECTOR.

DOES J 39200 DISPLAY RESISTANCE AS "OL" (INFINITE)?

NO → REPAIR SHORT TO GROUND IN CKT 885.

YES ↓

(12)
- THOROUGHLY RESPRAY THE LEFT REAR WHEEL SPEED SENSOR JUMPER HARNESS WITH THE SALT WATER SOLUTION.

IMPORTANT: DO NOT ALLOW WATER TO ENTER ANY WHEEL SPEED SENSOR CONNECTOR.

- DISCONNECT LEFT REAR WHEEL SPEED SENSOR DIRECTLY AT SENSOR.
- USING J 39200, MEASURE THE RESISTANCE BETWEEN GROUND AND TERMINAL "A," THEN "B" OF THE LEFT REAR WHEEL SPEED SENSOR HARNESS CONNECTOR.

DOES J 39200 DISPLAY RESISTANCE AS "OL" (INFINITE) FOR BOTH MEASUREMENTS?

NO → REPLACE LEFT REAR WHEEL SPEED SENSOR JUMPER HARNESS.

YES ↓

(13)
- USING J 39200, MEASURE THE RESISTANCE BETWEEN GROUND AND TERMINAL "A" OF THE LEFT REAR WHEEL SPEED SENSOR CONNECTOR.

DOES J 39200 DISPLAY RESISTANCE AS "OL" (INFINITE)?

NO → REPLACE WHEEL SPEED SENSOR.

YES ↓

MALFUNCTION IS INTERMITTENT. INSPECT ALL CONNECTORS AND HARNESSES FOR DAMAGE WHICH MAY RESULT IN A SHORT TO GROUND WHEN ALL COMPONENTS ARE CONNECTED. PAY PARTICULAR ATTENTION TO THE JUMPER HARNESS, SMALL NICKS OR CUTS CAN CAUSE A MALFUNCTION. REPAIR ANY DAMAGE IF FOUND. JUMPER HARNESS MUST BE REPLACED IF DAMAGED. REFER TO "DIAGNOSTIC AIDS" ON FACING PAGE FOR MORE INFORMATION.

DIAGNOSTIC CHART DTC 35 — 1994 SILHOUETTE, TRANS SPORT AND LUMINA APV

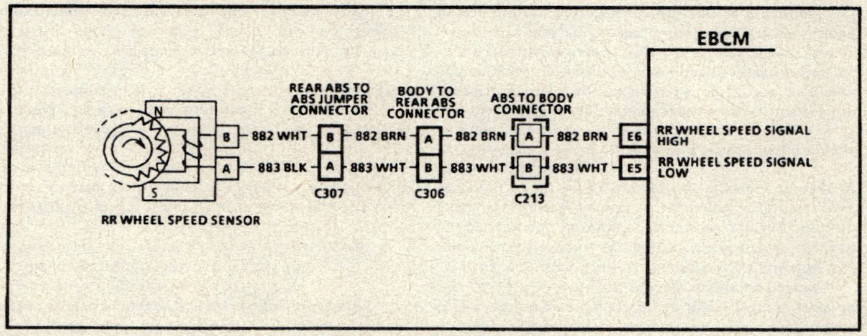

DTC 35
(Page 1 of 3)
RR WHEEL SPEED SNR CKT OPEN OR SHORT TO GND/BAT

Circuit Description:

As a toothed ring passes by the wheel speed sensor, changes in the electromagnetic field cause the wheel speed sensor to produce a sinusoidal (AC) voltage signal whose frequency is proportional to wheel speed. The magnitude of this signal is directly related to wheel speed and the proximity of the wheel speed sensor to the toothed ring, often referred to as the air gap.

DTC Will Set When: DTC 35 can be set anytime after initialization. If either of the right rear wheel speed circuits are shorted to voltage or ground, a malfunction exists.

Action Taken: A malfunction DTC is stored, ABS is disabled and the ABS warning lamp is turned "ON."

DTC Chart Test Description: Number(s) below refer to circled number(s) on the diagnostic chart.

1. Checks to see that the wheel speed sensor is within proper resistance values.
2. Checks for short to ground in wheel speed sensor wiring.
3. Checks for short to voltage in wheel speed sensor wiring.
4. Checks for a short to voltage in CKT 883.
5. Checks for a short to voltage in CKT 882.
6. Ensures malfunction was not due to physical damage of circuitry.

Diagnostic Aids: An "Intermittent" malfunction may be caused by a poor connection, rubbed through wire insulation, or a wire that is broken inside the insulation.

The frequency of the malfunction can be checked by using the enhanced diagnostic function of the Tech 1, as described in "Tech 1 Diagnostics" found in this section.

If the customer's comments reflect that the ABS warning lamp is "ON" only during moist environmental changes (rain, snow, vehicle wash), all wheel speed sensor circuitry should be thoroughly inspected for signs of water intrusion. If DTC is not current, clear DTCs and simulate the effects of water intrusion. Use the following procedure. Spray down the suspected area with a 5% salt water solution (two teaspoons of salt to 12 oz. of water). Test drive vehicle over various road surfaces (bumps, turns, etc.) above 24 km/h (15 mph) for at least 30 seconds. If DTC returns, replace suspected harness.

Any circuitry, that is suspected as causing the intermittent complaint, should be thoroughly checked for backed out terminals, improper mating, broken locks, improperly formed or damaged terminals, poor terminal to wiring connections or physical damage to the wiring harness.

Resistance of the wheel speed sensor will increase with an increase in sensor temperature.

When replacing a wheel speed sensor, inspect the sensor terminals and harness connector for corrosion and/or water intrusion. If evidence of corrosion or water intrusion exists, replace wheel speed sensor jumper harness. Likewise, if replacing a wheel speed sensor jumper harness, inspect sensor terminals. If evidence of corrosion or water intrusion exists, replace wheel speed sensor. Refer to "On-Vehicle Service" in this section.

DIAGNOSTIC CHART DTC 35 — 1994 SILHOUETTE, TRANS SPORT AND LUMINA APV

DTC 35
(Page 1 of 3)
RR WHEEL SPEED SNR CKT OPEN OR SHORT TO GND/BAT

Important

- WHEEL SPEED SENSOR INTERMITTENT MALFUNCTIONS MAY BE DIFFICULT TO LOCATE. CARE SHOULD BE TAKEN NOT TO DISTURB ANY ELECTRICAL CONNECTIONS PRIOR TO AN INDICATED STEP OF THIS CHART. THIS WILL ENSURE THAT AN INTERMITTENT CONNECTION WILL NOT BE CORRECTED BEFORE THE SOURCE OF THE MALFUNCTION IS FOUND.

(1)
- IGNITION "OFF."
- USING A 5% SALT WATER SOLUTION (TWO TEASPOONS OF SALT TO 12 OZ. WATER), THOROUGHLY SPRAY THE RIGHT REAR WHEEL SPEED SENSOR JUMPER HARNESS.
- DISCONNECT 32-WAY EBCM HARNESS CONNECTOR.
- USING J 39200, MEASURE THE RESISTANCE BETWEEN TERMINALS "E5" AND "E6" OF THE 32-WAY EBCM HARNESS CONNECTOR.

IS RESISTANCE BETWEEN 2059 AND 2353Ω WHEN SENSOR IS 20°C (68°F)? *(IF TEMPERATURE OF SENSOR IS SUSPECTED TO OTHER THAN 20°C (68°F), REFER TO THE TABLE ON PAGE 2 OF 2 OF THIS DTC CHART FOR APPLICABLE SENSOR RESISTANCE VALUES; THE VALUES IN THE TABLE ARE FOR TEMPERATURE OF THE SENSOR, NOT AIR TEMPERATURE.)*

- YES → (2)
- NO → DIAGNOSIS CONTINUED ON PAGE 2 OF 3 OF THIS DTC CHART.

(2)
- MEASURE THE RESISTANCE BETWEEN GROUND AND TERMINAL "E5" OF THE 32-WAY EBCM HARNESS CONNECTOR.

DOES J 39200 DISPLAY RESISTANCE AS "OL" (INFINITE)?

- YES → (3)
- NO → DIAGNOSIS CONTINUED ON PAGE 3 OF 3 OF THIS DTC CHART.

(3)
- IGNITION "ON."
- USING J 39200, MEASURE THE VOLTAGE BETWEEN GROUND AND TERMINAL "E5" OF THE 32-WAY EBCM HARNESS CONNECTOR.

IS VOLTAGE 1 VOLT OR LESS?

- NO → (4)
- YES → (6)

(4)
- IGNITION "OFF."
- DISCONNECT RIGHT REAR WHEEL SPEED SENSOR DIRECTLY AT SENSOR.
- IGNITION "ON."
- USING J 39200, MEASURE THE VOLTAGE BETWEEN GROUND AND TERMINAL "E5" OF THE 32-WAY EBCM HARNESS CONNECTOR.

IS VOLTAGE 1 VOLT OR LESS?

- YES → (5)
- NO → REPAIR SHORT TO VOLTAGE IN CKT 883.

(5)
- USING J 39200, MEASURE THE VOLTAGE BETWEEN GROUND AND TERMINAL "E6" OF THE 32-WAY EBCM HARNESS CONNECTOR.

IS VOLTAGE 1 VOLT OR LESS?

- YES → MALFUNCTION IS INTERMITTENT. INSPECT ALL CONNECTORS AND HARNESSES FOR DAMAGE WHICH MAY RESULT IN A SHORT TO VOLTAGE WHEN ALL COMPONENTS ARE CONNECTED. REPAIR ALL DAMAGE IF FOUND. REFER TO "DIAGNOSTIC AIDS" ON FACING PAGE FOR MORE INFORMATION.
- NO → REPAIR SHORT TO VOLTAGE IN CKT 882.

(6)
- IGNITION "OFF."
- INSPECT TERMINALS "E5" AND "E6" OF THE 32-WAY EBCM HARNESS CONNECTOR FOR POOR TERMINAL CONTACT AND CONTAMINATION. ALSO INSPECT CKTs 882 AND 883 FOR DAMAGE WHICH MAY RESULT IN AN OPEN CIRCUIT, SHORT TO GROUND, OR SHORT TO VOLTAGE. REPAIR ALL DAMAGE IF EVIDENT.
- RECONNECT ALL CONNECTORS.
- IGNITION "ON."

DOES DTC 35 SET AS A CURRENT DTC?

- YES → REPLACE EBCM.
- NO → MALFUNCTION IS INTERMITTENT. REFER TO "DIAGNOSTIC AIDS" ON FACING PAGE.

DIAGNOSTIC CHART DTC 35 — 1994 SILHOUETTE, TRANS SPORT AND LUMINA APV

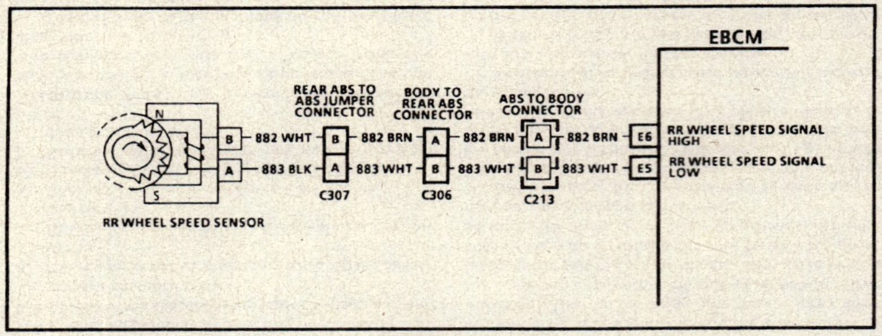

DTC 35
(Page 2 of 3)
RR WHEEL SPEED SNR CKT OPEN OR SHORT TO GND/BAT

Circuit Description:

As a toothed ring passes by the wheel speed sensor, changes in the electromagnetic field cause the wheel speed sensor to produce a sinusoidal (AC) voltage signal whose frequency is proportional to wheel speed. The magnitude of this signal is directly related to wheel speed and the proximity of the wheel speed sensor to the toothed ring, often referred to as the air gap.

DTC Will Set When: DTC 35 can be set anytime after initialization. If either of the right rear wheel speed circuits are shorted to voltage or ground, a malfunction exists.

Action Taken: A malfunction DTC is stored, ABS is disabled and the ABS warning lamp is turned "ON."

DTC Chart Test Description: Number(s) below refer to circled number(s) on the diagnostic chart.

7. Checks for high resistance in CKT 883.
8. Checks for high resistance in CKT 882.
9. Verifies proper wheel speed sensor resistance.

Diagnostic Aids: An "Intermittent" malfunction may be caused by a poor connection, rubbed through wire insulation, or a wire that is broken inside the insulation.

The frequency of the malfunction can be checked by using the enhanced diagnostic function of the Tech 1, as described in "Tech 1 Diagnostics" found in this section.

If the customer's comments reflect that the ABS warning lamp is "ON" only during moist environmental changes (rain, snow, vehicle wash), all wheel speed sensor circuitry should be thoroughly inspected for signs of water intrusion. If DTC is not current, clear DTCs and simulate the effects of water intrusion. Use the following procedure. Spray down the suspected area with a 5% salt water solution (two teaspoons of salt to 12 oz. of water).

Test drive vehicle over various road surfaces (bumps, turns, etc.) above 24 km/h (15 mph) for at least 30 seconds. If DTC returns, replace suspected harness.

Any circuitry, that is suspected as causing the intermittent complaint, should be thoroughly checked for backed out terminals, improper mating, broken locks, improperly formed or damaged terminals, poor terminal to wiring connections or physical damage to the wiring harness.

Resistance of the wheel speed sensor will increase with an increase in sensor temperature.

When replacing a wheel speed sensor, inspect the sensor terminals and harness connector for corrosion and/or water intrusion. If evidence of corrosion or water intrusion exists, replace wheel speed sensor jumper harness. Likewise, if replacing a wheel speed sensor jumper harness, inspect sensor terminals. If evidence of corrosion or water intrusion exists, replace wheel speed sensor. Refer to "On-Vehicle Service" in this section.

DIAGNOSTIC CHART DTC 35 — 1994 SILHOUETTE, TRANS SPORT AND LUMINA APV

DTC 35
(Page 2 of 3)
RR WHEEL SPEED SNR CKT OPEN OR SHORT TO GND/BAT

DIAGNOSIS CONTINUED FROM PAGE 1 OF 3 OF THIS DTC CHART.

(7)
- DISCONNECT RIGHT REAR WHEEL SPEED SENSOR DIRECTLY AT SENSOR.
- USING J 39200, MEASURE THE RESISTANCE BETWEEN TERMINAL "E5" OF THE 32-WAY EBCM HARNESS CONNECTOR AND TERMINAL "A" OF THE RIGHT REAR WHEEL SPEED SENSOR HARNESS CONNECTOR. IS RESISTANCE 2Ω OR LESS?

NO: REPAIR HIGH RESISTANCE IN CKT 883 (WHEEL SPEED SENSOR JUMPER MUST BE REPLACED IF DAMAGED).

YES:

(8)
- USING J 39200, MEASURE THE RESISTANCE BETWEEN TERMINAL "E6" OF THE 32-WAY EBCM HARNESS CONNECTOR AND TERMINAL "B" OF THE RIGHT REAR WHEEL SPEED SENSOR HARNESS CONNECTOR. IS RESISTANCE 2Ω OR LESS?

NO: REPAIR HIGH RESISTANCE IN CKT 882 (WHEEL SPEED SENSOR JUMPER MUST BE REPLACED IF DAMAGED).

YES:

(9)
- USING J 39200, MEASURE THE RESISTANCE BETWEEN TERMINAL "A" AND TERMINAL "B" OF THE RIGHT REAR WHEEL SPEED SENSOR CONNECTOR. IS RESISTANCE BETWEEN 2059 AND 2353Ω WHEN SENSOR IS 20°C (68°F)? *(IF TEMPERATURE OF SENSOR IS SUSPECTED TO OTHER THAN 20°C (68°F), REFER TO THE TABLE BELOW FOR APPLICABLE SENSOR RESISTANCE VALUES; THE VALUES IN THE TABLE ARE FOR TEMPERATURE OF THE SENSOR, NOT AIR TEMPERATURE.)*

NO: REPLACE WHEEL SPEED SENSOR.

YES: MALFUNCTION IS INTERMITTENT. INSPECT ALL CONNECTORS AND HARNESSES FOR DAMAGE WHICH MAY RESULT IN HIGH RESISTANCE WHEN ALL COMPONENTS ARE CONNECTED. REFER TO "DIAGNOSTIC AIDS" ON FACING PAGE FOR MORE INFORMATION. REPAIR ALL DAMAGE IF FOUND.

REAR WHEEL SPEED SENSOR

TEMPERATURE OF SENSOR VS. SENSOR RESISTANCE (APPROXIMATE)

TEMP. (°C)	TEMP. (°F)	RESISTANCE (OHMS)
-34 TO 4	-30 TO 40	1622 TO 2205
5 TO 43	41 TO 110	1930 TO 2566
44 TO 93	111 TO 200	2245 TO 3028

DIAGNOSTIC CHART DTC 35 — 1994 SILHOUETTE, TRANS SPORT AND LUMINA APV

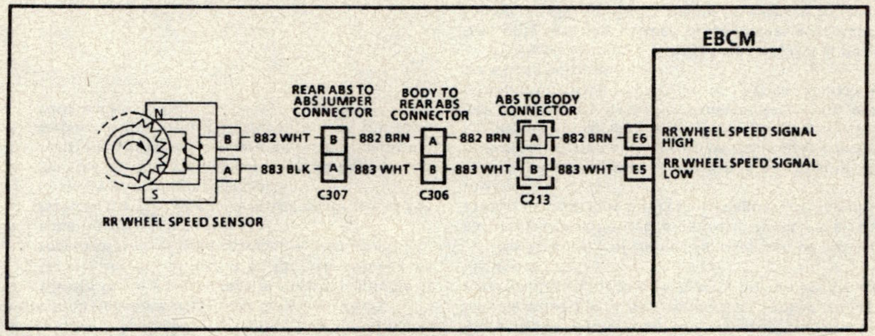

DTC 35
(Page 3 of 3)
RR WHEEL SPEED SNR CKT OPEN OR SHORT TO GND/BAT

Circuit Description:
As a toothed ring passes by the wheel speed sensor, changes in the electromagnetic field cause the wheel speed sensor to produce a sinusoidal (AC) voltage signal whose frequency is proportional to wheel speed. The magnitude of this signal is directly related to wheel speed and the proximity of the wheel speed sensor to the toothed ring, often referred to as the air gap.

DTC Will Set When: DTC 35 can be set anytime after initialization. If either of the right rear wheel speed circuits are shorted to voltage or ground, a malfunction exists.

Action Taken: A malfunction DTC is stored, ABS is disabled and the ABS warning lamp is turned "ON."

DTC Chart Test Description: Number(s) below refer to circled number(s) on the diagnostic chart.
10. Checks for short to ground in CKT 883.
11. Checks for short to ground in CKT 882.
12. Checks for short to ground in wheel speed sensor jumper harness.
13. Verifies that wheel speed sensor is not internally shorted to ground.

Diagnostic Aids: An "Intermittent" malfunction may be caused by a poor connection, rubbed through wire insulation, or a wire that is broken inside the insulation.

The frequency of the malfunction can be checked by using the enhanced diagnostic function of the Tech 1, as described in "Tech 1 Diagnostics" found in this section.

If the customer's comments reflect that the ABS warning lamp is "ON" only during moist environmental changes (rain, snow, vehicle wash), all wheel speed sensor circuitry should be thoroughly inspected for signs of water intrusion. If DTC is not current, clear DTCs and simulate the effects of water intrusion. Use the following procedure.

Spray down the suspected area with a 5% salt water solution (two teaspoons of salt to 12 oz. of water). Test drive vehicle over various road surfaces (bumps, turns, etc.) above 24 km/h (15 mph) for at least 30 seconds. If DTC returns, replace suspected harness.

Any circuitry, that is suspected as causing the intermittent complaint, should be thoroughly checked for backed out terminals, improper mating, broken locks, improperly formed or damaged terminals, poor terminal to wiring connections or physical damage to the wiring harness.

Resistance of the wheel speed sensor will increase with an increase in sensor temperature.

When replacing a wheel speed sensor, inspect the sensor terminals and harness connector for corrosion and/or water intrusion. If evidence of corrosion or water intrusion exists, replace wheel speed sensor jumper harness. Likewise, if replacing a wheel speed sensor jumper harness, inspect sensor terminals. If evidence of corrosion or water intrusion exists, replace wheel speed sensor. Refer to "On-Vehicle Service" in this section.

DIAGNOSTIC CHART DTC 35 — 1994 SILHOUETTE, TRANS SPORT AND LUMINA APV

DTC 35
(Page 3 of 3)
RR WHEEL SPEED SNR CKT OPEN OR SHORT TO GND/BAT

DIAGNOSIS CONTINUED FROM PAGE 1 OF 3 OF THIS DTC CHART.

(10)
- DISCONNECT CONNECTOR C307 (RH REAR UNDER BODY ABOVE FRONT SPRING HANGER).
- USING J 39200, MEASURE THE RESISTANCE BETWEEN GROUND AND TERMINAL "E5" OF THE 32-WAY EBCM HARNESS CONNECTOR.

DOES J 39200 DISPLAY RESISTANCE AS "OL" (INFINITE)?

- NO: REPAIR SHORT TO GROUND IN CKT 883.
- YES: go to (11)

(11)
- USING J 39200, MEASURE THE RESISTANCE BETWEEN GROUND AND TERMINAL "E6" OF THE 32-WAY EBCM HARNESS CONNECTOR.

DOES J 39200 DISPLAY RESISTANCE AS "OL" (INFINITE)?

- NO: REPAIR SHORT TO GROUND IN CKT 882.
- YES: go to (12)

(12)
- THOROUGHLY RESPRAY THE RIGHT REAR WHEEL SPEED SENSOR JUMPER HARNESS WITH THE SALT WATER SOLUTION.

IMPORTANT: DO NOT ALLOW WATER TO ENTER ANY WHEEL SPEED SENSOR CONNECTOR.

- DISCONNECT RIGHT REAR WHEEL SPEED SENSOR DIRECTLY AT SENSOR.
- USING J 39200, MEASURE THE RESISTANCE BETWEEN GROUND AND TERMINAL "A," THEN "B" OF THE RIGHT REAR WHEEL SPEED SENSOR HARNESS CONNECTOR.

DOES J 39200 DISPLAY RESISTANCE AS "OL" (INFINITE) FOR BOTH MEASUREMENTS?

- NO: REPLACE RIGHT REAR WHEEL SPEED SENSOR JUMPER HARNESS.
- YES: go to (13)

(13)
- USING J 39200, MEASURE THE RESISTANCE BETWEEN GROUND AND TERMINAL "A" OF THE RIGHT REAR WHEEL SPEED SENSOR CONNECTOR.

DOES J 39200 DISPLAY RESISTANCE AS "OL" (INFINITE)?

- NO: REPLACE WHEEL SPEED SENSOR.
- YES: MALFUNCTION IS INTERMITTENT. INSPECT ALL CONNECTORS AND HARNESSES FOR DAMAGE WHICH MAY RESULT IN A SHORT TO GROUND WHEN ALL COMPONENTS ARE CONNECTED. PAY PARTICULAR ATTENTION TO THE JUMPER HARNESS, SMALL NICKS OR CUTS CAN CAUSE A MALFUNCTION. REPAIR ANY DAMAGE IF FOUND. JUMPER HARNESS MUST BE REPLACED IF DAMAGED. REFER TO "DIAGNOSTIC AIDS" ON FACING PAGE FOR MORE INFORMATION.

DIAGNOSTIC CHART DTC 36 — 1994 SILHOUETTE, TRANS SPORT AND LUMINA APV

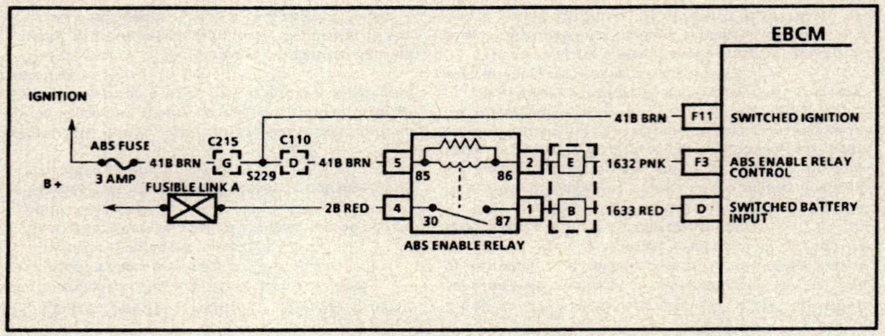

DTC 36

(Page 1 of 2)

LOW SYSTEM VOLTAGE

Circuit Description:

This DTC is used to monitor the voltage level available to the EBCM controller. If the voltage drops below 11 volts, full performance of the ABS system cannot be guaranteed. During ABS operation, there are several current requirements that will cause battery voltage to drop. Because of this, voltage is monitored prior to ABS to indicate good charging system condition and also during ABS when voltage may drop significantly.

DTC Will Set When: DTC 36 can only be set if the vehicle's speed is greater than 5 km/h (3 mph). If the switched battery voltage is less than 11.2 volts non-ABS or less than 9.6 volts ABS, a malfunction exists.

Action Taken: A malfunction DTC is stored, ABS is disabled and the ABS warning lamp is turned "ON."

DTC Chart Test Description: Number(s) below refer to circled number(s) on the diagnostic chart.

1. Checks to see what voltage is readily available at terminal "D" of the EBCM. If voltage is representative of good charging system condition, a malfunction is not present.
2. Isolates the EBCM to check the condition of the circuitry and charging system.
3. Isolates the low voltage condition to high circuit resistance, or improper charging system operation.
4. Isolates which circuit is the source of the high resistance condition.

Diagnostic Aids: An "Intermittent" malfunction may be caused by a poor connection, rubbed through wire insulation, or a wire that is broken inside the insulation.

The frequency of the malfunction can be checked by using the enhanced diagnostic function of the Tech 1, as described in "Tech 1 Diagnostics" found in this section.

Any circuitry, that is suspected as causing the intermittent complaint, should be thoroughly checked for backed out terminals, improper mating, broken locks, improperly formed or damaged terminals, poor terminal to wiring connections or physical damage to the wiring harness.

While performing a voltage load test, if it is noted that <u>only</u> ignition voltage drops below acceptable voltage levels, CKT 41B should be checked for high resistance or an open condition.

DIAGNOSTIC CHART DTC 36 — 1994 SILHOUETTE, TRANS SPORT AND LUMINA APV

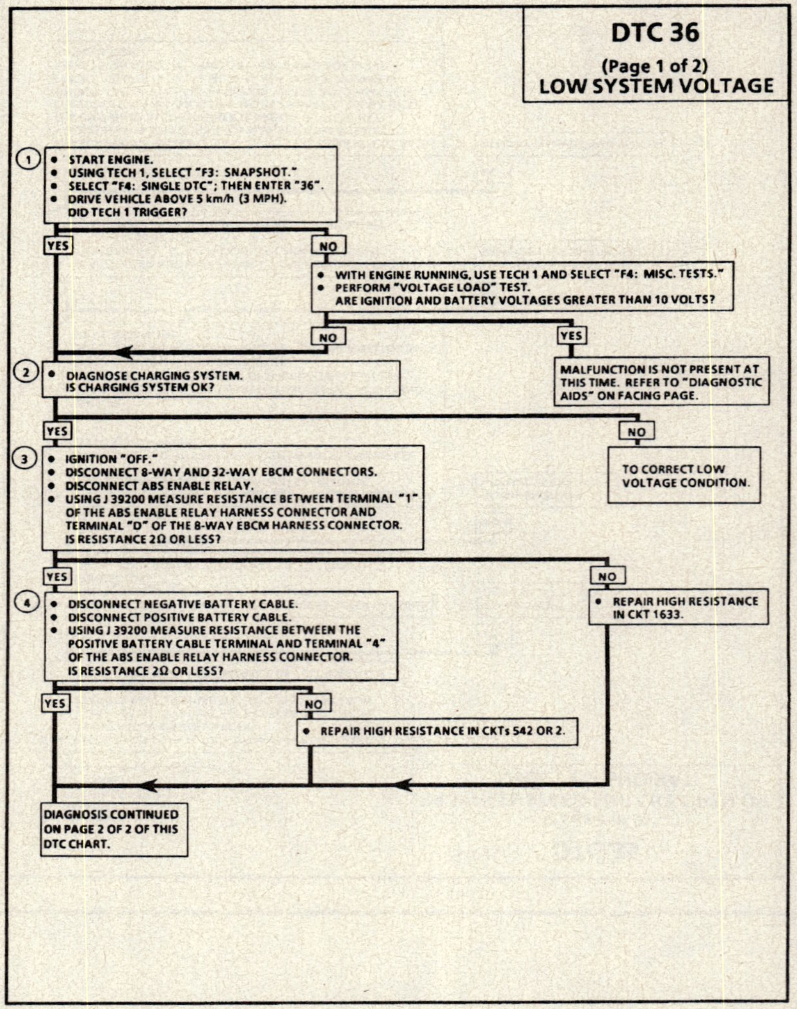

DIAGNOSTIC CHART DTC 36 — 1994 SILHOUETTE, TRANS SPORT AND LUMINA APV

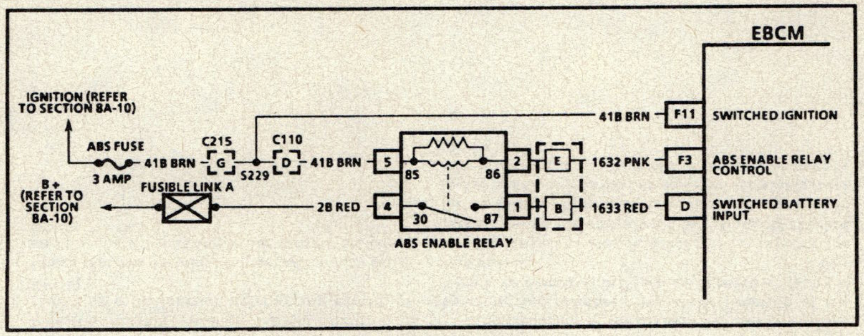

DTC 36
(Page 2 of 2)
LOW SYSTEM VOLTAGE

Circuit Description:

This DTC is used to monitor the voltage level available to the EBCM controller. If the voltage drops below 11 volts, full performance of the ABS system cannot be guaranteed. During ABS operation, there are several current requirements that will cause battery voltage to drop. Because of this, voltage is monitored prior to ABS to indicate good charging system condition and also during ABS when voltage may drop significantly.

DTC Will Set When: DTC 36 can only be set if the vehicle's speed is greater than 5 km/h (3 mph). If the switched battery voltage is less than 11.2 volts non-ABS or less than 9.6 volts ABS, a malfunction exists.

Action Taken: A malfunction DTC is stored, ABS is disabled and the ABS warning lamp is turned "ON."

DTC Chart Test Description: Number(s) below refer to circled number(s) on the diagnostic chart.

5. Checks for high resistance in CKT 41B.
6. Checks for high resistance in the ignition fuse feed circuit.
7. Checks for malfunctioning EBCM.

Diagnostic Aids: An "Intermittent" malfunction may be caused by a poor connection, rubbed through wire insulation, or a wire that is broken inside the insulation.

The frequency of the malfunction can be checked by using the enhanced diagnostic function of the Tech 1, as described in "Tech 1 Diagnostics" found in this section.

Any circuitry, that is suspected as causing the intermittent complaint, should be thoroughly checked for backed out terminals, improper mating, broken locks, improperly formed or damaged terminals, poor terminal to wiring connections or physical damage to the wiring harness.

While performing a voltage load test, if it is noted that <u>only</u> ignition voltage drops below acceptable voltage levels, CKT 41B should be checked for high resistance or an open condition.

DIAGNOSTIC CHART DTC 36 — 1994 SILHOUETTE, TRANS SPORT AND LUMINA APV

DTC 36
(Page 2 of 2)
LOW SYSTEM VOLTAGE

DIAGNOSIS CONTINUED FROM PAGE 1 OF 2 OF THIS DTC CHART.

(5)
- REMOVE ABS FUSE.
- USING J 39200 MEASURE RESISTANCE BETWEEN EACH ABS FUSE BLOCK TERMINAL AND TERMINAL "F11" OF THE 32-WAY EBCM HARNESS CONNECTOR.

IS EITHER MEASUREMENT 2Ω OR LESS?

NO: REPAIR HIGH RESISTANCE IN CKT 41.

YES:

(6)
- WITH BATTERY CABLES DISCONNECTED, TURN IGNITION "ON" (THIS PROVIDES IGNITION CIRCUIT CONTINUITY).
- USING J 39200, MEASURE RESISTANCE BETWEEN POSITIVE BATTERY CABLE TERMINAL AND EACH ABS FUSE BLOCK TERMINAL.

IS EITHER MEASUREMENT 2Ω OR LESS?

NO: REPAIR HIGH RESISTANCE IN THE "IGNITION SWITCH 'RUN' FUSE FEED" CIRCUIT.

YES:

(7)
- IGNITION "OFF."
- INSPECT ABS ENABLE RELAY AND ABS ENABLE RELAY HARNESS CONNECTOR FOR POOR TERMINAL CONTACT. INSPECT 8-WAY AND 32-WAY EBCM HARNESS CONNECTORS FOR A POOR CONNECTION TO THE EBCM. INSPECT POSITIVE BATTERY CABLE TERMINAL FOR POOR CONTACT TO THE BATTERY.

  REPLACE TERMINALS IF POOR CONTACT IS EVIDENT.
- RECONNECT ALL CONNECTORS.
- DRIVE VEHICLE ABOVE 5 km/h (3 MPH).

DOES DTC 36 SET AS A CURRENT DTC?

YES: REPLACE EBCM.

NO: MALFUNCTION IS MOST LIKELY DUE TO A POOR CONNECTION. REFER TO "DIAGNOSTIC AIDS" ON FACING PAGE.

DIAGNOSTIC CHART DTC 37 — 1994 SILHOUETTE, TRANS SPORT AND LUMINA APV

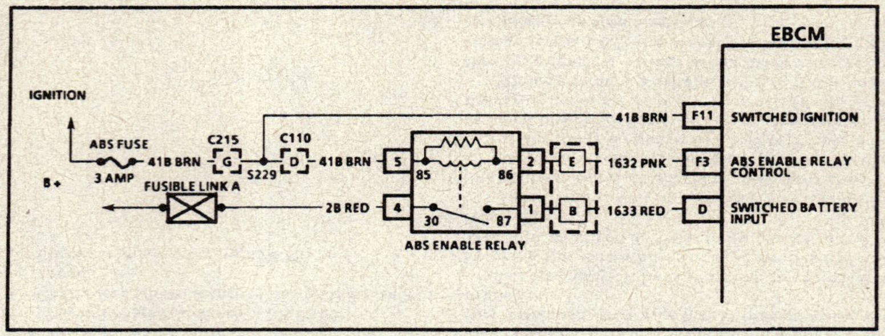

DTC 37
HIGH SYSTEM VOLTAGE

Circuit Description:
This DTC is designed to detect high vehicle voltage levels prior to any required motor movement (initialization or ABS). If excessive voltage exists, demagnetization of the motor magnets may occur, which would eventually affect or eliminate ABS performance.

DTC Will Set When: DTC 37 can only be set if the vehicle's speed is greater than 5 km/h (3 mph). If the switched battery voltage is greater than 17 volts, a malfunction exists.

Action Taken: A malfunction DTC is stored, ABS is disabled and the ABS warning lamp is turned "ON."

DTC Chart Test Description: Number(s) below refer to circled number(s) on the diagnostic chart.

1. Checks the voltage level being received by the EBCM.
2. Indicates whether the high voltage condition is caused by a malfunctioning charging system or EBCM.
3. Checks for poor terminal contact which may cause an intermittent malfunction or a malfunctioning EBCM.

Diagnostic Aids: An "Intermittent" malfunction may be caused by a poor connection, rubbed through wire insulation, or a wire that is broken inside the insulation.

The frequency of the malfunction can be checked by using the enhanced diagnostic function of the Tech 1, as described in "Tech 1 Diagnostics" found in this section.

Any circuitry, that is suspected as causing the intermittent complaint, should be thoroughly checked for backed out terminals, improper mating, broken locks, improperly formed or damaged terminals, poor terminal to wiring connections or physical damage to the wiring harness.

DIAGNOSTIC CHART DTC 37 — 1994 SILHOUETTE, TRANS SPORT AND LUMINA APV

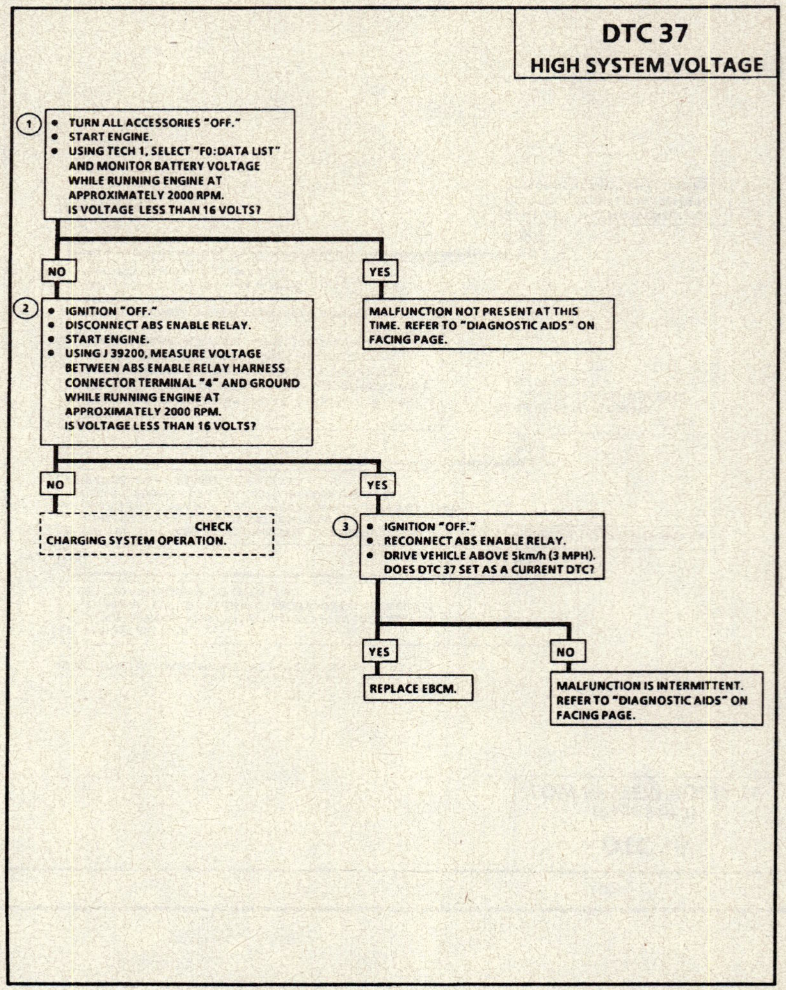

DIAGNOSTIC CHART DTC 38 — 1994 SILHOUETTE, TRANS SPORT AND LUMINA APV

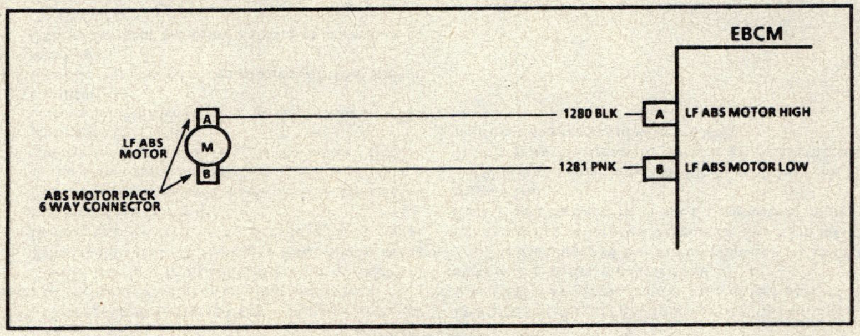

DTC 38

LEFT FRONT ESB WILL NOT HOLD MOTOR

Circuit Description:
This DTC is designed to detect a slipping left front ESB. During initialization and braking, the left front motor is re-homed. If the ESB slips, the motor/piston will move. During the next key "ON" initialization, a re-home of the motor verifies the motor/piston remained at the home position. If motor movement is detected, the ESB must be slipping.

DTC Will Set When: DTC 38 can be set during initialization. If the EBCM detects that the ESB could not hold the piston in the home position, a malfunction exists. DTC 86 is always set with DTC 38.

Action Taken: If an ESB cannot hold a piston in the home position, the piston may be backdriven when the brake pedal is applied, causing the brake pedal to drop. A malfunction DTC is stored, ABS is disabled, and the ABS warning lamp and the red "BRAKE" warning lamp are turned "ON."

DTC Chart Test Description: Number(s) below refer to circled number(s) on the diagnostic chart.

1. Checks the left front ESB. A broken or defective ESB would result in the left front piston being backdriven by hydraulic pressure and pedal movement would result.
2. Ensures that the proper motor circuits are located in the proper connector cavities.
3. Checks for high resistance in the motor "HIGH" circuitry.
4. Checks for high resistance in the motor "LOW" circuitry.
5. Checks for proper resistance of the motor windings.
6. Checks for poor terminal contact or corrosion in the connectors.
7. Checks for a malfunctioning EBCM.

Diagnostic Aids: An "Intermittent" malfunction in this DTC may result from a mechanical part of the system that sticks, binds, or slips.

The frequency of the malfunction can be checked by using the enhanced diagnostic function of the Tech 1, as described in "Tech 1 Diagnostics" found in this section.

The static modulator test function of the Tech 1 or T-100 may be used to locate an intermittent malfunction associated with the ESB.

DIAGNOSTIC CHART DTC 38 — 1994 SILHOUETTE, TRANS SPORT AND LUMINA APV

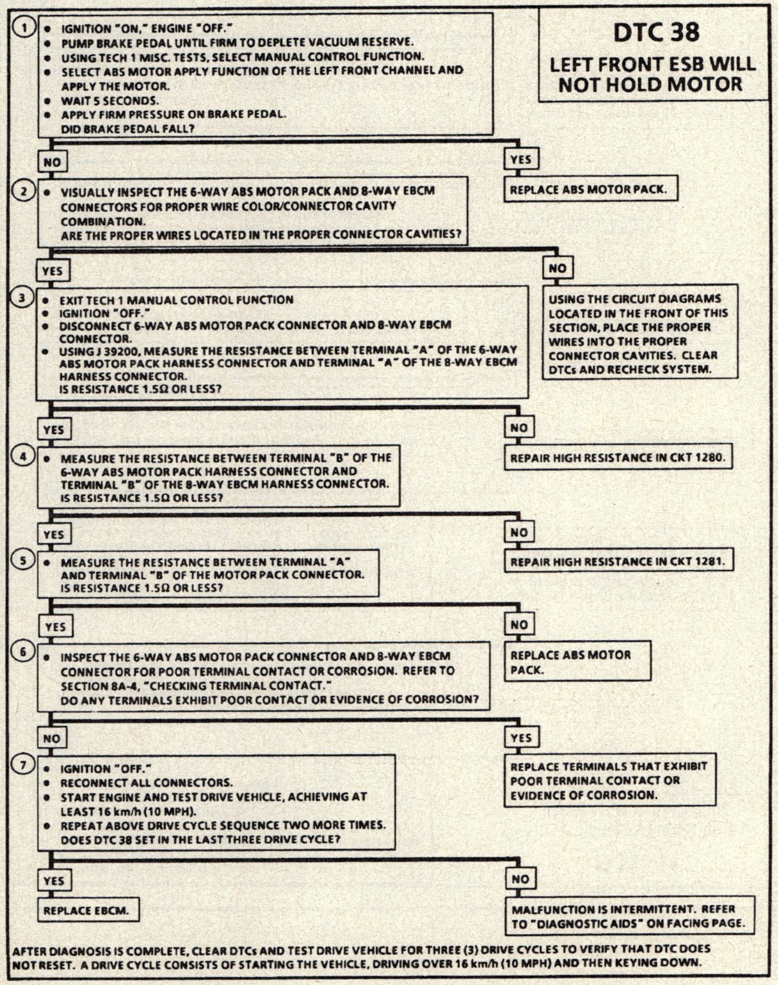

DIAGNOSTIC CHART DTC 41 — 1994 SILHOUETTE, TRANS SPORT AND LUMINA APV

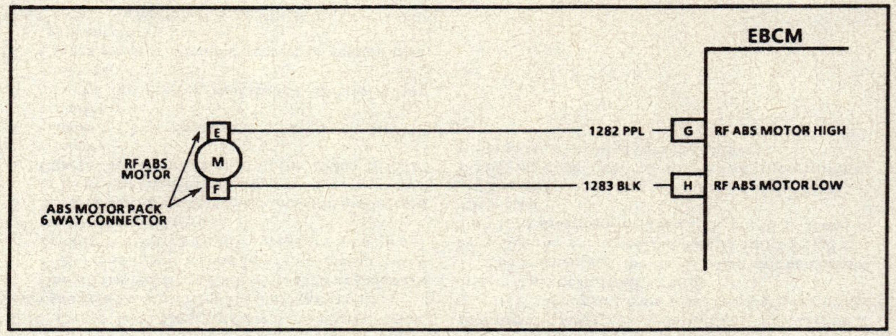

DTC 41

RIGHT FRONT ESB WILL NOT HOLD MOTOR

Circuit Description:

This DTC is designed to detect a slipping right front ESB. During initialization and braking, the right front motor is re-homed. If the ESB slips, the motor/piston will move. During the next key "ON" initialization, a re-home of the motor verifies the motor/piston remained at the home position. If motor movement is detected, the ESB must be slipping.

DTC Will Set When: DTC 41 can be set during initialization. If the EBCM detects that the ESB could not hold the piston in the home position, a malfunction exists. DTC 86 is always set with DTC 41.

Action Taken: If an ESB cannot hold a piston in the home position, the piston may be backdriven when the brake pedal is applied, causing the brake pedal to drop. A malfunction DTC is stored, ABS is disabled, and the ABS warning lamp and the red "BRAKE" warning lamp are turned "ON."

DTC Chart Test Description: Number(s) below refer to circled number(s) on the diagnostic chart.

1. Checks the right front ESB. A broken or defective ESB would result in the right front piston being backdriven by hydraulic pressure, and pedal movement would result.
2. Ensures that the proper motor circuits are located in the proper connector cavities.
3. Checks for high resistance in the motor "HIGH" circuitry.
4. Checks for high resistance in the motor "LOW" circuitry.
5. Checks for proper resistance of the motor windings.
6. Checks for poor terminal contact or corrosion in the connectors.
7. Checks for a malfunctioning EBCM.

Diagnostic Aids: An "Intermittent" malfunction in this DTC may result from a mechanical part of the system that sticks, binds, or slips.

The frequency of the malfunction can be checked by using the enhanced diagnostic function of the Tech 1, as described in "Tech 1 Diagnostics" found in this section.

The static modulator test function of the Tech 1 or T-100 may be used to locate an intermittent malfunction associated with the ESB.

DIAGNOSTIC CHART DTC 41 — 1994 SILHOUETTE, TRANS SPORT AND LUMINA APV

DTC 41
RIGHT FRONT ESB WILL NOT HOLD MOTOR

(1)
- IGNITION "ON," ENGINE "OFF."
- PUMP BRAKE PEDAL UNTIL FIRM TO DEPLETE VACUUM RESERVE.
- USING TECH 1 MISC. TESTS, SELECT MANUAL CONTROL FUNCTION.
- SELECT ABS MOTOR APPLY FUNCTION OF THE RIGHT FRONT CHANNEL AND APPLY THE MOTOR.
- WAIT 5 SECONDS.
- APPLY FIRM PRESSURE ON BRAKE PEDAL.

DID BRAKE PEDAL FALL?

- YES: REPLACE ABS MOTOR PACK.
- NO: go to (2)

(2)
- VISUALLY INSPECT THE 6-WAY ABS MOTOR PACK AND 8-WAY EBCM CONNECTORS FOR PROPER WIRE COLOR/CONNECTOR CAVITY COMBINATION.

ARE THE PROPER WIRES LOCATED IN THE PROPER CONNECTOR CAVITIES?

- NO: USING THE CIRCUIT DIAGRAMS LOCATED IN THE FRONT OF THIS SECTION, PLACE THE PROPER WIRES INTO THE PROPER CONNECTOR CAVITIES. CLEAR DTCs AND RECHECK SYSTEM.
- YES: go to (3)

(3)
- EXIT TECH 1 MANUAL CONTROL FUNCTION.
- IGNITION "OFF."
- DISCONNECT 6-WAY ABS MOTOR PACK CONNECTOR AND 8-WAY EBCM CONNECTOR.
- USING J 39200, MEASURE THE RESISTANCE BETWEEN TERMINAL "E" OF THE 6-WAY ABS MOTOR PACK HARNESS CONNECTOR AND TERMINAL "G" OF THE 8-WAY EBCM HARNESS CONNECTOR.

IS RESISTANCE 1.5Ω OR LESS?

- NO: REPAIR HIGH RESISTANCE IN CKT 1282.
- YES: go to (4)

(4)
- MEASURE THE RESISTANCE BETWEEN TERMINAL "F" OF THE 6-WAY ABS MOTOR PACK HARNESS CONNECTOR AND TERMINAL "H" OF THE 8-WAY EBCM HARNESS CONNECTOR.

IS RESISTANCE 1.5Ω OR LESS?

- NO: REPAIR HIGH RESISTANCE IN CKT 1283.
- YES: go to (5)

(5)
- MEASURE THE RESISTANCE BETWEEN TERMINAL "E" AND TERMINAL "F" OF THE ABS MOTOR PACK CONNECTOR.

IS RESISTANCE 1.5Ω OR LESS?

- NO: REPLACE ABS MOTOR PACK.
- YES: go to (6)

(6)
- INSPECT THE 6-WAY ABS MOTOR PACK CONNECTOR AND 8-WAY EBCM CONNECTOR FOR POOR TERMINAL CONTACT OR CORROSION.

DO ANY TERMINALS EXHIBIT POOR CONTACT OR EVIDENCE OF CORROSION?

- YES: REPLACE TERMINALS THAT EXHIBIT POOR TERMINAL CONTACT OR EVIDENCE OF CORROSION.
- NO: go to (7)

(7)
- IGNITION "OFF."
- RECONNECT ALL CONNECTORS.
- START ENGINE AND TEST DRIVE VEHICLE, ACHIEVING AT LEAST 16 km/h (10 MPH).
- REPEAT ABOVE DRIVE CYCLE SEQUENCE TWO MORE TIMES.

DOES DTC 41 SET IN THE LAST THREE DRIVE CYCLES?

- YES: REPLACE EBCM.
- NO: MALFUNCTION IS INTERMITTENT. REFER TO "DIAGNOSTIC AIDS" ON FACING PAGE.

AFTER DIAGNOSIS IS COMPLETE, CLEAR DTCs AND TEST DRIVE VEHICLE FOR THREE (3) DRIVE CYCLES TO VERIFY THAT DTC DOES NOT RESET. A DRIVE CYCLE CONSISTS OF STARTING THE VEHICLE, DRIVING OVER 16 KM/H (10 MPH) AND THEN KEYING DOWN.

DIAGNOSTIC CHART DTC 42 — 1994 SILHOUETTE, TRANS SPORT AND LUMINA APV

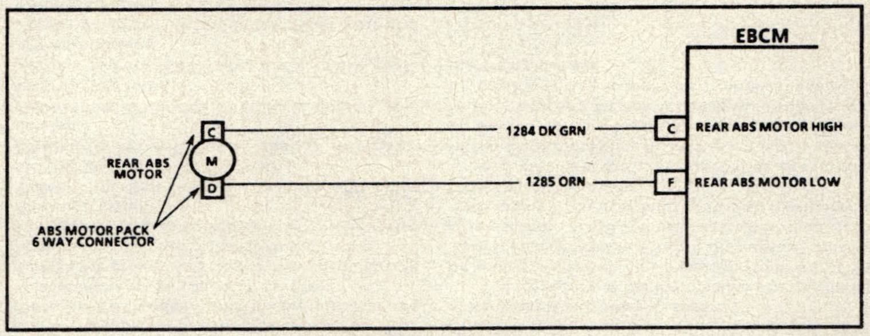

DTC 42

REAR ESB WILL NOT HOLD MOTOR

Circuit Description:
This DTC is designed to detect a slipping rear ESB. During initialization and braking, the rear motor is re-homed. If the ESB slips, the motor/piston will move. During the next key "ON" initialization, a re-home of the motor verifies the motor/piston remained at the home position. If motor movement is detected, the ESB must be slipping.

DTC Will Set When: DTC 42 can be set during initialization. If the EBCM detects that the ESB could not hold the piston in the home position, a malfunction exists. DTC 86 is always set with DTC 42.

Action Taken: If an ESB cannot hold a piston in the home position, the piston may be backdriven when the brake pedal is applied, causing the brake pedal to drop. A malfunction DTC is stored, ABS is disabled, and the ABS warning lamp and the red "BRAKE" warning lamp are turned "ON."

DTC Chart Test Description: Number(s) below refer to circled number(s) on the diagnostic chart.

1. Checks the rear ESB. A broken or defective ESB would result in the rear pistons being backdriven by hydraulic pressure and wheel movement would result.
2. Ensures that the proper motor circuits are located in the proper connector cavities.
3. Checks for high resistance in the motor "HIGH" circuitry.
4. Checks for high resistance in the motor "LOW" circuitry.
5. Checks for proper resistance of the motor windings.
6. Checks for poor terminal contact or corrosion in the connectors.
7. Checks for a malfunctioning EBCM.

Diagnostic Aids: An "Intermittent" malfunction in this DTC may result from a mechanical part of the system that sticks, binds, or slips.

The frequency of the malfunction can be checked by using the enhanced diagnostic function of the Tech 1, as described in "Tech 1 Diagnostics" found in this section.

The static modulator test function of the Tech 1 or T-100 may be used to locate an intermittent malfunction associated with the ESB.

DIAGNOSTIC CHART DTC 42 — 1994 SILHOUETTE, TRANS SPORT AND LUMINA APV

DTC 42
REAR ESB WILL NOT HOLD MOTOR

(1)
- RAISE THE VEHICLE SUCH THAT THE REAR WHEELS ARE APPROXIMATELY SIX INCHES OFF THE FLOOR.
- IGNITION "ON," ENGINE "OFF."
- PUMP BRAKE PEDAL UNTIL FIRM TO DEPLETE VACUUM RESERVE.
- APPLY FIRM PRESSURE ON BRAKE PEDAL AND RELEASE.
- REAPPLY VERY FIRM PRESSURE ON BRAKE PEDAL AND HAVE AN ASSISTANT TRY TO SPIN THE REAR WHEELS BY HAND.

COULD ASSISTANT SPIN THE REAR WHEELS?

YES: REPLACE ABS MOTOR PACK.

NO:

(2)
- VISUALLY INSPECT THE 6-WAY ABS MOTOR PACK AND 8-WAY EBCM CONNECTORS FOR PROPER WIRE COLOR/CONNECTOR CAVITY COMBINATION.

ARE THE PROPER WIRES LOCATED IN THE PROPER CONNECTOR CAVITIES?

NO: USING THE CIRCUIT DIAGRAMS LOCATED IN THE FRONT OF THIS SECTION, PLACE THE PROPER WIRES INTO THE PROPER CONNECTOR CAVITIES. CLEAR DTCs AND RECHECK SYSTEM.

YES:

(3)
- IGNITION "OFF."
- DISCONNECT 6-WAY ABS MOTOR PACK CONNECTOR AND 8-WAY EBCM CONNECTOR.
- USING J 39200, MEASURE THE RESISTANCE BETWEEN TERMINAL "C" OF THE 6-WAY ABS MOTOR PACK HARNESS CONNECTOR AND TERMINAL "C" OF THE 8-WAY EBCM HARNESS CONNECTOR.

IS RESISTANCE 1.5Ω OR LESS?

NO: REPAIR HIGH RESISTANCE IN CKT 1284.

YES:

(4)
- MEASURE THE RESISTANCE BETWEEN TERMINAL "D" OF THE 6-WAY ABS MOTOR PACK HARNESS CONNECTOR AND TERMINAL "F" OF THE 8-WAY EBCM HARNESS CONNECTOR.

IS RESISTANCE 1.5Ω OR LESS?

NO: REPAIR HIGH RESISTANCE IN CKT 1285.

YES:

(5)
- MEASURE THE RESISTANCE BETWEEN TERMINAL "C" AND TERMINAL "D" OF THE ABS MOTOR PACK CONNECTOR.

IS RESISTANCE 1.5Ω OR LESS?

NO: REPLACE ABS MOTOR PACK.

YES:

(6)
- INSPECT THE 6-WAY ABS MOTOR PACK CONNECTOR AND 8-WAY EBCM CONNECTOR FOR POOR TERMINAL CONTACT OR CORROSION.

DO ANY TERMINALS EXHIBIT POOR CONTACT OR EVIDENCE OF CORROSION?

YES: REPLACE TERMINALS THAT EXHIBIT POOR TERMINAL CONTACT OR EVIDENCE OF CORROSION.

NO:

(7)
- IGNITION "OFF."
- RECONNECT ALL CONNECTORS.
- START ENGINE AND TEST DRIVE VEHICLE, ACHIEVING AT LEAST 16 km/h (10 MPH).
- REPEAT ABOVE DRIVE CYCLE SEQUENCE TWO MORE TIMES.

DOES DTC 42 SET IN THE LAST THREE DRIVE CYCLES?

YES: REPLACE EBCM.

NO: MALFUNCTION WAS MOST LIKELY CAUSED BY A POOR CONNECTION. REFER TO "DIAGNOSTIC AIDS" ON FACING PAGE.

AFTER DIAGNOSIS IS COMPLETE, CLEAR DTCs AND TEST DRIVE VEHICLE FOR THREE (3) DRIVE CYCLES TO VERIFY DTC DOES NO RESET. A DRIVE CYCLE CONSISTS OF STARTING THE VEHICLE, DRIVING OVER 16 km/h (10 MPH) AND THEN KEYING DOWN.

DIAGNOSTIC CHART DTC 44 — 1994 SILHOUETTE, TRANS SPORT AND LUMINA APV

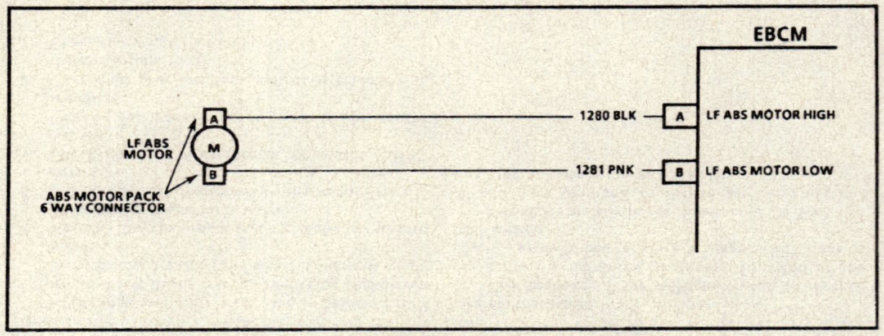

DTC 44

LEFT FRONT ABS CHANNEL WILL NOT MOVE

Circuit Description:

This DTC is designed to detect bound-up ESB, a stuck ABS motor, or a seized hydraulic modulator. When the release is commanded during initialization, the ESB should release the ABS motor, resulting in sensed current being less than commanded current (motor is spinning freely). If the ABS motor is not moving, sensed current will be equal to stall current.

DTC Will Set When: DTC 44 can be set during normal operation. If the EBCM detects a condition in which it cannot move the ABS motor in either direction, a malfunction exists.

Action Taken: This malfunction indicates the channel cannot be moved properly. A malfunction DTC is stored, ABS is disabled and the ABS warning lamp is turned "ON."

DTC Chart Test Description: Number(s) below refer to circled number(s) on the diagnostic chart.

1. Ensures that proper ABS motor circuits are located in the proper connector cavities.
2. Checks for proper ABS motor movement during apply and release commands from the Tech 1.
3. Compares EBCM command current to ABS motor feedback current.
4. Checks for a short circuit between the ABS motor "HIGH" and motor "LOW" circuits.
5. Checks for an ABS motor that is internally shorted.
6. Determines if the malfunction is due to poor terminal contact or the EBCM.
7. Verifies that the ABS motor can actually apply when commanded.
8. Checks for proper hydraulic modulator gear and piston movement.

Diagnostic Aids: An "Intermittent" malfunction in this DTC may result from a mechanical part of the system that sticks, binds, or slips.

The frequency of the malfunction can be checked by using the enhanced diagnostic function of the Tech 1, as described in "Tech 1 Diagnostics" found in this section. DTC 44 may set after modulator disassembly if the modulator pistons are positioned at the bottom of their bore.

Any circuitry, that is suspected as causing the intermittent complaint, should be thoroughly checked for backed out terminals, improper mating, broken locks, improperly formed or damaged terminals, poor terminal to wiring connections or physical damage to the wiring harness.

DIAGNOSTIC CHART DTC 44 — 1994 SILHOUETTE, TRANS SPORT AND LUMINA APV

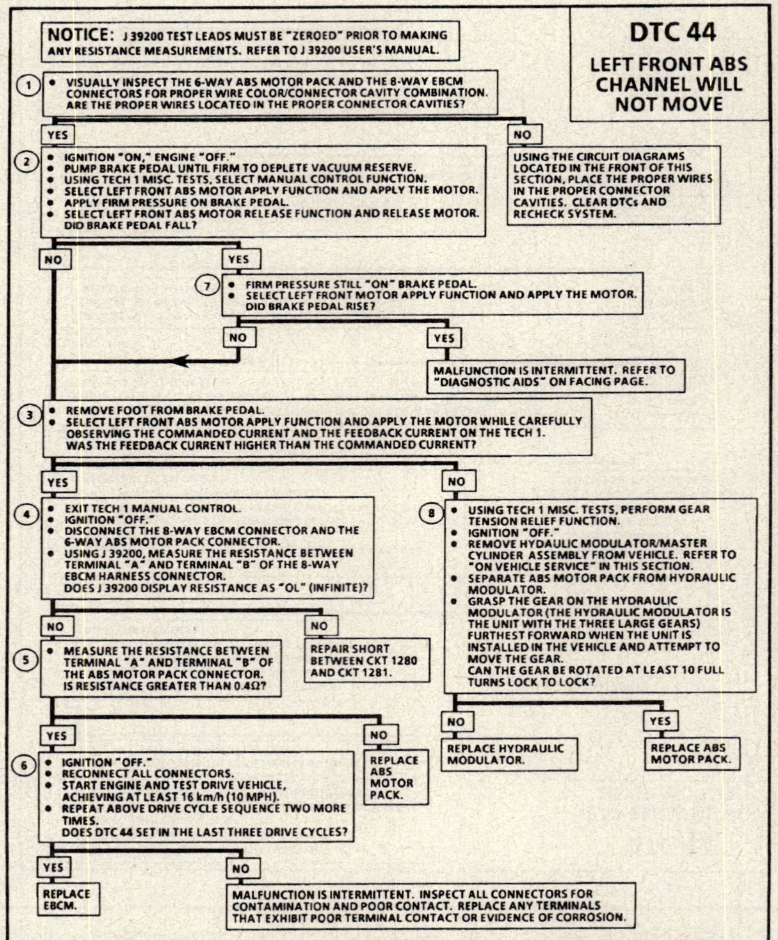

AFTER DIAGNOSIS IS COMPLETE, CLEAR DTCs AND TEST DRIVE VEHICLE FOR THREE (3) DRIVE CYCLES TO VERIFY THAT DTC DOES NOT RESET. A DRIVE CYCLE CONSISTS OF STARTING THE VEHICLE, DRIVING OVER 16 KM/H (10 MPH), AND THEN KEYING DOWN.

DIAGNOSTIC CHART DTC 45 — 1994 SILHOUETTE, TRANS SPORT AND LUMINA APV

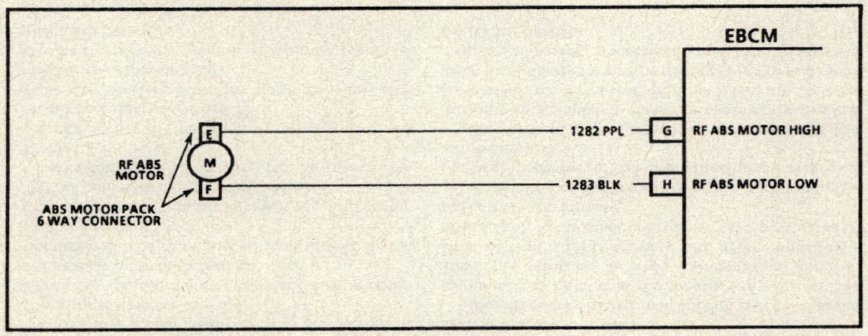

DTC 45

RIGHT FRONT ABS CHANNEL WILL NOT MOVE

Circuit Description:

This DTC is designed to detect bound-up ESB, a stuck ABS motor, or a seized hydraulic modulator. When the release is commanded during initialization, the ESB should release the ABS motor, resulting in sensed current being less than commanded current (ABS motor is spinning freely). If the ABS motor is not moving, sensed current will be equal to stall current.

DTC Will Set When: DTC 45 can be set during normal operation. If the EBCM detects a condition in which it cannot move the ABS motor in either direction, a malfunction exists.

Action Taken: This malfunction indicates the channel cannot be moved properly. A malfunction DTC is stored, ABS is disabled and the ABS warning lamp is turned "ON."

DTC Chart Test Description: Number(s) below refer to circled number(s) on the diagnostic chart.

1. Ensures that proper motor circuits are located in the proper connector cavities.
2. Checks for proper ABS motor movement during apply and release commands from the Tech 1.
3. Compares EBCM command current to ABS motor feedback current.
4. Checks for a short circuit between the ABS motor "HIGH" and motor "LOW" circuits.
5. Checks for an ABS motor that is internally shorted.
6. Checks to see if malfunction is due to poor terminal contact or the EBCM.
7. Verifies that the ABS motor can actually apply when commanded.
8. Checks for proper hydraulic modulator gear and piston movement.

Diagnostic Aids: An "Intermittent" malfunction in this DTC may result from a mechanical part of the system that sticks, binds, or slips.

The frequency of the malfunction can be checked by using the enhanced diagnostic function of the Tech 1, as described in "Tech 1 Diagnostics" found in this section. DTC 45 may set after modulator disassembly if the modulator pistons are positioned at the bottom of their bore.

Any circuitry, that is suspected as causing the intermittent complaint, should be thoroughly checked for backed out terminals, improper mating, broken locks, improperly formed or damaged terminals, poor terminal to wiring connections or physical damage to the wiring harness.

DIAGNOSTIC CHART DTC 45 — 1994 SILHOUETTE, TRANS SPORT AND LUMINA APV

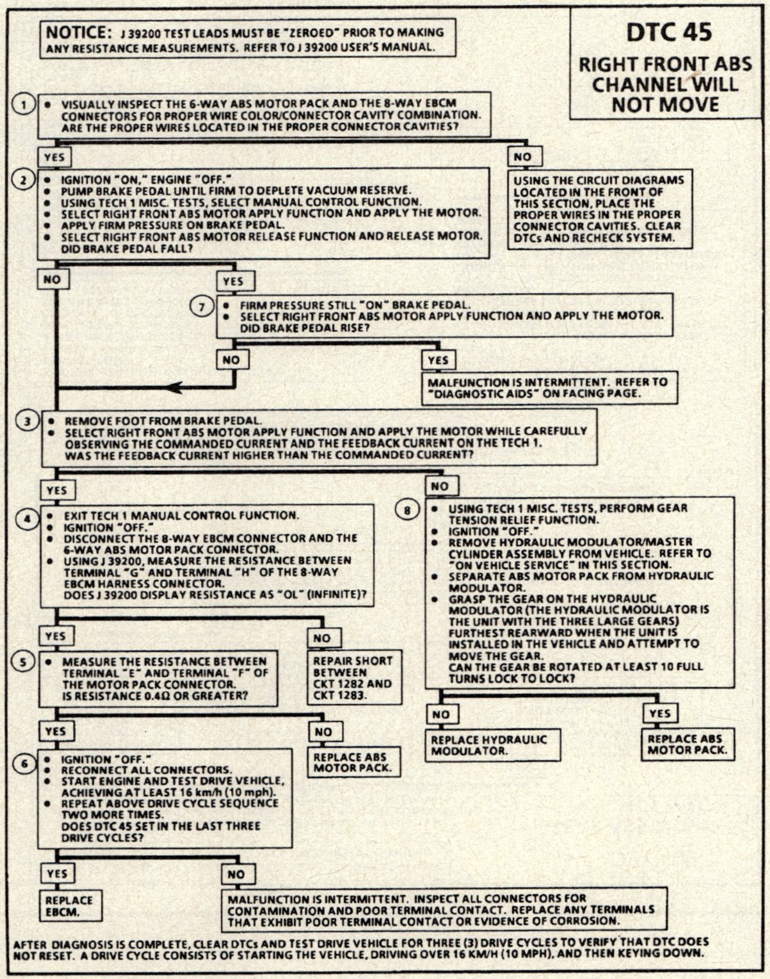

DIAGNOSTIC CHART DTC 46 — 1994 SILHOUETTE, TRANS SPORT AND LUMINA APV

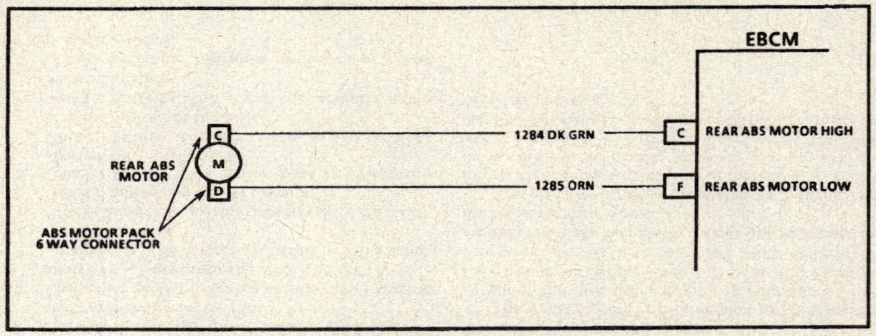

DTC 46

REAR ABS CHANNEL WILL NOT MOVE

Circuit Description:

This DTC is designed to detect bound-up ESB, a stuck ABS motor, or a seized hydraulic modulator. When the release is commanded during initialization, the ESB should release the ABS motor, resulting in sensed current being less than commanded current (motor is spinning freely). If the ABS motor is not moving, sensed current will be equal to stall current.

DTC Will Set When: DTC 46 can be set during normal operation. If the EBCM detects a condition in which it cannot move the ABS motor in either direction, a malfunction exists.

Action Taken: This malfunction indicates the channel cannot be moved properly. A malfunction DTC is stored, ABS is disabled and the ABS warning lamp is turned "ON."

DTC Chart Test Description: Number(s) below refer to circled number(s) on the diagnostic chart.

1. Ensures that proper motor circuits are located in the proper connector cavities.
2. Checks for proper motor movement during apply and release commands from the Tech 1.
3. Compares EBCM command current to ABS motor feedback current.
4. Checks for a short circuit between the ABS motor "HIGH" and ABS motor "LOW" circuits.
5. Checks for an ABS motor that is internally shorted.
6. Checks to see if malfunction is due to poor terminal contact or the EBCM.
7. This step verifies that the motor can actually release when commanded.
8. Checks for proper hydraulic modulator gear and piston movement.

Diagnostic Aids: An "Intermittent" malfunction in this DTC may result from a mechanical part of the system that sticks or binds.

The frequency of the malfunction can be checked by using the enhanced diagnostic function of the Tech 1, as described in "Tech 1 Diagnostics" found in this section. DTC 46 may set after modulator disassembly if the modulator pistons are positioned at the bottom of their bore.

Depending on the frequency of the malfunction, a physical inspection of the mechanical parts suspected may be necessary.

Any circuitry, that is suspected as causing the intermittent complaint, should be thoroughly checked for backed out terminals, improper mating, broken locks, improperly formed or damaged terminals, poor terminal to wiring connections or physical damage to the wiring harness.

DIAGNOSTIC CHART DTC 46 — 1994 SILHOUETTE, TRANS SPORT AND LUMINA APV

DTC 46
REAR ABS CHANNEL WILL NOT MOVE

NOTICE: J 39200 TEST LEADS MUST BE "ZEROED" PRIOR TO MAKING ANY RESISTANCE MEASUREMENTS. REFER TO J 39200 USER'S MANUAL.

(1)
- VISUALLY INSPECT THE 6-WAY ABS MOTOR PACK AND THE 8-WAY EBCM CONNECTORS FOR PROPER WIRE COLOR/CONNECTOR CAVITY COMBINATION. ARE THE PROPER WIRES LOCATED IN THE PROPER CONNECTOR CAVITIES?

NO: USING THE CIRCUIT DIAGRAMS LOCATED IN THE FRONT OF THIS SECTION, PLACE THE PROPER WIRES IN THE PROPER CONNECTOR CAVITIES. CLEAR DTCs AND RECHECK SYSTEM.

YES:

(2)
- RAISE AND SUPPORT THE VEHICLE SUCH THAT THE REAR WHEELS ARE APPROXIMATELY SIX INCHES OFF THE FLOOR.
- IGNITION "ON," ENGINE "OFF."
- PUMP BRAKE PEDAL UNTIL FIRM TO DEPLETE VACUUM RESERVE.
- APPLY FIRM PRESSURE ON BRAKE PEDAL.
- USING TECH 1 MISC. TESTS, SELECT MANUAL CONTROL FUNCTION.
- SELECT REAR AXLE ABS MOTOR APPLY FUNCTION AND APPLY THE MOTOR.
- HAVE AN ASSISTANT TRY TO SPIN THE REAR WHEELS BY HAND. COULD ASSISTANT SPIN THE REAR WHEELS?

NO:

(7)
- FIRM PRESSURE STILL "ON" BRAKE PEDAL.
- SELECT REAR AXLE ABS MOTOR RELEASE FUNCTION AND RELEASE THE MOTOR.
- HAVE AN ASSISTANT TRY TO SPIN THE REAR WHEELS BY HAND. COULD ASSISTANT SPIN THE REAR WHEELS?

YES: MALFUNCTION IS INTERMITTENT. REFER TO "DIAGNOSTIC AIDS" ON FACING PAGE.

NO: (go to 3)

YES (from 2):

(3)
- REMOVE FOOT FROM BRAKE PEDAL.
- SELECT REAR AXLE ABS MOTOR APPLY FUNCTION AND APPLY THE MOTOR WHILE CAREFULLY OBSERVING THE COMMANDED CURRENT AND THE FEEDBACK CURRENT ON THE TECH 1. WAS THE FEEDBACK CURRENT HIGHER THAN THE COMMANDED CURRENT?

YES:

(4)
- EXIT TECH 1 MANUAL CONTROL FUNCTION.
- IGNITION "OFF."
- DISCONNECT THE 8-WAY EBCM CONNECTOR AND THE 6-WAY ABS MOTOR PACK CONNECTOR.
- USING J 39200, MEASURE THE RESISTANCE BETWEEN TERMINAL "C" AND TERMINAL "F" OF THE 8-WAY EBCM HARNESS CONNECTOR. DOES J 39200 DISPLAY RESISTANCE AS "OL" (INFINITE)?

NO: REPAIR SHORT BETWEEN CKT 1284 AND CKT 1285.

YES:

(5)
- MEASURE THE RESISTANCE BETWEEN TERMINAL "C" AND TERMINAL "D" OF THE ABS MOTOR PACK CONNECTOR. IS RESISTANCE GREATER THAN 0.4Ω?

NO: REPLACE ABS MOTOR PACK.

YES:

(6)
- RECONNECT ALL CONNECTORS.
- IGNITION "OFF."
- START ENGINE AND TEST DRIVE VEHICLE, ACHIEVING AT LEAST 16 km/h (10 MPH).
- REPEAT ABOVE DRIVE CYCLE SEQUENCE TWO MORE TIMES. DOES DTC 46 SET IN THE LAST THREE DRIVE CYCLES?

YES: REPLACE EBCM.

NO: MALFUNCTION IS INTERMITTENT. INSPECT ALL CONNECTORS FOR CONTAMINATION AND POOR TERMINAL CONTACT. REPLACE ANY TERMINALS THAT EXHIBIT POOR TERMINAL CONTACT OR EVIDENCE OF CORROSION.

NO (from 3):

(8)
- USING TECH 1 MISC. TESTS, PERFORM GEAR TENSION RELIEF FUNCTION.
- IGNITION "OFF."
- REMOVE HYDRAULIC MODULATOR/MASTER CYLINDER ASSEMBLY FROM VEHICLE. REFER TO "ON VEHICLE SERVICE" IN THIS SECTION.
- SEPARATE ABS MOTOR PACK FROM HYDRAULIC MODULATOR.
- GRASP THE MIDDLE GEAR ON THE HYDRAULIC MODULATOR (THE HYDRAULIC MODULATOR IS THE UNIT WITH THE THREE LARGE GEARS) AND ATTEMPT TO MOVE THE GEAR. CAN THE GEAR BE ROTATED AT LEAST 7 FULL TURNS LOCK TO LOCK?

NO: REPLACE HYDRAULIC MODULATOR.

YES: REPLACE ABS MOTOR PACK.

AFTER DIAGNOSIS IS COMPLETE, CLEAR DTCs AND TEST DRIVE VEHICLE FOR THREE (3) DRIVE CYCLES TO VERIFY THAT DTC DOES NOT RESET. A DRIVE CYCLE CONSISTS OF STARTING THE VEHICLE, DRIVING OVER 16 KM/H (10 MPH), AND THEN KEYING DOWN.

DIAGNOSTIC CHART DTC 47 — 1994 SILHOUETTE, TRANS SPORT AND LUMINA APV

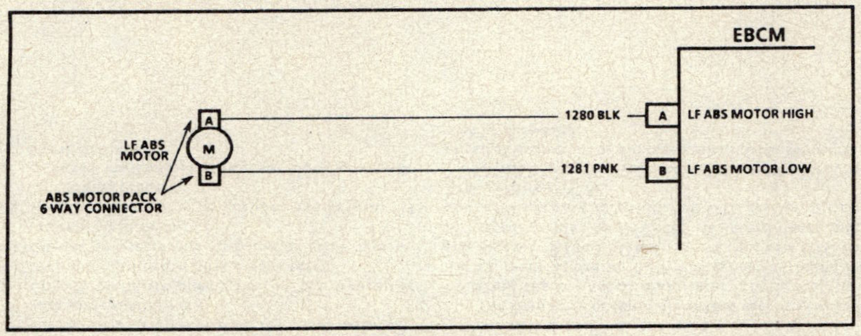

DTC 47
(Page 1 of 2)
LEFT FRONT ABS MOTOR FREE SPINS

Circuit Description:

This DTC is designed to detect a stripped nut or gear assembly during initialization. During the homing sequence, the piston should reach the top of the bore resulting in a stalled ABS motor. If this does not occur, the ABS motor must be spinning with little or no resistance, indicating a nut/screw or gear malfunction.

DTC Will Set When: DTC 47 can only be set during initialization. If the feedback current is less than the command current for a specified period of time, a malfunction exists.

Action Taken: A malfunction DTC is stored, ABS is disabled, and the ABS warning lamp is turned "ON."

DTC Chart Test Description: Number(s) below refer to circled number(s) on the diagnostic chart.

1. Checks to see if the corresponding open ABS motor DTC is also set.
2. Verifies that the ABS motor was actually applied as commanded by observing feedback current.
3. Verifies that the ABS motor can release.
4. This step verifies that the ABS motor can be applied by observing pedal movement.
5. Checks for an open in the ABS motor "HIGH" circuitry.
6. Checks for an open in the ABS motor "LOW" circuitry.
7. Checks for an ABS motor that is internally shorted.

Diagnostic Aids: An "Intermittent" malfunction in this DTC may result from a mechanical part of the system that sticks, binds, or slips.

The frequency of the malfunction can be checked by using the enhanced diagnostic function of the Tech 1, as described in "Tech 1 Diagnostics" found in this section. If the DTC only occurs once and DTC 56 also occurs, refer to DTC 56. If intermittent and enhanced diagnostics shows this DTC occurs during ABS, refer to DTC 56.

Depending on the frequency of the malfunction, a physical inspection of the mechanical parts suspected may be necessary.

DIAGNOSTIC CHART DTC 47 — 1994 SILHOUETTE, TRANS SPORT AND LUMINA APV

DTC 47
(Page 1 of 2)
LEFT FRONT ABS MOTOR FREE SPINS

1. IS DTC 56 ALSO CURRENTLY SET?
   - YES: PROCEED TO DTC 56 CHART FIRST.
   - NO: go to 2.
2.
   - IGNITION "ON," ENGINE "OFF."
   - PUMP BRAKE PEDAL UNTIL FIRM TO DEPLETE VACUUM RESERVE.
   - USING TECH 1 MISC. TESTS, SELECT MANUAL CONTROL FUNCTION.
   - SELECT LEFT FRONT ABS MOTOR APPLY FUNCTION AND APPLY THE MOTOR WHILE CAREFULLY OBSERVING THE COMMAND CURRENT AND FEEDBACK CURRENT ON THE TECH 1.

   WERE BOTH COMMAND AND FEEDBACK CURRENT 10 AMPS?
   - YES: go to 3.
   - NO: GO TO STEP 5
3.
   - APPLY FIRM PRESSURE ON BRAKE PEDAL.
   - SELECT LEFT FRONT ABS MOTOR RELEASE FUNCTION AND RELEASE THE MOTOR.

   DID THE BRAKE PEDAL FALL AND WERE BOTH COMMAND AND FEEDBACK CURRENT 6 AMPS?
   - YES: go to 4.
   - NO: GO TO STEP 5
4.
   - FIRM PRESSURE STILL ON BRAKE PEDAL.
   - SELECT LEFT FRONT ABS MOTOR APPLY FUNCTION AND APPLY THE MOTOR (THIS MOVES THE PISTON TO THE TOP OF ITS BORE.)

   DID BRAKE PEDAL RISE?
   - YES: MALFUNCTION IS INTERMITTENT. REFER TO "DIAGNOSTIC AIDS" ON FACING PAGE.
   - NO: go to 5.
5.
   - USING TECH 1 MISC. TESTS, PERFORM GEAR TENSION RELIEF FUNCTION.
   - IGNITION "OFF."
   - DISCONNECT 6-WAY ABS MOTOR PACK CONNECTOR AND 8-WAY EBCM CONNECTOR.
   - USING J 39200, MEASURE THE RESISTANCE BETWEEN TERMINAL "A" OF THE 6-WAY ABS MOTOR PACK HARNESS CONNECTOR AND TERMINAL "A" OF THE 8-WAY EBCM HARNESS CONNECTOR.

   IS RESISTANCE LESS THAN 1.5Ω?
   - YES: go to 6.
   - NO: REPAIR HIGH RESISTANCE IN CKT 1280.
6.
   - MEASURE THE RESISTANCE BETWEEN TERMINAL "B" OF THE 6-WAY ABS MOTOR PACK HARNESS CONNECTOR AND TERMINAL "B" OF THE 8-WAY EBCM HARNESS CONNECTOR.

   IS RESISTANCE LESS THAN 1.5Ω?
   - YES: go to 7.
   - NO: REPAIR HIGH RESISTANCE IN CKT 1281.
7.
   - MEASURE THE RESISTANCE BETWEEN TERMINAL "A" AND TERMINAL "B" OF THE ABS MOTOR PACK CONNECTOR.

   IS RESISTANCE LESS THAN 1.5Ω?
   - YES: DIAGNOSIS CONTINUED ON NEXT PAGE.
   - NO: REPLACE ABS MOTOR PACK.

AFTER DIAGNOSIS IS COMPLETE, CLEAR DTCs AND TEST DRIVE VEHICLE FOR THREE (3) DRIVE CYCLES TO VERIFY THAT DTC DOES NOT RESET. A DRIVE CYCLE CONSISTS OF STARTING THE VEHICLE, DRIVING OVER 16 KM/H (10 MPH), AND THEN KEYING DOWN.

DIAGNOSTIC CHART DTC 47 — 1994 SILHOUETTE, TRANS SPORT AND LUMINA APV

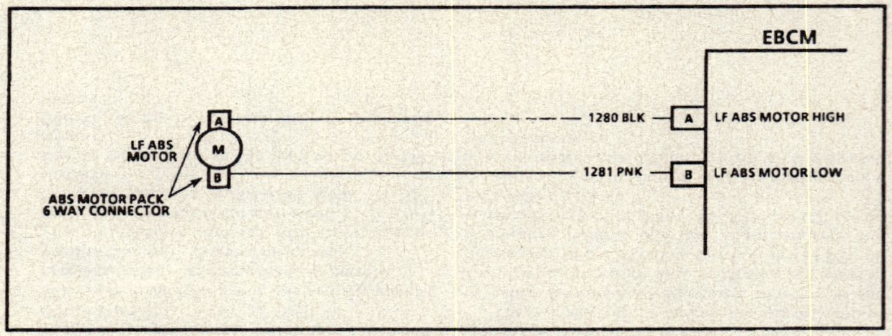

DTC 47
(Page 2 of 2)
LEFT FRONT ABS MOTOR FREE SPINS

Circuit Description:
This DTC is designed to detect a stripped nut or gear assembly during initialization. During the homing sequence, the piston should reach the top of the bore resulting in a stalled ABS motor. If this does not occur, the ABS motor must be spinning with little or no resistance, indicating a nut/screw or gear malfunction.

DTC Will Set When: DTC 47 can only be set during initialization. If the feedback current is less than the command current for a specified period of time, a malfunction exists.

Action Taken: A malfunction DTC is stored, ABS is disabled, and the ABS warning lamp is turned "ON."

DTC Chart Test Description: Number(s) below refer to circled number(s) on the diagnostic chart.

8. Checks for a stripped gear on the motor pack (unit with three small gears).
9. Checks for a stripped gear on the hydraulic modulator (unit with three large gears).
10. Checks to see if the ABS motor pack has an internal malfunction.
11. Verifies that the hydraulic modulator is functioning properly.
12. Ensures malfunction was not due to a poor terminal contact.

Diagnostic Aids: An "Intermittent" malfunction in this DTC may result from a mechanical part of the system that sticks, binds, or slips.

The frequency of the malfunction can be checked by using the enhanced diagnostic function of the Tech 1, as described in "Tech 1 Diagnostics" found in this section. If the DTC only occurs once and DTC 56 also occurs, refer to DTC 56. If intermittent and enhanced diagnostics shows this DTC occurs during ABS, refer to DTC 56.

Depending on the frequency of the malfunction, a physical inspection of the mechanical parts suspected may be necessary.

DIAGNOSTIC CHART DTC 47 — 1994 SILHOUETTE, TRANS SPORT AND LUMINA APV

DTC 47
(Page 2 of 2)
LEFT FRONT ABS MOTOR FREE SPINS

DIAGNOSIS CONTINUED FROM PREVIOUS PAGE.

(8)
- REMOVE HYDRAULIC MODULATOR/MASTER CYLINDER ASSEMBLY FROM VEHICLE. (REFER TO "ON-VEHICLE SERVICE" IN THIS SECTION.)
- REMOVE GEAR COVER.
- CHECK FOR STRIPPED GEAR ON ABS MOTOR PACK; ABS MOTOR PACK IS THE UNIT WITH THE THREE SMALL GEARS. THE LEFT FRONT GEAR IS THE GEAR FURTHEST FORWARD WHEN THE ASSEMBLY IS INSTALLED IN THE VEHICLE. IS THE ABS MOTOR PACK GEAR STRIPPED?

YES: REPLACE ABS MOTOR PACK.

NO:

(9)
- CHECK FOR A STRIPPED GEAR ON HYDRAULIC MODULATOR; MODULATOR IS THE UNIT WITH THE THREE LARGE GEARS. THE LEFT FRONT GEAR IS THE GEAR FURTHEST FORWARD WHEN THE UNIT IS INSTALLED IN THE VEHICLE. IS THE MODULATOR GEAR STRIPPED?

YES: REPLACE HYDRAULIC MODULATOR GEAR.

NO:

(10)
- RECONNECT ELECTRICAL CONNECTORS
- SECURELY POSITION HYDRAULIC MODULATOR/MASTER CYLINDER ASSEMBLY IN VEHICLE WITH GEAR COVER REMOVED SUCH THAT THE GEAR SET CAN BE OBSERVED. BE CAREFUL NOT TO DAMAGE GEAR SET OR ALLOW THE HYDRAULIC MODULATOR TO MOVE WHILE TESTING.
- IGNITION "ON."
- USING TECH 1 MISC. TESTS, SELECT MANUAL CONTROL FUNCTION.
- SELECT LEFT FRONT ABS MOTOR APPLY AND APPLY THE MOTOR WHILE OBSERVING THE GEAR SET FURTHEST FORWARD WHEN THE UNIT IS INSTALLED IN THE VEHICLE.
- SELECT LEFT ABS FRONT MOTOR RELEASE FUNCTION AND RELEASE THE MOTOR WHILE STILL OBSERVING THE GEAR SET.
- SELECT LEFT FRONT ABS MOTOR APPLY FUNCTION AND APPLY THE MOTOR. DID GEAR SET MOVE IN BOTH DIRECTIONS FOR AT LEAST ONE REVOLUTION?

NO: REPLACE ABS MOTOR PACK.

YES:

(11)
- USING TECH 1 MISC. TESTS, PERFORM GEAR TENSION RELIEF FUNCTION.
- IGNITION "OFF."
- SEPARATE THE ABS MOTOR PACK FROM THE HYDRAULIC MODULATOR.
- GRASP THE GEAR ON THE HYDRAULIC MODULATOR (THE HYDRAULIC MODULATOR IS THE UNIT WITH THE THREE LARGE GEARS) FURTHEST FORWARD WHEN THE UNIT IS INSTALLED IN THE VEHICLE AND ATTEMPT TO MOVE THE GEAR. CAN THE GEAR BE ROTATED MORE THAN 13 FULL TURNS LOCK TO LOCK?

YES: REPLACE THE HYDRAULIC MODULATOR.

NO:

(12)
- INSPECT ALL CONNECTORS AND TERMINALS FOR POOR TERMINAL CONTACT AND EVIDENCE OF CORROSION.
  ~~CONTACT~~/ REPLACE ALL TERMINALS THAT EXHIBIT SIGNS OF POOR TERMINAL CONTACT OR CORROSION.
- REINSTALL HYDRAULIC MODULATOR/MASTER CYLINDER ASSEMBLY IN VEHICLE.
- RECONNECT ALL CONNECTORS.
- IGNITION "OFF."
- START ENGINE AND TEST DRIVE VEHICLE, ACHIEVING AT LEAST 16 km/h (10 MPH).
- REPEAT ABOVE DRIVE CYCLE SEQUENCE TWO MORE TIMES. DOES DTC 47 SET IN THE LAST THREE DRIVE CYCLES?

YES: REPLACE EBCM.

NO: MALFUNCTION WAS MOSTLY LIKELY CAUSED BY A POOR CONNECTION. REFER TO "DIAGNOSTIC AIDS" ON FACING PAGE.

AFTER DIAGNOSIS IS COMPLETE, CLEAR DTCs AND TEST DRIVE VEHICLE FOR THREE (3) DRIVE CYCLES TO VERIFY THAT DTC DOES NOT RESET. A DRIVE CYCLE CONSISTS OF STARTING THE VEHICLE, DRIVING OVER 16 KM/H (10 MPH), AND THEN KEYING DOWN.

DIAGNOSTIC CHART DTC 48 — 1994 SILHOUETTE, TRANS SPORT AND LUMINA APV

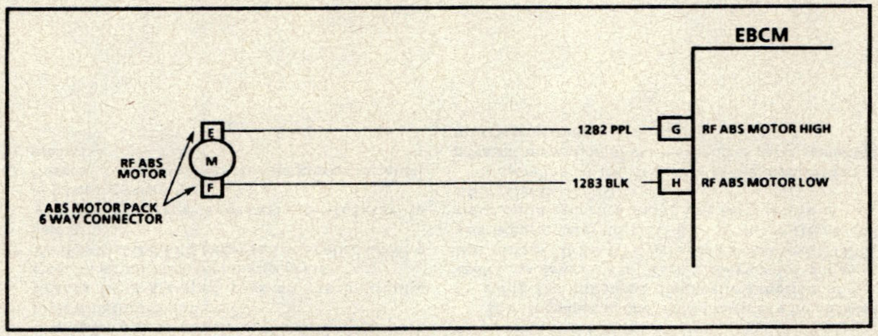

DTC 48
(Page 1 of 2)
RIGHT FRONT ABS MOTOR FREE SPINS

Circuit Description:
This DTC is designed to detect a stripped nut or gear assembly during initialization. During the homing sequence, the piston should reach the top of the bore resulting in a stalled ABS motor. If this does not occur, the ABS motor must be spinning with little or no resistance, indicating a nut/screw or gear malfunction.

DTC Will Set When: DTC 48 can only be set during initialization. If the feedback current is less than the command current for a specified period of time, a malfunction exists.

Action Taken: A malfunction DTC is stored, ABS is disabled, and the ABS warning lamp is turned "ON."

DTC Chart Test Description: Number(s) below refer to circled number(s) on the diagnostic chart.

1. Checks to see if the corresponding open ABS motor DTC is also set.
2. Verifies that the ABS motor was actually applied as commanded by observing feedback current.
3. Verifies that the ABS motor can release.
4. Verifies that the ABS motor can be applied by observing pedal movement.
5. Checks for a stripped gear on the ABS motor pack (unit with three small gears).
6. Checks for high resistance in the ABS motor "HIGH" circuitry.
6. Checks for high resistance in the ABS motor "LOW" circuitry.
7. Checks for proper resistance of the ABS motor windings.

Diagnostic Aids: An "Intermittent" malfunction in this DTC may result from a mechanical part of the system that sticks, binds, or slips.

The frequency of the malfunction can be checked by using the enhanced diagnostic function of the Tech 1, as described in "Tech 1 Diagnostics" found in this section. If the DTC only occurs once and DTC 61 also occurs, refer to DTC 61. If intermittent and enhanced diagnostics shows this DTC occurs during ABS, refer to DTC 61.

Depending on the frequency of the malfunction, a physical inspection of the mechanical parts suspected may be necessary.

DIAGNOSTIC CHART DTC 48 — 1994 SILHOUETTE, TRANS SPORT AND LUMINA APV

DTC 48
(Page 1 of 2)
RIGHT FRONT ABS MOTOR FREE SPINS

1. IS DTC 61 ALSO CURRENTLY SET?
   - YES: PROCEED TO DTC 61 CHART FIRST.
   - NO:
2. • IGNITION "ON," ENGINE "OFF."
   • PUMP BRAKE PEDAL UNTIL FIRM TO DEPLETE VACUUM RESERVE.
   • USING TECH 1 MISC. TESTS, SELECT MANUAL CONTROL FUNCTION.
   • SELECT RIGHT FRONT ABS MOTOR APPLY FUNCTION AND APPLY THE MOTOR WHILE CAREFULLY OBSERVING THE COMMAND CURRENT AND FEEDBACK CURRENT ON THE TECH 1.
   WERE BOTH COMMAND AND FEEDBACK CURRENT 10 AMPS?
   - NO: GO TO STEP 5
   - YES:
3. • APPLY FIRM PRESSURE ON BRAKE PEDAL.
   • SELECT RIGHT FRONT ABS MOTOR RELEASE FUNCTION AND RELEASE THE MOTOR.
   DID THE BRAKE PEDAL FALL AND WERE BOTH COMMAND AND FEEDBACK CURRENT 6 AMPS?
   - NO: GO TO STEP 5
   - YES:
4. • FIRM PRESSURE STILL ON BRAKE PEDAL.
   • SELECT RIGHT ABS MOTOR APPLY FUNCTION AND APPLY THE MOTOR (THIS MOVES THE PISTON TO THE TOP OF ITS BORE.)
   DID BRAKE PEDAL RISE?
   - YES: MALFUNCTION IS INTERMITTENT. REFER TO "DIAGNOSTIC AIDS" ON FACING PAGE.
   - NO:
5. • USING TECH 1 MISC. TESTS, PERFORM GEAR TENSION RELIEF FUNCTION.
   • IGNITION "OFF."
   • DISCONNECT 6-WAY ABS MOTOR PACK CONNECTOR AND 8-WAY EBCM CONNECTOR.
   • USING J 39200, MEASURE THE RESISTANCE BETWEEN TERMINAL "E" OF THE 6-WAY ABS MOTOR PACK HARNESS CONNECTOR AND TERMINAL "G" OF THE 8-WAY EBCM HARNESS CONNECTOR.
   IS RESISTANCE 1.5Ω OR LESS?
   - NO: REPAIR HIGH RESISTANCE IN CKT 1282.
   - YES:
6. • MEASURE THE RESISTANCE BETWEEN TERMINAL "F" OF THE 6-WAY ABS MOTOR PACK HARNESS CONNECTOR AND TERMINAL "H" OF THE 8-WAY EBCM HARNESS CONNECTOR.
   IS RESISTANCE 1.5Ω OR LESS?
   - NO: REPAIR HIGH RESISTANCE IN CKT 1283.
   - YES:
7. • MEASURE THE RESISTANCE BETWEEN TERMINAL "E" AND TERMINAL "F" OF THE ABS MOTOR PACK CONNECTOR.
   IS RESISTANCE 1.5Ω OR LESS?
   - NO: REPLACE ABS MOTOR PACK.
   - YES: DIAGNOSIS CONTINUED ON NEXT PAGE.

AFTER DIAGNOSIS IS COMPLETE, CLEAR DTCs AND TEST DRIVE VEHICLE FOR THREE (3) DRIVE CYCLES TO VERIFY THAT DTC DOES NOT RESET. A DRIVE CYCLE CONSISTS OF STARTING THE VEHICLE, DRIVING OVER 16 KM/H (10 MPH), AND THEN KEYING DOWN.

DIAGNOSTIC CHART DTC 48 — 1994 SILHOUETTE, TRANS SPORT AND LUMINA APV

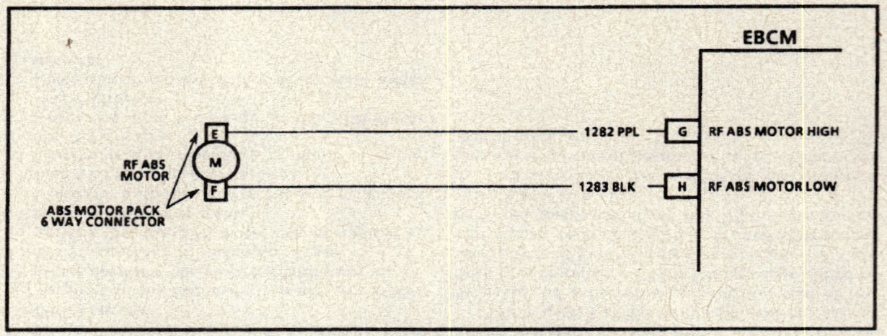

DTC 48
(Page 2 of 2)
RIGHT FRONT ABS MOTOR FREE SPINS

Circuit Description:
This DTC is designed to detect a stripped nut or gear assembly during initialization. During the homing sequence, the piston should reach the top of the bore resulting in a stalled ABS motor. If this does not occur, the ABS motor must be spinning with little or no resistance, indicating a nut/screw or gear malfunction.

DTC Will Set When: DTC 48 can only be set during initialization. If the feedback current is less than the command current for a specified period of time, a malfunction exists.

Action Taken: A malfunction DTC is stored, ABS is disabled, and the ABS warning lamp is turned "ON."

DTC Chart Test Description: Number(s) below refer to circled number(s) on the diagnostic chart.

8. Checks for a stripped gear on the ABS motor pack (unit with three small gears).
9. Checks for a stripped gear on the hydraulic modulator (unit with three large gears).
10. Verifies that the ABS motor can be applied under a load situation.
11. Verifies that the hydraulic modulator is functioning properly.
12. Ensures malfunction was not due to poor terminal contact.

Diagnostic Aids: An "Intermittent" malfunction in this DTC may result from a mechanical part of the system that sticks, binds, or slips.

The frequency of the malfunction can be checked by using the enhanced diagnostic function of the Tech 1, as described in "Tech 1 Diagnostics" found in this section. If the DTC only occurs once and DTC 61 also occurs, refer to DTC 61. If intermittent and enhanced diagnostics shows this DTC occurs during ABS, refer to DTC 61.

Depending on the frequency of the malfunction, a physical inspection of the mechanical parts suspected may be necessary.

DIAGNOSTIC CHART DTC 48 — 1994 SILHOUETTE, TRANS SPORT AND LUMINA APV

DTC 48
(Page 2 of 2)
RIGHT FRONT ABS MOTOR FREE SPINS

DIAGNOSIS CONTINUED FROM PREVIOUS PAGE.

(8)
- REMOVE HYDRAULIC MODULATOR/MASTER CYLINDER ASSEMBLY FROM VEHICLE. (REFER TO "ON-VEHICLE SERVICE" IN THIS SECTION.)
- REMOVE GEAR COVER.
- CHECK FOR STRIPPED GEAR ON ABS MOTOR PACK; ABS MOTOR PACK IS THE UNIT WITH THE THREE SMALL GEARS. THE RIGHT FRONT GEAR IS THE GEAR FURTHEST REARWARD WHEN THE ASSEMBLY IS INSTALLED IN THE VEHICLE. IS THE ABS MOTOR PACK GEAR STRIPPED?

YES: REPLACE ABS MOTOR PACK.

NO:

(9)
- CHECK FOR A STRIPPED GEAR ON HYDRAULIC MODULATOR; MODULATOR IS THE UNIT WITH THE THREE LARGE GEARS. THE RIGHT FRONT GEAR IS THE GEAR FURTHEST REARWARD WHEN THE UNIT IS INSTALLED IN THE VEHICLE. IS THE MODULATOR GEAR STRIPPED?

YES: REPLACE HYDRAULIC MODULATOR GEAR.

NO:

(10)
- RECONNECT ELECTRICAL CONNECTORS
- SECURELY POSITION HYDRAULIC MODULATOR/MASTER CYLINDER ASSEMBLY IN VEHICLE WITH GEAR COVER REMOVED SUCH THAT THE GEAR SET CAN BE OBSERVED. BE CAREFUL NOT TO DAMAGE GEAR SET OR ALLOW THE HYDRAULIC MODULATOR TO MOVE WHILE TESTING.
- IGNITION "ON."
- USING TECH 1 MISC. TESTS, SELECT MANUAL CONTROL FUNCTION.
- SELECT RIGHT FRONT ABS MOTOR APPLY AND APPLY THE MOTOR WHILE OBSERVING THE GEAR SET FURTHEST REARWARD WHEN THE UNIT IS INSTALLED IN THE VEHICLE.
- SELECT RIGHT FRONT ABS MOTOR RELEASE FUNCTION AND RELEASE THE MOTOR WHILE STILL OBSERVING THE GEAR SET.
- SELECT RIGHT FRONT ABS MOTOR APPLY FUNCTION AND APPLY THE MOTOR. DID GEAR SET MOVE IN BOTH DIRECTIONS FOR AT LEAST ONE REVOLUTION?

NO: REPLACE ABS MOTOR PACK.

YES:

(11)
- USING TECH 1 MISC. TESTS, PERFORM GEAR TENSION RELIEF FUNCTION.
- IGNITION "OFF."
- SEPARATE THE ABS MOTOR PACK FROM THE HYDRAULIC MODULATOR.
- GRASP THE GEAR ON THE HYDRAULIC MODULATOR (THE HYDRAULIC MODULATOR IS THE UNIT WITH THE THREE LARGE GEARS) FURTHEST REARWARD WHEN THE UNIT IS INSTALLED IN THE VEHICLE AND ATTEMPT TO MOVE THE GEAR. CAN THE GEAR BE ROTATED MORE THAN 13 FULL TURNS LOCK TO LOCK?

YES: REPLACE THE HYDRAULIC MODULATOR.

NO:

(12)
- INSPECT ALL CONNECTORS AND TERMINALS FOR POOR TERMINAL CONTACT AND EVIDENCE OF CORROSION.
  REPLACE ALL TERMINALS THAT EXHIBIT SIGNS OF POOR TERMINAL CONTACT OR CORROSION.
- REINSTALL HYDRAULIC MODULATOR/MASTER CYLINDER ASSEMBLY IN VEHICLE.
- RECONNECT ALL CONNECTORS.
- IGNITION "OFF."
- START ENGINE AND TEST DRIVE VEHICLE, ACHIEVING AT LEAST 16 km/h (10 MPH).
- REPEAT ABOVE DRIVE CYCLE SEQUENCE TWO MORE TIMES. DOES DTC 48 SET IN THE LAST THREE DRIVE CYCLES?

YES: REPLACE EBCM.

NO: MALFUNCTION WAS MOSTLY LIKELY CAUSED BY A POOR CONNECTION. REFER TO "DIAGNOSTIC AIDS" ON FACING PAGE.

AFTER DIAGNOSIS IS COMPLETE, CLEAR DTCs AND TEST DRIVE VEHICLE FOR THREE (3) DRIVE CYCLES TO VERIFY THAT DTC DOES NOT RESET. A DRIVE CYCLE CONSISTS OF STARTING THE VEHICLE, DRIVING OVER 16 KM/H (10 MPH), AND THEN KEYING DOWN.

DIAGNOSTIC CHART DTC 51 — 1994 SILHOUETTE, TRANS SPORT AND LUMINA APV

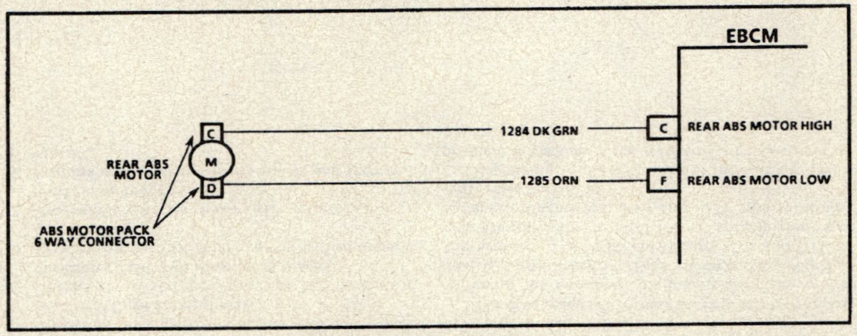

DTC 51
(Page 1 of 2)
REAR ABS MOTOR FREE SPINS

Circuit Description:
This DTC is designed to detect a stripped nut or gear assembly during initialization. During the homing sequence, the piston should reach the top of the bore resulting in a stalled ABS motor. If this does not occur, the ABS motor must be spinning with little or no resistance, indicating a nut/screw or gear malfunction.

DTC Will Set When: DTC 51 can be set during initialization. If the feedback current is less than the command current for a specified period of time, a malfunction exists.

Action Taken: A malfunction DTC is stored, ABS is disabled, and the ABS warning lamp is turned "ON."

DTC Chart Test Description: Number(s) below refer to circled number(s) on the diagnostic chart.

1. Checks to see if the corresponding open ABS motor DTC is also set.
2. Verifies that the ABS motor was actually applied as commanded by observing feedback current.
3. Verifies that the ABS motor can release.
4. Verifies that the ABS motor can be applied by observing pedal movement.
5. Checks for high resistance in the ABS motor "HIGH" circuitry.
6. Checks for high resistance in the ABS motor "LOW" circuitry.
7. Checks for proper resistance of the ABS motor windings.

Diagnostic Aids: An "Intermittent" malfunction in this DTC may result from a mechanical part of the system that sticks, binds, or slips.

The frequency of the malfunction can be checked by using the enhanced diagnostic function of the Tech 1, as described in "Tech 1 Diagnostics" found in this section. If the DTC only occurs once and DTC 64 also occurs, refer to DTC 64. If intermittent and enhanced diagnostics show this DTC occurs during ABS, refer to DTC 64.

Depending on the frequency of the malfunction, a physical inspection of the mechanical parts suspected may be necessary.

DIAGNOSTIC CHART DTC 51 — 1994 SILHOUETTE, TRANS SPORT AND LUMINA APV

DTC 51
(Page 1 of 2)
REAR ABS MOTOR FREE SPINS

1. IS DTC 64 ALSO CURRENTLY SET?
   - YES: PROCEED TO DTC 64 CHART FIRST.
   - NO: go to 2.
2. • IGNITION "ON," ENGINE "OFF."
   • PUMP BRAKE PEDAL UNTIL FIRM TO DEPLETE VACUUM RESERVE.
   • USING TECH 1 MISC. TESTS, SELECT MANUAL CONTROL FUNCTION.
   • SELECT REAR AXLE ABS MOTOR APPLY FUNCTION AND APPLY THE MOTOR WHILE CAREFULLY OBSERVING THE COMMAND CURRENT AND FEEDBACK CURRENT ON THE TECH 1.
   WERE BOTH COMMAND AND FEEDBACK CURRENT 10 AMPS?
   - YES: go to 3.
   - NO: GO TO STEP 5
3. • APPLY FIRM PRESSURE ON BRAKE PEDAL.
   • SELECT REAR AXLE ABS MOTOR RELEASE FUNCTION AND RELEASE THE MOTOR.
   DID THE BRAKE PEDAL FALL AND WERE BOTH COMMAND AND FEEDBACK CURRENT 6 AMPS?
   - YES: go to 4.
   - NO: GO TO STEP 5
4. • FIRM PRESSURE STILL ON BRAKE PEDAL.
   • SELECT REAR AXLE ABS MOTOR APPLY FUNCTION AND APPLY THE MOTOR (THIS MOVES THE PISTON TO THE TOP OF ITS BORE.)
   DID BRAKE PEDAL RISE?
   - YES: MALFUNCTION IS INTERMITTENT. REFER TO "DIAGNOSTIC AIDS" ON FACING PAGE.
   - NO: go to 5.
5. • USING TECH 1 MISC. TESTS, PERFORM GEAR TENSION RELIEF FUNCTION.
   • IGNITION "OFF."
   • DISCONNECT 6-WAY ABS MOTOR PACK CONNECTOR AND 8-WAY EBCM CONNECTOR.
   • USING J 39200, MEASURE THE RESISTANCE BETWEEN TERMINAL "C" OF THE 6-WAY ABS MOTOR PACK HARNESS CONNECTOR AND TERMINAL "C" OF THE 8-WAY EBCM HARNESS CONNECTOR.
   IS RESISTANCE 1.5Ω OR LESS?
   - YES: go to 6.
   - NO: REPAIR HIGH RESISTANCE IN CKT 1284.
6. • MEASURE THE RESISTANCE BETWEEN TERMINAL "D" OF THE 6-WAY ABS MOTOR PACK HARNESS CONNECTOR AND TERMINAL "F" OF THE 8-WAY EBCM HARNESS CONNECTOR.
   IS RESISTANCE 1.5Ω OR LESS?
   - YES: go to 7.
   - NO: REPAIR HIGH RESISTANCE IN CKT 1285.
7. • MEASURE THE RESISTANCE BETWEEN TERMINAL "C" AND TERMINAL "D" OF THE ABS MOTOR PACK CONNECTOR.
   IS RESISTANCE 1.5Ω OR LESS?
   - YES: DIAGNOSIS CONTINUED ON NEXT PAGE.
   - NO: REPLACE ABS MOTOR PACK.

AFTER DIAGNOSIS IS COMPLETE, CLEAR DTCs AND TEST DRIVE VEHICLE FOR THREE (3) DRIVE CYCLES TO VERIFY THAT DTC DOES NOT RESET. A DRIVE CYCLE CONSISTS OF STARTING THE VEHICLE, DRIVING OVER 16 KM/H (10 MPH), AND THEN KEYING DOWN.

DIAGNOSTIC CHART DTC 51 — 1994 SILHOUETTE, TRANS SPORT AND LUMINA APV

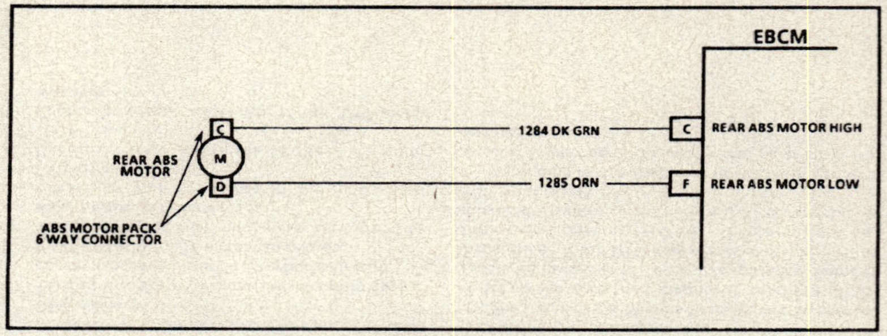

DTC 51
(Page 2 of 2)
REAR ABS MOTOR FREE SPINS

Circuit Description:
This DTC is designed to detect a stripped nut or gear assembly during initialization. During the homing sequence, the piston should reach the top of the bore resulting in a stalled ABS motor. If this does not occur, the ABS motor must be spinning with little or no resistance, indicating a nut/screw or gear malfunction.

DTC Will Set When: DTC 51 can be set during initialization. If the feedback current is less than the command current for a specified period of time, a malfunction exists.

Action Taken: A malfunction DTC is stored, ABS is disabled, and the ABS warning lamp is turned "ON."

DTC Chart Test Description: Number(s) below refer to circled number(s) on the diagnostic chart.

8. Checks for a stripped gear on the ABS motor pack (unit with three small gears).
9. Checks for a stripped gear on the hydraulic modulator (unit with three large gears).
10. Verifies that the ABS motor can be applied under a load situation.
11. Verifies that the hydraulic modulator is functioning properly.
12. Ensures malfunction was not due to poor terminal contact.

Diagnostic Aids: An "Intermittent" malfunction in this DTC may result from a mechanical part of the system that sticks, binds, or slips.

The frequency of the malfunction can be checked by using the enhanced diagnostic function of the Tech 1, as described in "Tech 1 Diagnostics" found in this section. If the DTC only occurs once and DTC 64 also occurs, refer to DTC 64. If intermittent and enhanced diagnostics show this DTC occurs during ABS, refer to DTC 64.

Depending on the frequency of the malfunction, a physical inspection of the mechanical parts suspected may be necessary.

DIAGNOSTIC CHART DTC 51 — 1994 SILHOUETTE, TRANS SPORT AND LUMINA APV

DTC 51
(Page 2 of 2)
REAR ABS MOTOR FREE SPINS

DIAGNOSIS CONTINUED FROM PREVIOUS PAGE.

(8)
- REMOVE HYDRAULIC MODULATOR/MASTER CYLINDER ASSEMBLY FROM VEHICLE.
- REMOVE GEAR COVER.
- CHECK FOR STRIPPED GEAR ON ABS MOTOR PACK; ABS MOTOR PACK IS THE UNIT WITH THE THREE SMALL GEARS. THE REAR AXLE GEAR IS THE MIDDLE GEAR OF THE THREE.

IS THE ABS MOTOR PACK GEAR STRIPPED?

YES: REPLACE ABS MOTOR PACK.

NO:

(9)
- CHECK FOR A STRIPPED GEAR ON HYDRAULIC MODULATOR; MODULATOR IS THE UNIT WITH THE THREE LARGE GEARS. THE REAR AXLE GEAR IS THE MIDDLE GEAR OF THE THREE.

IS THE MODULATOR GEAR STRIPPED?

YES: REPLACE HYDRAULIC MODULATOR GEAR.

NO:

(10)
- RECONNECT ELECTRICAL CONNECTORS
- SECURELY POSITION HYDRAULIC MODULATOR/MASTER CYLINDER ASSEMBLY IN VEHICLE WITH GEAR COVER REMOVED SUCH THAT THE GEAR SET CAN BE OBSERVED. BE CAREFUL NOT TO DAMAGE GEAR SET OR ALLOW THE HYDRAULIC MODULATOR TO MOVE WHILE TESTING.
- IGNITION "ON."
- USING TECH 1 MISC. TESTS, SELECT MANUAL CONTROL FUNCTION.
- SELECT REAR AXLE ABS MOTOR APPLY AND APPLY THE MOTOR WHILE OBSERVING THE MIDDLE GEAR SET.
- SELECT REAR AXLE ABS MOTOR RELEASE FUNCTION AND RELEASE THE MOTOR WHILE STILL OBSERVING THE GEAR SET.
- SELECT REAR AXLE ABS MOTOR APPLY FUNCTION AND APPLY THE MOTOR.

DID GEAR SET MOVE IN BOTH DIRECTIONS FOR AT LEAST ONE REVOLUTION?

NO: REPLACE ABS MOTOR PACK.

YES:

(11)
- USING TECH 1 MISC. TESTS, PERFORM GEAR TENSION RELIEF FUNCTION.
- IGNITION "OFF."
- SEPARATE THE ABS MOTOR PACK FROM THE HYDRAULIC MODULATOR.
- GRASP THE MIDDLE GEAR ON THE HYDRAULIC MODULATOR (THE HYDRAULIC MODULATOR IS THE UNIT WITH THE THREE LARGE GEARS) AND ATTEMPT TO MOVE THE GEAR.

CAN THE GEAR BE ROTATED MORE THAN 8 FULL TURNS LOCK TO LOCK?

YES: REPLACE THE HYDRAULIC MODULATOR.

NO:

(12)
- INSPECT ALL CONNECTORS AND TERMINALS FOR POOR TERMINAL CONTACT AND EVIDENCE OF CORROSION.
  REPLACE ALL TERMINALS THAT EXHIBIT SIGNS OF POOR TERMINAL CONTACT OR CORROSION.
- REINSTALL HYDRAULIC MODULATOR/MASTER CYLINDER ASSEMBLY IN VEHICLE.
- RECONNECT ALL CONNECTORS.
- IGNITION "OFF."
- START ENGINE AND TEST DRIVE VEHICLE, ACHIEVING AT LEAST 16 km/h (10 MPH).
- REPEAT ABOVE DRIVE CYCLE SEQUENCE TWO MORE TIMES.

DOES DTC 51 RESET?

YES: REPLACE EBCM.

NO: MALFUNCTION WAS MOSTLY LIKELY CAUSED BY A POOR CONNECTION. REFER TO "DIAGNOSTIC AIDS" ON FACING PAGE.

AFTER DIAGNOSIS IS COMPLETE, CLEAR DTCs AND TEST DRIVE VEHICLE FOR THREE (3) DRIVE CYCLES TO VERIFY THAT DTC DOES NOT RESET. A DRIVE CYCLE CONSISTS OF STARTING THE VEHICLE, DRIVING OVER 16 KM/H (10 MPH), AND THEN KEYING DOWN.

DIAGNOSTIC CHART DTC 52 — 1994 SILHOUETTE, TRANS SPORT AND LUMINA APV

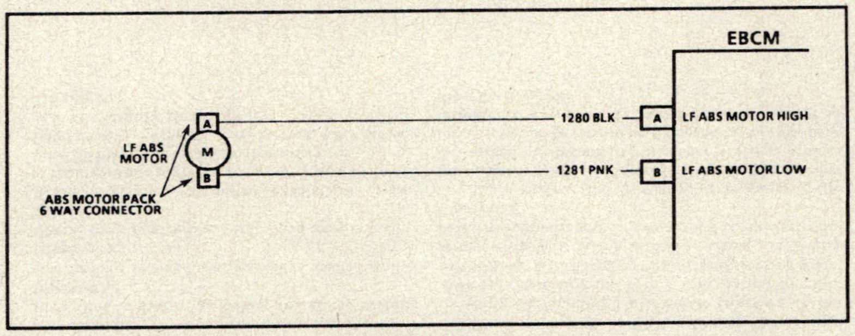

DTC 52

LEFT FRONT ABS CHANNEL IN RELEASE TOO LONG

Circuit Description:

This DTC will diagnose an ABS motor that is energized longer than expected. This could occur if a wheel speed sensor is malfunctioning, the ABS motor does not turn, the left front solenoid mechanically fails open, or the ABS motor wires are crossed.

DTC Will Set When: DTC 52 can be set only during an ABS stop. If the EBCM commands the left front ABS channel in release for three seconds, a malfunction exists.

Action Taken: A malfunction DTC is stored, ABS is disabled and the ABS warning lamp is turned "ON."

DTC Chart Test Description: Number(s) below refer to circled number(s) on the diagnostic chart.

1. Identifies a problem in a wheel speed sensor that may cause the system to be in release too long.
2. Identifies an ABS motor as being failed or wired incorrectly.
3. Checks for a solenoid that may have mechanically failed open.
4. Verifies proper solenoid operation.
5. If Step 3 has failed, this serves to isolate the cause of the hydraulic problem, to either the solenoid or the ABS hydraulic modulator assembly.
6. Determines whether a malfunctioning ABS motor pack or hydraulic modulator is the reason for DTC 52 being set.

Diagnostic Aids: An "Intermittent" malfunction may be caused by a mechanical part of the system that sticks or binds.

The frequency of the malfunction can be checked by using the enhanced diagnostic function of the Tech 1, as described in "Tech 1 Diagnostics" found in this section. DTC 52 may occur if on ice and steering wheel is turned to lock during braking. Using the Tech 1, perform hydraulic test to ensure total brake system is functional.

Any circuitry, that is suspected as causing the intermittent complaint, should be thoroughly checked for backed out terminals, improper mating, broken locks, improperly formed or damaged terminals, poor terminal to wiring connections or physical damage to the wiring harness.

DIAGNOSTIC CHART DTC 52 — 1994 SILHOUETTE, TRANS SPORT AND LUMINA APV

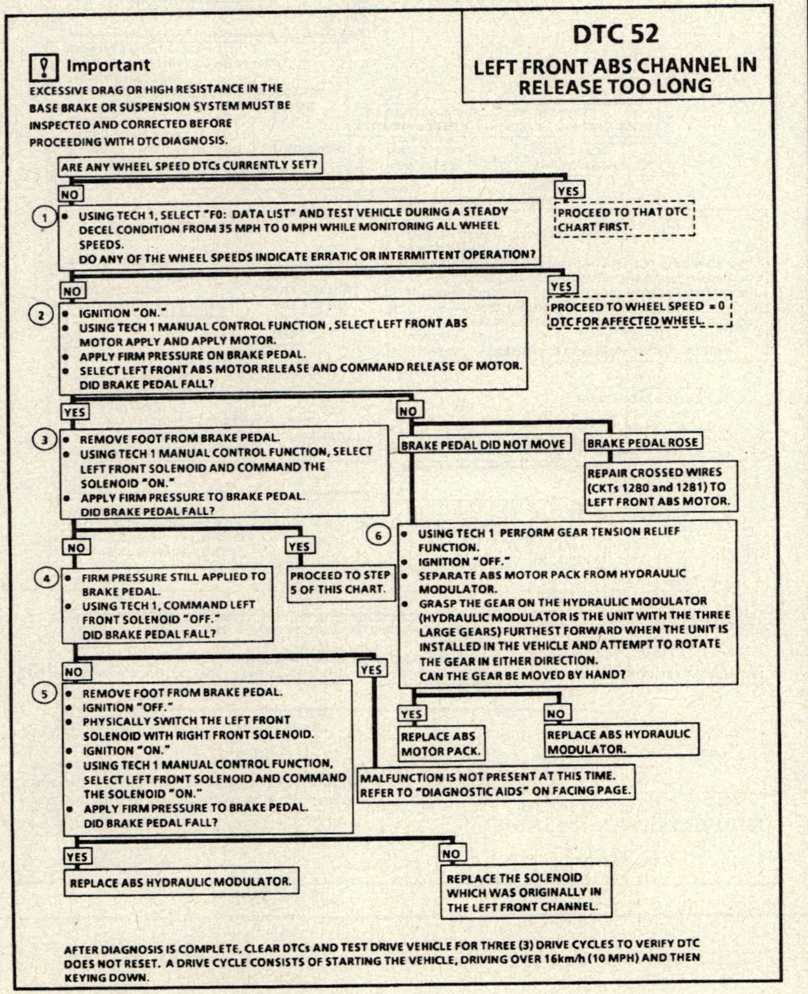

DIAGNOSTIC CHART DTC 53 — 1994 SILHOUETTE, TRANS SPORT AND LUMINA APV

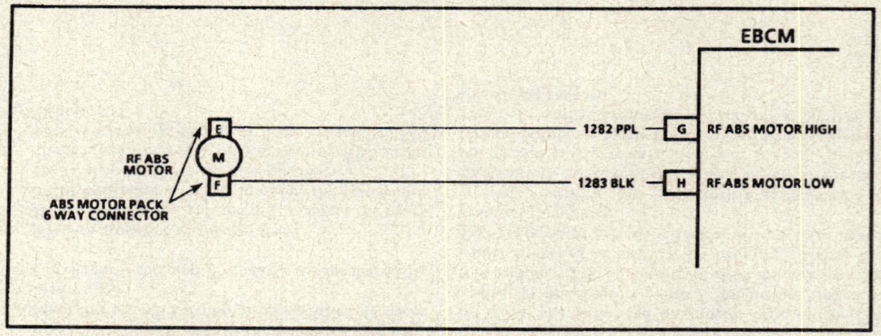

DTC 53

RIGHT FRONT ABS CHANNEL IN RELEASE TOO LONG

Circuit Description:

This DTC will diagnose an ABS motor that is energized longer than expected. This could occur if a wheel speed sensor is malfunctioning, the ABS motor does not turn, the right front solenoid mechanically fails open, or the ABS motor wires are crossed.

DTC Will Set When: DTC 53 can be set only during an ABS stop. If the EBCM commands the right front channel in release for three seconds, a malfunction exists.

Action Taken: A malfunction DTC is stored, ABS is disabled and the ABS warning lamp is turned "ON."

DTC Chart Test Description: Number(s) below refer to circled number(s) on the diagnostic chart.

1. Identifies a problem in a wheel speed sensor that may cause the system to be in release too long.
2. Identifies a motor as being failed or wired incorrectly.
3. Checks for a solenoid that may have mechanically failed open.
4. Checks for a solenoid that may have mechanically failed closed.
5. If Step 3 has failed, this serves to isolate the cause of the hydraulic problem, to either the solenoid or the ABS hydraulic modulator assembly.
6. Determines whether a malfunction ABS motor pack or hydraulic modulator is the reason for DTC 53 being set.

Diagnostic Aids: An "Intermittent" malfunction may be caused by a mechanical part of the system that sticks or binds.

The frequency of the malfunction can be checked by using the enhanced diagnostic function of the Tech 1, as described in "Tech 1 Diagnostics" found in this section. DTC 53 may occur on ice if steering wheel is turned to lock during braking. Using the Tech 1 perform hydraulic test to ensure total brake system is functional.

Any circuitry, that is suspected as causing the intermittent complaint, should be thoroughly checked for backed out terminals, improper mating, broken locks, improperly formed or damaged terminals, poor terminal to wiring connections or physical damage to the wiring harness.

DIAGNOSTIC CHART DTC 53 — 1994 SILHOUETTE, TRANS SPORT AND LUMINA APV

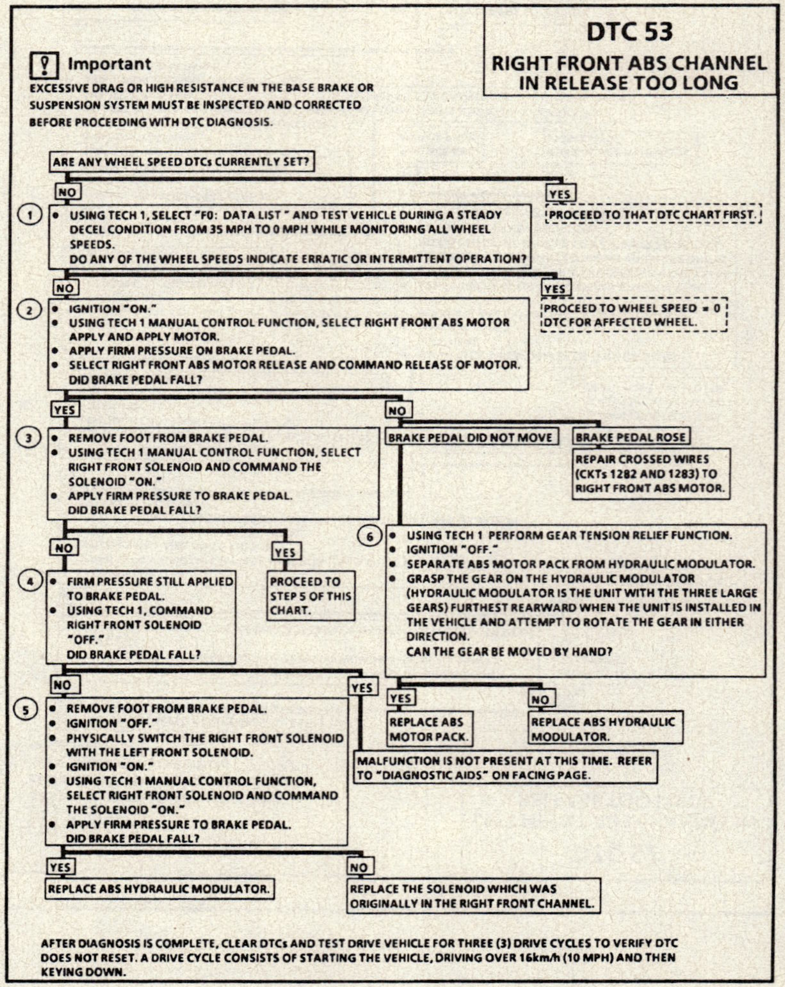

DIAGNOSTIC CHART DTC 54 — 1994 SILHOUETTE, TRANS SPORT AND LUMINA APV

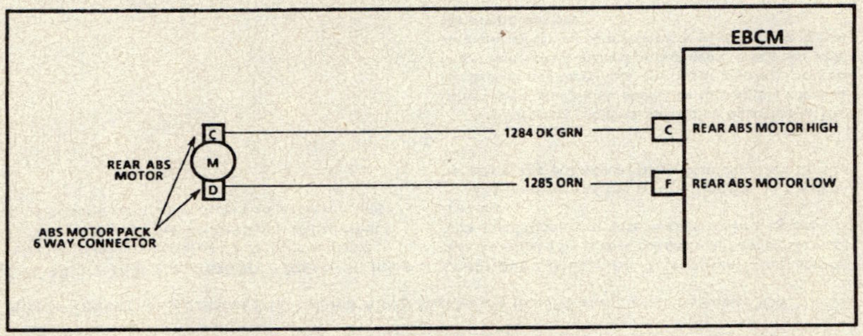

DTC 54

REAR ABS CHANNEL IN RELEASE TOO LONG

Circuit Description:
This DTC will diagnose an ABS motor that is energized longer than expected. This could occur if a wheel speed sensor is malfunctioning, the ABS motor does not turn, or the ABS motor wires are crossed.

DTC Will Set When: DTC 54 can be set only during an ABS stop. If the EBCM commands the rear ABS channel in release for three seconds, a malfunction exists.

Action Taken: A malfunction DTC is stored, ABS is disabled and the ABS warning lamp is turned "ON."

DTC Chart Test Description: Number(s) below refer to circled number(s) on the diagnostic chart.

1. Identifies a problem in a wheel speed sensor that may cause the system to be in release too long.
2. Checks for a wheel that may stick or bind because of a mechanical fault.
3. Checks to see if the ABS motor is capable of moving and applying the hydraulic piston for the rear wheels.
4. Ensures that the ABS motor wiring is not crossed.
5. Isolates the fault of a "no-apply" situation to either the ABS motor pack or ABS hydraulic modulator assembly.

Diagnostic Aids: An "Intermittent" malfunction may be caused by a mechanical part of the system that sticks or binds.

The frequency of the malfunction can be checked by using the enhanced diagnostic function of the Tech 1.

Using the Tech 1, perform hydraulic test to ensure total brake system is functional.

Any circuitry, that is suspected as causing the intermittent complaint, should be thoroughly checked for backed out terminals, improper mating, broken locks, improperly formed or damaged terminals, poor terminal to wiring connections or physical damage to the wiring harness.

DIAGNOSTIC CHART DTC 54 — 1994 SILHOUETTE, TRANS SPORT AND LUMINA APV

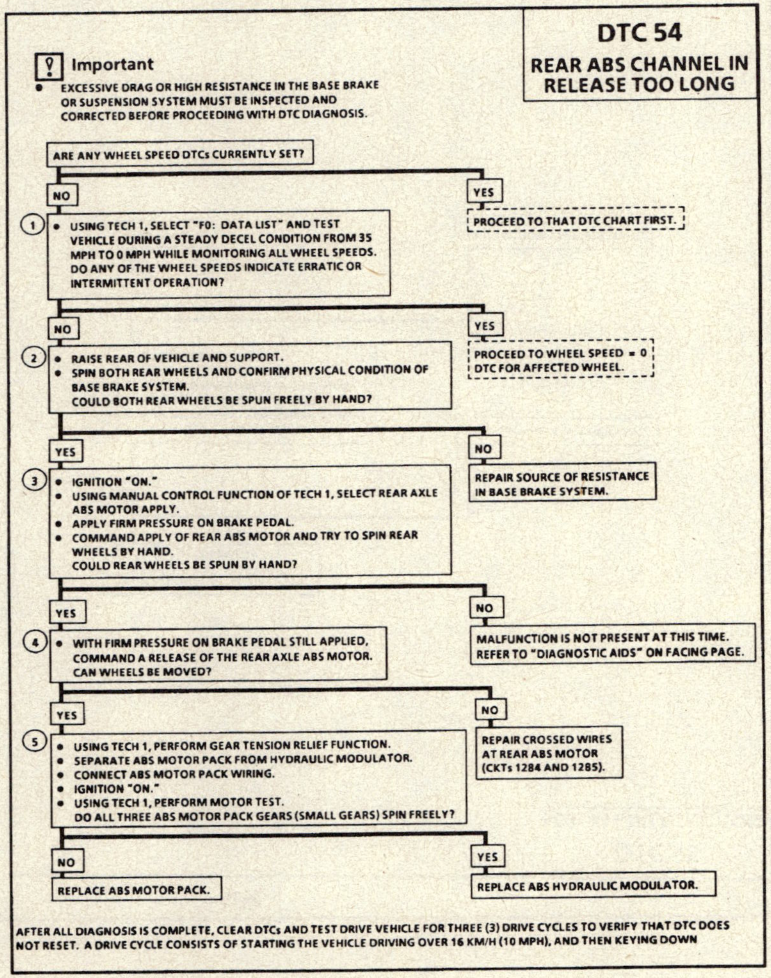

DIAGNOSTIC CHART DTC 55 — 1994 SILHOUETTE, TRANS SPORT AND LUMINA APV

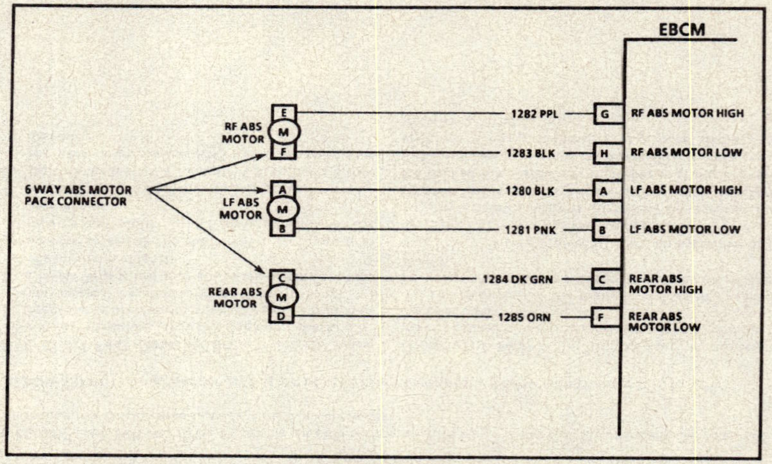

DTC 55
EBCM MALFUNCTION

Circuit Description:
This DTC identifies a malfunction detected by the MDI custom IC. It also ensures the cause of the malfunction is not a result of a problem with the ABS enable relay under a DTC 55 malfunction.

Action Taken: A malfunction DTC is stored, ABS is disabled and the ABS warning lamp is turned "ON."

DTC Chart Test Description: Number(s) below refer to circled number(s) on the diagnostic chart.
1. Checks to see if the malfunction is still present.
2. Checks to see if the malfunction is intermittent.

Diagnostic Aids: An "Intermittent" malfunction may be caused by a poor connection, rubbed through wire insulation, or a wire that is broken inside the insulation.

The frequency of the malfunction can be checked by using the enhanced diagnostic function of the Tech 1.

Any circuitry, that is suspected as causing the intermittent complaint, should be thoroughly checked for backed out terminals, improper mating, broken locks, improperly formed or damaged terminals, poor terminal to wiring connections or physical damage to the wiring harness.

DIAGNOSTIC CHART DTC 55 — 1994 SILHOUETTE, TRANS SPORT AND LUMINA APV

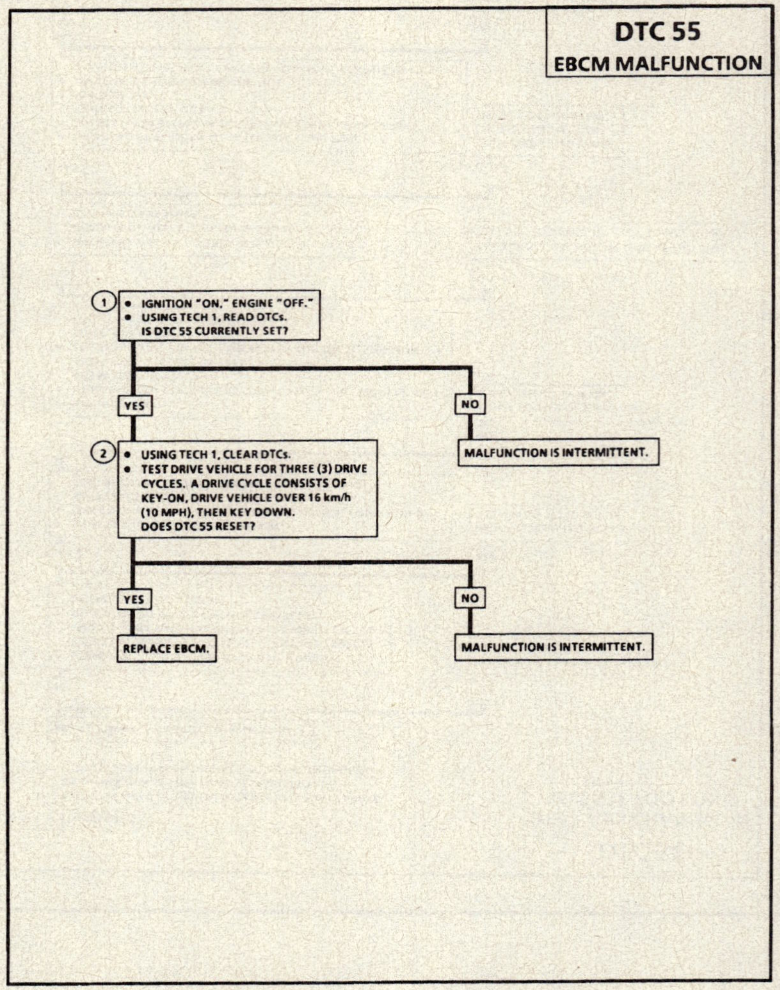

DIAGNOSTIC CHART DTC 56 — 1994 SILHOUETTE, TRANS SPORT AND LUMINA APV

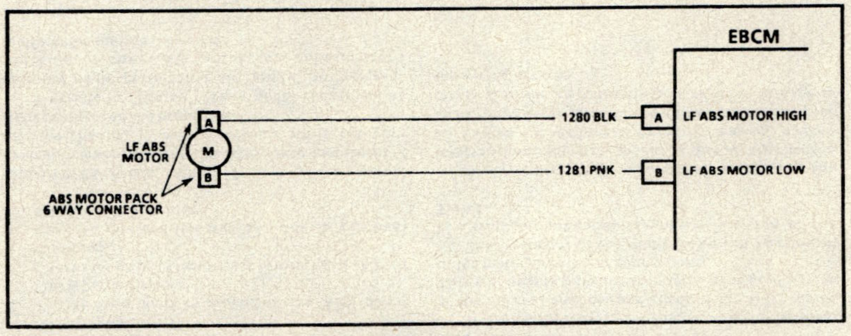

DTC 56

LEFT FRONT ABS MOTOR CIRCUIT OPEN

Circuit Description:
This DTC identifies an ABS motor that cannot be energized due to an open in its circuitry.

DTC Will Set When: DTC 56 can be set only when the motor is commanded "OFF." If the EBCM detects an out of range voltage on either of the left front ABS motor circuits indicating an open circuit, a malfunction exists.

Action Taken: An open ABS motor will not activate when requested. A malfunction DTC is stored, ABS is disabled and the ABS warning lamp is turned "ON."

DTC Chart Test Description: Number(s) below refer to circled number(s) on the diagnostic chart.
1. Checks for proper resistance of the ABS motor.
2. Checks for an open in the ABS motor "HIGH" circuitry.
3. Checks for an open in the ABS motor "LOW" circuitry.
4. Ensures malfunction was not due to a poor terminal contact.

Diagnostic Aids: Using Tech 1, select manual control function, and exercise ABS motor movement of affected channel in both directions while applying light pressure on the brake pedal.

If erratic or "jumpy" brake pedal movement is detected while performing an "apply" or "release" function of the ABS motor, an intermittent malfunction may be indicated.

An "Intermittent" malfunction may be caused by a poor connection, rubbed through wire insulation, or a wire that is broken inside the insulation.

If the malfunction is not current, wiggle the wires of the affected channel and check if the DTC resets. This will help to pinpoint an intermittent malfunction in the motor circuitry or connections.

The frequency of the malfunction can be checked by using the enhanced diagnostic function of the Tech 1,

Any circuitry, that is suspected as causing the intermittent complaint, should be thoroughly checked for backed out terminals, improper mating, broken locks, improperly formed or damaged terminals, poor terminal to wiring connections or physical damage to the wiring harness.

DIAGNOSTIC CHART DTC 56 — 1994 SILHOUETTE, TRANS SPORT AND LUMINA APV

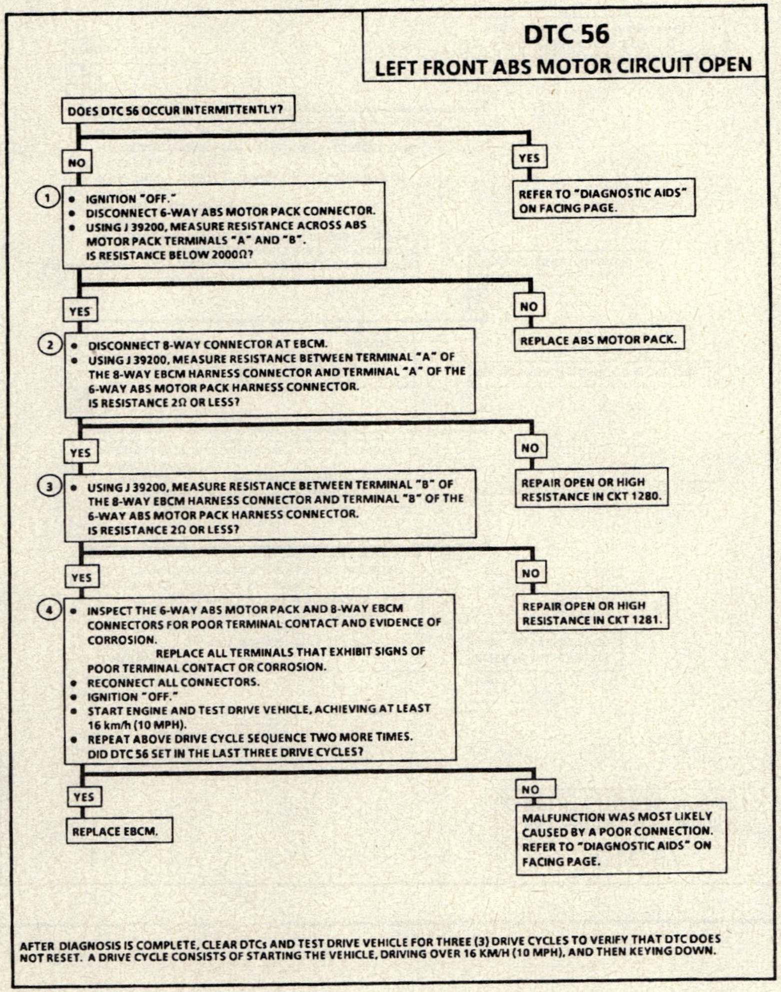

DIAGNOSTIC CHART DTC 57 — 1994 SILHOUETTE, TRANS SPORT AND LUMINA APV

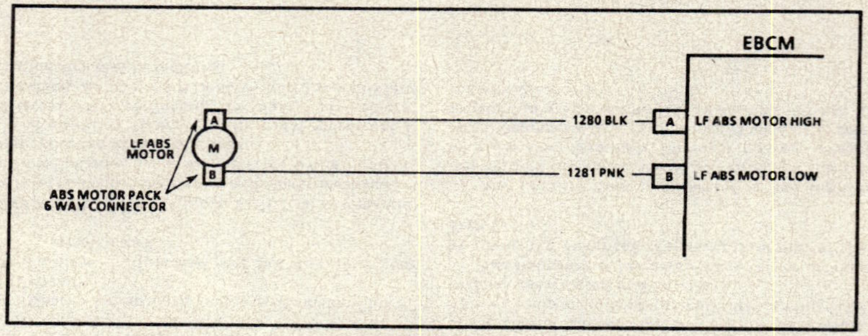

DTC 57

LEFT FRONT ABS MOTOR CIRCUIT SHORTED TO GND

Circuit Description:
This DTC identifies an ABS motor circuit that is shorted to ground. This malfunction will not allow the ABS motor to be controlled at the commanded current rate or will cause the driver circuit to allow current directly to ground.

DTC Will Set When: DTC 57 can be set anytime. If the EBCM detects an out of range voltage on either of the left front ABS motor circuits indicating a circuit shorted to ground, a malfunction exists.

Action Taken: A malfunction DTC is stored, ABS is disabled and the ABS warning lamp is turned "ON."

DTC Chart Test Description: Number(s) below refer to circled number(s) on the diagnostic chart.

1. Checks for a short to ground in the ABS motor "HIGH" circuitry.
2. Checks for a short to ground in the ABS motor "LOW" circuitry.
3. Checks for an ABS motor that is internally shorted to ground.
4. Ensures malfunction was not due to physical damage of the circuitry.

Diagnostic Aids: Using Tech 1, select manual control function, and exercise ABS motor movement of affected channel in both directions while applying light pressure on the brake pedal.

If erratic or "jumpy" brake pedal movement is detected while performing an "apply" or "release" function of the ABS motor, an intermittent malfunction may be indicated.

An "Intermittent" malfunction may be caused by a poor connection, rubbed through wire insulation, or a wire that is broken inside the insulation.

If the malfunction is not current, wiggle the wires of the affected channel and check if the DTC resets. This will help to pinpoint an intermittent malfunction in the motor circuitry or connections.

The frequency of the malfunction can be checked by using the enhanced diagnostic function of the Tech 1

Any circuitry, that is suspected as causing the intermittent complaint, should be thoroughly checked for backed out terminals, improper mating, broken locks, improperly formed or damaged terminals, poor terminal to wiring connections or physical damage to the wiring harness.

DIAGNOSTIC CHART DTC 57 — 1994 SILHOUETTE, TRANS SPORT AND LUMINA APV

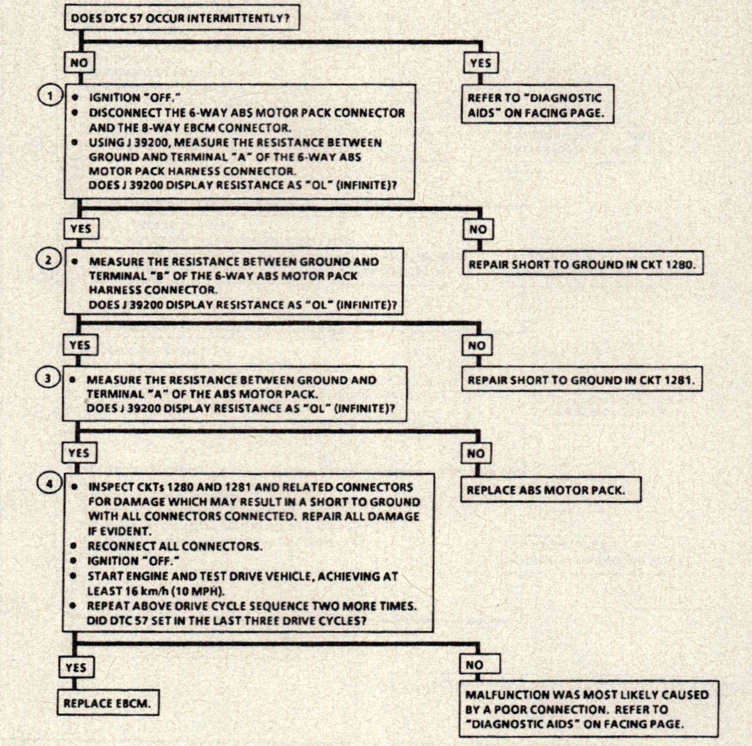

DIAGNOSTIC CHART DTC 58 — 1994 SILHOUETTE, TRANS SPORT AND LUMINA APV

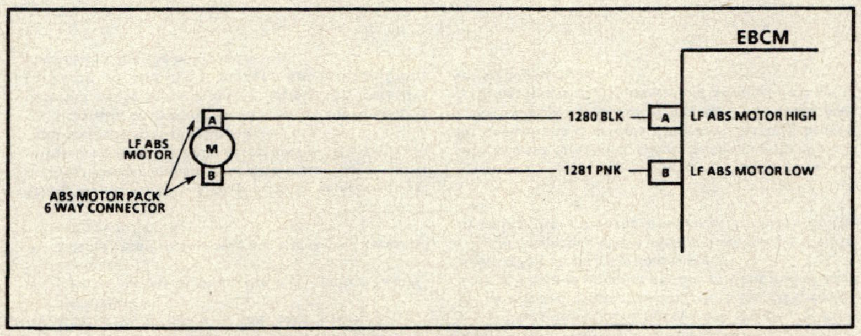

DTC 58

LEFT FRONT ABS MOTOR CIRCUIT SHORTED TO BATT

Circuit Description:
This DTC identifies an ABS motor circuit that is shorted to battery or an ABS motor that has low or no resistance. This malfunction will not allow the ABS motor to be controlled at the commanded current rate or will cause the ABS motor to turn in the opposite direction, or not at all.

DTC Will Set When: DTC 58 can be set only when the ABS motor is commanded "OFF." If the EBCM detects an out of range voltage on either of the left front ABS motor circuits indicating a circuit shorted to battery or an ABS motor shorted, a malfunction exists.

Action Taken: A malfunction DTC is stored, ABS is disabled and the ABS warning lamp is turned "ON."

DTC Chart Test Description: Number(s) below refer to circled number(s) on the diagnostic chart.

1. Checks for a short to voltage in the ABS motor "HIGH" circuitry.
2. Checks for a short to voltage in the ABS motor "LOW" circuitry.
3. Checks for a short between the "HIGH" and "LOW" motor circuits.
4. This step checks for an ABS motor that is internally shorted.
5. Ensures malfunction was not due to physical damage of the circuitry.

Diagnostic Aids: Using Tech 1, select manual control function, and exercise ABS motor movement of affected channel in both directions while applying light pressure on the brake pedal.

If erratic or "jumpy" brake pedal movement is detected while performing an "apply" or "release" function of the ABS motor, an intermittent malfunction may be indicated.

An "Intermittent" malfunction may be caused by a poor connection, rubbed through wire insulation, or a wire that is broken inside the insulation.

If the malfunction is not current, wiggle the wires of the affected channel and check if the DTC resets. This will help to pinpoint an intermittent malfunction in the motor circuitry or connections.

The frequency of the malfunction can be checked by using the enhanced diagnostic function of the Tech 1.

Any circuitry, that is suspected as causing the intermittent complaint, should be thoroughly checked for backed out terminals, improper mating, broken locks, improperly formed or damaged terminals, poor terminal to wiring connections or physical damage to the wiring harness.

DIAGNOSTIC CHART DTC 58 — 1994 SILHOUETTE, TRANS SPORT AND LUMINA APV

DTC 58

LEFT FRONT ABS MOTOR CIRCUIT SHORTED TO BATT

NOTICE: J 39200 TEST LEADS MUST BE "ZEROED" PRIOR TO MAKING ANY RESISTANCE MEASUREMENTS. REFER TO J 39200 USER'S MANUAL.

DOES DTC 58 OCCUR INTERMITTENTLY?

- YES: REFER TO "DIAGNOSTIC AIDS" ON FACING PAGE.
- NO: go to step 1.

(1)
- IGNITION "OFF."
- DISCONNECT THE 6-WAY ABS MOTOR PACK CONNECTOR AND THE 8-WAY EBCM CONNECTOR.
- IGNITION "ON."
- USING J 39200, MEASURE THE VOLTAGE BETWEEN GROUND AND TERMINAL "A" OF THE 6-WAY ABS MOTOR PACK HARNESS CONNECTOR. IS VOLTAGE 1 VOLT OR LESS?

NO: REPAIR SHORT TO VOLTAGE IN CKT 1280.
YES: go to step 2.

(2)
- MEASURE THE VOLTAGE BETWEEN GROUND AND TERMINAL "B" OF THE 6-WAY ABS MOTOR PACK HARNESS CONNECTOR. IS VOLTAGE 1 VOLT OR LESS?

NO: REPAIR SHORT TO VOLTAGE IN CKT 1281.
YES: go to step 3.

(3)
- IGNITION "OFF."
- USING J 39200, MEASURE THE RESISTANCE BETWEEN TERMINAL "A" AND TERMINAL "B" OF THE 6-WAY ABS MOTOR PACK HARNESS CONNECTOR. DOES J 39200 DISPLAY RESISTANCE AS "OL" (INFINITE)?

NO: REPAIR SHORT BETWEEN CKTs 1280 AND 1281.
YES: go to step 4.

(4)
- USING J 39200, MEASURE THE RESISTANCE BETWEEN TERMINAL "A" AND TERMINAL "B" OF THE ABS MOTOR PACK. IS RESISTANCE GREATER THAN 0.4Ω?

NO: REPLACE ABS MOTOR PACK.
YES: go to step 5.

(5)
- INSPECT CKTs 1280 AND 1281 AND RELATED CONNECTORS FOR DAMAGE WHICH MAY RESULT IN A SHORT TO VOLTAGE WITH ALL CONNECTORS CONNECTED. REPAIR ALL DAMAGE IF EVIDENT.
- RECONNECT ALL CONNECTORS.
- IGNITION "OFF."
- START ENGINE AND TEST DRIVE VEHICLE, ACHIEVING AT LEAST 16 km/h (10 MPH).
- REPEAT ABOVE DRIVE CYCLE SEQUENCE TWO MORE TIMES. DID DTC 58 SET IN THE LAST THREE DRIVE CYCLES?

YES: REPLACE EBCM.
NO: MALFUNCTION WAS MOST LIKELY CAUSED BY A POOR CONNECTION. REFER TO "DIAGNOSTIC AIDS" ON FACING PAGE.

AFTER DIAGNOSIS IS COMPLETE, CLEAR DTCs AND TEST DRIVE VEHICLE FOR THREE (3) DRIVE CYCLES TO VERIFY THAT DTC DOES NOT RESET. A DRIVE CYCLE CONSISTS OF STARTING THE VEHICLE, DRIVING OVER 16 KM/H (10 MPH), AND THEN KEYING DOWN.

DIAGNOSTIC CHART DTC 61 — 1994 SILHOUETTE, TRANS SPORT AND LUMINA APV

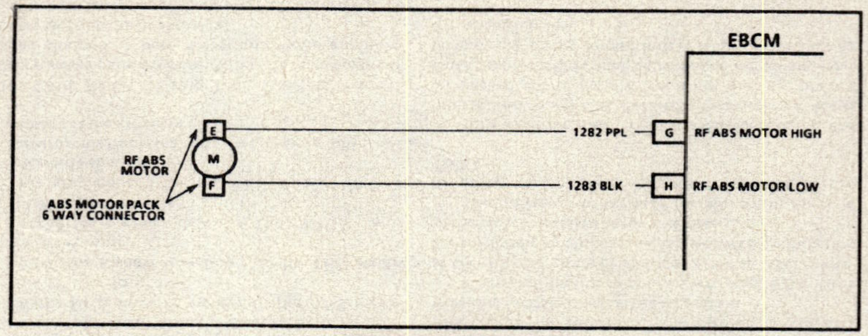

DTC 61

RIGHT FRONT ABS MOTOR CIRCUIT OPEN

Circuit Description:
This DTC identifies an ABS motor that cannot be energized due to an open in its circuitry. This malfunction will not allow proper front ABS operation.

DTC Will Set When: DTC 61 can be set only when the ABS motor is commanded "OFF." If the EBCM detects an out of range voltage on either of the right front ABS motor circuits indicating an open circuit, a malfunction exists.

Action Taken: An open ABS motor will not activate when requested. A malfunction DTC is stored, ABS is disabled and the ABS warning lamp is turned "ON."

DTC Chart Test Description: Number(s) below refer to circled number(s) on the diagnostic chart.

1. Checks for proper resistance of the ABS motor.
2. Checks for an open in the ABS motor "HIGH" circuitry.
3. Checks for an open in the ABS motor "LOW" circuitry.
4. Ensures malfunction was not due to poor terminal contact or EBCM.

Diagnostic Aids: Using Tech 1, select manual control function, and exercise ABS motor movement of affected channel in both directions while applying light pressure on the brake pedal.

If erratic or "jumpy" brake pedal movement is detected while performing an "apply" or "release" function of the ABS motor, an intermittent malfunction may be indicated.

An "Intermittent" malfunction may be caused by a poor connection, rubbed through wire insulation, or a wire that is broken inside the insulation.

If the malfunction is not current, wiggle the wires of the affected channel and check if the DTC resets. This will help to pinpoint an intermittent malfunction in the motor circuitry or connections.

The frequency of the malfunction can be checked by using the enhanced diagnostic function of the Tech 1

Any circuitry, that is suspected as causing the intermittent complaint, should be thoroughly checked for backed out terminals, improper mating, broken locks, improperly formed or damaged terminals, poor terminal to wiring connections or physical damage to the wiring harness.

DIAGNOSTIC CHART DTC 61 — 1994 SILHOUETTE, TRANS SPORT AND LUMINA APV

DTC 61

RIGHT FRONT ABS MOTOR CIRCUIT OPEN

DOES DTC 61 OCCUR INTERMITTENTLY?

- YES: REFER TO "DIAGNOSTIC AIDS" ON FACING PAGE.
- NO:

(1)
- IGNITION "OFF."
- DISCONNECT 6-WAY ABS MOTOR PACK CONNECTOR.
- USING J 39200, MEASURE RESISTANCE ACROSS ABS MOTOR PACK TERMINALS "E" AND "F".

IS RESISTANCE BELOW 2000Ω?

- NO: REPLACE ABS MOTOR PACK.
- YES:

(2)
- DISCONNECT 8-WAY CONNECTOR AT EBCM.
- USING J 39200, MEASURE RESISTANCE BETWEEN TERMINAL "G" OF THE 8-WAY EBCM HARNESS CONNECTOR AND TERMINAL "E" OF THE 6-WAY ABS MOTOR PACK HARNESS CONNECTOR.

IS RESISTANCE 2Ω OR LESS?

- NO: REPAIR OPEN OR HIGH RESISTANCE IN CKT 1282.
- YES:

(3)
- USING J 39200, MEASURE RESISTANCE BETWEEN TERMINAL "H" OF THE 8-WAY EBCM HARNESS CONNECTOR AND TERMINAL "F" OF THE 6-WAY ABS MOTOR PACK HARNESS CONNECTOR.

IS RESISTANCE 2Ω OR LESS?

- NO: REPAIR OPEN OR HIGH RESISTANCE IN CKT 1283.
- YES:

(4)
- INSPECT THE 6-WAY ABS MOTOR PACK AND 8-WAY EBCM CONNECTORS FOR POOR TERMINAL CONTACT AND EVIDENCE OF CORROSION.
  REPLACE ALL TERMINALS THAT EXHIBIT SIGNS OF POOR TERMINAL CONTACT OR CORROSION.
- RECONNECT ALL CONNECTORS.
- IGNITION "OFF."
- START ENGINE AND TEST DRIVE VEHICLE, ACHIEVING AT LEAST 16 km/h (10 MPH).
- REPEAT ABOVE DRIVE CYCLE SEQUENCE TWO MORE TIMES.

DID DTC 61 SET IN THE LAST THREE DRIVE CYCLES?

- YES: REPLACE EBCM.
- NO: MALFUNCTION WAS MOST LIKELY CAUSED BY A POOR CONNECTION. REFER TO "DIAGNOSTIC AIDS" ON FACING PAGE.

AFTER ALL DIAGNOSIS IS COMPLETE, CLEAR DTCs AND TEST DRIVE VEHICLE FOR THREE (3) DRIVE CYCLES TO VERIFY THAT DTC DOES NOT RESET. A DRIVE CYCLE CONSISTS OF STARTING THE VEHICLE, DRIVING OVER 16 km/h (10 MPH), AND THEN KEYING DOWN.

DIAGNOSTIC CHART DTC 62 — 1994 SILHOUETTE, TRANS SPORT AND LUMINA APV

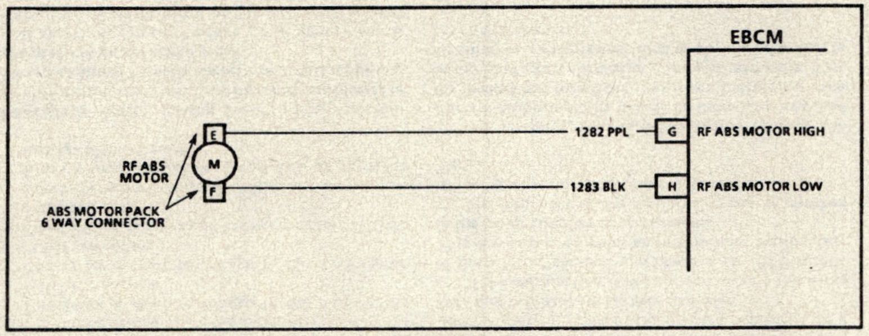

DTC 62
RIGHT FRONT ABS MOTOR CIRCUIT SHORTED TO GND

Circuit Description:
This DTC identifies an ABS motor circuit that is shorted to ground. This malfunction will not allow the ABS motor to be controlled at the commanded current rate or will cause the ABS motor driver circuit to allow current directly to ground.

DTC Will Set When: DTC 62 can be set anytime. If the EBCM detects an out of range voltage on either of the right front ABS motor circuits indicating a circuit shorted to ground, a malfunction exists.

Action Taken: A malfunction DTC is stored, ABS is disabled and the ABS warning lamp is turned "ON."

DTC Chart Test Description: Number(s) below refer to circled number(s) on the diagnostic chart.

1. Checks for a short to ground in the ABS motor "HIGH" circuitry.
2. Checks for a short to ground in the ABS motor "LOW" circuitry.
3. Checks for an ABS motor that is internally shorted to ground.
4. Ensures malfunction was not due to physical damage of the circuitry.

Diagnostic Aids: Using Tech 1, select manual control function, and exercise ABS motor movement of affected channel in both directions while applying light pressure on the brake pedal.

If erratic or "jumpy" brake pedal movement is detected while performing an "apply" or "release" function of the ABS motor, an intermittent malfunction may be indicated.

An "Intermittent" malfunction may be caused by a poor connection, rubbed through wire insulation, or a wire that is broken inside the insulation.

If the malfunction is not current, wiggle the wires of the affected channel and check if the DTC resets. This will help to pinpoint an intermittent malfunction in the motor circuitry or connections.

The frequency of the malfunction can be checked by using the enhanced diagnostic function of the Tech 1.

Any circuitry, that is suspected as causing the intermittent complaint, should be thoroughly checked for backed out terminals, improper mating, broken locks, improperly formed or damaged terminals, poor terminal to wiring connections or physical damage to the wiring harness.

DIAGNOSTIC CHART DTC 62 — 1994 SILHOUETTE, TRANS SPORT AND LUMINA APV

DTC 62
RIGHT FRONT ABS MOTOR CIRCUIT SHORTED TO GND

DOES DTC 62 OCCUR INTERMITTENTLY?

- YES → REFER TO "DIAGNOSTIC AIDS" ON FACING PAGE.
- NO →

(1)
- IGNITION "OFF."
- DISCONNECT 6-WAY ABS MOTOR PACK CONNECTOR AND THE 8-WAY EBCM CONNECTOR.
- USING J 39200, MEASURE THE RESISTANCE BETWEEN GROUND AND TERMINAL "E" OF THE 6-WAY ABS MOTOR PACK HARNESS CONNECTOR.

DOES J 39200 DISPLAY RESISTANCE AS "OL" (INFINITE)?

- NO → REPAIR SHORT TO GROUND IN CKT 1282.
- YES →

(2)
- MEASURE THE RESISTANCE BETWEEN GROUND AND TERMINAL "F" OF THE 6-WAY ABS MOTOR PACK HARNESS CONNECTOR.

DOES J 39200 DISPLAY RESISTANCE AS "OL" (INFINITE)?

- NO → REPAIR SHORT TO GROUND IN CKT 1283.
- YES →

(3)
- MEASURE THE RESISTANCE BETWEEN GROUND AND TERMINAL "E" OF THE ABS MOTOR PACK.

DOES J 39200 DISPLAY RESISTANCE AS "OL" (INFINITE)?

- NO → REPLACE ABS MOTOR PACK.
- YES →

(4)
- INSPECT CKTs 1282 AND 1283 AND RELATED CONNECTORS FOR DAMAGE WHICH MAY RESULT IN A SHORT TO GROUND WITH ALL CONNECTORS CONNECTED. REPAIR ALL DAMAGE IF EVIDENT.
- RECONNECT ALL CONNECTORS.
- IGNITION "OFF."
- START ENGINE AND TEST DRIVE VEHICLE, ACHIEVING AT LEAST 16 km/h (10 MPH).
- REPEAT ABOVE DRIVE CYCLE SEQUENCE TWO MORE TIMES.

DID DTC 62 SET IN THE LAST THREE DRIVE CYCLES?

- YES → REPLACE EBCM.
- NO → MALFUNCTION WAS MOST LIKELY CAUSED BY A POOR CONNECTION. REFER TO "DIAGNOSTIC AIDS" ON FACING PAGE.

AFTER ALL DIAGNOSIS IS COMPLETE, CLEAR DTCs AND TEST DRIVE VEHICLE FOR THREE (3) DRIVE CYCLES TO VERIFY THAT DTC DOES NOT RESET. A DRIVE CYCLE CONSISTS OF STARTING THE VEHICLE, DRIVING OVER 16 km/h (10 MPH), AND THEN KEYING DOWN.

DIAGNOSTIC CHART DTC 63 — 1994 SILHOUETTE, TRANS SPORT AND LUMINA APV

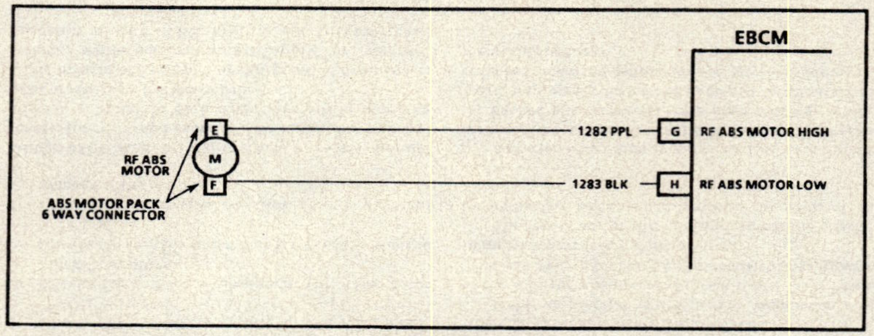

DTC 63

RIGHT FRONT ABS MOTOR CIRCUIT SHORTED TO BATT

Circuit Description:
This DTC identifies an ABS motor circuit that is shorted to battery or an ABS motor that has low or no resistance. This malfunction will not allow the ABS motor to be controlled at the commanded current rate or will cause the ABS motor to turn in the opposite direction or not at all.

DTC Will Set When: DTC 63 can be set only when the ABS motor is commanded "OFF." If the EBCM detects an out of range voltage on either of the right front ABS motor circuits indicating a circuit shorted to battery or an ABS motor shorted, a malfunction exists.

Action Taken: A malfunction DTC is stored, ABS is disabled and the ABS warning lamp is turned "ON."

DTC Chart Test Description: Number(s) below refer to circled number(s) on the diagnostic chart.

1. Checks for a short to voltage in the ABS motor "HIGH" circuitry.
2. Checks for a short to voltage in the ABS motor "LOW" circuitry.
3. Checks for a short circuit between motor "HIGH" and motor "LOW."
4. Checks for proper resistance of the ABS motor.
5. Ensures malfunction was not due to physical damage of the circuitry.

Diagnostic Aids: Using Tech 1, select manual control function, and exercise ABS motor movement of affected channel in both directions while applying light pressure on the brake pedal.

If erratic or "jumpy" brake pedal movement is detected while performing an "apply" or "release" function of the ABS motor, an intermittent malfunction may be indicated.

An "Intermittent" malfunction may be caused by a poor connection, rubbed through wire insulation, or a wire that is broken inside the insulation.

If the malfunction is not current, wiggle the wires of the affected channel and check if the DTC resets. This will help to pinpoint an intermittent malfunction in the motor circuitry or connections.

The frequency of the malfunction can be checked by using the enhanced diagnostic function of the Tech 1.

Any circuitry, that is suspected as causing the intermittent complaint, should be thoroughly checked for backed out terminals, improper mating, broken locks, improperly formed or damaged terminals, poor terminal to wiring connections or physical damage to the wiring harness.

DIAGNOSTIC CHART DTC 63 — 1994 SILHOUETTE, TRANS SPORT AND LUMINA APV

DTC 63

RIGHT FRONT ABS MOTOR CIRCUIT SHORTED TO BATT

NOTICE: J 39200 TEST LEADS MUST BE "ZEROED" PRIOR TO MAKING ANY RESISTANCE MEASUREMENTS. REFER TO J 39200 USER'S MANUAL.

DOES DTC 63 OCCUR INTERMITTENTLY?

- YES: REFER TO "DIAGNOSTIC AIDS" ON FACING PAGE.
- NO: go to 1.

(1)
- IGNITION "OFF."
- DISCONNECT THE 6-WAY ABS MOTOR PACK CONNECTOR AND THE 8-WAY EBCM CONNECTOR.
- IGNITION "ON."
- USING J 39200, MEASURE THE VOLTAGE BETWEEN GROUND AND TERMINAL "E" OF THE 6-WAY ABS MOTOR PACK HARNESS CONNECTOR.

IS VOLTAGE 1 VOLT OR LESS?

- NO: REPAIR SHORT TO VOLTAGE IN CKT 1282.
- YES: go to 2.

(2)
- MEASURE THE VOLTAGE BETWEEN GROUND AND TERMINAL "F" OF THE 6-WAY ABS MOTOR PACK HARNESS CONNECTOR.

IS VOLTAGE 1 VOLT OR LESS?

- NO: REPAIR SHORT TO VOLTAGE IN CKT 1283.
- YES: go to 3.

(3)
- IGNITION "OFF."
- USING J 39200, MEASURE THE RESISTANCE BETWEEN TERMINAL "E" AND TERMINAL "F" OF THE 6-WAY ABS MOTOR PACK HARNESS CONNECTOR.

DOES J 39200 DISPLAY RESISTANCE AS "OL" (INFINITE)?

- NO: REPAIR SHORT BETWEEN CKTs 1282 AND 1283.
- YES: go to 4.

(4)
- USING J 39200, MEASURE THE RESISTANCE BETWEEN TERMINAL "E" AND TERMINAL "F" OF THE ABS MOTOR PACK.

IS RESISTANCE GREATER THAN 0.4Ω?

- NO: REPLACE ABS MOTOR PACK.
- YES: go to 5.

(5)
- INSPECT CKTs 1282 AND 1283 AND RELATED CONNECTORS FOR DAMAGE WHICH MAY RESULT IN A SHORT TO VOLTAGE WITH ALL CONNECTORS CONNECTED. REPAIR ALL DAMAGE IF EVIDENT.
- RECONNECT ALL CONNECTORS.
- IGNITION "OFF."
- START ENGINE AND TEST DRIVE VEHICLE, ACHIEVING AT LEAST 16 km/h (10 MPH).
- REPEAT ABOVE DRIVE CYCLE SEQUENCE TWO MORE TIMES.

DID DTC 63 SET IN THE LAST THREE DRIVE CYCLES?

- YES: REPLACE EBCM.
- NO: MALFUNCTION WAS MOST LIKELY CAUSED BY A POOR CONNECTION. REFER TO "DIAGNOSTIC AIDS" ON FACING PAGE.

AFTER DIAGNOSIS IS COMPLETE, CLEAR DTCs AND TEST DRIVE VEHICLE FOR THREE (3) DRIVE CYCLES TO VERIFY THAT DTC DOES NOT RESET. A DRIVE CYCLE CONSISTS OF STARTING THE VEHICLE, DRIVING OVER 16 KM/H (10 MPH), AND THEN KEYING DOWN.

DIAGNOSTIC CHART DTC 64 — 1994 SILHOUETTE, TRANS SPORT AND LUMINA APV

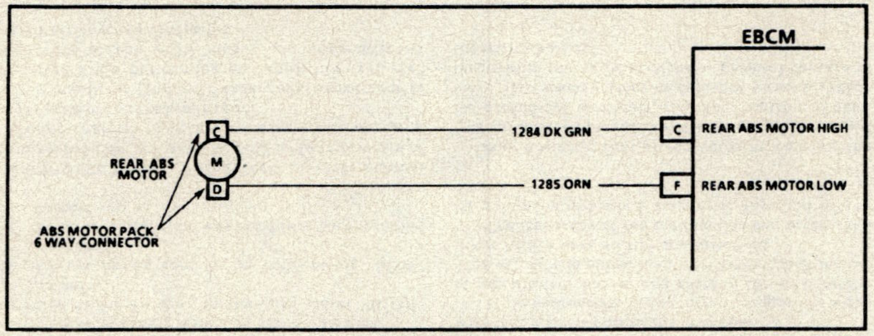

DTC 64
REAR ABS MOTOR CIRCUIT OPEN

Circuit Description:
This DTC identifies an ABS motor that cannot be energized due to an open in its circuitry. This malfunction will not allow proper rear ABS operation.

DTC Will Set When: DTC 64 can be set only when the ABS motor is commanded "OFF." If the EBCM detects an out of range voltage on either of the rear axle ABS motor circuits indicating an open circuit, a malfunction exists.

Action Taken: An open motor will not activate when requested. A malfunction DTC is stored, ABS is disabled and the ABS warning lamp is turned "ON."

DTC Chart Test Description: Number(s) below refer to circled number(s) on the diagnostic chart.

1. Checks for proper resistance of the ABS motor.
2. Checks for an open in the ABS motor "HIGH" circuitry.
3. Checks for an open in the ABS motor "LOW" circuitry.
4. Ensures malfunction was not due to poor terminal contact.

Diagnostic Aids: Using Tech 1, select manual control function, and exercise ABS motor movement of affected channel in both directions while applying light pressure on the brake pedal.

If erratic or "jumpy" brake pedal movement is detected while performing an "apply" or "release" function of the ABS motor, an intermittent malfunction may be indicated.

An "Intermittent" malfunction may be caused by a poor connection, rubbed through wire insulation, or a wire that is broken inside the insulation.

If the malfunction is not current, wiggle the wires of the affected channel and check if the DTC resets. This will help to pinpoint an intermittent malfunction in the motor circuitry or connections.

The frequency of the malfunction can be checked by using the enhanced diagnostic function of the Tech 1.

Any circuitry, that is suspected as causing the intermittent complaint, should be thoroughly checked for backed out terminals, improper mating, broken locks, improperly formed or damaged terminals, poor terminal to wiring connections or physical damage to the wiring harness.

DIAGNOSTIC CHART DTC 64 — 1994 SILHOUETTE, TRANS SPORT AND LUMINA APV

DTC 64
REAR ABS MOTOR CIRCUIT OPEN

DOES DTC 64 OCCUR INTERMITTENTLY?

- YES: REFER TO "DIAGNOSTIC AIDS" ON FACING PAGE.
- NO:

(1)
- IGNITION "OFF."
- DISCONNECT 6-WAY ABS MOTOR PACK CONNECTOR.
- USING J 39200, MEASURE RESISTANCE ACROSS ABS MOTOR PACK TERMINALS "C" AND "D".

IS RESISTANCE BELOW 2000Ω?

- NO: REPLACE ABS MOTOR PACK.
- YES:

(2)
- DISCONNECT 8-WAY CONNECTOR AT EBCM.
- USING J 39200, MEASURE RESISTANCE BETWEEN TERMINAL "C" OF THE 8-WAY EBCM HARNESS CONNECTOR AND TERMINAL "C" OF THE 6-WAY ABS MOTOR PACK HARNESS CONNECTOR.

IS RESISTANCE 2Ω OR LESS?

- NO: REPAIR OPEN OR HIGH RESISTANCE IN CKT 1284.
- YES:

(3)
- USING J 39200, MEASURE RESISTANCE BETWEEN TERMINAL "F" OF THE 8-WAY EBCM HARNESS CONNECTOR AND TERMINAL "D" OF THE 6-WAY ABS MOTOR PACK HARNESS CONNECTOR.

IS RESISTANCE 2Ω OR LESS?

- NO: REPAIR OPEN OR HIGH RESISTANCE IN CKT 1285.
- YES:

(4)
- INSPECT THE 6-WAY ABS MOTOR PACK AND 8-WAY EBCM CONNECTORS FOR POOR TERMINAL CONTACT AND EVIDENCE OF CORROSION.

REPLACE ALL TERMINALS THAT EXHIBIT SIGNS OF POOR TERMINAL CONTACT OR CORROSION.

- RECONNECT ALL CONNECTORS.
- IGNITION "OFF."
- START ENGINE AND TEST DRIVE VEHICLE, ACHIEVING AT LEAST 16 km/h (10 MPH).
- REPEAT ABOVE DRIVE CYCLE SEQUENCE TWO MORE TIMES.

DID DTC 64 SET IN THE LAST THREE DRIVE CYCLES?

- YES: REPLACE EBCM.
- NO: MALFUNCTION WAS MOST LIKELY CAUSED BY A POOR CONNECTION. REFER TO "DIAGNOSTIC AIDS" ON FACING PAGE.

AFTER ALL DIAGNOSIS IS COMPLETE, CLEAR DTCs AND TEST DRIVE VEHICLE FOR THREE (3) DRIVE CYCLES TO VERIFY THAT DTC DOES NOT RESET. A DRIVE CYCLE CONSISTS OF STARTING THE VEHICLE, DRIVING OVER 16 km/h (10 MPH), AND THEN KEYING DOWN.

DIAGNOSTIC CHART DTC 65 — 1994 SILHOUETTE, TRANS SPORT AND LUMINA APV

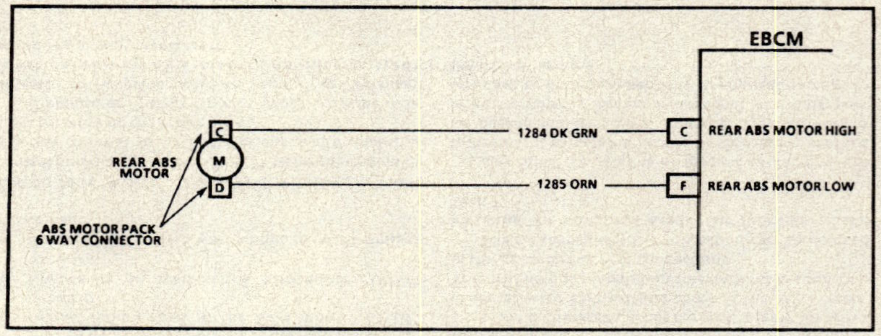

DTC 65
REAR ABS MOTOR CIRCUIT SHORTED TO GROUND

Circuit Description:
This DTC identifies an ABS motor circuit that is shorted to ground. This malfunction will not allow the ABS motor to be controlled at the commanded current rate or will cause the driver circuit to allow current directly to ground.

DTC Will Set When: DTC 65 can be set anytime. If the EBCM detects an out of range voltage on either of the rear ABS motor circuits indicating a circuit shorted to ground, a malfunction exists.

Action Taken: A malfunction DTC is stored, ABS is disabled and the ABS warning lamp is turned "ON."

DTC Chart Test Description: Number(s) below refer to circled number(s) on the diagnostic chart.

1. Checks for proper resistance of the ABS motor.
2. Checks for an open in the ABS motor "HIGH" circuitry.
3. Checks for an open in the ABS motor "LOW" circuitry.
4. Ensures malfunction was not due to poor terminal contact.

Diagnostic Aids: Using Tech 1, select manual control function, and exercise ABS motor movement of affected channel in both directions while applying light pressure on the brake pedal.

If erratic or "jumpy" brake pedal movement is detected while performing an "apply" or "release" function of the ABS motor, an intermittent malfunction may be indicated.

An "Intermittent" malfunction may be caused by a poor connection, rubbed through wire insulation, or a wire that is broken inside the insulation.

If the malfunction is not current, wiggle the wires of the affected channel and check if the DTC resets. This will help to pinpoint an intermittent malfunction in the ABS motor circuitry or connections.

The frequency of the malfunction can be checked by using the enhanced diagnostic function of the Tech 1.

Any circuitry, that is suspected as causing the intermittent complaint, should be thoroughly checked for backed out terminals, improper mating, broken locks, improperly formed or damaged terminals, poor terminal to wiring connections or physical damage to the wiring harness.

DIAGNOSTIC CHART DTC 65 — 1994 SILHOUETTE, TRANS SPORT AND LUMINA APV

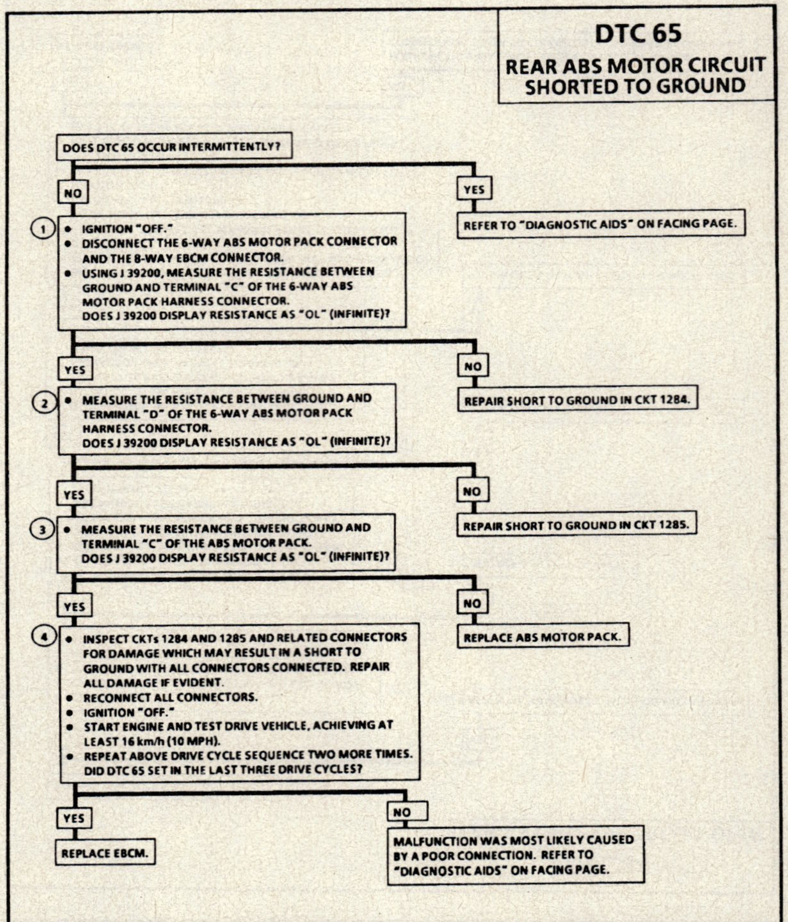

AFTER ALL DIAGNOSIS IS COMPLETE, CLEAR DTCs AND TEST DRIVE VEHICLE FOR THREE (3) DRIVE CYCLES TO VERIFY THAT DTC DOES NOT RESET. A DRIVE CYCLE CONSISTS OF STARTING THE VEHICLE, DRIVING OVER 16 km/h (10 MPH), AND THEN KEYING DOWN.

DIAGNOSTIC CHART DTC 66 — 1994 SILHOUETTE, TRANS SPORT AND LUMINA APV

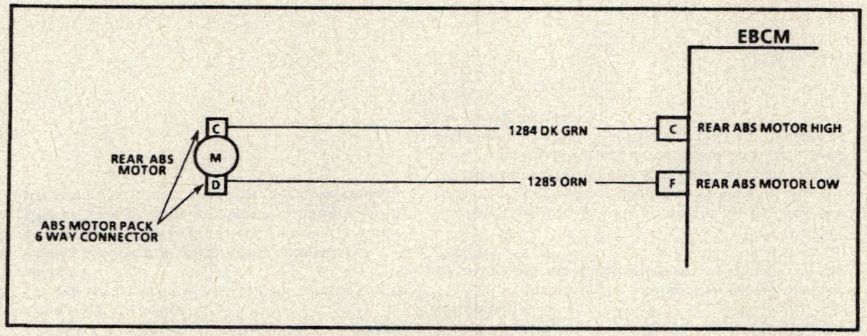

DTC 66
REAR ABS MOTOR CIRCUIT SHORTED TO BATTERY

Circuit Description:
This DTC identifies an ABS motor circuit that is shorted to battery or an ABS motor that has low or zero resistance. This malfunction will not allow the ABS motor to be controlled at the commanded current rate or will cause the ABS motor to turn in the opposite direction or not turn at all.

DTC Will Set When: DTC 66 can be set only when the ABS motor is commanded "OFF." If the EBCM detects an out of range voltage on either of the rear ABS motor circuits indicating a circuit shorted to battery or an ABS motor shorted, a malfunction exists.

Action Taken: A malfunction DTC is stored, ABS is disabled and the ABS warning lamp is turned "ON."

DTC Chart Test Description: Number(s) below refer to circled number(s) on the diagnostic chart.

1. Checks for a short to voltage in the ABS motor "HIGH" circuitry.
2. Checks for a short to voltage in the ABS motor "LOW" circuitry.
3. Checks for proper resistance of the ABS motor.
4. Checks for a short circuit between motor "HIGH" and motor "LOW."
5. Ensures malfunction was not due to physical damage of the circuitry.

Diagnostic Aids: Using Tech 1, select manual control function, and exercise ABS motor movement of affected channel in both directions while applying light pressure on the brake pedal.

If erratic or "jumpy" brake pedal movement is detected while performing an "apply" or "release" function of the ABS motor, an intermittent malfunction may be indicated.

An "Intermittent" malfunction may be caused by a poor connection, rubbed through wire insulation, or a wire that is broken inside the insulation.

If the malfunction is not current, wiggle the wires of the affected channel and check if the DTC resets. This will help to pinpoint an intermittent malfunction in the ABS motor circuitry or connections.

The frequency of the malfunction can be checked by using the enhanced diagnostic function of the Tech 1.

Any circuitry, that is suspected as causing the intermittent complaint, should be thoroughly checked for backed out terminals, improper mating, broken locks, improperly formed or damaged terminals, poor terminal to wiring connections or physical damage to the wiring harness.

DIAGNOSTIC CHART DTC 66 — 1994 SILHOUETTE, TRANS SPORT AND LUMINA APV

DTC 66
REAR ABS MOTOR CIRCUIT SHORTED TO BATTERY

NOTICE: J 39200 TEST LEADS MUST BE "ZEROED" PRIOR TO MAKING ANY RESISTANCE MEASUREMENTS. REFER TO J 39200 USER'S MANUAL.

DOES DTC 66 OCCUR INTERMITTENTLY?

- YES: REFER TO "DIAGNOSTIC AIDS" ON FACING PAGE.
- NO:

(1)
- IGNITION "OFF."
- DISCONNECT THE 6-WAY ABS MOTOR PACK CONNECTOR AND THE 8-WAY EBCM CONNECTOR.
- IGNITION "ON."
- USING J 39200, MEASURE THE VOLTAGE BETWEEN GROUND AND TERMINAL "C" OF THE 6-WAY ABS MOTOR PACK HARNESS CONNECTOR. IS VOLTAGE 1 VOLT OR LESS?

- NO: REPAIR SHORT TO VOLTAGE IN CKT 1284.
- YES:

(2)
- MEASURE THE VOLTAGE BETWEEN GROUND AND TERMINAL "D" OF THE 6-WAY ABS MOTOR PACK HARNESS CONNECTOR. IS VOLTAGE 1 VOLT OR LESS?

- NO: REPAIR SHORT TO VOLTAGE IN CKT 1285.
- YES:

(3)
- IGNITION "OFF."
- USING J 39200, MEASURE THE RESISTANCE BETWEEN TERMINAL "C" AND TERMINAL "D" OF THE 6-WAY ABS MOTOR PACK HARNESS CONNECTOR. DOES J 39200 DISPLAY RESISTANCE AS "OL" (INFINITE)?

- NO: REPAIR SHORT BETWEEN CKTs 1284 AND 1285.
- YES:

(4)
- USING J 39200, MEASURE THE RESISTANCE BETWEEN TERMINAL "C" AND TERMINAL "D" OF THE ABS MOTOR PACK. IS RESISTANCE GREATER THAN 0.4Ω?

- NO: REPLACE ABS MOTOR PACK
- YES:

(5)
- INSPECT CKTs 1284 AND 1285 AND RELATED CONNECTORS FOR DAMAGE WHICH MAY RESULT IN A SHORT TO VOLTAGE WITH ALL CONNECTORS CONNECTED. REPAIR ALL DAMAGE IF EVIDENT.
- RECONNECT ALL CONNECTORS.
- IGNITION "OFF."
- START ENGINE AND TEST DRIVE VEHICLE, ACHIEVING AT LEAST 16 km/h (10 MPH).
- REPEAT ABOVE DRIVE CYCLE SEQUENCE TWO MORE TIMES. DID DTC 66 SET IN THE LAST THREE DRIVE CYCLES?

- YES: REPLACE EBCM.
- NO: MALFUNCTION WAS MOST LIKELY CAUSED BY A POOR CONNECTION. REFER TO "DIAGNOSTIC AIDS" ON FACING PAGE.

AFTER DIAGNOSIS IS COMPLETE, CLEAR DTCs AND TEST DRIVE VEHICLE FOR THREE (3) DRIVE CYCLES TO VERIFY THAT DTC DOES NOT RESET. A DRIVE CYCLE CONSISTS OF STARTING THE VEHICLE, DRIVING OVER 16 KM/H (10 MPH), AND THEN KEYING DOWN.

DIAGNOSTIC CHART DTC 76 — 1994 SILHOUETTE, TRANS SPORT AND LUMINA APV

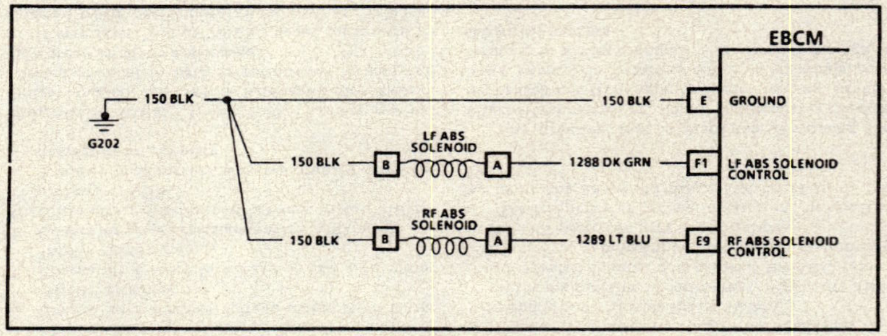

DTC 76

LEFT FRONT SOL CIRCUIT OPEN OR SHORTED TO BATT

Circuit Description:

This DTC identifies a solenoid that cannot be energized due to an open in its circuitry, or a solenoid that is always energized due to a short to battery in its circuitry between the driver and the solenoid. An open will not allow proper ABS operation, but the short to battery simply turns "ON" the solenoid. A path for base brakes is still allowed once the ABS motor re-homes and the check ball is lifted off its seat during key "ON" initialization.

DTC Will Set When: DTC 76 can be set only when the solenoid is commanded "OFF." If the EBCM detects an excessive voltage in the left front solenoid control circuit, a malfunction exists.

Action Taken: A malfunction DTC is stored, ABS is disabled and the ABS warning lamp is turned "ON."

DTC Chart Test Description: Number(s) below refer to circled number(s) on the diagnostic chart.

1. Checks for a short to voltage in the control circuitry of the solenoid.
2. Checks for an open in the control circuit of the solenoid.
3. Checks the solenoid coil for proper resistance.
4. Checks for an open in the solenoid ground circuit.
5. Checks for a possible intermittent malfunction in the solenoid circuitry due to poor terminal contact.

Diagnostic Aids: An "Intermittent" malfunction may be caused by a poor connection, rubbed through wire insulation, or a wire that is broken inside the insulation.

The frequency of the malfunction can be checked by using the enhanced diagnostic function of the Tech 1.

Any circuitry, that is suspected as causing the intermittent complaint, should be thoroughly checked for backed out terminals, improper mating, broken locks, improperly formed or damaged terminals, poor terminal to wiring connections or physical damage to the wiring harness.

DIAGNOSTIC CHART DTC 76 — 1994 SILHOUETTE, TRANS SPORT AND LUMINA APV

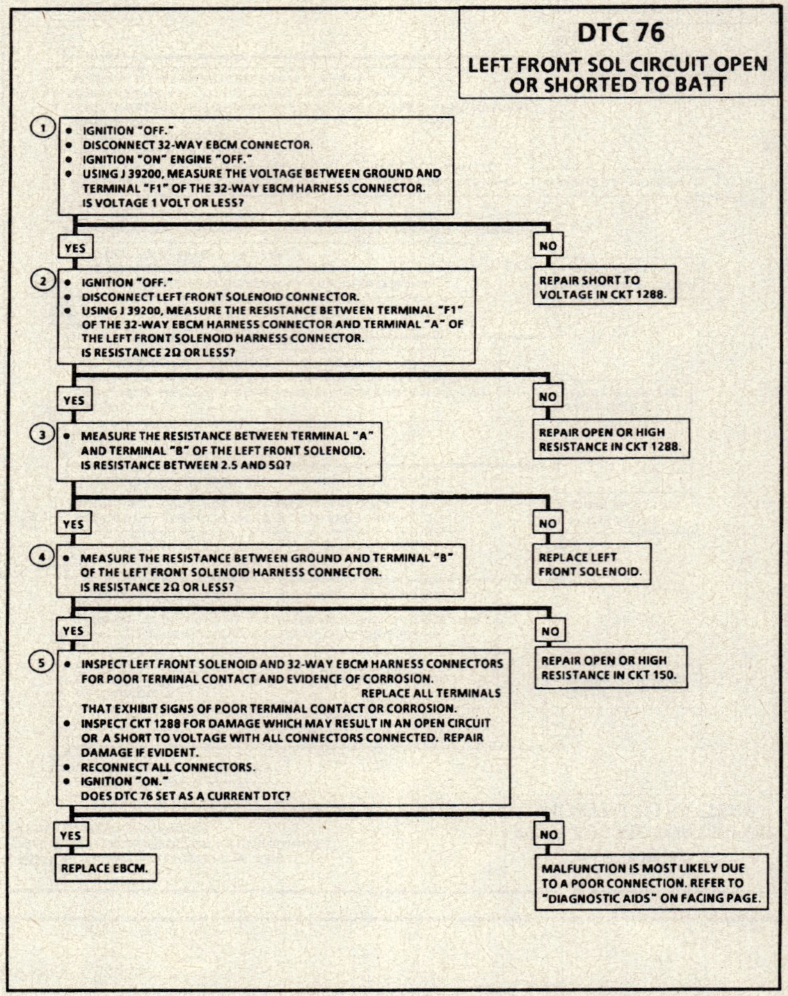

DIAGNOSTIC CHART DTC 77 — 1994 SILHOUETTE, TRANS SPORT AND LUMINA APV

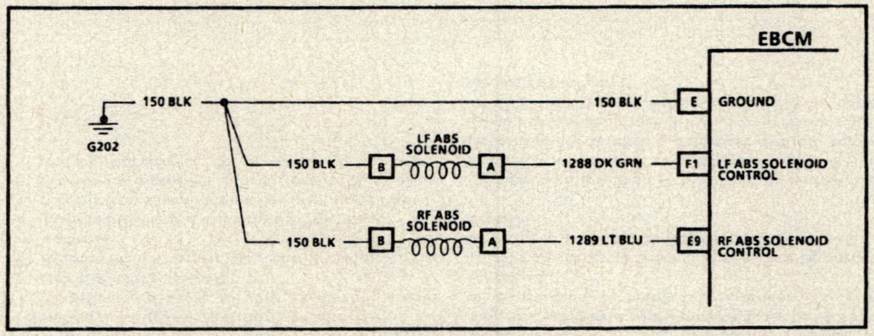

DTC 77

LEFT FRONT SOL CIRCUIT SHORTED TO GND

Circuit Description:

This DTC identifies a solenoid that cannot be energized due to an open in its driver circuitry, or a short to ground between the solenoid driver and the solenoid. These malfunctions can affect ABS operation since the flow of brake fluid to the wheel cylinder cannot be stopped, making ABS operation for that channel impossible.

DTC Will Set When: DTC 77 can be set only when the solenoid is commanded "ON." If the EBCM detects the left front solenoid control circuit voltage is out of specification, a malfunction exists.

Action Taken: A malfunction DTC is stored, ABS is disabled and the ABS warning lamp is turned "ON."

DTC Chart Test Description: Number(s) below refer to circled number(s) on the diagnostic chart.

1. Checks for a solenoid that is internally shorted to ground.
2. Checks for a solenoid that is not within proper resistance values.
3. Indicates if a short to ground exists in the solenoid circuitry.
4. Checks for a possible intermittent malfunction in the solenoid circuitry due to poor terminal contact.

Diagnostic Aids: An "Intermittent" malfunction may be caused by a poor connection, rubbed through wire insulation, or a wire that is broken inside the insulation.

The frequency of the malfunction can be checked by using the enhanced diagnostic function of the Tech 1,

Any circuitry, that is suspected as causing the intermittent complaint, should be thoroughly checked for backed out terminals, improper mating, broken locks, improperly formed or damaged terminals, poor terminal to wiring connections or physical damage to the wiring harness.

DIAGNOSTIC CHART DTC 77 — 1994 SILHOUETTE, TRANS SPORT AND LUMINA APV

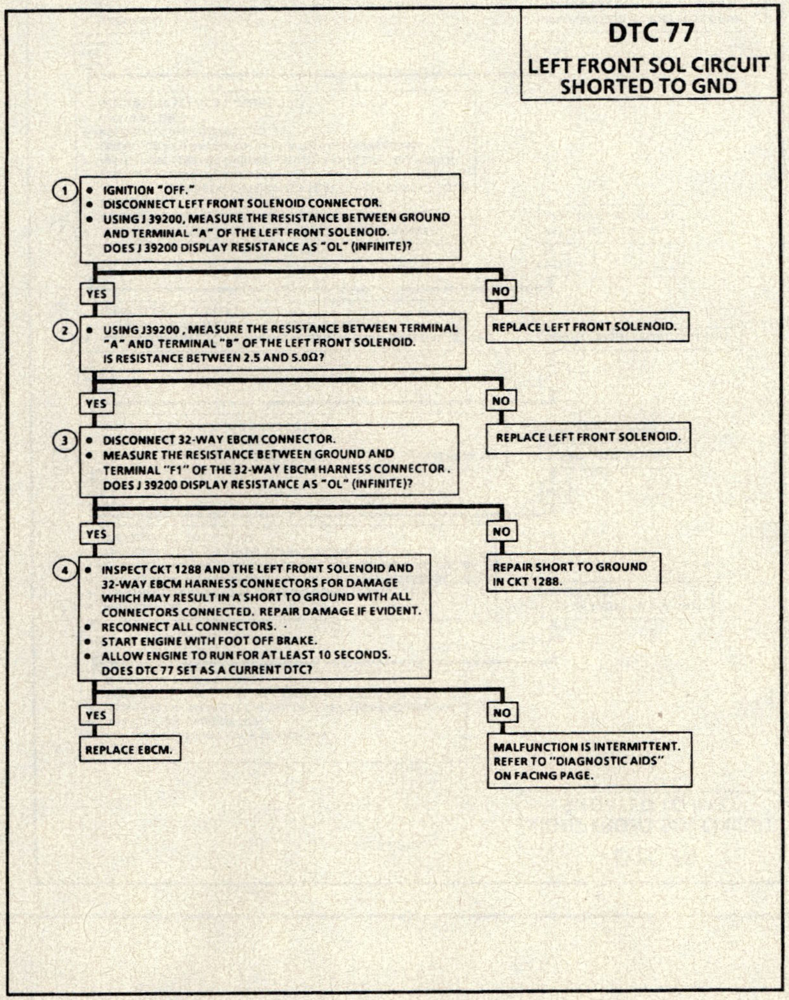

DIAGNOSTIC CHART DTC 78 — 1994 SILHOUETTE, TRANS SPORT AND LUMINA APV

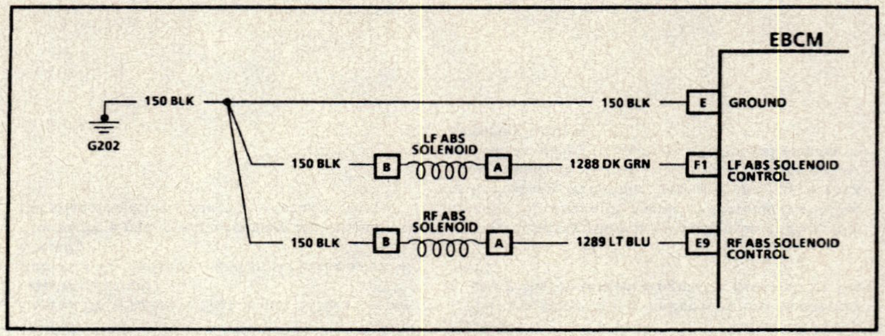

DTC 78
RIGHT FRONT SOL CIRCUIT SHORTED TO BATT

Circuit Description:

This DTC identifies a solenoid that cannot be energized due to an open in its circuitry, or a solenoid that is always energized due to a short to battery in its circuitry between the driver and the solenoid. An open will not allow proper ABS operation, but the short to battery simply turns "ON" the solenoid. A path for base brakes is still allowed once the ABS motor re-homes and the check ball is lifted off its seat during key "ON" initialization.

DTC Will Set When: DTC 78 can be set only when the solenoid is commanded "OFF." If the EBCM detects an excessive voltage in the right front solenoid control circuit, a malfunction exists.

Action Taken: A malfunction DTC is stored, ABS is disabled and the ABS warning lamp is turned "ON."

DTC Chart Test Description: Number(s) below refer to circled number(s) on the diagnostic chart.

1. Checks for a short to voltage in the control circuitry of the solenoid.
2. Checks for an open in the control circuit of the solenoid.
3. Checks the solenoid coil for proper resistance.
4. Checks for an open in the solenoid ground circuit.
5. Checks for a possible intermittent malfunction in the solenoid circuitry due to poor terminal contact.

Diagnostic Aids: An "Intermittent" malfunction may be caused by a poor connection, rubbed through wire insulation, or a wire that is broken inside the insulation.

The frequency of the malfunction can be checked by using the enhanced diagnostic function of the Tech 1

Any circuitry, that is suspected as causing the intermittent complaint, should be thoroughly checked for backed out terminals, improper mating, broken locks, improperly formed or damaged terminals, poor terminal to wiring connections or physical damage to the wiring harness.

DIAGNOSTIC CHART DTC 78 — 1994 SILHOUETTE, TRANS SPORT AND LUMINA APV

DTC 78
RIGHT FRONT SOL CIRCUIT SHORTED TO BATT

(1)
- IGNITION "OFF."
- DISCONNECT 32-WAY EBCM CONNECTOR.
- IGNITION "ON," ENGINE "OFF."
- USING J 39200, MEASURE THE VOLTAGE BETWEEN GROUND AND TERMINAL "E9" OF THE 32-WAY EBCM HARNESS CONNECTOR.

IS VOLTAGE 1 VOLT OR LESS?

NO: REPAIR SHORT TO VOLTAGE IN CKT 1289.

YES:

(2)
- IGNITION "OFF."
- DISCONNECT RIGHT FRONT SOLENOID CONNECTOR.
- USING J 39200, MEASURE THE RESISTANCE BETWEEN TERMINAL "E9" OF THE 32-WAY EBCM HARNESS CONNECTOR AND TERMINAL "A" OF THE RIGHT FRONT SOLENOID HARNESS CONNECTOR.

IS RESISTANCE 2Ω OR LESS?

NO: REPAIR OPEN OR HIGH RESISTANCE IN CKT 1289.

YES:

(3)
- MEASURE THE RESISTANCE BETWEEN TERMINAL "A" AND TERMINAL "B" OF THE RIGHT FRONT SOLENOID.

IS RESISTANCE BETWEEN 2.5 AND 5Ω?

NO: REPLACE RIGHT FRONT SOLENOID.

YES:

(4)
- MEASURE THE RESISTANCE BETWEEN GROUND AND TERMINAL "B" OF THE RIGHT FRONT SOLENOID HARNESS CONNECTOR.

IS RESISTANCE 2Ω OR LESS?

NO: REPAIR OPEN OR HIGH RESISTANCE IN CKT 150.

YES:

(5)
- INSPECT RIGHT FRONT SOLENOID AND 32-WAY EBCM HARNESS CONNECTORS FOR POOR TERMINAL CONTACT AND EVIDENCE OF CORROSION.

  REPLACE ALL TERMINALS THAT EXHIBIT SIGNS OF POOR TERMINAL CONTACT OR CORROSION.
- INSPECT CKT 1289 FOR DAMAGE WHICH MAY RESULT IN AN OPEN CIRCUIT OR A SHORT TO VOLTAGE WITH ALL CONNECTORS.
- RECONNECT ALL CONNECTORS.
- IGNITION "ON."

DOES DTC 78 SET AS A CURRENT DTC?

YES: REPLACE EBCM.

NO: MALFUNCTION IS MOST LIKELY DUE TO A POOR CONNECTION. REFER TO "DIAGNOSTIC AIDS" ON FACING PAGE.

DIAGNOSTIC CHART DTC 81 — 1994 SILHOUETTE, TRANS SPORT AND LUMINA APV

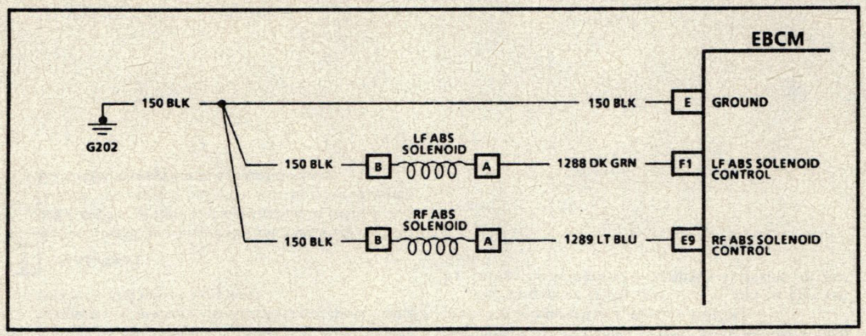

DTC 81
RIGHT FRONT SOL CIRCUIT SHORTED TO GND

Circuit Description:
This DTC identifies a solenoid that cannot be energized due to an open in its driver circuitry, or a short to ground between the solenoid driver and the solenoid. These malfunctions can affect ABS operation since the flow of brake fluid to the wheel cylinder cannot be stopped, making ABS operation for that channel impossible.

DTC Will Set When: DTC 81 can be set only when the solenoid is commanded "ON." If the EBCM detects the right front solenoid control circuit voltage is out of specification, a malfunction exists.

Action Taken: A malfunction DTC is stored, ABS is disabled and the ABS warning lamp is turned "ON."

DTC Chart Test Description: Number(s) below refer to circled number(s) on the diagnostic chart.
1. Checks for a solenoid that is internally shorted to ground.
2. Checks for a solenoid that is not within proper resistance values.
3. Indicates if a short to ground exists in the solenoid circuitry.
4. Checks for a possible intermittent malfunction in the solenoid circuitry due to poor terminal contact.

Diagnostic Aids: An "Intermittent" malfunction may be caused by a poor connection, rubbed through wire insulation, or a wire that is broken inside the insulation.

The frequency of the malfunction can be checked by using the enhanced diagnostic function of the Tech 1.

Any circuitry, that is suspected as causing the intermittent complaint, should be thoroughly checked for backed out terminals, improper mating, broken locks, improperly formed or damaged terminals, poor terminal to wiring connections or physical damage to the wiring harness.

DIAGNOSTIC CHART DTC 81 — 1994 SILHOUETTE, TRANS SPORT AND LUMINA APV

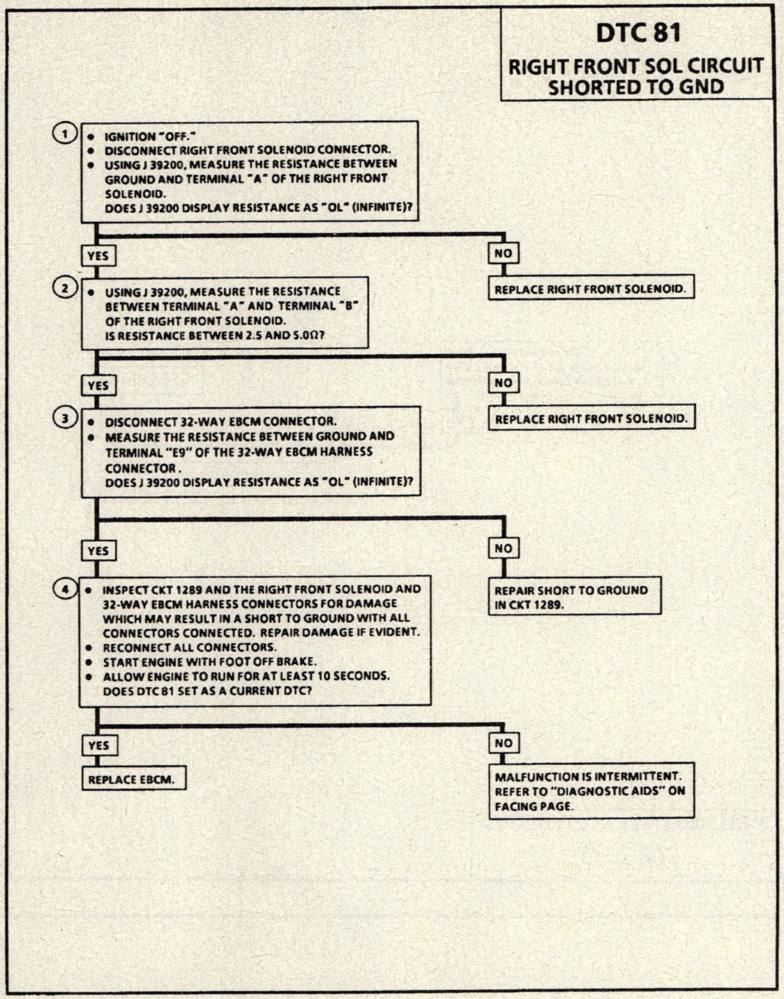

DIAGNOSTIC CHART DTC 82 — 1994 SILHOUETTE, TRANS SPORT AND LUMINA APV

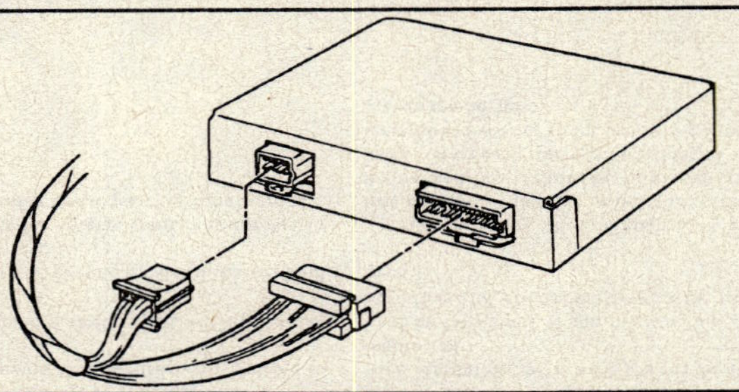

DTC 82
CALIBRATION MALFUNCTION

Circuit Description:
This DTC allows the EBCM to check for a calibration malfunction by comparing the calibration value to a known value stored in the EEPROM.
This DTC is also used as a security measure to prevent improper use of calibrations or changes to these calibrations that may alter the designed function of ABS.

DTC Will Set When: DTC 82 can be set at key-up. If the program identifier is incorrect or the memory checksum is incorrect, a malfunction exists.

Action Taken: A malfunction DTC is stored, ABS is disabled and the ABS warning lamp is turned "ON."

DTC Chart Test Description: Number(s) below refer to circled number(s) on the diagnostic chart.
1. Checks to see if the malfunction is present during diagnosis. If present, the EBCM is not functioning correctly and must be replaced.

[!] Important
- At the time of printing, 1994 vehicles with ABS VI were not being field reprogrammed to correct this failure. A service bulletin will be issued when EEPROM reprogramming is authorized.

Diagnostic Aids: An intermittent DTC 82 may be caused by a bad cell in the EEPROM that is sensitive to temperature changes. If DTC 82 failed more than once, but is intermittent, replace EBCM.
The frequency of the malfunction can be checked by using the enhanced diagnostic function of the Tech 1,

DIAGNOSTIC CHART DTC 82 — 1994 SILHOUETTE, TRANS SPORT AND LUMINA APV

DTC 82
CALIBRATION MALFUNCTION

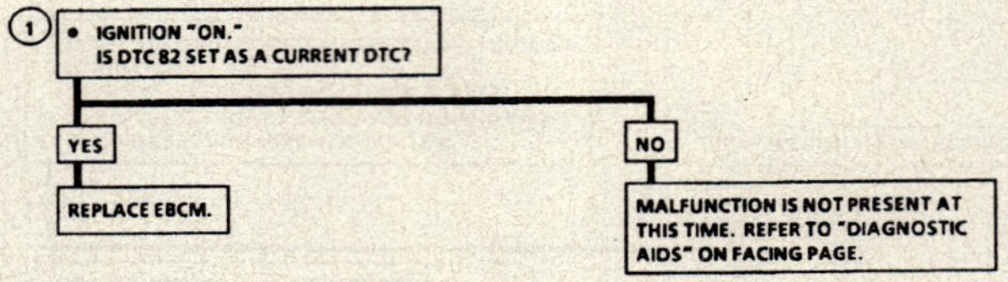

DIAGNOSTIC CHART DTC 86 — 1994 SILHOUETTE, TRANS SPORT AND LUMINA APV

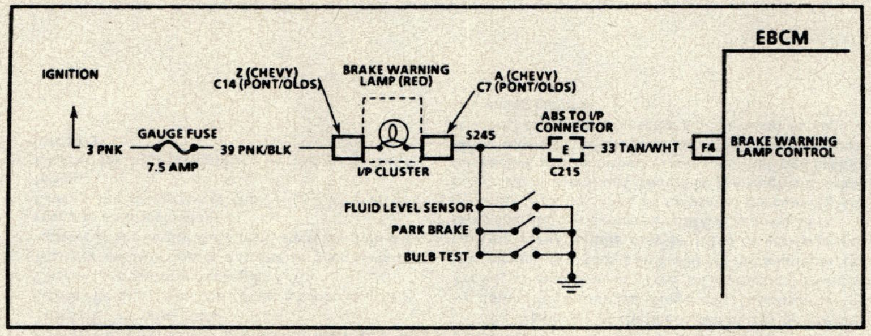

DTC 86

EBCM TURNED "ON" THE RED "BRAKE" WARNING LAMP

Circuit Description:

This DTC is provided as an information only DTC, and reflects the status of the command issued by the ABS EBCM to illuminate the red "BRAKE" warning lamp. If another DTC issues a command to illuminate the red "BRAKE" warning lamp, DTC 86 will be stored in EEPROM as a history DTC at the conclusion of the ignition cycle.

DTC Chart Test Description: Number(s) below refer to circled number(s) on the diagnostic chart.

1. This step serves to identify if a DTC other than DTC 86 commanded the red "BRAKE" warning lamp "ON."

Diagnostic Aids: Any ABS mechanical DTC that issues a command to illuminate the red "BRAKE" telltale will also result in DTC 86 being stored in EEPROM during shutdown. These DTCs are: 38, 41, 42, 46, and 51. If the motors are not in their home position, certain electrical DTCs will also command the red "BRAKE" telltale "ON." These DTCs are: 14, 16, 18, 55, 56, 57, 58, 61, 62, 63, 64, 65, and 66.

If any of these DTCs are indicated along with DTC 86, they must be corrected prior to addressing a DTC 86 malfunction.

DIAGNOSTIC CHART DTC 86 — 1994 SILHOUETTE, TRANS SPORT AND LUMINA APV

DTC 86

EBCM TURNED "ON" THE RED "BRAKE" WARNING LAMP

(1) ARE ANY DTCs CURRENTLY SET WITH DTC 86?

- NO: DTC 86 SET DUE TO VEHICLE TRANSIENT VOLTAGES; CLEAR DTC.
- YES: PROCEED TO "DIAGNOSTIC AIDS" ON FACING PAGE OF THIS CHART.

DIAGNOSTIC CHART DTC 87 — 1994 SILHOUETTE, TRANS SPORT AND LUMINA APV

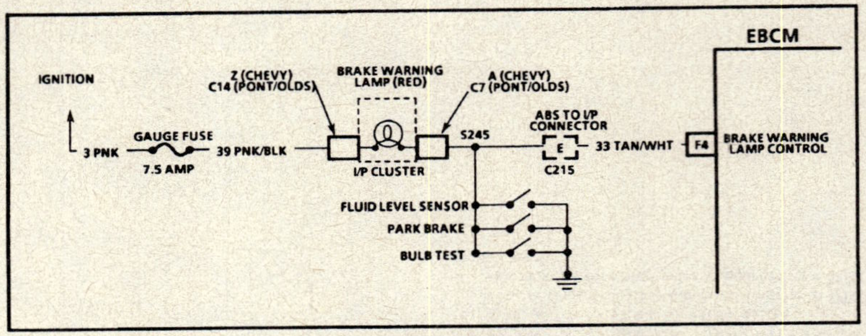

DTC 87

RED "BRAKE" WARNING LAMP CIRCUIT OPEN

Circuit Description:

This DTC is used to verify the EBCM has continuity to the red "BRAKE" warning lamp in case the EBCM must turn it "ON." This will only occur if an ABS malfunction is detected that may degrade base brake operation.

Because ABS is not the only device controlling the red "BRAKE" warning lamp (parking brake and low fluid sensor may also turn "ON" the lamp) a short to ground in this circuit cannot be detected.

DTC Will Set When: DTC 87 can be set after initialization is completed. If the red "BRAKE" warning lamp circuit voltage is out of specification indicating an open circuit, a malfunction exists.

Action Taken: A malfunction DTC is stored, ABS is not disabled.

DTC Chart Test Description: Number(s) below refer to circled number(s) on the diagnostic chart.

1. Checks the complete "BRAKE" warning lamp circuitry by attempting to manually turn "ON" the "BRAKE" warning lamp.
2. Checks for high resistance in park brake circuitry.
3. Checks for a malfunctioning EBCM.
4. Ensures malfunction is not due to poor terminal contact at the brake fluid level switch which may result in an intermittent.
5. Checks the integrity of the "BRAKE" warning lamp.
6. Checks for high resistance in the park brake circuitry.

Diagnostic Aids: An "Intermittent" malfunction may be caused by a poor connection, rubbed through wire insulation, or a wire that is broken inside the insulation.

The frequency of the malfunction can be checked by using the enhanced diagnostic function of the Tech 1, as described in "Tech 1 Diagnostics" found in this section. If a Tech 1 or T-100 are not available, the parking brake handle may be lifted to verify proper lamp operation and continuity of CKTs 3 and 39.

Any circuitry, that is suspected as causing the intermittent complaint, should be thoroughly checked for backed out terminals, improper mating, broken locks, improperly formed or damaged terminals, poor terminal to wiring connections or physical damage to the wiring harness.

DIAGNOSTIC CHART DTC 87 — 1994 SILHOUETTE, TRANS SPORT AND LUMINA APV

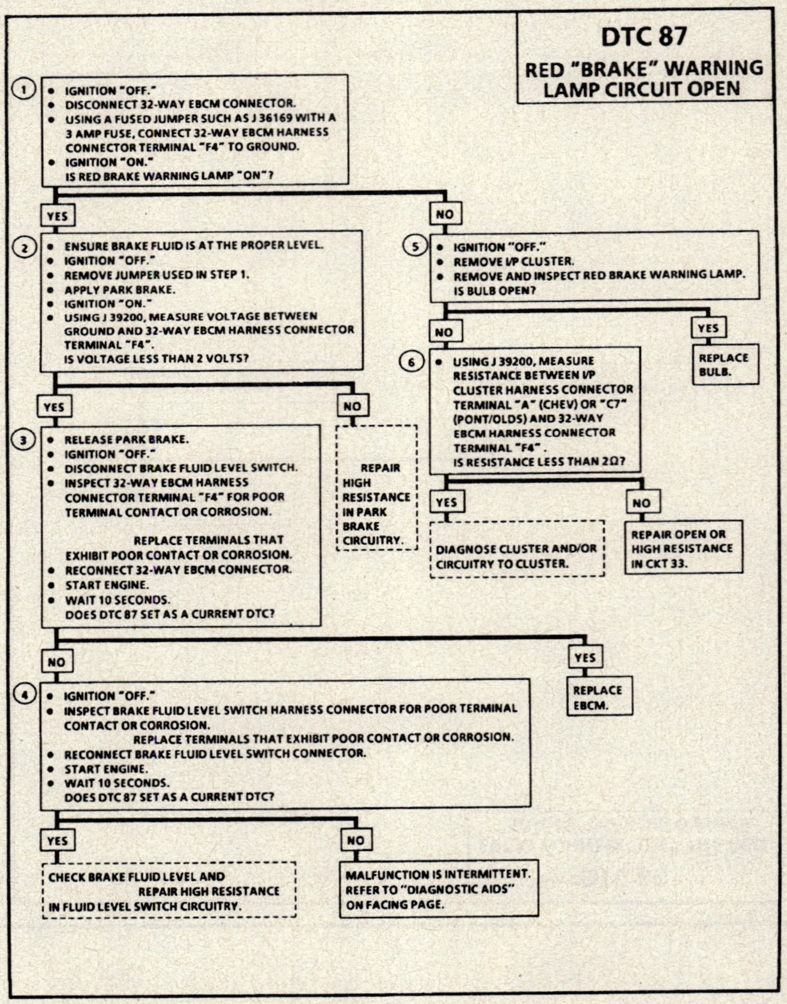

DIAGNOSTIC CHART DTC 88 — 1994 SILHOUETTE, TRANS SPORT AND LUMINA APV

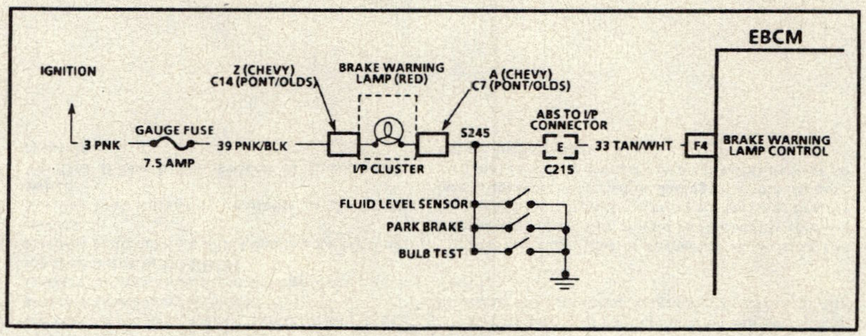

DTC 88

RED "BRAKE" WARNING LAMP CIRCUIT SHORTED TO BATT

Circuit Description:
This DTC identifies a short to battery between the EBCM and the red "BRAKE" warning lamp, or an open driver that does not allow the red "BRAKE" warning lamp to be illuminated by the EBCM. This will only occur if an ABS fault is detected that may degrade base brake operation.

DTC Will Set When: DTC 88 can be set only at the beginning of the three second bulb check. If the EBCM detects battery voltage on the red "BRAKE" warning lamp control circuit, a malfunction exists.

Action Taken: A malfunction DTC is stored. ABS is not disabled.

DTC Chart Test Description: Number(s) below refer to circled number(s) on the diagnostic chart.

1. Identifies if the ground circuit to the red "BRAKE" warning lamp is being completed by a source other than the EBCM.
2. Identifies if the malfunction is currently present.
3. By removing the 7.5 amps "GAUGES" fuse, the voltage source is eliminated. This test indicates if voltage is being supplied from a source other than the gauge fuse.
4. Ensures malfunction was not due to physical damage of the circuitry.
5. Checks for DTC 86 also being set.

Diagnostic Aids: An "Intermittent" malfunction may be caused by a poor connection, rubbed through wire insulation, or a wire that is broken inside the insulation.

The frequency of the malfunction can be checked by using the enhanced diagnostic function of the Tech 1.

Any circuitry, that is suspected as causing the intermittent complaint, should be thoroughly checked for backed out terminals, improper mating, broken locks, improperly formed or damaged terminals, poor terminal to wiring connections or physical damage to the wiring harness.

DIAGNOSTIC CHART DTC 88 — 1994 SILHOUETTE, TRANS SPORT AND LUMINA APV

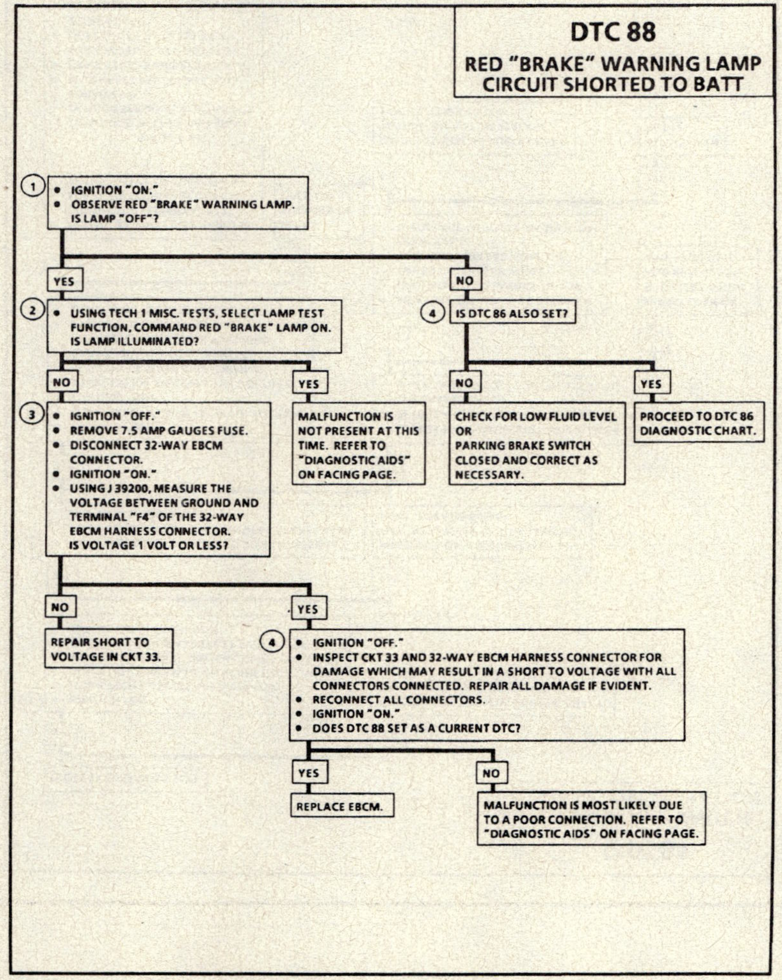

DIAGNOSTIC CHART DTC 91 — 1994 SILHOUETTE, TRANS SPORT AND LUMINA APV

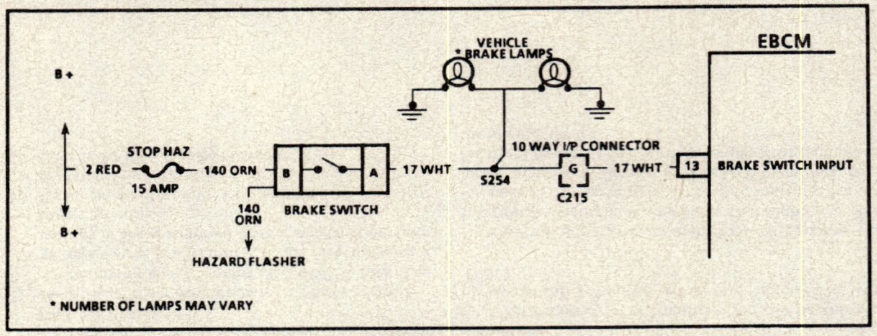

DTC 91

OPEN BRAKE SWITCH DURING DECELERATION

Circuit Description:
This DTC is used to detect an open brake switch in the non-ABS mode. The EBCM looks for deceleration rates that would indicate braking action and verifies this assumption by requiring several repeats of this detection method. In each case, ABS will not be available since no brake switch is seen.

DTC Will Set When: DTC 91 can be set if three deceleration cycles occur with the brake switch "OFF."

Action Taken: A malfunction DTC is stored, ABS is disabled and the ABS warning lamp is turned "ON."

DTC Chart Test Description: Number(s) below refer to circled number(s) on the diagnostic chart.

1. Indicates if the brake switch signal is being received by the EBCM.
2. Indicates if an open circuit exists in the brake switch or brake lamp circuitry.
3. Isolates the open circuit to either the brake switch input circuitry, or the EBCM.
4. Ensures malfunction was not due to poor terminal contact.
5. Verifies that voltage is available at the brake switch.
6. Verifies if the brake switch is functioning properly.

Diagnostic Aids: An "Intermittent" malfunction may be caused by a poor connection, rubbed through wire insulation, or a wire that is broken inside the insulation.

The frequency of the malfunction can be checked by using the enhanced diagnostic function of the Tech 1.

Any circuitry, that is suspected as causing the intermittent complaint, should be thoroughly checked for backed out terminals, improper mating, broken locks, improperly formed or damaged terminals, poor terminal to wiring connections or physical damage to the wiring harness.

DIAGNOSTIC CHART DTC 91 — 1994 SILHOUETTE, TRANS SPORT AND LUMINA APV

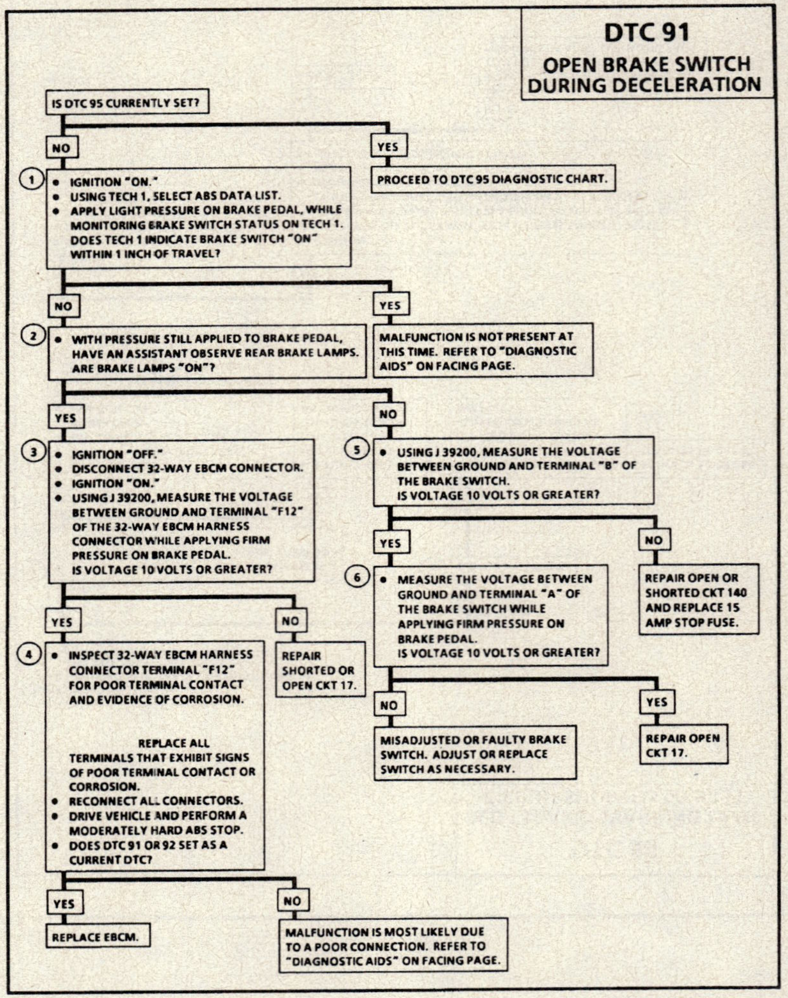

DIAGNOSTIC CHART DTC 92 — 1994 SILHOUETTE, TRANS SPORT AND LUMINA APV

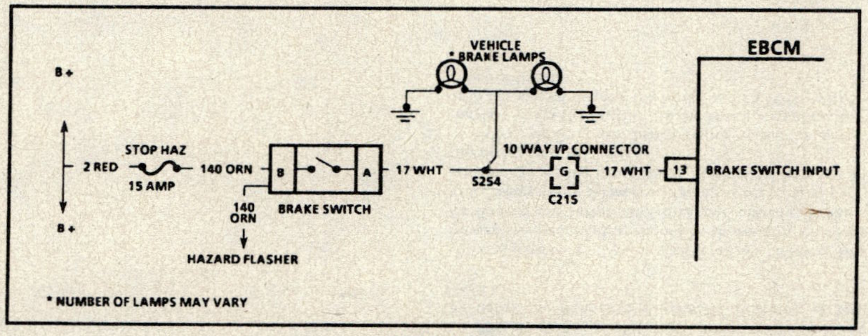

DTC 92

OPEN BRAKE SWITCH WHEN ABS WAS REQUIRED

Circuit Description:

This DTC is run to determine the proper operation of the brake switch. This is important because ABS is activated when the brake switch is "ON." If the brake switch is "OFF," ABS will never be activated. Since this malfunction is difficult to detect under normal braking conditions, this malfunction is only detected when ABS is required.

DTC Will Set When: DTC 92 can be set if the vehicle's speed is greater than 8 km/h (5 mph). If the brake was not "ON" and a release was required on two channels for 0.5 second, a malfunction exists.

Action Taken: A malfunction DTC is stored, ABS is disabled and the ABS warning lamp is turned "ON."

DTC Chart Test Description: Number(s) below refer to circled number(s) on the diagnostic chart.

1. Indicates if the brake switch signal is being received by the EBCM.
2. Indicates if an open circuit exists in the brake switch or brake lamp circuitry.
3. Isolates the open circuit to either the brake switch input circuitry, or the EBCM.
4. Ensures malfunction was not due to poor terminal contact.
5. Verifies the voltage is available at the brake switch.
6. Verifies if the brake switch is functioning properly.

Diagnostic Aids: An "Intermittent" malfunction may be caused by a poor connection, rubbed through wire insulation, or a wire that is broken inside the insulation.

The frequency of the malfunction can be checked by using the enhanced diagnostic function of the Tech 1.

Any circuitry, that is suspected as causing the intermittent complaint, should be thoroughly checked for backed out terminals, improper mating, broken locks, improperly formed or damaged terminals, poor terminal to wiring connections or physical damage to the wiring harness.

DIAGNOSTIC CHART DTC 92 — 1994 SILHOUETTE, TRANS SPORT AND LUMINA APV

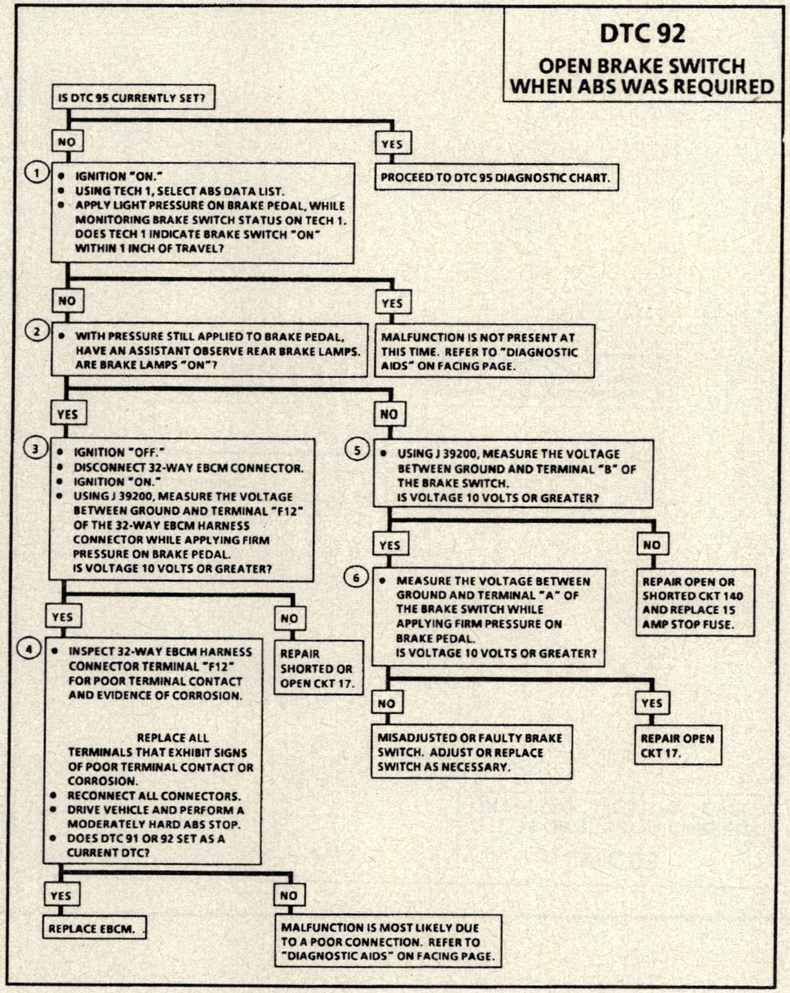

DIAGNOSTIC CHART DTC 93 — 1994 SILHOUETTE, TRANS SPORT AND LUMINA APV

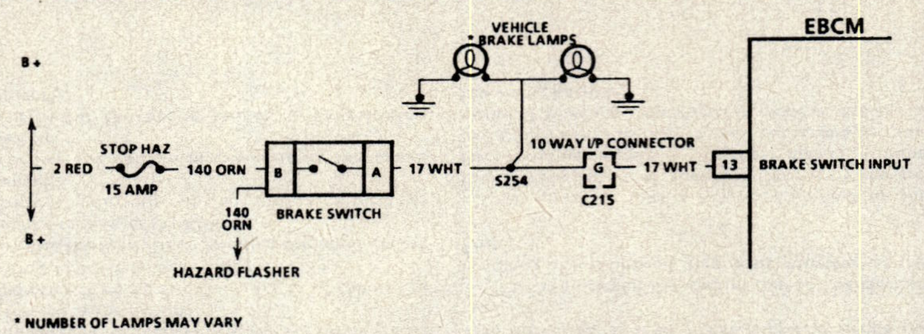

DTC 93

DTC 91 OR 92 SET IN CURRENT OR PREV IGN CYCLE

Circuit Description:

This DTC is the second portion of DTCs 91 and 92. If DTCs 91 or 92 occured during the last ignition cycle, DTC 93 becomes a current failure during the next ignition cycle, keeping ABS disabled until a brake switch "ON" state is seen. When a change is seen during an ignition cycle in which DTC 93 is a current malfunction, DTC 91 or 92 will clear itself at the end of the current ignition cycle, and ABS will enable itself at the start of the next ignition cycle. DTC 93 alone indicates 91 or 92 occurred previously, but is intermittent, or has been corrected.

DTC Chart Test Description: Number(s) below refer to circled number(s) on the diagnostic chart.

1. Indicates which DTC (either 91 or 92) caused DTC 93 to set.

Diagnostic Aids: An "Intermittent" malfunction may be caused by a poor connection, rubbed through wire insulation, or a wire that is broken inside the insulation.

The frequency of the malfunction can be checked by using the enhanced diagnostic function of the Tech 1,

Any circuitry, that is suspected as causing the intermittent complaint, should be thoroughly checked for backed out terminals, improper mating, broken locks, improperly formed or damaged terminals, poor terminal to wiring connections or physical damage to the wiring harness.

Also, verify proper brake switch operation using the data list of the Tech 1. As the brake is applied, the data list should display the brake switch "ON" within 1 inch of travel.

DIAGNOSTIC CHART DTC 93 — 1994 SILHOUETTE, TRANS SPORT AND LUMINA APV

DTC 93

DTC 91 OR 92 SET IN CURRENT OR PREVIOUS IGNITION CYCLE

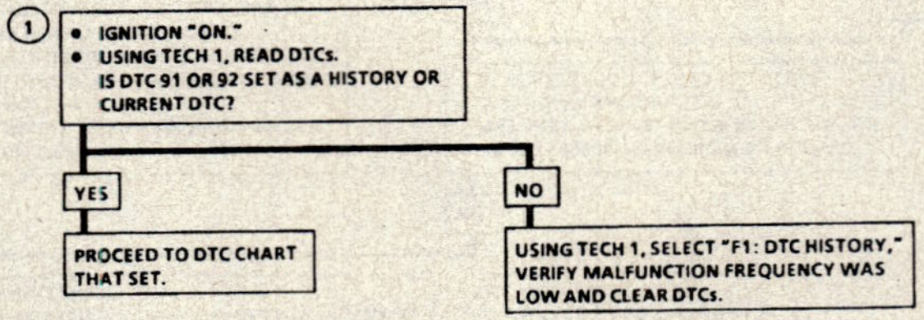

DIAGNOSTIC CHART DTC 94 — 1994 SILHOUETTE, TRANS SPORT AND LUMINA APV

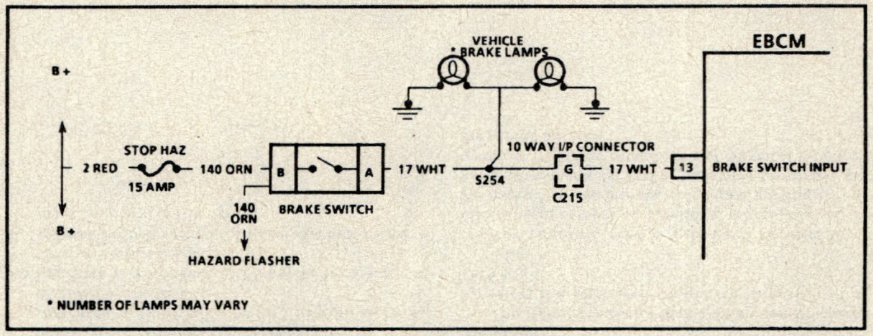

DTC 94

BRAKE SWITCH CONTACTS ALWAYS CLOSED

Circuit Description:

This DTC is run to determine the proper operation of the brake switch. This is important because ABS is activated when the brake switch is "ON" and turned "OFF" when the brake is "OFF." If the brake switch is always "ON," ABS operation will always be requested resulting in potential modulator cycling on rough roads. Additionally, this malfunction will most likely result in a dead battery (due to the brake lights staying "ON") if the driver is not informed of this malfunction.

DTC Will Set When: DTC 94 can be set when the vehicle reaches at least 40 km/h (25 mph). If the brake was never "OFF" during two consecutive drive cycles, a malfunction exists.

Action Taken: A malfunction DTC is stored. ABS is not disabled; however, the ABS warning will flash to indicate a malfunction exists.

DTC Chart Test Description: Number(s) below refer to circled number(s) on the diagnostic chart.

1. Checks if the malfunction is currently present.
2. Isolates the cause of the malfunction to either a malfunctioning or misadjusted brake switch, or a short to voltage in the brake switch circuitry.
3. Checks for unwanted voltage on the brake switch input circuit.
4. Ensures malfunction is not due to poor terminal contact or EBCM.

Diagnostic Aids: An "Intermittent" malfunction may be caused by a poor connection, rubbed through wire insulation, or a wire that is broken inside the insulation.

The frequency of the malfunction can be checked by using the enhanced diagnostic function of the Tech 1,

Any circuitry, that is suspected as causing the intermittent complaint, should be thoroughly checked for backed out terminals, improper mating, broken locks, improperly formed or damaged terminals, poor terminal to wiring connections or physical damage to the wiring harness.

DIAGNOSTIC CHART DTC 94 — 1994 SILHOUETTE, TRANS SPORT AND LUMINA APV

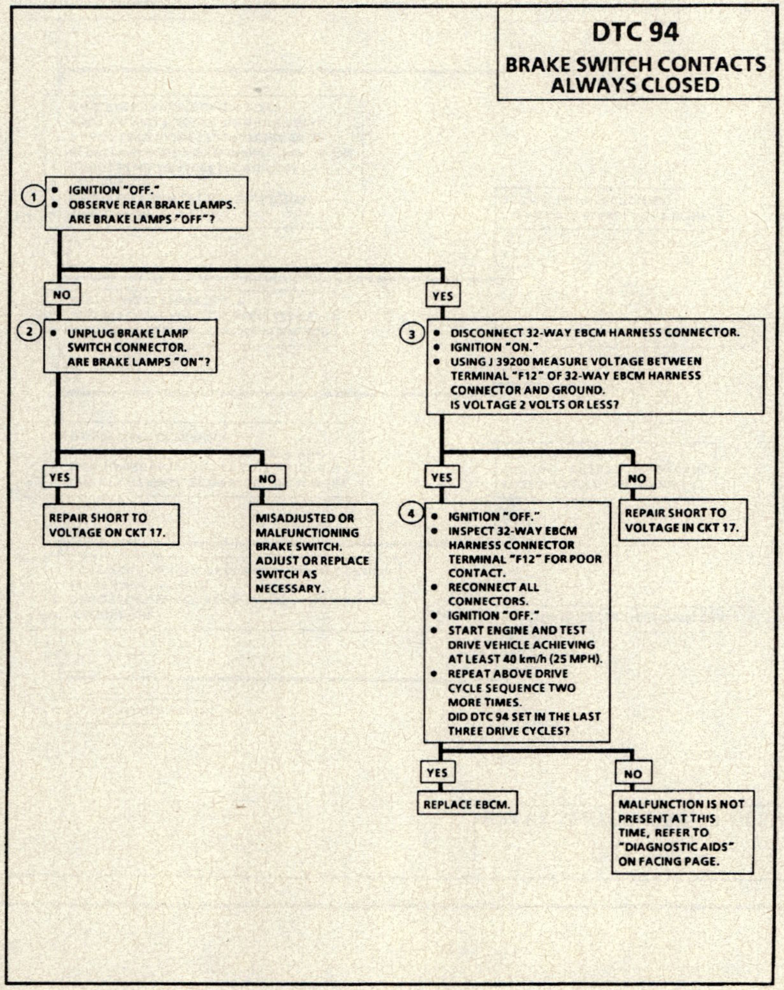

DIAGNOSTIC CHART DTC 95 — 1994 SILHOUETTE, TRANS SPORT AND LUMINA APV

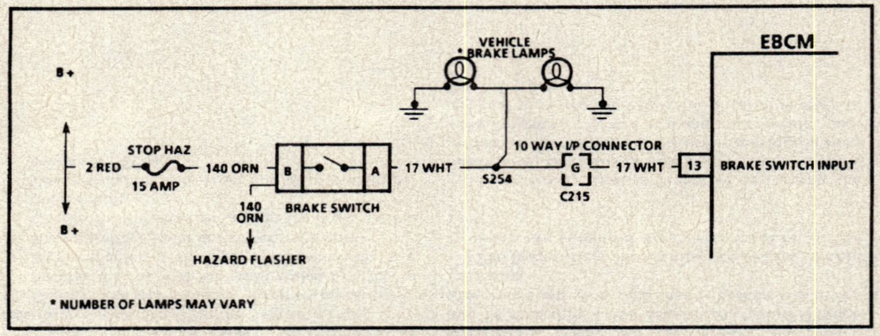

DTC 95

BRAKE SWITCH CIRCUIT OPEN

Circuit Description:
This DTC is used to identify open brake switch circuitry that prevents the brake switch input to the EBCM from changing states when the brake is applied. This DTC is used in conjunction with DTCs 91 and 92 to determine the cause of an open brake switch malfunction.

DTC Will Set When: DTC 95 can be set after initialization is completed. If the brake switch input voltage is out of specification for one second indicating an open circuit, a malfunction exists.

Action Taken: A malfunction DTC is stored, ABS is disabled and the ABS warning lamp is turned "ON."

DTC Chart Test Description: Number(s) below refer to circled number(s) on the diagnostic chart.

1. This step is used to confirm that an open in the brake switch circuitry currently exists.
2. Indicates if the brake switch signal is being received by the EBCM.
3. Indicates if an open circuit exists in the brake switch or brake lamp circuitry.
4. Isolates the open circuit to either the brake switch input circuitry, or the EBCM.

Diagnostic Aids: An "Intermittent" malfunction may be caused by a poor connection, rubbed through wire insulation, or a wire that is broken inside the insulation.

The frequency of the malfunction can be checked by using the enhanced diagnostic function of the Tech 1.

Any circuitry, that is suspected as causing the intermittent complaint, should be thoroughly checked for backed out terminals, improper mating, broken locks, improperly formed or damaged terminals, poor terminal to wiring connections or physical damage to the wiring harness.

DIAGNOSTIC CHART DTC 95 — 1994 SILHOUETTE, TRANS SPORT AND LUMINA APV

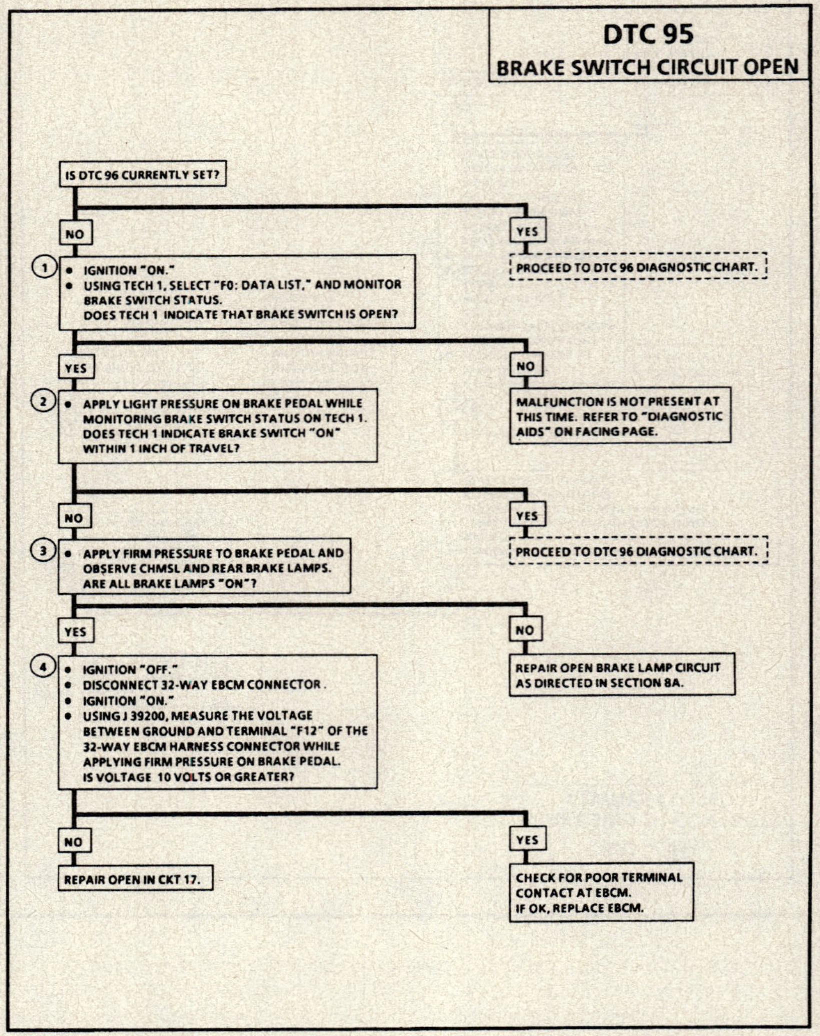

DIAGNOSTIC CHART DTC 96 — 1994 SILHOUETTE, TRANS SPORT AND LUMINA APV

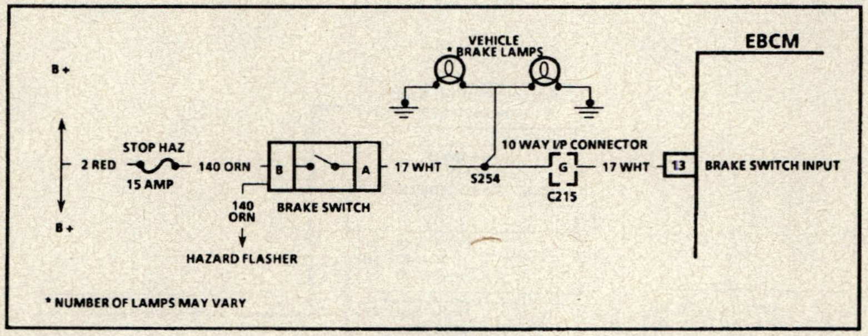

DTC 96

REAR BRAKE LAMP CIRCUIT OPEN

Circuit Description:

This DTC is designed to isolate the cause of a DTC 95 malfunction and indicate to the driver ABS is still available. If DTC 95 occurs with DTC 96, the brake lamps are open, the brake lamp grounds are open, or the Center High Mounted Stop Lamp (CHMSL) circuit is open during 4-way flasher use. The presence of battery voltage at the brake switch input indicates a valid brake switch input is still available.

DTC Will Set When: DTC 96 can be set only after DTC 95 has been set. If the EBCM detects battery voltage on the brake switch input circuit for 0.5 second, a malfunction exists.

Action Taken: A malfunction DTC is stored. ABS is not disabled.

DTC Chart Test Description: Number(s) below refer to circled number(s) on the diagnostic chart.

1. As a result of a malfunction of an additional brake switch circuit DTC, this DTC may be set. To ensure proper diagnosis, any additional brake switch DTCs must be repaired first.
2. Identifies if the malfunction is currently present in the brake circuit.

Diagnostic Aids: An "Intermittent" malfunction may be caused by a poor connection, rubbed through wire insulation, or a wire that is broken inside the insulation.

The frequency of the malfunction can be checked by using the enhanced diagnostic function of the Tech 1.

Any circuitry, that is suspected as causing the intermittent complaint, should be thoroughly checked for backed out terminals, improper mating, broken locks, improperly formed or damaged terminals, poor terminal to wiring connections or physical damage to the wiring harness.

DIAGNOSTIC CHART DTC 96 — 1994 SILHOUETTE, TRANS SPORT AND LUMINA APV

DTC 96

REAR BRAKE LAMP CIRCUIT OPEN

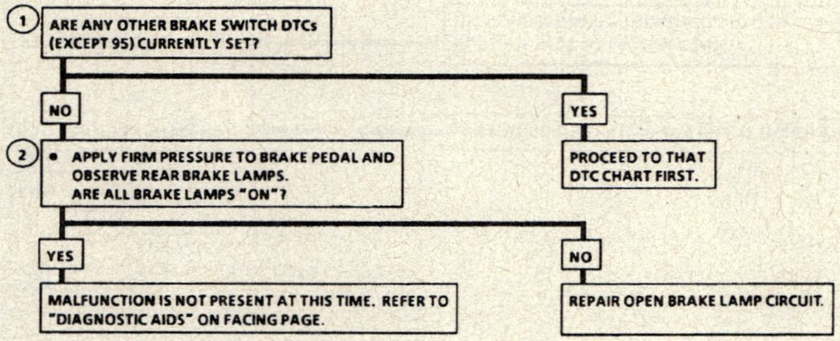

DIAGNOSTIC PROCESS CHART — 1994 SILHOUETTE, TRANS SPORT AND LUMINA APV

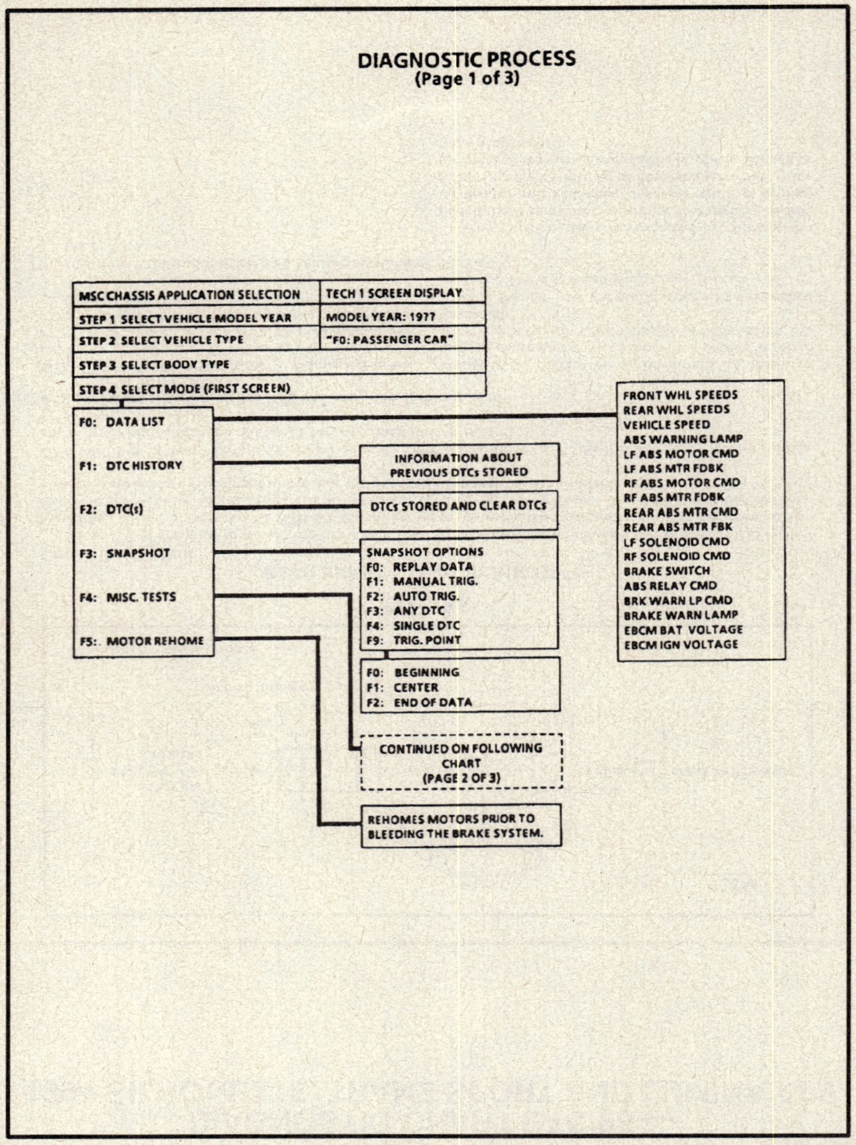

DIAGNOSTIC PROCESS CHART — 1994 SILHOUETTE, TRANS SPORT AND LUMINA APV

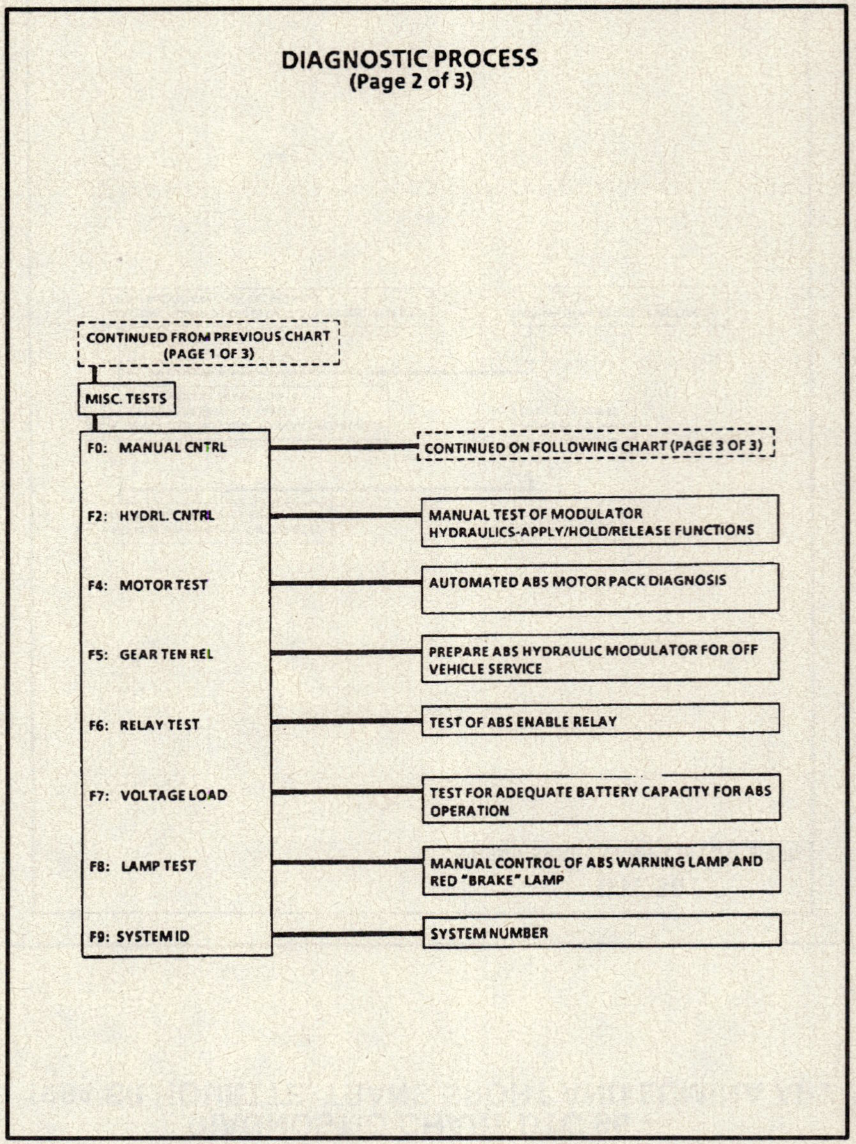

DIAGNOSTIC PROCESS CHART — 1994 SILHOUETTE, TRANS SPORT AND LUMINA APV

DIAGNOSTIC PROCESS
(Page 3 of 3)

CONTINUED FROM PREVIEWS CHART (PAGE 2 OF 3)

MANUAL CONTROL OF SOLENOIDS AND MOTORS.

↑ = TURN ON
↓ = TURN OFF
YES TO CONTINUE

SELECT FUNCTION
F0: LF APPLY
F1: LF RELEASE
F2: LF SOLENOID
F4: RF APPLY
F5: RF RELEASE
F6: RF SOLENOID
F8: REAR APPLY
F9: REAR RELEASE

MOTOR APPLY
COMMAND: XX AMPS
FEEDBACK: XX AMPS
ABS RELAY: ON/OFF

MOTOR RELEASE
COMMAND: XX AMPS
FEEDBACK: XX AMPS
ABS RELAY: ON/OFF

SOLENOID
COMMAND: ON/OFF
ABS RELAY: ON/OFF

DIAGNOSTIC CHART SYSTEM CHECK — 1994 SILHOUETTE, TRANS SPORT AND LUMINA APV

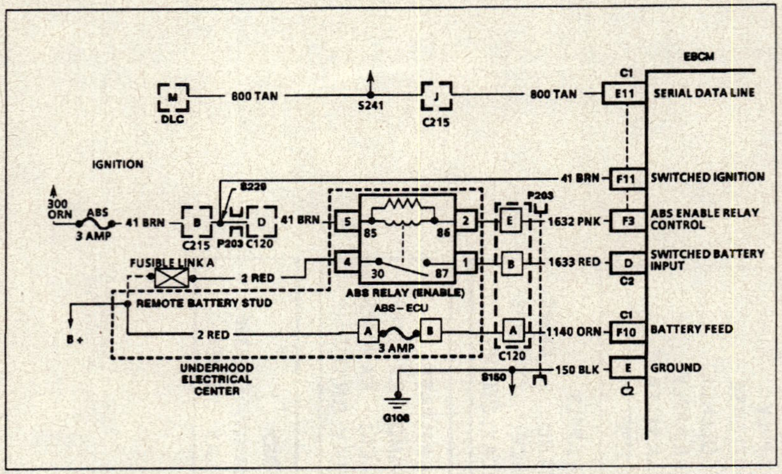

DIAGNOSTIC SYSTEM CHECK
(Page 1 of 4)

System Description:

The diagnostic system check is an organized approach to identifying a problem created by an Antilock Brake System (ABS) malfunction. It must be the starting point for any ABS complaint diagnosis because it directs the service technician to the next logical step in diagnosing the complaint.

Serial data is transmitted by the EBCM through terminal "E11". The EBCM is supplied switched ignition voltage through terminal "F11", battery voltage through terminal "F10" and switched battery voltage through terminal "D". The EBCM ground is provided through terminal "E".

Chart Test Description:

Number(s) below refer to circled number(s) on the diagnostic chart.

1. Indicates if EBCM is transmitting data.
2. Checks for a serial data line malfunction.
3. Checks for proper Tech 1 connection to DLC.
4. Checks for open ABS ECU 3 amp fuse.
5. Checks for open ABS 3 amp fuse.
6. Checks for high resistance in the EBCM ground circuit.
7. Checks for current DTCs.
8. Checks for proper ABS warning lamp operation.
9. Checks for history DTCs.

Diagnostic Aids:

Excessive resistance in the ground or power supply circuits will not allow communication with the EBCM. If communication with the EBCM is not possible, ensure the ABS ground connection is good and that there is no excessive resistance in any of the power supply circuits.

DIAGNOSTIC CHART SYSTEM CHECK — 1994 SILHOUETTE, TRANS SPORT AND LUMINA APV

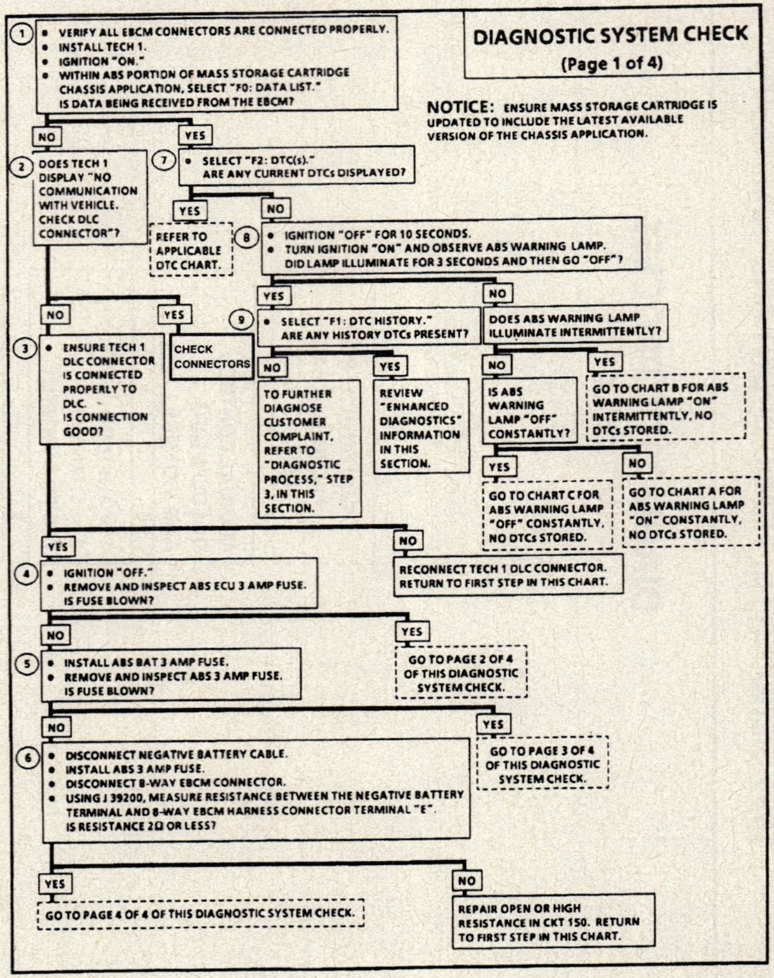

DIAGNOSTIC CHART SYSTEM CHECK — 1995 SILHOUETTE, TRANS SPORT AND LUMINA APV

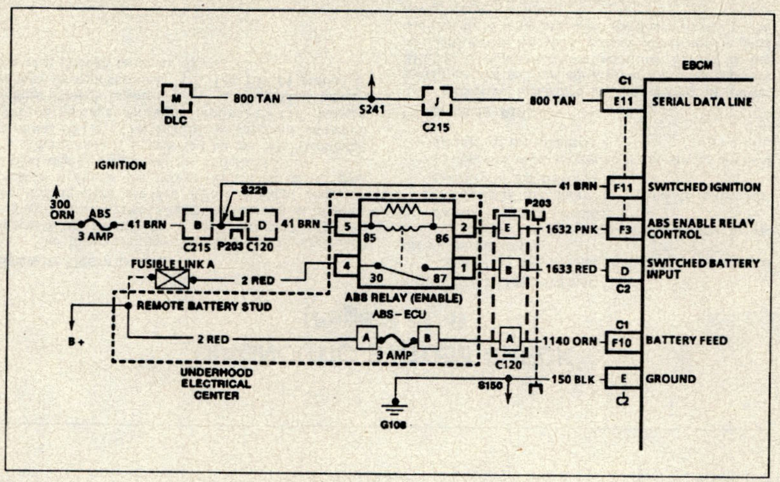

DIAGNOSTIC SYSTEM CHECK (Page 2 of 4)

System Description:

The diagnostic system check is an organized approach to identifying a problem created by an Antilock Brake System (ABS) malfunction. It must be the starting point for any ABS complaint diagnosis because it directs the service technician to the next logical step in diagnosing the complaint.

Serial data is transmitted by the EBCM through terminal "E11". The EBCM is supplied switched ignition voltage through terminal "F11", battery voltage through terminal "F10" and switched battery voltage through terminal "D". The EBCM ground is provided through terminal "E".

Chart Test Description:

Number(s) below refer to circled number(s) on the diagnostic chart.

10. Checks for possible short to ground in the battery feed circuitry.

11. Checks for a open ABS ECU 3 amp fuse

12. Ensures short to ground is not due to physical damage of the circuitry.

Diagnostic Aids:

Excessive resistance in the ground or power supply circuits will not allow communication with the EBCM. If communication with the EBCM is not possible, ensure the ABS ground connection is good and that there is no excessive resistance in any of the power supply circuits.

DIAGNOSTIC CHART SYSTEM CHECK — 1995 SILHOUETTE, TRANS SPORT AND LUMINA APV

DIAGNOSTIC SYSTEM CHECK (Page 2 of 4)

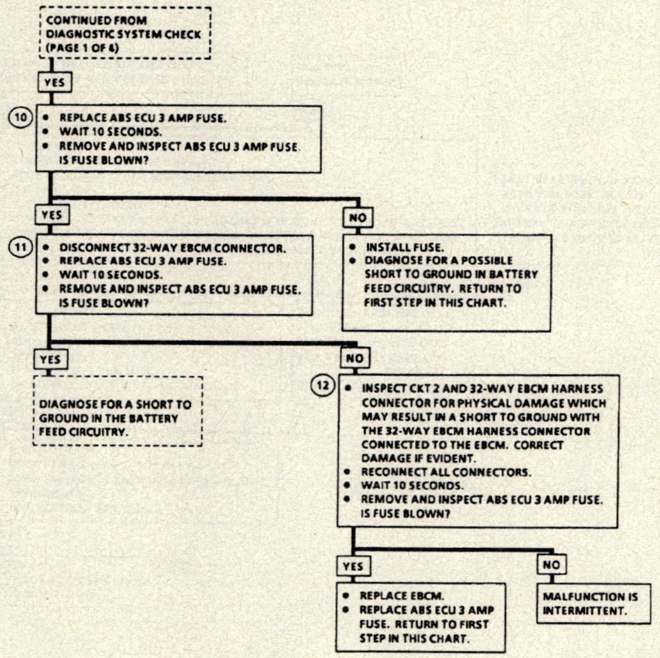

DIAGNOSTIC CHART SYSTEM CHECK — 1995 SILHOUETTE, TRANS SPORT AND LUMINA APV

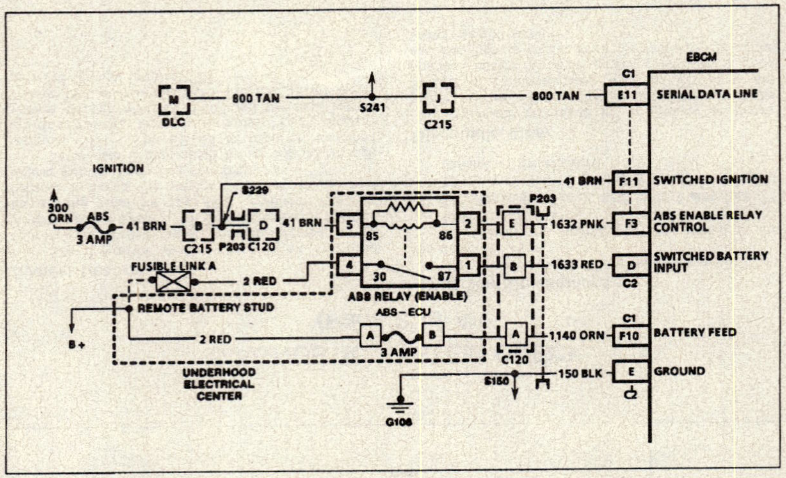

DIAGNOSTIC SYSTEM CHECK
(Page 3 of 4)

System Description:

The diagnostic system check is an organized approach to identifying a problem created by an Antilock Brake System (ABS) malfunction. It must be the starting point for any ABS complaint diagnosis because it directs the service technician to the next logical step in diagnosing the complaint.

Serial data is transmitted by the EBCM through terminal "E11". The EBCM is supplied switched ignition voltage through terminal "F11", battery voltage through terminal "F10" and switched battery voltage through terminal "D". The EBCM ground is provided through terminal "E".

Chart Test Description:

Number(s) below refer to circled number(s) on the diagnostic chart.

13. Checks for possible short to ground in the switched ignition circuitry.

14. Isolates the short to ground to either the circuitry or the EBCM.

15. Ensures short to ground is not due to physical damage of the circuitry.

Diagnostic Aids:

Excessive resistance in the ground or power supply circuits will not allow communication with the EBCM. If communication with the EBCM is not possible, ensure the ABS ground connection is good and that there is no excessive resistance in any of the power supply circuits.

DIAGNOSTIC CHART SYSTEM CHECK — 1995 SILHOUETTE, TRANS SPORT AND LUMINA APV

DIAGNOSTIC SYSTEM CHECK
(Page 3 of 4)

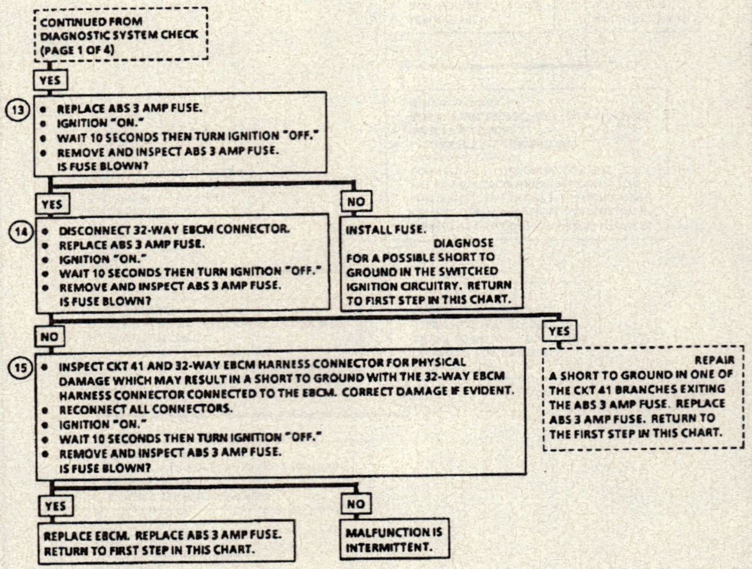

DIAGNOSTIC CHART SYSTEM CHECK — 1995 SILHOUETTE, TRANS SPORT AND LUMINA APV

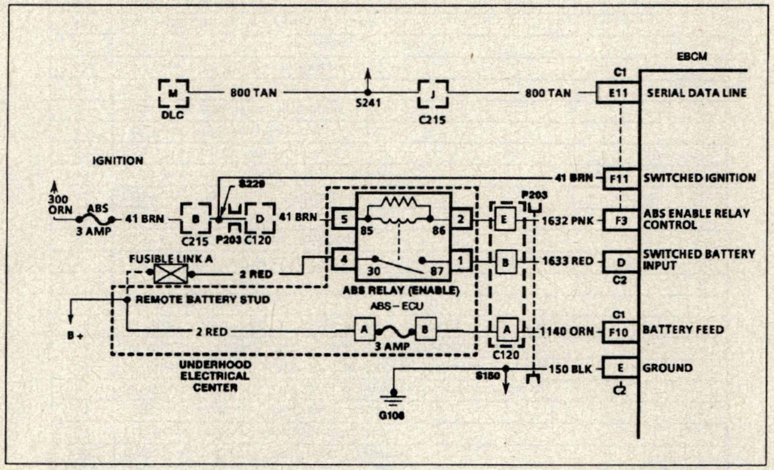

DIAGNOSTIC SYSTEM CHECK
(Page 4 of 4)

System Description:

The diagnostic system check is an organized approach to identifying a problem created by an Antilock Brake System (ABS) malfunction. It must be the starting point for any ABS complaint diagnosis because it directs the service technician to the next logical step in diagnosing the complaint.

Serial data is transmitted by the EBCM through terminal "E11". The EBCM is supplied switched ignition voltage through terminal "F11", battery voltage through terminal "F10" and switched battery voltage through terminal "D". The EBCM ground is provided through terminal "E".

Chart Test Description:

Number(s) below refer to circled number(s) on the diagnostic chart.

16. Checks for open in CKT 2.
17. Checks for high resistance in CKT 2.
18. Checks for high resistance in the switched ignition circuitry.
19. Ensures malfunction is not due to poor terminal contact.
20. Checks for high resistance in the switched ignition circuit between the fuse block and the EBCM.

Diagnostic Aids:

Excessive resistance in the ground or power supply circuits will not allow communication with the EBCM. If communication with the EBCM is not possible, ensure the ABS ground connection is good and that there is no excessive resistance in any of the power supply circuits.

DIAGNOSTIC CHART SYSTEM CHECK — 1995 SILHOUETTE, TRANS SPORT AND LUMINA APV

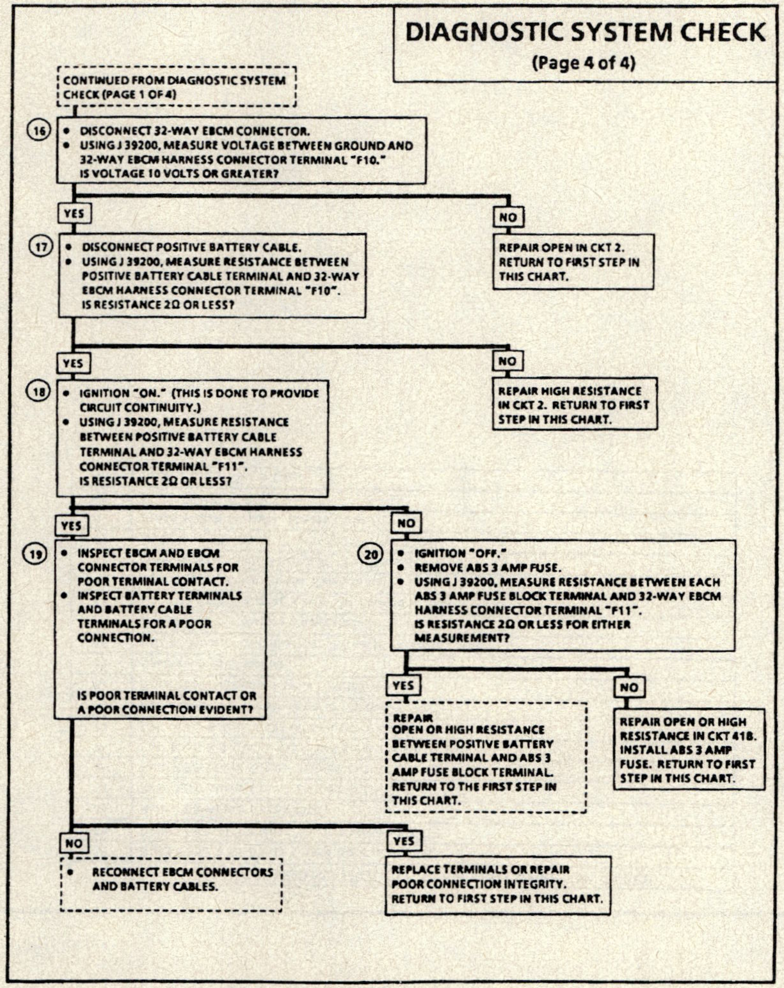

DIAGNOSTIC TROUBLE CODE CHART — 1995 SILHOUETTE, TRANS SPORT AND LUMINA APV

DIAGNOSTIC TROUBLE CODE AND SYMPTOM TABLE	
CHART	**SYMPTOM**
A	ABS (Amber) Warning Lamp "ON" Constantly, No DTCs Stored (Chev Only)
A	ABS (Amber) Warning Lamp "ON" Constantly, No DTCs Stored (Pont/Olds Only)
B	ABS (Amber) Warning Lamp "ON" Intermittently, No DTCs Stored
C	ABS (Amber) Warning Lamp "OFF" Constantly, No DTCs Stored (Chev Only)
D	Tech 1 Displays Undefined DTCs
DTC	**DESCRIPTION**
11	ABS Warning or Low Trac Lamp Circuit Malfunction (Chev Only)
11	ABS Warning or Low Trac Lamp Circuit Malfunction (Pont/Olds Only)
14	ABS Enable Relay Contact Circuit Open (1 of 3)
15	ABS Rly Contact CKT Short to Batt or Always Closed
16	ABS Enable Relay Coil Circuit Open
17	ABS Enable Relay Coil Circuit Shorted to Gnd
18	ABS Enable Relay Coil Circuit Shorted to Batt
21	Left Front Wheel Speed = 0
22	Right Front Wheel Speed = 0
23	Left Rear Wheel Speed = 0
24	Right Rear Wheel Speed = 0
25	Left Front Excessive Wheel Speed Variation
26	Right Front Excessive Wheel Speed Variation
27	Left Rear Excessive Wheel Speed Variation
28	Right Rear Excessive Wheel Speed Variation
32	LF Wheel Speed SNR CKT Open or Short to GND/BATT (1 of 3)
33	RF Wheel Speed SNR CKT Open or Short to GND/BATT (1 of 3)
34	LR Wheel Speed SNR CKT Open or Short to GND/BATT (1 of 3)
35	RR Wheel Speed SNR CKT Open or Short to GND/BATT (1 of 3)
36	Low System Voltage (1 of 2)
37	High System Voltage
38	Left Front ESB Will Not Hold Motor
41	Right Front ESB Will Not Hold Motor
42	Rear ESB Will Not Hold Motor
44	Left Front ABS Channel Will Not Move
45	Right Front ABS Channel Will Not Move
46	Rear ABS Channel Will Not Move
47	Left Front ABS Motor Free Spins (1 of 2)
48	Right Front ABS Motor Free Spins (1 of 2)
51	Rear ABS Motor Free Spins (1 of 2)
52	Left Front ABS Channel in Release Too Long
53	Right Front ABS Channel in Release Too Long
54	Rear ABS Channel in Release Too Long
55	EBCM Malfunction
56	Left Front ABS Motor Circuit Open
57	Left Front ABS Motor Circuit Shorted to GND
58	Left Front ABS Motor Circuit Shorted to BATT
61	Right Front ABS Motor Circuit Open

DIAGNOSTIC TROUBLE CODE CHART — 1995 SILHOUETTE, TRANS SPORT AND LUMINA APV

DIAGNOSTIC TROUBLE CODE AND SYMPTOM TABLE	
62	Right Front ABS Motor Circuit Shorted to GND
63	Right Front ABS Motor Circuit Shorted to BATT
64	Rear ABS Motor Circuit Open
65	Rear ABS Motor Circuit Shorted to GND
66	Rear ABS Motor Circuit Shorted to BATT
76	Left Front Sol Circuit Open or Shorted to BATT
77	Left Front Solenoid Circuit Shorted to GND
78	Right Front Sol Circuit Open or Shorted to BATT
81	Right Front Solenoid Circuit Shorted to GND
82	Calibration Malfunction
86	EBCM Turned "ON" the Red "BRAKE" Warning Lamp
87	Red "BRAKE" Warning Lamp Circuit Open
88	Red "BRAKE" Warning Lamp Ckt Shorted to Batt
91	Open Brake Switch During Deceleration
92	Open Brake Switch When ABS Was Required
93	DTC 91 or 92 Set in Current or Prev. Ign Cycle
94	Brake Switch Contacts Always Closed
95	Brake Switch Circuit Open
96	Brake Lamps Circuit Open

WIRING SCHEMATIC — 1995 SILHOUETTE, TRANS SPORT AND LUMINA APV, 2 OF 2

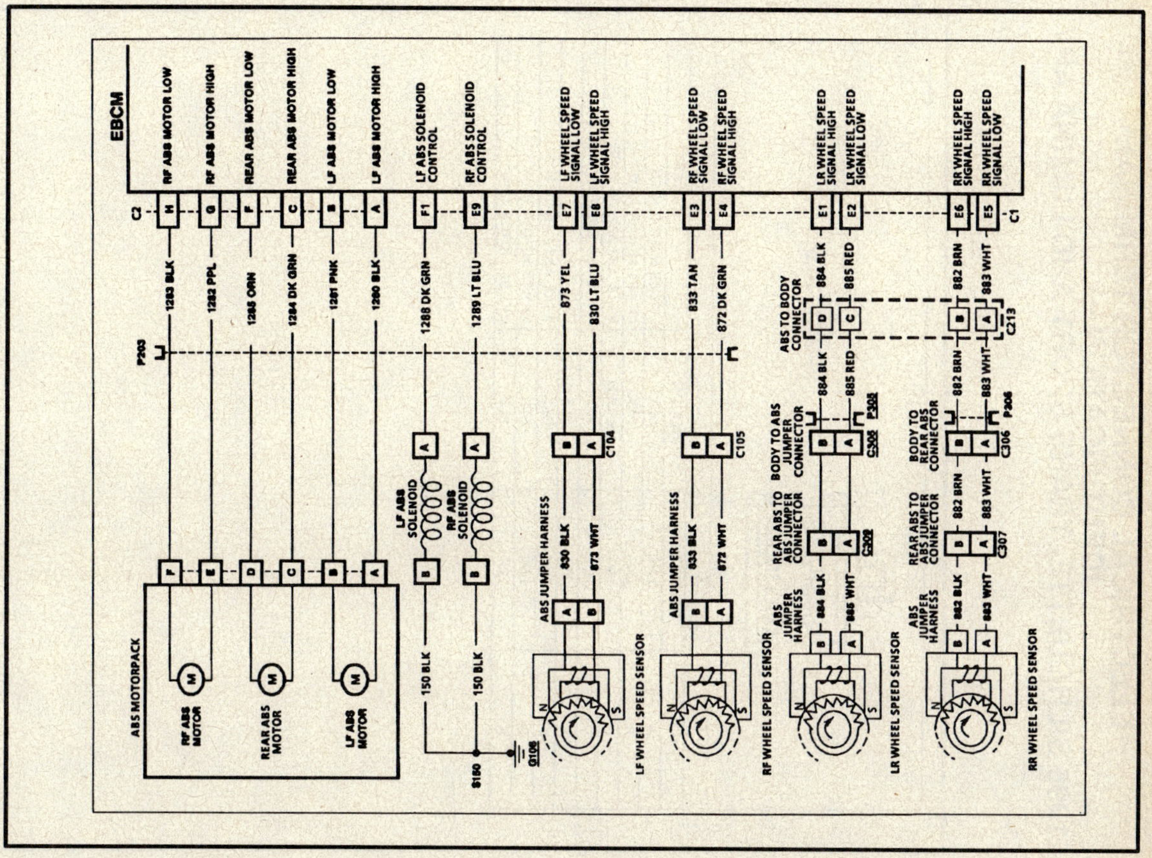

WIRING SCHEMATIC — 1995 SILHOUETTE, TRANS SPORT AND LUMINA APV, 1 OF 2

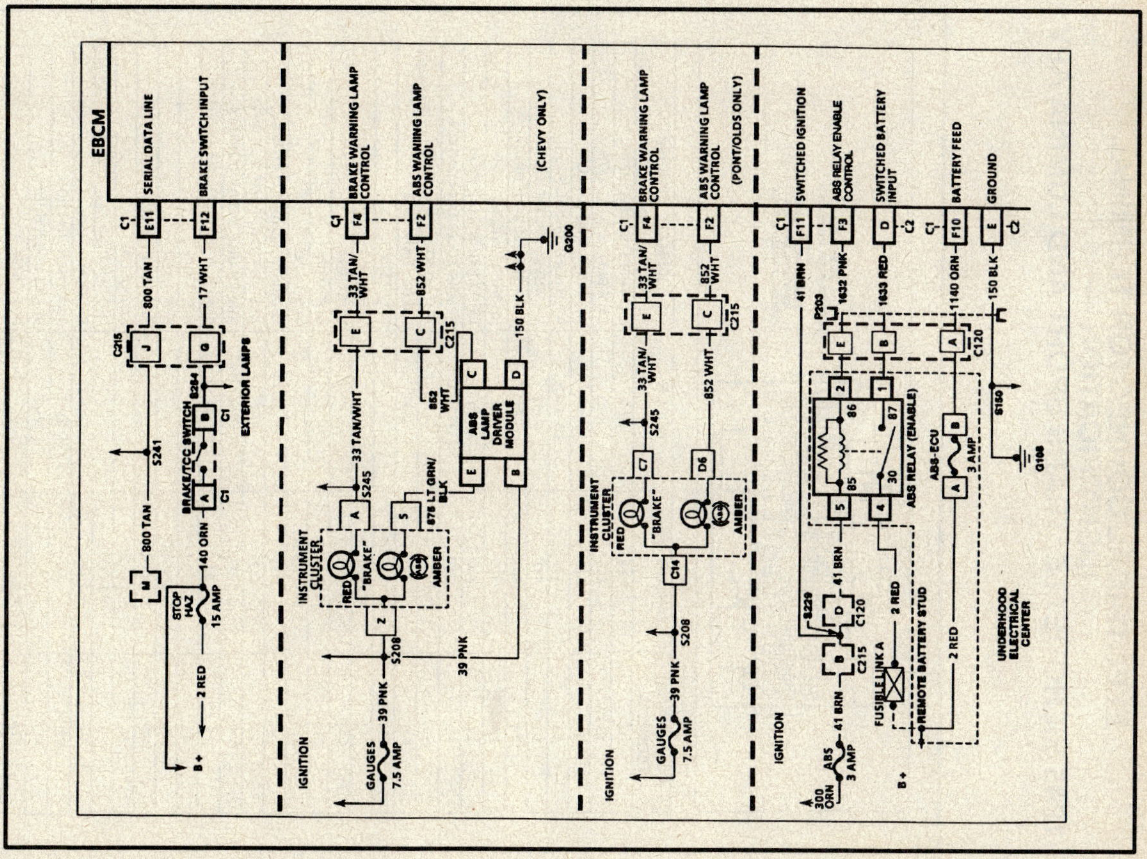

ELECTRICAL CONNECTOR TERMINAL IDENTIFICATION — 1995 SILHOUETTE, TRANS SPORT AND LUMINA APV

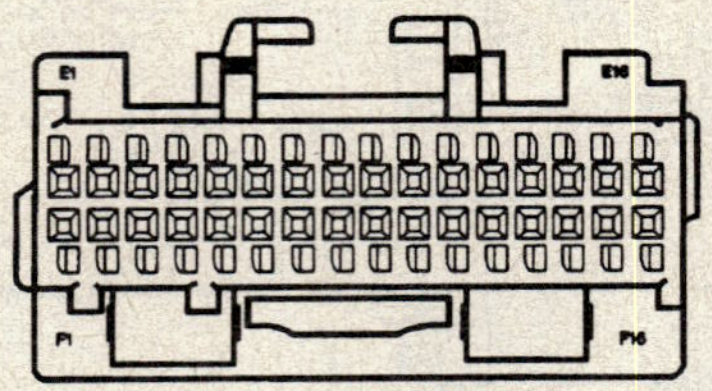

32-WAY F
MICRO-PACK 100 SERIES
BLU
ELECTRONIC BRAKE
CONTROL MODULE
(EBCM) CONNECTOR "C1"

CONNECTOR TERMINAL	CIRCUIT NO.	COLOR	CIRCUIT DESCRIPTION
E1	884	BLK	LR WHEEL SIGNAL HIGH
E2	885	RED	LR WHEEL SIGNAL LOW
E3	833	TAN	RF WHEEL SIGNAL LOW
E4	872	DK GRN	RF WHEEL SIGNAL HIGH
E5	883	WHT	RR WHEEL SIGNAL LOW
E6	882	BRN	RR WHEEL SIGNAL HIGH
E7	873	YEL	LF WHEEL SIGNAL LOW
E8	830	LT BLU	LF WHEEL SIGNAL HIGH
E9	1289	LT BLU	RF ABS SOLENOID CONTROL
E10	PLUGGED		NOT USED
E11	800	TAN	SERIAL DATA LINE
E12	PLUGGED		NOT USED
E13	PLUGGED		NOT USED
E14	PLUGGED		NOT USED
E15	PLUGGED		NOT USED
E16	PLUGGED		NOT USED
F1	1288	DK GRN	LF ABS SOLENOID CONTROL
F2	852	WHT	ABS WARNING LAMP CONTROL
F3	1632	PNK	ABS ENABLE RELAY CONTROL
F4	33	TAN/WHT	BRAKE WARNING LAMP CONTROL
F5	PLUGGED		NOT USED
F6	PLUGGED		NOT USED
F7	PLUGGED		NOT USED
F8	PLUGGED		NOT USED
F9	PLUGGED		NOT USED
F10	1040	ORN	BATTERY
F11	41	BRN	SWITCHED IGNITION
F12	17	WHT	BRAKE SWITCH INPUT
F13	PLUGGED		NOT USED
F14	PLUGGED		NOT USED
F15	PLUGGED		NOT USED
F16	PLUGGED		NOT USED

ELECTRICAL CONNECTOR TERMINAL IDENTIFICATION — 1995 SILHOUETTE, TRANS SPORT AND LUMINA APV

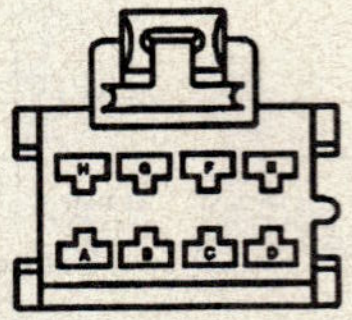

8-WAY F
METRI-PACK 280 SERIES
GRY CONNECTOR "C2"

PIN	CIRCUIT NO	COLOR	CIRCUIT
A	1280	BLK	LF ABS MOTOR HIGH
B	1281	PNK	LF ABS MOTOR LOW
C	1284	DK GRN	REAR ABS MOTOR HIGH
D	1633	RED	SWITCHED BATTERY INPUT
E	150	BLK	GROUND
F	1285	ORN	REAR ABS MOTOR LOW
G	1282	PPL	RF ABS MOTOR HIGH
H	1283	BLK	RF ABS MOTOR LOW

DIAGNOSTIC CHART A — 1995 LUMINA APV

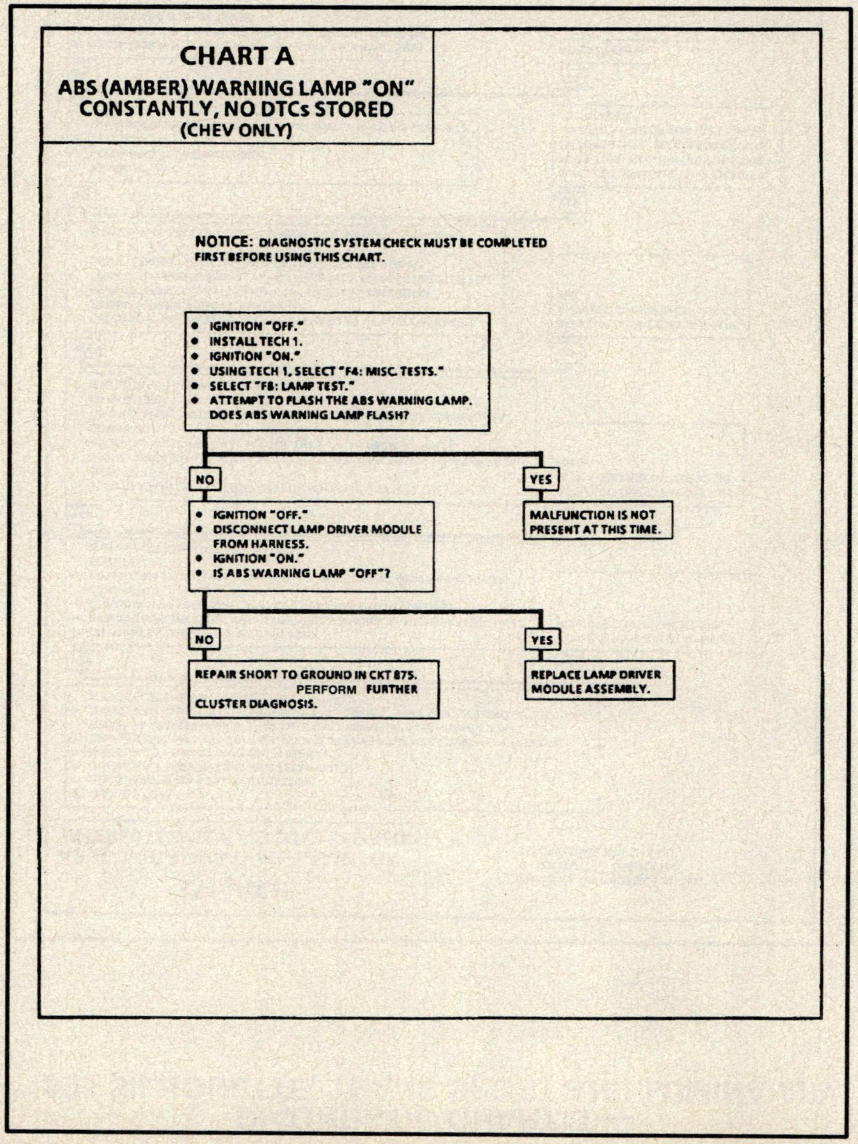

DIAGNOSTIC CHART A — 1995 SILHOUETTE AND TRANS SPORT

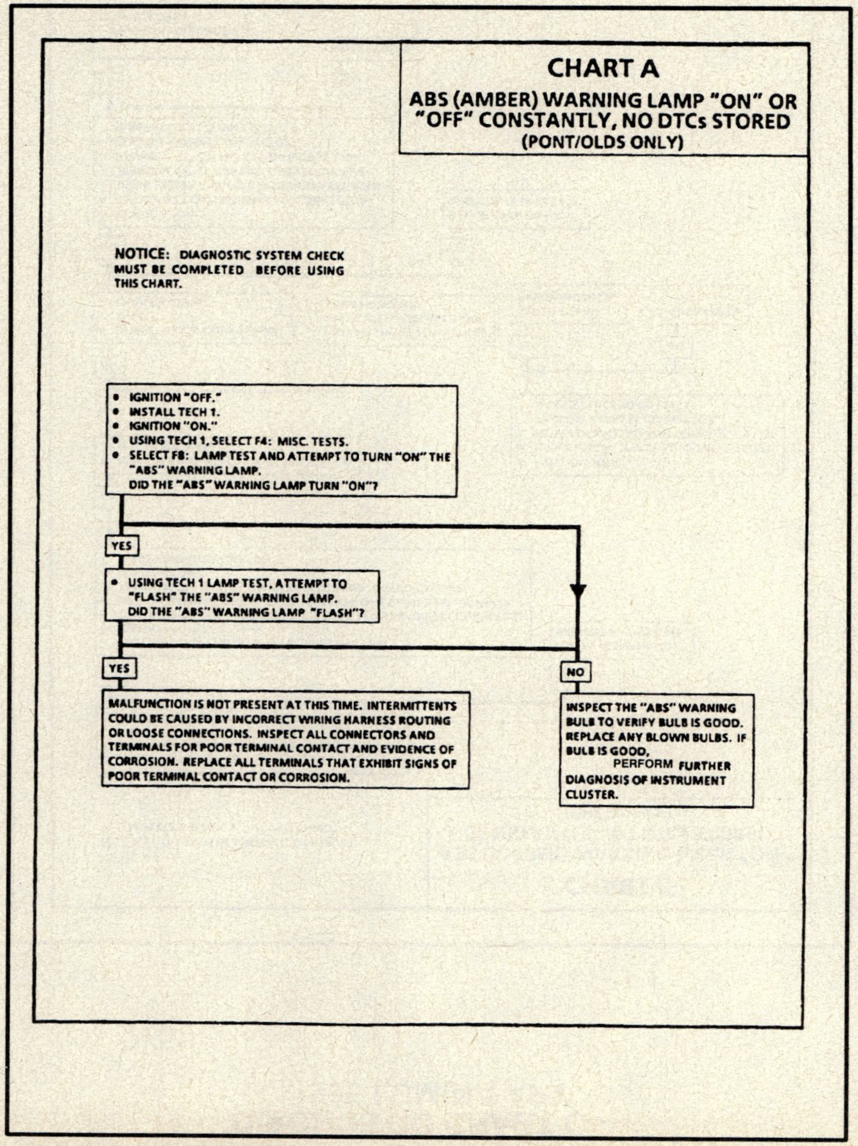

DIAGNOSTIC CHART B — 1995 SILHOUETTE, TRANS SPORT AND LUMINA APV

CHART B

ABS (AMBER) WARNING LAMP "ON" INTERMITTENTLY, NO DTCs STORED

NOTICE: DIAGNOSTIC SYSTEM CHECK MUST BE COMPLETED BEFORE USING THIS CHART.

- IGNITION "OFF."
- DISCONNECT 8-WAY EBCM CONNECTOR.
- DISCONNECT NEGATIVE BATTERY TERMINAL.
- DISCONNECT POSITIVE BATTERY TERMINAL.
- USING J 39200, MEASURE THE RESISTANCE BETWEEN THE NEGATIVE BATTERY CABLE TERMINAL AND TERMINAL "E" OF THE 8-WAY EBCM HARNESS CONNECTOR.
- OBSERVE MEASURED RESISTANCE WHILE "WIGGLING" WIRE HARNESS AND CONNECTORS.

IS RESISTANCE ALWAYS 2Ω OR LESS?

NO: REPAIR INTERMITTENT OPEN OR HIGH RESISTANCE IN CKT 150.

YES:

- DISCONNECT 32-WAY EBCM CONNECTOR.
- USING J 39200, MEASURE THE RESISTANCE BETWEEN POSITIVE BATTERY CABLE TERMINAL AND TERMINAL "F10" OF THE 32-WAY EBCM HARNESS CONNECTOR.
- OBSERVE MEASURED RESISTANCE WHILE "WIGGLING" WIRE HARNESS AND CONNECTORS.

IS RESISTANCE ALWAYS 2Ω OR LESS?

NO: REPAIR INTERMITTENT OPEN OR HIGH RESISTANCE IN CKTs 2 OR 1140.

YES:

- TURN IGNITION SWITCH "ON" (THIS IS DONE TO COMPLETE THE IGNITION CIRCUIT).
- USING J 39200, MEASURE THE RESISTANCE BETWEEN POSITIVE BATTERY CABLE TERMINAL AND TERMINAL "F11" OF THE 32-WAY EBCM HARNESS CONNECTOR.
- OBSERVE MEASURED RESISTANCE WHILE "WIGGLING" WIRE HARNESS AND CONNECTORS.

IS RESISTANCE ALWAYS 2Ω OR LESS?

NO: REPAIR INTERMITTENT OPEN OR HIGH RESISTANCE IN CKT 418 OR 300.

YES:

- CHECK POSITIVE BATTERY TERMINAL AND POSITIVE BATTERY CABLE TERMINAL FOR A POOR CONNECTION.
- INSPECT TERMINALS "F11" AND "F10" OF THE 32-WAY EBCM CONNECTOR AND TERMINAL "E" OF THE 8-WAY EBCM CONNECTOR FOR POOR TERMINAL CONTACT AND EVIDENCE OF CORROSION.

IS THERE EVIDENCE OF POOR TERMINAL CONTACT OR CORROSION?

YES: REPLACE TERMINALS THAT EXHIBIT POOR TERMINAL CONTACT OR SIGNS OF CORROSION. REPAIR CONNECTION INTEGRITY IF POOR CONTACT EXISTS AT BATTERY.

NO:

- IGNITION "OFF."
- RECONNECT ALL CONNECTORS.
- IGNITION "ON."

IS ABS WARNING LAMP "ON" INTERMITTENTLY WITH NO DTCs SET?

YES: REPLACE EBCM.

NO: MALFUNCTION IS NOT PRESENT AT THIS TIME. INTERMITTENTS MAY BE CAUSED BY INCORRECT WIRING HARNESS ROUTING OR LOOSE GROUND CONNECTIONS.

DIAGNOSTIC CHART C — 1995 LUMINA APV

CHART C

ABS (AMBER) WARNING LAMP "OFF" CONSTANTLY, NO DTCs STORED (CHEV ONLY)

NOTICE: DIAGNOSTIC SYSTEM CHECK MUST BE COMPLETED BEFORE USING THIS CHART.

- IGNITION "ON," ENGINE "OFF."
- USING TECH 1, SELECT LAMP TEST AND COMMAND THE ABS WARNING LAMP "ON."

IS ABS WARNING LAMP "ON"?

YES: MALFUNCTION IS NOT PRESENT AT THIS TIME.

NO:

NOTICE: DTC 11 MAY BE SET DURING THIS TEST.

- IGNITION "OFF."
- DISCONNECT LAMP DRIVER MODULE (LDM) FROM HARNESS CONNECTOR.
- WITH A FUSED JUMPER, SUCH AS J 36169 WITH A 3 AMP FUSE, CONNECT TERMINAL "E" OF THE LDM HARNESS CONNECTOR TO GROUND.
- IGNITION "ON."

IS ABS WARNING LAMP "ON"?

YES:

- IGNITION "OFF."
- USING J 39200, MEASURE THE RESISTANCE BETWEEN GROUND AND TERMINAL "D" OF THE LDM HARNESS CONNECTOR.

IS RESISTANCE 2Ω OR LESS?

- YES: REPLACE LAMP DRIVER MODULE.
- NO: REPAIR OPEN IN CKT 150.

NO:

- REMOVE AND INSPECT JUMPER WIRE FUSE.

IS FUSE BLOWN?

YES: REPAIR SHORT TO VOLTAGE IN CKT 875. REPLACE LAMP DRIVER MODULE.

NO:

- INSPECT ABS WARNING LAMP BULB.

IS BULB BLOWN?

YES: REPAIR OPEN OR HIGH RESISTANCE IN CKT 875.

NO:

- IGNITION "OFF."
- DISCONNECT CLUSTER HARNESS CONNECTOR.
- USING J 39200, MEASURE RESISTANCE BETWEEN TERMINAL "E" OF THE LDM CONNECTOR AND TERMINAL "S" ON THE ABS WARNING LAMP CLUSTER HARNESS CONNECTOR.

IS RESISTANCE 2Ω OR LESS?

- YES: REPLACE BULB.
- NO: PERFORM DIAGNOSIS OF I/P CLUSTER.

DIAGNOSTIC CHART D — 1995 SILHOUETTE, TRANS SPORT AND LUMINA APV

CHART D
TECH 1 DISPLAYS UNDEFINED DTCs

NOTICE: USE THIS CHART IF ANY OF THE FOLLOWING UNDEFINED DTCs ARE SET - 12, 13, 31, 43, 67, 68, 71, 72, 73, 74, 75, 83, 84, 85, 97, OR 98.

DIAGNOSTIC CHART DTC 11 — 1995 LUMINA APV

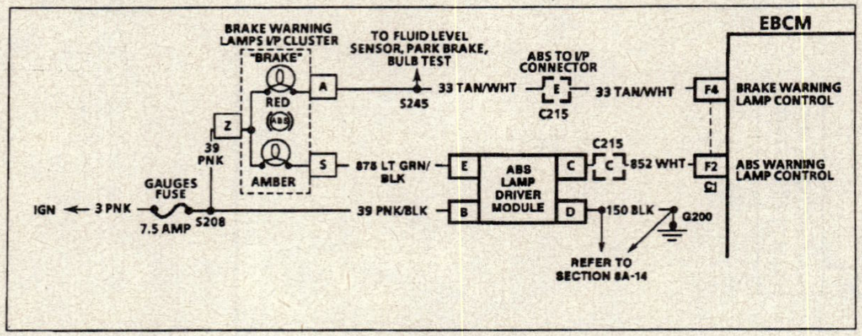

DTC 11
ABS WARNING LAMP CIRCUIT MALFUNCTION (CHEV ONLY)

Circuit Description:

This DTC checks the state of the amber ABS warning lamp to identify a situation in which the driver could not be warned of a system malfunction by the amber ABS warning lamp, or the warning lamp is always "ON." The ABS Lamp Driver Module (LDM) turns the amber ABS warning lamp "ON," unless the EBCM provides ground to turn it "OFF." Because of the circuitry in the LDM, only external malfunctions can be detected. As a result, the amber ABS warning lamp itself is not diagnosable, only the control line to the EBCM can be diagnosed. In the event of an open CKT 852, the amber ABS warning lamp will be "ON" at all times, due to the loss of ground at the LDM input. If the control line is shorted to ground, the amber ABS warning lamp is kept "OFF," due to the LDM input being grounded.

DTC Will Set When:

DTC 11 can be set only during the three second bulb check or when the amber ABS warning lamp is commanded "ON." If the EBCM cannot control the amber ABS warning lamp for two seconds, a malfunction exists.

Action Taken:

A malfunction DTC is stored, however, ABS is not disabled.

DTC Chart Test Description:

Number(s) below refer to circled number(s) on the diagnostic chart.

1. Determines if the ABS warning lamp can be properly controlled.
2. Determines if the ABS warning lamp is "ON" constantly.
3. Checks for other DTCs which could turn on the ABS warning lamp.
4. Determines if the ABS warning lamp can be turned "OFF" manually.
5. Checks the integrity of the ABS warning lamp circuitry.
6. Checks for a short to voltage on the ABS warning lamp circuitry.
7. Determines if a malfunctioning EBCM is the cause of DTC 11.
8. Verifies that the other I/P indicator lamps function properly.
9. Determines if the ABS warning lamp can be turned "ON" manually.
10. Checks for a short to ground in the ABS warning lamp circuitry.
11. Checks for a open between "GAUGES" fuse and LDM.
12. Verifies condition of the 7.5 amp "GAUGES" Fuse.
13. Determines if DTC 11 was set due to an EBCM malfunction.

Diagnostic Aids:

An "Intermittent" malfunction may be caused by a poor connection, rubbed through wire insulation, or a wire that is broken inside the insulation.

The lamp test function of the Tech 1 may be used to command the lamp "ON" while looking for an intermittent malfunction in the ABS warning lamp circuitry.

The frequency of the malfunction can be checked by using the enhanced diagnostic function of the Tech 1

Any circuitry that is suspected as causing the intermittent complaint should be thoroughly checked for backed out terminals, improper mating, broken locks, improperly formed or damaged terminals, poor terminal to wiring connections or physical damage to the wiring harness.

DIAGNOSTIC CHART DTC 11 — 1995 LUMINA APV

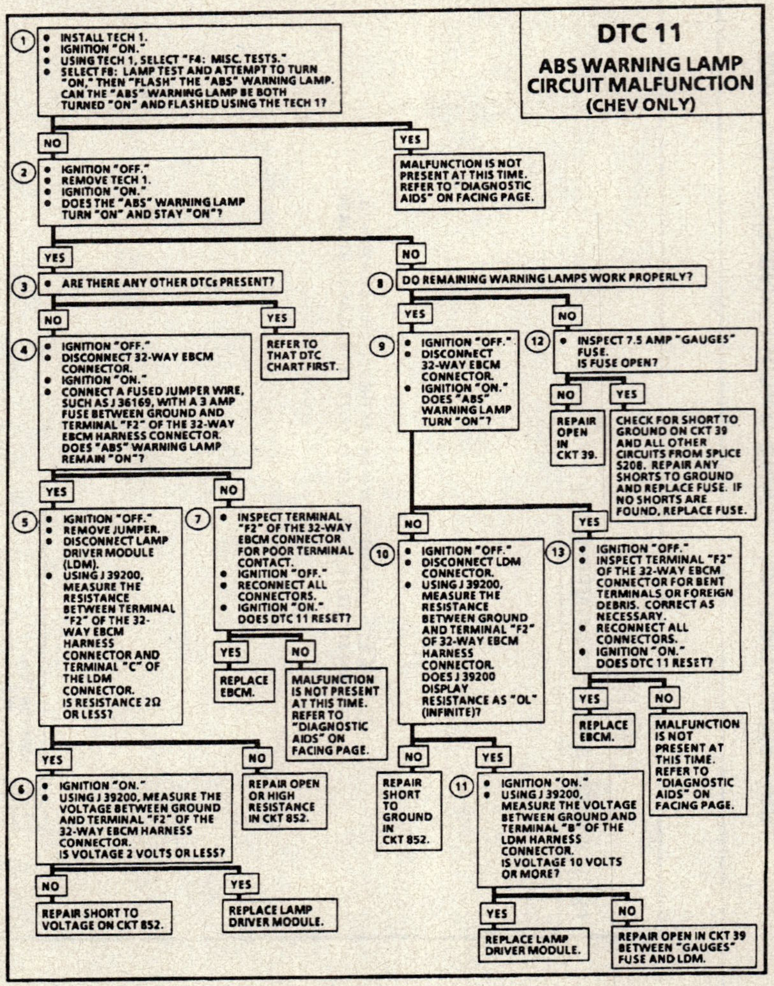

DIAGNOSTIC CHART DTC 11 — 1995 SILHOUETTE, TRANS SPORT AND LUMINA APV

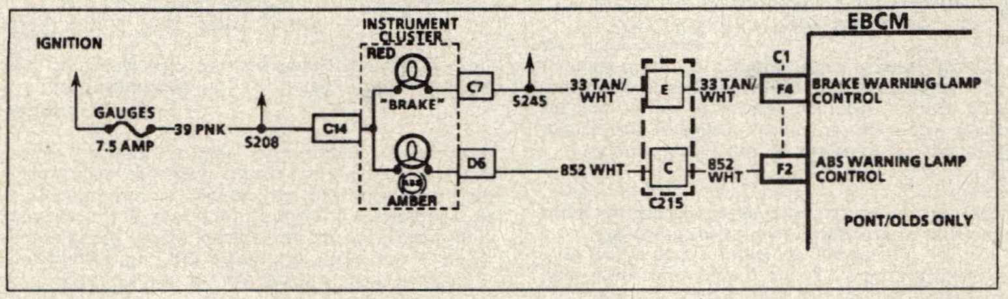

DTC 11
ABS WARNING LAMP CIRCUIT MALFUNCTION
(PONT/OLDS ONLY)

Circuit Description:

This DTC checks the state of both the amber ABS warning lamp and the red BRAKE warning lamp to identify a situation in which the driver could not be alerted to a system malfunction or of possible slippery driving conditions. Due to an integral lamp driver module within the instrument panel cluster, the EBCM must provide a ground to turn the amber ABS warning lamp "OFF." The red BRAKE warning lamp circuit is such that the EBCM provides a ground to turn "ON" the red BRAKE warning lamp.

DTC Will Set When:

DTC 11 can be set only during the three second bulb check or when the amber ABS warning lamp is commanded "ON." If the EBCM cannot control the amber ABS warning lamp for two seconds, a malfunction exists.

Action Taken:

A malfunction DTC is stored, however, ABS is not disabled.

DTC Chart Test Description:

Number(s) below refer to circled number(s) on the diagnostic chart.

1. Determines if the ABS warning lamp can be properly controlled.
2. Determines if the lamp is "ON" constantly.
3. Checks for other DTCs which could turn "ON" the ABS warning lamp.
4. Determines if the ABS warning lamp can be turned "OFF" manually.
5. Checks the integrity of the ABS warning lamp circuitry.
6. Checks for a short to voltage on the ABS warning lamp circuitry.
7. Determines if a malfunctioning EBCM is the cause of DTC 11.
8. Verifies that the other I/P indicator lamps function properly.
9. Determines if the ABS warning lamp can be turned "ON" manually.
10. Checks for a short to ground in the ABS warning lamp circuitry.
11. Verifies condition of the 7.5 amp "GAUGES" Fuse .
12. Determines if DTC 11 was set due to an EBCM malfunction.

Diagnostic Aids:

An "Intermittent" malfunction may be caused by a poor connection, rubbed through wire insulation, or a wire that is broken inside the insulation.

The lamp test function of the Tech 1 may be used to command the lamp "ON" while looking for an intermittent malfunction in the ABS warning lamp circuitry.

The frequency of the malfunction can be checked by using the enhanced diagnostic function of the Tech 1.

Any circuitry that is suspected of causing the intermittent complaint should be thoroughly checked for backed out terminals, improper mating, broken locks, improperly formed or damaged terminals, poor terminal to wiring connections or physical damage to the wiring harness.

DIAGNOSTIC CHART DTC 11 — 1995 SILHOUETTE, TRANS SPORT AND LUMINA APV

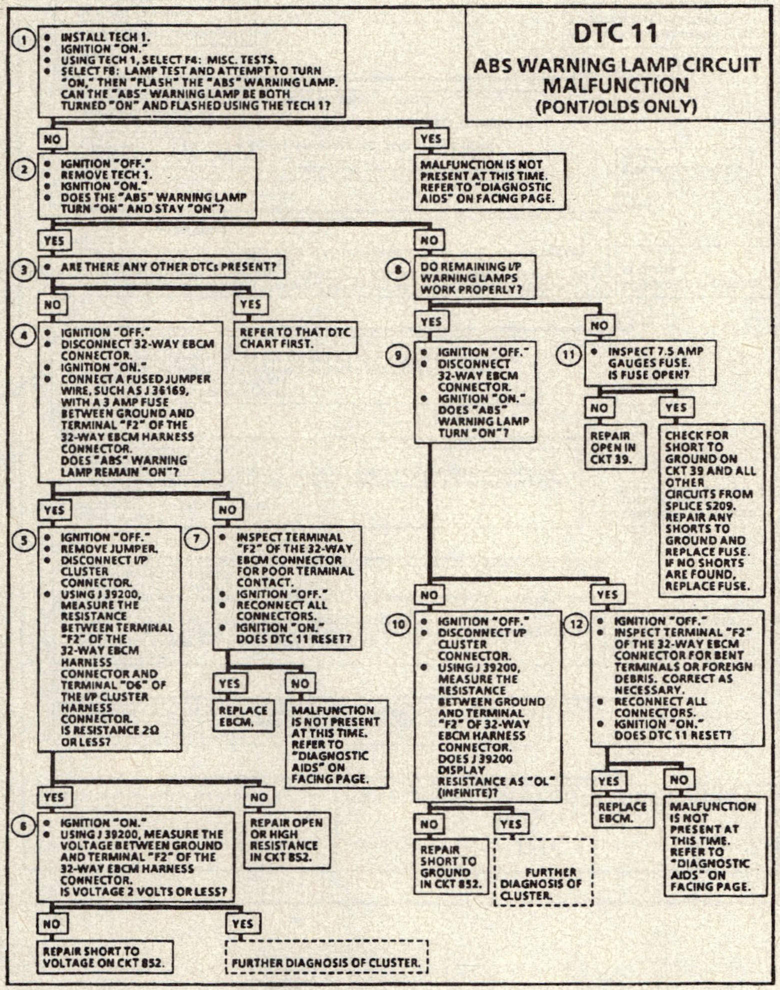

DIAGNOSTIC CHART DTC 14 — 1995 SILHOUETTE, TRANS SPORT AND LUMINA APV

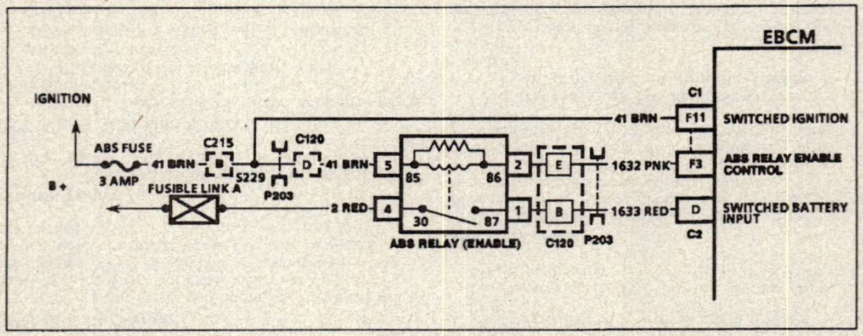

DTC 14
(Page 1 of 3)
ABS ENABLE RELAY CONTACT CIRCUIT OPEN

Circuit Description:

Ignition voltage is supplied through terminal "5" of the ABS relay. The EBCM then is able to energize the pull-in coil by completing the ground circuit at terminal "F3" of the EBCM. The magnetic field created closes the ABS relay contacts and allows battery voltage and current to be supplied to the EBCM, which supplies power to the motors and solenoids.

DTC Will Set When:

DTC 14 can be set anytime after the EBCM commands the ABS relay "ON" (the relay is first commanded "ON" during the three second bulb check). This test monitors the availability of current/voltage to the motors and solenoids. This malfunction indicates voltage is not available and would therefore not allow ABS operation if required.

Action Taken:

A malfunction DTC is stored, ABS is disabled and the amber ABS warning lamp is turned "ON."

DTC Chart Test Description:

Number(s) below refer to circled number(s) on the diagnostic chart.

1. Checks to see if DTC 16 is also set as current or history. If DTC 16 is also set, proceed to that chart first.
2. Verifies that the malfunction is currently present.
3. Verifies that the relay can be turned "ON" and provide voltage to the EBCM.
4. Checks for high resistance in the switched battery circuit from the relay to the EBCM.
5. Checks for high resistance in the switched battery circuit from the battery to the relay.
6. Verifies the condition of the relay.
7. Checks for poor terminal contact or corrosion at the connectors.
8. Determines if the EBCM is the cause of the malfunction.

Diagnostic Aids:

An "Intermittent" problem may be caused by a poor connection, rubbed through wire insulation, or a wire that is broken inside the insulation.

The frequency of the problem can be checked by using the enhanced diagnostic function of the Tech 1.

Any circuitry that is suspected of causing the intermittent complaint should be thoroughly checked for backed out terminals, improper mating, broken locks, improperly formed or damaged terminals, poor terminal to wiring connections or physical damage to the wiring harness.

VIBRATION, TEMPERATURE EFFECTS:

Check for vibration effects by performing the relay test function of the Tech 1. With the relay test commanded "ON," lightly tap the top and sides of the relay while monitoring relay voltage. If the relay voltage changes significantly, replace the relay.

If DTC 14 only sets when the vehicle is initially started in cold ambient conditions (temperature less than 32° F/0° C), replace the relay.

DIAGNOSTIC CHART DTC 14 — 1995 SILHOUETTE, TRANS SPORT AND LUMINA APV

DTC 14
(Page 1 of 3)
ABS ENABLE RELAY CONTACT CIRCUIT OPEN

① IS DTC 16 ALSO PRESENT IN HISTORY?
- YES: PROCEED TO DTC 16 CHART FIRST.
- NO:

②
- IGNITION "ON," ENGINE "OFF."
- USING TECH 1, SELECT DATA LIST AND OBSERVE EBCM BATTERY VOLTAGE.

DOES TECH 1 INDICATE THAT EBCM BATTERY VOLTAGE IS GREATER THAN 10 VOLTS?
- YES: MALFUNCTION IS NOT PRESENT AT THIS TIME. REFER TO "DIAGNOSTIC AIDS" ON FACING PAGE.
- NO:

③
- IGNITION "OFF."
- DISCONNECT 32-WAY AND 8-WAY EBCM CONNECTORS.
- CONNECT A FUSED JUMPER, SUCH AS J 36169 WITH A 3 AMP FUSE BETWEEN GROUND AND TERMINAL "F3" OF THE 32-WAY EBCM HARNESS CONNECTOR.
- IGNITION "ON."
- USING J 39200, MEASURE THE VOLTAGE BETWEEN GROUND AND TERMINAL "D" OF THE 8-WAY EBCM HARNESS CONNECTOR.

IS VOLTAGE 10 VOLTS OR GREATER?
- NO: DIAGNOSIS CONTINUED ON NEXT PAGE.
- YES:

④
- IGNITION "OFF."
- DISCONNECT JUMPER USED IN STEP 3.
- DISCONNECT ABS ENABLE RELAY.
- USING J 39200, MEASURE RESISTANCE BETWEEN TERMINAL "1" OF THE ABS ENABLE RELAY HARNESS CONNECTOR AND TERMINAL "D" OF THE 8-WAY EBCM HARNESS CONNECTOR.

IS RESISTANCE 2Ω OR LESS?
- NO: REPAIR HIGH RESISTANCE IN CKT 1633.
- YES:

⑤
- DISCONNECT NEGATIVE BATTERY CABLE.
- DISCONNECT POSITIVE BATTERY CABLE.
- USING J 39200, MEASURE THE RESISTANCE BETWEEN POSITIVE BATTERY CABLE TERMINAL AND TERMINAL "4" OF THE ABS ENABLE RELAY HARNESS CONNECTOR.

IS RESISTANCE 2Ω OR LESS?
- NO: REPAIR HIGH RESISTANCE IN CKT 2B.
- YES:

⑥
- USING A JUMPER WIRE, CONNECT TERMINAL "86" ON THE ABS ENABLE RELAY TO GROUND.
- USING A FUSED JUMPER WIRE, SUCH AS J 36169, WITH A 3 AMP FUSE, CONNECT TERMINAL "85" ON THE ABS ENABLE RELAY TO VOLTAGE.
- USING J 39200, MEASURE THE RESISTANCE BETWEEN TERMINALS "30" AND "87" ON THE ABS ENABLE RELAY.

IS RESISTANCE 2Ω OR LESS?
- NO: REPLACE ABS ENABLE RELAY.
- YES:

⑦
- INSPECT ABS ENABLE RELAY AND ABS ENABLE RELAY HARNESS CONNECTOR FOR A POOR CONNECTION OR POOR TERMINAL CONTACT.
- INSPECT EBCM AND EBCM CONNECTOR TERMINALS "F3" AND "D" FOR POOR TERMINAL CONTACT.
- INSPECT POSITIVE BATTERY TERMINAL AND POSITIVE BATTERY CABLE TERMINAL FOR A POOR CONNECTION.

IS POOR TERMINAL CONTACT OR A POOR CONNECTION EVIDENT?
- YES: REPLACE TERMINALS THAT EXHIBIT POOR TERMINAL CONTACT OR EVIDENCE OF CORROSION.
- NO:

⑧
- RECONNECT ALL CONNECTORS.
- IGNITION "ON."

DOES DTC 14 RESET?
- YES: REPLACE EBCM.
- NO: MALFUNCTION IS INTERMITTENT. REFER TO "DIAGNOSTIC AIDS" ON FACING PAGE.

DIAGNOSTIC CHART DTC 14 — 1995 SILHOUETTE, TRANS SPORT AND LUMINA APV

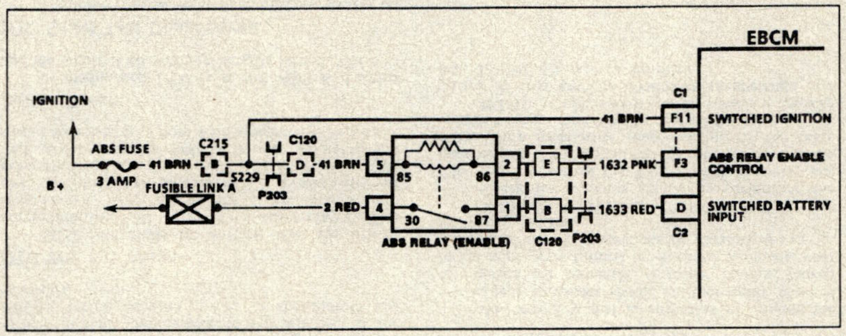

DTC 14
(Page 2 of 3)
ABS ENABLE RELAY CONTACT CIRCUIT OPEN

Circuit Description:

Ignition voltage is supplied through terminal "5" of the ABS relay. The EBCM then is able to energize the pull-in coil by completing the ground circuit at terminal "F3" of the EBCM. The magnetic field created closes the ABS relay contacts and allows battery voltage and current to be supplied to the EBCM, which supplies power to the motors and solenoids.

DTC Will Set When:

DTC 14 can be set anytime after the EBCM commands the ABS relay "ON" (the relay is first commanded "ON" during the three second bulb check). This test monitors the availability of current/voltage to the motors and solenoids. This malfunction indicates voltage is not available and would therefore not allow ABS operation if required.

Action Taken:

A malfunction DTC is stored, ABS is disabled and the amber ABS warning lamp is turned "ON."

DTC Chart Test Description:

Number(s) below refer to circled number(s) on the diagnostic chart.

9. Checks for high resistance in the entire relay coil circuit.
10. Verifies proper battery condition.
11. This step verifies proper voltage is available at the ABS relay.
12. Checks for an open in the switched battery circuit from the relay to the EBCM.
13. Verifies the condition of the ABS relay.

Diagnostic Aids:

An "Intermittent" problem may be caused by a poor connection, rubbed through wire insulation, or a wire that is broken inside the insulation.

The frequency of the problem can be checked by using the enhanced diagnostic function of the Tech 1,

Any circuitry that is suspected of causing the intermittent complaint should be thoroughly checked for backed out terminals, improper mating, broken locks, improperly formed or damaged terminals, poor terminal to wiring connections or physical damage to the wiring harness.

VIBRATION, TEMPERATURE EFFECTS:

Check for vibration effects by performing the relay test function of the Tech 1. With the relay test commanded "ON," lightly tap the top and sides of the relay while monitoring relay voltage. If the relay voltage changes significantly, replace the relay.

If DTC 14 only sets when the vehicle is initially started in cold ambient conditions (temperature less than 32° F/0° C), replace the relay.

DIAGNOSTIC CHART DTC 14 — 1995 SILHOUETTE, TRANS SPORT AND LUMINA APV

DTC 14
(Page 2 of 3)
ABS ENABLE RELAY CONTACT CIRCUIT OPEN

DIAGNOSIS CONTINUED FROM PREVIOUS PAGE.

(9)
- IGNITION "OFF."
- REMOVE JUMPER WIRE INSTALLED IN STEP #3 IF NOT ALREADY REMOVED.
- DISCONNECT NEGATIVE BATTERY CABLE.
- DISCONNECT POSITIVE BATTERY CABLE.
- WITH BATTERY STILL DISCONNECTED, TURN IGNITION SWITCH TO "RUN." THIS COMPLETES THE IGNITION CIRCUIT TO THE ABS ENABLE RELAY.
- USING J 39200, MEASURE THE RESISTANCE BETWEEN TERMINAL "F3" OF THE 32-WAY EBCM HARNESS CONNECTOR AND THE POSITIVE BATTERY CABLE TERMINAL.

IS RESISTANCE 95Ω OR LESS WHEN MEASURED AT APPROXIMATELY 20°C (68°F)?

NO → DIAGNOSIS CONTINUED ON NEXT PAGE.

YES ↓

(10)
- USING J 39200, MEASURE THE VOLTAGE BETWEEN THE NEGATIVE AND POSITIVE BATTERY TERMINALS.

IS VOLTAGE 10 VOLTS OR GREATER?

NO → CORRECT LOW VOLTAGE CONDITION.

YES ↓

(11)
- IGNITION "OFF."
- RECONNECT POSITIVE BATTERY CABLE.
- RECONNECT NEGATIVE BATTERY CABLE.
- DISCONNECT ABS ENABLE RELAY.
- MEASURE THE VOLTAGE BETWEEN TERMINAL "4" OF THE ABS ENABLE RELAY HARNESS CONNECTOR AND GROUND.

IS VOLTAGE 10 VOLTS OR GREATER?

NO → REPAIR OPEN OR HIGH RESISTANCE IN CKT 28.

YES ↓

(12)
- USING J 39200, MEASURE THE RESISTANCE BETWEEN TERMINAL "1" OF THE ABS ENABLE RELAY HARNESS CONNECTOR AND TERMINAL "D" OF THE 8-WAY EBCM HARNESS CONNECTOR.

IS RESISTANCE 2Ω OR LESS?

NO → REPAIR OPEN OR HIGH RESISTANCE IN CKT 1633.

YES ↓

(13)
- USING A JUMPER WIRE, CONNECT TERMINAL "86" ON THE ABS ENABLE RELAY TO GROUND.
- USING A FUSED JUMPER WIRE, SUCH AS J 36169, WITH A 3 AMP FUSE, CONNECT TERMINAL "85" ON THE ABS ENABLE RELAY TO B+.
- USING J 39200, MEASURE THE RESISTANCE BETWEEN TERMINALS "30" AND "87" ON THE ABS ENABLE RELAY.

IS RESISTANCE 2Ω OR LESS?

NO → REPLACE ABS ENABLE RELAY.

YES ↓

MALFUNCTION WAS MOST LIKELY DUE TO A POOR CONNECTION. INSPECT ALL CONNECTORS AND TERMINALS FOR POOR TERMINAL CONTACT AND EVIDENCE OF CORROSION.
REPLACE ALL TERMINALS THAT EXHIBIT EVIDENCE OF POOR TERMINAL CONTACT OR CORROSION.

DIAGNOSTIC CHART DTC 14 — 1995 SILHOUETTE, TRANS SPORT AND LUMINA APV

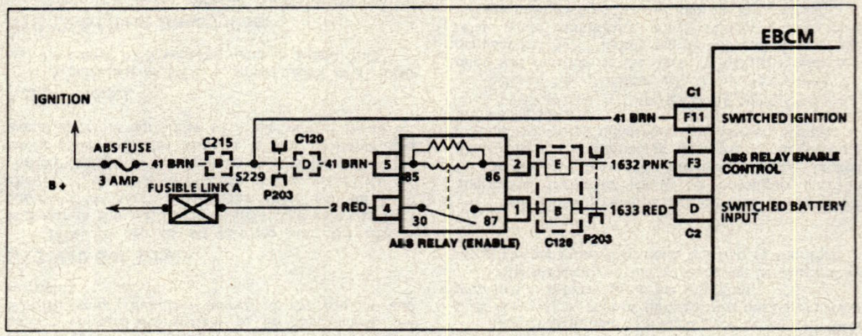

DTC 14
(Page 3 of 3)
ABS ENABLE RELAY CONTACT CIRCUIT OPEN

Circuit Description:

Ignition voltage is supplied through terminal "5" of the ABS relay. The EBCM then is able to energize the pull-in coil by completing the ground circuit at terminal "F3" of the EBCM. The magnetic field created closes the ABS relay contacts and allows battery voltage and current to be supplied to the EBCM, which supplies power to the motors and solenoids.

DTC Will Set When:

DTC 14 can be set anytime after the EBCM commands the ABS relay "ON" (the relay is first commanded "ON" during the three second bulb check). This test monitors the availability of current/voltage to the motors and solenoids. This malfunction indicates voltage is not available and would therefore not allow ABS operation if required

Action Taken:

A malfunction DTC is stored, ABS is disabled and the amber ABS warning lamp is turned "ON."

DTC Chart Test Description:

Number(s) below refer to circled number(s) on the diagnostic chart.

14. Checks for proper resistance of the relay coil.
15. Checks for high resistance in the ABS relay control circuit between the relay and the EBCM
16. Checks for high resistance in the ABS relay control circuit between the battery and the relay.

Diagnostic Aids:

An "Intermittent" problem may be caused by a poor connection, rubbed through wire insulation, or a wire that is broken inside the insulation.

The frequency of the problem can be checked by using the enhanced diagnostic function of the Tech 1,

Any circuitry that is suspected of causing the intermittent complaint should be thoroughly checked for backed out terminals, improper mating, broken locks, improperly formed or damaged terminals, poor terminal to wiring connections or physical damage to the wiring harness.

VIBRATION, TEMPERATURE EFFECTS:

Check for vibration effects by performing the relay test function of the Tech 1. With the relay test commanded "ON," lightly tap the top and sides of the relay while monitoring relay voltage. If the relay voltage changes significantly, replace the relay.

If DTC 14 only sets when the vehicle is initially started in cold ambient conditions (temperature less than 32° F/0° C), replace the relay.

DIAGNOSTIC CHART DTC 14 — 1995 SILHOUETTE, TRANS SPORT AND LUMINA APV

DTC 14
(Page 3 of 3)
ABS ENABLE RELAY CONTACT CIRCUIT OPEN

DIAGNOSIS CONTINUED FROM PREVIOUS PAGE.

(14)
- DISCONNECT ABS ENABLE RELAY.
- USING J 39200, MEASURE THE RESISTANCE BETWEEN TERMINALS "85" AND "86" ON THE ABS ENABLE RELAY. IS RESISTANCE 95Ω OR LESS WHEN MEASURED AT APPROXIMATELY 20°C (68°F)?

NO: REPLACE ABS ENABLE RELAY.

YES:

(15)
- MEASURE THE RESISTANCE BETWEEN TERMINAL "2" OF THE ABS ENABLE RELAY HARNESS CONNECTOR AND TERMINAL "F3" OF THE 32-WAY EBCM HARNESS CONNECTOR. IS RESISTANCE 2Ω OR LESS?

NO: REPAIR OPEN OR HIGH RESISTANCE IN CKT 1632.

YES:

(16)
- MEASURE THE RESISTANCE BETWEEN TERMINAL "5" OF THE ABS ENABLE RELAY HARNESS CONNECTOR AND THE POSITIVE BATTERY CABLE TERMINAL. IS RESISTANCE 2Ω OR LESS?

NO: REPAIR OPEN OR HIGH RESISTANCE IN CKT 41B, CKT 300, IGNITION SWITCH, OR BATTERY FEED CIRCUIT TO IGNITION SWITCH.

YES: MALFUNCTION WAS MOST LIKELY DUE TO A POOR CONNECTION. INSPECT ALL CONNECTORS AND TERMINALS FOR POOR TERMINAL CONTACT AND EVIDENCE OF CORROSION. REPLACE ALL TERMINALS THAT EXHIBIT EVIDENCE OF POOR TERMINAL CONTACT OR CORROSION.

DIAGNOSTIC CHART DTC 15 — 1995 SILHOUETTE, TRANS SPORT AND LUMINA APV

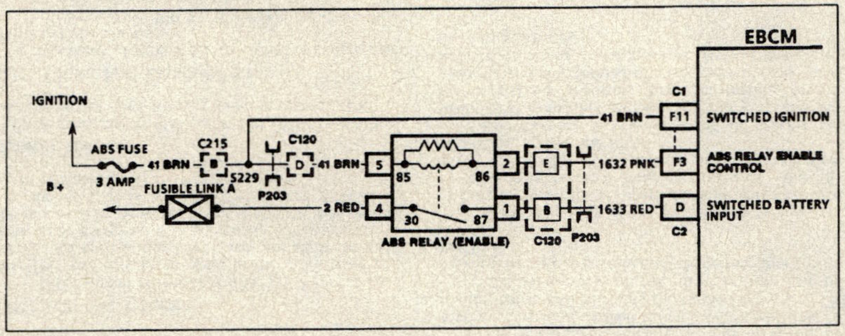

DTC 15
ABS ENABLE RLY CONTACT CKT SHORT TO BATT OR ALWAYS CLOSED

Circuit Description:

Ignition voltage is supplied through terminal "5" of the ABS relay. The EBCM then is able to energize the pull-in coil by completing the ground circuit at terminal "F3" of the EBCM. The magnetic field created closes the ABS relay contacts and allows battery voltage and current to be supplied to the EBCM, which supplies power to the motors and solenoids.

DTC Will Set When:

DTC 15 can be set only before the EBCM commands the relay "ON." This test determines if the ABS relay is energized when it should not be. This malfunction would not allow the ABS relay to remove power to the ABS system. If a second malfunction were to occur that requires the ABS relay to be turned "OFF," that malfunction cannot be removed if the relay cannot be controlled. The malfunction must be present for three consecutive drive cycles before the DTC is set.

Action Taken:

A malfunction DTC is stored. ABS is not disabled.

DTC Chart Test Description:

Number(s) below refer to circled number(s) on the diagnostic chart.

1. Determines if any other DTCs are present.
2. Indicates that the EBCM is capable of controlling the ABS relay as commanded.
3. Checks for a possible short to battery voltage in CKT 1633.
4. Determines whether the short to voltage is in CKT 1633 or internal to the EBCM.
5. Checks for a possible intermittent in wiring or connectors.

Diagnostic Aids:

An "Intermittent" malfunction may be caused by a poor connection, rubbed through wire insulation, or a wire that is broken inside the insulation.

The frequency of the malfunction can be checked by using the enhanced diagnostic function of the Tech 1.

Any circuitry that is suspected of causing the intermittent complaint should be thoroughly checked for backed out terminals, improper mating, broken locks, improperly formed or damaged terminals, poor terminal to wiring connections or physical damage to the wiring harness.

DIAGNOSTIC CHART DTC 15 — 1995 SILHOUETTE, TRANS SPORT AND LUMINA APV

DTC 15
ABS ENABLE RLY CONTACT CKT SHORT TO BATT OR ALWAYS CLOSED

(1) ARE ANY OTHER DTCs PRESENT?

- YES: PROCEED TO THAT DTC CHART FIRST.
- NO:

(2)
- IGNITION "ON," ENGINE "OFF."
- USING THE RELAY TEST FUNCTION OF THE TECH 1, COMMAND ABS ENABLE RELAY "OFF."

DOES TECH 1 INDICATE THAT ABS ENABLE RELAY IS "OFF" AND EBCM BATTERY VOLTAGE IS LESS THAN 5 VOLTS?

- YES: MALFUNCTION IS NOT PRESENT AT THIS TIME. REFER TO "DIAGNOSTIC AIDS" ON FACING PAGE.
- NO:

(3)
- IGNITION "OFF."
- DISCONNECT 32-WAY AND 8-WAY EBCM HARNESS CONNECTORS.
- IGNITION "ON."
- USING J 39200, MEASURE THE VOLTAGE BETWEEN GROUND AND TERMINAL "D" OF THE 8-WAY EBCM HARNESS CONNECTOR.

IS VOLTAGE 2 VOLTS OR LESS?

- NO:

(4)
- IGNITION "OFF."
- DISCONNECT ABS ENABLE RELAY.
- MEASURE THE VOLTAGE BETWEEN GROUND AND TERMINAL "D" OF THE 8-WAY EBCM HARNESS CONNECTOR.

IS VOLTAGE 2 VOLTS OR LESS?

- YES: REPLACE ABS ENABLE RELAY.
- NO: REPAIR SHORT TQ VOLTAGE IN CKT 1633.

- YES:

(5)
- IGNITION "OFF."
- INSERT CKT 1633 AND 8-WAY EBCM HARNESS CONNECT FOR DAMAGE WHICH MAY RESULT IN A SHORT TO VOLTAGE WHEN ALL CONNECTORS ARE CONNECTED. REPAIR DAMAGE IF EVIDENT.
- RECONNECT 8-WAY AND 32-WAY EBCM CONNECTORS.
- IGNITION "ON."
- USING RELAY TEST FUNCTION OF TECH 1, COMMAND ENABLE RELAY "OFF."

DOES TECH 1 INDICATE THAT ABS ENABLE RELAY IS "OFF" AND EBCM BATTERY VOLTAGE IS LESS THAN 5 VOLTS?

- NO: REPLACE EBCM.
- YES: MALFUNCTION IS INTERMITTENT. REFER TO "DIAGNOSTIC AIDS" ON FACING PAGE.

AFTER DIAGNOSIS IS COMPLETE, CLEAR DTCs, AND TEST DRIVE VEHICLE FOR 3 DRIVE CYCLES TO VERIFY THAT DTC DOES NOT RESET. A DRIVE CYCLE CONSISTS OF STARTING THE VEHICLE, DRIVING ABOVE 16 km/h (10 MPH), AND THEN KEYING DOWN.

DIAGNOSTIC CHART DTC 16 — 1995 SILHOUETTE, TRANS SPORT AND LUMINA APV

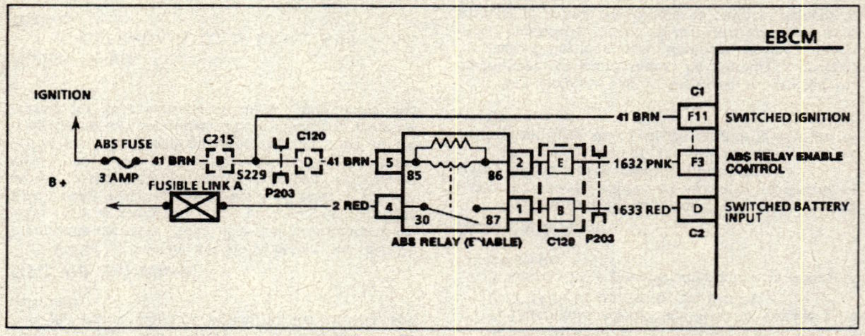

DTC 16
ABS ENABLE RELAY COIL CIRCUIT OPEN

Circuit Description:

Ignition voltage is supplied through terminal "5" of the ABS relay. The EBCM then is able to energize the pull-in coil by completing the ground circuit at terminal "F3" of the EBCM. The magnetic field created closes the ABS relay contacts and allows battery voltage and current to be supplied to the EBCM, which supplies power to the motors and solenoids.

DTC Will Set When:

DTC 16 can be set only after DTC 14 has been set. This test detects an open in the ABS relay coil circuit. An open in this circuit will not allow the ABS relay to be energized thus preventing voltage/current to the motors and solenoids. If this malfunction is present and the ignition is turned "OFF" before 5 km/h (3 mph) is reached, DTC 14 is set alone.

Action Taken:

A malfunction DTC is stored, ABS is disabled and the amber ABS warning lamp is turned "ON."

DTC Chart Test Description:

Number(s) below refer to circled number(s) on the diagnostic chart.

1. Indicates the EBCM is capable of controlling the ABS relay as commanded.
2. Checks to ensure that voltage is available to the pull-in coil of the ABS relay.
3. Ensures that there is continuity through the pull-in coil of the ABS relay.
4. Verifies the integrity of the relay control circuit.
5. Ensures malfunction was not due to poor terminal contact.

Diagnostic Aids:

An "Intermittent" malfunction may be caused by a poor connection, rubbed through wire insulation, or a wire that is broken inside the insulation.

The frequency of the malfunction can be checked by using the enhanced diagnostic function of the Tech 1.

If the frequency of the malfunction is high, but is currently intermittent, check for high coil resistance by measuring between relay terminals "2" and "5" using a J 39200. If resistance shows greater than 100 ohms, replace the relay.

Any circuitry that is suspected of causing the intermittent complaint should be thoroughly checked for backed out terminals, improper mating, broken locks, improperly formed or damaged terminals, poor terminal to wiring connections or physical damage to the wiring harness.

DIAGNOSTIC CHART DTC 16 — 1995 SILHOUETTE, TRANS SPORT AND LUMINA APV

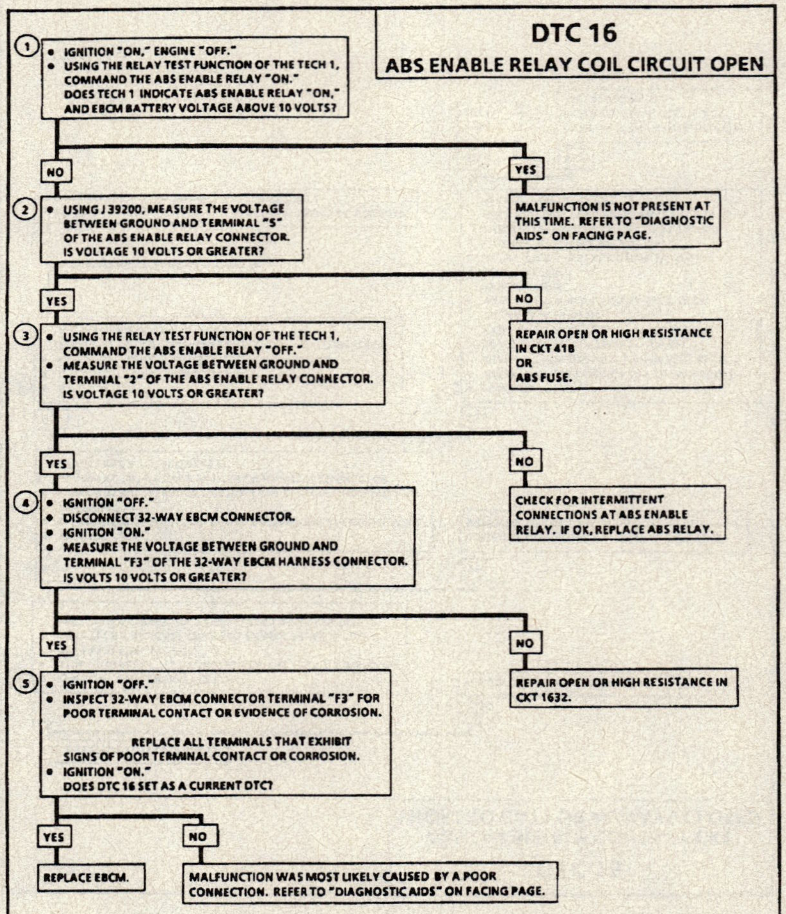

AFTER ALL DIAGNOSIS IS COMPLETE, CLEAR DTCs AND TEST DRIVE VEHICLE FOR THREE (3) DRIVE CYCLES TO VERIFY THAT DTC DOES NOT RESET. A DRIVE CYCLE CONSISTS OF STARTING THE VEHICLE, DRIVING OVER 16 km/h (10 MPH), AND THEN KEYING DOWN.

DIAGNOSTIC CHART DTC 17 — 1995 SILHOUETTE, TRANS SPORT AND LUMINA APV

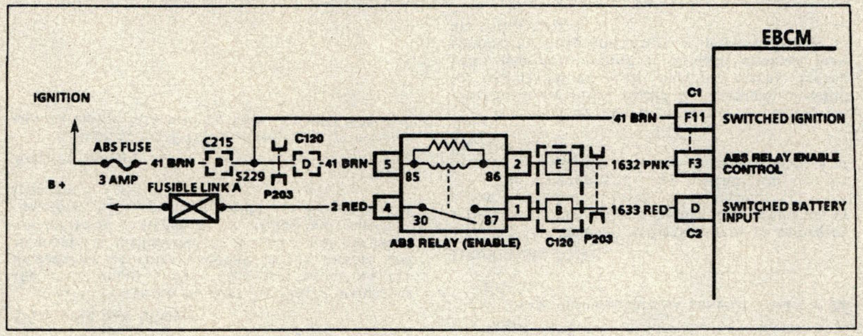

DTC 17

ABS ENABLE RELAY COIL CIRCUIT SHORTED TO GND

Circuit Description:

Ignition voltage is supplied through terminal "5" of the ABS relay. The EBCM then is able to energize the pull-in coil by completing the ground circuit at terminal "F3" of the EBCM. The magnetic field created closes the ABS relay contacts and allows battery voltage and current to be supplied to the EBCM, which supplies voltage to the motors and solenoids.

DTC Will Set When:

DTC 17 can be set before the EBCM commands the ABS relay "ON." This test determines if the ABS relay is energized when it should not be. This malfunction would not allow the ABS relay to remove power to the ABS system. If a second malfunction were to occur that requires the ABS relay to be turned "OFF," that malfunction cannot be removed if the relay cannot be controlled.

Action Taken:

A malfunction DTC is stored. ABS is not disabled.

DTC Chart Test Description:

Number(s) below refer to circled number(s) on the diagnostic chart.

1. Indicates that the EBCM is capable of controlling the ABS relay as commanded.
2. Checks to ensure that the ABS relay or control CKT 1632 are not shorted to ground.
3. Checks to see if the EBCM is internally shorted to ground. A grounded terminal "F3" would cause the ABS relay to be energized anytime the ignition was in the "RUN" position.
4. Checks for a short to ground in control CKT 1632 or a defective relay.

Diagnostic Aids:

An "Intermittent" malfunction may be caused by a poor connection, rubbed through wire insulation, or a wire that is broken inside the insulation.

The frequency of the malfunction can be checked by using the enhanced diagnostic function of the Tech 1.

Any circuitry that is suspected of causing the intermittent complaint should be thoroughly checked for backed out terminals, improper mating, broken locks, improperly formed or damaged terminals, poor terminal to wiring connections or physical damage to the wiring harness.

DIAGNOSTIC CHART DTC 17 — 1995 SILHOUETTE, TRANS SPORT AND LUMINA APV

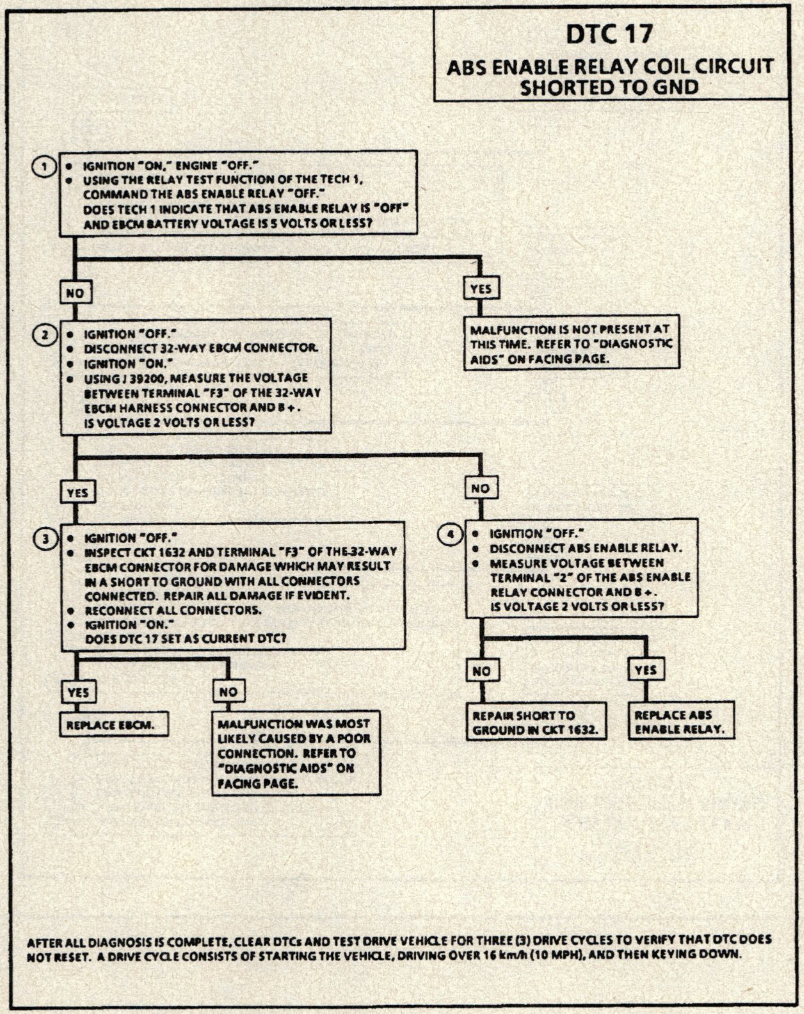

DIAGNOSTIC CHART DTC 18 — 1995 SILHOUETTE, TRANS SPORT AND LUMINA APV

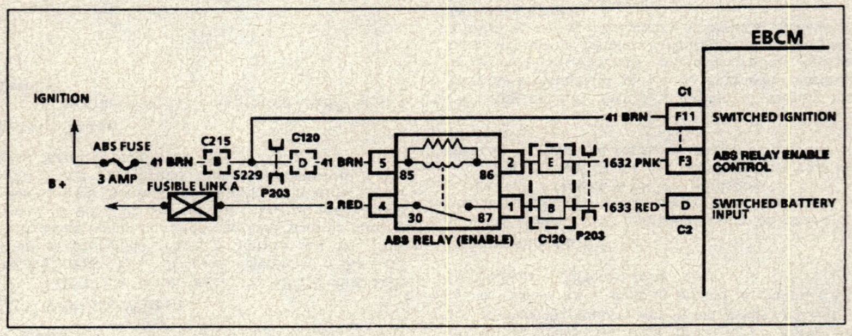

DTC 18

ABS ENABLE RELAY COIL CIRCUIT SHORTED TO BATT

Circuit Description:

Ignition voltage is supplied through terminal "5" of the ABS relay. The EBCM then is able to energize the pull-in coil by completing the ground circuit at terminal "F3" of the EBCM. The magnetic field created closes the ABS relay contacts and allows battery voltage and current to be supplied to the EBCM, which supplies voltage to the motors and solenoids.

DTC Will Set When:

DTC 18 can be set after the EBCM commands the ABS relay "ON." This test monitors the availability of current/voltage to the motors and solenoids. A malfunction of this type will not allow voltage to be available to the motors and solenoids, therefore, ABS operation would not be allowed if required.

Action Taken:

A malfunction DTC is stored, ABS is disabled and the amber ABS warning lamp is turned "ON."

DTC Chart Test Description:

Number(s) below refer to circled number(s) on the diagnostic chart.

1. Indicates that the EBCM is capable of controlling the ABS relay as commanded.
2. Checks for a short to voltage on the relay control circuit.
3. Verifies that the relay coil has the appropriate resistance.
4. Verifies that relay contacts have not failed closed.
5. Checks for possible poor terminal contact at the EBCM.

Diagnostic Aids:

An "Intermittent" malfunction may be caused by a poor connection, rubbed through wire insulation, or a wire that is broken inside the insulation.

The frequency of the malfunction can be checked by using the enhanced diagnostic function of the Tech 1,

Any circuitry that is suspected of causing the intermittent complaint should be thoroughly checked for backed out terminals, improper mating, broken locks, improperly formed or damaged terminals, poor terminal to wiring connections or physical damage to the wiring harness.

DIAGNOSTIC CHART DTC 18 — 1995 SILHOUETTE, TRANS SPORT AND LUMINA APV

DTC 18
ABS ENABLE RELAY COIL CIRCUIT SHORTED TO BATT

(1)
- IGNITION "ON," ENGINE "OFF."
- USING RELAY TEST FUNCTION OF TECH 1, COMMAND ABS ENABLE RELAY "ON."
- DOES TECH 1 INDICATE ABS ENABLE RELAY "ON" AND BATTERY VOLTAGE ABOVE 10 VOLTS?

YES: MALFUNCTION IS NOT PRESENT AT THIS TIME. REFER TO "DIAGNOSTIC AIDS" ON FACING PAGE.

NO:

(2)
- IGNITION "OFF."
- DISCONNECT ABS ENABLE RELAY.
- DISCONNECT 32-WAY EBCM CONNECTOR.
- IGNITION "ON."
- USING J 39200, MEASURE THE VOLTAGE BETWEEN GROUND AND TERMINAL "F3" OF THE 32-WAY EBCM HARNESS CONNECTOR. IS VOLTAGE 2 VOLTS OR LESS?

NO: REPAIR SHORT TO VOLTAGE CKT 1632.

YES:

(3)
- IGNITION "ON."
- USING J 39200, MEASURE THE RESISTANCE BETWEEN TERMINAL "2" AND TERMINAL "5" OF THE ABS ENABLE RELAY. IS RESISTANCE 40Ω OR GREATER?

NO: REPLACE ABS ENABLE RELAY.

YES:

(4)
- USING J 39200, MEASURE THE RESISTANCE BETWEEN TERMINALS "30" AND "87" OF THE ABS ENABLE RELAY. DOES J 39200 DISPLAY RESISTANCE AS "OL" (INFINITE)?

NO: REPLACE ABS ENABLE RELAY.

YES:

(5)
- INSPECT CKT 1632 AND TERMINAL "F3" OF THE 32-WAY EBCM CONNECTOR FOR DAMAGE WHICH MAY RESULT IN A SHORT TO VOLTAGE WITH ALL CONNECTORS CONNECTED. REPAIR ALL DAMAGE IF EVIDENT.
- RECONNECT ALL CONNECTORS.
- IGNITION "ON."
DID DTC 18 SET AS A CURRENT DTC?

YES: REPLACE EBCM.

NO: MALFUNCTION WAS MOST LIKELY CAUSED BY A POOR CONNECTION. REFER TO "DIAGNOSTIC AIDS" ON FACING PAGE.

AFTER ALL DIAGNOSIS IS COMPLETE, CLEAR DTCs AND TEST DRIVE VEHICLE FOR THREE (3) DRIVE CYCLES TO VERIFY THAT DTC DOES NOT RESET. A DRIVE CYCLE CONSISTS OF STARTING THE VEHICLE, DRIVING OVER 16 km/h (10 MPH), AND THEN KEYING DOWN.

DIAGNOSTIC CHART DTC 21 — 1995 SILHOUETTE, TRANS SPORT AND LUMINA APV

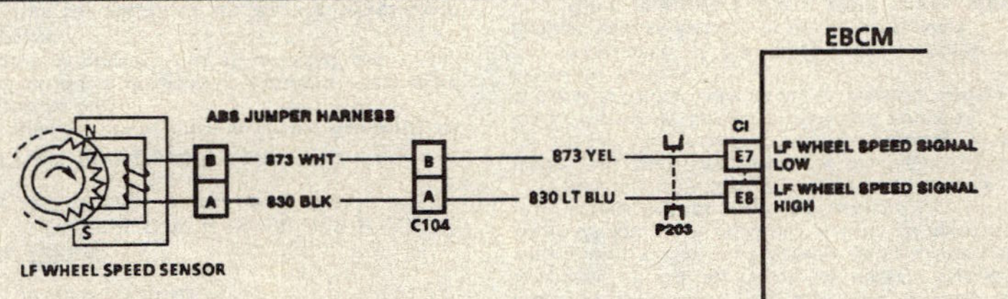

DTC 21
LEFT FRONT WHEEL SPEED = 0

Circuit Description:

As a toothed ring passes by the wheel speed sensor, changes in the electromagnetic field cause the wheel speed sensor to produce a sinusoidal (AC) voltage signal whose frequency is proportional to wheel speed. The magnitude of this signal is directly related to wheel speed and the proximity of the wheel speed sensor to the toothed ring, often referred to as the air gap.

DTC Will Set When:

DTC 21 can be set when the vehicle is not in an ABS stop. If the left front wheel speed = 0 and the vehicle's reference speed is greater than 8 km/h (5 mph), a malfunction exists.

Action Taken:

A malfunction DTC is stored, ABS is disabled and the amber ABS warning lamp is turned "ON."

DTC Chart Test Description:

Number(s) below refer to circled number(s) on the diagnostic chart.

1. This step is designed to identify a wheel speed sensor or circuitry that is damaged and visibly aparent.
2. Checks to see that the wheel speed sensor is within proper resistance values.
3. Verifies that the wheel speed sensor can generate an appropriate output signal.
4. Ensures that wheel speed sensor circuitry is not shorted together.
5. Ensures malfunction was not due to physical damage of the circuitry.

Diagnostic Aids:

An "Intermittent" malfunction may be caused by a poor connection, rubbed through wire insulation, or a wire that is broken inside the insulation.

The frequency of the malfunction can be checked by using the enhanced diagnostic function of the Tech 1.

If the customer's comments reflect that the ABS warning lamp is "ON" only during moist environmental changes (rain, snow, vehicle wash), all wheel speed sensor circuitry should be thoroughly inspected for signs of water intrusion. If DTC is not current, clear DTCs and simulate the effects of water intrusion. Use the following procedure. Spray the suspected area with a 5% salt water solution (two teaspoons of salt to 12 oz. of water). Test drive vehicle over various road surfaces (bumps, turns, etc.) above 24 km/h (15 mph) for at least 30 seconds. If DTC returns, replace suspected harness.

Any circuitry that is suspected of causing the intermittent complaint should be thoroughly checked for backed out terminals, improper mating, broken locks, improperly formed or damaged terminals, poor terminal to wiring connections or physical damage to the wiring harness.

Resistance of the wheel speed sensor will increase with an increase in sensor temperature.

When replacing a wheel speed sensor, inspect the sensor terminals and harness connector for corrosion and/or water intrusion. If evidence of corrosion or water intrusion exists, replace wheel speed sensor jumper harness. Likewise, if replacing a wheel speed sensor jumper harness, inspect sensor terminals. If evidence of corrosion or water intrusion exists, replace wheel speed sensor.

DIAGNOSTIC CHART DTC 21 — 1995 SILHOUETTE, TRANS SPORT AND LUMINA APV

DTC 21
LEFT FRONT WHEEL SPEED = 0

Important

- WHEEL SPEED SENSOR INTERMITTENT MALFUNCTIONS MAY BE DIFFICULT TO LOCATE. CARE SHOULD BE TAKEN NOT TO DISTURB ANY ELECTRICAL CONNECTIONS PRIOR TO AN INDICATED STEP OF THIS CHART. THIS WILL ENSURE THAT AN INTERMITTENT CONNECTION WILL NOT BE CORRECTED BEFORE THE SOURCE OF THE MALFUNCTION IS FOUND.

(1)
- IGNITION "OFF."
- PHYSICALLY INSPECT THE LEFT FRONT WHEEL SPEED SENSOR, LEFT FRONT WHEEL SPEED SENSOR JUMPER HARNESS, AND CONNECTORS FOR DAMAGE.

IS PHYSICAL DAMAGE INDICATED?

- YES: REPLACE WHEEL SPEED SENSOR AND/OR JUMPER HARNESS IF DAMAGED. DRIVE VEHICLE ABOVE 24 km/h (15 MPH) TO ENSURE DTC 21 DOES NOT SET.
- NO: go to step 2.

(2)
- DISCONNECT LEFT FRONT WHEEL SPEED SENSOR DIRECTLY AT SENSOR.
- USING J 39200, MEASURE THE RESISTANCE BETWEEN SENSOR TERMINALS "A" AND "B."

IS RESISTANCE BETWEEN 1020 AND 1137Ω WHEN SENSOR IS 20°C (68°F)? *(IF TEMPERATURE OF THE SENSOR IS SUSPECTED TO BE OTHER THAN 20°C (68°F), REFER TO THE TABLE BELOW FOR APPLICABLE SENSOR RESISTANCE VALUES; THE VALUES IN THE TABLE ARE FOR TEMPERATURE OF THE SENSOR, NOT AIR TEMPERATURE.)*

- NO: REPLACE WHEEL SPEED SENSOR.
- YES: go to step 3.

(3)
- WITH J 39200 STILL CONNECTED, SELECT A/C VOLTAGE SCALE.
- SPIN WHEEL BY HAND WHILE OBSERVING VOLTAGE.

IS VOLTAGE AT LEAST 100 mV (VOLTAGE WILL INCREASE AS WHEEL SPEED INCREASES)?

- NO: REPLACE WHEEL SPEED SENSOR.
- YES: go to step 4.

(4)
- IGNITION "OFF."
- DISCONNECT 32-WAY EBCM CONNECTOR.
- USING J 39200, MEASURE THE RESISTANCE BETWEEN TERMINAL "E7" AND "E8" OF THE 32-WAY EBCM HARNESS CONNECTOR.

DOES J 39200 DISPLAY RESISTANCE AS "OL" (INFINITE)?

- NO: REPAIR SHORT BETWEEN CKTs 830 AND 873.
- YES: go to step 5.

(5)
- INSPECT CKTs 830 AND 873 AND 32-WAY EBCM CONNECTOR FOR DAMAGE WHICH MAY RESULT IN A SHORT BETWEEN CKTs 830 AND 873 WITH ALL CONNECTORS CONNECTED. REPAIR ALL DAMAGE IF EVIDENT.
- RECONNECT ALL CONNECTORS.
- TEST DRIVE VEHICLE ABOVE 24 km/h (15 MPH) FOR AT LEAST 30 SECONDS.

DOES DTC 21 SET AS A CURRENT DTC?

- YES: REPLACE EBCM.
- NO: MALFUNCTION IS INTERMITTENT. REFER TO "DIAGNOSTIC AIDS" ON FACING PAGE.

FRONT WHEEL SPEED SENSOR
TEMPERATURE OF SENSOR VS. SENSOR RESISTANCE (APPROXIMATE)

TEMP. (°C)	TEMP. (°F)	RESISTANCE (OHMS)
-34 TO 4	-30 TO 40	804 TO 1066
5 TO 43	41 TO 110	956 TO 1240
44 TO 93	111 TO 200	1112 TO 1463

DIAGNOSTIC CHART DTC 22 — 1995 SILHOUETTE, TRANS SPORT AND LUMINA APV

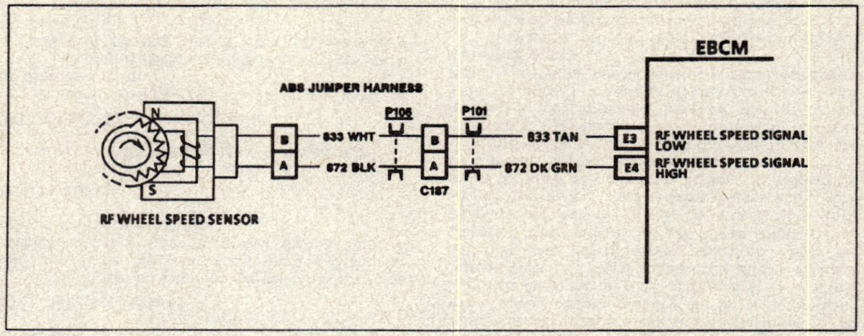

DTC 22
RIGHT FRONT WHEEL SPEED = 0

Circuit Description:

As a toothed ring passes by the wheel speed sensor, changes in the electromagnetic field cause the wheel speed sensor to produce a sinusoidal (AC) voltage signal whose frequency is proportional to wheel speed. The magnitude of this signal is directly related to wheel speed and the proximity of the wheel speed sensor to the toothed ring, often referred to as the air gap.

DTC Will Set When:

DTC 22 can be set when the vehicle is not in an ABS stop. If the right front wheel speed = 0 and the vehicle's reference speed is greater than 8 km/h (5 mph), a malfunction exists.

Action Taken:

A malfunction DTC is stored, ABS is disabled and the ABS warning lamp is turned "ON."

DTC Chart Test Description:

Number(s) below refer to circled number(s) on the diagnostic chart.

1. This step is designed to identify a wheel speed sensor or circuitry that is damaged and visibly apparent.
2. This test checks to see that the wheel speed sensor is within proper resistance values.
3. Verifies that the wheel speed sensor can generate an appropriate output signal.
4. Ensures that wheel speed sensor circuitry is not shorted together.
5. Ensures malfunction was not due to physical damage of the circuitry.

Diagnostic Aids:

An "Intermittent" malfunction may be caused by a poor connection, rubbed through wire insulation, or a wire that is broken inside the insulation.

The frequency of the malfunction can be checked by using the enhanced diagnostic function of the Tech 1,

If the customer's comments reflect that the ABS warning lamp is "ON" only during moist environmental changes (rain, snow, vehicle wash), all wheel speed sensor circuitry should be thoroughly inspected for signs of water intrusion. If DTC is not current, clear DTCs and simulate the effects of water intrusion. Use the following procedure. Spray the suspected area with a 5% salt water solution (two teaspoons of salt to 12 oz. of water). Test drive vehicle over various road surfaces (bumps, turns, etc.) above 24 km/h (15 mph) for at least 30 seconds. If DTC returns, replace suspected harness.

Any circuitry that is suspected of causing the intermittent complaint should be thoroughly checked for backed out terminals, improper mating, broken locks, improperly formed or damaged terminals, poor terminal to wiring connections or physical damage to the wiring harness.

Resistance of the wheel speed sensor will increase with an increase in sensor temperature.

When replacing a wheel speed sensor, inspect the sensor terminals and harness connector for corrosion and/or water intrusion. If evidence of corrosion or water intrusion exists, replace wheel speed sensor jumper harness. Likewise, if replacing a wheel speed sensor jumper harness, inspect sensor terminals. If evidence of corrosion or water intrusion exists, replace wheel speed sensor.

DIAGNOSTIC CHART DTC 22 — 1995 SILHOUETTE, TRANS SPORT AND LUMINA APV

DTC 22
RIGHT FRONT WHEEL SPEED = 0

Important

- WHEEL SPEED SENSOR INTERMITTENT MALFUNCTIONS MAY BE DIFFICULT TO LOCATE. CARE SHOULD BE TAKEN NOT TO DISTURB ANY ELECTRICAL CONNECTIONS PRIOR TO AN INDICATED STEP OF THIS CHART. THIS WILL ENSURE THAT AN INTERMITTENT CONNECTION WILL NOT BE CORRECTED BEFORE THE SOURCE OF THE MALFUNCTION IS FOUND.

(1)
- IGNITION "OFF."
- PHYSICALLY INSPECT THE RIGHT FRONT WHEEL SPEED SENSOR, RIGHT FRONT WHEEL SPEED SENSOR JUMPER HARNESS, AND CONNECTORS FOR DAMAGE.

IS PHYSICAL DAMAGE INDICATED?

YES: REPLACE WHEEL SPEED SENSOR AND/OR JUMPER HARNESS IF DAMAGED. DRIVE VEHICLE ABOVE 24 km/h (15 MPH) TO ENSURE DTC 22 DOES NOT SET.

NO: go to (2)

(2)
- DISCONNECT RIGHT FRONT WHEEL SPEED SENSOR DIRECTLY AT SENSOR.
- USING J 39200, MEASURE THE RESISTANCE BETWEEN SENSOR TERMINALS "A" AND "B."

IS RESISTANCE BETWEEN 1020 AND 1137Ω WHEN SENSOR IS 20°C (68°F)?
(IF TEMPERATURE OF THE SENSOR IS SUSPECTED TO BE OTHER THAN 20°C (68°F), REFER TO THE TABLE BELOW FOR APPLICABLE SENSOR RESISTANCE VALUES; THE VALUES IN THE TABLE ARE FOR TEMPERATURE OF THE SENSOR, NOT AIR TEMPERATURE.)

NO: REPLACE WHEEL SPEED SENSOR.

YES: go to (3)

(3)
- WITH J 39200 STILL CONNECTED, SELECT A/C VOLTAGE SCALE.
- SPIN WHEEL BY HAND WHILE OBSERVING VOLTAGE.

IS VOLTAGE AT LEAST 100 mV (VOLTAGE WILL INCREASE AS WHEEL SPEED INCREASES)?

NO: REPLACE WHEEL SPEED SENSOR.

YES: go to (4)

(4)
- IGNITION "OFF."
- DISCONNECT 32-WAY EBCM CONNECTOR.
- USING J 39200, MEASURE THE RESISTANCE BETWEEN TERMINAL "E3" AND "E4" OF THE 32-WAY EBCM HARNESS CONNECTOR.

DOES J 39200 DISPLAY RESISTANCE AS "OL" (INFINITE)?

NO: REPAIR SHORT BETWEEN CKTs 833 AND 872.

YES: go to (5)

(5)
- INSPECT CKTs 833 AND 872 AND 32-WAY EBCM CONNECTOR FOR DAMAGE WHICH MAY RESULT IN A SHORT BETWEEN CKTs 833 AND 872 WITH ALL CONNECTORS CONNECTED. REPAIR ALL DAMAGE IF EVIDENT.
- RECONNECT ALL CONNECTORS.
- TEST DRIVE VEHICLE ABOVE 24 km/h (15 MPH) FOR AT LEAST 30 SECONDS.

DOES DTC 22 SET AS A CURRENT DTC?

YES: REPLACE EBCM.

NO: MALFUNCTION IS INTERMITTENT. REFER TO "DIAGNOSTIC AIDS" ON FACING PAGE.

FRONT WHEEL SPEED SENSOR
TEMPERATURE OF SENSOR VS. SENSOR RESISTANCE (APPROXIMATE)

TEMP. (°C)	TEMP. (°F)	RESISTANCE (OHMS)
-34 TO 4	-30 TO 40	804 TO 1066
5 TO 43	41 TO 110	956 TO 1240
44 TO 93	111 TO 200	1112 TO 1463

DIAGNOSTIC CHART DTC 23 — 1995 SILHOUETTE, TRANS SPORT AND LUMINA APV

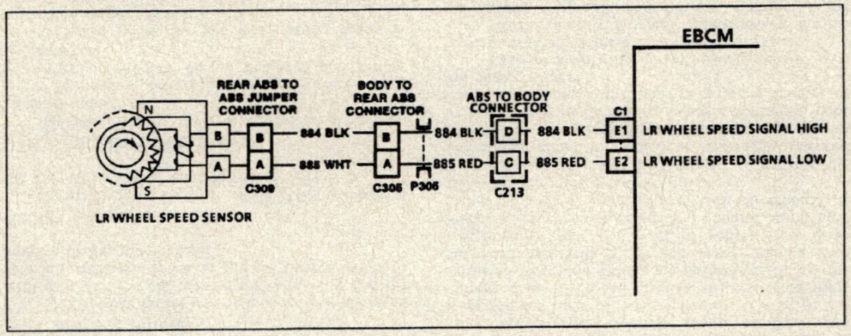

DTC 23
LEFT REAR WHEEL SPEED = 0

Circuit Description:

As a toothed ring passes by the wheel speed sensor, changes in the electromagnetic field cause the wheel speed sensor to produce a sinusoidal (AC) voltage signal whose frequency is proportional to wheel speed. The magnitude of this signal is directly related to wheel speed and the proximity of the wheel speed sensor to the toothed ring, often referred to as the air gap.

DTC Will Set When:

DTC 23 can be set when the vehicle is not in an ABS stop. If the left rear wheel speed = 0 and the vehicle's reference speed is greater than 8 km/h (5 mph), a malfunction exists.

Action Taken:

A malfunction DTC is stored, ABS is disabled and the ABS warning lamp is turned "ON."

DTC Chart Test Description:

Number(s) below refer to circled number(s) on the diagnostic chart.

1. This step is designed to identify a wheel speed sensor or circuitry that is damaged and visibly apparent.
2. Checks to see that the wheel speed sensor is within proper resistance values.
3. Verifies that the wheel speed sensor can generate an appropriate output signal.
4. Ensures that the wheel speed sensor circuitry is not shorted together.
5. Checks to see if malfunction was due to physical damage of the circuitry.
6. Checks for an incorrect or malfunctioning hub and bearing assembly.

Diagnostic Aids:

An "Intermittent" malfunction may be caused by a poor connection, rubbed through wire insulation, or a wire that is broken inside the insulation.

The frequency of the malfunction can be checked by using the enhanced diagnostic function of the Tech 1.

If the customer's comments reflect that the ABS warning lamp is "ON" only during moist environmental changes (rain, snow, vehicle wash), all wheel speed sensor circuitry should be thoroughly inspected for signs of water intrusion. If DTC is not current, clear DTCs and simulate the effects of water intrusion. Use the following procedure. Spray the suspected area with a 5% salt water solution (two teaspoons of salt to 12 oz. of water). Test drive vehicle over various road surfaces (bumps, turns, etc.) above 24 km/h (15 mph) for at least 30 seconds. If DTC returns, replace suspected harness.

Any circuitry that is suspected of causing the intermittent complaint should be thoroughly checked for backed out terminals, improper mating, broken locks, improperly formed or damaged terminals, poor terminal to wiring connections or physical damage to the wiring harness.

Resistance of the wheel speed sensor will increase with an increase in sensor temperature.

When replacing a wheel speed sensor, inspect the sensor terminals and harness connector for corrosion and/or water intrusion. If evidence of corrosion or water intrusion exists, replace wheel speed sensor jumper harness. Likewise, if replacing a wheel speed sensor jumper harness, inspect sensor terminals. If evidence of corrosion or water intrusion exists, replace wheel speed sensor.

DIAGNOSTIC CHART DTC 23 — 1995 SILHOUETTE, TRANS SPORT AND LUMINA APV

DTC 23
LEFT REAR WHEEL SPEED = 0

Important

- WHEEL SPEED SENSOR INTERMITTENT MALFUNCTIONS MAY BE DIFFICULT TO LOCATE. CARE SHOULD BE TAKEN NOT TO DISTURB ANY ELECTRICAL CONNECTIONS PRIOR TO AN INDICATED STEP OF THIS CHART. THIS WILL ENSURE THAT AN INTERMITTENT CONNECTION WILL NOT BE CORRECTED BEFORE THE SOURCE OF THE MALFUNCTION IS FOUND.

(1)
- IGNITION "OFF."
- PHYSICALLY INSPECT THE LEFT REAR WHEEL SPEED SENSOR, WHEEL SPEED SENSOR JUMPER HARNESS, AND CONNECTORS FOR DAMAGE.

IS PHYSICAL DAMAGE INDICATED?

YES: REPLACE WHEEL SPEED SENSOR AND/OR WHEEL SPEED SENSOR JUMPER HARNESS IF DAMAGED. DRIVE VEHICLE ABOVE 24 km/h (15 MPH) TO ENSURE DTC 23 DOES NOT SET.

NO:

(2)
- DISCONNECT LEFT REAR WHEEL SPEED SENSOR DIRECTLY AT SENSOR.
- USING J 39200, MEASURE THE RESISTANCE BETWEEN SENSOR TERMINALS "A" AND "B." IS RESISTANCE BETWEEN 2059 AND 2353Ω WHEN SENSOR IS 20°C (68°F)?

(IF TEMPERATURE OF THE SENSOR IS SUSPECTED TO BE OTHER THAN 20°C (68°F), REFER TO THE TABLE BELOW FOR APPLICABLE SENSOR RESISTANCE VALUES: THE VALUES IN THE TABLE ARE FOR TEMPERATURE OF THE SENSOR, NOT AIR TEMPERATURE.)

NO: REPLACE WHEEL SPEED SENSOR.

YES:

(3)
- WITH J 39200 STILL CONNECTED, SELECT A/C VOLTAGE SCALE.
- SPIN WHEEL BY HAND WHILE OBSERVING VOLTAGE.

IS VOLTAGE AT LEAST 100 mV (VOLTAGE WILL INCREASE AS WHEEL SPEED INCREASES)?

NO: REPLACE WHEEL SPEED SENSOR.

YES:

(4)
- IGNITION "OFF."
- DISCONNECT 32-WAY EBCM CONNECTOR.
- USING J 39200, MEASURE THE RESISTANCE BETWEEN TERMINAL "E1" AND "E2" OF THE 32-WAY EBCM HARNESS CONNECTOR.

DOES J 39200 DISPLAY RESISTANCE AS "OL" (INFINITE)?

NO: REPAIR SHORT BETWEEN CKTs 884 AND 885.

YES:

(5)
- INSPECT CKTs 884 AND 885 AND 32-WAY EBCM CONNECTOR FOR DAMAGE WHICH MAY RESULT IN A SHORT BETWEEN CKTs 884 AND 885 WITH ALL CONNECTORS CONNECTED. REPAIR ALL DAMAGE IF EVIDENT.
- RECONNECT ALL CONNECTORS.
- TEST DRIVE VEHICLE ABOVE 24 km/h (15 MPH) FOR AT LEAST 30 SECONDS.

DOES DTC 23 SET AS A CURRENT DTC?

NO: MALFUNCTION WAS MOST LIKELY CAUSED BY A POOR CONNECTION. REFER TO "DIAGNOSTIC AIDS" ON FACING PAGE.

YES:

(6)
- USING TECH 1, SELECT "F0: DATA LIST."
- MONITOR WHEEL SPEEDS WHILE SLOWLY ACCELERATING TO 65 km/h (40 MPH) AND THEN SLOWLY DECELERATING TO 0 km/h (0 MPH).

IS THE LEFT REAR WHEEL SPEED CONSTANTLY HIGHER THAN THE THREE REMAINING WHEEL SPEEDS?

NO: REPLACE EBCM.

YES: REPLACE INCORRECT OR MALFUNCTIONING LEFT WHEEL SPEED SENSOR/HUB AND BEARING ASSEMBLY WITH THE PROPER ASSEMBLY. REFER TO PARTS CATALOG FOR PROPER PART NUMBER.

REAR WHEEL SPEED SENSOR
TEMPERATURE OF SENSOR VS. SENSOR RESISTANCE (APPROXIMATE)

TEMP. (°C)	TEMP. (°F)	RESISTANCE (OHMS)
-34 TO 4	-30 TO 40	1622 TO 2205
5 TO 43	41 TO 110	1930 TO 2566
44 TO 93	111 TO 200	2245 TO 3028

DIAGNOSTIC CHART DTC 24 — 1995 SILHOUETTE, TRANS SPORT AND LUMINA APV

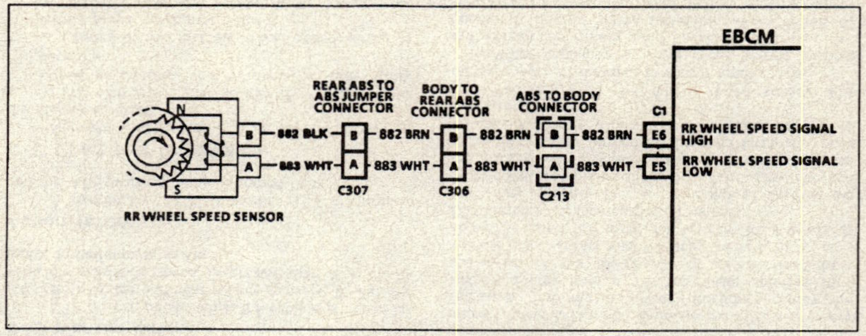

DTC 24
RIGHT REAR WHEEL SPEED = 0

Circuit Description:

As a toothed ring passes by the wheel speed sensor, changes in the electromagnetic field cause the wheel speed sensor to produce a sinusoidal (AC) voltage signal whose frequency is proportional to wheel speed. The magnitude of this signal is directly related to wheel speed and the proximity of the wheel speed sensor to the toothed ring, often referred to as the air gap.

DTC Will Set When:

DTC 24 can be set when the vehicle is not in an ABS stop. If the right rear wheel speed = 0 and the vehicle's reference speed is greater than 8 km/h (5 mph), a malfunction exists.

Action Taken:

A malfunction DTC is stored, ABS is disabled and the ABS warning lamp is turned "ON."

DTC Chart Test Description:

Number(s) below refer to circled number(s) on the diagnostic chart.

1. This step is designed to identify a wheel speed sensor or circuitry that is damaged and visibly apparent.
2. Checks to see that the wheel speed sensor is within proper resistance values.
3. Verifies that the wheel speed sensor can generate an appropriate output signal.
4. Ensures that the wheel speed sensor circuitry is not shorted together.
5. Checks to see if malfunction was due to physical damage of the circuitry.
6. Checks for an incorrect or malfunctioning hub and bearing assembly.

Diagnostic Aids:

An "Intermittent" malfunction may be caused by a poor connection, rubbed through wire insulation, or a wire that is broken inside the insulation.

The frequency of the malfunction can be checked by using the enhanced diagnostic function of the Tech 1.

If the customer's comments reflect that the ABS warning lamp is "ON" only during moist environmental changes (rain, snow, vehicle wash), all wheel speed sensor circuitry should be thoroughly inspected for signs of water intrusion. If DTC is not current, clear DTCs and simulate the effects of water intrusion. Use the following procedure. Spray the suspected area with a 5% salt water solution (two teaspoons of salt to 12 oz. of water). Test drive vehicle over various road surfaces (bumps, turns, etc.) above 24 km/h (15 mph) for at least 30 seconds. If DTC returns, replace suspected harness.

Any circuitry that is suspected of causing the intermittent complaint should be thoroughly checked for backed out terminals, improper mating, broken locks, improperly formed or damaged terminals, poor terminal to wiring connections or physical damage to the wiring harness.

Resistance of the wheel speed sensor will increase with an increase in sensor temperature.

When replacing a wheel speed sensor, inspect the sensor terminals and harness connector for corrosion and/or water intrusion. If evidence of corrosion or water intrusion exists, replace wheel speed sensor jumper harness. Likewise, if replacing a wheel speed sensor jumper harness, inspect sensor terminals. If evidence of corrosion or water intrusion exists, replace wheel speed sensor.

DIAGNOSTIC CHART DTC 24 — 1995 SILHOUETTE, TRANS SPORT AND LUMINA APV

DTC 24
RIGHT REAR WHEEL SPEED = 0

Important

- WHEEL SPEED SENSOR INTERMITTENT MALFUNCTIONS MAY BE DIFFICULT TO LOCATE. CARE SHOULD BE TAKEN NOT TO DISTURB ANY ELECTRICAL CONNECTIONS PRIOR TO AN INDICATED STEP OF THIS CHART. THIS WILL ENSURE THAT AN INTERMITTENT CONNECTION WILL NOT BE CORRECTED BEFORE THE SOURCE OF THE MALFUNCTION IS FOUND.

(1)
- IGNITION "OFF."
- PHYSICALLY INSPECT THE RIGHT REAR WHEEL SPEED SENSOR, WHEEL SPEED SENSOR JUMPER HARNESS, AND CONNECTORS FOR DAMAGE.

IS PHYSICAL DAMAGE INDICATED?

YES → REPLACE WHEEL SPEED SENSOR AND/OR WHEEL SPEED SENSOR JUMPER HARNESS IF DAMAGED. DRIVE VEHICLE ABOVE 24 km/h (15 MPH) TO ENSURE DTC 24 DOES NOT SET.

NO ↓

(2)
- DISCONNECT RIGHT REAR WHEEL SPEED SENSOR DIRECTLY AT SENSOR.
- USING J 39200, MEASURE THE RESISTANCE BETWEEN SENSOR TERMINALS "A" AND "B." IS RESISTANCE BETWEEN 2059 AND 2353Ω WHEN SENSOR IS 20°C (68°F)? *(IF TEMPERATURE OF THE SENSOR IS SUSPECTED TO BE OTHER THAN 20°C (68°F), REFER TO THE TABLE BELOW FOR APPLICABLE SENSOR RESISTANCE VALUES; THE VALUES IN THE TABLE ARE FOR TEMPERATURE OF THE SENSOR, NOT AIR TEMPERATURE.)*

NO → REPLACE WHEEL SPEED SENSOR.

YES ↓

(3)
- WITH J 39200 STILL CONNECTED, SELECT A/C VOLTAGE SCALE.
- SPIN WHEEL BY HAND WHILE OBSERVING VOLTAGE.

IS VOLTAGE AT LEAST 100 mV (VOLTAGE WILL INCREASE AS WHEEL SPEED INCREASES)?

NO → REPLACE WHEEL SPEED SENSOR.

YES ↓

(4)
- IGNITION "OFF."
- DISCONNECT 32-WAY EBCM CONNECTOR.
- USING J 39200, MEASURE THE RESISTANCE BETWEEN TERMINAL "E5" AND "E6" OF THE 32-WAY EBCM HARNESS CONNECTOR.

DOES J 39200 DISPLAY RESISTANCE AS "OL" (INFINITE)?

NO → REPAIR SHORT BETWEEN CKTs 882 AND 883.

YES ↓

(5)
- INSPECT CKTs 882 AND 883 AND 32-WAY EBCM CONNECTOR FOR DAMAGE WHICH MAY RESULT IN A SHORT BETWEEN CKTs 882 AND 883 WITH ALL CONNECTORS CONNECTED. REPAIR ALL DAMAGE IF EVIDENT.
- RECONNECT ALL CONNECTORS.
- TEST DRIVE VEHICLE ABOVE 24 km/h (15 MPH) FOR AT LEAST 30 SECONDS.

DOES DTC 24 SET AS A CURRENT DTC?

NO → MALFUNCTION WAS MOST LIKELY CAUSED BY A POOR CONNECTION. REFER TO "DIAGNOSTIC AIDS" ON FACING PAGE.

YES ↓

(6)
- USING TECH 1, SELECT "F0: DATA LIST."
- MONITOR WHEEL SPEEDS WHILE SLOWLY ACCELERATING TO 65 km/h (40 MPH) AND THEN SLOWLY DECELERATING TO 0 km/h (0 MPH).

IS THE RIGHT REAR WHEEL SPEED CONSTANTLY HIGHER THAN THE THREE REMAINING WHEEL SPEEDS?

NO → REPLACE EBCM.

YES → REPLACE INCORRECT OR MALFUNCTIONING RIGHT WHEEL SPEED SENSOR/HUB AND BEARING ASSEMBLY WITH THE PROPER ASSEMBLY. REFER TO PARTS CATALOG FOR PROPER PART NUMBER.

REAR WHEEL SPEED SENSOR
TEMPERATURE OF SENSOR VS. SENSOR RESISTANCE (APPROXIMATE)

TEMP. (°C)	TEMP. (°F)	RESISTANCE (OHMS)
-34 TO 4	-30 TO 40	1622 TO 2205
5 TO 43	41 TO 110	1930 TO 2566
44 TO 93	111 TO 200	2245 TO 3028

DIAGNOSTIC CHART DTC 25 — 1995 SILHOUETTE, TRANS SPORT AND LUMINA APV

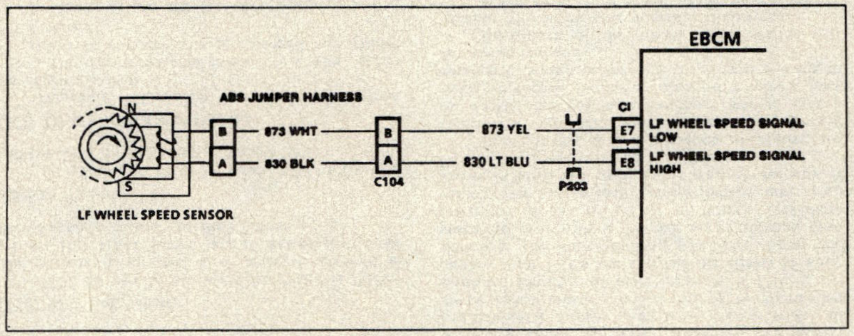

DTC 25
LEFT FRONT EXCESSIVE WHEEL SPEED VARIATION

Circuit Description:

As a toothed ring passes by the wheel speed sensor, changes in the electromagnetic field cause the wheel speed sensor to produce a sinusoidal (AC) voltage signal whose frequency is proportional to wheel speed. The magnitude of this signal is directly related to wheel speed and the proximity of the wheel speed sensor to the toothed ring, often referred to as the air gap.

DTC Will Set When:

DTC 25 can be set when the brake is "OFF." The purpose of this test is to detect a situation in which the left front wheel acceleration or deceleration is beyond specified limits.

Action Taken:

A malfunction DTC is stored, ABS is disabled and the ABS warning lamp is turned "ON."

DTC Chart Test Description:

Number(s) below refer to circled number(s) on the diagnostic chart.

1. This step is designed to identify a wheel speed sensor or circuitry that is damaged and visibly apparent.
2. Checks to see that the wheel speed sensor is within proper resistance values.
3. Verifies that the wheel speed sensor can generate an appropriate output signal.
4. Ensures that wheel speed sensor circuitry is not shorted together.
5. Ensures malfunction was not due to physical damage of the circuitry.

Diagnostic Aids:

An "Intermittent" malfunction may be caused by a poor connection, rubbed through wire insulation, or a wire that is broken inside the insulation.

The frequency of the malfunction can be checked by using the enhanced diagnostic function of the Tech 1,

If the customer's comments reflect that the ABS warning lamp is "ON" only during moist environmental changes (rain, snow, vehicle wash), all wheel speed sensor circuitry should be thoroughly inspected for signs of water intrusion. If DTC is not current, clear DTCs and simulate the effects of water intrusion. Use the following procedure. Spray the suspected area with a 5% salt water solution (two teaspoons of salt to 12 oz. of water). Test drive vehicle over various road surfaces (bumps, turns, etc.) above 24 km/h (15 mph) for at least 30 seconds. If DTC returns, replace suspected harness.

Any circuitry that is suspected of causing the intermittent complaint should be thoroughly checked for backed out terminals, improper mating, broken locks, improperly formed or damaged terminals, poor terminal to wiring connections or physical damage to the wiring harness.

Resistance of the wheel speed sensor will increase with an increase in sensor temperature.

When replacing a wheel speed sensor, inspect the sensor terminals and harness connector for corrosion and/or water intrusion. If evidence of corrosion or water intrusion exists, replace wheel speed sensor jumper harness. Likewise, if replacing a wheel speed sensor jumper harness, inspect sensor terminals. If evidence of corrosion or water intrusion exists, replace wheel speed sensor.

DIAGNOSTIC CHART DTC 25 — 1995 SILHOUETTE, TRANS SPORT AND LUMINA APV

DTC 25
LEFT FRONT EXCESSIVE WHEEL SPEED VARIATION

Important
- WHEEL SPEED SENSOR INTERMITTENT MALFUNCTIONS MAY BE DIFFICULT TO LOCATE. CARE SHOULD BE TAKEN NOT TO DISTURB ANY ELECTRICAL CONNECTIONS PRIOR TO AN INDICATED STEP OF THIS CHART. THIS WILL ENSURE THAT AN INTERMITTENT CONNECTION WILL NOT BE CORRECTED BEFORE THE SOURCE OF THE MALFUNCTION IS FOUND.

(1)
- IGNITION "OFF."
- PHYSICALLY INSPECT THE LEFT FRONT WHEEL SPEED SENSOR, LEFT FRONT WHEEL SPEED SENSOR JUMPER HARNESS, AND CONNECTORS FOR DAMAGE. IS PHYSICAL DAMAGE INDICATED?

YES: REPLACE WHEEL SPEED SENSOR AND/OR JUMPER HARNESS IF DAMAGED. DRIVE VEHICLE ABOVE 24 km/h (15 MPH) TO ENSURE DTC 25 DOES NOT SET.

NO:

(2)
- DISCONNECT LEFT FRONT WHEEL SPEED SENSOR DIRECTLY AT SENSOR.
- USING J 39200, MEASURE THE RESISTANCE BETWEEN SENSOR TERMINALS "A" AND "B."
  IS RESISTANCE BETWEEN 1020 AND 1137Ω WHEN SENSOR IS 20°C (68°F)?
  (IF TEMPERATURE OF THE SENSOR IS SUSPECTED TO BE OTHER THAN 20°C (68°F), REFER TO THE TABLE BELOW FOR APPLICABLE SENSOR RESISTANCE VALUES; THE VALUES IN THE TABLE ARE FOR TEMPERATURE OF THE SENSOR, NOT AIR TEMPERATURE.)

NO: REPLACE WHEEL SPEED SENSOR.

YES:

(3)
- WITH J 39200 STILL CONNECTED, SELECT A/C VOLTAGE SCALE.
- SPIN WHEEL BY HAND WHILE OBSERVING VOLTAGE.
  IS VOLTAGE AT LEAST 100 mV (VOLTAGE WILL INCREASE AS WHEEL SPEED INCREASES)?

NO: REPLACE WHEEL SPEED SENSOR.

YES:

(4)
- IGNITION "OFF."
- DISCONNECT 32-WAY EBCM CONNECTOR.
- USING J 39200, MEASURE THE RESISTANCE BETWEEN TERMINAL "E7" AND "E8" OF THE 32-WAY EBCM HARNESS CONNECTOR.
  DOES J 39200 DISPLAY RESISTANCE AS "OL" (INFINITE)?

NO: REPAIR SHORT BETWEEN CKTs 830 AND 873.

YES:

(5)
- INSPECT CKTs 830 AND 873 AND 32-WAY EBCM CONNECTOR FOR DAMAGE WHICH MAY RESULT IN A SHORT BETWEEN CKTs 830 AND 873 WITH ALL CONNECTORS CONNECTED. REPAIR ALL DAMAGE IF EVIDENT.
- RECONNECT ALL CONNECTORS.
- TEST DRIVE VEHICLE ABOVE 24 km/h (15 MPH) FOR AT LEAST 30 SECONDS. DOES DTC 25 SET AS A CURRENT DTC?

YES: REPLACE EBCM.

NO: MALFUNCTION IS INTERMITTENT. REFER TO "DIAGNOSTIC AIDS" ON FACING PAGE.

FRONT WHEEL SPEED SENSOR
TEMPERATURE OF SENSOR VS. SENSOR RESISTANCE (APPROXIMATE)

TEMP. (°C)	TEMP. (°F)	RESISTANCE (OHMS)
-34 TO 4	-30 TO 40	804 TO 1066
5 TO 43	41 TO 110	956 TO 1240
44 TO 93	111 TO 200	1112 TO 1463

DIAGNOSTIC CHART DTC 26 — 1995 SILHOUETTE, TRANS SPORT AND LUMINA APV

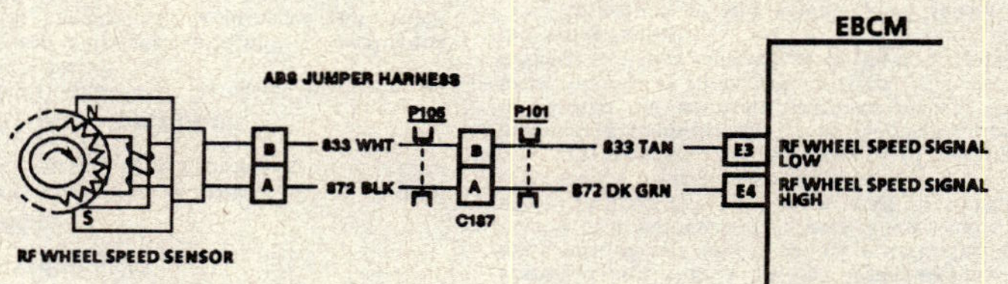

DTC 26
RIGHT FRONT EXCESSIVE WHEEL SPEED VARIATION

Circuit Description:

As a toothed ring passes by the wheel speed sensor, changes in the electromagnetic field cause the wheel speed sensor to produce a sinusoidal (AC) voltage signal whose frequency is proportional to wheel speed. The magnitude of this signal is directly related to wheel speed and the proximity of the wheel speed sensor to the toothed ring, often referred to as the air gap.

DTC Will Set When:

DTC 26 can be set when the brake is "OFF." The purpose of this test is to detect a situation in which the right front wheel acceleration or deceleration is beyond specified limits.

Action Taken:

A malfunction DTC is stored, ABS is disabled and the ABS warning lamp is turned "ON."

DTC Chart Test Description:

Number(s) below refer to circled number(s) on the diagnostic chart.

1. This step is designed to identify a wheel speed sensor or circuitry that is damaged and visibly apparent.
2. Checks to see that the wheel speed sensor is within proper resistance values.
3. Verifies that the wheel speed sensor can generate an appropriate output signal.
4. Ensures that wheel speed sensor circuitry is not shorted together.
5. Ensures malfunction was not due to physical damage of the circuitry.

Diagnostic Aids:

An "Intermittent" malfunction may be caused by a poor connection, rubbed through wire insulation, or a wire that is broken inside the insulation.

The frequency of the malfunction can be checked by using the enhanced diagnostic function of the Tech 1.

If the customer's comments reflect that the ABS warning lamp is "ON" only during moist environmental changes (rain, snow, vehicle wash), all wheel speed sensor circuitry should be thoroughly inspected for signs of water intrusion. If DTC is not current, clear DTCs and simulate the effects of water intrusion. Use the following procedure. Spray the suspected area with a 5% salt water solution (two teaspoons of salt to 12 oz. of water). Test drive vehicle over various road surfaces (bumps, turns, etc.) above 24 km/h (15 mph) for at least 30 seconds. If DTC returns, replace suspected harness.

Any circuitry that is suspected of causing the intermittent complaint should be thoroughly checked for backed out terminals, improper mating, broken locks, improperly formed or damaged terminals, poor terminal to wiring connections or physical damage to the wiring harness.

Resistance of the wheel speed sensor will increase with an increase in sensor temperature.

When replacing a wheel speed sensor, inspect the sensor terminals and harness connector for corrosion and/or water intrusion. If evidence of corrosion or water intrusion exists, replace wheel speed sensor jumper harness. Likewise, if replacing a wheel speed sensor jumper harness, inspect sensor terminals. If evidence of corrosion or water intrusion exists, replace wheel speed sensor.

DIAGNOSTIC CHART DTC 26 — 1995 SILHOUETTE, TRANS SPORT AND LUMINA APV

DTC 26
RIGHT FRONT EXCESSIVE WHEEL SPEED VARIATION

Important
- WHEEL SPEED SENSOR INTERMITTENT MALFUNCTIONS MAY BE DIFFICULT TO LOCATE. CARE SHOULD BE TAKEN NOT TO DISTURB ANY ELECTRICAL CONNECTIONS PRIOR TO AN INDICATED STEP OF THIS CHART. THIS WILL ENSURE THAT AN INTERMITTENT CONNECTION WILL NOT BE CORRECTED BEFORE THE SOURCE OF THE MALFUNCTION IS FOUND.

(1)
- IGNITION "OFF."
- PHYSICALLY INSPECT THE RIGHT FRONT WHEEL SPEED SENSOR, RIGHT FRONT WHEEL SPEED SENSOR JUMPER HARNESS, AND CONNECTORS FOR DAMAGE.

IS PHYSICAL DAMAGE INDICATED?

YES: REPLACE WHEEL SPEED SENSOR AND/OR JUMPER HARNESS IF DAMAGED. DRIVE VEHICLE ABOVE 24 km/h (15 MPH) TO ENSURE DTC 26 DOES NOT SET.

NO: (2)
- DISCONNECT RIGHT FRONT WHEEL SPEED SENSOR DIRECTLY AT SENSOR.
- USING J 39200, MEASURE THE RESISTANCE BETWEEN SENSOR TERMINALS "A" AND "B."

IS RESISTANCE BETWEEN 1020 AND 1137Ω WHEN SENSOR IS 20°C (68°F)?
(IF TEMPERATURE OF THE SENSOR IS SUSPECTED TO BE OTHER THAN 20°C (68°F), REFER TO THE TABLE BELOW FOR APPLICABLE SENSOR RESISTANCE VALUES; THE VALUES IN THE TABLE ARE FOR TEMPERATURE OF THE SENSOR, NOT AIR TEMPERATURE.)

NO: REPLACE WHEEL SPEED SENSOR.

YES: (3)
- WITH J 39200 STILL CONNECTED, SELECT A/C VOLTAGE SCALE.
- SPIN WHEEL BY HAND WHILE OBSERVING VOLTAGE.

IS VOLTAGE AT LEAST 100 mV (VOLTAGE WILL INCREASE AS WHEEL SPEED INCREASES)?

NO: REPLACE WHEEL SPEED SENSOR.

YES: (4)
- IGNITION "OFF."
- DISCONNECT 32-WAY EBCM CONNECTOR.
- USING J 39200, MEASURE THE RESISTANCE BETWEEN TERMINAL "E3" AND "E4" OF THE 32-WAY EBCM HARNESS CONNECTOR.

DOES J 39200 DISPLAY RESISTANCE AS "OL" (INFINITE)?

NO: REPAIR SHORT BETWEEN CKTs 833 AND 872.

YES: (5)
- INSPECT CKTs 833 AND 872 AND 32-WAY EBCM CONNECTOR FOR DAMAGE WHICH MAY RESULT IN A SHORT BETWEEN CKTs 833 AND 872 WITH ALL CONNECTORS CONNECTED. REPAIR ALL DAMAGE IF EVIDENT.
- RECONNECT ALL CONNECTORS.
- TEST DRIVE VEHICLE ABOVE 24 km/h (15 MPH) FOR AT LEAST 30 SECONDS.

DOES DTC 26 SET AS A CURRENT DTC?

YES: REPLACE EBCM.

NO: MALFUNCTION IS INTERMITTENT. REFER TO "DIAGNOSTIC AIDS" ON FACING PAGE.

FRONT WHEEL SPEED SENSOR
TEMPERATURE OF SENSOR VS. SENSOR RESISTANCE (APPROXIMATE)

TEMP. (°C)	TEMP. (°F)	RESISTANCE (OHMS)
-34 TO 4	-30 TO 40	804 TO 1066
5 TO 43	41 TO 110	956 TO 1240
44 TO 93	111 TO 200	1112 TO 1463

DIAGNOSTIC CHART DTC 27 — 1995 SILHOUETTE, TRANS SPORT AND LUMINA APV

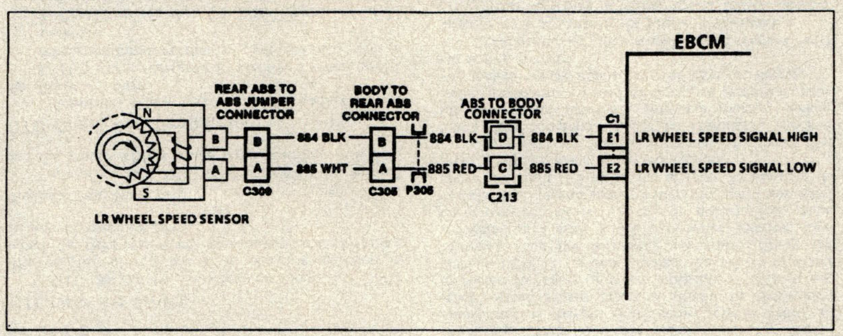

DTC 27
LEFT REAR EXCESSIVE WHEEL SPEED VARIATION

Circuit Description:

As a toothed ring passes by the wheel speed sensor, changes in the electromagnetic field cause the wheel speed sensor to produce a sinusoidal (AC) voltage signal whose frequency is proportional to wheel speed. The magnitude of this signal is directly related to wheel speed and the proximity of the wheel speed sensor to the toothed ring, often referred to as the air gap.

DTC Will Set When:

DTC 27 can be set when the brake is "OFF." The purpose of this test is to detect a situation in which the left rear wheel acceleration or deceleration is beyond specified limits.

Action Taken:

A malfunction DTC is stored, ABS is disabled and the ABS warning lamp is turned "ON."

DTC Chart Test Description:

Number(s) below refer to circled number(s) on the diagnostic chart.

1. This step is designed to identify a wheel speed sensor or circuitry that is damaged and visibly apparent.
2. Checks to see that the wheel speed sensor is within proper resistance values.
3. Verifies that the wheel speed sensor can generate an appropriate output signal.
4. Ensures that wheel speed sensor circuitry is not shorted together.
5. Ensures malfunction was not due to physical damage of the circuitry.

Diagnostic Aids:

An "Intermittent" malfunction may be caused by a poor connection, rubbed through wire insulation, or a wire that is broken inside the insulation.

The frequency of the malfunction can be checked by using the enhanced diagnostic function of the Tech 1,

If the customer's comments reflect that the ABS warning lamp is "ON" only during moist environmental changes (rain, snow, vehicle wash), all wheel speed sensor circuitry should be thoroughly inspected for signs of water intrusion. If DTC is not current, clear DTCs and simulate the effects of water intrusion. Use the following procedure. Spray the suspected area with a 5% salt water solution (two teaspoons of salt to 12 oz. of water). Test drive vehicle over various road surfaces (bumps, turns, etc.) above 24 km/h (15 mph) for at least 30 seconds. If DTC returns, replace suspected harness.

Any circuitry that is suspected of causing the intermittent complaint should be thoroughly checked for backed out terminals, improper mating, broken locks, improperly formed or damaged terminals, poor terminal to wiring connections or physical damage to the wiring harness.

Resistance of the wheel speed sensor will increase with an increase in sensor temperature.

When replacing a wheel speed sensor, inspect the sensor terminals and harness connector for corrosion and/or water intrusion. If evidence of corrosion or water intrusion exists, replace wheel speed sensor jumper harness. Likewise, if replacing a wheel speed sensor jumper harness, inspect sensor terminals. If evidence of corrosion or water intrusion exists, replace wheel speed sensor.

DIAGNOSTIC CHART DTC 27 — 1995 SILHOUETTE, TRANS SPORT AND LUMINA APV

DTC 27
LEFT REAR EXCESSIVE WHEEL SPEED VARIATION

Important
- WHEEL SPEED SENSOR INTERMITTENT MALFUNCTIONS MAY BE DIFFICULT TO LOCATE. CARE SHOULD BE TAKEN NOT TO DISTURB ANY ELECTRICAL CONNECTIONS PRIOR TO AN INDICATED STEP OF THIS CHART. THIS WILL ENSURE THAT AN INTERMITTENT CONNECTION WILL NOT BE CORRECTED BEFORE THE SOURCE OF THE MALFUNCTION IS FOUND.

(1)
- IGNITION "OFF."
- PHYSICALLY INSPECT THE LEFT REAR WHEEL SPEED SENSOR, WHEEL SPEED SENSOR JUMPER HARNESS, AND CONNECTORS FOR DAMAGE.

IS PHYSICAL DAMAGE INDICATED?

- YES: REPLACE WHEEL SPEED SENSOR AND/OR WHEEL SPEED SENSOR JUMPER HARNESS IF DAMAGED. DRIVE VEHICLE ABOVE 24 km/h (15 MPH) TO ENSURE DTC 27 DOES NOT SET.
- NO: go to (2)

(2)
- DISCONNECT LEFT REAR WHEEL SPEED SENSOR DIRECTLY AT SENSOR.
- USING J 39200, MEASURE THE RESISTANCE BETWEEN SENSOR TERMINALS "A" AND "B."

IS RESISTANCE BETWEEN 2059 AND 2353Ω WHEN SENSOR IS 20°C (68°F)?
(IF TEMPERATURE OF THE SENSOR IS SUSPECTED TO BE OTHER THAN 20°C (68°F), REFER TO THE TABLE BELOW FOR APPLICABLE SENSOR RESISTANCE VALUES; THE VALUES IN THE TABLE ARE FOR TEMPERATURE OF THE SENSOR, NOT AIR TEMPERATURE.)

- NO: REPLACE WHEEL SPEED SENSOR.
- YES: go to (3)

(3)
- WITH J 39200 STILL CONNECTED, SELECT A/C VOLTAGE SCALE.
- SPIN WHEEL BY HAND WHILE OBSERVING VOLTAGE.

IS VOLTAGE AT LEAST 100 mV (VOLTAGE WILL INCREASE AS WHEEL SPEED INCREASES)?

- NO: REPLACE WHEEL SPEED SENSOR.
- YES: go to (4)

(4)
- IGNITION "OFF."
- DISCONNECT 32-WAY EBCM CONNECTOR.
- USING J 39200, MEASURE THE RESISTANCE BETWEEN TERMINAL "E1" AND "E2" OF THE 32-WAY EBCM HARNESS CONNECTOR.

DOES J 39200 DISPLAY RESISTANCE AS "OL" (INFINITE)?

- NO: REPAIR SHORT BETWEEN CKTs 884 AND 885.
- YES: go to (5)

(5)
- INSPECT CKTs 884 AND 885 AND 32-WAY EBCM CONNECTOR FOR DAMAGE WHICH MAY RESULT IN A SHORT BETWEEN CKTs 884 AND 885 WITH ALL CONNECTORS CONNECTED. REPAIR ALL DAMAGE IF EVIDENT.
- RECONNECT ALL CONNECTORS.
- TEST DRIVE VEHICLE ABOVE 24 km/h (15 MPH) FOR AT LEAST 30 SECONDS.

DOES DTC 27 SET AS A CURRENT DTC?

- YES: REPLACE EBCM.
- NO: MALFUNCTION IS INTERMITTENT. REFER TO "DIAGNOSTIC AIDS" ON FACING PAGE.

REAR WHEEL SPEED SENSOR
TEMPERATURE OF SENSOR VS. SENSOR RESISTANCE (APPROXIMATE)

TEMP. (°C)	TEMP. (°F)	RESISTANCE (OHMS)
-34 TO 4	-30 TO 40	1622 TO 2205
5 TO 43	41 TO 110	1930 TO 2566
44 TO 93	111 TO 200	2245 TO 3028

DIAGNOSTIC CHART DTC 28 — 1995 SILHOUETTE, TRANS SPORT AND LUMINA APV

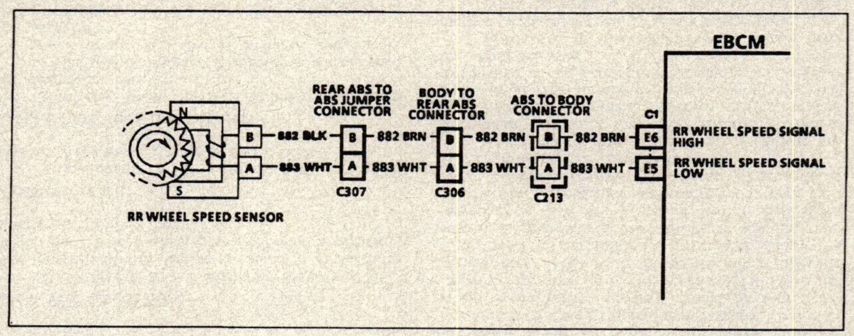

DTC 28
RIGHT REAR EXCESSIVE WHEEL SPEED VARIATION

Circuit Description:

As a toothed ring passes by the wheel speed sensor, changes in the electromagnetic field cause the wheel speed sensor to produce a sinusoidal (AC) voltage signal whose frequency is proportional to wheel speed. The magnitude of this signal is directly related to wheel speed and the proximity of the wheel speed sensor to the toothed ring, often referred to as the air gap.

DTC Will Set When:

DTC 28 can be set when the brake is "OFF." The purpose of this test is to detect a situation in which the right rear wheel acceleration or deceleration is beyond specified limits.

Action Taken:

A malfunction DTC is stored, ABS is disabled and the amber ABS warning lamp is turned "ON."

DTC Chart Test Description:

Number(s) below refer to circled number(s) on the diagnostic chart.

1. This step is designed to identify a wheel speed sensor or circuitry that is damaged and visibly apparent.
2. Checks to see that the wheel speed sensor is within proper resistance values.
3. Verifies that the wheel speed sensor can generate an appropriate output signal.
4. Ensures that wheel speed sensor circuitry is not shorted together.
5. Ensures malfunction was not due to physical damage of the circuitry.

Diagnostic Aids:

An "Intermittent" malfunction may be caused by a poor connection, rubbed through wire insulation, or a wire that is broken inside the insulation.

The frequency of the malfunction can be checked by using the enhanced diagnostic function of the Tech 1.

If the customer's comments reflect that the ABS warning lamp is "ON" only during moist environmental changes (rain, snow, vehicle wash), all wheel speed sensor circuitry should be thoroughly inspected for signs of water intrusion. If DTC is not current, clear DTCs and simulate the effects of water intrusion. Use the following procedure. Spray the suspected area with a 5% salt water solution (two teaspoons of salt to 12 oz. of water). Test drive vehicle over various road surfaces (bumps, turns, etc.) above 24 km/h (15 mph) for at least 30 seconds. If DTC returns, replace suspected harness.

Any circuitry that is suspected of causing the intermittent complaint should be thoroughly checked for backed out terminals, improper mating, broken locks, improperly formed or damaged terminals, poor terminal to wiring connections or physical damage to the wiring harness.

Resistance of the wheel speed sensor will increase with an increase in sensor temperature.

When replacing a wheel speed sensor, inspect the sensor terminals and harness connector for corrosion and/or water intrusion. If evidence of corrosion or water intrusion exists, replace wheel speed sensor jumper harness. Likewise, if replacing a wheel speed sensor jumper harness, inspect sensor terminals. If evidence of corrosion or water intrusion exists, replace wheel speed sensor.

DIAGNOSTIC CHART DTC 28 — 1995 SILHOUETTE, TRANS SPORT AND LUMINA APV

DTC 28
RIGHT REAR EXCESSIVE WHEEL SPEED VARIATION

Important

- WHEEL SPEED SENSOR INTERMITTENT MALFUNCTIONS MAY BE DIFFICULT TO LOCATE. CARE SHOULD BE TAKEN NOT TO DISTURB ANY ELECTRICAL CONNECTIONS PRIOR TO AN INDICATED STEP OF THIS CHART. THIS WILL ENSURE THAT AN INTERMITTENT CONNECTION WILL NOT BE CORRECTED BEFORE THE SOURCE OF THE MALFUNCTION IS FOUND.

(1)
- IGNITION "OFF."
- PHYSICALLY INSPECT THE RIGHT REAR WHEEL SPEED SENSOR, WHEEL SPEED SENSOR JUMPER HARNESS, AND CONNECTORS FOR DAMAGE.

IS PHYSICAL DAMAGE INDICATED?

YES: REPLACE WHEEL SPEED SENSOR AND/OR WHEEL SPEED SENSOR JUMPER HARNESS IF DAMAGED. DRIVE VEHICLE ABOVE 24 km/h (15 MPH) TO ENSURE DTC 28 DOES NOT SET.

NO:

(2)
- DISCONNECT RIGHT REAR WHEEL SPEED SENSOR DIRECTLY AT SENSOR.
- USING J 39200, MEASURE THE RESISTANCE BETWEEN SENSOR TERMINALS "A" AND "B."

IS RESISTANCE BETWEEN 2059 AND 2353Ω WHEN SENSOR IS 20°C (68°F)? *(IF TEMPERATURE OF THE SENSOR IS SUSPECTED TO BE OTHER THAN 20°C (68°F), REFER TO THE TABLE BELOW FOR APPLICABLE SENSOR RESISTANCE VALUES; THE VALUES IN THE TABLE ARE FOR TEMPERATURE OF THE SENSOR, NOT AIR TEMPERATURE.)*

NO: REPLACE WHEEL SPEED SENSOR.

YES:

(3)
- WITH J 39200 STILL CONNECTED, SELECT A/C VOLTAGE SCALE.
- SPIN WHEEL BY HAND WHILE OBSERVING VOLTAGE.

IS VOLTAGE AT LEAST 100 mV (VOLTAGE WILL INCREASE AS WHEEL SPEED INCREASES)?

NO: REPLACE WHEEL SPEED SENSOR.

YES:

(4)
- IGNITION "OFF."
- DISCONNECT 32-WAY EBCM CONNECTOR.
- USING J 39200, MEASURE THE RESISTANCE BETWEEN TERMINAL "E5" AND "E6" OF THE 32-WAY EBCM HARNESS CONNECTOR.

DOES J 39200 DISPLAY RESISTANCE AS "OL" (INFINITE)?

NO: REPAIR SHORT BETWEEN CKTs 882 AND 883.

YES:

(5)
- INSPECT CKTs 882 AND 883 AND 32-WAY EBCM CONNECTOR FOR DAMAGE WHICH MAY RESULT IN A SHORT BETWEEN CKTs 882 AND 883 WITH ALL CONNECTORS CONNECTED. REPAIR ALL DAMAGE IF EVIDENT.
- RECONNECT ALL CONNECTORS.
- TEST DRIVE VEHICLE ABOVE 24 km/h (15 MPH) FOR AT LEAST 30 SECONDS.

DOES DTC 28 SET AS A CURRENT DTC?

YES: REPLACE EBCM.

NO: MALFUNCTION IS INTERMITTENT. REFER TO "DIAGNOSTIC AIDS" ON FACING PAGE.

REAR WHEEL SPEED SENSOR

TEMPERATURE OF SENSOR VS. SENSOR RESISTANCE (APPROXIMATE)

TEMP. (°C)	TEMP. (°F)	RESISTANCE (OHMS)
-34 TO 4	-30 TO 40	1622 TO 2205
5 TO 43	41 TO 110	1930 TO 2566
44 TO 93	111 TO 200	2245 TO 3028

DIAGNOSTIC CHART DTC 32 — 1995 SILHOUETTE, TRANS SPORT AND LUMINA APV

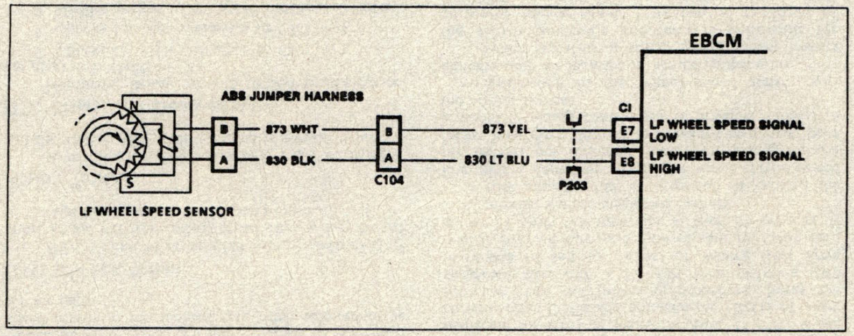

DTC 32
(Page 1 of 3)
LF WHEEL SPEED SNR CKT OPEN OR SHORT TO GND/BAT

Circuit Description:

As a toothed ring passes by the wheel speed sensor, changes in the electromagnetic field cause the wheel speed sensor to produce a sinusoidal (AC) voltage signal whose frequency is proportional to wheel speed. The magnitude of this signal is directly related to wheel speed and the proximity of the wheel speed sensor to the toothed ring often referred to as the air gap.

DTC Will Set When:

DTC 32 can be set anytime after initialization. If either of the left front wheel speed circuits are shorted to voltage or ground, a malfunction exists.

Action Taken:

A malfunction DTC is stored, ABS is disabled and the amber ABS warning lamp is turned "ON."

DTC Chart Test Description:

Number(s) below refer to circled number(s) on the diagnostic chart.

1. Checks to see that the wheel speed sensor is within proper resistance values.
2. Checks for short to ground in wheel speed sensor wiring.
3. Checks for short to voltage in wheel speed sensor wiring.
4. Checks for a short to voltage in CKT 873.
5. Checks for a short to voltage in CKT 830.
6. Ensures malfunction was not due to physical damage of circuitry.

Diagnostic Aids:

An "Intermittent" malfunction may be caused by a poor connection, rubbed through wire insulation, or a wire that is broken inside the insulation.

The frequency of the malfunction can be checked by using the enhanced diagnostic function of the Tech 1,

If the customer's comments reflect that the ABS warning lamp is "ON" only during moist environmental changes (rain, snow, vehicle wash), all wheel speed sensor circuitry should be thoroughly inspected for signs of water intrusion. If DTC is not current, clear DTCs and simulate the effects of water intrusion. Use the following procedure. Spray the suspected area with a 5% salt water solution (two teaspoons of salt to 12 oz. of water). Test drive vehicle over various road surfaces (bumps, turns, etc.) above 24 km/h (15 mph) for at least 30 seconds. If DTC returns, replace suspected harness.

Any circuitry that is suspected of causing the intermittent complaint should be thoroughly checked for backed out terminals, improper mating, broken locks, improperly formed or damaged terminals, poor terminal to wiring connections or physical damage to the wiring harness.

Resistance of the wheel speed sensor will increase with an increase in sensor temperature.

When replacing a wheel speed sensor, inspect the sensor terminals and harness connector for corrosion and/or water intrusion. If evidence of corrosion or water intrusion exists, replace wheel speed sensor jumper harness. Likewise, if replacing a wheel speed sensor jumper harness, inspect sensor terminals. If evidence of corrosion or water intrusion exists, replace wheel speed sensor.

DIAGNOSTIC CHART DTC 32 — 1995 SILHOUETTE, TRANS SPORT AND LUMINA APV

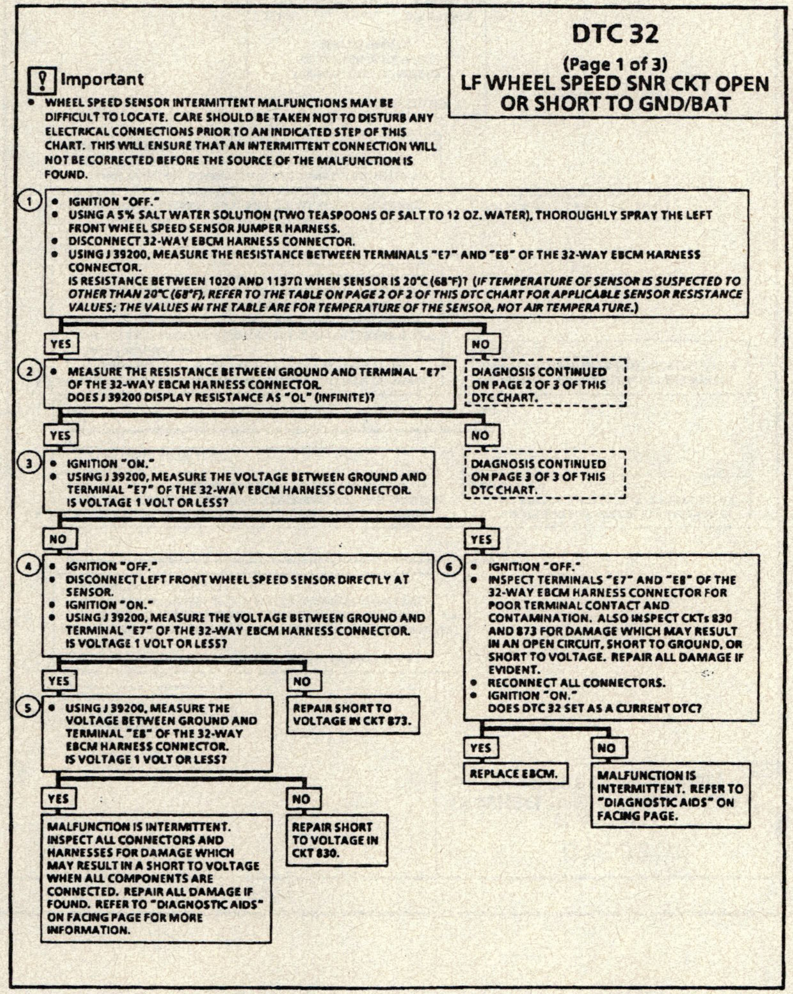

DIAGNOSTIC CHART DTC 32 — 1995 SILHOUETTE, TRANS SPORT AND LUMINA APV

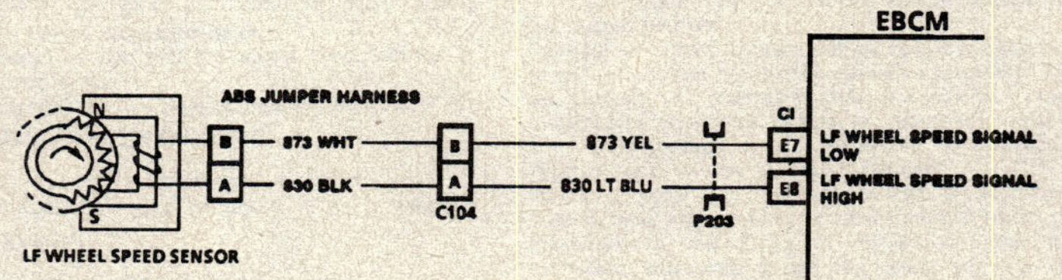

DTC 32
(Page 2 of 3)
LF WHEEL SPEED SNR CKT OPEN OR SHORT TO GND/BAT

Circuit Description:

As a toothed ring passes by the wheel speed sensor, changes in the electromagnetic field cause the wheel speed sensor to produce a sinusoidal (AC) voltage signal whose frequency is proportional to wheel speed. The magnitude of this signal is directly related to wheel speed and the proximity of the wheel speed sensor to the toothed ring often referred to as the air gap.

DTC Will Set When:

DTC 32 can be set anytime after initialization. If either of the left front wheel speed circuits are shorted to voltage or ground, a malfunction exists.

Action Taken:

A malfunction DTC is stored, ABS is disabled and the amber ABS warning lamp is turned "ON."

DTC Chart Test Description:

Number(s) below refer to circled number(s) on the diagnostic chart.

7. Checks for high resistance in CKT 873.
8. Checks for high resistance in CKT 830.
9. Verifies proper wheel speed sensor resistance.

Diagnostic Aids:

An "Intermittent" malfunction may be caused by a poor connection, rubbed through wire insulation, or a wire that is broken inside the insulation.

The frequency of the malfunction can be checked by using the enhanced diagnostic function of the Tech I.

If the customer's comments reflect that the ABS warning lamp is "ON" only during moist environmental changes (rain, snow, vehicle wash), all wheel speed sensor circuitry should be thoroughly inspected for signs of water intrusion. If DTC is not current, clear DTCs and simulate the effects of water intrusion. Use the following procedure. Spray the suspected area with a 5% salt water solution (two teaspoons of salt to 12 oz. of water). Test drive vehicle over various road surfaces (bumps, turns, etc.) above 24 km/h (15 mph) for at least 30 seconds. If DTC returns, replace suspected harness.

Any circuitry that is suspected of causing the intermittent complaint should be thoroughly checked for backed out terminals, improper mating, broken locks, improperly formed or damaged terminals, poor terminal to wiring connections or physical damage to the wiring harness.

Resistance of the wheel speed sensor will increase with an increase in sensor temperature.

When replacing a wheel speed sensor, inspect the sensor terminals and harness connector for corrosion and/or water intrusion. If evidence of corrosion or water intrusion exists, replace wheel speed sensor jumper harness. Likewise, if replacing a wheel speed sensor jumper harness, inspect sensor terminals. If evidence of corrosion or water intrusion exists, replace wheel speed sensor.

DIAGNOSTIC CHART DTC 32 — 1995 SILHOUETTE, TRANS SPORT AND LUMINA APV

DTC 32
(Page 2 of 3)
LF WHEEL SPEED SNR CKT OPEN OR SHORT TO GND/BATT

DIAGNOSIS CONTINUED FROM PAGE 1 OF 3 OF THIS DTC CHART.

(7)
- DISCONNECT LEFT FRONT WHEEL SPEED SENSOR DIRECTLY AT SENSOR.
- USING J 39200, MEASURE THE RESISTANCE BETWEEN TERMINAL "E7" OF THE 32-WAY EBCM HARNESS CONNECTOR AND TERMINAL "B" OF THE LEFT FRONT WHEEL SPEED SENSOR HARNESS CONNECTOR.

IS RESISTANCE 2Ω OR LESS?

NO: REPAIR HIGH RESISTANCE IN CKT 873 (WHEEL SPEED SENSOR JUMPER MUST BE REPLACED IF DAMAGED).

YES:

(8)
- USING J 39200, MEASURE THE RESISTANCE BETWEEN TERMINAL "E8" OF THE 32-WAY EBCM HARNESS CONNECTOR AND TERMINAL "A" OF THE LEFT FRONT WHEEL SPEED SENSOR HARNESS CONNECTOR.

IS RESISTANCE 2Ω OR LESS?

NO: REPAIR HIGH RESISTANCE IN CKT 830 (WHEEL SPEED SENSOR JUMPER MUST BE REPLACED IF DAMAGED).

YES:

(9)
- USING J 39200, MEASURE THE RESISTANCE BETWEEN TERMINAL "A" AND TERMINAL "B" OF THE LEFT FRONT WHEEL SPEED SENSOR CONNECTOR.

IS RESISTANCE BETWEEN 1020 AND 1137Ω WHEN SENSOR IS 20°C (68°F)? *(IF TEMPERATURE OF SENSOR IS SUSPECTED TO OTHER THAN 20°C (68°F), REFER TO THE TABLE BELOW FOR APPLICABLE SENSOR RESISTANCE VALUES; THE VALUES IN THE TABLE ARE FOR TEMPERATURE OF THE SENSOR, NOT AIR TEMPERATURE.)*

NO: REPLACE WHEEL SPEED SENSOR.

YES: MALFUNCTION IS INTERMITTENT. INSPECT ALL CONNECTORS AND HARNESSES FOR DAMAGE WHICH MAY RESULT IN HIGH RESISTANCE WHEN ALL COMPONENTS ARE CONNECTED. REFER TO "DIAGNOSTIC AIDS" ON FACING PAGE FOR MORE INFORMATION. REPAIR ALL DAMAGE IF FOUND.

FRONT WHEEL SPEED SENSOR

TEMPERATURE OF SENSOR VS. SENSOR RESISTANCE (APPROXIMATE)

TEMP. (°C)	TEMP. (°F)	RESISTANCE (OHMS)
-34 TO 4	-30 TO 40	804 TO 1066
5 TO 43	41 TO 110	956 TO 1240
44 TO 93	111 TO 200	1112 TO 1463

DIAGNOSTIC CHART DTC 32 — 1995 SILHOUETTE, TRANS SPORT AND LUMINA APV

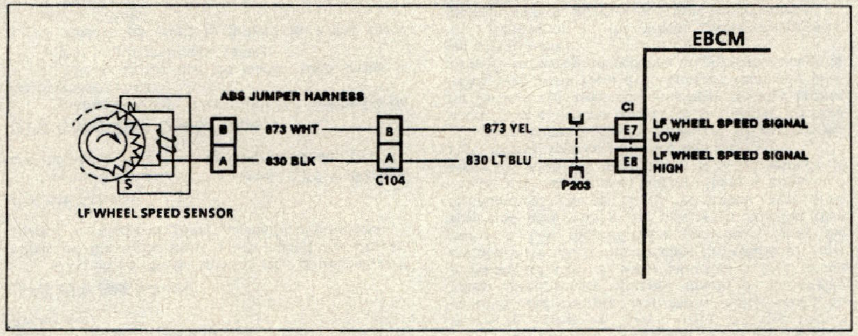

DTC 32
(Page 3 of 3)
LF WHEEL SPEED SNR CKT OPEN OR SHORT TO GND/BAT

Circuit Description:

As a toothed ring passes by the wheel speed sensor, changes in the electromagnetic field cause the wheel speed sensor to produce a sinusoidal (AC) voltage signal whose frequency is proportional to wheel speed. The magnitude of this signal is directly related to wheel speed and the proximity of the wheel speed sensor to the toothed ring often referred to as the air gap.

DTC Will Set When:

DTC 32 can be set anytime after initialization. If either of the left front wheel speed circuits are shorted to voltage or ground, a malfunction exists.

Action Taken:

A malfunction DTC is stored, ABS is disabled and the amber ABS warning lamp is turned "ON."

DTC Chart Test Description:

Number(s) below refer to circled number(s) on the diagnostic chart.

10. Checks for short to ground in CKT 873.
11. Checks for short to ground in CKT 830.
12. Checks for short to ground in wheel speed sensor jumper harness.
13. Verifies that wheel speed sensor is not internally shorted to ground.

Diagnostic Aids:

An "Intermittent" malfunction may be caused by a poor connection, rubbed through wire insulation, or a wire that is broken inside the insulation.

The frequency of the malfunction can be checked by using the enhanced diagnostic function of the Tech 1.

If the customer's comments reflect that the ABS warning lamp is "ON" only during moist environmental changes (rain, snow, vehicle wash), all wheel speed sensor circuitry should be thoroughly inspected for signs of water intrusion. If DTC is not current, clear DTCs and simulate the effects of water intrusion. Use the following procedure. Spray the suspected area with a 5% salt water solution (two teaspoons of salt to 12 oz. of water). Test drive vehicle over various road surfaces (bumps, turns, etc.) above 24 km/h (15 mph) for at least 30 seconds. If DTC returns, replace suspected harness.

Any circuitry that is suspected of causing the intermittent complaint should be thoroughly checked for backed out terminals, improper mating, broken locks, improperly formed or damaged terminals, poor terminal to wiring connections or physical damage to the wiring harness.

Resistance of the wheel speed sensor will increase with an increase in sensor temperature.

When replacing a wheel speed sensor, inspect the sensor terminals and harness connector for corrosion and/or water intrusion. If evidence of corrosion or water intrusion exists, replace wheel speed sensor jumper harness. Likewise, if replacing a wheel speed sensor jumper harness, inspect sensor terminals. If evidence of corrosion or water intrusion exists, replace wheel speed sensor.

DIAGNOSTIC CHART DTC 32 — 1995 SILHOUETTE, TRANS SPORT AND LUMINA APV

DTC 32
(Page 3 of 3)
LF WHEEL SPEED SNR CKT OPEN OR SHORT TO GND/BAT

DIAGNOSIS CONTINUED FROM PAGE 1 OF 3 OF THIS DTC CHART.

(10)
- DISCONNECT CONNECTOR C104 (MOUNTED ON LH SHOCK TOWER).
- USING J 39200, MEASURE THE RESISTANCE BETWEEN GROUND AND TERMINAL "E7" OF THE 32-WAY EBCM HARNESS CONNECTOR.

DOES J 39200 DISPLAY RESISTANCE AS "OL" (INFINITE)?

- NO: REPAIR SHORT TO GROUND IN CKT 873.
- YES:

(11)
- USING J 39200, MEASURE THE RESISTANCE BETWEEN GROUND AND TERMINAL "E8" OF THE 32-WAY EBCM HARNESS CONNECTOR.

DOES J 39200 DISPLAY RESISTANCE AS "OL" (INFINITE)?

- NO: REPAIR SHORT TO GROUND IN CKT 830.
- YES:

(12)
- THOROUGHLY RESPRAY THE LEFT FRONT WHEEL SPEED SENSOR JUMPER HARNESS WITH THE SALT WATER SOLUTION.

IMPORTANT: DO NOT ALLOW WATER TO ENTER ANY WHEEL SPEED SENSOR CONNECTOR.

- DISCONNECT LEFT FRONT WHEEL SPEED SENSOR DIRECTLY AT SENSOR.
- USING J 39200, MEASURE THE RESISTANCE BETWEEN GROUND AND TERMINAL "A," THEN "B" OF THE LEFT FRONT WHEEL SPEED SENSOR HARNESS CONNECTOR.

DOES J 39200 DISPLAY RESISTANCE AS "OL" (INFINITE) FOR BOTH MEASUREMENTS?

- NO: REPLACE LEFT FRONT WHEEL SPEED SENSOR JUMPER HARNESS.
- YES:

(13)
- USING J 39200, MEASURE THE RESISTANCE BETWEEN GROUND AND TERMINAL "A" OF THE LEFT FRONT WHEEL SPEED SENSOR CONNECTOR.

DOES J 39200 DISPLAY RESISTANCE AS "OL" (INFINITE)?

- NO: REPLACE WHEEL SPEED SENSOR.
- YES: MALFUNCTION IS INTERMITTENT. INSPECT ALL CONNECTORS AND HARNESSES FOR DAMAGE WHICH MAY RESULT IN A SHORT TO GROUND WHEN ALL COMPONENTS ARE CONNECTED. PAY PARTICULAR ATTENTION TO THE JUMPER HARNESS, SMALL NICKS OR CUTS CAN CAUSE A MALFUNCTION. REPAIR ANY DAMAGE IF FOUND. JUMPER HARNESS MUST BE REPLACED IF DAMAGED. REFER TO "DIAGNOSTIC AIDS" ON FACING PAGE FOR MORE INFORMATION.

DIAGNOSTIC CHART DTC 33 — 1995 SILHOUETTE, TRANS SPORT AND LUMINA APV

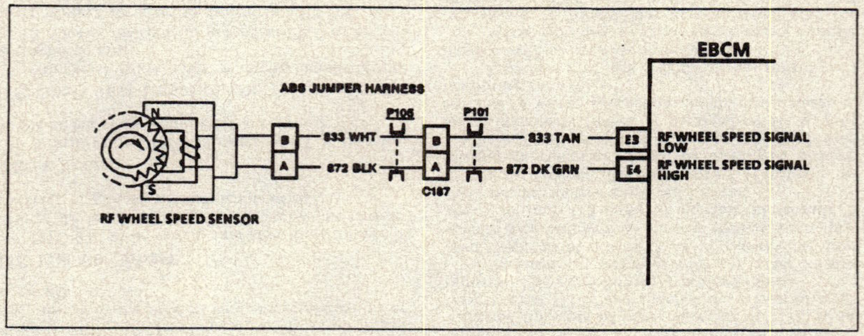

DTC 33
(Page 1 of 3)
RF WHEEL SPEED SNR CKT OPEN OR SHORT TO GND/BAT

Circuit Description:

As a toothed ring passes by the wheel speed sensor, changes in the electromagnetic field cause the wheel speed sensor to produce a sinusoidal (AC) voltage signal whose frequency is proportional to wheel speed. The magnitude of this signal is directly related to wheel speed and the proximity of the wheel speed sensor to the toothed ring, often referred to as the air gap.

DTC Will Set When:

DTC 33 can be set anytime after initialization. If either of the right front wheel speed circuits are shorted to voltage or ground, a malfunction exists.

Action Taken:

A malfunction DTC is stored, ABS is disabled and the amber ABS warning lamp is turned "ON."

DTC Chart Test Description:

Number(s) below refer to circled number(s) on the diagnostic chart.

1. Checks to see that the wheel speed sensor is within proper resistance values.
2. Checks for short to ground in wheel speed sensor wiring.
3. Checks for short to voltage in wheel speed sensor wiring.
4. Checks for a short to voltage in CKT 833.
5. Checks for a short to voltage in CKT 872.
6. Ensures malfunction was not due to physical damage of circuitry.

Diagnostic Aids:

An "Intermittent" malfunction may be caused by a poor connection, rubbed through wire insulation, or a wire that is broken inside the insulation.

The frequency of the malfunction can be checked by using the enhanced diagnostic function of the Tech 1,

If the customer's comments reflect that the ABS warning lamp is "ON" only during moist environmental changes (rain, snow, vehicle wash), all wheel speed sensor circuitry should be thoroughly inspected for signs of water intrusion. If DTC is not current, clear DTCs and simulate the effects of water intrusion. Use the following procedure. Spray the suspected area with a 5% salt water solution (two teaspoons of salt to 12 oz. of water). Test drive vehicle over various road surfaces (bumps, turns, etc.) above 24 km/h (15 mph) for at least 30 seconds. If DTC returns, replace suspected harness.

Any circuitry that is suspected of causing the intermittent complaint should be thoroughly checked for backed out terminals, improper mating, broken locks, improperly formed or damaged terminals, poor terminal to wiring connections or physical damage to the wiring harness.

Resistance of the wheel speed sensor will increase with an increase in sensor temperature.

When replacing a wheel speed sensor, inspect the sensor terminals and harness connector for corrosion and/or water intrusion. If evidence of corrosion or water intrusion exists, replace wheel speed sensor jumper harness. Likewise, if replacing a wheel speed sensor jumper harness, inspect sensor terminals. If evidence of corrosion or water intrusion exists, replace wheel speed sensor.

DIAGNOSTIC CHART DTC 33 — 1995 SILHOUETTE, TRANS SPORT AND LUMINA APV

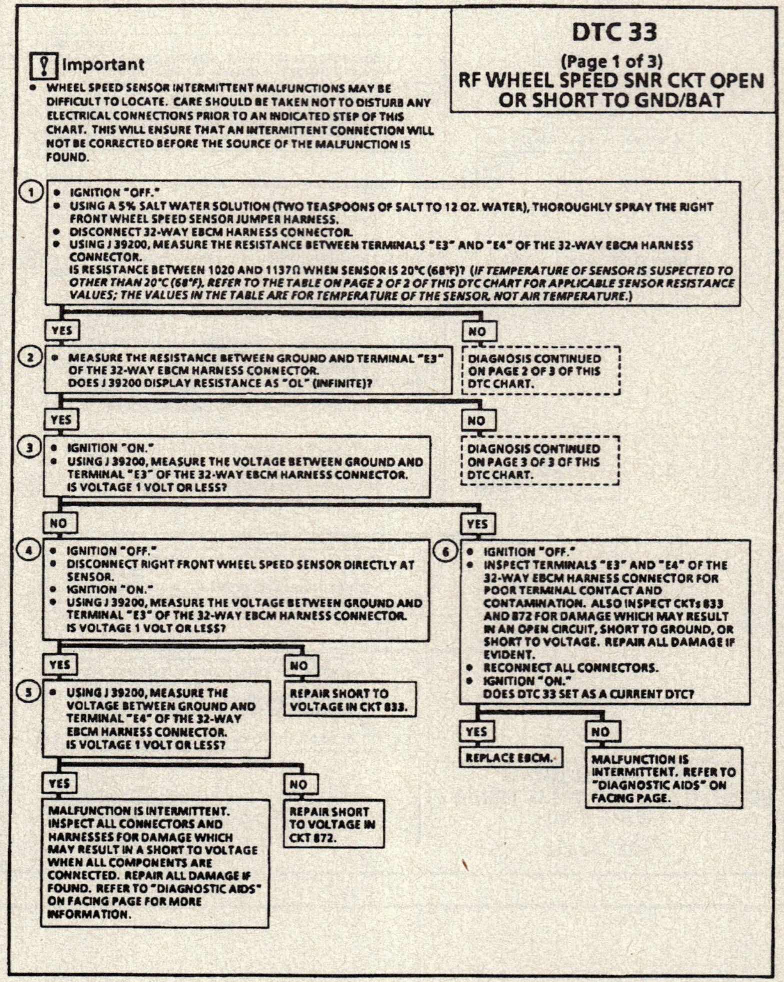

DIAGNOSTIC CHART DTC 33 — 1995 SILHOUETTE, TRANS SPORT AND LUMINA APV

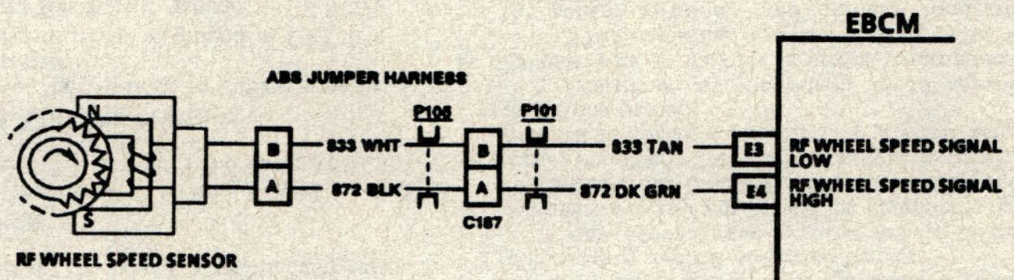

DTC 33
(Page 2 of 3)
RF WHEEL SPEED SNR CKT OPEN OR SHORT TO GND/BAT

Circuit Description:

As a toothed ring passes by the wheel speed sensor, changes in the electromagnetic field cause the wheel speed sensor to produce a sinusoidal (AC) voltage signal whose frequency is proportional to wheel speed. The magnitude of this signal is directly related to wheel speed and the proximity of the wheel speed sensor to the toothed ring, often referred to as the air gap.

DTC Will Set When:

DTC 33 can be set anytime after initialization. If either of the right front wheel speed circuits are shorted to voltage or ground, a malfunction exists.

Action Taken:

A malfunction DTC is stored, ABS is disabled and the amber ABS warning lamp is turned "ON."

DTC Chart Test Description:

Number(s) below refer to circled number(s) on the diagnostic chart.

7. Checks for high resistance in CKT 833.
8. Checks for high resistance in CKT 872.
9. Verifies proper wheel speed sensor resistance.

Diagnostic Aids:

An "Intermittent" malfunction may be caused by a poor connection, rubbed through wire insulation, or a wire that is broken inside the insulation.

The frequency of the malfunction can be checked by using the enhanced diagnostic function of the Tech 1.

If the customer's comments reflect that the ABS warning lamp is "ON" only during moist environmental changes (rain, snow, vehicle wash), all wheel speed sensor circuitry should be thoroughly inspected for signs of water intrusion. If DTC is not current, clear DTCs and simulate the effects of water intrusion. Use the following procedure. Spray the suspected area with a 5% salt water solution (two teaspoons of salt to 12 oz. of water). Test drive vehicle over various road surfaces (bumps, turns, etc.) above 24 km/h (15 mph) for at least 30 seconds. If DTC returns, replace suspected harness.

Any circuitry that is suspected of causing the intermittent complaint should be thoroughly checked for backed out terminals, improper mating, broken locks, improperly formed or damaged terminals, poor terminal to wiring connections or physical damage to the wiring harness.

Resistance of the wheel speed sensor will increase with an increase in sensor temperature.

When replacing a wheel speed sensor, inspect the sensor terminals and harness connector for corrosion and/or water intrusion. If evidence of corrosion or water intrusion exists, replace wheel speed sensor jumper harness. Likewise, if replacing a wheel speed sensor jumper harness, inspect sensor terminals. If evidence of corrosion or water intrusion exists, replace wheel speed sensor.

DIAGNOSTIC CHART DTC 33 — 1995 SILHOUETTE, TRANS SPORT AND LUMINA APV

DTC 33
(Page 2 of 3)
RF WHEEL SPEED SNR CKT SHORT TO GND/BAT

DIAGNOSIS CONTINUED FROM PAGE 1 OF 3 OF THIS DTC CHART.

7
- DISCONNECT RIGHT FRONT WHEEL SPEED SENSOR DIRECTLY AT SENSOR.
- USING J 39200, MEASURE THE RESISTANCE BETWEEN TERMINAL "E3" OF THE 32-WAY EBCM HARNESS CONNECTOR AND TERMINAL "B" OF THE RIGHT FRONT WHEEL SPEED SENSOR HARNESS CONNECTOR.

IS RESISTANCE 2Ω OR LESS?

NO: REPAIR HIGH RESISTANCE IN CKT 833 (WHEEL SPEED SENSOR JUMPER MUST BE REPLACED IF DAMAGED).

YES:

8
- USING J 39200, MEASURE THE RESISTANCE BETWEEN TERMINAL "E4" OF THE 32-WAY EBCM HARNESS CONNECTOR AND TERMINAL "A" OF THE RIGHT FRONT WHEEL SPEED SENSOR HARNESS CONNECTOR.

IS RESISTANCE 2Ω OR LESS?

NO: REPAIR HIGH RESISTANCE IN CKT 872 (WHEEL SPEED SENSOR JUMPER MUST BE REPLACED IF DAMAGED).

YES:

9
- USING J 39200, MEASURE THE RESISTANCE BETWEEN TERMINAL "A" AND TERMINAL "B" OF THE RIGHT FRONT WHEEL SPEED SENSOR CONNECTOR.

IS RESISTANCE BETWEEN 1020 AND 1137Ω WHEN SENSOR IS 20°C (68°F)? *(IF TEMPERATURE OF SENSOR IS SUSPECTED TO OTHER THAN 20°C (68°F), REFER TO THE TABLE BELOW FOR APPLICABLE SENSOR RESISTANCE VALUES; THE VALUES IN THE TABLE ARE FOR TEMPERATURE OF THE SENSOR, NOT AIR TEMPERATURE.)*

NO: REPLACE WHEEL SPEED SENSOR.

YES: MALFUNCTION IS INTERMITTENT. INSPECT ALL CONNECTORS AND HARNESSES FOR DAMAGE WHICH MAY RESULT IN HIGH RESISTANCE WHEN ALL COMPONENTS ARE CONNECTED. REFER TO "DIAGNOSTIC AIDS" ON FACING PAGE FOR MORE INFORMATION. REPAIR ALL DAMAGE IF FOUND.

FRONT WHEEL SPEED SENSOR

TEMPERATURE OF SENSOR VS. SENSOR RESISTANCE (APPROXIMATE)

TEMP. (°C)	TEMP. (°F)	RESISTANCE (OHMS)
-34 TO 4	-30 TO 40	804 TO 1066
5 TO 43	41 TO 110	956 TO 1240
44 TO 93	111 TO 200	1112 TO 1463

DIAGNOSTIC CHART DTC 33 — 1995 SILHOUETTE, TRANS SPORT AND LUMINA APV

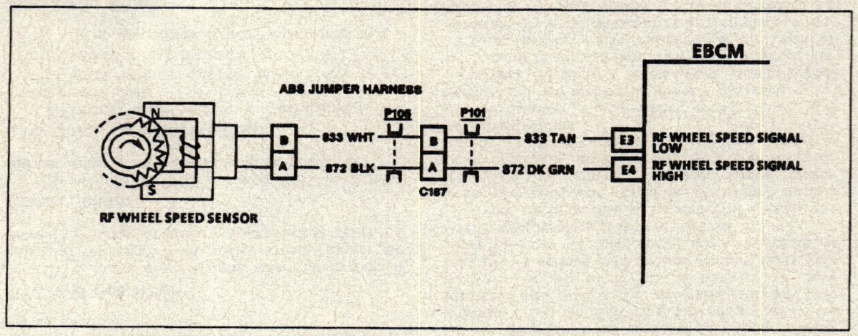

DTC 33
(Page 3 of 3)
RF WHEEL SPEED SNR CKT OPEN OR SHORT TO GND/BAT

Circuit Description:

As a toothed ring passes by the wheel speed sensor, changes in the electromagnetic field cause the wheel speed sensor to produce a sinusoidal (AC) voltage signal whose frequency is proportional to wheel speed. The magnitude of this signal is directly related to wheel speed and the proximity of the wheel speed sensor to the toothed ring, often referred to as the air gap.

DTC Will Set When:

DTC 33 can be set anytime after initialization. If either of the right front wheel speed circuits are shorted to voltage or ground, a malfunction exists.

Action Taken:

A malfunction DTC is stored, ABS is disabled and the amber ABS warning lamp is turned "ON."

DTC Chart Test Description:

Number(s) below refer to circled number(s) on the diagnostic chart.

10. Checks for short to ground in CKT 833.
11. Checks for short to ground in CKT 872.
12. Checks for short to ground in wheel speed sensor jumper harness.
13. Verifies that wheel speed sensor is not internally shorted to ground.

Diagnostic Aids:

An "Intermittent" malfunction may be caused by a poor connection, rubbed through wire insulation, or a wire that is broken inside the insulation.

The frequency of the malfunction can be checked by using the enhanced diagnostic function of the Tech 1,

If the customer's comments reflect that the ABS warning lamp is "ON" only during moist environmental changes (rain, snow, vehicle wash), all wheel speed sensor circuitry should be thoroughly inspected for signs of water intrusion. If DTC is not current, clear DTCs and simulate the effects of water intrusion. Use the following procedure. Spray the suspected area with a 5% salt water solution (two teaspoons of salt to 12 oz. of water). Test drive vehicle over various road surfaces (bumps, turns, etc.) above 24 km/h (15 mph) for at least 30 seconds. If DTC returns, replace suspected harness.

Any circuitry that is suspected of causing the intermittent complaint should be thoroughly checked for backed out terminals, improper mating, broken locks, improperly formed or damaged terminals, poor terminal to wiring connections or physical damage to the wiring harness.

Resistance of the wheel speed sensor will increase with an increase in sensor temperature.

When replacing a wheel speed sensor, inspect the sensor terminals and harness connector for corrosion and/or water intrusion. If evidence of corrosion or water intrusion exists, replace wheel speed sensor jumper harness. Likewise, if replacing a wheel speed sensor jumper harness, inspect sensor terminals. If evidence of corrosion or water intrusion exists, replace wheel speed sensor. Refer to "On-Vehicle Service" in this section.

DIAGNOSTIC CHART DTC 33 — 1995 SILHOUETTE, TRANS SPORT AND LUMINA APV

DTC 33
(Page 3 of 3)
RF WHEEL SPEED SNR CKT OPEN OR SHORT TO GND/BAT

DIAGNOSIS CONTINUED FROM PAGE 1 OF 3 OF THIS DTC CHART.

(10)
- DISCONNECT CONNECTOR C105 (MOUNTED ON RH SHOCK TOWER).
- USING J 39200, MEASURE THE RESISTANCE BETWEEN GROUND AND TERMINAL "E3" OF THE 32-WAY EBCM HARNESS CONNECTOR.

DOES J 39200 DISPLAY RESISTANCE AS "OL" (INFINITE)?

NO: REPAIR SHORT TO GROUND IN CKT 833.

YES:

(11)
- USING J 39200, MEASURE THE RESISTANCE BETWEEN GROUND AND TERMINAL "E4" OF THE 32-WAY EBCM HARNESS CONNECTOR.

DOES J 39200 DISPLAY RESISTANCE AS "OL" (INFINITE)?

NO: REPAIR SHORT TO GROUND IN CKT 872.

YES:

(12)
- THOROUGHLY RESPRAY THE RIGHT FRONT WHEEL SPEED SENSOR JUMPER HARNESS WITH THE SALT WATER SOLUTION.

IMPORTANT: DO NOT ALLOW WATER TO ENTER ANY WHEEL SPEED SENSOR CONNECTOR.

- DISCONNECT RIGHT FRONT WHEEL SPEED SENSOR DIRECTLY AT SENSOR.
- USING J 39200, MEASURE THE RESISTANCE BETWEEN GROUND AND TERMINAL "A," THEN "B" OF THE RIGHT FRONT WHEEL SPEED SENSOR HARNESS CONNECTOR.

DOES J 39200 DISPLAY RESISTANCE AS "OL" (INFINITE) FOR BOTH MEASUREMENTS?

NO: REPLACE RIGHT FRONT WHEEL SPEED SENSOR JUMPER HARNESS.

YES:

(13)
- USING J 39200, MEASURE THE RESISTANCE BETWEEN GROUND AND TERMINAL "A" OF THE RIGHT FRONT WHEEL SPEED SENSOR CONNECTOR.

DOES J 39200 DISPLAY RESISTANCE AS "OL" (INFINITE)?

NO: REPLACE WHEEL SPEED SENSOR.

YES: MALFUNCTION IS INTERMITTENT. INSPECT ALL CONNECTORS AND HARNESSES FOR DAMAGE WHICH MAY RESULT IN A SHORT TO GROUND WHEN ALL COMPONENTS ARE CONNECTED. PAY PARTICULAR ATTENTION TO THE JUMPER HARNESS, SMALL NICKS OR CUTS CAN CAUSE A MALFUNCTION. REPAIR ANY DAMAGE IF FOUND. JUMPER HARNESS MUST BE REPLACED IF DAMAGED. REFER TO "DIAGNOSTIC AIDS" ON FACING PAGE FOR MORE INFORMATION.

DIAGNOSTIC CHART DTC 34 — 1995 SILHOUETTE, TRANS SPORT AND LUMINA APV

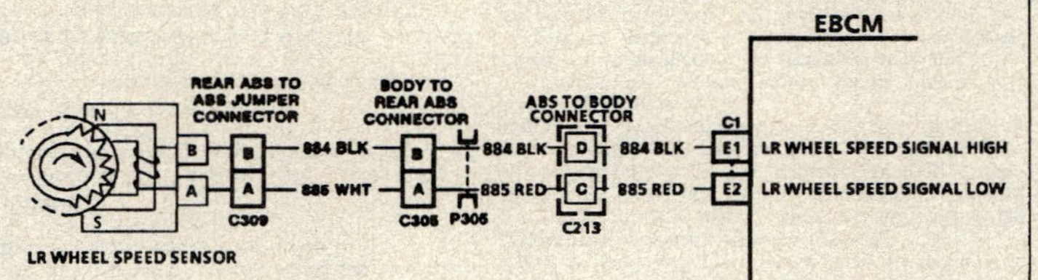

DTC 34
(Page 1 of 3)
LR WHEEL SPEED SNR CKT OPEN OR SHORT TO GND/BAT

Circuit Description:

As a toothed ring passes by the wheel speed sensor, changes in the electromagnetic field cause the wheel speed sensor to produce a sinusoidal (AC) voltage signal whose frequency is proportional to wheel speed. The magnitude of this signal is directly related to wheel speed and the proximity of the wheel speed sensor to the toothed ring, often referred to as the air gap.

DTC Will Set When:

DTC 34 can be set anytime after initialization. If either of the left rear wheel speed circuits are shorted to voltage or ground, a malfunction exists.

Action Taken:

A malfunction DTC is stored, ABS is disabled and the amber ABS warning lamp is turned "ON."

DTC Chart Test Description:

Number(s) below refer to circled number(s) on the diagnostic chart.

1. Checks to see that the wheel speed sensor is within proper resistance values.
2. Checks for short to ground in wheel speed sensor wiring.
3. Checks for short to voltage in wheel speed sensor wiring.
4. Checks for a short to voltage in CKT 884.
5. Checks for a short to voltage in CKT 885.
6. Ensures malfunction was not due to physical damage of circuitry.

Diagnostic Aids:

An "Intermittent" malfunction may be caused by a poor connection, rubbed through wire insulation, or a wire that is broken inside the insulation.

The frequency of the malfunction can be checked by using the enhanced diagnostic function of the Tech 1.

If the customer's comments reflect that the ABS warning lamp is "ON" only during moist environmental changes (rain, snow, vehicle wash), all wheel speed sensor circuitry should be thoroughly inspected for signs of water intrusion. If DTC is not current, clear DTCs and simulate the effects of water intrusion. Use the following procedure. Spray the suspected area with a 5% salt water solution (two teaspoons of salt to 12 oz. of water). Test drive vehicle over various road surfaces (bumps, turns, etc.) above 24 km/h (15 mph) for at least 30 seconds. If DTC returns, replace suspected harness.

Any circuitry that is suspected of causing the intermittent complaint should be thoroughly checked for backed out terminals, improper mating, broken locks, improperly formed or damaged terminals, poor terminal to wiring connections or physical damage to the wiring harness.

Resistance of the wheel speed sensor will increase with an increase in sensor temperature.

When replacing a wheel speed sensor, inspect the sensor terminals and harness connector for corrosion and/or water intrusion. If evidence of corrosion or water intrusion exists, replace wheel speed sensor jumper harness. Likewise, if replacing a wheel speed sensor jumper harness, inspect sensor terminals. If evidence of corrosion or water intrusion exists, replace wheel speed sensor.

DIAGNOSTIC CHART DTC 34 — 1995 SILHOUETTE, TRANS SPORT AND LUMINA APV

DTC 34
(Page 1 of 3)
LR WHEEL SPEED SNR CKT OPEN OR SHORT TO GND/BAT

Important

- WHEEL SPEED SENSOR INTERMITTENT MALFUNCTIONS MAY BE DIFFICULT TO LOCATE. CARE SHOULD BE TAKEN NOT TO DISTURB ANY ELECTRICAL CONNECTIONS PRIOR TO AN INDICATED STEP OF THIS CHART. THIS WILL ENSURE THAT AN INTERMITTENT CONNECTION WILL NOT BE CORRECTED BEFORE THE SOURCE OF THE MALFUNCTION IS FOUND.

(1)
- IGNITION "OFF."
- USING A 5% SALT WATER SOLUTION (TWO TEASPOONS OF SALT TO 12 OZ. WATER), THOROUGHLY SPRAY THE LEFT REAR WHEEL SPEED SENSOR JUMPER HARNESS.
- DISCONNECT 32-WAY EBCM HARNESS CONNECTOR.
- USING J 39200, MEASURE THE RESISTANCE BETWEEN TERMINALS "E1" AND "E2" OF THE 32-WAY EBCM HARNESS CONNECTOR.

IS RESISTANCE BETWEEN 2059 AND 2353Ω WHEN SENSOR IS 20°C (68°F)? *(IF TEMPERATURE OF SENSOR IS SUSPECTED TO OTHER THAN 20°C (68°F), REFER TO THE TABLE ON PAGE 2 OF 2 OF THIS DTC CHART FOR APPLICABLE SENSOR RESISTANCE VALUES; THE VALUES IN THE TABLE ARE FOR TEMPERATURE OF THE SENSOR, NOT AIR TEMPERATURE.)*

- NO → DIAGNOSIS CONTINUED ON PAGE 2 OF 3 OF THIS DTC CHART.
- YES →

(2)
- MEASURE THE RESISTANCE BETWEEN GROUND AND TERMINAL "E1" OF THE 32-WAY EBCM HARNESS CONNECTOR.

DOES J 39200 DISPLAY RESISTANCE AS "OL" (INFINITE)?

- NO → DIAGNOSIS CONTINUED ON PAGE 3 OF 3 OF THIS DTC CHART.
- YES →

(3)
- IGNITION "ON."
- USING J 39200, MEASURE THE VOLTAGE BETWEEN GROUND AND TERMINAL "E1" OF THE 32-WAY EBCM HARNESS CONNECTOR.

IS VOLTAGE 1 VOLT OR LESS?

- NO →

(4)
- IGNITION "OFF."
- DISCONNECT LEFT REAR WHEEL SPEED SENSOR DIRECTLY AT SENSOR.
- IGNITION "ON."
- USING J 39200, MEASURE THE VOLTAGE BETWEEN GROUND AND TERMINAL "E1" OF THE 32-WAY EBCM HARNESS CONNECTOR.

IS VOLTAGE 1 VOLT OR LESS?

- NO → REPAIR SHORT TO VOLTAGE IN CKT 884.
- YES →

(5)
- USING J 39200, MEASURE THE VOLTAGE BETWEEN GROUND AND TERMINAL "E2" OF THE 32-WAY EBCM HARNESS CONNECTOR.

IS VOLTAGE 1 VOLT OR LESS?

- YES → MALFUNCTION IS INTERMITTENT. INSPECT ALL CONNECTORS AND HARNESSES FOR DAMAGE WHICH MAY RESULT IN A SHORT TO VOLTAGE WHEN ALL COMPONENTS ARE CONNECTED. REPAIR ALL DAMAGE IF FOUND. REFER TO "DIAGNOSTIC AIDS" ON FACING PAGE FOR MORE INFORMATION.
- NO → REPAIR SHORT TO VOLTAGE IN CKT 885.

(3) YES →

(6)
- IGNITION "OFF."
- INSPECT TERMINALS "E1" AND "E2" OF THE 32-WAY EBCM HARNESS CONNECTOR FOR POOR TERMINAL CONTACT AND CONTAMINATION. ALSO INSPECT CKTs 884 AND 885 FOR DAMAGE WHICH MAY RESULT IN AN OPEN CIRCUIT, SHORT TO GROUND, OR SHORT TO VOLTAGE. REPAIR ALL DAMAGE IF EVIDENT.
- RECONNECT ALL CONNECTORS.
- IGNITION "ON."

DOES DTC 34 SET AS A CURRENT DTC?

- YES → REPLACE EBCM.
- NO → MALFUNCTION IS INTERMITTENT. REFER TO "DIAGNOSTIC AIDS" ON FACING PAGE.

DIAGNOSTIC CHART DTC 34 — 1995 SILHOUETTE, TRANS SPORT AND LUMINA APV

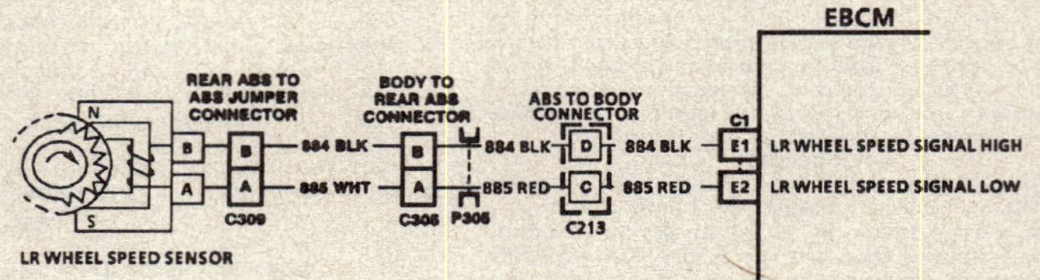

DTC 34
(Page 2 of 3)
LR WHEEL SPEED SNR CKT OPEN OR SHORT TO GND/BAT

Circuit Description:

As a toothed ring passes by the wheel speed sensor, changes in the electromagnetic field cause the wheel speed sensor to produce a sinusoidal (AC) voltage signal whose frequency is proportional to wheel speed. The magnitude of this signal is directly related to wheel speed and the proximity of the wheel speed sensor to the toothed ring, often referred to as the air gap.

DTC Will Set When:

DTC 34 can be set anytime after initialization. If either of the left rear wheel speed circuits are shorted to voltage or ground, a malfunction exists.

Action Taken:

A malfunction DTC is stored, ABS is disabled and the amber ABS warning lamp is turned "ON."

DTC Chart Test Description:

Number(s) below refer to circled number(s) on the diagnostic chart.

7. Checks for high resistance in CKT 884.
8. Checks for high resistance in CKT 885.
9. Verifies proper wheel speed sensor resistance.

Diagnostic Aids:

An "Intermittent" malfunction may be caused by a poor connection, rubbed through wire insulation, or a wire that is broken inside the insulation.

The frequency of the malfunction can be checked by using the enhanced diagnostic function of the Tech 1.

If the customer's comments reflect that the ABS warning lamp is "ON" only during moist environmental changes (rain, snow, vehicle wash), all wheel speed sensor circuitry should be thoroughly inspected for signs of water intrusion. If DTC is not current, clear DTCs and simulate the effects of water intrusion. Use the following procedure. Spray the suspected area with a 5% salt water solution (two teaspoons of salt to 12 oz. of water). Test drive vehicle over various road surfaces (bumps, turns, etc.) above 24 km/h (15 mph) for at least 30 seconds. If DTC returns, replace suspected harness.

Any circuitry that is suspected of causing the intermittent complaint should be thoroughly checked for backed out terminals, improper mating, broken locks, improperly formed or damaged terminals, poor terminal to wiring connections or physical damage to the wiring harness.

Resistance of the wheel speed sensor will increase with an increase in sensor temperature.

When replacing a wheel speed sensor, inspect the sensor terminals and harness connector for corrosion and/or water intrusion. If evidence of corrosion or water intrusion exists, replace wheel speed sensor jumper harness. Likewise, if replacing a wheel speed sensor jumper harness, inspect sensor terminals. If evidence of corrosion or water intrusion exists, replace wheel speed sensor.

DIAGNOSTIC CHART DTC 34 — 1995 SILHOUETTE, TRANS SPORT AND LUMINA APV

DTC 34
(Page 2 of 3)
LR WHEEL SPEED SNR CKT OPEN OR SHORT TO GND/BAT

DIAGNOSIS CONTINUED FROM PAGE 1 OF 3 OF THIS DTC CHART.

(7)
- DISCONNECT LEFT REAR WHEEL SPEED SENSOR DIRECTLY AT SENSOR.
- USING J 39200, MEASURE THE RESISTANCE BETWEEN TERMINAL "E1" OF THE 32-WAY EBCM HARNESS CONNECTOR AND TERMINAL "B" OF THE LEFT REAR WHEEL SPEED SENSOR HARNESS CONNECTOR.

IS RESISTANCE 2Ω OR LESS?

NO: REPAIR HIGH RESISTANCE IN CKT 884 (WHEEL SPEED SENSOR JUMPER MUST BE REPLACED IF DAMAGED).

YES:

(8)
- USING J 39200, MEASURE THE RESISTANCE BETWEEN TERMINAL "E2" OF THE 32-WAY EBCM HARNESS CONNECTOR AND TERMINAL "A" OF THE LEFT REAR WHEEL SPEED SENSOR HARNESS CONNECTOR.

IS RESISTANCE 2Ω OR LESS?

NO: REPAIR HIGH RESISTANCE IN CKT 885 (WHEEL SPEED SENSOR JUMPER MUST BE REPLACED IF DAMAGED).

YES:

(9)
- USING J 39200, MEASURE THE RESISTANCE BETWEEN TERMINAL "A" AND TERMINAL "B" OF THE LEFT REAR WHEEL SPEED SENSOR CONNECTOR.

IS RESISTANCE BETWEEN 2059 AND 2353Ω WHEN SENSOR IS 20°C (68°F)? *(IF TEMPERATURE OF SENSOR IS SUSPECTED TO OTHER THAN 20°C (68°F), REFER TO THE TABLE BELOW FOR APPLICABLE SENSOR RESISTANCE VALUES; THE VALUES IN THE TABLE ARE FOR TEMPERATURE OF THE SENSOR, NOT AIR TEMPERATURE.)*

NO: REPLACE WHEEL SPEED SENSOR.

YES: MALFUNCTION IS INTERMITTENT. INSPECT ALL CONNECTORS AND HARNESSES FOR DAMAGE WHICH MAY RESULT IN HIGH RESISTANCE WHEN ALL COMPONENTS ARE CONNECTED. REFER TO "DIAGNOSTIC AIDS" ON FACING PAGE FOR MORE INFORMATION. REPAIR ALL DAMAGE IF FOUND.

REAR WHEEL SPEED SENSOR

TEMPERATURE OF SENSOR VS. SENSOR RESISTANCE (APPROXIMATE)

TEMP. (°C)	TEMP. (°F)	RESISTANCE (OHMS)
-34 TO 4	-30 TO 40	1622 TO 2205
5 TO 43	41 TO 110	1930 TO 2566
44 TO 93	111 TO 200	2245 TO 3028

DIAGNOSTIC CHART DTC 34 — 1995 SILHOUETTE, TRANS SPORT AND LUMINA APV

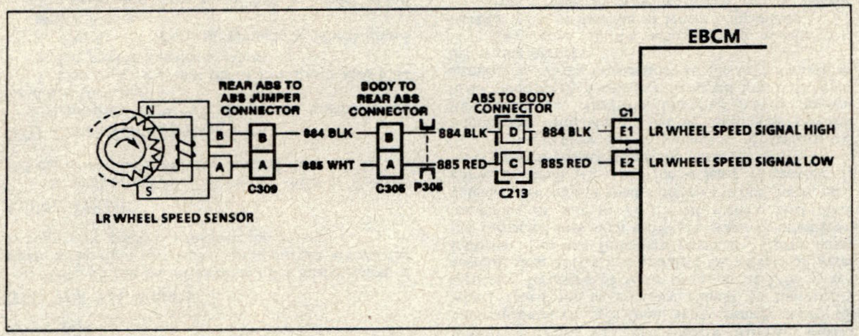

DTC 34
(Page 3 of 3)
LR WHEEL SPEED SNR CKT OPEN OR SHORT TO GND/BAT

Circuit Description:

As a toothed ring passes by the wheel speed sensor, changes in the electromagnetic field cause the wheel speed sensor to produce a sinusoidal (AC) voltage signal whose frequency is proportional to wheel speed. The magnitude of this signal is directly related to wheel speed and the proximity of the wheel speed sensor to the toothed ring, often referred to as the air gap.

DTC Will Set When:

DTC 34 can be set anytime after initialization. If either of the left rear wheel speed circuits are shorted to voltage or ground, a malfunction exists.

Action Taken:

A malfunction DTC is stored, ABS is disabled and the amber ABS warning lamp is turned "ON."

DTC Chart Test Description:

Number(s) below refer to circled number(s) on the diagnostic chart.

10. Checks for short to ground in CKT 884.
11. Checks for short to ground in CKT 885.
12. Checks for short to ground in wheel speed sensor jumper harness.
13. Verifies that wheel speed sensor is not internally shorted to ground.

Diagnostic Aids:

An "Intermittent" malfunction may be caused by a poor connection, rubbed through wire insulation, or a wire that is broken inside the insulation.

The frequency of the malfunction can be checked by using the enhanced diagnostic function of the Tech 1,

If the customer's comments reflect that the ABS warning lamp is "ON" only during moist environmental changes (rain, snow, vehicle wash), all wheel speed sensor circuitry should be thoroughly inspected for signs of water intrusion. If DTC is not current, clear DTCs and simulate the effects of water intrusion. Use the following procedure. Spray the suspected area with a 5% salt water solution (two teaspoons of salt to 12 oz. of water). Test drive vehicle over various road surfaces (bumps, turns, etc.) above 24 km/h (15 mph) for at least 30 seconds. If DTC returns, replace suspected harness.

Any circuitry that is suspected of causing the intermittent complaint should be thoroughly checked for backed out terminals, improper mating, broken locks, improperly formed or damaged terminals, poor terminal to wiring connections or physical damage to the wiring harness.

Resistance of the wheel speed sensor will increase with an increase in sensor temperature.

When replacing a wheel speed sensor, inspect the sensor terminals and harness connector for corrosion and/or water intrusion. If evidence of corrosion or water intrusion exists, replace wheel speed sensor jumper harness. Likewise, if replacing a wheel speed sensor jumper harness, inspect sensor terminals. If evidence of corrosion or water intrusion exists, replace wheel speed sensor.

DIAGNOSTIC CHART DTC 34 — 1995 SILHOUETTE, TRANS SPORT AND LUMINA APV

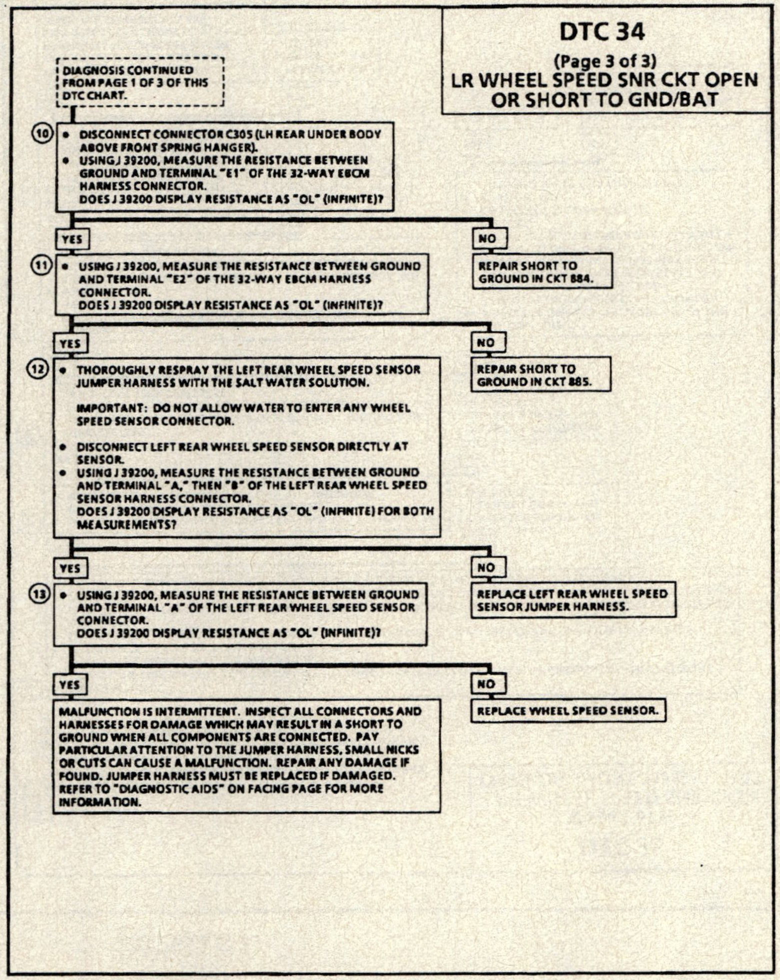

DIAGNOSTIC CHART DTC 35 — 1995 SILHOUETTE, TRANS SPORT AND LUMINA APV

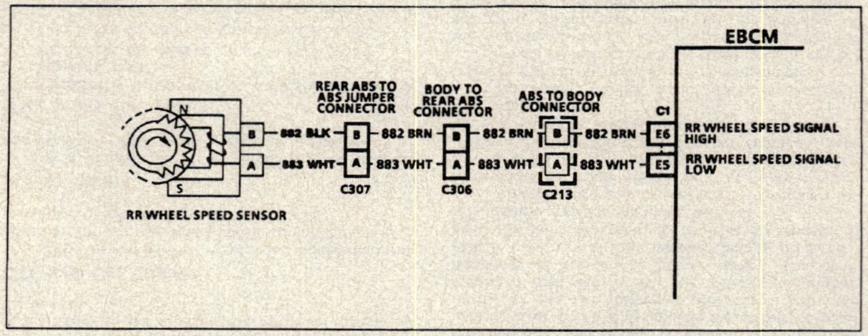

DTC 35
(Page 1 of 3)
RR WHEEL SPEED SNR CKT OPEN OR SHORT TO GND/BAT

Circuit Description:

As a toothed ring passes by the wheel speed sensor, changes in the electromagnetic field cause the wheel speed sensor to produce a sinusoidal (AC) voltage signal whose frequency is proportional to wheel speed. The magnitude of this signal is directly related to wheel speed and the proximity of the wheel speed sensor to the toothed ring, often referred to as the air gap.

DTC Will Set When:

DTC 35 can be set anytime after initialization. If either of the right rear wheel speed circuits are shorted to voltage or ground, a malfunction exists.

Action Taken:

A malfunction DTC is stored, ABS is disabled and the amber ABS warning lamp is turned "ON."

DTC Chart Test Description:

Number(s) below refer to circled number(s) on the diagnostic chart.

1. Checks to see that the wheel speed sensor is within proper resistance values.
2. Checks for short to ground in wheel speed sensor wiring.
3. Checks for short to voltage in wheel speed sensor wiring.
4. Checks for a short to voltage in CKT 883.
5. Checks for a short to voltage in CKT 882.
6. Ensures malfunction was not due to physical damage of circuitry.

Diagnostic Aids:

An "Intermittent" malfunction may be caused by a poor connection, rubbed through wire insulation, or a wire that is broken inside the insulation.

The frequency of the malfunction can be checked by using the enhanced diagnostic function of the Tech 1,

If the customer's comments reflect that the ABS warning lamp is "ON" only during moist environmental changes (rain, snow, vehicle wash), all wheel speed sensor circuitry should be thoroughly inspected for signs of water intrusion. If DTC is not current, clear DTCs and simulate the effects of water intrusion. Use the following procedure. Spray down the suspected area with a 5% salt water solution (two teaspoons of salt to 12 oz. of water). Test drive vehicle over various road surfaces (bumps, turns, etc.) above 24 km/h (15 mph) for at least 30 seconds. If DTC returns, replace suspected harness.

Any circuitry that is suspected of causing the intermittent complaint should be thoroughly checked for backed out terminals, improper mating, broken locks, improperly formed or damaged terminals, poor terminal to wiring connections or physical damage to the wiring harness.

Resistance of the wheel speed sensor will increase with an increase in sensor temperature.

When replacing a wheel speed sensor, inspect the sensor terminals and harness connector for corrosion and/or water intrusion. If evidence of corrosion or water intrusion exists, replace wheel speed sensor jumper harness. Likewise, if replacing a wheel speed sensor jumper harness, inspect sensor terminals. If evidence of corrosion or water intrusion exists, replace wheel speed sensor.

DIAGNOSTIC CHART DTC 35 — 1995 SILHOUETTE, TRANS SPORT AND LUMINA APV

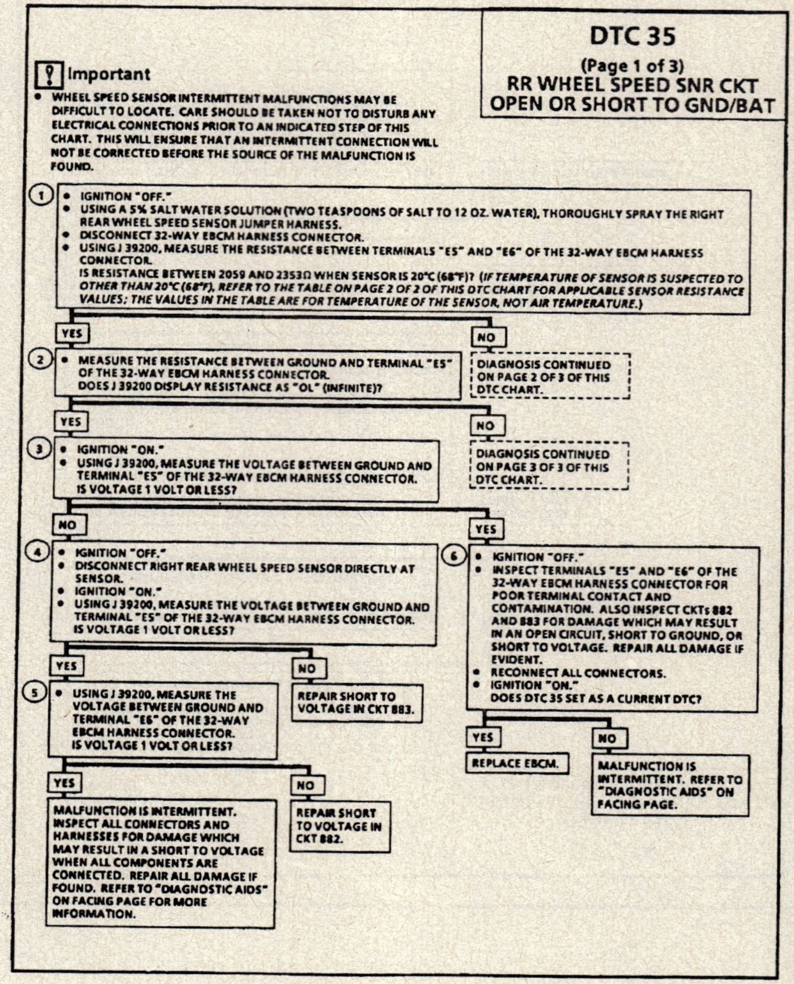

DIAGNOSTIC CHART DTC 35 — 1995 SILHOUETTE, TRANS SPORT AND LUMINA APV

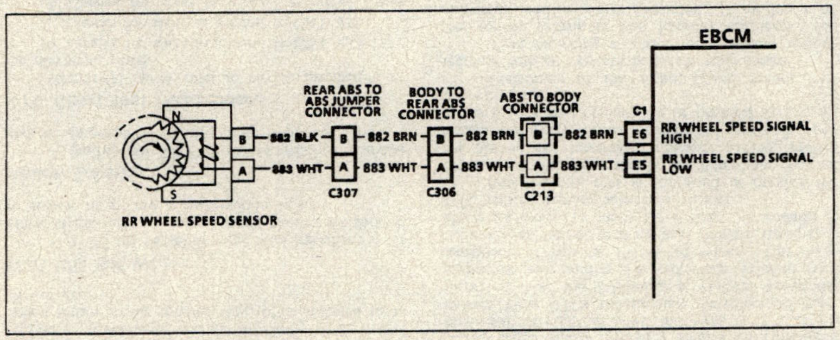

DTC 35
(Page 2 of 3)
RR WHEEL SPEED SNR CKT OPEN OR SHORT TO GND/BAT

Circuit Description:

As a toothed ring passes by the wheel speed sensor, changes in the electromagnetic field cause the wheel speed sensor to produce a sinusoidal (AC) voltage signal whose frequency is proportional to wheel speed. The magnitude of this signal is directly related to wheel speed and the proximity of the wheel speed sensor to the toothed ring, often referred to as the air gap.

DTC Will Set When:

DTC 35 can be set anytime after initialization. If either of the right rear wheel speed circuits are shorted to voltage or ground, a malfunction exists.

Action Taken:

A malfunction DTC is stored, ABS is disabled and the amber ABS warning lamp is turned "ON."

DTC Chart Test Description:

Number(s) below refer to circled number(s) on the diagnostic chart.

7. Checks for high resistance in CKT 883.
8. Checks for high resistance in CKT 882.
9. Verifies proper wheel speed sensor resistance.

Diagnostic Aids:

An "Intermittent" malfunction may be caused by a poor connection, rubbed through wire insulation, or a wire that is broken inside the insulation.

The frequency of the malfunction can be checked by using the enhanced diagnostic function of the Tech 1,

If the customer's comments reflect that the ABS warning lamp is "ON" only during moist environmental changes (rain, snow, vehicle wash), all wheel speed sensor circuitry should be thoroughly inspected for signs of water intrusion. If DTC is not current, clear DTCs and simulate the effects of water intrusion. Use the following procedure. Spray the suspected area with a 5% salt water solution (two teaspoons of salt to 12 oz. of water). Test drive vehicle over various road surfaces (bumps, turns, etc.) above 24 km/h (15 mph) for at least 30 seconds. If DTC returns, replace suspected harness.

Any circuitry that is suspected of causing the intermittent complaint should be thoroughly checked for backed out terminals, improper mating, broken locks, improperly formed or damaged terminals, poor terminal to wiring connections or physical damage to the wiring harness.

Resistance of the wheel speed sensor will increase with an increase in sensor temperature.

When replacing a wheel speed sensor, inspect the sensor terminals and harness connector for corrosion and/or water intrusion. If evidence of corrosion or water intrusion exists, replace wheel speed sensor jumper harness. Likewise, if replacing a wheel speed sensor jumper harness, inspect sensor terminals. If evidence of corrosion or water intrusion exists, replace wheel speed sensor.

DIAGNOSTIC CHART DTC 35 — 1995 SILHOUETTE, TRANS SPORT AND LUMINA APV

DTC 35
(Page 2 of 3)
RR WHEEL SPEED SNR CKT OPEN OR SHORT TO GND/BAT

DIAGNOSIS CONTINUED FROM PAGE 1 OF 3 OF THIS DTC CHART.

(7)
- DISCONNECT RIGHT REAR WHEEL SPEED SENSOR DIRECTLY AT SENSOR.
- USING J 39200, MEASURE THE RESISTANCE BETWEEN TERMINAL "E5" OF THE 32-WAY EBCM HARNESS CONNECTOR AND TERMINAL "A" OF THE RIGHT REAR WHEEL SPEED SENSOR HARNESS CONNECTOR. IS RESISTANCE 2Ω OR LESS?

NO: REPAIR HIGH RESISTANCE IN CKT 883 (WHEEL SPEED SENSOR JUMPER MUST BE REPLACED IF DAMAGED).

YES:

(8)
- USING J 39200, MEASURE THE RESISTANCE BETWEEN TERMINAL "E6" OF THE 32-WAY EBCM HARNESS CONNECTOR AND TERMINAL "B" OF THE RIGHT REAR WHEEL SPEED SENSOR HARNESS CONNECTOR. IS RESISTANCE 2Ω OR LESS?

NO: REPAIR HIGH RESISTANCE IN CKT 882 (WHEEL SPEED SENSOR JUMPER MUST BE REPLACED IF DAMAGED).

YES:

(9)
- USING J 39200, MEASURE THE RESISTANCE BETWEEN TERMINAL "A" AND TERMINAL "B" OF THE RIGHT REAR WHEEL SPEED SENSOR CONNECTOR. IS RESISTANCE BETWEEN 2059 AND 2353Ω WHEN SENSOR IS 20°C (68°F)? *(IF TEMPERATURE OF SENSOR IS SUSPECTED TO OTHER THAN 20°C (68°F), REFER TO THE TABLE BELOW FOR APPLICABLE SENSOR RESISTANCE VALUES; THE VALUES IN THE TABLE ARE FOR TEMPERATURE OF THE SENSOR, NOT AIR TEMPERATURE.)*

NO: REPLACE WHEEL SPEED SENSOR.

YES: MALFUNCTION IS INTERMITTENT. INSPECT ALL CONNECTORS AND HARNESSES FOR DAMAGE WHICH MAY RESULT IN HIGH RESISTANCE WHEN ALL COMPONENTS ARE CONNECTED. REFER TO "DIAGNOSTIC AIDS" ON FACING PAGE FOR MORE INFORMATION. REPAIR ALL DAMAGE IF FOUND.

REAR WHEEL SPEED SENSOR
TEMPERATURE OF SENSOR VS. SENSOR RESISTANCE (APPROXIMATE)

TEMP. (°C)	TEMP. (°F)	RESISTANCE (OHMS)
-34 TO 4	-30 TO 40	1622 TO 2205
5 TO 43	41 TO 110	1930 TO 2566
44 TO 93	111 TO 200	2245 TO 3028

DIAGNOSTIC CHART DTC 35 — 1995 SILHOUETTE, TRANS SPORT AND LUMINA APV

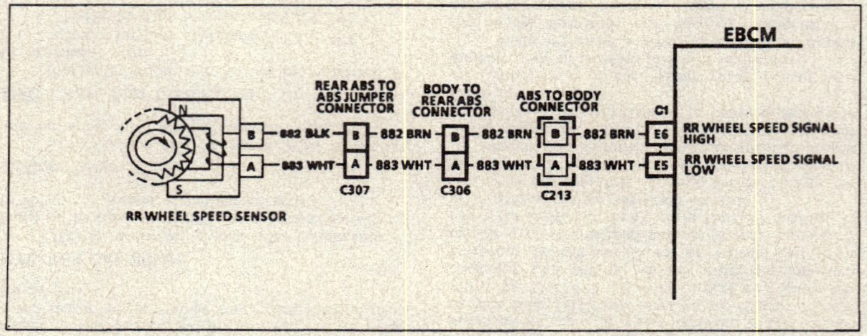

DTC 35
(Page 3 of 3)
RR WHEEL SPEED SNR CKT OPEN OR SHORT TO GND/BAT

Circuit Description:

As a toothed ring passes by the wheel speed sensor, changes in the electromagnetic field cause the wheel speed sensor to produce a sinusoidal (AC) voltage signal whose frequency is proportional to wheel speed. The magnitude of this signal is directly related to wheel speed and the proximity of the wheel speed sensor to the toothed ring, often referred to as the air gap.

DTC Will Set When:

DTC 35 can be set anytime after initialization. If either of the right rear wheel speed circuits are shorted to voltage or ground, a malfunction exists.

Action Taken:

A malfunction DTC is stored, ABS is disabled and the amber ABS warning lamp is turned "ON."

DTC Chart Test Description:

Number(s) below refer to circled number(s) on the diagnostic chart.

10. Checks for short to ground in CKT 883.
11. Checks for short to ground in CKT 882.
12. Checks for short to ground in wheel speed sensor jumper harness.
13. Verifies that wheel speed sensor is not internally shorted to ground.

Diagnostic Aids:

An "Intermittent" malfunction may be caused by a poor connection, rubbed through wire insulation, or a wire that is broken inside the insulation.

The frequency of the malfunction can be checked by using the enhanced diagnostic function of the Tech 1.

If the customer's comments reflect that the ABS warning lamp is "ON" only during moist environmental changes (rain, snow, vehicle wash), all wheel speed sensor circuitry should be thoroughly inspected for signs of water intrusion. If DTC is not current, clear DTCs and simulate the effects of water intrusion. Use the following procedure. Spray the suspected area with a 5% salt water solution (two teaspoons of salt to 12 oz. of water). Test drive vehicle over various road surfaces (bumps, turns, etc.) above 24 km/h (15 mph) for at least 30 seconds. If DTC returns, replace suspected harness.

Any circuitry that is suspected of causing the intermittent complaint should be thoroughly checked for backed out terminals, improper mating, broken locks, improperly formed or damaged terminals, poor terminal to wiring connections or physical damage to the wiring harness.

Resistance of the wheel speed sensor will increase with an increase in sensor temperature.

When replacing a wheel speed sensor, inspect the sensor terminals and harness connector for corrosion and/or water intrusion. If evidence of corrosion or water intrusion exists, replace wheel speed sensor jumper harness. Likewise, if replacing a wheel speed sensor jumper harness, inspect sensor terminals. If evidence of corrosion or water intrusion exists, replace wheel speed sensor.

DIAGNOSTIC CHART DTC 35 — 1995 SILHOUETTE, TRANS SPORT AND LUMINA APV

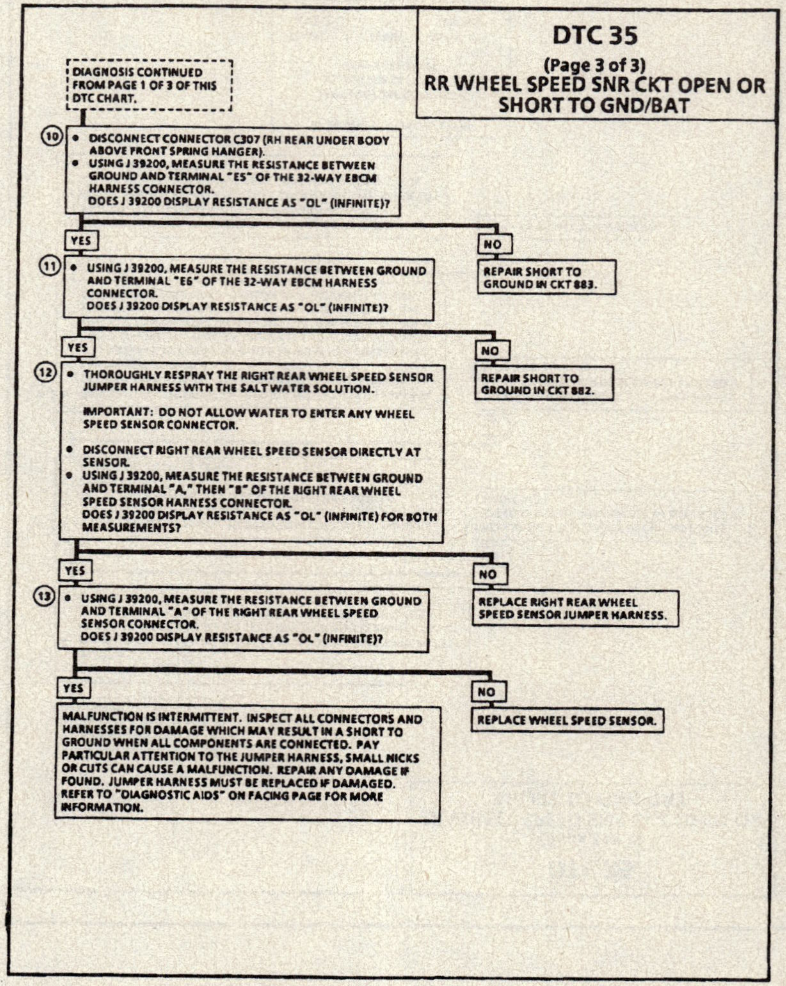

DIAGNOSTIC CHART DTC 36 — 1995 SILHOUETTE, TRANS SPORT AND LUMINA APV

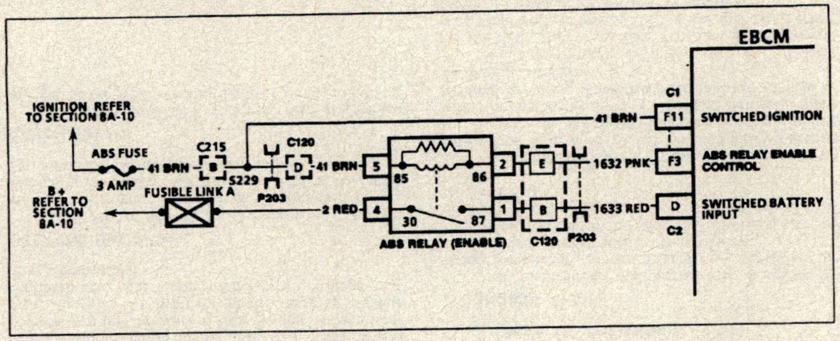

DTC 36
(Page 1 of 2)
LOW SYSTEM VOLTAGE

Circuit Description:

This DTC is used to monitor the voltage level available to the EBCM controller. If the voltage drops below 11 volts, full performance of the ABS system cannot be guaranteed. During ABS operation, there are several current requirements that will cause battery voltage to drop. Because of this, voltage is monitored prior to ABS to indicate good charging system condition and also during ABS when voltage may drop significantly.

DTC Will Set When:

DTC 36 can only be set if the vehicle's speed is greater than 5 km/h (3 mph). If the switched battery voltage is less than 11.2 volts non-ABS or less than 9.6 volts ABS, a malfunction exists.

Action Taken:

A malfunction DTC is stored, ABS is disabled and the amber ABS warning lamp is turned "ON."

DTC Chart Test Description:

Number(s) below refer to circled number(s) on the diagnostic chart.

1. Checks to see what voltage is readily available at terminal "D" of the EBCM. If voltage is representative of good charging system condition, a malfunction is not present.
2. Isolates the low voltage condition to high circuit resistance, or improper charging system operation.
3. Isolates which circuit is the source of the high resistance condition.
4. Checks for high resistance in CKTs 542 or 2.

Diagnostic Aids:

An "Intermittent" malfunction may be caused by a poor connection, rubbed through wire insulation, or a wire that is broken inside the insulation.

The frequency of the malfunction can be checked by using the enhanced diagnostic function of the Tech 1,

Any circuitry that is suspected of causing the intermittent complaint should be thoroughly checked for backed out terminals, improper mating, broken locks, improperly formed or damaged terminals, poor terminal to wiring connections or physical damage to the wiring harness.

While performing a voltage load test, if it is noted that only ignition voltage drops below acceptable voltage levels, CKT 41B should be checked for high resistance or an open condition.

DIAGNOSTIC CHART DTC 36 — 1995 SILHOUETTE, TRANS SPORT AND LUMINA APV

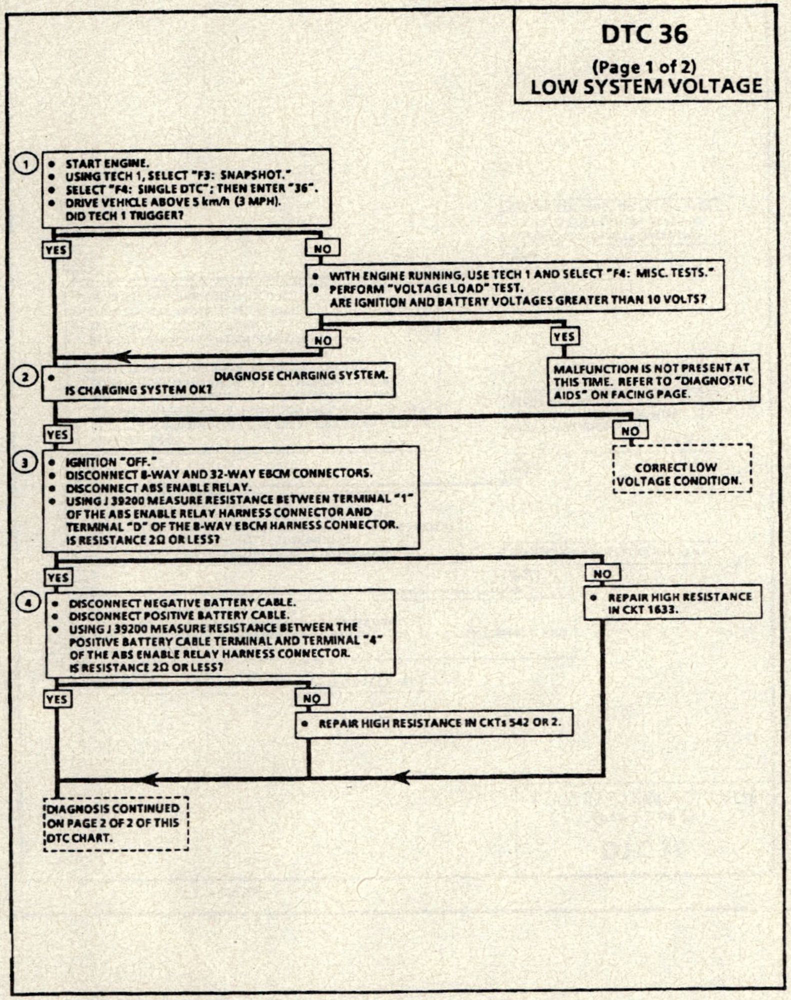

DIAGNOSTIC CHART DTC 36 — 1995 SILHOUETTE, TRANS SPORT AND LUMINA APV

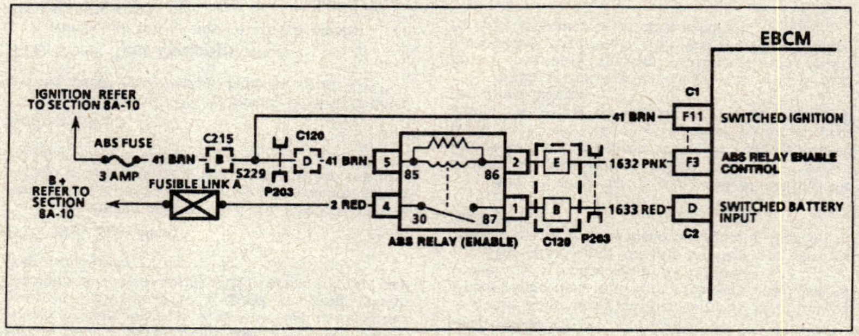

DTC 36
(Page 2 of 2)
LOW SYSTEM VOLTAGE

Circuit Description:

This DTC is used to monitor the voltage level available to the EBCM controller. If the voltage drops below 11 volts, full performance of the ABS system cannot be guaranteed. During ABS operation, there are several current requirements that will cause battery voltage to drop. Because of this, voltage is monitored prior to ABS to indicate good charging system condition and also during ABS when voltage may drop significantly.

DTC Will Set When:

DTC 36 can only be set if the vehicle's speed is greater than 5 km/h (3 mph). If the switched battery voltage is less than 11.2 volts non-ABS or less than 9.6 volts ABS, a malfunction exists.

Action Taken:

A malfunction DTC is stored, ABS is disabled and the amber ABS warning lamp is turned "ON."

DTC Chart Test Description:

Number(s) below refer to circled number(s) on the diagnostic chart.

5. Checks for high resistance in CKT 41.
6. Checks for high resistance in the ignition fuse feed circuit.
7. Checks for malfunctioning EBCM.

Diagnostic Aids:

An "Intermittent" malfunction may be caused by a poor connection, rubbed through wire insulation, or a wire that is broken inside the insulation.

The frequency of the malfunction can be checked by using the enhanced diagnostic function of the Tech 1.

Any circuitry that is suspected of causing the intermittent complaint should be thoroughly checked for backed out terminals, improper mating, broken locks, improperly formed or damaged terminals, poor terminal to wiring connections or physical damage to the wiring harness.

While performing a voltage load test, if it is noted that only ignition voltage drops below acceptable voltage levels, CKT 41 should be checked for high resistance or an open condition.

DIAGNOSTIC CHART DTC 36 — 1995 SILHOUETTE, TRANS SPORT AND LUMINA APV

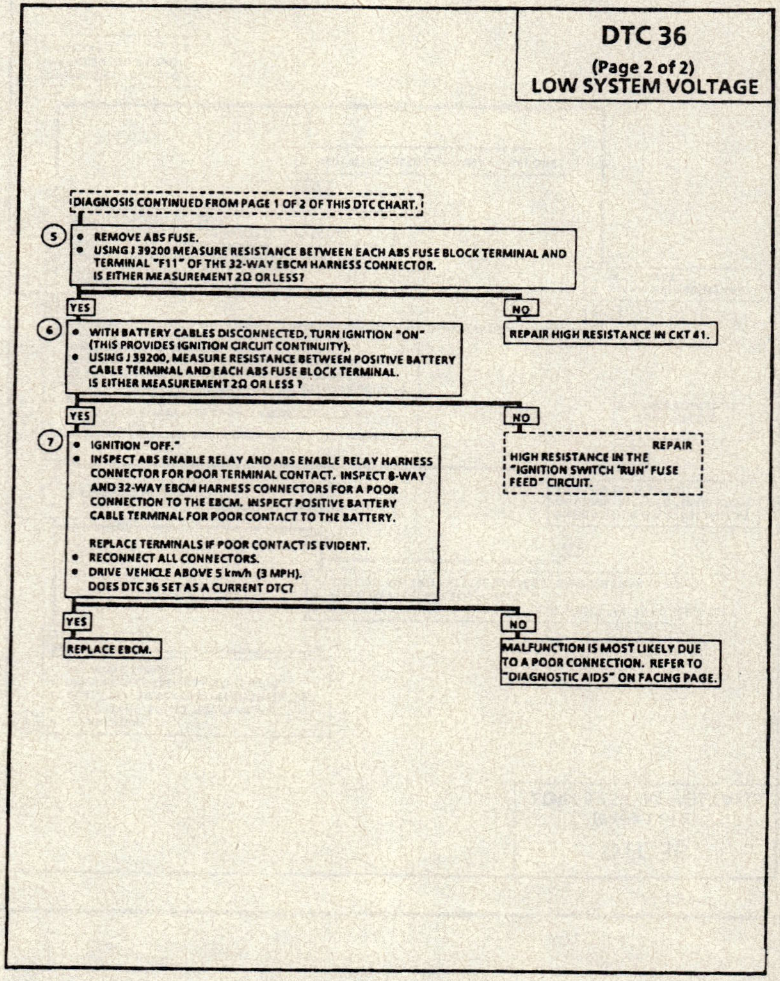

DIAGNOSTIC CHART DTC 37 — 1995 SILHOUETTE, TRANS SPORT AND LUMINA APV

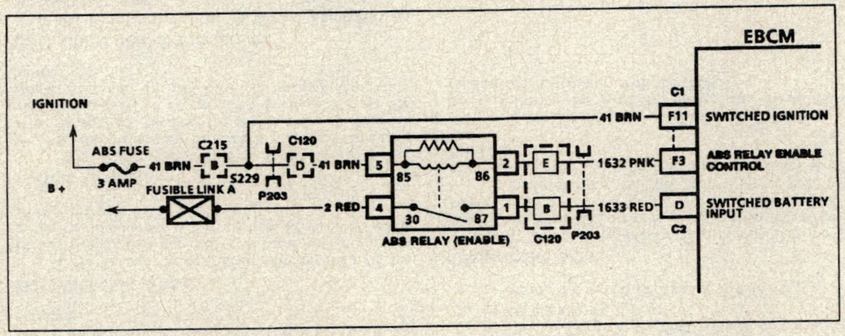

DTC 37
HIGH SYSTEM VOLTAGE

Circuit Description:

This DTC is designed to detect high vehicle voltage levels prior to any required motor movement (initialization or ABS). If excessive voltage exists, demagnetization of the motor magnets may occur, which would eventually affect or eliminate ABS performance.

DTC Will Set When:

DTC 37 can only be set if the vehicle's speed is greater than 5 km/h (3 mph). If the switched battery voltage is greater than 17 volts, a malfunction exists.

Action Taken:

A malfunction DTC is stored, ABS is disabled and the amber ABS warning lamp is turned "ON."

DTC Chart Test Description:

Number(s) below refer to circled number(s) on the diagnostic chart.

1. Checks the voltage level being received by the EBCM.
2. Indicates whether the high voltage condition is caused by a malfunctioning charging system or EBCM.
3. Checks for poor terminal contact which may cause an intermittent malfunction or a malfunctioning EBCM.

Diagnostic Aids:

An "Intermittent" malfunction may be caused by a poor connection, rubbed through wire insulation, or a wire that is broken inside the insulation.

The frequency of the malfunction can be checked by using the enhanced diagnostic function of the Tech 1,

Any circuitry that is suspected of causing the intermittent complaint should be thoroughly checked for backed out terminals, improper mating, broken locks, improperly formed or damaged terminals, poor terminal to wiring connections or physical damage to the wiring harness.

DIAGNOSTIC CHART DTC 37 — 1995 SILHOUETTE, TRANS SPORT AND LUMINA APV

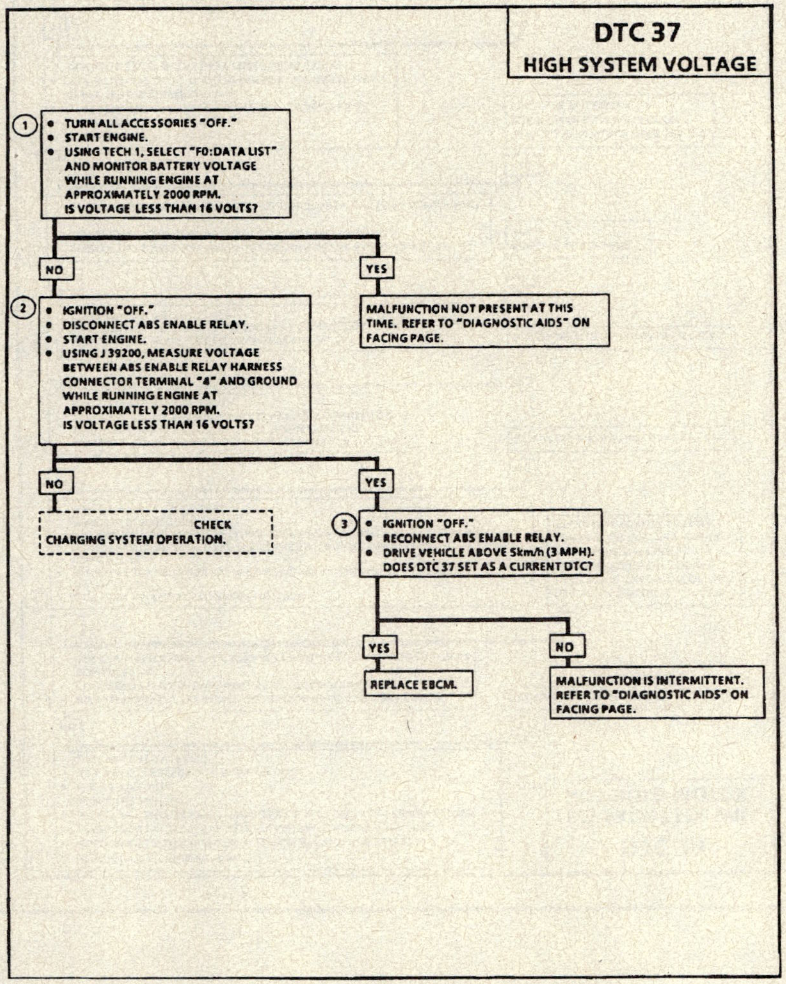

DIAGNOSTIC CHART DTC 38 — 1995 SILHOUETTE, TRANS SPORT AND LUMINA APV

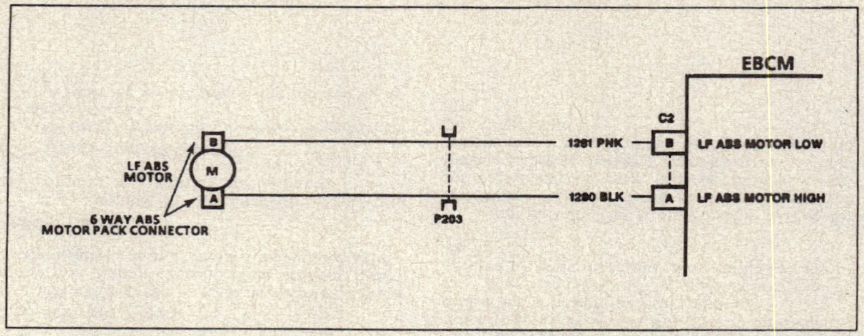

DTC 38
LEFT FRONT ESB WILL NOT HOLD MOTOR

Circuit Description:

This DTC is designed to detect a slipping left front ESB. During initialization and braking, the left front motor is re-homed. If the ESB slips, the motor/piston will move. During the next key to "RUN" initialization, a re-home of the motor verifies the motor/piston remained at the home position. If motor movement is detected, the ESB must be slipping.

DTC Will Set When:

DTC 38 can be set during initialization. If the EBCM detects that the ESB could not hold the piston in the home position, a malfunction exists. DTC 86 is always set with DTC 38.

Action Taken:

If an ESB cannot hold a piston in the home position, the piston may be backdriven when the brake pedal is applied, causing the brake pedal to drop. A malfunction DTC is stored, ABS is disabled, and the amber ABS and the red BRAKE warning lamps are turned "ON."

DTC Chart Test Description:

Number(s) below refer to circled number(s) on the diagnostic chart.

1. Checks the left front ESB. A broken or defective ESB would result in the left front piston being backdriven by hydraulic pressure and pedal movement would result.
2. Ensures that the proper motor circuits are located in the proper connector cavities.
3. Checks for high resistance in the motor "HIGH" circuitry.
4. Checks for high resistance in the motor "LOW" circuitry.
5. Checks for proper resistance of the motor windings.
6. Checks for poor terminal contact or corrosion in the connectors.
7. Checks for a malfunctioning EBCM.

Diagnostic Aids:

An "Intermittent" malfunction in this DTC may result from a mechanical part of the system that sticks, binds, or slips.

The frequency of the malfunction can be checked by using the enhanced diagnostic function of the Tech 1.

The static modulator test function of the Tech 1 or T-100 may be used to locate an intermittent malfunction associated with the ESB.

DIAGNOSTIC CHART DTC 38 — 1995 SILHOUETTE, TRANS SPORT AND LUMINA APV

DTC 38 — LEFT FRONT ESB WILL NOT HOLD MOTOR

(1)
- IGNITION "ON," ENGINE "OFF."
- PUMP BRAKE PEDAL UNTIL FIRM TO DEPLETE VACUUM RESERVE.
- USING TECH 1 MISC. TESTS, SELECT MANUAL CONTROL FUNCTION.
- SELECT ABS MOTOR APPLY FUNCTION OF THE LEFT FRONT CHANNEL AND APPLY THE MOTOR.
- WAIT 5 SECONDS.
- APPLY FIRM PRESSURE ON BRAKE PEDAL.

DID BRAKE PEDAL FALL?

- YES: REPLACE ABS MOTOR PACK.
- NO: go to (2)

(2)
- VISUALLY INSPECT THE 6-WAY ABS MOTOR PACK AND 8-WAY EBCM CONNECTORS FOR PROPER WIRE COLOR/CONNECTOR CAVITY COMBINATION.

ARE THE PROPER WIRES LOCATED IN THE PROPER CONNECTOR CAVITIES?

- NO: USING THE CIRCUIT DIAGRAMS LOCATED IN THE FRONT OF THIS SECTION, PLACE THE PROPER WIRES INTO THE PROPER CONNECTOR CAVITIES. CLEAR DTCs AND RECHECK SYSTEM.
- YES: go to (3)

(3)
- EXIT TECH 1 MANUAL CONTROL FUNCTION
- IGNITION "OFF."
- DISCONNECT 6-WAY ABS MOTOR PACK CONNECTOR AND 8-WAY EBCM CONNECTOR.
- USING J 39200, MEASURE THE RESISTANCE BETWEEN TERMINAL "A" OF THE 6-WAY ABS MOTOR PACK HARNESS CONNECTOR AND TERMINAL "A" OF THE 8-WAY EBCM HARNESS CONNECTOR.

IS RESISTANCE 1.5Ω OR LESS?

- NO: REPAIR HIGH RESISTANCE IN CKT 1280.
- YES: go to (4)

(4)
- MEASURE THE RESISTANCE BETWEEN TERMINAL "B" OF THE 6-WAY ABS MOTOR PACK HARNESS CONNECTOR AND TERMINAL "B" OF THE 8-WAY EBCM HARNESS CONNECTOR.

IS RESISTANCE 1.5Ω OR LESS?

- NO: REPAIR HIGH RESISTANCE IN CKT 1281.
- YES: go to (5)

(5)
- MEASURE THE RESISTANCE BETWEEN TERMINAL "A" AND TERMINAL "B" OF THE MOTOR PACK CONNECTOR.

IS RESISTANCE 1.5Ω OR LESS?

- NO: REPLACE ABS MOTOR PACK.
- YES: go to (6)

(6)
- INSPECT THE 6-WAY ABS MOTOR PACK CONNECTOR AND 8-WAY EBCM CONNECTOR FOR POOR TERMINAL CONTACT OR CORROSION.

DO ANY TERMINALS EXHIBIT POOR CONTACT OR EVIDENCE OF CORROSION?

- YES: REPLACE TERMINALS THAT EXHIBIT POOR TERMINAL CONTACT OR EVIDENCE OF CORROSION.
- NO: go to (7)

(7)
- IGNITION "OFF."
- RECONNECT ALL CONNECTORS.
- START ENGINE AND TEST DRIVE VEHICLE, ACHIEVING AT LEAST 16 km/h (10 MPH).
- REPEAT ABOVE DRIVE CYCLE SEQUENCE TWO MORE TIMES.

DOES DTC 38 SET IN THE LAST THREE DRIVE CYCLE?

- YES: REPLACE EBCM.
- NO: MALFUNCTION IS INTERMITTENT. REFER TO "DIAGNOSTIC AIDS" ON FACING PAGE.

AFTER DIAGNOSIS IS COMPLETE, CLEAR DTCs AND TEST DRIVE VEHICLE FOR THREE (3) DRIVE CYCLES TO VERIFY THAT DTC DOES NOT RESET. A DRIVE CYCLE CONSISTS OF STARTING THE VEHICLE, DRIVING OVER 16 km/h (10 MPH) AND THEN KEYING DOWN.

DIAGNOSTIC CHART DTC 41 — 1995 SILHOUETTE, TRANS SPORT AND LUMINA APV

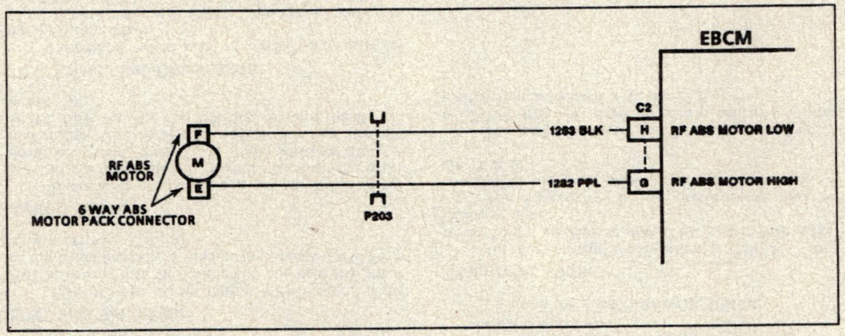

DTC 41
RIGHT FRONT ESB WILL NOT HOLD MOTOR

Circuit Description:

This DTC is designed to detect a slipping right front ESB. During initialization and braking, the right front motor is re-homed. If the ESB slips, the motor/piston will move. During the next key to "RUN" initialization, a re-home of the motor verifies the motor/piston remained at the home position. If motor movement is detected, the ESB must be slipping.

DTC Will Set When:

DTC 41 can be set during initialization. If the EBCM detects that the ESB could not hold the piston in the home position, a malfunction exists. DTC 86 is always set with DTC 41.

Action Taken:

If an ESB cannot hold a piston in the home position, the piston may be backdriven when the brake pedal is applied, causing the brake pedal to drop. A malfunction DTC is stored, ABS is disabled, and the amber ABS and the red BRAKE warning lamps are turned "ON."

DTC Chart Test Description:

Number(s) below refer to circled number(s) on the diagnostic chart.

1. Checks the right front ESB. A broken or defective ESB would result in the right front piston being backdriven by hydraulic pressure, and pedal movement would result.
2. Ensures that the proper motor circuits are located in the proper connector cavities.
3. Checks for high resistance in the motor "HIGH" circuitry.
4. Checks for high resistance in the motor "LOW" circuitry.
5. Checks for proper resistance of the motor windings.
6. Checks for poor terminal contact or corrosion in the connectors.
7. Checks for a malfunctioning EBCM.

Diagnostic Aids:

An "Intermittent" malfunction in this DTC may result from a mechanical part of the system that sticks, binds, or slips.

The frequency of the malfunction can be checked by using the enhanced diagnostic function of the Tech 1,

The static modulator test function of the Tech 1 or T-100 may be used to locate an intermittent malfunction associated with the ESB.

DIAGNOSTIC CHART DTC 41 — 1995 SILHOUETTE, TRANS SPORT AND LUMINA APV

DTC 41
RIGHT FRONT ESB WILL NOT HOLD MOTOR

(1)
- IGNITION "ON," ENGINE "OFF."
- PUMP BRAKE PEDAL UNTIL FIRM TO DEPLETE VACUUM RESERVE.
- USING TECH 1 MISC. TESTS, SELECT MANUAL CONTROL FUNCTION.
- SELECT ABS MOTOR APPLY FUNCTION OF THE RIGHT FRONT CHANNEL AND APPLY THE MOTOR.
- WAIT 5 SECONDS.
- APPLY FIRM PRESSURE ON BRAKE PEDAL.

DID BRAKE PEDAL FALL?

- YES: REPLACE ABS MOTOR PACK.
- NO: go to (2)

(2)
- VISUALLY INSPECT THE 6-WAY ABS MOTOR PACK AND 8-WAY EBCM CONNECTORS FOR PROPER WIRE COLOR/CONNECTOR CAVITY COMBINATION.

ARE THE PROPER WIRES LOCATED IN THE PROPER CONNECTOR CAVITIES?

- NO: USING THE CIRCUIT DIAGRAMS LOCATED IN THE FRONT OF THIS SECTION, PLACE THE PROPER WIRES INTO THE PROPER CONNECTOR CAVITIES. CLEAR DTCs AND RECHECK SYSTEM.
- YES: go to (3)

(3)
- EXIT TECH 1 MANUAL CONTROL FUNCTION.
- IGNITION "OFF."
- DISCONNECT 6-WAY ABS MOTOR PACK CONNECTOR AND 8-WAY EBCM CONNECTOR.
- USING J 39200, MEASURE THE RESISTANCE BETWEEN TERMINAL "E" OF THE 6-WAY ABS MOTOR PACK HARNESS CONNECTOR AND TERMINAL "G" OF THE 8-WAY EBCM HARNESS CONNECTOR.

IS RESISTANCE 1.5Ω OR LESS?

- NO: REPAIR HIGH RESISTANCE IN CKT 1282.
- YES: go to (4)

(4)
- MEASURE THE RESISTANCE BETWEEN TERMINAL "F" OF THE 6-WAY ABS MOTOR PACK HARNESS CONNECTOR AND TERMINAL "H" OF THE 8-WAY EBCM HARNESS CONNECTOR.

IS RESISTANCE 1.5Ω OR LESS?

- NO: REPAIR HIGH RESISTANCE IN CKT 1283.
- YES: go to (5)

(5)
- MEASURE THE RESISTANCE BETWEEN TERMINAL "E" AND TERMINAL "F" OF THE ABS MOTOR PACK CONNECTOR.

IS RESISTANCE 1.5Ω OR LESS?

- NO: REPLACE ABS MOTOR PACK.
- YES: go to (6)

(6)
- INSPECT THE 6-WAY ABS MOTOR PACK CONNECTOR AND 8-WAY EBCM CONNECTOR FOR POOR TERMINAL CONTACT OR CORROSION.

DO ANY TERMINALS EXHIBIT POOR CONTACT OR EVIDENCE OF CORROSION?

- YES: REPLACE TERMINALS THAT EXHIBIT POOR TERMINAL CONTACT OR EVIDENCE OF CORROSION.
- NO: go to (7)

(7)
- IGNITION "OFF."
- RECONNECT ALL CONNECTORS.
- START ENGINE AND TEST DRIVE VEHICLE, ACHIEVING AT LEAST 16 km/h (10 MPH).
- REPEAT ABOVE DRIVE CYCLE SEQUENCE TWO MORE TIMES.

DOES DTC 41 SET IN THE LAST THREE DRIVE CYCLES?

- YES: REPLACE EBCM.
- NO: MALFUNCTION IS INTERMITTENT. REFER TO "DIAGNOSTIC AIDS" ON FACING PAGE.

AFTER DIAGNOSIS IS COMPLETE, CLEAR DTCs AND TEST DRIVE VEHICLE FOR THREE (3) DRIVE CYCLES TO VERIFY THAT DTC DOES NOT RESET. A DRIVE CYCLE CONSISTS OF STARTING THE VEHICLE, DRIVING OVER 16 KM/H (10 MPH) AND THEN KEYING DOWN.

DIAGNOSTIC CHART DTC 42 — 1995 SILHOUETTE, TRANS SPORT AND LUMINA APV

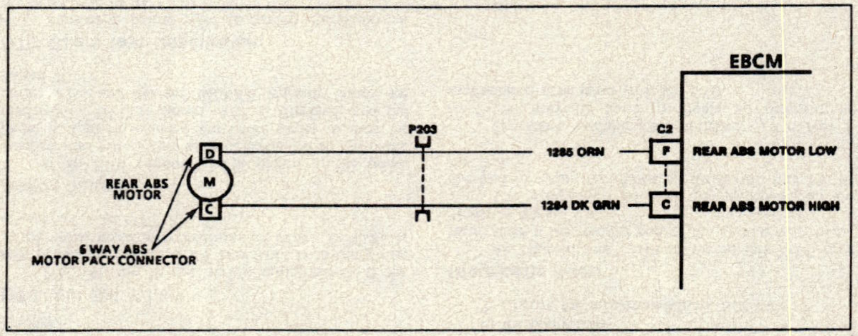

DTC 42
REAR ESB WILL NOT HOLD MOTOR

Circuit Description:

This DTC is designed to detect a slipping rear ESB. During initialization and braking, the rear motor is re-homed. If the ESB slips, the motor/piston will move. During the next key to "RUN," initialization, a re-home of the motor verifies the motor/piston remained at the home position. If motor movement is detected, the ESB must be slipping.

DTC Will Set When:

DTC 42 can be set during initialization. If the EBCM detects that the ESB could not hold the piston in the home position, a malfunction exists. DTC 86 is always set with DTC 42.

Action Taken:

If an ESB cannot hold a piston in the home position, the piston may be backdriven when the brake pedal is applied, causing the brake pedal to drop. A malfunction DTC is stored, ABS is disabled, and the amber ABS and the red BRAKE warning lamps are turned "ON."

DTC Chart Test Description:

Number(s) below refer to circled number(s) on the diagnostic chart.

1. Checks the rear ESB. A broken or defective ESB would result in the rear pistons being backdriven by hydraulic pressure and wheel movement would result.
2. Ensures that the proper motor circuits are located in the proper connector cavities.
3. Checks for high resistance in the motor "HIGH" circuitry.
4. Checks for high resistance in the motor "LOW" circuitry.
5. Checks for proper resistance of the motor windings.
6. Checks for poor terminal contact or corrosion in the connectors.
7. Checks for a malfunctioning EBCM.

Diagnostic Aids:

An "Intermittent" malfunction in this DTC may result from a mechanical part of the system that sticks, binds, or slips.

The frequency of the malfunction can be checked by using the enhanced diagnostic function of the Tech 1.

The static modulator test function of the Tech 1 or T-100 may be used to locate an intermittent malfunction associated with the ESB.

DIAGNOSTIC CHART DTC 42 — 1995 SILHOUETTE, TRANS SPORT AND LUMINA APV

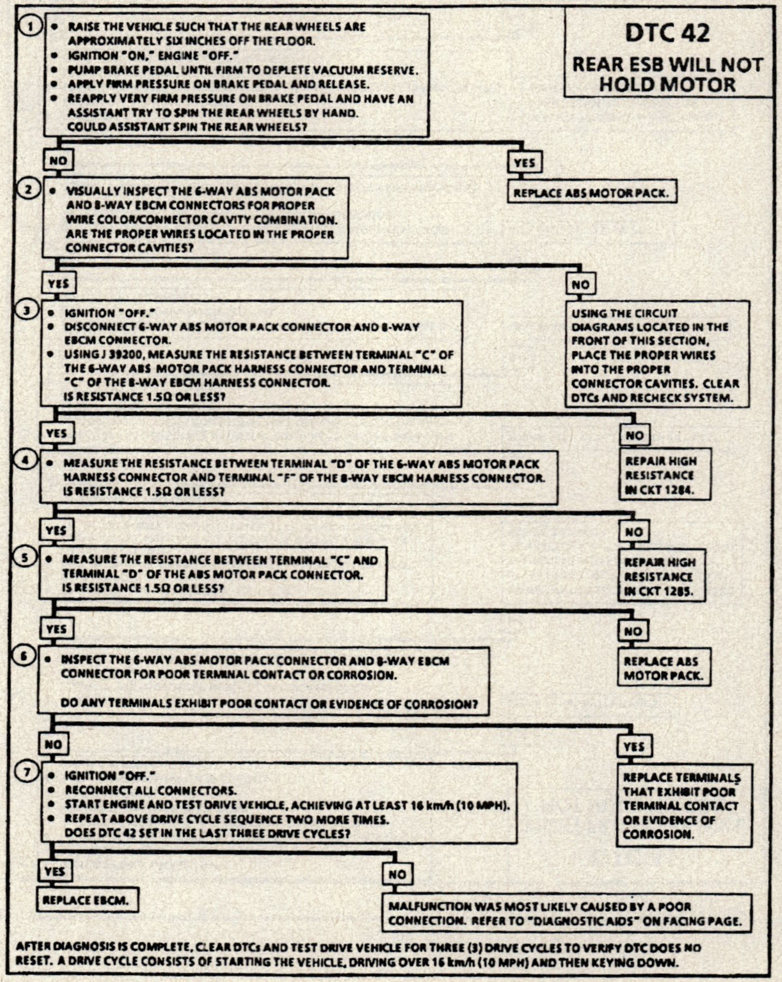

DIAGNOSTIC CHART DTC 44 — 1995 SILHOUETTE, TRANS SPORT AND LUMINA APV

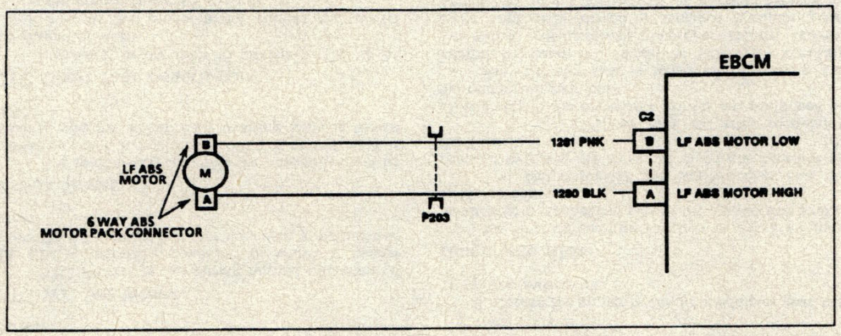

DTC 44
LEFT FRONT ABS CHANNEL WILL NOT MOVE

Circuit Description:

This DTC is designed to detect bound-up ESB, a stuck ABS motor, or a seized hydraulic modulator. When the release is commanded during initialization, the ESB should release the ABS motor, resulting in sensed current being less than commanded current (motor is spinning freely). If the ABS motor is not moving, sensed current will be equal to stall current.

DTC Will Set When:

DTC 44 can be set during normal operation. If the EBCM detects a condition in which it cannot move the ABS motor in either direction, a malfunction exists.

Action Taken:

This malfunction indicates the channel cannot be moved properly. A malfunction DTC is stored, ABS is disabled and the amber ABS warning lamp is turned "ON."

DTC Chart Test Description:

Number(s) below refer to circled number(s) on the diagnostic chart.

1. Ensures that proper ABS motor circuits are located in the proper connector cavities.
2. Checks for proper ABS motor movement during apply and release commands from the Tech 1.
3. Compares EBCM command current to ABS motor feedback current.
4. Checks for a short circuit between the ABS motor "HIGH" and motor "LOW" circuits.
5. Checks for an ABS motor that is internally shorted.
6. Determines if the malfunction is due to poor terminal contact or the EBCM.
7. Verifies that the ABS motor can actually apply when commanded.
8. Checks for proper hydraulic modulator gear and piston movement.

Diagnostic Aids:

An "Intermittent" malfunction in this DTC may result from a mechanical part of the system that sticks, binds, or slips.

The frequency of the malfunction can be checked by using the enhanced diagnostic function of the Tech 1,

DTC 44 may set after modulator disassembly if the modulator pistons are positioned at the bottom of their bore.

Any circuitry that is suspected of causing the intermittent complaint should be thoroughly checked for backed out terminals, improper mating, broken locks, improperly formed or damaged terminals, poor terminal to wiring connections or physical damage to the wiring harness.

DIAGNOSTIC CHART DTC 44 — 1995 SILHOUETTE, TRANS SPORT AND LUMINA APV

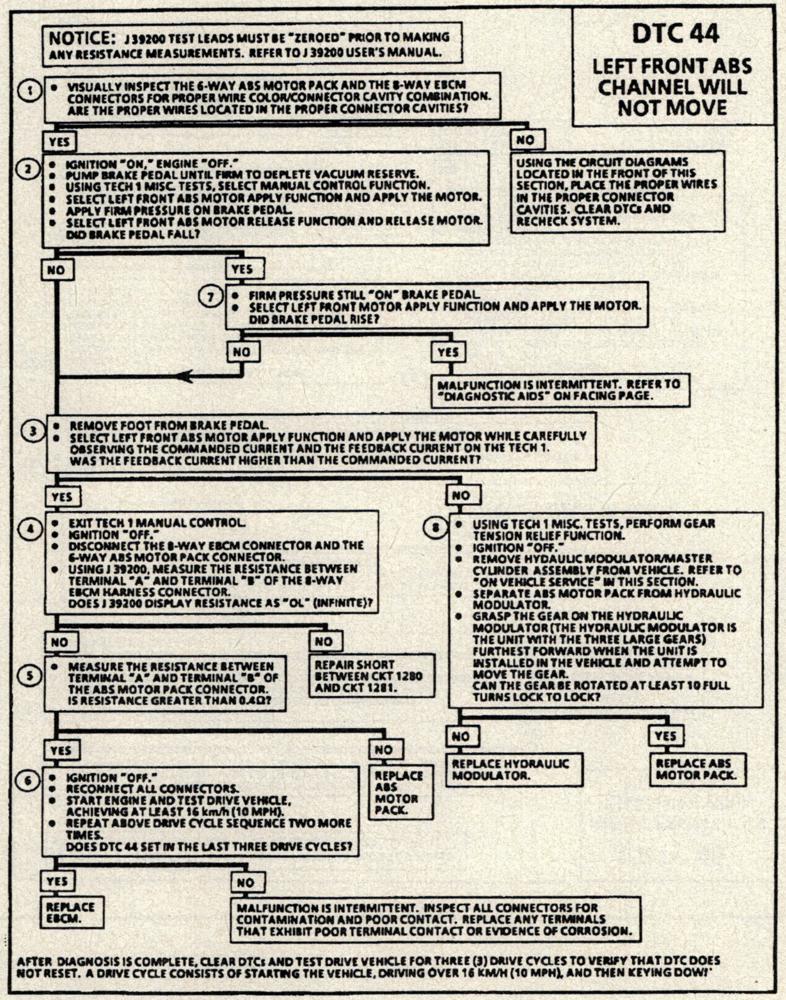

DIAGNOSTIC CHART DTC 45 — 1995 SILHOUETTE, TRANS SPORT AND LUMINA APV

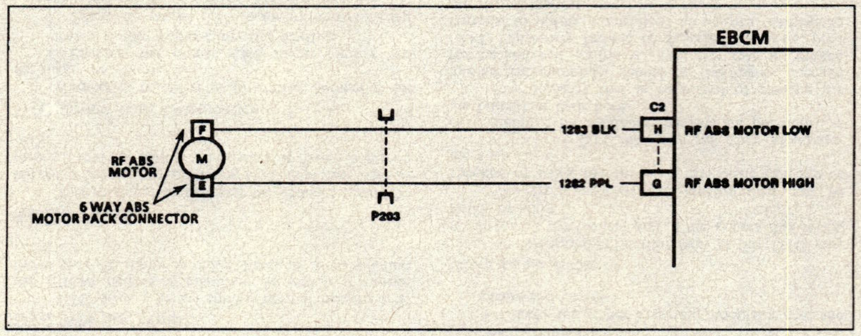

DTC 45
RIGHT FRONT ABS CHANNEL WILL NOT MOVE

Circuit Description:

This DTC is designed to detect bound-up ESB, a stuck ABS motor, or a seized hydraulic modulator. When the release is commanded during initialization, the ESB should release the ABS motor, resulting in sensed current being less than commanded current (ABS motor is spinning freely). If the ABS motor is not moving, sensed current will be equal to stall current.

DTC Will Set When:

DTC 45 can be set during normal operation. If the EBCM detects a condition in which it cannot move the ABS motor in either direction, a malfunction exists.

Action Taken:

This malfunction indicates the channel cannot be moved properly. A malfunction DTC is stored, ABS is disabled and the amber ABS warning lamp is turned "ON."

DTC Chart Test Description:

Number(s) below refer to circled number(s) on the diagnostic chart.

1. Ensures that proper motor circuits are located in the proper connector cavities.
2. Checks for proper ABS motor movement during apply and release commands from the Tech 1.
3. Compares EBCM command current to ABS motor feedback current.
4. Checks for a short circuit between the ABS motor "HIGH" and motor "LOW" circuits.
5. Checks for an ABS motor that is internally shorted.
6. Checks to see if malfunction is due to poor terminal contact or the EBCM.
7. Verifies that the ABS motor can actually apply when commanded.
8. Checks for proper hydraulic modulator gear and piston movement.

Diagnostic Aids:

An "Intermittent" malfunction in this DTC may result from a mechanical part of the system that sticks, binds, or slips.

The frequency of the malfunction can be checked by using the enhanced diagnostic function of the Tech 1,

DTC 45 may set after modulator disassembly if the modulator pistons are positioned at the bottom of their bore.

Any circuitry that is suspected of causing the intermittent complaint should be thoroughly checked for backed out terminals, improper mating, broken locks, improperly formed or damaged terminals, poor terminal to wiring connections or physical damage to the wiring harness.

DIAGNOSTIC CHART DTC 45 — 1995 SILHOUETTE, TRANS SPORT AND LUMINA APV

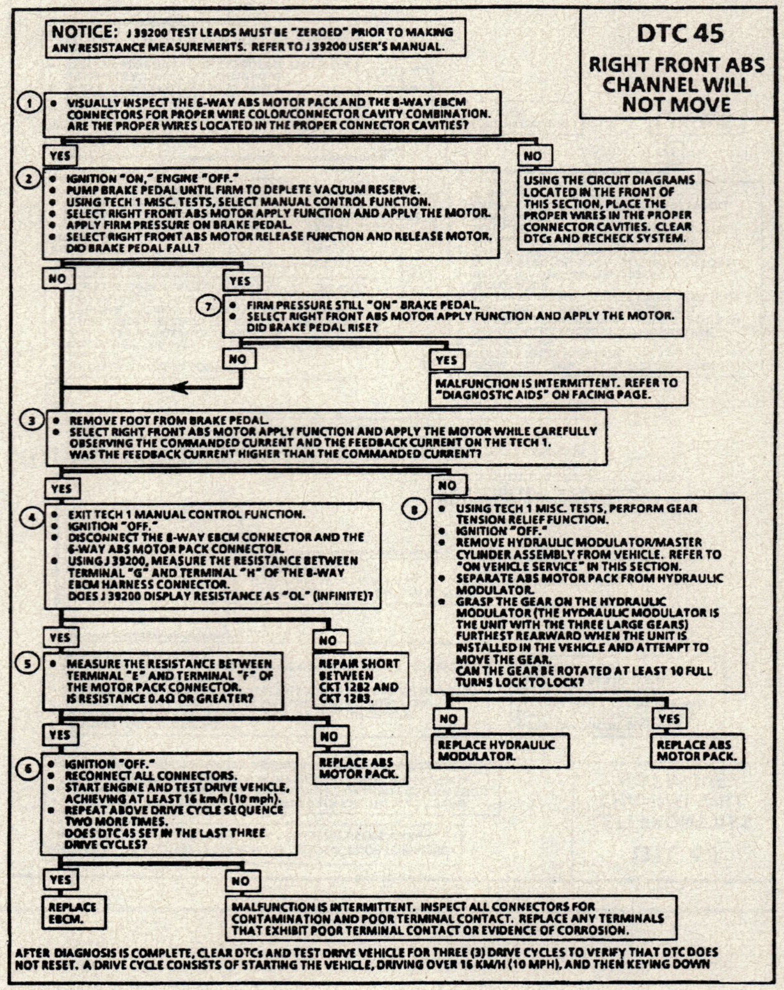

DIAGNOSTIC CHART DTC 46 — 1995 SILHOUETTE, TRANS SPORT AND LUMINA APV

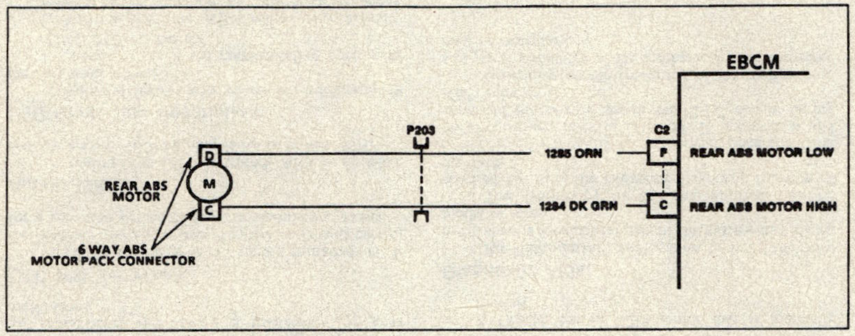

DTC 46
REAR ABS CHANNEL WILL NOT MOVE

Circuit Description:

This DTC is designed to detect bound-up ESB, a stuck ABS motor, or a seized hydraulic modulator. When the release is commanded during initialization, the ESB should release the ABS motor, resulting in sensed current being less than commanded current (motor is spinning freely). If the ABS motor is not moving, sensed current will be equal to stall current.

DTC Will Set When:

DTC 46 can be set during normal operation. If the EBCM detects a condition in which it cannot move the ABS motor in either direction, a malfunction exists.

Action Taken:

This malfunction indicates the channel cannot be moved properly. A malfunction DTC is stored, ABS is disabled and the amber ABS warning lamp is turned "ON."

DTC Chart Test Description:

Number(s) below refer to circled number(s) on the diagnostic chart.

1. Ensures that proper motor circuits are located in the proper connector cavities.
2. Checks for proper motor movement during apply and release commands from the Tech 1.
3. Compares EBCM command current to ABS motor feedback current.
4. Checks for a short circuit between the ABS motor "HIGH" and ABS motor "LOW" circuits.
5. Checks for an ABS motor that is internally shorted.
6. Checks to see if malfunction is due to poor terminal contact or the EBCM.
7. This step verifies that the motor can actually release when commanded.
8. Checks for proper hydraulic modulator gear and piston movement.

Diagnostic Aids:

An "Intermittent" malfunction in this DTC may result from a mechanical part of the system that sticks or binds.

The frequency of the malfunction can be checked by using the enhanced diagnostic function of the Tech 1,

DTC 46 may set after modulator disassembly if the modulator pistons are positioned at the bottom of their bore. Depending on the frequency of the malfunction, a physical inspection of the mechanical parts suspected may be necessary.

Any circuitry that is suspected of causing the intermittent complaint should be thoroughly checked for backed out terminals, improper mating, broken locks, improperly formed or damaged terminals, poor terminal to wiring connections or physical damage to the wiring harness.

DIAGNOSTIC CHART DTC 46 — 1995 SILHOUETTE, TRANS SPORT AND LUMINA APV

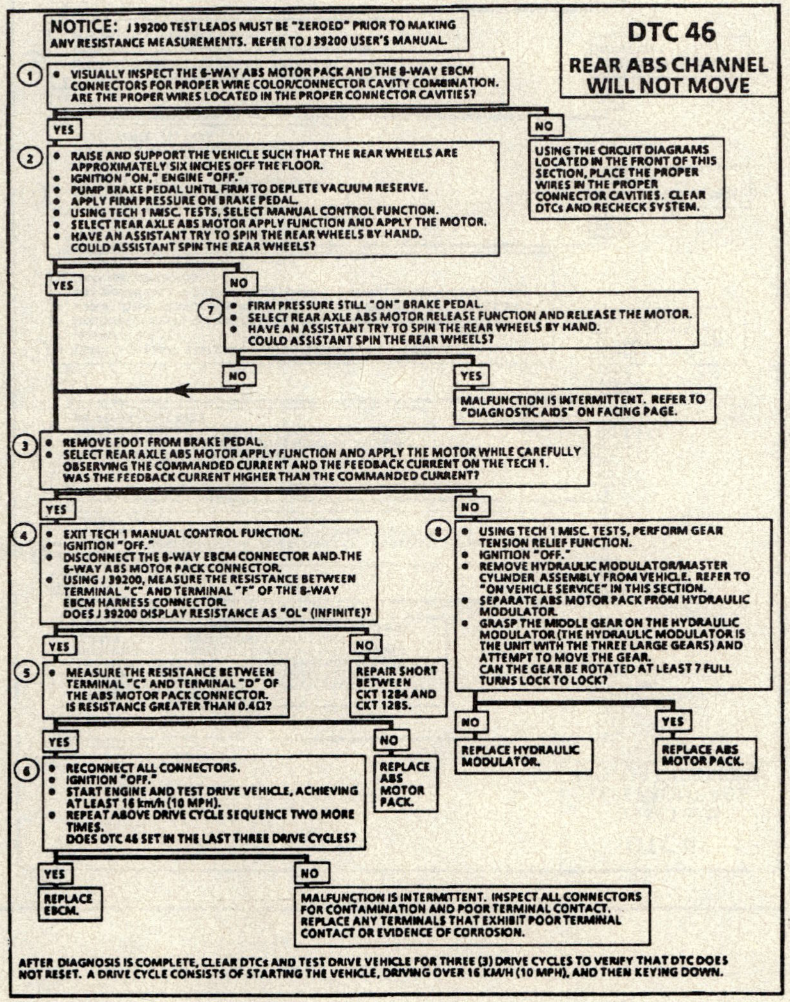

DIAGNOSTIC CHART DTC 47 — 1995 SILHOUETTE, TRANS SPORT AND LUMINA APV

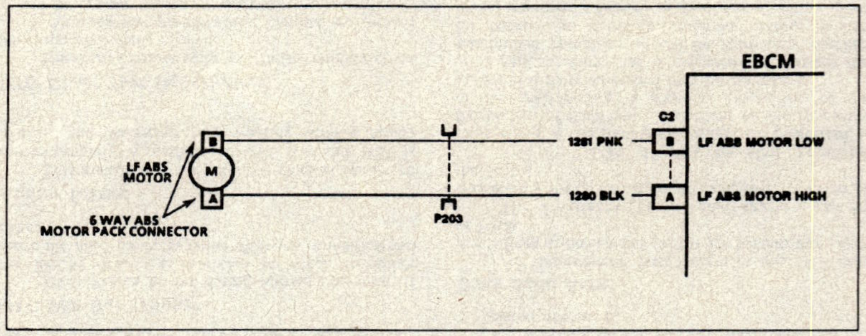

DTC 47
(Page 1 of 2)
LEFT FRONT ABS MOTOR FREE SPINS

Circuit Description:

This DTC is designed to detect a stripped nut or gear assembly during initialization. During the homing sequence, the piston should reach the top of the bore resulting in a stalled ABS motor. If this does not occur, the ABS motor must be spinning with little or no resistance, indicating a nut/screw or gear malfunction.

DTC Will Set When:

DTC 47 can only be set during initialization. If the feedback current is less than the command current for a specified period of time, a malfunction exists.

Action Taken:

A malfunction DTC is stored, ABS is disabled, and the amber ABS warning lamp is turned "ON."

DTC Chart Test Description:

Number(s) below refer to circled number(s) on the diagnostic chart.

1. Checks to see if the corresponding open ABS motor DTC is also set.
2. Verifies that the ABS motor was actually applied as commanded by observing feedback current.
3. Verifies that the ABS motor can release.
4. This step verifies that the ABS motor can be applied by observing pedal movement.
5. Checks for an open in the ABS motor "HIGH" circuitry.
6. Checks for an open in the ABS motor "LOW" circuitry.
7. Checks for an ABS motor that is internally shorted.

Diagnostic Aids:

An "Intermittent" malfunction in this DTC may result from a mechanical part of the system that sticks, binds, or slips.

The frequency of the malfunction can be checked by using the enhanced diagnostic function of the Tech 1,

If the DTC only occurs once and DTC 56 also occurs, refer to DTC 56. If intermittent and enhanced diagnostics shows this DTC occurs during ABS, refer to DTC 56.

Depending on the frequency of the malfunction, a physical inspection of the mechanical parts suspected may be necessary.

DIAGNOSTIC CHART DTC 47 — 1995 SILHOUETTE, TRANS SPORT AND LUMINA APV

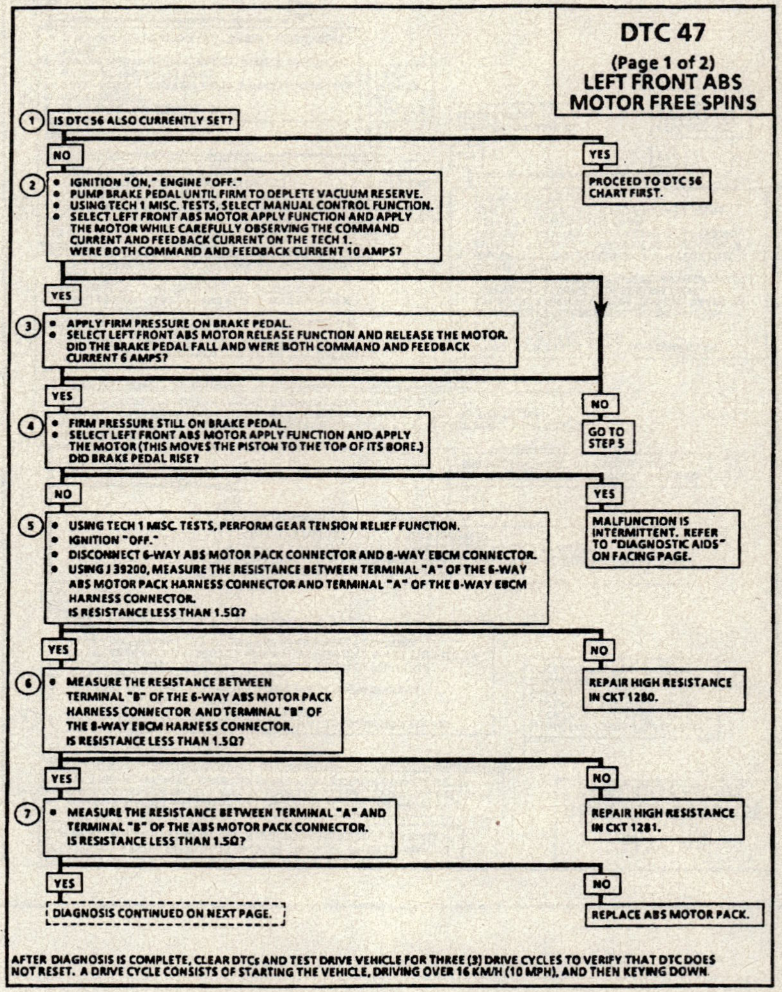

DIAGNOSTIC CHART DTC 47 — 1995 SILHOUETTE, TRANS SPORT AND LUMINA APV

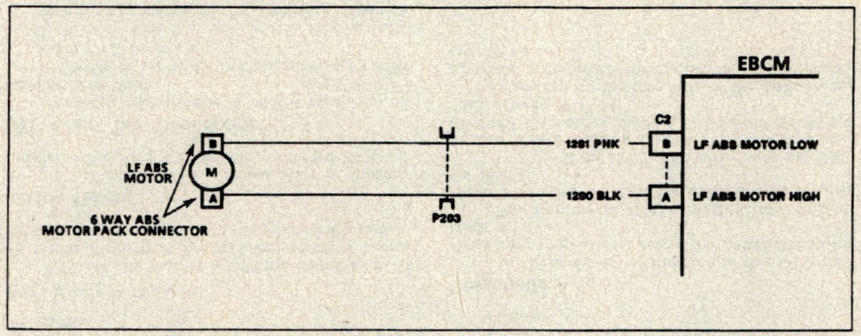

DTC 47
(Page 2 of 2)
LEFT FRONT ABS MOTOR FREE SPINS

Circuit Description:

This DTC is designed to detect a stripped nut or gear assembly during initialization. During the homing sequence, the piston should reach the top of the bore resulting in a stalled ABS motor. If this does not occur, the ABS motor must be spinning with little or no resistance, indicating a nut/screw or gear malfunction.

DTC Will Set When:

DTC 47 can only be set during initialization. If the feedback current is less than the command current for a specified period of time, a malfunction exists.

Action Taken:

A malfunction DTC is stored, ABS is disabled, and the amber ABS warning lamp is turned "ON."

DTC Chart Test Description:

Number(s) below refer to circled number(s) on the diagnostic chart.

8. Checks for a stripped gear on the motor pack (unit with three small gears).
9. Checks for a stripped gear on the hydraulic modulator (unit with three large gears).
10. Checks to see if the ABS motor pack has an internal malfunction.
11. Verifies that the hydraulic modulator is functioning properly.
12. Ensures malfunction was not due to a poor terminal contact.

Diagnostic Aids:

An "Intermittent" malfunction in this DTC may result from a mechanical part of the system that sticks, binds, or slips.

The frequency of the malfunction can be checked by using the enhanced diagnostic function of the Tech 1,

If the DTC only occurs once and DTC 56 also occurs, refer to DTC 56. If intermittent and enhanced diagnostics shows this DTC occurs during ABS, refer to DTC 56.

Depending on the frequency of the malfunction, a physical inspection of the mechanical parts suspected may be necessary.

DIAGNOSTIC CHART DTC 47 — 1995 SILHOUETTE, TRANS SPORT AND LUMINA APV

DTC 47 (Page 2 of 2) LEFT FRONT ABS MOTOR FREE SPINS

DIAGNOSIS CONTINUED FROM PREVIOUS PAGE.

(8)
- REMOVE HYDRAULIC MODULATOR/MASTER CYLINDER ASSEMBLY FROM VEHICLE.
- REMOVE GEAR COVER.
- CHECK FOR STRIPPED GEAR ON ABS MOTOR PACK; ABS MOTOR PACK IS THE UNIT WITH THE THREE SMALL GEARS. THE LEFT FRONT GEAR IS THE GEAR FURTHEST FORWARD WHEN THE ASSEMBLY IS INSTALLED IN THE VEHICLE. IS THE ABS MOTOR PACK GEAR STRIPPED?

YES: REPLACE ABS MOTOR PACK.

NO:

(9)
- CHECK FOR A STRIPPED GEAR ON HYDRAULIC MODULATOR; MODULATOR IS THE UNIT WITH THE THREE LARGE GEARS. THE LEFT FRONT GEAR IS THE GEAR FURTHEST FORWARD WHEN THE UNIT IS INSTALLED IN THE VEHICLE. IS THE MODULATOR GEAR STRIPPED?

YES: REPLACE HYDRAULIC MODULATOR GEAR.

NO:

(10)
- RECONNECT ELECTRICAL CONNECTORS
- SECURELY POSITION HYDRAULIC MODULATOR/MASTER CYLINDER ASSEMBLY IN VEHICLE WITH GEAR COVER REMOVED SUCH THAT THE GEAR SET CAN BE OBSERVED. BE CAREFUL NOT TO DAMAGE GEAR SET OR ALLOW THE HYDRAULIC MODULATOR TO MOVE WHILE TESTING.
- IGNITION "ON."
- USING TECH 1 MISC. TESTS, SELECT MANUAL CONTROL FUNCTION.
- SELECT LEFT FRONT ABS MOTOR APPLY AND APPLY THE MOTOR WHILE OBSERVING THE GEAR SET FURTHEST FORWARD WHEN THE UNIT IS INSTALLED IN THE VEHICLE.
- SELECT LEFT ABS FRONT MOTOR RELEASE FUNCTION AND RELEASE THE MOTOR WHILE STILL OBSERVING THE GEAR SET.
- SELECT LEFT FRONT ABS MOTOR APPLY FUNCTION AND APPLY THE MOTOR. DID GEAR SET MOVE IN BOTH DIRECTIONS FOR AT LEAST ONE REVOLUTION?

NO: REPLACE ABS MOTOR PACK.

YES:

(11)
- USING TECH 1 MISC. TESTS, PERFORM GEAR TENSION RELIEF FUNCTION.
- IGNITION "OFF."
- SEPARATE THE ABS MOTOR PACK FROM THE HYDRAULIC MODULATOR.
- GRASP THE GEAR ON THE HYDRAULIC MODULATOR (THE HYDRAULIC MODULATOR IS THE UNIT WITH THE THREE LARGE GEARS) FURTHEST FORWARD WHEN THE UNIT IS INSTALLED IN THE VEHICLE AND ATTEMPT TO MOVE THE GEAR. CAN THE GEAR BE ROTATED MORE THAN 13 FULL TURNS LOCK TO LOCK?

YES: REPLACE THE HYDRAULIC MODULATOR.

NO:

(12)
- INSPECT ALL CONNECTORS AND TERMINALS FOR POOR TERMINAL CONTACT AND EVIDENCE OF CORROSION.
  REPLACE ALL TERMINALS THAT EXHIBIT SIGNS OF POOR TERMINAL CONTACT OR CORROSION.
- REINSTALL HYDRAULIC MODULATOR/MASTER CYLINDER ASSEMBLY IN VEHICLE.
- RECONNECT ALL CONNECTORS.
- IGNITION "OFF."
- START ENGINE AND TEST DRIVE VEHICLE, ACHIEVING AT LEAST 16 km/h (10 MPH).
- REPEAT ABOVE DRIVE CYCLE SEQUENCE TWO MORE TIMES.
  DOES DTC 47 SET IN THE LAST THREE DRIVE CYCLES?

YES: REPLACE EBCM.

NO: MALFUNCTION WAS MOSTLY LIKELY CAUSED BY A POOR CONNECTION. REFER TO "DIAGNOSTIC AIDS" ON FACING PAGE.

AFTER DIAGNOSIS IS COMPLETE, CLEAR DTCs AND TEST DRIVE VEHICLE FOR THREE (3) DRIVE CYCLES TO VERIFY THAT DTC DOES NOT RESET. A DRIVE CYCLE CONSISTS OF STARTING THE VEHICLE, DRIVING OVER 16 KM/H (10 MPH), AND THEN KEYING DOWN

DIAGNOSTIC CHART DTC 48 — 1995 SILHOUETTE, TRANS SPORT AND LUMINA APV

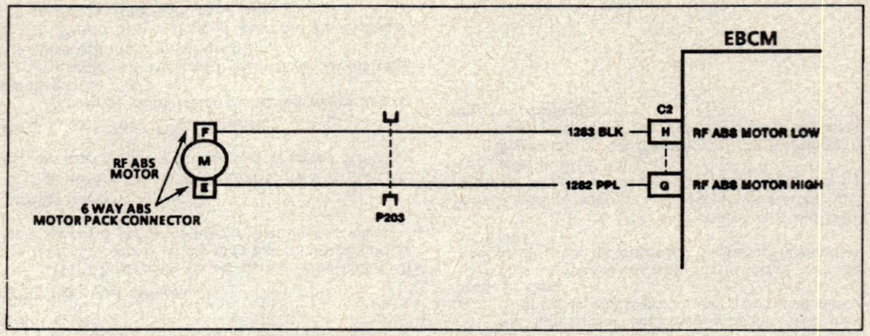

DTC 48
(Page 1 of 2)
RIGHT FRONT ABS MOTOR FREE SPINS

Circuit Description:

This DTC is designed to detect a stripped nut or gear assembly during initialization. During the homing sequence, the piston should reach the top of the bore resulting in a stalled ABS motor. If this does not occur, the ABS motor must be spinning with little or no resistance, indicating a nut/screw or gear malfunction.

DTC Will Set When:

DTC 48 can only be set during initialization. If the feedback current is less than the command current for a specified period of time, a malfunction exists.

Action Taken:

A malfunction DTC is stored, ABS is disabled, and the amber ABS warning lamp is turned "ON."

DTC Chart Test Description:

Number(s) below refer to circled number(s) on the diagnostic chart.

1. Checks to see if the corresponding open ABS motor DTC is also set.
2. Verifies that the ABS motor was actually applied as commanded by observing feedback current.
3. Verifies that the ABS motor can release.
4. Verifies that the ABS motor can be applied by observing pedal movement.
5. Checks for high resistance in the ABS motor "HIGH" circuitry.
6. Checks for high resistance in the ABS motor "LOW" circuitry.
7. Checks for proper resistance of the ABS motor windings.

Diagnostic Aids:

An "Intermittent" malfunction in this DTC may result from a mechanical part of the system that sticks, binds, or slips.

The frequency of the malfunction can be checked by using the enhanced diagnostic function of the Tech 1,

If the DTC only occurs once and DTC 61 also occurs, refer to DTC 61. If intermittent and enhanced diagnostics shows this DTC occurs during ABS, refer to DTC 61.

Depending on the frequency of the malfunction, a physical inspection of the mechanical parts suspected may be necessary.

DIAGNOSTIC CHART DTC 48 — 1995 SILHOUETTE, TRANS SPORT AND LUMINA APV

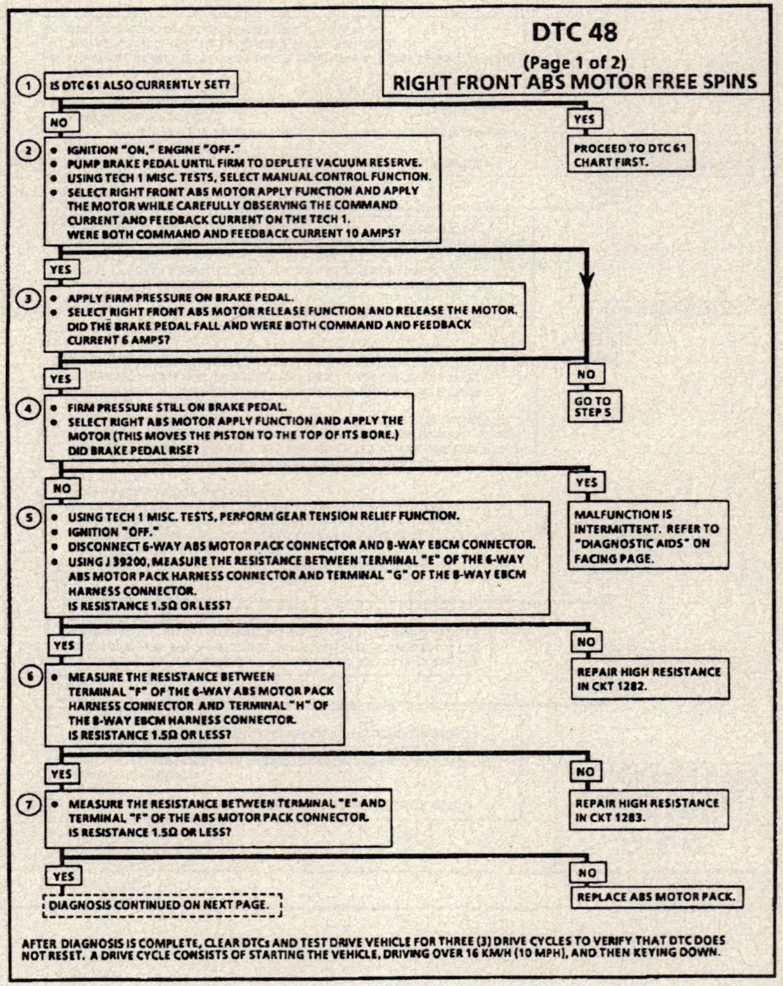

DIAGNOSTIC CHART DTC 48 — 1995 SILHOUETTE, TRANS SPORT AND LUMINA APV

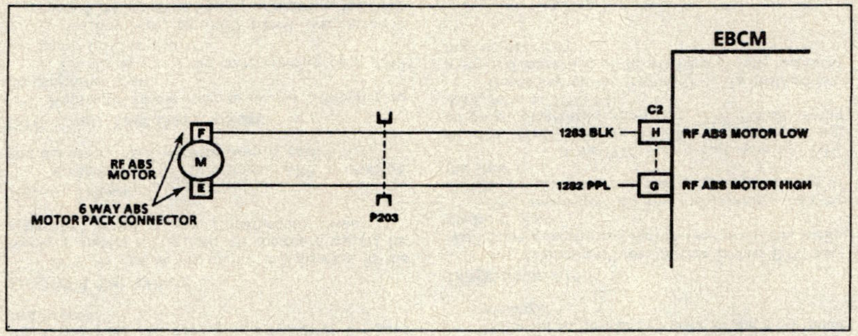

DTC 48
(Page 2 of 2)
RIGHT FRONT ABS MOTOR FREE SPINS

Circuit Description:

This DTC is designed to detect a stripped nut or gear assembly during initialization. During the homing sequence, the piston should reach the top of the bore resulting in a stalled ABS motor. If this does not occur, the ABS motor must be spinning with little or no resistance, indicating a nut/screw or gear malfunction.

DTC Will Set When:

DTC 48 can only be set during initialization. If the feedback current is less than the command current for a specified period of time, a malfunction exists.

Action Taken:

A malfunction DTC is stored, ABS is disabled, and the amber ABS warning lamp is turned "ON."

DTC Chart Test Description:

Number(s) below refer to circled number(s) on the diagnostic chart.

8. Checks for a stripped gear on the ABS motor pack (unit with three small gears).
9. Checks for a stripped gear on the hydraulic modulator (unit with three large gears).
10. Verifies that the ABS motor can be applied under a load situation.
11. Verifies that the hydraulic modulator is functioning properly.
12. Ensures malfunction was not due to poor terminal contact.

Diagnostic Aids:

An "Intermittent" malfunction in this DTC may result from a mechanical part of the system that sticks, binds, or slips.

The frequency of the malfunction can be checked by using the enhanced diagnostic function of the Tech 1,

If the DTC only occurs once and DTC 61 also occurs, refer to DTC 61. If intermittent and enhanced diagnostics shows this DTC occurs during ABS, refer to DTC 61.

Depending on the frequency of the malfunction, a physical inspection of the mechanical parts suspected may be necessary.

DIAGNOSTIC CHART DTC 48 — 1995 SILHOUETTE, TRANS SPORT AND LUMINA APV

DTC 48
(Page 2 of 2)
RIGHT FRONT ABS MOTOR FREE SPINS

DIAGNOSIS CONTINUED FROM PREVIOUS PAGE.

(8)
- REMOVE HYDRAULIC MODULATOR/MASTER CYLINDER ASSEMBLY FROM VEHICLE.
- REMOVE GEAR COVER.
- CHECK FOR STRIPPED GEAR ON ABS MOTOR PACK; ABS MOTOR PACK IS THE UNIT WITH THE THREE SMALL GEARS. THE RIGHT FRONT GEAR IS THE GEAR FURTHEST REARWARD WHEN THE ASSEMBLY IS INSTALLED IN THE VEHICLE.

IS THE ABS MOTOR PACK GEAR STRIPPED?

YES: REPLACE ABS MOTOR PACK.

NO:

(9)
- CHECK FOR A STRIPPED GEAR ON HYDRAULIC MODULATOR; MODULATOR IS THE UNIT WITH THE THREE LARGE GEARS. THE RIGHT FRONT GEAR IS THE GEAR FURTHEST REARWARD WHEN THE UNIT IS INSTALLED IN THE VEHICLE.

IS THE MODULATOR GEAR STRIPPED?

YES: REPLACE HYDRAULIC MODULATOR GEAR.

NO:

(10)
- RECONNECT ELECTRICAL CONNECTORS
- SECURELY POSITION HYDRAULIC MODULATOR/MASTER CYLINDER ASSEMBLY IN VEHICLE WITH GEAR COVER REMOVED SUCH THAT THE GEAR SET CAN BE OBSERVED. BE CAREFUL NOT TO DAMAGE GEAR SET OR ALLOW THE HYDRAULIC MODULATOR TO MOVE WHILE TESTING.
- IGNITION "ON."
- USING TECH 1 MISC. TESTS, SELECT MANUAL CONTROL FUNCTION.
- SELECT RIGHT FRONT ABS MOTOR APPLY AND APPLY THE MOTOR WHILE OBSERVING THE GEAR SET FURTHEST REARWARD WHEN THE UNIT IS INSTALLED IN THE VEHICLE.
- SELECT RIGHT FRONT ABS MOTOR RELEASE FUNCTION AND RELEASE THE MOTOR WHILE STILL OBSERVING THE GEAR SET.
- SELECT RIGHT FRONT ABS MOTOR APPLY FUNCTION AND APPLY THE MOTOR.

DID GEAR SET MOVE IN BOTH DIRECTIONS FOR AT LEAST ONE REVOLUTION?

NO: REPLACE ABS MOTOR PACK.

YES:

(11)
- USING TECH 1 MISC. TESTS, PERFORM GEAR TENSION RELIEF FUNCTION.
- IGNITION "OFF."
- SEPARATE THE ABS MOTOR PACK FROM THE HYDRAULIC MODULATOR.
- GRASP THE GEAR ON THE HYDRAULIC MODULATOR (THE HYDRAULIC MODULATOR IS THE UNIT WITH THE THREE LARGE GEARS) FURTHEST REARWARD WHEN THE UNIT IS INSTALLED IN THE VEHICLE AND ATTEMPT TO MOVE THE GEAR.

CAN THE GEAR BE ROTATED MORE THAN 13 FULL TURNS LOCK TO LOCK?

YES: REPLACE THE HYDRAULIC MODULATOR.

NO:

(12)
- INSPECT ALL CONNECTORS AND TERMINALS FOR POOR TERMINAL CONTACT AND EVIDENCE OF CORROSION.
REPLACE ALL TERMINALS THAT EXHIBIT SIGNS OF POOR TERMINAL CONTACT OR CORROSION.
- REINSTALL HYDRAULIC MODULATOR/MASTER CYLINDER ASSEMBLY IN VEHICLE.
- RECONNECT ALL CONNECTORS.
- IGNITION "OFF."
- START ENGINE AND TEST DRIVE VEHICLE, ACHIEVING AT LEAST 16 km/h (10 MPH).
- REPEAT ABOVE DRIVE CYCLE SEQUENCE TWO MORE TIMES.

DOES DTC 48 SET IN THE LAST THREE DRIVE CYCLES?

YES: REPLACE EBCM.

NO: MALFUNCTION WAS MOSTLY LIKELY CAUSED BY A POOR CONNECTION. REFER TO "DIAGNOSTIC AIDS" ON FACING PAGE.

AFTER DIAGNOSIS IS COMPLETE, CLEAR DTCs AND TEST DRIVE VEHICLE FOR THREE (3) DRIVE CYCLES TO VERIFY THAT DTC DOES NOT RESET. A DRIVE CYCLE CONSISTS OF STARTING THE VEHICLE, DRIVING OVER 16 KM/H (10 MPH), AND THEN KEYING DOWN

DIAGNOSTIC CHART DTC 51 — 1995 SILHOUETTE, TRANS SPORT AND LUMINA APV

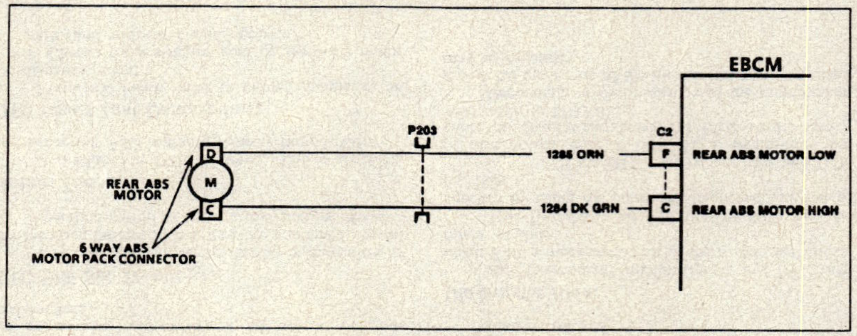

DTC 51
(Page 1 of 2)
REAR ABS MOTOR FREE SPINS

Circuit Description:

This DTC is designed to detect a stripped nut or gear assembly during initialization. During the homing sequence, the piston should reach the top of the bore resulting in a stalled ABS motor. If this does not occur, the ABS motor must be spinning with little or no resistance, indicating a nut/screw or gear malfunction.

DTC Will Set When:

DTC 51 can be set during initialization. If the feedback current is less than the command current for a specified period of time, a malfunction exists.

Action Taken:

A malfunction DTC is stored, ABS is disabled, and the amber ABS warning lamp is turned "ON."

DTC Chart Test Description:

Number(s) below refer to circled number(s) on the diagnostic chart.

1. Checks to see if the corresponding open ABS motor DTC is also set.
2. Verifies that the ABS motor was actually applied as commanded by observing feedback current.
3. Verifies that the ABS motor can release.
4. Verifies that the ABS motor can be applied by observing pedal movement.
5. Checks for high resistance in the ABS motor "HIGH" circuitry.
6. Checks for high resistance in the ABS motor "LOW" circuitry.
7. Checks for proper resistance of the ABS motor windings.

Diagnostic Aids:

An "Intermittent" malfunction in this DTC may result from a mechanical part of the system that sticks, binds, or slips.

The frequency of the malfunction can be checked by using the enhanced diagnostic function of the Tech 1.

If the DTC only occurs once and DTC 64 also occurs, refer to DTC 64. If intermittent and enhanced diagnostics show this DTC occurs during ABS, refer to DTC 64.

Depending on the frequency of the malfunction, a physical inspection of the mechanical parts suspected may be necessary.

DIAGNOSTIC CHART DTC 51 — 1995 SILHOUETTE, TRANS SPORT AND LUMINA APV

DTC 51
(Page 1 of 2)
REAR ABS MOTOR FREE SPINS

(1) IS DTC 64 ALSO CURRENTLY SET?
- YES: PROCEED TO DTC 64 CHART FIRST.
- NO: go to 2.

(2)
- IGNITION "ON," ENGINE "OFF."
- PUMP BRAKE PEDAL UNTIL FIRM TO DEPLETE VACUUM RESERVE.
- USING TECH 1 MISC. TESTS, SELECT MANUAL CONTROL FUNCTION.
- SELECT REAR AXLE ABS MOTOR APPLY FUNCTION AND APPLY THE MOTOR WHILE CAREFULLY OBSERVING THE COMMAND CURRENT AND FEEDBACK CURRENT ON THE TECH 1.

WERE BOTH COMMAND AND FEEDBACK CURRENT 10 AMPS?
- YES: go to 3.
- (NO: go to 5.)

(3)
- APPLY FIRM PRESSURE ON BRAKE PEDAL.
- SELECT REAR AXLE ABS MOTOR RELEASE FUNCTION AND RELEASE THE MOTOR.

DID THE BRAKE PEDAL FALL AND WERE BOTH COMMAND AND FEEDBACK CURRENT 6 AMPS?
- YES: go to 4.
- NO: GO TO STEP 5

(4)
- FIRM PRESSURE STILL ON BRAKE PEDAL.
- SELECT REAR AXLE ABS MOTOR APPLY FUNCTION AND APPLY THE MOTOR (THIS MOVES THE PISTON TO THE TOP OF ITS BORE.)

DID BRAKE PEDAL RISE?
- YES: MALFUNCTION IS INTERMITTENT. REFER TO "DIAGNOSTIC AIDS" ON FACING PAGE.
- NO: go to 5.

(5)
- USING TECH 1 MISC. TESTS, PERFORM GEAR TENSION RELIEF FUNCTION.
- IGNITION "OFF."
- DISCONNECT 6-WAY ABS MOTOR PACK CONNECTOR AND 8-WAY EBCM CONNECTOR.
- USING J 39200, MEASURE THE RESISTANCE BETWEEN TERMINAL "C" OF THE 6-WAY ABS MOTOR PACK HARNESS CONNECTOR AND TERMINAL "C" OF THE 8-WAY EBCM HARNESS CONNECTOR.

IS RESISTANCE 1.5Ω OR LESS?
- YES: go to 6.
- NO: REPAIR HIGH RESISTANCE IN CKT 1284.

(6)
- MEASURE THE RESISTANCE BETWEEN TERMINAL "D" OF THE 6-WAY ABS MOTOR PACK HARNESS CONNECTOR AND TERMINAL "F" OF THE 8-WAY EBCM HARNESS CONNECTOR.

IS RESISTANCE 1.5Ω OR LESS?
- YES: go to 7.
- NO: REPAIR HIGH RESISTANCE IN CKT 1285.

(7)
- MEASURE THE RESISTANCE BETWEEN TERMINAL "C" AND TERMINAL "D" OF THE ABS MOTOR PACK CONNECTOR.

IS RESISTANCE 1.5Ω OR LESS?
- YES: DIAGNOSIS CONTINUED ON NEXT PAGE.
- NO: REPLACE ABS MOTOR PACK.

AFTER DIAGNOSIS IS COMPLETE, CLEAR DTCs AND TEST DRIVE VEHICLE FOR THREE (3) DRIVE CYCLES TO VERIFY THAT DTC DOES NOT RESET. A DRIVE CYCLE CONSISTS OF STARTING THE VEHICLE, DRIVING OVER 16 KM/H (10 MPH), AND THEN KEYING DOWN.

DIAGNOSTIC CHART DTC 51 — 1995 SILHOUETTE, TRANS SPORT AND LUMINA APV

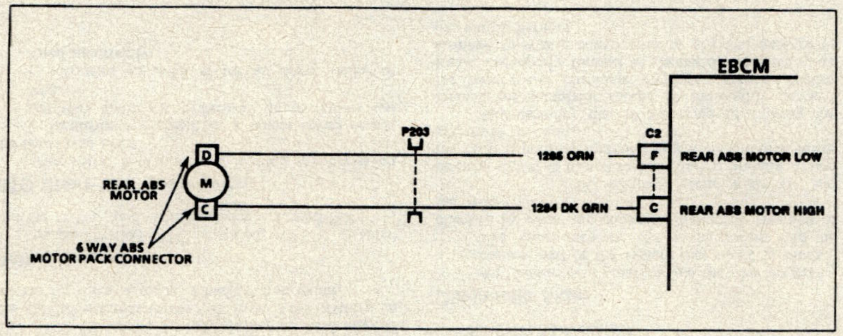

DTC 51
(Page 2 of 2)
REAR ABS MOTOR FREE SPINS

Circuit Description:

This DTC is designed to detect a stripped nut or gear assembly during initialization. During the homing sequence, the piston should reach the top of the bore resulting in a stalled ABS motor. If this does not occur, the ABS motor must be spinning with little or no resistance, indicating a nut/screw or gear malfunction.

DTC Will Set When:

DTC 51 can be set during initialization. If the feedback current is less than the command current for a specified period of time, a malfunction exists.

Action Taken:

A malfunction DTC is stored, ABS is disabled, and the amber ABS warning lamp is turned "ON."

DTC Chart Test Description:

Number(s) below refer to circled number(s) on the diagnostic chart.

8. Checks for a stripped gear on the ABS motor pack (unit with three small gears).
9. Checks for a stripped gear on the hydraulic modulator (unit with three large gears).
10. Verifies that the ABS motor can be applied under a load situation.
11. Verifies that the hydraulic modulator is functioning properly.
12. Ensures malfunction was not due to poor terminal contact.

Diagnostic Aids:

An "Intermittent" malfunction in this DTC may result from a mechanical part of the system that sticks, binds, or slips.

The frequency of the malfunction can be checked by using the enhanced diagnostic function of the Tech 1.

If the DTC only occurs once and DTC 64 also occurs, refer to DTC 64. If intermittent and enhanced diagnostics show this DTC occurs during ABS, refer to DTC 64.

Depending on the frequency of the malfunction, a physical inspection of the mechanical parts suspected may be necessary.

DIAGNOSTIC CHART DTC 51 — 1995 SILHOUETTE, TRANS SPORT AND LUMINA APV

DTC 51 (Page 2 of 2) REAR ABS MOTOR FREE SPINS

DIAGNOSIS CONTINUED FROM PREVIOUS PAGE.

(8)
- REMOVE HYDRAULIC MODULATOR/MASTER CYLINDER ASSEMBLY FROM VEHICLE.
- REMOVE GEAR COVER.
- CHECK FOR STRIPPED GEAR ON ABS MOTOR PACK; ABS MOTOR PACK IS THE UNIT WITH THE THREE SMALL GEARS. THE REAR AXLE GEAR IS THE MIDDLE GEAR OF THE THREE.

IS THE ABS MOTOR PACK GEAR STRIPPED?

YES → REPLACE ABS MOTOR PACK.

NO ↓

(9)
- CHECK FOR A STRIPPED GEAR ON HYDRAULIC MODULATOR; MODULATOR IS THE UNIT WITH THE THREE LARGE GEARS. THE REAR AXLE GEAR IS THE MIDDLE GEAR OF THE THREE.

IS THE MODULATOR GEAR STRIPPED?

YES → REPLACE HYDRAULIC MODULATOR GEAR.

NO ↓

(10)
- RECONNECT ELECTRICAL CONNECTORS
- SECURELY POSITION HYDRAULIC MODULATOR/MASTER CYLINDER ASSEMBLY IN VEHICLE WITH GEAR COVER REMOVED SUCH THAT THE GEAR SET CAN BE OBSERVED. BE CAREFUL NOT TO DAMAGE GEAR SET OR ALLOW THE HYDRAULIC MODULATOR TO MOVE WHILE TESTING.
- IGNITION "ON."
- USING TECH 1 MISC. TESTS, SELECT MANUAL CONTROL FUNCTION.
- SELECT REAR AXLE ABS MOTOR APPLY AND APPLY THE MOTOR WHILE OBSERVING THE MIDDLE GEAR SET.
- SELECT REAR AXLE ABS MOTOR RELEASE FUNCTION AND RELEASE THE MOTOR WHILE STILL OBSERVING THE GEAR SET.
- SELECT REAR AXLE ABS MOTOR APPLY FUNCTION AND APPLY THE MOTOR.

DID GEAR SET MOVE IN BOTH DIRECTIONS FOR AT LEAST ONE REVOLUTION?

NO → REPLACE ABS MOTOR PACK.

YES ↓

(11)
- USING TECH 1 MISC. TESTS, PERFORM GEAR TENSION RELIEF FUNCTION.
- IGNITION "OFF."
- SEPARATE THE ABS MOTOR PACK FROM THE HYDRAULIC MODULATOR.
- GRASP THE MIDDLE GEAR ON THE HYDRAULIC MODULATOR (THE HYDRAULIC MODULATOR IS THE UNIT WITH THE THREE LARGE GEARS) AND ATTEMPT TO MOVE THE GEAR.

CAN THE GEAR BE ROTATED MORE THAN 8 FULL TURNS LOCK TO LOCK?

YES → REPLACE THE HYDRAULIC MODULATOR.

NO ↓

(12)
- INSPECT ALL CONNECTORS AND TERMINALS FOR POOR TERMINAL CONTACT AND EVIDENCE OF CORROSION.
  REPLACE ALL TERMINALS THAT EXHIBIT SIGNS OF POOR TERMINAL CONTACT OR CORROSION.
- REINSTALL HYDRAULIC MODULATOR/MASTER CYLINDER ASSEMBLY IN VEHICLE.
- RECONNECT ALL CONNECTORS.
- IGNITION "OFF."
- START ENGINE AND TEST DRIVE VEHICLE, ACHIEVING AT LEAST 16 km/h (10 MPH).
- REPEAT ABOVE DRIVE CYCLE SEQUENCE TWO MORE TIMES.

DOES DTC 51 RESET?

YES → REPLACE EBCM.

NO → MALFUNCTION WAS MOSTLY LIKELY CAUSED BY A POOR CONNECTION. REFER TO "DIAGNOSTIC AIDS" ON FACING PAGE.

AFTER DIAGNOSIS IS COMPLETE, CLEAR DTCs AND TEST DRIVE VEHICLE FOR THREE (3) DRIVE CYCLES TO VERIFY THAT DTC DOES NOT RESET. A DRIVE CYCLE CONSISTS OF STARTING THE VEHICLE, DRIVING OVER 16 KM/H (10 MPH), AND THEN KEYING DOWN.

DIAGNOSTIC CHART DTC 52 — 1995 SILHOUETTE, TRANS SPORT AND LUMINA APV

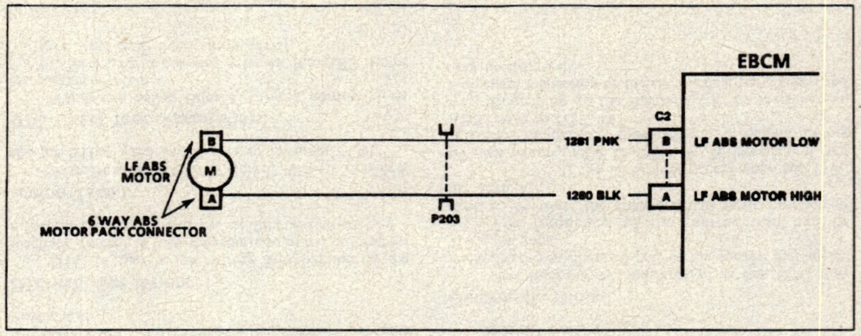

DTC 52
LEFT FRONT ABS CHANNEL IN RELEASE TOO LONG

Circuit Description:

This DTC will diagnose an ABS motor that is energized longer than expected. This could occur if a wheel speed sensor is malfunctioning, the ABS motor does not turn, the left front solenoid mechanically fails open, or the ABS motor wires are crossed.

DTC Will Set When:

DTC 52 can be set only during an ABS stop. If the EBCM commands the left front ABS channel in release for three seconds, a malfunction exists.

Action Taken:

A malfunction DTC is stored, ABS is disabled and the amber ABS warning lamp is turned "ON."

DTC Chart Test Description:

Number(s) below refer to circled number(s) on the diagnostic chart.

1. Identifies a problem in a wheel speed sensor that may cause the system to be in release too long.
2. Identifies an ABS motor as being failed or wired incorrectly.
3. Checks for a solenoid that may have mechanically failed open.
4. Verifies proper solenoid operation.
5. If Step 3 has failed, this serves to isolate the cause of the hydraulic problem, to either the solenoid or the ABS hydraulic modulator assembly.
6. Determines whether a malfunctioning ABS motor pack or hydraulic modulator is the reason for DTC 52 being set.

Diagnostic Aids:

An "Intermittent" malfunction may be caused by a mechanical part of the system that sticks or binds.

The frequency of the malfunction can be checked by using the enhanced diagnostic function of the Tech 1,

DTC 52 may occur if on ice and steering wheel is turned to lock during braking. Using the Tech 1, perform hydraulic test to ensure total brake system is functional.

Any circuitry that is suspected of causing the intermittent complaint should be thoroughly checked for backed out terminals, improper mating, broken locks, improperly formed or damaged terminals, poor terminal to wiring connections or physical damage to the wiring harness.

DIAGNOSTIC CHART DTC 52 — 1995 SILHOUETTE, TRANS SPORT AND LUMINA APV

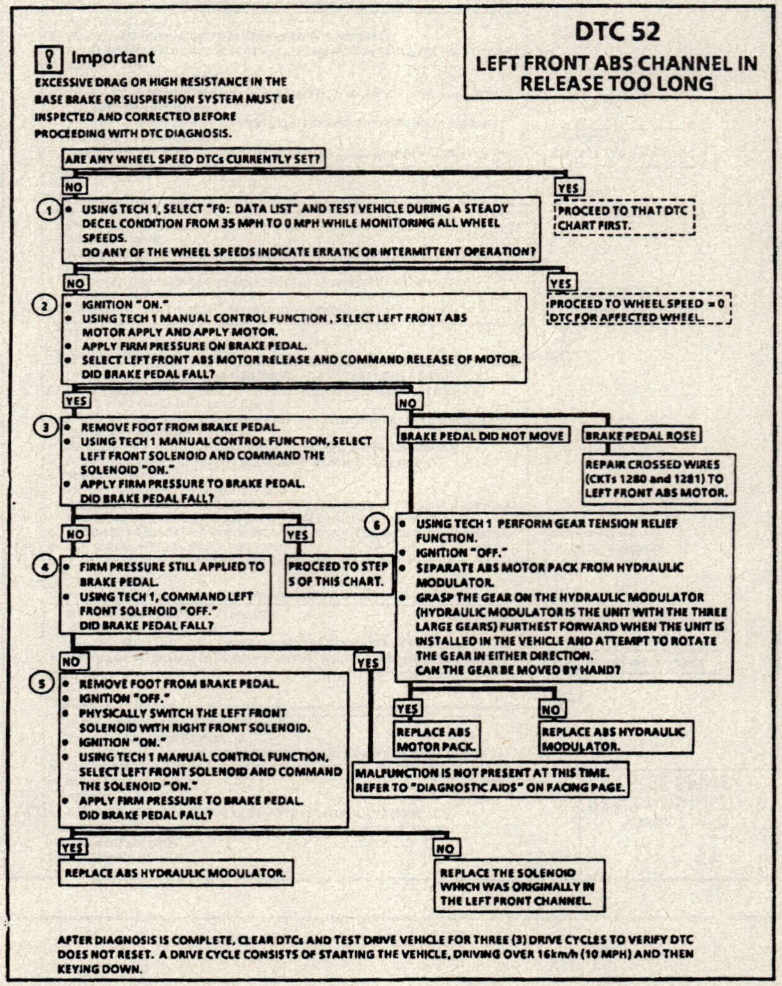

DIAGNOSTIC CHART DTC 53 — 1995 SILHOUETTE, TRANS SPORT AND LUMINA APV

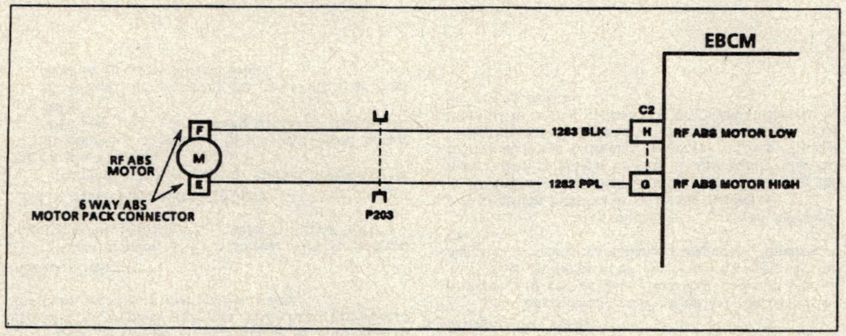

DTC 53
RIGHT FRONT ABS CHANNEL IN RELEASE TOO LONG

Circuit Description:

This DTC will diagnose an ABS motor that is energized longer than expected. This could occur if a wheel speed sensor is malfunctioning, the ABS motor does not turn, the right front solenoid mechanically fails open, or the ABS motor wires are crossed.

DTC Will Set When:

DTC 53 can be set only during an ABS stop. If the EBCM commands the right front channel in release for three seconds, a malfunction exists.

Action Taken:

A malfunction DTC is stored, ABS is disabled and the amber ABS warning lamp is turned "ON."

DTC Chart Test Description:

Number(s) below refer to circled number(s) on the diagnostic chart.

1. Identifies a problem in a wheel speed sensor that may cause the system to be in release too long.
2. Identifies a motor as being failed or wired incorrectly.
3. Checks for a solenoid that may have mechanically failed open.
4. Checks for a solenoid that may have mechanically failed closed.
5. If Step 3 has failed, this serves to isolate the cause of the hydraulic problem, to either the solenoid or the ABS hydraulic modulator assembly.
6. Determines whether a malfunction ABS motor pack or hydraulic modulator is the reason for DTC 53 being set.

Diagnostic Aids:

An "Intermittent" malfunction may be caused by a mechanical part of the system that sticks or binds.

The frequency of the malfunction can be checked by using the enhanced diagnostic function of the Tech 1.

DTC 53 may occur on ice if steering wheel is turned to lock during braking. Using the Tech 1 perform hydraulic test to ensure total brake system is functional.

Any circuitry that is suspected of causing the intermittent complaint should be thoroughly checked for backed out terminals, improper mating, broken locks, improperly formed or damaged terminals, poor terminal to wiring connections or physical damage to the wiring harness.

DIAGNOSTIC CHART DTC 53 — 1995 SILHOUETTE, TRANS SPORT AND LUMINA APV

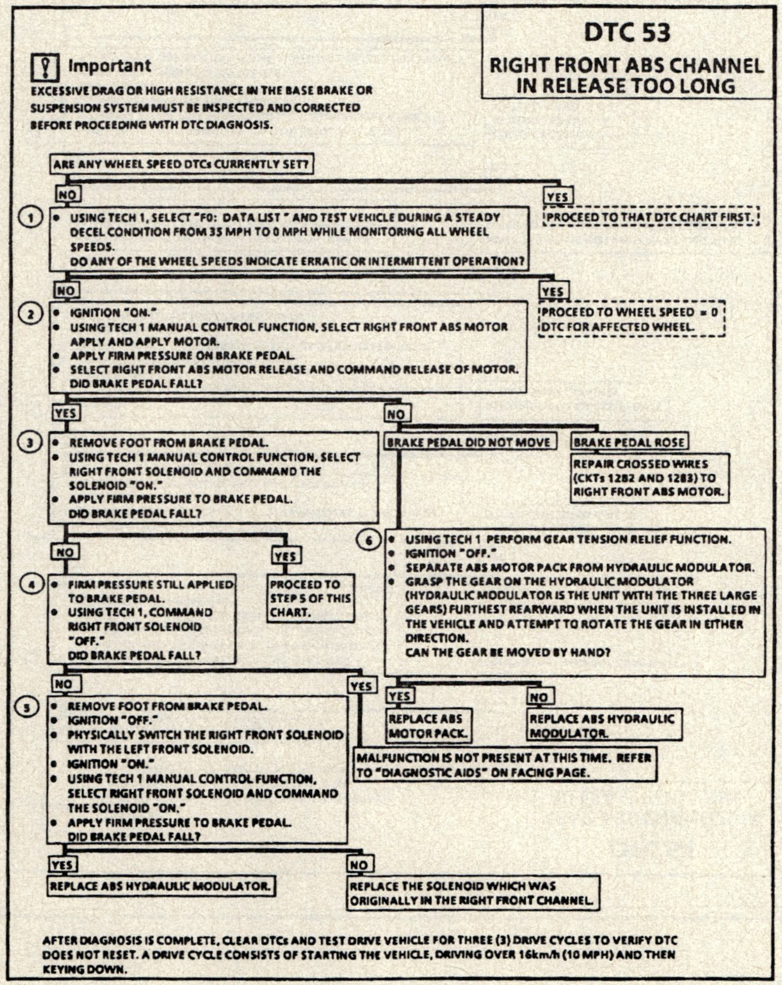

DIAGNOSTIC CHART DTC 54 — 1995 LUMINA APV

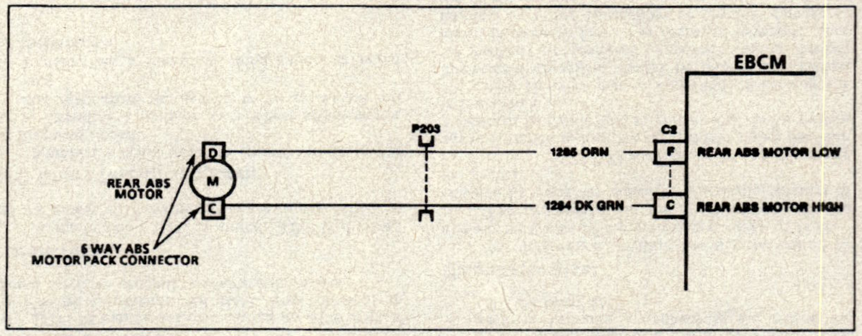

DTC 54
REAR ABS CHANNEL IN RELEASE TOO LONG

Circuit Description:

This DTC will diagnose an ABS motor that is energized longer than expected. This could occur if a wheel speed sensor is malfunctioning, the ABS motor does not turn, or the ABS motor wires are crossed.

DTC Will Set When:

DTC 54 can be set only during an ABS stop. If the EBCM commands the rear ABS channel in release for three seconds, a malfunction exists.

Action Taken:

A malfunction DTC is stored, ABS is disabled and the amber ABS warning lamp is turned "ON."

DTC Chart Test Description:

Number(s) below refer to circled number(s) on the diagnostic chart.

1. Identifies a problem in a wheel speed sensor that may cause the system to be in release too long.
2. Checks for a wheel that may stick or bind because of a mechanical fault.
3. Checks to see if the ABS motor is capable of moving and applying the hydraulic piston for the rear wheels.
4. Ensures that the ABS motor wiring is not crossed.
5. Isolates the fault of a "no-apply" situation to either the ABS motor pack or ABS hydraulic modulator assembly.

Diagnostic Aids:

An "Intermittent" malfunction may be caused by a mechanical part of the system that sticks or binds.

The frequency of the malfunction can be checked by using the enhanced diagnostic function of the Tech 1,

Using the Tech 1, perform hydraulic test to ensure total brake system is functional.

Any circuitry that is suspected of causing the intermittent complaint should be thoroughly checked for backed out terminals, improper mating, broken locks, improperly formed or damaged terminals, poor terminal to wiring connections or physical damage to the wiring harness.

DIAGNOSTIC CHART DTC 54 — 1995 LUMINA APV

DTC 54
REAR ABS CHANNEL IN RELEASE TOO LONG

Important
- EXCESSIVE DRAG OR HIGH RESISTANCE IN THE BASE BRAKE OR SUSPENSION SYSTEM MUST BE INSPECTED AND CORRECTED BEFORE PROCEEDING WITH DTC DIAGNOSIS.

ARE ANY WHEEL SPEED DTCs CURRENTLY SET?
- YES: PROCEED TO THAT DTC CHART FIRST.
- NO: go to 1.

(1)
- USING TECH 1, SELECT "F0: DATA LIST" AND TEST VEHICLE DURING A STEADY DECEL CONDITION FROM 35 MPH TO 0 MPH WHILE MONITORING ALL WHEEL SPEEDS. DO ANY OF THE WHEEL SPEEDS INDICATE ERRATIC OR INTERMITTENT OPERATION?
  - YES: PROCEED TO WHEEL SPEED = 0 DTC FOR AFFECTED WHEEL.
  - NO: go to 2.

(2)
- RAISE REAR OF VEHICLE AND SUPPORT.
- SPIN BOTH REAR WHEELS AND CONFIRM PHYSICAL CONDITION OF BASE BRAKE SYSTEM.
  COULD BOTH REAR WHEELS BE SPUN FREELY BY HAND?
  - NO: REPAIR SOURCE OF RESISTANCE IN BASE BRAKE SYSTEM.
  - YES: go to 3.

(3)
- IGNITION "ON."
- USING MANUAL CONTROL FUNCTION OF TECH 1, SELECT REAR AXLE ABS MOTOR APPLY.
- APPLY FIRM PRESSURE ON BRAKE PEDAL.
- COMMAND APPLY OF REAR ABS MOTOR AND TRY TO SPIN REAR WHEELS BY HAND.
  COULD REAR WHEELS BE SPUN BY HAND?
  - NO: MALFUNCTION IS NOT PRESENT AT THIS TIME. REFER TO "DIAGNOSTIC AIDS" ON FACING PAGE.
  - YES: go to 4.

(4)
- WITH FIRM PRESSURE ON BRAKE PEDAL STILL APPLIED, COMMAND A RELEASE OF THE REAR AXLE ABS MOTOR. CAN WHEELS BE MOVED?
  - NO: REPAIR CROSSED WIRES AT REAR ABS MOTOR (CKTs 1284 AND 1285).
  - YES: go to 5.

(5)
- USING TECH 1, PERFORM GEAR TENSION RELIEF FUNCTION.
- SEPARATE ABS MOTOR PACK FROM HYDRAULIC MODULATOR.
- CONNECT ABS MOTOR PACK WIRING.
- IGNITION "ON."
- USING TECH 1, PERFORM MOTOR TEST.
  DO ALL THREE ABS MOTOR PACK GEARS (SMALL GEARS) SPIN FREELY?
  - NO: REPLACE ABS MOTOR PACK.
  - YES: REPLACE ABS HYDRAULIC MODULATOR.

AFTER ALL DIAGNOSIS IS COMPLETE, CLEAR DTCs AND TEST DRIVE VEHICLE FOR THREE (3) DRIVE CYCLES TO VERIFY THAT DTC DOES NOT RESET. A DRIVE CYCLE CONSISTS OF STARTING THE VEHICLE DRIVING OVER 16 KM/H (10 MPH), AND THEN KEYING DOWN

DIAGNOSTIC CHART DTC 55 — 1995 LUMINA APV

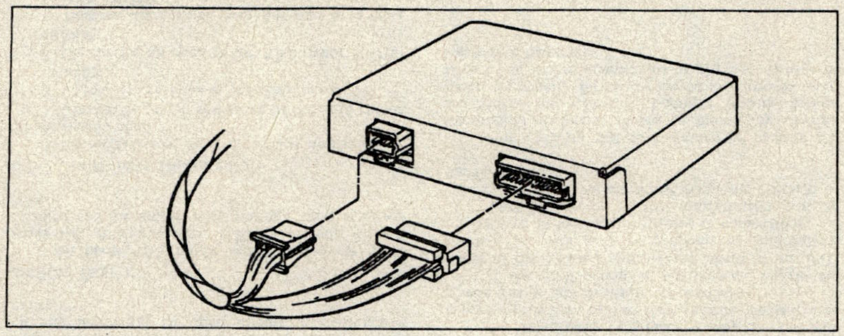

DTC 55
EBCM MALFUNCTION

Circuit Description:

This DTC identifies a malfunction detected by the MDI custom IC. It also ensures the cause of the malfunction is not a result of a problem with the ABS enable relay under a DTC 55 malfunction.

Action Taken:

A malfunction DTC is stored, ABS is disabled and the ABS warning lamp is turned "ON."

DTC Chart Test Description:

Number(s) below refer to circled number(s) on the diagnostic chart.

1. Checks to see if the malfunction is still present.
2. Checks to see if the malfunction is intermittent.

Diagnostic Aids:

An "Intermittent" malfunction may be caused by a poor connection, rubbed through wire insulation, or a wire that is broken inside the insulation.

The frequency of the malfunction can be checked by using the enhanced diagnostic function of the Tech 1.

Any circuitry that is suspected of causing the intermittent complaint should be thoroughly checked for backed out terminals, improper mating, broken locks, improperly formed or damaged terminals, poor terminal to wiring connections or physical damage to the wiring harness.

DIAGNOSTIC CHART DTC 55 — 1995 LUMINA APV

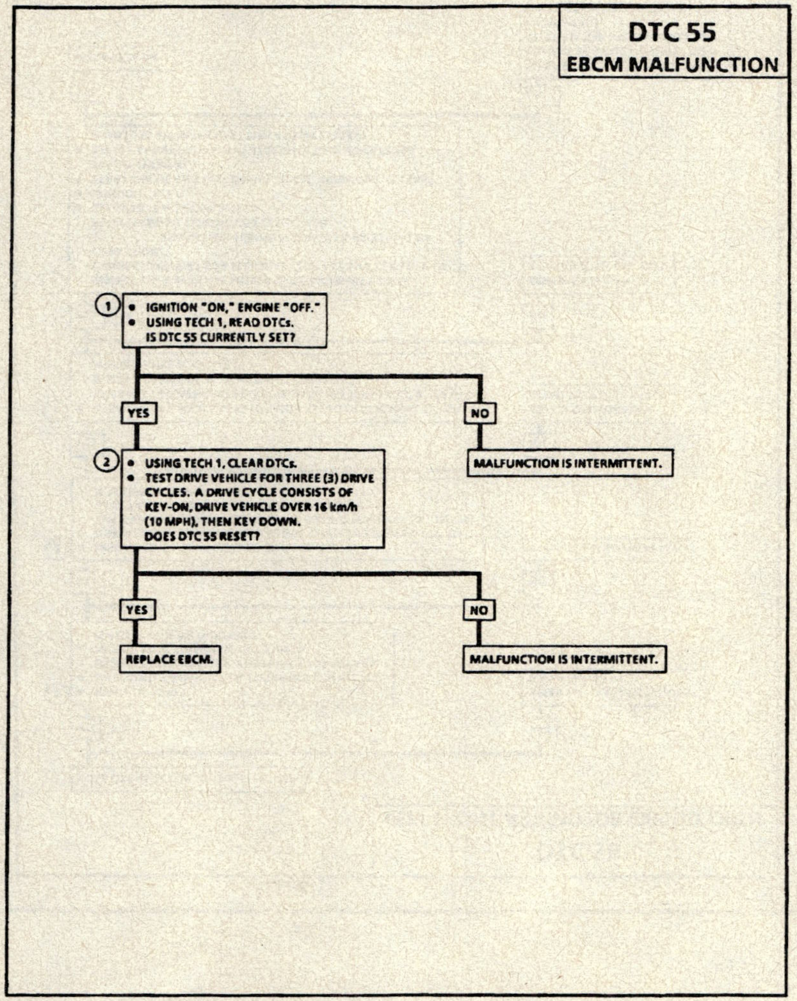

DIAGNOSTIC CHART DTC 56 — 1995 LUMINA APV

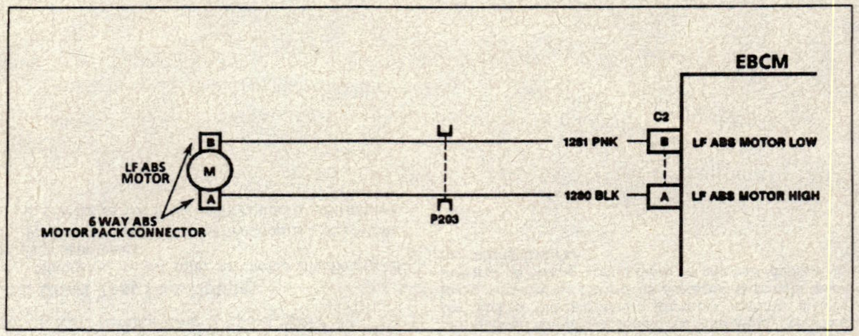

DTC 56
LEFT FRONT ABS MOTOR CIRCUIT OPEN

Circuit Description:

This DTC identifies an ABS motor that cannot be energized due to an open in its circuitry.

DTC Will Set When:

DTC 56 can be set only when the motor is commanded "OFF." If the EBCM detects an out of range voltage on either of the left front ABS motor circuits indicating an open circuit, a malfunction exists.

Action Taken:

An open ABS motor will not activate when requested. A malfunction DTC is stored, ABS is disabled and the amber ABS warning lamp is turned "ON."

DTC Chart Test Description:

Number(s) below refer to circled number(s) on the diagnostic chart.

1. Checks for proper resistance of the ABS motor.
2. Checks for an open in the ABS motor "HIGH" circuitry.
3. Checks for an open in the ABS motor "LOW" circuitry.
4. Ensures malfunction was not due to a poor terminal contact.

Diagnostic Aids:

Using Tech 1, select manual control function, and exercise ABS motor movement of affected channel in both directions while applying light pressure on the brake pedal.

If erratic or "jumpy" brake pedal movement is detected while performing an "apply" or "release" function of the ABS motor, an intermittent malfunction may be indicated.

An "Intermittent" malfunction may be caused by a poor connection, rubbed through wire insulation, or a wire that is broken inside the insulation.

If the malfunction is not current, wiggle the wires of the affected channel and check if the DTC resets. This will help to pinpoint an intermittent malfunction in the motor circuitry or connections.

The frequency of the malfunction can be checked by using the enhanced diagnostic function of the Tech 1,

Any circuitry that is suspected of causing the intermittent complaint should be thoroughly checked for backed out terminals, improper mating, broken locks, improperly formed or damaged terminals, poor terminal to wiring connections or physical damage to the wiring harness.

DIAGNOSTIC CHART DTC 56 — 1995 LUMINA APV

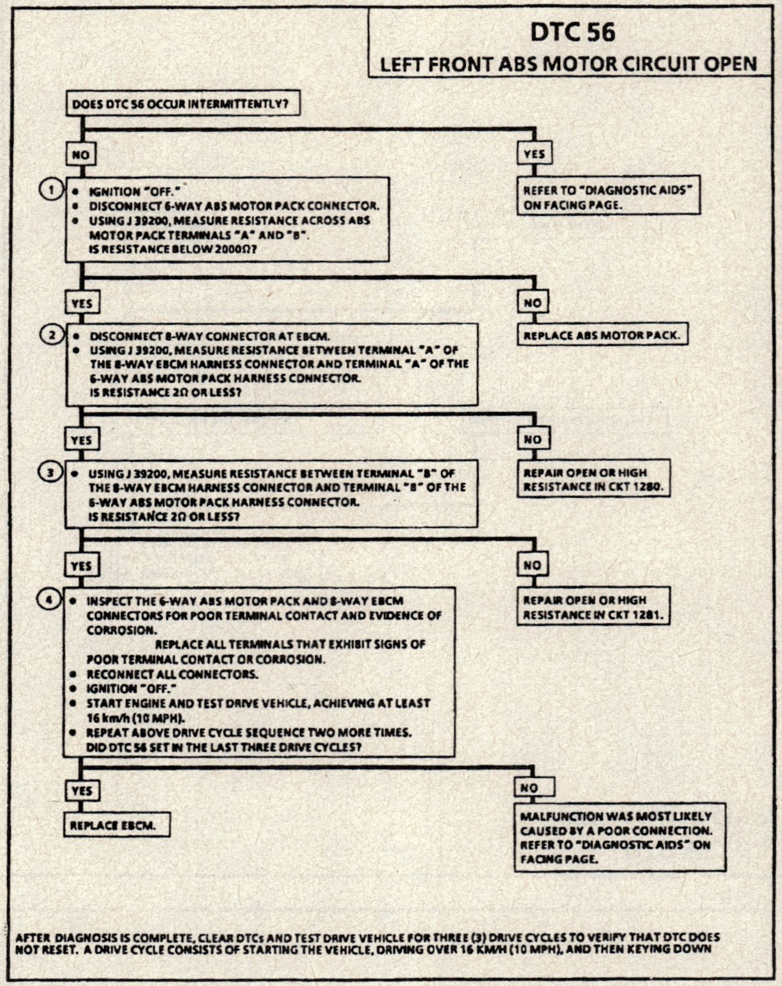

DIAGNOSTIC CHART DTC 57 — 1995 LUMINA APV

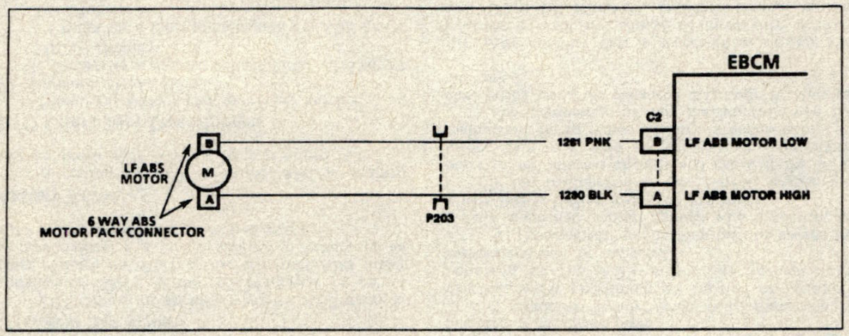

DTC 57
LEFT FRONT ABS MOTOR CIRCUIT SHORTED TO GND

Circuit Description:

This DTC identifies an ABS motor circuit that is shorted to ground. This malfunction will not allow the ABS motor to be controlled at the commanded current rate or will cause the driver circuit to allow current directly to ground.

DTC Will Set When:

DTC 57 can be set anytime. If the EBCM detects an out of range voltage on either of the left front ABS motor circuits indicating a circuit shorted to ground, a malfunction exists.

Action Taken:

A malfunction DTC is stored, ABS is disabled and the amber ABS warning lamp is turned "ON."

DTC Chart Test Description:

Number(s) below refer to circled number(s) on the diagnostic chart.

1. Checks for a short to ground in the ABS motor "HIGH" circuitry.
2. Checks for a short to ground in the ABS motor "LOW" circuitry.
3. Checks for an ABS motor that is internally shorted to ground.
4. Ensures malfunction was not due to physical damage of the circuitry.

Diagnostic Aids:

Using Tech 1, select manual control function, and exercise ABS motor movement of affected channel in both directions while applying light pressure on the brake pedal.

If erratic or "jumpy" brake pedal movement is detected while performing an "apply" or "release" function of the ABS motor, an intermittent malfunction may be indicated.

An "Intermittent" malfunction may be caused by a poor connection, rubbed through wire insulation, or a wire that is broken inside the insulation.

If the malfunction is not current, wiggle the wires of the affected channel and check if the DTC resets. This will help to pinpoint an intermittent malfunction in the motor circuitry or connections.

The frequency of the malfunction can be checked by using the enhanced diagnostic function of the Tech 1.

Any circuitry that is suspected of causing the intermittent complaint should be thoroughly checked for backed out terminals, improper mating, broken locks, improperly formed or damaged terminals, poor terminal to wiring connections or physical damage to the wiring harness.

DIAGNOSTIC CHART DTC 57 — 1995 LUMINA APV

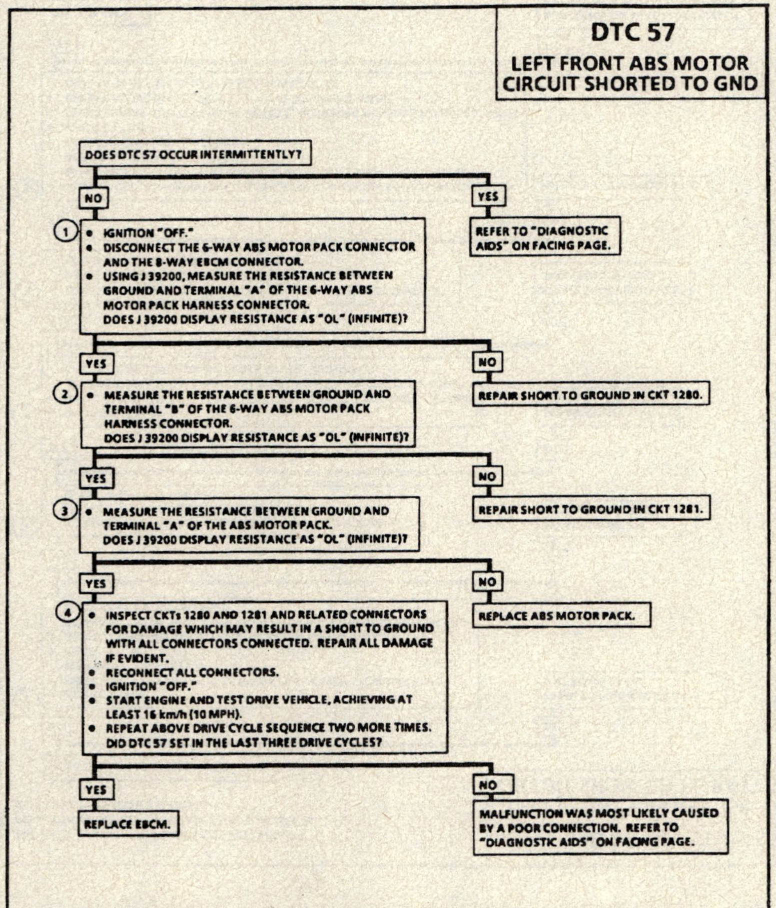

AFTER ALL DIAGNOSIS IS COMPLETE, CLEAR DTCs AND TEST DRIVE VEHICLE FOR THREE (3) DRIVE CYCLES TO VERIFY THAT DTC DOES NOT RESET. A DRIVE CYCLE CONSISTS OF STARTING THE VEHICLE, DRIVING OVER 16 km/h (10 MPH), AND THEN KEYING DOWN.

DIAGNOSTIC CHART DTC 58 — 1995 LUMINA APV

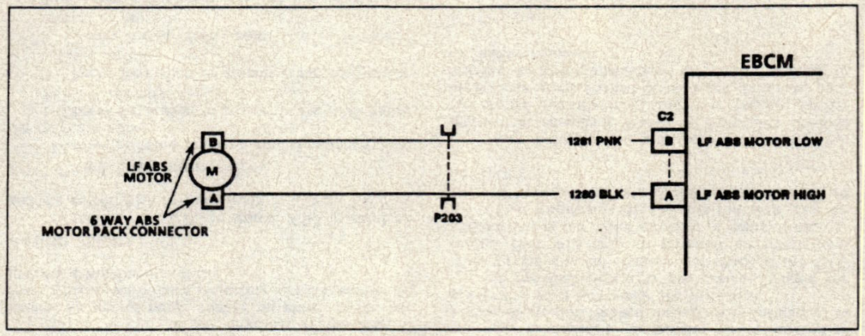

DTC 58
LEFT FRONT ABS MOTOR CIRCUIT SHORTED TO BATT

Circuit Description:

This DTC identifies an ABS motor circuit that is shorted to battery or an ABS motor that has low or no resistance. This malfunction will not allow the ABS motor to be controlled at the commanded current rate or will cause the ABS motor to turn in the opposite direction, or not at all.

DTC Will Set When:

DTC 58 can be set only when the ABS motor is commanded "OFF." If the EBCM detects an out of range voltage on either of the left front ABS motor circuits indicating a circuit shorted to battery or an ABS motor shorted, a malfunction exists.

Action Taken:

A malfunction DTC is stored, ABS is disabled and the amber ABS warning lamp is turned "ON."

DTC Chart Test Description:

Number(s) below refer to circled number(s) on the diagnostic chart.

1. Checks for a short to voltage in the ABS motor "HIGH" circuitry.
2. Checks for a short to voltage in the ABS motor "LOW" circuitry.
3. Checks for a short between the "HIGH" and "LOW" motor circuits.
4. This step checks for an ABS motor that is internally shorted.
5. Ensures malfunction was not due to physical damage of the circuitry.

Diagnostic Aids:

Using Tech 1, select manual control function, and exercise ABS motor movement of affected channel in both directions while applying light pressure on the brake pedal.

If erratic or "jumpy" brake pedal movement is detected while performing an "apply" or "release" function of the ABS motor, an intermittent malfunction may be indicated.

An "Intermittent" malfunction may be caused by a poor connection, rubbed through wire insulation, or a wire that is broken inside the insulation.

If the malfunction is not current, wiggle the wires of the affected channel and check if the DTC resets. This will help to pinpoint an intermittent malfunction in the motor circuitry or connections.

The frequency of the malfunction can be checked by using the enhanced diagnostic function of the Tech 1.

Any circuitry that is suspected of causing the intermittent complaint should be thoroughly checked for backed out terminals, improper mating, broken locks, improperly formed or damaged terminals, poor terminal to wiring connections or physical damage to the wiring harness.

DIAGNOSTIC CHART DTC 58 — 1995 LUMINA APV

DTC 58
LEFT FRONT ABS MOTOR CIRCUIT SHORTED TO BATT

NOTICE: J 39200 TEST LEADS MUST BE "ZEROED" PRIOR TO MAKING ANY RESISTANCE MEASUREMENTS. REFER TO J 39200 USER'S MANUAL.

DOES DTC 58 OCCUR INTERMITTENTLY?

- YES: REFER TO "DIAGNOSTIC AIDS" ON FACING PAGE.
- NO: go to step 1.

1
- IGNITION "OFF."
- DISCONNECT THE 6-WAY ABS MOTOR PACK CONNECTOR AND THE 8-WAY EBCM CONNECTOR.
- IGNITION "ON."
- USING J 39200, MEASURE THE VOLTAGE BETWEEN GROUND AND TERMINAL "A" OF THE 6-WAY ABS MOTOR PACK HARNESS CONNECTOR.

IS VOLTAGE 1 VOLT OR LESS?

- NO: REPAIR SHORT TO VOLTAGE IN CKT 1280.
- YES: go to step 2.

2
- MEASURE THE VOLTAGE BETWEEN GROUND AND TERMINAL "B" OF THE 6-WAY ABS MOTOR PACK HARNESS CONNECTOR.

IS VOLTAGE 1 VOLT OR LESS?

- NO: REPAIR SHORT TO VOLTAGE IN CKT 1281.
- YES: go to step 3.

3
- IGNITION "OFF."
- USING J 39200, MEASURE THE RESISTANCE BETWEEN TERMINAL "A" AND TERMINAL "B" OF THE 6-WAY ABS MOTOR PACK HARNESS CONNECTOR.

DOES J 39200 DISPLAY RESISTANCE AS "OL" (INFINITE)?

- NO: REPAIR SHORT BETWEEN CKTs 1280 AND 1281.
- YES: go to step 4.

4
- USING J 39200, MEASURE THE RESISTANCE BETWEEN TERMINAL "A" AND TERMINAL "B" OF THE ABS MOTOR PACK.

IS RESISTANCE GREATER THAN 0.4Ω?

- NO: REPLACE ABS MOTOR PACK.
- YES: go to step 5.

5
- INSPECT CKTs 1280 AND 1281 AND RELATED CONNECTORS FOR DAMAGE WHICH MAY RESULT IN A SHORT TO VOLTAGE WITH ALL CONNECTORS CONNECTED. REPAIR ALL DAMAGE IF EVIDENT.
- RECONNECT ALL CONNECTORS.
- IGNITION "OFF."
- START ENGINE AND TEST DRIVE VEHICLE, ACHIEVING AT LEAST 16 km/h (10 MPH).
- REPEAT ABOVE DRIVE CYCLE SEQUENCE TWO MORE TIMES.

DID DTC 58 SET IN THE LAST THREE DRIVE CYCLES?

- YES: REPLACE EBCM.
- NO: MALFUNCTION WAS MOST LIKELY CAUSED BY A POOR CONNECTION. REFER TO "DIAGNOSTIC AIDS" ON FACING PAGE.

AFTER DIAGNOSIS IS COMPLETE, CLEAR DTCs AND TEST DRIVE VEHICLE FOR THREE (3) DRIVE CYCLES TO VERIFY THAT DTC DOES NOT RESET. A DRIVE CYCLE CONSISTS OF STARTING THE VEHICLE, DRIVING OVER 16 KM/H (10 MPH), AND THEN KEYING DOWN.

DIAGNOSTIC CHART DTC 61 — 1995 LUMINA APV

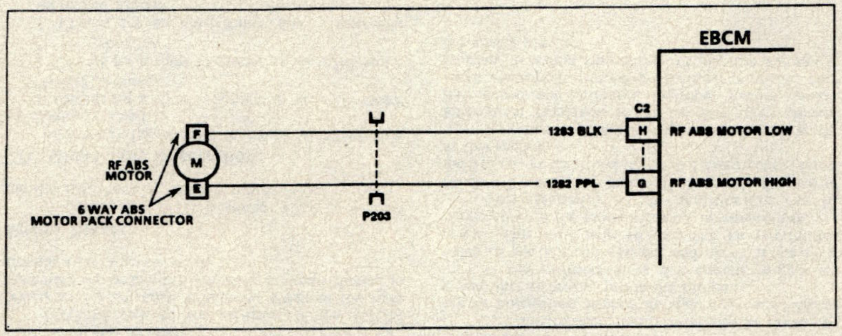

DTC 61
RIGHT FRONT ABS MOTOR CIRCUIT OPEN

Circuit Description:

This DTC identifies an ABS motor that cannot be energized due to an open in its circuitry. This malfunction will not allow proper front ABS operation.

DTC Will Set When:

DTC 61 can be set only when the ABS motor is commanded "OFF." If the EBCM detects an out of range voltage on either of the right front ABS motor circuits indicating an open circuit, a malfunction exists.

Action Taken:

An open ABS motor will not activate when requested. A malfunction DTC is stored, ABS is disabled and the amber ABS warning lamp is turned "ON."

DTC Chart Test Description:

Number(s) below refer to circled number(s) on the diagnostic chart.

1. Checks for proper resistance of the ABS motor.
2. Checks for an open in the ABS motor "HIGH" circuitry.
3. Checks for an open in the ABS motor "LOW" circuitry.
4. Ensures malfunction was not due to poor terminal contact or EBCM.

Diagnostic Aids:

Using Tech 1, select manual control function, and exercise ABS motor movement of affected channel in both directions while applying light pressure on the brake pedal.

If erratic or "jumpy" brake pedal movement is detected while performing an "apply" or "release" function of the ABS motor, an intermittent malfunction may be indicated.

An "Intermittent" malfunction may be caused by a poor connection, rubbed through wire insulation, or a wire that is broken inside the insulation.

If the malfunction is not current, wiggle the wires of the affected channel and check if the DTC resets. This will help to pinpoint an intermittent malfunction in the motor circuitry or connections.

The frequency of the malfunction can be checked by using the enhanced diagnostic function of the Tech 1.

Any circuitry that is suspected of causing the intermittent complaint should be thoroughly checked for backed out terminals, improper mating, broken locks, improperly formed or damaged terminals, poor terminal to wiring connections or physical damage to the wiring harness.

DIAGNOSTIC CHART DTC 61 — 1995 LUMINA APV

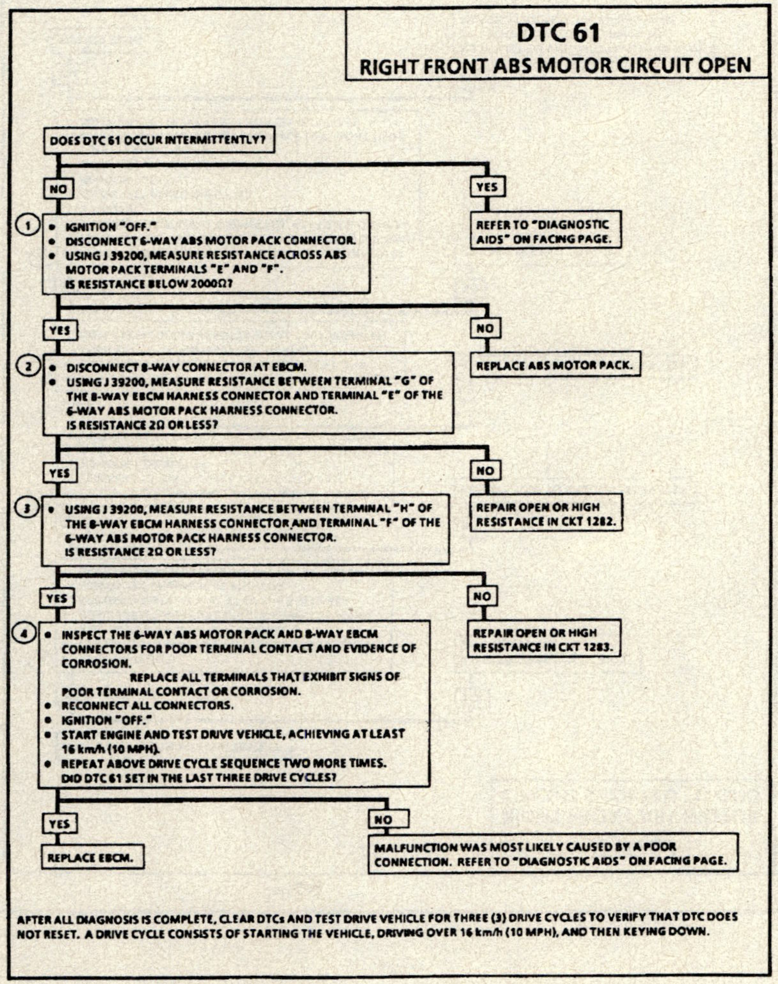

DIAGNOSTIC CHART DTC 62 — 1995 LUMINA APV

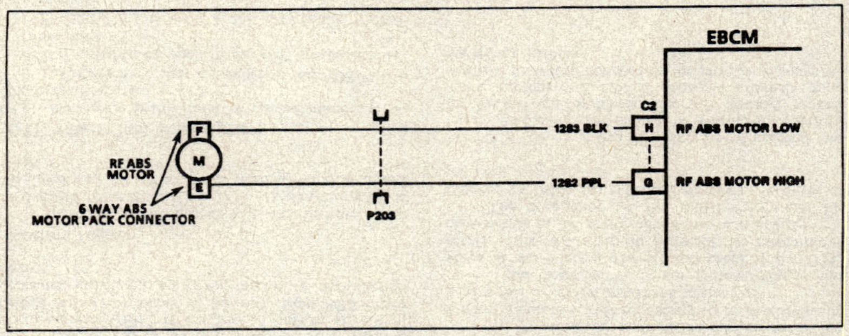

DTC 62
RIGHT FRONT ABS MOTOR CIRCUIT SHORTED TO GND

Circuit Description:

This DTC identifies an ABS motor circuit that is shorted to ground. This malfunction will not allow the ABS motor to be controlled at the commanded current rate or will cause the ABS motor driver circuit to allow current directly to ground.

DTC Will Set When:

DTC 62 can be set anytime. If the EBCM detects an out of range voltage on either of the right front ABS motor circuits indicating a circuit shorted to ground, a malfunction exists.

Action Taken:

A malfunction DTC is stored, ABS is disabled and the ABS warning lamp is turned "ON."

DTC Chart Test Description:

Number(s) below refer to circled number(s) on the diagnostic chart.

1. Checks for a short to ground in the ABS motor "HIGH" circuitry.
2. Checks for a short to ground in the ABS motor "LOW" circuitry.
3. Checks for an ABS motor that is internally shorted to ground.
4. Ensures malfunction was not due to physical damage of the circuitry.

Diagnostic Aids:

Using Tech 1, select manual control function, and exercise ABS motor movement of affected channel in both directions while applying light pressure on the brake pedal.

If errad or "jumpy" brake pedal movement is detected while performing an "apply" or "release" function of the ABS motor, an intermittent malfunction may be indicated.

An "Intermittent" malfunction may be caused by a poor connection, rubbed through wire insulation, or a wire that is broken inside the insulation.

If the malfunction is not current, wiggle the wires of the affected channel and check if the DTC resets. This will help to pinpoint an intermittent malfunction in the motor circuitry or connections.

The frequency of the malfunction can be checked by using the enhanced diagnostic function of the Tech 1, as described in "Tech 1 Diagnostics" found in this section.

Any circuitry that is suspected of causing the intermittent complaint should be thoroughly checked for backed out terminals, improper mating, broken locks, improperly formed or damaged terminals, poor terminal to wiring connections or physical damage to the wiring harness.

DIAGNOSTIC CHART DTC 62 — 1995 LUMINA APV

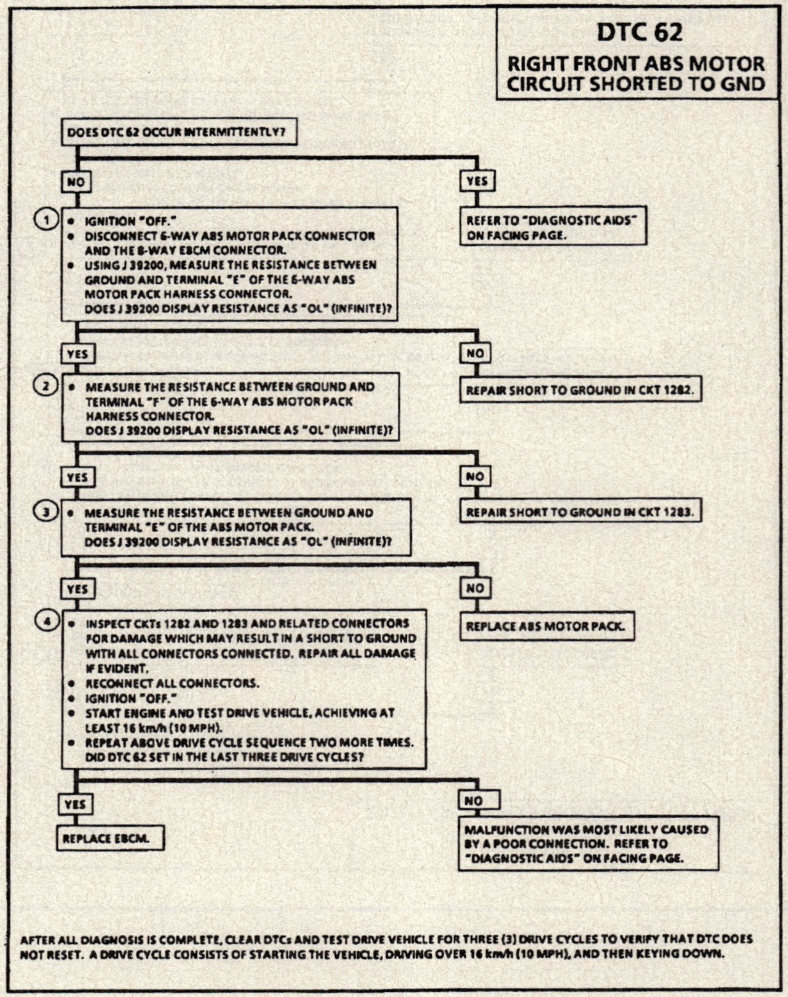

DIAGNOSTIC CHART DTC 63 — 1995 LUMINA APV

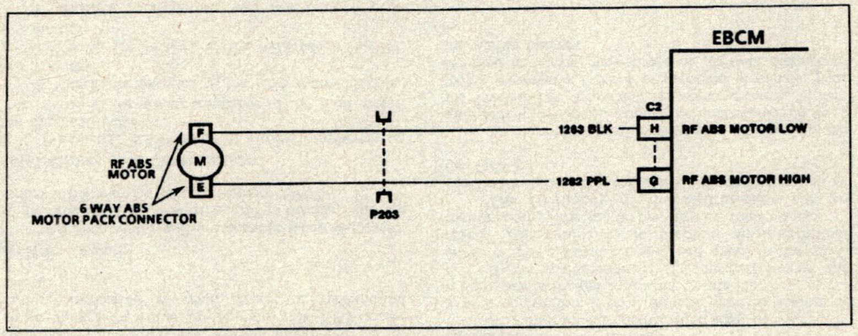

DTC 63
RIGHT FRONT ABS MOTOR CIRCUIT SHORTED TO BATT

Circuit Description:

This DTC identifies an ABS motor circuit that is shorted to battery or an ABS motor that has low or no resistance. This malfunction will not allow the ABS motor to be controlled at the commanded current rate or will cause the ABS motor to turn in the opposite direction or not at all.

DTC Will Set When:

DTC 63 can be set only when the ABS motor is commanded "OFF." If the EBCM detects an out of range voltage on either of the right front ABS motor circuits indicating a circuit shorted to battery or an ABS motor shorted, a malfunction exists.

Action Taken:

A malfunction DTC is stored, ABS is disabled and the amber ABS warning lamp is turned "ON."

DTC Chart Test Description:

Number(s) below refer to circled number(s) on the diagnostic chart.

1. Checks for a short to voltage in the ABS motor "HIGH" circuitry.
2. Checks for a short to voltage in the ABS motor "LOW" circuitry.
3. Checks for a short circuit between motor "HIGH" and motor "LOW."
4. Checks for proper resistance of the ABS motor.
5. Ensures malfunction was not due to physical damage of the circuitry.

Diagnostic Aids:

Using Tech 1, select manual control function, and exercise ABS motor movement of affected channel in both directions while applying light pressure on the brake pedal.

If erratic or "jumpy" brake pedal movement is detected while performing an "apply" or "release" function of the ABS motor, an intermittent malfunction may be indicated.

An "Intermittent" malfunction may be caused by a poor connection, rubbed through wire insulation, or a wire that is broken inside the insulation.

If the malfunction is not current, wiggle the wires of the affected channel and check if the DTC resets. This will help to pinpoint an intermittent malfunction in the motor circuitry or connections.

The frequency of the malfunction can be checked by using the enhanced diagnostic function of the Tech 1.

Any circuitry that is suspected of causing the intermittent complaint should be thoroughly checked for backed out terminals, improper mating, broken locks, improperly formed or damaged terminals, poor terminal to wiring connections or physical damage to the wiring harness.

DIAGNOSTIC CHART DTC 63 — 1995 LUMINA APV

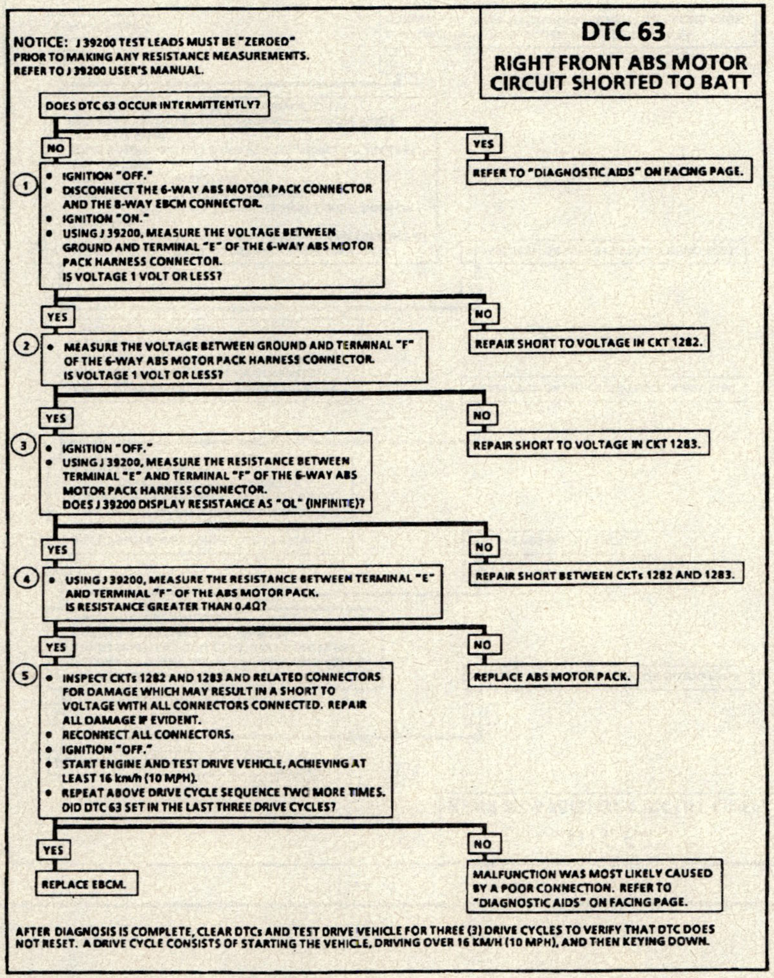

DIAGNOSTIC CHART DTC 64 — 1995 LUMINA APV

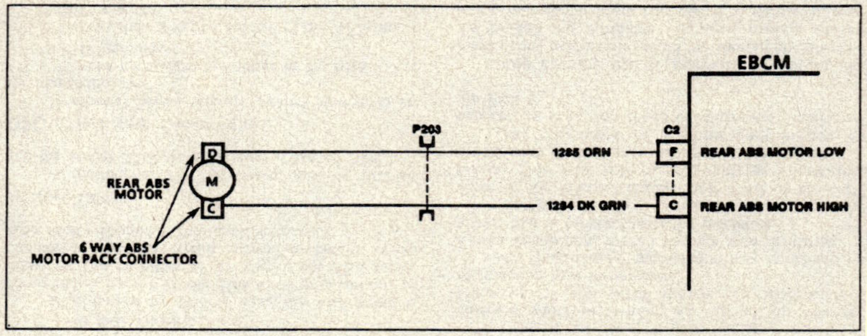

DTC 64
REAR ABS MOTOR CIRCUIT OPEN

Circuit Description:

This DTC identifies an ABS motor that cannot be energized due to an open in its circuitry. This malfunction will not allow proper rear ABS operation.

DTC Will Set When:

DTC 64 can be set only when the ABS motor is commanded "OFF." If the EBCM detects an out of range voltage on either of the rear axle ABS motor circuits indicating an open circuit, a malfunction exists.

Action Taken:

An open motor will not activate when requested. A malfunction DTC is stored, ABS is disabled and the amber ABS warning lamp is turned "ON."

DTC Chart Test Description:

Number(s) below refer to circled number(s) on the diagnostic chart.

1. Checks for proper resistance of the ABS motor.
2. Checks for an open in the ABS motor "HIGH" circuitry.
3. Checks for an open in the ABS motor "LOW" circuitry.
4. Ensures malfunction was not due to poor terminal contact.

Diagnostic Aids:

Using Tech 1, select manual control function, and exercise ABS motor movement of affected channel in both directions while applying light pressure on the brake pedal.

If erratic or "jumpy" brake pedal movement is detected while performing an "apply" or "release" function of the ABS motor, an intermittent malfunction may be indicated.

An "Intermittent" malfunction may be caused by a poor connection, rubbed through wire insulation, or a wire that is broken inside the insulation.

If the malfunction is not current, wiggle the wires of the affected channel and check if the DTC resets. This will help to pinpoint an intermittent malfunction in the motor circuitry or connections.

The frequency of the malfunction can be checked by using the enhanced diagnostic function of the Tech 1

Any circuitry that is suspected of causing the intermittent complaint should be thoroughly checked for backed out terminals, improper mating, broken locks, improperly formed or damaged terminals, poor terminal to wiring connections or physical damage to the wiring harness.

DIAGNOSTIC CHART DTC 64 — 1995 LUMINA APV

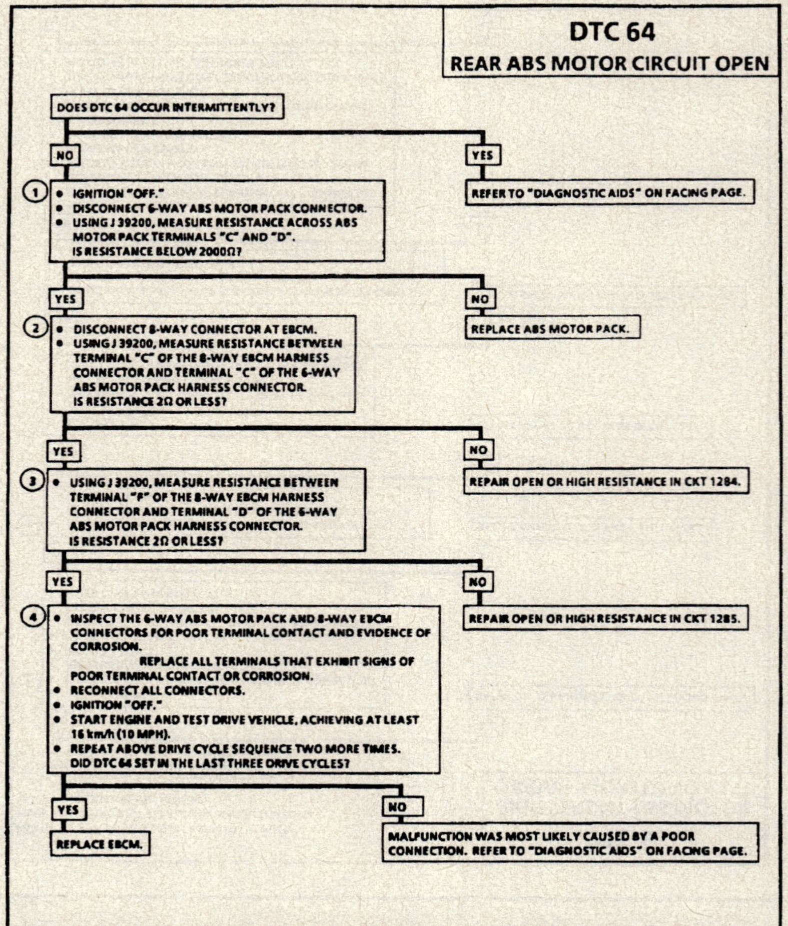

AFTER ALL DIAGNOSIS IS COMPLETE, CLEAR DTCs AND TEST DRIVE VEHICLE FOR THREE (3) DRIVE CYCLES TO VERIFY THAT DTC DOES NOT RESET. A DRIVE CYCLE CONSISTS OF STARTING THE VEHICLE, DRIVING OVER 16 km/h (10 MPH), AND THEN KEYING DOWN.

DIAGNOSTIC CHART DTC 65 — 1995 LUMINA APV

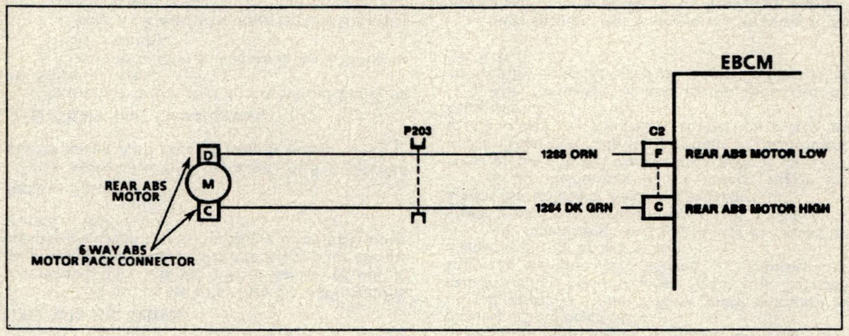

DTC 65
REAR ABS MOTOR CIRCUIT SHORTED TO GROUND

Circuit Description:

This DTC identifies an ABS motor circuit that is shorted to ground. This malfunction will not allow the ABS motor to be controlled at the commanded current rate or will cause the driver circuit to allow current directly to ground.

DTC Will Set When:

DTC 65 can be set anytime. If the EBCM detects an out of range voltage on either of the rear ABS motor circuits indicating a circuit shorted to ground, a malfunction exists.

Action Taken:

A malfunction DTC is stored, ABS is disabled and the amber ABS warning lamp is turned "ON."

DTC Chart Test Description:

Number(s) below refer to circled number(s) on the diagnostic chart.

1. Checks for a short to ground in the ABS motor "HIGH" circuitry.
2. Checks for a short to ground in the ABS motor "LOW" circuitry.
3. Checks for a short to ground in the ABS motor.
4. Ensures malfunction was not due to poor terminal contact.

Diagnostic Aids:

Using Tech 1, select manual control function, and exercise ABS motor movement of affected channel in both directions while applying light pressure on the brake pedal.

If erratic or "jumpy" brake pedal movement is detected while performing an "apply" or "release" function of the ABS motor, an intermittent malfunction may be indicated.

An "Intermittent" malfunction may be caused by a poor connection, rubbed through wire insulation, or a wire that is broken inside the insulation.

If the malfunction is not current, wiggle the wires of the affected channel and check if the DTC resets. This will help to pinpoint an intermittent malfunction in the ABS motor circuitry or connections.

The frequency of the malfunction can be checked by using the enhanced diagnostic function of the Tech 1

Any circuitry that is suspected of causing the intermittent complaint should be thoroughly checked for backed out terminals, improper mating, broken locks, improperly formed or damaged terminals, poor terminal to wiring connections or physical damage to the wiring harness.

DIAGNOSTIC CHART DTC 65 — 1995 LUMINA APV

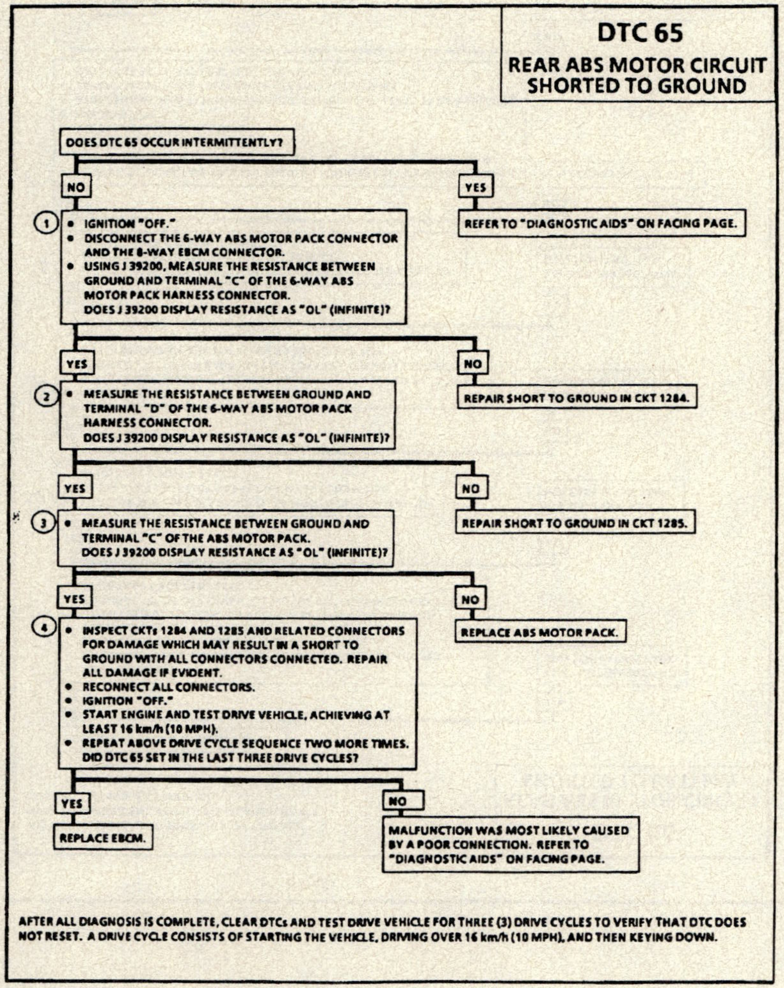

DIAGNOSTIC CHART DTC 66 — 1995 LUMINA APV

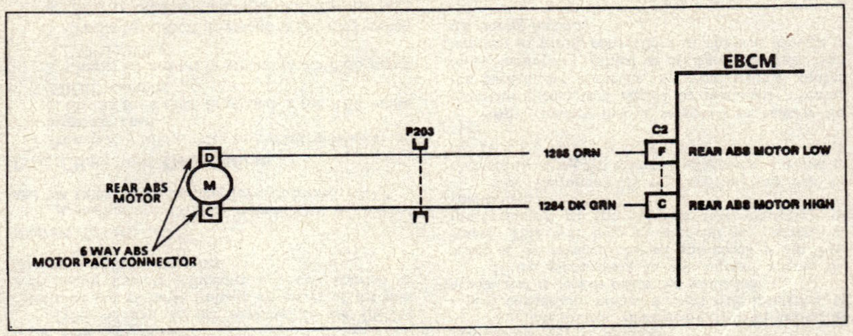

DTC 66
REAR ABS MOTOR CIRCUIT SHORTED TO BATTERY

Circuit Description:

This DTC identifies an ABS motor circuit that is shorted to battery or an ABS motor that has low or zero resistance. This malfunction will not allow the ABS motor to be controlled at the commanded current rate or will cause the ABS motor to turn in the opposite direction or not turn at all.

DTC Will Set When:

DTC 66 can be set only when the ABS motor is commanded "OFF." If the EBCM detects an out of range voltage on either of the rear ABS motor circuits indicating a circuit shorted to battery or an ABS motor shorted, a malfunction exists.

Action Taken:

A malfunction DTC is stored, ABS is disabled and the amber ABS warning lamp is turned "ON."

DTC Chart Test Description:

Number(s) below refer to circled number(s) on the diagnostic chart.

1. Checks for a short to voltage in the ABS motor "HIGH" circuitry.
2. Checks for a short to voltage in the ABS motor "LOW" circuitry.
3. Checks for a short between motor "HIGH" and motor "LOW" circuitry.
4. Checks for proper resistance of the ABS motor pack
5. Ensures malfunction was not due to physical damage of the circuitry.

Diagnostic Aids:

Using Tech 1, select manual control function, and exercise ABS motor movement of affected channel in both directions while applying light pressure on the brake pedal.

If erratic or "jumpy" brake pedal movement is detected while performing an "apply" or "release" function of the ABS motor, an intermittent malfunction may be indicated.

An "Intermittent" malfunction may be caused by a poor connection, rubbed through wire insulation, or a wire that is broken inside the insulation.

If the malfunction is not current, wiggle the wires of the affected channel and check if the DTC resets. This will help to pinpoint an intermittent malfunction in the ABS motor circuitry or connections.

The frequency of the malfunction can be checked by using the enhanced diagnostic function of the Tech 1,

Any circuitry that is suspected of causing the intermittent complaint should be thoroughly checked for backed out terminals, improper mating, broken locks, improperly formed or damaged terminals, poor terminal to wiring connections or physical damage to the wiring harness.

DIAGNOSTIC CHART DTC 66 — 1995 LUMINA APV

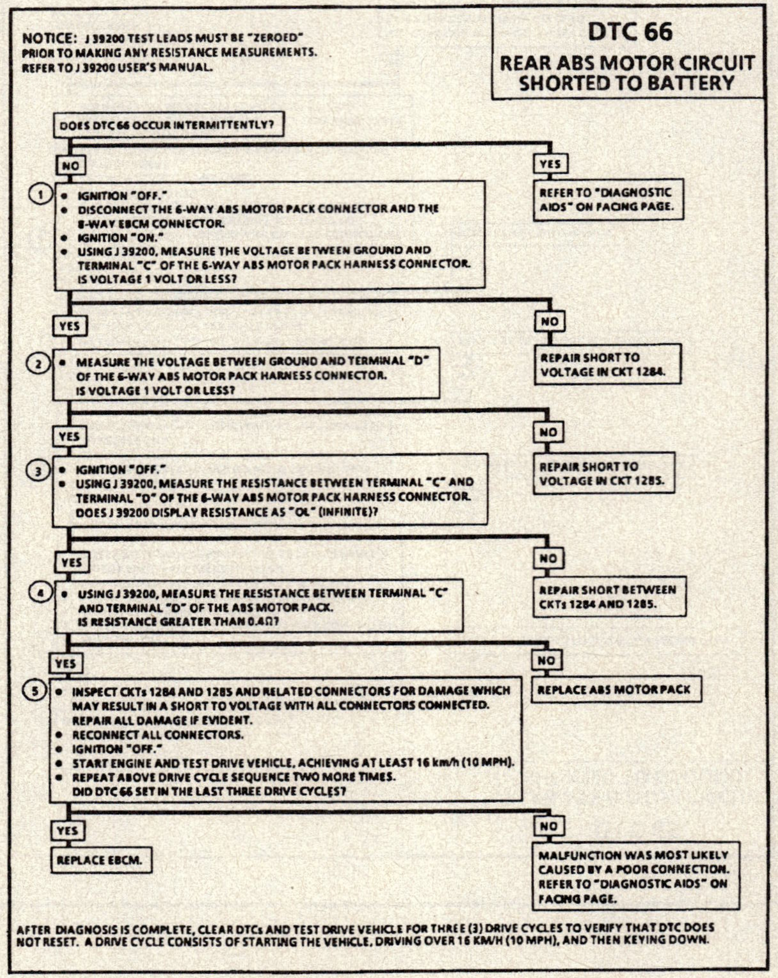

DIAGNOSTIC CHART DTC 76 — 1995 LUMINA APV

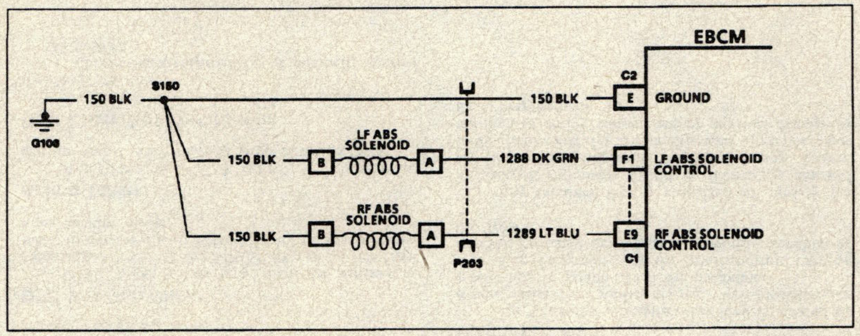

DTC 76
LEFT FRONT SOL CIRCUIT OPEN OR SHORTED TO BATT

Circuit Description:

This DTC identifies a solenoid that cannot be energized due to an open in its circuitry, or a solenoid that is always energized due to a short to battery in its circuitry between the driver and the solenoid. An open will not allow proper ABS operation, but the short to battery simply turns "ON" the solenoid. A path for base brakes is still allowed once the ABS motor re-homes and the check ball is lifted off its seat during key "ON" initialization.

DTC Will Set When:

DTC 76 can be set only when the solenoid is commanded "OFF." If the EBCM detects an excessive voltage in the left front solenoid control circuit, a malfunction exists.

Action Taken:

A malfunction DTC is stored, ABS is disabled and the amber ABS warning lamp is turned "ON."

DTC Chart Test Description:

Number(s) below refer to circled number(s) on the diagnostic chart.

1. Checks for a short to voltage in the control circuitry of the solenoid.
2. Checks for an open in the control circuit of the solenoid.
3. Checks the solenoid coil for proper resistance.
4. Checks for an open in the solenoid ground circuit.
5. Checks for a possible intermittent malfunction in the solenoid circuitry due to poor terminal contact.

Diagnostic Aids:

An "Intermittent" malfunction may be caused by a poor connection, rubbed through wire insulation, or a wire that is broken inside the insulation.

The frequency of the malfunction can be checked by using the enhanced diagnostic function of the Tech 1.

Any circuitry that is suspected of causing the intermittent complaint should be thoroughly checked for backed out terminals, improper mating, broken locks, improperly formed or damaged terminals, poor terminal to wiring connections or physical damage to the wiring harness.

DIAGNOSTIC CHART DTC 76 — 1995 LUMINA APV

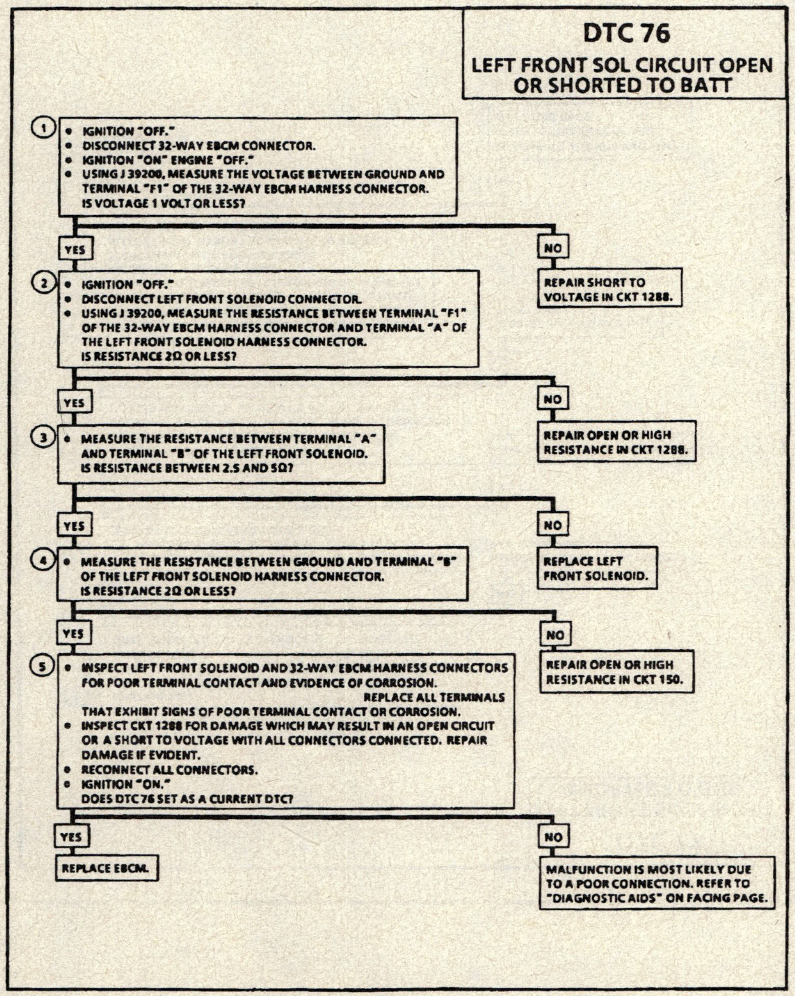

DIAGNOSTIC CHART DTC 77 — 1995 LUMINA APV

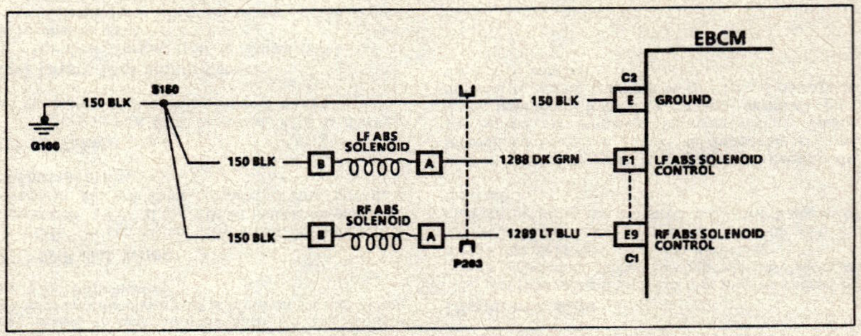

DTC 77
LEFT FRONT SOL CIRCUIT SHORTED TO GND

Circuit Description:

This DTC identifies a solenoid that cannot be energized due to an open in its driver circuitry, or a short to ground between the solenoid driver and the solenoid. These malfunctions can affect ABS operation since the flow of brake fluid to the wheel cylinder cannot be stopped, making ABS operation for that channel impossible.

DTC Will Set When:

DTC 77 can be set only when the solenoid is commanded "ON." If the EBCM detects the left front solenoid control circuit voltage is out of specification, a malfunction exists.

Action Taken:

A malfunction DTC is stored, ABS is disabled and the amber ABS warning lamp is turned "ON."

DTC Chart Test Description:

Number(s) below refer to circled number(s) on the diagnostic chart.

1. Checks for a solenoid that is internally shorted to ground.
2. Checks for a solenoid that is not within proper resistance values.
3. Indicates if a short to ground exists in the solenoid circuitry.
4. Checks for a possible intermittent malfunction in the solenoid circuitry due to poor terminal contact.

Diagnostic Aids:

An "Intermittent" malfunction may be caused by a poor connection, rubbed through wire insulation, or a wire that is broken inside the insulation.

The frequency of the malfunction can be checked by using the enhanced diagnostic function of the Tech 1,

Any circuitry that is suspected of causing the intermittent complaint should be thoroughly checked for backed out terminals, improper mating, broken locks, improperly formed or damaged terminals, poor terminal to wiring connections or physical damage to the wiring harness.

DIAGNOSTIC CHART DTC 77 — 1995 LUMINA APV

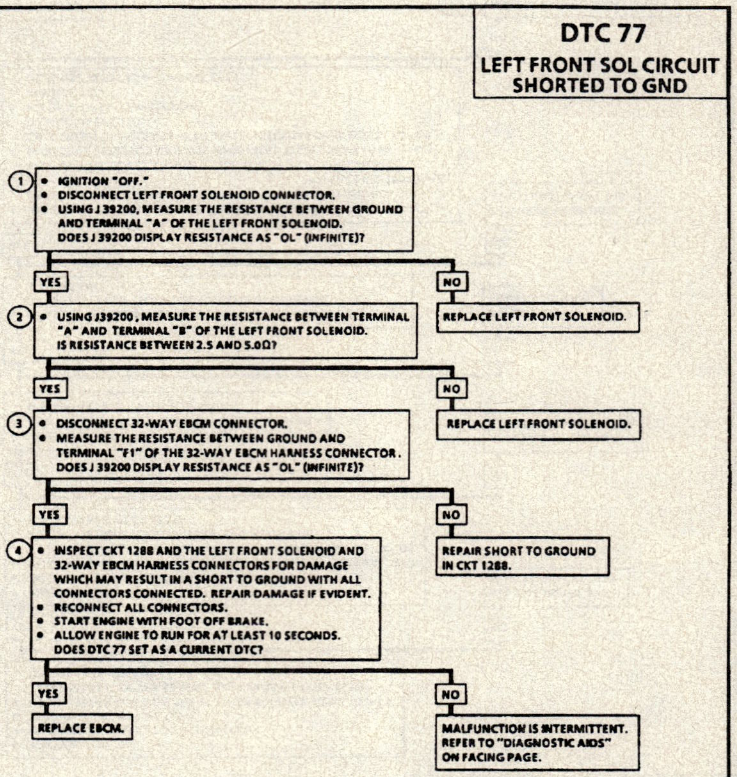

DIAGNOSTIC CHART DTC 78 — 1995 LUMINA APV

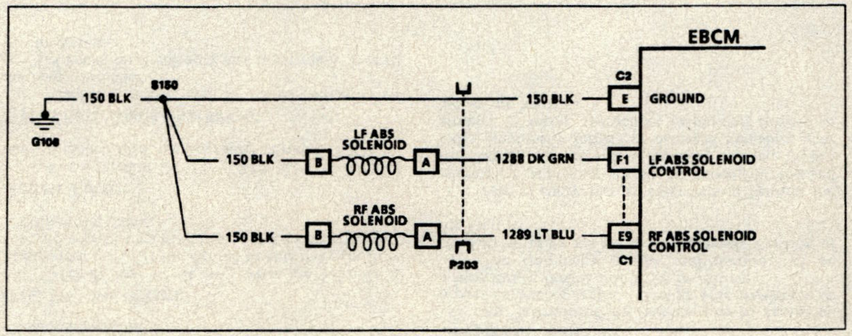

DTC 78

RIGHT FRONT SOL CIRCUIT SHORTED TO BATT

Circuit Description:

This DTC identifies a solenoid that cannot be energized due to an open in its circuitry, or a solenoid that is always energized due to a short to battery in its circuitry between the driver and the solenoid. An open will not allow proper ABS operation, but the short to battery simply turns "ON" the solenoid. A path for base brakes is still allowed once the ABS motor re-homes and the check ball is lifted off its seat during key "ON" initialization.

DTC Will Set When:

DTC 78 can be set only when the solenoid is commanded "OFF." If the EBCM detects an excessive voltage in the right front solenoid control circuit, a malfunction exists.

Action Taken:

A malfunction DTC is stored, ABS is disabled and the amber ABS warning lamp is turned "ON."

DTC Chart Test Description:

Number(s) below refer to circled number(s) on the diagnostic chart.

1. Checks for a short to voltage in the control circuitry of the solenoid.
2. Checks for an open in the control circuit of the solenoid.
3. Checks the solenoid coil for proper resistance.
4. Checks for an open in the solenoid ground circuit.
5. Checks for a possible intermittent malfunction in the solenoid circuitry due to poor terminal contact.

Diagnostic Aids:

An "Intermittent" malfunction may be caused by a poor connection, rubbed through wire insulation, or a wire that is broken inside the insulation.

The frequency of the malfunction can be checked by using the enhanced diagnostic function of the Tech 1.

Any circuitry that is suspected of causing the intermittent complaint should be thoroughly checked for backed out terminals, improper mating, broken locks, improperly formed or damaged terminals, poor terminal to wiring connections or physical damage to the wiring harness.

DIAGNOSTIC CHART DTC 78 — 1995 LUMINA APV

DTC 78
RIGHT FRONT SOL CIRCUIT SHORTED TO BATT

(1)
- IGNITION "OFF."
- DISCONNECT 32-WAY EBCM CONNECTOR.
- IGNITION "ON," ENGINE "OFF."
- USING J 39200, MEASURE THE VOLTAGE BETWEEN GROUND AND TERMINAL "E9" OF THE 32-WAY EBCM HARNESS CONNECTOR.

IS VOLTAGE 1 VOLT OR LESS?

NO: REPAIR SHORT TO VOLTAGE IN CKT 1289.

YES:

(2)
- IGNITION "OFF."
- DISCONNECT RIGHT FRONT SOLENOID CONNECTOR.
- USING J 39200, MEASURE THE RESISTANCE BETWEEN TERMINAL "E9" OF THE 32-WAY EBCM HARNESS CONNECTOR AND TERMINAL "A" OF THE RIGHT FRONT SOLENOID HARNESS CONNECTOR.

IS RESISTANCE 2Ω OR LESS?

NO: REPAIR OPEN OR HIGH RESISTANCE IN CKT 1289.

YES:

(3)
- MEASURE THE RESISTANCE BETWEEN TERMINAL "A" AND TERMINAL "B" OF THE RIGHT FRONT SOLENOID.

IS RESISTANCE BETWEEN 2.5 AND 5Ω?

NO: REPLACE RIGHT FRONT SOLENOID.

YES:

(4)
- MEASURE THE RESISTANCE BETWEEN GROUND AND TERMINAL "B" OF THE RIGHT FRONT SOLENOID HARNESS CONNECTOR.

IS RESISTANCE 2Ω OR LESS?

NO: REPAIR OPEN OR HIGH RESISTANCE IN CKT 150.

YES:

(5)
- INSPECT RIGHT FRONT SOLENOID AND 32-WAY EBCM HARNESS CONNECTORS FOR POOR TERMINAL CONTACT AND EVIDENCE OF CORROSION.
  REPLACE ALL TERMINALS THAT EXHIBIT SIGNS OF POOR TERMINAL CONTACT OR CORROSION.
- INSPECT CKT 1289 FOR DAMAGE WHICH MAY RESULT IN AN OPEN CIRCUIT OR A SHORT TO VOLTAGE WITH ALL CONNECTORS.
- RECONNECT ALL CONNECTORS.
- IGNITION "ON."

DOES DTC 78 SET AS A CURRENT DTC?

YES: REPLACE EBCM.

NO: MALFUNCTION IS MOST LIKELY DUE TO A POOR CONNECTION. REFER TO "DIAGNOSTIC AIDS" ON FACING PAGE.

DIAGNOSTIC CHART DTC 81 — 1995 LUMINA APV

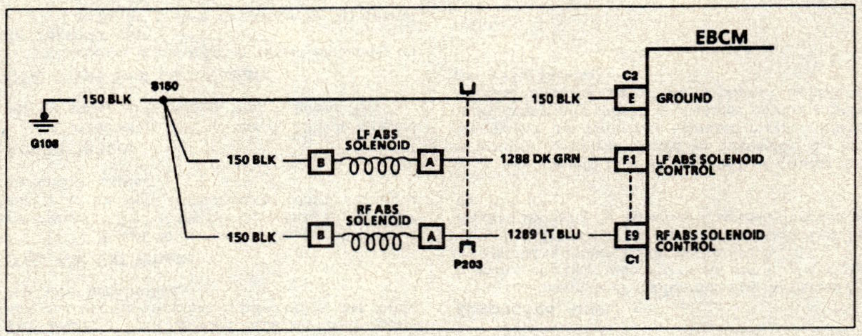

DTC 81
RIGHT FRONT SOL CIRCUIT SHORTED TO GND

Circuit Description:

This DTC identifies a solenoid that cannot be energized due to an open in its driver circuitry, or a short to ground between the solenoid driver and the solenoid. These malfunctions can affect ABS operation since the flow of brake fluid to the wheel cylinder cannot be stopped, making ABS operation for that channel impossible.

DTC Will Set When:

DTC 81 can be set only when the solenoid is commanded "ON." If the EBCM detects the right front solenoid control circuit voltage is out of specification, a malfunction exists.

Action Taken:

A malfunction DTC is stored, ABS is disabled and the amber ABS warning lamp is turned "ON."

DTC Chart Test Description:

Number(s) below refer to circled number(s) on the diagnostic chart.

1. Checks for a solenoid that is internally shorted to ground.
2. Checks for a solenoid that is not within proper resistance values.
3. Indicates if a short to ground exists in the solenoid circuitry.
4. Checks for a possible intermittent malfunction in the solenoid circuitry due to poor terminal contact.

Diagnostic Aids:

An "Intermittent" malfunction may be caused by a poor connection, rubbed through wire insulation, or a wire that is broken inside the insulation.

The frequency of the malfunction can be checked by using the enhanced diagnostic function of the Tech 1

Any circuitry that is suspected of causing the intermittent complaint should be thoroughly checked for backed out terminals, improper mating, broken locks, improperly formed or damaged terminals, poor terminal to wiring connections or physical damage to the wiring harness.

DIAGNOSTIC CHART DTC 81 — 1995 LUMINA APV

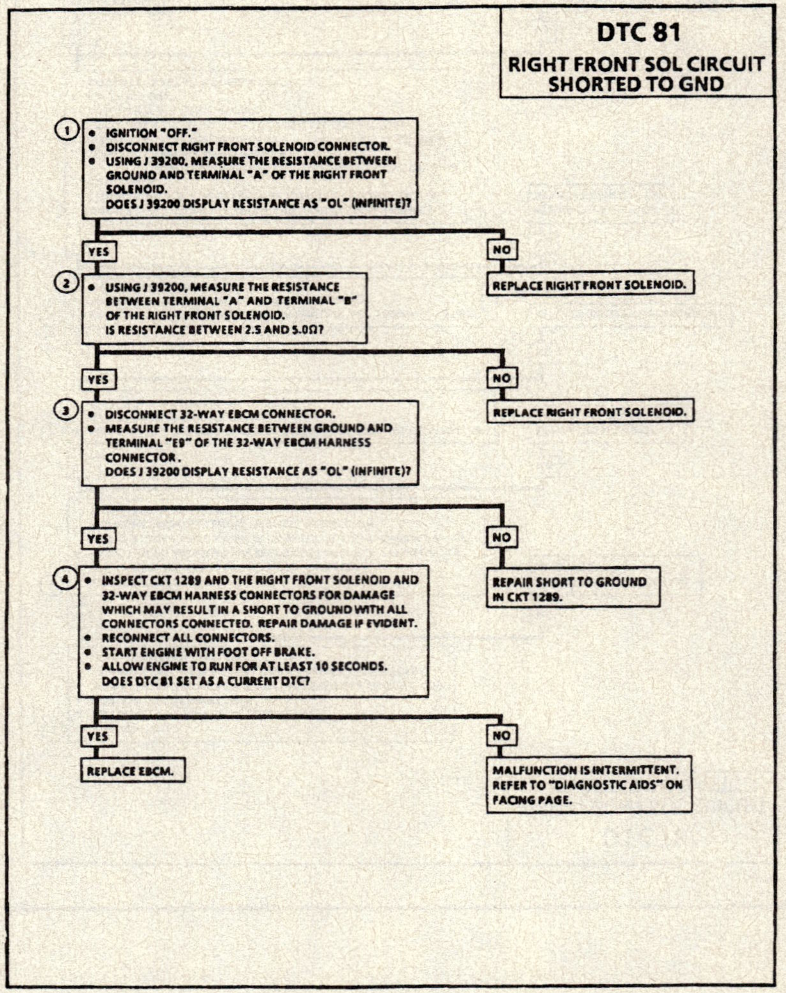

DIAGNOSTIC CHART DTC 82 — 1995 LUMINA APV

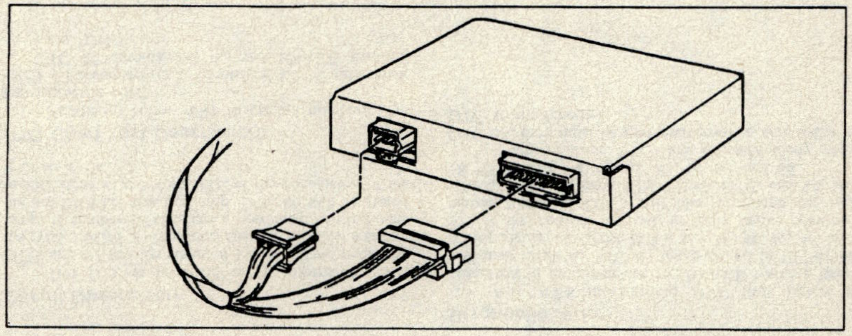

DTC 82
CALIBRATION MALFUNCTION

Circuit Description:

This DTC allows the EBCM to check for a calibration malfunction by comparing the calibration value to a known value stored in the EEPROM.

This DTC is also used as a security measure to prevent improper use of calibrations or changes to these calibrations that may alter the designed function of ABS.

DTC Will Set When:

DTC 82 can be set at key-up. If the program identifier is incorrect or the memory checksum is incorrect, a malfunction exists.

Action Taken:

A malfunction DTC is stored, ABS is disabled and the ABS warning lamp is turned "ON."

DTC Chart Test Description:

Number(s) below refer to circled number(s) on the diagnostic chart.

1. Checks to see if the malfunction is present during diagnosis. If present, the EBCM is not functioning correctly and must be replaced.

Diagnostic Aids:

An intermittent DTC 82 may be caused by a bad cell in the EEPROM that is sensitive to temperature changes. If DTC 82 failed more than once, but is intermittent, replace EBCM.

The frequency of the malfunction can be checked by using the enhanced diagnostic function of the Tech 1

DIAGNOSTIC CHART DTC 82 — 1995 LUMINA APV

DTC 82
CALIBRATION MALFUNCTION

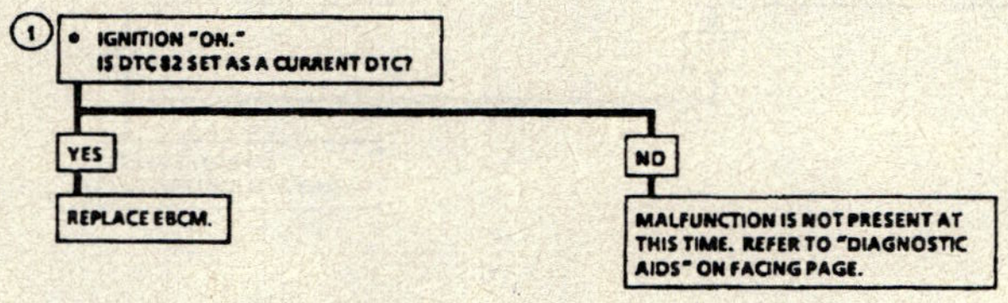

DIAGNOSTIC CHART DTC 86 — 1995 LUMINA APV

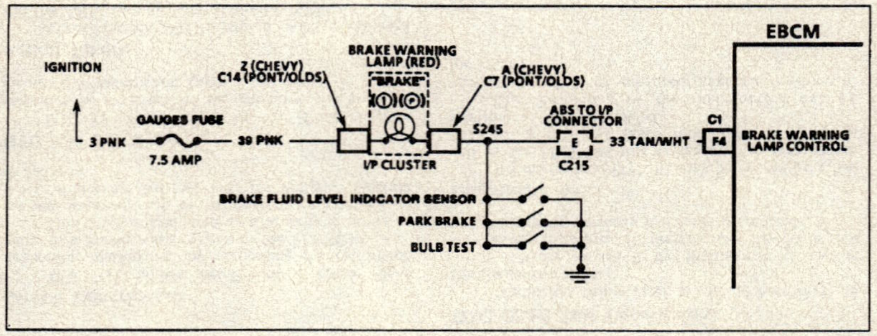

DTC 86
EBCM TURNED "ON" THE RED BRAKE WARNING LAMP

Circuit Description:

This DTC is provided as an information only DTC, and reflects the status of the command issued by the ABS EBCM to illuminate the red BRAKE warning lamp. If another DTC issues a command to illuminate the red BRAKE warning lamp, DTC 86 will be stored in EEPROM as a history DTC at the conclusion of the ignition cycle.

DTC Chart Test Description:

Number(s) below refer to circled number(s) on the diagnostic chart.

1. This step serves to identify if a DTC other than DTC 86 commanded the red BRAKE warning lamp "ON."

Diagnostic Aids:

Any ABS mechanical DTC that issues a command to illuminate the red BRAKE warning lamp will also result in DTC 86 being stored in EEPROM during shutdown. These DTCs are: 38, 41, 42, 46, and 51. If the motors are not in their home position, certain electrical DTCs will also command the red BRAKE warning lamp "ON." These DTCs are: 14, 16, 18, 55, 56, 57, 58, 61, 62, 63, 64, 65, and 66.

If any of these DTCs are indicated along with DTC 86, they must be corrected prior to addressing a DTC 86 malfunction.

DIAGNOSTIC CHART DTC 86 — 1995 LUMINA APV

DTC 86
EBCM TURNED "ON" THE RED "BRAKE" WARNING LAMP

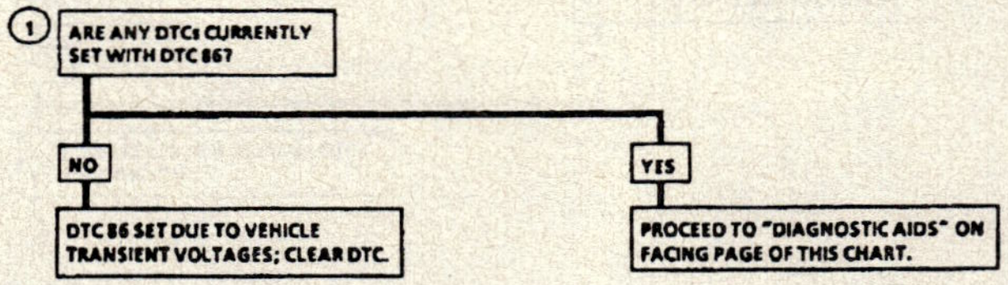

DIAGNOSTIC CHART DTC 87 — 1995 LUMINA APV

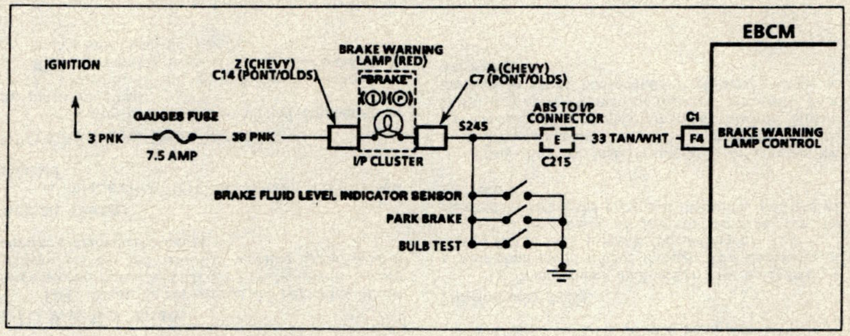

DTC 87
RED BRAKE WARNING LAMP CIRCUIT OPEN

Circuit Description:

This DTC is used to verify the EBCM has continuity to the red BRAKE warning lamp in case the EBCM must turn it "ON." This will only occur if an ABS malfunction is detected that may degrade base brake operation.

Because ABS is not the only device controlling the red BRAKE warning lamp (parking brake and brake fluid level switch may also turn "ON" the lamp) a short to ground in this circuit cannot be detected.

DTC Will Set When:

DTC 87 can be set after initialization is completed. If the red BRAKE warning lamp circuit voltage is out of specification indicating an open circuit, a malfunction exists.

Action Taken:

A malfunction DTC is stored, ABS is not disabled.

DTC Chart Test Description:

Number(s) below refer to circled number(s) on the diagnostic chart.

1. Checks the complete BRAKE warning lamp circuitry by attempting to manually turn "ON" the BRAKE warning lamp.
2. Checks for voltage in park brake circuitry.
3. Checks for a malfunctioning EBCM or poor terminal contact.
4. Ensures malfunction is not due to poor terminal contact at the brake fluid level switch which may result in an intermittent.
5. Checks the integrity of the BRAKE warning lamp.
6. Checks for high resistance in the park brake circuitry.

Diagnostic Aids:

An "Intermittent" malfunction may be caused by a poor connection, rubbed through wire insulation, or a wire that is broken inside the insulation.

The frequency of the malfunction can be checked by using the enhanced diagnostic function of the Tech 1.

If a Tech 1 or T-100 are not available, the parking brake handle may be lifted to verify proper lamp operation and continuity of CKTs 3 and 39.

Any circuitry that is suspected of causing the intermittent complaint should be thoroughly checked for backed out terminals, improper mating, broken locks, improperly formed or damaged terminals, poor terminal to wiring connections or physical damage to the wiring harness.

DIAGNOSTIC CHART DTC 87 — 1995 LUMINA APV

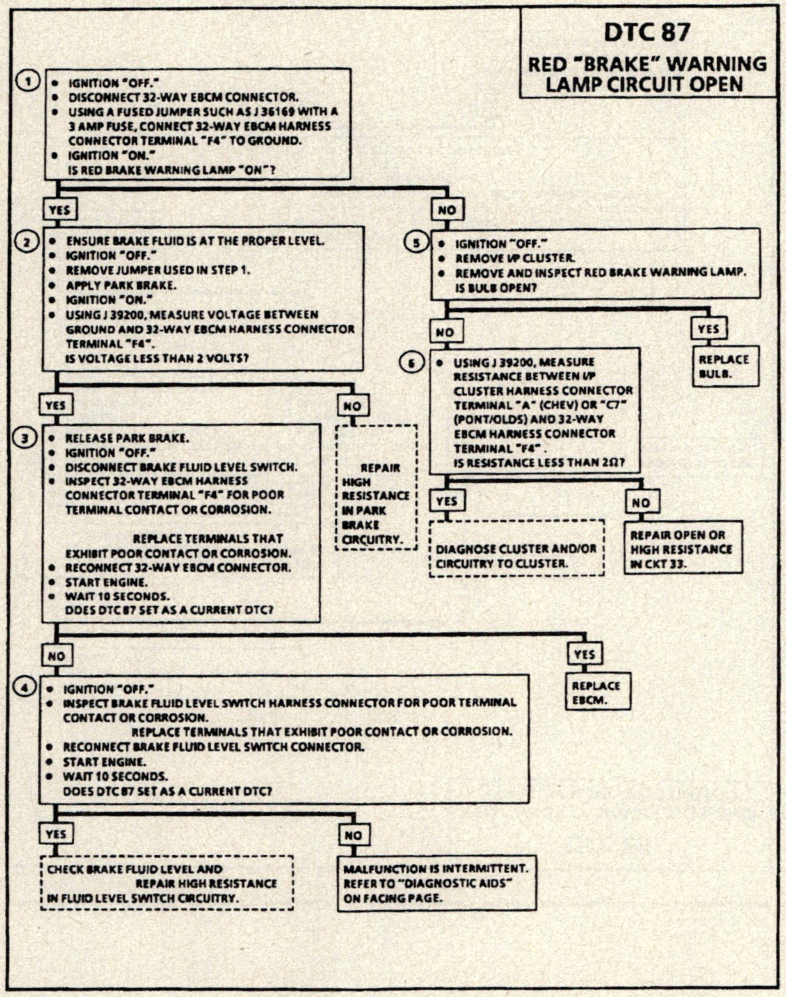

DIAGNOSTIC CHART DTC 88 — 1995 LUMINA APV

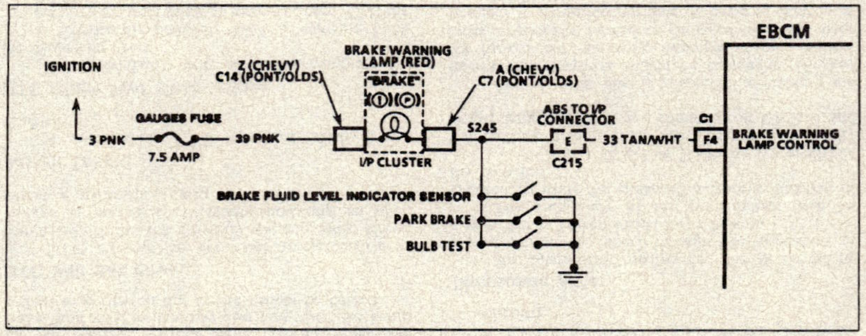

DTC 88
RED BRAKE WARNING LAMP CIRCUIT SHORTED TO BATT

Circuit Description:

This DTC identifies a short to battery between the EBCM and the red BRAKE warning lamp, or an open driver that does not allow the red BRAKE warning lamp to be illuminated by the EBCM. This will only occur if an ABS fault is detected that may degrade base brake operation.

DTC Will Set When:

DTC 88 can be set only at the beginning of the three second bulb check. If the EBCM detects battery voltage on the red BRAKE warning lamp control circuit, a malfunction exists.

Action Taken:

A malfunction DTC is stored. ABS is not disabled.

DTC Chart Test Description:

Number(s) below refer to circled number(s) on the diagnostic chart.

1. Identifies if the ground circuit to the red BRAKE warning lamp is being completed by a source other than the EBCM.
2. Identifies if the malfunction is currently present.
3. By removing the "GAUGES" 7.5 amp fuse, the voltage source is eliminated. This test indicates if voltage is being supplied from a source other than the fuse block.
4. Ensures malfunction was not due to physical damage of the circuitry.
5. Checks for DTC 86 also being set.

Diagnostic Aids:

An "Intermittent" malfunction may be caused by a poor connection, rubbed through wire insulation, or a wire that is broken inside the insulation.

The frequency of the malfunction can be checked by using the enhanced diagnostic function of the Tech 1

Any circuitry that is suspected of causing the intermittent complaint should be thoroughly checked for backed out terminals, improper mating, broken locks, improperly formed or damaged terminals, poor terminal to wiring connections or physical damage to the wiring harness.

DIAGNOSTIC CHART DTC 88 — 1995 LUMINA APV

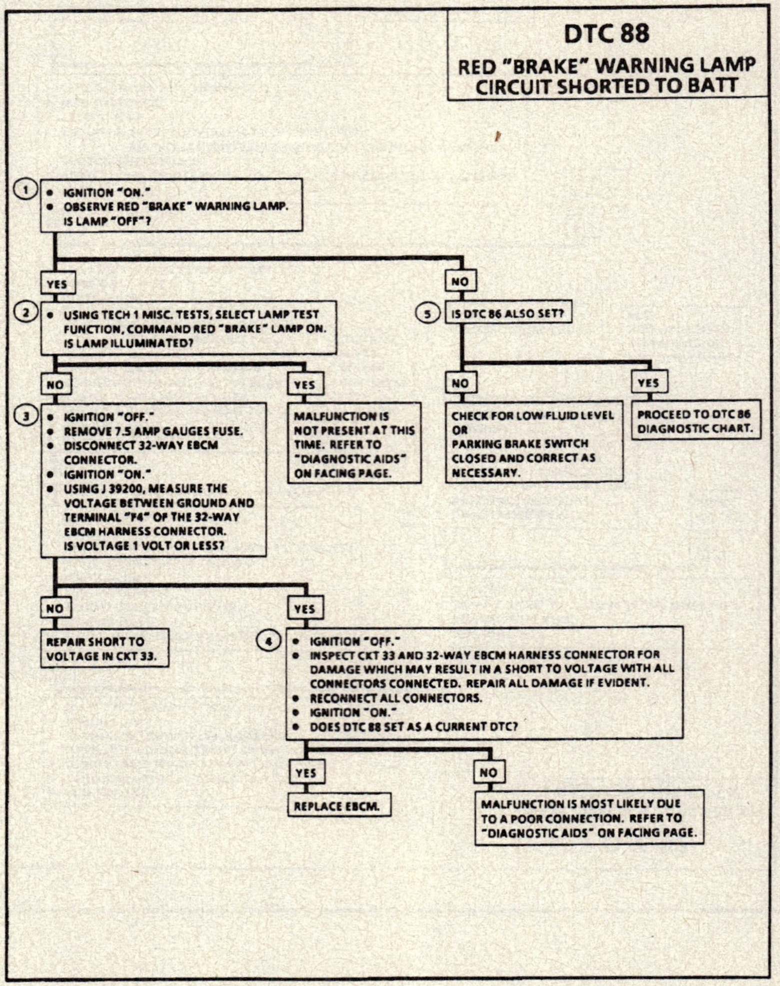

DIAGNOSTIC CHART DTC 91 — 1995 LUMINA APV

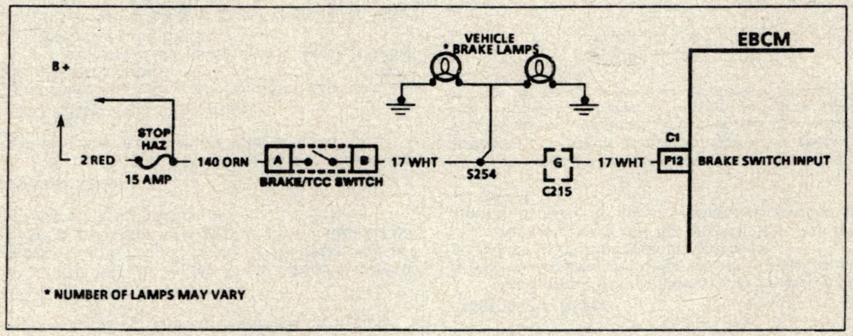

DTC 91
OPEN BRAKE SWITCH DURING DECELERATION

Circuit Description:

This DTC is used to detect an open brake switch in the non-ABS mode. The EBCM looks for deceleration rates that would indicate braking action and verifies this assumption by requiring several repeats of this detection method. In each case, ABS will not be available since no brake switch is seen.

DTC Will Set When:

DTC 91 can be set if three deceleration cycles occur with the brake switch "OFF."

Action Taken:

A malfunction DTC is stored, ABS is disabled and the amber ABS warning lamp is turned "ON."

DTC Chart Test Description:

Number(s) below refer to circled number(s) on the diagnostic chart.

1. Indicates if the brake switch signal is being received by the EBCM.
2. Indicates if an open circuit exists in the brake switch or brake lamp circuitry.
3. Isolates the open circuit to either the brake switch input circuitry, or the EBCM.
4. Ensures malfunction was not due to poor terminal contact.
5. Verifies that voltage is available at the brake switch.
6. Verifies if the brake switch is functioning properly.

Diagnostic Aids:

An "Intermittent" malfunction may be caused by a poor connection, rubbed through wire insulation, or a wire that is broken inside the insulation.

The frequency of the malfunction can be checked by using the enhanced diagnostic function of the Tech 1

Any circuitry that is suspected of causing the intermittent complaint should be thoroughly checked for backed out terminals, improper mating, broken locks, improperly formed or damaged terminals, poor terminal to wiring connections or physical damage to the wiring harness.

DIAGNOSTIC CHART DTC 91 — 1995 LUMINA APV

ANTILOCK BRAKE SYSTEM 5E1-165

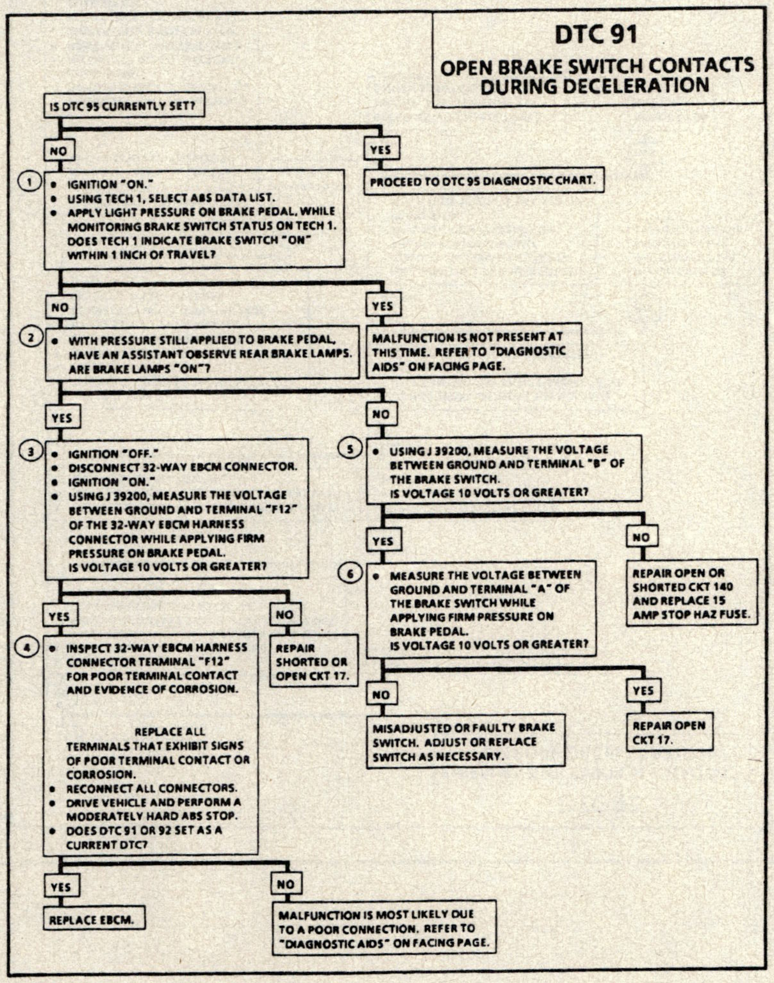

DIAGNOSTIC CHART DTC 92 — 1995 LUMINA APV

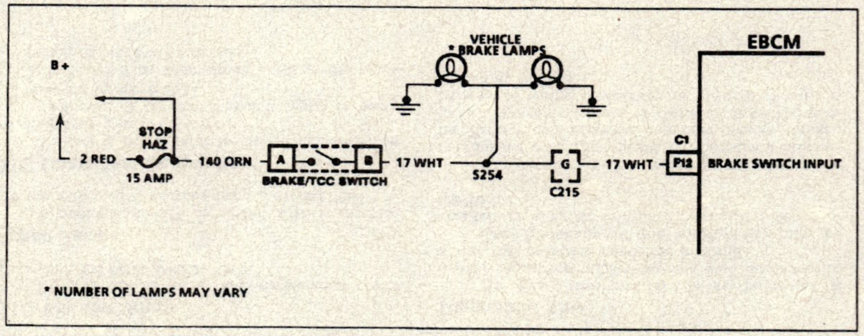

DTC 92
OPEN BRAKE SWITCH WHEN ABS WAS REQUIRED

Circuit Description:

This DTC is run to determine the proper operation of the brake switch. This is important because ABS is activated when the brake switch is "ON." If the brake switch is "OFF," ABS will never be activated. Since this malfunction is difficult to detect under normal braking conditions, this malfunction is only detected when ABS is required.

DTC Will Set When:

DTC 92 can be set if the vehicle's speed is greater than 8 km/h (5 mph). If the brake was not "ON" and a release was required on two channels for 0.5 second, a malfunction exists.

Action Taken:

A malfunction DTC is stored, ABS is disabled and the amber ABS warning lamp is turned "ON."

DTC Chart Test Description:

Number(s) below refer to circled number(s) on the diagnostic chart.

1. Indicates if the brake switch signal is being received by the EBCM.
2. Indicates if an open circuit exists in the brake switch or brake lamp circuitry.
3. Isolates the open circuit to either the brake switch input circuitry, or the EBCM.
4. Ensures malfunction was not due to poor terminal contact.
5. Verifies the voltage is available at the brake switch.
6. Verifies if the brake switch is functioning properly.

Diagnostic Aids:

An "Intermittent" malfunction may be caused by a poor connection, rubbed through wire insulation, or a wire that is broken inside the insulation.

The frequency of the malfunction can be checked by using the enhanced diagnostic function of the Tech 1,

Any circuitry that is suspected of causing the intermittent complaint should be thoroughly checked for backed out terminals, improper mating, broken locks, improperly formed or damaged terminals, poor terminal to wiring connections or physical damage to the wiring harness.

DIAGNOSTIC CHART DTC 92 — 1995 LUMINA APV

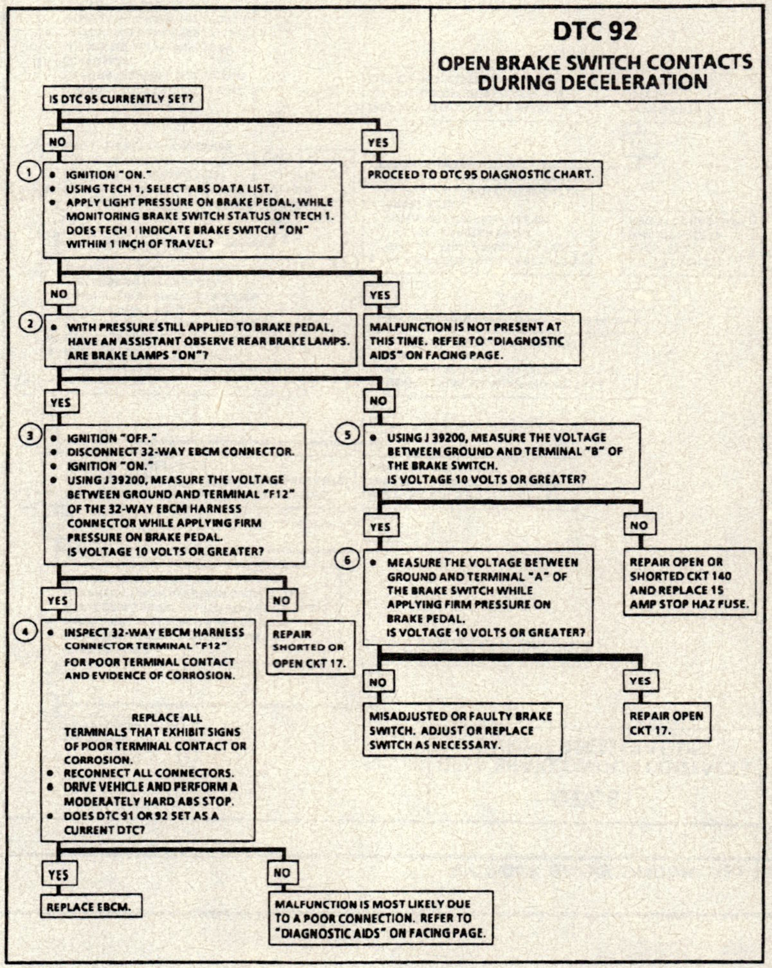

DIAGNOSTIC CHART DTC 93 — 1995 LUMINA APV

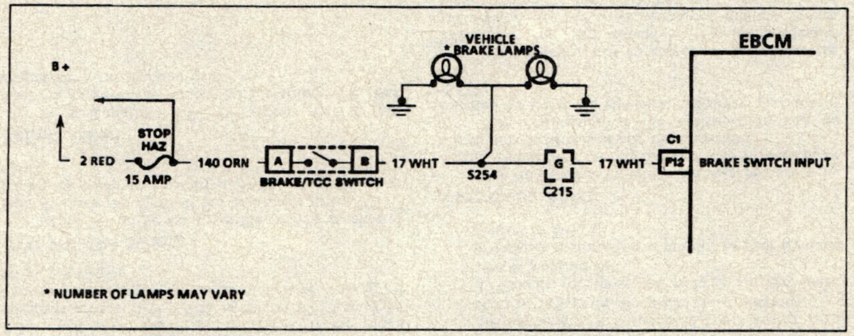

DTC 93
DTC 91 OR 92 SET IN CURRENT OR PREV IGN CYCLE

Circuit Description:

This DTC is the second portion of DTCs 91 and 92. If DTCs 91 or 92 occurred during the last ignition cycle, DTC 93 becomes a current failure during the next ignition cycle, keeping ABS disabled until a brake switch "ON" state is seen. When a change is seen during an ignition cycle in which DTC 93 is a current malfunction, DTC 91 or 92 will clear itself at the end of the current ignition cycle, and ABS will enable itself at the start of the next ignition cycle. DTC 93 alone indicates 91 or 92 occurred previously, but is intermittent, or has been corrected.

DTC Chart Test Description:

Number(s) below refer to circled number(s) on the diagnostic chart.

1. Indicates which DTC (either 91 or 92) caused DTC 93 to set.

Diagnostic Aids:

An "Intermittent" malfunction may be caused by a poor connection, rubbed through wire insulation, or a wire that is broken inside the insulation.

The frequency of the malfunction can be checked by using the enhanced diagnostic function of the Tech 1

Any circuitry that is suspected of causing the intermittent complaint should be thoroughly checked for backed out terminals, improper mating, broken locks, improperly formed or damaged terminals, poor terminal to wiring connections or physical damage to the wiring harness.

Also, verify proper brake switch operation using the data list of the Tech 1. As the brake is applied, the data list should display the brake switch "ON" within 1 inch of travel.

DIAGNOSTIC CHART DTC 93 — 1995 LUMINA APV

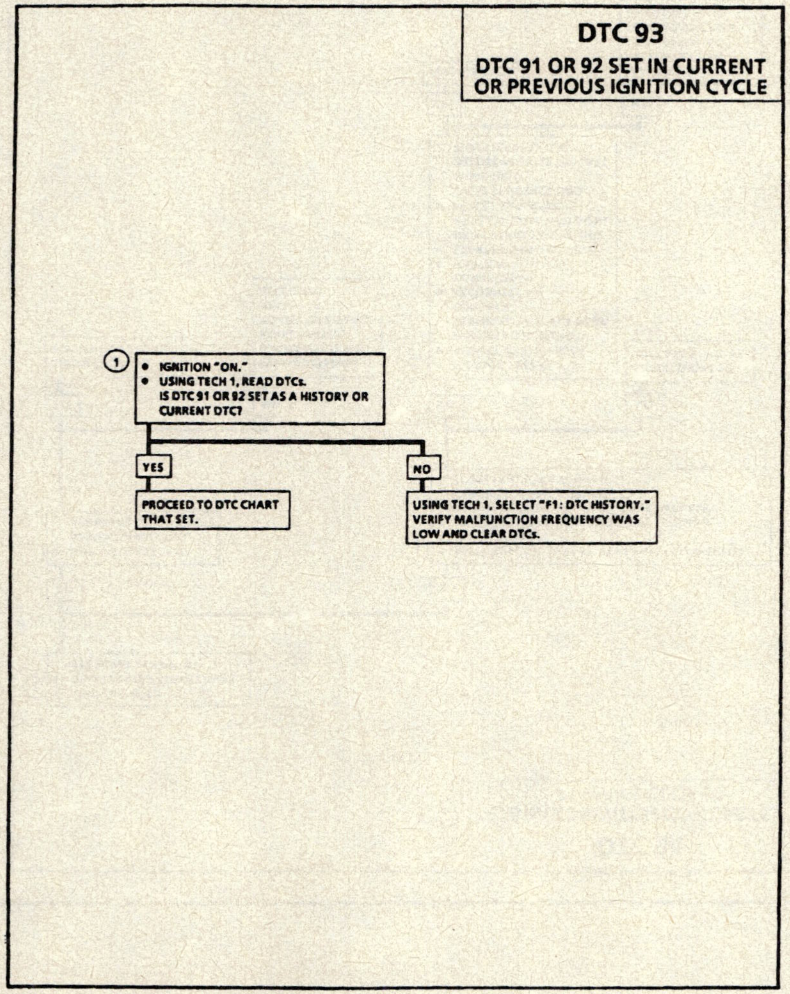

DIAGNOSTIC CHART DTC 94 — 1995 LUMINA APV

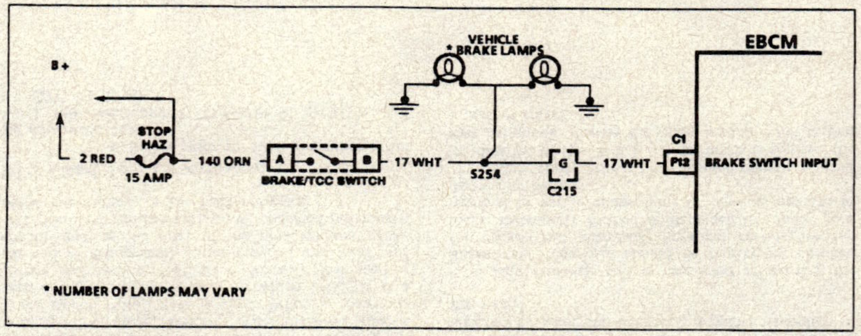

DTC 94
BRAKE SWITCH CONTACTS ALWAYS CLOSED

Circuit Description:

This DTC is run to determine the proper operation of the brake switch. This is important as ABS is activated when the brake switch is "ON" and deactivated when the brake is "OFF." If the brake switch is always "ON," ABS operation will always be requested resulting in potential modulator cycling on rough roads.

DTC Will Set When:

DTC 94 can be set when the vehicle reaches at least 40 km/h (25 mph). If the brake was never "OFF" during two consecutive drive cycles, a malfunction exists.

Action Taken:

A malfunction DTC is stored. ABS is not disabled, however, the amber ABS warning will flash to indicate a malfunction exists.

DTC Chart Test Description:

Number(s) below refer to circled number(s) on the diagnostic chart.

1. Checks if the malfunction is currently present.
2. Isolates the cause of the malfunction to either a malfunctioning or misadjusted brake switch, or a short to voltage in the brake switch circuitry.
3. Checks for unwanted voltage on the brake switch input circuit.
4. Ensures malfunction is not due to poor terminal contact or EBCM.

Diagnostic Aids:

An "Intermittent" malfunction may be caused by a poor connection, rubbed through wire insulation, or a wire that is broken inside the insulation.

The frequency of the malfunction can be checked by using the enhanced diagnostic function of the Tech 1

Any circuitry that is suspected of causing the intermittent complaint should be thoroughly checked for backed out terminals, improper mating, broken locks, improperly formed or damaged terminals, poor terminal to wiring connections or physical damage to the wiring harness.

DIAGNOSTIC CHART DTC 94 — 1995 LUMINA APV

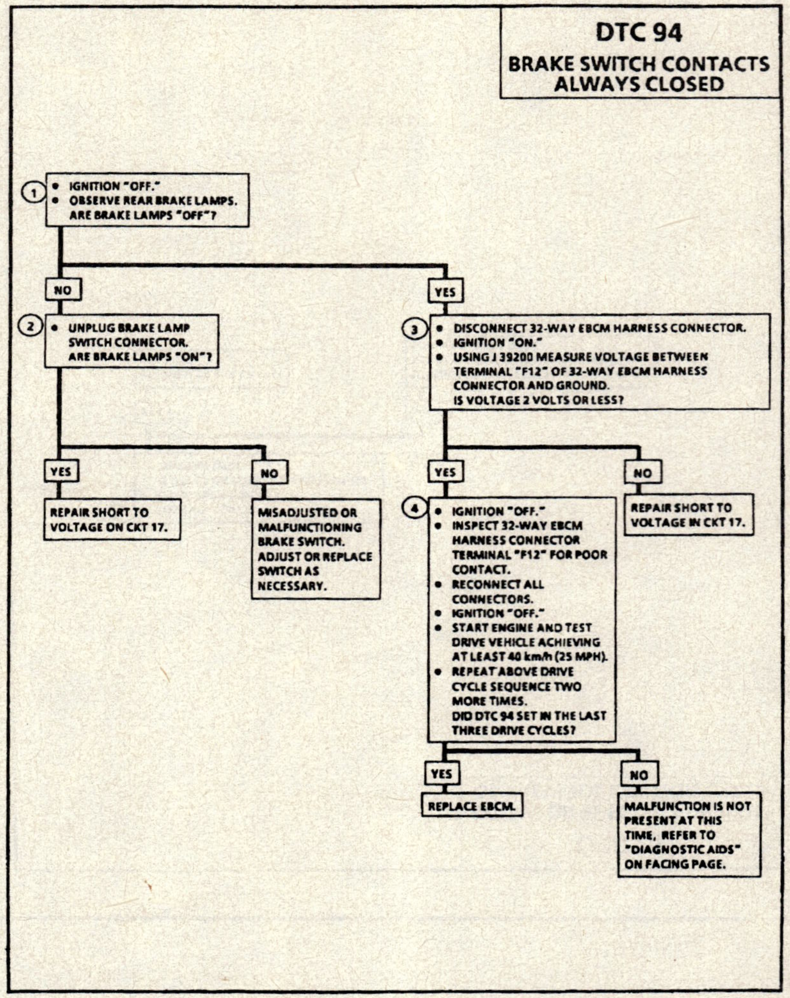

DIAGNOSTIC CHART DTC 95 — 1995 LUMINA APV

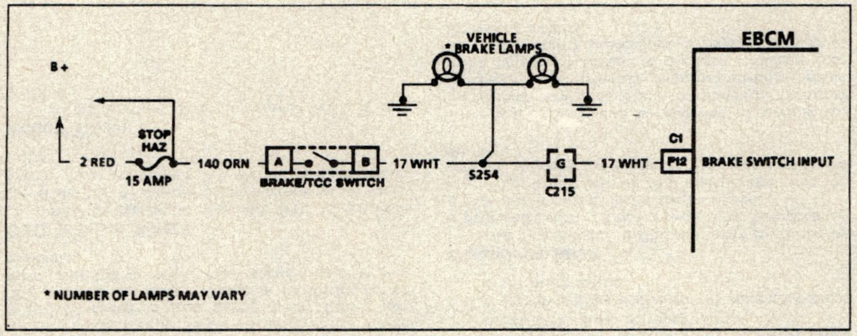

DTC 95
BRAKE SWITCH CIRCUIT OPEN

Circuit Description:

This DTC is used to identify open brake switch circuitry that prevents the brake switch input to the EBCM from changing states when the brake is applied. This DTC is used in conjunction with DTCs 91 and 92 to determine the cause of an open brake switch malfunction.

DTC Will Set When:

DTC 95 can be set after initialization is completed. If the brake switch input voltage is out of specification for one second indicating an open circuit, a malfunction exists.

Action Taken:

A malfunction DTC is stored, ABS is disabled and the amber ABS warning lamp is turned "ON."

DTC Chart Test Description:

Number(s) below refer to circled number(s) on the diagnostic chart.

1. This step is used to confirm that an open in the brake switch circuitry currently exists.
2. Indicates if the brake switch signal is being received by the EBCM.
3. Indicates if an open circuit exists in the brake switch or brake lamp circuitry.
4. Isolates the open circuit to either the brake switch input circuitry, or the EBCM.

Diagnostic Aids:

An "Intermittent" malfunction may be caused by a poor connection, rubbed through wire insulation, or a wire that is broken inside the insulation.

The frequency of the malfunction can be checked by using the enhanced diagnostic function of the Tech 1

Any circuitry that is suspected of causing the intermittent complaint should be thoroughly checked for backed out terminals, improper mating, broken locks, improperly formed or damaged terminals, poor terminal to wiring connections or physical damage to the wiring harness.

DIAGNOSTIC CHART DTC 95 — 1995 LUMINA APV

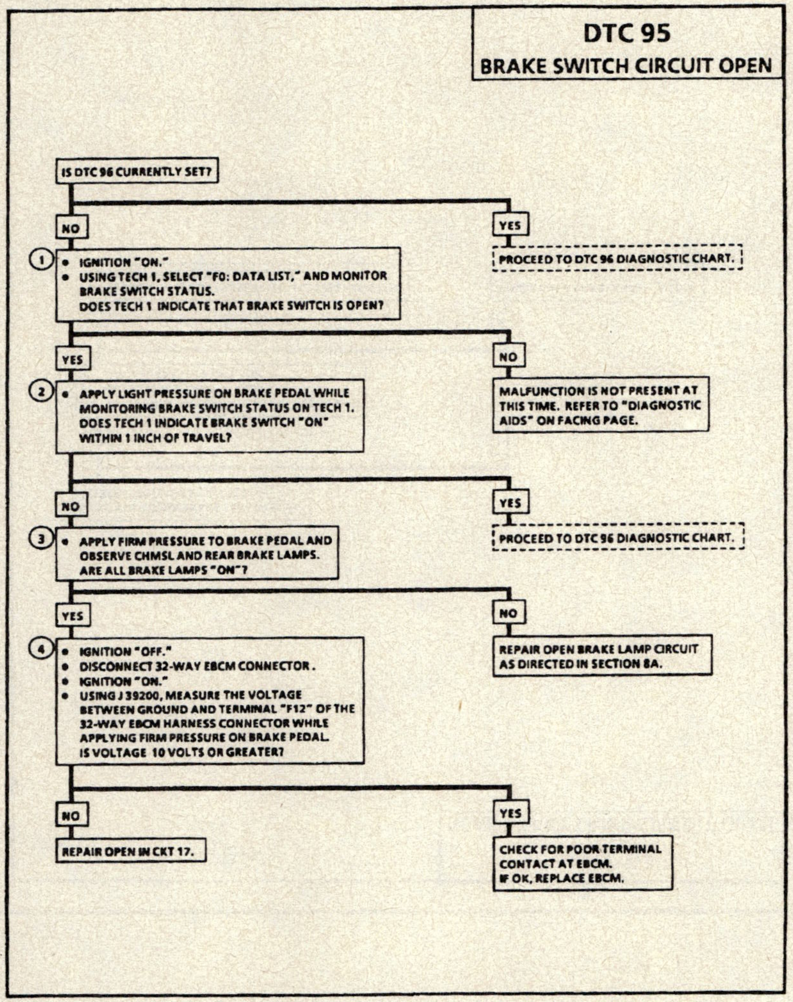

DIAGNOSTIC CHART DTC 96 — 1995 LUMINA APV

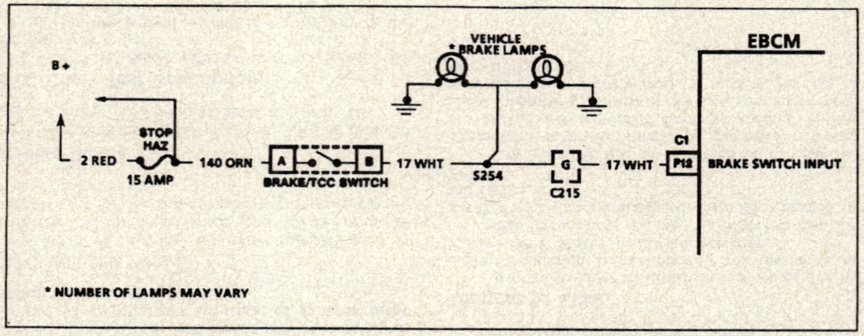

DTC 96
REAR BRAKE LAMP CIRCUIT OPEN

Circuit Description:

This DTC is designed to isolate the cause of a DTC 95 malfunction and indicate to the driver ABS is still available. If DTC 95 occurs with DTC 96, the brake lamps are open, the brake lamp grounds are open, or the Center High Mounted Stop Lamp (CHMSL) circuit is open during 4-way flasher use. The presence of battery voltage at the brake switch input indicates a valid brake switch input is still available.

DTC Will Set When:

DTC 96 can be set only after DTC 95 has been set. If the EBCM detects battery voltage on the brake switch input circuit for 0.5 second, a malfunction exists.

Action Taken:

A malfunction DTC is stored. ABS is not disabled.

DTC Chart Test Description:

Number(s) below refer to circled number(s) on the diagnostic chart.

1. As a result of a malfunction of an additional brake switch circuit DTC, this DTC may be set. To ensure proper diagnosis, any additional brake switch DTCs must be repaired first.
2. Identifies if the malfunction is currently present in the brake circuit.

Diagnostic Aids:

An "Intermittent" malfunction may be caused by a poor connection, rubbed through wire insulation, or a wire that is broken inside the insulation.

The frequency of the malfunction can be checked by using the enhanced diagnostic function of the Tech 1

Any circuitry that is suspected of causing the intermittent complaint should be thoroughly checked for backed out terminals, improper mating, broken locks, improperly formed or damaged terminals, poor terminal to wiring connections or physical damage to the wiring harness.

DIAGNOSTIC CHART DTC 96 — 1995 LUMINA APV

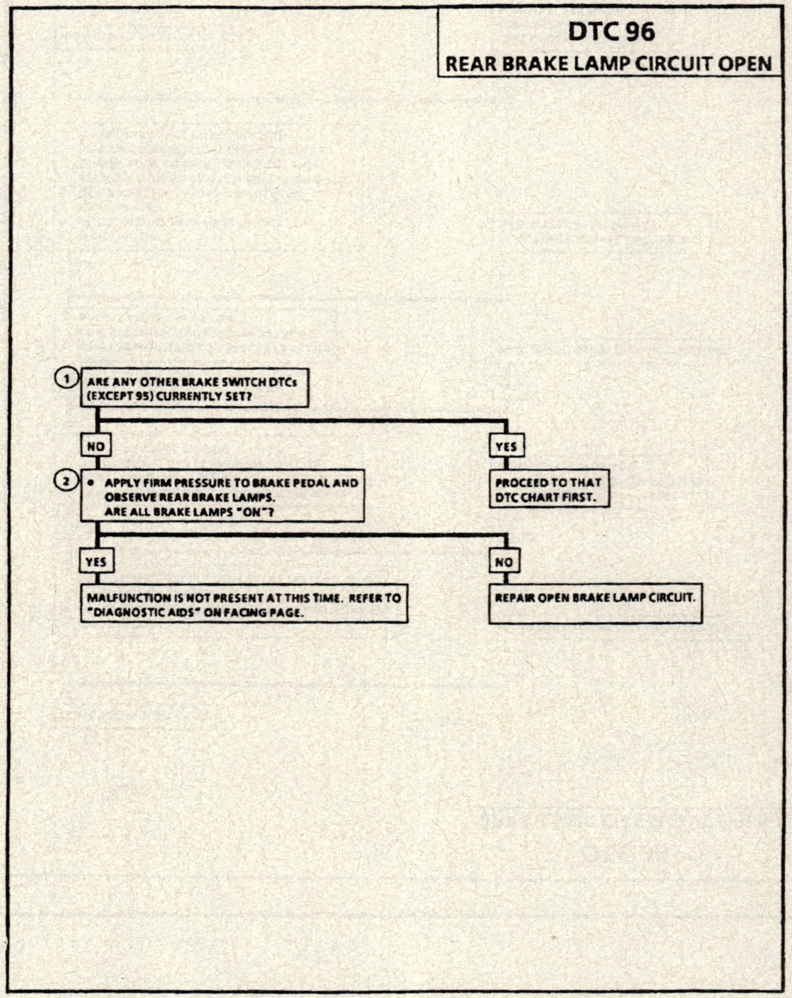

DIAGNOSTIC PROCESS CHART — 1994 CENTURY, CUTLASS CIERA AND CUTLASS CRUISER

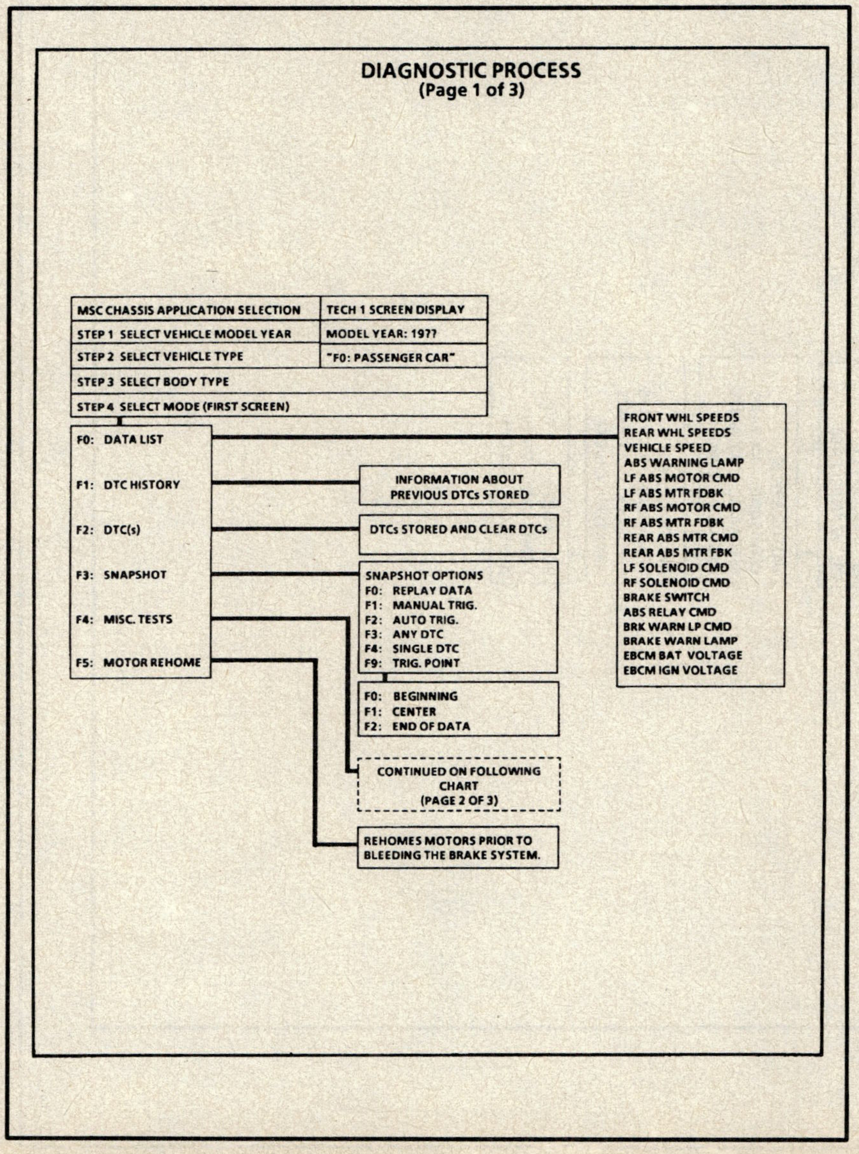

DIAGNOSTIC PROCESS CHART — 1994 CENTURY, CUTLASS CIERA AND CUTLASS CRUISER

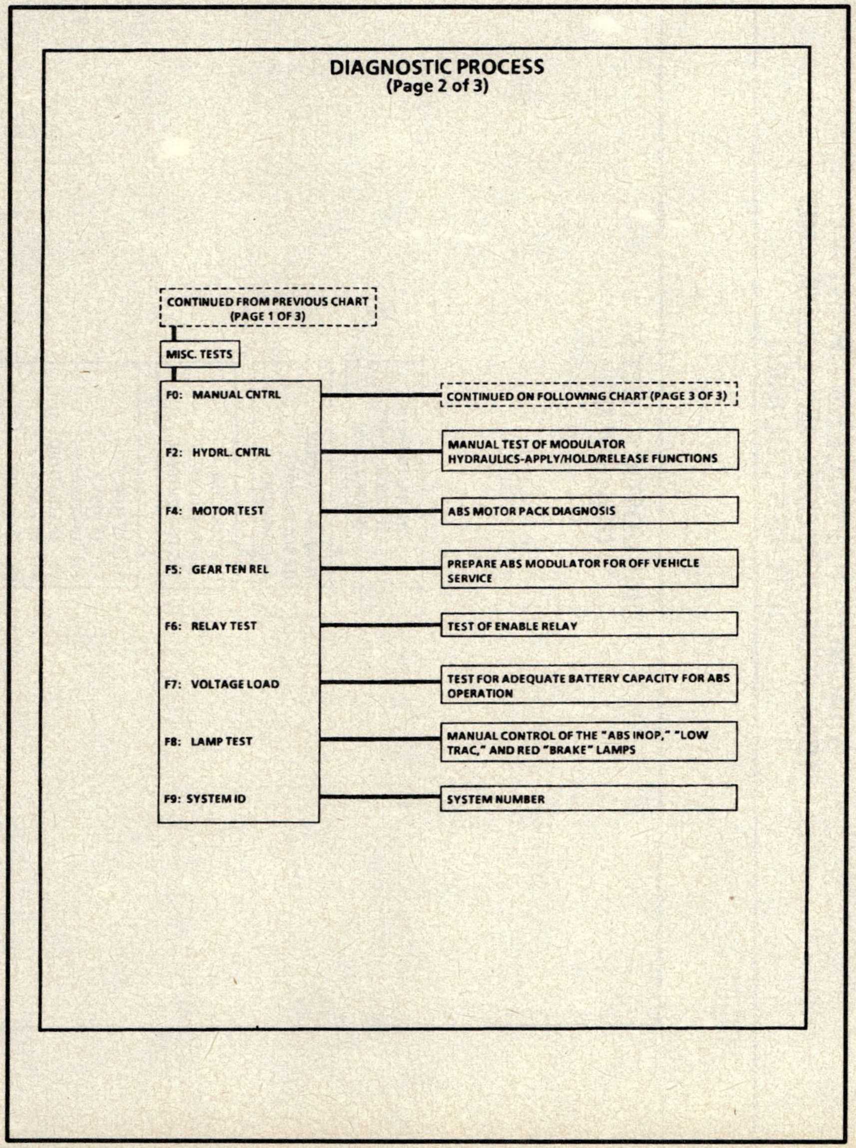

DIAGNOSTIC PROCESS CHART — 1994 CENTURY, CUTLASS CIERA AND CUTLASS CRUISER

DIAGNOSTIC PROCESS
(Page 3 of 3)

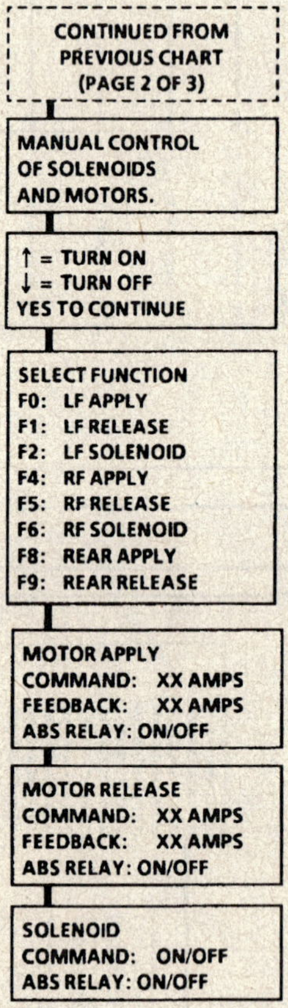

DIAGNOSTIC CHART SYSTEM CHECK — 1994 CENTURY, CUTLASS CIERA AND CUTLASS CRUISER

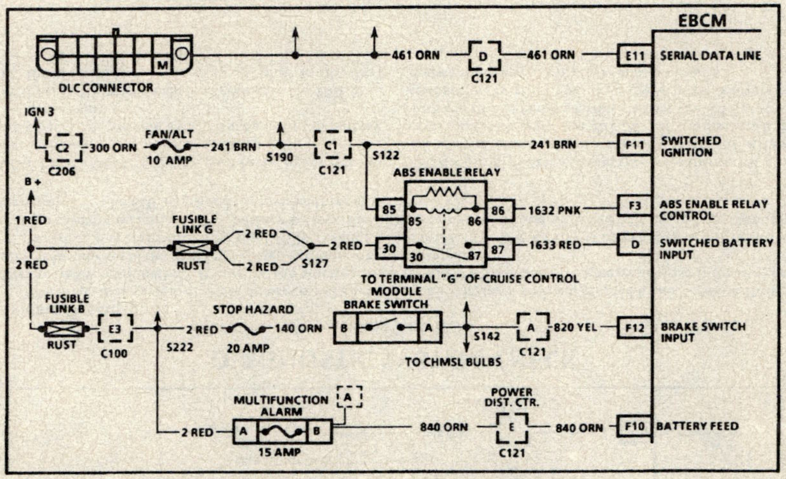

DIAGNOSTIC SYSTEM CHECK

(Page 1 of 4)

System Description:

The diagnostic system check is an organized approach to identifying a problem created by an Antilock Brake System (ABS) malfunction. It must be the starting point for any ABS complaint diagnosis because it directs the service technician to the next logical step in diagnosing the complaint.

Serial data is transmitted by the EBCM through terminal "E11". The EBCM is supplied switched ignition voltage through terminal "F11", battery voltage through terminal "F10" and switched battery voltage through terminal "D". The EBCM ground is provided through terminal "E".

Chart Test Description: Number(s) below refer to circled number(s) on the diagnostic chart.

1. Indicates if EBCM is transmitting data.
2. Checks for a serial data line malfunction.
3. Checks for proper Tech 1 connection to DLC.
4. Checks for open multifunction alarm 15 amp.
5. Checks for open fan/alt 10 amp fuse.
6. Checks for high resistance in the EBCM ground circuit.
7. Checks for current DTCs.
8. Checks for proper ABS warning lamp operation.
9. Checks for history DTCs.

Diagnostic Aids: Excessive resistance in the ground or power supply circuits will not allow communication with the EBCM. If communication with the EBCM is not possible, ensure the ABS ground connection is good and that there is no excessive resistance in any of the power supply circuits.

DIAGNOSTIC CHART SYSTEM CHECK — 1994 CENTURY, CUTLASS CIERA AND CUTLASS CRUISER

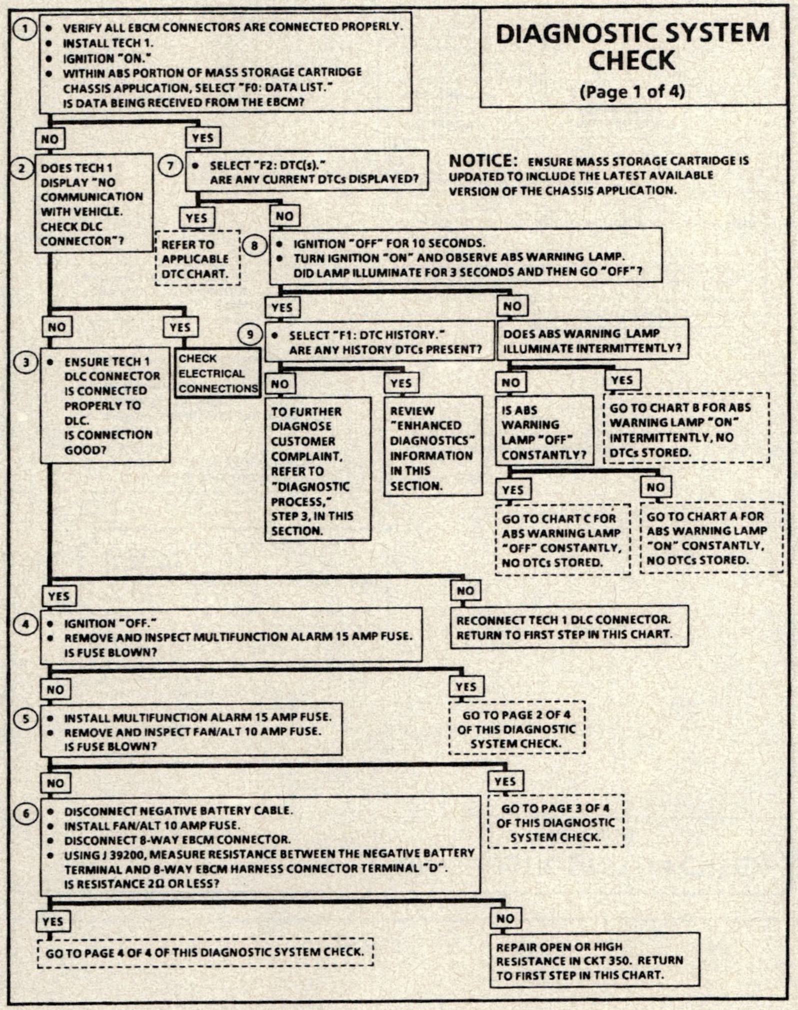

DIAGNOSTIC CHART SYSTEM CHECK — 1994 CENTURY, CUTLASS CIERA AND CUTLASS CRUISER

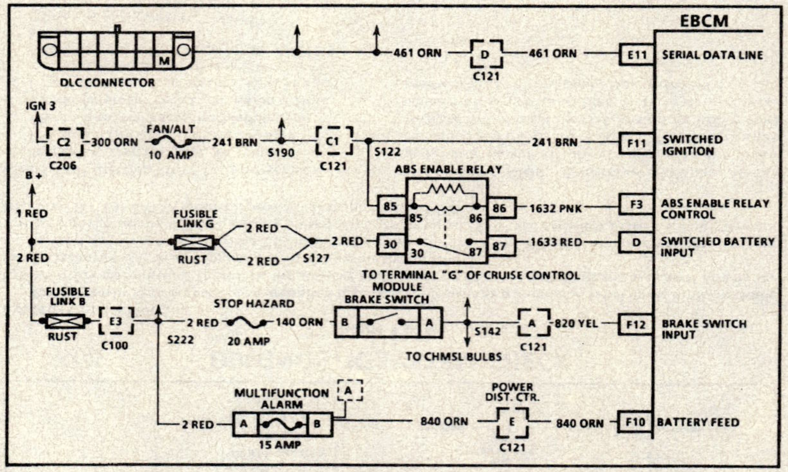

DIAGNOSTIC SYSTEM CHECK

(Page 2 of 4)

System Description:

The diagnostic system check is an organized approach to identifying a problem created by an Antilock Brake System (ABS) malfunction. It must be the starting point for any ABS complaint diagnosis because it directs the service technician to the next logical step in diagnosing the complaint.

Serial data is transmitted by the EBCM through terminal "E11". The EBCM is supplied switched ignition voltage through terminal "F11", battery voltage through terminal "F10" and switched battery voltage through terminal "D". The EBCM ground is provided through terminal "E".

Chart Test Description: Number(s) below refer to circled number(s) on the diagnostic chart.

10. Checks for possible short to ground in the battery feed circuitry.
11. Checks for possible short to ground in the EBCM.
12. Ensures short to ground is not due to physical damage of the circuitry.

Diagnostic Aids: Excessive resistance in the ground or power supply circuits will not allow communication with the EBCM. If communication with the EBCM is not possible, ensure the ABS ground connection is good and that there is no excessive resistance in any of the power supply circuits.

DIAGNOSTIC CHART SYSTEM CHECK — 1994 CENTURY, CUTLASS CIERA AND CUTLASS CRUISER

DIAGNOSTIC SYSTEM CHECK

(Page 2 of 4)

CONTINUED FROM DIAGNOSTIC SYSTEM CHECK (PAGE 1 OF 4)

YES

(10)
- REPLACE MULTIFUNCTION ALARM 15 AMP FUSE.
- WAIT 10 SECONDS.
- REMOVE AND INSPECT MULTIFUNCTION ALARM 15 AMP FUSE.

IS FUSE BLOWN?

NO:
- INSTALL FUSE.
- DIAGNOSE FOR A POSSIBLE SHORT TO GROUND IN BATTERY FEED CIRCUITRY. RETURN TO FIRST STEP IN THIS CHART.

YES:

(11)
- DISCONNECT 32-WAY EBCM CONNECTOR.
- REPLACE MULTIFUNCTION ALARM 15 AMP FUSE.
- WAIT 10 SECONDS.
- REMOVE AND INSPECT MULTIFUNCTION ALARM 15 AMP FUSE.

IS FUSE BLOWN?

YES: DIAGNOSE FOR A SHORT TO GROUND IN THE BATTERY FEED CIRCUITRY.

NO:

(12)
- INSPECT CKT 840 AND 32-WAY EBCM HARNESS CONNECTOR FOR PHYSICAL DAMAGE WHICH MAY RESULT IN A SHORT TO GROUND WITH THE 32-WAY EBCM HARNESS CONNECTOR CONNECTED TO THE EBCM. CORRECT DAMAGE IF EVIDENT.
- RECONNECT ALL CONNECTORS
- WAIT 10 SECONDS.
- REMOVE AND INSPECT MULTIFUNCTION ALARM 15 AMP FUSE.

IS FUSE BLOWN?

YES:
- REPLACE EBCM.
- REPLACE MULTIFUNCTION ALARM 15 AMP FUSE. RETURN TO FIRST STEP IN THIS CHART.

NO: MALFUNCTION IS INTERMITTENT.

DIAGNOSTIC CHART SYSTEM CHECK — 1994 CENTURY, CUTLASS CIERA AND CUTLASS CRUISER

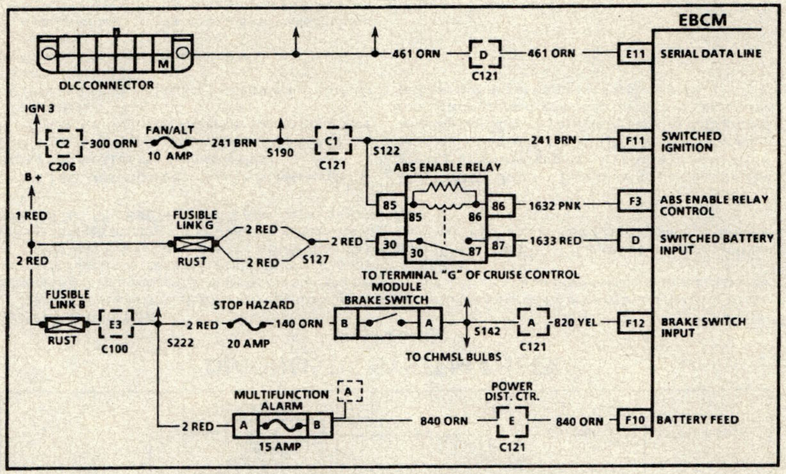

DIAGNOSTIC SYSTEM CHECK

(Page 3 of 4)

System Description:

The diagnostic system check is an organized approach to identifying a problem created by an Antilock Brake System (ABS) malfunction. It must be the starting point for any ABS complaint diagnosis because it directs the service technician to the next logical step in diagnosing the complaint.

Serial data is transmitted by the EBCM through terminal "E11". The EBCM is supplied switched ignition voltage through terminal "F11", battery voltage through terminal "F10" and switched battery voltage through terminal "D". The EBCM ground is provided through terminal "E".

Chart Test Description: Number(s) below refer to circled number(s) on the diagnostic chart.

13. Checks for possible short to ground in the switched ignition circuitry.
14. Isolates the short to ground to either the circuitry or the EBCM.
15. Ensures short to ground is not due to physical damage of the circuitry.

Diagnostic Aids: Excessive resistance in the ground or power supply circuits will not allow communication with the EBCM. If communication with the EBCM is not possible, ensure the ABS ground connection is good and that there is no excessive resistance in any of the power supply circuits.

DIAGNOSTIC CHART SYSTEM CHECK — 1994 CENTURY, CUTLASS CIERA AND CUTLASS CRUISER

DIAGNOSTIC SYSTEM CHECK

(Page 3 of 4)

CONTINUED FROM DIAGNOSTIC SYSTEM CHECK (PAGE 1 OF 4)

YES

(13)
- REPLACE FAN/ALT 10 AMP FUSE.
- IGNITION "ON."
- WAIT 10 SECONDS THEN TURN IGNITION "OFF."
- REMOVE AND INSPECT FAN/ALT 10 AMP FUSE.

IS FUSE BLOWN?

YES →

(14)
- DISCONNECT 32-WAY EBCM CONNECTOR.
- REPLACE FAN/ALT 10 AMP FUSE.
- IGNITION "ON."
- WAIT 10 SECONDS THEN TURN IGNITION "OFF."
- REMOVE AND INSPECT FAN/ALT 10 AMP FUSE.

IS FUSE BLOWN?

NO → INSTALL FAN/ALT 10 AMP FUSE. REFER TO SECTION 8A TO DIAGNOSE FOR A POSSIBLE SHORT TO GROUND IN THE SWITCHED IGNITION CIRCUITRY. RETURN TO FIRST STEP IN THIS CHART.

(From 14) YES → REPAIR A SHORT TO GROUND IN ONE OF THE CKT 241 BRANCHES EXITING THE FAN/ALT 10 AMP FUSE. REPLACE FAN/ALT 10 AMP FUSE. RETURN TO THE FIRST STEP IN THIS CHART.

(From 14) NO →

(15)
- INSPECT CKT 241 AND 32-WAY EBCM HARNESS CONNECTOR FOR PHYSICAL DAMAGE WHICH MAY RESULT IN A SHORT TO GROUND WITH THE 32-WAY EBCM HARNESS CONNECTOR CONNECTED TO THE EBCM. CORRECT DAMAGE IF EVIDENT.
- RECONNECT ALL CONNECTORS.
- IGNITION "ON."
- WAIT 10 SECONDS THEN TURN IGNITION "OFF."
- REMOVE AND INSPECT FAN/ALT 10 AMP FUSE.

IS FUSE BLOWN?

YES → REPLACE EBCM. REPLACE FAN/ALT 10 AMP FUSE. RETURN TO FIRST STEP IN THIS CHART.

NO → MALFUNCTION IS INTERMITTENT.

DIAGNOSTIC CHART SYSTEM CHECK — 1994 CENTURY, CUTLASS CIERA AND CUTLASS CRUISER

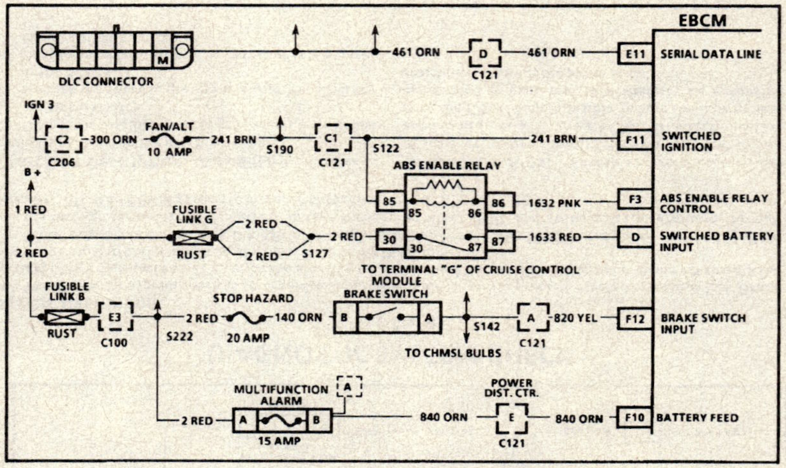

DIAGNOSTIC SYSTEM CHECK
(Page 4 of 4)

System Description:
The diagnostic system check is an organized approach to identifying a problem created by an Antilock Brake System (ABS) malfunction. It must be the starting point for any ABS complaint diagnosis because it directs the service technician to the next logical step in diagnosing the complaint.
Serial data is transmitted by the EBCM through terminal "E11". The EBCM is supplied switched ignition voltage through terminal "F11", battery voltage through terminal "F10" and switched battery voltage through terminal "D". The EBCM ground is provided through terminal "E".

Chart Test Description: Number(s) below refer to circled number(s) on the diagnostic chart.
16. Checks for proper battery voltage at the EBCM.
17. Checks for high resistance in the battery feed circuitry.
18. Checks for high resistance in the switched ignition circuitry.
19. Ensures malfunction is not due to poor terminal contact.
20. Checks for high resistance in the switched ignition circuit between the fuse block and the EBCM.

Diagnostic Aids: Excessive resistance in the ground or power supply circuits will not allow communication with the EBCM. If communication with the EBCM is not possible, ensure the ABS ground connection is good and that there is no excessive resistance in any of the power supply circuits.

DIAGNOSTIC CHART SYSTEM CHECK — 1994 CENTURY, CUTLASS CIERA AND CUTLASS CRUISER

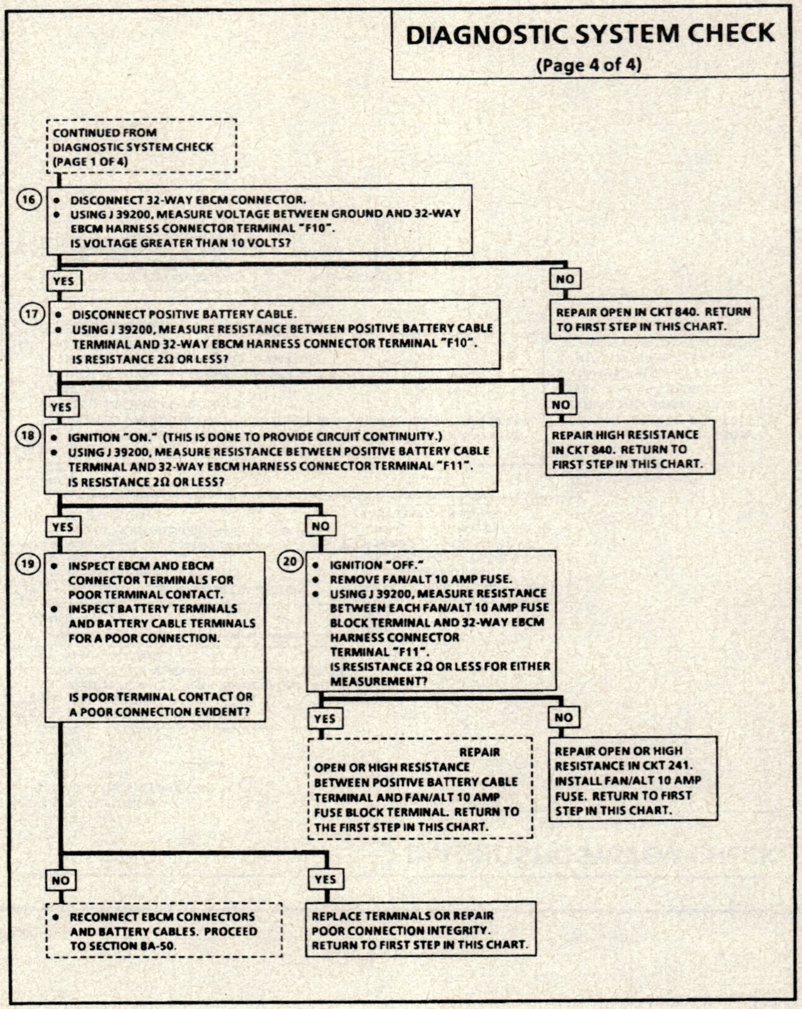

DIAGNOSTIC TROUBLE CODE CHART — 1994 CENTURY, CUTLASS CIERA AND CUTLASS CRUISER

DIAGNOSTIC TROUBLE CODE AND SYMPTOM TABLE

CHART	SYMPTOM
A	ABS (Amber) Warning Lamp "ON" Constantly, No DTCs Stored
B	ABS (Amber) Warning Lamp "ON" Intermittently, No DTCs Stored
C	ABS (Amber) Warning Lamp "OFF" Constantly, No DTCs Stored
D	"Low Traction" Warning Lamp "ON" with Ignition "OFF," No DTCs Stored
E	Tech 1 Displays Undefined DTCs
DIAGNOSTIC TROUBLE CODE	**DESCRIPTION**
11	ABS Warning Lamp Circuit Malfunction (1 of 2)
14	ABS Enable Relay Contact Circuit Open (1 of 3)
15	ABS Enable Rly. Contact Ckt Short to Batt or Always Closed
16	ABS Enable Relay Coil Circuit Open
17	ABS Enable Relay Coil Circuit Shorted to Gnd
18	ABS Enable Relay Coil Circuit Shorted to Batt
21	Left Front Wheel Speed = 0
22	Right Front Wheel Speed = 0
23	Left Rear Wheel Speed = 0
24	Right Rear Wheel Speed = 0
25	Left Front Excessive Wheel Speed Variation
26	Right Front Excessive Wheel Speed Variation
27	Left Rear Excessive Wheel Speed Variation
28	Right Rear Excessive Wheel Speed Variation
32	LF Wheel Speed Snr Ckt Open or Short to Gnd/Batt (1 of 3)
33	RF Wheel Speed Snr Ckt Open or Short to Gnd/Batt (1 of 3)
34	LR Wheel Speed Snr Ckt Open or Short to Gnd/Batt (1 of 3)
35	RR Wheel Speed Snr Ckt Open or Short to Gnd/Batt (1 of 3)
36	Low System Voltage (1 of 2)
37	High System Voltage
38	Left Front ESB Will Not Hold Motor
41	Right Front ESB Will Not Hold Motor
42	Rear ESB Will Not Hold Motor
44	Left Front Channel Will Not Move
45	Right Front Channel Will Not Move
46	Rear Channel Will Not Move
47	Left Front ABS Motor Free Spins (1 of 2)

DIAGNOSTIC TROUBLE CODE CHART — 1994 CENTURY, CUTLASS CIERA AND CUTLASS CRUISER

DIAGNOSTIC TROUBLE CODE AND SYMPTOM TABLE

DIAGNOSTIC TROUBLE CODE	DESCRIPTION
48	Right Front ABS Motor Free Spins (1 of 2)
51	Rear ABS Motor Free Spins (1 of 2)
52	Left Front Channel in Release Too Long
53	Right Front Channel in Release Too Long
54	Rear Channel in Release Too Long
55	EBCM Malfunction
56	Left Front ABS Motor Circuit Open
57	Left Front ABS Motor Circuit Shorted to Gnd
58	Left Front ABS Motor Circuit Shorted to Batt
61	Right Front ABS Motor Circuit Open
62	Right Front ABS Motor Circuit Shorted to Gnd
63	Right Front ABS Motor Circuit Shorted to Batt
64	Rear ABS Motor Circuit Open
65	Rear ABS Motor Circuit Shorted to Ground
66	Rear ABS Motor Circuit Shorted to Battery
76	Left Front Sol. Circuit Open or Shorted to Batt
77	Left Front Solenoid Circuit Shorted to Gnd
78	Right Front Sol. Circuit Open or Shorted to Batt
81	Right Front Solenoid Circuit Shorted to Gnd
82	Calibration Malfunction
86	EBCM Turned "ON" the Red "Brake" Warning Lamp
87	Red "Brake" Warning Lamp Circuit Open
88	Red "Brake" Warning Lamp Ckt Shorted to Batt
91	Open Brake Switch During Deceleration
92	Open Brake Switch When ABS Was Required
93	DTC 91 or 92 Set in Current or Prev. Ign Cycle
94	Brake Switch Contacts Always Closed
95	Brake Switch Circuit Open
96	Rear Brake Lamp Circuit Open

ELECTRICAL CONNECTOR TERMINAL IDENTIFICATION — 1994 CENTURY, CUTLASS CIERA AND CUTLASS CRUISER

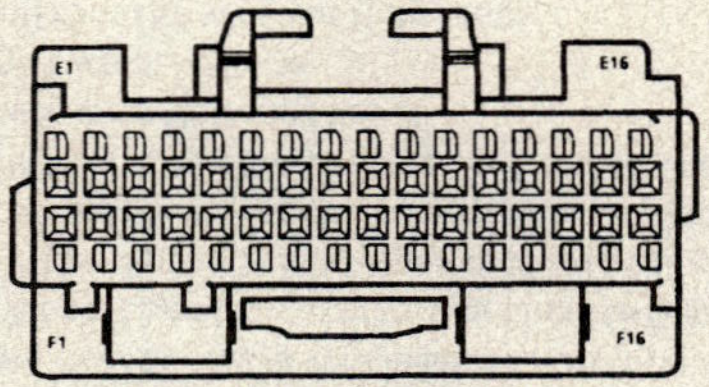

EBCM 32-WAY CONNECTOR

CONNECTOR TERMINAL	CIRCUIT NO.	COLOR	CIRCUIT DESCRIPTION
E1	884	BLK	LR WHEEL SIGNAL HIGH
E2	885	RED	LR WHEEL SIGNAL LOW
E3	833	TAN	RF WHEEL SIGNAL LOW
E4	872	DK GRN	RF WHEEL SIGNAL HIGH
E5	883	WHT	RR WHEEL SIGNAL LOW
E6	882	BRN	RR WHEEL SIGNAL HIGH
E7	873	YEL	LF WHEEL SIGNAL LOW
E8	830	LT BLU	LF WHEEL SIGNAL HIGH
E9	1289	LT BLU	RF ABS SOLENOID CONTROL
E10	PLUGGED		NOT USED
E11	461	ORN	SERIAL DATA LINE
E12	PLUGGED		NOT USED
E13	PLUGGED		NOT USED
E14	PLUGGED		NOT USED
E15	PLUGGED		NOT USED
E16	PLUGGED		NOT USED
F1	1288	DK GRN	LF ABS SOLENOID CONTROL
F2	852	WHT	ABS WARNING LAMP CONTROL
F3	1632	PNK	ABS ENABLE RELAY CONTROL
F4	33	TAN/WHT	BRAKE WARNING LAMP CONTROL
F5	1537	DK BLU	ABS ACTIVE LAMP CONTROL
F6	PLUGGED		NOT USED
F7	PLUGGED		NOT USED
F8	PLUGGED		NOT USED
F9	PLUGGED		NOT USED
F10	840	ORN	BATTERY FEED
F11	241	BRN	SWITCHED IGNITION
F12	820	YEL	BRAKE SWITCH INPUT
F13	PLUGGED		NOT USED
F14	PLUGGED		NOT USED
F15	PLUGGED		NOT USED
F16	PLUGGED		NOT USED

ELECTRICAL CONNECTOR TERMINAL IDENTIFICATION — 1994 CENTURY, CUTLASS CIERA AND CUTLASS CRUISER

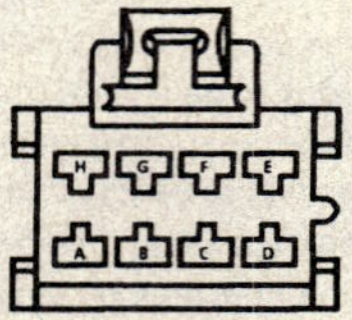

EBCM 8-WAY CONNECTOR

PIN	CIRCUIT NO.	COLOR	CIRCUIT
A	1280	BLK	LF ABS MOTOR HIGH
B	1281	PNK	LF ABS MOTOR LOW
C	1284	DK GRN	REAR ABS MOTOR HIGH
D	1633	RED	SWITCHED BATTERY INPUT
E	350	BLK	GROUND
F	1285	ORN	REAR ABS MOTOR LOW
G	1282	PPL	RF ABS MOTOR HIGH
H	1283	BLK	RF ABS MOTOR LOW

DIAGNOSTIC CHART A — 1994 CENTURY, CUTLASS CIERA AND CUTLASS CRUISER

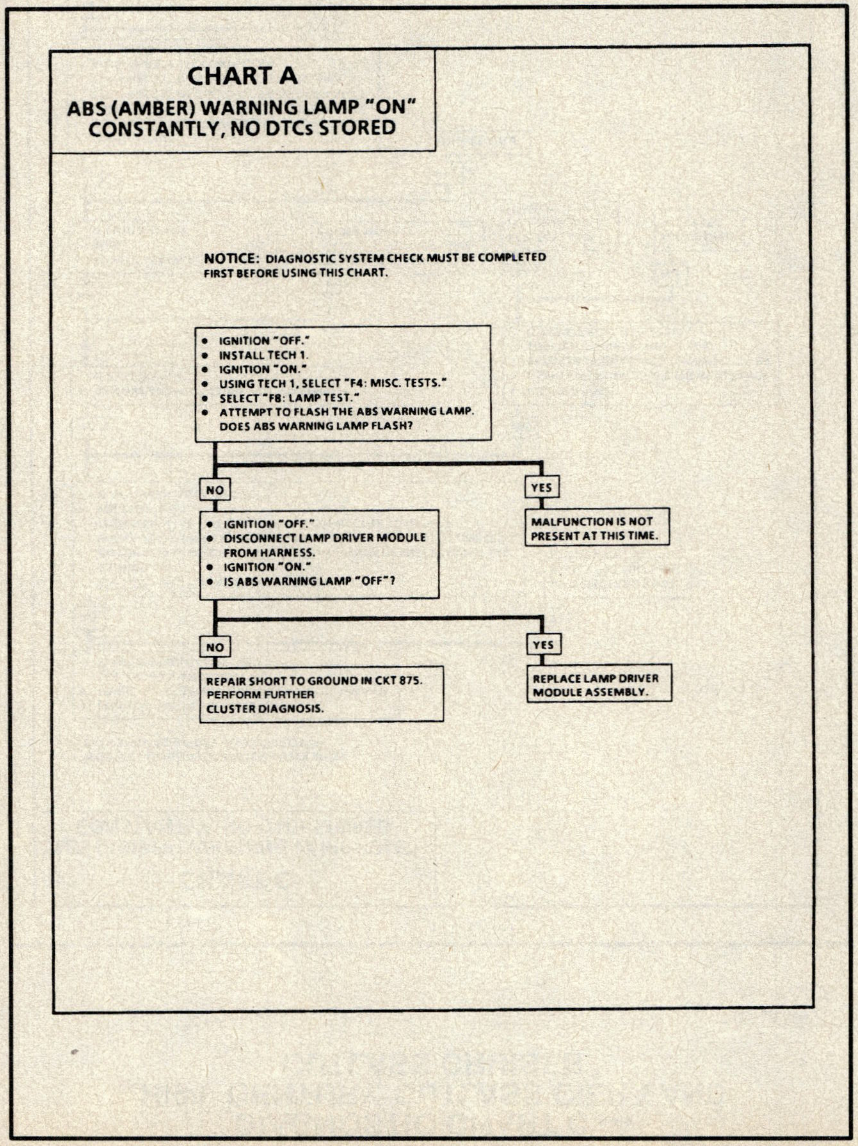

DIAGNOSTIC CHART B — 1994 CENTURY, CUTLASS CIERA AND CUTLASS CRUISER

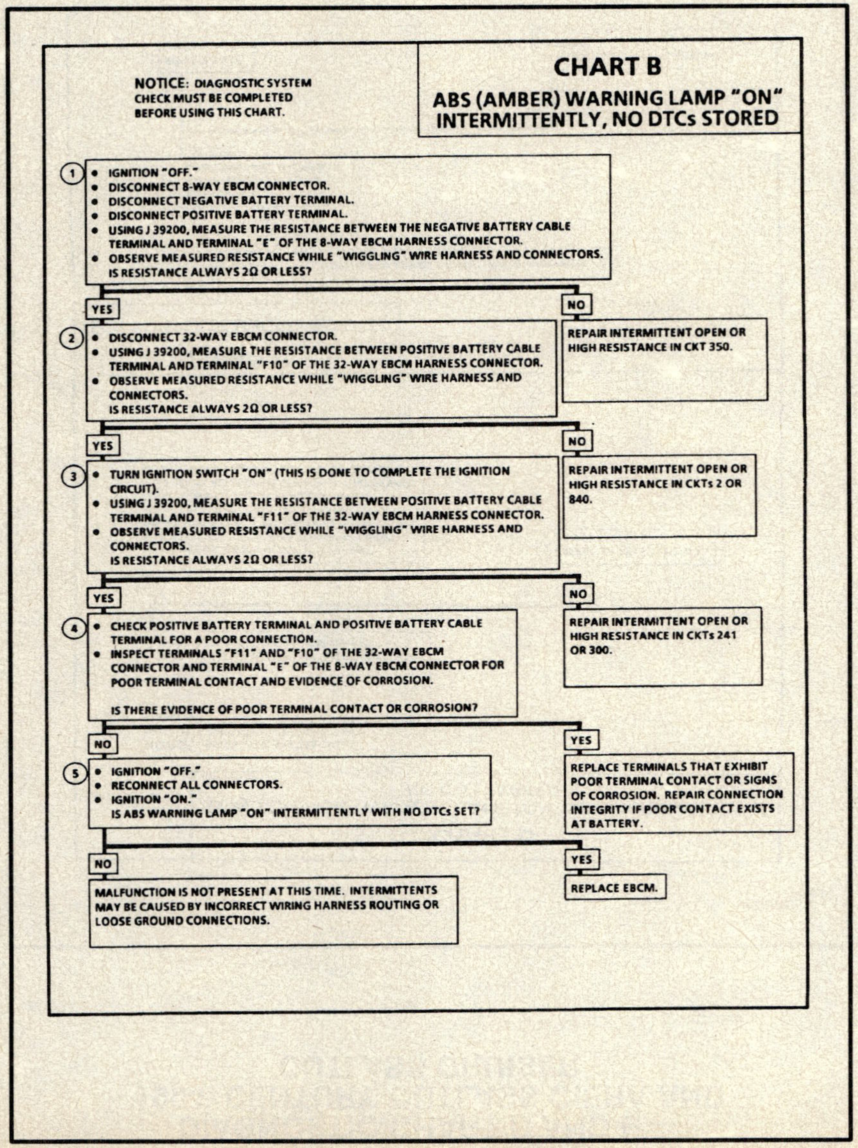

DIAGNOSTIC CHART C — 1994 CENTURY, CUTLASS CIERA AND CUTLASS CRUISER

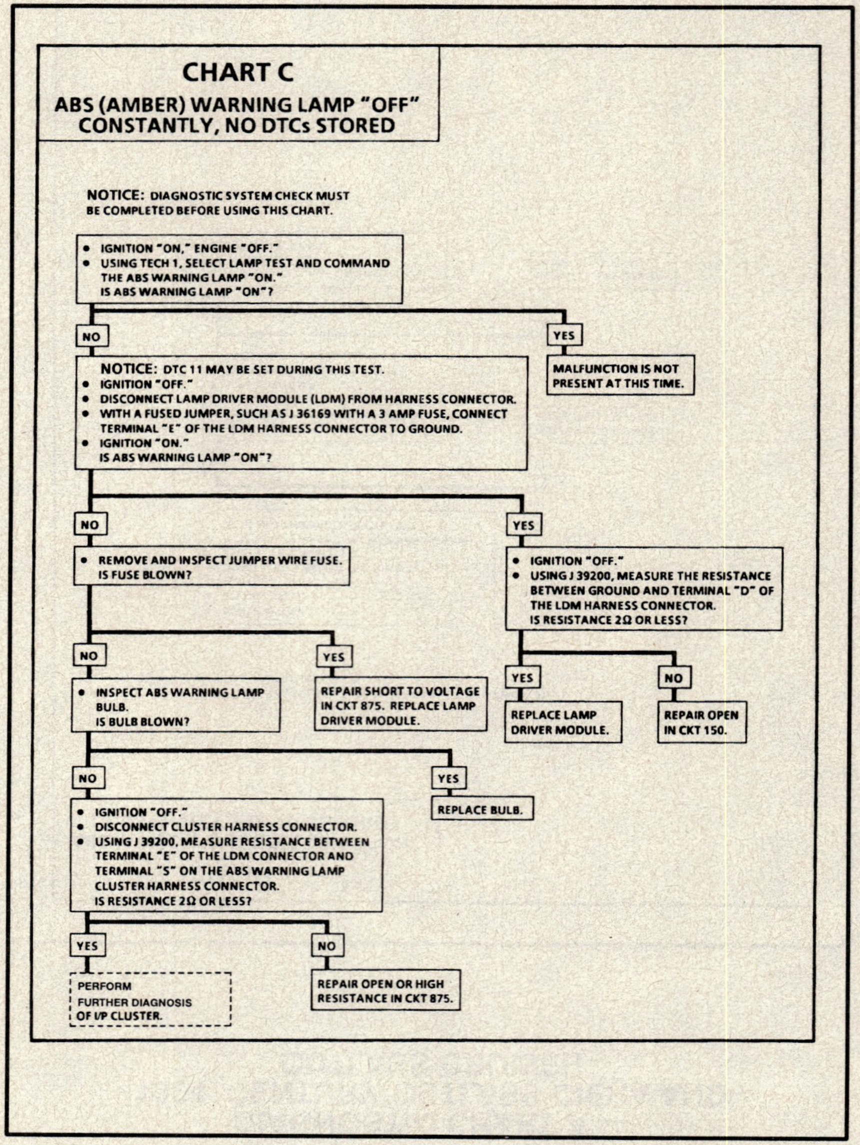

DIAGNOSTIC CHART D AND E — 1994 CENTURY, CUTLASS CIERA AND CUTLASS CRUISER

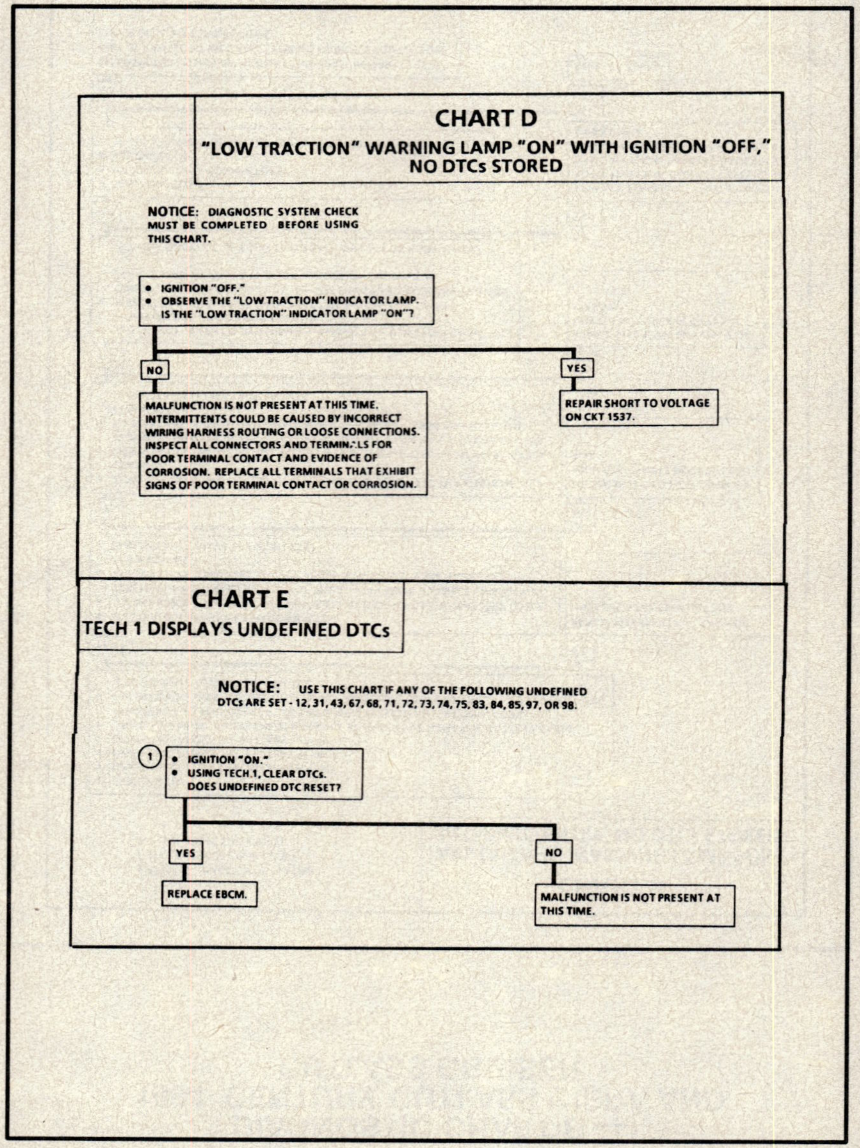

DIAGNOSTIC CHART DTC 11 — 1994 CENTURY, CUTLASS CIERA AND CUTLASS CRUISER

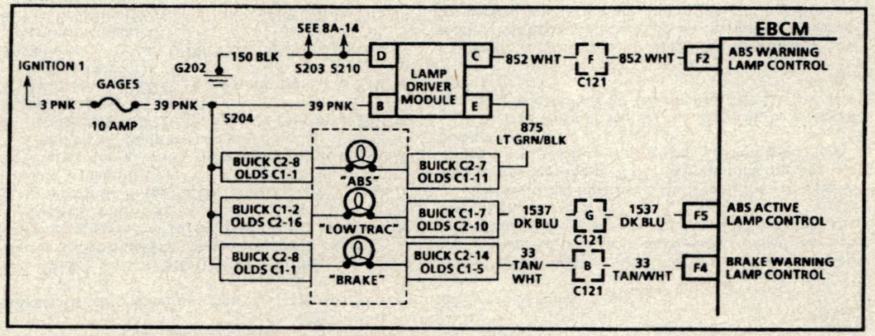

DTC 11 (Page 1 of 2)

ABS WARNING LAMP CIRCUIT MALFUNCTION

Circuit Description:

This DTC checks the state of the ABS warning lamp to identify a situation in which the driver could not be warned of a system malfunction by the ABS warning lamp, or the warning lamp is always "ON." The Lamp Driver Module (LDM) turns the ABS warning lamp "ON," unless the EBCM provides ground to turn it "OFF." Because of the circuitry in the LDM, only external malfunctions can be detected. As a result, the ABS warning lamp itself is not diagnosable, only the control line to the EBCM can be diagnosed. In the event of an open CKT 852, the ABS warning lamp will be "ON" at all times, due to the loss of ground at the LDM input. If the control line is shorted to ground, the ABS warning lamp is kept "OFF," due to the LDM input being grounded.

DTC Will Set When: DTC 11 can be set only during the three second bulb check or when the ABS warning lamp is commanded "ON." If the EBCM cannot control the ABS warning lamp for two seconds, a malfunction exists.

Action Taken: A malfunction DTC is stored; however, ABS is not disabled.

DTC Chart Test Description: Number(s) below refer to circled number(s) on the diagnostic chart.

1. Checks to see if ABS lamp functions properly. Normal operation would indicate that the problem is not present.
2. Isolates the cause of the lamp remaining "ON" to a short to voltage, a short in the EBCM or an open in CKT 39 which supplies voltage to the lamp driver module.
3. Determines if any other DTCs are present or if DTC 11 is set falsely.
4. Ensures that the circuitry from the EBCM to the lamp driver module is complete.
5. Determines whether the lamp remains "ON" due to a lamp driver malfunction or an open in CKT 852.
6. Checks to see if there is a short to voltage in CKT 852.
7. Checks for a malfunctioning EBCM.
8. Ensures that the ignition circuit to the lamp driver module is functional. An open indicator fuse or open CKT 3, 39, or 875 would result in inoperative warning lamp.
9. Isolates the cause of the ABS lamp remaining "OFF" by disconnecting the 32-way EBCM connector, CKT 852 should be open and cause the ABS lamp to be "ON."
10. Isolates the cause of the lamp remaining "OFF" to a short to ground in CKT 852 or a malfunctioning lamp driver module.
11. Checks for an open in CKT 39.
12. Determines if the "GAGES" fuse or CKT 39 is open, causing the I/P warning lamps to remain "OFF."
13. Checks for malfunctioning EBCM.

Diagnostic Aids: An "Intermittent" malfunction may be caused by a poor connection, rubbed through wire insulation, or a wire that is broken inside the insulation.

The lamp test function of the Tech 1 may be used to command the lamp "ON" while looking for an intermittent malfunction in the ABS warning lamp circuitry.

The frequency of the malfunction can be checked by using the enhanced diagnostic function of the Tech 1.

Any circuitry that is suspected as causing the intermittent complaint should be thoroughly checked for backed out terminals, improper mating, broken locks, improperly formed or damaged terminals, poor terminal to wiring connections or physical damage to the wiring harness.

DIAGNOSTIC CHART DTC 11 — 1994 CENTURY, CUTLASS CIERA AND CUTLASS CRUISER

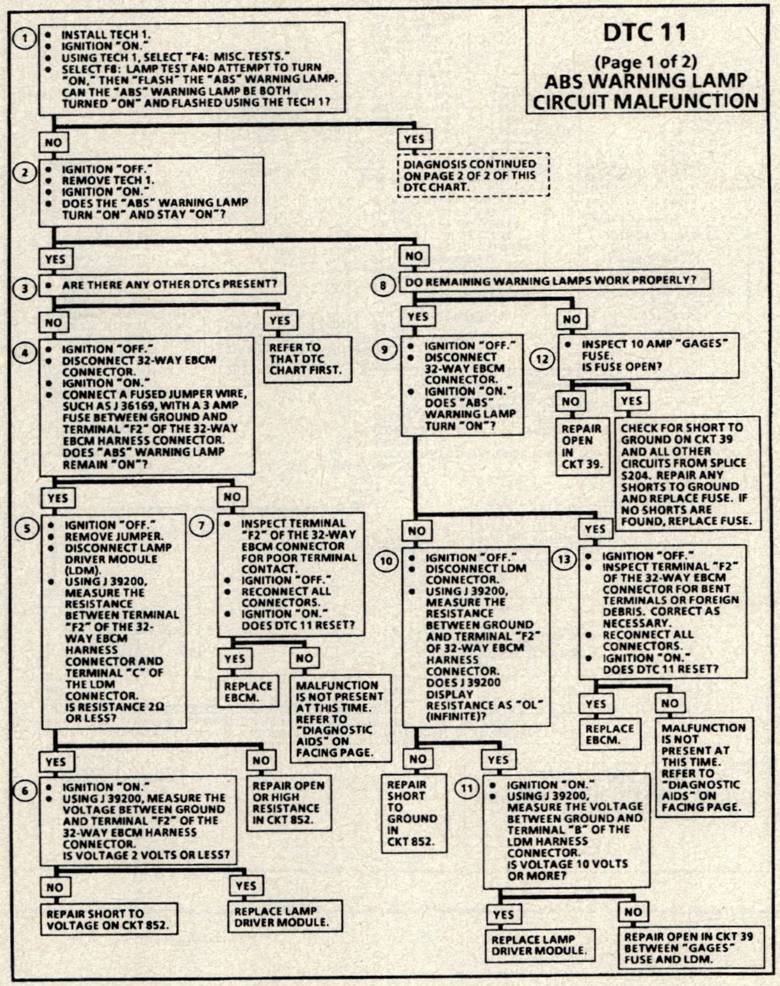

DIAGNOSTIC CHART DTC 11 — 1994 CENTURY, CUTLASS CIERA AND CUTLASS CRUISER

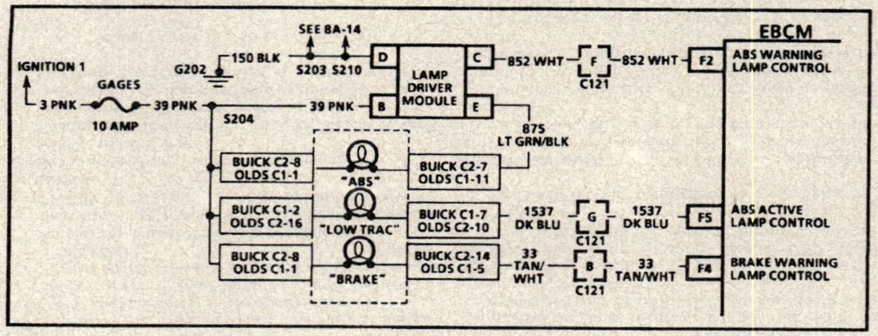

DTC 11

(Page 2 of 2)

ABS WARNING LAMP CIRCUIT MALFUNCTION

Circuit Description:

This DTC checks the state of the ABS warning lamp to identify a situation in which the driver could not be warned of a system malfunction by the ABS warning lamp, or the warning lamp is always "ON." The Lamp Driver Module (LDM) turns the ABS warning lamp "ON," unless the EBCM provides ground to turn it "OFF." Because of the circuitry in the LDM, only external malfunctions can be detected. As a result, the ABS warning lamp itself is not diagnosable, only the control line to the EBCM can be diagnosed. In the event of an open CKT 852, the ABS warning lamp will be "ON" at all times, due to the loss of ground at the LDM input. If the control line is shorted to ground, the ABS warning lamp is kept "OFF," due to the LDM input being grounded.

DTC Will Set When: DTC 11 only can be set during the three second bulb check or when either of the lamps are commanded "ON." If the EBCM cannot control either the ABS warning lamp or the "LOW TRACTION" lamp for two seconds, a malfunction exists.

Action Taken: A malfunction DTC is stored; however, ABS is not disabled.

DTC Chart Test Description: Number(s) below refer to circled number(s) on the diagnostic chart.

14. Determines if the "LOW TRACTION" lamp can be properly controlled.
15. Determines if the "LOW TRACTION" lamp is "ON" constantly.
16. Checks for a short to ground in the "LOW TRACTION" lamp circuitry.
17. Determines if DTC 11 was set due to a EBCM malfunction.
18. Determines if DTC 11 was set due to a EBCM malfunction.
19. Determines if the "LOW TRACTION" lamp can be controlled manually.
20. Verifies the integrity of the "LOW TRACTION" lamp circuitry.
21. Determines if there is a short to voltage in the "LOW TRACTION" lamp circuitry.
22. Determines if DTC 11 was set due to an EBCM malfunction.

Diagnostic Aids: An "Intermittent" malfunction may be caused by a poor connection, rubbed through wire insulation, or a wire that is broken inside the insulation.

The lamp test function of the Tech 1 may be used to command the lamp "ON" while looking for an intermittent malfunction in the "LOW TRACTION" lamp circuitry.

The frequency of the malfunction can be checked by using the enhanced diagnostic function of the Tech 1

Any circuitry that is suspected as causing the intermittent complaint should be thoroughly checked for backed out terminals, improper mating, broken locks, improperly formed or damaged terminals, poor terminal to wiring connections or physical damage to the wiring harness.

DIAGNOSTIC CHART DTC 11 — 1994 CENTURY, CUTLASS CIERA AND CUTLASS CRUISER

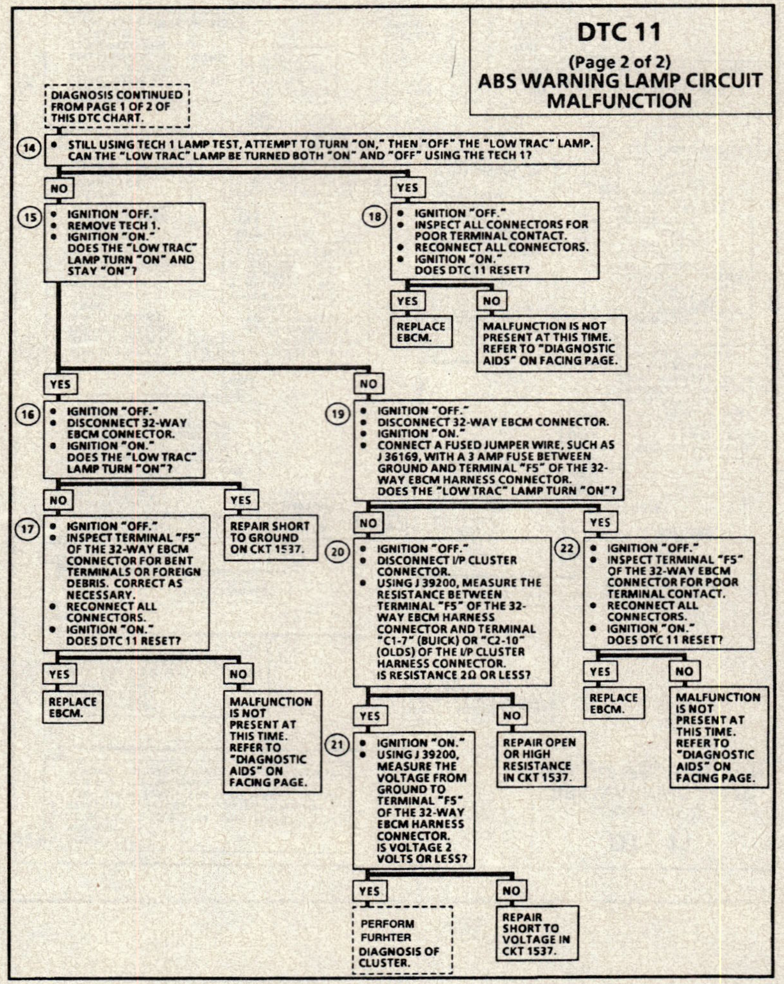

DIAGNOSTIC CHART DTC 14 — 1994 CENTURY, CUTLASS CIERA AND CUTLASS CRUISER

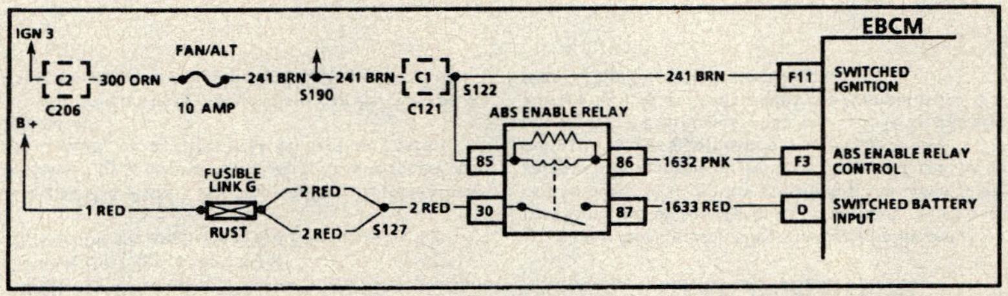

DTC 14
(Page 1 of 3)
ABS ENABLE RELAY CONTACT CIRCUIT OPEN

Circuit Description:
Ignition voltage is supplied through terminal "85" of the ABS enable relay. The EBCM then is able to energize the pull-in coil by completing the ground circuit at terminal "F3" of the EBCM. The magnetic field created closes the ABS enable relay contacts and allows battery voltage and current to be supplied to the EBCM, which supplies power to the motors and solenoids.

DTC Will Set When: DTC 14 can be set anytime after the EBCM commands the ABS enable relay "ON" (the relay is first commanded "ON" during the three second bulb check). This test monitors the availability of current/voltage to the motors and solenoids. This malfunction indicates voltage is not available and would therefore not allow ABS operation if required.

Action Taken: A malfunction DTC is stored, ABS is disabled and the ABS warning lamp is turned "ON."

DTC Chart Test Description: Number(s) below refer to circled number(s) on the diagnostic chart.

1. Checks to see if DTC 16 is also set as current or history. If DTC 16 is also set, proceed to that chart first.
2. Verifies that the malfunction is currently present.
3. Verifies that the relay can be turned "ON" and provide voltage to the EBCM.
4. Checks for high resistance in the switched battery circuit from the relay to the EBCM.
5. Checks for high resistance in the switched battery circuit from the battery to the relay.
6. Verifies the condition of the ABS enable relay.
7. Checks for poor terminal contact or corrosion at the connectors.
8. Determines if the EBCM is the cause of the malfunction.

Diagnostic Aids: An "Intermittent" problem may be caused by a poor connection, rubbed through wire insulation, or a wire that is broken inside the insulation.

The frequency of the problem can be checked by using the enhanced diagnostic function of the Tech 1.

Any circuitry that is suspected as causing the "Intermittent" complaint should be thoroughly checked for backed out terminals, improper mating, broken locks, improperly formed or damaged terminals, poor terminal to wiring connections or physical damage to the wiring harness.

VIBRATION, TEMPERATURE EFFECTS:

Check for vibration effects by performing the relay test function of the Tech 1. With the relay test commanded "ON," lightly tap the top and sides of the relay while monitoring relay voltage. If the relay voltage changes significantly, replace the relay.

If DTC 14 only sets when the vehicle is initially started in cold ambient conditions (temperature less than 32°F-0°C), replace the relay.

DIAGNOSTIC CHART DTC 14 — 1994 CENTURY, CUTLASS CIERA AND CUTLASS CRUISER

DTC 14
(Page 1 of 3)
ABS ENABLE RELAY CONTACT CIRCUIT OPEN

(1) IS DTC 16 ALSO PRESENT OR IN HISTORY?
- YES → PROCEED TO DTC 16 CHART FIRST.
- NO ↓

(2)
- IGNITION "ON," ENGINE "OFF."
- USING TECH 1 SELECT DATA LIST AND OBSERVE EBCM BATTERY VOLTAGE. DOES TECH 1 INDICATE THAT EBCM BATTERY VOLTAGE IS GREATER THAN 10 VOLTS?

- YES → MALFUNCTION IS NOT PRESENT AT THIS TIME. REFER TO "DIAGNOSTIC AIDS" ON FACING PAGE.
- NO ↓

(3)
- IGNITION "OFF."
- DISCONNECT 32-WAY AND 8-WAY EBCM CONNECTORS.
- CONNECT A FUSED JUMPER, SUCH AS J 36169 WITH A 3 AMP FUSE BETWEEN GROUND AND TERMINAL "F3" OF THE 32-WAY EBCM HARNESS CONNECTOR.
- IGNITION "ON."
- USING J 39200, MEASURE THE VOLTAGE BETWEEN GROUND AND TERMINAL "D" OF THE 8-WAY EBCM HARNESS CONNECTOR. IS VOLTAGE 10 VOLTS OR GREATER?

- NO → DIAGNOSIS CONTINUED ON NEXT PAGE.
- YES ↓

(4)
- IGNITION "OFF."
- DISCONNECT JUMPER USED IN STEP 3.
- DISCONNECT ABS ENABLE RELAY.
- USING J 39200, MEASURE RESISTANCE BETWEEN TERMINAL "87" OF THE ABS ENABLE RELAY HARNESS CONNECTOR AND TERMINAL "D" OF THE 8-WAY EBCM HARNESS CONNECTOR. IS RESISTANCE 2Ω OR LESS?

- NO → REPAIR HIGH RESISTANCE IN CKT 1633.
- YES ↓

(5)
- DISCONNECT NEGATIVE BATTERY CABLE.
- DISCONNECT POSITIVE BATTERY CABLE.
- USING J 39200, MEASURE THE RESISTANCE BETWEEN POSITIVE BATTERY CABLE TERMINAL AND TERMINAL "30" OF THE ABS ENABLE RELAY HARNESS CONNECTOR. IS RESISTANCE 2Ω OR LESS?

- NO → REPAIR HIGH RESISTANCE IN CKT 1 OR 2. REFER TO SECTION 8A.
- YES ↓

(6)
- USING A JUMPER WIRE, CONNECT TERMINAL "86" ON THE ABS ENABLE RELAY TO GROUND.
- USING A FUSED JUMPER WIRE, SUCH AS J 36169, WITH A 3 AMP FUSE, CONNECT TERMINAL "85" ON THE ABS ENABLE RELAY TO B+.
- USING J 39200, MEASURE THE RESISTANCE BETWEEN TERMINALS "30" AND "87" ON THE ABS ENABLE RELAY. IS RESISTANCE 2Ω OR LESS?

- NO → REPLACE ABS ENABLE RELAY.
- YES ↓

(7)
- INSPECT ABS ENABLE RELAY AND ABS ENABLE RELAY HARNESS CONNECTOR FOR A POOR CONNECTION OR POOR TERMINAL CONTACT.
- INSPECT EBCM AND EBCM CONNECTOR TERMINALS "F3" AND "D" FOR POOR TERMINAL CONTACT.
- INSPECT POSITIVE BATTERY TERMINAL AND POSITIVE BATTERY CABLE TERMINAL FOR A POOR CONNECTION. REFER TO SECTION 8A-4, "CHECKING TERMINAL CONTACT." IS POOR TERMINAL CONTACT OR A POOR CONNECTION EVIDENT?

- YES → REPLACE TERMINALS THAT EXHIBIT POOR TERMINAL CONTACT OR EVIDENCE OF CORROSION.
- NO ↓

(8)
- RECONNECT ALL CONNECTORS.
- IGNITION "ON." DOES DTC 14 RESET?

- YES → REPLACE EBCM.
- NO → MALFUNCTION IS MOST LIKELY DUE TO A POOR CONNECTION. REFER TO "DIAGNOSTIC AIDS" ON FACING PAGE.

DIAGNOSTIC CHART DTC 14 — 1994 CENTURY, CUTLASS CIERA AND CUTLASS CRUISER

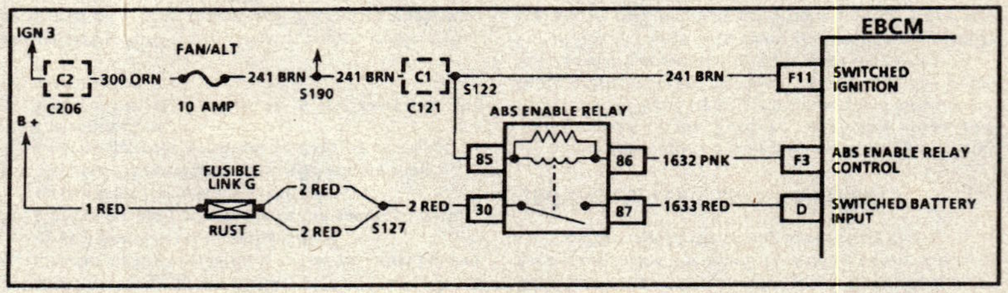

DTC 14

(Page 2 of 3)

ABS ENABLE RELAY CONTACT CIRCUIT OPEN

Circuit Description:

Ignition voltage is supplied through terminal "85" of the ABS enable relay. The EBCM then is able to energize the pull-in coil by completing the ground circuit at terminal "F3" of the EBCM. The magnetic field created closes the ABS enable relay contacts and allows battery voltage and current to be supplied to the EBCM, which supplies power to the motors and solenoids.

DTC Will Set When: DTC 14 can be set anytime after the EBCM commands the ABS enable relay "ON" (the relay is first commanded "ON" during the three second bulb check). This test monitors the availability of current/voltage to the motors and solenoids. This malfunction indicates voltage is not available and would therefore not allow ABS operation if required.

Action Taken: A malfunction DTC is stored, ABS is disabled and the ABS warning lamp is turned "ON."

DTC Chart Test Description: Number(s) below refer to circled number(s) on the diagnostic chart.

9. Checks for high resistance in the entire relay coil circuit.
10. Verifies proper battery condition.
11. This step verifies proper voltage is available at the ABS enable relay.
12. Checks for an open in the switched battery circuit from the relay to the EBCM.
13. Verifies the condition of the ABS enable relay.

Diagnostic Aids: An "Intermittent" problem may be caused by a poor connection, rubbed through wire insulation, or a wire that is broken inside the insulation.

The frequency of the problem can be checked by using the enhanced diagnostic function of the Tech 1.

Any circuitry that is suspected as causing the "Intermittent" complaint should be thoroughly checked for backed out terminals, improper mating, broken locks, improperly formed or damaged terminals, poor terminal to wiring connections or physical damage to the wiring harness.

VIBRATION, TEMPERATURE EFFECTS:

Check for vibration effects by performing the relay test function of the Tech 1. With the relay test commanded "ON," lightly tap the top and sides of the relay while monitoring relay voltage. If the relay voltage changes significantly, replace the relay.

If DTC 14 only sets when the vehicle is initially started in cold ambient conditions (temperature less than 32°F-0°C), replace the relay.

DIAGNOSTIC CHART DTC 14 — 1994 CENTURY, CUTLASS CIERA AND CUTLASS CRUISER

DTC 14

(Page 2 of 3)

ABS ENABLE RELAY CONTACT CIRCUIT OPEN

DIAGNOSIS CONTINUED FROM PREVIOUS PAGE.

(9)
- IGNITION "OFF."
- REMOVE JUMPER WIRE INSTALLED IN STEP #3, IF NOT ALREADY REMOVED.
- DISCONNECT NEGATIVE BATTERY CABLE.
- DISCONNECT POSITIVE BATTERY CABLE.
- WITH BATTERY STILL DISCONNECTED, TURN IGNITION SWITCH TO "RUN." THIS COMPLETES THE IGNITION CIRCUIT TO THE ABS ENABLE RELAY.
- USING J 39200, MEASURE THE RESISTANCE BETWEEN TERMINAL "F3" OF THE 32-WAY EBCM HARNESS CONNECTOR AND THE POSITIVE BATTERY CABLE TERMINAL. IS RESISTANCE 95Ω OR LESS WHEN MEASURED AT APPROXIMATELY 20°C (68°F)?

NO: DIAGNOSIS CONTINUED ON NEXT PAGE.

YES:

(10)
- USING J 39200, MEASURE THE VOLTAGE BETWEEN THE NEGATIVE AND POSITIVE BATTERY TERMINALS. IS VOLTAGE 10 VOLTS OR GREATER?

NO: CORRECT LOW VOLTAGE CONDITION.

YES:

(11)
- IGNITION "OFF."
- RECONNECT POSITIVE BATTERY CABLE.
- RECONNECT NEGATIVE BATTERY CABLE.
- DISCONNECT ABS ENABLE RELAY.
- MEASURE THE VOLTAGE BETWEEN TERMINAL "30" OF THE ABS ENABLE RELAY HARNESS CONNECTOR AND GROUND. IS VOLTAGE 10 VOLTS OR GREATER?

NO: REPAIR OPEN OR HIGH RESISTANCE IN CKT 1 OR 2.

YES:

(12)
- USING J 39200, MEASURE THE RESISTANCE BETWEEN TERMINAL "87" OF THE ABS ENABLE RELAY HARNESS CONNECTOR AND TERMINAL "D" OF THE 8-WAY EBCM HARNESS CONNECTOR. IS RESISTANCE 2Ω OR LESS?

NO: REPAIR OPEN OR HIGH RESISTANCE IN CKT 1633.

YES:

(13)
- USING A JUMPER WIRE, CONNECT TERMINAL "86" ON THE ABS ENABLE RELAY TO GROUND.
- USING A FUSED JUMPER WIRE, SUCH AS J 36169, WITH A 3 AMP FUSE, CONNECT TERMINAL "85" ON THE ABS ENABLE RELAY TO B+.
- USING J 39200, MEASURE THE RESISTANCE BETWEEN TERMINALS "30" AND "87" ON THE ABS ENABLE RELAY. IS RESISTANCE 2Ω OR LESS?

NO: REPLACE ABS ENABLE RELAY.

YES: MALFUNCTION WAS MOST LIKELY DUE TO A POOR CONNECTION. INSPECT ALL CONNECTORS AND TERMINALS FOR POOR TERMINAL CONTACT AND EVIDENCE OF CORROSION. REPLACE ALL TERMINALS THAT EXHIBIT EVIDENCE OF POOR TERMINAL CONTACT OR CORROSION.

DIAGNOSTIC CHART DTC 14 — 1994 CENTURY, CUTLASS CIERA AND CUTLASS CRUISER

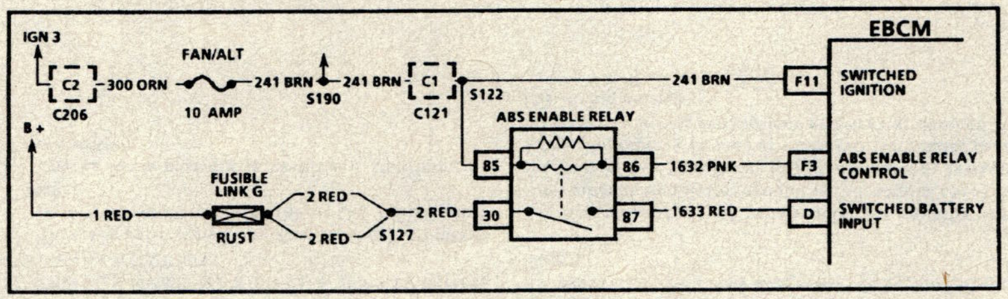

DTC 14

(Page 3 of 3)

ABS ENABLE RELAY CONTACT CIRCUIT OPEN

Circuit Description:

Ignition voltage is supplied through terminal "85" of the ABS enable relay. The EBCM then is able to energize the pull-in coil by completing the ground circuit at terminal "F3" of the EBCM. The magnetic field created closes the ABS enable relay contacts and allows battery voltage and current to be supplied to the EBCM, which supplies power to the motors and solenoids.

DTC Will Set When: DTC 14 can be set anytime after the EBCM commands the ABS enable relay "ON" (the relay is first commanded "ON" during the three second bulb check). This test monitors the availability of current/voltage to the motors and solenoids. This malfunction indicates voltage is not available and would therefore not allow ABS operation if required.

Action Taken: A malfunction DTC is stored, ABS is disabled and the ABS warning lamp is turned "ON."

DTC Chart Test Description: Number(s) below refer to circled number(s) on the diagnostic chart.

14. Checks for proper resistance of the relay coil.
15. Checks for high resistance in the ABS enable relay control circuit between the relay and the EBCM.
16. Checks for high resistance in the ABS enable relay control circuit between the battery and the relay.

Diagnostic Aids: An "Intermittent" problem may be caused by a poor connection, rubbed through wire insulation, or a wire that is broken inside the insulation.

The frequency of the problem can be checked by using the enhanced diagnostic function of the Tech 1.

Any circuitry that is suspected as causing the "Intermittent" complaint should be thoroughly checked for backed out terminals, improper mating, broken locks, improperly formed or damaged terminals, poor terminal to wiring connections or physical damage to the wiring harness.

VIBRATION, TEMPERATURE EFFECTS:

Check for vibration effects by performing the relay test function of the Tech 1. With the relay test commanded "ON," lightly tap the top and sides of the relay while monitoring relay voltage. If the relay voltage changes significantly, replace the relay.

If DTC 14 only sets when the vehicle is initially started in cold ambient conditions (temperature less than 32°F-0°C), replace the relay.

DIAGNOSTIC CHART DTC 14 — 1994 CENTURY, CUTLASS CIERA AND CUTLASS CRUISER

DTC 14

(Page 3 of 3)

ABS ENABLE RELAY CONTACT CIRCUIT OPEN

DIAGNOSIS CONTINUED FROM PREVIOUS PAGE.

(14)
- DISCONNECT ABS ENABLE RELAY.
- USING J 39200, MEASURE THE RESISTANCE BETWEEN TERMINALS "85" AND "86" ON THE ABS ENABLE RELAY. IS RESISTANCE 95Ω OR LESS WHEN MEASURED AT APPROXIMATELY 20°C (68°F)?

NO: REPLACE ABS ENABLE RELAY.

YES:

(15)
- MEASURE THE RESISTANCE BETWEEN TERMINAL "86" OF THE ABS ENABLE RELAY HARNESS CONNECTOR AND TERMINAL "F3" OF THE 32-WAY EBCM HARNESS CONNECTOR. IS RESISTANCE 2Ω OR LESS?

NO: REPAIR OPEN OR HIGH RESISTANCE IN CKT 1632.

YES:

(16)
- MEASURE THE RESISTANCE BETWEEN TERMINAL "85" OF THE ABS ENABLE RELAY HARNESS CONNECTOR AND THE POSITIVE BATTERY CABLE TERMINAL. IS RESISTANCE 2Ω OR LESS?

NO: REPAIR OPEN OR HIGH RESISTANCE IN CKT 241, CKT 300, IGNITION SWITCH, OR BATTERY FEED CIRCUIT TO IGNITION SWITCH. REFER TO SECTION 8A FOR FURTHER INFORMATION.

YES: MALFUNCTION WAS MOST LIKELY DUE TO A POOR CONNECTION. INSPECT ALL CONNECTORS AND TERMINALS FOR POOR TERMINAL CONTACT AND EVIDENCE OF CORROSION. REPLACE ALL TERMINALS THAT EXHIBIT EVIDENCE OF POOR TERMINAL CONTACT OR CORROSION.

DIAGNOSTIC CHART DTC 15 — 1994 CENTURY, CUTLASS CIERA AND CUTLASS CRUISER

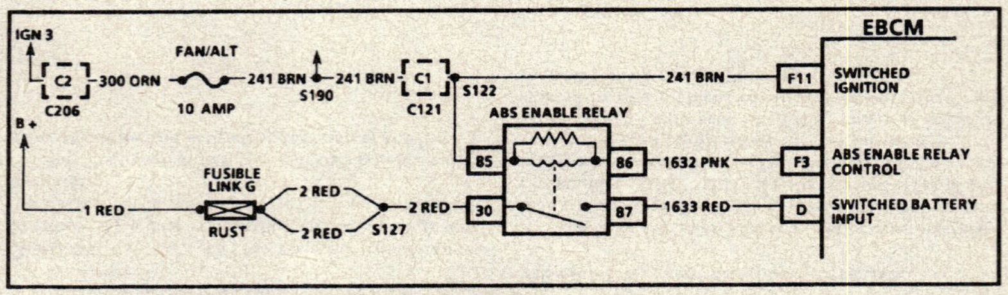

DTC 15

ABS ENABLE RLY. CONTACT CKT SHORT TO BATT OR ALWAYS CLOSED

Circuit Description:

Ignition voltage is supplied through terminal "85" of the ABS enable relay. The EBCM then is able to energize the pull-in coil by completing the ground circuit at terminal "F3" of the EBCM. The magnetic field created closes the ABS enable relay contacts and allows battery voltage and current to be supplied to the EBCM, which supplies power to the motors and solenoids.

DTC Will Set When: DTC 15 can be set only before the EBCM commands the relay "ON." This test determines if the ABS enable relay is energized when it should not be. This malfunction would not allow the ABS enable relay to remove power to the ABS system. If a second malfunction were to occur that requires the ABS enable relay to be turned "OFF," that malfunction cannot be removed if the relay cannot be controlled. The malfunction must be present for three consecutive drive cycles before the DTC is set.

Action Taken: A malfunction DTC is stored. ABS is not disabled.

DTC Chart Test Description: Number(s) below refer to circled number(s) on the diagnostic chart.

1. Determines if any other DTCs are present.
2. Indicates that the EBCM is capable of controlling the ABS enable relay as commanded.
3. This step checks for a possible short to battery voltage in CKT 1633.
4. This step determines whether the short to voltage is in CKT 1633 or internal to the ABS enable relay.
5. Checks for a possible intermittent in wiring or connectors.

Diagnostic Aids: An "Intermittent" malfunction may be caused by a poor connection, rubbed through wire insulation, or a wire that is broken inside the insulation.

The frequency of the malfunction can be checked by using the enhanced diagnostic function of the Tech 1

Any circuitry, that is suspected as causing the intermittent complaint, should be thoroughly checked for backed out terminals, improper mating, broken locks, improperly formed or damaged terminals, poor terminal to wiring connections or physical damage to the wiring harness.

DIAGNOSTIC CHART DTC 15 — 1994 CENTURY, CUTLASS CIERA AND CUTLASS CRUISER

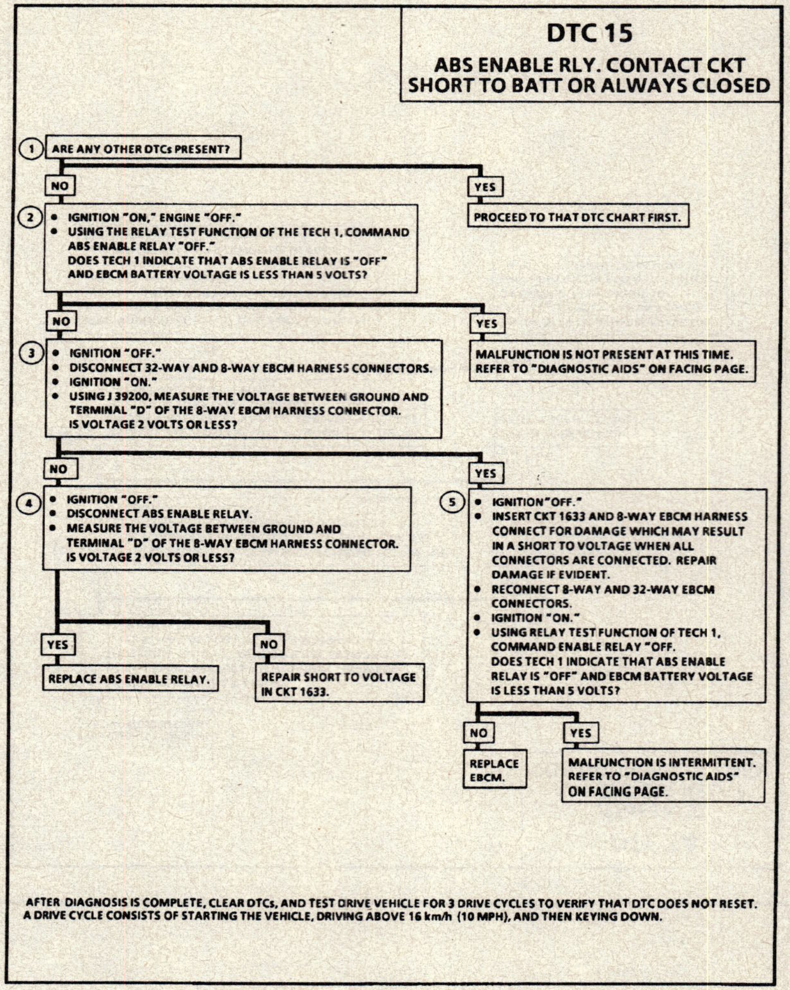

DIAGNOSTIC CHART DTC 16 — 1994 CENTURY, CUTLASS CIERA AND CUTLASS CRUISER

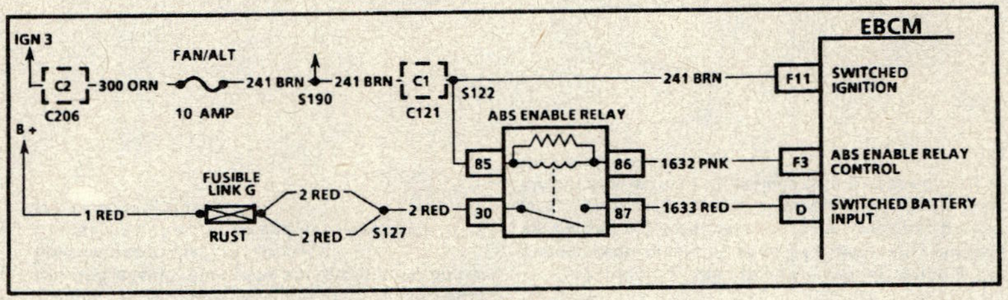

DTC 16

ABS ENABLE RELAY COIL CIRCUIT OPEN

Circuit Description:
Ignition voltage is supplied through terminal "85" of the ABS enable relay. The EBCM then is able to energize the pull-in coil by completing the ground circuit at terminal "F3" of the EBCM. The magnetic field created closes the ABS enable relay contacts and allows battery voltage and current to be supplied to the EBCM, which supplies power to the motors and solenoids.

DTC Will Set When: DTC 16 can be set only after DTC 14 has been set. This test detects an open in the ABS enable relay coil circuit. An open in this circuit will not allow the ABS enable relay to be energized thus preventing voltage/current to the motors and solenoids. If this malfunction is present and the ignition is turned "OFF" before 5 km/h (3 mph) is reached, DTC 14 is set alone.

Action Taken: A malfunction DTC is stored, ABS is disabled and the ABS warning lamp is turned "ON."

DTC Chart Test Description: Number(s) below refer to circled number(s) on the diagnostic chart.
1. Indicates the EBCM is capable of controlling the ABS enable relay as commanded.
2. Checks to ensure that voltage is available to the pull-in coil of the ABS enable relay.
3. Ensures that there is continuity through the pull-in coil of the ABS enable relay.
4. Verifies the integrity of the relay control circuit.
5. Ensures malfunction was not due to poor terminal contact.

Diagnostic Aids: An "Intermittent" malfunction may be caused by a poor connection, rubbed through wire insulation, or a wire that is broken inside the insulation.

The frequency of the malfunction can be checked by using the enhanced diagnostic function of the Tech 1

If the frequency of the malfunction is high, but is currently intermittent, check for high coil resistance by measuring between relay terminals "85" and "86" using a J 39200. If resistance shows greater than 100Ω, replace the relay.

Any circuitry, that is suspected as causing the intermittent complaint, should be thoroughly checked for backed out terminals, improper mating, broken locks, improperly formed or damaged terminals, poor terminal to wiring connections or physical damage to the wiring harness.

DIAGNOSTIC CHART DTC 16 — 1994 CENTURY, CUTLASS CIERA AND CUTLASS CRUISER

DTC 16

ABS ENABLE RELAY COIL CIRCUIT OPEN

(1)
- IGNITION "ON," ENGINE "OFF."
- USING THE RELAY TEST FUNCTION OF THE TECH 1, COMMAND THE ABS ENABLE RELAY "ON."

DOES TECH 1 INDICATE ABS ENABLE RELAY "ON," AND EBCM BATTERY VOLTAGE ABOVE 10 VOLTS?

YES: MALFUNCTION IS NOT PRESENT AT THIS TIME. REFER TO "DIAGNOSTIC AIDS" ON FACING PAGE.

NO:

(2)
- USING J 39200, MEASURE THE VOLTAGE BETWEEN GROUND AND TERMINAL "85" OF THE ABS ENABLE RELAY CONNECTOR.

IS VOLTAGE 10 VOLTS OR GREATER?

NO: REPAIR OPEN OR HIGH RESISTANCE IN CKT 241 OR FAN/ALT FUSE.

YES:

(3)
- USING THE RELAY TEST FUNCTION OF THE TECH 1, COMMAND THE ABS ENABLE RELAY "OFF."
- MEASURE THE VOLTAGE BETWEEN GROUND AND TERMINAL "86" OF THE ABS ENABLE RELAY CONNECTOR.

IS VOLTAGE 10 VOLTS OR GREATER?

NO: CHECK FOR INTERMITTENT CONNECTIONS AT ABS ENABLE RELAY. IF OK, REPLACE ABS RELAY.

YES:

(4)
- IGNITION "OFF."
- DISCONNECT 32-WAY EBCM CONNECTOR.
- IGNITION "ON."
- MEASURE THE VOLTAGE BETWEEN GROUND AND TERMINAL "F3" OF THE 32-WAY EBCM HARNESS CONNECTOR.

IS VOLTAGE 10 VOLTS OR GREATER?

NO: REPAIR OPEN OR HIGH RESISTANCE IN CKT 1632.

YES:

(5)
- IGNITION "OFF."
- INSPECT 32-WAY EBCM CONNECTOR TERMINAL "F3" FOR POOR TERMINAL CONTACT OR EVIDENCE OF CORROSION.

REPLACE ALL TERMINALS THAT EXHIBIT SIGNS OF POOR TERMINAL CONTACT OR CORROSION.
- RECONNECT ALL CONNECTORS.
- IGNITION "ON."

DOES DTC 16 SET AS A CURRENT DTC?

YES: REPLACE EBCM.

NO: MALFUNCTION WAS MOST LIKELY CAUSED BY A POOR CONNECTION. REFER TO "DIAGNOSTIC AIDS" ON FACING PAGE.

AFTER ALL DIAGNOSIS IS COMPLETE, CLEAR DTCs AND TEST DRIVE VEHICLE FOR THREE (3) DRIVE CYCLES TO VERIFY THAT DTC DOES NOT RESET. A DRIVE CYCLE CONSISTS OF STARTING THE VEHICLE, DRIVING OVER 16 km/h (10 MPH), AND THEN KEYING DOWN.

DIAGNOSTIC CHART DTC 17 — 1994 CENTURY, CUTLASS CIERA AND CUTLASS CRUISER

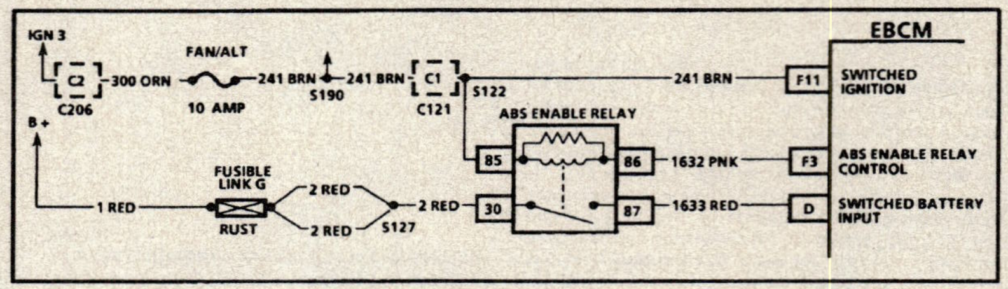

DTC 17

ABS ENABLE RELAY COIL CIRCUIT SHORTED TO GND

Circuit Description:

Ignition voltage is supplied through terminal "85" of the ABS enable relay. The EBCM then is able to energize the pull-in coil by completing the ground circuit at terminal "F3" of the EBCM. The magnetic field created closes the ABS enable relay contacts and allows battery voltage and current to be supplied to the EBCM, which supplies voltage to the motors and solenoids.

DTC Will Set When: DTC 17 can be set before the EBCM commands the ABS enable relay "ON." This test determines if the ABS enable relay is energized when it should not be. This malfunction would not allow the ABS enable relay to remove power to the ABS system. If a second malfunction were to occur that requires the ABS enable relay to be turned "OFF," that malfunction cannot be removed if the relay cannot be controlled.

Action Taken: A malfunction DTC is stored. ABS is not disabled.

DTC Chart Test Description: Number(s) below refer to circled number(s) on the diagnostic chart.

1. Indicates that the EBCM is capable of controlling the ABS enable relay as commanded.
2. Checks to ensure that the ABS enable relay or control CKT 1632 are not shorted to ground.
3. Checks to see if the EBCM is internally shorted to ground. A grounded terminal "F3" would cause the ABS enable relay to be energized anytime the ignition was in the "ON" position.
4. Checks for a short to ground in control CKT 1632 or a defective relay.

Diagnostic Aids: An "Intermittent" malfunction may be caused by a poor connection, rubbed through wire insulation, or a wire that is broken inside the insulation.

The frequency of the malfunction can be checked by using the enhanced diagnostic function of the Tech 1.

Any circuitry, that is suspected as causing the intermittent complaint, should be thoroughly checked for backed out terminals, improper mating, broken locks, improperly formed or damaged terminals, poor terminal to wiring connections or physical damage to the wiring harness.

DIAGNOSTIC CHART DTC 17 — 1994 CENTURY, CUTLASS CIERA AND CUTLASS CRUISER

DTC 17

ABS ENABLE RELAY COIL CIRCUIT SHORTED TO GND

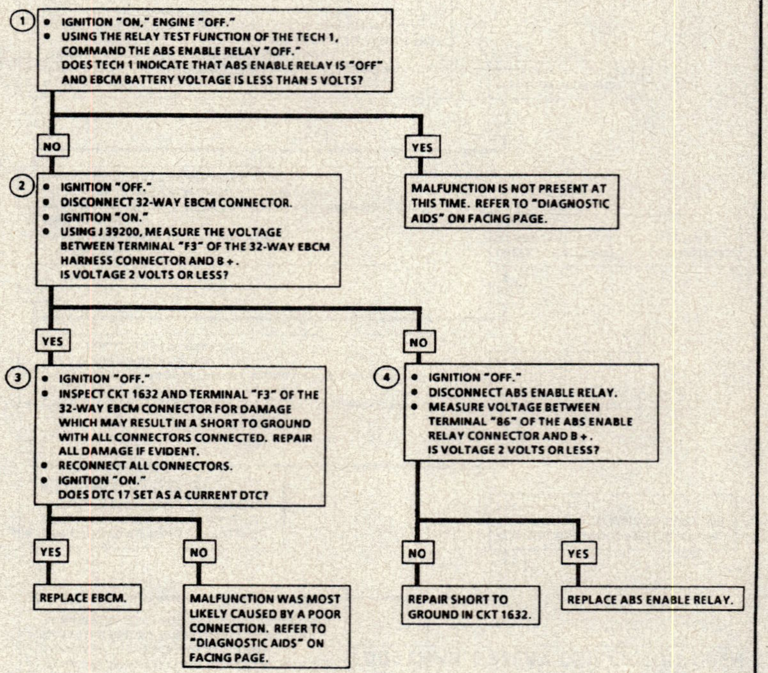

AFTER ALL DIAGNOSIS IS COMPLETE, CLEAR DTCs AND TEST DRIVE VEHICLE FOR THREE (3) DRIVE CYCLES TO VERIFY THAT DTC DOES NOT RESET. A DRIVE CYCLE CONSISTS OF STARTING THE VEHICLE, DRIVING OVER 16 km/h (10 MPH), AND THEN KEYING DOWN.

DIAGNOSTIC CHART DTC 18 — 1994 CENTURY, CUTLASS CIERA AND CUTLASS CRUISER

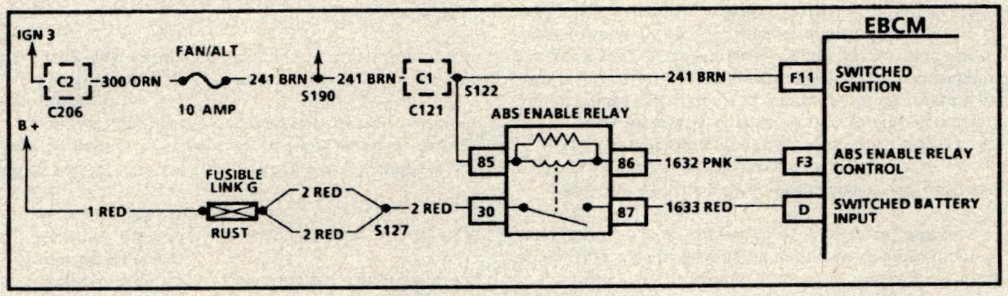

DTC 18

ABS ENABLE RELAY COIL CIRCUIT SHORTED TO BATTERY

Circuit Description:
Ignition voltage is supplied through terminal "85" of the ABS enable relay. The EBCM then is able to energize the pull-in coil by completing the ground circuit at terminal "F3" of the EBCM. The magnetic field created closes the ABS enable relay contacts and allows battery voltage and current to be supplied to the EBCM, which supplies voltage to the motors and solenoids.

DTC Will Set When: DTC 18 can be set after the EBCM commands the enable relay "ON." This test monitors the availability of current/voltage to the motors and solenoids. A malfunction of this type will not allow voltage to be available to the motors and solenoids; therefore, ABS operation would not be allowed if required.

Action Taken: A malfunction DTC is stored, ABS is disabled and the ABS warning lamp is turned "ON."

DTC Chart Test Description: Number(s) below refer to circled number(s) on the diagnostic chart.
1. Indicates that the EBCM is capable of controlling the ABS enable relay as commanded.
2. Checks for a short to voltage on the relay control circuit.
3. Verifies that the relay coil has the appropriate resistance.
4. Verifies that relay contacts are not failed closed.
5. Checks for possible poor terminal contact at the EBCM.

Diagnostic Aids: An "Intermittent" malfunction may be caused by a poor connection, rubbed through wire insulation, or a wire that is broken inside the insulation.

The frequency of the malfunction can be checked by using the enhanced diagnostic function of the Tech 1

Any circuitry, that is suspected as causing the intermittent complaint, should be thoroughly checked for backed out terminals, improper mating, broken locks, improperly formed or damaged terminals, poor terminal to wiring connections or physical damage to the wiring harness.

DIAGNOSTIC CHART DTC 18 — 1994 CENTURY, CUTLASS CIERA AND CUTLASS CRUISER

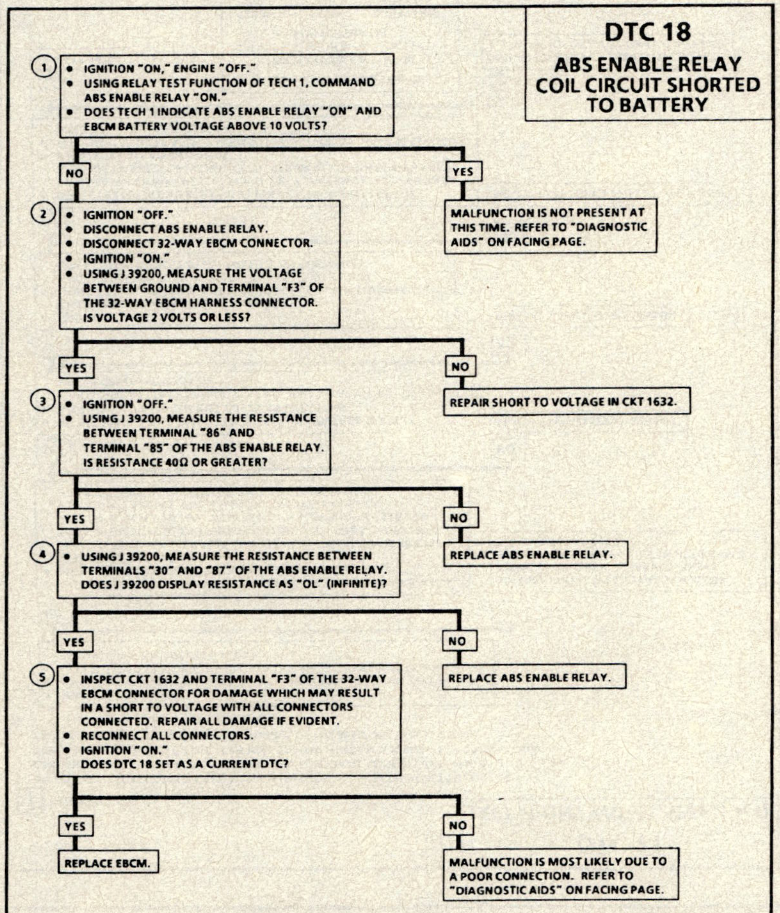

AFTER ALL DIAGNOSIS IS COMPLETE, CLEAR DTCs AND TEST DRIVE VEHICLE FOR THREE (3) DRIVE CYCLES TO VERIFY THAT DTC DOES NOT RESET. A DRIVE CYCLE CONSISTS OF STARTING THE VEHICLE, DRIVING OVER 16 km/h (10 MPH), AND THEN KEYING DOWN.

DIAGNOSTIC CHART DTC 21 — 1994 CENTURY, CUTLASS CIERA AND CUTLASS CRUISER

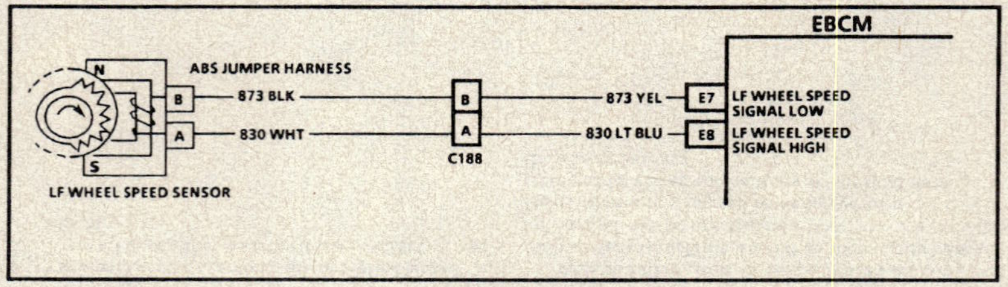

DTC 21

LEFT FRONT WHEEL SPEED = 0

Circuit Description:

As a toothed ring passes by the wheel speed sensor, changes in the electromagnetic field cause the wheel speed sensor to produce a sinusoidal (AC) voltage signal whose frequency is proportional to wheel speed. The magnitude of this signal is directly related to wheel speed and the proximity of the wheel speed sensor to the toothed ring, often referred to as the air gap.

DTC Will Set When: DTC 21 can be set when the vehicle is not in an ABS stop. If the left front wheel speed = 0 and the vehicle's reference speed is greater than 8 km/h (5 mph), a malfunction exists.

Action Taken: A malfunction DTC is stored, ABS is disabled and the ABS warning lamp is turned "ON."

DTC Chart Test Description: Number(s) below refer to circled number(s) on the diagnostic chart.

1. This step is designed to identify a wheel speed sensor or circuitry that is damaged and visibly apparent.
2. Checks to see that the wheel speed sensor is within proper resistance values.
3. Verifies that the wheel speed sensor can generate an appropriate output signal.
4. Ensures that wheel speed sensor circuitry is not shorted together.
5. Ensures malfunction was not due to physical damage of the circuitry.

Diagnostic Aids: An "Intermittent" malfunction may be caused by a poor connection, rubbed through wire insulation, or a wire that is broken inside the insulation.

The frequency of the malfunction can be checked by using the enhanced diagnostic function of the Tech I,

If the customer's comments reflect that the ABS warning lamp is "ON" only during moist environmental changes (rain, snow, vehicle wash), all wheel speed sensor circuitry should be thoroughly inspected for signs of water intrusion.

If DTC is not current, clear DTCs and simulate the effects of water intrusion. Use the following procedure. Spray down the suspected area with a 5% salt water solution (two teaspoons of salt to 12 oz. of water). Test drive vehicle over various road surfaces (bumps, turns, etc.) above 24 km/h (15 mph) for at least 30 seconds. If DTC returns, replace suspected harness.

Any circuitry, that is suspected as causing the intermittent complaint, should be thoroughly checked for backed out terminals, improper mating, broken locks, improperly formed or damaged terminals, poor terminal to wiring connections or physical damage to the wiring harness.

Resistance of the wheel speed sensor will increase with an increase in sensor temperature.

When replacing a wheel speed sensor, inspect the sensor terminals and harness connector for corrosion and/or water intrusion. If evidence of corrosion or water intrusion exists, replace wheel speed sensor jumper harness. Likewise, if replacing a wheel speed sensor jumper harness, inspect sensor terminals. If evidence of corrosion or water intrusion exists, replace wheel speed sensor.

DIAGNOSTIC CHART DTC 21 — 1994 CENTURY, CUTLASS CIERA AND CUTLASS CRUISER

DTC 21

LEFT FRONT WHEEL SPEED = 0

Important

- WHEEL SPEED SENSOR INTERMITTENT MALFUNCTIONS MAY BE DIFFICULT TO LOCATE. CARE SHOULD BE TAKEN NOT TO DISTURB ANY ELECTRICAL CONNECTIONS PRIOR TO AN INDICATED STEP OF THIS CHART. THIS WILL ENSURE THAT AN INTERMITTENT CONNECTION WILL NOT BE CORRECTED BEFORE THE SOURCE OF THE MALFUNCTION IS FOUND.

(1)
- IGNITION "OFF."
- PHYSICALLY INSPECT THE LEFT FRONT WHEEL SPEED SENSOR, LEFT FRONT WHEEL SPEED SENSOR JUMPER HARNESS, AND CONNECTORS FOR DAMAGE. IS PHYSICAL DAMAGE INDICATED?

YES: REPLACE WHEEL SPEED SENSOR AND/OR JUMPER HARNESS IF DAMAGED. DRIVE VEHICLE ABOVE 24 km/h (15 MPH) TO ENSURE DTC 21 DOES NOT SET.

NO:

(2)
- DISCONNECT LEFT FRONT WHEEL SPEED SENSOR DIRECTLY AT SENSOR.
- USING J 39200, MEASURE THE RESISTANCE BETWEEN SENSOR TERMINALS "A" AND "B." IS RESISTANCE BETWEEN 1020 AND 1137Ω WHEN SENSOR IS 20°C (68°F)? *(IF TEMPERATURE OF THE SENSOR IS SUSPECTED TO BE OTHER THAN 20°C (68°F), REFER TO THE TABLE BELOW FOR APPLICABLE SENSOR RESISTANCE VALUES; THE VALUES IN THE TABLE ARE FOR TEMPERATURE OF THE SENSOR, NOT AIR TEMPERATURE.)*

NO: REPLACE WHEEL SPEED SENSOR.

YES:

(3)
- WITH J 39200 STILL CONNECTED, SELECT A/C VOLTAGE SCALE.
- SPIN WHEEL BY HAND WHILE OBSERVING VOLTAGE. IS VOLTAGE AT LEAST 100 mV (VOLTAGE WILL INCREASE AS WHEEL SPEED INCREASES)?

NO: REPLACE WHEEL SPEED SENSOR.

YES:

(4)
- IGNITION "OFF."
- DISCONNECT 32-WAY EBCM CONNECTOR.
- USING J 39200, MEASURE THE RESISTANCE BETWEEN TERMINAL "E7" AND "E8" OF THE 32-WAY EBCM HARNESS CONNECTOR. DOES J 39200 DISPLAY RESISTANCE AS "OL" (INFINITE)?

NO: REPAIR SHORT BETWEEN CKTs 830 AND 873.

YES:

(5)
- INSPECT CKTs 830 AND 873 AND 32-WAY EBCM CONNECTOR FOR DAMAGE WHICH MAY RESULT IN A SHORT BETWEEN CKTs 830 AND 873 WITH ALL CONNECTORS CONNECTED. REPAIR ALL DAMAGE IF EVIDENT.
- RECONNECT ALL CONNECTORS.
- TEST DRIVE VEHICLE ABOVE 24 km/h (15 MPH) FOR AT LEAST 30 SECONDS. DOES DTC 21 SET AS A CURRENT DTC?

YES: REPLACE EBCM.

NO: MALFUNCTION IS INTERMITTENT. REFER TO "DIAGNOSTIC AIDS" ON FACING PAGE.

FRONT WHEEL SPEED SENSOR
TEMPERATURE OF SENSOR VS. SENSOR RESISTANCE (APPROXIMATE)

TEMP. (°C)	TEMP. (°F)	RESISTANCE (OHMS)
-34 TO 4	-30 TO 40	804 TO 1066
5 TO 43	41 TO 110	956 TO 1240
44 TO 93	111 TO 200	1112 TO 1463

DIAGNOSTIC CHART DTC 22 — 1994 CENTURY, CUTLASS CIERA AND CUTLASS CRUISER

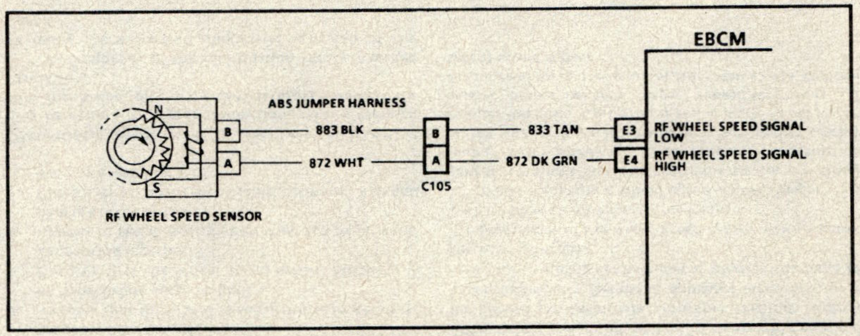

DTC 22

RIGHT FRONT WHEEL SPEED = 0

Circuit Description:

As a toothed ring passes by the wheel speed sensor, changes in the electromagnetic field cause the wheel speed sensor to produce a sinusoidal (AC) voltage signal whose frequency is proportional to wheel speed. The magnitude of this signal is directly related to wheel speed and the proximity of the wheel speed sensor to the toothed ring, often referred to as the air gap.

DTC Will Set When: DTC 22 can be set when the vehicle is not in an ABS stop. If the right front wheel speed = 0 and the vehicle's reference speed is greater than 8 km/h (5 mph), a malfunction exists.

Action Taken: A malfunction DTC is stored, ABS is disabled and the ABS warning lamp is turned "ON."

DTC Chart Test Description: Number(s) below refer to circled number(s) on the diagnostic chart.

1. This step is designed to identify a wheel speed sensor or circuitry that is damaged and visibly apparent.
2. This test checks to see that the wheel speed sensor is within proper resistance values.
3. Verifies that the wheel speed sensor can generate an appropriate output signal.
4. Ensures that wheel speed sensor circuitry is not shorted together.
5. Ensures malfunction was not due to physical damage of the circuitry.

Diagnostic Aids: An "Intermittent" malfunction may be caused by a poor connection, rubbed through wire insulation, or a wire that is broken inside the insulation.

The frequency of the malfunction can be checked by using the enhanced diagnostic function of the Tech 1.

If the customer's comments reflect that the ABS warning lamp is "ON" only during moist environmental changes (rain, snow, vehicle wash), all wheel speed sensor circuitry should be thoroughly inspected for signs of water intrusion. If DTC is not current, clear DTCs and simulate the effects of water intrusion. Use the following procedure. Spray down the suspected area with a 5% salt water solution (two teaspoons of salt to 12 oz. of water). Test drive vehicle over various road surfaces (bumps, turns, etc.) above 24 km/h (15 mph) for at least 30 seconds. If DTC returns, replace suspected harness.

Any circuitry, that is suspected as causing the intermittent complaint, should be thoroughly checked for backed out terminals, improper mating, broken locks, improperly formed or damaged terminals, poor terminal to wiring connections or physical damage to the wiring harness.

Resistance of the wheel speed sensor will increase with an increase in sensor temperature.

When replacing a wheel speed sensor, inspect the sensor terminals and harness connector for corrosion and/or water intrusion. If evidence of corrosion or water intrusion exists, replace wheel speed sensor jumper harness. Likewise, if replacing a wheel speed sensor jumper harness, inspect sensor terminals. If evidence of corrosion or water intrusion exists, replace wheel speed sensor.

DIAGNOSTIC CHART DTC 22 — 1994 CENTURY, CUTLASS CIERA AND CUTLASS CRUISER

DTC 22

RIGHT FRONT WHEEL SPEED = 0

Important

- WHEEL SPEED SENSOR INTERMITTENT MALFUNCTIONS MAY BE DIFFICULT TO LOCATE. CARE SHOULD BE TAKEN NOT TO DISTURB ANY ELECTRICAL CONNECTIONS PRIOR TO AN INDICATED STEP OF THIS CHART. THIS WILL ENSURE THAT AN INTERMITTENT CONNECTION WILL NOT BE CORRECTED BEFORE THE SOURCE OF THE MALFUNCTION IS FOUND.

(1)
- IGNITION "OFF."
- PHYSICALLY INSPECT THE RIGHT FRONT WHEEL SPEED SENSOR, RIGHT FRONT WHEEL SPEED SENSOR JUMPER HARNESS, AND CONNECTORS FOR DAMAGE. IS PHYSICAL DAMAGE INDICATED?

YES: REPLACE WHEEL SPEED SENSOR AND/OR JUMPER HARNESS IF DAMAGED. DRIVE VEHICLE ABOVE 24 km/h (15 MPH) TO ENSURE DTC 22 DOES NOT SET.

NO:

(2)
- DISCONNECT RIGHT FRONT WHEEL SPEED SENSOR DIRECTLY AT SENSOR.
- USING J 39200, MEASURE THE RESISTANCE BETWEEN SENSOR TERMINALS "A" AND "B." IS RESISTANCE BETWEEN 1020 AND 1137Ω WHEN SENSOR IS 20°C (68°F)? *(IF TEMPERATURE OF THE SENSOR IS SUSPECTED TO BE OTHER THAN 20°C (68°F), REFER TO THE TABLE BELOW FOR APPLICABLE SENSOR RESISTANCE VALUES; THE VALUES IN THE TABLE ARE FOR TEMPERATURE OF THE SENSOR, NOT AIR TEMPERATURE.)*

NO: REPLACE WHEEL SPEED SENSOR.

YES:

(3)
- WITH J 39200 STILL CONNECTED, SELECT A/C VOLTAGE SCALE.
- SPIN WHEEL BY HAND WHILE OBSERVING VOLTAGE. IS VOLTAGE AT LEAST 100 mV (VOLTAGE WILL INCREASE AS WHEEL SPEED INCREASES)?

NO: REPLACE WHEEL SPEED SENSOR.

YES:

(4)
- IGNITION "OFF."
- DISCONNECT 32-WAY EBCM CONNECTOR.
- USING J 39200, MEASURE THE RESISTANCE BETWEEN TERMINAL "E3" AND "E4" OF THE 32-WAY EBCM HARNESS CONNECTOR. DOES J 39200 DISPLAY RESISTANCE AS "OL" (INFINITE)?

NO: REPAIR SHORT BETWEEN CKTs 833 AND 872.

YES:

(5)
- INSPECT CKTs 833 AND 872 AND 32-WAY EBCM CONNECTOR FOR DAMAGE WHICH MAY RESULT IN A SHORT BETWEEN CKTs 833 AND 872 WITH ALL CONNECTORS CONNECTED. REPAIR ALL DAMAGE IF EVIDENT.
- RECONNECT ALL CONNECTORS.
- TEST DRIVE VEHICLE ABOVE 24 km/h (15 MPH) FOR AT LEAST 30 SECONDS. DOES DTC 22 SET AS A CURRENT DTC?

YES: REPLACE EBCM.

NO: MALFUNCTION IS INTERMITTENT. REFER TO "DIAGNOSTIC AIDS" ON FACING PAGE.

FRONT WHEEL SPEED SENSOR
TEMPERATURE OF SENSOR VS. SENSOR RESISTANCE (APPROXIMATE)

TEMP. (°C)	TEMP. (°F)	RESISTANCE (OHMS)
-34 TO 4	-30 TO 40	804 TO 1066
5 TO 43	41 TO 110	956 TO 1240
44 TO 93	111 TO 200	1112 TO 1463

DIAGNOSTIC CHART DTC 23 — 1994 CENTURY, CUTLASS CIERA AND CUTLASS CRUISER

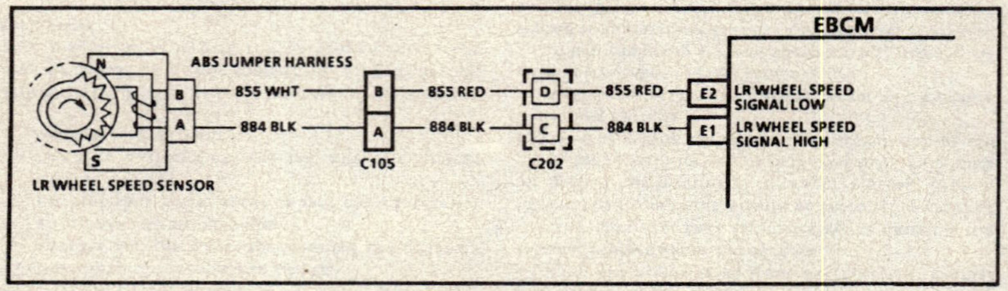

DTC 23

LEFT REAR WHEEL SPEED = 0

Circuit Description:

As a toothed ring passes by the wheel speed sensor, changes in the electromagnetic field cause the wheel speed sensor to produce a sinusoidal (AC) voltage signal whose frequency is proportional to wheel speed. The magnitude of this signal is directly related to wheel speed and the proximity of the wheel speed sensor to the toothed ring, often referred to as the air gap.

DTC Will Set When: DTC 23 can be set when the vehicle is not in an ABS stop. If the left rear wheel speed = 0 and the vehicle's reference speed is greater than 8 km/h (5 mph), a malfunction exists.

Action Taken: A malfunction DTC is stored, ABS is disabled and the ABS warning lamp is turned "ON."

DTC Chart Test Description: Number(s) below refer to circled number(s) on the diagnostic chart.

1. This step is designed to identify a wheel speed sensor or circuitry that is damaged and visibly apparent.
2. Checks to see that the wheel speed sensor is within proper resistance values.
3. Verifies that the wheel speed sensor can generate an appropriate output signal.
4. Ensures that the wheel speed sensor circuitry is not shorted together.
5. Checks to see if malfunction was due to physical damage of the circuitry.
6. Checks for an incorrect or malfunctioning hub and bearing assembly.

Diagnostic Aids: An "Intermittent" malfunction may be caused by a poor connection, rubbed through wire insulation, or a wire that is broken inside the insulation.

The frequency of the malfunction can be checked by using the enhanced diagnostic function of the Tech 1.

If the customer's comments reflect that the ABS warning lamp is "ON" only during moist environmental changes (rain, snow, vehicle wash), all wheel speed sensor circuitry should be thoroughly inspected for signs of water intrusion. If DTC is not current, clear DTCs and simulate the effects of water intrusion. Use the following procedure. Spray down the suspected area with a 5% salt water solution (two teaspoons of salt to 12 oz. of water). Test drive vehicle over various road surfaces (bumps, turns, etc.) above 24 km/h (15 mph) for at least 30 seconds. If DTC returns, replace suspected harness.

Any circuitry, that is suspected as causing the intermittent complaint, should be thoroughly checked for backed out terminals, improper mating, broken locks, improperly formed or damaged terminals, poor terminal to wiring connections or physical damage to the wiring harness.

Resistance of the wheel speed sensor will increase with an increase in sensor temperature.

When replacing a wheel speed sensor, inspect the sensor terminals and harness connector for corrosion and/or water intrusion. If evidence of corrosion or water intrusion exists, replace wheel speed sensor jumper harness. Likewise, if replacing a wheel speed sensor jumper harness, inspect sensor terminals. If evidence of corrosion or water intrusion exists, replace wheel speed sensor.

DIAGNOSTIC CHART DTC 23 — 1994 CENTURY, CUTLASS CIERA AND CUTLASS CRUISER

DTC 23
LEFT REAR WHEEL SPEED = 0

Important

- WHEEL SPEED SENSOR INTERMITTENT MALFUNCTIONS MAY BE DIFFICULT TO LOCATE. CARE SHOULD BE TAKEN NOT TO DISTURB ANY ELECTRICAL CONNECTIONS PRIOR TO AN INDICATED STEP OF THIS CHART. THIS WILL ENSURE THAT AN INTERMITTENT CONNECTION WILL NOT BE CORRECTED BEFORE THE SOURCE OF THE MALFUNCTION IS FOUND.

(1)
- IGNITION "OFF."
- PHYSICALLY INSPECT THE LEFT REAR WHEEL SPEED SENSOR, WHEEL SPEED SENSOR JUMPER HARNESS, AND CONNECTORS FOR DAMAGE.

IS PHYSICAL DAMAGE INDICATED?

- YES: REPLACE WHEEL SPEED SENSOR AND/OR WHEEL SPEED SENSOR JUMPER HARNESS IF DAMAGED. DRIVE VEHICLE ABOVE 24 km/h (15 MPH) TO ENSURE DTC 23 DOES NOT SET.
- NO: go to step 2.

(2)
- DISCONNECT LEFT REAR WHEEL SPEED SENSOR DIRECTLY AT SENSOR.
- USING J 39200, MEASURE THE RESISTANCE BETWEEN SENSOR TERMINALS "A" AND "B."

IS RESISTANCE BETWEEN 2059 AND 2353Ω WHEN SENSOR IS 20°C (68°F)?

(IF TEMPERATURE OF THE SENSOR IS SUSPECTED TO BE OTHER THAN 20°C (68°F), REFER TO THE TABLE BELOW FOR APPLICABLE SENSOR RESISTANCE VALUES; THE VALUES IN THE TABLE ARE FOR TEMPERATURE OF THE SENSOR, NOT AIR TEMPERATURE.)

- NO: REPLACE WHEEL SPEED SENSOR.
- YES: go to step 3.

(3)
- WITH J 39200 STILL CONNECTED, SELECT A/C VOLTAGE SCALE.
- SPIN WHEEL BY HAND WHILE OBSERVING VOLTAGE.

IS VOLTAGE AT LEAST 100 mV (VOLTAGE WILL INCREASE AS WHEEL SPEED INCREASES)?

- NO: REPLACE WHEEL SPEED SENSOR.
- YES: go to step 4.

(4)
- IGNITION "OFF."
- DISCONNECT 32-WAY EBCM CONNECTOR.
- USING J 39200, MEASURE THE RESISTANCE BETWEEN TERMINAL "E1" AND "E2" OF THE 32-WAY EBCM HARNESS CONNECTOR.

DOES J 39200 DISPLAY RESISTANCE AS "OL" (INFINITE)?

- NO: REPAIR SHORT BETWEEN CKTs 884 AND 885.
- YES: go to step 5.

(5)
- INSPECT CKTs 884 AND 885 AND 32-WAY EBCM CONNECTOR FOR DAMAGE WHICH MAY RESULT IN A SHORT BETWEEN CKTs 884 AND 885 WITH ALL CONNECTORS CONNECTED. REPAIR ALL DAMAGE IF EVIDENT.
- RECONNECT ALL CONNECTORS.
- TEST DRIVE VEHICLE ABOVE 24 km/h (15 MPH) FOR AT LEAST 30 SECONDS.

DOES DTC 23 SET AS A CURRENT DTC?

- NO: MALFUNCTION WAS MOST LIKELY CAUSED BY A POOR CONNECTION. REFER TO "DIAGNOSTIC AIDS" ON FACING PAGE.
- YES: go to step 6.

(6)
- USING TECH 1, SELECT "F0: DATA LIST."
- MONITOR WHEEL SPEEDS WHILE SLOWLY ACCELERATING TO 65 km/h (40 MPH) AND THEN SLOWLY DECELERATING TO 0 km/h (0 MPH).

IS THE LEFT REAR WHEEL SPEED CONSTANTLY HIGHER THAN THE THREE REMAINING WHEEL SPEEDS?

- YES: REPLACE INCORRECT OR MALFUNCTIONING LEFT WHEEL SPEED SENSOR/HUB AND BEARING ASSEMBLY WITH THE PROPER ASSEMBLY. REFER TO PARTS CATALOG FOR PROPER PART NUMBER.
- NO: REPLACE EBCM.

REAR WHEEL SPEED SENSOR
TEMPERATURE OF SENSOR VS. SENSOR RESISTANCE (APPROXIMATE)

TEMP. (°C)	TEMP. (°F)	RESISTANCE (OHMS)
-34 TO 4	-30 TO 40	1622 TO 2205
5 TO 43	41 TO 110	1930 TO 2566
44 TO 93	111 TO 200	2245 TO 3028

DIAGNOSTIC CHART DTC 24 — 1994 CENTURY, CUTLASS CIERA AND CUTLASS CRUISER

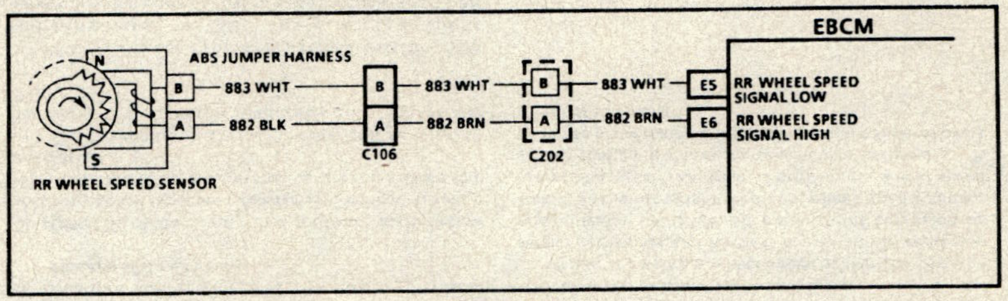

DTC 24

RIGHT REAR WHEEL SPEED = 0

Circuit Description:

As a toothed ring passes by the wheel speed sensor, changes in the electromagnetic field cause the wheel speed sensor to produce a sinusoidal (AC) voltage signal whose frequency is proportional to wheel speed. The magnitude of this signal is directly related to wheel speed and the proximity of the wheel speed sensor to the toothed ring, often referred to as the air gap.

DTC Will Set When: DTC 24 can be set when the vehicle is not in an ABS stop. If the right rear wheel speed = 0 and the vehicle's reference speed is greater than 8 km/h (5 mph), a malfunction exists.

Action Taken: A malfunction DTC is stored, ABS is disabled and the ABS warning lamp is turned "ON."

DTC Chart Test Description: Number(s) below refer to circled number(s) on the diagnostic chart.

1. This step is designed to identify a wheel speed sensor or circuitry that is damaged and visibly apparent.
2. Checks to see that the wheel speed sensor is within proper resistance values.
3. Verifies that the wheel speed sensor can generate an appropriate output signal.
4. Ensures that the wheel speed sensor circuitry is not shorted together.
5. Checks to see if malfunction was due to physical damage of the circuitry.
6. Checks for an incorrect or malfunctioning hub and bearing assembly.

Diagnostic Aids: An "Intermittent" malfunction may be caused by a poor connection, rubbed through wire insulation, or a wire that is broken inside the insulation.

The frequency of the malfunction can be checked by using the enhanced diagnostic function of the Tech 1.

If the customer's comments reflect that the ABS warning lamp is "ON" only during moist environmental changes (rain, snow, vehicle wash), all wheel speed sensor circuitry should be thoroughly inspected for signs of water intrusion. If DTC is not current, clear DTCs and simulate the effects of water intrusion. Use the following procedure. Spray down the suspected area with a 5% salt water solution (two teaspoons of salt to 12 oz. of water). Test drive vehicle over various road surfaces (bumps, turns, etc.) above 24 km/h (15 mph) for at least 30 seconds. If DTC returns, replace suspected harness.

Any circuitry, that is suspected as causing the intermittent complaint, should be thoroughly checked for backed out terminals, improper mating, broken locks, improperly formed or damaged terminals, poor terminal to wiring connections or physical damage to the wiring harness.

Resistance of the wheel speed sensor will increase with an increase in sensor temperature.

When replacing a wheel speed sensor, inspect the sensor terminals and harness connector for corrosion and/or water intrusion. If evidence of corrosion or water intrusion exists, replace wheel speed sensor jumper harness. Likewise, if replacing a wheel speed sensor jumper harness, inspect sensor terminals. If evidence of corrosion or water intrusion exists, replace wheel speed sensor.

DIAGNOSTIC CHART DTC 24 — 1994 CENTURY, CUTLASS CIERA AND CUTLASS CRUISER

DTC 24
RIGHT REAR WHEEL SPEED = 0

Important
- WHEEL SPEED SENSOR INTERMITTENT MALFUNCTIONS MAY BE DIFFICULT TO LOCATE. CARE SHOULD BE TAKEN NOT TO DISTURB ANY ELECTRICAL CONNECTIONS PRIOR TO AN INDICATED STEP OF THIS CHART. THIS WILL ENSURE THAT AN INTERMITTENT CONNECTION WILL NOT BE CORRECTED BEFORE THE SOURCE OF THE MALFUNCTION IS FOUND.

(1)
- IGNITION "OFF."
- PHYSICALLY INSPECT THE RIGHT REAR WHEEL SPEED SENSOR, WHEEL SPEED SENSOR JUMPER HARNESS, AND CONNECTORS FOR DAMAGE. IS PHYSICAL DAMAGE INDICATED?

YES → REPLACE WHEEL SPEED SENSOR AND/OR WHEEL SPEED SENSOR JUMPER HARNESS IF DAMAGED. DRIVE VEHICLE ABOVE 24 km/h (15 MPH) TO ENSURE DTC 24 DOES NOT SET.

NO ↓

(2)
- DISCONNECT RIGHT REAR WHEEL SPEED SENSOR DIRECTLY AT SENSOR.
- USING J 39200, MEASURE THE RESISTANCE BETWEEN SENSOR TERMINALS "A" AND "B." IS RESISTANCE BETWEEN 2059 AND 2353Ω WHEN SENSOR IS 20°C (68°F)? *(IF TEMPERATURE OF THE SENSOR IS SUSPECTED TO BE OTHER THAN 20°C (68°F), REFER TO THE TABLE BELOW FOR APPLICABLE SENSOR RESISTANCE VALUES; THE VALUES IN THE TABLE ARE FOR TEMPERATURE OF THE SENSOR, NOT AIR TEMPERATURE.)*

NO → REPLACE WHEEL SPEED SENSOR.

YES ↓

(3)
- WITH J 39200 STILL CONNECTED, SELECT A/C VOLTAGE SCALE.
- SPIN WHEEL BY HAND WHILE OBSERVING VOLTAGE. IS VOLTAGE AT LEAST 100 mV (VOLTAGE WILL INCREASE AS WHEEL SPEED INCREASES)?

NO → REPLACE WHEEL SPEED SENSOR.

YES ↓

(4)
- IGNITION "OFF."
- DISCONNECT 32-WAY EBCM CONNECTOR.
- USING J 39200, MEASURE THE RESISTANCE BETWEEN TERMINAL "E5" AND "E6" OF THE 32-WAY EBCM HARNESS CONNECTOR. DOES J 39200 DISPLAY RESISTANCE AS "OL" (INFINITE)?

NO → REPAIR SHORT BETWEEN CKTs 883 AND 883.

YES ↓

(5)
- INSPECT CKTs 882 AND 883 AND 32-WAY EBCM CONNECTOR FOR DAMAGE WHICH MAY RESULT IN A SHORT BETWEEN CKTs 882 AND 883 WITH ALL CONNECTORS CONNECTED. REPAIR ALL DAMAGE IF EVIDENT.
- RECONNECT ALL CONNECTORS.
- TEST DRIVE VEHICLE ABOVE 24 km/h (15 MPH) FOR AT LEAST 30 SECONDS. DOES DTC 24 SET AS A CURRENT DTC?

NO → MALFUNCTION WAS MOST LIKELY CAUSED BY A POOR CONNECTION. REFER TO "DIAGNOSTIC AIDS" ON FACING PAGE.

YES ↓

(6)
- USING TECH 1, SELECT "F0: DATA LIST."
- MONITOR WHEEL SPEEDS WHILE SLOWLY ACCELERATING TO 65 km/h (40 MPH) AND THEN SLOWLY DECELERATING TO 0 km/h (0 MPH). IS THE RIGHT REAR WHEEL SPEED CONSTANTLY HIGHER THAN THE THREE REMAINING WHEEL SPEEDS?

NO → REPLACE EBCM.

YES → REPLACE INCORRECT OR MALFUNCTIONING RIGHT WHEEL SPEED SENSOR/HUB AND BEARING ASSEMBLY WITH THE PROPER ASSEMBLY. REFER TO PARTS CATALOG FOR PROPER PART NUMBER.

REAR WHEEL SPEED SENSOR
TEMPERATURE OF SENSOR VS. SENSOR RESISTANCE (APPROXIMATE)

TEMP. (°C)	TEMP. (°F)	RESISTANCE (OHMS)
-34 TO 4	-30 TO 40	1622 TO 2205
5 TO 43	41 TO 110	1930 TO 2566
44 TO 93	111 TO 200	2245 TO 3028

DIAGNOSTIC CHART DTC 25 — 1994 CENTURY, CUTLASS CIERA AND CUTLASS CRUISER

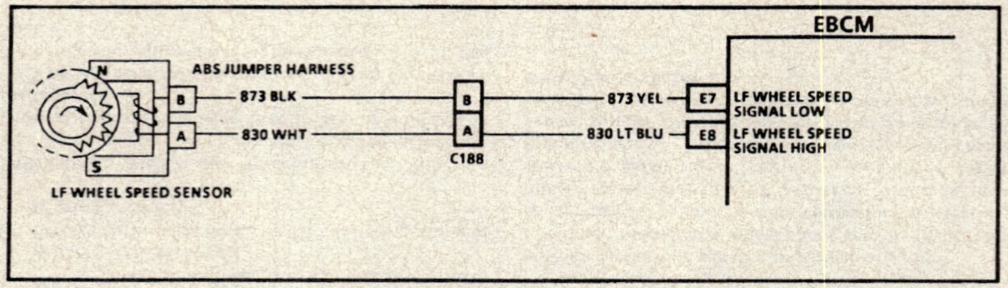

DTC 25

LEFT FRONT EXCESSIVE WHEEL SPEED VARIATION

Circuit Description:
As a toothed ring passes by the wheel speed sensor, changes in the electromagnetic field cause the wheel speed sensor to produce a sinusoidal (AC) voltage signal whose frequency is proportional to wheel speed. The magnitude of this signal is directly related to wheel speed and the proximity of the wheel speed sensor to the toothed ring, often referred to as the air gap.

DTC Will Set When: DTC 25 can be set when the brake is "OFF." The purpose of this test is to detect a situation in which the left front wheel acceleration or deceleration is beyond specified limits.

Action Taken: A malfunction DTC is stored, ABS is disabled and the ABS warning lamp is turned "ON."

DTC Chart Test Description: Number(s) below refer to circled number(s) on the diagnostic chart.

1. This step is designed to identify a wheel speed sensor or circuitry that is damaged and visibly apparent.
2. Checks to see that the wheel speed sensor is within proper resistance values.
3. Verifies that the wheel speed sensor can generate an appropriate output signal.
4. Ensures that wheel speed sensor circuitry is not shorted together.
5. Ensures malfunction was not due to physical damage of the circuitry.

Diagnostic Aids: An "Intermittent" malfunction may be caused by a poor connection, rubbed through wire insulation, or a wire that is broken inside the insulation.

The frequency of the malfunction can be checked by using the enhanced diagnostic function of the Tech 1.

If the customer's comments reflect that the ABS warning lamp is "ON" only during moist environmental changes (rain, snow, vehicle wash), all wheel speed sensor circuitry should be thoroughly inspected for signs of water intrusion. If DTC is not current, clear DTCs and simulate the effects of water intrusion. Use the following procedure. Spray down the suspected area with a 5% salt water solution (two teaspoons of salt to 12 oz. of water). Test drive vehicle over various road surfaces (bumps, turns, etc.) above 24 km/h (15 mph) for at least 30 seconds. If DTC returns, replace suspected harness.

Any circuitry, that is suspected as causing the intermittent complaint, should be thoroughly checked for backed out terminals, improper mating, broken locks, improperly formed or damaged terminals, poor terminal to wiring connections or physical damage to the wiring harness.

Resistance of the wheel speed sensor will increase with an increase in sensor temperature.

When replacing a wheel speed sensor, inspect the sensor terminals and harness connector for corrosion and/or water intrusion. If evidence of corrosion or water intrusion exists, replace wheel speed sensor jumper harness. Likewise, if replacing a wheel speed sensor jumper harness, inspect sensor terminals. If evidence of corrosion or water intrusion exists, replace wheel speed sensor.

DIAGNOSTIC CHART DTC 25 — 1994 CENTURY, CUTLASS CIERA AND CUTLASS CRUISER

DTC 25
LEFT FRONT EXCESSIVE WHEEL SPEED VARIATION

Important
- WHEEL SPEED SENSOR INTERMITTENT MALFUNCTIONS MAY BE DIFFICULT TO LOCATE. CARE SHOULD BE TAKEN NOT TO DISTURB ANY ELECTRICAL CONNECTIONS PRIOR TO AN INDICATED STEP OF THIS CHART. THIS WILL ENSURE THAT AN INTERMITTENT CONNECTION WILL NOT BE CORRECTED BEFORE THE SOURCE OF THE MALFUNCTION IS FOUND.

(1)
- IGNITION "OFF."
- PHYSICALLY INSPECT THE LEFT FRONT WHEEL SPEED SENSOR, LEFT FRONT WHEEL SPEED SENSOR JUMPER HARNESS, AND CONNECTORS FOR DAMAGE. IS PHYSICAL DAMAGE INDICATED?

YES: REPLACE WHEEL SPEED SENSOR AND/OR JUMPER HARNESS IF DAMAGED. DRIVE VEHICLE ABOVE 24 km/h (15 MPH) TO ENSURE DTC 25 DOES NOT SET.

NO:

(2)
- DISCONNECT LEFT FRONT WHEEL SPEED SENSOR DIRECTLY AT SENSOR.
- USING J 39200, MEASURE THE RESISTANCE BETWEEN SENSOR TERMINALS "A" AND "B."
IS RESISTANCE BETWEEN 1020 AND 1137Ω WHEN SENSOR IS 20°C (68°F)? *(IF TEMPERATURE OF THE SENSOR IS SUSPECTED TO BE OTHER THAN 20°C (68°F), REFER TO THE TABLE BELOW FOR APPLICABLE SENSOR RESISTANCE VALUES; THE VALUES IN THE TABLE ARE FOR TEMPERATURE OF THE SENSOR, NOT AIR TEMPERATURE.)*

NO: REPLACE WHEEL SPEED SENSOR.

YES:

(3)
- WITH J 39200 STILL CONNECTED, SELECT A/C VOLTAGE SCALE.
- SPIN WHEEL BY HAND WHILE OBSERVING VOLTAGE.
IS VOLTAGE AT LEAST 100 mV (VOLTAGE WILL INCREASE AS WHEEL SPEED INCREASES)?

NO: REPLACE WHEEL SPEED SENSOR.

YES:

(4)
- IGNITION "OFF."
- DISCONNECT 32-WAY EBCM CONNECTOR.
- USING J 39200, MEASURE THE RESISTANCE BETWEEN TERMINAL "E7" AND "E8" OF THE 32-WAY EBCM HARNESS CONNECTOR.
DOES J 39200 DISPLAY RESISTANCE AS "OL" (INFINITE)?

NO: REPAIR SHORT BETWEEN CKTs 830 AND 873.

YES:

(5)
- INSPECT CKTs 830 AND 873 AND 32-WAY EBCM CONNECTOR FOR DAMAGE WHICH MAY RESULT IN A SHORT BETWEEN CKTs 830 AND 873 WITH ALL CONNECTORS CONNECTED. REPAIR ALL DAMAGE IF EVIDENT.
- RECONNECT ALL CONNECTORS.
- TEST DRIVE VEHICLE ABOVE 24 km/h (15 MPH) FOR AT LEAST 30 SECONDS.
DOES DTC 25 SET AS A CURRENT DTC?

YES: REPLACE EBCM.

NO: MALFUNCTION IS INTERMITTENT. REFER TO "DIAGNOSTIC AIDS" ON FACING PAGE.

FRONT WHEEL SPEED SENSOR
TEMPERATURE OF SENSOR VS. SENSOR RESISTANCE (APPROXIMATE)

TEMP. (°C)	TEMP. (°F)	RESISTANCE (OHMS)
-34 TO 4	-30 TO 40	804 TO 1066
5 TO 43	41 TO 110	956 TO 1240
44 TO 93	111 TO 200	1112 TO 1463

DIAGNOSTIC CHART DTC 26 — 1994 CENTURY, CUTLASS CIERA AND CUTLASS CRUISER

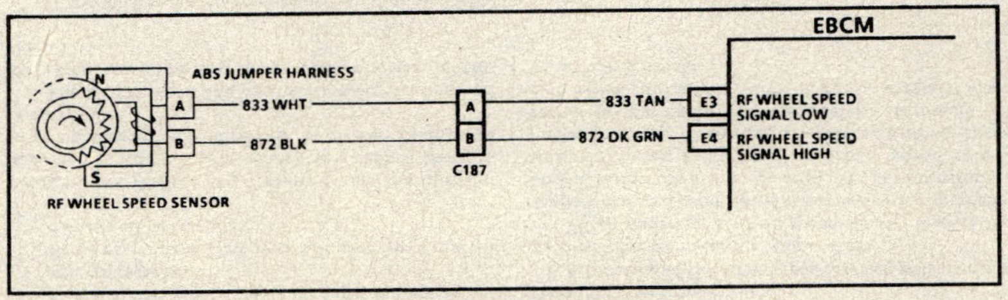

DTC 26

RIGHT FRONT EXCESSIVE WHEEL SPEED VARIATION

Circuit Description:

As a toothed ring passes by the wheel speed sensor, changes in the electromagnetic field cause the wheel speed sensor to produce a sinusoidal (AC) voltage signal whose frequency is proportional to wheel speed. The magnitude of this signal is directly related to wheel speed and the proximity of the wheel speed sensor to the toothed ring, often referred to as the air gap.

DTC Will Set When: DTC 26 can be set when the brake is "OFF." The purpose of this test is to detect a situation in which the right front wheel acceleration or deceleration is beyond specified limits.

Action Taken: A malfunction DTC is stored, ABS is disabled and the ABS warning lamp is turned "ON."

DTC Chart Test Description: Number(s) below refer to circled number(s) on the diagnostic chart.

1. This step is designed to identify a wheel speed sensor or circuitry that is damaged and visibly apparent.
2. Checks to see that the wheel speed sensor is within proper resistance values.
3. Verifies that the wheel speed sensor can generate an appropriate output signal.
4. Ensures that wheel speed sensor circuitry is not shorted together.
5. Ensures malfunction was not due to physical damage of the circuitry.

Diagnostic Aids: An "Intermittent" malfunction may be caused by a poor connection, rubbed through wire insulation, or a wire that is broken inside the insulation.

The frequency of the malfunction can be checked by using the enhanced diagnostic function of the Tech 1.

If the customer's comments reflect that the ABS warning lamp is "ON" only during moist environmental changes (rain, snow, vehicle wash), all wheel speed sensor circuitry should be thoroughly inspected for signs of water intrusion. If DTC is not current, clear DTCs and simulate the effects of water intrusion. Use the following procedure. Spray down the suspected area with a 5% salt water solution (two teaspoons of salt to 12 oz. of water). Test drive vehicle over various road surfaces (bumps, turns, etc.) above 24 km/h (15 mph) for at least 30 seconds. If DTC returns, replace suspected harness.

Any circuitry, that is suspected as causing the intermittent complaint, should be thoroughly checked for backed out terminals, improper mating, broken locks, improperly formed or damaged terminals, poor terminal to wiring connections or physical damage to the wiring harness.

Resistance of the wheel speed sensor will increase with an increase in sensor temperature.

When replacing a wheel speed sensor, inspect the sensor terminals and harness connector for corrosion and/or water intrusion. If evidence of corrosion or water intrusion exists, replace wheel speed sensor jumper harness. Likewise, if replacing a wheel speed sensor jumper harness, inspect sensor terminals. If evidence of corrosion or water intrusion exists, replace wheel speed sensor.

DIAGNOSTIC CHART DTC 26 — 1994 CENTURY, CUTLASS CIERA AND CUTLASS CRUISER

DTC 26 RIGHT FRONT EXCESSIVE WHEEL SPEED VARIATION

Important

- WHEEL SPEED SENSOR INTERMITTENT MALFUNCTIONS MAY BE DIFFICULT TO LOCATE. CARE SHOULD BE TAKEN NOT TO DISTURB ANY ELECTRICAL CONNECTIONS PRIOR TO AN INDICATED STEP OF THIS CHART. THIS WILL ENSURE THAT AN INTERMITTENT CONNECTION WILL NOT BE CORRECTED BEFORE THE SOURCE OF THE MALFUNCTION IS FOUND.

(1)
- IGNITION "OFF."
- PHYSICALLY INSPECT THE RIGHT FRONT WHEEL SPEED SENSOR, RIGHT FRONT WHEEL SPEED SENSOR JUMPER HARNESS, AND CONNECTORS FOR DAMAGE.

IS PHYSICAL DAMAGE INDICATED?

YES → REPLACE WHEEL SPEED SENSOR AND/OR JUMPER HARNESS IF DAMAGED. DRIVE VEHICLE ABOVE 24 km/h (15 MPH) TO ENSURE DTC 26 DOES NOT SET.

NO ↓

(2)
- DISCONNECT RIGHT FRONT WHEEL SPEED SENSOR DIRECTLY AT SENSOR.
- USING J 39200, MEASURE THE RESISTANCE BETWEEN SENSOR TERMINALS "A" AND "B."

IS RESISTANCE BETWEEN 1020 AND 1137Ω WHEN SENSOR IS 20°C (68°F)? *(IF TEMPERATURE OF THE SENSOR IS SUSPECTED TO BE OTHER THAN 20°C (68°F), REFER TO THE TABLE BELOW FOR APPLICABLE SENSOR RESISTANCE VALUES; THE VALUES IN THE TABLE ARE FOR TEMPERATURE OF THE SENSOR, NOT AIR TEMPERATURE.)*

NO → REPLACE WHEEL SPEED SENSOR.

YES ↓

(3)
- WITH J 39200 STILL CONNECTED, SELECT A/C VOLTAGE SCALE.
- SPIN WHEEL BY HAND WHILE OBSERVING VOLTAGE.

IS VOLTAGE AT LEAST 100 mV (VOLTAGE WILL INCREASE AS WHEEL SPEED INCREASES)?

NO → REPLACE WHEEL SPEED SENSOR.

YES ↓

(4)
- IGNITION "OFF."
- DISCONNECT 32-WAY EBCM CONNECTOR.
- USING J 39200, MEASURE THE RESISTANCE BETWEEN TERMINAL "E3" AND "E4" OF THE 32-WAY EBCM HARNESS CONNECTOR.

DOES J 39200 DISPLAY RESISTANCE AS "OL" (INFINITE)?

NO → REPAIR SHORT BETWEEN CKTs 833 AND 872.

YES ↓

(5)
- INSPECT CKTs 833 AND 872 AND 32-WAY EBCM CONNECTOR FOR DAMAGE WHICH MAY RESULT IN A SHORT BETWEEN CKTs 833 AND 872 WITH ALL CONNECTORS CONNECTED. REPAIR ALL DAMAGE IF EVIDENT.
- RECONNECT ALL CONNECTORS.
- TEST DRIVE VEHICLE ABOVE 24 km/h (15 MPH) FOR AT LEAST 30 SECONDS.

DOES DTC 26 SET AS A CURRENT DTC?

YES → REPLACE EBCM.

NO → MALFUNCTION IS INTERMITTENT. REFER TO "DIAGNOSTIC AIDS" ON FACING PAGE.

FRONT WHEEL SPEED SENSOR
TEMPERATURE OF SENSOR VS. SENSOR RESISTANCE (APPROXIMATE)

TEMP. (°C)	TEMP. (°F)	RESISTANCE (OHMS)
-34 TO 4	-30 TO 40	804 TO 1066
5 TO 43	41 TO 110	956 TO 1240
44 TO 93	111 TO 200	1112 TO 1463

DIAGNOSTIC CHART DTC 27 — 1994 CENTURY, CUTLASS CIERA AND CUTLASS CRUISER

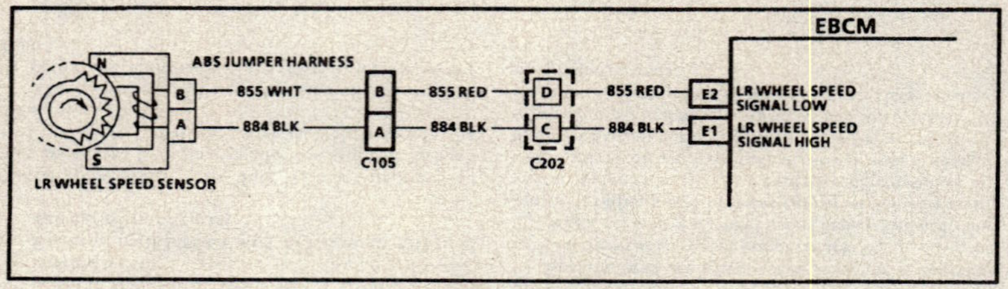

DTC 27

LEFT REAR EXCESSIVE WHEEL SPEED VARIATION

Circuit Description:

As a toothed ring passes by the wheel speed sensor, changes in the electromagnetic field cause the wheel speed sensor to produce a sinusoidal (AC) voltage signal whose frequency is proportional to wheel speed. The magnitude of this signal is directly related to wheel speed and the proximity of the wheel speed sensor to the toothed ring, often referred to as the air gap.

DTC Will Set When: DTC 27 can be set when the brake is "OFF." The purpose of this test is to detect a situation in which the left rear wheel acceleration or deceleration is beyond specified limits.

Action Taken: A malfunction DTC is stored, ABS is disabled and the ABS warning lamp is turned "ON."

DTC Chart Test Description: Number(s) below refer to circled number(s) on the diagnostic chart.

1. This step is designed to identify a wheel speed sensor or circuitry that is damaged and visibly apparent.
2. Checks to see that the wheel speed sensor is within proper resistance values.
3. Verifies that the wheel speed sensor can generate an appropriate output signal.
4. Ensures that wheel speed sensor circuitry is not shorted together.
5. Ensures malfunction was not due to physical damage of the circuitry.

Diagnostic Aids: An "Intermittent" malfunction may be caused by a poor connection, rubbed through wire insulation, or a wire that is broken inside the insulation.

The frequency of the malfunction can be checked by using the enhanced diagnostic function of the Tech 1

If the customer's comments reflect that the ABS warning lamp is "ON" only during moist environmental changes (rain, snow, vehicle wash), all wheel speed sensor circuitry should be thoroughly inspected for signs of water intrusion. If DTC is not current, clear DTCs and simulate the effects of water intrusion. Use the following procedure. Spray down the suspected area with a 5% salt water solution (two teaspoons of salt to 12 oz. of water). Test drive vehicle over various road surfaces (bumps, turns, etc.) above 24 km/h (15 mph) for at least 30 seconds. If DTC returns, replace suspected harness.

Any circuitry, that is suspected as causing the intermittent complaint, should be thoroughly checked for backed out terminals, improper mating, broken locks, improperly formed or damaged terminals, poor terminal to wiring connections or physical damage to the wiring harness.

Resistance of the wheel speed sensor will increase with an increase in sensor temperature.

When replacing a wheel speed sensor, inspect the sensor terminals and harness connector for corrosion and/or water intrusion. If evidence of corrosion or water intrusion exists, replace wheel speed sensor jumper harness. Likewise, if replacing a wheel speed sensor jumper harness, inspect sensor terminals. If evidence of corrosion or water intrusion exists, replace wheel speed sensor.

DIAGNOSTIC CHART DTC 27 — 1994 CENTURY, CUTLASS CIERA AND CUTLASS CRUISER

DTC 27
LEFT REAR EXCESSIVE WHEEL SPEED VARIATION

Important

- WHEEL SPEED SENSOR INTERMITTENT MALFUNCTIONS MAY BE DIFFICULT TO LOCATE. CARE SHOULD BE TAKEN NOT TO DISTURB ANY ELECTRICAL CONNECTIONS PRIOR TO AN INDICATED STEP OF THIS CHART. THIS WILL ENSURE THAT AN INTERMITTENT CONNECTION WILL NOT BE CORRECTED BEFORE THE SOURCE OF THE MALFUNCTION IS FOUND.

(1)
- IGNITION "OFF."
- PHYSICALLY INSPECT THE LEFT REAR WHEEL SPEED SENSOR, WHEEL SPEED SENSOR JUMPER HARNESS, AND CONNECTORS FOR DAMAGE. IS PHYSICAL DAMAGE INDICATED?

YES: REPLACE WHEEL SPEED SENSOR AND/OR WHEEL SPEED SENSOR JUMPER HARNESS IF DAMAGED. DRIVE VEHICLE ABOVE 24 km/h (15 MPH) TO ENSURE DTC 27 DOES NOT SET.

NO:

(2)
- DISCONNECT LEFT REAR WHEEL SPEED SENSOR DIRECTLY AT SENSOR.
- USING J 39200, MEASURE THE RESISTANCE BETWEEN SENSOR TERMINALS "A" AND "B."

IS RESISTANCE BETWEEN 2059 AND 2353Ω WHEN SENSOR IS 20°C (68°F)? *(IF TEMPERATURE OF THE SENSOR IS SUSPECTED TO BE OTHER THAN 20°C (68°F), REFER TO THE TABLE BELOW FOR APPLICABLE SENSOR RESISTANCE VALUES; THE VALUES IN THE TABLE ARE FOR TEMPERATURE OF THE SENSOR, NOT AIR TEMPERATURE.)*

NO: REPLACE WHEEL SPEED SENSOR.

YES:

(3)
- WITH J 39200 STILL CONNECTED, SELECT A/C VOLTAGE SCALE.
- SPIN WHEEL BY HAND WHILE OBSERVING VOLTAGE.

IS VOLTAGE AT LEAST 100 mV (VOLTAGE WILL INCREASE AS WHEEL SPEED INCREASES)?

NO: REPLACE WHEEL SPEED SENSOR.

YES:

(4)
- IGNITION "OFF."
- DISCONNECT 32-WAY EBCM CONNECTOR.
- USING J 39200, MEASURE THE RESISTANCE BETWEEN TERMINAL "E1" AND "E2" OF THE 32-WAY EBCM HARNESS CONNECTOR.

DOES J 39200 DISPLAY RESISTANCE AS "OL" (INFINITE)?

NO: REPAIR SHORT BETWEEN CKTs 884 AND 885.

YES:

(5)
- INSPECT CKTs 884 AND 885 AND 32-WAY EBCM CONNECTOR FOR DAMAGE WHICH MAY RESULT IN A SHORT BETWEEN CKTs 884 AND 885 WITH ALL CONNECTORS CONNECTED. REPAIR ALL DAMAGE IF EVIDENT.
- RECONNECT ALL CONNECTORS.
- TEST DRIVE VEHICLE ABOVE 24 km/h (15 MPH) FOR AT LEAST 30 SECONDS.

DOES DTC 27 SET AS A CURRENT DTC?

YES: REPLACE EBCM.

NO: MALFUNCTION IS INTERMITTENT. REFER TO "DIAGNOSTIC AIDS" ON FACING PAGE.

REAR WHEEL SPEED SENSOR
TEMPERATURE OF SENSOR VS. SENSOR RESISTANCE (APPROXIMATE)

TEMP. (°C)	TEMP. (°F)	RESISTANCE (OHMS)
-34 TO 4	-30 TO 40	1622 TO 2205
5 TO 43	41 TO 110	1930 TO 2566
44 TO 93	111 TO 200	2245 TO 3028

DIAGNOSTIC CHART DTC 28 — 1994 CENTURY, CUTLASS CIERA AND CUTLASS CRUISER

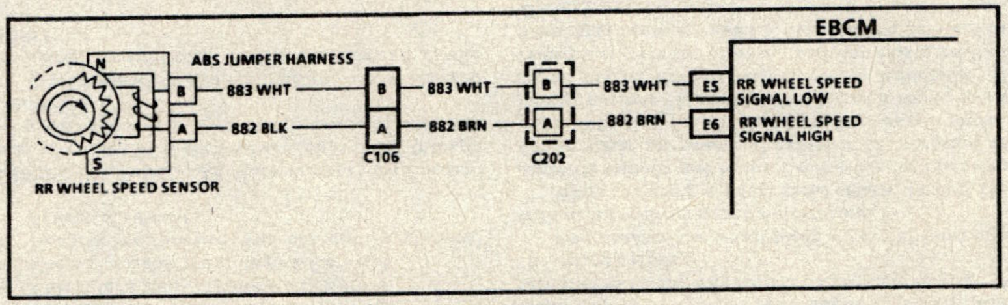

DTC 28

RIGHT REAR EXCESSIVE WHEEL SPEED VARIATION

Circuit Description:
As a toothed ring passes by the wheel speed sensor, changes in the electromagnetic field cause the wheel speed sensor to produce a sinusoidal (AC) voltage signal whose frequency is proportional to wheel speed. The magnitude of this signal is directly related to wheel speed and the proximity of the wheel speed sensor to the toothed ring, often referred to as the air gap.

DTC Will Set When: DTC 28 can be set when the brake is "OFF." The purpose of this test is to detect a situation in which the right rear wheel acceleration or deceleration is beyond specified limits.

Action Taken: A malfunction DTC is stored, ABS is disabled and the ABS warning lamp is turned "ON."

DTC Chart Test Description: Number(s) below refer to circled number(s) on the diagnostic chart.

1. This step is designed to identify a wheel speed sensor or circuitry that is damaged and visibly apparent.
2. Checks to see that the wheel speed sensor is within proper resistance values.
3. Verifies that the wheel speed sensor can generate an appropriate output signal.
4. Ensures that wheel speed sensor circuitry is not shorted together.
5. Ensures malfunction was not due to physical damage of the circuitry.

Diagnostic Aids: An "Intermittent" malfunction may be caused by a poor connection, rubbed through wire insulation, or a wire that is broken inside the insulation.

The frequency of the malfunction can be checked by using the enhanced diagnostic function of the Tech 1.

If the customer's comments reflect that the ABS warning lamp is "ON" only during moist environmental changes (rain, snow, vehicle wash), all wheel speed sensor circuitry should be thoroughly inspected for signs of water intrusion. If DTC is not current, clear DTCs and simulate the effects of water intrusion. Use the following procedure. Spray down the suspected area with a 5% salt water solution (two teaspoons of salt to 12 oz. of water). Test drive vehicle over various road surfaces (bumps, turns, etc.) above 24 km/h (15 mph) for at least 30 seconds. If DTC returns, replace suspected harness.

Any circuitry, that is suspected as causing the intermittent complaint, should be thoroughly checked for backed out terminals, improper mating, broken locks, improperly formed or damaged terminals, poor terminal to wiring connections or physical damage to the wiring harness.

Resistance of the wheel speed sensor will increase with an increase in sensor temperature.

When replacing a wheel speed sensor, inspect the sensor terminals and harness connector for corrosion and/or water intrusion. If evidence of corrosion or water intrusion exists, replace wheel speed sensor jumper harness. Likewise, if replacing a wheel speed sensor jumper harness, inspect sensor terminals. If evidence of corrosion or water intrusion exists, replace wheel speed sensor.

DIAGNOSTIC CHART DTC 28 — 1994 CENTURY, CUTLASS CIERA AND CUTLASS CRUISER

DTC 28
RIGHT REAR EXCESSIVE WHEEL SPEED VARIATION

Important

- WHEEL SPEED SENSOR INTERMITTENT MALFUNCTIONS MAY BE DIFFICULT TO LOCATE. CARE SHOULD BE TAKEN NOT TO DISTURB ANY ELECTRICAL CONNECTIONS PRIOR TO AN INDICATED STEP OF THIS CHART. THIS WILL ENSURE THAT AN INTERMITTENT CONNECTION WILL NOT BE CORRECTED BEFORE THE SOURCE OF THE MALFUNCTION IS FOUND.

(1)
- IGNITION "OFF."
- PHYSICALLY INSPECT THE RIGHT REAR WHEEL SPEED SENSOR, WHEEL SPEED SENSOR JUMPER HARNESS, AND CONNECTORS FOR DAMAGE. IS PHYSICAL DAMAGE INDICATED?

YES → REPLACE WHEEL SPEED SENSOR AND/OR WHEEL SPEED SENSOR JUMPER HARNESS IF DAMAGED. DRIVE VEHICLE ABOVE 24 km/h (15 MPH) TO ENSURE DTC 28 DOES NOT SET.

NO →

(2)
- DISCONNECT RIGHT REAR WHEEL SPEED SENSOR DIRECTLY AT SENSOR.
- USING J 39200, MEASURE THE RESISTANCE BETWEEN SENSOR TERMINALS "A" AND "B."
IS RESISTANCE BETWEEN 2059 AND 2353Ω WHEN SENSOR IS 20°C (68°F)?
(IF TEMPERATURE OF THE SENSOR IS SUSPECTED TO BE OTHER THAN 20°C (68°F), REFER TO THE TABLE BELOW FOR APPLICABLE SENSOR RESISTANCE VALUES; THE VALUES IN THE TABLE ARE FOR TEMPERATURE OF THE SENSOR, NOT AIR TEMPERATURE.)

NO → REPLACE WHEEL SPEED SENSOR.

YES →

(3)
- WITH J 39200 STILL CONNECTED, SELECT A/C VOLTAGE SCALE.
- SPIN WHEEL BY HAND WHILE OBSERVING VOLTAGE.
IS VOLTAGE AT LEAST 100 mV (VOLTAGE WILL INCREASE AS WHEEL SPEED INCREASES)?

NO → REPLACE WHEEL SPEED SENSOR.

YES →

(4)
- IGNITION "OFF."
- DISCONNECT 32-WAY EBCM CONNECTOR.
- USING J 39200, MEASURE THE RESISTANCE BETWEEN TERMINAL "E5" AND "E6" OF THE 32-WAY EBCM HARNESS CONNECTOR.
DOES J 39200 DISPLAY RESISTANCE AS "OL" (INFINITE)?

NO → REPAIR SHORT BETWEEN CKTs 882 AND 883.

YES →

(5)
- INSPECT CKTs 882 AND 883 AND 32-WAY EBCM CONNECTOR FOR DAMAGE WHICH MAY RESULT IN A SHORT BETWEEN CKTs 882 AND 883 WITH ALL CONNECTORS CONNECTED. REPAIR ALL DAMAGE IF EVIDENT.
- RECONNECT ALL CONNECTORS.
- TEST DRIVE VEHICLE ABOVE 24 km/h (15 MPH) FOR AT LEAST 30 SECONDS.
DOES DTC 28 SET AS A CURRENT DTC?

YES → REPLACE EBCM.

NO → MALFUNCTION IS INTERMITTENT. REFER TO "DIAGNOSTIC AIDS" ON FACING PAGE.

REAR WHEEL SPEED SENSOR
TEMPERATURE OF SENSOR VS. SENSOR RESISTANCE (APPROXIMATE)

TEMP. (°C)	TEMP. (°F)	RESISTANCE (OHMS)
-34 TO 4	-30 TO 40	1622 TO 2205
5 TO 43	41 TO 110	1930 TO 2566
44 TO 93	111 TO 200	2245 TO 3028

DIAGNOSTIC CHART DTC 32 — 1994 CENTURY, CUTLASS CIERA AND CUTLASS CRUISER

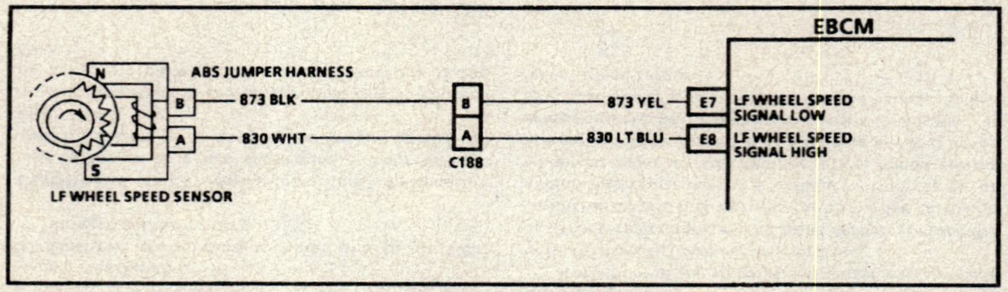

DTC 32

(Page 1 of 3)

LF WHEEL SPEED SNR CKT OPEN OR SHORT TO GND/BATT

Circuit Description:

As a toothed ring passes by the wheel speed sensor, changes in the electromagnetic field cause the wheel speed sensor to produce a sinusoidal (AC) voltage signal whose frequency is proportional to wheel speed. The magnitude of this signal is directly related to wheel speed and the proximity of the wheel speed sensor to the toothed ring often referred to as the air gap.

DTC Will Set When: DTC 32 can be set anytime after initialization. If either of the left front wheel speed circuits are shorted to voltage or ground, a malfunction exists.

Action Taken: A malfunction DTC is stored, ABS is disabled and the ABS warning lamp is turned "ON."

DTC Chart Test Description: Number(s) below refer to circled number(s) on the diagnostic chart.

1. Checks to see that the wheel speed sensor is within proper resistance values.
2. Checks for short to ground in wheel speed sensor wiring.
3. Checks for short to voltage in wheel speed sensor wiring.
4. Checks for a short to voltage in CKT 873.
5. Checks for a short to voltage in CKT 830.
6. Ensures malfunction was not due to physical damage of circuitry.

Diagnostic Aids: An "Intermittent" malfunction may be caused by a poor connection, rubbed through wire insulation, or a wire that is broken inside the insulation.

The frequency of the malfunction can be checked by using the enhanced diagnostic function of the Tech 1

If the customer's comments reflect that the ABS warning lamp is "ON" only during moist environmental changes (rain, snow, vehicle wash), all wheel speed sensor circuitry should be thoroughly inspected for signs of water intrusion. If DTC is not current, clear DTCs and simulate the effects of water intrusion. Use the following procedure. Spray down the suspected area with a 5% salt water solution (two teaspoons of salt to 12 oz. of water). Test drive vehicle over various road surfaces (bumps, turns, etc.) above 24 km/h (15 mph) for at least 30 seconds. If DTC returns, replace suspected harness.

Any circuitry, that is suspected as causing the intermittent complaint, should be thoroughly checked for backed out terminals, improper mating, broken locks, improperly formed or damaged terminals, poor terminal to wiring connections or physical damage to the wiring harness.

Resistance of the wheel speed sensor will increase with an increase in sensor temperature.

When replacing a wheel speed sensor, inspect the sensor terminals and harness connector for corrosion and/or water intrusion. If evidence of corrosion or water intrusion exists, replace wheel speed sensor jumper harness. Likewise, if replacing a wheel speed sensor jumper harness, inspect sensor terminals. If evidence of corrosion or water intrusion exists, replace wheel speed sensor. Refer to "On-Vehicle Service" in this section.

DIAGNOSTIC CHART DTC 32 — 1994 CENTURY, CUTLASS CIERA AND CUTLASS CRUISER

DTC 32

(Page 1 of 3)

LF WHEEL SPEED SNR CKT OPEN OR SHORT TO GND/BATT

Important

- WHEEL SPEED SENSOR INTERMITTENT MALFUNCTIONS MAY BE DIFFICULT TO LOCATE. CARE SHOULD BE TAKEN NOT TO DISTURB ANY ELECTRICAL CONNECTIONS PRIOR TO AN INDICATED STEP OF THIS CHART. THIS WILL ENSURE THAT AN INTERMITTENT CONNECTION WILL NOT BE CORRECTED BEFORE THE SOURCE OF THE MALFUNCTION IS FOUND.

(1)
- IGNITION "OFF."
- USING A 5% SALT WATER SOLUTION (TWO TEASPOONS OF SALT TO 12 OZ. WATER), THOROUGHLY SPRAY THE LEFT FRONT WHEEL SPEED SENSOR JUMPER HARNESS.
- DISCONNECT 32-WAY EBCM HARNESS CONNECTOR.
- USING J 39200, MEASURE THE RESISTANCE BETWEEN TERMINALS "E7" AND "E8" OF THE 32-WAY EBCM HARNESS CONNECTOR.

IS RESISTANCE BETWEEN 1020 AND 1137Ω WHEN SENSOR IS 20°C (68°F)? *(IF TEMPERATURE OF SENSOR IS SUSPECTED TO OTHER THAN 20°C (68°F), REFER TO THE TABLE ON PAGE 2 OF 2 OF THIS DTC CHART FOR APPLICABLE SENSOR RESISTANCE VALUES; THE VALUES IN THE TABLE ARE FOR TEMPERATURE OF THE SENSOR, NOT AIR TEMPERATURE.)*

- YES → (2)
- NO → DIAGNOSIS CONTINUED ON PAGE 2 OF 3 OF THIS DTC CHART.

(2)
- MEASURE THE RESISTANCE BETWEEN GROUND AND TERMINAL "E7" OF THE 32-WAY EBCM HARNESS CONNECTOR.

DOES J 39200 DISPLAY RESISTANCE AS "OL" (INFINITE)?

- YES → (3)
- NO → DIAGNOSIS CONTINUED ON PAGE 3 OF 3 OF THIS DTC CHART.

(3)
- IGNITION "ON."
- USING J 39200, MEASURE THE VOLTAGE BETWEEN GROUND AND TERMINAL "E7" OF THE 32-WAY EBCM HARNESS CONNECTOR.

IS VOLTAGE 1 VOLT OR LESS?

- NO → (4)
- YES → (6)

(4)
- IGNITION "OFF."
- DISCONNECT LEFT FRONT WHEEL SPEED SENSOR DIRECTLY AT SENSOR.
- IGNITION "ON."
- USING J 39200, MEASURE THE VOLTAGE BETWEEN GROUND AND TERMINAL "E7" OF THE 32-WAY EBCM HARNESS CONNECTOR.

IS VOLTAGE 1 VOLT OR LESS?

- YES → (5)
- NO → REPAIR SHORT TO VOLTAGE IN CKT 873.

(5)
- USING J 39200, MEASURE THE VOLTAGE BETWEEN GROUND AND TERMINAL "E8" OF THE 32-WAY EBCM HARNESS CONNECTOR.

IS VOLTAGE 1 VOLT OR LESS?

- YES → MALFUNCTION IS INTERMITTENT. INSPECT ALL CONNECTORS AND HARNESSES FOR DAMAGE WHICH MAY RESULT IN A SHORT TO VOLTAGE WHEN ALL COMPONENTS ARE CONNECTED. REPAIR ALL DAMAGE IF FOUND. REFER TO "DIAGNOSTIC AIDS" ON FACING PAGE FOR MORE INFORMATION.
- NO → REPAIR SHORT TO VOLTAGE IN CKT 830.

(6)
- IGNITION "OFF."
- INSPECT TERMINALS "E7" AND "E8" OF THE 32-WAY EBCM HARNESS CONNECTOR FOR POOR TERMINAL CONTACT AND CONTAMINATION. ALSO INSPECT CKTs 830 AND 873 FOR DAMAGE WHICH MAY RESULT IN AN OPEN CIRCUIT, SHORT TO GROUND, OR SHORT TO VOLTAGE. REPAIR ALL DAMAGE IF EVIDENT.
- RECONNECT ALL CONNECTORS.
- IGNITION "ON."

DOES DTC 32 SET AS A CURRENT DTC?

- YES → REPLACE EBCM.
- NO → MALFUNCTION IS INTERMITTENT. REFER TO "DIAGNOSTIC AIDS" ON FACING PAGE.

DIAGNOSTIC CHART DTC 32 — 1994 CENTURY, CUTLASS CIERA AND CUTLASS CRUISER

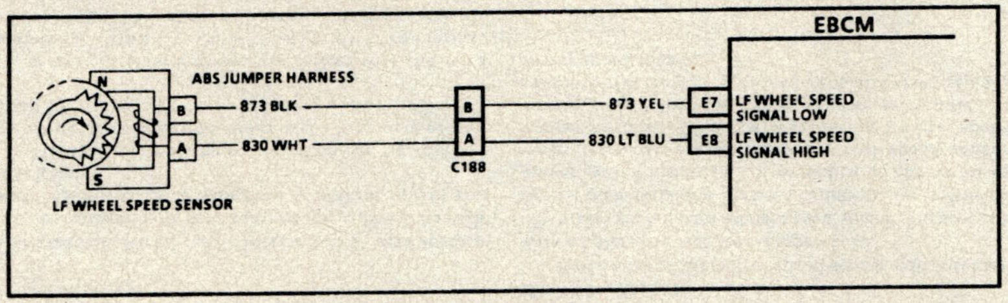

DTC 32

(Page 2 of 3)

LF WHEEL SPEED SNR CKT OPEN OR SHORT TO GND/BATT

Circuit Description:

As a toothed ring passes by the wheel speed sensor, changes in the electromagnetic field cause the wheel speed sensor to produce a sinusoidal (AC) voltage signal whose frequency is proportional to wheel speed. The magnitude of this signal is directly related to wheel speed and the proximity of the wheel speed sensor to the toothed ring often referred to as the air gap.

DTC Will Set When: DTC 32 can be set anytime after initialization. If either of the left front wheel speed circuits are shorted to voltage or ground, a malfunction exists.

Action Taken: A malfunction DTC is stored, ABS is disabled and the ABS warning lamp is turned "ON."

DTC Chart Test Description: Number(s) below refer to circled number(s) on the diagnostic chart.

7. Checks for high resistance in CKT 873.
8. Checks for high resistance in CKT 830.
9. Verifies proper resistance value of the wheel speed sensor.

Diagnostic Aids: An "Intermittent" malfunction may be caused by a poor connection, rubbed through wire insulation, or a wire that is broken inside the insulation.

The frequency of the malfunction can be checked by using the enhanced diagnostic function of the Tech 1.

If the customer's comments reflect that the ABS warning lamp is "ON" only during moist environmental changes (rain, snow, vehicle wash), all wheel speed sensor circuitry should be thoroughly inspected for signs of water intrusion. If DTC is not current, clear DTCs and simulate the effects of water intrusion. Use the following procedure. Spray down the suspected area with a 5% salt water solution (two teaspoons of salt to 12 oz. of water). Test drive vehicle over various road surfaces (bumps, turns, etc.) above 24 km/h (15 mph) for at least 30 seconds. If DTC returns, replace suspected harness.

Any circuitry, that is suspected as causing the intermittent complaint, should be thoroughly checked for backed out terminals, improper mating, broken locks, improperly formed or damaged terminals, poor terminal to wiring connections or physical damage to the wiring harness.

Resistance of the wheel speed sensor will increase with an increase in sensor temperature.

When replacing a wheel speed sensor, inspect the sensor terminals and harness connector for corrosion and/or water intrusion. If evidence of corrosion or water intrusion exists, replace wheel speed sensor jumper harness. Likewise, if replacing a wheel speed sensor jumper harness, inspect sensor terminals. If evidence of corrosion or water intrusion exists, replace wheel speed sensor. Refer to "On-Vehicle Service" in this section.

DIAGNOSTIC CHART DTC 32 — 1994 CENTURY, CUTLASS CIERA AND CUTLASS CRUISER

DTC 32

(Page 2 of 3)

LF WHEEL SPEED SNR CKT OPEN OR SHORT TO GND/BATT

DIAGNOSIS CONTINUED FROM PAGE 1 OF 3 OF THIS DTC CHART.

(7)
- DISCONNECT LEFT FRONT WHEEL SPEED SENSOR DIRECTLY AT SENSOR.
- USING J 39200, MEASURE THE RESISTANCE BETWEEN TERMINAL "E7" OF THE 32-WAY EBCM HARNESS CONNECTOR AND TERMINAL "B" OF THE LEFT FRONT WHEEL SPEED SENSOR HARNESS CONNECTOR.

IS RESISTANCE 2Ω OR LESS?

NO: REPAIR HIGH RESISTANCE IN CKT 873 (WHEEL SPEED SENSOR JUMPER MUST BE REPLACED IF DAMAGED).

YES:

(8)
- USING J 39200, MEASURE THE RESISTANCE BETWEEN TERMINAL "E8" OF THE 32-WAY EBCM HARNESS CONNECTOR AND TERMINAL "A" OF THE LEFT FRONT WHEEL SPEED SENSOR HARNESS CONNECTOR.

IS RESISTANCE 2Ω OR LESS?

NO: REPAIR HIGH RESISTANCE IN CKT 830 (WHEEL SPEED SENSOR JUMPER MUST BE REPLACED IF DAMAGED).

YES:

(9)
- USING J 39200, MEASURE THE RESISTANCE BETWEEN TERMINAL "A" AND TERMINAL "B" OF THE LEFT FRONT WHEEL SPEED SENSOR CONNECTOR.

IS RESISTANCE BETWEEN 1020 AND 1137Ω WHEN SENSOR IS 20°C (68°F)? *(IF TEMPERATURE OF SENSOR IS SUSPECTED TO OTHER THAN 20°C (68°F), REFER TO THE TABLE BELOW FOR APPLICABLE SENSOR RESISTANCE VALUES; THE VALUES IN THE TABLE ARE FOR TEMPERATURE OF THE SENSOR, NOT AIR TEMPERATURE.)*

NO: REPLACE WHEEL SPEED SENSOR.

YES: MALFUNCTION IS INTERMITTENT. INSPECT ALL CONNECTORS AND HARNESSES FOR DAMAGE WHICH MAY RESULT IN HIGH RESISTANCE WHEN ALL COMPONENTS ARE CONNECTED. REFER TO "DIAGNOSTIC AIDS" ON FACING PAGE FOR MORE INFORMATION. REPAIR ALL DAMAGE IF FOUND.

FRONT WHEEL SPEED SENSOR

TEMPERATURE OF SENSOR VS. SENSOR RESISTANCE (APPROXIMATE)

TEMP. (°C)	TEMP. (°F)	RESISTANCE (OHMS)
-34 TO 4	-30 TO 40	804 TO 1066
5 TO 43	41 TO 110	956 TO 1240
44 TO 93	111 TO 200	1112 TO 1463

DIAGNOSTIC CHART DTC 32 — 1994 CENTURY, CUTLASS CIERA AND CUTLASS CRUISER

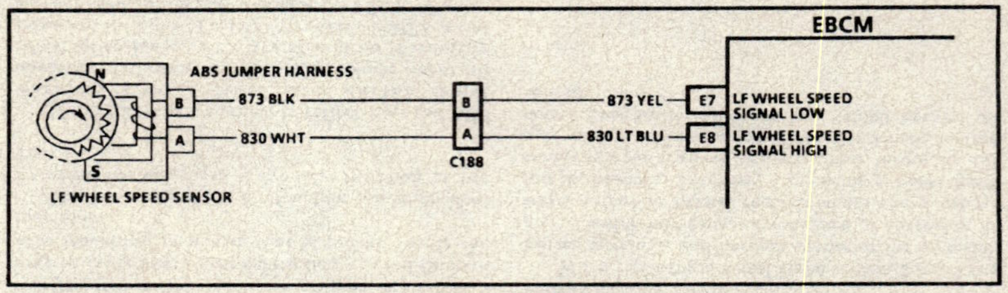

DTC 32

(Page 3 of 3)
LF WHEEL SPEED SNR CKT OPEN OR SHORT TO GND/BATT

Circuit Description:
As a toothed ring passes by the wheel speed sensor, changes in the electromagnetic field cause the wheel speed sensor to produce a sinusoidal (AC) voltage signal whose frequency is proportional to wheel speed. The magnitude of this signal is directly related to wheel speed and the proximity of the wheel speed sensor to the toothed ring often referred to as the air gap.

DTC Will Set When: DTC 32 can be set anytime after initialization. If either of the left front wheel speed circuits are shorted to voltage or ground, a malfunction exists.

Action Taken: A malfunction DTC is stored, ABS is disabled and the ABS warning lamp is turned "ON."

DTC Chart Test Description: Number(s) below refer to circled number(s) on the diagnostic chart.

10. Checks for short to ground in CKT 873.
11. Checks for short to ground in CKT 830.
12. Checks for short to ground in wheel speed sensor jumper harness.
13. Checks for an internal short to ground in the wheel speed sensor.

Diagnostic Aids: An "Intermittent" malfunction may be caused by a poor connection, rubbed through wire insulation, or a wire that is broken inside the insulation.

The frequency of the malfunction can be checked by using the enhanced diagnostic function of the Tech 1

If the customer's comments reflect that the ABS warning lamp is "ON" only during moist environmental changes (rain, snow, vehicle wash), all wheel speed sensor circuitry should be thoroughly inspected for signs of water intrusion. If DTC is not current, clear DTCs and simulate the effects of water intrusion. Use the following procedure. Spray down the suspected area with a 5% salt water solution (two teaspoons of salt to 12 oz. of water). Test drive vehicle over various road surfaces (bumps, turns, etc.) above 24 km/h (15 mph) for at least 30 seconds. If DTC returns, replace suspected harness.

Any circuitry, that is suspected as causing the intermittent complaint, should be thoroughly checked for backed out terminals, improper mating, broken locks, improperly formed or damaged terminals, poor terminal to wiring connections or physical damage to the wiring harness.

Resistance of the wheel speed sensor will increase with an increase in sensor temperature.

When replacing a wheel speed sensor, inspect the sensor terminals and harness connector for corrosion and/or water intrusion. If evidence of corrosion or water intrusion exists, replace wheel speed sensor jumper harness. Likewise, if replacing a wheel speed sensor jumper harness, inspect sensor terminals. If evidence of corrosion or water intrusion exists, replace wheel speed sensor.

DIAGNOSTIC CHART DTC 32 — 1994 CENTURY, CUTLASS CIERA AND CUTLASS CRUISER

DTC 32

(Page 3 of 3)
LF WHEEL SPEED SNR CKT OPEN OR SHORT TO GND/BATT

DIAGNOSIS CONTINUED FROM PAGE 1 OF 3 OF THIS DTC CHART.

(10)
- DISCONNECT CONNECTOR C188 (MOUNTED ON LH SHOCK TOWER).
- USING J 39200, MEASURE THE RESISTANCE BETWEEN GROUND AND TERMINAL "E7" OF THE 32-WAY EBCM HARNESS CONNECTOR.

DOES J 39200 DISPLAY RESISTANCE AS "OL" (INFINITE)?

NO → REPAIR SHORT TO GROUND IN CKT 873.

YES ↓

(11)
- USING J 39200, MEASURE THE RESISTANCE BETWEEN GROUND AND TERMINAL "E8" OF THE 32-WAY EBCM HARNESS CONNECTOR.

DOES J 39200 DISPLAY RESISTANCE AS "OL" (INFINITE)?

NO → REPAIR SHORT TO GROUND IN CKT 830.

YES ↓

(12)
- THOROUGHLY RESPRAY THE LEFT FRONT WHEEL SPEED SENSOR JUMPER HARNESS WITH THE SALT WATER SOLUTION.

IMPORTANT: DO NOT ALLOW WATER TO ENTER ANY WHEEL SPEED SENSOR CONNECTOR.

- DISCONNECT LEFT FRONT WHEEL SPEED SENSOR DIRECTLY AT SENSOR.
- USING J 39200, MEASURE THE RESISTANCE BETWEEN GROUND AND TERMINAL "A," THEN "B" OF THE LEFT FRONT WHEEL SPEED SENSOR HARNESS CONNECTOR.

DOES J 39200 DISPLAY RESISTANCE AS "OL" (INFINITE) FOR BOTH MEASUREMENTS?

NO → REPLACE LEFT FRONT WHEEL SPEED SENSOR JUMPER HARNESS.

YES ↓

(13)
- USING J 39200, MEASURE THE RESISTANCE BETWEEN GROUND AND TERMINAL "A" OF THE LEFT FRONT WHEEL SPEED SENSOR CONNECTOR.

DOES J 39200 DISPLAY RESISTANCE AS "OL" (INFINITE)?

NO → REPLACE WHEEL SPEED SENSOR.

YES → MALFUNCTION IS INTERMITTENT. INSPECT ALL CONNECTORS AND HARNESSES FOR DAMAGE WHICH MAY RESULT IN A SHORT TO GROUND WHEN ALL COMPONENTS ARE CONNECTED. PAY PARTICULAR ATTENTION TO THE JUMPER HARNESS, SMALL NICKS OR CUTS CAN CAUSE A MALFUNCTION. REPAIR ANY DAMAGE IF FOUND. JUMPER HARNESS MUST BE REPLACED IF DAMAGED. REFER TO "DIAGNOSTIC AIDS" ON FACING PAGE FOR MORE INFORMATION.

DIAGNOSTIC CHART DTC 33 — 1994 CENTURY, CUTLASS CIERA AND CUTLASS CRUISER

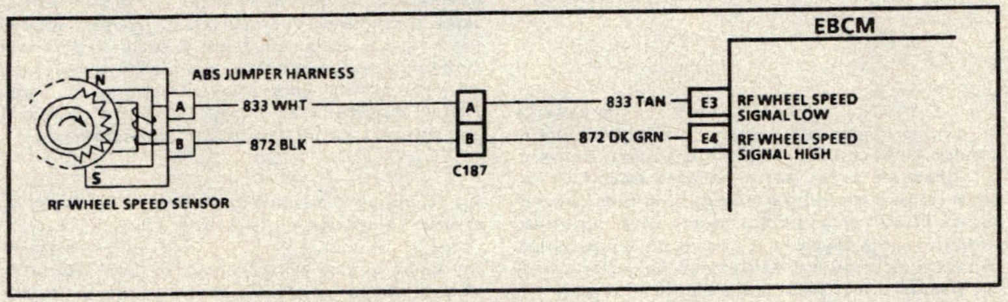

DTC 33

(Page 1 of 3)
RF WHEEL SPEED SNR CKT OPEN OR SHORT TO GND/BATT

Circuit Description:

As a toothed ring passes by the wheel speed sensor, changes in the electromagnetic field cause the wheel speed sensor to produce a sinusoidal (AC) voltage signal whose frequency is proportional to wheel speed. The magnitude of this signal is directly related to wheel speed and the proximity of the wheel speed sensor to the toothed ring, often referred to as the air gap.

DTC Will Set When: DTC 33 can be set anytime after initialization. If either of the right front wheel speed circuits are shorted to voltage or ground, a malfunction exists.

Action Taken: A malfunction DTC is stored, ABS is disabled and the ABS warning lamp is turned "ON."

DTC Chart Test Description: Number(s) below refer to circled number(s) on the diagnostic chart.

1. Checks to see that the wheel speed sensor is within proper resistance values.
2. Checks for short to ground in wheel speed sensor wiring.
3. Checks for short to voltage in wheel speed sensor wiring.
4. Checks for a short to voltage in CKT 833.
5. Checks for a short to voltage in CKT 872.
6. Ensures malfunction was not due to physical damage of circuitry.

Diagnostic Aids: An "Intermittent" malfunction may be caused by a poor connection, rubbed through wire insulation, or a wire that is broken inside the insulation.

The frequency of the malfunction can be checked by using the enhanced diagnostic function of the Tech 1.

If the customer's comments reflect that the ABS warning lamp is "ON" only during moist environmental changes (rain, snow, vehicle wash), all wheel speed sensor circuitry should be thoroughly inspected for signs of water intrusion. If DTC is not current, clear DTCs and simulate the effects of water intrusion. Use the following procedure. Spray down the suspected area with a 5% salt water solution (two teaspoons of salt to 12 oz. of water). Test drive vehicle over various road surfaces (bumps, turns, etc.) above 24 km/h (15 mph) for at least 30 seconds. If DTC returns, replace suspected harness.

Any circuitry, that is suspected as causing the intermittent complaint, should be thoroughly checked for backed out terminals, improper mating, broken locks, improperly formed or damaged terminals, poor terminal to wiring connections or physical damage to the wiring harness.

Resistance of the wheel speed sensor will increase with an increase in sensor temperature.

When replacing a wheel speed sensor, inspect the sensor terminals and harness connector for corrosion and/or water intrusion. If evidence of corrosion or water intrusion exists, replace wheel speed sensor jumper harness. Likewise, if replacing a wheel speed sensor jumper harness, inspect sensor terminals. If evidence of corrosion or water intrusion exists, replace wheel speed sensor. Refer to "On-Vehicle Service" in this section.

DIAGNOSTIC CHART DTC 33 — 1994 CENTURY, CUTLASS CIERA AND CUTLASS CRUISER

DTC 33

(Page 1 of 3)
RF WHEEL SPEED SNR CKT OPEN OR SHORT TO GND/BATT

Important

- WHEEL SPEED SENSOR INTERMITTENT MALFUNCTIONS MAY BE DIFFICULT TO LOCATE. CARE SHOULD BE TAKEN NOT TO DISTURB ANY ELECTRICAL CONNECTIONS PRIOR TO AN INDICATED STEP OF THIS CHART. THIS WILL ENSURE THAT AN INTERMITTENT CONNECTION WILL NOT BE CORRECTED BEFORE THE SOURCE OF THE MALFUNCTION IS FOUND.

(1)
- IGNITION "OFF."
- USING A 5% SALT WATER SOLUTION (TWO TEASPOONS OF SALT TO 12 OZ. WATER), THOROUGHLY SPRAY THE RIGHT FRONT WHEEL SPEED SENSOR JUMPER HARNESS.
- DISCONNECT 32-WAY EBCM HARNESS CONNECTOR.
- USING J 39200, MEASURE THE RESISTANCE BETWEEN TERMINALS "E3" AND "E4" OF THE 32-WAY EBCM HARNESS CONNECTOR.

IS RESISTANCE BETWEEN 1020 AND 1137Ω WHEN SENSOR IS 20°C (68°F)? *(IF TEMPERATURE OF SENSOR IS SUSPECTED TO OTHER THAN 20°C (68°F), REFER TO THE TABLE ON PAGE 2 OF 2 OF THIS DTC CHART FOR APPLICABLE SENSOR RESISTANCE VALUES; THE VALUES IN THE TABLE ARE FOR TEMPERATURE OF THE SENSOR, NOT AIR TEMPERATURE.)*

- NO → DIAGNOSIS CONTINUED ON PAGE 2 OF 3 OF THIS DTC CHART.
- YES → (2)

(2)
- MEASURE THE RESISTANCE BETWEEN GROUND AND TERMINAL "E3" OF THE 32-WAY EBCM HARNESS CONNECTOR.

DOES J 39200 DISPLAY RESISTANCE AS "OL" (INFINITE)?

- NO → DIAGNOSIS CONTINUED ON PAGE 3 OF 3 OF THIS DTC CHART.
- YES → (3)

(3)
- IGNITION "ON."
- USING J 39200, MEASURE THE VOLTAGE BETWEEN GROUND AND TERMINAL "E3" OF THE 32-WAY EBCM HARNESS CONNECTOR.

IS VOLTAGE 1 VOLT OR LESS?

- NO → (4)
- YES → (6)

(4)
- IGNITION "OFF."
- DISCONNECT RIGHT FRONT WHEEL SPEED SENSOR DIRECTLY AT SENSOR.
- IGNITION "ON."
- USING J 39200, MEASURE THE VOLTAGE BETWEEN GROUND AND TERMINAL "E3" OF THE 32-WAY EBCM HARNESS CONNECTOR.

IS VOLTAGE 1 VOLT OR LESS?

- YES → (5)
- NO → REPAIR SHORT TO VOLTAGE IN CKT 833.

(5)
- USING J 39200, MEASURE THE VOLTAGE BETWEEN GROUND AND TERMINAL "E4" OF THE 32-WAY EBCM HARNESS CONNECTOR.

IS VOLTAGE 1 VOLT OR LESS?

- YES → MALFUNCTION IS INTERMITTENT. INSPECT ALL CONNECTORS AND HARNESSES FOR DAMAGE WHICH MAY RESULT IN A SHORT TO VOLTAGE WHEN ALL COMPONENTS ARE CONNECTED. REPAIR ALL DAMAGE IF FOUND. REFER TO "DIAGNOSTIC AIDS" ON FACING PAGE FOR MORE INFORMATION.
- NO → REPAIR SHORT TO VOLTAGE IN CKT 872.

(6)
- IGNITION "OFF."
- INSPECT TERMINALS "E3" AND "E4" OF THE 32-WAY EBCM HARNESS CONNECTOR FOR POOR TERMINAL CONTACT AND CONTAMINATION. ALSO INSPECT CKTs 833 AND 872 FOR DAMAGE WHICH MAY RESULT IN AN OPEN CIRCUIT, SHORT TO GROUND, OR SHORT TO VOLTAGE. REPAIR ALL DAMAGE IF EVIDENT.
- RECONNECT ALL CONNECTORS.
- IGNITION "ON."

DOES DTC 33 SET AS A CURRENT DTC?

- YES → REPLACE EBCM.
- NO → MALFUNCTION IS INTERMITTENT. REFER TO "DIAGNOSTIC AIDS" ON FACING PAGE.

DIAGNOSTIC CHART DTC 33 — 1994 CENTURY, CUTLASS CIERA AND CUTLASS CRUISER

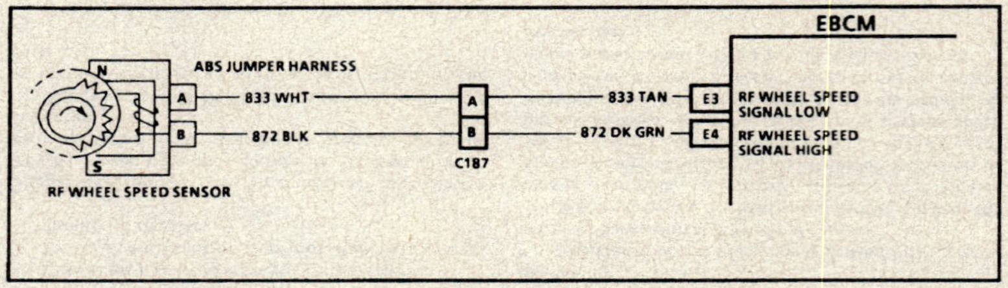

DTC 33

(Page 2 of 3)

RF WHEEL SPEED SNR CKT OPEN OR SHORT TO GND/BATT

Circuit Description:

As a toothed ring passes by the wheel speed sensor, changes in the electromagnetic field cause the wheel speed sensor to produce a sinusoidal (AC) voltage signal whose frequency is proportional to wheel speed. The magnitude of this signal is directly related to wheel speed and the proximity of the wheel speed sensor to the toothed ring, often referred to as the air gap.

DTC Will Set When: DTC 33 can be set anytime after initialization. If either of the right front wheel speed circuits are shorted to voltage or ground, a malfunction exists.

Action Taken: A malfunction DTC is stored, ABS is disabled and the ABS warning lamp is turned "ON."

DTC Chart Test Description: Number(s) below refer to circled number(s) on the diagnostic chart.

7. Checks for high resistance in CKT 833.
8. Checks for high resistance in CKT 872.
9. Verifies proper resistance value of the wheel speed sensor.

Diagnostic Aids: An "Intermittent" malfunction may be caused by a poor connection, rubbed through wire insulation, or a wire that is broken inside the insulation.

The frequency of the malfunction can be checked by using the enhanced diagnostic function of the Tech I.

If the customer's comments reflect that the ABS warning lamp is "ON" only during moist environmental changes (rain, snow, vehicle wash), all wheel speed sensor circuitry should be thoroughly inspected for signs of water intrusion. If DTC is not current, clear DTCs and simulate the effects of water intrusion. Use the following procedure. Spray down the suspected area with a 5% salt water solution (two teaspoons of salt to 12 oz. of water). Test drive vehicle over various road surfaces (bumps, turns, etc.) above 24 km/h (15 mph) for at least 30 seconds. If DTC returns, replace suspected harness.

Any circuitry, that is suspected as causing the intermittent complaint, should be thoroughly checked for backed out terminals, improper mating, broken locks, improperly formed or damaged terminals, poor terminal to wiring connections or physical damage to the wiring harness.

Resistance of the wheel speed sensor will increase with an increase in sensor temperature.

When replacing a wheel speed sensor, inspect the sensor terminals and harness connector for corrosion and/or water intrusion. If evidence of corrosion or water intrusion exists, replace wheel speed sensor jumper harness. Likewise, if replacing a wheel speed sensor jumper harness, inspect sensor terminals. If evidence of corrosion or water intrusion exists, replace wheel speed sensor. Refer to "On-Vehicle Service" in this section.

DIAGNOSTIC CHART DTC 33 — 1994 CENTURY, CUTLASS CIERA AND CUTLASS CRUISER

DTC 33

(Page 2 of 3)

RF WHEEL SPEED SNR CKT OPEN OR SHORT TO GND/BATT

DIAGNOSIS CONTINUED FROM PAGE 1 OF 3 OF THIS DTC CHART.

(7)
- DISCONNECT RIGHT FRONT WHEEL SPEED SENSOR DIRECTLY AT SENSOR.
- USING J 39200, MEASURE THE RESISTANCE BETWEEN TERMINAL "E3" OF THE 32-WAY EBCM HARNESS CONNECTOR AND TERMINAL "A" OF THE RIGHT FRONT WHEEL SPEED SENSOR HARNESS CONNECTOR.

IS RESISTANCE 2Ω OR LESS?

NO: REPAIR HIGH RESISTANCE IN CKT 833 (WHEEL SPEED SENSOR JUMPER MUST BE REPLACED IF DAMAGED).

YES:

(8)
- USING J 39200, MEASURE THE RESISTANCE BETWEEN TERMINAL "E4" OF THE 32-WAY EBCM HARNESS CONNECTOR AND TERMINAL "B" OF THE RIGHT FRONT WHEEL SPEED SENSOR HARNESS CONNECTOR.

IS RESISTANCE 2Ω OR LESS?

NO: REPAIR HIGH RESISTANCE IN CKT 872 (WHEEL SPEED SENSOR JUMPER MUST BE REPLACED IF DAMAGED).

YES:

(9)
- USING J 39200, MEASURE THE RESISTANCE BETWEEN TERMINAL "A" AND TERMINAL "B" OF THE RIGHT FRONT WHEEL SPEED SENSOR CONNECTOR.

IS RESISTANCE BETWEEN 1020 AND 1137Ω WHEN SENSOR IS 20°C (68°F)? *(IF TEMPERATURE OF SENSOR IS SUSPECTED TO OTHER THAN 20°C (68°F), REFER TO THE TABLE BELOW FOR APPLICABLE SENSOR RESISTANCE VALUES; THE VALUES IN THE TABLE ARE FOR TEMPERATURE OF THE SENSOR, NOT AIR TEMPERATURE.)*

NO: REPLACE WHEEL SPEED SENSOR.

YES: MALFUNCTION IS INTERMITTENT. INSPECT ALL CONNECTORS AND HARNESSES FOR DAMAGE WHICH MAY RESULT IN HIGH RESISTANCE WHEN ALL COMPONENTS ARE CONNECTED. REFER TO "DIAGNOSTIC AIDS" ON FACING PAGE FOR MORE INFORMATION. REPAIR ALL DAMAGE IF FOUND.

FRONT WHEEL SPEED SENSOR

TEMPERATURE OF SENSOR VS. SENSOR RESISTANCE (APPROXIMATE)

TEMP. (°C)	TEMP. (°F)	RESISTANCE (OHMS)
-34 TO 4	-30 TO 40	804 TO 1066
5 TO 43	41 TO 110	956 TO 1240
44 TO 93	111 TO 200	1112 TO 1463

DIAGNOSTIC CHART DTC 33 — 1994 CENTURY, CUTLASS CIERA AND CUTLASS CRUISER

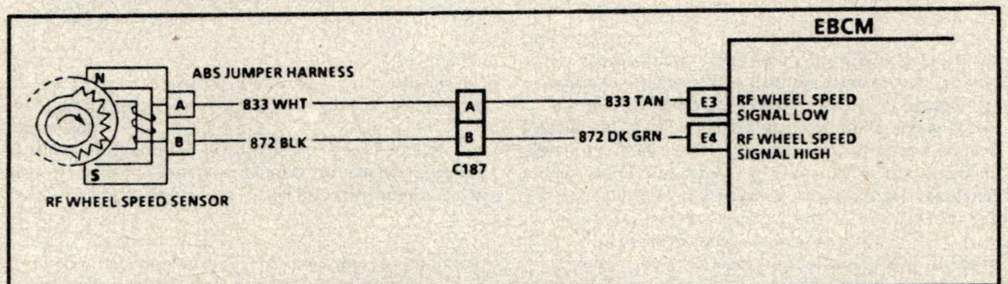

DTC 33

(Page 3 of 3)
RF WHEEL SPEED SNR CKT OPEN OR SHORT TO GND/BATT

Circuit Description:

As a toothed ring passes by the wheel speed sensor, changes in the electromagnetic field cause the wheel speed sensor to produce a sinusoidal (AC) voltage signal whose frequency is proportional to wheel speed. The magnitude of this signal is directly related to wheel speed and the proximity of the wheel speed sensor to the toothed ring, often referred to as the air gap.

DTC Will Set When: DTC 33 can be set anytime after initialization. If either of the right front wheel speed circuits are shorted to voltage or ground, a malfunction exists.

Action Taken: A malfunction DTC is stored, ABS is disabled and the ABS warning lamp is turned "ON."

DTC Chart Test Description: Number(s) below refer to circled number(s) on the diagnostic chart.

10. Checks for short to ground in CKT 833.
11. Checks for short to ground in CKT 872.
12. Checks for short to ground in wheel speed sensor jumper harness.
13. Checks for an internal short to ground in the wheel speed sensor.

Diagnostic Aids: An "Intermittent" malfunction may be caused by a poor connection, rubbed through wire insulation, or a wire that is broken inside the insulation.

The frequency of the malfunction can be checked by using the enhanced diagnostic function of the Tech 1.

If the customer's comments reflect that the ABS warning lamp is "ON" only during moist environmental changes (rain, snow, vehicle wash), all wheel speed sensor circuitry should be thoroughly inspected for signs of water intrusion. If DTC is not current, clear DTCs and simulate the effects of water intrusion. Use the following procedure. Spray down the suspected area with a 5% salt water solution (two teaspoons of salt to 12 oz. of water). Test drive vehicle over various road surfaces (bumps, turns, etc.) above 24 km/h (15 mph) for at least 30 seconds. If DTC returns, replace suspected harness.

Any circuitry, that is suspected as causing the intermittent complaint, should be thoroughly checked for backed out terminals, improper mating, broken locks, improperly formed or damaged terminals, poor terminal to wiring connections or physical damage to the wiring harness.

Resistance of the wheel speed sensor will increase with an increase in sensor temperature.

When replacing a wheel speed sensor, inspect the sensor terminals and harness connector for corrosion and/or water intrusion. If evidence of corrosion or water intrusion exists, replace wheel speed sensor jumper harness. Likewise, if replacing a wheel speed sensor jumper harness, inspect sensor terminals. If evidence of corrosion or water intrusion exists, replace wheel speed sensor. Refer to "On-Vehicle Service" in this section.

DIAGNOSTIC CHART DTC 33 — 1994 CENTURY, CUTLASS CIERA AND CUTLASS CRUISER

DTC 33

(Page 3 of 3)
RF WHEEL SPEED SNR CKT OPEN OR SHORT TO GND/BATT

DIAGNOSIS CONTINUED FROM PAGE 1 OF 3 OF THIS DTC CHART.

(10)
- DISCONNECT CONNECTOR C187 (MOUNTED ON RH SHOCK TOWER).
- USING J 39200, MEASURE THE RESISTANCE BETWEEN GROUND AND TERMINAL "E3" OF THE 32-WAY EBCM HARNESS CONNECTOR.

DOES J 39200 DISPLAY RESISTANCE AS "OL" (INFINITE)?

NO → REPAIR SHORT TO GROUND IN CKT 833.

YES ↓

(11)
- USING J 39200, MEASURE THE RESISTANCE BETWEEN GROUND AND TERMINAL "E4" OF THE 32-WAY EBCM HARNESS CONNECTOR.

DOES J 39200 DISPLAY RESISTANCE AS "OL" (INFINITE)?

NO → REPAIR SHORT TO GROUND IN CKT 872.

YES ↓

(12)
- THOROUGHLY RESPRAY THE RIGHT FRONT WHEEL SPEED SENSOR JUMPER HARNESS WITH THE SALT WATER SOLUTION.

IMPORTANT: DO NOT ALLOW WATER TO ENTER ANY WHEEL SPEED SENSOR CONNECTOR.

- DISCONNECT RIGHT FRONT WHEEL SPEED SENSOR DIRECTLY AT SENSOR.
- USING J 39200, MEASURE THE RESISTANCE BETWEEN GROUND AND TERMINAL "A," THEN "B" OF THE RIGHT FRONT WHEEL SPEED SENSOR HARNESS CONNECTOR.

DOES J 39200 DISPLAY RESISTANCE AS "OL" (INFINITE) FOR BOTH MEASUREMENTS?

NO → REPLACE RIGHT FRONT WHEEL SPEED SENSOR JUMPER HARNESS.

YES ↓

(13)
- USING J 39200, MEASURE THE RESISTANCE BETWEEN GROUND AND TERMINAL "A" OF THE RIGHT FRONT WHEEL SPEED SENSOR CONNECTOR.

DOES J 39200 DISPLAY RESISTANCE AS "OL" (INFINITE)?

NO → REPLACE WHEEL SPEED SENSOR.

YES ↓

MALFUNCTION IS INTERMITTENT. INSPECT ALL CONNECTORS AND HARNESSES FOR DAMAGE WHICH MAY RESULT IN A SHORT TO GROUND WHEN ALL COMPONENTS ARE CONNECTED. PAY PARTICULAR ATTENTION TO THE JUMPER HARNESS, SMALL NICKS OR CUTS CAN CAUSE A MALFUNCTION. REPAIR ANY DAMAGE IF FOUND. JUMPER HARNESS MUST BE REPLACED IF DAMAGED. REFER TO "DIAGNOSTIC AIDS" ON FACING PAGE FOR MORE INFORMATION.

DIAGNOSTIC CHART DTC 34 — 1994 CENTURY, CUTLASS CIERA AND CUTLASS CRUISER

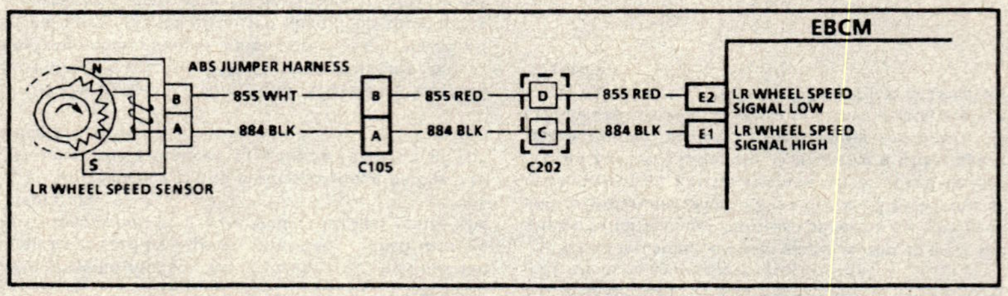

DTC 34

(Page 1 of 3)

LR WHEEL SPEED SNR CKT OPEN OR SHORT TO GND/BATT

Circuit Description:

As a toothed ring passes by the wheel speed sensor, changes in the electromagnetic field cause the wheel speed sensor to produce a sinusoidal (AC) voltage signal whose frequency is proportional to wheel speed. The magnitude of this signal is directly related to wheel speed and the proximity of the wheel speed sensor to the toothed ring, often referred to as the air gap.

DTC Will Set When: DTC 34 can be set anytime after initialization. If either of the left rear wheel speed circuits are shorted to voltage or ground, a malfunction exists.

Action Taken: A malfunction DTC is stored, ABS is disabled and the ABS warning lamp is turned "ON."

DTC Chart Test Description: Number(s) below refer to circled number(s) on the diagnostic chart.

1. Checks to see that the wheel speed sensor is within proper resistance values.
2. Checks for short to ground in wheel speed sensor wiring.
3. Checks for short to voltage in wheel speed sensor wiring.
4. Checks for a short to voltage in CKT 884.
5. Checks for a short to voltage in CKT 885.
6. Ensures malfunction was not due to physical damage of circuitry.

Diagnostic Aids: An "Intermittent" malfunction may be caused by a poor connection, rubbed through wire insulation, or a wire that is broken inside the insulation.

The frequency of the malfunction can be checked by using the enhanced diagnostic function of the Tech 1.

If the customer's comments reflect that the ABS warning lamp is "ON" only during moist environmental changes (rain, snow, vehicle wash), all wheel speed sensor circuitry should be thoroughly inspected for signs of water intrusion. If DTC is not current, clear DTCs and simulate the effects of water intrusion. Use the following procedure. Spray down the suspected area with a 5% salt water solution (two teaspoons of salt to 12 oz. of water). Test drive vehicle over various road surfaces (bumps, turns, etc.) above 24 km/h (15 mph) for at least 30 seconds. If DTC returns, replace suspected harness.

Any circuitry, that is suspected as causing the intermittent complaint, should be thoroughly checked for backed out terminals, improper mating, broken locks, improperly formed or damaged terminals, poor terminal to wiring connections or physical damage to the wiring harness.

Resistance of the wheel speed sensor will increase with an increase in sensor temperature.

When replacing a wheel speed sensor, inspect the sensor terminals and harness connector for corrosion and/or water intrusion. If evidence of corrosion or water intrusion exists, replace wheel speed sensor jumper harness. Likewise, if replacing a wheel speed sensor jumper harness, inspect sensor terminals. If evidence of corrosion or water intrusion exists, replace wheel speed sensor. Refer to "On-Vehicle Service" in this section.

DIAGNOSTIC CHART DTC 34 — 1994 CENTURY, CUTLASS CIERA AND CUTLASS CRUISER

DTC 34

(Page 1 of 3)

LR WHEEL SPEED SNR CKT OPEN OR SHORT TO GND/BATT

Important

- WHEEL SPEED SENSOR INTERMITTENT MALFUNCTIONS MAY BE DIFFICULT TO LOCATE. CARE SHOULD BE TAKEN NOT TO DISTURB ANY ELECTRICAL CONNECTIONS PRIOR TO AN INDICATED STEP OF THIS CHART. THIS WILL ENSURE THAT AN INTERMITTENT CONNECTION WILL NOT BE CORRECTED BEFORE THE SOURCE OF THE MALFUNCTION IS FOUND.

(1)
- IGNITION "OFF."
- USING A 5% SALT WATER SOLUTION (TWO TEASPOONS OF SALT TO 12 OZ. WATER), THOROUGHLY SPRAY THE LEFT REAR WHEEL SPEED SENSOR JUMPER HARNESS.
- DISCONNECT 32-WAY EBCM HARNESS CONNECTOR.
- USING J 39200, MEASURE THE RESISTANCE BETWEEN TERMINALS "E1" AND "E2" OF THE 32-WAY EBCM HARNESS CONNECTOR.

IS RESISTANCE BETWEEN 2059 AND 2353Ω WHEN SENSOR IS 20°C (68°F)? *(IF TEMPERATURE OF SENSOR IS SUSPECTED TO OTHER THAN 20°C (68°F), REFER TO THE TABLE ON PAGE 2 OF 2 OF THIS DTC CHART FOR APPLICABLE SENSOR RESISTANCE VALUES; THE VALUES IN THE TABLE ARE FOR TEMPERATURE OF THE SENSOR, NOT AIR TEMPERATURE.)*

NO → DIAGNOSIS CONTINUED ON PAGE 2 OF 3 OF THIS DTC CHART.

YES ↓

(2)
- MEASURE THE RESISTANCE BETWEEN GROUND AND TERMINAL "E1" OF THE 32-WAY EBCM HARNESS CONNECTOR.

DOES J 39200 DISPLAY RESISTANCE AS "OL" (INFINITE)?

NO → DIAGNOSIS CONTINUED ON PAGE 3 OF 3 OF THIS DTC CHART.

YES ↓

(3)
- IGNITION "ON."
- USING J 39200, MEASURE THE VOLTAGE BETWEEN GROUND AND TERMINAL "E1" OF THE 32-WAY EBCM HARNESS CONNECTOR.

IS VOLTAGE 1 VOLT OR LESS?

NO ↓

(4)
- IGNITION "OFF."
- DISCONNECT LEFT REAR WHEEL SPEED SENSOR DIRECTLY AT SENSOR.
- IGNITION "ON."
- USING J 39200, MEASURE THE VOLTAGE BETWEEN GROUND AND TERMINAL "E1" OF THE 32-WAY EBCM HARNESS CONNECTOR.

IS VOLTAGE 1 VOLT OR LESS?

NO → REPAIR SHORT TO VOLTAGE IN CKT 884.

YES ↓

(5)
- USING J 39200, MEASURE THE VOLTAGE BETWEEN GROUND AND TERMINAL "E2" OF THE 32-WAY EBCM HARNESS CONNECTOR.

IS VOLTAGE 1 VOLT OR LESS?

NO → REPAIR SHORT TO VOLTAGE IN CKT 885.

YES → MALFUNCTION IS INTERMITTENT. INSPECT ALL CONNECTORS AND HARNESSES FOR DAMAGE WHICH MAY RESULT IN A SHORT TO VOLTAGE WHEN ALL COMPONENTS ARE CONNECTED. REPAIR ALL DAMAGE IF FOUND. REFER TO "DIAGNOSTIC AIDS" ON FACING PAGE FOR MORE INFORMATION.

(3) YES ↓

(6)
- IGNITION "OFF."
- INSPECT TERMINALS "E1" AND "E2" OF THE 32-WAY EBCM HARNESS CONNECTOR FOR POOR TERMINAL CONTACT AND CONTAMINATION. ALSO INSPECT CKTs 884 AND 885 FOR DAMAGE WHICH MAY RESULT IN AN OPEN CIRCUIT, SHORT TO GROUND, OR SHORT TO VOLTAGE. REPAIR ALL DAMAGE IF EVIDENT.
- RECONNECT ALL CONNECTORS.
- IGNITION "ON."

DOES DTC 34 SET AS A CURRENT DTC?

YES → REPLACE EBCM.

NO → MALFUNCTION IS INTERMITTENT. REFER TO "DIAGNOSTIC AIDS" ON FACING PAGE.

DIAGNOSTIC CHART DTC 34 — 1994 CENTURY, CUTLASS CIERA AND CUTLASS CRUISER

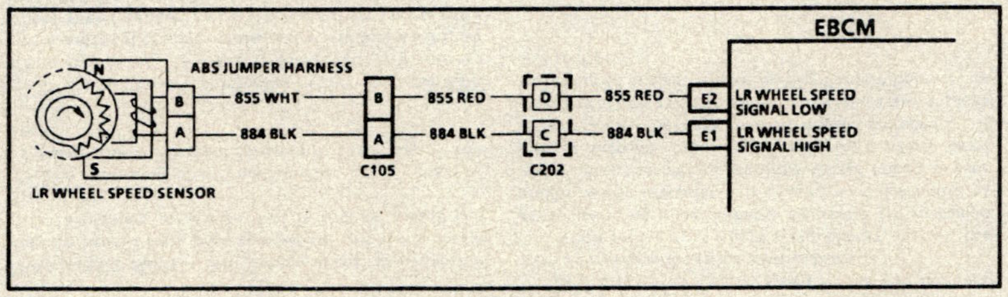

DTC 34

(Page 2 of 3)
LR WHEEL SPEED SNR CKT OPEN OR SHORT TO GND/BATT

Circuit Description:

As a toothed ring passes by the wheel speed sensor, changes in the electromagnetic field cause the wheel speed sensor to produce a sinusoidal (AC) voltage signal whose frequency is proportional to wheel speed. The magnitude of this signal is directly related to wheel speed and the proximity of the wheel speed sensor to the toothed ring, often referred to as the air gap.

DTC Will Set When: DTC 34 can be set anytime after initialization. If either of the left rear wheel speed circuits are shorted to voltage or ground, a malfunction exists.

Action Taken: A malfunction DTC is stored, ABS is disabled and the ABS warning lamp is turned "ON."

DTC Chart Test Description: Number(s) below refer to circled number(s) on the diagnostic chart.

7. Checks for high resistance in CKT 884.
8. Checks for high resistance in CKT 885.
9. Verifies proper resistance value of the wheel speed sensor.

Diagnostic Aids: An "Intermittent" malfunction may be caused by a poor connection, rubbed through wire insulation, or a wire that is broken inside the insulation.

The frequency of the malfunction can be checked by using the enhanced diagnostic function of the Tech 1.

If the customer's comments reflect that the ABS warning lamp is "ON" only during moist environmental changes (rain, snow, vehicle wash), all wheel speed sensor circuitry should be thoroughly inspected for signs of water intrusion. If DTC is not current, clear DTCs and simulate the effects of water intrusion. Use the following procedure. Spray down the suspected area with a 5% salt water solution (two teaspoons of salt to 12 oz. of water). Test drive vehicle over various road surfaces (bumps, turns, etc.) above 24 km/h (15 mph) for at least 30 seconds. If DTC returns, replace suspected harness.

Any circuitry, that is suspected as causing the intermittent complaint, should be thoroughly checked for backed out terminals, improper mating, broken locks, improperly formed or damaged terminals, poor terminal to wiring connections or physical damage to the wiring harness.

Resistance of the wheel speed sensor will increase with an increase in sensor temperature.

When replacing a wheel speed sensor, inspect the sensor terminals and harness connector for corrosion and/or water intrusion. If evidence of corrosion or water intrusion exists, replace wheel speed sensor jumper harness. Likewise, if replacing a wheel speed sensor jumper harness, inspect sensor terminals. If evidence of corrosion or water intrusion exists, replace wheel speed sensor. Refer to "On-Vehicle Service" in this section.

DIAGNOSTIC CHART DTC 34 — 1994 CENTURY, CUTLASS CIERA AND CUTLASS CRUISER

DTC 34

(Page 2 of 3)
LR WHEEL SPEED SNR CKT OPEN OR SHORT TO GND/BATT

DIAGNOSIS CONTINUED FROM PAGE 1 OF 3 OF THIS DTC CHART.

(7)
- DISCONNECT LEFT REAR WHEEL SPEED SENSOR DIRECTLY AT SENSOR.
- USING J 39200, MEASURE THE RESISTANCE BETWEEN TERMINAL "E1" OF THE 32-WAY EBCM HARNESS CONNECTOR AND TERMINAL "B" OF THE LEFT REAR WHEEL SPEED SENSOR HARNESS CONNECTOR.

IS RESISTANCE 2Ω OR LESS?

NO: REPAIR HIGH RESISTANCE IN CKT 884 (WHEEL SPEED SENSOR JUMPER MUST BE REPLACED IF DAMAGED).

YES:

(8)
- USING J 39200, MEASURE THE RESISTANCE BETWEEN TERMINAL "E2" OF THE 32-WAY EBCM HARNESS CONNECTOR AND TERMINAL "A" OF THE LEFT REAR WHEEL SPEED SENSOR HARNESS CONNECTOR.

IS RESISTANCE 2Ω OR LESS?

NO: REPAIR HIGH RESISTANCE IN CKT 885 (WHEEL SPEED SENSOR JUMPER MUST BE REPLACED IF DAMAGED).

YES:

(9)
- USING J 39200, MEASURE THE RESISTANCE BETWEEN TERMINAL "A" AND TERMINAL "B" OF THE LEFT REAR WHEEL SPEED SENSOR CONNECTOR.

IS RESISTANCE BETWEEN 2059 AND 2353Ω WHEN SENSOR IS 20°C (68°F)? *(IF TEMPERATURE OF SENSOR IS SUSPECTED TO OTHER THAN 20°C (68°F), REFER TO THE TABLE BELOW FOR APPLICABLE SENSOR RESISTANCE VALUES; THE VALUES IN THE TABLE ARE FOR TEMPERATURE OF THE SENSOR, NOT AIR TEMPERATURE.)*

NO: REPLACE WHEEL SPEED SENSOR.

YES: MALFUNCTION IS INTERMITTENT. INSPECT ALL CONNECTORS AND HARNESSES FOR DAMAGE WHICH MAY RESULT IN HIGH RESISTANCE WHEN ALL COMPONENTS ARE CONNECTED. REFER TO "DIAGNOSTIC AIDS" ON FACING PAGE FOR MORE INFORMATION. REPAIR ALL DAMAGE IF FOUND.

REAR WHEEL SPEED SENSOR

TEMPERATURE OF SENSOR VS. SENSOR RESISTANCE (APPROXIMATE)

TEMP. (°C)	TEMP. (°F)	RESISTANCE (OHMS)
-34 TO 4	-30 TO 40	1622 TO 2205
5 TO 43	41 TO 110	1930 TO 2566
44 TO 93	111 TO 200	2245 TO 3028

DIAGNOSTIC CHART DTC 34 — 1994 CENTURY, CUTLASS CIERA AND CUTLASS CRUISER

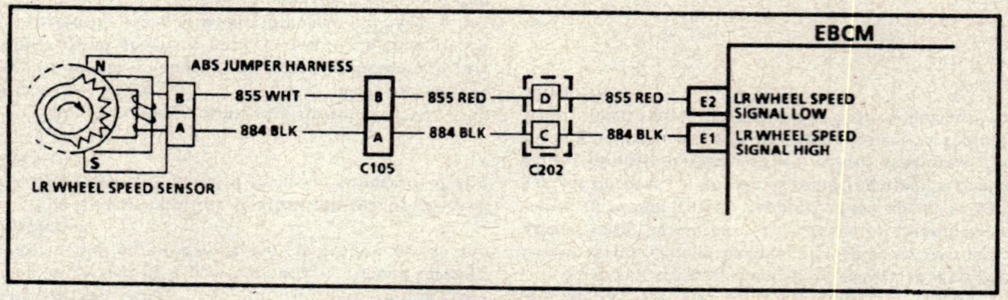

DTC 34

(Page 3 of 3)
LR WHEEL SPEED SNR CKT OPEN OR SHORT TO GND/BATT

Circuit Description:

As a toothed ring passes by the wheel speed sensor, changes in the electromagnetic field cause the wheel speed sensor to produce a sinusoidal (AC) voltage signal whose frequency is proportional to wheel speed. The magnitude of this signal is directly related to wheel speed and the proximity of the wheel speed sensor to the toothed ring, often referred to as the air gap.

DTC Will Set When: DTC 34 can be set anytime after initialization. If either of the left rear wheel speed circuits are shorted to voltage or ground, a malfunction exists.

Action Taken: A malfunction DTC is stored, ABS is disabled and the ABS warning lamp is turned "ON."

DTC Chart Test Description: Number(s) below refer to circled number(s) on the diagnostic chart.

10. Checks for short to ground in CKT 884.
11. Checks for short to ground in CKT 885.
12. Checks for short to ground in wheel speed sensor jumper harness.
13. Checks for an internal short to ground in the wheel speed sensor.

Diagnostic Aids: An "Intermittent" malfunction may be caused by a poor connection, rubbed through wire insulation, or a wire that is broken inside the insulation.

The frequency of the malfunction can be checked by using the enhanced diagnostic function of the Tech 1,

If the customer's comments reflect that the ABS warning lamp is "ON" only during moist environmental changes (rain, snow, vehicle wash), all wheel speed sensor circuitry should be thoroughly inspected for signs of water intrusion. If DTC is not current, clear DTCs and simulate the effects of water intrusion. Use the following procedure. Spray down the suspected area with a 5% salt water solution (two teaspoons of salt to 12 oz. of water). Test drive vehicle over various road surfaces (bumps, turns, etc.) above 24 km/h (15 mph) for at least 30 seconds. If DTC returns, replace suspected harness.

Any circuitry, that is suspected as causing the intermittent complaint, should be thoroughly checked for backed out terminals, improper mating, broken locks, improperly formed or damaged terminals, poor terminal to wiring connections or physical damage to the wiring harness.

Resistance of the wheel speed sensor will increase with an increase in sensor temperature.

When replacing a wheel speed sensor, inspect the sensor terminals and harness connector for corrosion and/or water intrusion. If evidence of corrosion or water intrusion exists, replace wheel speed sensor jumper harness. Likewise, if replacing a wheel speed sensor jumper harness, inspect sensor terminals. If evidence of corrosion or water intrusion exists, replace wheel speed sensor. Refer to "On-Vehicle Service" in this section.

DIAGNOSTIC CHART DTC 34 — 1994 CENTURY, CUTLASS CIERA AND CUTLASS CRUISER

DTC 34

(Page 3 of 3)
LR WHEEL SPEED SNR CKT OPEN OR SHORT TO GND/BATT

DIAGNOSIS CONTINUED FROM PAGE 1 OF 3 OF THIS DTC CHART.

(10)
- DISCONNECT CONNECTOR C105 (LH REAR UNDER BODY NEAR FRAME RAIL).
- USING J 39200, MEASURE THE RESISTANCE BETWEEN GROUND AND TERMINAL "E1" OF THE 32-WAY EBCM HARNESS CONNECTOR.

DOES J 39200 DISPLAY RESISTANCE AS "OL" (INFINITE)?

- NO → REPAIR SHORT TO GROUND IN CKT 884.
- YES ↓

(11)
- USING J 39200, MEASURE THE RESISTANCE BETWEEN GROUND AND TERMINAL "E2" OF THE 32-WAY EBCM HARNESS CONNECTOR.

DOES J 39200 DISPLAY RESISTANCE AS "OL" (INFINITE)?

- NO → REPAIR SHORT TO GROUND IN CKT 885.
- YES ↓

(12)
- THOROUGHLY RESPRAY THE LEFT REAR WHEEL SPEED SENSOR JUMPER HARNESS WITH THE SALT WATER SOLUTION.

IMPORTANT: DO NOT ALLOW WATER TO ENTER ANY WHEEL SPEED SENSOR CONNECTOR.

- DISCONNECT LEFT REAR WHEEL SPEED SENSOR DIRECTLY AT SENSOR.
- USING J 39200, MEASURE THE RESISTANCE BETWEEN GROUND AND TERMINAL "A," THEN "B" OF THE LEFT REAR WHEEL SPEED SENSOR HARNESS CONNECTOR.

DOES J 39200 DISPLAY RESISTANCE AS "OL" (INFINITE) FOR BOTH MEASUREMENTS?

- NO → REPLACE LEFT REAR WHEEL SPEED SENSOR JUMPER HARNESS.
- YES ↓

(13)
- USING J 39200, MEASURE THE RESISTANCE BETWEEN GROUND AND TERMINAL "A" OF THE LEFT REAR WHEEL SPEED SENSOR CONNECTOR.

DOES J 39200 DISPLAY RESISTANCE AS "OL" (INFINITE)?

- NO → REPLACE WHEEL SPEED SENSOR.
- YES ↓

MALFUNCTION IS INTERMITTENT. INSPECT ALL CONNECTORS AND HARNESSES FOR DAMAGE WHICH MAY RESULT IN A SHORT TO GROUND WHEN ALL COMPONENTS ARE CONNECTED. PAY PARTICULAR ATTENTION TO THE JUMPER HARNESS, SMALL NICKS OR CUTS CAN CAUSE A MALFUNCTION. REPAIR ANY DAMAGE IF FOUND. JUMPER HARNESS MUST BE REPLACED IF DAMAGED. REFER TO "DIAGNOSTIC AIDS" ON FACING PAGE FOR MORE INFORMATION.

DIAGNOSTIC CHART DTC 35 — 1994 CENTURY, CUTLASS CIERA AND CUTLASS CRUISER

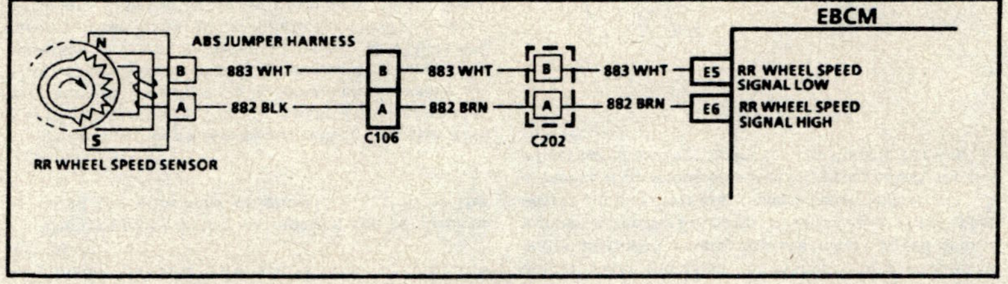

DTC 35

(Page 1 of 3)
RR WHEEL SPEED SNR CKT OPEN OR SHORT TO GND/BATT

Circuit Description:
As a toothed ring passes by the wheel speed sensor, changes in the electromagnetic field cause the wheel speed sensor to produce a sinusoidal (AC) voltage signal whose frequency is proportional to wheel speed. The magnitude of this signal is directly related to wheel speed and the proximity of the wheel speed sensor to the toothed ring, often referred to as the air gap.

DTC Will Set When: DTC 35 can be set anytime after initialization. If either of the right rear wheel speed circuits are shorted to voltage or ground, a malfunction exists.

Action Taken: A malfunction DTC is stored, ABS is disabled and the ABS warning lamp is turned "ON."

DTC Chart Test Description: Number(s) below refer to circled number(s) on the diagnostic chart.

1. Checks to see that the wheel speed sensor is within proper resistance values.
2. Checks for short to ground in wheel speed sensor wiring.
3. Checks for short to voltage in wheel speed sensor wiring.
4. Checks for a short to voltage in CKT 883.
5. Checks for a short to voltage in CKT 882.
6. Ensures malfunction was not due to physical damage of circuitry.

Diagnostic Aids: An "Intermittent" malfunction may be caused by a poor connection, rubbed through wire insulation, or a wire that is broken inside the insulation.

The frequency of the malfunction can be checked by using the enhanced diagnostic function of the Tech 1.

If the customer's comments reflect that the ABS warning lamp is "ON" only during moist environmental changes (rain, snow, vehicle wash), all wheel speed sensor circuitry should be thoroughly inspected for signs of water intrusion. If DTC is not current, clear DTCs and simulate the effects of water intrusion. Use the following procedure. Spray down the suspected area with a 5% salt water solution (two teaspoons of salt to 12 oz. of water). Test drive vehicle over various road surfaces (bumps, turns, etc.) above 24 km/h (15 mph) for at least 30 seconds. If DTC returns, replace suspected harness.

Any circuitry, that is suspected as causing the intermittent complaint, should be thoroughly checked for backed out terminals, improper mating, broken locks, improperly formed or damaged terminals, poor terminal to wiring connections or physical damage to the wiring harness.

Resistance of the wheel speed sensor will increase with an increase in sensor temperature.

When replacing a wheel speed sensor, inspect the sensor terminals and harness connector for corrosion and/or water intrusion. If evidence of corrosion or water intrusion exists, replace wheel speed sensor jumper harness. Likewise, if replacing a wheel speed sensor jumper harness, inspect sensor terminals. If evidence of corrosion or water intrusion exists, replace wheel speed sensor. Refer to "On-Vehicle Service" in this section.

DIAGNOSTIC CHART DTC 35 — 1994 CENTURY, CUTLASS CIERA AND CUTLASS CRUISER

DTC 35

(Page 1 of 3)
RR WHEEL SPEED SNR CKT OPEN OR SHORT TO GND/BATT

Important
- WHEEL SPEED SENSOR INTERMITTENT MALFUNCTIONS MAY BE DIFFICULT TO LOCATE. CARE SHOULD BE TAKEN NOT TO DISTURB ANY ELECTRICAL CONNECTIONS PRIOR TO AN INDICATED STEP OF THIS CHART. THIS WILL ENSURE THAT AN INTERMITTENT CONNECTION WILL NOT BE CORRECTED BEFORE THE SOURCE OF THE MALFUNCTION IS FOUND.

(1)
- IGNITION "OFF."
- USING A 5% SALT WATER SOLUTION (TWO TEASPOONS OF SALT TO 12 OZ. WATER), THOROUGHLY SPRAY THE RIGHT REAR WHEEL SPEED SENSOR JUMPER HARNESS.
- DISCONNECT 32-WAY EBCM HARNESS CONNECTOR.
- USING J 39200, MEASURE THE RESISTANCE BETWEEN TERMINALS "E5" AND "E6" OF THE 32-WAY EBCM HARNESS CONNECTOR.

IS RESISTANCE BETWEEN 2059 AND 2353Ω WHEN SENSOR IS 20°C (68°F)? *(IF TEMPERATURE OF SENSOR IS SUSPECTED TO OTHER THAN 20°C (68°F), REFER TO THE TABLE ON PAGE 2 OF 2 OF THIS DTC CHART FOR APPLICABLE SENSOR RESISTANCE VALUES; THE VALUES IN THE TABLE ARE FOR TEMPERATURE OF THE SENSOR, NOT AIR TEMPERATURE.)*

- YES → (2)
- NO → DIAGNOSIS CONTINUED ON PAGE 2 OF 3 OF THIS DTC CHART.

(2)
- MEASURE THE RESISTANCE BETWEEN GROUND AND TERMINAL "E5" OF THE 32-WAY EBCM HARNESS CONNECTOR.

DOES J 39200 DISPLAY RESISTANCE AS "OL" (INFINITE)?

- YES → (3)
- NO → DIAGNOSIS CONTINUED ON PAGE 3 OF 3 OF THIS DTC CHART.

(3)
- IGNITION "ON."
- USING J 39200, MEASURE THE VOLTAGE BETWEEN GROUND AND TERMINAL "E5" OF THE 32-WAY EBCM HARNESS CONNECTOR.

IS VOLTAGE 1 VOLT OR LESS?

- NO → (4)
- YES → (6)

(4)
- IGNITION "OFF."
- DISCONNECT RIGHT REAR WHEEL SPEED SENSOR DIRECTLY AT SENSOR.
- IGNITION "ON."
- USING J 39200, MEASURE THE VOLTAGE BETWEEN GROUND AND TERMINAL "E5" OF THE 32-WAY EBCM HARNESS CONNECTOR.

IS VOLTAGE 1 VOLT OR LESS?

- YES → (5)
- NO → REPAIR SHORT TO VOLTAGE IN CKT 883.

(5)
- USING J 39200, MEASURE THE VOLTAGE BETWEEN GROUND AND TERMINAL "E6" OF THE 32-WAY EBCM HARNESS CONNECTOR.

IS VOLTAGE 1 VOLT OR LESS?

- YES → MALFUNCTION IS INTERMITTENT. INSPECT ALL CONNECTORS AND HARNESSES FOR DAMAGE WHICH MAY RESULT IN A SHORT TO VOLTAGE WHEN ALL COMPONENTS ARE CONNECTED. REPAIR ALL DAMAGE IF FOUND. REFER TO "DIAGNOSTIC AIDS" ON FACING PAGE FOR MORE INFORMATION.
- NO → REPAIR SHORT TO VOLTAGE IN CKT 882.

(6)
- IGNITION "OFF."
- INSPECT TERMINALS "E5" AND "E6" OF THE 32-WAY EBCM HARNESS CONNECTOR FOR POOR TERMINAL CONTACT AND CONTAMINATION. ALSO INSPECT CKTs 882 AND 883 FOR DAMAGE WHICH MAY RESULT IN AN OPEN CIRCUIT, SHORT TO GROUND, OR SHORT TO VOLTAGE. REPAIR ALL DAMAGE IF EVIDENT.
- RECONNECT ALL CONNECTORS.
- IGNITION "ON."

DOES DTC 35 SET AS A CURRENT DTC?

- YES → REPLACE EBCM.
- NO → MALFUNCTION IS INTERMITTENT. REFER TO "DIAGNOSTIC AIDS" ON FACING PAGE.

DIAGNOSTIC CHART DTC 35 — 1994 CENTURY, CUTLASS CIERA AND CUTLASS CRUISER

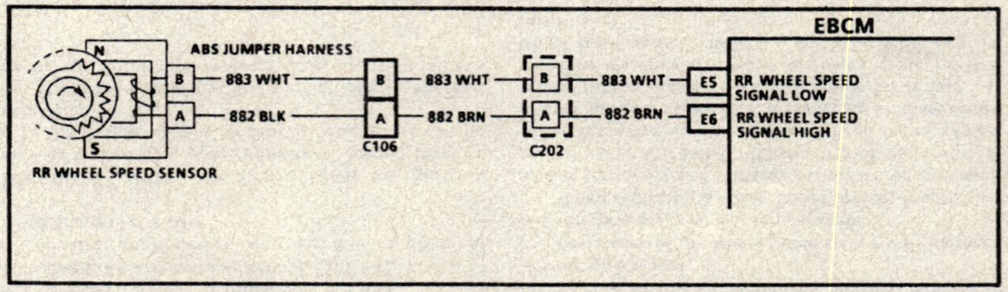

DTC 35

(Page 2 of 3)

RR WHEEL SPEED SNR CKT OPEN OR SHORT TO GND/BATT

Circuit Description:

As a toothed ring passes by the wheel speed sensor, changes in the electromagnetic field cause the wheel speed sensor to produce a sinusoidal (AC) voltage signal whose frequency is proportional to wheel speed. The magnitude of this signal is directly related to wheel speed and the proximity of the wheel speed sensor to the toothed ring, often referred to as the air gap.

DTC Will Set When: DTC 35 can be set anytime after initialization. If either of the right rear wheel speed circuits are shorted to voltage or ground, a malfunction exists.

Action Taken: A malfunction DTC is stored, ABS is disabled and the ABS warning lamp is turned "ON."

DTC Chart Test Description: Number(s) below refer to circled number(s) on the diagnostic chart.

7. Checks for high resistance in CKT 883.
8. Checks for high resistance in CKT 882.
9. Verifies the proper resistance value of the sensor.

Diagnostic Aids: An "Intermittent" malfunction may be caused by a poor connection, rubbed through wire insulation, or a wire that is broken inside the insulation.

The frequency of the malfunction can be checked by using the enhanced diagnostic function of the Tech 1.

If the customer's comments reflect that the ABS warning lamp is "ON" only during moist environmental changes (rain, snow, vehicle wash), all wheel speed sensor circuitry should be thoroughly inspected for signs of water intrusion. If DTC is not current, clear DTCs and simulate the effects of water intrusion. Use the following procedure. Spray down the suspected area with a 5% salt water solution (two teaspoons of salt to 12 oz. of water). Test drive vehicle over various road surfaces (bumps, turns, etc.) above 24 km/h (15 mph) for at least 30 seconds. If DTC returns, replace suspected harness.

Any circuitry, that is suspected as causing the intermittent complaint, should be thoroughly checked for backed out terminals, improper mating, broken locks, improperly formed or damaged terminals, poor terminal to wiring connections or physical damage to the wiring harness.

Resistance of the wheel speed sensor will increase with an increase in sensor temperature.

When replacing a wheel speed sensor, inspect the sensor terminals and harness connector for corrosion and/or water intrusion. If evidence of corrosion or water intrusion exists, replace wheel speed sensor jumper harness. Likewise, if replacing a wheel speed sensor jumper harness, inspect sensor terminals. If evidence of corrosion or water intrusion exists, replace wheel speed sensor. Refer to "On-Vehicle Service" in this section.

DIAGNOSTIC CHART DTC 35 — 1994 CENTURY, CUTLASS CIERA AND CUTLASS CRUISER

DTC 35

(Page 2 of 3)

RR WHEEL SPEED SNR CKT OPEN OR SHORT TO GND/BATT

DIAGNOSIS CONTINUED FROM PAGE 1 OF 3 OF THIS DTC CHART.

(7)
- DISCONNECT RIGHT REAR WHEEL SPEED SENSOR DIRECTLY AT SENSOR.
- USING J 39200, MEASURE THE RESISTANCE BETWEEN TERMINAL "E5" OF THE 32-WAY EBCM HARNESS CONNECTOR AND TERMINAL "B" OF THE RIGHT REAR WHEEL SPEED SENSOR HARNESS CONNECTOR.

IS RESISTANCE 2Ω OR LESS?

NO: REPAIR HIGH RESISTANCE IN CKT 883 (WHEEL SPEED SENSOR JUMPER MUST BE REPLACED IF DAMAGED).

YES:

(8)
- USING J 39200, MEASURE THE RESISTANCE BETWEEN TERMINAL "E6" OF THE 32-WAY EBCM HARNESS CONNECTOR AND TERMINAL "A" OF THE RIGHT REAR WHEEL SPEED SENSOR HARNESS CONNECTOR.

IS RESISTANCE 2Ω OR LESS?

NO: REPAIR HIGH RESISTANCE IN CKT 882 (WHEEL SPEED SENSOR JUMPER MUST BE REPLACED IF DAMAGED).

YES:

(9)
- USING J 39200, MEASURE THE RESISTANCE BETWEEN TERMINAL "A" AND TERMINAL "B" OF THE RIGHT REAR WHEEL SPEED SENSOR CONNECTOR.

IS RESISTANCE BETWEEN 2059 AND 2353Ω WHEN SENSOR IS 20°C (68°F)? *(IF TEMPERATURE OF SENSOR IS SUSPECTED TO OTHER THAN 20°C (68°F), REFER TO THE TABLE BELOW FOR APPLICABLE SENSOR RESISTANCE VALUES; THE VALUES IN THE TABLE ARE FOR TEMPERATURE OF THE SENSOR, NOT AIR TEMPERATURE.)*

YES: MALFUNCTION IS INTERMITTENT. INSPECT ALL CONNECTORS AND HARNESSES FOR DAMAGE WHICH MAY RESULT IN HIGH RESISTANCE WHEN ALL COMPONENTS ARE CONNECTED. REFER TO "DIAGNOSTIC AIDS" ON FACING PAGE FOR MORE INFORMATION. REPAIR ALL DAMAGE IF FOUND.

NO: REPLACE WHEEL SPEED SENSOR.

REAR WHEEL SPEED SENSOR

TEMPERATURE OF SENSOR VS. SENSOR RESISTANCE (APPROXIMATE)

TEMP. (°C)	TEMP. (°F)	RESISTANCE (OHMS)
-34 TO 4	-30 TO 40	1622 TO 2205
5 TO 43	41 TO 110	1930 TO 2566
44 TO 93	111 TO 200	2245 TO 3028

DIAGNOSTIC CHART DTC 35 — 1994 CENTURY, CUTLASS CIERA AND CUTLASS CRUISER

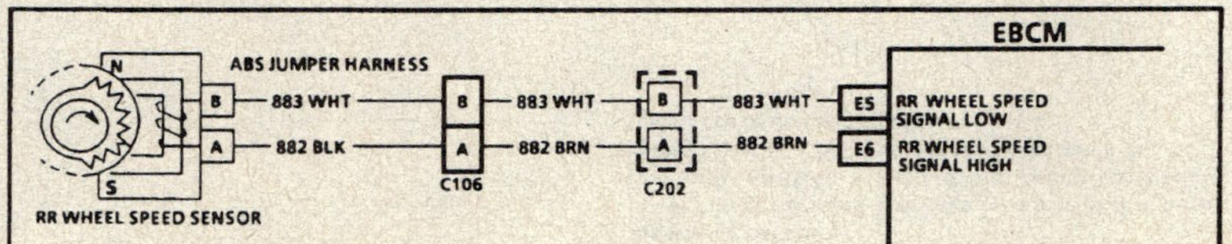

DTC 35

(Page 3 of 3)
RR WHEEL SPEED SNR CKT OPEN OR SHORT TO GND/BATT

Circuit Description:

As a toothed ring passes by the wheel speed sensor, changes in the electromagnetic field cause the wheel speed sensor to produce a sinusoidal (AC) voltage signal whose frequency is proportional to wheel speed. The magnitude of this signal is directly related to wheel speed and the proximity of the wheel speed sensor to the toothed ring, often referred to as the air gap.

DTC Will Set When: DTC 35 can be set anytime after initialization. If either of the right rear wheel speed circuits are shorted to voltage or ground, a malfunction exists.

Action Taken: A malfunction DTC is stored, ABS is disabled and the ABS warning lamp is turned "ON."

DTC Chart Test Description: Number(s) below refer to circled number(s) on the diagnostic chart.

10. Checks for short to ground in CKT 883.
11. Checks for short to ground in CKT 882.
12. Checks for short to ground in wheel speed sensor jumper harness.
13. Checks for an internal short to ground in the wheel speed sensor.

Diagnostic Aids: An "Intermittent" malfunction may be caused by a poor connection, rubbed through wire insulation, or a wire that is broken inside the insulation.

The frequency of the malfunction can be checked by using the enhanced diagnostic function of the Tech 1.

If the customer's comments reflect that the ABS warning lamp is "ON" only during moist environmental changes (rain, snow, vehicle wash), all wheel speed sensor circuitry should be thoroughly inspected for signs of water intrusion. If DTC is not current, clear DTCs and simulate the effects of water intrusion. Use the following procedure. Spray down the suspected area with a 5% salt water solution (two teaspoons of salt to 12 oz. of water). Test drive vehicle over various road surfaces (bumps, turns, etc.) above 24 km/h (15 mph) for at least 30 seconds. If DTC returns, replace suspected harness.

Any circuitry, that is suspected as causing the intermittent complaint, should be thoroughly checked for backed out terminals, improper mating, broken locks, improperly formed or damaged terminals, poor terminal to wiring connections or physical damage to the wiring harness.

Resistance of the wheel speed sensor will increase with an increase in sensor temperature.

When replacing a wheel speed sensor, inspect the sensor terminals and harness connector for corrosion and/or water intrusion. If evidence of corrosion or water intrusion exists, replace wheel speed sensor jumper harness. Likewise, if replacing a wheel speed sensor jumper harness, inspect sensor terminals. If evidence of corrosion or water intrusion exists, replace wheel speed sensor. Refer to "On-Vehicle Service" in this section.

DIAGNOSTIC CHART DTC 35 — 1994 CENTURY, CUTLASS CIERA AND CUTLASS CRUISER

DTC 35

(Page 3 of 3)
RR WHEEL SPEED SNR CKT OPEN OR SHORT TO GND/BATT

DIAGNOSIS CONTINUED FROM PAGE 1 OF 3 OF THIS DTC CHART.

(10)
- DISCONNECT CONNECTOR C106 (RH REAR UNDER BODY NEAR FRAME RAIL).
- USING J 39200, MEASURE THE RESISTANCE BETWEEN GROUND AND TERMINAL "E5" OF THE 32-WAY EBCM HARNESS CONNECTOR.
DOES J 39200 DISPLAY RESISTANCE AS "OL" (INFINITE)?

NO: REPAIR SHORT TO GROUND IN CKT 883.

YES:

(11)
- USING J 39200, MEASURE THE RESISTANCE BETWEEN GROUND AND TERMINAL "E6" OF THE 32-WAY EBCM HARNESS CONNECTOR.
DOES J 39200 DISPLAY RESISTANCE AS "OL" (INFINITE)?

NO: REPAIR SHORT TO GROUND IN CKT 882.

YES:

(12)
- THOROUGHLY RESPRAY THE RIGHT REAR WHEEL SPEED SENSOR JUMPER HARNESS WITH THE SALT WATER SOLUTION.

IMPORTANT: DO NOT ALLOW WATER TO ENTER ANY WHEEL SPEED SENSOR CONNECTOR.

- DISCONNECT RIGHT REAR WHEEL SPEED SENSOR DIRECTLY AT SENSOR.
- USING J 39200, MEASURE THE RESISTANCE BETWEEN GROUND AND TERMINAL "A," THEN "B" OF THE RIGHT REAR WHEEL SPEED SENSOR HARNESS CONNECTOR.
DOES J 39200 DISPLAY RESISTANCE AS "OL" (INFINITE) FOR BOTH MEASUREMENTS?

NO: REPLACE RIGHT REAR WHEEL SPEED SENSOR JUMPER HARNESS.

YES:

(13)
- USING J 39200, MEASURE THE RESISTANCE BETWEEN GROUND AND TERMINAL "A" OF THE RIGHT REAR WHEEL SPEED SENSOR CONNECTOR.
DOES J 39200 DISPLAY RESISTANCE AS "OL" (INFINITE)?

NO: REPLACE WHEEL SPEED SENSOR.

YES: MALFUNCTION IS INTERMITTENT. INSPECT ALL CONNECTORS AND HARNESSES FOR DAMAGE WHICH MAY RESULT IN A SHORT TO GROUND WHEN ALL COMPONENTS ARE CONNECTED. PAY PARTICULAR ATTENTION TO THE JUMPER HARNESS, SMALL NICKS OR CUTS CAN CAUSE A MALFUNCTION. REPAIR ANY DAMAGE IF FOUND. JUMPER HARNESS MUST BE REPLACED IF DAMAGED. REFER TO "DIAGNOSTIC AIDS" ON FACING PAGE FOR MORE INFORMATION.

DIAGNOSTIC CHART DTC 36 — 1994 CENTURY, CUTLASS CIERA AND CUTLASS CRUISER

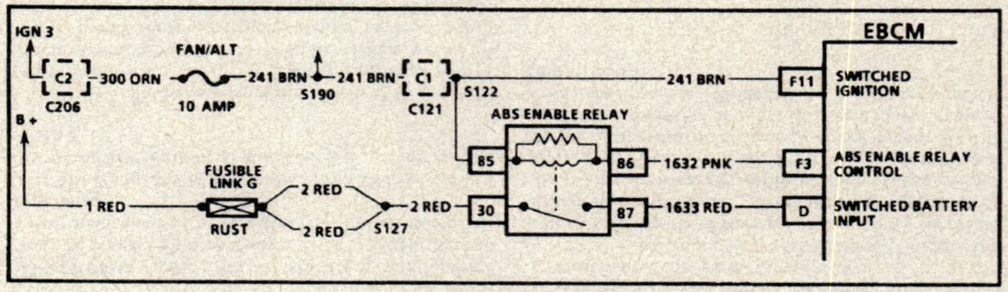

DTC 36

(Page 1 of 2)
LOW SYSTEM VOLTAGE

Circuit Description:

This DTC is used to monitor the voltage level available to the EBCM controller. If the voltage drops below 11 volts, full performance of the ABS system cannot be guaranteed. During ABS operation, there are several current requirements that will cause battery voltage to drop. Because of this, voltage is monitored prior to ABS to indicate good charging system condition and also during ABS when voltage may drop significantly.

DTC Will Set When: DTC 36 can only be set if the vehicle's speed is greater than 5 km/h (3 mph). If the switched battery voltage is less than 11.2 volts non-ABS or less than 9.6 volts ABS, a malfunction exists.

Action Taken: A malfunction DTC is stored, ABS is disabled and the ABS warning lamp is turned "ON."

DTC Chart Test Description: Number(s) below refer to circled number(s) on the diagnostic chart.

1. Checks to see what voltage is readily available at terminal "D" of the EBCM. If voltage is representative of good charging system condition, a malfunction is not present.
2. Isolates the low voltage condition to high circuit resistance, or improper charging system operation.
3. Isolates which circuit is the source of the high resistance condition.
4. Checks for high resistance in CKT 2 or 1.

Diagnostic Aids: An "Intermittent" malfunction may be caused by a poor connection, rubbed through wire insulation, or a wire that is broken inside the insulation.

The frequency of the malfunction can be checked by using the enhanced diagnostic function of the Tech 1.

Any circuitry, that is suspected as causing the intermittent complaint, should be thoroughly checked for backed out terminals, improper mating, broken locks, improperly formed or damaged terminals, poor terminal to wiring connections or physical damage to the wiring harness.

While performing a voltage load test, if it is noted that <u>only</u> ignition voltage drops below acceptable voltage levels, CKT 241 should be checked for high resistance or an open condition.

DIAGNOSTIC CHART DTC 36 — 1994 CENTURY, CUTLASS CIERA AND CUTLASS CRUISER

DTC 36

(Page 1 of 2)
LOW SYSTEM VOLTAGE

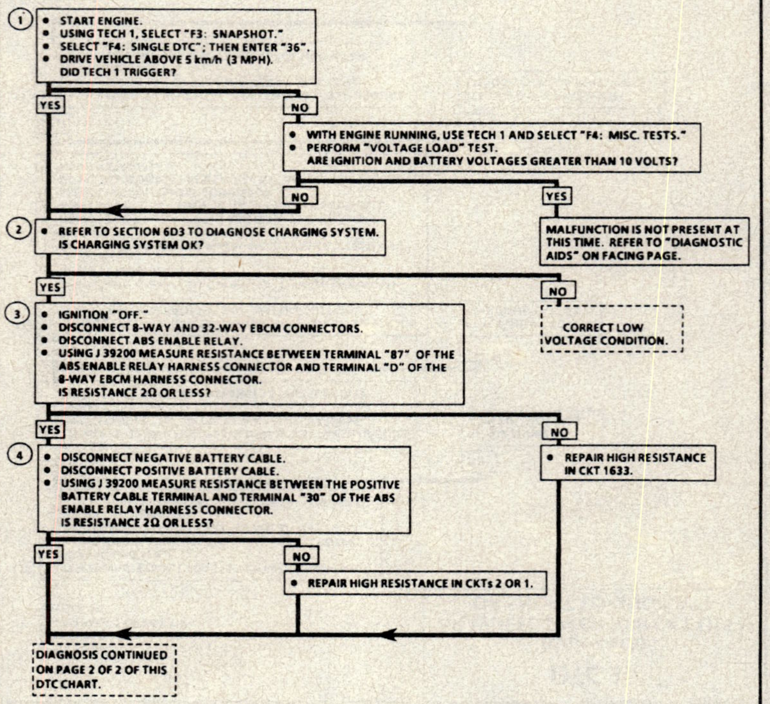

DIAGNOSTIC CHART DTC 36 — 1994 CENTURY, CUTLASS CIERA AND CUTLASS CRUISER

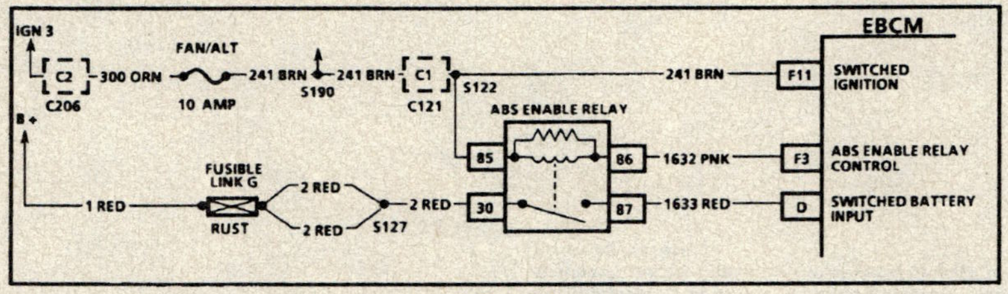

DTC 36
(Page 2 of 2)
LOW SYSTEM VOLTAGE

Circuit Description:

This DTC is used to monitor the voltage level available to the EBCM controller. If the voltage drops below 11 volts, full performance of the ABS system cannot be guaranteed. During ABS operation, there are several current requirements that will cause battery voltage to drop. Because of this, voltage is monitored prior to ABS to indicate good charging system condition and also during ABS when voltage may drop significantly.

DTC Will Set When: DTC 36 can only be set if the vehicle's speed is greater than 5 km/h (3 mph). If the switched battery voltage is less than 11.2 volts non-ABS or less than 9.6 volts ABS, a malfunction exists.

Action Taken: A malfunction DTC is stored, ABS is disabled and the ABS warning lamp is turned "ON."

DTC Chart Test Description: Number(s) below refer to circled number(s) on the diagnostic chart.

5. Checks for high resistance in CKT 241.
6. Checks for high resistance in the ignition fuse feed circuit.
7. Checks for poor terminal contact which may cause an intermittent malfunction or a malfunctioning EBCM.

Diagnostic Aids: An "Intermittent" malfunction may be caused by a poor connection, rubbed through wire insulation, or a wire that is broken inside the insulation.

The frequency of the malfunction can be checked by using the enhanced diagnostic function of the Tech 1.

Any circuitry, that is suspected as causing the intermittent complaint, should be thoroughly checked for backed out terminals, improper mating, broken locks, improperly formed or damaged terminals, poor terminal to wiring connections or physical damage to the wiring harness.

While performing a voltage load test, if it is noted that <u>only</u> ignition voltage drops below acceptable voltage levels, CKT 241 should be checked for high resistance or an open condition.

DIAGNOSTIC CHART DTC 36 — 1994 CENTURY, CUTLASS CIERA AND CUTLASS CRUISER

DTC 36
(Page 2 of 2)
LOW SYSTEM VOLTAGE

DIAGNOSIS CONTINUED FROM PAGE 1 OF 2 OF THIS DTC CHART.

(5)
- REMOVE FAN/ALT FUSE.
- USING J 39200 MEASURE RESISTANCE BETWEEN EACH FAN/ALT FUSE BLOCK TERMINAL AND TERMINAL "F11" OF THE 32-WAY EBCM HARNESS CONNECTOR.

IS EITHER MEASUREMENT 2Ω OR LESS?

NO → REPAIR HIGH RESISTANCE IN CKT 241.

YES ↓

(6)
- WITH BATTERY CABLES DISCONNECTED, TURN IGNITION "ON" (THIS PROVIDES IGNITION CIRCUIT CONTINUITY).
- USING J 39200, MEASURE RESISTANCE BETWEEN POSITIVE BATTERY CABLE TERMINAL AND EACH FAN/ALT FUSE BLOCK TERMINAL.

IS EITHER MEASUREMENT 2Ω OR LESS ?

NO → REPAIR HIGH RESISTANCE IN THE "IGNITION SWITCH 'RUN' FUSE FEED" CIRCUIT.

YES ↓

(7)
- IGNITION "OFF."
- INSPECT ABS ENABLE RELAY AND ABS ENABLE RELAY HARNESS CONNECTOR FOR POOR TERMINAL CONTACT. INSPECT 8-WAY AND 32-WAY EBCM HARNESS CONNECTORS FOR A POOR CONNECTION TO THE EBCM. INSPECT POSITIVE BATTERY CABLE TERMINAL FOR POOR CONTACT TO THE BATTERY.

REPLACE TERMINALS IF POOR CONTACT IS EVIDENT.
- RECONNECT ALL CONNECTORS.
- DRIVE VEHICLE ABOVE 5 km/h (3 MPH).

DOES DTC 36 SET AS A CURRENT DTC?

YES → REPLACE EBCM.

NO → MALFUNCTION IS MOST LIKELY DUE TO A POOR CONNECTION. REFER TO "DIAGNOSTIC AIDS" ON FACING PAGE.

DIAGNOSTIC CHART DTC 37 — 1994 CENTURY, CUTLASS CIERA AND CUTLASS CRUISER

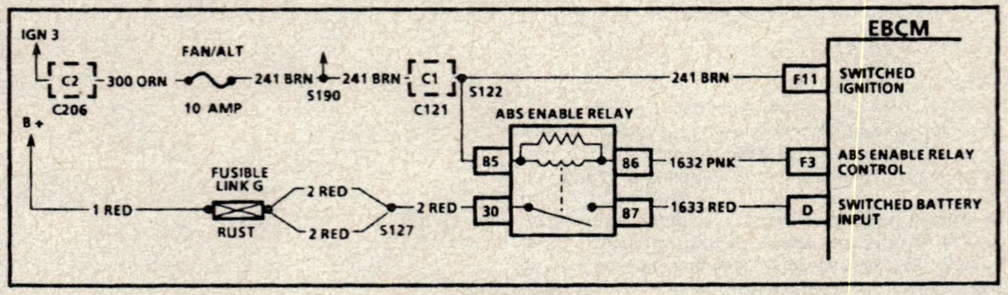

DTC 37

HIGH SYSTEM VOLTAGE

Circuit Description:

This DTC is designed to detect high vehicle voltage levels prior to any required motor movement (initialization or ABS). If excessive voltage exists, demagnetization of the motor magnets may occur, which would eventually affect or eliminate ABS performance.

DTC Will Set When: DTC 37 can only be set if the vehicle's speed is greater than 5 km/h (3 mph). If the switched battery voltage is greater than 17 volts, a malfunction exists.

Action Taken: A malfunction DTC is stored, ABS is disabled and the ABS warning lamp is turned "ON."

DTC Chart Test Description: Number(s) below refer to circled number(s) on the diagnostic chart.

1. Checks the voltage level being received by the EBCM.
2. Indicates whether the high voltage condition is caused by a malfunctioning charging system or EBCM.
3. Checks for poor terminal contact which may cause an intermittent malfunction or a malfunctioning EBCM.

Diagnostic Aids: An "Intermittent" malfunction may be caused by a poor connection, rubbed through wire insulation, or a wire that is broken inside the insulation.

The frequency of the malfunction can be checked by using the enhanced diagnostic function of the Tech 1

Any circuitry, that is suspected as causing the intermittent complaint, should be thoroughly checked for backed out terminals, improper mating, broken locks, improperly formed or damaged terminals, poor terminal to wiring connections or physical damage to the wiring harness.

DIAGNOSTIC CHART DTC 37 — 1994 CENTURY, CUTLASS CIERA AND CUTLASS CRUISER

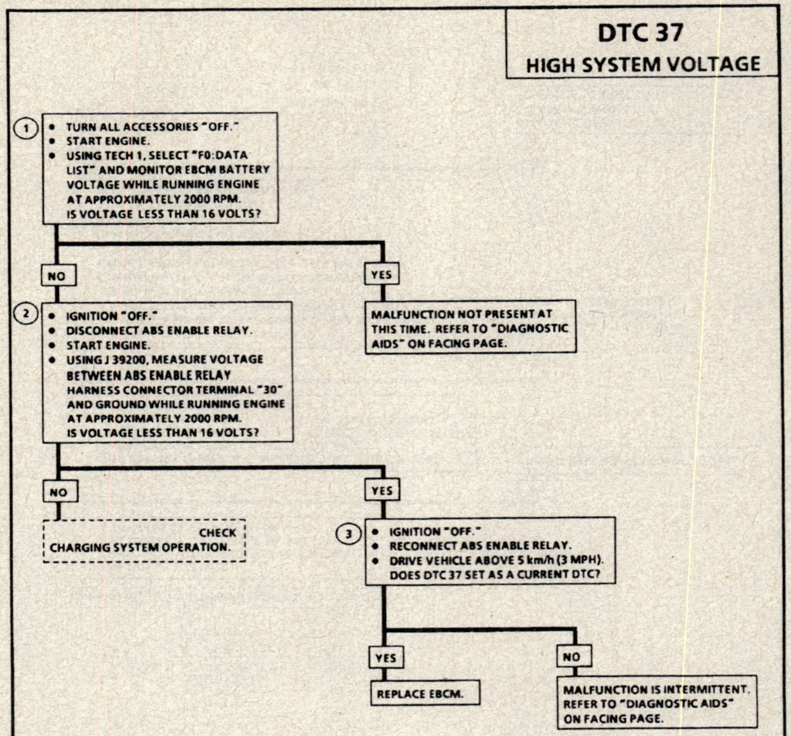

DIAGNOSTIC CHART DTC 38 — 1994 CENTURY, CUTLASS CIERA AND CUTLASS CRUISER

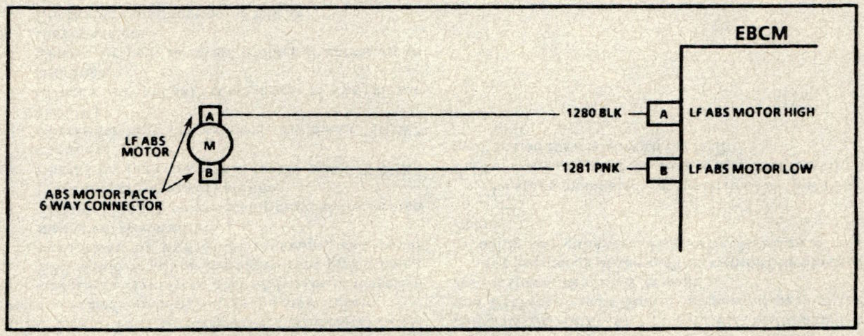

DTC 38

LEFT FRONT ESB WILL NOT HOLD MOTOR

Circuit Description:
This DTC is designed to detect a slipping left front ESB. During initialization and braking, the left front motor is rehomed. If the ESB slips, the motor/piston will move. During the next key "ON" initialization, a rehome of the motor verifies the motor/piston remained at the home position. If motor movement is detected, the ESB must be slipping.

DTC Will Set When: DTC 38 can be set during initialization. If the EBCM detects that the ESB could not hold the piston in the home position, a malfunction exists. DTC 86 is always set with DTC 38.

Action Taken: If an ESB cannot hold a piston in the home position, the piston may be backdriven when the brake pedal is applied, causing the brake pedal to drop. A malfunction DTC is stored, ABS is disabled, and the ABS warning lamp and the red "BRAKE" warning lamp are turned "ON."

DTC Chart Test Description: Number(s) below refer to circled number(s) on the diagnostic chart.

1. Checks the left front ESB. A broken or defective ESB would result in the left front piston being backdriven by hydraulic pressure and pedal movement would result.
2. Ensures that the proper motor circuits are located in the proper connector cavities.
3. Checks for high resistance in the motor "HIGH" circuitry.
4. Checks for high resistance in the motor "LOW" circuitry.
5. Checks for proper resistance of the motor windings.
6. Checks for poor terminal contact or corrosion in the connectors.
7. Checks for a malfunctioning EBCM.

Diagnostic Aids: An "Intermittent" malfunction in this DTC may result from a mechanical part of the system that sticks, binds, or slips.

The frequency of the malfunction can be checked by using the enhanced diagnostic function of the Tech 1.

The static modulator test function of the Tech 1 or T-100 may be used to locate an intermittent malfunction associated with the ESB.

DIAGNOSTIC CHART DTC 38 — 1994 CENTURY, CUTLASS CIERA AND CUTLASS CRUISER

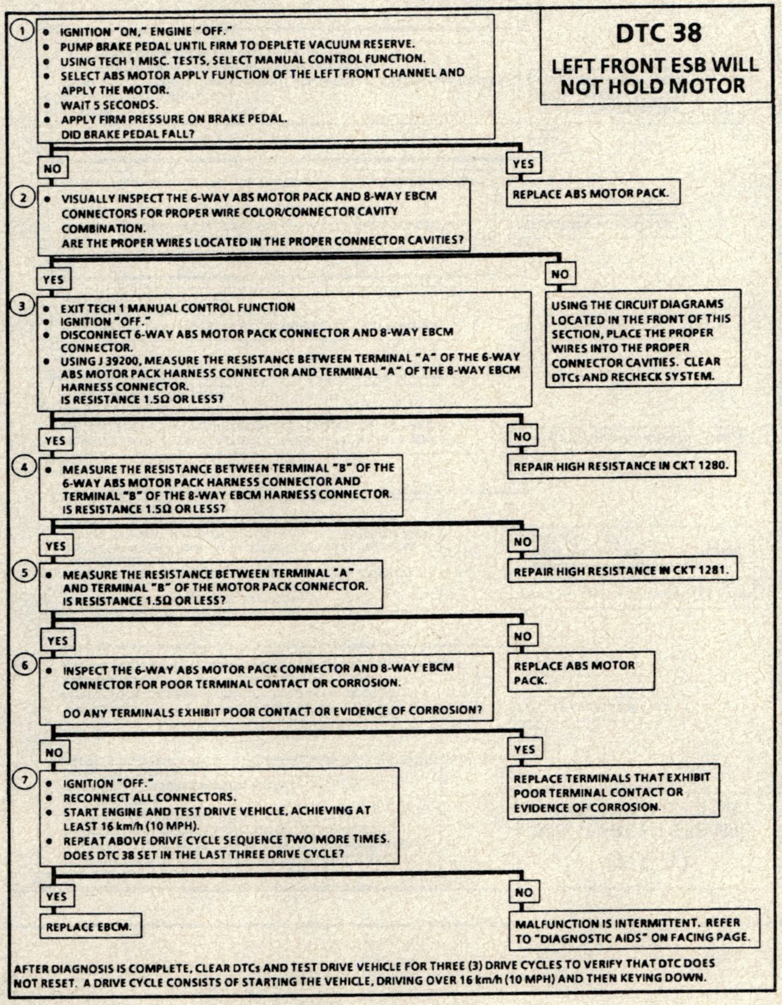

DIAGNOSTIC CHART DTC 41 — 1994 CENTURY, CUTLASS CIERA AND CUTLASS CRUISER

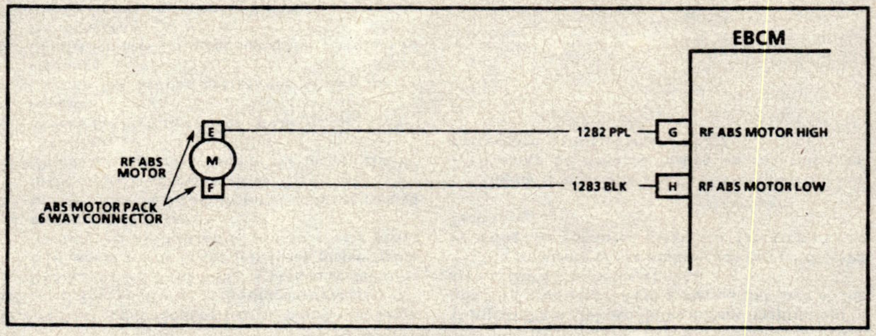

DTC 41

RIGHT FRONT ESB WILL NOT HOLD MOTOR

Circuit Description:
This DTC is designed to detect a slipping right front ESB. During initialization and braking, the right front motor is rehomed. If the ESB slips, the motor/piston will move. During the next key "ON" initialization, a rehome of the motor verifies the motor/piston remained at the home position. If motor movement is detected, the ESB must be slipping.

DTC Will Set When: DTC 41 can be set during initialization. If the EBCM detects that the ESB could not hold the piston in the home position, a malfunction exists. DTC 86 is always set with DTC 41.

Action Taken: If an ESB cannot hold a piston in the home position, the piston may be backdriven when the brake pedal is applied, causing the brake pedal to drop. A malfunction DTC is stored, ABS is disabled, and the ABS warning lamp and the red "BRAKE" warning lamp are turned "ON."

DTC Chart Test Description: Number(s) below refer to circled number(s) on the diagnostic chart.

1. Checks the right front ESB. A broken or defective ESB would result in the right front piston being backdriven by hydraulic pressure, and pedal movement would result.
2. Ensures that the proper motor circuits are located in the proper connector cavities.
3. Checks for high resistance in the motor "HIGH" circuitry.
4. Checks for high resistance in the motor "LOW" circuitry.
5. Checks for proper resistance of the motor windings.
6. Checks for poor terminal contact or corrosion in the connectors.
7. Checks for a malfunctioning EBCM.

Diagnostic Aids: An "Intermittent" malfunction in this DTC may result from a mechanical part of the system that sticks, binds, or slips.

The frequency of the malfunction can be checked by using the enhanced diagnostic function of the Tech 1

The static modulator test function of the Tech 1 or T-100 may be used to locate an intermittent malfunction associated with the ESB.

DIAGNOSTIC CHART DTC 41 — 1994 CENTURY, CUTLASS CIERA AND CUTLASS CRUISER

DTC 41
RIGHT FRONT ESB WILL NOT HOLD MOTOR

(1)
- IGNITION "ON," ENGINE "OFF."
- PUMP BRAKE PEDAL UNTIL FIRM TO DEPLETE VACUUM RESERVE.
- USING TECH 1 MISC. TESTS, SELECT MANUAL CONTROL FUNCTION.
- SELECT ABS MOTOR APPLY FUNCTION OF THE RIGHT FRONT CHANNEL AND APPLY THE MOTOR.
- WAIT 5 SECONDS.
- APPLY FIRM PRESSURE ON BRAKE PEDAL.

DID BRAKE PEDAL FALL?

YES: REPLACE ABS MOTOR PACK.

NO: (2)
- VISUALLY INSPECT THE 6-WAY ABS MOTOR PACK AND 8-WAY EBCM CONNECTORS FOR PROPER WIRE COLOR/CONNECTOR CAVITY COMBINATION.

ARE THE PROPER WIRES LOCATED IN THE PROPER CONNECTOR CAVITIES?

NO: USING THE CIRCUIT DIAGRAMS LOCATED IN THE FRONT OF THIS SECTION, PLACE THE PROPER WIRES INTO THE PROPER CONNECTOR CAVITIES. CLEAR DTCs AND RECHECK SYSTEM.

YES: (3)
- EXIT TECH 1 MANUAL CONTROL FUNCTION.
- IGNITION "OFF."
- DISCONNECT 6-WAY ABS MOTOR PACK CONNECTOR AND 8-WAY EBCM CONNECTOR.
- USING J 39200, MEASURE THE RESISTANCE BETWEEN TERMINAL "E" OF THE 6-WAY ABS MOTOR PACK HARNESS CONNECTOR AND TERMINAL "G" OF THE 8-WAY EBCM HARNESS CONNECTOR.

IS RESISTANCE 1.5Ω OR LESS?

NO: REPAIR HIGH RESISTANCE IN CKT 1282.

YES: (4)
- MEASURE THE RESISTANCE BETWEEN TERMINAL "F" OF THE 6-WAY ABS MOTOR PACK HARNESS CONNECTOR AND TERMINAL "H" OF THE 8-WAY EBCM HARNESS CONNECTOR.

IS RESISTANCE 1.5Ω OR LESS?

NO: REPAIR HIGH RESISTANCE IN CKT 1283.

YES: (5)
- MEASURE THE RESISTANCE BETWEEN TERMINAL "E" AND TERMINAL "F" OF THE ABS MOTOR PACK CONNECTOR.

IS RESISTANCE 1.5Ω OR LESS?

NO: REPLACE ABS MOTOR PACK.

YES: (6)
- INSPECT THE 6-WAY ABS MOTOR PACK CONNECTOR AND 8-WAY EBCM CONNECTOR FOR POOR TERMINAL CONTACT OR CORROSION.

DO ANY TERMINALS EXHIBIT POOR CONTACT OR EVIDENCE OF CORROSION?

YES: REPLACE TERMINALS THAT EXHIBIT POOR TERMINAL CONTACT OR EVIDENCE OF CORROSION.

NO: (7)
- IGNITION "OFF."
- RECONNECT ALL CONNECTORS.
- START ENGINE AND TEST DRIVE VEHICLE, ACHIEVING AT LEAST 16 km/h (10 MPH).
- REPEAT ABOVE DRIVE CYCLE SEQUENCE TWO MORE TIMES.

DOES DTC 41 SET IN THE LAST THREE DRIVE CYCLES?

YES: REPLACE EBCM.

NO: MALFUNCTION IS INTERMITTENT. REFER TO "DIAGNOSTIC AIDS" ON FACING PAGE.

AFTER DIAGNOSIS IS COMPLETE, CLEAR DTCs AND TEST DRIVE VEHICLE FOR THREE (3) DRIVE CYCLES TO VERIFY THAT DTC DOES NOT RESET. A DRIVE CYCLE CONSISTS OF STARTING THE VEHICLE, DRIVING OVER 16 KM/H (10 MPH) AND THEN KEYING DOWN.

DIAGNOSTIC CHART DTC 42 — 1994 CENTURY, CUTLASS CIERA AND CUTLASS CRUISER

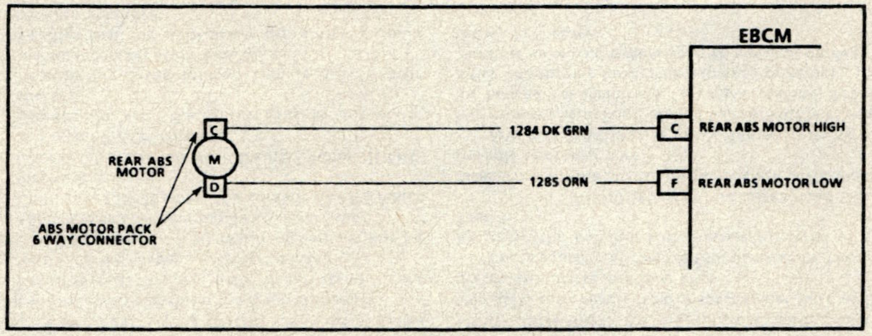

DTC 42

REAR ESB WILL NOT HOLD MOTOR

Circuit Description:

This DTC is designed to detect a slipping rear ESB. During initialization and braking, the rear motor is rehomed. If the ESB slips, the motor/piston will move. During the next key "ON" initialization, a rehome of the motor verifies the motor/piston remained at the home position. If motor movement is detected, the ESB must be slipping.

DTC Will Set When: DTC 42 can be set during initialization. If the EBCM detects that the ESB could not hold the piston in the home position, a malfunction exists. DTC 86 is always set with DTC 42.

Action Taken: If an ESB cannot hold a piston in the home position, the piston may be backdriven when the brake pedal is applied, causing the brake pedal to drop. A malfunction DTC is stored, ABS is disabled, and the ABS warning lamp and the red "BRAKE" warning lamp are turned "ON."

DTC Chart Test Description: Number(s) below refer to circled number(s) on the diagnostic chart.

1. Checks the rear ESB. A broken or defective ESB would result in the rear pistons being backdriven by hydraulic pressure and wheel movement would result.
2. Ensures that the proper motor circuits are located in the proper connector cavities.
3. Checks for high resistance in the motor "HIGH" circuitry.
4. Checks for high resistance in the motor "LOW" circuitry.
5. Checks for proper resistance of the motor windings.
6. Checks for poor terminal contact or corrosion in the connectors.
7. Checks for a malfunctioning EBCM.

Diagnostic Aids: An "Intermittent" malfunction in this DTC may result from a mechanical part of the system that sticks, binds, or slips.

The frequency of the malfunction can be checked by using the enhanced diagnostic function of the Tech 1

The static modulator test function of the Tech 1 or T-100 may be used to locate an intermittent malfunction associated with the ESB.

DIAGNOSTIC CHART DTC 42 — 1994 CENTURY, CUTLASS CIERA AND CUTLASS CRUISER

DTC 42

REAR ESB WILL NOT HOLD MOTOR

(1)
- RAISE THE VEHICLE SUCH THAT THE REAR WHEELS ARE APPROXIMATELY SIX INCHES OFF THE FLOOR.
- IGNITION "ON," ENGINE "OFF."
- PUMP BRAKE PEDAL UNTIL FIRM TO DEPLETE VACUUM RESERVE.
- APPLY FIRM PRESSURE ON BRAKE PEDAL AND RELEASE.
- REAPPLY VERY FIRM PRESSURE ON BRAKE PEDAL AND HAVE AN ASSISTANT TRY TO SPIN THE REAR WHEELS BY HAND. COULD ASSISTANT SPIN THE REAR WHEELS?

YES: REPLACE ABS MOTOR PACK.

NO:

(2)
- VISUALLY INSPECT THE 6-WAY ABS MOTOR PACK AND 8-WAY EBCM CONNECTORS FOR PROPER WIRE COLOR/CONNECTOR CAVITY COMBINATION. ARE THE PROPER WIRES LOCATED IN THE PROPER CONNECTOR CAVITIES?

NO: USING THE CIRCUIT DIAGRAMS LOCATED IN THE FRONT OF THIS SECTION, PLACE THE PROPER WIRES INTO THE PROPER CONNECTOR CAVITIES. CLEAR DTCs AND RECHECK SYSTEM.

YES:

(3)
- IGNITION "OFF."
- DISCONNECT 6-WAY ABS MOTOR PACK CONNECTOR AND 8-WAY EBCM CONNECTOR.
- USING J 39200, MEASURE THE RESISTANCE BETWEEN TERMINAL "C" OF THE 6-WAY ABS MOTOR PACK HARNESS CONNECTOR AND TERMINAL "C" OF THE 8-WAY EBCM HARNESS CONNECTOR. IS RESISTANCE 1.5Ω OR LESS?

NO: REPAIR HIGH RESISTANCE IN CKT 1284.

YES:

(4)
- MEASURE THE RESISTANCE BETWEEN TERMINAL "D" OF THE 6-WAY ABS MOTOR PACK HARNESS CONNECTOR AND TERMINAL "F" OF THE 8-WAY EBCM HARNESS CONNECTOR. IS RESISTANCE 1.5Ω OR LESS?

NO: REPAIR HIGH RESISTANCE IN CKT 1285.

YES:

(5)
- MEASURE THE RESISTANCE BETWEEN TERMINAL "C" AND TERMINAL "D" OF THE ABS MOTOR PACK CONNECTOR. IS RESISTANCE 1.5Ω OR LESS?

NO: REPLACE ABS MOTOR PACK.

YES:

(6)
- INSPECT THE 6-WAY ABS MOTOR PACK CONNECTOR AND 8-WAY EBCM CONNECTOR FOR POOR TERMINAL CONTACT OR CORROSION.

DO ANY TERMINALS EXHIBIT POOR CONTACT OR EVIDENCE OF CORROSION?

YES: REPLACE TERMINALS THAT EXHIBIT POOR TERMINAL CONTACT OR EVIDENCE OF CORROSION.

NO:

(7)
- IGNITION "OFF."
- RECONNECT ALL CONNECTORS.
- START ENGINE AND TEST DRIVE VEHICLE, ACHIEVING AT LEAST 16 km/h (10 MPH).
- REPEAT ABOVE DRIVE CYCLE SEQUENCE TWO MORE TIMES. DOES DTC 42 SET IN THE LAST THREE DRIVE CYCLES?

YES: REPLACE EBCM.

NO: MALFUNCTION WAS MOST LIKELY CAUSED BY A POOR CONNECTION. REFER TO "DIAGNOSTIC AIDS" ON FACING PAGE.

AFTER DIAGNOSIS IS COMPLETE, CLEAR DTCs AND TEST DRIVE VEHICLE FOR THREE (3) DRIVE CYCLES TO VERIFY DTC DOES NO RESET. A DRIVE CYCLE CONSISTS OF STARTING THE VEHICLE, DRIVING OVER 16 km/h (10 MPH) AND THEN KEYING DOWN.

DIAGNOSTIC CHART DTC 44 — 1994 CENTURY, CUTLASS CIERA AND CUTLASS CRUISER

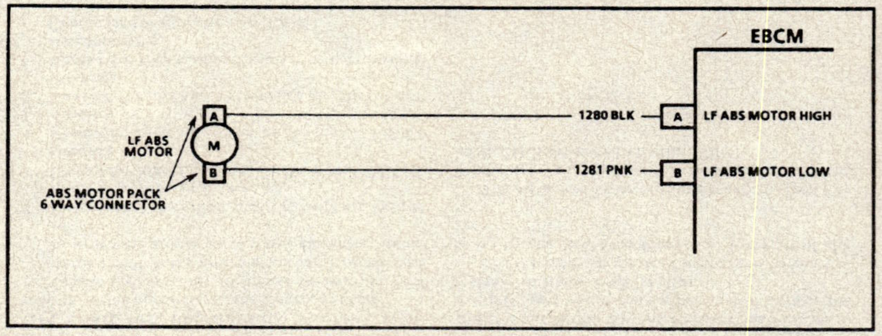

DTC 44

LEFT FRONT CHANNEL WILL NOT MOVE

Circuit Description:

This DTC is designed to detect bound-up ESB, a stuck ABS motor, or a seized hydraulic modulator. When the release is commanded during initialization, the ESB should release the ABS motor, resulting in sensed current being less than commanded current (motor is spinning freely). If the ABS motor is not moving, sensed current will be equal to stall current.

DTC Will Set When: DTC 44 can be set during normal operation. If the EBCM detects a condition in which it cannot move the ABS motor in either direction, a malfunction exists.

Action Taken: This malfunction indicates the channel cannot be moved properly. A malfunction DTC is stored, ABS is disabled and the ABS warning lamp is turned "ON."

DTC Chart Test Description: Number(s) below refer to circled number(s) on the diagnostic chart.

1. Ensures that proper ABS motor circuits are located in the proper connector cavities.
2. Checks for proper ABS motor movement during apply and release commands from the Tech 1.
3. Compares EBCM command current to ABS motor feedback current.
4. Checks for a short between the ABS motor "HIGH" and motor "LOW" circuits.
5. Checks for an ABS motor that is internally shorted.
6. Determines if the malfunction is due to poor terminal contact or the EBCM.
7. Verifies that the ABS motor can actually apply when commanded.
8. Checks for proper hydraulic modulator gear and piston movement.

Diagnostic Aids: An "Intermittent" malfunction in this DTC may result from a mechanical part of the system that sticks, binds, or slips.

The frequency of the malfunction can be checked by using the enhanced diagnostic function of the Tech 1.

DTC 44 may set after modulator disassembly if the modulator pistons are positioned at the bottom of their bore.

Any circuitry, that is suspected as causing the intermittent complaint, should be thoroughly checked for backed out terminals, improper mating, broken locks, improperly formed or damaged terminals, poor terminal to wiring connections or physical damage to the wiring harness.

DIAGNOSTIC CHART DTC 44 — 1994 CENTURY, CUTLASS CIERA AND CUTLASS CRUISER

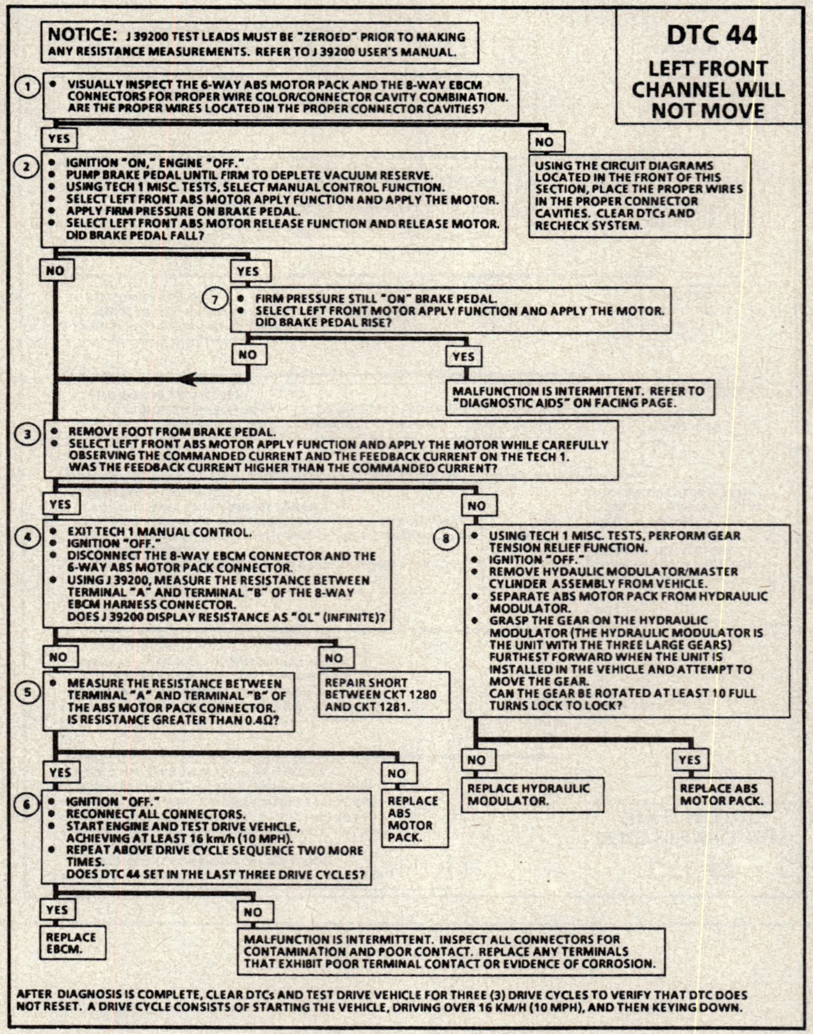

DIAGNOSTIC CHART DTC 45 — 1994 CENTURY, CUTLASS CIERA AND CUTLASS CRUISER

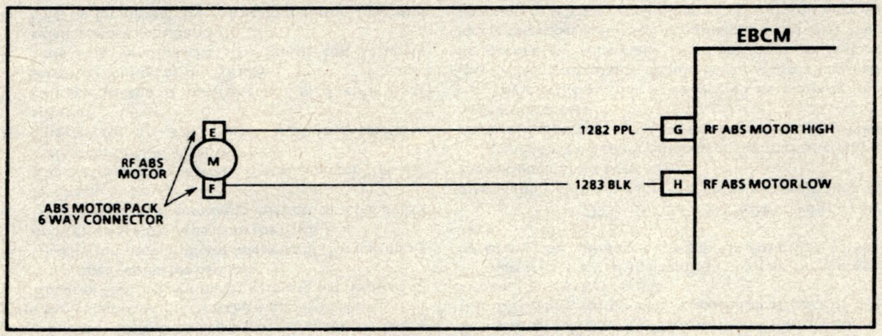

DTC 45

RIGHT FRONT CHANNEL WILL NOT MOVE

Circuit Description:

This DTC is designed to detect bound-up ESB, a stuck ABS motor, or a seized hydraulic modulator. When the release is commanded during initialization, the ESB should release the ABS motor, resulting in sensed current being less than commanded current (ABS motor is spinning freely). If the ABS motor is not moving, sensed current will be equal to stall current.

DTC Will Set When: DTC 45 can be set during normal operation. If the EBCM detects a condition in which it cannot move the ABS motor in either direction, a malfunction exists.

Action Taken: This malfunction indicates the channel cannot be moved properly. A malfunction DTC is stored, ABS is disabled and the ABS warning lamp is turned "ON."

DTC Chart Test Description: Number(s) below refer to circled number(s) on the diagnostic chart.

1. Ensures that proper motor circuits are located in the proper connector cavities.
2. Checks for proper ABS motor movement during apply and release commands from the Tech 1.
3. Compares EBCM command current to ABS motor feedback current.
4. Checks for a short between the ABS motor "HIGH" and motor "LOW" circuits.
5. Checks for an ABS motor that is internally shorted.
6. Checks to see if malfunction is due to poor terminal contact or the EBCM.
7. Verifies that the ABS motor can actually apply when commanded.
8. Checks for proper hydraulic modulator gear and piston movement.

Diagnostic Aids: An "Intermittent" malfunction in this DTC may result from a mechanical part of the system that sticks, binds, or slips.

The frequency of the malfunction can be checked by using the enhanced diagnostic function of the Tech 1.

DTC 45 may set after modulator disassembly if the modulator pistons are positioned at the bottom of their bore.

Any circuitry, that is suspected as causing the intermittent complaint, should be thoroughly checked for backed out terminals, improper mating, broken locks, improperly formed or damaged terminals, poor terminal to wiring connections or physical damage to the wiring harness.

DIAGNOSTIC CHART DTC 45 — 1994 CENTURY, CUTLASS CIERA AND CUTLASS CRUISER

DTC 45
RIGHT FRONT CHANNEL WILL NOT MOVE

NOTICE: J 39200 TEST LEADS MUST BE "ZEROED" PRIOR TO MAKING ANY RESISTANCE MEASUREMENTS. REFER TO J 39200 USER'S MANUAL.

(1)
- VISUALLY INSPECT THE 6-WAY ABS MOTOR PACK AND THE 8-WAY EBCM CONNECTORS FOR PROPER WIRE COLOR/CONNECTOR CAVITY COMBINATION. ARE THE PROPER WIRES LOCATED IN THE PROPER CONNECTOR CAVITIES?

NO: USING THE CIRCUIT DIAGRAMS LOCATED IN THE FRONT OF THIS SECTION, PLACE THE PROPER WIRES IN THE PROPER CONNECTOR CAVITIES. CLEAR DTCs AND RECHECK SYSTEM.

YES:

(2)
- IGNITION "ON," ENGINE "OFF."
- PUMP BRAKE PEDAL UNTIL FIRM TO DEPLETE VACUUM RESERVE.
- USING TECH 1 MISC. TESTS, SELECT MANUAL CONTROL FUNCTION.
- SELECT RIGHT FRONT ABS MOTOR APPLY FUNCTION AND APPLY THE MOTOR.
- APPLY FIRM PRESSURE ON BRAKE PEDAL.
- SELECT RIGHT FRONT ABS MOTOR RELEASE FUNCTION AND RELEASE MOTOR. DID BRAKE PEDAL FALL?

YES:

(7)
- FIRM PRESSURE STILL "ON" BRAKE PEDAL.
- SELECT RIGHT FRONT ABS MOTOR APPLY FUNCTION AND APPLY THE MOTOR. DID BRAKE PEDAL RISE?

YES: MALFUNCTION IS INTERMITTENT. REFER TO "DIAGNOSTIC AIDS".

NO: (go to 3)

NO:

(3)
- REMOVE FOOT FROM BRAKE PEDAL.
- SELECT RIGHT FRONT ABS MOTOR APPLY FUNCTION AND APPLY THE MOTOR WHILE CAREFULLY OBSERVING THE COMMANDED CURRENT AND THE FEEDBACK CURRENT ON THE TECH 1. WAS THE FEEDBACK CURRENT HIGHER THAN THE COMMANDED CURRENT?

YES:

(4)
- EXIT TECH 1 MANUAL CONTROL FUNCTION.
- IGNITION "OFF."
- DISCONNECT THE 8-WAY EBCM CONNECTOR AND THE 6-WAY ABS MOTOR PACK CONNECTOR.
- USING J 39200, MEASURE THE RESISTANCE BETWEEN TERMINAL "G" AND TERMINAL "H" OF THE 8-WAY EBCM HARNESS CONNECTOR. DOES J 39200 DISPLAY RESISTANCE AS "OL" (INFINITE)?

NO: REPAIR SHORT BETWEEN CKT 1282 AND CKT 1283.

YES:

(5)
- MEASURE THE RESISTANCE BETWEEN TERMINAL "E" AND TERMINAL "F" OF THE MOTOR PACK CONNECTOR. IS RESISTANCE 0.4Ω OR GREATER?

NO: REPLACE ABS MOTOR PACK.

YES:

(6)
- IGNITION "OFF."
- RECONNECT ALL CONNECTORS.
- START ENGINE AND TEST DRIVE VEHICLE, ACHIEVING AT LEAST 16 km/h (10 mph).
- REPEAT ABOVE DRIVE CYCLE SEQUENCE TWO MORE TIMES. DOES DTC 45 SET IN THE LAST THREE DRIVE CYCLES?

YES: REPLACE EBCM.

NO: MALFUNCTION IS INTERMITTENT. INSPECT ALL CONNECTORS FOR CONTAMINATION AND POOR TERMINAL CONTACT. REPLACE ANY TERMINALS THAT EXHIBIT POOR TERMINAL CONTACT OR EVIDENCE OF CORROSION.

NO:

(8)
- USING TECH 1 MISC. TESTS, PERFORM GEAR TENSION RELIEF FUNCTION.
- IGNITION "OFF."
- REMOVE HYDRAULIC MODULATOR/MASTER CYLINDER ASSEMBLY FROM VEHICLE.
- SEPARATE ABS MOTOR PACK FROM HYDRAULIC MODULATOR.
- GRASP THE GEAR ON THE HYDRAULIC MODULATOR (THE HYDRAULIC MODULATOR IS THE UNIT WITH THE THREE LARGE GEARS) FURTHEST REARWARD WHEN THE UNIT IS INSTALLED IN THE VEHICLE AND ATTEMPT TO MOVE THE GEAR. CAN THE GEAR BE ROTATED AT LEAST 10 FULL TURNS LOCK TO LOCK?

NO: REPLACE HYDRAULIC MODULATOR.

YES: REPLACE ABS MOTOR PACK.

AFTER DIAGNOSIS IS COMPLETE, CLEAR DTCs AND TEST DRIVE VEHICLE FOR THREE (3) DRIVE CYCLES TO VERIFY THAT DTC DOES NOT RESET. A DRIVE CYCLE CONSISTS OF STARTING THE VEHICLE, DRIVING OVER 16 KM/H (10 MPH), AND THEN KEYING DOWN.

DIAGNOSTIC CHART DTC 46 — 1994 CENTURY, CUTLASS CIERA AND CUTLASS CRUISER

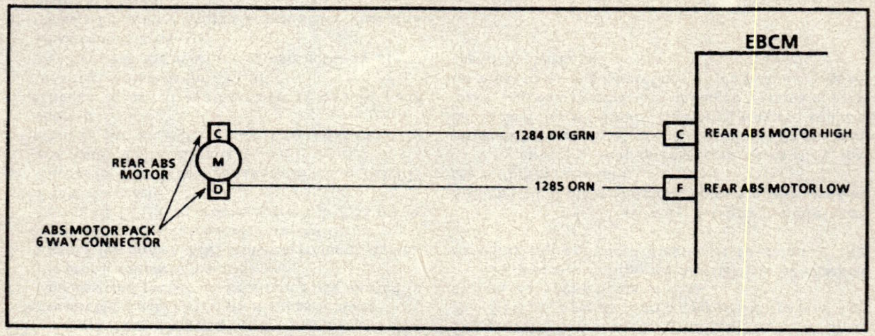

DTC 46

REAR CHANNEL WILL NOT MOVE

Circuit Description:

This DTC is designed to detect bound-up ESB, a stuck ABS motor, or a seized hydraulic modulator. When the release is commanded during initialization, the ESB should release the ABS motor, resulting in sensed current being less than commanded current (motor is spinning freely). If the ABS motor is not moving, sensed current will be equal to stall current.

DTC Will Set When: DTC 46 can be set during normal operation. If the EBCM detects a condition in which it cannot move the ABS motor in either direction, a malfunction exists.

Action Taken: This malfunction indicates the channel cannot be moved properly. A malfunction DTC is stored, ABS is disabled and the ABS warning lamp is turned "ON."

DTC Chart Test Description: Number(s) below refer to circled number(s) on the diagnostic chart.

1. Ensures that proper motor circuits are located in the proper connector cavities.
2. Checks for proper motor movement during apply and release commands from the Tech 1.
3. Compares EBCM command current to ABS motor feedback current.
4. Checks for a short between the ABS motor "HIGH" and motor "LOW" circuits.
5. Checks for an ABS motor that is internally shorted.
6. Checks to see if malfunction is due to poor terminal contact or the EBCM.
7. This step verifies that the motor can actually release when commanded.
8. Checks for proper hydraulic modulator gear and piston movement.

Diagnostic Aids: An "Intermittent" malfunction in this DTC may result from a mechanical part of the system that sticks or binds.

The frequency of the malfunction can be checked by using the enhanced diagnostic function of the Tech 1.

DTC 46 may set after modulator disassembly if the modulator pistons are positioned at the bottom of their bore.

Depending on the frequency of the malfunction, a physical inspection of the mechanical parts suspected may be necessary.

Any circuitry, that is suspected as causing the intermittent complaint, should be thoroughly checked for backed out terminals, improper mating, broken locks, improperly formed or damaged terminals, poor terminal to wiring connections or physical damage to the wiring harness.

DIAGNOSTIC CHART DTC 46 — 1994 CENTURY, CUTLASS CIERA AND CUTLASS CRUISER

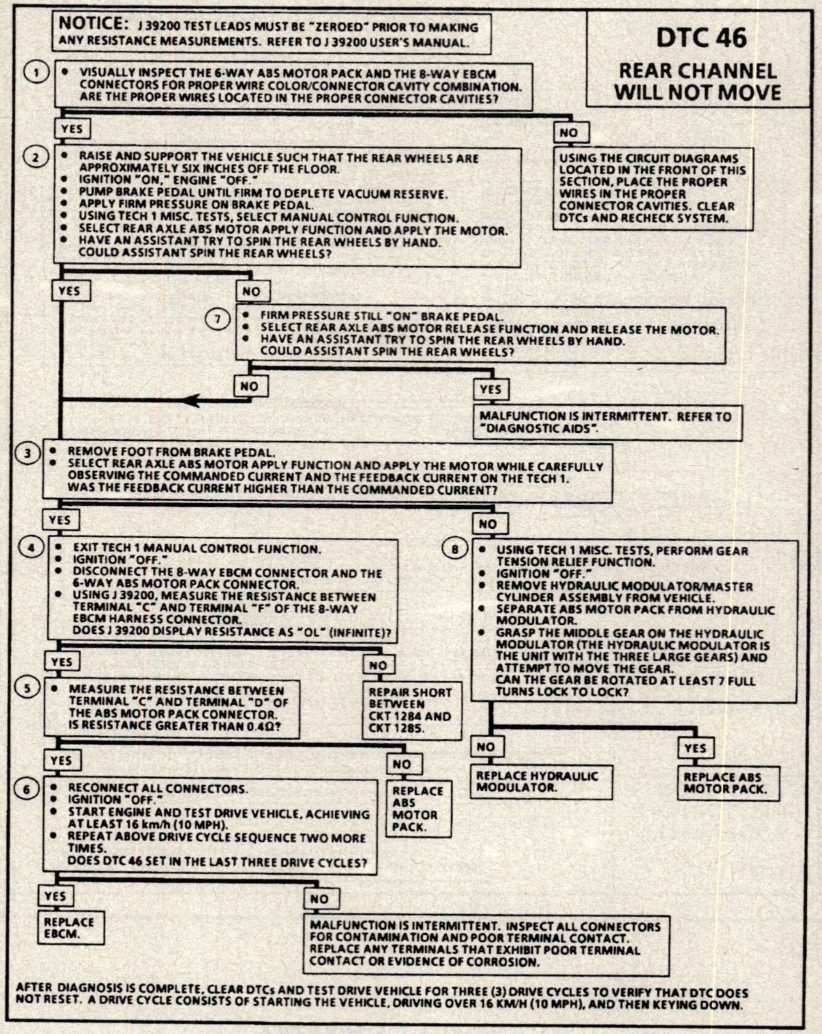

DIAGNOSTIC CHART DTC 47 — 1994 CENTURY, CUTLASS CIERA AND CUTLASS CRUISER

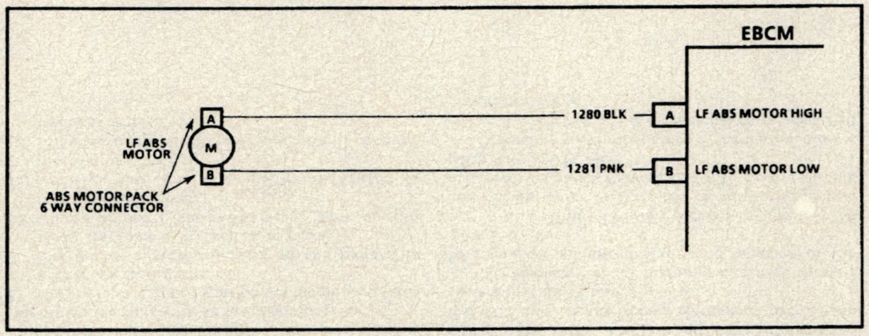

DTC 47

LEFT FRONT ABS MOTOR FREE SPINS

Circuit Description:
This DTC is designed to detect a stripped nut or gear assembly during initialization. During the homing sequence, the piston should reach the top of the bore resulting in a stalled ABS motor. If this does not occur, the ABS motor must be spinning with little or no resistance, indicating a nut/screw or gear malfunction.

DTC Will Set When: DTC 47 can only be set during initialization. If the feedback current is less than the command current for a specified period of time, a malfunction exists.

Action Taken: A malfunction DTC is stored, ABS is disabled, and the ABS warning lamp is turned "ON."

DTC Chart Test Description: Number(s) below refer to circled number(s) on the diagnostic chart.

1. Checks to see if the corresponding open ABS motor DTC 56 is also set.
2. Verifies that the ABS motor was actually applied as commanded by observing feedback current.
3. Verifies that the ABS motor can release.
4. This step verifies that the ABS motor can be applied by observing pedal movement.
5. Checks for an open in the ABS motor "HIGH" circuitry.
6. Checks for an open in the ABS motor "LOW" circuitry.
7. Checks for an ABS motor that is internally shorted.

Diagnostic Aids: An "Intermittent" malfunction in this DTC may result from a mechanical part of the system that sticks, binds, or slips.

The frequency of the malfunction can be checked by using the enhanced diagnostic function of the Tech 1. If the DTC only occurs once and DTC 56 also occurs, refer to DTC 56. If intermittent and enhanced diagnostics shows this DTC occurs during ABS, refer to DTC 56.

Depending on the frequency of the malfunction, a physical inspection of the mechanical parts suspected may be necessary.

DIAGNOSTIC CHART DTC 47 — 1994 CENTURY, CUTLASS CIERA AND CUTLASS CRUISER

DTC 47
(Page 1 of 2)
LEFT FRONT ABS MOTOR FREE SPINS

1. IS DTC 56 ALSO CURRENTLY SET?
   - YES: PROCEED TO DTC 56 CHART FIRST.
   - NO: go to step 2.
2.
   - IGNITION "ON," ENGINE "OFF."
   - PUMP BRAKE PEDAL UNTIL FIRM TO DEPLETE VACUUM RESERVE.
   - USING TECH 1 MISC. TESTS, SELECT MANUAL CONTROL FUNCTION.
   - SELECT LEFT FRONT ABS MOTOR APPLY FUNCTION AND APPLY THE MOTOR WHILE CAREFULLY OBSERVING THE COMMAND CURRENT AND FEEDBACK CURRENT ON THE TECH 1.

   WERE BOTH COMMAND AND FEEDBACK CURRENT 10 AMPS?
   - YES: go to step 3.
   - NO: go to step 5.
3.
   - APPLY FIRM PRESSURE ON BRAKE PEDAL.
   - SELECT LEFT FRONT ABS MOTOR RELEASE FUNCTION AND RELEASE THE MOTOR.

   DID THE BRAKE PEDAL FALL AND WERE BOTH COMMAND AND FEEDBACK CURRENT 6 AMPS?
   - YES: go to step 4.
   - NO: GO TO STEP 5
4.
   - FIRM PRESSURE STILL ON BRAKE PEDAL.
   - SELECT LEFT FRONT ABS MOTOR APPLY FUNCTION AND APPLY THE MOTOR (THIS MOVES THE PISTON TO THE TOP OF ITS BORE.)

   DID BRAKE PEDAL RISE?
   - YES: MALFUNCTION IS INTERMITTENT. REFER TO "DIAGNOSTIC AIDS".
   - NO: go to step 5.
5.
   - USING TECH 1 MISC. TESTS, PERFORM GEAR TENSION RELIEF FUNCTION.
   - IGNITION "OFF."
   - DISCONNECT 6-WAY ABS MOTOR PACK CONNECTOR AND 8-WAY EBCM CONNECTOR.
   - USING J 39200, MEASURE THE RESISTANCE BETWEEN TERMINAL "A" OF THE 6-WAY ABS MOTOR PACK HARNESS CONNECTOR AND TERMINAL "A" OF THE 8-WAY EBCM HARNESS CONNECTOR.

   IS RESISTANCE LESS THAN 1.5Ω?
   - YES: go to step 6.
   - NO: REPAIR HIGH RESISTANCE IN CKT 1280.
6.
   - MEASURE THE RESISTANCE BETWEEN TERMINAL "B" OF THE 6-WAY ABS MOTOR PACK HARNESS CONNECTOR AND TERMINAL "B" OF THE 8-WAY EBCM HARNESS CONNECTOR.

   IS RESISTANCE LESS THAN 1.5Ω?
   - YES: go to step 7.
   - NO: REPAIR HIGH RESISTANCE IN CKT 1281.
7.
   - MEASURE THE RESISTANCE BETWEEN TERMINAL "A" AND TERMINAL "B" OF THE ABS MOTOR PACK CONNECTOR.

   IS RESISTANCE LESS THAN 1.5Ω?
   - YES: DIAGNOSIS CONTINUED ON NEXT PAGE.
   - NO: REPLACE ABS MOTOR PACK.

AFTER DIAGNOSIS IS COMPLETE, CLEAR DTCs AND TEST DRIVE VEHICLE FOR THREE (3) DRIVE CYCLES TO VERIFY THAT DTC DOES NOT RESET. A DRIVE CYCLE CONSISTS OF STARTING THE VEHICLE, DRIVING OVER 16 KM/H (10 MPH), AND THEN KEYING DOWN.

DIAGNOSTIC CHART DTC 47 — 1994 CENTURY, CUTLASS CIERA AND CUTLASS CRUISER

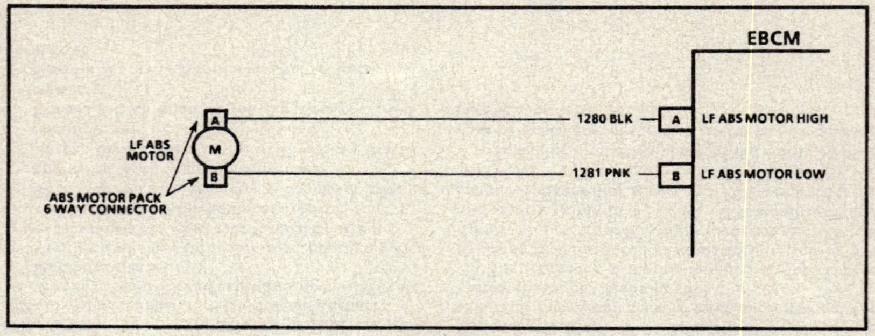

DTC 47

(Page 2 of 2)

LEFT FRONT ABS MOTOR FREE SPINS

Circuit Description:
This DTC is designed to detect a stripped nut or gear assembly during initialization. During the homing sequence, the piston should reach the top of the bore resulting in a stalled ABS motor. If this does not occur, the ABS motor must be spinning with little or no resistance, indicating a nut/screw or gear malfunction.

DTC Will Set When: DTC 47 can only be set during initialization. If the feedback current is less than the command current for a specified period of time, a malfunction exists.

Action Taken: A malfunction DTC is stored, ABS is disabled, and the ABS warning lamp is turned "ON."

DTC Chart Test Description: Number(s) below refer to circled number(s) on the diagnostic chart.

8. Checks for a stripped gear on the motor pack (unit with three small gears).
9. Checks for a stripped gear on the hydraulic modulator (unit with three large gears).
10. Checks to see if the ABS motor pack has an internal malfunction.
11. Verifies that the hydraulic modulator is functioning properly.
12. Ensures malfunction was not due to a poor terminal contact.

Diagnostic Aids: An "Intermittent" malfunction in this DTC may result from a mechanical part of the system that sticks, binds, or slips.

The frequency of the malfunction can be checked by using the enhanced diagnostic function of the Tech 1.

If the DTC only occurs once and DTC 56 also occurs, refer to DTC 56. If intermittent and enhanced diagnostics shows this DTC occurs during ABS, refer to DTC 56.

Depending on the frequency of the malfunction, a physical inspection of the mechanical parts suspected may be necessary.

DIAGNOSTIC CHART DTC 47 — 1994 CENTURY, CUTLASS CIERA AND CUTLASS CRUISER

DTC 47
(Page 2 of 2)
LEFT FRONT ABS MOTOR FREE SPINS

DIAGNOSIS CONTINUED FROM PREVIOUS PAGE.

(8)
- REMOVE HYDRAULIC MODULATOR/MASTER CYLINDER ASSEMBLY FROM VEHICLE.
- REMOVE GEAR COVER.
- CHECK FOR STRIPPED GEAR ON ABS MOTOR PACK; ABS MOTOR PACK IS THE UNIT WITH THE THREE SMALL GEARS. THE LEFT FRONT GEAR IS THE GEAR FURTHEST FORWARD WHEN THE ASSEMBLY IS INSTALLED IN THE VEHICLE. IS THE ABS MOTOR PACK GEAR STRIPPED?

YES: REPLACE ABS MOTOR PACK.

NO:

(9)
- CHECK FOR A STRIPPED GEAR ON HYDRAULIC MODULATOR; MODULATOR IS THE UNIT WITH THE THREE LARGE GEARS. THE LEFT FRONT GEAR IS THE GEAR FURTHEST FORWARD WHEN THE UNIT IS INSTALLED IN THE VEHICLE. IS THE MODULATOR GEAR STRIPPED?

YES: REPLACE HYDRAULIC MODULATOR GEAR.

NO:

(10)
- RECONNECT ELECTRICAL CONNECTORS
- SECURELY POSITION HYDRAULIC MODULATOR/MASTER CYLINDER ASSEMBLY IN VEHICLE WITH GEAR COVER REMOVED SUCH THAT THE GEAR SET CAN BE OBSERVED. BE CAREFUL NOT TO DAMAGE GEAR SET OR ALLOW THE HYDRAULIC MODULATOR TO MOVE WHILE TESTING.
- IGNITION "ON."
- USING TECH 1 MISC. TESTS, SELECT MANUAL CONTROL FUNCTION.
- SELECT LEFT FRONT ABS MOTOR APPLY AND APPLY THE MOTOR WHILE OBSERVING THE GEAR SET FURTHEST FORWARD WHEN THE UNIT IS INSTALLED IN THE VEHICLE.
- SELECT LEFT ABS FRONT MOTOR RELEASE FUNCTION AND RELEASE THE MOTOR WHILE STILL OBSERVING THE GEAR SET.
- SELECT LEFT FRONT ABS MOTOR APPLY FUNCTION AND APPLY THE MOTOR. DID GEAR SET MOVE IN BOTH DIRECTIONS FOR AT LEAST ONE REVOLUTION?

NO: REPLACE ABS MOTOR PACK.

YES:

(11)
- USING TECH 1 MISC. TESTS, PERFORM GEAR TENSION RELIEF FUNCTION.
- IGNITION "OFF."
- SEPARATE THE ABS MOTOR PACK FROM THE HYDRAULIC MODULATOR.
- GRASP THE GEAR ON THE HYDRAULIC MODULATOR (THE HYDRAULIC MODULATOR IS THE UNIT WITH THE THREE LARGE GEARS) FURTHEST FORWARD WHEN THE UNIT IS INSTALLED IN THE VEHICLE AND ATTEMPT TO MOVE THE GEAR. CAN THE GEAR BE ROTATED MORE THAN 13 FULL TURNS LOCK TO LOCK?

YES: REPLACE THE HYDRAULIC MODULATOR.

NO:

(12)
- INSPECT ALL CONNECTORS AND TERMINALS FOR POOR TERMINAL CONTACT AND EVIDENCE OF CORROSION. REPLACE ALL TERMINALS THAT EXHIBIT SIGNS OF POOR TERMINAL
- REINSTALL HYDRAULIC MODULATOR/MASTER CYLINDER ASSEMBLY IN VEHICLE.
- RECONNECT ALL CONNECTORS.
- IGNITION "OFF."
- START ENGINE AND TEST DRIVE VEHICLE, ACHIEVING AT LEAST 16 km/h (10 MPH).
- REPEAT ABOVE DRIVE CYCLE SEQUENCE TWO MORE TIMES. DOES DTC 47 SET IN THE LAST THREE DRIVE CYCLES?

YES: REPLACE EBCM.

NO: MALFUNCTION WAS MOSTLY LIKELY CAUSED BY A POOR CONNECTION.

AFTER DIAGNOSIS IS COMPLETE, CLEAR DTCs AND TEST DRIVE VEHICLE FOR THREE (3) DRIVE CYCLES TO VERIFY THAT DTC DOES NOT RESET. A DRIVE CYCLE CONSISTS OF STARTING THE VEHICLE, DRIVING OVER 16 KM/H (10 MPH), AND THEN KEYING DOWN.

DIAGNOSTIC CHART DTC 48 — 1994 CENTURY, CUTLASS CIERA AND CUTLASS CRUISER

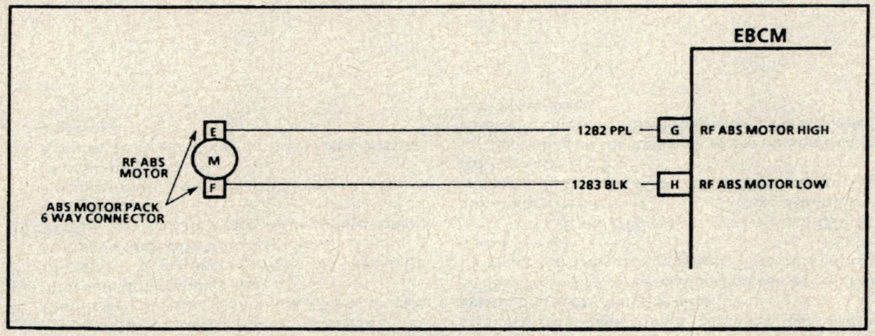

DTC 48

(Page 1 of 2)
RIGHT FRONT ABS MOTOR FREE SPINS

Circuit Description:

This DTC is designed to detect a stripped nut or gear assembly during initialization. During the homing sequence, the piston should reach the top of the bore resulting in a stalled ABS motor. If this does not occur, the ABS motor must be spinning with little or no resistance, indicating a nut/screw or gear malfunction.

DTC Will Set When: DTC 48 can only be set during initialization. If the feedback current is less than the command current for a specified period of time, a malfunction exists.

Action Taken: A malfunction DTC is stored, ABS is disabled, and the ABS warning lamp is turned "ON."

DTC Chart Test Description: Number(s) below refer to circled number(s) on the diagnostic chart.

1. Checks to see if the corresponding open ABS motor DTC 61 is also set.
2. Verifies that the ABS motor was actually applied as commanded by observing feedback current.
3. Verifies that the ABS motor can release.
4. Verifies that the ABS motor can be applied by observing pedal movement.
5. Checks for high resistance in the ABS motor "HIGH" circuitry.
6. Checks for high resistance in the ABS motor "LOW" circuitry.
7. Checks for proper resistance of the ABS motor windings.

Diagnostic Aids: An "Intermittent" malfunction in this DTC may result from a mechanical part of the system that sticks, binds, or slips.

The frequency of the malfunction can be checked by using the enhanced diagnostic function of the Tech 1.

If the DTC only occurs once and DTC 61 also occurs, refer to DTC 61. If intermittent and enhanced diagnostics shows this DTC occurs during ABS, refer to DTC 61.

Depending on the frequency of the malfunction, a physical inspection of the mechanical parts suspected may be necessary.

DIAGNOSTIC CHART DTC 48 — 1994 CENTURY, CUTLASS CIERA AND CUTLASS CRUISER

DTC 48

(Page 1 of 2)
RIGHT FRONT ABS MOTOR FREE SPINS

(1) IS DTC 61 ALSO CURRENTLY SET?

- YES: PROCEED TO DTC 61 CHART FIRST.
- NO:

(2)
- IGNITION "ON," ENGINE "OFF."
- PUMP BRAKE PEDAL UNTIL FIRM TO DEPLETE VACUUM RESERVE.
- USING TECH 1 MISC. TESTS, SELECT MANUAL CONTROL FUNCTION.
- SELECT RIGHT FRONT ABS MOTOR APPLY FUNCTION AND APPLY THE MOTOR WHILE CAREFULLY OBSERVING THE COMMAND CURRENT AND FEEDBACK CURRENT ON THE TECH 1.

WERE BOTH COMMAND AND FEEDBACK CURRENT 10 AMPS?

- YES:

(3)
- APPLY FIRM PRESSURE ON BRAKE PEDAL.
- SELECT RIGHT FRONT ABS MOTOR RELEASE FUNCTION AND RELEASE THE MOTOR.

DID THE BRAKE PEDAL FALL AND WERE BOTH COMMAND AND FEEDBACK CURRENT 6 AMPS?

- NO (from step 2 or step 3): GO TO STEP 5
- YES:

(4)
- FIRM PRESSURE STILL ON BRAKE PEDAL.
- SELECT RIGHT ABS MOTOR APPLY FUNCTION AND APPLY THE MOTOR (THIS MOVES THE PISTON TO THE TOP OF ITS BORE.)

DID BRAKE PEDAL RISE?

- YES: MALFUNCTION IS INTERMITTENT. REFER TO "DIAGNOSTIC AIDS".
- NO:

(5)
- USING TECH 1 MISC. TESTS, PERFORM GEAR TENSION RELIEF FUNCTION.
- IGNITION "OFF."
- DISCONNECT 6-WAY ABS MOTOR PACK CONNECTOR AND 8-WAY EBCM CONNECTOR.
- USING J 39200, MEASURE THE RESISTANCE BETWEEN TERMINAL "E" OF THE 6-WAY ABS MOTOR PACK HARNESS CONNECTOR AND TERMINAL "G" OF THE 8-WAY EBCM HARNESS CONNECTOR.

IS RESISTANCE 1.5Ω OR LESS?

- NO: REPAIR HIGH RESISTANCE IN CKT 1282.
- YES:

(6)
- MEASURE THE RESISTANCE BETWEEN TERMINAL "F" OF THE 6-WAY ABS MOTOR PACK HARNESS CONNECTOR AND TERMINAL "H" OF THE 8-WAY EBCM HARNESS CONNECTOR.

IS RESISTANCE 1.5Ω OR LESS?

- NO: REPAIR HIGH RESISTANCE IN CKT 1283.
- YES:

(7)
- MEASURE THE RESISTANCE BETWEEN TERMINAL "E" AND TERMINAL "F" OF THE ABS MOTOR PACK CONNECTOR.

IS RESISTANCE 1.5Ω OR LESS?

- NO: REPLACE ABS MOTOR PACK.
- YES: DIAGNOSIS CONTINUED ON NEXT PAGE.

AFTER DIAGNOSIS IS COMPLETE, CLEAR DTCs AND TEST DRIVE VEHICLE FOR THREE (3) DRIVE CYCLES TO VERIFY THAT DTC DOES NOT RESET. A DRIVE CYCLE CONSISTS OF STARTING THE VEHICLE, DRIVING OVER 16 KM/H (10 MPH), AND THEN KEYING DOWN.

DIAGNOSTIC CHART DTC 48 — 1994 CENTURY, CUTLASS CIERA AND CUTLASS CRUISER

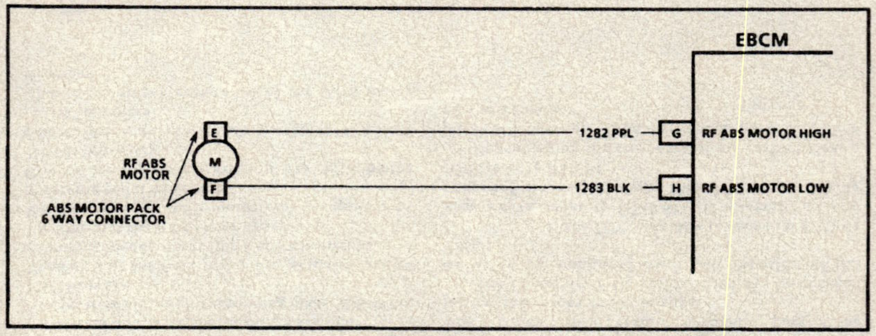

DTC 48
(Page 2 of 2)
RIGHT FRONT ABS MOTOR FREE SPINS

Circuit Description:
This DTC is designed to detect a stripped nut or gear assembly during initialization. During the homing sequence, the piston should reach the top of the bore resulting in a stalled ABS motor. If this does not occur, the ABS motor must be spinning with little or no resistance, indicating a nut/screw or gear malfunction.

DTC Will Set When: DTC 48 can only be set during initialization. If the feedback current is less than the command current for a specified period of time, a malfunction exists.

Action Taken: A malfunction DTC is stored, ABS is disabled, and the ABS warning lamp is turned "ON."

DTC Chart Test Description: Number(s) below refer to circled number(s) on the diagnostic chart.
8. Checks for a stripped gear on the ABS motor pack (unit with three small gears).
9. Checks for a stripped gear on the hydraulic modulator (unit with three large gears).
10. Verifies that the ABS motor can be applied under a load situation.
11. Verifies that the hydraulic modulator is functioning properly.
12. Ensures malfunction was not due to poor terminal contact.

Diagnostic Aids: An "Intermittent" malfunction in this DTC may result from a mechanical part of the system that sticks, binds, or slips.

The frequency of the malfunction can be checked by using the enhanced diagnostic function of the Tech 1.

If the DTC only occurs once and DTC 61 also occurs, refer to DTC 61. If intermittent and enhanced diagnostics shows this DTC occurs during ABS, refer to DTC 61.

Depending on the frequency of the malfunction, a physical inspection of the mechanical parts suspected may be necessary.

DIAGNOSTIC CHART DTC 48 — 1994 CENTURY, CUTLASS CIERA AND CUTLASS CRUISER

DTC 48
(Page 2 of 2)
RIGHT FRONT ABS MOTOR FREE SPINS

DIAGNOSIS CONTINUED FROM PREVIOUS PAGE.

(8)
- REMOVE HYDRAULIC MODULATOR/MASTER CYLINDER ASSEMBLY FROM VEHICLE.
- REMOVE GEAR COVER.
- CHECK FOR STRIPPED GEAR ON ABS MOTOR PACK; ABS MOTOR PACK IS THE UNIT WITH THE THREE SMALL GEARS. THE RIGHT FRONT GEAR IS THE GEAR FURTHEST REARWARD WHEN THE ASSEMBLY IS INSTALLED IN THE VEHICLE. IS THE ABS MOTOR PACK GEAR STRIPPED?

YES: REPLACE ABS MOTOR PACK.

NO:

(9)
- CHECK FOR A STRIPPED GEAR ON HYDRAULIC MODULATOR; MODULATOR IS THE UNIT WITH THE THREE LARGE GEARS. THE RIGHT FRONT GEAR IS THE GEAR FURTHEST REARWARD WHEN THE UNIT IS INSTALLED IN THE VEHICLE. IS THE MODULATOR GEAR STRIPPED?

YES: REPLACE HYDRAULIC MODULATOR GEAR.

NO:

(10)
- RECONNECT ELECTRICAL CONNECTORS
- SECURELY POSITION HYDRAULIC MODULATOR/MASTER CYLINDER ASSEMBLY IN VEHICLE WITH GEAR COVER REMOVED SUCH THAT THE GEAR SET CAN BE OBSERVED. BE CAREFUL NOT TO DAMAGE GEAR SET OR ALLOW THE HYDRAULIC MODULATOR TO MOVE WHILE TESTING.
- IGNITION "ON."
- USING TECH 1 MISC. TESTS, SELECT MANUAL CONTROL FUNCTION.
- SELECT RIGHT FRONT ABS MOTOR APPLY AND APPLY THE MOTOR WHILE OBSERVING THE GEAR SET FURTHEST REARWARD WHEN THE UNIT IS INSTALLED IN THE VEHICLE.
- SELECT RIGHT FRONT ABS MOTOR RELEASE FUNCTION AND RELEASE THE MOTOR WHILE STILL OBSERVING THE GEAR SET.
- SELECT RIGHT FRONT ABS MOTOR APPLY FUNCTION AND APPLY THE MOTOR. DID GEAR SET MOVE IN BOTH DIRECTIONS FOR AT LEAST ONE REVOLUTION?

NO: REPLACE ABS MOTOR PACK.

YES:

(11)
- USING TECH 1 MISC. TESTS, PERFORM GEAR TENSION RELIEF FUNCTION.
- IGNITION "OFF."
- SEPARATE THE ABS MOTOR PACK FROM THE HYDRAULIC MODULATOR.
- GRASP THE GEAR ON THE HYDRAULIC MODULATOR (THE HYDRAULIC MODULATOR IS THE UNIT WITH THE THREE LARGE GEARS) FURTHEST REARWARD WHEN THE UNIT IS INSTALLED IN THE VEHICLE AND ATTEMPT TO MOVE THE GEAR. CAN THE GEAR BE ROTATED MORE THAN 13 FULL TURNS LOCK TO LOCK?

YES: REPLACE THE HYDRAULIC MODULATOR.

NO:

(12)
- INSPECT ALL CONNECTORS AND TERMINALS FOR POOR TERMINAL CONTACT AND EVIDENCE OF CORROSION. REPLACE ALL TERMINALS THAT EXHIBIT SIGNS OF POOR TERMINAL CONTACT OR CORROSION.
- REINSTALL HYDRAULIC MODULATOR/MASTER CYLINDER ASSEMBLY IN VEHICLE.
- RECONNECT ALL CONNECTORS.
- IGNITION "OFF."
- START ENGINE AND TEST DRIVE VEHICLE, ACHIEVING AT LEAST 16 km/h (10 MPH).
- REPEAT ABOVE DRIVE CYCLE SEQUENCE TWO MORE TIMES. DOES DTC 48 SET IN THE LAST THREE DRIVE CYCLES?

YES: REPLACE EBCM.

NO: MALFUNCTION WAS MOSTLY LIKELY CAUSED BY A POOR CONNECTION. REFER TO "DIAGNOSTIC AIDS".

AFTER DIAGNOSIS IS COMPLETE, CLEAR DTCs AND TEST DRIVE VEHICLE FOR THREE (3) DRIVE CYCLES TO VERIFY THAT DTC DOES NOT RESET. A DRIVE CYCLE CONSISTS OF STARTING THE VEHICLE, DRIVING OVER 16 KM/H (10 MPH), AND THEN KEYING DOWN.

DIAGNOSTIC CHART DTC 51 — 1994 CENTURY, CUTLASS CIERA AND CUTLASS CRUISER

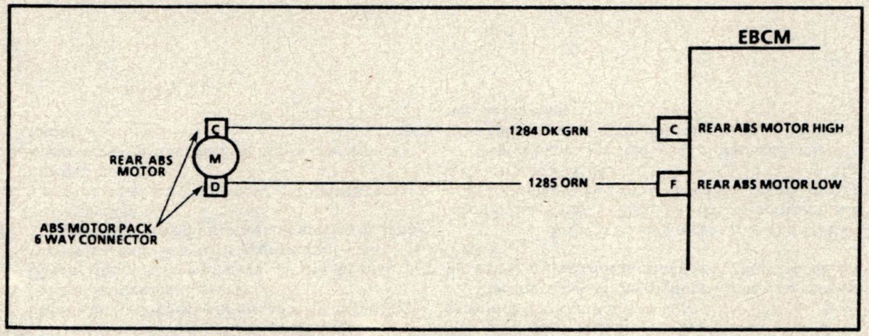

DTC 51
(Page 1 of 2)
REAR ABS MOTOR FREE SPINS

Circuit Description:

This DTC is designed to detect a stripped nut or gear assembly during initialization. During the homing sequence, the piston should reach the top of the bore resulting in a stalled ABS motor. If this does not occur, the ABS motor must be spinning with little or no resistance, indicating a nut/screw or gear malfunction.

DTC Will Set When: DTC 51 can be set during initialization. If the feedback current is less than the command current for a specified period of time, a malfunction exists.

Action Taken: A malfunction DTC is stored, ABS is disabled, and the ABS warning lamp is turned "ON."

DTC Chart Test Description: Number(s) below refer to circled number(s) on the diagnostic chart.

1. Checks to see if the corresponding open ABS motor DTC 64 is also set.
2. Verifies that the ABS motor was actually applied as commanded by observing feedback current.
3. Verifies that the ABS motor can release.
4. Verifies that the ABS motor can be applied by observing pedal movement.
5. Checks for high resistance in the ABS motor "HIGH" circuitry.
6. Checks for high resistance in the ABS motor "LOW" circuitry.
7. Checks for proper resistance of the ABS motor windings.

Diagnostic Aids: An "Intermittent" malfunction in this DTC may result from a mechanical part of the system that sticks, binds, or slips.

The frequency of the malfunction can be checked by using the enhanced diagnostic function of the Tech 1.

If the DTC only occurs once and DTC 64 also occurs, refer to DTC 64. If intermittent and enhanced diagnostics show this DTC occurs during ABS, refer to DTC 64.

Depending on the frequency of the malfunction, a physical inspection of the mechanical parts suspected may be necessary.

DIAGNOSTIC CHART DTC 51 — 1994 CENTURY, CUTLASS CIERA AND CUTLASS CRUISER

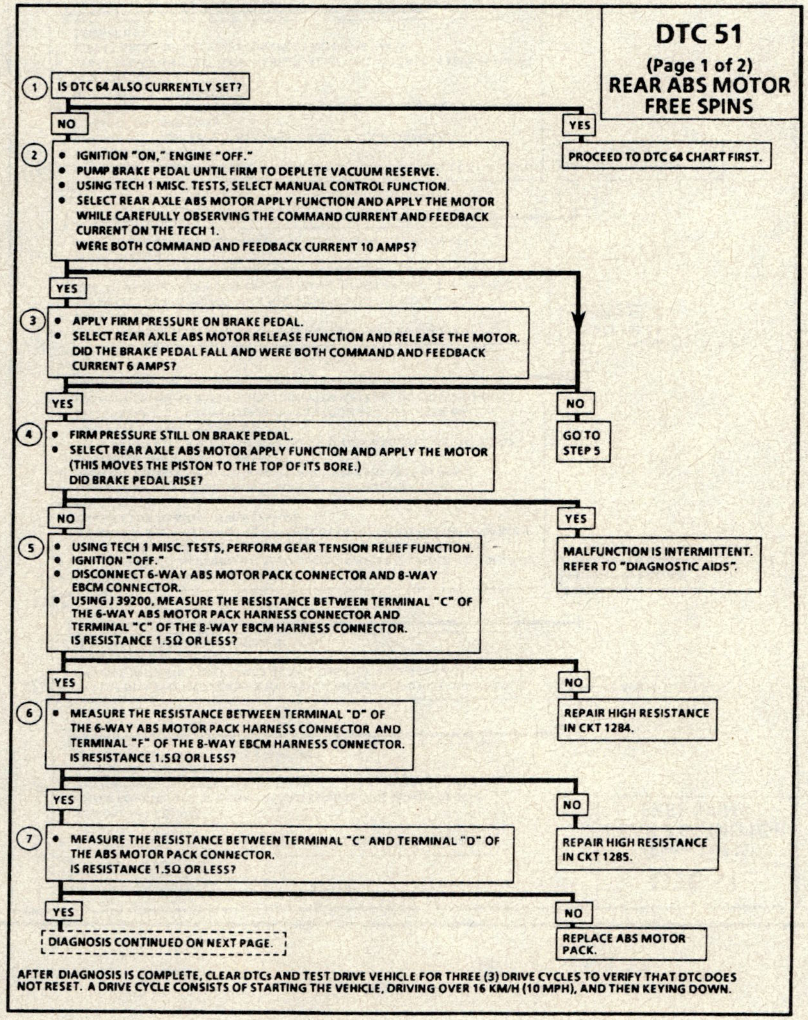

DIAGNOSTIC CHART DTC 51 — 1994 CENTURY, CUTLASS CIERA AND CUTLASS CRUISER

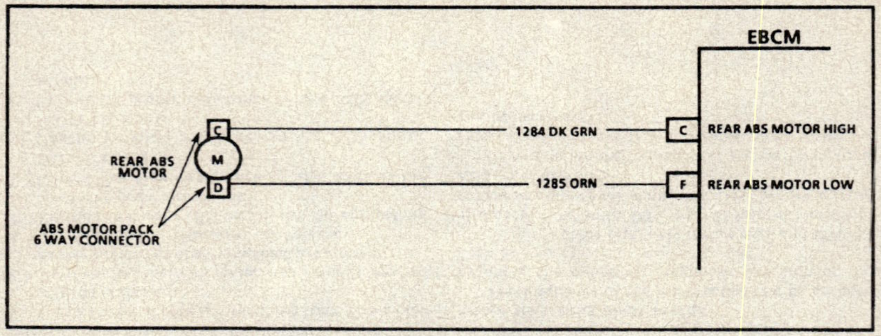

DTC 51
(Page 2 of 2)
REAR ABS MOTOR FREE SPINS

Circuit Description:
This DTC is designed to detect a stripped nut or gear assembly during initialization. During the homing sequence, the piston should reach the top of the bore resulting in a stalled ABS motor. If this does not occur, the ABS motor must be spinning with little or no resistance, indicating a nut/screw or gear malfunction.

DTC Will Set When: DTC 51 can be set during initialization. If the feedback current is less than the command current for a specified period of time, a malfunction exists.

Action Taken: A malfunction DTC is stored, ABS is disabled, and the ABS warning lamp is turned "ON."

DTC Chart Test Description: Number(s) below refer to circled number(s) on the diagnostic chart.
8. Checks for a stripped gear on the ABS motor pack (unit with three small gears).
9. Checks for a stripped gear on the hydraulic modulator (unit with three large gears).
10. Verifies that the ABS motor can be applied under a load situation.
11. Verifies that the hydraulic modulator is functioning properly.
12. Ensures malfunction was not due to poor terminal contact.

Diagnostic Aids: An "Intermittent" malfunction in this DTC may result from a mechanical part of the system that sticks, binds, or slips.

The frequency of the malfunction can be checked by using the enhanced diagnostic function of the Tech 1.

If the DTC only occurs once and DTC 64 also occurs, refer to DTC 64. If intermittent and enhanced diagnostics show this DTC occurs during ABS, refer to DTC 64.

Depending on the frequency of the malfunction, a physical inspection of the mechanical parts suspected may be necessary.

DIAGNOSTIC CHART DTC 51 — 1994 CENTURY, CUTLASS CIERA AND CUTLASS CRUISER

DTC 51
(Page 2 of 2)
REAR ABS MOTOR FREE SPINS

DIAGNOSIS CONTINUED FROM PREVIOUS PAGE.

(8)
- REMOVE HYDRAULIC MODULATOR/MASTER CYLINDER ASSEMBLY FROM VEHICLE.
- REMOVE GEAR COVER.
- CHECK FOR STRIPPED GEAR ON ABS MOTOR PACK; ABS MOTOR PACK IS THE UNIT WITH THE THREE SMALL GEARS. THE REAR AXLE GEAR IS THE MIDDLE GEAR OF THE THREE.

IS THE ABS MOTOR PACK GEAR STRIPPED?

YES: REPLACE ABS MOTOR PACK.

NO:

(9)
- CHECK FOR A STRIPPED GEAR ON HYDRAULIC MODULATOR; MODULATOR IS THE UNIT WITH THE THREE LARGE GEARS. THE REAR AXLE GEAR IS THE MIDDLE GEAR OF THE THREE.

IS THE MODULATOR GEAR STRIPPED?

YES: REPLACE HYDRAULIC MODULATOR GEAR.

NO:

(10)
- RECONNECT ELECTRICAL CONNECTORS
- SECURELY POSITION HYDRAULIC MODULATOR/MASTER CYLINDER ASSEMBLY IN VEHICLE WITH GEAR COVER REMOVED SUCH THAT THE GEAR SET CAN BE OBSERVED. BE CAREFUL NOT TO DAMAGE GEAR SET OR ALLOW THE HYDRAULIC MODULATOR TO MOVE WHILE TESTING.
- IGNITION "ON."
- USING TECH 1 MISC. TESTS, SELECT MANUAL CONTROL FUNCTION.
- SELECT REAR AXLE ABS MOTOR APPLY AND APPLY THE MOTOR WHILE OBSERVING THE MIDDLE GEAR SET.
- SELECT REAR AXLE ABS MOTOR RELEASE FUNCTION AND RELEASE THE MOTOR WHILE STILL OBSERVING THE GEAR SET.
- SELECT REAR AXLE ABS MOTOR APPLY FUNCTION AND APPLY THE MOTOR.

DID GEAR SET MOVE IN BOTH DIRECTIONS FOR AT LEAST ONE REVOLUTION?

NO: REPLACE ABS MOTOR PACK.

YES:

(11)
- USING TECH 1 MISC. TESTS, PERFORM GEAR TENSION RELIEF FUNCTION.
- IGNITION "OFF."
- SEPARATE THE ABS MOTOR PACK FROM THE HYDRAULIC MODULATOR.
- GRASP THE MIDDLE GEAR ON THE HYDRAULIC MODULATOR (THE HYDRAULIC MODULATOR IS THE UNIT WITH THE THREE LARGE GEARS) AND ATTEMPT TO MOVE THE GEAR.

CAN THE GEAR BE ROTATED MORE THAN 8 FULL TURNS LOCK TO LOCK?

YES: REPLACE THE HYDRAULIC MODULATOR.

NO:

(12)
- INSPECT ALL CONNECTORS AND TERMINALS FOR POOR TERMINAL CONTACT AND EVIDENCE OF CORROSION.
  REPLACE ALL TERMINALS THAT EXHIBIT SIGNS OF POOR TERMINAL CONTACT OR CORROSION.
- REINSTALL HYDRAULIC MODULATOR/MASTER CYLINDER ASSEMBLY IN VEHICLE.
- RECONNECT ALL CONNECTORS.
- IGNITION "OFF."
- START ENGINE AND TEST DRIVE VEHICLE, ACHIEVING AT LEAST 16 km/h (10 MPH).
- REPEAT ABOVE DRIVE CYCLE SEQUENCE TWO MORE TIMES.

DOES DTC 51 RESET?

YES: REPLACE EBCM.

NO: MALFUNCTION WAS MOSTLY LIKELY CAUSED BY A POOR CONNECTION. REFER TO "DIAGNOSTIC AIDS".

AFTER DIAGNOSIS IS COMPLETE, CLEAR DTCs AND TEST DRIVE VEHICLE FOR THREE (3) DRIVE CYCLES TO VERIFY THAT DTC DOES NOT RESET. A DRIVE CYCLE CONSISTS OF STARTING THE VEHICLE, DRIVING OVER 16 KM/H (10 MPH), AND THEN KEYING DOWN.

DIAGNOSTIC CHART DTC 52 — 1994 CENTURY, CUTLASS CIERA AND CUTLASS CRUISER

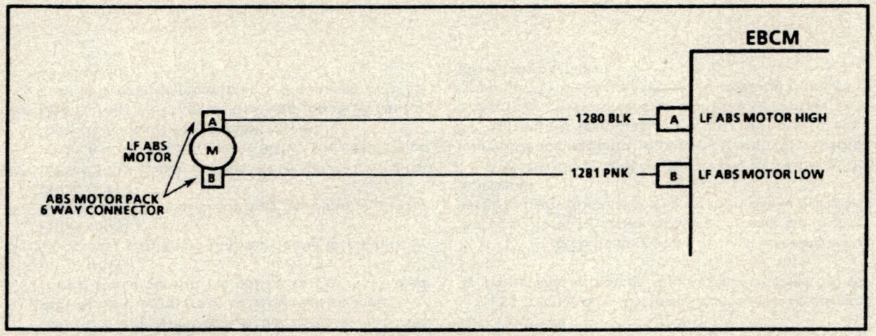

DTC 52

LEFT FRONT CHANNEL IN RELEASE TOO LONG

Circuit Description:

This DTC will diagnose an ABS motor that is energized longer than expected. This could occur if a wheel speed sensor is malfunctioning, the ABS motor does not turn, the left front solenoid mechanically fails open, or the ABS motor wires are crossed.

DTC Will Set When: DTC 52 can be set only during an ABS stop. If the EBCM commands the left front ABS channel in release for three seconds, a malfunction exists.

Action Taken: A malfunction DTC is stored, ABS is disabled and the ABS warning lamp is turned "ON."

DTC Chart Test Description: Number(s) below refer to circled number(s) on the diagnostic chart.

1. Identifies a problem in a wheel speed sensor that may cause the system to be in release too long.
2. Identifies an ABS motor as being failed or wired incorrectly.
3. Checks for a solenoid that may have mechanically failed open.
4. Verifies proper solenoid operation.
5. If Step 3 has failed, this serves to isolate the cause of the hydraulic problem, to either the solenoid or the ABS hydraulic modulator assembly.
6. Determines whether a malfunctioning ABS motor pack or hydraulic modulator is the reason for DTC 52 being set.

Diagnostic Aids: An "Intermittent" malfunction may be caused by a mechanical part of the system that sticks or binds.

The frequency of the malfunction can be checked by using the enhanced diagnostic function of the Tech 1.

DTC 52 may occur if on ice and steering wheel is turned to lock during braking. Using the Tech 1, perform hydraulic test to ensure total brake system is functional.

Any circuitry, that is suspected as causing the intermittent complaint, should be thoroughly checked for backed out terminals, improper mating, broken locks, improperly formed or damaged terminals, poor terminal to wiring connections or physical damage to the wiring harness.

DIAGNOSTIC CHART DTC 52 — 1994 CENTURY, CUTLASS CIERA AND CUTLASS CRUISER

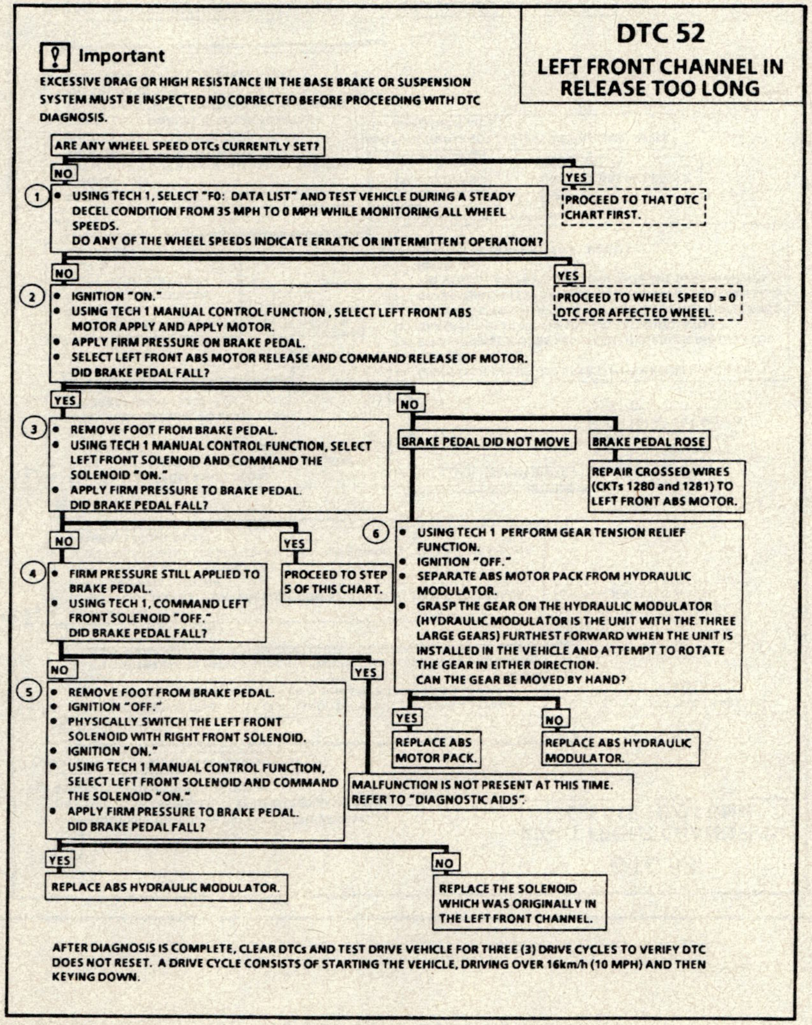

DIAGNOSTIC CHART DTC 53 — 1994 CENTURY, CUTLASS CIERA AND CUTLASS CRUISER

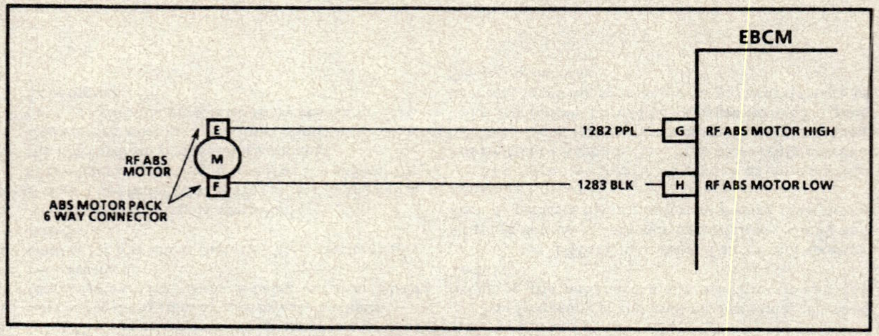

DTC 53

RIGHT FRONT CHANNEL IN RELEASE TOO LONG

Circuit Description:

This DTC will diagnose an ABS motor that is energized longer than expected. This could occur if a wheel speed sensor is malfunctioning, the ABS motor does not turn, the right front solenoid mechanically fails open, or the ABS motor wires are crossed.

DTC Will Set When: DTC 53 can be set only during an ABS stop. If the EBCM commands the right front channel in release for three seconds, a malfunction exists.

Action Taken: A malfunction DTC is stored, ABS is disabled and the ABS warning lamp is turned "ON."

DTC Chart Test Description: Number(s) below refer to circled number(s) on the diagnostic chart.

1. Identifies a problem in a wheel speed sensor that may cause the system to be in release too long.
2. Identifies a motor as being failed or wired incorrectly.
3. Checks for a solenoid that may have mechanically failed open.
4. Checks for a solenoid that may have mechanically failed closed.
5. If Step 3 has failed, this serves to isolate the cause of the hydraulic problem, to either the solenoid or the ABS hydraulic modulator assembly.
6. Determines whether a malfunction ABS motor pack or hydraulic modulator is the reason for DTC 53 being set.

Diagnostic Aids: An "Intermittent" malfunction may be caused by a mechanical part of the system that sticks or binds.

The frequency of the malfunction can be checked by using the enhanced diagnostic function of the Tech 1.

DTC 53 may occur on ice if steering wheel is turned to lock during braking. Using the Tech 1 perform hydraulic test to ensure total brake system is functional.

Any circuitry, that is suspected as causing the intermittent complaint, should be thoroughly checked for backed out terminals, improper mating, broken locks, improperly formed or damaged terminals, poor terminal to wiring connections or physical damage to the wiring harness.

DIAGNOSTIC CHART DTC 53 — 1994 CENTURY, CUTLASS CIERA AND CUTLASS CRUISER

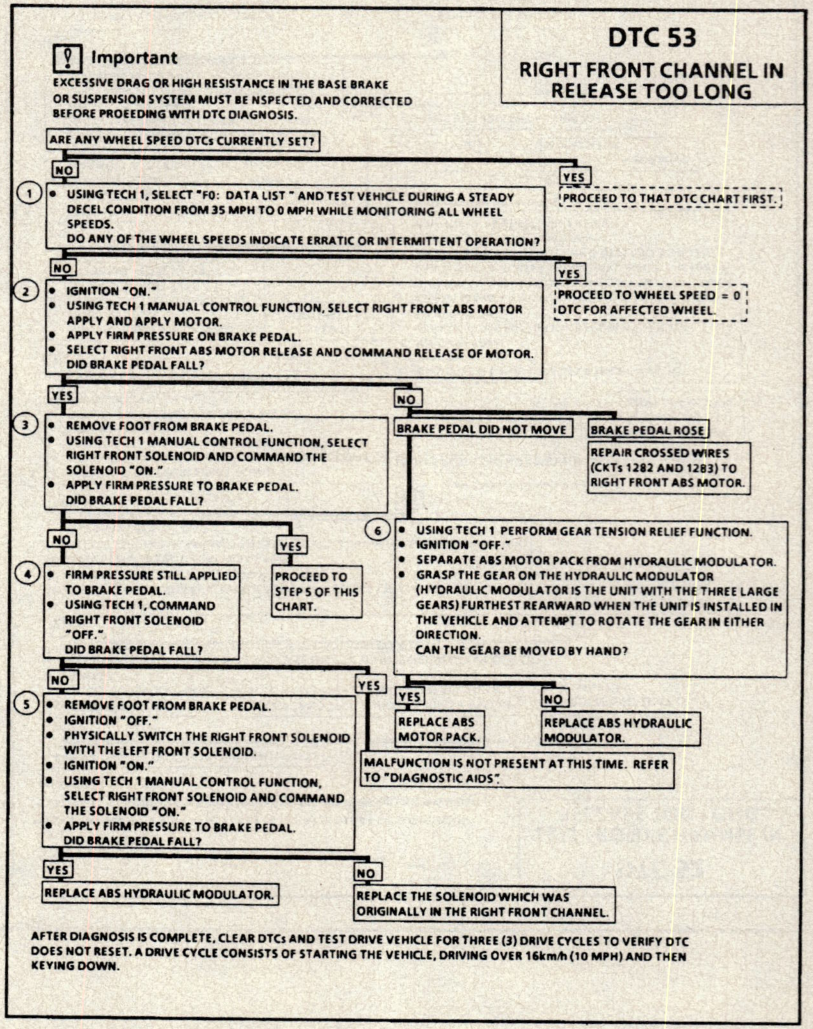

DIAGNOSTIC CHART DTC 54 — 1994 CENTURY, CUTLASS CIERA AND CUTLASS CRUISER

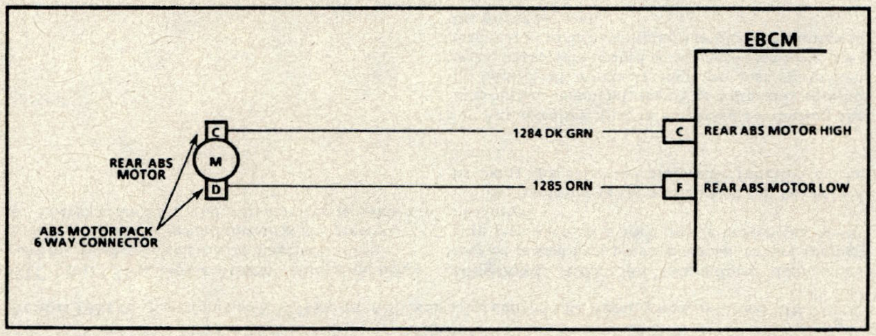

DTC 54

REAR CHANNEL IN RELEASE TOO LONG

Circuit Description:

This DTC will diagnose an ABS motor that is energized longer than expected. This could occur if a wheel speed sensor is malfunctioning, the ABS motor does not turn, or the ABS motor wires are crossed.

DTC Will Set When: DTC 54 can be set only during an ABS stop. If the EBCM commands the rear ABS channel in release for three seconds, a malfunction exists.

Action Taken: A malfunction DTC is stored, ABS is disabled and the ABS warning lamp is turned "ON."

DTC Chart Test Description: Number(s) below refer to circled number(s) on the diagnostic chart.

1. Identifies a problem in a wheel speed sensor that may cause the system to be in release too long.
2. Checks for a wheel that may stick or bind because of a mechanical fault.
3. Checks to see if the ABS motor is capable of moving and applying the hydraulic piston for the rear wheels.
4. Ensures that the ABS motor wiring is not crossed.
5. Isolates the fault of a "no-apply" situation to either the ABS motor pack or ABS hydraulic modulator assembly.

Diagnostic Aids: An "Intermittent" malfunction may be caused by a mechanical part of the system that sticks or binds.

The frequency of the malfunction can be checked by using the enhanced diagnostic function of the Tech 1.

Using the Tech 1, perform hydraulic test to ensure total brake system is functional.

Any circuitry, that is suspected as causing the intermittent complaint, should be thoroughly checked for backed out terminals, improper mating, broken locks, improperly formed or damaged terminals, poor terminal to wiring connections or physical damage to the wiring harness.

DIAGNOSTIC CHART DTC 54 — 1994 CENTURY, CUTLASS CIERA AND CUTLASS CRUISER

DTC 54
REAR CHANNEL IN RELEASE TOO LONG

Important
- EXCESSIVE DRAG OR HIGH RESISTANCE IN THE BASE BRAKE OR SUSPENSION SYSTEM MUST BE INSPECTED AND CORRECTED BEFORE PROCEEDING WITH DTC DIAGNOSIS.

ARE ANY WHEEL SPEED DTCs CURRENTLY SET?
- YES: PROCEED TO THAT DTC CHART FIRST.
- NO:

(1)
- USING TECH 1, SELECT "F0: DATA LIST" AND TEST VEHICLE DURING A STEADY DECEL CONDITION FROM 35 MPH TO 0 MPH WHILE MONITORING ALL WHEEL SPEEDS.
DO ANY OF THE WHEEL SPEEDS INDICATE ERRATIC OR INTERMITTENT OPERATION?
- YES: PROCEED TO WHEEL SPEED = 0 DTC FOR AFFECTED WHEEL.
- NO:

(2)
- RAISE REAR OF VEHICLE AND SUPPORT
- SPIN BOTH REAR WHEELS AND CONFIRM PHYSICAL CONDITION OF BASE BRAKE SYSTEM.
COULD BOTH REAR WHEELS BE SPUN FREELY BY HAND?
- NO: REPAIR SOURCE OF RESISTANCE IN BASE BRAKE SYSTEM.
- YES:

(3)
- IGNITION "ON."
- USING MANUAL CONTROL FUNCTION OF TECH 1, SELECT REAR AXLE ABS MOTOR APPLY.
- APPLY FIRM PRESSURE ON BRAKE PEDAL.
- COMMAND APPLY OF REAR ABS MOTOR AND TRY TO SPIN REAR WHEELS BY HAND.
COULD REAR WHEELS BE SPUN BY HAND?
- NO: MALFUNCTION IS NOT PRESENT AT THIS TIME. REFER TO "DIAGNOSTIC AIDS".
- YES:

(4)
- WITH FIRM PRESSURE ON BRAKE PEDAL STILL APPLIED, COMMAND A RELEASE OF THE REAR AXLE ABS MOTOR.
CAN WHEELS BE MOVED?
- NO: REPAIR CROSSED WIRES AT REAR ABS MOTOR (CKTs 1284 AND 1285).
- YES:

(5)
- USING TECH 1, PERFORM GEAR TENSION RELIEF FUNCTION.
- SEPARATE ABS MOTOR PACK FROM HYDRAULIC MODULATOR.
- CONNECT ABS MOTOR PACK WIRING.
- IGNITION "ON."
- USING TECH 1, PERFORM MOTOR TEST.
DO ALL THREE ABS MOTOR PACK GEARS (SMALL GEARS) SPIN FREELY?
- NO: REPLACE ABS MOTOR PACK.
- YES: REPLACE ABS HYDRAULIC MODULATOR.

AFTER ALL DIAGNOSIS IS COMPLETE, CLEAR DTCs AND TEST DRIVE VEHICLE FOR THREE (3) DRIVE CYCLES TO VERIFY THAT DTC DOES NOT RESET. A DRIVE CYCLE CONSISTS OF STARTING THE VEHICLE DRIVING OVER 16 KM/H (10 MPH), AND THEN KEYING DOWN.

DIAGNOSTIC CHART DTC 55 — 1994 CENTURY, CUTLASS CIERA AND CUTLASS CRUISER

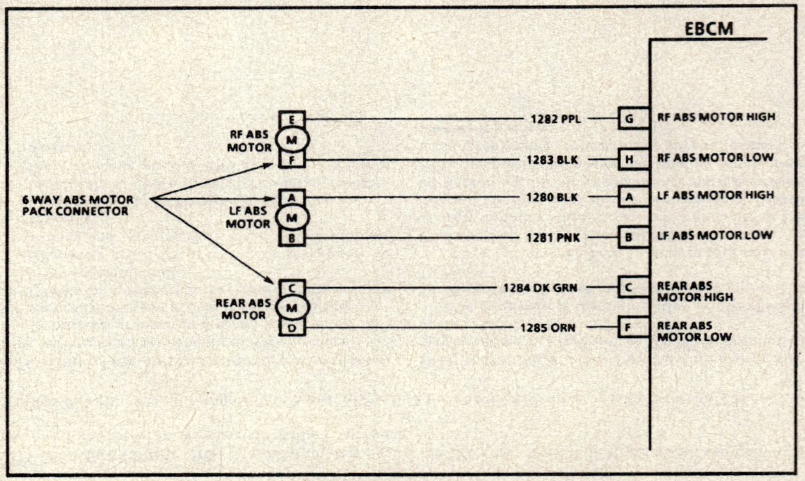

DTC 55

EBCM MALFUNCTION

Circuit Description:

This DTC identifies a malfunction detected by the MDI custom IC. It also ensures the cause of the malfunction is not a result of a problem with the ABS enable relay under a DTC 55 malfunction.

Action Taken: A malfunction DTC is stored, ABS is disabled and the ABS warning lamp is turned "ON."

DTC Chart Test Description: Number(s) below refer to circled number(s) on the diagnostic chart.

1. Checks to see if the malfunction is still present.
2. Checks to see if the malfunction is intermittent.

Diagnostic Aids: An "Intermittent" malfunction may be caused by a poor connection, rubbed through wire insulation, or a wire that is broken inside the insulation.

The frequency of the malfunction can be checked by using the enhanced diagnostic function of the Tech 1.

Any circuitry, that is suspected as causing the intermittent complaint, should be thoroughly checked for backed out terminals, improper mating, broken locks, improperly formed or damaged terminals, poor terminal to wiring connections or physical damage to the wiring harness.

DIAGNOSTIC CHART DTC 55 — 1994 CENTURY, CUTLASS CIERA AND CUTLASS CRUISER

DTC 55

EBCM MALFUNCTION

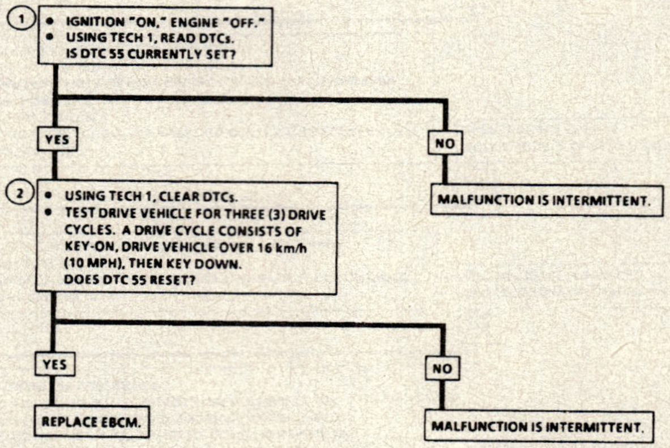

DIAGNOSTIC CHART DTC 56 — 1994 CENTURY, CUTLASS CIERA AND CUTLASS CRUISER

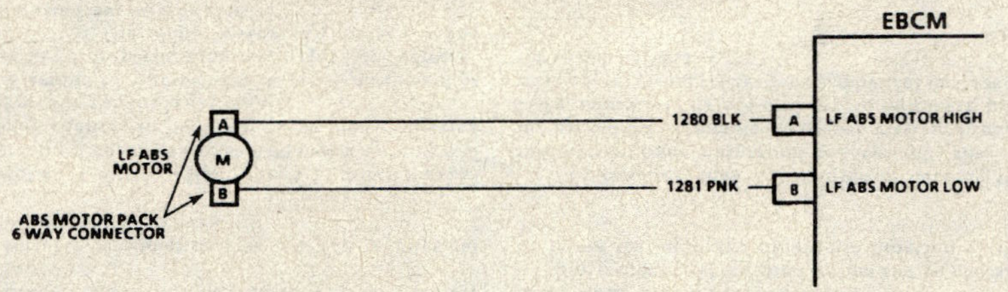

DTC 56

LEFT FRONT ABS MOTOR CIRCUIT OPEN

Circuit Description:
This DTC identifies an ABS motor that cannot be energized due to an open in its circuitry.

DTC Will Set When: DTC 56 can be set only when the motor is commanded "OFF." If the EBCM detects an out of range voltage on either of the left front ABS motor circuits indicating an open circuit, a malfunction exists.

Action Taken: An open ABS motor will not activate when requested. A malfunction DTC is stored, ABS is disabled and the ABS warning lamp is turned "ON."

DTC Chart Test Description: Number(s) below refer to circled number(s) on the diagnostic chart.

1. Checks for proper resistance of the ABS motor.
2. Checks for an open in the ABS motor "HIGH" circuitry.
3. Checks for an open in the ABS motor "LOW" circuitry.
4. Ensures malfunction was not due to a poor terminal contact.

Diagnostic Aids: Using Tech 1, select manual control function, and exercise ABS motor movement of affected channel in both directions while applying light pressure on the brake pedal.

If erratic or "jumpy" brake pedal movement is detected while performing an "apply" or "release" function of the ABS motor, an intermittent malfunction may be indicated.

An "Intermittent" malfunction may be caused by a poor connection, rubbed through wire insulation, or a wire that is broken inside the insulation.

If the malfunction is not current, wiggle the wires of the affected channel and check if the DTC resets. This will help to pinpoint an intermittent malfunction in the motor circuitry or connections.

The frequency of the malfunction can be checked by using the enhanced diagnostic function of the Tech 1.

Any circuitry, that is suspected as causing the intermittent complaint, should be thoroughly checked for backed out terminals, improper mating, broken locks, improperly formed or damaged terminals, poor terminal to wiring connections or physical damage to the wiring harness.

DIAGNOSTIC CHART DTC 56 — 1994 CENTURY, CUTLASS CIERA AND CUTLASS CRUISER

DTC 56

LEFT FRONT ABS MOTOR CIRCUIT OPEN

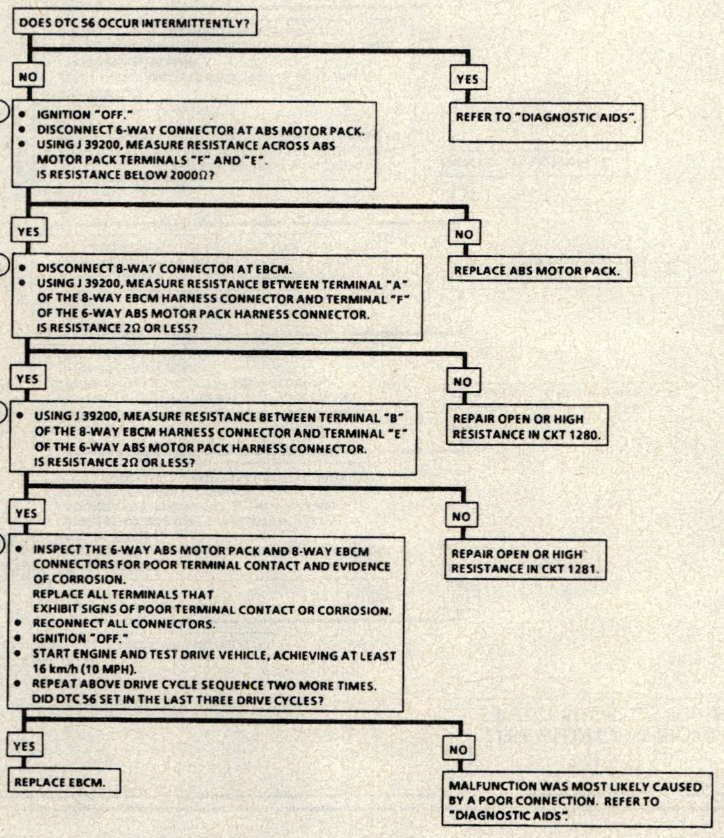

AFTER ALL DIAGNOSIS IS COMPLETE, CLEAR DTCs AND TEST DRIVE VEHICLE FOR THREE (3) DRIVE CYCLES TO VERIFY THAT DTC DOES NOT RESET. A DRIVE CYCLE CONSISTS OF STARTING THE VEHICLE, DRIVING OVER 16 KM/H (10 MPH), AND THEN KEYING DOWN.

DIAGNOSTIC CHART DTC 57 — 1994 CENTURY, CUTLASS CIERA AND CUTLASS CRUISER

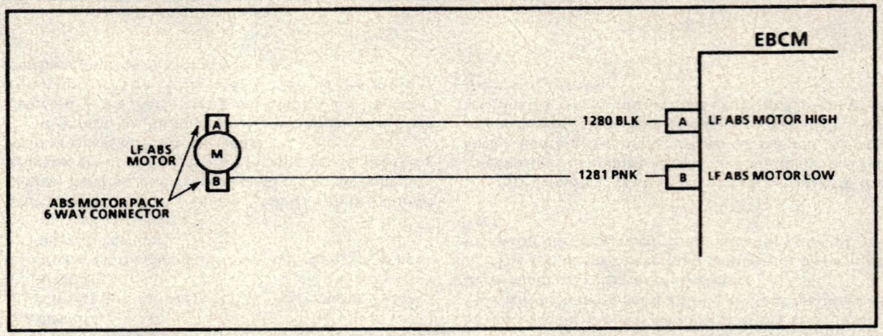

DTC 57

LEFT FRONT ABS MOTOR CIRCUIT SHORTED TO GND

Circuit Description:
This DTC identifies an ABS motor circuit that is shorted to ground. This malfunction will not allow the ABS motor to be controlled at the commanded current rate or will cause the driver circuit to allow current directly to ground.

DTC Will Set When: DTC 57 can be set anytime. If the EBCM detects an out of range voltage on either of the left front ABS motor circuits indicating a circuit shorted to ground, a malfunction exists.

Action Taken: A malfunction DTC is stored, ABS is disabled and the ABS warning lamp is turned "ON."

DTC Chart Test Description: Number(s) below refer to circled number(s) on the diagnostic chart.

1. Checks for a short to ground in the ABS motor "HIGH" circuitry.
2. Checks for a short to ground in the ABS motor "LOW" circuitry.
3. Checks for an ABS motor that is internally shorted to ground.
4. Ensures malfunction was not due to physical damage of the circuitry.

Diagnostic Aids: Using Tech 1, select manual control function, and exercise ABS motor movement of affected channel in both directions while applying light pressure on the brake pedal.

If erratic or "jumpy" brake pedal movement is detected while performing an "apply" or "release" function of the ABS motor, an intermittent malfunction may be indicated.

An "Intermittent" malfunction may be caused by a poor connection, rubbed through wire insulation, or a wire that is broken inside the insulation.

If the malfunction is not current, wiggle the wires of the affected channel and check if the DTC resets. This will help to pinpoint an intermittent malfunction in the motor circuitry or connections.

The frequency of the malfunction can be checked by using the enhanced diagnostic function of the Tech 1.

Any circuitry, that is suspected as causing the intermittent complaint, should be thoroughly checked for backed out terminals, improper mating, broken locks, improperly formed or damaged terminals, poor terminal to wiring connections or physical damage to the wiring harness.

DIAGNOSTIC CHART DTC 57 — 1994 CENTURY, CUTLASS CIERA AND CUTLASS CRUISER

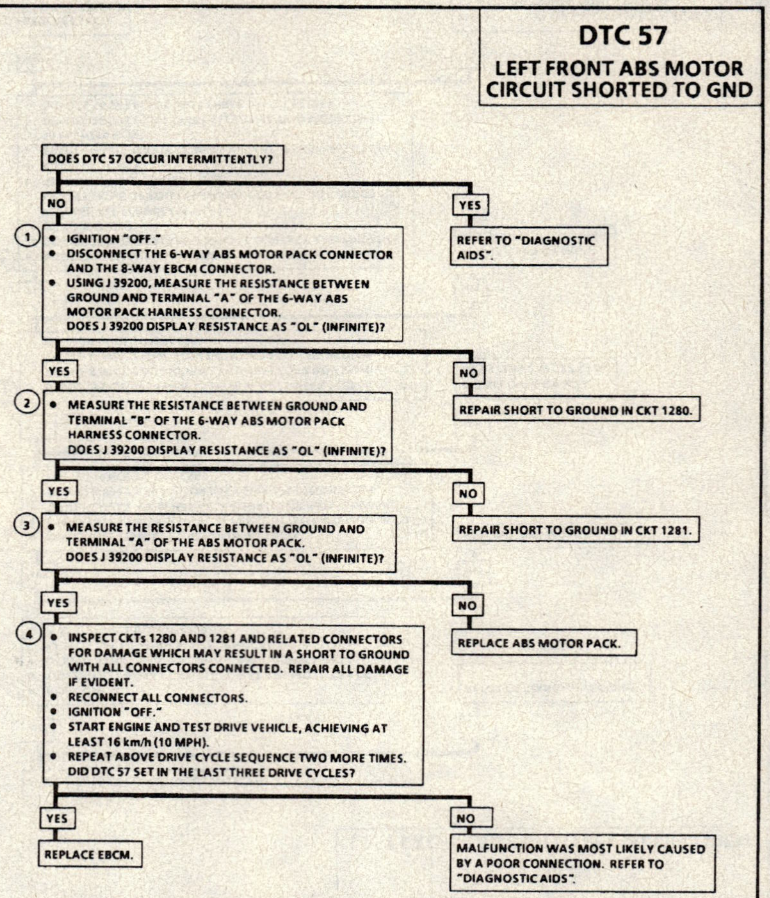

AFTER ALL DIAGNOSIS IS COMPLETE, CLEAR DTCs AND TEST DRIVE VEHICLE FOR THREE (3) DRIVE CYCLES TO VERIFY THAT DTC DOES NOT RESET. A DRIVE CYCLE CONSISTS OF STARTING THE VEHICLE, DRIVING OVER 16 km/h (10 MPH), AND THEN KEYING DOWN.

DIAGNOSTIC CHART DTC 58 — 1994 CENTURY, CUTLASS CIERA AND CUTLASS CRUISER

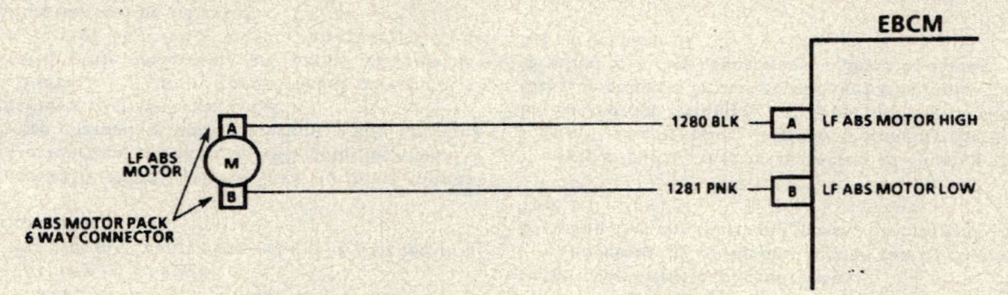

DTC 58

LEFT FRONT ABS MOTOR CIRCUIT SHORTED TO BATT

Circuit Description:
This DTC identifies an ABS motor circuit that is shorted to battery or an ABS motor that has low or no resistance. This malfunction will not allow the ABS motor to be controlled at the commanded current rate or will cause the ABS motor to turn in the opposite direction, or not at all.

DTC Will Set When: DTC 58 can be set only when the ABS motor is commanded "OFF." If the EBCM detects an out of range voltage on either of the left front ABS motor circuits indicating a circuit shorted to battery or an ABS motor shorted, a malfunction exists.

Action Taken: A malfunction DTC is stored, ABS is disabled and the ABS warning lamp is turned "ON."

DTC Chart Test Description: Number(s) below refer to circled number(s) on the diagnostic chart.

1. Checks for a short to voltage in the ABS motor "HIGH" circuitry.
2. Checks for a short to voltage in the ABS motor "LOW" circuitry.
3. Checks for a short between the "HIGH" and "LOW" ABS motor circuits.
4. Checks for an ABS motor that is internally shorted.
5. Ensures malfunction was not due to physical damage of the circuitry.

Diagnostic Aids: Using Tech 1, select manual control function, and exercise ABS motor movement of affected channel in both directions while applying light pressure on the brake pedal.

If erratic or "jumpy" brake pedal movement is detected while performing an "apply" or "release" function of the ABS motor, an intermittent malfunction may be indicated.

An "Intermittent" malfunction may be caused by a poor connection, rubbed through wire insulation, or a wire that is broken inside the insulation.

If the malfunction is not current, wiggle the wires of the affected channel and check if the DTC resets. This will help to pinpoint an intermittent malfunction in the motor circuitry or connections.

The frequency of the malfunction can be checked by using the enhanced diagnostic function of the Tech 1.

Any circuitry, that is suspected as causing the intermittent complaint, should be thoroughly checked for backed out terminals, improper mating, broken locks, improperly formed or damaged terminals, poor terminal to wiring connections or physical damage to the wiring harness.

DIAGNOSTIC CHART DTC 58 — 1994 CENTURY, CUTLASS CIERA AND CUTLASS CRUISER

DTC 58

LEFT FRONT ABS MOTOR CIRCUIT SHORTED TO BATT

NOTICE: J 39200 TEST LEADS MUST BE "ZEROED" PRIOR TO MAKING ANY RESISTANCE MEASUREMENTS. REFER TO J 39200 USER'S MANUAL.

DOES DTC 58 OCCUR INTERMITTENTLY?

- YES: REFER TO "DIAGNOSTIC AIDS".
- NO:

(1)
- IGNITION "OFF."
- DISCONNECT 6-WAY ABS MOTOR PACK CONNECTOR AND 8-WAY EBCM CONNECTOR.
- IGNITION "ON."
- USING J 39200, MEASURE THE VOLTAGE BETWEEN GROUND AND TERMINAL "F" OF THE 6-WAY ABS MOTOR PACK HARNESS CONNECTOR. IS VOLTAGE 1 VOLT OR LESS?

- NO: REPAIR SHORT TO VOLTAGE IN CKT 1280.
- YES:

(2)
- MEASURE THE VOLTAGE BETWEEN GROUND AND TERMINAL "E" OF THE 6-WAY ABS MOTOR PACK HARNESS CONNECTOR. IS VOLTAGE 1 VOLT OR LESS?

- NO: REPAIR SHORT TO VOLTAGE IN CKT 1281.
- YES:

(3)
- IGNITION "OFF."
- USING J 39200, MEASURE THE RESISTANCE BETWEEN TERMINAL "E" AND TERMINAL "F" OF THE 6-WAY ABS MOTOR PACK HARNESS CONNECTOR. DOES J 39200 DISPLAY RESISTANCE AS "OL" (INFINITE)?

- NO: REPAIR SHORT BETWEEN CKTs 1280 AND 1281.
- YES:

(4)
- USING J 39200, MEASURE THE RESISTANCE BETWEEN TERMINAL "E" AND TERMINAL "F" OF THE ABS MOTOR PACK. IS RESISTANCE GREATER THAN 0.4Ω?

- NO: REPLACE ABS MOTOR PACK.
- YES:

(5)
- INSPECT CKTs 1280 AND 1281 AND RELATED CONNECTORS FOR DAMAGE WHICH MAY RESULT IN A SHORT TO GROUND WITH ALL CONNECTORS CONNECTED. REPAIR ALL DAMAGE IF EVIDENT.
- RECONNECT ALL CONNECTORS.
- IGNITION "OFF."
- START ENGINE AND TEST DRIVE VEHICLE, ACHIEVING AT LEAST 16 km/h (10 MPH).
- REPEAT ABOVE DRIVE CYCLE SEQUENCE TWO MORE TIMES. DID DTC 58 SET IN THE LAST THREE DRIVE CYCLES?

- YES: REPLACE EBCM.
- NO: MALFUNCTION WAS MOST LIKELY CAUSED BY A POOR CONNECTION. REFER TO "DIAGNOSTIC AIDS".

AFTER ALL DIAGNOSIS IS COMPLETE, CLEAR DTCs AND TEST DRIVE VEHICLE FOR THREE (3) DRIVE CYCLES TO VERIFY THAT DTC DOES NOT RESET. A DRIVE CYCLE CONSISTS OF STARTING THE VEHICLE, DRIVING OVER 16 km/h (10 MPH), AND THEN KEYING DOWN.

DIAGNOSTIC CHART DTC 61 — 1994 CENTURY, CUTLASS CIERA AND CUTLASS CRUISER

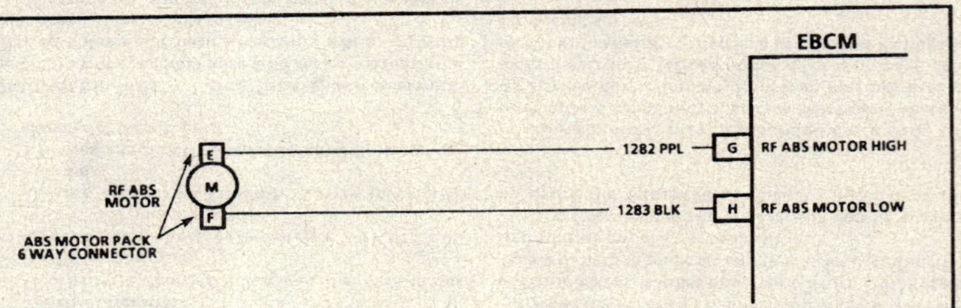

DTC 61

RIGHT FRONT ABS MOTOR CIRCUIT OPEN

Circuit Description:
This DTC identifies an ABS motor that cannot be energized due to an open in its circuitry. This malfunction will not allow proper front ABS operation.

DTC Will Set When: DTC 61 can be set only when the ABS motor is commanded "OFF." If the EBCM detects an out of range voltage on either of the right front ABS motor circuits indicating an open circuit, a malfunction exists.

Action Taken: An open ABS motor will not activate when requested. A malfunction DTC is stored, ABS is disabled and the ABS warning lamp is turned "ON."

DTC Chart Test Description: Number(s) below refer to circled number(s) on the diagnostic chart.

1. Checks for proper resistance of the ABS motor.
2. Checks for an open in the ABS motor "HIGH" circuitry.
3. Checks for an open in the ABS motor "LOW" circuitry.
4. Ensures malfunction was not due to poor terminal contact or EBCM.

Diagnostic Aids: Using Tech 1, select manual control function, and exercise ABS motor movement of affected channel in both directions while applying light pressure on the brake pedal.

If erratic or "jumpy" brake pedal movement is detected while performing an "apply" or "release" function of the ABS motor, an intermittent malfunction may be indicated.

An "Intermittent" malfunction may be caused by a poor connection, rubbed through wire insulation, or a wire that is broken inside the insulation.

If the malfunction is not current, wiggle the wires of the affected channel and check if the DTC resets. This will help to pinpoint an intermittent malfunction in the motor circuitry or connections.

The frequency of the malfunction can be checked by using the enhanced diagnostic function of the Tech 1.

Any circuitry, that is suspected as causing the intermittent complaint, should be thoroughly checked for backed out terminals, improper mating, broken locks, improperly formed or damaged terminals, poor terminal to wiring connections or physical damage to the wiring harness.

DIAGNOSTIC CHART DTC 61 — 1994 CENTURY, CUTLASS CIERA AND CUTLASS CRUISER

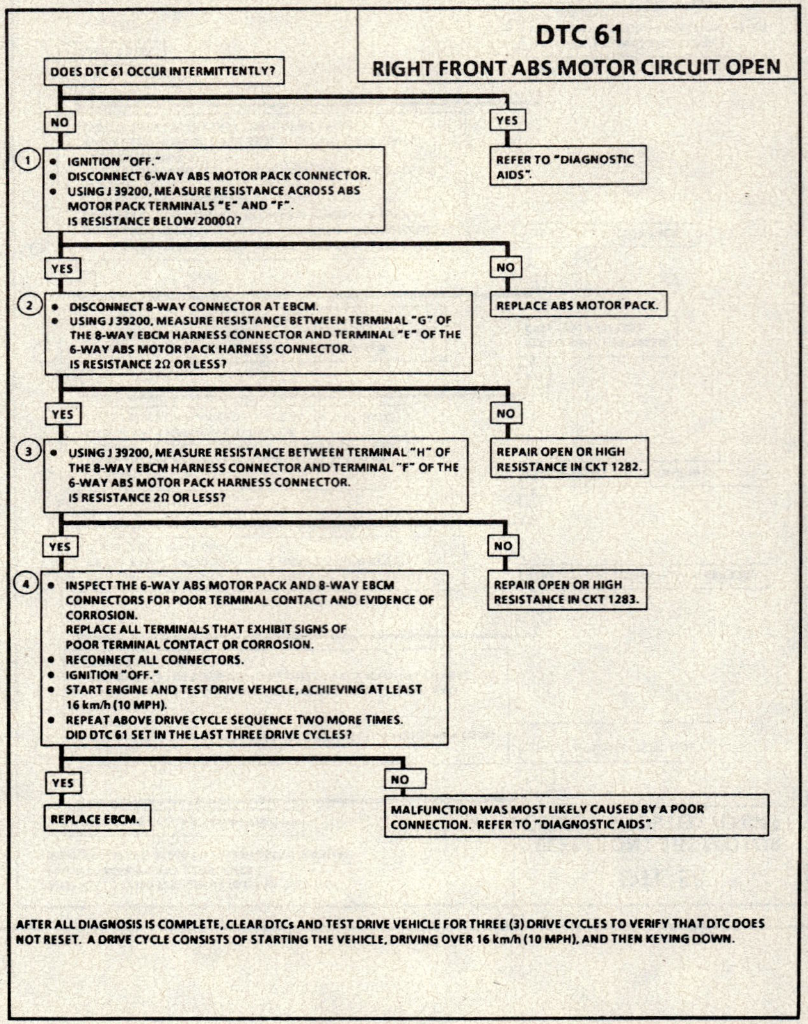

DIAGNOSTIC CHART DTC 62 — 1994 CENTURY, CUTLASS CIERA AND CUTLASS CRUISER

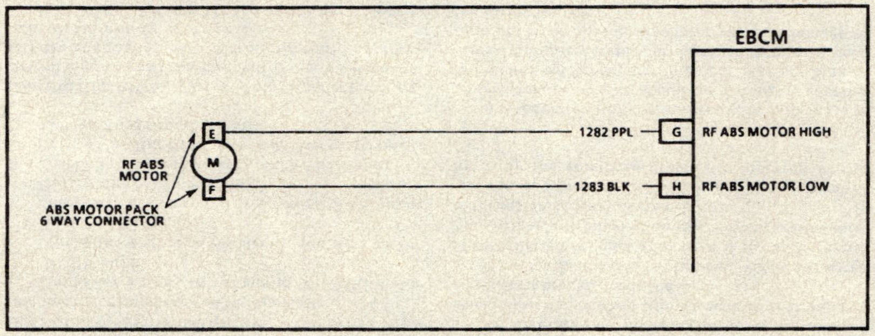

DTC 62

RIGHT FRONT ABS MOTOR CIRCUIT SHORTED TO GND

Circuit Description:
This DTC identifies an ABS motor circuit that is shorted to ground. This malfunction will not allow the ABS motor to be controlled at the commanded current rate or will cause the ABS motor driver circuit to allow current directly to ground.

DTC Will Set When: DTC 62 can be set anytime. If the EBCM detects an out of range voltage on either of the right front ABS motor circuits indicating a circuit shorted to ground, a malfunction exists.

Action Taken: A malfunction DTC is stored, ABS is disabled and the ABS warning lamp is turned "ON."

DTC Chart Test Description: Number(s) below refer to circled number(s) on the diagnostic chart.

1. Checks for a short to ground in the ABS motor "HIGH" circuitry.
2. Checks for a short to ground in the ABS motor "LOW" circuitry.
3. Checks for an ABS motor that is internally shorted to ground.
4. Ensures malfunction was not due to physical damage of the circuitry.

Diagnostic Aids: Using Tech 1, select manual control function, and exercise ABS motor movement of affected channel in both directions while applying light pressure on the brake pedal.

If erratic or "jumpy" brake pedal movement is detected while performing an "apply" or "release" function of the ABS motor, an intermittent malfunction may be indicated.

An "Intermittent" malfunction may be caused by a poor connection, rubbed through wire insulation, or a wire that is broken inside the insulation.

If the malfunction is not current, wiggle the wires of the affected channel and check if the DTC resets. This will help to pinpoint an intermittent malfunction in the motor circuitry or connections.

The frequency of the malfunction can be checked by using the enhanced diagnostic function of the Tech 1.

Any circuitry, that is suspected as causing the intermittent complaint, should be thoroughly checked for backed out terminals, improper mating, broken locks, improperly formed or damaged terminals, poor terminal to wiring connections or physical damage to the wiring harness.

DIAGNOSTIC CHART DTC 62 — 1994 CENTURY, CUTLASS CIERA AND CUTLASS CRUISER

DTC 62

RIGHT FRONT ABS MOTOR CIRCUIT SHORTED TO GND

DOES DTC 62 OCCUR INTERMITTENTLY?

- YES: REFER TO "DIAGNOSTIC AIDS".
- NO:

(1)
- IGNITION "OFF."
- DISCONNECT 6-WAY ABS MOTOR PACK CONNECTOR AND THE 8-WAY EBCM CONNECTOR.
- USING J 39200, MEASURE THE RESISTANCE BETWEEN GROUND AND TERMINAL "E" OF THE 6-WAY ABS MOTOR PACK HARNESS CONNECTOR.

DOES J 39200 DISPLAY RESISTANCE AS "OL" (INFINITE)?

- NO: REPAIR SHORT TO GROUND IN CKT 1282.
- YES:

(2)
- MEASURE THE RESISTANCE BETWEEN GROUND AND TERMINAL "F" OF THE 6-WAY ABS MOTOR PACK HARNESS CONNECTOR.

DOES J 39200 DISPLAY RESISTANCE AS "OL" (INFINITE)?

- NO: REPAIR SHORT TO GROUND IN CKT 1283.
- YES:

(3)
- MEASURE THE RESISTANCE BETWEEN GROUND AND TERMINAL "E" OF THE ABS MOTOR PACK.

DOES J 39200 DISPLAY RESISTANCE AS "OL" (INFINITE)?

- NO: REPLACE ABS MOTOR PACK.
- YES:

(4)
- INSPECT CKTs 1282 AND 1283 AND RELATED CONNECTORS FOR DAMAGE WHICH MAY RESULT IN A SHORT TO GROUND WITH ALL CONNECTORS CONNECTED. REPAIR ALL DAMAGE IF EVIDENT.
- RECONNECT ALL CONNECTORS.
- IGNITION "OFF."
- START ENGINE AND TEST DRIVE VEHICLE, ACHIEVING AT LEAST 16 km/h (10 MPH).
- REPEAT ABOVE DRIVE CYCLE SEQUENCE TWO MORE TIMES.

DID DTC 62 SET IN THE LAST THREE DRIVE CYCLES?

- YES: REPLACE EBCM.
- NO: MALFUNCTION WAS MOST LIKELY CAUSED BY A POOR CONNECTION. REFER TO "DIAGNOSTIC AIDS".

AFTER ALL DIAGNOSIS IS COMPLETE, CLEAR DTCs AND TEST DRIVE VEHICLE FOR THREE (3) DRIVE CYCLES TO VERIFY THAT DTC DOES NOT RESET. A DRIVE CYCLE CONSISTS OF STARTING THE VEHICLE, DRIVING OVER 16 km/h (10 MPH), AND THEN KEYING DOWN.

DIAGNOSTIC CHART DTC 63 — 1994 CENTURY, CUTLASS CIERA AND CUTLASS CRUISER

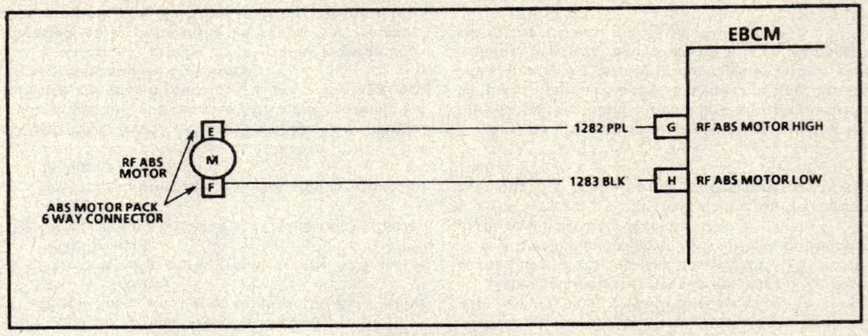

DTC 63

RIGHT FRONT ABS MOTOR CIRCUIT SHORTED TO BATT

Circuit Description:
This DTC identifies an ABS motor circuit that is shorted to battery or an ABS motor that has low or no resistance. This malfunction will not allow the ABS motor to be controlled at the commanded current rate or will cause the ABS motor to turn in the opposite direction or not at all.

DTC Will Set When: DTC 63 can be set only when the ABS motor is commanded "OFF." If the EBCM detects an out of range voltage on either of the right front ABS motor circuits indicating a circuit shorted to battery or an ABS motor shorted, a malfunction exists.

Action Taken: A malfunction DTC is stored, ABS is disabled and the ABS warning lamp is turned "ON."

DTC Chart Test Description: Number(s) below refer to circled number(s) on the diagnostic chart.
1. Checks for a short to voltage in the ABS motor "HIGH" circuitry.
2. Checks for a short to voltage in the ABS motor "LOW" circuitry.
3. Checks for a short circuit between ABS motor "HIGH" and ABS motor "LOW."
4. Checks for proper resistance of the ABS motor.
5. Ensures malfunction was not due to physical damage of the circuitry.

Diagnostic Aids: Using Tech 1, select manual control function, and exercise ABS motor movement of affected channel in both directions while applying light pressure on the brake pedal.
If erratic or "jumpy" brake pedal movement is detected while performing an "apply" or "release" function of the ABS motor, an intermittent malfunction may be indicated.

An "Intermittent" malfunction may be caused by a poor connection, rubbed through wire insulation, or a wire that is broken inside the insulation.
If the malfunction is not current, wiggle the wires of the affected channel and check if the DTC resets. This will help to pinpoint an intermittent malfunction in the motor circuitry or connections.
The frequency of the malfunction can be checked by using the enhanced diagnostic function of the Tech 1.

Any circuitry, that is suspected as causing the intermittent complaint, should be thoroughly checked for backed out terminals, improper mating, broken locks, improperly formed or damaged terminals, poor terminal to wiring connections or physical damage to the wiring harness.

DIAGNOSTIC CHART DTC 63 — 1994 CENTURY, CUTLASS CIERA AND CUTLASS CRUISER

DTC 63
RIGHT FRONT ABS MOTOR CIRCUIT SHORTED TO BATT

NOTICE: J 39200 TEST LEADS MUST BE "ZEROED" PRIOR TO MAKING ANY RESISTANCE MEASUREMENTS. REFER TO J 39200 USER'S MANUAL.

DOES DTC 63 OCCUR INTERMITTENTLY?
- YES: REFER TO "DIAGNOSTIC AIDS".
- NO: go to step 1.

(1)
- IGNITION "OFF."
- DISCONNECT 6-WAY ABS MOTOR PACK CONNECTOR AND 8-WAY EBCM CONNECTOR.
- IGNITION "ON."
- USING J 39200, MEASURE THE VOLTAGE BETWEEN GROUND AND TERMINAL "B" OF THE 6-WAY ABS MOTOR PACK HARNESS CONNECTOR.

IS VOLTAGE 1 VOLT OR LESS?
- NO: REPAIR SHORT TO VOLTAGE IN CKT 1282.
- YES: go to step 2.

(2)
- MEASURE THE VOLTAGE BETWEEN GROUND AND TERMINAL "A" OF THE 6-WAY ABS MOTOR PACK HARNESS CONNECTOR.

IS VOLTAGE 1 VOLT OR LESS?
- NO: REPAIR SHORT TO VOLTAGE IN CKT 1283.
- YES: go to step 3.

(3)
- IGNITION "OFF."
- USING J 39200, MEASURE THE RESISTANCE BETWEEN TERMINAL "A" AND TERMINAL "B" OF THE 6-WAY ABS MOTOR PACK HARNESS CONNECTOR.

DOES J 39200 DISPLAY RESISTANCE AS "OL" (INFINITE)?
- NO: REPAIR SHORT BETWEEN CKTs 1282 AND 1283.
- YES: go to step 4.

(4)
- USING J 39200, MEASURE THE RESISTANCE BETWEEN TERMINAL "A" AND TERMINAL "B" OF THE ABS MOTOR PACK.

IS RESISTANCE GREATER THAN 0.4Ω?
- NO: REPLACE ABS MOTOR PACK.
- YES: go to step 5.

(5)
- INSPECT CKTs 1282 AND 1283 AND RELATED CONNECTORS FOR DAMAGE WHICH MAY RESULT IN A SHORT TO GROUND WITH ALL CONNECTORS CONNECTED. REPAIR ALL DAMAGE IF EVIDENT.
- RECONNECT ALL CONNECTORS.
- IGNITION "OFF."
- START ENGINE AND TEST DRIVE VEHICLE, ACHIEVING AT LEAST 16 km/h (10 MPH).
- REPEAT ABOVE DRIVE CYCLE SEQUENCE TWO MORE TIMES.

DID DTC 63 SET IN THE LAST THREE DRIVE CYCLES?
- YES: REPLACE EBCM.
- NO: MALFUNCTION WAS MOST LIKELY CAUSED BY A POOR CONNECTION. REFER TO "DIAGNOSTIC AIDS".

AFTER ALL DIAGNOSIS IS COMPLETE, CLEAR DTCs AND TEST DRIVE VEHICLE FOR THREE (3) DRIVE CYCLES TO VERIFY THAT DTC DOES NOT RESET. A DRIVE CYCLE CONSISTS OF STARTING THE VEHICLE, DRIVING OVER 16 km/h (10 MPH), AND THEN KEYING DOWN

DIAGNOSTIC CHART DTC 64 — 1994 CENTURY, CUTLASS CIERA AND CUTLASS CRUISER

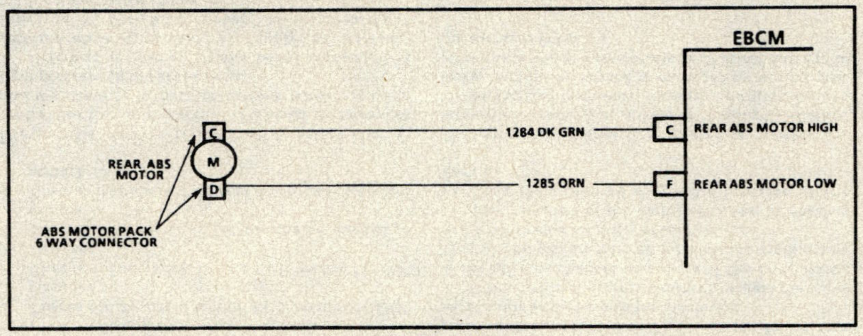

DTC 64

REAR ABS MOTOR CIRCUIT OPEN

Circuit Description:
This DTC identifies an ABS motor that cannot be energized due to an open in its circuitry. This malfunction will not allow proper rear ABS operation.

DTC Will Set When: DTC 64 can be set only when the ABS motor is commanded "OFF." If the EBCM detects an out of range voltage on either of the rear axle ABS motor circuits indicating an open circuit, a malfunction exists.

Action Taken: An open motor will not activate when requested. A malfunction DTC is stored, ABS is disabled and the ABS warning lamp is turned "ON."

DTC Chart Test Description: Number(s) below refer to circled number(s) on the diagnostic chart.

1. Checks for proper resistance of the ABS motor.
2. Checks for an open in the ABS motor "HIGH" circuitry.
3. Checks for an open in the ABS motor "LOW" circuitry.
4. Ensures malfunction was not due to poor terminal contact.

Diagnostic Aids: Using Tech 1, select manual control function, and exercise ABS motor movement of affected channel in both directions while applying light pressure on the brake pedal.

If erratic or "jumpy" brake pedal movement is detected while performing an "apply" or "release" function of the ABS motor, an intermittent malfunction may be indicated.

An "Intermittent" malfunction may be caused by a poor connection, rubbed through wire insulation, or a wire that is broken inside the insulation.

If the malfunction is not current, wiggle the wires of the affected channel and check if the DTC resets. This will help to pinpoint an intermittent malfunction in the motor circuitry or connections.

The frequency of the malfunction can be checked by using the enhanced diagnostic function of the Tech 1.

Any circuitry, that is suspected as causing the intermittent complaint, should be thoroughly checked for backed out terminals, improper mating, broken locks, improperly formed or damaged terminals, poor terminal to wiring connections or physical damage to the wiring harness.

DIAGNOSTIC CHART DTC 64 — 1994 CENTURY, CUTLASS CIERA AND CUTLASS CRUISER

DTC 64

REAR ABS MOTOR CIRCUIT OPEN

DOES DTC 64 OCCUR INTERMITTENTLY?

- YES: REFER TO "DIAGNOSTIC AIDS".
- NO:

(1)
- IGNITION "OFF."
- DISCONNECT 6-WAY ABS MOTOR PACK CONNECTOR.
- USING J 39200, MEASURE RESISTANCE ACROSS ABS MOTOR PACK TERMINALS "C" AND "D".

IS RESISTANCE BELOW 2000Ω?

- NO: REPLACE ABS MOTOR PACK.
- YES:

(2)
- DISCONNECT 8-WAY CONNECTOR AT EBCM.
- USING J 39200, MEASURE RESISTANCE BETWEEN TERMINAL "C" OF THE 8-WAY EBCM HARNESS CONNECTOR AND TERMINAL "C" OF THE 6-WAY ABS MOTOR PACK HARNESS CONNECTOR.

IS RESISTANCE 2Ω OR LESS?

- NO: REPAIR OPEN OR HIGH RESISTANCE IN CKT 1284.
- YES:

(3)
- USING J 39200, MEASURE RESISTANCE BETWEEN TERMINAL "F" OF THE 8-WAY EBCM HARNESS CONNECTOR AND TERMINAL "D" OF THE 6-WAY ABS MOTOR PACK HARNESS CONNECTOR.

IS RESISTANCE 2Ω OR LESS?

- NO: REPAIR OPEN OR HIGH RESISTANCE IN CKT 1285.
- YES:

(4)
- INSPECT THE 6-WAY ABS MOTOR PACK AND 8-WAY EBCM CONNECTORS FOR POOR TERMINAL CONTACT AND EVIDENCE OF CORROSION.
  REPLACE ALL TERMINALS THAT EXHIBIT SIGNS OF POOR TERMINAL CONTACT OR CORROSION.
- RECONNECT ALL CONNECTORS.
- IGNITION "OFF."
- START ENGINE AND TEST DRIVE VEHICLE, ACHIEVING AT LEAST 16 km/h (10 MPH).
- REPEAT ABOVE DRIVE CYCLE SEQUENCE TWO MORE TIMES.

DID DTC 64 SET IN THE LAST THREE DRIVE CYCLES?

- YES: REPLACE EBCM.
- NO: MALFUNCTION WAS MOST LIKELY CAUSED BY A POOR CONNECTION. REFER TO "DIAGNOSTIC AIDS".

AFTER ALL DIAGNOSIS IS COMPLETE, CLEAR DTCs AND TEST DRIVE VEHICLE FOR THREE (3) DRIVE CYCLES TO VERIFY THAT DTC DOES NOT RESET. A DRIVE CYCLE CONSISTS OF STARTING THE VEHICLE, DRIVING OVER 16 km/h (10 MPH), AND THEN KEYING DOWN.

DIAGNOSTIC CHART DTC 65 — 1994 CENTURY, CUTLASS CIERA AND CUTLASS CRUISER

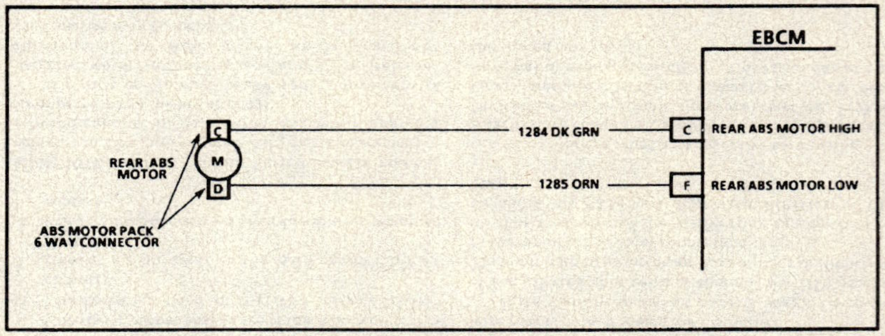

DTC 65

REAR ABS MOTOR CIRCUIT SHORTED TO GROUND

Circuit Description:
This DTC identifies an ABS motor circuit that is shorted to ground. This malfunction will not allow the ABS motor to be controlled at the commanded current rate or will cause the driver circuit to allow current directly to ground.

DTC Will Set When: DTC 65 can be set anytime. If the EBCM detects an out of range voltage on either of the rear ABS motor circuits indicating a circuit shorted to ground, a malfunction exists.

Action Taken: A malfunction DTC is stored, ABS is disabled and the ABS warning lamp is turned "ON."

DTC Chart Test Description: Number(s) below refer to circled number(s) on the diagnostic chart.

1. Checks for an open in the ABS motor "HIGH" circuitry.
2. Checks for an open in the ABS motor "LOW" circuitry.
3. Ensures ABS motor is not internally shorted to ground.
4. Ensures malfunction was not due to poor terminal contact.

Diagnostic Aids: Using Tech 1, select manual control function, and exercise ABS motor movement of affected channel in both directions while applying light pressure on the brake pedal.

If erratic or "jumpy" brake pedal movement is detected while performing an "apply" or "release" function of the ABS motor, an intermittent malfunction may be indicated.

An "Intermittent" malfunction may be caused by a poor connection, rubbed through wire insulation, or a wire that is broken inside the insulation.

If the malfunction is not current, wiggle the wires of the affected channel and check if the DTC resets. This will help to pinpoint an intermittent malfunction in the ABS motor circuitry or connections.

The frequency of the malfunction can be checked by using the enhanced diagnostic function of the Tech 1.

Any circuitry, that is suspected as causing the intermittent complaint, should be thoroughly checked for backed out terminals, improper mating, broken locks, improperly formed or damaged terminals, poor terminal to wiring connections or physical damage to the wiring harness.

DIAGNOSTIC CHART DTC 65 — 1994 CENTURY, CUTLASS CIERA AND CUTLASS CRUISER

DTC 65

REAR ABS MOTOR CIRCUIT SHORTED TO GROUND.

DOES DTC 65 OCCUR INTERMITTENTLY?

- YES: REFER TO "DIAGNOSTIC AIDS".
- NO:

(1)
- IGNITION "OFF."
- DISCONNECT THE 6-WAY ABS MOTOR PACK CONNECTOR AND THE 8-WAY EBCM CONNECTOR.
- USING J 39200, MEASURE THE RESISTANCE BETWEEN GROUND AND TERMINAL "C" OF THE 6-WAY ABS MOTOR PACK HARNESS CONNECTOR.

DOES J 39200 DISPLAY RESISTANCE AS "OL" (INFINITE)?

- NO: REPAIR SHORT TO GROUND IN CKT 1284.
- YES:

(2)
- MEASURE THE RESISTANCE BETWEEN GROUND AND TERMINAL "D" OF THE 6-WAY ABS MOTOR PACK HARNESS CONNECTOR.

DOES J 39200 DISPLAY RESISTANCE AS "OL" (INFINITE)?

- NO: REPAIR SHORT TO GROUND IN CKT 1285.
- YES:

(3)
- MEASURE THE RESISTANCE BETWEEN GROUND AND TERMINAL "C" OF THE ABS MOTOR PACK.

DOES J 39200 DISPLAY RESISTANCE AS "OL" (INFINITE)?

- NO: REPLACE ABS MOTOR PACK.
- YES:

(4)
- INSPECT CKTs 1284 AND 1285 AND RELATED CONNECTORS FOR DAMAGE WHICH MAY RESULT IN A SHORT TO GROUND WITH ALL CONNECTORS CONNECTED. REPAIR ALL DAMAGE IF EVIDENT.
- RECONNECT ALL CONNECTORS.
- IGNITION "OFF."
- START ENGINE AND TEST DRIVE VEHICLE, ACHIEVING AT LEAST 16 km/h (10 MPH).
- REPEAT ABOVE DRIVE CYCLE SEQUENCE TWO MORE TIMES.

DID DTC 65 SET IN THE LAST THREE DRIVE CYCLES?

- YES: REPLACE EBCM.
- NO: MALFUNCTION WAS MOST LIKELY CAUSED BY A POOR CONNECTION. REFER TO "DIAGNOSTIC AIDS".

AFTER ALL DIAGNOSIS IS COMPLETE, CLEAR DTCs AND TEST DRIVE VEHICLE FOR THREE (3) DRIVE CYCLES TO VERIFY THAT DTC DOES NOT RESET. A DRIVE CYCLE CONSISTS OF STARTING THE VEHICLE, DRIVING OVER 16 km/h (10 MPH), AND THEN KEYING DOWN.

DIAGNOSTIC CHART DTC 66 — 1994 CENTURY, CUTLASS CIERA AND CUTLASS CRUISER

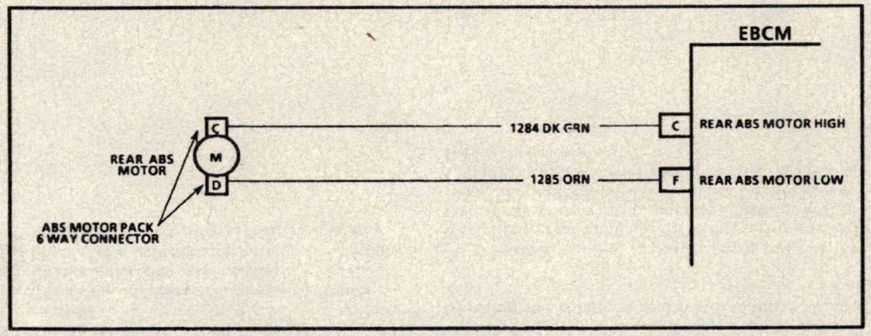

DTC 66

REAR ABS MOTOR CIRCUIT SHORTED TO BATTERY

Circuit Description:
This DTC identifies an ABS motor circuit that is shorted to battery or an ABS motor that has low or zero resistance. This malfunction will not allow the ABS motor to be controlled at the commanded current rate or will cause the ABS motor to turn in the opposite direction or not turn at all.

DTC Will Set When: DTC 66 can be set only when the ABS motor is commanded "OFF." If the EBCM detects an out of range voltage on either of the rear ABS motor circuits indicating a circuit shorted to battery or an ABS motor shorted, a malfunction exists.

Action Taken: A malfunction DTC is stored, ABS is disabled and the ABS warning lamp is turned "ON."

DTC Chart Test Description: Number(s) below refer to circled number(s) on the diagnostic chart.

1. Checks for a short to voltage in the ABS motor "HIGH" circuitry.
2. Checks for a short to voltage in the ABS motor "LOW" circuitry.
3. Checks for a short between ABS motor "HIGH" and ABS motor "LOW" circuits.
4. Checks for proper resistance of the ABS motor.
5. Ensures malfunction was not due to physical damage of the circuitry.

Diagnostic Aids: Using Tech 1, select manual control function, and exercise ABS motor movement of affected channel in both directions while applying light pressure on the brake pedal.

If erratic or "jumpy" brake pedal movement is detected while performing an "apply" or "release" function of the ABS motor, an intermittent malfunction may be indicated.

An "Intermittent" malfunction may be caused by a poor connection, rubbed through wire insulation, or a wire that is broken inside the insulation.

If the malfunction is not current, wiggle the wires of the affected channel and check if the DTC resets. This will help to pinpoint an intermittent malfunction in the ABS motor circuitry or connections.

The frequency of the malfunction can be checked by using the enhanced diagnostic function of the Tech 1.

Any circuitry, that is suspected as causing the intermittent complaint, should be thoroughly checked for backed out terminals, improper mating, broken locks, improperly formed or damaged terminals, poor terminal to wiring connections or physical damage to the wiring harness.

DIAGNOSTIC CHART DTC 66 — 1994 CENTURY, CUTLASS CIERA AND CUTLASS CRUISER

DTC 66

REAR ABS MOTOR CIRCUIT SHORTED TO BATTERY

NOTICE: J 39200 TEST LEADS MUST BE "ZEROED" PRIOR TO MAKING ANY RESISTANCE MEASUREMENTS. REFER TO J 39200 USER'S MANUAL.

DOES DTC 66 OCCUR INTERMITTENTLY?
- YES → REFER TO "DIAGNOSTIC AIDS".
- NO ↓

(1)
- IGNITION "OFF."
- DISCONNECT THE 6-WAY ABS MOTOR PACK CONNECTOR AND THE 8-WAY EBCM CONNECTOR.
- IGNITION "ON."
- USING J 39200, MEASURE THE VOLTAGE BETWEEN GROUND AND TERMINAL "C" OF THE 6-WAY ABS MOTOR PACK HARNESS CONNECTOR. IS VOLTAGE 1 VOLT OR LESS?

- NO → REPAIR SHORT TO VOLTAGE IN CKT 1284.
- YES ↓

(2)
- MEASURE THE VOLTAGE BETWEEN GROUND AND TERMINAL "D" OF THE 6-WAY ABS MOTOR PACK HARNESS CONNECTOR. IS VOLTAGE 1 VOLT OR LESS?

- NO → REPAIR SHORT TO VOLTAGE IN CKT 1285.
- YES ↓

(3)
- IGNITION "OFF."
- USING J 39200, MEASURE THE RESISTANCE BETWEEN TERMINAL "C" AND TERMINAL "D" OF THE 6-WAY ABS MOTOR PACK HARNESS CONNECTOR. DOES J 39200 DISPLAY RESISTANCE AS "OL" (INFINITE)?

- NO → REPAIR SHORT BETWEEN CKTs 1284 AND 1285.
- YES ↓

(4)
- USING J 39200, MEASURE THE RESISTANCE BETWEEN TERMINAL "C" AND TERMINAL "D" OF THE ABS MOTOR PACK. IS RESISTANCE GREATER THAN 0.4Ω?

- NO → REPLACE ABS MOTOR PACK
- YES ↓

(5)
- INSPECT CKTs 1284 AND 1285 AND RELATED CONNECTORS FOR DAMAGE WHICH MAY RESULT IN A SHORT TO VOLTAGE WITH ALL CONNECTORS CONNECTED. REPAIR ALL DAMAGE IF EVIDENT.
- RECONNECT ALL CONNECTORS.
- IGNITION "OFF."
- START ENGINE AND TEST DRIVE VEHICLE, ACHIEVING AT LEAST 16 km/h (10 MPH).
- REPEAT ABOVE DRIVE CYCLE SEQUENCE TWO MORE TIMES. DID DTC 66 SET IN THE LAST THREE DRIVE CYCLES?

- YES → REPLACE EBCM.
- NO → MALFUNCTION WAS MOST LIKELY CAUSED BY A POOR CONNECTION. REFER TO "DIAGNOSTIC AIDS".

AFTER DIAGNOSIS IS COMPLETE, CLEAR DTCs AND TEST DRIVE VEHICLE FOR THREE (3) DRIVE CYCLES TO VERIFY THAT DTC DOES NOT RESET. A DRIVE CYCLE CONSISTS OF STARTING THE VEHICLE, DRIVING OVER 16 KM/H (10 MPH), AND THEN KEYING DOWN.

DIAGNOSTIC CHART DTC 76 — 1994 CENTURY, CUTLASS CIERA AND CUTLASS CRUISER

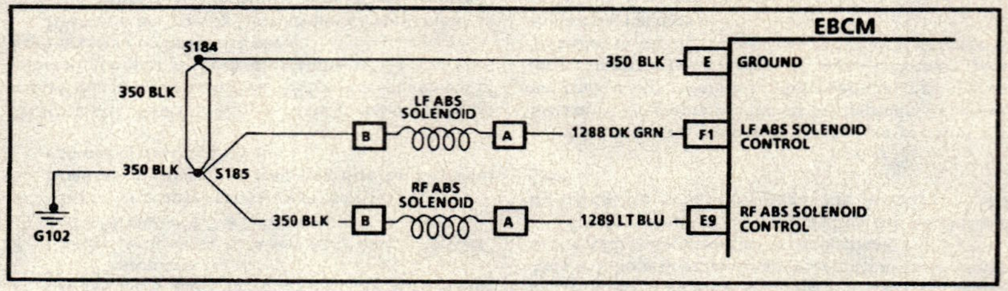

DTC 76

LEFT FRONT SOL. CIRCUIT OPEN OR SHORTED TO BATT

Circuit Description:

This DTC identifies a solenoid that cannot be energized due to an open in its driver circuitry, or a solenoid that is always energized due to a short to battery in its circuitry between the driver and the solenoid. An open will not allow proper ABS operation, but the short to battery simply turns "ON" the solenoid. A path for base brakes is still allowed once the ABS motor rehomes and the check ball is lifted off its seat during key "ON" initialization.

DTC Will Set When: DTC 76 can be set only when the solenoid is commanded "OFF." If the EBCM detects an excessive voltage in the left front solenoid control circuit, a malfunction exists.

Action Taken: A malfunction DTC is stored, ABS is disabled and the ABS warning lamp is turned "ON."

DTC Chart Test Description: Number(s) below refer to circled number(s) on the diagnostic chart.

1. Checks for a short to voltage in the control circuitry of the solenoid.
2. Checks for an open in the control circuit of the solenoid.
3. Checks the solenoid coil for proper resistance.
4. Checks for an open in the solenoid ground circuit.
5. Checks for a possible intermittent malfunction in the solenoid circuitry due to poor terminal contact.

Diagnostic Aids: An "Intermittent" malfunction may be caused by a poor connection, rubbed through wire insulation, or a wire that is broken inside the insulation.

The frequency of the malfunction can be checked by using the enhanced diagnostic function of the Tech 1.

Any circuitry, that is suspected as causing the intermittent complaint, should be thoroughly checked for backed out terminals, improper mating, broken locks, improperly formed or damaged terminals, poor terminal to wiring connections or physical damage to the wiring harness.

DIAGNOSTIC CHART DTC 76 — 1994 CENTURY, CUTLASS CIERA AND CUTLASS CRUISER

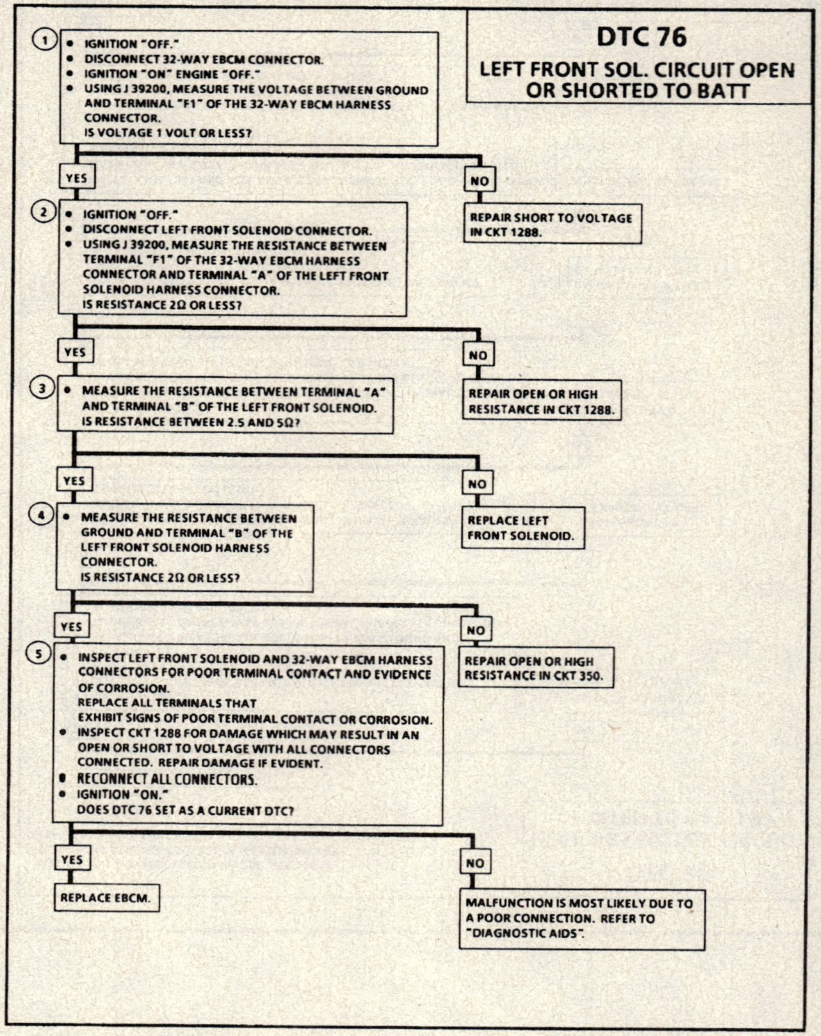

DIAGNOSTIC CHART DTC 77 — 1994 CENTURY, CUTLASS CIERA AND CUTLASS CRUISER

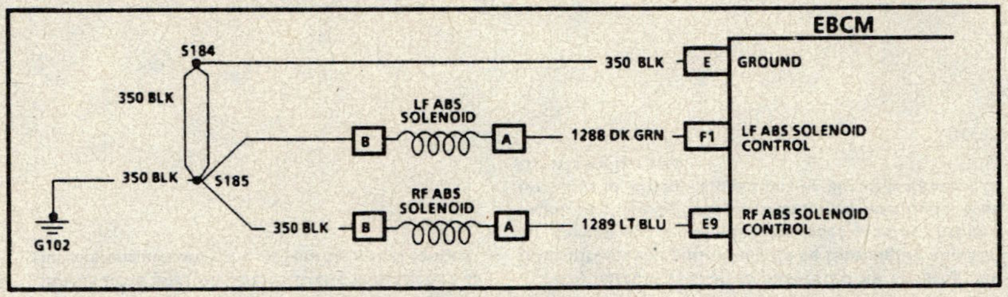

DTC 77

LEFT FRONT SOLENOID CIRCUIT SHORTED TO GND

Circuit Description:

This DTC identifies a solenoid that cannot be energized due to an open in its driver circuitry, or a short to ground between the solenoid driver and the solenoid. These malfunctions can affect ABS operation since the flow of brake fluid to the wheel cylinder cannot be stopped, making ABS operation for that channel impossible.

DTC Will Set When: DTC 77 can be set only when the solenoid is commanded "ON." If the EBCM detects the left front solenoid control circuit voltage is out of specification, a malfunction exists.

Action Taken: A malfunction DTC is stored, ABS is disabled and the ABS warning lamp is turned "ON."

DTC Chart Test Description: Number(s) below refer to circled number(s) on the diagnostic chart.

1. Checks for a solenoid that is internally shorted to ground.
2. Checks for a solenoid that is not within proper resistance values.
3. Indicates if a short to ground exists in the solenoid circuitry.
4. Checks for a possible intermittent malfunction in the solenoid circuitry due to poor terminal contact.

Diagnostic Aids: An "Intermittent" malfunction may be caused by a poor connection, rubbed through wire insulation, or a wire that is broken inside the insulation.

The frequency of the malfunction can be checked by using the enhanced diagnostic function of the Tech 1.

Any circuitry, that is suspected as causing the intermittent complaint, should be thoroughly checked for backed out terminals, improper mating, broken locks, improperly formed or damaged terminals, poor terminal to wiring connections or physical damage to the wiring harness.

DIAGNOSTIC CHART DTC 77 — 1994 CENTURY, CUTLASS CIERA AND CUTLASS CRUISER

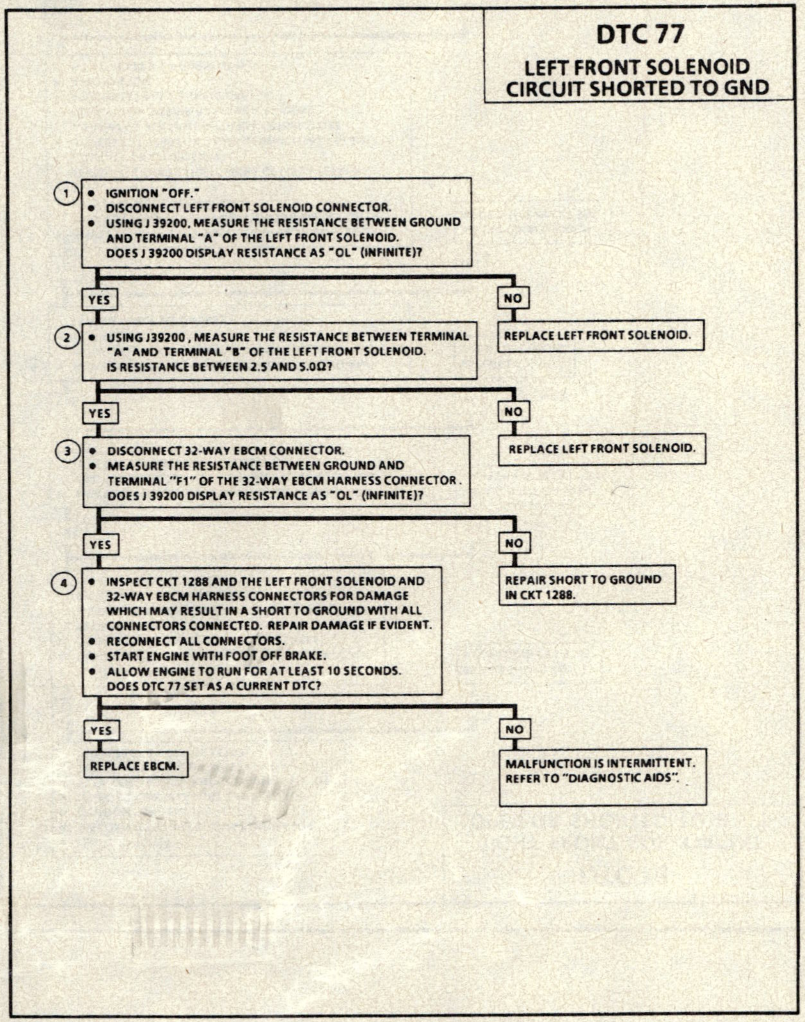

DIAGNOSTIC CHART DTC 78 — 1994 CENTURY, CUTLASS CIERA AND CUTLASS CRUISER

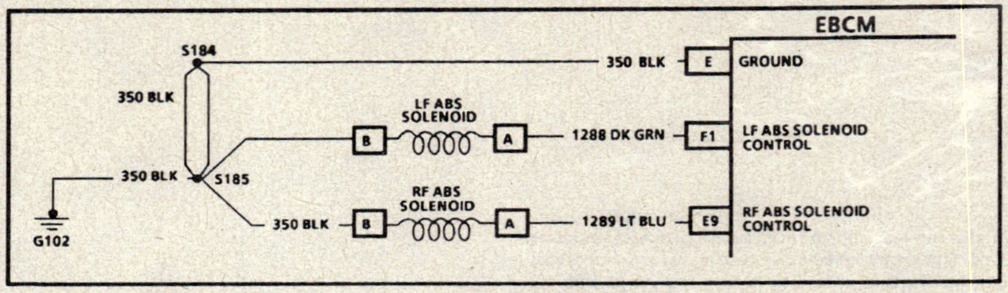

DTC 78

RIGHT FRONT SOL. CIRCUIT OPEN OR SHORTED TO BATT

Circuit Description:
This DTC identifies a solenoid that cannot be energized due to an open in its circuitry, or a solenoid that is always energized due to a short to battery in its circuitry between the driver and the solenoid. An open will not allow proper ABS operation, but the short to battery simply turns "ON" the solenoid. A path for base brakes is still allowed once the ABS motor rehomes and the check ball is lifted off its seat during key "ON" initialization.

DTC Will Set When: DTC 78 can be set only when the solenoid is commanded "OFF." If the EBCM detects an excessive voltage in the right front solenoid control circuit, a malfunction exists.

Action Taken: A malfunction DTC is stored, ABS is disabled and the ABS warning lamp is turned "ON."

DTC Chart Test Description: Number(s) below refer to circled number(s) on the diagnostic chart.

1. Checks for a short to voltage in the control circuitry of the solenoid.
2. Checks for an open in the control circuit of the solenoid.
3. Checks the solenoid coil for proper resistance.
4. Checks for an open in the solenoid ground circuit.
5. Checks for a possible intermittent malfunction in the solenoid circuitry due to poor terminal contact.

Diagnostic Aids: An "Intermittent" malfunction may be caused by a poor connection, rubbed through wire insulation, or a wire that is broken inside the insulation.

The frequency of the malfunction can be checked by using the enhanced diagnostic function of the Tech 1.

Any circuitry, that is suspected as causing the intermittent complaint, should be thoroughly checked for backed out terminals, improper mating, broken locks, improperly formed or damaged terminals, poor terminal to wiring connections or physical damage to the wiring harness.

DIAGNOSTIC CHART DTC 78 — 1994 CENTURY, CUTLASS CIERA AND CUTLASS CRUISER

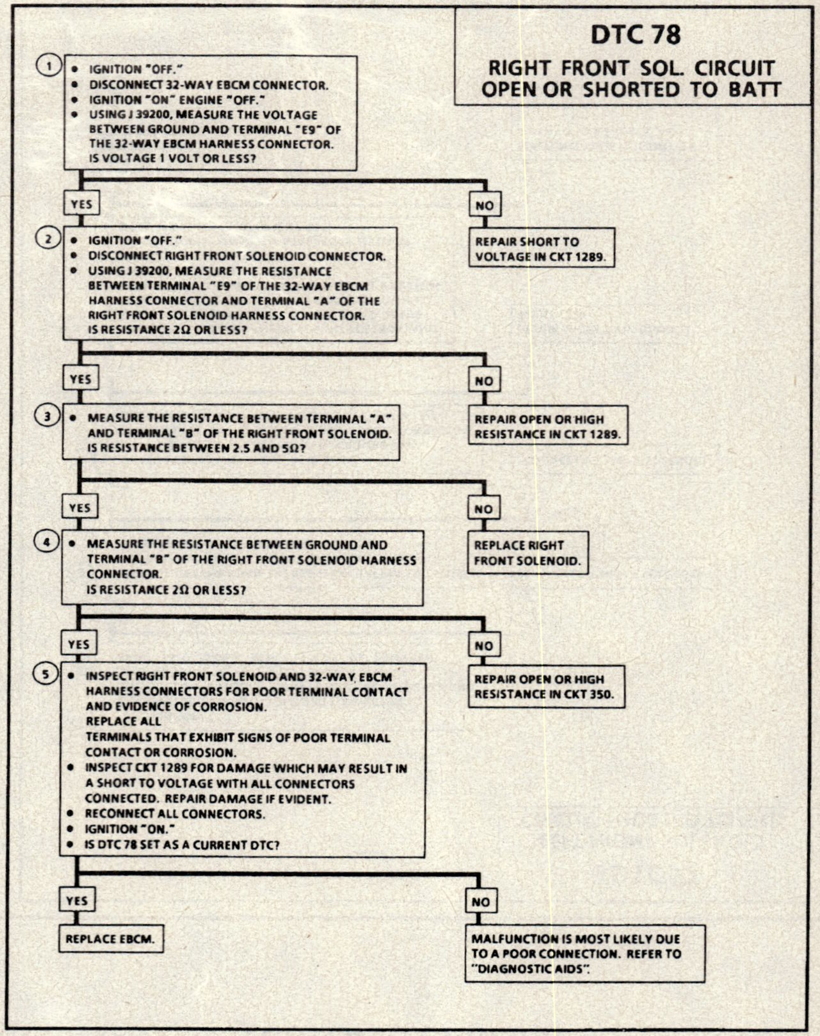

DIAGNOSTIC CHART DTC 81 — 1994 CENTURY, CUTLASS CIERA AND CUTLASS CRUISER

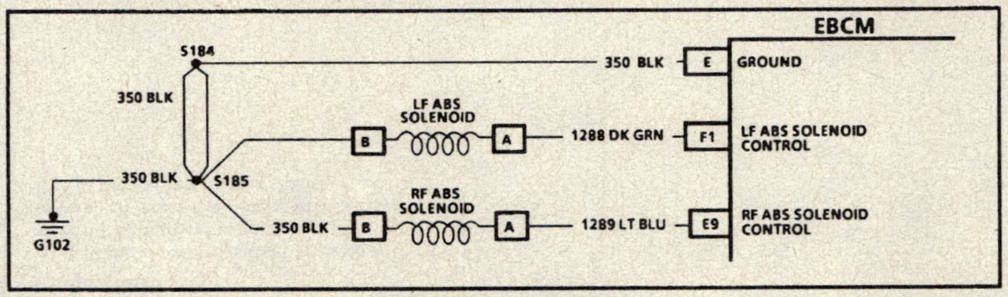

DTC 81

RIGHT FRONT SOLENOID CIRCUIT SHORTED TO GND

Circuit Description:
This DTC identifies a solenoid that cannot be energized due to an open in its driver circuitry, or a short to ground between the solenoid driver and the solenoid. These malfunctions can affect ABS operation since the flow of brake fluid to the wheel cylinder cannot be stopped, making ABS operation for that channel impossible.

DTC Will Set When: DTC 81 can be set only when the solenoid is commanded "ON." If the EBCM detects the right front solenoid control circuit voltage is out of specification, a malfunction exists.

Action Taken: A malfunction DTC is stored, ABS is disabled and the ABS warning lamp is turned "ON."

DTC Chart Test Description: Number(s) below refer to circled number(s) on the diagnostic chart.

1. Checks for a solenoid that is internally shorted to ground.
2. Checks for a solenoid that is not within proper resistance values.
3. Indicates if a short to ground exists in the solenoid circuitry.
4. Checks for a possible intermittent malfunction in the solenoid circuitry due to poor terminal contact.

Diagnostic Aids: An "Intermittent" malfunction may be caused by a poor connection, rubbed through wire insulation, or a wire that is broken inside the insulation.

The frequency of the malfunction can be checked by using the enhanced diagnostic function of the Tech 1.

Any circuitry, that is suspected as causing the intermittent complaint, should be thoroughly checked for backed out terminals, improper mating, broken locks, improperly formed or damaged terminals, poor terminal to wiring connections or physical damage to the wiring harness.

DIAGNOSTIC CHART DTC 81 — 1994 CENTURY, CUTLASS CIERA AND CUTLASS CRUISER

DTC 81
RIGHT FRONT SOLENOID CIRCUIT SHORTED TO GND

(1)
- IGNITION "OFF."
- DISCONNECT RIGHT FRONT SOLENOID CONNECTOR.
- USING J 39200, MEASURE THE RESISTANCE BETWEEN GROUND AND TERMINAL "A" OF THE RIGHT FRONT SOLENOID.

DOES J 39200 DISPLAY RESISTANCE AS "OL" (INFINITE)?

NO: REPLACE RIGHT FRONT SOLENOID.

YES:

(2)
- USING J 39200, MEASURE THE RESISTANCE BETWEEN TERMINAL "A" AND TERMINAL "B" OF THE RIGHT FRONT SOLENOID.

IS RESISTANCE BETWEEN 2.5 AND 5.0Ω?

NO: REPLACE RIGHT FRONT SOLENOID.

YES:

(3)
- DISCONNECT 32-WAY EBCM CONNECTOR.
- MEASURE THE RESISTANCE BETWEEN GROUND AND TERMINAL "E9" OF THE 32-WAY EBCM HARNESS CONNECTOR.

DOES J 39200 DISPLAY RESISTANCE AS "OL" (INFINITE)?

NO: REPAIR SHORT TO GROUND IN CKT 1289.

YES:

(4)
- INSPECT CKT 1289 AND THE RIGHT FRONT SOLENOID AND 32-WAY EBCM HARNESS CONNECTORS FOR DAMAGE WHICH MAY RESULT IN A SHORT TO GROUND WITH ALL CONNECTORS CONNECTED. REPAIR DAMAGE IF EVIDENT.
- RECONNECT ALL CONNECTORS.
- START ENGINE WITH FOOT OFF BRAKE.
- ALLOW ENGINE TO RUN FOR AT LEAST 10 SECONDS.

DOES DTC 81 SET AS A CURRENT DTC?

YES: REPLACE EBCM.

NO: MALFUNCTION IS INTERMITTENT. REFER TO "DIAGNOSTIC AIDS".

DIAGNOSTIC CHART DTC 82 — 1994 CENTURY, CUTLASS CIERA AND CUTLASS CRUISER

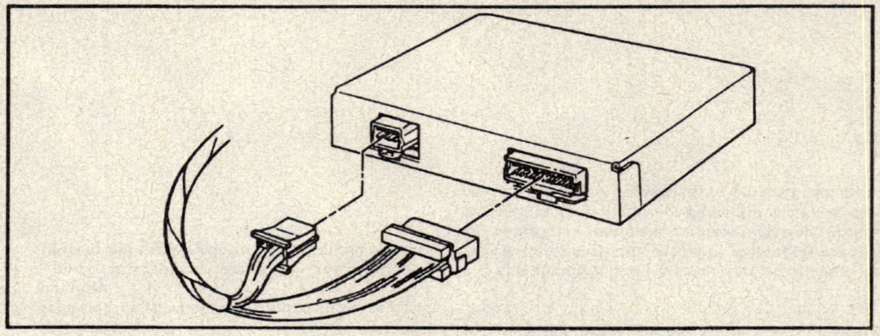

DTC 82

CALIBRATION MALFUNCTION

Circuit Description:

This DTC allows the EBCM to check for a calibration malfunction by comparing the calibration value to a known value stored in the EEPROM.

This DTC is also used as a security measure to prevent improper use of calibrations or changes to these calibrations that may alter the designed function of ABS.

DTC Will Set When: DTC 82 can be set at key-up. If the program identifier is incorrect or the memory checksum is incorrect, a malfunction exists.

Action Taken: A malfunction DTC is stored, ABS is disabled and the ABS warning lamp is turned "ON."

DTC Chart Test Description: Number(s) below refer to circled number(s) on the diagnostic chart.

1. Checks to see if the malfunction is present during diagnosis. If present, the EBCM is not functioning correctly and must be replaced.

Important

- At the time of printing, 1994 vehicles with ABS VI were not being field reprogrammed to correct this failure. A service bulletin will be issued when EEPROM reprogramming is authorized.

Diagnostic Aids: An intermittent DTC 82 may be caused by a bad cell in the EEPROM that is sensitive to temperature changes. If DTC 82 failed more than once, but is intermittent, replace EBCM.

The frequency of the malfunction can be checked by using the enhanced diagnostic function of the Tech 1.

DIAGNOSTIC CHART DTC 82 — 1994 CENTURY, CUTLASS CIERA AND CUTLASS CRUISER

DTC 82

CALIBRATION MALFUNCTION

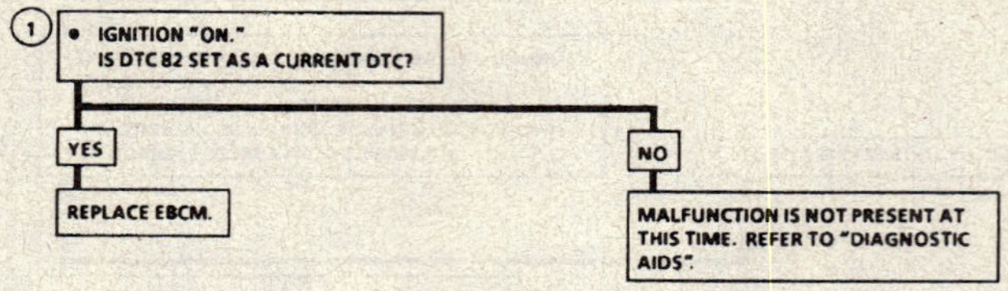

DIAGNOSTIC CHART DTC 86 — 1994 CENTURY, CUTLASS CIERA AND CUTLASS CRUISER

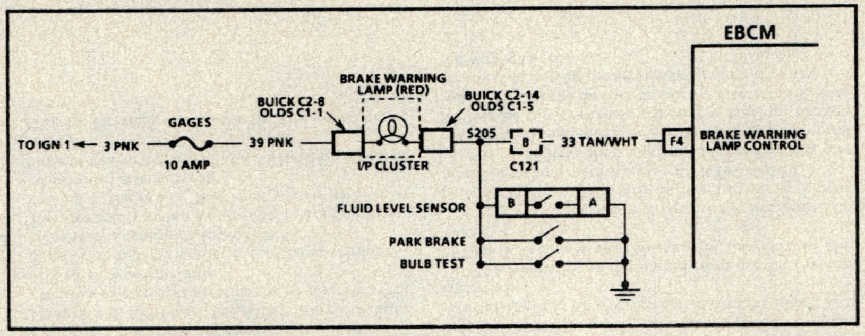

DTC 86

EBCM TURNED "ON" THE RED "BRAKE" WARNING LAMP

Circuit Description:

This DTC is provided as an information only DTC, and reflects the status of the command issued by the ABS EBCM to illuminate the red "BRAKE" warning lamp. If another DTC issues a command to illuminate the red "BRAKE" warning lamp, DTC 86 will be stored in EEPROM as a history DTC at the conclusion of the ignition cycle.

DTC Chart Test Description: Number(s) below refer to circled number(s) on the diagnostic chart.

1. This step serves to identify if a DTC other than DTC 86 commanded the red "BRAKE" warning lamp "ON."

Diagnostic Aids: Any ABS mechanical DTC that issues a command to illuminate the red "BRAKE" telltale will also result in DTC 86 being stored in EEPROM during shutdown. These DTCs are: 38, 41, 42, 46, and 51. If the motors are not in their home position, certain electrical DTCs will also command the red "BRAKE" telltale "ON." These DTCs are: 14, 16, 18, 55, 56, 57, 58, 61, 62, 63, 64, 65, and 66.

If any of these DTCs are indicated along with DTC 86, they must be corrected prior to addressing a DTC 86 malfunction.

DIAGNOSTIC CHART DTC 86 — 1994 CENTURY, CUTLASS CIERA AND CUTLASS CRUISER

DTC 86

EBCM TURNED "ON" THE RED "BRAKE" WARNING LAMP

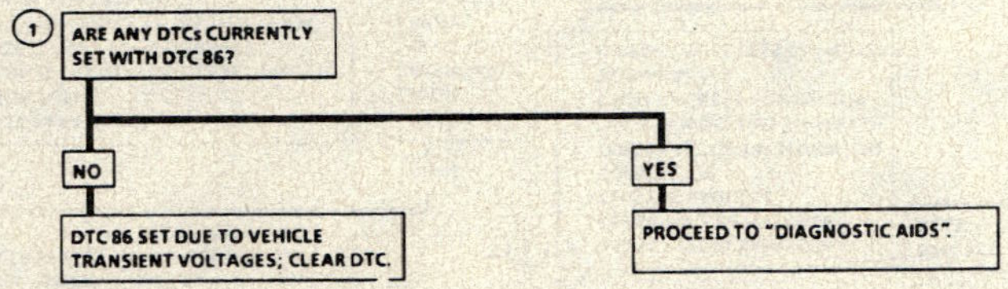

DIAGNOSTIC CHART DTC 87 — 1994 CENTURY, CUTLASS CIERA AND CUTLASS CRUISER

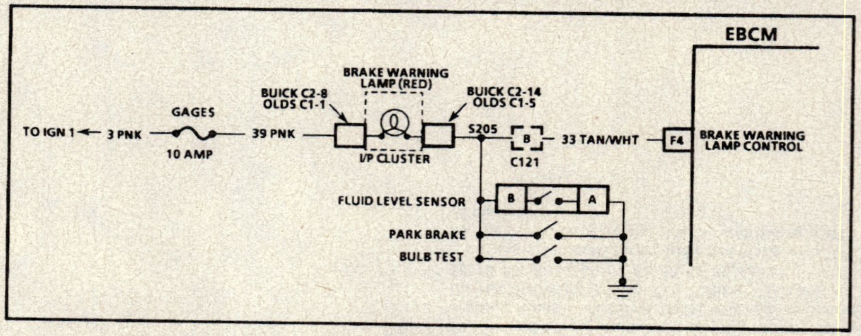

DTC 87

RED "BRAKE" WARNING LAMP CIRCUIT OPEN

Circuit Description:

This DTC is used to verify the EBCM has continuity to the red "BRAKE" warning lamp in case the EBCM must turn it "ON." This will only occur if an ABS malfunction is detected that may degrade base brake operation.

Because ABS is not the only device controlling the red "BRAKE" warning lamp (parking brake and low fluid sensor may also turn "ON" the lamp) a short to ground in this circuit cannot be detected.

DTC Will Set When: DTC 87 can be set after initialization is completed. If the red "BRAKE" warning lamp circuit voltage is out of specification indicating an open circuit, a malfunction exists.

Action Taken: A malfunction DTC is stored, ABS is not disabled.

DTC Chart Test Description: Number(s) below refer to circled number(s) on the diagnostic chart.

1. Checks the complete "BRAKE" warning lamp circuitry by attempting to manually turn "ON" the "BRAKE" warning lamp.
2. Checks for high resistance in park brake circuitry.
3. Checks for a malfunctioning EBCM.
4. Ensures malfunction is not due to poor terminal contact at the brake fluid level switch which may result in an intermittent.
5. Checks the integrity of the "BRAKE" warning lamp.
6. Checks for high resistance in the park brake circuitry.

Diagnostic Aids: An "Intermittent" malfunction may be caused by a poor connection, rubbed through wire insulation, or a wire that is broken inside the insulation.

The frequency of the malfunction can be checked by using the enhanced diagnostic function of the Tech 1.

If a Tech 1 or T-100 are not available, the parking brake handle may be lifted to verify proper lamp operation and continuity of CKTs 3 and 39.

Any circuitry, that is suspected as causing the intermittent complaint, should be thoroughly checked for backed out terminals, improper mating, broken locks, improperly formed or damaged terminals, poor terminal to wiring connections or physical damage to the wiring harness.

DIAGNOSTIC CHART DTC 87 — 1994 CENTURY, CUTLASS CIERA AND CUTLASS CRUISER

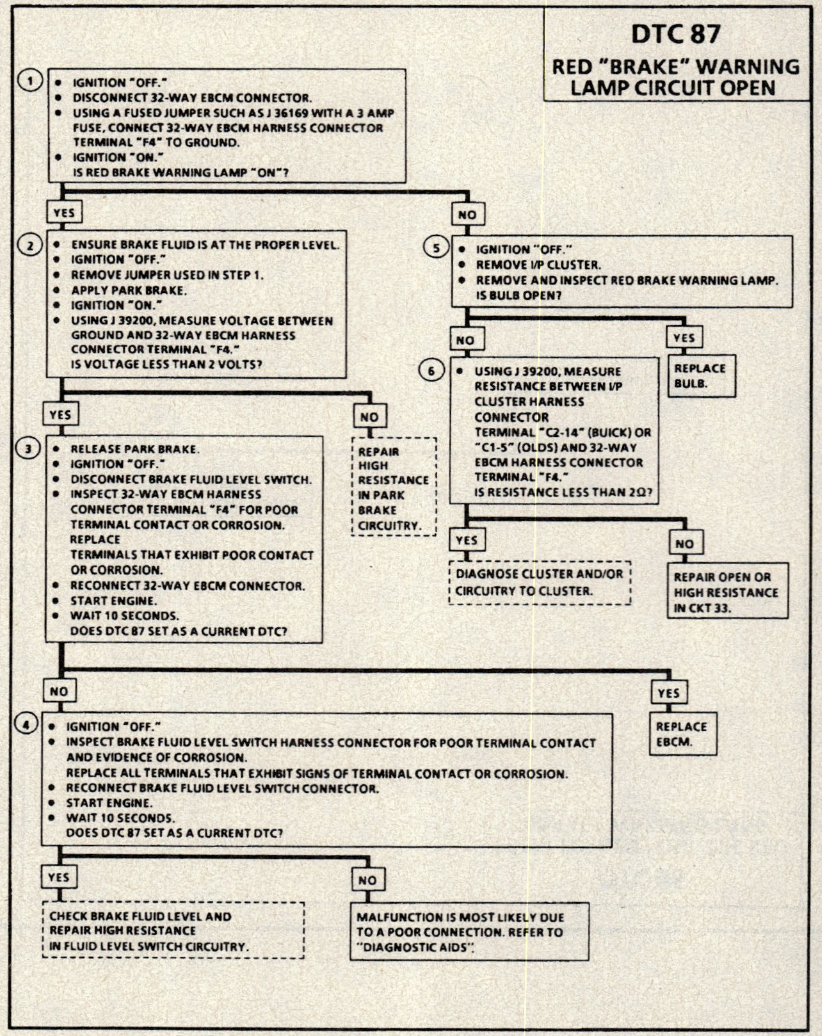

DIAGNOSTIC CHART DTC 88 — 1994 CENTURY, CUTLASS CIERA AND CUTLASS CRUISER

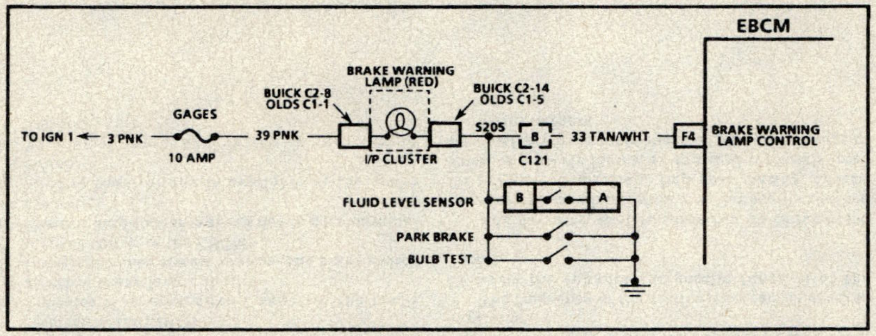

DTC 88

RED "BRAKE" WARNING LAMP CKT SHORTED TO BATT

Circuit Description:

This DTC identifies a short to battery between the EBCM and the red "BRAKE" warning lamp, or an open driver that does not allow the red "BRAKE" warning lamp to be illuminated by the EBCM. This will only occur if an ABS fault is detected that may degrade base brake operation.

DTC Will Set When: DTC 88 can be set only at the beginning of the three second bulb check. If the EBCM detects battery voltage on the red "BRAKE" warning lamp control circuit, a malfunction exists.

Action Taken: A malfunction DTC is stored. ABS is not disabled.

DTC Chart Test Description: Number(s) below refer to circled number(s) on the diagnostic chart.

1. Identifies if a malfunction is currently present.
2. With the Tech 1 connected, attempts to turn "ON" the "BRAKE" warning lamp.
3. Checks for a short to voltage in CKT 33.
4. Ensures malfunction was not due to physical damage of the circuitry.
5. Checks for DTC 86 also being set.

Diagnostic Aids: An "Intermittent" malfunction may be caused by a poor connection, rubbed through wire insulation, or a wire that is broken inside the insulation.

The frequency of the malfunction can be checked by using the enhanced diagnostic function of the Tech 1.

Any circuitry, that is suspected as causing the intermittent complaint, should be thoroughly checked for backed out terminals, improper mating, broken locks, improperly formed or damaged terminals, poor terminal to wiring connections or physical damage to the wiring harness.

DIAGNOSTIC CHART DTC 88 — 1994 CENTURY, CUTLASS CIERA AND CUTLASS CRUISER

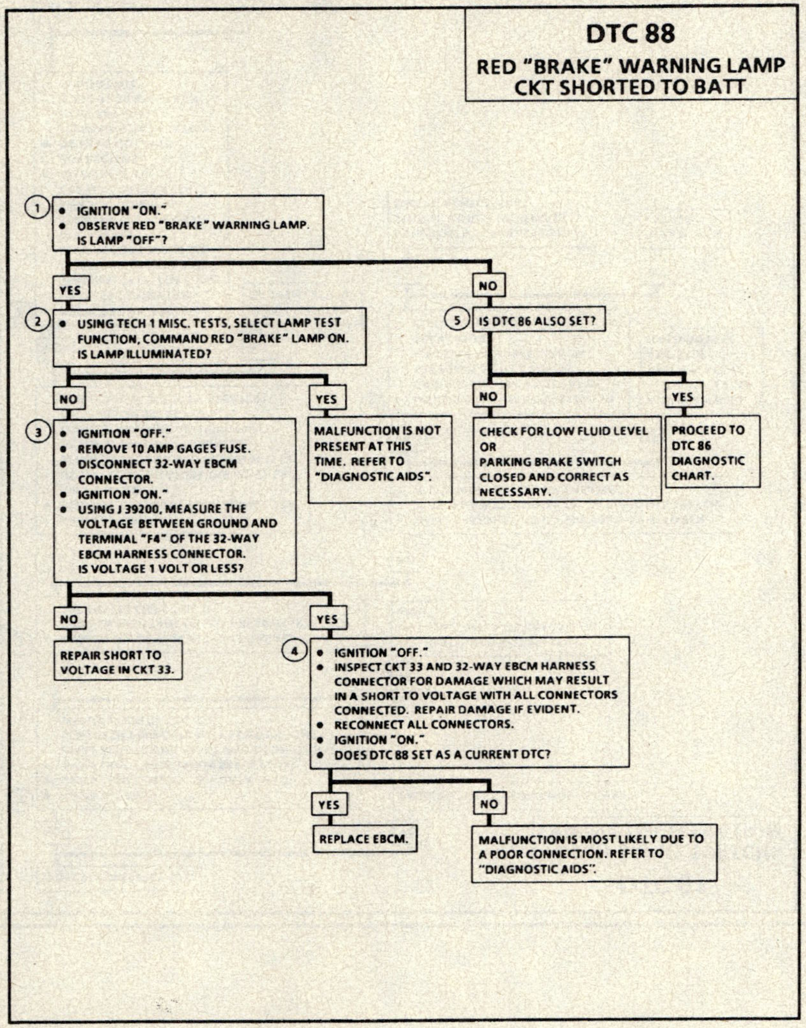

DIAGNOSTIC CHART DTC 91 — 1994 CENTURY, CUTLASS CIERA AND CUTLASS CRUISER

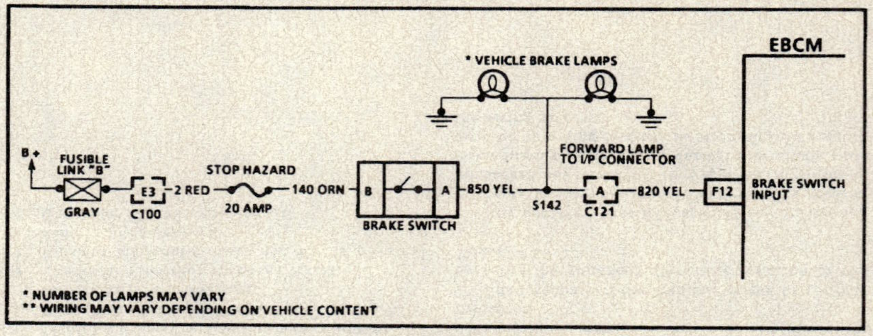

DTC 91

OPEN BRAKE SWITCH DURING DECELERATION

Circuit Description:

This DTC is used to detect an open brake switch in the non-ABS mode. The EBCM looks for deceleration rates that would indicate braking action and verifies this assumption by requiring several repeats of this detection method. In each case, ABS will not be available since no brake switch is seen.

DTC Will Set When: DTC 91 can be set if three deceleration cycles occur with the brake switch "OFF."

Action Taken: A malfunction DTC is stored, ABS is disabled and the ABS warning lamp is turned "ON."

DTC Chart Test Description: Number(s) below refer to circled number(s) on the diagnostic chart.

1. Indicates if the brake switch signal is being received by the EBCM.
2. Indicates if an open circuit exists in the brake switch or brake lamp circuitry.
3. Isolates the open circuit to either the brake switch input circuitry, or the EBCM.
4. Ensures malfunction was not due to poor terminal contact.
5. Verifies that voltage is available at the brake switch.
6. Verifies if the brake switch is functioning properly.

Diagnostic Aids: An "Intermittent" malfunction may be caused by a poor connection, rubbed through wire insulation, or a wire that is broken inside the insulation.

The frequency of the malfunction can be checked by using the enhanced diagnostic function of the Tech 1.

Any circuitry, that is suspected as causing the intermittent complaint, should be thoroughly checked for backed out terminals, improper mating, broken locks, improperly formed or damaged terminals, poor terminal to wiring connections or physical damage to the wiring harness.

DIAGNOSTIC CHART DTC 91 — 1994 CENTURY, CUTLASS CIERA AND CUTLASS CRUISER

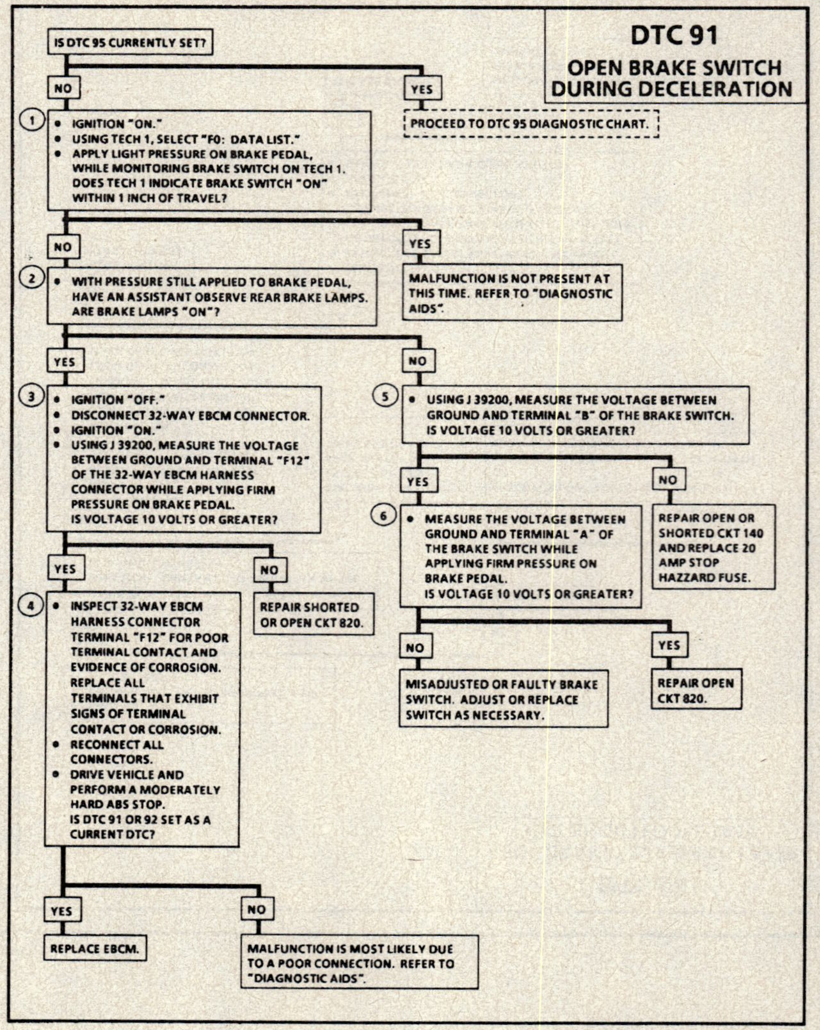

DIAGNOSTIC CHART DTC 92 — 1994 CENTURY, CUTLASS CIERA AND CUTLASS CRUISER

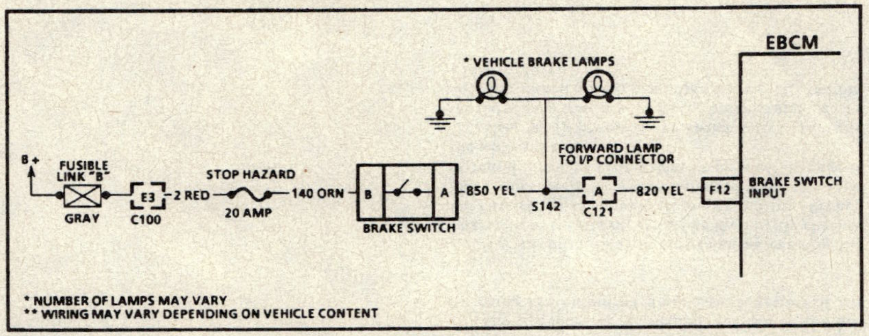

DTC 92

OPEN BRAKE SWITCH WHEN ABS WAS REQUIRED

Circuit Description:

This DTC is run to determine the proper operation of the brake switch. This is important because ABS is activated when the brake switch is "ON." If the brake switch is "OFF," ABS will never be activated. Since this malfunction is difficult to detect under normal braking conditions, this malfunction is only detected when ABS is required.

DTC Will Set When: DTC 92 can be set if the vehicle's speed is greater than 8 km/h (5 mph). If the brake was not "ON" and a release was required on two channels for 0.5 second, a malfunction exists.

Action Taken: A malfunction DTC is stored, ABS is disabled and the ABS warning lamp is turned "ON."

DTC Chart Test Description: Number(s) below refer to circled number(s) on the diagnostic chart.

1. Indicates if the brake switch signal is being received by the EBCM.
2. Indicates if an open circuit exists in the brake switch or brake lamp circuitry.
3. Isolates the open circuit to either the brake switch input circuitry, or the EBCM.
4. Ensures malfunction was not due to poor terminal contact.
5. Verifies the voltage is available at the brake switch.
6. Verifies if the brake switch is functioning properly.

Diagnostic Aids: An "Intermittent" malfunction may be caused by a poor connection, rubbed through wire insulation, or a wire that is broken inside the insulation.

The frequency of the malfunction can be checked by using the enhanced diagnostic function of the Tech 1.

Any circuitry, that is suspected as causing the intermittent complaint, should be thoroughly checked for backed out terminals, improper mating, broken locks, improperly formed or damaged terminals, poor terminal to wiring connections or physical damage to the wiring harness.

DIAGNOSTIC CHART DTC 92 — 1994 CENTURY, CUTLASS CIERA AND CUTLASS CRUISER

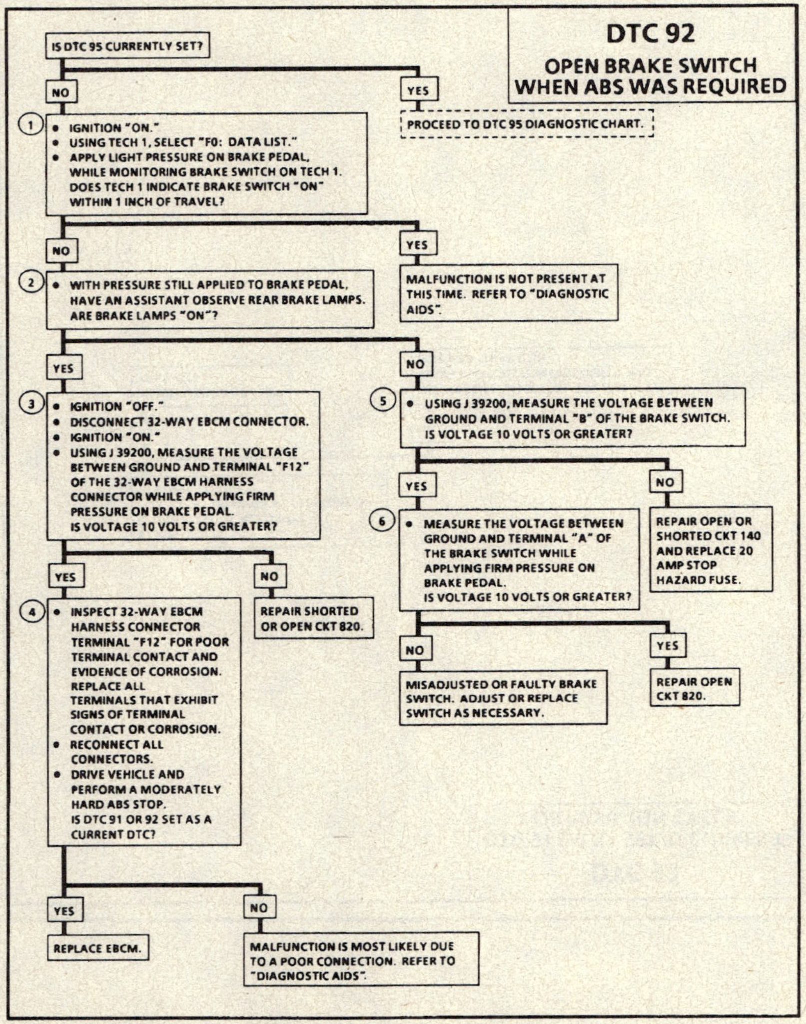

DIAGNOSTIC CHART DTC 93 — 1994 CENTURY, CUTLASS CIERA AND CUTLASS CRUISER

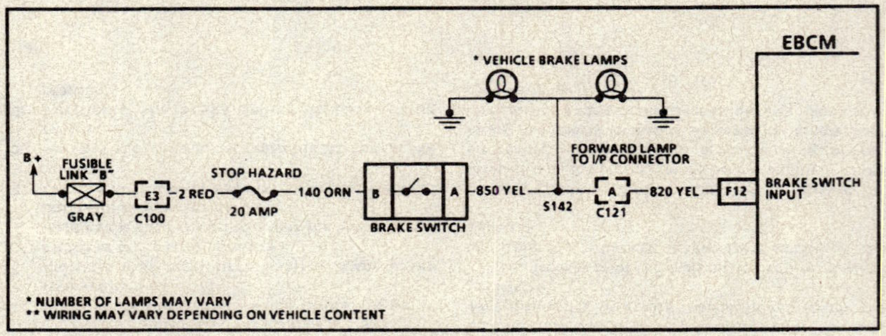

DTC 93

DTC 91 OR 92 SET IN CURRENT OR PREV. IGN CYCLE

Circuit Description:

This DTC is the second portion of DTCs 91 and 92. If DTCs 91 or 92 occured during the last ignition cycle, DTC 93 becomes a current failure during the next ignition cycle, keeping ABS disabled until a brake switch "ON" state is seen. When a change is seen during an ignition cycle in which DTC 93 is a current malfunction, DTC 91 or 92 will clear itself at the end of the current ignition cycle, and ABS will enable itself at the start of the next ignition cycle. DTC 93 alone indicates 91 or 92 occurred previously, but is intermittent, or has been corrected.

DTC Chart Test Description: Number(s) below refer to circled number(s) on the diagnostic chart.

1. Indicates which DTC (either 91 or 92) caused DTC 93 to set.

Diagnostic Aids: An "Intermittent" malfunction may be caused by a poor connection, rubbed through wire insulation, or a wire that is broken inside the insulation.

The frequency of the malfunction can be checked by using the enhanced diagnostic function of the Tech 1.

Any circuitry, that is suspected as causing the intermittent complaint, should be thoroughly checked for backed out terminals, improper mating, broken locks, improperly formed or damaged terminals, poor terminal to wiring connections or physical damage to the wiring harness.

Also, verify proper brake switch operation using the data list of the Tech 1. As the brake is applied, the data list should display the brake switch "ON" within 1 inch of travel.

DIAGNOSTIC CHART DTC 93 — 1994 CENTURY, CUTLASS CIERA AND CUTLASS CRUISER

DTC 93

DTC 91 OR 92 SET IN CURRENT OR PREV. IGN CYCLE

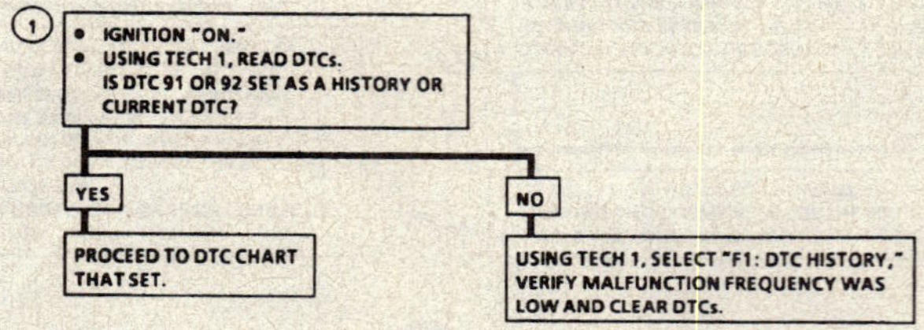

DIAGNOSTIC CHART DTC 94 — 1994 CENTURY, CUTLASS CIERA AND CUTLASS CRUISER

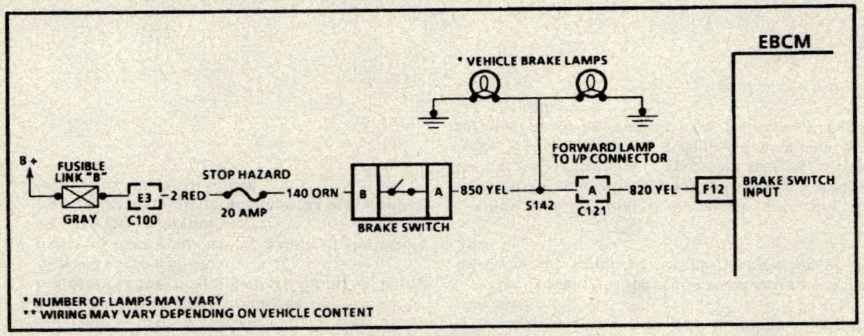

DTC 94

BRAKE SWITCH CONTACTS ALWAYS CLOSED

Circuit Description:

This DTC is run to determine the proper operation of the brake switch. This is important because ABS is activated when the brake switch is "ON" and turned "OFF" when the brake is "OFF." If the brake switch is always "ON," ABS operation will always be requested resulting in potential modulator cycling on rough roads. Additionally, this malfunction will most likely result in a dead battery (due to the brake lights staying "ON") if the driver is not informed of this malfunction.

DTC Will Set When: DTC 94 can be set when the vehicle reaches at least 40 km/h (25 mph). If the brake was never "OFF" during two consecutive drive cycles, a malfunction exists.

Action Taken: A malfunction DTC is stored. ABS is not disabled; however, the ABS warning will flash to indicate a malfunction exists.

DTC Chart Test Description: Number(s) below refer to circled number(s) on the diagnostic chart.

1. Checks if the malfunction is currently present.
2. Isolates the cause of the malfunction to either a malfunctioning or misadjusted brake switch, or a short to voltage in the brake switch circuitry.
3. Checks for unwanted voltage on the brake switch input circuit.
4. Ensures malfunction is not due to poor terminal contact or EBCM.

Diagnostic Aids: An "Intermittent" malfunction may be caused by a poor connection, rubbed through wire insulation, or a wire that is broken inside the insulation.

The frequency of the malfunction can be checked by using the enhanced diagnostic function of the Tech 1.

Any circuitry, that is suspected as causing the intermittent complaint, should be thoroughly checked for backed out terminals, improper mating, broken locks, improperly formed or damaged terminals, poor terminal to wiring connections or physical damage to the wiring harness.

DIAGNOSTIC CHART DTC 94 — 1994 CENTURY, CUTLASS CIERA AND CUTLASS CRUISER

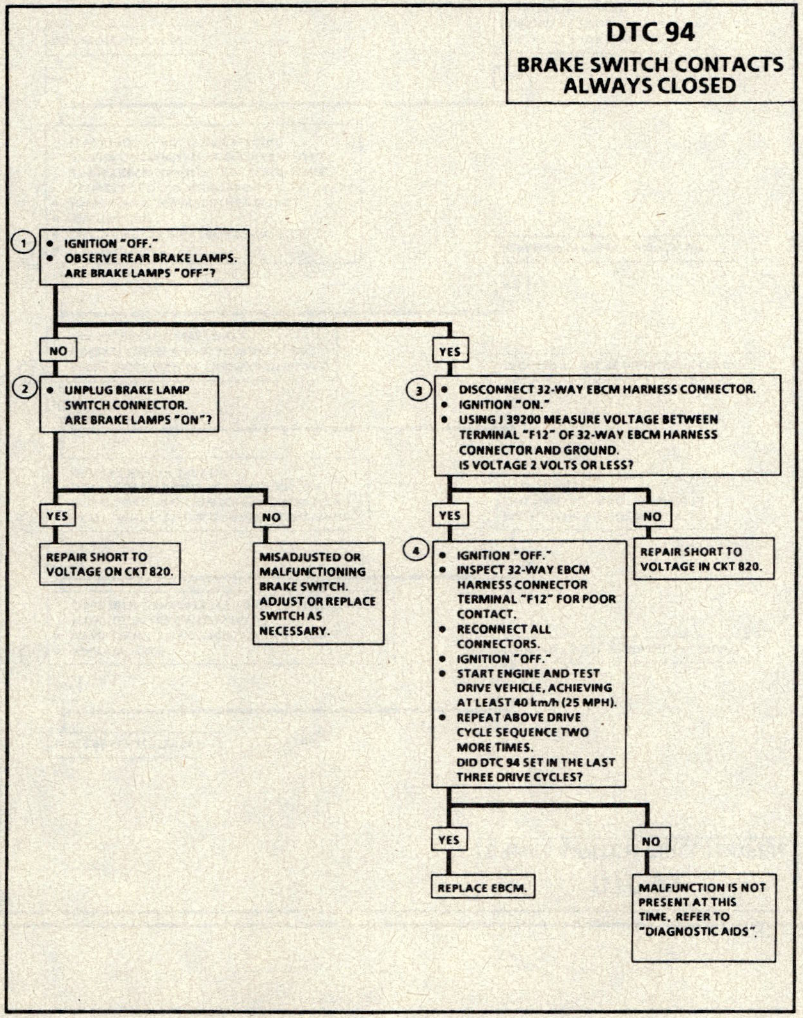

DIAGNOSTIC CHART DTC 95 — 1994 CENTURY, CUTLASS CIERA AND CUTLASS CRUISER

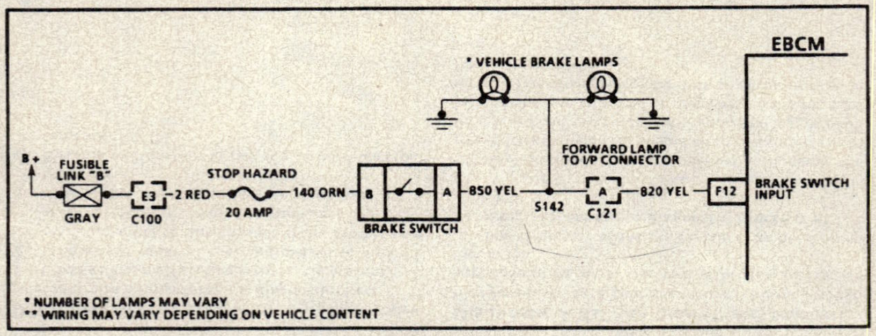

DTC 95

BRAKE SWITCH CIRCUIT OPEN

Circuit Description:
This DTC is used to identify open brake switch circuitry that prevents the brake switch input to the EBCM from changing states when the brake is applied. This DTC is used in conjunction with DTCs 91 and 92 to determine the cause of an open brake switch malfunction.

DTC Will Set When: DTC 95 can be set after initialization is completed. If the brake switch input voltage is out of specification for one second indicating an open circuit, a malfunction exists.

Action Taken: A malfunction DTC is stored, ABS is disabled and the ABS warning lamp is turned "ON."

DTC Chart Test Description: Number(s) below refer to circled number(s) on the diagnostic chart.

1. This step is used to confirm that an open in the brake switch circuitry currently exists.
2. Indicates if the brake switch signal is being received by the EBCM.
3. Indicates if an open circuit exists in the brake switch or brake lamp circuitry.
4. Isolates the open circuit to either the brake switch input circuitry, or the EBCM.

Diagnostic Aids: An "Intermittent" malfunction may be caused by a poor connection, rubbed through wire insulation, or a wire that is broken inside the insulation.

The frequency of the malfunction can be checked by using the enhanced diagnostic function of the Tech 1.

Any circuitry, that is suspected as causing the intermittent complaint, should be thoroughly checked for backed out terminals, improper mating, broken locks, improperly formed or damaged terminals, poor terminal to wiring connections or physical damage to the wiring harness.

DIAGNOSTIC CHART DTC 95 — 1994 CENTURY, CUTLASS CIERA AND CUTLASS CRUISER

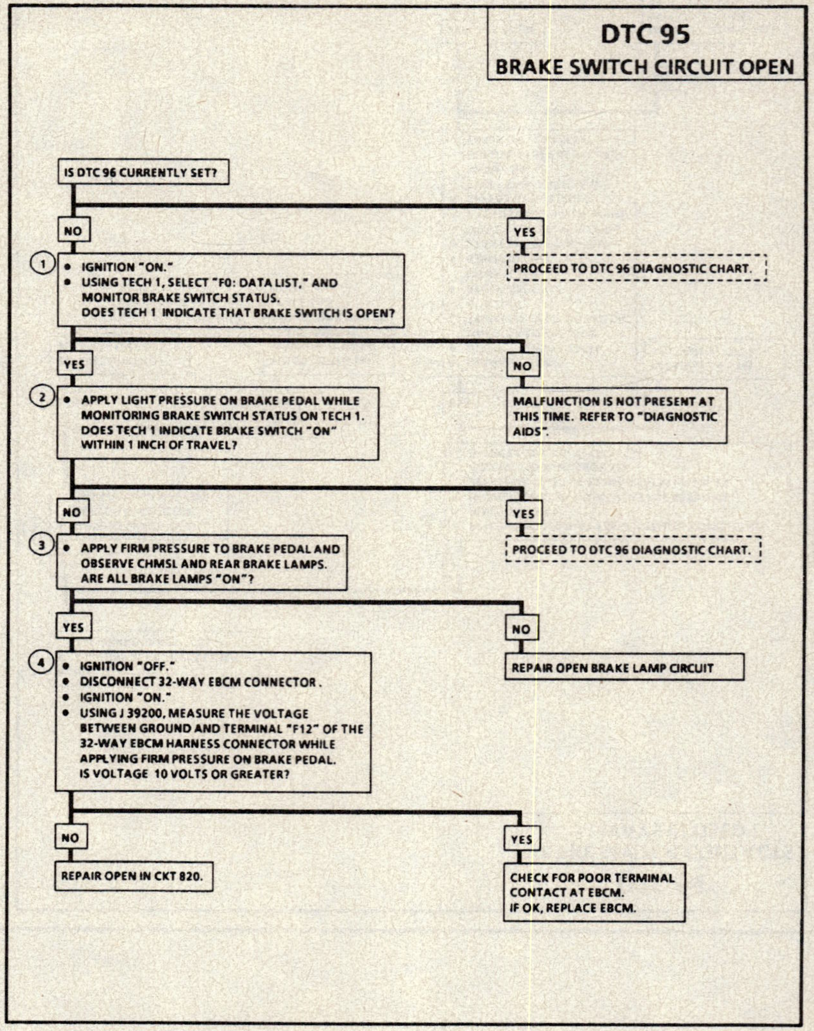

DIAGNOSTIC CHART DTC 96 — 1994 CENTURY, CUTLASS CIERA AND CUTLASS CRUISER

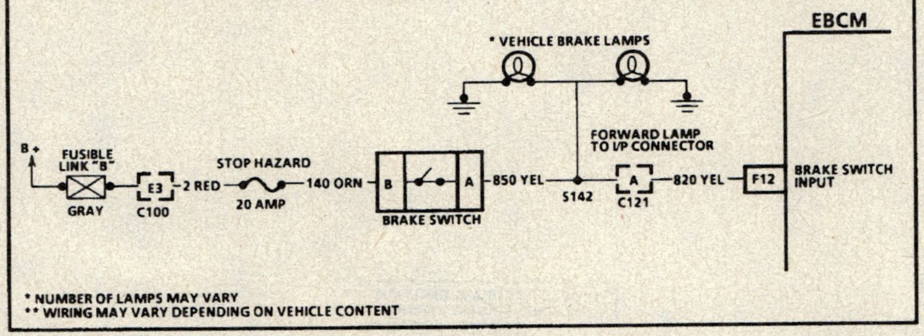

DTC 96
REAR BRAKE LAMP CIRCUIT OPEN

Circuit Description:
This DTC is designed to isolate the cause of a DTC 95 malfunction and indicate to the driver ABS is still available. If DTC 95 occurs with DTC 96, the brake lamps are open, the brake lamp grounds are open, or the Center High Mounted Stop Lamp (CHMSL) circuit is open during 4-way flasher use. The presence of battery voltage at the brake switch input indicates a valid brake switch input is still available.

DTC Will Set When: DTC 96 can be set only after DTC 95 has been set. If the EBCM detects battery voltage on the brake switch input circuit for 0.5 second, a malfunction exists.

Action Taken: A malfunction DTC is stored. ABS is not disabled.

DTC Chart Test Description: Number(s) below refer to circled number(s) on the diagnostic chart.

1. As a result of a malfunction of an additional brake switch circuit DTC, this DTC may be set. To ensure proper diagnosis, any additional brake switch DTCs must be repaired first.
2. Identifies if the malfunction is currently present in the brake circuit.

Diagnostic Aids: An "Intermittent" malfunction may be caused by a poor connection, rubbed through wire insulation, or a wire that is broken inside the insulation.

The frequency of the malfunction can be checked by using the enhanced diagnostic function of the Tech 1.

Any circuitry, that is suspected as causing the intermittent complaint, should be thoroughly checked for backed out terminals, improper mating, broken locks, improperly formed or damaged terminals, poor terminal to wiring connections or physical damage to the wiring harness.

DIAGNOSTIC CHART DTC 96 — 1994 CENTURY, CUTLASS CIERA AND CUTLASS CRUISER

DTC 96
REAR BRAKE LAMP CIRCUIT OPEN

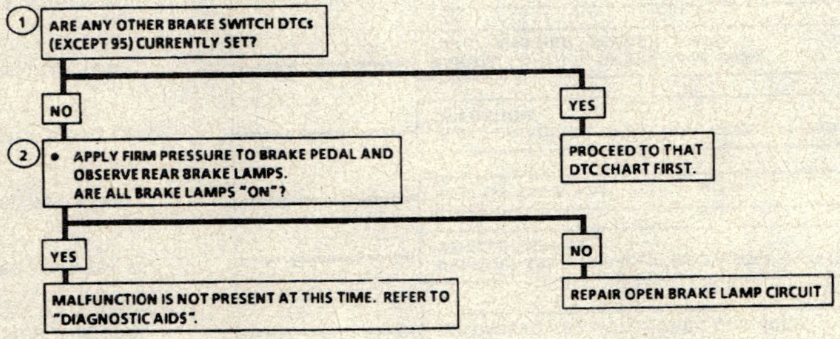

DIAGNOSTIC PROCESS CHART — 1995 CENTURY, CUTLASS CIERA AND CUTLASS CRUISER

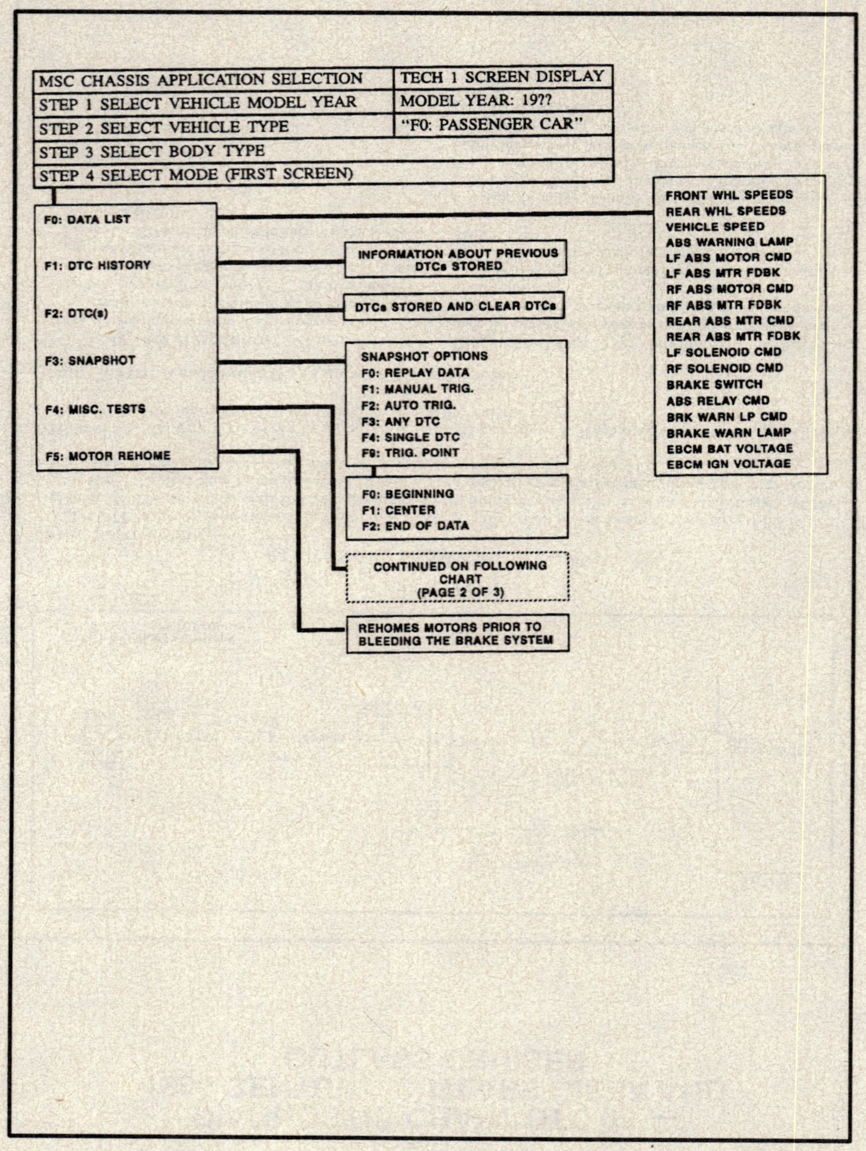

DIAGNOSTIC PROCESS CHART — 1995 CENTURY, CUTLASS CIERA AND CUTLASS CRUISER

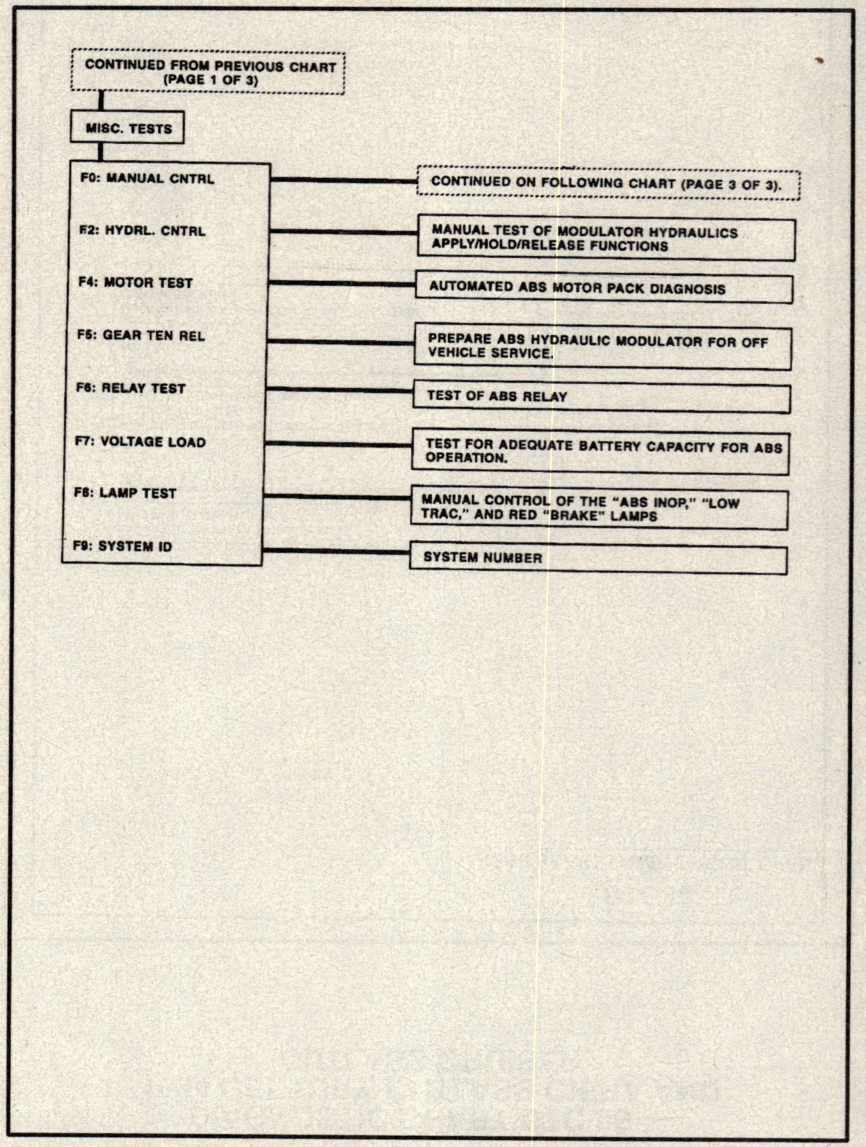

DIAGNOSTIC PROCESS CHART — 1995 CENTURY, CUTLASS CIERA AND CUTLASS CRUISER

CONTINUED FROM PREVIOUS CHART (PAGE 2 OF 3)

MANUAL CONTROL OF SOLENOIDS AND MOTORS.

↑ = TURN ON
↓ = TURN OFF
YES TO CONTINUE

SELECT FUNCTION
F0: LF APPLY
F1: LF RELEASE
F2: LF SOLENOID
F4: RF APPLY
F5: RF RELEASE
F6: RF SOLENOID
F8: REAR APPLY
F9: REAR RELEASE

MOTOR APPLY
COMMAND: XX AMPS
FEEDBACK: XX AMPS
ABS RELAY: ON/OFF

MOTOR RELEASE
COMMAND: XX AMPS
FEEDBACK: XX AMPS
ABS RELAY: ON/OFF

SOLENOID
COMMAND: ON/OFF
ABS RELAY: ON/OFF

DIAGNOSTIC CHART SYSTEM CHECK — 1995 CENTURY, CUTLASS CIERA AND CUTLASS CRUISER

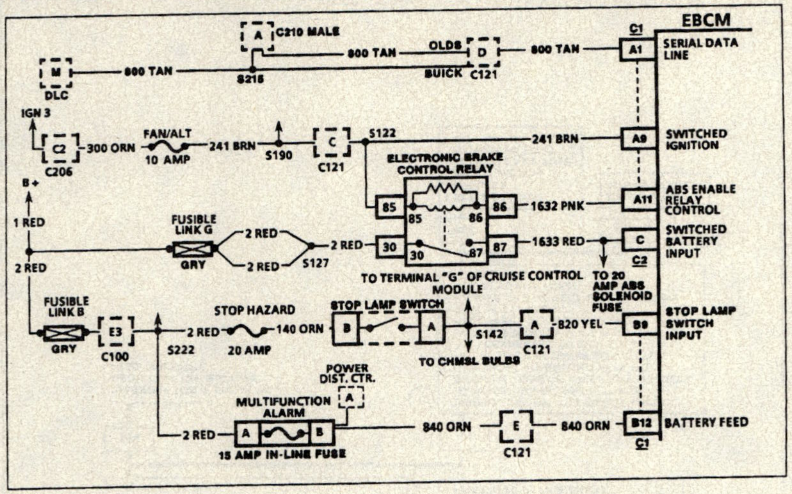

DIAGNOSTIC SYSTEM CHECK

System Description:

The diagnostic system check is an organized approach to identifying a problem created by an Antilock Brake System (ABS) malfunction. It must be the starting point for any ABS complaint diagnosis because it directs the service technician to the next logical step in diagnosing the complaint.

Serial data is transmitted by the EBCM through terminal "A1". The EBCM is supplied switched ignition voltage through terminal "A9", battery voltage through terminal "B12", and switched battery voltage through terminal "C". The EBCM ground is provided through terminal "D".

Chart Test Description:

Number(s) below refer to circled number(s) on the diagnostic chart.

1. Indicates if EBCM is transmitting data.
2. Checks for a serial data line malfunction.
3. Checks for proper Tech 1 connection to DLC.
4. Checks for open multifunction alarm fuse 15 amp.
5. Checks for open fan/alt 10 amp fuse.
6. Checks for high resistance in the EBCM ground circuit.
7. Checks for current DTCs.
8. Checks for proper ABS warning lamp operation.
9. Checks for history DTCs.
10. Checks for possible short to ground in the battery feed circuitry.
11. Checks for possible short to ground in the EBCM.
12. Ensures short to ground is not due to physical damage of the circuitry.

Diagnostic Aids:

Excessive resistance in the ground or power supply circuits will not allow communication with the EBCM. If communication with the EBCM is not possible, ensure the ABS ground connection is good and that there is no excessive resistance in any of the power supply circuits.

DIAGNOSTIC CHART SYSTEM CHECK — 1995 CENTURY, CUTLASS CIERA AND CUTLASS CRUISER

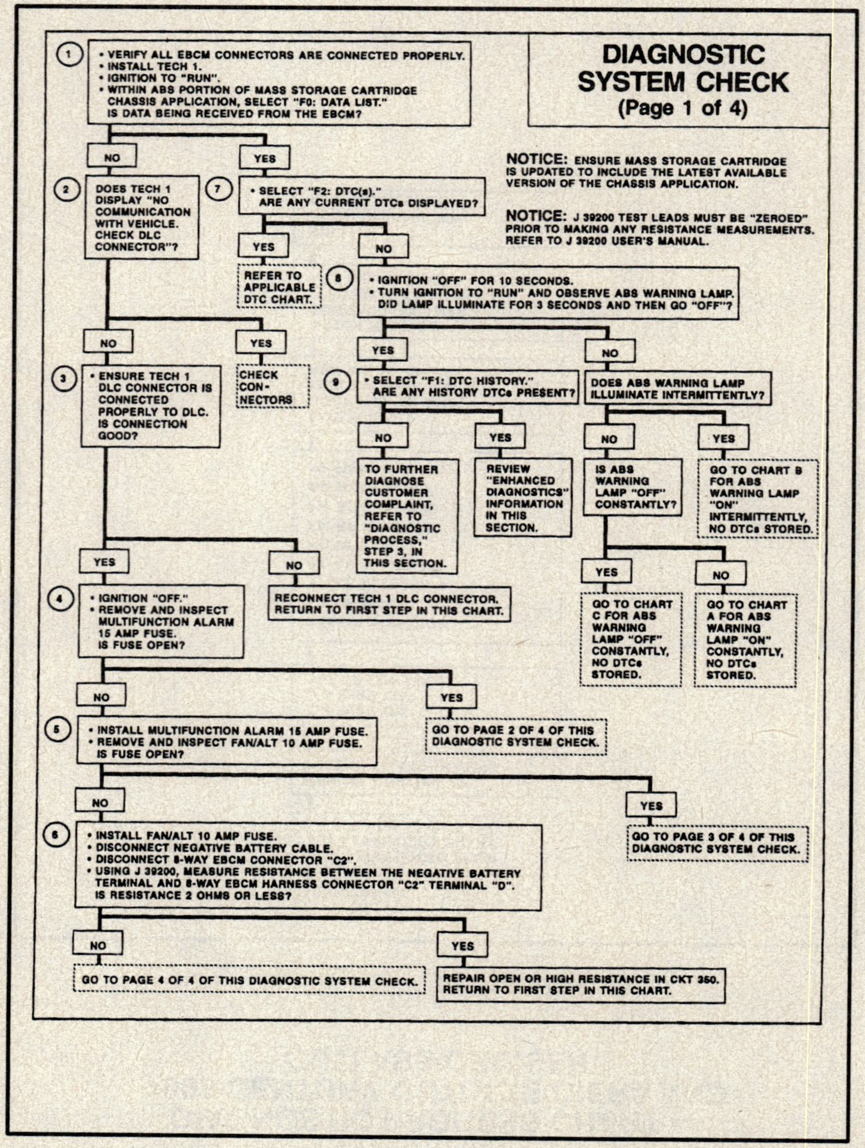

DIAGNOSTIC CHART SYSTEM CHECK — 1995 CENTURY, CUTLASS CIERA AND CUTLASS CRUISER

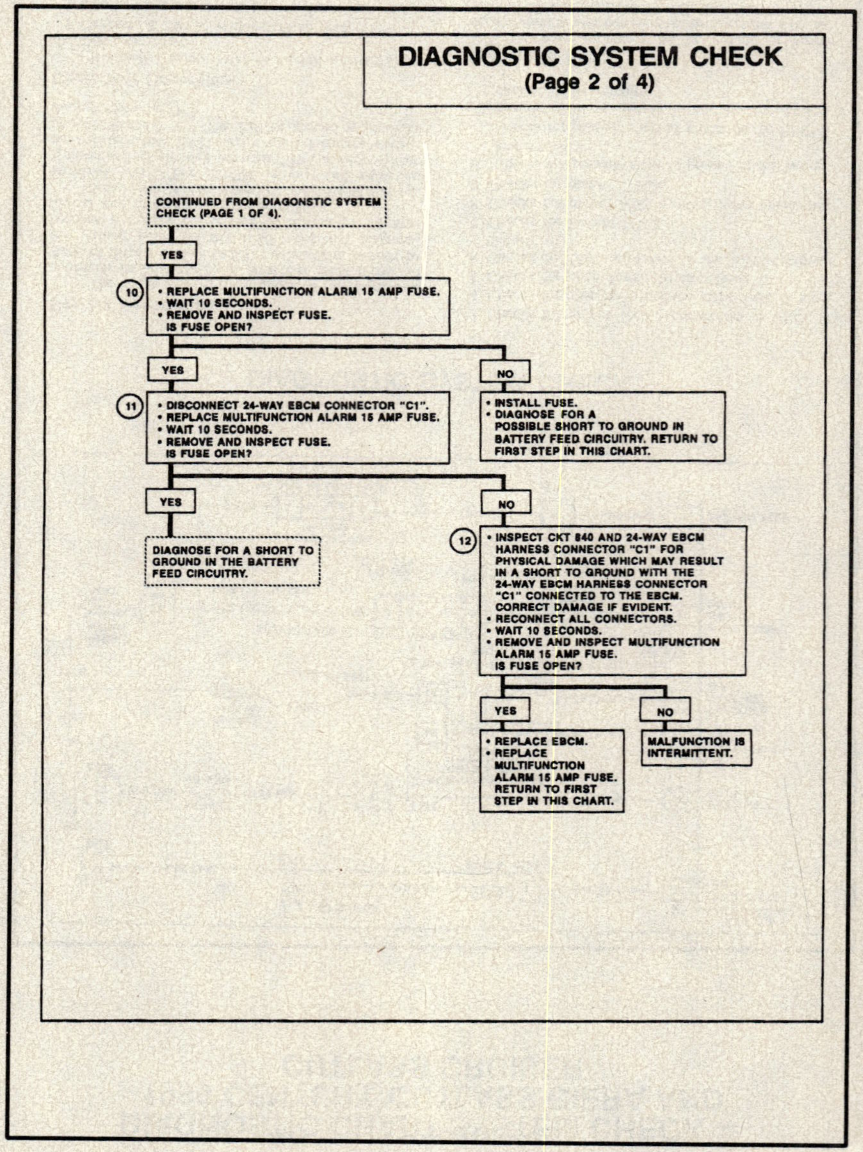

DIAGNOSTIC CHART SYSTEM CHECK — 1995 CENTURY, CUTLASS CIERA AND CUTLASS CRUISER

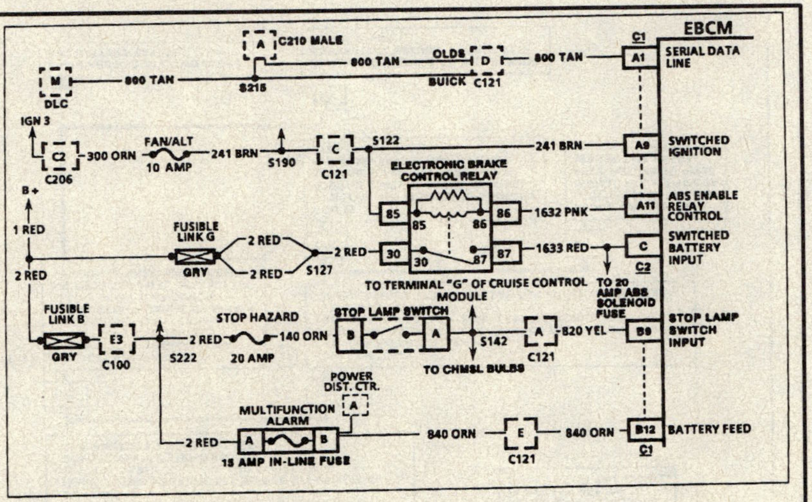

DIAGNOSTIC SYSTEM CHECK

System Description:

The diagnostic system check is an organized approach to identifying a problem created by an Antilock Brake System (ABS) malfunction. It must be the starting point for any ABS complaint diagnosis because it directs the service technician to the next logical step in diagnosing the complaint.

Serial data is transmitted by the EBCM through terminal "A1". The EBCM is supplied switched ignition voltage through terminal "A9", battery voltage through terminal "B12", and switched battery voltage through terminal "C". The EBCM ground is provided through terminal "D".

Chart Test Description:

Number(s) below refer to circled number(s) on the diagnostic chart.

13. Checks for possible short to ground in the switched ignition circuitry.
14. Isolates the short to ground to either the circuitry or the EBCM.
15. Ensures short to ground is not due to physical damage of the circuitry.
16. Checks for proper battery voltage at the EBCM.
17. Checks for high resistance in the battery feed circuitry.
18. Checks for high resistance in the switched ignition circuitry.
19. Ensures malfunction is not due to poor terminal contact.
20. Checks for high resistance in the switched ignition circuit between the fuse block and the EBCM.

Diagnostic Aids:

Excessive resistance in the ground or power supply circuits will not allow communication with the EBCM. If communication with the EBCM is not possible, ensure the ABS ground connection is good and that there is no excessive resistance in any of the power supply circuits.

DIAGNOSTIC CHART SYSTEM CHECK — 1995 CENTURY, CUTLASS CIERA AND CUTLASS CRUISER

DIAGNOSTIC SYSTEM CHECK
(Page 3 of 4)

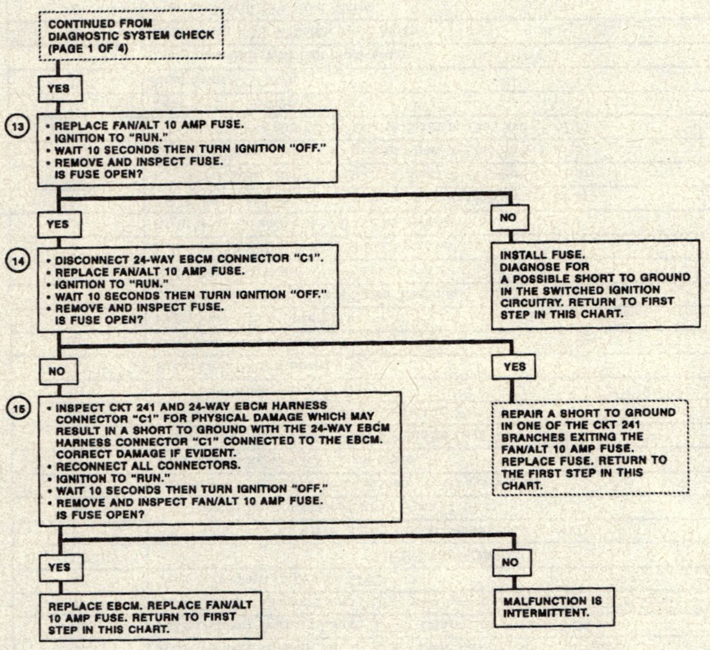

DIAGNOSTIC CHART SYSTEM CHECK — 1995 CENTURY, CUTLASS CIERA AND CUTLASS CRUISER

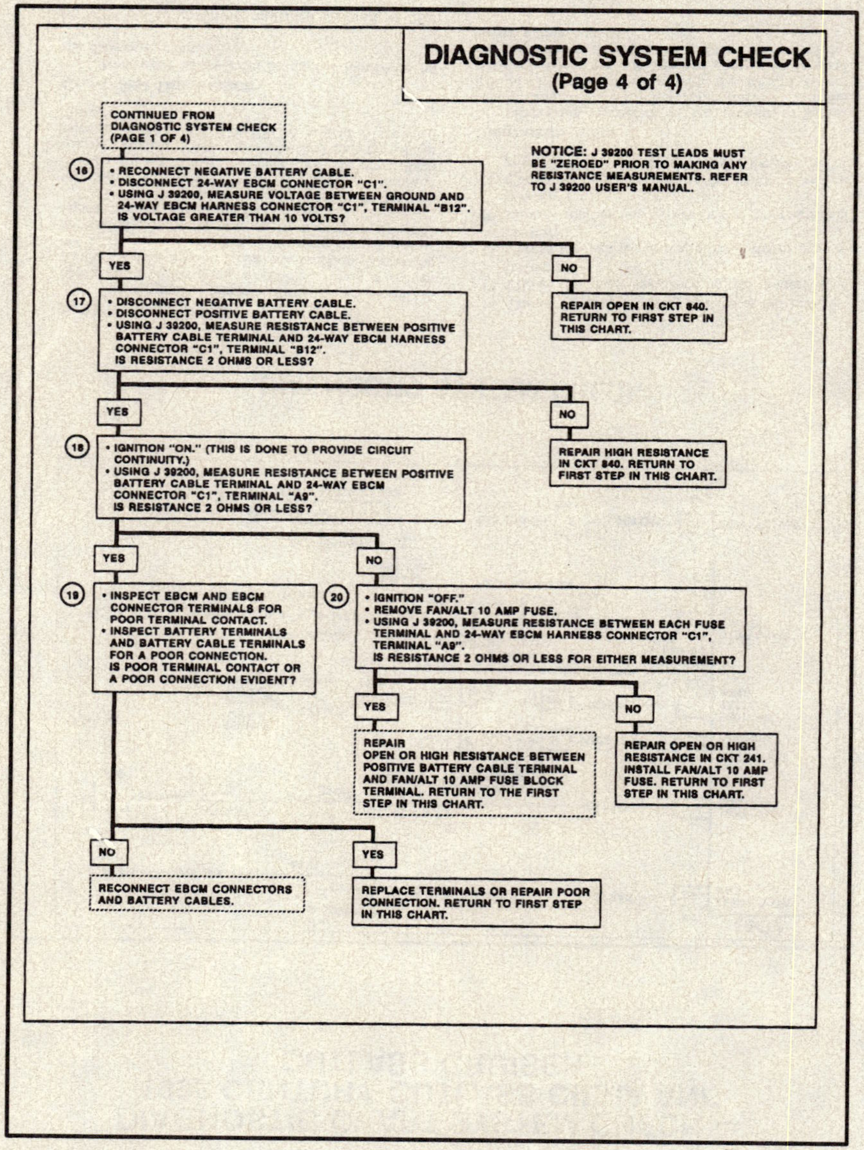

DIAGNOSTIC TROUBLE CODE CHART — 1995 CENTURY, CUTLASS CIERA AND CUTLASS CRUISER

DIAGNOSTIC TROUBLE CODE AND SYMPTOM TABLE	
CHART	**SYMPTOM**
A	ABS (Amber) Warning Lamp "ON" Constantly, No DTCs Stored
B	ABS (Amber) Warning Lamp "ON" Intermittently, No DTCs Stored
C	ABS (Amber) Warning Lamp "OFF" Constantly, No DTCs Stored
D	LOW TRAC Indicator Lamp "ON" with Ignition "OFF," No DTCs Stored
E	Tech 1 Displays Undefined DTCs
DTC	**DESCRIPTION**
11	ABS Warning or Low Trac Lamp Circuit Malfunction (1 of 2)
14	ABS Enable Relay Contact Circuit Open (1 of 3)
15	ABS Rly Contact CKT Short to Batt or Always Closed
16	ABS Enable Relay Coil Circuit Open
17	ABS Enable Relay Coil Circuit Shorted to Gnd
18	ABS Enable Relay Coil Circuit Shorted to Batt
21	Left Front Wheel Speed = 0
22	Right Front Wheel Speed = 0
23	Left Rear Wheel Speed = 0
24	Right Rear Wheel Speed = 0
25	Left Front Excessive Wheel Speed Variation
26	Right Front Excessive Wheel Speed Variation
27	Left Rear Excessive Wheel Speed Variation
28	Right Rear Excessive Wheel Speed Variation
32	LF Wheel Speed Snr CKT Open or Short to Gnd/Batt (1 of 3)
33	RF Wheel Speed Snr CKT Open or Short to Gnd/Batt (1 of 3)
34	LR Wheel Speed Snr CKT Open or Short to Gnd/Batt (1 of 3)
35	RR Wheel Speed Snr CKT Open or Short to Gnd/Batt (1 of 3)
36	Low System Voltage (1 of 2)
37	High System Voltage
38	Left Front ESB Will Not Hold Motor
41	Right Front ESB Will Not Hold Motor
42	Rear ESB Will Not Hold Motor
44	Left Front Channel Will Not Move
45	Right Front Channel Will Not Move
46	Rear Channel Will Not Move
47	Left Front Motor Free Spins (1 of 2)
48	Right Front Motor Free Spins (1 of 2)
51	Rear ABS Motor Free Spins (1 of 2)
52	Left Front Channel in Release Too Long
53	Right Front Channel in Release Too Long
54	Rear Channel in Release Too Long
55	EBCM Malfunction
56	Left Front ABS Motor Circuit Open
57	Left Front ABS Motor Circuit Shorted to Ground
58	Left Front ABS Motor Circuit Shorted to Battery
61	Right Front ABS Motor Circuit Open
62	Right Front ABS Motor Circuit Shorted to Gnd

DIAGNOSTIC TROUBLE CODE CHART — 1995 CENTURY, CUTLASS CIERA AND CUTLASS CRUISER

DIAGNOSTIC TROUBLE CODE AND SYMPTOM TABLE	
63	Right Front ABS Motor Circuit Shorted to Batt
64	Rear ABS Motor Circuit Open
65	Rear ABS Motor Circuit Shorted to Ground
66	Rear ABS Motor Circuit Shorted to Battery
76	Left Front Sol Circuit Open or Shorted to Gnd
77	Left Front Solenoid Circuit Shorted to Batt
78	Right Front Sol Circuit Open or Shorted to Gnd
81	Right Front Solenoid Circuit Shorted to Batt
82	Calibration Malfunction
86	EBCM Turned "ON" the Red "BRAKE" Warning Lamp
87	Red "BRAKE" Warning Lamp Circuit Open
91	Open Brake Switch During Deceleration
92	Open Brake Switch When ABS Was Required
93	DTC 91 or 92 Set in Current or Prev. Ign Cycle
94	Brake Switch Contacts Always Closed
95	Brake Switch Circuit Open
96	Brake Lamps Circuit Open

ELECTRICAL CONNECTOR TERMINAL IDENTIFICATION — 1995 CENTURY, CUTLASS CIERA AND CUTLASS CRUISER

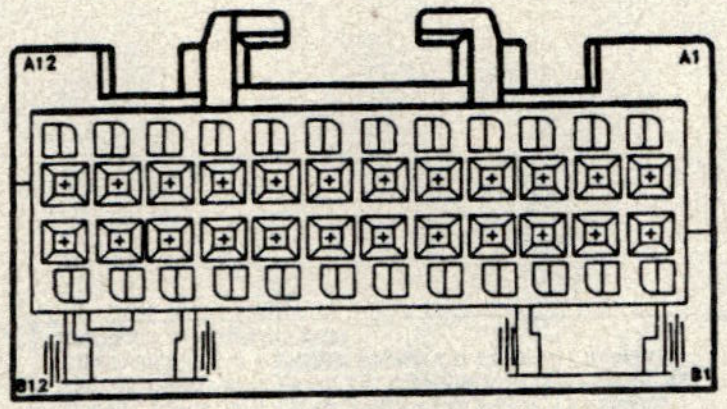

12110088
24-WAY F MICRO-PACK 100 SERIES
LT GRY
ELECTRONIC BRAKE CONTROL MODULE (EBCM) CONNECTOR "C1"

CONNECTOR TERMINAL	CIRCUIT NO.	COLOR	CIRCUIT DESCRIPTION
A1	800	TAN	SERIAL DATA LINE
A2	882	BRN	RR WHEEL SIGNAL HIGH
A3	884	BLK	LR WHEEL SIGNAL HIGH
A4	872	DK GRN	RF WHEEL SIGNAL HIGH
A5	830	LT BLU	LF WHEEL SIGNAL HIGH
A6	1537	DK BLU	ABS ACTIVE LAMP CONTROL
A7	867	LT GRN	ABS WARNING LAMP CONTROL
A8	PLUGGED		NOT USED
A9	241	BRN	SWITCHED IGNITION
A10	1288	DK GRN	LF ABS SOLENOID CONTROL
A11	1632	PNK	ABS ENABLE RELAY CONTROL
A12	PLUGGED		NOT USED
B1	883	WHT	RR WHEEL SIGNAL LOW
B2	885	RED	LR WHEEL SIGNAL LOW
B3	833	TAN	RF WHEEL SIGNAL LOW
B4	873	YEL	LF WHEEL SIGNAL LOW
B5	33	TAN/WHT	BRAKE WARNING LAMP CONTROL
B6	1289	LT BLU	RF ABS SOLENOID CONTROL
B7	PLUGGED		NOT USED
B8	PLUGGED		NOT USED
B9	820	YEL	BRAKE SWITCH INPUT
B10	PLUGGED		NOT USED
B11	PLUGGED		NOT USED
B12	840	ORN	BATTERY FEED

ELECTRICAL CONNECTOR TERMINAL IDENTIFICATION — 1995 CENTURY, CUTLASS CIERA AND CUTLASS CRUISER

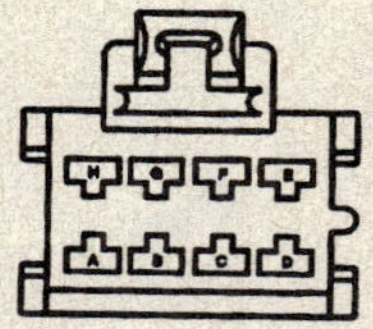

8-WAY F
METRI-PACK 280 SERIES
GRY CONNECTOR "C2"

PIN	CIRCUIT NO.	COLOR	CIRCUIT
A	1282	PPL	RF ABS MOTOR HIGH
B	1283	BLK	RF ABS MOTOR LOW
C	1633	RED	SWITCHED BATTERY INPUT
D	350	BLK	GROUND
E	1285	ORN	REAR ABS MOTOR LOW
F	1284	DK GRN	REAR ABS MOTOR HIGH
G	1280	BLK	LF ABS MOTOR HIGH
H	1281	PNK	LF ABS MOTOR LOW

DIAGNOSTIC CHART A — 1995 CENTURY, CUTLASS CIERA AND CUTLASS CRUISER

CHART A
ABS (AMBER) WARNING LAMP "ON" CONSTANTLY, NO DTCs STORED

NOTICE: DIAGNOSTIC SYSTEM CHECK MUST BE COMPLETED BEFORE USING THIS CHART.

- IGNITION "OFF."
- INSTALL TECH 1.
- IGNITION TO "RUN."
- USING TECH 1, SELECT "F4: MISC. TESTS."
- SELECT "F8: LAMP TEST" AND ATTEMPT TO TURN "ON" THE AMBER ABS WARNING LAMP.

DID THE AMBER ABS WARNING LAMP TURN "ON"?

YES

- USING TECH 1 LAMP TEST, ATTEMPT TO "FLASH" THE ABS WARNING LAMP.

DID THE ABS WARNING LAMP "FLASH"?

YES

MALFUNCTION IS NOT PRESENT AT THIS TIME. INTEMITTENTS COULD BE CAUSED BY INCORRECT WIRING HARNESS ROUTING OR LOOSE CONNECTIONS. INSPECT ALL CONNECTORS AND TERMINALS FOR POOR TERMINAL CONTACT AND EVIDENCE OF CORROSION. REPLACE ALL TERMINALS THAT EXHIBIT SIGNS OF POOR TERMINAL CONTACT OR CORROSION.

NO

INSPECT THE ABS WARNING LAMP BULB TO VERIFY BULB IS NOT OPEN. REPLACE ANY OPEN BULBS. IF BULB IS GOOD, REPAIR SHORT TO GROUND IN CKT 875.
PERFORM FURTHER DIAGNOSIS OF INSTRUMENT CLUSTER.

DIAGNOSTIC CHART B — 1995 CENTURY, CUTLASS CIERA AND CUTLASS CRUISER

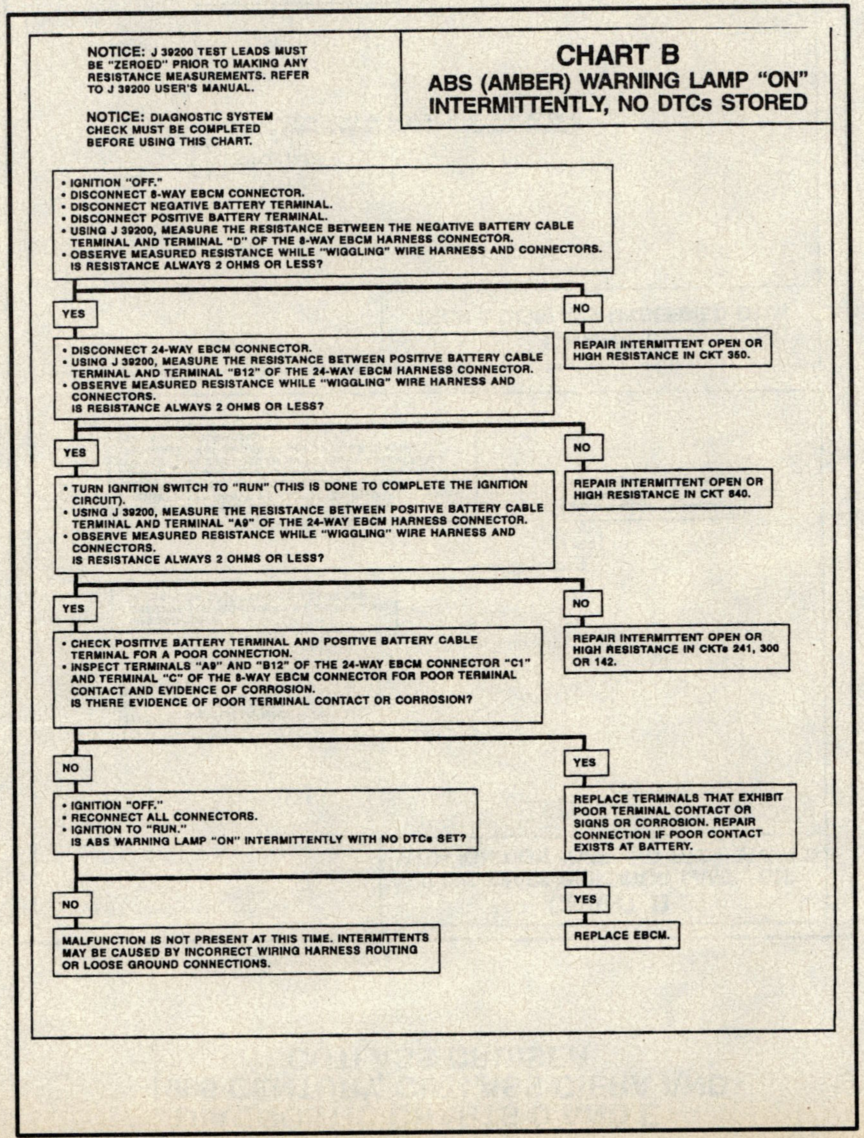

DIAGNOSTIC CHART C — 1995 CENTURY, CUTLASS CIERA AND CUTLASS CRUISER

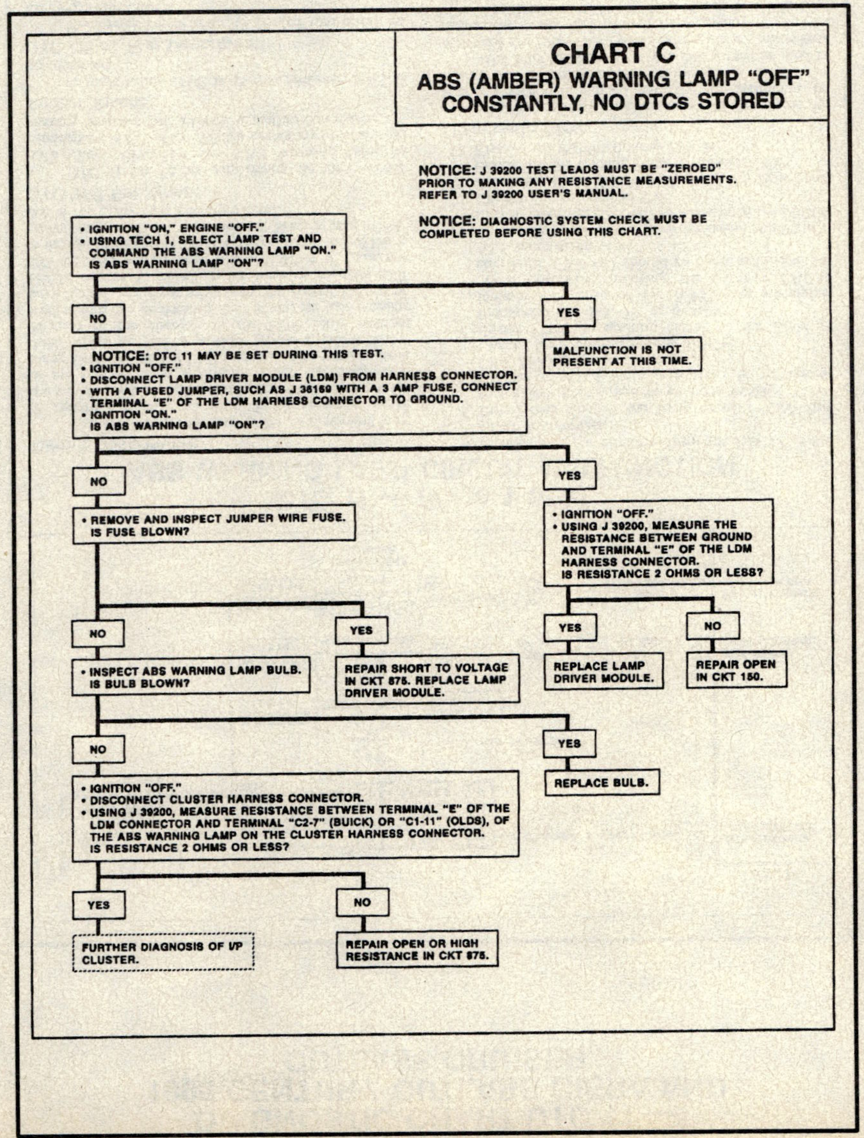

DIAGNOSTIC CHARTS D AND E — 1995 CENTURY, CUTLASS CIERA AND CUTLASS CRUISER

CHART D
"LOW TRAC" INDICATOR LAMP "ON" WITH IGNITION "OFF," NO DTCs STORED

NOTICE: DIAGNOSTIC SYSTEM CHECK MUST BE COMPLETED BEFORE USING THIS CHART.

- IGNITION "OFF."
- OBSERVE THE "LOW TRAC" WARNING LAMP. IS THE "LOW TRAC" WARNING LAMP "ON"?

NO: MALFUNCTION IS NOT PRESENT AT THIS TIME. INTERMITTENTS COULD BE CAUSED BY INCORRECT WIRING HARNESS ROUTING OR LOOSE CONNECTIONS. INSPECT ALL CONNECTORS AND TERMINALS FOR POOR TERMINAL CONTACT AND EVIDENCE OF CORROSION. REPLACE ALL TERMINALS THAT EXHIBIT SIGNS OF POOR TERMINAL CONTACT OR CORROSION.

YES: REPAIR SHORT TO VOLTAGE ON CKT 1537.

CHART E
TECH 1 DISPLAYS UNDEFINED DTCs

IMPORTANT
- USE THIS CHART IF ANY OF THE FOLLOWING UNDEFINED DTCs ARE SET - 12, 13, 31, 43, 57, 68, 71, 72, 73, 75, 83, 84, 97, OR 98.

- IGNITION TO "RUN."
- USING TECH 1 , CLEAR DTCs. DOES UNDEFINED DTC RESET?

YES: REPLACE EBCM.

NO: MALFUNCTION IS NOT PRESENT AT THIS TIME.

DIAGNOSTIC CHART DTC 11 — 1995 CENTURY, CUTLASS CIERA AND CUTLASS CRUISER

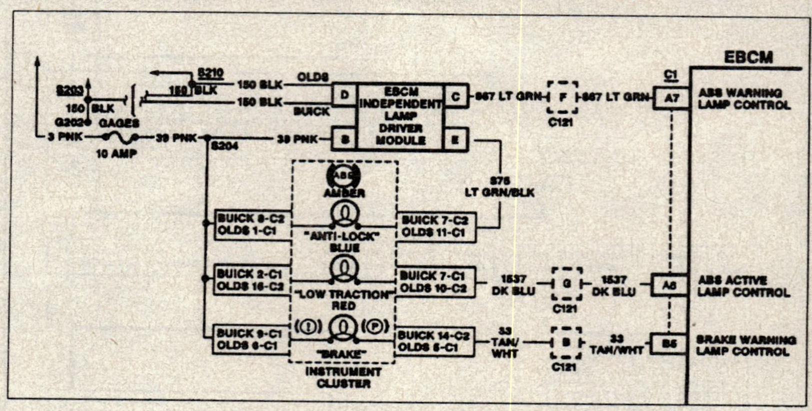

DTC 11 — (Page 1 of 2)
ABS WARNING LAMP CIRCUIT MALFUNCTION

Circuit Description:

This DTC checks the state of the ABS warning lamp to identify a situation in which the driver could not be warned of a system malfunction by the ABS warning lamp, or the warning lamp is always "ON." The Independent Lamp Driver Module (LDM) turns the ABS warning lamp "ON," unless the EBCM provides ground to turn it "OFF." Because of the circuitry in the LDM, only external malfunctions can be detected. As a result, the ABS warning lamp itself is not diagnosable, only the control line to the EBCM can be diagnosed. In the event of an open CKT 867, the ABS warning lamp will be "ON" at all times, due to the loss of ground at the LDM input. If the control line is shorted to ground, the ABS warning lamp is kept "OFF," due to the LDM input being grounded.

DTC Will Set When:

DTC 11 can be set only during the three second bulb check or when the ABS warning lamp is commanded "ON." If the EBCM cannot control the ABS warning lamp for two seconds, a malfunction exists.

Action Taken:

A malfunction DTC is stored, however, ABS is not disabled.

DTC Chart Test Description:

Number(s) below refer to circled number(s) on the diagnostic chart.

1. Checks to see if ABS lamp functions properly. Normal operation would indicate that the problem is not present.
2. Isolates the cause of the lamp remaining "ON" to a short to voltage, a short in the EBCM or an open in CKT 39 which supplies voltage to the LDM.
3. Determines if any other DTCs are present or if DTC 11 is set falsely.
4. Ensures that the circuitry from the EBCM to the LDM is complete.
5. Determines whether the lamp remains "ON" due to a LDM malfunction or an open in CKT 867.
6. Checks to see if there is a short to voltage in CKT 867.
7. Checks for a malfunctioning EBCM.
8. Ensures that the ignition circuit to the LDM is functional. CKTs 3, 39, or 875.
9. Isolates the cause of the ABS lamp remaining "OFF" by disconnecting the 32-way EBCM connector, CKT 867 should be open and cause the ABS lamp to be "ON."
10. Isolates the cause of the lamp remaining "OFF" to a short to ground in CKT 867 or a malfunctioning LDM.
11. Checks for an open in CKT 39.
12. Determines if the "GAGES" fuse or CKT 39 is open, causing the I/P warning lamps to remain "OFF."
13. Checks for malfunctioning EBCM.

Diagnostic Aids:

An "Intermittent" malfunction may be caused by a poor connection, rubbed through wire insulation, or a wire that is broken inside the insulation.

The lamp test function of the Tech 1 may be used to command the lamp "ON" while looking for an intermittent malfunction in the ABS warning lamp circuitry.

The frequency of the malfunction can be checked by using the enhanced diagnostic function of the Tech 1.

Any circuitry that is suspected as causing the intermittent complaint should be thoroughly checked for backed out terminals, improper mating, broken locks, improperly formed or damaged terminals, poor terminal to wiring connections or physical damage to the wiring harness.

DIAGNOSTIC CHART DTC 11 — 1995 CENTURY, CUTLASS CIERA AND CUTLASS CRUISER

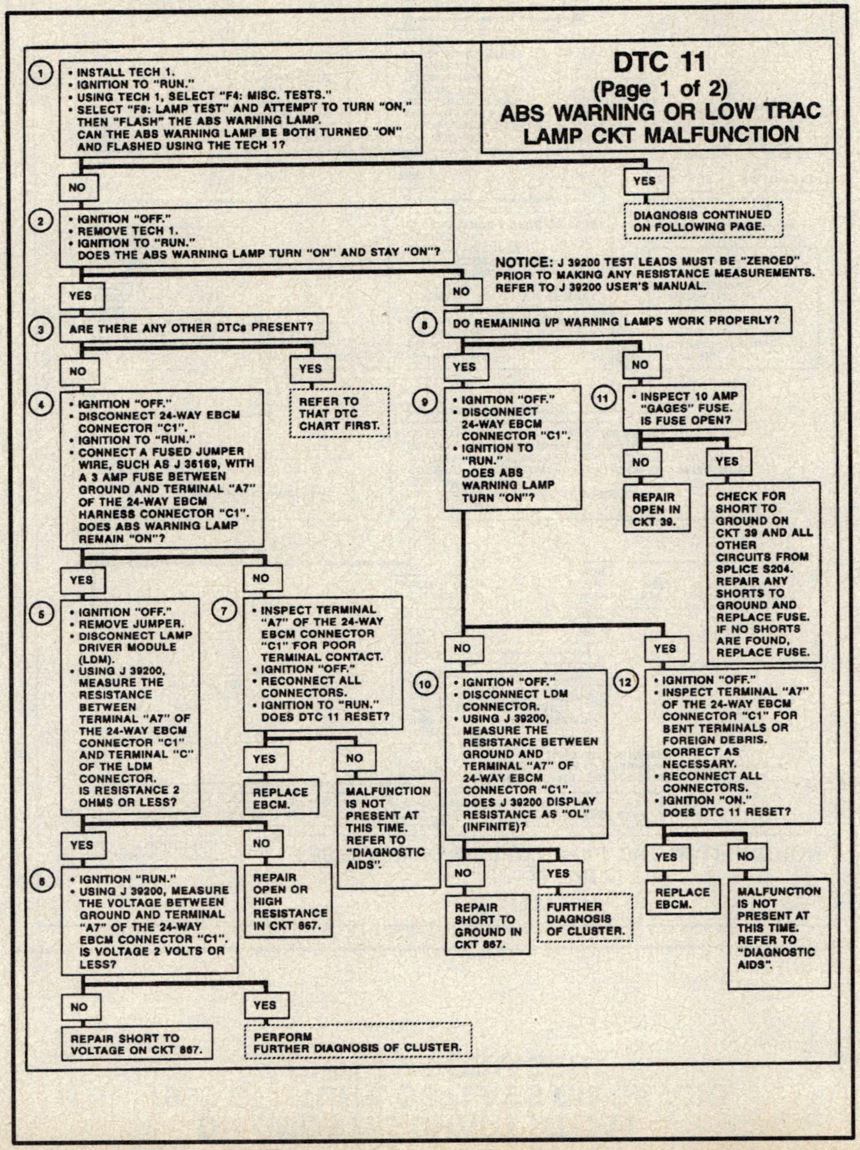

DIAGNOSTIC CHART DTC 11 — 1995 CENTURY, CUTLASS CIERA AND CUTLASS CRUISER

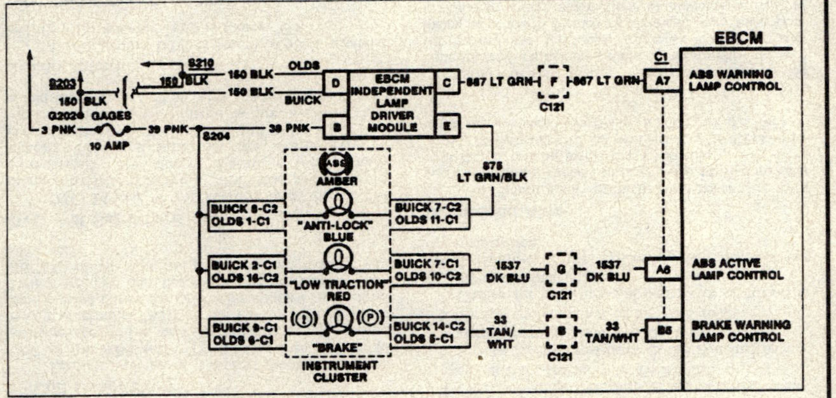

DTC 11 — (Page 2 of 2)
ABS WARNING LAMP CIRCUIT MALFUNCTION

Circuit Description:

This DTC checks the state of the ABS warning lamp to identify a situation in which the driver could not be warned of a system malfunction by the ABS warning lamp, or the warning lamp is always "ON." The Independent Lamp Driver Module (LDM) turns the ABS warning lamp "ON," unless the EBCM provides ground to turn it "OFF." Because of the circuitry in the LDM, only external malfunctions can be detected. As a result, the ABS warning lamp itself is not diagnosable, only the control line to the EBCM can be diagnosed. In the event of an open CKT 867, the ABS warning lamp will be "ON" at all times, due to the loss of ground at the LDM input. If the control line is shorted to ground, the ABS warning lamp is kept "OFF," due to the LDM input being grounded.

DTC Will Set When:

DTC 11 only can be set during the three second bulb check or when either of the lamps are commanded "ON." If the EBCM cannot control either the ABS warning lamp or the "LOW TRACTION" lamp for two seconds, a malfunction exists.

Action Taken:

A malfunction DTC is stored, however, ABS is not disabled.

DTC Chart Test Description:

Number(s) below refer to circled number(s) on the diagnostic chart.

14. Determines if the "LOW TRACTION" lamp can be properly controlled.
15. Determines if the "LOW TRACTION" lamp is "ON" constantly.
16. Checks for a short to ground in the "LOW TRACTION" lamp circuitry.
17. Determines if DTC 11 was set due to a EBCM malfunction.
18. Determines if DTC 11 was set due to a EBCM malfunction.
19. Determines if the "LOW TRACTION" lamp can be controlled manually.
20. Verifies the integrity of the "LOW TRACTION" lamp circuitry.
21. Determines if there is a short to voltage in the "LOW TRACTION" lamp circuitry.
22. Determines if DTC 11 was set due to an EBCM malfunction.

Diagnostic Aids:

An "Intermittent" malfunction may be caused by a poor connection, rubbed through wire insulation, or a wire that is broken inside the insulation.

The lamp test function of the Tech 1 may be used to command the lamp "ON" while looking for an intermittent malfunction in the "LOW TRACTION" lamp circuitry.

The frequency of the malfunction can be checked by using the enhanced diagnostic function of the Tech 1.

Any circuitry that is suspected as causing the intermittent complaint should be thoroughly checked for backed out terminals, improper mating, broken locks, improperly formed or damaged terminals, poor terminal to wiring connections or physical damage to the wiring harness.

DIAGNOSTIC CHART DTC 11 — 1995 CENTURY, CUTLASS CIERA AND CUTLASS CRUISER

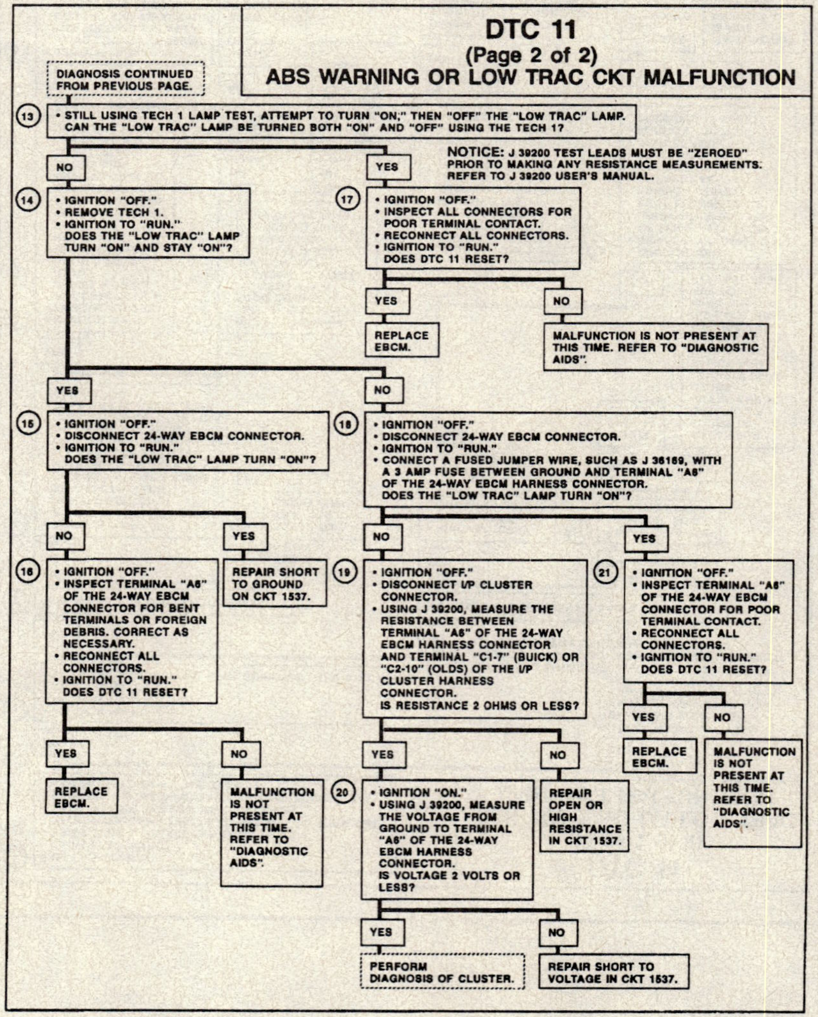

DIAGNOSTIC CHART DTC 14 — 1995 CENTURY, CUTLASS CIERA AND CUTLASS CRUISER

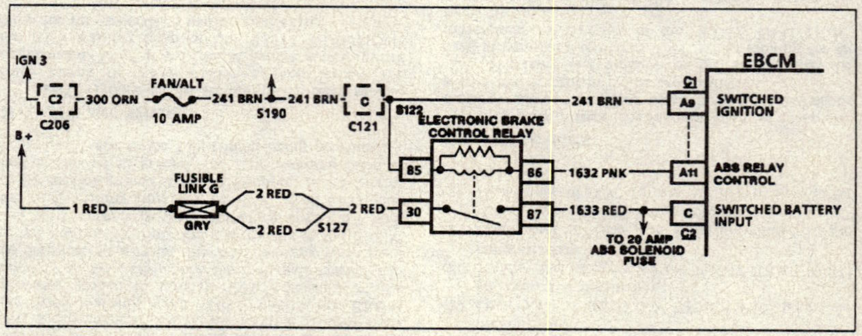

DTC 14

ABS ENABLE RELAY CONTACT CIRCUIT OPEN

Circuit Description:

Ignition voltage is supplied through terminal "85" of the ABS relay. The EBCM then is able to energize the pull-in coil by completing the ground circuit at terminal "A11" of the EBCM. The magnetic field created closes the ABS relay contacts and allows battery voltage and current to be supplied to the EBCM, which supplies power to the motors and solenoids.

DTC Will Set When:

DTC 14 can be set anytime after the EBCM commands the ABS relay "ON" (the relay is first commanded "ON" during the three second bulb check). This test monitors the availability of current/voltage to the motors and solenoids. This malfunction indicates voltage is not available and would therefore not allow ABS operation if required.

Action Taken:

A malfunction DTC is stored, ABS is disabled and the ABS warning lamp is turned "ON."

DTC Chart Test Description:

Number(s) below refer to circled number(s) on the diagnostic chart.

1. Checks to see if DTC 16 is also set as current or history. If DTC 16 is also set, proceed to that chart first.
2. Verifies that the malfunction is currently present.
3. Verifies that the relay can be turned "ON" and provide voltage to the EBCM.
4. Checks for high resistance in the switched battery circuit from the relay to the EBCM.
5. Checks for high resistance in the switched battery circuit from the battery to the relay.
6. Verifies the condition of the ABS relay.
7. Checks for poor terminal contact or corrosion at the connectors.
8. Determines if the EBCM is the cause of the malfunction.

Diagnostic Aids:

An "Intermittent" problem may be caused by a poor connection, rubbed through wire insulation, or a wire that is broken inside the insulation.

The frequency of the problem can be checked by using the enhanced diagnostic function of the Tech 1.

Any circuitry that is suspected as causing the "Intermittent" complaint should be thoroughly checked for backed out terminals, improper mating, broken locks, improperly formed or damaged terminals, poor terminal to wiring connections or physical damage to the wiring harness.

VIBRATION, TEMPERATURE EFFECTS:

Check for vibration effects by performing the relay test function of the Tech 1. With the relay test commanded "ON," lightly tap the top and sides of the relay while monitoring relay voltage. If the relay voltage changes significantly, replace the relay.

If DTC 14 only sets when the vehicle is initially started in cold ambient conditions (temperature less than 32° F-0° C), replace the relay.

DIAGNOSTIC CHART DTC 14 — 1995 CENTURY, CUTLASS CIERA AND CUTLASS CRUISER

DTC 14 (Page 1 of 3) ABS ENABLE RELAY CONTACT CIRCUIT OPEN

NOTICE: J 39200 TEST LEADS MUST BE "ZEROED" PRIOR TO MAKING ANY RESISTANCE MEASUREMENTS. REFER TO J 39200 USER'S MANUAL.

1. IS DTC 16 ALSO PRESENT OR IN HISTORY?
   - YES: PROCEED TO DTC 16 CHART FIRST.
   - NO: go to 2.
2. • IGNITION TO "RUN," ENGINE "OFF."
   • USING TECH 1, SELECT DATA LIST AND OBSERVE EBCM BATTERY VOLTAGE.
   DOES TECH 1 INDICATE THAT EBCM BATTERY VOLTAGE IS GREATER THAN 10 VOLTS?
   - YES: MALFUNCTION IS NOT PRESENT AT THIS TIME. REFER TO "DIAGNOSTIC AIDS" ON FACING PAGE.
   - NO: go to 3.
3. • IGNITION "OFF."
   • DISCONNECT 24-WAY AND 8-WAY EBCM CONNECTORS.
   • CONNECT A FUSED JUMPER SUCH AS J 36169 WITH A 3 AMP FUSE BETWEEN GROUND AND TERMINAL "A11" OF THE 24-WAY EBCM HARNESS CONNECTOR.
   • IGNITION TO "RUN."
   • USING J 39200, MEASURE THE VOLTAGE BETWEEN GROUND AND TERMINAL "C" OF THE 8-WAY EBCM HARNESS CONNECTOR.
   IS VOLTAGE 10 VOLTS OR GREATER?
   - NO: DIAGNOSIS CONTINUED ON NEXT PAGE.
   - YES: go to 4.
4. • IGNITION "OFF."
   • DISCONNECT JUMPER USED IN STEP 3.
   • DISCONNECT ABS RELAY.
   • USING J 39200, MEASURE RESISTANCE BETWEEN TERMINAL "87" OF THE ABS RELAY HARNESS CONNECTOR AND TERMINAL "C" OF THE 8-WAY EBCM HARNESS CONNECTOR.
   IS RESISTANCE 2 OHMS OR LESS?
   - NO: REPAIR HIGH RESISTANCE IN CKT 1633.
   - YES: go to 5.
5. • DISCONNECT NEGATIVE BATTERY CABLE.
   • DISCONNECT POSITIVE BATTERY CABLE.
   • USING J 39200, MEASURE THE RESISTANCE BETWEEN POSITIVE BATTERY CABLE TERMINAL AND TERMINAL "30" OF THE ABS RELAY HARNESS CONNECTOR.
   IS RESISTANCE 2 OHMS OR LESS?
   - NO: REPAIR HIGH RESISTANCE IN CKT 1 OR 2.
   - YES: go to 6.
6. • USING A JUMPER WIRE, CONNECT TERMINAL "86" ON THE ABS RELAY TO GROUND.
   • USING A FUSED JUMPER WIRE, SUCH AS J 36169, WITH A 3 AMP FUSE, CONNECT TERMINAL "85" ON THE ABS RELAY TO B+.
   • USING J 39200, MEASURE THE RESISTANCE BETWEEN TERMINALS "30" AND "87" ON THE ABS RELAY.
   IS RESISTANCE 2 OHMS OR LESS?
   - NO: REPLACE ABS RELAY.
   - YES: go to 7.
7. • INSPECT ABS RELAY AND ABS RELAY HARNESS CONNECTOR FOR A POOR CONNECTION OR POOR TERMINAL CONTACT.
   • INSPECT EBCM AND EBCM CONNECTOR TERMINALS "A11" AND "C" FOR POOR TERMINAL CONTACT.
   • INSPECT POSITIVE BATTERY TERMINAL AND POSITIVE BATTERY CABLE TERMINAL FOR A POOR CONNECTION.
   IS POOR TERMINAL CONTACT OR A POOR CONNECTION EVIDENT?
   - YES: REPLACE TERMINALS THAT EXHIBIT POOR TERMINAL CONTACT OR EVIDENCE OF CORROSION.
   - NO: go to 8.
8. • RECONNECT ALL CONNECTORS.
   • IGNITION TO "RUN."
   DOES DTC 14 RESET?
   - YES: REPLACE EBCM.
   - NO: MALFUNCTION IS INTERMITTENT. REFER TO "DIAGNOSTIC AIDS".

DIAGNOSTIC CHART DTC 14 — 1995 CENTURY, CUTLASS CIERA AND CUTLASS CRUISER

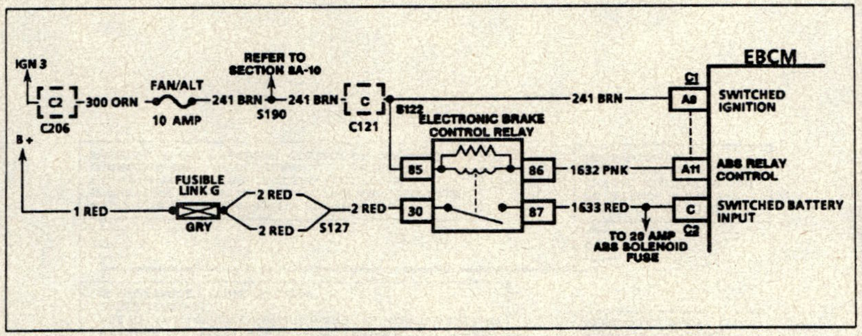

DTC 14

ABS ENABLE RELAY CONTACT CIRCUIT OPEN

Circuit Description:

Ignition voltage is supplied through terminal "85" of the ABS relay. The EBCM then is able to energize the pull-in coil by completing the ground circuit at terminal "A11" of the EBCM. The magnetic field created closes the ABS relay contacts and allows battery voltage and current to be supplied to the EBCM, which supplies power to the motors and solenoids.

DTC Will Set When:

DTC 14 can be set anytime after the EBCM commands the ABS enable relay "ON" (the relay is first commanded "ON" during the three second bulb check). This test monitors the availability of current/voltage to the motors and solenoids. This malfunction indicates voltage is not available and would therefore not allow ABS operation if required.

Action Taken:

A malfunction DTC is stored, ABS is disabled and the ABS warning lamp is turned "ON."

DTC Chart Test Description:

Number(s) below refer to circled number(s) on the diagnostic chart.

9. Checks for high resistance in the entire relay coil circuit.
10. Verifies proper battery condition.
11. This step verifies proper voltage is available at the ABS relay.
12. Checks for an open in the switched battery circuit from the relay to the EBCM.
13. Verifies the condition of the ABS relay.
14. Checks for proper resistance of the relay coil.
15. Checks for high resistance in the ABS relay control circuit between the relay and the EBCM.
16. Checks for high resistance in the ABS relay control circuit between the battery and the relay.

Diagnostic Aids:

An "Intermittent" problem may be caused by a poor connection, rubbed through wire insulation, or a wire that is broken inside the insulation.

The frequency of the problem can be checked by using the enhanced diagnostic function of the Tech 1

Any circuitry that is suspected as causing the "Intermittent" complaint should be thoroughly checked for backed out terminals, improper mating, broken locks, improperly formed or damaged terminals, poor terminal to wiring connections or physical damage to the wiring harness.

VIBRATION, TEMPERATURE EFFECTS:

Check for vibration effects by performing the relay test function of the Tech 1. With the relay test commanded "ON," lightly tap the top and sides of the relay while monitoring relay voltage. If the relay voltage changes significantly, replace the relay.

If DTC 14 only sets when the vehicle is initially started in cold ambient conditions (temperature less than 32° F-0° C), replace the relay.

DIAGNOSTIC CHART DTC 14 — 1995 CENTURY, CUTLASS CIERA AND CUTLASS CRUISER

DTC 14
(Page 2 of 3)
ABS ENABLE RELAY CONTACT CIRCUIT OPEN

NOTICE: J 39200 TEST LEADS MUST BE "ZEROED" PRIOR TO MAKING ANY RESISTANCE MEASUREMENTS. REFER TO J 39200 USER'S MANUAL.

DIAGNOSIS CONTINUED FROM PREVIOUS PAGE.

(9)
- IGNITION "OFF."
- REMOVE JUMPER WIRE INSTALLED IN STEP #3, IF NOT ALREADY REMOVED.
- DISCONNECT NEGATIVE BATTERY CABLE.
- DISCONNECT POSITIVE BATTERY CABLE.
- WITH BATTERY STILL DISCONNECTED, TURN IGNITION SWITCH TO "RUN." THIS COMPLETES THE IGNITION CIRCUIT TO THE ABS RELAY.
- USING J 39200, MEASURE THE RESISTANCE BETWEEN TERMINAL "A11" OF THE 24-WAY EBCM HARNESS CONNECTOR AND THE POSITIVE BATTERY CABLE TERMINAL. IS RESISTANCE 95 OHMS OR LESS WHEN MEASURED AT APPROXIMATELY 20°C (68°F)?

YES → go to step 10. NO → DIAGNOSIS CONTINUED ON NEXT PAGE.

(10)
- USING J 39200, MEASURE THE VOLTAGE BETWEEN THE NEGATIVE AND POSITIVE BATTERY TERMINALS. IS VOLTAGE 10 VOLTS OR GREATER?

YES → go to step 11. NO → CORRECT LOW VOLTAGE CONDITION.

(11)
- IGNITION "OFF."
- RECONNECT POSITIVE BATTERY CABLE.
- RECONNECT NEGATIVE BATTERY CABLE.
- DISCONNECT ABS RELAY.
- MEASURE THE VOLTAGE BETWEEN TERMINAL "30" OF THE ABS RELAY HARNESS CONNECTOR AND GROUND. IS VOLTAGE 10 VOLTS OR GREATER?

YES → go to step 12. NO → REPAIR OPEN OR HIGH RESISTANCE IN CKT 1 OR 2.

(12)
- USING J 39200, MEASURE THE RESISTANCE BETWEEN TERMINAL "87" OF THE ABS RELAY HARNESS CONNECTOR AND TERMINAL "C" OF THE 8-WAY EBCM HARNESS CONNECTOR. IS RESISTANCE 2 OHMS OR LESS?

YES → go to step 13. NO → REPAIR OPEN OR HIGH RESISTANCE IN CKT 1633.

(13)
- USING A JUMPER WIRE, CONNECT TERMINAL "86" ON THE ABS RELAY TO GROUND.
- USING A FUSED JUMPER WIRE, SUCH AS J 38169, WITH A 3 AMP FUSE, CONNECT TERMINAL "85" ON THE ABS RELAY TO B+.
- USING J 39200, MEASURE THE RESISTANCE BETWEEN TERMINALS "30" AND "87" ON THE ABS RELAY. IS RESISTANCE 2 OHMS OR LESS?

YES → MALFUNCTION WAS MOST LIKELY DUE TO A POOR CONNECTION. INSPECT ALL CONNECTORS AND TERMINALS FOR POOR TERMINAL CONTACT AND EVIDENCE OF CORROSION. REPLACE ALL TERMINALS THAT EXHIBIT EVIDENCE OF POOR TERMINAL CONTACT OR CORROSION.

NO → REPLACE ABS RELAY.

DIAGNOSTIC CHART DTC 14 — 1995 CENTURY, CUTLASS CIERA AND CUTLASS CRUISER

DTC 14
(Page 3 of 3)
ABS ENABLE RELAY CONTACT CIRCUIT OPEN

NOTICE: J 39200 TEST LEADS MUST BE "ZEROED" PRIOR TO MAKING ANY RESISTANCE MEASUREMENTS. REFER TO J 39200 USER'S MANUAL.

DIAGNOSIS CONTINUED FROM PREVIOUS PAGE.

(14)
- DISCONNECT ABS RELAY.
- USING J 39200, MEASURE THE RESISTANCE BETWEEN TERMINALS "85" AND "86" ON THE ABS RELAY. IS RESISTANCE 95 OHMS OR LESS WHEN MEASURED AT APPROXIMATELY 20°C (68°F)?

YES → go to step 15. NO → REPLACE ABS RELAY.

(15)
- MEASURE THE RESISTANCE BETWEEN TERMINAL "86" OF THE ABS RELAY HARNESS CONNECTOR AND TERMINAL "A11" OF THE 24-WAY EBCM HARNESS CONNECTOR. IS RESISTANCE 2 OHMS OR LESS?

YES → go to step 16. NO → REPAIR OPEN OR HIGH RESISTANCE IN CKT 1632.

(16)
- MEASURE THE RESISTANCE BETWEEN TERMINAL "85" OF THE ABS RELAY HARNESS CONNECTOR AND THE POSITIVE BATTERY CABLE TERMINAL. IS RESISTANCE 2 OHMS OR LESS?

YES → MALFUNCTION WAS MOST LIKELY DUE TO A POOR CONNECTION. INSPECT ALL CONNECTORS AND TERMINALS FOR POOR TERMINAL CONTACT AND EVIDENCE OF CORROSION. REPLACE ALL TERMINALS THAT EXHIBIT EVIDENCE OF POOR TERMINAL CONTACT OR CORROSION.

NO → REPAIR OPEN OR HIGH RESISTANCE IN CKT 241, CKT 300, IGNITION SWITCH, OR BATTERY FEED CIRCUIT TO IGNITION SWITCH.

DIAGNOSTIC CHART DTC 15 — 1995 CENTURY, CUTLASS CIERA AND CUTLASS CRUISER

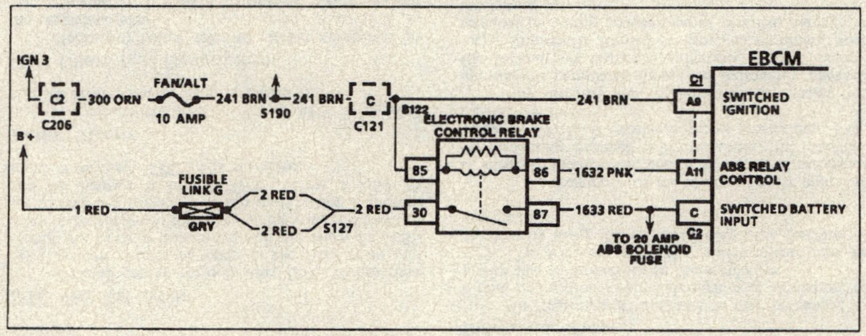

DTC 15

ABS ENABLE RLY. CONTACT CKT SHORT TO BATT OR ALWAYS CLOSED

Circuit Description:

Ignition voltage is supplied through terminal "85" of the ABS relay. The EBCM then is able to energize the pull-in coil by completing the ground circuit at terminal "A11" of the EBCM. The magnetic field created closes the ABS relay contacts and allows battery voltage and current to be supplied to the EBCM, which supplies power to the motors and solenoids.

DTC Will Set When:

DTC 15 can be set only before the EBCM commands the relay "ON." This test determines if the ABS relay is energized when it should not be. This malfunction would not allow the ABS relay to remove power to the ABS system. If a second malfunction were to occur that requires the ABS relay to be turned "OFF," that malfunction cannot be removed if the relay cannot be controlled. The malfunction must be present for three consecutive drive cycles before the DTC is set.

Action Taken:

A malfunction DTC is stored. ABS is not disabled.

DTC Chart Test Description:

Number(s) below refer to circled number(s) on the diagnostic chart.

1. Determines if any other DTCs are present.
2. Indicates that the EBCM is capable of controlling the ABS relay as commanded.
3. This step checks for a possible short to battery voltage in CKT 1633.
4. This step determines whether the short to voltage is in CKT 1633 or internal to the ABS relay.
5. Checks for a possible intermittent in wiring or connectors.

Diagnostic Aids:

An "Intermittent" malfunction may be caused by a poor connection, rubbed through wire insulation, or a wire that is broken inside the insulation.

The frequency of the malfunction can be checked by using the enhanced diagnostic function of the Tech 1.

Any circuitry that is suspected as causing the intermittent complaint should be thoroughly checked for backed out terminals, improper mating, broken locks, improperly formed or damaged terminals, poor terminal to wiring connections or physical damage to the wiring harness.

DIAGNOSTIC CHART DTC 15 — 1995 CENTURY, CUTLASS CIERA AND CUTLASS CRUISER

DTC 15
ABS RLY CONTACT SHORT TO BATT OR ALWAYS CLOSED

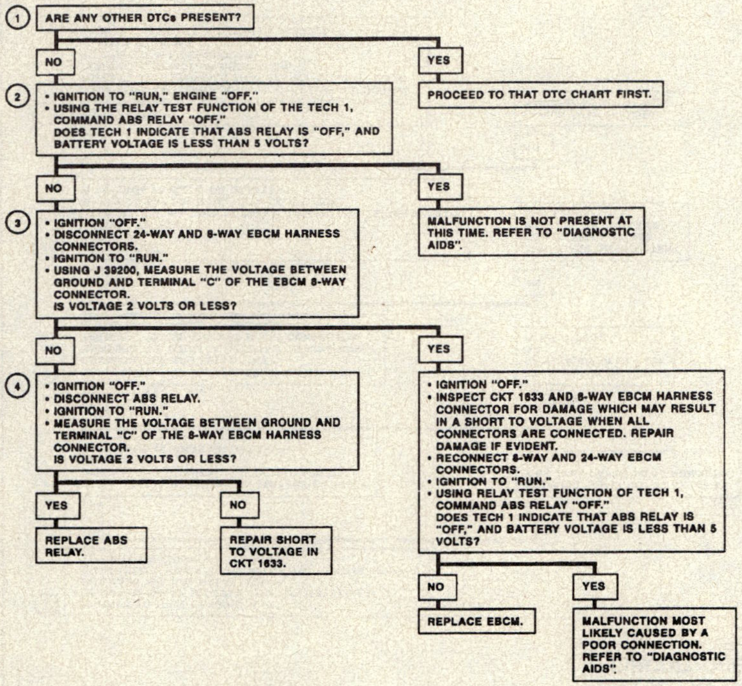

AFTER ALL DIAGNOSIS IS COMPLETE, CLEAR DTCs, AND TEST DRIVE VEHICLE FOR 3 DRIVE CYCLES TO VERIFY THAT DTC DOES NOT RESET. A DRIVE CYCLE CONSISTS OF STARTING THE VEHICLE, DRIVING ABOVE 16 km/h (10 mph), AND THEN IGNITION TO "OFF."

DIAGNOSTIC CHART DTC 16 — 1995 CENTURY, CUTLASS CIERA AND CUTLASS CRUISER

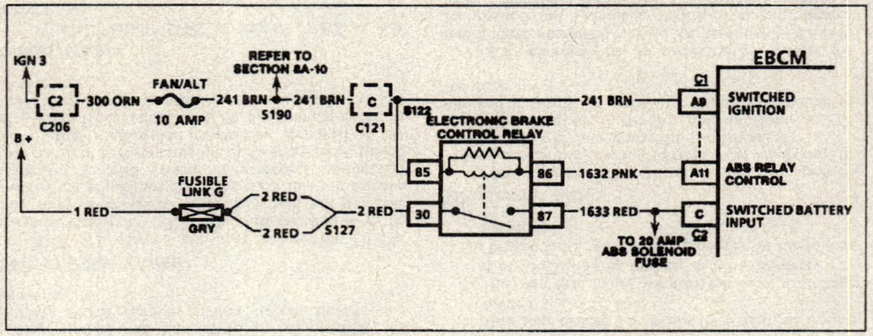

DTC 16
ABS ENABLE RELAY COIL CIRCUIT OPEN

Circuit Description:

Ignition voltage is supplied through terminal "85" of the ABS relay. The EBCM then is able to energize the pull-in coil by completing the ground circuit at terminal "A11" of the EBCM. The magnetic field created closes the ABS relay contacts and allows battery voltage and current to be supplied to the EBCM, which supplies power to the motors and solenoids.

DTC Will Set When:

DTC 16 can be set only after DTC 14 has been set. This test detects an open in the ABS relay coil circuit. An open in this circuit will not allow the ABS relay to be energized thus preventing voltage/current to the motors and solenoids. If this malfunction is present and the ignition is turned "OFF" before 5 km/h (3 mph) is reached, DTC 14 is set alone.

Action Taken:

A malfunction DTC is stored, ABS is disabled and the ABS warning lamp is turned "ON."

DTC Chart Test Description:

Number(s) below refer to circled number(s) on the diagnostic chart.

1. Indicates the EBCM is capable of controlling the ABS relay as commanded.
2. Checks to ensure that voltage is available to the pull-in coil of the ABS relay.
3. Ensures that there is continuity through the pull-in coil of the ABS relay.
4. Verifies the integrity of the relay control circuit.
5. Ensures malfunction was not due to poor terminal contact.

Diagnostic Aids:

An "Intermittent" malfunction may be caused by a poor connection, rubbed through wire insulation, or a wire that is broken inside the insulation.

The frequency of the malfunction can be checked by using the enhanced diagnostic function of the Tech 1.

If the frequency of the malfunction is high, but is currently intermittent, check for high coil resistance by measuring between relay terminals "85" and"86" using a J 39200. If resistance shows greater than 100(, replace the relay.

Any circuitry that is suspected as causing the intermittent complaint should be thoroughly checked for backed out terminals, improper mating, broken locks, improperly formed or damaged terminals, poor terminal to wiring connections or physical damage to the wiring harness.

DIAGNOSTIC CHART DTC 16 — 1995 CENTURY, CUTLASS CIERA AND CUTLASS CRUISER

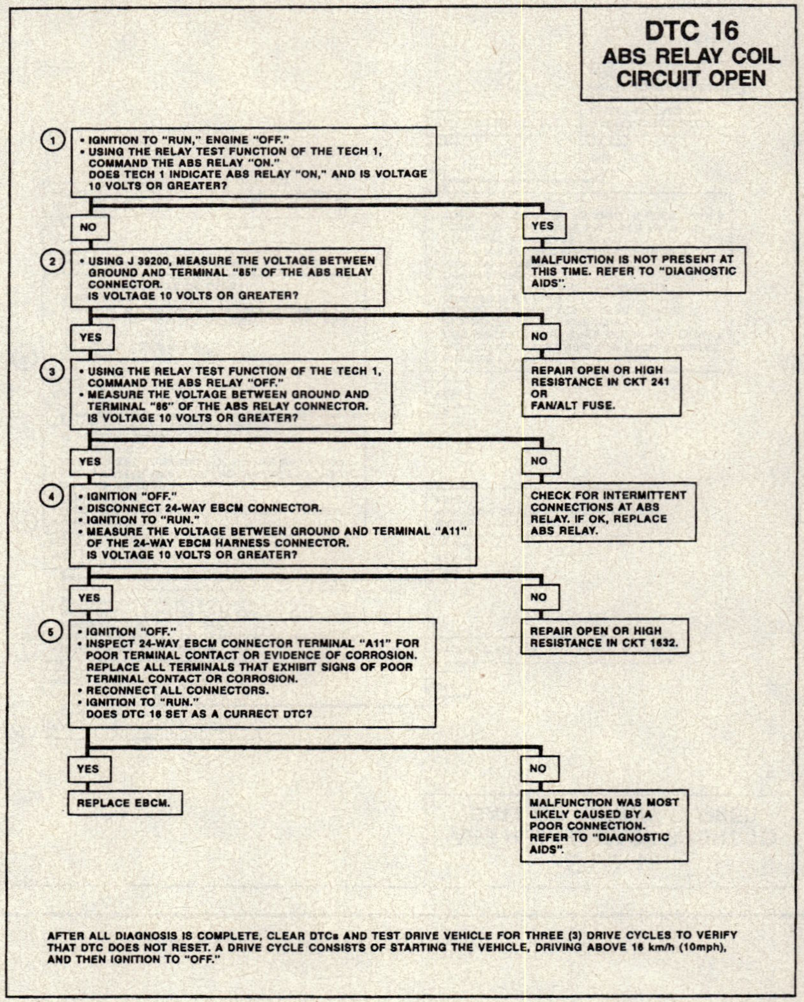

DIAGNOSTIC CHART DTC 17 — 1995 CENTURY, CUTLASS CIERA AND CUTLASS CRUISER

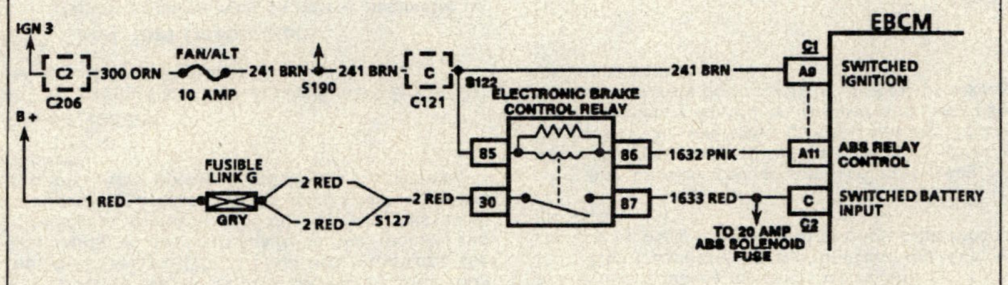

DTC 17
ABS ENABLE RELAY COIL CIRCUIT SHORTED TO GND

Circuit Description:

Ignition voltage is supplied through terminal "85" of the ABS relay. The EBCM then is able to energize the pull-in coil by completing the ground circuit at terminal "A11" of the EBCM. The magnetic field created closes the ABS relay contacts and allows battery voltage and current to be supplied to the EBCM, which supplies voltage to the motors and solenoids.

DTC Will Set When:

DTC 17 can be set before the EBCM commands the ABS relay "ON." This test determines if the ABS relay is energized when it should not be. This malfunction would not allow the ABS relay to remove power to the ABS system. If a second malfunction were to occur that requires the ABS relay to be turned "OFF," that malfunction cannot be removed if the relay cannot be controlled.

Action Taken:

A malfunction DTC is stored. ABS is not disabled.

DTC Chart Test Description:

Number(s) below refer to circled number(s) on the diagnostic chart.

1. Indicates that the EBCM is capable of controlling the ABS relay as commanded.
2. Checks to ensure that the ABS relay or control CKT 1632 are not shorted to ground.
3. Checks to see if the EBCM is internally shorted to ground. A grounded terminal "A11" would cause the ABS relay to be energized anytime the ignition was in the "ON" position.
4. Checks for a short to ground in control CKT 1632 or a defective relay.

Diagnostic Aids:

An "Intermittent" malfunction may be caused by a poor connection, rubbed through wire insulation, or a wire that is broken inside the insulation.

The frequency of the malfunction can be checked by using the enhanced diagnostic function of the Tech 1.

Any circuitry that is suspected as causing the intermittent complaint should be thoroughly checked for backed out terminals, improper mating, broken locks, improperly formed or damaged terminals, poor terminal to wiring connections or physical damage to the wiring harness.

DIAGNOSTIC CHART DTC 17 — 1995 CENTURY, CUTLASS CIERA AND CUTLASS CRUISER

DTC 17
ABS RELAY COIL CIRCUIT SHORTED TO GND

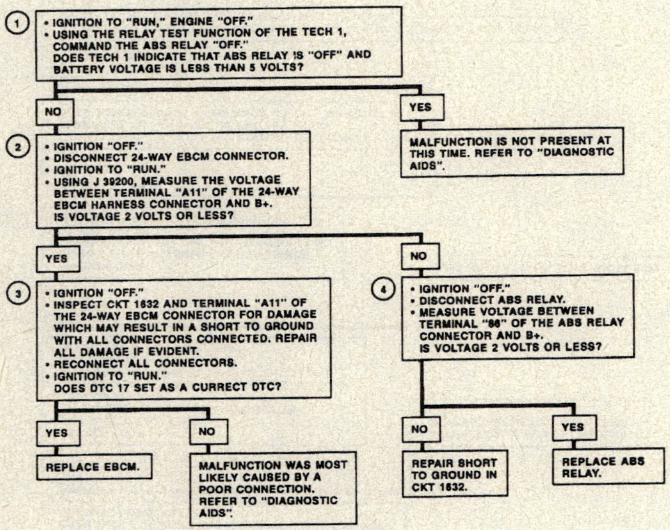

AFTER ALL DIAGNOSIS IS COMPLETE, CLEAR DTCs AND TEST DRIVE VEHICLE FOR THREE (3) DRIVE CYCLES TO VERIFY THAT DTC DOES NOT RESET. A DRIVE CYCLE CONSISTS OF STARTING THE VEHICLE, DRIVING OVER 16 km/h (10 mph), AND THEN IGNITION TO "OFF."

DIAGNOSTIC CHART DTC 18 — 1995 CENTURY, CUTLASS CIERA AND CUTLASS CRUISER

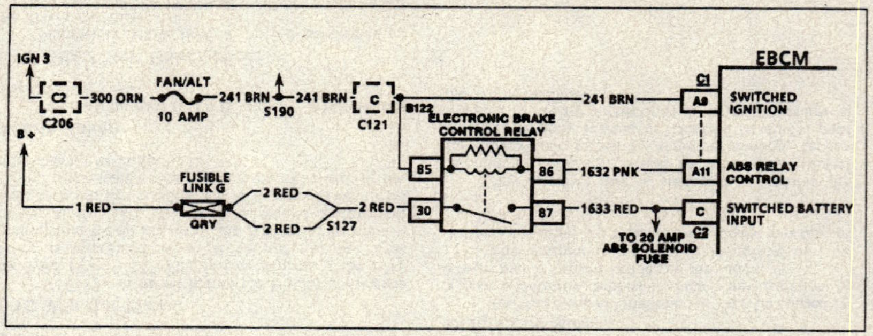

DTC 18

ABS ENABLE RELAY COIL CIRCUIT SHORTED TO BATTERY

Circuit Description:

Ignition voltage is supplied through terminal "85" of the ABS relay. The EBCM then is able to energize the pull-in coil by completing the ground circuit at terminal "A11" of the EBCM. The magnetic field created closes the ABS relay contacts and allows battery voltage and current to be supplied to the EBCM, which supplies voltage to the motors and solenoids.

DTC Will Set When:

DTC 18 can be set after the EBCM commands the ABS relay "ON." This test monitors the availability of current/voltage to the motors and solenoids. A malfunction of this type will not allow voltage to be available to the motors and solenoids, therefore, ABS operation would not be allowed if required.

Action Taken:

A malfunction DTC is stored, ABS is disabled and the ABS warning lamp is turned "ON."

DTC Chart Test Description:

Number(s) below refer to circled number(s) on the diagnostic chart.

1. Indicates that the EBCM is capable of controlling the ABS relay as commanded.
2. Checks for a short to voltage on the relay control circuit.
3. Verifies that the relay coil has the appropriate resistance.
4. Verifies that relay contacts are not failed closed.
5. Checks for possible poor terminal contact at the EBCM.

Diagnostic Aids:

An "Intermittent" malfunction may be caused by a poor connection, rubbed through wire insulation, or a wire that is broken inside the insulation.

The frequency of the malfunction can be checked by using the enhanced diagnostic function of the Tech 1.

Any circuitry that is suspected as causing the intermittent complaint should be thoroughly checked for backed out terminals, improper mating, broken locks, improperly formed or damaged terminals, poor terminal to wiring connections or physical damage to the wiring harness.

DIAGNOSTIC CHART DTC 18 — 1995 CENTURY, CUTLASS CIERA AND CUTLASS CRUISER

DTC 18
ABS RELAY COIL CIRCUIT SHORTED TO BATT

(1)
- IGNITION TO "RUN," ENGINE "OFF."
- USING THE RELAY TEST FUNCTION OF TECH 1, COMMAND ABS RELAY "ON."

DOES TECH 1 INDICATE ABS RELAY "ON" AND BATTERY VOLTAGE ABOVE 10 VOLTS?

YES: MALFUNCTION IS NOT PRESENT AT THIS TIME. REFER TO "DIAGNOSTIC AIDS".

NO:

(2)
- IGNITION "OFF."
- DISCONNECT ABS RELAY.
- DISCONNECT 24-WAY EBCM CONNECTOR.
- IGNITION TO "RUN."
- USING J 39200, MEASURE THE VOLTAGE BETWEEN GROUND AND TERMINAL "A11" OF THE 24-WAY EBCM HARNESS CONNECTOR.

IS VOLTAGE 2 VOLTS OR LESS?

NO: REPAIR SHORT TO VOLTAGE IN CKT 1632.

YES:

(3)
- IGNITION "OFF."
- USING J 39200, MEASURE THE RESISTANCE BETWEEN TERMINAL "86" AND TERMINAL "85" OF THE ABS RELAY.

IS RESISTANCE 40 OHMS OR GREATER?

NO: REPLACE ABS RELAY.

YES:

(4)
- USING J 39200, MEASURE THE RESISTANCE BETWEEN TERMINALS "30" AND "87" OF THE ABS RELAY.

DOES J 39200 DISPLAY RESISTANCE AS "OL" (INFINITE)?

NO: REPLACE ABS RELAY.

YES:

(5)
- INSPECT CKT 1632 AND TERMINAL "A11" OF THE 24-WAY EBCM CONNECTOR FOR DAMAGE WHICH MAY RESULT IN A SHORT TO VOLTAGE WITH ALL CONNECTORS CONNECTED. REPAIR ALL DAMAGE IF EVIDENT.
- RECONNECT ALL CONNECTORS.
- IGNITION TO "RUN."

DOES DTC 18 SET AS A CURRENT DTC?

YES: REPLACE EBCM.

NO: MALFUNCTION IS MOST LIKELY DUE TO A POOR CONNECTION. REFER TO "DIAGNOSTIC AIDS".

AFTER ALL DIAGNOSIS IS COMPLETE, CLEAR DTCs AND TEST DRIVE VEHICLE FOR THREE (3) DRIVE CYCLES TO VERIFY THAT DTC DOES NOT RESET. A DRIVE CYCLE CONSISTS OF STARTING THE VEHICLE, DRIVING OVER 16 km/h (10 mph), AND THEN IGNITION TO "OFF."

DIAGNOSTIC CHART DTC 21 — 1995 CENTURY, CUTLASS CIERA AND CUTLASS CRUISER

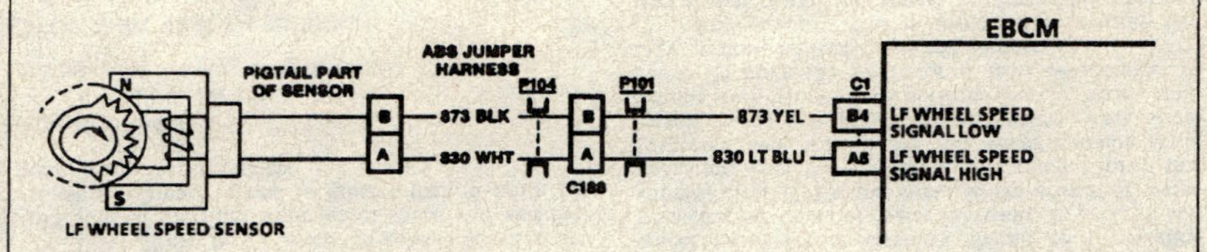

DTC 21
LEFT FRONT WHEEL SPEED = 0

Circuit Description:

As a toothed ring passes by the wheel speed sensor, changes in the electromagnetic field cause the wheel speed sensor to produce a sinusoidal (AC) voltage signal whose frequency is proportional to wheel speed. The magnitude of this signal is directly related to wheel speed and the proximity of the wheel speed sensor to the toothed ring, often referred to as the air gap.

DTC Will Set When:

DTC 21 can be set when the vehicle is not in an ABS stop. If the left front wheel speed = 0 and the vehicle's reference speed is greater than 8 km/h (5 mph), a malfunction exists.

Action Taken:

A malfunction DTC is stored, ABS is disabled and the ABS warning lamp is turned "ON."

DTC Chart Test Description:

Number(s) below refer to circled number(s) on the diagnostic chart.

1. This step is designed to identify a wheel speed sensor or circuitry that is damaged and visibly apparent.
2. Checks to see that the wheel speed sensor is within proper resistance values.
3. Verifies that the wheel speed sensor can generate an appropriate output signal.
4. Ensures that wheel speed sensor circuitry is not shorted together.
5. Ensures malfunction was not due to physical damage of the circuitry.

Diagnostic Aids:

An "Intermittent" malfunction may be caused by a poor connection, rubbed through wire insulation, or a wire that is broken inside the insulation.

The frequency of the malfunction can be checked by using the enhanced diagnostic function of the Tech 1.

If the customer's comments reflect that the ABS warning lamp is "ON" only during moist environmental changes (rain, snow, vehicle wash), all wheel speed sensor circuitry should be thoroughly inspected for signs of water intrusion. If DTC is not current, clear DTCs and simulate the effects of water intrusion. Use the following procedure. Spray the suspected area with a 5% salt water solution (two teaspoons of salt to 12 oz. of water). Test drive vehicle over various road surfaces (bumps, turns, etc.) above 24 km/h (15 mph) for at least 30 seconds. If DTC returns, replace suspected harness.

Any circuitry that is suspected as causing the intermittent complaint should be thoroughly checked for backed out terminals, improper mating, broken locks, improperly formed or damaged terminals, poor terminal to wiring connections or physical damage to the wiring harness.

Resistance of the wheel speed sensor will increase with an increase in sensor temperature.

When replacing a wheel speed sensor, inspect the sensor terminals and harness connector for corrosion and/or water intrusion. If evidence of corrosion or water intrusion exists, replace wheel speed sensor jumper harness. Likewise, if replacing a wheel speed sensor jumper harness, inspect sensor terminals. If evidence of corrosion or water intrusion exists, replace wheel speed sensor.

DIAGNOSTIC CHART DTC 21 — 1995 CENTURY, CUTLASS CIERA AND CUTLASS CRUISER

DTC 21
LEFT FRONT WHEEL SPEED = 0

IMPORTANT:
- WHEEL SPEED SENSOR INTERMITTENT MALFUNCTIONS MAY BE DIFFICULT TO LOCATE. CARE SHOULD BE TAKEN NOT TO DISTURB ANY ELECTRICAL CONNECTIONS PRIOR TO AN INDICATED STEP OF THIS CHART. THIS WILL ENSURE THAT AN INTERMITTENT CONNECTION WILL NOT BE CORRECTED BEFORE THE SOURCE OF THE MALFUNCTION IS FOUND.

(1)
- IGNITION "OFF."
- PHYSICALLY INSPECT THE LEFT FRONT WHEEL SPEED SENSOR, LEFT FRONT WHEEL SPEED SENSOR JUMPER HARNESS, AND CONNECTORS FOR DAMAGE. INSPECT THE WHEEL SPEED SENSOR AND TOOTHED RING FOR LOOSENESS OR DAMAGE. PAY PARTICULAR ATTENTION TO THE TOOTHED RING. ANY DEVIATION WILL AFFECT THE WHEEL SPEED SENSOR OUTPUT SIGNAL.

IS PHYSICAL DAMAGE INDICATED?

YES: REPLACE WHEEL SPEED SENSOR AND/OR JUMPER HARNESS IF DAMAGED. REPAIR ANY SENSOR OR TOOTHED RING LOOSENESS. DRIVE VEHICLE ABOVE 24 km/h (15 mph) TO ENSURE DTC 21 DOES NOT SET.

NO:

(2)
- DISCONNECT LEFT FRONT WHEEL SPEED SENSOR DIRECTLY AT SENSOR.
- USING J 39200, MEASURE THE RESISTANCE BETWEEN SENSOR TERMINALS "A" AND "B".

IS RESISTANCE BETWEEN 1661 AND 1929 OHMS WHEN SENSOR IS 20°C (68°F)?
(IF TEMPERATURE OF THE SENSOR IS SUSPECTED TO BE OTHER THAN 20°C (68°F), REFER TO THE TABLE BELOW FOR APPLICABLE SENSOR RESISTANCE VALUES; THE VALUES IN THE TABLE ARE FOR TEMPERATURE OF THE SENSOR, NOT AIR TEMPERATURE.)

NO: REPLACE WHEEL SPEED SENSOR.

YES:

(3)
- WITH J 39200 STILL CONNECTED, SELECT A/C VOLTAGE SCALE.
- SPIN WHEEL BY HAND WHILE OBSERVING VOLTAGE.

IS VOLTAGE AT LEAST 100 mV (VOLTAGE WILL INCREASE AS WHEEL SPEED INCREASES)?

NO: VERIFY THAT TOOTHED RING IS INTACT. IF OK, REPLACE WHEEL SPEED SENSOR

YES:

(4)
- DISCONNECT 24-WAY EBCM CONNECTOR.
- USING J 39200, MEASURE THE RESISTANCE BETWEEN TERMINAL "B4" AND "A8" OF THE 24-WAY EBCM HARNESS CONNECTOR.

DOES J 39200 DISPLAY RESISTANCE AS "OL" (INFINITE)?

NO: REPAIR SHORT BETWEEN CKTs 830 AND 873.

YES:

(5)
- INSPECT CKTs 830 AND 873 AND 24-WAY EBCM CONNECTOR FOR DAMAGE WHICH MAY RESULT IN A SHORT BETWEEN CKTs 830 AND 873 WITH ALL CONNECTORS CONNECTED. REPAIR ALL DAMAGE IF EVIDENT.
- RECONNECT ALL CONNECTORS.
- TEST DRIVE VEHICLE ABOVE 24 km/h (15 mph) FOR AT LEAST 30 SECONDS.

DOES DTC 21 SET AS A CURRENT DTC?

YES: REPLACE EBCM.

NO: MALFUNCTION IS INTERMITTENT. REFER TO "DIAGNOSTIC AIDS".

FRONT WHEEL SPEED SENSOR
TEMPERATURE OF SENSOR VS. SENSOR RESISTANCE (APPROXIMATE)

TEMP. (°C)	TEMP. (°F)	RESISTANCE (OHMS)
-34 TO 4	-30 TO 40	1258 TO 1608
5 TO 43	41 TO 110	1502 TO 2103
44 TO 93	111 TO 200	1747 TO 2482

DIAGNOSTIC CHART DTC 22 — 1995 CENTURY, CUTLASS CIERA AND CUTLASS CRUISER

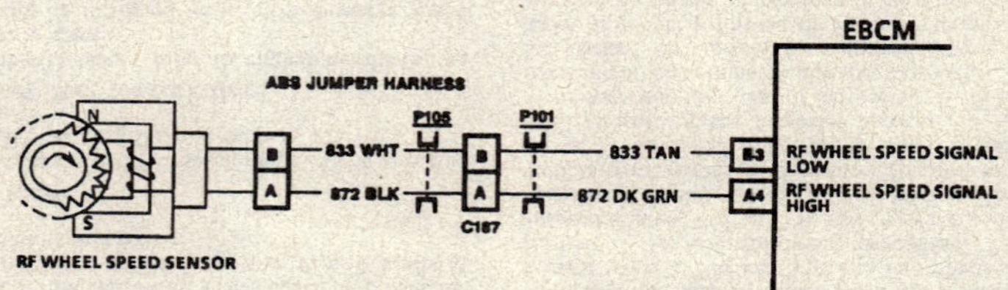

DTC 22
RIGHT FRONT WHEEL SPEED = 0

Circuit Description:

As a toothed ring passes by the wheel speed sensor, changes in the electromagnetic field cause the wheel speed sensor to produce a sinusoidal (AC) voltage signal whose frequency is proportional to wheel speed. The magnitude of this signal is directly related to wheel speed and the proximity of the wheel speed sensor to the toothed ring, often referred to as the air gap.

DTC Will Set When:

DTC 22 can be set when the vehicle is not in an ABS stop. If the right front wheel speed = 0 and the vehicle's reference speed is greater than 8 km/h (5 mph), a malfunction exists.

Action Taken:

A malfunction DTC is stored, ABS is disabled and the ABS warning lamp is turned "ON."

DTC Chart Test Description:

Number(s) below refer to circled number(s) on the diagnostic chart.

1. This step is designed to identify a wheel speed sensor or circuitry that is damaged and visibly apparent.
2. This test checks to see that the wheel speed sensor is within proper resistance values.
3. Verifies that the wheel speed sensor can generate an appropriate output signal.
4. Ensures that wheel speed sensor circuitry is not shorted together.
5. Ensures malfunction was not due to physical damage of the circuitry.

Diagnostic Aids:

An "Intermittent" malfunction may be caused by a poor connection, rubbed through wire insulation, or a wire that is broken inside the insulation.

The frequency of the malfunction can be checked by using the enhanced diagnostic function of the Tech 1.

If the customer's comments reflect that the ABS warning lamp is "ON" only during moist environmental changes (rain, snow, vehicle wash), all wheel speed sensor circuitry should be thoroughly inspected for signs of water intrusion. If DTC is not current, clear DTCs and simulate the effects of water intrusion. Use the following procedure. Spray the suspected area with a 5% salt water solution (two teaspoons of salt to 12 oz. of water). Test drive vehicle over various road surfaces (bumps, turns, etc.) above 24 km/h (15 mph) for at least 30 seconds. If DTC returns, replace suspected harness.

Any circuitry, that is suspected as causing the intermittent complaint, should be thoroughly checked for backed out terminals, improper mating, broken locks, improperly formed or damaged terminals, poor terminal to wiring connections or physical damage to the wiring harness.

Resistance of the wheel speed sensor will increase with an increase in sensor temperature.

When replacing a wheel speed sensor, inspect the sensor terminals and harness connector for corrosion and/or water intrusion. If evidence of corrosion or water intrusion exists, replace wheel speed sensor jumper harness. Likewise, if replacing a wheel speed sensor jumper harness, inspect sensor terminals. If evidence of corrosion or water intrusion exists, replace wheel speed sensor.

DIAGNOSTIC CHART DTC 22 — 1995 CENTURY, CUTLASS CIERA AND CUTLASS CRUISER

DTC 22
RIGHT FRONT WHEEL SPEED = 0

IMPORTANT:
- WHEEL SPEED SENSOR INTERMITTENT MALFUNCTIONS MAY BE DIFFICULT TO LOCATE. CARE SHOULD BE TAKEN NOT TO DISTURB ANY ELECTRICAL CONNECTIONS PRIOR TO AN INDICATED STEP OF THIS CHART. THIS WILL ENSURE THAT AN INTERMITTENT CONNECTION WILL NOT BE CORRECTED BEFORE THE SOURCE OF THE MALFUNCTION IS FOUND.

(1)
- IGNITION "OFF."
- PHYSICALLY INSPECT THE RIGHT FRONT WHEEL SPEED SENSOR, RIGHT FRONT WHEEL SPEED SENSOR JUMPER HARNESS, AND CONNECTORS FOR DAMAGE. INSPECT THE WHEEL SPEED SENSOR AND TOOTHED RING FOR LOOSENESS OR DAMAGE. PAY PARTICULAR ATTENTION TO THE TOOTHED RING. ANY DEVIATION WILL AFFECT THE WHEEL SPEED SENSOR OUTPUT SIGNAL.

IS PHYSICAL DAMAGE INDICATED?

- YES: REPLACE WHEEL SPEED SENSOR AND/OR JUMPER HARNESS IF DAMAGED. REPAIR ANY SENSOR OR TOOTHED RING LOOSENESS. DRIVE VEHICLE ABOVE 24 km/h (15 mph) TO ENSURE DTC 22 DOES NOT SET.
- NO: go to (2)

(2)
- DISCONNECT RIGHT FRONT WHEEL SPEED SENSOR DIRECTLY AT SENSOR.
- USING J 39200, MEASURE THE RESISTANCE BETWEEN SENSOR TERMINALS "A" AND "B."

IS RESISTANCE BETWEEN 1661 AND 1929 OHMS WHEN SENSOR IS 20°C (68°F)?

(IF TEMPERATURE OF THE SENSOR IS SUSPECTED TO BE OTHER THAN 20°C (68°F), REFER TO THE TABLE BELOW FOR APPLICABLE SENSOR RESISTANCE VALUES; THE VALUES IN THE TABLE ARE FOR TEMPERATURE OF THE SENSOR, NOT AIR TEMPERATURE.)

- NO: REPLACE WHEEL SPEED SENSOR.
- YES: go to (3)

(3)
- WITH J 39200 STILL CONNECTED, SELECT A/C VOLTAGE SCALE.
- SPIN WHEEL BY HAND WHILE OBSERVING VOLTAGE.

IS VOLTAGE AT LEAST 100 mV (VOLTAGE WILL INCREASE AS WHEEL SPEED INCREASES)?

- NO: VERIFY THAT TOOTHED RING IS INTACT. IF OK, REPLACE WHEEL SPEED SENSOR.
- YES: go to (4)

(4)
- DISCONNECT 24-WAY EBCM CONNECTOR.
- USING J 39200, MEASURE THE RESISTANCE BETWEEN TERMINAL "B3" AND "A4" OF THE 24-WAY EBCM HARNESS CONNECTOR.

DOES J 39200 DISPLAY RESISTANCE AS "OL" (INFINITE)?

- NO: REPAIR SHORT BETWEEN CKTs 833 AND 872.
- YES: go to (5)

(5)
- INSPECT CKTs 833 AND 872 AND 24-WAY EBCM CONNECTOR FOR DAMAGE WHICH MAY RESULT IN A SHORT BETWEEN CKTs 833 AND 872 WITH ALL CONNECTORS CONNECTED. REPAIR ALL DAMAGE IF EVIDENT.
- RECONNECT ALL CONNECTORS.
- TEST DRIVE VEHICLE ABOVE 24 km/h (15 mph) FOR AT LEAST 30 SECONDS.

DOES DTC 22 SET AS A CURRENT DTC?

- YES: REPLACE EBCM.
- NO: MALFUNCTION IS INTERMITTENT. REFER TO "DIAGNOSTIC AIDS".

FRONT WHEEL SPEED SENSOR

TEMPERATURE OF SENSOR VS. SENSOR RESISTANCE (APPROXIMATE)

TEMP. (°C)	TEMP. (°F)	RESISTANCE (OHMS)
-34 TO 4	-30 TO 40	1258 TO 1808
5 TO 43	41 TO 110	1502 TO 2103
44 TO 93	111 TO 200	1747 TO 2482

DIAGNOSTIC CHART DTC 23 — 1995 CENTURY, CUTLASS CIERA AND CUTLASS CRUISER

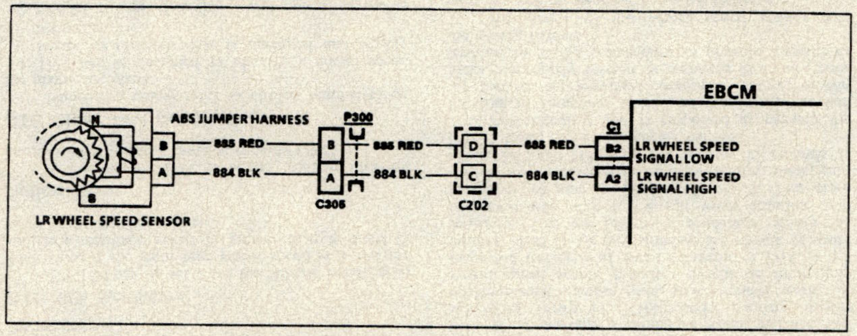

DTC 23
LEFT REAR WHEEL SPEED = 0

Circuit Description:

As a toothed ring passes by the wheel speed sensor, changes in the electromagnetic field cause the wheel speed sensor to produce a sinusoidal (AC) voltage signal whose frequency is proportional to wheel speed. The magnitude of this signal is directly related to wheel speed and the proximity of the wheel speed sensor to the toothed ring, often referred to as the air gap.

DTC Will Set When:

DTC 23 can be set when the vehicle is not in an ABS stop. If the left rear wheel speed = 0 and the vehicle's reference speed is greater than 8 km/h (5 mph), a malfunction exists.

Action Taken:

A malfunction DTC is stored, ABS is disabled and the ABS warning lamp is turned "ON."

DTC Chart Test Description:

Number(s) below refer to circled number(s) on the diagnostic chart.

1. This step is designed to identify a wheel speed sensor or circuitry that is damaged and visibly apparent.
2. Checks to see that the wheel speed sensor is within proper resistance values.
3. Verifies that the wheel speed sensor can generate an appropriate output signal.
4. Ensures that the wheel speed sensor circuitry is not shorted together.
5. Checks to see if malfunction was due to physical damage of the circuitry.
6. Checks for an incorrect or malfunctioning hub and bearing assembly.

Diagnostic Aids:

An "Intermittent" malfunction may be caused by a poor connection, rubbed through wire insulation, or a wire that is broken inside the insulation.

The frequency of the malfunction can be checked by using the enhanced diagnostic function of the Tech 1.

If the customer's comments reflect that the ABS warning lamp is "ON" only during moist environmental changes (rain, snow, vehicle wash), all wheel speed sensor circuitry should be thoroughly inspected for signs of water intrusion. If DTC is not current, clear DTCs and simulate the effects of water intrusion. Use the following procedure. Spray the suspected area with a 5% salt water solution (two teaspoons of salt to 12 oz. of water). Test drive vehicle over various road surfaces (bumps, turns, etc.) above 24 km/h (15 mph) for at least 30 seconds. If DTC returns, replace suspected harness.

Any circuitry that is suspected as causing the intermittent complaint should be thoroughly checked for backed out terminals, improper mating, broken locks, improperly formed or damaged terminals, poor terminal to wiring connections or physical damage to the wiring harness.

Resistance of the wheel speed sensor will increase with an increase in sensor temperature.

When replacing a wheel speed sensor, inspect the sensor terminals and harness connector for corrosion and/or water intrusion. If evidence of corrosion or water intrusion exists, replace wheel speed sensor jumper harness. Likewise, if replacing a wheel speed sensor jumper harness, inspect sensor terminals. If evidence of corrosion or water intrusion exists, replace wheel speed sensor.

DIAGNOSTIC CHART DTC 23 — 1995 CENTURY, CUTLASS CIERA AND CUTLASS CRUISER

DTC 23
LEFT REAR WHEEL WHEEL SPEED = 0

IMPORTANT:

- WHEEL SPEED SENSOR INTERMITTENT MALFUNCTIONS MAY BE DIFFICULT TO LOCATE. CARE SHOULD BE TAKEN NOT TO DISTURB ANY ELECTRICAL CONNECTIONS PRIOR TO AN INDICATED STEP OF THIS CHART. THIS WILL ENSURE THAT AN INTERMITTENT CONNECTION WILL NOT BE CORRECTED BEFORE THE SOURCE OF THE MALFUNCTION IS FOUND.

(1)
- IGNITION "OFF."
- PHYSICALLY INSPECT THE LEFT REAR WHEEL SPEED SENSOR, WHEEL SPEED SENSOR JUMPER HARNESS, AND CONNECTORS FOR DAMAGE.

IS PHYSICAL DAMAGE INDICATED?

YES → REPLACE WHEEL SPEED SENSOR AND/OR WHEEL SPEED SENSOR JUMPER HARNESS IF DAMAGED. DRIVE VEHICLE ABOVE 24 km/h (15 mph) TO ENSURE DTC 23 DOES NOT SET.

NO ↓

(2)
- DISCONNECT LEFT REAR WHEEL SPEED SENSOR DIRECTLY AT SENSOR.
- USING J 39200, MEASURE THE RESISTANCE BETWEEN SENSOR TERMINALS "A" AND "B."

IS RESISTANCE BETWEEN 2059 AND 2353 OHMS WHEN SENSOR IS 20°C (68°F)? *(IF TEMPERATURE OF THE SENSOR IS SUSPECTED TO BE OTHER THAN 20°C (68°F), REFER TO THE TABLE BELOW FOR APPLICABLE SENSOR RESISTANCE VALUES; THE VALUES IN THE TABLE ARE FOR TEMPERATURE OF THE SENSOR, NOT AIR TEMPERATURE.)*

NO → REPLACE WHEEL SPEED SENSOR.

YES ↓

(3)
- WITH J 39200 STILL CONNECTED, SELECT A/C VOLTAGE SCALE.
- SPIN WHEEL BY HAND WHILE OBSERVING VOLTAGE.

IS VOLTAGE AT LEAST 100 mV (VOLTAGE WILL INCREASE AS WHEEL SPEED INCREASES)?

NO → REPLACE WHEEL SPEED SENSOR.

YES ↓

(4)
- DISCONNECT 24-WAY EBCM CONNECTOR.
- USING J 39200, MEASURE THE RESISTANCE BETWEEN TERMINAL "A3" AND "B2" OF THE 24-WAY EBCM HARNESS CONNECTOR.

DOES J 39200 DISPLAY RESISTANCE AS "OL" (INFINITE)?

NO → REPAIR SHORT BETWEEN CKTs 884 AND 885.

YES ↓

(5)
- INSPECT CKTs 884 AND 885 AND 24-WAY EBCM CONNECTOR FOR DAMAGE WHICH MAY RESULT IN A SHORT BETWEEN CKTs 884 AND 885 WITH ALL CONNECTORS CONNECTED. REPAIR ALL DAMAGE IF EVIDENT.
- RECONNECT ALL CONNECTORS.
- TEST DRIVE VEHICLE ABOVE 24 km/h (15 mph) FOR AT LEAST 30 SECONDS.

DOES DTC 23 SET AS A CURRENT DTC?

NO → MALFUNCTION WAS MOST LIKELY CAUSED BY A POOR CONNECTION. REFER TO "DIAGNOSTIC AIDS".

YES ↓

(6)
- USING TECH 1, SELECT "F0: DATA LIST."
- MONITOR WHEEL SPEEDS WHILE SLOWLY ACCELERATING TO 65 km/h (40 mph) AND THEN SLOWLY DECELERATING TO 0 km/h (0 mph).

IS THE LEFT REAR WHEEL SPEED CONSTANTLY HIGHER THAN THE THREE REMAINING WHEEL SPEEDS?

NO → REPLACE EBCM.

YES → REPLACE INCORRECT OR MALFUNCTIONING LEFT WHEEL SPEED SENSOR/HUB AND BEARING ASSEMBLY WITH THE PROPER ASSEMBLY. REFER TO PARTS CATALOG FOR PROPER PART NUMBER.

REAR WHEEL SPEED SENSOR
TEMPERATURE OF SENSOR VS. SENSOR RESISTANCE (APPROXIMATE)

TEMP. (°C)	TEMP. (°F)	RESISTANCE (OHMS)
-34 TO 4	-30 TO 40	1622 TO 2205
5 TO 43	41 TO 110	1930 TO 2566
44 TO 93	111 TO 200	2245 TO 3028

DIAGNOSTIC CHART DTC 24 — 1995 CENTURY, CUTLASS CIERA AND CUTLASS CRUISER

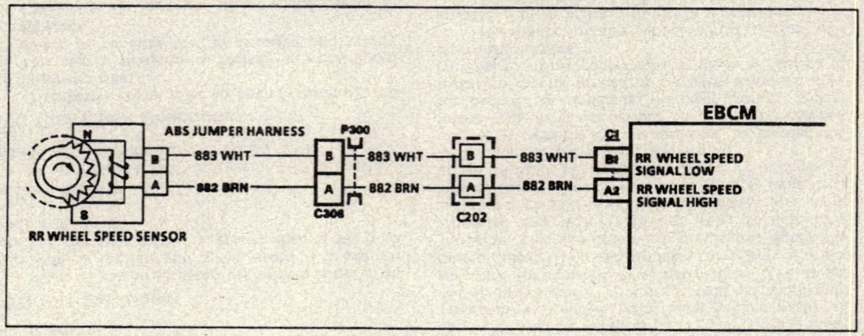

DTC 24
RIGHT REAR WHEEL SPEED = 0

Circuit Description:

As a toothed ring passes by the wheel speed sensor, changes in the electromagnetic field cause the wheel speed sensor to produce a sinusoidal (AC) voltage signal whose frequency is proportional to wheel speed. The magnitude of this signal is directly related to wheel speed and the proximity of the wheel speed sensor to the toothed ring, often referred to as the air gap.

DTC Will Set When:

DTC 24 can be set when the vehicle is not in an ABS stop. If the right rear wheel speed = 0 and the vehicle's reference speed is greater than 8 km/h (5 mph), a malfunction exists.

Action Taken:

A malfunction DTC is stored, ABS is disabled and the ABS warning lamp is turned "ON."

DTC Chart Test Description:

Number(s) below refer to circled number(s) on the diagnostic chart.

1. This step is designed to identify a wheel speed sensor or circuitry that is damaged and visibly apparent.
2. Checks to see that the wheel speed sensor is within proper resistance values.
3. Verifies that the wheel speed sensor can generate an appropriate output signal.
4. Ensures that the wheel speed sensor circuitry is not shorted together.
5. Checks to see if malfunction was due to physical damage of the circuitry.
6. Checks for an incorrect or malfunctioning hub and bearing assembly.

Diagnostic Aids:

An "Intermittent" malfunction may be caused by a poor connection, rubbed through wire insulation, or a wire that is broken inside the insulation.

The frequency of the malfunction can be checked by using the enhanced diagnostic function of the Tech 1.

If the customer's comments reflect that the ABS warning lamp is "ON" only during moist environmental changes (rain, snow, vehicle wash), all wheel speed sensor circuitry should be thoroughly inspected for signs of water intrusion. If DTC is not current, clear DTCs and simulate the effects of water intrusion. Use the following procedure. Spray the suspected area with a 5% salt water solution (two teaspoons of salt to 12 oz. of water). Test drive vehicle over various road surfaces (bumps, turns, etc.) above 24 km/h (15 mph) for at least 30 seconds. If DTC returns, replace suspected harness.

Any circuitry that is suspected as causing the intermittent complaint should be thoroughly checked for backed out terminals, improper mating, broken locks, improperly formed or damaged terminals, poor terminal to wiring connections or physical damage to the wiring harness.

Resistance of the wheel speed sensor will increase with an increase in sensor temperature.

When replacing a wheel speed sensor, inspect the sensor terminals and harness connector for corrosion and/or water intrusion. If evidence of corrosion or water intrusion exists, replace wheel speed sensor jumper harness. Likewise, if replacing a wheel speed sensor jumper harness, inspect sensor terminals. If evidence of corrosion or water intrusion exists, replace wheel speed sensor.

DIAGNOSTIC CHART DTC 24 — 1995 CENTURY, CUTLASS CIERA AND CUTLASS CRUISER

DTC 24
RIGHT REAR WHEEL SPEED = 0

IMPORTANT:

- WHEEL SPEED SENSOR INTERMITTENT MALFUNCTIONS MAY BE DIFFICULT TO LOCATE. CARE SHOULD BE TAKEN NOT TO DISTURB ANY ELECTRICAL CONNECTIONS PRIOR TO AN INDICATED STEP OF THIS CHART. THIS WILL ENSURE THAT AN INTERMITTENT CONNECTION WILL NOT BE CORRECTED BEFORE THE SOURCE OF THE MALFUNCTION IS FOUND.

(1)
- IGNITION "OFF."
- PHYSICALLY INSPECT THE RIGHT REAR WHEEL SPEED SENSOR, WHEEL SPEED SENSOR JUMPER HARNESS, AND CONNECTORS FOR DAMAGE.

IS PHYSICAL DAMAGE INDICATED?

YES: REPLACE WHEEL SPEED SENSOR AND/OR WHEEL SPEED SENSOR JUMPER HARNESS IF DAMAGED. DRIVE VEHICLE ABOVE 24 km/h (15 mph) TO ENSURE DTC 24 DOES NOT SET.

NO: go to (2).

(2)
- DISCONNECT RIGHT REAR WHEEL SPEED SENSOR DIRECTLY AT SENSOR.
- USING J 39200, MEASURE THE RESISTANCE BETWEEN SENSOR TERMINALS "A" AND "B."

IS RESISTANCE BETWEEN 2059 AND 2353 OHMS WHEN SENSOR IS 20°C (68°F)? *(IF TEMPERATURE OF THE SENSOR IS SUSPECTED TO BE OTHER THAN 20°C (68°F), REFER TO THE TABLE BELOW FOR APPLICABLE SENSOR RESISTANCE VALUES; THE VALUES IN THE TABLE ARE FOR TEMPERATURE OF THE SENSOR, NOT AIR TEMPERATURE.)*

NO: REPLACE WHEEL SPEED SENSOR.

YES: go to (3).

(3)
- WITH J 39200 STILL CONNECTED, SELECT A/C VOLTAGE SCALE.
- SPIN WHEEL BY HAND WHILE OBSERVING VOLTAGE.

IS VOLTAGE AT LEAST 100 mV (VOLTAGE WILL INCREASE AS WHEEL SPEED INCREASES)?

NO: REPLACE WHEEL SPEED SENSOR.

YES: go to (4).

(4)
- DISCONNECT 24-WAY EBCM CONNECTOR.
- USING J 39200, MEASURE THE RESISTANCE BETWEEN TERMINAL "B1" AND "A2" OF THE 24-WAY EBCM HARNESS CONNECTOR.

DOES J 39200 DISPLAY RESISTANCE AS "OL" (INFINITE)?

NO: REPAIR SHORT BETWEEN CKTs 882 AND 883.

YES: go to (5).

(5)
- INSPECT CKTs 882 AND 883 AND 24-WAY EBCM CONNECTOR FOR DAMAGE WHICH MAY RESULT IN A SHORT BETWEEN CKTs 882 AND 883 WITH ALL CONNECTORS CONNECTED. REPAIR ALL DAMAGE IF EVIDENT.
- RECONNECT ALL CONNECTORS.
- TEST DRIVE VEHICLE ABOVE 24 km/h (15 mph) FOR AT LEAST 30 SECONDS.

DOES DTC 24 SET AS A CURRENT DTC?

NO: MALFUNCTION WAS MOST LIKELY CAUSED BY A POOR CONNECTION. REFER TO "DIAGNOSTIC AIDS".

YES: go to (6).

(6)
- USING TECH 1, SELECT "F0: DATA LIST."
- MONITOR WHEEL SPEEDS WHILE SLOWLY ACCELERATING TO 65 km/h (40 mph) AND THEN SLOWLY DECELERATING TO 0 km/h (0 mph).

IS THE RIGHT REAR WHEEL SPEED CONSTANTLY HIGHER THAN THE THREE REMAINING WHEEL SPEEDS?

NO: REPLACE EBCM.

YES: REPLACE INCORRECT OR MALFUNCTIONING RIGHT WHEEL SPEED SENSOR/HUB AND BEARING ASSEMBLY WITH THE PROPER ASSEMBLY. REFER TO PARTS CATALOG FOR PROPER PART NUMBER.

REAR WHEEL SPEED SENSOR
TEMPERATURE OF SENSOR VS. SENSOR RESISTANCE (APPROXIMATE)

TEMP. (°C)	TEMP. (°F)	RESISTANCE (OHMS)
-34 TO 4	-30 TO 40	1822 TO 2206
5 TO 43	41 TO 110	1930 TO 2566
44 TO 93	111 TO 200	2246 TO 3026

DIAGNOSTIC CHART DTC 25 — 1995 CENTURY, CUTLASS CIERA AND CUTLASS CRUISER

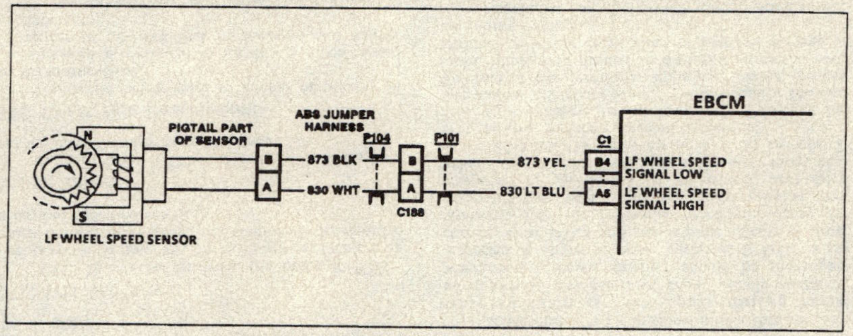

DTC 25
LEFT FRONT EXCESSIVE WHEEL SPEED VARIATION

Circuit Description:

As a toothed ring passes by the wheel speed sensor, changes in the electromagnetic field cause the wheel speed sensor to produce a sinusoidal (AC) voltage signal whose frequency is proportional to wheel speed. The magnitude of this signal is directly related to wheel speed and the proximity of the wheel speed sensor to the toothed ring, often referred to as the air gap.

DTC Will Set When:

DTC 25 can be set when the brake is "OFF." The purpose of this test is to detect a situation in which the left front wheel acceleration or deceleration is beyond specified limits.

Action Taken:

A malfunction DTC is stored, ABS is disabled and the ABS warning lamp is turned "ON."

DTC Chart Test Description:

Number(s) below refer to circled number(s) on the diagnostic chart.

1. This step is designed to identify a wheel speed sensor or circuitry that is damaged and visibly apparent.
2. Checks to see that the wheel speed sensor is within proper resistance values.
3. Verifies that the wheel speed sensor can generate an appropriate output signal.
4. Ensures that wheel speed sensor circuitry is not shorted together.
5. Ensures malfunction was not due to physical damage of the circuitry.

Diagnostic Aids:

An "Intermittent" malfunction may be caused by a poor connection, rubbed through wire insulation, or a wire that is broken inside the insulation.

The frequency of the malfunction can be checked by using the enhanced diagnostic function of the Tech 1.

If the customer's comments reflect that the ABS warning lamp is "ON" only during moist environmental changes (rain, snow, vehicle wash), all wheel speed sensor circuitry should be thoroughly inspected for signs of water intrusion. If DTC is not current, clear DTCs and simulate the effects of water intrusion. Use the following procedure. Spray the suspected area with a 5% salt water solution (two teaspoons of salt to 12 oz. of water). Test drive vehicle over various road surfaces (bumps, turns, etc.) above 24 km/h (15 mph) for at least 30 seconds. If DTC returns, replace suspected harness.

Any circuitry that is suspected of causing the intermittent complaint should be thoroughly checked for backed out terminals, improper mating, broken locks, improperly formed or damaged terminals, poor terminal to wiring connections or physical damage to the wiring harness.

Resistance of the wheel speed sensor will increase with an increase in sensor temperature.

When replacing a wheel speed sensor, inspect the sensor terminals and harness connector for corrosion and/or water intrusion. If evidence of corrosion or water intrusion exists, replace wheel speed sensor jumper harness. Likewise, if replacing a wheel speed sensor jumper harness, inspect sensor terminals. If evidence of corrosion or water intrusion exists, replace wheel speed sensor.

DIAGNOSTIC CHART DTC 25 — 1995 CENTURY, CUTLASS CIERA AND CUTLASS CRUISER

DTC 25
LEFT FRONT EXCESSIVE WHEEL SPEED VARIATION

IMPORTANT:
- WHEEL SPEED SENSOR INTERMITTENT MALFUNCTIONS MAY BE DIFFICULT TO LOCATE. CARE SHOULD BE TAKEN NOT TO DISTURB ANY ELECTRICAL CONNECTIONS PRIOR TO AN INDICATED STEP OF THIS CHART. THIS WILL ENSURE THAT AN INTERMITTENT CONNECTION WILL NOT BE CORRECTED BEFORE THE SOURCE OF THE MALFUNCTION IS FOUND.

(1)
- IGNITION "OFF."
- PHYSICALLY INSPECT THE LEFT FRONT WHEEL SPEED SENSOR, LEFT FRONT WHEEL SPEED SENSOR JUMPER HARNESS, AND CONNECTORS FOR DAMAGE. INSPECT THE WHEEL SPEED SENSOR AND TOOTHED RING FOR LOOSENESS OR DAMAGE. PAY PARTICULAR ATTENTION TO THE TOOTHED RING. ANY DEVIATION WILL AFFECT THE WHEEL SPEED SENSOR OUTPUT SIGNAL.

IS PHYSICAL DAMAGE INDICATED?

- YES: REPLACE WHEEL SPEED SENSOR AND/OR JUMPER HARNESS IF DAMAGED. REPAIR ANY SENSOR OR TOOTHED RING LOOSENESS. DRIVE VEHICLE ABOVE 24 km/h (15 mph) TO ENSURE DTC 25 DOES NOT SET.
- NO: go to (2)

(2)
- DISCONNECT LEFT FRONT WHEEL SPEED SENSOR DIRECTLY AT SENSOR.
- USING J 39200, MEASURE THE RESISTANCE BETWEEN SENSOR TERMINALS "A" AND "B."

IS RESISTANCE BETWEEN 1651 AND 1929 OHMS WHEN SENSOR IS 20°C (68°F)? *(IF TEMPERATURE OF THE SENSOR IS SUSPECTED TO BE OTHER THAN 20°C (68°F), REFER TO THE TABLE BELOW FOR APPLICABLE SENSOR RESISTANCE VALUES; THE VALUES IN THE TABLE ARE FOR TEMPERATURE OF THE SENSOR, NOT AIR TEMPERATURE.)*

- NO: REPLACE WHEEL SPEED SENSOR.
- YES: go to (3)

(3)
- WITH J 39200 STILL CONNECTED, SELECT A/C VOLTAGE SCALE.
- SPIN WHEEL BY HAND WHILE OBSERVING VOLTAGE.

IS VOLTAGE AT LEAST 100 mV (VOLTAGE WILL INCREASE AS WHEEL SPEED INCREASES)?

- NO: VERIFY THAT TOOTHED RING IS INTACT. IF OK, REPLACE WHEEL SPEED SENSOR.
- YES: go to (4)

(4)
- DISCONNECT 24-WAY EBCM CONNECTOR.
- USING J 39200, MEASURE THE RESISTANCE BETWEEN TERMINAL "B4" AND "A5" OF THE 24-WAY EBCM HARNESS CONNECTOR.

DOES J 39200 DISPLAY RESISTANCE AS "OL" (INFINITE)?

- NO: REPAIR SHORT BETWEEN CKTs 830 AND 873.
- YES: go to (5)

(5)
- INSPECT CKTs 830 AND 873 AND 24-WAY EBCM CONNECTOR FOR DAMAGE WHICH MAY RESULT IN A SHORT BETWEEN CKTs 830 AND 873 WITH ALL CONNECTORS CONNECTED. REPAIR ALL DAMAGE IF EVIDENT.
- RECONNECT ALL CONNECTORS.
- TEST DRIVE VEHICLE ABOVE 24 km/h (15 mph) FOR AT LEAST 30 SECONDS.

DOES DTC 25 SET AS A CURRENT DTC?

- YES: REPLACE EBCM.
- NO: MALFUNCTION IS INTERMITTENT. REFER TO "DIAGNOSTIC AIDS".

FRONT WHEEL SPEED SENSOR

TEMPERATURE OF SENSOR VS. SENSOR RESISTANCE (APPROXIMATE)

TEMP. (°C)	TEMP. (°F)	RESISTANCE (OHMS)
-34 TO 4	-30 TO 40	1258 TO 1808
5 TO 43	41 TO 110	1502 TO 2103
44 TO 93	111 TO 200	1747 TO 2482

DIAGNOSTIC CHART DTC 26 — 1995 CENTURY, CUTLASS CIERA AND CUTLASS CRUISER

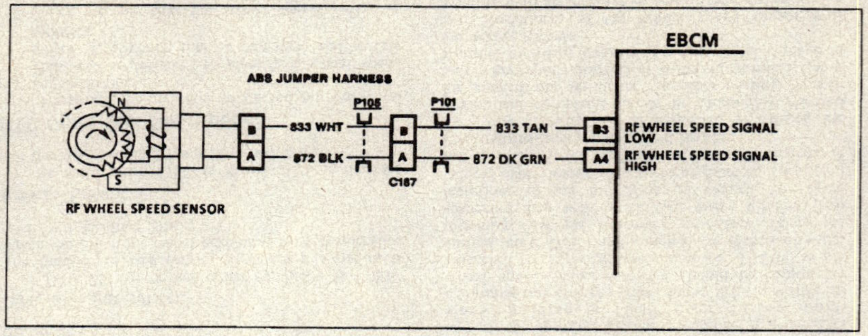

DTC 26
RIGHT FRONT EXCESSIVE WHEEL SPEED VARIATION

Circuit Description:

As a toothed ring passes by the wheel speed sensor, changes in the electromagnetic field cause the wheel speed sensor to produce a sinusoidal (AC) voltage signal whose frequency is proportional to wheel speed. The magnitude of this signal is directly related to wheel speed and the proximity of the wheel speed sensor to the toothed ring, often referred to as the air gap.

DTC Will Set When:

DTC 26 can be set when the brake is "OFF." The purpose of this test is to detect a situation in which the left front wheel acceleration or deceleration is beyond specified limits.

Action Taken:

A malfunction DTC is stored, ABS is disabled and the ABS warning lamp is turned "ON."

DTC Chart Test Description:

Number(s) below refer to circled number(s) on the diagnostic chart.

1. This step is designed to identify a wheel speed sensor or circuitry that is damaged and visibly apparent.
2. Checks to see that the wheel speed sensor is within proper resistance values.
3. Verifies that the wheel speed sensor can generate an appropriate output signal.
4. Ensures that wheel speed sensor circuitry is not shorted together.
5. Ensures malfunction was not due to physical damage of the circuitry.

Diagnostic Aids:

An "Intermittent" malfunction may be caused by a poor connection, rubbed through wire insulation, or a wire that is broken inside the insulation.

The frequency of the malfunction can be checked by using the enhanced diagnostic function of the Tech 1.

If the customer's comments reflect that the ABS warning lamp is "ON" only during moist environmental changes (rain, snow, vehicle wash), all wheel speed sensor circuitry should be thoroughly inspected for signs of water intrusion. If DTC is not current, clear DTCs and simulate the effects of water intrusion. Use the following procedure. Spray the suspected area with a 5% salt water solution (two teaspoons of salt to 12 oz. of water). Test drive vehicle over various road surfaces (bumps, turns, etc.) above 24 km/h (15 mph) for at least 30 seconds. If DTC returns, replace suspected harness.

Any circuitry that is suspected of causing the intermittent complaint should be thoroughly checked for backed out terminals, improper mating, broken locks, improperly formed or damaged terminals, poor terminal to wiring connections or physical damage to the wiring harness.

Resistance of the wheel speed sensor will increase with an increase in sensor temperature.

When replacing a wheel speed sensor, inspect the sensor terminals and harness connector for corrosion and/or water intrusion. If evidence of corrosion or water intrusion exists, replace wheel speed sensor jumper harness. Likewise, if replacing a wheel speed sensor jumper harness, inspect sensor terminals. If evidence of corrosion or water intrusion exists, replace wheel speed sensor.

DIAGNOSTIC CHART DTC 26 — 1995 CENTURY, CUTLASS CIERA AND CUTLASS CRUISER

DTC 26
RIGHT FRONT EXCESSIVE WHEEL SPEED VARIATION

IMPORTANT:
- WHEEL SPEED SENSOR INTERMITTENT MALFUNCTIONS MAY BE DIFFICULT TO LOCATE. CARE SHOULD BE TAKEN NOT TO DISTURB ANY ELECTRICAL CONNECTIONS PRIOR TO AN INDICATED STEP OF THIS CHART. THIS WILL ENSURE THAT AN INTERMITTENT CONNECTION WILL NOT BE CORRECTED BEFORE THE SOURCE OF THE MALFUNCTION IS FOUND.

(1)
- IGNITION "OFF."
- PHYSICALLY INSPECT THE RIGHT FRONT WHEEL SPEED SENSOR, RIGHT FRONT WHEEL SPEED SENSOR JUMPER HARNESS, AND CONNECTORS FOR DAMAGE. INSPECT THE WHEEL SPEED SENSOR AND TOOTHED RING FOR LOOSENESS OR DAMAGE. PAY PARTICULAR ATTENTION TO THE TOOTHED RING. ANY DEVIATION WILL AFFECT THE WHEEL SPEED SENSOR OUTPUT SIGNAL.

IS PHYSICAL DAMAGE INDICATED?

- YES: REPLACE WHEEL SPEED SENSOR AND/OR JUMPER HARNESS IF DAMAGED. REPAIR ANY SENSOR OR TOOTHED RING LOOSENESS. DRIVE VEHICLE ABOVE 24 km/h (15 mph) TO ENSURE DTC 26 DOES NOT SET.
- NO: go to step 2.

(2)
- DISCONNECT RIGHT FRONT WHEEL SPEED SENSOR DIRECTLY AT SENSOR.
- USING J 39200, MEASURE THE RESISTANCE BETWEEN SENSOR TERMINALS "A" AND "B."

IS RESISTANCE BETWEEN 1681 AND 1929 OHMS WHEN SENSOR IS 20°C (68°F)? *(IF TEMPERATURE OF THE SENSOR IS SUSPECTED TO BE OTHER THAN 20°C (68°F), REFER TO THE TABLE BELOW FOR APPLICABLE SENSOR RESISTANCE VALUES; THE VALUES IN THE TABLE ARE FOR TEMPERATURE OF THE SENSOR, NOT AIR TEMPERATURE.)*

- NO: REPLACE WHEEL SPEED SENSOR.
- YES: go to step 3.

(3)
- WITH J 39200 STILL CONNECTED, SELECT A/C VOLTAGE SCALE.
- SPIN WHEEL BY HAND WHILE OBSERVING VOLTAGE.

IS VOLTAGE AT LEAST 100 mV (VOLTAGE WILL INCREASE AS WHEEL SPEED INCREASES)?

- NO: VERIFY THAT TOOTHED RING IS INTACT. IF OK, REPLACE WHEEL SPEED SENSOR.
- YES: go to step 4.

(4)
- DISCONNECT 24-WAY EBCM CONNECTOR.
- USING J 39200, MEASURE THE RESISTANCE BETWEEN TERMINAL "B3" AND "A4" OF THE 24-WAY EBCM HARNESS CONNECTOR.

DOES J 39200 DISPLAY RESISTANCE AS "OL" (INFINITE)?

- NO: REPAIR SHORT BETWEEN CKTs 833 AND 872.
- YES: go to step 5.

(5)
- INSPECT CKTs 833 AND 872 AND 24-WAY EBCM CONNECTOR FOR DAMAGE WHICH MAY RESULT IN A SHORT BETWEEN CKTs 833 AND 872 WITH ALL CONNECTORS CONNECTED. REPAIR ALL DAMAGE IF EVIDENT.
- RECONNECT ALL CONNECTORS.
- TEST DRIVE VEHICLE ABOVE 24 km/h (15 mph) FOR AT LEAST 30 SECONDS.

DOES DTC 26 SET AS A CURRENT DTC?

- YES: REPLACE EBCM.
- NO: MALFUNCTION IS INTERMITTENT. REFER TO "DIAGNOSTIC AIDS".

FRONT WHEEL SPEED SENSOR

TEMPERATURE OF SENSOR VS. SENSOR RESISTANCE (APPROXIMATE)

TEMP. (°C)	TEMP. (°F)	RESISTANCE (OHMS)
-34 TO 4	-30 TO 40	1258 TO 1808
5 TO 43	41 TO 110	1502 TO 2103
44 TO 93	111 TO 200	1747 TO 2482

DIAGNOSTIC CHART DTC 27 — 1995 CENTURY, CUTLASS CIERA AND CUTLASS CRUISER

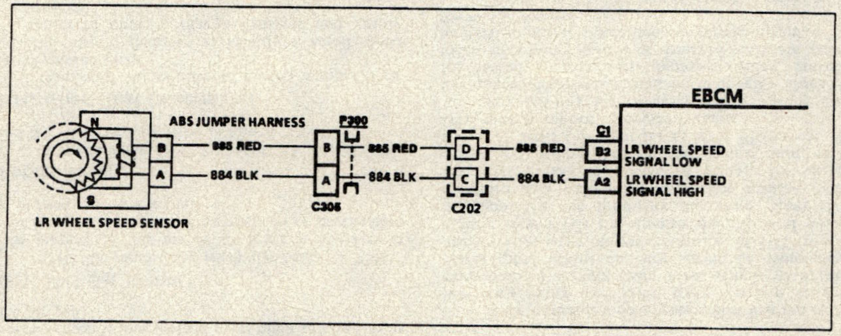

DTC 27
LEFT REAR EXCESSIVE WHEEL SPEED VARIATION

Circuit Description:

As a toothed ring passes by the wheel speed sensor, changes in the electromagnetic field cause the wheel speed sensor to produce a sinusoidal (AC) voltage signal whose frequency is proportional to wheel speed. The magnitude of this signal is directly related to wheel speed and the proximity of the wheel speed sensor to the toothed ring, often referred to as the air gap.

DTC Will Set When:

DTC 27 can be set when the brake is "OFF." The purpose of this test is to detect a situation in which the left rear wheel acceleration or deceleration is beyond specified limits.

Action Taken:

A malfunction DTC is stored, ABS is disabled and the ABS warning lamp is turned "ON."

DTC Chart Test Description:

Number(s) below refer to circled number(s) on the diagnostic chart.

1. This step is designed to identify a wheel speed sensor or circuitry that is damaged and visibly apparent.
2. Checks to see that the wheel speed sensor is within proper resistance values.
3. Verifies that the wheel speed sensor can generate an appropriate output signal.
4. Ensures that wheel speed sensor circuitry is not shorted together.
5. Ensures malfunction was not due to physical damage of the circuitry.

Diagnostic Aids:

An "Intermittent" malfunction may be caused by a poor connection, rubbed through wire insulation, or a wire that is broken inside the insulation.

The frequency of the malfunction can be checked by using the enhanced diagnostic function of the Tech I.

If the customer's comments reflect that the ABS warning lamp is "ON" only during moist environmental changes (rain, snow, vehicle wash), all wheel speed sensor circuitry should be thoroughly inspected for signs of water intrusion. If DTC is not current, clear DTCs and simulate the effects of water intrusion. Use the following procedure. Spray the suspected area with a 5% salt water solution (two teaspoons of salt to 12 oz. of water). Test drive vehicle over various road surfaces (bumps, turns, etc.) above 24 km/h (15 mph) for at least 30 seconds. If DTC returns, replace suspected harness.

Any circuitry that is suspected of causing the intermittent complaint should be thoroughly checked for backed out terminals, improper mating, broken locks, improperly formed or damaged terminals, poor terminal to wiring connections or physical damage to the wiring harness.

Resistance of the wheel speed sensor will increase with an increase in sensor temperature.

When replacing a wheel speed sensor, inspect the sensor terminals and harness connector for corrosion and/or water intrusion. If evidence of corrosion or water intrusion exists, replace wheel speed sensor jumper harness. Likewise, if replacing a wheel speed sensor jumper harness, inspect sensor terminals. If evidence of corrosion or water intrusion exists, replace wheel speed sensor.

DIAGNOSTIC CHART DTC 27 — 1995 CENTURY, CUTLASS CIERA AND CUTLASS CRUISER

DTC 27
LEFT REAR EXCESSIVE WHEEL SPEED VARIATION

IMPORTANT:
- WHEEL SPEED SENSOR INTERMITTENT MALFUNCTIONS MAY BE DIFFICULT TO LOCATE. CARE SHOULD BE TAKEN NOT TO DISTURB ANY ELECTRICAL CONNECTIONS PRIOR TO AN INDICATED STEP OF THIS CHART. THIS WILL ENSURE THAT AN INTERMITTENT CONNECTION WILL NOT BE CORRECTED BEFORE THE SOURCE OF THE MALFUNCTION IS FOUND.

(1)
- IGNITION "OFF."
- PHYSICALLY INSPECT THE LEFT REAR WHEEL SPEED SENSOR, WHEEL SPEED SENSOR JUMPER HARNESS, AND CONNECTORS FOR DAMAGE.

IS PHYSICAL DAMAGE INDICATED?

- YES: REPLACE WHEEL SPEED SENSOR AND/OR WHEEL SPEED SENSOR JUMPER HARNESS IF DAMAGED. DRIVE VEHICLE ABOVE 24 km/h (15 mph) TO ENSURE DTC 27 DOES NOT SET.
- NO: go to (2)

(2)
- DISCONNECT LEFT FRONT WHEEL SPEED SENSOR DIRECTLY AT SENSOR.
- USING J 39200, MEASURE THE RESISTANCE BETWEEN SENSOR TERMINALS "A" AND "B."

IS RESISTANCE BETWEEN 2059 AND 2353 OHMS WHEN SENSOR IS 20°C (68°F)? *(IF TEMPERATURE OF THE SENSOR IS SUSPECTED TO BE OTHER THAN 20°C (68°F), REFER TO THE TABLE BELOW FOR APPLICABLE SENSOR RESISTANCE VALUES: THE VALUES IN THE TABLE ARE FOR TEMPERATURE OF THE SENSOR, NOT AIR TEMPERATURE.)*

- NO: REPLACE WHEEL SPEED SENSOR.
- YES: go to (3)

(3)
- WITH J 39200 STILL CONNECTED, SELECT A/C VOLTAGE SCALE.
- SPIN WHEEL BY HAND WHILE OBSERVING VOLTAGE.

IS VOLTAGE AT LEAST 100 mV (VOLTAGE WILL INCREASE AS WHEEL SPEED INCREASES)?

- NO: REPLACE WHEEL SPEED SENSOR.
- YES: go to (4)

(4)
- DISCONNECT 24-WAY EBCM CONNECTOR.
- USING J 39200, MEASURE THE RESISTANCE BETWEEN TERMINAL "A3" AND "B2" OF THE 24-WAY EBCM HARNESS CONNECTOR.

DOES J 39200 DISPLAY RESISTANCE AS "OL" (INFINITE)?

- NO: REPAIR SHORT BETWEEN CKTs 884 AND 885.
- YES: go to (5)

(5)
- INSPECT CKTs 884 AND 885 AND 24-WAY EBCM CONNECTOR FOR DAMAGE WHICH MAY RESULT IN A SHORT BETWEEN CKTs 884 AND 885 WITH ALL CONNECTORS CONNECTED. REPAIR ALL DAMAGE IF EVIDENT.
- RECONNECT ALL CONNECTORS.
- TEST DRIVE VEHICLE ABOVE 24 km/h (15 mph) FOR AT LEAST 30 SECONDS.

DOES DTC 27 SET AS A CURRENT DTC?

- YES: REPLACE EBCM.
- NO: MALFUNCTION IS INTERMITTENT. REFER TO "DIAGNOSTIC AIDS".

REAR WHEEL SPEED SENSOR

TEMPERATURE OF SENSOR VS. SENSOR RESISTANCE (APPROXIMATE)

TEMP. (°C)	TEMP. (°F)	RESISTANCE (OHMS)
-34 TO 4	-30 TO 40	1822 TO 2205
5 TO 43	41 TO 110	1930 TO 2558
44 TO 93	111 TO 200	2245 TO 3028

DIAGNOSTIC CHART DTC 28 — 1995 CENTURY, CUTLASS CIERA AND CUTLASS CRUISER

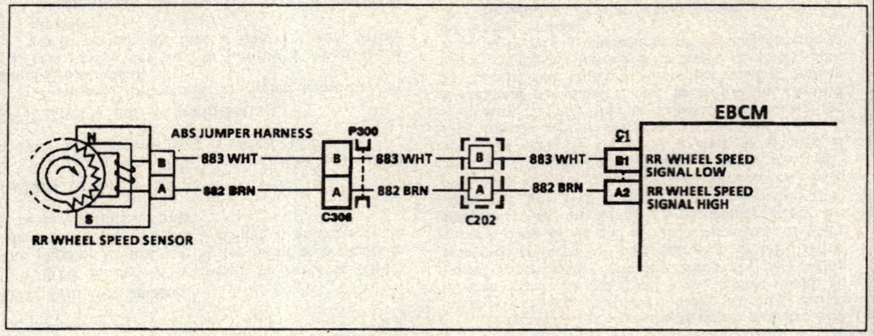

DTC 28
RIGHT REAR EXCESSIVE WHEEL SPEED VARIATION

Circuit Description:

As a toothed ring passes by the wheel speed sensor, changes in the electromagnetic field cause the wheel speed sensor to produce a sinusoidal (AC) voltage signal whose frequency is proportional to wheel speed. The magnitude of this signal is directly related to wheel speed and the proximity of the wheel speed sensor to the toothed ring, often referred to as the air gap.

DTC Will Set When:

DTC 28 can be set when the brake is "OFF." The purpose of this test is to detect a situation in which the right rear wheel acceleration or deceleration is beyond specified limits.

Action Taken:

A malfunction DTC is stored, ABS is disabled and the ABS warning lamp is turned "ON."

DTC Chart Test Description:

Number(s) below refer to circled number(s) on the diagnostic chart.

1. This step is designed to identify a wheel speed sensor or circuitry that is damaged and visibly apparent.
2. Checks to see that the wheel speed sensor is within proper resistance values.
3. Verifies that the wheel speed sensor can generate an appropriate output signal.
4. Ensures that wheel speed sensor circuitry is not shorted together.
5. Ensures malfunction was not due to physical damage of the circuitry.

Diagnostic Aids:

An "Intermittent" malfunction may be caused by a poor connection, rubbed through wire insulation, or a wire that is broken inside the insulation.

The frequency of the malfunction can be checked by using the enhanced diagnostic function of the Tech 1.

If the customer's comments reflect that the ABS warning lamp is "ON" only during moist environmental changes (rain, snow, vehicle wash), all wheel speed sensor circuitry should be thoroughly inspected for signs of water intrusion. If DTC is not current, clear DTCs and simulate the effects of water intrusion. Use the following procedure. Spray the suspected area with a 5% salt water solution (two teaspoons of salt to 12 oz. of water). Test drive vehicle over various road surfaces (bumps, turns, etc.) above 24 km/h (15 mph) for at least 30 seconds. If DTC returns, replace suspected harness.

Any circuitry that is suspected of causing the intermittent complaint should be thoroughly checked for backed out terminals, improper mating, broken locks, improperly formed or damaged terminals, poor terminal to wiring connections or physical damage to the wiring harness.

Resistance of the wheel speed sensor will increase with an increase in sensor temperature.

When replacing a wheel speed sensor, inspect the sensor terminals and harness connector for corrosion and/or water intrusion. If evidence of corrosion or water intrusion exists, replace wheel speed sensor jumper harness. Likewise, if replacing a wheel speed sensor jumper harness, inspect sensor terminals. If evidence of corrosion or water intrusion exists, replace wheel speed sensor.

DIAGNOSTIC CHART DTC 28 — 1995 CENTURY, CUTLASS CIERA AND CUTLASS CRUISER

DTC 28
RIGHT REAR EXCESSIVE WHEEL SPEED VARIATION

IMPORTANT:
- WHEEL SPEED SENSOR INTERMITTENT MALFUNCTIONS MAY BE DIFFICULT TO LOCATE. CARE SHOULD BE TAKEN NOT TO DISTURB ANY ELECTRICAL CONNECTIONS PRIOR TO AN INDICATED STEP OF THIS CHART. THIS WILL ENSURE THAT AN INTERMITTENT CONNECTION WILL NOT BE CORRECTED BEFORE THE SOURCE OF THE MALFUNCTION IS FOUND.

(1)
- IGNITION "OFF."
- PHYSICALLY INSPECT THE RIGHT REAR WHEEL SPEED SENSOR, RIGHT REAR WHEEL SPEED SENSOR JUMPER HARNESS, AND CONNECTORS FOR DAMAGE.

IS PHYSICAL DAMAGE INDICATED?

YES: REPLACE WHEEL SPEED SENSOR AND/OR WHEEL SPEED SENSOR JUMPER HARNESS IF DAMAGED. DRIVE VEHICLE ABOVE 24 km/h (15 mph) TO ENSURE DTC 28 DOES NOT SET.

NO:

(2)
- DISCONNECT RIGHT REAR WHEEL SPEED SENSOR DIRECTLY AT SENSOR.
- USING J 39200, MEASURE THE RESISTANCE BETWEEN SENSOR TERMINALS "A" AND "B."

IS RESISTANCE BETWEEN 2059 AND 2363 OHMS WHEN SENSOR IS 20°C (68°F)? *(IF TEMPERATURE OF THE SENSOR IS SUSPECTED TO BE OTHER THAN 20°C (68°F), REFER TO THE TABLE BELOW FOR APPLICABLE SENSOR RESISTANCE VALUES: THE VALUES IN THE TABLE ARE FOR TEMPERATURE OF THE SENSOR, NOT AIR TEMPERATURE.)*

NO: REPLACE WHEEL SPEED SENSOR.

YES:

(3)
- WITH J 39200 STILL CONNECTED, SELECT A/C VOLTAGE SCALE.
- SPIN WHEEL BY HAND WHILE OBSERVING VOLTAGE.

IS VOLTAGE AT LEAST 100 mV (VOLTAGE WILL INCREASE AS WHEEL SPEED INCREASES)?

NO: REPLACE WHEEL SPEED SENSOR.

YES:

(4)
- DISCONNECT 24-WAY EBCM CONNECTOR.
- USING J 39200, MEASURE THE RESISTANCE BETWEEN TERMINAL "B1" AND "A2" OF THE 24-WAY EBCM HARNESS CONNECTOR.

DOES J 39200 DISPLAY RESISTANCE AS "OL" (INFINITE)?

NO: REPAIR SHORT BETWEEN CKTs 882 AND 883.

YES:

(5)
- INSPECT CKTs 882 AND 883 AND 24-WAY EBCM CONNECTOR FOR DAMAGE WHICH MAY RESULT IN A SHORT BETWEEN CKTs 882 AND 883 WITH ALL CONNECTORS CONNECTED. REPAIR ALL DAMAGE IF EVIDENT.
- RECONNECT ALL CONNECTORS.
- TEST DRIVE VEHICLE ABOVE 24 km/h (15 mph) FOR AT LEAST 30 SECONDS.

DOES DTC 28 SET AS A CURRENT DTC?

YES: REPLACE EBCM.

NO: MALFUNCTION IS INTERMITTENT. REFER TO "DIAGNOSTIC AIDS".

REAR WHEEL SPEED SENSOR

TEMPERATURE OF SENSOR VS. SENSOR RESISTANCE (APPROXIMATE)

TEMP. (°C)	TEMP. (°F)	RESISTANCE (OHMS)
-34 TO 4	-30 TO 40	1622 TO 2205
5 TO 43	41 TO 110	1930 TO 2586
44 TO 93	111 TO 200	2245 TO 3028

DIAGNOSTIC CHART DTC 32 — 1995 CENTURY, CUTLASS CIERA AND CUTLASS CRUISER

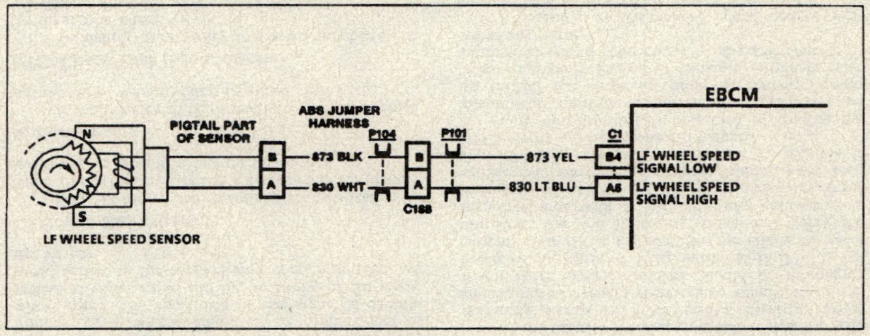

DTC 32
(Page 1 of 3)
LF WHEEL SPEED SNR CKT OPEN OR SHORT TO GND/BATT

Circuit Description:

As a toothed ring passes by the wheel speed sensor, changes in the electromagnetic field cause the wheel speed sensor to produce a sinusoidal (AC) voltage signal whose frequency is proportional to wheel speed. The magnitude of this signal is directly related to wheel speed and the proximity of the wheel speed sensor to the toothed ring often referred to as the air gap.

DTC Will Set When:

DTC 32 can be set anytime after initialization. If either of the left front wheel speed circuits are shorted to voltage or ground, a malfunction exists.

Action Taken:

A malfunction DTC is stored, ABS is disabled and the ABS warning lamp is turned "ON."

DTC Chart Test Description:

Number(s) below refer to circled number(s) on the diagnostic chart.

1. Checks to see that the wheel speed sensor is within proper resistance values.
2. Checks for short to ground in wheel speed sensor wiring.
3. Checks for short to voltage in wheel speed sensor wiring.
4. Checks for a short to voltage in CKT 873.
5. Checks for a short to voltage in CKT 830.
6. Ensures malfunction was not due to physical damage of circuitry.

Diagnostic Aids:

An "Intermittent" malfunction may be caused by a poor connection, rubbed through wire insulation, or a wire that is broken inside the insulation.

The frequency of the malfunction can be checked by using the enhanced diagnostic function of the Tech 1.

If the customer's comments reflect that the ABS warning lamp is "ON" only during moist environmental changes (rain, snow, vehicle wash), all wheel speed sensor circuitry should be thoroughly inspected for signs of water intrusion. If DTC is not current, clear DTCs and simulate the effects of water intrusion. Use the following procedure. Spray the suspected area with a 5% salt water solution (two teaspoons of salt to 12 oz. of water). Test drive vehicle over various road surfaces (bumps, turns, etc.) above 24 km/h (15 mph) for at least 30 seconds. If DTC returns, replace suspected harness.

Any circuitry that is suspected of causing the intermittent complaint should be thoroughly checked for backed out terminals, improper mating, broken locks, improperly formed or damaged terminals, poor terminal to wiring connections or physical damage to the wiring harness.

Resistance of the wheel speed sensor will increase with an increase in sensor temperature.

When replacing a wheel speed sensor, inspect the sensor terminals and harness connector for corrosion and/or water intrusion. If evidence of corrosion or water intrusion exists, replace wheel speed sensor jumper harness. Likewise, if replacing a wheel speed sensor jumper harness, inspect sensor terminals. If evidence of corrosion or water intrusion exists, replace wheel speed sensor.

DIAGNOSTIC CHART DTC 32 — 1995 CENTURY, CUTLASS CIERA AND CUTLASS CRUISER

DTC 32
(Page 1 of 3)
LF WHEEL SPEED SNR CKT OPEN OR SHORT TO GND/BATT

IMPORTANT:

- WHEEL SPEED SENSOR INTERMITTENT MALFUNCTIONS MAY BE DIFFICULT TO LOCATE. CARE SHOULD BE TAKEN NOT TO DISTURB ANY ELECTRICAL CONNECTIONS PRIOR TO AN INDICATED STEP OF THIS CHART. THIS WILL ENSURE THAT AN INTERMITTENT CONNECTION WILL NOT BE CORRECTED BEFORE THE SOURCE OF THE MALFUNCTION IS FOUND.

(1)
- IGNITION "OFF."
- USING A 5% SALT WATER SOLUTION (TWO TEASPOONS OF SALT TO 12 OZ. WATER), THOROUGHLY SPRAY THE LEFT FRONT WHEEL SPEED SENSOR JUMPER HARNESS.
- DISCONNECT 24-WAY EBCM HARNESS CONNECTOR.
- USING J 39200, MEASURE THE RESISTANCE BETWEEN TERMINALS "B4" AND "A5" OF THE 24-WAY EBCM HARNESS CONNECTOR.

IS RESISTANCE BETWEEN 1661 AND 1929 OHMS WHEN SENSOR IS 20°C (68°F)? *(IF TEMPERATURE OF SENSOR IS SUSPECTED TO OTHER THAN 20°C (68°F), REFER TO THE TABLE ON PAGE 2 OF 2 OF THIS DTC CHART FOR APPLICABLE SENSOR RESISTANCE VALUES: THE VALUES IN THE TABLE ARE FOR TEMPERATURE OF THE SENSOR, NOT AIR TEMPERATURE.)*

NO → DIAGNOSIS CONTINUED ON PAGE 2 OF 3 OF THIS DTC CHART.

YES →

(2)
- MEASURE THE RESISTANCE BETWEEN GROUND AND TERMINAL "B4" OF THE 24-WAY EBCM HARNESS CONNECTOR.

DOES J 39200 DISPLAY RESISTANCE AS "OL" (INFINITE)?

NO → DIAGNOSIS CONTINUED ON PAGE 3 OF 3 OF THIS DTC CHART.

YES →

(3)
- IGNITION TO "RUN."
- USING J 39200, MEASURE THE VOLTAGE BETWEEN GROUND AND TERMINAL "B4" OF THE 24-WAY EBCM HARNESS CONNECTOR.

IS VOLTAGE 1 VOLT OR LESS?

NO →

(4)
- IGNITION "OFF."
- DISCONNECT LEFT FRONT WHEEL SPEED SENSOR DIRECTLY AT SENSOR.
- IGNITION TO "RUN."
- USING J 39200, MEASURE THE VOLTAGE BETWEEN GROUND AND TERMINAL "B4" OF THE 24-WAY EBCM HARNESS CONNECTOR.

IS VOLTAGE 1 VOLT OR LESS?

NO → REPAIR SHORT TO VOLTAGE IN CKT 873.

YES →

(5)
- USING J 39200, MEASURE THE VOLTAGE BETWEEN GROUND AND TERMINAL "A5" OF THE 24-WAY EBCM HARNESS CONNECTOR.

IS VOLTAGE 1 VOLT OR LESS?

NO → REPAIR SHORT TO VOLTAGE IN CKT 830.

YES → MALFUNCTION IS INTERMITTENT. INSPECT ALL CONNECTORS AND HARNESSES FOR DAMAGE WHICH MAY RESULT IN A SHORT TO VOLTAGE WHEN ALL COMPONENTS ARE CONNECTED. REPAIR ALL DAMAGE IF FOUND. REFER TO "DIAGNOSIS AIDS" ON FACING PAGE FOR MORE INFORMATION.

YES (from 3) →

(6)
- IGNITION "OFF."
- INSPECT TERMINALS "B4" AND "A5" OF THE 24-WAY EBCM HARNESS CONNECTOR FOR POOR TERMINAL CONTACT AND CONTAMINATION. ALSO INSPECT CKTs 830 AND 873 FOR DAMAGE WHICH MAY RESULT IN AN OPEN CIRCUIT, SHORT TO GROUND, OR SHORT TO VOLTAGE. REPAIR ALL DAMAGE IF EVIDENT.
- RECONNECT ALL CONNECTORS.
- IGNITION TO "RUN."

DOES DTC 32 SET AS A CURRENT DTC?

YES → REPLACE EBCM.

NO → MALFUNCTION IS INTERMITTENT. REFER TO "DIAGNOSTIC AIDS".

DIAGNOSTIC CHART DTC 32 — 1995 CENTURY, CUTLASS CIERA AND CUTLASS CRUISER

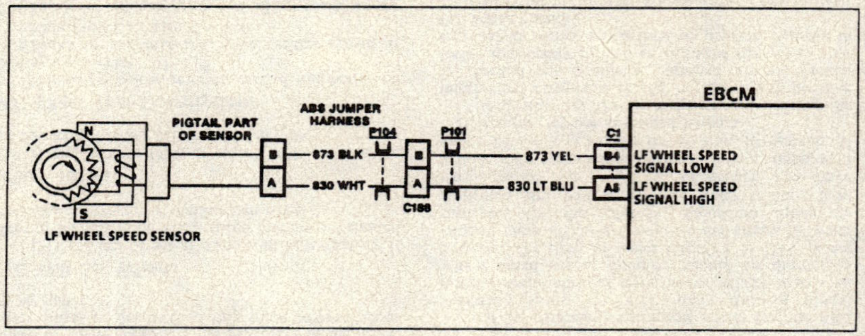

DTC 32
(Page 2 of 3)
LF WHEEL SPEED SNR CKT OPEN OR SHORT TO GND/BATT

Circuit Description:

As a toothed ring passes by the wheel speed sensor, changes in the electromagnetic field cause the wheel speed sensor to produce a sinusoidal (AC) voltage signal whose frequency is proportional to wheel speed. The magnitude of this signal is directly related to wheel speed and the proximity of the wheel speed sensor to the toothed ring often referred to as the air gap.

DTC Will Set When:

DTC 32 can be set anytime after initialization. If either of the left front wheel speed circuits are shorted to voltage or ground, a malfunction exists.

Action Taken:

A malfunction DTC is stored, ABS is disabled and the ABS warning lamp is turned "ON."

DTC Chart Test Description:

Number(s) below refer to circled number(s) on the diagnostic chart.

7. Checks for high resistance in CKT 873.
8. Checks for high resistance in CKT 830.
9. Verifies proper resistance value of the wheel speed sensor.

Diagnostic Aids:

An "Intermittent" malfunction may be caused by a poor connection, rubbed through wire insulation, or a wire that is broken inside the insulation.

The frequency of the malfunction can be checked by using the enhanced diagnostic function of the Tech 1.

If the customer's comments reflect that the ABS warning lamp is "ON" only during moist environmental changes (rain, snow, vehicle wash), all wheel speed sensor circuitry should be thoroughly inspected for signs of water intrusion. If DTC is not current, clear DTCs and simulate the effects of water intrusion. Use the following procedure. Spray the suspected area with a 5% salt water solution (two teaspoons of salt to 12 oz. of water). Test drive vehicle over various road surfaces (bumps, turns, etc.) above 24 km/h (15 mph) for at least 30 seconds. If DTC returns, replace suspected harness.

Any circuitry that is suspected of causing the intermittent complaint should be thoroughly checked for backed out terminals, improper mating, broken locks, improperly formed or damaged terminals, poor terminal to wiring connections or physical damage to the wiring harness.

Resistance of the wheel speed sensor will increase with an increase in sensor temperature.

When replacing a wheel speed sensor, inspect the sensor terminals and harness connector for corrosion and/or water intrusion. If evidence of corrosion or water intrusion exists, replace wheel speed sensor jumper harness. Likewise, if replacing a wheel speed sensor jumper harness, inspect sensor terminals. If evidence of corrosion or water intrusion exists, replace wheel speed sensor.

DIAGNOSTIC CHART DTC 32 — 1995 CENTURY, CUTLASS CIERA AND CUTLASS CRUISER

DTC 32
(Page 2 of 3)
LF WHEEL SPEED SNR CKT OPEN OR SHORT TO GND/BATT

DIAGNOSIS CONTINUED FROM PAGE 1 OF 3 OF THIS DTC CHART.

(7)
- DISCONNECT LEFT FRONT WHEEL SPEED SENSOR DIRECTLY AT SENSOR.
- USING J 39200, MEASURE THE RESISTANCE BETWEEN TERMINAL "B4" OF THE 24-WAY EBCM HARNESS CONNECTOR AND TERMINAL "B" OF THE LEFT FRONT WHEEL SPEED SENSOR HARNESS CONNECTOR. IS RESISTANCE 2 OHMS OR LESS?

NO: REPAIR HIGH RESISTANCE IN CKT 873 (WHEEL SPEED SENSOR JUMPER MUST BE REPLACED IF DAMAGED).

YES:

(8)
- USING J 39200, MEASURE THE RESISTANCE BETWEEN TERMINAL "A5" OF THE 24-WAY EBCM HARNESS CONNECTOR AND TERMINAL "A" OF THE LEFT FRONT WHEEL SPEED SENSOR HARNESS CONNECTOR. IS RESISTANCE 2 OHMS OR LESS?

NO: REPAIR HIGH RESISTANCE IN CKT 830 (WHEEL SPEED SENSOR JUMPER MUST BE REPLACED IF DAMAGED).

YES:

(9)
- USING J 39200, MEASURE THE RESISTANCE BETWEEN TERMINAL "A" AND TERMINAL "B" OF THE LEFT FRONT WHEEL SPEED SENSOR CONNECTOR. IS RESISTANCE BETWEEN 1681 AND 1929 OHMS WHEN SENSOR IS 20°C (68°F)? *(IF TEMPERATURE OF SENSOR IS SUSPECTED TO OTHER THAN 20°C (68°F), REFER TO THE TABLE BELOW FOR APPLICABLE SENSOR RESISTANCE VALUES: THE VALUES IN THE TABLE ARE FOR TEMPERATURE OF THE SENSOR, NOT AIR TEMPERATURE.)*

NO: REPLACE WHEEL SPEED SENSOR.

YES: MALFUNCTION IS INTERMITTENT. INSPECT ALL CONNECTORS AND HARNESSES FOR DAMAGE WHICH MAY RESULT IN HIGH RESISTANCE WHEN ALL COMPONENTS ARE CONNECTED. REFER TO "DIAGNOSTIC AIDS" ON FACING PAGE FOR MORE INFORMATION. REPAIR ALL DAMAGE IF FOUND.

FRONT WHEEL SPEED SENSOR

TEMPERATURE OF SENSOR VS. SENSOR RESISTANCE (APPROXIMATE)

TEMP. (°C)	TEMP. (°F)	RESISTANCE (OHMS)
-34 TO 4	-30 TO 40	1258 TO 1806
5 TO 43	41 TO 110	1502 TO 2103
44 TO 93	111 TO 200	1747 TO 2482

DIAGNOSTIC CHART DTC 32 — 1995 CENTURY, CUTLASS CIERA AND CUTLASS CRUISER

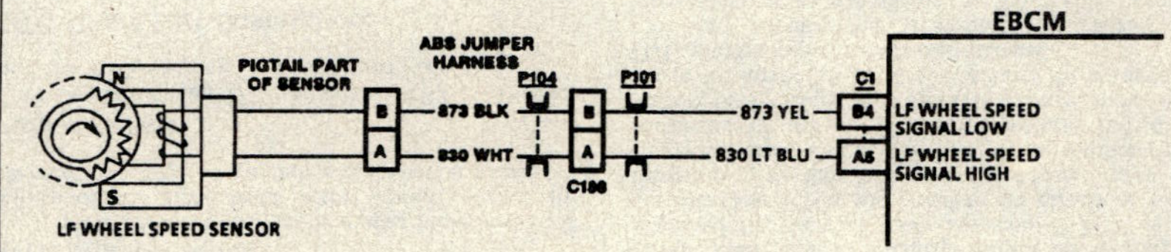

DTC 32

LF WHEEL SPEED SNR CKT OPEN OR SHORT TO GND/BATT

Circuit Description:

As a toothed ring passes by the wheel speed sensor, changes in the electromagnetic field cause the wheel speed sensor to produce a sinusoidal (AC) voltage signal whose frequency is proportional to wheel speed. The magnitude of this signal is directly related to wheel speed and the proximity of the wheel speed sensor to the toothed ring often referred to as the air gap.

DTC Will Set When:

DTC 32 can be set anytime after initialization. If either of the left front wheel speed circuits are shorted to voltage or ground, a malfunction exists.

Action Taken:

A malfunction DTC is stored, ABS is disabled and the ABS warning lamp is turned "ON."

DTC Chart Test Description:

Number(s) below refer to circled number(s) on the diagnostic chart.

10. Checks for short to ground in CKT 873.
11. Checks for short to ground in CKT 830.
12. Checks for short to ground in wheel speed sensor jumper harness.
13. Checks for an internal short to ground in the wheel speed sensor.

Diagnostic Aids:

An "Intermittent" malfunction may be caused by a poor connection, rubbed through wire insulation, or a wire that is broken inside the insulation.

The frequency of the malfunction can be checked by using the enhanced diagnostic function of the Tech 1.

If the customer's comments reflect that the ABS warning lamp is "ON" only during moist environmental changes (rain, snow, vehicle wash), all wheel speed sensor circuitry should be thoroughly inspected for signs of water intrusion. If DTC is not current, clear DTCs and simulate the effects of water intrusion. Use the following procedure. Spray the suspected area with a 5% salt water solution (two teaspoons of salt to 12 oz. of water). Test drive vehicle over various road surfaces (bumps, turns, etc.) above 24 km/h (15 mph) for at least 30 seconds. If DTC returns, replace suspected harness.

Any circuitry that is suspected of causing the intermittent complaint should be thoroughly checked for backed out terminals, improper mating, broken locks, improperly formed or damaged terminals, poor terminal to wiring connections or physical damage to the wiring harness.

Resistance of the wheel speed sensor will increase with an increase in sensor temperature.

When replacing a wheel speed sensor, inspect the sensor terminals and harness connector for corrosion and/or water intrusion. If evidence of corrosion or water intrusion exists, replace wheel speed sensor jumper harness. Likewise, if replacing a wheel speed sensor jumper harness, inspect sensor terminals. If evidence of corrosion or water intrusion exists, replace wheel speed sensor.

DIAGNOSTIC CHART DTC 32 — 1995 CENTURY, CUTLASS CIERA AND CUTLASS CRUISER

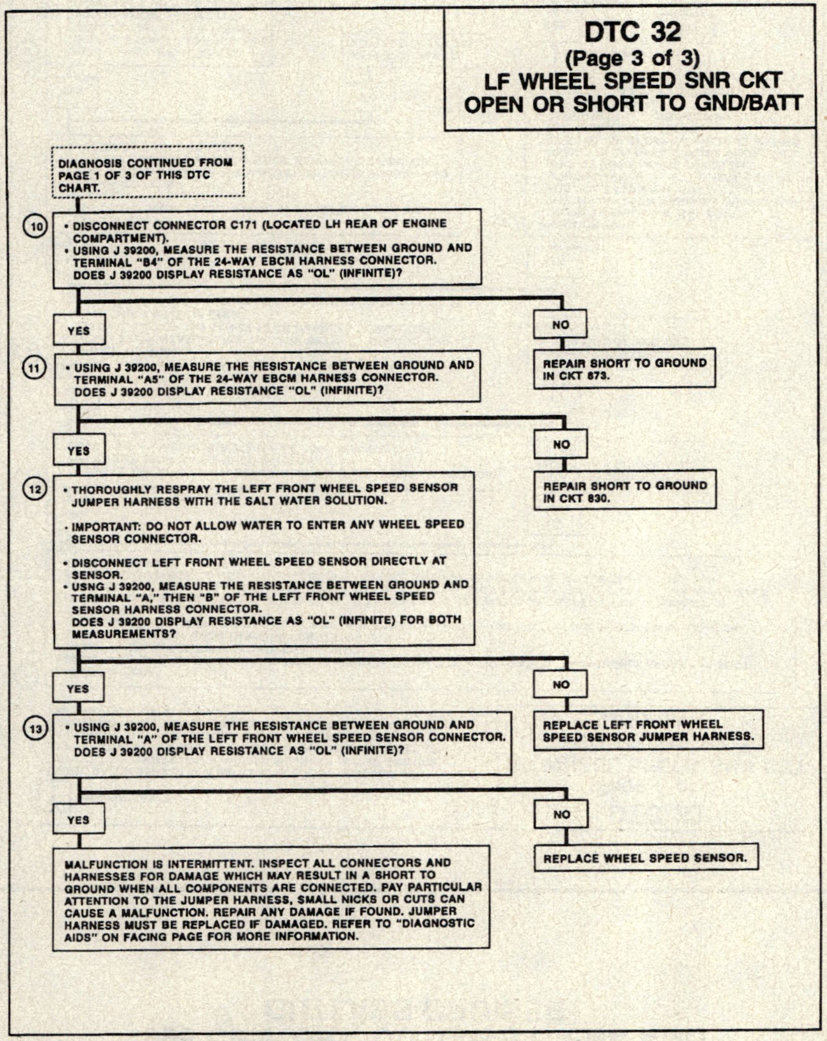

DIAGNOSTIC CHART DTC 33 — 1995 CENTURY, CUTLASS CIERA AND CUTLASS CRUISER

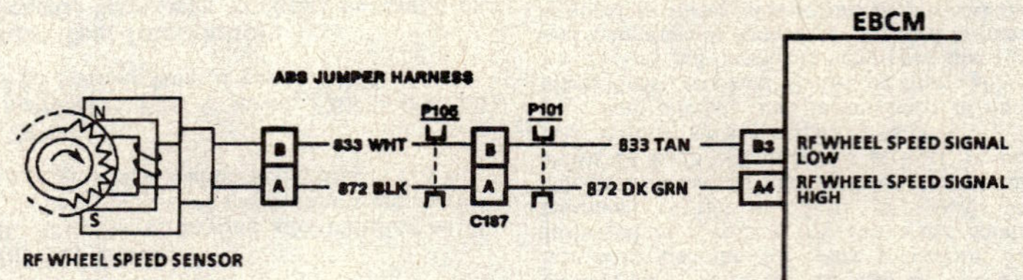

DTC 33
(Page 1 of 3)
RF WHEEL SPEED SNR CKT OPEN OR SHORT TO GND/BATT

Circuit Description:

As a toothed ring passes by the wheel speed sensor, changes in the electromagnetic field cause the wheel speed sensor to produce a sinusoidal (AC) voltage signal whose frequency is proportional to wheel speed. The magnitude of this signal is directly related to wheel speed and the proximity of the wheel speed sensor to the toothed ring, often referred to as the air gap.

DTC Will Set When:

DTC 33 can be set anytime after initialization. If either of the right front wheel speed circuits are shorted to voltage or ground, a malfunction exists.

Action Taken:

A malfunction DTC is stored, ABS is disabled and the ABS warning lamp is turned "ON."

DTC Chart Test Description:

Number(s) below refer to circled number(s) on the diagnostic chart.

1. Checks to see that the wheel speed sensor is within proper resistance values.
2. Checks for short to ground in wheel speed sensor wiring.
3. Checks for short to voltage in wheel speed sensor wiring.
4. Checks for a short to voltage in CKT 833.
5. Checks for a short to voltage in CKT 872.
6. Ensures malfunction was not due to physical damage of circuitry.

Diagnostic Aids:

An "Intermittent" malfunction may be caused by a poor connection, rubbed through wire insulation, or a wire that is broken inside the insulation.

The frequency of the malfunction can be checked by using the enhanced diagnostic function of the Tech 1.

If the customer's comments reflect that the ABS warning lamp is "ON" only during moist environmental changes (rain, snow, vehicle wash), all wheel speed sensor circuitry should be thoroughly inspected for signs of water intrusion. If DTC is not current, clear DTCs and simulate the effects of water intrusion. Use the following procedure. Spray the suspected area with a 5% salt water solution (two teaspoons of salt to 12 oz. of water). Test drive vehicle over various road surfaces (bumps, turns, etc.) above 24 km/h (15 mph) for at least 30 seconds. If DTC returns, replace suspected harness.

Any circuitry that is suspected of causing the intermittent complaint should be thoroughly checked for backed out terminals, improper mating, broken locks, improperly formed or damaged terminals, poor terminal to wiring connections or physical damage to the wiring harness.

Resistance of the wheel speed sensor will increase with an increase in sensor temperature.

When replacing a wheel speed sensor, inspect the sensor terminals and harness connector for corrosion and/or water intrusion. If evidence of corrosion or water intrusion exists, replace wheel speed sensor jumper harness. Likewise, if replacing a wheel speed sensor jumper harness, inspect sensor terminals. If evidence of corrosion or water intrusion exists, replace wheel speed sensor.

DIAGNOSTIC CHART DTC 33 — 1995 CENTURY, CUTLASS CIERA AND CUTLASS CRUISER

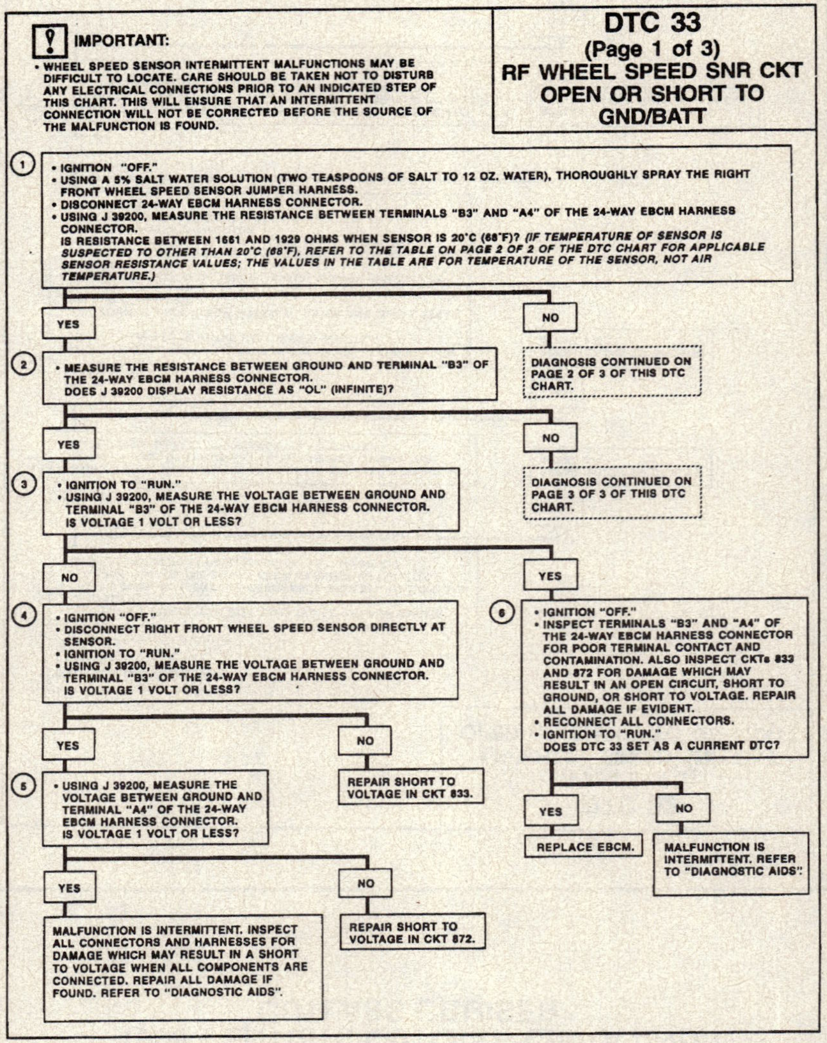

DIAGNOSTIC CHART DTC 33 — 1995 CENTURY, CUTLASS CIERA AND CUTLASS CRUISER

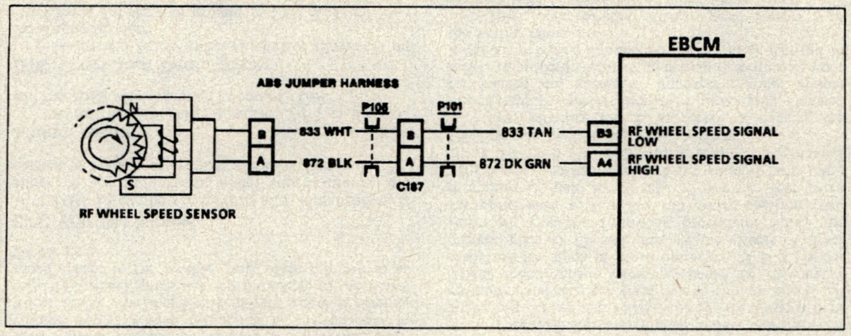

DTC 33
(Page 2 of 3)
RF WHEEL SPEED SNR CKT OPEN OR SHORT TO GND/BATT

Circuit Description:

As a toothed ring passes by the wheel speed sensor, changes in the electromagnetic field cause the wheel speed sensor to produce a sinusoidal (AC) voltage signal whose frequency is proportional to wheel speed. The magnitude of this signal is directly related to wheel speed and the proximity of the wheel speed sensor to the toothed ring, often referred to as the air gap.

DTC Will Set When:

DTC 33 can be set anytime after initialization. If either of the right front wheel speed circuits are shorted to voltage or ground, a malfunction exists.

Action Taken:

A malfunction DTC is stored, ABS is disabled and the ABS warning lamp is turned "ON."

DTC Chart Test Description:

Number(s) below refer to circled number(s) on the diagnostic chart.

7. Checks for high resistance in CKT 833.
8. Checks for high resistance in CKT 872.
9. Verifies proper resistance value of the wheel speed sensor.

Diagnostic Aids:

An "Intermittent" malfunction may be caused by a poor connection, rubbed through wire insulation, or a wire that is broken inside the insulation.

The frequency of the malfunction can be checked by using the enhanced diagnostic function of the Tech 1.

If the customer's comments reflect that the ABS warning lamp is "ON" only during moist environmental changes (rain, snow, vehicle wash), all wheel speed sensor circuitry should be thoroughly inspected for signs of water intrusion. If DTC is not current, clear DTCs and simulate the effects of water intrusion. Use the following procedure. Spray the suspected area with a 5% salt water solution (two teaspoons of salt to 12 oz. of water). Test drive vehicle over various road surfaces (bumps, turns, etc.) above 24 km/h (15 mph) for at least 30 seconds. If DTC returns, replace suspected harness.

Any circuitry that is suspected of causing the intermittent complaint should be thoroughly checked for backed out terminals, improper mating, broken locks, improperly formed or damaged terminals, poor terminal to wiring connections or physical damage to the wiring harness.

Resistance of the wheel speed sensor will increase with an increase in sensor temperature.

When replacing a wheel speed sensor, inspect the sensor terminals and harness connector for corrosion and/or water intrusion. If evidence of corrosion or water intrusion exists, replace wheel speed sensor jumper harness. Likewise, if replacing a wheel speed sensor jumper harness, inspect sensor terminals. If evidence of corrosion or water intrusion exists, replace wheel speed sensor.

DIAGNOSTIC CHART DTC 33 — 1995 CENTURY, CUTLASS CIERA AND CUTLASS CRUISER

DTC 33
(Page 2 of 3) RF WHEEL SPEED SNR CKT OPEN OR SHORT TO GND/BATT

DIAGNOSIS CONTINUED FROM PAGE 1 OF 3 OF THIS DTC CHART.

(7)
- DISCONNECT RIGHT FRONT WHEEL SPEED SENSOR DIRECTLY AT SENSOR
- USING J 39200, MEASURE THE RESISTANCE BETWEEN TERMINAL "B3" OF THE 24-WAY EBCM HARNESS CONNECTOR AND TERMINAL "B" OF THE RIGHT FRONT WHEEL SPEED SENSOR HARNESS CONNECTOR.

IS RESISTANCE 2 OHMS OR LESS?

- NO: REPAIR HIGH RESISTANCE IN CKT 833 (WHEEL SPEED SENSOR JUMPER MUST BE REPLACED IF DAMAGED).
- YES:

(8)
- USING J 39200, MEASURE THE RESISTANCE BETWEEN TERMINAL "A4" OF THE 24-WAY EBCM HARNESS CONNECTOR AND TERMINAL "A" OF THE RIGHT FRONT WHEEL SPEED SENSOR HARNESS CONNECTOR.

IS RESISTANCE 2 OHMS OR LESS?

- NO: REPAIR HIGH RESISTANCE IN CKT 872 (WHEEL SPEED SENSOR JUMPER MUST BE REPLACED IF DAMAGED).
- YES:

(9)
- USING J 39200, MEASURE THE RESISTANCE BETWEEN TERMINAL "A" AND TERMINAL "B" OF THE RIGHT FRONT WHEEL SPEED SENSOR CONNECTOR.

IS RESISTANCE BETWEEN 1661 AND 1929 OHMS WHEN SENSOR IS 20°C (68°F)? *(IF TEMPERATURE OF SENSOR IS SUSPECTED TO OTHER THAN 20°C (68°F), REFER TO THE TABLE BELOW FOR APPLICABLE SENSOR RESISTANCE VALUES; THE VALUES IN THE TABLE FOR TEMPERATURE OF THE SENSOR, NOT AIR TEMPERATURE.)*

- NO: REPLACE WHEEL SPEED SENSOR.
- YES: MALFUNCTION IS INTERMITTENT. INSPECT ALL CONNECTORS AND HARNESSES FOR DAMAGE WHICH MAY RESULT IN HIGH RESISTANCE WHEN ALL COMPONENTS ARE CONNECTED. REFER TO "DIAGNOSTIC AIDS" ON FACING PAGE FOR MORE INFORMATION. REPAIR ALL DAMAGE IF FOUND.

FRONT WHEEL SPEED SENSOR

TEMPERATURE OF SENSOR VS. SENSOR RESISTANCE (APPROXIMATE)

TEMP. (°C)	TEMP. (°F)	RESISTANCE (OHMS)
-34 TO 4	-30 TO 40	1258 TO 1808
5 TO 43	41 TO 110	1502 TO 2103
44 TO 93	111 TO 200	1747 TO 2482

DIAGNOSTIC CHART DTC 33 — 1995 CENTURY, CUTLASS CIERA AND CUTLASS CRUISER

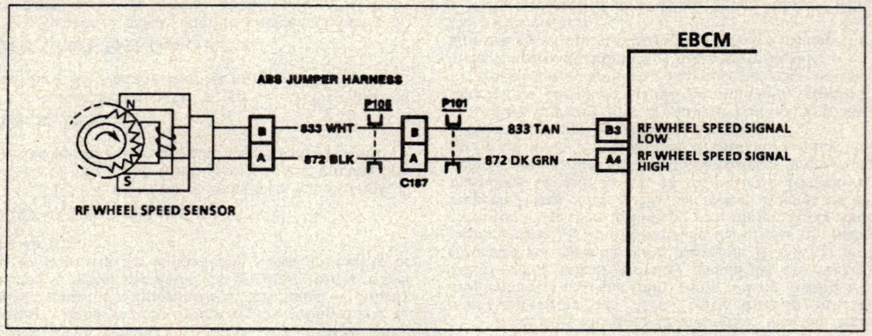

DTC 33
(Page 3 of 3)
RF WHEEL SPEED SNR CKT OPEN OR SHORT TO GND/BATT

Circuit Description:

As a toothed ring passes by the wheel speed sensor, changes in the electromagnetic field cause the wheel speed sensor to produce a sinusoidal (AC) voltage signal whose frequency is proportional to wheel speed. The magnitude of this signal is directly related to wheel speed and the proximity of the wheel speed sensor to the toothed ring, often referred to as the air gap.

DTC Will Set When:

DTC 33 can be set anytime after initialization. If either of the right front wheel speed circuits are shorted to voltage or ground, a malfunction exists.

Action Taken:

A malfunction DTC is stored, ABS is disabled and the ABS warning lamp is turned "ON."

DTC Chart Test Description:

Number(s) below refer to circled number(s) on the diagnostic chart.

10. Checks for short to ground in CKT 833.
11. Checks for short to ground in CKT 872.
12. Checks for short to ground in wheel speed sensor jumper harness.
13. Checks for an internal short to ground in the wheel speed sensor.

Diagnostic Aids:

An "Intermittent" malfunction may be caused by a poor connection, rubbed through wire insulation, or a wire that is broken inside the insulation.

The frequency of the malfunction can be checked by using the enhanced diagnostic function of the Tech 1.

If the customer's comments reflect that the ABS warning lamp is "ON" only during moist environmental changes (rain, snow, vehicle wash), all wheel speed sensor circuitry should be thoroughly inspected for signs of water intrusion. If DTC is not current, clear DTCs and simulate the effects of water intrusion. Use the following procedure. Spray the suspected area with a 5% salt water solution (two teaspoons of salt to 12 oz. of water). Test drive vehicle over various road surfaces (bumps, turns, etc.) above 24 km/h (15 mph) for at least 30 seconds. If DTC returns, replace suspected harness.

Any circuitry that is suspected of causing the intermittent complaint should be thoroughly checked for backed out terminals, improper mating, broken locks, improperly formed or damaged terminals, poor terminal to wiring connections or physical damage to the wiring harness.

Resistance of the wheel speed sensor will increase with an increase in sensor temperature.

When replacing a wheel speed sensor, inspect the sensor terminals and harness connector for corrosion and/or water intrusion. If evidence of corrosion or water intrusion exists, replace wheel speed sensor jumper harness. Likewise, if replacing a wheel speed sensor jumper harness, inspect sensor terminals. If evidence of corrosion or water intrusion exists, replace wheel speed sensor.

DIAGNOSTIC CHART DTC 33 — 1995 CENTURY, CUTLASS CIERA AND CUTLASS CRUISER

DTC 33
(Page 3 of 3)RF WHEEL SPEED SNR CKT OPEN OR SHORT TO GND/BATT

DIAGNOSIS CONTINUED FROM PAGE 1 OF 3 OF THIS DTC CHART.

(10)
- DISCONNECT CONNECTOR C172 (LOCATED RH REAR OF ENGINE COMPARTMENT).
- USING J 39200, MEASURE THE RESISTANCE BETWEEN GROUND AND TERMINAL "B3" OF THE 24-WAY EBCM HARNESS CONNECTOR. DOES J 39200 DISPLAY RESISTANCE AS "OL" (INFINITE)?

NO → REPAIR SHORT TO GROUND IN CKT 833.

YES ↓

(11)
- USING J 39200, MEASURE THE RESISTANCE BETWEEN GROUND AND TERMINAL "A4" OF THE 24-WAY EBCM HARNESS CONNECTOR. DOES J 39200 DISPLAY RESISTANCE AS "OL" (INFINITE)?

NO → REPAIR SHORT TO GROUND IN CKT 872.

YES ↓

(12)
- THOROUGHLY RESPRAY THE RIGHT FRONT WHEEL SPEED SENSOR JUMPER HARNESS WITH THE SALT WATER SOLUTION.
- IMPORTANT: DO NOT ALLOW WATER TO ENTER ANY WHEEL SPEED SENSOR CONNECTOR.
- DISCONNECT RIGHT FRONT WHEEL SPEED SENSOR DIRECTY AT SENSOR.
- USING J 39200, MEASURE THE RESISTANCE BETWEEN GROUND AND TERMINAL "A," THEN "B" OF THE RIGHT FRONT WHEEL SPEED SENSOR HARNESS CONNECTOR. DOES J 39200 DISPLAY RESISTANCE AS "OL" (INFINITE) FOR BOTH MEASUREMENTS?

NO → REPLACE RIGHT FRONT WHEEL SPEED SENSOR JUMPER HARNESS.

YES ↓

(13)
- USING J 39200, MEASURE THE RESISTANCE BETWEEN GROUND AND TERMINAL "A" OF THE RIGHT FRONT WHEEL SPEED SENSOR CONNECTOR. DOES J 39200 DISPLAY RESISTANCE AS "OL" (INFINITE)?

NO → REPLACE WHEEL SPEED SENSOR.

YES ↓

MALFUNCTION IS INTERMITTENT. INSPECT ALL CONNECTORS AND HARNESSES FOR DAMAGE WHICH MAY RESULT IN A SHORT TO GROUND WHEN ALL COMPONENTS ARE CONNECTED. PAY PARTICULAR ATTENTION TO THE JUMPER HARNESS, SMALL NICKS OR CUTS CAN CAUSE A MALFUNCTION. REPAIR ANY DAMAGE IF FOUND. JUMPER HARNESS MUST BE REPLACED IF DAMAGED. REFER TO "DIAGNOSTIC AIDS".

DIAGNOSTIC CHART DTC 34 — 1995 CENTURY, CUTLASS CIERA AND CUTLASS CRUISER

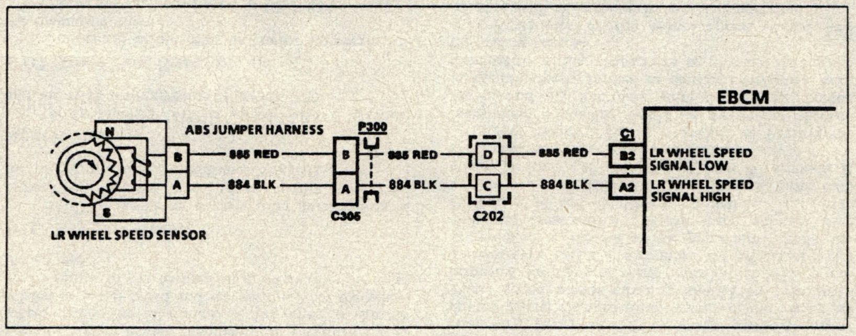

DTC 34
(Page 1 of 3)
LR WHEEL SPEED SNR CKT OPEN OR SHORT TO GND/BATT

Circuit Description:

As a toothed ring passes by the wheel speed sensor, changes in the electromagnetic field cause the wheel speed sensor to produce a sinusoidal (AC) voltage signal whose frequency is proportional to wheel speed. The magnitude of this signal is directly related to wheel speed and the proximity of the wheel speed sensor to the toothed ring, often referred to as the air gap.

DTC Will Set When:

DTC 34 can be set anytime after initialization. If either of the left rear wheel speed circuits are shorted to voltage or ground, a malfunction exists.

Action Taken:

A malfunction DTC is stored, ABS is disabled and the ABS warning lamp is turned "ON."

DTC Chart Test Description:

Number(s) below refer to circled number(s) on the diagnostic chart.

1. Checks to see that the wheel speed sensor is within proper resistance values.
2. Checks for short to ground in wheel speed sensor wiring.
3. Checks for short to voltage in wheel speed sensor wiring.
4. Checks for a short to voltage in CKT 884.
5. Checks for a short to voltage in CKT 885.
6. Ensures malfunction was not due to physical damage of circuitry.

Diagnostic Aids:

An "Intermittent" malfunction may be caused by a poor connection, rubbed through wire insulation, or a wire that is broken inside the insulation.

The frequency of the malfunction can be checked by using the enhanced diagnostic function of the Tech 1.

If the customer's comments reflect that the ABS warning lamp is "ON" only during moist environmental changes (rain, snow, vehicle wash), all wheel speed sensor circuitry should be thoroughly inspected for signs of water intrusion. If DTC is not current, clear DTCs and simulate the effects of water intrusion. Use the following procedure. Spray the suspected area with a 5% salt water solution (two teaspoons of salt to 12 oz. of water). Test drive vehicle over various road surfaces (bumps, turns, etc.) above 24 km/h (15 mph) for at least 30 seconds. If DTC returns, replace suspected harness.

Any circuitry that is suspected of causing the intermittent complaint should be thoroughly checked for backed out terminals, improper mating, broken locks, improperly formed or damaged terminals, poor terminal to wiring connections or physical damage to the wiring harness.

Resistance of the wheel speed sensor will increase with an increase in sensor temperature.

When replacing a wheel speed sensor, inspect the sensor terminals and harness connector for corrosion and/or water intrusion. If evidence of corrosion or water intrusion exists, replace wheel speed sensor jumper harness. Likewise, if replacing a wheel speed sensor jumper harness, inspect sensor terminals. If evidence of corrosion or water intrusion exists, replace wheel speed sensor.

DIAGNOSTIC CHART DTC 34 — 1995 CENTURY, CUTLASS CIERA AND CUTLASS CRUISER

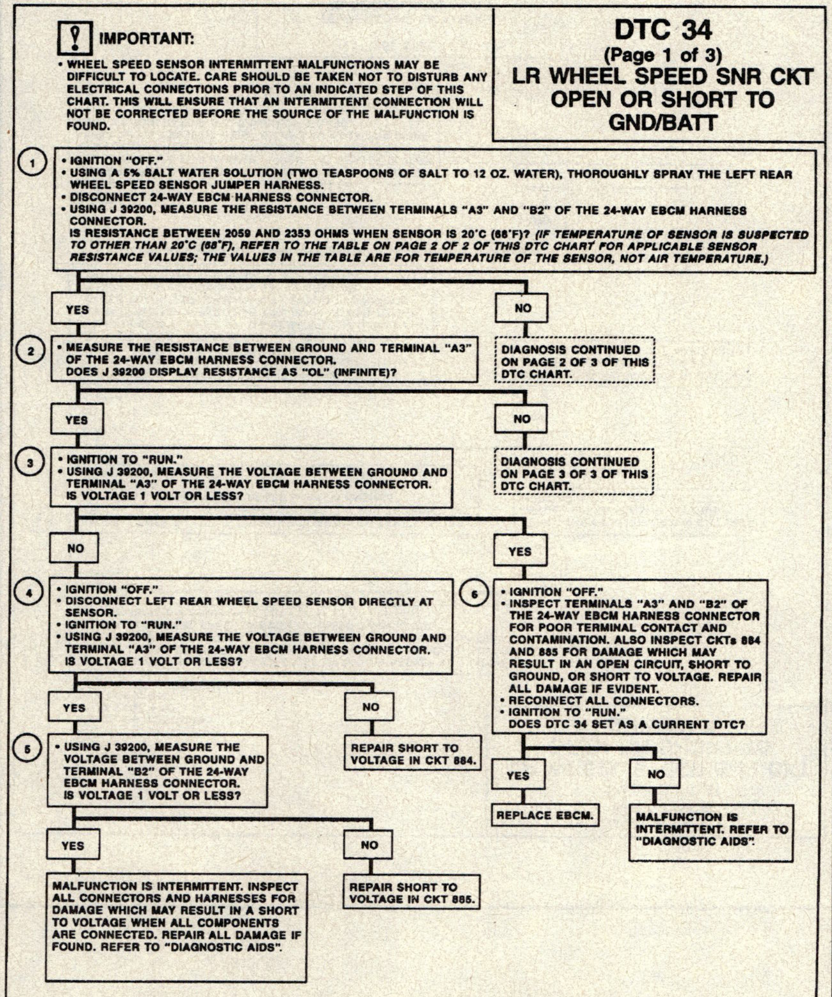

DIAGNOSTIC CHART DTC 34 — 1995 CENTURY, CUTLASS CIERA AND CUTLASS CRUISER

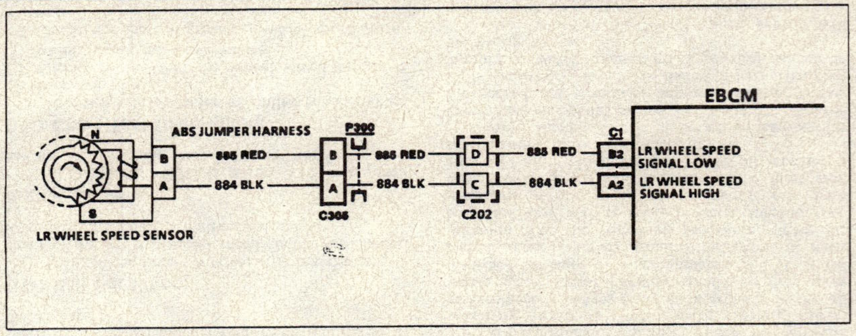

DTC 34
(Page 2 of 3)
LR WHEEL SPEED SNR CKT OPEN OR SHORT TO GND/BATT

Circuit Description:

As a toothed ring passes by the wheel speed sensor, changes in the electromagnetic field cause the wheel speed sensor to produce a sinusoidal (AC) voltage signal whose frequency is proportional to wheel speed. The magnitude of this signal is directly related to wheel speed and the proximity of the wheel speed sensor to the toothed ring, often referred to as the air gap.

DTC Will Set When:

DTC 34 can be set anytime after initialization. If either of the left rear wheel speed circuits are shorted to voltage or ground, a malfunction exists.

Action Taken:

A malfunction DTC is stored, ABS is disabled and the ABS warning lamp is turned "ON."

DTC Chart Test Description:

Number(s) below refer to circled number(s) on the diagnostic chart.

7. Checks for high resistance in CKT 884.
8. Checks for high resistance in CKT 885.
9. Verifies proper resistance value of the wheel speed sensor.

Diagnostic Aids:

An "Intermittent" malfunction may be caused by a poor connection, rubbed through wire insulation, or a wire that is broken inside the insulation.

The frequency of the malfunction can be checked by using the enhanced diagnostic function of the Tech 1.

If the customer's comments reflect that the ABS warning lamp is "ON" only during moist environmental changes (rain, snow, vehicle wash), all wheel speed sensor circuitry should be thoroughly inspected for signs of water intrusion. If DTC is not current, clear DTCs and simulate the effects of water intrusion. Use the following procedure. Spray the suspected area with a 5% salt water solution (two teaspoons of salt to 12 oz. of water). Test drive vehicle over various road surfaces (bumps, turns, etc.) above 24 km/h (15 mph) for at least 30 seconds. If DTC returns, replace suspected harness.

Any circuitry that is suspected of causing the intermittent complaint should be thoroughly checked for backed out terminals, improper mating, broken locks, improperly formed or damaged terminals, poor terminal to wiring connections or physical damage to the wiring harness.

Resistance of the wheel speed sensor will increase with an increase in sensor temperature.

When replacing a wheel speed sensor, inspect the sensor terminals and harness connector for corrosion and/or water intrusion. If evidence of corrosion or water intrusion exists, replace wheel speed sensor jumper harness. Likewise, if replacing a wheel speed sensor jumper harness, inspect sensor terminals. If evidence of corrosion or water intrusion exists, replace wheel speed sensor.

DIAGNOSTIC CHART DTC 34 — 1995 CENTURY, CUTLASS CIERA AND CUTLASS CRUISER

DTC 34
(Page 2 of 3)
LR WHEEL SPEED SNR CKT OPEN OR SHORT TO GND/BATT

DIAGNOSIS CONTINUED FROM PAGE 1 OF 3 OF THIS DTC CHART.

(7)
- DISCONNECT LEFT REAR WHEEL SPEED SENSOR DIRECTLY AT SENSOR.
- USING J 39200, MEASURE THE RESISTANCE BETWEEN TERMINAL "A3" OF THE 24-WAY EBCM HARNESS CONNECTOR AND TERMINAL "B" OF THE LEFT REAR WHEEL SPEED SENSOR HARNESS CONNECTOR.
IS RESISTANCE 2 OHMS OR LESS?

NO → REPAIR HIGH RESISTANCE IN CKT 884 (WHEEL SPEED SENSOR JUMPER MUST BE REPLACED IF DAMAGED).

YES ↓

(8)
- USING J 39200, MEASURE RESISTANCE BETWEEN TERMINAL "B2" OF THE 24-WAY EBCM HARNESS CONNECTOR AND TERMINAL "A" OF THE LEFT REAR WHEEL SPEED SENSOR HARNESS CONNECTOR.
IS RESISTANCE 2 OHMS OR LESS?

NO → REPAIR HIGH RESISTANCE IN CKT 885 (WHEEL SPEED SENSOR JUMPER MUST BE REPLACED IF DAMAGED).

YES ↓

(9)
- USING J 39200, MEASURE THE RESISTANCE BETWEEN TERMINAL "A" AND TERMINAL "B" OF THE LEFT REAR WHEEL SPEED SENSOR CONNECTOR.
IS RESISTANCE BETWEEN 2059 AND 2353 OHMS WHEN SENSOR IS 20°C (68°F)? *(IF TEMPERATURE OF SENSOR IS SUSPECTED TO OTHER THAN 20°C (68°F), REFER TO THE TABLE BELOW FOR APPLICABLE SENSOR RESISTANCE VALUES; THE VALUES IN THE TABLE ARE FOR TEMPERATURE OF THE SENSOR, NOT AIR TEMPERATURE.)*

NO → REPLACE WHEEL SPEED SENSOR.

YES → MALFUNCTION IS INTERMITTENT. INSPECT ALL CONNECTORS AND HARNESSES FOR DAMAGE WHICH MAY RESULT IN HIGH RESISTANCE WHEN ALL COMPONENTS ARE CONNECTED. REFER TO "DIAGNOSTIC AIDS" ON FACING PAGE FOR MORE INFORMATION. REPAIR ALL DAMAGE IF FOUND.

REAR WHEEL SPEED SENSOR

TEMPERATURE OF SENSOR VS. SENSOR RESISTANCE (APPROXIMATE)

TEMP. (°C)	TEMP. (°F)	RESISTANCE (OHMS)
-34 TO 4	-30 TO 40	1822 TO 2205
5 TO 43	41 TO 110	1930 TO 2566
44 TO 93	111 TO 200	2245 TO 3028

DIAGNOSTIC CHART DTC 34 — 1995 CENTURY, CUTLASS CIERA AND CUTLASS CRUISER

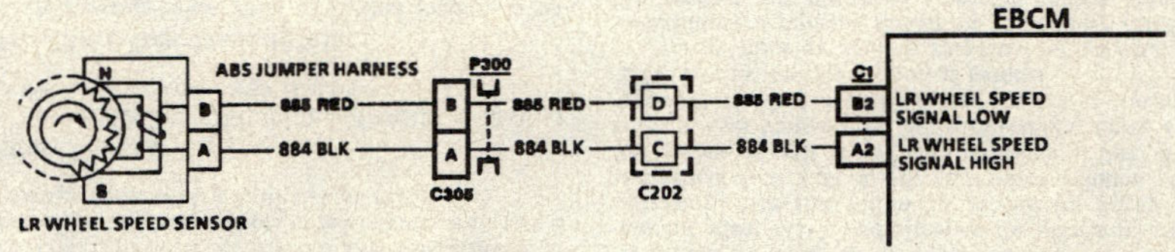

DTC 34
(Page 3 of 3)
LR WHEEL SPEED SNR CKT OPEN OR SHORT TO GND/BATT

Circuit Description:

As a toothed ring passes by the wheel speed sensor, changes in the electromagnetic field cause the wheel speed sensor to produce a sinusoidal (AC) voltage signal whose frequency is proportional to wheel speed. The magnitude of this signal is directly related to wheel speed and the proximity of the wheel speed sensor to the toothed ring, often referred to as the air gap.

DTC Will Set When:

DTC 34 can be set anytime after initialization. If either of the left rear wheel speed circuits are shorted to voltage or ground, a malfunction exists.

Action Taken:

A malfunction DTC is stored, ABS is disabled and the ABS warning lamp is turned "ON."

DTC Chart Test Description:

Number(s) below refer to circled number(s) on the diagnostic chart.

10. Checks for short to ground in CKT 884.
11. Checks for short to ground in CKT 885.
12. Checks for short to ground in wheel speed sensor jumper harness.
13. Checks for an internal short to ground in the wheel speed sensor.

Diagnostic Aids:

An "Intermittent" malfunction may be caused by a poor connection, rubbed through wire insulation, or a wire that is broken inside the insulation.

The frequency of the malfunction can be checked by using the enhanced diagnostic function of the Tech 1.

If the customer's comments reflect that the ABS warning lamp is "ON" only during moist environmental changes (rain, snow, vehicle wash), all wheel speed sensor circuitry should be thoroughly inspected for signs of water intrusion. If DTC is not current, clear DTCs and simulate the effects of water intrusion. Use the following procedure. Spray the suspected area with a 5% salt water solution (two teaspoons of salt to 12 oz. of water). Test drive vehicle over various road surfaces (bumps, turns, etc.) above 24 km/h (15 mph) for at least 30 seconds. If DTC returns, replace suspected harness.

Any circuitry that is suspected of causing the intermittent complaint should be thoroughly checked for backed out terminals, improper mating, broken locks, improperly formed or damaged terminals, poor terminal to wiring connections or physical damage to the wiring harness.

Resistance of the wheel speed sensor will increase with an increase in sensor temperature.

When replacing a wheel speed sensor, inspect the sensor terminals and harness connector for corrosion and/or water intrusion. If evidence of corrosion or water intrusion exists, replace wheel speed sensor jumper harness. Likewise, if replacing a wheel speed sensor jumper harness, inspect sensor terminals. If evidence of corrosion or water intrusion exists, replace wheel speed sensor.

DIAGNOSTIC CHART DTC 34 — 1995 CENTURY, CUTLASS CIERA AND CUTLASS CRUISER

DTC 34
(Page 3 of 3)
LR WHEEL SPEED SNR CKT OPEN OR SHORT TO GND/BATT

DIAGNOSIS CONTINUED FROM PAGE 1 OF 3 OF THIS DTC CHART.

(10)
- DISCONNECT CONNECTOR C371 (LOCATED BEHIND LH SIDE OF REAR SEAT).
- USING J 39200, MEASURE THE RESISTANCE BETWEEN GROUND AND TERMINAL "10" OF THE 24-WAY EBCM HARNESS CONNECTOR.

DOES J 39200 DISPLAY RESISTANCE AS "OL" (INFINITE)?

- NO: REPAIR SHORT TO GROUND IN CKT 884.
- YES: go to (11).

(11)
- USING J 39200, MEASURE THE RESISTANCE BETWEEN GROUND AND TERMINAL "23" OF THE 24-WAY EBCM HARNESS CONNECTOR.

DOES J 39200 DISPLAY RESISTANCE AS "OL" (INFINITE)?

- NO: REPAIR SHORT TO GROUND IN CKT 885.
- YES: go to (12).

(12)
- THOROUGHLY RESPRAY THE LEFT REAR WHEEL SPEED SENSOR JUMPER HARNESS WITH THE SALT WATER SOLUTION.

IMPORTANT: DO NOT ALLOW WATER TO ENTER ANY WHEEL SPEED SENSOR CONNECTOR.

- DISCONNECT LEFT REAR WHEEL SPEED SENSOR DIRECTLY AT SENSOR.
- USING J 39200, MEASURE THE RESISTANCE BETWEEN GROUND AND TERMINAL "A", THEN "B" OF THE LEFT REAR WHEEL SPEED SENSOR HARNESS CONNECTOR.

DOES J 39200 DISPLAY RESISTANCE AS "OL" (INFINITE) FOR BOTH MEASUREMENTS?

- NO: REPLACE LEFT REAR WHEEL SPEED SENSOR JUMPER HARNESS.
- YES: go to (13).

(13)
- USING J 39200, MEASURE THE RESISTANCE BETWEEN GROUND AND TERMINAL "A" OF THE LEFT REAR WHEEL SPEED SENSOR CONNECTOR.

DOES J 39200 DISPLAY RESISTANCE AS "OL" (INFINITE)?

- NO: REPLACE WHEEL SPEED SENSOR.
- YES: MALFUNCTION IS INTERMITTENT. INSPECT ALL CONNECTORS AND HARNESSES FOR DAMAGE WHICH MAY RESULT IN A SHORT TO GROUND WHEN ALL COMPONENTS ARE CONNECTED. PAY PARTICULAR ATTENTION TO THE JUMPER HARNESS. SMALL NICKS OR CUTS CAN CAUSE A MALFUNCTION. REPAIR ANY DAMAGE IF FOUND. JUMPER HARNESS MUST BE REPLACED IF DAMAGED. REFER TO "DIAGNOSTIC AIDS".

DIAGNOSTIC CHART DTC 35 — 1995 CENTURY, CUTLASS CIERA AND CUTLASS CRUISER

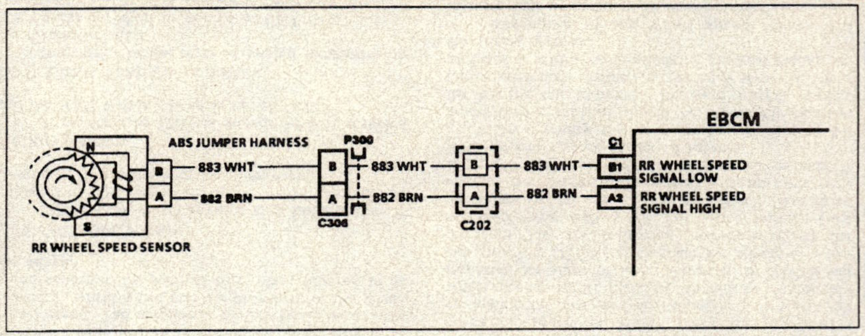

DTC 35 (Page 1 of 3) RR WHEEL SPEED SNR CKT OPEN OR SHORT TO GND/BATT

Circuit Description:

As a toothed ring passes by the wheel speed sensor, changes in the electromagnetic field cause the wheel speed sensor to produce a sinusoidal (AC) voltage signal whose frequency is proportional to wheel speed. The magnitude of this signal is directly related to wheel speed and the proximity of the wheel speed sensor to the toothed ring, often referred to as the air gap.

DTC Will Set When:

DTC 35 can be set anytime after initialization. If either of the right rear wheel speed circuits are shorted to voltage or ground, a malfunction exists.

Action Taken:

A malfunction DTC is stored, ABS is disabled and the ABS warning lamp is turned "ON."

DTC Chart Test Description:

Number(s) below refer to circled number(s) on the diagnostic chart.

1. Checks to see that the wheel speed sensor is within proper resistance values.
2. Checks for short to ground in wheel speed sensor wiring.
3. Checks for short to voltage in wheel speed sensor wiring.
4. Checks for a short to voltage in CKT 883.
5. Checks for a short to voltage in CKT 882.
6. Ensures malfunction was not due to physical damage of circuitry.

Diagnostic Aids:

An "Intermittent" malfunction may be caused by a poor connection, rubbed through wire insulation, or a wire that is broken inside the insulation.

The frequency of the malfunction can be checked by using the enhanced diagnostic function of the Tech 1.

If the customer's comments reflect that the ABS warning lamp is "ON" only during moist environmental changes (rain, snow, vehicle wash), all wheel speed sensor circuitry should be thoroughly inspected for signs of water intrusion. If DTC is not current, clear DTCs and simulate the effects of water intrusion. Use the following procedure. Spray the suspected area with a 5% salt water solution (two teaspoons of salt to 12 oz. of water). Test drive vehicle over various road surfaces (bumps, turns, etc.) above 24 km/h (15 mph) for at least 30 seconds. If DTC returns, replace suspected harness.

Any circuitry that is suspected of causing the intermittent complaint should be thoroughly checked for backed out terminals, improper mating, broken locks, improperly formed or damaged terminals, poor terminal to wiring connections or physical damage to the wiring harness.

Resistance of the wheel speed sensor will increase with an increase in sensor temperature.

When replacing a wheel speed sensor, inspect the sensor terminals and harness connector for corrosion and/or water intrusion. If evidence of corrosion or water intrusion exists, replace wheel speed sensor jumper harness. Likewise, if replacing a wheel speed sensor jumper harness, inspect sensor terminals. If evidence of corrosion or water intrusion exists, replace wheel speed sensor.

DIAGNOSTIC CHART DTC 35 — 1995 CENTURY, CUTLASS CIERA AND CUTLASS CRUISER

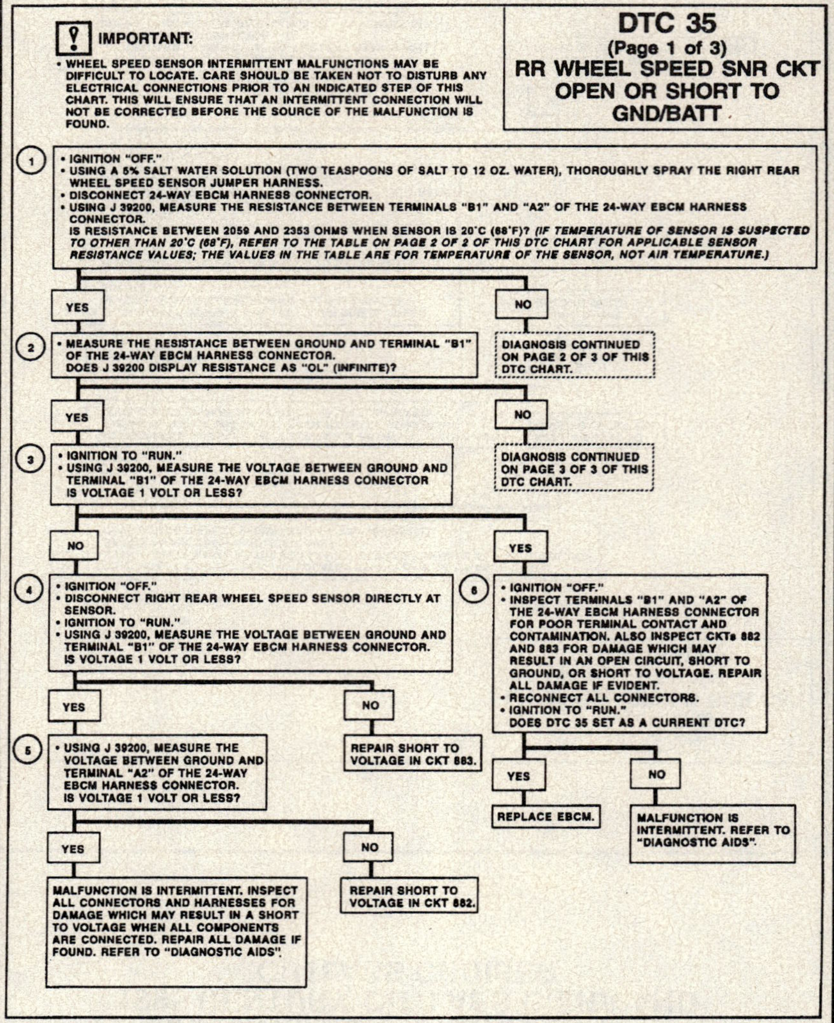

DIAGNOSTIC CHART DTC 35 — 1995 CENTURY, CUTLASS CIERA AND CUTLASS CRUISER

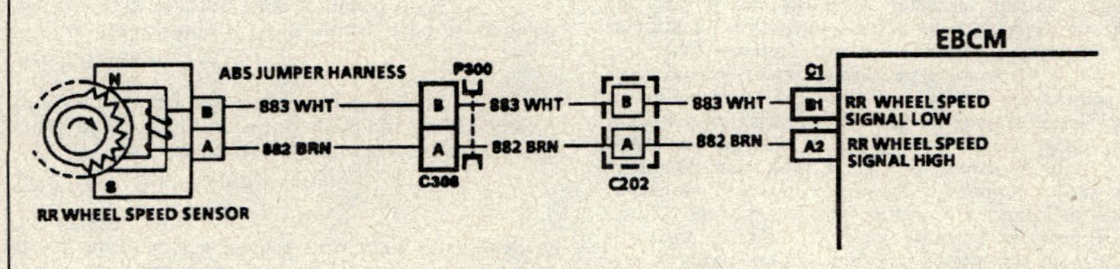

DTC 35
(Page 2 of 3)
RR WHEEL SPEED SNR CKT OPEN OR SHORT TO GND/BATT

Circuit Description:

As a toothed ring passes by the wheel speed sensor, changes in the electromagnetic field cause the wheel speed sensor to produce a sinusoidal (AC) voltage signal whose frequency is proportional to wheel speed. The magnitude of this signal is directly related to wheel speed and the proximity of the wheel speed sensor to the toothed ring, often referred to as the air gap.

DTC Will Set When:

DTC 35 can be set anytime after initialization. If either of the right rear wheel speed circuits are shorted to voltage or ground, a malfunction exists.

Action Taken:

A malfunction DTC is stored, ABS is disabled and the ABS warning lamp is turned "ON."

DTC Chart Test Description:

Number(s) below refer to circled number(s) on the diagnostic chart.

7. Checks for high resistance in CKT 883.
8. Checks for high resistance in CKT 882.
9. Verifies the proper resistance value of the sensor.

Diagnostic Aids:

An "Intermittent" malfunction may be caused by a poor connection, rubbed through wire insulation, or a wire that is broken inside the insulation.

The frequency of the malfunction can be checked by using the enhanced diagnostic function of the Tech 1.

If the customer's comments reflect that the ABS warning lamp is "ON" only during moist environmental changes (rain, snow, vehicle wash), all wheel speed sensor circuitry should be thoroughly inspected for signs of water intrusion. If DTC is not current, clear DTCs and simulate the effects of water intrusion. Use the following procedure. Spray the suspected area with a 5% salt water solution (two teaspoons of salt to 12 oz. of water). Test drive vehicle over various road surfaces (bumps, turns, etc.) above 24 km/h (15 mph) for at least 30 seconds. If DTC returns, replace suspected harness.

Any circuitry that is suspected of causing the intermittent complaint should be thoroughly checked for backed out terminals, improper mating, broken locks, improperly formed or damaged terminals, poor terminal to wiring connections or physical damage to the wiring harness.

Resistance of the wheel speed sensor will increase with an increase in sensor temperature.

When replacing a wheel speed sensor, inspect the sensor terminals and harness connector for corrosion and/or water intrusion. If evidence of corrosion or water intrusion exists, replace wheel speed sensor jumper harness. Likewise, if replacing a wheel speed sensor jumper harness, inspect sensor terminals. If evidence of corrosion or water intrusion exists, replace wheel speed sensor.

DIAGNOSTIC CHART DTC 35 — 1995 CENTURY, CUTLASS CIERA AND CUTLASS CRUISER

DTC 35
(Page 2 of 3) RR WHEEL SPEED SNR CKT OPEN OR SHORT TO GND/BATT

DIAGNOSIS CONTINUED FROM PAGE 1 OF 3 OF THIS DTC CHART.

(7)
- DISCONNECT RIGHT REAR WHEEL SPEED SENSOR DIRECTLY AT SENSOR.
- USING J 39200, MEASURE THE RESISTANCE BETWEEN TERMINAL "B1" OF THE 24-WAY EBCM HARNESS CONNECTOR AND TERMINAL "A" OF THE RIGHT REAR WHEEL SPEED SENSOR HARNESS CONNECTOR.

IS RESISTANCE 2 OHMS OR LESS?

NO: REPAIR HIGH RESISTANCE IN CKT 883 (WHEEL SPEED SENSOR JUMPER MUST BE REPLACED IF DAMAGED).

YES:

(8)
- USING J 39200, MEASURE THE RESISTANCE BETWEEN TERMINAL "A2" OF THE 24-WAY EBCM HARNESS CONNECTOR AND TERMINAL "B" OF THE RIGHT REAR WHEEL SPEED SENSOR HARNESS CONNECTOR.

IS RESISTANCE 2 OHMS OR LESS?

NO: REPAIR HIGH RESISTANCE IN CKT 882 (WHEEL SPEED SENSOR JUMPER MUST BE REPLACED IF DAMAGED).

YES:

(9)
- USING J 39200, MEASURE THE RESISTANCE BETWEEN TERMINAL "A" AND TERMINAL "B" OF THE RIGHT REAR WHEEL SPEED SENSOR CONNECTOR.

IS RESISTANCE BETWEEN 2059 AND 2353 OHMS WHEN SENSOR IS 20°C (68°F)? *(IF TEMPERATURE OF SENSOR IS SUSPECTED TO OTHER THAN 20°C (68°F), REFER TO THE TABLE BELOW FOR APPLICABLE SENSOR RESISTANCE VALUES; THE VALUES IN THE TABLE FOR TEMPERATURE OF THE SENSOR, NOT AIR TEMPERATURE.)*

YES: MALFUNCTION IS INTERMITTENT. INSPECT ALL CONNECTORS AND HARNESSES FOR DAMAGE WHICH MAY RESULT IN HIGH RESISTANCE WHEN ALL COMPONENTS ARE CONNECTED. REFER TO "DIAGNOSTIC AIDS" ON FACING PAGE FOR MORE INFORMATION. REPAIR ALL DAMAGE IF FOUND.

NO: REPLACE WHEEL SPEED SENSOR.

REAR WHEEL SPEED SENSOR

TEMPERATURE OF SENSOR VS. SENSOR RESISTANCE (APPROXIMATE)

TEMP. (°C)	TEMP. (°F)	RESISTANCE (OHMS)
-34 TO 4	-30 TO 40	1622 TO 2205
5 TO 43	41 TO 110	1930 TO 2566
44 TO 93	111 TO 200	2245 TO 3028

DIAGNOSTIC CHART DTC 35 — 1995 CENTURY, CUTLASS CIERA AND CUTLASS CRUISER

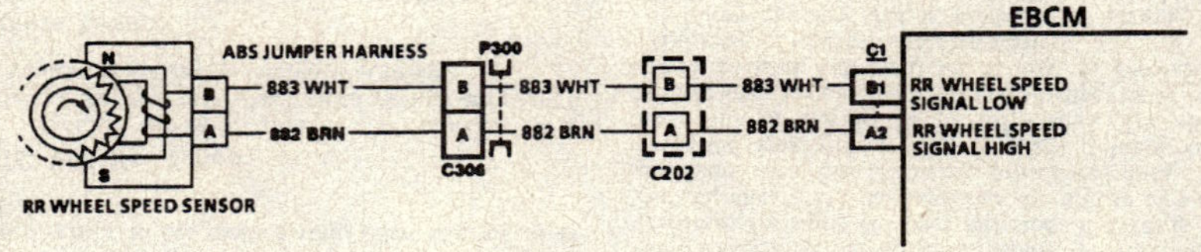

DTC 35
(Page 3 of 3)
RR WHEEL SPEED SNR CKT OPEN OR SHORT TO GND/BATT

Circuit Description:

As a toothed ring passes by the wheel speed sensor, changes in the electromagnetic field cause the wheel speed sensor to produce a sinusoidal (AC) voltage signal whose frequency is proportional to wheel speed. The magnitude of this signal is directly related to wheel speed and the proximity of the wheel speed sensor to the toothed ring, often referred to as the air gap.

DTC Will Set When:

DTC 35 can be set anytime after initialization. If either of the right rear wheel speed circuits are shorted to voltage or ground, a malfunction exists.

Action Taken:

A malfunction DTC is stored, ABS is disabled and the ABS warning lamp is turned "ON."

DTC Chart Test Description:

Number(s) below refer to circled number(s):

10. Checks for short to ground in CKT 883.
11. Checks for short to ground in CKT 882.
12. Checks for short to ground in wheel speed sensor jumper harness.
13. Checks for an internal short to ground in the wheel speed sensor.

Diagnostic Aids:

An "Intermittent" malfunction may be caused by a poor connection, rubbed through wire insulation, or a wire that is broken inside the insulation.

The frequency of the malfunction can be checked by using the enhanced diagnostic function of the Tech 1.

If the customer's comments reflect that the ABS warning lamp is "ON" only during moist environmental changes (rain, snow, vehicle wash), all wheel speed sensor circuitry should be thoroughly inspected for signs of water intrusion. If DTC is not current, clear DTCs and simulate the effects of water intrusion. Use the following procedure. Spray the suspected area with a 5% salt water solution (two teaspoons of salt to 12 oz. of water). Test drive vehicle over various road surfaces (bumps, turns, etc.) above 24 km/h (15 mph) for at least 30 seconds. If DTC returns, replace suspected harness.

Any circuitry that is suspected of causing the intermittent complaint should be thoroughly checked for backed out terminals, improper mating, broken locks, improperly formed or damaged terminals, poor terminal to wiring connections or physical damage to the wiring harness.

Resistance of the wheel speed sensor will increase with an increase in sensor temperature.

When replacing a wheel speed sensor, inspect the sensor terminals and harness connector for corrosion and/or water intrusion. If evidence of corrosion or water intrusion exists, replace wheel speed sensor jumper harness. Likewise, if replacing a wheel speed sensor jumper harness, inspect sensor terminals. If evidence of corrosion or water intrusion exists, replace wheel speed sensor.

DIAGNOSTIC CHART DTC 35 — 1995 CENTURY, CUTLASS CIERA AND CUTLASS CRUISER

DTC 35
(Page 3 of 3)
RR WHEEL SPEED SNR CKT OPEN OR SHORT TO GND/BATT

DIAGNOSIS CONTINUED FROM PAGE 1 OF 3 OF THIS DTC CHART.

(10)
- DISCONNECT CONNECTOR C372 (LOCATED BEHIND RH SIDE OF REAR SEAT).
- USING J 39200, MEASURE THE RESISTANCE BETWEEN GROUND AND TERMINAL "B1" OF THE 24-WAY EBCM HARNESS CONNECTOR. DOES J 39200 DISPLAY RESISTANCE AS "OL" (INFINITE)?

NO → REPAIR SHORT TO GROUND IN CKT 883.

YES →

(11)
- USING J 39200, MEASURE THE RESISTANCE BETWEEN GROUND AND TERMINAL "A2" OF THE 24-WAY EBCM HARNESS CONNECTOR. DOES J 39200 DISPLAY RESISTANCE AS "OL" (INFINITE)?

NO → REPAIR SHORT TO GROUND IN CKT 882.

YES →

(12)
- THOROUGHLY RESPRAY THE RIGHT REAR WHEEL SPEED SENSOR JUMPER HARNESS WITH THE SALT WATER SOLUTION.
- IMPORTANT: DO NOT ALLOW WATER TO ENTER ANY WHEEL SPEED SENSOR CONNECTOR.
- DISCONNECT RIGHT REAR WHEEL SPEED SENSOR DIRECTLY AT SENSOR.
- USING J 39200, MEASURE THE RESISTANCE BETWEEN GROUND AND TERMINAL "A," THEN "B" OF THE RIGHT REAR WHEEL SPEED SENSOR HARNESS CONNECTOR. DOES J 39200 DISPLAY RESISTANCE AS "OL" (INFINITE) FOR BOTH MEASUREMENTS?

NO → REPLACE RIGHT REAR WHEEL SPEED SENSOR JUMPER HARNESS.

YES →

(13)
- USING J 39200, MEASURE THE RESISTANCE BETWEEN GROUND AND TERMINAL "A" OF THE RIGHT REAR WHEEL SPEED SENSOR CONNECTOR. DOES J 39200 DISPLAY RESISTANCE AS "OL" (INFINITE)?

NO → REPLACE WHEEL SPEED SENSOR.

YES → MALFUNCTION IS INTERMITTENT. INSPECT ALL CONNECTORS AND HARNESSES FOR DAMAGE WHICH MAY RESULT IN A SHORT TO GROUND WHEN ALL COMPONENTS ARE CONNECTED. PAY PARTICULAR ATTENTION TO THE JUMPER HARNESS. SMALL NICKS OR CUTS CAN CAUSE A MALFUNCTION. REPAIR ANY DAMAGE IF FOUND. JUMPER HARNESS MUST BE REPLACED IF DAMAGED. REFER TO "DIAGNOSTIC AIDS".

DIAGNOSTIC CHART DTC 36 — 1995 CENTURY, CUTLASS CIERA AND CUTLASS CRUISER

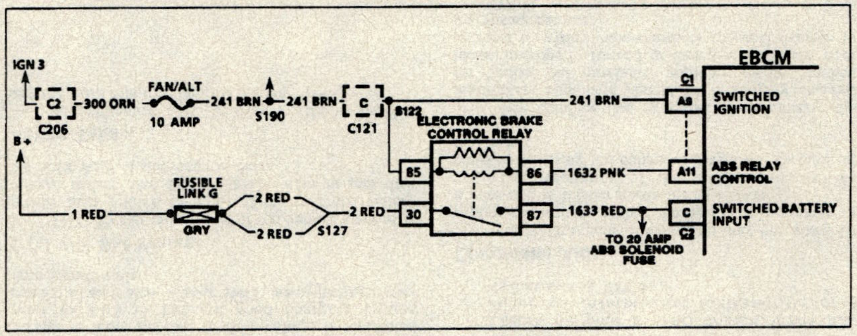

DTC 36
(Page 1 of 2)
LOW SYSTEM VOLTAGE

Circuit Description:

This DTC is used to monitor the voltage level available to the EBCM controller. If the voltage drops below 11 volts, full performance of the ABS system cannot be guaranteed. During ABS operation, there are several current requirements that will cause battery voltage to drop. Because of this, voltage is monitored prior to ABS to indicate good charging system condition and also during ABS when voltage may drop significantly.

DTC Will Set When:

DTC 36 can only be set if the vehicle's speed is greater than 5 km/h (3 mph). If the switched battery voltage is less than 11.2 volts non-ABS or less than 9.6 volts ABS, a malfunction exists.

Action Taken:

A malfunction DTC is stored, ABS is disabled and the ABS warning lamp is turned "ON."

DTC Chart Test Description:

Number(s) below refer to circled number(s) on the diagnostic chart.

1. Checks to see what voltage is readily available at terminal "C" of the EBCM. If voltage is representative of good charging system condition, a malfunction is not present.
2. Isolates the low voltage condition to high circuit resistance, or improper charging system operation.
3. Isolates which circuit is the source of the high resistance condition.
4. Checks for high resistance in CKT 2 or 1.

Diagnostic Aids:

An "Intermittent" malfunction may be caused by a poor connection, rubbed through wire insulation, or a wire that is broken inside the insulation.

The frequency of the malfunction can be checked by using the enhanced diagnostic function of the Tech 1.

Any circuitry that is suspected of causing the intermittent complaint should be thoroughly checked for backed out terminals, improper mating, broken locks, improperly formed or damaged terminals, poor terminal to wiring connections or physical damage to the wiring harness.

While performing a voltage load test, if it is noted that only ignition voltage drops below acceptable voltage levels, CKT 241 should be checked for high resistance or an open condition.

DIAGNOSTIC CHART DTC 36 — 1995 CENTURY, CUTLASS CIERA AND CUTLASS CRUISER

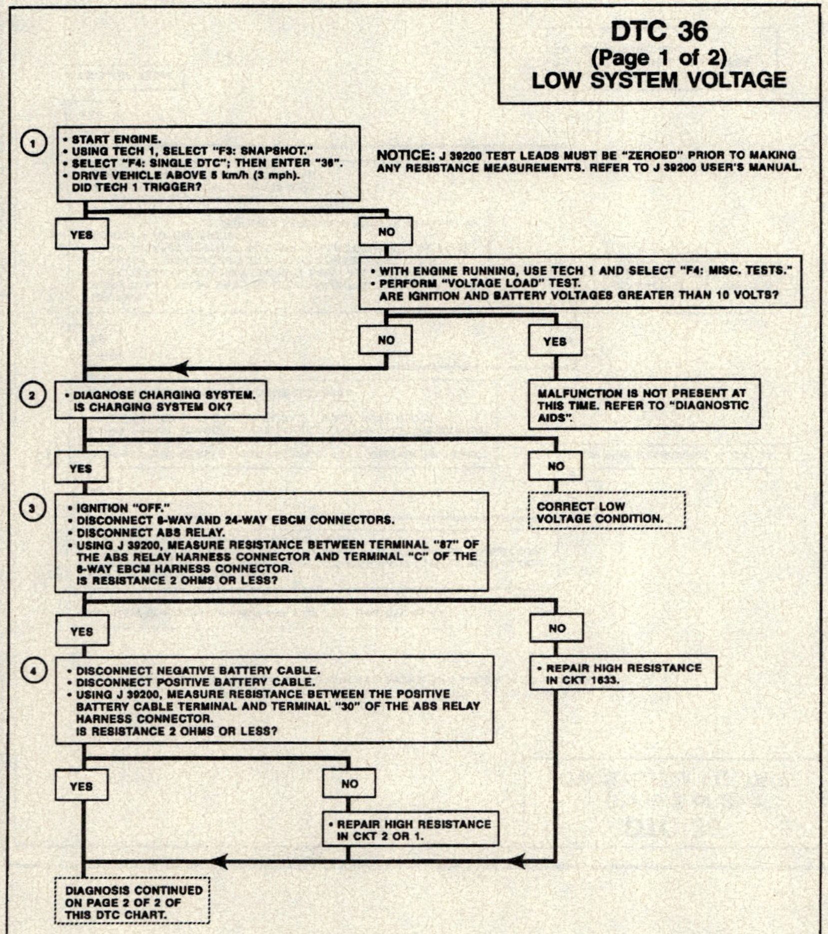

DIAGNOSTIC CHART DTC 36 — 1995 CENTURY, CUTLASS CIERA AND CUTLASS CRUISER

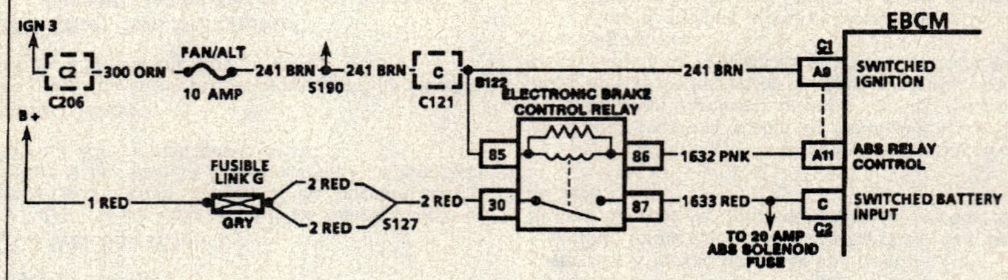

DTC 36
(Page 2 of 2)
LOW SYSTEM VOLTAGE

Circuit Description:

This DTC is used to monitor the voltage level available to the EBCM controller. If the voltage drops below 11 volts, full performance of the ABS system cannot be guaranteed. During ABS operation, there are several current requirements that will cause battery voltage to drop. Because of this, voltage is monitored prior to ABS to indicate good charging system condition and also during ABS when voltage may drop significantly.

DTC Will Set When:

DTC 36 can only be set if the vehicle's speed is greater than 5 km/h (3 mph). If the switched battery voltage is less than 11.2 volts non-ABS or less than 9.6 volts ABS, a malfunction exists.

Action Taken:

A malfunction DTC is stored, ABS is disabled and the ABS warning lamp is turned "ON."

DTC Chart Test Description:

Number(s) below refer to circled number(s) on the diagnostic chart.

5. Checks for high resistance in CKT 241.
6. Checks for high resistance in the ignition fuse feed circuit.
7. Checks for poor terminal contact which may cause an intermittent malfunction or a malfunctioning EBCM.

Diagnostic Aids:

An "Intermittent" malfunction may be caused by a poor connection, rubbed through wire insulation, or a wire that is broken inside the insulation.

The frequency of the malfunction can be checked by using the enhanced diagnostic function of the Tech 1.

Any circuitry that is suspected of causing the intermittent complaint should be thoroughly checked for backed out terminals, improper mating, broken locks, improperly formed or damaged terminals, poor terminal to wiring connections or physical damage to the wiring harness.

While performing a voltage load test, if it is noted that only ignition voltage drops below acceptable voltage levels, CKT 241 should be checked for high resistance or an open condition.

DIAGNOSTIC CHART DTC 36 — 1995 CENTURY, CUTLASS CIERA AND CUTLASS CRUISER

DTC 36
(Page 2 of 2)
LOW SYSTEM VOLTAGE

NOTICE: J 39200 TEST LEADS MUST BE "ZEROED" PRIOR TO MAKING ANY RESISTANCE MEASUREMENTS. REFER TO J 39200 USER'S MANUAL.

DIAGNOSIS CONTINUED FROM PAGE 1 OF 2 OF THIS DTC CHART.

(5)
- REMOVE FAN/ALT 10 AMP FUSE.
- USING J 39200, MEASURE RESISTANCE BETWEEN EACH FAN/ALT FUSE BLOCK TERMINAL AND TERMINAL "A9" OF THE 24-WAY EBCM HARNESS CONNECTOR. IS EITHER MEASUREMENT 2 OHMS OR LESS?

NO: REPAIR HIGH RESISTANCE IN CKT 241.

YES:

(6)
- WITH BATTERY CABLES DISCONNECTED, TURN IGNITION TO "RUN." (THIS PROVIDES IGNITION CIRCUIT CONTINUITY).
- USING J 39200, MEASURE RESISTANCE BETWEEN POSITIVE BATTERY CABLE TERMINAL AND EACH FAN/ALT 10 AMP FUSE BLOCK TERMINAL. IS EITHER MEASUREMENT 2 OHMS OR LESS?

NO: REPAIR HIGH RESISTANCE IN THE "IGNITION SWITCH 'RUN' FUSE FEED" CIRCUIT.

YES:

(7)
- IGNITION "OFF."
- INSPECT ABS RELAY AND ABS RELAY HARNESS CONNECTOR FOR POOR TERMINAL CONTACT. INSPECT 8-WAY AND 24-WAY EBCM HARNESS CONNECTORS FOR A POOR CONNECTION TO THE EBCM. INSPECT POSITIVE BATTERY CABLE TERMINAL FOR POOR CONTACT TO THE BATTERY. REPLACE TERMINALS IF POOR CONTACT IS EVIDENT.
- RECONNECT ALL CONNECTORS.
- DRIVE VEHICLE ABOVE 5 km/h (3 mph). DOES DTC 36 SET AS A CURRENT DTC?

YES: REPLACE EBCM.

NO: MALFUNCTION IS MOST LIKELY DUE TO A POOR CONNECTION. REFER TO "DIAGNOSTIC AIDS".

DIAGNOSTIC CHART DTC 37 — 1995 CENTURY, CUTLASS CIERA AND CUTLASS CRUISER

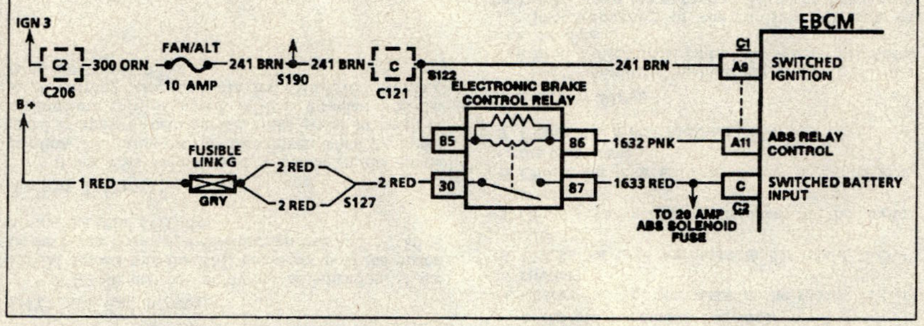

DTC 37
HIGH SYSTEM VOLTAGE

Circuit Description:

This DTC is designed to detect high vehicle voltage levels prior to any required motor movement (initialization or ABS). If excessive voltage exists, demagnetization of the motor magnets may occur, which would eventually affect or eliminate ABS performance.

DTC Will Set When:

DTC 37 can only be set if the vehicle's speed is greater than 5 km/h (3 mph). If the switched battery voltage is greater than 17 volts, a malfunction exists.

Action Taken:

A malfunction DTC is stored, ABS is disabled and the ABS warning lamp is turned "ON."

DTC Chart Test Description:

Number(s) below refer to circled number(s) on the diagnostic chart.

1. Checks the voltage level being received by the EBCM.
2. Indicates whether the high voltage condition is caused by a malfunctioning charging system or EBCM.
3. Checks for poor terminal contact which may cause an intermittent malfunction or a malfunctioning EBCM.

Diagnostic Aids:

An "Intermittent" malfunction may be caused by a poor connection, rubbed through wire insulation, or a wire that is broken inside the insulation.

The frequency of the malfunction can be checked by using the enhanced diagnostic function of the Tech 1.

Any circuitry that is suspected of causing the intermittent complaint should be thoroughly checked for backed out terminals, improper mating, broken locks, improperly formed or damaged terminals, poor terminal to wiring connections or physical damage to the wiring harness.

DIAGNOSTIC CHART DTC 37 — 1995 CENTURY, CUTLASS CIERA AND CUTLASS CRUISER

DTC 37
HIGH SYSTEM VOLTAGE

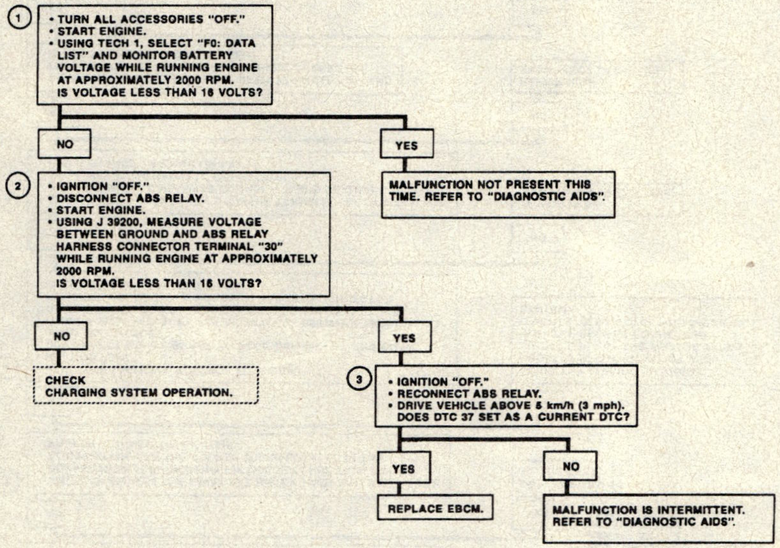

DIAGNOSTIC CHART DTC 38 — 1995 CENTURY, CUTLASS CIERA AND CUTLASS CRUISER

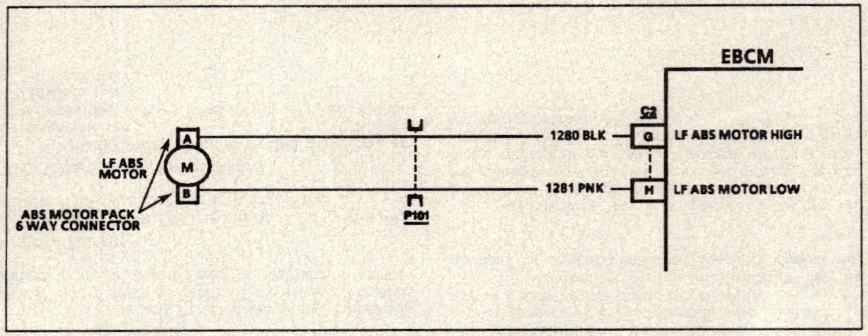

DTC 38
LEFT FRONT ESB WILL NOT HOLD MOTOR

Circuit Description:

This DTC is designed to detect a slipping left front ESB. During initialization and braking, the left front motor is rehomed. If the ESB slips, the motor/piston will move. During the next key "ON" initialization, a rehome of the motor verifies the motor/piston remained at the home position. If motor movement is detected, the ESB must be slipping.

DTC Will Set When:

DTC 38 can be set during initialization. If the EBCM detects that the ESB could not hold the piston in the home position, a malfunction exists. DTC 86 is always set with DTC 38.

Action Taken:

If an ESB cannot hold a piston in the home position, the piston may be backdriven when the brake pedal is applied, causing the brake pedal to drop. A malfunction DTC is stored, ABS is disabled, and the ABS warning lamp and the red "BRAKE" warning lamp are turned "ON."

DTC Chart Test Description:

Number(s) below refer to circled number(s) on the diagnostic chart.

1. Checks the left front ESB. A broken or defective ESB would result in the left front piston being backdriven by hydraulic pressure and pedal movement would result.
2. Ensures that the proper motor circuits are located in the proper connector cavities.
3. Checks for high resistance in the motor "HIGH" circuitry.
4. Checks for high resistance in the motor "LOW" circuitry.
5. Checks for proper resistance of the motor windings.
6. Checks for poor terminal contact or corrosion in the connectors.
7. Checks for a malfunctioning EBCM.

Diagnostic Aids:

An "Intermittent" malfunction in this DTC may result from a mechanical part of the system that sticks, binds, or slips.

The frequency of the malfunction can be checked by using the enhanced diagnostic function of the Tech 1.

The static modulator test function of the Tech 1 or T-100 may be used to locate an intermittent malfunction associated with the ESB.

DIAGNOSTIC CHART DTC 38 — 1995 CENTURY, CUTLASS CIERA AND CUTLASS CRUISER

DTC 38
LEFT FRONT ESB WILL NOT HOLD MOTOR

NOTICE: J 39200 TEST LEADS MUST BE "ZEROED" PRIOR TO MAKING ANY RESISTANCE MEASUREMENTS. REFER TO J 39200 USER'S MANUAL.

(1)
- IGNITION TO "RUN," ENGINE "OFF."
- PUMP BRAKE PEDAL UNTIL FIRM TO DEPLETE VACUUM RESERVE.
- USING TECH 1 MISC TESTS, SELECT MANUAL CONTROL FUNCTION.
- SELECT ABS MOTOR APPLY FUNCTION OF THE LEFT FRONT ABS CHANNEL AND APPLY THE MOTOR.
- WAIT 5 SECONDS.
- APPLY FIRM PRESSURE ON BRAKE PEDAL.

DID BRAKE PEDAL FALL?

- YES → REPLACE ABS MOTOR PACK.
- NO → (2)

(2)
- VISUALLY INSPECT THE 6-WAY ABS MOTOR PACK AND 6-WAY EBCM CONNECTORS FOR PROPER WIRE COLOR/CONNECTOR CAVITY COMBINATION. ARE THE PROPER WIRES LOCATED IN THE PROPER CONNECTOR CAVITIES?

- NO → USING THE CIRCUIT DIAGRAMS LOCATED IN THE FRONT OF THIS SECTION, PLACE THE PROPER WIRES INTO THE PROPER CONNECTOR CAVITIES. CLEAR DTCs AND RECHECK SYSTEM.
- YES → (3)

(3)
- EXIT TECH 1 MANUAL CONTROL FUNCTION.
- IGNITION "OFF."
- DISCONNECT 6-WAY ABS MOTOR PACK CONNECTOR AND 6-WAY EBCM CONNECTOR.
- USING J 39200, MEASURE THE RESISTANCE BETWEEN TERMINAL "A" OF THE 6-WAY ABS MOTOR PACK HARNESS CONNECTOR AND TERMINAL "G" OF THE 6-WAY EBCM HARNESS CONNECTOR.

IS RESISTANCE 1.5 OHMS OR LESS?

- NO → REPAIR HIGH RESISTANCE IN CKT 1280.
- YES → (4)

(4)
- MEASURE THE RESISTANCE BETWEEN TERMINAL "B" OF THE 6-WAY ABS MOTOR PACK HARNESS CONNECTOR AND TERMINAL "H" OF THE 6-WAY EBCM HARNESS CONNECTOR.

IS RESISTANCE 1.5 OHMS OR LESS?

- NO → REPAIR HIGH RESISTANCE IN CKT 1281.
- YES → (5)

(5)
- MEASURE THE RESISTANCE BETWEEN TERMINAL "A" AND TERMINAL "B" OF THE ABS MOTOR PACK CONNECTOR.

IS RESISTANCE 1.5 OHMS OR LESS?

- NO → REPLACE ABS MOTOR PACK.
- YES → (6)

(6)
- INSPECT THE 6-WAY ABS MOTOR PACK CONNECTOR AND 6-WAY EBCM CONNECTOR FOR POOR TERMINAL CONTACT OR CORROSION. DO ANY TERMINALS EXHIBIT POOR CONTACT OR EVIDENCE OF CORROSION?

- YES → REPLACE TERMINALS THAT EXHIBIT POOR TERMINAL CONTACT OR EVIDENCE OF CORROSION.
- NO → (7)

(7)
- IGNITION "OFF."
- RECONNECT ALL CONNECTORS.
- START ENGINE WITH FOOT OFF BRAKE.
- ALLOW ENGINE TO RUN FOR AT LEAST 10 SECONDS.
- REPEAT ABOVE IGNITION CYCLE SEQUENCE TWO MORE TIMES.

DID DTC 38 SET IN THE LAST THREE IGNITION CYCLES?

- YES → REPLACE EBCM.
- NO → MALFUNCTION IS INTERMITTENT. REFER TO "DIAGNOSTIC AIDS".

AFTER DIAGNOSIS IS COMPLETE, CLEAR DTCs AND TEST DRIVE VEHICLE FOR THREE (3) DRIVE CYCLES TO VERIFY THAT DTC DOES NOT RESET. A DRIVE CYCLE CONSISTS OF STARTING THE VEHICLE, DRIVING OVER 16 km/h (10 mph) AND THEN IGNITION "OFF."

DIAGNOSTIC CHART DTC 41 — 1995 CENTURY, CUTLASS CIERA AND CUTLASS CRUISER

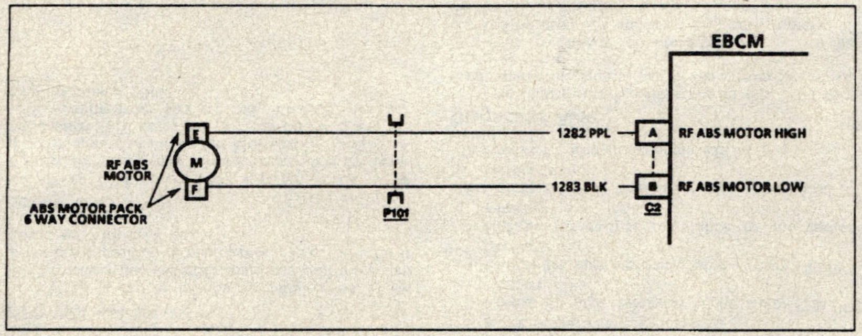

DTC 41
RIGHT FRONT ESB WILL NOT HOLD MOTOR

Circuit Description:

This DTC is designed to detect a slipping right front ESB. During initialization and braking, the right front motor is rehomed. If the ESB slips, the motor/piston will move. During the next key "ON" initialization, a rehome of the motor verifies the motor/piston remained at the home position. If motor movement is detected, the ESB must be slipping.

DTC Will Set When:

DTC 41 can be set during initialization. If the EBCM detects that the ESB could not hold the piston in the home position, a malfunction exists. DTC 86 is always set with DTC 41.

Action Taken:

If an ESB cannot hold a piston in the home position, the piston may be backdriven when the brake pedal is applied, causing the brake pedal to drop. A malfunction DTC is stored, ABS is disabled, and the ABS warning lamp and the red "BRAKE" warning lamp are turned "ON."

DTC Chart Test Description:

Number(s) below refer to circled number(s) on the diagnostic chart.

1. Checks the right front ESB. A broken or defective ESB would result in the right front piston being backdriven by hydraulic pressure, and pedal movement would result.
2. Ensures that the proper motor circuits are located in the proper connector cavities.
3. Checks for high resistance in the motor "HIGH" circuitry.
4. Checks for high resistance in the motor "LOW" circuitry.
5. Checks for proper resistance of the motor windings.
6. Checks for poor terminal contact or corrosion in the connectors.
7. Checks for a malfunctioning EBCM.

Diagnostic Aids:

An "Intermittent" malfunction in this DTC may result from a mechanical part of the system that sticks, binds, or slips.

The frequency of the malfunction can be checked by using the enhanced diagnostic function of the Tech 1.

The static modulator test function of the Tech 1 or T-100 may be used to locate an intermittent malfunction associated with the ESB.

DIAGNOSTIC CHART DTC 41 — 1995 CENTURY, CUTLASS CIERA AND CUTLASS CRUISER

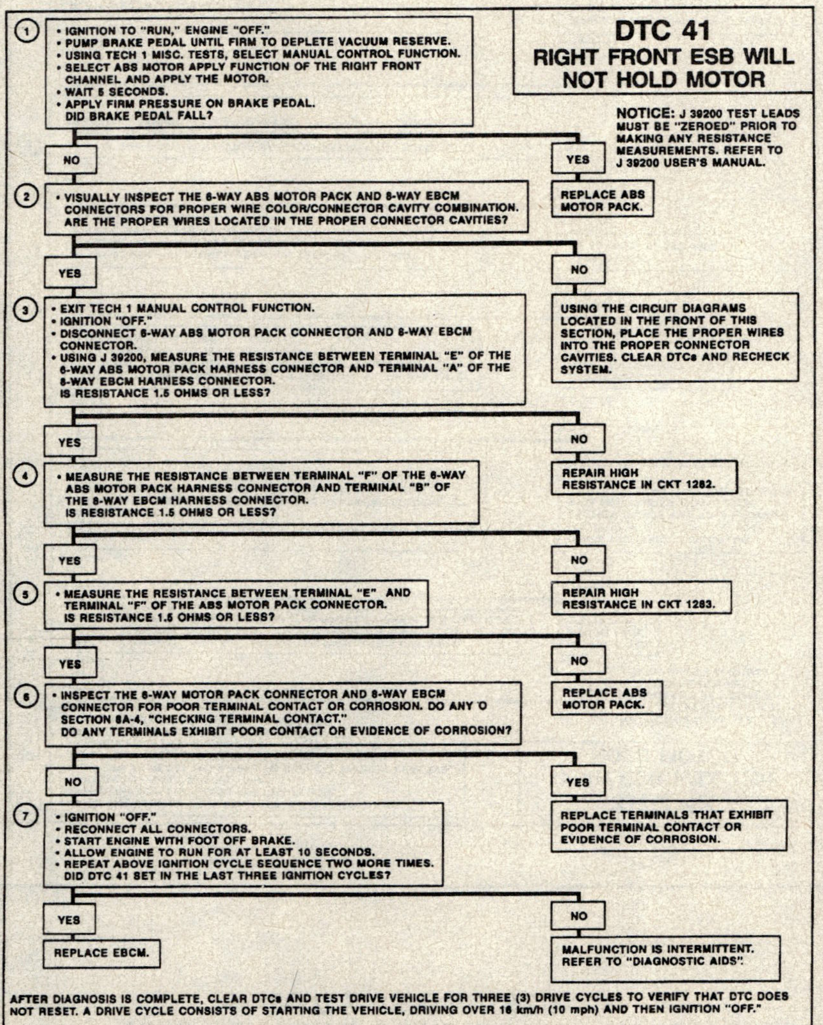

DIAGNOSTIC CHART DTC 42 — 1995 CENTURY, CUTLASS CIERA AND CUTLASS CRUISER

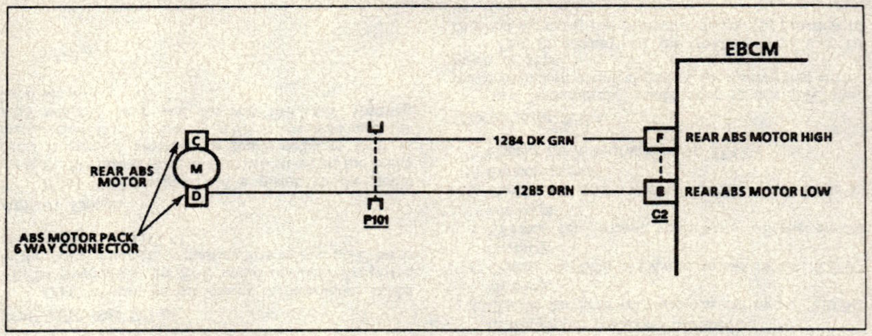

DTC 42
REAR ESB WILL NOT HOLD MOTOR

Circuit Description:

This DTC is designed to detect a slipping rear ESB. During initialization and braking, the rear motor is rehomed. If the ESB slips, the motor/piston will move. During the next key "ON" initialization, a rehome of the motor verifies the motor/piston remained at the home position. If motor movement is detected, the ESB must be slipping.

DTC Will Set When:

DTC 42 can be set during initialization. If the EBCM detects that the ESB could not hold the piston in the home position, a malfunction exists. DTC 86 is always set with DTC 42.

Action Taken:

If an ESB cannot hold a piston in the home position, the piston may be backdriven when the brake pedal is applied, causing the brake pedal to drop. A malfunction DTC is stored, ABS is disabled, and the ABS warning lamp and the red "BRAKE" warning lamp are turned "ON."

DTC Chart Test Description:

Number(s) below refer to circled number(s) on the diagnostic chart.

1. Checks the rear ESB. A broken or defective ESB would result in the rear pistons being backdriven by hydraulic pressure and wheel movement would result.
2. Ensures that the proper motor circuits are located in the proper connector cavities.
3. Checks for high resistance in the motor "HIGH" circuitry.
4. Checks for high resistance in the motor "LOW" circuitry.
5. Checks for proper resistance of the motor windings.
6. Checks for poor terminal contact or corrosion in the connectors.
7. Checks for a malfunctioning EBCM.

Diagnostic Aids:

An "Intermittent" malfunction in this DTC may result from a mechanical part of the system that sticks, binds, or slips.

The frequency of the malfunction can be checked by using the enhanced diagnostic function of the Tech 1.

The static modulator test function of the Tech 1 or T-100 may be used to locate an intermittent malfunction associated with the ESB.

DIAGNOSTIC CHART DTC 42 — 1995 CENTURY, CUTLASS CIERA AND CUTLASS CRUISER

DTC 42
REAR ESB WILL NOT HOLD MOTOR

NOTICE: J 39200 TEST LEADS MUST BE "ZEROED" PRIOR TO MAKING ANY RESISTANCE MEASUREMENTS. REFER TO J 39200 USER'S MANUAL.

(1)
- RAISE THE VEHICLE SUCH THAT THE REAR WHEELS ARE APPROXIMATELY SIX INCHES OFF THE FLOOR.
- IGNITION TO "RUN," ENGINE "OFF."
- PUMP BRAKE PEDAL UNTIL FIRM TO DEPLETE VACUUM RESERVE.
- APPLY FIRM PRESSURE ON BRAKE PEDAL AND RELEASE.
- REAPPLY VERY FIRM PRESSURE ON BRAKE PEDAL AND HAVE AN ASSISTANT TRY TO SPIN THE REAR WHEELS BY HAND. COULD ASSISTANT SPIN THE REAR WHEELS?

NO → (2)
YES → REPLACE ABS MOTOR PACK.

(2)
- VISUALLY INSPECT THE 6-WAY ABS MOTOR PACK AND 6-WAY EBCM CONNECTORS FOR PROPER WIRE COLOR/CONNECTOR CAVITY COMBINATION. ARE THE PROPER WIRES LOCATED IN THE PROPER CONNECTOR CAVITIES?

YES → (3)
NO → USING THE CIRCUIT DIAGRAMS LOCATED IN THE FRONT OF THIS SECTION, PLACE THE PROPER WIRES INTO THE PROPER CONNECTOR CAVITIES. CLEAR DTCs AND RECHECK SYSTEM.

(3)
- IGNITION "OFF."
- DISCONNECT 6-WAY ABS MOTOR PACK CONNECTOR AND 6-WAY EBCM CONNECTOR.
- USING J 39200, MEASURE THE RESISTANCE BETWEEN TERMINAL "C" OF THE 6-WAY ABS MOTOR PACK HARNESS CONNECTOR AND TERMINAL "F" OF THE 6-WAY EBCM HARNESS CONNECTOR. IS RESISTANCE 1.5 OHMS OR LESS?

YES → (4)
NO → REPAIR HIGH RESISTANCE IN CKT 1284.

(4)
- MEASURE THE RESISTANCE BETWEEN TERMINAL "D" OF THE 6-WAY ABS MOTOR PACK HARNESS CONNECTOR AND TERMINAL "E" OF THE 6-WAY EBCM HARNESS CONNECTOR "C2." IS RESISTANCE 1.5 OHMS OR LESS?

YES → (5)
NO → REPAIR HIGH RESISTANCE IN CKT 1285.

(5)
- MEASURE THE RESISTANCE BETWEEN TERMINAL "C" AND TERMINAL "D" OF THE ABS MOTOR PACK CONNECTOR. IS RESISTANCE 1.5 OHMS OR LESS?

YES → (6)
NO → REPLACE ABS MOTOR PACK.

(6)
- INSPECT THE 6-WAY ABS MOTOR PACK CONNECTOR AND 6-WAY EBCM CONNECTOR FOR POOR TERMINAL CONTACT OR CORROSION. DO ANY TERMINALS EXHIBIT POOR CONTACT OR EVIDENCE OF CORROSION?

NO → (7)
YES → REPLACE TERMINALS THAT EXHIBIT POOR TERMINAL CONTACT OR EVIDENCE OF CORROSION.

(7)
- IGNITION "OFF."
- RECONNECT ALL CONNECTORS.
- START ENGINE WITH FOOT OFF BRAKE.
- ALLOW ENGINE TO RUN FOR AT LEAST 10 SECONDS.
- REPEAT ABOVE IGNITION CYCLE SEQUENCE TWO MORE TIMES. DID DTC 42 SET IN THE LAST THREE IGNITION CYCLES?

YES → REPLACE EBCM.
NO → MALFUNCTION IS INTERMITTENT. REFER TO "DIAGNOSTIC AIDS".

AFTER DIAGNOSIS IS COMPLETE, CLEAR DTCs AND TEST DRIVE VEHICLE FOR THREE (3) DRIVE CYCLES TO VERIFY DTC DOES NOT RESET. A DRIVE CYCLE CONSISTS OF STARTING THE VEHICLE, DRIVING OVER 16 km/h (10 mph) AND THEN IGNITION "OFF."

DIAGNOSTIC CHART DTC 44 — 1995 CENTURY, CUTLASS CIERA AND CUTLASS CRUISER

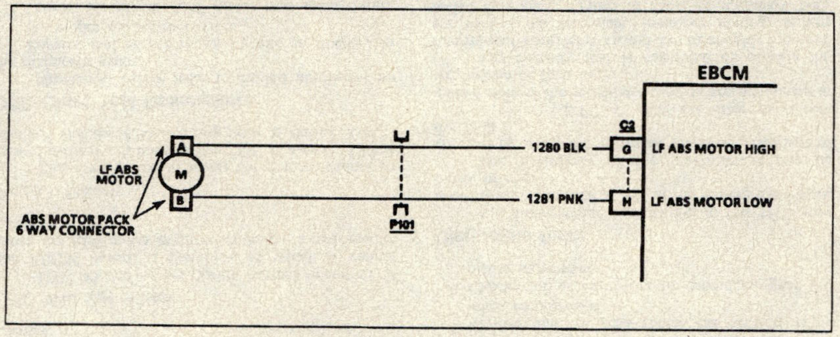

DTC 44
LEFT FRONT CHANNEL WILL NOT MOVE

Circuit Description:

This DTC is designed to detect a bound-up ESB, a stuck ABS motor, or a seized hydraulic modulator. When the release is commanded during initialization, the ESB should release the ABS motor, resulting in sensed current being less than commanded current (motor is spinning freely). If the ABS motor is not moving, sensed current will be equal to stall current.

DTC Will Set When:

DTC 44 can be set during normal operation. If the EBCM detects a condition in which it cannot move the ABS motor in either direction, a malfunction exists.

Action Taken:

This malfunction indicates the channel cannot be moved properly. A malfunction DTC is stored, ABS is disabled and the ABS warning lamp is turned "ON."

DTC Chart Test Description:

Number(s) below refer to circled number(s) on the diagnostic chart.

1. Ensures that proper ABS motor circuits are located in the proper connector cavities.
2. Checks for proper ABS motor movement during apply and release commands from the Tech 1.
3. Compares EBCM command current to ABS motor feedback current.
4. Checks for a short between the ABS motor "HIGH" and motor "LOW" circuits.
5. Checks for an ABS motor that is internally shorted.
6. Determines if the malfunction is due to poor terminal contact or the EBCM.
7. Verifies that the ABS motor can actually apply when commanded.
8. Checks for proper hydraulic modulator gear and piston movement.

Diagnostic Aids:

An "Intermittent" malfunction in this DTC may result from a mechanical part of the system that sticks, binds, or slips.

The frequency of the malfunction can be checked by using the enhanced diagnostic function of the Tech 1.

DTC 44 may set after modulator disassembly if the modulator pistons are positioned at the bottom of their bore.

Any circuitry that is suspected of causing the intermittent complaint should be thoroughly checked for backed out terminals, improper mating, broken locks, improperly formed or damaged terminals, poor terminal to wiring connections or physical damage to the wiring harness.

DIAGNOSTIC CHART DTC 44 — 1995 CENTURY, CUTLASS CIERA AND CUTLASS CRUISER

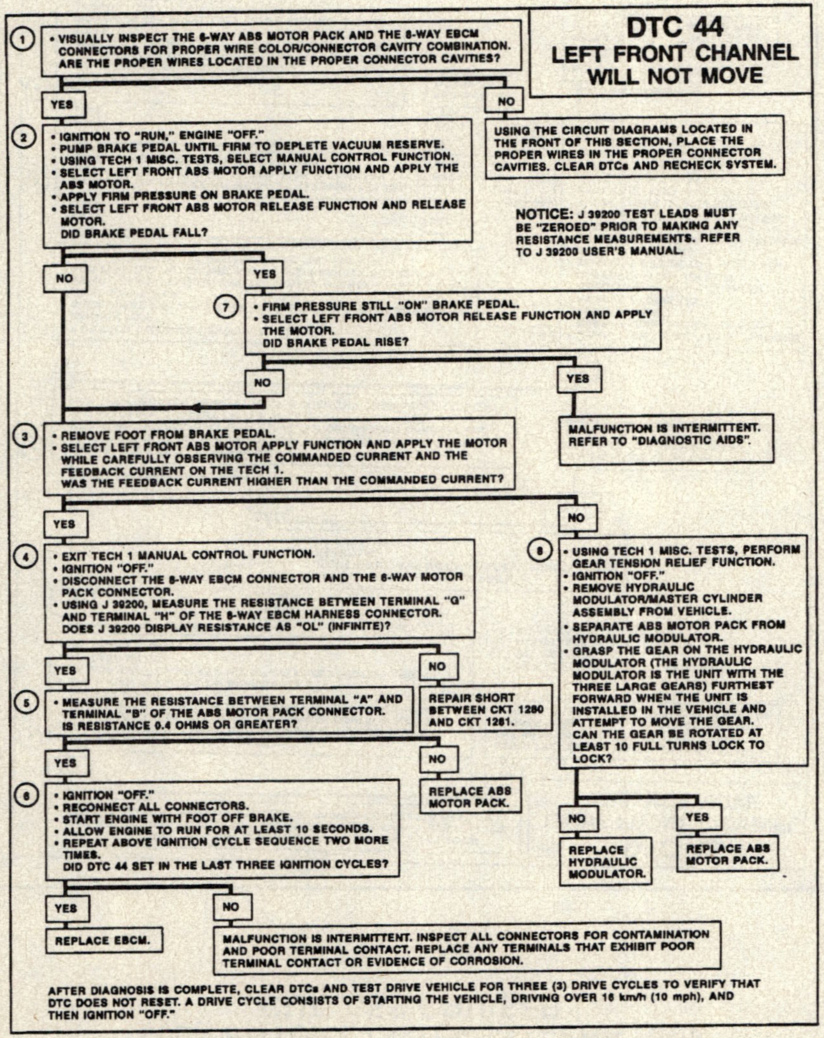

DIAGNOSTIC CHART DTC 45 — 1995 CENTURY, CUTLASS CIERA AND CUTLASS CRUISER

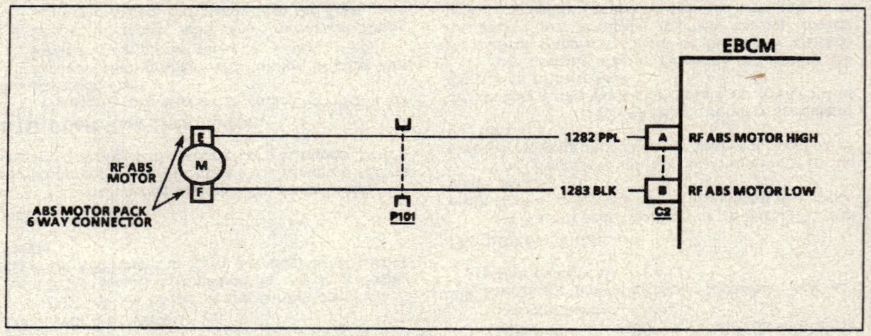

DTC 45
RIGHT FRONT CHANNEL WILL NOT MOVE

Circuit Description:

This DTC is designed to detect bound-up ESB, a stuck ABS motor, or a seized hydraulic modulator. When the release is commanded during initialization, the ESB should release the ABS motor, resulting in sensed current being less than commanded current (ABS motor is spinning freely). If the ABS motor is not moving, sensed current will be equal to stall current.

DTC Will Set When:

DTC 45 can be set during normal operation. If the EBCM detects a condition in which it cannot move the ABS motor in either direction, a malfunction exists.

Action Taken:

This malfunction indicates the channel cannot be moved properly. A malfunction DTC is stored, ABS is disabled and the ABS warning lamp is turned "ON."

DTC Chart Test Description:

Number(s) below refer to circled number(s) on the diagnostic chart.

1. Ensures that proper motor circuits are located in the proper connector cavities.
2. Checks for proper ABS motor movement during apply and release commands from the Tech 1.
3. Compares EBCM command current to ABS motor feedback current.
4. Checks for a short between the ABS motor "HIGH" and motor "LOW" circuits.
5. Checks for an ABS motor that is internally shorted.
6. Checks to see if malfunction is due to poor terminal contact or the EBCM.
7. Verifies that the ABS motor can actually apply when commanded.
8. Checks for proper hydraulic modulator gear and piston movement.

Diagnostic Aids:

An "Intermittent" malfunction in this DTC may result from a mechanical part of the system that sticks, binds, or slips.

The frequency of the malfunction can be checked by using the enhanced diagnostic function of the Tech 1.

DTC 45 may set after modulator disassembly if the modulator pistons are positioned at the bottom of their bore.

Any circuitry that is suspected of causing the intermittent complaint should be thoroughly checked for backed out terminals, improper mating, broken locks, improperly formed or damaged terminals, poor terminal to wiring connections or physical damage to the wiring harness.

DIAGNOSTIC CHART DTC 45 — 1995 CENTURY, CUTLASS CIERA AND CUTLASS CRUISER

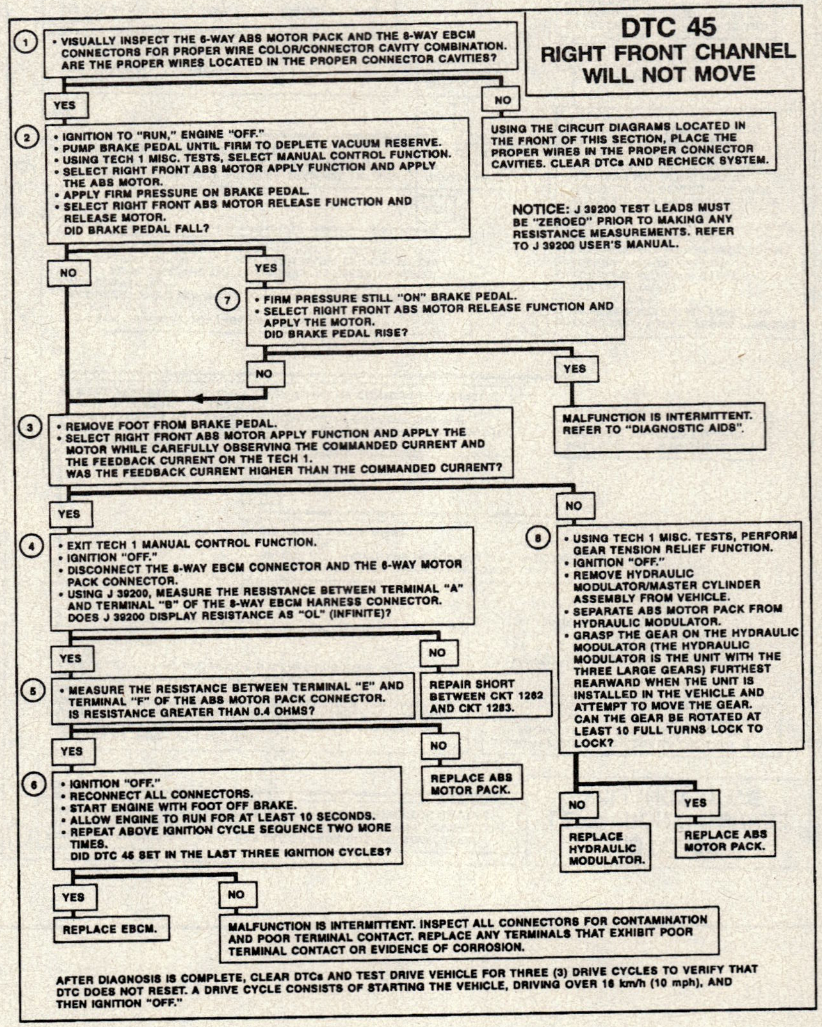

DIAGNOSTIC CHART DTC 46 — 1995 CENTURY, CUTLASS CIERA AND CUTLASS CRUISER

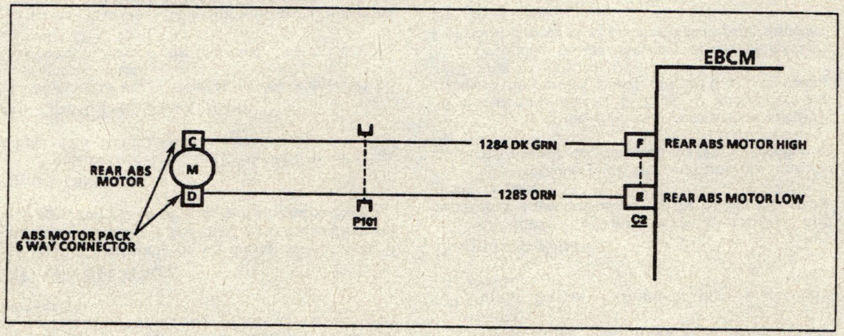

DTC 46 REAR CHANNEL WILL NOT MOVE

Circuit Description:

This DTC is designed to detect a bound-up ESB, a stuck ABS motor, or a seized hydraulic modulator. When the release is commanded during initialization, the ESB should release the ABS motor, resulting in sensed current being less than commanded current (motor is spinning freely). If the ABS motor is not moving, sensed current will be equal to stall current.

DTC Will Set When:

DTC 46 can be set during normal operation. If the EBCM detects a condition in which it cannot move the ABS motor in either direction, a malfunction exists.

Action Taken:

This malfunction indicates the channel cannot be moved properly. A malfunction DTC is stored, ABS is disabled and the ABS warning lamp is turned "ON."

DTC Chart Test Description:

Number(s) below refer to circled number(s) on the diagnostic chart.

1. Ensures that proper motor circuits are located in the proper connector cavities.
2. Checks for proper motor movement during apply and release commands from the Tech 1.
3. Compares EBCM command current to ABS motor feedback current.
4. Checks for a short between the ABS motor "HIGH" and motor "LOW" circuits.
5. Checks for an ABS motor that is internally shorted.
6. Checks to see if malfunction is due to poor terminal contact or the EBCM.
7. This step verifies that the motor can actually release when commanded.
8. Checks for proper hydraulic modulator gear and piston movement.

Diagnostic Aids:

An "Intermittent" malfunction in this DTC may result from a mechanical part of the system that sticks or binds.

The frequency of the malfunction can be checked by using the enhanced diagnostic function of the Tech 1.

DTC 46 may set after modulator disassembly if the modulator pistons are positioned at the bottom of their bore. Depending on the frequency of the malfunction, a physical inspection of the mechanical parts suspected may be necessary.

Any circuitry that is suspected of causing the intermittent complaint should be thoroughly checked for backed out terminals, improper mating, broken locks, improperly formed or damaged terminals, poor terminal to wiring connections or physical damage to the wiring harness.

DIAGNOSTIC CHART DTC 46 — 1995 CENTURY, CUTLASS CIERA AND CUTLASS CRUISER

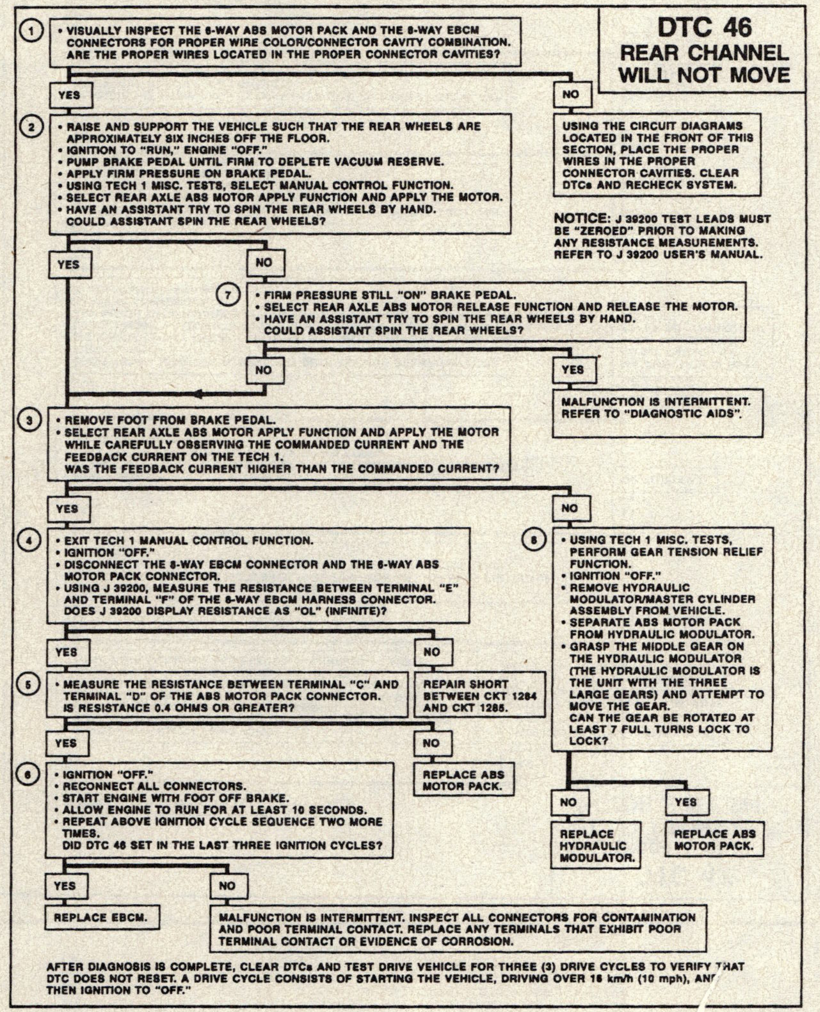

DIAGNOSTIC CHART DTC 47 — 1995 CENTURY, CUTLASS CIERA AND CUTLASS CRUISER

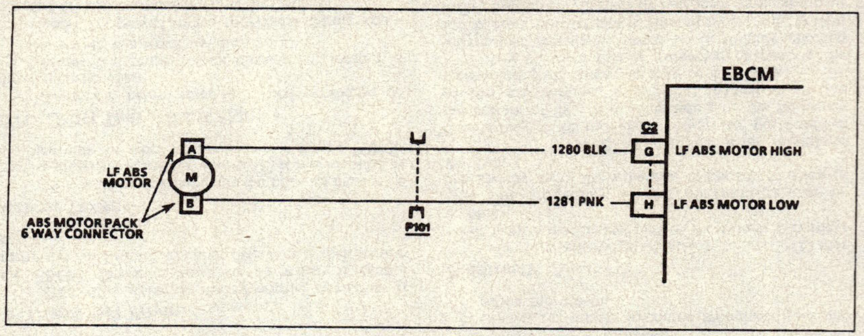

DTC 47
(Page 1 of 2)
LEFT FRONT ABS MOTOR FREE SPINS

Circuit Description:

This DTC is designed to detect a stripped nut or gear assembly during initialization. During the homing sequence, the piston should reach the top of the bore resulting in a stalled ABS motor. If this does not occur, the ABS motor must be spinning with little or no resistance, indicating a nut/screw or gear malfunction.

DTC Will Set When:

DTC 47 can only be set during initialization. If the feedback current is less than the command current for a specified period of time, a malfunction exists.

Action Taken:

A malfunction DTC is stored, ABS is disabled, and the ABS warning lamp is turned "ON."

DTC Chart Test Description:

Number(s) below refer to circled number(s) on the diagnostic chart.

1. Checks to see if the corresponding open ABS motor DTC 56 is also set.
2. Verifies that the ABS motor was actually applied as commanded by observing feedback current.
3. Verifies that the ABS motor can release.
4. This step verifies that the ABS motor can be applied by observing pedal movement.
5. Checks for an open in the ABS motor "HIGH" circuitry.
6. Checks for an open in the ABS motor "LOW" circuitry.
7. Checks for an ABS motor that is internally shorted.

Diagnostic Aids:

An "Intermittent" malfunction in this DTC may result from a mechanical part of the system that sticks, binds, or slips.

The frequency of the malfunction can be checked by using the enhanced diagnostic function of the Tech 1.

If the DTC only occurs once and DTC 56 also occurs, refer to DTC 56. If intermittent and enhanced diagnostics shows this DTC occurs during ABS, refer to DTC 56.

Depending on the frequency of the malfunction, a physical inspection of the mechanical parts suspected may be necessary.

DIAGNOSTIC CHART DTC 47 — 1995 CENTURY, CUTLASS CIERA AND CUTLASS CRUISER

DTC 47 (Page 1 of 2) LEFT FRONT ABS MOTOR FREE SPINS

① IS DTC 56 ALSO CURRENTLY SET?

- YES: PROCEED TO DTC 56 CHART FIRST.
  NOTICE: J 39200 TEST LEADS MUST BE "ZEROED" PRIOR TO MAKING ANY RESISTANCE MEASUREMENTS. REFER TO J 39200 USER'S MANUAL.
- NO:

② • IGNITION TO "RUN," ENGINE "OFF."
• PUMP BRAKE PEDAL UNTIL FIRM TO DEPLETE VACUUM RESERVE.
• USING TECH 1 MISC. TESTS, SELECT MANUAL CONTROL FUNCTION.
• SELECT LEFT FRONT ABS MOTOR APPLY FUNCTION AND APPLY THE MOTOR WHILE CAREFULLY OBSERVING THE COMMAND CURRENT AND FEEDBACK CURRENT ON THE TECH 1.
WERE BOTH COMMAND AND FEEDBACK CURRENT 10 AMPS?

- NO: GO TO STEP 5.
- YES:

③ • APPLY FIRM PRESSURE ON BRAKE PEDAL.
• SELECT LEFT FRONT ABS MOTOR RELEASE FUNCTION AND RELEASE THE MOTOR.
DID THE BRAKE PEDAL FALL AND WERE BOTH COMMAND AND FEEDBACK CURRENT 6 AMPS?

- NO: GO TO STEP 5.
- YES:

④ • FIRM PRESSURE STILL ON BRAKE PEDAL.
• SELECT LEFT FRONT ABS MOTOR APPLY FUNCTION AND APPLY THE MOTOR (THIS MOVES THE PISTON TO THE TOP OF ITS BORE).
DID BRAKE PEDAL RISE?

- YES: MALFUNCTION IS MOST LIKELY CAUSED BY POOR CONNECTION. REFER TO "DIAGNOSTIC AIDS".
- NO:

⑤ • USING TECH 1 MISC. TESTS, PERFORM GEAR TENSION RELIEF FUNCTION.
• IGNITION "OFF."
• DISCONNECT 6-WAY ABS MOTOR PACK CONNECTOR AND 6-WAY EBCM CONNECTOR.
• USING J 39200, MEASURE THE RESISTANCE BETWEEN TERMINAL "A" OF THE 6-WAY ABS MOTOR PACK HARNESS CONNECTOR AND TERMINAL "G" OF THE 6-WAY EBCM HARNESS CONNECTOR.
IS RESISTANCE 1.5 OHMS OR LESS?

- NO: REPAIR HIGH RESISTANCE IN CKT 1280.
- YES:

⑥ • MEASURE THE RESISTANCE BETWEEN TERMINAL "B" OF THE 6-WAY ABS MOTOR PACK HARNESS CONNECTOR AND TERMINAL "H" OF THE 6-WAY EBCM HARNESS CONNECTOR.
IS RESISTANCE 1.5 OHMS OR LESS?

- NO: REPAIR HIGH RESISTANCE IN CKT 1281.
- YES:

⑦ • MEASURE THE RESISTANCE BETWEEN TERMINAL "A" AND TERMINAL "B" OF THE ABS MOTOR PACK CONNECTOR.
IS RESISTANCE 1.5 OHMS OR LESS?

- NO: REPLACE ABS MOTOR PACK.
- YES: DIAGNOSIS CONTINUED ON NEXT PAGE.

AFTER DIAGNOSIS IS COMPLETE, CLEAR DTCs AND TEST DRIVE VEHICLE FOR THREE (3) DRIVE CYCLES TO VERIFY THAT DTC DOES NOT RESET. A DRIVE CYCLE CONSISTS OF STARTING THE VEHICLE, DRIVING OVER 16 km/h (10 mph), AND THEN IGNITION TO "OFF."

DIAGNOSTIC CHART DTC 47 — 1995 CENTURY, CUTLASS CIERA AND CUTLASS CRUISER

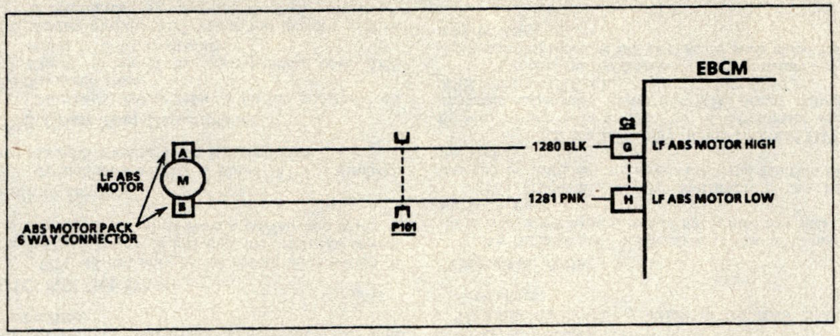

DTC 47
(Page 2 of 2)
LEFT FRONT ABS MOTOR FREE SPINS

Circuit Description:

This DTC is designed to detect a stripped nut or gear assembly during initialization. During the homing sequence, the piston should reach the top of the bore resulting in a stalled ABS motor. If this does not occur, the ABS motor must be spinning with little or no resistance, indicating a nut/screw or gear malfunction.

DTC Will Set When:

DTC 47 can only be set during initialization. If the feedback current is less than the command current for a specified period of time, a malfunction exists.

Action Taken:

A malfunction DTC is stored, ABS is disabled, and the ABS warning lamp is turned "ON."

DTC Chart Test Description:

Number(s) below refer to circled number(s) on the diagnostic chart.

8. Checks for a stripped gear on the motor pack (unit with three small gears).
9. Checks for a stripped gear on the hydraulic modulator (unit with three large gears).
10. Checks to see if the ABS motor pack has an internal malfunction.
11. Verifies that the hydraulic modulator is functioning properly.
12. Ensures malfunction was not due to a poor terminal contact.

Diagnostic Aids:

An "Intermittent" malfunction in this DTC may result from a mechanical part of the system that sticks, binds, or slips.

The frequency of the malfunction can be checked by using the enhanced diagnostic function of the Tech 1.

If the DTC only occurs once and DTC 56 also occurs, refer to DTC 56. If intermittent and enhanced diagnostics shows this DTC occurs during ABS, refer to DTC 56.

Depending on the frequency of the malfunction, a physical inspection of the mechanical parts suspected may be necessary.

DIAGNOSTIC CHART DTC 47 — 1995 CENTURY, CUTLASS CIERA AND CUTLASS CRUISER

DTC 47 (Page 2 of 2) LEFT FRONT ABS MOTOR FREE SPINS

DIAGNOSIS CONTINUED FROM PREVIOUS PAGE.

(8)
- REMOVE HYDRAULIC MODULATOR/MASTER CYLINDER ASSEMBLY FROM VEHICLE (REFER TO "ON-VEHICLE SERVICE" IN THIS SECTION).
- REMOVE GEAR COVER.
- CHECK FOR STRIPPED GEAR ON ABS MOTOR PACK; ABS MOTOR PACK IS THE UNIT WITH THE THREE SMALL GEARS. THE LEFT FRONT GEAR IS THE GEAR FURTHEST FORWARD WHEN THE ASSEMBLY IS INSTALLED IN THE VEHICLE.

IS THE ABS MOTOR PACK GEAR STRIPPED?

- YES: REPLACE ABS MOTOR PACK.
- NO: go to (9).

(9)
- CHECK FOR A STRIPPED GEAR ON HYDRAULIC MODULATOR; MODULATOR IS THE UNIT WITH THE THREE LARGE GEARS. THE LEFT FRONT GEAR IS THE GEAR FURTHEST FORWARD WHEN THE UNIT IS INSTALLED IN THE VEHICLE.

IS THE MODULATOR GEAR STRIPPED?

- YES: REPLACE HYDRAULIC MODULATOR GEAR.
- NO: go to (10).

(10)
- RECONNECT ELECTRICAL CONNECTORS.
- SECURELY POSITION HYDRAULIC MODULATOR/MASTER CYLINDER ASSEMBLY IN VEHICLE WITH GEAR COVER REMOVED SUCH THAT THE GEAR SET CAN BE OBSERVED. BE CAREFUL NOT TO DAMAGE GEAR SET OR ALLOW THE HYDRAULIC MODULATOR TO MOVE WHILE TESTING.
- IGNITION TO "RUN."
- USING TECH 1 SELECT MANUAL CONTROL FUNCTION.
- SELECT LEFT FRONT ABS MOTOR APPLY AND APPLY THE MOTOR WHILE OBSERVING THE GEAR SET FURTHEST FORWARD WHEN THE UNIT IS INSTALLED IN THE VEHICLE.
- SELECT LEFT FRONT ABS MOTOR RELEASE FUNCTION AND RELEASE THE MOTOR WHILE STILL OBSERVING THE GEAR SET.
- SELECT LEFT FRONT ABS MOTOR APPLY FUNCTION AND APPLY THE MOTOR.

DID GEAR SET MOVE IN BOTH DIRECTIONS FOR AT LEAST ONE REVOLUTION?

- NO: REPLACE ABS MOTOR PACK.
- YES: go to (11).

(11)
- USING TECH 1 PERFORM GEAR TENSION RELIEF FUNCTION.
- IGNITION "OFF."
- SEPARATE THE ABS MOTOR PACK FROM THE HYDRAULIC MODULATOR.
- GRASP THE GEAR ON THE HYDRAULIC MODULATOR (THE HYDRAULIC MODULATOR IS THE UNIT WITH THE THREE LARGE GEARS) FURTHEST FORWARD WHEN THE UNIT IS INSTALLED IN THE VEHICLE AND ATTEMPT TO MOVE THE GEAR.

CAN THE GEAR BE ROTATED MORE THAN 13 FULL TURNS LOCK TO LOCK?

- YES: REPLACE THE HYDRAULIC MODULATOR.
- NO: go to (12).

(12)
- INSPECT ALL CONNECTORS AND TERMINALS FOR POOR TERMINAL CONTACT AND EVIDENCE OF CORROSION.
  REPLACE ALL TERMINALS THAT EXHIBIT SIGNS OF POOR TERMINAL CONTACT OR CORROSION.
- REINSTALL HYDRAULIC MODULATOR/MASTER CYLINDER ASSEMBLY IN VEHICLE.
- RECONNECT ALL CONNECTORS.
- IGNITION "OFF."
- START ENGINE WITH FOOT OFF BRAKE.
- ALLOW ENGINE TO RUN FOR AT LEAST 10 SECONDS.
- REPEAT ABOVE IGNITION CYCLE SEQUENCE TWO MORE TIMES.

DID DTC 47 SET IN THE LAST THREE IGNITION CYCLES?

- YES: REPLACE EBCM.
- NO: MALFUNCTION IS MOST LIKELY CAUSED BY A POOR CONNECTION. REFER TO "DIAGNOSTIC AIDS".

AFTER DIAGNOSIS IS COMPLETE, CLEAR DTCs AND TEST DRIVE VEHICLE FOR THREE (3) DRIVE CYCLES TO VERIFY THAT DTC DOES NOT RESET. A DRIVE CYCLE CONSISTS OF STARTING THE VEHICLE, DRIVING OVER 16 km/h (10 mph), AND THEN IGNITION TO "OFF."

DIAGNOSTIC CHART DTC 48 — 1995 CENTURY, CUTLASS CIERA AND CUTLASS CRUISER

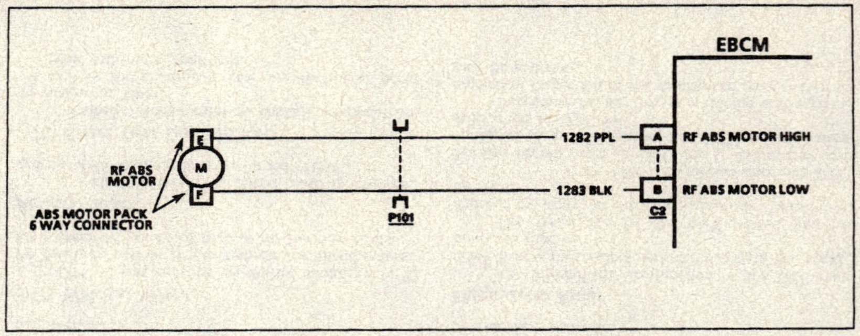

DTC 48
(Page 1 of 2)
RIGHT FRONT ABS MOTOR FREE SPINS

Circuit Description:

This DTC is designed to detect a stripped nut or gear assembly during initialization. During the homing sequence, the piston should reach the top of the bore resulting in a stalled ABS motor. If this does not occur, the ABS motor must be spinning with little or no resistance, indicating a nut/screw or gear malfunction.

DTC Will Set When:

DTC 48 can only be set during initialization. If the feedback current is less than the command current for a specified period of time, a malfunction exists.

Action Taken:

A malfunction DTC is stored, ABS is disabled, and the ABS warning lamp is turned "ON."

DTC Chart Test Description:

Number(s) below refer to circled number(s) on the diagnostic chart.

1. Checks to see if the corresponding open ABS motor DTC 61 is also set.
2. Verifies that the ABS motor was actually applied as commanded by observing feedback current.
3. Verifies that the ABS motor can release.
4. Verifies that the ABS motor can be applied by observing pedal movement.
5. Checks for high resistance in the ABS motor "HIGH" circuitry.
6. Checks for high resistance in the ABS motor "LOW" circuitry.
7. Checks for proper resistance of the ABS motor windings.

Diagnostic Aids:

An "Intermittent" malfunction in this DTC may result from a mechanical part of the system that sticks, binds, or slips.

The frequency of the malfunction can be checked by using the enhanced diagnostic function of the Tech 1.

If the DTC only occurs once and DTC 61 also occurs, refer to DTC 61. If intermittent and enhanced diagnostics shows this DTC occurs during ABS, refer to DTC 61.

Depending on the frequency of the malfunction, a physical inspection of the mechanical parts suspected may be necessary.

DIAGNOSTIC CHART DTC 48 — 1995 CENTURY, CUTLASS CIERA AND CUTLASS CRUISER

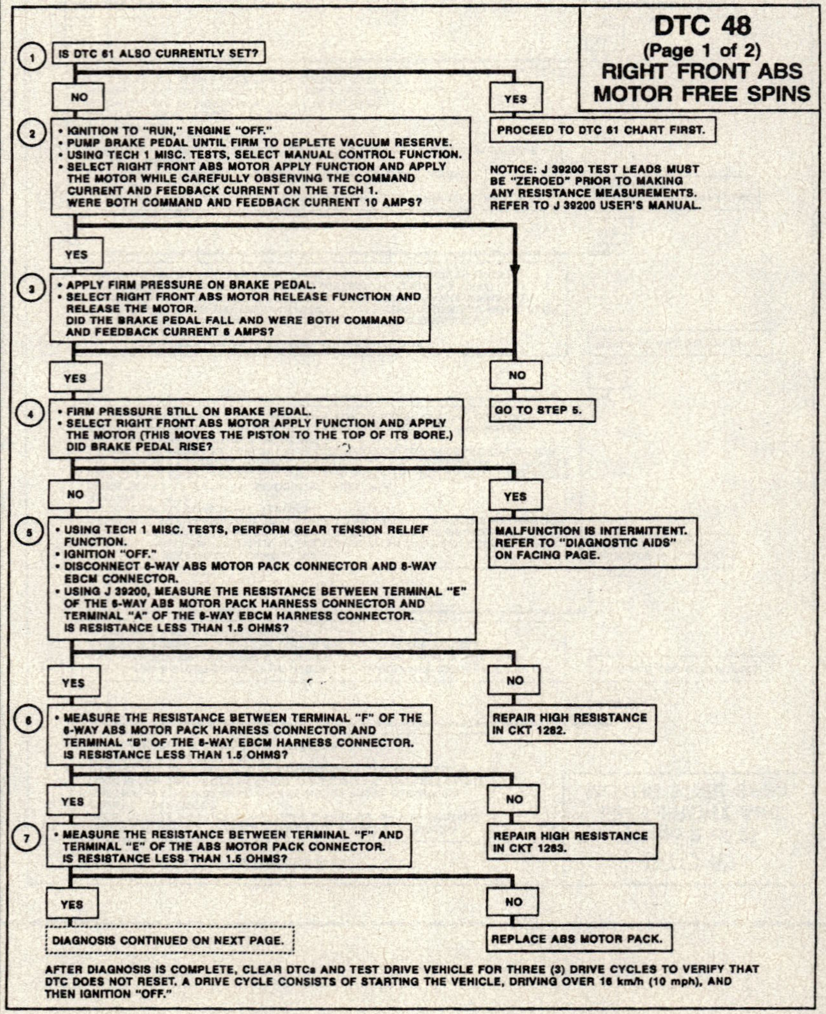

DIAGNOSTIC CHART DTC 48 — 1995 CENTURY, CUTLASS CIERA AND CUTLASS CRUISER

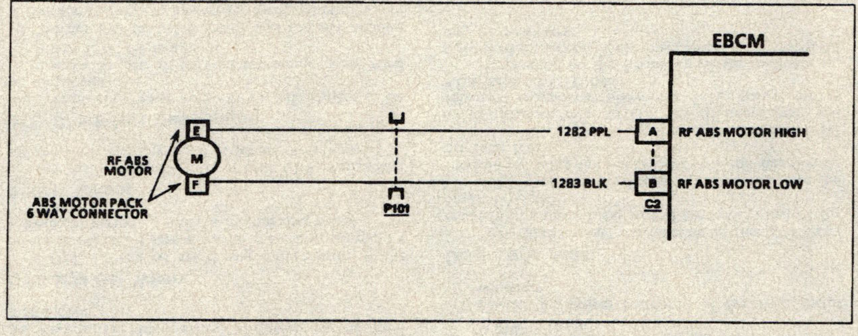

DTC 48
(Page 2 of 2)
RIGHT FRONT ABS MOTOR FREE SPINS

Circuit Description:

This DTC is designed to detect a stripped nut or gear assembly during initialization. During the homing sequence, the piston should reach the top of the bore resulting in a stalled ABS motor. If this does not occur, the ABS motor must be spinning with little or no resistance, indicating a nut/screw or gear malfunction.

DTC Will Set When:

DTC 48 can only be set during initialization. If the feedback current is less than the command current for a specified period of time, a malfunction exists.

Action Taken:

A malfunction DTC is stored, ABS is disabled, and the ABS warning lamp is turned "ON."

DTC Chart Test Description:

Number(s) below refer to circled number(s) on the diagnostic chart.

8. Checks for a stripped gear on the ABS motor pack (unit with three small gears).
9. Checks for a stripped gear on the hydraulic modulator (unit with three large gears).
10. Verifies that the ABS motor can be applied under a load situation.
11. Verifies that the hydraulic modulator is functioning properly.
12. Ensures malfunction was not due to poor terminal contact.

Diagnostic Aids:

An "Intermittent" malfunction in this DTC may result from a mechanical part of the system that sticks, binds, or slips.

The frequency of the malfunction can be checked by using the enhanced diagnostic function of the Tech 1.

If the DTC only occurs once and DTC 61 also occurs, refer to DTC 61. If intermittent and enhanced diagnostics shows this DTC occurs during ABS, refer to DTC 61.

Depending on the frequency of the malfunction, a physical inspection of the mechanical parts suspected may be necessary.

DIAGNOSTIC CHART DTC 48 — 1995 CENTURY, CUTLASS CIERA AND CUTLASS CRUISER

DTC 48
(Page 2 of 2)
RIGHT FRONT ABS MOTOR FREE SPINS

DIAGNOSIS CONTINUED FROM PREVIOUS PAGE.

(8)
- REMOVE HYDRAULIC MODULATOR/MASTER CYLINDER ASSEMBLY FROM VEHICLE.
- REMOVE GEAR COVER.
- CHECK FOR STRIPPED GEAR ON ABS MOTOR PACK; ABS MOTOR PACK IS THE UNIT WITH THE THREE SMALL GEARS. THE RIGHT FRONT GEAR IS THE GEAR FURTHEST REARWARD WHEN THE ASSEMBLY IS INSTALLED IN THE VEHICLE. IS THE ABS MOTOR PACK GEAR STRIPPED?

YES → REPLACE ABS MOTOR PACK.

NO ↓

(9)
- CHECK FOR A STRIPPED GEAR ON HYDRAULIC MODULATOR; MODULATOR IS THE UNIT WITH THE THREE LARGE GEARS. THE RIGHT FRONT GEAR IS THE GEAR FURTHEST REARWARD WHEN THE UNIT IS INSTALLED IN THE VEHICLE. IS THE MODULATOR GEAR STRIPPED?

YES → REPLACE HYDRAULIC MODULATOR GEAR.

NO ↓

(10)
- RECONNECT ELECTRICAL CONNECTORS.
- SECURELY POSITION HYDRAULIC MODULATOR/MASTER CYLINDER ASSEMBLY IN VEHICLE WITH GEAR COVER REMOVED SUCH THAT THE GEAR SET CAN BE OBSERVED. BE CAREFUL NOT TO DAMAGE GEAR SET OR ALLOW THE HYDRAULIC MODULATOR TO MOVE WHILE TESTING.
- IGNITION TO "RUN."
- USING TECH 1, SELECT MANUAL CONTROL FUNCTION.
- SELECT RIGHT FRONT ABS MOTOR APPLY AND APPLY THE MOTOR WHILE OBSERVING THE GEAR SET FURTHEST REARWARD WHEN THE UNIT IS INSTALLED IN THE VEHICLE.
- SELECT RIGHT FRONT ABS MOTOR RELEASE FUNCTION AND RELEASE THE MOTOR WHILE STILL OBSERVING THE GEAR SET.
- SELECT RIGHT FRONT ABS MOTOR APPLY FUNCTION AND APPLY THE MOTOR. DID GEAR SET MOVE IN BOTH DIRECTIONS FOR AT LEAST ONE REVOLUTION?

NO → REPLACE ABS MOTOR PACK.

YES ↓

(11)
- USING TECH 1, PERFORM GEAR TENSION RELIEF FUNCTION.
- IGNITION "OFF."
- SEPARATE THE ABS MOTOR PACK FROM THE HYDRAULIC MODULATOR.
- GRASP THE GEAR ON THE HYDRAULIC MODULATOR (THE HYDRAULIC MODULATOR IS THE UNIT WITH THE THREE LARGE GEARS) FURTHEST REARWARD WHEN THE UNIT IS INSTALLED IN THE VEHICLE AND ATTEMPT TO MOVE THE GEAR.

CAN THE GEAR BE ROTATED MORE THAN 13 FULL TURNS LOCK TO LOCK?

YES → REPLACE THE HYDRAULIC MODULATOR.

NO ↓

(12)
- INSPECT ALL CONNECTORS AND TERMINALS FOR POOR TERMINAL CONTACT AND EVIDENCE OF CORROSION.

REPLACE ALL TERMINALS THAT EXHIBIT SIGNS OF POOR TERMINAL CONTACT OR CORROSION.
- REINSTALL HYDRAULIC MODULATOR/MASTER CYLINDER ASSEMBLY IN VEHICLE.
- RECONNECT ALL CONNECTORS.
- IGNITION "OFF."
- START ENGINE WITH FOOT OFF BRAKE.
- ALLOW ENGINE TO RUN FOR AT LEAST 10 SECONDS.
- REPEAT ABOVE IGNITION CYCLE SEQUENCE TWO MORE TIMES.

DID DTC 48 SET IN THE LAST THREE IGNITION CYCLES?

YES → REPLACE EBCM.

NO → MALFUNCTION WAS MOST LIKELY CAUSED BY A POOR CONNECTION. REFER TO "DIAGNOSTIC AIDS".

AFTER DIAGNOSIS IS COMPLETE, CLEAR DTCs AND TEST DRIVE VEHICLE FOR THREE (3) DRIVE CYCLES TO VERIFY THAT DTC DOES NOT RESET. A DRIVE CYCLE CONSISTS OF STARTING THE VEHICLE, DRIVING OVER 16 km/h (10 mph), AND THEN IGNITION "OFF."

DIAGNOSTIC CHART DTC 51 — 1995 CENTURY, CUTLASS CIERA AND CUTLASS CRUISER

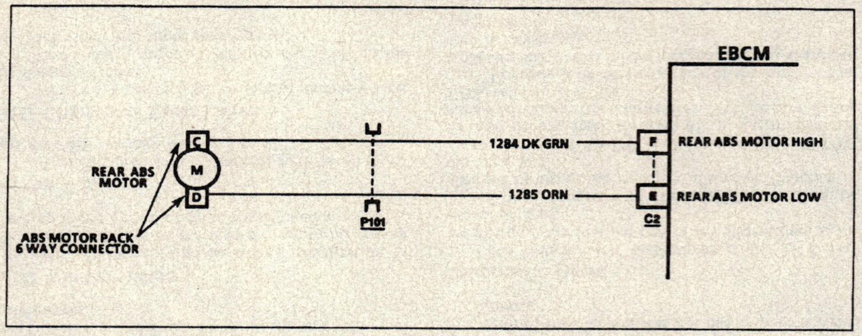

DTC 51
(Page 1 of 2)
REAR ABS MOTOR FREE SPINS

Circuit Description:

This DTC is designed to detect a stripped nut or gear assembly during initialization. During the homing sequence, the piston should reach the top of the bore resulting in a stalled ABS motor. If this does not occur, the ABS motor must be spinning with little or no resistance, indicating a nut/screw or gear malfunction.

DTC Will Set When:

DTC 51 can be set during initialization. If the feedback current is less than the command current for a specified period of time, a malfunction exists.

Action Taken:

A malfunction DTC is stored, ABS is disabled, and the ABS warning lamp is turned "ON."

DTC Chart Test Description:

Number(s) below refer to circled number(s) on the diagnostic chart.

1. Checks to see if the corresponding open ABS motor DTC 64 is also set.
2. Verifies that the ABS motor was actually applied as commanded by observing feedback current.
3. Verifies that the ABS motor can release.
4. Verifies that the ABS motor can be applied by observing pedal movement.
5. Checks for high resistance in the ABS motor "HIGH" circuitry.
6. Checks for high resistance in the ABS motor "LOW" circuitry.
7. Checks for proper resistance of the ABS motor windings.

Diagnostic Aids:

An "Intermittent" malfunction in this DTC may result from a mechanical part of the system that sticks, binds, or slips.

The frequency of the malfunction can be checked by using the enhanced diagnostic function of the Tech 1.

If the DTC only occurs once and DTC 64 also occurs, refer to DTC 64. If intermittent and enhanced diagnostics show this DTC occurs during ABS, refer to DTC 64.

Depending on the frequency of the malfunction, a physical inspection of the mechanical parts suspected may be necessary.

DIAGNOSTIC CHART DTC 51 — 1995 CENTURY, CUTLASS CIERA AND CUTLASS CRUISER

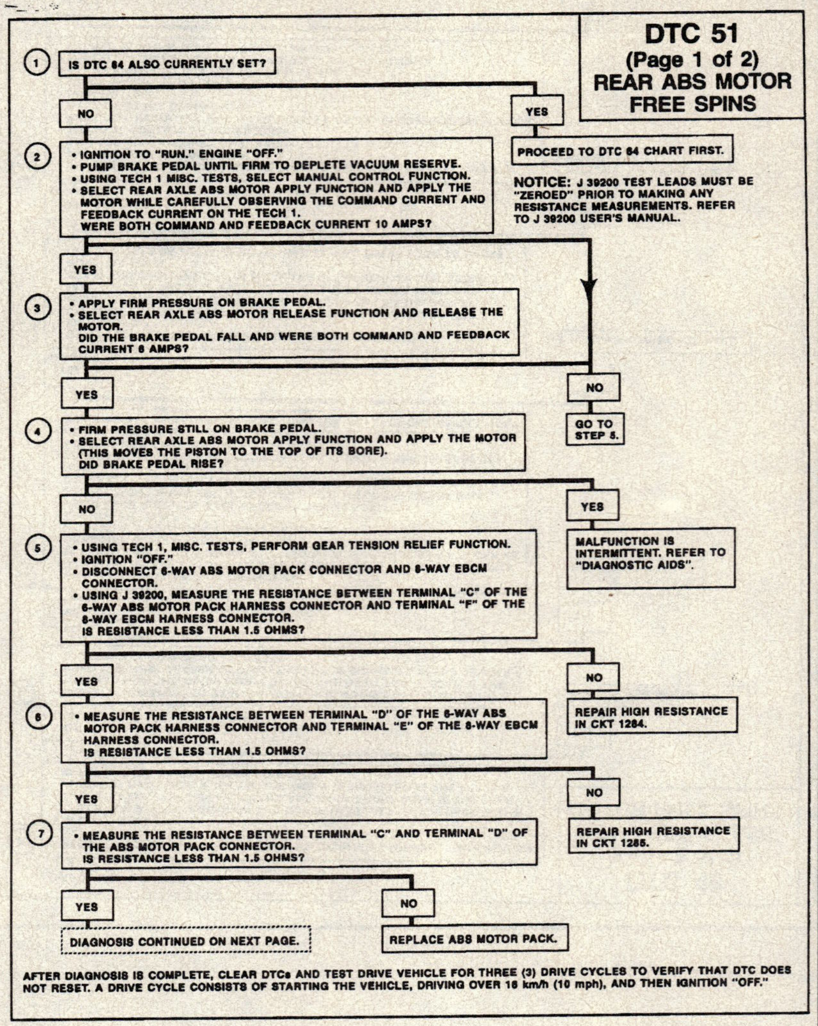

DIAGNOSTIC CHART DTC 51 — 1995 CENTURY, CUTLASS CIERA AND CUTLASS CRUISER

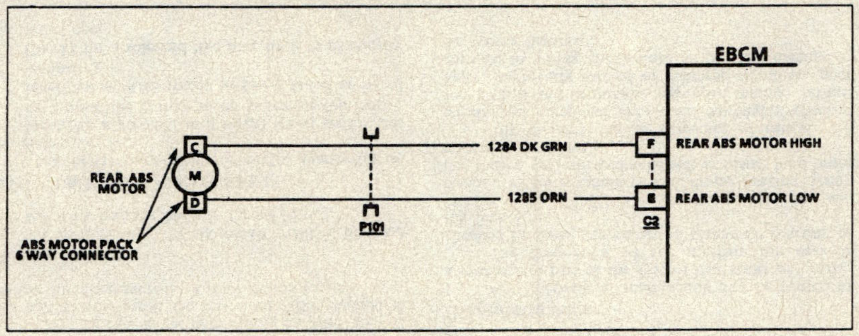

DTC 51
(Page 2 of 2)
REAR ABS MOTOR FREE SPINS

Circuit Description:

This DTC is designed to detect a stripped nut or gear assembly during initialization. During the homing sequence, the piston should reach the top of the bore resulting in a stalled ABS motor. If this does not occur, the ABS motor must be spinning with little or no resistance, indicating a nut/screw or gear malfunction.

DTC Will Set When:

DTC 51 can be set during initialization. If the feedback current is less than the command current for a specified period of time, a malfunction exists.

Action Taken:

A malfunction DTC is stored, ABS is disabled, and the ABS warning lamp is turned "ON."

DTC Chart Test Description:

Number(s) below refer to circled number(s) on the diagnostic chart.

8. Checks for a stripped gear on the ABS motor pack (unit with three small gears).
9. Checks for a stripped gear on the hydraulic modulator (unit with three large gears).
10. Verifies that the ABS motor can be applied under a load situation.
11. Verifies that the hydraulic modulator is functioning properly.
12. Ensures malfunction was not due to poor terminal contact.

Diagnostic Aids:

An "Intermittent" malfunction in this DTC may result from a mechanical part of the system that sticks, binds, or slips.

The frequency of the malfunction can be checked by using the enhanced diagnostic function of the Tech 1.

If the DTC only occurs once and DTC 64 also occurs, refer to DTC 64. If intermittent and enhanced diagnostics show this DTC occurs during ABS, refer to DTC 64.

Depending on the frequency of the malfunction, a physical inspection of the mechanical parts suspected may be necessary.

DIAGNOSTIC CHART DTC 51 — 1995 CENTURY, CUTLASS CIERA AND CUTLASS CRUISER

DTC 51 (Page 2 of 2) REAR ABS MOTOR FREE SPINS

DIAGNOSIS CONTINUED FROM PREVIOUS PAGE.

(8)
- REMOVE HYDRAULIC MODULATOR/MASTER CYLINDER ASSEMBLY FROM VEHICLE
- REMOVE GEAR COVER.
- CHECK FOR STRIPPED GEAR ON ABS MOTOR PACK; ABS MOTOR PACK IS THE UNIT WITH THE THREE SMALL GEARS. THE REAR AXLE GEAR IS THE MIDDLE GEAR OF THE THREE.

IS THE ABS MOTOR PACK GEAR STRIPPED?

- YES → REPLACE ABS MOTOR PACK.
- NO → (9)

(9)
- CHECK FOR A STRIPPED GEAR ON HYDRAULIC MODULATOR; MODULATOR IS THE UNIT WITH THE THREE LARGE GEARS. THE REAR AXLE GEAR IS THE MIDDLE GEAR OF THE THREE.

IS THE MODULATOR GEAR STRIPPED?

- YES → REPLACE HYDRAULIC MODULATOR GEAR.
- NO → (10)

(10)
- RECONNECT ELECTRICAL CONNECTORS.
- SECURELY POSITION HYDRAULIC MODULATOR/MASTER CYLINDER ASSEMBLY IN VEHICLE WITH GEAR COVER REMOVED SUCH THAT THE GEAR SET CAN BE OBSERVED. BE CAREFUL NOT TO DAMAGE GEAR SET OR ALLOW THE HYDRAULIC MODULATOR TO MOVE WHILE TESTING.
- IGNITION TO "RUN."
- USING TECH 1, SELECT MANUAL CONTROL FUNCTION.
- SELECT REAR AXLE ABS MOTOR APPLY AND APPLY THE MOTOR WHILE OBSERVING THE MIDDLE GEAR SET.
- SELECT REAR AXLE ABS MOTOR RELEASE FUNCTION AND RELEASE THE MOTOR WHILE OBSERVING THE GEAR SET.
- SELECT REAR AXLE ABS MOTOR APPLY FUNCTION AND APPLY THE MOTOR.

DID GEAR SET MOVE IN BOTH DIRECTIONS FOR AT LEAST ONE REVOLUTION?

- NO → REPLACE ABS MOTOR PACK.
- YES → (11)

(11)
- USING TECH 1, PERFORM GEAR TENSION RELIEF FUNCTION.
- IGNITION "OFF."
- SEPARATE THE ABS MOTOR PACK FROM THE HYDRAULIC MODULATOR.
- GRASP THE MIDDLE GEAR ON THE HYDRAULIC MODULATOR (THE HYDRAULIC MODULATOR IS THE UNIT WITH THE THREE LARGE GEARS) AND ATTEMPT TO MOVE THE GEAR.

CAN THE GEAR BE ROTATED MORE THAN 8 FULL TURNS LOCK TO LOCK?

- YES → REPLACE THE HYDRAULIC MODULATOR.
- NO → (12)

(12)
- INSPECT ALL CONNECTORS AND TERMINALS FOR POOR TERMINAL CONTACT AND EVIDENCE OF CORROSION.
- REPLACE ALL TERMINALS THAT EXHIBIT SIGNS OF POOR TERMINAL CONTACT OR CORROSION.
- REINSTALL HYDRAULIC MODULATOR/MASTER CYLINDER ASSEMBLY IN VEHICLE.
- RECONNECT ALL CONNECTORS.
- IGNITION "OFF."
- START ENGINE WITH FOOT OFF BRAKE.
- ALLOW ENGINE TO RUN FOR AT LEAST 10 SECONDS.
- REPEAT ABOVE IGNITION CYCLE SEQUENCE TWO MORE TIMES.

DID DTC 51 SET IN THE LAST THREE IGNITION CYCLES?

- YES → REPLACE EBCM.
- NO → MALFUNCTION WAS MOST LIKELY CAUSED BY A POOR CONNECTION. REFER TO "DIAGNOSTIC AIDS".

AFTER DIAGNOSIS IS COMPLETE, CLEAR DTCs AND TEST DRIVE VEHICLE FOR THREE (3) DRIVE CYCLES TO VERIFY THAT DTC DOES NOT RESET. A DRIVE CYCLE CONSISTS OF STARTING THE VEHICLE, DRIVING OVER 16 km/h (10 mph), AND THEN IGNITION "OFF."

DIAGNOSTIC CHART DTC 52 — 1995 CENTURY, CUTLASS CIERA AND CUTLASS CRUISER

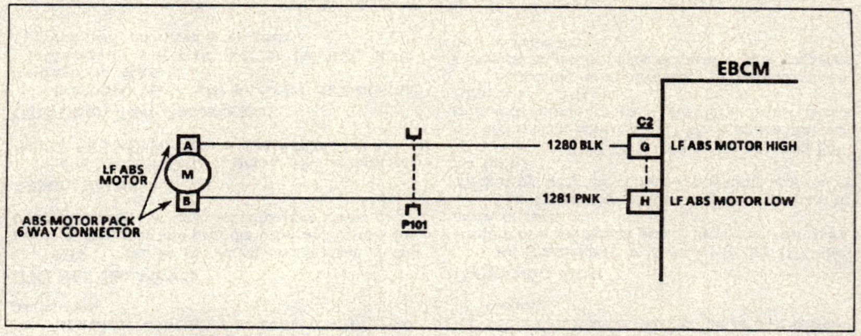

DTC 52
LEFT FRONT CHANNEL IN RELEASE TOO LONG

Circuit Description:

This DTC will diagnose an ABS motor that is energized longer than expected. This could occur if a wheel speed sensor is malfunctioning, the ABS motor does not turn, the left front solenoid mechanically fails open, or the ABS motor wires are crossed.

DTC Will Set When:

DTC 52 can be set only during an ABS stop. If the EBCM commands the left front ABS channel in release for three seconds, a malfunction exists.

Action Taken:

A malfunction DTC is stored, ABS is disabled and the ABS warning lamp is turned "ON."

DTC Chart Test Description:

Number(s) below refer to circled number(s) on the diagnostic chart.

1. Identifies a problem in a wheel speed sensor that may cause the system to be in release too long.
2. Identifies an ABS motor as being failed or wired incorrectly.
3. Checks for a solenoid that may have mechanically failed open.
4. Verifies proper solenoid operation.
5. If Step 3 has failed, this serves to isolate the cause of the hydraulic problem, to either the solenoid or the ABS hydraulic modulator assembly.
6. Determines whether a malfunctioning ABS motor pack or hydraulic modulator is the reason for DTC 52 being set.

Diagnostic Aids:

An "Intermittent" malfunction may be caused by a mechanical part of the system that sticks or binds.

The frequency of the malfunction can be checked by using the enhanced diagnostic function of the Tech 1.

DTC 52 may occur if on ice and steering wheel is turned to lock during braking. Using the Tech 1, perform hydraulic test to ensure total brake system is functional.

Any circuitry that is suspected of causing the intermittent complaint should be thoroughly checked for backed out terminals, improper mating, broken locks, improperly formed or damaged terminals, poor terminal to wiring connections or physical damage to the wiring harness.

DIAGNOSTIC CHART DTC 52 — 1995 CENTURY, CUTLASS CIERA AND CUTLASS CRUISER

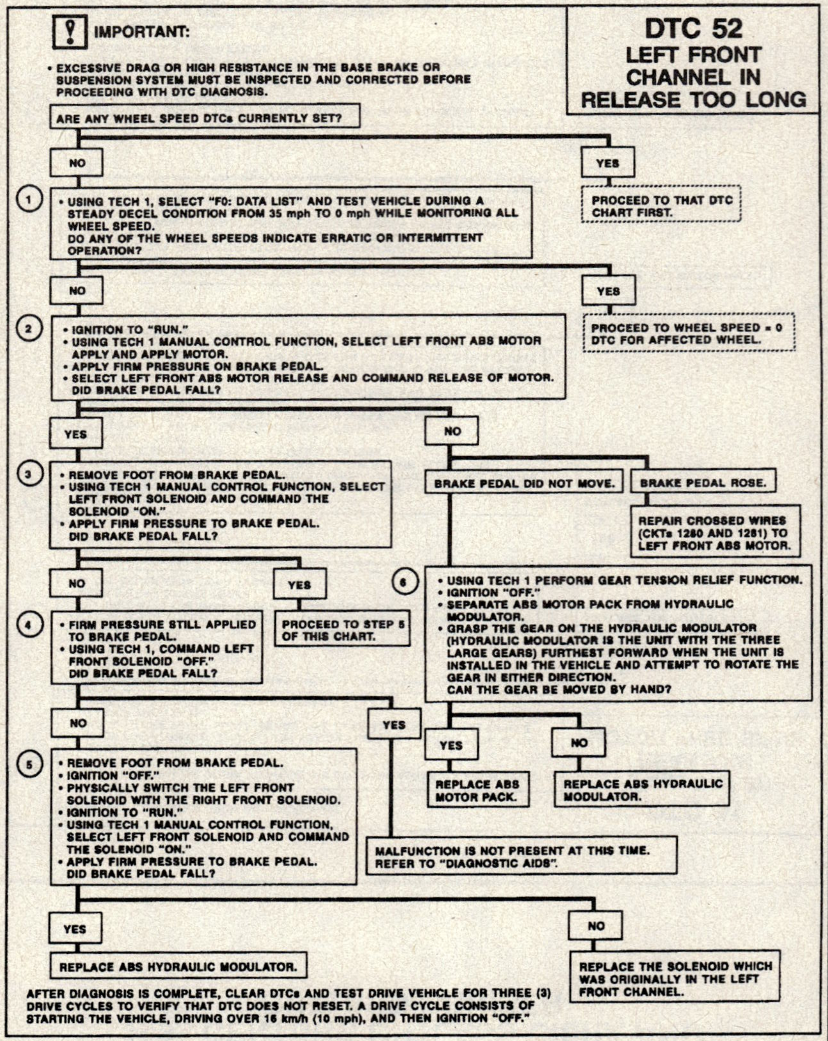

DIAGNOSTIC CHART DTC 53 — 1995 CENTURY, CUTLASS CIERA AND CUTLASS CRUISER

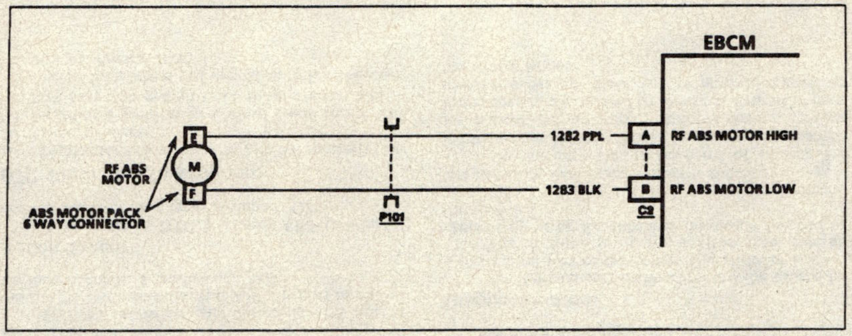

DTC 53
RIGHT FRONT CHANNEL IN RELEASE TOO LONG

Circuit Description:

This DTC will diagnose an ABS motor that is energized longer than expected. This could occur if a wheel speed sensor is malfunctioning, the ABS motor does not turn, the right front solenoid mechanically fails open, or the ABS motor wires are crossed.

DTC Will Set When:

DTC 53 can be set only during an ABS stop. If the EBCM commands the right front channel in release for three seconds, a malfunction exists.

Action Taken:

A malfunction DTC is stored, ABS is disabled and the ABS warning lamp is turned "ON."

DTC Chart Test Description:

Number(s) below refer to circled number(s) on the diagnostic chart.

1. Identifies a problem in a wheel speed sensor that may cause the system to be in release too long.
2. Identifies a motor as being failed or wired incorrectly.
3. Checks for a solenoid that may have mechanically failed open.
4. Checks for a solenoid that may have mechanically failed closed.
5. If Step 3 has failed, this serves to isolate the cause of the hydraulic problem, to either the solenoid or the ABS hydraulic modulator assembly.
6. Determines whether a malfunction ABS motor pack or hydraulic modulator is the reason for DTC 53 being set.

Diagnostic Aids:

An "Intermittent" malfunction may be caused by a mechanical part of the system that sticks or binds.

The frequency of the malfunction can be checked by using the enhanced diagnostic function of the Tech 1.

DTC 53 may occur on ice if steering wheel is turned to lock during braking. Using the Tech 1 perform hydraulic test to ensure total brake system is functional.

Any circuitry that is suspected of causing the intermittent complaint should be thoroughly checked for backed out terminals, improper mating, broken locks, improperly formed or damaged terminals, poor terminal to wiring connections or physical damage to the wiring harness.

DIAGNOSTIC CHART DTC 53 — 1995 CENTURY, CUTLASS CIERA AND CUTLASS CRUISER

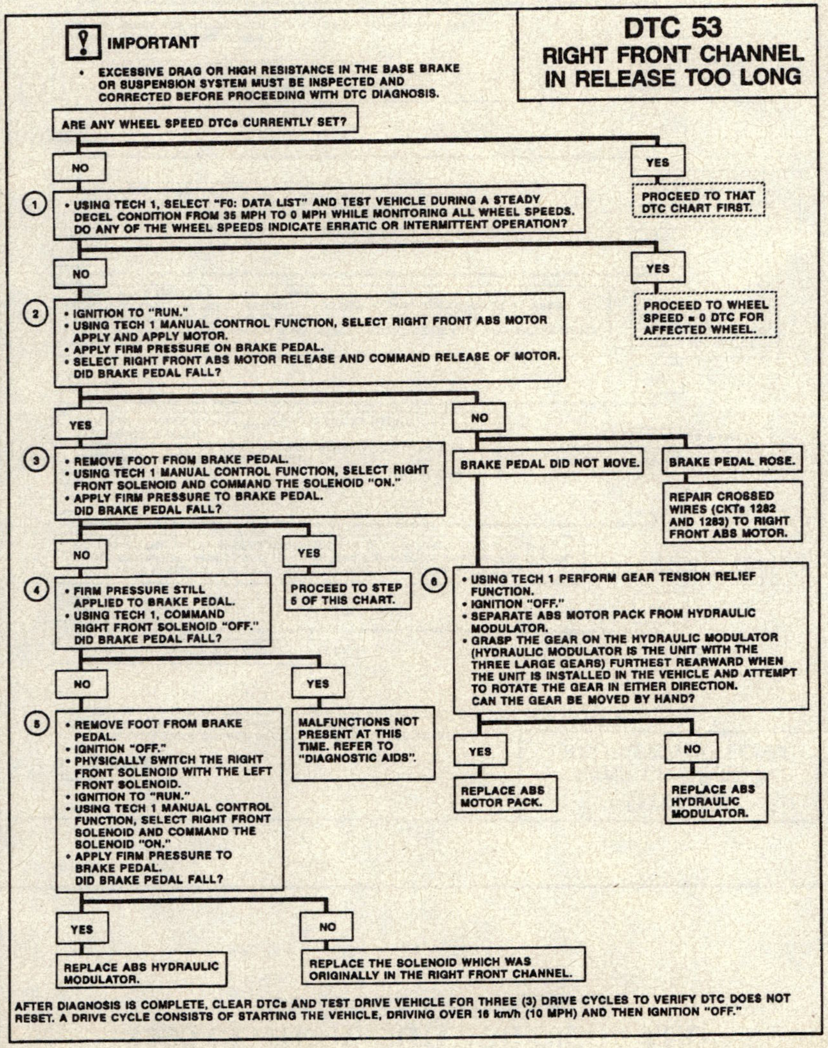

DIAGNOSTIC CHART DTC 54 — 1995 CENTURY, CUTLASS CIERA AND CUTLASS CRUISER

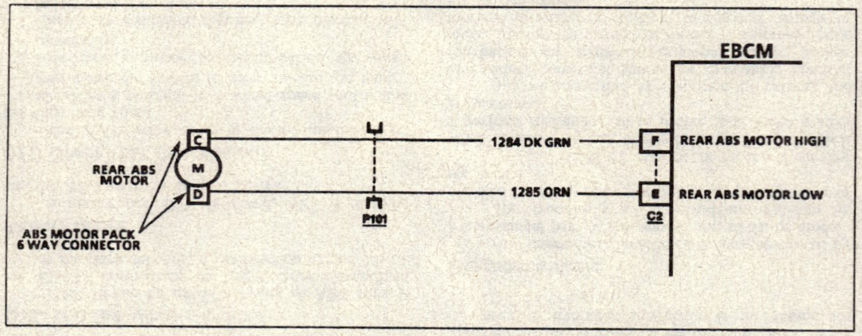

DTC 54
REAR CHANNEL IN RELEASE TOO LONG

Circuit Description:

This DTC will diagnose an ABS motor that is energized longer than expected. This could occur if a wheel speed sensor is malfunctioning, the ABS motor does not turn, or the ABS motor wires are crossed.

DTC Will Set When:

DTC 54 can be set only during an ABS stop. If the EBCM commands the rear ABS channel in release for three seconds, a malfunction exists.

Action Taken:

A malfunction DTC is stored, ABS is disabled and the ABS warning lamp is turned "ON."

DTC Chart Test Description:

Number(s) below refer to circled number(s) on the diagnostic chart.

1. Identifies a problem in a wheel speed sensor that may cause the system to be in release too long.
2. Checks for a wheel that may stick or bind because of a mechanical fault.
3. Checks to see if the ABS motor is capable of moving and applying the hydraulic piston for the rear wheels.
4. Ensures that the ABS motor wiring is not crossed.
5. Isolates the fault of a "no-apply" situation to either the ABS motor pack or ABS hydraulic modulator assembly.

Diagnostic Aids:

An "Intermittent" malfunction may be caused by a mechanical part of the system that sticks or binds.

The frequency of the malfunction can be checked by using the enhanced diagnostic function of the Tech 1.

Using the Tech 1, perform hydraulic test to ensure total brake system is functional.

Any circuitry that is suspected of causing the intermittent complaint should be thoroughly checked for backed out terminals, improper mating, broken locks, improperly formed or damaged terminals, poor terminal to wiring connections or physical damage to the wiring harness.

DIAGNOSTIC CHART DTC 54 — 1995 CENTURY, CUTLASS CIERA AND CUTLASS CRUISER

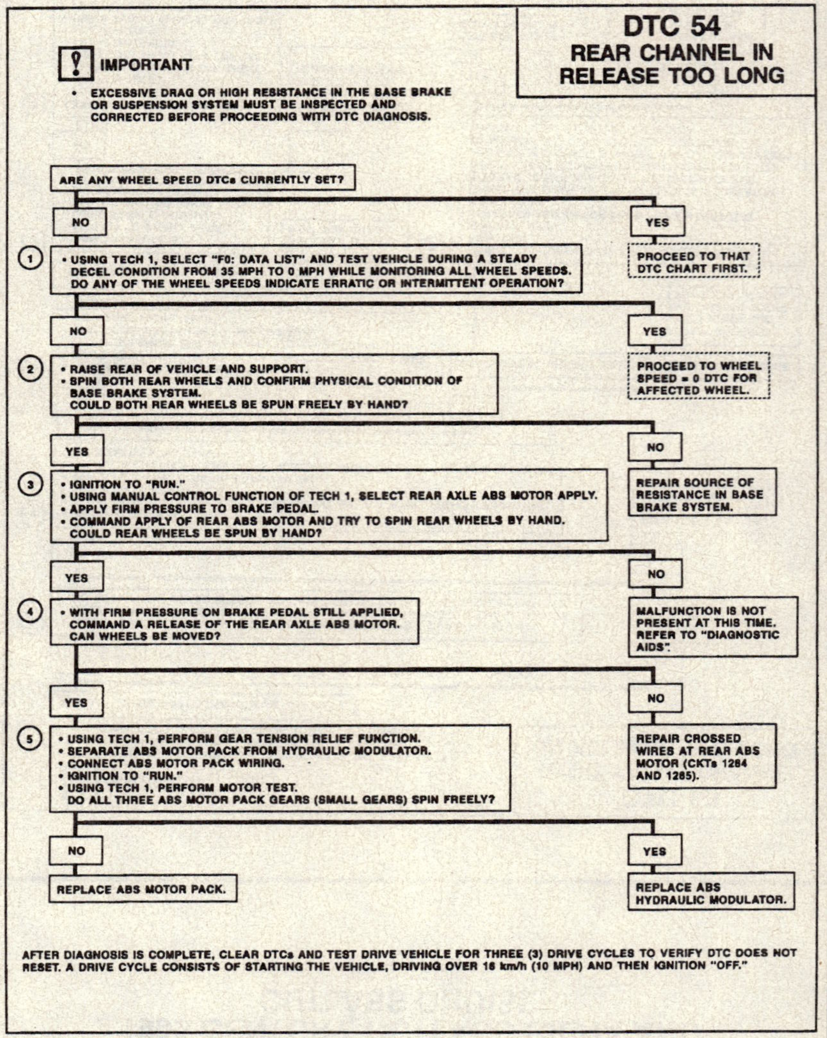

DIAGNOSTIC CHART DTC 55 — 1995 CENTURY, CUTLASS CIERA AND CUTLASS CRUISER

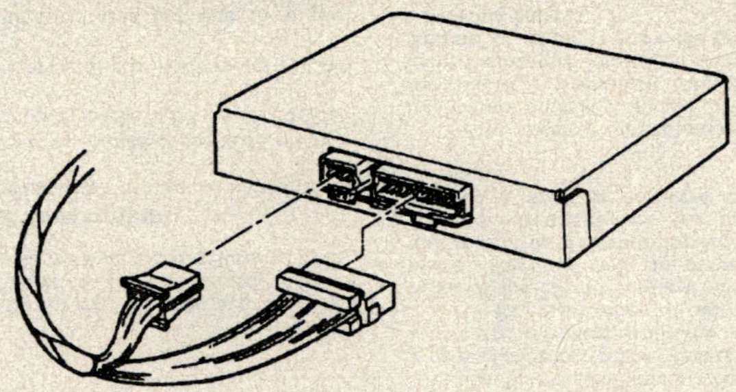

DTC 55
EBCM MALFUNCTION

Circuit Description:

This DTC identifies a malfunction detected by the MDI custom IC. It also ensures the cause of the malfunction is not a result of a problem with the ABS enable relay under a DTC 55 malfunction.

Action Taken:

A malfunction DTC is stored, ABS is disabled and the ABS warning lamp is turned "ON."

DTC Chart Test Description:

Number(s) below refer to circled number(s) on the diagnostic chart.

1. Checks to see if the malfunction is still present.
2. Checks to see if the malfunction is intermittent.

Diagnostic Aids:

An "Intermittent" malfunction may be caused by a poor connection, rubbed through wire insulation, or a wire that is broken inside the insulation.

The frequency of the malfunction can be checked by using the enhanced diagnostic function of the Tech 1.

Any circuitry that is suspected of causing the intermittent complaint should be thoroughly checked for backed out terminals, improper mating, broken locks, improperly formed or damaged terminals, poor terminal to wiring connections or physical damage to the wiring harness.

DIAGNOSTIC CHART DTC 55 — 1995 CENTURY, CUTLASS CIERA AND CUTLASS CRUISER

DTC 55
EBCM MALFUNCTION

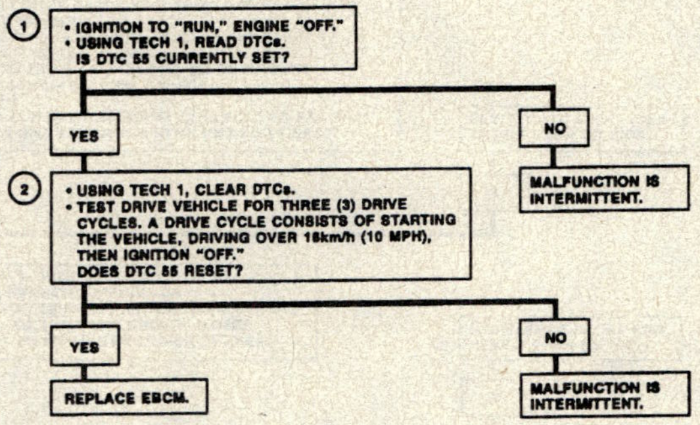

DIAGNOSTIC CHART DTC 56 — 1995 CENTURY, CUTLASS CIERA AND CUTLASS CRUISER

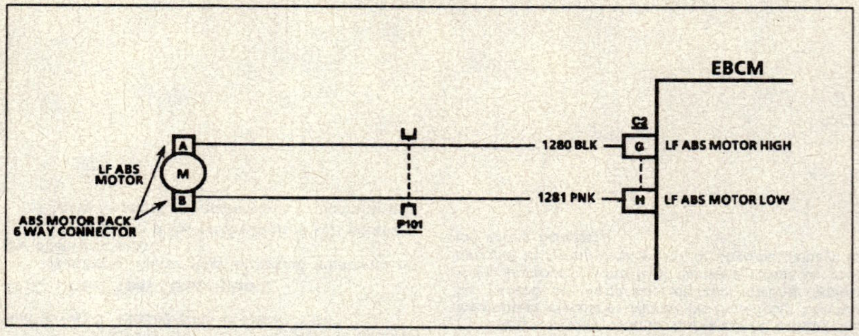

DTC 56
LEFT FRONT ABS MOTOR CIRCUIT OPEN

Circuit Description:

This DTC identifies an ABS motor that cannot be energized due to an open in its circuitry.

DTC Will Set When:

DTC 56 can be set only when the motor is commanded "OFF." If the EBCM detects an out of range voltage on either of the left front ABS motor circuits indicating an open circuit, a malfunction exists.

Action Taken:

An open ABS motor will not activate when requested. A malfunction DTC is stored, ABS is disabled and the ABS warning lamp is turned "ON."

DTC Chart Test Description:

Number(s) below refer to circled number(s) on the diagnostic chart.

1. Checks for proper resistance of the ABS motor.
2. Checks for an open in the ABS motor "HIGH" circuitry.
3. Checks for an open in the ABS motor "LOW" circuitry.
4. Ensures malfunction was not due to a poor terminal contact.

Diagnostic Aids:

Using Tech 1, select manual control function, and exercise ABS motor movement of affected channel in both directions while applying light pressure on the brake pedal.

If erratic or "jumpy" brake pedal movement is detected while performing an "apply" or "release" function of the ABS motor, an intermittent malfunction may be indicated.

An "Intermittent" malfunction may be caused by a poor connection, rubbed through wire insulation, or a wire that is broken inside the insulation.

If the malfunction is not current, wiggle the wires of the affected channel and check if the DTC resets. This will help to pinpoint an intermittent malfunction in the motor circuitry or connections.

The frequency of the malfunction can be checked by using the enhanced diagnostic function of the Tech 1.

Any circuitry that is suspected of causing the intermittent complaint should be thoroughly checked for backed out terminals, improper mating, broken locks, improperly formed or damaged terminals, poor terminal to wiring connections or physical damage to the wiring harness.

DIAGNOSTIC CHART DTC 56 — 1995 CENTURY, CUTLASS CIERA AND CUTLASS CRUISER

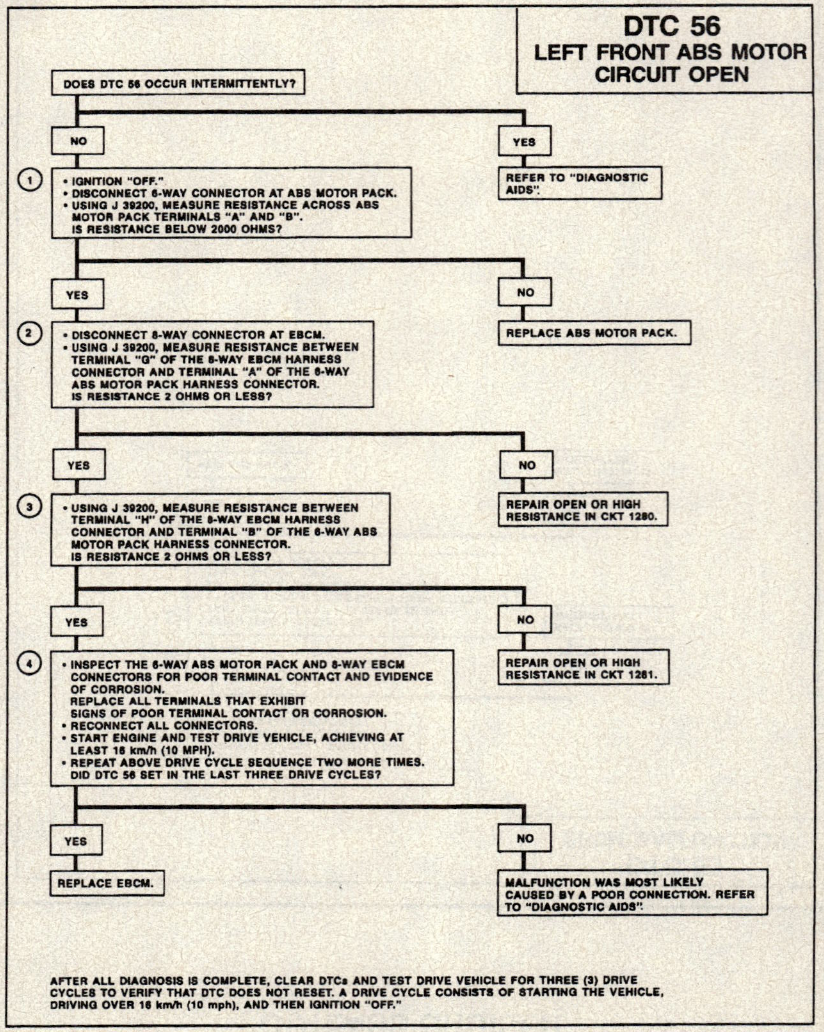

DIAGNOSTIC CHART DTC 57 — 1995 CENTURY, CUTLASS CIERA AND CUTLASS CRUISER

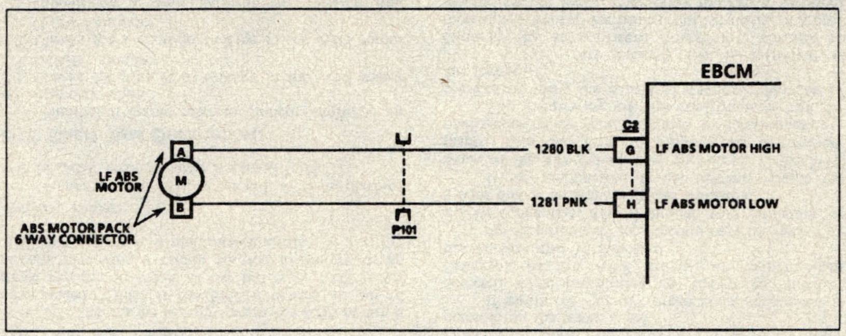

DTC 57
LEFT FRONT ABS MOTOR CIRCUIT SHORTED TO GND

Circuit Description:

This DTC identifies an ABS motor circuit that is shorted to ground. This malfunction will not allow the ABS motor to be controlled at the commanded current rate or will cause the driver circuit to allow current directly to ground.

DTC Will Set When:

DTC 57 can be set anytime. If the EBCM detects an out of range voltage on either of the left front ABS motor circuits indicating a circuit shorted to ground, a malfunction exists.

Action Taken:

A malfunction DTC is stored, ABS is disabled and the ABS warning lamp is turned "ON."

DTC Chart Test Description:

Number(s) below refer to circled number(s) on the diagnostic chart.

1. Checks for a short to ground in the ABS motor "HIGH" circuitry.
2. Checks for a short to ground in the ABS motor "LOW" circuitry.
3. Checks for an ABS motor that is internally shorted to ground.
4. Ensures malfunction was not due to physical damage of the circuitry.

Diagnostic Aids:

Using Tech 1, select manual control function, and exercise ABS motor movement of affected channel in both directions while applying light pressure on the brake pedal.

If erratic or "jumpy" brake pedal movement is detected while performing an "apply" or "release" function of the ABS motor, an intermittent malfunction may be indicated.

An "Intermittent" malfunction may be caused by a poor connection, rubbed through wire insulation, or a wire that is broken inside the insulation.

If the malfunction is not current, wiggle the wires of the affected channel and check if the DTC resets. This will help to pinpoint an intermittent malfunction in the motor circuitry or connections.

The frequency of the malfunction can be checked by using the enhanced diagnostic function of the Tech 1.

Any circuitry that is suspected of causing the intermittent complaint should be thoroughly checked for backed out terminals, improper mating, broken locks, improperly formed or damaged terminals, poor terminal to wiring connections or physical damage to the wiring harness.

DIAGNOSTIC CHART DTC 57 — 1995 CENTURY, CUTLASS CIERA AND CUTLASS CRUISER

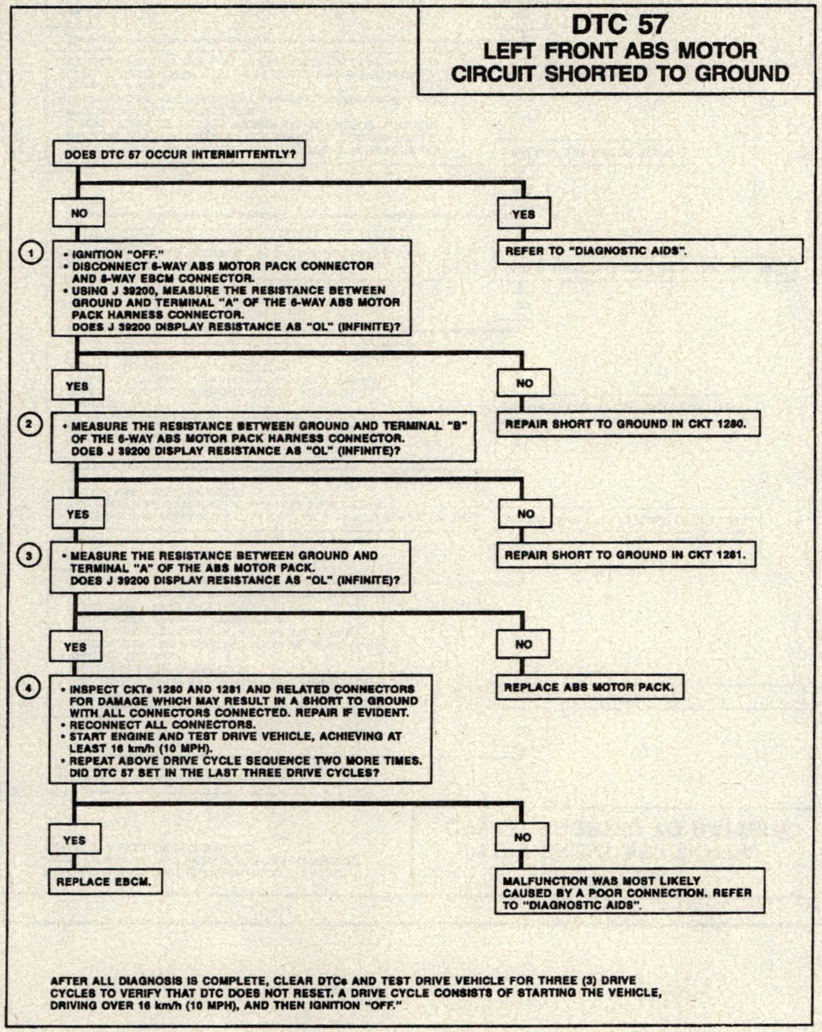

DIAGNOSTIC CHART DTC 58 — 1995 CENTURY, CUTLASS CIERA AND CUTLASS CRUISER

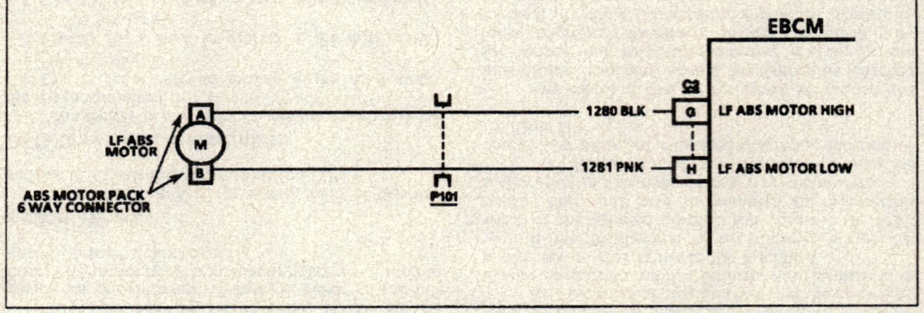

DTC 58
LEFT FRONT ABS MOTOR CIRCUIT SHORTED TO BATT

Circuit Description:

This DTC identifies an ABS motor circuit that is shorted to battery or an ABS motor that has low or no resistance. This malfunction will not allow the ABS motor to be controlled at the commanded current rate or will cause the ABS motor to turn in the opposite direction, or not at all.

DTC Will Set When:

DTC 58 can be set only when the ABS motor is commanded "OFF." If the EBCM detects an out of range voltage on either of the left front ABS motor circuits indicating a circuit shorted to battery or an ABS motor shorted, a malfunction exists.

Action Taken:

A malfunction DTC is stored, ABS is disabled and the ABS warning lamp is turned "ON."

DTC Chart Test Description:

Number(s) below refer to circled number(s) on the diagnostic chart.

1. Checks for a short to voltage in the ABS motor "HIGH" circuitry.
2. Checks for a short to voltage in the ABS motor "LOW" circuitry.
3. Checks for a short between the "HIGH" and "LOW" ABS motor circuits.
4. Checks for an ABS motor that is internally shorted.
5. Ensures malfunction was not due to physical damage of the circuitry.

Diagnostic Aids:

Using Tech 1, select manual control function, and exercise ABS motor movement of affected channel in both directions while applying light pressure on the brake pedal.

If erratic or "jumpy" brake pedal movement is detected while performing an "apply" or "release" function of the ABS motor, an intermittent malfunction may be indicated.

An "Intermittent" malfunction may be caused by a poor connection, rubbed through wire insulation, or a wire that is broken inside the insulation.

If the malfunction is not current, wiggle the wires of the affected channel and check if the DTC resets. This will help to pinpoint an intermittent malfunction in the motor circuitry or connections.

The frequency of the malfunction can be checked by using the enhanced diagnostic function of the Tech 1.

Any circuitry, that is suspected as causing the intermittent complaint, should be thoroughly checked for backed out terminals, improper mating, broken locks, improperly formed or damaged terminals, poor terminal to wiring connections or physical damage to the wiring harness.

DIAGNOSTIC CHART DTC 58 — 1995 CENTURY, CUTLASS CIERA AND CUTLASS CRUISER

NOTICE: J 39200 TEST LEADS MUST BE "ZEROED" PRIOR TO MAKING ANY RESISTANCE MEASUREMENTS. REFER TO J 39200 USER'S MANUAL.

DTC 58 LEFT FRONT ABS MOTOR CIRCUIT SHORTED TO BATTERY

DOES DTC 58 OCCUR INTERMITTENTLY?

- YES: REFER TO "DIAGNOSTIC AIDS".
- NO:

(1)
- IGNITION "OFF."
- DISCONNECT 6-WAY ABS MOTOR PACK CONNECTOR AND 8-WAY EBCM CONNECTOR.
- IGNITION TO "RUN."
- USING J 39200, MEASURE THE VOLTAGE BETWEEN GROUND AND TERMINAL "A" OF THE 6-WAY ABS MOTOR PACK HARNESS CONNECTOR.

IS VOLTAGE 1 VOLT OR LESS?

- NO: REPAIR SHORT TO VOLTAGE IN CKT 1280.
- YES:

(2)
- MEASURE THE VOLTAGE BETWEEN GROUND AND TERMINAL "B" OF THE 6-WAY ABS MOTOR PACK HARNESS CONNECTOR.

IS VOLTAGE 1 VOLT OR LESS?

- NO: REPAIR SHORT TO GROUND IN CKT 1281.
- YES:

(3)
- IGNITION "OFF."
- USING J 39200, MEASURE THE RESISTANCE BETWEEN TERMINAL "B" AND TERMINAL "A" OF THE 6-WAY ABS MOTOR PACK HARNESS CONNECTOR.

DOES J 39200 DISPLAY RESISTANCE AS "OL" (INFINITE)?

- NO: REPAIR SHORT BETWEEN CKTs 1280 AND 1281.
- YES:

(4)
- USING J 39200, MEASURE THE RESISTANCE BETWEEN TERMINAL "B" AND TERMINAL "A" OF THE ABS MOTOR PACK.

IS RESISTANCE GREATER THAN 0.4 OHMS?

- NO: REPLACE ABS MOTOR PACK.
- YES:

(5)
- INSPECT CKTs 1280 AND 1281 AND RELATED CONNECTORS FOR DAMAGE WHICH MAY RESULT IN A SHORT TO GROUND WITH ALL CONNECTORS CONNECTED. REPAIR ALL DAMAGE IF EVIDENT.
- RECONNECT ALL CONNECTORS.
- START ENGINE AND TEST DRIVE VEHICLE, ACHIEVING AT LEAST 16 km/h (10 MPH).
- REPEAT ABOVE DRIVE CYCLE SEQUENCE TWO MORE TIMES.

DID DTC 58 SET IN THE LAST THREE DRIVE CYCLES?

- YES: REPLACE EBCM.
- NO: MALFUNCTION WAS MOST LIKELY CAUSED BY A POOR CONNECTION. REFER TO "DIAGNOSTIC AIDS".

AFTER ALL DIAGNOSIS IS COMPLETE, CLEAR DTCs AND TEST DRIVE VEHICLE FOR THREE (3) DRIVE CYCLES TO VERIFY THAT DTC DOES NOT RESET. A DRIVE CYCLE CONSISTS OF STARTING THE VEHICLE, DRIVING OVER 16 km/h (10 MPH), AND THEN IGNITION "OFF."

DIAGNOSTIC CHART DTC 61 — 1995 CENTURY, CUTLASS CIERA AND CUTLASS CRUISER

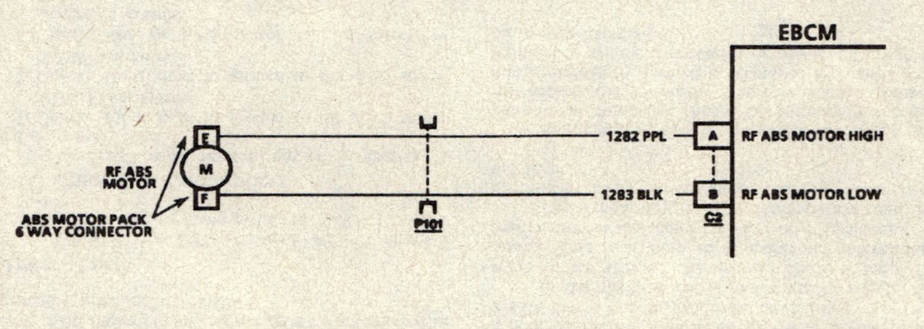

DTC 61
RIGHT FRONT ABS MOTOR CIRCUIT OPEN

Circuit Description:

This DTC identifies an ABS motor that cannot be energized due to an open in its circuitry. This malfunction will not allow proper front ABS operation.

DTC Will Set When:

DTC 61 can be set only when the ABS motor is commanded "OFF." If the EBCM detects an out of range voltage on either of the right front ABS motor circuits indicating an open circuit, a malfunction exists.

Action Taken:

An open ABS motor will not activate when requested. A malfunction DTC is stored, ABS is disabled and the ABS warning lamp is turned "ON."

DTC Chart Test Description:

Number(s) below refer to circled number(s) on the diagnostic chart.

1. Checks for proper resistance of the ABS motor.
2. Checks for an open in the ABS motor "HIGH" circuitry.
3. Checks for an open in the ABS motor "LOW" circuitry.
4. Ensures malfunction was not due to poor terminal contact or EBCM.

Diagnostic Aids:

Using Tech 1, select manual control function, and exercise ABS motor movement of affected channel in both directions while applying light pressure on the brake pedal.

If erratic or "jumpy" brake pedal movement is detected while performing an "apply" or "release" function of the ABS motor, an intermittent malfunction may be indicated.

An "Intermittent" malfunction may be caused by a poor connection, rubbed through wire insulation, or a wire that is broken inside the insulation.

If the malfunction is not current, wiggle the wires of the affected channel and check if the DTC resets. This will help to pinpoint an intermittent malfunction in the motor circuitry or connections.

The frequency of the malfunction can be checked by using the enhanced diagnostic function of the Tech 1.

Any circuitry that is suspected of causing the intermittent complaint should be thoroughly checked for backed out terminals, improper mating, broken locks, improperly formed or damaged terminals, poor terminal to wiring connections or physical damage to the wiring harness.

DIAGNOSTIC CHART DTC 61 — 1995 CENTURY, CUTLASS CIERA AND CUTLASS CRUISER

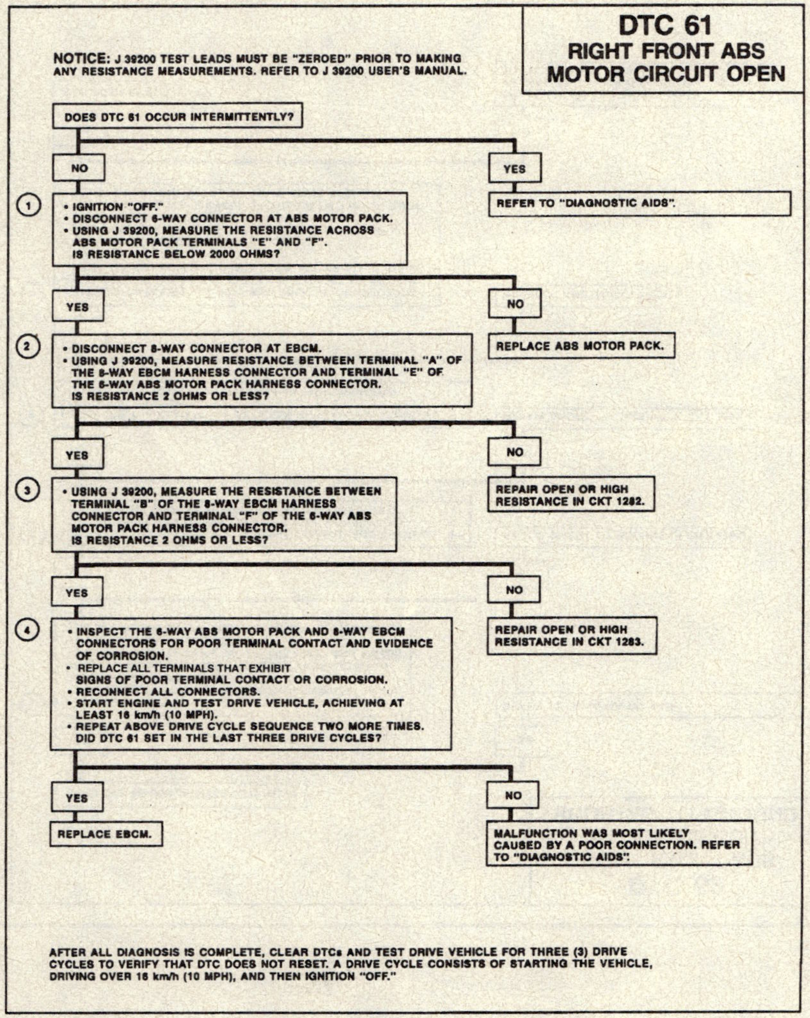

DIAGNOSTIC CHART DTC 62 — 1995 CENTURY, CUTLASS CIERA AND CUTLASS CRUISER

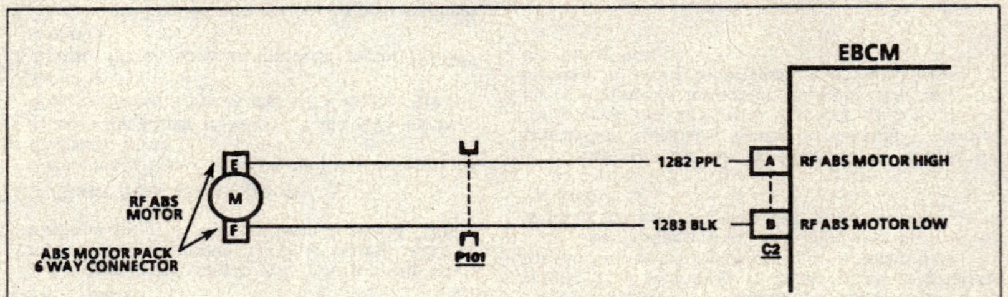

DTC 62
RIGHT FRONT ABS MOTOR CIRCUIT SHORTED TO GND

Circuit Description:

This DTC identifies an ABS motor circuit that is shorted to ground. This malfunction will not allow the ABS motor to be controlled at the commanded current rate or will cause the ABS motor driver circuit to allow current directly to ground.

DTC Will Set When:

DTC 62 can be set anytime. If the EBCM detects an out of range voltage on either of the right front ABS motor circuits indicating a circuit shorted to ground, a malfunction exists.

Action Taken:

A malfunction DTC is stored, ABS is disabled and the ABS warning lamp is turned "ON."

DTC Chart Test Description:

Number(s) below refer to circled number(s) on the diagnostic chart.

1. Checks for a short to ground in the ABS motor "HIGH" circuitry.
2. Checks for a short to ground in the ABS motor "LOW" circuitry.
3. Checks for an ABS motor that is internally shorted to ground.
4. Ensures malfunction was not due to physical damage of the circuitry.

Diagnostic Aids:

Using Tech 1, select manual control function, and exercise ABS motor movement of affected channel in both directions while applying light pressure on the brake pedal.

If erratic or "jumpy" brake pedal movement is detected while performing an "apply" or "release" function of the ABS motor, an intermittent malfunction may be indicated.

An "Intermittent" malfunction may be caused by a poor connection, rubbed through wire insulation, or a wire that is broken inside the insulation.

If the malfunction is not current, wiggle the wires of the affected channel and check if the DTC resets. This will help to pinpoint an intermittent malfunction in the motor circuitry or connections.

The frequency of the malfunction can be checked by using the enhanced diagnostic function of the Tech 1.

Any circuitry that is suspected of causing the intermittent complaint should be thoroughly checked for backed out terminals, improper mating, broken locks, improperly formed or damaged terminals, poor terminal to wiring connections or physical damage to the wiring harness.

DIAGNOSTIC CHART DTC 62 — 1995 CENTURY, CUTLASS CIERA AND CUTLASS CRUISER

DTC 62
RIGHT FRONT ABS MOTOR CIRCUIT SHORTED TO GROUND

DOES DTC 62 OCCUR INTERMITTENTLY?

- YES: REFER TO "DIAGNOSTIC AIDS".
- NO: go to step 1.

①
- IGNITION "OFF."
- DISCONNECT 6-WAY ABS MOTOR PACK CONNECTOR AND 6-WAY EBCM CONNECTOR.
- USING J 39200, MEASURE THE RESISTANCE BETWEEN GROUND AND TERMINAL "E" OF THE 6-WAY ABS MOTOR PACK HARNESS CONNECTOR.

DOES J 39200 DISPLAY RESISTANCE AS "OL" (INFINITE)?

- NO: REPAIR SHORT TO GROUND IN CKT 1282.
- YES: go to step 2.

②
- MEASURE THE RESISTANCE BETWEEN GROUND AND TERMINAL "F" OF THE 6-WAY ABS MOTOR PACK HARNESS CONNECTOR.

DOES J 39200 DISPLAY RESISTANCE AS "OL" (INFINITE)?

- NO: REPAIR SHORT TO GROUND IN CKT 1283.
- YES: go to step 3.

③
- MEASURE THE RESISTANCE BETWEEN GROUND AND TERMINAL "E" OF THE ABS MOTOR PACK.

DOES J 39200 DISPLAY RESISTANCE AS "OL" (INFINITE)?

- NO: REPLACE ABS MOTOR PACK.
- YES: go to step 4.

④
- INSPECT CKTs 1282 AND 1283 AND RELATED CONNECTORS FOR DAMAGE WHICH MAY RESULT IN A SHORT TO GROUND WITH ALL CONNECTORS CONNECTED. REPAIR ALL DAMAGE IF EVIDENT.
- RECONNECT ALL CONNECTORS.
- START ENGINE AND TEST DRIVE VEHICLE, ACHIEVING AT LEAST 16 km/h (10 MPH).
- REPEAT ABOVE DRIVE CYCLE SEQUENCE TWO MORE TIMES.

DID DTC 62 SET IN THE LAST THREE DRIVE CYCLES?

- YES: REPLACE EBCM.
- NO: MALFUNCTION WAS MOST LIKELY CAUSED BY A POOR CONNECTION. REFER TO "DIAGNOSTIC AIDS".

AFTER ALL DIAGNOSIS IS COMPLETE, CLEAR DTCs AND TEST DRIVE VEHICLE FOR THREE (3) DRIVE CYCLES TO VERIFY THAT DTC DOES NOT RESET. A DRIVE CYCLE CONSISTS OF STARTING THE VEHICLE, DRIVING OVER 16 km/h (10 MPH), AND THEN IGNITION "OFF."

DIAGNOSTIC CHART DTC 63 — 1995 CENTURY, CUTLASS CIERA AND CUTLASS CRUISER

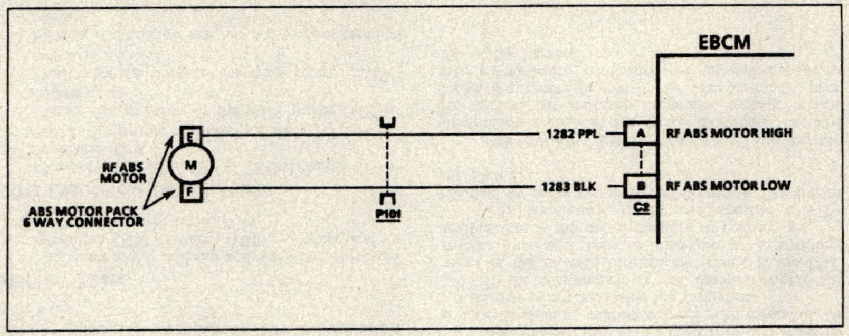

DTC 63

RIGHT FRONT ABS MOTOR CIRCUIT SHORTED TO BATT

Circuit Description:

This DTC identifies an ABS motor circuit that is shorted to battery or an ABS motor that has low or no resistance. This malfunction will not allow the ABS motor to be controlled at the commanded current rate or will cause the ABS motor to turn in the opposite direction or not at all.

DTC Will Set When:

DTC 63 can be set only when the ABS motor is commanded "OFF." If the EBCM detects an out of range voltage on either of the right front ABS motor circuits indicating a circuit shorted to battery or an ABS motor shorted, a malfunction exists.

Action Taken:

A malfunction DTC is stored, ABS is disabled and the ABS warning lamp is turned "ON."

DTC Chart Test Description:

Number(s) below refer to circled number(s) on the diagnostic chart.

1. Checks for a short to voltage in the ABS motor "HIGH" circuitry.
2. Checks for a short to voltage in the ABS motor "LOW" circuitry.
3. Checks for a short circuit between ABS motor "HIGH" and ABS motor "LOW."
4. Checks for proper resistance of the ABS motor.
5. Ensures malfunction was not due to physical damage of the circuitry.

Diagnostic Aids:

Using Tech 1, select manual control function, and exercise ABS motor movement of affected channel in both directions while applying light pressure on the brake pedal.

If erratic or "jumpy" brake pedal movement is detected while performing an "apply" or "release" function of the ABS motor, an intermittent malfunction may be indicated.

An "Intermittent" malfunction may be caused by a poor connection, rubbed through wire insulation, or a wire that is broken inside the insulation.

If the malfunction is not current, wiggle the wires of the affected channel and check if the DTC resets. This will help to pinpoint an intermittent malfunction in the motor circuitry or connections.

The frequency of the malfunction can be checked by using the enhanced diagnostic function of the Tech 1.

Any circuitry that is suspected of causing the intermittent complaint should be thoroughly checked for backed out terminals, improper mating, broken locks, improperly formed or damaged terminals, poor terminal to wiring connections or physical damage to the wiring harness.

DIAGNOSTIC CHART DTC 63 — 1995 CENTURY, CUTLASS CIERA AND CUTLASS CRUISER

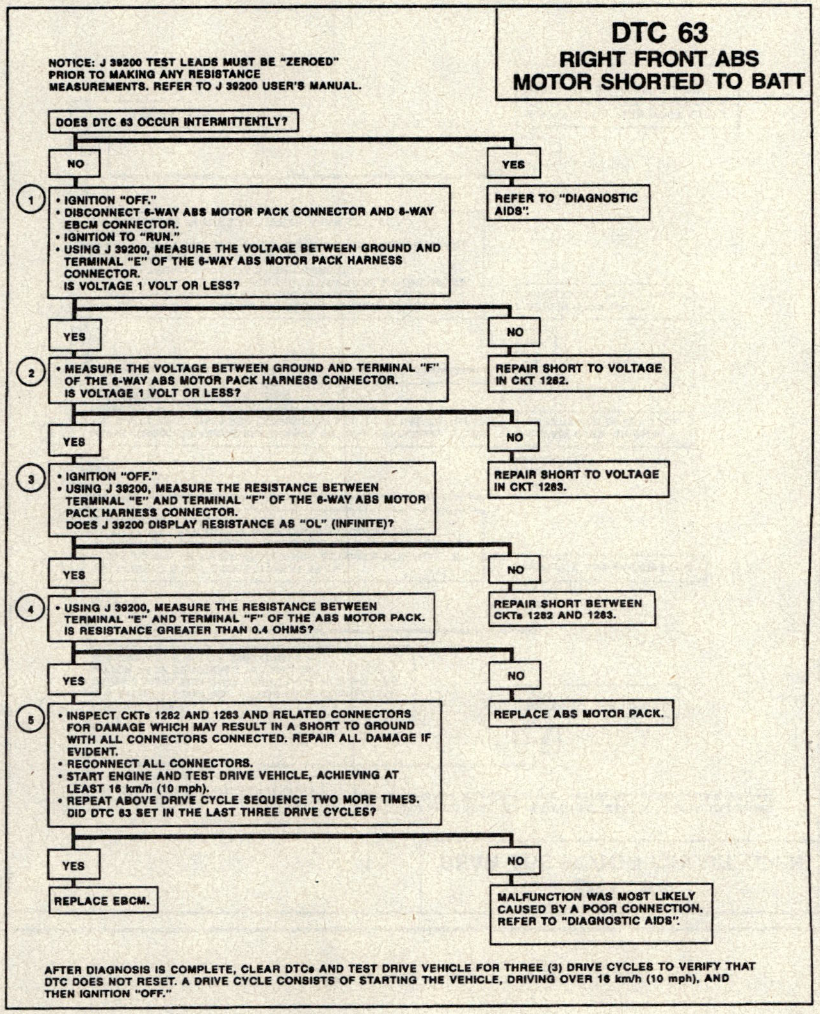

DIAGNOSTIC CHART DTC 64 — 1995 CENTURY, CUTLASS CIERA AND CUTLASS CRUISER

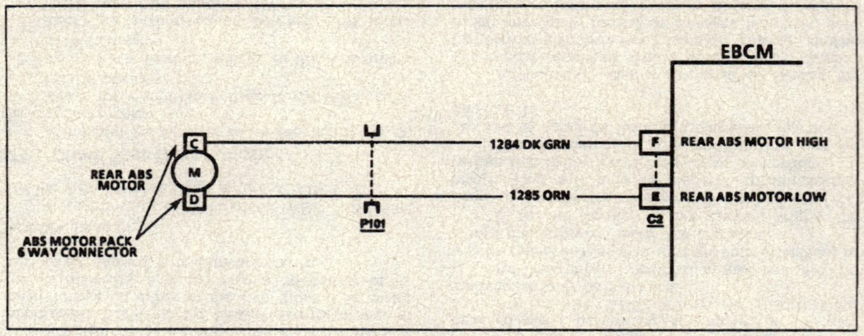

DTC 64
REAR ABS MOTOR CIRCUIT OPEN

Circuit Description:

This DTC identifies an ABS motor that cannot be energized due to an open in its circuitry. This malfunction will not allow proper rear ABS operation.

DTC Will Set When:

DTC 64 can be set only when the ABS motor is commanded "OFF." If the EBCM detects an out of range voltage on either of the rear axle ABS motor circuits indicating an open circuit, a malfunction exists.

Action Taken:

An open motor will not activate when requested. A malfunction DTC is stored, ABS is disabled and the ABS warning lamp is turned "ON."

DTC Chart Test Description:

Number(s) below refer to circled number(s) on the diagnostic chart.

1. Checks for proper resistance of the ABS motor.
2. Checks for an open in the ABS motor "HIGH" circuitry.
3. Checks for an open in the ABS motor "LOW" circuitry.
4. Ensures malfunction was not due to poor terminal contact.

Diagnostic Aids:

Using Tech 1, select manual control function, and exercise ABS motor movement of affected channel in both directions while applying light pressure on the brake pedal.

If erratic or "jumpy" brake pedal movement is detected while performing an "apply" or "release" function of the ABS motor, an intermittent malfunction may be indicated.

An "Intermittent" malfunction may be caused by a poor connection, rubbed through wire insulation, or a wire that is broken inside the insulation.

If the malfunction is not current, wiggle the wires of the affected channel and check if the DTC resets. This will help to pinpoint an intermittent malfunction in the motor circuitry or connections.

The frequency of the malfunction can be checked by using the enhanced diagnostic function of the Tech 1.

Any circuitry that is suspected of causing the intermittent complaint should be thoroughly checked for backed out terminals, improper mating, broken locks, improperly formed or damaged terminals, poor terminal to wiring connections or physical damage to the wiring harness.

DIAGNOSTIC CHART DTC 64 — 1995 CENTURY, CUTLASS CIERA AND CUTLASS CRUISER

DTC 64
REAR ABS MOTOR CIRCUIT OPEN

NOTICE: J 39200 TEST LEADS MUST BE "ZEROED" PRIOR TO MAKING ANY RESISTANCE MEASUREMENTS. REFER TO J 39200 USER'S MANUAL.

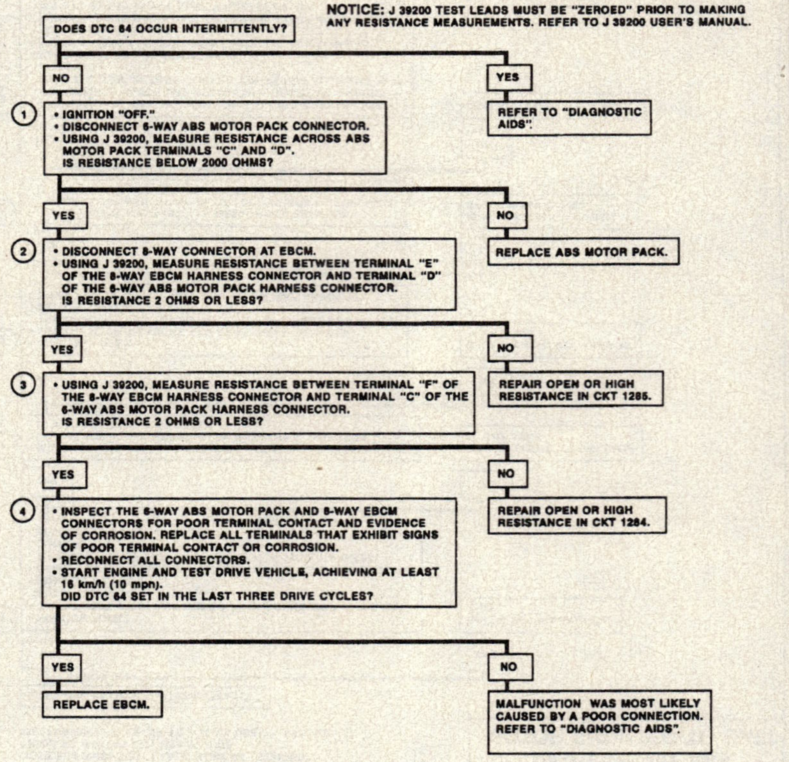

DIAGNOSTIC CHART DTC 65 — 1995 CENTURY, CUTLASS CIERA AND CUTLASS CRUISER

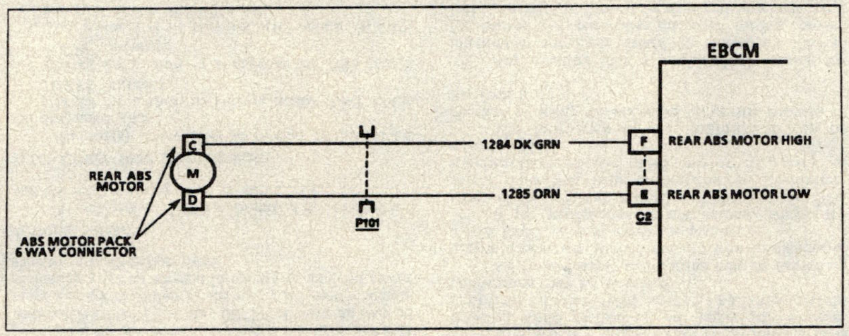

DTC 65
REAR ABS MOTOR CIRCUIT SHORTED TO GROUND

Circuit Description:

This DTC identifies an ABS motor circuit that is shorted to ground. This malfunction will not allow the ABS motor to be controlled at the commanded current rate or will cause the driver circuit to allow current directly to ground.

DTC Will Set When:

DTC 65 can be set anytime. If the EBCM detects an out of range voltage on either of the rear ABS motor circuits indicating a circuit shorted to ground, a malfunction exists.

Action Taken:

A malfunction DTC is stored, ABS is disabled and the ABS warning lamp is turned "ON."

DTC Chart Test Description:

Number(s) below refer to circled number(s) on the diagnostic chart.

1. Checks for an open in the ABS motor "HIGH" circuitry.
2. Checks for an open in the ABS motor "LOW" circuitry.
3. Ensures ABS motor is not internally shorted to ground.
4. Ensures malfunction was not due to poor terminal contact.

Diagnostic Aids:

Using Tech 1, select manual control function, and exercise ABS motor movement of affected channel in both directions while applying light pressure on the brake pedal.

If erratic or "jumpy" brake pedal movement is detected while performing an "apply" or "release" function of the ABS motor, an intermittent malfunction may be indicated.

An "Intermittent" malfunction may be caused by a poor connection, rubbed through wire insulation, or a wire that is broken inside the insulation.

If the malfunction is not current, wiggle the wires of the affected channel and check if the DTC resets. This will help to pinpoint an intermittent malfunction in the ABS motor circuitry or connections.

The frequency of the malfunction can be checked by using the enhanced diagnostic function of the Tech 1.

Any circuitry that is suspected of causing the intermittent complaint should be thoroughly checked for backed out terminals, improper mating, broken locks, improperly formed or damaged terminals, poor terminal to wiring connections or physical damage to the wiring harness.

DIAGNOSTIC CHART DTC 65 — 1995 CENTURY, CUTLASS CIERA AND CUTLASS CRUISER

DTC 65
REAR ABS MOTOR CIRCUIT SHORTED TO GROUND

DOES DTC 65 OCCUR INTERMITTENTLY?

- YES → REFER TO "DIAGNOSTIC AIDS".
- NO →

(1)
- IGNITION "OFF."
- DISCONNECT 6-WAY ABS MOTOR PACK CONNECTOR AND 8-WAY EBCM CONNECTOR.
- USING A J 39200, MEASURE THE RESISTANCE BETWEEN GROUND AND TERMINAL "C" OF THE 6-WAY ABS MOTOR PACK HARNESS CONNECTOR.

DOES J 39200 DISPLAY RESISTANCE AS "OL" (INFINITE)?

- NO → REPAIR SHORT TO GROUND IN CKT 1284.
- YES →

(2)
- MEASURE THE RESISTANCE BETWEEN GROUND AND TERMINAL "D" OF THE 6-WAY ABS MOTOR PACK HARNESS CONNECTOR.

DOES J 39200 DISPLAY RESISTANCE AS "OL" (INFINITE)?

- NO → REPAIR SHORT TO GROUND IN CKT 1285.
- YES →

(3)
- MEASURE THE RESISTANCE BETWEEN GROUND AND TERMINAL "C" OF THE ABS MOTOR PACK.

DOES J 39200 DISPLAY RESISTANCE AS "OL" (INFINITE)?

- NO → REPLACE ABS MOTOR PACK.
- YES →

(4)
- INSPECT CKTs 1284 AND 1285 AND RELATED CONNECTORS FOR DAMAGE WHICH MAY RESULT IN A SHORT TO GROUND WITH ALL CONNECTORS CONNECTED. REPAIR ALL DAMAGE IF EVIDENT.
- RECONNECT ALL CONNECTORS.
- START ENGINE AND TEST DRIVE VEHICLE, ACHIEVING AT LEAST 16 km/h (10 mph).
- REPEAT ABOVE DRIVE CYCLE SEQUENCE TWO MORE TIMES.

DID DTC 65 SET IN THE LAST THREE DRIVE CYCLES?

- YES → REPLACE EBCM.
- NO → MALFUNCTION WAS MOST LIKELY CAUSED BY A POOR CONNECTION. REFER TO "DIAGNOSTIC AIDS".

AFTER ALL DIAGNOSIS IS COMPLETE, CLEAR DTCs AND TEST DRIVE VEHICLE FOR THREE (3) DRIVE CYCLES TO VERIFY THAT DTC DOES NOT RESET. A DRIVE CYCLE CONSISTS OF STARTING THE VEHICLE, DRIVING OVER 16 km/h (10 mph), AND THEN IGNITION "OFF."

DIAGNOSTIC CHART DTC 66 — 1995 CENTURY, CUTLASS CIERA AND CUTLASS CRUISER

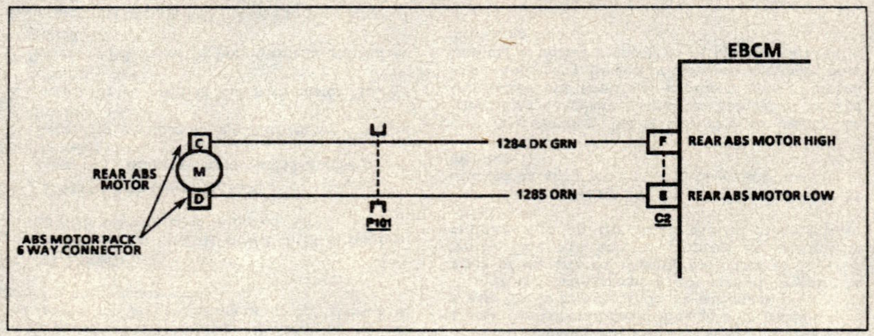

DTC 66
REAR ABS MOTOR CIRCUIT SHORTED TO BATTERY

Circuit Description:

This DTC identifies an ABS motor circuit that is shorted to battery or an ABS motor that has low or zero resistance. This malfunction will not allow the ABS motor to be controlled at the commanded current rate or will cause the ABS motor to turn in the opposite direction or not turn at all.

DTC Will Set When:

DTC 66 can be set only when the ABS motor is commanded "OFF." If the EBCM detects an out of range voltage on either of the rear ABS motor circuits indicating a circuit shorted to battery or an ABS motor shorted, a malfunction exists.

Action Taken:

A malfunction DTC is stored, ABS is disabled and the ABS warning lamp is turned "ON."

DTC Chart Test Description:

Number(s) below refer to circled number(s) on the diagnostic chart.

1. Checks for a short to voltage in the ABS motor "HIGH" circuitry.
2. Checks for a short to voltage in the ABS motor "LOW" circuitry.
3. Checks for a short between ABS motor "HIGH" and ABS motor "LOW" circuits.
4. Checks for proper resistance of the ABS motor.
5. Ensures malfunction was not due to physical damage of the circuitry.

Diagnostic Aids:

Using Tech 1, select manual control function, and exercise ABS motor movement of affected channel in both directions while applying light pressure on the brake pedal.

If erratic or "jumpy" brake pedal movement is detected while performing an "apply" or "release" function of the ABS motor, an intermittent malfunction may be indicated.

An "Intermittent" malfunction may be caused by a poor connection, rubbed through wire insulation, or a wire that is broken inside the insulation.

If the malfunction is not current, wiggle the wires of the affected channel and check if the DTC resets. This will help to pinpoint an intermittent malfunction in the ABS motor circuitry or connections.

The frequency of the malfunction can be checked by using the enhanced diagnostic function of the Tech 1.

Any circuitry that is suspected of causing the intermittent complaint should be thoroughly checked for backed out terminals, improper mating, broken locks, improperly formed or damaged terminals, poor terminal to wiring connections or physical damage to the wiring harness.

DIAGNOSTIC CHART DTC 66 — 1995 CENTURY, CUTLASS CIERA AND CUTLASS CRUISER

DTC 66
REAR ABS MOTOR CIRCUIT SHORTED TO BATTERY

NOTICE: J 39200 TEST LEADS MUST BE "ZEROED" PRIOR TO MAKING ANY RESISTANCE MEASUREMENTS. REFER TO J 39200 USER'S MANUAL.

DOES DTC 66 OCCUR INTERMITTENTLY?

- YES: REFER TO "DIAGNOSTIC AIDS".
- NO:

(1)
- IGNITION "OFF."
- DISCONNECT 6-WAY ABS MOTOR PACK CONNECTOR AND 8-WAY EBCM CONNECTOR.
- IGNITION TO "RUN."
- USING A J 39200, MEASURE THE VOLTAGE BETWEEN GROUND AND TERMINAL "C" OF THE 6-WAY ABS MOTOR PACK HARNESS CONNECTOR.

IS VOLTAGE 1 VOLT OR LESS?

- NO: REPAIR SHORT TO VOLTAGE IN CKT 1284.
- YES:

(2)
- MEASURE THE VOLTAGE BETWEEN GROUND AND TERMINAL "D" OF THE 6-WAY ABS MOTOR PACK HARNESS CONNECTOR.

IS VOLTAGE 1 VOLT OR LESS?

- NO: REPAIR SHORT TO VOLTAGE IN CKT 1285.
- YES:

(3)
- IGNITION "OFF."
- USING J 39200, MEASURE THE RESISTANCE BETWEEN TERMINAL "C" AND TERMINAL "D" OF THE 6-WAY ABS MOTOR PACK HARNESS CONNECTOR.

DOES J 39200 DISPLAY RESISTANCE AS "OL" (INFINITE)?

- NO: REPAIR SHORT BETWEEN CKTs 1284 AND 1285.
- YES:

(4)
- USING J 39200, MEASURE THE RESISTANCE BETWEEN TERMINAL "C" AND TERMINAL "D" OF THE ABS MOTOR PACK.

IS RESISTANCE GREATER THAN 0.4 OHMS?

- NO: REPLACE ABS MOTOR PACK.
- YES:

(5)
- INSPECT CKTs 1284 AND 1285 AND RELATED CONNECTORS FOR DAMAGE WHICH MAY RESULT IN A SHORT TO VOLTAGE WITH ALL CONNECTORS CONNECTED. REPAIR ALL DAMAGE IF EVIDENT.
- RECONNECT ALL CONNECTORS.
- START ENGINE AND TEST DRIVE VEHICLE, ACHIEVING AT LEAST 16 km/h (10 mph).
- REPEAT ABOVE DRIVE CYCLE SEQUENCE TWO MORE TIMES.

DID DTC 66 SET IN THE LAST THREE DRIVE CYCLES?

- YES: REPLACE EBCM.
- NO: MALFUNCTION WAS MOSTLY CAUSED BY A POOR CONNECTION. REFER TO "DIAGNOSTIC AIDS".

AFTER ALL DIAGNOSIS IS COMPLETE, CLEAR DTCs AND TEST DRIVE VEHICLE FOR THREE (3) DRIVE CYCLES TO VERIFY THAT DTC DOES NOT RESET. A DRIVE CYCLE CONSISTS OF STARTING THE VEHICLE, DRIVING OVER 16 km/h (10 mph), AND THEN IGNITION "OFF."

DIAGNOSTIC CHART DTC 76 — 1995 CENTURY, CUTLASS CIERA AND CUTLASS CRUISER

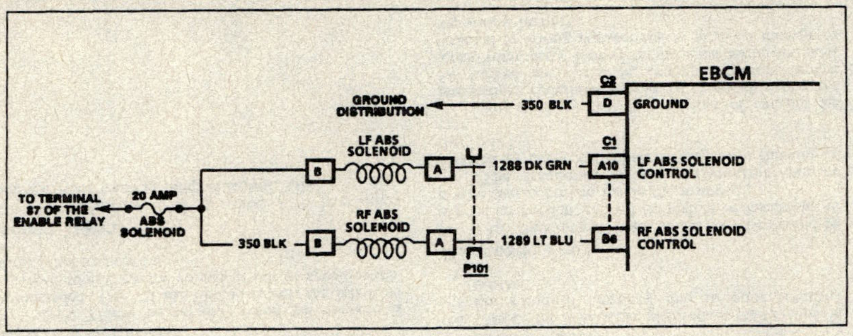

DTC 76
LEFT FRONT SOL. CIRCUIT OPEN OR SHORTED TO BATT

Circuit Description:

This DTC identifies a solenoid that cannot be energized due to an open in its driver circuitry, or a solenoid that is always energized due to a short to battery in its circuitry between the driver and the solenoid. An open will not allow proper ABS operation, but the short to battery simply turns "ON" the solenoid. A path for base brakes is still allowed once the ABS motor rehomes and the check ball is lifted off its seat during key "ON" initialization.

DTC Will Set When:

DTC 76 can be set only when the solenoid is commanded "OFF." If the EBCM detects an excessive voltage in the left front solenoid control circuit, a malfunction exists.

Action Taken:

A malfunction DTC is stored, ABS is disabled and the ABS warning lamp is turned "ON."

DTC Chart Test Description:

Number(s) below refer to circled number(s) on the diagnostic chart.

1. Checks for a short to voltage in the control circuitry of the solenoid.
2. Checks for an open in the control circuit of the solenoid.
3. Checks the solenoid coil for proper resistance.
4. Checks for an open in the solenoid ground circuit.
5. Checks for a possible intermittent malfunction in the solenoid circuitry due to poor terminal contact.

Diagnostic Aids:

An "Intermittent" malfunction may be caused by a poor connection, rubbed through wire insulation, or a wire that is broken inside the insulation.

The frequency of the malfunction can be checked by using the enhanced diagnostic function of the Tech 1.

Any circuitry that is suspected of causing the intermittent complaint should be thoroughly checked for backed out terminals, improper mating, broken locks, improperly formed or damaged terminals, poor terminal to wiring connections or physical damage to the wiring harness.

DIAGNOSTIC CHART DTC 76 — 1995 CENTURY, CUTLASS CIERA AND CUTLASS CRUISER

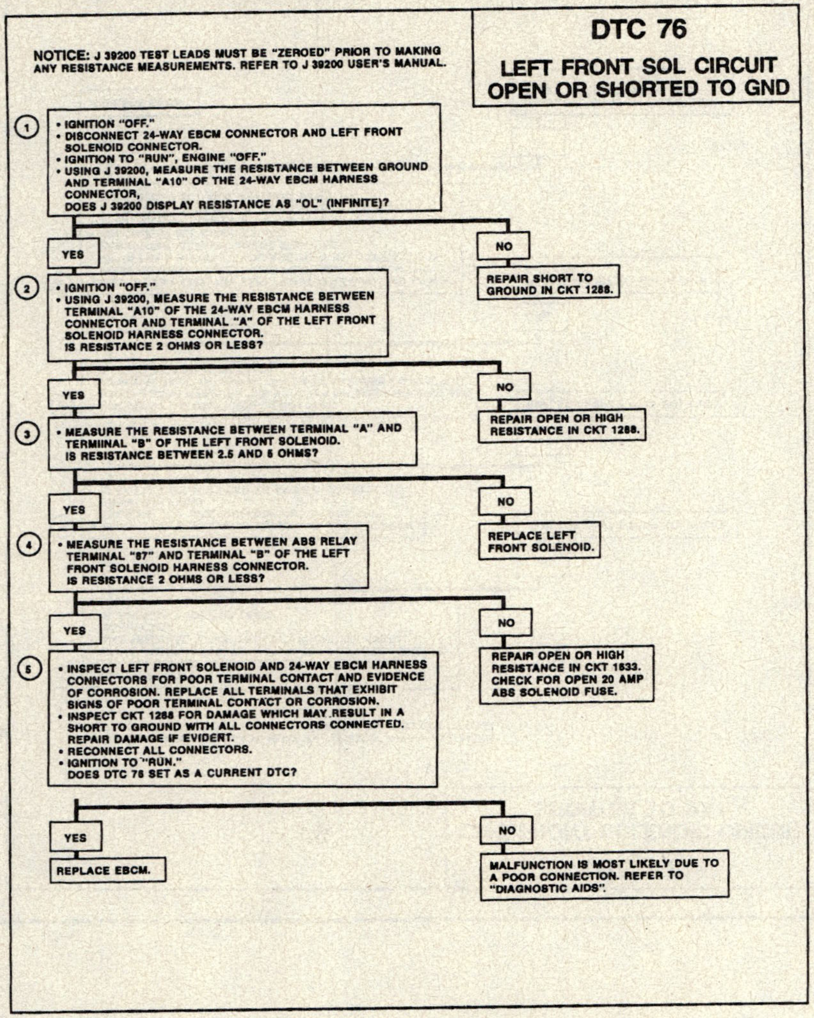

DIAGNOSTIC CHART DTC 77 — 1995 CENTURY, CUTLASS CIERA AND CUTLASS CRUISER

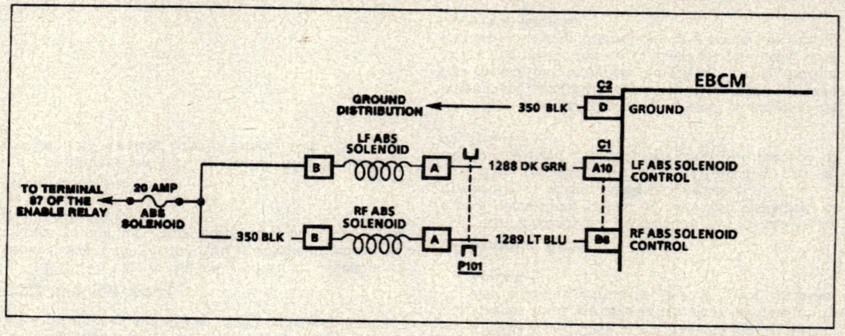

DTC 77
LEFT FRONT SOLENOID CIRCUIT SHORTED TO GND

Circuit Description:

This DTC identifies a solenoid that cannot be energized due to an open in its driver circuitry, or a short to ground between the solenoid driver and the solenoid. These malfunctions can affect ABS operation since the flow of brake fluid to the wheel cylinder cannot be stopped, making ABS operation for that channel impossible.

DTC Will Set When:

DTC 77 can be set only when the solenoid is commanded "ON." If the EBCM detects the left front solenoid control circuit voltage is out of specification, a malfunction exists.

Action Taken:

A malfunction DTC is stored, ABS is disabled and the ABS warning lamp is turned "ON."

DTC Chart Test Description:

Number(s) below refer to circled number(s) on the diagnostic chart.

1. Checks for a solenoid that is internally shorted to ground.
2. Checks for a solenoid that is not within proper resistance values.
3. Indicates if a short to ground exists in the solenoid circuitry.
4. Checks for a possible intermittent malfunction in the solenoid circuitry due to poor terminal contact.

Diagnostic Aids:

An "Intermittent" malfunction may be caused by a poor connection, rubbed through wire insulation, or a wire that is broken inside the insulation.

The frequency of the malfunction can be checked by using the enhanced diagnostic function of the Tech 1.

Any circuitry that is suspected of causing the intermittent complaint should be thoroughly checked for backed out terminals, improper mating, broken locks, improperly formed or damaged terminals, poor terminal to wiring connections or physical damage to the wiring harness.

DIAGNOSTIC CHART DTC 77 — 1995 CENTURY, CUTLASS CIERA AND CUTLASS CRUISER

DTC 77
LEFT FRONT SOLENOID CIRCUIT SHORTED TO BAT

NOTICE: J 39200 TEST LEADS MUST BE "ZEROED" PRIOR TO MAKING ANY RESISTANCE MEASUREMENTS. REFER TO J 39200 USER'S MANUAL.

(1)
- IGNITION "OFF."
- DISCONNECT LEFT FRONT SOLENOID CONNECTOR.
- USING J 39200, MEASURE THE VOLTAGE BETWEEN GROUND AND TERMINAL "A" OF THE LEFT FRONT SOLENOID CONNECTOR.

IS VOLTAGE 1 VOLT OR LESS?

- NO: REPLACE LEFT FRONT SOLENOID.
- YES: go to (2)

(2)
- USING J 39200, MEASURE THE RESISTANCE BETWEEN TERMINAL "A" AND TERMINAL "B" OF THE LEFT FRONT SOLENOID.

IS RESISTANCE BETWEEN 2.5 AND 5.0 OHMS?

- NO: REPLACE LEFT FRONT SOLENOID.
- YES: go to (3)

(3)
- DISCONNECT 24-WAY EBCM CONNECTOR.
- USING J 39200 MEASURE THE VOLTAGE BETWEEN GROUND AND TERMINAL "A10" OF THE 24-WAY EBCM HARNESS CONNECTOR.

IS VOLTAGE 1 VOLT OR LESS?

- NO: REPAIR SHORT TO VOLTAGE IN CKT 1288.
- YES: go to (4)

(4)
- INSPECT CKT 1288, THE LEFT FRONT SOLENOID AND 24-WAY EBCM HARNESS CONNECTORS FOR DAMAGE WHICH MAY RESULT IN A SHORT TO VOLTAGE WITH ALL CONNECTORS CONNECTED. REPAIR DAMAGE IF EVIDENT.
- RECONNECT ALL CONNECTORS.
- START ENGINE WITH FOOT OFF BRAKE.
- ALLOW ENGINE TO RUN FOR AT LEAST 10 SECONDS.

DOES DTC 77 SET AS A CURRENT DTC?

- YES: REPLACE EBCM.
- NO: MALFUNCTION IS INTERMITTENT. REFER TO "DIAGNOSTIC AIDS".

DIAGNOSTIC CHART DTC 78 — 1995 CENTURY, CUTLASS CIERA AND CUTLASS CRUISER

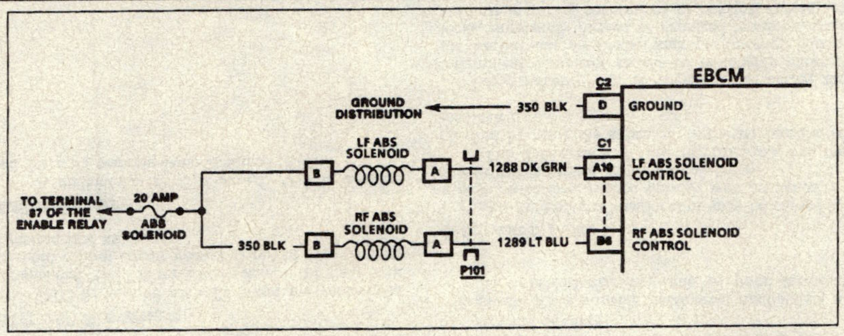

DTC 78
RIGHT FRONT SOL. CIRCUIT OPEN OR SHORTED TO BATT

Circuit Description:

This DTC identifies a solenoid that cannot be energized due to an open in its circuitry, or a solenoid that is always energized due to a short to battery in its circuitry between the driver and the solenoid. An open will not allow proper ABS operation, but the short to battery simply turns "ON" the solenoid. A path for base brakes is still allowed once the ABS motor rehomes and the check ball is lifted off its seat during key "ON" initialization.

DTC Will Set When:

DTC 78 can be set only when the solenoid is commanded "OFF." If the EBCM detects an excessive voltage in the right front solenoid control circuit, a malfunction exists.

Action Taken:

A malfunction DTC is stored, ABS is disabled and the ABS warning lamp is turned "ON."

DTC Chart Test Description:

Number(s) below refer to circled number(s) on the diagnostic chart.

1. Checks for a short to voltage in the control circuitry of the solenoid.
2. Checks for an open in the control circuit of the solenoid.
3. Checks the solenoid coil for proper resistance.
4. Checks for an open in the solenoid ground circuit.
5. Checks for a possible intermittent malfunction in the solenoid circuitry due to poor terminal contact.

Diagnostic Aids:

An "Intermittent" malfunction may be caused by a poor connection, rubbed through wire insulation, or a wire that is broken inside the insulation.

The frequency of the malfunction can be checked by using the enhanced diagnostic function of the Tech 1.

Any circuitry, that is suspected as causing the intermittent complaint, should be thoroughly checked for backed out terminals, improper mating, broken locks, improperly formed or damaged terminals, poor terminal to wiring connections or physical damage to the wiring harness.

DIAGNOSTIC CHART DTC 78 — 1995 CENTURY, CUTLASS CIERA AND CUTLASS CRUISER

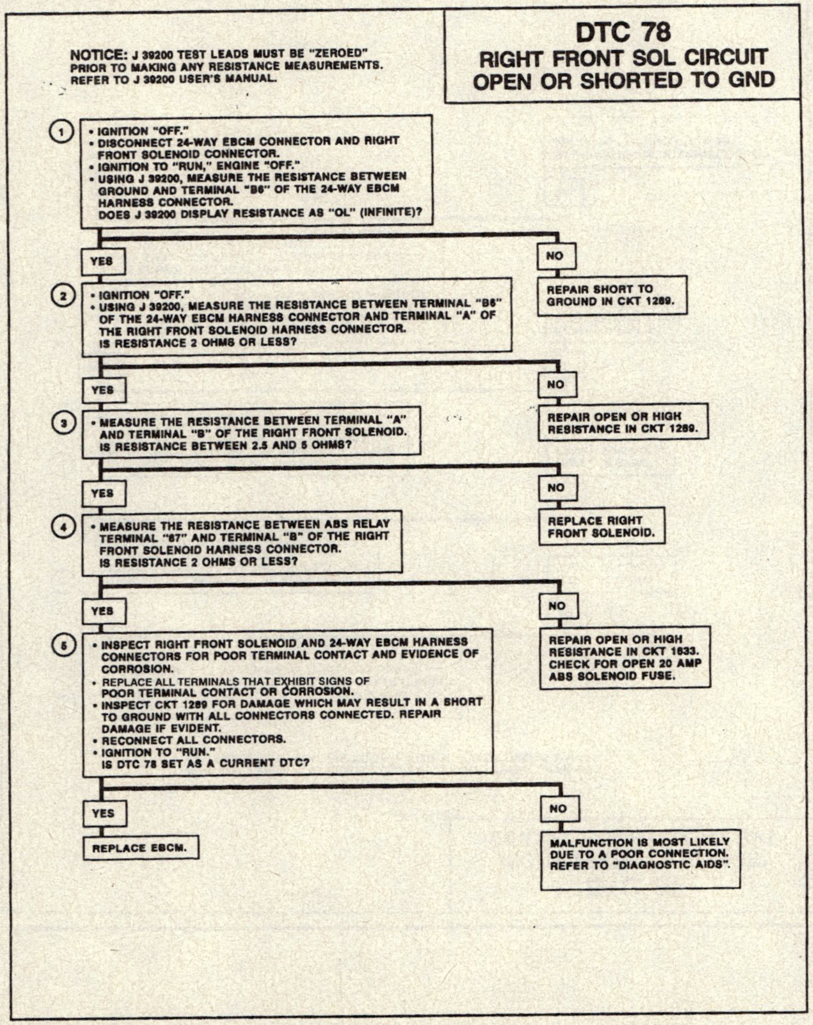

DIAGNOSTIC CHART DTC 81 — 1995 CENTURY, CUTLASS CIERA AND CUTLASS CRUISER

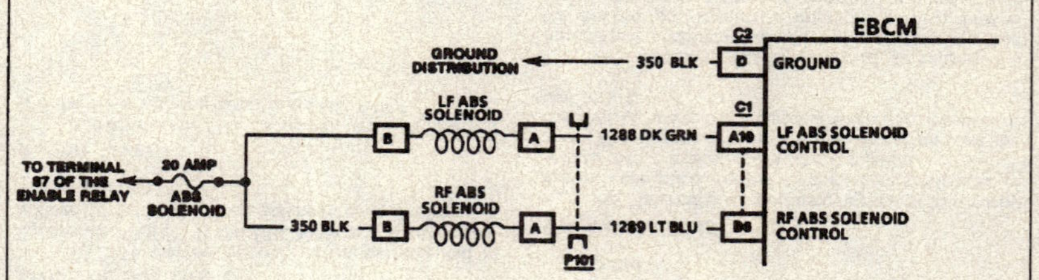

DTC 81
RIGHT FRONT SOLENOID CIRCUIT SHORTED TO GND

Circuit Description:

This DTC identifies a solenoid that cannot be energized due to an open in its driver circuitry, or a short to ground between the solenoid driver and the solenoid. These malfunctions can affect ABS operation since the flow of brake fluid to the wheel cylinder cannot be stopped, making ABS operation for that channel impossible.

DTC Will Set When:

DTC 81 can be set only when the solenoid is commanded "ON." If the EBCM detects the right front solenoid control circuit voltage is out of specification, a malfunction exists.

Action Taken:

A malfunction DTC is stored, ABS is disabled and the ABS warning lamp is turned "ON."

DTC Chart Test Description:

Number(s) below refer to circled number(s) on the diagnostic chart.

1. Checks for a solenoid that is internally shorted to ground.
2. Checks for a solenoid that is not within proper resistance values.
3. Indicates if a short to ground exists in the solenoid circuitry.
4. Checks for a possible intermittent malfunction in the solenoid circuitry due to poor terminal contact.

Diagnostic Aids:

An "Intermittent" malfunction may be caused by a poor connection, rubbed through wire insulation, or a wire that is broken inside the insulation.

The frequency of the malfunction can be checked by using the enhanced diagnostic function of the Tech 1.

Any circuitry that is suspected of causing the intermittent complaint should be thoroughly checked for backed out terminals, improper mating, broken locks, improperly formed or damaged terminals, poor terminal to wiring connections or physical damage to the wiring harness.

DIAGNOSTIC CHART DTC 81 — 1995 CENTURY, CUTLASS CIERA AND CUTLASS CRUISER

DTC 81
RIGHT FRONT SOLENOID CIRCUIT SHORTED TO BAT

NOTICE: J 39200 TEST LEADS MUST BE "ZEROED" PRIOR TO MAKING ANY RESISTANCE MEASUREMENTS. REFER TO J 39200 USER'S MANUAL.

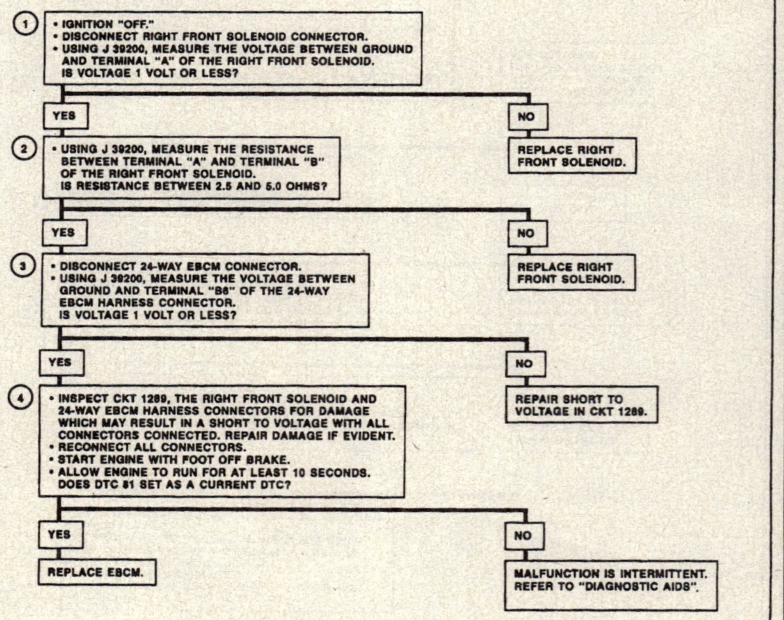

DIAGNOSTIC CHART DTC 82 — 1995 CENTURY, CUTLASS CIERA AND CUTLASS CRUISER

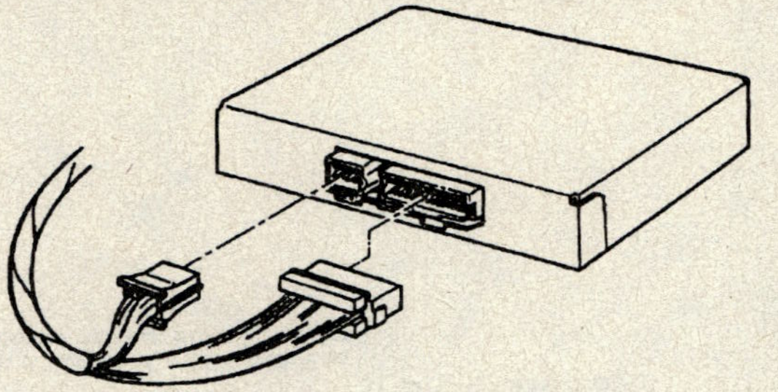

DTC 82
CALIBRATION MALFUNCTION

Circuit Description:

This DTC allows the EBCM to check for a calibration malfunction by comparing the calibration value to a known value stored in the EEPROM.

This DTC is also used as a security measure to prevent improper use of calibrations or changes to these calibrations that may alter the designed function of ABS.

DTC Will Set When:

DTC 82 can be set at key-up. If the program identifier is incorrect or the memory checksum is incorrect, a malfunction exists.

Action Taken:

A malfunction DTC is stored, ABS is disabled and the ABS warning lamp is turned "ON."

DTC Chart Test Description:

Number(s) below refer to circled number(s) on the diagnostic chart.

1. Checks to see if the malfunction is present during diagnosis. If present, the EBCM is not functioning correctly and must be replaced.

Diagnostic Aids:

An intermittent DTC 82 may be caused by a bad cell in the EEPROM that is sensitive to temperature changes. If DTC 82 failed more than once, but is intermittent, replace EBCM.

The frequency of the malfunction can be checked by using the enhanced diagnostic function of the Tech 1.

DIAGNOSTIC CHART DTC 82 — 1995 CENTURY, CUTLASS CIERA AND CUTLASS CRUISER

DTC 82
CALIBRATION MALFUNCTION

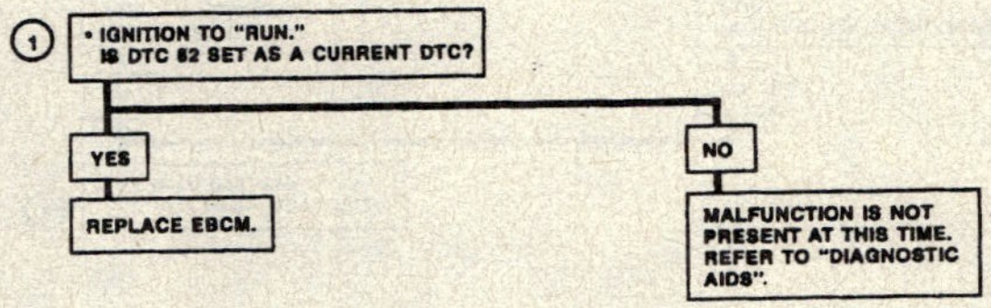

DIAGNOSTIC CHART DTC 86 — 1995 CENTURY, CUTLASS CIERA AND CUTLASS CRUISER

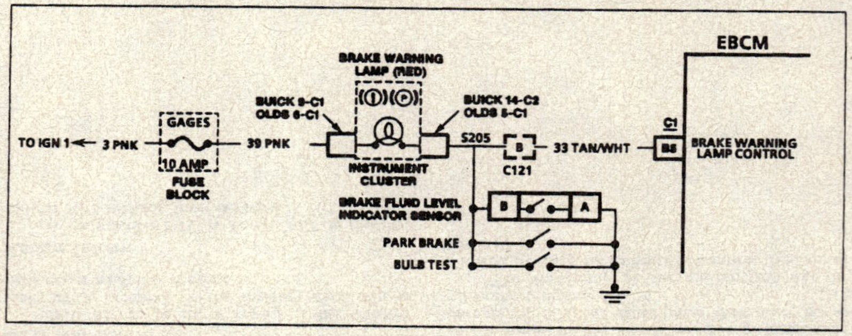

DTC 86
EBCM TURNED "ON" THE RED "BRAKE" WARNING LAMP

Circuit Description:

This DTC is provided as an information only DTC, and reflects the status of the command issued by the ABS EBCM to illuminate the red BRAKE warning lamp. If another DTC issues a command to illuminate the red BRAKE warning lamp, DTC 86 will be stored in EEPROM as a history DTC at the conclusion of the ignition cycle.

DTC Chart Test Description:

Number(s) below refer to circled number(s) on the diagnostic chart.

1. This step serves to identify if a DTC other than DTC 86 commanded the red BRAKE warning lamp "ON."

Diagnostic Aids:

Any ABS mechanical DTC that issues a command to illuminate the red BRAKE warning lamp will also result in DTC 86 being stored in EEPROM during shutdown. These DTCs are: 38, 41, 42, 46, and 51. If the motors are not in their home position, certain electrical DTCs will also command the red BRAKE warning lamp "ON." These DTCs are: 14, 16, 18, 55, 56, 57, 58, 61, 62, 63, 64, 65, and 66.

If any of these DTCs are indicated along with DTC 86, they must be corrected prior to addressing a DTC 86 malfunction.

DIAGNOSTIC CHART DTC 86 — 1995 CENTURY, CUTLASS CIERA AND CUTLASS CRUISER

DTC 86
EBCM TURNED "ON" THE RED BRAKE WARNING LAMP

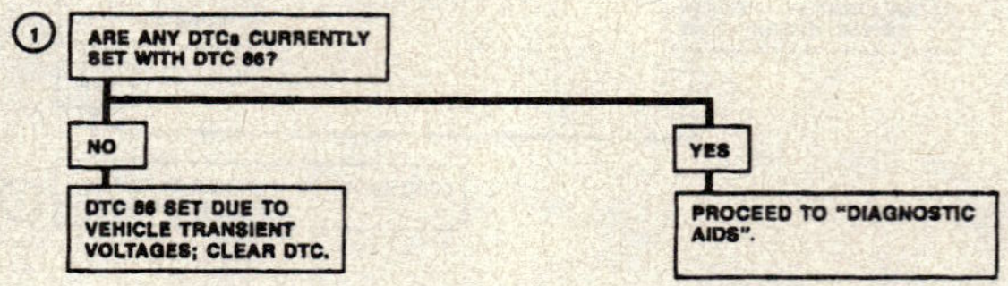

DIAGNOSTIC CHART DTC 87 — 1995 CENTURY, CUTLASS CIERA AND CUTLASS CRUISER

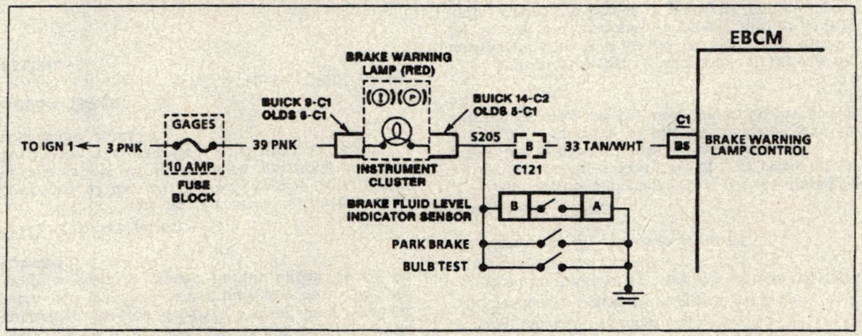

DTC 87
RED "BRAKE" WARNING LAMP CIRCUIT OPEN OR SHORTED TO BATT.
(Page 1 of 2)

Circuit Description:

This DTC is used to verify the EBCM has continuity to the red BRAKE warning lamp in case the EBCM must turn it "ON." This will only occur if an ABS malfunction is detected that may degrade base brake operation. Because ABS is not the only device controlling the red BRAKE warning lamp (parking brake and low fluid sensor may also turn "ON" the lamp) a short to ground in this circuit cannot be detected.

DTC Will Set When:

DTC 87 can be set after initialization is completed. If the red BRAKE warning lamp circuit voltage is out of specification indicating an open circuit, or a circuit that is shorted to battery, a malfunction exists.

Action Taken:

A malfunction DTC is stored. ABS is not disabled.

DTC Chart Test Description:

Number(s) below refer to circled number(s) on the diagnostic chart.

1. Checks whether the BRAKE warning lamp circuitry is open or shorted to battery.
2. Checks the complete BRAKE warning lamp circuitry by attempting to manually turn "ON" the BRAKE warning lamp.
3. Checks for high resistance in park brake circuitry.
4. Checks for a malfunctioning EBCM.
5. Ensures malfunction is not due to poor terminal contact at the brake fluid level switch which may result in an intermittent.
6. Checks the integrity of the "BRAKE" warning lamp.
7. Checks for high resistance in the park brake circuitry.

Diagnostic Aids:

An "Intermittent" malfunction may be caused by a poor connection, rubbed through wire insulation, or a wire that is broken inside the insulation.

The frequency of the malfunction can be checked by using the enhanced diagnostic function of the Tech 1.

If a Tech 1 or T-100 are not available, the parking brake handle may be lifted to verify proper lamp operation and continuity of CKTs 3 and 39.

Any circuitry that is suspected of causing the intermittent complaint should be thoroughly checked for backed out terminals, improper mating, broken locks, improperly formed or damaged terminals, poor terminal to wiring connections or physical damage to the wiring harness.

DIAGNOSTIC CHART DTC 87 — 1995 CENTURY, CUTLASS CIERA AND CUTLASS CRUISER

DTC 87 (Page 1 of 2) RED "BRAKE" WARNING LAMP CIRCUIT OPEN OR SHORTED TO BATT.

(1)
- IGNITION "OFF."
- DISCONNECT 24-WAY EBCM HARNESS CONNECTOR.
- DISCONNECT INSTRUMENT CLUSTER HARNESS CONNECTOR.
- TURN IGNITION TO "RUN."
- USING J 39200, MEASURE VOLTAGE FROM GROUND TO TERMINAL "B5" OF THE 24-WAY EBCM HARNESS CONNECTOR.

IS VOLTAGE GREATER THAN 1 VOLT?

- YES: GO TO DTC 87 PAGE 2 OF 2.
- NO: go to step 2.

(2)
- IGNITION "OFF."
- CONNECT INSTRUMENT CLUSTER HARNESS CONNECTOR.
- USING A FUSED JUMPER SUCH AS J 36169 WITH A 3 AMP FUSE, CONNECT 24-WAY EBCM HARNESS CONNECTOR TERMINAL "B5" TO GROUND.
- IGNITION "ON."

IS RED BRAKE WARNING LAMP "ON"?

- YES: go to step 3.
- NO: go to step 6.

(3)
- ENSURE BRAKE FLUID IS AT THE PROPER LEVEL.
- IGNITION "OFF."
- REMOVE JUMPER USED IN STEP 1.
- APPLY PARK BRAKE.
- IGNITION "ON."
- USING J 39200, MEASURE VOLTAGE BETWEEN GROUND AND 24-WAY EBCM HARNESS CONNECTOR TERMINAL "B5".

IS VOLTAGE LESS THAN 2 VOLTS?

- YES: go to step 4.
- NO: REPAIR HIGH RESISTANCE IN PARK BRAKE CIRCUITRY.

(4)
- RELEASE PARK BRAKE.
- IGNITION "OFF."
- DISCONNECT BRAKE FLUID LEVEL SWITCH.
- INSPECT 24-WAY EBCM HARNESS CONNECTOR TERMINAL "B5" FOR POOR TERMINAL CONTACT OR CORROSION.
- REPLACe TERMINALS THAT EXHIBIT POOR CONTACT OR CORROSION.
- RECONNECT 24-WAY EBCM CONNECTOR.
- START ENGINE.
- WAIT 10 SECONDS.

DOES DTC 87 SET AS A CURRENT DTC?

- YES: REPLACE EBCM.
- NO: go to step 5.

(5)
- IGNITION "OFF."
- INSPECT BRAKE FLUID LEVEL SWITCH HARNESS CONNECTOR FOR POOR TERMINAL CONTACT AND EVIDENCE OF CORROSION. REPLACE ALL TERMINALS THAT EXHIBIT SIGNS OF TERMINAL CONTACT OR CORROSION.
- RECONNECT BRAKE FLUID LEVEL SWITCH CONNECTOR.
- START ENGINE.
- WAIT 10 SECONDS.

DOES DTC 87 SET AS A CURRENT DTC?

- YES:
  - CHECK BRAKE FLUID LEVEL.
  - REPAIR HIGH RESISTANCE IN FLUID LEVEL SWITCH CIRCUITRY.
- NO: MALFUNCTION IS MOST LIKELY DUE TO A POOR CONNECTION. REFER TO "DIAGNOSTIC AIDS".

(6)
- IGNITION "OFF."
- REMOVE I/P CLUSTER.
- REMOVE AND INSPECT RED BRAKE WARNING LAMP.

IS BULB OPEN?

- YES: REPLACE BULB.
- NO: go to step 7.

(7)
- USING J 39200, MEASURE RESISTANCE BETWEEN I/P CLUSTER HARNESS CONNECTOR TERMINAL "C2-14" (BUICK) OR "C1-5" (OLDS) AND 24-WAY EBCM HARNESS CONNECTOR TERMINAL "B5".

IS RESISTANCE LESS THAN 2 OHMS?

- YES: DIAGNOSE CLUSTER AND/OR CIRCUITRY TO CLUSTER.
- NO: REPAIR OPEN OR HIGH RESISTANCE IN CKT 33.

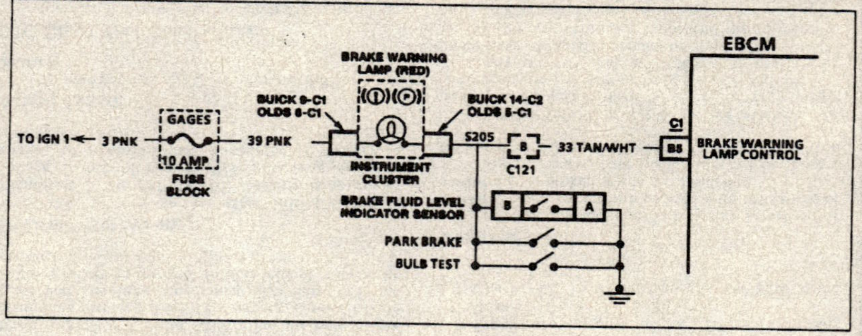

DTC 87
RED "BRAKE" WARNING LAMP CIRCUIT OPEN OR SHORTED TO BATT.

Circuit Description:

This DTC is used to verify the EBCM has continuity to the red BRAKE warning lamp in case the EBCM must turn it "ON." This will only occur if an ABS malfunction is detected that may degrade base brake operation. Because ABS is not the only device controlling the red BRAKE warning lamp (parking brake and low fluid sensor may also turn "ON" the lamp) a short to ground in this circuit cannot be detected.

DTC Will Set When:

DTC 87 can be set after initialization is completed. If the red BRAKE warning lamp circuit voltage is out of specification indicating an open circuit, or a circuit that is shorted to battery, a malfunction exists.

Action Taken:

A malfunction DTC is stored. ABS is not disabled.

DTC Chart Test Description:

Number(s) below refer to circled number(s) on the diagnostic chart.

8. Identifies if a malfunction is currently present.
9. With the Tech 1 connected, attempts to turn "ON" the "BRAKE" warning lamp.
10. Checks for a short to voltage in CKT 33.
11. Ensures malfunction was not due to physical damage of the circuitry.
12. Checks for DTC 86 also being set.

Diagnostic Aids:

An "Intermittent" malfunction may be caused by a poor connection, rubbed through wire insulation, or a wire that is broken inside the insulation.

The frequency of the malfunction can be checked by using the enhanced diagnostic function of the Tech 1.

Any circuitry that is suspected of causing the intermittent complaint should be thoroughly checked for backed out terminals, improper mating, broken locks, improperly formed or damaged terminals, poor terminal to wiring connections or physical damage to the wiring harness.

DIAGNOSTIC CHART DTC 87 —
1995 CENTURY, CUTLASS CIERA AND
CUTLASS CRUISER

DTC 87
(Page 2 of 2)
RED "BRAKE" WARNING LAMP CIRCUIT OPEN OR SHORTED TO BATT.

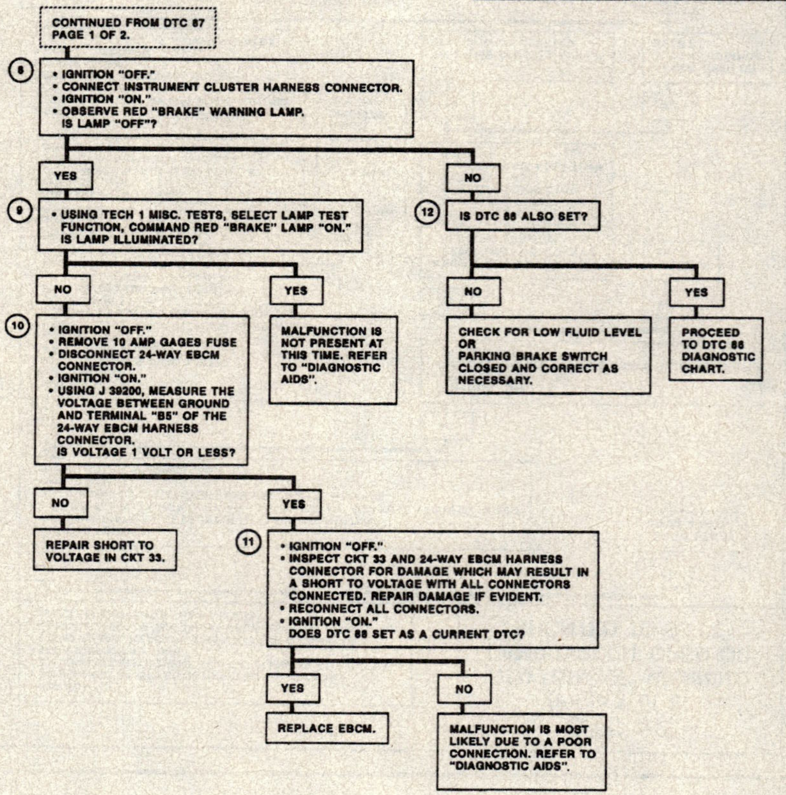

DIAGNOSTIC CHART DTC 91 — 1995 CENTURY, CUTLASS CIERA AND CUTLASS CRUISER

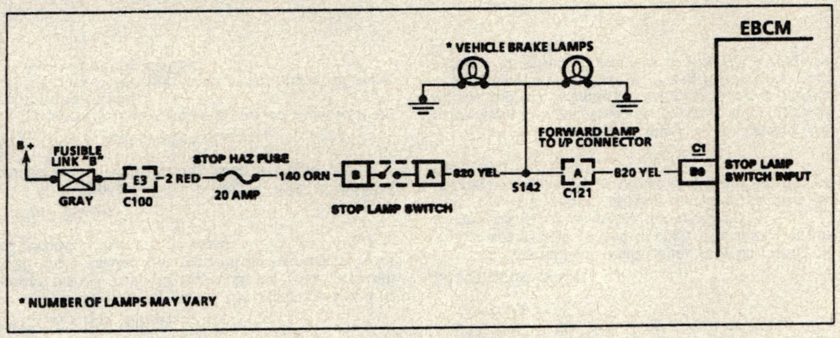

DTC 91
OPEN BRAKE SWITCH DURING DECELERATION

Circuit Description:

This DTC is used to detect an open stop lamp switch in the non-ABS mode. The EBCM looks for deceleration rates that would indicate braking action and verifies this assumption by requiring several repeats of this detection method. In each case, ABS will not be available since the stop lamp switch signal is not seen.

DTC Will Set When:

DTC 91 can be set if three deceleration cycles occur with the stop lamp switch "OFF."

Action Taken:

A malfunction DTC is stored, ABS is disabled and the amber ABS warning lamp is turned "ON."

DTC Chart Test Description:

Number(s) below refer to circled number(s) on the diagnostic chart.

1. Indicates if the stop lamp switch signal is being received by the EBCM.
2. Indicates if an open circuit exists in the stop lamp switch or stop lamp circuitry.
3. Isolates the open circuit to either the stop lamp switch input circuitry, or the EBCM.
4. Ensures malfunction was not due to poor terminal contact.
5. Verifies that voltage is available at the stop lamp switch.
6. Verifies if the stop lamp switch is functioning properly.

Diagnostic Aids:

An "Intermittent" malfunction may be caused by a poor connection, rubbed through wire insulation, or a wire that is broken inside the insulation.

The frequency of the malfunction can be checked by using the enhanced diagnostic function of the Tech 1.

Any circuitry that is suspected of causing the intermittent complaint should be thoroughly checked for backed out terminals, improper mating, broken locks, improperly formed or damaged terminals, poor terminal to wiring connections or physical damage to the wiring harness.

DIAGNOSTIC CHART DTC 91 — 1995 CENTURY, CUTLASS CIERA AND CUTLASS CRUISER

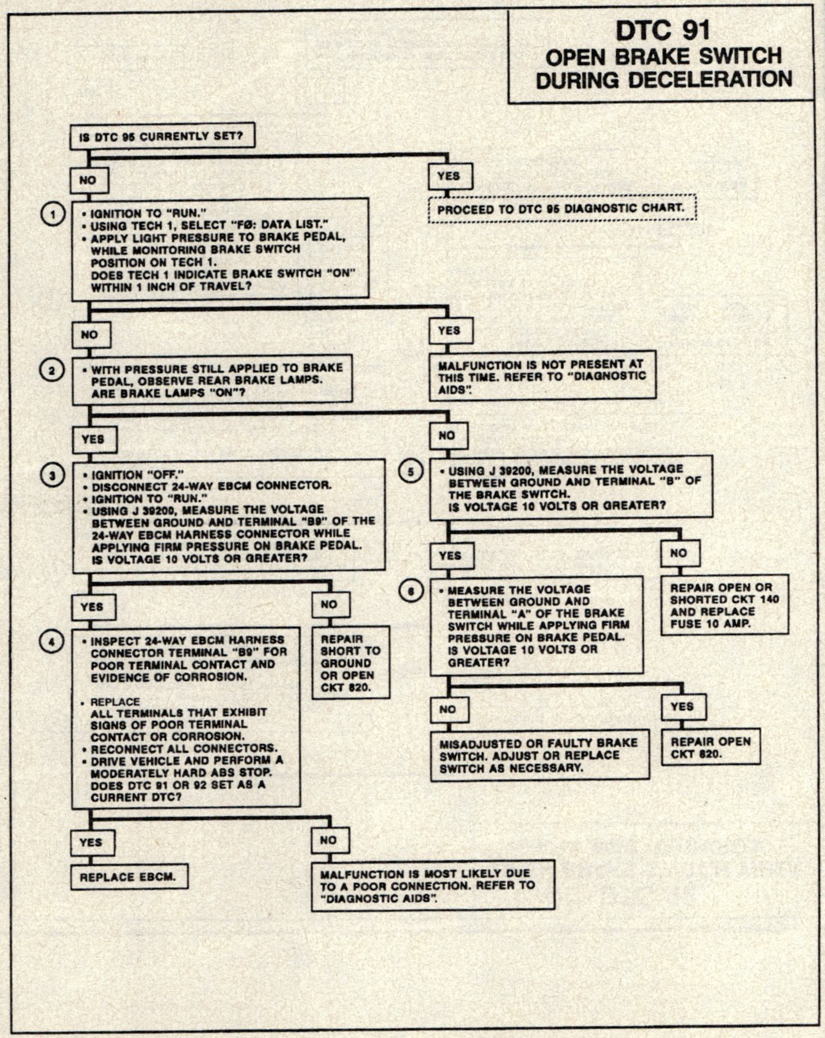

DIAGNOSTIC CHART DTC 92 — 1995 CENTURY, CUTLASS CIERA AND CUTLASS CRUISER

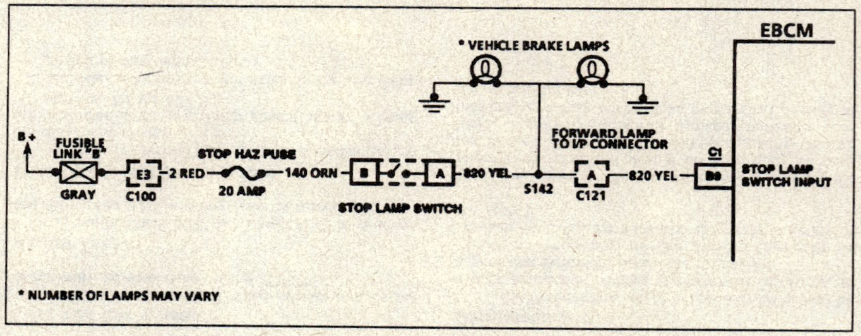

DTC 92
OPEN BRAKE SWITCH WHEN ABS WAS REQUIRED

Circuit Description:

This DTC is run to determine the proper operation of the stop lamp switch. This is important because ABS is activated when the stop lamp switch is "ON." If the stop lamp switch is "OFF," ABS will never be activated. Since this malfunction is difficult to detect under normal braking conditions, this malfunction is only detected when ABS is required.

DTC Will Set When:

DTC 92 can be set if the vehicle's speed is greater than 8 km/h (5 mph). If the brake was not "ON" and a release was required on two channels for 0.5 second, a malfunction exists.

Action Taken:

A malfunction DTC is stored, ABS is disabled and the amber ABS warning lamp is turned "ON."

DTC Chart Test Description:

Number(s) below refer to circled number(s) on the diagnostic chart.

1. Indicates if the stop lamp switch signal is being received by the EBCM.
2. Indicates if an open circuit exists in the stop lamp switch or brake lamp circuitry.
3. Isolates the open circuit to either the stop lamp switch input circuitry, or the EBCM.
4. Ensures malfunction was not due to poor terminal contact.
5. Verifies the voltage is available at the stop lamp switch.
6. Verifies if the stop lamp switch is functioning properly.

Diagnostic Aids:

An "Intermittent" malfunction may be caused by a poor connection, rubbed through wire insulation, or a wire that is broken inside the insulation.

The frequency of the malfunction can be checked by using the enhanced diagnostic function of the Tech 1.

Any circuitry that is suspected of causing the intermittent complaint should be thoroughly checked for backed out terminals, improper mating, broken locks, improperly formed or damaged terminals, poor terminal to wiring connections or physical damage to the wiring harness.

DIAGNOSTIC CHART DTC 92 — 1995 CENTURY, CUTLASS CIERA AND CUTLASS CRUISER

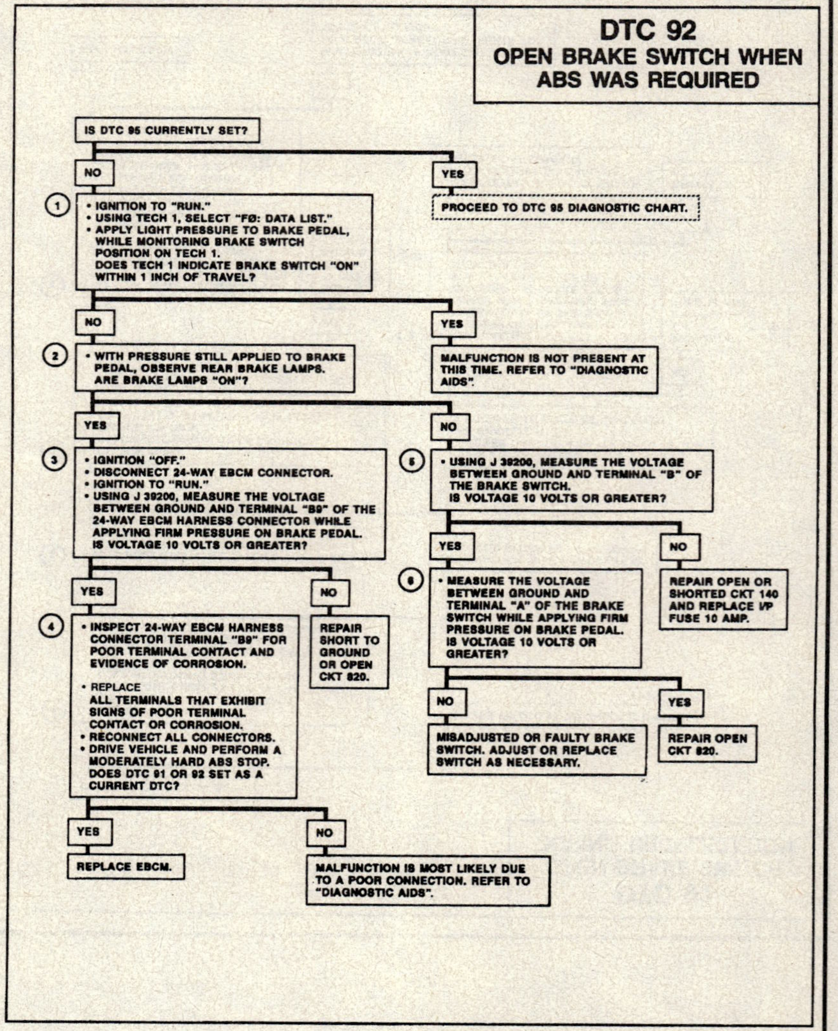

DIAGNOSTIC CHART DTC 93 — 1995 CENTURY, CUTLASS CIERA AND CUTLASS CRUISER

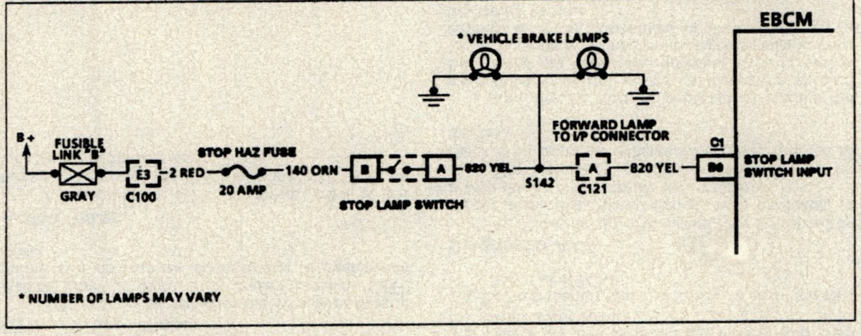

DTC 93
DTC 91 OR 92 SET IN CURRENT OR PREV. IGN CYCLE

Circuit Description:

This DTC is the second portion of DTCs 91 and 92. If DTCs 91 or 92 occurred during the last ignition cycle, DTC 93 becomes a current failure during the next ignition cycle, keeping ABS disabled until a stop lamp switch "ON" state is seen. When a change is seen during an ignition cycle in which DTC 93 is a current malfunction, DTC 91 or 92 will clear itself at the end of the current ignition cycle, and ABS will enable itself at the start of the next ignition cycle. DTC 93 alone indicates 91 or 92 occurred previously, but is intermittent, or has been corrected.

DTC Chart Test Description:

Number(s) below refer to circled number(s) on the diagnostic chart.

1. Indicates which DTC (either 91 or 92) caused DTC 93 to set.

Diagnostic Aids:

An "Intermittent" malfunction may be caused by a poor connection, rubbed through wire insulation, or a wire that is broken inside the insulation.

The frequency of the malfunction can be checked by using the enhanced diagnostic function of the Tech 1.

Any circuitry that is suspected of causing the intermittent complaint should be thoroughly checked for backed out terminals, improper mating, broken locks, improperly formed or damaged terminals, poor terminal to wiring connections or physical damage to the wiring harness.

Also, verify proper stop lamp switch operation using the data list of the Tech 1. As the brake is applied, the data list should display the stop lamp switch "ON" within 1 inch of travel.

DIAGNOSTIC CHART DTC 93 — 1995 CENTURY, CUTLASS CIERA AND CUTLASS CRUISER

DTC 93
DTC 91 OR 92 SET IN CURRENT OR PREV. IGN CYCLE

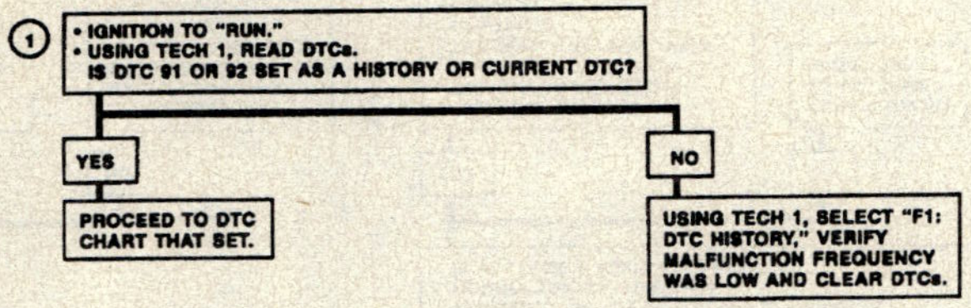

DIAGNOSTIC CHART DTC 94 — 1995 CENTURY, CUTLASS CIERA AND CUTLASS CRUISER

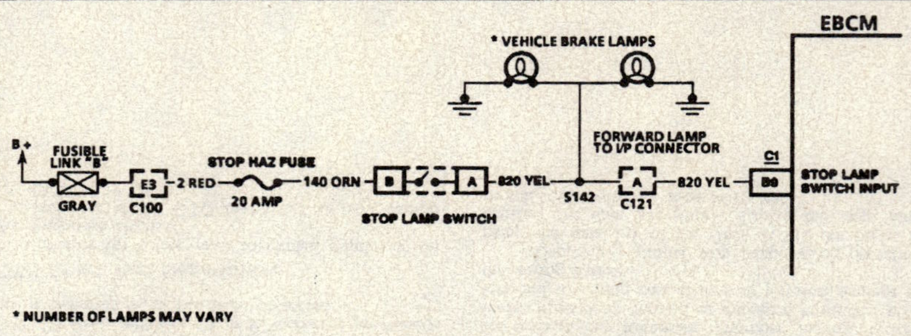

DTC 94
BRAKE SWITCH CONTACTS ALWAYS CLOSED

Circuit Description:

This DTC is run to determine the proper operation of the stop lamp switch. This is important because ABS is activated when the stop lamp switch is "ON" and turned "OFF" when the stop lamp switch is "OFF." If the stop lamp switch is always "ON," ABS operation will always be requested resulting in potential modulator cycling on rough roads.

DTC Will Set When:

DTC 94 can be set when the vehicle reaches at least 40 km/h (25 mph). If the brake was never "OFF" during two consecutive drive cycles, a malfunction exists.

Action Taken:

A malfunction DTC is stored. ABS is not disabled, however, the ABS warning will flash to indicate a malfunction exists.

DTC Chart Test Description:

Number(s) below refer to circled number(s) on the diagnostic chart.

1. Checks if the malfunction is currently present.
2. Isolates the cause of the malfunction to either a malfunctioning or misadjusted stop lamp switch, or a short to voltage in the stop lamp switch circuitry.
3. Checks for unwanted voltage on the stop lamp switch input circuit.
4. Ensures malfunction is not due to poor terminal contact or EBCM.

Diagnostic Aids:

An "Intermittent" malfunction may be caused by a poor connection, rubbed through wire insulation, or a wire that is broken inside the insulation.

The frequency of the malfunction can be checked by using the enhanced diagnostic function of the Tech 1.

Any circuitry that is suspected of causing the intermittent complaint should be thoroughly checked for backed out terminals, improper mating, broken locks, improperly formed or damaged terminals, poor terminal to wiring connections or physical damage to the wiring harness.

DIAGNOSTIC CHART DTC 94 — 1995 CENTURY, CUTLASS CIERA AND CUTLASS CRUISER

DTC 94
BRAKE SWITCH CONTACTS ALWAYS CLOSED

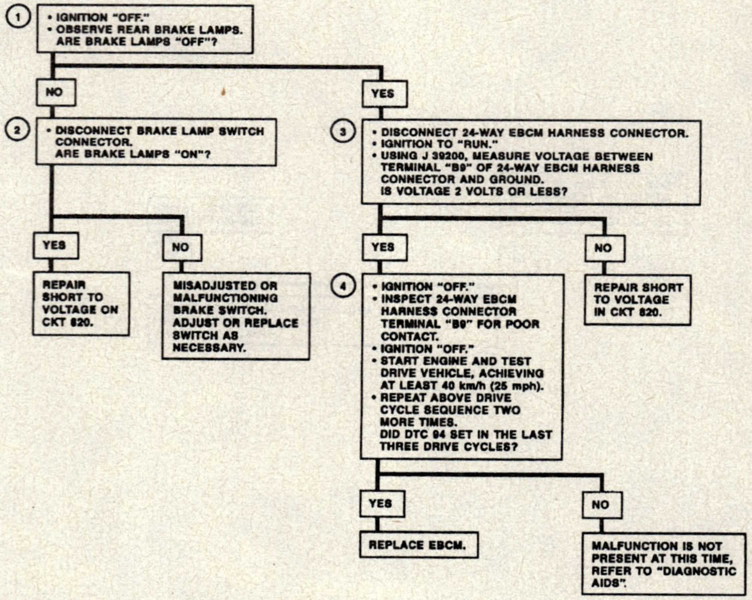

DIAGNOSTIC CHART DTC 95 — 1995 CENTURY, CUTLASS CIERA AND CUTLASS CRUISER

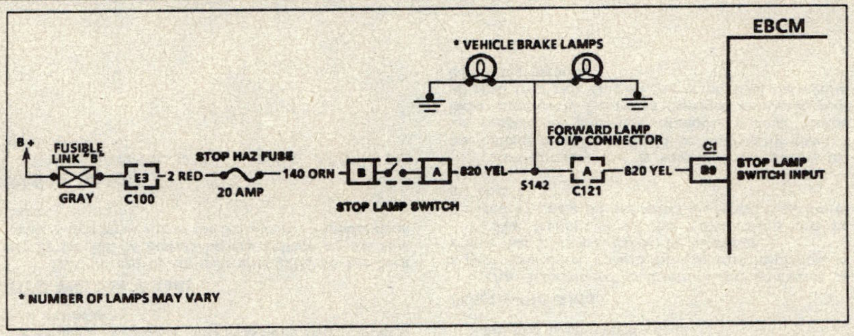

DTC 95
BRAKE SWITCH CIRCUIT OPEN

Circuit Description:
This DTC is used to identify open stop lamp switch circuitry that prevents the stop lamp switch input to the EBCM from changing states when the brake is applied. This DTC is used in conjunction with DTCs 91 and 92 to determine the cause of an open stop lamp switch malfunction.

DTC Will Set When:
DTC 95 can be set after initialization is completed. If the stop lamp switch input voltage is out of specification for one second indicating an open circuit, a malfunction exists.

Action Taken:
A malfunction DTC is stored, ABS is disabled and the amber ABS warning lamp is turned "ON."

DTC Chart Test Description:
Number(s) below refer to circled number(s) on the diagnostic chart.
1. This step is used to confirm that an open in the stop lamp switch circuitry currently exists.
2. Indicates if the stop lamp switch signal is being received by the EBCM.
3. Indicates if an open circuit exists in the stop lamp switch or stop lamp circuitry.
4. Isolates the open circuit to either the stop lamp switch input circuitry, or the EBCM.

Diagnostic Aids:
An "Intermittent" malfunction may be caused by a poor connection, rubbed through wire insulation, or a wire that is broken inside the insulation.
The frequency of the malfunction can be checked by using the enhanced diagnostic function of the Tech 1.

Any circuitry that is suspected of causing the intermittent complaint should be thoroughly checked for backed out terminals, improper mating, broken locks, improperly formed or damaged terminals, poor terminal to wiring connections or physical damage to the wiring harness.

DIAGNOSTIC CHART DTC 95 — 1995 CENTURY, CUTLASS CIERA AND CUTLASS CRUISER

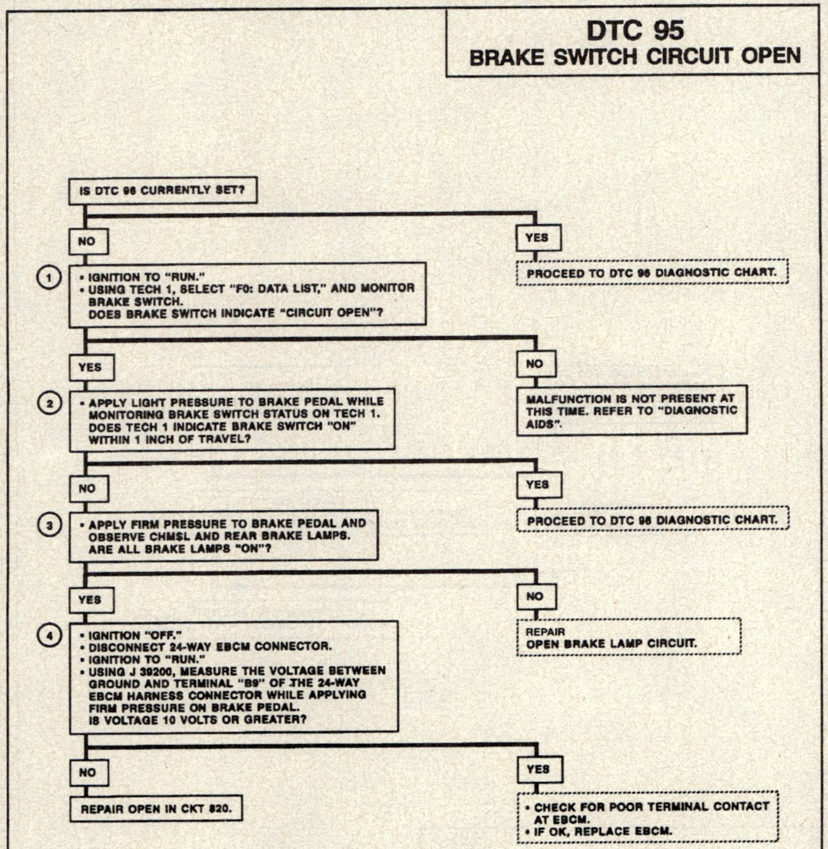

DIAGNOSTIC CHART DTC 96 — 1995 CENTURY, CUTLASS CIERA AND CUTLASS CRUISER

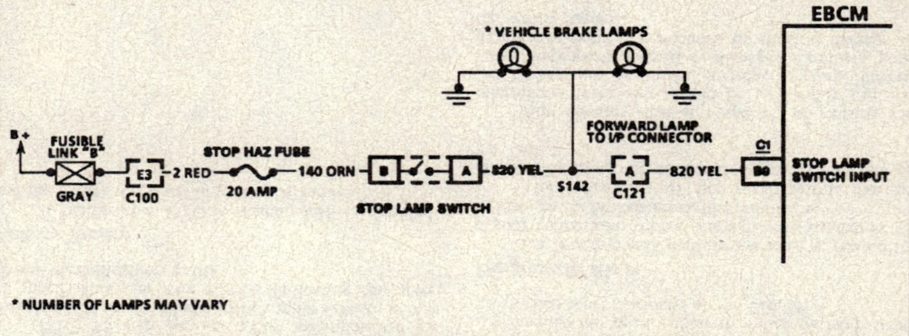

DTC 96
REAR BRAKE LAMP CIRCUIT OPEN

Circuit Description:

This DTC is designed to isolate the cause of a DTC 95 malfunction and indicate to the driver ABS is still available. If DTC 95 occurs with DTC 96, the brake lamps are open, the brake lamp grounds are open, or the Center High Mounted Stop Lamp (CHMSL) circuit is open during 4-way flasher use. The presence of battery voltage at the stop lamp switch input indicates a valid stop lamp switch input is still available.

DTC Will Set When:

DTC 96 can be set only after DTC 95 has been set. If the EBCM detects battery voltage on the stop lamp switch input circuit for 0.5 second, a malfunction exists.

Action Taken:

A malfunction DTC is stored. ABS is not disabled.

DTC Chart Test Description:

Number(s) below refer to circled number(s) on the diagnostic chart.

1. As a result of a malfunction of an additional stop lamp switch circuit DTC, this DTC may be set. To ensure proper diagnosis, any additional stop lamp switch DTCs must be repaired first.
2. Identifies if the malfunction is currently present in the stop lamp switch circuit.

Diagnostic Aids:

An "Intermittent" malfunction may be caused by a poor connection, rubbed through wire insulation, or a wire that is broken inside the insulation.

The frequency of the malfunction can be checked by using the enhanced diagnostic function of the Tech 1.

Any circuitry that is suspected of causing the intermittent complaint should be thoroughly checked for backed out terminals, improper mating, broken locks, improperly formed or damaged terminals, poor terminal to wiring connections or physical damage to the wiring harness.

DIAGNOSTIC CHART DTC 96 — 1995 CENTURY, CUTLASS CIERA AND CUTLASS CRUISER

DTC 96
BRAKE LAMPS CIRCUIT OPEN

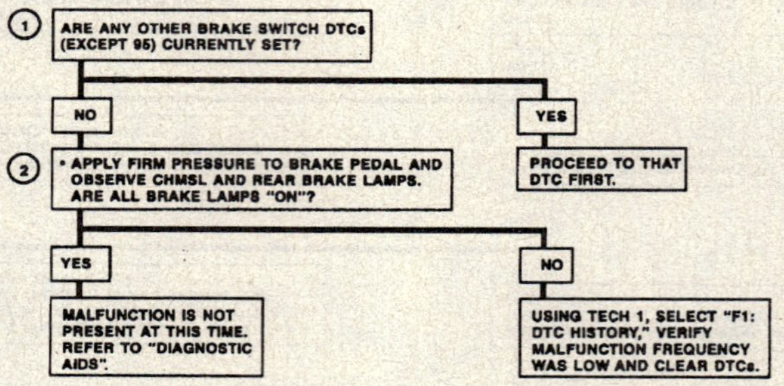

DIAGNOSTIC CHART DTC 96 — 1995 CENTURY, CUTLASS CIERA AND CUTLASS CRUISER

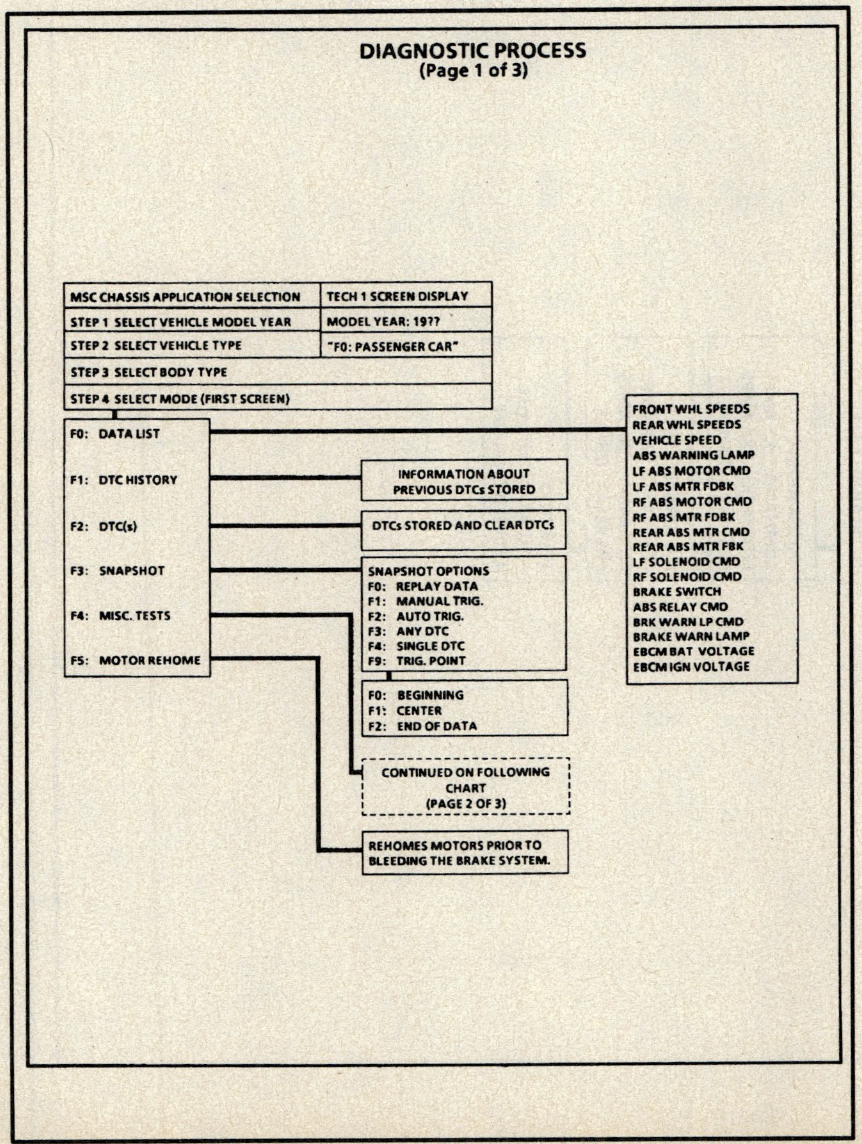

DIAGNOSTIC CHART DTC 96 — 1995 CENTURY, CUTLASS CIERA AND CUTLASS CRUISER

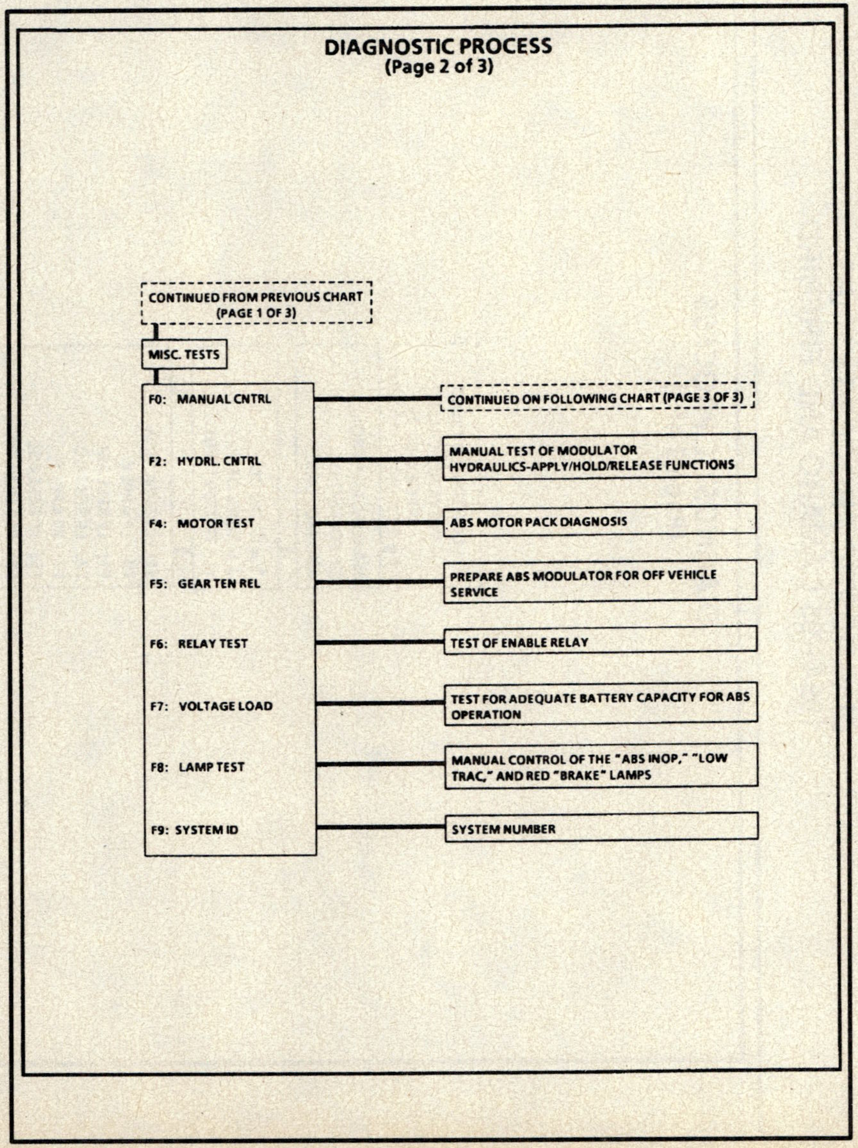

DIAGNOSTIC PROCESS CHART — 1994–95 CAMARO AND FIREBIRD

DIAGNOSTIC PROCESS
(Page 3 of 3)

CONTINUED FROM PREVIOUS CHART (PAGE 2 OF 3)

MANUAL CONTROL OF SOLENOIDS AND MOTORS.

↑ = TURN ON
↓ = TURN OFF
YES TO CONTINUE

SELECT FUNCTION
F0: LF APPLY
F1: LF RELEASE
F2: LF SOLENOID
F4: RF APPLY
F5: RF RELEASE
F6: RF SOLENOID
F8: REAR APPLY
F9: REAR RELEASE

MOTOR APPLY
COMMAND: XX AMPS
FEEDBACK: XX AMPS
ABS RELAY: ON/OFF

MOTOR RELEASE
COMMAND: XX AMPS
FEEDBACK: XX AMPS
ABS RELAY: ON/OFF

SOLENOID
COMMAND: ON/OFF
ABS RELAY: ON/OFF

DIAGNOSTIC CHART SYSTEM CHECK — 1994–95 CAMARO AND FIREBIRD

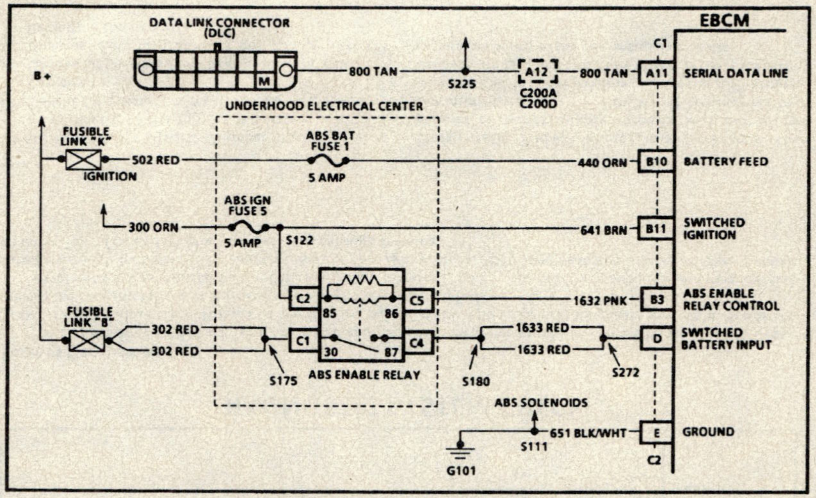

DIAGNOSTIC SYSTEM CHECK

System Description:

The diagnostic system check is an organized approach to identifying a problem created by an Antilock Brake System (ABS) malfunction. It must be the starting point for any ABS complaint diagnosis because it directs the service technician to the next logical step in diagnosing the complaint.

Serial data is transmitted by the EBCM through terminal "A11". The EBCM is supplied switched ignition voltage through terminal "B11", battery voltage through terminal "B10" and switched battery voltage through terminal "D". The EBCM ground is provided through terminal "E".

Chart Test Description: Number(s) below refer to circled number(s) on the diagnostic chart.

1. Indicates if EBCM is transmitting data.
2. Checks for a serial data line malfunction.
3. Checks for proper Tech 1 connection to DLC.
4. Checks for open ABS BAT 5 amp Fuse 1.
5. Checks for open ABS IGN 5 amp Fuse 5.
6. Checks for high resistance in the EBCM ground circuit.
7. Checks for current DTCs.
8. Checks for proper ABS warning lamp operation.
9. Checks for history DTCs.

Diagnostic Aids: Excessive resistance in the ground or power supply circuits will not allow communication with the EBCM. If communication with the EBCM is not possible, ensure the ABS ground connection is good and that there is no excessive resistance in any of the power supply circuits.

DIAGNOSTIC CHART SYSTEM CHECK — 1994–95 CAMARO AND FIREBIRD

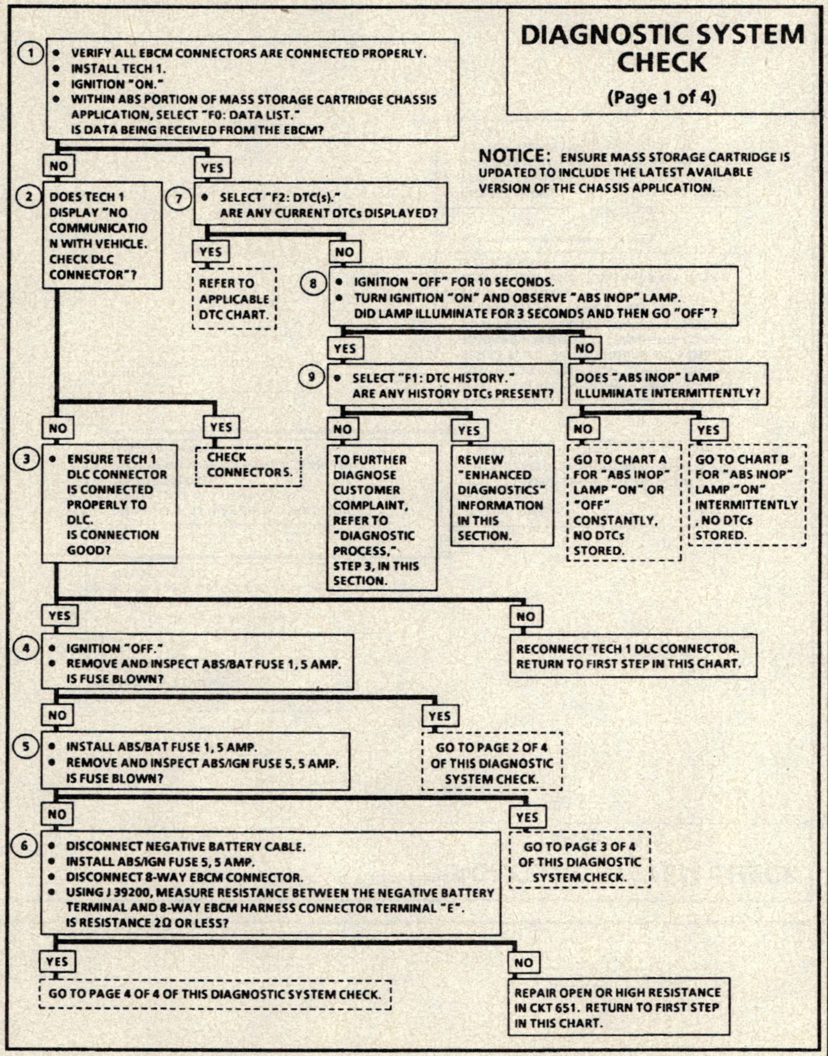

DIAGNOSTIC CHART SYSTEM CHECK — 1994–95 CAMARO AND FIREBIRD

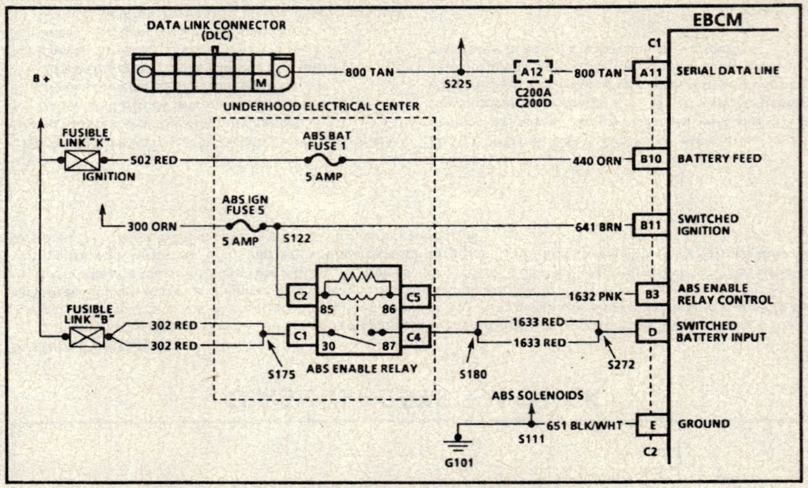

DIAGNOSTIC SYSTEM CHECK

System Description:

The diagnostic system check is an organized approach to identifying a problem created by an Antilock Brake System (ABS) malfunction. It must be the starting point for any ABS complaint diagnosis because it directs the service technician to the next logical step in diagnosing the complaint.

Serial data is transmitted by the EBCM through terminal "A11". The EBCM is supplied switched ignition voltage through terminal "B11", battery voltage through terminal "B10" and switched battery voltage through terminal "D". The EBCM ground is provided through terminal "E".

Chart Test Description: Number(s) below refer to circled number(s) on the diagnostic chart.

10. Checks for possible short to ground in the battery feed circuitry.
11. Checks for possible short to ground in the EBCM.
12. Ensures short to ground is not due to physical damage of the circuitry.

Diagnostic Aids: Excessive resistance in the ground or power supply circuits will not allow communication with the EBCM. If communication with the EBCM is not possible, ensure the ABS ground connection is good and that there is no excessive resistance in any of the power supply circuits.

DIAGNOSTIC CHART SYSTEM CHECK — 1994–95 CAMARO AND FIREBIRD

DIAGNOSTIC SYSTEM CHECK

(Page 2 of 4)

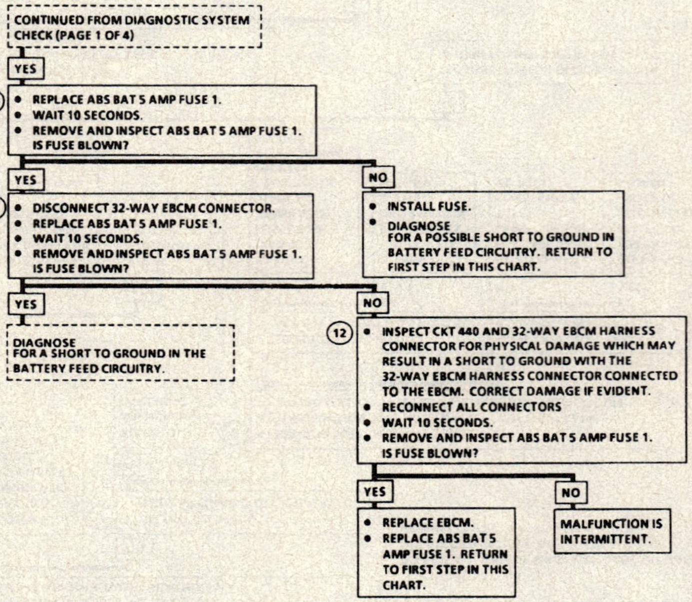

DIAGNOSTIC CHART SYSTEM CHECK — 1994–95 CAMARO AND FIREBIRD

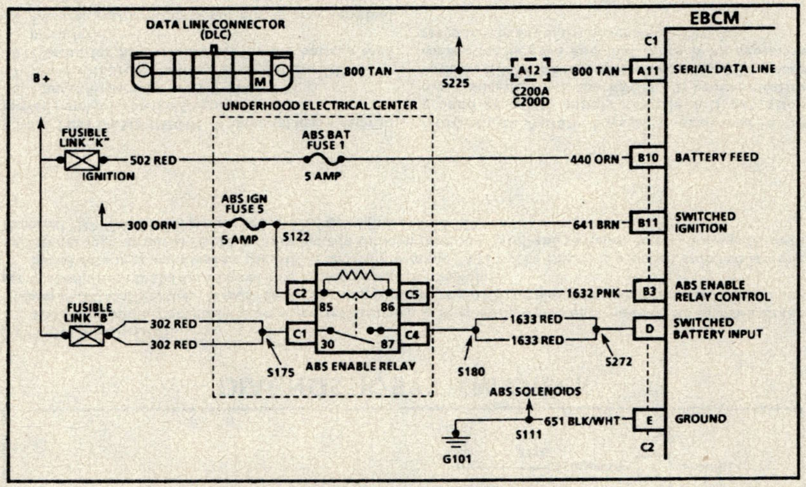

DIAGNOSTIC SYSTEM CHECK

System Description:
The diagnostic system check is an organized approach to identifying a problem created by an Antilock Brake System (ABS) malfunction. It must be the starting point for any ABS complaint diagnosis because it directs the service technician to the next logical step in diagnosing the complaint.

Serial data is transmitted by the EBCM through terminal "A11". The EBCM is supplied switched ignition voltage through terminal "B11", battery voltage through terminal "B10" and switched battery voltage through terminal "D". The EBCM ground is provided through terminal "E".

Chart Test Description: Number(s) below refer to circled number(s) on the diagnostic chart.

13. Checks for possible short to ground in the switched ignition circuitry.
14. Isolates the short to ground to either the circuitry or the EBCM.
15. Ensures short to ground is not due to physical damage of the circuitry.

Diagnostic Aids: Excessive resistance in the ground or power supply circuits will not allow communication with the EBCM. If communication with the EBCM is not possible, ensure the ABS ground connection is good and that there is no excessive resistance in any of the power supply circuits.

DIAGNOSTIC CHART SYSTEM CHECK — 1994–95 CAMARO AND FIREBIRD

DIAGNOSTIC SYSTEM CHECK

(Page 3 of 4)

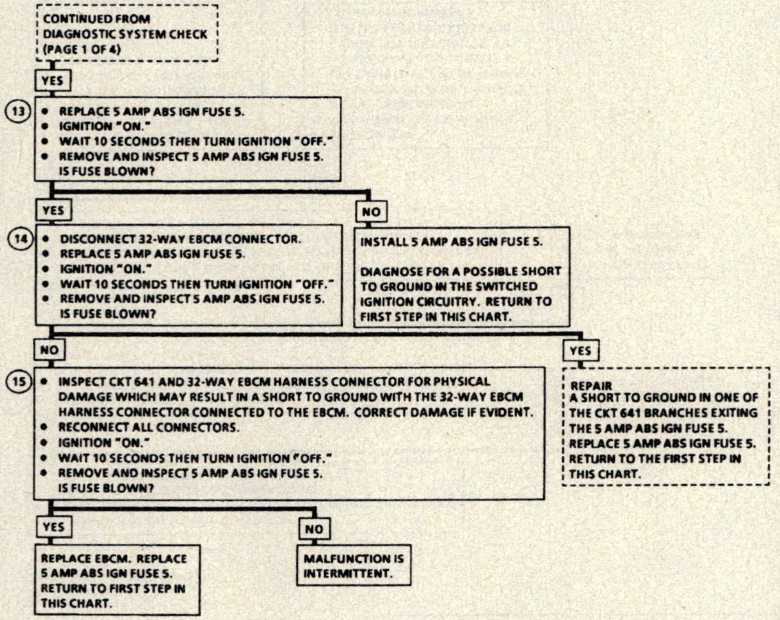

DIAGNOSTIC CHART SYSTEM CHECK — 1994–95 CAMARO AND FIREBIRD

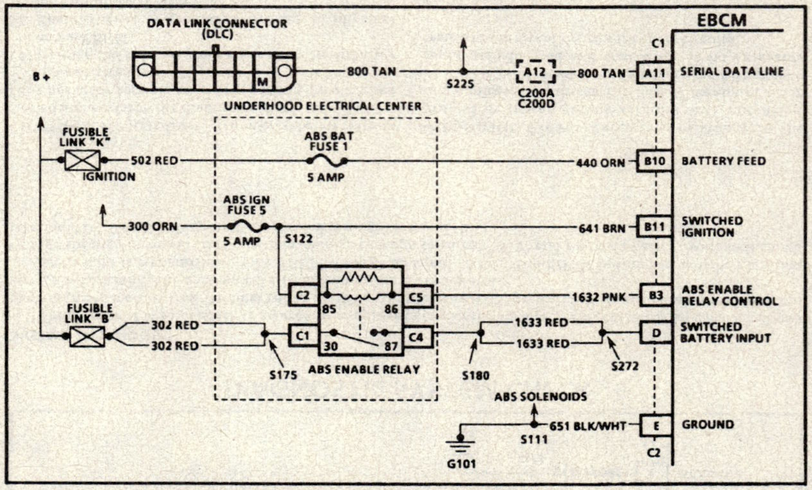

DIAGNOSTIC SYSTEM CHECK

System Description:

The diagnostic system check is an organized approach to identifying a problem created by an Antilock Brake System (ABS) malfunction. It must be the starting point for any ABS complaint diagnosis because it directs the service technician to the next logical step in diagnosing the complaint.

Serial data is transmitted by the EBCM through terminal "A11". The EBCM is supplied switched ignition voltage through terminal "B11", battery voltage through terminal "B10" and switched battery voltage through terminal "D". The EBCM ground is provided through terminal "E".

Chart Test Description: Number(s) below refer to circled number(s) on the diagnostic chart.

16. Checks for an open in CKT 440 or 502.
17. Checks for proper battery voltage at the EBCM.
18. Checks for high resistance in the battery feed circuitry.
19. Ensures malfunction is not due to poor terminal contact.
20. Checks for high resistance in the switched ignition circuit between the fuse block and the EBCM.

Diagnostic Aids: Excessive resistance in the ground or power supply circuits will not allow communication with the EBCM. If communication with the EBCM is not possible, ensure the ABS ground connection is good and that there is no excessive resistance in any of the power supply circuits.

DIAGNOSTIC CHART SYSTEM CHECK — 1994–95 CAMARO AND FIREBIRD

DIAGNOSTIC SYSTEM CHECK

(Page 4 of 4)

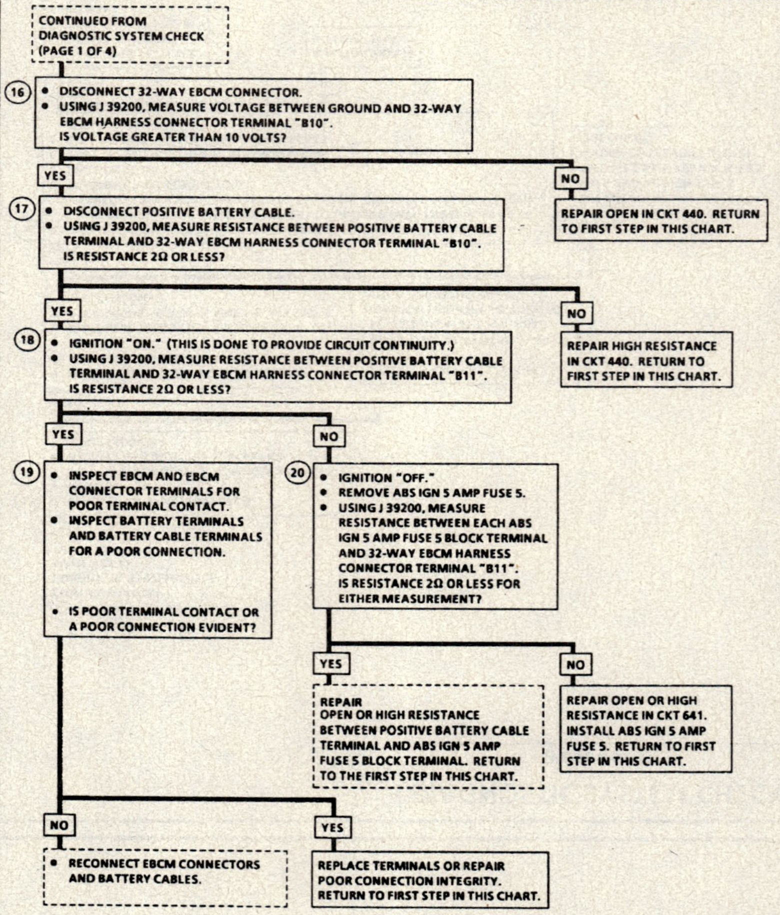

DIAGNOSTIC TROUBLE CODE CHART — 1994–95 CAMARO AND FIREBIRD

DIAGNOSTIC TROUBLE CODE AND SYMPTOM TABLE

CHART	SYMPTOM
A	"ABS INOP" Lamp "ON" or "OFF" Constantly, No DTCs Stored
B	"ABS INOP" Lamp "ON" Intermittently, No DTCs Stored
C	"Low TRAC" Indicator Lamp "ON" With Ignition "OFF," No DTCs Stored
D	Tech 1 Displays Undefined DTCs
DIAGNOSTIC TROUBLE CODE	**DESCRIPTION**
11	ABS Warning or Low Trac Lamp Ckt Malfunction (1 of 2)
14	ABS Enable Relay Contact Circuit Open (1 of 3)
15	ABS Rly Contact Ckt Short to Batt or Always Closed
16	ABS Enable Relay Coil Circuit Open
17	ABS Enable Relay Coil Circuit Shorted to Gnd
18	ABS Enable Relay Coil Circuit Shorted to Batt
21	Left Front Wheel Speed = 0
22	Right Front Wheel Speed = 0
24	Rear Wheel Speed = 0
25	Left Front Excessive Wheel Speed Variation
26	Right Front Excessive Wheel Speed Variation
28	Rear Excessive Wheel Speed Variation
32	LF Wheel Speed Snr Ckt Open or Short to Gnd/Batt (1 fo 3)
33	RF Wheel Speed Snr Ckt Open or Short to Gnd/Batt (1 of 3)
35	RR Wheel Speed Snr Ckt Open or Short to Gnd/Batt (1 of 3)
36	Low System Voltage (1 of 2)
37	High System Voltage
38	Left Front ESB Will Not Hold Motor
41	Right Front ESB Will Not Hold Motor
42	Rear ESB Will Not Hold Motor
44	Left Front ABS Channel Will Not Move
45	Right Front ABS Channel Will Not Move
46	Rear ABS Channel Will Not Move
47	Left Front ABS Motor Free Spins (1 of 2)
48	Right Front ABS Motor Free Spins (1 of 2)
51	Rear ABS Motor Free Spins (1 of 2)
52	Left Front ABS Channel in Release Too Long
53	Right Front ABS Channel In Release Too Long
54	Rear ABS Channel in Release Too Long
55	EBCM Malfunction
56	Left Front ABS Motor Circuit Open
57	Left Front ABS Motor Circuit Shorted to Gnd
58	Left Front ABS Motor Circuit Shorted to Batt

DIAGNOSTIC TROUBLE CODE CHART — 1994–95 CAMARO AND FIREBIRD

DIAGNOSTIC TROUBLE CODE AND SYMPTOM TABLE

DIAGNOSTIC TROUBLE CODE	DESCRIPTION
61	Right Front ABS Motor Circuit Open
62	Right Front ABS Motor Circuit Shorted to Gnd
63	Right Front ABS Motor Circuit Shorted to Batt
64	Rear ABS Motor Circuit Open
65	Rear ABS Motor Circuit Shorted to Ground
66	Rear ABS Motor Circuit Shorted to Battery
76	Left Front Sol Circuit Open or Shorted to Batt
77	Left Front Solenoid Circuit Shorted to Gnd
78	Right Front Sol Circuit Open or Shorted to Batt
81	Right Front Solenoid Circuit Shorted to Gnd
82	Calibration Malfunction
86	EBCM Turned "ON" the Red "BRAKE" Warning Lamp
87	Red "BRAKE" Warning Lamp Circuit Open
88	Red "BRAKE" Warning Lamp Ckt Shorted to Batt
91	Open Brake Switch During Deceleration
92	Open Brake Switch When ABS Was Required
93	DTC 91 or 92 Set in Current or Prev. Ign. Cycle
94	Brake Switch Contacts Always Closed
95	Brake Switch Circuit Open
96	Rear Brake Lamp Circuit Open

ELECTRICAL CONNECTOR TERMINAL IDENTIFICATION — 1994–95 CAMARO AND FIREBIRD

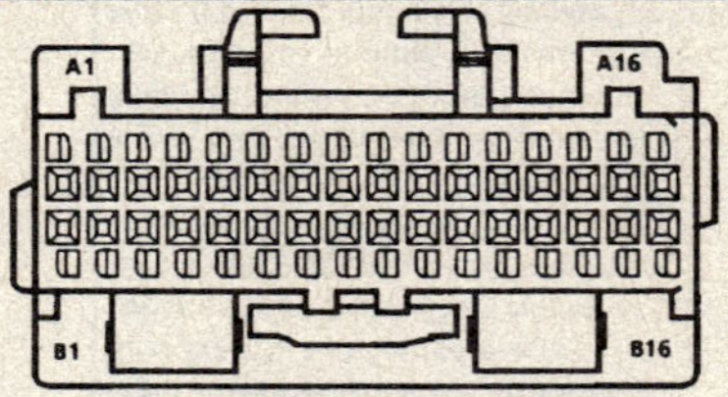

EBCM 32-WAY CONNECTOR

CONNECTOR TERMINAL	CIRCUIT NO.	COLOR	CIRCUIT DESCRIPTION
A1	PLUGGED		NOT USED
A2	PLUGGED		NOT USED
A3	833	TAN	RF WHEEL SPEED SIGNAL LOW
A4	872	DK GRN	RF WHEEL SPEED SIGNAL HIGH
A5	883	WHT	REAR WHEEL SPEED SIGNAL LOW
A6	882	BRN	REAR WHEEL SPEED SIGNAL HIGH
A7	873	YEL	LF WHEEL SPEED SIGNAL LOW
A8	830	LT BLU	LF WHEEL SPEED SIGNAL HIGH
A9	1289	LT BLU	RF ABS SOLENOID CONTROL
A10	PLUGGED		NOT USED
A11	800	TAN	SERIAL DATA LINE
A12	PLUGGED		NOT USED
A13	PLUGGED		NOT USED
A14	PLUGGED		NOT USED
A15	PLUGGED		NOT USED
A16	PLUGGED		NOT USED
B1	1288	DK GRN	LF ABS SOLENOID CONTROL
B2	867	LT GRN	ABS WARNING LAMP CONTROL
B3	1632	PNK	ABS ENABLE RELAY CONTROL
B4	33	TAN/WHT	BRAKE WARNING LAMP CONTROL
B5	1537	DK BLU	ABS ACTIVE LAMP CONTROL
B6	PLUGGED		NOT USED
B7	PLUGGED		NOT USED
B8	PLUGGED		NOT USED
B9	PLUGGED		NOT USED
B10	440	ORN	BATTERY FEED
B11	641	BRN	SWITCHED IGNITION
B12	820	YEL	BRAKE SWITCH INPUT
B13	PLUGGED		NOT USED
B14	PLUGGED		NOT USED
B15	PLUGGED		NOT USED
B16	PLUGGED		NOT USED

ELECTRICAL CONNECTOR TERMINAL IDENTIFICATION — 1994–95 CAMARO AND FIREBIRD

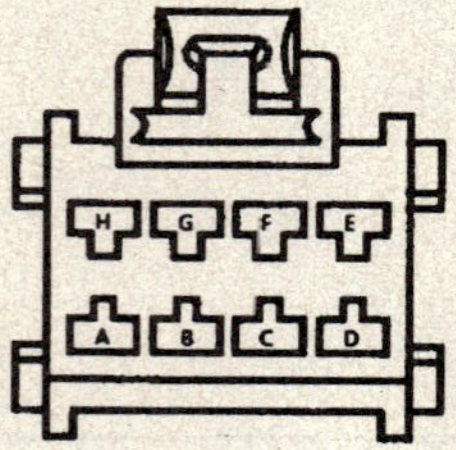

EBCM 8-WAY CONNECTOR

PIN	CIRCUIT NO.	COLOR	CIRCUIT
A	1280	BLK	LF ABS MOTOR HIGH
B	1281	PNK	LF ABS MOTOR LOW
C	1284	DK GRN	REAR ABS MOTOR HIGH
D	1633	RED	SWITCHED BATTERY
E	651	BLK/WHT	GROUND
F	1285	ORN	REAR ABS MOTOR LOW
G	1282	PPL	RF ABS MOTOR HIGH
H	1283	BLK	RF ABS MOTOR LOW

DIAGNOSTIC CHART A — 1994–95 CAMARO AND FIREBIRD

CHART A
"ABS INOP" LAMP "ON" OR "OFF" CONSTANTLY, NO DTCs STORED

NOTICE: DIAGNOSTIC SYSTEM CHECK MUST BE COMPLETED BEFORE USING THIS CHART.

- IGNITION "OFF."
- INSTALL TECH 1.
- IGNITION "ON."
- USING TECH 1, SELECT "F4: MISC. TESTS."
- SELECT "F8: LAMP TEST" AND ATTEMPT TO TURN "ON" THE "ABS INOP" LAMP.

DID THE "ABS INOP" LAMP TURN "ON"?

- **YES:**
  - USING TECH 1 LAMP TEST, ATTEMPT TO "FLASH" THE "ABS INOP" LAMP.

  DID THE "ABS INOP" LAMP "FLASH"?
  - **YES:** MALFUNCTION IS NOT PRESENT AT THIS TIME. INTERMITTENTS COULD BE CAUSED BY INCORRECT WIRING HARNESS ROUTING OR LOOSE CONNECTIONS. INSPECT ALL CONNECTORS AND TERMINALS FOR POOR TERMINAL CONTACT AND EVIDENCE OF CORROSION. REPLACE ALL TERMINALS THAT EXHIBIT SIGNS OF POOR TERMINAL CONTACT OR CORROSION.
  - **NO:** INSPECT THE "ABS INOP" BULB TO VERIFY BULB IS GOOD. REPLACE ANY BLOWN BULBS. IF BULB IS GOOD, PERFORM FURTHER DIAGNOSIS OF INSTRUMENT CLUSTER.
- **NO:** INSPECT THE "ABS INOP" BULB TO VERIFY BULB IS GOOD. REPLACE ANY BLOWN BULBS. IF BULB IS GOOD, PERFORM FURTHER DIAGNOSIS OF INSTRUMENT CLUSTER.

DIAGNOSTIC CHART B — 1994–95 CAMARO AND FIREBIRD

CHART B
"ABS INOP" LAMP "ON" INTERMITTENTLY, NO DTCs STORED

NOTICE: DIAGNOSTIC SYSTEM CHECK MUST BE COMPLETED BEFORE USING THIS CHART.

1
- IGNITION "OFF."
- DISCONNECT 8-WAY EBCM CONNECTOR.
- DISCONNECT NEGATIVE BATTERY TERMINAL.
- DISCONNECT POSITIVE BATTERY TERMINAL.
- USING J 39200, MEASURE THE RESISTANCE BETWEEN THE NEGATIVE BATTERY CABLE TERMINAL AND TERMINAL "E" OF THE 8-WAY EBCM HARNESS CONNECTOR.
- OBSERVE MEASURED RESISTANCE WHILE "WIGGLING" WIRE HARNESS AND CONNECTORS.

IS RESISTANCE ALWAYS 2Ω OR LESS?

- **NO:** REPAIR INTERMITTENT OPEN OR HIGH RESISTANCE IN CKT 651.
- **YES:** Go to step 2.

2
- DISCONNECT 32-WAY EBCM CONNECTOR.
- USING J 39200, MEASURE THE RESISTANCE BETWEEN POSITIVE BATTERY CABLE TERMINAL AND TERMINAL "B10" OF THE 32-WAY EBCM HARNESS CONNECTOR.
- OBSERVE MEASURED RESISTANCE WHILE "WIGGLING" WIRE HARNESS AND CONNECTORS.

IS RESISTANCE ALWAYS 2Ω OR LESS?

- **NO:** REPAIR INTERMITTENT OPEN OR HIGH RESISTANCE IN CKTs 440 OR 502.
- **YES:** Go to step 3.

3
- TURN IGNITION SWITCH "ON" (THIS IS DONE TO COMPLETE THE IGNITION CIRCUIT).
- USING J 39200, MEASURE THE RESISTANCE BETWEEN POSITIVE BATTERY CABLE TERMINAL AND TERMINAL "B11" OF THE 32-WAY EBCM HARNESS CONNECTOR.
- OBSERVE MEASURED RESISTANCE WHILE "WIGGLING" WIRE HARNESS AND CONNECTORS.

IS RESISTANCE ALWAYS 2Ω OR LESS?

- **NO:** REPAIR INTERMITTENT OPEN OR HIGH RESISTANCE IN CKTs 641 OR 300.
- **YES:** Go to step 4.

4
- CHECK POSITIVE BATTERY TERMINAL AND POSITIVE BATTERY CABLE TERMINAL FOR A POOR CONNECTION.
- INSPECT TERMINALS "B11" AND "B10" OF THE 32-WAY EBCM CONNECTOR AND TERMINAL "E" OF THE 8-WAY EBCM CONNECTOR FOR POOR TERMINAL CONTACT AND EVIDENCE OF CORROSION.

IS THERE EVIDENCE OF POOR TERMINAL CONTACT OR CORROSION?

- **YES:** REPLACE TERMINALS THAT EXHIBIT POOR TERMINAL CONTACT OR SIGNS OF CORROSION. REPAIR CONNECTION INTEGRITY IF POOR CONTACT EXISTS AT BATTERY.
- **NO:** Go to step 5.

5
- IGNITION "OFF."
- RECONNECT ALL CONNECTORS.
- IGNITION "ON."

IS "ABS INOP" LAMP "ON" INTERMITTENTLY WITH NO DTCs SET?

- **YES:** REPLACE EBCM.
- **NO:** MALFUNCTION IS NOT PRESENT AT THIS TIME. INTERMITTENTS MAY BE CAUSED BY INCORRECT WIRING HARNESS ROUTING OR LOOSE GROUND CONNECTIONS.

DIAGNOSTIC CHART C — 1994–95 CAMARO AND FIREBIRD

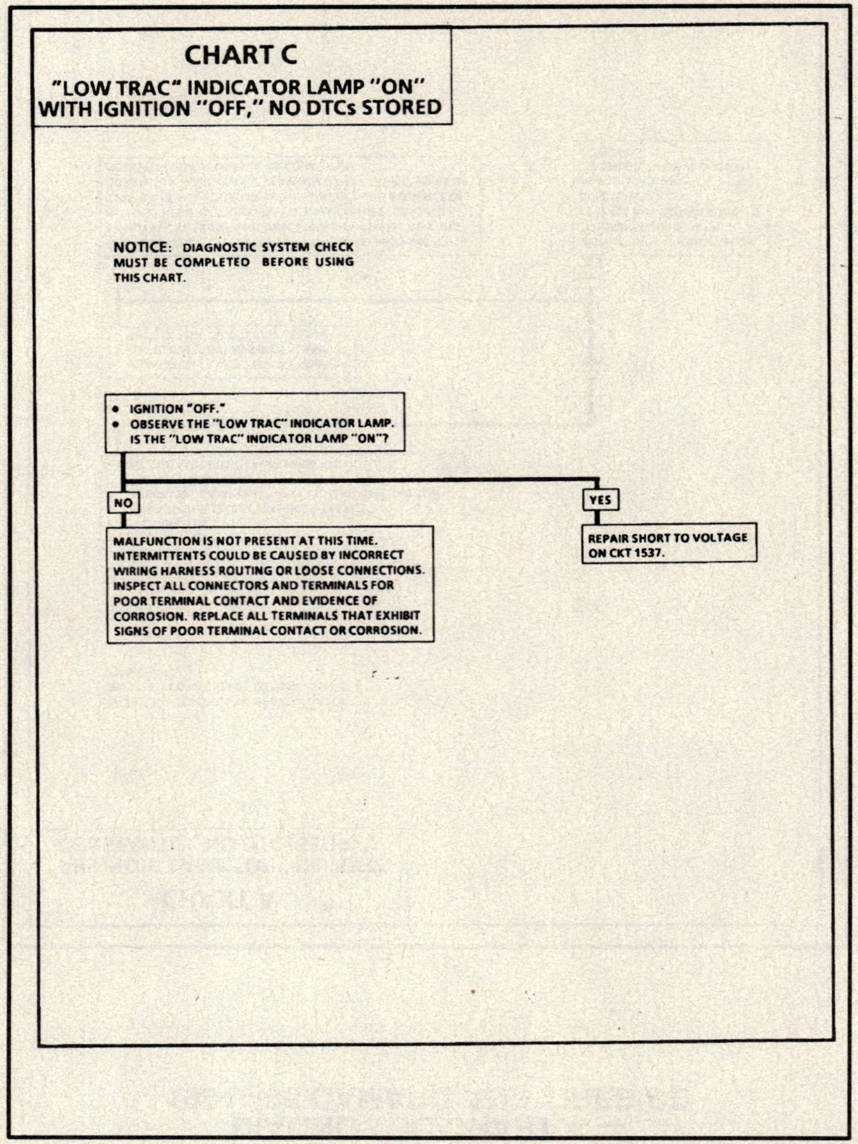

DIAGNOSTIC CHART D — 1994–95 CAMARO AND FIREBIRD

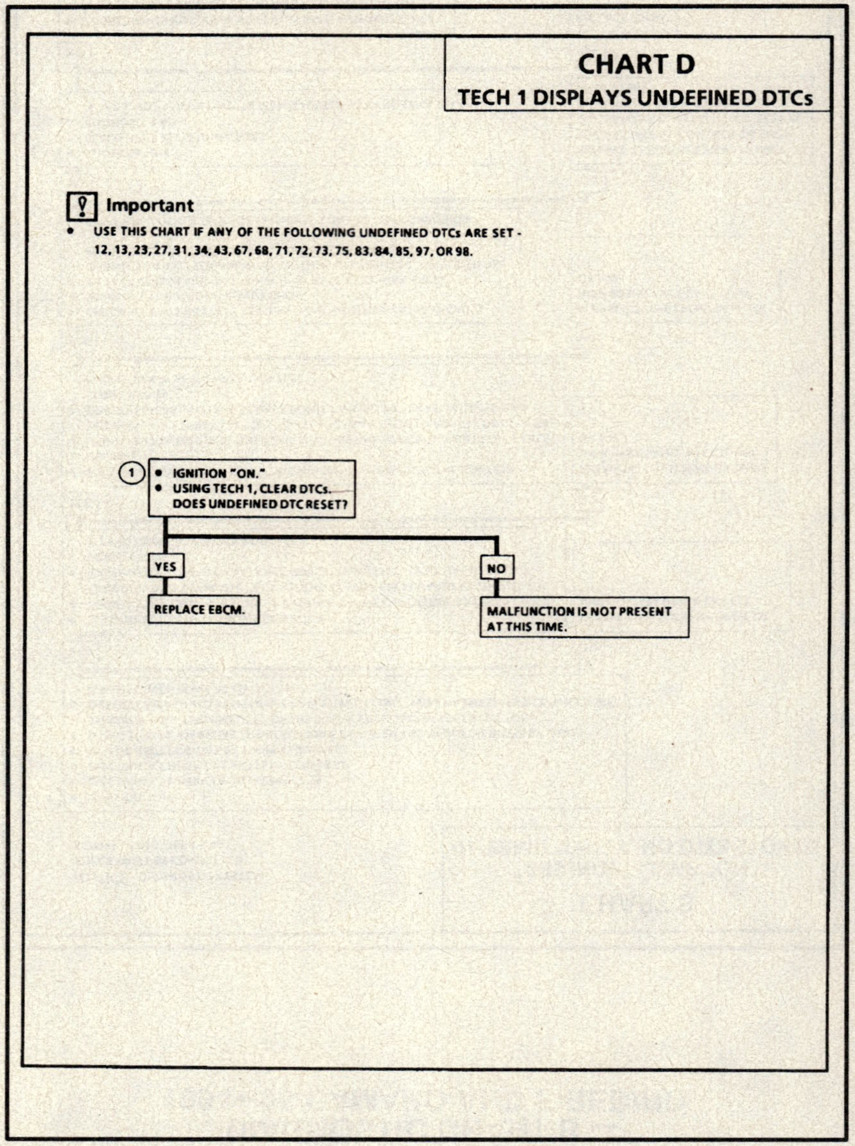

DIAGNOSTIC CHART DTC 11 — 1994–95 CAMARO AND FIREBIRD

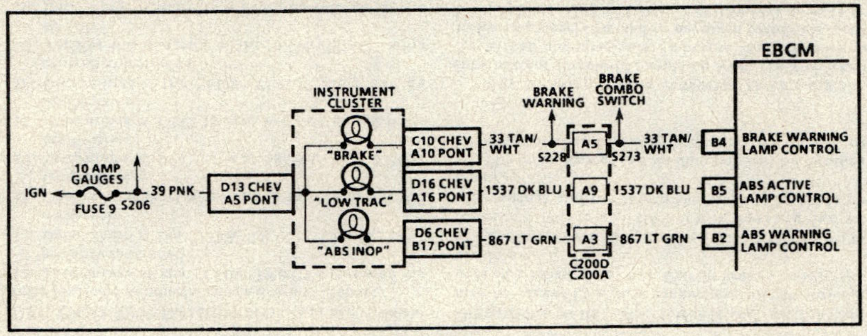

DTC 11
(Page 1 of 2)
ABS WARNING OR LOW TRAC LAMP CKT MALFUNCTION

Circuit Description:

This DTC checks the state of both the "ABS INOP" lamp and the "LOW TRAC" lamp to identify a situation in which the driver could not be alerted to a system malfunction or of possible slippery driving conditions. Due to an integral lamp driver module within the instrument panel cluster, the EBCM must provide a ground to turn the "ABS INOP" lamp "OFF." The "LOW TRAC" lamp circuit is such that the EBCM provides a ground to turn "ON" the "LOW TRAC" lamp.

DTC Will Set When: DTC 11 can only be set during the three second bulb check or when either of the lamps are commanded "ON." If the EBCM cannot control either the "ABS INOP" lamp or the "LOW TRAC" lamp for two seconds, a malfunction exists.

Action Taken: A malfunction DTC is stored; however, ABS is not disabled.

DTC Chart Test Description: Number(s) below refer to circled number(s) on the diagnostic chart.

1. Determines if the "ABS INOP" lamp can be properly controlled.
2. Determines if the lamp is "ON" constantly.
3. Checks for other DTCs which could turn "ON" the "ABS INOP" lamp.
4. Determines if the "ABS INOP" lamp can be turned "OFF" manually.
5. Checks the integrity of the "ABS INOP" lamp circuitry.
6. Checks for a short to voltage on the "ABS INOP" lamp circuitry.
7. Determines if a malfunctioning EBCM is the cause of DTC 11.
8. Verifies that the other I/P indicator lamps function properly.
9. Determines if the "ABS INOP" lamp can be turned "ON" manually.
10. Checks for a short to ground in the "ABS INOP" lamp circuitry.
11. Verifies condition of the 10 amp "GAGES" Fuse 9.
12. Determines if DTC 11 was set due to an EBCM malfunction.

Diagnostic Aids: An "Intermittent" malfunction may be caused by a poor connection, rubbed through wire insulation, or a wire that is broken inside the insulation.

The lamp test function of the Tech 1 may be used to command the lamp "ON" while looking for an intermittent malfunction in the "ABS INOP" lamp circuitry.

The frequency of the malfunction can be checked by using the enhanced diagnostic function of the Tech 1.

Any circuitry that is suspected as causing the intermittent complaint should be thoroughly checked for backed out terminals, improper mating, broken locks, improperly formed or damaged terminals, poor terminal to wiring connections or physical damage to the wiring harness.

DIAGNOSTIC CHART DTC 11 — 1994–95 CAMARO AND FIREBIRD

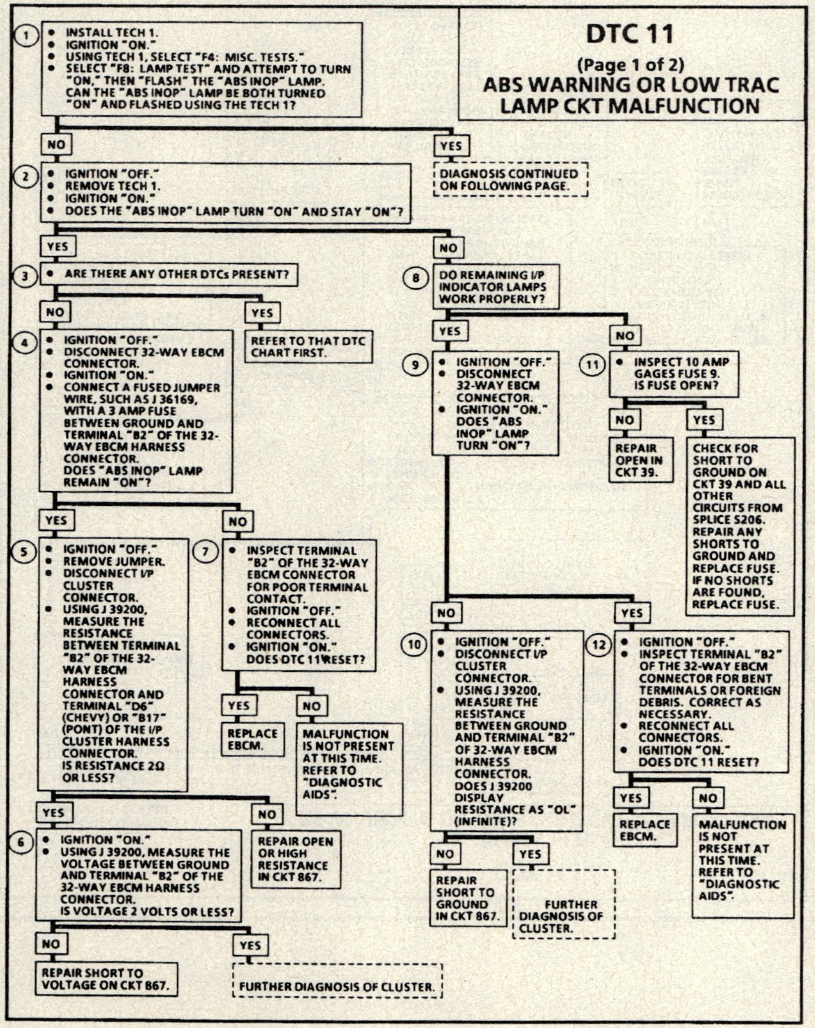

DIAGNOSTIC CHART DTC 11 — 1994–95 CAMARO AND FIREBIRD

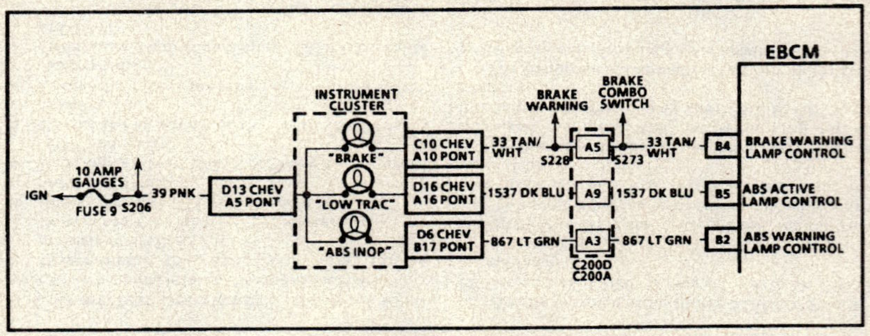

DTC 11

(Page 2 of 2)

ABS WARNING OR LOW TRAC LAMP CKT MALFUNCTION

Circuit Description:

This DTC checks the state of both the "ABS INOP" lamp and the "LOW TRAC" lamp to identify a situation in which the driver could not be alerted to a system malfunction or of possible slippery driving conditions. Due to an integral lamp driver module within the instrument panel cluster, the EBCM must provide a ground to turn the "ABS INOP" lamp "OFF." The "LOW TRAC" lamp circuit is such that the EBCM provides a ground to turn "ON" the "LOW TRAC" lamp.

DTC Will Set When: DTC 11 can only be set during the three second bulb check or when either of the lamps are commanded "ON." If the EBCM cannot control either the "ABS INOP" lamp or the "LOW TRAC" lamp for two seconds, a malfunction exists.

Action Taken: A malfunction DTC is stored; however, ABS is not disabled.

DTC Chart Test Description: Number(s) below refer to circled number(s) on the diagnostic chart.

13. Determines if the "LOW TRAC" lamp can be properly controlled.
14. Determines if the "LOW TRAC" lamp is "ON" constantly.
15. Checks for a short to ground in the "LOW TRAC" lamp circuitry.
16. Determines if DTC 11 was set due to a EBCM malfunction.
17. Determines if DTC 11 was set due to a EBCM malfunction.
18. Determines if the "LOW TRAC" lamp can be controlled manually.
19. Verifies the integrity of the "LOW TRAC" lamp circuitry.
20. Determines if there is a short to voltage in the "LOW TRAC" lamp circuitry.
21. Determines if DTC 11 was set due to an EBCM malfunction.

Diagnostic Aids: An "Intermittent" malfunction may be caused by a poor connection, rubbed through wire insulation, or a wire that is broken inside the insulation.

The lamp test function of the Tech 1 may be used to command the lamp "ON" while looking for an intermittent malfunction in the "LOW TRAC" lamp circuitry.

The frequency of the malfunction can be checked by using the enhanced diagnostic function of the Tech 1.

Any circuitry that is suspected as causing the intermittent complaint should be thoroughly checked for backed out terminals, improper mating, broken locks, improperly formed or damaged terminals, poor terminal to wiring connections or physical damage to the wiring harness.

DIAGNOSTIC CHART DTC 11 — 1994–95 CAMARO AND FIREBIRD

DTC 11

(Page 2 of 2)

ABS WARNING OR LOW TRAC CKT MALFUNCTION

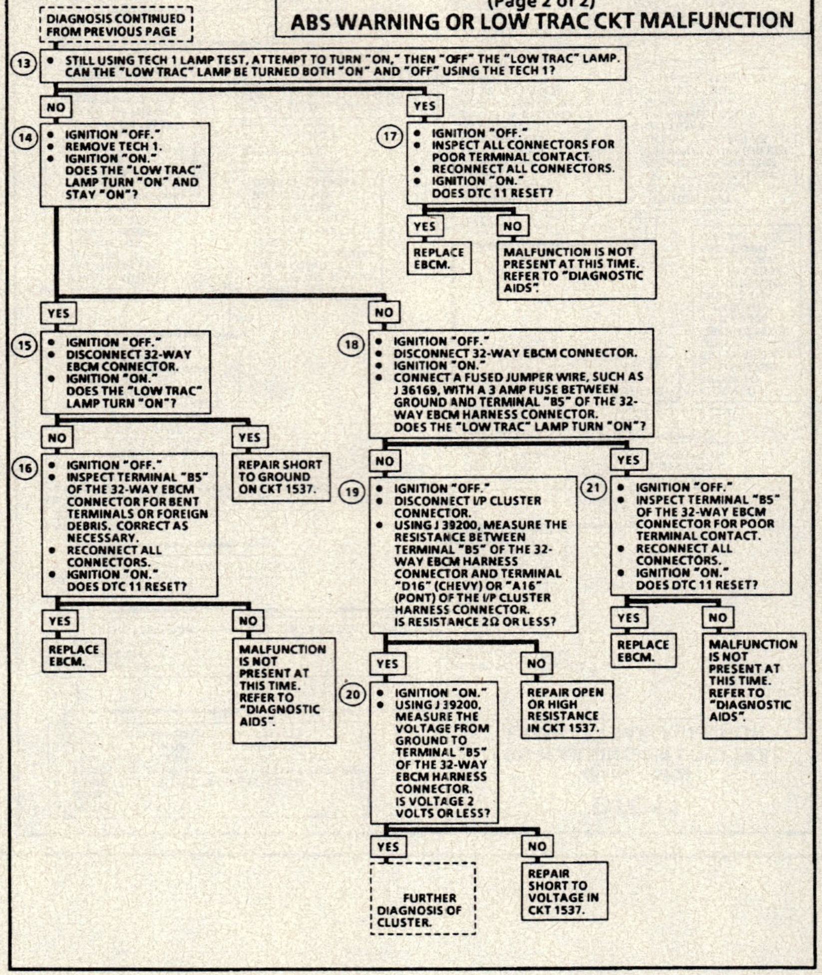

DIAGNOSTIC CHART DTC 14 — 1994–95 CAMARO AND FIREBIRD

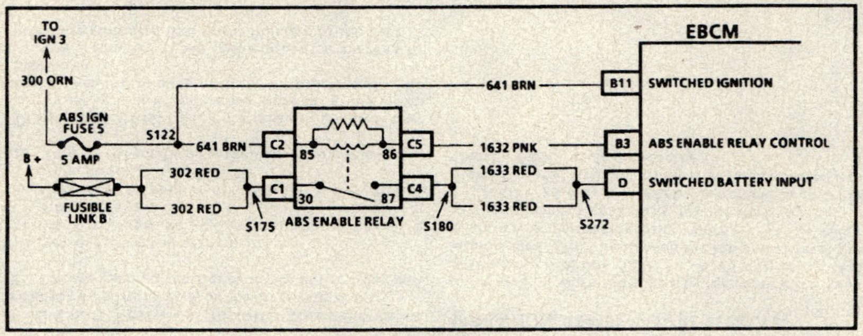

DTC 14

(Page 1 of 3)

ABS ENABLE RELAY CONTACT CIRCUIT OPEN

Circuit Description:
Ignition voltage is supplied through terminal "C2" of the ABS enable relay. The EBCM then is able to energize the pull-in coil by completing the ground circuit at terminal "B3" of the EBCM. The magnetic field created closes the ABS enable relay contacts and allows battery voltage and current to be supplied to the EBCM, which supplies power to the motors and solenoids.

DTC Will Set When: DTC 14 can be set anytime after the EBCM commands the ABS enable relay "ON" (the relay is first commanded "ON" during the three second bulb check). This test monitors the availability of current/voltage to the motors and solenoids. This malfunction indicates voltage is not available and would therefore not allow ABS operation if required.

Action Taken: A malfunction DTC is stored, ABS is disabled and the "ABS INOP" lamp is turned "ON."

DTC Chart Test Description: Number(s) below refer to circled number(s) on the diagnostic chart.

1. Checks to see if DTC 16 is also set as current or history. If DTC 16 is also set, proceed to that chart first.
2. Verifies that the malfunction is currently present.
3. Verifies that the relay can be turned "ON" and provide voltage to the EBCM.
4. Checks for high resistance in the switched battery circuit from the relay to the EBCM.
5. Checks for high resistance in the switched battery circuit from the battery to the relay.
6. Checks for high resistance in the relay contacts.
7. Checks for poor terminal contact or corrosion at the connectors.
8. Determines if the EBCM is the cause of the malfunction.

Diagnostic Aids: An "Intermittent" problem may be caused by a poor connection, rubbed through wire insulation, or a wire that is broken inside the insulation.

The frequency of the problem can be checked by using the enhanced diagnostic function of the Tech 1.

Any circuitry that is suspected as causing the "Intermittent" complaint should be thoroughly checked for backed out terminals, improper mating, broken locks, improperly formed or damaged terminals, poor terminal to wiring connections or physical damage to the wiring harness.

VIBRATION, TEMPERATURE EFFECTS:

Check for vibration effects by performing the relay test function of the Tech 1. With the relay test commanded "ON," lightly tap the top and sides of the relay while monitoring relay voltage. If the relay voltage changes significantly, replace the relay.

If DTC 14 only sets when the vehicle is initially started in cold ambient conditions (temperature less than 32°F-0°C), replace the relay.

DIAGNOSTIC CHART DTC 14 — 1994–95 CAMARO AND FIREBIRD

DTC 14

(Page 1 of 3)

ABS ENABLE RELAY CONTACT CIRCUIT OPEN

(1) IS DTC 16 ALSO PRESENT OR IN HISTORY?

- YES: PROCEED TO DTC 16 CHART FIRST.
- NO: go to 2.

(2)
- IGNITION "ON," ENGINE "OFF."
- USING TECH 1, SELECT DATA LIST AND OBSERVE EBCM BATTERY VOLTAGE.

DOES TECH 1 INDICATE THAT EBCM BATTERY VOLTAGE IS GREATER THAN 10 VOLTS?

- YES: MALFUNCTION IS NOT PRESENT AT THIS TIME. REFER TO "DIAGNOSTIC AIDS".
- NO: go to 3.

(3)
- IGNITION "OFF."
- DISCONNECT 32-WAY AND 8-WAY EBCM CONNECTORS.
- CONNECT A FUSED JUMPER, SUCH AS J 36169 WITH A 3 AMP FUSE BETWEEN GROUND AND TERMINAL "B3" OF THE 32-WAY EBCM HARNESS CONNECTOR.
- IGNITION "ON."
- USING J 39200, MEASURE THE VOLTAGE BETWEEN GROUND AND TERMINAL "D" OF THE 8-WAY EBCM HARNESS CONNECTOR.

IS VOLTAGE 10 VOLTS OR GREATER?

- NO: DIAGNOSIS CONTINUED ON NEXT PAGE.
- YES: go to 4.

(4)
- IGNITION "OFF."
- DISCONNECT JUMPER USED IN STEP 3.
- DISCONNECT ABS ENABLE RELAY.
- USING J 39200, MEASURE RESISTANCE BETWEEN TERMINAL "C4" OF THE ABS ENABLE RELAY HARNESS CONNECTOR AND TERMINAL "D" OF THE 8-WAY EBCM HARNESS CONNECTOR.

IS RESISTANCE 2Ω OR LESS?

- NO: REPAIR HIGH RESISTANCE IN CKT 1633.
- YES: go to 5.

(5)
- DISCONNECT NEGATIVE BATTERY CABLE.
- DISCONNECT POSITIVE BATTERY CABLE.
- USING J 39200, MEASURE THE RESISTANCE BETWEEN POSITIVE BATTERY CABLE TERMINAL AND TERMINAL "C1" OF THE ABS ENABLE RELAY HARNESS CONNECTOR.

IS RESISTANCE 2Ω OR LESS?

- NO: REPAIR HIGH RESISTANCE IN CKT 302.
- YES: go to 6.

(6)
- USING A JUMPER WIRE, CONNECT TERMINAL "86" ON THE ABS ENABLE RELAY TO GROUND.
- USING A FUSED JUMPER WIRE, SUCH AS J 36169, WITH A 3 AMP FUSE, CONNECT TERMINAL "85" ON THE ABS ENABLE RELAY TO B+.
- USING J 39200, MEASURE THE RESISTANCE BETWEEN TERMINALS "30" AND "87" ON THE ABS ENABLE RELAY.

IS RESISTANCE 2Ω OR LESS?

- NO: REPLACE ABS ENABLE RELAY.
- YES: go to 7.

(7)
- INSPECT ABS ENABLE RELAY AND ABS ENABLE RELAY HARNESS CONNECTOR FOR A POOR CONNECTION OR POOR TERMINAL CONTACT.
- INSPECT EBCM AND EBCM CONNECTOR TERMINALS "B3" AND "D" FOR POOR TERMINAL CONTACT.
- INSPECT POSITIVE BATTERY TERMINAL AND POSITIVE BATTERY CABLE TERMINAL FOR A POOR CONNECTION.

IS POOR TERMINAL CONTACT OR A POOR CONNECTION EVIDENT?

- YES: REPLACE TERMINALS THAT EXHIBIT POOR TERMINAL CONTACT OR EVIDENCE OF CORROSION.
- NO: go to 8.

(8)
- RECONNECT ALL CONNECTORS.
- IGNITION "ON."

DOES DTC 14 RESET?

- YES: REPLACE EBCM.
- NO: MALFUNCTION IS INTERMITTENT. REFER TO "DIAGNOSTIC AIDS".

DIAGNOSTIC CHART DTC 14 — 1994–95 CAMARO AND FIREBIRD

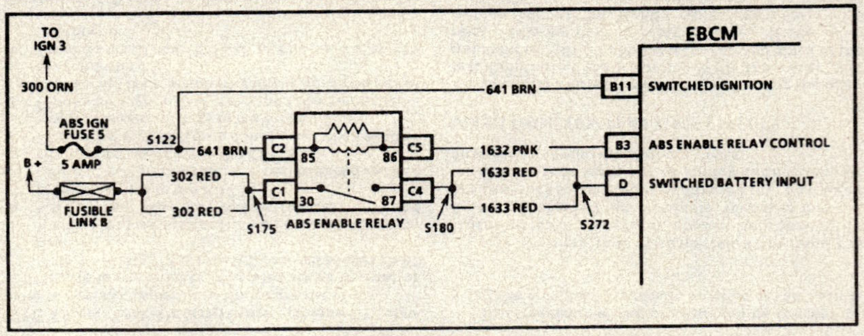

DTC 14

(Page 2 of 3)

ABS ENABLE RELAY CONTACT CIRCUIT OPEN

Circuit Description:

Ignition voltage is supplied through terminal "C2" of the ABS enable relay. The EBCM then is able to energize the pull-in coil by completing the ground circuit at terminal "B3" of the EBCM. The magnetic field created closes the ABS enable relay contacts and allows battery voltage and current to be supplied to the EBCM, which supplies power to the motors and solenoids.

DTC Will Set When: DTC 14 can be set anytime after the EBCM commands the ABS enable relay "ON" (the relay is first commanded "ON" during the three second bulb check). This test monitors the availability of current/voltage to the motors and solenoids. This malfunction indicates voltage is not available and would therefore not allow ABS operation if required.

Action Taken: A malfunction DTC is stored, ABS is disabled and the "ABS INOP" lamp is turned "ON."

DTC Chart Test Description: Number(s) below refer to circled number(s) on the diagnostic chart.

9. Checks for high resistance in the entire relay coil circuit.
10. Verifies proper battery condition.
11. Verifies proper voltage is available at the ABS enable relay.
12. Checks for an open in the switched battery circuit from the relay to the EBCM.
13. Verifies the condition of the ABS enable relay.

Diagnostic Aids: An "Intermittent" problem may be caused by a poor connection, rubbed through wire insulation, or a wire that is broken inside the insulation.

The frequency of the problem can be checked by using the enhanced diagnostic function of the Tech 1.

Any circuitry that is suspected as causing the "Intermittent" complaint should be thoroughly checked for backed out terminals, improper mating, broken locks, improperly formed or damaged terminals, poor terminal to wiring connections or physical damage to the wiring harness.

VIBRATION, TEMPERATURE EFFECTS:

Check for vibration effects by performing the relay test function of the Tech 1. With the relay test commanded "ON," lightly tap the top and sides of the relay while monitoring relay voltage. If the relay voltage changes significantly, replace the relay.

If DTC 14 only sets when the vehicle is initially started in cold ambient conditions (temperature less than 32°F-0°C), replace the relay.

DIAGNOSTIC CHART DTC 14 — 1994–95 CAMARO AND FIREBIRD

DTC 14

(Page 2 of 3)

ABS ENABLE RELAY CONTACT CIRCUIT OPEN

DIAGNOSIS CONTINUED FROM PREVIOUS PAGE.

(9)
- IGNITION "OFF."
- REMOVE JUMPER WIRE INSTALLED IN STEP 3, IF NOT ALREADY REMOVED.
- DISCONNECT NEGATIVE BATTERY CABLE.
- DISCONNECT POSITIVE BATTERY CABLE.
- WITH BATTERY STILL DISCONNECTED, TURN IGNITION SWITCH TO "RUN." THIS COMPLETES THE IGNITION CIRCUIT TO THE ABS ENABLE RELAY.
- USING J 39200, MEASURE THE RESISTANCE BETWEEN TERMINAL "B3" OF THE 32-WAY EBCM HARNESS CONNECTOR AND THE POSITIVE BATTERY CABLE TERMINAL. IS RESISTANCE 95Ω OR LESS WHEN MEASURED AT APPROXIMATELY 20 °C (68°F)?

NO → DIAGNOSIS CONTINUED ON NEXT PAGE.

YES ↓

(10)
- USING J 39200, MEASURE THE VOLTAGE BETWEEN THE NEGATIVE AND POSITIVE BATTERY TERMINALS. IS VOLTAGE 10 VOLTS OR GREATER?

NO → CORRECT LOW VOLTAGE CONDITION.

YES ↓

(11)
- IGNITION "OFF."
- RECONNECT POSITIVE BATTERY CABLE.
- RECONNECT NEGATIVE BATTERY CABLE.
- DISCONNECT ABS ENABLE RELAY.
- MEASURE THE VOLTAGE BETWEEN TERMINAL "C1" OF THE ABS ENABLE RELAY HARNESS CONNECTOR AND GROUND. IS VOLTAGE 10 VOLTS OR GREATER?

NO → REPAIR OPEN OR HIGH RESISTANCE IN CKT 302.

YES ↓

(12)
- USING J 39200, MEASURE THE RESISTANCE BETWEEN TERMINAL "C4" OF THE ABS ENABLE RELAY HARNESS CONNECTOR AND TERMINAL "D" OF THE 8-WAY EBCM HARNESS CONNECTOR. IS RESISTANCE 2Ω OR LESS?

NO → REPAIR OPEN OR HIGH RESISTANCE IN CKT 1633.

YES ↓

(13)
- USING A JUMPER WIRE, CONNECT TERMINAL "86" ON THE ABS ENABLE RELAY TO GROUND.
- USING A FUSED JUMPER WIRE, SUCH AS J 36169, WITH A 3 AMP FUSE, CONNECT TERMINAL "85" ON THE ABS ENABLE RELAY TO VOLTAGE.
- USING J 39200, MEASURE THE RESISTANCE BETWEEN TERMINALS "30" AND "87" ON THE ABS ENABLE RELAY. IS RESISTANCE 2Ω OR LESS?

NO → REPLACE ABS ENABLE RELAY.

YES → MALFUNCTION WAS MOST LIKELY DUE TO A POOR CONNECTION. INSPECT ALL CONNECTORS AND TERMINALS FOR POOR TERMINAL CONTACT AND EVIDENCE OF CORROSION. REPLACE ALL TERMINALS THAT EXHIBIT EVIDENCE OF POOR TERMINAL CONTACT OR CORROSION.

DIAGNOSTIC CHART DTC 14 — 1994–95 CAMARO AND FIREBIRD

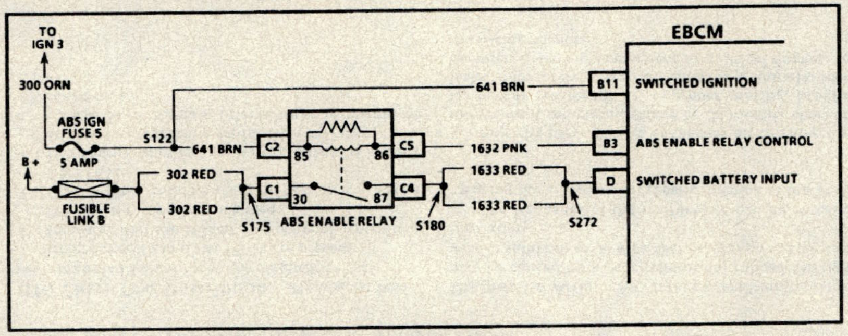

DTC 14

ABS ENABLE RELAY CONTACT CIRCUIT OPEN

Circuit Description:
Ignition voltage is supplied through terminal "C2" of the ABS enable relay. The EBCM then is able to energize the pull-in coil by completing the ground circuit at terminal "B3" of the EBCM. The magnetic field created closes the ABS enable relay contacts and allows battery voltage and current to be supplied to the EBCM, which supplies power to the motors and solenoids.

DTC Will Set When: DTC 14 can be set anytime after the EBCM commands the ABS enable relay "ON" (the relay is first commanded "ON" during the three second bulb check). This test monitors the availability of current/voltage to the motors and solenoids. This malfunction indicates voltage is not available and would therefore not allow ABS operation if required.

Action Taken: A malfunction DTC is stored, ABS is disabled and the "ABS INOP" lamp is turned "ON."

DTC Chart Test Description: Number(s) below refer to circled number(s) on the diagnostic chart.
14. Checks for proper resistance of the relay coil.
15. Checks for high resistance in the ABS enable relay control circuit between the relay and the EBCM.
16. Checks for high resistance in the ABS enable relay control circuit between the battery and the relay.

Diagnostic Aids: An "Intermittent" problem may be caused by a poor connection, rubbed through wire insulation, or a wire that is broken inside the insulation.

The frequency of the problem can be checked by using the enhanced diagnostic function of the Tech 1.

Any circuitry that is suspected as causing the "Intermittent" complaint should be thoroughly checked for backed out terminals, improper mating, broken locks, improperly formed or damaged terminals, poor terminal to wiring connections or physical damage to the wiring harness.

VIBRATION, TEMPERATURE EFFECTS:

Check for vibration effects by performing the relay test function of the Tech 1. With the relay test commanded "ON," lightly tap the top and sides of the relay while monitoring relay voltage. If the relay voltage changes significantly, replace the relay.

If DTC 14 only sets when the vehicle is initially started in cold ambient conditions (temperature less than 32°F-0°C), replace the relay.

DIAGNOSTIC CHART DTC 14 — 1994–95 CAMARO AND FIREBIRD

DTC 14
(Page 3 of 3)
ABS ENABLE RELAY CONTACT CIRCUIT OPEN

DIAGNOSIS CONTINUED FROM PREVIOUS PAGE.

(14)
- DISCONNECT ABS ENABLE RELAY.
- USING J 39200, MEASURE THE RESISTANCE BETWEEN TERMINALS "85" AND "86" ON THE ABS ENABLE RELAY. IS RESISTANCE 95Ω OR LESS WHEN MEASURED AT APPROXIMATELY 24°C (75°F)?

NO → REPLACE ABS ENABLE RELAY.

YES ↓

(15)
- MEASURE THE RESISTANCE BETWEEN TERMINAL "C5" OF THE ABS ENABLE RELAY HARNESS CONNECTOR AND TERMINAL "B3" OF THE 32-WAY EBCM HARNESS CONNECTOR. IS RESISTANCE 2Ω OR LESS?

NO → REPAIR OPEN OR HIGH RESISTANCE IN CKT 1632.

YES ↓

(16)
- MEASURE THE RESISTANCE BETWEEN TERMINAL "C2" OF THE ABS ENABLE RELAY HARNESS CONNECTOR AND THE POSITIVE BATTERY CABLE TERMINAL. IS RESISTANCE 2Ω OR LESS?

NO → REPAIR OPEN OR HIGH RESISTANCE IN CKT 641, CKT 300, IGNITION SWITCH, OR BATTERY FEED CIRCUIT TO IGNITION SWITCH.

YES → MALFUNCTION WAS MOST LIKELY DUE TO A POOR CONNECTION. INSPECT ALL CONNECTORS AND TERMINALS FOR POOR TERMINAL CONTACT AND EVIDENCE OF CORROSION.
REPLACE
ALL TERMINALS THAT EXHIBIT EVIDENCE OF POOR TERMINAL CONTACT OR CORROSION.

DIAGNOSTIC CHART DTC 15 — 1994–95 CAMARO AND FIREBIRD

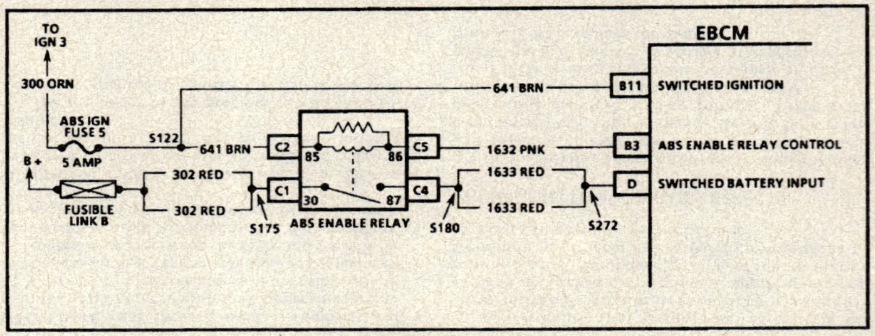

DTC 15

ABS RLY CONTACT CKT SHORT TO BATT OR ALWAYS CLOSED

Circuit Description:

Ignition voltage is supplied through terminal "C2" of the ABS enable relay. The EBCM then is able to energize the pull-in coil by completing the ground circuit at terminal "B3" of the EBCM. The magnetic field created closes the ABS enable relay contacts and allows battery voltage and current to be supplied to the EBCM, which supplies power to the motors and solenoids.

DTC Will Set When: DTC 15 can be set only before the EBCM commands the relay "ON." This test determines if the ABS enable relay is energized when it should not be. This malfunction would not allow the ABS enable relay to remove power to the ABS system. If a second malfunction were to occur that requires the ABS enable relay to be turned "OFF," that malfunction cannot be removed if the relay cannot be controlled. The malfunction must be present for three consecutive drive cycles before the DTC is set.

Action Taken: A malfunction DTC is stored. ABS is not disabled.

DTC Chart Test Description: Number(s) below refer to circled number(s) on the diagnostic chart.

1. Determines if any other DTCs are present.
2. Indicates that the EBCM is capable of controlling the enable relay as commanded.
3. Checks for a possible short to battery voltage in CKT 1633.
4. Determines whether the short to voltage is in CKT 1633 or as a result of a malfunctioning relay.
5. Checks for a possible intermittent in wiring or connectors.

Diagnostic Aids: An "Intermittent" malfunction may be caused by a poor connection, rubbed through wire insulation, or a wire that is broken inside the insulation.

The frequency of the malfunction can be checked by using the enhanced diagnostic function of the Tech 1.

Any circuitry, that is suspected as causing the intermittent complaint, should be thoroughly checked for backed out terminals, improper mating, broken locks, improperly formed or damaged terminals, poor terminal to wiring connections or physical damage to the wiring harness.

DIAGNOSTIC CHART DTC 15 — 1994–95 CAMARO AND FIREBIRD

DTC 15

ABS RLY CONTACT CKT SHORT TO BATT OR ALWAYS CLOSED

(1) ARE ANY OTHER DTCs SET?

- YES: PROCEED TO THAT DTC CHART FIRST.
- NO:

(2)
- IGNITION "ON," ENGINE "OFF."
- USING THE RELAY TEST FUNCTION OF THE TECH 1, COMMAND ABS ENABLE RELAY "OFF."

DOES TECH 1 INDICATE THAT ABS ENABLE RELAY IS "OFF" AND BATTERY VOLTAGE IS LESS THAN 5 VOLTS?

- YES: MALFUNCTION IS NOT PRESENT AT THIS TIME. REFER TO "DIAGNOSTIC AIDS".
- NO:

(3)
- IGNITION "OFF."
- DISCONNECT 32-WAY AND 8-WAY EBCM HARNESS CONNECTORS.
- IGNITION "ON."
- USING J 39200, MEASURE THE VOLTAGE BETWEEN GROUND AND TERMINAL "D" OF THE 8-WAY EBCM CONNECTOR.

IS VOLTAGE 2 VOLTS OR LESS?

- YES:

(5)
- IGNITION "OFF."
- INSPECT CKT 1633 AND THE 8-WAY EBCM HARNESS CONNECTOR FOR DAMAGE WHICH MAY RESULT IN A SHORT TO VOLTAGE WHEN ALL CONNECTORS ARE CONNECTED. REPAIR DAMAGE IF EVIDENT.
- RECONNECT 8-WAY AND 32-WAY EBCM CONNECTORS.
- IGNITION "ON."
- USING RELAY TEST FUNCTION OF TECH 1, COMMAND ABS ENABLE RELAY "OFF."

DOES TECH 1 INDICATE THAT ABS ENABLE RELAY IS "OFF" AND BATTERY VOLTAGE IS LESS THAN 5 VOLTS?

- NO: REPLACE EBCM.
- YES: MALFUNCTION IS INTERMITTENT. REFER TO "DIAGNOSTIC AIDS".

- NO:

(4)
- IGNITION "OFF."
- DISCONNECT ABS ENABLE RELAY.
- IGNITION "ON."
- MEASURE THE VOLTAGE BETWEEN GROUND AND TERMINAL "D" OF THE 8-WAY EBCM HARNESS CONNECTOR.

IS VOLTAGE 2 VOLTS OR LESS?

- YES: REPLACE ABS ENABLE RELAY.
- NO: REPAIR SHORT TO VOLTAGE IN CKT 1633.

AFTER ALL DIAGNOSIS IS COMPLETE, CLEAR DTCs, AND TEST DRIVE VEHICLE FOR 3 DRIVE CYCLES TO VERIFY THAT DTC DOES NOT RESET. A DRIVE CYCLE CONSISTS OF STARTING THE VEHICLE, DRIVING ABOVE 16 km/h (10 MPH), AND THEN KEYING DOWN.

DIAGNOSTIC CHART DTC 16 — 1994–95 CAMARO AND FIREBIRD

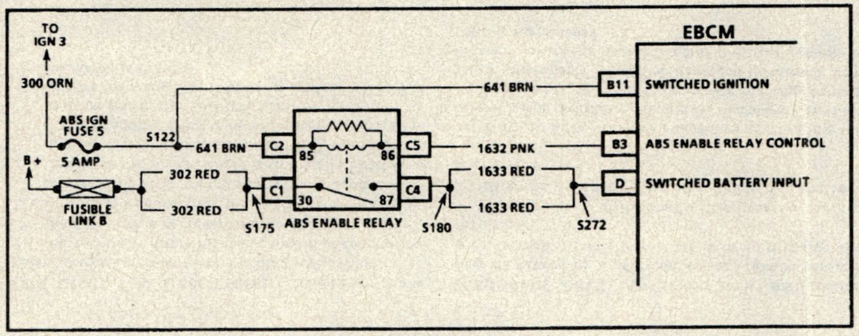

DTC 16

ABS ENABLE RELAY COIL CIRCUIT OPEN

Circuit Description:

Ignition voltage is supplied through terminal "C2" of the ABS enable relay. The EBCM then is able to energize the pull-in coil by completing the ground circuit at terminal "B3" of the EBCM. The magnetic field created closes the ABS enable relay contacts and allows battery voltage and current to be supplied to the EBCM, which supplies power to the motors and solenoids.

DTC Will Set When: DTC 16 can be set only after DTC 14 has been set. This test detects an open in the ABS enable relay coil circuit. An open in this circuit will not allow the ABS enable relay to be energized thus preventing voltage/current to the motors and solenoids. If this malfunction is present and the ignition is turned "OFF" before 5 km/h (3 mph) is reached, DTC 14 is set alone.

Action Taken: A malfunction DTC is stored, ABS is disabled and the "ABS INOP" lamp is turned "ON."

DTC Chart Test Description: Number(s) below refer to circled number(s) on the diagnostic chart.

1. Indicates the EBCM is capable of controlling the ABS enable relay as commanded.
2. Checks to ensure that voltage is available to the pull-in coil of the ABS enable relay.
3. Ensures that there is continuity through the pull-in coil of the ABS enable relay.
4. Verifies the integrity of the relay control circuit.
5. Ensures malfunction was not due to poor terminal contact.

Diagnostic Aids: An "Intermittent" malfunction may be caused by a poor connection, rubbed through wire insulation, or a wire that is broken inside the insulation.

The frequency of the malfunction can be checked by using the enhanced diagnostic function of the Tech 1.

If the frequency of the malfunction is high, but is currently intermittent, check for high coil resistance by measuring between relay terminals "C4" and "C5" using a J 39200. If resistance shows greater than 100Ω, replace the relay.

Any circuitry, that is suspected as causing the intermittent complaint, should be thoroughly checked for backed out terminals, improper mating, broken locks, improperly formed or damaged terminals, poor terminal to wiring connections or physical damage to the wiring harness.

DIAGNOSTIC CHART DTC 16 — 1994–95 CAMARO AND FIREBIRD

DTC 16
ABS ENABLE RELAY COIL CIRCUIT OPEN

(1)
- IGNITION "ON," ENGINE "OFF."
- USING THE RELAY TEST FUNCTION OF THE TECH 1, COMMAND THE ABS ENABLE RELAY "ON."
DOES TECH 1 INDICATE ABS ENABLE RELAY "ON" AND BATTERY VOLTAGE ABOVE 10 VOLTS?

YES → MALFUNCTION IS NOT PRESENT AT THIS TIME. REFER TO "DIAGNOSTIC AIDS".

NO → (2)
- USING J 39200, MEASURE THE VOLTAGE BETWEEN GROUND AND TERMINAL "C2" OF THE ABS ENABLE RELAY CONNECTOR.
IS VOLTAGE 10 VOLTS OR GREATER?

NO → REPAIR OPEN OR HIGH RESISTANCE IN CKT 641 OR ABS IGN FUSE 5.

YES → (3)
- USING THE RELAY TEST FUNCTION OF THE TECH 1, COMMAND THE ABS ENABLE RELAY "OFF."
- MEASURE THE VOLTAGE BETWEEN GROUND AND TERMINAL "C5" OF THE ABS ENABLE RELAY CONNECTOR.
IS VOLTAGE 10 VOLTS OR GREATER?

NO → CHECK FOR INTERMITTENT CONNECTIONS AT ABS ENABLE RELAY. IF OK, REPLACE ABS RELAY.

YES → (4)
- IGNITION "OFF."
- DISCONNECT 32-WAY EBCM CONNECTOR.
- IGNITION "ON."
- MEASURE THE VOLTAGE BETWEEN GROUND AND TERMINAL "B3" OF THE 32-WAY EBCM HARNESS CONNECTOR.
IS VOLTAGE 10 VOLTS OR GREATER?

NO → REPAIR OPEN OR HIGH RESISTANCE IN CKT 1632.

YES → (5)
- IGNITION "OFF."
- INSPECT 32-WAY EBCM CONNECTOR TERMINAL "B3" FOR POOR TERMINAL CONTACT OR EVIDENCE OF CORROSION.
- REPLACE ALL TERMINALS THAT EXHIBIT SIGNS OF POOR TERMINAL CONTACT OR CORROSION.
- RECONNECT ALL CONNECTORS.
- IGNITION "ON."
DOES DTC 16 SET AS A CURRENT DTC?

YES → REPLACE EBCM.

NO → MALFUNCTION WAS MOST LIKELY CAUSED BY A POOR CONNECTION. REFER TO "DIAGNOSTIC AIDS".

AFTER ALL DIAGNOSIS IS COMPLETE, CLEAR DTCs AND TEST DRIVE VEHICLE FOR THREE (3) DRIVE CYCLES TO VERIFY THAT DTC DOES NOT RESET. A DRIVE CYCLE CONSISTS OF STARTING THE VEHICLE, DRIVING OVER 16 km/h (10 MPH), AND THEN KEYING DOWN.

DIAGNOSTIC CHART DTC 17 — 1994–95 CAMARO AND FIREBIRD

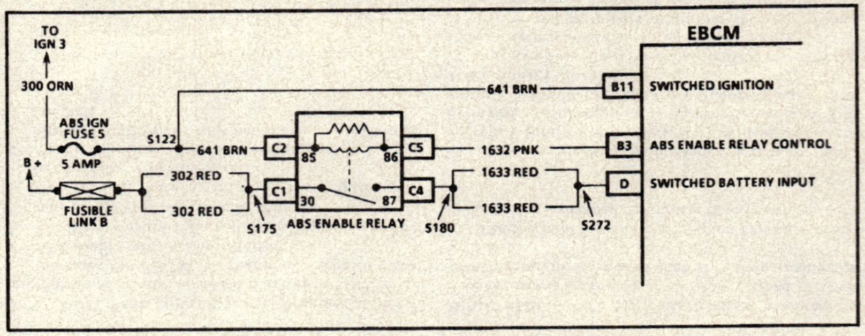

DTC 17

ABS ENABLE RELAY COIL CIRCUIT SHORTED TO GND

Circuit Description:

Ignition voltage is supplied through terminal "C2" of the ABS enable relay. The EBCM then is able to energize the pull-in coil by completing the ground circuit at terminal "B3" of the EBCM. The magnetic field created closes the ABS enable relay contacts and allows battery voltage and current to be supplied to the EBCM, which supplies voltage to the motors and solenoids.

DTC Will Set When: DTC 17 can be set before the EBCM commands the ABS enable relay "ON." This test determines if the ABS enable relay is energized when it should not be. This malfunction would not allow the ABS enable relay to remove power to the ABS system. If a second malfunction were to occur that requires the ABS enable relay to be turned "OFF," that malfunction cannot be removed if the relay cannot be controlled.

Action Taken: A malfunction DTC is stored. ABS is not disabled.

DTC Chart Test Description: Number(s) below refer to circled number(s) on the diagnostic chart.

1. Indicates that the EBCM is capable of controlling the ABS enable relay as commanded.
2. Checks to ensure that the ABS enable relay or control CKT 1632 are not shorted to ground.
3. Checks to see if the EBCM is internally shorted to ground. A grounded terminal "B3" would cause the ABS enable relay to be energized any time the ignition was in the "ON" position.
4. Checks for a short to ground in control CKT 1632 or a defective relay.

Diagnostic Aids: An "Intermittent" malfunction may be caused by a poor connection, rubbed through wire insulation, or a wire that is broken inside the insulation.

The frequency of the malfunction can be checked by using the enhanced diagnostic function of the Tech 1.

Any circuitry, that is suspected as causing the intermittent complaint, should be thoroughly checked for backed out terminals, improper mating, broken locks, improperly formed or damaged terminals, poor terminal to wiring connections or physical damage to the wiring harness.

DIAGNOSTIC CHART DTC 17 — 1994–95 CAMARO AND FIREBIRD

DTC 17

ABS ENABLE RELAY COIL CIRCUIT SHORTED TO GND

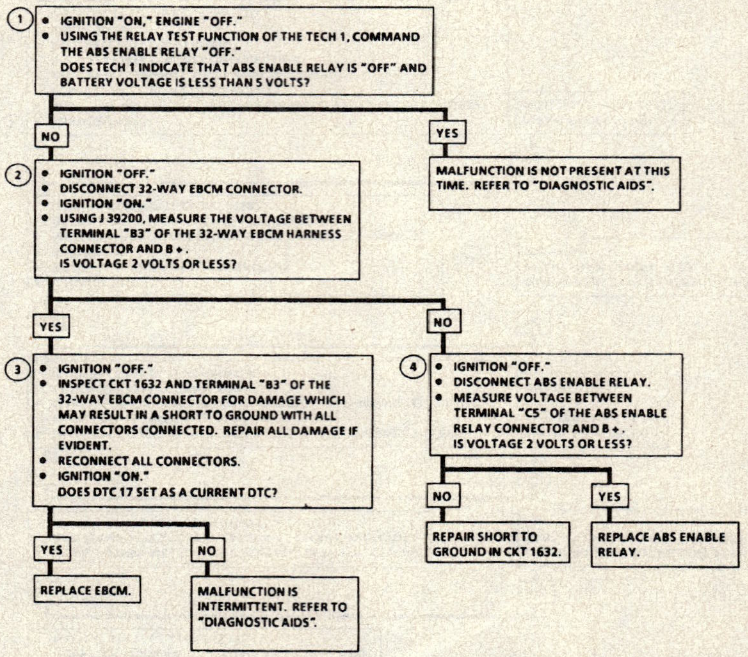

AFTER ALL DIAGNOSIS IS COMPLETE, CLEAR DTCs AND TEST DRIVE VEHICLE FOR THREE (3) DRIVE CYCLES TO VERIFY THAT DTC DOES NOT RESET. A DRIVE CYCLE CONSISTS OF STARTING THE VEHICLE, DRIVING OVER 16 km/h (10 MPH), AND THEN KEYING DOWN.

DIAGNOSTIC CHART DTC 18 — 1994–95 CAMARO AND FIREBIRD

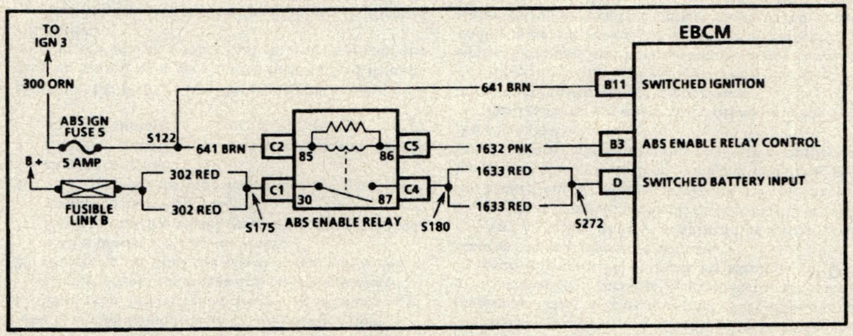

DTC 18

ABS ENABLE RELAY COIL CIRCUIT SHORTED TO BATT

Circuit Description:
Ignition voltage is supplied through terminal "C2" of the ABS enable relay. The EBCM then is able to energize the pull-in coil by completing the ground circuit at terminal "B3" of the EBCM. The magnetic field created closes the ABS enable relay contacts and allows battery voltage and current to be supplied to the EBCM, which supplies voltage to the motors and solenoids.

DTC Will Set When: DTC 18 can be set after the EBCM commands the ABS enable relay "ON." This test monitors the availability of current/voltage to the motors and solenoids. A malfunction of this type will not allow voltage to be available to the motors and solenoids; therefore, ABS operation would not be allowed if required.

Action Taken: A malfunction DTC is stored, ABS is disabled and the "ABS INOP" lamp is turned "ON."

DTC Chart Test Description: Number(s) below refer to circled number(s) on the diagnostic chart.
1. Indicates that the EBCM is capable of controlling the ABS enable relay as commanded.
2. Checks for a short to voltage on the relay control circuit.
3. Verifies that the relay coil has the appropriate resistance.
4. Checks for relay contacts that are failed closed.
5. Checks for possible poor terminal contact at the EBCM.

Diagnostic Aids: An "Intermittent" malfunction may be caused by a poor connection, rubbed through wire insulation, or a wire that is broken inside the insulation.

The frequency of the malfunction can be checked by using the enhanced diagnostic function of the Tech 1.

Any circuitry, that is suspected as causing the intermittent complaint, should be thoroughly checked for backed out terminals, improper mating, broken locks, improperly formed or damaged terminals, poor terminal to wiring connections or physical damage to the wiring harness.

DIAGNOSTIC CHART DTC 18 — 1994–95 CAMARO AND FIREBIRD

DTC 18
ABS ENABLE RELAY COIL CIRCUIT SHORTED TO BATT

1.
- IGNITION "ON," ENGINE "OFF."
- USING RELAY TEST FUNCTION OF TECH 1, COMMAND ABS ENABLE RELAY "ON." DOES TECH 1 INDICATE ABS ENABLE RELAY "ON" AND BATTERY VOLTAGE ABOVE 10 VOLTS?

YES: MALFUNCTION IS NOT PRESENT AT THIS TIME. REFER TO "DIAGNOSTIC AIDS".

NO:

2.
- IGNITION "OFF."
- DISCONNECT ABS ENABLE RELAY.
- DISCONNECT 32-WAY EBCM CONNECTOR.
- IGNITION "ON."
- USING J 39200, MEASURE THE VOLTAGE BETWEEN GROUND AND TERMINAL "B3" OF THE 32-WAY EBCM HARNESS CONNECTOR. IS VOLTAGE 2 VOLTS OR LESS?

NO: REPAIR SHORT TO VOLTAGE IN CKT 1632.

YES:

3.
- IGNITION "OFF."
- USING J 39200, MEASURE THE RESISTANCE BETWEEN TERMINAL "C2" AND TERMINAL "C5" OF THE ABS ENABLE RELAY. IS RESISTANCE 40Ω OR GREATER?

NO: REPLACE ABS ENABLE RELAY.

YES:

4.
- USING J 39200, MEASURE THE RESISTANCE BETWEEN TERMINALS "30" AND "87" OF THE ABS ENABLE RELAY. DOES J 39200 DISPLAY RESISTANCE AS "OL" (INFINITE)?

NO: REPLACE ABS ENABLE RELAY.

YES:

5.
- INSPECT CKT 1632 AND TERMINAL "B3" OF THE 32-WAY EBCM CONNECTOR FOR DAMAGE WHICH MAY RESULT IN A SHORT TO VOLTAGE WITH ALL CONNECTORS CONNECTED. REPAIR ALL DAMAGE IF EVIDENT.
- RECONNECT ALL CONNECTORS.
- IGNITION "ON." DOES DTC 18 SET AS A CURRENT DTC?

YES: REPLACE EBCM.

NO: MALFUNCTION IS INTERMITTENT. REFER TO "DIAGNOSTIC AIDS".

AFTER ALL DIAGNOSIS IS COMPLETE, CLEAR DTCs AND TEST DRIVE VEHICLE FOR THREE (3) DRIVE CYCLES TO VERIFY THAT DTC DOES NOT RESET. A DRIVE CYCLE CONSISTS OF STARTING THE VEHICLE, DRIVING OVER 16 km/h (10 MPH), AND THEN KEYING DOWN.

DIAGNOSTIC CHART DTC 21 — 1994–95 CAMARO AND FIREBIRD

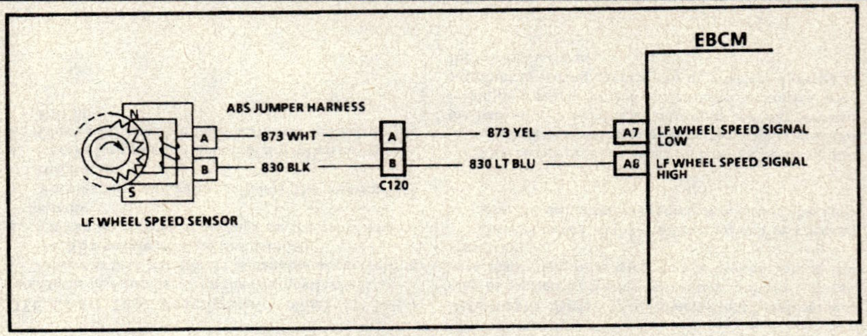

DTC 21

LEFT FRONT WHEEL SPEED = 0

Circuit Description:
As a toothed ring passes by the wheel speed sensor, changes in the electromagnetic field cause the wheel speed sensor to produce a sinusoidal (AC) voltage signal whose frequency is proportional to wheel speed. The magnitude of this signal is directly related to wheel speed and the proximity of the wheel speed sensor to the toothed ring, often referred to as the air gap.

DTC Will Set When: DTC 21 can be set when the vehicle is not in an ABS stop. If the left front wheel speed = 0 and the vehicle's reference speed is greater than 8 km/h (5 mph), a malfunction exists.

Action Taken: A malfunction DTC is stored, ABS is disabled and the ABS warning lamp is turned "ON."

DTC Chart Test Description: Number(s) below refer to circled number(s) on the diagnostic chart.

1. This step identifies a wheel speed sensor or circuitry that is damaged and visibly apparent.
2. Checks to see that the wheel speed sensor coil is within proper resistance values.
3. Verifies that the wheel speed sensor can generate an appropriate output signal.
4. This ensures that the wheel speed circuitry is not shorted together.
5. Ensures that the malfunction was not due to physical damage.

Diagnostic Aids: An "Intermittent" malfunction may be caused by a poor connection, rubbed through wire insulation, or a wire that is broken inside the insulation.

The frequency of the malfunction can be checked by using the enhanced diagnostic function of the Tech 1.

If the customer's comments reflect that the ABS warning lamp is "ON" only during moist environmental changes (rain, snow, vehicle wash), all wheel speed sensor circuitry should be thoroughly inspected for signs of water intrusion. If DTC is not current, clear DTCs and simulate the effects of water intrusion. Use the following procedure. Spray down the suspected area with a 5% salt water solution (two teaspoons of salt to 12 oz. of water). Test drive vehicle over various road surfaces (bumps, turns, etc.) above 24 km/h (15 mph) for at least 30 seconds. If DTC returns, replace suspected harness.

Any circuitry, that is suspected as causing the intermittent complaint, should be thoroughly checked for backed out terminals, improper mating, broken locks, improperly formed or damaged terminals, poor terminal to wiring connections or physical damage to the wiring harness.

Resistance of the wheel speed sensor will increase with an increase in sensor temperature.

When replacing a wheel speed sensor, inspect the sensor terminals and harness connector for corrosion and/or water intrusion. If evidence of corrosion or water intrusion exists, replace wheel speed sensor jumper harness. Likewise, if replacing a wheel speed sensor jumper harness, inspect sensor terminals. If evidence of corrosion or water intrusion exists, replace wheel speed sensor.

DIAGNOSTIC CHART DTC 21 — 1994–95 CAMARO AND FIREBIRD

DTC 21
LEFT FRONT WHEEL SPEED = 0

Important

- WHEEL SPEED SENSOR INTERMITTENT MALFUNCTIONS MAY BE DIFFICULT TO LOCATE. CARE SHOULD BE TAKEN NOT TO DISTURB ANY ELECTRICAL CONNECTIONS PRIOR TO AN INDICATED STEP OF THIS CHART. THIS WILL ENSURE THAT AN INTERMITTENT CONNECTION WILL NOT BE CORRECTED BEFORE THE SOURCE OF THE MALFUNCTION IS FOUND.

(1)
- IGNITION "OFF."
- PHYSICALLY INSPECT THE LEFT FRONT WHEEL SPEED SENSOR, LEFT FRONT WHEEL SPEED SENSOR JUMPER HARNESS, AND CONNECTORS FOR DAMAGE. IS PHYSICAL DAMAGE INDICATED?

YES: REPLACE WHEEL SPEED SENSOR AND/OR JUMPER HARNESS IF DAMAGED. DRIVE VEHICLE ABOVE 24 km/h (15 MPH) TO ENSURE DTC 21 DOES NOT SET.

NO:

(2)
- DISCONNECT LEFT FRONT WHEEL SPEED SENSOR DIRECTLY AT SENSOR.
- USING J 39200, MEASURE THE RESISTANCE BETWEEN SENSOR TERMINALS "A" AND "B." IS RESISTANCE BETWEEN 1020 AND 1137Ω WHEN SENSOR IS 20°C (68°F)? *(IF TEMPERATURE OF THE SENSOR IS SUSPECTED TO BE OTHER THAN 20°C (68°F), REFER TO THE TABLE BELOW FOR APPLICABLE SENSOR RESISTANCE VALUES; THE VALUES IN THE TABLE ARE FOR TEMPERATURE OF THE SENSOR, NOT AIR TEMPERATURE.)*

NO: REPLACE WHEEL SPEED SENSOR.

YES:

(3)
- WITH J 39200 STILL CONNECTED, SELECT A/C VOLTAGE SCALE.
- SPIN WHEEL BY HAND WHILE OBSERVING VOLTAGE. IS VOLTAGE AT LEAST 100 mV (VOLTAGE WILL INCREASE AS WHEEL SPEED INCREASES)?

NO: REPLACE WHEEL SPEED SENSOR.

YES:

(4)
- IGNITION "OFF."
- DISCONNECT 32-WAY EBCM CONNECTOR.
- USING J 39200, MEASURE THE RESISTANCE BETWEEN TERMINAL "A7" AND "A8" OF THE 32-WAY EBCM HARNESS CONNECTOR. DOES J 39200 DISPLAY RESISTANCE AS "OL" (INFINITE)?

NO: REPAIR SHORT BETWEEN CKTs 830 AND 873.

YES:

(5)
- INSPECT CKTs 830 AND 873 AND 32-WAY EBCM CONNECTOR FOR DAMAGE WHICH MAY RESULT IN A SHORT BETWEEN CKTs 830 AND 873 WITH ALL CONNECTORS CONNECTED. REPAIR ALL DAMAGE IF EVIDENT.
- RECONNECT ALL CONNECTORS.
- TEST DRIVE VEHICLE ABOVE 24 km/h (15 MPH) FOR AT LEAST 30 SECONDS. DOES DTC 21 SET AS A CURRENT DTC?

YES: REPLACE EBCM.

NO: MALFUNCTION IS INTERMITTENT. REFER TO "DIAGNOSTIC AIDS."

FRONT WHEEL SPEED SENSOR
TEMPERATURE OF SENSOR VS. SENSOR RESISTANCE (APPROXIMATE)

TEMP. (°C)	TEMP. (°F)	RESISTANCE (OHMS)
-34 TO 4	-30 TO 40	804 TO 1066
5 TO 43	41 TO 110	956 TO 1240
44 TO 93	111 TO 200	1112 TO 1463

DIAGNOSTIC CHART DTC 22 — 1994–95 CAMARO AND FIREBIRD

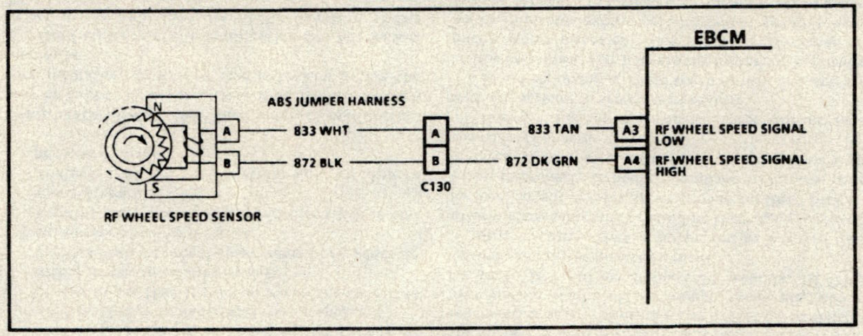

DTC 22

RIGHT FRONT WHEEL SPEED = 0

Circuit Description:

As a toothed ring passes by the wheel speed sensor, changes in the electromagnetic field cause the wheel speed sensor to produce a sinusoidal (AC) voltage signal whose frequency is proportional to wheel speed. The magnitude of this signal is directly related to wheel speed and the proximity of the wheel speed sensor to the toothed ring, often referred to as the air gap.

DTC Will Set When: DTC 22 can be set when the vehicle is not in an ABS stop. If the right front wheel speed = 0 and the vehicle's reference speed is greater than 8 km/h (5 mph), a malfunction exists.

Action Taken: A malfunction DTC is stored, ABS is disabled and the ABS warning lamp is turned "ON."

DTC Chart Test Description: Number(s) below refer to circled number(s) on the diagnostic chart.

1. This step identifies a wheel speed sensor or circuitry that is damaged and visibly apparent.
2. Checks to see that the wheel speed sensor coil is within proper resistance values.
3. Verifies that the wheel speed sensor can generate an appropriate output signal.
4. This ensures that the wheel speed circuitry is not shorted together.
5. Ensures that the malfunction was not due to physical damage.

Diagnostic Aids: An "Intermittent" malfunction may be caused by a poor connection, rubbed through wire insulation, or a wire that is broken inside the insulation.

The frequency of the malfunction can be checked by using the enhanced diagnostic function of the Tech 1.

If the customer's comments reflect that the ABS warning lamp is "ON" only during moist environmental changes (rain, snow, vehicle wash), all wheel speed sensor circuitry should be thoroughly inspected for signs of water intrusion. If DTC is not current, clear DTCs and simulate the effects of water intrusion. Use the following procedure. Spray down the suspected area with a 5% salt water solution (two teaspoons of salt to 12 oz. of water). Test drive vehicle over various road surfaces (bumps, turns, etc.) above 24 km/h (15 mph) for at least 30 seconds. If DTC returns, replace suspected harness.

Any circuitry, that is suspected as causing the intermittent complaint, should be thoroughly checked for backed out terminals, improper mating, broken locks, improperly formed or damaged terminals, poor terminal to wiring connections or physical damage to the wiring harness.

Resistance of the wheel speed sensor will increase with an increase in sensor temperature.

When replacing a wheel speed sensor, inspect the sensor terminals and harness connector for corrosion and/or water intrusion. If evidence of corrosion or water intrusion exists, replace wheel speed sensor jumper harness. Likewise, if replacing a wheel speed sensor jumper harness, inspect sensor terminals. If evidence of corrosion or water intrusion exists, replace wheel speed sensor.

DIAGNOSTIC CHART DTC 22 — 1994–95 CAMARO AND FIREBIRD

DTC 22
RIGHT FRONT WHEEL SPEED = 0

Important
- WHEEL SPEED SENSOR INTERMITTENT MALFUNCTIONS MAY BE DIFFICULT TO LOCATE. CARE SHOULD BE TAKEN NOT TO DISTURB ANY ELECTRICAL CONNECTIONS PRIOR TO AN INDICATED STEP OF THIS CHART. THIS WILL ENSURE THAT AN INTERMITTENT CONNECTION WILL NOT BE CORRECTED BEFORE THE SOURCE OF THE MALFUNCTION IS FOUND.

(1)
- IGNITION "OFF."
- PHYSICALLY INSPECT THE RIGHT FRONT WHEEL SPEED SENSOR, RIGHT FRONT WHEEL SPEED SENSOR JUMPER HARNESS, AND CONNECTORS FOR DAMAGE. IS PHYSICAL DAMAGE INDICATED?

NO → (2) | YES → REPLACE WHEEL SPEED SENSOR AND/OR JUMPER HARNESS IF DAMAGED. DRIVE VEHICLE ABOVE 24 km/h (15 MPH) TO ENSURE DTC 22 DOES NOT SET.

(2)
- DISCONNECT RIGHT FRONT WHEEL SPEED SENSOR DIRECTLY AT SENSOR.
- USING J 39200, MEASURE THE RESISTANCE BETWEEN SENSOR TERMINALS "A" AND "B."
  IS RESISTANCE BETWEEN 1020 AND 1137Ω WHEN SENSOR IS 20°C (68°F)?
  (IF TEMPERATURE OF THE SENSOR IS SUSPECTED TO BE OTHER THAN 20°C (68°F), REFER TO THE TABLE BELOW FOR APPLICABLE SENSOR RESISTANCE VALUES; THE VALUES IN THE TABLE ARE FOR TEMPERATURE OF THE SENSOR, NOT AIR TEMPERATURE.)

YES → (3) | NO → REPLACE WHEEL SPEED SENSOR.

(3)
- WITH J 39200 STILL CONNECTED, SELECT A/C VOLTAGE SCALE.
- SPIN WHEEL BY HAND WHILE OBSERVING VOLTAGE. IS VOLTAGE AT LEAST 100 mV (VOLTAGE WILL INCREASE AS WHEEL SPEED INCREASES)?

YES → (4) | NO → REPLACE WHEEL SPEED SENSOR.

(4)
- IGNITION "OFF."
- DISCONNECT 32-WAY EBCM CONNECTOR.
- USING J 39200, MEASURE THE RESISTANCE BETWEEN TERMINAL "A3" AND "A4" OF THE 32-WAY EBCM HARNESS CONNECTOR. DOES J 39200 DISPLAY RESISTANCE AS "OL" (INFINITE)?

YES → (5) | NO → REPAIR SHORT BETWEEN CKTs 833 AND 872.

(5)
- INSPECT CKTs 833 AND 872 AND 32-WAY EBCM CONNECTOR FOR DAMAGE WHICH MAY RESULT IN A SHORT BETWEEN CKTs 833 AND 872 WITH ALL CONNECTORS CONNECTED. REPAIR ALL DAMAGE IF EVIDENT.
- RECONNECT ALL CONNECTORS.
- TEST DRIVE VEHICLE ABOVE 24 km/h (15 MPH) FOR AT LEAST 30 SECONDS. DOES DTC 22 SET AS A CURRENT DTC?

YES → REPLACE EBCM. | NO → MALFUNCTION IS INTERMITTENT. REFER TO "DIAGNOSTIC AIDS".

FRONT WHEEL SPEED SENSOR
TEMPERATURE OF SENSOR VS. SENSOR RESISTANCE (APPROXIMATE)

TEMP. (°C)	TEMP. (°F)	RESISTANCE (OHMS)
-34 TO 4	-30 TO 40	804 TO 1066
5 TO 43	41 TO 110	956 TO 1240
44 TO 93	111 TO 200	1112 TO 1463

DIAGNOSTIC CHART DTC 24 — 1994–95 CAMARO AND FIREBIRD

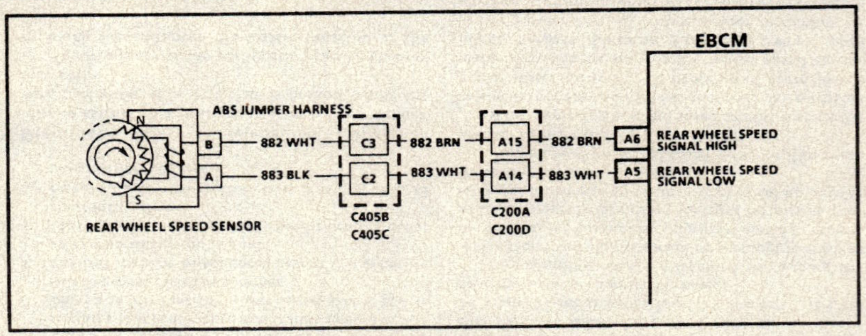

DTC 24

REAR WHEEL SPEED = 0

Circuit Description:

As a toothed ring passes by the wheel speed sensor, changes in the electromagnetic field cause the wheel speed sensor to produce a sinusoidal (AC) voltage signal whose frequency is proportional to wheel speed. The magnitude of this signal is directly related to wheel speed and the proximity of the wheel speed sensor to the toothed ring, often referred to as the air gap.

DTC Will Set When: DTC 24 can be set when the vehicle is not in an ABS stop. If the rear wheel speed = 0 and the vehicle's reference speed is greater than 8 km/h (5 mph), a malfunction exists.

Action Taken: A malfunction DTC is stored, ABS is disabled and the ABS warning lamp is turned "ON."

DTC Chart Test Description: Number(s) below refer to circled number(s) on the diagnostic chart.

1. This step identifies a wheel speed sensor or circuitry that is damaged and visibly apparent.
2. Checks to see that the wheel speed sensor coil is within proper resistance values.
3. Verifies that the wheel speed sensor can generate an appropriate output signal.
4. Verifies that the wheel speed circuitry is not shorted together.
5. Ensures that the malfunction was not due to physical damage.

Diagnostic Aids: An "Intermittent" malfunction may be caused by a poor connection, rubbed through wire insulation, or a wire that is broken inside the insulation.

The frequency of the malfunction can be checked by using the enhanced diagnostic function of the Tech 1.

If the customer's comments reflect that the ABS warning lamp is "ON" only during moist environmental changes (rain, snow, vehicle wash), all wheel speed sensor circuitry should be thoroughly inspected for signs of water intrusion. If DTC is not current, clear DTCs and simulate the effects of water intrusion. Use the following procedure. Spray down the suspected area with a 5% salt water solution (two teaspoons of salt to 12 oz. of water). Test drive vehicle over various road surfaces (bumps, turns, etc.) above 24 km/h (15 mph) for at least 30 seconds. If DTC returns, replace suspected harness.

Any circuitry, that is suspected as causing the intermittent complaint, should be thoroughly checked for backed out terminals, improper mating, broken locks, improperly formed or damaged terminals, poor terminal to wiring connections or physical damage to the wiring harness.

Resistance of the wheel speed sensor will increase with an increase in sensor temperature.

When replacing a wheel speed sensor, inspect the sensor terminals and harness connector for corrosion and/or water intrusion. If evidence of corrosion or water intrusion exists, replace wheel speed sensor jumper harness. Likewise, if replacing a wheel speed sensor jumper harness, inspect sensor terminals. If evidence of corrosion or water intrusion exists, replace wheel speed sensor.

DIAGNOSTIC CHART DTC 24 — 1994–95 CAMARO AND FIREBIRD

DTC 24
REAR WHEEL SPEED = 0

Important

- WHEEL SPEED SENSOR INTERMITTENT MALFUNCTIONS MAY BE DIFFICULT TO LOCATE. CARE SHOULD BE TAKEN NOT TO DISTURB ANY ELECTRICAL CONNECTIONS PRIOR TO AN INDICATED STEP OF THIS CHART. THIS WILL ENSURE THAT AN INTERMITTENT CONNECTION WILL NOT BE CORRECTED BEFORE THE SOURCE OF THE MALFUNCTION IS FOUND.

(1)
- IGNITION "OFF."
- PHYSICALLY INSPECT THE REAR WHEEL SPEED SENSOR, WHEEL SPEED SENSOR JUMPER HARNESS, AND CONNECTORS FOR DAMAGE.

IS PHYSICAL DAMAGE INDICATED?

YES: REPLACE WHEEL SPEED SENSOR AND/OR WHEEL SPEED SENSOR JUMPER HARNESS IF DAMAGED. DRIVE VEHICLE ABOVE 24 km/h (15 MPH) TO ENSURE DTC 24 DOES NOT SET.

NO:

(2)
- DISCONNECT REAR WHEEL SPEED SENSOR DIRECTLY AT SENSOR.
- USING J 39200, MEASURE THE RESISTANCE BETWEEN SENSOR TERMINALS "A" AND "B."

IS RESISTANCE BETWEEN 969 AND 1185Ω WHEN SENSOR IS 20°C (68°F)? *(IF TEMPERATURE OF THE SENSOR IS SUSPECTED TO BE OTHER THAN 20°C (68°F), REFER TO THE TABLE BELOW FOR APPLICABLE SENSOR RESISTANCE VALUES; THE VALUES IN THE TABLE ARE FOR TEMPERATURE OF THE SENSOR, NOT AIR TEMPERATURE.)*

NO: REPLACE WHEEL SPEED SENSOR.

YES:

(3)
- WITH J 39200 STILL CONNECTED, SELECT A/C VOLTAGE SCALE.
- VEHICLE TRANSMISSION IN NEUTRAL.
- SPIN WHEEL BY HAND WHILE OBSERVING VOLTAGE.

IS VOLTAGE AT LEAST 100 mV (VOLTAGE WILL INCREASE AS WHEEL SPEED INCREASES)?

NO: REPLACE WHEEL SPEED SENSOR.

YES:

(4)
- IGNITION "OFF."
- DISCONNECT 32-WAY EBCM CONNECTOR.
- USING J 39200, MEASURE THE RESISTANCE BETWEEN TERMINAL "A5" AND "A6" OF THE 32-WAY EBCM HARNESS CONNECTOR.

DOES J 39200 DISPLAY RESISTANCE AS "OL" (INFINITE)?

NO: REPAIR SHORT BETWEEN CKTs 882 AND 883.

YES:

(5)
- INSPECT CKTs 882 AND 883 AND 32-WAY EBCM CONNECTOR FOR DAMAGE WHICH MAY RESULT IN A SHORT BETWEEN CKTs 882 AND 883 WITH ALL CONNECTORS CONNECTED. REPAIR ALL DAMAGE IF EVIDENT.
- RECONNECT ALL CONNECTORS.
- TEST DRIVE VEHICLE ABOVE 24 km/h (15 MPH) FOR AT LEAST 30 SECONDS.

DOES DTC 24 SET AS A CURRENT DTC?

YES: REPLACE EBCM.

NO: MALFUNCTION IS INTERMITTENT. REFER TO "DIAGNOSTIC AIDS".

REAR WHEEL SPEED SENSOR

TEMPERATURE OF SENSOR VS. SENSOR RESISTANCE (APPROXIMATE)

TEMP. (°C)	TEMP. (°F)	RESISTANCE (OHMS)
-34 TO 4	-30 TO 40	763 TO 1110
5 TO 43	41 TO 110	908 TO 1292
44 TO 93	111 TO 200	1057 TO 1525

DIAGNOSTIC CHART DTC 25 — 1994–95 CAMARO AND FIREBIRD

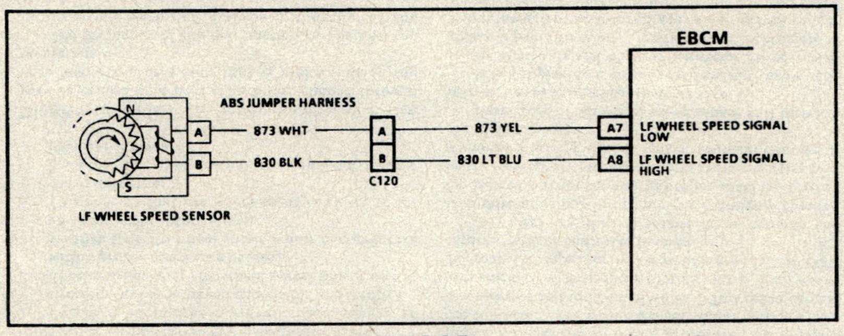

DTC 25

LEFT FRONT EXCESSIVE WHEEL SPEED VARIATION

Circuit Description:

As a toothed ring passes by the wheel speed sensor, changes in the electromagnetic field cause the wheel speed sensor to produce a sinusoidal (AC) voltage signal whose frequency is proportional to wheel speed. The magnitude of this signal is directly related to wheel speed and the proximity of the wheel speed sensor to the toothed ring, often referred to as the air gap.

DTC Will Set When: DTC 25 can be set when the brake is "OFF." The purpose of this test is to detect a situation in which the left front wheel acceleration or deceleration is beyond specified limits.

Action Taken: A malfunction DTC is stored, ABS is disabled and the ABS warning lamp is turned "ON."

DTC Chart Test Description: Number(s) below refer to circled number(s) on the diagnostic chart.

1. This step identifies a wheel speed sensor or circuitry that is damaged and visibly apparent.
2. Checks to see that the wheel speed sensor coil is within proper resistance values.
3. Verifies that the wheel speed sensor can generate an appropriate output signal.
4. This ensures that the wheel speed circuitry is not shorted together.
5. Ensures that the malfunction was not due to physical damage.

Diagnostic Aids: An "Intermittent" malfunction may be caused by a poor connection, rubbed through wire insulation, or a wire that is broken inside the insulation.

The frequency of the malfunction can be checked by using the enhanced diagnostic function of the Tech 1.

If the customer's comments reflect that the ABS warning lamp is "ON" only during moist environmental changes (rain, snow, vehicle wash), all wheel speed sensor circuitry should be thoroughly inspected for signs of water intrusion. If DTC is not current, clear DTCs and simulate the effects of water intrusion. Use the following procedure. Spray down the suspected area with a 5% salt water solution (two teaspoons of salt to 12 oz. of water). Test drive vehicle over various road surfaces (bumps, turns, etc.) above 24 km/h (15 mph) for at least 30 seconds. If DTC returns, replace suspected harness.

Any circuitry, that is suspected as causing the intermittent complaint, should be thoroughly checked for backed out terminals, improper mating, broken locks, improperly formed or damaged terminals, poor terminal to wiring connections or physical damage to the wiring harness.

Resistance of the wheel speed sensor will increase with an increase in sensor temperature.

When replacing a wheel speed sensor, inspect the sensor terminals and harness connector for corrosion and/or water intrusion. If evidence of corrosion or water intrusion exists, replace wheel speed sensor jumper harness. Likewise, if replacing a wheel speed sensor jumper harness, inspect sensor terminals. If evidence of corrosion or water intrusion exists, replace wheel speed sensor.

DIAGNOSTIC CHART DTC 25 — 1994–95 CAMARO AND FIREBIRD

DTC 25
LEFT FRONT EXCESSIVE WHEEL SPEED VARIATION

Important

- WHEEL SPEED SENSOR INTERMITTENT MALFUNCTIONS MAY BE DIFFICULT TO LOCATE. CARE SHOULD BE TAKEN NOT TO DISTURB ANY ELECTRICAL CONNECTIONS PRIOR TO AN INDICATED STEP OF THIS CHART. THIS WILL ENSURE THAT AN INTERMITTENT CONNECTION WILL NOT BE CORRECTED BEFORE THE SOURCE OF THE MALFUNCTION IS FOUND.

(1)
- IGNITION "OFF."
- PHYSICALLY INSPECT THE LEFT FRONT WHEEL SPEED SENSOR, LEFT FRONT WHEEL SPEED SENSOR JUMPER HARNESS, AND CONNECTORS FOR DAMAGE. IS PHYSICAL DAMAGE INDICATED?

YES → REPLACE WHEEL SPEED SENSOR AND/OR JUMPER HARNESS IF DAMAGED. DRIVE VEHICLE ABOVE 24 km/h (15 MPH) TO ENSURE DTC 25 DOES NOT SET.

NO ↓

(2)
- DISCONNECT LEFT FRONT WHEEL SPEED SENSOR DIRECTLY AT SENSOR.
- USING J 39200, MEASURE THE RESISTANCE BETWEEN SENSOR TERMINALS "A" AND "B."

IS RESISTANCE BETWEEN 1020 AND 1137Ω WHEN SENSOR IS 20°C (68°F)?
(IF TEMPERATURE OF THE SENSOR IS SUSPECTED TO BE OTHER THAN 20°C (68°F), REFER TO THE TABLE BELOW FOR APPLICABLE SENSOR RESISTANCE VALUES; THE VALUES IN THE TABLE ARE FOR TEMPERATURE OF THE SENSOR, NOT AIR TEMPERATURE.)

NO → REPLACE WHEEL SPEED SENSOR.

YES ↓

(3)
- WITH J 39200 STILL CONNECTED, SELECT A/C VOLTAGE SCALE.
- SPIN WHEEL BY HAND WHILE OBSERVING VOLTAGE.

IS VOLTAGE AT LEAST 100 mV (VOLTAGE WILL INCREASE AS WHEEL SPEED INCREASES)?

NO → REPLACE WHEEL SPEED SENSOR.

YES ↓

(4)
- IGNITION "OFF."
- DISCONNECT 32-WAY EBCM CONNECTOR.
- USING J 39200, MEASURE THE RESISTANCE BETWEEN TERMINAL "A7" AND "A8" OF THE 32-WAY EBCM HARNESS CONNECTOR. DOES J 39200 DISPLAY RESISTANCE AS "OL" (INFINITE)?

NO → REPAIR SHORT BETWEEN CKTs 830 AND 873.

YES ↓

(5)
- INSPECT CKTs 830 AND 873 AND 32-WAY EBCM CONNECTOR FOR DAMAGE WHICH MAY RESULT IN A SHORT BETWEEN CKTs 830 AND 873 WITH ALL CONNECTORS CONNECTED. REPAIR ALL DAMAGE IF EVIDENT.
- RECONNECT ALL CONNECTORS.
- TEST DRIVE VEHICLE ABOVE 24 km/h (15 MPH) FOR AT LEAST 30 SECONDS. DOES DTC 25 SET AS A CURRENT DTC?

YES → REPLACE EBCM.

NO → MALFUNCTION IS INTERMITTENT. REFER TO "DIAGNOSTIC AIDS".

FRONT WHEEL SPEED SENSOR
TEMPERATURE OF SENSOR VS. SENSOR RESISTANCE (APPROXIMATE)

TEMP. (°C)	TEMP. (°F)	RESISTANCE (OHMS)
-34 TO 4	-30 TO 40	804 TO 1066
5 TO 43	41 TO 110	956 TO 1240
44 TO 93	111 TO 200	1112 TO 1463

DIAGNOSTIC CHART DTC 26 — 1994–95 CAMARO AND FIREBIRD

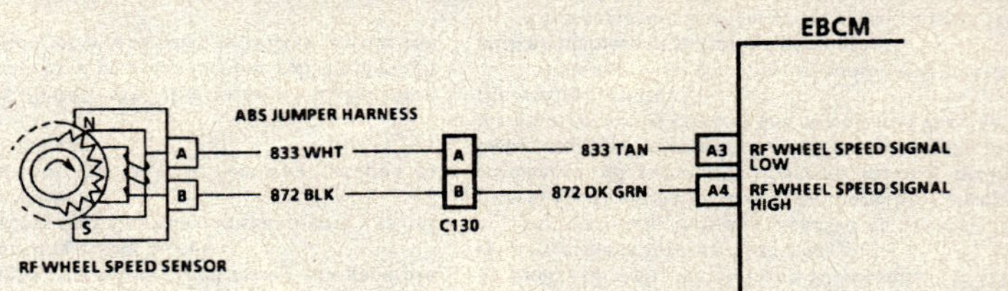

DTC 26

RIGHT FRONT EXCESSIVE WHEEL SPEED VARIATION

Circuit Description:
As a toothed ring passes by the wheel speed sensor, changes in the electromagnetic field cause the wheel speed sensor to produce a sinusoidal (AC) voltage signal whose frequency is proportional to wheel speed. The magnitude of this signal is directly related to wheel speed and the proximity of the wheel speed sensor to the toothed ring, often referred to as the air gap.

DTC Will Set When: DTC 26 can be set when the brake is "OFF." The purpose of this test is to detect a situation in which the right front wheel acceleration or deceleration is beyond specified limits.

Action Taken: A malfunction DTC is stored, ABS is disabled and the ABS warning lamp is turned "ON."

DTC Chart Test Description: Number(s) below refer to circled number(s) on the diagnostic chart.
1. This step identifies a wheel speed sensor or circuitry that is damaged and visibly apparent.
2. Checks to see that the wheel speed sensor coil is within proper resistance values.
3. Verifies that the wheel speed sensor can generate an appropriate output signal.
4. This ensures that the wheel speed circuitry is not shorted together.
5. Ensures that the malfunction was not due to physical damage.

Diagnostic Aids: An "Intermittent" malfunction may be caused by a poor connection, rubbed through wire insulation, or a wire that is broken inside the insulation.

The frequency of the malfunction can be checked by using the enhanced diagnostic function of the Tech 1.

If the customer's comments reflect that the ABS warning lamp is "ON" only during moist environmental changes (rain, snow, vehicle wash), all wheel speed sensor circuitry should be thoroughly inspected for signs of water intrusion. If DTC is not current, clear DTCs and simulate the effects of water intrusion. Use the following procedure. Spray down the suspected area with a 5% salt water solution (two teaspoons of salt to 12 oz. of water). Test drive vehicle over various road surfaces (bumps, turns, etc.) above 24 km/h (15 mph) for at least 30 seconds. If DTC returns, replace suspected harness.

Any circuitry, that is suspected as causing the intermittent complaint, should be thoroughly checked for backed out terminals, improper mating, broken locks, improperly formed or damaged terminals, poor terminal to wiring connections or physical damage to the wiring harness.

Resistance of the wheel speed sensor will increase with an increase in sensor temperature.

When replacing a wheel speed sensor, inspect the sensor terminals and harness connector for corrosion and/or water intrusion. If evidence of corrosion or water intrusion exists, replace wheel speed sensor jumper harness. Likewise, if replacing a wheel speed sensor jumper harness, inspect sensor terminals. If evidence of corrosion or water intrusion exists, replace wheel speed sensor.

DIAGNOSTIC CHART DTC 26 — 1994–95 CAMARO AND FIREBIRD

DTC 26

RIGHT FRONT EXCESSIVE WHEEL SPEED VARIATION

Important
- WHEEL SPEED SENSOR INTERMITTENT MALFUNCTIONS MAY BE DIFFICULT TO LOCATE. CARE SHOULD BE TAKEN NOT TO DISTURB ANY ELECTRICAL CONNECTIONS PRIOR TO AN INDICATED STEP OF THIS CHART. THIS WILL ENSURE THAT AN INTERMITTENT CONNECTION WILL NOT BE CORRECTED BEFORE THE SOURCE OF THE MALFUNCTION IS FOUND.

(1)
- IGNITION "OFF."
- PHYSICALLY INSPECT THE RIGHT FRONT WHEEL SPEED SENSOR, RIGHT FRONT WHEEL SPEED SENSOR JUMPER HARNESS, AND CONNECTORS FOR DAMAGE.

IS PHYSICAL DAMAGE INDICATED?

YES: REPLACE WHEEL SPEED SENSOR AND/OR JUMPER HARNESS IF DAMAGED. DRIVE VEHICLE ABOVE 24 km/h (15 MPH) TO ENSURE DTC 26 DOES NOT SET.

NO:

(2)
- DISCONNECT RIGHT FRONT WHEEL SPEED SENSOR DIRECTLY AT SENSOR.
- USING J 39200, MEASURE THE RESISTANCE BETWEEN SENSOR TERMINALS "A" AND "B."

IS RESISTANCE BETWEEN 1020 AND 1137Ω WHEN SENSOR IS 20°C (68°F)?
(IF TEMPERATURE OF THE SENSOR IS SUSPECTED TO BE OTHER THAN 20°C (68°F), REFER TO THE TABLE BELOW FOR APPLICABLE SENSOR RESISTANCE VALUES; THE VALUES IN THE TABLE ARE FOR TEMPERATURE OF THE SENSOR, NOT AIR TEMPERATURE.)

NO: REPLACE WHEEL SPEED SENSOR.

YES:

(3)
- WITH J 39200 STILL CONNECTED, SELECT A/C VOLTAGE SCALE.
- SPIN WHEEL BY HAND WHILE OBSERVING VOLTAGE.

IS VOLTAGE AT LEAST 100 mV (VOLTAGE WILL INCREASE AS WHEEL SPEED INCREASES)?

NO: REPLACE WHEEL SPEED SENSOR.

YES:

(4)
- IGNITION "OFF."
- DISCONNECT 32-WAY EBCM CONNECTOR.
- USING J 39200, MEASURE THE RESISTANCE BETWEEN TERMINAL "A3" AND "A4" OF THE 32-WAY EBCM HARNESS CONNECTOR.

DOES J 39200 DISPLAY RESISTANCE AS "OL" (INFINITE)?

NO: REPAIR SHORT BETWEEN CKTs 833 AND 872.

YES:

(5)
- INSPECT CKTs 833 AND 872 AND 32-WAY EBCM CONNECTOR FOR DAMAGE WHICH MAY RESULT IN A SHORT BETWEEN CKTs 833 AND 872 WITH ALL CONNECTORS CONNECTED. REPAIR ALL DAMAGE IF EVIDENT.
- RECONNECT ALL CONNECTORS.
- TEST DRIVE VEHICLE ABOVE 24 km/h (15 MPH) FOR AT LEAST 30 SECONDS.

DOES DTC 26 SET AS A CURRENT DTC?

YES: REPLACE EBCM.

NO: MALFUNCTION IS INTERMITTENT. REFER TO "DIAGNOSTIC AIDS".

FRONT WHEEL SPEED SENSOR
TEMPERATURE OF SENSOR VS. SENSOR RESISTANCE (APPROXIMATE)

TEMP. (°C)	TEMP. (°F)	RESISTANCE (OHMS)
-34 TO 4	-30 TO 40	804 TO 1066
5 TO 43	41 TO 110	956 TO 1240
44 TO 93	111 TO 200	1112 TO 1463

DIAGNOSTIC CHART DTC 28 — 1994–95 CAMARO AND FIREBIRD

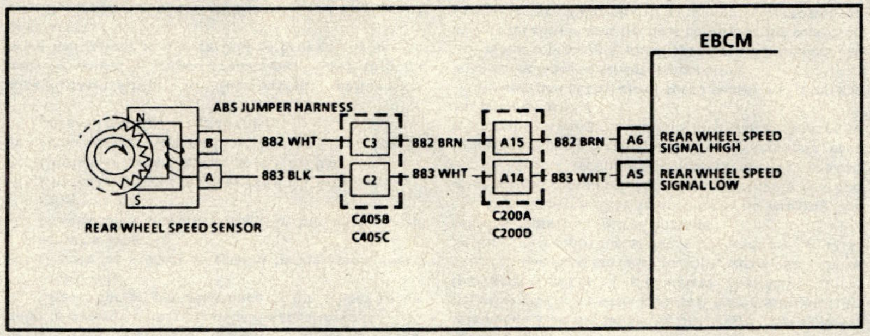

DTC 28

REAR EXCESSIVE WHEEL SPEED VARIATION

Circuit Description:

As a toothed ring passes by the wheel speed sensor, changes in the electromagnetic field cause the wheel speed sensor to produce a sinusoidal (AC) voltage signal whose frequency is proportional to wheel speed. The magnitude of this signal is directly related to wheel speed and the proximity of the wheel speed sensor to the toothed ring, often referred to as the air gap.

DTC Will Set When: DTC 28 can be set when the brake is "OFF." The purpose of this test is to detect a situation in which the right rear wheel acceleration or deceleration is beyond specified limits.

Action Taken: A malfunction DTC is stored, ABS is disabled and the ABS warning lamp is turned "ON."

DTC Chart Test Description: Number(s) below refer to circled number(s) on the diagnostic chart.

1. This step identifies a wheel speed sensor or circuitry that is damaged and visibly apparent.
2. Checks to see that the wheel speed sensor coil is within proper resistance values.
3. Verifies that the wheel speed sensor can generate an appropriate output signal.
4. This ensures that the wheel speed circuitry is not shorted together.
5. Ensures that the malfunction was not due to physical damage.
6. Checks rear sensor toothed ring for physical damage.

Diagnostic Aids: An "Intermittent" malfunction may be caused by a poor connection, rubbed through wire insulation, or a wire that is broken inside the insulation.

The frequency of the malfunction can be checked by using the enhanced diagnostic function of the Tech 1.

If the customer's comments reflect that the ABS warning lamp is "ON" only during moist environmental changes (rain, snow, vehicle wash), all wheel speed sensor circuitry should be thoroughly inspected for signs of water intrusion. If DTC is not current, clear DTCs and simulate the effects of water intrusion. Use the following procedure. Spray down the suspected area with a 5% salt water solution (two teaspoons of salt to 12 oz. of water). Test drive vehicle over various road surfaces (bumps, turns, etc.) above 24 km/h (15 mph) for at least 30 seconds. If DTC returns, replace suspected harness.

Any circuitry, that is suspected as causing the intermittent complaint, should be thoroughly checked for backed out terminals, improper mating, broken locks, improperly formed or damaged terminals, poor terminal to wiring connections or physical damage to the wiring harness.

Resistance of the wheel speed sensor will increase with an increase in sensor temperature.

When replacing a wheel speed sensor, inspect the sensor terminals and harness connector for corrosion and/or water intrusion. If evidence of corrosion or water intrusion exists, replace wheel speed sensor jumper harness. Likewise, if replacing a wheel speed sensor jumper harness, inspect sensor terminals. If evidence of corrosion or water intrusion exists, replace wheel speed sensor.

DIAGNOSTIC CHART DTC 28 — 1994–95 CAMARO AND FIREBIRD

DTC 28
REAR EXCESSIVE WHEEL SPEED VARIATION

Important

- WHEEL SPEED SENSOR INTERMITTENT MALFUNCTIONS MAY BE DIFFICULT TO LOCATE. CARE SHOULD BE TAKEN NOT TO DISTURB ANY ELECTRICAL CONNECTIONS PRIOR TO AN INDICATED STEP OF THIS CHART. THIS WILL ENSURE THAT AN INTERMITTENT CONNECTION WILL NOT BE CORRECTED BEFORE THE SOURCE OF THE MALFUNCTION IS FOUND.

(1)
- IGNITION "OFF."
- PHYSICALLY INSPECT THE REAR WHEEL SPEED SENSOR, WHEEL SPEED SENSOR JUMPER HARNESS, AND CONNECTORS FOR DAMAGE.

IS PHYSICAL DAMAGE INDICATED?

- YES: REPLACE WHEEL SPEED SENSOR AND/OR WHEEL SPEED SENSOR JUMPER HARNESS IF DAMAGED. DRIVE VEHICLE ABOVE 24 km/h (15 MPH) TO ENSURE DTC 28 DOES NOT SET.
- NO: go to (2).

(2)
- DISCONNECT REAR WHEEL SPEED SENSOR DIRECTLY AT SENSOR.
- USING J 39200, MEASURE THE RESISTANCE BETWEEN SENSOR TERMINALS "A" AND "B."

IS RESISTANCE BETWEEN 969 AND 1185Ω WHEN SENSOR IS 20°C (68°F)?
(IF TEMPERATURE OF THE SENSOR IS SUSPECTED TO BE OTHER THAN 20°C (68°F), REFER TO THE TABLE BELOW FOR APPLICABLE SENSOR RESISTANCE VALUES; THE VALUES IN THE TABLE ARE FOR TEMPERATURE OF THE SENSOR, NOT AIR TEMPERATURE.)

- NO: REPLACE WHEEL SPEED SENSOR.
- YES: go to (3).

(3)
- WITH J 39200 STILL CONNECTED, SELECT A/C VOLTAGE SCALE.
- VEHICLE TRANSMISSION IN NEUTRAL.
- SPIN WHEEL BY HAND WHILE OBSERVING VOLTAGE.

IS VOLTAGE AT LEAST 100 mV (VOLTAGE WILL INCREASE AS WHEEL SPEED INCREASES)?

- NO: REPLACE WHEEL SPEED SENSOR.
- YES: go to (4).

(4)
- IGNITION "OFF."
- DISCONNECT 32-WAY EBCM CONNECTOR.
- USING J 39200, MEASURE THE RESISTANCE BETWEEN TERMINAL "A5" AND "A6" OF THE 32-WAY EBCM HARNESS CONNECTOR.

DOES J 39200 DISPLAY RESISTANCE AS "OL" (INFINITE)?

- NO: REPAIR SHORT BETWEEN CKTs 882 AND 883.
- YES: go to (5).

(5)
- INSPECT CKTs 882 AND 883 AND 32-WAY EBCM CONNECTOR FOR DAMAGE WHICH MAY RESULT IN A SHORT BETWEEN CKTs 882 AND 883 WITH ALL CONNECTORS CONNECTED. REPAIR ALL DAMAGE IF EVIDENT.
- RECONNECT ALL CONNECTORS.
- TEST DRIVE VEHICLE ABOVE 24 km/h (15 MPH) FOR AT LEAST 30 SECONDS.

DOES DTC 28 SET AS A CURRENT DTC?

- NO: MALFUNCTION IS INTERMITTENT. REFER TO "DIAGNOSTIC AIDS".
- YES: go to (6).

(6)
- INSPECT REAR SENSOR TOOTHED RING IN REAR AXLE DIFFERENTIAL FOR MISSING TEETH OR OTHER PHYSICAL DAMAGE.

IS ANY PHYSICAL DAMAGE INDICATED?

- NO: REPLACE EBCM.
- YES: REPLACE TOOTHED RING.

REAR WHEEL SPEED SENSOR
TEMPERATURE OF SENSOR VS. SENSOR RESISTANCE (APPROXIMATE)

TEMP. (°C)	TEMP. (°F)	RESISTANCE (OHMS)
-34 TO 4	-30 TO 40	763 TO 1110
5 TO 43	41 TO 110	908 TO 1292
44 TO 93	111 TO 200	1057 TO 1525

DIAGNOSTIC CHART DTC 32 — 1994–95 CAMARO AND FIREBIRD

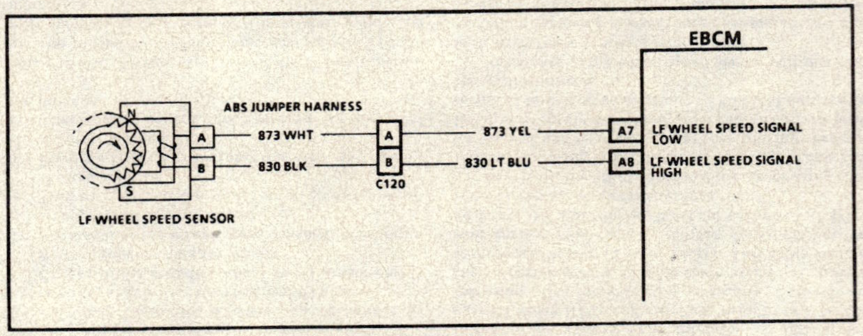

DTC 32

(Page 1 of 3)

LF WHEEL SPEED SNR CKT OPEN OR SHORT TO GND/BATT

Circuit Description:

As a toothed ring passes by the wheel speed sensor, changes in the electromagnetic field cause the wheel speed sensor to produce a sinusoidal (AC) voltage signal whose frequency is proportional to wheel speed. The magnitude of this signal is directly related to wheel speed and the proximity of the wheel speed sensor to the toothed ring often referred to as the air gap.

DTC Will Set When: DTC 32 can be set anytime after initialization. If either of the left front wheel speed circuits are open or shorted to voltage or ground, a malfunction exists.

Action Taken: A malfunction DTC is stored, ABS is disabled and the ABS warning lamp is turned "ON."

DTC Chart Test Description: Number(s) below refer to circled number(s) on the diagnostic chart.

1. Checks for proper resistance of the wheel speed sensor coil.
2. Checks for a short to ground in the wheel speed sensor wiring.
3. Checks for a short to voltage in the wheel speed sensor wiring.
4. Checks for a short to voltage in CKT 873.
5. Checks for a short to voltage in CKT 830.
6. Ensures that the malfunction was not due to physical damage of the circuitry.

Diagnostic Aids: An "Intermittent" malfunction may be caused by a poor connection, rubbed through wire insulation, or a wire that is broken inside the insulation.

The frequency of the malfunction can be checked by using the enhanced diagnostic function of the Tech 1.

If the customer's comments reflect that the ABS warning lamp is "ON" only during moist environmental changes (rain, snow, vehicle wash), all wheel speed sensor circuitry should be thoroughly inspected for signs of water intrusion. If DTC is not current, clear DTCs and simulate the effects of water intrusion. Use the following procedure. Spray down the suspected area with a 5% salt water solution (two teaspoons of salt to 12 oz. of water). Test drive vehicle over various road surfaces (bumps, turns, etc.) above 24 km/h (15 mph) for at least 30 seconds. If DTC returns, replace suspected harness.

Any circuitry, that is suspected as causing the intermittent complaint, should be thoroughly checked for backed out terminals, improper mating, broken locks, improperly formed or damaged terminals, poor terminal to wiring connections or physical damage to the wiring harness.

Resistance of the wheel speed sensor will increase with an increase in sensor temperature.

When replacing a wheel speed sensor, inspect the sensor terminals and harness connector for corrosion and/or water intrusion. If evidence of corrosion or water intrusion exists, replace wheel speed sensor jumper harness. Likewise, if replacing a wheel speed sensor jumper harness, inspect sensor terminals. If evidence of corrosion or water intrusion exists, replace wheel speed sensor.

DIAGNOSTIC CHART DTC 32 — 1994–95 CAMARO AND FIREBIRD

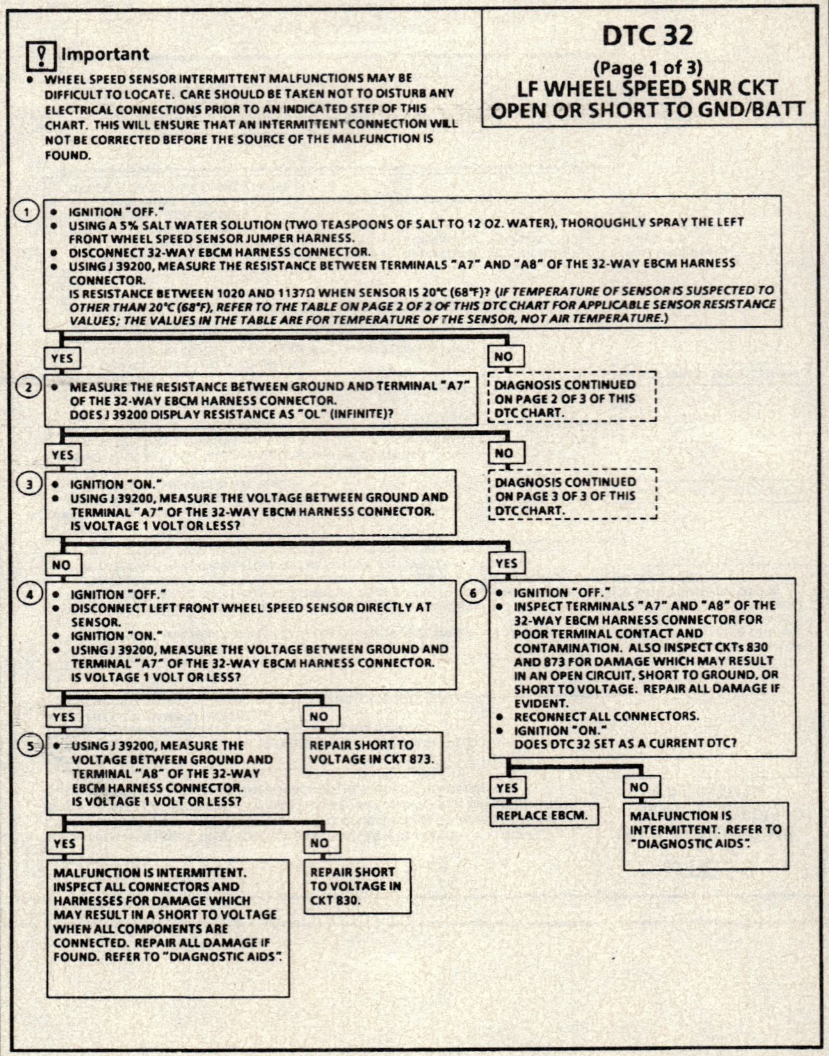

DIAGNOSTIC CHART DTC 32 — 1994–95 CAMARO AND FIREBIRD

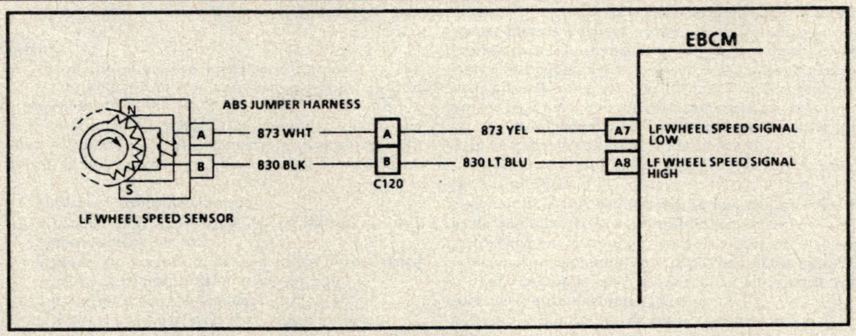

DTC 32
(Page 2 of 3)
LF WHEEL SPEED SNR CKT OPEN OR SHORT TO GND/BATT

Circuit Description:
As a toothed ring passes by the wheel speed sensor, changes in the electromagnetic field cause the wheel speed sensor to produce a sinusoidal (AC) voltage signal whose frequency is proportional to wheel speed. The magnitude of this signal is directly related to wheel speed and the proximity of the wheel speed sensor to the toothed ring often referred to as the air gap.

DTC Will Set When: DTC 32 can be set anytime after initialization. If either of the left front wheel speed circuits are open or shorted to voltage or ground, a malfunction exists.

Action Taken: A malfunction DTC is stored, ABS is disabled and the ABS warning lamp is turned "ON."

DTC Chart Test Description: Number(s) below refer to circled number(s) on the diagnostic chart.
7. Checks for high resistance in CKT 873.
8. Checks for high resistance in CKT 830.
9. Verifies proper resistance of the wheel speed sensor coil

Diagnostic Aids: An "Intermittent" malfunction may be caused by a poor connection, rubbed through wire insulation, or a wire that is broken inside the insulation.

The frequency of the malfunction can be checked by using the enhanced diagnostic function of the Tech 1.

If the customer's comments reflect that the ABS warning lamp is "ON" only during moist environmental changes (rain, snow, vehicle wash), all wheel speed sensor circuitry should be thoroughly inspected for signs of water intrusion. If DTC is not current, clear DTCs and simulate the effects of water intrusion. Use the following procedure. Spray down the suspected area with a 5% salt water solution (two teaspoons of salt to 12 oz. of water). Test drive vehicle over various road surfaces (bumps, turns, etc.) above 24 km/h (15 mph) for at least 30 seconds. If DTC returns, replace suspected harness.

Any circuitry, that is suspected as causing the intermittent complaint, should be thoroughly checked for backed out terminals, improper mating, broken locks, improperly formed or damaged terminals, poor terminal to wiring connections or physical damage to the wiring harness.

Resistance of the wheel speed sensor will increase with an increase in sensor temperature.

When replacing a wheel speed sensor, inspect the sensor terminals and harness connector for corrosion and/or water intrusion. If evidence of corrosion or water intrusion exists, replace wheel speed sensor jumper harness. Likewise, if replacing a wheel speed sensor jumper harness, inspect sensor terminals. If evidence of corrosion or water intrusion exists, replace wheel speed sensor.

DIAGNOSTIC CHART DTC 32 — 1994–95 CAMARO AND FIREBIRD

DTC 32
(Page 2 of 3)
LF WHEEL SPEED SNR CKT OPEN OR SHORT TO GND/BATT

DIAGNOSIS CONTINUED FROM PAGE 1 OF 3 OF THIS DTC CHART.

(7)
- DISCONNECT LEFT FRONT WHEEL SPEED SENSOR DIRECTLY AT SENSOR.
- USING J 39200, MEASURE THE RESISTANCE BETWEEN TERMINAL "A7" OF THE 32-WAY EBCM HARNESS CONNECTOR AND TERMINAL "B" OF THE LEFT FRONT WHEEL SPEED SENSOR HARNESS CONNECTOR. IS RESISTANCE 2Ω OR LESS?

NO: REPAIR HIGH RESISTANCE IN CKT 873 (WHEEL SPEED SENSOR JUMPER MUST BE REPLACED IF DAMAGED).

YES:

(8)
- USING J 39200, MEASURE THE RESISTANCE BETWEEN TERMINAL "A8" OF THE 32-WAY EBCM HARNESS CONNECTOR AND TERMINAL "A" OF THE LEFT FRONT WHEEL SPEED SENSOR HARNESS CONNECTOR. IS RESISTANCE 2Ω OR LESS?

NO: REPAIR HIGH RESISTANCE IN CKT 830 (WHEEL SPEED SENSOR JUMPER MUST BE REPLACED IF DAMAGED).

YES:

(9)
- USING J 39200, MEASURE THE RESISTANCE BETWEEN TERMINAL "A" AND TERMINAL "B" OF THE LEFT FRONT WHEEL SPEED SENSOR CONNECTOR. IS RESISTANCE BETWEEN 1020 AND 1137Ω WHEN SENSOR IS 20°C (68°F)? *(IF TEMPERATURE OF SENSOR IS SUSPECTED TO OTHER THAN 20°C (68°F), REFER TO THE TABLE BELOW FOR APPLICABLE SENSOR RESISTANCE VALUES; THE VALUES IN THE TABLE ARE FOR TEMPERATURE OF THE SENSOR, NOT AIR TEMPERATURE.)*

NO: REPLACE WHEEL SPEED SENSOR.

YES: MALFUNCTION IS INTERMITTENT. INSPECT ALL CONNECTORS AND HARNESSES FOR DAMAGE WHICH MAY RESULT IN HIGH RESISTANCE WHEN ALL COMPONENTS ARE CONNECTED. REFER TO "DIAGNOSTIC AIDS". REPAIR ALL DAMAGE IF FOUND.

FRONT WHEEL SPEED SENSOR

TEMPERATURE OF SENSOR VS. SENSOR RESISTANCE (APPROXIMATE)

TEMP. (°C)	TEMP. (°F)	RESISTANCE (OHMS)
-34 TO 4	-30 TO 40	804 TO 1066
5 TO 43	41 TO 110	956 TO 1240
44 TO 93	111 TO 200	1112 TO 1463

DIAGNOSTIC CHART DTC 32 — 1994–95 CAMARO AND FIREBIRD

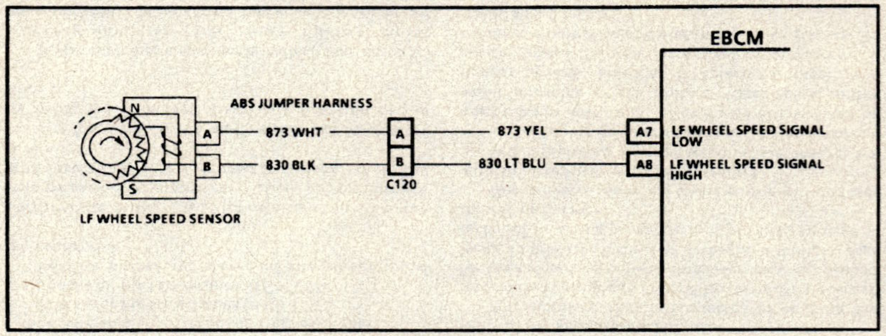

DTC 32

LF WHEEL SPEED SNR CKT OPEN OR SHORT TO GND/BATT

Circuit Description:

As a toothed ring passes by the wheel speed sensor, changes in the electromagnetic field cause the wheel speed sensor to produce a sinusoidal (AC) voltage signal whose frequency is proportional to wheel speed. The magnitude of this signal is directly related to wheel speed and the proximity of the wheel speed sensor to the toothed ring often referred to as the air gap.

DTC Will Set When: DTC 32 can be set anytime after initialization. If either of the left front wheel speed circuits are open or shorted to voltage or ground, a malfunction exists.

Action Taken: A malfunction DTC is stored, ABS is disabled and the ABS warning lamp is turned "ON."

DTC Chart Test Description: Number(s) below refer to circled number(s) on the diagnostic chart.

10. Checks for a short to ground in CKT 873.
11. Checks for a short to ground in CKT 830.
12. Checks for a short to ground in the wheel speed sensor jumper harness.
13. Verifies that the wheel speed sensor is not internally shorted to ground.

Diagnostic Aids: An "Intermittent" malfunction may be caused by a poor connection, rubbed through wire insulation, or a wire that is broken inside the insulation.

The frequency of the malfunction can be checked by using the enhanced diagnostic function of the Tech 1.

If the customer's comments reflect that the ABS warning lamp is "ON" only during moist environmental changes (rain, snow, vehicle wash), all wheel speed sensor circuitry should be thoroughly inspected for signs of water intrusion. If DTC is not current, clear DTCs and simulate the effects of water intrusion. Use the following procedure. Spray down the suspected area with a 5% salt water solution (two teaspoons of salt to 12 oz. of water). Test drive vehicle over various road surfaces (bumps, turns, etc.) above 24 km/h (15 mph) for at least 30 seconds. If DTC returns, replace suspected harness.

Any circuitry, that is suspected as causing the intermittent complaint, should be thoroughly checked for backed out terminals, improper mating, broken locks, improperly formed or damaged terminals, poor terminal to wiring connections or physical damage to the wiring harness.

Resistance of the wheel speed sensor will increase with an increase in sensor temperature.

When replacing a wheel speed sensor, inspect the sensor terminals and harness connector for corrosion and/or water intrusion. If evidence of corrosion or water intrusion exists, replace wheel speed sensor jumper harness. Likewise, if replacing a wheel speed sensor jumper harness, inspect sensor terminals. If evidence of corrosion or water intrusion exists, replace wheel speed sensor.

DIAGNOSTIC CHART DTC 32 — 1994–95 CAMARO AND FIREBIRD

DTC 32

(Page 3 of 3)

LF WHEEL SPEED SNR CKT OPEN OR SHORT TO GND/BATT

DIAGNOSIS CONTINUED FROM PAGE 1 OF 3 OF THIS DTC CHART.

(10)
- DISCONNECT CONNECTOR C120 (LOCATED LH INNER FRAME RAIL).
- USING J 39200, MEASURE THE RESISTANCE BETWEEN GROUND AND TERMINAL "A7" OF THE 32-WAY EBCM HARNESS CONNECTOR.

DOES J 39200 DISPLAY RESISTANCE AS "OL" (INFINITE)?

- NO: REPAIR SHORT TO GROUND IN CKT 873.
- YES:

(11)
- USING J 39200, MEASURE THE RESISTANCE BETWEEN GROUND AND TERMINAL "A8" OF THE 32-WAY EBCM HARNESS CONNECTOR.

DOES J 39200 DISPLAY RESISTANCE AS "OL" (INFINITE)?

- NO: REPAIR SHORT TO GROUND IN CKT 830.
- YES:

(12)
- THOROUGHLY RESPRAY THE LEFT FRONT WHEEL SPEED SENSOR JUMPER HARNESS WITH THE SALT WATER SOLUTION.
- IMPORTANT: DO NOT ALLOW WATER TO ENTER ANY WHEEL SPEED SENSOR CONNECTOR.
- DISCONNECT LEFT FRONT WHEEL SPEED SENSOR DIRECTLY AT SENSOR.
- USING J 39200, MEASURE THE RESISTANCE BETWEEN GROUND AND TERMINAL "A," THEN "B" OF THE LEFT FRONT WHEEL SPEED SENSOR HARNESS CONNECTOR.

DOES J 39200 DISPLAY RESISTANCE AS "OL" (INFINITE) FOR BOTH MEASUREMENTS?

- NO: REPLACE LEFT FRONT WHEEL SPEED SENSOR JUMPER HARNESS.
- YES:

(13)
- USING J 39200, MEASURE THE RESISTANCE BETWEEN GROUND AND TERMINAL "A" OF THE LEFT FRONT WHEEL SPEED SENSOR CONNECTOR.

DOES J 39200 DISPLAY RESISTANCE AS "OL" (INFINITE)?

- NO: REPLACE WHEEL SPEED SENSOR.
- YES: MALFUNCTION IS INTERMITTENT. INSPECT ALL CONNECTORS AND HARNESSES FOR DAMAGE WHICH MAY RESULT IN A SHORT TO GROUND WHEN ALL COMPONENTS ARE CONNECTED. PAY PARTICULAR ATTENTION TO THE JUMPER HARNESS, SMALL NICKS OR CUTS CAN CAUSE A MALFUNCTION. REPAIR ANY DAMAGE IF FOUND. JUMPER HARNESS MUST BE REPLACED IF DAMAGED. REFER TO "DIAGNOSTIC AIDS".

DIAGNOSTIC CHART DTC 33 — 1994–95 CAMARO AND FIREBIRD

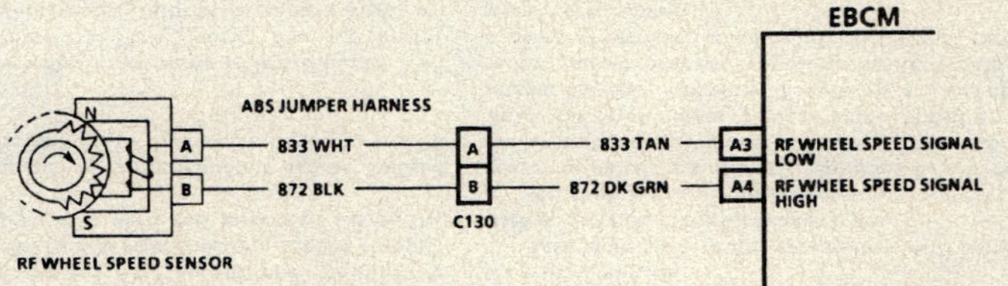

DTC 33

RF WHEEL SPEED SNR CKT OPEN OR SHORT TO GND/BATT

Circuit Description:

As a toothed ring passes by the wheel speed sensor, changes in the electromagnetic field cause the wheel speed sensor to produce a sinusoidal (AC) voltage signal whose frequency is proportional to wheel speed. The magnitude of this signal is directly related to wheel speed and the proximity of the wheel speed sensor to the toothed ring, often referred to as the air gap.

DTC Will Set When: DTC 33 can be set anytime after initialization. If either of the right front wheel speed circuits are open or shorted to voltage or ground, a malfunction exists.

Action Taken: A malfunction DTC is stored, ABS is disabled and the ABS warning lamp is turned "ON."

DTC Chart Test Description: Number(s) below refer to circled number(s) on the diagnostic chart.

1. Checks for proper resistance of the wheel speed sensor coil.
2. Checks for a short to ground in the wheel speed sensor wiring.
3. Checks for a short to voltage in the wheel speed sensor wiring.
4. Checks for a short to voltage in CKT 833.
5. Checks for a short to voltage in CKT 872.
6. Ensures that the malfunction was not due to physical damage of the circuitry.

Diagnostic Aids: An "Intermittent" malfunction may be caused by a poor connection, rubbed through wire insulation, or a wire that is broken inside the insulation.

The frequency of the malfunction can be checked by using the enhanced diagnostic function of the Tech 1.

If the customer's comments reflect that the ABS warning lamp is "ON" only during moist environmental changes (rain, snow, vehicle wash), all wheel speed sensor circuitry should be thoroughly inspected for signs of water intrusion. If DTC is not current, clear DTCs and simulate the effects of water intrusion. Use the following procedure. Spray down the suspected area with a 5% salt water solution (two teaspoons of salt to 12 oz. of water). Test drive vehicle over various road surfaces (bumps, turns, etc.) above 24 km/h (15 mph) for at least 30 seconds. If DTC returns, replace suspected harness.

Any circuitry, that is suspected as causing the intermittent complaint, should be thoroughly checked for backed out terminals, improper mating, broken locks, improperly formed or damaged terminals, poor terminal to wiring connections or physical damage to the wiring harness.

Resistance of the wheel speed sensor will increase with an increase in sensor temperature.

When replacing a wheel speed sensor, inspect the sensor terminals and harness connector for corrosion and/or water intrusion. If evidence of corrosion or water intrusion exists, replace wheel speed sensor jumper harness. Likewise, if replacing a wheel speed sensor jumper harness, inspect sensor terminals. If evidence of corrosion or water intrusion exists, replace wheel speed sensor.

DIAGNOSTIC CHART DTC 33 — 1994–95 CAMARO AND FIREBIRD

DTC 33
(Page 1 of 3)
RF WHEEL SPEED SNR CKT OPEN OR SHORT TO GND/BATT

Important

- WHEEL SPEED SENSOR INTERMITTENT MALFUNCTIONS MAY BE DIFFICULT TO LOCATE. CARE SHOULD BE TAKEN NOT TO DISTURB ANY ELECTRICAL CONNECTIONS PRIOR TO AN INDICATED STEP OF THIS CHART. THIS WILL ENSURE THAT AN INTERMITTENT CONNECTION WILL NOT BE CORRECTED BEFORE THE SOURCE OF THE MALFUNCTION IS FOUND.

(1)
- IGNITION "OFF."
- USING A 5% SALT WATER SOLUTION (TWO TEASPOONS OF SALT TO 12 OZ. WATER), THOROUGHLY SPRAY THE RIGHT FRONT WHEEL SPEED SENSOR JUMPER HARNESS.
- DISCONNECT 32-WAY EBCM HARNESS CONNECTOR.
- USING J 39200, MEASURE THE RESISTANCE BETWEEN TERMINALS "A3" AND "A4" OF THE 32-WAY EBCM HARNESS CONNECTOR.

IS RESISTANCE BETWEEN 1020 AND 1137Ω WHEN SENSOR IS 20°C (68°F)? *(IF TEMPERATURE OF SENSOR IS SUSPECTED TO OTHER THAN 20°C (68°F), REFER TO THE TABLE ON PAGE 2 OF 2 OF THIS DTC CHART FOR APPLICABLE SENSOR RESISTANCE VALUES; THE VALUES IN THE TABLE ARE FOR TEMPERATURE OF THE SENSOR, NOT AIR TEMPERATURE.)*

- YES → (2)
- NO → DIAGNOSIS CONTINUED ON PAGE 2 OF 3 OF THIS DTC CHART.

(2)
- MEASURE THE RESISTANCE BETWEEN GROUND AND TERMINAL "A3" OF THE 32-WAY EBCM HARNESS CONNECTOR.

DOES J 39200 DISPLAY RESISTANCE AS "OL" (INFINITE)?

- YES → (3)
- NO → DIAGNOSIS CONTINUED ON PAGE 3 OF 3 OF THIS DTC CHART.

(3)
- IGNITION "ON."
- USING J 39200, MEASURE THE VOLTAGE BETWEEN GROUND AND TERMINAL "A3" OF THE 32-WAY EBCM HARNESS CONNECTOR.

IS VOLTAGE 1 VOLT OR LESS?

- NO → (4)
- YES → (6)

(4)
- IGNITION "OFF."
- DISCONNECT RIGHT FRONT WHEEL SPEED SENSOR DIRECTLY AT SENSOR.
- IGNITION "ON."
- USING J 39200, MEASURE THE VOLTAGE BETWEEN GROUND AND TERMINAL "A3" OF THE 32-WAY EBCM HARNESS CONNECTOR.

IS VOLTAGE 1 VOLT OR LESS?

- YES → (5)
- NO → REPAIR SHORT TO VOLTAGE IN CKT 833.

(5)
- USING J 39200, MEASURE THE VOLTAGE BETWEEN GROUND AND TERMINAL "A4" OF THE 32-WAY EBCM HARNESS CONNECTOR.

IS VOLTAGE 1 VOLT OR LESS?

- YES → MALFUNCTION IS INTERMITTENT. INSPECT ALL CONNECTORS AND HARNESSES FOR DAMAGE WHICH MAY RESULT IN A SHORT TO VOLTAGE WHEN ALL COMPONENTS ARE CONNECTED. REPAIR ALL DAMAGE IF FOUND. REFER TO "DIAGNOSTIC AIDS."
- NO → REPAIR SHORT TO VOLTAGE IN CKT 872.

(6)
- IGNITION "OFF."
- INSPECT TERMINALS "A3" AND "A4" OF THE 32-WAY EBCM HARNESS CONNECTOR FOR POOR TERMINAL CONTACT AND CONTAMINATION. ALSO INSPECT CKTs 833 AND 872 FOR DAMAGE WHICH MAY RESULT IN AN OPEN CIRCUIT, SHORT TO GROUND, OR SHORT TO VOLTAGE. REPAIR ALL DAMAGE IF EVIDENT.
- RECONNECT ALL CONNECTORS.
- IGNITION "ON."

DOES DTC 33 SET AS A CURRENT DTC?

- YES → REPLACE EBCM.
- NO → MALFUNCTION IS INTERMITTENT. REFER TO "DIAGNOSTIC AIDS".

DIAGNOSTIC CHART DTC 33 — 1994–95 CAMARO AND FIREBIRD

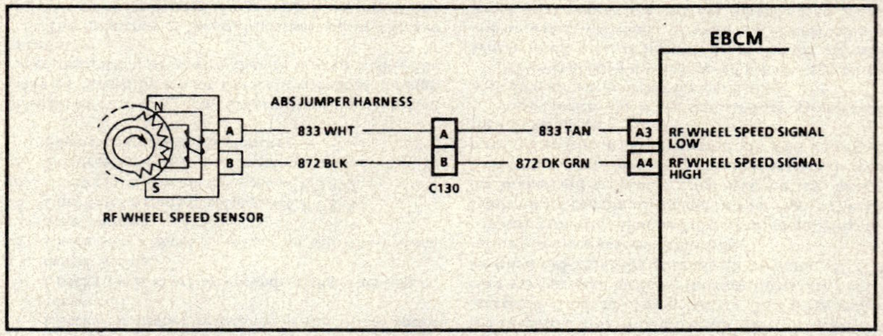

DTC 33

RF WHEEL SPEED SNR CKT OPEN OR SHORT TO GND/BATT

Circuit Description:

As a toothed ring passes by the wheel speed sensor, changes in the electromagnetic field cause the wheel speed sensor to produce a sinusoidal (AC) voltage signal whose frequency is proportional to wheel speed. The magnitude of this signal is directly related to wheel speed and the proximity of the wheel speed sensor to the toothed ring, often referred to as the air gap.

DTC Will Set When: DTC 33 can be set anytime after initialization. If either of the right front wheel speed circuits are open or shorted to voltage or ground, a malfunction exists.

Action Taken: A malfunction DTC is stored, ABS is disabled and the ABS warning lamp is turned "ON."

DTC Chart Test Description: Number(s) below refer to circled number(s) on the diagnostic chart.

7. Checks for high resistance in CKT 833.
8. Checks for high resistance in CKT 872.
9. Verifies proper resistance of the wheel speed sensor coil.

Diagnostic Aids: An "Intermittent" malfunction may be caused by a poor connection, rubbed through wire insulation, or a wire that is broken inside the insulation.

The frequency of the malfunction can be checked by using the enhanced diagnostic function of the Tech 1.

If the customer's comments reflect that the ABS warning lamp is "ON" only during moist environmental changes (rain, snow, vehicle wash), all wheel speed sensor circuitry should be thoroughly inspected for signs of water intrusion. If DTC is not current, clear DTCs and simulate the effects of water intrusion. Use the following procedure. Spray down the suspected area with a 5% salt water solution (two teaspoons of salt to 12 oz. of water). Test drive vehicle over various road surfaces (bumps, turns, etc.) above 24 km/h (15 mph) for at least 30 seconds. If DTC returns, replace suspected harness.

Any circuitry, that is suspected as causing the intermittent complaint, should be thoroughly checked for backed out terminals, improper mating, broken locks, improperly formed or damaged terminals, poor terminal to wiring connections or physical damage to the wiring harness.

Resistance of the wheel speed sensor will increase with an increase in sensor temperature.

When replacing a wheel speed sensor, inspect the sensor terminals and harness connector for corrosion and/or water intrusion. If evidence of corrosion or water intrusion exists, replace wheel speed sensor jumper harness. Likewise, if replacing a wheel speed sensor jumper harness, inspect sensor terminals. If evidence of corrosion or water intrusion exists, replace wheel speed sensor.

DIAGNOSTIC CHART DTC 33 — 1994–95 CAMARO AND FIREBIRD

DTC 33
(Page 2 of 3)
RF WHEEL SPEED SNR CKT OPEN OR SHORT TO GND/BATT

DIAGNOSIS CONTINUED FROM PAGE 1 OF 3 OF THIS DTC CHART.

(7)
- DISCONNECT RIGHT FRONT WHEEL SPEED SENSOR DIRECTLY AT SENSOR.
- USING J 39200, MEASURE THE RESISTANCE BETWEEN TERMINAL "A3" OF THE 32-WAY EBCM HARNESS CONNECTOR AND TERMINAL "A" OF THE RIGHT FRONT WHEEL SPEED SENSOR HARNESS CONNECTOR.

IS RESISTANCE 2Ω OR LESS?

NO: REPAIR HIGH RESISTANCE IN CKT 833 (WHEEL SPEED SENSOR JUMPER MUST BE REPLACED IF DAMAGED).

YES:

(8)
- USING J 39200, MEASURE THE RESISTANCE BETWEEN TERMINAL "A4" OF THE 32-WAY EBCM HARNESS CONNECTOR AND TERMINAL "B" OF THE RIGHT FRONT WHEEL SPEED SENSOR HARNESS CONNECTOR.

IS RESISTANCE 2Ω OR LESS?

NO: REPAIR HIGH RESISTANCE IN CKT 872 (WHEEL SPEED SENSOR JUMPER MUST BE REPLACED IF DAMAGED).

YES:

(9)
- USING J 39200, MEASURE THE RESISTANCE BETWEEN TERMINAL "A" AND TERMINAL "B" OF THE RIGHT FRONT WHEEL SPEED SENSOR CONNECTOR.

IS RESISTANCE BETWEEN 1020 AND 1137Ω WHEN SENSOR IS 20°C (68°F)? *(IF TEMPERATURE OF SENSOR IS SUSPECTED TO OTHER THAN 20°C (68°F), REFER TO THE TABLE BELOW FOR APPLICABLE SENSOR RESISTANCE VALUES; THE VALUES IN THE TABLE ARE FOR TEMPERATURE OF THE SENSOR, NOT AIR TEMPERATURE.)*

YES: MALFUNCTION IS INTERMITTENT. INSPECT ALL CONNECTORS AND HARNESSES FOR DAMAGE WHICH MAY RESULT IN HIGH RESISTANCE WHEN ALL COMPONENTS ARE CONNECTED. REFER TO "DIAGNOSTIC AIDS." REPAIR ALL DAMAGE IF FOUND.

NO: REPLACE WHEEL SPEED SENSOR.

FRONT WHEEL SPEED SENSOR

TEMPERATURE OF SENSOR VS. SENSOR RESISTANCE (APPROXIMATE)

TEMP. (°C)	TEMP. (°F)	RESISTANCE (OHMS)
-34 TO 4	-30 TO 40	804 TO 1066
5 TO 43	41 TO 110	956 TO 1240
44 TO 93	111 TO 200	1112 TO 1463

DIAGNOSTIC CHART DTC 33 — 1994–95 CAMARO AND FIREBIRD

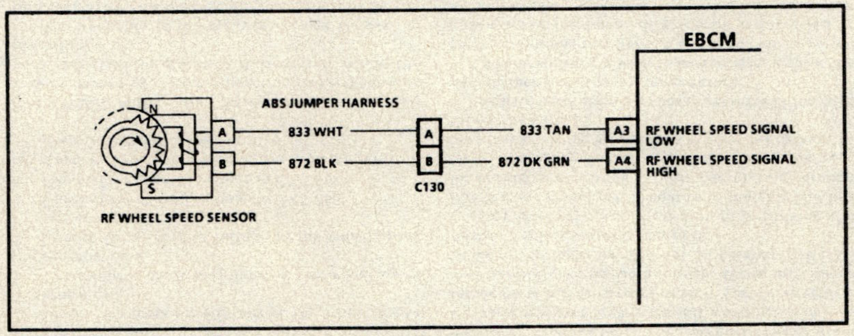

DTC 33

RF WHEEL SPEED SNR CKT OPEN OR SHORT TO GND/BATT

Circuit Description:

As a toothed ring passes by the wheel speed sensor, changes in the electromagnetic field cause the wheel speed sensor to produce a sinusoidal (AC) voltage signal whose frequency is proportional to wheel speed. The magnitude of this signal is directly related to wheel speed and the proximity of the wheel speed sensor to the toothed ring, often referred to as the air gap.

DTC Will Set When: DTC 33 can be set anytime after initialization. If either of the right front wheel speed circuits are open or shorted to voltage or ground, a malfunction exists.

Action Taken: A malfunction DTC is stored, ABS is disabled and the ABS warning lamp is turned "ON."

DTC Chart Test Description: Number(s) below refer to circled number(s) on the diagnostic chart.

10. Checks for a short to ground in CKT 833.
11. Checks for a short to ground in CKT 872.
12. Checks for a short to ground in the wheel speed sensor jumper harness.
13. Verifies that the wheel speed sensor is not internally shorted to ground.

Diagnostic Aids: An "Intermittent" malfunction may be caused by a poor connection, rubbed through wire insulation, or a wire that is broken inside the insulation.

The frequency of the malfunction can be checked by using the enhanced diagnostic function of the Tech 1.

If the customer's comments reflect that the ABS warning lamp is "ON" only during moist environmental changes (rain, snow, vehicle wash), all wheel speed sensor circuitry should be thoroughly inspected for signs of water intrusion. If DTC is not current, clear DTCs and simulate the effects of water intrusion. Use the following procedure. Spray down the suspected area with a 5% salt water solution (two teaspoons of salt to 12 oz. of water). Test drive vehicle over various road surfaces (bumps, turns, etc.) above 24 km/h (15 mph) for at least 30 seconds. If DTC returns, replace suspected harness.

Any circuitry, that is suspected as causing the intermittent complaint, should be thoroughly checked for backed out terminals, improper mating, broken locks, improperly formed or damaged terminals, poor terminal to wiring connections or physical damage to the wiring harness.

Resistance of the wheel speed sensor will increase with an increase in sensor temperature.

When replacing a wheel speed sensor, inspect the sensor terminals and harness connector for corrosion and/or water intrusion. If evidence of corrosion or water intrusion exists, replace wheel speed sensor jumper harness. Likewise, if replacing a wheel speed sensor jumper harness, inspect sensor terminals. If evidence of corrosion or water intrusion exists, replace wheel speed sensor.

DIAGNOSTIC CHART DTC 33 — 1994–95 CAMARO AND FIREBIRD

DTC 33

(Page 3 of 3)

RF WHEEL SPEED SNR CKT OPEN OR SHORT TO GND/BATT

DIAGNOSIS CONTINUED FROM PAGE 1 OF 3 OF THIS DTC CHART.

(10)
- DISCONNECT CONNECTOR C130 (LOCATED RH INNER FRAME RAIL).
- USING J 39200, MEASURE THE RESISTANCE BETWEEN GROUND AND TERMINAL "A3" OF THE 32-WAY EBCM HARNESS CONNECTOR.

DOES J 39200 DISPLAY RESISTANCE AS "OL" (INFINITE)?

NO: REPAIR SHORT TO GROUND IN CKT 833.

YES:

(11)
- USING J 39200, MEASURE THE RESISTANCE BETWEEN GROUND AND TERMINAL "A4" OF THE 32-WAY EBCM HARNESS CONNECTOR.

DOES J 39200 DISPLAY RESISTANCE AS "OL" (INFINITE)?

NO: REPAIR SHORT TO GROUND IN CKT 872.

YES:

(12)
- THOROUGHLY RESPRAY THE RIGHT FRONT WHEEL SPEED SENSOR JUMPER HARNESS WITH THE SALT WATER SOLUTION.
- IMPORTANT: DO NOT ALLOW WATER TO ENTER ANY WHEEL SPEED SENSOR CONNECTOR.
- DISCONNECT RIGHT FRONT WHEEL SPEED SENSOR DIRECTLY AT SENSOR.
- USING J 39200, MEASURE THE RESISTANCE BETWEEN GROUND AND TERMINAL "A," THEN "B" OF THE RIGHT FRONT WHEEL SPEED SENSOR HARNESS CONNECTOR.

DOES J 39200 DISPLAY RESISTANCE AS "OL" (INFINITE) FOR BOTH MEASUREMENTS?

NO: REPLACE RIGHT FRONT WHEEL SPEED SENSOR.

YES:

(13)
- USING J 39200, MEASURE THE RESISTANCE BETWEEN GROUND AND TERMINAL "A" OF THE RIGHT FRONT WHEEL SPEED SENSOR CONNECTOR.

DOES J 39200 DISPLAY RESISTANCE AS "OL" (INFINITE)?

NO: REPLACE WHEEL SPEED SENSOR.

YES: MALFUNCTION IS INTERMITTENT. INSPECT ALL CONNECTORS AND HARNESSES FOR DAMAGE WHICH MAY RESULT IN A SHORT TO GROUND WHEN ALL COMPONENTS ARE CONNECTED. PAY PARTICULAR ATTENTION TO THE JUMPER HARNESS, SMALL NICKS OR CUTS CAN CAUSE A MALFUNCTION. REPAIR ANY DAMAGE IF FOUND. JUMPER HARNESS MUST BE REPLACED IF DAMAGED. REFER TO "DIAGNOSTIC AIDS".

DIAGNOSTIC CHART DTC 35 — 1994–95 CAMARO AND FIREBIRD

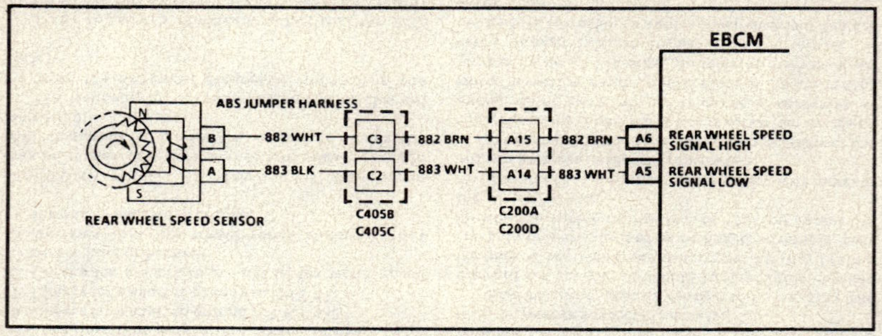

DTC 35

RR WHEEL SPEED SNR CKT OPEN OR SHORT TO GND/BATT

Circuit Description:

As a toothed ring passes by the wheel speed sensor, changes in the electromagnetic field cause the wheel speed sensor to produce a sinusoidal (AC) voltage signal whose frequency is proportional to wheel speed. The magnitude of this signal is directly related to wheel speed and the proximity of the wheel speed sensor to the toothed ring, often referred to as the air gap.

DTC Will Set When: DTC 35 can be set anytime after initialization. If either of the right rear wheel speed circuits are open or shorted to voltage or ground, a malfunction exists.

Action Taken: A malfunction DTC is stored, ABS is disabled and the ABS warning lamp is turned "ON."

DTC Chart Test Description: Number(s) below refer to circled number(s) on the diagnostic chart.

1. Checks for proper resistance of the wheel speed sensor coil.
2. Checks for a short to ground in the wheel speed sensor wiring.
3. Checks for a short to voltage in the wheel speed sensor wiring.
4. Checks for a short to voltage in CKT 883.
5. Checks for a short to voltage in CKT 882.
6. Ensures that the malfunction was not due to physical damage of the circuitry.

Diagnostic Aids: An "Intermittent" malfunction may be caused by a poor connection, rubbed through wire insulation, or a wire that is broken inside the insulation.

The frequency of the malfunction can be checked by using the enhanced diagnostic function of the Tech 1.

If the customer's comments reflect that the ABS warning lamp is "ON" only during moist environmental changes (rain, snow, vehicle wash), all wheel speed sensor circuitry should be thoroughly inspected for signs of water intrusion. If DTC is not current, clear DTCs and simulate the effects of water intrusion. Use the following procedure. Spray down the suspected area with a 5% salt water solution (two teaspoons of salt to 12 oz. of water). Test drive vehicle over various road surfaces (bumps, turns, etc.) above 24 km/h (15 mph) for at least 30 seconds. If DTC returns, replace suspected harness.

Any circuitry, that is suspected as causing the intermittent complaint, should be thoroughly checked for backed out terminals, improper mating, broken locks, improperly formed or damaged terminals, poor terminal to wiring connections or physical damage to the wiring harness.

Resistance of the wheel speed sensor will increase with an increase in sensor temperature.

When replacing a wheel speed sensor, inspect the sensor terminals and harness connector for corrosion and/or water intrusion. If evidence of corrosion or water intrusion exists, replace wheel speed sensor jumper harness. Likewise, if replacing a wheel speed sensor jumper harness, inspect sensor terminals. If evidence of corrosion or water intrusion exists, replace wheel speed sensor.

DIAGNOSTIC CHART DTC 35 — 1994–95 CAMARO AND FIREBIRD

DTC 35

(Page 1 of 3)

RR WHEEL SPEED SNR CKT OPEN OR SHORT TO GND/BATT

Important

- WHEEL SPEED SENSOR INTERMITTENT MALFUNCTIONS MAY BE DIFFICULT TO LOCATE. CARE SHOULD BE TAKEN NOT TO DISTURB ANY ELECTRICAL CONNECTIONS PRIOR TO AN INDICATED STEP OF THIS CHART. THIS WILL ENSURE THAT AN INTERMITTENT CONNECTION WILL NOT BE CORRECTED BEFORE THE SOURCE OF THE MALFUNCTION IS FOUND.

(1)
- IGNITION "OFF."
- USING A 5% SALT WATER SOLUTION (TWO TEASPOONS OF SALT TO 12 OZ. WATER), THOROUGHLY SPRAY THE REAR WHEEL SPEED SENSOR JUMPER HARNESS.
- DISCONNECT 32-WAY EBCM HARNESS CONNECTOR.
- USING J 39200, MEASURE THE RESISTANCE BETWEEN TERMINALS "A5" AND "A6" OF THE 32-WAY EBCM HARNESS CONNECTOR.

IS RESISTANCE BETWEEN 969 AND 1185Ω WHEN SENSOR IS 20°C (68°F)? *(IF TEMPERATURE OF SENSOR IS SUSPECTED TO OTHER THAN 20°C (68°F), REFER TO THE TABLE ON PAGE 2 OF 2 OF THIS DTC CHART FOR APPLICABLE SENSOR RESISTANCE VALUES; THE VALUES IN THE TABLE ARE FOR TEMPERATURE OF THE SENSOR, NOT AIR TEMPERATURE.)*

- NO → DIAGNOSIS CONTINUED ON PAGE 2 OF 3 OF THIS DTC CHART.
- YES →

(2)
- MEASURE THE RESISTANCE BETWEEN GROUND AND TERMINAL "A5" OF THE 32-WAY EBCM HARNESS CONNECTOR.

DOES J 39200 DISPLAY RESISTANCE AS "OL" (INFINITE)?

- NO → DIAGNOSIS CONTINUED ON PAGE 3 OF 3 OF THIS DTC CHART.
- YES →

(3)
- IGNITION "ON."
- USING J 39200, MEASURE THE VOLTAGE BETWEEN GROUND AND TERMINAL "A5" OF THE 32-WAY EBCM HARNESS CONNECTOR.

IS VOLTAGE 1 VOLT OR LESS?

- NO →

(4)
- IGNITION "OFF."
- DISCONNECT REAR WHEEL SPEED SENSOR DIRECTLY AT SENSOR.
- IGNITION "ON."
- USING J 39200, MEASURE THE VOLTAGE BETWEEN GROUND AND TERMINAL "A5" OF THE 32-WAY EBCM HARNESS CONNECTOR.

IS VOLTAGE 1 VOLT OR LESS?

- NO → REPAIR SHORT TO VOLTAGE IN CKT 883.
- YES →

(5)
- USING J 39200, MEASURE THE VOLTAGE BETWEEN GROUND AND TERMINAL "A6" OF THE 32-WAY EBCM HARNESS CONNECTOR.

IS VOLTAGE 1 VOLT OR LESS?

- YES → MALFUNCTION IS INTERMITTENT. INSPECT ALL CONNECTORS AND HARNESSES FOR DAMAGE WHICH MAY RESULT IN A SHORT TO VOLTAGE WHEN ALL COMPONENTS ARE CONNECTED. REPAIR ALL DAMAGE IF FOUND. REFER TO "DIAGNOSTIC AIDS".
- NO → REPAIR SHORT TO VOLTAGE IN CKT 882.

(From step 3) YES →

(6)
- IGNITION "OFF."
- INSPECT TERMINALS "A5" AND "A6" OF THE 32-WAY EBCM HARNESS CONNECTOR FOR POOR TERMINAL CONTACT AND CONTAMINATION. ALSO INSPECT CKTs 882 AND 883 FOR DAMAGE WHICH MAY RESULT IN AN OPEN CIRCUIT, SHORT TO GROUND, OR SHORT TO VOLTAGE. REPAIR ALL DAMAGE IF EVIDENT.
- RECONNECT ALL CONNECTORS.
- IGNITION "ON."

DOES DTC 35 SET AS A CURRENT DTC?

- YES → REPLACE EBCM.
- NO → MALFUNCTION IS INTERMITTENT. REFER TO "DIAGNOSTIC AIDS".

DIAGNOSTIC CHART DTC 35 — 1994–95 CAMARO AND FIREBIRD

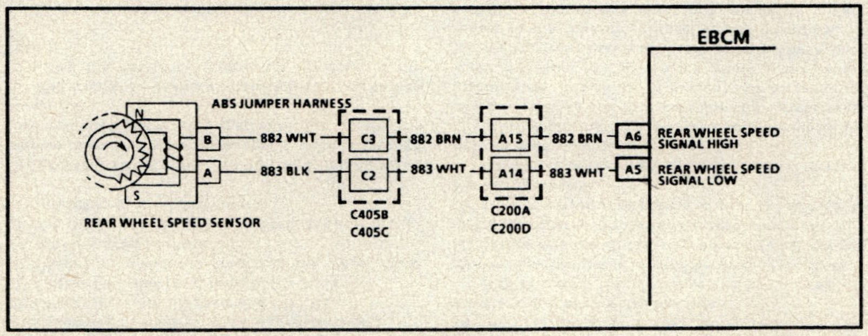

DTC 35

RR WHEEL SPEED SNR CKT OPEN OR SHORT TO GND/BATT

Circuit Description:

As a toothed ring passes by the wheel speed sensor, changes in the electromagnetic field cause the wheel speed sensor to produce a sinusoidal (AC) voltage signal whose frequency is proportional to wheel speed. The magnitude of this signal is directly related to wheel speed and the proximity of the wheel speed sensor to the toothed ring, often referred to as the air gap.

DTC Will Set When: DTC 35 can be set anytime after initialization. If either of the right rear wheel speed circuits are open or shorted to voltage or ground, a malfunction exists.

Action Taken: A malfunction DTC is stored, ABS is disabled and the ABS warning lamp is turned "ON."

DTC Chart Test Description: Number(s) below refer to circled number(s) on the diagnostic chart.

7. Checks for high resistance in CKT 833.
8. Checks for high resistance in CKT 882.
9. Verifies proper resistance of the wheel speed sensor coil.

Diagnostic Aids: An "Intermittent" malfunction may be caused by a poor connection, rubbed through wire insulation, or a wire that is broken inside the insulation.

The frequency of the malfunction can be checked by using the enhanced diagnostic function of the Tech 1.

If the customer's comments reflect that the ABS warning lamp is "ON" only during moist environmental changes (rain, snow, vehicle wash), all wheel speed sensor circuitry should be thoroughly inspected for signs of water intrusion. If DTC is not current, clear DTCs and simulate the effects of water intrusion. Use the following procedure. Spray down the suspected area with a 5% salt water solution (two teaspoons of salt to 12 oz. of water). Test drive vehicle over various road surfaces (bumps, turns, etc.) above 24 km/h (15 mph) for at least 30 seconds. If DTC returns, replace suspected harness.

Any circuitry, that is suspected as causing the intermittent complaint, should be thoroughly checked for backed out terminals, improper mating, broken locks, improperly formed or damaged terminals, poor terminal to wiring connections or physical damage to the wiring harness.

Resistance of the wheel speed sensor will increase with an increase in sensor temperature.

When replacing a wheel speed sensor, inspect the sensor terminals and harness connector for corrosion and/or water intrusion. If evidence of corrosion or water intrusion exists, replace wheel speed sensor jumper harness. Likewise, if replacing a wheel speed sensor jumper harness, inspect sensor terminals. If evidence of corrosion or water intrusion exists, replace wheel speed sensor.

DIAGNOSTIC CHART DTC 35 — 1994–95 CAMARO AND FIREBIRD

DTC 35
(Page 2 of 3)
RR WHEEL SPEED SNR CKT OPEN OR SHORT TO GND/BATT

DIAGNOSIS CONTINUED FROM PAGE 1 OF 3 OF THIS DTC CHART.

(7)
- DISCONNECT REAR WHEEL SPEED SENSOR DIRECTLY AT SENSOR.
- USING J 39200, MEASURE THE RESISTANCE BETWEEN TERMINAL "A5" OF THE 32-WAY EBCM HARNESS CONNECTOR AND TERMINAL "A" OF THE REAR WHEEL SPEED SENSOR HARNESS CONNECTOR.

IS RESISTANCE 2Ω OR LESS?

- NO: REPAIR HIGH RESISTANCE IN CKT 883 (WHEEL SPEED SENSOR JUMPER MUST BE REPLACED IF DAMAGED).
- YES:

(8)
- USING J 39200, MEASURE THE RESISTANCE BETWEEN TERMINAL "A6" OF THE 32-WAY EBCM HARNESS CONNECTOR AND TERMINAL "B" OF THE REAR WHEEL SPEED SENSOR HARNESS CONNECTOR.

IS RESISTANCE 2Ω OR LESS?

- NO: REPAIR HIGH RESISTANCE IN CKT 882 (WHEEL SPEED SENSOR JUMPER MUST BE REPLACED IF DAMAGED).
- YES:

(9)
- USING J 39200, MEASURE THE RESISTANCE BETWEEN TERMINAL "A" AND TERMINAL "B" OF THE REAR WHEEL SPEED SENSOR CONNECTOR.

IS RESISTANCE BETWEEN 969 AND 1185Ω WHEN SENSOR IS 20°C (68°F)? *(IF TEMPERATURE OF SENSOR IS SUSPECTED TO OTHER THAN 20°C (68°F), REFER TO THE TABLE BELOW FOR APPLICABLE SENSOR RESISTANCE VALUES; THE VALUES IN THE TABLE ARE FOR TEMPERATURE OF THE SENSOR, NOT AIR TEMPERATURE.)*

- NO: REPLACE WHEEL SPEED SENSOR.
- YES: MALFUNCTION IS INTERMITTENT. INSPECT ALL CONNECTORS AND HARNESSES FOR DAMAGE WHICH MAY RESULT IN HIGH RESISTANCE WHEN ALL COMPONENTS ARE CONNECTED. REFER TO "DIAGNOSTIC AIDS". REPAIR ALL DAMAGE IF FOUND.

REAR WHEEL SPEED SENSOR

TEMPERATURE OF SENSOR VS. SENSOR RESISTANCE (APPROXIMATE)

TEMP. (°C)	TEMP. (°F)	RESISTANCE (OHMS)
-34 TO 4	-30 TO 40	763 TO 1110
5 TO 43	41 TO 110	908 TO 1292
44 TO 93	111 TO 200	1057 TO 1525

DIAGNOSTIC CHART DTC 35 — 1994–95 CAMARO AND FIREBIRD

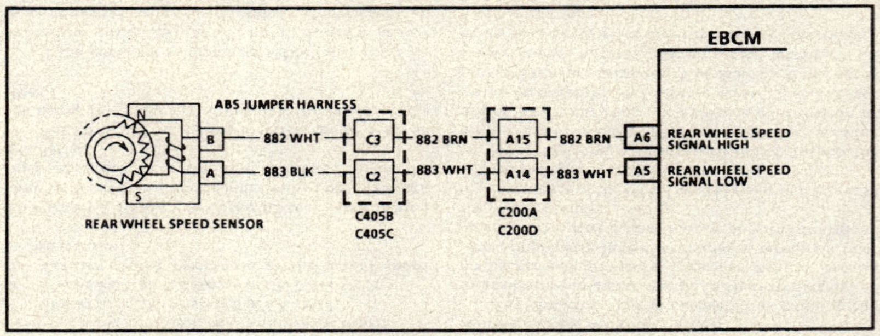

DTC 35

RR WHEEL SPEED SNR CKT OPEN OR SHORT TO GND/BATT

Circuit Description:

As a toothed ring passes by the wheel speed sensor, changes in the electromagnetic field cause the wheel speed sensor to produce a sinusoidal (AC) voltage signal whose frequency is proportional to wheel speed. The magnitude of this signal is directly related to wheel speed and the proximity of the wheel speed sensor to the toothed ring, often referred to as the air gap.

DTC Will Set When: DTC 35 can be set anytime after initialization. If either of the right rear wheel speed circuits are open or shorted to voltage or ground, a malfunction exists.

Action Taken: A malfunction DTC is stored, ABS is disabled and the ABS warning lamp is turned "ON."

DTC Chart Test Description: Number(s) below refer to circled number(s) on the diagnostic chart.

10. Checks for a short to ground in CKT 883.
11. Checks for a short to ground in CKT 882.
12. Checks for a short to ground in the wheel speed sensor jumper harness.
13. Verifies that the wheel speed sensor is not internally shorted to ground.

Diagnostic Aids: An "Intermittent" malfunction may be caused by a poor connection, rubbed through wire insulation, or a wire that is broken inside the insulation.

The frequency of the malfunction can be checked by using the enhanced diagnostic function of the Tech 1.

If the customer's comments reflect that the ABS warning lamp is "ON" only during moist environmental changes (rain, snow, vehicle wash), all wheel speed sensor circuitry should be thoroughly inspected for signs of water intrusion. If DTC is not current, clear DTCs and simulate the effects of water intrusion. Use the following procedure. Spray down the suspected area with a 5% salt water solution (two teaspoons of salt to 12 oz. of water). Test drive vehicle over various road surfaces (bumps, turns, etc.) above 24 km/h (15 mph) for at least 30 seconds. If DTC returns, replace suspected harness.

Any circuitry, that is suspected as causing the intermittent complaint, should be thoroughly checked for backed out terminals, improper mating, broken locks, improperly formed or damaged terminals, poor terminal to wiring connections or physical damage to the wiring harness.

Resistance of the wheel speed sensor will increase with an increase in sensor temperature.

When replacing a wheel speed sensor, inspect the sensor terminals and harness connector for corrosion and/or water intrusion. If evidence of corrosion or water intrusion exists, replace wheel speed sensor jumper harness. Likewise, if replacing a wheel speed sensor jumper harness, inspect sensor terminals. If evidence of corrosion or water intrusion exists, replace wheel speed sensor.

DIAGNOSTIC CHART DTC 35 — 1994–95 CAMARO AND FIREBIRD

DTC 35
(Page 3 of 3)
RR WHEEL SPEED SNR CKT OPEN OR SHORT TO GND/BATT

DIAGNOSIS CONTINUED FROM PAGE 1 OF 3 OF THIS DTC CHART.

(10)
- DISCONNECT CONNECTOR C405B, C405C (LOCATED ON THE LH SIDE REAR UNDER BODY, BEHIND REAR SEAT).
- USING J 39200, MEASURE THE RESISTANCE BETWEEN GROUND AND TERMINAL "A5" OF THE 32-WAY EBCM HARNESS CONNECTOR.

DOES J 39200 DISPLAY RESISTANCE AS "OL" (INFINITE)?

NO: REPAIR SHORT TO GROUND IN CKT 883.

YES:

(11)
- USING J 39200, MEASURE THE RESISTANCE BETWEEN GROUND AND TERMINAL "A6" OF THE 32-WAY EBCM HARNESS CONNECTOR.

DOES J 39200 DISPLAY RESISTANCE AS "OL" (INFINITE)?

NO: REPAIR SHORT TO GROUND IN CKT 882.

YES:

(12)
- THOROUGHLY RESPRAY THE REAR WHEEL SPEED SENSOR JUMPER HARNESS WITH THE SALT WATER SOLUTION.
- IMPORTANT: DO NOT ALLOW WATER TO ENTER ANY WHEEL SPEED SENSOR CONNECTOR.
- DISCONNECT REAR WHEEL SPEED SENSOR DIRECTLY AT SENSOR.
- USING J 39200, MEASURE THE RESISTANCE BETWEEN GROUND AND TERMINAL "A," THEN "B" OF THE REAR WHEEL SPEED SENSOR HARNESS CONNECTOR.

DOES J 39200 DISPLAY RESISTANCE AS "OL" (INFINITE) FOR BOTH MEASUREMENTS?

NO: REPLACE REAR WHEEL SPEED SENSOR JUMPER HARNESS.

YES:

(13)
- USING J 39200, MEASURE THE RESISTANCE BETWEEN GROUND AND TERMINAL "A" OF THE REAR WHEEL SPEED SENSOR CONNECTOR.

DOES J 39200 DISPLAY RESISTANCE AS "OL" (INFINITE)?

NO: REPLACE WHEEL SPEED SENSOR.

YES: MALFUNCTION IS INTERMITTENT. INSPECT ALL CONNECTORS AND HARNESSES FOR DAMAGE WHICH MAY RESULT IN A SHORT TO GROUND WHEN ALL COMPONENTS ARE CONNECTED. PAY PARTICULAR ATTENTION TO THE JUMPER HARNESS, SMALL NICKS OR CUTS CAN CAUSE A MALFUNCTION. REPAIR ANY DAMAGE IF FOUND. JUMPER HARNESS MUST BE REPLACED IF DAMAGED. REFER TO "DIAGNOSTIC AIDS".

DIAGNOSTIC CHART DTC 36 — 1994–95 CAMARO AND FIREBIRD

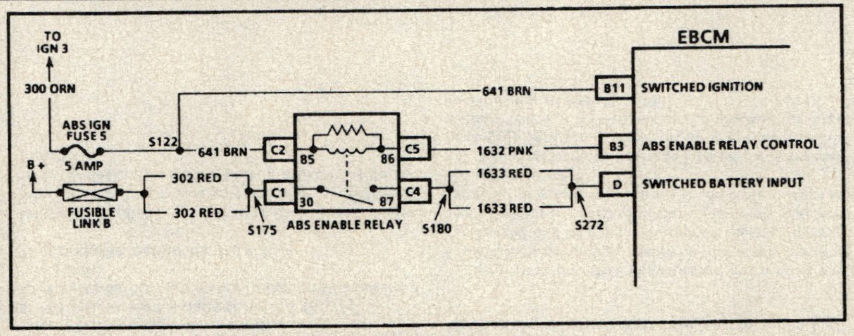

DTC 36
(Page 1 of 2)
LOW SYSTEM VOLTAGE

Circuit Description:
This DTC is used to monitor the voltage level available to the EBCM controller. If the voltage drops below 11.5 volts, full performance of the ABS system cannot be guaranteed. During ABS operation, there are several current requirements that will cause battery voltage to drop. Because of this, voltage is monitored prior to ABS to indicate good charging system condition and also during ABS when voltage may drop significantly.

DTC Will Set When: DTC 36 can only be set if the vehicle's speed is greater than 5 km/h (3 mph). If the switched battery voltage is less than 11.5 volts non-ABS or less than 9.0 volts ABS, a malfunction exists.

Action Taken: A malfunction DTC is stored, ABS is disabled and the "ABS INOP" lamp is turned "ON."

DTC Chart Test Description: Number(s) below refer to circled number(s) on the diagnostic chart.

1. Checks to see what voltage is readily available at terminal "D" of the EBCM. If voltage is representative of good charging system condition, a malfunction is not present.
2. Isolates the condition to high circuit resistance, or improper charging system operation.
3. Isolates which circuit is the source of the low voltage condition.
4. Checks for high resistance in CKT 302.

Diagnostic Aids: An "Intermittent" malfunction may be caused by a poor connection, rubbed through wire insulation, or a wire that is broken inside the insulation.

The frequency of the malfunction can be checked by using the enhanced diagnostic function of the Tech 1.

Any circuitry, that is suspected as causing the intermittent complaint, should be thoroughly checked for backed out terminals, improper mating, broken locks, improperly formed or damaged terminals, poor terminal to wiring connections or physical damage to the wiring harness.

While performing a voltage load test, if it is noted that <u>only</u> ignition voltage drops below acceptable voltage levels, CKT 641 should be checked for high resistance or an open condition.

DIAGNOSTIC CHART DTC 36 — 1994–95 CAMARO AND FIREBIRD

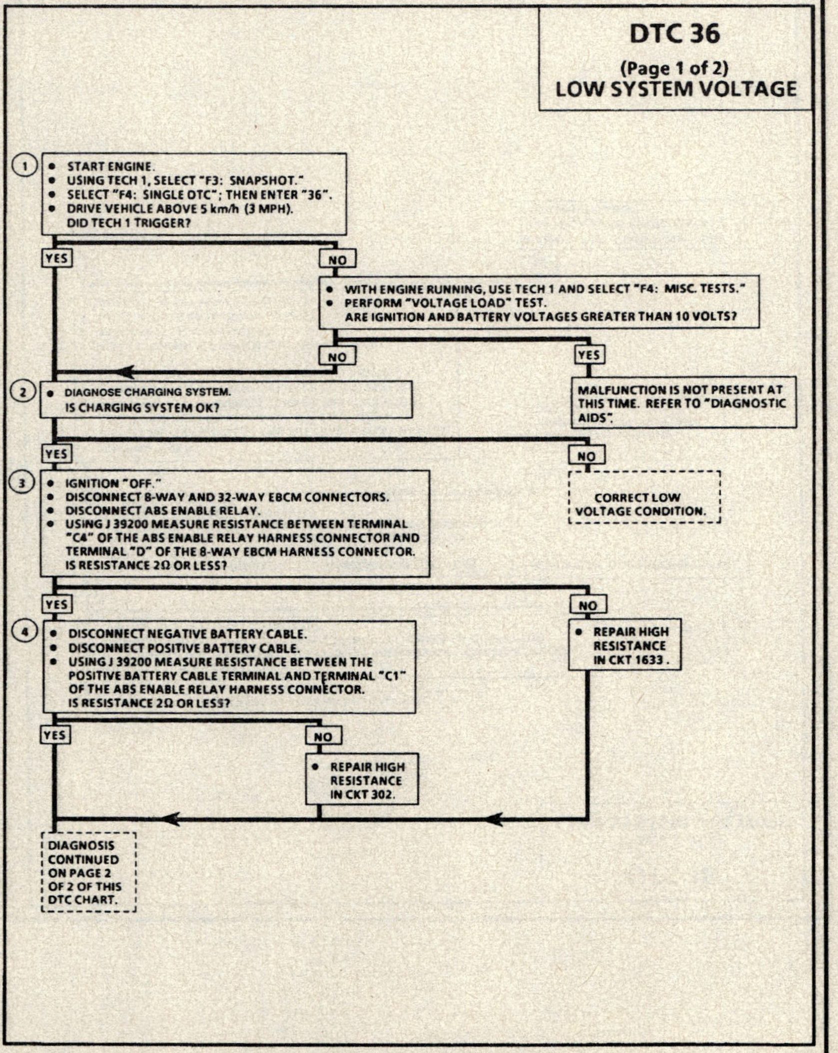

DIAGNOSTIC CHART DTC 36 — 1994–95 CAMARO AND FIREBIRD

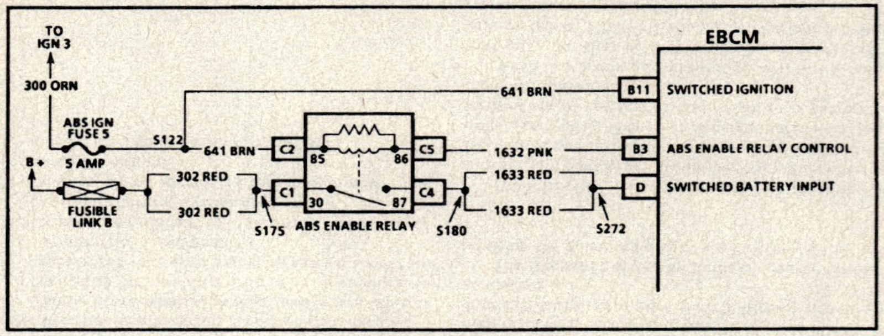

DTC 36

LOW SYSTEM VOLTAGE

Circuit Description:

This DTC is used to monitor the voltage level available to the EBCM controller. If the voltage drops below 11.5 volts, full performance of the ABS system cannot be guaranteed. During ABS operation, there are several current requirements that will cause battery voltage to drop. Because of this, voltage is monitored prior to ABS to indicate good charging system condition and also during ABS when voltage may drop significantly.

DTC Will Set When: DTC 36 can only be set if the vehicle's speed is greater than 5 km/h (3 mph). If the switched battery voltage is less than 11.5 volts non-ABS or less than 9.0 volts ABS, a malfunction exists.

Action Taken: A malfunction DTC is stored, ABS is disabled and the "ABS INOP" lamp is turned "ON."

DTC Chart Test Description: Number(s) below refer to circled number(s) on the diagnostic chart.

5. Checks for high resistance in CKT 641.
6. Checks for high resistance in the ignition fuse feed circuit.
7. Checks for malfunctioning EBCM.

Diagnostic Aids: An "Intermittent" malfunction may be caused by a poor connection, rubbed through wire insulation, or a wire that is broken inside the insulation.

The frequency of the malfunction can be checked by using the enhanced diagnostic function of the Tech 1.

Any circuitry, that is suspected as causing the intermittent complaint, should be thoroughly checked for backed out terminals, improper mating, broken locks, improperly formed or damaged terminals, poor terminal to wiring connections or physical damage to the wiring harness.

While performing a voltage load test, if it is noted that only ignition voltage drops below acceptable voltage levels, CKT 641 should be checked for high resistance or an open condition.

DIAGNOSTIC CHART DTC 36 — 1994–95 CAMARO AND FIREBIRD

DTC 36
(Page 2 of 2)
LOW SYSTEM VOLTAGE

DIAGNOSIS CONTINUED FROM PAGE 1 OF 2 OF THIS DTC CHART.

(5)
- REMOVE ABS IGN 5 AMP FUSE 5.
- USING J 39200 MEASURE RESISTANCE BETWEEN EACH ABS IGN 5 AMP FUSE 5 BLOCK TERMINAL AND TERMINAL "B11" OF THE 32-WAY EBCM HARNESS CONNECTOR. IS EITHER MEASUREMENT 2Ω OR LESS?

YES → (6) | NO → REPAIR HIGH RESISTANCE IN CKT 641.

(6)
- WITH BATTERY CABLES DISCONNECTED, TURN IGNITION "ON" (THIS PROVIDES IGNITION CIRCUIT CONTINUITY).
- USING J 39200, MEASURE RESISTANCE BETWEEN POSITIVE BATTERY CABLE TERMINAL AND EACH ABS IGN 5 AMP FUSE 5 BLOCK TERMINAL. IS EITHER MEASUREMENT 2Ω OR LESS ?

YES → (7) | NO → REPAIR HIGH RESISTANCE IN THE "IGNITION SWITCH 'RUN' FUSE FEED" CIRCUIT.

(7)
- IGNITION "OFF."
- INSPECT ABS ENABLE RELAY AND ABS ENABLE RELAY HARNESS CONNECTOR FOR POOR TERMINAL CONTACT. INSPECT 8-WAY AND 32-WAY EBCM HARNESS CONNECTORS FOR A POOR CONNECTION TO THE EBCM. INSPECT POSITIVE BATTERY CABLE TERMINAL FOR POOR CONTACT TO THE BATTERY.
- REPLACE TERMINALS IF POOR CONTACT IS EVIDENT.
- RECONNECT ALL CONNECTORS.
- DRIVE VEHICLE ABOVE 5 km/h (3 MPH). DOES DTC 36 SET AS A CURRENT DTC?

YES → REPLACE EBCM. | NO → MALFUNCTION IS MOST LIKELY DUE TO A POOR CONNECTION. REFER TO "DIAGNOSTIC AIDS".

DIAGNOSTIC CHART DTC 37 — 1994–95 CAMARO AND FIREBIRD

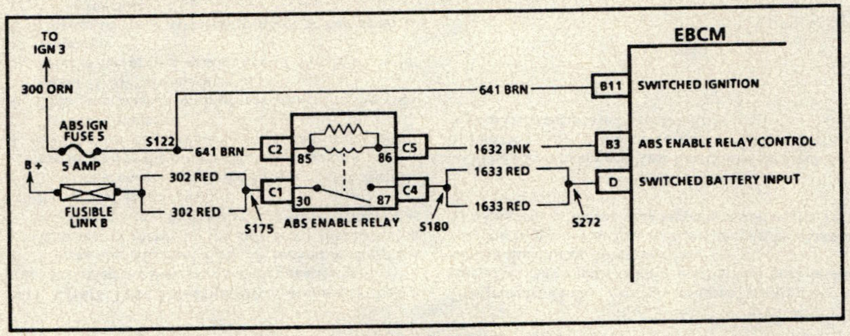

DTC 37
HIGH SYSTEM VOLTAGE

Circuit Description:
This DTC is designed to detect high vehicle voltage levels prior to any required motor movement (initialization or ABS). If excessive voltage exists, demagnetization of the motor magnets may occur, which would eventually affect or eliminate ABS performance.

DTC Will Set When: DTC 37 can only be set if the vehicle's speed is greater than 5 km/h (3 mph). If the switched battery voltage is greater than 17 volts, a malfunction exists.

Action Taken: A malfunction DTC is stored, ABS is disabled and the "ABS INOP" lamp is turned "ON."

DTC Chart Test Description: Number(s) below refer to circled number(s) on the diagnostic chart.
1. Checks the voltage level being received by the EBCM.
2. Checks to see if high voltage condition is due to charging system malfunction.
3. Checks for poor terminal contact which may cause an intermittent malfunctioning, or a malfunctioning EBCM.

Diagnostic Aids: An "Intermittent" malfunction may be caused by a poor connection, rubbed through wire insulation, or a wire that is broken inside the insulation.

The frequency of the malfunction can be checked by using the enhanced diagnostic function of the Tech 1.

Any circuitry, that is suspected as causing the intermittent complaint, should be thoroughly checked for backed out terminals, improper mating, broken locks, improperly formed or damaged terminals, poor terminal to wiring connections or physical damage to the wiring harness.

DIAGNOSTIC CHART DTC 37 — 1994–95 CAMARO AND FIREBIRD

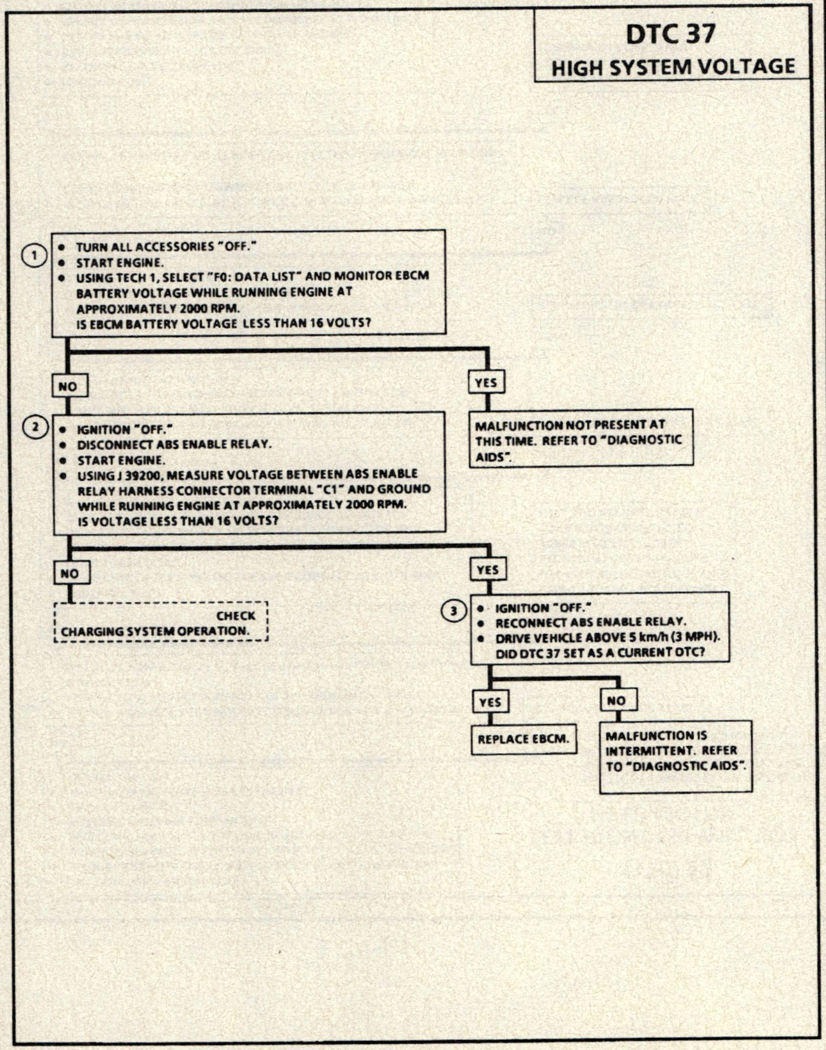

DIAGNOSTIC CHART DTC 38 — 1994–95 CAMARO AND FIREBIRD

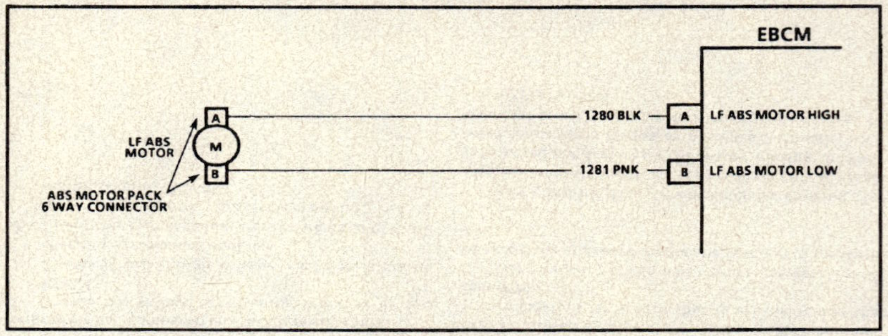

DTC 38

LEFT FRONT ESB WILL NOT HOLD MOTOR

Circuit Description:

This DTC is designed to detect a slipping left front ESB. During initialization and braking, the left front motor is re-homed. If the ESB slips, the motor/piston will move. During the next key "ON" initialization, a re-home of the motor verifies the motor/piston remained at the home position. If motor movement is detected, the ESB must be slipping.

DTC Will Set When: DTC 38 can be set during initialization. If the EBCM detects that the ESB could not hold the piston in the home position, a malfunction exists. DTC 86 is always set with DTC 38.

Action Taken: If an ESB cannot hold a piston in the home position, the piston may be backdriven when the brake pedal is applied, causing the brake pedal to drop. A malfunction DTC is stored, ABS is disabled, and the "ABS INOP" lamp and the red "BRAKE" warning lamp are turned "ON."

DTC Chart Test Description: Number(s) below refer to circled number(s) on the diagnostic chart.

1. Checks the left front ESB. A broken or defective ESB would result in the left front piston being backdriven by hydraulic pressure and pedal movement would result.
2. This step ensures that the proper motor circuits are located in the proper connector cavities.
3. Checks for high resistance in the ABS motor "HIGH" circuitry.
4. This step checks for high resistance in the ABS motor "LOW" circuitry.
5. Checks for proper resistance of the ABS motor windings.
6. Checks for poor terminal contact or corrosion in the connectors.
7. Checks for a malfunctioning EBCM.

Diagnostic Aids: An "Intermittent" malfunction in this DTC may result from a mechanical part of the system that sticks, binds, or slips.

The frequency of the malfunction can be checked by using the enhanced diagnostic function of the Tech 1.

The static modulator test function of the Tech 1 or T-100 may be used to locate an intermittent malfunction associated with the ESB.

DIAGNOSTIC CHART DTC 38 — 1994–95 CAMARO AND FIREBIRD

DTC 38

LEFT FRONT ESB WILL NOT HOLD MOTOR

NOTICE: J 39200 TEST LEADS MUST BE "ZEROED" PRIOR TO MAKING ANY RESISTANCE MEASUREMENTS. REFER TO J 39200 USER'S MANUAL.

(1)
- IGNITION "ON," ENGINE "OFF."
- PUMP BRAKE PEDAL UNTIL FIRM TO DEPLETE VACUUM RESERVE.
- USING TECH 1 MISC. TESTS, SELECT MANUAL CONTROL FUNCTION.
- SELECT MOTOR APPLY FUNCTION OF THE LEFT FRONT ABS CHANNEL AND APPLY THE MOTOR.
- WAIT 5 SECONDS.
- APPLY FIRM PRESSURE ON BRAKE PEDAL.

DID BRAKE PEDAL FALL?

- YES: REPLACE ABS MOTOR PACK.
- NO: go to step 2.

(2)
- VISUALLY INSPECT THE 6-WAY ABS MOTOR PACK AND 8-WAY EBCM CONNECTORS FOR PROPER WIRE COLOR/CONNECTOR CAVITY COMBINATION.

ARE THE PROPER WIRES LOCATED IN THE PROPER CONNECTOR CAVITIES?

- NO: USING THE CIRCUIT DIAGRAMS LOCATED IN THE FRONT OF THIS SECTION, PLACE THE PROPER WIRES INTO THE PROPER CONNECTOR CAVITIES. CLEAR DTCs AND RECHECK SYSTEM.
- YES: go to step 3.

(3)
- IGNITION "OFF."
- DISCONNECT 6-WAY ABS MOTOR PACK CONNECTOR AND 8-WAY EBCM CONNECTOR.
- USING J 39200, MEASURE THE RESISTANCE BETWEEN TERMINAL "A" OF THE 6-WAY ABS MOTOR PACK HARNESS CONNECTOR AND TERMINAL "A" OF THE 8-WAY EBCM HARNESS CONNECTOR.

IS RESISTANCE 1.5Ω OR LESS?

- NO: REPAIR HIGH RESISTANCE IN CKT 1280.
- YES: go to step 4.

(4)
- MEASURE THE RESISTANCE BETWEEN TERMINAL "B" OF THE 6-WAY ABS MOTOR PACK HARNESS CONNECTOR AND TERMINAL "B" OF THE 8-WAY EBCM HARNESS CONNECTOR.

IS RESISTANCE 1.5Ω OR LESS?

- NO: REPAIR HIGH RESISTANCE IN CKT 1281.
- YES: go to step 5.

(5)
- MEASURE THE RESISTANCE BETWEEN TERMINAL "A" AND TERMINAL "B" OF THE ABS MOTOR PACK CONNECTOR.

IS RESISTANCE 1.5Ω OR LESS?

- NO: REPLACE ABS MOTOR PACK.
- YES: go to step 6.

(6)
- INSPECT THE 6-WAY ABS MOTOR PACK CONNECTOR AND 8-WAY EBCM CONNECTOR FOR POOR TERMINAL CONTACT OR CORROSION.
- DO ANY TERMINALS EXHIBIT POOR CONTACT OR EVIDENCE OF CORROSION?

- YES: REPLACE TERMINALS THAT EXHIBIT POOR TERMINAL CONTACT OR EVIDENCE OF CORROSION.
- NO: go to step 7.

(7)
- IGNITION "OFF."
- RECONNECT ALL CONNECTORS.
- START ENGINE WITH FOOT OFF BRAKE.
- ALLOW ENGINE TO RUN FOR AT LEAST 10 SECONDS.
- REPEAT ABOVE IGNITION CYCLE SEQUENCE TWO MORE TIMES.

DID DTC 38 SET IN THE LAST THREE IGNITION CYCLES?

- YES: REPLACE EBCM.
- NO: MALFUNCTION IS INTERMITTENT. REFER TO "DIAGNOSTIC AIDS".

AFTER DIAGNOSIS IS COMPLETE, CLEAR DTCs AND TEST DRIVE VEHICLE FOR THREE (3) DRIVE CYCLES TO VERIFY THAT DTC DOES NOT RESET. A DRIVE CYCLE CONSISTS OF STARTING THE VEHICLE, DRIVING OVER 16 km/h (10 MPH) AND THEN KEYING DOWN.

DIAGNOSTIC CHART DTC 41 — 1994–95 CAMARO AND FIREBIRD

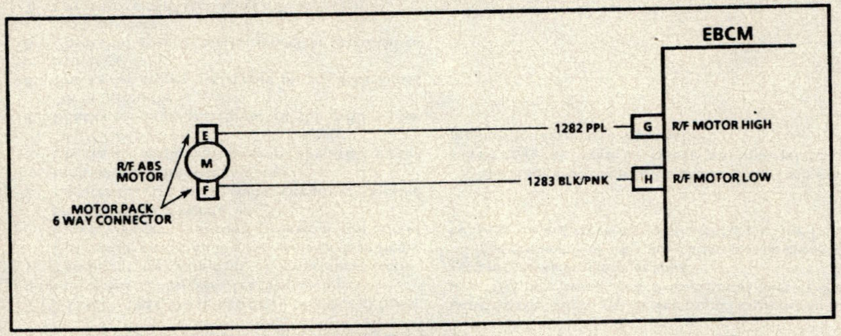

DTC 41

RIGHT FRONT ESB WILL NOT HOLD MOTOR

Circuit Description:
This DTC is designed to detect a slipping right front ESB. During initialization and braking, the right front motor is re-homed. If the ESB slips, the motor/piston will move. During the next key "ON" initialization, a re-home of the motor verifies the motor/piston remained at the home position. If motor movement is detected, the ESB must be slipping.

DTC Will Set When: DTC 41 can be set during initialization. If the EBCM detects that the ESB could not hold the piston in the home position, a malfunction exists. DTC 86 is always set with DTC 41.

Action Taken: If an ESB cannot hold a piston in the home position, the piston may be backdriven when the brake pedal is applied, causing the brake pedal to drop. A malfunction DTC is stored, ABS is disabled, and the "ABS INOP" lamp and the red "BRAKE" warning lamp are turned "ON."

DTC Chart Test Description: Number(s) below refer to circled number(s) on the diagnostic chart.

1. Checks the right front ESB. A broken or defective ESB would result in the right front piston being backdriven by hydraulic pressure, and pedal movement would result.
2. Ensures that the proper motor circuits are located in the proper connector cavities.
3. Checks for high resistance in the ABS motor "HIGH" circuitry.
4. This step checks for high resistance in the ABS motor "LOW" circuitry.
5. Checks for proper resistance of the ABS motor windings.
6. Checks for poor terminal contact or corrosion in the connectors.
7. Checks for a malfunctioning EBCM.

Diagnostic Aids: An "Intermittent" malfunction in this DTC may result from a mechanical part of the system that sticks, binds, or slips.

The frequency of the malfunction can be checked by using the enhanced diagnostic function of the Tech 1.

The static modulator test function of the Tech 1 or T-100 may be used to locate an intermittent malfunction associated with the ESB.

DIAGNOSTIC CHART DTC 41 — 1994–95 CAMARO AND FIREBIRD

DTC 41

RIGHT FRONT ESB WILL NOT HOLD MOTOR

NOTICE: J 39200 TEST LEADS MUST BE "ZEROED" PRIOR TO MAKING ANY RESISTANCE MEASUREMENTS. REFER TO J 39200 USER'S MANUAL.

(1)
- IGNITION "ON," ENGINE "OFF."
- PUMP BRAKE PEDAL UNTIL FIRM TO DEPLETE VACUUM RESERVE.
- USING TECH 1 MISC. TESTS, SELECT MANUAL CONTROL FUNCTION.
- SELECT MOTOR APPLY FUNCTION OF THE RIGHT FRONT ABS CHANNEL AND APPLY THE MOTOR.
- WAIT 5 SECONDS.
- APPLY FIRM PRESSURE ON BRAKE PEDAL.

DID BRAKE PEDAL FALL?

- YES: REPLACE ABS MOTOR PACK.
- NO: go to (2)

(2)
- VISUALLY INSPECT THE 6-WAY ABS MOTOR PACK AND 8-WAY EBCM CONNECTORS FOR PROPER WIRE COLOR/CONNECTOR CAVITY COMBINATION.

ARE THE PROPER WIRES LOCATED IN THE PROPER CONNECTOR CAVITIES?

- NO: USING THE CIRCUIT DIAGRAMS LOCATED IN THE FRONT OF THIS SECTION, PLACE THE PROPER WIRES INTO THE PROPER CONNECTOR CAVITIES. CLEAR DTCs AND RECHECK SYSTEM.
- YES: go to (3)

(3)
- IGNITION "OFF."
- DISCONNECT 6-WAY ABS MOTOR PACK CONNECTOR AND 8-WAY EBCM CONNECTOR.
- USING J 39200, MEASURE THE RESISTANCE BETWEEN TERMINAL "E" OF THE 6-WAY ABS MOTOR PACK HARNESS CONNECTOR AND TERMINAL "G" OF THE 8-WAY EBCM HARNESS CONNECTOR.

IS RESISTANCE 1.5Ω OR LESS?

- NO: REPAIR HIGH RESISTANCE IN CKT 1282.
- YES: go to (4)

(4)
- MEASURE THE RESISTANCE BETWEEN TERMINAL "F" OF THE 6-WAY ABS MOTOR PACK HARNESS CONNECTOR AND TERMINAL "H" OF THE 8-WAY EBCM HARNESS CONNECTOR.

IS RESISTANCE 1.5Ω OR LESS?

- NO: REPAIR HIGH RESISTANCE IN CKT 1283.
- YES: go to (5)

(5)
- MEASURE THE RESISTANCE BETWEEN TERMINAL "E" AND TERMINAL "F" OF THE ABS MOTOR PACK CONNECTOR.

IS RESISTANCE 1.5Ω OR LESS?

- NO: REPLACE ABS MOTOR PACK.
- YES: go to (6)

(6)
- INSPECT THE 6-WAY ABS MOTOR PACK CONNECTOR AND 8-WAY EBCM CONNECTOR FOR POOR TERMINAL CONTACT OR CORROSION.
- DO ANY TERMINALS EXHIBIT POOR CONTACT OR EVIDENCE OF CORROSION?

- YES: REPLACE TERMINALS THAT EXHIBIT POOR TERMINAL CONTACT OR EVIDENCE OF CORROSION.
- NO: go to (7)

(7)
- IGNITION "OFF."
- RECONNECT ALL CONNECTORS.
- START ENGINE WITH FOOT OFF BRAKE.
- ALLOW ENGINE TO RUN FOR AT LEAST 10 SECONDS.
- REPEAT ABOVE IGNITION CYCLE SEQUENCE TWO MORE TIMES.

DID DTC 41 SET IN THE LAST THREE IGNITION CYCLES?

- YES: REPLACE EBCM.
- NO: MALFUNCTION IS INTERMITTENT. REFER TO "DIAGNOSTIC AIDS".

AFTER DIAGNOSIS IS COMPLETE, CLEAR DTCs AND TEST DRIVE VEHICLE FOR THREE (3) DRIVE CYCLES TO VERIFY THAT DTC DOES NOT RESET. A DRIVE CYCLE CONSISTS OF STARTING THE VEHICLE, DRIVING OVER 16 KM/H (10 MPH) AND THEN KEYING DOWN.

DIAGNOSTIC CHART DTC 42 — 1994–95 CAMARO AND FIREBIRD

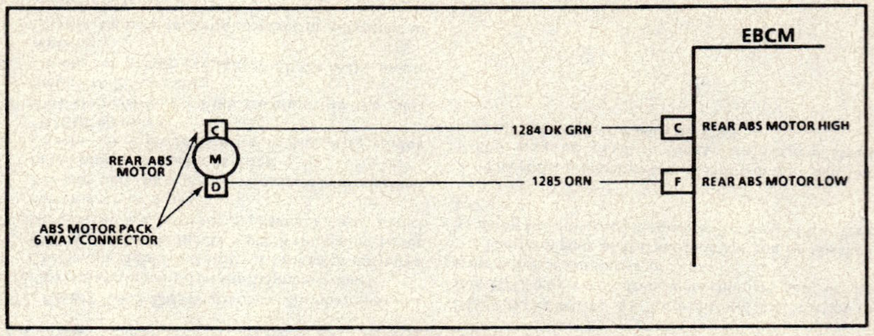

DTC 42
REAR ESB WILL NOT HOLD MOTOR

Circuit Description:
This DTC is designed to detect a slipping rear axle ESB. During initialization and braking, the rear motor is re-homed. If the ESB slips, the motor/piston will move. During the next key "ON" initialization, a re-home of the motor verifies the motor/piston remained at the home position. If motor movement is detected, the ESB must be slipping.

DTC Will Set When: DTC 42 can be set during initialization. If the EBCM detects that the ESB could not hold the piston in the home position, a malfunction exists. DTC 86 is always set with DTC 42.

Action Taken: If an ESB can hold a piston in the home position, the piston may be backdriven when the brake pedal is applied, causing the brake pedal to drop. A malfunction DTC is stored, ABS is disabled, and the "ABS INOP" lamp and the red "BRAKE" warning lamp are turned "ON."

DTC Chart Test Description: Number(s) below refer to circled number(s) on the diagnostic chart.

1. Checks the rear axle ESB. A broken or defective ESB would result in the rear axle pistons being backdriven by hydraulic pressure and wheel movement would result.
2. Ensures that the proper motor circuits are located in the proper connector cavities.
3. Checks for high resistance in the ABS motor "HIGH" circuitry.
4. Checks for high resistance in the ABS motor "LOW" circuitry.
5. Checks for proper resistance of the ABS motor windings.
6. Checks for poor terminal contact or corrosion in the connectors.
7. Checks for a malfunctioning EBCM.

Diagnostic Aids: An "Intermittent" malfunction in this DTC may result from a mechanical part of the system that sticks, binds, or slips.

The frequency of the malfunction can be checked by using the enhanced diagnostic function of the Tech 1.

The static modulator test function of the Tech 1 or T-100 may be used to locate an intermittent malfunction associated with the ESB.

DIAGNOSTIC CHART DTC 42 — 1994–95 CAMARO AND FIREBIRD

DTC 42
REAR ESB WILL NOT HOLD MOTOR

NOTICE: J 39200 TEST LEADS MUST BE "ZEROED" PRIOR TO MAKING ANY RESISTANCE MEASUREMENTS. REFER TO J 39200 USER'S MANUAL.

(1)
- RAISE THE VEHICLE SUCH THAT THE REAR WHEELS ARE APPROXIMATELY SIX INCHES OFF THE FLOOR.
- IGNITION "ON," ENGINE "OFF."
- PUMP BRAKE PEDAL UNTIL FIRM TO DEPLETE VACUUM RESERVE.
- APPLY FIRM PRESSURE ON BRAKE PEDAL AND RELEASE.
- REAPPLY VERY FIRM PRESSURE ON BRAKE PEDAL AND HAVE AN ASSISTANT TRY TO SPIN THE REAR WHEELS BY HAND.

COULD ASSISTANT SPIN THE REAR WHEELS?

YES: REPLACE ABS MOTOR PACK.

NO: go to (2).

(2)
- VISUALLY INSPECT THE 6-WAY ABS MOTOR PACK AND 8-WAY EBCM CONNECTORS FOR PROPER WIRE COLOR/CONNECTOR CAVITY COMBINATION.

ARE THE PROPER WIRES LOCATED IN THE PROPER CONNECTOR CAVITIES?

NO: USING THE CIRCUIT DIAGRAMS LOCATED IN THE FRONT OF THIS SECTION, PLACE THE PROPER WIRES INTO THE PROPER CONNECTOR CAVITIES. CLEAR DTCs AND RECHECK SYSTEM.

YES: go to (3).

(3)
- IGNITION "OFF."
- DISCONNECT 6-WAY ABS MOTOR PACK CONNECTOR AND 8-WAY EBCM CONNECTOR.
- USING J 39200, MEASURE THE RESISTANCE BETWEEN TERMINAL "C" OF THE 6-WAY ABS MOTOR PACK HARNESS CONNECTOR AND TERMINAL "C" OF THE 8-WAY EBCM HARNESS CONNECTOR.

IS RESISTANCE 1.5Ω OR LESS?

NO: REPAIR HIGH RESISTANCE IN CKT 1284.

YES: go to (4).

(4)
- MEASURE THE RESISTANCE BETWEEN TERMINAL "D" OF THE 6-WAY ABS MOTOR PACK HARNESS CONNECTOR AND TERMINAL "F" OF THE 8-WAY EBCM HARNESS CONNECTOR.

IS RESISTANCE 1.5Ω OR LESS?

NO: REPAIR HIGH RESISTANCE IN CKT 1285.

YES: go to (5).

(5)
- MEASURE THE RESISTANCE BETWEEN TERMINAL "C" AND TERMINAL "D" OF THE ABS MOTOR PACK CONNECTOR.

IS RESISTANCE 1.5Ω OR LESS?

NO: REPLACE ABS MOTOR PACK.

YES: go to (6).

(6)
- INSPECT THE 6-WAY ABS MOTOR PACK CONNECTOR AND 8-WAY EBCM CONNECTOR FOR POOR TERMINAL CONTACT OR CORROSION.

DO ANY TERMINALS EXHIBIT POOR CONTACT OR EVIDENCE OF CORROSION?

YES: REPLACE TERMINALS THAT EXHIBIT POOR TERMINAL CONTACT OR EVIDENCE OF CORROSION.

NO: go to (7).

(7)
- IGNITION "OFF."
- RECONNECT ALL CONNECTORS.
- START ENGINE WITH FOOT OFF BRAKE.
- ALLOW ENGINE TO RUN FOR AT LEAST 10 SECONDS.
- REPEAT ABOVE IGNITION CYCLE SEQUENCE TWO MORE TIMES.

DID DTC 42 SET IN THE LAST THREE IGNITION CYCLES?

YES: REPLACE EBCM.

NO: MALFUNCTION IS INTERMITTENT. REFER TO "DIAGNOSTIC AIDS" ON FACING PAGE.

AFTER DIAGNOSIS IS COMPLETE, CLEAR DTCs AND TEST DRIVE VEHICLE FOR THREE (3) DRIVE CYCLES TO VERIFY DTC DOES NOT RESET. A DRIVE CYCLE CONSISTS OF STARTING THE VEHICLE, DRIVING OVER 16 km/h (10 MPH) AND THEN KEYING DOWN.

DIAGNOSTIC CHART DTC 44 — 1994–95 CAMARO AND FIREBIRD

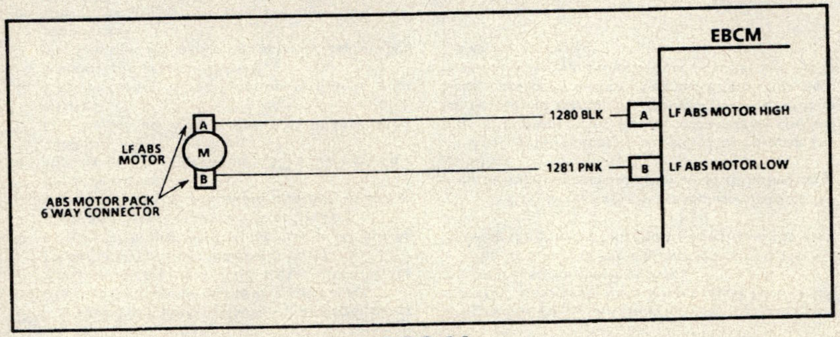

DTC 44

LEFT FRONT ABS CHANNEL WILL NOT MOVE

Circuit Description:

This DTC is designed to detect bound-up ESB, a stuck ABS motor, or a seized hydraulic modulator. When the release is commanded during initialization, the ESB should release the ABS motor, resulting in sensed current being less than commanded current (motor is spinning freely). If the ABS motor is not moving, sensed current will be equal to stall current.

DTC Will Set When: DTC 44 can be set during normal operation. If the EBCM detects a condition in which it cannot move the motor in either direction, a malfunction exists.

Action Taken: This malfunction indicates the channel cannot be moved properly. A malfunction DTC is stored, ABS is disabled and the "ABS INOP" lamp is turned "ON."

DTC Chart Test Description: Number(s) below refer to circled number(s) on the diagnostic chart.

1. Ensures that proper motor circuits are located in the proper connector cavities.
2. Checks for proper ABS motor movement during apply and release commands from the Tech 1.
3. Compares EBCM command current to motor feedback current.
4. Checks for a short circuit between the ABS motor "HIGH" and ABS motor "LOW" circuits.
5. Checks for an ABS motor that is internally shorted.
6. Determines if the malfunction is caused by a defective EBCM or poor terminal contact.
7. Verifies that the ABS motor can actually apply when commanded.
8. Checks for proper hydraulic modulator gear and piston movement.

Diagnostic Aids: An "Intermittent" malfunction in this DTC may result from a mechanical part of the system that sticks, binds, or slips.

The frequency of the malfunction can be checked by using the enhanced diagnostic function of the Tech 1.

DTC 44 may set after modulator disassembly if the modulator pistons are positioned at the bottom of their bore.

Any circuitry, that is suspected as causing the intermittent complaint, should be thoroughly checked for backed out terminals, improper mating, broken locks, improperly formed or damaged terminals, poor terminal to wiring connections or physical damage to the wiring harness.

DIAGNOSTIC CHART DTC 44 — 1994–95 CAMARO AND FIREBIRD

DTC 44
LEFT FRONT ABS CHANNEL WILL NOT MOVE

(1)
- VISUALLY INSPECT THE 6-WAY ABS MOTOR PACK AND THE 8-WAY EBCM CONNECTORS FOR PROPER WIRE COLOR/CONNECTOR CAVITY COMBINATION.

ARE THE PROPER WIRES LOCATED IN THE PROPER CONNECTOR CAVITIES?

- YES → (2)
- NO → USING THE CIRCUIT DIAGRAMS LOCATED IN THE FRONT OF THIS SECTION, PLACE THE PROPER WIRES IN THE PROPER CONNECTOR CAVITIES. CLEAR DTCs AND RECHECK SYSTEM.

(2)
- IGNITION "ON," ENGINE "OFF."
- PUMP BRAKE PEDAL UNTIL FIRM TO DEPLETE VACUUM RESERVE.
- USING TECH 1 MISC. TESTS, SELECT MANUAL CONTROL FUNCTION.
- SELECT LEFT FRONT ABS MOTOR APPLY FUNCTION AND APPLY THE MOTOR.
- APPLY FIRM PRESSURE ON BRAKE PEDAL.
- SELECT LEFT FRONT ABS MOTOR RELEASE FUNCTION AND RELEASE MOTOR.

DID BRAKE PEDAL FALL?

- NO → (3)
- YES → (7)

(7)
- FIRM PRESSURE STILL "ON" BRAKE PEDAL.
- SELECT LEFT FRONT ABS MOTOR APPLY FUNCTION AND APPLY THE MOTOR.

DID BRAKE PEDAL RISE?

- NO → (3)
- YES → MALFUNCTION IS INTERMITTENT. REFER TO "DIAGNOSTIC AIDS".

(3)
- REMOVE FOOT FROM BRAKE PEDAL.
- SELECT LEFT FRONT ABS MOTOR APPLY FUNCTION AND APPLY THE MOTOR WHILE CAREFULLY OBSERVING THE COMMANDED CURRENT AND THE FEEDBACK CURRENT ON THE TECH 1.

WAS THE FEEDBACK CURRENT HIGHER THAN THE COMMANDED CURRENT?

- YES → (4)
- NO → (8)

(4)
- EXIT TECH 1 MANUAL CONTROL FUNCTION.
- IGNITION "OFF."
- DISCONNECT THE 8-WAY EBCM CONNECTOR AND THE 6-WAY ABS MOTOR PACK CONNECTOR.
- USING J 39200, MEASURE THE RESISTANCE BETWEEN TERMINAL "A" AND TERMINAL "B" OF THE 8-WAY EBCM HARNESS CONNECTOR.

DOES J 39200 DISPLAY RESISTANCE AS "OL" (INFINITE)?

- YES → (5)
- NO → REPAIR SHORT BETWEEN CKT 1280 AND CKT 1281.

(5)
- MEASURE THE RESISTANCE BETWEEN TERMINAL "A" AND TERMINAL "B" OF THE ABS MOTOR PACK CONNECTOR.

IS RESISTANCE 0.4Ω OR GREATER?

- YES → (6)
- NO → REPLACE ABS MOTOR PACK.

(6)
- RECONNECT ALL CONNECTORS.
- IGNITION "OFF."
- START ENGINE WITH FOOT OFF BRAKE.
- ALLOW ENGINE TO RUN FOR AT LEAST 10 SECONDS.
- REPEAT ABOVE IGNITION CYCLE SEQUENCE TWO MORE TIMES.

DID DTC 44 SET IN THE LAST THREE IGNITION CYCLES?

- YES → REPLACE EBCM.
- NO → MALFUNCTION IS INTERMITTENT. INSPECT ALL CONNECTORS FOR CONTAMINATION AND POOR TERMINAL CONTACT. REPLACE ANY TERMINALS THAT EXHIBIT POOR TERMINAL CONTACT OR EVIDENCE OF CORROSION.

(8)
- USING TECH 1 MISC. TESTS, PERFORM GEAR TENSION RELIEF FUNCTION.
- IGNITION "OFF."
- REMOVE HYDRAULIC MODULATOR/MOTOR PACK ASSEMBLY FROM VEHICLE.
- SEPARATE ABS MOTOR PACK FROM HYDRAULIC MODULATOR.
- GRASP THE GEAR ON THE HYDRAULIC MODULATOR (THE HYDRAULIC MODULATOR IS THE UNIT WITH THE THREE LARGE GEARS) FURTHEST FORWARD WHEN THE UNIT IS INSTALLED IN THE VEHICLE AND ATTEMPT TO MOVE THE GEAR.

CAN THE GEAR BE ROTATED AT LEAST 10 FULL TURNS LOCK TO LOCK?

- NO → REPLACE HYDRAULIC MODULATOR.
- YES → REPLACE ABS MOTOR PACK.

* AFTER DIAGNOSIS IS COMPLETE, CLEAR DTCs AND TEST DRIVE VEHICLE FOR THREE (3) DRIVE CYCLES TO VERIFY THAT DTC DOES NOT RESET. A DRIVE CYCLE CONSISTS OF STARTING THE VEHICLE, DRIVING OVER 16 KM/H (10 MPH), AND THEN KEYING DOWN. AND THEN KEYING DOWN.

DIAGNOSTIC CHART DTC 45 — 1994–95 CAMARO AND FIREBIRD

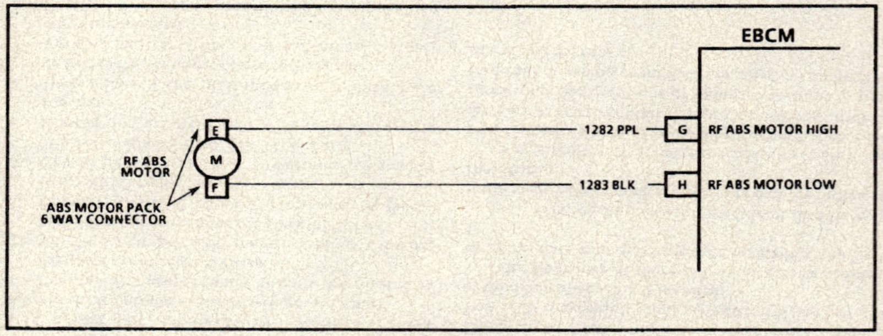

DTC 45

RIGHT FRONT ABS CHANNEL WILL NOT MOVE

Circuit Description:
This DTC is designed to detect bound-up ESB, a stuck ABS motor, or a seized hydraulic modulator. When the release is commanded during initialization, the ESB should release the ABS motor, resulting in sensed current being less than commanded current (motor is spinning freely). If the ABS motor is not moving, sensed current will be equal to stall current.

DTC Will Set When: DTC 45 can be set during normal operation. If the EBCM detects a condition in which it cannot move the ABS motor in either direction, a malfunction exists.

Action Taken: This malfunction indicates the channel cannot be moved properly. A malfunction DTC is stored, ABS is disabled and the "ABS INOP" lamp is turned "ON."

DTC Chart Test Description: Number(s) below refer to circled number(s) on the diagnostic chart.

1. Ensures that proper ABS motor circuits are located in the proper connector cavities.
2. Checks for proper ABS motor movement during apply and release commands from the Tech 1.
3. Compares EBCM command current to motor feedback current.
4. Checks for short between CKT 1282 and CKT 1283.
5. Checks for an ABS motor that is internally shorted.
6. Checks to see if malfunction is due to poor terminal contact or the EBCM.
7. Verifies that the motor can actually apply when commanded.
8. Checks for proper hydraulic modulator gear and piston movement.

Diagnostic Aids: An "Intermittent" malfunction in this DTC may result from a mechanical part of the system that sticks, binds, or slips.

The frequency of the malfunction can be checked by using the enhanced diagnostic function of the Tech 1.

DTC 45 may set after modulator disassembly if the modulator pistons are positioned at the bottom of their bore.

Any circuitry, that is suspected as causing the intermittent complaint, should be thoroughly checked for backed out terminals, improper mating, broken locks, improperly formed or damaged terminals, poor terminal to wiring connections or physical damage to the wiring harness.

DIAGNOSTIC CHART DTC 45 — 1994–95 CAMARO AND FIREBIRD

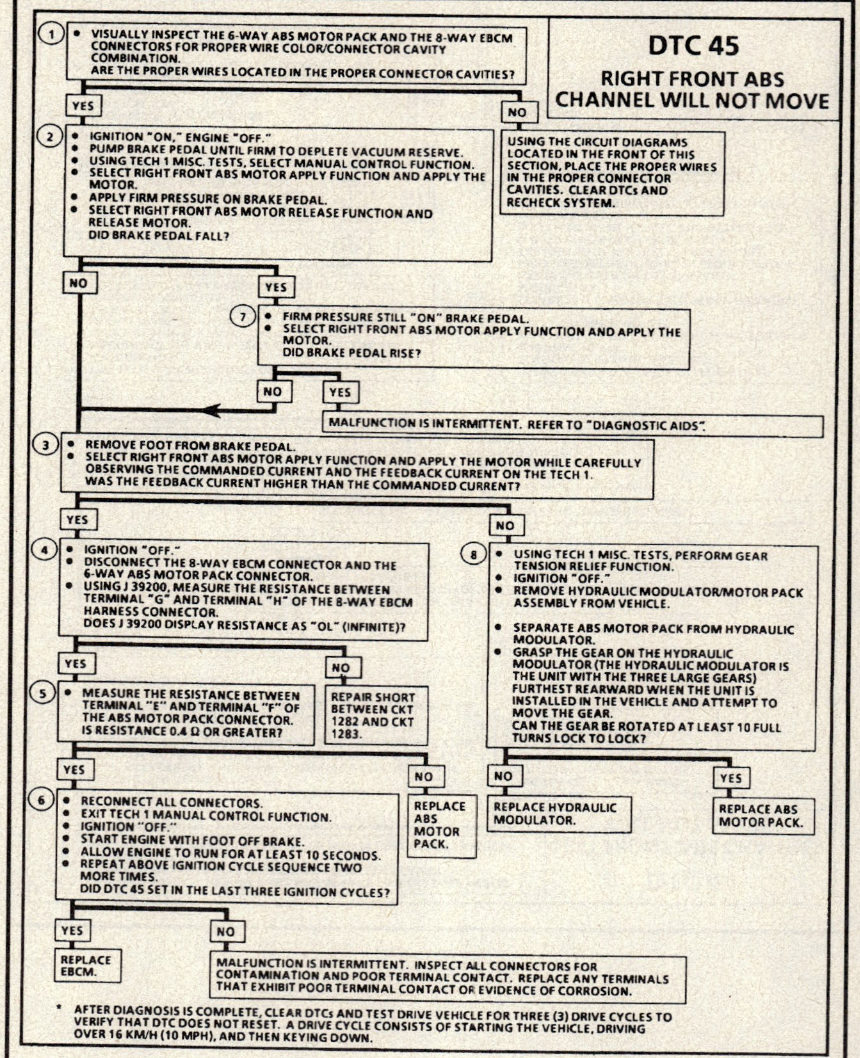

DIAGNOSTIC CHART DTC 46 — 1994–95 CAMARO AND FIREBIRD

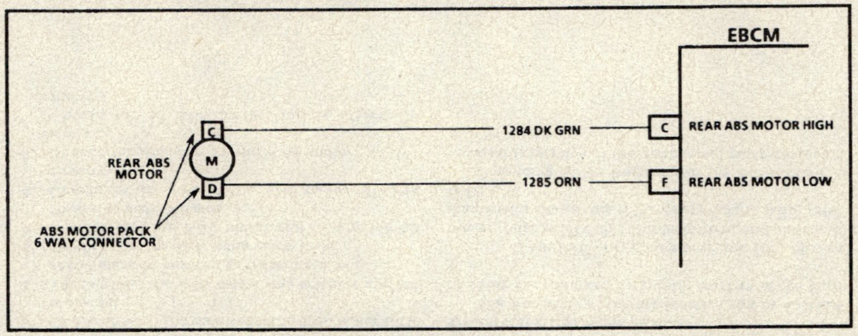

DTC 46

REAR ABS CHANNEL WILL NOT MOVE

Circuit Description:

This DTC is designed to detect bound-up ESB, a stuck ABS motor, or a seized hydraulic modulator. When the release is commanded during initialization, the ESB should release the ABS motor, resulting in sensed current being less than commanded current (motor is spinning freely). If the ABS motor is not moving, sensed current will be equal to stall current.

DTC Will Set When: DTC 46 can be set during normal operation. If the EBCM detects a condition in which it cannot move the motor in either direction, a malfunction exists.

Action Taken: This malfunction indicates the channel cannot be moved properly. A malfunction DTC is stored, ABS is disabled and the "ABS INOP" lamp is turned "ON."

DTC Chart Test Description: Number(s) below refer to circled number(s) on the diagnostic chart.

1. Ensures that proper ABS motor circuits are located in the proper connector cavities.
2. Checks for proper ABS motor movement during apply and release commands from the Tech 1.
3. Compares EBCM command current to ABS motor feedback current.
4. Checks for a short circuit between the ABS motor "HIGH" and ABS motor "LOW" circuits.
5. Checks to see if malfunction is due to poor terminal contact or the EBCM.
6. Verifies that the motor can actually release when commanded.
7. Checks for proper hydraulic modulator gear and piston movement.

Diagnostic Aids: An "Intermittent" malfunction in this DTC may result from a mechanical part of the system that sticks or binds.

The frequency of the malfunction can be checked by using the enhanced diagnostic function of the Tech 1.

DTC 46 may set after modulator disassembly if the modulator pistons are positioned at the bottom of their bore.

Depending on the frequency of the malfunction, a physical inspection of the mechanical parts suspected may be necessary.

Any circuitry, that is suspected as causing the intermittent complaint, should be thoroughly checked for backed out terminals, improper mating, broken locks, improperly formed or damaged terminals, poor terminal to wiring connections or physical damage to the wiring harness.

DIAGNOSTIC CHART DTC 46 — 1994–95 CAMARO AND FIREBIRD

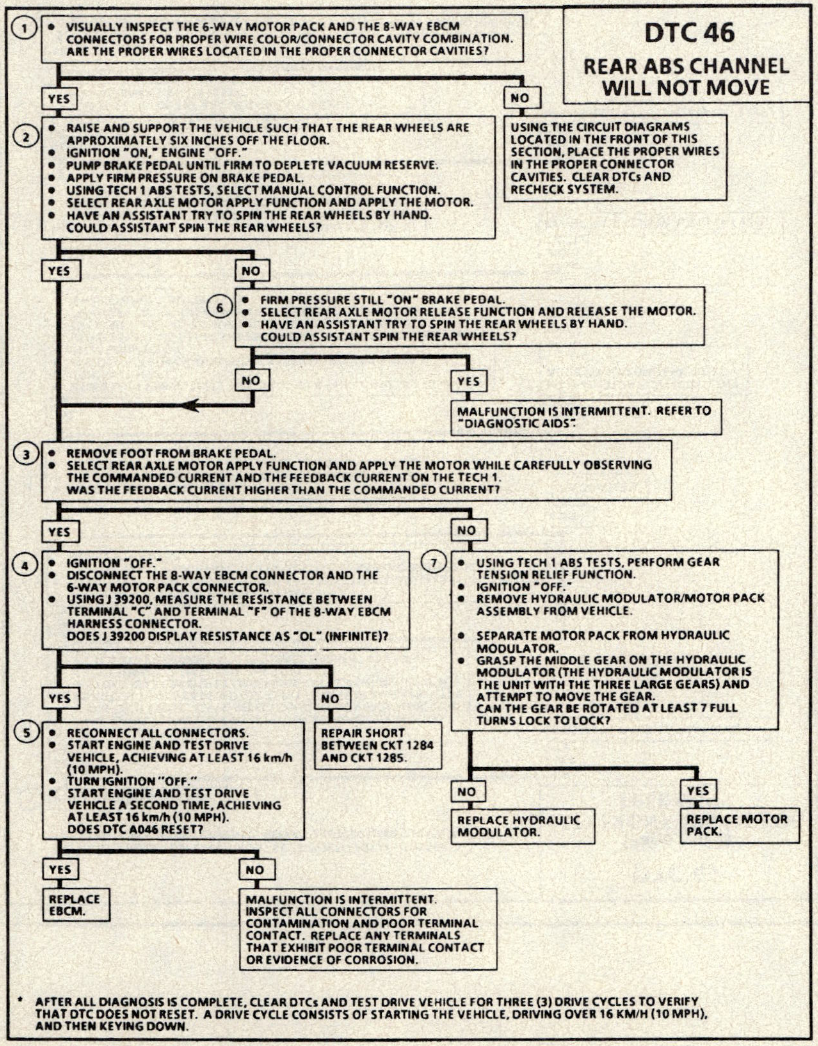

DIAGNOSTIC CHART DTC 47 — 1994–95 CAMARO AND FIREBIRD

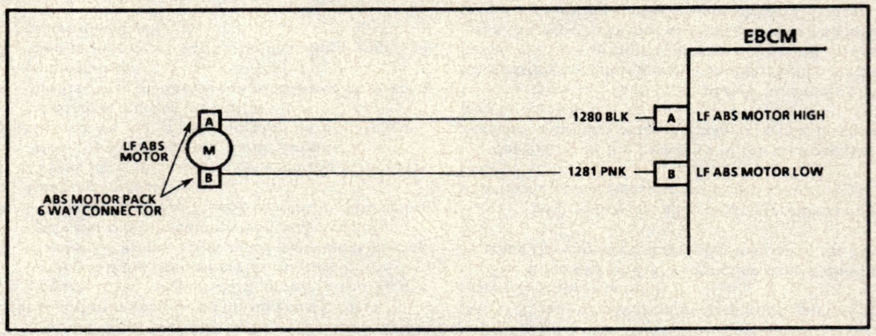

DTC 47

LEFT FRONT ABS MOTOR FREE SPINS

Circuit Description:
This DTC is designed to detect a stripped nut or gear assembly during initialization. During the homing sequence, the piston should reach the top of the bore resulting in a stalled ABS motor. If this does not occur, the ABS motor must be spinning with little or no resistance, indicating a nut/screw or gear malfunction.

DTC Will Set When: DTC 47 can only be set during initialization. If the feedback current is less than the command current for a specified period of time, a malfunction exists.

Action Taken: A malfunction DTC is stored, ABS is disabled, and the "ABS INOP" lamp is turned "ON."

DTC Chart Test Description: Number(s) below refer to circled number(s) on the diagnostic chart.

1. Checks to see if the corresponding open motor DTC is also set.
2. Verifies that the ABS motor was actually applied as command by observing feedback current.
3. Verifies that the ABS motor can release.
4. Verifies that the ABS motor can be applied by observing pedal movement.
5. Checks for an open in the ABS motor "HIGH" circuitry.
6. Checks for an open in the ABS motor "LOW" circuitry.
7. Checks for an ABS motor that is internally shorted.

Diagnostic Aids: An "Intermittent" malfunction in this DTC may result from a mechanical part of the system that sticks, binds, or slips.

The frequency of the malfunction can be checked by using the enhanced diagnostic function of the Tech 1.

If the DTC only occurs once and DTC 56 also occurs, refer to DTC 56. If intermittent and enhanced diagnostics shows this DTC occurs during ABS, refer to DTC 56.

Depending on the frequency of the malfunction, a physical inspection of the mechanical parts suspected may be necessary.

DIAGNOSTIC CHART DTC 47 — 1994–95 CAMARO AND FIREBIRD

DTC 47
(Page 1 of 2)
LEFT FRONT ABS MOTOR FREE SPINS

NOTICE: J 39200 TEST LEADS MUST BE "ZEROED" PRIOR TO MAKING ANY RESISTANCE MEASUREMENTS. REFER TO J 39200 USER'S MANUAL.

(1) IS DTC 56 ALSO CURRENTLY SET?
- YES: PROCEED TO DTC 56 CHART FIRST.
- NO: go to (2)

(2)
- IGNITION "ON," ENGINE "OFF."
- PUMP BRAKE PEDAL UNTIL FIRM TO DEPLETE VACUUM RESERVE.
- USING TECH 1 MISC. TESTS, SELECT MANUAL CONTROL FUNCTION.
- SELECT LEFT FRONT ABS MOTOR APPLY FUNCTION AND APPLY THE MOTOR WHILE CAREFULLY OBSERVING THE COMMAND CURRENT AND FEEDBACK CURRENT ON THE TECH 1.
WERE BOTH COMMAND AND FEEDBACK CURRENT 10 AMPS?
- YES: go to (3)
- NO: GO TO STEP 5.

(3)
- APPLY FIRM PRESSURE ON BRAKE PEDAL.
- SELECT LEFT FRONT ABS MOTOR RELEASE FUNCTION AND RELEASE THE MOTOR.
DID THE BRAKE PEDAL FALL AND WERE BOTH COMMAND AND FEEDBACK CURRENT 6 AMPS?
- YES: go to (4)
- NO: GO TO STEP 5.

(4)
- FIRM PRESSURE STILL ON BRAKE PEDAL.
- SELECT LEFT ABS MOTOR APPLY FUNCTION AND APPLY THE MOTOR (THIS MOVES THE PISTON TO THE TOP OF ITS BORE.)
DID BRAKE PEDAL RISE?
- YES: MALFUNCTION IS INTERMITTENT. REFER TO "DIAGNOSTIC AIDS".
- NO: go to (5)

(5)
- USING TECH 1 MISC. TESTS, PERFORM GEAR TENSION RELIEF FUNCTION.
- IGNITION "OFF."
- DISCONNECT 6-WAY ABS MOTOR PACK CONNECTOR AND 8-WAY EBCM CONNECTOR.
- USING J 39200, MEASURE THE RESISTANCE BETWEEN TERMINAL "A" OF THE 6-WAY ABS MOTOR PACK HARNESS CONNECTOR AND TERMINAL "A" OF THE 8-WAY EBCM HARNESS CONNECTOR.
IS RESISTANCE 1.5Ω OR LESS?
- YES: go to (6)
- NO: REPAIR HIGH RESISTANCE IN CKT 1280.

(6)
- MEASURE THE RESISTANCE BETWEEN TERMINAL "B" OF THE 6-WAY ABS MOTOR PARK HARNESS CONNECTOR AND TERMINAL "B" OF THE 8-WAY EBCM HARNESS CONNECTOR.
IS RESISTANCE 1.5Ω OR LESS?
- YES: go to (7)
- NO: REPAIR HIGH RESISTANCE IN CKT 1281.

(7)
- MEASURE THE RESISTANCE BETWEEN TERMINAL "A" AND TERMINAL "B" OF THE ABS MOTOR PACK CONNECTOR.
IS RESISTANCE 1.5Ω OR LESS?
- YES: DIAGNOSIS CONTINUED ON NEXT PAGE.
- NO: REPLACE ABS MOTOR PACK.

AFTER DIAGNOSIS IS COMPLETE, CLEAR DTCs AND TEST DRIVE VEHICLE FOR THREE (3) DRIVE CYCLES TO VERIFY THAT DTC DOES NOT RESET. A DRIVE CYCLE CONSISTS OF STARTING THE VEHICLE, DRIVING OVER 16 km/h (10 MPH), AND THEN KEYING DOWN.

DIAGNOSTIC CHART DTC 47 — 1994–95 CAMARO AND FIREBIRD

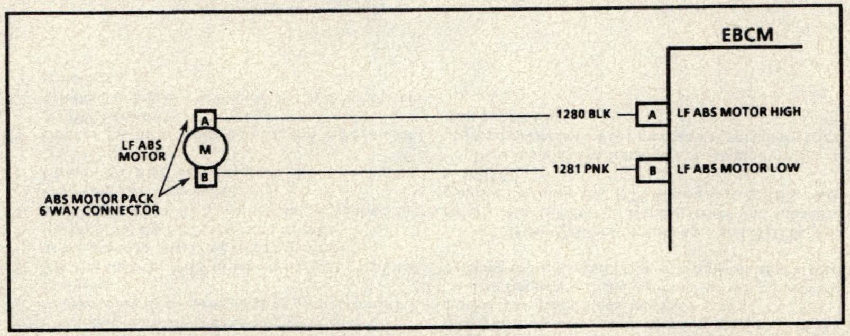

DTC 47
(Page 2 of 2)
LEFT FRONT ABS MOTOR FREE SPINS

Circuit Description:
This DTC is designed to detect a stripped nut or gear assembly during initialization. During the homing sequence, the piston should reach the top of the bore resulting in a stalled motor. If this does not occur, the motor must be spinning with little or no resistance, indicating a nut/screw or gear malfunction.

DTC Will Set When: DTC 47 can only be set during initialization. If the feedback current is less than the command current for a specified period of time, a malfunction exists.

Action Taken: A malfunction DTC is stored, ABS is disabled, and the "ABS INOP" lamp is turned "ON."

DTC Chart Test Description: Number(s) below refer to circled number(s) on the diagnostic chart.

8. Checks for a stripped gear on the ABS motor pack (unit with three small gears).
9. Checks for a stripped gear on the hydraulic modulator (unit with three large gears).
10. Checks to see if the ABS motor pack has an internal malfunction.
11. Checks for a malfunctioning hydraulic modulator.
12. Ensures malfunction was not due to poor terminal contact.

Diagnostic Aids: An "Intermittent" malfunction in this DTC may result from a mechanical part of the system that sticks, binds, or slips.

The frequency of the malfunction can be checked by using the enhanced diagnostic function of the Tech 1.

If the DTC only occurs once and DTC 56 also occurs, refer to DTC 56. If intermittent and enhanced diagnostics shows this DTC occurs during ABS, refer to DTC 56.

Depending on the frequency of the malfunction, a physical inspection of the mechanical parts suspected may be necessary.

DIAGNOSTIC CHART DTC 47 — 1994–95 CAMARO AND FIREBIRD

DTC 47
(Page 2 of 2)
LEFT FRONT ABS MOTOR FREE SPINS

DIAGNOSIS CONTINUED FROM PREVIOUS PAGE.

(8)
- REMOVE HYDRAULIC MODULATOR/MOTOR PACK ASSEMBLY FROM VEHICLE
- REMOVE GEAR COVER.
- CHECK FOR STRIPPED GEAR ON ABS MOTOR PACK; ABS MOTOR PACK IS THE UNIT WITH THE THREE SMALL GEARS. THE LEFT FRONT GEAR IS THE GEAR FURTHEST FORWARD WHEN THE ASSEMBLY IS INSTALLED IN THE VEHICLE. IS THE ABS MOTOR PACK GEAR STRIPPED?

YES: REPLACE ABS MOTOR PACK.

NO:

(9)
- CHECK FOR A STRIPPED GEAR ON HYDRAULIC MODULATOR; MODULATOR IS THE UNIT WITH THE THREE LARGE GEARS. THE LEFT FRONT GEAR IS THE GEAR FURTHEST FORWARD WHEN THE UNIT IS INSTALLED IN THE VEHICLE. IS THE MODULATOR GEAR STRIPPED?

YES: REPLACE HYDRAULIC MODULATOR GEAR.

NO:

(10)
- RECONNECT ELECTRICAL CONNECTORS
- SECURELY POSITION HYDRAULIC MODULATOR/MOTOR PACK ASSEMBLY IN VEHICLE WITH GEAR COVER REMOVED SUCH THAT THE GEAR SET CAN BE OBSERVED. BE CAREFUL NOT TO DAMAGE GEAR SET OR ALLOW THE HYDRAULIC MODULATOR TO MOVE WHILE TESTING.
- IGNITION "ON."
- USING TECH 1 MISC. TESTS SELECT MANUAL CONTROL FUNCTION.
- SELECT LEFT FRONT ABS MOTOR APPLY AND APPLY THE MOTOR WHILE OBSERVING THE GEAR SET FURTHEST FORWARD WHEN THE UNIT IS INSTALLED IN THE VEHICLE.
- SELECT LEFT FRONT ABS MOTOR RELEASE FUNCTION AND RELEASE THE MOTOR WHILE STILL OBSERVING THE GEAR SET.
- SELECT LEFT FRONT ABS MOTOR APPLY FUNCTION AND APPLY THE MOTOR. DID GEAR SET MOVE IN BOTH DIRECTIONS FOR AT LEAST ONE REVOLUTION?

NO: REPLACE ABS MOTOR PACK.

YES:

(11)
- USING TECH 1 MISC. TESTS, PERFORM GEAR TENSION RELIEF FUNCTION.
- IGNITION "OFF."
- SEPARATE THE ABS MOTOR PACK FROM THE HYDRAULIC MODULATOR.
- GRASP THE GEAR ON THE HYDRAULIC MODULATOR (THE HYDRAULIC MODULATOR IS THE UNIT WITH THE THREE LARGE GEARS) FURTHEST FORWARD WHEN THE UNIT IS INSTALLED IN THE VEHICLE AND ATTEMPT TO MOVE THE GEAR. CAN THE GEAR BE ROTATED MORE THAN 13 FULL TURNS LOCK TO LOCK?

YES: REPLACE THE HYDRAULIC MODULATOR.

NO:

(12)
- INSPECT ALL CONNECTORS AND TERMINALS FOR POOR TERMINAL CONTACT AND EVIDENCE OF CORROSION.
- REPLACE ALL TERMINALS THAT EXHIBIT SIGNS OF POORTERMINAL CONTACT OR CORROSION.
- REINSTALL HYDRAULIC MODULATOR/MOTOR PACK ASSEMBLY IN VEHICLE.
- RECONNECT ALL CONNECTORS.
- IGNITION "OFF."
- START ENGINE WITH FOOT OFF BRAKE.
- ALLOW ENGINE TO RUN FOR AT LEAST 10 SECONDS.
- REPEAT ABOVE IGNITION CYCLE SEQUENCE TWO MORE TIMES. DID DTC 47 SET IN THE LAST THREE IGNITION CYCLES?

YES: REPLACE EBCM.

NO: MALFUNCTION IS MOST LIKELY CAUSED BY A POOR CONNECTION. REFER TO "DIAGNOSTIC AIDS".

AFTER DIAGNOSIS IS COMPLETE, CLEAR DTCs AND TEST DRIVE VEHICLE FOR THREE (3) DRIVE CYCLES TO VERIFY THAT DTC DOES NOT RESET. A DRIVE CYCLE CONSISTS OF STARTING THE VEHICLE, DRIVING OVER 16 km/h (10 MPH), AND THEN KEYING DOWN.

DIAGNOSTIC CHART DTC 48 — 1994–95 CAMARO AND FIREBIRD

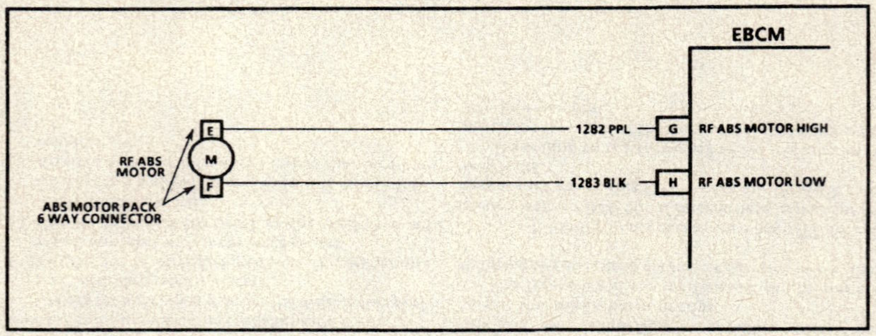

DTC 48

RIGHT FRONT ABS MOTOR FREE SPINS

Circuit Description:

This DTC is designed to detect a stripped nut or gear assembly during initialization. During the homing sequence, the piston should reach the top of the bore resulting in a stalled ABS motor. If this does not occur, the ABS motor must be spinning with little or no resistance, indicating a nut/screw or gear malfunction.

DTC Will Set When: DTC 48 can only be set during initialization. If the feedback current is less than the command current for a specified period of time, a malfunction exists.

Action Taken: A malfunction DTC is stored, ABS is disabled, and the "ABS INOP" lamp is turned "ON."

DTC Chart Test Description: Number(s) below refer to circled number(s) on the diagnostic chart.

1. Checks to see if the corresponding open motor DTC is also set.
2. Verifies that the ABS motor was actually applied as command by observing feedback current.
3. Verifies that the ABS motor can release.
4. Verifies that the ABS motor can be applied by observing pedal movement.
5. Checks for high resistance in the ABS motor "HIGH" circuitry.
6. Checks for high resistance in the ABS motor "LOW" circuitry.
7. Checks for proper resistance of the ABS motor windings.

Diagnostic Aids: An "Intermittent" malfunction in this DTC may result from a mechanical part of the system that sticks, binds, or slips.

The frequency of the malfunction can be checked by using the enhanced diagnostic function of the Tech 1.

If the DTC only occurs once and DTC 61 also occurs, refer to DTC 61. If intermittent and enhanced diagnostics shows this DTC occurs during ABS, refer to DTC 61.

Depending on the frequency of the malfunction, a physical inspection of the mechanical parts suspected may be necessary.

DIAGNOSTIC CHART DTC 48 — 1994–95 CAMARO AND FIREBIRD

DTC 48
(Page 1 of 2)
RIGHT FRONT ABS MOTOR FREE SPINS

NOTICE: J 39200 TEST LEADS MUST BE "ZEROED" PRIOR TO MAKING ANY RESISTANCE MEASUREMENTS. REFER TO J 39200 USER'S MANUAL.

(1) IS DTC 61 ALSO CURRENTLY SET?
- YES: PROCEED TO DTC 61 CHART FIRST.
- NO: go to (2).

(2)
- IGNITION "ON," ENGINE "OFF."
- PUMP BRAKE PEDAL UNTIL FIRM TO DEPLETE VACUUM RESERVE.
- USING TECH 1 MISC. TESTS, SELECT MANUAL CONTROL FUNCTION.
- SELECT RIGHT FRONT ABS MOTOR APPLY FUNCTION AND APPLY THE MOTOR WHILE CAREFULLY OBSERVING THE COMMAND CURRENT AND FEEDBACK CURRENT ON THE TECH 1.

WERE BOTH COMMAND AND FEEDBACK CURRENT 10 AMPS?
- YES: go to (3).
- NO: go to step 5.

(3)
- APPLY FIRM PRESSURE ON BRAKE PEDAL.
- SELECT RIGHT FRONT ABS MOTOR RELEASE FUNCTION AND RELEASE THE MOTOR.

DID THE BRAKE PEDAL FALL AND WERE BOTH COMMAND AND FEEDBACK CURRENT 6 AMPS?
- YES: go to (4).
- NO: GO TO STEP 5.

(4)
- FIRM PRESSURE STILL ON BRAKE PEDAL.
- SELECT RIGHT ABS MOTOR APPLY FUNCTION AND APPLY THE MOTOR (THIS MOVES THE PISTON TO THE TOP OF ITS BORE.)

DID BRAKE PEDAL RISE?
- YES: MALFUNCTION IS INTERMITTENT. REFER TO "DIAGNOSTIC AIDS".
- NO: go to (5).

(5)
- USING TECH 1 MISC. TESTS, PERFORM GEAR TENSION RELIEF FUNCTION.
- IGNITION "OFF."
- DISCONNECT 6-WAY ABS MOTOR PACK CONNECTOR AND 8-WAY EBCM CONNECTOR.
- USING J 39200, MEASURE THE RESISTANCE BETWEEN TERMINAL "E" OF THE 6-WAY ABS MOTOR PACK HARNESS CONNECTOR AND TERMINAL "G" OF THE 8-WAY EBCM HARNESS CONNECTOR.

IS RESISTANCE 1.5Ω OR LESS?
- YES: go to (6).
- NO: REPAIR HIGH RESISTANCE IN CKT 1282.

(6)
- MEASURE THE RESISTANCE BETWEEN TERMINAL "F" OF THE 6-WAY ABS MOTOR PARK HARNESS CONNECTOR AND TERMINAL "H" OF THE 8-WAY EBCM HARNESS CONNECTOR.

IS RESISTANCE 1.5Ω OR LESS?
- YES: go to (7).
- NO: REPAIR HIGH RESISTANCE IN CKT 1283.

(7)
- MEASURE THE RESISTANCE BETWEEN TERMINAL "E" AND TERMINAL "F" OF THE ABS MOTOR PACK CONNECTOR.

IS RESISTANCE 1.5Ω OR LESS?
- YES: DIAGNOSIS CONTINUED ON NEXT PAGE.
- NO: REPLACE ABS MOTOR PACK.

AFTER DIAGNOSIS IS COMPLETE, CLEAR DTCs AND TEST DRIVE VEHICLE FOR THREE (3) DRIVE CYCLES TO VERIFY THAT DTC DOES NOT RESET. A DRIVE CYCLE CONSISTS OF STARTING THE VEHICLE, DRIVING OVER 16 km/h (10 MPH), AND THEN KEYING DOWN.

DIAGNOSTIC CHART DTC 48 — 1994–95 CAMARO AND FIREBIRD

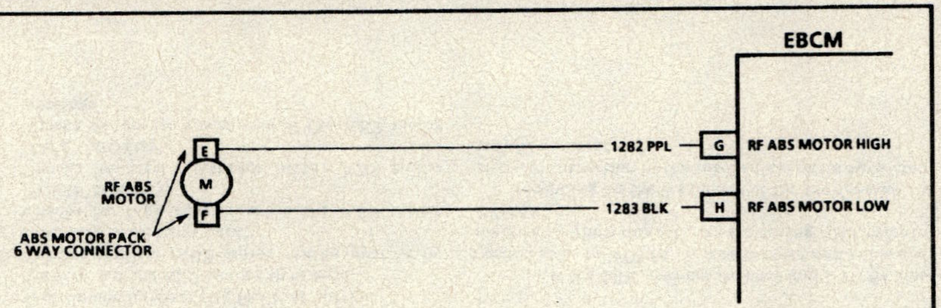

DTC 48

RIGHT FRONT ABS MOTOR FREE SPINS

Circuit Description:
This DTC is designed to detect a stripped nut or gear assembly during initialization. During the homing sequence, the piston should reach the top of the bore resulting in a stalled ABS motor. If this does not occur, the ABS motor must be spinning with little or no resistance, indicating a nut/screw or gear malfunction.

DTC Will Set When: DTC 48 can only be set during initialization. If the feedback current is less than the command current for a specified period of time, a malfunction exists.

Action Taken: A malfunction DTC is stored, ABS is disabled, and the "ABS INOP" lamp is turned "ON."

DTC Chart Test Description: Number(s) below refer to circled number(s) on the diagnostic chart.

8. Checks for a stripped gear on the ABS motor pack (unit with three small gears).
9. Checks for a stripped gear on the hydraulic modulator (unit with three large gears).
10. Verifies that the ABS motor can be applied under a load situation.
11. Verifies that the hydraulic modulator is functioning properly.
12. Ensures malfunction was not due to poor terminal contact.

Diagnostic Aids: An "Intermittent" malfunction in this DTC may result from a mechanical part of the system that sticks, binds, or slips.

The frequency of the malfunction can be checked by using the enhanced diagnostic function of the Tech 1, as described in "Tech 1 Diagnostics" found in this section. If the DTC only occurs once and DTC 61 also occurs, refer to DTC 61. If intermittent and enhanced diagnostics shows this DTC occurs during ABS, refer to DTC 61.

Depending on the frequency of the malfunction, a physical inspection of the mechanical parts suspected may be necessary.

DIAGNOSTIC CHART DTC 48 — 1994–95 CAMARO AND FIREBIRD

DTC 48
(Page 2 of 2)
RIGHT FRONT ABS MOTOR FREE SPINS

DIAGNOSIS CONTINUED FROM PREVIOUS PAGE.

(8)
- REMOVE HYDRAULIC MODULATOR/MOTOR PACK ASSEMBLY FROM VEHICLE.
- REMOVE GEAR COVER.
- CHECK FOR STRIPPED GEAR ON ABS MOTOR PACK; ABS MOTOR PACK IS THE UNIT WITH THE THREE SMALL GEARS. THE RIGHT FRONT GEAR IS THE GEAR FURTHEST REARWARD WHEN THE ASSEMBLY IS INSTALLED IN THE VEHICLE. IS THE ABS MOTOR PACK GEAR STRIPPED?

YES → REPLACE ABS MOTOR PACK.

NO ↓

(9)
- CHECK FOR A STRIPPED GEAR ON HYDRAULIC MODULATOR; MODULATOR IS THE UNIT WITH THE THREE LARGE GEARS. THE RIGHT FRONT GEAR IS THE GEAR FURTHEST REARWARD WHEN THE UNIT IS INSTALLED IN THE VEHICLE. IS THE MODULATOR GEAR STRIPPED?

YES → REPLACE HYDRAULIC MODULATOR GEAR.

NO ↓

(10)
- RECONNECT ELECTRICAL CONNECTORS.
- SECURELY POSITION HYDRAULIC MODULATOR/MOTOR PACK ASSEMBLY IN VEHICLE WITH GEAR COVER REMOVED SUCH THAT THE GEAR SET CAN BE OBSERVED. BE CAREFUL NOT TO DAMAGE GEAR SET OR ALLOW THE HYDRAULIC MODULATOR TO MOVE WHILE TESTING.
- IGNITION "ON."
- USING TECH 1 MISC. TESTS, SELECT MANUAL CONTROL FUNCTION.
- SELECT RIGHT FRONT ABS MOTOR APPLY AND APPLY THE MOTOR WHILE OBSERVING THE GEAR SET FURTHEST REARWARD WHEN THE UNIT IS INSTALLED IN THE VEHICLE.
- SELECT RIGHT FRONT ABS MOTOR RELEASE FUNCTION AND RELEASE THE MOTOR WHILE STILL OBSERVING THE GEAR SET.
- SELECT RIGHT FRONT ABS MOTOR APPLY FUNCTION AND APPLY THE MOTOR. DID GEAR SET MOVE IN BOTH DIRECTIONS FOR AT LEAST ONE REVOLUTION?

NO → REPLACE ABS MOTOR PACK.

YES ↓

(11)
- USING TECH 1 MISC. TESTS, PERFORM GEAR TENSION RELIEF FUNCTION.
- IGNITION "OFF."
- SEPARATE THE ABS MOTOR PACK FROM THE HYDRAULIC MODULATOR.
- GRASP THE GEAR ON THE HYDRAULIC MODULATOR (THE HYDRAULIC MODULATOR IS THE UNIT WITH THE THREE LARGE GEARS) FURTHEST REARWARD WHEN THE UNIT IS INSTALLED IN THE VEHICLE AND ATTEMPT TO MOVE THE GEAR. CAN THE GEAR BE ROTATED MORE THAN 1½ FULL TURNS LOCK TO LOCK?

YES → REPLACE THE HYDRAULIC MODULATOR.

NO ↓

(12)
- INSPECT ALL CONNECTORS AND TERMINALS FOR POOR TERMINAL CONTACT AND EVIDENCE OF CORROSION.
- REPLACE ALL TERMINALS THAT EXHIBIT SIGNS OF POOR TERMINAL CONTACT OR CORROSION.
- REINSTALL HYDRAULIC MODULATOR/MOTOR PACK ASSEMBLY IN VEHICLE.
- RECONNECT ALL CONNECTORS.
- IGNITION "OFF."
- START ENGINE WITH FOOT OFF BRAKE.
- ALLOW ENGINE TO RUN FOR AT LEAST 10 SECONDS.
- REPEAT ABOVE IGNITION CYCLE SEQUENCE TWO MORE TIMES. DID DTC 48 SET IN THE LAST THREE IGNITION CYCLES?

YES → REPLACE EBCM.

NO → MALFUNCTION WAS MOST LIKELY CAUSED BY A POOR CONNECTION. REFER TO "DIAGNOSTIC AIDS".

AFTER DIAGNOSIS IS COMPLETE, CLEAR DTCs AND TEST DRIVE VEHICLE FOR THREE (3) DRIVE CYCLES TO VERIFY THAT DTC DOES NOT RESET. A DRIVE CYCLE CONSISTS OF STARTING THE VEHICLE, DRIVING OVER 16 km/h (10 MPH), AND THEN KEYING DOWN.

DIAGNOSTIC CHART DTC 51 — 1994–95 CAMARO AND FIREBIRD

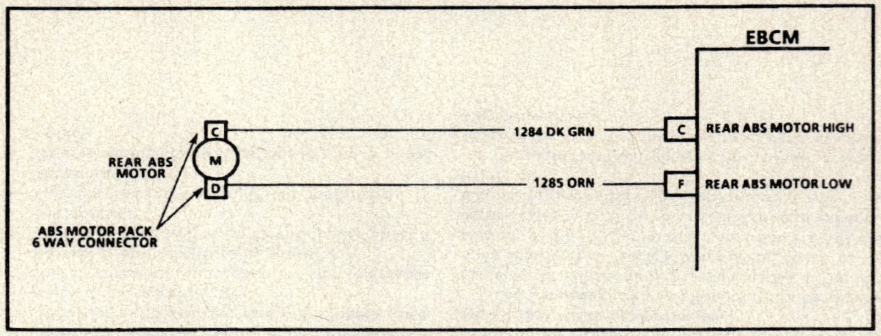

DTC 51

REAR ABS MOTOR FREE SPINS

Circuit Description:
This DTC is designed to detect a stripped nut or gear assembly during initialization. During the homing sequence, the piston should reach the top of the bore resulting in a stalled ABS motor. If this does not occur, the ABS motor must be spinning with little or no resistance, indicating a nut/screw or gear malfunction.

DTC Will Set When: DTC 51 can be set during initialization. If the feedback current is less than the command current for a specified period of time, a malfunction exists.

Action Taken: A malfunction DTC is stored, ABS is disabled, and the "ABS INOP" lamp is turned "ON."

DTC Chart Test Description: Number(s) below refer to circled number(s) on the diagnostic chart.

1. Checks to see if the corresponding open ABS motor DTC is also set.
2. Verifies that the ABS motor was actually applied as command by observing feedback current.
3. Verifies that the ABS motor can release.
4. Verifies that the ABS motor can be applied by observing pedal movement.
5. Checks for high resistance in the ABS motor "HIGH" circuitry.
6. Checks for high resistance in the ABS motor "LOW" circuitry.
7. Checks for proper resistance of the ABS motor windings.

Diagnostic Aids: An "Intermittent" malfunction in this DTC may result from a mechanical part of the system that sticks, binds, or slips.

The frequency of the malfunction can be checked by using the enhanced diagnostic function of the Tech 1.

If the DTC only occurs once and DTC 64 also occurs, refer to DTC 64. If intermittent and enhanced diagnostics show this DTC occurs during ABS, refer to DTC 64.

Depending on the frequency of the malfunction, a physical inspection of the mechanical parts suspected may be necessary.

DIAGNOSTIC CHART DTC 51 — 1994–95 CAMARO AND FIREBIRD

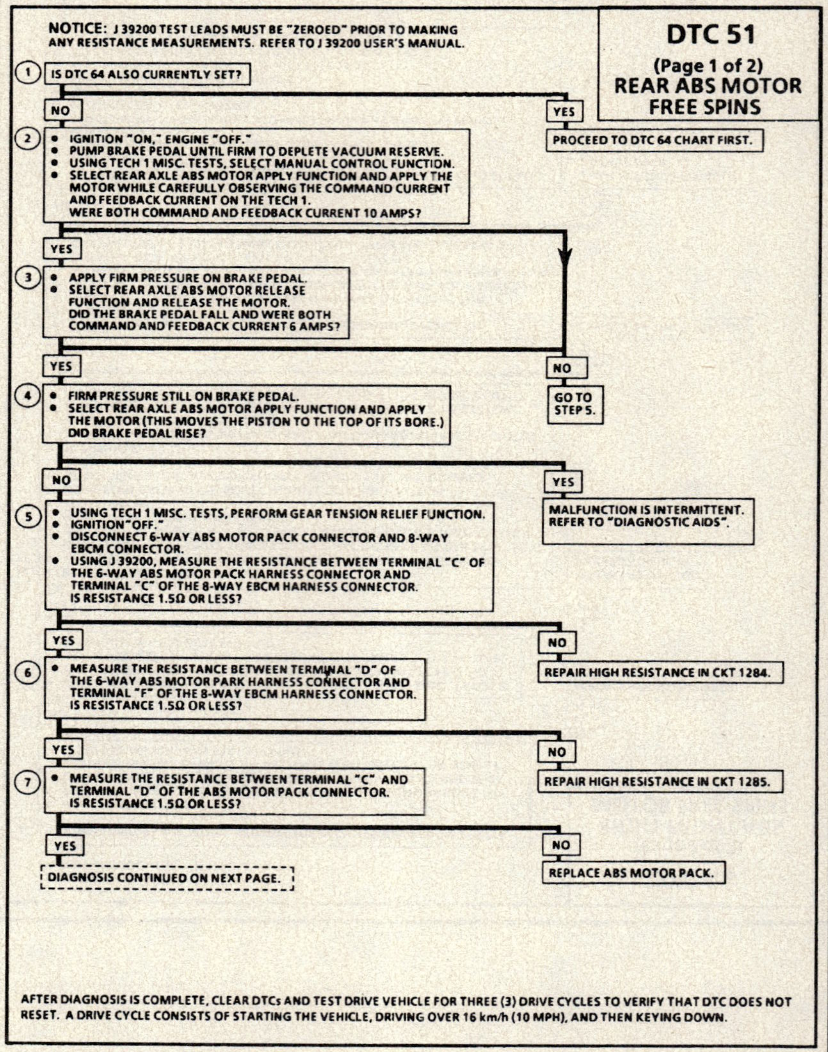

DIAGNOSTIC CHART DTC 51 — 1994–95 CAMARO AND FIREBIRD

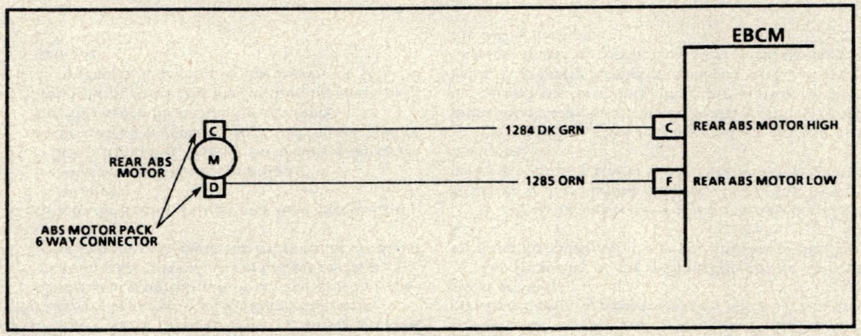

DTC 51

REAR ABS MOTOR FREE SPINS

Circuit Description:
This DTC is designed to detect a stripped nut or gear assembly during initialization. During the homing sequence, the piston should reach the top of the bore resulting in a stalled motor. If this does not occur, the ABS motor must be spinning with little or no resistance, indicating a nut/screw or gear malfunction.

DTC Will Set When: DTC 51 can be set during initialization. If the feedback current is less than the command current for a specified period of time, a malfunction exists.

Action Taken: A malfunction DTC is stored, ABS is disabled, and the "ABS INOP" lamp is turned "ON."

DTC Chart Test Description: Number(s) below refer to circled number(s) on the diagnostic chart.

8. Checks for a stripped gear on the ABS motor pack (unit with three small gears).
9. Checks for a stripped gear on the hydraulic modulator (unit with three large gears).
10. Verifies that the motor can be applied under a load situation.
11. Verifies that the hydraulic modulator is functioning properly.
12. Ensures malfunction was not due to poor terminal contact.

Diagnostic Aids: An "Intermittent" malfunction in this DTC may result from a mechanical part of the system that sticks, binds, or slips.

The frequency of the malfunction can be checked by using the enhanced diagnostic function of the Tech 1.

If the DTC only occurs once and DTC 64 also occurs, refer to DTC 64. If intermittent and enhanced diagnostics show this DTC occurs during ABS, refer to DTC 64.

Depending on the frequency of the malfunction, a physical inspection of the mechanical parts suspected may be necessary.

DIAGNOSTIC CHART DTC 51 — 1994–95 CAMARO AND FIREBIRD

DTC 51
(Page 2 of 2)
REAR ABS MOTOR FREE SPINS

DIAGNOSIS CONTINUED FROM PREVIOUS PAGE.

(8)
- REMOVE HYDRAULIC MODULATOR/MOTOR PACK ASSEMBLY FROM VEHICLE
- REMOVE GEAR COVER.
- CHECK FOR STRIPPED GEAR ON ABS MOTOR PACK; ABS MOTOR PACK IS THE UNIT WITH THE THREE SMALL GEARS. THE REAR AXLE GEAR IS THE MIDDLE GEAR OF THE THREE.
IS THE ABS MOTOR PACK GEAR STRIPPED?

YES: REPLACE ABS MOTOR PACK.

NO:

(9)
- CHECK FOR A STRIPPED GEAR ON HYDRAULIC MODULATOR; MODULATOR IS THE UNIT WITH THE THREE LARGE GEARS. THE REAR AXLE GEAR IS THE MIDDLE GEAR OF THE THREE.
IS THE MODULATOR GEAR STRIPPED?

YES: REPLACE HYDRAULIC MODULATOR GEAR.

NO:

(10)
- RECONNECT ELECTRICAL CONNECTORS
- SECURELY POSITION HYDRAULIC MODULATOR/MOTOR PACK ASSEMBLY IN VEHICLE WITH GEAR COVER REMOVED SUCH THAT THE GEAR SET CAN BE OBSERVED. BE CAREFUL NOT TO DAMAGE GEAR SET OR ALLOW THE HYDRAULIC MODULATOR TO MOVE WHILE TESTING.
- IGNITION "ON."
- USING TECH 1 MISC. TESTS, SELECT MANUAL CONTROL FUNCTION.
- SELECT REAR AXLE ABS MOTOR APPLY AND APPLY THE MOTOR WHILE OBSERVING THE MIDDLE GEAR SET.
- SELECT REAR AXLE ABS MOTOR RELEASE FUNCTION AND RELEASE THE MOTOR WHILE STILL OBSERVING THE GEAR SET.
- SELECT REAR AXLE ABS MOTOR APPLY FUNCTION AND APPLY THE MOTOR.
DID GEAR SET MOVE IN BOTH DIRECTIONS FOR AT LEAST ONE REVOLUTION?

NO: REPLACE ABS MOTOR PACK.

YES:

(11)
- USING TECH 1 MISC. TESTS, PERFORM GEAR TENSION RELIEF FUNCTION.
- IGNITION "OFF."
- SEPARATE THE ABS MOTOR PACK FROM THE HYDRAULIC MODULATOR.
- GRASP THE MIDDLE GEAR ON THE HYDRAULIC MODULATOR (THE HYDRAULIC MODULATOR IS THE UNIT WITH THE THREE LARGE GEARS) AND ATTEMPT TO MOVE THE GEAR.
CAN THE GEAR BE ROTATED MORE THAN 8 FULL TURNS LOCK TO LOCK?

YES: REPLACE THE HYDRAULIC MODULATOR.

NO:

(12)
- INSPECT ALL CONNECTORS AND TERMINALS FOR POOR TERMINAL CONTACT AND EVIDENCE OF CORROSION.
- REPLACE ALL TERMINALS THAT EXHIBIT SIGNS OF POOR TERMINAL CONTACT OR CORROSION.
- REINSTALL HYDRAULIC MODULATOR/MOTOR PACK ASSEMBLY IN VEHICLE.
- RECONNECT ALL CONNECTORS.
- IGNITION "OFF."
- START ENGINE WITH FOOT OFF BRAKE.
- ALLOW ENGINE TO RUN FOR AT LEAST 10 SECONDS.
- REPEAT ABOVE IGNITION CYCLE SEQUENCE TWO MORE TIMES.
DID DTC 51 SET IN THE LAST THREE IGNITION CYCLES?

YES: REPLACE EBCM.

NO: MALFUNCTION WAS MOST LIKELY CAUSED BY A POOR CONNECTION. REFER TO "DIAGNOSTIC AIDS".

AFTER DIAGNOSIS IS COMPLETE, CLEAR DTCs AND TEST DRIVE VEHICLE FOR THREE (3) DRIVE CYCLES TO VERIFY THAT DTC DOES NOT RESET. A DRIVE CYCLE CONSISTS OF STARTING THE VEHICLE, DRIVING OVER 16 km/h (10 MPH), AND THEN KEYING DOWN.

DIAGNOSTIC CHART DTC 52 — 1994–95 CAMARO AND FIREBIRD

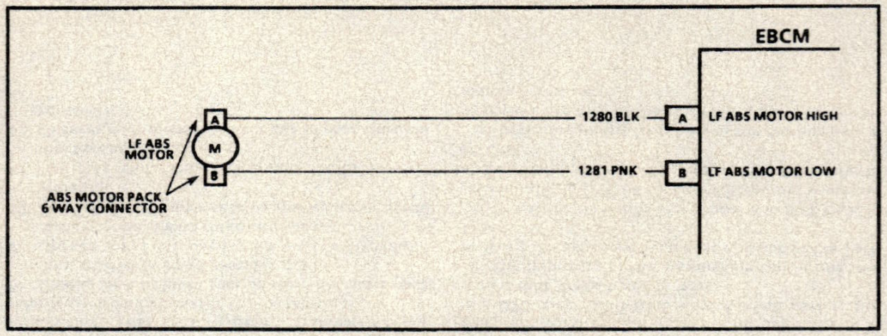

DTC 52

LEFT FRONT ABS CHANNEL IN RELEASE TOO LONG

Circuit Description:

This DTC will diagnose an ABS motor that is energized longer than expected. This could occur if a wheel speed sensor is malfunctioning, the ABS motor does not turn, the left front solenoid mechanically fails open, or the ABS motor wires are crossed.

DTC Will Set When: DTC 52 can be set only during an ABS stop. If the EBCM commands the left front ABS channel in release for three seconds, a malfunction exists.

Action Taken: A malfunction DTC is stored, ABS is disabled and the "ABS INOP" lamp is turned "ON."

DTC Chart Test Description: Number(s) below refer to circled number(s) on the diagnostic chart.

1. Identifies a problem in a wheel speed sensor that may cause the system to be in release too long.
2. Identifies an ABS motor as being failed or wired incorrectly.
3. Checks for a solenoid that may have mechanically failed open.
4. Verifies proper solenoid operation.
5. If Step 3 has failed, this serves to isolate the cause of the hydraulic problem, to either the solenoid or the ABS hydraulic modulator assembly.
6. Determines whether a malfunctioning motor pack or hydraulic modulator is the reason for DTC 52 being set.

Diagnostic Aids: An "Intermittent" malfunction may be caused by a mechanical part of the system that sticks or binds.

The frequency of the malfunction can be checked by using the enhanced diagnostic function of the Tech 1.

DTC 52 may occur if on ice and steering wheel is turned to lock during braking. Using the Tech 1, perform hydraulic test to ensure total brake system is functional.

Any circuitry, that is suspected as causing the intermittent complaint, should be thoroughly checked for backed out terminals, improper mating, broken locks, improperly formed or damaged terminals, poor terminal to wiring connections or physical damage to the wiring harness.

DIAGNOSTIC CHART DTC 52 — 1994–95 CAMARO AND FIREBIRD

DTC 52

LEFT FRONT ABS CHANNEL IN RELEASE TOO LONG

Important

EXCESSIVE DRAG OR HIGH RESISTANCE IN THE BASE BRAKE OR SUSPENSION SYSTEM MUST BE INSPECTED AND CORRECTED BEFORE PROCEEDING WITH DTC DIAGNOSIS.

ARE ANY WHEEL SPEED DTCs CURRENTLY SET?

- YES: PROCEED TO THAT DTC CHART FIRST.
- NO: go to 1.

(1)
- USING TECH 1, SELECT "F0: DATA LIST" AND TEST VEHICLE DURING A STEADY DECEL CONDITION FROM 35 MPH TO 0 MPH WHILE MONITORING ALL WHEEL SPEEDS.
- DO ANY OF THE WHEEL SPEEDS INDICATE ERRATIC OR INTERMITTENT OPERATION?

YES: PROCEED TO WHEEL SPEED = 0 DTC FOR AFFECTED WHEEL.
NO: go to 2.

(2)
- IGNITION "ON."
- USING TECH 1 MANUAL CONTROL FUNCTION, SELECT LEFT FRONT ABS MOTOR APPLY AND APPLY MOTOR.
- APPLY FIRM PRESSURE ON BRAKE PEDAL.
- SELECT LEFT FRONT ABS MOTOR RELEASE AND COMMAND RELEASE OF MOTOR.

DID BRAKE PEDAL FALL?

YES: go to 3.
NO:
- BRAKE PEDAL DID NOT MOVE: go to 6.
- BRAKE PEDAL ROSE: REPAIR CROSSED WIRES (CKTs 1280 and 1281) TO LEFT FRONT ABS MOTOR.

(3)
- REMOVE FOOT FROM BRAKE PEDAL.
- USING TECH 1 MANUAL CONTROL FUNCTION, SELECT LEFT FRONT SOLENOID AND COMMAND THE SOLENOID "ON."
- APPLY FIRM PRESSURE TO BRAKE PEDAL.

DID BRAKE PEDAL FALL?

YES: PROCEED TO STEP 5 OF THIS CHART.
NO: go to 4.

(4)
- FIRM PRESSURE STILL APPLIED TO BRAKE PEDAL.
- USING TECH 1, COMMAND LEFT FRONT SOLENOID "OFF."

DID BRAKE PEDAL FALL?

YES: MALFUNCTION IS NOT PRESENT AT THIS TIME. REFER TO "DIAGNOSTIC AIDS" ON FACING PAGE.
NO: go to 5.

(5)
- REMOVE FOOT FROM BRAKE PEDAL.
- IGNITION "OFF."
- PHYSICALLY SWITCH THE LEFT FRONT SOLENOID WITH RIGHT FRONT SOLENOID.
- IGNITION "ON."
- USING TECH 1 MANUAL CONTROL FUNCTION, SELECT LEFT FRONT SOLENOID AND COMMAND THE SOLENOID "ON."
- APPLY FIRM PRESSURE TO BRAKE PEDAL.

DID BRAKE PEDAL FALL?

YES: REPLACE ABS HYDRAULIC MODULATOR.
NO: REPLACE THE SOLENOID WHICH WAS ORIGINALLY IN THE LEFT FRONT CHANNEL.

(6)
- USING TECH 1 PERFORM GEAR TENSION RELIEF FUNCTION.
- IGNITION "OFF."
- REMOVE HYDRAULIC MODULATOR/MOTOR PACK ASSEMBLY FROM VEHICLE.
- SEPARATE ABS MOTOR PACK FROM HYDRAULIC MODULATOR.
- GRASP THE GEAR ON THE HYDRAULIC MODULATOR (HYDRAULIC MODULATOR IS THE UNIT WITH THE THREE LARGE GEARS) FURTHEST FORWARD WHEN THE UNIT IS INSTALLED IN THE VEHICLE AND ATTEMPT TO ROTATE THE GEAR IN EITHER DIRECTION.

CAN THE GEAR BE MOVED BY HAND?

YES: REPLACE ABS MOTOR PACK.
NO: REPLACE ABS HYDRAULIC MODULATOR.

AFTER DIAGNOSIS IS COMPLETE, CLEAR DTCs AND TEST DRIVE VEHICLE FOR THREE (3) DRIVE CYCLES TO VERIFY DTC DOES NOT RESET. A DRIVE CYCLE CONSISTS OF STARTING THE VEHICLE, DRIVING OVER 16km/h (10 MPH) AND THEN KEYING DOWN.

DIAGNOSTIC CHART DTC 53 — 1994–95 CAMARO AND FIREBIRD

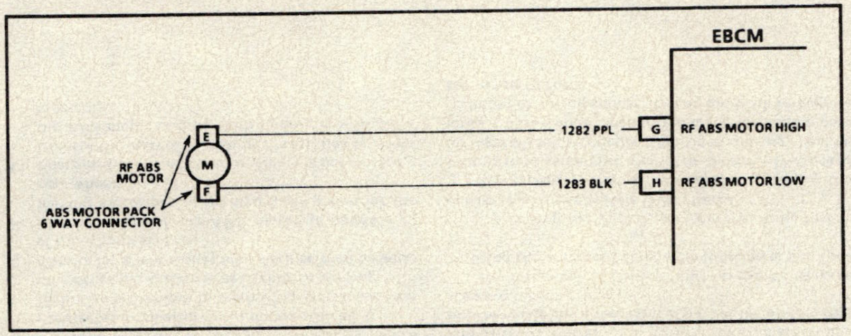

DTC 53

RIGHT FRONT ABS CHANNEL IN RELEASE TOO LONG

Circuit Description:

This DTC will diagnose a motor that is energized longer than expected. This could occur if a wheel speed sensor is malfunctioning, the motor does not turn, the right front solenoid mechanically fails open, or the ABS motor wires are crossed.

DTC Will Set When: DTC 53 can be set only during an ABS stop. If the EBCM commands the right front channel in release for three seconds, a malfunction exists.

Action Taken: A malfunction DTC is stored, ABS is disabled and the "ABS INOP" lamp is turned "ON."

DTC Chart Test Description: Number(s) below refer to circled number(s) on the diagnostic chart.

1. Identifies a problem in a wheel speed sensor that may cause the system to be in release too long.
2. Identifies a motor as being failed or wired incorrectly.
3. Checks for a solenoid that may have mechanically failed open.
4. If Step 3 has failed, this serves to isolate the cause of the hydraulic problem, to either the solenoid or the ABS hydraulic modulator assembly.
5. Checks for a solenoid that may have mechanically failed closed.
6. Determines whether a malfunction motor pack or hydraulic modulator is the reason for DTC 53 being set.

Diagnostic Aids: An "Intermittent" malfunction may be caused by a mechanical part of the system that sticks or binds.

The frequency of the malfunction can be checked by using the enhanced diagnostic function of the Tech 1,

DTC 53 may occur on ice if steering wheel is turned to lock during braking. Using the Tech 1 perform hydraulic test to ensure total brake system is functional.

Any circuitry, that is suspected as causing the intermittent complaint, should be thoroughly checked for backed out terminals, improper mating, broken locks, improperly formed or damaged terminals, poor terminal to wiring connections or physical damage to the wiring harness.

DIAGNOSTIC CHART DTC 53 — 1994–95 CAMARO AND FIREBIRD

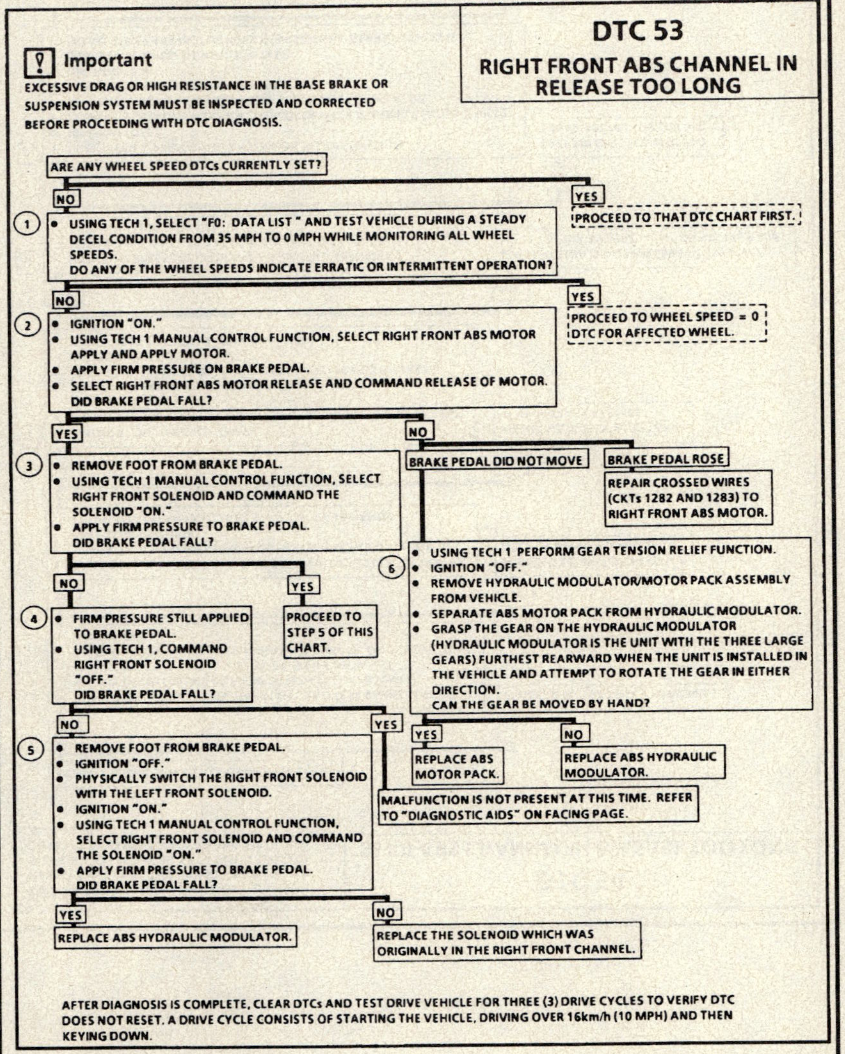

DIAGNOSTIC CHART DTC 54 — 1994–95 CAMARO AND FIREBIRD

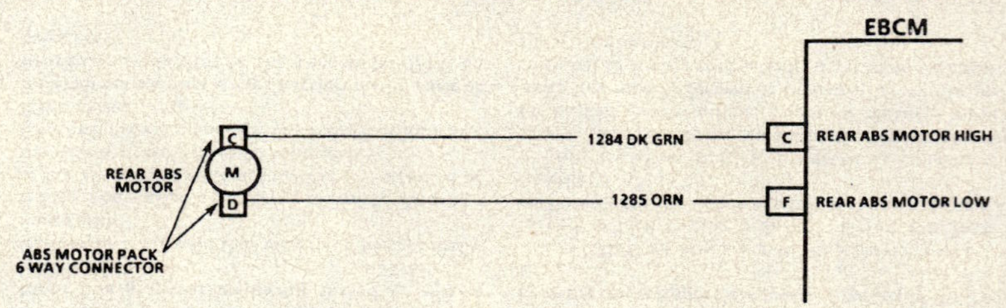

DTC 54

REAR ABS CHANNEL IN RELEASE TOO LONG

Circuit Description:
This DTC will diagnose an ABS motor that is energized longer than expected. This could occur if a wheel speed sensor is malfunctioning, the ABS motor does not turn, or the ABS motor wires are crossed.

DTC Will Set When: DTC 54 can be set only during an ABS stop. If the EBCM commands the rear axle channel in release for three seconds, a malfunction exists.

Action Taken: A malfunction DTC is stored, ABS is disabled and the "ABS INOP" lamp is turned "ON."

DTC Chart Test Description: Number(s) below refer to circled number(s) on the diagnostic chart.

1. Identifies a problem in a wheel speed sensor that may cause the system to be in release too long.
2. Checks for a wheel that may stick or bind because of a mechanical fault.
3. Checks to see if the ABS motor is capable of moving and applying the hydraulic piston for the rear wheels.
4. Ensures that the ABS motor wiring is not crossed.
5. Isolates the fault of a "no-apply" situation to either the ABS motor pack or ABS hydraulic modulator assembly.

Diagnostic Aids: An "Intermittent" malfunction may be caused by a mechanical part of the system that sticks or binds.

The frequency of the malfunction can be checked by using the enhanced diagnostic function of the Tech 1,

Using the Tech 1, perform hydraulic test to ensure total brake system is functional.

Any circuitry, that is suspected as causing the intermittent complaint, should be thoroughly checked for backed out terminals, improper mating, broken locks, improperly formed or damaged terminals, poor terminal to wiring connections or physical damage to the wiring harness.

DIAGNOSTIC CHART DTC 54 — 1994–95 CAMARO AND FIREBIRD

IMPORTANT: EXCESSIVE DRAG OR HIGH RESISTANCE IN THE BASE BRAKE OR SUSPENSION SYSTEM MUST BE INSPECTED AND CORRECTED BEFORE PROCEEDING WITH DTC DIAGNOSIS.

DTC 54

REAR ABS CHANNEL IN RELEASE TOO LONG

ARE ANY WHEEL SPEED DTCs CURRENTLY SET?

- YES: PROCEED TO THAT DTC CHART FIRST.
- NO:

(1)
- USING TECH 1, SELECT "F0: DATA LIST" AND TEST VEHICLE DURING A STEADY DECEL CONDITION FROM 35 MPH TO 0 MPH WHILE MONITORING ALL WHEEL SPEEDS ON TECH 1 DATA LIST.

DO ANY OF THE WHEEL SPEEDS INDICATE ERRATIC OR INTERMITTENT OPERATION?

- YES: PROCEED TO WHEEL SPEED = 0 DTC FOR AFFECTED WHEEL.
- NO:

(2)
- RAISE REAR OF VEHICLE AND SUPPORT AS INDICATED IN SECTION 0A.
- SPIN BOTH REAR WHEELS AND CONFIRM PHYSICAL CONDITION OF BASE BRAKE SYSTEM.

COULD BOTH REAR WHEELS BE SPUN FREELY BY HAND?

- NO: REPAIR SOURCE OF RESISTANCE IN BASE BRAKE SYSTEM.
- YES:

(3)
- IGNITION "ON," ENGINE "OFF."
- USING MANUAL CONTROL FUNCTION OF TECH 1, SELECT REAR AXLE ABS MOTOR APPLY.
- APPLY FIRM PRESSURE ON BRAKE PEDAL.
- COMMAND APPLY OF REAR ABS MOTOR AND TRY TO SPIN REAR WHEELS BY HAND.

COULD REAR WHEELS BE SPUN BY HAND?

- NO: MALFUNCTION IS NOT PRESENT AT THIS TIME. REFER TO "DIAGNOSTIC AIDS" ON FACING PAGE.
- YES:

(4)
- WITH FIRM PRESSURE ON BRAKE PEDAL STILL APPLIED, COMMAND A RELEASE OF THE REAR AXLE ABS MOTOR.

CAN WHEELS BE MOVED?

- NO: REPAIR CROSSED WIRES AT REAR ABS MOTOR CIRCUIT.
- YES:

(5)
- USING TECH 1, PERFORM GEAR TENSION RELIEF FUNCTION.
- IGNITION "OFF."
- REMOVE HYDRAULIC MODULATOR/MOTOR PACK ASSEMBLY FROM VEHICLE.
- SEPARATE ABS MOTOR PACK FROM HYDRAULIC MODULATOR.
- CONNECT ABS MOTOR PACK WIRING.
- IGNITION "ON."
- USING TECH 1, PERFORM MOTOR TEST.

DO ALL THREE ABS MOTOR PACK GEARS (SMALL GEARS) SPIN FREELY?

- NO: REPLACE ABS MOTOR PACK.
- YES: REPLACE ABS HYDRAULIC MODULATOR.

AFTER ALL DIAGNOSIS IS COMPLETE, CLEAR DTCs AND TEST DRIVE VEHICLE FOR THREE (3) DRIVE CYCLES TO VERIFY THAT DTC DOES NOT RESET. A DRIVE CYCLE CONSISTS OF STARTING THE VEHICLE DRIVING OVER 16 KM/H (10 MPH), AND THEN KEYING DOWN.

DIAGNOSTIC CHART DTC 55 — 1994–95 CAMARO AND FIREBIRD

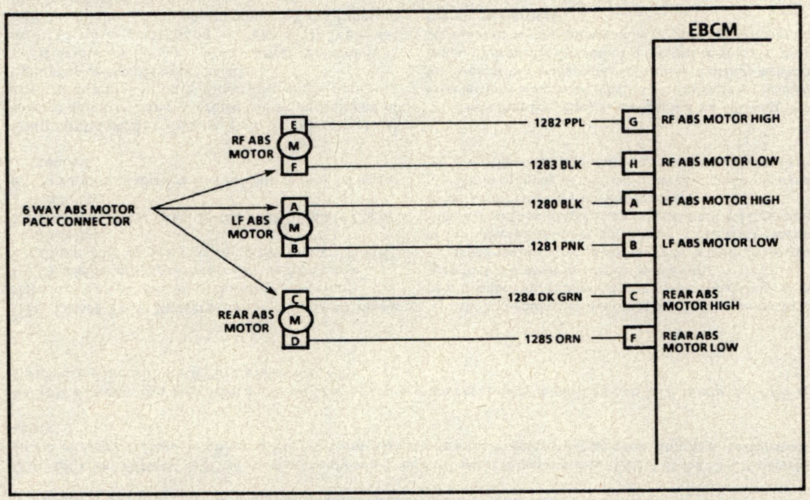

DTC 55
EBCM MALFUNCTION

Circuit Description:
This DTC identifies a malfunction detected by the MDI custom IC. It also ensures the cause of the malfunction is not a result of a problem with the ABS enable relay under a DTC 55 malfunction.

Action Taken: A malfunction DTC is stored, ABS is disabled and the "ABS INOP" lamp is turned "ON."

DTC Chart Test Description: Number(s) below refer to circled number(s) on the diagnostic chart.
1. Checks to see if the malfunction is still present.
2. Checks to see if the malfunction is intermittent.

DIAGNOSTIC CHART DTC 55 — 1994–95 CAMARO AND FIREBIRD

DTC 55
EBCM MALFUNCTION

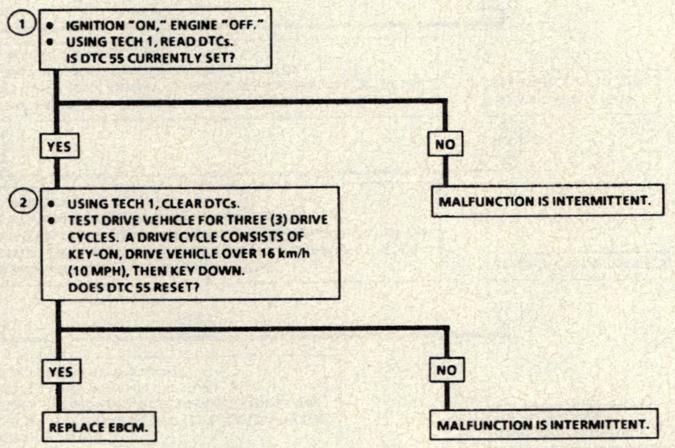

DIAGNOSTIC CHART DTC 56 — 1994–95 CAMARO AND FIREBIRD

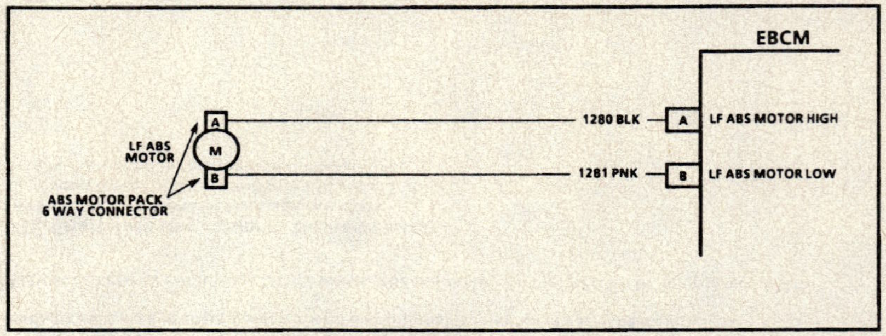

DTC 56
LEFT FRONT ABS MOTOR CIRCUIT OPEN

Circuit Description:
This DTC identifies an ABS motor that cannot be energized due to an open in its circuitry.

DTC Will Set When: DTC 56 can be set only when the ABS motor is commanded "OFF." If the EBCM detects an out of range voltage on either of the left front ABS motor circuits indicating an open circuit, a malfunction exists.

Action Taken: An open ABS motor will not activate when requested. A malfunction DTC is stored, "ABS is disabled and the ABS INOP" lamp is turned "ON."

DTC Chart Test Description: Number(s) below refer to circled number(s) on the diagnostic chart.

1. Checks for proper resistance of the ABS motor.
2. Checks for an open in the ABS motor "HIGH" circuitry.
3. Checks for an open in the ABS motor "LOW" circuitry.
4. Ensures malfunction was not due to poor terminal contact.

Diagnostic Aids: Using Tech 1, select manual control function, and exercise motor movement of affected channel in both directions while applying light pressure on the brake pedal.

If erratic or "jumpy" brake pedal movement is detected while performing an "apply" or "release" function of the motor, an intermittent malfunction may be indicated.

An "Intermittent" malfunction may be caused by a poor connection, rubbed through wire insulation, or a wire that is broken inside the insulation.

If the malfunction is not current, wiggle the wires of the affected channel and check if the DTC resets. This will help to pinpoint an intermittent malfunction in the motor circuitry or connections.

The frequency of the malfunction can be checked by using the enhanced diagnostic function of the Tech 1.

Any circuitry, that is suspected as causing the intermittent complaint, should be thoroughly checked for backed out terminals, improper mating, broken locks, improperly formed or damaged terminals, poor terminal to wiring connections or physical damage to the wiring harness.

DIAGNOSTIC CHART DTC 56 — 1994–95 CAMARO AND FIREBIRD

DTC 56
LEFT FRONT ABS MOTOR CIRCUIT OPEN

DOES DTC 56 OCCUR INTERMITTENTLY?

- YES: REFER TO "DIAGNOSTIC AIDS" ON FACING PAGE.
- NO:

(1)
- IGNITION "OFF."
- DISCONNECT 6-WAY ABS MOTOR PACK CONNECTOR.
- USING J 39200, MEASURE RESISTANCE ACROSS ABS MOTOR PACK TERMINALS "A" AND "B".

IS RESISTANCE BELOW 2000Ω?

- NO: REPLACE ABS MOTOR PACK.
- YES:

(2)
- DISCONNECT 8-WAY CONNECTOR AT EBCM.
- USING J 39200, MEASURE RESISTANCE BETWEEN TERMINAL "A" OF THE 8-WAY EBCM HARNESS CONNECTOR AND TERMINAL "A" OF THE 6-WAY ABS MOTOR PACK HARNESS CONNECTOR.

IS RESISTANCE 2Ω OR LESS?

- NO: REPAIR OPEN OR HIGH RESISTANCE IN CKT 1280.
- YES:

(3)
- USING J 39200, MEASURE RESISTANCE BETWEEN TERMINAL "B" OF THE 8-WAY EBCM HARNESS CONNECTOR AND TERMINAL "B" OF THE 6-WAY ABS MOTOR PACK HARNESS CONNECTOR.

IS RESISTANCE 2Ω OR LESS?

- NO: REPAIR OPEN OR HIGH RESISTANCE IN CKT 1281.
- YES:

(4)
- INSPECT THE 6-WAY ABS MOTOR PACK AND 8-WAY EBCM CONNECTORS FOR POOR TERMINAL CONTACT AND EVIDENCE OF CORROSION.

REPLACE ALL TERMINALS THAT EXHIBIT SIGNS OF POOR TERMINAL CONTACT OR CORROSION.
- RECONNECT ALL CONNECTORS.
- IGNITION "OFF."
- START ENGINE AND TEST DRIVE VEHICLE, ACHIEVING AT LEAST 16 km/h (10 MPH).
- REPEAT ABOVE DRIVE CYCLE SEQUENCE TWO MORE TIMES.

DID DTC 56 SET IN THE LAST THREE DRIVE CYCLES?

- YES: REPLACE EBCM.
- NO: MALFUNCTION WAS MOST LIKELY CAUSED BY A POOR CONNECTION. REFER TO "DIAGNOSTIC AIDS" ON FACING PAGE.

AFTER DIAGNOSIS IS COMPLETE, CLEAR DTCs AND TEST DRIVE VEHICLE FOR THREE (3) DRIVE CYCLES TO VERIFY THAT DTC DOES NOT RESET. A DRIVE CYCLE CONSISTS OF STARTING THE VEHICLE, DRIVING OVER 16 KM/H (10 MPH), AND THEN KEYING DOWN.

DIAGNOSTIC CHART DTC 57 — 1994–95 CAMARO AND FIREBIRD

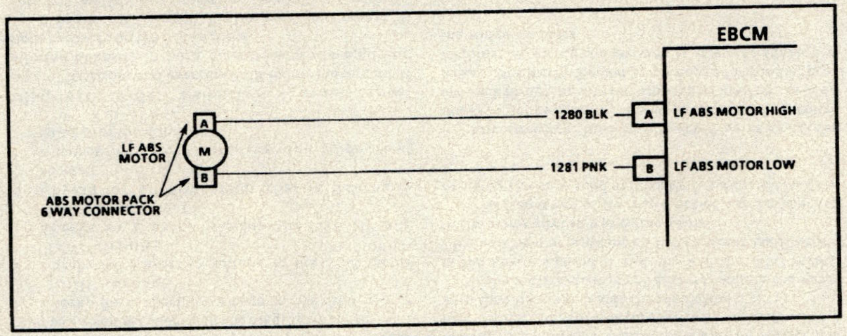

DTC 57

LEFT FRONT ABS MOTOR CIRCUIT SHORTED TO GND

Circuit Description:
This DTC identifies an ABS motor circuit that is shorted to ground. This malfunction will not allow the ABS motor to be controlled at the commanded current rate or will cause the driver circuit to allow current directly to ground.

DTC Will Set When: DTC 57 can be set anytime. If the EBCM detects an out of range voltage on either of the left front ABS motor circuits indicating a circuit shorted to ground, a malfunction exists.

Action Taken: A malfunction DTC is stored, ABS is disabled and the "ABS INOP" lamp is turned "ON."

DTC Chart Test Description: Number(s) below refer to circled number(s) on the diagnostic chart.

1. Checks for a short to ground in the ABS motor "HIGH" circuitry.
2. Checks for a short to ground in the ABS motor "LOW" circuitry.
3. Checks for an ABS motor that is internally shorted to ground.
4. Ensures malfunction was not due to physical damage of the circuitry.

Diagnostic Aids: Using Tech 1, select manual control function, and exercise motor movement of affected channel in both directions while applying light pressure on the brake pedal.

If erratic or "jumpy" brake pedal movement is detected while performing an "apply" or "release" function of the motor, an intermittent malfunction may be indicated.

An "Intermittent" malfunction may be caused by a poor connection, rubbed through wire insulation, or a wire that is broken inside the insulation.

If the malfunction is not current, wiggle the wires of the affected channel and check if the DTC resets. This will help to pinpoint an intermittent malfunction in the motor circuitry or connections.

The frequency of the malfunction can be checked by using the enhanced diagnostic function of the Tech 1,

Any circuitry, that is suspected as causing the intermittent complaint, should be thoroughly checked for backed out terminals, improper mating, broken locks, improperly formed or damaged terminals, poor terminal to wiring connections or physical damage to the wiring harness.

DIAGNOSTIC CHART DTC 57 — 1994–95 CAMARO AND FIREBIRD

DTC 57

LEFT FRONT ABS MOTOR CIRCUIT SHORTED TO GND

DOES DTC 57 OCCUR INTERMITTENTLY?

- YES: REFER TO "DIAGNOSTIC AIDS" ON FACING PAGE.
- NO:

(1)
- IGNITION "OFF."
- DISCONNECT THE 6-WAY ABS MOTOR PACK CONNECTOR AND THE 8-WAY EBCM CONNECTOR.
- USING J 39200, MEASURE THE RESISTANCE BETWEEN GROUND AND TERMINAL "A" OF THE 6-WAY ABS MOTOR PACK HARNESS CONNECTOR.

DOES J 39200 DISPLAY RESISTANCE AS "OL" (INFINITE)?

- NO: REPAIR SHORT TO GROUND IN CKT 1280.
- YES:

(2)
- MEASURE THE RESISTANCE BETWEEN GROUND AND TERMINAL "B" OF THE 6-WAY ABS MOTOR PACK HARNESS CONNECTOR.

DOES J 39200 DISPLAY RESISTANCE AS "OL" (INFINITE)?

- NO: REPAIR SHORT TO GROUND IN CKT 1281.
- YES:

(3)
- MEASURE THE RESISTANCE BETWEEN GROUND AND TERMINAL "A" OF THE ABS MOTOR PACK.

DOES J 39200 DISPLAY RESISTANCE AS "OL" (INFINITE)?

- NO: REPLACE ABS MOTOR PACK.
- YES:

(4)
- INSPECT CKTs 1280 AND 1281 AND RELATED CONNECTORS FOR DAMAGE WHICH MAY RESULT IN A SHORT TO GROUND WITH ALL CONNECTORS CONNECTED. REPAIR ALL DAMAGE IF EVIDENT.
- RECONNECT ALL CONNECTORS.
- IGNITION "OFF."
- START ENGINE AND TEST DRIVE VEHICLE, ACHIEVING AT LEAST 16 km/h (10 MPH).
- REPEAT ABOVE DRIVE CYCLE SEQUENCE TWO MORE TIMES.

DID DTC 57 SET IN THE LAST THREE DRIVE CYCLES?

- YES: REPLACE EBCM.
- NO: MALFUNCTION WAS MOST LIKELY CAUSED BY A POOR CONNECTION. REFER TO "DIAGNOSTIC AIDS" ON FACING PAGE.

AFTER ALL DIAGNOSIS IS COMPLETE, CLEAR DTCs AND TEST DRIVE VEHICLE FOR THREE (3) DRIVE CYCLES TO VERIFY THAT DTC DOES NOT RESET. A DRIVE CYCLE CONSISTS OF STARTING THE VEHICLE, DRIVING OVER 16 km/h (10 MPH), AND THEN KEYING DOWN.

DIAGNOSTIC CHART DTC 58 — 1994–95 CAMARO AND FIREBIRD

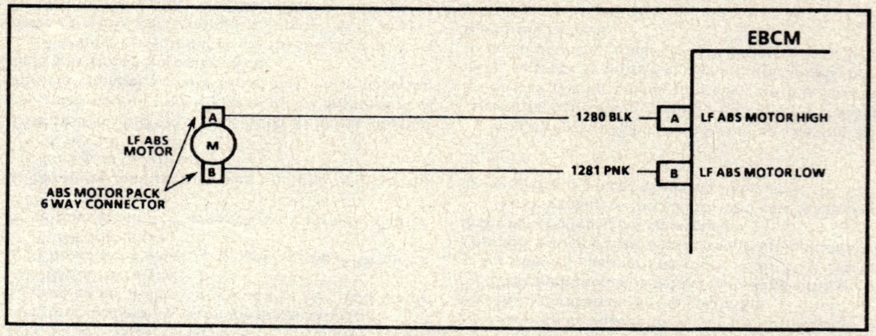

DTC 58

LEFT FRONT ABS MOTOR CIRCUIT SHORTED TO BATT

Circuit Description:

This DTC identifies an ABS motor circuit that is shorted to battery or an ABS motor that has low or no resistance. This malfunction will not allow the ABS motor to be controlled at the commanded current rate or will cause the ABS motor to turn in the opposite direction, or not at all.

DTC Will Set When: DTC 58 can be set only when the ABS motor is commanded "OFF." If the EBCM detects an out of range voltage on either of the left front ABS motor circuits indicating a circuit shorted to battery or an ABS motor shorted, a malfunction exists.

Action Taken: A malfunction DTC is stored, ABS is disabled and the "ABS INOP" lamp is turned "ON."

DTC Chart Test Description: Number(s) below refer to circled number(s) on the diagnostic chart.

1. Checks for a short to voltage in the ABS motor "HIGH" circuitry.
2. Checks for a short to voltage in the ABS motor "LOW" circuitry.
3. Checks for a short between the "HIGH" and "LOW" motor circuits.
4. Checks for an ABS motor that is internally shorted.
5. Ensures malfunction was not due to physical damage of the circuitry.

Diagnostic Aids: Using Tech 1, select manual control function, and exercise ABS motor movement of affected channel in both directions while applying light pressure on the brake pedal.

If erratic or "jumpy" brake pedal movement is detected while performing an "apply" or "release" function of the ABS motor, an intermittent malfunction may be indicated.

An "Intermittent" malfunction may be caused by a poor connection, rubbed through wire insulation, or a wire that is broken inside the insulation.

If the malfunction is not current, wiggle the wires of the affected channel and check if the DTC resets. This will help to pinpoint an intermittent malfunction in the motor circuitry or connections.

The frequency of the malfunction can be checked by using the enhanced diagnostic function of the Tech 1,

Any circuitry, that is suspected as causing the intermittent complaint, should be thoroughly checked for backed out terminals, improper mating, broken locks, improperly formed or damaged terminals, poor terminal to wiring connections or physical damage to the wiring harness.

DIAGNOSTIC CHART DTC 58 — 1994–95 CAMARO AND FIREBIRD

DTC 58

LEFT FRONT ABS MOTOR CIRCUIT SHORTED TO BATT

NOTICE: J 39200 TEST LEADS MUST BE "ZEROED" PRIOR TO MAKING ANY RESISTANCE MEASUREMENTS. REFER TO J 39200 USER'S MANUAL.

DOES DTC 58 OCCUR INTERMITTENTLY?

- YES: REFER TO "DIAGNOSTIC AIDS" ON FACING PAGE.
- NO:

(1)
- IGNITION "OFF."
- DISCONNECT THE 6-WAY ABS MOTOR PACK CONNECTOR AND THE 8-WAY EBCM CONNECTOR.
- IGNITION "ON."
- USING J 39200, MEASURE THE VOLTAGE BETWEEN GROUND AND TERMINAL "A" OF THE 6-WAY ABS MOTOR PACK HARNESS CONNECTOR.

IS VOLTAGE 1 VOLT OR LESS?

- NO: REPAIR SHORT TO VOLTAGE IN CKT 1280.
- YES:

(2)
- MEASURE THE VOLTAGE BETWEEN GROUND AND TERMINAL "B" OF THE 6-WAY ABS MOTOR PACK HARNESS CONNECTOR.

IS VOLTAGE 1 VOLT OR LESS?

- NO: REPAIR SHORT TO VOLTAGE IN CKT 1281.
- YES:

(3)
- IGNITION "OFF."
- USING J 39200, MEASURE THE RESISTANCE BETWEEN TERMINAL "A" AND TERMINAL "B" OF THE 6-WAY ABS MOTOR PACK HARNESS CONNECTOR.

DOES J 39200 DISPLAY RESISTANCE AS "OL" (INFINITE)?

- NO: REPAIR SHORT BETWEEN CKTs 1280 AND 1281.
- YES:

(4)
- USING J 39200, MEASURE THE RESISTANCE BETWEEN TERMINAL "A" AND TERMINAL "B" OF THE ABS MOTOR PACK.

IS RESISTANCE GREATER THAN 0.4Ω?

- NO: REPLACE ABS MOTOR PACK.
- YES:

(5)
- INSPECT CKTs 1280 AND 1281 AND RELATED CONNECTORS FOR DAMAGE WHICH MAY RESULT IN A SHORT TO VOLTAGE WITH ALL CONNECTORS CONNECTED. REPAIR ALL DAMAGE IF EVIDENT.
- RECONNECT ALL CONNECTORS.
- IGNITION "OFF."
- START ENGINE AND TEST DRIVE VEHICLE, ACHIEVING AT LEAST 16 km/h (10 MPH).
- REPEAT ABOVE DRIVE CYCLE SEQUENCE TWO MORE TIMES.

DID DTC 58 SET IN THE LAST THREE DRIVE CYCLES?

- YES: REPLACE EBCM.
- NO: MALFUNCTION WAS MOST LIKELY CAUSED BY A POOR CONNECTION. REFER TO "DIAGNOSTIC AIDS" ON FACING PAGE.

AFTER DIAGNOSIS IS COMPLETE, CLEAR DTCs AND TEST DRIVE VEHICLE FOR THREE (3) DRIVE CYCLES TO VERIFY THAT DTC DOES NOT RESET. A DRIVE CYCLE CONSISTS OF STARTING THE VEHICLE, DRIVING OVER 16 KM/H (10 MPH), AND THEN KEYING DOWN.

DIAGNOSTIC CHART DTC 61 — 1994–95 CAMARO AND FIREBIRD

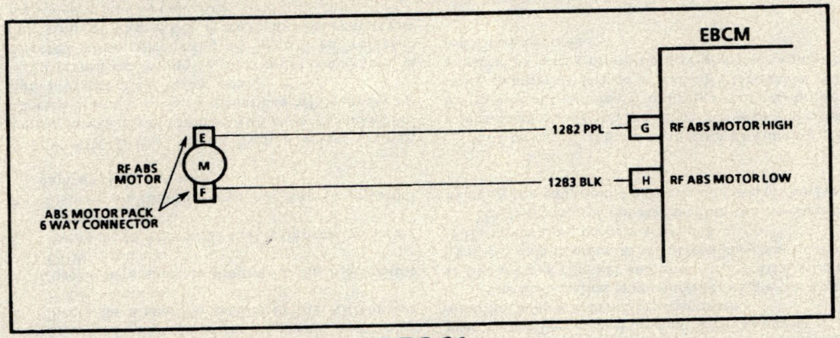

DTC 61

RIGHT FRONT ABS MOTOR CIRCUIT OPEN

Circuit Description:
This DTC identifies an ABS motor that cannot be energized due to an open in its circuitry. This malfunction will not allow proper front ABS operation.

DTC Will Set When: DTC 61 can be set only when the ABS motor is commanded "OFF." If the EBCM detects an out of range voltage on either of the right front ABS motor circuits indicating an open circuit, a malfunction exists.

Action Taken: An open motor will not activate when requested. A malfunction DTC is stored, ABS is disabled and the "ABS INOP" lamp is turned "ON."

DTC Chart Test Description: Number(s) below refer to circled number(s) on the diagnostic chart.

1. Checks for proper resistance of the ABS motor.
2. Checks for an open in the ABS motor "HIGH" circuitry.
3. Checks for an open in the ABS motor "LOW" circuitry.
4. Ensures malfunction was not due to physical damage of the circuitry.

Diagnostic Aids: Using Tech 1, select manual control function, and exercise ABS motor movement of affected channel in both directions while applying light pressure on the brake pedal.

If erratic or "jumpy" brake pedal movement is detected while performing an "apply" or "release" function of the ABS motor, an intermittent malfunction may be indicated.

An "Intermittent" malfunction may be caused by a poor connection, rubbed through wire insulation, or a wire that is broken inside the insulation.

If the malfunction is not current, wiggle the wires of the affected channel and check if the DTC resets. This will help to pinpoint an intermittent malfunction in the motor circuitry or connections.

The frequency of the malfunction can be checked by using the enhanced diagnostic function of the Tech 1,

Any circuitry, that is suspected as causing the intermittent complaint, should be thoroughly checked for backed out terminals, improper mating, broken locks, improperly formed or damaged terminals, poor terminal to wiring connections or physical damage to the wiring harness.

DIAGNOSTIC CHART DTC 61 — 1994–95 CAMARO AND FIREBIRD

DTC 61

RIGHT FRONT ABS MOTOR CIRCUIT OPEN

DOES DTC 61 OCCUR INTERMITTENTLY?

- YES: REFER TO "DIAGNOSTIC AIDS" ON FACING PAGE.
- NO:

(1)
- IGNITION "OFF."
- DISCONNECT 6-WAY ABS MOTOR PACK CONNECTOR.
- USING J 39200, MEASURE RESISTANCE ACROSS ABS MOTOR PACK TERMINALS "E" AND "F".

IS RESISTANCE BELOW 2000Ω?

- NO: REPLACE ABS MOTOR PACK.
- YES:

(2)
- DISCONNECT 8-WAY CONNECTOR AT EBCM.
- USING J 39200, MEASURE RESISTANCE BETWEEN TERMINAL "G" OF THE 8-WAY EBCM HARNESS CONNECTOR AND TERMINAL "E" OF THE 6-WAY ABS MOTOR PACK HARNESS CONNECTOR.

IS RESISTANCE 2Ω OR LESS?

- NO: REPAIR OPEN OR HIGH RESISTANCE IN CKT 1282.
- YES:

(3)
- USING J 39200, MEASURE RESISTANCE BETWEEN TERMINAL "H" OF THE 8-WAY EBCM HARNESS CONNECTOR AND TERMINAL "F" OF THE 6-WAY ABS MOTOR PACK HARNESS CONNECTOR.

IS RESISTANCE 2Ω OR LESS?

- NO: REPAIR OPEN OR HIGH RESISTANCE IN CKT 1283.
- YES:

(4)
- INSPECT THE 6-WAY ABS MOTOR PACK AND 8-WAY EBCM CONNECTORS FOR POOR TERMINAL CONTACT AND EVIDENCE OF CORROSION.
  REPLACE ALL TERMINALS THAT EXHIBIT SIGNS OF POOR TERMINAL CONTACT OR CORROSION.
- RECONNECT ALL CONNECTORS.
- IGNITION "OFF."
- START ENGINE AND TEST DRIVE VEHICLE, ACHIEVING AT LEAST 16 km/h (10 MPH).
- REPEAT ABOVE DRIVE CYCLE SEQUENCE TWO MORE TIMES.

DID DTC 61 SET IN THE LAST THREE DRIVE CYCLES?

- YES: REPLACE EBCM.
- NO: MALFUNCTION WAS MOST LIKELY CAUSED BY A POOR CONNECTION. REFER TO "DIAGNOSTIC AIDS" ON FACING PAGE.

AFTER ALL DIAGNOSIS IS COMPLETE, CLEAR DTCs AND TEST DRIVE VEHICLE FOR THREE (3) DRIVE CYCLES TO VERIFY THAT DTC DOES NOT RESET. A DRIVE CYCLE CONSISTS OF STARTING THE VEHICLE, DRIVING OVER 16 km/h (10 MPH), AND THEN KEYING DOWN.

DIAGNOSTIC CHART DTC 62 — 1994–95 CAMARO AND FIREBIRD

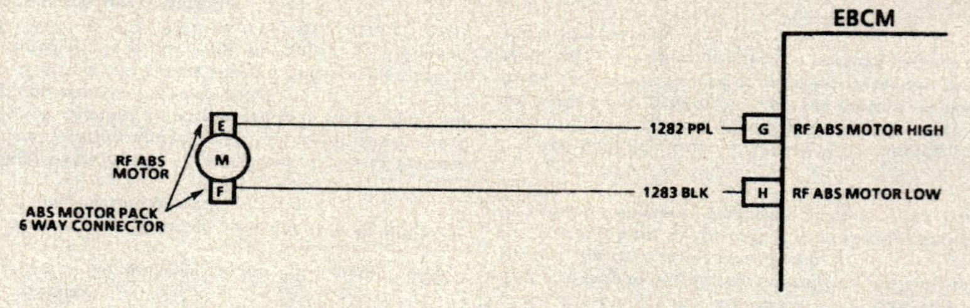

DTC 62

RIGHT FRONT ABS MOTOR CIRCUIT SHORTED TO GND

Circuit Description:
This DTC identifies an ABS motor circuit that is shorted to ground. This malfunction will not allow the ABS motor to be controlled at the commanded current rate or will cause the ABS motor driver circuit to allow current directly to ground.

DTC Will Set When: DTC 62 can be set anytime. If the EBCM detects an out of range voltage on either of the right front ABS motor circuits indicating a circuit shorted to ground, a malfunction exists.

Action Taken: A malfunction DTC is stored, ABS is disabled and the "ABS INOP" lamp is turned "ON."

DTC Chart Test Description: Number(s) below refer to circled number(s) on the diagnostic chart.

1. Checks for a short to ground in the ABS motor "HIGH" circuitry.
2. Checks for a short to ground in the ABS motor "LOW" circuitry.
3. Checks for an ABS motor that is internally shorted to ground.
4. Ensures malfunction was not due to physical damage of the circuitry.

Diagnostic Aids: Using Tech 1, select manual control function, and exercise ABS motor movement of affected channel in both directions while applying light pressure on the brake pedal.

If erratic or "jumpy" brake pedal movement is detected while performing an "apply" or "release" function of the ABS motor, an intermittent malfunction may be indicated.

An "Intermittent" malfunction may be caused by a poor connection, rubbed through wire insulation, or a wire that is broken inside the insulation.

If the malfunction is not current, wiggle the wires of the affected channel and check if the DTC resets. This will help to pinpoint an intermittent malfunction in the motor circuitry or connections.

The frequency of the malfunction can be checked by using the enhanced diagnostic function of the Tech 1.

Any circuitry, that is suspected as causing the intermittent complaint, should be thoroughly checked for backed out terminals, improper mating, broken locks, improperly formed or damaged terminals, poor terminal to wiring connections or physical damage to the wiring harness.

DIAGNOSTIC CHART DTC 62 — 1994–95 CAMARO AND FIREBIRD

DTC 62

RIGHT FRONT ABS MOTOR CIRCUIT SHORTED TO GND

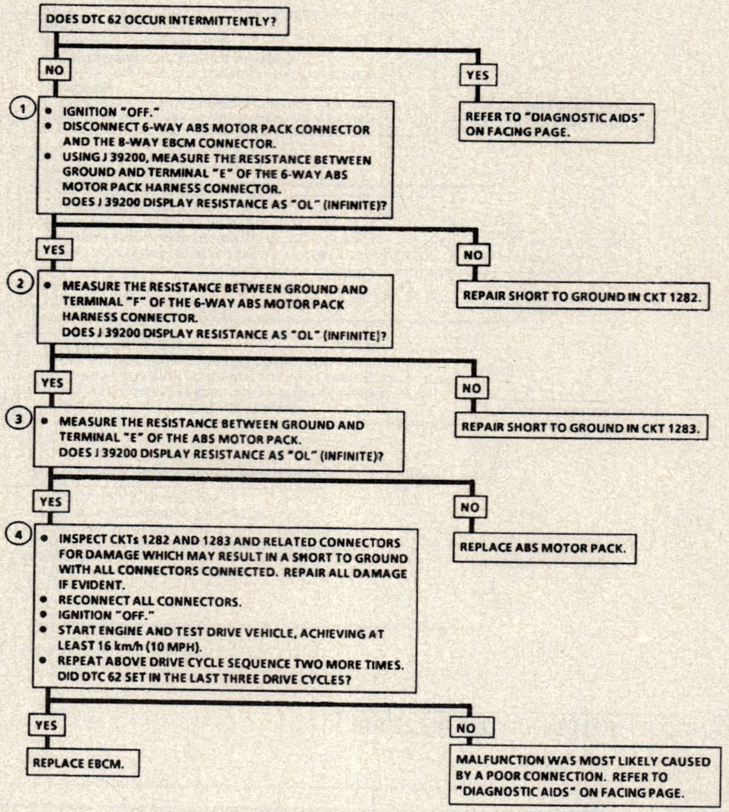

AFTER ALL DIAGNOSIS IS COMPLETE, CLEAR DTCs AND TEST DRIVE VEHICLE FOR THREE (3) DRIVE CYCLES TO VERIFY THAT DTC DOES NOT RESET. A DRIVE CYCLE CONSISTS OF STARTING THE VEHICLE, DRIVING OVER 16 km/h (10 MPH), AND THEN KEYING DOWN.

DIAGNOSTIC CHART DTC 63 — 1994–95 CAMARO AND FIREBIRD

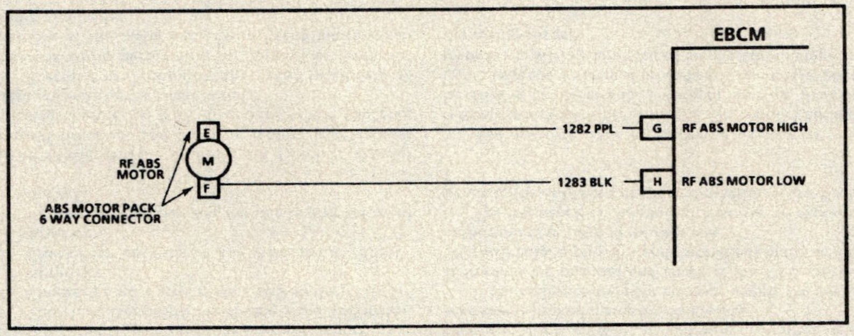

DTC 63

RIGHT FRONT ABS MOTOR CIRCUIT SHORTED TO BATT

Circuit Description:

This DTC identifies an ABS motor circuit that is shorted to battery or an ABS motor that has low or no resistance. This malfunction will not allow the ABS motor to be controlled at the commanded current rate or will cause the ABS motor to turn in the opposite direction or not at all.

DTC Will Set When: DTC 63 can be set only when the ABS motor is commanded "OFF." If the EBCM detects an out of range voltage on either of the right front ABS motor circuits indicating a circuit shorted to battery or an ABS motor shorted, a malfunction exists.

Action Taken: A malfunction DTC is stored, ABS is disabled and the "ABS INOP" lamp is turned "ON."

DTC Chart Test Description: Number(s) below refer to circled number(s) on the diagnostic chart.

1. Checks for a short to voltage in the ABS motor "HIGH" circuitry.
2. Checks for a short to voltage in the ABS motor "LOW" circuitry.
3. Checks for short circuit between ABS motor "HIGH" and motor "LOW".
4. Checks for proper resistance of the ABS motor.
5. Ensures malfunction was not due to physical damage of the circuitry.

Diagnostic Aids: Using Tech 1, select manual control function, and exercise motor movement of affected channel in both directions while applying light pressure on the brake pedal.

If erratic or "jumpy" brake pedal movement is detected while performing an "apply" or "release" function of the motor, an intermittent malfunction may be indicated.

An "Intermittent" malfunction may be caused by a poor connection, rubbed through wire insulation, or a wire that is broken inside the insulation.

If the malfunction is not current, wiggle the wires of the affected channel and check if the DTC resets. This will help to pinpoint an intermittent malfunction in the motor circuitry or connections.

The frequency of the malfunction can be checked by using the enhanced diagnostic function of the Tech 1.

Any circuitry, that is suspected as causing the intermittent complaint, should be thoroughly checked for backed out terminals, improper mating, broken locks, improperly formed or damaged terminals, poor terminal to wiring connections or physical damage to the wiring harness.

DIAGNOSTIC CHART DTC 63 — 1994–95 CAMARO AND FIREBIRD

DTC 63
RIGHT FRONT MOTOR CIRCUIT SHORTED TO BATT

NOTICE: J 39200 TEST LEADS MUST BE "ZEROED" PRIOR TO MAKING ANY RESISTANCE MEASUREMENTS. REFER TO J 39200 USER'S MANUAL.

DOES DTC 63 OCCUR INTERMITTENTLY?

- YES → REFER TO "DIAGNOSTIC AIDS" ON FACING PAGE.
- NO →

(1)
- IGNITION "OFF."
- DISCONNECT THE 6-WAY ABS MOTOR PACK CONNECTOR AND THE 8-WAY EBCM CONNECTOR.
- IGNITION "ON."
- USING J 39200, MEASURE THE VOLTAGE BETWEEN GROUND AND TERMINAL "E" OF THE 6-WAY ABS MOTOR PACK HARNESS CONNECTOR. IS VOLTAGE 1 VOLT OR LESS?

- NO → REPAIR SHORT TO VOLTAGE IN CKT 1282.
- YES →

(2)
- MEASURE THE VOLTAGE BETWEEN GROUND AND TERMINAL "F" OF THE 6-WAY ABS MOTOR PACK HARNESS CONNECTOR. IS VOLTAGE 1 VOLT OR LESS?

- NO → REPAIR SHORT TO VOLTAGE IN CKT 1283.
- YES →

(3)
- IGNITION "OFF."
- USING J 39200, MEASURE THE RESISTANCE BETWEEN TERMINAL "E" AND TERMINAL "F" OF THE 6-WAY ABS MOTOR PACK HARNESS CONNECTOR. DOES J 39200 DISPLAY RESISTANCE AS "OL" (INFINITE)?

- NO → REPAIR SHORT BETWEEN CKTs 1282 AND 1283.
- YES →

(4)
- USING J 39200, MEASURE THE RESISTANCE BETWEEN TERMINAL "E" AND TERMINAL "F" OF THE ABS MOTOR PACK. IS RESISTANCE GREATER THAN 0.4Ω?

- NO → REPLACE ABS MOTOR PACK.
- YES →

(5)
- INSPECT CKTs 1282 AND 1283 AND RELATED CONNECTORS FOR DAMAGE WHICH MAY RESULT IN A SHORT TO VOLTAGE WITH ALL CONNECTORS CONNECTED. REPAIR ALL DAMAGE IF EVIDENT.
- RECONNECT ALL CONNECTORS.
- IGNITION "OFF."
- START ENGINE AND TEST DRIVE VEHICLE, ACHIEVING AT LEAST 16 km/h (10 MPH).
- REPEAT ABOVE DRIVE CYCLE SEQUENCE TWO MORE TIMES. DID DTC 63 SET IN THE LAST THREE DRIVE CYCLES?

- YES → REPLACE EBCM.
- NO → MALFUNCTION WAS MOST LIKELY CAUSED BY A POOR CONNECTION. REFER TO "DIAGNOSTIC AIDS" ON FACING PAGE.

AFTER DIAGNOSIS IS COMPLETE, CLEAR DTCs AND TEST DRIVE VEHICLE FOR THREE (3) DRIVE CYCLES TO VERIFY THAT DTC DOES NOT RESET. A DRIVE CYCLE CONSISTS OF STARTING THE VEHICLE, DRIVING OVER 16 KM/H (10 MPH), AND THEN KEYING DOWN.

DIAGNOSTIC CHART DTC 64 — 1994–95 CAMARO AND FIREBIRD

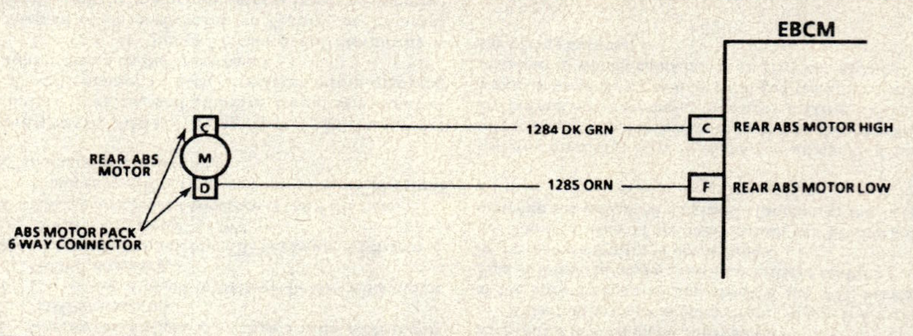

DTC 64

REAR ABS MOTOR CIRCUIT OPEN

Circuit Description:

This DTC identifies an ABS motor that cannot be energized due to an open in its circuitry. This malfunction will not allow proper rear ABS operation.

DTC Will Set When: DTC 64 can be set only when the ABS motor is commanded "OFF." If the EBCM detects an out of range voltage on either of the rear axle ABS motor circuits indicating an open circuit, a malfunction exists.

Action Taken: An open motor will not activate when requested. A malfunction DTC is stored, ABS is disabled and the "ABS INOP" lamp is turned "ON."

DTC Chart Test Description: Number(s) below refer to circled number(s) on the diagnostic chart.

1. This checks for proper resistance of the ABS motor.
2. Checks for an open in the ABS motor "HIGH" circuitry.
3. Checks for an open in the ABS motor "LOW" circuitry.
4. Ensures malfunction was not due to poor terminal contact.

Diagnostic Aids: Using Tech 1, select manual control function, and exercise motor movement of affected channel in both directions while applying light pressure on the brake pedal.

If erratic or "jumpy" brake pedal movement is detected while performing an "apply" or "release" function of the motor, an intermittent malfunction may be indicated.

An "Intermittent" malfunction may be caused by a poor connection, rubbed through wire insulation, or a wire that is broken inside the insulation.

If the malfunction is not current, wiggle the wires of the affected channel and check if the DTC resets. This will help to pinpoint an intermittent malfunction in the motor circuitry or connections.

The frequency of the malfunction can be checked by using the enhanced diagnostic function of the Tech 1,

Any circuitry, that is suspected as causing the intermittent complaint, should be thoroughly checked for backed out terminals, improper mating, broken locks, improperly formed or damaged terminals, poor terminal to wiring connections or physical damage to the wiring harness.

DIAGNOSTIC CHART DTC 64 — 1994–95 CAMARO AND FIREBIRD

DTC 64

REAR ABS MOTOR CIRCUIT OPEN

DOES DTC 64 OCCUR INTERMITTENTLY?

- YES → REFER TO "DIAGNOSTIC AIDS" ON FACING PAGE.
- NO ↓

(1)
- IGNITION "OFF."
- DISCONNECT 6-WAY ABS MOTOR PACK CONNECTOR.
- USING J 39200, MEASURE RESISTANCE ACROSS ABS MOTOR PACK TERMINALS "C" AND "D".

IS RESISTANCE BELOW 2000Ω?

- NO → REPLACE ABS MOTOR PACK.
- YES ↓

(2)
- DISCONNECT 8-WAY CONNECTOR AT EBCM.
- USING J 39200, MEASURE RESISTANCE BETWEEN TERMINAL "C" OF THE 8-WAY EBCM HARNESS CONNECTOR AND TERMINAL "C" OF THE 6-WAY ABS MOTOR PACK HARNESS CONNECTOR.

IS RESISTANCE 2Ω OR LESS?

- NO → REPAIR OPEN OR HIGH RESISTANCE IN CKT 1284.
- YES ↓

(3)
- USING J 39200, MEASURE RESISTANCE BETWEEN TERMINAL "F" OF THE 8-WAY EBCM HARNESS CONNECTOR AND TERMINAL "D" OF THE 6-WAY ABS MOTOR PACK HARNESS CONNECTOR.

IS RESISTANCE 2Ω OR LESS?

- NO → REPAIR OPEN OR HIGH RESISTANCE IN CKT 1285.
- YES ↓

(4)
- INSPECT THE 6-WAY ABS MOTOR PACK AND 8-WAY EBCM CONNECTORS FOR POOR TERMINAL CONTACT AND EVIDENCE OF CORROSION.

REPLACE ALL TERMINALS THAT EXHIBIT SIGNS OF POOR TERMINAL CONTACT OR CORROSION.

- RECONNECT ALL CONNECTORS.
- IGNITION "OFF."
- START ENGINE AND TEST DRIVE VEHICLE, ACHIEVING AT LEAST 16 km/h (10 MPH).
- REPEAT ABOVE DRIVE CYCLE SEQUENCE TWO MORE TIMES.

DID DTC 64 SET IN THE LAST THREE DRIVE CYCLES?

- YES → REPLACE EBCM.
- NO → MALFUNCTION WAS MOST LIKELY CAUSED BY A POOR CONNECTION. REFER TO "DIAGNOSTIC AIDS" ON FACING PAGE.

AFTER ALL DIAGNOSIS IS COMPLETE, CLEAR DTCs AND TEST DRIVE VEHICLE FOR THREE (3) DRIVE CYCLES TO VERIFY THAT DTC DOES NOT RESET. A DRIVE CYCLE CONSISTS OF STARTING THE VEHICLE, DRIVING OVER 16 km/h (10 MPH), AND THEN KEYING DOWN.

DIAGNOSTIC CHART DTC 65 — 1994–95 CAMARO AND FIREBIRD

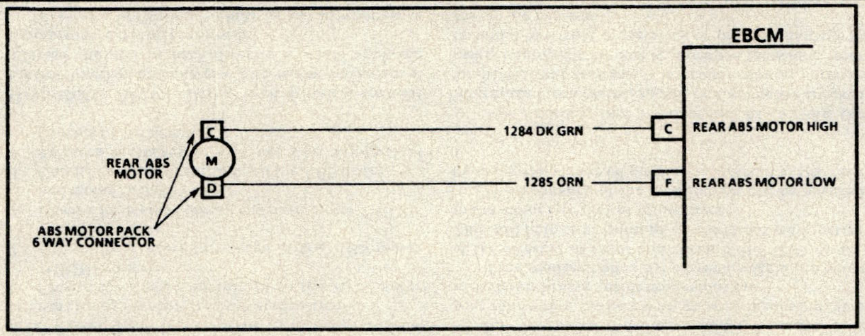

DTC 65

REAR ABS MOTOR CIRCUIT SHORTED TO GROUND

Circuit Description:
This DTC identifies an ABS motor circuit that is shorted to ground. This malfunction will not allow the ABS motor to be controlled at the commanded current rate or will cause the driver circuit to allow current directly to ground.

DTC Will Set When: DTC 65 can be set anytime. If the EBCM detects an out of range voltage on either of the rear axle ABS motor circuits indicating a circuit shorted to ground, a malfunction exists.

Action Taken: A malfunction DTC is stored, ABS is disabled and the "ABS INOP" lamp is turned "ON."

DTC Chart Test Description: Number(s) below refer to circled number(s) on the diagnostic chart.

1. Checks for a short to ground in the ABS motor "HIGH" circuitry.
2. Checks for a short to ground in the ABS motor "LOW" circuitry.
3. Checks for an ABS motor that is internally shorted to ground.
4. Ensures malfunction was not due to physical damage of the circuitry.

Diagnostic Aids: Using Tech 1, select manual control function, and exercise ABS motor movement of affected channel in both directions while applying light pressure on the brake pedal.

If erratic or "jumpy" brake pedal movement is detected while performing an "apply" or "release" function of the ABS motor, an intermittent malfunction may be indicated.

An "Intermittent" malfunction may be caused by a poor connection, rubbed through wire insulation, or a wire that is broken inside the insulation.

If the malfunction is not current, wiggle the wires of the affected channel and check if the DTC resets. This will help to pinpoint an intermittent malfunction in the motor circuitry or connections.

The frequency of the malfunction can be checked by using the enhanced diagnostic function of the Tech 1,

Any circuitry, that is suspected as causing the intermittent complaint, should be thoroughly checked for backed out terminals, improper mating, broken locks, improperly formed or damaged terminals, poor terminal to wiring connections or physical damage to the wiring harness.

DIAGNOSTIC CHART DTC 65 — 1994–95 CAMARO AND FIREBIRD

DTC 65

REAR ABS MOTOR CIRCUIT SHORTED TO GROUND

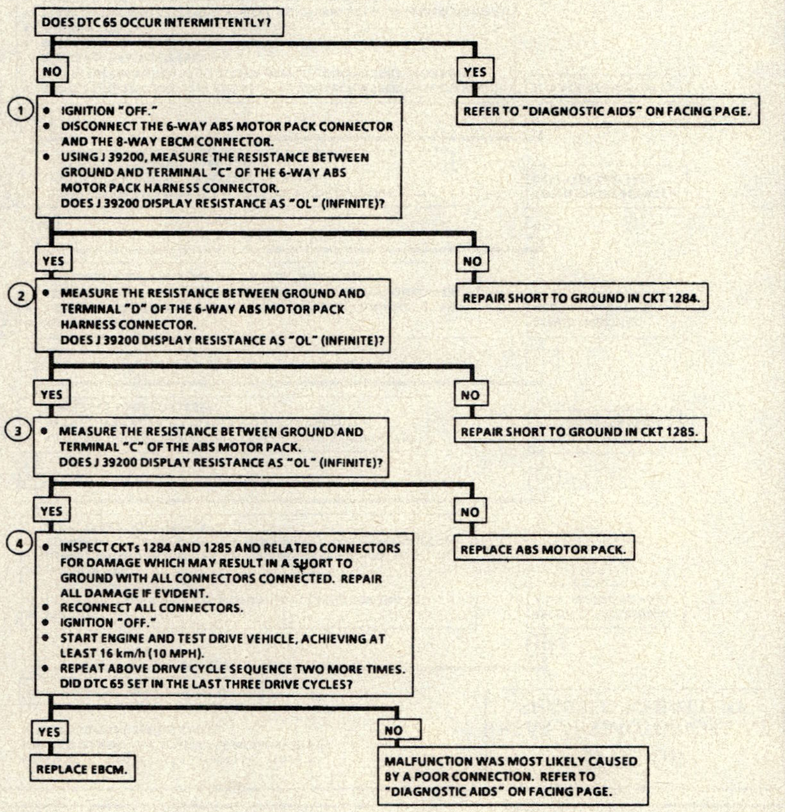

AFTER ALL DIAGNOSIS IS COMPLETE, CLEAR DTCs AND TEST DRIVE VEHICLE FOR THREE (3) DRIVE CYCLES TO VERIFY THAT DTC DOES NOT RESET. A DRIVE CYCLE CONSISTS OF STARTING THE VEHICLE, DRIVING OVER 16 km/h (10 MPH), AND THEN KEYING DOWN.

DIAGNOSTIC CHART DTC 66 — 1994–95 CAMARO AND FIREBIRD

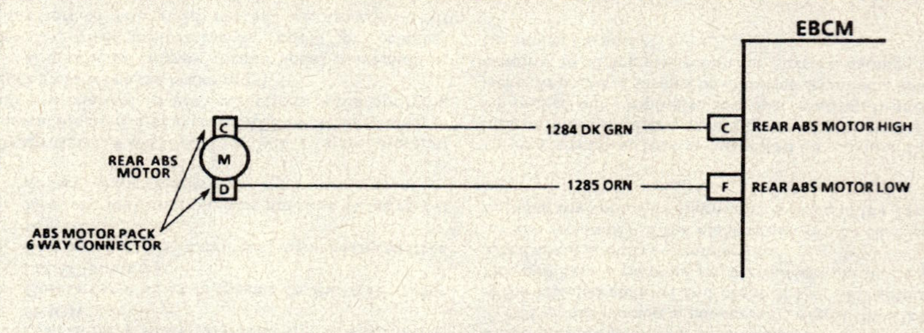

DTC 66

REAR ABS MOTOR CIRCUIT SHORTED TO BATTERY

Circuit Description:
This DTC identifies an ABS motor circuit that is shorted to battery or an ABS motor that has low or zero resistance. This malfunction will not allow the ABS motor to be controlled at the commanded current rate or will cause the ABS motor to turn in the opposite direction or not turn at all.

DTC Will Set When: DTC 66 can be set only when the ABS motor is commanded "OFF." If the EBCM detects an out of range voltage on either of the rear ABS axle motor circuits indicating a circuit shorted to battery or an ABS motor shorted, a malfunction exists.

Action Taken: A malfunction DTC is stored, ABS is disabled and the "ABS INOP" lamp is turned "ON."

DTC Chart Test Description: Number(s) below refer to circled number(s) on the diagnostic chart.

1. Checks for a short to voltage in the ABS motor "HIGH" circuitry.
2. Checks for a short to voltage in the ABS motor "LOW" circuitry.
3. Checks for a short circuit between motor "HIGH" and motor "LOW".
4. Checks for proper resistance of the ABS motor.
5. Ensures malfunction was not due to physical damage of the circuitry.

Diagnostic Aids: Using Tech 1, select manual control function, and exercise ABS motor movement of affected channel in both directions while applying light pressure on the brake pedal.

If erratic or "jumpy" brake pedal movement is detected while performing an "apply" or "release" function of the ABS motor, an intermittent malfunction may be indicated.

An "Intermittent" malfunction may be caused by a poor connection, rubbed through wire insulation, or a wire that is broken inside the insulation.

If the malfunction is not current, wiggle the wires of the affected channel and check if the DTC resets. This will help to pinpoint an intermittent malfunction in the motor circuitry or connections.

The frequency of the malfunction can be checked by using the enhanced diagnostic function of the Tech 1.

Any circuitry, that is suspected as causing the intermittent complaint, should be thoroughly checked for backed out terminals, improper mating, broken locks, improperly formed or damaged terminals, poor terminal to wiring connections or physical damage to the wiring harness.

DIAGNOSTIC CHART DTC 66 — 1994–95 CAMARO AND FIREBIRD, CONTINUED

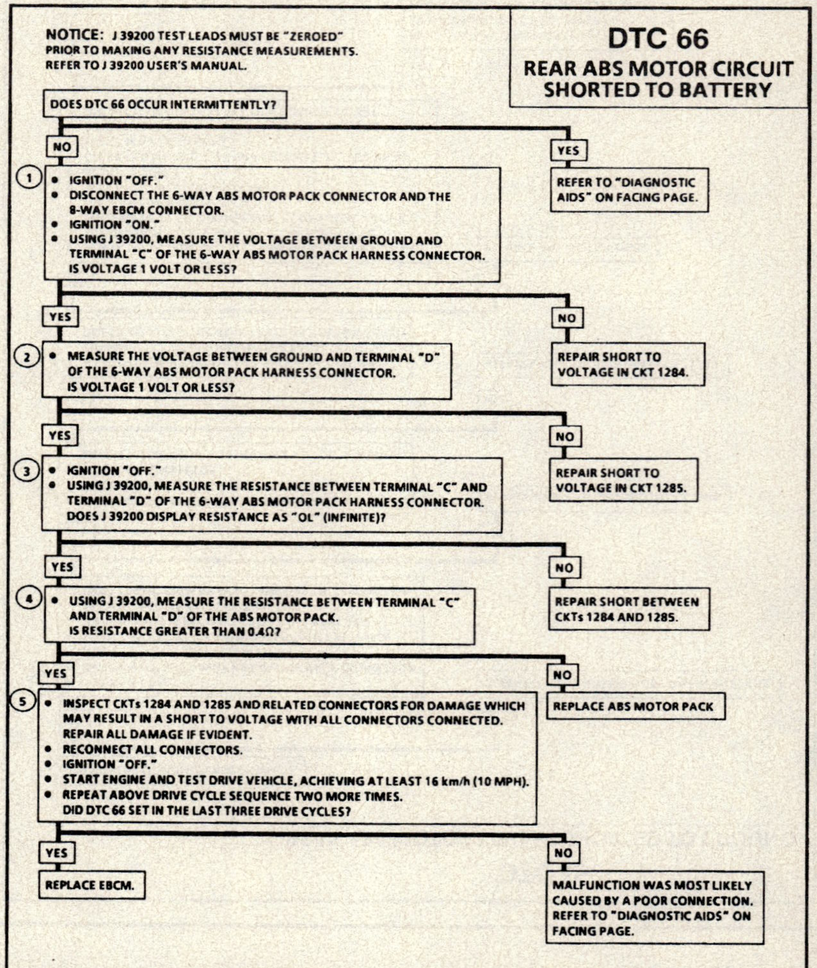

AFTER DIAGNOSIS IS COMPLETE, CLEAR DTCs AND TEST DRIVE VEHICLE FOR THREE (3) DRIVE CYCLES TO VERIFY THAT DTC DOES NOT RESET. A DRIVE CYCLE CONSISTS OF STARTING THE VEHICLE, DRIVING OVER 16 KM/H (10 MPH), AND THEN KEYING DOWN.

DIAGNOSTIC CHART DTC 76 — 1994–95 CAMARO AND FIREBIRD

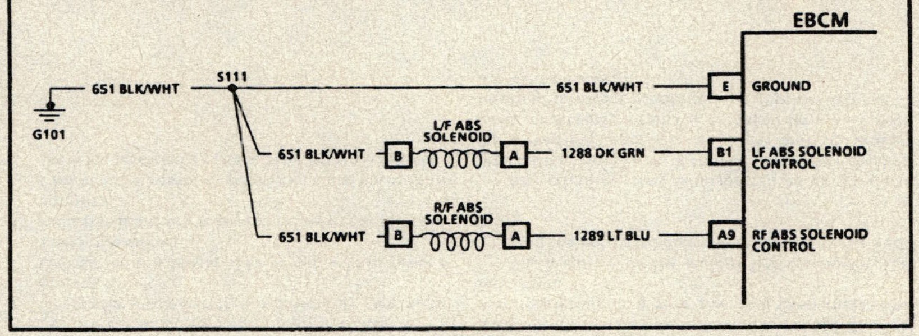

DTC 76

LEFT FRONT SOL CIRCUIT OPEN OR SHORTED TO BATT

Circuit Description:
This DTC identifies a solenoid that cannot be energized due to an open in its circuitry, or a solenoid that is always energized due to a short to battery in its circuitry between the driver and the solenoid. An open will not allow proper ABS operation, but the short to battery simply turns "ON" the solenoid. A path for base brakes is still allowed once the motor re-homes and the check ball is lifted off its seat during key "ON" initialization.

DTC Will Set When: DTC 76 can be set only when the solenoid is commanded "OFF." If the EBCM detects an excessive voltage in the left front solenoid control circuit, a malfunction exists.

Action Taken: A malfunction DTC is stored, ABS is disabled and the "ABS INOP" lamp is turned "ON."

DTC Chart Test Description: Number(s) below refer to circled number(s) on the diagnostic chart.

1. Checks for a short to voltage in the control circuitry of the solenoid.
2. Checks for an open in the control circuit of the solenoid.
3. Checks the solenoid coil for proper resistance.
4. Checks for an open in the solenoid ground circuit.
5. Checks for a possible intermittent malfunction in the solenoid circuitry due to poor terminal contact.

Diagnostic Aids: An "Intermittent" malfunction may be caused by a poor connection, rubbed through wire insulation, or a wire that is broken inside the insulation.

The frequency of the malfunction can be checked by using the enhanced diagnostic function of the Tech 1,

Any circuitry, that is suspected as causing the intermittent complaint, should be thoroughly checked for backed out terminals, improper mating, broken locks, improperly formed or damaged terminals, poor terminal to wiring connections or physical damage to the wiring harness.

DIAGNOSTIC CHART DTC 76 — 1994–95 CAMARO AND FIREBIRD, CONTINUED

DTC 76
LEFT FRONT SOL CIRCUIT OPEN OR SHORTED TO BATT

(1)
- IGNITION "OFF."
- DISCONNECT 32-WAY EBCM CONNECTOR.
- IGNITION "ON," ENGINE "OFF."
- USING J 39200, MEASURE THE VOLTAGE BETWEEN GROUND AND TERMINAL "B1" OF THE 32-WAY EBCM HARNESS CONNECTOR.

IS VOLTAGE 1 VOLT OR LESS?

NO → REPAIR SHORT TO VOLTAGE IN CKT 1288.

YES ↓

(2)
- IGNITION "OFF."
- DISCONNECT LEFT FRONT SOLENOID CONNECTOR.
- USING J 39200, MEASURE THE RESISTANCE BETWEEN TERMINAL "B1" OF THE 32-WAY EBCM HARNESS CONNECTOR AND TERMINAL "A" OF THE LEFT FRONT SOLENOID HARNESS CONNECTOR.

IS RESISTANCE 2Ω OR LESS?

NO → REPAIR OPEN OR HIGH RESISTANCE IN CKT 1288.

YES ↓

(3)
- MEASURE THE RESISTANCE BETWEEN TERMINAL "A" AND TERMINAL "B" OF THE LEFT FRONT SOLENOID.

IS RESISTANCE BETWEEN 2.5 AND 5Ω?

NO → REPLACE LEFT FRONT SOLENOID.

YES ↓

(4)
- MEASURE THE RESISTANCE BETWEEN GROUND AND TERMINAL "B" OF THE LEFT FRONT SOLENOID HARNESS CONNECTOR.

IS RESISTANCE 2Ω OR LESS?

NO → REPAIR OPEN OR HIGH RESISTANCE IN CKT 651.

YES ↓

(5)
- INSPECT LEFT FRONT SOLENOID AND 32-WAY EBCM HARNESS CONNECTORS FOR POOR TERMINAL CONTACT AND EVIDENCE OF CORROSION.

REPLACE ALL TERMINALS THAT EXHIBIT SIGNS OF POOR TERMINAL CONTACT OR CORROSION.
- INSPECT CKT 1288 FOR DAMAGE WHICH MAY RESULT IN A SHORT TO VOLTAGE WITH ALL CONNECTORS CONNECTED. REPAIR DAMAGE IF EVIDENT.
- RECONNECT ALL CONNECTORS.
- IGNITION "ON."

DOES DTC 76 SET AS A CURRENT DTC?

YES → REPLACE EBCM.

NO → MALFUNCTION IS INTERMITTENT. REFER TO "DIAGNOSTIC AIDS" ON FACING PAGE.

DIAGNOSTIC CHART DTC 77 — 1994–95 CAMARO AND FIREBIRD

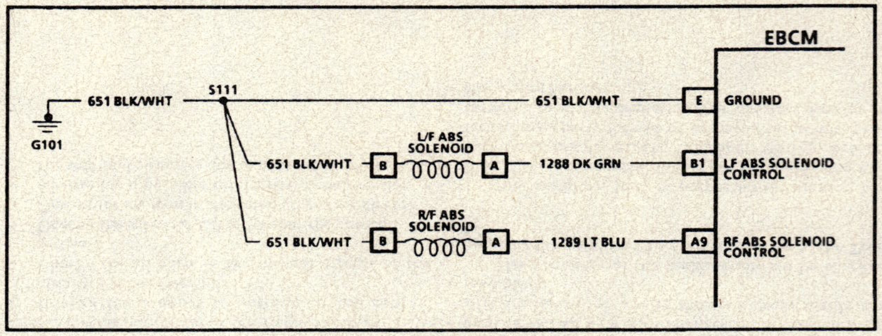

DTC 77

LEFT FRONT SOLENOID CIRCUIT SHORTED TO GND

Circuit Description:

This DTC identifies a solenoid that cannot be energized due to an open in its driver circuitry, or a short to ground between the solenoid driver and the solenoid. These malfunctions can affect ABS operation since the flow of brake fluid to the wheel cylinder cannot be stopped, making ABS operation for that channel impossible.

DTC Will Set When: DTC 77 can be set only when the solenoid is commanded "ON." If the EBCM detects the left front solenoid control circuit voltage is out of specification, a malfunction exists.

Action Taken: A malfunction DTC is stored, ABS is disabled and the "ABS INOP" lamp is turned "ON."

DTC Chart Test Description: Number(s) below refer to circled number(s) on the diagnostic chart.

1. Checks for a solenoid that is internally shorted to ground.
2. Checks for a solenoid that is not within proper resistance values.
3. Indicates if a short to ground exists in the solenoid circuitry.
4. Checks for a possible intermittent malfunction in the solenoid circuitry due to poor terminal contact.

Diagnostic Aids: An "Intermittent" malfunction may be caused by a poor connection, rubbed through wire insulation, or a wire that is broken inside the insulation.

The frequency of the malfunction can be checked by using the enhanced diagnostic function of the Tech 1.

Any circuitry, that is suspected as causing the intermittent complaint, should be thoroughly checked for backed out terminals, improper mating, broken locks, improperly formed or damaged terminals, poor terminal to wiring connections or physical damage to the wiring harness.

DIAGNOSTIC CHART DTC 77 — 1994–95 CAMARO AND FIREBIRD, CONTINUED

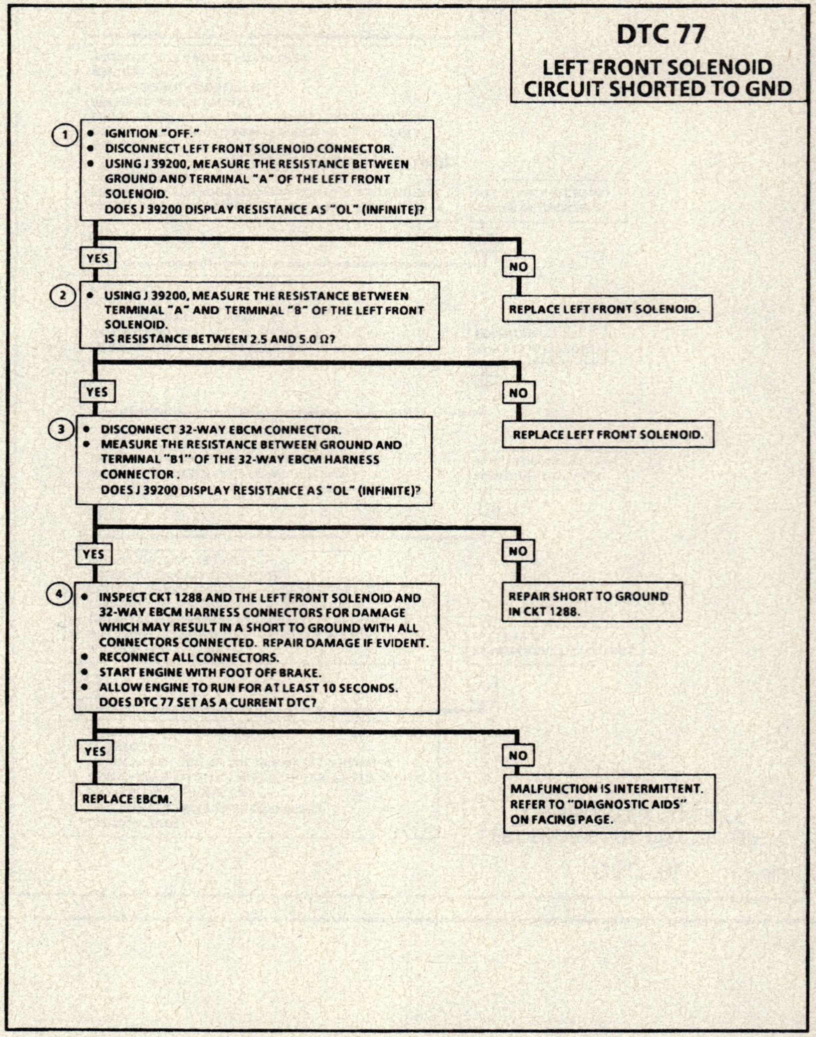

DIAGNOSTIC CHART DTC 78 — 1994–95 CAMARO AND FIREBIRD

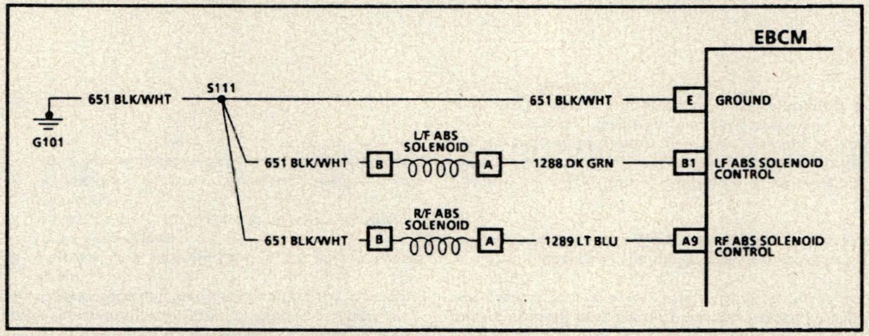

DTC 78

RIGHT FRONT SOL CIRCUIT OPEN OR SHORTED TO BATT

Circuit Description:

This DTC identifies a solenoid that cannot be energized due to an open in its circuitry, or a solenoid that is always energized due to a short to battery in its circuitry between the driver and the solenoid. An open will not allow proper ABS operation, but the short to battery simply turns "ON" the solenoid. A path for base brakes is still allowed once the motor re-homes and the check ball is lifted off its seat during key "ON" initialization.

DTC Will Set When: DTC 78 can be set only when the solenoid is commanded "OFF." If the EBCM detects an excessive voltage in the right front solenoid control circuit, a malfunction exists.

Action Taken: A malfunction DTC is stored, ABS is disabled and the "ABS INOP" lamp is turned "ON."

DTC Chart Test Description: Number(s) below refer to circled number(s) on the diagnostic chart.

1. Checks for a short to voltage in the control circuitry of the solenoid.
2. Checks for an open in the control circuit of the solenoid.
3. Checks the solenoid coil for proper resistance.
4. Checks for an open in the solenoid ground circuit.
5. Checks for a possible intermittent malfunction in the solenoid circuitry due to poor terminal contact.

Diagnostic Aids: An "Intermittent" malfunction may be caused by a poor connection, rubbed through wire insulation, or a wire that is broken inside the insulation.

The frequency of the malfunction can be checked by using the enhanced diagnostic function of the Tech 1.

Any circuitry, that is suspected as causing the intermittent complaint, should be thoroughly checked for backed out terminals, improper mating, broken locks, improperly formed or damaged terminals, poor terminal to wiring connections or physical damage to the wiring harness.

DIAGNOSTIC CHART DTC 78 — 1994–95 CAMARO AND FIREBIRD, CONTINUED

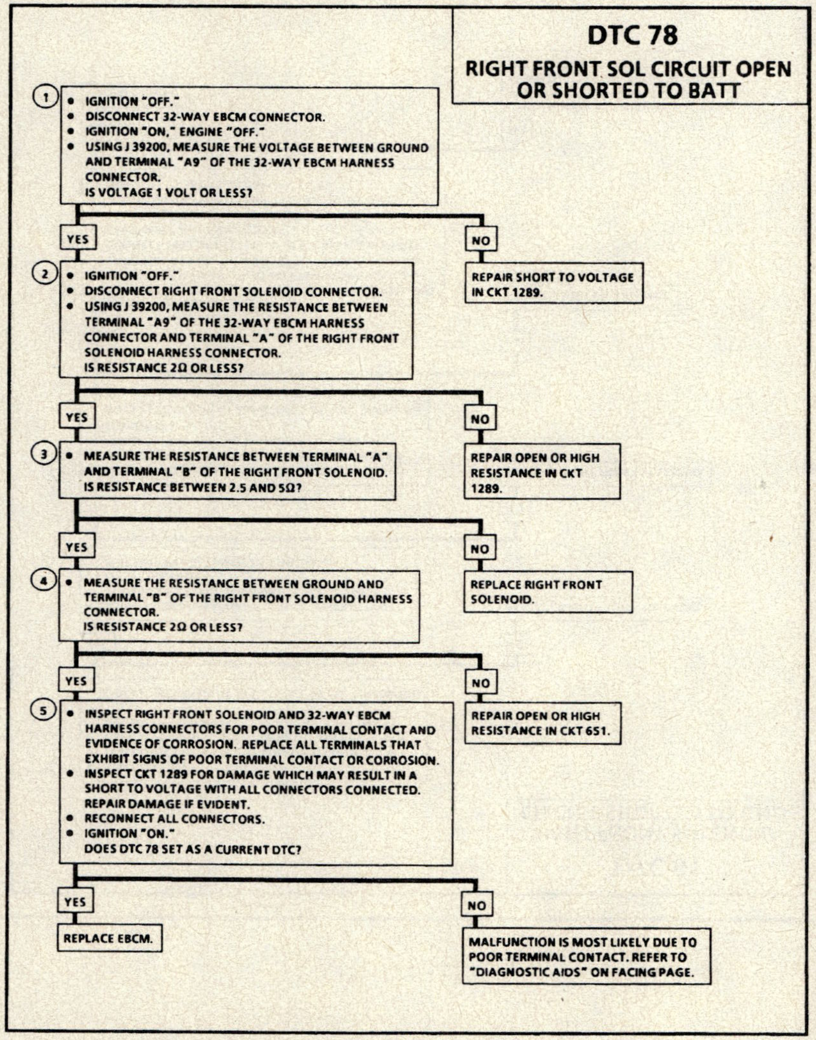

DIAGNOSTIC CHART DTC 81 — 1994–95 CAMARO AND FIREBIRD

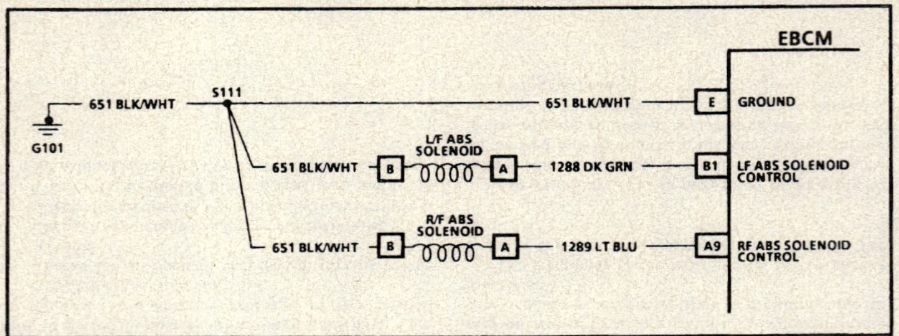

DTC 81

RIGHT FRONT SOLENOID CIRCUIT SHORTED TO GND

Circuit Description:

This DTC identifies a solenoid that cannot be energized due to an open in its driver circuitry, or a short to ground between the solenoid driver and the solenoid. These malfunctions can affect ABS operation since the flow of brake fluid to the wheel cylinder cannot be stopped, making ABS operation for that channel impossible.

DTC Will Set When: DTC 81 can be set only when the solenoid is commanded "ON." If the EBCM detects the right front solenoid control circuit voltage is out of specification, a malfunction exists.

Action Taken: A malfunction DTC is stored, ABS is disabled and the "ABS INOP" lamp is turned "ON."

DTC Chart Test Description: Number(s) below refer to circled number(s) on the diagnostic chart.

1. Checks for a solenoid that is internally shorted to ground.
2. Checks for a solenoid that is not within proper resistance values.
3. Indicates if a short to ground exists in the solenoid circuitry.
4. Checks for a possible intermittent malfunction in the solenoid circuitry due to poor terminal contact.

Diagnostic Aids: An "Intermittent" malfunction may be caused by a poor connection, rubbed through wire insulation, or a wire that is broken inside the insulation.

The frequency of the malfunction can be checked by using the enhanced diagnostic function of the Tech 1,

Any circuitry, that is suspected as causing the intermittent complaint, should be thoroughly checked for backed out terminals, improper mating, broken locks, improperly formed or damaged terminals, poor terminal to wiring connections or physical damage to the wiring harness.

DIAGNOSTIC CHART DTC 81 — 1994–95 CAMARO AND FIREBIRD, CONTINUED

DTC 81
RIGHT FRONT SOLENOID CIRCUIT SHORTED TO GND

(1)
- IGNITION "OFF."
- DISCONNECT RIGHT FRONT SOLENOID CONNECTOR.
- USING J 39200, MEASURE THE RESISTANCE BETWEEN GROUND AND TERMINAL "A" OF THE RIGHT FRONT SOLENOID.

DOES J 39200 DISPLAY RESISTANCE AS "OL" (INFINITE)?

- NO: REPLACE RIGHT FRONT SOLENOID.
- YES: go to (2)

(2)
- USING J 39200, MEASURE THE RESISTANCE BETWEEN TERMINAL "A" AND TERMINAL "B" OF THE RIGHT FRONT SOLENOID.

IS RESISTANCE BETWEEN 2.5 AND 5.0 Ω?

- NO: REPLACE RIGHT FRONT SOLENOID.
- YES: go to (3)

(3)
- DISCONNECT 32-WAY EBCM CONNECTOR.
- MEASURE THE RESISTANCE BETWEEN GROUND AND TERMINAL "A9" OF THE 32-WAY EBCM HARNESS CONNECTOR.

DOES J 39200 DISPLAY RESISTANCE AS "OL" (INFINITE)?

- NO: REPAIR SHORT TO GROUND IN CKT 1289.
- YES: go to (4)

(4)
- INSPECT CKT 1289 AND THE RIGHT FRONT SOLENOID AND 32-WAY EBCM HARNESS CONNECTORS FOR DAMAGE WHICH MAY RESULT IN A SHORT TO GROUND WITH ALL CONNECTORS CONNECTED. REPAIR DAMAGE IF EVIDENT.
- RECONNECT ALL CONNECTORS.
- START ENGINE WITH FOOT OFF BRAKE.
- ALLOW ENGINE TO RUN FOR AT LEAST 10 SECONDS.

DOES DTC 81 SET AS A CURRENT DTC?

- YES: REPLACE EBCM.
- NO: MALFUNCTION IS INTERMITTENT. REFER TO "DIAGNOSTIC AIDS" ON FACING PAGE.

DIAGNOSTIC CHART DTC 82 — 1994–95 CAMARO AND FIREBIRD

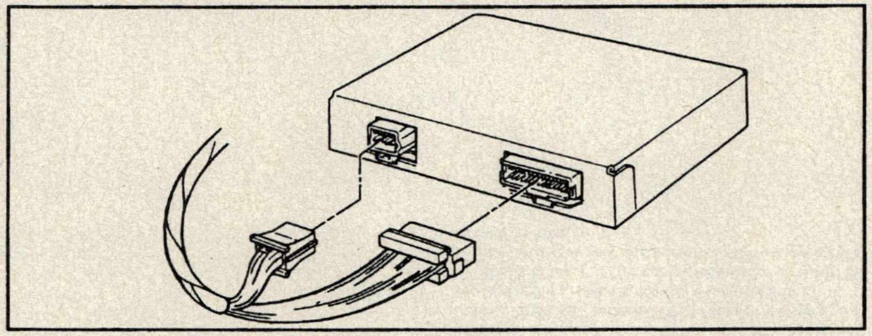

DTC 82
CALIBRATION MALFUNCTION

Circuit Description:
This DTC allows the EBCM to check for a calibration malfunction by comparing the calibration value to a known value stored in the EEPROM.
This DTC is also used as a security measure to prevent improper use of calibrations or changes to these calibrations that may alter the designed function of ABS.

DTC Will Set When: DTC 82 can be set at key-up. If the program identifier is incorrect or the memory checksum is incorrect, a malfunction exists.

Action Taken: A malfunction DTC is stored, ABS is disabled and the "ABS INOP" lamp is turned "ON."

DTC Chart Test Description: Number(s) below refer to circled number(s) on the diagnostic chart.
1. Checks to see if the malfunction is present during diagnosis. If present, the EBCM is not functioning correctly and must be replaced.

Important
- At the time of printing, 1994 vehicles with ABS-VI were not being field reprogrammed to correct this failure. A service bulletin will be issued when EEPROM reprogramming is authorized.

Diagnostic Aids: An intermittent DTC 82 may be caused by a bad cell in the EEPROM that is sensitive to temperature changes. If DTC 82 failed more than once, but is intermittent, replace EBCM.
The frequency of the malfunction can be checked by using the enhanced diagnostic function of the Tech 1,

DIAGNOSTIC CHART DTC 82 — 1994–95 CAMARO AND FIREBIRD, CONTINUED

DTC 82
CALIBRATION MALFUNCTION

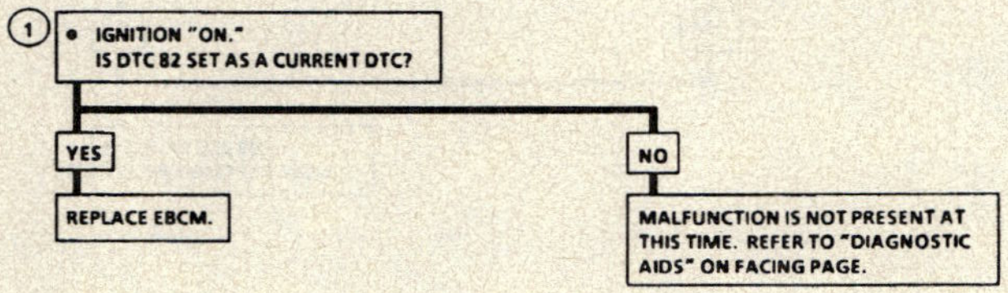

DIAGNOSTIC CHART DTC 86 — 1994–95 CAMARO AND FIREBIRD

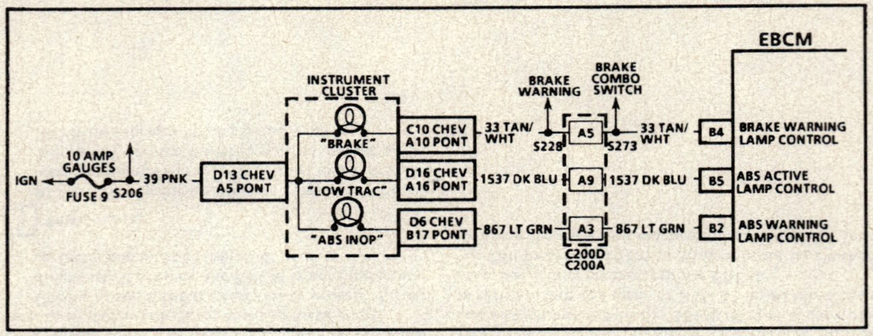

DTC 86

EBCM TURNED "ON" THE RED "BRAKE" WARNING LAMP

Circuit Description:

This DTC is provided as an information only DTC, and reflects the status of the command issued by the ABS EBCM to illuminate the red brake warning lamp. If another DTC issues a command to illuminate the red brake warning lamp, DTC 86 will be stored in EEPROM as a history DTC at the conclusion of the ignition cycle.

DTC Chart Test Description: Number(s) below refer to circled number(s) on the diagnostic chart.

1. Serves to identify if a DTC other than DTC 86 commanded the red brake warning lamp "ON."

Diagnostic Aids: Any ABS mechanical DTC that issues a command to illuminate the red "BRAKE" telltale will also result in DTC 86 being stored in EEPROM during shut down. These DTCs are: 38, 41, 42, 46, and 51. If the motors are not in their home position, certain electrical DTCs will also command the red "BRAKE" telltale "ON." These DTCs are: 14, 16, 18, 55, 56, 57, 58, 61, 62, 63, 64, 65, and 66.

If any of these DTCs are indicated along with DTC 86, they must be corrected prior to addressing a DTC 86 malfunction.

DIAGNOSTIC CHART DTC 86 — 1994–95 CAMARO AND FIREBIRD, CONTINUED

DTC 86

EBCM TURNED "ON" THE RED "BRAKE" WARNING LAMP

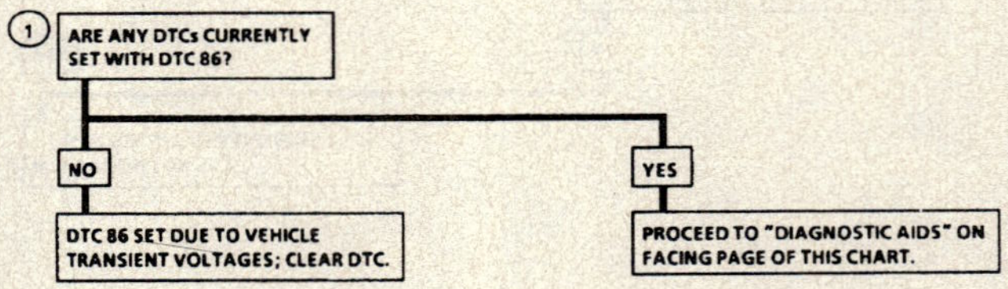

DIAGNOSTIC CHART DTC 87 — 1994–95 CAMARO AND FIREBIRD

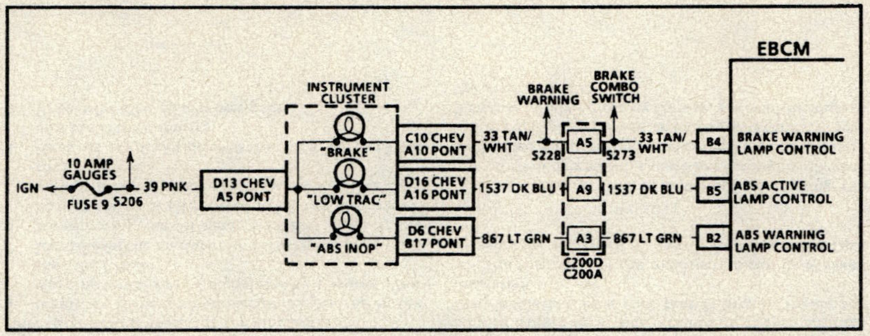

DTC 87

RED "BRAKE" WARNING LAMP CIRCUIT OPEN

Circuit Description:

This DTC is used to verify the EBCM has continuity to the red brake warning lamp in case the EBCM must turn it "ON." This will only occur if an ABS malfunction is detected that may degrade base brake operation.

Because ABS is not the only device controlling the red brake warning lamp (parking brake and low fluid sensor may also turn "ON" the lamp) a short to ground in this circuit cannot be detected.

DTC Will Set When: DTC 87 can be set after initialization is completed. If the red brake warning lamp circuit voltage is out of specification indicating an open circuit, a malfunction exists.

Action Taken: A malfunction DTC is stored, ABS is not disabled.

DTC Chart Test Description: Number(s) below refer to circled number(s) on the diagnostic chart.

1. Identifies if the ground circuit to the red brake warning lamp is completed by a source other than the EBCM.
2. Checks for high resistance in park brake circuitry.
3. Checks for malfunctioning EBCM.
4. Checks for high resistance in fluid level switch circuitry.
5. Checks for malfunctioning warning lamp bulb.
6. Checks for open or high resistance in CKT 33.

Diagnostic Aids: An "Intermittent" malfunction may be caused by a poor connection, rubbed through wire insulation, or a wire that is broken inside the insulation.

The frequency of the malfunction can be checked by using the enhanced diagnostic function of the Tech 1.

If a Tech 1 or T-100 are not available, the parking brake handle may be lifted to verify proper lamp operation and continuity of CKTs 3 and 39.

Any circuitry, that is suspected as causing the intermittent complaint, should be thoroughly checked for backed out terminals, improper mating, broken locks, improperly formed or damaged terminals, poor terminal to wiring connections or physical damage to the wiring harness.

DIAGNOSTIC CHART DTC 87 — 1994–95 CAMARO AND FIREBIRD, CONTINUED

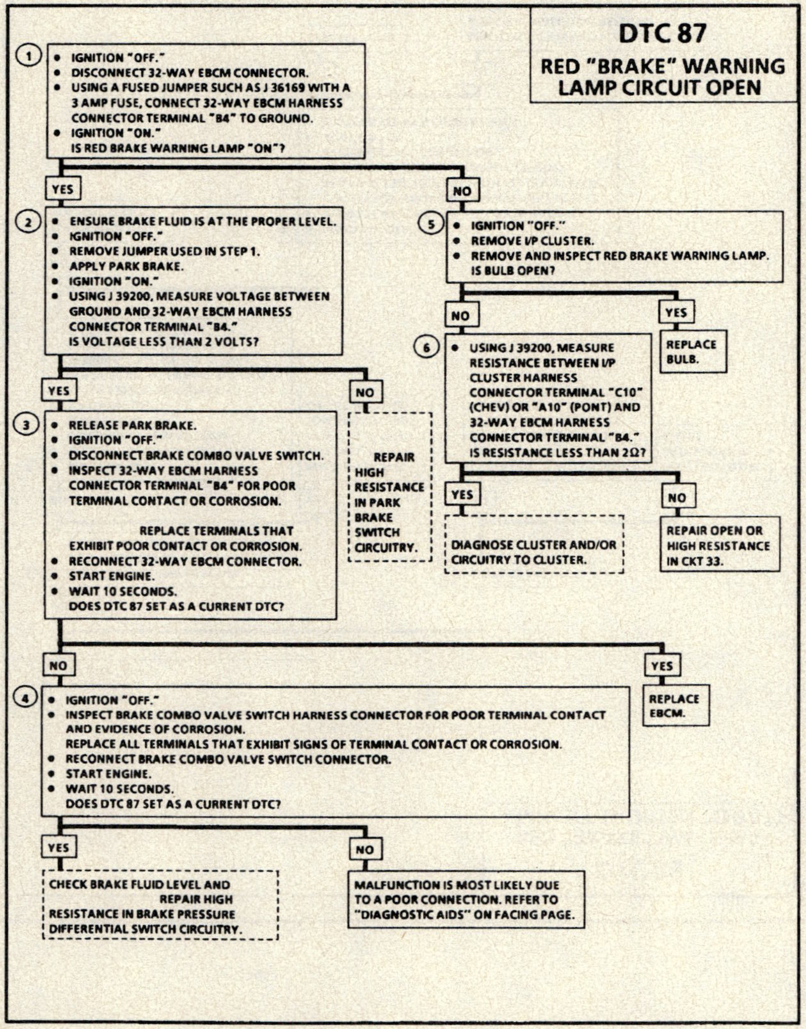

DIAGNOSTIC CHART DTC 88 — 1994–95 CAMARO AND FIREBIRD

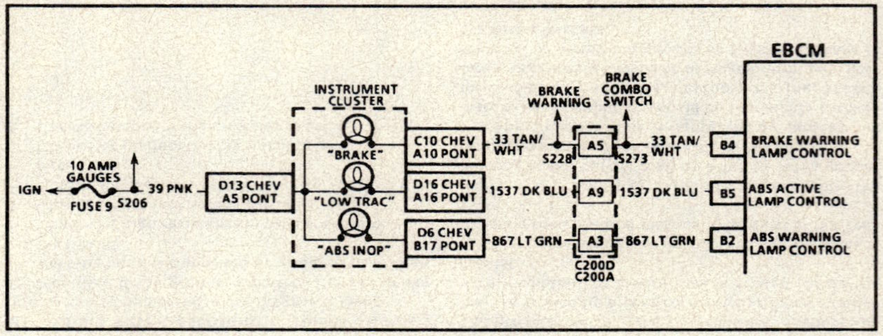

DTC 88

RED "BRAKE" WARNING LAMP CKT SHORTED TO BATT

Circuit Description:

This DTC identifies a short to battery between the EBCM and the red brake warning lamp, or an open driver that does not allow the red brake warning lamp to be illuminated by the EBCM. This will only occur if an ABS fault is detected that may degrade base brake operation.

DTC Will Set When: DTC 88 can be set only at the beginning of the three second bulb check. If the EBCM detects battery voltage on the red brake warning lamp control circuit, a malfunction exists.

Action Taken: A malfunction DTC is stored. ABS is not disabled; however, the "ABS INOP" lamp will flash to indicate a malfunction exists.

DTC Chart Test Description: Number(s) below refer to circled number(s) on the diagnostic chart.

1. Identifies if the ground circuit to the red brake warning lamp is being completed by a source other than the EBCM.
2. Identifies if the malfunction is currently present.
3. By removing the 10 amp "GAGES" Fuse 9, the voltage source is eliminated. This test indicates if voltage is being supplied from a source other than the gauge fuse.
4. Ensures malfunction was not due to physical damage of the circuitry.
5. Checks for DTC 86 also being set.

Diagnostic Aids: An "Intermittent" malfunction may be caused by a poor connection, rubbed through wire insulation, or a wire that is broken inside the insulation.

The frequency of the malfunction can be checked by using the enhanced diagnostic function of the Tech 1,

Any circuitry, that is suspected as causing the intermittent complaint, should be thoroughly checked for backed out terminals, improper mating, broken locks, improperly formed or damaged terminals, poor terminal to wiring connections or physical damage to the wiring harness.

DIAGNOSTIC CHART DTC 88 — 1994–95 CAMARO AND FIREBIRD, CONTINUED

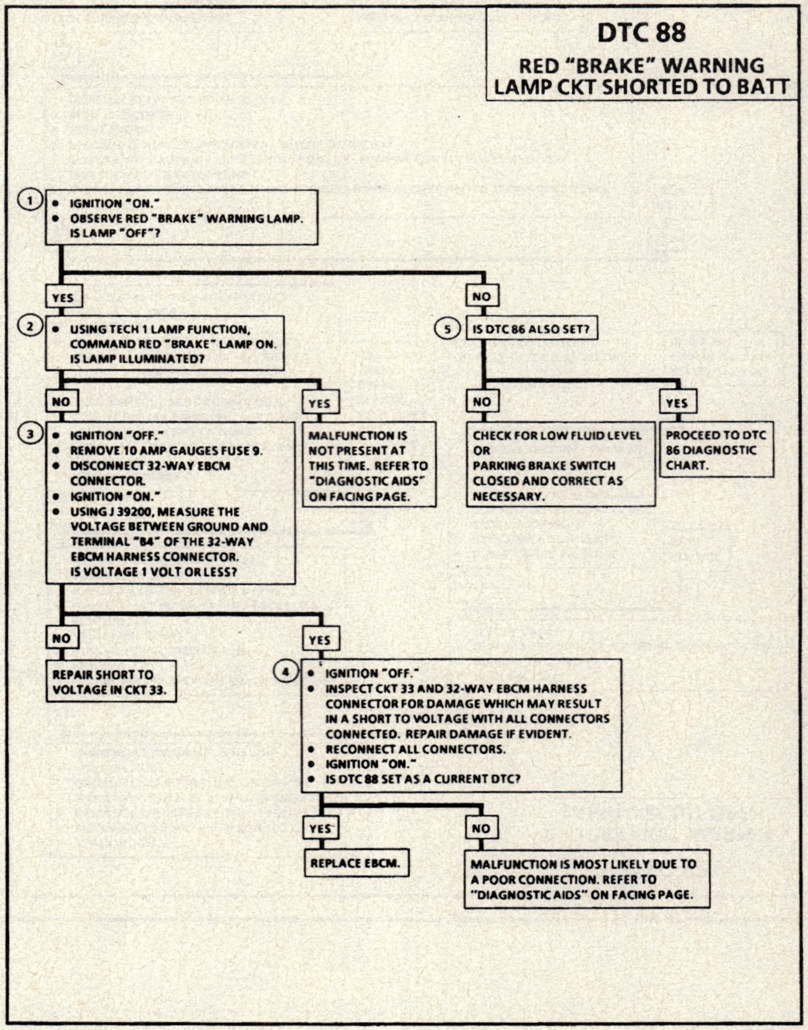

DIAGNOSTIC CHART DTC 91 — 1994–95 CAMARO AND FIREBIRD

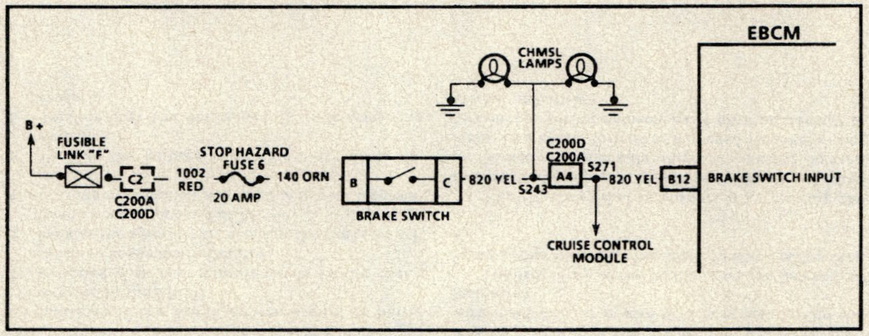

DTC 91

OPEN BRAKE SWITCH DURING DECELERATION

Circuit Description:
This DTC is used to detect an open brake switch in the non-ABS mode. The EBCM looks for deceleration rates that would indicate braking action and verifies this assumption by requiring several repeats of this detection method. In each case, ABS will not be available since no brake switch is seen.

DTC Will Set When: DTC 91 can be set if three deceleration cycles occur with the brake switch "OFF."

Action Taken: A malfunction DTC is stored, ABS is disabled and the "ABS INOP" lamp is turned "ON."

DTC Chart Test Description: Number(s) below refer to circled number(s) on the diagnostic chart.

1. Indicates if the brake switch signal is being received by the EBCM.
2. Indicates if an open circuit exists in the brake switch or brake lamp circuitry.
3. Isolates the open circuit to either the brake switch input circuitry, or the EBCM.
4. Ensures malfunction was not due to poor terminal contact.
5. Verifies that voltage is available at the brake switch.
6. Verifies that the brake switch is functioning properly.

Diagnostic Aids: An "Intermittent" malfunction may be caused by a poor connection, rubbed through wire insulation, or a wire that is broken inside the insulation.

The frequency of the malfunction can be checked by using the enhanced diagnostic function of the Tech 1,

Any circuitry, that is suspected as causing the intermittent complaint, should be thoroughly checked for backed out terminals, improper mating, broken locks, improperly formed or damaged terminals, poor terminal to wiring connections or physical damage to the wiring harness.

DIAGNOSTIC CHART DTC 91 — 1994–95 CAMARO AND FIREBIRD, CONTINUED

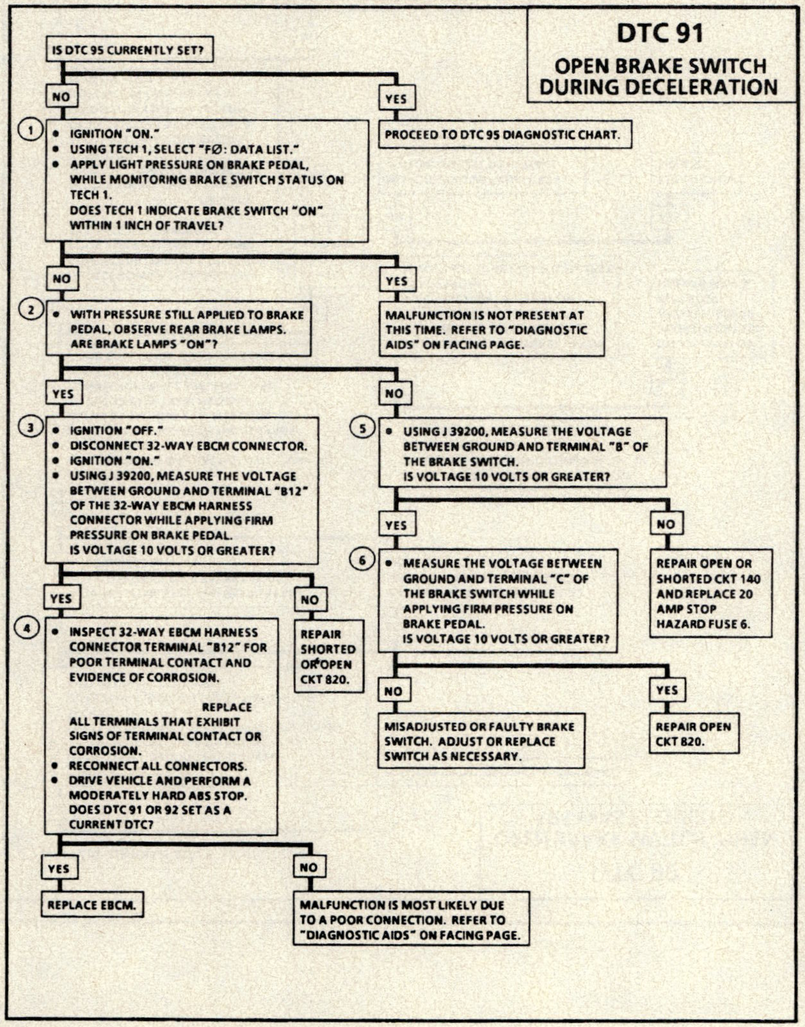

DIAGNOSTIC CHART DTC 92 — 1994–95 CAMARO AND FIREBIRD

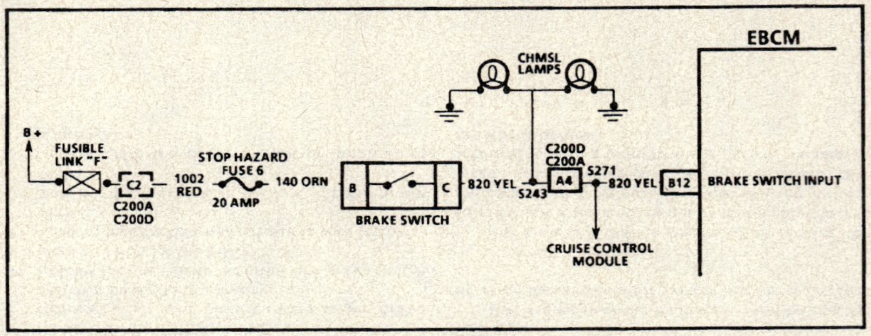

DTC 92

OPEN BRAKE SWITCH WHEN ABS WAS REQUIRED

Circuit Description:

This DTC is run to determine the proper operation of the brake switch. This is important because ABS is activated when the brake switch is "ON." If the brake switch is "OFF," ABS will never be activated. Since this malfunction is difficult to detect under normal braking conditions, this malfunction is only detected when ABS is required.

DTC Will Set When: DTC 92 can be set if the vehicle's speed is greater than 8 km/h (5 mph). If the brake was not "ON" and a release was required on two channels for 0.5 seconds, a malfunction exists.

Action Taken: A malfunction DTC is stored, ABS is disabled and the "ABS INOP" lamp is turned "ON."

DTC Chart Test Description: Number(s) below refer to circled number(s) on the diagnostic chart.

1. Indicates if the brake switch signal is being received by the EBCM.
2. Indicates if an open circuit exists in the brake switch or brake lamp circuitry.
3. Isolates the open circuit to either the brake switch input circuitry, or the EBCM.
4. Ensures malfunction was not due to poor terminal contact.
5. Verifies the voltage is available at the brake switch.
6. Verifies that the brake switch is functioning properly.

Diagnostic Aids: An "Intermittent" malfunction may be caused by a poor connection, rubbed through wire insulation, or a wire that is broken inside the insulation.

The frequency of the malfunction can be checked by using the enhanced diagnostic function of the Tech 1,

Any circuitry, that is suspected as causing the intermittent complaint, should be thoroughly checked for backed out terminals, improper mating, broken locks, improperly formed or damaged terminals, poor terminal to wiring connections or physical damage to the wiring harness.

DIAGNOSTIC CHART DTC 92 — 1994–95 CAMARO AND FIREBIRD, CONTINUED

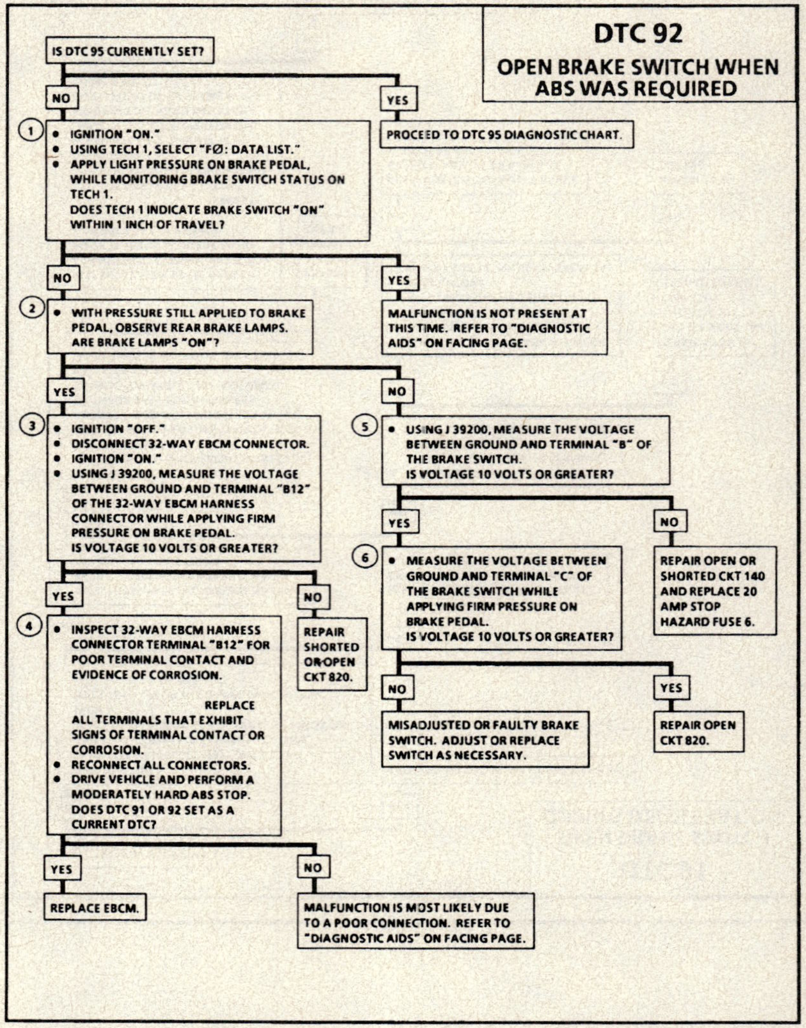

DIAGNOSTIC CHART DTC 93 — 1994–95 CAMARO AND FIREBIRD

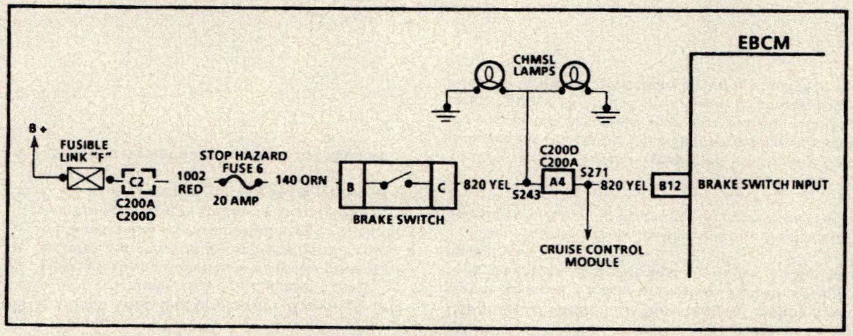

DTC 93

DTC 91 OR 92 SET IN CURRENT OR PREV. IGN CYCLE

Circuit Description:

This DTC is the second portion of DTCs 91 and 92. If DTCs 91 or 92 failed during the last ignition cycle, DTC 93 becomes a current failure during the next ignition cycle, keeping ABS disabled until a brake switch "ON" state is seen. When a change is seen during an ignition cycle in which DTC 93 is a current malfunction, DTC 91 or 92 will clear itself at the end of the current ignition cycle, and ABS will enable itself at the start of the next ignition cycle. DTC 93 alone indicates 91 or 92 failed previously, but is intermittent, or has been corrected.

DTC Chart Test Description: Number(s) below refer to circled number(s) on the diagnostic chart.

1. Indicates which DTC (either 91 or 92) caused DTC 93 to set.

Diagnostic Aids: An "Intermittent" malfunction may be caused by a poor connection, rubbed through wire insulation, or a wire that is broken inside the insulation.

The frequency of the malfunction can be checked by using the enhanced diagnostic function of the Tech 1, as described in "Tech 1 Diagnostics" found in this section.

Any circuitry, that is suspected as causing the intermittent complaint, should be thoroughly checked for backed out terminals, improper mating, broken locks, improperly formed or damaged terminals, poor terminal to wiring connections or physical damage to the wiring harness.

Also, verify proper brake switch operation using the data list of the Tech 1. As the brake is applied, the data list should display the brake switch "ON" within 1 inch of travel.

DIAGNOSTIC CHART DTC 93 — 1994–95 CAMARO AND FIREBIRD, CONTINUED

DTC 93

DTC 91 OR 92 SET IN CURRENT OR PREV. IGN CYCLE

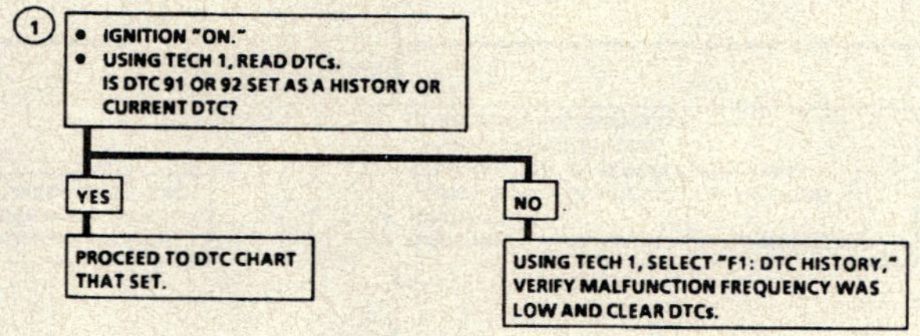

DIAGNOSTIC CHART DTC 94 — 1994–95 CAMARO AND FIREBIRD

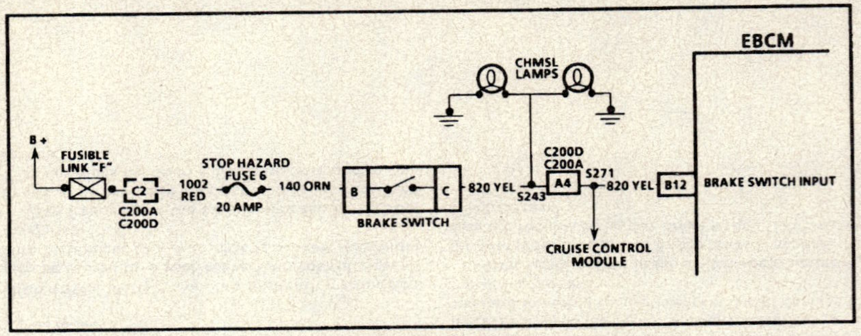

DTC 94

BRAKE SWITCH CONTACTS ALWAYS CLOSED

Circuit Description:
This DTC is run to determine the proper operation of the brake switch. This is important because ABS is activated when the brake switch is "ON" and turned "OFF" when the brake is "OFF." If the brake switch is always "ON," ABS operation will always be requested resulting in potential modulator cycling on rough roads. Additionally, this malfunction will most likely result in a dead battery (due to the brake lights staying "ON") if the driver is not informed of this malfunction.

DTC Will Set When: DTC 94 can be set when the vehicle reaches at least 40 km/h (25 mph). If the brake was never "OFF" during two consecutive drive cycles, a malfunction exists.

Action Taken: A malfunction DTC is stored. ABS is not disabled; however.

DTC Chart Test Description: Number(s) below refer to circled number(s) on the diagnostic chart.

1. Checks if the malfunction is currently present.
2. Isolates the cause of the malfunction to either a malfunctioning or misadjusted brake switch, or a short to voltage in the brake switch circuitry.
3. Checks for unwanted voltage on the brake switch input circuit.
4. Checks for a possible intermittent malfunction.

Diagnostic Aids: An "Intermittent" malfunction may be caused by a poor connection, rubbed through wire insulation, or a wire that is broken inside the insulation.

The frequency of the malfunction can be checked by using the enhanced diagnostic function of the Tech 1,

Any circuitry, that is suspected as causing the intermittent complaint, should be thoroughly checked for backed out terminals, improper mating, broken locks, improperly formed or damaged terminals, poor terminal to wiring connections or physical damage to the wiring harness.

DIAGNOSTIC CHART DTC 94 — 1994–95 CAMARO AND FIREBIRD, CONTINUED

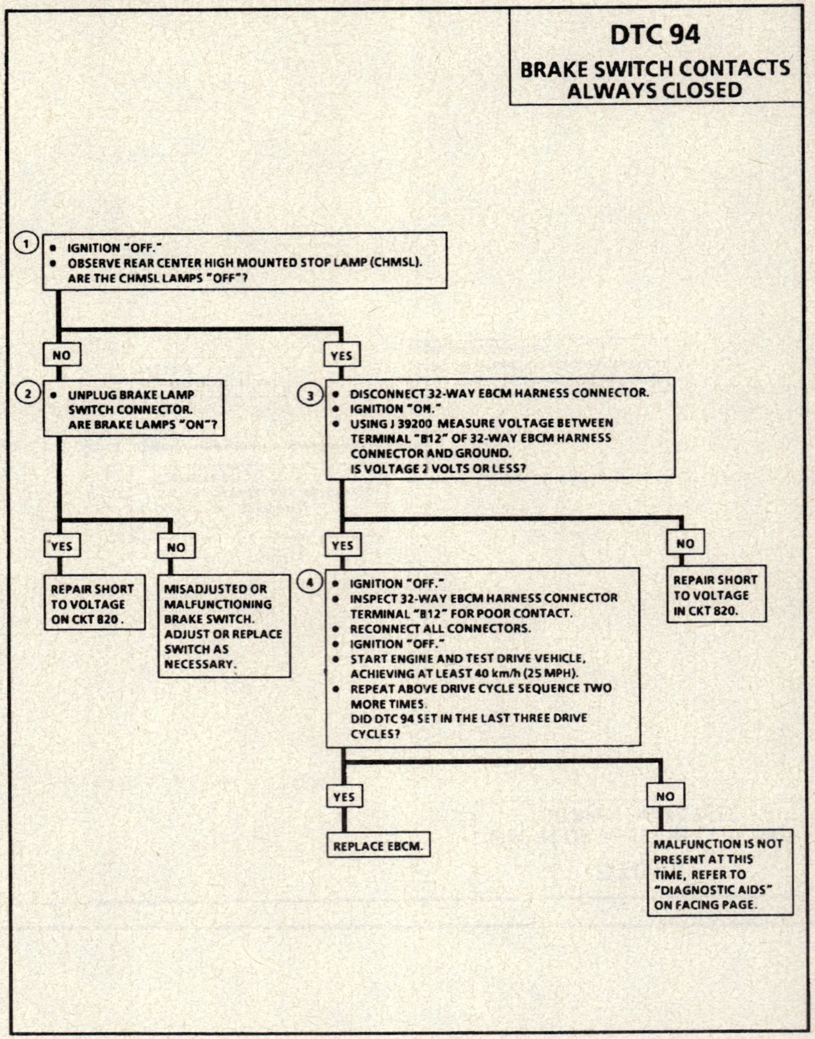

DIAGNOSTIC CHART DTC 95 — 1994–95 CAMARO AND FIREBIRD

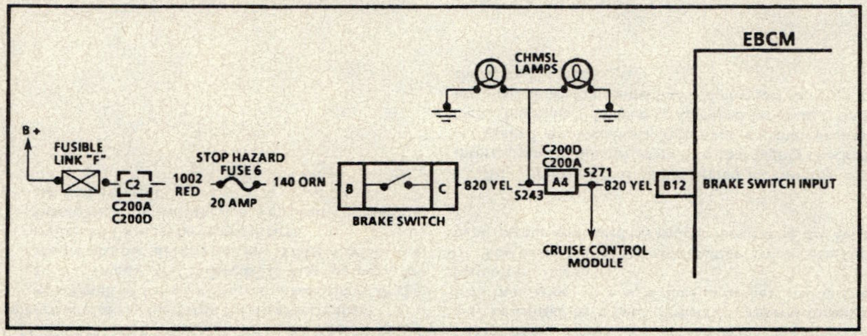

DTC 95

BRAKE SWITCH CIRCUIT OPEN

Circuit Description:

This DTC is used to identify open brake switch circuitry that prevents the brake switch input to the EBCM from changing states when the brake is applied. This DTC is used in conjunction with DTCs 91 and 92 to determine the cause of an open brake switch malfunction.

DTC Will Set When: DTC 95 can be set after initialization is completed. If the brake switch input voltage is out of specification for one second indicating an open circuit, a malfunction exists.

Action Taken: A malfunction DTC is stored, ABS is disabled and the "ABS INOP" lamp is turned "ON."

DTC Chart Test Description: Number(s) below refer to circled number(s) on the diagnostic chart.

1. Is used to confirm that an open in the brake switch circuitry currently exists.
2. Indicates if the brake switch signal is being received by the EBCM.
3. Indicates if an open circuit exists in the brake switch or brake lamp circuitry.
4. Isolates the open circuit to either the brake switch input circuitry, or the EBCM.

Diagnostic Aids: An "Intermittent" malfunction may be caused by a poor connection, rubbed through wire insulation, or a wire that is broken inside the insulation.

The frequency of the malfunction can be checked by using the enhanced diagnostic function of the Tech 1,

Any circuitry, that is suspected as causing the intermittent complaint, should be thoroughly checked for backed out terminals, improper mating, broken locks, improperly formed or damaged terminals, poor terminal to wiring connections or physical damage to the wiring harness.

DIAGNOSTIC CHART DTC 95 — 1994–95 CAMARO AND FIREBIRD, CONTINUED

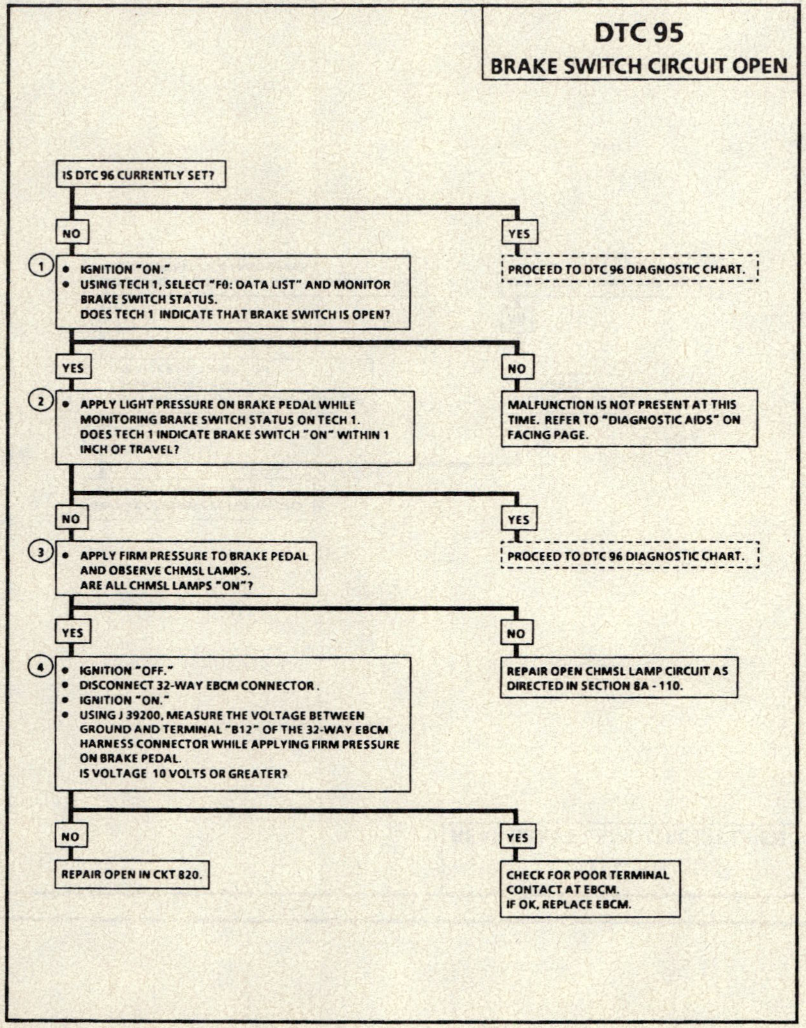

DIAGNOSTIC CHART DTC 96 — 1994–95 CAMARO AND FIREBIRD

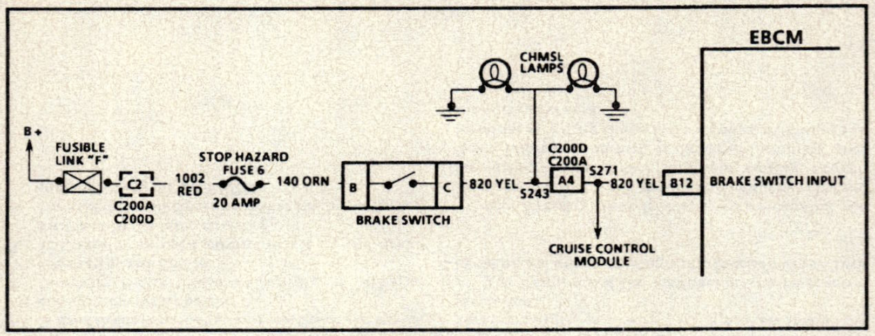

DTC 96

REAR BRAKE LAMP CIRCUIT OPEN

Circuit Description:

This DTC is designed to isolate the cause of a DTC 95 malfunction and indicate to the driver ABS is still available. If DTC 95 fails with DTC 96, the Center High Mounted Stop Lamp (CHMSL) circuit is open. The presence of battery voltage at the brake switch input indicates a valid brake switch input is still available.

DTC Will Set When: DTC 96 can be set only after DTC 95 has been set. If the EBCM detects battery voltage on the brake switch input circuit for 0.5 seconds, a malfunction exists.

Action Taken: A malfunction DTC is stored. ABS is not disabled; however.

DTC Chart Test Description: Number(s) below refer to circled number(s) on the diagnostic chart.

1. As a result of a malfunction of an additional brake switch circuit DTC, this DTC may be set., To insure proper diagnosis, any additional brake switch DTCs must be repaired first.
2. Identifies if the malfunction is currently present in the brake circuit.

Diagnostic Aids: An "Intermittent" malfunction may be caused by a poor connection, rubbed through wire insulation, or a wire that is broken inside the insulation.

The frequency of the malfunction can be checked by using the enhanced diagnostic function of the Tech 1,

Any circuitry, that is suspected as causing the intermittent complaint, should be thoroughly checked for backed out terminals, improper mating, broken locks, improperly formed or damaged terminals, poor terminal to wiring connections or physical damage to the wiring harness.

DIAGNOSTIC CHART DTC 96 — 1994–95 CAMARO AND FIREBIRD, CONTINUED

DTC 96

REAR BRAKE LAMP CIRCUIT OPEN

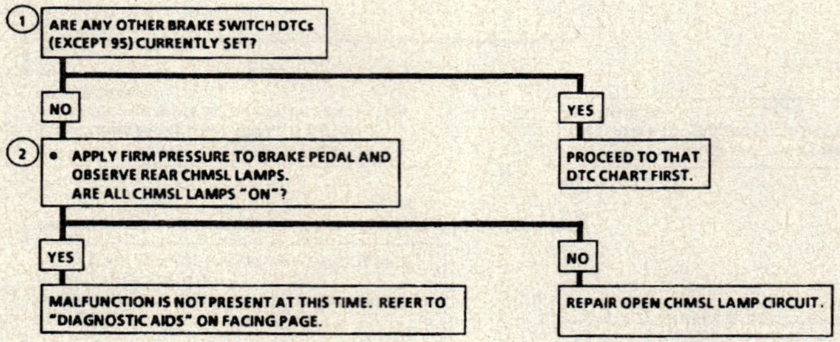

DIAGNOSTIC PROCESS CHART — 1995 CAMARO AND FIREBIRD WITH TCS SYSTEM

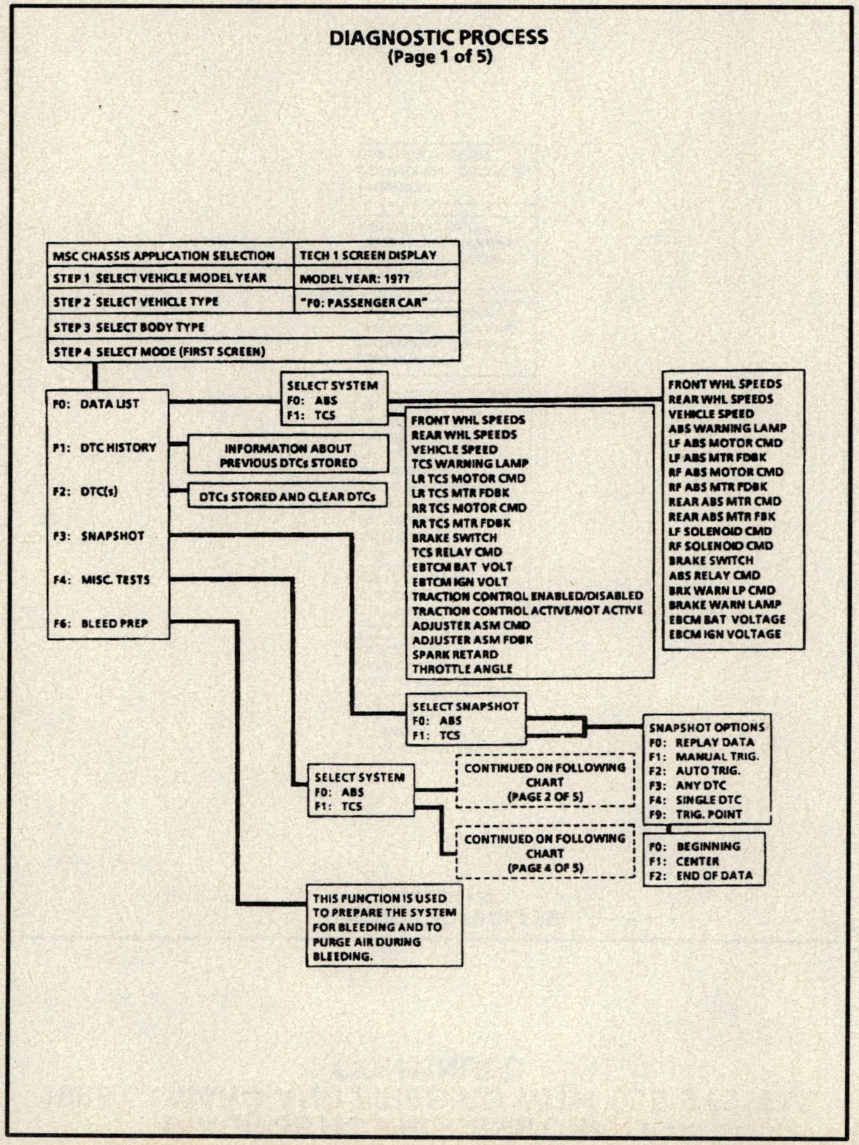

DIAGNOSTIC PROCESS CHART — 1995 CAMARO AND FIREBIRD WITH TCS SYSTEM, CONTINUED

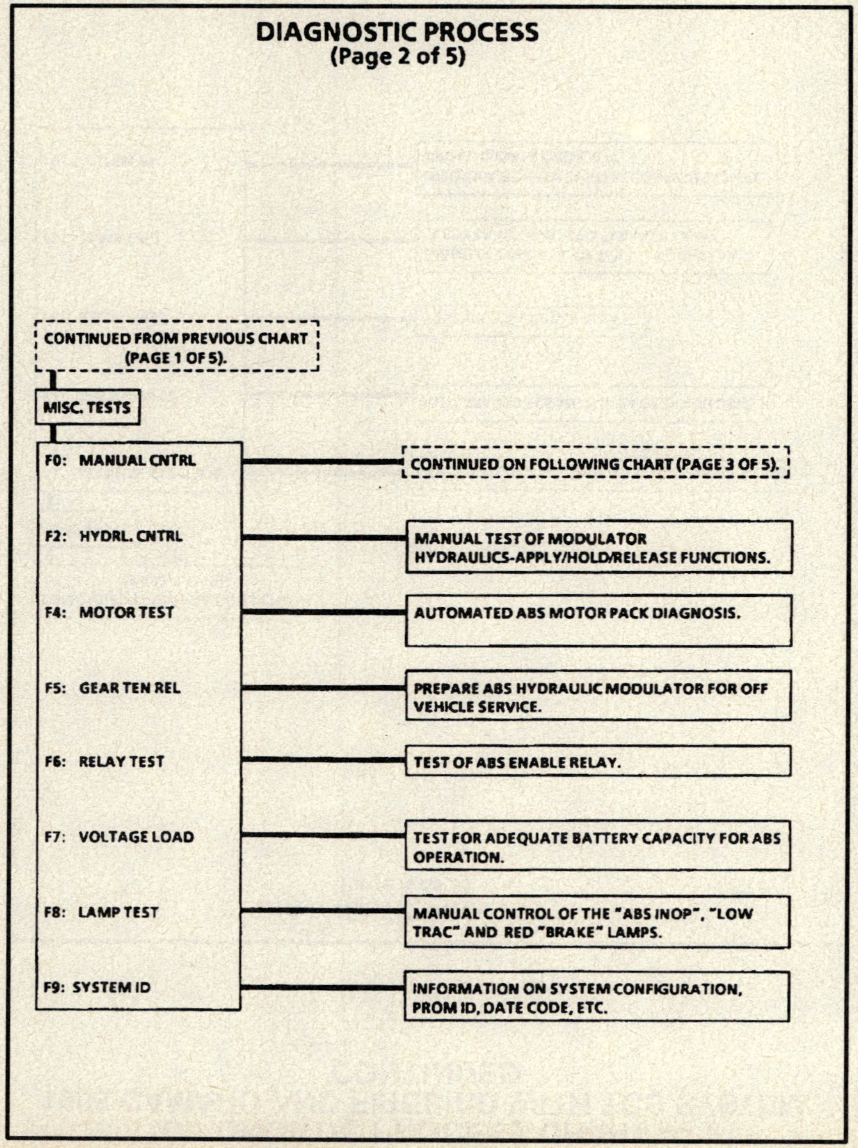

DIAGNOSTIC PROCESS CHART — 1995 CAMARO AND FIREBIRD WITH TCS SYSTEM, CONTINUED

DIAGNOSTIC PROCESS
(Page 3 of 5)

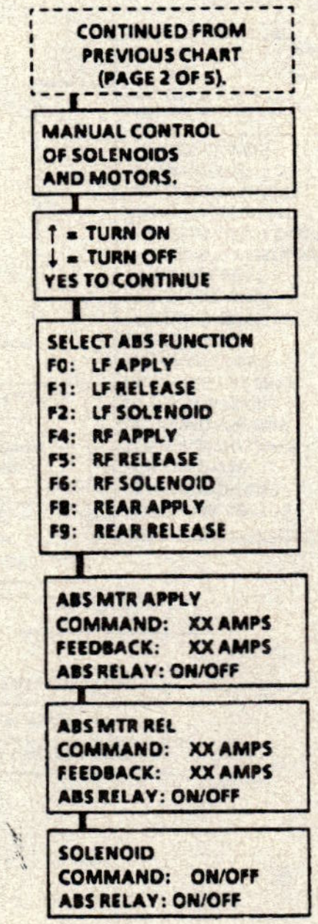

DIAGNOSTIC PROCESS CHART — 1995 CAMARO AND FIREBIRD WITH TCS SYSTEM, CONTINUED

DIAGNOSTIC PROCESS
(Page 4 of 5)

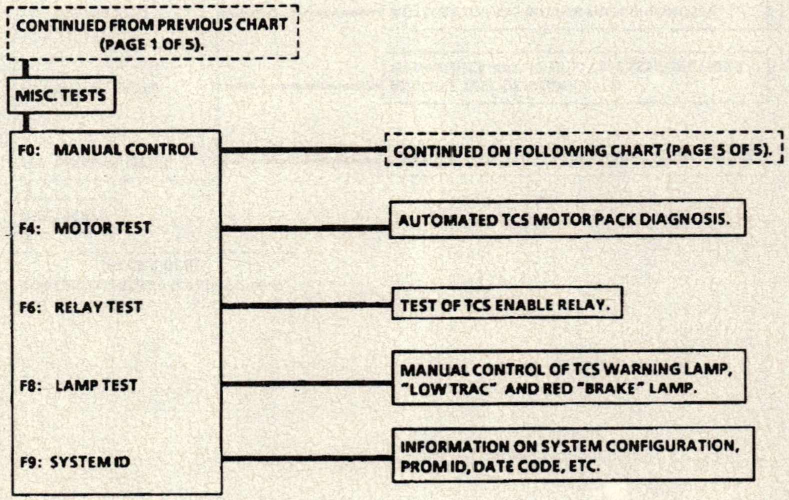

DIAGNOSTIC PROCESS CHART — 1995 CAMARO AND FIREBIRD WITH TCS SYSTEM, CONTINUED

DIAGNOSTIC PROCESS
(Page 5 of 5)

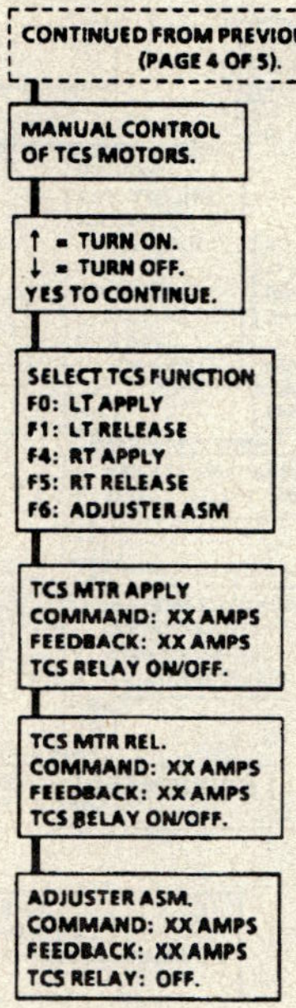

DIAGNOSTIC CHART SYSTEM CHECK — 1995 CAMARO AND FIREBIRD WITH TCS

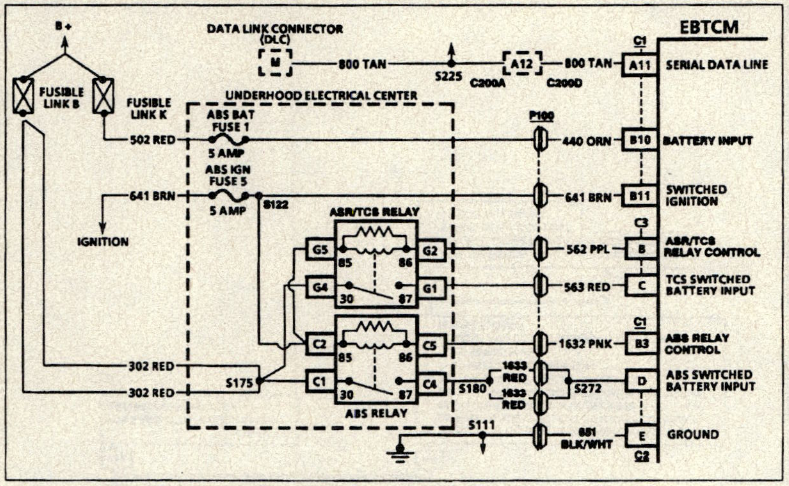

DIAGNOSTIC SYSTEM CHECK

System Description:

The diagnostic system check is an organized approach to identifying a problem created by an Antilock Brake Traction Control System (ABS VI/TCS) malfunction. It must be the starting point for any ABS VI/TCS complaint diagnosis because it directs the service technician to the next logical step in diagnosing the complaint.

Serial data is transmitted by the EBTCM through connector "C1", terminal "A11". The EBTCM is supplied switched ignition voltage through connector "C1", terminal "B11", battery voltage through connector "C1", terminal "B10" and switched battery voltage through connector "C2", terminal" D" and terminal "C". The EBTCM ground is provided through connector "C2", terminal "E".

Chart Test Description:

Number(s) below refer to circled number(s) on the diagnostic chart.

1. Indicates if EBTCM is transmitting data.
2. Checks for a serial data line malfunction.
3. Checks for proper Tech 1 connection to DLC.
4. Checks for open ABS BAT 5 amp Fuse 1.
5. Checks for open ABS IGN 5 amp Fuse 5.
6. Checks for high resistance in the EBTCM ground circuit.
7. Checks for current DTCs.
8. Checks for proper ABS and TCS warning lamp operation.
9. Checks for history DTCs.
10. Checks for possible short to ground in the battery feed circuitry.
11. Checks for possible short to ground in the EBTCM.
12. Ensures short to ground is not due to physical damage of the circuitry.
13. Checks for possible short to ground in the switched ignition circuitry.
14. Isolates the short to ground to either the circuitry or the EBTCM.
15. Ensures short to ground is not due to physical damage of the circuitry.
16. Checks for an open in CKT 440 or 502.
17. Checks for high resistance in CKT 440 or 502.
18. Checks for high resistance in the switched ignition feed circuitry.
19. Ensures malfunction is not due to poor terminal contact.
20. Checks for high resistance in the switched ignition circuit between the fuse block and the EBTCM.

Diagnostic Aids:

Excessive resistance in the ground or power supply circuits will not allow communication with the EBTCM. If communication with the EBTCM is not possible, ensure the ABS VI/TCS ground connection is good and that there is no excessive resistance in any of the power supply circuits.

DIAGNOSTIC CHART SYSTEM CHECK — 1995 CAMARO AND FIREBIRD WITH TCS, CONTINUED

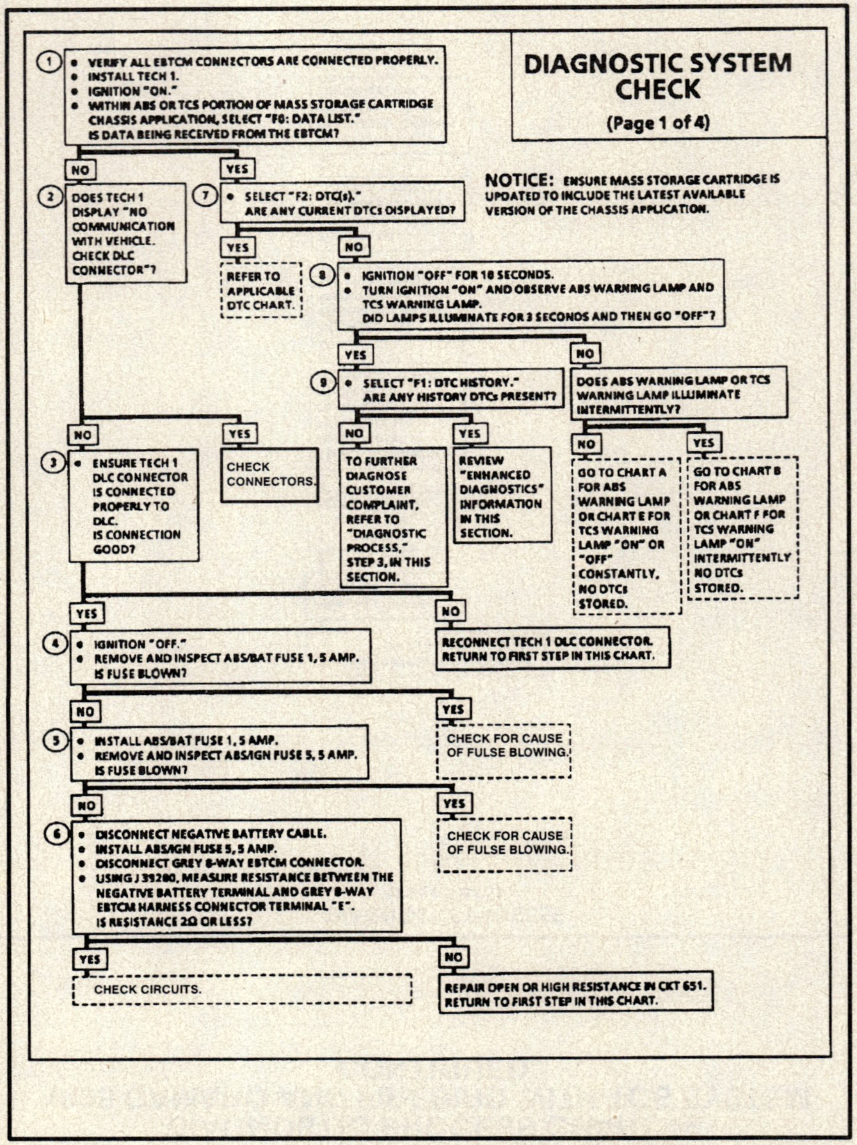

DIAGNOSTIC CHART SYSTEM CHECK — 1995 CAMARO AND FIREBIRD WITH TCS, CONTINUED

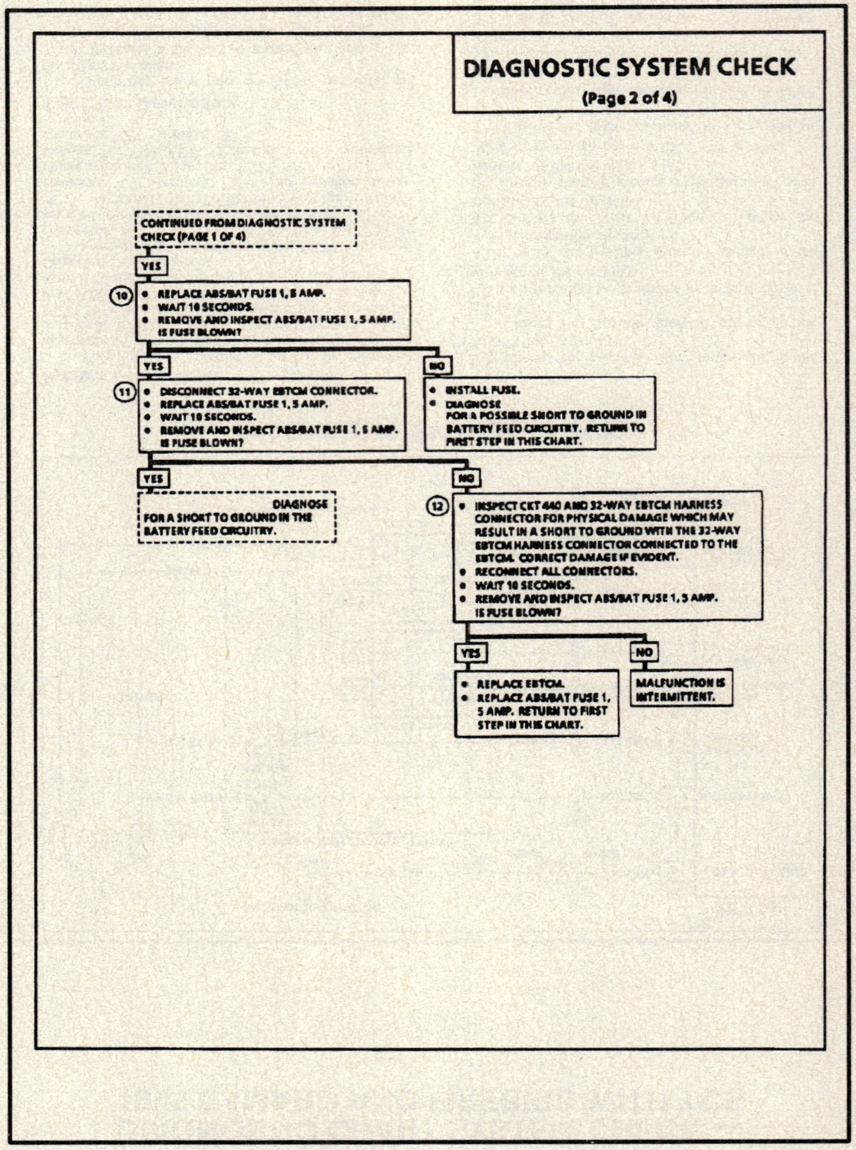

DIAGNOSTIC CHART SYSTEM CHECK — 1995 CAMARO AND FIREBIRD WITH TCS, CONTINUED

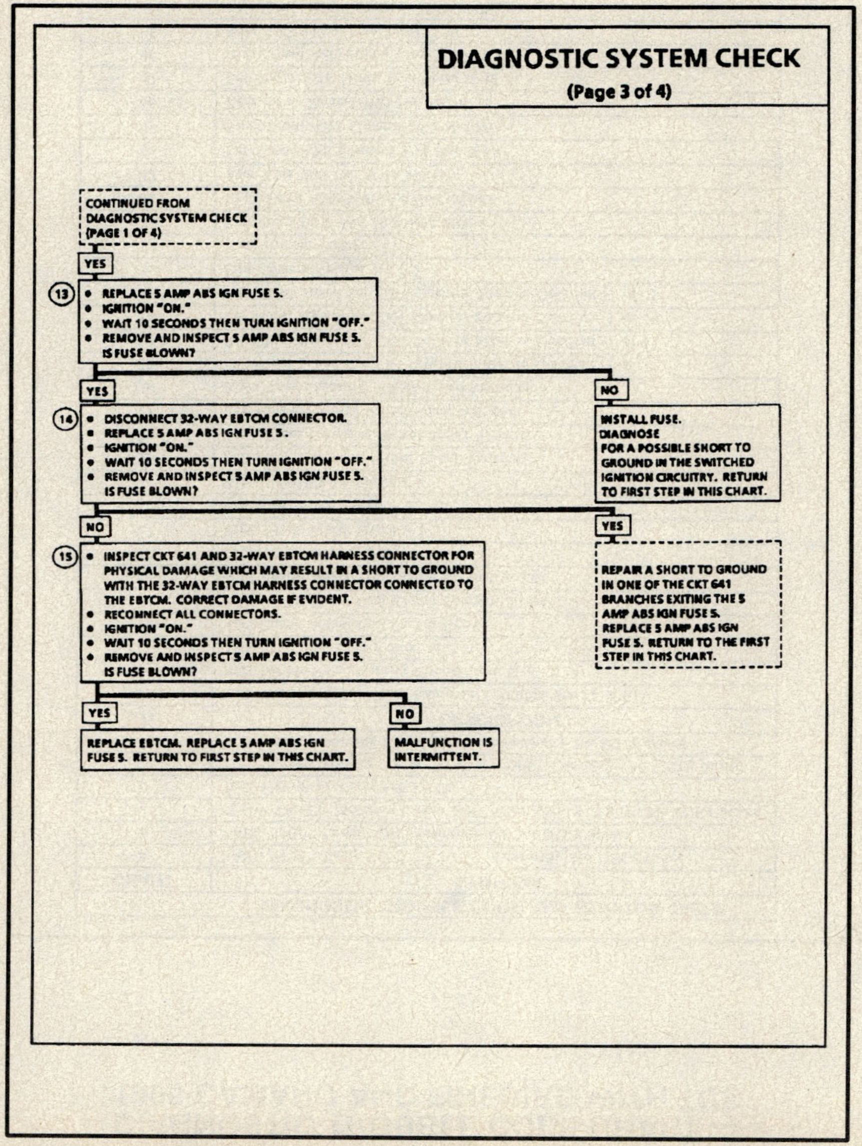

DIAGNOSTIC CHART SYSTEM CHECK — 1995 CAMARO AND FIREBIRD WITH TCS, CONTINUED

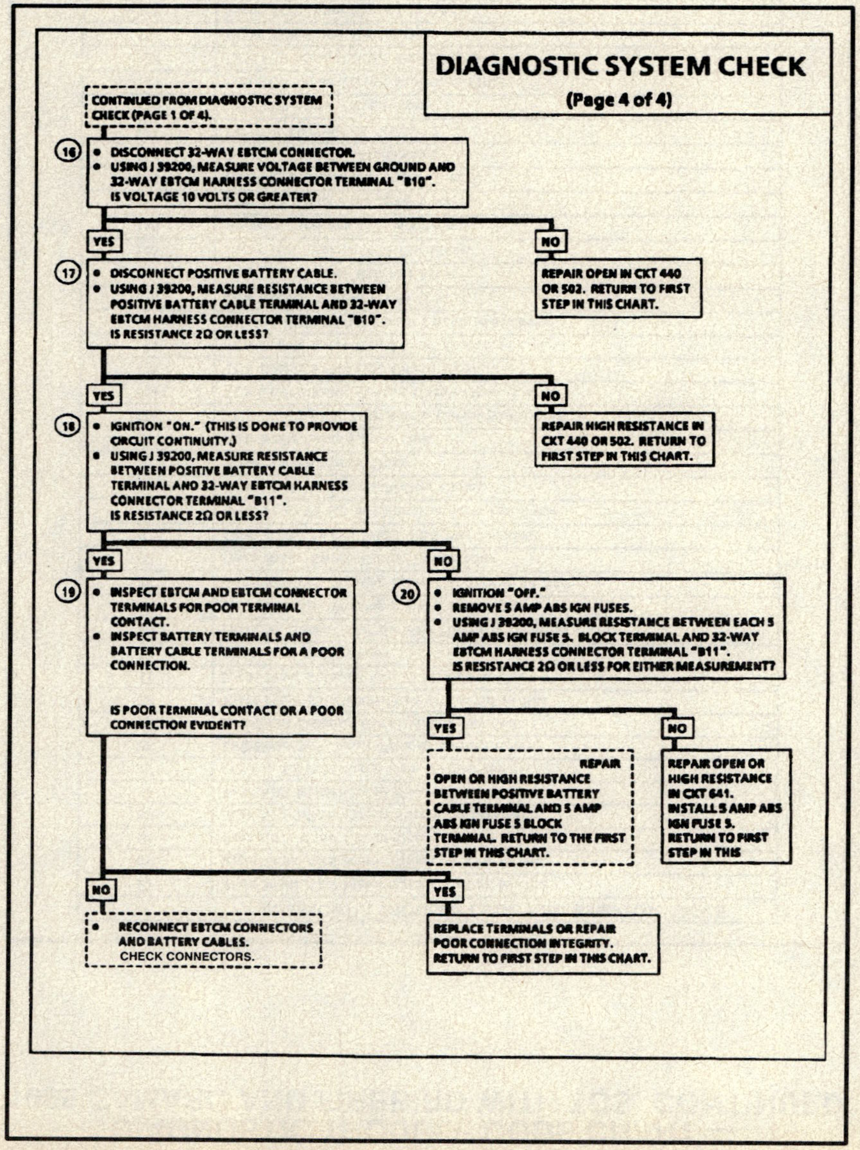

DIAGNOSTIC TROUBLE CODE CHART — 1995 CAMARO AND FIREBIRD WITH TCS

DIAGNOSTIC TROUBLE CODE AND SYMPTOM TABLE	
CHART	**SYMPTOM**
A	"ABS INOP" Lamp "ON" or "OFF" Constantly, No DTCs Stored
B	"ABS INOP" Lamp "ON" Intermittently, No DTCs Stored
C	"LOW TRAC" Indicator Lamp "ON" With Ignition "OFF", No DTCs Stored
D	Tech 1 Displays Undefined DTCs
E	TCS (Amber) Warning Lamp "ON" or "OFF" Constantly, No DTCs Stored
F	TCS (Amber) Warning Lamp "ON" Intermittently, No DTCs Stored
DTC	**DESCRIPTION**
11	ABS Warning or Low Trac Lamp CKT Malfunction (1 of 2)
14	ABS Enable Relay Contact Circuit Open (1 of 3)
15	ABS Rly Contact CKT Short to Batt or Always Closed
16	ABS Enable Relay Coil Circuit Open
17	ABS Enable Relay Coil Circuit Shorted to Gnd
18	ABS Enable Relay Coil Circuit Shorted to Batt
21	Left Front Wheel Speed = 0
22	Right Front Wheel Speed = 0
23	Left Rear Wheel Speed = 0
24	Right Rear Wheel Speed = 0
25	Left Front Excessive Wheel Speed Variation
26	Right Front Excessive Wheel Speed Variation
27	Left Rear Excessive Wheel Speed Variation
28	Right Rear Excessive Wheel Speed Variation
32	LF Wheel Speed Snr CKT Open or Short to Gnd/Batt (1 of 3)
33	RF Wheel Speed Snr CKT Open or Short to Gnd/Batt (1 of 3)
34	LR Wheel Speed Snr CKT Open or Short to Gnd/Batt (1 of 3)
35	RR Wheel Speed Snr CKT Open or Short to Gnd/Batt (1 of 3)
36	Low System Voltage (1 of 2)
37	High System Voltage
38	Left Front ESB Will Not Hold Motor
41	Right Front ESB Will Not Hold Motor
42	Rear ESB Will Not Hold Motor
44	Left Front ABS Channel Will Not Move
45	Right Front ABS Channel Will Not Move
46	Rear ABS Channel Will Not Move
47	Left Front ABS Motor Free Spins (1 of 2)
48	Right Front ABS Motor Free Spins (1 of 2)
51	Rear ABS Motor Free Spins (1 of 2)
52	Left Front ABS Channel in Release Too Long
53	Right Front ABS Channel in Release Too Long
54	Rear ABS Channel in Release Too Long
55	EBTCM Malfunction (ABS)
56	Left Front ABS Motor Circuit Open
57	Left Front ABS Motor Circuit Shorted to Gnd
58	Left Front ABS Motor Circuit Shorted to Batt
61	Right Front ABS Motor Circuit Open

DIAGNOSTIC TROUBLE CODE CHART — 1995 CAMARO AND FIREBIRD WITH TCS, CONTINUED

DIAGNOSTIC TROUBLE CODE AND SYMPTOM TABLE	
62	Right Front ABS Motor Circuit Shorted to Gnd
63	Right Front ABS Motor Circuit Shorted to Batt
64	Rear ABS Motor Circuit Open
65	Rear ABS Motor Circuit Shorted to Ground
66	Rear ABS Motor Circuit Shorted to Battery
75	PCM to EBTCM Serial Data Malfunction
76	Left Front Sol Circuit Open or Shorted to Batt
77	Left Front Solenoid Circuit Shorted to Gnd
78	Right Front Sol Circuit Open or shorted to Batt
81	Right Front Solenoid Circuit Shorted to Gnd
82	Calibration Malfunction
86	EBTCM Turned "ON" the Red "BRAKE" Warning Lamp
87	Red "BRAKE" Warning Lamp Circuit Open
88	Red "BRAKE" Warning Lamp Ckt Shorted to Batt
91	Open Brake Switch During Deceleration
92	Open Brake Switch When ABS Was Required
93	DTC 91 or 92 Set in Current or Prev. Ign. Cycle
94	Brake Switch Contacts Always Closed
95	Brake Switch Circuit Open
96	Rear Brake Lamp Circuit Open
111	TCS Warning Lamp or Active Circuit Malfunction
114	TCS Enable Relay Contact Circuit Open (1 of 3)
116	TCS Enable Relay Coil Circuit Open
118	TCS Enable Relay Coil Circuit Shorted to Batt
123	PCM to EBTCM TP Sensor Circuit Malfunction
136	Low System Voltage (TCS) (1 of 2)
144	Left TCS Channel Will Not Move
145	Right TCS Channel Will Not Move
146	Adjuster Assembly Will Not Move
147	Left TCS Motor Free Spins (1 of 2)
148	Right TCS Motor Free Spins (1 of 2)
154	Adjuster Assembly Malfunction
155	EBTCM Malfunction (TCS)
156	Left TCS Motor Circuit Open
157	Left TCS Motor Circuit Shorted to GND
158	Left TCS Motor Circuit Shorted to BATT
161	Right TCS Motor Circuit Open
162	Right TCS Motor Shorted to GND
163	Right TCS Motor Shorted to BATT
164	Adjuster Assembly Motor Circuit Open
165	Adjusted Assembly Motor Circuit Shorted to Gnd
166	Adjuster Assembly Motor Circuit Shorted to Batt

ELECTRICAL CONNECTOR TERMINAL — 1995 CAMARO AND FIREBIRD WITH TCS

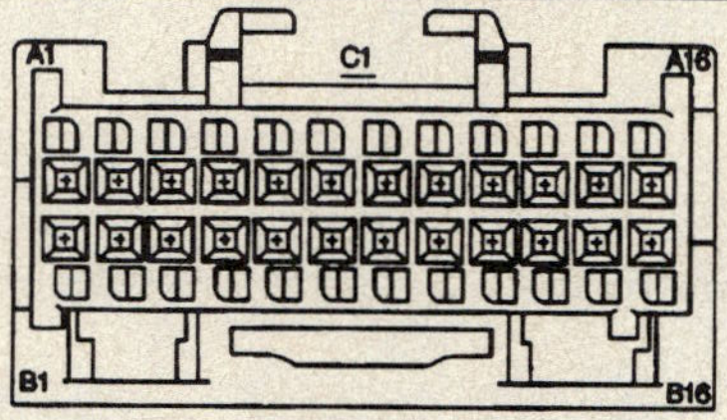

32 – WAY F MICRO – PACK 100 SERIES
DK RED

ELECTRONIC BRAKE/TRACTION CONTROL MODULE (EBTCM)

CONNECTOR TERMINAL	CIRCUIT NO.	COLOR	CIRCUIT DESCRIPTION
A1	884	BLK	LR WHEEL SPEED SIGNAL HIGH
A2	885	RED	LR WHEEL SPEED SIGNAL LOW
A3	883	TAN	RF WHEEL SPEED SIGNAL LOW
A4	872	DK GRN	RF WHEEL SPEED SIGNAL HIGH
A5	883	WHT	RR WHEEL SPEED SIGNAL LOW
A6	882	BRN	RR WHEEL SPEED SIGNAL HIGH
A7	873	YEL	LF WHEEL SPEED SIGNAL LOW
A8	830	LT BLU	LF WHEEL SPEED SIGNAL HIGH
A9	1289	LT BLU	RF ABS SOLENOID CONTROL
A10	PLUGGED		NOT USED
A11	800	TAN	SERIAL DATA LINE
A12	PLUGGED		NOT USED
A13	PLUGGED		NOT USED
A14	464	TAN/BLK	RECEIVED THROTTLE POSITION
A15	1571	BRN/WHT	TCS CONTROL SWITCH SIGNAL
A16	832	LT BLU	TRACTION CONTROL ACTIVE SIGNAL
B1	1288	DK GRN	LF ABS SOLENOID CONTROL
B2	867	LT GRN	ABS WARNING LAMP CONTROL
B3	1632	PNK	ABS RELAY CONTROL
B4	33	TAN/WHT	BRAKE WARNING LAMP CONTROL
B5	1537	DK BLU	ABS/TCS ACTIVE LAMP CONTROL
B6	PLUGGED		NOT USED
B7	PLUGGED		NOT USED
B8	PLUGGED		NOT USED
B9	PLUGGED		NOT USED
B10	440	ORN	BATTERY FEED
B11	641	BRN	SWITCHED IGNITION
B12	820	YEL	BRAKE SWITCH INPUT
B13	PLUGGED		NOT USED
B14	463	ORN/BLK	SPARK RETARD REQUEST
B15	1572	PPL/WHT	TCS WARNING LAMP CONTROL
B16	PLUGGED		NOT USED

ELECTRICAL CONNECTOR TERMINAL — 1995 CAMARO AND FIREBIRD WITH TCS, CONTINUED

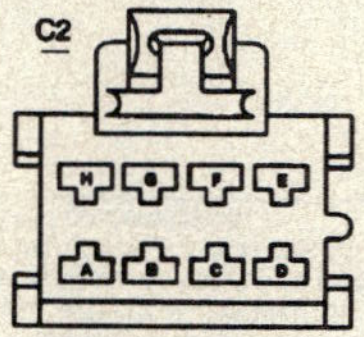

ELECTRONIC BRAKE/TRACTION CONTROL MODULE (EBTCM)

8 – WAY F METRI – PACK 280 SERIES
LR GRY

PIN	CIRCUIT NO.	COLOR	CIRCUIT
A	1280	BLK	LF ABS MOTOR HIGH
B	1281	PNK	LF ABS MOTOR LOW
C	1284	DK GRN	REAR ABS MOTOR HIGH
D	1633	RED	ABS SWITCHED BATTERY INPUT
E	851	BLK/WHT	GROUND
F	1285	ORN	REAR ABS MOTOR LOW
G	1282	PPL	RF ABS MOTOR HIGH
H	1283	BLK	RF ABS MOTOR LOW

ELECTRICAL CONNECTOR TERMINAL IDENTIFICATION — 1995 CAMARO AND FIREBIRD WITH TCS, CONTINUED

C3

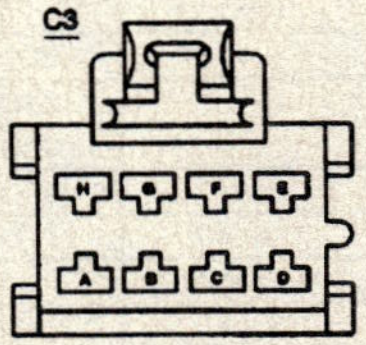

ELECTRONIC BRAKE/TRACTION CONTROL MODULE (EBTCM)

8-WAY F METRI-PACK 280 SERIES
BLK

PIN	CIRCUIT NO.	COLOR	CIRCUIT
A	566	LT GRN	RR TCS MOTOR HIGH
B	562	PPL	TCS ENABLE RELAY CONTROL
C	563	RED	TCS SWITCHED BATTERY INPUT
D	561	GRY	TCS ADJUSTER ASSEMBLY MOTOR LOW
E	560	WHT	TCS ADJUSTER ASSEMBLY MOTOR HIGH
F	564	LT BLU	LR TCS MOTOR HIGH
G	565	DK BLU	LR TCS MOTOR LOW
H	567	DK GRN	RR TCS MOTOR LOW

DIAGNOSTIC CHART A — 1995 CAMARO AND FIREBIRD WITH TCS

CHART A

ABS (AMBER) WARNING LAMP "ON" OR "OFF" CONSTANTLY, NO DTCs STORED

NOTICE: DIAGNOSTIC SYSTEM CHECK MUST BE COMPLETED BEFORE USING THIS CHART.

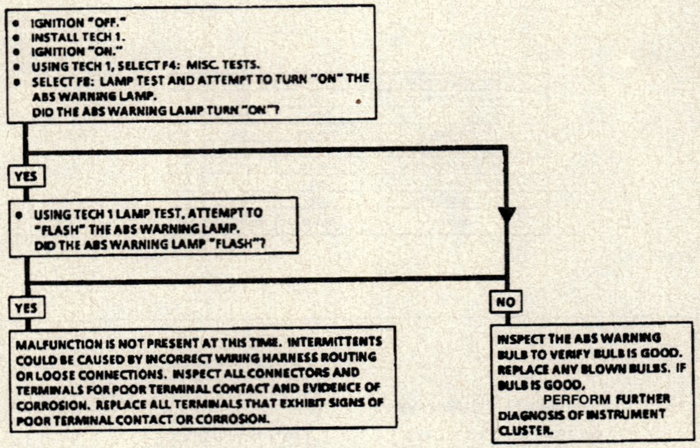

DIAGNOSTIC CHART B — 1995 CAMARO AND FIREBIRD WITH TCS

CHART B
ABS (AMBER) WARNING LAMP "ON" INTERMITTENTLY, NO DTCs STORED

NOTICE: DIAGNOSTIC SYSTEM CHECK MUST BE COMPLETED BEFORE USING THIS CHART.

- IGNITION "OFF."
- DISCONNECT GREY 8-WAY EBTCM CONNECTOR.
- DISCONNECT NEGATIVE BATTERY TERMINAL.
- DISCONNECT POSITIVE BATTERY TERMINAL.
- USING J 39200, MEASURE THE RESISTANCE BETWEEN THE NEGATIVE BATTERY CABLE TERMINAL AND TERMINAL "E" OF THE GRAY 8-WAY EBTCM HARNESS CONNECTOR.
- OBSERVE MEASURED RESISTANCE WHILE "WIGGLING" WIRE HARNESS AND CONNECTORS.

IS RESISTANCE ALWAYS 2Ω OR LESS?

NO: REPAIR INTERMITTENT OPEN OR HIGH RESISTANCE IN CKT 851.

YES:
- DISCONNECT 32-WAY EBTCM CONNECTOR.
- USING J 39200, MEASURE THE RESISTANCE BETWEEN POSITIVE BATTERY CABLE TERMINAL AND TERMINAL "B10" OF THE 32-WAY EBTCM HARNESS CONNECTOR.
- OBSERVE MEASURED RESISTANCE WHILE "WIGGLING" WIRE HARNESS AND CONNECTORS.

IS RESISTANCE ALWAYS 2Ω OR LESS?

NO: REPAIR INTERMITTENT OPEN OR HIGH RESISTANCE IN CKTs 502 OR 440.

YES:
- TURN IGNITION SWITCH "ON" (THIS IS DONE TO COMPLETE THE IGNITION CIRCUIT).
- USING J 39200, MEASURE THE RESISTANCE BETWEEN POSITIVE BATTERY CABLE TERMINAL AND TERMINAL "B11" OF THE 32-WAY EBTCM HARNESS CONNECTOR.
- OBSERVE MEASURED RESISTANCE WHILE "WIGGLING" WIRE HARNESS AND CONNECTORS.

IS RESISTANCE ALWAYS 2Ω OR LESS?

NO: REPAIR INTERMITTENT OPEN OR HIGH RESISTANCE IN CKT 641.

YES:
- CHECK POSITIVE BATTERY TERMINAL AND POSITIVE BATTERY CABLE TERMINAL FOR A POOR CONNECTION.
- INSPECT TERMINALS "B11" AND "B10" OF THE 32-WAY EBTCM CONNECTOR AND TERMINAL "E" OF THE GREY 8-WAY EBTCM CONNECTOR FOR POOR TERMINAL CONTACT AND EVIDENCE OF CORROSION.

IS THERE EVIDENCE OF POOR TERMINAL CONTACT OR CORROSION?

YES: REPLACE TERMINALS THAT EXHIBIT POOR TERMINAL CONTACT OR SIGNS OF CORROSION. REPAIR CONNECTION INTEGRITY IF POOR CONTACT EXISTS AT BATTERY.

NO:
- IGNITION "OFF."
- RECONNECT ALL CONNECTORS.
- IGNITION "ON."

IS ABS WARNING LAMP "ON" INTERMITTENTLY WITH NO DTCs SET?

YES: REPLACE EBTCM.

NO: MALFUNCTION IS NOT PRESENT AT THIS TIME. INTERMITTENTS MAY BE CAUSED BY INCORRECT WIRING HARNESS ROUTING OR LOOSE GROUND CONNECTIONS.

DIAGNOSTIC CHART C — 1995 CAMARO AND FIREBIRD WITH TCS

CHART C
ABS/TCS ACTIVE INDICATOR LAMP "ON" WITH IGNITION "OFF," NO DTCs STORED

NOTICE: DIAGNOSTIC SYSTEM CHECK MUST BE COMPLETED BEFORE USING THIS CHART.

- IGNITION "OFF."
- OBSERVE THE ABS/TCS ACTIVE WARNING LAMP.

IS THE ABS/TCS ACTIVE WARNING LAMP "ON"?

NO: MALFUNCTION IS NOT PRESENT AT THIS TIME. INTERMITTENTS COULD BE CAUSED BY INCORRECT WIRING HARNESS ROUTING OR LOOSE CONNECTIONS. INSPECT ALL CONNECTORS AND TERMINALS FOR POOR TERMINAL CONTACT AND EVIDENCE OF CORROSION. REPLACE ALL TERMINALS THAT EXHIBIT SIGNS OF POOR TERMINAL CONTACT OR CORROSION.

YES: REPAIR SHORT TO VOLTAGE ON CKT 1537.

DIAGNOSTIC CHART D — 1995 CAMARO AND FIREBIRD WITH TCS

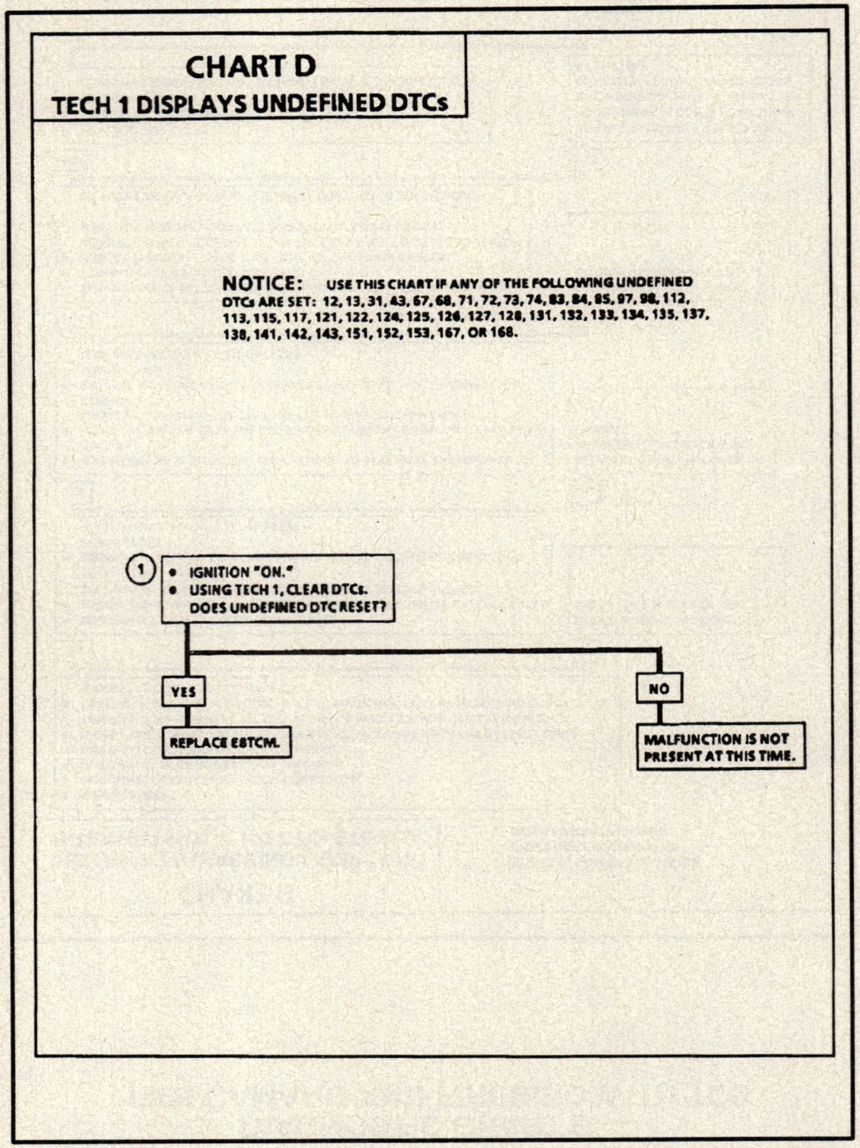

DIAGNOSTIC CHART E — 1995 CAMARO AND FIREBIRD WITH TCS

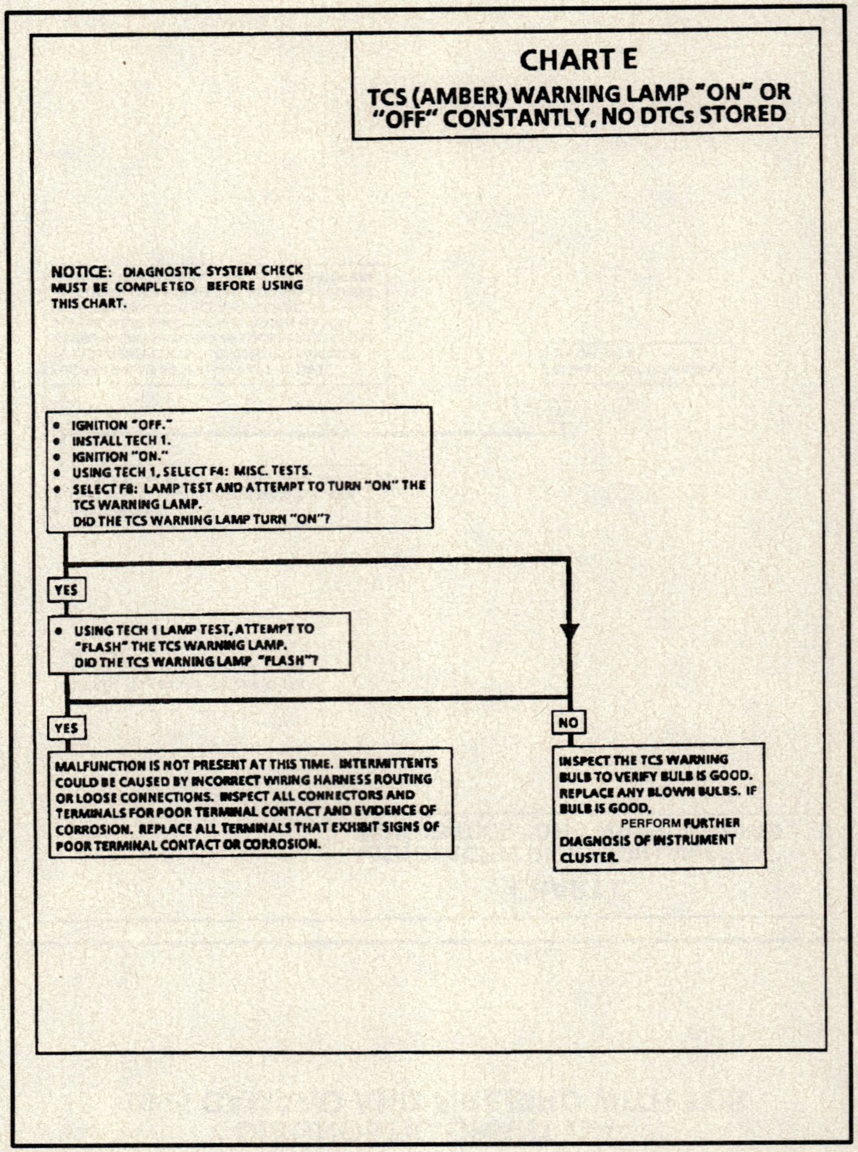

DIAGNOSTIC CHART F — 1995 CAMARO AND FIREBIRD WITH TCS

CHART F

TCS (AMBER) WARNING LAMP "ON" INTERMITTENTLY, NO DTCs STORED

NOTICE: DIAGNOSTIC SYSTEM CHECK MUST BE COMPLETED BEFORE USING THIS CHART.

- IGNITION "OFF."
- DISCONNECT GREY 8-WAY EBTCM CONNECTOR.
- DISCONNECT NEGATIVE BATTERY TERMINAL.
- DISCONNECT POSITIVE BATTERY TERMINAL.
- USING J 39200, MEASURE THE RESISTANCE BETWEEN THE NEGATIVE BATTERY CABLE TERMINAL AND TERMINAL "E" OF THE GREY 8-WAY EBTCM HARNESS CONNECTOR.
- OBSERVE MEASURED RESISTANCE WHILE "WIGGLING" WIRE HARNESS AND CONNECTORS.

IS RESISTANCE ALWAYS 2Ω OR LESS?

NO: REPAIR INTERMITTENT OPEN OR HIGH RESISTANCE IN CKT 651.

YES:

- DISCONNECT 32-WAY EBTCM CONNECTOR.
- USING J 39200, MEASURE THE RESISTANCE BETWEEN POSITIVE BATTERY CABLE TERMINAL AND TERMINAL "B10" OF THE 32-WAY EBTCM HARNESS CONNECTOR.
- OBSERVE MEASURED RESISTANCE WHILE "WIGGLING" WIRE HARNESS AND CONNECTORS.

IS RESISTANCE ALWAYS 2Ω OR LESS?

NO: REPAIR INTERMITTENT OPEN OR HIGH RESISTANCE IN CKTs 502 OR 440.

YES:

- TURN IGNITION SWITCH "ON" (THIS IS DONE TO COMPLETE THE IGNITION CIRCUIT).
- USING J 39200, MEASURE THE RESISTANCE BETWEEN POSITIVE BATTERY CABLE TERMINAL AND TERMINAL "B11" OF THE 32-WAY EBTCM HARNESS CONNECTOR.
- OBSERVE MEASURED RESISTANCE WHILE "WIGGLING" WIRE HARNESS AND CONNECTORS.

IS RESISTANCE ALWAYS 2Ω OR LESS?

NO: REPAIR INTERMITTENT OPEN OR HIGH RESISTANCE IN CKT 641.

YES:

- CHECK POSITIVE BATTERY TERMINAL AND POSITIVE BATTERY CABLE TERMINAL FOR A POOR CONNECTION.
- INSPECT TERMINALS "B11" AND "B10" OF THE 32-WAY EBTCM CONNECTOR AND TERMINAL "E" OF THE GREY 8-WAY EBTCM CONNECTOR FOR POOR TERMINAL CONTACT AND EVIDENCE OF CORROSION.

IS THERE EVIDENCE OF POOR TERMINAL CONTACT OR CORROSION?

YES: REPLACE TERMINALS THAT EXHIBIT POOR TERMINAL CONTACT OR SIGNS OF CORROSION. REPAIR CONNECTION INTEGRITY IF POOR CONTACT EXISTS AT BATTERY.

NO:

- IGNITION "OFF."
- RECONNECT ALL CONNECTORS.
- IGNITION "ON."

IS TCS WARNING LAMP "ON" INTERMITTENTLY WITH NO DTCs SET?

YES: REPLACE EBTCM.

NO: MALFUNCTION IS NOT PRESENT AT THIS TIME. INTERMITTENTS MAY BE CAUSED BY INCORRECT WIRING HARNESS ROUTING OR LOOSE GROUND CONNECTIONS.

DIAGNOSTIC CHART DTC 11 — 1995 CAMARO AND FIREBIRD WITH TCS

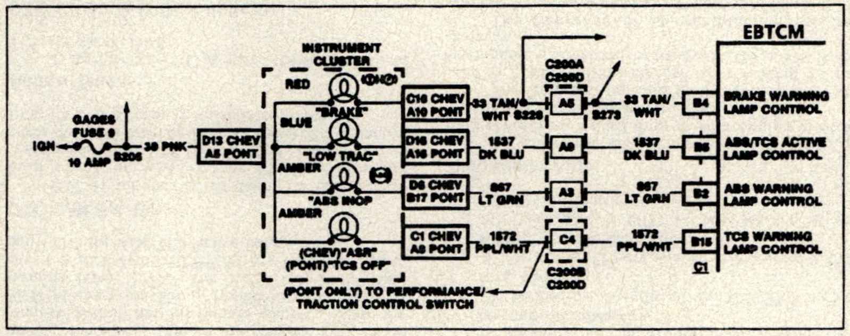

DTC 11
(Page 1 of 2)
ABS WARNING OR LOW TRAC LAMP CKT MALFUNCTION

Circuit Description:

This DTC checks the state of both the ABS warning lamp and the ABS/TCS active lamp to identify a situation in which the driver could not be alerted to a system malfunction or of possible slippery driving conditions. Due to an integral lamp driver module within the instrument panel cluster, the EBTCM must provide a ground to turn the ABS warning lamp "OFF." The ABS/TCS active lamp circuit is such that the EBTCM provides a ground to turn "ON" the ABS/TCS active lamp.

DTC Will Set When:

DTC 11 can only be set during the three second bulb check or when either of the lamps are commanded "ON." If the EBTCM cannot control either the ABS warning lamp or the ABS/TCS active lamp for two seconds, a malfunction exists.

Action Taken:

A malfunction DTC is stored; however, ABS or TCS are not disabled.

DTC Chart Test Description:

Number(s) below refer to circled number(s) on the diagnostic chart.

1. Determines if the ABS warning lamp can be properly controlled.
2. Determines if the lamp is "ON" constantly.
3. Checks for other DTCs which could turn "ON" the ABS warning lamp.
4. Determines if the ABS warning lamp can be turned "OFF" manually.
5. Checks the integrity of the ABS warning lamp circuitry.
6. Checks for a short to voltage on the ABS warning lamp circuitry.
7. Determines if a malfunctioning EBTCM is the cause of DTC 11.
8. Verifies that the other I/P indicator lamps function properly.
9. Determines if the ABS warning lamp can be turned "ON" manually.
10. Verifies condition of Fuse 9.
11. Determines if DTC 11 was set due to an EBTCM malfunction.
12. Checks for a short to ground in the ABS warning lamp circuitry.

Diagnostic Aids:

An intermittent malfunction may be caused by a poor connection, rubbed through wire insulation, or a wire that is broken inside the insulation.

The lamp test function of the Tech 1 may be used to command the lamp on" while looking for an intermittent malfunction in the ABS warning lamp circuitry.

The frequency of the malfunction can be checked by using the enhanced diagnostic function of the Tech 1.

Any circuitry that is suspected of causing the intermittent complaint should be thoroughly checked for backed out terminals, improper mating, broken locks, improperly formed or damaged terminals, poor terminal to wiring connections or physical damage to the wiring harness.

DIAGNOSTIC CHART DTC 11 — 1995 CAMARO AND FIREBIRD WITH TCS, CONTINUED

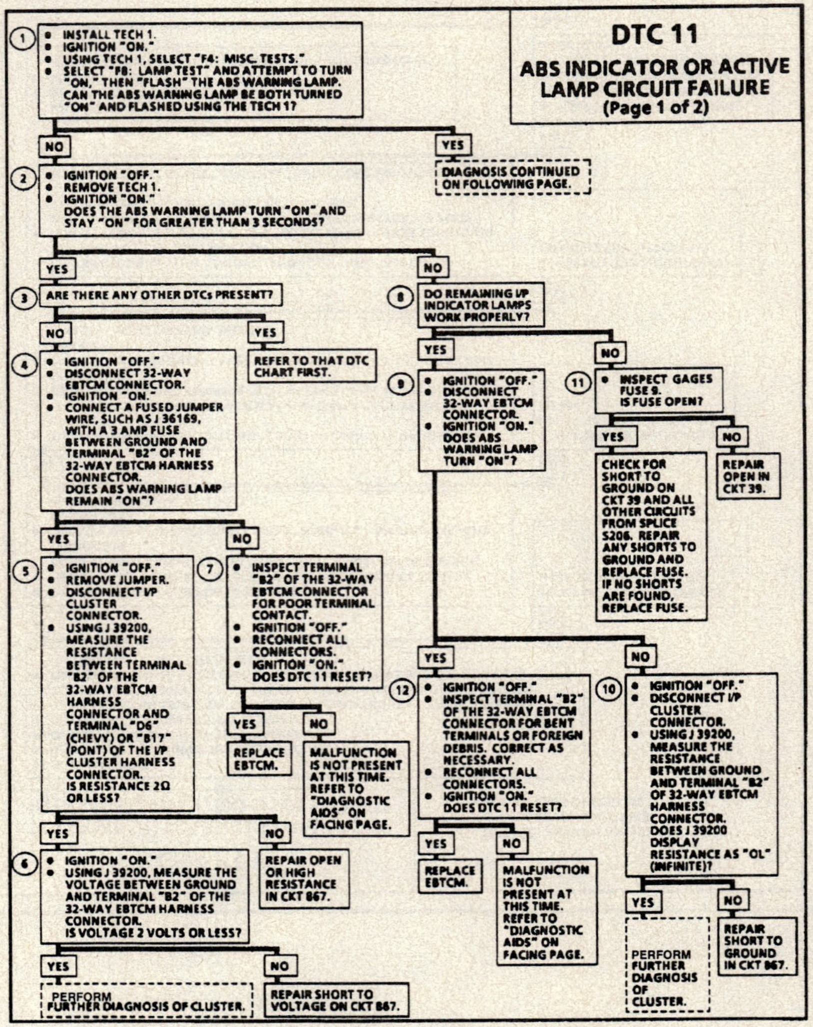

DIAGNOSTIC CHART DTC 11 — 1995 CAMARO AND FIREBIRD WITH TCS, CONTINUED

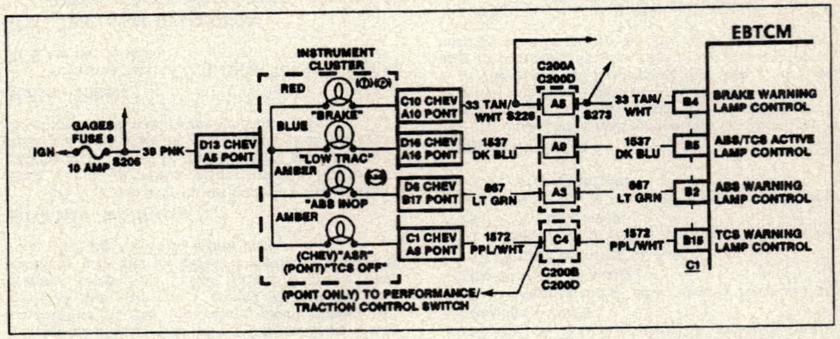

DTC 11
(Page 2 of 2)
ABS WARNING OR LOW TRAC LAMP CKT MALFUNCTION

Circuit Description:

This DTC checks the state of both the ABS warning lamp and the ABS/TCS active lamp to identify a situation in which the driver could not be alerted to a system malfunction or of possible slippery driving conditions. Due to an integral lamp driver module within the instrument panel cluster, the EBTCM must provide a ground to turn the ABS warning lamp "OFF." The ABS/TCS active lamp circuit is such that the EBTCM provides a ground to turn "ON" the ABS/TCS active lamp.

DTC Will Set When:

DTC 11 can only be set during the three second bulb check or when either of the lamps are commanded "ON." If the EBTCM cannot control either the ABS warning lamp or the ABS/TCS active lamp for two seconds, a malfunction exists.

Action Taken:

A malfunction DTC is stored; however, ABS or TCS are not disabled.

DTC Chart Test Description:

Number(s) below refer to circled number(s) on the diagnostic chart.

13. Determines if the ABS/TCS active lamp can be properly controlled.
14. Determines if the ABS/TCS active lamp is "ON" constantly.
15. Checks for a short to ground in the ABS/TCS active lamp circuitry.
16. Determines if DTC 11 was set due to a EBTCM malfunction.
17. Determines if DTC 11 was set due to a EBTCM malfunction.
18. Determines if the ABS/TCS active lamp can be controlled manually.
19. Verifies the integrity of the ABS/TCS active lamp circuitry.
20. Determines if there is a short to voltage in the ABS/TCS active lamp circuitry.
21. Determines if DTC 11 was set due to an EBTCM malfunction.

Diagnostic Aids:

An intermittent malfunction may be caused by a poor connection, rubbed through wire insulation, or a wire that is broken inside the insulation.

The lamp test function of the Tech 1 may be used to command the lamp on" while looking for an intermittent malfunction in the ABS/TCS active lamp circuitry.

The frequency of the malfunction can be checked by using the enhanced diagnostic function of the Tech 1.

Any circuitry that is suspected of causing the intermittent complaint should be thoroughly checked for backed out terminals, improper mating, broken locks, improperly formed or damaged terminals, poor terminal to wiring connections or physical damage to the wiring harness.

DIAGNOSTIC CHART DTC 11 — 1995 CAMARO AND FIREBIRD WITH TCS, CONTINUED

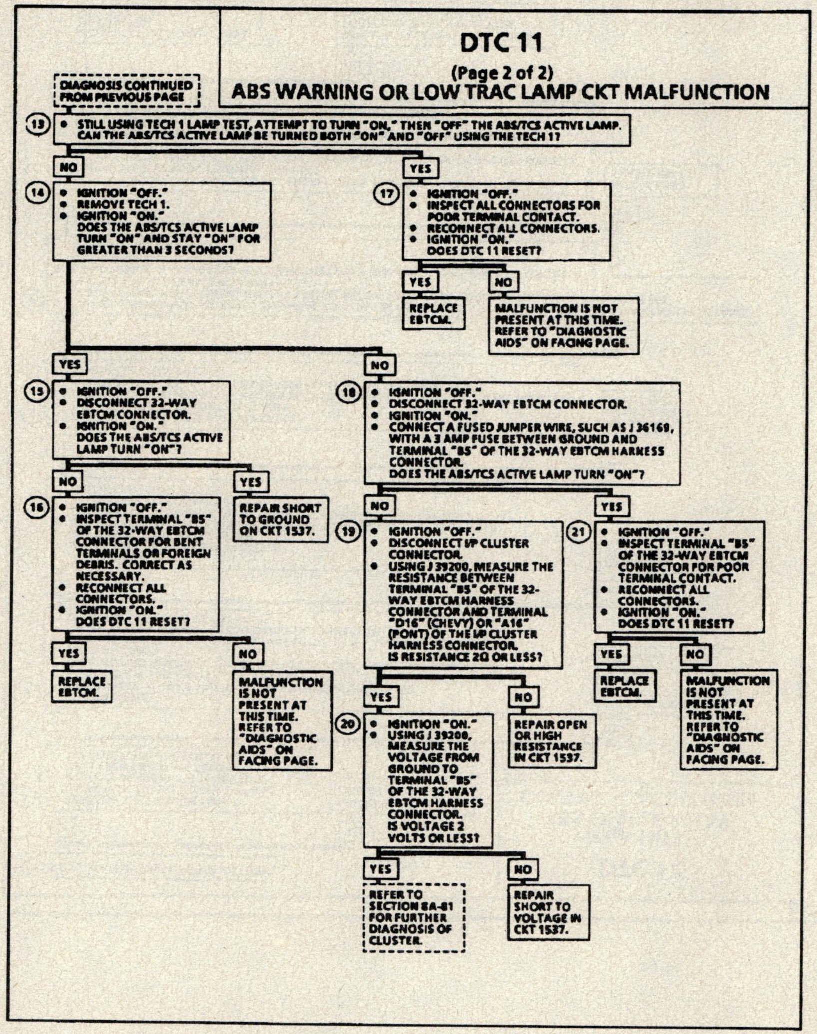

DIAGNOSTIC CHART DTC 14 — 1995 CAMARO AND FIREBIRD WITH TCS

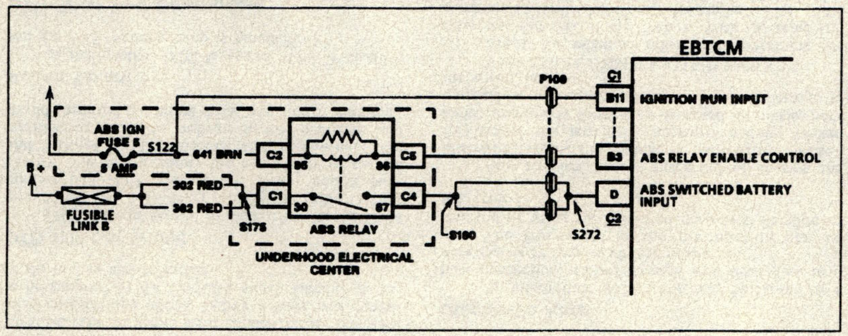

DTC 14
(Page 1 of 3)
ABS ENABLE RELAY CONTACT CIRCUIT OPEN

Circuit Description:

Ignition voltage is supplied through terminal "C2" of the ABS relay. The EBTCM then is able to energize the pull-in coil by completing the ground circuit at connector "C1", terminal "B3" of the EBTCM. The magnetic field created closes the ABS relay contacts and allows battery voltage and current to be supplied to the EBTCM, which supplies power to the motors and solenoids.

DTC Will Set When:

DTC 14 can be set anytime after the EBTCM commands the ABS relay "ON" (the relay is first commanded "ON" during the three second bulb check). This test monitors the availability of current and voltage to the motors and solenoids. This malfunction indicates voltage is not available and would therefore not allow ABS operation if required.

Action Taken:

A malfunction DTC is stored, ABS is disabled and the ABS warning lamp is turned "ON."

DTC Chart Test Description:

Number(s) below refer to circled number(s) on the diagnostic chart.

1. Checks to see if DTC 16 is also set as current or history. If DTC 16 is also set, proceed to that chart first.
2. Verifies that the malfunction is currently present.
3. Verifies that the relay can be turned "ON" and provide voltage to the EBTCM.
4. Checks for high resistance in the switched battery circuit from the relay to the EBTCM.
5. Checks for high resistance in the switched battery circuit from the battery to the relay.
6. Checks for high resistance in the relay contacts.
7. Checks for poor terminal contact or corrosion at the connectors.
8. Determines if the EBTCM is the cause of the malfunction.

Diagnostic Aids:

An intermittent malfunction may be caused by a poor connection, rubbed through wire insulation, or a wire that is broken inside the insulation.

The frequency of the malfunction can be checked by using the enhanced diagnostic function of the Tech 1.

Any circuitry that is suspected of causing the intermittent complaint should be thoroughly checked for backed out terminals, improper mating, broken locks, improperly formed or damaged terminals, poor terminal to wiring connections or physical damage to the wiring harness.

VIBRATION, TEMPERATURE EFFECTS:

Check for vibration effects by performing the relay test function of the Tech 1. With the relay test commanded "ON," lightly tap the top and sides of the relay while monitoring relay voltage. If the relay voltage changes significantly, replace the relay.

If DTC 14 only sets when the vehicle is initially started in cold ambient conditions (temperature less than 32° F-0° C), replace the relay.

DIAGNOSTIC CHART DTC 14 — 1995 CAMARO AND FIREBIRD WITH TCS, CONTINUED

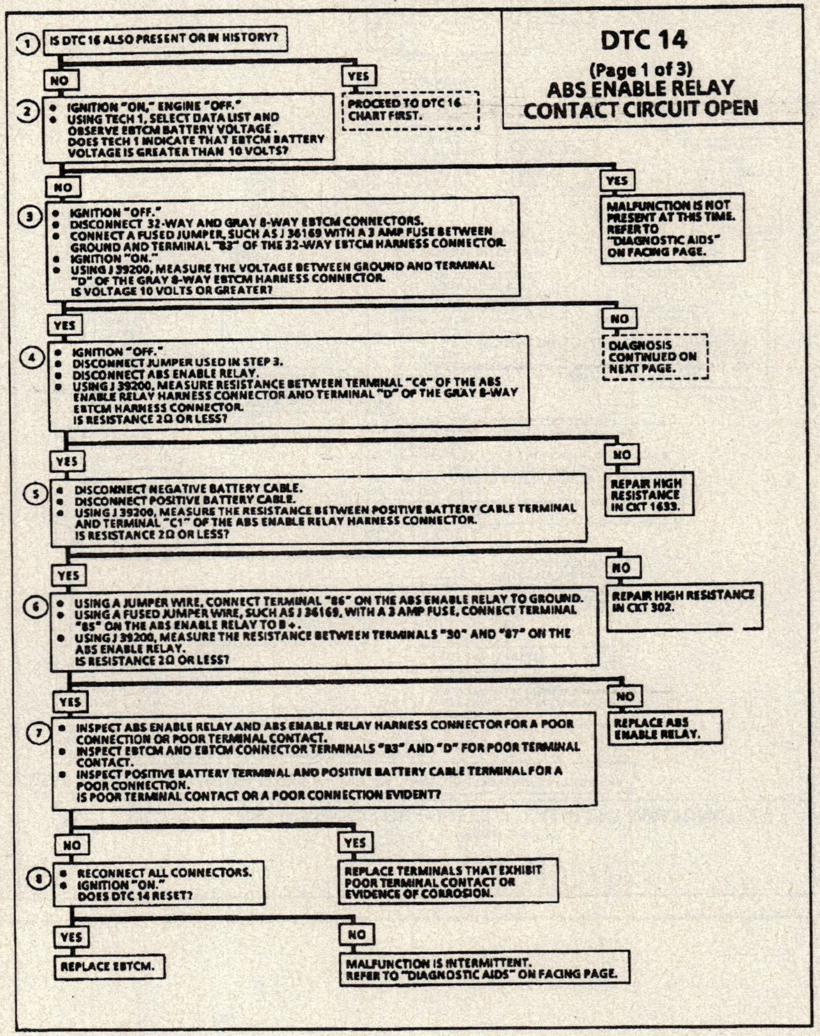

DIAGNOSTIC CHART DTC 14 — 1995 CAMARO AND FIREBIRD WITH TCS, CONTINUED

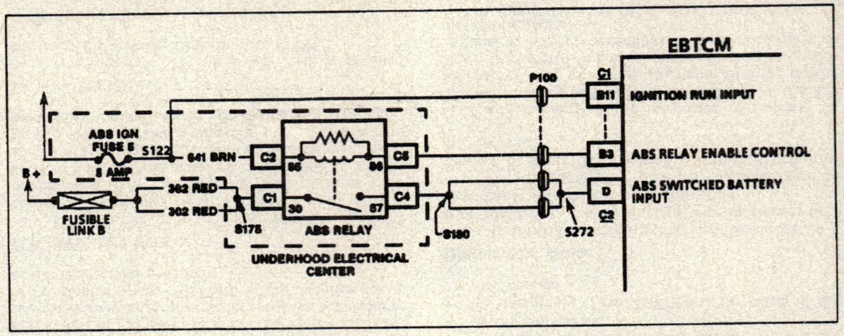

DTC 14
(Page 2 of 3)
ABS ENABLE RELAY CONTACT CIRCUIT OPEN

Circuit Description:

Ignition voltage is supplied through terminal "C2" of the ABS relay. The EBTCM then is able to energize the pull-in coil by completing the ground circuit at connector "C1", terminal "B3" of the EBTCM. The magnetic field created closes the ABS relay contacts and allows battery voltage and current to be supplied to the EBTCM, which supplies power to the motors and solenoids.

DTC Will Set When:

DTC 14 can be set anytime after the EBTCM commands the ABS relay "ON" (the relay is first commanded "ON" during the three second bulb check). This test monitors the availability of current and voltage to the motors and solenoids. This malfunction indicates voltage is not available and would therefore not allow ABS operation if required.

Action Taken:

A malfunction DTC is stored, ABS is disabled and the ABS warning lamp is turned "ON."

DTC Chart Test Description:

Number(s) below refer to circled number(s) on the diagnostic chart.

9. Checks for high resistance in the entire relay coil circuit.
10. Verifies proper battery condition.
11. Verifies proper voltage is available at the ABS relay.
12. Checks for an open in the switched battery circuit from the relay to the EBTCM.
13. Verifies the condition of the ABS relay.

Diagnostic Aids:

An intermittent malfunction may be caused by a poor connection, rubbed through wire insulation, or a wire that is broken inside the insulation.

The frequency of the malfunction can be checked by using the enhanced diagnostic function of the Tech 1,

Any circuitry that is suspected of causing the intermittent complaint should be thoroughly checked for backed out terminals, improper mating, broken locks, improperly formed or damaged terminals, poor terminal to wiring connections or physical damage to the wiring harness.

VIBRATION, TEMPERATURE EFFECTS:

Check for vibration effects by performing the relay test function of the Tech 1. With the relay test commanded "ON," lightly tap the top and sides of the relay while monitoring relay voltage. If the relay voltage changes significantly, replace the relay.

If DTC 14 only sets when the vehicle is initially started in cold ambient conditions (temperature less than 32° F-0° C), replace the relay.

DIAGNOSTIC CHART DTC 14 — 1995 CAMARO AND FIREBIRD WITH TCS, CONTINUED

DTC 14
(Page 2 of 3)
ABS ENABLE RELAY CONTACT CIRCUIT OPEN

DIAGNOSIS CONTINUED FROM PREVIOUS PAGE.

(9)
- IGNITION "OFF."
- REMOVE JUMPER WIRE INSTALLED IN STEP 3, IF NOT ALREADY REMOVED.
- DISCONNECT NEGATIVE BATTERY CABLE.
- DISCONNECT POSITIVE BATTERY CABLE.
- WITH BATTERY STILL DISCONNECTED, TURN IGNITION SWITCH TO "RUN." THIS COMPLETES THE IGNITION CIRCUIT TO THE ABS ENABLE RELAY.
- USING J 39200, MEASURE THE RESISTANCE BETWEEN TERMINAL "B3" OF THE 32-WAY EBTCM HARNESS CONNECTOR AND THE POSITIVE BATTERY CABLE TERMINAL. IS RESISTANCE 95Ω OR LESS WHEN MEASURED AT APPROXIMATELY 20 °C (68°F)?

NO: DIAGNOSIS CONTINUED ON NEXT PAGE.

YES:

(10)
- USING J 39200, MEASURE THE VOLTAGE BETWEEN THE NEGATIVE AND POSITIVE BATTERY TERMINALS. IS VOLTAGE 10 VOLTS OR GREATER?

NO: REFER TO SECTION 6D3 TO CORRECT LOW VOLTAGE CONDITION.

YES:

(11)
- IGNITION "OFF."
- RECONNECT POSITIVE BATTERY CABLE.
- RECONNECT NEGATIVE BATTERY CABLE.
- DISCONNECT ABS ENABLE RELAY.
- USING J 39200, MEASURE THE VOLTAGE BETWEEN TERMINAL "C1" OF THE ABS ENABLE RELAY HARNESS CONNECTOR AND GROUND. IS VOLTAGE 10 VOLTS OR GREATER?

NO: REPAIR OPEN OR HIGH RESISTANCE IN CKT 302.

YES:

(12)
- USING J 39200, MEASURE THE RESISTANCE BETWEEN TERMINAL "C4" OF THE ABS ENABLE RELAY HARNESS CONNECTOR AND TERMINAL "D" OF THE GRAY 8-WAY EBTCM HARNESS CONNECTOR. IS RESISTANCE 2Ω OR LESS?

NO: REPAIR OPEN OR HIGH RESISTANCE IN CKT 1633.

YES:

(13)
- USING A JUMPER WIRE, CONNECT TERMINAL "86" ON THE ABS ENABLE RELAY TO GROUND.
- USING A FUSED JUMPER WIRE, SUCH AS J 36169, WITH A 3 AMP FUSE, CONNECT TERMINAL "85" ON THE ABS ENABLE RELAY TO B+.
- USING J 39200, MEASURE THE RESISTANCE BETWEEN TERMINALS "30" AND "87" ON THE ABS ENABLE RELAY. IS RESISTANCE 2Ω OR LESS?

NO: REPLACE ABS ENABLE RELAY.

YES: MALFUNCTION WAS MOST LIKELY DUE TO A POOR CONNECTION. INSPECT ALL CONNECTORS AND TERMINALS FOR POOR TERMINAL CONTACT AND EVIDENCE OF CORROSION. REPLACE ALL TERMINALS THAT EXHIBIT EVIDENCE OF POOR TERMINAL CONTACT OR CORROSION.

DIAGNOSTIC CHART DTC 14 — 1995 CAMARO AND FIREBIRD WITH TCS, CONTINUED

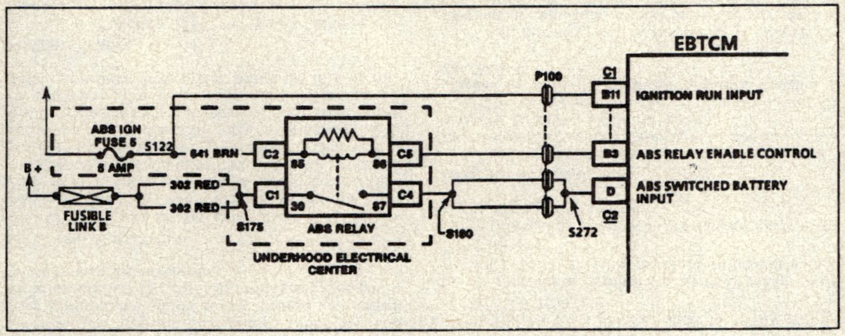

DTC 14
(Page 3 of 3)
ABS ENABLE RELAY CONTACT CIRCUIT OPEN

Circuit Description:

Ignition voltage is supplied through terminal "C2" of the ABS relay. The EBTCM then is able to energize the pull-in coil by completing the ground circuit at connector "C1", terminal "B3" of the EBTCM. The magnetic field created closes the ABS relay contacts and allows battery voltage and current to be supplied to the EBTCM, which supplies power to the motors and solenoids.

DTC Will Set When:

DTC 14 can be set anytime after the EBTCM commands the ABS relay "ON" (the relay is first commanded "ON" during the three second bulb check). This test monitors the availability of current and voltage to the motors and solenoids. This malfunction indicates voltage is not available and would therefore not allow ABS operation if required.

Action Taken:

A malfunction DTC is stored, ABS is disabled and the ABS warning lamp is turned "ON."

DTC Chart Test Description:

Number(s) below refer to circled number(s) on the diagnostic chart.

14. Checks for proper resistance of the relay coil.
15. Checks for high resistance in the ABS enable relay control circuit between the relay and the EBTCM.
16. Checks for high resistance in the ABS enable relay control circuit between the battery and the relay.

Diagnostic Aids:

An intermittent malfunction may be caused by a poor connection, rubbed through wire insulation, or a wire that is broken inside the insulation.

The frequency of the malfunction can be checked by using the enhanced diagnostic function of the Tech 1,

Any circuitry that is suspected of causing the intermittent complaint should be thoroughly checked for backed out terminals, improper mating, broken locks, improperly formed or damaged terminals, poor terminal to wiring connections or physical damage to the wiring harness.

VIBRATION, TEMPERATURE EFFECTS:

Check for vibration effects by performing the relay test function of the Tech 1. With the relay test commanded "ON," lightly tap the top and sides of the relay while monitoring relay voltage. If the relay voltage changes significantly, replace the relay.

If DTC 14 only sets when the vehicle is initially started in cold ambient conditions (temperature less than 32° F-0° C), replace the relay.

DIAGNOSTIC CHART DTC 14 — 1995 CAMARO AND FIREBIRD WITH TCS, CONTINUED

DTC 14
(Page 3 of 3)
ABS ENABLE RELAY CONTACT CIRCUIT OPEN

DIAGNOSIS CONTINUED FROM PREVIOUS PAGE.

(14)
- DISCONNECT ABS ENABLE RELAY.
- USING J 39200, MEASURE THE RESISTANCE BETWEEN TERMINALS "85" AND "86" ON THE ABS ENABLE RELAY. IS RESISTANCE 95Ω OR LESS WHEN MEASURED AT APPROXIMATELY 24°C (75°F)?

NO → REPLACE ABS ENABLE RELAY.

YES ↓

(15)
- USING J 39200, MEASURE THE RESISTANCE BETWEEN TERMINAL "C5" OF THE ABS ENABLE RELAY HARNESS CONNECTOR AND TERMINAL "B3" OF THE 32-WAY EBTCM HARNESS CONNECTOR. IS RESISTANCE 2Ω OR LESS?

NO → REPAIR OPEN OR HIGH RESISTANCE IN CKT 1632.

YES ↓

(16)
- USING J 39200, MEASURE THE RESISTANCE BETWEEN TERMINAL "C2" OF THE ABS ENABLE RELAY HARNESS CONNECTOR AND THE POSITIVE BATTERY CABLE TERMINAL. IS RESISTANCE 2Ω OR LESS?

NO → REPAIR OPEN OR HIGH RESISTANCE IN CKT 641, CKT 300, IGNITION SWITCH, OR BATTERY FEED CIRCUIT TO IGNITION SWITCH. REFER TO SECTION 8A FOR FURTHER INFORMATION.

YES ↓

MALFUNCTION WAS MOST LIKELY DUE TO A POOR CONNECTION. INSPECT ALL CONNECTORS AND TERMINALS FOR POOR TERMINAL CONTACT AND EVIDENCE OF CORROSION.
REPLACE ALL TERMINALS THAT EXHIBIT EVIDENCE OF POOR TERMINAL CONTACT OR CORROSION.

DIAGNOSTIC CHART DTC 15 — 1995 CAMARO AND FIREBIRD WITH TCS

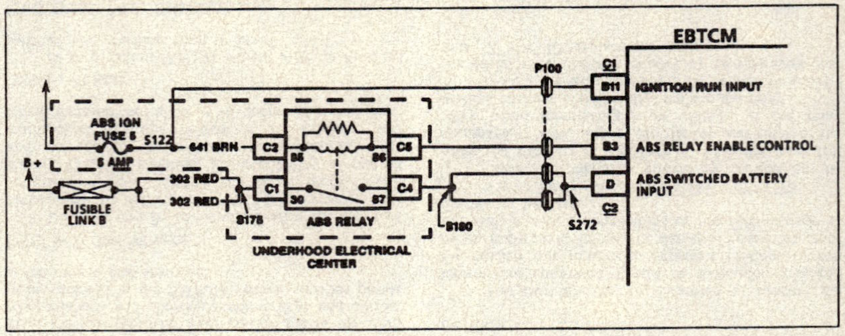

DTC 15
ABS RLY CONTACT CKT SHORT TO BATT OR ALWAYS CLOSED

Circuit Description:

Ignition voltage is supplied through terminal "C2" of the ABS relay. The EBTCM then is able to energize the pull-in coil by completing the ground circuit at connector "C1", terminal "B3" of the EBTCM. The magnetic field created closes the ABS relay contacts and allows battery voltage and current to be supplied to the EBTCM, which supplies power to the motors and solenoids.

DTC Will Set When:

DTC 15 can be set only before the EBTCM commands the relay "ON." This test determines if the ABS relay is energized when it should not be. This malfunction would not allow the ABS relay to remove power to the EBTCM. If a second malfunction were to occur that requires the ABS relay to be turned "OFF," that malfunction cannot be removed if the relay cannot be controlled. The malfunction must be present for three consecutive drive cycles before the DTC is set.

Action Taken:

A malfunction DTC is stored. ABS is not disabled.

DTC Chart Test Description:

Number(s) below refer to circled number(s) on the diagnostic chart.

1. Determines if any other DTCs are present.
2. Indicates that the EBTCM is capable of controlling the ABS relay as commanded.
3. Checks for a possible short to battery voltage in CKT 1633.
4. Determines whether the short to voltage is in CKT 1633 or as a result of a malfunctioning relay.
5. Checks for a possible intermittent malfunction in wiring or connectors.

Diagnostic Aids:

An intermittent malfunction may be caused by a poor connection, rubbed through wire insulation, or a wire that is broken inside the insulation.

The frequency of the malfunction can be checked by using the enhanced diagnostic function of the Tech 1.

Any circuitry, that is suspected of causing the intermittent complaint, should be thoroughly checked for backed out terminals, improper mating, broken locks, improperly formed or damaged terminals, poor terminal to wiring connections or physical damage to the wiring harness.

DIAGNOSTIC CHART DTC 15 — 1995 CAMARO AND FIREBIRD WITH TCS, CONTINUED

DTC 15
ABS RLY CONTACT CKT SHORT TO BATT OR ALWAYS CLOSED

(1) ARE ANY OTHER DTCs SET?

- YES: PROCEED TO THAT DTC CHART FIRST.
- NO:

(2)
- IGNITION "ON," ENGINE "OFF."
- USING THE RELAY TEST FUNCTION OF THE TECH 1, COMMAND ABS ENABLE RELAY "OFF." DOES TECH 1 INDICATE THAT ABS ENABLE RELAY IS "OFF" AND BATTERY VOLTAGE IS LESS THAN 5 VOLTS?

- YES: MALFUNCTION IS NOT PRESENT AT THIS TIME. REFER TO "DIAGNOSTIC AIDS" ON FACING PAGE.
- NO:

(3)
- IGNITION "OFF."
- DISCONNECT 32-WAY AND GRAY 8-WAY EBTCM HARNESS CONNECTORS.
- IGNITION "ON."
- USING J 39200, MEASURE THE VOLTAGE BETWEEN GROUND AND TERMINAL "D" OF THE GRAY 8-WAY EBTCM CONNECTOR. IS VOLTAGE 2 VOLTS OR LESS?

- YES:

(5)
- IGNITION "OFF."
- INSPECT CKT 1633 AND THE GRAY 8-WAY EBTCM HARNESS CONNECTOR FOR DAMAGE WHICH MAY RESULT IN A SHORT TO VOLTAGE WHEN ALL CONNECTORS ARE CONNECTED. REPAIR DAMAGE IF EVIDENT.
- RECONNECT GRAY 8-WAY AND 32-WAY EBTCM CONNECTORS.
- IGNITION "ON."
- USING RELAY TEST FUNCTION OF TECH 1, COMMAND ABS ENABLE RELAY "OFF." DOES TECH 1 INDICATE THAT ABS ENABLE RELAY IS "OFF" AND BATTERY VOLTAGE IS LESS THAN 5 VOLTS?

  - NO: REPLACE EBTCM.
  - YES: MALFUNCTION IS INTERMITTENT. REFER TO "DIAGNOSTIC AIDS" ON FACING PAGE.

- NO:

(4)
- IGNITION "OFF."
- DISCONNECT ABS ENABLE RELAY.
- IGNITION "ON."
- USING J 39200, MEASURE THE VOLTAGE BETWEEN GROUND AND TERMINAL "D" OF THE GRAY 8-WAY EBTCM HARNESS CONNECTOR. IS VOLTAGE 2 VOLTS OR LESS?

  - YES: REPLACE ABS ENABLE RELAY.
  - NO: REPAIR SHORT TO VOLTAGE IN CKT 1633.

AFTER ALL DIAGNOSIS IS COMPLETE, CLEAR DTCs, AND TEST DRIVE VEHICLE FOR 3 DRIVE CYCLES TO VERIFY THAT DTC DOES NOT RESET. A DRIVE CYCLE CONSISTS OF STARTING THE VEHICLE, DRIVING ABOVE 16 km/h (10 MPH), AND THEN TURNING KEY "OFF."

DIAGNOSTIC CHART DTC 16 — 1995 CAMARO AND FIREBIRD WITH TCS

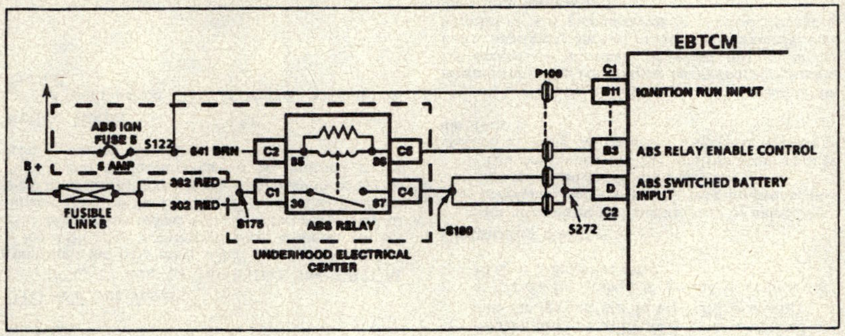

DTC 16
ABS ENABLE RELAY COIL CIRCUIT OPEN

Circuit Description:

Ignition voltage is supplied through terminal "C2" of the ABS relay. The EBTCM then is able to energize the pull-in coil by completing the ground circuit at connector "C1", terminal "B3" of the EBTCM. The magnetic field created closes the ABS relay contacts and allows battery voltage and current to be supplied to the EBTCM, which supplies power to the motors and solenoids.

DTC Will Set When:

DTC 16 can be set only after DTC 14 has been set. This test detects an open in the ABS enable relay coil circuit. An open in this circuit will not allow the ABS relay to be energized thus preventing voltage and current to the motors and solenoids. If this malfunction is present and the ignition is turned "OFF" before 5 km/h (3 mph) is reached, DTC 14 is set alone.

Action Taken:

A malfunction DTC is stored, ABS is disabled and the ABS warning lamp is turned "ON."

DTC Chart Test Description:

Number(s) below refer to circled number(s) on the diagnostic chart.

1. Indicates the EBTCM is capable of controlling the ABS relay as commanded.
2. Checks to ensure that voltage is available to the pull-in coil of the ABS relay.
3. Ensures that there is continuity through the pull-in coil of the ABS relay.
4. Verifies the integrity of the relay control circuit.
5. Ensures malfunction was not due to poor terminal contact.

Diagnostic Aids:

An intermittent malfunction may be caused by a poor connection, rubbed through wire insulation, or a wire that is broken inside the insulation.

The frequency of the malfunction can be checked by using the enhanced diagnostic function of the Tech 1,

If the frequency of the malfunction is high, but is currently intermittent, check for high coil resistance by measuring between relay terminals "C2" and" C5" using a J 39200. If resistance shows greater than 100 ohms, replace the relay.

Any circuitry, that is suspected of causing the intermittent complaint, should be thoroughly checked for backed out terminals, improper mating, broken locks, improperly formed or damaged terminals, poor terminal to wiring connections or physical damage to the wiring harness.

DIAGNOSTIC CHART DTC 16 — 1995 CAMARO AND FIREBIRD WITH TCS, CONTINUED

DTC 16
ABS ENABLE RELAY COIL CIRCUIT OPEN

1. • IGNITION "ON," ENGINE "OFF."
   • USING THE RELAY TEST FUNCTION OF THE TECH 1, COMMAND THE ABS ENABLE RELAY "ON."
   DOES TECH 1 INDICATE ABS ENABLE RELAY "ON" AND BATTERY VOLTAGE ABOVE 10 VOLTS?
   - YES: MALFUNCTION IS NOT PRESENT AT THIS TIME. REFER TO "DIAGNOSTIC AIDS" ON FACING PAGE.
   - NO: go to 2.

2. • USING J 39200, MEASURE THE VOLTAGE BY BACKPROBING BETWEEN GROUND AND TERMINAL "C2" OF THE ABS ENABLE RELAY CONNECTOR.
   IS VOLTAGE 10 VOLTS OR GREATER?
   - NO: REPAIR OPEN OR HIGH RESISTANCE IN CKT 641 OR ABS IGN FUSE 5.
   - YES: go to 3.

3. • USING THE RELAY TEST FUNCTION OF THE TECH 1, COMMAND THE ABS ENABLE RELAY "OFF."
   • USING J 39200, MEASURE THE VOLTAGE BY BACKPROBING BETWEEN GROUND AND TERMINAL "C5" OF THE ABS ENABLE RELAY CONNECTOR.
   IS VOLTAGE 10 VOLTS OR GREATER?
   - NO: CHECK FOR INTERMITTENT CONNECTIONS AT ABS ENABLE RELAY. IF OK, REPLACE ABS RELAY.
   - YES: go to 4.

4. • IGNITION "OFF."
   • DISCONNECT 32-WAY EBTCM CONNECTOR.
   • IGNITION "ON."
   • USING J 39200, MEASURE THE VOLTAGE BETWEEN GROUND AND TERMINAL "B3" OF THE 32-WAY EBTCM HARNESS CONNECTOR.
   IS VOLTAGE 10 VOLTS OR GREATER?
   - NO: REPAIR OPEN OR HIGH RESISTANCE IN CKT 1632.
   - YES: go to 5.

5. • IGNITION "OFF."
   • INSPECT 32-WAY EBTCM CONNECTOR TERMINAL "B3" FOR POOR TERMINAL CONTACT OR EVIDENCE OF CORROSION.
   REPLACE ALL TERMINALS THAT EXHIBIT SIGNS OF POOR TERMINAL CONTACT OR CORROSION.
   • RECONNECT ALL CONNECTORS.
   • IGNITION "ON."
   DOES DTC 16 SET AS A CURRENT DTC?
   - YES: REPLACE EBTCM.
   - NO: MALFUNCTION WAS MOST LIKELY CAUSED BY A POOR CONNECTION. REFER TO "DIAGNOSTIC AIDS" ON FACING PAGE.

AFTER ALL DIAGNOSIS IS COMPLETE, CLEAR DTCs AND TEST DRIVE VEHICLE FOR THREE (3) DRIVE CYCLES TO VERIFY THAT DTC DOES NOT RESET. A DRIVE CYCLE CONSISTS OF STARTING THE VEHICLE, DRIVING OVER 16 km/h (10 MPH), AND THEN TURNING KEY "OFF."

DIAGNOSTIC CHART DTC 17 — 1995 CAMARO AND FIREBIRD WITH TCS

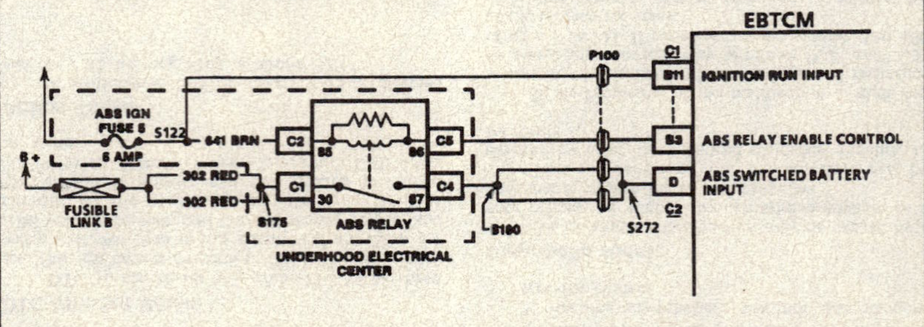

DTC 17
ABS ENABLE RELAY COIL CIRCUIT SHORTED TO GND

Circuit Description:

Ignition voltage is supplied through terminal "C2" of the ABS relay. The EBTCM then is able to energize the pull-in coil by completing the ground circuit at terminal "B3" of the EBTCM. The magnetic field created closes the ABS relay contacts and allows battery voltage and current to be supplied to the EBTCM, which supplies voltage to the motors and solenoids.

DTC Will Set When:

DTC 17 can be set before the EBTCM commands the ABS relay "ON." This test determines if the ABS relay is energized when it should not be. This malfunction would not allow the ABS relay to remove power to the EBTCM. If a second malfunction were to occur that requires the ABS relay to be turned "OFF," that malfunction cannot be removed if the relay cannot be controlled.

Action Taken:

A malfunction DTC is stored. ABS is not disabled.

DTC Chart Test Description:

Number(s) below refer to circled number(s) on the diagnostic chart.

1. Indicates that the EBTCM is capable of controlling the ABS relay as commanded.
2. Checks to ensure that the ABS relay or control CKT 1632 are not shorted to ground.
3. Checks to see if the EBTCM is internally shorted to ground. A grounded terminal "B3" would cause the ABS relay to be energized any time the ignition was in the "ON" position.
4. Checks for a short to ground in control CKT 1632 or a defective relay.

Diagnostic Aids:

An intermittent malfunction may be caused by a poor connection, rubbed through wire insulation, or a wire that is broken inside the insulation.

The frequency of the malfunction can be checked by using the enhanced diagnostic function of the Tech 1.

Any circuitry, that is suspected of causing the intermittent complaint, should be thoroughly checked for backed out terminals, improper mating, broken locks, improperly formed or damaged terminals, poor terminal to wiring connections or physical damage to the wiring harness.

DIAGNOSTIC CHART DTC 17 — 1995 CAMARO AND FIREBIRD WITH TCS, CONTINUED

DTC 17
ABS ENABLE RELAY COIL CIRCUIT SHORTED TO GROUND

(1)
- IGNITION "ON," ENGINE "OFF."
- USING THE RELAY TEST FUNCTION OF THE TECH 1, COMMAND THE ABS ENABLE RELAY "OFF."

DOES TECH 1 INDICATE THAT ABS ENABLE RELAY IS "OFF" AND BATTERY VOLTAGE IS LESS THAN 5 VOLTS?

- YES → MALFUNCTION IS NOT PRESENT AT THIS TIME. REFER TO "DIAGNOSTIC AIDS" ON FACING PAGE.
- NO → (2)

(2)
- IGNITION "OFF."
- DISCONNECT 32-WAY EBTCM CONNECTOR.
- IGNITION "ON."
- USING J 39200, MEASURE THE VOLTAGE BETWEEN TERMINAL "B3" OF THE 32-WAY EBTCM HARNESS CONNECTOR AND B+.

IS VOLTAGE 2 VOLTS OR LESS?

- YES → (3)
- NO → (4)

(3)
- IGNITION "OFF."
- INSPECT CKT 1632 AND TERMINAL "B3" OF THE 32-WAY EBTCM CONNECTOR FOR DAMAGE WHICH MAY RESULT IN A SHORT TO GROUND WITH ALL CONNECTORS CONNECTED. REPAIR ALL DAMAGE IF EVIDENT.
- RECONNECT ALL CONNECTORS.
- IGNITION "ON."

DOES DTC 17 SET AS A CURRENT DTC?

- YES → REPLACE EBTCM.
- NO → MALFUNCTION IS INTERMITTENT. REFER TO "DIAGNOSTIC AIDS" ON FACING PAGE.

(4)
- IGNITION "OFF."
- DISCONNECT ABS ENABLE RELAY.
- USING J 39200, MEASURE VOLTAGE BETWEEN TERMINAL "C5" OF THE ABS ENABLE RELAY CONNECTOR AND B+.

IS VOLTAGE 2 VOLTS OR LESS?

- NO → REPAIR SHORT TO GROUND IN CKT 1632.
- YES → REPLACE ABS ENABLE RELAY.

AFTER ALL DIAGNOSIS IS COMPLETE, CLEAR DTCs AND TEST DRIVE VEHICLE FOR THREE (3) DRIVE CYCLES TO VERIFY THAT DTC DOES NOT RESET. A DRIVE CYCLE CONSISTS OF STARTING THE VEHICLE, DRIVING OVER 16 km/h (10 MPH), AND THEN TURNING KEY "OFF."

DIAGNOSTIC CHART DTC 18 — 1995 CAMARO AND FIREBIRD WITH TCS

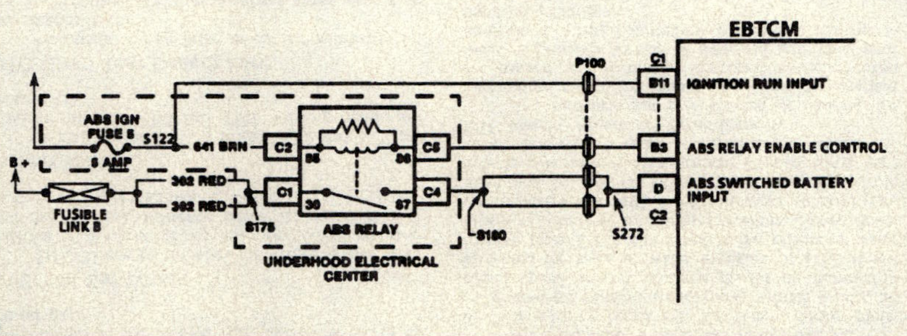

DTC 18
ABS ENABLE RELAY COIL CIRCUIT SHORTED TO BATT

Circuit Description:

Ignition voltage is supplied through terminal "C2" of the ABS relay. The EBTCM then is able to energize the pull-in coil by completing the ground circuit at connector "C1", terminal "B3" of the EBTCM. The magnetic field created closes the ABS relay contacts and allows battery voltage and current to be supplied to the EBTCM, which supplies voltage to the motors and solenoids.

DTC Will Set When:

DTC 18 can be set after the EBTCM commands the ABS relay "ON." This test monitors the availability of current and voltage to the motors and solenoids. A malfunction of this type will not allow voltage to be available to the motors and solenoids; therefore, ABS operation would not be allowed if required.

Action Taken:

A malfunction DTC is stored, ABS is disabled and the ABS warning lamp is turned "ON." DTCs 14, 15, 16, 17 ,18, 36, 37

DTC Chart Test Description:

Number(s) below refer to circled number(s) on the diagnostic chart.

1. Indicates that the EBTCM is capable of controlling the ABS relay as commanded.
2. Checks for a short to voltage on the relay control circuit.
3. Verifies that the relay coil has the appropriate resistance.
4. Checks for relay contacts that are failed closed.
5. Checks for possible poor terminal contact at the EBTCM.

Diagnostic Aids:

An intermittent malfunction may be caused by a poor connection, rubbed through wire insulation, or a wire that is broken inside the insulation.

The frequency of the malfunction can be checked by using the enhanced diagnostic function of the Tech 1,

Any circuitry, that is suspected of causing the intermittent complaint, should be thoroughly checked for backed out terminals, improper mating, broken locks, improperly formed or damaged terminals, poor terminal to wiring connections or physical damage to the wiring harness.

DIAGNOSTIC CHART DTC 18 — 1995 CAMARO AND FIREBIRD WITH TCS, CONTINUED

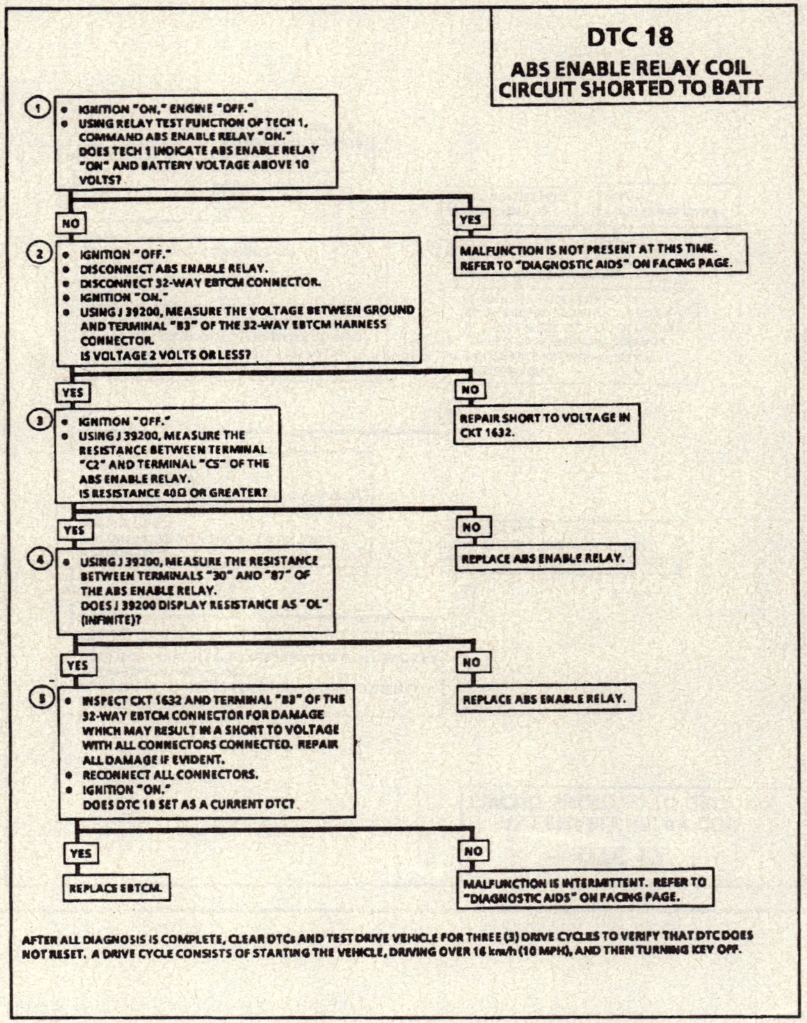

DIAGNOSTIC CHART DTC 21 — 1995 CAMARO AND FIREBIRD WITH TCS

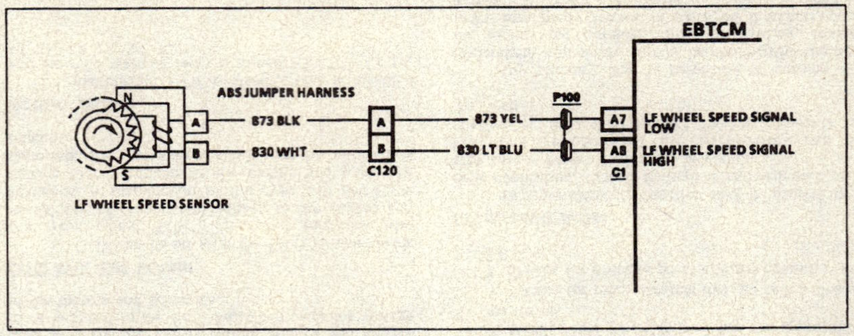

DTC 21
LEFT FRONT WHEEL SPEED = 0

Circuit Description:

As a toothed ring passes by the wheel speed sensor, changes in the electromagnetic field cause the wheel speed sensor to produce a sinusoidal(AC) voltage signal whose frequency is proportional to wheel speed. The magnitude of this signal is directly related to wheel speed and the proximity of the wheel speed sensor to the toothed ring, often referred to as the air gap.

DTC Will Set When:

DTC 21 can be set when the vehicle is not in an ABS or TCS event. If the left front wheel speed=0 and the vehicle's reference speed is greater than 8 km/h (5 mph), a malfunction exists.

Action Taken:

A malfunction DTC is stored, ABS and TCS are disabled and the ABS and TCS warning lamps are turned "ON."

DTC Chart Test Description:

Number(s) below refer to circled number(s) on the diagnostic chart.

1. This step identifies a wheel speed sensor or circuitry that is damaged and visibly apparent.
2. Checks to see that the wheel speed sensor coil is within proper resistance values.
3. Verifies that the wheel speed sensor can generate an appropriate output signal.
4. This ensures that the wheel speed circuitry is not shorted together.
5. Ensures that the malfunction was not due to physical damage.

Diagnostic Aids:

An intermittent malfunction may be caused by a poor connection, rubbed through wire insulation, or a wire that is broken inside the insulation.

The frequency of the malfunction can be checked by using the enhanced diagnostic function of the Tech 1,

If the customer's comments reflect that the ABS or TCS warning lamps are "ON" only during moist environmental changes (rain, snow, vehicle wash), all wheel speed sensor circuitry should be thoroughly inspected for signs of water intrusion. If DTC is not current, clear DTCs and simulate the effects of water intrusion. Use the following procedure. Spray down the suspected area with a 5% salt water solution (two teaspoons of salt to 12 oz. of water). Test drive vehicle over various road surfaces (bumps, turns, etc.) above 24km/h (15 mph) for at least 30 seconds. If DTC returns, replace suspected harness.

Any circuitry that is suspected of causing the intermittent complaint should be thoroughly checked for backed out terminals, improper mating, broken locks, improperly formed or damaged terminals, poor terminal to wiring connections or physical damage to the wiring harness.

Resistance of the wheel speed sensor will increase with an increase in sensor temperature.

When replacing a wheel speed sensor, inspect the sensor terminals and harness connector for corrosion and/or water intrusion. If evidence of corrosion or water intrusion exists, replace wheel speed sensor jumper harness. Likewise, if replacing a wheel speed sensor jumper harness, inspect sensor terminals. If evidence of corrosion or water intrusion exists, replace wheel speed sensor.

DIAGNOSTIC CHART DTC 21 — 1995 CAMARO AND FIREBIRD WITH TCS, CONTINUED

DTC 21
LEFT FRONT WHEEL SPEED = 0

Important
- WHEEL SPEED SENSOR INTERMITTENT MALFUNCTIONS MAY BE DIFFICULT TO LOCATE. CARE SHOULD BE TAKEN NOT TO DISTURB ANY ELECTRICAL CONNECTIONS PRIOR TO AN INDICATED STEP OF THIS CHART. THIS WILL ENSURE THAT AN INTERMITTENT CONNECTION WILL NOT BE CORRECTED BEFORE THE SOURCE OF THE MALFUNCTION IS FOUND.

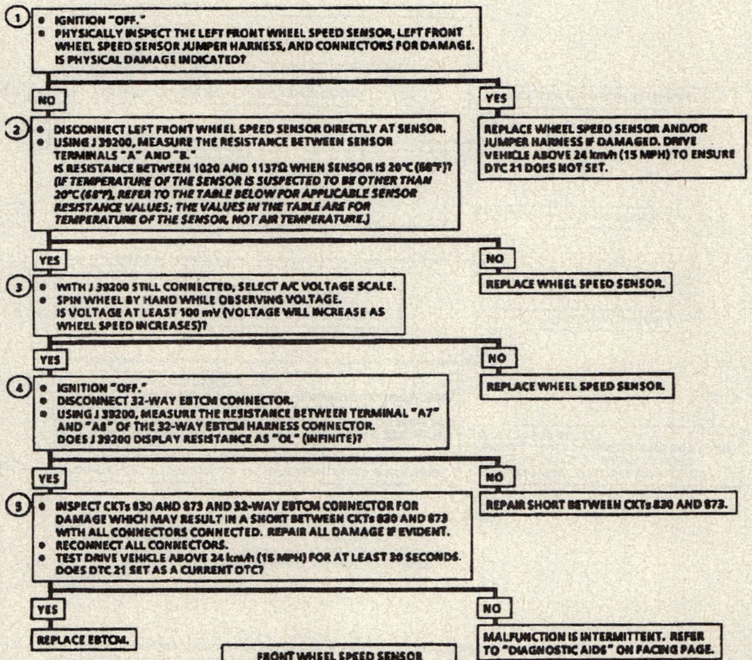

FRONT WHEEL SPEED SENSOR
TEMPERATURE OF SENSOR VS. SENSOR RESISTANCE (APPROXIMATE)

TEMP. (°C)	TEMP. (°F)	RESISTANCE (OHMS)
-34 TO 4	-30 TO 40	804 TO 1066
5 TO 43	41 TO 110	956 TO 1240
44 TO 93	111 TO 200	1112 TO 1463

DIAGNOSTIC CHART DTC 22 — 1995 CAMARO AND FIREBIRD WITH TCS

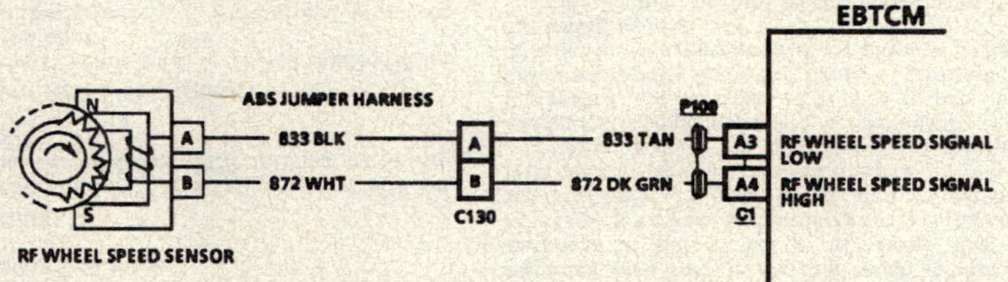

DTC 22
RIGHT FRONT WHEEL SPEED = 0

Circuit Description:

As a toothed ring passes by the wheel speed sensor, changes in the electromagnetic field cause the wheel speed sensor to produce a sinusoidal(AC) voltage signal whose frequency is proportional to wheel speed. The magnitude of this signal is directly related to wheel speed and the proximity of the wheel speed sensor to the toothed ring, often referred to as the air gap.

DTC Will Set When:

DTC 22 can be set when the vehicle is not in an ABS or TCS event. If the right front wheel speed=0 and the vehicle's reference speed is greater than 8 km/h (5 mph), a malfunction exists.

Action Taken:

A malfunction DTC is stored, ABS and TCS are disabled and the ABS and TCS warning lamps are turned "ON."

DTC Chart Test Description:

Number(s) below refer to circled number(s) on the diagnostic chart.

1. This step identifies a wheel speed sensor or circuitry that is damaged and visibly apparent.
2. Checks to see that the wheel speed sensor coil is within proper resistance values.
3. Verifies that the wheel speed sensor can generate an appropriate output signal.
4. This ensures that the wheel speed circuitry is not shorted together.
5. Ensures that the malfunction was not due to physical damage.

Diagnostic Aids:

An intermittent malfunction may be caused by a poor connection, rubbed through wire insulation, or a wire that is broken inside the insulation.

The frequency of the malfunction can be checked by using the enhanced diagnostic function of the Tech 1.

If the customer's comments reflect that the ABS or TCS warning lamps are "ON" only during moist environmental changes (rain, snow, vehicle wash), all wheel speed sensor circuitry should be thoroughly inspected for signs of water intrusion. If DTC is not current, clear DTCs and simulate the effects of water intrusion. Use the following procedure. Spray the suspected area with a 5% salt water solution (two teaspoons of salt to 12 oz. of water). Test drive vehicle over various road surfaces (bumps, turns, etc.) above 24km/h (15 mph) for at least 30 seconds. If DTC returns, replace suspected harness.

Any circuitry that is suspected of causing the intermittent complaint should be thoroughly checked for backed out terminals, improper mating, broken locks, improperly formed or damaged terminals, poor terminal to wiring connections or physical damage to the wiring harness.

Resistance of the wheel speed sensor will increase with an increase in sensor temperature.

When replacing a wheel speed sensor, inspect the sensor terminals and harness connector for corrosion and/or water intrusion. If evidence of corrosion or water intrusion exists, replace wheel speed sensor jumper harness. Likewise, if replacing a wheel speed sensor jumper harness, inspect sensor terminals. If evidence of corrosion or water intrusion exists, replace wheel speed sensor.

DIAGNOSTIC CHART DTC 22 — 1995 CAMARO AND FIREBIRD WITH TCS, CONTINUED

DTC 22
RIGHT FRONT WHEEL SPEED = 0

Important

- WHEEL SPEED SENSOR INTERMITTENT MALFUNCTIONS MAY BE DIFFICULT TO LOCATE. CARE SHOULD BE TAKEN NOT TO DISTURB ANY ELECTRICAL CONNECTIONS PRIOR TO AN INDICATED STEP OF THIS CHART. THIS WILL ENSURE THAT AN INTERMITTENT CONNECTION WILL NOT BE CORRECTED BEFORE THE SOURCE OF THE MALFUNCTION IS FOUND.

(1)
- IGNITION "OFF."
- PHYSICALLY INSPECT THE RIGHT FRONT WHEEL SPEED SENSOR, RIGHT FRONT WHEEL SPEED SENSOR JUMPER HARNESS, AND CONNECTORS FOR DAMAGE.

IS PHYSICAL DAMAGE INDICATED?

YES: REPLACE WHEEL SPEED SENSOR AND/OR JUMPER HARNESS IF DAMAGED. DRIVE VEHICLE ABOVE 24 km/h (15 MPH) TO ENSURE DTC 22 DOES NOT SET.

NO:

(2)
- DISCONNECT RIGHT FRONT WHEEL SPEED SENSOR DIRECTLY AT SENSOR.
- USING J 39200, MEASURE THE RESISTANCE BETWEEN SENSOR TERMINALS "A" AND "B."

IS RESISTANCE BETWEEN 1020 AND 1137Ω WHEN SENSOR IS 20°C (68°F)?
(IF TEMPERATURE OF THE SENSOR IS SUSPECTED TO BE OTHER THAN 20°C (68°F), REFER TO THE TABLE BELOW FOR APPLICABLE SENSOR RESISTANCE VALUES; THE VALUES IN THE TABLE ARE FOR TEMPERATURE OF THE SENSOR, NOT AIR TEMPERATURE.)

NO: REPLACE WHEEL SPEED SENSOR.

YES:

(3)
- WITH J 39200 STILL CONNECTED, SELECT A/C VOLTAGE SCALE.
- SPIN WHEEL BY HAND WHILE OBSERVING VOLTAGE.

IS VOLTAGE AT LEAST 100 mV (VOLTAGE WILL INCREASE AS WHEEL SPEED INCREASES)?

NO: REPLACE WHEEL SPEED SENSOR.

YES:

(4)
- IGNITION "OFF."
- DISCONNECT 32-WAY EBTCM CONNECTOR.
- USING J 39200, MEASURE THE RESISTANCE BETWEEN TERMINAL "A3" AND "A4" OF THE 32-WAY EBTCM HARNESS CONNECTOR.

DOES J 39200 DISPLAY RESISTANCE AS "OL" (INFINITE)?

NO: REPAIR SHORT BETWEEN CKTs 833 AND 872.

YES:

(5)
- INSPECT CKTs 833 AND 872 AND 32-WAY EBTCM CONNECTOR FOR DAMAGE WHICH MAY RESULT IN A SHORT BETWEEN CKTs 833 AND 872 WITH ALL CONNECTORS CONNECTED. REPAIR ALL DAMAGE IF EVIDENT.
- RECONNECT ALL CONNECTORS.
- TEST DRIVE VEHICLE ABOVE 24 km/h (15 MPH) FOR AT LEAST 30 SECONDS.

DOES DTC 22 SET AS A CURRENT DTC?

YES: REPLACE EBTCM.

NO: MALFUNCTION IS INTERMITTENT. REFER TO "DIAGNOSTIC AIDS" ON FACING PAGE.

FRONT WHEEL SPEED SENSOR
TEMPERATURE OF SENSOR VS. SENSOR RESISTANCE (APPROXIMATE)

TEMP. (°C)	TEMP. (°F)	RESISTANCE (OHMS)
-34 TO 4	-30 TO 40	804 TO 1066
5 TO 43	41 TO 110	956 TO 1240
44 TO 93	111 TO 200	1112 TO 1463

DIAGNOSTIC CHART DTC 23 — 1995 CAMARO AND FIREBIRD WITH TCS

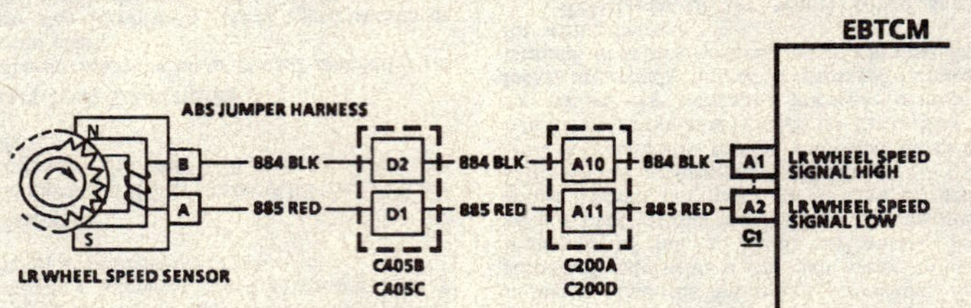

DTC 23
LEFT REAR WHEEL SPEED = 0

Circuit Description:

As a toothed ring passes by the wheel speed sensor, changes in the electromagnetic field cause the wheel speed sensor to produce a sinusoidal(AC) voltage signal whose frequency is proportional to wheel speed. The magnitude of this signal is directly related to wheel speed and the proximity of the wheel speed sensor to the toothed ring, often referred to as the air gap.

DTC Will Set When:

DTC 23 can be set when the vehicle is not in an ABS or TCS event. If the left rear wheel speed=0 and the vehicle's reference speed is greater than 8 km/h (5 mph), a malfunction exists.

Action Taken:

A malfunction DTC is stored, ABS and TCS are disabled and the ABS and TCS warning lamps are turned "ON."

DTC Chart Test Description:

Number(s) below refer to circled number(s) on the diagnostic chart.

1. This step identifies a wheel speed sensor or circuitry that is damaged and visibly apparent.
2. Checks to see that the wheel speed sensor coil is within proper resistance values.
3. Verifies that the wheel speed sensor can generate an appropriate output signal.
4. Verifies that the wheel speed circuitry is not shorted together.
5. Ensures that the malfunction was not due to physical damage.

Diagnostic Aids:

An intermittent malfunction may be caused by a poor connection, rubbed through wire insulation, or a wire that is broken inside the insulation.

The frequency of the malfunction can be checked by using the enhanced diagnostic function of the Tech 1,

If the customer's comments reflect that the ABS or TCS warning lamp is "ON" only during moist environmental changes (rain, snow, vehicle wash),all wheel speed sensor circuitry should be thoroughly inspected for signs of water intrusion. If DTC is not current, clear DTCs and simulate the effects of water intrusion. Use the following procedure. Spray the suspected area with a 5% salt water solution (two teaspoons of salt to 12 oz. of water). Test drive vehicle over various road surfaces (bumps, turns, etc.) above 24km/h (15 mph) for at least 30 seconds. If DTC returns, replace suspected harness.

Any circuitry that is suspected of causing the intermittent complaint should be thoroughly checked for backed out terminals, improper mating, broken locks, improperly formed or damaged terminals, poor terminal to wiring connections or physical damage to the wiring harness.

Resistance of the wheel speed sensor will increase with an increase in sensor temperature.

When replacing a wheel speed sensor, inspect the sensor terminals and harness connector for corrosion and/or water intrusion. If evidence of corrosion or water intrusion exists, replace wheel speed sensor jumper harness. Likewise, if replacing a wheel speed sensor jumper harness, inspect sensor terminals. If evidence of corrosion or water intrusion exists, replace wheel speed sensor.

DIAGNOSTIC CHART DTC 23 — 1995 CAMARO AND FIREBIRD WITH TCS, CONTINUED

DTC 23
LEFT REAR WHEEL SPEED = 0

Important

- WHEEL SPEED SENSOR INTERMITTENT MALFUNCTIONS MAY BE DIFFICULT TO LOCATE. CARE SHOULD BE TAKEN NOT TO DISTURB ANY ELECTRICAL CONNECTIONS PRIOR TO AN INDICATED STEP OF THIS CHART. THIS WILL ENSURE THAT AN INTERMITTENT CONNECTION WILL NOT BE CORRECTED BEFORE THE SOURCE OF THE MALFUNCTION IS FOUND.

(1)
- IGNITION "OFF."
- PHYSICALLY INSPECT THE LEFT REAR WHEEL SPEED SENSOR, WHEEL SPEED SENSOR JUMPER HARNESS, AND CONNECTORS FOR DAMAGE.

IS PHYSICAL DAMAGE INDICATED?

YES: REPLACE WHEEL SPEED SENSOR AND/OR WHEEL SPEED SENSOR JUMPER HARNESS IF DAMAGED. DRIVE VEHICLE ABOVE 24 km/h (15 MPH) TO ENSURE DTC 23 DOES NOT SET.

NO:

(2)
- DISCONNECT LEFT REAR WHEEL SPEED SENSOR DIRECTLY AT SENSOR.
- USING J 39200, MEASURE THE RESISTANCE BETWEEN SENSOR TERMINALS "A" AND "B."

IS RESISTANCE BETWEEN 2579 AND 3315Ω WHEN SENSOR IS 20°C (68°F)?
(IF TEMPERATURE OF THE SENSOR IS SUSPECTED TO BE OTHER THAN 20°C (68°F), REFER TO THE TABLE BELOW FOR APPLICABLE SENSOR RESISTANCE VALUES; THE VALUES IN THE TABLE ARE FOR TEMPERATURE OF THE SENSOR, NOT AIR TEMPERATURE.)

NO: REPLACE WHEEL SPEED SENSOR.

YES:

(3)
- WITH J 39200 STILL CONNECTED, SELECT A/C VOLTAGE SCALE.
- VEHICLE TRANSMISSION IN NEUTRAL.
- SPIN WHEEL BY HAND WHILE OBSERVING VOLTAGE.

IS VOLTAGE AT LEAST 100 mV (VOLTAGE WILL INCREASE AS WHEEL SPEED INCREASES)?

NO: REPLACE WHEEL SPEED SENSOR.

YES:

(4)
- IGNITION "OFF."
- DISCONNECT 32-WAY EBTCM CONNECTOR.
- USING J 39200, MEASURE THE RESISTANCE BETWEEN TERMINAL "A1" AND "A2" OF THE 32-WAY EBTCM HARNESS CONNECTOR.

DOES J 39200 DISPLAY RESISTANCE AS "OL" (INFINITE)?

NO: REPAIR SHORT BETWEEN CKTs 884 AND 885.

YES:

(5)
- INSPECT CKTs 884 AND 885 AND 32-WAY EBTCM CONNECTOR FOR DAMAGE WHICH MAY RESULT IN A SHORT BETWEEN CKTs 884 AND 885 WITH ALL CONNECTORS CONNECTED. REPAIR ALL DAMAGE IF EVIDENT.
- RECONNECT ALL CONNECTORS.
- TEST DRIVE VEHICLE ABOVE 24 km/h (15 MPH) FOR AT LEAST 30 SECONDS.

DOES DTC 23 SET AS A CURRENT DTC?

YES: REPLACE EBTCM.

NO: MALFUNCTION IS INTERMITTENT. REFER TO "DIAGNOSTIC AIDS" ON FACING PAGE.

REAR WHEEL SPEED SENSOR
TEMPERATURE OF SENSOR VS. SENSOR RESISTANCE (APPROXIMATE)

TEMP. (°C)	TEMP. (°F)	RESISTANCE (OHMS)
-34 TO 4	-30 TO 40	2032 TO 3107
5 TO 43	41 TO 110	2417 TO 3615
44 TO 93	111 TO 200	2812 TO 4266

DIAGNOSTIC CHART DTC 24 — 1995 CAMARO AND FIREBIRD WITH TCS

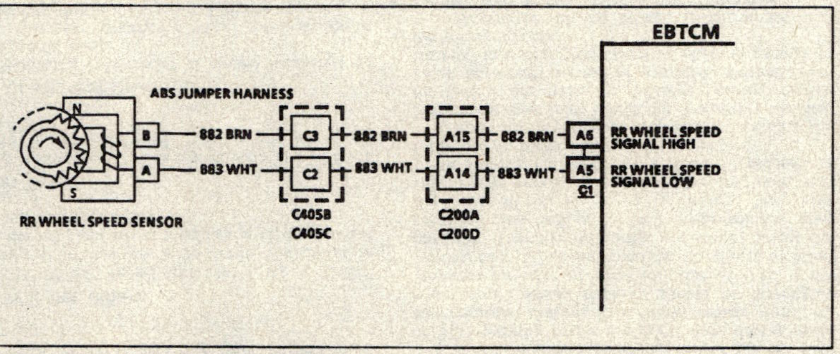

DTC 24
RIGHT REAR WHEEL SPEED = 0

Circuit Description:

As a toothed ring passes by the wheel speed sensor, changes in the electromagnetic field cause the wheel speed sensor to produce a sinusoidal(AC) voltage signal whose frequency is proportional to wheel speed. The magnitude of this signal is directly related to wheel speed and the proximity of the wheel speed sensor to the toothed ring, often referred to as the air gap.

DTC Will Set When:

DTC 24 can be set when the vehicle is not in an ABS or TCS event. If the right rear wheel speed=0 and the vehicle's reference speed is greater than 8 km/h (5 mph), a malfunction exists.

Action Taken:

A malfunction DTC is stored, ABS and TCS are disabled and the ABS and TCS warning lamps are turned "ON."

DTC Chart Test Description:

Number(s) below refer to circled number(s) on the diagnostic chart.

1. This step identifies a wheel speed sensor or circuitry that is damaged and visibly apparent.
2. Checks to see that the wheel speed sensor coil is within proper resistance values.
3. Verifies that the wheel speed sensor can generate an appropriate output signal.
4. Verifies that the wheel speed circuitry is not shorted together.
5. Ensures that the malfunction was not due to physical damage.

Diagnostic Aids:

An intermittent malfunction may be caused by a poor connection, rubbed through wire insulation, or a wire that is broken inside the insulation.

The frequency of the malfunction can be checked by using the enhanced diagnostic function of the Tech 1,

If the customer's comments reflect that the ABS or TCS warning lamps are "ON" only during moist environmental changes (rain, snow, vehicle wash), all wheel speed sensor circuitry should be thoroughly inspected for signs of water intrusion. If DTC is not current, clear DTCs and simulate the effects of water intrusion. Use the following procedure. Spray the suspected area with a 5% salt water solution (two teaspoons of salt to 12 oz. of water). Test drive vehicle over various road surfaces (bumps, turns, etc.) above 24km/h (15 mph) for at least 30 seconds. If DTC returns, replace suspected harness.

Any circuitry that is suspected of causing the intermittent complaint should be thoroughly checked for backed out terminals, improper mating, broken locks, improperly formed or damaged terminals, poor terminal to wiring connections or physical damage to the wiring harness.

Resistance of the wheel speed sensor will increase with an increase in sensor temperature.

When replacing a wheel speed sensor, inspect the sensor terminals and harness connector for corrosion and/or water intrusion. If evidence of corrosion or water intrusion exists, replace wheel speed sensor jumper harness. Likewise, if replacing a wheel speed sensor jumper harness, inspect sensor terminals. If evidence of corrosion or water intrusion exists, replace wheel speed sensor.

DIAGNOSTIC CHART DTC 24 — 1995 CAMARO AND FIREBIRD WITH TCS, CONTINUED

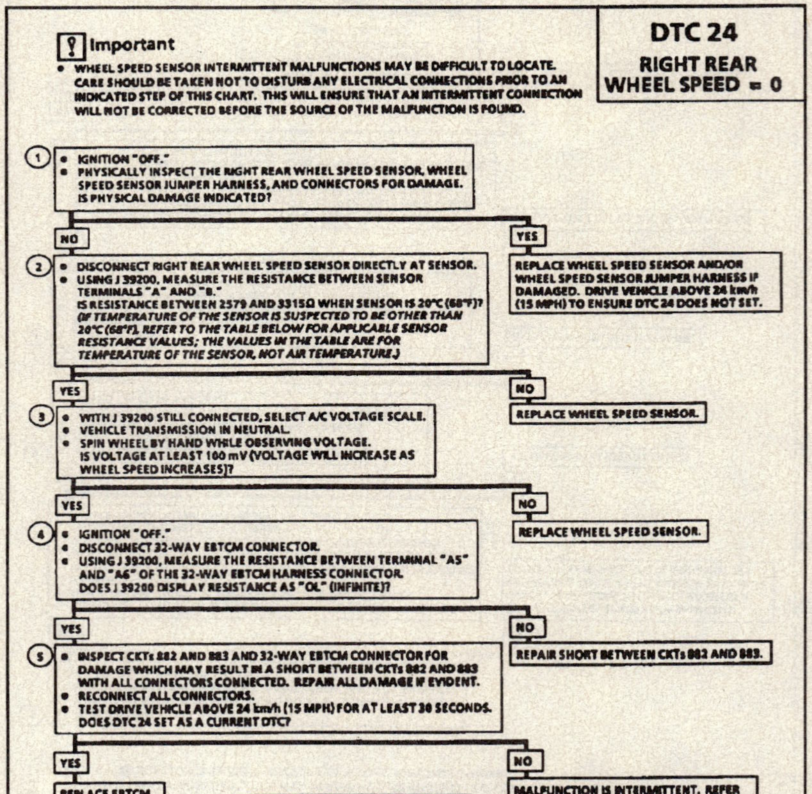

REAR WHEEL SPEED SENSOR
TEMPERATURE OF SENSOR VS. SENSOR RESISTANCE (APPROXIMATE)

TEMP. (°C)	TEMP. (°F)	RESISTANCE (OHMS)
-34 TO 4	-30 TO 40	2032 TO 3107
5 TO 43	41 TO 110	2417 TO 3615
44 TO 93	111 TO 200	2812 TO 4266

DIAGNOSTIC CHART DTC 25 — 1995 CAMARO AND FIREBIRD WITH TCS

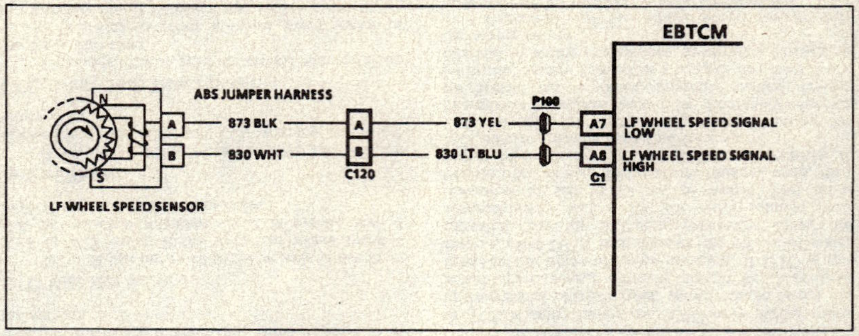

DTC 25
LEFT FRONT EXCESSIVE WHEEL SPEED VARIATION

Circuit Description:

As a toothed ring passes by the wheel speed sensor, changes in the electromagnetic field cause the wheel speed sensor to produce a sinusoidal(AC) voltage signal whose frequency is proportional to wheel speed. The magnitude of this signal is directly related to wheel speed and the proximity of the wheel speed sensor to the toothed ring, often referred to as the air gap.

DTC Will Set When:

DTC 25 can be set when the brake is "OFF." The purpose of this test is to detect a situation in which the left front wheel acceleration or deceleration is beyond specified limits.

Action Taken:

A malfunction DTC is stored, ABS and TCS are disabled and the ABS and TCS warning lamps are turned "ON."

DTC Chart Test Description:

Number(s) below refer to circled number(s) on the diagnostic chart.

1. This step identifies a wheel speed sensor or circuitry that is damaged and visibly apparent.
2. Checks to see that the wheel speed sensor coil is within proper resistance values.
3. Verifies that the wheel speed sensor can generate an appropriate output signal.
4. This ensures that the wheel speed circuitry is not shorted together.
5. Ensures that the malfunction was not due to physical damage.

Diagnostic Aids:

An intermittent malfunction may be caused by a poor connection, rubbed through wire insulation, or a wire that is broken inside the insulation.

The frequency of the malfunction can be checked by using the enhanced diagnostic function of the Tech 1,

If the customer's comments reflect that the ABS or TCS warning lamps are "ON" only during moist environmental changes (rain, snow, vehicle wash), all wheel speed sensor circuitry should be thoroughly inspected for signs of water intrusion. If DTC is not current, clear DTCs and simulate the effects of water intrusion. Use the following procedure. Spray the suspected area with a 5% salt water solution (two teaspoons of salt to 12 oz. of water). Test drive vehicle over various road surfaces (bumps, turns, etc.) above 24km/h (15 mph) for at least 30 seconds. If DTC returns, replace suspected harness.

Any circuitry that is suspected of causing the intermittent complaint should be thoroughly checked for backed out terminals, improper mating, broken locks, improperly formed or damaged terminals, poor terminal to wiring connections or physical damage to the wiring harness.

Resistance of the wheel speed sensor will increase with an increase in sensor temperature.

When replacing a wheel speed sensor, inspect the sensor terminals and harness connector for corrosion and/or water intrusion. If evidence of corrosion or water intrusion exists, replace wheel speed sensor jumper harness. Likewise, if replacing a wheel speed sensor jumper harness, inspect sensor terminals. If evidence of corrosion or water intrusion exists, replace wheel speed sensor.

DIAGNOSTIC CHART DTC 25 — 1995 CAMARO AND FIREBIRD WITH TCS, CONTINUED

DTC 25
LEFT FRONT EXCESSIVE WHEEL SPEED VARIATION

Important
- WHEEL SPEED SENSOR INTERMITTENT MALFUNCTIONS MAY BE DIFFICULT TO LOCATE. CARE SHOULD BE TAKEN NOT TO DISTURB ANY ELECTRICAL CONNECTIONS PRIOR TO AN INDICATED STEP OF THIS CHART. THIS WILL ENSURE THAT AN INTERMITTENT CONNECTION WILL NOT BE CORRECTED BEFORE THE SOURCE OF THE MALFUNCTION IS FOUND.

(1)
- IGNITION "OFF."
- PHYSICALLY INSPECT THE LEFT FRONT WHEEL SPEED SENSOR, LEFT FRONT WHEEL SPEED SENSOR JUMPER HARNESS, AND CONNECTORS FOR DAMAGE. IS PHYSICAL DAMAGE INDICATED?

YES → REPLACE WHEEL SPEED SENSOR AND/OR JUMPER HARNESS IF DAMAGED. DRIVE VEHICLE ABOVE 24 km/h (15 MPH) TO ENSURE DTC 25 DOES NOT SET.

NO ↓

(2)
- DISCONNECT LEFT FRONT WHEEL SPEED SENSOR DIRECTLY AT SENSOR.
- USING J 39200, MEASURE THE RESISTANCE BETWEEN SENSOR TERMINALS "A" AND "B."

IS RESISTANCE BETWEEN 1020 AND 1137Ω WHEN SENSOR IS 20°C (68°F)? *(IF TEMPERATURE OF THE SENSOR IS SUSPECTED TO BE OTHER THAN 20°C (68°F), REFER TO THE TABLE BELOW FOR APPLICABLE SENSOR RESISTANCE VALUES; THE VALUES IN THE TABLE ARE FOR TEMPERATURE OF THE SENSOR, NOT AIR TEMPERATURE.)*

NO → REPLACE WHEEL SPEED SENSOR.

YES ↓

(3)
- WITH J 39200 STILL CONNECTED, SELECT A/C VOLTAGE SCALE.
- SPIN WHEEL BY HAND WHILE OBSERVING VOLTAGE.

IS VOLTAGE AT LEAST 100 mV (VOLTAGE WILL INCREASE AS WHEEL SPEED INCREASES)?

NO → REPLACE WHEEL SPEED SENSOR.

YES ↓

(4)
- IGNITION "OFF."
- DISCONNECT 32-WAY EBTCM CONNECTOR.
- USING J 39200, MEASURE THE RESISTANCE BETWEEN TERMINAL "A7" AND "A8" OF THE 32-WAY EBTCM HARNESS CONNECTOR.

DOES J 39200 DISPLAY RESISTANCE AS "OL" (INFINITE)?

NO → REPAIR SHORT BETWEEN CKTs 830 AND 873.

YES ↓

(5)
- INSPECT CKTs 830 AND 873 AND 32-WAY EBTCM CONNECTOR FOR DAMAGE WHICH MAY RESULT IN A SHORT BETWEEN CKTs 830 AND 873 WITH ALL CONNECTORS CONNECTED. REPAIR ALL DAMAGE IF EVIDENT.
- RECONNECT ALL CONNECTORS.
- TEST DRIVE VEHICLE ABOVE 24 km/h (15 MPH) FOR AT LEAST 30 SECONDS. DOES DTC 25 SET AS A CURRENT DTC?

YES → REPLACE EBTCM.

NO → MALFUNCTION IS INTERMITTENT. REFER TO "DIAGNOSTIC AIDS" ON FACING PAGE.

FRONT WHEEL SPEED SENSOR
TEMPERATURE OF SENSOR VS. SENSOR RESISTANCE (APPROXIMATE)

TEMP. (°C)	TEMP. (°F)	RESISTANCE (OHMS)
-34 TO 4	-30 TO 40	804 TO 1066
5 TO 43	41 TO 110	956 TO 1240
44 TO 93	111 TO 200	1112 TO 1463

DIAGNOSTIC CHART DTC 26 — 1995 CAMARO AND FIREBIRD WITH TCS

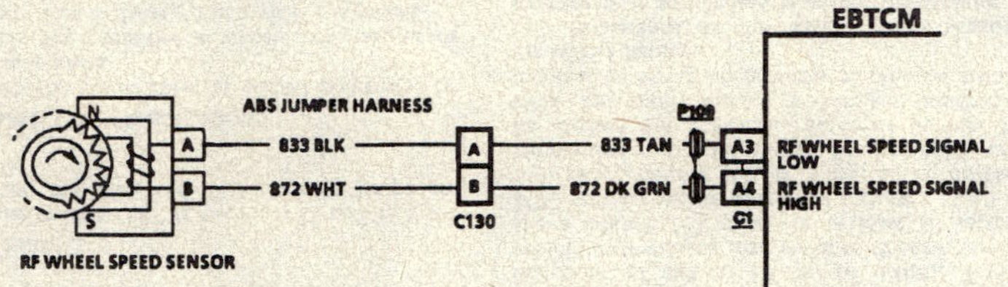

DTC 26
RIGHT FRONT EXCESSIVE WHEEL SPEED VARIATION

Circuit Description:

As a toothed ring passes by the wheel speed sensor, changes in the electromagnetic field cause the wheel speed sensor to produce a sinusoidal(AC) voltage signal whose frequency is proportional to wheel speed. The magnitude of this signal is directly related to wheel speed and the proximity of the wheel speed sensor to the toothed ring, often referred to as the air gap.

DTC Will Set When:

DTC 26 can be set when the brake is "OFF." The purpose of this test is to detect a situation in which the right front wheel acceleration or deceleration is beyond specified limits.

Action Taken:

A malfunction DTC is stored, ABS and TCS are disabled and the ABS and TCS warning lamps are turned "ON."

DTC Chart Test Description:

Number(s) below refer to circled number(s) on the diagnostic chart.

1. This step identifies a wheel speed sensor or circuitry that is damaged and visibly apparent.
2. Checks to see that the wheel speed sensor coil is within proper resistance values.
3. Verifies that the wheel speed sensor can generate an appropriate output signal.
4. This ensures that the wheel speed circuitry is not shorted together.
5. Ensures that the malfunction was not due to physical damage.

Diagnostic Aids:

An intermittent malfunction may be caused by a poor connection, rubbed through wire insulation, or a wire that is broken inside the insulation.

The frequency of the malfunction can be checked by using the enhanced diagnostic function of the Tech 1,

If the customer's comments reflect that the ABS or TCS warning lamps are "ON" only during moist environmental changes (rain, snow, vehicle wash), all wheel speed sensor circuitry should be thoroughly inspected for signs of water intrusion. If DTC is not current, clear DTCs and simulate the effects of water intrusion. Use the following procedure. Spray the suspected area with a 5% salt water solution (two teaspoons of salt to 12 oz. of water). Test drive vehicle over various road surfaces (bumps, turns, etc.) above 24km/h (15 mph) for at least 30 seconds. If DTC returns, replace suspected harness.

Any circuitry that is suspected of causing the intermittent complaint should be thoroughly checked for backed out terminals, improper mating, broken locks, improperly formed or damaged terminals, poor terminal to wiring connections or physical damage to the wiring harness.

Resistance of the wheel speed sensor will increase with an increase in sensor temperature.

When replacing a wheel speed sensor, inspect the sensor terminals and harness connector for corrosion and/or water intrusion. If evidence of corrosion or water intrusion exists, replace wheel speed sensor jumper harness. Likewise, if replacing a wheel speed sensor jumper harness, inspect sensor terminals. If evidence of corrosion or water intrusion exists, replace wheel speed sensor.

DIAGNOSTIC CHART DTC 26 — 1995 CAMARO AND FIREBIRD WITH TCS, CONTINUED

DTC 26 RIGHT FRONT EXCESSIVE WHEEL SPEED VARIATION

Important

- WHEEL SPEED SENSOR INTERMITTENT MALFUNCTIONS MAY BE DIFFICULT TO LOCATE. CARE SHOULD BE TAKEN NOT TO DISTURB ANY ELECTRICAL CONNECTIONS PRIOR TO AN INDICATED STEP OF THIS CHART. THIS WILL ENSURE THAT AN INTERMITTENT CONNECTION WILL NOT BE CORRECTED BEFORE THE SOURCE OF THE MALFUNCTION IS FOUND.

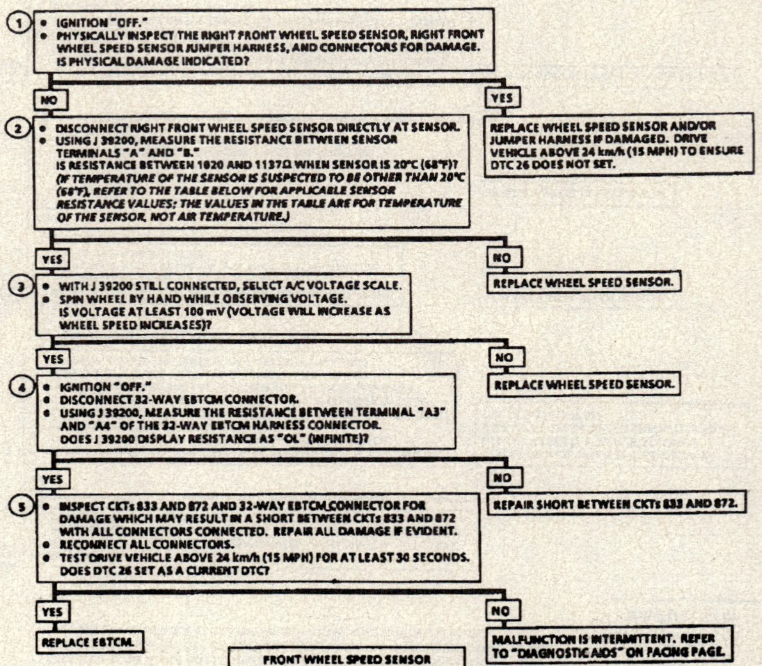

FRONT WHEEL SPEED SENSOR

TEMPERATURE OF SENSOR VS. SENSOR RESISTANCE (APPROXIMATE)

TEMP. (°C)	TEMP. (°F)	RESISTANCE (OHMS)
-34 TO 4	-30 TO 40	804 TO 1066
5 TO 43	41 TO 110	956 TO 1240
44 TO 93	111 TO 200	1112 TO 1463

DIAGNOSTIC CHART DTC 27 — 1995 CAMARO AND FIREBIRD WITH TCS

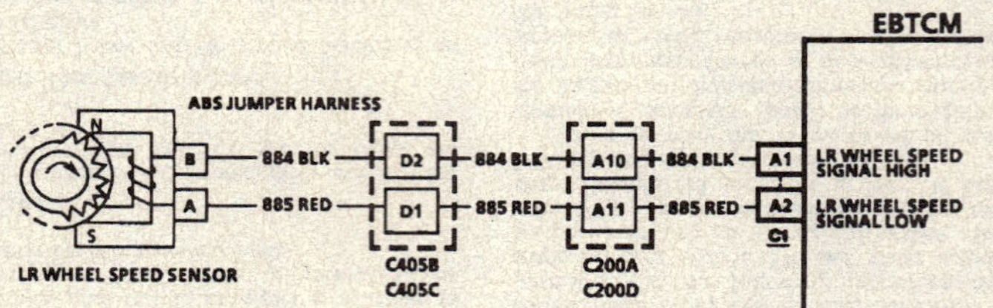

DTC 27
LEFT REAR EXCESSIVE WHEEL SPEED VARIATION

Circuit Description:

As a toothed ring passes by the wheel speed sensor, changes in the electromagnetic field cause the wheel speed sensor to produce a sinusoidal(AC) voltage signal whose frequency is proportional to wheel speed. The magnitude of this signal is directly related to wheel speed and the proximity of the wheel speed sensor to the toothed ring, often referred to as the air gap.

DTC Will Set When:

DTC 27 can be set when the brake is "OFF." The purpose of this test is to detect a situation in which the left rear wheel acceleration or deceleration is beyond specified limits.

Action Taken:

A malfunction DTC is stored, ABS and TCS are disabled and the ABS and TCS warning lamps are turned "ON."

DTC Chart Test Description:

Number(s) below refer to circled number(s) on the diagnostic chart.

1. This step identifies a wheel speed sensor or circuitry that is damaged and visibly apparent.
2. Checks to see that the wheel speed sensor coil is within proper resistance values.
3. Verifies that the wheel speed sensor can generate an appropriate output signal.
4. This ensures that the wheel speed circuitry is not shorted together.
5. Ensures that the malfunction was not due to physical damage.
6. Checks rear sensor toothed ring for physical damage.

Diagnostic Aids:

An intermittent malfunction may be caused by a poor connection, rubbed through wire insulation, or a wire that is broken inside the insulation.

The frequency of the malfunction can be checked by using the enhanced diagnostic function of the Tech 1.

If the customer's comments reflect that the ABS or TCS warning lamps are "ON" only during moist environmental changes (rain, snow, vehicle wash), all wheel speed sensor circuitry should be thoroughly inspected for signs of water intrusion. If DTC is not current, clear DTCs and simulate the effects of water intrusion. Use the following procedure. Spray the suspected area with a 5% salt water solution (two teaspoons of salt to 12 oz. of water). Test drive vehicle over various road surfaces (bumps, turns, etc.) above 24km/h (15 mph) for at least 30 seconds. If DTC returns, replace suspected harness.

Any circuitry that is suspected of causing the intermittent complaint should be thoroughly checked for backed out terminals, improper mating, broken locks, improperly formed or damaged terminals, poor terminal to wiring connections or physical damage to the wiring harness.

Resistance of the wheel speed sensor will increase with an increase in sensor temperature.

When replacing a wheel speed sensor, inspect the sensor terminals and harness connector for corrosion and/or water intrusion. If evidence of corrosion or water intrusion exists, replace wheel speed sensor jumper harness. Likewise, if replacing a wheel speed sensor jumper harness, inspect sensor terminals. If evidence of corrosion or water intrusion exists, replace wheel speed sensor.

DIAGNOSTIC CHART DTC 27 — 1995 CAMARO AND FIREBIRD WITH TCS, CONTINUED

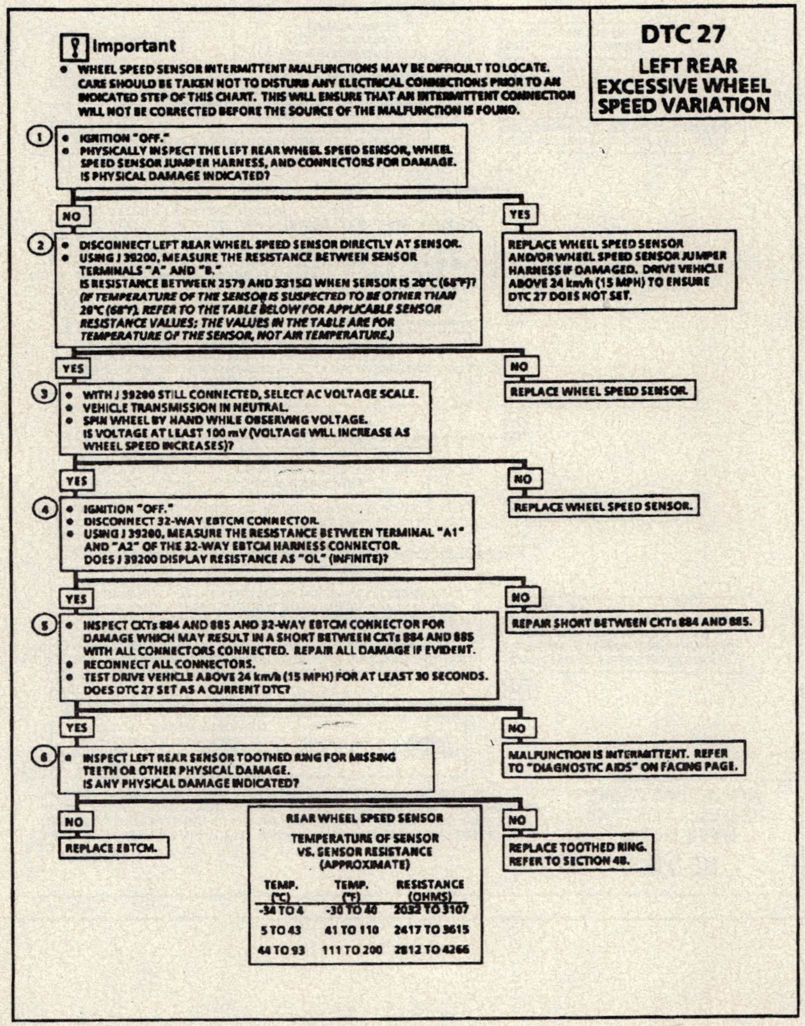

REAR WHEEL SPEED SENSOR
TEMPERATURE OF SENSOR VS. SENSOR RESISTANCE (APPROXIMATE)

TEMP. (°C)	TEMP. (°F)	RESISTANCE (OHMS)
-34 TO 4	-30 TO 40	2032 TO 3107
5 TO 43	41 TO 110	2417 TO 3615
44 TO 93	111 TO 200	2812 TO 4266

DIAGNOSTIC CHART DTC 28 — 1995 CAMARO AND FIREBIRD WITH TCS

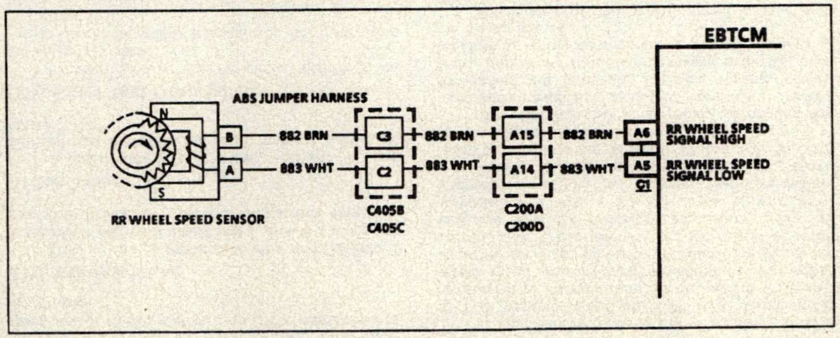

DTC 28
RIGHT REAR EXCESSIVE WHEEL SPEED VARIATION

Circuit Description:

As a toothed ring passes by the wheel speed sensor, changes in the electromagnetic field cause the wheel speed sensor to produce a sinusoidal(AC) voltage signal whose frequency is proportional to wheel speed. The magnitude of this signal is directly related to wheel speed and the proximity of the wheel speed sensor to the toothed ring, often referred to as the air gap.

DTC Will Set When:

DTC 28 can be set when the brake is "OFF." The purpose of this test is to detect a situation in which the right rear wheel acceleration or deceleration is beyond specified limits.

Action Taken:

A malfunction DTC is stored, ABS and TCS are disabled and the ABS and TCS warning lamps are turned "ON."

DTC Chart Test Description:

Number(s) below refer to circled number(s) on the diagnostic chart.

1. This step identifies a wheel speed sensor or circuitry that is damaged and visibly apparent.
2. Checks to see that the wheel speed sensor coil is within proper resistance values.
3. Verifies that the wheel speed sensor can generate an appropriate output signal.
4. This ensures that the wheel speed circuitry is not shorted together.
5. Ensures that the malfunction was not due to physical damage.
6. Checks rear sensor toothed ring for physical damage.

Diagnostic Aids:

An intermittent malfunction may be caused by a poor connection, rubbed through wire insulation, or a wire that is broken inside the insulation.

The frequency of the malfunction can be checked by using the enhanced diagnostic function of the Tech 1,

If the customer's comments reflect that the ABS or TCS warning lamps are "ON" only during moist environmental changes (rain, snow, vehicle wash), all wheel speed sensor circuitry should be thoroughly inspected for signs of water intrusion. If DTC is not current, clear DTCs and simulate the effects of water intrusion. Use the following procedure. Spray the suspected area with a 5% salt water solution (two teaspoons of salt to 12 oz. of water).Test drive vehicle over various road surfaces (bumps, turns, etc.) above 24km/h (15 mph) for at least 30 seconds. If DTC returns, replace suspected harness.

Any circuitry that is suspected of causing the intermittent complaint should be thoroughly checked for backed out terminals, improper mating, broken locks, improperly formed or damaged terminals, poor terminal to wiring connections or physical damage to the wiring harness.

Resistance of the wheel speed sensor will increase with an increase in sensor temperature.

When replacing a wheel speed sensor, inspect the sensor terminals and harness connector for corrosion and/or water intrusion. If evidence of corrosion or water intrusion exists, replace wheel speed sensor jumper harness. Likewise, if replacing a wheel speed sensor jumper harness, inspect sensor terminals. If evidence of corrosion or water intrusion exists, replace wheel speed sensor.

DIAGNOSTIC CHART DTC 28 — 1995 CAMARO AND FIREBIRD WITH TCS, CONTINUED

DTC 28
RIGHT REAR EXCESSIVE WHEEL SPEED VARIATION

Important
- WHEEL SPEED SENSOR INTERMITTENT MALFUNCTIONS MAY BE DIFFICULT TO LOCATE. CARE SHOULD BE TAKEN NOT TO DISTURB ANY ELECTRICAL CONNECTIONS PRIOR TO AN INDICATED STEP OF THIS CHART. THIS WILL ENSURE THAT AN INTERMITTENT CONNECTION WILL NOT BE CORRECTED BEFORE THE SOURCE OF THE MALFUNCTION IS FOUND.

(1)
- IGNITION "OFF."
- PHYSICALLY INSPECT THE RIGHT REAR WHEEL SPEED SENSOR, WHEEL SPEED SENSOR JUMPER HARNESS, AND CONNECTORS FOR DAMAGE. IS PHYSICAL DAMAGE INDICATED?

YES: REPLACE WHEEL SPEED SENSOR AND/OR WHEEL SPEED SENSOR JUMPER HARNESS IF DAMAGED. DRIVE VEHICLE ABOVE 24 km/h (15 MPH) TO ENSURE DTC 28 DOES NOT SET.

NO:

(2)
- DISCONNECT RIGHT REAR WHEEL SPEED SENSOR DIRECTLY AT SENSOR.
- USING J 39200, MEASURE THE RESISTANCE BETWEEN SENSOR TERMINALS "A" AND "B." IS RESISTANCE BETWEEN 2579 AND 3315Ω WHEN SENSOR IS 20°C (68°F)? *(IF TEMPERATURE OF THE SENSOR IS SUSPECTED TO BE OTHER THAN 20°C (68°F), REFER TO THE TABLE BELOW FOR APPLICABLE SENSOR RESISTANCE VALUES; THE VALUES IN THE TABLE ARE FOR TEMPERATURE OF THE SENSOR, NOT AIR TEMPERATURE.)*

NO: REPLACE WHEEL SPEED SENSOR.

YES:

(3)
- WITH J 39200 STILL CONNECTED, SELECT A/C VOLTAGE SCALE.
- VEHICLE TRANSMISSION IN NEUTRAL.
- SPIN WHEEL BY HAND WHILE OBSERVING VOLTAGE. IS VOLTAGE AT LEAST 100 mV (VOLTAGE WILL INCREASE AS WHEEL SPEED INCREASES)?

NO: REPLACE WHEEL SPEED SENSOR.

YES:

(4)
- IGNITION "OFF."
- DISCONNECT 32-WAY EBTCM CONNECTOR.
- USING J 39200, MEASURE THE RESISTANCE BETWEEN TERMINAL "A5" AND "A6" OF THE 32-WAY EBTCM HARNESS CONNECTOR. DOES J 39200 DISPLAY RESISTANCE AS "OL" (INFINITE)?

NO: REPAIR SHORT BETWEEN CKTs 882 AND 883.

YES:

(5)
- INSPECT CKTs 882 AND 883 AND 32-WAY EBTCM CONNECTOR FOR DAMAGE WHICH MAY RESULT IN A SHORT BETWEEN CKTs 882 AND 883 WITH ALL CONNECTORS CONNECTED. REPAIR ALL DAMAGE IF EVIDENT.
- RECONNECT ALL CONNECTORS.
- TEST DRIVE VEHICLE ABOVE 24 km/h (15 MPH) FOR AT LEAST 30 SECONDS. DOES DTC 28 SET AS A CURRENT DTC?

NO: MALFUNCTION IS INTERMITTENT. REFER TO "DIAGNOSTIC AIDS" ON FACING PAGE.

(6)
- INSPECT RIGHT REAR SENSOR TOOTHED RING FOR MISSING TEETH OR OTHER PHYSICAL DAMAGE. REFER TO SECTION 4B. IS ANY PHYSICAL DAMAGE INDICATED?

NO: REPLACE EBTCM.

YES: REPLACE TOOTHED RING. REFER TO SECTION 4B.

REAR WHEEL SPEED SENSOR
TEMPERATURE OF SENSOR VS. SENSOR RESISTANCE (APPROXIMATE)

TEMP. (°C)	TEMP. (°F)	RESISTANCE (OHMS)
-34 TO 4	-30 TO 40	2032 TO 3107
5 TO 43	41 TO 110	2417 TO 3615
44 TO 93	111 TO 200	2812 TO 4266

DIAGNOSTIC CHART DTC 32 — 1995 CAMARO AND FIREBIRD WITH TCS

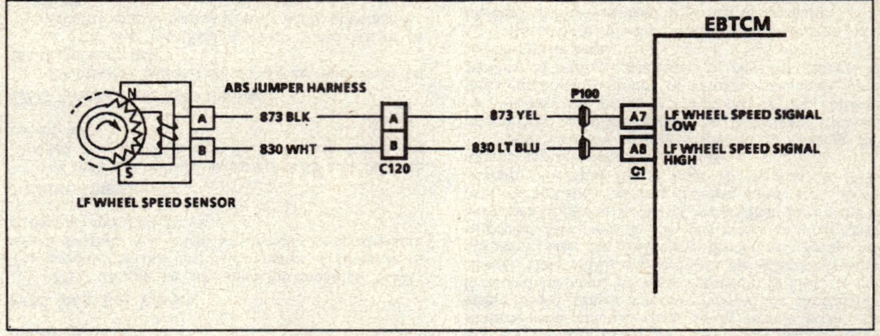

DTC 32
(Page 1 of 3)
LF WHEEL SPEED SNR CKT OPEN OR SHORT TO GND/BATT

Circuit Description:

As a toothed ring passes by the wheel speed sensor, changes in the electromagnetic field cause the wheel speed sensor to produce a sinusoidal(AC) voltage signal whose frequency is proportional to wheel speed. The magnitude of this signal is directly related to wheel speed and the proximity of the wheel speed sensor to the toothed ring often referred to as the air gap.

DTC Will Set When:

DTC 32 can be set anytime after initialization. If either of the left front wheel speed circuits are open or shorted to voltage or ground, a malfunction exists.

Action Taken:

A malfunction DTC is stored, ABS and TCS are disabled and the ABS and TCS warning lamps are turned "ON."

DTC Chart Test Description:

Number(s) below refer to circled number(s) on the diagnostic chart.

1. Checks for proper resistance of the wheel speed sensor coil.
2. Checks for a short to ground in the wheel speed sensor wiring.
3. Checks for a short to voltage in the wheel speed sensor wiring.
4. Checks for a short to voltage in CKT 873.
5. Checks for a short to voltage in CKT 830.
6. Ensures that the malfunction was not due to physical damage of the circuitry.

Diagnostic Aids:

An intermittent malfunction may be caused by a poor connection, rubbed through wire insulation, or a wire that is broken inside the insulation.

The frequency of the malfunction can be checked by using the enhanced diagnostic function of the Tech 1.

If the customer's comments reflect that the ABS or TCS warning lamps are "ON" only during moist environmental changes (rain, snow, vehicle wash), all wheel speed sensor circuitry should be thoroughly inspected for signs of water intrusion. If DTC is not current, clear DTCs and simulate the effects of water intrusion. Use the following procedure. Spray the suspected area with a 5% salt water solution (two teaspoons of salt to 12 oz. of water).Test drive vehicle over various road surfaces (bumps, turns, etc.) above 24km/h (15 mph) for at least 30 seconds. If DTC returns, replace suspected harness.

Any circuitry that is suspected of causing the intermittent complaint should be thoroughly checked for backed out terminals, improper mating, broken locks, improperly formed or damaged terminals, poor terminal to wiring connections or physical damage to the wiring harness.

Resistance of the wheel speed sensor will increase with an increase in sensor temperature.

When replacing a wheel speed sensor, inspect the sensor terminals and harness connector for corrosion and/or water intrusion. If evidence of corrosion or water intrusion exists, replace wheel speed sensor jumper harness. Likewise, if replacing a wheel speed sensor jumper harness, inspect sensor terminals. If evidence of corrosion or water intrusion exists, replace wheel speed sensor.

DIAGNOSTIC CHART DTC 32 — 1995 CAMARO AND FIREBIRD WITH TCS, CONTINUED

DTC 32
(Page 1 of 3)
LF WHEEL SPEED SNR CKT OPEN OR SHORT TO GND/BATT

Important
- WHEEL SPEED SENSOR INTERMITTENT MALFUNCTIONS MAY BE DIFFICULT TO LOCATE. CARE SHOULD BE TAKEN NOT TO DISTURB ANY ELECTRICAL CONNECTIONS PRIOR TO AN INDICATED STEP OF THIS CHART. THIS WILL ENSURE THAT AN INTERMITTENT CONNECTION WILL NOT BE CORRECTED BEFORE THE SOURCE OF THE MALFUNCTION IS FOUND.

(1)
- IGNITION "OFF."
- USING A 5% SALT WATER SOLUTION (TWO TEASPOONS OF SALT TO 12 OZ. WATER), THOROUGHLY SPRAY THE LEFT FRONT WHEEL SPEED SENSOR JUMPER HARNESS.
- DISCONNECT 32-WAY EBTCM HARNESS CONNECTOR.
- USING J 39200, MEASURE THE RESISTANCE BETWEEN TERMINALS "A7" AND "A8" OF THE 32-WAY EBTCM HARNESS CONNECTOR.

IS RESISTANCE BETWEEN 1020 AND 1137Ω WHEN SENSOR IS 20°C (68°F)? *(IF TEMPERATURE OF SENSOR IS SUSPECTED TO BE OTHER THAN 20°C (68°F), REFER TO THE TABLE ON PAGE 2 OF 2 OF THIS DTC CHART FOR APPLICABLE SENSOR RESISTANCE VALUES; THE VALUES IN THE TABLE ARE FOR TEMPERATURE OF THE SENSOR, NOT AIR TEMPERATURE.)*

- YES → (2)
- NO → DIAGNOSIS CONTINUED ON PAGE 2 OF 3 OF THIS DTC CHART.

(2)
- USING J 39200, MEASURE THE RESISTANCE BETWEEN GROUND AND TERMINAL "A7" OF THE 32-WAY EBTCM HARNESS CONNECTOR.

DOES J 39200 DISPLAY RESISTANCE AS "OL" (INFINITE)?

- YES → (3)
- NO → DIAGNOSIS CONTINUED ON PAGE 3 OF 3 OF THIS DTC CHART.

(3)
- IGNITION "ON."
- USING J 39200, MEASURE THE VOLTAGE BETWEEN GROUND AND TERMINAL "A7" OF THE 32-WAY EBTCM HARNESS CONNECTOR.

IS VOLTAGE 1 VOLT OR LESS?

- NO → (4)
- YES → (6)

(4)
- IGNITION "OFF."
- DISCONNECT LEFT FRONT WHEEL SPEED SENSOR DIRECTLY AT SENSOR.
- IGNITION "ON."
- USING J 39200, MEASURE THE VOLTAGE BETWEEN GROUND AND TERMINAL "A7" OF THE 32-WAY EBTCM HARNESS CONNECTOR.

IS VOLTAGE 1 VOLT OR LESS?

- YES → (5)
- NO → REPAIR SHORT TO VOLTAGE IN CKT 873.

(5)
- USING J 39200, MEASURE THE VOLTAGE BETWEEN GROUND AND TERMINAL "A8" OF THE 32-WAY EBTCM HARNESS CONNECTOR.

IS VOLTAGE 1 VOLT OR LESS?

- YES → MALFUNCTION IS INTERMITTENT. INSPECT ALL CONNECTORS AND HARNESSES FOR DAMAGE WHICH MAY RESULT IN A SHORT TO VOLTAGE WHEN ALL COMPONENTS ARE CONNECTED. REPAIR ALL DAMAGE IF FOUND. REFER TO "DIAGNOSTIC AIDS" ON FACING PAGE FOR MORE INFORMATION.
- NO → REPAIR SHORT TO VOLTAGE IN CKT 830.

(6)
- IGNITION "OFF."
- INSPECT TERMINALS "A7" AND "A8" OF THE 32-WAY EBTCM HARNESS CONNECTOR FOR POOR TERMINAL CONTACT AND CONTAMINATION. ALSO INSPECT CKTs 830 AND 873 FOR DAMAGE WHICH MAY RESULT IN AN OPEN CIRCUIT, SHORT TO GROUND, OR SHORT TO VOLTAGE. REPAIR ALL DAMAGE IF EVIDENT.
- RECONNECT ALL CONNECTORS.
- IGNITION "ON."

DOES DTC 32 SET AS A CURRENT DTC?

- YES → REPLACE EBTCM.
- NO → MALFUNCTION IS INTERMITTENT. REFER TO "DIAGNOSTIC AIDS" ON FACING PAGE.

DIAGNOSTIC CHART DTC 32 — 1995 CAMARO AND FIREBIRD WITH TCS, CONTINUED

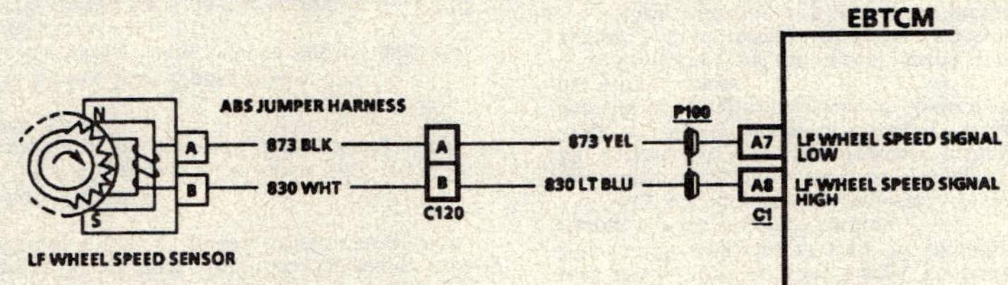

DTC 32
(Page 2 of 3)
LF WHEEL SPEED SNR CKT OPEN OR SHORT TO GND/BATT

Circuit Description:

As a toothed ring passes by the wheel speed sensor, changes in the electromagnetic field cause the wheel speed sensor to produce a sinusoidal(AC) voltage signal whose frequency is proportional to wheel speed. The magnitude of this signal is directly related to wheel speed and the proximity of the wheel speed sensor to the toothed ring, often referred to as the air gap.

DTC Will Set When:

DTC 32 can be set anytime after initialization. If either of the left front wheel speed circuits are open or shorted to voltage or ground, a malfunction exists.

Action Taken:

A malfunction DTC is stored, ABS and TCS are disabled and the ABS and TCS warning lamps are turned "ON."

DTC Chart Test Description:

Number(s) below refer to circled number(s) on the diagnostic chart.

7. Checks for high resistance in CKT 873.
8. Checks for high resistance in CKT 830.
9. Verifies proper resistance of the wheel speed sensor coil.

Diagnostic Aids:

An intermittent malfunction may be caused by a poor connection, rubbed through wire insulation, or a wire that is broken inside the insulation.

The frequency of the malfunction can be checked by using the enhanced diagnostic function of the Tech 1.

If the customer's comments reflect that the ABS or TCS warning lamps are "ON" only during moist environmental changes (rain, snow, vehicle wash), all wheel speed sensor circuitry should be thoroughly inspected for signs of water intrusion. If DTC is not current, clear DTCs and simulate the effects of water intrusion. Use the following procedure. Spray the suspected area with a 5% salt water solution (two teaspoons of salt to 12 oz. of water).Test drive vehicle over various road surfaces (bumps, turns, etc.) above 24km/h (15 mph) for at least 30 seconds. If DTC returns, replace suspected harness.

Any circuitry that is suspected of causing the intermittent complaint should be thoroughly checked for backed out terminals, improper mating, broken locks, improperly formed or damaged terminals, poor terminal to wiring connections or physical damage to the wiring harness.

Resistance of the wheel speed sensor will increase with an increase in sensor temperature.

When replacing a wheel speed sensor, inspect the sensor terminals and harness connector for corrosion and/or water intrusion. If evidence of corrosion or water intrusion exists, replace wheel speed sensor jumper harness. Likewise, if replacing a wheel speed sensor jumper harness, inspect sensor terminals. If evidence of corrosion or water intrusion exists, replace wheel speed sensor.

DIAGNOSTIC CHART DTC 32 — 1995 CAMARO AND FIREBIRD WITH TCS, CONTINUED

DTC 32
(Page 2 of 3)
LF WHEEL SPEED SNR CKT OPEN OR SHORT TO GND/BATT

DIAGNOSIS CONTINUED FROM PAGE 1 OF 3 OF THIS DTC CHART.

(7)
- DISCONNECT LEFT FRONT WHEEL SPEED SENSOR DIRECTLY AT SENSOR.
- USING J 39200, MEASURE THE RESISTANCE BETWEEN TERMINAL "A7" OF THE 32-WAY EBTCM HARNESS CONNECTOR AND TERMINAL "A" OF THE LEFT FRONT WHEEL SPEED SENSOR HARNESS CONNECTOR. IS RESISTANCE 2Ω OR LESS?

NO: REPAIR HIGH RESISTANCE IN CKT 873 (WHEEL SPEED SENSOR JUMPER MUST BE REPLACED IF DAMAGED).

YES:

(8)
- USING J 39200, MEASURE THE RESISTANCE BETWEEN TERMINAL "A8" OF THE 32-WAY EBTCM HARNESS CONNECTOR AND TERMINAL "B" OF THE LEFT FRONT WHEEL SPEED SENSOR HARNESS CONNECTOR. IS RESISTANCE 2Ω OR LESS?

NO: REPAIR HIGH RESISTANCE IN CKT 830 (WHEEL SPEED SENSOR JUMPER MUST BE REPLACED IF DAMAGED).

YES:

(9)
- USING J 39200, MEASURE THE RESISTANCE BETWEEN TERMINAL "A" AND TERMINAL "B" OF THE LEFT FRONT WHEEL SPEED SENSOR CONNECTOR. IS RESISTANCE BETWEEN 1020 AND 1137Ω WHEN SENSOR IS 20°C (68°F)? *(IF TEMPERATURE OF SENSOR IS SUSPECTED TO OTHER THAN 20°C (68°F), REFER TO THE TABLE BELOW FOR APPLICABLE SENSOR RESISTANCE VALUES; THE VALUES IN THE TABLE ARE FOR TEMPERATURE OF THE SENSOR, NOT AIR TEMPERATURE.)*

NO: REPLACE WHEEL SPEED SENSOR.

YES: MALFUNCTION IS INTERMITTENT. INSPECT ALL CONNECTORS AND HARNESSES FOR DAMAGE WHICH MAY RESULT IN HIGH RESISTANCE WHEN ALL COMPONENTS ARE CONNECTED. REFER TO "DIAGNOSTIC AIDS" ON FACING PAGE FOR MORE INFORMATION. REPAIR ALL DAMAGE IF FOUND.

FRONT WHEEL SPEED SENSOR
TEMPERATURE OF SENSOR VS. SENSOR RESISTANCE (APPROXIMATE)

TEMP. (°C)	TEMP. (°F)	RESISTANCE (OHMS)
-34 TO 4	-30 TO 40	804 TO 1066
5 TO 43	41 TO 110	956 TO 1240
44 TO 93	111 TO 200	1112 TO 1463

DIAGNOSTIC CHART DTC 32 — 1995 CAMARO AND FIREBIRD WITH TCS, CONTINUED

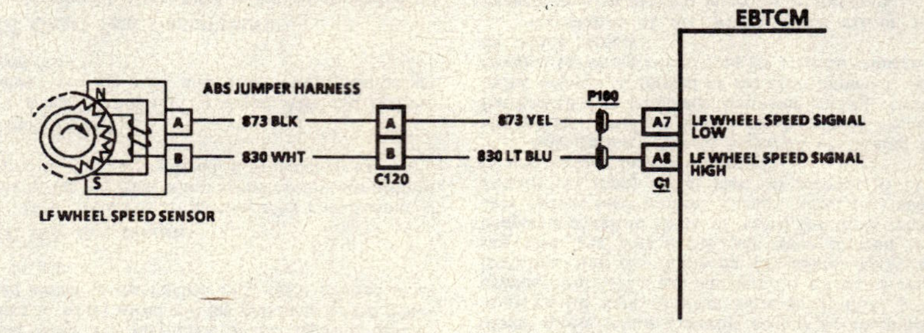

DTC 32
(Page 3 of 3)
LF WHEEL SPEED SNR CKT OPEN OR SHORT TO GND/BATT

Circuit Description:

As a toothed ring passes by the wheel speed sensor, changes in the electromagnetic field cause the wheel speed sensor to produce a sinusoidal(AC) voltage signal whose frequency is proportional to wheel speed. The magnitude of this signal is directly related to wheel speed and the proximity of the wheel speed sensor to the toothed ring, often referred to as the air gap.

DTC Will Set When:

DTC 32 can be set anytime after initialization. If either of the left front wheel speed circuits are open or shorted to voltage or ground, a malfunction exists.

Action Taken:

A malfunction DTC is stored, ABS and TCS are disabled and the ABS and TCS warning lamps are turned "ON."

DTC Chart Test Description:

Number(s) below refer to circled number(s) on the diagnostic chart.

10. Checks for a short to ground in CKT 873.
11. Checks for a short to ground in CKT 830.
12. Checks for a short to ground in the wheel speed sensor jumper harness.
13. Verifies that the wheel speed sensor is not internally shorted to ground.

Diagnostic Aids:

An intermittent malfunction may be caused by a poor connection, rubbed through wire insulation, or a wire that is broken inside the insulation.

The frequency of the malfunction can be checked by using the enhanced diagnostic function of the Tech 1.

If the customer's comments reflect that the ABS or TCS warning lamps are "ON" only during moist environmental changes (rain, snow, vehicle wash), all wheel speed sensor circuitry should be thoroughly inspected for signs of water intrusion. If DTC is not current, clear DTCs and simulate the effects of water intrusion. Use the following procedure. Spray the suspected area with a 5% salt water solution (two teaspoons of salt to 12 oz. of water).Test drive vehicle over various road surfaces (bumps, turns, etc.) above 24km/h (15 mph) for at least 30 seconds. If DTC returns, replace suspected harness.

Any circuitry that is suspected of causing the intermittent complaint should be thoroughly checked for backed out terminals, improper mating, broken locks, improperly formed or damaged terminals, poor terminal to wiring connections or physical damage to the wiring harness.

Resistance of the wheel speed sensor will increase with an increase in sensor temperature.

When replacing a wheel speed sensor, inspect the sensor terminals and harness connector for corrosion and/or water intrusion. If evidence of corrosion or water intrusion exists, replace wheel speed sensor jumper harness. Likewise, if replacing a wheel speed sensor jumper harness, inspect sensor terminals. If evidence of corrosion or water intrusion exists, replace wheel speed sensor.

DIAGNOSTIC CHART DTC 32 — 1995 CAMARO AND FIREBIRD WITH TCS, CONTINUED

DTC 32
(Page 3 of 3)
LF WHEEL SPEED SNR CKT OPEN OR SHORT TO GND/BATT

DIAGNOSIS CONTINUED FROM PAGE 1 OF 3 OF THIS DTC CHART.

(10)
- DISCONNECT CONNECTOR C120 (LOCATED LH INNER FRAME RAIL).
- USING J 39200, MEASURE THE RESISTANCE BETWEEN GROUND AND TERMINAL "A7" OF THE 32-WAY EBCM HARNESS CONNECTOR.

DOES J 39200 DISPLAY RESISTANCE AS "OL" (INFINITE)?

NO → REPAIR SHORT TO GROUND IN CKT 873.

YES ↓

(11)
- USING J 39200, MEASURE THE RESISTANCE BETWEEN GROUND AND TERMINAL "A8" OF THE 32-WAY EBCM HARNESS CONNECTOR.

DOES J 39200 DISPLAY RESISTANCE AS "OL" (INFINITE)?

NO → REPAIR SHORT TO GROUND IN CKT 830.

YES ↓

(12)
- THOROUGHLY RESPRAY THE LEFT FRONT WHEEL SPEED SENSOR JUMPER HARNESS WITH THE SALT WATER SOLUTION.

IMPORTANT: DO NOT ALLOW WATER TO ENTER ANY WHEEL SPEED SENSOR CONNECTOR.

- DISCONNECT LEFT FRONT WHEEL SPEED SENSOR DIRECTLY AT SENSOR.
- USING J 39200, MEASURE THE RESISTANCE BETWEEN GROUND AND TERMINAL "A," THEN "B" OF THE LEFT FRONT WHEEL SPEED SENSOR HARNESS CONNECTOR.

DOES J 39200 DISPLAY RESISTANCE AS "OL" (INFINITE) FOR BOTH MEASUREMENTS?

NO → REPLACE LEFT FRONT WHEEL SPEED SENSOR JUMPER HARNESS.

YES ↓

(13)
- USING J 39200, MEASURE THE RESISTANCE BETWEEN GROUND AND TERMINAL "A" OF THE LEFT FRONT WHEEL SPEED SENSOR CONNECTOR.

DOES J 39200 DISPLAY RESISTANCE AS "OL" (INFINITE)?

NO → REPLACE WHEEL SPEED SENSOR.

YES ↓

MALFUNCTION IS INTERMITTENT. INSPECT ALL CONNECTORS AND HARNESSES FOR DAMAGE WHICH MAY RESULT IN A SHORT TO GROUND WHEN ALL COMPONENTS ARE CONNECTED. PAY PARTICULAR ATTENTION TO THE JUMPER HARNESS, SMALL NICKS OR CUTS CAN CAUSE A MALFUNCTION. REPAIR ANY DAMAGE IF FOUND. JUMPER HARNESS MUST BE REPLACED IF DAMAGED. REFER TO "DIAGNOSTIC AIDS" ON FACING PAGE FOR MORE INFORMATION.

DIAGNOSTIC CHART DTC 33 — 1995 CAMARO AND FIREBIRD WITH TCS

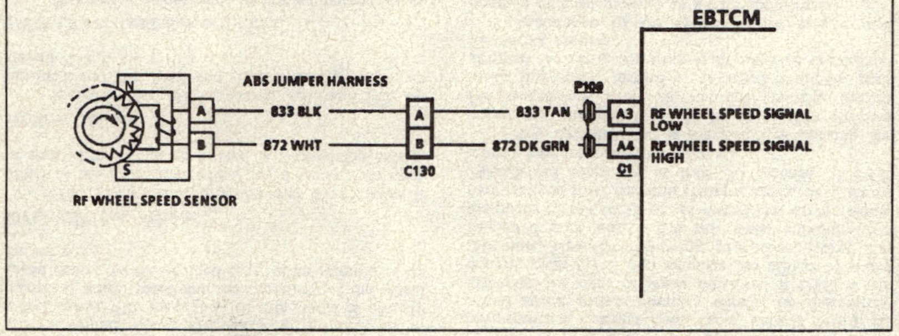

DTC 33
(Page 1 of 3)
RF WHEEL SPEED SNR CKT OPEN OR SHORT TO GND/BATT

Circuit Description:
As a toothed ring passes by the wheel speed sensor, changes in the electromagnetic field cause the wheel speed sensor to produce a sinusoidal(AC) voltage signal whose frequency is proportional to wheel speed. The magnitude of this signal is directly related to wheel speed and the proximity of the wheel speed sensor to the toothed ring, often referred to as the air gap.

DTC Will Set When:
DTC 33 can be set anytime after initialization. If either of the right front wheel speed circuits are open or shorted to voltage or ground, a malfunction exists.

Action Taken:
A malfunction DTC is stored, ABS and TCS are disabled and the ABS and TCS warning lamps are turned "ON."

DTC Chart Test Description:
Number(s) below refer to circled number(s) on the diagnostic chart.
1. Checks for proper resistance of the wheel speed sensor coil.
2. Checks for a short to ground in the wheel speed sensor wiring.
3. Checks for a short to voltage in the wheel speed sensor wiring.
4. Checks for a short to voltage in CKT 833.
5. Checks for a short to voltage in CKT 872.
6. Ensures that the malfunction was not due to physical damage of the circuitry.

Diagnostic Aids:
An intermittent malfunction may be caused by a poor connection, rubbed through wire insulation, or a wire that is broken inside the insulation.

The frequency of the malfunction can be checked by using the enhanced diagnostic function of the Tech 1,

If the customer's comments reflect that the ABS or TCS warning lamps are "ON" only during moist environmental changes (rain, snow, vehicle wash), all wheel speed sensor circuitry should be thoroughly inspected for signs of water intrusion. If DTC is not current, clear DTCs and simulate the effects of water intrusion. Use the following procedure. Spray the suspected area with a 5% salt water solution (two teaspoons of salt to 12 oz. of water).Test drive vehicle over various road surfaces (bumps, turns, etc.) above 24km/h (15 mph) for at least 30 seconds. If DTC returns, replace suspected harness.

Any circuitry that is suspected of causing the intermittent complaint should be thoroughly checked for backed out terminals, improper mating, broken locks, improperly formed or damaged terminals, poor terminal to wiring connections or physical damage to the wiring harness.

Resistance of the wheel speed sensor will increase with an increase in sensor temperature.

When replacing a wheel speed sensor, inspect the sensor terminals and harness connector for corrosion and/or water intrusion. If evidence of corrosion or water intrusion exists, replace wheel speed sensor jumper harness. Likewise, if replacing a wheel speed sensor jumper harness, inspect sensor terminals. If evidence of corrosion or water intrusion exists, replace wheel speed sensor.

DIAGNOSTIC CHART DTC 33 — 1995 CAMARO AND FIREBIRD WITH TCS, CONTINUED

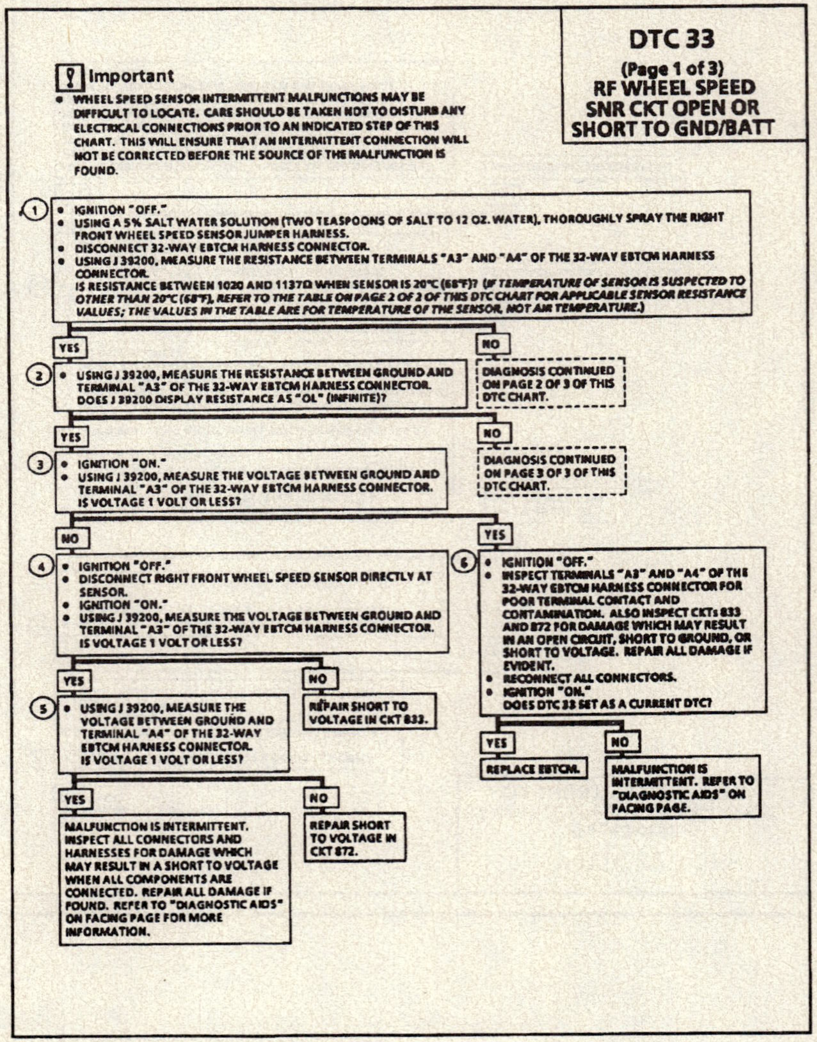

DIAGNOSTIC CHART DTC 33 — 1995 CAMARO AND FIREBIRD WITH TCS, CONTINUED

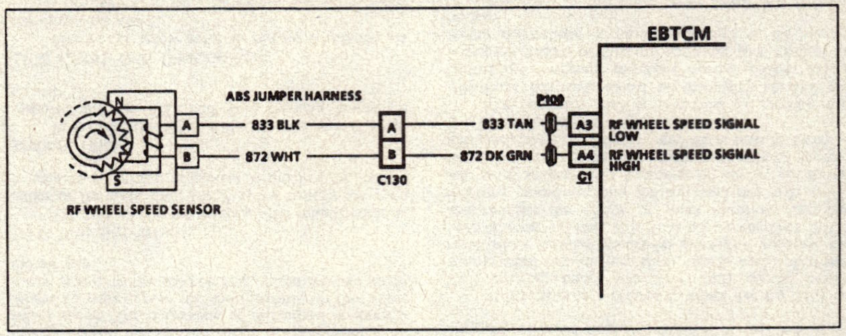

DTC 33
(Page 2 of 3)
RF WHEEL SPEED SNR CKT OPEN OR SHORT TO GND/BATT

Circuit Description:

As a toothed ring passes by the wheel speed sensor, changes in the electromagnetic field cause the wheel speed sensor to produce a sinusoidal(AC) voltage signal whose frequency is proportional to wheel speed. The magnitude of this signal is directly related to wheel speed and the proximity of the wheel speed sensor to the toothed ring, often referred to as the air gap.

DTC Will Set When:

DTC 33 can be set anytime after initialization. If either of the right front wheel speed circuits are open or shorted to voltage or ground, a malfunction exists.

Action Taken:

A malfunction DTC is stored, ABS and TCS are disabled and the ABS and TCS warning lamps are turned "ON."

DTC Chart Test Description:

Number(s) below refer to circled number(s) on the diagnostic chart.

7. Checks for high resistance in CKT 833.
8. Checks for high resistance in CKT 872.
9. Verifies proper resistance of the wheel speed sensor coil.

Diagnostic Aids:

An intermittent malfunction may be caused by a poor connection, rubbed through wire insulation, or a wire that is broken inside the insulation.

The frequency of the malfunction can be checked by using the enhanced diagnostic function of the Tech 1.

If the customer's comments reflect that the ABS or TCS warning lamps are "ON" only during moist environmental changes (rain, snow, vehicle wash), all wheel speed sensor circuitry should be thoroughly inspected for signs of water intrusion. If DTC is not current, clear DTCs and simulate the effects of water intrusion. Use the following procedure. Spray the suspected area with a 5% salt water solution (two teaspoons of salt to 12 oz. of water).Test drive vehicle over various road surfaces (bumps, turns, etc.) above 24km/h (15 mph) for at least 30 seconds. If DTC returns, replace suspected harness.

Any circuitry that is suspected of causing the intermittent complaint should be thoroughly checked for backed out terminals, improper mating, broken locks, improperly formed or damaged terminals, poor terminal to wiring connections or physical damage to the wiring harness.

Resistance of the wheel speed sensor will increase with an increase in sensor temperature.

When replacing a wheel speed sensor, inspect the sensor terminals and harness connector for corrosion and/or water intrusion. If evidence of corrosion or water intrusion exists, replace wheel speed sensor jumper harness. Likewise, if replacing a wheel speed sensor jumper harness, inspect sensor terminals. If evidence of corrosion or water intrusion exists, replace wheel speed sensor.

DIAGNOSTIC CHART DTC 33 — 1995 CAMARO AND FIREBIRD WITH TCS, CONTINUED

DTC 33
(Page 2 of 3)
RF WHEEL SPEED SNR CKT OPEN OR SHORT TO GND/BATT

DIAGNOSIS CONTINUED FROM PAGE 1 OF 3 OF THIS DTC CHART.

(7)
- DISCONNECT RIGHT FRONT WHEEL SPEED SENSOR DIRECTLY AT SENSOR.
- USING J 39200, MEASURE THE RESISTANCE BETWEEN TERMINAL "A3" OF THE 32-WAY EBTCM HARNESS CONNECTOR AND TERMINAL "A" OF THE RIGHT FRONT WHEEL SPEED SENSOR HARNESS CONNECTOR.

IS RESISTANCE 2Ω OR LESS?

- NO: REPAIR HIGH RESISTANCE IN CKT 833 (WHEEL SPEED SENSOR JUMPER MUST BE REPLACED IF DAMAGED).
- YES:

(8)
- USING J 39200, MEASURE THE RESISTANCE BETWEEN TERMINAL "A4" OF THE 32-WAY EBTCM HARNESS CONNECTOR AND TERMINAL "B" OF THE RIGHT FRONT WHEEL SPEED SENSOR HARNESS CONNECTOR.

IS RESISTANCE 2Ω OR LESS?

- NO: REPAIR HIGH RESISTANCE IN CKT 872 (WHEEL SPEED SENSOR JUMPER MUST BE REPLACED IF DAMAGED).
- YES:

(9)
- USING J 39200, MEASURE THE RESISTANCE BETWEEN TERMINAL "A" AND TERMINAL "B" OF THE RIGHT FRONT WHEEL SPEED SENSOR CONNECTOR.

IS RESISTANCE BETWEEN 1020 AND 1137Ω WHEN SENSOR IS 20°C (68°F)? *(IF TEMPERATURE OF SENSOR IS SUSPECTED TO OTHER THAN 20°C (68°F), REFER TO THE TABLE BELOW FOR APPLICABLE SENSOR RESISTANCE VALUES; THE VALUES IN THE TABLE ARE FOR TEMPERATURE OF THE SENSOR, NOT AIR TEMPERATURE.)*

- NO: REPLACE WHEEL SPEED SENSOR.
- YES: MALFUNCTION IS INTERMITTENT. INSPECT ALL CONNECTORS AND HARNESSES FOR DAMAGE WHICH MAY RESULT IN HIGH RESISTANCE WHEN ALL COMPONENTS ARE CONNECTED. REFER TO "DIAGNOSTIC AIDS" ON FACING PAGE FOR MORE INFORMATION. REPAIR ALL DAMAGE IF FOUND.

FRONT WHEEL SPEED SENSOR

TEMPERATURE OF SENSOR VS. SENSOR RESISTANCE (APPROXIMATE)

TEMP. (°C)	TEMP. (°F)	RESISTANCE (OHMS)
-34 TO 4	-30 TO 40	804 TO 1066
5 TO 43	41 TO 110	956 TO 1240
44 TO 93	111 TO 200	1112 TO 1463

DIAGNOSTIC CHART DTC 33 — 1995 CAMARO AND FIREBIRD WITH TCS, CONTINUED

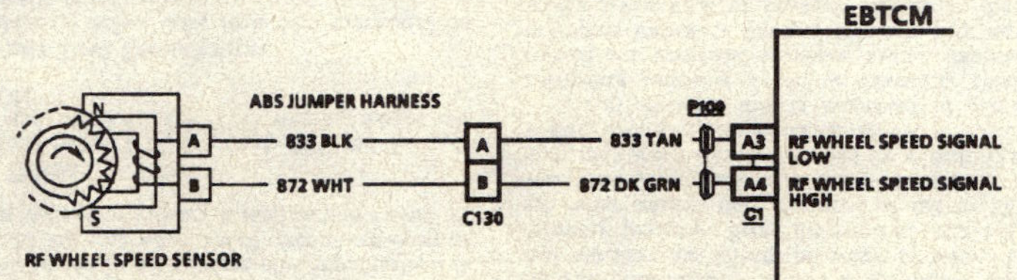

DTC 33
(Page 3 of 3)
RF WHEEL SPEED SNR CKT OPEN OR SHORT TO GND/BATT

Circuit Description:

As a toothed ring passes by the wheel speed sensor, changes in the electromagnetic field cause the wheel speed sensor to produce a sinusoidal(AC) voltage signal whose frequency is proportional to wheel speed. The magnitude of this signal is directly related to wheel speed and the proximity of the wheel speed sensor to the toothed ring, often referred to as the air gap.

DTC Will Set When:

DTC 33 can be set anytime after initialization. If either of the right front wheel speed circuits are open or shorted to voltage or ground, a malfunction exists.

Action Taken:

A malfunction DTC is stored, ABS and TCS are disabled and the ABS and TCS warning lamps are turned "ON."

DTC Chart Test Description:

Number(s) below refer to circled number(s) on the diagnostic chart.

10. Checks for a short to ground in CKT 833.
11. Checks for a short to ground in CKT 872.
12. Checks for a short to ground in the wheel speed sensor jumper harness.
13. Verifies that the wheel speed sensor is not internally shorted to ground.

Diagnostic Aids:

An intermittent malfunction may be caused by a poor connection, rubbed through wire insulation, or a wire that is broken inside the insulation.

The frequency of the malfunction can be checked by using the enhanced diagnostic function of the Tech 1,

If the customer's comments reflect that the ABS or TCS warning lamps are "ON" only during moist environmental changes (rain, snow, vehicle wash), all wheel speed sensor circuitry should be thoroughly inspected for signs of water intrusion. If DTC is not current, clear DTCs and simulate the effects of water intrusion. Use the following procedure. Spray the suspected area with a 5% salt water solution (two teaspoons of salt to 12 oz. of water).Test drive vehicle over various road surfaces (bumps, turns, etc.) above 24km/h (15 mph) for at least 30 seconds. If DTC returns, replace suspected harness.

Any circuitry that is suspected of causing the intermittent complaint should be thoroughly checked for backed out terminals, improper mating, broken locks, improperly formed or damaged terminals, poor terminal to wiring connections or physical damage to the wiring harness.

Resistance of the wheel speed sensor will increase with an increase in sensor temperature.

When replacing a wheel speed sensor, inspect the sensor terminals and harness connector for corrosion and/or water intrusion. If evidence of corrosion or water intrusion exists, replace wheel speed sensor jumper harness. Likewise, if replacing a wheel speed sensor jumper harness, inspect sensor terminals. If evidence of corrosion or water intrusion exists, replace wheel speed sensor.

DIAGNOSTIC CHART DTC 33 — 1995 CAMARO AND FIREBIRD WITH TCS, CONTINUED

DTC 33
(Page 3 of 3)
RF WHEEL SPEED SNR CKT OPEN OR SHORT TO GND/BATT

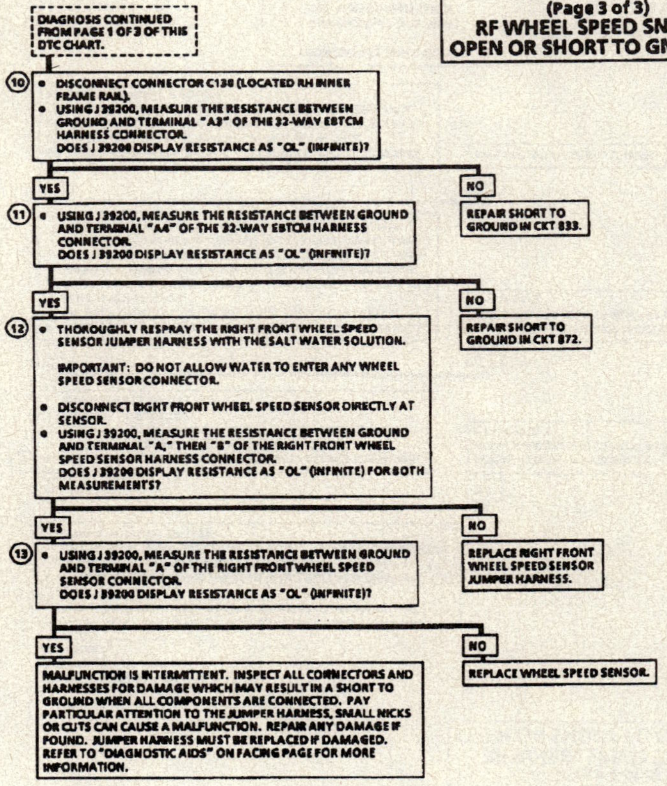

DIAGNOSTIC CHART DTC 34 — 1995 CAMARO AND FIREBIRD WITH TCS

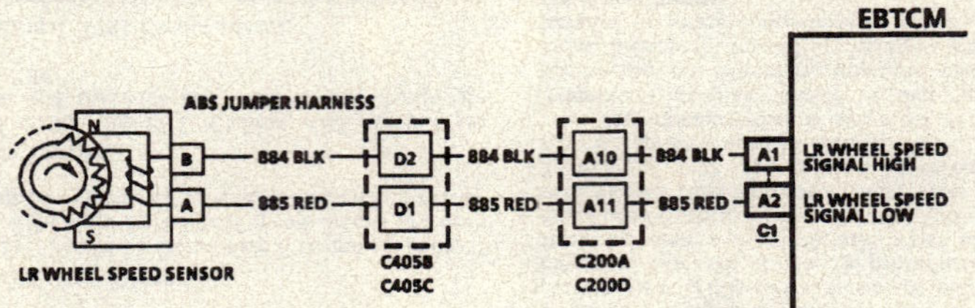

DTC 34
(Page 1 of 3)
LR WHEEL SPEED SNR CKT OPEN OR SHORT TO GND/BATT

Circuit Description:

As a toothed ring passes by the wheel speed sensor, changes in the electromagnetic field cause the wheel speed sensor to produce a sinusoidal(AC) voltage signal whose frequency is proportional to wheel speed. The magnitude of this signal is directly related to wheel speed and the proximity of the wheel speed sensor to the toothed ring, often referred to as the air gap.

DTC Will Set When:

DTC 34 can be set anytime after initialization. If either of the left rear wheel speed circuits are open or shorted to voltage or ground, a malfunction exists.

Action Taken:

A malfunction DTC is stored, ABS and TCS are disabled and the ABS and TCS warning lamps are turned "ON."

DTC Chart Test Description:

Number(s) below refer to circled number(s) on the diagnostic chart.

1. Checks for proper resistance of the wheel speed sensor coil.
2. Checks for a short to ground in the wheel speed sensor wiring.
3. Checks for a short to voltage in the wheel speed sensor wiring.
4. Checks for a short to voltage in CKT 885.
5. Checks for a short to voltage in CKT 884.
6. Ensures that the malfunction was not due to physical damage of the circuitry.

Diagnostic Aids:

An intermittent malfunction may be caused by a poor connection, rubbed through wire insulation, or a wire that is broken inside the insulation.

The frequency of the malfunction can be checked by using the enhanced diagnostic function of the Tech 1,

If the customer's comments reflect that the ABS or TCS warning lamps are "ON" only during moist environmental changes (rain, snow, vehicle wash), all wheel speed sensor circuitry should be thoroughly inspected for signs of water intrusion. If DTC is not current, clear DTCs and simulate the effects of water intrusion. Use the following procedure. Spray the suspected area with a 5% salt water solution (two teaspoons of salt to 12 oz. of water).Test drive vehicle over various road surfaces (bumps, turns, etc.) above 24km/h (15 mph) for at least 30 seconds. If DTC returns, replace suspected harness.

Any circuitry that is suspected of causing the intermittent complaint should be thoroughly checked for backed out terminals, improper mating, broken locks, improperly formed or damaged terminals, poor terminal to wiring connections or physical damage to the wiring harness.

Resistance of the wheel speed sensor will increase with an increase in sensor temperature.

When replacing a wheel speed sensor, inspect the sensor terminals and harness connector for corrosion and/or water intrusion. If evidence of corrosion or water intrusion exists, replace wheel speed sensor jumper harness. Likewise, if replacing a wheel speed sensor jumper harness, inspect sensor terminals. If evidence of corrosion or water intrusion exists, replace wheel speed sensor.

DIAGNOSTIC CHART DTC 34 — 1995 CAMARO AND FIREBIRD WITH TCS, CONTINUED

DTC 34 (Page 1 of 3) LR WHEEL SPEED SNR CKT OPEN OR SHORT TO GND/BATT

Important

- WHEEL SPEED SENSOR INTERMITTENT MALFUNCTIONS MAY BE DIFFICULT TO LOCATE. CARE SHOULD BE TAKEN NOT TO DISTURB ANY ELECTRICAL CONNECTIONS PRIOR TO AN INDICATED STEP OF THIS CHART. THIS WILL ENSURE THAT AN INTERMITTENT CONNECTION WILL NOT BE CORRECTED BEFORE THE SOURCE OF THE MALFUNCTION IS FOUND.

(1)
- IGNITION "OFF."
- USING A 5% SALT WATER SOLUTION (TWO TEASPOONS OF SALT TO 12 OZ. WATER), THOROUGHLY SPRAY THE LEFT REAR WHEEL SPEED SENSOR JUMPER HARNESS.
- DISCONNECT 32-WAY EBTCM HARNESS CONNECTOR.
- USING J 39200, MEASURE THE RESISTANCE BETWEEN TERMINALS "A1" AND "A2" OF THE 32-WAY EBTCM HARNESS CONNECTOR.

IS RESISTANCE BETWEEN 2579 AND 3315Ω WHEN SENSOR IS 20°C (68°F)? *(IF TEMPERATURE OF SENSOR IS SUSPECTED TO BE OTHER THAN 20°C (68°F), REFER TO THE TABLE ON PAGE 2 OF 2 OF THIS DTC CHART FOR APPLICABLE SENSOR RESISTANCE VALUES; THE VALUES IN THE TABLE ARE FOR TEMPERATURE OF THE SENSOR, NOT AIR TEMPERATURE.)*

- YES → (2)
- NO → DIAGNOSIS CONTINUED ON PAGE 2 OF 3 OF THIS DTC CHART.

(2)
- USING J 39200, MEASURE THE RESISTANCE BETWEEN GROUND AND TERMINAL "A2" OF THE 32-WAY EBTCM HARNESS CONNECTOR. DOES J 39200 DISPLAY RESISTANCE AS "OL" (INFINITE)?

- YES → (3)
- NO → DIAGNOSIS CONTINUED ON PAGE 3 OF 3 OF THIS DTC CHART.

(3)
- IGNITION "ON."
- USING J 39200, MEASURE THE VOLTAGE BETWEEN GROUND AND TERMINAL "A2" OF THE 32-WAY EBTCM HARNESS CONNECTOR. IS VOLTAGE 1 VOLT OR LESS?

- NO → (4)
- YES → (6)

(4)
- IGNITION "OFF."
- DISCONNECT LEFT REAR WHEEL SPEED SENSOR DIRECTLY AT SENSOR.
- IGNITION "ON."
- USING J 39200, MEASURE THE VOLTAGE BETWEEN GROUND AND TERMINAL "A2" OF THE 32-WAY EBTCM HARNESS CONNECTOR.

IS VOLTAGE 1 VOLT OR LESS?

- YES → (5)
- NO → REPAIR SHORT TO VOLTAGE IN CKT 885.

(5)
- USING J 39200, MEASURE THE VOLTAGE BETWEEN GROUND AND TERMINAL "A1" OF THE 32-WAY EBTCM HARNESS CONNECTOR. IS VOLTAGE 1 VOLT OR LESS?

- YES → MALFUNCTION IS INTERMITTENT. INSPECT ALL CONNECTORS AND HARNESSES FOR DAMAGE WHICH MAY RESULT IN A SHORT TO VOLTAGE WHEN ALL COMPONENTS ARE CONNECTED. REPAIR ALL DAMAGE IF FOUND. REFER TO "DIAGNOSTIC AIDS" ON FACING PAGE FOR MORE INFORMATION.
- NO → REPAIR SHORT TO VOLTAGE IN CKT 884.

(6)
- IGNITION "OFF."
- INSPECT TERMINALS "A2" AND "A1" OF THE 32-WAY EBTCM HARNESS CONNECTOR FOR POOR TERMINAL CONTACT AND CONTAMINATION. ALSO INSPECT CKTs 884 AND 885 FOR DAMAGE WHICH MAY RESULT IN AN OPEN CIRCUIT, SHORT TO GROUND, OR SHORT TO VOLTAGE. REPAIR ALL DAMAGE IF EVIDENT.
- RECONNECT ALL CONNECTORS.
- IGNITION "ON."

DOES DTC 34 SET AS A CURRENT DTC?

- YES → REPLACE EBTCM.
- NO → MALFUNCTION IS INTERMITTENT. REFER TO "DIAGNOSTIC AIDS" ON FACING PAGE.

DIAGNOSTIC CHART DTC 34 — 1995 CAMARO AND FIREBIRD WITH TCS, CONTINUED

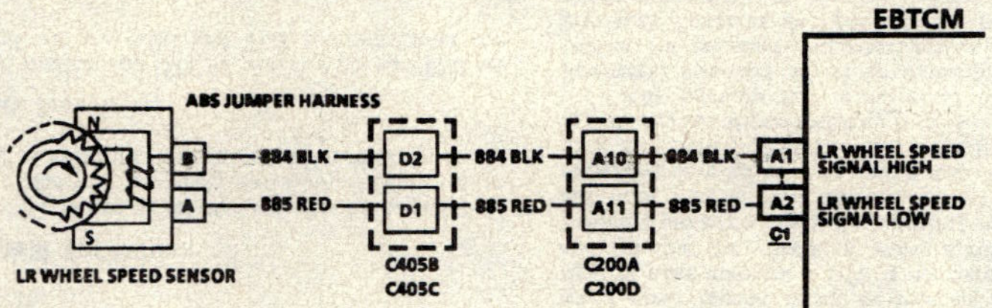

DTC 34 (Page 2 of 3) LR WHEEL SPEED SNR CKT OPEN OR SHORT TO GND/BATT

Circuit Description:

As a toothed ring passes by the wheel speed sensor, changes in the electromagnetic field cause the wheel speed sensor to produce a sinusoidal(AC) voltage signal whose frequency is proportional to wheel speed. The magnitude of this signal is directly related to wheel speed and the proximity of the wheel speed sensor to the toothed ring, often referred to as the air gap.

DTC Will Set When:

DTC 34 can be set anytime after initialization. If either of the left rear wheel speed circuits are open or shorted to voltage or ground, a malfunction exists.

Action Taken:

A malfunction DTC is stored, ABS and TCS are disabled and the ABS and TCS warning lamps are turned "ON."

DTC Chart Test Description:

Number(s) below refer to circled number(s) on the diagnostic chart.

7. Checks for high resistance in CKT 885.
8. Checks for high resistance in CKT 884.
9. Verifies proper resistance of the wheel speed sensor coil.

Diagnostic Aids:

An intermittent malfunction may be caused by a poor connection, rubbed through wire insulation, or a wire that is broken inside the insulation.

The frequency of the malfunction can be checked by using the enhanced diagnostic function of the Tech 1,

If the customer's comments reflect that the ABS or TCS warning lamps are "ON" only during moist environmental changes (rain, snow, vehicle wash), all wheel speed sensor circuitry should be thoroughly inspected for signs of water intrusion. If DTC is not current, clear DTCs and simulate the effects of water intrusion. Use the following procedure. Spray the suspected area with a 5% salt water solution (two teaspoons of salt to 12 oz. of water).Test drive vehicle over various road surfaces (bumps, turns, etc.) above 24km/h (15 mph) for at least 30 seconds. If DTC returns, replace suspected harness.

Any circuitry that is suspected of causing the intermittent complaint should be thoroughly checked for backed out terminals, improper mating, broken locks, improperly formed or damaged terminals, poor terminal to wiring connections or physical damage to the wiring harness.

Resistance of the wheel speed sensor will increase with an increase in sensor temperature.

When replacing a wheel speed sensor, inspect the sensor terminals and harness connector for corrosion and/or water intrusion. If evidence of corrosion or water intrusion exists, replace wheel speed sensor jumper harness. Likewise, if replacing a wheel speed sensor jumper harness, inspect sensor terminals. If evidence of corrosion or water intrusion exists, replace wheel speed sensor.

DIAGNOSTIC CHART DTC 34 — 1995 CAMARO AND FIREBIRD WITH TCS, CONTINUED

DTC 34
(Page 2 of 3)
LR WHEEL SPEED SNR CKT OPEN OR SHORT TO GND/BATT

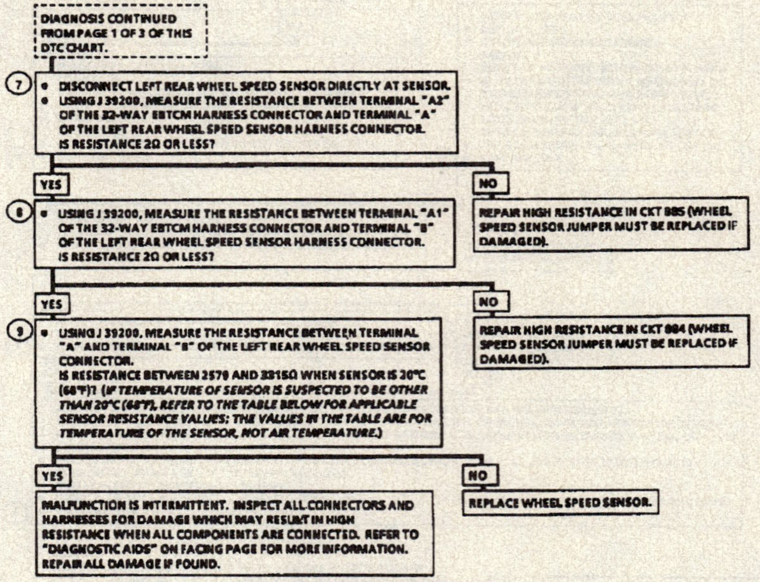

REAR WHEEL SPEED SENSOR

TEMPERATURE OF SENSOR VS. SENSOR RESISTANCE (APPROXIMATE)

TEMP. (°C)	TEMP. (°F)	RESISTANCE (OHMS)
-34 TO 4	-30 TO 40	2032 TO 3107
5 TO 43	41 TO 110	2417 TO 3615
44 TO 93	111 TO 200	2812 TO 4256

DIAGNOSTIC CHART DTC 34 — 1995 CAMARO AND FIREBIRD WITH TCS, CONTINUED

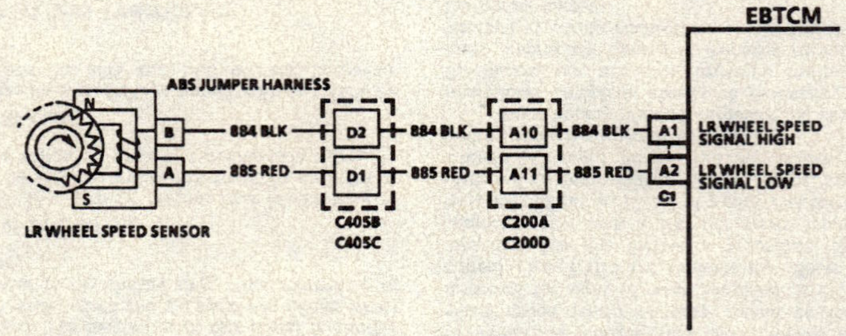

DTC 34
(Page 3 of 3)
LR WHEEL SPEED SNR CKT OPEN OR SHORT TO GND/BATT

Circuit Description:

As a toothed ring passes by the wheel speed sensor, changes in the electromagnetic field cause the wheel speed sensor to produce a sinusoidal(AC) voltage signal whose frequency is proportional to wheel speed. The magnitude of this signal is directly related to wheel speed and the proximity of the wheel speed sensor to the toothed ring, often referred to as the air gap.

DTC Will Set When:

DTC 34 can be set anytime after initialization. If either of the left rear wheel speed circuits are open or shorted to voltage or ground, a malfunction exists.

Action Taken:

A malfunction DTC is stored, ABS and TCS are disabled and the ABS and TCS warning lamps are turned "ON."

DTC Chart Test Description:

Number(s) below refer to circled number(s) on the diagnostic chart.

10. Checks for a short to ground in CKT 885.
11. Checks for a short to ground in CKT 884.
12. Checks for a short to ground in the wheel speed sensor jumper harness.
13. Verifies that the wheel speed sensor is not internally shorted to ground.

Diagnostic Aids:

An intermittent malfunction may be caused by a poor connection, rubbed through wire insulation, or a wire that is broken inside the insulation.

The frequency of the malfunction can be checked by using the enhanced diagnostic function of the Tech 1,

If the customer's comments reflect that the ABS or TCS warning lamps are "ON" only during moist environmental changes (rain, snow, vehicle wash), all wheel speed sensor circuitry should be thoroughly inspected for signs of water intrusion. If DTC is not current, clear DTCs and simulate the effects of water intrusion. Use the following procedure. Spray the suspected area with a 5% salt water solution (two teaspoons of salt to 12 oz. of water).Test drive vehicle over various road surfaces (bumps, turns, etc.) above 24km/h (15 mph) for at least 30 seconds. If DTC returns, replace suspected harness.

Any circuitry that is suspected of causing the intermittent complaint should be thoroughly checked for backed out terminals, improper mating, broken locks, improperly formed or damaged terminals, poor terminal to wiring connections or physical damage to the wiring harness.

Resistance of the wheel speed sensor will increase with an increase in sensor temperature.

When replacing a wheel speed sensor, inspect the sensor terminals and harness connector for corrosion and/or water intrusion. If evidence of corrosion or water intrusion exists, replace wheel speed sensor jumper harness. Likewise, if replacing a wheel speed sensor jumper harness, inspect sensor terminals. If evidence of corrosion or water intrusion exists, replace wheel speed sensor.

DIAGNOSTIC CHART DTC 34 — 1995 CAMARO AND FIREBIRD WITH TCS, CONTINUED

DTC 34
(Page 3 of 3)
LR WHEEL SPEED SNR CKT OPEN OR SHORT TO GND/BATT

DIAGNOSIS CONTINUED FROM PAGE 1 OF 3 OF THIS DTC CHART.

(10)
- DISCONNECT CONNECTOR C405B, C405C (LOCATED ON THE LH SIDE REAR UNDER BODY, BEHIND REAR SEAT).
- USING J 39200, MEASURE THE RESISTANCE BETWEEN GROUND AND TERMINAL "A2" OF THE 32-WAY EBTCM HARNESS CONNECTOR.

DOES J 39200 DISPLAY RESISTANCE AS "OL" (INFINITE)?

- NO → REPAIR SHORT TO GROUND IN CKT 885.
- YES ↓

(11)
- USING J 39200, MEASURE THE RESISTANCE BETWEEN GROUND AND TERMINAL "A1" OF THE 32-WAY EBTCM HARNESS CONNECTOR.

DOES J 39200 DISPLAY RESISTANCE AS "OL" (INFINITE)?

- NO → REPAIR SHORT TO GROUND IN CKT 884.
- YES ↓

(12)
- THOROUGHLY RESPRAY THE LEFT REAR WHEEL SPEED SENSOR JUMPER HARNESS WITH THE SALT WATER SOLUTION.

IMPORTANT: DO NOT ALLOW WATER TO ENTER ANY WHEEL SPEED SENSOR CONNECTOR.

- DISCONNECT LEFT REAR WHEEL SPEED SENSOR DIRECTLY AT SENSOR.
- USING J 39200, MEASURE THE RESISTANCE BETWEEN GROUND AND TERMINAL "A," THEN "B" OF THE LEFT REAR WHEEL SPEED SENSOR HARNESS CONNECTOR.

DOES J 39200 DISPLAY RESISTANCE AS "OL" (INFINITE) FOR BOTH MEASUREMENTS?

- NO → REPLACE REAR WHEEL SPEED SENSOR JUMPER HARNESS.
- YES ↓

(13)
- USING J 39200, MEASURE THE RESISTANCE BETWEEN GROUND AND TERMINAL "A" OF THE LEFT REAR WHEEL SPEED SENSOR CONNECTOR.

DOES J 39200 DISPLAY RESISTANCE AS "OL" (INFINITE)?

- NO → REPLACE WHEEL SPEED SENSOR.
- YES → MALFUNCTION IS INTERMITTENT. INSPECT ALL CONNECTORS AND HARNESSES FOR DAMAGE WHICH MAY RESULT IN A SHORT TO GROUND WHEN ALL COMPONENTS ARE CONNECTED. PAY PARTICULAR ATTENTION TO THE JUMPER HARNESS, SMALL NICKS OR CUTS CAN CAUSE A MALFUNCTION. REPAIR ANY DAMAGE IF FOUND. JUMPER HARNESS MUST BE REPLACED IF DAMAGED. REFER TO "DIAGNOSTIC AIDS" ON FACING PAGE FOR MORE INFORMATION.

DIAGNOSTIC CHART DTC 35 — 1995 CAMARO AND FIREBIRD WITH TCS

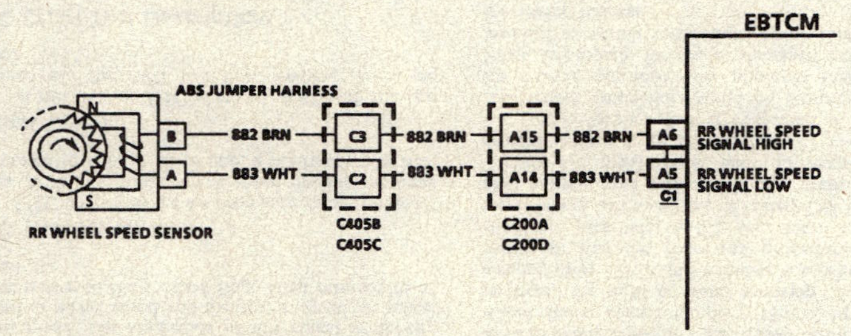

DTC 35
(Page 1 of 3)
RR WHEEL SPEED SNR CKT OPEN OR SHORT TO GND/BATT

Circuit Description:

As a toothed ring passes by the wheel speed sensor, changes in the electromagnetic field cause the wheel speed sensor to produce a sinusoidal(AC) voltage signal whose frequency is proportional to wheel speed. The magnitude of this signal is directly related to wheel speed and the proximity of the wheel speed sensor to the toothed ring, often referred to as the air gap.

DTC Will Set When:

DTC 35 can be set anytime after initialization. If either of the right rear wheel speed circuits are open or shorted to voltage or ground, a malfunction exists.

Action Taken:

A malfunction DTC is stored, ABS and TCS are disabled and the ABS and TCS warning lamps are turned "ON."

DTC Chart Test Description:

Number(s) below refer to circled number(s) on the diagnostic chart.

1. Checks for proper resistance of the wheel speed sensor coil.
2. Checks for a short to ground in the wheel speed sensor wiring.
3. Checks for a short to voltage in the wheel speed sensor wiring.
4. Checks for a short to voltage in CKT 883.
5. Checks for a short to voltage in CKT 882.
6. Ensures that the malfunction was not due to physical damage of the circuitry.

Diagnostic Aids:

An intermittent malfunction may be caused by a poor connection, rubbed through wire insulation, or a wire that is broken inside the insulation.

The frequency of the malfunction can be checked by using the enhanced diagnostic function of the Tech 1,

If the customer's comments reflect that the ABS or TCS warning lamps are "ON" only during moist environmental changes (rain, snow, vehicle wash), all wheel speed sensor circuitry should be thoroughly inspected for signs of water intrusion. If DTC is not current, clear DTCs and simulate the effects of water intrusion. Use the following procedure. Spray the suspected area with a 5% salt water solution (two teaspoons of salt to 12 oz. of water).Test drive vehicle over various road surfaces (bumps, turns, etc.) above 24km/h (15 mph) for at least 30 seconds. If DTC returns, replace suspected harness.

Any circuitry that is suspected of causing the intermittent complaint should be thoroughly checked for backed out terminals, improper mating, broken locks, improperly formed or damaged terminals, poor terminal to wiring connections or physical damage to the wiring harness.

Resistance of the wheel speed sensor will increase with an increase in sensor temperature.

When replacing a wheel speed sensor, inspect the sensor terminals and harness connector for corrosion and/or water intrusion. If evidence of corrosion or water intrusion exists, replace wheel speed sensor jumper harness. Likewise, if replacing a wheel speed sensor jumper harness, inspect sensor terminals. If evidence of corrosion or water intrusion exists, replace wheel speed sensor.

DIAGNOSTIC CHART DTC 35 — 1995 CAMARO AND FIREBIRD WITH TCS, CONTINUED

DTC 35
(Page 1 of 3)
RR WHEEL SPEED SNR CKT OPEN OR SHORT TO GND/BATT

Important
- WHEEL SPEED SENSOR INTERMITTENT MALFUNCTIONS MAY BE DIFFICULT TO LOCATE. CARE SHOULD BE TAKEN NOT TO DISTURB ANY ELECTRICAL CONNECTIONS PRIOR TO AN INDICATED STEP OF THIS CHART. THIS WILL ENSURE THAT AN INTERMITTENT CONNECTION WILL NOT BE CORRECTED BEFORE THE SOURCE OF THE MALFUNCTION IS FOUND.

(1)
- IGNITION "OFF."
- USING A 5% SALT WATER SOLUTION (TWO TEASPOONS OF SALT TO 12 OZ. WATER), THOROUGHLY SPRAY THE RIGHT REAR WHEEL SPEED SENSOR JUMPER HARNESS.
- DISCONNECT 32-WAY EBTCM HARNESS CONNECTOR.
- USING J 39200, MEASURE THE RESISTANCE BETWEEN TERMINALS "A5" AND "A6" OF THE 32-WAY EBTCM HARNESS CONNECTOR.

IS RESISTANCE BETWEEN 2579 AND 3315Ω WHEN SENSOR IS 20°C (68°F)? *(IF TEMPERATURE OF SENSOR IS SUSPECTED TO BE OTHER THAN 20°C (68°F), REFER TO THE TABLE ON PAGE 2 OF 2 OF THIS DTC CHART FOR APPLICABLE SENSOR RESISTANCE VALUES; THE VALUES IN THE TABLE ARE FOR TEMPERATURE OF THE SENSOR, NOT AIR TEMPERATURE.)*

YES → (2)
NO → DIAGNOSIS CONTINUED ON PAGE 2 OF 3 OF THIS DTC CHART.

(2)
- USING J 39200 MEASURE THE RESISTANCE BETWEEN GROUND AND TERMINAL "A5" OF THE 32-WAY EBTCM HARNESS CONNECTOR.

DOES J 39200 DISPLAY RESISTANCE AS "OL" (INFINITE)?

YES → (3)
NO → DIAGNOSIS CONTINUED ON PAGE 3 OF 3 OF THIS DTC CHART.

(3)
- IGNITION "ON."
- USING J 39200, MEASURE THE VOLTAGE BETWEEN GROUND AND TERMINAL "A5" OF THE 32-WAY EBTCM HARNESS CONNECTOR.

IS VOLTAGE 1 VOLT OR LESS?

NO → (4)
YES → (6)

(4)
- IGNITION "OFF."
- DISCONNECT RIGHT REAR WHEEL SPEED SENSOR DIRECTLY AT SENSOR.
- IGNITION "ON."
- USING J 39200, MEASURE THE VOLTAGE BETWEEN GROUND AND TERMINAL "A5" OF THE 32-WAY EBTCM HARNESS CONNECTOR.

IS VOLTAGE 1 VOLT OR LESS?

YES → (5)
NO → REPAIR SHORT TO VOLTAGE IN CKT 883.

(5)
- USING J 39200, MEASURE THE VOLTAGE BETWEEN GROUND AND TERMINAL "A6" OF THE 32-WAY EBTCM HARNESS CONNECTOR.

IS VOLTAGE 1 VOLT OR LESS?

YES → MALFUNCTION IS INTERMITTENT. INSPECT ALL CONNECTORS AND HARNESSES FOR DAMAGE WHICH MAY RESULT IN A SHORT TO VOLTAGE WHEN ALL COMPONENTS ARE CONNECTED. REPAIR ALL DAMAGE IF FOUND. REFER TO "DIAGNOSTIC AIDS" ON FACING PAGE FOR MORE INFORMATION.
NO → REPAIR SHORT TO VOLTAGE IN CKT 882.

(6)
- IGNITION "OFF."
- INSPECT TERMINALS "A5" AND "A6" OF THE 32-WAY EBTCM HARNESS CONNECTOR FOR POOR TERMINAL CONTACT AND CONTAMINATION. ALSO INSPECT CKTs 882 AND 883 FOR DAMAGE WHICH MAY RESULT IN AN OPEN CIRCUIT, SHORT TO GROUND, OR SHORT TO VOLTAGE. REPAIR ALL DAMAGE IF EVIDENT.
- RECONNECT ALL CONNECTORS.
- IGNITION "ON."

DOES DTC 35 SET AS A CURRENT DTC?

YES → REPLACE EBTCM.
NO → MALFUNCTION IS INTERMITTENT. REFER TO "DIAGNOSTIC AIDS" ON FACING PAGE.

DIAGNOSTIC CHART DTC 35 — 1995 CAMARO AND FIREBIRD WITH TCS, CONTINUED

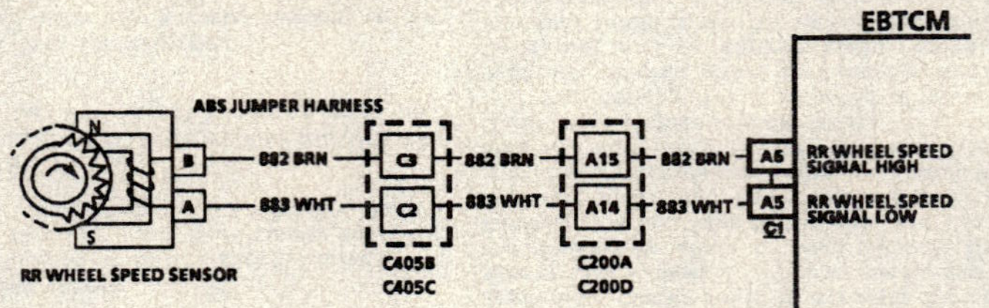

DTC 35
(Page 2 of 3)
RR WHEEL SPEED SNR CKT OPEN OR SHORT TO GND/BATT

Circuit Description:

As a toothed ring passes by the wheel speed sensor, changes in the electromagnetic field cause the wheel speed sensor to produce a sinusoidal (AC) voltage signal whose frequency is proportional to wheel speed. The magnitude of this signal is directly related to wheel speed and the proximity of the wheel speed sensor to the toothed ring, often referred to as the air gap.

DTC Will Set When:

DTC 35 can be set anytime after initialization. If either of the right rear wheel speed circuits are open or shorted to voltage or ground, a malfunction exists.

Action Taken:

A malfunction DTC is stored, ABS and TCS are disabled and the ABS and TCS warning lamps are turned "ON."

DTC Chart Test Description:

Number(s) below refer to circled number(s) on the diagnostic chart.

7. Checks for high resistance in CKT 883.
8. Checks for high resistance in CKT 882.
9. Verifies proper resistance of the wheel speed sensor coil.

Diagnostic Aids:

An intermittent malfunction may be caused by a poor connection, rubbed through wire insulation, or a wire that is broken inside the insulation.

The frequency of the malfunction can be checked by using the enhanced diagnostic function of the Tech 1,

If the customer's comments reflect that the ABS or TCS warning lamps are "ON" only during moist environmental changes (rain, snow, vehicle wash), all wheel speed sensor circuitry should be thoroughly inspected for signs of water intrusion. If DTC is not current, clear DTCs and simulate the effects of water intrusion. Use the following procedure. Spray the suspected area with a 5% salt water solution (two teaspoons of salt to 12 oz. of water).Test drive vehicle over various road surfaces (bumps, turns, etc.) above 24km/h (15 mph) for at least 30 seconds. If DTC returns, replace suspected harness.

Any circuitry that is suspected of causing the intermittent complaint should be thoroughly checked for backed out terminals, improper mating, broken locks, improperly formed or damaged terminals, poor terminal to wiring connections or physical damage to the wiring harness.

Resistance of the wheel speed sensor will increase with an increase in sensor temperature.

When replacing a wheel speed sensor, inspect the sensor terminals and harness connector for corrosion and/or water intrusion. If evidence of corrosion or water intrusion exists, replace wheel speed sensor jumper harness. Likewise, if replacing a wheel speed sensor jumper harness, inspect sensor terminals. If evidence of corrosion or water intrusion exists, replace wheel speed sensor.

DIAGNOSTIC CHART DTC 35 — 1995 CAMARO AND FIREBIRD WITH TCS, CONTINUED

DTC 35
(Page 2 of 3)
RR WHEEL SPEED SNR CKT OPEN OR SHORT TO GND/BATT

DIAGNOSIS CONTINUED FROM PAGE 1 OF 3 OF THIS DTC CHART.

(7)
- DISCONNECT RIGHT REAR WHEEL SPEED SENSOR DIRECTLY AT SENSOR.
- USING J 39200, MEASURE THE RESISTANCE BETWEEN TERMINAL "A5" OF THE 32-WAY EBTCM HARNESS CONNECTOR AND TERMINAL "A" OF THE RIGHT REAR WHEEL SPEED SENSOR HARNESS CONNECTOR.

IS RESISTANCE 2Ω OR LESS?

NO: REPAIR HIGH RESISTANCE IN CKT 883 (WHEEL SPEED SENSOR JUMPER MUST BE REPLACED IF DAMAGED).

YES:

(8)
- USING J 39200, MEASURE THE RESISTANCE BETWEEN TERMINAL "A6" OF THE 32-WAY EBTCM HARNESS CONNECTOR AND TERMINAL "B" OF THE RIGHT REAR WHEEL SPEED SENSOR HARNESS CONNECTOR.

IS RESISTANCE 2Ω OR LESS?

NO: REPAIR HIGH RESISTANCE IN CKT 882 (WHEEL SPEED SENSOR JUMPER MUST BE REPLACED IF DAMAGED).

YES:

(9)
- USING J 39200, MEASURE THE RESISTANCE BETWEEN TERMINAL "A" AND TERMINAL "B" OF THE RIGHT REAR WHEEL SPEED SENSOR CONNECTOR.

IS RESISTANCE BETWEEN 2579 AND 3315Ω WHEN SENSOR IS 20°C (68°F)? *(IF TEMPERATURE OF SENSOR IS SUSPECTED TO BE OTHER THAN 20°C (68°F), REFER TO THE TABLE BELOW FOR APPLICABLE SENSOR RESISTANCE VALUES; THE VALUES IN THE TABLE ARE FOR TEMPERATURE OF THE SENSOR, NOT AIR TEMPERATURE.)*

NO: REPLACE WHEEL SPEED SENSOR.

YES: MALFUNCTION IS INTERMITTENT. INSPECT ALL CONNECTORS AND HARNESSES FOR DAMAGE WHICH MAY RESULT IN HIGH RESISTANCE WHEN ALL COMPONENTS ARE CONNECTED.

REPAIR ALL DAMAGE IF FOUND.

REAR WHEEL SPEED SENSOR

TEMPERATURE OF SENSOR VS. SENSOR RESISTANCE (APPROXIMATE)

TEMP. (°C)	TEMP. (°F)	RESISTANCE (OHMS)
-34 TO 4	-30 TO 40	2032 TO 3187
5 TO 43	41 TO 110	2417 TO 3615
44 TO 93	111 TO 200	2812 TO 4266

DIAGNOSTIC CHART DTC 35 — 1995 CAMARO AND FIREBIRD WITH TCS, CONTINUED

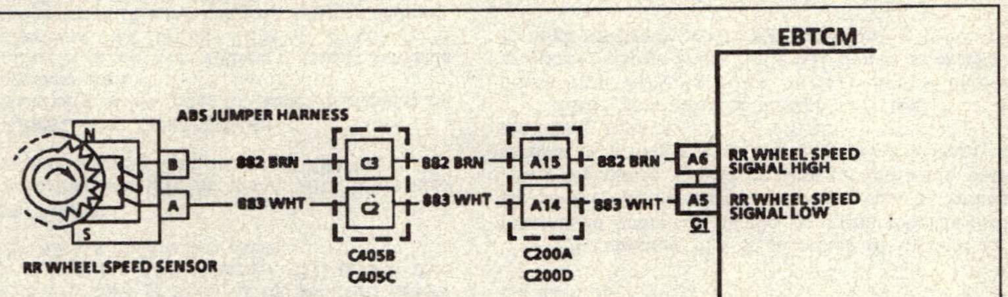

DTC 35
(Page 3 of 3)
RR WHEEL SPEED SNR CKT OPEN OR SHORT TO GND/BATT

Circuit Description:

As a toothed ring passes by the wheel speed sensor, changes in the electromagnetic field cause the wheel speed sensor to produce a sinusoidal(AC) voltage signal whose frequency is proportional to wheel speed. The magnitude of this signal is directly related to wheel speed and the proximity of the wheel speed sensor to the toothed ring, often referred to as the air gap.

DTC Will Set When:

DTC 35 can be set anytime after initialization. If either of the right rear wheel speed circuits are open or shorted to voltage or ground, a malfunction exists.

Action Taken:

A malfunction DTC is stored, ABS and TCS are disabled and the ABS and TCS warning lamps are turned "ON."

DTC Chart Test Description:

Number(s) below refer to circled number(s) on the diagnostic chart.

10. Checks for a short to ground in CKT 883.
11. Checks for a short to ground in CKT 882.
12. Checks for a short to ground in the wheel speed sensor jumper harness.
13. Verifies that the wheel speed sensor is not internally shorted to ground.

Diagnostic Aids:

An intermittent malfunction may be caused by a poor connection, rubbed through wire insulation, or a wire that is broken inside the insulation.

The frequency of the malfunction can be checked by using the enhanced diagnostic function of the Tech 1,

If the customer's comments reflect that the ABS or TCS warning lamps are "ON" only during moist environmental changes (rain, snow, vehicle wash), all wheel speed sensor circuitry should be thoroughly inspected for signs of water intrusion. If DTC is not current, clear DTCs and simulate the effects of water intrusion. Use the following procedure. Spray the suspected area with a 5% salt water solution (two teaspoons of salt to 12 oz. of water). Test drive vehicle over various road surfaces (bumps, turns, etc.) above 24 km/h (15 mph) for at least 30 seconds. If DTC returns, replace suspected harness.

Any circuitry that is suspected of causing the intermittent complaint should be thoroughly checked for backed out terminals, improper mating, broken locks, improperly formed or damaged terminals, poor terminal to wiring connections or physical damage to the wiring harness.

Resistance of the wheel speed sensor will increase with an increase in sensor temperature.

When replacing a wheel speed sensor, inspect the sensor terminals and harness connector for corrosion and/or water intrusion. If evidence of corrosion or water intrusion exists, replace wheel speed sensor jumper harness. Likewise, if replacing a wheel speed sensor jumper harness, inspect sensor terminals. If evidence of corrosion or water intrusion exists, replace wheel speed sensor.

DIAGNOSTIC CHART DTC 35 — 1995 CAMARO AND FIREBIRD WITH TCS, CONTINUED

DTC 35
(Page 3 of 3)
RR WHEEL SPEED SNR CKT OPEN OR SHORT TO GND/BATT

DIAGNOSIS CONTINUED FROM PAGE 1 OF 3 OF THIS DTC CHART.

(10)
- DISCONNECT CONNECTOR C405B, C405C (LOCATED ON THE LH SIDE REAR UNDER BODY, BEHIND REAR SEAT).
- USING J 39200, MEASURE THE RESISTANCE BETWEEN GROUND AND TERMINAL "A5" OF THE 32-WAY EBTCM HARNESS CONNECTOR.

DOES J 39200 DISPLAY RESISTANCE AS "OL" (INFINITE)?

NO → REPAIR SHORT TO GROUND IN CKT 883.

YES →

(11)
- USING J 39200, MEASURE THE RESISTANCE BETWEEN GROUND AND TERMINAL "A6" OF THE 32-WAY EBTCM HARNESS CONNECTOR.

DOES J 39200 DISPLAY RESISTANCE AS "OL" (INFINITE)?

NO → REPAIR SHORT TO GROUND IN CKT 882.

YES →

(12)
- THOROUGHLY RESPRAY THE RIGHT REAR WHEEL SPEED SENSOR JUMPER HARNESS WITH THE SALT WATER SOLUTION.

IMPORTANT: DO NOT ALLOW WATER TO ENTER ANY WHEEL SPEED SENSOR CONNECTOR.

- DISCONNECT RIGHT REAR WHEEL SPEED SENSOR DIRECTLY AT SENSOR.
- USING J 39200, MEASURE THE RESISTANCE BETWEEN GROUND AND TERMINAL "A," THEN "B" OF THE RIGHT REAR WHEEL SPEED SENSOR HARNESS CONNECTOR.

DOES J 39200 DISPLAY RESISTANCE AS "OL" (INFINITE) FOR BOTH MEASUREMENTS?

NO → REPLACE REAR WHEEL SPEED SENSOR JUMPER HARNESS.

YES →

(13)
- USING J 39200, MEASURE THE RESISTANCE BETWEEN GROUND AND TERMINAL "A" OF THE RIGHT REAR WHEEL SPEED SENSOR CONNECTOR.

DOES J 39200 DISPLAY RESISTANCE AS "OL" (INFINITE)?

NO → REPLACE WHEEL SPEED SENSOR.

YES → MALFUNCTION IS INTERMITTENT. INSPECT ALL CONNECTORS AND HARNESSES FOR DAMAGE WHICH MAY RESULT IN A SHORT TO GROUND WHEN ALL COMPONENTS ARE CONNECTED. PAY PARTICULAR ATTENTION TO THE JUMPER HARNESS, SMALL NICKS OR CUTS CAN CAUSE A MALFUNCTION. REPAIR ANY DAMAGE IF FOUND. JUMPER HARNESS MUST BE REPLACED IF DAMAGED. REFER TO "DIAGNOSTIC AIDS" ON FACING PAGE FOR MORE INFORMATION.

DIAGNOSTIC CHART DTC 36 — 1995 CAMARO AND FIREBIRD WITH TCS

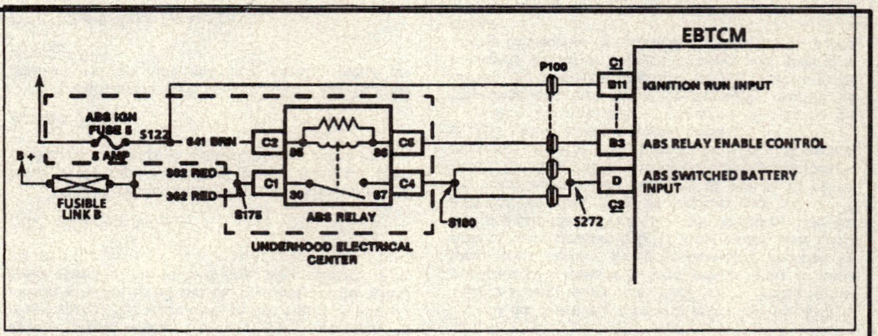

DTC 36
(Page 1 of 2)
LOW SYSTEM VOLTAGE

Circuit Description:
This DTC is used to monitor the voltage level available to the EBTCM. If the voltage drops below 11.5 volts, full performance of the ABS system cannot be guaranteed. During ABS operation, there are several current requirements that will cause battery voltage to drop. Because of this, voltage is monitored prior to ABS to indicate good charging system condition and also during ABS when voltage may drop significantly.

DTC Will Set When:
DTC 36 can only be set if the vehicle's speed is greater than 5 km/h (3 mph). If the switched battery voltage is less than 11.5 volts non-ABS or less than 9.0 volts ABS, a malfunction exists.

Action Taken:
A malfunction DTC is stored, ABS is disabled and the ABS warning lamp is turned "ON."

DTC Chart Test Description:
Number(s) below refer to circled number(s) on the diagnostic chart.

1. Checks to see what voltage is readily available at terminal "D" of the EBTCM. If voltage is representative of good charging system condition, a malfunction is not present.
2. Isolates the condition to high circuit resistance, or improper charging system operation.
3. Isolates which circuit is the source of the low voltage condition.
4. Checks for high resistance in CKT 302.

Diagnostic Aids:
An intermittent malfunction may be caused by a poor connection, rubbed through wire insulation, or a wire that is broken inside the insulation.

The frequency of the malfunction can be checked by using the enhanced diagnostic function of the Tech 1,

Any circuitry that is suspected of causing the intermittent complaint should be thoroughly checked for backed out terminals, improper mating, broken locks, improperly formed or damaged terminals, poor terminal to wiring connections or physical damage to the wiring harness.

While performing a voltage load test, if it is noted that only ignition voltage drops below acceptable voltage levels, CKT 641 should be checked for high resistance or an open condition.

DIAGNOSTIC CHART DTC 36 — 1995 CAMARO AND FIREBIRD WITH TCS, CONTINUED

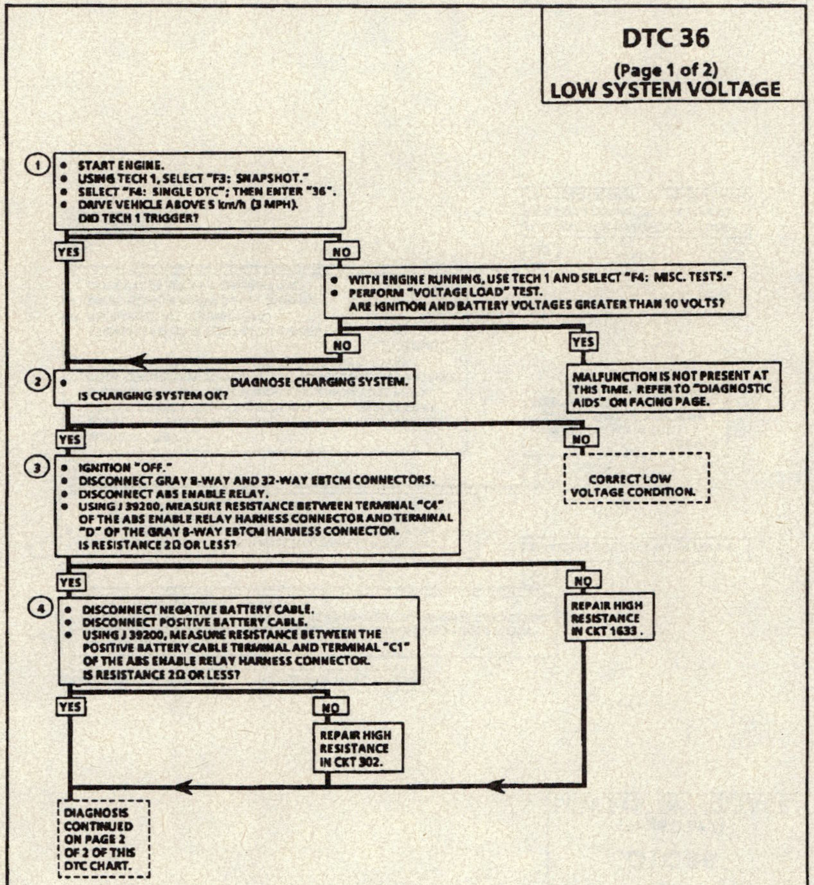

DIAGNOSTIC CHART DTC 36 — 1995 CAMARO AND FIREBIRD WITH TCS, CONTINUED

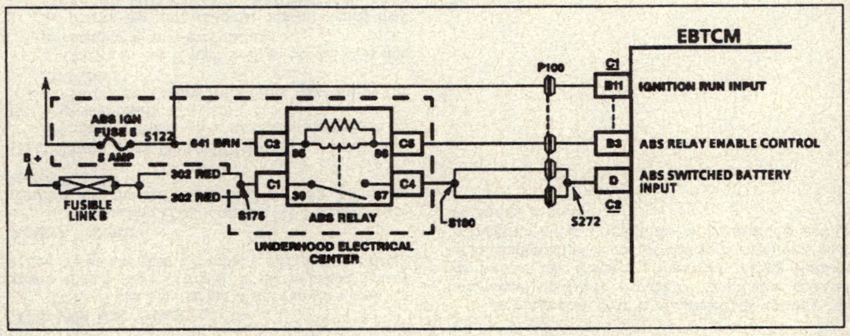

DTC 36 (Page 2 of 2) LOW SYSTEM VOLTAGE

Circuit Description:

This DTC is used to monitor the voltage level available to the EBTCM controller. If the voltage drops below 11.5 volts, full performance of the ABS system cannot be guaranteed. During ABS operation, there are several current requirements that will cause battery voltage to drop. Because of this, voltage is monitored prior to ABS to indicate good charging system condition and also during ABS when voltage may drop significantly.

DTC Will Set When:

DTC 36 can only be set if the vehicle's speed is greater than 5 km/h (3 mph). If the switched battery voltage is less than 11.5 volts non-ABS or less than 9.0 volts ABS, a malfunction exists.

Action Taken:

A malfunction DTC is stored, ABS is disabled and the ABS warning lamp is turned "ON."

DTC Chart Test Description:

Number(s) below refer to circled number(s) on the diagnostic chart.

5. Checks for high resistance in CKT 641.
6. Checks for high resistance in the ignition fused feed circuit.
7. Checks for malfunctioning EBTCM.

Diagnostic Aids:

An intermittent malfunction may be caused by a poor connection, rubbed through wire insulation, or a wire that is broken inside the insulation.

The frequency of the malfunction can be checked by using the enhanced diagnostic function of the Tech 1,

Any circuitry that is suspected of causing the intermittent complaint should be thoroughly checked for backed out terminals, improper mating, broken locks, improperly formed or damaged terminals, poor terminal to wiring connections or physical damage to the wiring harness.

While performing a voltage load test, if it is noted that only ignition voltage drops below acceptable voltage levels, CKT 641 should be checked for high resistance or an open condition.

DIAGNOSTIC CHART DTC 36 — 1995 CAMARO AND FIREBIRD WITH TCS, CONTINUED

DTC 36
(Page 2 of 2)
LOW SYSTEM VOLTAGE

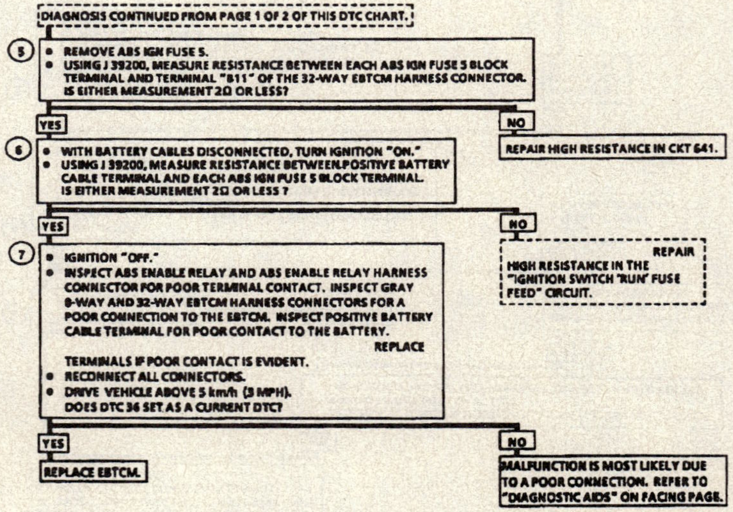

DIAGNOSTIC CHART DTC 37 — 1995 CAMARO AND FIREBIRD WITH TCS

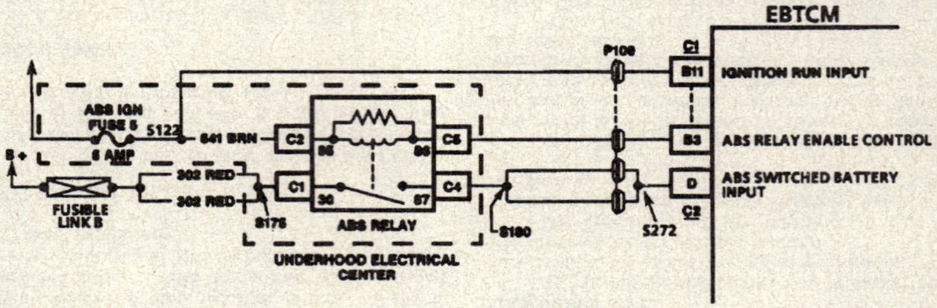

DTC 37
HIGH SYSTEM VOLTAGE

Circuit Description:
This DTC is designed to detect high vehicle voltage levels prior to any required motor movement (initialization or ABS/TCS). If excessive voltage exists, demagnetization of the motor magnets may occur, which would eventually affect or eliminate ABS/TCS performance.

DTC Will Set When:
DTC 37 can only be set if the vehicle's speed is greater than 5 km/h (3 mph). If the switched battery voltage is greater than 17 volts, a malfunction exists.

Action Taken:
A malfunction DTC is stored, ABS and TCS are disabled and the ABS and TCS warning lamps are turned "ON."

DTC Chart Test Description:
Number(s) below refer to circled number(s) on the diagnostic chart.
1. Checks the voltage level being received by the EBTCM.
2. Checks to see if high voltage condition is due to charging system malfunction.
3. Checks for poor terminal contact which may cause an intermittent malfunction, or a malfunctioning EBTCM.

Diagnostic Aids:
An intermittent malfunction may be caused by a poor connection, rubbed through wire insulation, or a wire that is broken inside the insulation.
The frequency of the malfunction can be checked by using the enhanced diagnostic function of the Tech 1,

Any circuitry that is suspected of causing the intermittent complaint should be thoroughly checked for backed out terminals, improper mating, broken locks, improperly formed or damaged terminals, poor terminal to wiring connections or physical damage to the wiring harness.

DIAGNOSTIC CHART DTC 37 — 1995 CAMARO AND FIREBIRD WITH TCS, CONTINUED

DTC 37
HIGH SYSTEM VOLTAGE

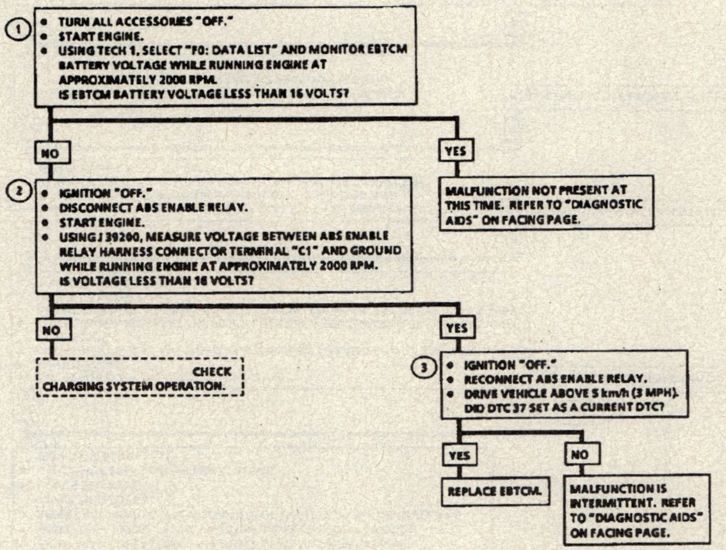

DIAGNOSTIC CHART DTC 38 — 1995 CAMARO AND FIREBIRD WITH TCS

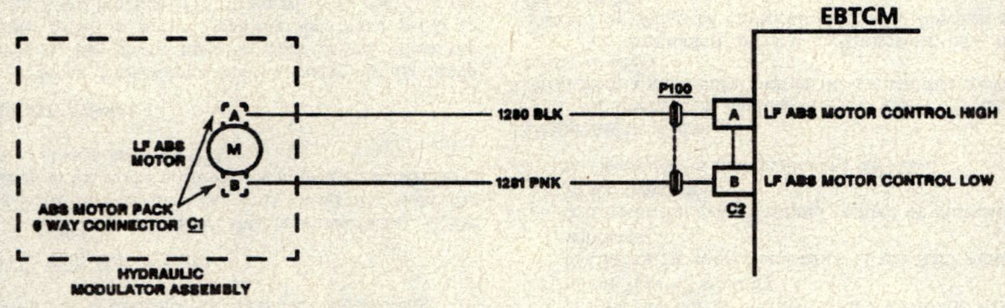

DTC 38
LEFT FRONT ESB WILL NOT HOLD MOTOR

Circuit Description:

This DTC is designed to detect a slipping left front ESB. During initialization and braking, the left front motor is re-homed. If the ESB slips, the motor/piston will move. During the next key to "RUN" initialization, a re-home of the motor verifies the motor/piston remained at the home position. If motor movement is detected, the ESB must be slipping.

DTC Will Set When:

DTC 38 can be set during initialization. If the EBTCM detects that the ESB could not hold the piston in the home position, a malfunction exists. DTC 86 is always set with DTC 38.

Action Taken:

If an ESB cannot hold a piston in the home position, the piston may be back driven when the brake pedal is applied, causing the brake pedal to drop. A malfunction DTC is stored, ABS and TCS are disabled. The amber ABS and TCS warning lamps and the red BRAKE warning lamp are turned "ON."

DTC Chart Test Description:

Number(s) below refer to circled number(s) on the diagnostic chart.

1. Checks the left front ESB. A broken or defective ESB would result in the left front piston being back driven by hydraulic pressure and pedal movement would result.
2. Checks for high resistance in the ABS motor "HIGH" circuitry.
3. This step checks for high resistance in the ABS motor "LOW" circuitry.
4. Checks for proper resistance of the ABS motor windings.
5. Checks for poor terminal contact or corrosion in the connectors.
6. Checks for a malfunctioning EBTCM.

Diagnostic Aids:

An intermittent malfunction in this DTC may result from a mechanical part of the system that sticks, binds, or slips.

The frequency of the malfunction can be checked by using the enhanced diagnostic function of the Tech 1,

The static modulator test function of the Tech 1 or T-100 may be used to locate an intermittent malfunction associated with the ESB.

DIAGNOSTIC CHART DTC 38 — 1995 CAMARO AND FIREBIRD WITH TCS, CONTINUED

DTC 38
LEFT FRONT ESB WILL NOT HOLD MOTOR

(1)
- IGNITION "ON," ENGINE "OFF."
- PUMP BRAKE PEDAL UNTIL FIRM TO DEPLETE VACUUM RESERVE.
- USING TECH 1 MISC. TESTS, SELECT MANUAL CONTROL FUNCTION.
- SELECT ABS MOTOR APPLY FUNCTION OF THE LEFT FRONT CHANNEL AND APPLY THE MOTOR.
- WAIT 5 SECONDS.
- APPLY FIRM PRESSURE ON BRAKE PEDAL.

DID BRAKE PEDAL FALL?

YES: REPLACE ABS MOTOR PACK.

NO:

(2)
- EXIT TECH 1 MANUAL CONTROL FUNCTION
- IGNITION "OFF."
- DISCONNECT 6-WAY ABS MOTOR PACK CONNECTOR AND GRAY 8-WAY EBTCM CONNECTOR.
- USING J 39200, MEASURE THE RESISTANCE BETWEEN TERMINAL "A" OF THE 6-WAY ABS MOTOR PACK HARNESS CONNECTOR AND TERMINAL "A" OF THE GRAY 8-WAY EBTCM HARNESS CONNECTOR.

IS RESISTANCE 1.5Ω OR LESS?

NO: REPAIR HIGH RESISTANCE IN CKT 1280.

YES:

(3)
- USING J 39200, MEASURE THE RESISTANCE BETWEEN TERMINAL "B" OF THE 6-WAY ABS MOTOR PACK HARNESS CONNECTOR AND TERMINAL "B" OF THE GRAY 8-WAY EBTCM HARNESS CONNECTOR.

IS RESISTANCE 1.5Ω OR LESS?

NO: REPAIR HIGH RESISTANCE IN CKT 1281.

YES:

(4)
- USING J 39200, MEASURE THE RESISTANCE BETWEEN TERMINAL "A" AND TERMINAL "B" OF THE MOTOR PACK CONNECTOR.

IS RESISTANCE 1.5Ω OR LESS?

NO: REPLACE ABS MOTOR PACK.

YES:

(5)
- INSPECT THE 6-WAY ABS MOTOR PACK CONNECTOR AND GRAY 8-WAY EBTCM CONNECTOR FOR POOR TERMINAL CONTACT OR CORROSION.

DO ANY TERMINALS EXHIBIT POOR CONTACT OR EVIDENCE OF CORROSION?

YES: REPLACE TERMINALS THAT EXHIBIT POOR TERMINAL CONTACT OR EVIDENCE OF CORROSION.

NO:

(6)
- IGNITION "OFF."
- RECONNECT ALL CONNECTORS.
- START ENGINE AND TEST DRIVE VEHICLE, ACHIEVING AT LEAST 16 km/h (10 MPH).
- REPEAT ABOVE DRIVE CYCLE SEQUENCE TWO MORE TIMES.

DOES DTC 38 SET IN THE LAST THREE DRIVE CYCLE?

YES: REPLACE EBTCM.

NO: MALFUNCTION IS INTERMITTENT. REFER TO "DIAGNOSTIC AIDS" ON FACING PAGE.

AFTER DIAGNOSIS IS COMPLETE, CLEAR DTCs AND TEST DRIVE VEHICLE FOR THREE (3) DRIVE CYCLES TO VERIFY THAT DTC DOES NOT RESET. A DRIVE CYCLE CONSISTS OF STARTING THE VEHICLE, DRIVING OVER 16 km/h (10 MPH), AND THEN TURNING KEY OFF.

DIAGNOSTIC CHART DTC 41 — 1995 CAMARO AND FIREBIRD WITH TCS

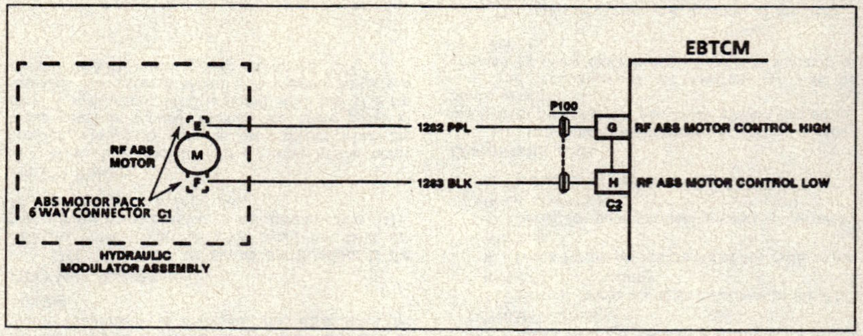

DTC 41
RIGHT FRONT ESB WILL NOT HOLD MOTOR

Circuit Description:

This DTC is designed to detect a slipping right front ESB. During initialization and braking, the right front motor is re-homed. If the ESB slips, the motor/piston will move. During the next key "ON" initialization, a re-home of the motor verifies the motor/piston remained at the home position. If motor movement is detected, the ESB must be slipping.

DTC Will Set When:

DTC 41 can be set during initialization. If the EBTCM detects that the ESB could not hold the piston in the home position, a malfunction exists. DTC 86 is always set with DTC 41.

Action Taken:

If an ESB cannot hold a piston in the home position, the piston may be back driven when the brake pedal is applied, causing the brake pedal to drop. A malfunction DTC is stored, ABS and TCS are disabled. The amber ABS and TCS warning lamps and the red BRAKE warning lamp are turned "ON."

DTC Chart Test Description:

Number(s) below refer to circled number(s) on the diagnostic chart.

1. Checks the right front ESB. A broken or defective ESB would result in the right front piston being back driven by hydraulic pressure, and pedal movement would result.
2. Checks for high resistance in the ABS motor "HIGH" circuitry.
3. This step checks for high resistance in the ABS motor "LOW" circuitry.
4. Checks for proper resistance of the ABS motor windings.
5. Checks for poor terminal contact or corrosion in the connectors.
6. Checks for a malfunctioning EBTCM.

Diagnostic Aids:

An intermittent malfunction in this DTC may result from a mechanical part of the system that sticks, binds, or slips.

The frequency of the malfunction can be checked by using the enhanced diagnostic function of the Tech 1,

The static modulator test function of the Tech 1 or T-100 may be used to locate an intermittent malfunction associated with the ESB.

DIAGNOSTIC CHART DTC 41 — 1995 CAMARO AND FIREBIRD WITH TCS, CONTINUED

DTC 41
RIGHT FRONT ESB WILL NOT HOLD MOTOR

1.
   - IGNITION "ON," ENGINE "OFF."
   - PUMP BRAKE PEDAL UNTIL FIRM TO DEPLETE VACUUM RESERVE.
   - USING TECH 1 MISC. TESTS, SELECT MANUAL CONTROL FUNCTION.
   - SELECT ABS MOTOR APPLY FUNCTION OF THE RIGHT FRONT CHANNEL AND APPLY THE MOTOR.
   - WAIT 5 SECONDS.
   - APPLY FIRM PRESSURE ON BRAKE PEDAL.

   DID BRAKE PEDAL FALL?
   - YES: REPLACE ABS MOTOR PACK.
   - NO: go to 2.

2.
   - EXIT TECH 1 MANUAL CONTROL FUNCTION.
   - IGNITION "OFF."
   - DISCONNECT 6-WAY ABS MOTOR PACK CONNECTOR AND GRAY 8-WAY EBTCM CONNECTOR.
   - USING J 39200, MEASURE THE RESISTANCE BETWEEN TERMINAL "E" OF THE 6-WAY ABS MOTOR PACK HARNESS CONNECTOR AND TERMINAL "G" OF THE GRAY 8-WAY EBTCM HARNESS CONNECTOR.

   IS RESISTANCE 1.5Ω OR LESS?
   - NO: REPAIR HIGH RESISTANCE IN CKT 1282.
   - YES: go to 3.

3.
   - USING J 39200, MEASURE THE RESISTANCE BETWEEN TERMINAL "F" OF THE 6-WAY ABS MOTOR PACK HARNESS CONNECTOR AND TERMINAL "H" OF THE GRAY 8-WAY EBTCM HARNESS CONNECTOR.

   IS RESISTANCE 1.5Ω OR LESS?
   - NO: REPAIR HIGH RESISTANCE IN CKT 1283.
   - YES: go to 4.

4.
   - USING J 39200, MEASURE THE RESISTANCE BETWEEN TERMINAL "E" AND TERMINAL "F" OF THE ABS MOTOR PACK CONNECTOR.

   IS RESISTANCE 1.5Ω OR LESS?
   - NO: REPLACE ABS MOTOR PACK.
   - YES: go to 5.

5.
   - INSPECT THE 6-WAY ABS MOTOR PACK CONNECTOR AND GRAY 8-WAY EBTCM CONNECTOR FOR POOR TERMINAL CONTACT OR CORROSION.

   DO ANY TERMINALS EXHIBIT POOR CONTACT OR EVIDENCE OF CORROSION?
   - YES: REPLACE TERMINALS THAT EXHIBIT POOR TERMINAL CONTACT OR EVIDENCE OF CORROSION.
   - NO: go to 6.

6.
   - IGNITION "OFF."
   - RECONNECT ALL CONNECTORS.
   - START ENGINE AND TEST DRIVE VEHICLE, ACHIEVING AT LEAST 16 km/h (10 MPH).
   - REPEAT ABOVE DRIVE CYCLE SEQUENCE TWO MORE TIMES.

   DOES DTC41 SET IN THE LAST THREE DRIVE CYCLES?
   - YES: REPLACE EBTCM.
   - NO: MALFUNCTION IS INTERMITTENT. REFER TO "DIAGNOSTIC AIDS" ON FACING PAGE.

AFTER DIAGNOSIS IS COMPLETE, CLEAR DTCs AND TEST DRIVE VEHICLE FOR THREE (3) DRIVE CYCLES TO VERIFY THAT DTC DOES NOT RESET. A DRIVE CYCLE CONSISTS OF STARTING THE VEHICLE, DRIVING OVER 16 km/h (10 MPH), AND THEN TURNING KEY OFF.

DIAGNOSTIC CHART DTC 42 — 1995 CAMARO AND FIREBIRD WITH TCS

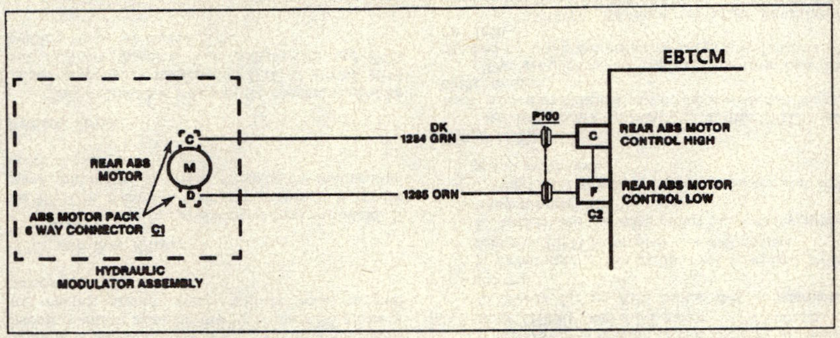

DTC 42
REAR ESB WILL NOT HOLD MOTOR

Circuit Description:
This DTC is designed to detect a slipping rear axle ESB. During initialization and braking, the rear motor is re-homed. If the ESB slips, the motor/piston will move. During the next key "ON" initialization, a re-home of the motor verifies the motor/piston remained at the home position. If motor movement is detected, the ESB must be slipping.

DTC Will Set When:
DTC 42 can be set during initialization. If the EBTCM detects that the ESB could not hold the piston in the home position, a malfunction exists. DTC 86 is always set with DTC 42.

Action Taken:
If an ESB can hold a piston in the home position, the piston may be back driven when the brake pedal is applied, causing the brake pedal to drop. A malfunction DTC is stored, ABS and TCS are disabled. The amber ABS and TCS warning lamps and the red BRAKE warning lamp are turned "ON."

DTC Chart Test Description:
Number(s) below refer to circled number(s) on the diagnostic chart.
1. Checks the rear axle ESB. A broken or defective ESB would result in the rear axle pistons being back driven by hydraulic pressure and wheel movement would result.
2. Checks for high resistance in the ABS motor "HIGH" circuitry.
3. Checks for high resistance in the ABS motor "LOW" circuitry.
4. Checks for proper resistance of the ABS motor windings.
5. Checks for poor terminal contact or corrosion in the connectors.
6. Checks for a malfunctioning EBTCM.

Diagnostic Aids:
An intermittent malfunction in this DTC may result from a mechanical part of the system that sticks, binds, or slips.
The frequency of the malfunction can be checked by using the enhanced diagnostic function of the Tech 1,

The static modulator test function of the Tech 1 or T-100 may be used to locate an intermittent malfunction associated with the ESB.

DIAGNOSTIC CHART DTC 42 — 1995 CAMARO AND FIREBIRD WITH TCS, CONTINUED

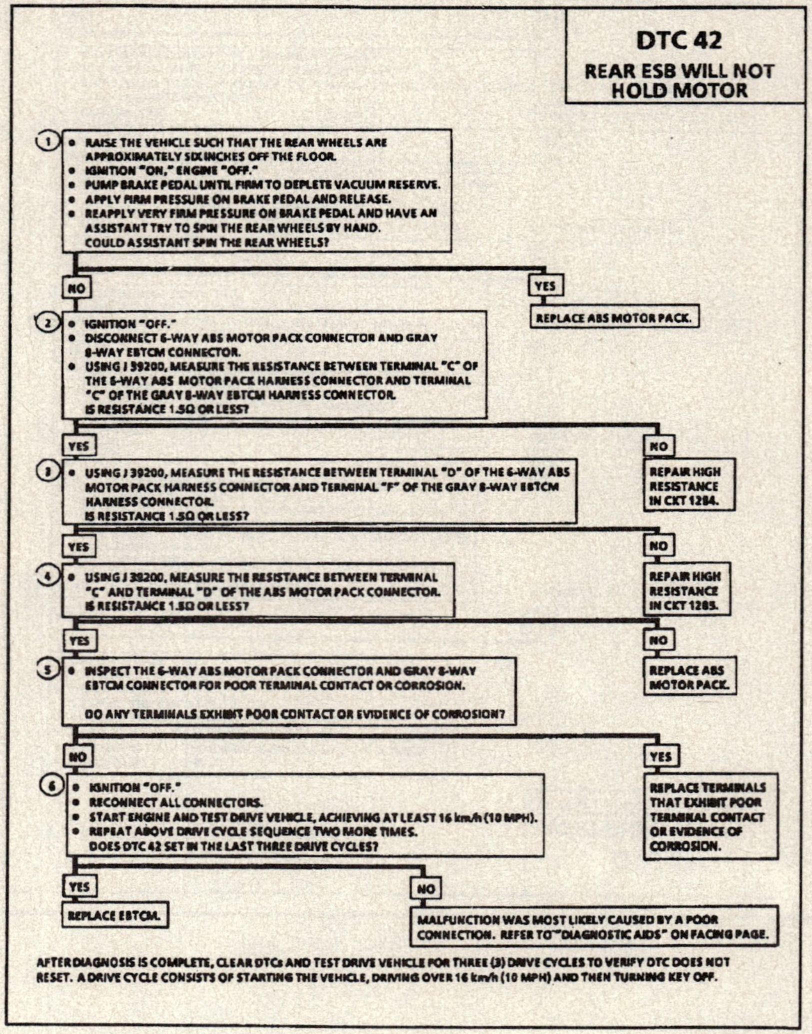

DIAGNOSTIC CHART DTC 44 — 1995 CAMARO AND FIREBIRD WITH TCS

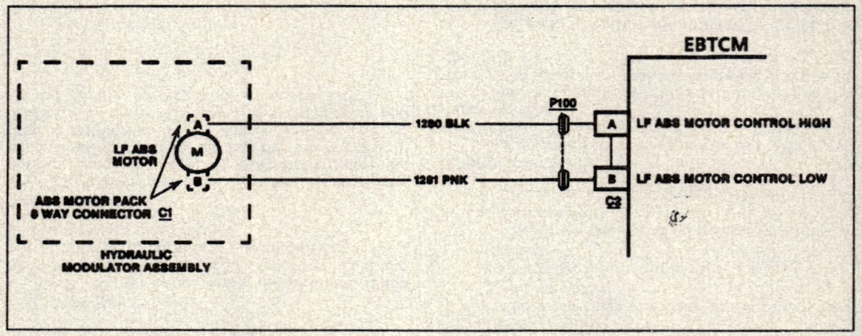

DTC 44
LEFT FRONT ABS CHANNEL WILL NOT MOVE

Circuit Description:

This DTC is designed to detect bound-up ESB, a stuck ABS motor, or a seized ABS hydraulic modulator. When the release is commanded during initialization, the ESB should release the ABS motor, resulting in sensed current being less than commanded current (motor is spinning freely). If the ABS motor is not moving, sensed current will be equal to stall current.

DTC Will Set When:

DTC 44 can be set during normal operation. If the EBTCM detects a condition in which it cannot move the motor in either direction, a malfunction exists.

Action Taken:

This malfunction indicates the channel cannot be moved properly. A malfunction DTC is stored, ABS and TCS are disabled. The amber ABS and TCS warning lamps are turned "ON."

DTC Chart Test Description:

Number(s) below refer to circled number(s) on the diagnostic chart.

1. Checks for proper ABS motor movement during apply and release commands from the Tech 1.
2. Compares EBTCM command current to motor feedback current.
3. Checks for a shorted circuit between the ABS motor "HIGH" and ABS motor "LOW" circuits.
4. Checks for an ABS motor that is internally shorted.
5. Determines if the malfunction is caused by a defective EBTCM or poor terminal contact.
6. Verifies that the ABS motor can actually apply when commanded.
7. Checks for proper hydraulic modulator gear and piston movement.

Diagnostic Aids:

An intermittent malfunction in this DTC may result from a mechanical part of the system that sticks, binds, or slips.

The frequency of the malfunction can be checked by using the enhanced diagnostic function of the Tech 1,

DTC 44 may set after modulator disassembly if the modulator pistons are positioned at the bottom of their bore.

Any circuitry that is suspected of causing the intermittent complaint should be thoroughly checked for backed out terminals, improper mating, broken locks, improperly formed or damaged terminals, poor terminal to wiring connections or physical damage to the wiring harness.

DIAGNOSTIC CHART DTC 44 — 1995 CAMARO AND FIREBIRD WITH TCS, CONTINUED

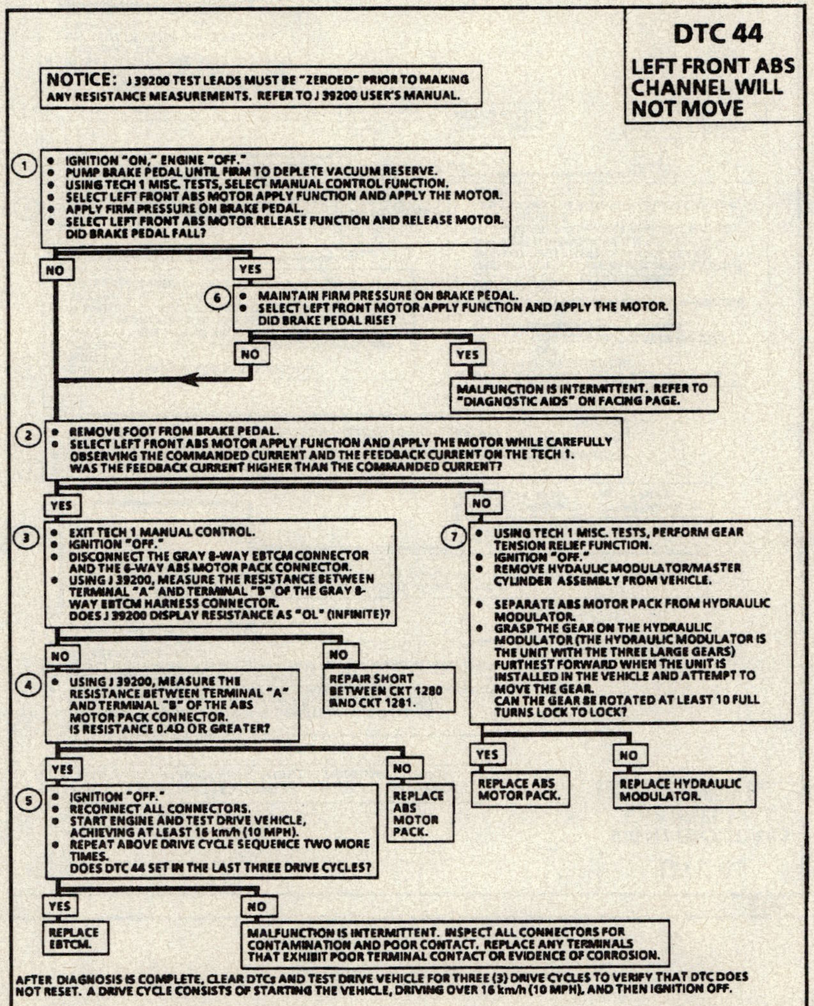

DIAGNOSTIC CHART DTC 45 — 1995 CAMARO AND FIREBIRD WITH TCS

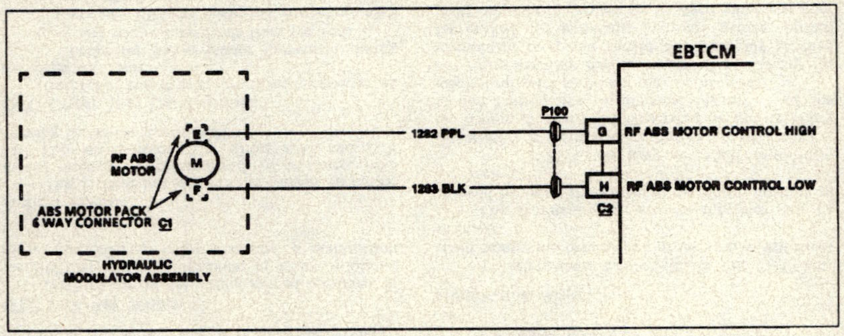

DTC 45
RIGHT FRONT ABS CHANNEL WILL NOT MOVE

Circuit Description:

This DTC is designed to detect bound-up ESB, a stuck ABS motor, or a seized ABS hydraulic modulator. When the release is commanded during initialization, the ESB should release the ABS motor, resulting in sensed current being less than commanded current (motor is spinning freely). If the ABS motor is not moving, sensed current will be equal to stall current.

DTC Will Set When:

DTC 45 can be set during normal operation. If the EBTCM detects a condition in which it cannot move the ABS motor in either direction, a malfunction exists.

Action Taken:

This malfunction indicates the channel cannot be moved properly. A malfunction DTC is stored, ABS and TCS are disabled. The amber ABS and TCS warning lamps are turned "ON."

DTC Chart Test Description:

Number(s) below refer to circled number(s) on the diagnostic chart.

1. Checks for proper motor movement during apply and release commands from the Tech 1.
2. Compares EBTCM command current to motor feedback current.
3. Checks for shorted circuit between the ABS motor "HIGH" and ABS motor "LOW" circuits.
4. Checks for an ABS motor that is internally shorted.
5. Checks to see if malfunction is due to poor terminal contact or the EBTCM.
6. Verifies that the motor can actually apply when commanded.
7. Checks for proper hydraulic modulator gear and piston movement.

Diagnostic Aids:

An intermittent malfunction in this DTC may result from a mechanical part of the system that sticks, binds, or slips.

The frequency of the malfunction can be checked by using the enhanced diagnostic function of the Tech 1,

DTC 45 may set after modulator disassembly if the modulator pistons are positioned at the bottom of their bore.

Any circuitry that is suspected of causing the intermittent complaint should be thoroughly checked for backed out terminals, improper mating, broken locks, improperly formed or damaged terminals, poor terminal to wiring connections or physical damage to the wiring harness.

DIAGNOSTIC CHART DTC 45 — 1995 CAMARO AND FIREBIRD WITH TCS, CONTINUED

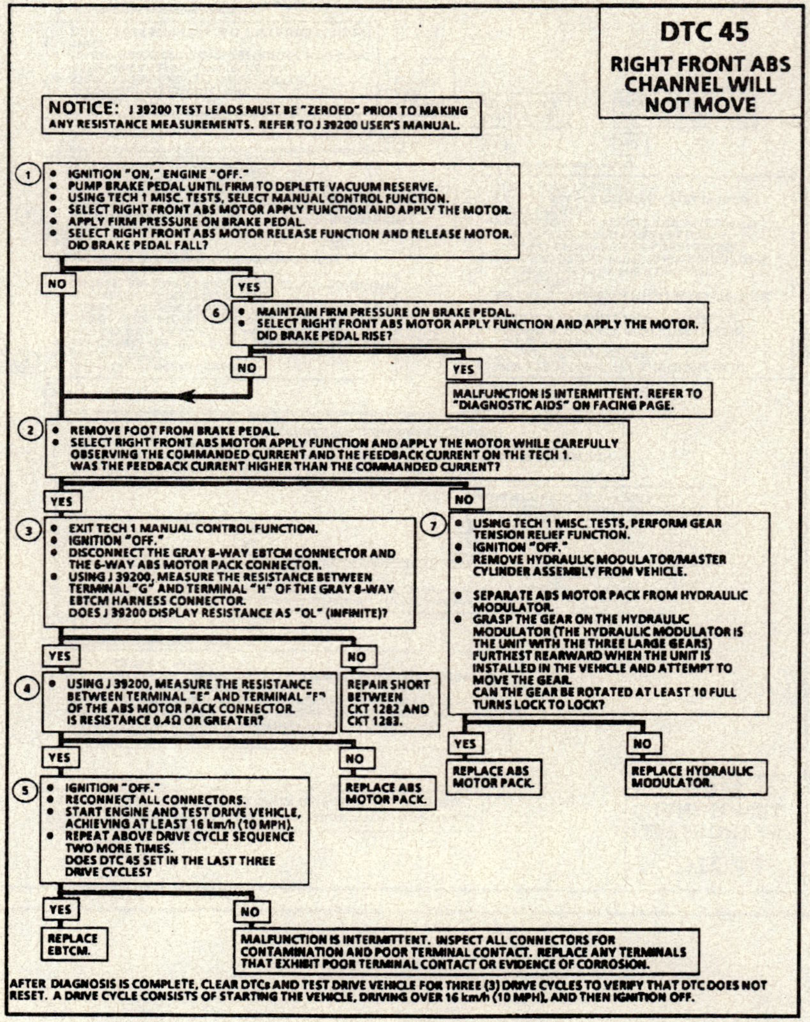

DIAGNOSTIC CHART DTC 46 — 1995 CAMARO AND FIREBIRD WITH TCS

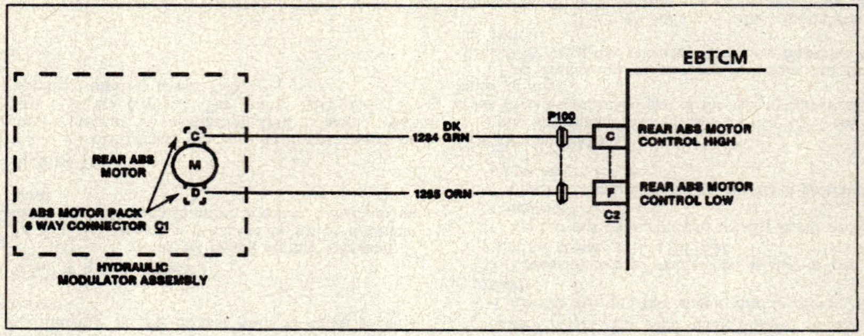

DTC 46
REAR ABS CHANNEL WILL NOT MOVE

Circuit Description:

This DTC is designed to detect bound-up ESB, a stuck ABS motor, or a seized hydraulic modulator. When the release is commanded during initialization, the ESB should release the ABS motor, resulting in sensed current being less than commanded current (motor is spinning freely). If the ABS motor is not moving, sensed current will be equal to stall current.

DTC Will Set When:

DTC 46 can be set during normal operation. If the EBTCM detects a condition in which it cannot move the motor in either direction, a malfunction exists.

Action Taken:

This malfunction indicates the channel cannot be moved properly. A malfunction DTC is stored, ABS and TCS are disabled. The amber ABS and TCS warning lamps are turned "ON."

DTC Chart Test Description:

Number(s) below refer to circled number(s) on the diagnostic chart.

1. Checks for proper motor movement during apply and release commands from the Tech 1.
2. Compares EBTCM command current to ABS motor feedback current.
3. Checks for a short circuit between the ABS motor "HIGH" and ABS motor "LOW" circuits.
4. Checks for an ABS motor that is internally shorted.
5. Checks to see if malfunction is due to poor terminal contact or the EBTCM.
6. Verifies that the motor can actually release when commanded.
7. Checks for proper hydraulic modulator gear and piston movement.

Diagnostic Aids:

An intermittent malfunction in this DTC may result from a mechanical part of the system that sticks or binds.

The frequency of the malfunction can be checked by using the enhanced diagnostic function of the Tech 1,

DTC 46 may set after modulator disassembly if the modulator pistons are positioned at the bottom of their bore. Depending on the frequency of the malfunction, a physical inspection of the mechanical parts suspected may be necessary.

Any circuitry that is suspected of causing the intermittent complaint should be thoroughly checked for backed out terminals, improper mating, broken locks, improperly formed or damaged terminals, poor terminal to wiring connections or physical damage to the wiring harness.

DIAGNOSTIC CHART DTC 46 — 1995 CAMARO AND FIREBIRD WITH TCS, CONTINUED

DTC 46
REAR ABS CHANNEL WILL NOT MOVE

NOTICE: J 39200 TEST LEADS MUST BE "ZEROED" PRIOR TO MAKING ANY RESISTANCE MEASUREMENTS. REFER TO J 39200 USER'S MANUAL.

(1)
- RAISE AND SUPPORT THE VEHICLE SUCH THAT THE REAR WHEELS ARE APPROXIMATELY SIX INCHES OFF THE FLOOR.
- IGNITION "ON," ENGINE "OFF."
- PUMP BRAKE PEDAL UNTIL FIRM TO DEPLETE VACUUM RESERVE.
- APPLY FIRM PRESSURE ON BRAKE PEDAL.
- USING TECH 1 MISC. TESTS, SELECT MANUAL CONTROL FUNCTION.
- SELECT REAR AXLE ABS MOTOR APPLY FUNCTION AND APPLY THE MOTOR.
- HAVE AN ASSISTANT TRY TO SPIN THE REAR WHEELS BY HAND.

COULD ASSISTANT SPIN THE REAR WHEELS?

YES → (2); NO → (6)

(6)
- MAINTAIN FIRM PRESSURE ON BRAKE PEDAL.
- SELECT REAR AXLE ABS MOTOR RELEASE FUNCTION AND RELEASE THE MOTOR.
- HAVE AN ASSISTANT TRY TO SPIN THE REAR WHEELS BY HAND.

COULD ASSISTANT SPIN THE REAR WHEELS?

NO → (2); YES → MALFUNCTION IS INTERMITTENT. REFER TO "DIAGNOSTIC AIDS" ON FACING PAGE.

(2)
- REMOVE FOOT FROM BRAKE PEDAL.
- SELECT REAR AXLE ABS MOTOR APPLY FUNCTION AND APPLY THE MOTOR WHILE CAREFULLY OBSERVING THE COMMANDED CURRENT AND THE FEEDBACK CURRENT ON THE TECH 1.

WAS THE FEEDBACK CURRENT HIGHER THAN THE COMMANDED CURRENT?

YES → (3); NO → (7)

(3)
- EXIT TECH 1 MANUAL CONTROL FUNCTION.
- IGNITION "OFF."
- DISCONNECT THE GRAY 8-WAY EBTCM CONNECTOR AND THE 6-WAY ABS MOTOR PACK CONNECTOR.
- USING J 39200, MEASURE THE RESISTANCE BETWEEN TERMINAL "C" AND TERMINAL "F" OF THE GRAY 8-WAY EBTCM HARNESS CONNECTOR.

DOES J 39200 DISPLAY RESISTANCE AS "OL" (INFINITE)?

YES → (4); NO → REPAIR SHORT BETWEEN CKT 1284 AND CKT 1285.

(4)
- USING J 39200, MEASURE THE RESISTANCE BETWEEN TERMINAL "C" AND TERMINAL "D" OF THE ABS MOTOR PACK CONNECTOR.

IS RESISTANCE GREATER THAN 0.4Ω?

YES → (5); NO → REPLACE ABS MOTOR PACK.

(5)
- RECONNECT ALL CONNECTORS.
- IGNITION "OFF."
- START ENGINE AND TEST DRIVE VEHICLE, ACHIEVING AT LEAST 16 km/h (10 MPH).
- REPEAT ABOVE DRIVE CYCLE SEQUENCE TWO MORE TIMES.

DOES DTC 46 SET IN THE LAST THREE DRIVE CYCLES?

YES → REPLACE EBTCM.
NO → MALFUNCTION IS INTERMITTENT. INSPECT ALL CONNECTORS FOR CONTAMINATION AND POOR TERMINAL CONTACT. REPLACE ANY TERMINALS THAT EXHIBIT POOR TERMINAL CONTACT OR EVIDENCE OF CORROSION.

(7)
- USING TECH 1 MISC. TESTS, PERFORM GEAR TENSION RELIEF FUNCTION.
- IGNITION "OFF."
- REMOVE HYDRAULIC MODULATOR/MASTER CYLINDER ASSEMBLY FROM VEHICLE.
- SEPARATE ABS MOTOR PACK FROM HYDRAULIC MODULATOR.
- GRASP THE MIDDLE GEAR ON THE HYDRAULIC MODULATOR (THE HYDRAULIC MODULATOR IS THE UNIT WITH THE THREE LARGE GEARS) AND ATTEMPT TO MOVE THE GEAR.

CAN THE GEAR BE ROTATED AT LEAST 7 FULL TURNS LOCK TO LOCK?

YES → REPLACE ABS MOTOR PACK.
NO → REPLACE HYDRAULIC MODULATOR.

AFTER DIAGNOSIS IS COMPLETE, CLEAR DTCs AND TEST DRIVE VEHICLE FOR THREE (3) DRIVE CYCLES TO VERIFY THAT DTC DOES NOT RESET. A DRIVE CYCLE CONSISTS OF STARTING THE VEHICLE, DRIVING OVER 16 km/h (10 MPH), AND THEN IGNITION OFF.

DIAGNOSTIC CHART DTC 47 — 1995 CAMARO AND FIREBIRD WITH TCS

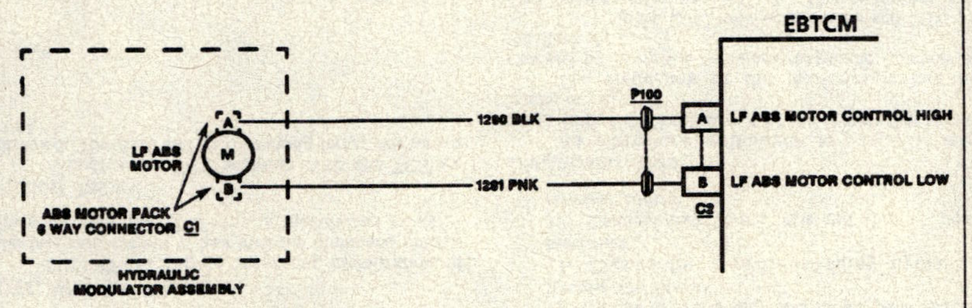

DTC 47
(Page 1 of 2)
LEFT FRONT ABS MOTOR FREE SPINS

Circuit Description:

This DTC is designed to detect a stripped nut or gear assembly during initialization. During the homing sequence, the piston should reach the top of the bore resulting in a stalled ABS motor. If this does not occur, the ABS motor must be spinning with little or no resistance, indicating a nut/screw or gear malfunction.

DTC Will Set When:

DTC 47 can only be set during initialization. If the feedback current is less than the command current for a specified period of time, a malfunction exists.

Action Taken:

A malfunction DTC is stored, ABS and TCS are disabled. The amber ABS and TCS warning lamps are turned "ON."

DTC Chart Test Description:

Number(s) below refer to circled number(s) on the diagnostic chart.

1. Checks to see if the corresponding open motor DTC is also set.
2. Verifies that the ABS motor was actually applied as command by observing feedback current.
3. Verifies that the ABS motor can release.
4. Verifies that the ABS motor can be applied by observing pedal movement.
5. Checks for an open in the ABS motor "HIGH" circuitry.
6. Checks for an open in the ABS motor "LOW" circuitry.
7. Checks for an ABS motor that is internally shorted.

Diagnostic Aids:

An intermittent malfunction in this DTC may result from a mechanical part of the system that sticks, binds, or slips.

The frequency of the malfunction can be checked by using the enhanced diagnostic function of the Tech 1,

If the DTC only occurs once and DTC 56 also occurs, refer to DTC 56. If intermittent and enhanced diagnostics shows this DTC occurs during ABS, refer to DTC 56.

Depending on the frequency of the malfunction, a physical inspection of the mechanical parts suspected may be necessary.

DIAGNOSTIC CHART DTC 47 — 1995 CAMARO AND FIREBIRD WITH TCS, CONTINUED

DTC 47
(1 OF 2)
LEFT FRONT ABS MOTOR FREE SPINS

NOTICE: J 39200 TEST LEADS MUST BE "ZEROED" PRIOR TO MAKING ANY RESISTANCE MEASUREMENTS. REFER TO J 39200 USER'S MANUAL.

1. IS DTC 56 ALSO CURRENTLY SET?
   - YES: PROCEED TO DTC 56 CHART FIRST.
   - NO: Go to step 2.
2. • IGNITION "ON," ENGINE "OFF."
   • PUMP BRAKE PEDAL UNTIL FIRM TO DEPLETE VACUUM RESERVE.
   • USING TECH 1 MISC. TESTS, SELECT MANUAL CONTROL FUNCTION.
   • SELECT LEFT FRONT ABS MOTOR APPLY FUNCTION AND APPLY THE MOTOR WHILE CAREFULLY OBSERVING THE COMMAND CURRENT AND FEEDBACK CURRENT ON THE TECH 1.
   WERE BOTH COMMAND AND FEEDBACK CURRENT APPROXIMATELY 10 AMPS?
   - YES: Go to step 3.
   - NO: Go to step 5.
3. • APPLY FIRM PRESSURE ON BRAKE PEDAL.
   • SELECT LEFT FRONT ABS MOTOR RELEASE FUNCTION AND RELEASE THE MOTOR.
   DID THE BRAKE PEDAL FALL AND WERE BOTH COMMAND AND FEEDBACK CURRENT APPROXIMATELY 6 AMPS?
   - YES: Go to step 4.
   - NO: GO TO STEP 5.
4. • MAINTAIN FIRM PRESSURE ON BRAKE PEDAL.
   • SELECT LEFT ABS MOTOR APPLY FUNCTION AND APPLY THE MOTOR (THIS MOVES THE PISTON TO THE TOP OF ITS BORE.)
   DID BRAKE PEDAL RISE?
   - YES: MALFUNCTION IS INTERMITTENT. REFER TO "DIAGNOSTIC AIDS" ON FACING PAGE.
   - NO: Go to step 5.
5. • USING TECH 1 MISC. TESTS, PERFORM GEAR TENSION RELIEF FUNCTION.
   • IGNITION "OFF."
   • DISCONNECT 6-WAY ABS MOTOR PACK CONNECTOR AND GRAY 8-WAY EBTCM CONNECTOR.
   • USING J 39200, MEASURE THE RESISTANCE BETWEEN TERMINAL "A" OF THE 6-WAY ABS MOTOR PACK HARNESS CONNECTOR AND TERMINAL "A" OF THE GRAY 8-WAY EBTCM HARNESS CONNECTOR.
   IS RESISTANCE 1.5Ω OR LESS?
   - YES: Go to step 6.
   - NO: REPAIR HIGH RESISTANCE IN CKT 1280.
6. • USING J 39200, MEASURE THE RESISTANCE BETWEEN TERMINAL "B" OF THE 6-WAY ABS MOTOR PACK HARNESS CONNECTOR AND TERMINAL "B" OF THE GRAY 8-WAY EBTCM HARNESS CONNECTOR.
   IS RESISTANCE 1.5Ω OR LESS?
   - YES: Go to step 7.
   - NO: REPAIR HIGH RESISTANCE IN CKT 1281.
7. • USING J 39200, MEASURE THE RESISTANCE BETWEEN TERMINAL "A" AND TERMINAL "B" OF THE ABS MOTOR PACK CONNECTOR.
   IS RESISTANCE 1.5Ω OR LESS?
   - YES: DIAGNOSIS CONTINUED ON NEXT PAGE.
   - NO: REPLACE ABS MOTOR PACK.

AFTER DIAGNOSIS IS COMPLETE, CLEAR DTCs AND TEST DRIVE VEHICLE FOR THREE (3) DRIVE CYCLES TO VERIFY THAT DTC DOES NOT RESET. A DRIVE CYCLE CONSISTS OF STARTING THE VEHICLE, DRIVING OVER 16 km/h (10 MPH) AND THEN IGNITION "OFF."

DIAGNOSTIC CHART DTC 47 — 1995 CAMARO AND FIREBIRD WITH TCS, CONTINUED

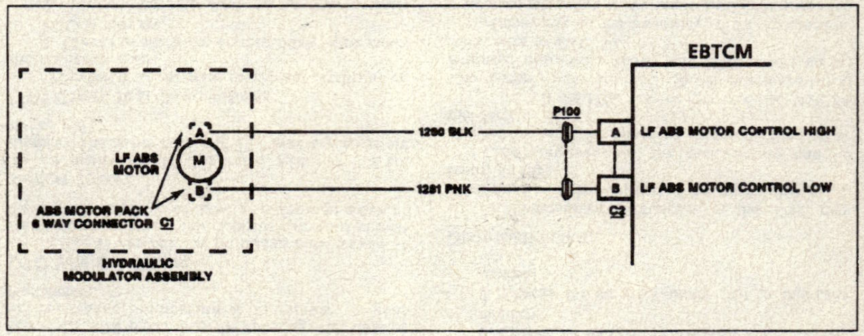

DTC 47
(Page 2 of 2)
LEFT FRONT ABS MOTOR FREE SPINS

Circuit Description:

This DTC is designed to detect a stripped nut or gear assembly during initialization. During the homing sequence, the piston should reach the top of the bore resulting in a stalled motor. If this does not occur, the motor must be spinning with little or no resistance, indicating a nut/screw or gear malfunction.

DTC Will Set When:

DTC 47 can only be set during initialization. If the feedback current is less than the command current for a specified period of time, a malfunction exists.

Action Taken:

A malfunction DTC is stored, ABS and TCS are disabled. The ABS and TCS warning lamps are turned "ON."

DTC Chart Test Description:

Number(s) below refer to circled number(s) on the diagnostic chart.

8. Checks for a stripped gear on the ABS motor pack (unit with three small gears).
9. Checks for a stripped gear on the hydraulic modulator (unit with three large gears).
10. Checks to see if the ABS motor pack has an internal malfunction.
11. Checks for a malfunctioning hydraulic modulator.
12. Ensures malfunction was not due to poor terminal contact.

Diagnostic Aids:

An intermittent malfunction in this DTC may result from a mechanical part of the system that sticks, binds, or slips.

The frequency of the malfunction can be checked by using the enhanced diagnostic function of the Tech 1.

If the DTC only occurs once and DTC 56 also occurs, refer to DTC 56. If intermittent and enhanced diagnostics shows this DTC occurs during ABS, refer to DTC 56.

Depending on the frequency of the malfunction, a physical inspection of the mechanical parts suspected may be necessary.

DIAGNOSTIC CHART DTC 47 — 1995 CAMARO AND FIREBIRD WITH TCS, CONTINUED

DTC 47
(Page 2 of 2)
LEFT FRONT ABS MOTOR FREE SPINS

DIAGNOSIS CONTINUED FROM PREVIOUS PAGE.

(8)
- REMOVE HYDRAULIC MODULATOR/MOTOR PACK ASSEMBLY FROM VEHICLE
- REMOVE GEAR COVER.
- CHECK FOR STRIPPED GEAR ON ABS MOTOR PACK; ABS MOTOR PACK IS THE UNIT WITH THE THREE SMALL GEARS. THE LEFT FRONT GEAR IS THE GEAR FURTHEST FORWARD WHEN THE ASSEMBLY IS INSTALLED IN THE VEHICLE. IS THE ABS MOTOR PACK GEAR STRIPPED?

YES: REPLACE ABS MOTOR PACK.

NO:

(9)
- CHECK FOR A STRIPPED GEAR ON HYDRAULIC MODULATOR; MODULATOR IS THE UNIT WITH THE THREE LARGE GEARS. THE LEFT FRONT GEAR IS THE GEAR FURTHEST FORWARD WHEN THE UNIT IS INSTALLED IN THE VEHICLE. IS THE MODULATOR GEAR STRIPPED?

YES: REPLACE HYDRAULIC MODULATOR GEAR.

NO:

(10)
- RECONNECT ELECTRICAL CONNECTORS.
- SECURELY POSITION HYDRAULIC MODULATOR/MOTOR PACK ASSEMBLY IN VEHICLE WITH GEAR COVER REMOVED SUCH THAT THE GEAR SET CAN BE OBSERVED. BE CAREFUL NOT TO DAMAGE GEAR SET OR ALLOW THE HYDRAULIC MODULATOR TO MOVE WHILE TESTING.
- IGNITION "ON."
- USING TECH 1 MISC. TESTS SELECT MANUAL CONTROL FUNCTION.
- SELECT LEFT FRONT ABS MOTOR APPLY AND APPLY THE MOTOR WHILE OBSERVING THE GEAR SET FURTHEST FORWARD WHEN THE UNIT IS INSTALLED IN THE VEHICLE.
- SELECT LEFT FRONT ABS MOTOR RELEASE FUNCTION AND RELEASE THE MOTOR WHILE STILL OBSERVING THE GEAR SET.
- SELECT LEFT FRONT ABS MOTOR APPLY FUNCTION AND APPLY THE MOTOR. DID GEAR SET MOVE IN BOTH DIRECTIONS FOR AT LEAST ONE REVOLUTION?

NO: REPLACE ABS MOTOR PACK.

YES:

(11)
- USING TECH 1 MISC. TESTS, PERFORM GEAR TENSION RELIEF FUNCTION.
- IGNITION "OFF."
- SEPARATE THE ABS MOTOR PACK FROM THE HYDRAULIC MODULATOR.
- GRASP THE GEAR ON THE HYDRAULIC MODULATOR (THE HYDRAULIC MODULATOR IS THE UNIT WITH THE THREE LARGE GEARS) FURTHEST FORWARD WHEN THE UNIT IS INSTALLED IN THE VEHICLE AND ATTEMPT TO MOVE THE GEAR. CAN THE GEAR BE ROTATED MORE THAN 13 FULL TURNS LOCK TO LOCK?

YES: REPLACE THE HYDRAULIC MODULATOR.

NO:

(12)
- INSPECT ALL CONNECTORS AND TERMINALS FOR POOR TERMINAL CONTACT AND EVIDENCE OF CORROSION. REPLACE ALL TERMINALS THAT EXHIBIT SIGNS OF POOR TERMINAL CONTACT OR CORROSION.
- REINSTALL HYDRAULIC MODULATOR/MOTOR PACK ASSEMBLY IN VEHICLE.
- RECONNECT ALL CONNECTORS.
- IGNITION "OFF."
- START ENGINE WITH FOOT OFF BRAKE.
- ALLOW ENGINE TO RUN FOR AT LEAST 10 SECONDS.
- REPEAT ABOVE IGNITION CYCLE SEQUENCE TWO MORE TIMES. DID DTC 47 SET IN THE LAST THREE IGNITION CYCLES?

YES: REPLACE EBTCM.

NO: MALFUNCTION IS MOST LIKELY CAUSED BY A POOR CONNECTION. REFER TO "DIAGNOSTIC AIDS" ON FACING PAGE.

AFTER DIAGNOSIS IS COMPLETE, CLEAR DTCs AND TEST DRIVE VEHICLE FOR THREE (3) DRIVE CYCLES TO VERIFY THAT DTC DOES NOT RESET. A DRIVE CYCLE CONSISTS OF STARTING THE VEHICLE, DRIVING OVER 16 km/h (10 MPH), AND THEN TURNING KEY OFF.

DIAGNOSTIC CHART DTC 48 — 1995 CAMARO AND FIREBIRD WITH TCS

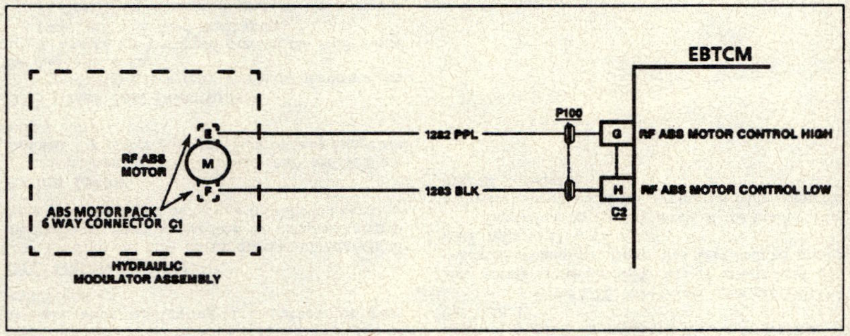

DTC 48
(Page 1 of 2)
RIGHT FRONT ABS MOTOR FREE SPINS

Circuit Description:

This DTC is designed to detect a stripped nut or gear assembly during initialization. During the homing sequence, the piston should reach the top of the bore resulting in a stalled ABS motor. If this does not occur, the ABS motor must be spinning with little or no resistance, indicating a nut/screw or gear malfunction.

DTC Will Set When:

DTC 48 can only be set during initialization. If the feedback current is less than the command current for a specified period of time, a malfunction exists.

Action Taken:

A malfunction DTC is stored, ABS and TCS are disabled. The amber ABS and TCS warning lamps are turned "ON."

DTC Chart Test Description:

Number(s) below refer to circled number(s) on the diagnostic chart.

1. Checks to see if the corresponding open motor DTC is also set.
2. Verifies that the ABS motor was actually applied as command by observing feedback current.
3. Verifies that the ABS motor can release.
4. Verifies that the ABS motor can be applied by observing pedal movement.
5. Checks for high resistance in the ABS motor "HIGH" circuitry.
6. Checks for high resistance in the ABS motor "LOW" circuitry.
7. Checks for proper resistance of the ABS motor windings.

Diagnostic Aids:

An intermittent malfunction in this DTC may result from a mechanical part of the system that sticks, binds, or slips.

The frequency of the malfunction can be checked by using the enhanced diagnostic function of the Tech 1,

If the DTC only occurs once and DTC 61 also occurs, refer to DTC 61. If intermittent and enhanced diagnostics shows this DTC occurs during ABS, refer to DTC 61.

Depending on the frequency of the malfunction, a physical inspection of the mechanical parts suspected may be necessary.

DIAGNOSTIC CHART DTC 48 — 1995 CAMARO AND FIREBIRD WITH TCS, CONTINUED

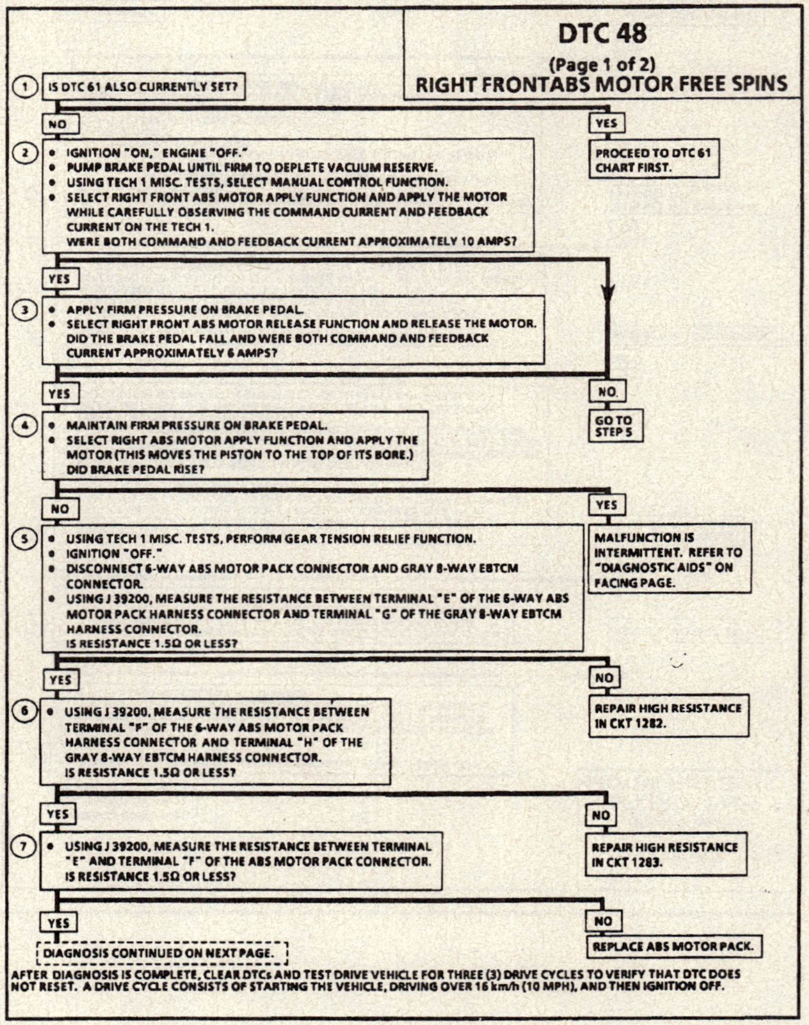

DIAGNOSTIC CHART DTC 48 — 1995 CAMARO AND FIREBIRD WITH TCS, CONTINUED

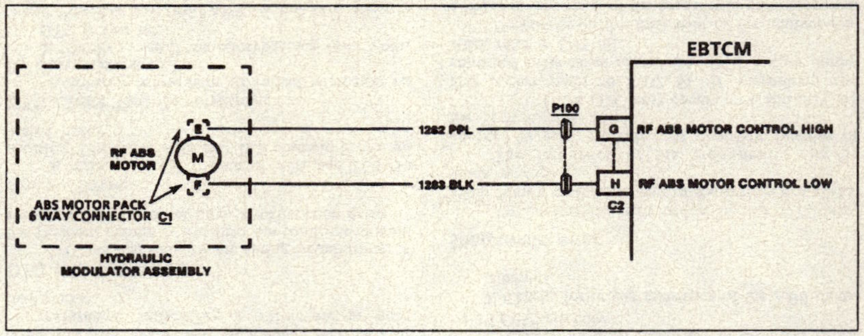

DTC 48
(Page 2 of 2)
RIGHT FRONT ABS MOTOR FREE SPINS

Circuit Description:

This DTC is designed to detect a stripped nut or gear assembly during initialization. During the homing sequence, the piston should reach the top of the bore resulting in a stalled ABS motor. If this does not occur, the ABS motor must be spinning with little or no resistance, indicating a nut/screw or gear malfunction.

DTC Will Set When:

DTC 48 can only be set during initialization. If the feedback current is less than the command current for a specified period of time, a malfunction exists.

Action Taken:

A malfunction DTC is stored, ABS and TCS are disabled. The amber ABS and TCS warning lamps are turned "ON."

DTC Chart Test Description:

Number(s) below refer to circled number(s) on the diagnostic chart.

8. Checks for a stripped gear on the ABS motor pack (unit with three small gears).
9. Checks for a stripped gear on the hydraulic modulator (unit with three large gears).
10. Checks to see if the ABS motor pack has an internal malfunction.
11. Checks for a malfunctioning hydraulic modulator.
12. Ensures malfunction was not due to poor terminal contact.

Diagnostic Aids:

An intermittent malfunction in this DTC may result from a mechanical part of the system that sticks, binds, or slips.

The frequency of the malfunction can be checked by using the enhanced diagnostic function of the Tech 1,

If the DTC only occurs once and DTC 61 also occurs, refer to DTC 61. If intermittent and enhanced diagnostics shows this DTC occurs during ABS, refer to DTC 61.

Depending on the frequency of the malfunction, a physical inspection of the mechanical parts suspected may be necessary.

DIAGNOSTIC CHART DTC 48 — 1995 CAMARO AND FIREBIRD WITH TCS, CONTINUED

DTC 48
(Page 2 of 2)
RIGHT FRONT ABS MOTOR FREE SPINS

DIAGNOSIS CONTINUED FROM PREVIOUS PAGE.

(8)
- REMOVE HYDRAULIC MODULATOR/MASTER CYLINDER ASSEMBLY FROM VEHICLE.
- REMOVE GEAR COVER.
- CHECK FOR STRIPPED GEAR ON ABS MOTOR PACK; ABS MOTOR PACK IS THE UNIT WITH THE THREE SMALL GEARS. THE RIGHT FRONT GEAR IS THE GEAR FURTHEST REARWARD WHEN THE ASSEMBLY IS INSTALLED IN THE VEHICLE. IS THE ABS MOTOR PACK GEAR STRIPPED?

YES: REPLACE ABS MOTOR PACK.

NO:

(9)
- CHECK FOR A STRIPPED GEAR ON HYDRAULIC MODULATOR; MODULATOR IS THE UNIT WITH THE THREE LARGE GEARS. THE RIGHT FRONT GEAR IS THE GEAR FURTHEST REARWARD WHEN THE UNIT IS INSTALLED IN THE VEHICLE. IS THE MODULATOR GEAR STRIPPED?

YES: REPLACE HYDRAULIC MODULATOR GEAR.

NO:

(10)
- RECONNECT ELECTRICAL CONNECTORS
- SECURELY POSITION HYDRAULIC MODULATOR/MASTER CYLINDER ASSEMBLY IN VEHICLE WITH GEAR COVER REMOVED SUCH THAT THE GEAR SET CAN BE OBSERVED. BE CAREFUL NOT TO DAMAGE GEAR SET OR ALLOW THE HYDRAULIC MODULATOR TO MOVE WHILE TESTING.
- IGNITION "ON."
- USING TECH 1 MISC. TESTS, SELECT MANUAL CONTROL FUNCTION.
- SELECT RIGHT FRONT ABS MOTOR APPLY AND APPLY THE MOTOR WHILE OBSERVING THE GEAR SET FURTHEST REARWARD WHEN THE UNIT IS INSTALLED IN THE VEHICLE.
- SELECT RIGHT FRONT ABS MOTOR RELEASE FUNCTION AND RELEASE THE MOTOR WHILE STILL OBSERVING THE GEAR SET.
- SELECT RIGHT FRONT ABS MOTOR APPLY FUNCTION AND APPLY THE MOTOR. DID GEAR SET MOVE IN BOTH DIRECTIONS FOR AT LEAST ONE REVOLUTION?

NO: REPLACE ABS MOTOR PACK.

YES:

(11)
- USING TECH 1 MISC. TESTS, PERFORM GEAR TENSION RELIEF FUNCTION.
- IGNITION "OFF."
- SEPARATE THE ABS MOTOR PACK FROM THE HYDRAULIC MODULATOR.
- GRASP THE GEAR ON THE HYDRAULIC MODULATOR (THE HYDRAULIC MODULATOR IS THE UNIT WITH THE THREE LARGE GEARS) FURTHEST REARWARD WHEN THE UNIT IS INSTALLED IN THE VEHICLE AND ATTEMPT TO MOVE THE GEAR. CAN THE GEAR BE ROTATED MORE THAN 13 FULL TURNS LOCK TO LOCK?

YES: REPLACE THE HYDRAULIC MODULATOR.

NO:

(12)
- INSPECT ALL CONNECTORS AND TERMINALS FOR POOR TERMINAL CONTACT AND EVIDENCE OF CORROSION. REPLACE ALL TERMINALS THAT EXHIBIT SIGNS OF POOR TERMINAL CONTACT OR CORROSION.
- REINSTALL HYDRAULIC MODULATOR/MASTER CYLINDER ASSEMBLY IN VEHICLE.
- RECONNECT ALL CONNECTORS.
- IGNITION "OFF."
- START ENGINE AND TEST DRIVE VEHICLE, ACHIEVING AT LEAST 16 km/h (10 MPH).
- REPEAT ABOVE DRIVE CYCLE SEQUENCE TWO MORE TIMES. DOES DTC 48 SET IN THE LAST THREE DRIVE CYCLES?

YES: REPLACE EBTCM.

NO: MALFUNCTION WAS MOST LIKELY CAUSED BY A POOR CONNECTION. REFER TO "DIAGNOSTIC AIDS" ON FACING PAGE.

AFTER DIAGNOSIS IS COMPLETE, CLEAR DTCs AND TEST DRIVE VEHICLE FOR THREE (3) DRIVE CYCLES TO VERIFY THAT DTC DOES NOT RESET. A DRIVE CYCLE CONSISTS OF STARTING THE VEHICLE, DRIVING OVER 16 KM/H (10 MPH), AND THEN TURNING KEY OFF.

DIAGNOSTIC CHART DTC 51 — 1995 CAMARO AND FIREBIRD WITH TCS

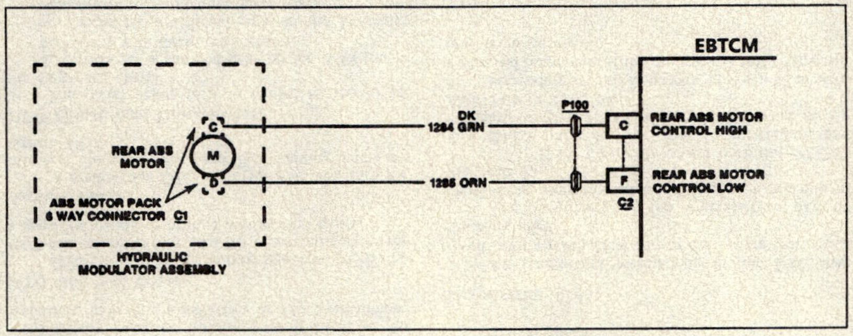

DTC 51
(Page 1 of 2)
REAR ABS MOTOR FREE SPINS

Circuit Description:

This DTC is designed to detect a stripped nut or gear assembly during initialization. During the homing sequence, the piston should reach the top of the bore resulting in a stalled ABS motor. If this does not occur, the ABS motor must be spinning with little or no resistance, indicating a nut/screw or gear malfunction.

DTC Will Set When:

DTC 51 can be set during initialization. If the feedback current is less than the command current for a specified period of time, a malfunction exists.

Action Taken:

A malfunction DTC is stored, ABS and TCS are disabled. The amber ABS and TCS warning lamps are turned "ON."

DTC Chart Test Description:

Number(s) below refer to circled number(s) on the diagnostic chart.

1. Checks to see if the corresponding open ABS motor DTC is also set.
2. Verifies that the ABS motor was actually applied as command by observing feedback current.
3. Verifies that the ABS motor can release.
4. Verifies that the ABS motor can be applied by observing pedal movement.
5. Checks for high resistance in the ABS motor "HIGH" circuitry.
6. Checks for high resistance in the ABS motor "LOW" circuitry.
7. Checks for proper resistance of the ABS motor windings.

Diagnostic Aids:

An intermittent malfunction in this DTC may result from a mechanical part of the system that sticks, binds, or slips.

The frequency of the malfunction can be checked by using the enhanced diagnostic function of the Tech 1.

If the DTC only occurs once and DTC 64 also occurs, refer to DTC 64. If intermittent and enhanced diagnostics show this DTC occurs during ABS, refer to DTC 64.

Depending on the frequency of the malfunction, a physical inspection of the mechanical parts suspected may be necessary.

DIAGNOSTIC CHART DTC 51 — 1995 CAMARO AND FIREBIRD WITH TCS, CONTINUED

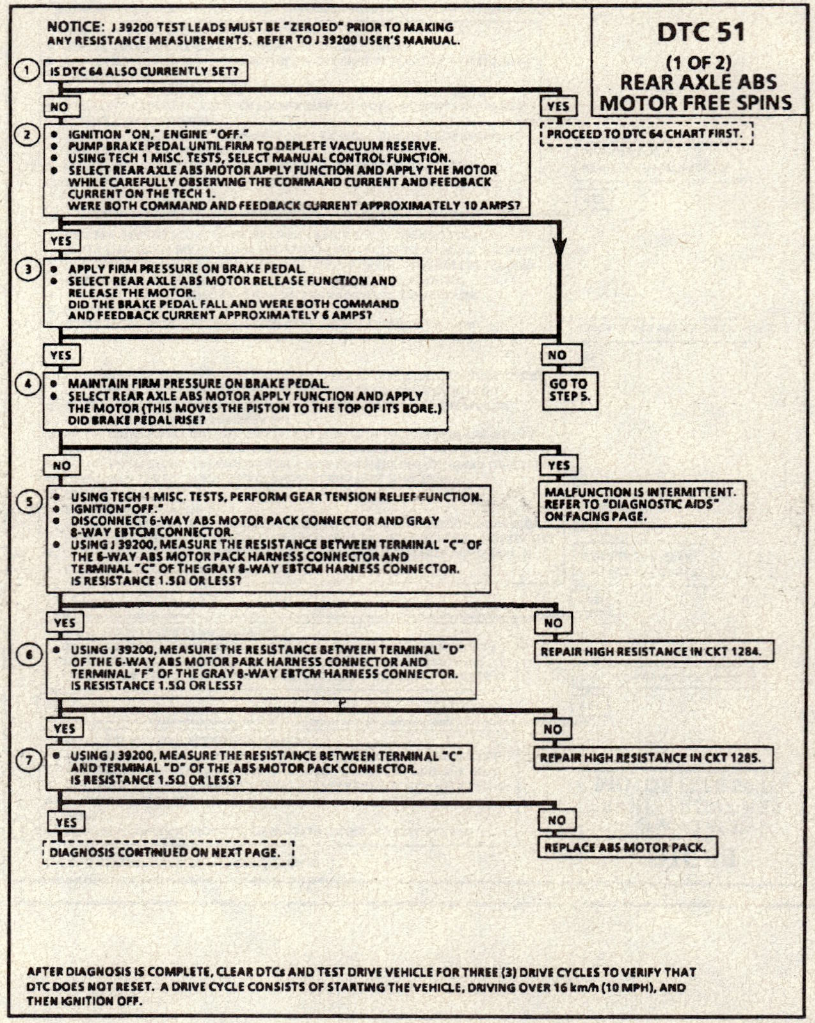

DIAGNOSTIC CHART DTC 51 — 1995 CAMARO AND FIREBIRD WITH TCS, CONTINUED

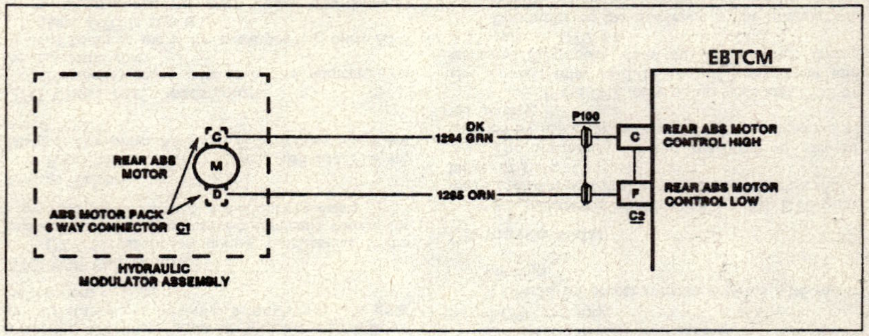

DTC 51
(Page 2 of 2)
REAR ABS MOTOR FREE SPINS

Circuit Description:

This DTC is designed to detect a stripped nut or gear assembly during initialization. During the homing sequence, the piston should reach the top of the bore resulting in a stalled motor. If this does not occur, the ABS motor must be spinning with little or no resistance, indicating a nut/screw or gear malfunction.

DTC Will Set When:

DTC 51 can be set during initialization. If the feedback current is less than the command current for a specified period of time, a malfunction exists.

Action Taken:

A malfunction DTC is stored, ABS and TCS are disabled. The amber ABS and TCS warning lamps are turned "ON."

DTC Chart Test Description:

Number(s) below refer to circled number(s) on the diagnostic chart.

8. Checks for a stripped gear on the ABS motor pack (unit with three small gears).
9. Checks for a stripped gear on the hydraulic modulator (unit with three large gears).
10. Checks to see if the ABS motor pack has an internal malfunction.
11. Checks for a malfunctioning hydraulic modulator.
12. Ensures malfunction was not due to poor terminal contact.

Diagnostic Aids:

An intermittent malfunction in this DTC may result from a mechanical part of the system that sticks, binds, or slips.

The frequency of the malfunction can be checked by using the enhanced diagnostic function of the Tech 1,

If the DTC only occurs once and DTC 64 also occurs, refer to DTC 64. If intermittent and enhanced diagnostics show this DTC occurs during ABS, refer to DTC 64.

Depending on the frequency of the malfunction, a physical inspection of the mechanical parts suspected may be necessary.

DIAGNOSTIC CHART DTC 51 — 1995 CAMARO AND FIREBIRD WITH TCS, CONTINUED

DTC 51
(Page 2 of 2)
REAR ABS MOTOR FREE SPINS

DIAGNOSIS CONTINUED FROM PREVIOUS PAGE.

(8)
- REMOVE HYDRAULIC MODULATOR/MOTOR PACK ASSEMBLY FROM VEHICLE.
- REMOVE GEAR COVER.
- CHECK FOR STRIPPED GEAR ON ABS MOTOR PACK; ABS MOTOR PACK IS THE UNIT WITH THE THREE SMALL GEARS. THE REAR AXLE GEAR IS THE MIDDLE GEAR OF THE THREE.

IS THE ABS MOTOR PACK GEAR STRIPPED?

YES: REPLACE ABS MOTOR PACK.

NO:

(9)
- CHECK FOR A STRIPPED GEAR ON HYDRAULIC MODULATOR; MODULATOR IS THE UNIT WITH THE THREE LARGE GEARS. THE REAR AXLE GEAR IS THE MIDDLE GEAR OF THE THREE.

IS THE MODULATOR GEAR STRIPPED?

YES: REPLACE HYDRAULIC MODULATOR GEAR.

NO:

(10)
- RECONNECT ELECTRICAL CONNECTORS.
- SECURELY POSITION HYDRAULIC MODULATOR/MOTOR PACK ASSEMBLY IN VEHICLE WITH GEAR COVER REMOVED SUCH THAT THE GEAR SET CAN BE OBSERVED. BE CAREFUL NOT TO DAMAGE GEAR SET OR ALLOW THE HYDRAULIC MODULATOR TO MOVE WHILE TESTING.
- IGNITION "ON."
- USING TECH 1 MISC. TESTS, SELECT MANUAL CONTROL FUNCTION.
- SELECT REAR AXLE ABS MOTOR APPLY AND APPLY THE MOTOR WHILE OBSERVING THE MIDDLE GEAR SET.
- SELECT REAR AXLE ABS MOTOR RELEASE FUNCTION AND RELEASE THE MOTOR WHILE STILL OBSERVING THE GEAR SET.
- SELECT REAR AXLE ABS MOTOR APPLY FUNCTION AND APPLY THE MOTOR.

DID GEAR SET MOVE IN BOTH DIRECTIONS FOR AT LEAST ONE REVOLUTION?

NO: REPLACE ABS MOTOR PACK.

YES:

(11)
- USING TECH 1 MISC. TESTS, PERFORM GEAR TENSION RELIEF FUNCTION.
- IGNITION "OFF."
- SEPARATE THE ABS MOTOR PACK FROM THE HYDRAULIC MODULATOR.
- GRASP THE MIDDLE GEAR ON THE HYDRAULIC MODULATOR (THE HYDRAULIC MODULATOR IS THE UNIT WITH THE THREE LARGE GEARS) AND ATTEMPT TO MOVE THE GEAR.

CAN THE GEAR BE ROTATED MORE THAN 8 FULL TURNS LOCK TO LOCK?

YES: REPLACE THE HYDRAULIC MODULATOR.

NO:

(12)
- INSPECT ALL CONNECTORS AND TERMINALS FOR POOR TERMINAL CONTACT AND EVIDENCE OF CORROSION.
  REPLACE ALL TERMINALS THAT EXHIBIT SIGNS OF POOR TERMINAL CONTACT OR CORROSION.
- REINSTALL HYDRAULIC MODULATOR/MOTOR PACK ASSEMBLY IN VEHICLE.
- RECONNECT ALL CONNECTORS.
- IGNITION "OFF."
- START ENGINE WITH FOOT OFF BRAKE.
- ALLOW ENGINE TO RUN FOR AT LEAST 10 SECONDS.
- REPEAT ABOVE IGNITION CYCLE SEQUENCE TWO MORE TIMES.

DID DTC 51 SET IN THE LAST THREE IGNITION CYCLES?

YES: REPLACE EBTCM.

NO: MALFUNCTION WAS MOST LIKELY CAUSED BY A POOR CONNECTION. REFER TO "DIAGNOSTIC AIDS" ON FACING PAGE.

AFTER DIAGNOSIS IS COMPLETE, CLEAR DTCs AND TEST DRIVE VEHICLE FOR THREE (3) DRIVE CYCLES TO VERIFY THAT DTC DOES NOT RESET. A DRIVE CYCLE CONSISTS OF STARTING THE VEHICLE, DRIVING OVER 16 km/h (10 MPH), AND THEN TURNING KEY OFF.

DIAGNOSTIC CHART DTC 52 — 1995 CAMARO AND FIREBIRD WITH TCS

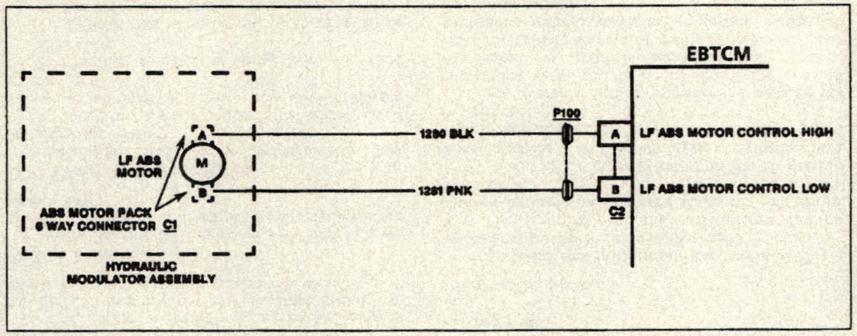

DTC 52
LEFT FRONT ABS CHANNEL IN RELEASE TOO LONG

Circuit Description:

This DTC will diagnose an ABS motor that is energized longer than expected. This could occur if a wheel speed sensor is malfunctioning, the ABS motor does not turn, the left front solenoid mechanically fails open, or the ABS motor wires are crossed.

DTC Will Set When:

DTC 52 can be set only during an ABS stop. If the EBTCM commands the left front ABS channel in release for three seconds, a malfunction exists.

Action Taken:

A malfunction DTC is stored, ABS and TCS are disabled. The amber ABS and TCS warning lamps are turned "ON."

DTC Chart Test Description:

Number(s) below refer to circled number(s) on the diagnostic chart.

1. Identifies a malfunction in a wheel speed sensor that may cause the system to be in release too long.
2. Identifies an ABS motor as being failed or wired incorrectly.
3. Checks for a solenoid that may have mechanically failed open.
4. Verifies proper solenoid operation.
5. If Step 3 has failed, this serves to isolate the cause of the hydraulic malfunction, to either the solenoid or the ABS hydraulic modulator assembly.
6. Determines whether a malfunctioning motor pack or hydraulic modulator is the reason for DTC 52 being set.

Diagnostic Aids:

An intermittent malfunction may be caused by a mechanical part of the system that sticks or binds.

The frequency of the malfunction can be checked by using the enhanced diagnostic function of the Tech 1.

DTC 52 may occur if on ice and steering wheel is turned to lock during braking. Using the Tech 1, perform hydraulic test to ensure total brake system is functional.

Any circuitry that is suspected of causing the intermittent complaint should be thoroughly checked for backed out terminals, improper mating, broken locks, improperly formed or damaged terminals, poor terminal to wiring connections or physical damage to the wiring harness.

DIAGNOSTIC CHART DTC 52 — 1995 CAMARO AND FIREBIRD WITH TCS, CONTINUED

DTC 52
LEFT FRONT ABS CHANNEL IN RELEASE TOO LONG

Important
EXCESSIVE DRAG OR HIGH RESISTANCE IN THE BASE BRAKE OR SUSPENSION SYSTEM MUST BE INSPECTED AND CORRECTED BEFORE PROCEEDING WITH DTC DIAGNOSIS.

ARE ANY WHEEL SPEED DTCs CURRENTLY SET?
- YES: PROCEED TO THAT DTC CHART FIRST.
- NO: go to step 1.

(1)
- USING TECH 1, SELECT "F0: DATA LIST" AND TEST VEHICLE DURING A STEADY DECEL CONDITION FROM 35 MPH TO 0 MPH WHILE MONITORING ALL WHEEL SPEEDS.

DO ANY OF THE WHEEL SPEEDS INDICATE ERRATIC OR INTERMITTENT OPERATION?
- YES: PROCEED TO WHEEL SPEED = 0 DTC FOR AFFECTED WHEEL.
- NO: go to step 2.

(2)
- IGNITION "ON."
- USING TECH 1 MANUAL CONTROL FUNCTION, SELECT LEFT FRONT ABS MOTOR APPLY AND APPLY MOTOR.
- APPLY FIRM PRESSURE ON BRAKE PEDAL.
- SELECT LEFT FRONT ABS MOTOR RELEASE AND COMMAND RELEASE OF MOTOR.

DID BRAKE PEDAL FALL?
- YES: go to step 3.
- NO:
  - BRAKE PEDAL DID NOT MOVE: go to step 6.
  - BRAKE PEDAL ROSE: REPAIR CROSSED WIRES (CKTs 1280 and 1281) TO LEFT FRONT ABS MOTOR.

(3)
- REMOVE FOOT FROM BRAKE PEDAL.
- USING TECH 1 MANUAL CONTROL FUNCTION, SELECT LEFT FRONT SOLENOID AND COMMAND THE SOLENOID "ON."
- APPLY FIRM PRESSURE TO BRAKE PEDAL.

DID BRAKE PEDAL FALL?
- YES: PROCEED TO STEP 5 OF THIS CHART.
- NO: go to step 4.

(4)
- FIRM PRESSURE STILL APPLIED TO BRAKE PEDAL.
- USING TECH 1, COMMAND LEFT FRONT SOLENOID "OFF."

DID BRAKE PEDAL FALL?
- YES: MALFUNCTION IS NOT PRESENT AT THIS TIME. REFER TO "DIAGNOSTIC AIDS" ON FACING PAGE.
- NO: go to step 5.

(5)
- REMOVE FOOT FROM BRAKE PEDAL.
- IGNITION "OFF."
- PHYSICALLY SWITCH THE LEFT FRONT SOLENOID WITH RIGHT FRONT SOLENOID.
- IGNITION "ON."
- USING TECH 1 MANUAL CONTROL FUNCTION, SELECT LEFT FRONT SOLENOID AND COMMAND THE SOLENOID "ON."
- APPLY FIRM PRESSURE TO BRAKE PEDAL.

DID BRAKE PEDAL FALL?
- YES: REPLACE ABS HYDRAULIC MODULATOR.
- NO: REPLACE THE SOLENOID WHICH WAS ORIGINALLY IN THE LEFT FRONT CHANNEL.

(6)
- USING TECH 1 PERFORM GEAR TENSION RELIEF FUNCTION.
- IGNITION "OFF."
- SEPARATE ABS MOTOR PACK FROM HYDRAULIC MODULATOR.
- GRASP THE GEAR ON THE HYDRAULIC MODULATOR (HYDRAULIC MODULATOR IS THE UNIT WITH THE THREE LARGE GEARS) FURTHEST FORWARD WHEN THE UNIT IS INSTALLED IN THE VEHICLE AND ATTEMPT TO ROTATE THE GEAR IN EITHER DIRECTION.

CAN THE GEAR BE MOVED BY HAND?
- YES: REPLACE ABS MOTOR PACK.
- NO: REPLACE ABS HYDRAULIC MODULATOR.

AFTER DIAGNOSIS IS COMPLETE, CLEAR DTCs AND TEST DRIVE VEHICLE FOR THREE (3) DRIVE CYCLES TO VERIFY DTC DOES NOT RESET. A DRIVE CYCLE CONSISTS OF STARTING THE VEHICLE, DRIVING OVER 16km/h (10 MPH) AND THEN TURNING KEY OFF.

DIAGNOSTIC CHART DTC 53 — 1995 CAMARO AND FIREBIRD WITH TCS

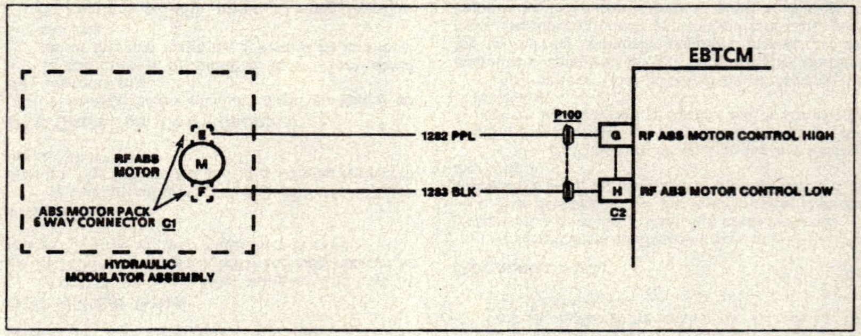

DTC 53
RIGHT FRONT ABS CHANNEL IN RELEASE TOO LONG

Circuit Description:
This DTC will diagnose a motor that is energized longer than expected. This could occur if a wheel speed sensor is malfunctioning, the motor does not turn, the right front solenoid mechanically fails open, or the ABS motor wires are crossed.

DTC Will Set When:
DTC 53 can be set only during an ABS stop. If the EBTCM commands the right front channel in release for three seconds, a malfunction exists.

Action Taken:
A malfunction DTC is stored, ABS and TCS are disabled. The amber ABS and TCS warning lamps are turned "ON."

DTC Chart Test Description:
Number(s) below refer to circled number(s) on the diagnostic chart.

1. Identifies a malfunction in a wheel speed sensor that may cause the system to be in release too long.
2. Identifies a motor as being failed or wired incorrectly.
3. Checks for a solenoid that may have mechanically failed open.
4. If Step 3 has failed, this serves to isolate the cause of the hydraulic malfunction, to either the solenoid or the ABS hydraulic modulator assembly.
5. Checks for a solenoid that may have mechanically failed closed.
6. Determines whether a malfunction motor pack or hydraulic modulator is the reason for DTC 53 being set.

Diagnostic Aids:
An intermittent malfunction may be caused by a mechanical part of the system that sticks or binds.

The frequency of the malfunction can be checked by using the enhanced diagnostic function of the Tech 1.

DTC 53 may occur on ice if steering wheel is turned to lock during braking. Using the Tech 1 perform hydraulic test to ensure total brake system is functional.

Any circuitry that is suspected of causing the intermittent complaint should be thoroughly checked for backed out terminals, improper mating, broken locks, improperly formed or damaged terminals, poor terminal to wiring connections or physical damage to the wiring harness.

DIAGNOSTIC CHART DTC 53 — 1995 CAMARO AND FIREBIRD WITH TCS, CONTINUED

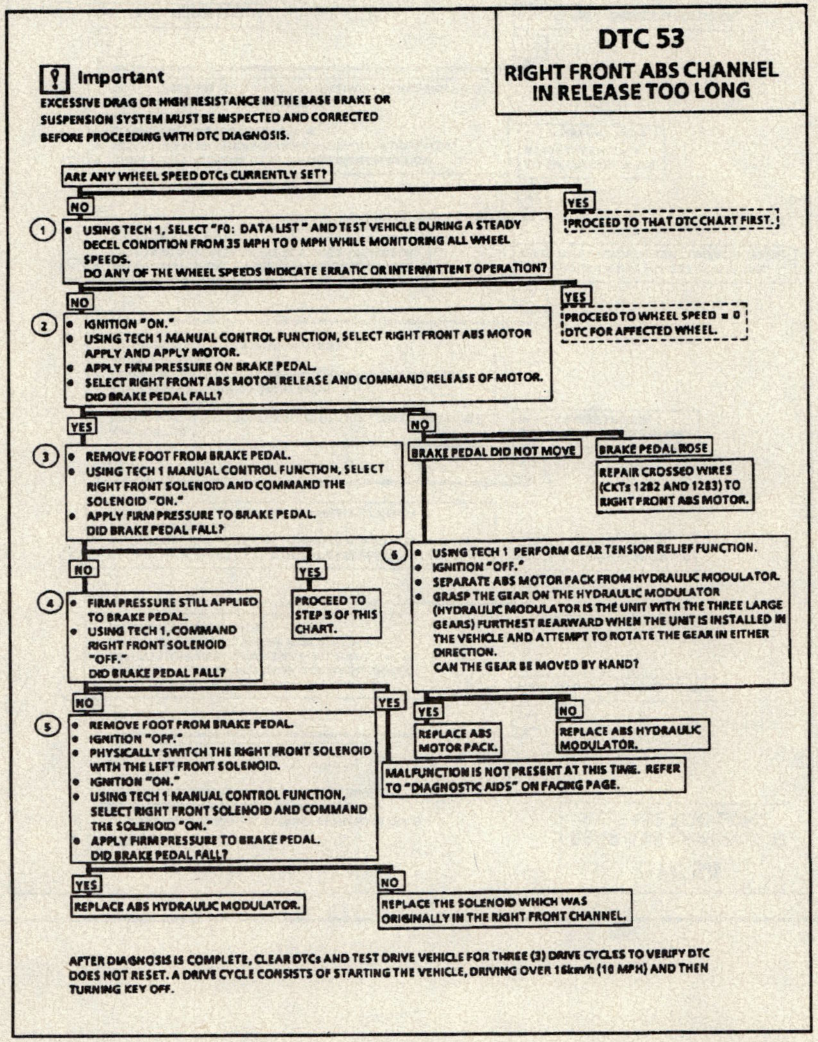

DIAGNOSTIC CHART DTC 54 — 1995 CAMARO AND FIREBIRD WITH TCS

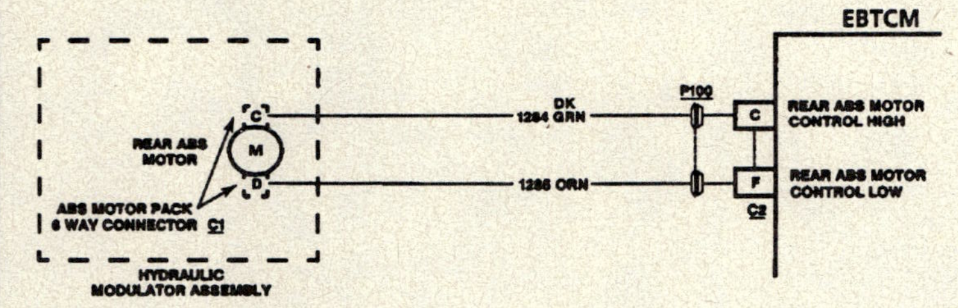

DTC 54
REAR ABS CHANNEL IN RELEASE TOO LONG

Circuit Description:

This DTC will diagnose an ABS motor that is energized longer than expected. This could occur if a wheel speed sensor is malfunctioning, the ABS motor does not turn, or the ABS motor wires are crossed.

DTC Will Set When:

DTC 54 can be set only during an ABS stop. If the EBTCM commands the rear axle channel in release for three seconds, a malfunction exists.

Action Taken:

A malfunction DTC is stored, ABS and TCS are disabled. The amber ABS and TCS warning lamps are turned "ON."

DTC Chart Test Description:

Number(s) below refer to circled number(s) on the diagnostic chart.

1. Identifies a malfunction in a wheel speed sensor that may cause the system to be in release too long.
2. Checks for a wheel that may stick or bind because of a mechanical malfunction.
3. Checks to see if the ABS motor is capable of moving and applying the hydraulic piston for the rear wheels.
4. Ensures that the ABS motor wiring is not crossed.
5. Isolates the fault of a "no-apply" situation to either the ABS motor pack or ABS hydraulic modulator assembly.

Diagnostic Aids:

An intermittent malfunction may be caused by a mechanical part of the system that sticks or binds.

The frequency of the malfunction can be checked by using the enhanced diagnostic function of the Tech 1.

Using the Tech 1, perform hydraulic test to ensure total brake system is functional.

Any circuitry that is suspected of causing the intermittent complaint should be thoroughly checked for backed out terminals, improper mating, broken locks, improperly formed or damaged terminals, poor terminal to wiring connections or physical damage to the wiring harness.

DIAGNOSTIC CHART DTC 54 — 1995 CAMARO AND FIREBIRD WITH TCS, CONTINUED

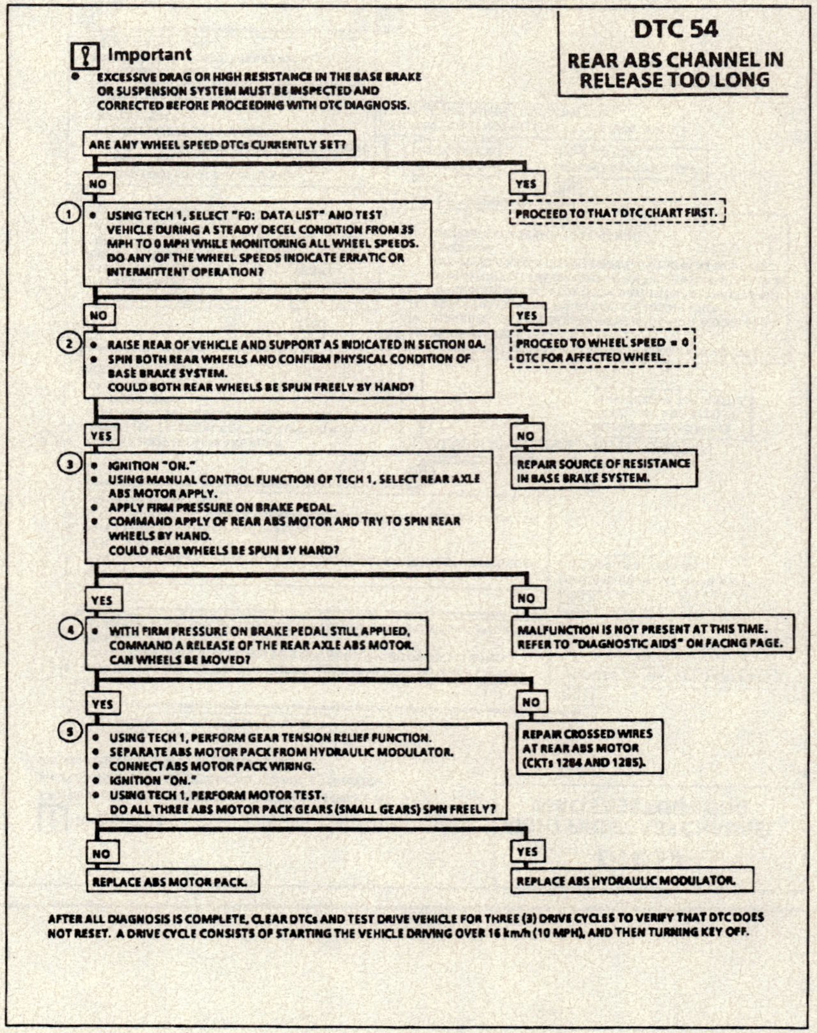

DIAGNOSTIC CHART DTC 55 — 1995 CAMARO AND FIREBIRD WITH TCS

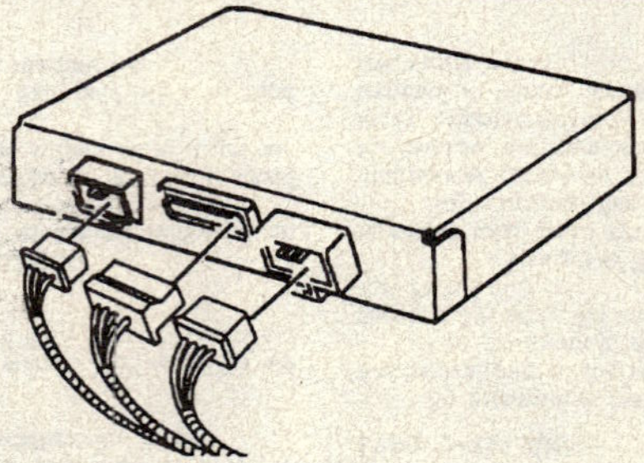

DTC 55
EBTCM MALFUNCTION (ABS)

Circuit Description:
This DTC identifies an internal malfunction detected by the EBTCM ABS motor driver interface custom integrated circuit.

Action Taken:
A malfunction DTC is stored, ABS and TCS are disabled. The amber ABS and TCS warning lamps are turned "ON."

DTC Chart Test Description:
Number(s) below refer to circled number(s) on the diagnostic chart.
1. Checks to see if the malfunction is still present.
2. Checks to see if the malfunction is intermittent.

DIAGNOSTIC CHART DTC 55 — 1995 CAMARO AND FIREBIRD WITH TCS, CONTINUED

DTC 55
EBTCM MALFUNCTION (ABS)

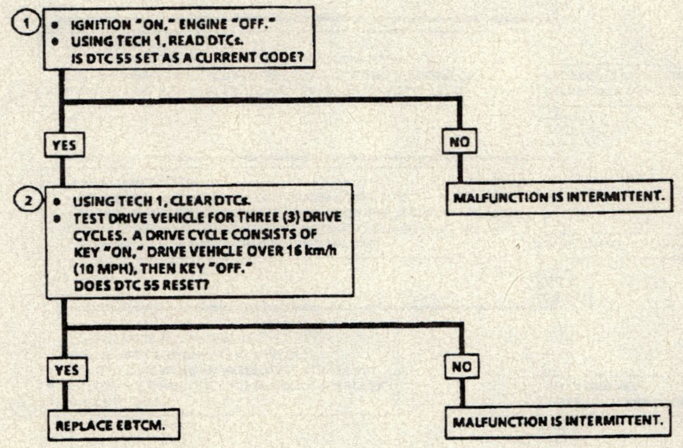

DIAGNOSTIC CHART DTC 56 — 1995 CAMARO AND FIREBIRD WITH TCS

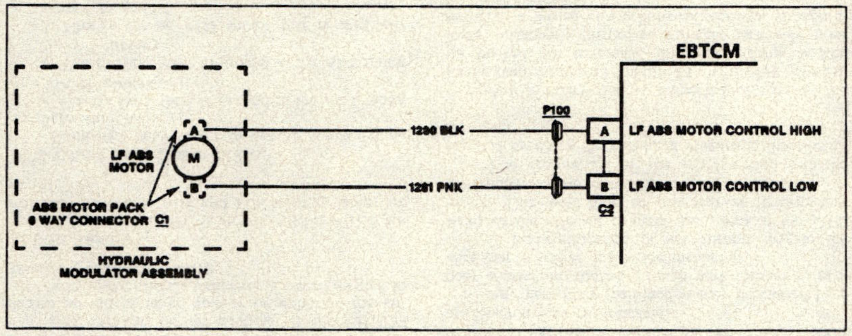

DTC 56
LEFT FRONT ABS MOTOR CIRCUIT OPEN

Circuit Description:

This DTC identifies an ABS motor that cannot be energized due to an open in its circuitry. This malfunction will not allow proper ABS operation.

DTC Will Set When:

DTC 56 can be set only when the ABS motor is commanded "OFF." If the EBTCM detects an out of range voltage on either of the left front ABS motor circuits indicating an open circuit, a malfunction exists.

Action Taken:

An open ABS motor will not activate when requested. A malfunction DTC is stored, ABS and TCS are disabled. The amber ABS and TCS warning lamps are turned "ON."

DTC Chart Test Description:

Number(s) below refer to circled number(s) on the diagnostic chart.

1. Checks for proper resistance of the ABS motor.
2. Checks for an open in the ABS motor "HIGH" circuitry.
3. Checks for an open in the ABS motor "LOW" circuitry.
4. Ensures malfunction was not due to poor terminal contact.

Diagnostic Aids:

Using Tech 1, select manual control function, and exercise ABS motor movement of affected channel in both directions while applying light pressure on the brake pedal.

If erratic or "jumpy" brake pedal movement is detected while performing an "apply" or "release" function of the ABS motor, an intermittent malfunction may be indicated.

An intermittent malfunction may be caused by a poor connection, rubbed through wire insulation, or a wire that is broken inside the insulation.

If the malfunction is not current, wiggle the wires of the affected channel and check if the DTC resets. This will help to pinpoint an intermittent malfunction in the motor circuitry or connections.

The frequency of the malfunction can be checked by using the enhanced diagnostic function of the Tech 1,

Any circuitry that is suspected of causing the intermittent complaint should be thoroughly checked for backed out terminals, improper mating, broken locks, improperly formed or damaged terminals, poor terminal to wiring connections or physical damage to the wiring harness.

DIAGNOSTIC CHART DTC 56 — 1995 CAMARO AND FIREBIRD WITH TCS, CONTINUED

DTC 56
LEFT FRONT ABS MOTOR CIRCUIT OPEN

DOES DTC 56 OCCUR INTERMITTENTLY?

- YES: REFER TO "DIAGNOSTIC AIDS" ON FACING PAGE.
- NO:

(1)
- IGNITION "OFF."
- DISCONNECT 6-WAY ABS MOTOR PACK CONNECTOR.
- USING J 39200, MEASURE RESISTANCE ACROSS ABS MOTOR PACK TERMINALS "A" AND "B".

IS RESISTANCE BELOW 2000Ω?

- NO: REPLACE ABS MOTOR PACK.
- YES:

(2)
- DISCONNECT GRAY 8-WAY CONNECTOR AT EBTCM.
- USING J 39200, MEASURE RESISTANCE BETWEEN TERMINAL "A" OF THE GRAY 8-WAY EBTCM HARNESS CONNECTOR AND TERMINAL "A" OF THE 6-WAY ABS MOTOR PACK HARNESS CONNECTOR.

IS RESISTANCE 2Ω OR LESS?

- NO: REPAIR OPEN OR HIGH RESISTANCE IN CKT 1280.
- YES:

(3)
- USING J 39200, MEASURE RESISTANCE BETWEEN TERMINAL "B" OF THE GRAY 8-WAY EBTCM HARNESS CONNECTOR AND TERMINAL "B" OF THE 6-WAY ABS MOTOR PACK HARNESS CONNECTOR.

IS RESISTANCE 2Ω OR LESS?

- NO: REPAIR OPEN OR HIGH RESISTANCE IN CKT 1281.
- YES:

(4)
- INSPECT THE 6-WAY ABS MOTOR PACK AND GRAY 8-WAY EBTCM CONNECTORS FOR POOR TERMINAL CONTACT AND EVIDENCE OF CORROSION.
  REPLACE ALL TERMINALS THAT EXHIBIT SIGNS OF POOR TERMINAL CONTACT OR CORROSION.
- IGNITION "OFF."
- RECONNECT ALL CONNECTORS.
- START ENGINE AND TEST DRIVE VEHICLE, ACHIEVING AT LEAST 16 km/h (10 MPH).
- REPEAT ABOVE DRIVE CYCLE SEQUENCE TWO MORE TIMES.

DID DTC 56 SET IN THE LAST THREE DRIVE CYCLES?

- YES: REPLACE EBTCM.
- NO: MALFUNCTION WAS MOST LIKELY CAUSED BY A POOR CONNECTION. REFER TO "DIAGNOSTIC AIDS" ON FACING PAGE.

AFTER ALL DIAGNOSIS IS COMPLETE, CLEAR DTCs AND TEST DRIVE VEHICLE FOR THREE (3) DRIVE CYCLES TO VERIFY THAT DTC DOES NOT RESET. A DRIVE CYCLE CONSISTS OF STARTING THE VEHICLE, DRIVING OVER 16 km/h (10 MPH), AND THEN IGNITION OFF.

DIAGNOSTIC CHART DTC 57 — 1995 CAMARO AND FIREBIRD WITH TCS

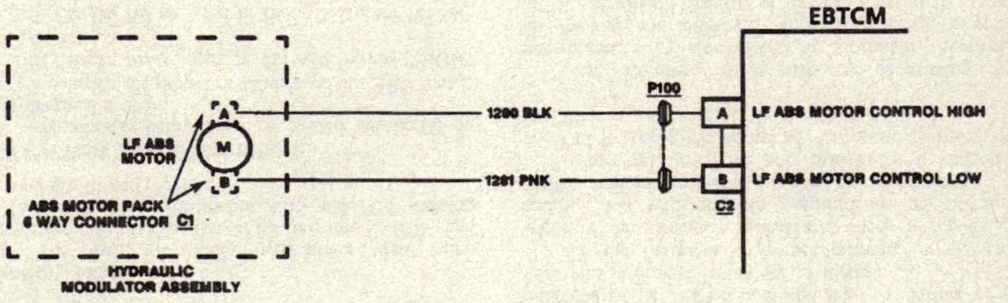

DTC 57
LEFT FRONT ABS MOTOR CIRCUIT SHORTED TO GND

Circuit Description:

This DTC identifies an ABS motor circuit that is shorted to ground. This malfunction will not allow the ABS motor to be controlled at the commanded current rate or will cause the ABS motor driver circuit to allow current directly to ground.

DTC Will Set When:

DTC 57 can be set anytime. If the EBTCM detects an out of range voltage on either of the left front ABS motor circuits indicating a circuit shorted to ground, a malfunction exists.

Action Taken:

A malfunction DTC is stored, ABS and TCS are disabled. The amber ABS and TCS warning lamps are turned "ON."

DTC Chart Test Description:

Number(s) below refer to circled number(s) on the diagnostic chart.

1. Checks for a short to ground in the ABS motor "HIGH" circuitry.
2. Checks for a short to ground in the ABS motor "LOW" circuitry.
3. Checks for an ABS motor that is internally shorted to ground.
4. Ensures malfunction was not due to physical damage of the circuitry.

Diagnostic Aids:

Using Tech 1, select manual control function, and exercise ABS motor movement of affected channel in both directions while applying light pressure on the brake pedal.

If erratic or "jumpy" brake pedal movement is detected while performing an "apply" or "release" function of the ABS motor, an intermittent malfunction may be indicated.

An intermittent malfunction may be caused by a poor connection, rubbed through wire insulation, or a wire that is broken inside the insulation.

If the malfunction is not current, wiggle the wires of the affected channel and check if the DTC resets. This will help to pinpoint an intermittent malfunction in the motor circuitry or connections.

The frequency of the malfunction can be checked by using the enhanced diagnostic function of the Tech 1.

Any circuitry that is suspected of causing the intermittent complaint should be thoroughly checked for backed out terminals, improper mating, broken locks, improperly formed or damaged terminals, poor terminal to wiring connections or physical damage to the wiring harness.

DIAGNOSTIC CHART DTC 57 — 1995 CAMARO AND FIREBIRD WITH TCS, CONTINUED

DTC 57
LEFT FRONT ABS MOTOR CIRCUIT SHORTED TO GND

DOES DTC 57 OCCUR INTERMITTENTLY?

- YES: REFER TO "DIAGNOSTIC AIDS" ON FACING PAGE.
- NO:

(1)
- IGNITION "OFF."
- DISCONNECT THE 6-WAY ABS MOTOR PACK CONNECTOR AND THE GRAY 8-WAY EBTCM CONNECTOR.
- USING J 39200, MEASURE THE RESISTANCE BETWEEN GROUND AND TERMINAL "A" OF THE 6-WAY ABS MOTOR PACK HARNESS CONNECTOR.

DOES J 39200 DISPLAY RESISTANCE AS "OL" (INFINITE)?

- NO: REPAIR SHORT TO GROUND IN CKT 1280.
- YES:

(2)
- USING J 39200, MEASURE THE RESISTANCE BETWEEN GROUND AND TERMINAL "B" OF THE 6-WAY ABS MOTOR PACK HARNESS CONNECTOR.

DOES J 39200 DISPLAY RESISTANCE AS "OL" (INFINITE)?

- NO: REPAIR SHORT TO GROUND IN CKT 1281.
- YES:

(3)
- USING J 39200, MEASURE THE RESISTANCE BETWEEN GROUND AND TERMINAL "A" OF THE ABS MOTOR PACK.

DOES J 39200 DISPLAY RESISTANCE AS "OL" (INFINITE)?

- NO: REPLACE ABS MOTOR PACK.
- YES:

(4)
- INSPECT CKTs 1280 AND 1281 AND RELATED CONNECTORS FOR DAMAGE WHICH MAY RESULT IN A SHORT TO GROUND WITH ALL CONNECTORS CONNECTED. REPAIR ALL DAMAGE IF EVIDENT.
- IGNITION "OFF."
- RECONNECT ALL CONNECTORS.
- START ENGINE AND TEST DRIVE VEHICLE, ACHIEVING AT LEAST 16 km/h (10 MPH).
- REPEAT ABOVE DRIVE CYCLE SEQUENCE TWO MORE TIMES.

DID DTC 57 SET IN THE LAST THREE DRIVE CYCLES?

- YES: REPLACE EBTCM.
- NO: MALFUNCTION WAS MOST LIKELY CAUSED BY A POOR CONNECTION. REFER TO "DIAGNOSTIC AIDS" ON FACING PAGE.

AFTER ALL DIAGNOSIS IS COMPLETE, CLEAR DTCs AND TEST DRIVE VEHICLE FOR THREE (3) DRIVE CYCLES TO VERIFY THAT DTC DOES NOT RESET. A DRIVE CYCLE CONSISTS OF STARTING THE VEHICLE, DRIVING OVER 16 km/h (10 MPH), AND THEN IGNITION OFF.

DIAGNOSTIC CHART DTC 58 — 1995 CAMARO AND FIREBIRD WITH TCS

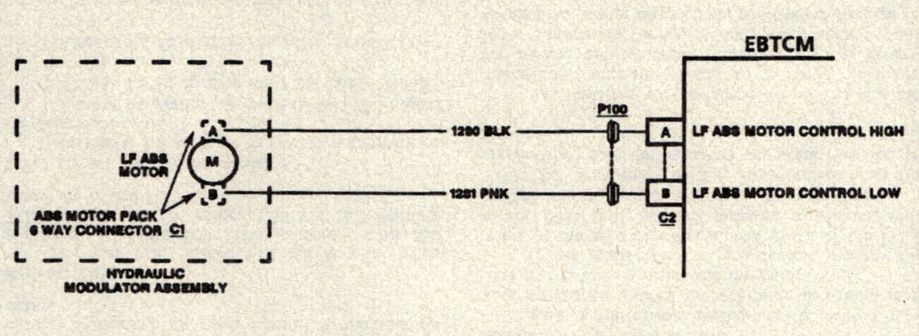

DTC 58
LEFT FRONT ABS MOTOR CIRCUIT SHORTED TO BATT

Circuit Description:

This DTC identifies an ABS motor circuit that is shorted to battery or an ABS motor that has low or no resistance. This malfunction will not allow the ABS motor to be controlled at the commanded current rate or will cause the ABS motor to turn in the opposite direction, or not at all.

DTC Will Set When:

DTC 58 can be set only when the ABS motor is commanded "OFF." If the EBTCM detects an out of range voltage on either of the left front ABS motor circuits indicating a circuit shorted to battery or an ABS motor shorted, a malfunction exists.

Action Taken:

A malfunction DTC is stored, ABS and TCS are disabled. The amber ABS and TCS warning lamps are turned "ON."

DTC Chart Test Description:

Number(s) below refer to circled number(s) on the diagnostic chart.

1. Checks for a short to voltage in the ABS motor "HIGH" circuitry.
2. Checks for a short to voltage in the ABS motor "LOW" circuitry.
3. Checks for a short between the "HIGH" and "LOW" motor circuits.
4. Checks for an ABS motor that is internally shorted.
5. Ensures malfunction was not due to physical damage of the circuitry.

Diagnostic Aids:

Using Tech 1, select manual control function, and exercise ABS motor movement of affected channel in both directions while applying light pressure on the brake pedal.

If erratic or "jumpy" brake pedal movement is detected while performing an "apply" or "release" function of the ABS motor, an intermittent malfunction may be indicated.

An intermittent malfunction may be caused by a poor connection, rubbed through wire insulation, or a wire that is broken inside the insulation.

If the malfunction is not current, wiggle the wires of the affected channel and check if the DTC resets. This will help to pinpoint an intermittent malfunction in the motor circuitry or connections.

The frequency of the malfunction can be checked by using the enhanced diagnostic function of the Tech 1.

Any circuitry that is suspected of causing the intermittent complaint should be thoroughly checked for backed out terminals, improper mating, broken locks, improperly formed or damaged terminals, poor terminal to wiring connections or physical damage to the wiring harness.

DIAGNOSTIC CHART DTC 58 — 1995 CAMARO AND FIREBIRD WITH TCS, CONTINUED

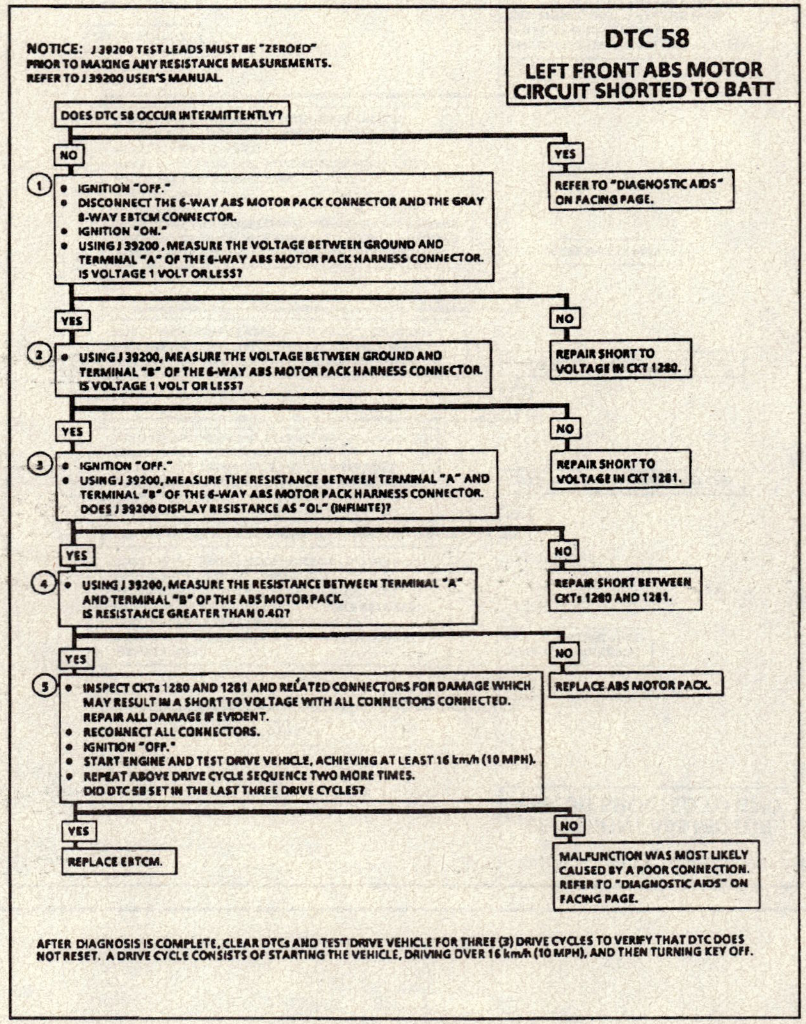

DIAGNOSTIC CHART DTC 61 — 1995 CAMARO AND FIREBIRD WITH TCS

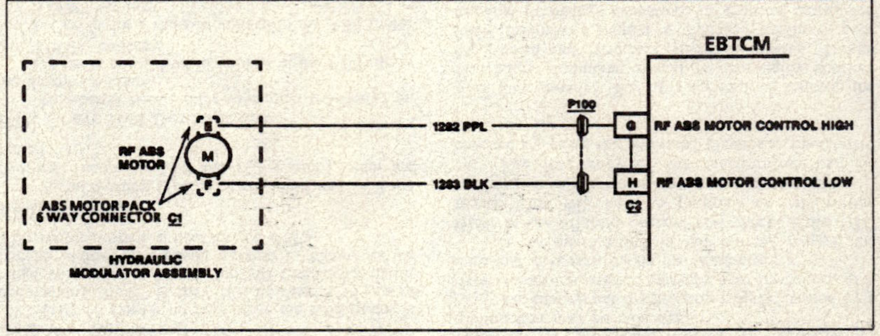

DTC 61
RIGHT FRONT ABS MOTOR CIRCUIT OPEN

Circuit Description:

This DTC identifies an ABS motor that cannot be energized due to an open in its circuitry. This malfunction will not allow proper ABS operation.

DTC Will Set When:

DTC 61 can be set only when the ABS motor is commanded "OFF." If the EBTCM detects an out of range voltage on either of the right front ABS motor circuits indicating an open circuit, a malfunction exists.

Action Taken:

An open ABS motor will not activate when requested. A malfunction DTC is stored, ABS and TCS are disabled. The amber ABS and TCS warning lamps are turned "ON."

DTC Chart Test Description:

Number(s) below refer to circled number(s) on the diagnostic chart.

1. Checks for proper resistance of the ABS motor.
2. Checks for an open in the ABS motor "HIGH" circuitry.
3. Checks for an open in the ABS motor "LOW" circuitry.
4. Ensures malfunction was not due to poor terminal contact.

Diagnostic Aids:

Using Tech 1, select manual control function, and exercise ABS motor movement of affected channel in both directions while applying light pressure on the brake pedal.

If erratic or "jumpy" brake pedal movement is detected while performing an "apply" or "release" function of the ABS motor, an intermittent malfunction may be indicated.

An intermittent malfunction may be caused by a poor connection, rubbed through wire insulation, or a wire that is broken inside the insulation.

If the malfunction is not current, wiggle the wires of the affected channel and check if the DTC resets. This will help to pinpoint an intermittent malfunction in the motor circuitry or connections.

The frequency of the malfunction can be checked by using the enhanced diagnostic function of the Tech 1.

Any circuitry that is suspected of causing the intermittent complaint should be thoroughly checked for backed out terminals, improper mating, broken locks, improperly formed or damaged terminals, poor terminal to wiring connections or physical damage to the wiring harness.

DIAGNOSTIC CHART DTC 61 — 1995 CAMARO AND FIREBIRD WITH TCS, CONTINUED

DTC 61
RIGHT FRONT ABS MOTOR CIRCUIT OPEN

DOES DTC 61 OCCUR INTERMITTENTLY?

- YES: REFER TO "DIAGNOSTIC AIDS" ON FACING PAGE.
- NO:

(1)
- IGNITION "OFF."
- DISCONNECT 6-WAY ABS MOTOR PACK CONNECTOR.
- USING J 39200, MEASURE RESISTANCE ACROSS ABS MOTOR PACK TERMINALS "E" AND "F".
  IS RESISTANCE BELOW 2000Ω?

- NO: REPLACE ABS MOTOR PACK.
- YES:

(2)
- DISCONNECT GRAY 8-WAY CONNECTOR AT EBTCM.
- USING J 39200, MEASURE RESISTANCE BETWEEN TERMINAL "G" OF THE GRAY 8-WAY EBTCM HARNESS CONNECTOR AND TERMINAL "E" OF THE 6-WAY ABS MOTOR PACK HARNESS CONNECTOR.
  IS RESISTANCE 2Ω OR LESS?

- NO: REPAIR OPEN OR HIGH RESISTANCE IN CKT 1282.
- YES:

(3)
- USING J 39200, MEASURE RESISTANCE BETWEEN TERMINAL "H" OF THE GRAY 8-WAY EBTCM HARNESS CONNECTOR AND TERMINAL "F" OF THE 6-WAY ABS MOTOR PACK HARNESS CONNECTOR.
  IS RESISTANCE 2Ω OR LESS?

- NO: REPAIR OPEN OR HIGH RESISTANCE IN CKT 1283.
- YES:

(4)
- INSPECT THE 6-WAY ABS MOTOR PACK AND GRAY 8-WAY EBTCM CONNECTORS FOR POOR TERMINAL CONTACT AND EVIDENCE OF CORROSION.
  REPLACE ALL TERMINALS THAT EXHIBIT SIGNS OF POOR TERMINAL CONTACT OR CORROSION.
- IGNITION "OFF."
- RECONNECT ALL CONNECTORS.
- START ENGINE AND TEST DRIVE VEHICLE, ACHIEVING AT LEAST 16 km/h (10 MPH).
- REPEAT ABOVE DRIVE CYCLE SEQUENCE TWO MORE TIMES.
  DID DTC 61 SET IN THE LAST THREE DRIVE CYCLES?

- YES: REPLACE EBTCM.
- NO: MALFUNCTION WAS MOST LIKELY CAUSED BY A POOR CONNECTION. REFER TO "DIAGNOSTIC AIDS" ON FACING PAGE.

AFTER ALL DIAGNOSIS IS COMPLETE, CLEAR DTCs AND TEST DRIVE VEHICLE FOR THREE (3) DRIVE CYCLES TO VERIFY THAT DTC DOES NOT RESET. A DRIVE CYCLE CONSISTS OF STARTING THE VEHICLE, DRIVING OVER 16 km/h (10 MPH), AND THEN TURNING KEY OFF.

DIAGNOSTIC CHART DTC 62 — 1995 CAMARO AND FIREBIRD WITH TCS

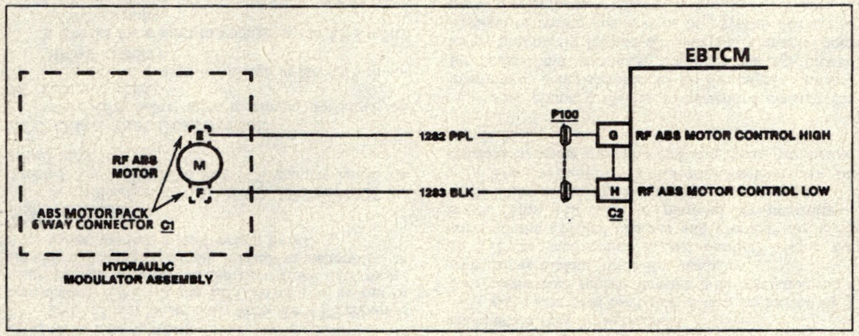

DTC 62
RIGHT FRONT ABS MOTOR CIRCUIT SHORTED TO GND

Circuit Description:
This DTC identifies an ABS motor circuit that is shorted to ground. This malfunction will not allow the ABS motor to be controlled at the commanded current rate or will cause the ABS motor driver circuit to allow current directly to ground.

DTC Will Set When:
DTC 62 can be set anytime. If the EBTCM detects an out of range voltage on either of the right front ABS motor circuits indicating a circuit shorted to ground, a malfunction exists.

Action Taken:
A malfunction DTC is stored, ABS and TCS are disabled. The amber ABS and TCS warning lamps are turned "ON."

DTC Chart Test Description:
Number(s) below refer to circled number(s) on the diagnostic chart.
1. Checks for a short to ground in the ABS motor "HIGH" circuitry.
2. Checks for a short to ground in the ABS motor "LOW" circuitry.
3. Checks for an ABS motor that is internally shorted to ground.
4. Ensures malfunction was not due to physical damage of the circuitry.

Diagnostic Aids:
Using Tech 1, select manual control function, and exercise ABS motor movement of affected channel in both directions while applying light pressure on the brake pedal.

If erratic or "jumpy" brake pedal movement is detected while performing an "apply" or "release" function of the ABS motor, an intermittent malfunction may be indicated.

An intermittent malfunction may be caused by a poor connection, rubbed through wire insulation, or a wire that is broken inside the insulation.

If the malfunction is not current, wiggle the wires of the affected channel and check if the DTC resets. This will help to pinpoint an intermittent malfunction in the motor circuitry or connections.

The frequency of the malfunction can be checked by using the enhanced diagnostic function of the Tech 1.

Any circuitry, that is suspected as causing the intermittent complaint, should be thoroughly checked for backed out terminals, improper mating, broken locks, improperly formed or damaged terminals, poor terminal to wiring connections or physical damage to the wiring harness.

DIAGNOSTIC CHART DTC 62 — 1995 CAMARO AND FIREBIRD WITH TCS, CONTINUED

DTC 62
RIGHT FRONT ABS MOTOR CIRCUIT SHORTED TO GND

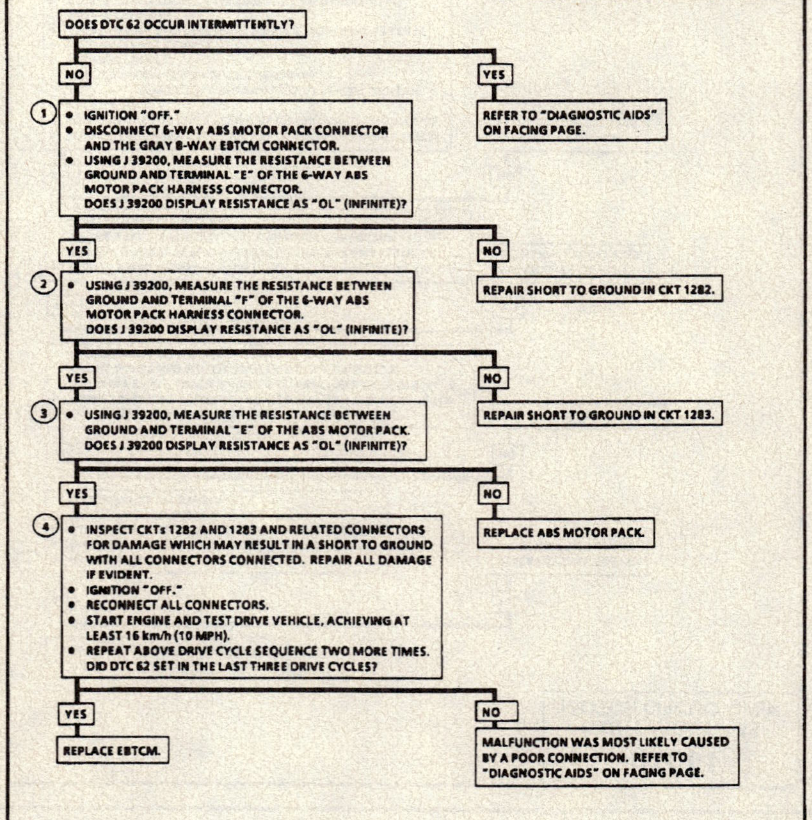

AFTER ALL DIAGNOSIS IS COMPLETE, CLEAR DTCs AND TEST DRIVE VEHICLE FOR THREE (3) DRIVE CYCLES TO VERIFY THAT DTC DOES NOT RESET. A DRIVE CYCLE CONSISTS OF STARTING THE VEHICLE, DRIVING OVER 16 km/h (10 MPH), AND THEN TURNING KEY OFF.

DIAGNOSTIC CHART DTC 63 — 1995 CAMARO AND FIREBIRD WITH TCS

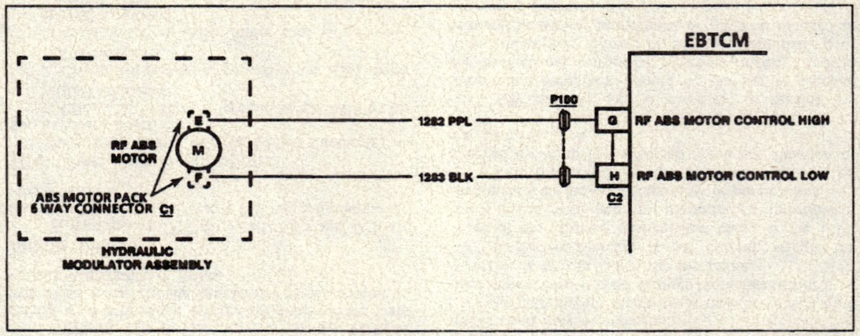

DTC 63
RIGHT FRONT ABS MOTOR CIRCUIT SHORTED TO BATT

Circuit Description:

This DTC identifies an ABS motor circuit that is shorted to battery or an ABS motor that has low or no resistance. This malfunction will not allow the ABS motor to be controlled at the commanded current rate or will cause the ABS motor to turn in the opposite direction, or not at all.

DTC Will Set When:

DTC 63 can be set only when the ABS motor is commanded "OFF." If the EBTCM detects an out of range voltage on either of the right front ABS motor circuits indicating a circuit shorted to battery or an ABS motor shorted, a malfunction exists.

Action Taken:

A malfunction DTC is stored, ABS and TCS are disabled. The amber ABS and TCS warning lamps are turned "ON."

DTC Chart Test Description:

Number(s) below refer to circled number(s) on the diagnostic chart.

1. Checks for a short to voltage in the ABS motor "HIGH" circuitry.
2. Checks for a short to voltage in the ABS motor "LOW" circuitry.
3. Checks for short circuit between ABS motor "HIGH" and motor "LOW".
4. Checks for an ABS motor that is internally shorted.
5. Ensures malfunction was not due to physical damage of the circuitry.

Diagnostic Aids:

Using Tech 1, select manual control function, and exercise ABS motor movement of affected channel in both directions while applying light pressure on the brake pedal.

If erratic or "jumpy" brake pedal movement is detected while performing an "apply" or "release" function of the ABS motor, an intermittent malfunction may be indicated.

An intermittent malfunction may be caused by a poor connection, rubbed through wire insulation, or a wire that is broken inside the insulation.

If the malfunction is not current, wiggle the wires of the affected channel and check if the DTC resets. This will help to pinpoint an intermittent malfunction in the motor circuitry or connections.

The frequency of the malfunction can be checked by using the enhanced diagnostic function of the Tech 1,

Any circuitry that is suspected of causing the intermittent complaint should be thoroughly checked for backed out terminals, improper mating, broken locks, improperly formed or damaged terminals, poor terminal to wiring connections or physical damage to the wiring harness.

DIAGNOSTIC CHART DTC 63 — 1995 CAMARO AND FIREBIRD WITH TCS, CONTINUED

DTC 63
RIGHT FRONT ABS MOTOR CIRCUIT SHORTED TO BATT

NOTICE: J 39200 TEST LEADS MUST BE "ZEROED" PRIOR TO MAKING ANY RESISTANCE MEASUREMENTS. REFER TO J 39200 USER'S MANUAL.

DOES DTC 63 OCCUR INTERMITTENTLY?

- YES: REFER TO "DIAGNOSTIC AIDS" ON FACING PAGE.
- NO:

(1)
- IGNITION "OFF."
- DISCONNECT THE 6-WAY ABS MOTOR PACK CONNECTOR AND THE GRAY 8-WAY EBTCM CONNECTOR.
- IGNITION "ON."
- USING J 39200, MEASURE THE VOLTAGE BETWEEN GROUND AND TERMINAL "E" OF THE 6-WAY ABS MOTOR PACK HARNESS CONNECTOR. IS VOLTAGE 1 VOLT OR LESS?

- NO: REPAIR SHORT TO VOLTAGE IN CKT 1282.
- YES:

(2)
- USING J 39200, MEASURE THE VOLTAGE BETWEEN GROUND AND TERMINAL "F" OF THE 6-WAY ABS MOTOR PACK HARNESS CONNECTOR. IS VOLTAGE 1 VOLT OR LESS?

- NO: REPAIR SHORT TO VOLTAGE IN CKT 1283.
- YES:

(3)
- IGNITION "OFF."
- USING J 39200, MEASURE THE RESISTANCE BETWEEN TERMINAL "E" AND TERMINAL "F" OF THE 6-WAY ABS MOTOR PACK HARNESS CONNECTOR. DOES J 39200 DISPLAY RESISTANCE AS "OL" (INFINITE)?

- NO: REPAIR SHORT BETWEEN CKTs 1282 AND 1283.
- YES:

(4)
- USING J 39200, MEASURE THE RESISTANCE BETWEEN TERMINAL "E" AND TERMINAL "F" OF THE ABS MOTOR PACK. IS RESISTANCE GREATER THAN 0.4Ω?

- NO: REPLACE ABS MOTOR PACK.
- YES:

(5)
- INSPECT CKTs 1282 AND 1283 AND RELATED CONNECTORS FOR DAMAGE WHICH MAY RESULT IN A SHORT TO VOLTAGE WITH ALL CONNECTORS CONNECTED. REPAIR ALL DAMAGE IF EVIDENT.
- IGNITION "OFF."
- RECONNECT ALL CONNECTORS.
- START ENGINE AND TEST DRIVE VEHICLE, ACHIEVING AT LEAST 16 km/h (10 MPH).
- REPEAT ABOVE DRIVE CYCLE SEQUENCE TWO MORE TIMES. DID DTC 63 SET IN THE LAST THREE DRIVE CYCLES?

- YES: REPLACE EBTCM.
- NO: MALFUNCTION WAS MOST LIKELY CAUSED BY A POOR CONNECTION. REFER TO "DIAGNOSTIC AIDS" ON FACING PAGE.

AFTER DIAGNOSIS IS COMPLETE, CLEAR DTCs AND TEST DRIVE VEHICLE FOR THREE (3) DRIVE CYCLES TO VERIFY THAT DTC DOES NOT RESET. A DRIVE CYCLE CONSISTS OF STARTING THE VEHICLE, DRIVING OVER 16 km/h (10 MPH), AND THEN TURNING KEY OFF.

DIAGNOSTIC CHART DTC 64 — 1995 CAMARO AND FIREBIRD WITH TCS

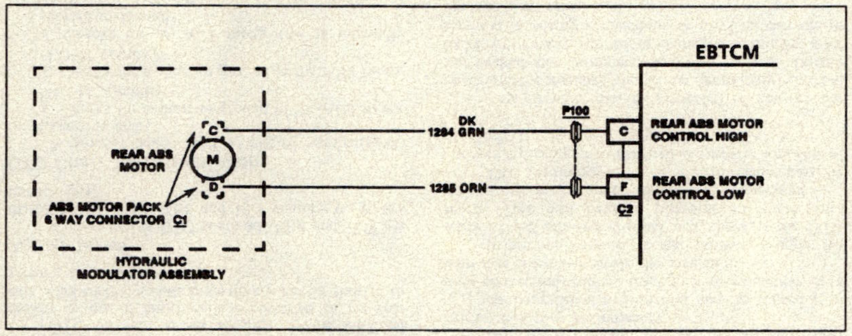

DTC 64
REAR ABS MOTOR CIRCUIT OPEN

Circuit Description:

This DTC identifies an ABS motor that cannot be energized due to an open in its circuitry. This malfunction will not allow proper ABS operation.

DTC Will Set When:

DTC 64 can be set only when the ABS motor is commanded "OFF." If the EBTCM detects an out of range voltage on either of the rear axle ABS motor circuits indicating an open circuit, a malfunction exists.

Action Taken:

An open ABS motor will not activate when requested. A malfunction DTC is stored, ABS and TCS are disabled. The amber ABS and TCS warning lamps are turned "ON."

DTC Chart Test Description:

Number(s) below refer to circled number(s) on the diagnostic chart.

1. This checks for proper resistance of the ABS motor.
2. Checks for an open in the ABS motor "HIGH" circuitry.
3. Checks for an open in the ABS motor "LOW" circuitry.
4. Ensures malfunction was not due to poor terminal contact.

Diagnostic Aids:

Using Tech 1, select manual control function, and exercise ABS motor movement of affected channel in both directions while applying light pressure on the brake pedal.

If erratic or "jumpy" brake pedal movement is detected while performing an "apply" or "release" function of the ABS motor, an intermittent malfunction may be indicated.

An intermittent malfunction may be caused by a poor connection, rubbed through wire insulation, or a wire that is broken inside the insulation.

If the malfunction is not current, wiggle the wires of the affected channel and check if the DTC resets. This will help to pinpoint an intermittent malfunction in the motor circuitry or connections.

The frequency of the malfunction can be checked by using the enhanced diagnostic function of the Tech 1,

Any circuitry that is suspected of causing the intermittent complaint should be thoroughly checked for backed out terminals, improper mating, broken locks, improperly formed or damaged terminals, poor terminal to wiring connections or physical damage to the wiring harness.

DIAGNOSTIC CHART DTC 64 — 1995 CAMARO AND FIREBIRD WITH TCS, CONTINUED

DTC 64
REAR ABS MOTOR CIRCUIT OPEN

DOES DTC 64 OCCUR INTERMITTENTLY?

- YES → REFER TO "DIAGNOSTIC AIDS" ON FACING PAGE.
- NO →

(1)
- IGNITION "OFF."
- DISCONNECT 6-WAY ABS MOTOR PACK CONNECTOR.
- USING J 39200, MEASURE RESISTANCE ACROSS ABS MOTOR PACK TERMINALS "C" AND "D".

IS RESISTANCE BELOW 2000Ω?

- NO → REPLACE ABS MOTOR PACK.
- YES →

(2)
- DISCONNECT GRAY 8-WAY CONNECTOR AT EBTCM.
- USING J 39200, MEASURE RESISTANCE BETWEEN TERMINAL "C" OF THE GRAY 8-WAY EBTCM HARNESS CONNECTOR AND TERMINAL "C" OF THE 6-WAY ABS MOTOR PACK HARNESS CONNECTOR.

IS RESISTANCE 2Ω OR LESS?

- NO → REPAIR OPEN OR HIGH RESISTANCE IN CKT 1284.
- YES →

(3)
- USING J 39200, MEASURE RESISTANCE BETWEEN TERMINAL "F" OF THE GRAY 8-WAY EBTCM HARNESS CONNECTOR AND TERMINAL "D" OF THE 6-WAY ABS MOTOR PACK HARNESS CONNECTOR.

IS RESISTANCE 2Ω OR LESS?

- NO → REPAIR OPEN OR HIGH RESISTANCE IN CKT 1285.
- YES →

(4)
- INSPECT THE 6-WAY ABS MOTOR PACK AND GRAY 8-WAY EBTCM CONNECTORS FOR POOR TERMINAL CONTACT AND EVIDENCE OF CORROSION.
  REPLACE ALL TERMINALS THAT EXHIBIT SIGNS OF POOR TERMINAL CONTACT OR CORROSION.
- IGNITION "OFF."
- RECONNECT ALL CONNECTORS.
- START ENGINE AND TEST DRIVE VEHICLE, ACHIEVING AT LEAST 16 km/h (10 MPH).
- REPEAT ABOVE DRIVE CYCLE SEQUENCE TWO MORE TIMES.

DID DTC 64 SET IN THE LAST THREE DRIVE CYCLES?

- YES → REPLACE EBTCM.
- NO → MALFUNCTION WAS MOST LIKELY CAUSED BY A POOR CONNECTION. REFER TO "DIAGNOSTIC AIDS" ON FACING PAGE.

AFTER ALL DIAGNOSIS IS COMPLETE, CLEAR DTCs AND TEST DRIVE VEHICLE FOR THREE (3) DRIVE CYCLES TO VERIFY THAT DTC DOES NOT RESET. A DRIVE CYCLE CONSISTS OF STARTING THE VEHICLE, DRIVING OVER 16 km/h (10 MPH), AND THEN TURNING KEY OFF.

DIAGNOSTIC CHART DTC 65 — 1995 CAMARO AND FIREBIRD WITH TCS

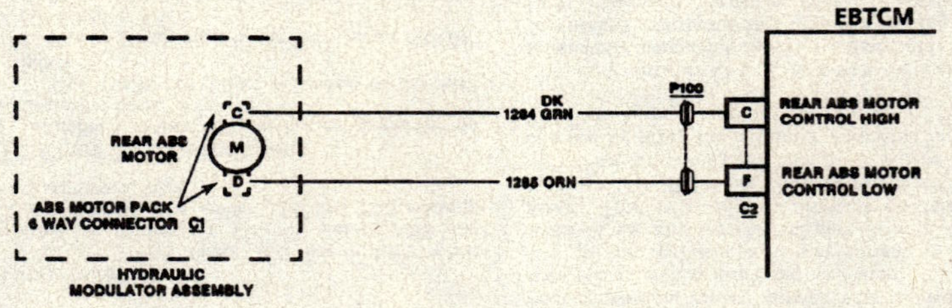

DTC 65
REAR ABS MOTOR CIRCUIT SHORTED TO GROUND

Circuit Description:

This DTC identifies an ABS motor circuit that is shorted to ground. This malfunction will not allow the ABS motor to be controlled at the commanded current rate or will cause the ABS motor driver circuit to allow current directly to ground.

DTC Will Set When:

DTC 65 can be set anytime. If the EBTCM detects an out of range voltage on either of the rear axle ABS motor circuits indicating a circuit shorted to ground, a malfunction exists.

Action Taken:

A malfunction DTC is stored, ABS and TCS are disabled. The amber ABS and TCS warning lamps are turned "ON."

DTC Chart Test Description:

Number(s) below refer to circled number(s) on the diagnostic chart.

1. Checks for a short to ground in the ABS motor "HIGH" circuitry.
2. Checks for a short to ground in the ABS motor "LOW" circuitry.
3. Checks for an ABS motor that is internally shorted to ground.
4. Ensures malfunction was not due to physical damage of the circuitry.

Diagnostic Aids:

Using Tech 1, select manual control function, and exercise ABS motor movement of affected channel in both directions while applying light pressure on the brake pedal.

If erratic or "jumpy" brake pedal movement is detected while performing an "apply" or "release" function of the ABS motor, an intermittent malfunction may be indicated.

An intermittent malfunction may be caused by a poor connection, rubbed through wire insulation, or a wire that is broken inside the insulation.

If the malfunction is not current, wiggle the wires of the affected channel and check if the DTC resets. This will help to pinpoint an intermittent malfunction in the motor circuitry or connections.

The frequency of the malfunction can be checked by using the enhanced diagnostic function of the Tech 1.

Any circuitry that is suspected of causing the intermittent complaint should be thoroughly checked for backed out terminals, improper mating, broken locks, improperly formed or damaged terminals, poor terminal to wiring connections or physical damage to the wiring harness.

DIAGNOSTIC CHART DTC 65 — 1995 CAMARO AND FIREBIRD WITH TCS, CONTINUED

DTC 65
REAR ABS MOTOR CIRCUIT SHORTED TO GROUND

DOES DTC 65 OCCUR INTERMITTENTLY?

- YES: REFER TO "DIAGNOSTIC AIDS" ON FACING PAGE.
- NO:

(1)
- IGNITION "OFF."
- DISCONNECT THE 6-WAY ABS MOTOR PACK CONNECTOR AND THE GRAY 8-WAY EBTCM CONNECTOR.
- USING J 39200, MEASURE THE RESISTANCE BETWEEN GROUND AND TERMINAL "C" OF THE 6-WAY ABS MOTOR PACK HARNESS CONNECTOR.
DOES J 39200 DISPLAY RESISTANCE AS "OL" (INFINITE)?

- NO: REPAIR SHORT TO GROUND IN CKT 1284.
- YES:

(2)
- USING J 39200, MEASURE THE RESISTANCE BETWEEN GROUND AND TERMINAL "D" OF THE 6-WAY ABS MOTOR PACK HARNESS CONNECTOR.
DOES J 39200 DISPLAY RESISTANCE AS "OL" (INFINITE)?

- NO: REPAIR SHORT TO GROUND IN CKT 1285.
- YES:

(3)
- USING J 39200, MEASURE THE RESISTANCE BETWEEN GROUND AND TERMINAL "C" OF THE ABS MOTOR PACK.
DOES J 39200 DISPLAY RESISTANCE AS "OL" (INFINITE)?

- NO: REPLACE ABS MOTOR PACK.
- YES:

(4)
- INSPECT CKTs 1284 AND 1285 AND RELATED CONNECTORS FOR DAMAGE WHICH MAY RESULT IN A SHORT TO GROUND WITH ALL CONNECTORS CONNECTED. REPAIR ALL DAMAGE IF EVIDENT.
- IGNITION "OFF."
- RECONNECT ALL CONNECTORS.
- START ENGINE AND TEST DRIVE VEHICLE, ACHIEVING AT LEAST 16 km/h (10 MPH).
- REPEAT ABOVE DRIVE CYCLE SEQUENCE TWO MORE TIMES.
DID DTC 65 SET IN THE LAST THREE DRIVE CYCLES?

- YES: REPLACE EBTCM.
- NO: MALFUNCTION WAS MOST LIKELY CAUSED BY A POOR CONNECTION. REFER TO "DIAGNOSTIC AIDS" ON FACING PAGE.

AFTER ALL DIAGNOSIS IS COMPLETE, CLEAR DTCs AND TEST DRIVE VEHICLE FOR THREE (3) DRIVE CYCLES TO VERIFY THAT DTC DOES NOT RESET. A DRIVE CYCLE CONSISTS OF STARTING THE VEHICLE, DRIVING OVER 16 km/h (10 MPH), AND THEN TURNING KEY OFF.

DIAGNOSTIC CHART DTC 66 — 1995 CAMARO AND FIREBIRD WITH TCS

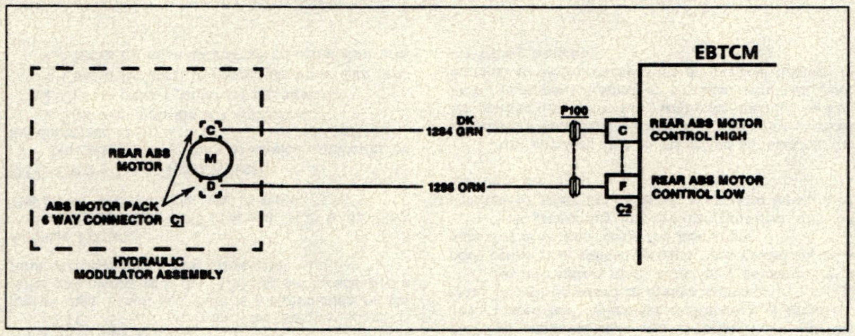

DTC 66
REAR ABS MOTOR CIRCUIT SHORTED TO BATTERY

Circuit Description:
This DTC identifies an ABS motor circuit that is shorted to battery or an ABS motor that has low or no resistance. This malfunction will not allow the ABS motor to be controlled at the commanded current rate or will cause the ABS motor to turn in the opposite direction, or not turn at all.

DTC Will Set When:
DTC 66 can be set only when the ABS motor is commanded "OFF." If the EBTCM detects an out of range voltage on either of the rear ABS axle motor circuits indicating a circuit shorted to battery or an ABS motor shorted, a malfunction exists.

Action Taken:
A malfunction DTC is stored, ABS and TCS are disabled. The amber ABS and TCS warning lamps are turned "ON."

DTC Chart Test Description:
Number(s) below refer to circled number(s) on the diagnostic chart.
1. Checks for a short to voltage in the ABS motor "HIGH" circuitry.
2. Checks for a short to voltage in the ABS motor "LOW" circuitry.
3. Checks for a short circuit between motor "HIGH" and motor "LOW".
4. Checks for an ABS motor that is internally shorted.
5. Ensures malfunction was not due to physical damage of the circuitry.

Diagnostic Aids:
Using Tech 1, select manual control function, and exercise ABS motor movement of affected channel in both directions while applying light pressure on the brake pedal.

If erratic or "jumpy" brake pedal movement is detected while performing an "apply" or "release" function of the ABS motor, an intermittent malfunction may be indicated.

An intermittent malfunction may be caused by a poor connection, rubbed through wire insulation, or a wire that is broken inside the insulation.

If the malfunction is not current, wiggle the wires of the affected channel and check if the DTC resets. This will help to pinpoint an intermittent malfunction in the motor circuitry or connections.

The frequency of the malfunction can be checked by using the enhanced diagnostic function of the Tech 1.

Any circuitry that is suspected of causing the intermittent complaint should be thoroughly checked for backed out terminals, improper mating, broken locks, improperly formed or damaged terminals, poor terminal to wiring connections or physical damage to the wiring harness.

DIAGNOSTIC CHART DTC 66 — 1995 CAMARO AND FIREBIRD WITH TCS, CONTINUED

DTC 66
REAR ABS MOTOR CIRCUIT SHORTED TO BATTERY

NOTICE: J 39200 TEST LEADS MUST BE "ZEROED" PRIOR TO MAKING ANY RESISTANCE MEASUREMENTS. REFER TO J 39200 USER'S MANUAL.

DOES DTC 66 OCCUR INTERMITTENTLY?

- YES: REFER TO "DIAGNOSTIC AIDS" ON FACING PAGE.
- NO:

(1)
- IGNITION "OFF."
- DISCONNECT THE 6-WAY ABS MOTOR PACK CONNECTOR AND THE GRAY 8-WAY EBTCM CONNECTOR.
- IGNITION "ON."
- USING J 39200, MEASURE THE VOLTAGE BETWEEN GROUND AND TERMINAL "C" OF THE 6-WAY ABS MOTOR PACK HARNESS CONNECTOR. IS VOLTAGE 1 VOLT OR LESS?

- NO: REPAIR SHORT TO VOLTAGE IN CKT 1284.
- YES:

(2)
- USING J 39200, MEASURE THE VOLTAGE BETWEEN GROUND AND TERMINAL "D" OF THE 6-WAY ABS MOTOR PACK HARNESS CONNECTOR. IS VOLTAGE 1 VOLT OR LESS?

- NO: REPAIR SHORT TO VOLTAGE IN CKT 1285.
- YES:

(3)
- IGNITION "OFF."
- USING J 39200, MEASURE THE RESISTANCE BETWEEN TERMINAL "C" AND TERMINAL "D" OF THE 6-WAY ABS MOTOR PACK HARNESS CONNECTOR. DOES J 39200 DISPLAY RESISTANCE AS "OL" (INFINITE)?

- NO: REPAIR SHORT BETWEEN CKTs 1284 AND 1285.
- YES:

(4)
- USING J 39200, MEASURE THE RESISTANCE BETWEEN TERMINAL "C" AND TERMINAL "D" OF THE ABS MOTOR PACK. IS RESISTANCE GREATER THAN 0.4Ω?

- NO: REPLACE ABS MOTOR PACK
- YES:

(5)
- INSPECT CKTs 1284 AND 1285 AND RELATED CONNECTORS FOR DAMAGE WHICH MAY RESULT IN A SHORT TO VOLTAGE WITH ALL CONNECTORS CONNECTED. REPAIR ALL DAMAGE IF EVIDENT.
- IGNITION "OFF."
- RECONNECT ALL CONNECTORS.
- START ENGINE AND TEST DRIVE VEHICLE, ACHIEVING AT LEAST 16 km/h (10 MPH).
- REPEAT ABOVE DRIVE CYCLE SEQUENCE TWO MORE TIMES. DID DTC 66 SET IN THE LAST THREE DRIVE CYCLES?

- YES: REPLACE EBTCM.
- NO: MALFUNCTION WAS MOST LIKELY CAUSED BY A POOR CONNECTION. REFER TO "DIAGNOSTIC AIDS" ON FACING PAGE.

AFTER DIAGNOSIS IS COMPLETE, CLEAR DTCs AND TEST DRIVE VEHICLE FOR THREE (3) DRIVE CYCLES TO VERIFY THAT DTC DOES NOT RESET. A DRIVE CYCLE CONSISTS OF STARTING THE VEHICLE, DRIVING OVER 16 km/h (10 MPH), AND THEN TURNING KEY OFF.

DIAGNOSTIC CHART DTC 75 — 1995 CAMARO AND FIREBIRD WITH TCS

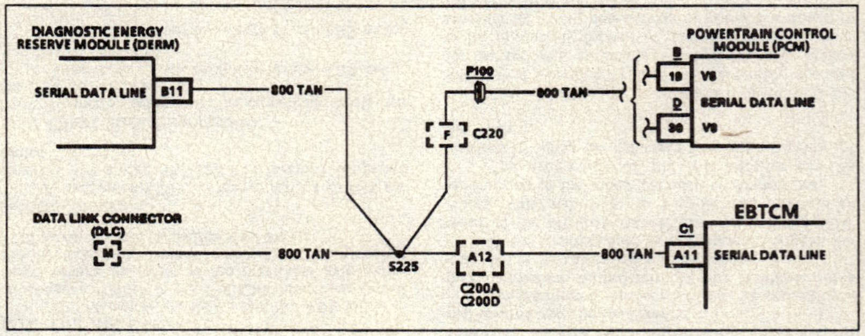

DTC 75
PCM TO EBTCM SERIAL DATA MALFUNCTION

Circuit Description:

The serial data communications circuit is used to transmit messages from one electronic control unit to another when interface is necessary between systems. The ABS VI/TCS systems depends on information from the PCM for proper operation.

DTC Will Set When:

DTC 75 will set when the vehicle speed is greater than 3 mph and there is a malfunction on the serial data circuit, or if the PCM has not broadcasted a serial message for at least 10 seconds.

Action Taken:

A malfunction DTC is stored, TCS is disabled. The amber TCS warning lamp is turned "ON."

DTC Chart Test Description:

Number(s) below refer to circled number(s) on the diagnostic chart.

1. This step tests the test equipment.
2. Checks other systems on the circuit.
3. Checks for short to ground on serial data line.
4. Checks for short to voltage on serial data line.
5. Checks for poor terminal contact or corrosion which can cause malfunction.
6. Attempts to establish communication with modules.

Diagnostic Aids:

When DTC 75 is current (malfunction in CKT 800), no communication can be established with the Tech 1 scan tool. When the malfunction is removed, DTC 75 will be stored in system memory.

An intermittent malfunction may be caused by a poor connection, rubbed through wire insulation, or a wire that is broken inside the insulation.

The frequency of the malfunction can be checked by using the enhanced diagnostic function of the Tech 1.

Any circuitry that is suspected of causing the intermittent complaint should be thoroughly checked for backed out terminals, improper mating, broken locks, improperly formed or damaged terminals, poor terminal to wiring connections or physical damage to the wiring harness.

DIAGNOSTIC CHART DTC 75 — 1995 CAMARO AND FIREBIRD WITH TCS, CONTINUED

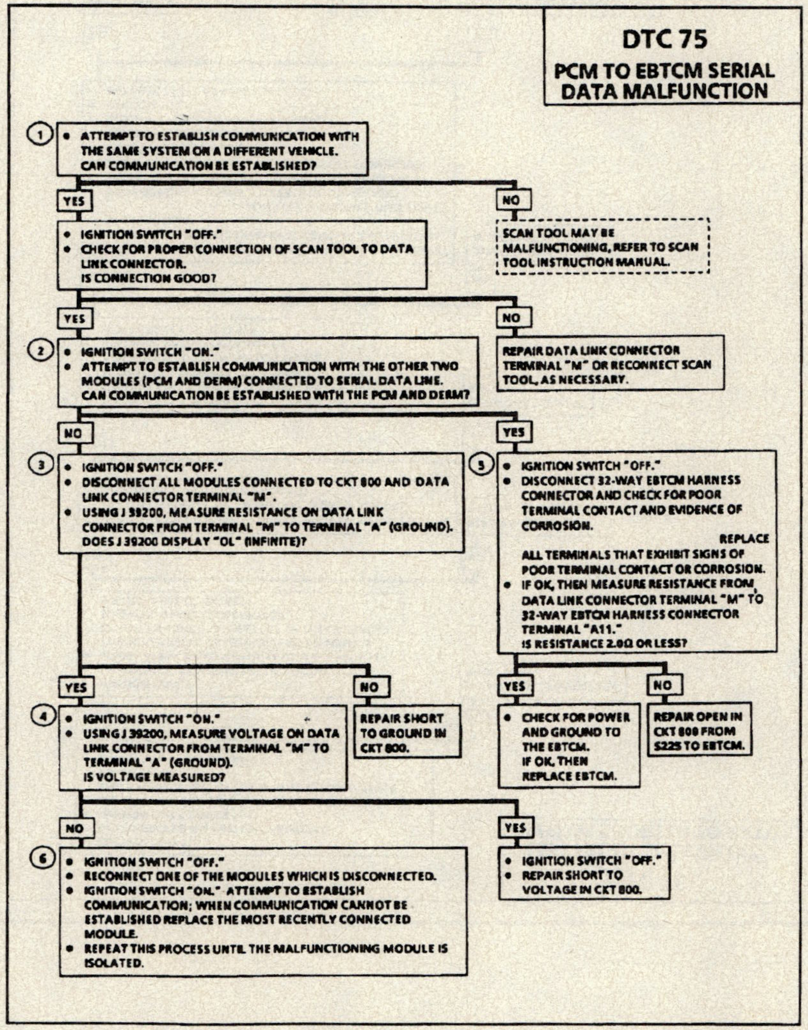

DIAGNOSTIC CHART DTC 76 — 1995 CAMARO AND FIREBIRD WITH TCS

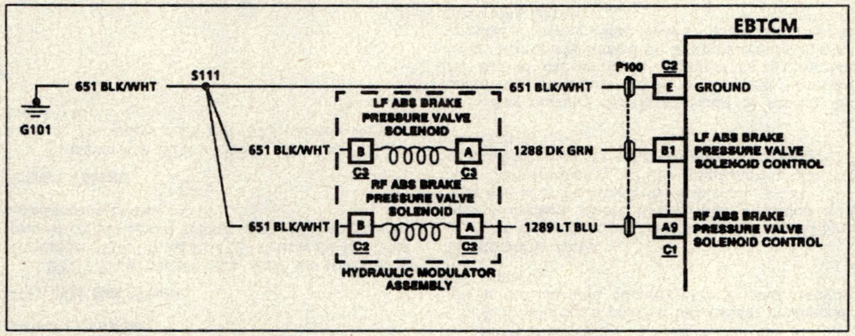

DTC 76
LEFT FRONT SOL CIRCUIT OPEN OR SHORTED TO BATT

Circuit Description:
This DTC identifies a solenoid that cannot be energized due to an open in its circuitry, or a solenoid that is always energized due to a short to battery in its circuitry between the driver and the solenoid. An open will not allow proper ABS operation, but the short to battery simply turns "ON" the solenoid. A path for base brakes is still allowed once the motor re-homes and the check ball is lifted off its seat during key "ON" initialization.

DTC Will Set When:
DTC 76 can be set only when the solenoid is commanded "OFF." If the EBTCM detects an excessive voltage in the left front solenoid control circuit, a malfunction exists.

Action Taken:
A malfunction DTC is stored, ABS and TCS are disabled. The amber ABS and TCS warning lamps are turned "ON."

DTC Chart Test Description:
Number(s) below refer to circled number(s) on the diagnostic chart.
1. Checks for a short to voltage in the control circuitry of the solenoid.
2. Checks for an open in the control circuit of the solenoid.
3. Checks the solenoid coil for proper resistance.
4. Checks for an open in the solenoid ground circuit.
5. Checks for a possible intermittent malfunction in the solenoid circuitry due to poor terminal contact.

Diagnostic Aids:
An intermittent malfunction may be caused by a poor connection, rubbed through wire insulation, or a wire that is broken inside the insulation.
The frequency of the malfunction can be checked by using the enhanced diagnostic function of the Tech 1.

Any circuitry that is suspected of causing the intermittent complaint should be thoroughly checked for backed out terminals, improper mating, broken locks, improperly formed or damaged terminals, poor terminal to wiring connections or physical damage to the wiring harness.

DIAGNOSTIC CHART DTC 76 — 1995 CAMARO AND FIREBIRD WITH TCS, CONTINUED

DTC 76
LEFT FRONT SOL CIRCUIT OPEN OR SHORTED TO BATT

1.
   - IGNITION "OFF."
   - DISCONNECT 32-WAY EBTCM CONNECTOR.
   - IGNITION "ON," ENGINE "OFF."
   - USING J 39200, MEASURE THE VOLTAGE BETWEEN GROUND AND TERMINAL "B1" OF THE 32-WAY EBTCM HARNESS CONNECTOR.

   IS VOLTAGE 1 VOLT OR LESS?

   - YES: go to 2.
   - NO: REPAIR SHORT TO VOLTAGE IN CKT 1288.

2.
   - IGNITION "OFF."
   - DISCONNECT LEFT FRONT SOLENOID CONNECTOR.
   - USING J 39200, MEASURE THE RESISTANCE BETWEEN TERMINAL "B1" OF THE 32-WAY EBTCM HARNESS CONNECTOR AND TERMINAL "A" OF THE LEFT FRONT SOLENOID HARNESS CONNECTOR.

   IS RESISTANCE 2Ω OR LESS?

   - YES: go to 3.
   - NO: REPAIR OPEN OR HIGH RESISTANCE IN CKT 1288.

3.
   - USING J 39200, MEASURE THE RESISTANCE BETWEEN TERMINAL "A" AND TERMINAL "B" OF THE LEFT FRONT SOLENOID.

   IS RESISTANCE BETWEEN 2.5 AND 5Ω?

   - YES: go to 4.
   - NO: REPLACE LEFT FRONT SOLENOID.

4.
   - DISCONNECT NEGATIVE BATTERY CABLE.
   - USING J 39200, MEASURE THE RESISTANCE BETWEEN NEGATIVE BATTERY TERMINAL AND TERMINAL "B" OF THE LEFT FRONT SOLENOID HARNESS CONNECTOR.

   IS RESISTANCE 2Ω OR LESS?

   - YES: go to 5.
   - NO: REPAIR OPEN OR HIGH RESISTANCE IN CKT 651.

5.
   - INSPECT LEFT FRONT SOLENOID AND 32-WAY EBTCM HARNESS CONNECTORS FOR POOR TERMINAL CONTACT AND EVIDENCE OF CORROSION.
     REPLACE ALL TERMINALS THAT EXHIBIT SIGNS OF POOR TERMINAL CONTACT OR CORROSION.
   - INSPECT CKT 1288 FOR DAMAGE WHICH MAY RESULT IN A SHORT TO VOLTAGE WITH ALL CONNECTORS CONNECTED. REPAIR DAMAGE IF EVIDENT.
   - RECONNECT ALL CONNECTORS.
   - RECONNECT NEGATIVE BATTERY CABLE.
   - IGNITION "ON."

   DOES DTC 76 SET AS A CURRENT DTC?

   - YES: REPLACE EBTCM.
   - NO: MALFUNCTION IS INTERMITTENT. REFER TO "DIAGNOSTIC AIDS" ON FACING PAGE.

DIAGNOSTIC CHART DTC 77 — 1995 CAMARO AND FIREBIRD WITH TCS

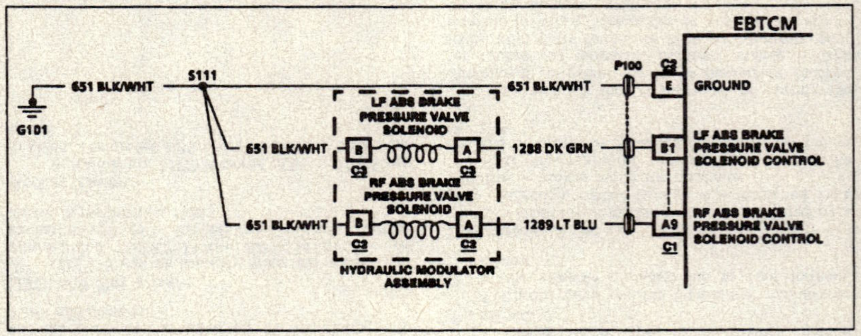

DTC 77
LEFT FRONT SOLENOID CIRCUIT SHORTED TO GND

Circuit Description:

This DTC identifies a solenoid that cannot be energized due to an open in its driver circuitry, or a short to ground between the solenoid driver and the solenoid. These malfunctions can affect ABS operation since the flow of brake fluid to the wheel cylinder cannot be stopped, making ABS operation for that channel impossible.

DTC Will Set When:

DTC 77 can be set only when the solenoid is commanded "ON." If the EBTCM detects the left front solenoid control circuit voltage is out of specification, a malfunction exists.

Action Taken:

A malfunction DTC is stored, ABS and TCS are disabled. The amber ABS and TCS warning lamp is turned "ON."

DTC Chart Test Description:

Number(s) below refer to circled number(s) on the diagnostic chart.

1. Checks for a solenoid that is internally shorted to ground.
2. Checks for a solenoid that is not within proper resistance values.
3. Indicates if a short to ground exists in the solenoid circuitry.
4. Checks for a possible intermittent malfunction in the solenoid circuitry due to poor terminal contact.

Diagnostic Aids:

An intermittent malfunction may be caused by a poor connection, rubbed through wire insulation, or a wire that is broken inside the insulation.

The frequency of the malfunction can be checked by using the enhanced diagnostic function of the Tech 1,

Any circuitry that is suspected of causing the intermittent complaint should be thoroughly checked for backed out terminals, improper mating, broken locks, improperly formed or damaged terminals, poor terminal to wiring connections or physical damage to the wiring harness.

DIAGNOSTIC CHART DTC 77 — 1995 CAMARO AND FIREBIRD WITH TCS, CONTINUED

DTC 77
LEFT FRONT SOLENOID CIRCUIT SHORTED TO GND

(1)
- IGNITION "OFF."
- DISCONNECT LEFT FRONT SOLENOID CONNECTOR.
- USING J 39200, MEASURE THE RESISTANCE BETWEEN GROUND AND TERMINAL "A" OF THE LEFT FRONT SOLENOID.

DOES J 39200 DISPLAY RESISTANCE AS "OL" (INFINITE)?

- YES → go to step 2
- NO → REPLACE LEFT FRONT SOLENOID.

(2)
- USING J 39200, MEASURE THE RESISTANCE BETWEEN TERMINAL "A" AND TERMINAL "B" OF THE LEFT FRONT SOLENOID.

IS RESISTANCE BETWEEN 2.5 AND 5.0Ω?

- YES → go to step 3
- NO → REPLACE LEFT FRONT SOLENOID.

(3)
- DISCONNECT 32-WAY EBTCM CONNECTOR.
- USING J 39200, MEASURE THE RESISTANCE BETWEEN GROUND AND TERMINAL "B1" OF THE 32-WAY EBTCM HARNESS CONNECTOR.

DOES J 39200 DISPLAY RESISTANCE AS "OL" (INFINITE)?

- YES → go to step 4
- NO → REPAIR SHORT TO GROUND IN CKT 1288.

(4)
- INSPECT CKT 1288 AND THE LEFT FRONT SOLENOID AND 32-WAY EBTCM HARNESS CONNECTORS FOR DAMAGE WHICH MAY RESULT IN A SHORT TO GROUND WITH ALL CONNECTORS CONNECTED. REPAIR DAMAGE IF EVIDENT.
- RECONNECT ALL CONNECTORS.
- START ENGINE WITH FOOT OFF BRAKE.
- ALLOW ENGINE TO RUN FOR AT LEAST 10 SECONDS.

DOES DTC 77 SET AS A CURRENT DTC?

- YES → REPLACE EBTCM.
- NO → MALFUNCTION IS INTERMITTENT. REFER TO "DIAGNOSTIC AIDS" ON FACING PAGE.

DIAGNOSTIC CHART DTC 78 — 1995 CAMARO AND FIREBIRD WITH TCS

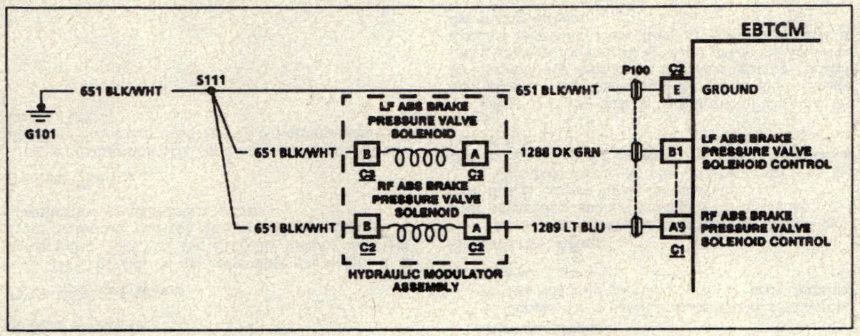

DTC 78
RIGHT FRONT SOL CIRCUIT OPEN OR SHORTED TO BATT

Circuit Description:

This DTC identifies a solenoid that cannot be energized due to an open in its circuitry, or a solenoid that is always energized due to a short to battery in its circuitry between the driver and the solenoid. An open will not allow proper ABS operation, but the short to battery simply turns "ON" the solenoid. A path for base brakes is still allowed once the motor re-homes and the check ball is lifted off its seat during key "ON" initialization.

DTC Will Set When:

DTC 78 can be set only when the solenoid is commanded "OFF." If the EBTCM detects an excessive voltage in the right front solenoid control circuit, a malfunction exists.

Action Taken:

A malfunction DTC is stored, ABS and TCS are disabled. The amber ABS and TCS warning lamps are turned "ON."

DTC Chart Test Description:

Number(s) below refer to circled number(s) on the diagnostic chart.

1. Checks for a short to voltage in the control circuitry of the solenoid.
2. Checks for an open in the control circuit of the solenoid.
3. Checks the solenoid coil for proper resistance.
4. Checks for an open in the solenoid ground circuit.
5. Checks for a possible intermittent malfunction in the solenoid circuitry due to poor terminal contact.

Diagnostic Aids:

An intermittent malfunction may be caused by a poor connection, rubbed through wire insulation, or a wire that is broken inside the insulation.

The frequency of the malfunction can be checked by using the enhanced diagnostic function of the Tech 1.

Any circuitry that is suspected of causing the intermittent complaint should be thoroughly checked for backed out terminals, improper mating, broken locks, improperly formed or damaged terminals, poor terminal to wiring connections or physical damage to the wiring harness.

DIAGNOSTIC CHART DTC 78 — 1995 CAMARO AND FIREBIRD WITH TCS, CONTINUED

DTC 78
RIGHT FRONT SOL CIRCUIT OPEN OR SHORTED TO BATT

(1)
- IGNITION "OFF."
- DISCONNECT 32-WAY EBTCM CONNECTOR.
- IGNITION "ON," ENGINE "OFF."
- USING J 39200, MEASURE THE VOLTAGE BETWEEN GROUND AND TERMINAL "A9" OF THE 32-WAY EBTCM HARNESS CONNECTOR.

IS VOLTAGE 1 VOLT OR LESS?

- YES: go to (2)
- NO: REPAIR SHORT TO VOLTAGE IN CKT 1289.

(2)
- IGNITION "OFF."
- DISCONNECT RIGHT FRONT SOLENOID CONNECTOR.
- USING J 39200, MEASURE THE RESISTANCE BETWEEN TERMINAL "A9" OF THE 32-WAY EBTCM HARNESS CONNECTOR AND TERMINAL "A" OF THE RIGHT FRONT SOLENOID HARNESS CONNECTOR.

IS RESISTANCE 2Ω OR LESS?

- YES: go to (3)
- NO: REPAIR OPEN OR HIGH RESISTANCE IN CKT 1289.

(3)
- USING J 39200, MEASURE THE RESISTANCE BETWEEN TERMINAL "A" AND TERMINAL "B" OF THE RIGHT FRONT SOLENOID.

IS RESISTANCE BETWEEN 2.5 AND 5Ω?

- YES: go to (4)
- NO: REPLACE RIGHT FRONT SOLENOID.

(4)
- DISCONNECT NEGATIVE BATTERY CABLE.
- USING J 39200, MEASURE THE RESISTANCE BETWEEN NEGATIVE BATTERY TERMINAL AND TERMINAL "B" OF THE RIGHT FRONT SOLENOID HARNESS CONNECTOR.

IS RESISTANCE 2Ω OR LESS?

- YES: go to (5)
- NO: REPAIR OPEN OR HIGH RESISTANCE IN CKT 651.

(5)
- INSPECT RIGHT FRONT SOLENOID AND 32-WAY EBTCM HARNESS CONNECTORS FOR POOR TERMINAL CONTACT AND EVIDENCE OF CORROSION. REPLACE ALL TERMINALS THAT EXHIBIT SIGNS OF POOR TERMINAL CONTACT OR CORROSION.
- INSPECT CKT 1289 FOR DAMAGE WHICH MAY RESULT IN A SHORT TO VOLTAGE WITH ALL CONNECTORS CONNECTED. REPAIR DAMAGE IF EVIDENT.
- RECONNECT ALL CONNECTORS.
- RECONNECT NEGATIVE BATTERY CABLE.
- IGNITION "ON."

DOES DTC 78 SET AS A CURRENT DTC?

- YES: REPLACE EBTCM.
- NO: MALFUNCTION IS MOST LIKELY DUE TO POOR TERMINAL CONTACT. REFER TO "DIAGNOSTIC AIDS" ON FACING PAGE.

DIAGNOSTIC CHART DTC 81 — 1995 CAMARO AND FIREBIRD WITH TCS

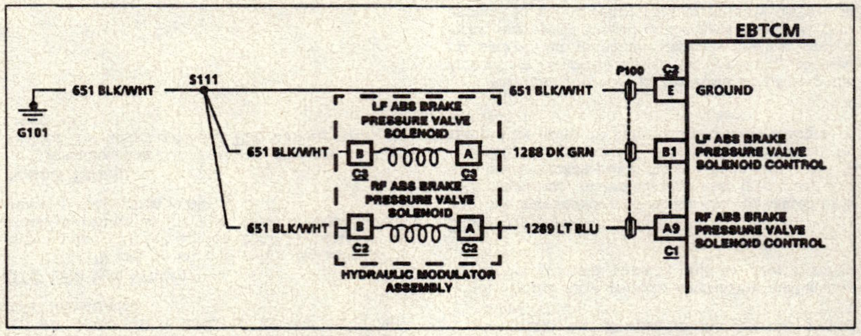

DTC 81
RIGHT FRONT SOLENOID CIRCUIT SHORTED TO GND

Circuit Description:

This DTC identifies a solenoid that cannot be energized due to an open in its driver circuitry, or a short to ground between the solenoid driver and the solenoid. These malfunctions can affect ABS operation since the flow of brake fluid to the wheel cylinder cannot be stopped, making ABS operation for that channel impossible.

DTC Will Set When:

DTC 81 can be set only when the solenoid is commanded "ON." If the EBTCM detects the right front solenoid control circuit voltage is out of specification, a malfunction exists.

Action Taken:

A malfunction DTC is stored, ABS and TCS are disabled. The amber ABS and TCS warning lamps are turned "ON."

DTC Chart Test Description:

Number(s) below refer to circled number(s) on the diagnostic chart.

1. Checks for a solenoid that is internally shorted to ground.
2. Checks for a solenoid that is not within proper resistance values.
3. Indicates if a short to ground exists in the solenoid circuitry.
4. Checks for a possible intermittent malfunction in the solenoid circuitry due to poor terminal contact.

Diagnostic Aids:

An intermittent malfunction may be caused by a poor connection, rubbed through wire insulation, or a wire that is broken inside the insulation.

The frequency of the malfunction can be checked by using the enhanced diagnostic function of the Tech 1,

Any circuitry that is suspected of causing the intermittent complaint should be thoroughly checked for backed out terminals, improper mating, broken locks, improperly formed or damaged terminals, poor terminal to wiring connections or physical damage to the wiring harness.

DIAGNOSTIC CHART DTC 81 — 1995 CAMARO AND FIREBIRD WITH TCS, CONTINUED

DTC 81
RIGHT FRONT SOLENOID CIRCUIT SHORTED TO GND

1.
   - IGNITION "OFF."
   - DISCONNECT RIGHT FRONT SOLENOID CONNECTOR.
   - USING J 39200, MEASURE THE RESISTANCE BETWEEN GROUND AND TERMINAL "A" OF THE RIGHT FRONT SOLENOID.

   DOES J 39200 DISPLAY RESISTANCE AS "OL" (INFINITE)?
   - NO: REPLACE RIGHT FRONT SOLENOID.
   - YES: go to 2.

2.
   - USING J 39200, MEASURE THE RESISTANCE BETWEEN TERMINAL "A" AND TERMINAL "B" OF THE RIGHT FRONT SOLENOID.

   IS RESISTANCE BETWEEN 2.5 AND 5.0Ω?
   - NO: REPLACE RIGHT FRONT SOLENOID.
   - YES: go to 3.

3.
   - DISCONNECT 32-WAY EBTCM CONNECTOR.
   - USING J 39200, MEASURE THE RESISTANCE BETWEEN GROUND AND TERMINAL "E9" OF THE 32-WAY EBTCM HARNESS CONNECTOR.

   DOES J 39200 DISPLAY RESISTANCE AS "OL" (INFINITE)?
   - NO: REPAIR SHORT TO GROUND IN CKT 1289.
   - YES: go to 4.

4.
   - INSPECT CKT 1289 AND THE RIGHT FRONT SOLENOID AND 32-WAY EBTCM HARNESS CONNECTORS FOR DAMAGE WHICH MAY RESULT IN A SHORT TO GROUND WITH ALL CONNECTORS CONNECTED. REPAIR DAMAGE IF EVIDENT.
   - RECONNECT ALL CONNECTORS.
   - START ENGINE WITH FOOT OFF BRAKE.
   - ALLOW ENGINE TO RUN FOR AT LEAST 10 SECONDS.

   DOES DTC 81 SET AS A CURRENT DTC?
   - YES: REPLACE EBTCM.
   - NO: MALFUNCTION IS INTERMITTENT. REFER TO "DIAGNOSTIC AIDS" ON FACING PAGE.

DIAGNOSTIC CHART DTC 82 — 1995 CAMARO AND FIREBIRD WITH TCS

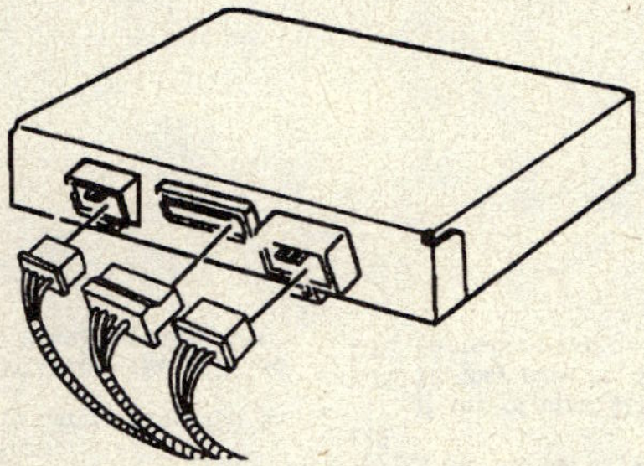

DTC 82
CALIBRATION MALFUNCTION

Circuit Description:

This DTC allows the EBTCM to check for a calibration malfunction by comparing the calibration value to a known value stored in the EEPROM.

This DTC is also used as a security measure to prevent improper use of calibrations or changes to these calibrations that may alter the designed function of ABS and TCS.

DTC Will Set When:

DTC 82 can be set at key-up. If the program identifier is incorrect or the memory checksum is incorrect, a malfunction exists.

Action Taken:

A malfunction DTC is stored, ABS and TCS are disabled. The amber ABS and TCS warning lamps are turned "ON."

DTC Chart Test Description:

Number(s) below refer to circled number(s) on the diagnostic chart.

1. Checks to see if the malfunction is present during diagnosis. If present, the EBTCM is not functioning correctly and must be replaced.

Diagnostic Aids:

An intermittent DTC 82 may be caused by a bad cell in the EEPROM that is sensitive to temperature changes. If DTC 82 failed more than once, but is intermittent, replace EBTCM.

The frequency of the malfunction can be checked by using the enhanced diagnostic function of the Tech 1.

DIAGNOSTIC CHART DTC 82 — 1995 CAMARO AND FIREBIRD WITH TCS, CONTINUED

DTC 82
CALIBRATION MALFUNCTION

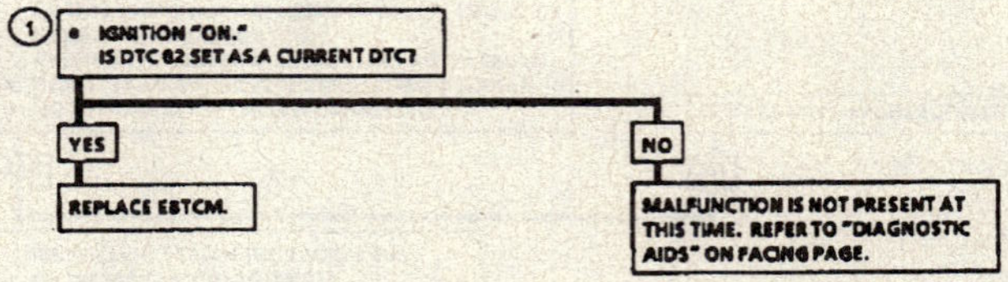

THIS CHART ASSUMES THAT A CURRENT DTC IS STORED INDICATING THAT THIS MALFUNCTION IS PRESENT.

DIAGNOSTIC CHART DTC 86 — 1995 CAMARO AND FIREBIRD WITH TCS

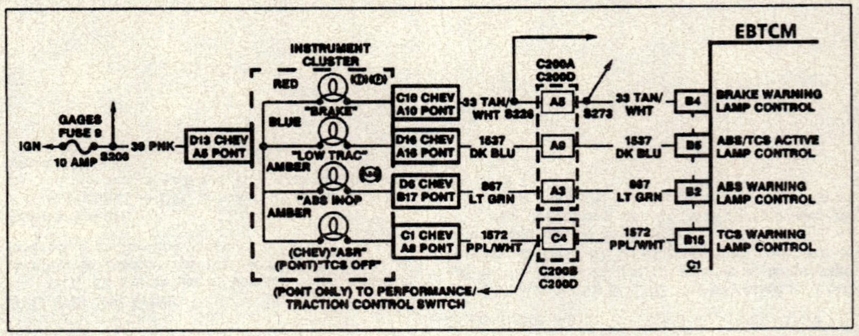

DTC 86

EBTCM TURNED "ON" THE RED BRAKE WARNING LAMP

Circuit Description:

This DTC is provided as an information only DTC, and reflects the status of the command issued by the ABS EBTCM to illuminate the red BRAKE warning lamp. If another DTC issues a command to illuminate the red BRAKE warning lamp, DTC 86 will be stored in EEPROM as a history DTC at the conclusion of the ignition cycle.

DTC Chart Test Description:

Number(s) below refer to circled number(s) on the diagnostic chart.

1. Serves to identify if a DTC other than DTC 86 commanded the red BRAKE warning lamp "ON."

Diagnostic Aids:

Any ABS mechanical DTC that issues a command to illuminate the red BRAKE warning lamp will also result in DTC 86 being stored in EEPROM during shut down. These DTCs are: 38, 41, 42, 46, and 51. If the motor car not in their home position, certain electrical DTCs will also command the red BRAKE warning lamp "ON." These DTCs are: 14, 16, 18, 55, 56, 57, 58, 61, 62, 63, 64, 65, 66, 144, and 145.

If any of these DTCs are indicated along with DTC 86, they must be corrected prior to addressing a DTC 86 malfunction.

DIAGNOSTIC CHART DTC 86 — 1995 CAMARO AND FIREBIRD WITH TCS, CONTINUED

DTC 86
EBTCM TURNED "ON" THE RED "BRAKE" WARNING LAMP

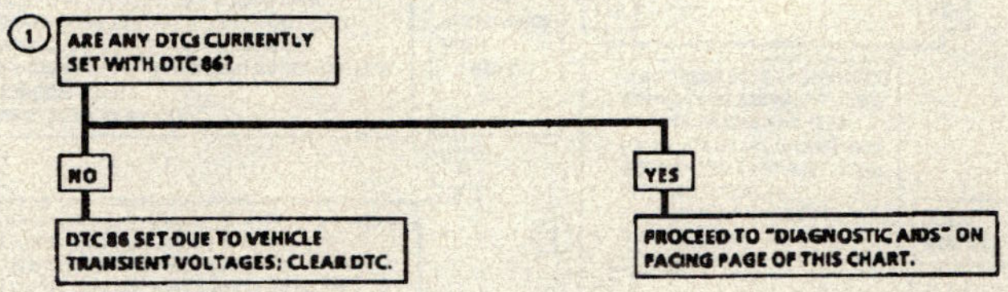

DIAGNOSTIC CHART DTC 87 — 1995 CAMARO AND FIREBIRD WITH TCS

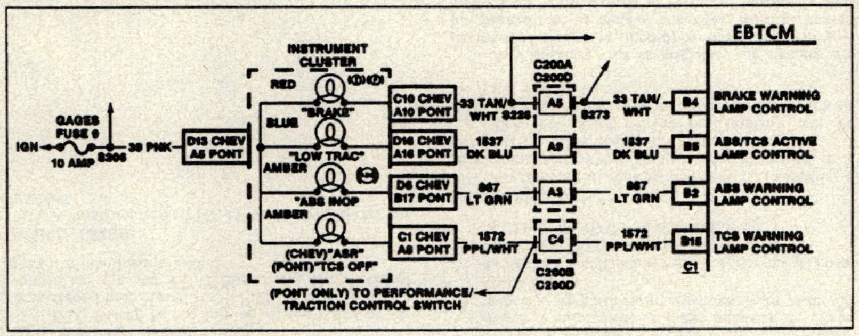

DTC 87
RED BRAKE WARNING LAMP CIRCUIT OPEN

Circuit Description:

This DTC is used to verify the EBTCM has continuity to the red BRAKE warning lamp in case the EBTCM must turn it "ON." This will only occur if an ABS malfunction is detected that may degrade base brake operation. Because ABS is not the only device controlling the red BRAKE warning lamp (parking brake and low fluid sensor may also turn "ON" the lamp) a short to ground in this circuit cannot be detected.

DTC Will Set When:

DTC 87 can be set after initialization is completed. If the red BRAKE warning lamp circuit voltage is out of specification indicating an open circuit, a malfunction exists.

Action Taken:

A malfunction DTC is stored, ABS is not disabled.

DTC Chart Test Description:

Number(s) below refer to circled number(s) on the diagnostic chart.

1. Identifies if the ground circuit to the red BRAKE warning lamp is completed by a source other than the EBTCM.
2. Checks for high resistance in park brake circuitry.
3. Checks for malfunctioning EBTCM.
4. Checks for high resistance in brake pressure differential switch circuitry.
5. Checks for malfunctioning warning lamp bulb.
6. Checks for open or high resistance in CKT 33.

Diagnostic Aids:

An intermittent malfunction may be caused by a poor connection, rubbed through wire insulation, or a wire that is broken inside the insulation.

The frequency of the malfunction can be checked by using the enhanced diagnostic function of the Tech 1,

If a Tech 1 or T-100 are not available, the parking brake handle may be lifted to verify proper lamp operation and continuity of CKTs 3 and 39.

Any circuitry that is suspected of causing the intermittent complaint should be thoroughly checked for backed out terminals, improper mating, broken locks, improperly formed or damaged terminals, poor terminal to wiring connections or physical damage to the wiring harness.

DIAGNOSTIC CHART DTC 87 — 1995 CAMARO AND FIREBIRD WITH TCS, CONTINUED

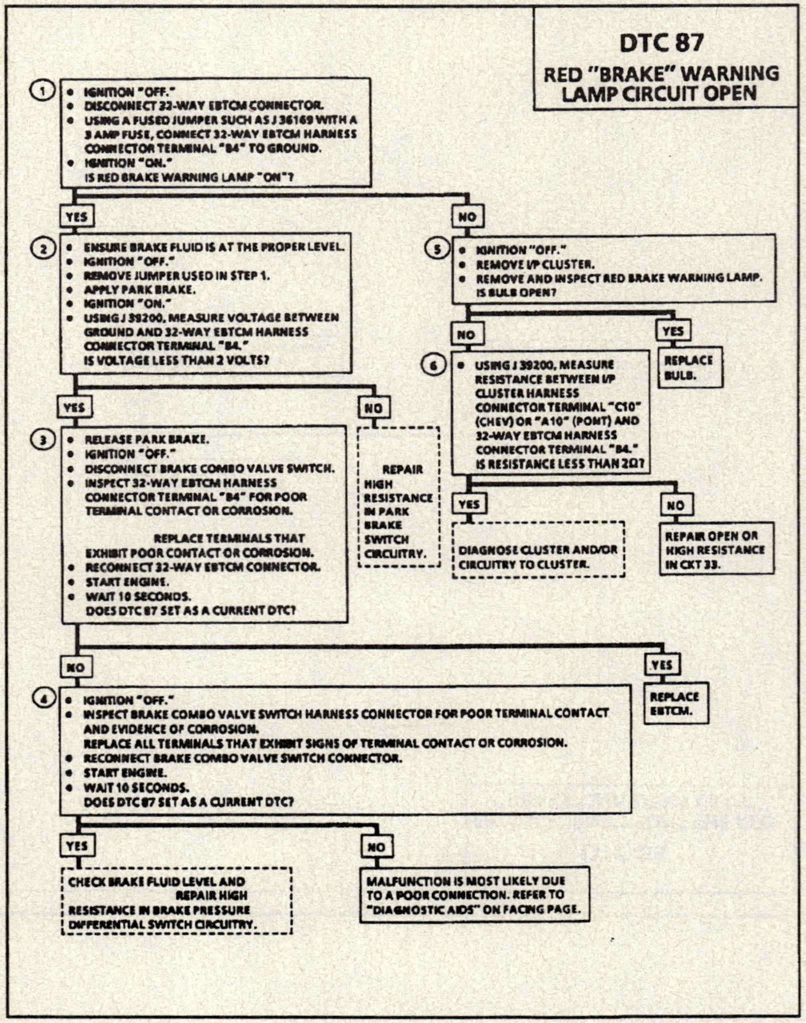

DIAGNOSTIC CHART DTC 88 — 1995 CAMARO AND FIREBIRD WITH TCS

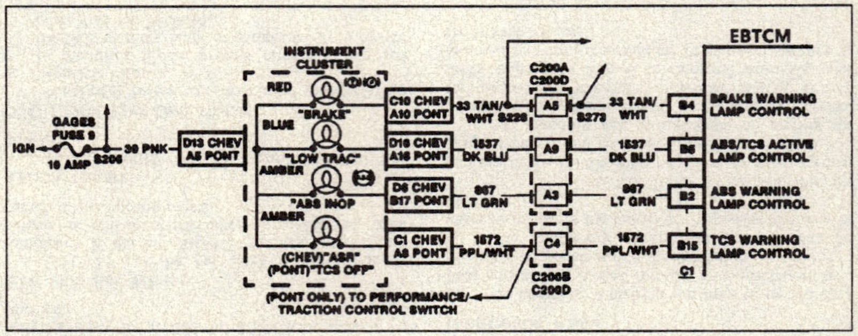

DTC 88
RED BRAKE WARNING LAMP CKT SHORTED TO BATT

Circuit Description:
This DTC identifies a short to battery between the EBTCM and the red BRAKE warning lamp, or an open driver that does not allow the red BRAKE warning lamp to be illuminated by the EBTCM. This will only occur if an ABS malfunction is detected that may degrade base brake operation.

DTC Will Set When:
DTC 88 can be set only at the beginning of the three second bulb check. If the EBTCM detects battery voltage on the red BRAKE warning lamp control circuit, a malfunction exists.

Action Taken:
A malfunction DTC is stored. ABS is not disabled.

DTC Chart Test Description:
Number(s) below refer to circled number(s) on the diagnostic chart.

1. Identifies if the ground circuit to the red BRAKE warning lamp is being completed by a source other than the EBTCM.
2. Identifies if the malfunction is currently present.
3. By removing the "GAGES" Fuse 9, the voltage source is eliminated. This test indicates if voltage is being supplied from a source other than the gauge fuse.
4. Ensures malfunction was not due to physical damage of the circuitry.
5. Checks for DTC 86 also being set.

Diagnostic Aids:
An intermittent malfunction may be caused by a poor connection, rubbed through wire insulation, or a wire that is broken inside the insulation.

The frequency of the malfunction can be checked by using the enhanced diagnostic function of the Tech 1.

Any circuitry that is suspected of causing the intermittent complaint should be thoroughly checked for backed out terminals, improper mating, broken locks, improperly formed or damaged terminals, poor terminal to wiring connections or physical damage to the wiring harness.

DIAGNOSTIC CHART DTC 88 — 1995 CAMARO AND FIREBIRD WITH TCS, CONTINUED

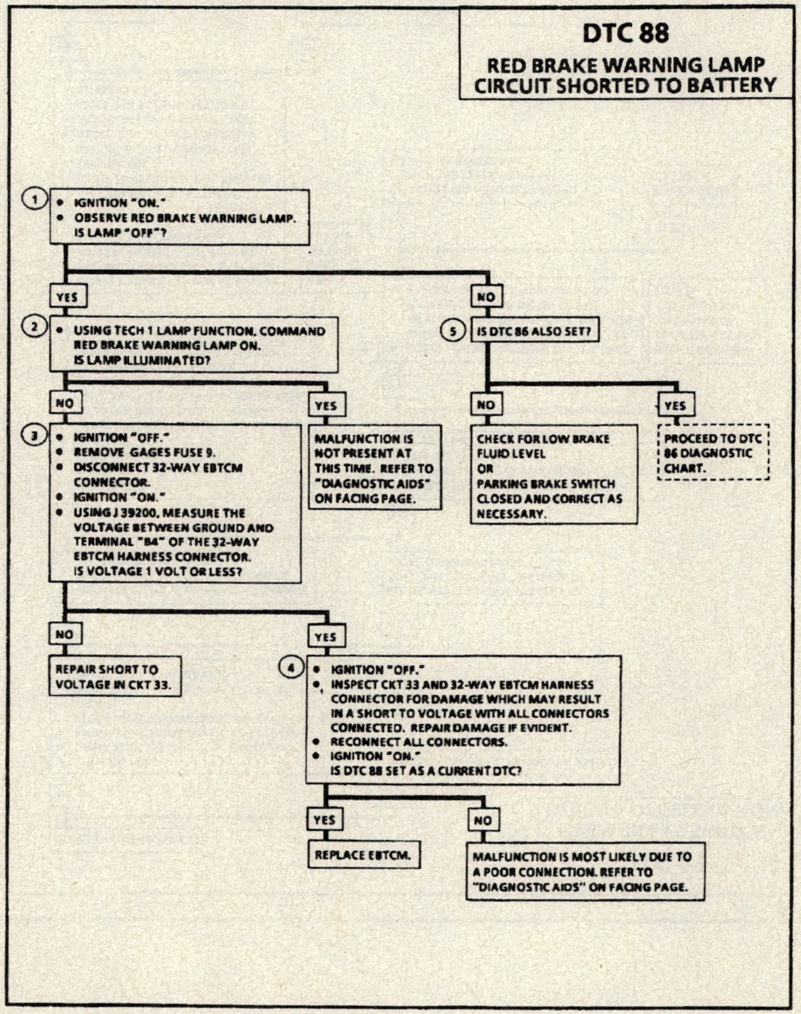

DIAGNOSTIC CHART DTC 91 — 1995 CAMARO AND FIREBIRD WITH TCS

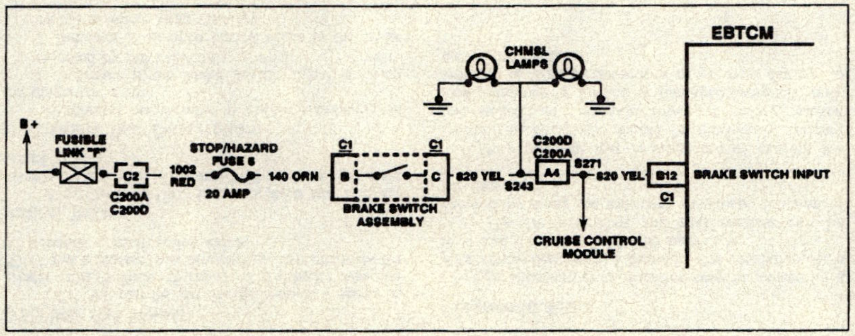

DTC 91
OPEN BRAKE SWITCH DURING DECELERATION

Circuit Description:

This DTC is used to detect an open brake switch in the non-ABS mode. The EBTCM looks for deceleration rates that would indicate braking action and verifies this assumption by requiring several repeats of this detection method. In each case, ABS will not be available since no brake switch is seen.

DTC Will Set When:

DTC 91 can be set if three deceleration cycles occur with the brake switch "OFF."

Action Taken:

A malfunction DTC is stored, ABS and TCS are disabled. The amber ABS and TCS warning lamps are turned "ON."

DTC Chart Test Description:

Number(s) below refer to circled number(s) on the diagnostic chart.

1. Indicates if the brake switch signal is being received by the EBTCM.
2. Indicates if an open circuit exists in the brake switch or brake lamp circuitry.
3. Isolates the open circuit to either the brake switch input circuitry, or the EBTCM.
4. Ensures malfunction was not due to poor terminal contact.
5. Verifies that voltage is available at the brake switch.
6. Verifies that the brake switch is functioning properly.

Diagnostic Aids:

An intermittent malfunction may be caused by a poor connection, rubbed through wire insulation, or a wire that is broken inside the insulation.

The frequency of the malfunction can be checked by using the enhanced diagnostic function of the Tech 1.

Any circuitry that is suspected of causing the intermittent complaint should be thoroughly checked for backed out terminals, improper mating, broken locks, improperly formed or damaged terminals, poor terminal to wiring connections or physical damage to the wiring harness.

DIAGNOSTIC CHART DTC 91 — 1995 CAMARO AND FIREBIRD WITH TCS, CONTINUED

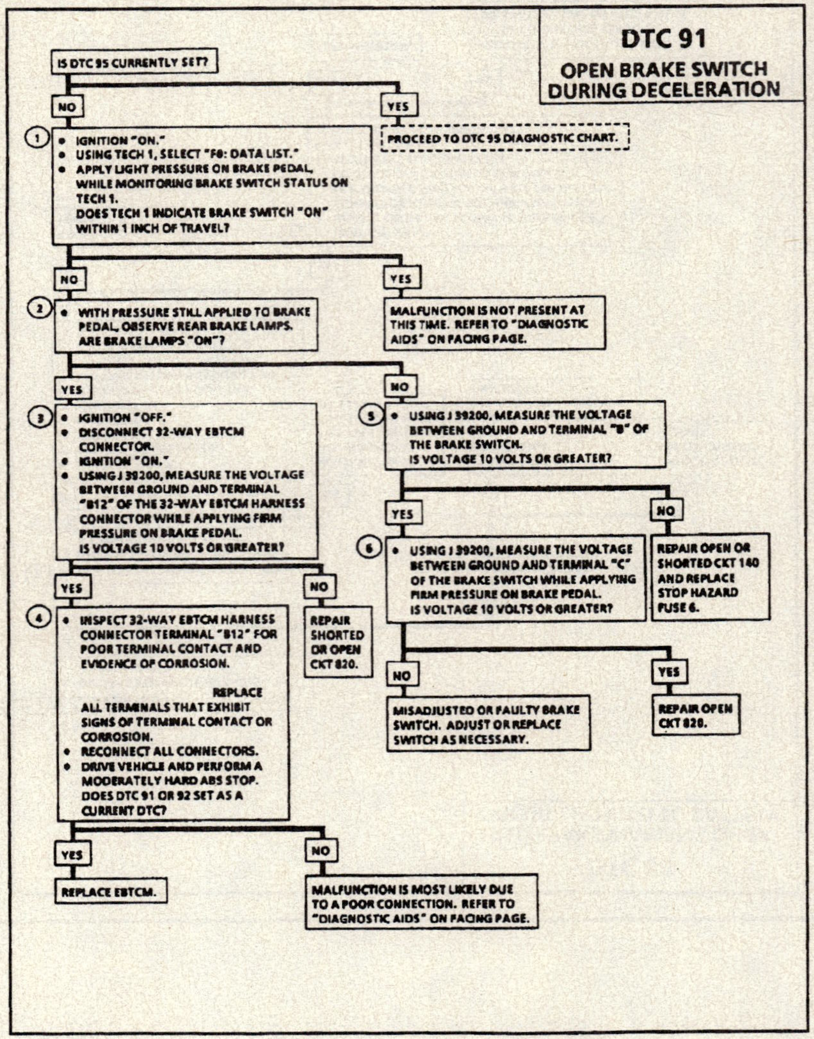

DIAGNOSTIC CHART DTC 92 — 1995 CAMARO AND FIREBIRD WITH TCS

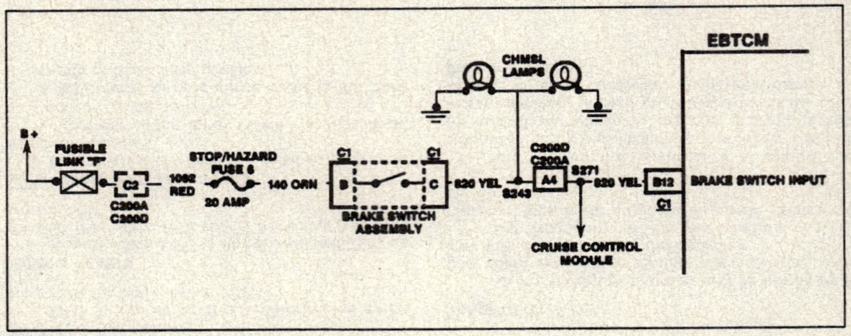

DTC 92
OPEN BRAKE SWITCH WHEN ABS WAS REQUIRED

Circuit Description:

This DTC is run to determine the proper operation of the brake switch. This is important because ABS is activated when the brake switch is "ON." If the brake switch is "OFF," ABS will never be activated. Since this malfunction is difficult to detect under normal braking conditions, this malfunction is only detected when ABS is required.

DTC Will Set When:

DTC 92 can be set if the vehicle's speed is greater than 8 km/h (5 mph). If the brake was not "ON" and a release was required on two channels for 0.5 seconds, a malfunction exists.

Action Taken:

A malfunction DTC is stored, ABS and TCS are disabled. The amber ABS and TCS warning lamps are turned "ON."

DTC Chart Test Description:

Number(s) below refer to circled number(s) on the diagnostic chart.

1. Indicates if the brake switch signal is being received by the EBTCM.
2. Indicates if an open circuit exists in the brake switch or brake lamp circuitry.
3. Isolates the open circuit to either the brake switch input circuitry, or the EBTCM.
4. Ensures malfunction was not due to poor terminal contact.
5. Verifies the voltage is available at the brake switch.
6. Verifies that the brake switch is functioning properly.

Diagnostic Aids:

An intermittent malfunction may be caused by a poor connection, rubbed through wire insulation, or a wire that is broken inside the insulation.

The frequency of the malfunction can be checked by using the enhanced diagnostic function of the Tech 1,

Any circuitry that is suspected of causing the intermittent complaint should be thoroughly checked for backed out terminals, improper mating, broken locks, improperly formed or damaged terminals, poor terminal to wiring connections or physical damage to the wiring harness.

DIAGNOSTIC CHART DTC 92 — 1995 CAMARO AND FIREBIRD WITH TCS, CONTINUED

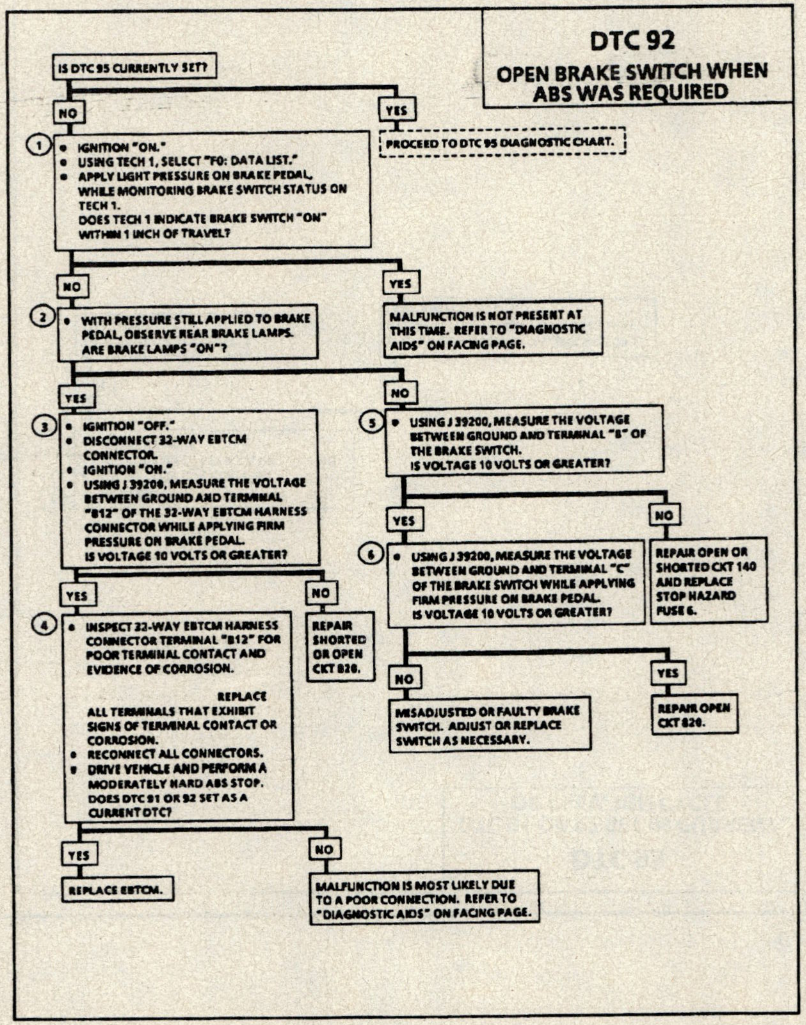

DIAGNOSTIC CHART DTC 93 — 1995 CAMARO AND FIREBIRD WITH TCS

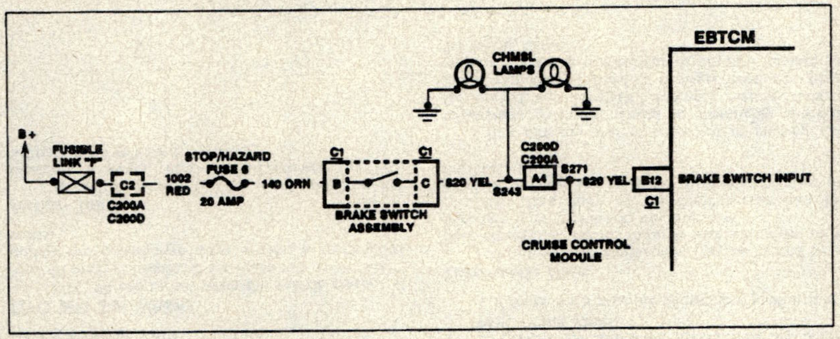

DTC 93
DTC 91 OR 92 SET IN CURRENT OR PREV. IGN CYCLE

Circuit Description:

This DTC is the second portion of DTCs 91 and 92. If DTCs 91 or 92 failed during the last ignition cycle, DTC 93 becomes a current failure during the next ignition cycle, keeping ABS and TCS disabled until a brake switch "ON" state is seen. When a change is seen during an ignition cycle in which DTC 93 is a current malfunction, DTC 91 or 92 will clear itself at the end of the current ignition cycle, and ABS and will be enabled at the start of the next ignition cycle. DTC 93 alone indicates 91 or 92 failed previously, but is intermittent, or has been corrected.

DTC Chart Test Description:

Number(s) below refer to circled number(s) on the diagnostic chart.

1. Indicates which DTC (either 91 or 92) caused DTC 93 to set.

Diagnostic Aids:

An intermittent malfunction may be caused by a poor connection, rubbed through wire insulation, or a wire that is broken inside the insulation.

The frequency of the malfunction can be checked by using the enhanced diagnostic function of the Tech 1,

Any circuitry that is suspected of causing the intermittent complaint should be thoroughly checked for backed out terminals, improper mating, broken locks, improperly formed or damaged terminals, poor terminal to wiring connections or physical damage to the wiring harness.

Also, verify proper brake switch operation using the data list of the Tech 1. As the brake is applied, the data list should display the brake switch "ON" within 1 inch of travel.

DIAGNOSTIC CHART DTC 93 — 1995 CAMARO AND FIREBIRD WITH TCS, CONTINUED

DTC 93

DTC 91 OR 92 SET IN CURRENT OR PREV. IGN CYCLE

(1)
- IGNITION "ON."
- USING TECH 1, READ DTCs. IS DTC 91 OR 92 SET AS A HISTORY OR CURRENT DTC?

YES → PROCEED TO DTC CHART THAT SET.

NO → USING TECH 1, SELECT "F1: DTC HISTORY." VERIFY MALFUNCTION FREQUENCY WAS LOW AND CLEAR DTCs.

DIAGNOSTIC CHART DTC 94 — 1995 CAMARO AND FIREBIRD WITH TCS

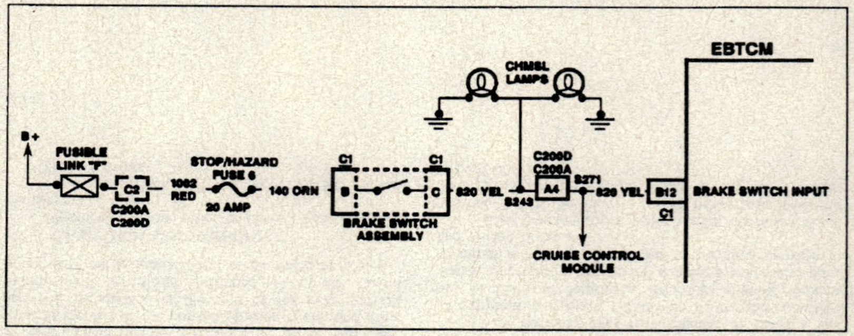

DTC 94
BRAKE SWITCH CONTACTS ALWAYS CLOSED

Circuit Description:

This DTC is run to determine the proper operation of the brake switch. This is important because ABS is activated when the brake switch is "ON" and turned "OFF" when the brake is "OFF." If the brake switch is always "ON," ABS operation will always be requested resulting in potential modulator cycling on rough roads.

DTC Will Set When:

DTC 94 can be set when the vehicle reaches at least 40 km/h (25 mph). If the brake was never "OFF" during two consecutive drive cycles, a malfunction exists.

Action Taken:

A malfunction DTC is stored. ABS is not disabled; however, TCS is disabled and the TCS warning lamp is turned "ON."

DTC Chart Test Description:

Number(s) below refer to circled number(s) on the diagnostic chart.

1. Checks if the malfunction is currently present.
2. Isolates the cause of the malfunction to either a malfunctioning or misadjusted brake switch, or a short to voltage in the brake switch circuitry.
3. Checks for unwanted voltage on the brake switch input circuit.
4. Checks for a possible intermittent malfunction.

Diagnostic Aids:

An intermittent malfunction may be caused by a poor connection, rubbed through wire insulation, or a wire that is broken inside the insulation.

The frequency of the malfunction can be checked by using the enhanced diagnostic function of the Tech 1.

Any circuitry that is suspected of causing the intermittent complaint should be thoroughly checked for backed out terminals, improper mating, broken locks, improperly formed or damaged terminals, poor terminal to wiring connections or physical damage to the wiring harness.

DIAGNOSTIC CHART DTC 94 — 1995 CAMARO AND FIREBIRD WITH TCS, CONTINUED

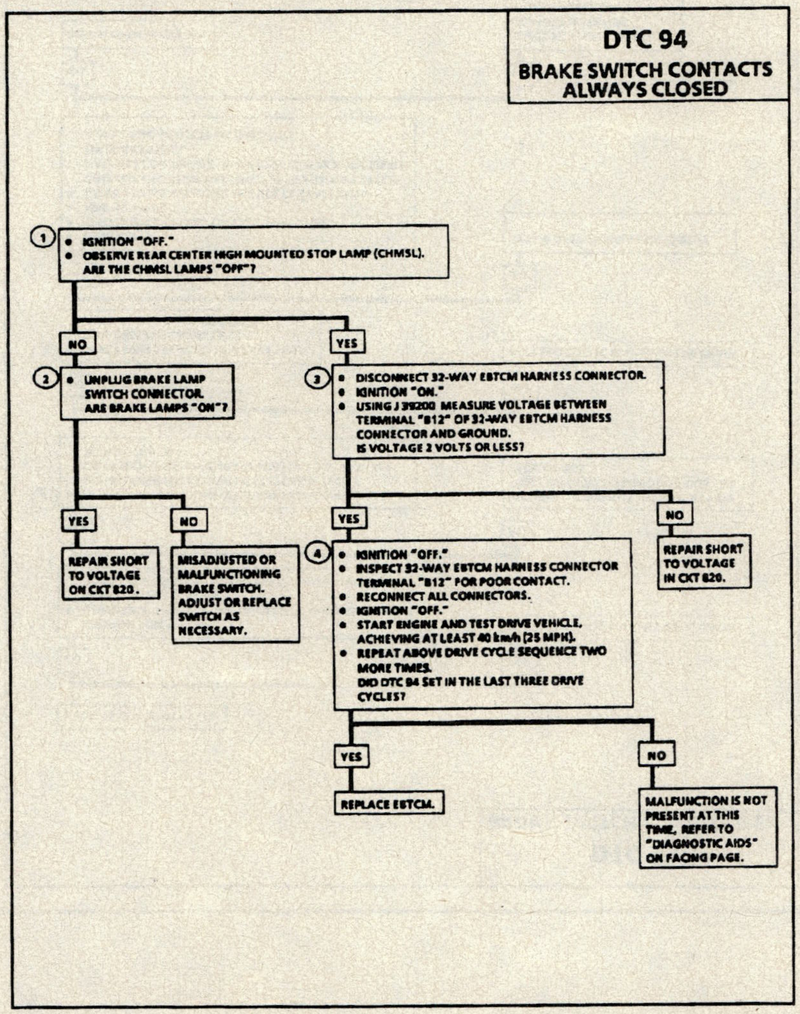

DIAGNOSTIC CHART DTC 95 — 1995 CAMARO AND FIREBIRD WITH TCS

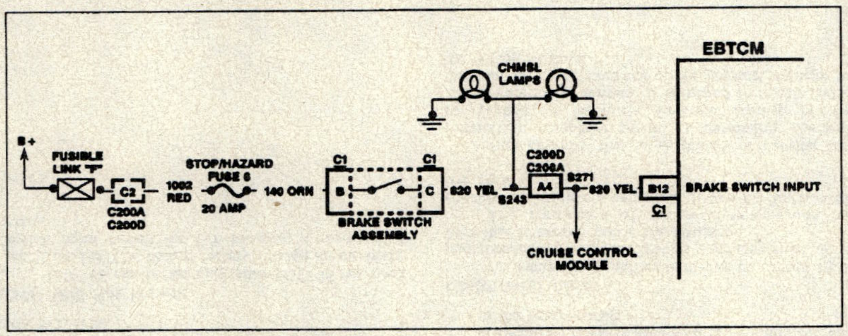

DTC 95
BRAKE SWITCH CIRCUIT OPEN

Circuit Description:
This DTC is used to identify open brake switch circuitry that prevents the brake switch input to the EBTCM from changing states when the brake is applied. This DTC is used in conjunction with DTCs 91 and 92 to determine the cause of an open brake switch malfunction.

DTC Will Set When:
DTC 95 can be set after initialization is completed. If the brake switch input voltage is out of specification for one second indicating an open circuit, a malfunction exists.

Action Taken:
A malfunction DTC is stored, ABS and TCS are disabled. The amber ABS and TCS warning lamps are turned "ON."

DTC Chart Test Description:
Number(s) below refer to circled number(s) on the diagnostic chart.
1. Is used to confirm that an open in the brake switch circuitry currently exists.
2. Indicates if the brake switch signal is being received by the EBTCM.
3. Indicates if an open circuit exists in the brake switch or brake lamp circuitry.
4. Isolates the open circuit to either the brake switch input circuitry, or the EBTCM.

Diagnostic Aids:
An intermittent malfunction may be caused by a poor connection, rubbed through wire insulation, or a wire that is broken inside the insulation.
The frequency of the malfunction can be checked by using the enhanced diagnostic function of the Tech 1.

Any circuitry that is suspected of causing the intermittent complaint should be thoroughly checked for backed out terminals, improper mating, broken locks, improperly formed or damaged terminals, poor terminal to wiring connections or physical damage to the wiring harness.

DIAGNOSTIC CHART DTC 95 — 1995 CAMARO AND FIREBIRD WITH TCS, CONTINUED

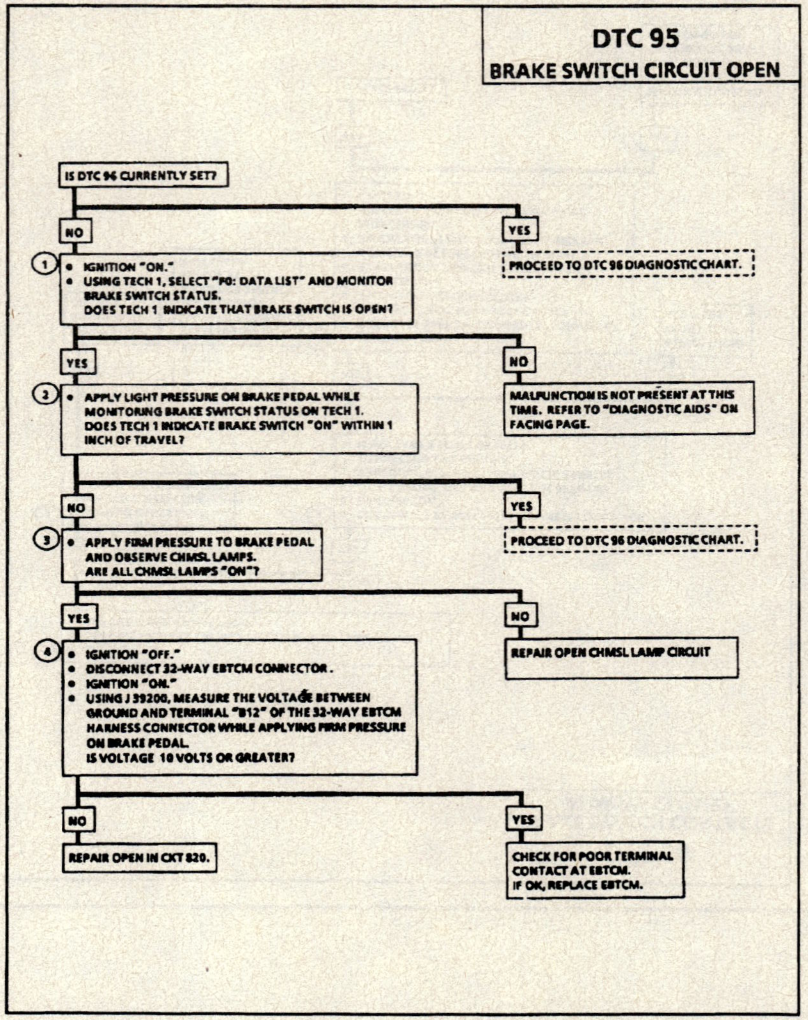

DIAGNOSTIC CHART DTC 96 — 1995 CAMARO AND FIREBIRD WITH TCS

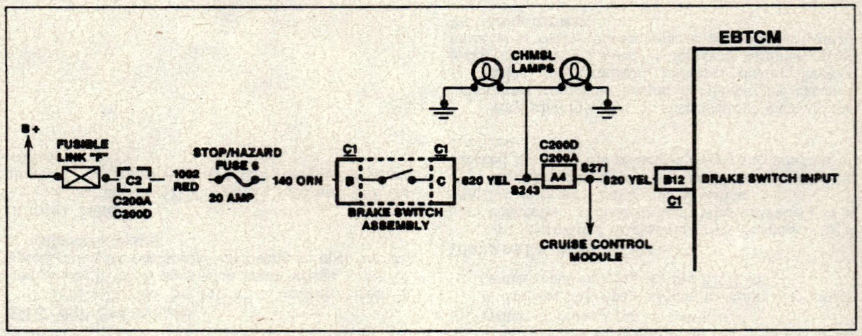

DTC 96
REAR BRAKE LAMP CIRCUIT OPEN

Circuit Description:

This DTC is designed to isolate the cause of a DTC 95 malfunction and indicate to the driver ABS is still available. If DTC 95 fails with DTC 96, the Center High Mounted Stop Lamp (CHMSL) circuit is open. The presence of battery voltage at the brake switch input indicates a valid brake switch input is still available.

DTC Will Set When:

DTC 96 can be set only after DTC 95 has been set. If the EBTCM detects battery voltage on the brake switch input circuit for 0.5 seconds, a malfunction exists.

DTC Chart Test Description:

Number(s) below refer to circled number(s) on the diagnostic chart.

1. As a result of a malfunction of an additional brake switch circuit DTC, this DTC may be set. To insure proper diagnosis, any additional brake switch DTCs must be repaired first.
2. Identifies if the malfunction is currently present in the brake circuit.

Diagnostic Aids:

An intermittent malfunction may be caused by a poor connection, rubbed through wire insulation, or a wire that is broken inside the insulation.

The frequency of the malfunction can be checked by using the enhanced diagnostic function of the Tech 1.

Any circuitry that is suspected of causing the intermittent complaint should be thoroughly checked for backed out terminals, improper mating, broken locks, improperly formed or damaged terminals, poor terminal to wiring connections or physical damage to the wiring harness.

DIAGNOSTIC CHART DTC 96 — 1995 CAMARO AND FIREBIRD WITH TCS, CONTINUED

DTC 96
REAR BRAKE LAMP CIRCUIT OPEN

1. ARE ANY OTHER BRAKE SWITCH DTCs (EXCEPT 95) CURRENTLY SET?
   - YES: PROCEED TO THAT DTC CHART FIRST.
   - NO: go to 2.
2. • APPLY FIRM PRESSURE TO BRAKE PEDAL AND OBSERVE REAR CHMSL LAMPS. ARE ALL CHMSL LAMPS "ON"?
   - YES: MALFUNCTION IS NOT PRESENT AT THIS TIME. REFER TO "DIAGNOSTIC AIDS" ON FACING PAGE.
   - NO: REPAIR OPEN CHMSL LAMP CIRCUIT.

DIAGNOSTIC CHART DTC 111 — 1995 CAMARO AND FIREBIRD WITH TCS

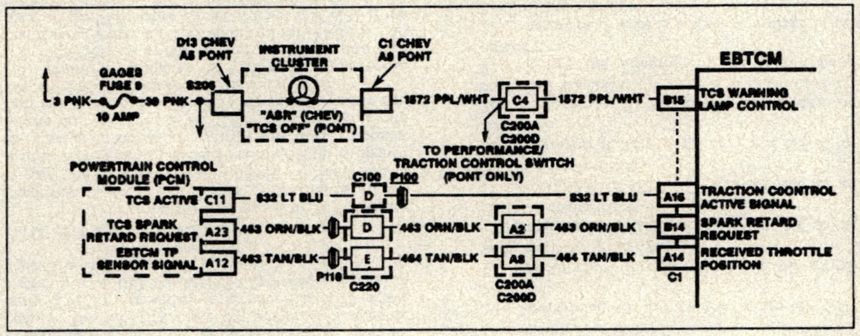

DTC 111
(Page 1 of 3)
TCS WARNING LAMP OR ACTIVE CIRCUIT MALFUNCTION

Circuit Description:

This DTC checks the state of both the TCS warning lamp and the TCS active circuit to identify a situation in which the driver could not be alerted to a system malfunction or the availability of traction control. The EBTCM turns the amber TCS warning lamp "ON," by providing a ground. The TCS active circuit is such that the EBTCM provides a ground to signal the PCM that TCS is active.

DTC Will Set When:

DTC 111 can be set any time. The TCS warning lamp will set DTC 111 if open or shorted to ground during the 3 second bulb check. In addition a malfunction exists if the TCS warning lamp is opened any time after the bulb check for more than 2 seconds. Shorts to voltage on the TCS warning lamp are detected anytime after the bulb check is complete and the EBTCM is commanding the lamp "OFF."

Shorts to voltage on the TCS active circuit when the EBTCM is commanding it "ON" and the malfunction exists for 2 seconds will set DTC 111. An open on the TCS active circuit will set DTC 111 if the malfunction exists for 2 seconds. An open on the spark retard line while the TCS is not active will set DTC 111 if the malfunction is present in excess of 2 seconds. A short to ground malfunction exists if the TCS active circuit is commanded "OFF" by the EBTCM and the short is present for 2 seconds.

Action Taken:

A malfunction DTC is stored, however ABS and TCS are not disabled.

DTC Chart Test Description:

Number(s) below refer to circled number(s) on the diagnostic chart.

1. Determines if the TCS warning lamp can be properly controlled.
2. determines if the lamp is "ON" constantly.
3. Checks for other DTCs which could turn "ON" the TCS warning lamp.
4. Determines if the TCS warning lamp can be turned "OFF" manually.
5. Checks the integrity of the TCS warning lamp circuitry.
6. Checks for a short to voltage on the TCS warning lamp circuitry.
7. Determines if a malfunctioning EBTCM is the cause of DTC 111.
8. Verifies that the other I/P indicator lamps function properly.
9. Determines if the TCS warning lamp can be turned "ON" manually.
10. Checks for a short to ground in the TCS warning lamp circuitry.
11. Checks for a open between the "Gages" fuse and the I/P cluster.
12. Verifies condition of the "Gages" fuse.
13. Determines if DTC 111 was set due to a EBTCM malfunction.

Diagnostic Aids:

Refer to Page 3 of 3 of this chart for diagnostic aids.

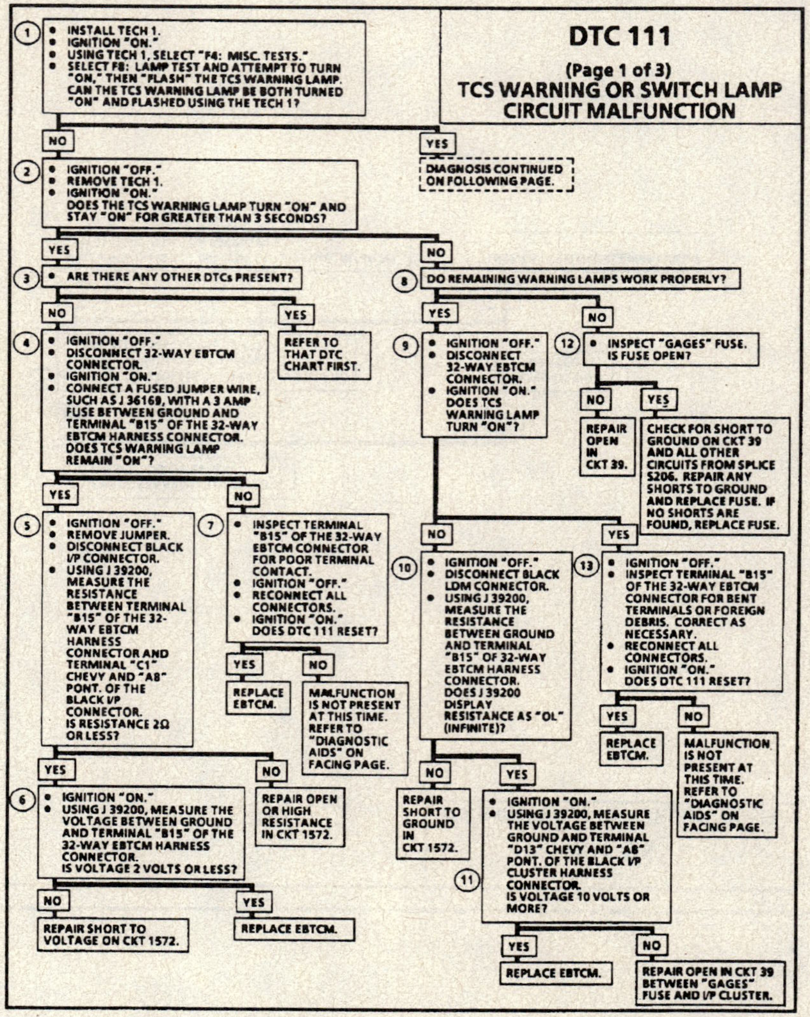

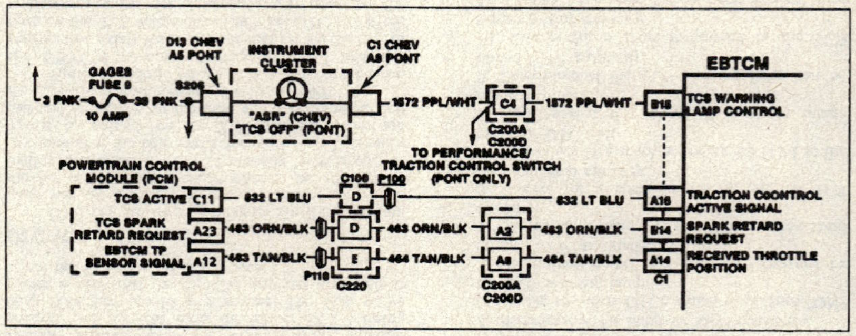

DTC 111
(Page 2 of 3)
TCS WARNING LAMP OR ACTIVE CIRCUIT MALFUNCTION

Circuit Description:

This DTC checks the state of both the TCS warning lamp and the TCS active circuit to identify a situation in which the driver could not be alerted to a system malfunction or the availability of traction control. The EBTCM turns the amber TCS warning lamp "ON," by providing a ground. The TCS active circuit is such that the EBTCM provides a ground to signal the PCM that TCS is active.

DTC Will Set When:

DTC 111 can be set any time. The TCS warning lamp will set DTC 111 if open or shorted to ground during the 3 second bulb check. In addition a malfunction exists if the TCS warning lamp is opened any time after the bulb check for more than 2 seconds. Shorts to voltage on the TCS warning lamp are detected anytime after the bulb check is complete and the EBTCM is commanding the lamp "OFF."

Shorts to voltage on the TCS active circuit when the EBTCM is commanding it "ON" and the malfunction exists for 2 seconds will set DTC 111. An open on the TCS active circuit will set DTC 111 if the malfunction exists for 2 seconds. An open on the spark retard line while the TCS is not active will set DTC 111 if the malfunction is present in excess of 2 seconds. A short to ground malfunction exists if the TCS active circuit is commanded "OFF" by the EBTCM and the short is present for 2 seconds.

Action Taken:

A malfunction DTC is stored, however ABS and TCS are not disabled.

DTC Chart Test Description:

Number(s) below refer to circled number(s) on the diagnostic chart.

14. Determines if the switch lamp can be properly controlled. (Pont Only)
15. Determines id the switch lamp is "ON" constantly.
16. Checks for a short to ground in the switch lamp circuitry (Pont Only)
17. Determines if DTC 111 was set due to an EBTCM malfunction.
18. Determines if DTC 111 was set due to an EBTCM malfunction.
19. Determines if there is an open in CKT 41 or ABS fuse.
20. Determines if the switch lamp can be controlled manually.
21. Verifies the integrity of the switch lamp circuitry.
22. Determines if there is a short to voltage in the switch lamp circuitry.
23. Determines if DTC 111 was set due to an EBTCM malfunction.

Diagnostic Aids:

Refer to page 3 of 3 of this chart for diagnostic aids.

DIAGNOSTIC CHART DTC 111 — 1995 CAMARO AND FIREBIRD WITH TCS, CONTINUED

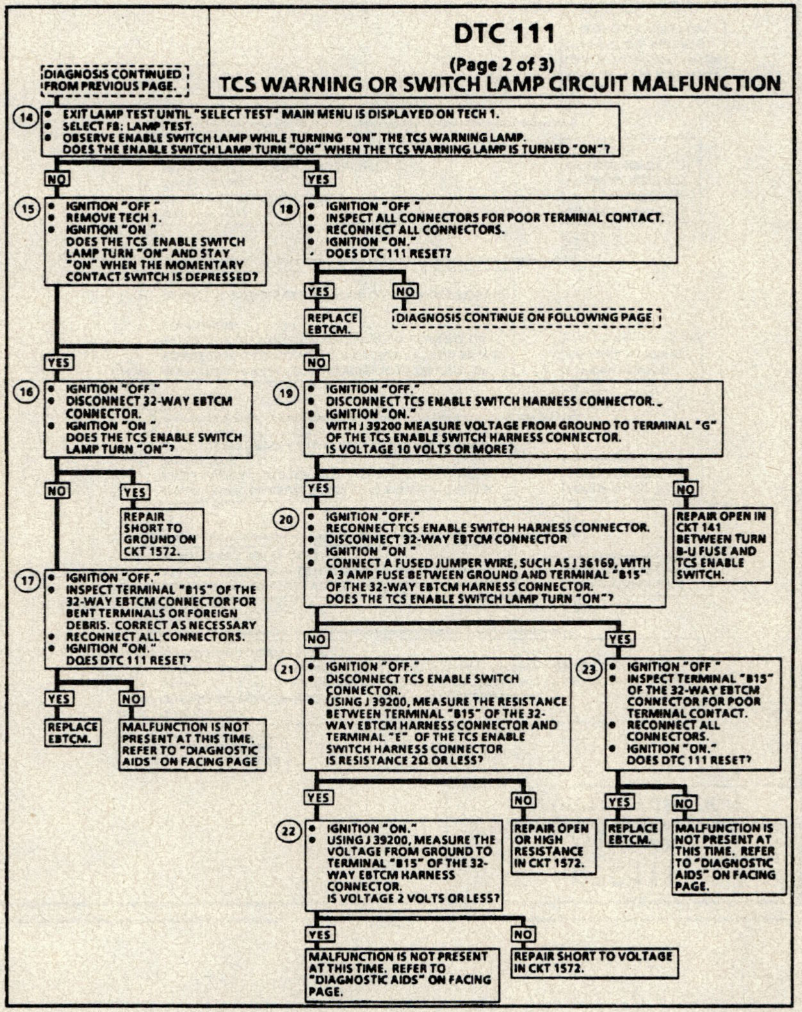

DIAGNOSTIC CHART DTC 111 — 1995 CAMARO AND FIREBIRD WITH TCS, CONTINUED

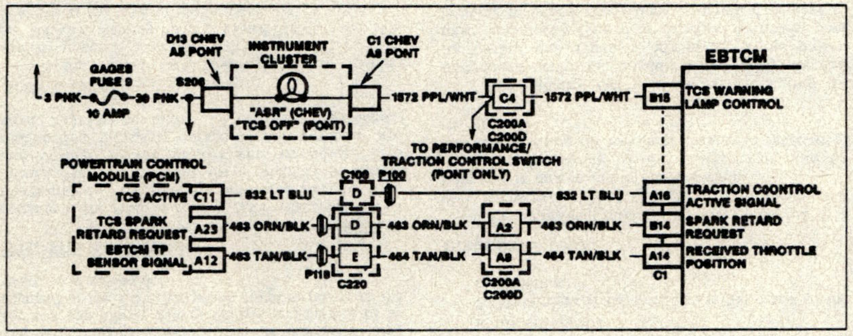

DTC 111
(Page 3 of 3)
TCS WARNING LAMP OR ACTIVE CIRCUIT MALFUNCTION

Circuit Description:

This DTC checks the state of both the TCS warning lamp and the TCS active circuit to identify a situation in which the driver could not be alerted to a system malfunction or the availability of traction control. The EBTCM turns the amber TCS warning lamp "ON," by providing a ground. The TCS active circuit is such that the EBTCM provides a ground to signal the PCM that TCS is active.

DTC Will Set When:

DTC 111 can be set any time. The TCS warning lamp will set DTC 111 if open or shorted to ground during the 3 second bulb check. In addition a malfunction exists if the TCS warning lamp is opened any time after the bulb check for more than 2 seconds. Shorts to voltage on the TCS warning lamp are detected anytime after the bulb check is complete and the EBTCM is commanding the lamp "OFF."

Shorts to voltage on the TCS active circuit when the EBTCM is commanding it "ON" and the malfunction exists for 2 seconds will set DTC 111. An open on the TCS active circuit will set DTC 111 if the malfunction exists for 2 seconds. An open on the spark retard line while the TCS is not active will set DTC 111 if the malfunction is present in excess of 2 seconds. A short to ground malfunction exists if the TCS active circuit is commanded "OFF" by the EBTCM and the short is present for 2 seconds.

Action Taken:

A malfunction DTC is stored, however ABS and TCS are not disabled.

DTC Chart Test Description:

Number(s) below refer to circled number(s) on the diagnostic chart.

24. Checks for or high resistance in spark retard request line (CKT 463) between PCM and EBTCM.

25. Check for short to voltage in spark retard request line (CKT 463) between PCM and EBTCM.

26. Checks for short to ground in spark retard request line (CKT 463) between PCM and EBTCM.

27. Checks for poor terminal contact or corrosion in the connectors in CKT 463.

Diagnostic Aids:

An intermittent malfunction may be caused by a poor connection, rubbed through wire insulation, or a wire that is broken inside the insulation.

The frequency of the malfunction can be checked by using the enhanced diagnostic function of the Tech 1.

Any circuitry that is suspected of causing the intermittent complaint should be thoroughly checked for backed out terminals, improper mating, broken locks, improperly formed or damaged terminals, poor terminal to wiring connections or physical damage to the wiring harness.

DIAGNOSTIC CHART DTC 111 — 1995 CAMARO AND FIREBIRD WITH TCS, CONTINUED

DTC 111
(Page 3 of 3)
TCS WARNING LAMP OR ACTIVE CIRCUIT MALFUNCTION

DIAGNOSIS CONTINUED FROM PREVIOUS PAGE.

(24)
- IGNITION "OFF."
- DISCONNECT PCM CONNECTOR "A" 32-WAY EBTCM CONNECTOR.
- USING J 39200, MEASURE THE RESISTANCE BETWEEN TERMINAL "23" OF THE PCM HARNESS CONNECTOR "A" AND TERMINAL "B14" OF THE 32-WAY EBTCM HARNESS CONNECTOR. IS RESISTANCE 2Ω OR LESS.?

NO: REPAIR OPEN OR HIGH RESISTANCE IN CKT 463.

YES:

(25)
- USING J 39200 MEASURE VOLTAGE BETWEEN TERMINAL "B14" OF THE 32-WAY EBTCM CONNECTOR AND GROUND. IS VOLTAGE 2 VOLTS OR LESS?

NO: REPAIR SHORT TO VOLTAGE IN CKT 463.

YES:

(26)
- USING J 39200 MEASURE RESISTANCE BETWEEN GROUND AND TERMINAL "B14" OF THE 32-WAY EBTCM HARNESS CONNECTOR. DOES J 39200 DISPLAY RESISTANCE AS "OL" (INFINITE)?

NO: REPAIR SHORT TO GROUND IN CKT 463.

YES:

(27)
- INSPECT TERMINALS "D" OF CONNECTOR C220, "A2" OF CONNECTOR C200 A/D, PCM CONNECTOR "A" AND 32-WAY EBTCM CONNECTOR FOR POOR TERMINAL CONTACT OR CORROSION.

DO ANY TERMINALS EXHIBIT POOR CONTACT OR EVIDENCE OF CORROSION?

YES: REPLACE TERMINALS THAT EXHIBIT POOR TERMINAL CONTACT OR EVIDENCE OF CORROSION.

NO:
- IGNITION "OFF."
- RECONNECT ALL CONNECTORS.
- START ENGINE.

DOES DTC 111 SET AS A CURRENT DTC?

YES: REPLACE EBTCM.

NO: MALFUNCTION WAS MOST LIKELY CAUSED BY A POOR CONNECTION. REFER TO "DIAGNOSTIC AIDS" ON FACING PAGE.

DIAGNOSTIC CHART DTC 114 — 1995 CAMARO AND FIREBIRD WITH TCS

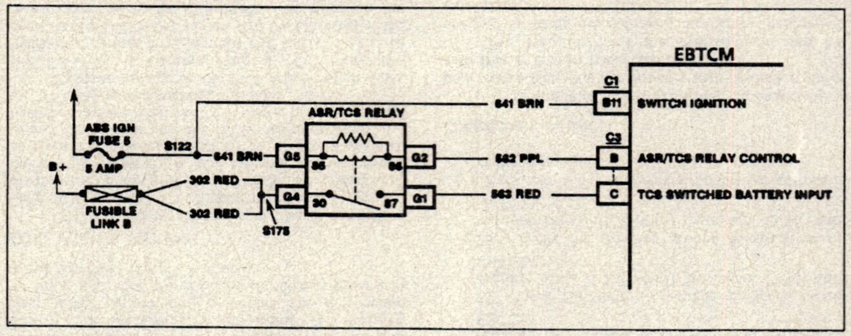

DTC 114
(Page 1 of 3)
TCS ENABLE RELAY CONTACT CIRCUIT OPEN

Circuit Description:

Ignition voltage is supplied through terminal "G5" of the ASR/TCS relay. The EBTCM then is able to energize the pull-in coil by completing the ground circuit at connector "C3",terminal "B" of the EBTCM. The magnetic field created closes the ASR/TCS relay contacts and allows battery voltage and current to be supplied to the EBTCM, which supplies power to the motors and solenoids.

DTC Will Set When:

DTC 114 can be set anytime after the EBTCM commands the ASR/TCS relay "ON" (the relay is first commanded "ON" during the three second bulb check). This test monitors the availability of current and voltage to the motors and solenoids. This malfunction indicates voltage is not available and would therefore not allow TCS operation if required.

Action Taken:

A malfunction DTC is stored, TCS is disabled and the amber TCS warning lamp is turned "ON." The red BRAKE warning lamp is turned "ON" if the TCS is active at the time of malfunction.

DTC Chart Test Description:

Number(s) below refer to circled number(s) on the diagnostic chart.

1. Checks to see if DTC 116 is also set as current or history. If DTC 116 is also set, proceed to that chart first.
2. Verifies that the malfunction is currently present. 3. Verifies that the relay can be turned "ON" and provide voltage to the EBTCM.
4. Checks for high resistance in the switched battery circuit from the relay to the EBTCM.
5. Checks for high resistance in the switched battery circuit from the battery to the relay.
6. Checks for high resistance in the relay contacts.
7. Checks for poor terminal contact or corrosion at the connectors.
8. Determines if the EBTCM is the cause of the malfunction.

Diagnostic Aids:

An intermittent malfunction may be caused by a poor connection, rubbed through wire insulation, or a wire that is broken inside the insulation.

The frequency of the malfunction can be checked by using the enhanced diagnostic function of the Tech 1,

Any circuitry that is suspected of causing the intermittent complaint should be thoroughly checked for backed out terminals, improper mating, broken locks, improperly formed or damaged terminals, poor terminal to wiring connections or physical damage to the wiring harness.

VIBRATION, TEMPERATURE EFFECTS:

Check for vibration effects by performing the relay test function of the Tech 1. With the relay test commanded "ON," lightly tap the top and sides of the relay while monitoring relay voltage. If the relay voltage changes significantly, replace the relay.

If DTC 114 only sets when the vehicle is initially started in cold ambient conditions (temperature less than 32° F-0° C), replace the relay.

DIAGNOSTIC CHART DTC 114 — 1995 CAMARO AND FIREBIRD WITH TCS, CONTINUED

DTC 114
(Page 1 of 3)
TCS RELAY CONTACT CIRCUIT OPEN

(1) IS DTC 116 ALSO PRESENT OR IN HISTORY?
- YES: PROCEED TO DTC 116 CHART FIRST.
- NO: go to step 2.

(2)
- IGNITION "ON," ENGINE "OFF."
- USING TECH 1, SELECT DATA LIST AND OBSERVE EBTCM BATTERY VOLTAGE.

DOES TECH 1 INDICATE THAT EBTCM BATTERY VOLTAGE IS GREATER THAN 10 VOLTS?
- YES: MALFUNCTION IS NOT PRESENT AT THIS TIME. REFER TO "DIAGNOSTIC AIDS" ON FACING PAGE.
- NO: go to step 3.

(3)
- IGNITION "OFF."
- DISCONNECT BLACK 8-WAY EBTCM CONNECTOR.
- CONNECT A FUSED JUMPER, SUCH AS J 36169 WITH A 3 AMP FUSE BETWEEN GROUND AND TERMINAL "B" OF THE BLACK 8-WAY EBTCM HARNESS CONNECTOR.
- IGNITION "ON."
- USING J 39200, MEASURE THE VOLTAGE BETWEEN GROUND AND TERMINAL "C" OF THE BLACK 8-WAY EBTCM HARNESS CONNECTOR.

IS VOLTAGE 10 VOLTS OR GREATER?
- NO: DIAGNOSIS CONTINUED ON NEXT PAGE.
- YES: go to step 4.

(4)
- IGNITION "OFF."
- DISCONNECT JUMPER USED IN STEP 3.
- DISCONNECT TCS RELAY.
- USING J 39200, MEASURE RESISTANCE BETWEEN TERMINAL "G1" OF THE TCS RELAY CONNECTOR AND TERMINAL "C" OF THE BLACK 8-WAY EBTCM HARNESS CONNECTOR.

IS RESISTANCE 2Ω OR LESS?
- NO: REPAIR HIGH RESISTANCE IN CKT 563.
- YES: go to step 5.

(5)
- DISCONNECT NEGATIVE BATTERY CABLE.
- DISCONNECT POSITIVE BATTERY CABLE.
- USING J 39200, MEASURE THE RESISTANCE BETWEEN POSITIVE BATTERY CABLE TERMINAL AND TERMINAL "G4" OF THE TCS RELAY CONNECTOR.

IS RESISTANCE 2Ω OR LESS?
- NO: REPAIR HIGH RESISTANCE IN CKT 302.
- YES: go to step 6.

(6)
- USING A JUMPER WIRE, CONNECT TERMINAL "86" ON THE TCS RELAY TO GROUND.
- USING A FUSED JUMPER WIRE, SUCH AS J 36169, WITH A 3 AMP FUSE, CONNECT TERMINAL "85" ON THE TCS RELAY TO B+.
- USING J 39200, MEASURE THE RESISTANCE BETWEEN TERMINALS "30" AND "87" ON THE TCS RELAY.

IS RESISTANCE 2Ω OR LESS?
- NO: REPLACE TCS RELAY.
- YES: go to step 7.

(7)
- INSPECT TCS RELAY AND TCS RELAY CONNECTOR FOR A POOR CONNECTION OR POOR TERMINAL CONTACT.
- INSPECT EBTCM AND EBTCM CONNECTOR TERMINALS "B" AND "C" FOR POOR TERMINAL CONTACT.
- INSPECT POSITIVE BATTERY TERMINAL AND POSITIVE BATTERY CABLE TERMINAL FOR A POOR CONNECTION.

IS POOR TERMINAL CONTACT OR A POOR CONNECTION EVIDENT?
- YES: REPLACE TERMINALS THAT EXHIBIT POOR TERMINAL CONTACT OR EVIDENCE OF CORROSION.
- NO: go to step 8.

(8)
- RECONNECT ALL CONNECTORS.
- IGNITION "ON."

DOES DTC 114 RESET?
- YES: REPLACE EBTCM.
- NO: MALFUNCTION IS INTERMITTENT. REFER TO "DIAGNOSTIC AIDS" ON FACING PAGE.

DIAGNOSTIC CHART DTC 114 — 1995 CAMARO AND FIREBIRD WITH TCS, CONTINUED

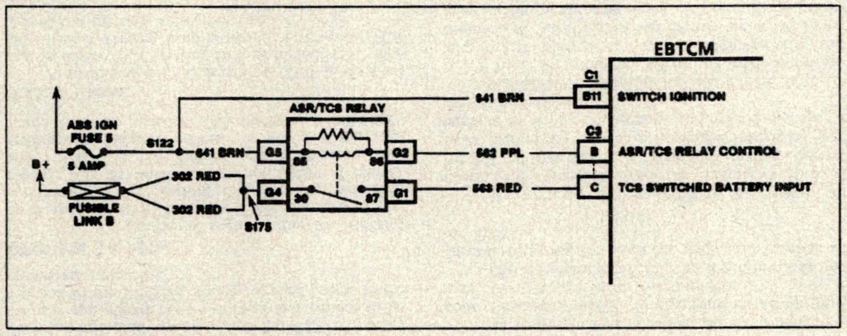

DTC 114
(Page 2 of 3)
TCS ENABLE RELAY CONTACT CIRCUIT OPEN

Circuit Description:

Ignition voltage is supplied through terminal "G5" of the ASR/TCS relay. The EBTCM then is able to energize the pull-in coil by completing the ground circuit at connector "C3",terminal "B" of the EBTCM. The magnetic field created closes the ASR/TCS relay contacts and allows battery voltage and current to be supplied to the EBTCM, which supplies power to the motors and solenoids.

DTC Will Set When:

DTC 114 can be set anytime after the EBTCM commands the ASR/TCS relay "ON" (the relay is first commanded "ON" during the three second bulb check). This test monitors the availability of current and voltage to the motors and solenoids. This malfunction indicates voltage is not available and would therefore not allow TCS operation if required.

Action Taken:

A malfunction DTC is stored, TCS is disabled and the amber TCS warning lamp is turned "ON." The red BRAKE warning lamp is turned "ON" if the TCS is active at the time of malfunction.

DTC Chart Test Description:

Number(s) below refer to circled number(s) on the diagnostic chart.

9. Checks for high resistance in the entire relay coil circuit.
10. Verifies proper battery condition.
11. Verifies proper voltage is available at the ASR/TCS relay.
12. Checks for an open in the switched battery circuit from the relay to the EBTCM.
13. Verifies the condition of the ASR/TCS relay.

Diagnostic Aids:

An intermittent malfunction may be caused by a poor connection, rubbed through wire insulation, or a wire that is broken inside the insulation.

The frequency of the malfunction can be checked by using the enhanced diagnostic function of the Tech 1,

Any circuitry that is suspected of causing the intermittent complaint should be thoroughly checked for backed out terminals, improper mating, broken locks, improperly formed or damaged terminals, poor terminal to wiring connections or physical damage to the wiring harness.

VIBRATION, TEMPERATURE EFFECTS:

Check for vibration effects by performing the relay test function of the Tech 1. With the relay test commanded "ON," lightly tap the top and sides of the relay while monitoring relay voltage. If the relay voltage changes significantly, replace the relay.

If DTC 114 only sets when the vehicle is initially started in cold ambient conditions (temperature less than 32° F-0° C), replace the relay.

DIAGNOSTIC CHART DTC 114 — 1995 CAMARO AND FIREBIRD WITH TCS, CONTINUED

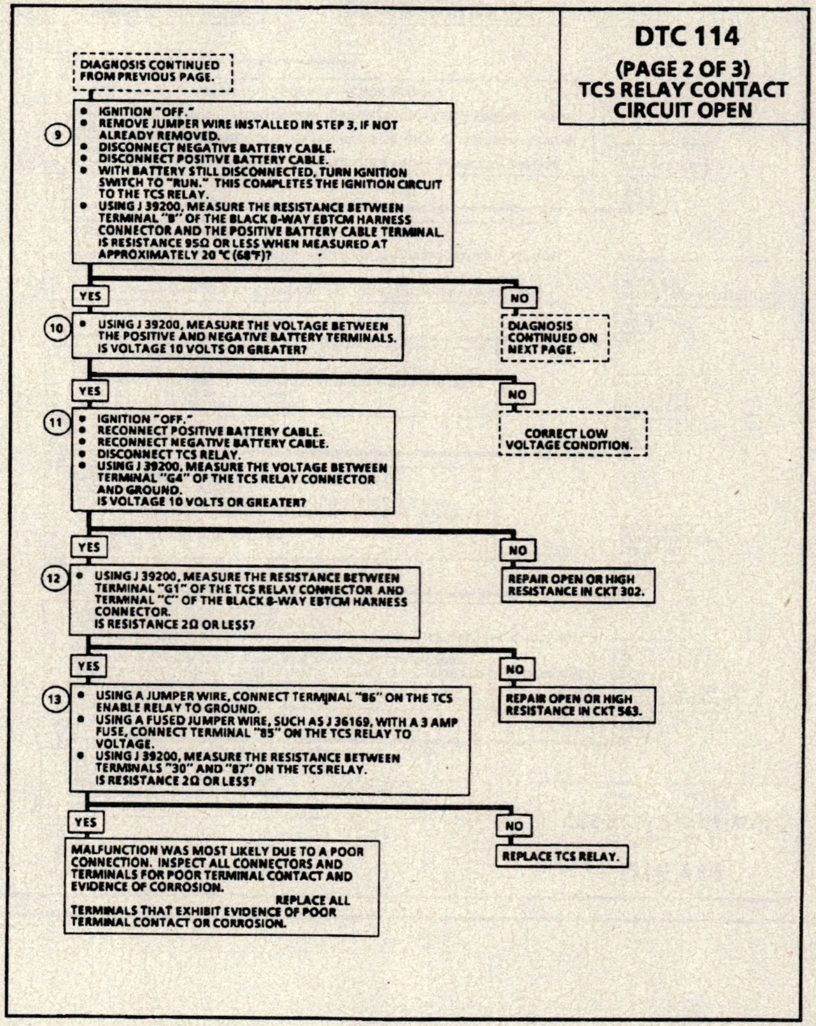

DIAGNOSTIC CHART DTC 114 — 1995 CAMARO AND FIREBIRD WITH TCS, CONTINUED

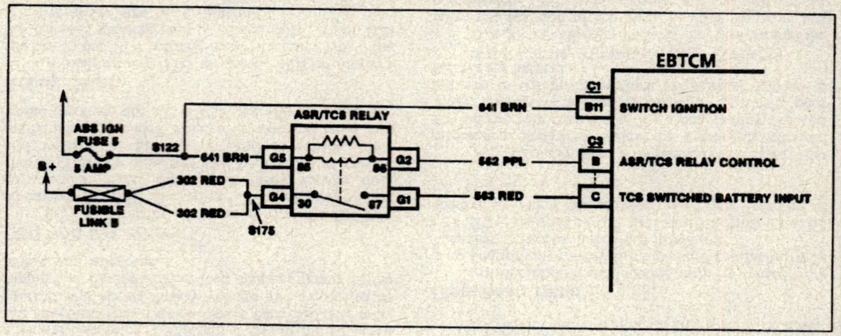

DTC 114
(Page 3 of 3)
TCS ENABLE RELAY CONTACT CIRCUIT OPEN

Circuit Description:

Ignition voltage is supplied through terminal "G5" of the ASR/TCS relay. The EBTCM then is able to energize the pull-in coil by completing the ground circuit at connector "C3",terminal "B" of the EBTCM. The magnetic field created closes the ASR/TCS relay contacts and allows battery voltage and current to be supplied to the EBTCM, which supplies power to the motors and solenoids.

DTC Will Set When:

DTC 114 can be set anytime after the EBTCM commands the ASR/TCS relay "ON" (the relay is first commanded "ON" during the three second bulb check). This test monitors the availability of current and voltage to the motors and solenoids. This malfunction indicates voltage is not available and would therefore not allow TCS operation if required.

Action Taken:

A malfunction DTC is stored, TCS is disabled and the amber TCS warning lamp is turned "ON." The red BRAKE warning lamp is turned "ON" if the TCS is active at the time of malfunction.

DTC Chart Test Description:

Number(s) below refer to circled number(s) on the diagnostic chart.

14. Checks for proper resistance of the relay coil.

15. Checks for high resistance in the ASR/TCS relay control circuit between the relay and the EBTCM.

16. Checks for high resistance in the ASR/TCS relay control circuit between the battery and the relay.

Diagnostic Aids:

An intermittent malfunction may be caused by a poor connection, rubbed through wire insulation, or a wire that is broken inside the insulation.

The frequency of the malfunction can be checked by using the enhanced diagnostic function of the Tech 1,

Any circuitry that is suspected of causing the intermittent complaint should be thoroughly checked for backed out terminals, improper mating, broken locks, improperly formed or damaged terminals, poor terminal to wiring connections or physical damage to the wiring harness.

VIBRATION, TEMPERATURE EFFECTS:

Check for vibration effects by performing the relay test function of the Tech 1. With the relay test commanded "ON," lightly tap the top and sides of the relay while monitoring relay voltage. If the relay voltage changes significantly, replace the relay.

If DTC 114 only sets when the vehicle is initially started in cold ambient conditions (temperature less than 32° F-0° C), replace the relay.

DIAGNOSTIC CHART DTC 114 — 1995 CAMARO AND FIREBIRD WITH TCS, CONTINUED

DTC 114
(Page 3 of 3)
TCS RELAY CONTACT CIRCUIT OPEN

DIAGNOSIS CONTINUED FROM PREVIOUS PAGE.

(14)
- DISCONNECT TCS RELAY.
- USING J 39200, MEASURE THE RESISTANCE BETWEEN TERMINALS "85" AND "86" ON THE TCS RELAY. IS RESISTANCE 95Ω OR LESS WHEN MEASURED AT APPROXIMATELY 24°C (75°F)?

NO: REPLACE TCS RELAY.

YES:

(15)
- USING J 39200, MEASURE THE RESISTANCE BETWEEN TERMINAL "G2" OF THE TCS RELAY CONNECTOR AND TERMINAL "B" OF THE BLACK 8-WAY EBTCM HARNESS CONNECTOR. IS RESISTANCE 2Ω OR LESS?

NO: REPAIR OPEN OR HIGH RESISTANCE IN CKT 562.

YES:

(16)
- USING J 39200, MEASURE THE RESISTANCE BETWEEN TERMINAL "G5" OF THE TCS RELAY CONNECTOR AND THE POSITIVE BATTERY CABLE TERMINAL. IS RESISTANCE 2Ω OR LESS?

NO: REPAIR OPEN OR HIGH RESISTANCE IN CKT 641, IGNITION SWITCH, OR BATTERY FEED CIRCUIT TO IGNITION SWITCH.

YES: MALFUNCTION WAS MOST LIKELY DUE TO A POOR CONNECTION. INSPECT ALL CONNECTORS AND TERMINALS FOR POOR TERMINAL CONTACT AND EVIDENCE OF CORROSION. REPLACE ALL TERMINALS THAT EXHIBIT EVIDENCE OF POOR TERMINAL CONTACT OR CORROSION.

DIAGNOSTIC CHART DTC 116 — 1995 CAMARO AND FIREBIRD WITH TCS

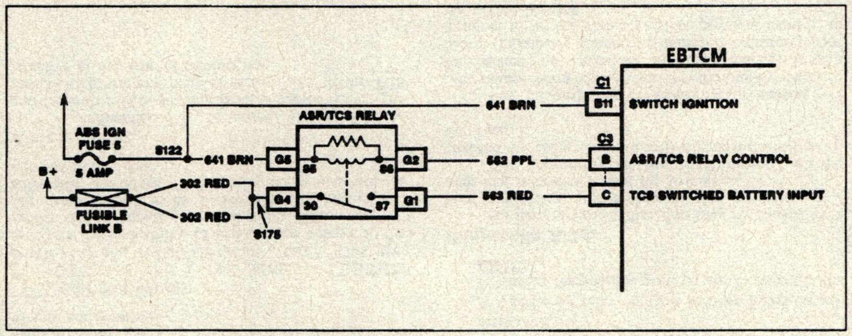

DTC 116
TCS ENABLE RELAY COIL CIRCUIT OPEN

Circuit Description:
Ignition voltage is supplied through terminal "G5" of the ASR/TCS relay. The EBTCM then is able to energize the pull-in coil by completing the ground circuit at connector "C3",terminal "B" of the EBTCM. The magnetic field created closes the ASR/TCS relay contacts and allows battery voltage and current to be supplied to the EBTCM, which supplies power to the motors and solenoids.

DTC Will Set When:
DTC 116 can be set only after DTC 114 has been set. This test detects an open in the ASR/TCS relay coil circuit. An open in this circuit will not allow the ASR/TCS relay to be energized thus preventing voltage and current to the motors and solenoids. If this malfunction is present and the ignition is turned "OFF" before 5 km/h (3 mph) is reached, DTC 114 is set alone.

Action Taken:
A malfunction DTC is stored, TCS is disabled and the amber TCS warning lamp is turned "ON." The red BRAKE warning lamp is turned "ON" if the TCS is active at the time of malfunction.

DTC Chart Test Description:
Number(s) below refer to circled number(s) on the diagnostic chart.
1. Indicates the EBTCM is capable of controlling the ASR/TCS relay as commanded.
2. Checks to ensure that voltage is available to the pull-in coil of the ASR/TCS relay.
3. Ensures that there is continuity through the pull-in coil of the ASR/TCS relay.
4. Verifies the integrity of the relay control circuit.
5. Ensures malfunction was not due to poor terminal contact.

Diagnostic Aids:
An intermittent malfunction may be caused by a poor connection, rubbed through wire insulation, or a wire that is broken inside the insulation.

The frequency of the malfunction can be checked by using the enhanced diagnostic function of the Tech 1,

If the frequency of the malfunction is high, but is currently intermittent, check for high coil resistance by measuring between relay terminals "G1" and "G2" using a J 39200. If resistance shows greater than 100 ohms, replace the relay.

Any circuitry that is suspected of causing the intermittent complaint should be thoroughly checked for backed out terminals, improper mating, broken locks, improperly formed or damaged terminals, poor terminal to wiring connections or physical damage to the wiring harness.

DIAGNOSTIC CHART DTC 116 — 1995 CAMARO AND FIREBIRD WITH TCS, CONTINUED

DTC 116
TCS RELAY COIL CIRCUIT OPEN

(1)
- IGNITION "ON," ENGINE "OFF."
- USING THE RELAY TEST FUNCTION OF THE TECH 1, COMMAND THE TCS RELAY "ON."

DOES TECH 1 INDICATE TCS RELAY "ON" AND BATTERY VOLTAGE ABOVE 10 VOLTS?

YES: MALFUNCTION IS NOT PRESENT AT THIS TIME. REFER TO "DIAGNOSTIC AIDS" ON FACING PAGE.

NO:

(2)
- USING J 39200, MEASURE THE VOLTAGE BY BACKPROBING BETWEEN GROUND AND TERMINAL "G5" OF THE TCS RELAY CONNECTOR.

IS VOLTAGE 10 VOLTS OR GREATER?

NO: REPAIR OPEN OR HIGH RESISTANCE IN CKT 641 OR ABS IGN FUSE 5.

YES:

(3)
- USING THE RELAY TEST FUNCTION OF THE TECH 1, COMMAND THE TCS RELAY "OFF."
- USING J 39200, MEASURE THE VOLTAGE BY BACKPROBING BETWEEN GROUND AND TERMINAL "G2" OF THE TCS RELAY CONNECTOR.

IS VOLTAGE 10 VOLTS OR GREATER?

NO: CHECK FOR INTERMITTENT CONNECTIONS AT TCS RELAY. IF OK, REPLACE TCS RELAY.

YES:

(4)
- IGNITION "OFF."
- DISCONNECT BLACK 8-WAY EBTCM CONNECTOR.
- IGNITION "ON."
- MEASURE THE VOLTAGE BETWEEN GROUND AND TERMINAL "B" OF THE BLACK 8-WAY EBTCM HARNESS CONNECTOR.

IS VOLTAGE 10 VOLTS OR GREATER?

NO: REPAIR OPEN OR HIGH RESISTANCE IN CKT 562.

YES:

(5)
- IGNITION "OFF."
- INSPECT BLACK 8-WAY EBTCM CONNECTOR TERMINAL "B" FOR POOR TERMINAL CONTACT OR EVIDENCE OF CORROSION. REPLACE ALL TERMINALS THAT EXHIBIT SIGNS OF POOR TERMINAL CONTACT OR CORROSION.
- RECONNECT ALL CONNECTORS.
- IGNITION "ON."

DOES DTC 116 SET AS A CURRENT DTC?

YES: REPLACE EBTCM.

NO: MALFUNCTION WAS MOST LIKELY CAUSED BY A POOR CONNECTION. REFER TO "DIAGNOSTIC AIDS" ON FACING PAGE.

AFTER ALL DIAGNOSIS IS COMPLETE, CLEAR DTCs AND TEST DRIVE VEHICLE FOR THREE (3) DRIVE CYCLES TO VERIFY THAT DTC DOES NOT RESET. A DRIVE CYCLE CONSISTS OF STARTING THE VEHICLE, DRIVING OVER 16 km/h (10 MPH), AND THEN IGNITION "OFF."

DIAGNOSTIC CHART DTC 118 — 1995 CAMARO AND FIREBIRD WITH TCS

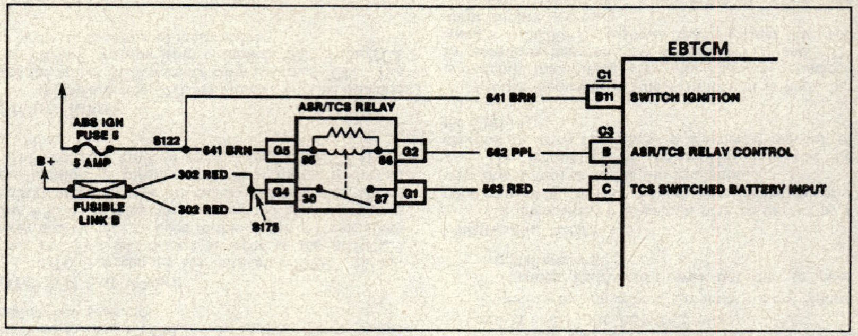

DTC 118
TCS ENABLE RELAY COIL CIRCUIT SHORTED TO BATT

Circuit Description:

Ignition voltage is supplied through terminal "G5" of the ASR/TCS relay. The EBTCM then is able to energize the pull-in coil by completing the ground circuit at connector "C3",terminal "B" of the EBTCM. The magnetic field created closes the ASR/TCS relay contacts and allows battery voltage and current to be supplied to the EBTCM, which supplies voltage to the motors and solenoids.

DTC Will Set When:

DTC 118 can be set after the EBTCM commands the ASR/TCS relay "ON." This test monitors the availability of current and voltage to the motors and solenoids. A malfunction of this type will not allow voltage to be available to the motors and solenoids; therefore, TCS operation would not be allowed if required.

Action Taken:

A malfunction DTC is stored, TCS is disabled and the amber TCS warning lamp is turned "ON." The red BRAKE warning lamp is turned "ON" if the TCS is active at the time of malfunction.

DTC Chart Test Description:

Number(s) below refer to circled number(s) on the diagnostic chart.

1. Indicates that the EBTCM is capable of controlling the ASR/TCS relay as commanded.
2. Checks for a short to voltage on the relay control circuit.
3. Verifies that the relay coil has the appropriate resistance.
4. Checks for relay contacts that are failed closed.
5. Checks for possible poor terminal contact at the EBTCM.

Diagnostic Aids:

An intermittent malfunction may be caused by a poor connection, rubbed through wire insulation, or a wire that is broken inside the insulation.

The frequency of the malfunction can be checked by using the enhanced diagnostic function of the Tech 1,

Any circuitry that is suspected of causing the intermittent complaint should be thoroughly checked for backed out terminals, improper mating, broken locks, improperly formed or damaged terminals, poor terminal to wiring connections or physical damage to the wiring harness.

DIAGNOSTIC CHART DTC 118 — 1995 CAMARO AND FIREBIRD WITH TCS, CONTINUED

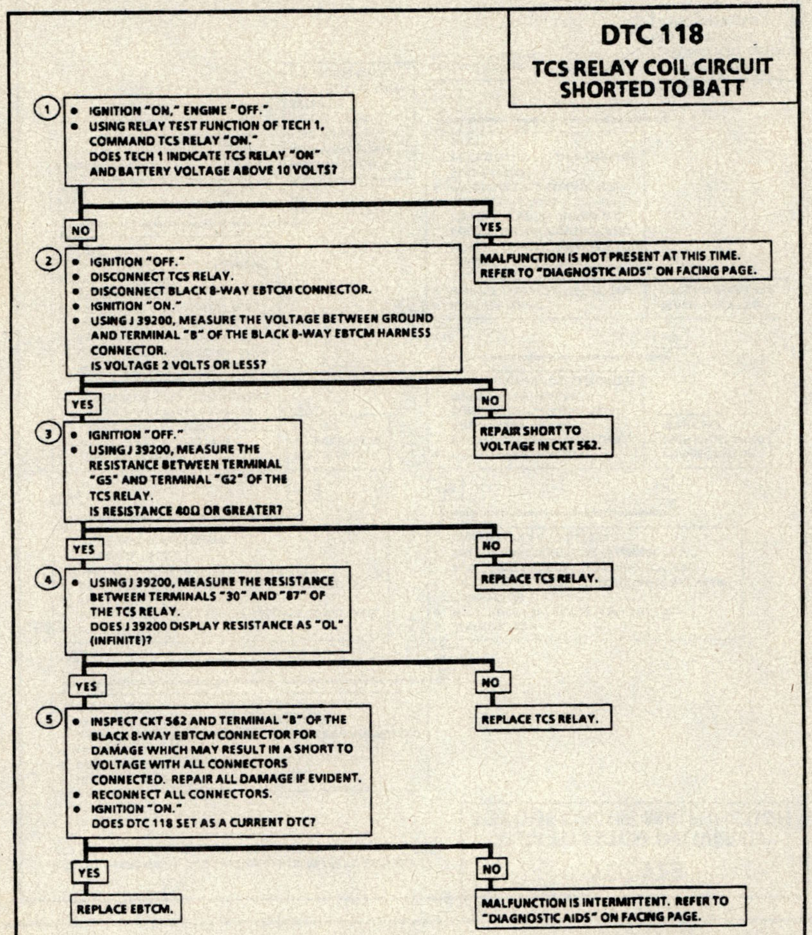

AFTER ALL DIAGNOSIS IS COMPLETE, CLEAR DTCs AND TEST DRIVE VEHICLE FOR THREE (3) DRIVE CYCLES TO VERIFY THAT DTC DOES NOT RESET. A DRIVE CYCLE CONSISTS OF STARTING THE VEHICLE, DRIVING OVER 16 km/h (10 MPH), AND THEN IGNITION "OFF."

DIAGNOSTIC CHART DTC 123 — 1995 CAMARO AND FIREBIRD WITH TCS

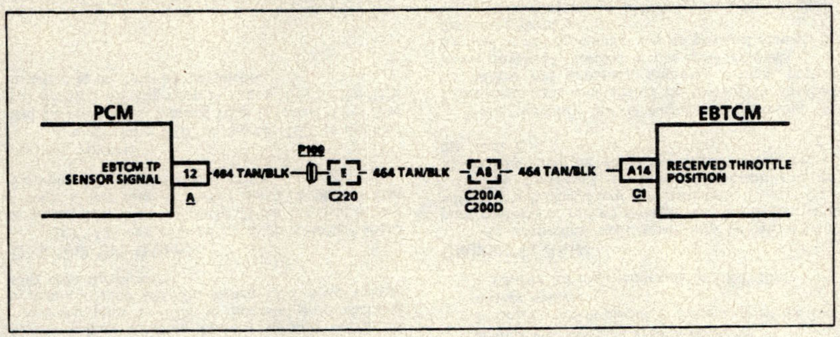

DTC 123
PCM TO EBTCM TP SENSOR CIRCUIT MALFUNCTION

Circuit Description:
The traction control system uses the antilock brake system in conjunction with the PCM fuel and ignition controls to limit drive wheel slippage during acceleration. The PCM controlled portion of the traction control system reduces engine torque by altering spark timing and controlling throttle position. The EBTCM sends a spark retard request to the PCM via a Pulse Width Modulated (PWM) signal requesting the desired spark level for proper traction control system operation (CKT 463). The PCM then provides a throttle position PWM signal (CKT 464) informing the EBTCM what the actual throttle position of the engine is.

DTC Will Set When:
DTC 123 will set when there is a malfunction in the throttle position circuit (CKT 464) or if the signal is out of range.

Action Taken:
A malfunction DTC is stored, TCS is disabled at the end of the current TCS cycle and the amber TCS warning lamp is turned "ON."

DTC Chart Test Description:
Number(s) below refer to circled number(s) on the diagnostic chart,
1. Checks for proper voltage at the PCM terminal.
2. Checks for short to voltage in CKT 464.
3. Ensures malfunction was not due to physical damage of circuitry.
4. Checks for open in CKT 464.
5. Checks for short to ground in CKT 464.
6. Checks for possible poor terminal contact at the EBTCM.

Diagnostic Aids:
An intermittent malfunction may be caused by a poor connection, rubbed through wire insulation, or a wire that is broken inside the insulation.
The frequency of the malfunction can be checked by using the enhanced diagnostic function of the Tech 1,

Any circuitry that is suspected of causing the intermittent complaints should be thoroughly checked for backed out terminals, improper mating, broken locks, improperly formed or damaged terminals, poor terminal to wiring connections or physical damage to the wiring harness.
Before beginning diagnostics perform a "OBD" system check. If OK begin chart.

DIAGNOSTIC CHART DTC 123 — 1995 CAMARO AND FIREBIRD WITH TCS, CONTINUED

DTC 123
PCM TO EBTCM DELIVERED TORQUE CIRCUIT MALFUNCTION

NOTE: PERFORM AN OBD SYSTEM CHECK, REFER TO SECTION 6E3 BEFORE BEGINNING CHART. IF OK, BEGIN CHART.

(1)
- IGNITION "OFF."
- DISCONNECT PCM CONNECTOR "A".
- IGNITION "ON."
- USING J 39200, MEASURE VOLTAGE BETWEEN GROUND AND TERMINAL "12" OF THE PCM CONNECTOR "A".

IS VOLTAGE APPROXIMATELY 12 VOLTS?

YES → (2) / NO → (4)

(2)
- IGNITION "OFF."
- DISCONNECT 32-WAY EBTCM HARNESS CONNECTOR.
- IGNITION "ON."
- USING J 39200, MEASURE VOLTAGE BETWEEN GROUND AND 32-WAY EBTCM CONNECTOR TERMINAL "A14".

IS VOLTAGE ZERO VOLTS?

YES → (3) / NO → REPAIR SHORT TO VOLTAGE IN CKT 464.

(3)
- IGNITION "OFF."
- INSPECT CKT 464 AND PCM CONNECTOR "A", TERMINAL "12" AND 32-WAY EBTCM HARNESS CONNECTOR "A14" FOR DAMAGE WHICH MAY RESULT IN AN INTERMITTENT OPEN, SHORT TO GROUND, OR SHORT TO B+ WITH ALL CONNECTORS CONNECTED. REPAIR DAMAGE IF EVIDENT.
- RECONNECT ALL CONNECTORS.
- START ENGINE.
- WAIT 10 SECONDS.

DOES DTC 123 SET AS A CURRENT DTC?

YES → REPLACE PCM. / NO → MALFUNCTION IS MOST LIKELY DUE TO A POOR CONNECTION. REFER TO "DIAGNOSTIC AIDS" ON FACING PAGE.

(4)
- IGNITION "OFF."
- DISCONNECT 32-WAY EBTCM HARNESS CONNECTOR.
- USING J 39200, MEASURE RESISTANCE BETWEEN PCM CONNECTOR "A" TERMINAL "12" AND 32-WAY EBTCM CONNECTOR TERMINAL "A14".

IS RESISTANCE 2Ω OR LESS?

YES → (5) / NO → REPAIR OPEN OR HIGH RESISTANCE IN CKT 464.

(5)
- USING J 39200 MEASURE RESISTANCE BETWEEN GROUND AND 32-WAY EBTCM CONNECTOR TERMINAL "A14".

DOES J 39200 DISPLAY RESISTANCE AS "OL" (INFINITE)?

YES → (6) / NO → REPAIR SHORT TO GROUND IN CKT 464.

(6)
- IGNITION "OFF."
- INSPECT TERMINAL "12" OF THE PCM CONNECTOR "A" AND TERMINAL "A14" OF THE 32-WAY EBTCM CONNECTOR FOR POOR TERMINAL CONTACT AND CONTAMINATION. REPAIR ALL DAMAGE IF EVIDENT.
- RECONNECT ALL CONNECTORS.
- START ENGINE.

DOES DTC 123 SET AS A CURRENT DTC?

YES → REPLACE EBTCM. / NO → MALFUNCTION IS INTERMITTENT. REFER TO "DIAGNOSTIC AIDS" ON FACING PAGE.

DIAGNOSTIC CHART DTC 136 — 1995 CAMARO AND FIREBIRD WITH TCS

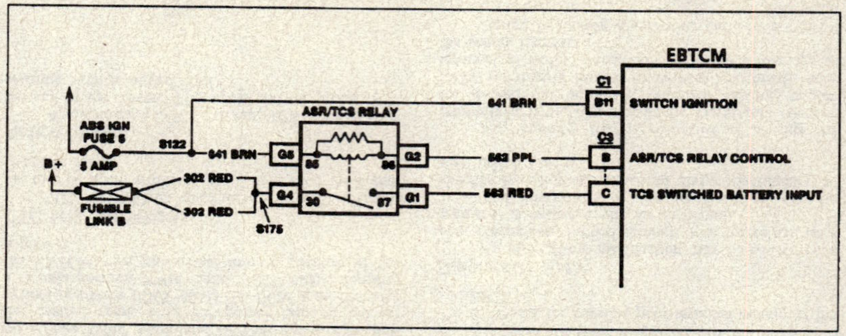

DTC 136
(Page 1 of 2)
LOW SYSTEM VOLTAGE (TCS)

Circuit Description:
This DTC is used to monitor the voltage level available to the EBTCM controller. If the voltage drops below 11.5 volts, full performance of the TCS system cannot be guaranteed. During TCS operation, there are several current requirements that will cause battery voltage to drop. Because of this, voltage is monitored prior to TCS to indicate good charging system condition and also during TCS when voltage may drop significantly.

DTC Will Set When:
DTC 136 can only be set if the vehicle's speed is greater than 5 km/h (3 mph). If the switched battery voltage is less than 11.5 volts non-TCS or less than 9.0 volts TCS, a malfunction exists.

Action Taken:
A malfunction DTC is stored, TCS is disabled and the amber TCS warning lamp is turned "ON." The red BRAKE warning lamp is turned "ON" if the TCS is active at the time of malfunction.

DTC Chart Test Description:
Number(s) below refer to circled number(s) on the diagnostic chart.
1. Checks to see what voltage is readily available at connector "C3",terminal "C" of the EBTCM
2. Isolates the condition to high circuit resistance, or improper charging system operation.
3. Isolates which circuit is the source of the low voltage condition.
4. Checks for high resistance in CKT 302.

Diagnostic Aids:
An intermittent malfunction may be caused by a poor connection, rubbed through wire insulation, or a wire that is broken inside the insulation.

The frequency of the malfunction can be checked by using the enhanced diagnostic function of the Tech 1,

Any circuitry that is suspected of causing the intermittent complaint should be thoroughly checked for backed out terminals, improper mating, broken locks, improperly formed or damaged terminals, poor terminal to wiring connections or physical damage to the wiring harness.

While performing a voltage load test, if it is noted that only ignition voltage drops below acceptable voltage levels, CKT 641 should be checked for high resistance or an open condition.

DIAGNOSTIC CHART DTC 136 — 1995 CAMARO AND FIREBIRD WITH TCS, CONTINUED

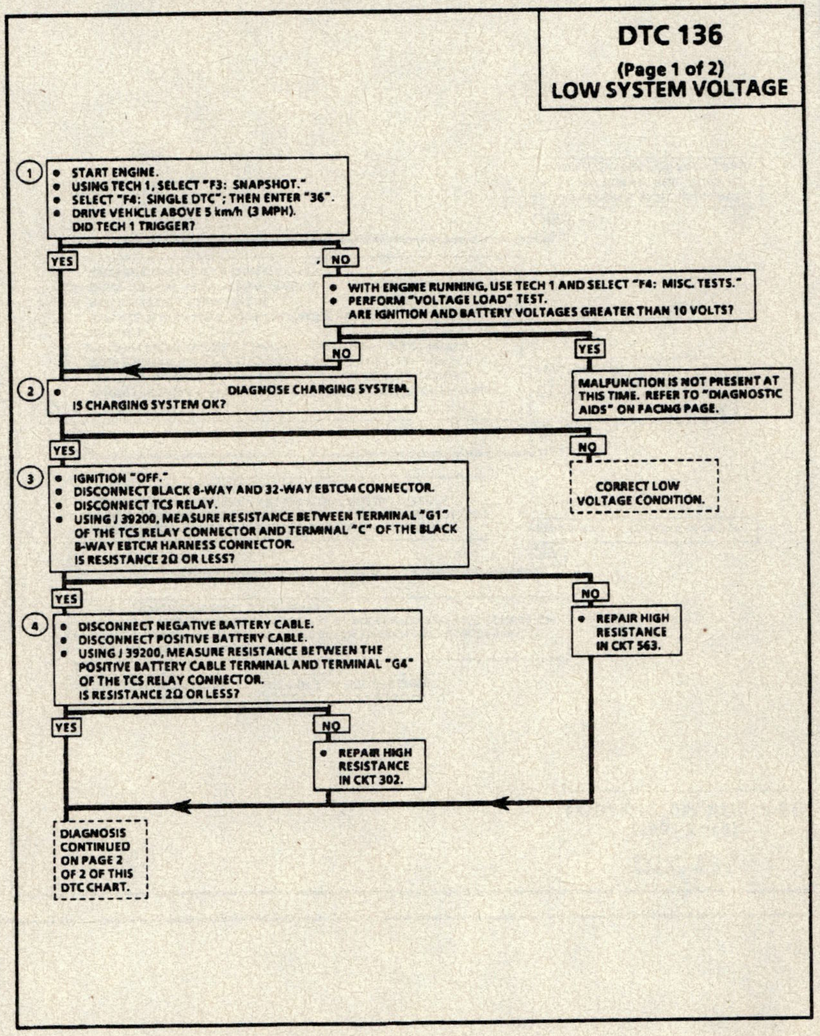

DIAGNOSTIC CHART DTC 136 — 1995 CAMARO AND FIREBIRD WITH TCS, CONTINUED

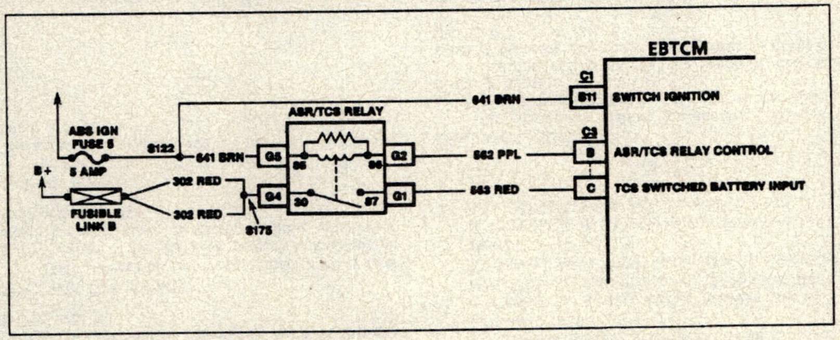

DTC 136
(Page 2 of 2)
LOW SYSTEM VOLTAGE (TCS)

Circuit Description:

This DTC is used to monitor the voltage level available to the EBTCM controller. If the voltage drops below 11.5 volts, full performance of the TCS system cannot be guaranteed. During TCS operation, there are several current requirements that will cause battery voltage to drop. Because of this, voltage is monitored prior to TCS to indicate good charging system condition and also during TCS when voltage may drop significantly.

DTC Will Set When:

DTC 136 can only be set if the vehicle's speed is greater than 5 km/h (3 mph). If the switched battery voltage is less than 11.5 volts non-TCS or less than 9.0 volts TCS, a malfunction exists.

Action Taken:

A malfunction DTC is stored, TCS is disabled and the amber TCS warning lamp is turned "ON." The red BRAKE warning lamp is turned "ON" if the TCS is active at the time of malfunction.

DTC Chart Test Description:

Number(s) below refer to circled number(s) on the diagnostic chart.

5. Checks for high resistance in CKT 641.
6. Checks for high resistance in the ignition fuse feed circuit.
7. Checks for malfunctioning EBTCM.

Diagnostic Aids:

An intermittent malfunction may be caused by a poor connection, rubbed through wire insulation, or a wire that is broken inside the insulation.

The frequency of the malfunction can be checked by using the enhanced diagnostic function of the Tech 1.

Any circuitry that is suspected of causing the intermittent complaint should be thoroughly checked for backed out terminals, improper mating, broken locks, improperly formed or damaged terminals, poor terminal to wiring connections or physical damage to the wiring harness.

While performing a voltage load test, if it is noted that only ignition voltage drops below acceptable voltage levels, CKT 641 should be checked for high resistance or an open condition.

DIAGNOSTIC CHART DTC 136 — 1995 CAMARO AND FIREBIRD WITH TCS, CONTINUED

DTC 136
(Page 2 of 2)
LOW SYSTEM VOLTAGE

DIAGNOSIS CONTINUED FROM PAGE 1 OF 2 OF THIS DTC CHART.

(5)
- REMOVE ABS IGN FUSE 5.
- USING J 39200, MEASURE RESISTANCE BETWEEN EACH ABS IGN FUSE 5 BLOCK TERMINAL AND TERMINAL "B11" OF THE 32-WAY EBTCM HARNESS CONNECTOR. IS EITHER MEASUREMENT 2Ω OR LESS?

NO: REPAIR HIGH RESISTANCE IN CKT 641.

YES:

(6)
- WITH BATTERY CABLES DISCONNECTED, TURN IGNITION "ON."
- USING J 39200, MEASURE RESISTANCE BETWEEN POSITIVE BATTERY CABLE TERMINAL AND EACH ABS IGN FUSE 5 BLOCK TERMINAL. IS EITHER MEASUREMENT 2Ω OR LESS ?

NO: REPAIR HIGH RESISTANCE IN THE "IGNITION SWITCH 'RUN' FUSE FEED" CIRCUIT.

YES:

(7)
- IGNITION "OFF."
- INSPECT TCS RELAY AND TCS RELAY CONNECTOR FOR POOR TERMINAL CONTACT. INSPECT BLACK 8-WAY AND 32-WAY EBTCM HARNESS CONNECTORS FOR A POOR CONNECTION TO THE EBTCM. INSPECT POSITIVE BATTERY CABLE TERMINAL FOR POOR CONTACT TO THE BATTERY.
  REPLACE TERMINALS IF POOR CONTACT IS EVIDENT.
- RECONNECT ALL CONNECTORS.
- DRIVE VEHICLE ABOVE 5 km/h (3 MPH).
  DOES DTC 136 SET AS A CURRENT DTC?

YES: REPLACE EBTCM.

NO: MALFUNCTION IS MOST LIKELY DUE TO A POOR CONNECTION. REFER TO "DIAGNOSTIC AIDS" ON FACING PAGE.

DIAGNOSTIC CHART DTC 144 — 1995 CAMARO AND FIREBIRD WITH TCS

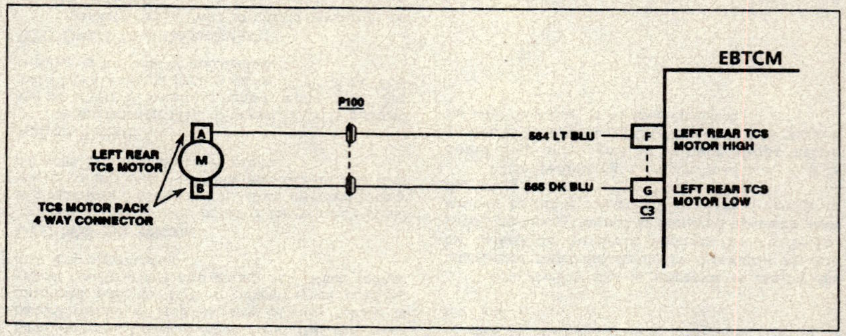

DTC 144
LEFT TCS CHANNEL WILL NOT MOVE

Circuit Description:
This DTC is designed to detect a stuck TCS motor or a seized hydraulic modulator. During initialization or after a TCS cycle, the EBTCM commands the TCS motor in the reverse direction and the forward direction to ensure that the motor moves properly. If the TCS motor does not move, improper current feedback will be detected by the EBTCM, and a DTC will set.

DTC Will Set When:
DTC 144 can be set during initialization or after a TCS cycle. If the EBTCM detects a condition in which it cannot move the motor in either direction, a malfunction exists.

Action Taken:
A malfunction DTC is stored, TCS is disabled. The amber TCS warning and the red BRAKE warning lamps are turned "ON."

DTC Chart Test Description:
Number(s) below refer to circled number(s) on the diagnostic chart.
1. Checks for proper TCS motor movement during apply and release commands from the Tech 1.
2. Verifies that the TCS motor can actually apply when commanded.
3. Compares EBTCM command current to motor feedback current.
4. Checks for a short circuit between the TCS motor "HIGH" and TCS motor "LOW" circuits.
5. Checks for a TCS motor that is internally shorted.
6. Determines if the malfunction is caused by a defective EBTCM or poor terminal contact.

Diagnostic Aids:
An intermittent malfunction in this DTC may result from a mechanical part of the system that sticks, binds, or slips.
The frequency of the malfunction can be checked by using the enhanced diagnostic function of the Tech 1,

Any circuitry that is suspected of causing the intermittent complaint should be thoroughly checked for backed out terminals, improper mating, broken locks, improperly formed or damaged terminals, poor terminal to wiring connections or physical damage to the wiring harness.

DIAGNOSTIC CHART DTC 144 — 1995 CAMARO AND FIREBIRD WITH TCS, CONTINUED

DTC 144 LEFT TCS CHANNEL WILL NOT MOVE

NOTICE: J 39200 TEST LEADS MUST BE "ZEROED" PRIOR TO MAKING ANY RESISTANCE MEASUREMENTS. REFER TO J 39200 USER'S MANUAL.

(1)
- RAISE VEHICLE.
- IGNITION TO "RUN." ENGINE "OFF."
- PUMP BRAKE PEDAL UNTIL FIRM TO DEPLETE VACUUM RESERVE.
- USING TECH 1 MISC. TEST, SELECT MANUAL CONTROL FUNCTION.
- FOOT OFF BRAKE PEDAL.
- SELECT LEFT REAR TCS MOTOR RELEASE FUNCTION AND RELEASE THE MOTOR.

DOES THE LEFT REAR WHEEL ROTATE?

NO → go to (3). YES → go to (2).

(2)
- USING TECH 1 MISC. TEST, SELECT MANUAL CONTROL FUNCTION.
- FOOT OFF BRAKE PEDAL.
- SELECT LEFT REAR TCS MOTOR APPLY FUNCTION AND APPLY THE MOTOR.

DOES THE LEFT REAR WHEEL ROTATE?

YES → go to (3).
NO → MALFUNCTION IS INTERMITTENT. REFER TO "DIAGNOSTIC AIDS" ON FACING PAGE.

(3)
- FOOT OFF BRAKE PEDAL.
- SELECT LEFT REAR TCS MOTOR RELEASE FUNCTION AND RELEASE THE TCS MOTOR WHILE CAREFULLY OBSERVING THE COMMANDED CURRENT AND THE FEEDBACK CURRENT ON THE TECH 1.

WAS FEEDBACK CURRENT HIGHER THAN THE COMMANDED CURRENT?

YES → go to (4).
NO → REPLACE TCS HYDRAULIC MODULATOR/MOTOR PACK ASSEMBLY.

(4)
- EXIT TECH 1 MANUAL CONTROL.
- IGNITION "OFF."
- DISCONNECT THE BLACK 8-WAY EBTCM CONNECTOR AND THE 4-WAY TCS MOTOR PACK CONNECTOR.
- USING J 39200, MEASURE THE RESISTANCE BETWEEN TERMINAL "F" AND TERMINAL "G" OF THE BLACK 8-WAY EBTCM HARNESS CONNECTOR.

DOES J 39200 DISPLAY RESISTANCE AS "OL" (INFINITE)?

YES → go to (5).
NO → REPAIR SHORT BETWEEN CKT 564 AND CKT 565.

(5)
- USING J 39200, MEASURE THE RESISTANCE BETWEEN TERMINAL "A" AND TERMINAL "B" OF THE TCS MOTOR PACK CONNECTOR.

IS RESISTANCE GREATER THAN 0.4 OHMS?

YES → go to (6).
NO → REPLACE TCS HYDRAULIC MODULATOR/MOTOR PACK ASSEMBLY.

(6)
- RECONNECT ALL CONNECTORS.
- START ENGINE AND TEST DRIVE VEHICLE, ACHIEVING AT LEAST 16 km/h (10 MPH).
- REPEAT ABOVE DRIVE CYCLE SEQUENCE TWO MORE TIMES.

DOES DTC 144 SET IN THE LAST THREE DRIVE CYCLES?

YES → REPLACE EBTCM.
NO → MALFUNCTION IS INTERMITTENT. INSPECT ALL CONNECTORS FOR CONTAMINATION AND POOR CONTACT. REPLACE ANY TERMINALS THAT EXHIBIT POOR TERMINAL CONTACT OR EVIDENCE OF CORROSION.

AFTER DIAGNOSIS IS COMPLETE, CLEAR DTCs AND TEST DRIVE VEHICLE FOR THREE (3) DRIVE CYCLES TO VERIFY THAT DTC DOES NOT RESET. A DRIVE CYCLE CONSISTS OF STARTING THE VEHICLE, DRIVING OVER 16 km/h (10 MPH), AND THEN IGNITION "OFF."

DIAGNOSTIC CHART DTC 145 — 1995 CAMARO AND FIREBIRD WITH TCS

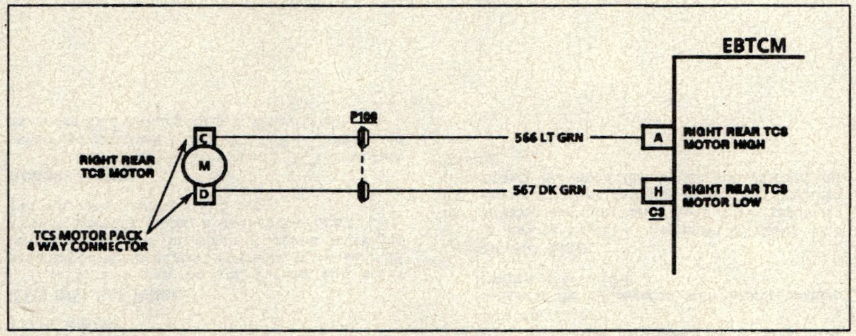

DTC 145
RIGHT TCS CHANNEL WILL NOT MOVE

Circuit Description:

This DTC is designed to detect a stuck TCS motor or a seized hydraulic modulator. During initialization or after a TCS event, the EBTCM commands the TCS motor in the reverse direction and the forward direction to ensure that the motor moves properly. If the TCS motor does not move, improper current feedback will be detected by the EBTCM, and a DTC will set.

DTC Will Set When:

DTC 145 can be set during initialization or after a TCS cycle. If the EBTCM detects a condition in which it cannot move the motor in either direction, a malfunction exists.

Action Taken:

A malfunction DTC is stored, TCS is disabled. The amber TCS warning and the red BRAKE warning lamps are turned "ON."

DTC Chart Test Description:

Number(s) below refer to circled number(s) on the diagnostic chart.

1. Checks for proper TCS motor movement during apply and release commands from the Tech 1.
2. Verifies that the motor can actually apply when commanded.
3. Compares EBTCM command current to motor feedback current.
4. Checks for short circuit between TCS motor "HIGH" and TCS motor "LOW" circuits.
5. Checks for a TCS motor that is internally shorted.
6. Checks to see if malfunction is due to poor terminal contact or the EBTCM.

Diagnostic Aids:

An intermittent malfunction in this DTC may result from a mechanical part of the system that sticks, binds, or slips.

The frequency of the malfunction can be checked by using the enhanced diagnostic function of the Tech 1.

Any circuitry that is suspected of causing the intermittent complaint should be thoroughly checked for backed out terminals, improper mating, broken locks, improperly formed or damaged terminals, poor terminal to wiring connections or physical damage to the wiring harness.

DIAGNOSTIC CHART DTC 145 — 1995 CAMARO AND FIREBIRD WITH TCS, CONTINUED

DTC 145
RIGHT TCS CHANNEL WILL NOT MOVE

NOTICE: J 39200 TEST LEADS MUST BE "ZEROED" PRIOR TO MAKING ANY RESISTANCE MEASUREMENTS. REFER TO J 39200 USER'S MANUAL.

1
- RAISE VEHICLE.
- IGNITION "ON," ENGINE "OFF."
- PUMP BRAKE PEDAL UNTIL FIRM TO DEPLETE VACUUM RESERVE.
- USING TECH 1 MISC. TEST, SELECT MANUAL CONTROL FUNCTION.
- FOOT OFF BRAKE PEDAL.
- SELECT RIGHT REAR TCS MOTOR RELEASE FUNCTION AND RELEASE THE MOTOR. DOES THE RIGHT REAR WHEEL ROTATE?

NO → 3; YES → 2

2
- USING TECH 1 MISC. TEST, SELECT MANUAL CONTROL FUNCTION.
- FOOT OFF BRAKE PEDAL.
- SELECT RIGHT REAR TCS MOTOR APPLY FUNCTION AND APPLY THE MOTOR.
- DOES THE RIGHT REAR WHEEL ROTATE?

YES → 3; NO → MALFUNCTION IS INTERMITTENT. REFER TO "DIAGNOSTIC AIDS" ON FACING PAGE.

3
- FOOT OFF BRAKE PEDAL.
- SELECT RIGHT REAR TCS MOTOR RELEASE FUNCTION AND RELEASE THE TCS MOTOR WHILE CAREFULLY OBSERVING THE COMMANDED CURRENT AND THE FEEDBACK CURRENT ON THE TECH 1. WAS FEEDBACK CURRENT HIGHER THAN THE COMMANDED CURRENT?

YES → 4; NO → 7

4
- EXIT TECH 1 MANUAL CONTROL.
- IGNITION "OFF."
- DISCONNECT THE BLACK 8-WAY EBTCM CONNECTOR AND THE 4-WAY TCS MOTOR PACK CONNECTOR.
- USING J 39200, MEASURE THE RESISTANCE BETWEEN TERMINAL "A" AND TERMINAL "H" OF THE BLACK 8-WAY EBTCM HARNESS CONNECTOR. DOES J 39200 DISPLAY RESISTANCE AS "OL" (INFINITE)?

YES → 5; NO → REPAIR SHORT BETWEEN CKT 566 AND CKT 567.

5
- USING J 39200, MEASURE THE RESISTANCE BETWEEN TERMINAL "C" AND TERMINAL "D" OF THE TCS MOTOR PACK CONNECTOR. IS RESISTANCE GREATER THAN 0.4Ω?

YES → 6; NO → REPLACE TCS MOTOR PACK.

6
- IGNITION "OFF."
- RECONNECT ALL CONNECTORS.
- START ENGINE AND TEST DRIVE VEHICLE, ACHIEVING AT LEAST 16 km/h (10 MPH).
- REPEAT ABOVE DRIVE CYCLE SEQUENCE TWO MORE TIMES. DOES DTC 145 SET IN THE LAST THREE DRIVE CYCLES?

YES → REPLACE EBTCM.; NO → MALFUNCTION IS INTERMITTENT. INSPECT ALL CONNECTORS FOR CONTAMINATION AND POOR CONTACT. REPLACE ANY TERMINALS THAT EXHIBIT POOR TERMINAL CONTACT OR EVIDENCE OF CORROSION.

7
- IGNITION "OFF."
- REMOVE TCS MODULATOR/MOTOR PACK ASSEMBLY FROM VEHICLE, REFER TO "ON VEHICLE SERVICE" IN THIS SECTION.
- SEPARATE TCS MOTOR PACK FROM HYDRAULIC MODULATOR.
- GRASP THE GEAR ON THE HYDRAULIC MODULATOR. (THE RIGHT TCS CHANNEL IS FURTHEST FORWARD AS THE MODULATOR/MOTOR PACK SITS IN VEHICLE). CAN THE GEAR BE ROTATED AT LEAST 7 FULL TURNS LOCK TO LOCK?

NO → REPLACE HYDRAULIC MODULATOR.; YES → REPLACE TCS MOTOR PACK.

AFTER DIAGNOSIS IS COMPLETE, CLEAR DTCs AND TEST DRIVE VEHICLE FOR THREE (3) DRIVE CYCLES TO VERIFY THAT DTC DOES NOT RESET. A DRIVE CYCLE CONSISTS OF STARTING THE VEHICLE, DRIVING OVER 16 km/h (10 MPH), AND THEN IGNITION OFF.

DIAGNOSTIC CHART DTC 146 — 1995 CAMARO AND FIREBIRD WITH TCS

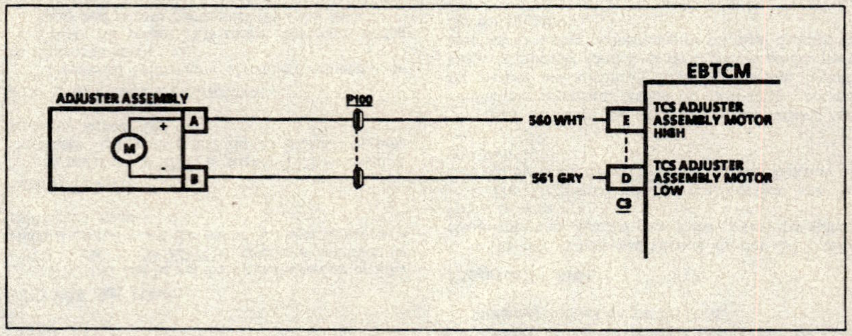

DTC 146
ADJUSTER ASSEMBLY WILL NOT MOVE

Circuit Description:
This DTC is designed to detect a stalled or frozen TCS adjuster assembly motor. During initialization, the adjuster assembly motor is commanded through a test sequence to determine if it can be moved for proper TCS operation. If the adjuster assembly motor cannot be moved, improper current feedback will be detected by the EBTCM, and a DTC will set.

DTC Will Set When:
DTC 146 can be set during and after initialization of the adjuster assembly. If the EBTCM detects a condition in which it cannot move the adjuster assembly motor, a malfunction exists. DTC 146 will not set during a TCS event.

Action Taken:
A malfunction DTC is stored, TCS modulator motors are commanded "HOME", TCS is disabled and the amber TCS warning lamp is turned "ON."

DTC Chart Test Description:
Number(s) below refer to circled number(s) on the diagnostic chart.
1. Checks to see if DTC's 154, 164,165 or 166 are set.
2. Checks for any mechanical binding of the adjuster assembly, attached cables, and components to which the cables are attached, which could cause excessive load on the adjuster assembly.
3. Checks for initialization with throttle position between 10% to 60%.

Diagnostic Aids:
It may be helpful to disconnect all cables from the adjuster assembly and move them manually throughout their entire operating range. Also check the components to which the cables are attached for binding.

DIAGNOSTIC CHART DTC 146 — 1995 CAMARO AND FIREBIRD WITH TCS, CONTINUED

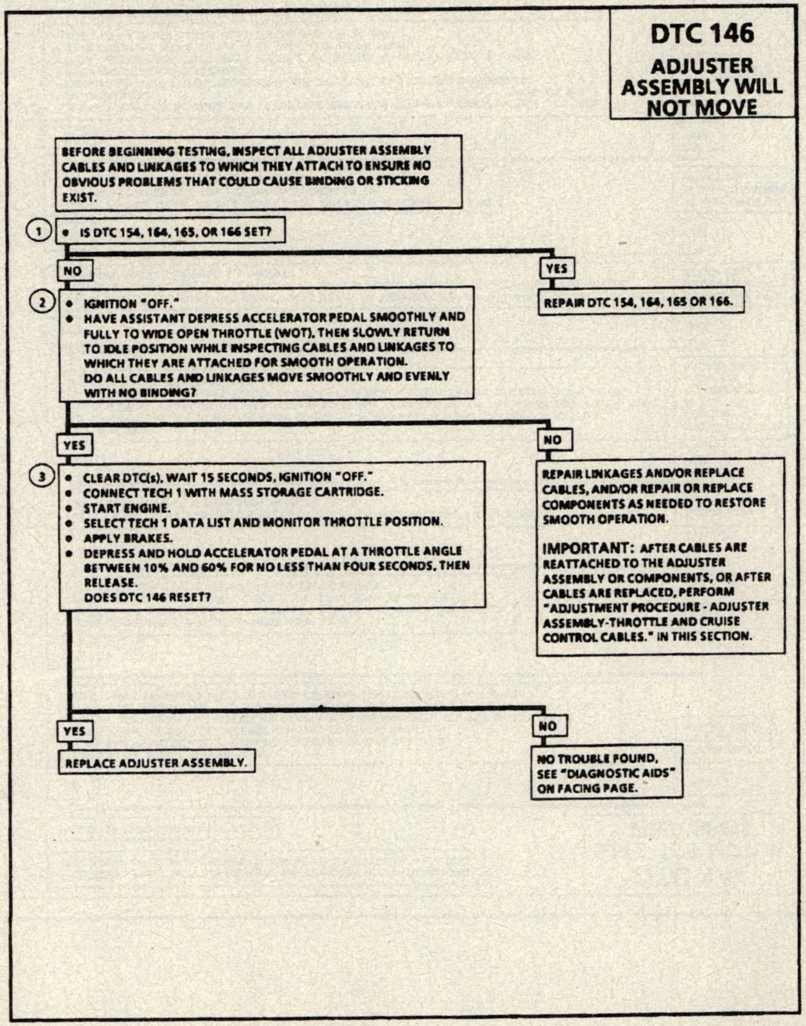

DIAGNOSTIC CHART DTC 147 — 1995 CAMARO AND FIREBIRD WITH TCS

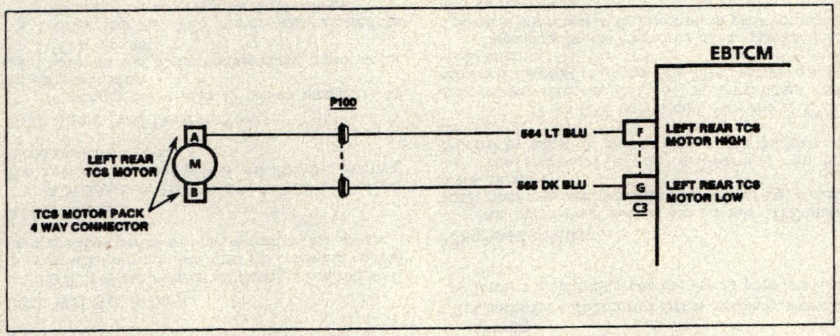

DTC 147
LEFT TCS MOTOR FREE SPINS

Circuit Description:

This DTC is designed to detect a stripped nut or gear assembly during initialization. During the homing sequence, the piston should reach the bottom of the bore resulting in a stalled TCS motor. If this does not occur, the TCS motor must be spinning with little or no resistance, indicating a nut/screw or gear malfunction.

DTC Will Set When:

DTC 147 can only be set during initialization. If the feedback current is less than the command current for a specified period of time, a malfunction exists.

Action Taken:

A malfunction DTC is stored, TCS is disabled. The amber TCS warning and red BRAKE warning lamps are turned "ON."

DTC Chart Test Description:

Number(s) below refer to circled number(s) on the diagnostic chart.

1. Checks to see if the corresponding open motor DTC is also set.
2. Verifies that the TCS motor was released as commanded by observing feedback current.
3. Verifies that the TCS motor can be applied as commanded by observing feedback current.
4. Verifies that the TCS motor can be released by observing wheel movement.
5. Checks for an open in the TCS motor "HIGH" circuitry.
6. Checks for an open in the TCS motor "LOW" circuitry.
7. Checks for a TCS motor that is internally shorted.
8. Ensures malfunction was not due to poor terminal contact.

Diagnostic Aids:

An intermittent malfunction in this DTC may result from a mechanical part of the system that sticks, binds, or slips.

The frequency of the malfunction can be checked by using the enhanced diagnostic function of the Tech 1.

If the DTC only occurs once and DTC 56 also occurs, refer to DTC 56. If intermittent and enhanced diagnostics shows this DTC occurs during ABS, refer to DTC 56.

Depending on the frequency of the malfunction, a physical inspection of the mechanical parts suspected may be necessary.

DIAGNOSTIC CHART DTC 147 — 1995 CAMARO AND FIREBIRD WITH TCS, CONTINUED

DTC 147 LEFT TCS MOTOR FREE SPINS

NOTICE: J 39200 TEST LEADS MUST BE "ZEROED" PRIOR TO MAKING ANY RESISTANCE MEASUREMENTS. REFER TO J 39200 USER'S MANUAL.

1. IS DTC 156 ALSO CURRENTLY SET?
   - YES: PROCEED TO DTC 156 CHART FIRST.
   - NO: go to step 2.
2. • IGNITION TO "RUN," ENGINE "OFF."
   • PUMP BRAKE PEDAL UNTIL FIRM TO DEPLETE VACUUM RESERVE.
   • USING TECH 1 MISC. TEST, SELECT MANUAL CONTROL FUNCTION.
   • SELECT LEFT REAR TCS MOTOR RELEASE FUNCTION AND RELEASE THE MOTOR WHILE CAREFULLY OBSERVING THE COMMAND CURRENT AND FEEDBACK CURRENT ON THE TECH 1. WERE BOTH COMMAND AND FEEDBACK CURRENT APPROXIMATELY 10 AMPS?
   - YES: go to step 3.
   - NO: GO TO STEP 5.
3. • RAISE DRIVE WHEELS AND SUPPORT VEHICLE.
   • SELECT LEFT REAR TCS MOTOR, APPLY FUNCTION AND APPLY THE MOTOR. DOES THE LEFT REAR WHEEL LOCK, AND WERE BOTH COMMAND AND FEEDBACK CURRENTS APPROXIMATELY 8 AMPS?
   - YES: go to step 4.
   - NO: GO TO STEP 5.
4. • SELECT LEFT REAR TCS MOTOR RELEASE FUNCTION AND RELEASE THE MOTOR. CAN THE LEFT REAR WHEEL SPIN FREELY?
   - YES: MALFUNCTION IS INTERMITTENT. REFER TO "DIAGNOSTIC AIDS" ON FACING PAGE.
   - NO: go to step 5.
5. • IGNITION "OFF."
   • DISCONNECT 4-WAY TCS MOTOR PACK CONNECTOR AND BLACK 8-WAY EBTCM CONNECTOR.
   • USING J 39200, MEASURE THE RESISTANCE BETWEEN TERMINAL "A" OF THE 4-WAY TCS MOTOR PACK HARNESS CONNECTOR AND TERMINAL "F" OF THE BLACK 8-WAY EBTCM HARNESS CONNECTOR. IS RESISTANCE LESS THAN 1.5 OHMS?
   - YES: go to step 6.
   - NO: REPAIR OPEN OR HIGH RESISTANCE IN CKT 564.
6. • USING J 39200, MEASURE THE RESISTANCE BETWEEN TERMINAL "B" OF THE 4-WAY TCS MOTOR PACK HARNESS CONNECTOR AND TERMINAL "G" OF THE BLACK 8-WAY EBTCM HARNESS CONNECTOR. IS RESISTANCE LESS THAN 1.5 OHMS?
   - YES: go to step 7.
   - NO: REPAIR OPEN OR HIGH RESISTANCE IN CKT 565.
7. • USING J 39200, MEASURE THE RESISTANCE BETWEEN TERMINAL "A" AND TERMINAL "B" OF THE TCS MOTOR PACK CONNECTOR. IS RESISTANCE LESS THAN 1.5 OHMS?
   - YES: go to step 8.
   - NO: REPLACE TCS HYDRAULIC MODULATOR/MOTOR PACK ASSEMBLY.
8. • INSPECT ALL CONNECTORS AND TERMINALS FOR POOR TERMINAL CONTACT AND EVIDENCE OF CORROSION. REPLACE ALL TERMINALS THAT EXHIBIT SIGNS OF POOR TERMINAL CONTACT OR CORROSION.
   • RECONNECT ALL CONNECTORS.
   • START ENGINE AND TEST DRIVE VEHICLE, ACHIEVING AT LEAST 16 km/h (10 MPH).
   • REPEAT ABOVE DRIVE CYCLE SEQUENCE TWO MORE TIMES. DOES DTC 147 SET IN THE LAST THREE DRIVE CYCLES?
   - YES: REPLACE EBTCM.
   - NO: MALFUNCTION WAS MOST LIKELY CAUSED BY A POOR CONNECTION. REFER TO "DIAGNOSTIC AIDS" ON FACING PAGE.

AFTER DIAGNOSIS IS COMPLETE, CLEAR DTCs AND TEST DRIVE VEHICLE FOR THREE (3) DRIVE CYCLES TO VERIFY THAT DTC DOES NOT RESET. A DRIVE CYCLE CONSISTS OF STARTING THE VEHICLE, DRIVING OVER 16 km/h (10 MPH), AND THEN IGNITION "OFF".

DIAGNOSTIC CHART DTC 148 — 1995 CAMARO AND FIREBIRD WITH TCS

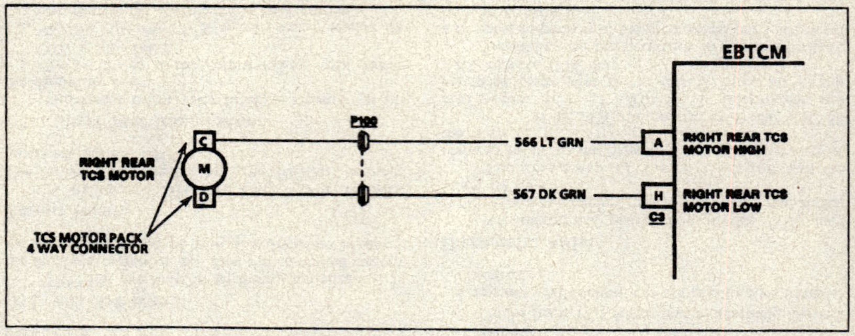

DTC 148
RIGHT TCS MOTOR FREE SPINS

Circuit Description:

This DTC is designed to detect a stripped nut or gear assembly during initialization. During the homing sequence, the piston should reach the bottom of the bore resulting in a stalled TCS motor. If this does not occur, the TCS motor must be spinning with little or no resistance, indicating a nut/screw or gear malfunction.

DTC Will Set When:

DTC 148 can only be set during initialization. If the feedback current is less than the command current for a specified period of time, a malfunction exists.

Action Taken:

A malfunction DTC is stored, TCS is disabled. The amber TCS warning and red BRAKE warning lamps are turned "ON."

DTC Chart Test Description:

Number(s) below refer to circled number(s) on the diagnostic chart.

1. Checks to see if the corresponding open motor DTC is also set.
2. Verifies that the TCS motor was released as commanded by observing feedback current.
3. Verifies that the TCS motor can be applied as commanded by observing feedback current.
4. Verifies that the TCS motor can be released by observing wheel movement.
5. Checks for an open in the TCS motor "HIGH" circuitry.
6. Checks for an open in the TCS motor "LOW" circuitry.
7. Checks for a TCS motor that is internally shorted.
8. Ensures malfunction was not due to poor terminal contact.

Diagnostic Aids:

An intermittent malfunction in this DTC may result from a mechanical part of the system that sticks, binds, or slips.

The frequency of the malfunction can be checked by using the enhanced diagnostic function of the Tech 1.

If the DTC only occurs once and DTC 61 also occurs, refer to DTC 61. If intermittent and enhanced diagnostics shows this DTC occurs during ABS, refer to DTC 61.

Depending on the frequency of the malfunction, a physical inspection of the mechanical parts suspected may be necessary.

DIAGNOSTIC CHART DTC 148 — 1995 CAMARO AND FIREBIRD WITH TCS, CONTINUED

DTC 148
RIGHT TCS MOTOR FREE SPINS

NOTICE: J 39200 TEST LEADS MUST BE "ZEROED" PRIOR TO MAKING ANY RESISTANCE MEASUREMENTS. REFER TO J 39200 USER'S MANUAL.

(1) IS DTC 161 ALSO CURRENTLY SET?
- YES: PROCEED TO DTC 161 CHART FIRST.
- NO: go to 2.

(2)
- IGNITION TO "RUN," ENGINE "OFF."
- PUMP BRAKE PEDAL UNTIL FIRM TO DEPLETE VACUUM RESERVE.
- USING TECH 1 MISC. TEST, SELECT MANUAL CONTROL FUNCTION.
- SELECT RIGHT REAR TCS MOTOR RELEASE FUNCTION AND RELEASE THE MOTOR WHILE CAREFULLY OBSERVING THE COMMAND CURRENT AND FEEDBACK CURRENT ON THE TECH 1. WERE BOTH COMMAND AND FEEDBACK CURRENT APPROXIMATELY 10 AMPS?
- YES: go to 3.

(3)
- RAISE DRIVE WHEELS AND SUPPORT VEHICLE.
- SELECT RIGHT REAR TCS MOTOR, APPLY FUNCTION AND APPLY THE MOTOR. DOES THE RIGHT REAR WHEEL LOCK, AND WERE BOTH COMMAND AND FEEDBACK CURRENTS APPROXIMATELY 8 AMPS?
- NO: GO TO STEP 5.
- YES: go to 4.

(4)
- SELECT RIGHT REAR TCS MOTOR RELEASE FUNCTION AND RELEASE THE MOTOR. CAN THE RIGHT REAR WHEEL SPIN FREELY?
- YES: MALFUNCTION IS INTERMITTENT. REFER TO "DIAGNOSTIC AIDS" ON FACING PAGE.
- NO: go to 5.

(5)
- IGNITION "OFF."
- DISCONNECT 4-WAY TCS MOTOR PACK CONNECTOR AND BLACK 8-WAY EBTCM CONNECTOR.
- USING J 39200, MEASURE THE RESISTANCE BETWEEN TERMINAL "C" OF THE 4-WAY TCS MOTOR PACK HARNESS CONNECTOR AND TERMINAL "A" OF THE BLACK 8-WAY EBTCM HARNESS CONNECTOR. IS RESISTANCE LESS THAN 1.5 OHMS?
- NO: REPAIR OPEN OR HIGH RESISTANCE IN CKT 566.
- YES: go to 6.

(6)
- USING J 39200, MEASURE THE RESISTANCE BETWEEN TERMINAL "D" OF THE 4-WAY TCS MOTOR PACK HARNESS CONNECTOR AND TERMINAL "H" OF THE BLACK 8-WAY EBTCM HARNESS CONNECTOR. IS RESISTANCE LESS THAN 1.5 OHMS?
- NO: REPAIR OPEN OR HIGH RESISTANCE IN CKT 567.
- YES: go to 7.

(7)
- USING J 39200, MEASURE THE RESISTANCE BETWEEN TERMINAL "C" AND TERMINAL "D" OF THE TCS MOTOR PACK CONNECTOR. IS RESISTANCE LESS THAN 1.5 OHMS?
- NO: REPLACE TCS HYDRAULIC MODULATOR/ MOTOR PACK ASSEMBLY.
- YES: go to 8.

(8)
- INSPECT ALL CONNECTORS AND TERMINALS FOR POOR TERMINAL CONTACT AND EVIDENCE OF CORROSION. REPLACE ALL TERMINALS THAT EXHIBIT SIGNS OF POOR TERMINAL CONTACT OR CORROSION.
- RECONNECT ALL CONNECTORS.
- START ENGINE AND TEST DRIVE VEHICLE, ACHIEVING AT LEAST 16 km/h (10 MPH).
- REPEAT ABOVE DRIVE CYCLE SEQUENCE TWO MORE TIMES. DOES DTC 148 SET IN THE LAST THREE DRIVE CYCLES?
- YES: REPLACE EBTCM.
- YES: MALFUNCTION WAS MOST LIKELY CAUSED BY A POOR CONNECTION. REFER TO "DIAGNOSTIC AIDS" ON FACING PAGE.

AFTER DIAGNOSIS IS COMPLETE, CLEAR DTCs AND TEST DRIVE VEHICLE FOR THREE (3) DRIVE CYCLES TO VERIFY THAT DTC DOES NOT RESET. A DRIVE CYCLE CONSISTS OF STARTING THE VEHICLE, DRIVING OVER 16 km/h (10 MPH), AND THEN IGNITION "OFF".

DIAGNOSTIC CHART DTC 154 — 1995 CAMARO AND FIREBIRD WITH TCS

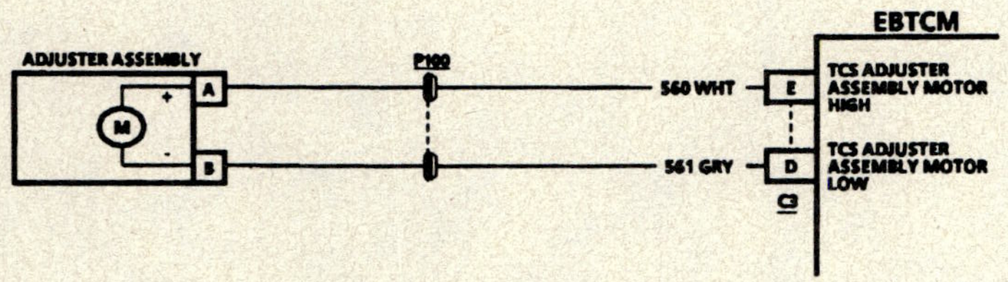

DTC 154
ADJUSTER ASSEMBLY MALFUNCTION

Circuit Description:

This DTC is designed to detect malfunctions of the adjuster assembly that can not be detected by the adjuster assembly electrical tests or the initialization movement test, it is necessary to monitor the performance of the adjuster assembly during a TCS event. The TPS monitors the actual position of the adjuster assembly while the EBTCM commands a target position. A malfunction exists is the difference between the two positions becomes larger then a calibration value indicative of a non-responding adjuster assembly.

DTC Will Set When:

If (during an TCS event) the EBTCM delivers current greater than 16 amps to the adjuster assembly for more than three seconds without the adjuster assembly reaching it's desired position.

DTC Chart Test Description:

Number(s) below refer to circled number(s) on the diagnostic chart.

1. Checks to see if DTC 146 set, since DTC 146 conditions might affect DTC 154.
2. Checks for any mechanical binding of the adjuster assembly, attached cables, and components to which the cables are attached, which could cause excessive load on the adjuster assembly.
3. Since DTC 154 only sets during an TCS, this check raises the vehicle on a hoist such that the rear wheels will spin while the fronts do not, inducing the system into attempting TCS control. This chart ensures the throttle angle must be greater than 10% and less than 60%, and the system must attempt TCS control for greater than 3 seconds to ensure that the DTC setting throttle and timing parameters are met when checking for a DTC 154.

Diagnostic Aids:

It may be helpful to disconnect all cables from the adjuster assembly and move them manually throughout their entire operating range. Also check the components to which the cables are attached for binding.

DIAGNOSTIC CHART DTC 154 — 1995 CAMARO AND FIREBIRD WITH TCS, CONTINUED

DTC 154
ADJUSTER ASSEMBLY MALFUNCTION

BEFORE BEGINNING TESTING, INSPECT ALL ADJUSTER ASSEMBLY CABLES AND LINKAGES TO WHICH THEY ATTACH TO ENSURE NO OBVIOUS PROBLEMS THAT COULD CAUSE BINDING OR STICKING EXIST.

(1)
- IS DTC 146 SET?

YES → REPAIR DTC 146.

NO →

(2)
- IGNITION "OFF."
- HAVE ASSISTANT DEPRESS ACCELERATOR PEDAL SMOOTHLY AND FULLY TO WIDE OPEN THROTTLE (WOT), THEN SLOWLY RETURN TO IDLE POSITION WHILE INSPECTING CABLES AND LINKAGES TO WHICH THEY ARE ATTACHED FOR SMOOTH OPERATION.

DO ALL CABLES AND LINKAGES MOVE SMOOTHLY AND EVENLY WITH NO BINDING?

NO → REPAIR LINKAGES AND/OR REPLACE CABLES, AND/OR REPAIR OR REPLACE COMPONENTS AS NEEDED TO RESTORE SMOOTH OPERATION.

IMPORTANT: AFTER CABLES ARE REATTACHED TO THE ADJUSTER ASSEMBLY OR COMPONENTS, OR AFTER CABLES ARE REPLACED, PERFORM "ADJUSTMENT PROCEDURE - ADJUSTER ASSEMBLY-THROTTLE AND CRUISE CONTROL CABLES." IN THIS SECTION.

IF A NEW TP SENSOR OR THROTTLE BODY IS INSTALLED, BE SURE TO PERFORM "TP SENSOR LEARN PROCEDURE" IN THIS SECTION.

YES →

(3)
- CLEAR DTC(s), WAIT 15 SECONDS, IGNITION "OFF."
- CONNECT TECH 1 WITH MASS STORAGE CARTRIDGE.
- START ENGINE.
- SELECT TECH 1 DATA LIST AND MONITOR THROTTLE POSITION.
- APPLY BRAKES.
- PUT VEHICLE IN GEAR.
- RELEASE BRAKES (AND CLUTCH, IF MANUAL TRANS).
- DEPRESS AND HOLD ACCELERATOR PEDAL AT A THROTTLE ANGLE BETWEEN 15% AND 45% FOR NO LESS THAN FOUR SECONDS, THEN RELEASE.

DOES DTC 154 RESET?

YES → REPLACE ADJUSTER ASSEMBLY.

NO → NO TROUBLE FOUND. SEE "DIAGNOSTIC AIDS" ON FACING PAGE.

DIAGNOSTIC CHART DTC 155 — 1995 CAMARO AND FIREBIRD WITH TCS

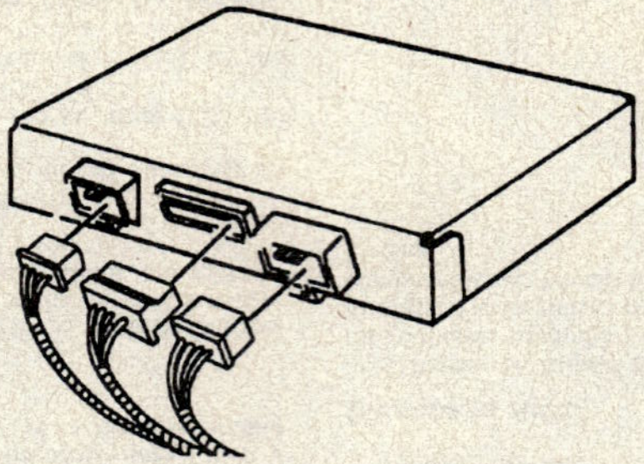

DTC 155
EBTCM MALFUNCTION (TCS)

Circuit Description:
This DTC identifies an internal malfunction detected by the EBTCM TCS motor driver interface custom integrated circuit.

Action Taken:
A malfunction DTC is stored, TCS is disabled and the amber TCS warning lamp is turned "ON." The red BRAKE warning lamp is turned "ON" if the rear ABS motor is not in the home position.

DTC Chart Test Description:
Number(s) below refer to circled number(s) on the diagnostic chart.
1. Checks to see if the malfunction is still present.
2. Checks to see if the malfunction is intermittent.

DIAGNOSTIC CHART DTC 155 — 1995 CAMARO AND FIREBIRD WITH TCS, CONTINUED

DTC 155
EBTCM MALFUNCTION

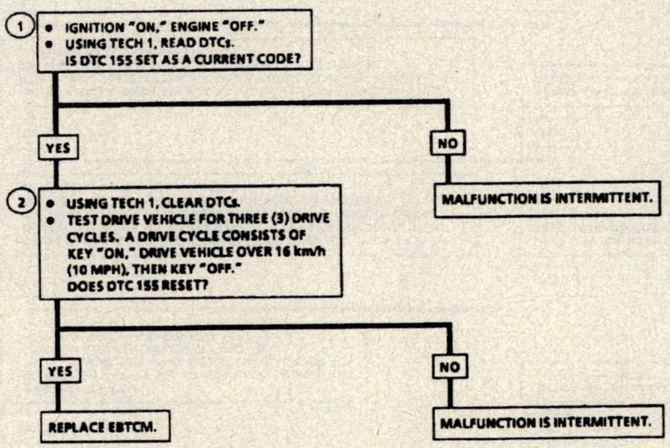

AFTER DIAGNOSIS IS COMPLETE, CLEAR DTCs AND TEST DRIVE VEHICLE FOR THREE (3) DRIVE CYCLES TO VERIFY THAT DTC DOES NOT RESET. A DRIVE CYCLE CONSISTS OF STARTING THE VEHICLE, DRIVING OVER 16 km/h (10 MPH), AND THEN IGNITION OFF.

DIAGNOSTIC CHART DTC 156 — 1995 CAMARO AND FIREBIRD WITH TCS

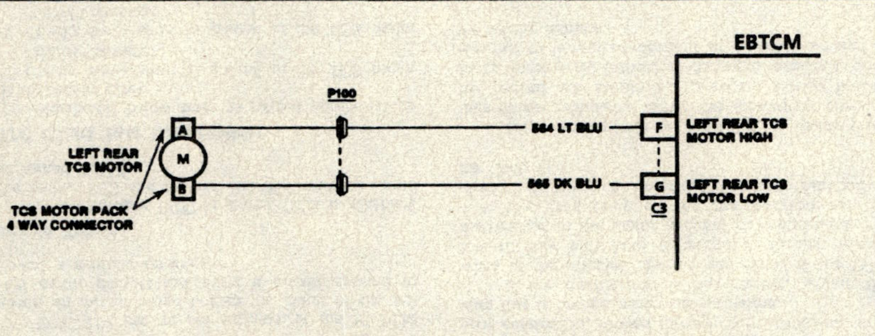

DTC 156
LEFT TCS MOTOR CIRCUIT OPEN

Circuit Description:

This DTC identifies a TCS motor that cannot be energized due to an open in its circuitry. This malfunction will not allow proper TCS operation.

DTC Will Set When:

DTC 156 can be set only when the TCS motor is commanded "OFF." If the EBTCM detects an out of range voltage on either of the left TCS motor circuits indicating an open circuit, a malfunction exists.

Action Taken:

An open TCS motor will not activate when requested. A malfunction DTC is stored, TCS is disabled and the amber TCS warning lamp is turned "ON."

DTC Chart Test Description:

Number(s) below refer to circled number(s) on the diagnostic chart.

1. Checks for proper resistance of the TCS motor.
2. Checks for an open in the TCS motor "HIGH" circuitry.
3. Checks for an open in the TCS motor "LOW" circuitry.
4. Ensures malfunction was not due to poor terminal contact.

Diagnostic Aids:

An intermittent malfunction may be caused by a poor connection, rubbed through wire insulation, or a wire that is broken inside the insulation.

If the malfunction is not current, wiggle the wires of the affected channel and check if the DTC resets. This will help to pinpoint an intermittent malfunction in the motor circuitry or connections.

The frequency of the malfunction can be checked by using the enhanced diagnostic function of the Tech 1,

Any circuitry that is suspected of causing the intermittent complaint should be thoroughly checked for backed out terminals, improper mating, broken locks, improperly formed or damaged terminals, poor terminal to wiring connections or physical damage to the wiring harness.

DIAGNOSTIC CHART DTC 156 — 1995 CAMARO AND FIREBIRD WITH TCS, CONTINUED

DTC 156
LEFT TCS MOTOR CIRCUIT OPEN

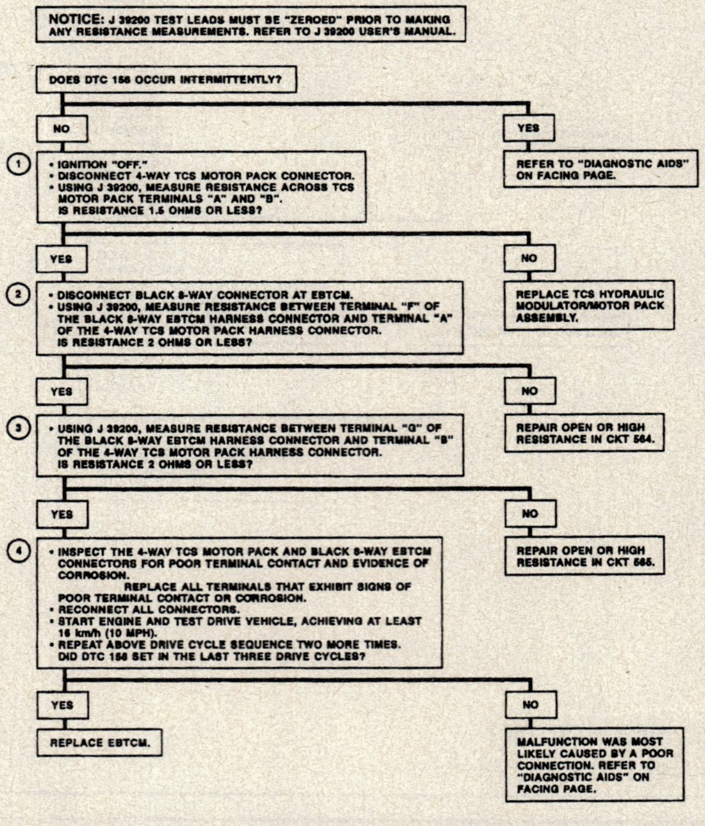

AFTER DIAGNOSIS IS COMPLETE, CLEAR DTCs AND TEST DRIVE VEHICLE FOR THREE (3) DRIVE CYCLES TO VERIFY THAT DTC DOES NOT RESET. A DRIVE CYCLE CONSISTS OF STARTING THE VEHICLE, DRIVING OVER 16 km/h (10 MPH), AND THEN TURNING KEY "OFF."

DIAGNOSTIC CHART DTC 157 — 1995 CAMARO AND FIREBIRD WITH TCS

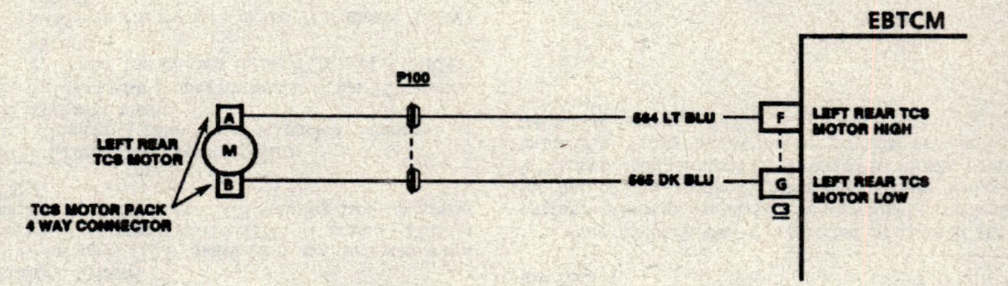

DTC 157
LEFT TCS MOTOR CIRCUIT SHORTED TO GND

Circuit Description:

This DTC identifies an TCS motor circuit that is shorted to ground. This malfunction will not allow the TCS motor to be controlled at the commanded current rate or will cause the driver circuit to allow current directly to ground.

DTC Will Set When:

DTC 157 can be set anytime. If the EBTCM detects an out of range voltage on either of the left TCS motor circuits indicating a circuit shorted to ground, a malfunction exists.

Action Taken:

A malfunction DTC is stored, TCS is disabled. The amber TCS and the red BRAKE warning lamps are turned "ON."

DTC Chart Test Description:

Number(s) below refer to circled number(s) on the diagnostic chart.

1. Checks for a short to ground in the TCS motor "HIGH" circuitry.
2. Checks for a short to ground in the TCS motor "LOW" circuitry.
3. Checks for an TCS motor that is internally shorted to ground.
4. Ensures malfunction was not due to physical damage of the circuitry.

Diagnostic Aids:

An intermittent malfunction may be caused by a poor connection, rubbed through wire insulation, or a wire that is broken inside the insulation.

If the malfunction is not current, wiggle the wires of the affected channel and check if the DTC resets. This will help to pinpoint an intermittent malfunction in the motor circuitry or connections.

The frequency of the malfunction can be checked by using the enhanced diagnostic function of the Tech 1.

Any circuitry that is suspected of causing the intermittent complaint should be thoroughly checked for backed out terminals, improper mating, broken locks, improperly formed or damaged terminals, poor terminal to wiring connections or physical damage to the wiring harness.

DIAGNOSTIC CHART DTC 157 — 1995 CAMARO AND FIREBIRD WITH TCS, CONTINUED

DTC 157
LEFT TCS MOTOR CIRCUIT SHORTED TO GND

NOTICE: J 39200 TEST LEADS MUST BE "ZEROED" PRIOR TO MAKING ANY RESISTANCE MEASUREMENTS. REFER TO J 39200 USER'S MANUAL.

DOES DTC 157 OCCUR INTERMITTENTLY?

- YES: REFER TO "DIAGNOSTIC AIDS" ON FACING PAGE.
- NO:

(1)
- IGNITION "OFF."
- DISCONNECT THE 4-WAY TCS MOTOR PACK CONNECTOR AND THE BLACK 8-WAY EBTCM CONNECTOR.
- USING J 39200, MEASURE THE RESISTANCE BETWEEN GROUND AND TERMINAL "A" OF THE 4-WAY TCS MOTOR PACK HARNESS CONNECTOR.

DOES J 39200 DISPLAY RESISTANCE AS "OL" (INFINITE)?

- NO: REPAIR SHORT TO GROUND IN CKT 564.
- YES:

(2)
- USING J 39200, MEASURE THE RESISTANCE BETWEEN GROUND AND TERMINAL "B" OF THE 4-WAY TCS MOTOR PACK HARNESS CONNECTOR.

DOES J 39200 DISPLAY RESISTANCE AS "OL" (INFINITE)?

- NO: REPAIR SHORT TO GROUND IN CKT 565.
- YES:

(3)
- USING J 39200, MEASURE RESISTANCE BETWEEN GROUND AND TERMINAL "A" OF THE TCS MOTOR PACK CONNECTOR.

DOES J 39200 DISPLAY RESISTANCE AS "OL" (INFINITE)?

- NO: REPLACE TCS HYDRAULIC MODULATOR/MOTOR PACK ASSEMBLY.
- YES:

(4)
- INSPECT CKTs 564 AND 565 AND RELATED CONNECTORS FOR DAMAGE WHICH MAY RESULT IN A SHORT TO GROUND WITH ALL CONNECTORS CONNECTED. REPAIR ALL DAMAGE IF EVIDENT.
- RECONNECT ALL CONNECTORS.
- START ENGINE AND TEST DRIVE VEHICLE, ACHIEVING AT LEAST 16 km/h (10 MPH).
- REPEAT ABOVE DRIVE CYCLE SEQUENCE TWO MORE TIMES.

DID DTC 157 SET IN THE LAST THREE DRIVE CYCLES?

- YES: REPLACE EBTCM.
- NO: MALFUNCTION WAS MOST LIKELY CAUSED BY A POOR CONNECTION. REFER TO "DIAGNOSTIC AIDS" ON FACING PAGE.

AFTER ALL DIAGNOSIS IS COMPLETE, CLEAR DTCs AND TEST DRIVE VEHICLE FOR THREE (3) DRIVE CYCLES TO VERIFY THAT DTC DOES NOT RESET. A DRIVE CYCLE CONSISTS OF STARTING THE VEHICLE, DRIVING OVER 16 km/h (10 MPH), AND THEN TURNING KEY "OFF."

DIAGNOSTIC CHART DTC 158 — 1995 CAMARO AND FIREBIRD WITH TCS

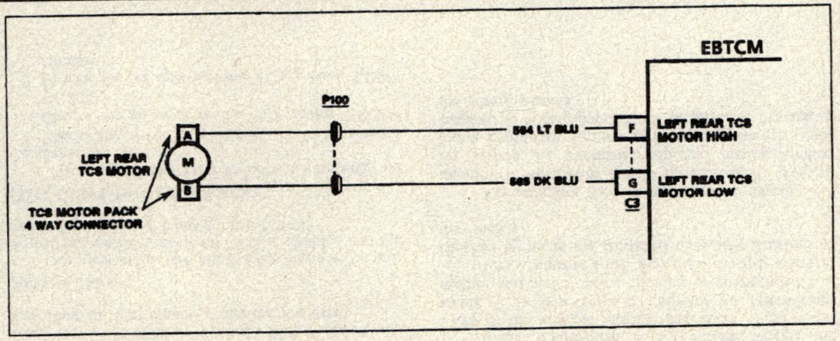

DTC 158
LEFT TCS MOTOR CIRCUIT SHORTED TO BATT

Circuit Description:

This DTC identifies a TCS motor circuit that is shorted to battery or a TCS motor that has low or no resistance. This malfunction will not allow the TCS motor to be controlled at the commanded current rate or will cause the TCS motor to turn in the opposite direction, or not at all.

DTC Will Set When:

DTC 158 can be set only when the TCS motor is commanded "OFF." If the EBTCM detects an out of range voltage on either of the left TCS motor circuits indicating a circuit shorted to battery or a TCS motor shorted, a malfunction exists.

Action Taken:

A malfunction DTC is stored, TCS is disabled. The amber TCS and the red BRAKE warning lamps are turned "ON."

DTC Chart Test Description:

Number(s) below refer to circled number(s) on the diagnostic chart.

1. Checks for a short to voltage in the TCS motor "HIGH" circuitry.
2. Checks for a short to voltage in the TCS motor "LOW" circuitry.
3. Checks for a short between the "HIGH" and "LOW" motor circuits.
4. Checks for a TCS motor that is internally shorted.
5. Ensures malfunction was not due to physical damage of the circuitry.

Diagnostic Aids:

An intermittent malfunction may be caused by a poor connection, rubbed through wire insulation, or a wire that is broken inside the insulation.

If the malfunction is not current, wiggle the wires of the affected channel and check if the DTC resets. This will help to pinpoint an intermittent malfunction in the motor circuitry or connections.

The frequency of the malfunction can be checked by using the enhanced diagnostic function of the Tech 1

Any circuitry that is suspected of causing the intermittent complaint should be thoroughly checked for backed out terminals, improper mating, broken locks, improperly formed or damaged terminals, poor terminal to wiring connections or physical damage to the wiring harness.

DIAGNOSTIC CHART DTC 158 — 1995 CAMARO AND FIREBIRD WITH TCS, CONTINUED

DTC 158
LEFT TCS MOTOR CIRCUIT SHORTED TO BATT

NOTICE: J 39200 TEST LEADS MUST BE "ZEROED" PRIOR TO MAKING ANY RESISTANCE MEASUREMENTS. REFER TO J 39200 USER'S MANUAL.

DOES DTC 158 OCCUR INTERMITTENTLY?

- YES → REFER TO "DIAGNOSTIC AIDS" ON FACING PAGE.
- NO →

(1)
- IGNITION "OFF."
- DISCONNECT THE 4-WAY TCS MOTOR PACK CONNECTOR AND THE BLACK 8-WAY EBTCM CONNECTOR.
- IGNITION TO "RUN."
- USING J 39200, MEASURE THE VOLTAGE BETWEEN GROUND AND TERMINAL "A" OF THE 4-WAY TCS MOTOR PACK HARNESS CONNECTOR. IS VOLTAGE 1 VOLT OR LESS?

- NO → REPAIR SHORT TO VOLTAGE IN CKT 564.
- YES →

(2)
- USING J 39200, MEASURE THE VOLTAGE BETWEEN GROUND AND TERMINAL "B" OF THE 4-WAY TCS MOTOR PACK HARNESS CONNECTOR. IS VOLTAGE 1 VOLT OR LESS?

- NO → REPAIR SHORT TO VOLTAGE IN CKT 565.
- YES →

(3)
- IGNITION "OFF."
- USING J 39200, MEASURE THE RESISTANCE BETWEEN TERMINAL "A" AND TERMINAL "B" OF THE 4-WAY TCS MOTOR PACK HARNESS CONNECTOR. DOES J 39200 DISPLAY RESISTANCE AS "OL" (INFINITE)?

- NO → REPAIR SHORT BETWEEN CKTs 564 AND 565.
- YES →

(4)
- USING J 39200, MEASURE THE RESISTANCE BETWEEN TERMINAL "A" AND TERMINAL "B" OF THE TCS MOTOR PACK CONNECTOR. IS RESISTANCE GREATER THAN 0.4 OHMS?

- NO → REPLACE TCS HYDRAULIC MODULATOR/MOTOR PACK ASSEMBLY.
- YES →

(5)
- INSPECT CKTs 564 AND 565 AND RELATED CONNECTORS FOR DAMAGE WHICH MAY RESULT IN A SHORT TO VOLTAGE WITH ALL CONNECTORS CONNECTED. REPAIR ALL DAMAGE IF EVIDENT.
- RECONNECT ALL CONNECTORS.
- START ENGINE AND TEST DRIVE VEHICLE, ACHIEVING AT LEAST 16 km/h (10 MPH).
- REPEAT ABOVE DRIVE CYCLE SEQUENCE TWO MORE TIMES. DID DTC 158 SET IN THE LAST THREE DRIVE CYCLES?

- YES → REPLACE EBTCM.
- NO → MALFUNCTION WAS MOST LIKELY CAUSED BY A POOR CONNECTION. REFER TO "DIAGNOSTIC AIDS" ON FACING PAGE.

AFTER DIAGNOSIS IS COMPLETE, CLEAR DTCs AND TEST DRIVE VEHICLE FOR THREE (3) DRIVE CYCLES TO VERIFY THAT DTC DOES NOT RESET. A DRIVE CYCLE CONSISTS OF STARTING THE VEHICLE, DRIVING OVER 16 km/h (10 MPH), AND THEN TURNING KEY "OFF."

DIAGNOSTIC CHART DTC 161 — 1995 CAMARO AND FIREBIRD WITH TCS

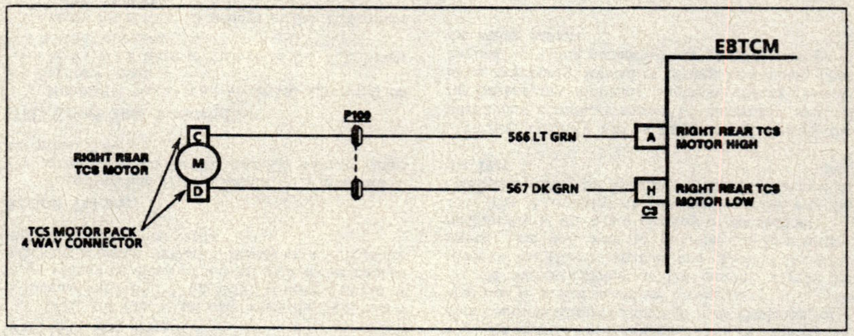

DTC 161
RIGHT TCS MOTOR CIRCUIT OPEN

Circuit Description:

This DTC identifies a TCS motor that cannot be energized due to an open in its circuitry. This malfunction will not allow proper TCS operation.

DTC Will Set When:

DTC 161 can be set only when the TCS motor is commanded "OFF." If the EBTCM detects an out of range voltage on either of the right TCS motor circuits indicating an open circuit, a malfunction exists.

Action Taken:

An open motor will not activate when requested. A malfunction DTC is stored, TCS is disabled and the amber TCS warning lamp is turned "ON."

DTC Chart Test Description:

Number(s) below refer to circled number(s) on the diagnostic chart.

1. Checks for proper resistance of the TCS motor.
2. Checks for an open in the TCS motor "HIGH" circuitry.
3. Checks for an open in the TCS motor "LOW" circuitry.
4. Ensures malfunction was not due to physical damage of the circuitry.

Diagnostic Aids:

An intermittent malfunction may be caused by a poor connection, rubbed through wire insulation, or a wire that is broken inside the insulation.

If the malfunction is not current, wiggle the wires of the affected channel and check if the DTC resets. This will help to pinpoint an intermittent malfunction in the motor circuitry or connections.

The frequency of the malfunction can be checked by using the enhanced diagnostic function of the Tech 1.

Any circuitry that is suspected of causing the intermittent complaint should be thoroughly checked for backed out terminals, improper mating, broken locks, improperly formed or damaged terminals, poor terminal to wiring connections or physical damage to the wiring harness.

DIAGNOSTIC CHART DTC 161 — 1995 CAMARO AND FIREBIRD WITH TCS, CONTINUED

DTC 161
RIGHT TCS MOTOR CIRCUIT OPEN

NOTICE: J 39200 TEST LEADS MUST BE "ZEROED" PRIOR TO MAKING ANY RESISTANCE MEASUREMENTS. REFER TO J 39200 USER'S MANUAL.

DOES DTC 161 OCCUR INTERMITTENTLY?

YES: REFER TO "DIAGNOSTIC AIDS" ON FACING PAGE.

NO:

(1)
- IGNITION "OFF."
- DISCONNECT 4-WAY TCS MOTOR PACK CONNECTOR.
- USING J 39200, MEASURE RESISTANCE ACROSS TCS MOTOR PACK TERMINALS "C" AND "D".

IS RESISTANCE 1.5 OHMS OR LESS?

NO: REPLACE TCS HYDRAULIC MODULATOR/MOTOR PACK ASSEMBLY.

YES:

(2)
- DISCONNECT BLACK 8-WAY CONNECTOR AT EBTCM.
- USING J 39200, MEASURE RESISTANCE BETWEEN TERMINAL "A" OF THE BLACK 8-WAY EBTCM HARNESS CONNECTOR AND TERMINAL "C" OF THE 4-WAY TCS MOTOR PACK HARNESS CONNECTOR.

IS RESISTANCE 2 OHMS OR LESS?

NO: REPAIR OPEN OR HIGH RESISTANCE IN CKT 566.

YES:

(3)
- USING J 39200, MEASURE RESISTANCE BETWEEN TERMINAL "H" OF THE BLACK 8-WAY EBTCM HARNESS CONNECTOR AND TERMINAL "D" OF THE 4-WAY TCS MOTOR PACK HARNESS CONNECTOR.

IS RESISTANCE 2 OHMS OR LESS?

NO: REPAIR OPEN OR HIGH RESISTANCE IN CKT 567.

YES:

(4)
- INSPECT THE 4-WAY TCS MOTOR PACK AND BLACK 8-WAY EBTCM CONNECTORS FOR POOR TERMINAL CONTACT AND EVIDENCE OF CORROSION.

REPLACE ALL TERMINALS THAT EXHIBIT SIGNS OF POOR TERMINAL CONTACT OR CORROSION.
- RECONNECT ALL CONNECTORS.
- START ENGINE AND TEST DRIVE VEHICLE, ACHIEVING AT LEAST 16 km/h (10 MPH).
- REPEAT ABOVE DRIVE CYCLE SEQUENCE TWO MORE TIMES.

DID DTC 161 SET IN THE LAST THREE DRIVE CYCLES?

YES: REPLACE EBTCM.

NO: MALFUNCTION WAS MOST LIKELY CAUSED BY A POOR CONNECTION. REFER TO "DIAGNOSTIC AIDS" ON FACING PAGE.

AFTER ALL DIAGNOSIS IS COMPLETE, CLEAR DTCs AND TEST DRIVE VEHICLE FOR THREE (3) DRIVE CYCLES TO VERIFY THAT DTC DOES NOT RESET. A DRIVE CYCLE CONSISTS OF STARTING THE VEHICLE, DRIVING OVER 16 km/h (10 MPH), AND THEN TURNING KEY "OFF."

DIAGNOSTIC CHART DTC 162 — 1995 CAMARO AND FIREBIRD WITH TCS

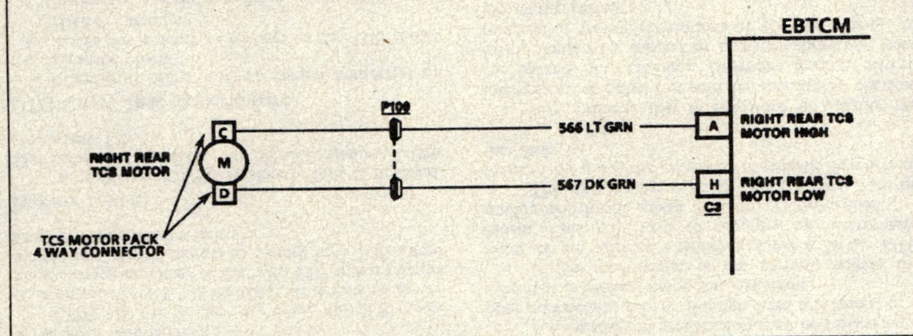

DTC 162
RIGHT TCS MOTOR CIRCUIT SHORTED TO GND

Circuit Description:

This DTC identifies a TCS motor circuit that is shorted to ground. This malfunction will not allow the TCS motor to be controlled at the commanded current rate or will cause the TCS motor driver circuit to allow current directly to ground.

DTC Will Set When:

DTC 162 can be set anytime. If the EBTCM detects an out of range voltage on either of the right TCS motor circuits indicating a circuit shorted to ground, a malfunction exists.

Action Taken:

A malfunction DTC is stored, TCS is disabled. The amber TCS and the red BRAKE warning lamps are turned "ON."

DTC Chart Test Description:

Number(s) below refer to circled number(s) on the diagnostic chart.

1. Checks for a short to ground in the TCS motor "HIGH" circuitry.
2. Checks for a short to ground in the TCS motor "LOW" circuitry.
3. Checks for a TCS motor that is internally shorted to ground.
4. Ensures malfunction was not due to physical damage of the circuitry.

Diagnostic Aids:

An intermittent malfunction may be caused by a poor connection, rubbed through wire insulation, or a wire that is broken inside the insulation.

If the malfunction is not current, wiggle the wires of the affected channel and check if the DTC resets. This will help to pinpoint an intermittent malfunction in the motor circuitry or connections.

The frequency of the malfunction can be checked by using the enhanced diagnostic function of the Tech 1.

Any circuitry that is suspected of causing the intermittent complaint should be thoroughly checked for backed out terminals, improper mating, broken locks, improperly formed or damaged terminals, poor terminal to wiring connections or physical damage to the wiring harness.

DIAGNOSTIC CHART DTC 162 — 1995 CAMARO AND FIREBIRD WITH TCS, CONTINUED

DTC 162
RIGHT TCS MOTOR CIRCUIT SHORTED TO GND

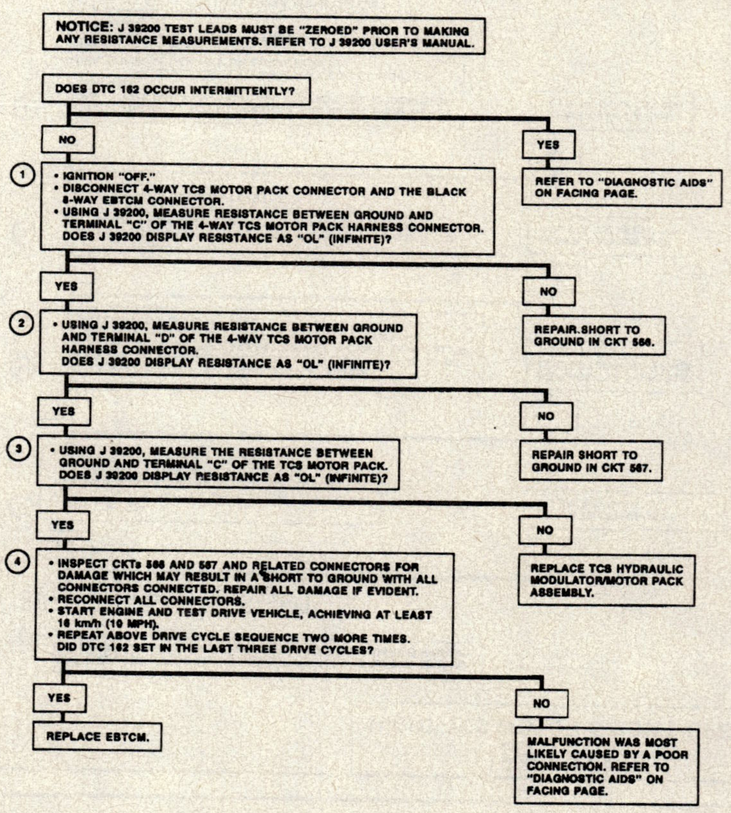

AFTER ALL DIAGNOSIS IS COMPLETE, CLEAR DTCs AND TEST DRIVE VEHICLE FOR THREE (3) DRIVE CYCLES TO VERIFY THAT DTC DOES NOT RESET. A DRIVE CYCLE CONSISTS OF STARTING THE VEHICLE, DRIVING OVER 16 km/h (10 MPH), AND THEN TURNING KEY "OFF."

DIAGNOSTIC CHART DTC 163 — 1995 CAMARO AND FIREBIRD WITH TCS

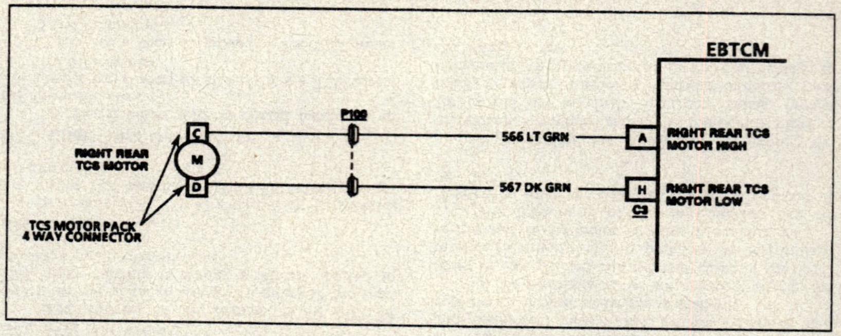

DTC 163
RIGHT TCS MOTOR CIRCUIT SHORTED TO BATT

Circuit Description:

This DTC identifies a TCS motor circuit that is shorted to battery or a TCS motor that has low or no resistance. This malfunction will not allow the TCS motor to be controlled at the commanded current rate or will cause the TCS motor to turn in the opposite direction or not at all.

DTC Will Set When:

DTC 163 can be set only when the TCS motor is commanded "OFF." If the EBTCM detects an out of range voltage on either of the right TCS motor circuits indicating a circuit shorted to battery or a TCS motor shorted, a malfunction exists.

Action Taken:

A malfunction DTC is stored, TCS is disabled. The amber TCS and the red BRAKE warning lamps are turned "ON."

DTC Chart Test Description:

Number(s) below refer to circled number(s) on the diagnostic chart.

1. Checks for a short to voltage in the TCS motor "HIGH" circuitry.
2. Checks for a short to voltage in the TCS motor "LOW" circuitry.
3. Checks for short circuit between TCS motor "HIGH" and motor "LOW".
4. Checks for proper resistance of the TCS motor.
5. Ensures malfunction was not due to physical damage of the circuitry.

Diagnostic Aids:

An intermittent malfunction may be caused by a poor connection, rubbed through wire insulation, or a wire that is broken inside the insulation.

If the malfunction is not current, wiggle the wires of the affected channel and check if the DTC resets. This will help to pinpoint an intermittent malfunction in the motor circuitry or connections.

The frequency of the malfunction can be checked by using the enhanced diagnostic function of the Tech 1.

Any circuitry that is suspected of causing the intermittent complaint should be thoroughly checked for backed out terminals, improper mating, broken locks, improperly formed or damaged terminals, poor terminal to wiring connections or physical damage to the wiring harness.

DIAGNOSTIC CHART DTC 163 — 1995 CAMARO AND FIREBIRD WITH TCS, CONTINUED

DTC 163
RIGHT TCS MOTOR CIRCUIT SHORTED TO BATT

NOTICE: J 39200 TEST LEADS MUST BE "ZEROED" PRIOR TO MAKING ANY RESISTANCE MEASUREMENTS. REFER TO J 39200 USER'S MANUAL.

DOES DTC 163 OCCUR INTERMITTENTLY?

- YES: REFER TO "DIAGNOSTIC AIDS" ON FACING PAGE.
- NO:

(1)
- IGNITION "OFF."
- DISCONNECT THE 4-WAY TCS MOTOR PACK CONNECTOR AND THE BLACK 8-WAY EBTCM CONNECTOR.
- IGNITION TO "RUN."
- USING J 39200, MEASURE THE VOLTAGE BETWEEN GROUND AND TERMINAL "C" OF THE 4-WAY TCS MOTOR PACK HARNESS CONNECTOR. IS VOLTAGE 1 VOLT OR LESS?

- NO: REPAIR SHORT TO VOLTAGE IN CKT 566.
- YES:

(2)
- USING J 39200, MEASURE THE VOLTAGE BETWEEN GROUND AND TERMINAL "D" OF THE 4-WAY TCS MOTOR PACK HARNESS CONNECTOR. IS VOLTAGE 1 VOLT OR LESS?

- NO: REPAIR SHORT TO VOLTAGE IN CKT 567.
- YES:

(3)
- IGNITION "OFF."
- USING J 39200, MEASURE THE RESISTANCE BETWEEN TERMINAL "C" AND TERMINAL "D" OF THE 4-WAY TCS MOTOR PACK HARNESS CONNECTOR. DOES J 39200 DISPLAY RESISTANCE AS "OL" (INFINITE)?

- NO: REPAIR SHORT BETWEEN CKTs 566 AND 567.
- YES:

(4)
- USING J 39200, MEASURE THE RESISTANCE BETWEEN TERMINAL "C" AND TERMINAL "D" OF THE TCS MOTOR PACK CONNECTOR. IS RESISTANCE GREATER THAN 0.4 OHMS?

- NO: REPLACE TCS HYDRAULIC MODULATOR/MOTOR PACK ASSEMBLY.
- YES:

(5)
- INSPECT CKTs 566 AND 567 AND RELATED CONNECTORS FOR DAMAGE WHICH MAY RESULT IN A SHORT TO VOLTAGE WITH ALL CONNECTORS CONNECTED. REPAIR ALL DAMAGE IF EVIDENT.
- RECONNECT ALL CONNECTORS.
- START ENGINE AND TEST DRIVE VEHICLE, ACHIEVING AT LEAST 16 km/h (10 MPH).
- REPEAT ABOVE DRIVE CYCLE SEQUENCE TWO MORE TIMES. DID DTC 163 SET IN THE LAST THREE DRIVE CYCLES?

- YES: REPLACE EBTCM.
- NO: MALFUNCTION WAS MOST LIKELY CAUSED BY A POOR CONNECTION. REFER TO "DIAGNOSTIC AIDS" ON FACING PAGE.

AFTER DIAGNOSIS IS COMPLETE, CLEAR DTCs AND TEST DRIVE VEHICLE FOR THREE (3) DRIVE CYCLES TO VERIFY THAT DTC DOES NOT RESET. A DRIVE CYCLE CONSISTS OF STARTING THE VEHICLE, DRIVING OVER 16 km/h (10 MPH), AND THEN TURNING KEY "OFF."

DIAGNOSTIC CHART DTC 164 — 1995 CAMARO AND FIREBIRD WITH TCS

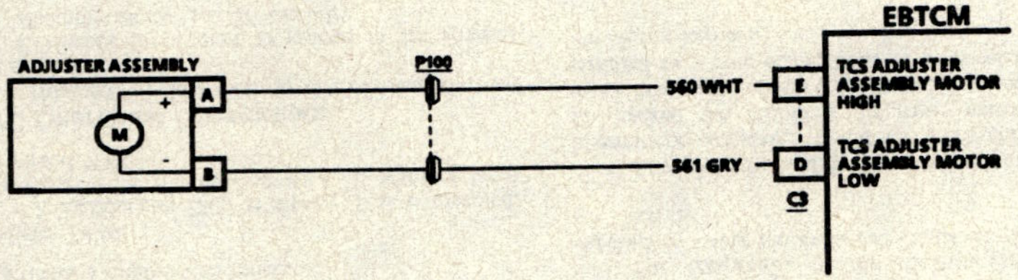

DTC 164
ADJUSTER ASSEMBLY MOTOR CIRCUIT OPEN

Circuit Description:
This DTC identifies a TCS adjuster assembly motor that cannot be energized due to an open in its circuitry.

DTC Will Set When:
DTC 164 can be set only when the adjuster assembly motor is commanded "OFF." If the EBTCM detects an out of range voltage on either of the adjuster assembly circuits indicating an open circuit, a malfunction exists.

Action Taken:
An open adjuster assembly motor will not activate when requested. A malfunction DTC is stored, TCS is disabled, and the amber TCS warning lamp is turned "ON."

DTC Chart Test Description:
Number(s) below refer to circled number(s) on the diagnostic chart.

1. Checks for proper resistance of the adjuster assembly motor.
2. Checks for an open in the adjuster assembly motor "HIGH" circuitry.
3. Checks for an open in the adjuster assembly motor "LOW" circuitry.
4. Ensures malfunction was not due to poor terminal contact.

Diagnostic Aids:
An intermittent malfunction may be caused by a poor connection, rubbed through wire insulation, or a wire that is broken inside the insulation.

The frequency of the malfunction can be checked by using the enhanced diagnostic function of the Tech 1,

Any circuitry that is suspected of causing the intermittent complaint should be thoroughly checked for backed out terminals, improper mating, broken locks, improperly formed or damaged terminals, poor terminal to wiring connections or physical damage to the wiring harness.

DIAGNOSTIC CHART DTC 164 — 1995 CAMARO AND FIREBIRD WITH TCS, CONTINUED

DTC 164
ADJUSTER ASSEMBLY MOTOR CIRCUIT OPEN

1.
- IGNITION "OFF."
- DISCONNECT THE ADJUSTER ASSEMBLY MOTOR CONNECTOR.
- USING J 39200, MEASURE THE RESISTANCE ACROSS ADJUSTER ASSEMBLY MOTOR TERMINALS "A" AND "B".

IS RESISTANCE LESS THAN 15Ω?

- NO: REPLACE ADJUSTER ASSEMBLY MOTOR.
- YES: go to 2.

2.
- DISCONNECT BLACK 8-WAY CONNECTOR AT EBTCM.
- USING J 39200, MEASURE RESISTANCE BETWEEN TERMINAL "E" OF THE BLACK 8-WAY EBTCM HARNESS CONNECTOR AND TERMINAL "A" OF THE ADJUSTER ASSEMBLY HARNESS CONNECTOR.

IS RESISTANCE 2Ω OR LESS?

- NO: REPAIR OPEN OR HIGH RESISTANCE IN CKT 560.
- YES: go to 3.

3.
- USING J 39200, MEASURE RESISTANCE BETWEEN TERMINAL "D" OF THE BLACK 8-WAY EBTCM HARNESS CONNECTOR AND TERMINAL "B" OF THE ADJUSTER ASSEMBLY HARNESS CONNECTOR.

IS RESISTANCE 2Ω OR LESS?

- NO: REPLACE OPEN OR HIGH RESISTANCE IN CKT 561.
- YES: go to 4.

4.
- INSPECT THE ADJUSTER ASSEMBLY MOTOR AND BLACK 8-WAY EBTCM CONNECTORS FOR POOR TERMINAL CONTACT AND EVIDENCE OF CORROSION.
  REPLACE ALL TERMINALS THAT EXHIBIT SIGNS OF POOR TERMINAL CONTACT OR CORROSION.
- RECONNECT ALL CONNECTORS.
- IGNITION "OFF."
- START ENGINE AND TEST DRIVE VEHICLE, ACHIEVING AT LEAST 16 km/h (10 MPH).
- REPEAT ABOVE DRIVE CYCLE SEQUENCE TWO MORE TIMES.

DID DTC 164 SET IN THE LAST THREE DRIVE CYCLES?

- YES: REPLACE EBTCM.
- NO: MALFUNCTION WAS MOST LIKELY CAUSED BY A POOR CONNECTION. REFER TO "DIAGNOSTIC AIDS" ON FACING PAGE.

AFTER DIAGNOSIS IS COMPLETE, CLEAR DTCs AND TEST DRIVE VEHICLE FOR THREE (3) DRIVE CYCLES TO VERIFY THAT DTC DOES NOT RESET. A DRIVE CYCLE CONSISTS OF STARTING THE VEHICLE, DRIVING OVER 16 km/h (10 MPH), AND THEN TURNING KEY "OFF."

DIAGNOSTIC CHART DTC 165 — 1995 CAMARO AND FIREBIRD WITH TCS

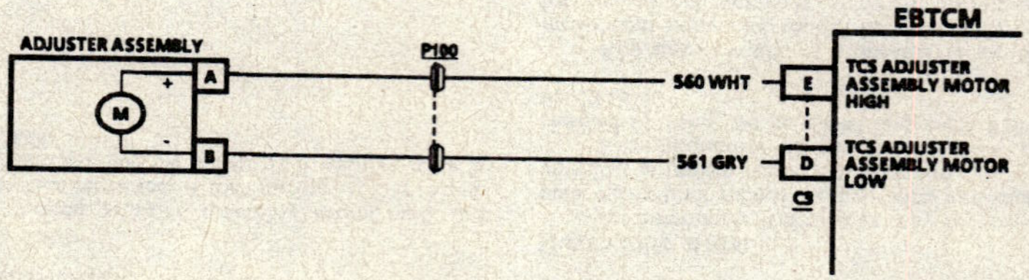

DTC 165
ADJUSTER ASSEMBLY MOTOR CIRCUIT SHORTED TO GND

Circuit Description:

This DTC identifies a TCS adjuster assembly motor circuit that is shorted to ground. This malfunction will not allow the adjuster assembly motor to be controlled at the commanded current rate or will cause the driver circuit to allow current directly to ground.

DTC Will Set When:

DTC 165 can be set anytime. If the EBTCM detects an out of range voltage on either of the adjuster assembly circuits indicating a circuit shorted to ground, a malfunction exists.

Action Taken:

A malfunction DTC is stored, TCS is disabled. The amber TCS and the red BRAKE warning lamp are turned "ON."

DTC Chart Test Description:

Number(s) below refer to circled number(s) on the diagnostic chart.

1. Checks for a short to ground in the adjuster assembly motor "HIGH" circuitry.
2. Checks for a short to ground in the adjuster assembly motor "LOW" circuitry.
3. Checks for an adjuster assembly motor that is internally shorted to ground.
4. Ensures malfunction was not due to physical damage of the circuitry.

Diagnostic Aids:

An intermittent malfunction may be caused by a poor connection, rubbed through wire insulation, or a wire that is broken inside the insulation.

The frequency of the malfunction can be checked by using the enhanced diagnostic function of the Tech 1.

Any circuitry that is suspected of causing the intermittent complaint should be thoroughly checked for backed out terminals, improper mating, broken locks, improperly formed or damaged terminals, poor terminal to wiring connections or physical damage to the wiring harness.

DIAGNOSTIC CHART DTC 165 — 1995 CAMARO AND FIREBIRD WITH TCS, CONTINUED

DTC 165
ADJUSTER ASSEMBLY MOTOR CIRCUIT SHORTED TO GND

(1)
- IGNITION "OFF."
- DISCONNECT THE ADJUSTER ASSEMBLY MOTOR CONNECTOR AND THE BLACK 8-WAY EBTCM CONNECTOR.
- USING J 39200, MEASURE THE RESISTANCE BETWEEN GROUND AND TERMINAL "A" OF THE ADJUSTER ASSEMBLY HARNESS CONNECTOR. DOES J 39200 DISPLAY RESISTANCE AS "OL" (INFINITE)?

NO: REPAIR SHORT TO GROUND IN CKT 560.

YES:

(2)
- USING J 39200, MEASURE THE RESISTANCE BETWEEN GROUND AND TERMINAL "B" OF THE ADJUSTER ASSEMBLY HARNESS CONNECTOR. DOES J 39200 DISPLAY RESISTANCE AS "OL" (INFINITE)?

NO: REPAIR SHORT TO GROUND IN CKT 561.

YES:

(3)
- USING J 39200, MEASURE THE RESISTANCE BETWEEN GROUND AND TERMINAL "A" OF THE ADJUSTER ASSEMBLY. DOES J 39200 DISPLAY RESISTANCE AS "OL" (INFINITE)?

NO: REPLACE ADJUSTER ASSEMBLY MOTOR.

YES:

(4)
- INSPECT CKTs 560 AND 561 AND RELATED CONNECTORS FOR DAMAGE WHICH MAY RESULT IN A SHORT TO GROUND WITH ALL CONNECTORS CONNECTED. REPAIR ALL DAMAGE IF EVIDENT.
- RECONNECT ALL CONNECTORS.
- IGNITION "OFF."
- START ENGINE AND TEST DRIVE VEHICLE, ACHIEVING AT LEAST 16 km/h (10 MPH).
- REPEAT ABOVE DRIVE CYCLE SEQUENCE TWO MORE TIMES. DID DTC 165 SET IN THE LAST THREE DRIVE CYCLES?

YES: REPLACE EBTCM.

NO: MALFUNCTION WAS MOST LIKELY CAUSED BY A POOR CONNECTION. REFER TO "DIAGNOSTIC AIDS" ON FACING PAGE.

AFTER DIAGNOSIS IS COMPLETE, CLEAR DTCs AND TEST DRIVE VEHICLE FOR THREE (3) DRIVE CYCLES TO VERIFY THAT DTC DOES NOT RESET. A DRIVE CYCLE CONSISTS OF STARTING THE VEHICLE, DRIVING OVER 16 km/h (10 MPH), AND THEN TURNING KEY "OFF."

DIAGNOSTIC CHART DTC 166 — 1995 CAMARO AND FIREBIRD WITH TCS

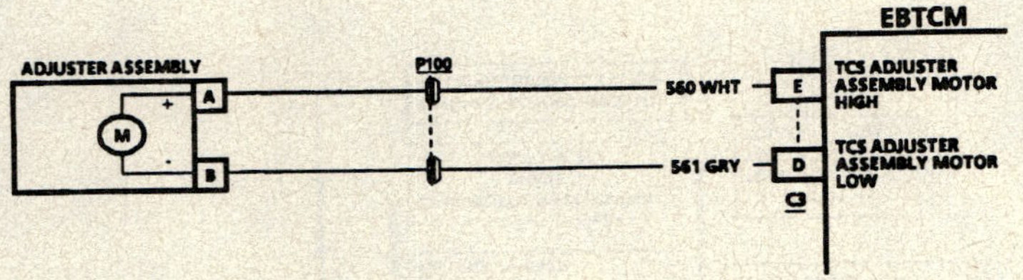

DTC 166
ADJUSTER ASSEMBLY MOTOR CIRCUIT SHORTED TO BATT

Circuit Description:

This DTC identifies an adjuster assembly motor circuit that is shorted to battery or an adjuster assembly motor that has low or no resistance. This malfunction will not allow the adjuster assembly motor to be controlled at the commanded current rate or will cause the adjuster assembly motor to turn in the opposite direction, or not at all.

DTC Will Set When:

DTC 166 can be set at any time. If the EBTCM detects an out of range voltage on either of the adjuster assembly circuits indicating a circuit shorted to battery or an adjuster assembly motor shorted, a malfunction exists.

Action Taken:

A malfunction DTC is stored, TCS is disabled. The amber TCS and the red BRAKE warning lamp are turned "ON."

DTC Chart Test Description:

Number(s) below refer to circled number(s) on the diagnostic chart.

1. Checks for a short to voltage in the adjuster assembly motor "HIGH" circuitry.
2. Checks for a short to voltage in the adjuster assembly motor "LOW" circuitry.
3. Checks for a short between the "HIGH" and "LOW" motor circuits..
4. Checks for an adjuster assembly motor that is internally shorted.
5. Ensures malfunction was not due to physical damage of the circuitry.

Diagnostic Aids:

An intermittent malfunction may be caused by a poor connection, rubbed through wire insulation, or a wire that is broken inside the insulation.

The frequency of the malfunction can be checked by using the enhanced diagnostic function of the Tech 1,

Any circuitry that is suspected of causing the intermittent complaint should be thoroughly checked for backed out terminals, improper mating, broken locks, improperly formed or damaged terminals, poor terminal to wiring connections or physical damage to the wiring harness.

DIAGNOSTIC CHART DTC 166 — 1995 CAMARO AND FIREBIRD WITH TCS, CONTINUED

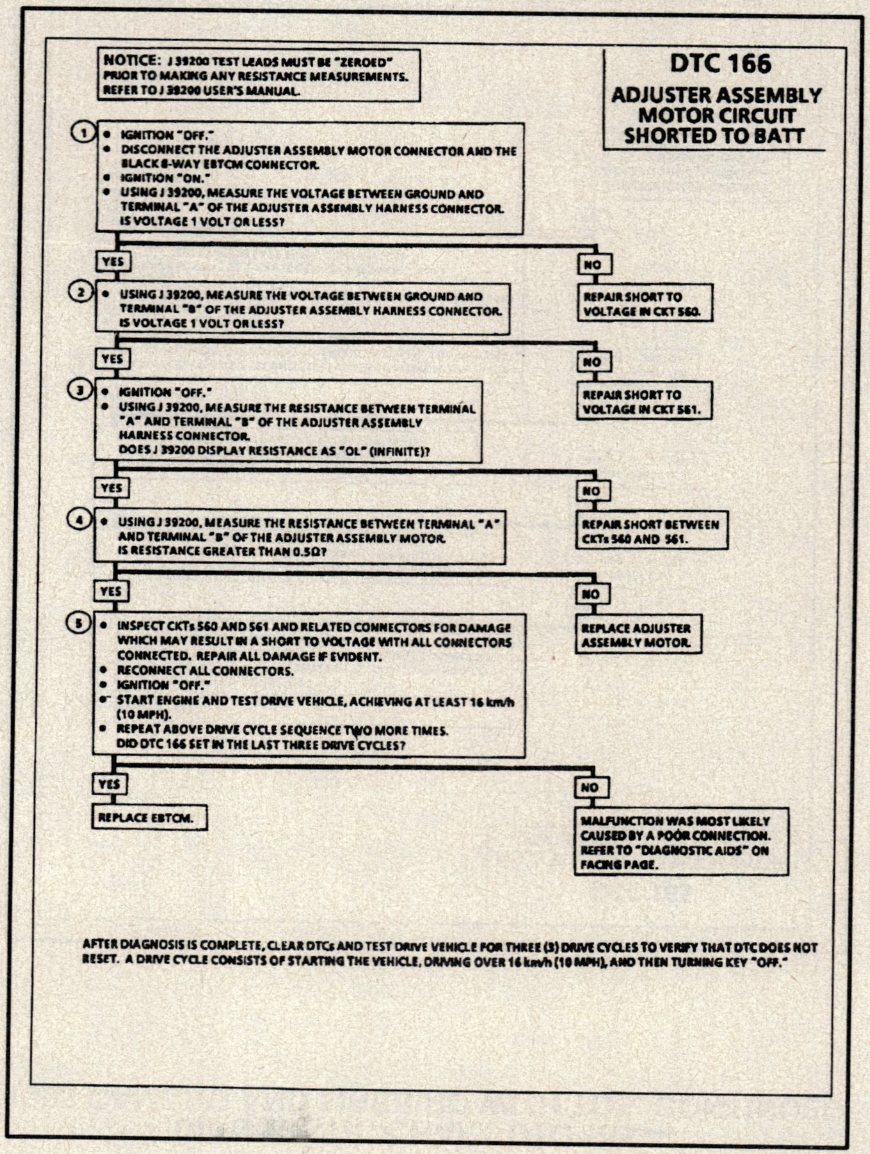

DIAGNOSTIC PROCESS CHART — 1995 CAVALIER AND SUNFIRE WITH TCS

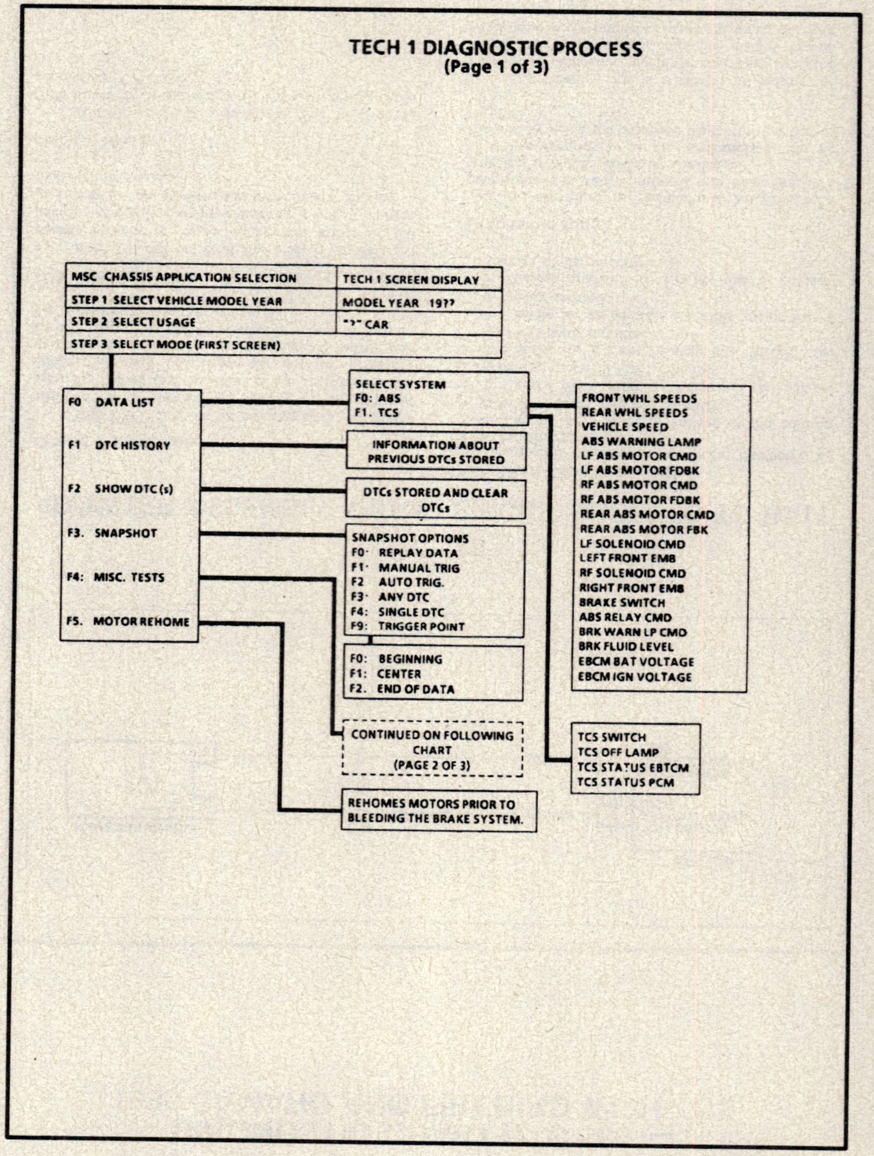

DIAGNOSTIC PROCESS CHART — 1995 CAVALIER AND SUNFIRE WITH TCS, CONTINUED

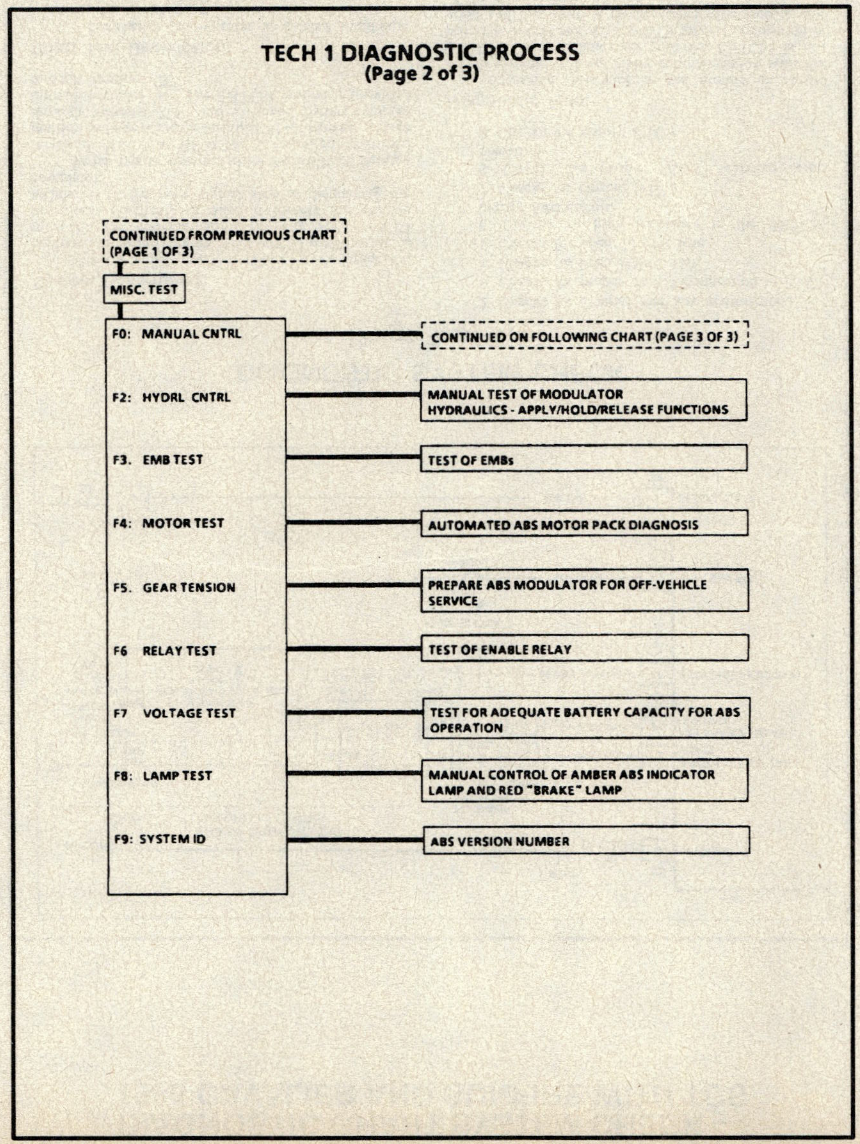

DIAGNOSTIC PROCESS CHART — 1995 CAVALIER AND SUNFIRE WITH TCS, CONTINUED

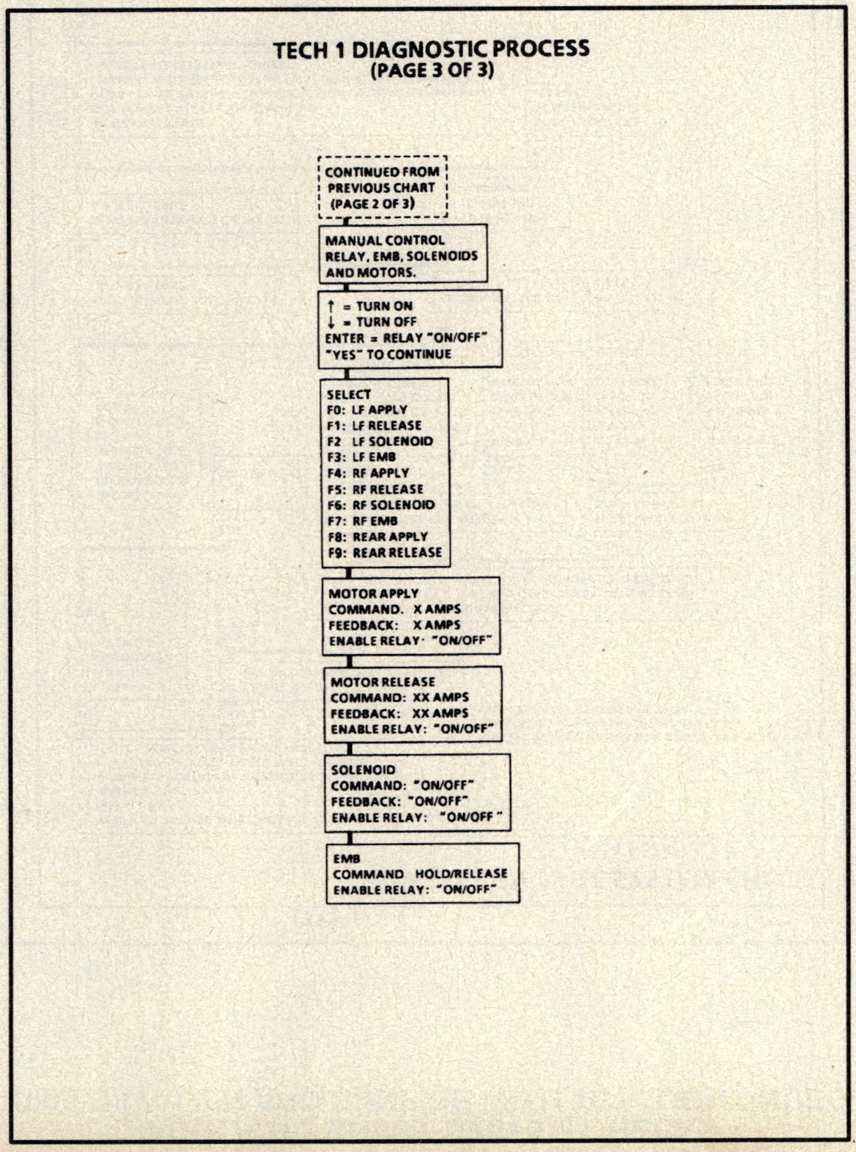

DIAGNOSTIC CHART SYSTEM CHECK — 1995 CAVALIER AND SUNFIRE WITH TCS

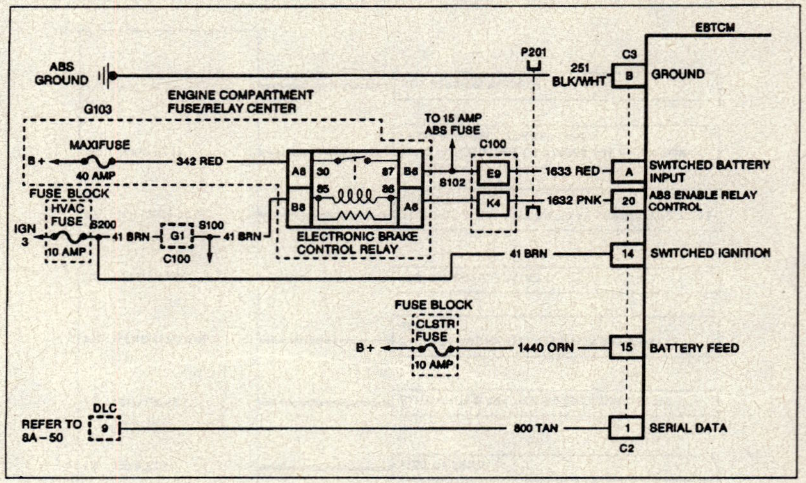

DIAGNOSTIC SYSTEM CHECK
(Page 1 of 4)

System Description:

The diagnostic system check is an organized approach to identifying an ABS w/ETS malfunction. It must be the starting point for any ABS w/ETS complaint diagnosis because it directs the service technician to the next logical step in diagnosing the complaint.

Serial data is transmitted by the EBTCM through terminal "1." The EBTCM is supplied switched ignition voltage through terminal "14," battery voltage through terminal "15" and switched battery voltage through terminal "A." The EBTCM ground is provided through terminal "B."

Chart Test Description:

Number(s) below refer to circled number(s) on the diagnostic chart.

1. Indicates if EBTCM is transmitting data.
2. Checks for a serial data link malfunction.
3. Checks for proper Tech 1 connection to DLC.
4. Checks for open HVAC fuse.
5. Checks for open CLSTR fuse.
6. Checks for high resistance in the EBTCM battery feed circuit.
7. Checks for current DTC's.
8. Checks for proper "ABS" warning lamp operation.
9. Checks for history DTC's.

Diagnostic Aids:

Excessive resistance in the ground or power supply circuits will not allow communication with the EBTCM. If communication with the EBTCM is not possible, ensure the ABS w/ETS ground connection is good and that there is no excessive resistance in any of the power supply circuits.

DIAGNOSTIC CHART SYSTEM CHECK — 1995 CAVALIER AND SUNFIRE WITH TCS, CONTINUED

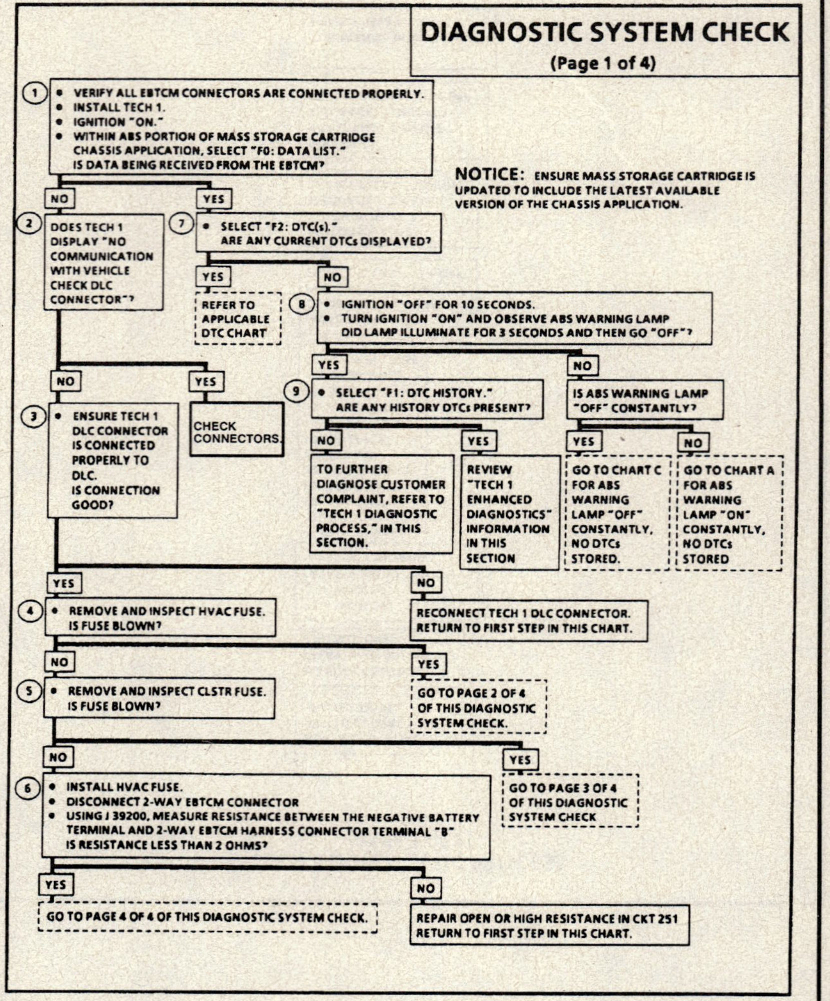

DIAGNOSTIC CHART SYSTEM CHECK — 1995 CAVALIER AND SUNFIRE WITH TCS, CONTINUED

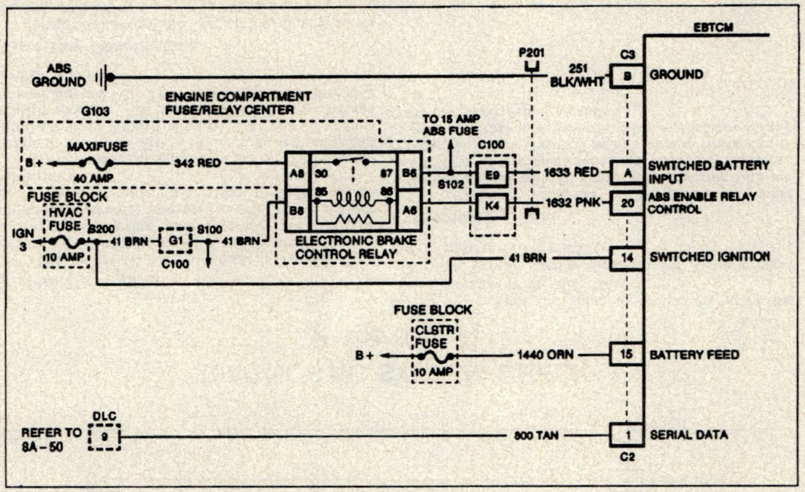

DIAGNOSTIC SYSTEM CHECK
(Page 2 of 4)

System Description:

The diagnostic system check is an organized approach to identifying an ABS w/ETS malfunction. It must be the starting point for any ABS w/ETS complaint diagnosis because it directs the service technician to the next logical step in diagnosing the complaint.

Serial data is transmitted by the EBTCM through terminal "1." The EBTCM is supplied switched ignition voltage through terminal "14," battery voltage through terminal "15" and switched battery voltage through terminal "A." The EBTCM ground is provided through terminal "B."

Chart Test Description:

Number(s) below refer to circled number(s) on the diagnostic chart.

10. Checks for possible short to ground in the switched ignition circuitry.

11. Isolates the short to ground to either the circuitry or the EBTCM.

12. Ensures short to ground is not due to physical damage of the circuitry.

Diagnostic Aids:

Excessive resistance in the ground or power supply circuits will not allow communication with the EBTCM. If communication with the EBTCM is not possible, ensure the ABS w/ETS ground connection is good and that there is no excessive resistance in any of the power supply circuits.

DIAGNOSTIC CHART SYSTEM CHECK — 1995 CAVALIER AND SUNFIRE WITH TCS, CONTINUED

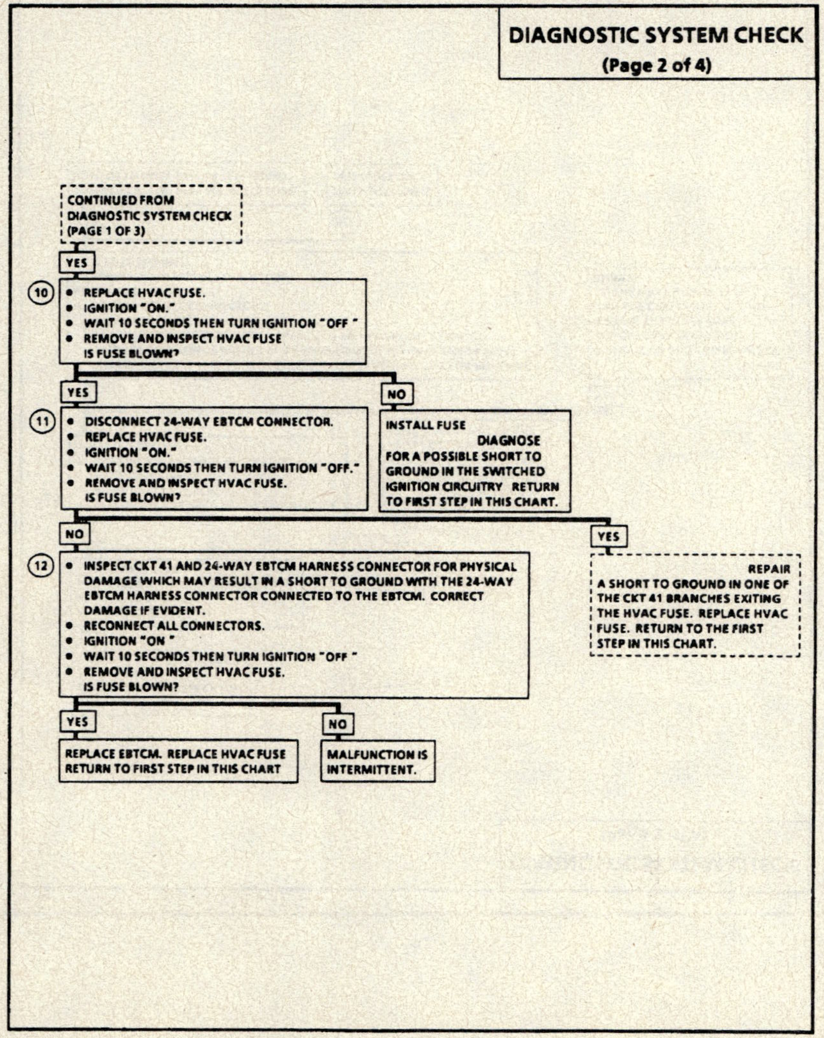

DIAGNOSTIC CHART SYSTEM CHECK — 1995 CAVALIER AND SUNFIRE WITH TCS, CONTINUED

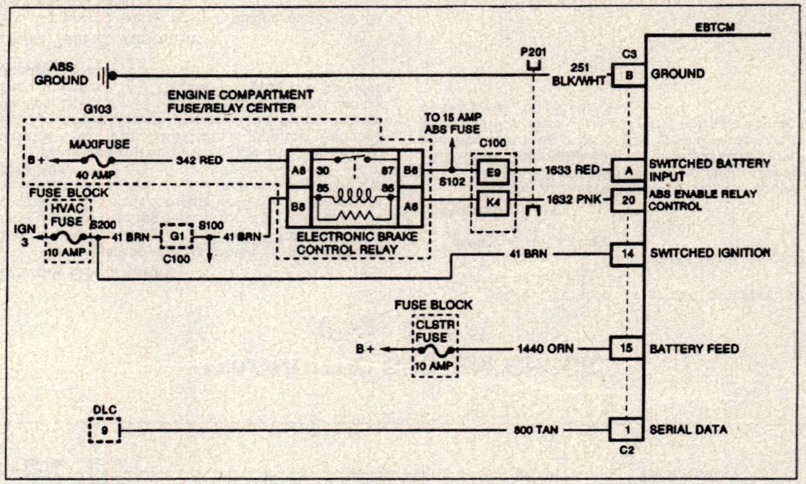

DIAGNOSTIC SYSTEM CHECK
(Page 3 of 4)

System Description:

The diagnostic system check is an organized approach to identifying an ABS w/ETS malfunction. It must be the starting point for any ABS w/ETS complaint diagnosis because it directs the service technician to the next logical step in diagnosing the complaint.

Serial data is transmitted by the EBTCM through terminal "1." The EBTCM is supplied switched ignition voltage through terminal "14," battery voltage through terminal "15" and switched battery voltage through terminal "A". The EBTCM ground is provided through terminal "B."

Chart Test Description:

Number(s) below refer to circled number(s) on the diagnostic chart.

13. Checks for possible short to ground in the battery feed circuitry.

14. Isolates the short to ground to either the circuitry or the EBTCM.

15. Ensures short to ground is not due to physical damage of the circuitry.

Diagnostic Aids:

Excessive resistance in the ground or power supply circuits will not allow communication with the EBTCM. If communication with the EBTCM is not possible, ensure the ABS w/ETS ground connection is good and that there is no excessive resistance in any of the power supply circuits.

DIAGNOSTIC CHART SYSTEM CHECK — 1995 CAVALIER AND SUNFIRE WITH TCS, CONTINUED

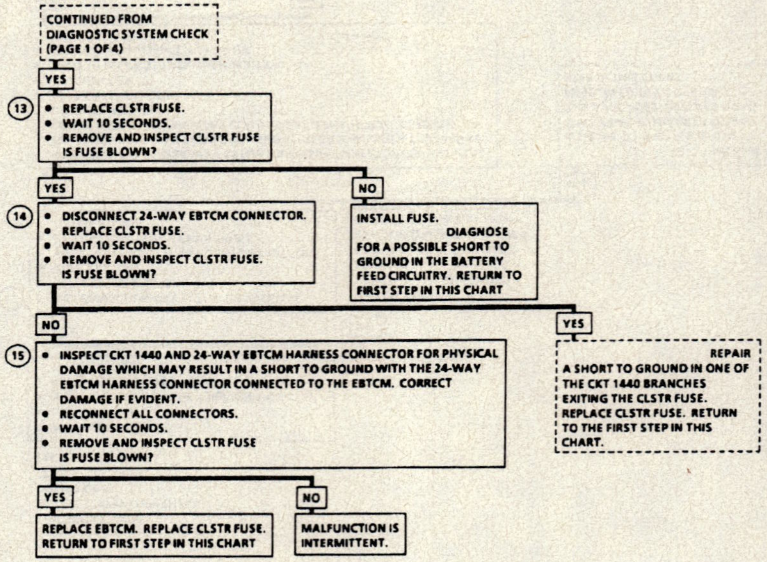

DIAGNOSTIC CHART SYSTEM CHECK — 1995 CAVALIER AND SUNFIRE WITH TCS, CONTINUED

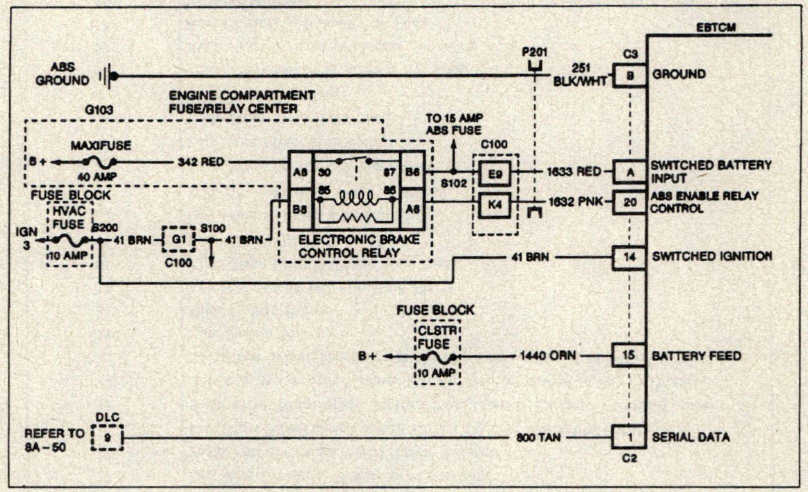

DIAGNOSTIC SYSTEM CHECK
(Page 4 of 4)

System Description:

The diagnostic system check is an organized approach to identifying an ABS w/ETS malfunction. It must be the starting point for any ABS w/ETS complaint diagnosis because it directs the service technician to the next logical step in diagnosing the complaint.

Serial data is transmitted by the EBTCM through terminal "1." The EBTCM is supplied switched ignition voltage through terminal "14," battery voltage through terminal "15" and switched battery voltage through terminal "A". The EBTCM ground is provided through terminal "B."

Chart Test Description:

Number(s) below refer to circled number(s) on the diagnostic chart.

16. Checks for proper battery feed voltage at the EBTCM.
17. Checks for proper ignition feed voltage at the EBTCM.
18. Checks for high resistance in the battery feed circuit and the switched ignition feed circuit.
19. Ensures malfunction is not due to poor terminal contact.
20. Checks for an open circuit in the battery feed circuit between the fuse block and the EBTCM.
21. Checks for an open circuit in the switched ignition feed circuit between the fuse block and the EBTCM.
22. Isolates the high resistance circuit to either the battery feed circuit or the switched ignition feed circuit.
23. Isolates the high resistance in the battery feed circuit between the fuse block and the EBTCM.
24. Isolates the high resistance in the ignition feed circuit between the fuse block and the EBTCM.

Diagnostic Aids:

Excessive resistance in the ground or power supply circuits will not allow communication with the EBTCM. If communication with the EBTCM is not possible, ensure the ABS w/ETS ground connection is good and that there is no excessive resistance in any of the power supply circuits.

DIAGNOSTIC CHART SYSTEM CHECK — 1995 CAVALIER AND SUNFIRE WITH TCS, CONTINUED

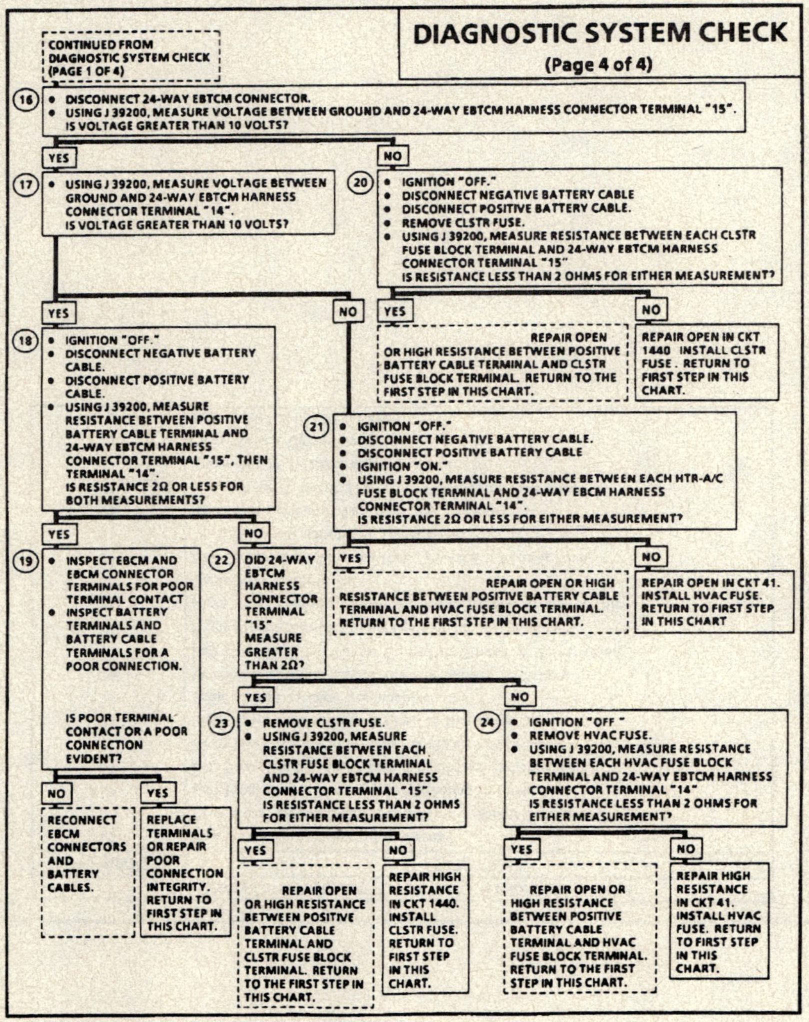

DIAGNOSTIC TROUBLE CODE CHART — 1995 CAVALIER AND SUNFIRE WITH TCS

DIAGNOSTIC TROUBLE CODE AND SYMPTOM TABLE	
CHART	SYMPTOM
A	ABS (Amber) Indicator Lamp "ON" Constantly, No DTCs Stored
B	Tech 1 Displays Undefined DTCs
C	ABS (Amber) Indicator Lamp "OFF" Constantly, No DTCs Stored
DIAGNOSTIC TROUBLE CODE	DESCRIPTION
14	Electronic Brake Control Relay Contacts Circuit Open
15	Electronic Brake Control Relay Contacts Circuit Shorted to Battery
16	Electronic Brake Control Relay Coil Circuit Open
17	Electronic Brake Control Relay Coil Circuit Shorted to Ground
18	Electronic Brake Control Relay Coil Circuit Shorted to Battery or Coil Shorted
21	Left Front Wheel Speed = 0
22	Right Front Wheel Speed = 0
23	Left Rear Wheel Speed = 0
24	Right Rear Wheel Speed = 0
25	Left Front Excessive Wheel Speed Variation
26	Right Front Excessive Wheel Speed Variation
27	Left Rear Excessive Wheel Speed Variation
28	Right Rear Excessive Wheel Speed Variation
32	Left Front Wheel Speed Sensor Circuit Open or Shorted to Battery/Ground
33	Right Front Wheel Speed Sensor Circuit Open or Shorted to Battery/Ground
34	Left Rear Wheel Speed Sensor Circuit Open or Shorted to Battery/Ground
35	Right Rear Wheel Speed Sensor Circuit Open or Shorted to Battery/Ground
36	Low System Voltage
37	High System Voltage
38	Left Front EMB Will Not Hold Motor
41	Right Front EMB Will Not Hold Motor
42	Rear ESB Will Not Hold Motor
44	Left Front Channel Will Not Move
45	Right Front Channel Will Not Move
46	Rear Axle Channel Will Not Move
47	Left Front Motor Free Spins
48	Right Front Motor Free Spins
51	Rear Motor Free Spins
52	Left Front Channel in Release Too Long
53	Right Front Channel in Release Too Long
54	Rear Channel in Release Too Long
55	EBTCM Malfunction
56	Left Front Motor Circuit Open
57	Left Front Motor Circuit Shorted to Ground
58	Left Front Motor Circuit Shorted to Battery or Motor Shorted
61	Right Front Motor Circuit Open
62	Right Front Motor Circuit Shorted to Ground
63	Right Front Motor Circuit Shorted to Battery or Motor Shorted
64	Rear Motor Circuit Open

DIAGNOSTIC TROUBLE CODE CHART — 1995 CAVALIER AND SUNFIRE WITH TCS, CONTINUED

DIAGNOSTIC TROUBLE CODE AND SYMPTOM TABLE	
CHART	SYMPTOM
65	Rear Motor Circuit Shorted to Ground
66	Rear Motor Circuit Shorted to Battery or Motor Shorted
67	Left Front EMB Circuit Open or Shorted to Ground
68	Left Front EMB Circuit Shorted to Battery or Driver Open
71	Right Front EMB Circuit Open or Shorted to Ground
72	Right Front EMB Circuit Shorted to Battery or Driver Open
75	Serial Communication Malfunction
76	Left Front Solenoid Circuit Open or Shorted to Battery
77	Left Front Solenoid Circuit Shorted to Ground or Driver Open
78	Right Front Solenoid Circuit Open or Shorted to Battery
81	Right Front Solenoid Circuit Shorted to Ground or Driver Open
82	Calibration Malfunction
86	EBTCM Turned "ON" the Red "BRAKE" Warning Lamp
91	Open Brake Switch During Deceleration
92	Open Brake Switch When ABS Was Required
93	DTCs 91 or 92 Set in Current or Previous Ignition Cycle
94	Brake Switch Contacts Always Closed
95	Brake Switch Circuit Open
96	Rear Brake Lamp Circuit Open

ELECTRICAL CONNECTOR TERMINAL IDENTIFICATION — 1995 CAVALIER AND SUNFIRE WITH TCS

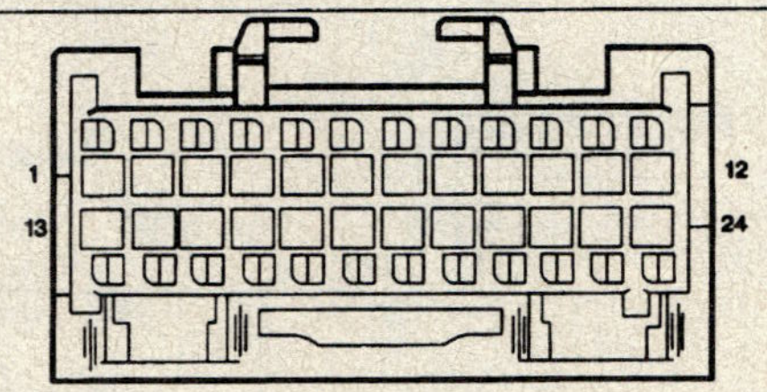

PIN	CIRCUIT NO.	COLOR	CIRCUIT
1	800	TAN	SERIAL DATA
2	1289	LT BLU	R/F ABS SOLENOID CONTROL
3	830	LT BLU	L/F WHEEL SIGNAL HIGH
4	873	YEL	L/F WHEEL SIGNAL LOW
5	882	BRN	R/R WHEEL SIGNAL HIGH
6	883	WHT	R/R WHEEL SIGNAL LOW
7	872	DK GRN	R/F WHEEL SIGNAL HIGH
8	833	TAN	R/F WHEEL SIGNAL LOW
9	884	BLK	L/R WHEEL SIGNAL HIGH
10	885	RED	L/R WHEEL SIGNAL LOW
11	OPEN		NOT USED
12	OPEN		NOT USED
13	17	WHT	BRAKE SWITCH INPUT
14	41	BRN	SWITCHED IGNITION
15	1440	ORN	BATTERY FEED
16	OPEN		NOT USED
17	1286	LT GRN	L/F EMB CONTROL
18	1287	GRY	R/F EMB CONTROL
19	849	PNK	BRAKE FLUID LEVEL SWITCH INPUT
20	1632	PNK	ENABLE RELAY CONTROL
21	OPEN		NOT USED
22	1288	DK GRN	L/F ABS SOLENOID CONTROL
23	OPEN		NOT USED
24	OPEN		NOT USED

ELECTRICAL CONNECTOR TERMINAL IDENTIFICATION — 1995 CAVALIER AND SUNFIRE WITH TCS, CONTINUED

6 WAY EBTCM HARNESS CONNECTOR

PIN	CIRCUIT NO.	COLOR	CIRCUIT
C	1284	DK GRN	REAR MOTOR HIGH
D	1285	ORN	REAR MOTOR LOW
E	1281	PNK	L/F MOTOR LOW
F	1280	BLK	L/F MOTOR HIGH
G	1283	BLK	R/F MOTOR LOW
H	1282	PPL	R/F MOTOR HIGH

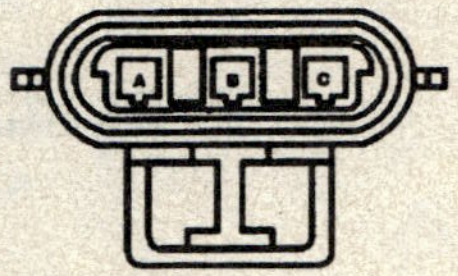

2 WAY EBTCM HARNESS CONNECTOR

PIN	CIRCUIT NO.	COLOR	CIRCUIT
A	1633	RED	SWITCHED BATTERY INPUT
B	251	BLK/WHT	GROUND

DIAGNOSTIC CHART A AND B — 1995 CAVALIER AND SUNFIRE WITH TCS

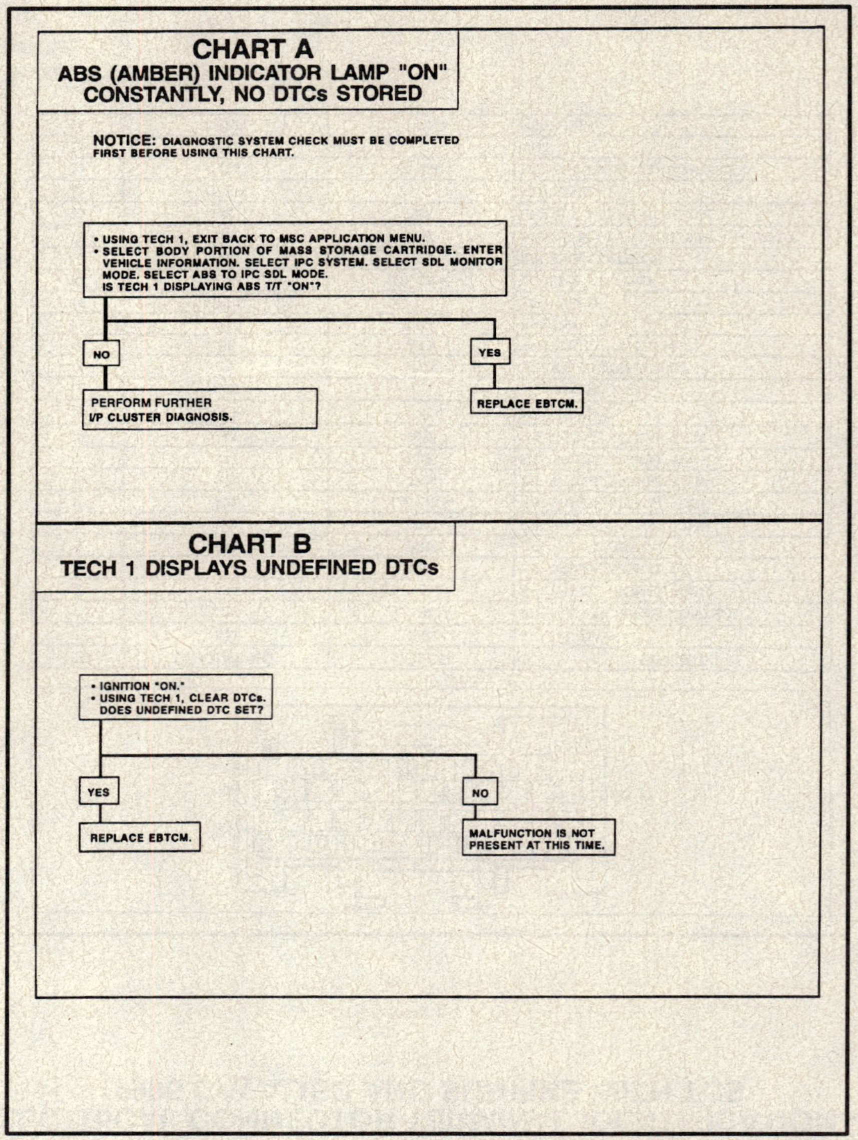

DIAGNOSTIC CHART C — 1995 CAVALIER AND SUNFIRE WITH TCS

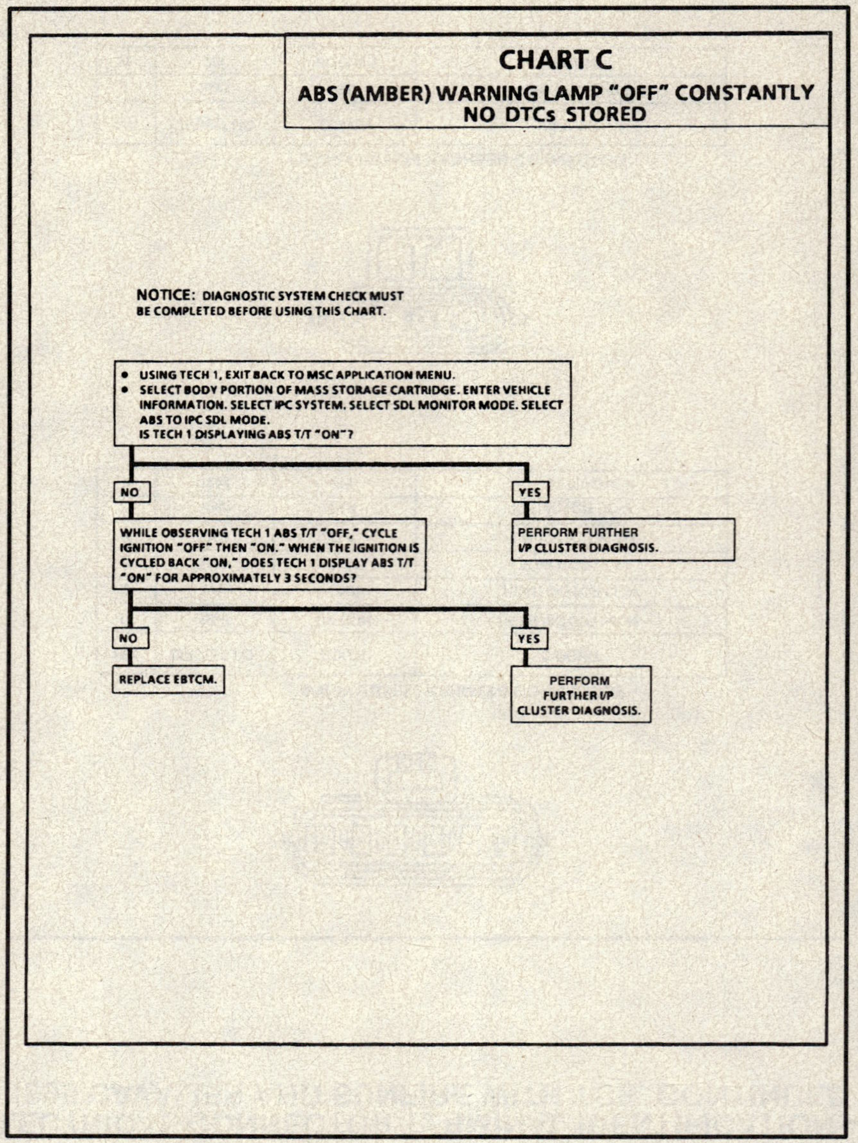

DIAGNOSTIC CHART DTC 14 — 1995 CAVALIER AND SUNFIRE WITH TCS

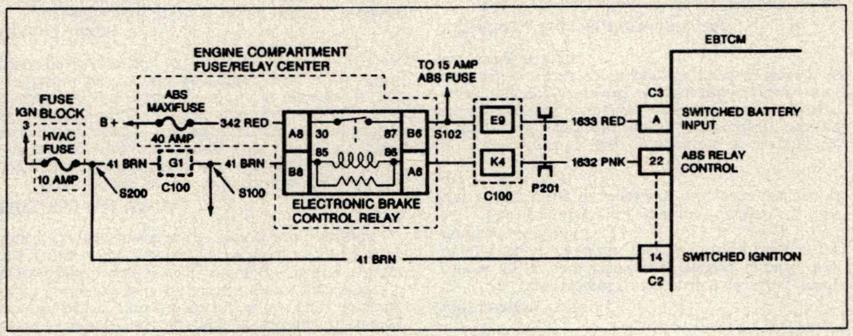

DTC 14
(Page 1 of 3)
ELECTRONIC BRAKE CONTROL RELAY CONTACT CIRCUIT OPEN

Circuit Description:

Ignition voltage is supplied through terminal "85" of the electronic brake control relay. The EBTCM is then able to energize the pull-in coil by completing the ground circuit at EBTCM terminal "18." The magnetic field created closes the electronic brake control relay contacts and allows battery voltage and current to be supplied to the EMB's and EBTCM, which supplies power to the motors and solenoids.

DTC Will Set When:

DTC 14 can be set anytime after the EBTCM commands the electronic brake control relay "ON" (the relay is first commanded "ON" during the three second bulb check). This test monitors the availability of current/voltage to the motors, solenoids, and EMB's. This malfunction indicates voltage is not available and would therefore not allow ABS operation if required.

Action Taken:

A malfunction DTC is stored, ABS is disabled and the "ABS" warning lamp is turned "ON."

DTC Chart Test Description: Number(s) below refer to circled number(s) on the diagnostic chart.

1. Checks to see if the EBTCM is sensing battery voltage at terminal "A".
2. Checks for proper operation of the electronic brake control relay and related circuitry.
3. Checks for excessive resistance in the switched battery input circuitry.
4. Checks for excessive resistance in the battery feed circuit.
5. Checks for excessive resistance in the electronic brake control relay contacts.
6. Determines if the malfunction is due to poor terminal contact.
7. Determines if the malfunction is due to the EBTCM.

Diagnostic Aids:

An "intermittent" malfunction is most likely caused by a poor connection, rubbed through wire insulation, or a wire that is broken inside the insulation.

The frequency of the malfunction is most likely checked by using the enhanced diagnostic function of the Tech 1.

Any circuitry that is suspected as causing the "intermittent" complaint should be thoroughly checked for backed out terminals, improper mating, broken locks, improperly formed or damaged terminals, poor terminal to wiring connections or physical damage to the wiring harness.

Vibration, Temperature Effects:

Check for vibration effects by performing the relay test function of the Tech 1. With the relay commanded "ON," lightly tap the top and sides of the relay while monitoring battery voltage. If the battery voltage changes significantly, replace the relay.

If DTC 14 only sets when the vehicle is initially started in cold ambient conditions (temperature less than 32°F/0°C), replace the electronic brake control relay.

DIAGNOSTIC CHART DTC 14 — 1995 CAVALIER AND SUNFIRE WITH TCS, CONTINUED

DTC 14
(Page 1 of 3)
ELECTRONIC BRAKE CONTROL RELAY CONTACTS CIRCUIT OPEN

IS DTC 16 ALSO PRESENT OR IN HISTORY?
- YES → PROCEED TO THAT DTC CHART FIRST.
- NO ↓

(1)
- IGNITION "ON," ENGINE "OFF."
- USING TECH 1, SELECT DATA LIST AND OBSERVE ABS BATTERY VOLTAGE. DOES TECH 1 INDICATE THAT ABS BATTERY VOLTAGE IS GREATER THAN 10 VOLTS?

- YES → MALFUNCTION IS NOT PRESENT AT THIS TIME. REFER TO "DIAGNOSTIC AIDS" ON FACING PAGE.
- NO ↓

(2)
- IGNITION "OFF."
- DISCONNECT 24-WAY EBTCM CONNECTOR.
- DISCONNECT 2-WAY EBTCM CONNECTOR.
- CONNECT A FUSED JUMPER WIRE SUCH AS J 36169 WITH A 3 AMP FUSE BETWEEN GROUND AND 24-WAY EBTCM HARNESS CONNECTOR TERMINAL "20".
- IGNITION "ON."
- USING J 39200, MEASURE VOLTAGE BETWEEN GROUND AND 2-WAY EBTCM HARNESS CONNECTOR TERMINAL "A".

IS VOLTAGE GREATER THAN 10 VOLTS?

- NO → REFER TO PAGE 2 OF 3 OF THIS DTC CHART.
- YES ↓

(3)
- IGNITION "OFF."
- DISCONNECT JUMPER WIRE USED IN STEP 2.
- DISCONNECT ELECTRONIC BRAKE CONTROL RELAY.
- USING J 39200, MEASURE RESISTANCE BETWEEN 2-WAY EBTCM HARNESS CONNECTOR TERMINAL "A" AND ENGINE COMPARTMENT FUSE/RELAY CENTER TERMINAL "B6".

IS RESISTANCE LESS THAN 2 OHMS?

- NO → REPAIR HIGH RESISTANCE IN CKT 1633.
- YES ↓

(4)
- DISCONNECT NEGATIVE BATTERY CABLE.
- DISCONNECT POSITIVE BATTERY CABLE.
- USING J 39200, MEASURE RESISTANCE BETWEEN POSITIVE BATTERY CABLE TERMINAL AND ENGINE COMPARTMENT FUSE/RELAY CENTER TERMINAL "A8".

IS RESISTANCE LESS THAN 2 OHMS?

- NO → REPAIR HIGH RESISTANCE IN B+ FEED CIRCUIT.
- YES ↓

(5)
- USING A JUMPER WIRE, CONNECT ELECTRONIC BRAKE CONTROL RELAY TERMINAL "85" TO GROUND. THEN, USING A FUSED JUMPER WIRE SUCH AS J 36169 WITH A 3 AMP FUSE, CONNECT ELECTRONIC BRAKE CONTROL RELAY TERMINAL "86" TO B+.
- USING J 39200, MEASURE THE RESISTANCE BETWEEN ENABLE RELAY TERMINALS "30" AND "87".

IS RESISTANCE LESS THAN 2 OHMS?

- NO → REPLACE ELECTRONIC BRAKE CONTROL RELAY.
- YES ↓

(6)
- INSPECT ELECTRONIC BRAKE CONTROL RELAY AND ENGINE COMPARTMENT FUSE/RELAY CENTER FOR A POOR CONNECTION OR POOR TERMINAL CONTACT. INSPECT EBTCM AND EBTCM CONNECTOR TERMINALS "20" AND "A" FOR POOR TERMINAL CONTACT. INSPECT POSITIVE BATTERY TERMINAL AND POSITIVE BATTERY CABLE TERMINAL FOR A POOR CONNECTION.

IS POOR TERMINAL CONTACT OR A POOR CONNECTION EVIDENT?

- YES → REPLACE TERMINALS OR REPAIR CONNECTION INTEGRITY (IF POOR CONTACT EXISTS AT BATTERY TERMINAL) THAT EXHIBIT(S) POOR CONTACT.
- NO ↓

(7)
- RECONNECT ALL CONNECTORS.
- IGNITION "ON."

IS DTC 14 SET AS A CURRENT DTC?

- YES → REPLACE EBTCM.
- NO → MALFUNCTION IS INTERMITTENT. REFER TO "DIAGNOSTIC AIDS" ON FACING PAGE.

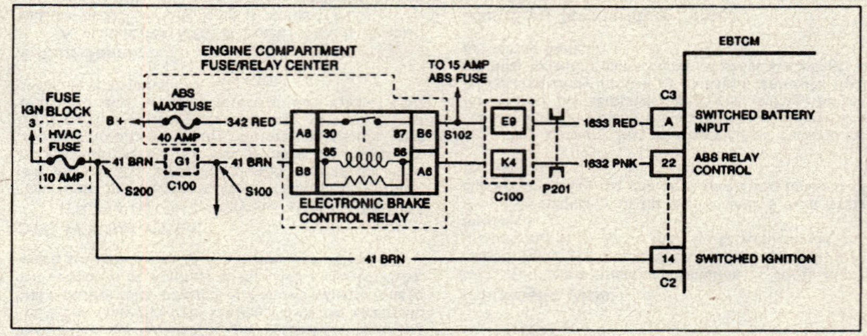

DTC 14
(Page 2 of 3)
ELECTRONIC BRAKE CONTROL RELAY CONTACT CIRCUIT OPEN

Circuit Description:

Ignition voltage is supplied through terminal "85" of the electronic brake control relay. The EBTCM is then able to energize the pull-in coil by completing the ground circuit at EBTCM terminal "20." The magnetic field created closes the electronic brake control relay contacts and allows battery voltage and current to be supplied to the EMB's and EBTCM, which supplies power to the motors and solenoids.

DTC Will Set When:

DTC 14 can be set anytime after the EBTCM commands the electronic brake control relay "ON" (the relay is first commanded "ON" during the three second bulb check). This test monitors the availability of current/voltage to the motors, solenoids, and EMB's. This malfunction indicates voltage is not available and would therefore not allow ABS operation if required.

Action Taken:

A malfunction DTC is stored, ABS is disabled and the "ABS" warning lamp is turned "ON."

DTC Chart Test Description:

Number(s) below refer to circled number(s) on the diagnostic chart.

8. Checks for excessive resistance in the electronic brake control relay control circuitry.
9. Checks for proper battery voltage.
10. Checks for an open in the BATTERY feed circuit.
11. Checks for an open in the switched battery input circuit.
12. Checks for proper operation of the electronic brake control relay.

Diagnostic Aids:

An "intermittent" malfunction is most likely caused by a poor connection, rubbed through wire insulation, or a wire that is broken inside the insulation.

The frequency of the malfunction can be checked by using the enhanced diagnostic function of the Tech 1.

Any circuitry that is suspected as causing the "intermittent" complaint should be thoroughly checked for backed out terminals, improper mating, broken locks, improperly formed or damaged terminals, poor terminal to wiring connections or physical damage to the wiring harness.

Vibration, Temperature Effects:

Check for vibration effects by performing the relay test function of the Tech 1. With the relay commanded "ON," lightly tap the top and sides of the relay while monitoring battery voltage. If the battery voltage changes significantly, replace the relay.

If DTC 14 only sets when the vehicle is initially started in cold ambient conditions (temperature less than 32°F/0°C), replace the electronic brake control relay.

DTC 14
(Page 2 of 3)
ELECTRONIC BRAKE CONTROL RELAY CONTACT CIRCUITS OPEN

CONTINUED FROM PAGE 1 OF 3 OF THIS CHART.

(8)
- IGNITION "OFF."
- DISCONNECT JUMPER WIRE USED IN STEP 2.
- DISCONNECT NEGATIVE BATTERY CABLE FROM BATTERY.
- DISCONNECT POSITIVE BATTERY CABLE FROM BATTERY.
- TURN IGNITION "ON" (THIS IS DONE TO PROVIDE CIRCUIT CONTINUITY).
- USING J 39200, MEASURE RESISTANCE BETWEEN POSITIVE BATTERY CABLE TERMINAL AND 24-WAY EBTCM HARNESS CONNECTOR TERMINAL "20". IS RESISTANCE LESS THAN 95 OHMS WHEN RELAY IS APPROXIMATELY 25°C (77°F)?

NO: REFER TO PAGE 3 OF 3 OF THIS DTC CHART.

YES:

(9)
- USING J 39200, MEASURE VOLTAGE BETWEEN THE NEGATIVE AND POSITIVE BATTERY TERMINALS. IS VOLTAGE GREATER THAN 10 VOLTS?

NO: CORRECT LOW VOLTAGE CONDITION.

YES:

(10)
- IGNITION "OFF."
- RECONNECT BATTERY CABLES.
- DISCONNECT ELECTRONIC BRAKE CONTROL RELAY.
- USING J 39200, MEASURE VOLTAGE BETWEEN GROUND AND ENGINE COMPARTMENT FUSE/RELAY CENTER TERMINAL "A8". IS VOLTAGE GREATER THAN 10 VOLTS?

NO: REPAIR OPEN IN B+ FEED CIRCUIT.

YES:

(11)
- USING J 39200, MEASURE RESISTANCE BETWEEN 2-WAY EBTCM HARNESS CONNECTOR TERMINAL "A" AND ENGINE COMPARTMENT FUSE/RELAY CENTER TERMINAL "B6". IS RESISTANCE LESS THAN 2 OHMS?

NO: INSPECT IP TO ENGINE HARNESS CONNECTOR FOR A POOR CONNECTION OR POOR TERMINAL CONTACT. IS POOR TERMINAL CONTACT OR A POOR CONNECTION EVIDENT?
- NO: REPAIR OPEN IN CKT 1633.
- YES: REPLACE TERMINALS OR REPAIR CONNECTION INTEGRITY THAT EXHIBITS POOR CONTACT.

YES:

(12)
- USING A JUMPER WIRE, CONNECT ELECTRONIC BRAKE CONTROL RELAY TERMINAL "86" TO GROUND. THEN, USING A FUSED JUMPER WIRE SUCH AS J 36169 WITH A 3 AMP FUSE, CONNECT ENABLE RELAY TERMINAL "85" TO B+.
- USING J 39200, MEASURE RESISTANCE BETWEEN ENABLE RELAY TERMINALS "30" AND "87". IS RESISTANCE LESS THAN 2 OHMS?

NO: REPLACE ELECTRONIC BRAKE CONTROL RELAY.

YES: MALFUNCTION WAS DUE TO A POOR CONNECTION. REFER TO SECTION 8A, "CHECKING TERMINAL CONTACT" TO TEST FOR POOR TERMINAL CONTACT BETWEEN ELECTRONIC BRAKE CONTROL RELAY AND ENGINE COMPARTMENT FUSE/RELAY CENTER AND/OR AT POSITIVE BATTERY TERMINAL. REPLACE OR REPAIR TERMINALS THAT EXHIBIT POOR TERMINAL CONTACT.

DIAGNOSTIC CHART DTC 14 — 1995 CAVALIER AND SUNFIRE WITH TCS, CONTINUED

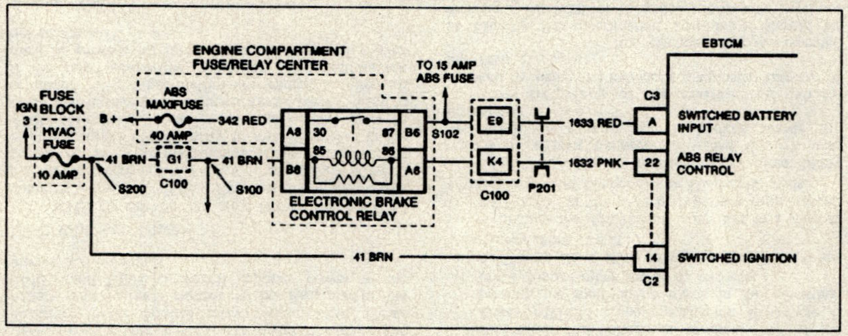

DTC 14
(Page 3 of 3)
ELECTRONIC BRAKE CONTROL RELAY CONTACT CIRCUIT OPEN

Circuit Description:

Ignition voltage is supplied through terminal "85" of the electronic brake control relay. The EBTCM is then able to energize the pull-in coil by completing the ground circuit at EBTCM terminal "20". The magnetic field created closes the electronic brake control relay contacts and allows battery voltage and current to be supplied to the EMB's and EBTCM, which supplies power to the motors and solenoids.

DTC Will Set When:

DTC 14 can be set anytime after the EBTCM commands the electronic brake control relay "ON" (the relay is first commanded "ON" during the three second bulb check). This test monitors the availability of current/voltage to the motors, solenoids, and EMB's. This malfunction indicates voltage is not available and would therefore not allow ABS operation if required.

Action Taken:

A malfunction DTC is stored, ABS is disabled and the "ABS" warning lamp is turned "ON."

DTC Chart Test Description:

Number(s) below refer to circled number(s) on the diagnostic chart.

13. Checks for excessive resistance in the electronic brake control relay coil.

14. Checks for excessive resistance in the electronic brake control relay control circuit between the relay and the EBTCM.

15. Checks for excessive resistance in the ignition switch "run" fuse feed circuit.

Diagnostic Aids:

An "intermittent" malfunction is most likely caused by a poor connection, rubbed through wire insulation, or a wire that is broken inside the insulation.

The frequency of the malfunction can be checked by using the enhanced diagnostic function of the Tech 1,

Any circuitry that is suspected as causing the "intermittent" complaint should be thoroughly checked for backed out terminals, improper mating, broken locks, improperly formed or damaged terminals, poor terminal to wiring connections or physical damage to the wiring harness.

Vibration, Temperature Effects:

Check for vibration effects by performing the relay test function of the Tech 1. With the relay commanded "ON," lightly tap the top and sides of the relay while monitoring battery voltage. If the battery voltage changes significantly, replace the relay.

If DTC 14 only sets when the vehicle is initially started in cold ambient conditions (temperature less than 32°F-0°C),replace the electronic brake control relay.

DIAGNOSTIC CHART DTC 14 — 1995 CAVALIER AND SUNFIRE WITH TCS, CONTINUED

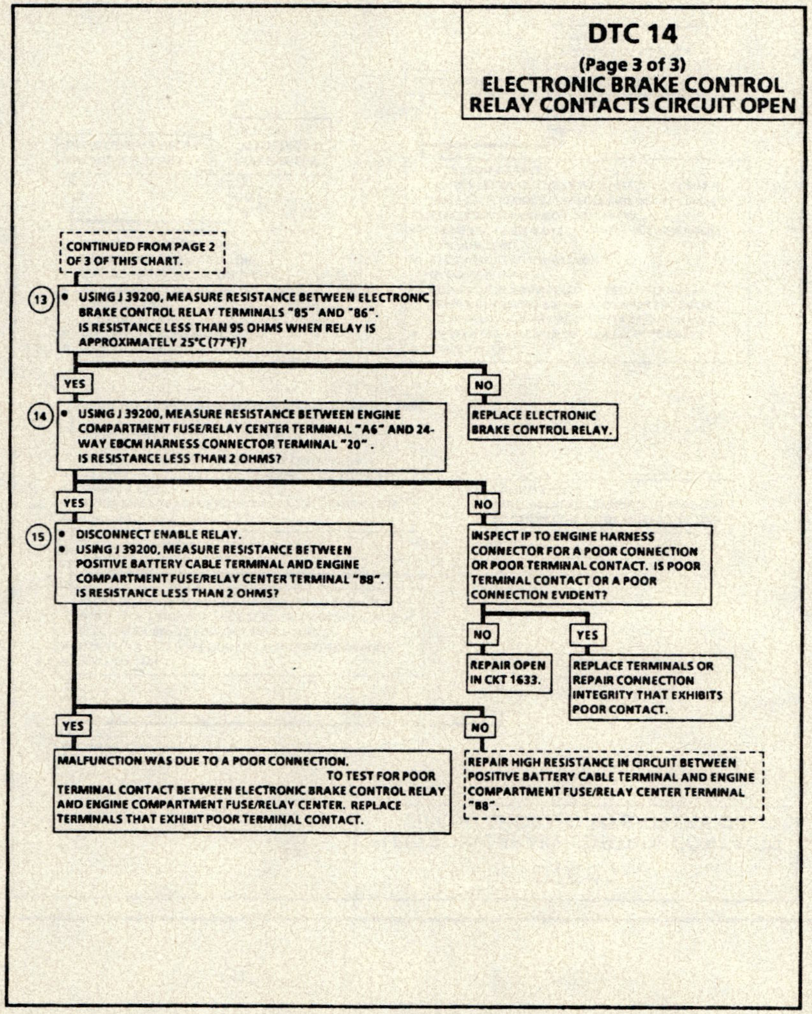

DIAGNOSTIC CHART DTC 15 — 1995 CAVALIER AND SUNFIRE WITH TCS

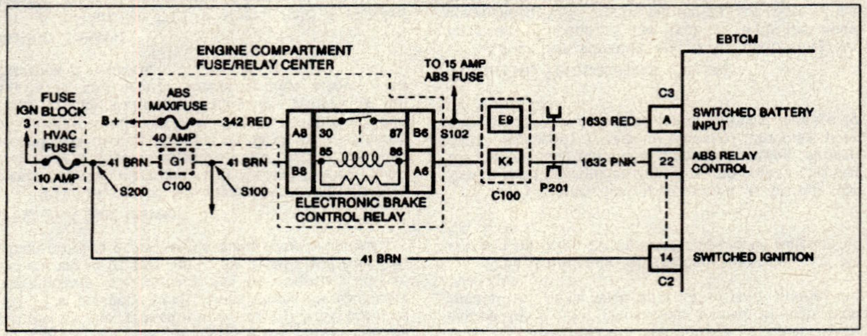

DTC 15

ELECTRONIC BRAKE CONTROL RELAY CONTACT CIRCUIT SHORTED TO BATTERY OR ALWAYS CLOSED

Circuit Description:

Ignition voltage is supplied through terminal "85" of the Electronic brake control relay. The EBTCM is then able to energize the pull-in coil by completing the ground circuit at terminal "20" of the EBTCM. The magnetic field created closes the electronic brake control relay contacts and allows battery voltage and current to be supplied to the EMB's and EBTCM, which supplies power to the motors and solenoids.

DTC Will Set When:

DTC 15 can be set only before the EBTCM commands the relay "ON." This test determines if the electronic brake control relay is energized when it should not be. This malfunction would not allow the electronic brake control relay to remove power to the ABS motor pack. If a second malfunction were to occur that requires the electronic brake control relay to be turned "OFF," that malfunction cannot be removed if the relay cannot be controlled. The malfunction must be present for three consecutive ignition cycles before the DTC is set.

Action Taken:

A malfunction DTC is stored. ABS is not disabled and the "ABS" warning lamp will remain "OFF."

DTC Chart Test Description:

Number(s) below refer to circled number(s) on the diagnostic chart.

1. Indicates if the EBTCM is capable of controlling the electronic brake control relay as commanded.
2. Checks for voltage at terminal "A" of the 2-way EBTCM harness connector. If voltage is present, the malfunction exists in the electronic brake control relay and/or its circuitry.
3. Checks for a short to voltage in the switched battery input circuit.
4. Ensures the malfunction is not due to a short to BATTERY in the switched battery input circuit with EBTCM connected. sh Diagnostic Aids:

An "intermittent" malfunction is most likely caused by a poor connection, rubbed through wire insulation, or a wire that is broken inside the insulation.

The frequency of the malfunction can be checked by using the enhanced diagnostic function of the Tech 1,

Any circuitry, that is suspected as causing the intermittent complaint, should be thoroughly checked for backed out terminals, improper mating, broken locks, improperly formed or damaged terminals, poor terminal to wiring connections or physical damage to the wiring harness.

DIAGNOSTIC CHART DTC 15 — 1995 CAVALIER AND SUNFIRE WITH TCS, CONTINUED

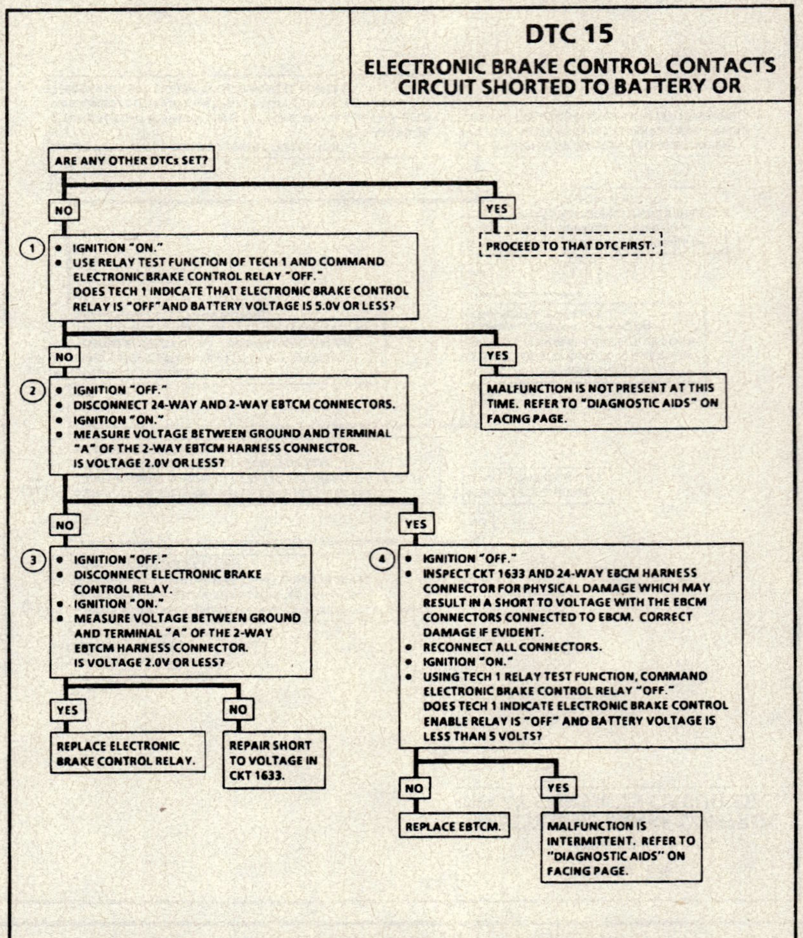

• AFTER DIAGNOSIS IS COMPLETE, CLEAR DTCs AND TEST DRIVE VEHICLE FOR THREE (3) DRIVE CYCLES TO VERIFY DTC DOES NOT RESET. A DRIVE CYCLE CONSISTS OF STARTING THE VEHICLE, DRIVING ABOVE 16 km/h (10 MPH) AND THEN TURNING IGNITION "OFF."

DIAGNOSTIC CHART DTC 16 — 1995 CAVALIER AND SUNFIRE WITH TCS

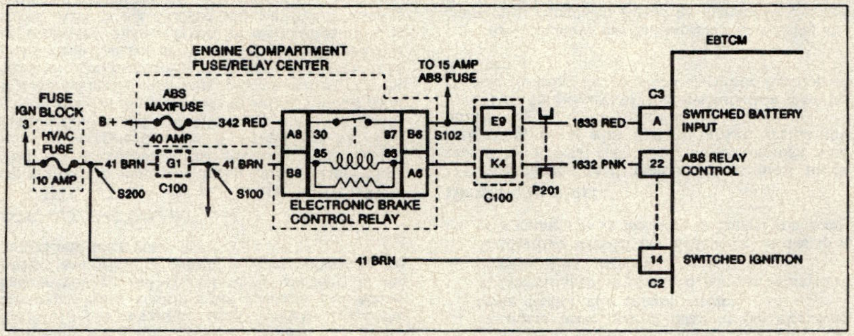

DTC 16
ELECTRONIC BRAKE CONTROL RELAY COIL CIRCUIT OPEN

Circuit Description:

Ignition voltage is supplied through terminal "85" of the electronic brake control relay. The EBTCM is then able to energize the pull-in coil by completing the ground circuit at terminal "20" of the EBTCM. The magnetic field created closes the electronic brake control relay contacts and allows battery voltage and current to be supplied to the EMB's and EBTCM, which supplies power to the motors and solenoids.

DTC Will Set When:

DTC 16 can be set anytime after the electronic brake control relay is commanded "ON." This test detects an open in the electronic brake control relay coil circuit. An open in this circuit will not allow the electronic brake control relay to be energized thus preventing power to the motors, EMB's and solenoids. A code 86 will be set with code 16 if the rear channel is not expected to be in the home position.

Action Taken:

A malfunction DTC is stored, ABS is disabled and the "ABS" warning lamp is turned "ON."

DTC Chart Test Description:

Number(s) below refer to circled number(s) on the diagnostic chart.

1. Indicates if the EBTCM is capable of controlling the electronic brake control relay as commanded.
2. Ensures that there is continuity through the pull-in coil of the electronic brake control relay.
3. Checks to ensure that voltage is available to the pull-in coil of the electronic brake control relay.
4. Checks continuity in the electronic brake control relay control circuit.
5. Ensures DTC was not set due to a poor connection.
6. Identifies whether the malfunction is due to a poor connection or the EBTCM.
7. Identifies whether the malfunction is due to a poor connection in the Dash to Engine connector.

Diagnostic Aids:

An "intermittent" malfunction is most likely caused by a poor connection, rubbed through wire insulation, or a wire that is broken inside the insulation.

The frequency of the malfunction can be checked by using the enhanced diagnostic function of the Tech 1,

If the frequency of the malfunction is high, but is currently intermittent, check for high coil resistance by measuring resistance between relay terminals "86" and "85". If resistance is greater than 100 ohms, replace the relay.

Any circuitry, that is suspected as causing the intermittent complaint, should be thoroughly checked for backed out terminals, improper mating, broken locks, improperly formed or damaged terminals, poor terminal to wiring connections or physical damage to the wiring harness.

DIAGNOSTIC CHART DTC 16 — 1995 CAVALIER AND SUNFIRE WITH TCS, CONTINUED

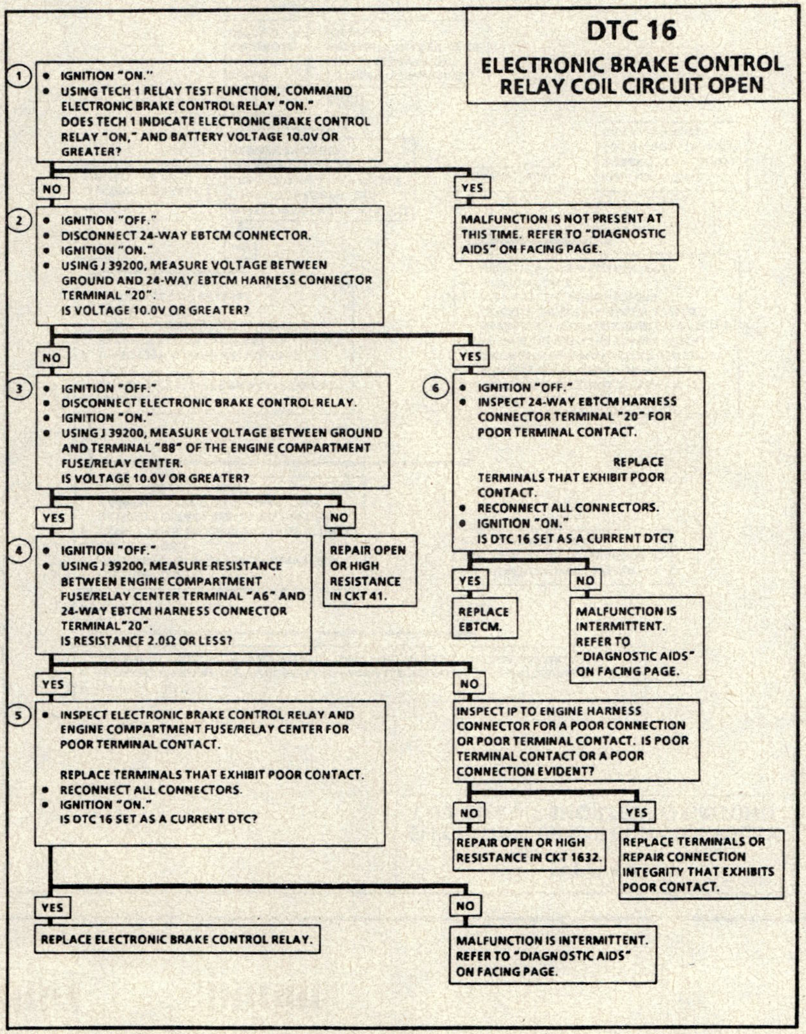

DIAGNOSTIC CHART DTC 17 — 1995 CAVALIER AND SUNFIRE WITH TCS

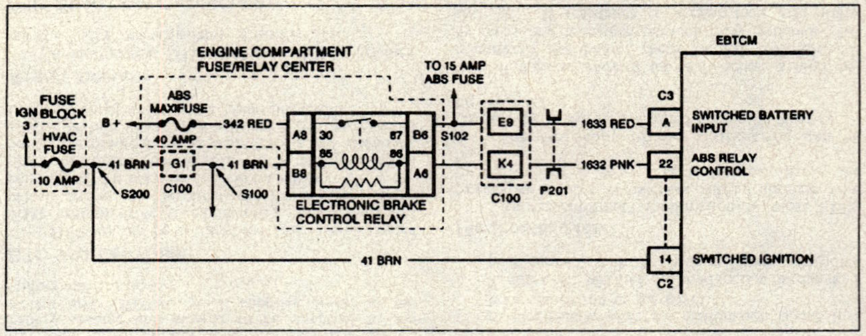

DTC 17
ELECTRONIC BRAKE CONTROL RELAY COIL CIRCUIT SHORTED TO GROUND

Circuit Description:

Ignition voltage is supplied through terminal "85" of the electronic brake control relay. The EBTCM is then able to energize the pull-in coil by completing the ground circuit at terminal "20" of the EBTCM. The magnetic field created closes the electronic brake control relay contacts and allows battery voltage and current to be supplied to the EMB's and EBTCM, which supplies voltage to the motors and solenoids.

DTC Will Set When:

DTC 17 can be set before the EBTCM commands the electronic brake control relay "ON." This test determines if the electronic brake control relay is energized when it should not be. This malfunction would not allow the electronic brake control relay to remove power to the ABS motor pack. If a second malfunction were to occur that requires the electronic brake control relay to be turned "OFF," that malfunction can not be removed if the relay can not be controlled. Action Taken: A malfunction DTC is stored. ABS is not disabled and the "ABS" warning lamp will remain "OFF."

DTC Chart Test Description:

Number(s) below refer to circled number(s) on the diagnostic chart.

1. Indicates if the EBTCM is capable of controlling the electronic brake control relay as commanded.
2. Ensures the electronic brake control relay or control circuit are not shorted to ground.
3. This step isolates the malfunction to either the electronic brake control relay or the electronic brake control relay control circuit.
4. Ensures short to ground is not due to physical damage of the circuitry.
5. Identifies whether the malfunction is due to a poor connection in the Dash to Engine connector.

Diagnostic Aids:

An "intermittent" malfunction is most likely caused by a poor connection, rubbed through wire insulation, or a wire that is broken inside the insulation.

The frequency of the malfunction can be checked by using the enhanced diagnostic function of the Tech 1,

Any circuitry, that is suspected as causing the intermittent complaint, should be thoroughly checked for backed out terminals, improper mating, broken locks, improperly formed or damaged terminals, poor terminal to wiring connections or physical damage to the wiring harness.

DIAGNOSTIC CHART DTC 17 — 1995 CAVALIER AND SUNFIRE WITH TCS, CONTINUED

DTC 17
ELECTRONIC BRAKE CONTROL RELAY COIL CIRCUIT SHORTED TO GROUND

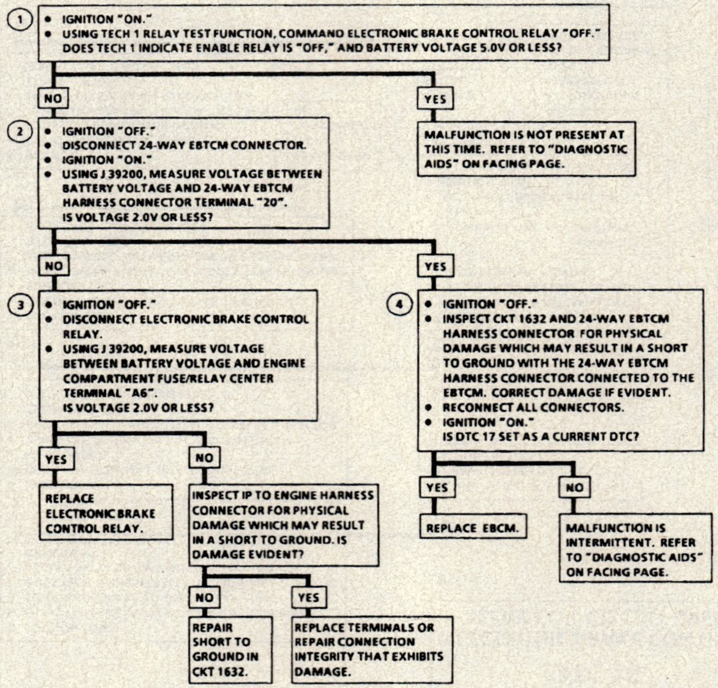

DIAGNOSTIC CHART DTC 18 — 1995 CAVALIER AND SUNFIRE WITH TCS

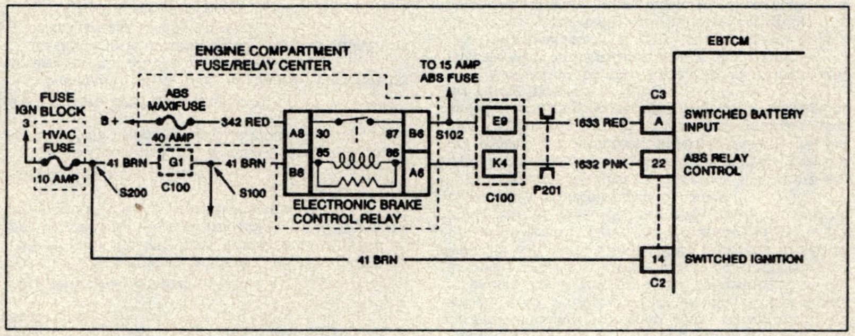

DTC 18
ELECTRONIC BRAKE CONTROL RELAY COIL CIRCUIT SHORTED TO BATTERY OR COIL SHORTED

Circuit Description:

Ignition voltage is supplied through terminal "85" of the electronic brake control relay. The EBTCM is then able to energize the pull-in coil by completing the ground circuit at terminal "20" of the EBTCM. The magnetic field created closes the electronic brake control relay contacts and allows battery voltage and current to be supplied to the EMB's and EBTCM, which supplies voltage to the motors and solenoids.

DTC Will Set When:

DTC 18 can be set after the EBTCM commands the electronic brake control relay "ON." This test monitors the availability of current/voltage to the motors, solenoids and EMB's. A malfunction of this type will not allow voltage to be available to the motors, solenoids and EMB's; therefore, ABS operation would not be allowed if required. DTC 86 is set with DTC 18 if the rear channel is not expected to be in the home position.

Action Taken:

A malfunction DTC is stored, ABS is disabled and the "ABS" warning lamp is turned "ON."

DTC Chart Test Description:

Number(s) below refer to circled number(s) on the diagnostic chart.

1. Indicates if the EBTCM is capable of controlling the electronic brake control relay as commanded.
2. With the electronic brake control relay removed, voltage should not be available at terminal "86". Any voltage at this point would indicate that the electronic brake control relay control circuit was shorted to a voltage source.
3. Checks for a short across the electronic brake control relay coil.
4. Checks for an internal short in the electronic brake control relay between the BATTERY feed terminal "30" and the relay pull-in coil.
5. Ensures short to BATTERY was not due to physical damage of the electronic brake control relay control circuit.
6. Identifies whether the malfunction is due to a shorted connection in the Dash to Engine connector.

Diagnostic Aids:

An "intermittent" malfunction is most likely caused by a poor connection, rubbed through wire insulation, or a wire that is broken inside the insulation.

The frequency of the malfunction can be checked by using the enhanced diagnostic function of the Tech 1.

Any circuitry, that is suspected as causing the intermittent complaint, should be thoroughly checked for backed out terminals, improper mating, broken locks, improperly formed or damaged terminals, poor terminal to wiring connections or physical damage to the wiring harness.

DIAGNOSTIC CHART DTC 18 — 1995 CAVALIER AND SUNFIRE WITH TCS, CONTINUED

DTC 18
ELECTRONIC BRAKE CONTROL RELAY COIL CIRCUIT SHORTED TO BATTERY OR COIL SHORTED

(1)
- IGNITION "ON."
- USING RELAY TEST FUNCTION OF TECH 1, COMMAND ELECTRONIC BRAKE CONTROL RELAY "ON." DOES TECH 1 INDICATE ELECTRONIC BRAKE CONTROL RELAY "ON" AND BATTERY VOLTAGE 10.0V OR GREATER?

YES: MALFUNCTION IS NOT PRESENT AT THIS TIME. REFER TO "DIAGNOSTIC AIDS" ON FACING PAGE.

NO:

(2)
- IGNITION "OFF."
- DISCONNECT ELECTRONIC BRAKE CONTROL RELAY.
- DISCONNECT 24-WAY EBTCM CONNECTOR.
- IGNITION "ON."
- USING J 39200, MEASURE VOLTAGE BETWEEN GROUND AND ENGINE COMPARTMENT FUSE/RELAY CENTER TERMINAL "A6". IS VOLTAGE 2.0V OR LESS?

NO: INSPECT IP TO ENGINE HARNESS CONNECTOR FOR PHYSICAL DAMAGE WHICH MAY RESULT IN A SHORT TO BATTERY. IS DAMAGE EVIDENT?
- NO: REPAIR SHORT TO GROUND IN CKT 1632.
- YES: REPLACE TERMINALS OR REPAIR CONNECTION INTEGRITY THAT EXHIBITS DAMAGE.

YES:

(3)
- USING J 39200, MEASURE RESISTANCE BETWEEN ELECTRONIC BRAKE CONTROL RELAY TERMINALS "85" AND "86". IS RESISTANCE 40.0Ω OR GREATER?

NO: REPLACE ELECTRONIC BRAKE CONTROL RELAY.

YES:

(4)
- USING J 39200, MEASURE RESISTANCE BETWEEN ELECTRONIC BRAKE CONTROL RELAY TERMINALS "86" AND "30". DOES J 39200 DISPLAY RESISTANCE AS "OL" (INFINITE)?

NO: REPLACE ELECTRONIC BRAKE CONTROL RELAY.

YES:

(5)
- IGNITION "OFF."
- INSPECT CKT 1632 AND 24-WAY EBTCM HARNESS CONNECTOR FOR PHYSICAL DAMAGE WHICH MAY RESULT IN A SHORT TO BATTERY WITH THE 24-WAY EBTCM HARNESS CONNECTOR CONNECTED TO THE EBTCM. CORRECT DAMAGE IF EVIDENT.
- RECONNECT ALL CONNECTORS.
- IGNITION "ON."

IS DTC 18 SET AS A CURRENT DTC?

YES: REPLACE EBTCM.

NO: MALFUNCTION IS INTERMITTENT. REFER TO "DIAGNOSTIC AIDS" ON FACING PAGE.

DIAGNOSTIC CHART DTC 21 — 1995 CAVALIER AND SUNFIRE WITH TCS

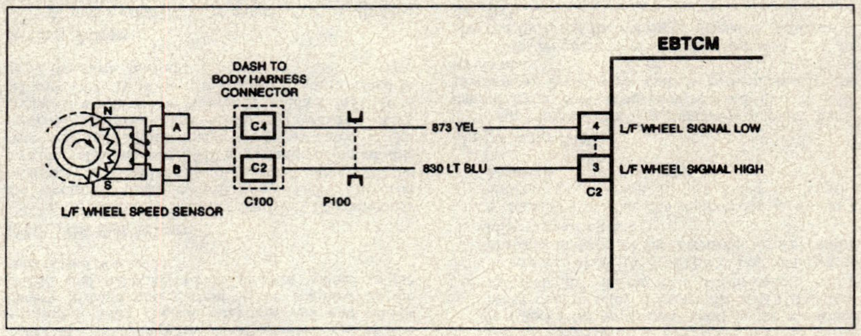

DTC 21
LEFT FRONT WHEEL SPEED = 0

Circuit Description:

As a toothed ring passes by the wheel speed sensor, changes in the electromagnetic field cause the wheel speed sensor to produce a sinusoidal (AC) voltage signal whose frequency is proportional to wheel speed. The magnitude of this signal is directly related to wheel speed and the proximity of the wheel speed sensor to the toothed ring, often referred to as the air gap.

DTC Will Set When:

DTC 21 can be set when the vehicle is not in an ABS stop. If the left front wheel speed is equal to zero for two seconds and the vehicle speed is greater than 8 km/h (5 mph), a malfunction exists.

Action Taken:

A malfunction DTC is stored, ABS is disabled and the "ABS" warning lamp is turned "ON."

DTC Chart Test Description:

Number(s) below refer to circled number(s) on the diagnostic chart.

1. Identifies if a wheel speed sensor or circuitry is damaged and visibly apparent.
2. Checks wheel speed sensor for proper resistance value.
3. Ensures the wheel speed sensor and toothed ring generate the proper voltage.
4. Tests for a short between the wheel speed signal high and low circuits.
5. Identifies if a wheel speed circuit could be shorted when the EBTCM is connected.

Diagnostic Aids:

An "intermittent" malfunction is most likely caused by a poor connection, rubbed through wire insulation, or a wire that is broken inside the insulation. The frequency of the malfunction can be checked by using the enhanced diagnostic function of the Tech 1.

If the customer's comments reflect that the "ABS" warning lamp is "ON" only during humid conditions (rain, snow, vehicle wash), all wheel speed sensor circuitry should be thoroughly inspected for signs of water intrusion. Use the following procedure. Spray down the suspected area with 5% salt water solution (two teaspoons of salt to 12 oz. of water). Drive vehicle above 24 km/h (15 mph) for at least 30 seconds. If DTC returns immediately, replace suspected harness.

Any circuitry, that is suspected as causing the intermittent complaint, should be thoroughly checked for backed out terminals, improper mating, broken locks, improperly formed or damaged terminals, poor terminal to wiring connections or physical damage to the wiring harness.

Resistance of the wheel speed sensor will increase with an increase in sensor temperature.

When replacing a wheel speed sensor, inspect the sensor terminals and harness connector for corrosion and/or water intrusion. If evidence of corrosion or water intrusion exists, replace wheel speed sensor harness.

If wheel speed sensor harness shows evidence of water intrusion, inspect wheel speed sensor for water intrusion. If water intrusion is evident, replace wheel speed sensor.

DIAGNOSTIC CHART DTC 21 — 1995 CAVALIER AND SUNFIRE WITH TCS, CONTINUED

DTC 21
LEFT FRONT WHEEL SPEED = 0

Important

- WHEEL SPEED SENSOR INTERMITTENT MALFUNCTIONS MAY BE DIFFICULT TO LOCATE. CARE SHOULD BE TAKEN NOT TO DISTURB ANY ELECTRICAL CONNECTIONS PRIOR TO AN INDICATED STEP OF THIS CHART. THIS WILL ENSURE THAT AN INTERMITTENT CONNECTION WILL NOT BE CORRECTED BEFORE THE SOURCE OF THE MALFUNCTION IS FOUND.

(1)
- IGNITION "OFF."
- INSPECT WHEEL SPEED SENSOR WIRING AND CONNECTORS FOR DAMAGE. INSPECT WHEEL SPEED SENSOR AND TOOTHED RING FOR LOOSENESS OR DAMAGE. PAY PARTICULAR ATTENTION TO THE TOOTHED RING. ANY SIGNIFICANT DAMAGE WILL AFFECT THE WHEEL SPEED INPUT SIGNAL. IS PHYSICAL DAMAGE OR SENSOR/TOOTHED RING LOOSENESS EVIDENT?

YES: CORRECT PHYSICAL DAMAGE; CLEAR DTCs AND RECHECK.
NO: go to (2).

(2)
- DISCONNECT CONNECTOR FROM LEFT FRONT WHEEL SPEED SENSOR.
- USING J 39200, MEASURE RESISTANCE BETWEEN SENSOR TERMINALS "A" AND "B". IS RESISTANCE BETWEEN 1628 AND 1889 Ω WHEN SENSOR IS 25°C (77°F)? *(IF TEMPERATURE OF THE SENSOR IS SUSPECTED TO BE OTHER THAN 25°C (77°F), REFER TO THE TABLE BELOW FOR THE APPLICABLE SENSOR RESISTANCE VALUES; THE VALUES IN THE TABLE ARE FOR TEMPERATURE OF THE SENSOR NOT AIR TEMPERATURE.)*

YES: go to (3).
NO: REPLACE WHEEL SPEED SENSOR.

(3)
- WITH J 39200 STILL CONNECTED, SELECT AC VOLTAGE SCALE.
- SPIN WHEEL BY HAND WHILE MONITORING VOLTAGE. IS VOLTAGE 100 mV OR GREATER (VOLTAGE WILL INCREASE AS WHEEL SPEED INCREASES)?

YES: go to (4).
NO: REPLACE WHEEL SPEED SENSOR.

(4)
- DISCONNECT 24-WAY EBTCM CONNECTOR.
- USING J 39200, MEASURE RESISTANCE BETWEEN 24-WAY EBTCM HARNESS CONNECTOR TERMINALS "3" AND "4". DOES J 39200 DISPLAY RESISTANCE AS "OL" (INFINITE)?

YES: go to (5).
NO: REPAIR SHORT BETWEEN CKTs 830 AND 873.

(5)
- INSPECT CKT 830 AND 873 AND 24-WAY EBTCM HARNESS CONNECTOR FOR PHYSICAL DAMAGE WHICH MAY RESULT IN A SHORT BETWEEN CKT 830 AND 873 WITH THE 24-WAY EBTCM HARNESS CONNECTOR CONNECTED TO THE EBTCM. CORRECT DAMAGE IF EVIDENT.
- RECONNECT ALL CONNECTORS.
- DRIVE VEHICLE ABOVE 24 km/h (15 MPH) FOR AT LEAST 30 SECONDS. IS DTC 21 SET AS A CURRENT DTC?

YES: REPLACE EBTCM.
NO: MALFUNCTION IS INTERMITTENT. REFER TO "DIAGNOSTIC AIDS" ON FACING PAGE.

FRONT WHEEL SPEED SENSOR
TEMPERATURE OF SENSOR VS. SENSOR RESISTANCE (APPROXIMATE)

TEMP. (°C)	TEMP. (°F)	RESISTANCE (OHMS)
-34 TO 4	-30 TO 40	1258 TO 1739
5 TO 43	41 TO 110	1503 TO 2023
44 TO 93	111 TO 200	1747 TO 2384

DIAGNOSTIC CHART DTC 22 — 1995 CAVALIER AND SUNFIRE WITH TCS

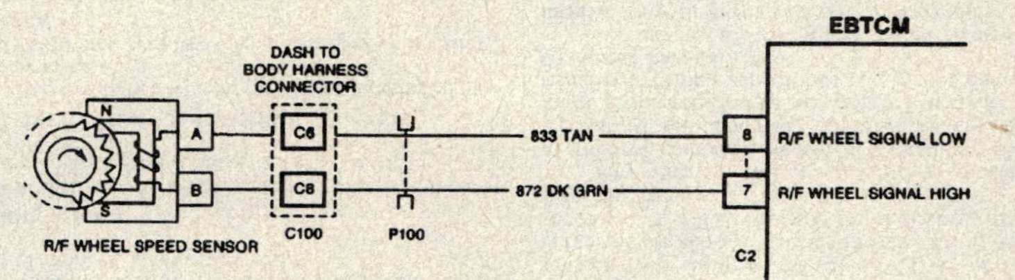

DTC 22
RIGHT FRONT WHEEL SPEED = 0

Circuit Description:

As a toothed ring passes by the wheel speed sensor, changes in the electromagnetic field cause the wheel speed sensor to produce a sinusoidal (AC) voltage signal whose frequency is proportional to wheel speed. The magnitude of this signal is directly related to wheel speed and the proximity of the wheel speed sensor to the toothed ring, often referred to as the air gap.

DTC Will Set When:

DTC 22 can be set when the vehicle is not in an ABS stop. If the right front wheel speed is equal to zero for two seconds and the vehicle speed is greater than 8 km/h (5 mph), a malfunction exists.

Action Taken:

A malfunction DTC is stored, ABS is disabled and the "ABS" warning lamp is turned "ON."

DTC Chart Test Description:

Number(s) below refer to circled number(s) on the diagnostic chart.

1. Identifies if a wheel speed sensor or circuitry is damaged and visibly apparent.
2. Checks wheel speed sensor for proper resistance value.
3. Ensures the wheel speed sensor and toothed ring generate the proper voltage.
4. Tests for a short between the wheel speed signal high and low circuits.
5. Identifies if a wheel speed circuit could be shorted when the EBTCM is connected.

Diagnostic Aids:

An "intermittent" malfunction is most likely caused by a poor connection, rubbed through wire insulation, or a wire that is broken inside the insulation. The frequency of the malfunction can be checked by using the enhanced diagnostic function of the Tech 1.

If the customer's comments reflect that the "ABS" warning lamp is "ON" only during humid conditions (rain, snow, vehicle wash), all wheel speed sensor circuitry should be thoroughly inspected for signs of water intrusion. Use the following procedure. Spray down the suspected area with 5% salt water solution (two teaspoons of salt to 12 oz. of water). Drive vehicle above 24 km/h (15 mph) for at least 30 seconds. If DTC returns immediately, replace suspected harness.

Any circuitry, that is suspected as causing the intermittent complaint, should be thoroughly checked for backed out terminals, improper mating, broken locks, improperly formed or damaged terminals, poor terminal to wiring connections or physical damage to the wiring harness.

Resistance of the wheel speed sensor will increase with an increase in sensor temperature.

When replacing a wheel speed sensor, inspect the sensor terminals and harness connector for corrosion and/or water intrusion. If evidence of corrosion or water intrusion exists, replace wheel speed sensor harness.

If wheel speed sensor harness shows evidence of water intrusion, inspect wheel speed sensor for water intrusion. If water intrusion is evident, replace wheel speed sensor.

DIAGNOSTIC CHART DTC 22 — 1995 CAVALIER AND SUNFIRE WITH TCS, CONTINUED

DTC 22
RIGHT FRONT WHEEL SPEED = 0

Important

- WHEEL SPEED SENSOR INTERMITTENT MALFUNCTIONS MAY BE DIFFICULT TO LOCATE. CARE SHOULD BE TAKEN NOT TO DISTURB ANY ELECTRICAL CONNECTIONS PRIOR TO AN INDICATED STEP OF THIS CHART. THIS WILL ENSURE THAT AN INTERMITTENT CONNECTION WILL NOT BE CORRECTED BEFORE THE SOURCE OF THE MALFUNCTION IS FOUND.

(1)
- IGNITION "OFF."
- INSPECT WHEEL SPEED SENSOR WIRING AND CONNECTORS FOR DAMAGE. INSPECT WHEEL SPEED SENSOR AND TOOTHED RING FOR LOOSENESS OR DAMAGE. PAY PARTICULAR ATTENTION TO THE TOOTHED RING. ANY SIGNIFICANT DAMAGE WILL AFFECT THE WHEEL SPEED INPUT SIGNAL.

IS PHYSICAL DAMAGE OR SENSOR/TOOTHED RING LOOSENESS EVIDENT?

YES → CORRECT PHYSICAL DAMAGE; CLEAR DTCs AND RECHECK.

NO ↓

(2)
- DISCONNECT CONNECTOR FROM RIGHT FRONT WHEEL SPEED SENSOR.
- USING J 39200, MEASURE RESISTANCE BETWEEN SENSOR TERMINALS "A" AND "B". IS RESISTANCE BETWEEN 1628Ω AND 1889Ω WHEN SENSOR IS 25°C (77°F)? *[IF TEMPERATURE OF THE SENSOR IS SUSPECTED TO BE OTHER THAN 25°C (77°F), REFER TO THE TABLE BELOW FOR THE APPLICABLE SENSOR RESISTANCE VALUES; THE VALUES IN THE TABLE ARE FOR TEMPERATURE OF THE SENSOR NOT AIR TEMPERATURE.]*

NO → REPLACE WHEEL SPEED SENSOR.

YES ↓

(3)
- WITH J 39200 STILL CONNECTED, SELECT AC VOLTAGE SCALE.
- SPIN WHEEL BY HAND WHILE MONITORING VOLTAGE.

IS VOLTAGE 100 mV OR GREATER (VOLTAGE WILL INCREASE AS WHEEL SPEED INCREASES)?

NO → REPLACE WHEEL SPEED SENSOR.

YES ↓

(4)
- DISCONNECT 24-WAY EBTCM CONNECTOR.
- USING J 39200, MEASURE RESISTANCE BETWEEN 24-WAY EBTCM HARNESS CONNECTOR TERMINALS "7" AND "8".

DOES J 39200 DISPLAY RESISTANCE AS "OL" (INFINITE)?

NO → REPAIR SHORT BETWEEN CKTs 833 AND 872.

YES ↓

(5)
- INSPECT CKT 833 AND 872 AND 24-WAY EBTCM HARNESS CONNECTOR FOR PHYSICAL DAMAGE WHICH MAY RESULT IN A SHORT BETWEEN CKT 833 AND 872 WITH THE 24-WAY EBTCM HARNESS CONNECTOR CONNECTED TO THE EBTCM. CORRECT DAMAGE IF EVIDENT.
- RECONNECT ALL CONNECTORS.
- DRIVE VEHICLE ABOVE 24 km/h (15 MPH) FOR AT LEAST 30 SECONDS.

IS DTC 22 SET AS A CURRENT DTC?

YES → REPLACE EBTCM.

NO → MALFUNCTION IS INTERMITTENT. REFER TO "DIAGNOSTIC AIDS" ON FACING PAGE.

FRONT WHEEL SPEED SENSOR
TEMPERATURE OF SENSOR VS. SENSOR RESISTANCE (APPROXIMATE)

TEMP. (°C)	TEMP. (°F)	RESISTANCE (OHMS)
-34 TO 4	-30 TO 40	1258 TO 1739
5 TO 43	41 TO 110	1503 TO 2023
44 TO 93	111 TO 200	1747 TO 2384

DIAGNOSTIC CHART DTC 23 — 1995 CAVALIER AND SUNFIRE WITH TCS

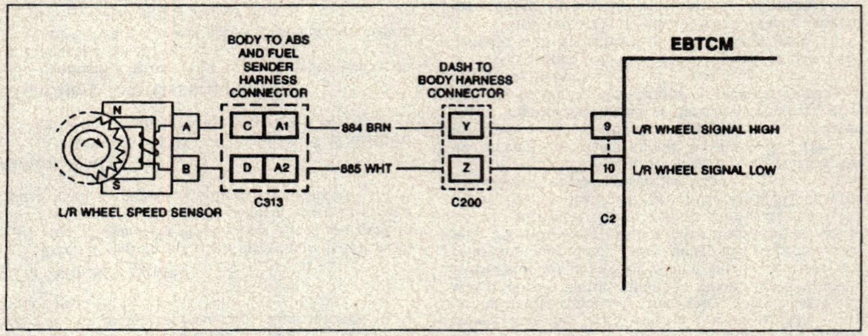

DTC 23
(Page 1 of 2)
LEFT REAR WHEEL SPEED = 0

Circuit Description:

As a toothed ring passes by the wheel speed sensor, changes in the electromagnetic field cause the wheel speed sensor to produce a sinusoidal (AC) voltage signal whose frequency is proportional to wheel speed. The magnitude of this signal is directly related to wheel speed and the proximity of the wheel speed sensor to the toothed ring, often referred to as the air gap.

DTC Will Set When:

DTC 23 can be set when the vehicle is not in an ABS stop. If the left rear wheel speed is equal to zero for two seconds and the vehicle speed is greater than 8 km/h (5 mph), a malfunction exists.

Action Taken:

A malfunction DTC is stored, ABS is disabled and the "ABS" warning lamp is turned "ON."

DTC Chart Test Description:

Number(s) below refer to circled number(s) on the diagnostic chart.

1. Identifies circuitry that is damaged and visibly apparent.
2. Checks for proper resistance at the ABS rear axle harness connector.
3. Checks for a short between the wheel speed signal high and low circuits in the ABS rear axle harness.
4. Checks wheel speed sensor for proper resistance value.

Diagnostic Aids:

An "intermittent" malfunction is most likely caused by a poor connection, rubbed through wire insulation, or a wire that is broken inside the insulation.

The frequency of the malfunction can be checked by using the enhanced diagnostic function of the Tech 1.

If the customer's comments reflect that the "ABS" warning lamp is "ON" only during humid conditions (rain, snow, vehicle wash), all wheel speed sensor circuitry should be thoroughly inspected for signs of water intrusion. Use the following procedure. Spray down the suspected area with 5% salt water solution (two teaspoons of salt to 12 oz. of water). Drive vehicle above 24 km/h (15 mph) for at least 30 seconds. If DTC returns immediately, replace suspected harness.

Any circuitry, that is suspected as causing the intermittent complaint, should be thoroughly checked for backed out terminals, improper mating, broken locks, improperly formed or damaged terminals, poor terminal to wiring connections or physical damage to the wiring harness.

Resistance of the wheel speed sensor will increase with an increase in sensor temperature.

When replacing a wheel speed sensor, inspect the sensor terminals and harness connector for corrosion and/or water intrusion. If evidence or corrosion or water intrusion exists, replace wheel speed sensor harness.

If wheel speed sensor harness shows evidence of water intrusion, inspect wheel speed sensor for water intrusion. If water intrusion is evident, replace wheel speed sensor.

DIAGNOSTIC CHART DTC 23 — 1995 CAVALIER AND SUNFIRE WITH TCS, CONTINUED

DTC 23
(Page 1 of 2)
LEFT REAR WHEEL SPEED = 0

Important

- WHEEL SPEED SENSOR INTERMITTENT MALFUNCTIONS MAY BE DIFFICULT TO LOCATE. CARE SHOULD BE TAKEN NOT TO DISTURB ANY ELECTRICAL CONNECTIONS PRIOR TO AN INDICATED STEP OF THIS CHART. THIS WILL ENSURE THAT AN INTERMITTENT CONNECTION WILL NOT BE CORRECTED BEFORE THE SOURCE OF THE MALFUNCTION IS FOUND.

(1)
- IGNITION "OFF."
- INSPECT WHEEL SPEED SENSOR WIRING AND CONNECTORS FOR DAMAGE. IS DAMAGE EVIDENT?

YES → REPAIR DAMAGE. DRIVE VEHICLE ABOVE 24 km/h (15 MPH) FOR AT LEAST 30 SECONDS TO ENSURE DTC 23 DOES NOT SET.

NO ↓

(2)
- DISCONNECT ABS REAR AXLE HARNESS FROM BODY TO ABS AND FUEL SENDER CONNECTOR.
- CONNECT J 39200, MEASURE RESISTANCE BETWEEN ABS REAR AXLE HARNESS CONNECTOR TERMINALS "A" AND "B".

IS RESISTANCE BETWEEN 2100 AND 2400 Ω WHEN SENSOR IS 25°C (77°F)? *[IF TEMPERATURE OF THE SENSOR IS SUSPECTED TO BE OTHER THAN 25°C (77°F), REFER TO THE TABLE BELOW FOR THE APPLICABLE SENSOR RESISTANCE VALUES; THE VALUES IN THE TABLE ARE FOR TEMPERATURE OF THE SENSOR, NOT AIR TEMPERATURE.]*

YES → GO TO PAGE 2 OF 2 OF THIS CHART.

NO ↓

(3)
- REMOVE LEFT REAR INTEGRAL BEARING AND SPEED SENSOR ASSEMBLY (TUBULAR AXLE ONLY).
- DISCONNECT CONNECTOR FROM LEFT REAR WHEEL SPEED SENSOR.
- USING J 39200, MEASURE RESISTANCE BETWEEN ABS REAR AXLE HARNESS CONNECTOR TERMINALS "A" AND "B".

DOES J 39200 DISPLAY RESISTANCE AS "OL" (INFINITE)?

NO → REPLACE ABS REAR AXLE HARNESS.

YES ↓

(4)
- USING J 39200, MEASURE RESISTANCE BETWEEN WHEEL SPEED SENSOR TERMINALS "A" AND "B." IS RESISTANCE BETWEEN 2100 AND 2400 Ω WHEN SENSOR IS 25°C (77°F)? *[IF TEMPERATURE OF THE SENSOR IS SUSPECTED TO BE OTHER THAN 25°C (77°F), REFER TO THE TABLE BELOW FOR THE APPLICABLE SENSOR RESISTANCE VALUES; THE VALUES IN THE TABLE ARE FOR TEMPERATURE OF THE SENSOR, NOT AIR TEMPERATURE.]*

NO → REPLACE INTEGRAL BEARING AND SPEED SENSOR ASSEMBLY.

YES → MALFUNCTION WAS DUE TO AN INTERMITTENT SHORT BETWEEN CKTs 884 AND 885. REPLACE ABS REAR AXLE HARNESS.

REAR WHEEL SPEED SENSOR

TEMPERATURE OF SENSOR VS. SENSOR RESISTANCE (APPROXIMATE)

TEMP. (°C)	TEMP. (°F)	RESISTANCE (OHMS)
-34 TO 4	-30 TO 40	1623 TO 2210
5 TO 43	41 TO 110	1939 TO 2570
44 TO 93	111 TO 200	2254 TO 3029

DIAGNOSTIC CHART DTC 23 — 1995 CAVALIER AND SUNFIRE WITH TCS, CONTINUED

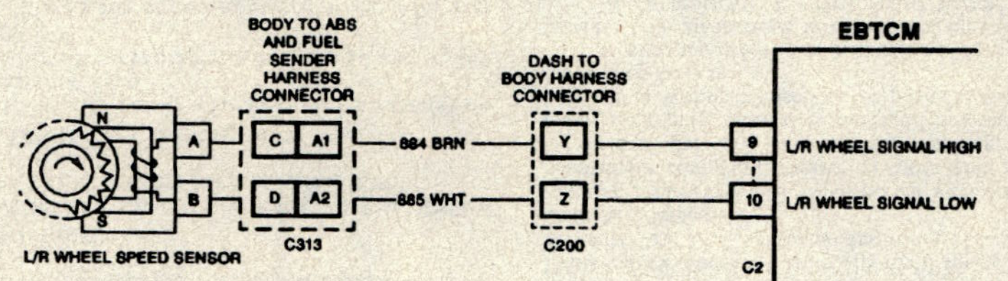

DTC 23
(Page 2 of 2)
LEFT REAR WHEEL SPEED = 0

Circuit Description:

As a toothed ring passes by the wheel speed sensor, changes in the electromagnetic field cause the wheel speed sensor to produce a sinusoidal (AC) voltage signal whose frequency is proportional to wheel speed. The magnitude of this signal is directly related to wheel speed and the proximity of the wheel speed sensor to the toothed ring, often referred to as the air gap.

DTC Will Set When:

DTC 23 can be set when the vehicle is not in an ABS stop. If the left rear wheel speed is equal to zero for two seconds and the vehicle speed is greater than 8 km/h (5 mph), a malfunction exists.

Action Taken:

A malfunction DTC is stored, ABS is disabled and the "ABS" warning lamp is turned "ON."

DTC Chart Test Description:

Number(s) below refer to circled number(s) on the diagnostic chart.

5. This test ensures the wheel speed sensor and sensor ring generate the proper voltage.
6. Checks for short between the wheel speed signal high and low circuits in the dash or body harness.
7. Ensures malfunction was not due to physical damage of the circuitry.

Diagnostic Aids:

An "intermittent" malfunction is most likely caused by a poor connection, rubbed through wire insulation, or a wire that is broken inside the insulation.

The frequency of the malfunction can be checked by using the enhanced diagnostic function of the Tech 1.

If the customer's comments reflect that the "ABS" warning lamp is "ON" only during humid conditions (rain, snow, vehicle wash), all wheel speed sensor circuitry should be thoroughly inspected for signs of water intrusion. Use the following procedure. Spray down the suspected area with 5% salt water solution (two teaspoons of salt to 12 oz. of water). Drive vehicle above 24 km/h (15 mph) for at least 30 seconds. If DTC returns immediately, replace suspected harness.

Any circuitry, that is suspected as causing the intermittent complaint, should be thoroughly checked for backed out terminals, improper mating, broken locks, improperly formed or damaged terminals, poor terminal to wiring connections or physical damage to the wiring harness.

Resistance of the wheel speed sensor will increase with an increase in sensor temperature. When replacing a wheel speed sensor, inspect the sensor terminals and harness connector for corrosion and/or water intrusion. If evidence or corrosion or water intrusion exists, replace wheel speed sensor harness.

If wheel speed sensor harness shows evidence of water intrusion, inspect wheel speed sensor for water intrusion. If water intrusion is evident, replace wheel speed sensor.

DIAGNOSTIC CHART DTC 23 — 1995 CAVALIER AND SUNFIRE WITH TCS, CONTINUED

DTC 23
(Page 2 of 2)
LEFT REAR WHEEL SPEED = 0
(W/ETS OR VES ONLY)

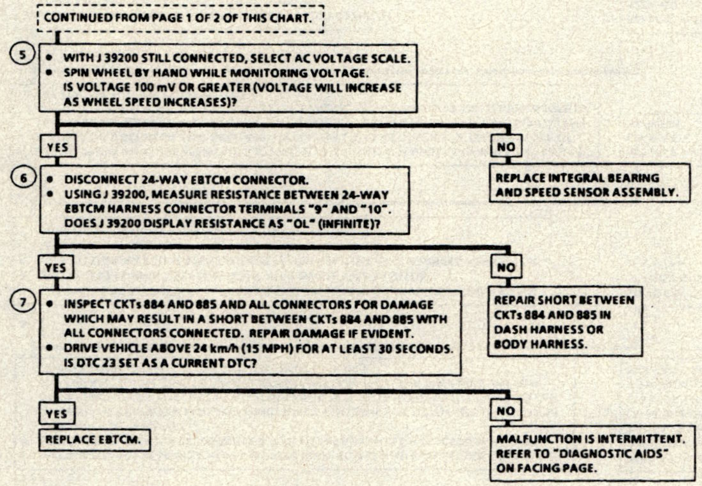

DIAGNOSTIC CHART DTC 24 — 1995 CAVALIER AND SUNFIRE WITH TCS

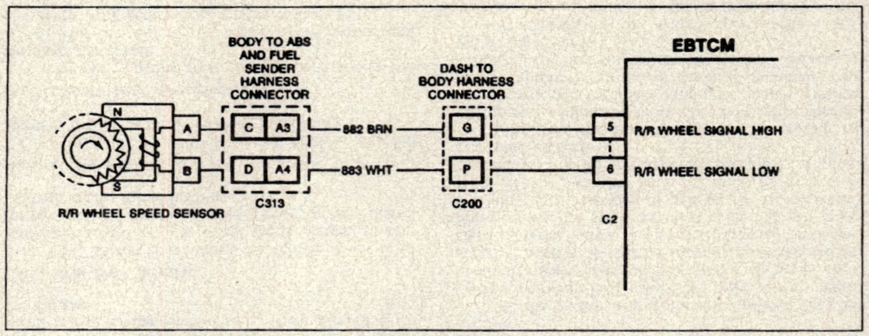

DTC 24
(Page 1 of 2)
RIGHT REAR WHEEL SPEED = 0

Circuit Description:

As a toothed ring passes by the wheel speed sensor, changes in the electromagnetic field cause the wheel speed sensor to produce a sinusoidal (AC) voltage signal whose frequency is proportional to wheel speed. The magnitude of this signal is directly related to wheel speed and the proximity of the wheel speed sensor to the toothed ring, often referred to as the air gap.

DTC Will Set When:

DTC 24 can be set when the vehicle is not in an ABS stop. If the right rear wheel speed is equal to zero for two seconds and the vehicle speed is greater than 8 km/h (5 mph), a malfunction exists.

Action Taken:

A malfunction DTC is stored, ABS is disabled and the "ABS" warning lamp is turned "ON."

DTC Chart Test Description:

Number(s) below refer to circled number(s) on the diagnostic chart.

1. Identifies circuitry that is damaged and visibly apparent.
2. Checks for proper resistance at the ABS rear axle harness connector.
3. Checks for a short between the wheel speed signal high and low circuits in the ABS rear axle harness.
4. Checks wheel speed sensor for proper resistance value.

Diagnostic Aids:

An "intermittent" malfunction is most likely caused by a poor connection, rubbed through wire insulation, or a wire that is broken inside the insulation.

The frequency of the malfunction can be checked by using the enhanced diagnostic function of the Tech 1.

If the customer's comments reflect that the "ABS" warning lamp is "ON" only during humid conditions (rain, snow, vehicle wash), all wheel speed sensor circuitry should be thoroughly inspected for signs of water intrusion. Use the following procedure. Spray down the suspected area with 5% salt water solution (two teaspoons of salt to 12 oz. of water). Drive vehicle above 24 km/h (15 mph) for at least 30 seconds. If DTC returns immediately, replace suspected harness.

Any circuitry, that is suspected as causing the intermittent complaint, should be thoroughly checked for backed out terminals, improper mating, broken locks, improperly formed or damaged terminals, poor terminal to wiring connections or physical damage to the wiring harness.

Resistance of the wheel speed sensor will increase with an increase in sensor temperature.

When replacing a wheel speed sensor, inspect the sensor terminals and harness connector for corrosion and/or water intrusion. If evidence or corrosion or water intrusion exists, replace wheel speed sensor harness.

If wheel speed sensor harness shows evidence of water intrusion, inspect wheel speed sensor for water intrusion. If water intrusion is evident, replace wheel speed sensor.

DIAGNOSTIC CHART DTC 24 — 1995 CAVALIER AND SUNFIRE WITH TCS, CONTINUED

Important

- WHEEL SPEED SENSOR INTERMITTENT MALFUNCTIONS MAY BE DIFFICULT TO LOCATE. CARE SHOULD BE TAKEN NOT TO DISTURB ANY ELECTRICAL CONNECTIONS PRIOR TO AN INDICATED STEP OF THIS CHART. THIS WILL ENSURE THAT AN INTERMITTENT CONNECTION WILL NOT BE CORRECTED BEFORE THE SOURCE OF THE MALFUNCTION IS FOUND.

DTC 24
(Page 1 of 2)
RIGHT REAR WHEEL SPEED = 0

(1)
- IGNITION "OFF."
- INSPECT WHEEL SPEED SENSOR WIRING AND CONNECTORS FOR DAMAGE. IS DAMAGE EVIDENT?

YES: REPAIR DAMAGE. DRIVE VEHICLE ABOVE 24 km/h (15 MPH) FOR AT LEAST 30 SECONDS TO ENSURE DTC 24 DOES NOT SET.

NO:

(2)
- DISCONNECT ABS REAR AXLE HARNESS FROM BODY TO ABS AND FUEL SENDER CONNECTOR.
- USING J 39200, MEASURE RESISTANCE BETWEEN ABS REAR AXLE HARNESS CONNECTOR TERMINALS "C" AND "D."

IS RESISTANCE BETWEEN 2100 AND 2400Ω WHEN SENSOR IS 25°C (77°F)? *[IF TEMPERATURE OF THE SENSOR IS SUSPECTED TO BE OTHER THAN 25°C (77°F), REFER TO THE TABLE BELOW FOR THE APPLICABLE SENSOR RESISTANCE VALUES; THE VALUES IN THE TABLE ARE FOR TEMPERATURE OF THE SENSOR, NOT AIR TEMPERATURE.]*

YES: GO TO PAGE 2 OF 2 OF THIS CHART.

NO:

(3)
- REMOVE RIGHT REAR INTEGRAL BEARING AND SPEED SENSOR ASSEMBLY (TUBULAR AXLE ONLY).
- DISCONNECT CONNECTOR FROM RIGHT REAR WHEEL SPEED SENSOR.
- USING J 39200, MEASURE RESISTANCE BETWEEN ABS REAR AXLE HARNESS CONNECTOR TERMINALS "C" AND "D."

DOES J 39200 DISPLAY RESISTANCE AS "OL" (INFINITE)?

NO: REPLACE ABS REAR AXLE HARNESS.

YES:

(4)
- USING J 39200, MEASURE RESISTANCE BETWEEN WHEEL SPEED SENSOR TERMINALS "A" AND "B."

IS RESISTANCE BETWEEN 2100 AND 2400Ω WHEN SENSOR IS 25°C (77°F)? *[IF TEMPERATURE OF THE SENSOR IS SUSPECTED TO BE OTHER THAN 25°C (77°F), REFER TO THE TABLE BELOW FOR THE APPLICABLE SENSOR RESISTANCE VALUES; THE VALUES IN THE TABLE ARE FOR TEMPERATURE OF THE SENSOR, NOT AIR TEMPERATURE.]*

YES: MALFUNCTION WAS DUE TO AN INTERMITTENT SHORT BETWEEN CKTs 882 AND 883. REPLACE ABS REAR AXLE HARNESS.

NO: REPLACE INTEGRAL BEARING AND SPEED SENSOR ASSEMBLY.

REAR WHEEL SPEED SENSOR

TEMPERATURE OF SENSOR VS. SENSOR RESISTANCE (APPROXIMATE)

TEMP. (°C)	TEMP. (°F)	RESISTANCE (OHMS)
-34 TO 4	-30 TO 40	1623 TO 2210
5 TO 43	41 TO 110	1939 TO 2570
44 TO 93	111 TO 200	2254 TO 3029

DIAGNOSTIC CHART DTC 24 — 1995 CAVALIER AND SUNFIRE WITH TCS, CONTINUED

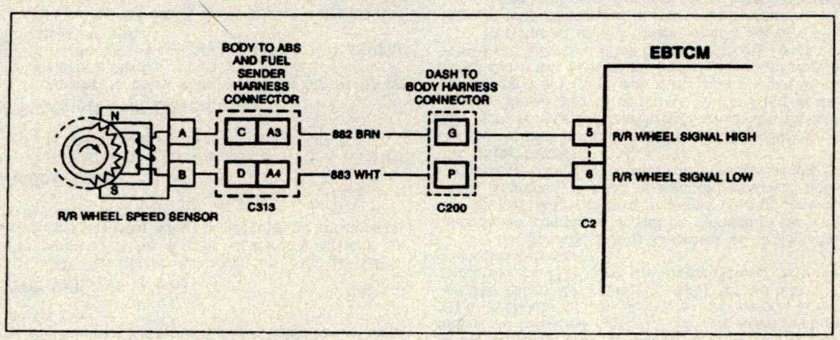

DTC 24
(Page 2 of 2)
RIGHT REAR WHEEL SPEED = 0

Circuit Description:

As a toothed ring passes by the wheel speed sensor, changes in the electromagnetic field cause the wheel speed sensor to produce a sinusoidal (AC) voltage signal whose frequency is proportional to wheel speed. The magnitude of this signal is directly related to wheel speed and the proximity of the wheel speed sensor to the toothed ring, often referred to as the air gap.

DTC Will Set When:

DTC 24 can be set when the vehicle is not in an ABS stop. If the right rear wheel speed is equal to zero for two seconds and the vehicle speed is greater than 8 km/h (5 mph), a malfunction exists.

Action Taken:

A malfunction DTC is stored, ABS is disabled and the "ABS" warning lamp is turned "ON."

DTC Chart Test Description:

Number(s) below refer to circled number(s) on the diagnostic chart.

5. This test ensures the wheel speed sensor and sensor ring generate the proper voltage.
6. Checks for short between the wheel speed signal high and low circuits in the dash or body harness.
7. Ensures malfunction was not due to physical damage of the circuitry.

Diagnostic Aids:

An "intermittent" malfunction is most likely caused by a poor connection, rubbed through wire insulation, or a wire that is broken inside the insulation.

The frequency of the malfunction can be checked by using the enhanced diagnostic function of the Tech 1.

If the customer's comments reflect that the "ABS" warning lamp is "ON" only during humid conditions (rain, snow, vehicle wash), all wheel speed sensor circuitry should be thoroughly inspected for signs of water intrusion. Use the following procedure. Spray down the suspected area with 5% salt water solution (two teaspoons of salt to 12 oz. of water). Drive vehicle above 24 km/h (15 mph) for at least 30 seconds. If DTC returns immediately, replace suspected harness.

Any circuitry, that is suspected as causing the intermittent complaint, should be thoroughly checked for backed out terminals, improper mating, broken locks, improperly formed or damaged terminals, poor terminal to wiring connections or physical damage to the wiring harness.

Resistance of the wheel speed sensor will increase with an increase in sensor temperature. When replacing a wheel speed sensor, inspect the sensor terminals and harness connector for corrosion and/or water intrusion. If evidence or corrosion or water intrusion exists, replace wheel speed sensor harness.

If wheel speed sensor harness shows evidence of water intrusion, inspect wheel speed sensor for water intrusion. If water intrusion is evident, replace wheel speed sensor.

DIAGNOSTIC CHART DTC 24 — 1995 CAVALIER AND SUNFIRE WITH TCS, CONTINUED

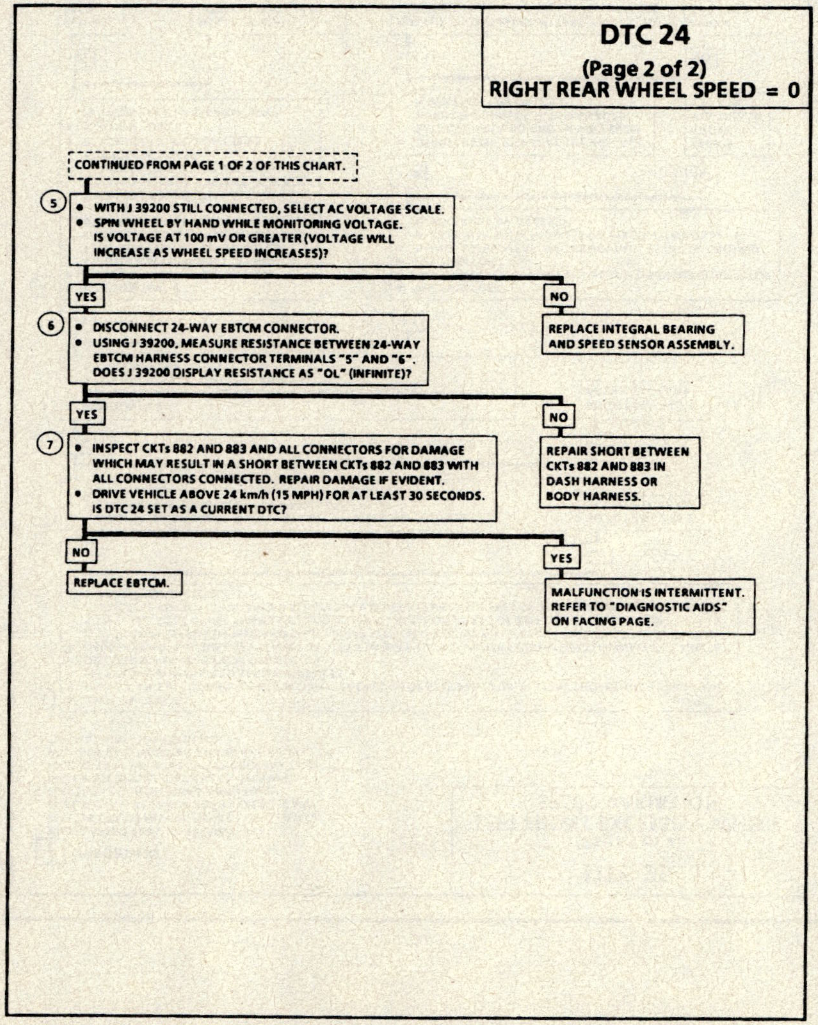

DIAGNOSTIC CHART DTC 25 — 1995 CAVALIER AND SUNFIRE WITH TCS

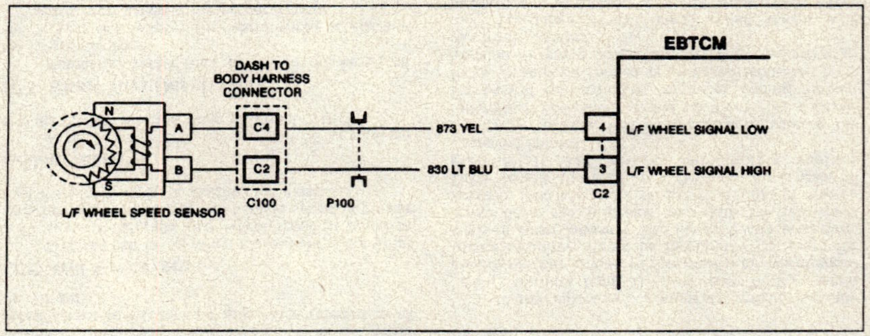

DTC 25

(Page 1 of 3)

LEFT FRONT EXCESSIVE WHEEL SPEED VARIATION

Circuit Description:

As a toothed ring passes by the wheel speed sensor, changes in the electromagnetic field cause the wheel speed sensor to produce a sinusoidal (AC) voltage signal whose frequency is proportional to wheel speed. The magnitude of this signal is directly related to wheel speed and the proximity of the wheel speed sensor to the toothed ring, often referred to as the air gap.

DTC Will Set When:

DTC 25 can be set when the brake is "OFF." The purpose of this test is to detect a situation in which the left front wheel acceleration or deceleration is beyond specified limits.

Action Taken:

A malfunction DTC is stored, ABS is disabled and the "ABS" warning lamp is turned "ON."

DTC Chart Test Description:

Number(s) below refer to circled number(s) on the diagnostic chart.

1. Checks wheel speed sensor for proper resistance value.
2. Ensures the wheel speed sensor is not shorted to ground.
3. Ensures the wheel speed sensor is not shorted to voltage.
4. Ensures that DTC 25 was not set due to a poor connection at the EBTCM or physical damage of the circuitry.
5. Checks for proper voltage at the wheel signal low speed sensor harness connector.
6. Checks for proper voltage at the wheel signal high speed sensor harness connector.

The frequency of the malfunction can be checked by using the enhanced diagnostic function of the Tech 1.

If the customer's comments reflect that the "ABS" warning lamp is "ON" only during humid conditions (rain, snow, vehicle wash), all wheel speed sensor circuitry should be thoroughly inspected for water intrusion. Use the following procedure. Spray down the suspected area with 5% salt water solution (two teaspoons of salt to 12 oz. of water). Drive vehicle above 24 km/h (15 mph) for at least 30 seconds. If DTC returns immediately, replace suspected harness.

Any circuitry, that is suspected as causing the intermittent complaint, should be thoroughly checked for backed out terminals, improper mating, broken locks, improperly formed or damaged terminals, poor terminal to wiring connections or physical damage to the wiring harness.

A careful visual inspection of the toothed ring, CV joint, bearing and wheel speed sensor for physical damage should also be performed. If DTC sets at the same mph every time and wheel speed variation is noted above this mph on the Tech 1, the speed ring is most likely damaged. This can occur at any speed.

Resistance of the wheel speed sensor will increase with an increase in sensor temperature.

When replacing a wheel speed sensor, inspect the sensor terminals and harness connector for corrosion and/or water intrusion. If evidence of corrosion or water intrusion exists, replace wheel speed sensor harness.

If wheel speed sensor harness shows evidence of water intrusion, inspect wheel speed sensor for water intrusion. If water intrusion is evident, replace wheel speed sensor.

DIAGNOSTIC CHART DTC 25 — 1995 CAVALIER AND SUNFIRE WITH TCS, CONTINUED

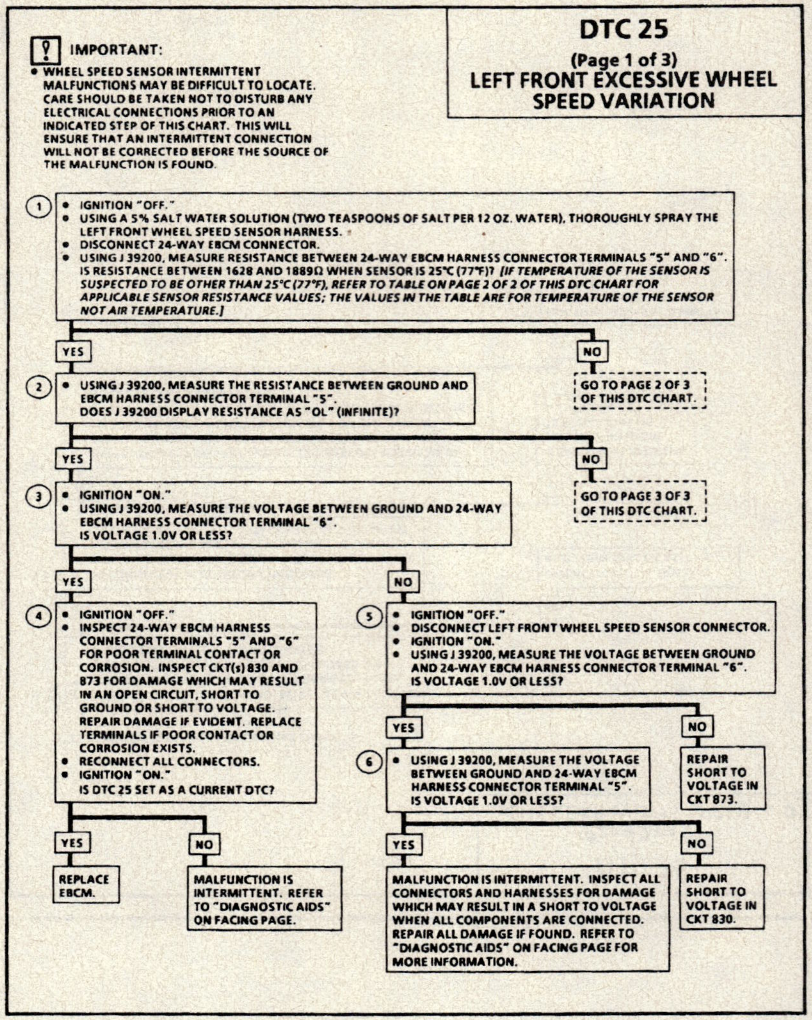

DIAGNOSTIC CHART DTC 25 — 1995 CAVALIER AND SUNFIRE WITH TCS, CONTINUED

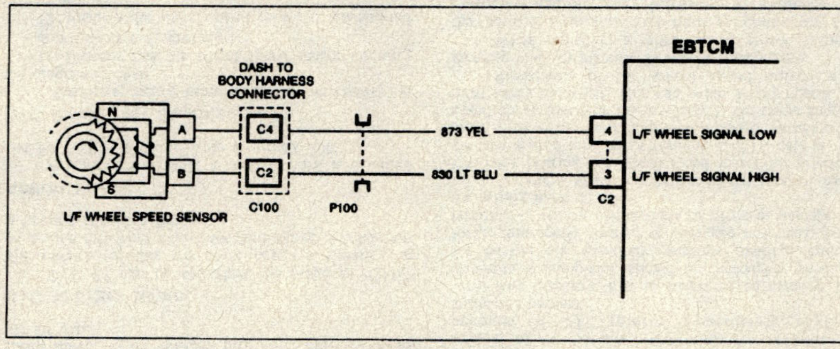

DTC 25
(Page 2 of 3)
LEFT FRONT EXCESSIVE WHEEL SPEED VARIATION

Circuit Description:

As a toothed ring passes by the wheel speed sensor, changes in the electromagnetic field cause the wheel speed sensor to produce a sinusoidal (AC) voltage signal whose frequency is proportional to wheel speed. The magnitude of this signal is directly related to wheel speed and the proximity of the wheel speed sensor to the toothed ring, often referred to as the air gap.

DTC Will Set When:

DTC 25 can be set when the brake is "OFF." The purpose of this test is to detect a situation in which the left front wheel acceleration or deceleration is beyond specified limits.

Action Taken:

A malfunction DTC is stored, ABS is disabled and the "ABS" warning lamp is turned "ON."

DTC Chart Test Description:

Number(s) below refer to circled number(s) on the diagnostic chart.

7. Tests for a short between the wheel signal high and wheel signal low circuits.
8. Checks for an open in the wheel signal low circuit.
9. Checks for an open in the wheel signal high circuit.
10. Checks for proper resistance of the wheel speed sensor.

Diagnostic Aids:

An "intermittent" malfunction is most likely caused by a poor connection, rubbed through wire insulation, or a wire that is broken inside the insulation.

The frequency of the malfunction can be checked by using the enhanced diagnostic function of the Tech 1.

If the customer's comments reflect that the "ABS" warning lamp is "ON" only during humid conditions (rain, snow, vehicle wash), all wheel speed sensor circuitry should be thoroughly inspected for water intrusion. Use the following procedure. Spray down the suspected area with 5% salt water solution (two teaspoons of salt to 12 oz. of water). Drive vehicle above 24 km/h (15 mph) for at least 30 seconds. If DTC returns immediately, replace suspected harness.

Any circuitry, that is suspected as causing the intermittent complaint, should be thoroughly checked for backed out terminals, improper mating, broken locks, improperly formed or damaged terminals, poor terminal to wiring connections or physical damage to the wiring harness.

A careful visual inspection of the toothed ring, CV joint, bearing and wheel speed sensor for physical damage should also be performed. If DTC sets at the same mph every time and wheel speed variation is noted above this mph on the Tech 1, the speed ring is most likely damaged. This can occur at any speed.

Resistance of the wheel speed sensor will increase with an increase in sensor temperature.

When replacing a wheel speed sensor, inspect the sensor terminals and harness connector for corrosion and/or water intrusion. If evidence of corrosion or water intrusion exists, replace wheel speed sensor harness.

If wheel speed sensor harness shows evidence of water intrusion, inspect wheel speed sensor for water intrusion. If water intrusion is evident, replace wheel speed sensor.

DIAGNOSTIC CHART DTC 25 — 1995 CAVALIER AND SUNFIRE WITH TCS, CONTINUED

DTC 25
(Page 2 of 3)
LEFT FRONT EXCESSIVE WHEEL SPEED VARIATION

CONTINUED FROM PAGE 1 OF 3 OF THIS DTC CHART.

(7)
- DISCONNECT LEFT FRONT WHEEL SPEED SENSOR CONNECTOR.
- USING J 39200, MEASURE THE RESISTANCE BETWEEN 24-WAY EBCM HARNESS CONNECTOR TERMINALS "5" AND "6".

DOES J 39200 DISPLAY RESISTANCE AS "OL" (INFINITE)?

NO: REPAIR SHORT BETWEEN CKTs 830 AND 873.

YES:

(8)
- USING J 39200, MEASURE THE RESISTANCE BETWEEN 24-WAY EBCM HARNESS CONNECTOR TERMINAL "6" AND LEFT FRONT WHEEL SPEED SENSOR HARNESS CONNECTOR TERMINAL "A."

IS RESISTANCE 2.0Ω OR LESS?

NO: REPAIR OPEN OR HIGH RESISTANCE IN CKT 873.

YES:

(9)
- USING J 39200, MEASURE THE RESISTANCE BETWEEN 24-WAY EBCM HARNESS CONNECTOR TERMINAL "5" AND LEFT FRONT WHEEL SPEED SENSOR HARNESS CONNECTOR TERMINAL "B."

IS RESISTANCE 2.0Ω OR LESS?

NO: REPAIR OPEN OR HIGH RESISTANCE IN CKT 830.

YES:

(10)
- USING J 39200, MEASURE THE RESISTANCE BETWEEN LEFT FRONT WHEEL SPEED SENSOR TERMINALS "A" AND "B."

IS RESISTANCE BETWEEN 1628 AND 1889Ω WHEN SENSOR IS 25°C (77°F)?
[IF TEMPERATURE OF THE SENSOR IS SUSPECTED TO BE OTHER THAN 25°C (77°F), REFER TO THE TABLE BELOW FOR APPLICABLE SENSOR RESISTANCE VALUES; THE VALUES IN THE TABLE ARE FOR TEMPERATURE OF THE SENSOR NOT AIR TEMPERATURE.]

NO: REPLACE WHEEL SPEED SENSOR.

YES: MALFUNCTION IS INTERMITTENT. INSPECT ALL CONNECTORS AND HARNESSES FOR DAMAGES WHICH MAY RESULT IN HIGH RESISTANCE WHEN ALL COMPONENTS ARE CONNECTED. REFER TO "DIAGNOSTIC AIDS" ON FACING PAGE FOR MORE INFORMATION. REPAIR ALL DAMAGE IF FOUND.

FRONT WHEEL SPEED SENSOR
TEMPERATURE OF SENSOR VS. SENSOR RESISTANCE (APPROXIMATE)

TEMP. (°C)	TEMP. (°F)	RESISTANCE (OHMS)
-34 TO 4	-30 TO 40	1258 TO 1739
5 TO 43	41 TO 110	1503 TO 2023
44 TO 93	111 TO 200	1747 TO 2384

DIAGNOSTIC CHART DTC 25 — 1995 CAVALIER AND SUNFIRE WITH TCS, CONTINUED

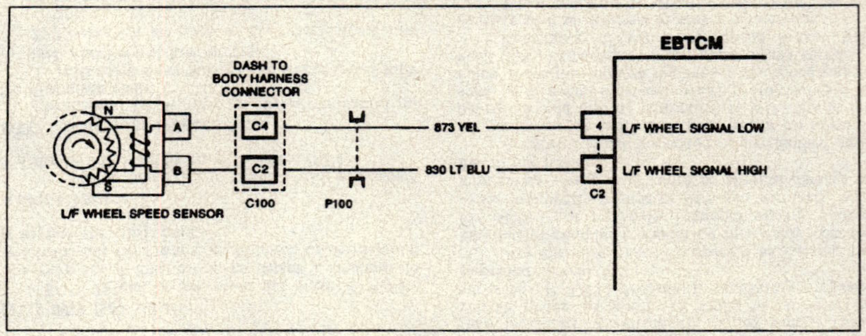

DTC 25
(Page 3 of 3)
LEFT FRONT EXCESSIVE WHEEL SPEED VARIATION

Circuit Description:

As a toothed ring passes by the wheel speed sensor, changes in the electromagnetic field cause the wheel speed sensor to produce a sinusoidal (AC) voltage signal whose frequency is proportional to wheel speed. The magnitude of this signal is directly related to wheel speed and the proximity of the wheel speed sensor to the toothed ring, often referred to as the air gap.

DTC Will Set When:

DTC 25 can be set when the brake is "OFF." The purpose of this test is to detect a situation in which the left front wheel acceleration or deceleration is beyond specified limits.

Action Taken:

A malfunction DTC is stored, ABS is disabled and the "ABS" warning lamp is turned "ON."

DTC Chart Test Description:

Number(s) below refer to circled number(s) on the diagnostic chart.

11. Ensures that the wheel speed sensor circuitry is not shorted to ground.

12. Ensures the wheel speed sensor is not shorted to ground.

Diagnostic Aids:

An "intermittent" malfunction is most likely caused by a poor connection, rubbed through wire insulation, or a wire that is broken inside the insulation.

The frequency of the malfunction can be checked by using the enhanced diagnostic function of the Tech 1.

If the customer's comments reflect that the "ABS" warning lamp is "ON" only during humid conditions (rain, snow, vehicle wash), all wheel speed sensor circuitry should be thoroughly inspected for water intrusion. Use the following procedure. Spray down the suspected area with 5% salt water solution (two teaspoons of salt to 12 oz. of water). Drive vehicle above 24 km/h (15 mph) for at least 30 seconds. If DTC returns immediately, replace suspected harness.

Any circuitry, that is suspected as causing the intermittent complaint, should be thoroughly checked for backed out terminals, improper mating, broken locks, improperly formed or damaged terminals, poor terminal to wiring connections or physical damage to the wiring harness.

A careful visual inspection of the toothed ring, CV joint, bearing and wheel speed sensor for physical damage should also be performed. If DTC sets at the same mph every time and wheel speed variation is noted above this mph on the Tech 1, the speed ring is most likely damaged. This can occur at any speed.

Resistance of the wheel speed sensor will increase with an increase in sensor temperature.

When replacing a wheel speed sensor, inspect the sensor terminals and harness connector for corrosion and/or water intrusion. If evidence of corrosion or water intrusion exists, replace wheel speed sensor harness.

If wheel speed sensor harness shows evidence of water intrusion, inspect wheel speed sensor for water intrusion. If water intrusion is evident, replace wheel speed sensor.

DIAGNOSTIC CHART DTC 25 — 1995 CAVALIER AND SUNFIRE WITH TCS, CONTINUED

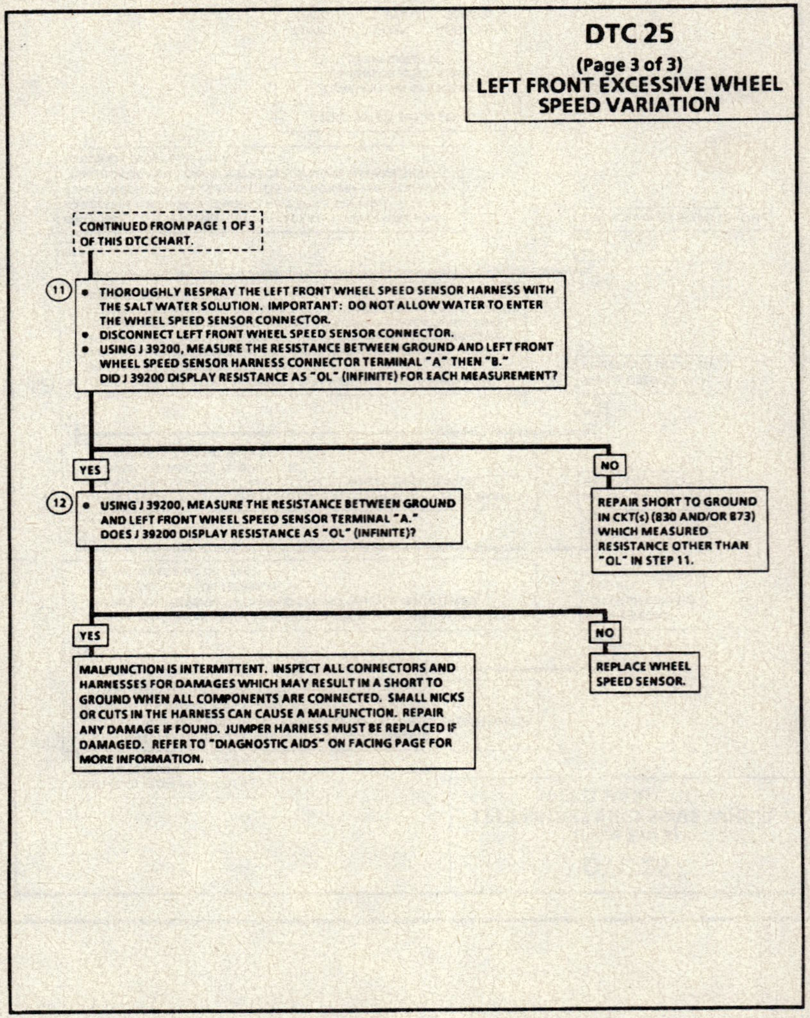

DIAGNOSTIC CHART DTC 26 — 1995 CAVALIER AND SUNFIRE WITH TCS

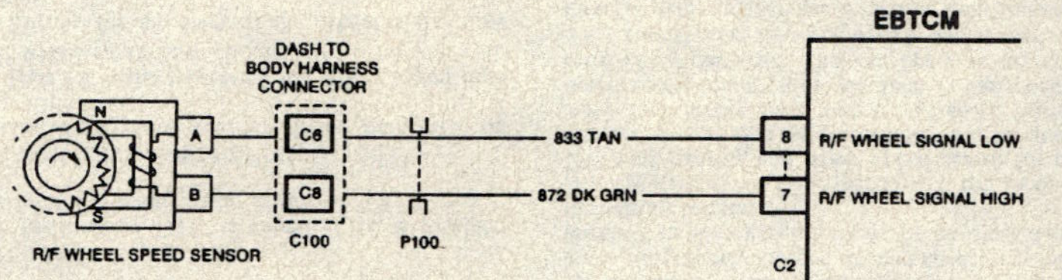

DTC 26
(Page 1 of 3)
RIGHT FRONT EXCESSIVE WHEEL SPEED VARIATION

Circuit Description:

As a toothed ring passes by the wheel speed sensor, changes in the electromagnetic field cause the wheel speed sensor to produce a sinusoidal (AC) voltage signal whose frequency is proportional to wheel speed. The magnitude of this signal is directly related to wheel speed and the proximity of the wheel speed sensor to the toothed ring, often referred to as the air gap.

DTC Will Set When:

DTC 26 can be set when the brake is "OFF." The purpose of this test is to detect a situation in which the right front wheel acceleration or deceleration is beyond specified limits.

Action Taken:

A malfunction DTC is stored, ABS is disabled and the "ABS" warning lamp is turned "ON."

DTC Chart Test Description:

Number(s) below refer to circled number(s) on the diagnostic chart.

1. Checks wheel speed sensor for proper resistance value.
2. Ensures the wheel speed sensor is not shorted to ground.
3. Ensures the wheel speed sensor is not shorted to voltage.
4. Ensures that DTC 26 was not set due to a poor connection at the EBTCM or physical damage of the circuitry.
5. Checks for proper voltage at the wheel signal low speed sensor harness connector.
6. Checks for proper voltage at the wheel signal high speed sensor harness connector.

The frequency of the malfunction can be checked by using the enhanced diagnostic function of the Tech 1, as described in "Enhanced Diagnostics," found in this section.

If the customer's comments reflect that the "ABS" warning lamp is "ON" only during humid conditions (rain, snow, vehicle wash), all wheel speed sensor circuitry should be thoroughly inspected for water intrusion. Use the following procedure. Spray down the suspected area with 5% salt water solution (two teaspoons of salt to 12 oz. of water). Drive vehicle above 24 km/h (15 mph) for at least 30 seconds. If DTC returns immediately, replace suspected harness.

Any circuitry, that is suspected as causing the intermittent complaint, should be thoroughly checked for backed out terminals, improper mating, broken locks, improperly formed or damaged terminals, poor terminal to wiring connections or physical damage to the wiring harness.

A careful visual inspection of the toothed ring, CV joint, bearing and wheel speed sensor for physical damage should also be performed. If DTC sets at the same mph every time and wheel speed variation is noted above this mph on the Tech 1, the speed ring is most likely damaged. This can occur at any speed.

Resistance of the wheel speed sensor will increase with an increase in sensor temperature.

When replacing a wheel speed sensor, inspect the sensor terminals and harness connector for corrosion and/or water intrusion. If evidence of corrosion or water intrusion exists, replace wheel speed sensor harness.

If wheel speed sensor harness shows evidence of water intrusion, inspect wheel speed sensor for water intrusion. If water intrusion is evident, replace wheel speed sensor.

DIAGNOSTIC CHART DTC 26 — 1995 CAVALIER AND SUNFIRE WITH TCS, CONTINUED

DTC 26
(Page 1 of 3)
RIGHT FRONT EXCESSIVE WHEEL SPEED VARIATION

IMPORTANT:
- WHEEL SPEED SENSOR INTERMITTENT MALFUNCTIONS MAY BE DIFFICULT TO LOCATE. CARE SHOULD BE TAKEN NOT TO DISTURB ANY ELECTRICAL CONNECTIONS PRIOR TO AN INDICATED STEP OF THIS CHART. THIS WILL ENSURE THAT AN INTERMITTENT CONNECTION WILL NOT BE CORRECTED BEFORE THE SOURCE OF THE MALFUNCTION IS FOUND.

(1)
- IGNITION "OFF."
- USING A 5% SALT WATER SOLUTION (TWO TEASPOONS OF SALT PER 12 OZ. WATER), THOROUGHLY SPRAY THE RIGHT FRONT WHEEL SPEED SENSOR HARNESS.
- DISCONNECT 24-WAY EBCM CONNECTOR.
- USING J 39200, MEASURE RESISTANCE BETWEEN 24-WAY EBCM HARNESS CONNECTOR TERMINALS "9" AND "10". IS RESISTANCE BETWEEN 1628 AND 1889Ω WHEN THE SENSOR IS 25°C (77°F)? *[IF TEMPERATURE OF THE SENSOR IS SUSPECTED TO BE OTHER THAN 25°C (77°F), REFER TO TABLE ON PAGE 2 OF 2 OF THIS DTC CHART FOR APPLICABLE SENSOR RESISTANCE VALUES; THE VALUES IN THE TABLE ARE FOR TEMPERATURE OF THE SENSOR NOT AIR TEMPERATURE.]*

YES → (2)
NO → GO TO PAGE 2 OF 3 OF THIS DTC CHART.

(2)
- USING J 39200, MEASURE THE RESISTANCE BETWEEN GROUND AND EBCM HARNESS CONNECTOR TERMINAL "9".
DOES J 39200 DISPLAY RESISTANCE AS "OL" (INFINITE)?

YES → (3)
NO → GO TO PAGE 3 OF 3 OF THIS DTC CHART.

(3)
- IGNITION "ON."
- USING J 39200, MEASURE THE VOLTAGE BETWEEN GROUND AND 24-WAY EBCM HARNESS CONNECTOR TERMINAL "10".
IS VOLTAGE 1.0V OR LESS?

YES → (4)
NO → (5)

(4)
- IGNITION "OFF."
- INSPECT 24-WAY EBCM HARNESS CONNECTOR TERMINALS "9" AND "10" FOR POOR TERMINAL CONTACT OR CORROSION. INSPECT CKT(s) 833 AND 872 FOR DAMAGE WHICH MAY RESULT IN AN OPEN CIRCUIT, SHORT TO GROUND OR SHORT TO VOLTAGE. REPAIR DAMAGE IF EVIDENT. REPLACE TERMINALS IF POOR CONTACT OR CORROSION EXISTS.
- RECONNECT ALL CONNECTORS.
- IGNITION "ON."
IS DTC 26 SET AS A CURRENT DTC?

YES → REPLACE EBCM.
NO → MALFUNCTION IS INTERMITTENT. REFER TO "DIAGNOSTIC AIDS" ON FACING PAGE.

(5)
- IGNITION "OFF."
- DISCONNECT RIGHT FRONT WHEEL SPEED SENSOR CONNECTOR.
- IGNITION "ON."
- USING J 39200, MEASURE THE VOLTAGE BETWEEN GROUND AND 24-WAY EBCM HARNESS CONNECTOR TERMINAL "10".
IS VOLTAGE 1.0V OR LESS?

YES → (6)
NO → REPAIR SHORT TO VOLTAGE IN CKT 833.

(6)
- USING J 39200, MEASURE THE VOLTAGE BETWEEN GROUND AND 24-WAY EBCM HARNESS CONNECTOR TERMINAL "9".
IS VOLTAGE 1.0V OR LESS?

YES → MALFUNCTION IS INTERMITTENT. INSPECT ALL CONNECTORS AND HARNESSES FOR DAMAGE WHICH MAY RESULT IN A SHORT TO VOLTAGE WHEN ALL COMPONENTS ARE CONNECTED. REPAIR ALL DAMAGE IF FOUND. REFER TO "DIAGNOSTIC AIDS" ON FACING PAGE FOR MORE INFORMATION.
NO → REPAIR SHORT TO VOLTAGE IN CKT 872.

DIAGNOSTIC CHART DTC 26 — 1995 CAVALIER AND SUNFIRE WITH TCS, CONTINUED

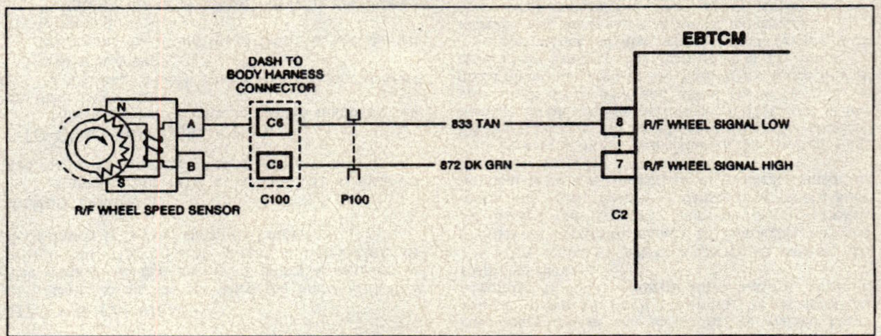

DTC 26
(Page 2 of 3)
RIGHT FRONT EXCESSIVE WHEEL SPEED VARIATION

Circuit Description:

As a toothed ring passes by the wheel speed sensor, changes in the electromagnetic field cause the wheel speed sensor to produce a sinusoidal (AC) voltage signal whose frequency is proportional to wheel speed. The magnitude of this signal is directly related to wheel speed and the proximity of the wheel speed sensor to the toothed ring, often referred to as the air gap.

DTC Will Set When:

DTC 26 can be set when the brake is "OFF." The purpose of this test is to detect a situation in which the right front wheel acceleration or deceleration is beyond specified limits.

Action Taken:

A malfunction DTC is stored, ABS is disabled and the "ABS" warning lamp is turned "ON."

DTC Chart Test Description:

Number(s) below refer to circled number(s) on the diagnostic chart.

7. Tests for a short between the wheel signal high and wheel signal low circuits.
8. Checks for an open in the wheel signal low circuit.
9. Checks for an open in the wheel signal high circuit.
10. Checks for proper resistance of the wheel speed sensor.

Diagnostic Aids:

An "intermittent" malfunction is most likely caused by a poor connection, rubbed through wire insulation, or a wire that is broken inside the insulation.

The frequency of the malfunction can be checked by using the enhanced diagnostic function of the Tech 1.

If the customer's comments reflect that the "ABS" warning lamp is "ON" only during humid conditions (rain, snow, vehicle wash), all wheel speed sensor circuitry should be thoroughly inspected for water intrusion. Use the following procedure. Spray down the suspected area with 5% salt water solution (two teaspoons of salt to 12 oz. of water). Drive vehicle above 24 km/h (15 mph) for at least 30 seconds. If DTC returns immediately, replace suspected harness.

Any circuitry, that is suspected as causing the intermittent complaint, should be thoroughly checked for backed out terminals, improper mating, broken locks, improperly formed or damaged terminals, poor terminal to wiring connections or physical damage to the wiring harness.

A careful visual inspection of the toothed ring, CV joint, bearing and wheel speed sensor for physical damage should also be performed. If DTC sets at the same mph every time and wheel speed variation is noted above this mph on the Tech 1, the speed ring is most likely damaged. This can occur at any speed.

Resistance of the wheel speed sensor will increase with an increase in sensor temperature.

When replacing a wheel speed sensor, inspect the sensor terminals and harness connector for corrosion and/or water intrusion. If evidence of corrosion or water intrusion exists, replace wheel speed sensor harness.

If wheel speed sensor harness shows evidence of water intrusion, inspect wheel speed sensor for water intrusion. If water intrusion is evident, replace wheel speed sensor.

DIAGNOSTIC CHART DTC 26 — 1995 CAVALIER AND SUNFIRE WITH TCS, CONTINUED

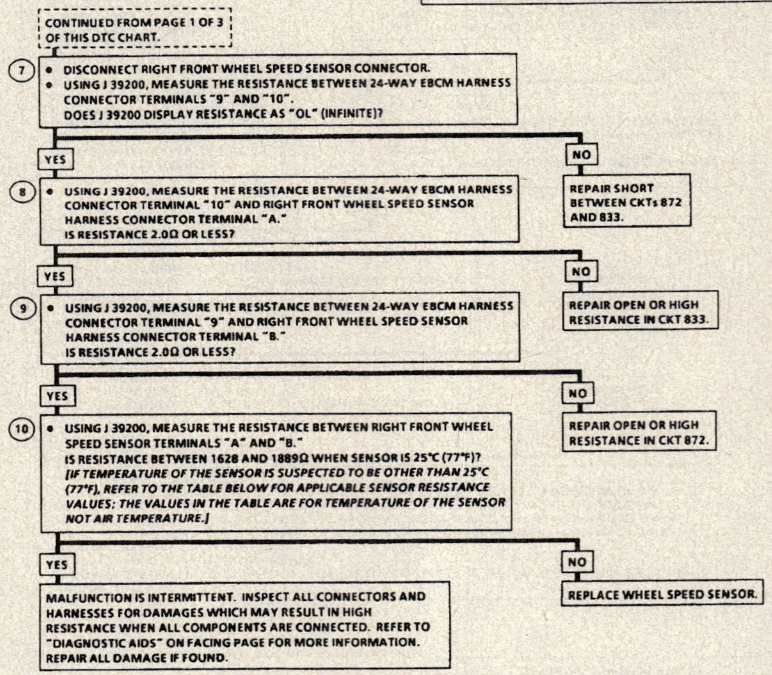

FRONT WHEEL SPEED SENSOR

TEMPERATURE OF SENSOR VS. SENSOR RESISTANCE (APPROXIMATE)

TEMP. (°C)	TEMP. (°F)	RESISTANCE (OHMS)
-34 TO 4	-30 TO 40	1258 TO 1739
5 TO 43	41 TO 110	1503 TO 2023
44 TO 93	111 TO 200	1747 TO 2384

DIAGNOSTIC CHART DTC 26 — 1995 CAVALIER AND SUNFIRE WITH TCS, CONTINUED

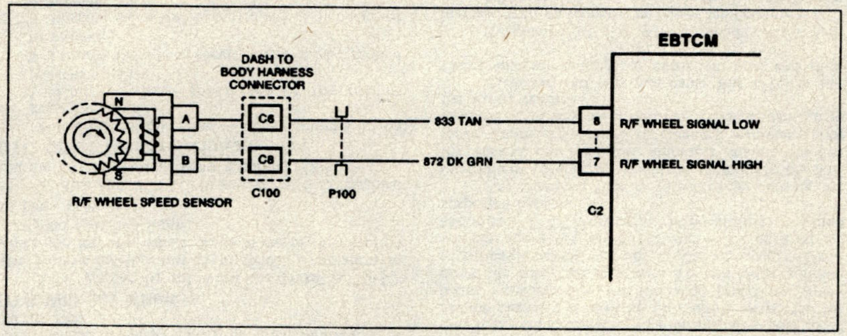

DTC 26
(Page 3 of 3)
RIGHT FRONT EXCESSIVE WHEEL SPEED VARIATION

Circuit Description:

As a toothed ring passes by the wheel speed sensor, changes in the electromagnetic field cause the wheel speed sensor to produce a sinusoidal (AC) voltage signal whose frequency is proportional to wheel speed. The magnitude of this signal is directly related to wheel speed and the proximity of the wheel speed sensor to the toothed ring, often referred to as the air gap.

DTC Will Set When:

DTC 26 can be set when the brake is "OFF." The purpose of this test is to detect a situation in which the right front wheel acceleration or deceleration is beyond specified limits.

Action Taken:

A malfunction DTC is stored, ABS is disabled and the "ABS" warning lamp is turned "ON."

DTC Chart Test Description:

Number(s) below refer to circled number(s) on the diagnostic chart.

11. Ensures that the wheel speed sensor circuitry is not shorted to ground.
12. Ensures the wheel speed sensor is not shorted to ground.

Diagnostic Aids:

An "intermittent" malfunction is most likely caused by a poor connection, rubbed through wire insulation, or a wire that is broken inside the insulation.

The frequency of the malfunction can be checked by using the enhanced diagnostic function of the Tech 1.

If the customer's comments reflect that the "ABS" warning lamp is "ON" only during humid conditions (rain, snow, vehicle wash), all wheel speed sensor circuitry should be thoroughly inspected for water intrusion. Use the following procedure. Spray down the suspected area with 5% salt water solution (two teaspoons of salt to 12 oz. of water). Drive vehicle above 24 km/h (15 mph) for at least 30 seconds. If DTC returns immediately, replace suspected harness.

Any circuitry, that is suspected as causing the intermittent complaint, should be thoroughly checked for backed out terminals, improper mating, broken locks, improperly formed or damaged terminals, poor terminal to wiring connections or physical damage to the wiring harness.

A careful visual inspection of the toothed ring, CV joint, bearing and wheel speed sensor for physical damage should also be performed. If DTC sets at the same mph every time and wheel speed variation is noted above this mph on the Tech 1, the speed ring is most likely damaged. This can occur at any speed.

Resistance of the wheel speed sensor will increase with an increase in sensor temperature.

When replacing a wheel speed sensor, inspect the sensor terminals and harness connector for corrosion and/or water intrusion. If evidence of corrosion or water intrusion exists, replace wheel speed sensor harness.

If wheel speed sensor harness shows evidence of water intrusion, inspect wheel speed sensor for water intrusion. If water intrusion is evident, replace wheel speed sensor.

DIAGNOSTIC CHART DTC 26 — 1995 CAVALIER AND SUNFIRE WITH TCS, CONTINUED

DTC 26
(Page 3 of 3)
RIGHT FRONT EXCESSIVE WHEEL SPEED VARIATION

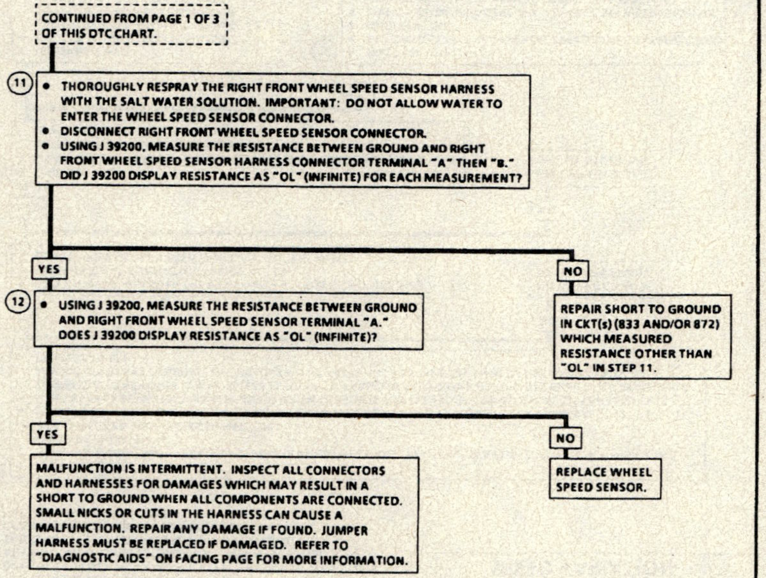

DIAGNOSTIC CHART DTC 27 — 1995 CAVALIER AND SUNFIRE WITH TCS

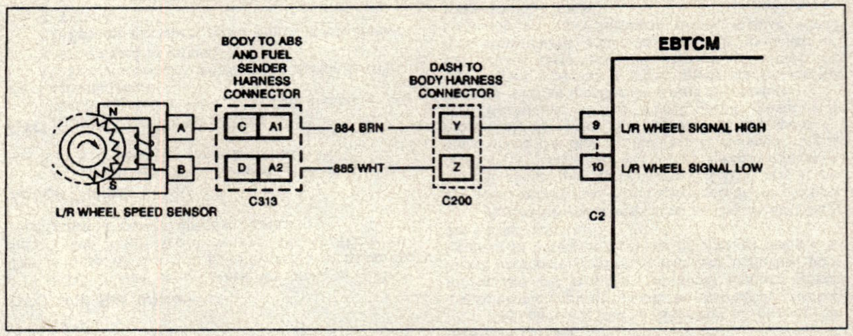

DTC 27
(Page 1 of 3)
LEFT REAR EXCESSIVE WHEEL SPEED VARIATION

Circuit Description:

As a toothed ring passes by the wheel speed sensor, changes in the electromagnetic field cause the wheel speed sensor to produce a sinusoidal (AC) voltage signal whose frequency is proportional to wheel speed. The magnitude of this signal is directly related to wheel speed and the proximity of the wheel speed sensor to the toothed ring, often referred to as the air gap.

DTC Will Set When:

DTC 27 can be set when the brake is "OFF." The purpose of this test is to detect a situation in which the left rear wheel acceleration or deceleration is beyond specified limits.

Action Taken:

A malfunction DTC is stored, ABS is disabled and the "ABS" warning lamp is turned "ON."

DTC Chart Test Description:

Number(s) below refer to circled number(s) on the diagnostic chart.

1. Checks wheel speed sensor for proper resistance value.
2. Ensures the wheel speed sensor is not shorted to ground.
3. Ensures the wheel speed sensor is not shorted to voltage.
4. Ensures that DTC 27 was not set due to a poor connection at the EBTCM or physical damage of the circuitry.
5. Checks for proper voltage at the wheel signal low speed sensor harness connector.
6. Checks for proper voltage at the wheel signal high speed sensor harness connector.

Diagnostic Aids:

An "intermittent" malfunction is most likely caused by a poor connection, rubbed through wire insulation, or a wire that is broken inside the insulation.

The frequency of the malfunction can be checked by using the enhanced diagnostic function of the Tech 1

If the customer's comments reflect that the "ABS" warning lamp is "ON" only during humid conditions (rain, snow, vehicle wash), all wheel speed sensor circuitry should be thoroughly inspected for water intrusion. Use the following procedure. Spray down the suspected area with 5% salt water solution (two teaspoons of salt to 12 oz. of water). Drive vehicle above 24 km/h (15 mph) for at least 30 seconds. If DTC returns immediately, replace suspected harness.

Any circuitry, that is suspected as causing the intermittent complaint, should be thoroughly checked for backed out terminals, improper mating, broken locks, improperly formed or damaged terminals, poor terminal to wiring connections or physical damage to the wiring harness.

Excessive end play can cause this DTC to set. Check integral bearing for excessive axial end play.

Resistance of the wheel speed sensor will increase with an increase in sensor temperature.

When replacing a wheel speed sensor, inspect the sensor terminals and harness connector for corrosion and/or water intrusion. If evidence of corrosion or water intrusion exists, replace wheel speed sensor harness.

If wheel speed sensor harness shows evidence of water intrusion, inspect wheel speed sensor for water intrusion. If water intrusion is evident, replace wheel speed sensor.

DIAGNOSTIC CHART DTC 27 — 1995 CAVALIER AND SUNFIRE WITH TCS, CONTINUED

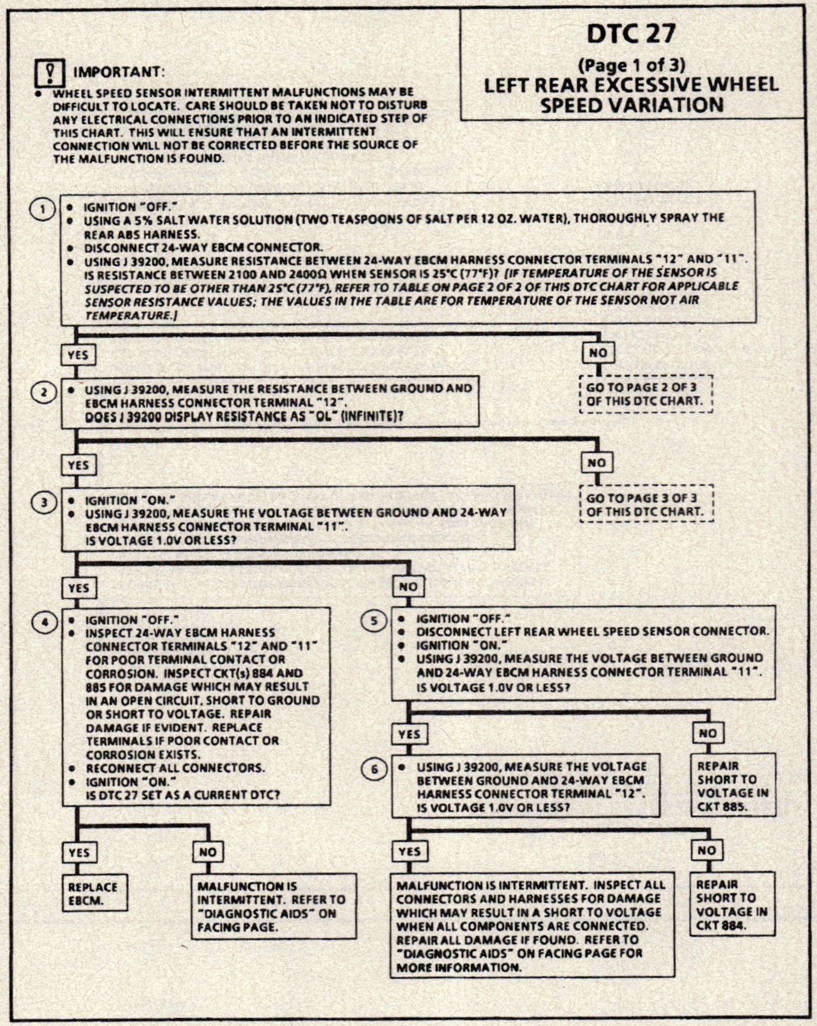

DIAGNOSTIC CHART DTC 27 — 1995 CAVALIER AND SUNFIRE WITH TCS, CONTINUED

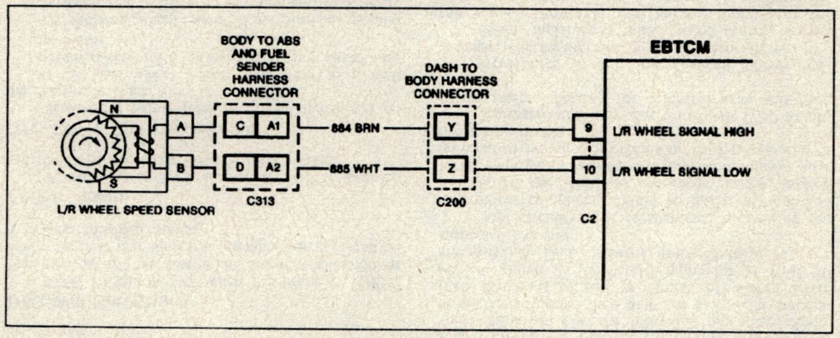

DTC 27
(Page 2 of 3)
LEFT REAR EXCESSIVE WHEEL SPEED VARIATION

Circuit Description:
As a toothed ring passes by the wheel speed sensor, changes in the electromagnetic field cause the wheel speed sensor to produce a sinusoidal (AC) voltage signal whose frequency is proportional to wheel speed. The magnitude of this signal is directly related to wheel speed and the proximity of the wheel speed sensor to the toothed ring, often referred to as the air gap.

DTC Will Set When:
DTC 27 can be set when the brake is "OFF." The purpose of this test is to detect a situation in which the left rear wheel acceleration or deceleration is beyond specified limits.

Action Taken:
A malfunction DTC is stored, ABS is disabled and the "ABS" warning lamp is turned "ON."

DTC Chart Test Description:
Number(s) below refer to circled number(s) on the diagnostic chart.

7. Tests for a short between the wheel signal high and wheel signal low circuits.
8. Checks for an open in the wheel signal low circuit.
9. Checks for an open in the wheel signal high circuit.
10. Checks for proper resistance of the wheel speed sensor.

Diagnostic Aids:
An "intermittent" malfunction is most likely caused by a poor connection, rubbed through wire insulation, or a wire that is broken inside the insulation.

The frequency of the malfunction can be checked by using the enhanced diagnostic function of the Tech 1.

If the customer's comments reflect that the "ABS" warning lamp is "ON" only during humid conditions (rain, snow, vehicle wash), all wheel speed sensor circuitry should be thoroughly inspected for water intrusion. Use the following procedure. Spray down the suspected area with 5% salt water solution (two teaspoons of salt to 12 oz. of water). Drive vehicle above 24 km/h (15 mph) for at least 30 seconds. If DTC returns immediately, replace suspected harness.

Any circuitry, that is suspected as causing the intermittent complaint, should be thoroughly checked for backed out terminals, improper mating, broken locks, improperly formed or damaged terminals, poor terminal to wiring connections or physical damage to the wiring harness.

Excessive end play can cause this DTC to set. Check integral bearing for excessive axial end play.

Resistance of the wheel speed sensor will increase with an increase in sensor temperature.

When replacing a wheel speed sensor, inspect the sensor terminals and harness connector for corrosion and/or water intrusion. If evidence of corrosion or water intrusion exists, replace wheel speed sensor harness.

If wheel speed sensor harness shows evidence of water intrusion, inspect wheel speed sensor for water intrusion. If water intrusion is evident, replace wheel speed sensor.

DIAGNOSTIC CHART DTC 27 — 1995 CAVALIER AND SUNFIRE WITH TCS, CONTINUED

DTC 27
(Page 2 of 3)
LEFT REAR EXCESSIVE WHEEL SPEED VARIATION

CONTINUED FROM PAGE 1 OF 3 OF THIS DTC CHART.

(7)
- DISCONNECT LEFT REAR WHEEL SPEED SENSOR CONNECTOR.
- USING J 39200, MEASURE THE RESISTANCE BETWEEN 24-WAY EBCM HARNESS CONNECTOR TERMINALS "11" AND "12". DOES J 39200 DISPLAY RESISTANCE AS "OL" (INFINITE)?

NO: REPAIR SHORT BETWEEN CKTs 884 AND 885.

YES:

(8)
- USING J 39200, MEASURE THE RESISTANCE BETWEEN 24-WAY EBCM HARNESS CONNECTOR TERMINAL "12" AND LEFT REAR WHEEL SPEED SENSOR HARNESS CONNECTOR TERMINAL "A." IS RESISTANCE 2.0Ω OR LESS?

NO: REPAIR OPEN OR HIGH RESISTANCE IN CKT 884.

YES:

(9)
- USING J 39200, MEASURE THE RESISTANCE BETWEEN 24-WAY EBCM HARNESS CONNECTOR TERMINAL "11" AND LEFT REAR WHEEL SPEED SENSOR HARNESS CONNECTOR TERMINAL "B." IS RESISTANCE 2.0Ω OR LESS?

NO: REPAIR OPEN OR HIGH RESISTANCE IN CKT 885.

YES:

(10)
- USING J 39200, MEASURE THE RESISTANCE BETWEEN LEFT REAR WHEEL SPEED SENSOR TERMINALS "A" AND "B." IS RESISTANCE BETWEEN 2100 AND 2400Ω WHEN SENSOR IS 25°C (77°F)? *[IF TEMPERATURE OF THE SENSOR IS SUSPECTED TO BE OTHER THAN 25°C (77°F), REFER TO THE TABLE BELOW FOR APPLICABLE SENSOR RESISTANCE VALUES; THE VALUES IN THE TABLE ARE FOR TEMPERATURE OF THE SENSOR NOT AIR TEMPERATURE.]*

NO: REPLACE INTEGRAL BEARING AND SPEED SENSOR ASSEMBLY.

YES: MALFUNCTION IS INTERMITTENT. INSPECT ALL CONNECTORS AND HARNESSES FOR DAMAGES WHICH MAY RESULT IN HIGH RESISTANCE WHEN ALL COMPONENTS ARE CONNECTED. REFER TO "DIAGNOSTIC AIDS" ON FACING PAGE FOR MORE INFORMATION. REPAIR ALL DAMAGE IF FOUND.

REAR WHEEL SPEED SENSOR

TEMPERATURE OF SENSOR VS. SENSOR RESISTANCE (APPROXIMATE)

TEMP. (°C)	TEMP. (°F)	RESISTANCE (OHMS)
-34 TO 4	-30 TO 40	1623 TO 2210
5 TO 43	41 TO 110	1939 TO 2570
44 TO 93	111 TO 200	2254 TO 3029

DIAGNOSTIC CHART DTC 27 — 1995 CAVALIER AND SUNFIRE WITH TCS, CONTINUED

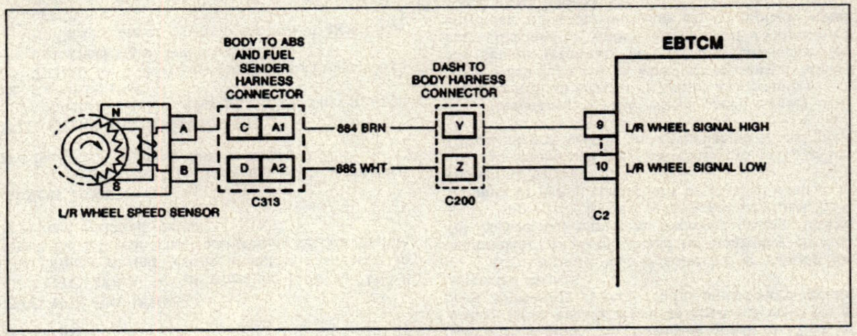

DTC 27
(Page 3 of 3)
LEFT REAR EXCESSIVE WHEEL SPEED VARIATION

Circuit Description:
As a toothed ring passes by the wheel speed sensor, changes in the electromagnetic field cause the wheel speed sensor to produce a sinusoidal (AC) voltage signal whose frequency is proportional to wheel speed. The magnitude of this signal is directly related to wheel speed and the proximity of the wheel speed sensor to the toothed ring, often referred to as the air gap.

DTC Will Set When:
DTC 27 can be set when the brake is "OFF." The purpose of this test is to detect a situation in which the left rear wheel acceleration or deceleration is beyond specified limits.

Action Taken:
A malfunction DTC is stored, ABS is disabled and the "ABS" warning lamp is turned "ON."

DTC Chart Test Description:
Number(s) below refer to circled number(s) on the diagnostic chart.
11. Isolates wheel signal high short to ground between rear ABS axle harness and dash/body harnesses.
12. Isolates wheel signal low short to ground between rear ABS axle harness and dash/body harnesses.
13. Ensures that the wheel speed sensor harness is not shorted to ground.
14. Ensures that the wheel speed sensor is not shorted to ground.

Diagnostic Aids:
An "intermittent" malfunction is most likely caused by a poor connection, rubbed through wire insulation, or a wire that is broken inside the insulation.
The frequency of the malfunction can be checked by using the enhanced diagnostic function of the Tech 1.

If the customer's comments reflect that the "ABS" warning lamp is "ON" only during humid conditions (rain, snow, vehicle wash), all wheel speed sensor circuitry should be thoroughly inspected for water intrusion. Use the following procedure. Spray down the suspected area with 5% salt water solution (two teaspoons of salt to 12 oz. of water). Drive vehicle above 24 km/h (15 mph) for at least 30 seconds. If DTC returns immediately, replace suspected harness.
Any circuitry, that is suspected as causing the intermittent complaint, should be thoroughly checked for backed out terminals, improper mating, broken locks, improperly formed or damaged terminals, poor terminal to wiring connections or physical damage to the wiring harness.
Excessive end play can cause this DTC to set. Check integral bearing for excessive axial end play.

Resistance of the wheel speed sensor will increase with an increase in sensor temperature.
When replacing a wheel speed sensor, inspect the sensor terminals and harness connector for corrosion and/or water intrusion. If evidence of corrosion or water intrusion exists, replace wheel speed sensor harness.

If wheel speed sensor harness shows evidence of water intrusion, inspect wheel speed sensor for water intrusion. If water intrusion is evident, replace wheel speed sensor.

DIAGNOSTIC CHART DTC 27 — 1995 CAVALIER AND SUNFIRE WITH TCS, CONTINUED

DTC 27
(Page 3 of 3)
LEFT REAR EXCESSIVE WHEEL SPEED VARIATION

CONTINUED FROM PAGE 1 OF 3 OF THIS DTC CHART.

(11)
- DISCONNECT REAR ABS HARNESS FROM BODY TO ABS AND FUEL SENDER HARNESS CONNECTOR.
- USING J 39200, MEASURE THE RESISTANCE BETWEEN GROUND AND 24-WAY EBCM HARNESS CONNECTOR TERMINAL "12".

DOES J 39200 DISPLAY RESISTANCE AS "OL" (INFINITE)?

NO → REPAIR SHORT TO GROUND IN CKT 884.

YES ↓

(12)
- USING J 39200, MEASURE THE RESISTANCE BETWEEN GROUND AND 24-WAY EBCM HARNESS CONNECTOR TERMINAL "11".

DOES J 39200 DISPLAY RESISTANCE AS "OL" (INFINITE)?

NO → REPAIR SHORT TO GROUND IN CKT 885.

YES ↓

(13)
- THOROUGHLY RESPRAY THE REAR ABS HARNESS WITH THE SALT WATER SOLUTION. IMPORTANT: DO NOT ALLOW WATER TO ENTER THE WHEEL SPEED SENSOR CONNECTOR.
- DISCONNECT LEFT REAR WHEEL SPEED SENSOR CONNECTOR.
- USING J 39200, MEASURE THE RESISTANCE BETWEEN GROUND AND LEFT REAR WHEEL SPEED SENSOR HARNESS CONNECTOR TERMINAL "A" THEN "B."

DID J 39200 DISPLAY RESISTANCE AS "OL" (INFINITE) FOR EACH MEASUREMENT?

NO → REPLACE REAR ABS HARNESS.

YES ↓

(14)
- USING J 39200, MEASURE THE RESISTANCE BETWEEN SENSOR TERMINAL "A" AND INTEGRAL BEARING AND SPEED SENSOR ASSEMBLY MOUNTING FLANGE.

DOES J 39200 DISPLAY RESISTANCE AS "OL" (INFINITE)?

NO → REPLACE INTEGRAL BEARING AND SPEED SENSOR ASSEMBLY.

YES ↓

MALFUNCTION IS INTERMITTENT. INSPECT ALL CONNECTORS AND HARNESSES FOR DAMAGES WHICH MAY RESULT IN A SHORT TO GROUND WHEN ALL COMPONENTS ARE CONNECTED. SMALL NICKS OR CUTS IN THE HARNESS CAN CAUSE A MALFUNCTION. REPAIR ANY DAMAGE IF FOUND. JUMPER HARNESS MUST BE REPLACED IF DAMAGED. REFER TO "DIAGNOSTIC AIDS" ON FACING PAGE FOR MORE INFORMATION.

DIAGNOSTIC CHART DTC 28 — 1995 CAVALIER AND SUNFIRE WITH TCS

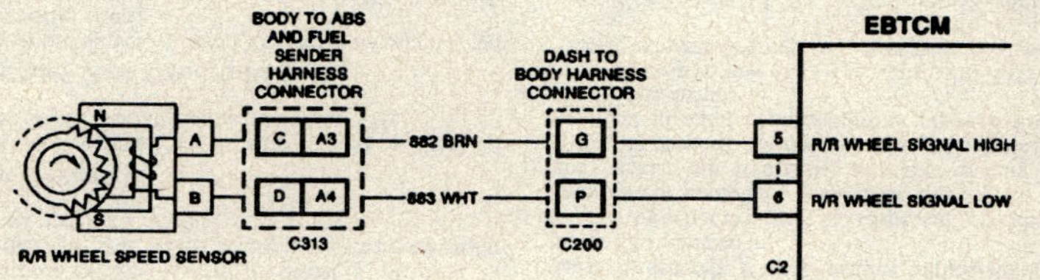

DTC 28
(Page 1 of 3)
RIGHT REAR EXCESSIVE WHEEL SPEED VARIATION

Circuit Description:

As a toothed ring passes by the wheel speed sensor, changes in the electromagnetic field cause the wheel speed sensor to produce a sinusoidal (AC) voltage signal whose frequency is proportional to wheel speed. The magnitude of this signal is directly related to wheel speed and the proximity of the wheel speed sensor to the toothed ring, often referred to as the air gap.

DTC Will Set When:

DTC 28 can be set when the brake is "OFF." The purpose of this test is to detect a situation in which the right rear wheel acceleration or deceleration is beyond specified limits.

Action Taken:

A malfunction DTC is stored, ABS is disabled and the "ABS" warning lamp is turned "ON."

DTC Chart Test Description:

Number(s) below refer to circled number(s) on the diagnostic chart.

1. Checks wheel speed sensor for proper resistance value.
2. Ensures the wheel speed sensor is not shorted to ground.
3. Ensures the wheel speed sensor is not shorted to voltage.
4. Ensures that DTC 28 was not set due to a poor connection at the EBTCM or physical damage of the circuitry.
5. Checks for proper voltage at the wheel signal low speed sensor harness connector.
6. Checks for proper voltage at the wheel signal high speed sensor harness connector.

Diagnostic Aids:

An "intermittent" malfunction is most likely caused by a poor connection, rubbed through wire insulation, or a wire that is broken inside the insulation.

The frequency of the malfunction can be checked by using the enhanced diagnostic function of the Tech 1.

If the customer's comments reflect that the "ABS" warning lamp is "ON" only during humid conditions (rain, snow, vehicle wash), all wheel speed sensor circuitry should be thoroughly inspected for water intrusion. Use the following procedure. Spray down the suspected area with 5% salt water solution (two teaspoons of salt to 12 oz. of water). Drive vehicle above 24 km/h (15 mph) for at least 30 seconds. If DTC returns immediately, replace suspected harness.

Any circuitry, that is suspected as causing the intermittent complaint, should be thoroughly checked for backed out terminals, improper mating, broken locks, improperly formed or damaged terminals, poor terminal to wiring connections or physical damage to the wiring harness.

Excessive end play can cause this DTC to set. Check integral bearing for excessive axial end play.

Resistance of the wheel speed sensor will increase with an increase in sensor temperature.

When replacing a wheel speed sensor, inspect the sensor terminals and harness connector for corrosion and/or water intrusion. If evidence of corrosion or water intrusion exists, replace wheel speed sensor harness.

If wheel speed sensor harness shows evidence of water intrusion, inspect wheel speed sensor for water intrusion. If water intrusion is evident, replace wheel speed sensor.

DIAGNOSTIC CHART DTC 28 — 1995 CAVALIER AND SUNFIRE WITH TCS, CONTINUED

DTC 28
(Page 1 of 3)
RIGHT REAR EXCESSIVE WHEEL SPEED VARIATION

IMPORTANT:
- WHEEL SPEED SENSOR INTERMITTENT MALFUNCTIONS MAY BE DIFFICULT TO LOCATE. CARE SHOULD BE TAKEN NOT TO DISTURB ANY ELECTRICAL CONNECTIONS PRIOR TO AN INDICATED STEP OF THIS CHART. THIS WILL ENSURE THAT AN INTERMITTENT CONNECTION WILL NOT BE CORRECTED BEFORE THE SOURCE OF THE MALFUNCTION IS FOUND.

(1)
- IGNITION "OFF."
- USING A 5% SALT WATER SOLUTION (TWO TEASPOONS OF SALT PER 12 OZ. WATER), THOROUGHLY SPRAY THE REAR ABS HARNESS.
- DISCONNECT 24-WAY EBCM CONNECTOR.
- USING J 39200, MEASURE RESISTANCE BETWEEN 24-WAY EBCM HARNESS CONNECTOR TERMINALS "7" AND "8". IS RESISTANCE BETWEEN 2100 AND 2400 Ω WHEN SENSOR IS 25°C (77°F)? *[IF TEMPERATURE OF THE SENSOR IS SUSPECTED TO BE OTHER THAN 25°C (77°F), REFER TO TABLE ON PAGE 2 OF 2 OF THIS DTC CHART FOR APPLICABLE SENSOR RESISTANCE VALUES; THE VALUES IN THE TABLE ARE FOR TEMPERATURE OF THE SENSOR NOT AIR TEMPERATURE.]*

YES → (2); NO → GO TO PAGE 2 OF 3 OF THIS DTC CHART.

(2)
- USING J 39200, MEASURE THE RESISTANCE BETWEEN GROUND AND EBCM HARNESS CONNECTOR TERMINAL "7". DOES J 39200 DISPLAY RESISTANCE AS "OL" (INFINITE)?

YES → (3); NO → GO TO PAGE 3 OF 3 OF THIS DTC CHART.

(3)
- IGNITION "ON."
- USING J 39200, MEASURE THE VOLTAGE BETWEEN GROUND AND 24-WAY EBCM HARNESS CONNECTOR TERMINAL "8". IS VOLTAGE 1.0V OR LESS?

YES → (4); NO → (5)

(4)
- IGNITION "OFF."
- INSPECT 24-WAY EBCM HARNESS CONNECTOR TERMINALS "7" AND "8" FOR POOR TERMINAL CONTACT OR CORROSION. INSPECT CKT(s) 882 AND 883 FOR DAMAGE WHICH MAY RESULT IN AN OPEN CIRCUIT, SHORT TO GROUND OR SHORT TO VOLTAGE. REPAIR DAMAGE IF EVIDENT. REPLACE TERMINALS IF POOR CONTACT OR CORROSION EXISTS.
- RECONNECT ALL CONNECTORS.
- IGNITION "ON."
- IS DTC 28 SET AS A CURRENT DTC?

YES → REPLACE EBCM.
NO → MALFUNCTION IS INTERMITTENT. REFER TO "DIAGNOSTIC AIDS" ON FACING PAGE.

(5)
- IGNITION "OFF."
- DISCONNECT RIGHT REAR WHEEL SPEED SENSOR CONNECTOR.
- IGNITION "ON."
- USING J 39200, MEASURE THE VOLTAGE BETWEEN GROUND AND 24-WAY EBCM HARNESS CONNECTOR TERMINAL "8". IS VOLTAGE 1.0V OR LESS?

YES → (6); NO → REPAIR SHORT TO VOLTAGE IN CKT 883.

(6)
- USING J 39200, MEASURE THE VOLTAGE BETWEEN GROUND AND 24-WAY EBCM HARNESS CONNECTOR TERMINAL "7". IS VOLTAGE 1.0V OR LESS?

YES → MALFUNCTION IS INTERMITTENT. INSPECT ALL CONNECTORS AND HARNESSES FOR DAMAGE WHICH MAY RESULT IN A SHORT TO VOLTAGE WHEN ALL COMPONENTS ARE CONNECTED. REPAIR ALL DAMAGE IF FOUND. REFER TO "DIAGNOSTIC AIDS" ON FACING PAGE FOR MORE INFORMATION.
NO → REPAIR SHORT TO VOLTAGE IN CKT 882.

DIAGNOSTIC CHART DTC 28 — 1995 CAVALIER AND SUNFIRE WITH TCS, CONTINUED

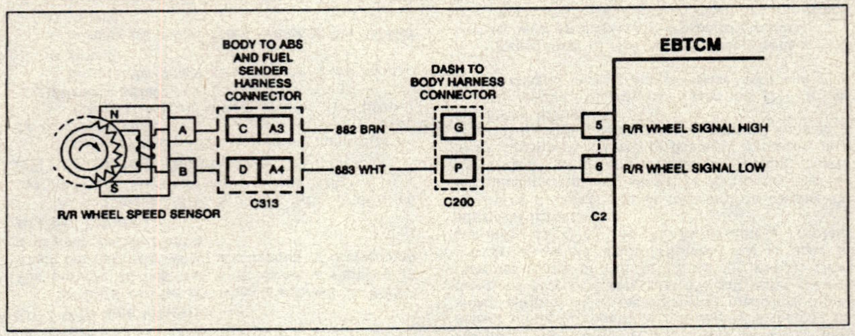

DTC 28
(Page 2 of 3)
RIGHT REAR EXCESSIVE WHEEL SPEED VARIATION

Circuit Description:

As a toothed ring passes by the wheel speed sensor, changes in the electromagnetic field cause the wheel speed sensor to produce a sinusoidal (AC) voltage signal whose frequency is proportional to wheel speed. The magnitude of this signal is directly related to wheel speed and the proximity of the wheel speed sensor to the toothed ring, often referred to as the air gap.

DTC Will Set When:

DTC 28 can be set when the brake is "OFF." The purpose of this test is to detect a situation in which the right rear wheel acceleration or deceleration is beyond specified limits.

Action Taken:

A malfunction DTC is stored, ABS is disabled and the "ABS" warning lamp is turned "ON."

DTC Chart Test Description:

Number(s) below refer to circled number(s) on the diagnostic chart.

7. Tests for a short between the wheel signal high and wheel signal low circuits.
8. Checks for an open in the wheel signal low circuit.
9. Checks for an open in the wheel signal high circuit.
10. Checks for proper resistance of the wheel speed sensor.

Diagnostic Aids:

An "intermittent" malfunction is most likely caused by a poor connection, rubbed through wire insulation, or a wire that is broken inside the insulation.

The frequency of the malfunction can be checked by using the enhanced diagnostic function of the Tech 1.

If the customer's comments reflect that the "ABS" warning lamp is "ON" only during humid conditions (rain, snow, vehicle wash), all wheel speed sensor circuitry should be thoroughly inspected for water intrusion. Use the following procedure. Spray down the suspected area with 5% salt water solution (two teaspoons of salt to 12 oz. of water). Drive vehicle above 24 km/h (15 mph) for at least 30 seconds. If DTC returns immediately, replace suspected harness.

Any circuitry, that is suspected as causing the intermittent complaint, should be thoroughly checked for backed out terminals, improper mating, broken locks, improperly formed or damaged terminals, poor terminal to wiring connections or physical damage to the wiring harness.

Excessive end play can cause this DTC to set. Check integral bearing for excessive axial end play.

Resistance of the wheel speed sensor will increase with an increase in sensor temperature.

When replacing a wheel speed sensor, inspect the sensor terminals and harness connector for corrosion and/or water intrusion. If evidence of corrosion or water intrusion exists, replace wheel speed sensor harness.

If wheel speed sensor harness shows evidence of water intrusion, inspect wheel speed sensor for water intrusion. If water intrusion is evident, replace wheel speed sensor.

DTC 28
(Page 2 of 3)
RIGHT REAR EXCESSIVE WHEEL SPEED VARIATION

CONTINUED FROM PAGE 1 OF 3 OF THIS DTC CHART.

(7)
- DISCONNECT RIGHT REAR WHEEL SPEED SENSOR CONNECTOR.
- USING J 39200, MEASURE THE RESISTANCE BETWEEN 24-WAY EBCM HARNESS CONNECTOR TERMINALS "7" AND "8".

DOES J 39200 DISPLAY RESISTANCE AS "OL" (INFINITE)?

NO: REPAIR SHORT BETWEEN CKTs 882 AND 883.

YES:

(8)
- USING J 39200, MEASURE THE RESISTANCE BETWEEN 24-WAY EBCM HARNESS CONNECTOR TERMINAL "7" AND RIGHT REAR WHEEL SPEED SENSOR HARNESS CONNECTOR TERMINAL "A."

IS RESISTANCE 2.0Ω OR LESS?

NO: REPAIR OPEN OR HIGH RESISTANCE IN CKT 882.

YES:

(9)
- USING J 39200, MEASURE THE RESISTANCE BETWEEN 24-WAY EBCM HARNESS CONNECTOR TERMINAL "8" AND RIGHT REAR WHEEL SPEED SENSOR HARNESS CONNECTOR TERMINAL "B."

IS RESISTANCE 2.0Ω OR LESS?

NO: REPAIR OPEN OR HIGH RESISTANCE IN CKT 883.

YES:

(10)
- USING J 39200, MEASURE THE RESISTANCE BETWEEN RIGHT REAR WHEEL SPEED SENSOR TERMINALS "A" AND "B."

IS RESISTANCE BETWEEN 2100 AND 2400Ω WHEN SENSOR IS 25°C (77°F)?
[IF TEMPERATURE OF THE SENSOR IS SUSPECTED TO BE OTHER THAN 25°C (77°F), REFER TO THE TABLE BELOW FOR APPLICABLE SENSOR RESISTANCE VALUES; THE VALUES IN THE TABLE ARE FOR TEMPERATURE OF THE SENSOR NOT AIR TEMPERATURE.]

NO: REPLACE INTEGRAL BEARING AND SPEED SENSOR ASSEMBLY.

YES: MALFUNCTION IS INTERMITTENT. INSPECT ALL CONNECTORS AND HARNESSES FOR DAMAGES WHICH MAY RESULT IN HIGH RESISTANCE WHEN ALL COMPONENTS ARE CONNECTED. REFER TO "DIAGNOSTIC AIDS" ON FACING PAGE FOR MORE INFORMATION. REPAIR ALL DAMAGE IF FOUND.

REAR WHEEL SPEED SENSOR

TEMPERATURE OF SENSOR VS. SENSOR RESISTANCE (APPROXIMATE)

TEMP. (°C)	TEMP. (°F)	RESISTANCE (OHMS)
-34 TO 4	-30 TO 40	1623 TO 2210
5 TO 43	41 TO 110	1939 TO 2570
44 TO 93	111 TO 200	2254 TO 3029

DIAGNOSTIC CHART DTC 28 — 1995 CAVALIER AND SUNFIRE WITH TCS, CONTINUED

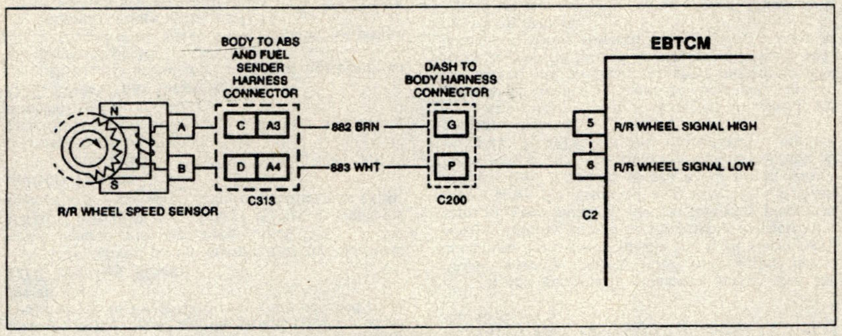

DTC 28
(Page 3 of 3)
RIGHT REAR EXCESSIVE WHEEL SPEED VARIATION

Circuit Description:

As a toothed ring passes by the wheel speed sensor, changes in the electromagnetic field cause the wheel speed sensor to produce a sinusoidal (AC) voltage signal whose frequency is proportional to wheel speed. The magnitude of this signal is directly related to wheel speed and the proximity of the wheel speed sensor to the toothed ring, often referred to as the air gap.

DTC Will Set When:

DTC 28 can be set when the brake is "OFF." The purpose of this test is to detect a situation in which the right rear wheel acceleration or deceleration is beyond specified limits.

Action Taken:

A malfunction DTC is stored, ABS is disabled and the "ABS" warning lamp is turned "ON."

DTC Chart Test Description:

Number(s) below refer to circled number(s) on the diagnostic chart.

11. Isolates wheel signal high short to ground between rear ABS axle harness and dash/body harnesses.
12. Isolates wheel signal low short to ground between rear ABS axle harness and dash/body harnesses.
13. Ensures that the wheel speed sensor harness is not shorted to ground.
14. Ensures that the wheel speed sensor is not shorted to ground.

Diagnostic Aids:

An "intermittent" malfunction is most likely caused by a poor connection, rubbed through wire insulation, or a wire that is broken inside the insulation.

The frequency of the malfunction can be checked by using the enhanced diagnostic function of the Tech 1.

If the customer's comments reflect that the "ABS" warning lamp is "ON" only during humid conditions (rain, snow, vehicle wash), all wheel speed sensor circuitry should be thoroughly inspected for water intrusion. Use the following procedure. Spray down the suspected area with 5% salt water solution (two teaspoons of salt to 12 oz. of water). Drive vehicle above 24 km/h (15 mph) for at least 30 seconds. If DTC returns immediately, replace suspected harness.

Any circuitry, that is suspected as causing the intermittent complaint, should be thoroughly checked for backed out terminals, improper mating, broken locks, improperly formed or damaged terminals, poor terminal to wiring connections or physical damage to the wiring harness.

Excessive end play can cause this DTC to set. Check integral bearing for excessive axial end play.

Resistance of the wheel speed sensor will increase with an increase in sensor temperature.

When replacing a wheel speed sensor, inspect the sensor terminals and harness connector for corrosion and/or water intrusion. If evidence of corrosion or water intrusion exists, replace wheel speed sensor harness.

If wheel speed sensor harness shows evidence of water intrusion, inspect wheel speed sensor for water intrusion. If water intrusion is evident, replace wheel speed sensor.

DIAGNOSTIC CHART DTC 28 — 1995 CAVALIER AND SUNFIRE WITH TCS, CONTINUED

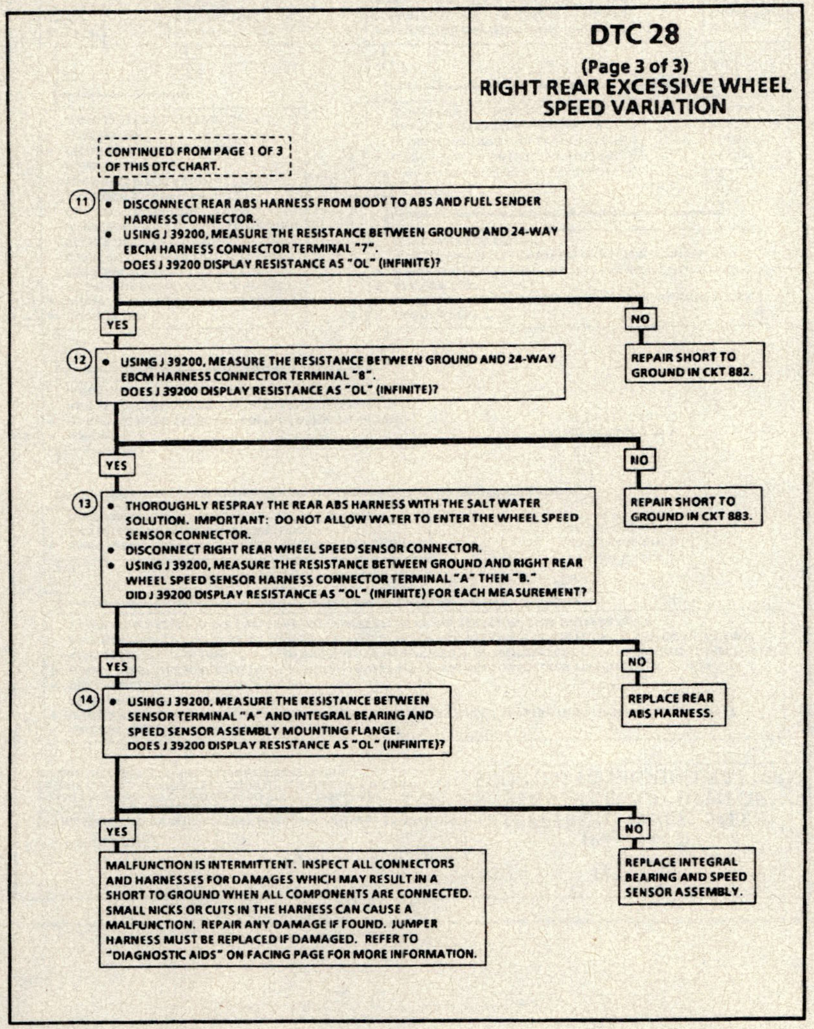

DIAGNOSTIC CHART DTC 32 — 1995 CAVALIER AND SUNFIRE WITH TCS

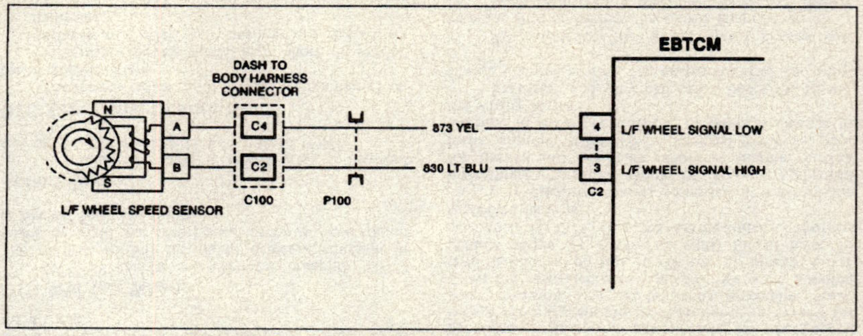

DTC 32
(Page 1 of 3)
LEFT FRONT WHEEL SPEED SENSOR CIRCUIT OPEN OR SHORTED TO GROUND/BATTERY

Circuit Description:

As a toothed ring passes by the wheel speed sensor, changes in the electromagnetic field cause the wheel speed sensor to produce a sinusoidal (AC) voltage signal whose frequency is proportional to wheel speed. The magnitude of this signal is directly related to wheel speed and the proximity of the wheel speed sensor to the toothed ring, often referred to as the air gap.

DTC Will Set When:

DTC 32 can be set anytime after the electronic brake control relay has been turned "ON." If the EBTCM detects an out of range voltage in either the high or low wheel speed circuits, a malfunction exists.

Action Taken:

A malfunction DTC is stored, the "ABS" warning lamp is turned "ON," the "ETS OFF" indicator is turned "ON," ABS is disabled, and ETS is disabled.

DTC Chart Test Description:

Number(s) below refer to circled number(s) on the diagnostic chart.

1. Checks wheel speed sensor circuitry for proper resistance values at the EBTCM.
2. Ensures that the wheel speed sensor signal high circuitry is not shorted to ground.
3. Checks for a short to voltage in either the high or low wheel signal circuits.
4. Identifies a wheel speed sensor or circuitry that is damaged and visibly apparent.
5. Determines if wheel speed signal low is shorted to voltage.
6. Determines if wheel speed signal high is shorted to voltage.

Diagnostic Aids:

An "intermittent" malfunction is most likely caused by a poor connection, rubbed through wire insulation, or a wire that is broken inside the insulation.

The frequency of the malfunction can be checked by using the enhanced diagnostic function of the Tech 1.

If the customer's comments reflect that the "ABS" warning lamp is "ON" only during humid conditions (rain, snow, vehicle wash) all wheel speed sensor circuitry should be thoroughly inspected for signs of water intrusion. Use the following procedure. Spray down the suspected area with 5% salt water solution (two teaspoons of salt to 12 oz. of water). Drive vehicle above 24 km/h (15 mph) for at least 30 seconds. If DTC returns immediately, replace suspected harness.

Any circuitry, that is suspected as causing the intermittent complaint, should be thoroughly checked for backed out terminals, improper mating, broken locks, improperly formed or damaged terminals, poor terminal to wiring connections or physical damage to the wiring harness.

Resistance of the wheel speed sensor will increase with an increase in sensor temperature. When replacing a wheel speed sensor, inspect the sensor terminals and harness connector for corrosion and/or water intrusion. If evidence of corrosion or water intrusion exists, replace wheel speed sensor harness.

If wheel speed sensor harness shows evidence of water intrusion, inspect wheel speed sensor for water intrusion. If water intrusion is evident, replace wheel speed sensor.

DIAGNOSTIC CHART DTC 32 — 1995 CAVALIER AND SUNFIRE WITH TCS, CONTINUED

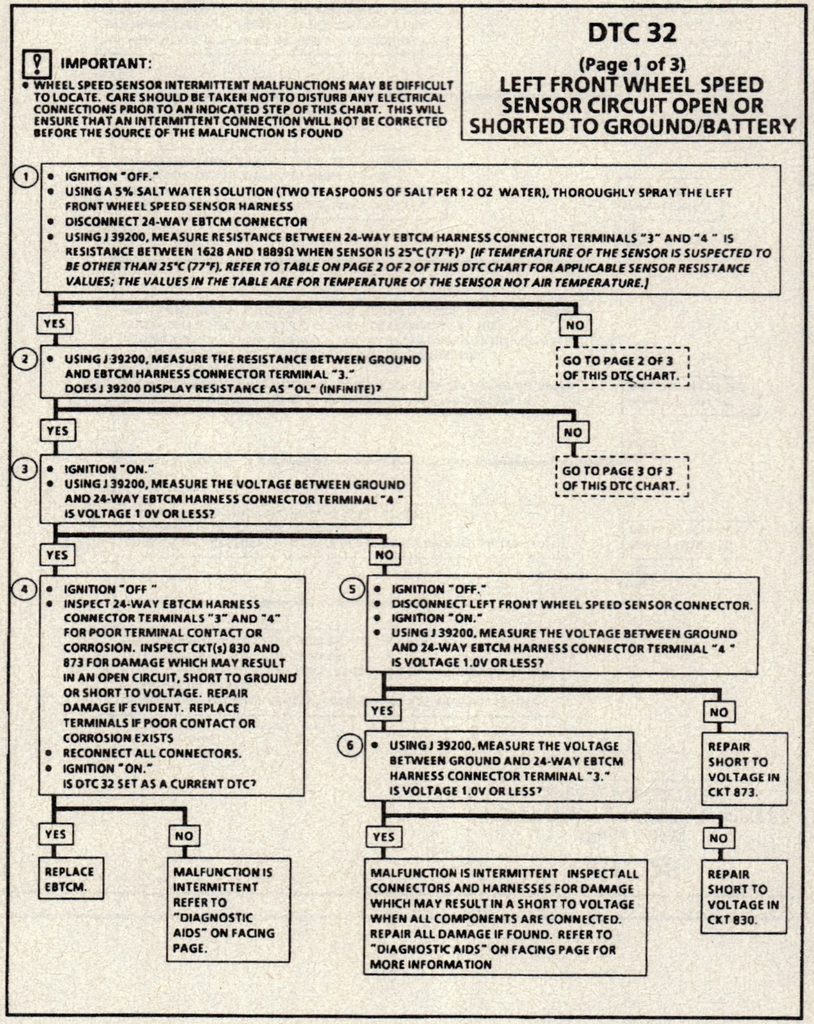

DIAGNOSTIC CHART DTC 32 — 1995 CAVALIER AND SUNFIRE WITH TCS, CONTINUED

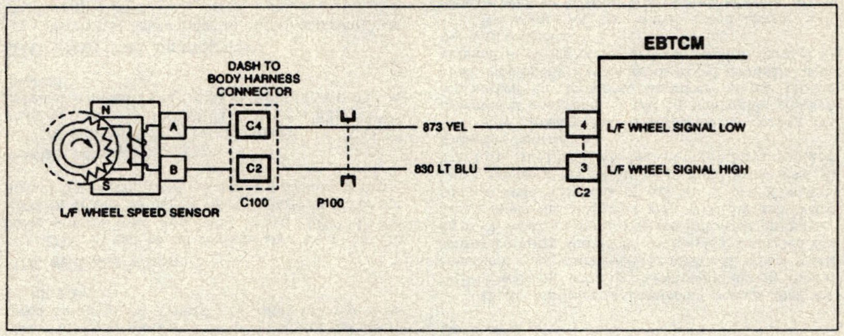

DTC 32
(Page 2 of 3)
LEFT FRONT WHEEL SPEED SENSOR CIRCUIT OPEN OR SHORTED TO GROUND/BATTERY

Circuit Description:

As a toothed ring passes by the wheel speed sensor, changes in the electromagnetic field cause the wheel speed sensor to produce a sinusoidal (AC) voltage signal whose frequency is proportional to wheel speed. The magnitude of this signal is directly related to wheel speed and the proximity of the wheel speed sensor to the toothed ring, often referred to as the air gap.

DTC Will Set When:

DTC 32 can be set anytime after the electronic brake control relay has been turned "ON." If the EBTCM detects an out of range voltage in either the high or low wheel speed circuits, a malfunction exists.

Action Taken:

A malfunction DTC is stored, the "ABS" warning lamp is turned "ON," the "ETS OFF" indicator is turned "ON," ABS is disabled, and ETS is disabled.

DTC Chart Test Description:

Number(s) below refer to circled number(s) on the diagnostic chart.

7. Checks for open or high resistance in wheel speed signal low circuitry.
8. Checks for open or high resistance in wheel speed signal high circuitry.
9. Checks wheel speed sensor for proper resistance value.

Diagnostic Aids:

An "intermittent" malfunction is most likely caused by a poor connection, rubbed through wire insulation, or a wire that is broken inside the insulation.

The frequency of the malfunction can be checked by using the enhanced diagnostic function of the Tech 1.

If the customer's comments reflect that the "ABS" warning lamp is "ON" only during humid conditions (rain, snow, vehicle wash), all wheel speed sensor circuitry should be thoroughly inspected for signs of water intrusion. Use the following procedure. Spray down the suspected area with 5% salt water solution (two teaspoons of salt to 12 oz. of water). Drive vehicle above 24 km/h (15 mph) for at least 30 seconds. If DTC returns immediately, replace suspected harness.

Any circuitry, that is suspected as causing the intermittent complaint, should be thoroughly checked for backed out terminals, improper mating, broken locks, improperly formed or damaged terminals, poor terminal to wiring connections or physical damage to the wiring harness.

Resistance of the wheel speed sensor will increase with an increase in sensor temperature. When replacing a wheel speed sensor, inspect the sensor terminals and harness connector for corrosion and/or water intrusion. If evidence of corrosion or water intrusion exists, replace wheel speed sensor harness.

If wheel speed sensor harness shows evidence of water intrusion, inspect wheel speed sensor for water intrusion. If water intrusion is evident, replace wheel speed sensor.

DIAGNOSTIC CHART DTC 32 — 1995 CAVALIER AND SUNFIRE WITH TCS, CONTINUED

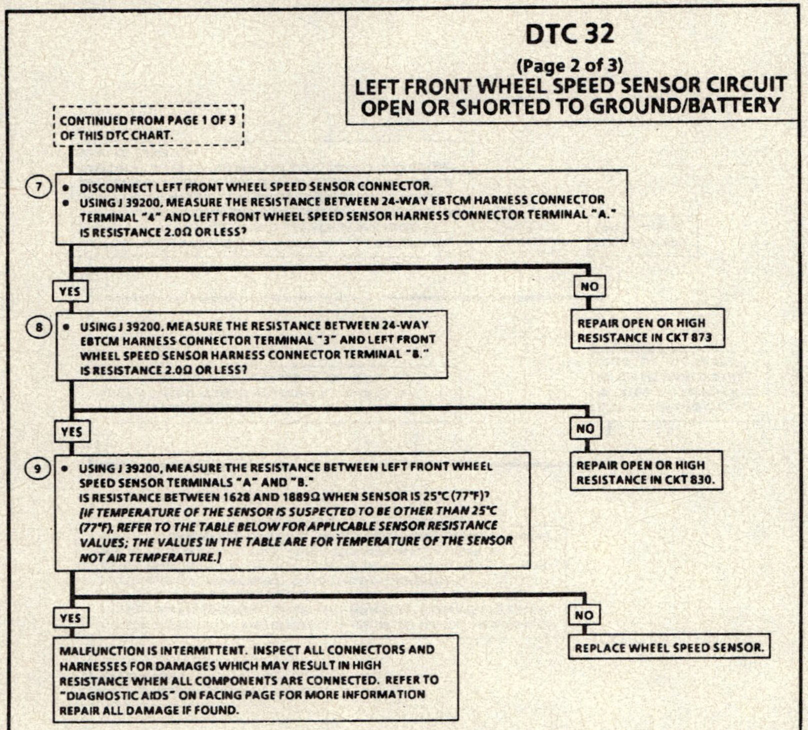

FRONT WHEEL SPEED SENSOR

TEMPERATURE OF SENSOR VS. SENSOR RESISTANCE (APPROXIMATE)

TEMP. (°C)	TEMP (°F)	RESISTANCE (OHMS)
-34 TO 4	-30 TO 40	1258 TO 1739
5 TO 43	41 TO 110	1503 TO 2023
44 TO 93	111 TO 200	1747 TO 2384

DIAGNOSTIC CHART DTC 32 — 1995 CAVALIER AND SUNFIRE WITH TCS, CONTINUED

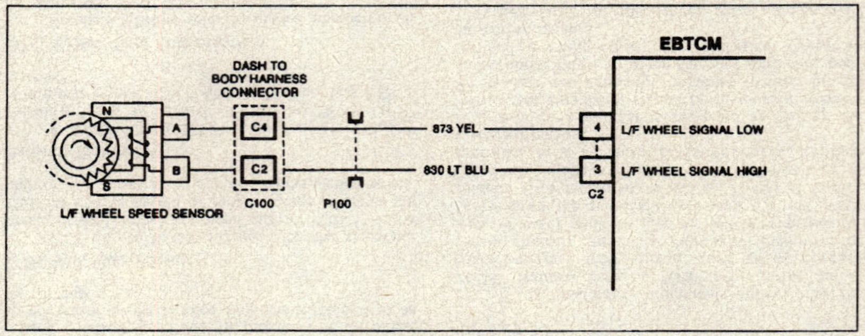

DTC 32
(Page 3 of 3)
LEFT FRONT WHEEL SPEED SENSOR CIRCUIT OPEN OR SHORTED TO GROUND/BATTERY

Circuit Description:

As a toothed ring passes by the wheel speed sensor, changes in the electromagnetic field cause the wheel speed sensor to produce a sinusoidal (AC) voltage signal whose frequency is proportional to wheel speed. The magnitude of this signal is directly related to wheel speed and the proximity of the wheel speed sensor to the toothed ring, often referred to as the air gap.

DTC Will Set When:

DTC 32 can be set anytime after the electronic brake control relay has been turned "ON." If the EBTCM detects an out of range voltage in either the high or low wheel speed circuits, a malfunction exists.

Action Taken:

A malfunction DTC is stored, the "ABS" warning lamp is turned "ON," the "ETS OFF" indicator is turned "ON," ABS is disabled, and ETS is disabled.

DTC Chart Test Description:

Number(s) below refer to circled number(s) on the diagnostic chart.

10. Checks for intermittent short in wheel speed signal circuitry.

11. Checks for short to ground in wheel speed sensor.

Diagnostic Aids:

An "intermittent" malfunction is most likely caused by a poor connection, rubbed through wire insulation, or a wire that is broken inside the insulation.

The frequency of the malfunction can be checked by using the enhanced diagnostic function of the Tech 1

If the customer's comments reflect that the "ABS" warning lamp is "ON" only during humid conditions (rain, snow, vehicle wash), all wheel speed sensor circuitry should be thoroughly inspected for signs of water intrusion. Use the following procedure. Spray down the suspected area with 5% salt water solution (two teaspoons of salt to 12 oz. of water). Drive vehicle above 24 km/h (15 mph) for at least 30 seconds. If DTC returns immediately, replace suspected harness.

Any circuitry, that is suspected as causing the intermittent complaint, should be thoroughly checked for backed out terminals, improper mating, broken locks, improperly formed or damaged terminals, poor terminal to wiring connections or physical damage to the wiring harness.

Resistance of the wheel speed sensor will increase with an increase in sensor temperature. When replacing a wheel speed sensor, inspect the sensor terminals and harness connector for corrosion and/or water intrusion. If evidence of corrosion or water intrusion exists, replace wheel speed sensor harness.

If wheel speed sensor harness shows evidence of water intrusion, inspect wheel speed sensor for water intrusion. If water intrusion is evident, replace wheel speed sensor.

DIAGNOSTIC CHART DTC 32 — 1995 CAVALIER AND SUNFIRE WITH TCS, CONTINUED

DTC 32
(Page 3 of 3)
LEFT FRONT WHEEL SPEED SENSOR CIRCUIT OPEN OR SHORTED TO GROUND/BATTERY

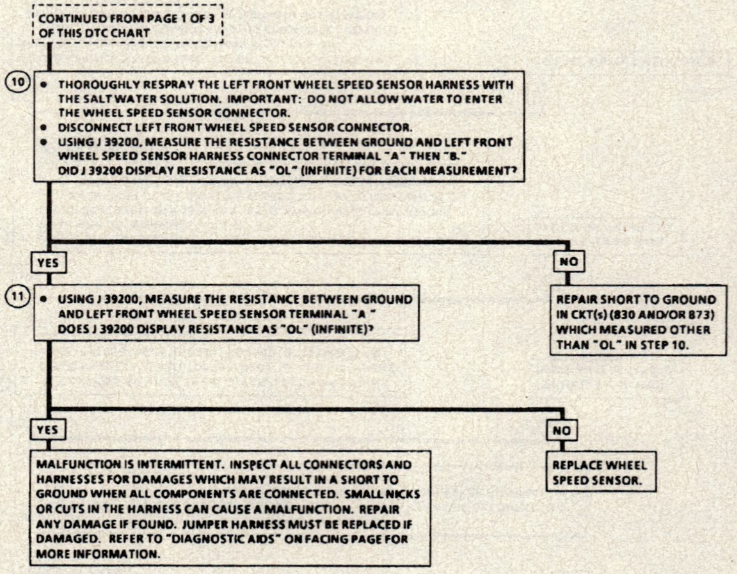

DIAGNOSTIC CHART DTC 33 — 1995 CAVALIER AND SUNFIRE WITH TCS

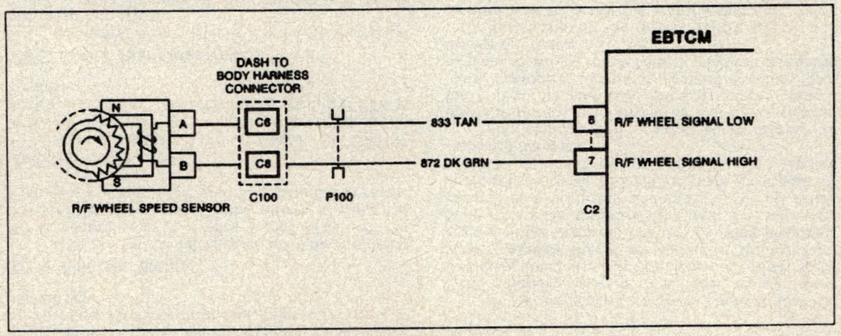

DTC 33
(Page 1 of 3)
RIGHT FRONT WHEEL SPEED SENSOR CIRCUIT OPEN OR SHORTED TO GROUND/BATTERY

Circuit Description:

As a toothed ring passes by the wheel speed sensor, changes in the electromagnetic field cause the wheel speed sensor to produce a sinusoidal (AC) voltage signal whose frequency is proportional to wheel speed. The magnitude of this signal is directly related to wheel speed and the proximity of the wheel speed sensor to the toothed ring, often referred to as the air gap.

DTC Will Set When:

DTC 33 can be set anytime after the electronic brake control relay has been turned "ON." If the EBTCM detects an out of range voltage in either the high or low wheel speed circuits, a malfunction exists.

Action Taken:

A malfunction DTC is stored, the "ABS" warning lamp is turned "ON," the "ETS OFF" indicator is turned "ON," ABS is disabled, and ETS is disabled.

DTC Chart Test Description:

Number(s) below refer to circled number(s) on the diagnostic chart.

1. Checks wheel speed sensor circuitry for proper resistance values at the EBTCM.
2. Ensures that the wheel speed sensor signal high circuitry is not shorted to ground.
3. Checks for a short to voltage in either the high or low wheel signal circuits.
4. Identifies a wheel speed sensor or circuitry that is damaged and visibly apparent.
5. Determines if wheel speed signal low is shorted to voltage.
6. Determines if wheel speed signal high is shorted to voltage.

Diagnostic Aids:

An "intermittent" malfunction is most likely caused by a poor connection, rubbed through wire insulation, or a wire that is broken inside the insulation.

The frequency of the malfunction can be checked by using the enhanced diagnostic function of the Tech 1.

If the customer's comments reflect that the "ABS" warning lamp is "ON" only during humid conditions (rain, snow, vehicle wash) all wheel speed sensor circuitry should be thoroughly inspected for signs of water intrusion. Use the following procedure. Spray down the suspected area with 5% salt water solution (two teaspoons of salt to 12 oz. of water). Drive vehicle above 24 km/h (15 mph) for at least 30 seconds. If DTC returns immediately, replace suspected harness.

Any circuitry, that is suspected as causing the intermittent complaint, should be thoroughly checked for backed out terminals, improper mating, broken locks, improperly formed or damaged terminals, poor terminal to wiring connections or physical damage to the wiring harness.

Resistance of the wheel speed sensor will increase with an increase in sensor temperature. When replacing a wheel speed sensor, inspect the sensor terminals and harness connector for corrosion and/or water intrusion. If evidence of corrosion or water intrusion exists, replace wheel speed sensor harness.

If wheel speed sensor harness shows evidence of water intrusion, inspect wheel speed sensor for water intrusion. If water intrusion is evident, replace wheel speed sensor.

DIAGNOSTIC CHART DTC 33 — 1995 CAVALIER AND SUNFIRE WITH TCS, CONTINUED

IMPORTANT:

- WHEEL SPEED SENSOR INTERMITTENT MALFUNCTIONS MAY BE DIFFICULT TO LOCATE. CARE SHOULD BE TAKEN NOT TO DISTURB ANY ELECTRICAL CONNECTIONS PRIOR TO AN INDICATED STEP OF THIS CHART. THIS WILL ENSURE THAT AN INTERMITTENT CONNECTION WILL NOT BE CORRECTED BEFORE THE SOURCE OF THE MALFUNCTION IS FOUND.

DTC 33
(Page 1 of 3)
RIGHT FRONT WHEEL SPEED SENSOR CIRCUIT OPEN OR SHORTED TO GROUND/BATTERY

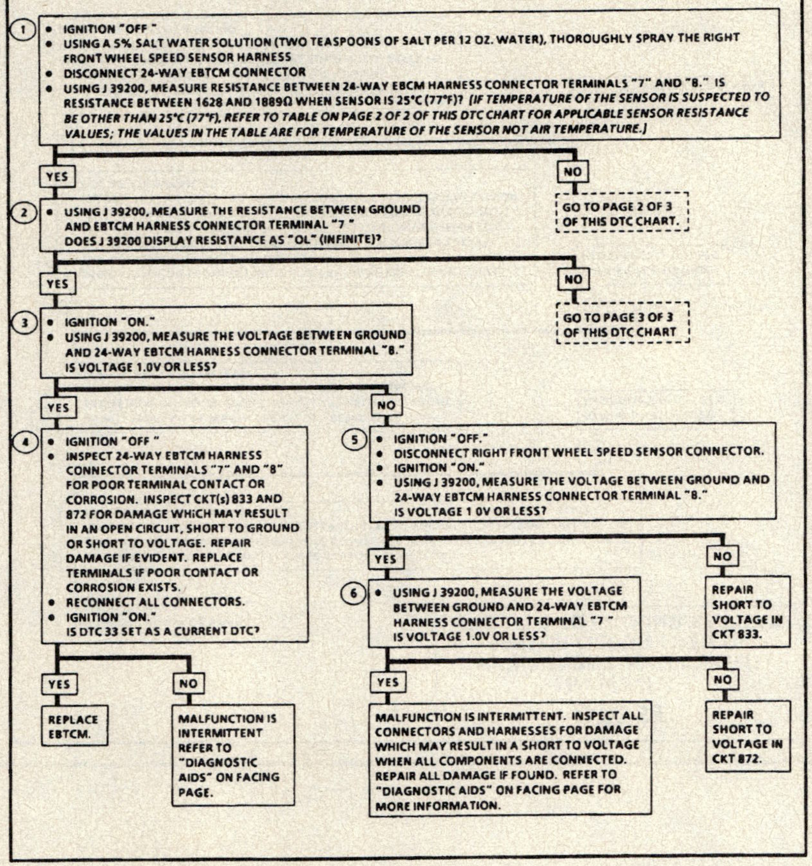

DIAGNOSTIC CHART DTC 33 — 1995 CAVALIER AND SUNFIRE WITH TCS, CONTINUED

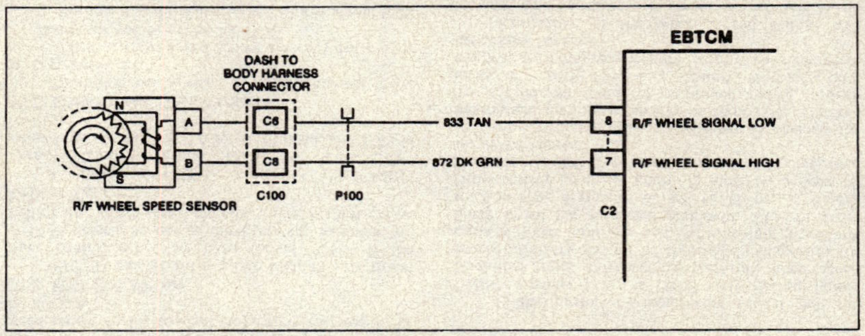

DTC 33
(Page 2 of 3)
RIGHT FRONT WHEEL SPEED SENSOR CIRCUIT OPEN OR SHORTED TO GROUND/BATTERY

Circuit Description:

As a toothed ring passes by the wheel speed sensor, changes in the electromagnetic field cause the wheel speed sensor to produce a sinusoidal (AC) voltage signal whose frequency is proportional to wheel speed. The magnitude of this signal is directly related to wheel speed and the proximity of the wheel speed sensor to the toothed ring, often referred to as the air gap.

DTC Will Set When:

DTC 33 can be set anytime after the electronic brake control relay has been turned "ON." If the EBTCM detects an out of range voltage in either the high or low wheel speed circuits, a malfunction exists.

Action Taken:

A malfunction DTC is stored, the "ABS" warning lamp is turned "ON," the "ETS OFF" indicator is turned "ON," ABS is disabled, and ETS is disabled.

DTC Chart Test Description:

Number(s) below refer to circled number(s) on the diagnostic chart.

7. Checks for open or high resistance in wheel speed signal low circuitry.
8. Checks for open or high resistance in wheel speed signal high circuitry.
9. Checks wheel speed sensor for proper resistance value.

Diagnostic Aids:

An "intermittent" malfunction is most likely caused by a poor connection, rubbed through wire insulation, or a wire that is broken inside the insulation.

The frequency of the malfunction can be checked by using the enhanced diagnostic function of the Tech 1.

If the customer's comments reflect that the "ABS" warning lamp is "ON" only during humid conditions (rain, snow, vehicle wash), all wheel speed sensor circuitry should be thoroughly inspected for signs of water intrusion. Use the following procedure. Spray down the suspected area with 5% salt water solution (two teaspoons of salt to 12 oz. of water). Drive vehicle above 24 km/h (15 mph) for at least 30 seconds. If DTC returns immediately, replace suspected harness.

Any circuitry, that is suspected as causing the intermittent complaint, should be thoroughly checked for backed out terminals, improper mating, broken locks, improperly formed or damaged terminals, poor terminal to wiring connections or physical damage to the wiring harness.

Resistance of the wheel speed sensor will increase with an increase in sensor temperature. When replacing a wheel speed sensor, inspect the sensor terminals and harness connector for corrosion and/or water intrusion. If evidence of corrosion or water intrusion exists, replace wheel speed sensor harness.

If wheel speed sensor harness shows evidence of water intrusion, inspect wheel speed sensor for water intrusion. If water intrusion is evident, replace wheel speed sensor.

DIAGNOSTIC CHART DTC 33 — 1995 CAVALIER AND SUNFIRE WITH TCS, CONTINUED

DTC 33
(Page 2 of 3)
RIGHT FRONT WHEEL SPEED SENSOR CIRCUIT OPEN OR SHORTED TO GROUND/BATTERY

CONTINUED FROM PAGE 1 OF 3 OF THIS DTC CHART.

(7)
- DISCONNECT RIGHT FRONT WHEEL SPEED SENSOR CONNECTOR.
- USING J 39200, MEASURE THE RESISTANCE BETWEEN 24-WAY EBTCM HARNESS CONNECTOR TERMINAL "8" AND RIGHT FRONT WHEEL SPEED SENSOR HARNESS CONNECTOR TERMINAL "A." IS RESISTANCE 2.0Ω OR LESS?

NO: REPAIR OPEN OR HIGH RESISTANCE IN CKT 833.

YES:

(8)
- USING J 39200, MEASURE THE RESISTANCE BETWEEN 24-WAY EBTCM HARNESS CONNECTOR TERMINAL "7" AND RIGHT FRONT WHEEL SPEED SENSOR HARNESS CONNECTOR TERMINAL "B." IS RESISTANCE 2.0Ω OR LESS?

NO: REPAIR OPEN OR HIGH RESISTANCE IN CKT 872

YES:

(9)
- USING J 39200, MEASURE THE RESISTANCE BETWEEN RIGHT FRONT WHEEL SPEED SENSOR TERMINALS "A" AND "B."
  IS RESISTANCE BETWEEN 1628 AND 1889Ω WHEN SENSOR IS 25°C (77°F)?
  (IF TEMPERATURE OF THE SENSOR IS SUSPECTED TO BE OTHER THAN 25°C (77°F), REFER TO THE TABLE BELOW FOR APPLICABLE SENSOR RESISTANCE VALUES; THE VALUES IN THE TABLE ARE FOR TEMPERATURE OF THE SENSOR NOT AIR TEMPERATURE.)

NO: REPLACE WHEEL SPEED SENSOR.

YES: MALFUNCTION IS INTERMITTENT. INSPECT ALL CONNECTORS AND HARNESSES FOR DAMAGES WHICH MAY RESULT IN HIGH RESISTANCE WHEN ALL COMPONENTS ARE CONNECTED. REFER TO "DIAGNOSTIC AIDS" ON FACING PAGE FOR MORE INFORMATION. REPAIR ALL DAMAGE IF FOUND.

FRONT WHEEL SPEED SENSOR

TEMPERATURE OF SENSOR VS. SENSOR RESISTANCE (APPROXIMATE)

TEMP. (°C)	TEMP (°F)	RESISTANCE (OHMS)
-34 TO 4	-30 TO 40	1258 TO 1739
5 TO 43	41 TO 110	1503 TO 2023
44 TO 93	111 TO 200	1747 TO 2384

DIAGNOSTIC CHART DTC 33 — 1995 CAVALIER AND SUNFIRE WITH TCS, CONTINUED

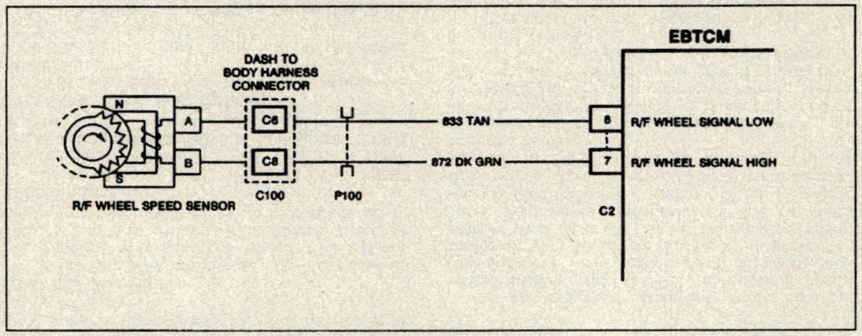

DTC 33
(Page 3 of 3)
RIGHT FRONT WHEEL SPEED SENSOR CIRCUIT OPEN OR SHORTED TO GROUND/BATTERY

Circuit Description:

As a toothed ring passes by the wheel speed sensor, changes in the electromagnetic field cause the wheel speed sensor to produce a sinusoidal (AC) voltage signal whose frequency is proportional to wheel speed. The magnitude of this signal is directly related to wheel speed and the proximity of the wheel speed sensor to the toothed ring, often referred to as the air gap.

DTC Will Set When:

DTC 33 can be set anytime after the electronic brake control relay has been turned "ON." If the EBTCM detects an out of range voltage in either the high or low wheel speed circuits, a malfunction exists.

Action Taken:

A malfunction DTC is stored, the "ABS" warning lamp is turned "ON," the "ETS OFF" indicator is turned "ON," ABS is disabled, and ETS is disabled.

DTC Chart Test Description:

Number(s) below refer to circled number(s) on the diagnostic chart.

10. Checks for intermittent short in wheel speed signal circuitry.

11. Checks for short to ground in wheel speed sensor.

Diagnostic Aids:

An "intermittent" malfunction is most likely caused by a poor connection, rubbed through wire insulation, or a wire that is broken inside the insulation.

The frequency of the malfunction can be checked by using the enhanced diagnostic function of the Tech 1.

If the customer's comments reflect that the "ABS" warning lamp is "ON" only during humid conditions (rain, snow, vehicle wash), all wheel speed sensor circuitry should be thoroughly inspected for signs of water intrusion. Use the following procedure. Spray down the suspected area with 5% salt water solution (two teaspoons of salt to 12 oz. of water). Drive vehicle above 24 km/h (15 mph) for at least 30 seconds. If DTC returns immediately, replace suspected harness.

Any circuitry, that is suspected as causing the intermittent complaint, should be thoroughly checked for backed out terminals, improper mating, broken locks, improperly formed or damaged terminals, poor terminal to wiring connections or physical damage to the wiring harness.

Resistance of the wheel speed sensor will increase with an increase in sensor temperature. When replacing a wheel speed sensor, inspect the sensor terminals and harness connector for corrosion and/or water intrusion. If evidence of corrosion or water intrusion exists, replace wheel speed sensor harness.

If wheel speed sensor harness shows evidence of water intrusion, inspect wheel speed sensor for water intrusion. If water intrusion is evident, replace wheel speed sensor.

DIAGNOSTIC CHART DTC 33 — 1995 CAVALIER AND SUNFIRE WITH TCS, CONTINUED

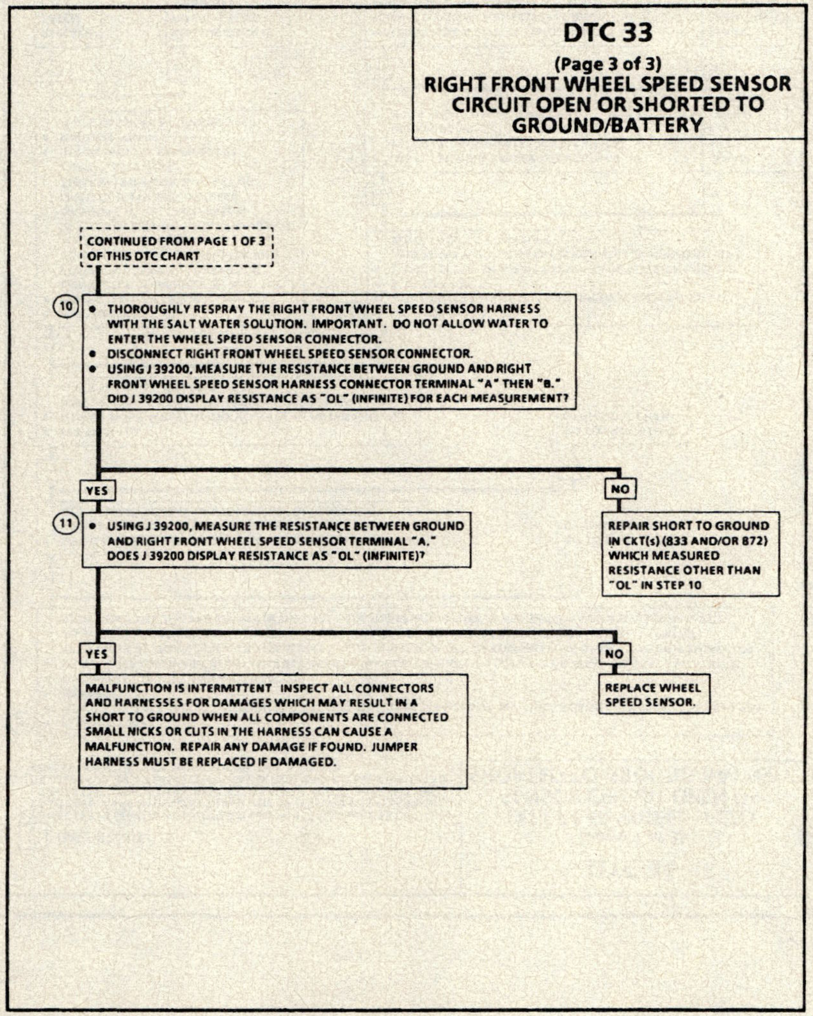

DIAGNOSTIC CHART DTC 34 — 1995 CAVALIER AND SUNFIRE WITH TCS

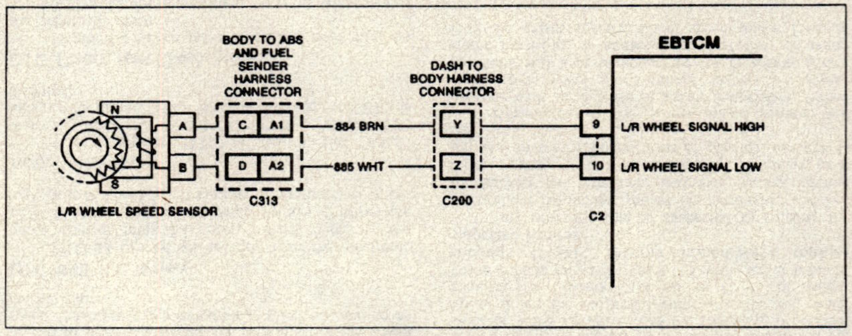

DTC 34
(Page 1 of 3)
LEFT REAR WHEEL SPEED SENSOR CIRCUIT OPEN OR SHORTED TO GROUND/BATTERY

Circuit Description:

As a toothed ring passes by the wheel speed sensor, changes in the electromagnetic field cause the wheel speed sensor to produce a sinusoidal (AC) voltage signal whose frequency is proportional to wheel speed. The magnitude of this signal is directly related to wheel speed and the proximity of the wheel speed sensor to the toothed ring, often referred to as the air gap.

DTC Will Set When:

DTC 34 can be set anytime after the electronic brake control relay has been turned "ON." If the EBTCM detects an out of range voltage in either the high or low wheel speed circuits, a malfunction exists.

Action Taken:

A malfunction DTC is stored, the "ABS" warning lamp is turned "ON," the "ETS OFF" indicator is turned "ON," ABS is disabled, and ETS is disabled.

DTC Chart Test Description:

Number(s) below refer to circled number(s) on the diagnostic chart.

1. Checks wheel speed sensor circuitry for proper resistance values at the EBTCM.
2. Ensures that the wheel speed sensor signal high circuitry is not shorted to ground.
3. Checks for a short to voltage in either the high or low wheel signal circuits.
4. Identifies a wheel speed sensor or circuitry that is damaged and visibly apparent.
5. Determines if wheel speed signal low is shorted to voltage.
6. Determines if wheel speed signal high is shorted to voltage.

Diagnostic Aids:

An "intermittent" malfunction is most likely caused by a poor connection, rubbed through wire insulation, or a wire that is broken inside the insulation.

The frequency of the malfunction can be checked by using the enhanced diagnostic function of the Tech 1.

If the customer's comments reflect that the "ABS" warning lamp is "ON" only during humid conditions (rain, snow, vehicle wash) all wheel speed sensor circuitry should be thoroughly inspected for signs of water intrusion. Use the following procedure. Spray down the suspected area with 5% salt water solution (two teaspoons of salt to 12 oz. of water). Drive vehicle above 24 km/h (15 mph) for at least 30 seconds. If DTC returns immediately, replace suspected harness.

Any circuitry, that is suspected as causing the intermittent complaint, should be thoroughly checked for backed out terminals, improper mating, broken locks, improperly formed or damaged terminals, poor terminal to wiring connections or physical damage to the wiring harness.

Resistance of the wheel speed sensor will increase with an increase in sensor temperature. When replacing a wheel speed sensor, inspect the sensor terminals and harness connector for corrosion and/or water intrusion. If evidence of corrosion or water intrusion exists, replace wheel speed sensor harness.

If wheel speed sensor harness shows evidence of water intrusion, inspect wheel speed sensor for water intrusion. If water intrusion is evident, replace wheel speed sensor.

DIAGNOSTIC CHART DTC 34 — 1995 CAVALIER AND SUNFIRE WITH TCS, CONTINUED

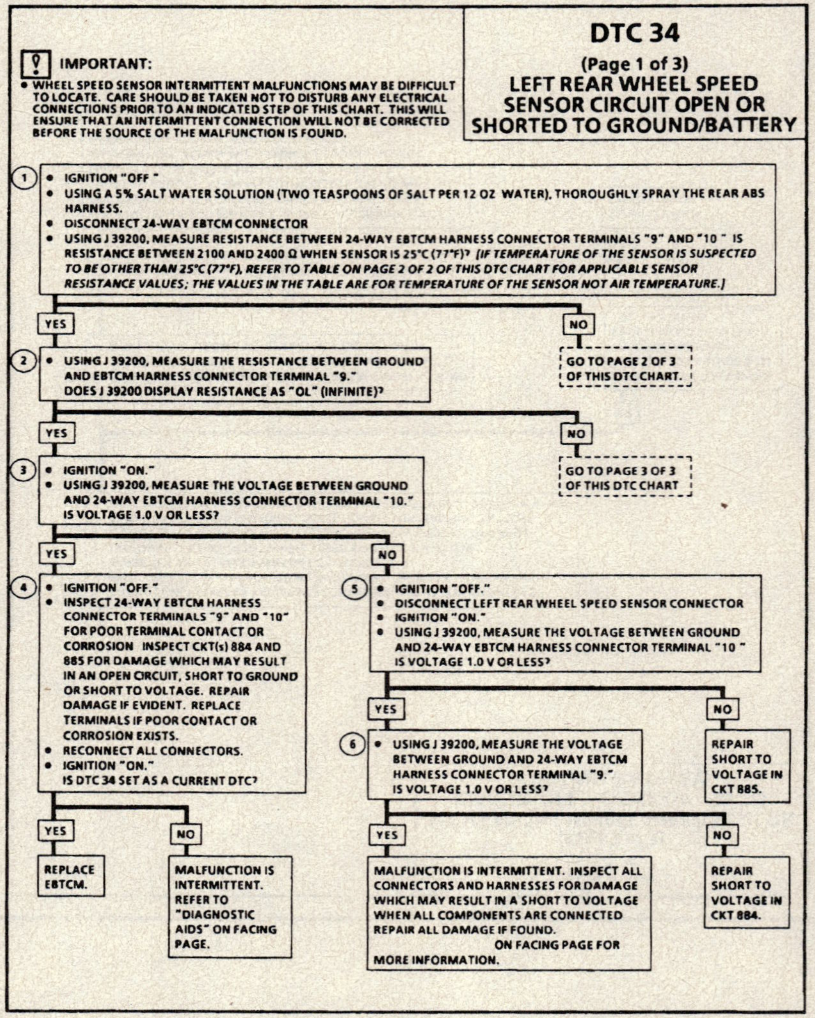

DIAGNOSTIC CHART DTC 34 — 1995 CAVALIER AND SUNFIRE WITH TCS, CONTINUED

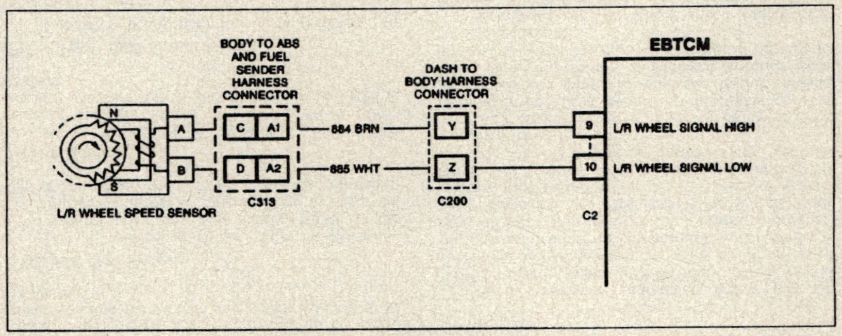

DTC 34
(Page 2 of 3)
LEFT REAR WHEEL SPEED SENSOR CIRCUIT OPEN OR SHORTED TO GROUND/BATTERY

Circuit Description:

As a toothed ring passes by the wheel speed sensor, changes in the electromagnetic field cause the wheel speed sensor to produce a sinusoidal (AC) voltage signal whose frequency is proportional to wheel speed. The magnitude of this signal is directly related to wheel speed and the proximity of the wheel speed sensor to the toothed ring, often referred to as the air gap.

DTC Will Set When:

DTC 34 can be set anytime after the electronic brake control relay has been turned "ON." If the EBTCM detects an out of range voltage in either the high or low wheel speed circuits, a malfunction exists.

Action Taken:

A malfunction DTC is stored, the "ABS" warning lamp is turned "ON," the "ETS OFF" indicator is turned "ON," ABS is disabled, and ETS is disabled.

DTC Chart Test Description:

Number(s) below refer to circled number(s) on the diagnostic chart.

7. Checks for open or high resistance in wheel speed signal high circuitry.
8. Checks for open or high resistance in wheel speed signal low circuitry.
9. Checks wheel speed sensor for proper resistance value.

Diagnostic Aids:

An "intermittent" malfunction is most likely caused by a poor connection, rubbed through wire insulation, or a wire that is broken inside the insulation.

The frequency of the malfunction can be checked by using the enhanced diagnostic function of the Tech 1.

If the customer's comments reflect that the "ABS" warning lamp is "ON" only during humid conditions (rain, snow, vehicle wash), all wheel speed sensor circuitry should be thoroughly inspected for signs of water intrusion. Use the following procedure. Spray down the suspected area with 5% salt water solution (two teaspoons of salt to 12 oz. of water). Drive vehicle above 24 km/h (15 mph) for at least 30 seconds. If DTC returns immediately, replace suspected harness.

Any circuitry, that is suspected as causing the intermittent complaint, should be thoroughly checked for backed out terminals, improper mating, broken locks, improperly formed or damaged terminals, poor terminal to wiring connections or physical damage to the wiring harness.

Resistance of the wheel speed sensor will increase with an increase in sensor temperature. When replacing a wheel speed sensor, inspect the sensor terminals and harness connector for corrosion and/or water intrusion. If evidence of corrosion or water intrusion exists, replace wheel speed sensor harness.

If wheel speed sensor harness shows evidence of water intrusion, inspect wheel speed sensor for water intrusion. If water intrusion is evident, replace wheel speed sensor.

DIAGNOSTIC CHART DTC 34 — 1995 CAVALIER AND SUNFIRE WITH TCS, CONTINUED

DTC 34
(Page 2 of 3)
LEFT REAR WHEEL SPEED SENSOR CIRCUIT OPEN OR SHORTED TO GROUND/BATTERY

CONTINUED FROM PAGE 1 OF 3 OF THIS DTC CHART.

(7)
- DISCONNECT LEFT REAR WHEEL SPEED SENSOR CONNECTOR.
- USING J 39200, MEASURE THE RESISTANCE BETWEEN 24-WAY EBTCM HARNESS CONNECTOR TERMINAL "9" AND LEFT REAR WHEEL SPEED SENSOR HARNESS CONNECTOR TERMINAL "A". IS RESISTANCE LESS THAN 2.0Ω OR LESS?

NO: REPAIR OPEN OR HIGH RESISTANCE IN CKT 884.

YES:

(8)
- USING J 39200, MEASURE THE RESISTANCE BETWEEN 24-WAY EBTCM HARNESS CONNECTOR TERMINAL "10" AND LEFT REAR WHEEL SPEED SENSOR HARNESS CONNECTOR TERMINAL "B". IS RESISTANCE LESS THAN 2.0Ω OR LESS?

NO: REPAIR OPEN OR HIGH RESISTANCE IN CKT 885.

YES:

(9)
- USING J 39200, MEASURE THE RESISTANCE BETWEEN LEFT REAR WHEEL SPEED SENSOR TERMINALS "A" AND "B". IS RESISTANCE BETWEEN 2100 AND 2400Ω WHEN SENSOR IS 25°C (77°F)? *[IF TEMPERATURE OF THE SENSOR IS SUSPECTED TO BE OTHER THAN 25°C (77°F), REFER TO THE TABLE BELOW FOR APPLICABLE SENSOR RESISTANCE VALUES; THE VALUES IN THE TABLE ARE FOR TEMPERATURE OF THE SENSOR NOT AIR TEMPERATURE.]*

NO: REPLACE INTEGRAL BEARING AND SPEED SENSOR ASSEMBLY.

YES: MALFUNCTION IS INTERMITTENT. INSPECT ALL CONNECTORS AND HARNESSES FOR DAMAGES WHICH MAY RESULT IN HIGH RESISTANCE WHEN ALL COMPONENTS ARE CONNECTED. REFER TO "DIAGNOSTIC AIDS" ON FACING PAGE FOR MORE INFORMATION. REPAIR ALL DAMAGE IF FOUND.

REAR WHEEL SPEED SENSOR

TEMPERATURE OF SENSOR VS. SENSOR RESISTANCE (APPROXIMATE)

TEMP (°C)	TEMP (°F)	RESISTANCE (OHMS)
-34 TO 4	-30 TO 40	1623 TO 2210
5 TO 43	41 TO 110	1939 TO 2570
44 TO 93	111 TO 200	2254 TO 3029

DIAGNOSTIC CHART DTC 34 — 1995 CAVALIER AND SUNFIRE WITH TCS, CONTINUED

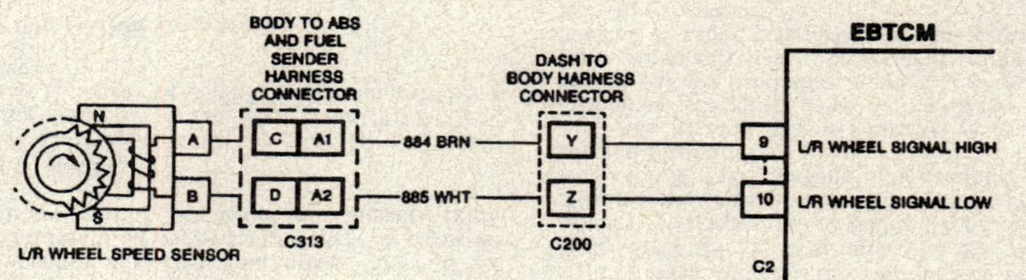

DTC 34
(Page 3 of 3)
LEFT REAR WHEEL SPEED SENSOR CIRCUIT OPEN OR SHORTED TO GROUND/BATTERY

Circuit Description:

As a toothed ring passes by the wheel speed sensor, changes in the electromagnetic field cause the wheel speed sensor to produce a sinusoidal (AC) voltage signal whose frequency is proportional to wheel speed. The magnitude of this signal is directly related to wheel speed and the proximity of the wheel speed sensor to the toothed ring, often referred to as the air gap.

DTC Will Set When:

DTC 34 can be set anytime after the electronic brake control relay has been turned "ON." If the EBTCM detects an out of range voltage in either the high or low wheel speed circuits, a malfunction exists.

Action Taken:

A malfunction DTC is stored, the "ABS" warning lamp is turned "ON," the "ETS OFF" indicator is turned "ON," ABS is disabled, and ETS is disabled.

DTC Chart Test Description:

Number(s) below refer to circled number(s) on the diagnostic chart.

10. Isolates short to ground in wheel sensor signal high to dash or body harness.
11. Isolates short to ground in wheel sensor signal low to dash or body harness.
12. Checks for intermittent short in rear ABS axle harness.
13. Checks for short to ground in wheel speed sensor.

Diagnostic Aids:

An "intermittent" malfunction is most likely caused by a poor connection, rubbed through wire insulation, or a wire that is broken inside the insulation.

The frequency of the malfunction can be checked by using the enhanced diagnostic function of the Tech 1.

If the customer's comments reflect that the "ABS" warning lamp is "ON" only during humid conditions (rain, snow, vehicle wash), all wheel speed sensor circuitry should be thoroughly inspected for signs of water intrusion. Use the following procedure. Spray down the suspected area with 5% salt water solution (two teaspoons of salt to 12 oz. of water). Drive vehicle above 24 km/h (15 mph) for at least 30 seconds. If DTC returns immediately, replace suspected harness.

Any circuitry, that is suspected as causing the intermittent complaint, should be thoroughly checked for backed out terminals, improper mating, broken locks, improperly formed or damaged terminals, poor terminal to wiring connections or physical damage to the wiring harness.

Resistance of the wheel speed sensor will increase with an increase in sensor temperature. When replacing a wheel speed sensor, inspect the sensor terminals and harness connector for corrosion and/or water intrusion. If evidence of corrosion or water intrusion exists, replace wheel speed sensor harness.

If wheel speed sensor harness shows evidence of water intrusion, inspect wheel speed sensor for water intrusion. If water intrusion is evident, replace wheel speed sensor.

DIAGNOSTIC CHART DTC 34 — 1995 CAVALIER AND SUNFIRE WITH TCS, CONTINUED

DTC 34
(Page 3 of 3)
LEFT REAR WHEEL SPEED SENSOR CIRCUIT OPEN OR SHORTED TO GROUND/BATTERY

CONTINUED FROM PAGE 1 OF 3 OF THIS DTC CHART

(10)
- DISCONNECT REAR ABS HARNESS FROM BODY TO ABS AND FUEL SENDER HARNESS CONNECTOR.
- USING J 39200, MEASURE THE RESISTANCE BETWEEN GROUND AND 24-WAY EBTCM HARNESS CONNECTOR TERMINAL "9."

DOES J 39200 DISPLAY RESISTANCE AS "OL" (INFINITE)?

NO: REPAIR SHORT TO GROUND IN CKT 884.

YES:

(11)
- USING J 39200, MEASURE THE RESISTANCE BETWEEN GROUND AND 24-WAY EBTCM HARNESS CONNECTOR TERMINAL "10."

DOES J 39200 DISPLAY RESISTANCE AS "OL" (INFINITE)?

NO: REPAIR SHORT TO GROUND IN CKT 885

YES:

(12)
- THOROUGHLY RESPRAY THE REAR ABS HARNESS WITH THE SALT WATER SOLUTION. IMPORTANT: DO NOT ALLOW WATER TO ENTER THE WHEEL SPEED SENSOR CONNECTOR.
- DISCONNECT LEFT REAR WHEEL SPEED SENSOR CONNECTOR.
- USING J 39200, MEASURE THE RESISTANCE BETWEEN GROUND AND LEFT REAR WHEEL SPEED SENSOR HARNESS CONNECTOR TERMINAL "A" THEN "B"

DID J 39200 DISPLAY RESISTANCE AS "OL" (INFINITE) FOR EACH MEASUREMENT?

NO: REPLACE REAR ABS HARNESS

YES:

(13)
- USING J 39200, MEASURE THE RESISTANCE BETWEEN SENSOR TERMINAL "A" AND INTEGRAL BEARING AND SPEED SENSOR ASSEMBLY MOUNTING FLANGE.

DOES J 39200 DISPLAY RESISTANCE AS "OL" (INFINITE)?

NO: REPLACE INTEGRAL BEARING AND SPEED SENSOR ASSEMBLY.

YES: MALFUNCTION IS INTERMITTENT. INSPECT ALL CONNECTORS AND HARNESSES FOR DAMAGES WHICH MAY RESULT IN A SHORT TO GROUND WHEN ALL COMPONENTS ARE CONNECTED. SMALL NICKS OR CUTS IN THE HARNESS CAN CAUSE A MALFUNCTION. REPAIR ANY DAMAGE IF FOUND. JUMPER HARNESS MUST BE REPLACED IF DAMAGED. REFER TO "DIAGNOSTIC AIDS" ON FACING PAGE FOR MORE INFORMATION.

DIAGNOSTIC CHART DTC 35 — 1995 CAVALIER AND SUNFIRE WITH TCS

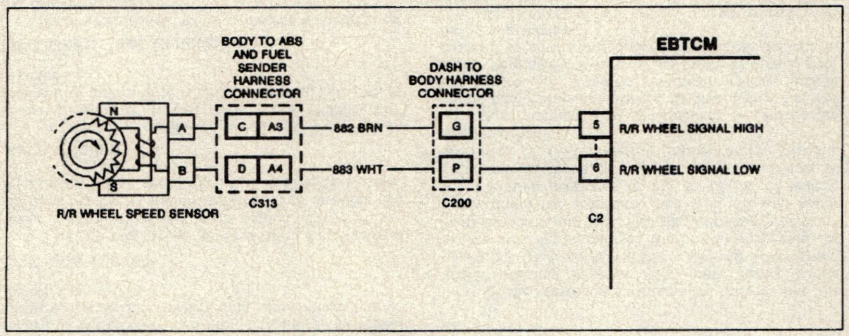

DTC 35
(Page 1 of 3)
RIGHT REAR WHEEL SPEED SENSOR CIRCUIT OPEN OR SHORTED TO GROUND/BATTERY

Circuit Descripton:

As a toothed ring passes by the wheel speed sensor, changes in the electromagnetic field cause the wheel speed sensor to produce a sinusoidal (AC) voltage signal whose frequency is proportional to wheel speed. The magnitude of this signal is directly related to wheel speed and the proximity of the wheel speed sensor to the toothed ring, often referred to as the air gap.

DTC Will Set When:

DTC 35 can be set anytime after the electronic brake control relay has been turned "ON." If the EBTCM detects an out of range voltage in either the high or low wheel speed circuits, a malfunction exists.

Action Taken:

A malfunction DTC is stored, the "ABS" warning lamp is turned "ON," the "ETS OFF" indicator is turned "ON," ABS is disabled, and ETS is disabled.

DTC Chart Test Description:

Number(s) below refer to circled number(s) on the diagnostic chart.

1. Checks wheel speed sensor circuitry for proper resistance values at the EBTCM.
2. Ensures that the wheel speed sensor signal high circuitry is not shorted to ground.
3. Checks for a short to voltage in either the high or low wheel signal circuits.
4. Identifies a wheel speed sensor or circuitry that is damaged and visibly apparent.
5. Determines if wheel speed signal low is shorted to voltage.
6. Determines if wheel speed signal high is shorted to voltage.

Diagnostic Aids:

An "intermittent" malfunction is most likely caused by a poor connection, rubbed through wire insulation, or a wire that is broken inside the insulation.

The frequency of the malfunction can be checked by using the enhanced diagnostic function of the Tech 1.

If the customer's comments reflect that the "ABS" warning lamp is "ON" only during humid conditions (rain, snow, vehicle wash) all wheel speed sensor circuitry should be thoroughly inspected for signs of water intrusion. Use the following procedure. Spray down the suspected area with 5% salt water solution (two teaspoons of salt to 12 oz. of water). Drive vehicle above 24 km/h (15 mph) for at least 30 seconds. If DTC returns immediately, replace suspected harness.

Any circuitry, that is suspected as causing the intermittent complaint, should be thoroughly checked for backed out terminals, improper mating, broken locks, improperly formed or damaged terminals, poor terminal to wiring connections or physical damage to the wiring harness.

Resistance of the wheel speed sensor will increase with an increase in sensor temperature. When replacing a wheel speed sensor, inspect the sensor terminals and harness connector for corrosion and/or water intrusion. If evidence of corrosion or water intrusion exists, replace wheel speed sensor harness.

If wheel speed sensor harness shows evidence of water intrusion, inspect wheel speed sensor for water intrusion. If water intrusion is evident, replace wheel speed sensor.

DIAGNOSTIC CHART DTC 35 — 1995 CAVALIER AND SUNFIRE WITH TCS, CONTINUED

DTC 35
(Page 1 of 3)
RIGHT REAR WHEEL SPEED SENSOR CIRCUIT OPEN OR SHORTED TO GROUND/BATTERY

IMPORTANT:
- WHEEL SPEED SENSOR INTERMITTENT MALFUNCTIONS MAY BE DIFFICULT TO LOCATE. CARE SHOULD BE TAKEN NOT TO DISTURB ANY ELECTRICAL CONNECTIONS PRIOR TO AN INDICATED STEP OF THIS CHART. THIS WILL ENSURE THAT AN INTERMITTENT CONNECTION WILL NOT BE CORRECTED BEFORE THE SOURCE OF THE MALFUNCTION IS FOUND.

(1)
- IGNITION "OFF."
- USING A 5% SALT WATER SOLUTION (TWO TEASPOONS OF SALT PER 12 OZ. WATER), THOROUGHLY SPRAY THE REAR ABS HARNESS.
- DISCONNECT 24-WAY EBTCM CONNECTOR
- USING J 39200, MEASURE RESISTANCE BETWEEN 24-WAY EBTCM HARNESS CONNECTOR TERMINALS "5" AND "6" IS RESISTANCE BETWEEN 2100 AND 2400 Ω WHEN SENSOR IS 25°C (77°F)? *[IF TEMPERATURE OF THE SENSOR IS SUSPECTED TO BE OTHER THAN 25°C (77°F), REFER TO TABLE ON PAGE 2 OF 2 OF THIS DTC CHART FOR APPLICABLE SENSOR RESISTANCE VALUES; THE VALUES IN THE TABLE ARE FOR TEMPERATURE OF THE SENSOR NOT AIR TEMPERATURE.]*

YES → (2)
NO → GO TO PAGE 2 OF 3 OF THIS DTC CHART.

(2)
- USING J 39200, MEASURE THE RESISTANCE BETWEEN GROUND AND EBTCM HARNESS CONNECTOR TERMINAL "5." DOES J 39200 DISPLAY RESISTANCE AS "OL" (INFINITE)?

YES → (3)
NO → GO TO PAGE 3 OF 3 OF THIS DTC CHART

(3)
- IGNITION "ON."
- USING J 39200, MEASURE THE VOLTAGE BETWEEN GROUND AND 24-WAY EBTCM HARNESS CONNECTOR TERMINAL "6." IS VOLTAGE 1.0 V OR LESS?

YES → (4)
NO → (5)

(4)
- IGNITION "OFF."
- INSPECT 24-WAY EBTCM HARNESS CONNECTOR TERMINALS "5" AND "6" FOR POOR TERMINAL CONTACT OR CORROSION. INSPECT CKT(s) 882 AND 883 FOR DAMAGE WHICH MAY RESULT IN AN OPEN CIRCUIT, SHORT TO GROUND OR SHORT TO VOLTAGE. REPAIR DAMAGE IF EVIDENT. REPLACE TERMINALS IF POOR CONTACT OR CORROSION EXISTS.
- RECONNECT ALL CONNECTORS.
- IGNITION "ON."

IS DTC 35 SET AS A CURRENT DTC?

YES → REPLACE EBTCM.
NO → MALFUNCTION IS INTERMITTENT. REFER TO "DIAGNOSTIC AIDS" ON FACING PAGE.

(5)
- IGNITION "OFF."
- DISCONNECT RIGHT REAR WHEEL SPEED SENSOR CONNECTOR.
- IGNITION "ON."
- USING J 39200, MEASURE THE VOLTAGE BETWEEN GROUND AND 24-WAY EBTCM HARNESS CONNECTOR TERMINAL "6." IS VOLTAGE 1.0 V OR LESS?

YES → (6)
NO → REPAIR SHORT TO VOLTAGE IN CKT 883.

(6)
- USING J 39200, MEASURE THE VOLTAGE BETWEEN GROUND AND 24-WAY EBTCM HARNESS CONNECTOR TERMINAL "5." IS VOLTAGE 1.0 V OR LESS?

YES → MALFUNCTION IS INTERMITTENT. INSPECT ALL CONNECTORS AND HARNESSES FOR DAMAGE WHICH MAY RESULT IN A SHORT TO VOLTAGE WHEN ALL COMPONENTS ARE CONNECTED. REPAIR ALL DAMAGE IF FOUND. REFER TO "DIAGNOSTIC AIDS" ON FACING PAGE FOR MORE INFORMATION.
NO → REPAIR SHORT TO VOLTAGE IN CKT 882.

DIAGNOSTIC CHART DTC 35 — 1995 CAVALIER AND SUNFIRE WITH TCS, CONTINUED

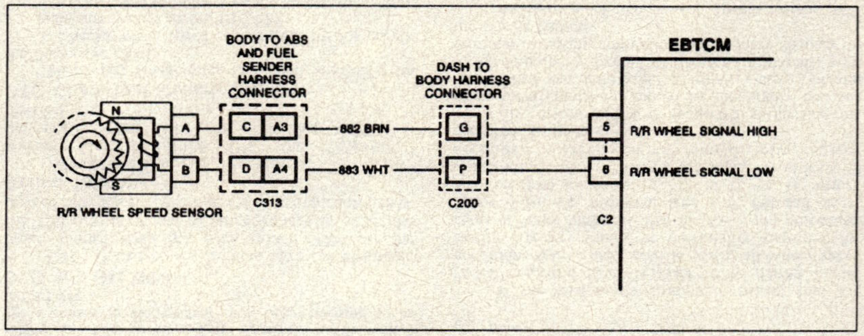

DTC 35
(Page 2 of 3)
RIGHT REAR WHEEL SPEED SENSOR CIRCUIT OPEN OR SHORTED TO GROUND/BATTERY

Circuit Description:

As a toothed ring passes by the wheel speed sensor, changes in the electromagnetic field cause the wheel speed sensor to produce a sinusoidal (AC) voltage signal whose frequency is proportional to wheel speed. The magnitude of this signal is directly related to wheel speed and the proximity of the wheel speed sensor to the toothed ring, often referred to as the air gap.

DTC Will Set When:

DTC 35 can be set anytime after the electronic brake control relay has been turned "ON." If the EBTCM detects an out of range voltage in either the high or low wheel speed circuits, a malfunction exists.

Action Taken:

A malfunction DTC is stored, the "ABS" warning lamp is turned "ON," the "ETS OFF" indicator is turned "ON," ABS is disabled, and ETS is disabled.

DTC Chart Test Description:

Number(s) below refer to circled number(s) on the diagnostic chart.

7. Checks for open or high resistance in wheel speed signal high circuitry.
8. Checks for open or high resistance in wheel speed signal low circuitry.
9. Checks wheel speed sensor for proper resistance value.

Diagnostic Aids:

An "intermittent" malfunction is most likely caused by a poor connection, rubbed through wire insulation, or a wire that is broken inside the insulation.

The frequency of the malfunction can be checked by using the enhanced diagnostic function of the Tech 1.

If the customer's comments reflect that the "ABS" warning lamp is "ON" only during humid conditions (rain, snow, vehicle wash), all wheel speed sensor circuitry should be thoroughly inspected for signs of water intrusion. Use the following procedure. Spray down the suspected area with 5% salt water solution (two teaspoons of salt to 12 oz. of water). Drive vehicle above 24 km/h (15 mph) for at least 30 seconds. If DTC returns immediately, replace suspected harness.

Any circuitry, that is suspected as causing the intermittent complaint, should be thoroughly checked for backed out terminals, improper mating, broken locks, improperly formed or damaged terminals, poor terminal to wiring connections or physical damage to the wiring harness.

Resistance of the wheel speed sensor will increase with an increase in sensor temperature. When replacing a wheel speed sensor, inspect the sensor terminals and harness connector for corrosion and/or water intrusion. If evidence of corrosion or water intrusion exists, replace wheel speed sensor harness.

If wheel speed sensor harness shows evidence of water intrusion, inspect wheel speed sensor for water intrusion. If water intrusion is evident, replace wheel speed sensor.

DIAGNOSTIC CHART DTC 35 — 1995 CAVALIER AND SUNFIRE WITH TCS, CONTINUED

DTC 35
(Page 2 of 3)
RIGHT REAR WHEEL SPEED SENSOR CIRCUIT OPEN OR SHORTED TO GROUND/BATTERY

CONTINUED FROM PAGE 1 OF 3 OF THIS DTC CHART.

(7)
- DISCONNECT RIGHT REAR WHEEL SPEED SENSOR CONNECTOR
- USING J 39200, MEASURE THE RESISTANCE BETWEEN 24-WAY EBTCM HARNESS CONNECTOR TERMINAL "5" AND RIGHT REAR WHEEL SPEED SENSOR HARNESS CONNECTOR TERMINAL "A." IS RESISTANCE 2.0Ω OR LESS?

NO → REPAIR OPEN OR HIGH RESISTANCE IN CKT 882.

YES ↓

(8)
- USING J 39200, MEASURE THE RESISTANCE BETWEEN 24-WAY EBTCM HARNESS CONNECTOR TERMINAL "6" AND RIGHT REAR WHEEL SPEED SENSOR HARNESS CONNECTOR TERMINAL "B." IS RESISTANCE 2.0Ω OR LESS?

NO → REPAIR OPEN OR HIGH RESISTANCE IN CKT 883.

YES ↓

(9)
- USING J 39200, MEASURE THE RESISTANCE BETWEEN RIGHT REAR WHEEL SPEED SENSOR TERMINALS "A" AND "B." IS RESISTANCE BETWEEN 2100 AND 2400Ω WHEN SENSOR IS 25°C (77°F)? *(IF TEMPERATURE OF THE SENSOR IS SUSPECTED TO BE OTHER THAN 25°C (77°F), REFER TO THE TABLE BELOW FOR APPLICABLE SENSOR RESISTANCE VALUES; THE VALUES IN THE TABLE ARE FOR TEMPERATURE OF THE SENSOR NOT AIR TEMPERATURE.)*

NO → REPLACE INTEGRAL BEARING AND SPEED SENSOR ASSEMBLY

YES → MALFUNCTION IS INTERMITTENT. INSPECT ALL CONNECTORS AND HARNESSES FOR DAMAGES WHICH MAY RESULT IN HIGH RESISTANCE WHEN ALL COMPONENTS ARE CONNECTED. REFER TO "DIAGNOSTIC AIDS" ON FACING PAGE FOR MORE INFORMATION. REPAIR ALL DAMAGE IF FOUND.

REAR WHEEL SPEED SENSOR

TEMPERATURE OF SENSOR VS. SENSOR RESISTANCE (APPROXIMATE)

TEMP (°C)	TEMP. (°F)	RESISTANCE (OHMS)
-34 TO 4	-30 TO 40	1623 TO 2210
5 TO 43	41 TO 110	1939 TO 2570
44 TO 93	111 TO 200	2254 TO 3029

DIAGNOSTIC CHART DTC 35 — 1995 CAVALIER AND SUNFIRE WITH TCS, CONTINUED

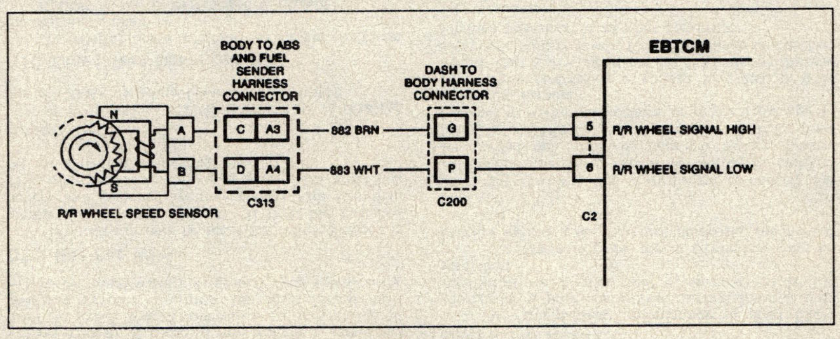

DTC 35
(Page 3 of 3)
RIGHT REAR WHEEL SPEED SENSOR CIRCUIT OPEN OR SHORTED TO GROUND/BATTERY

Circuit Description:

As a toothed ring passes by the wheel speed sensor, changes in the electromagnetic field cause the wheel speed sensor to produce a sinusoidal (AC) voltage signal whose frequency is proportional to wheel speed. The magnitude of this signal is directly related to wheel speed and the proximity of the wheel speed sensor to the toothed ring, often referred to as the air gap.

DTC Will Set When:

DTC 35 can be set anytime after the electronic brake control relay has been turned "ON." If the EBTCM detects an out of range voltage in either the high or low wheel speed circuits, a malfunction exists.

Action Taken:

A malfunction DTC is stored, the "ABS" warning lamp is turned "ON," the "ETS OFF" indicator is turned "ON," ABS is disabled, and ETS is disabled.

DTC Chart Test Description:

Number(s) below refer to circled number(s) on the diagnostic chart.

10. Isolates short to ground in wheel sensor signal high to dash or body harness.
11. Isolates short to ground in wheel sensor signal low to dash or body harness.
12. Checks for intermittent short in rear ABS axle harness.
13. Checks for short to ground in wheel speed sensor.

Diagnostic Aids:

An "intermittent" malfunction is most likely caused by a poor connection, rubbed through wire insulation, or a wire that is broken inside the insulation.

The frequency of the malfunction can be checked by using the enhanced diagnostic function of the Tech 1.

If the customer's comments reflect that the "ABS" warning lamp is "ON" only during humid conditions (rain, snow, vehicle wash), all wheel speed sensor circuitry should be thoroughly inspected for signs of water intrusion. Use the following procedure. Spray down the suspected area with 5% salt water solution (two teaspoons of salt to 12 oz. of water). Drive vehicle above 24 km/h (15 mph) for at least 30 seconds. If DTC returns immediately, replace suspected harness.

Any circuitry, that is suspected as causing the intermittent complaint, should be thoroughly checked for backed out terminals, improper mating, broken locks, improperly formed or damaged terminals, poor terminal to wiring connections or physical damage to the wiring harness.

Resistance of the wheel speed sensor will increase with an increase in sensor temperature. When replacing a wheel speed sensor, inspect the sensor terminals and harness connector for corrosion and/or water intrusion. If evidence of corrosion or water intrusion exists, replace wheel speed sensor harness.

If wheel speed sensor harness shows evidence of water intrusion, inspect wheel speed sensor for water intrusion. If water intrusion is evident, replace wheel speed sensor.

DTC 35
(Page 3 of 3)
RIGHT REAR WHEEL SPEED SENSOR CIRCUIT OPEN OR SHORTED TO GROUND/BATTERY

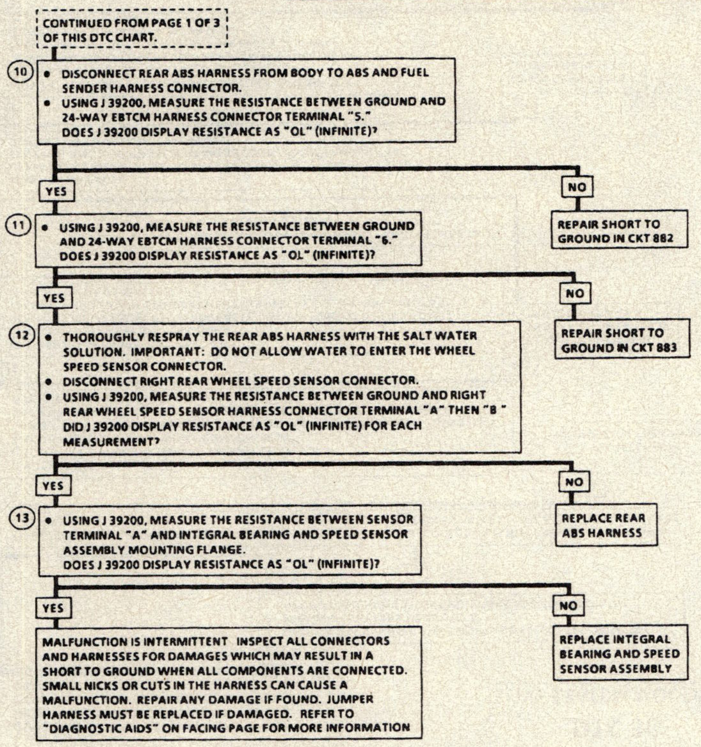

DIAGNOSTIC CHART DTC 36 — 1995 CAVALIER AND SUNFIRE WITH TCS

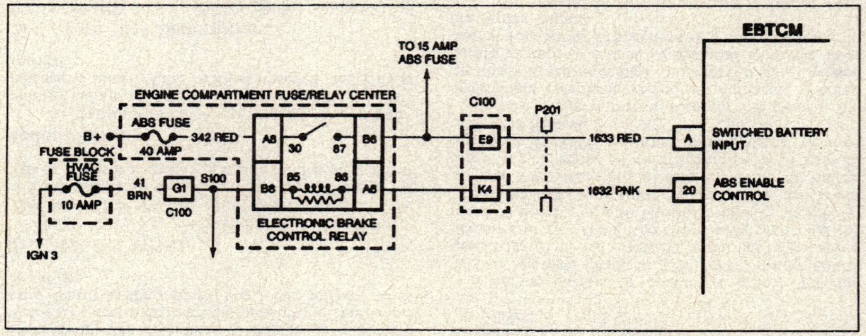

DTC 36
LOW SYSTEM VOLTAGE

Circuit Description:

This DTC is used to monitor the voltage level available to the EBTCM. If the voltage drops below 11.0 volts, full performance of the ABS w/ETS cannot be guaranteed. During ABS w/ETS operation, there are several current requirements that will cause battery voltage to drop. Because of this, voltage is monitored prior to ABS w/ETS operation to indicate good charging system condition and also during ABS w/ETS operation when voltage may drop significantly.

DTC Will Set When:

DTC 36 can only be set if the vehicle's speed is greater than 5 km/h (3 mph). If both the switched battery and switched ignition voltages are less than 11.2 volts (less than 9.6 volts when ABS w/ETS is active), a malfunction exists.

Action Taken:

A malfunction DTC is stored, ABS is disabled and the "ABS" warning lamp is turned "ON."

DTC Chart Test Description:

Number(s) below refer to circled number(s) on the diagnostic chart.

1. Tests to see if malfunction is present.
2. Checks charging system integrity.
3. Checks for excessive resistance in the switched battery input circuit.
4. Tests for high resistance in the electronic brake control relay circuit.
5. Ensures low voltage condition was not due to a poor connection.

Diagnostic Aids:

An "intermittent" malfunction is most likely caused by a poor connection, rubbed through wire insulation, or a wire that is broken inside the insulation.

The frequency of the malfunction can be checked by using the enhanced diagnostic function of the Tech 1

Any circuitry, that is suspected as causing the intermittent complaint, should be thoroughly checked for backed out terminals, improper mating, broken locks, improperly formed or damaged terminals, poor terminal to wiring connections or physical damage to the wiring harness.

While performing a voltage load test, if it is noted that only ignition voltage drops below acceptable voltage levels, CKT 650 should be checked for high resistance or an open condition.

DIAGNOSTIC CHART DTC 36 — 1995 CAVALIER AND SUNFIRE WITH TCS, CONTINUED

DTC 36
LOW SYSTEM VOLTAGE

1
- START ENGINE.
- USING TECH 1, SELECT "F3: SNAPSHOT."
- SELECT "F4: SINGLE DTC"; THEN ENTER "36".
- DRIVE VEHICLE ABOVE 5 km/h (3 MPH).

DID TECH 1 TRIGGER?

- YES → go to step 2
- NO →
  - WITH ENGINE RUNNING, USE TECH 1 AND SELECT "F4: MISC. TESTS."
  - PERFORM "VOLTAGE LOAD" TEST.

  ARE IGNITION AND BATTERY VOLTAGES 10.0V OR GREATER?
  - NO → go to step 2
  - YES → MALFUNCTION IS NOT PRESENT AT THIS TIME. REFER TO "DIAGNOSTIC AIDS" ON FACING PAGE.

2
- REFER TO SECTION 6D3 TO DIAGNOSE CHARGING SYSTEM.

IS CHARGING SYSTEM OK?

- NO → CORRECT LOW VOLTAGE CONDITION.
- YES → go to step 3

3
- IGNITION "OFF."
- DISCONNECT 2-WAY AND 24-WAY EBTCM CONNECTORS
- DISCONNECT ELECTRONIC BRAKE CONTROL RELAY
- USING J 39200 MEASURE RESISTANCE BETWEEN ENGINE COMPARTMENT FUSE/RELAY CENTER TERMINAL "B6" AND 2-WAY EBTCM HARNESS CONNECTOR TERMINAL "A".

IS RESISTANCE 2 0Ω OR LESS?

- NO →
  - REPAIR OPEN OR HIGH RESISTANCE IN CKT 1633.
  - PROCEED TO STEP 4.
- YES →
  - DISCONNECT NEGATIVE BATTERY CABLE.
  - DISCONNECT POSITIVE BATTERY CABLE.
  - USING J 39200 MEASURE RESISTANCE BETWEEN THE POSITIVE BATTERY CABLE TERMINAL AND ENGINE COMPARTMENT FUSE/RELAY CENTER TERMINAL "A8".

  IS RESISTANCE 2 0Ω OR LESS?
  - NO →
    - REPAIR OPEN OR HIGH RESISTANCE IN CKT 342.
    - PROCEED TO STEP 4.
  - YES → go to step 4

4
- REMOVE HVAC FUSE.
- USING J 39200 MEASURE RESISTANCE BETWEEN EACH HVAC FUSE BLOCK TERMINAL AND 24-WAY EBCM HARNESS CONNECTOR TERMINAL "14"

IS EITHER MEASUREMENT 2 0Ω OR LESS?

- NO → REPAIR OPEN OR HIGH RESISTANCE IN CKT 41
- YES →
  - WITH BATTERY CABLES DISCONNECTED, TURN IGNITION "ON" (THIS PROVIDES CIRCUIT CONTINUITY).
  - USING J 39200, MEASURE RESISTANCE BETWEEN POSITIVE BATTERY CABLE TERMINAL AND EACH HVAC FUSE BLOCK TERMINAL.

  IS EITHER MEASUREMENT 2 0Ω OR LESS?
  - NO → REPAIR HIGH RESISTANCE IN THE "IGNITION SWITCH 'RUN' FUSE FEED" CIRCUIT.
  - YES → go to step 5

5
- IGNITION "OFF"
- INSPECT ELECTRONIC BRAKE CONTROL RELAY AND ENGINE COMPARTMENT FUSE/RELAY CENTER FOR POOR TERMINAL CONTACT. INSPECT 2-WAY AND 24-WAY EBTCM HARNESS CONNECTORS FOR A POOR CONNECTION TO THE EBTCM. INSPECT POSITIVE BATTERY CABLE TERMINAL FOR POOR CONTACT TO THE BATTERY.
  REPLACE TERMINALS IF POOR CONTACT IS EVIDENT
- RECONNECT ALL CONNECTORS.
- DRIVE VEHICLE ABOVE 5 km/h (3 MPH).

IS DTC 36 SET AS A CURRENT DTC?

- YES → REPLACE EBTCM.
- NO → MALFUNCTION IS INTERMITTENT. REFER TO "DIAGNOSTIC AIDS" ON FACING PAGE

DIAGNOSTIC CHART DTC 37 — 1995 CAVALIER AND SUNFIRE WITH TCS

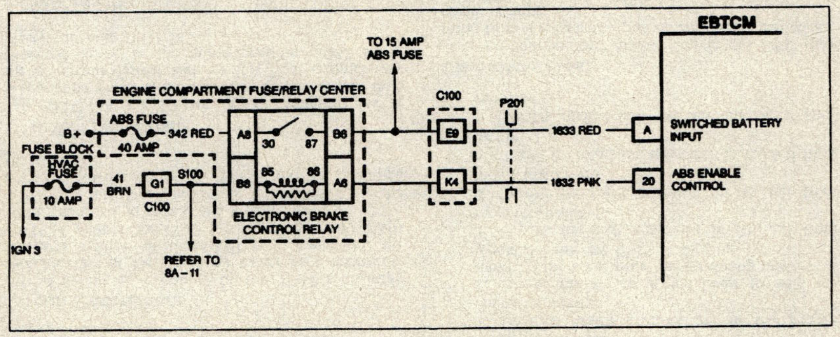

DTC 37
HIGH SYSTEM VOLTAGE

Circuit Description:

This DTC is designed to detect high vehicle voltage levels prior to any required motor movement (initialization or ABS). If excessive voltage exists, demagnetization of the motor magnets may occur, which would eventually affect or eliminate ABS w/ETS performance.

DTC Will Set When:

DTC 37 can only be set if the vehicle's speed is greater than 5 km/h (3 mph). If the switched battery voltage is greater than 17 volts, a malfunction exists.

Action Taken:

A malfunction DTC is stored, ABS is disabled and the "ABS" warning lamp is turned "ON."

DTC Chart Test Description:

Number(s) below refer to circled number(s) on the diagnostic chart.

1. Checks the voltage level being received by the EBTCM.
2. Isolates the malfunction to either the EBTCM or the charging system.

Diagnostic Aids:

An "intermittent" malfunction is most likely caused by a poor connection, rubbed through wire insulation, or a wire that is broken inside the insulation.

The frequency of the malfunction can be checked by using the enhanced diagnostic function of the Tech 1 tool

Any circuitry, that is suspected as causing the intermittent malfunction, should be thoroughly checked for backed out terminals, improper mating, broken locks, improperly formed or damaged terminals, poor terminal to wiring connections or physical damage to the wiring harness.

DIAGNOSTIC CHART DTC 37 — 1995 CAVALIER AND SUNFIRE WITH TCS, CONTINUED

DTC 37
HIGH SYSTEM VOLTAGE

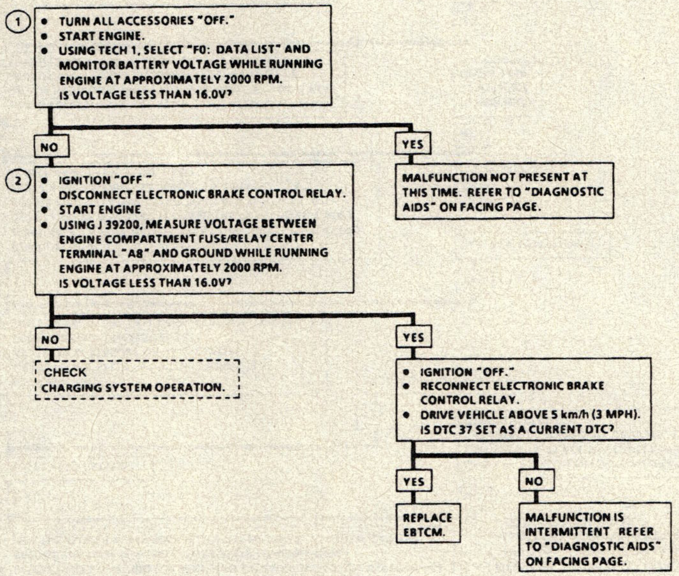

DIAGNOSTIC CHART DTC 38 — 1995 CAVALIER AND SUNFIRE WITH TCS

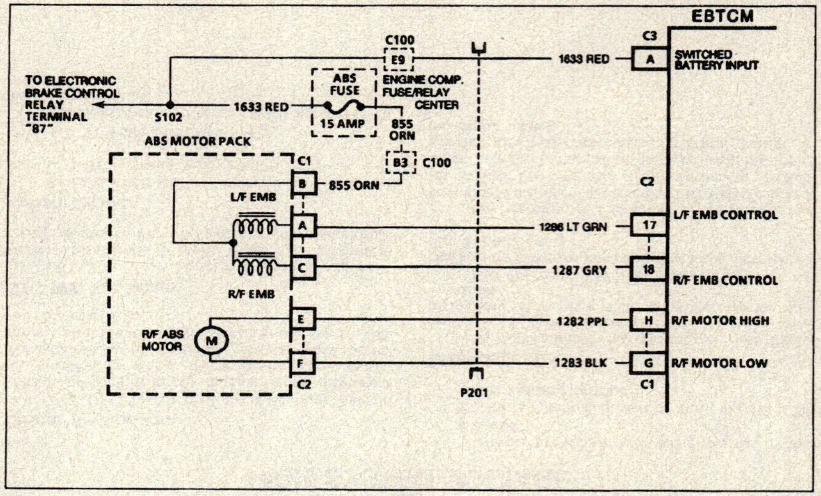

DTC 38
LEFT FRONT EMB WILL NOT HOLD MOTOR

Circuit Description:

Current is supplied to the EMB's through terminal "B" of the three-way motor pack connector. When a release of the EMB is commanded, the EBTCM grounds the EMB input and allows the EMB to be energized to release the motor.

The left front motor is commanded in the forward direction while the EMB is released (energized).

DTC Will Set When:

DTC 38 can be set during initialization. A malfunction exists if the feedback current is less than the command current and the left front channel was expected to be in the home position. DTC 86 is always set with DTC 38.

Action Taken:

If an EMB can not hold a motor in the home position, the motor may be back driven when the brake pedal is applied causing the brake pedal to drop. A malfunction DTC is stored, ABS is disabled and the "ABS" warning lamp is turned "ON."

DTC Chart Test Description:

Number(s) below refer to circled number(s) on the diagnostic chart.

1. Ensures motor circuits are in the proper connector cavity.
2. Checks the ability of the EMB to hold the motor. If the brake pedal moves during this test, it would indicate the EMB is slipping.
3. Checks for high resistance in the left front motor high circuit.
4. Checks for high resistance in the left front motor low circuit.
5. Checks for excessive resistance in the left front motor.
6. Ensures malfunction was not due to a poor connection.

Diagnostic Aids:

An "intermittent" malfunction in this DTC may result from a mechanical part of the system that sticks, binds, or slips.

The frequency of the malfunction can be checked by using the enhanced diagnostic function of the Tech 1

NOTICE: This test may fail if low voltage (10 volts or less) was present when initialization occurred. Ensure battery/charging system is functioning properly.

DIAGNOSTIC CHART DTC 38 — 1995 CAVALIER AND SUNFIRE WITH TCS, CONTINUED

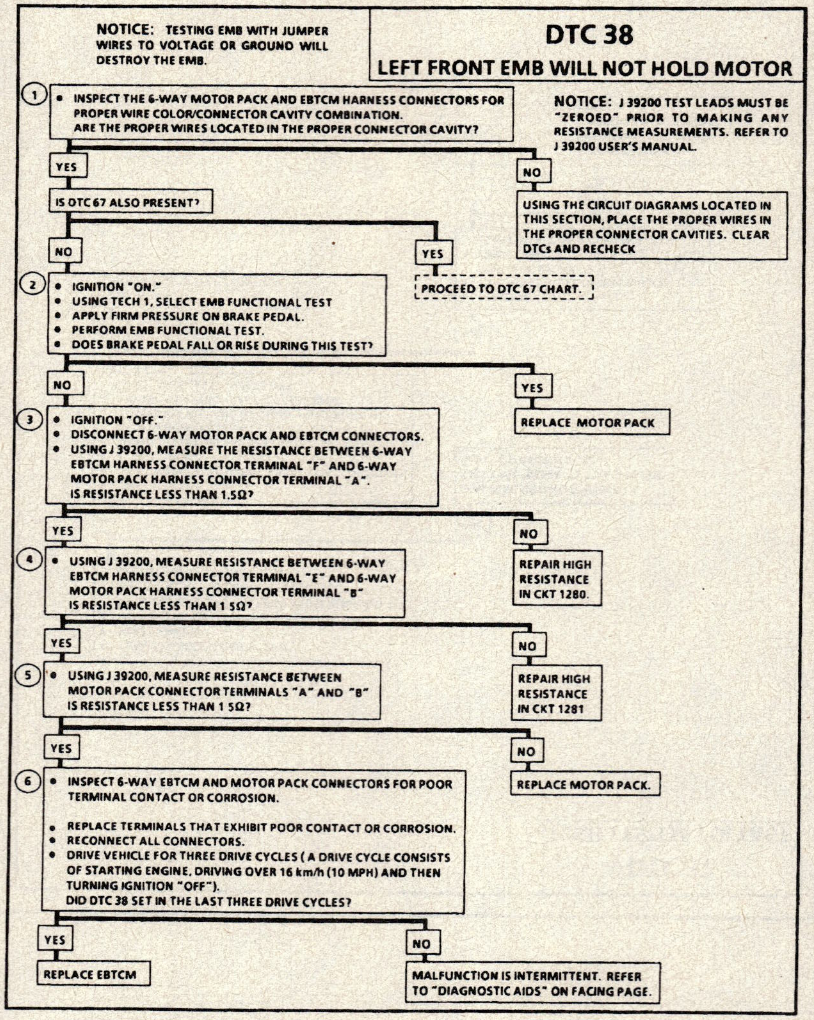

DIAGNOSTIC CHART DTC 41 — 1995 CAVALIER AND SUNFIRE WITH TCS

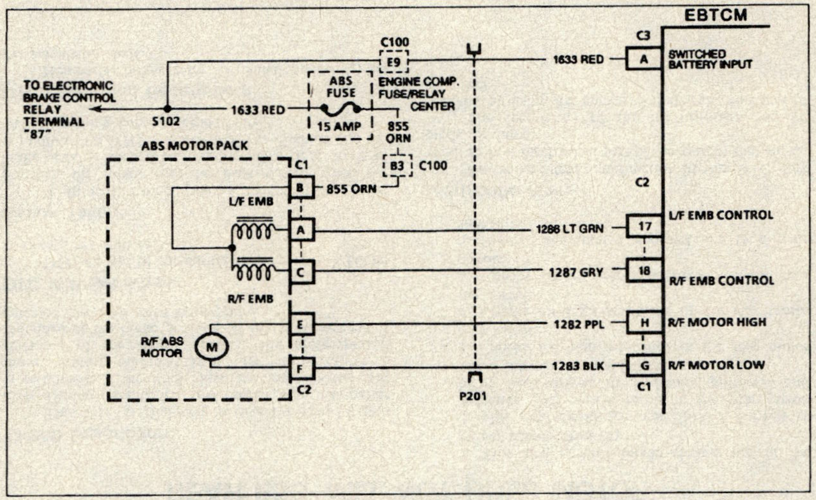

DTC 41

RIGHT FRONT EMB WILL NOT HOLD MOTOR

Circuit Description:

Current is supplied to the EMB's through terminal "B" of the three-way motor pack connector. When a release of the EMB is commanded, the EBTCM grounds the EMB input and allows the EMB to be energized to release the motor.

The right front motor is commanded in the forward direction while the EMB is released (energized).

DTC Will Set When:

DTC 41 can be set during initialization. A malfunction exists if the feedback current is less than the command current and the right front channel was expected to be in the home position. DTC 86 is always set with DTC 41.

Action Taken:

If an EMB can not hold a motor in the home position, the motor may be back driven when the brake pedal is applied causing the brake pedal to drop. A malfunction DTC is stored, ABS is disabled and the "ABS" warning lamp is turned "ON."

DTC Chart Test Description:

Number(s) below refer to circled number(s) on the diagnostic chart.

1. Ensures motor circuits are in the proper connector cavity.
2. Checks the ability of the EMB to hold the motor. If the brake pedal moves during this test, it would indicate the EMB is slipping.
3. Checks for high resistance in the right front motor high circuit.
4. Checks for high resistance in the right front motor low circuit.
5. Checks for excessive resistance in the right front motor.
6. Ensures malfunction was not due to a poor connection.

Diagnostic Aids:

An "intermittent" malfunction in this DTC may result from a mechanical part of the system that sticks, binds, or slips.

The frequency of the malfunction can be checked by using the enhanced diagnostic function of the Tech 1 tool

NOTICE: This DTC may fail if low voltage (10 volts or less) was present when initialization occurred. Ensure battery/charging system is functioning properly.

DIAGNOSTIC CHART DTC 41 — 1995 CAVALIER AND SUNFIRE WITH TCS, CONTINUED

DTC 41

RIGHT FRONT EMB WILL NOT HOLD MOTOR

NOTICE: TESTING EMB WITH JUMPER WIRES TO VOLTAGE OR GROUND WILL DESTROY THE EMB.

NOTICE: J 39200 TEST LEADS MUST BE "ZEROED" PRIOR TO MAKING ANY RESISTANCE MEASUREMENTS. REFER TO J 39200 USER'S MANUAL

(1)
- INSPECT THE 6-WAY MOTOR PACK AND EBTCM HARNESS CONNECTORS FOR PROPER WIRE COLOR/CONNECTOR CAVITY COMBINATION. ARE THE PROPER WIRES LOCATED IN THE PROPER CONNECTOR CAVITY?

NO → USING THE CIRCUIT DIAGRAMS LOCATED IN THIS SECTION, PLACE THE PROPER WIRES IN THE PROPER CONNECTOR CAVITIES. CLEAR DTCs AND RECHECK.

YES → IS DTC 71 ALSO PRESENT?

YES → PROCEED TO DTC 71 CHART.

NO ↓

(2)
- IGNITION "ON."
- USING TECH 1, SELECT EMB FUNCTIONAL TEST.
- APPLY FIRM PRESSURE ON BRAKE PEDAL.
- PERFORM EMB FUNCTIONAL TEST. DOES BRAKE PEDAL FALL OR RISE DURING THIS TEST?

YES → REPLACE MOTOR PACK.

NO ↓

(3)
- IGNITION "OFF."
- DISCONNECT 6-WAY MOTOR PACK AND EBTCM CONNECTORS.
- USING J 39200, MEASURE THE RESISTANCE BETWEEN 6-WAY EBTCM HARNESS CONNECTOR TERMINAL "H" AND 6-WAY MOTOR PACK HARNESS CONNECTOR TERMINAL "E". IS RESISTANCE LESS THAN 1.5Ω?

NO → REPAIR HIGH RESISTANCE IN CKT 1282

YES ↓

(4)
- USING J 39200, MEASURE RESISTANCE BETWEEN 6-WAY EBCM HARNESS CONNECTOR TERMINAL "G" AND 6-WAY MOTOR PACK HARNESS CONNECTOR TERMINAL "F". IS RESISTANCE LESS THAN 1.5Ω?

NO → REPAIR HIGH RESISTANCE IN CKT 1283.

YES ↓

(5)
- USING J 39200, MEASURE RESISTANCE BETWEEN MOTOR PACK CONNECTOR TERMINALS "E" AND "F". IS RESISTANCE LESS THAN 1.5Ω?

NO → REPLACE MOTOR PACK.

YES ↓

(6)
- INSPECT 6-WAY EBTCM AND MOTOR PACK CONNECTORS FOR POOR TERMINAL CONTACT OR CORROSION.
- REPLACE TERMINALS THAT EXHIBIT POOR CONTACT OR CORROSION
- RECONNECT ALL CONNECTORS.
- DRIVE VEHICLE FOR THREE DRIVE CYCLES (A DRIVE CYCLE CONSISTS OF STARTING ENGINE, DRIVING OVER 16 km/h (10 MPH) AND THEN TURNING IGNITION "OFF"). DID DTC 41 SET IN THE LAST THREE DRIVE CYCLES?

YES → REPLACE EBTCM.

NO → MALFUNCTION IS INTERMITTENT. REFER TO "DIAGNOSTIC AIDS" ON FACING PAGE.

DIAGNOSTIC CHART DTC 42 — 1995 CAVALIER AND SUNFIRE WITH TCS

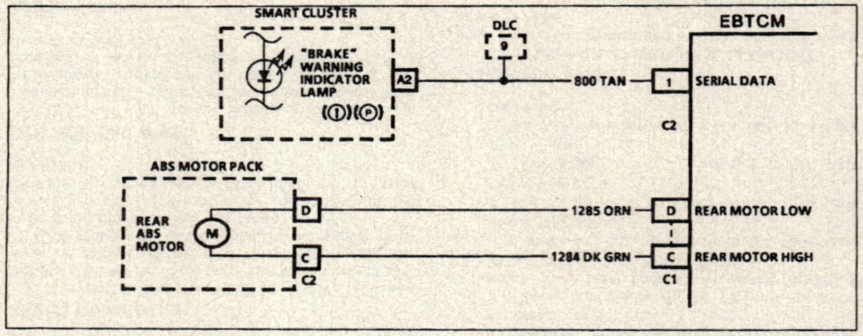

DTC 42
REAR ESB WILL NOT HOLD MOTOR

Circuit Description:

This DTC is designed to detect a slipping rear ESB. During initialization and braking, the rear motor is re-homed. If the ESB slips, the motor/piston will move. During the next key "ON" initialization, a re-home of the motor verifies the motor/piston remained at the home position. If motor movement is detected, the ESB must be slipping.

DTC Will Set When:

DTC 42 can be set during initialization. DTC 86 is always set with DTC 42.

Action Taken:

If an ESB cannot hold a motor in the home position, the motor may be back driven when the brake pedal is applied causing the brake pedal to drop. A malfunction DTC is stored, ABS is disabled and the "ABS" warning lamp is turned "ON."

DTC Chart Test Description:

Number(s) below refer to circled number(s) on the diagnostic chart.

1. This step ensures motor circuits are in the proper connector cavity.
2. This step checks the rear ESB. A broken or defective ESB would result in the rear piston being back driven by hydraulic pressure, and wheel movement would result.
3. Checks for high resistance in the rear motor high circuit.
4. Checks for high resistance in the rear motor low circuit.
5. Checks for excessive resistance in the rear motor.
6. Ensures malfunction was not due to a poor connection.

Diagnostic Aids:

An "intermittent" malfunction in this DTC may result from a mechanical part of the system that sticks, binds, or slips.

The frequency of the malfunction can be checked by using the enhanced diagnostic function of the Tech 1

DIAGNOSTIC CHART DTC 42 — 1995 CAVALIER AND SUNFIRE WITH TCS, CONTINUED

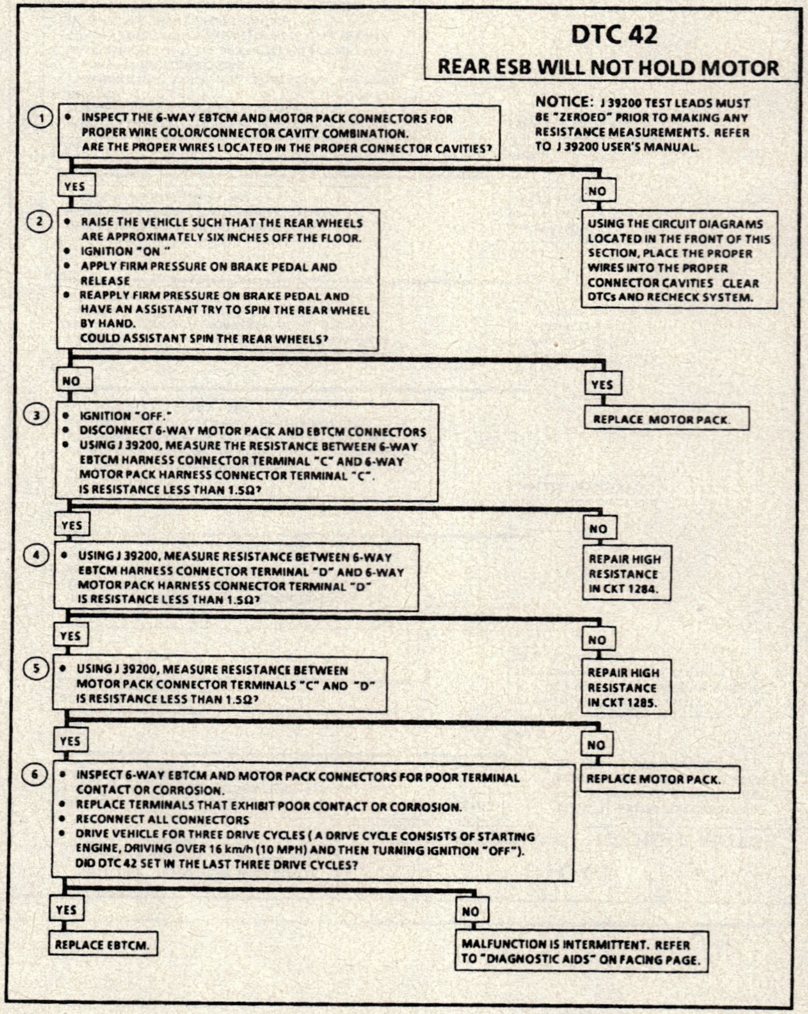

DIAGNOSTIC CHART DTC 44 — 1995 CAVALIER AND SUNFIRE WITH TCS

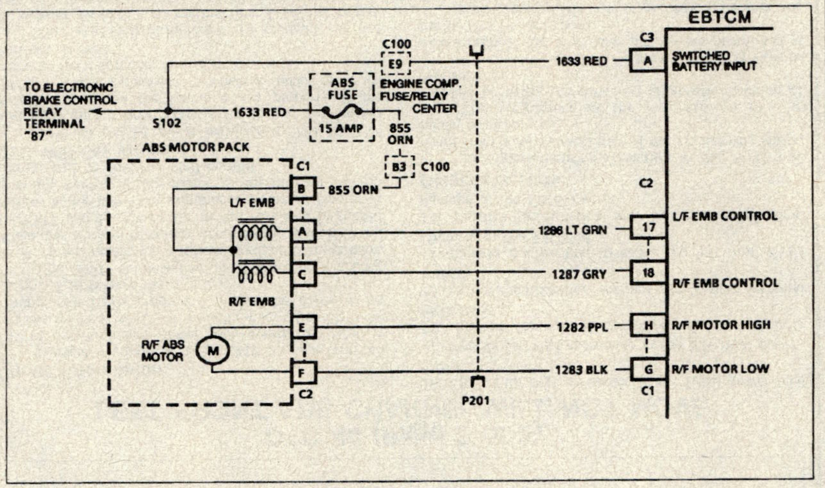

DTC 44 (Page 1 of 2)
LEFT FRONT ABS CHANNEL WILL NOT MOVE

Circuit Description:

Current is supplied to the EMB through terminal "B" of the three-way motor pack connector. When a release of the EMB is commanded, the EBTCM grounds the EMB input and allows the EMB to be energized to release the motor.

This DTC is designed to detect non-actuating EMB, stuck motor or a seized hydraulic modulator. When the release is commanded during initialization, the EMB should release the motor, resulting in sensed current being less than commanded current (motor is spinning freely). If the motor is not moving, sensed current will be equal to stall current.

DTC Will Set When:

DTC 44 can be set at initialization. The EBTCM commands the left front motor in the reverse and forward direction to ensure the motor moves properly. If the EBTCM detects a condition in which it cannot move the motor in either direction, a malfunction exists.

Action Taken:

This malfunction indicates the channel can not be moved properly. A malfunction DTC is stored, ABS is disabled and the "ABS" warning lamp is turned "ON."

DTC Chart Test Description:

Number(s) below refer to circled number(s) on the diagnostic chart.

1. Ensures motor circuits are in the proper connector cavity.
2. Checks for motor movement during release.
3. Compares feedback current to command current during apply.
4. Checks for a short between motor high and low circuits.
5. Checks for a shorted motor.
6. Ensures malfunction was not due to physical damage of the circuitry.
7. Checks if motor can apply properly under load.

Diagnostic Aids:

An "intermittent" malfunction in this DTC may result from a mechanical part of the system that sticks, binds, or slips.

The frequency of the malfunction can be checked by using the enhanced diagnostic function of the Tech 1

DTC 44 may set after modulator disassembly if the modulator pistons are positioned at the bottom of their bore.

Any circuitry, that is suspected as causing the intermittent complaint, should be thoroughly checked for backed out terminals, improper mating, broken locks, improperly formed or damaged terminals, poor terminal to wiring connections or physical damage to the wiring harness.

NOTICE: This DTC may fail if low voltage (10 volts or less) was present when initialization occurred. Ensure battery/charging system is functioning properly.

DIAGNOSTIC CHART DTC 44 — 1995 CAVALIER AND SUNFIRE WITH TCS, CONTINUED

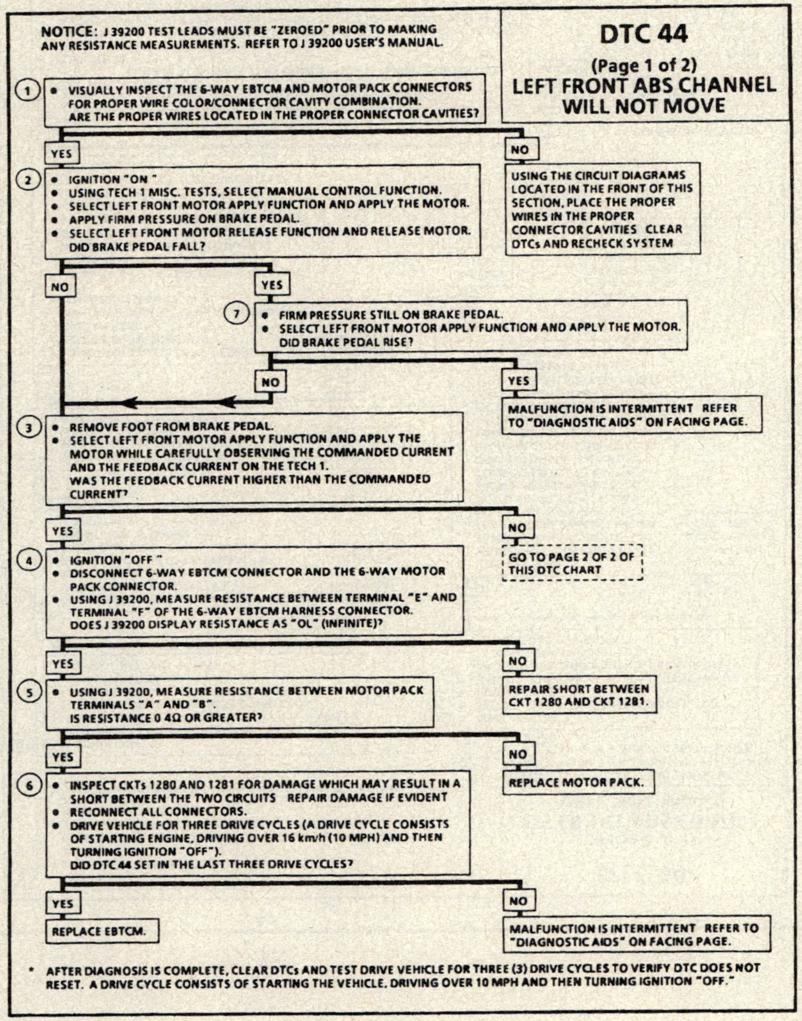

DIAGNOSTIC CHART DTC 44 — 1995 CAVALIER AND SUNFIRE WITH TCS, CONTINUED

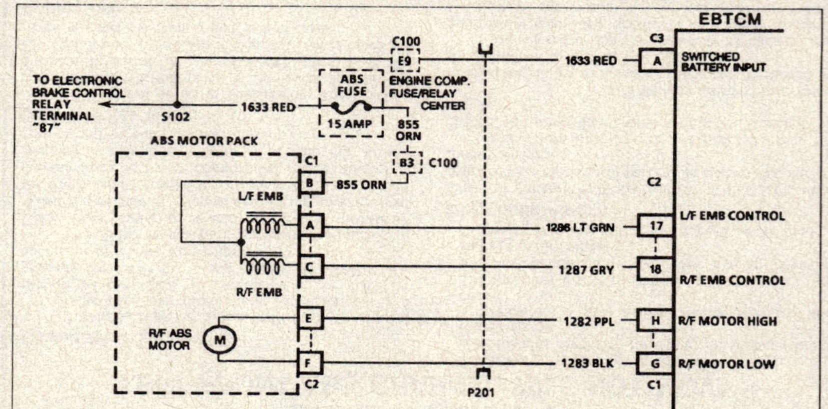

DTC 44 (Page 2 of 2)
LEFT FRONT ABS CHANNEL WILL NOT MOVE

Circuit Description:

Current is supplied to the EMB through terminal "B" of the three-way motor pack connector. When a release of the EMB is commanded, the EBTCM grounds the EMB input and allows the EMB to be energized to release the motor.

This DTC is designed to detect non-actuating EMB, stuck motor or a seized hydraulic modulator. When the release is commanded during initialization, the EMB should release the motor, resulting in sensed current being less than commanded current (motor is spinning freely). If the motor is not moving, sensed current will be equal to stall current.

DTC Will Set When:

DTC 44 can be set at initialization. The EBTCM commands the left front motor in the reverse and forward direction to ensure the motor moves properly. If the EBTCM detects a condition in which it cannot move the motor in either direction, a malfunction exists.

Action Taken:

This malfunction indicates the channel can not be moved properly. A malfunction DTC is stored, ABS is disabled and the "ABS" warning lamp is turned "ON."

DTC Chart Test Description:

Number(s) below refer to circled number(s) on the diagnostic chart.

8. Checks for a short to BATTERY in the left front EMB control circuit.
9. Checks for high resistance in the left front EMB control circuit.
10. Checks for high resistance in the EMB power feed circuit.
11. Checks for high resistance in the left front EMB.
12. Checks for proper operation of the hydraulic modulator.
13. Ensures malfunction was not due to poor terminal contact.
14. Isolates malfunction to either the left front EMB circuitry or the EBTCM.
15. Ensures malfunction was not due to physical damage of the EMB circuit.

Diagnostic Aids:

An "intermittent" malfunction in this DTC may result from a mechanical part of the system that sticks, binds, or slips.

The frequency of the malfunction can be checked by using the enhanced diagnostic function of the Tech 1 . DTC 44 may set after modulator disassembly if the modulator pistons are positioned at the bottom of their bore.

Any circuitry, that is suspected as causing the intermittent complaint, should be thoroughly checked for backed out terminals, improper mating, broken locks, improperly formed or damaged terminals, poor terminal to wiring connections or physical damage to the wiring harness.

NOTICE: This DTC may fail if low voltage (10 volts or less) was present when initialization occurred. Ensure battery/charging system is functioning properly.

DIAGNOSTIC CHART DTC 44 — 1995 CAVALIER AND SUNFIRE WITH TCS, CONTINUED

DTC 44
(Page 2 of 2)
LEFT FRONT ABS CHANNEL WILL NOT MOVE

NOTICE: J 39200 TEST LEADS MUST BE "ZEROED" PRIOR TO MAKING ANY RESISTANCE MEASUREMENTS. REFER TO J 39200 USER'S MANUAL.

(8)
- IGNITION "OFF."
- DISCONNECT 3-WAY MOTOR PACK CONNECTOR.
- IGNITION "ON."
- USING J 39200, MEASURE VOLTAGE BETWEEN GROUND AND 3-WAY MOTOR PACK HARNESS CONNECTOR TERMINAL "A."

IS VOLTAGE 1.0V OR LESS?

YES → (9); NO → (14)

(9)
- IGNITION "OFF."
- DISCONNECT 24-WAY EBTCM CONNECTOR.
- USING J 39200, MEASURE THE RESISTANCE BETWEEN 3-WAY MOTOR PACK HARNESS CONNECTOR TERMINAL "A" AND 24-WAY EBTCM HARNESS CONNECTOR TERMINAL "19".

IS RESISTANCE 2 0Ω OR LESS?

YES → (10); NO → REPAIR OPEN OR HIGH RESISTANCE IN CKT 1286.

(10)
- DISCONNECT 2-WAY EBTCM CONNECTOR
- USING J 39200, MEASURE THE RESISTANCE BETWEEN 2-WAY EBTCM HARNESS CONNECTOR TERMINAL "A" AND 3-WAY MOTOR PACK HARNESS CONNECTOR TERMINAL "B."

IS RESISTANCE 2 0Ω OR LESS?

YES → (11); NO → REPAIR OPEN OR HIGH RESISTANCE IN CKT 855 OR CKT 1633

(11)
- USING J 39200, MEASURE THE RESISTANCE BETWEEN MOTOR PACK TERMINALS "A" AND "B."

IS RESISTANCE 2 0Ω OR LESS?

YES → (12); NO → REPLACE MOTOR PACK.

(12)
- USING TECH 1 MISC. TESTS, PERFORM GEAR TENSION RELIEF FUNCTION.
- IGNITION "OFF."
- REMOVE HYDRAULIC MODULATOR/MASTER CYLINDER ASSEMBLY.
- SEPARATE MOTOR PACK FROM HYDRAULIC MODULATOR.
- GRASP THE GEAR ON THE HYDRAULIC MODULATOR (THE HYDRAULIC MODULATOR IS THE UNIT WITH THE THREE LARGE GEARS) FURTHEST FROM THE BRAKE BOOSTER AND ATTEMPT TO MOVE THE GEAR BY HAND

CAN GEAR BE ROTATED AT LEAST 8 5 FULL TURNS LOCK TO LOCK?

YES → (13); NO → REPLACE HYDRAULIC MODULATOR.

(13)
- INSPECT MOTOR PACK AND EBTCM CONNECTORS FOR POOR TERMINAL CONTACT OR CORROSION.
- REPLACE TERMINALS THAT EXHIBIT POOR TERMINAL CONTACT OR CORROSION.
- RECONNECT ALL CONNECTORS.
- DRIVE VEHICLE FOR THREE DRIVE CYCLES (A DRIVE CYCLE CONSISTS OF STARTING ENGINE, DRIVING OVER 16 km/h (10 MPH) AND THEN TURNING IGNITION "OFF")

DID DTC 44 SET IN THE LAST THREE DRIVE CYCLES?

YES → REPLACE MOTOR PACK.; NO → MALFUNCTION IS INTERMITTENT. REFER TO "DIAGNOSTIC AIDS" ON FACING PAGE.

(14)
- IGNITION "OFF."
- DISCONNECT 24-WAY EBTCM CONNECTOR
- IGNITION "ON."
- USING J 39200, MEASURE VOLTAGE BETWEEN GROUND AND 3-WAY MOTOR PACK HARNESS CONNECTOR TERMINAL "A."

IS VOLTAGE 1 0V OR LESS?

YES → (15); NO → REPAIR SHORT TO VOLTAGE IN CKT 1286.

(15)
- IGNITION "OFF."
- INSPECT CKT 1286 AND 24-WAY EBTCM HARNESS CONNECTOR FOR PHYSICAL DAMAGE WHICH MAY RESULT IN A SHORT TO VOLTAGE WITH THE 24-WAY EBCM HARNESS CONNECTOR CONNECTED TO THE EBTCM. CORRECT DAMAGE IF EVIDENT
- RECONNECT ALL CONNECTORS.
- DRIVE VEHICLE FOR THREE DRIVE CYCLES (A DRIVE CYCLE CONSISTS OF STARTING ENGINE, DRIVING OVER 16 km/h (10 MPH) AND THEN TURNING IGNITION "OFF").

DID DTC 44 SET IN THE LAST THREE DRIVE CYCLES?

YES → REPLACE EBTCM.; NO → MALFUNCTION IS INTERMITTENT REFER TO "DIAGNOSTIC AIDS" ON FACING PAGE

*AFTER DIAGNOSIS IS COMPLETE, CLEAR DTCs AND TEST DRIVE VEHICLE FOR THREE (3) DRIVE CYCLES TO VERIFY DTC DOES NOT RESET. A DRIVE CYCLE CONSISTS OF STARTING THE VEHICLE, DRIVING OVER 10 MPH AND THEN KEYING DOWN.

DIAGNOSTIC CHART DTC 45 — 1995 CAVALIER AND SUNFIRE WITH TCS

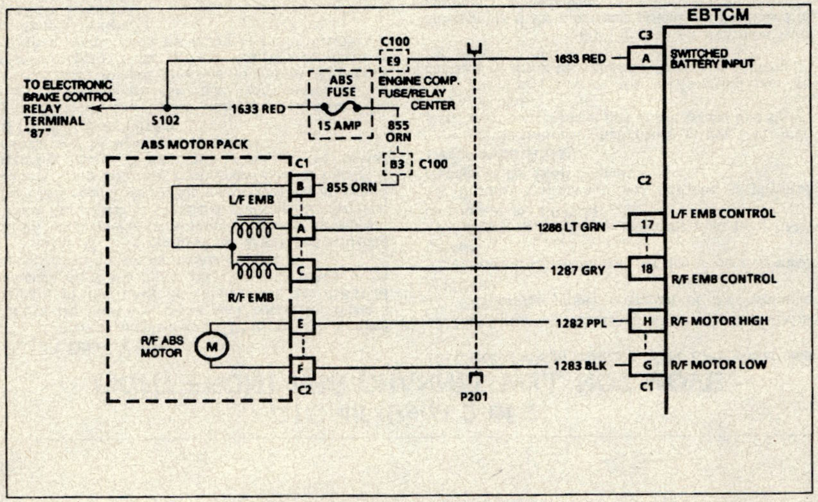

DTC 45 (Page 1 of 2) RIGHT FRONT ABS CHANNEL WILL NOT MOVE

Circuit Description:

Current is supplied to the EMB through terminal "B" of the three-way motor pack connector. When a release of the EMB is commanded, the EBTCM grounds the EMB input and allows the EMB to be energized to release the motor.

This DTC is designed to detect non-actuating EMB, stuck motor or a seized hydraulic modulator. When the release is commanded during initialization, the EMB should release the motor, resulting in sensed current being less than commanded current (motor is spinning freely). If the motor is not moving, sensed current will be equal to stall current.

DTC Will Set When:

DTC 45 can be set during initialization. The EBTCM commands the right front motor in the reverse and forward direction to ensure the motor moves properly. If the EBTCM detects a condition in which it cannot move the motor in either direction, a malfunction exists.

Action Taken:

This malfunction indicates the channel can not be moved properly. A malfunction DTC is stored, ABS is disabled and the "ABS" warning lamp is turned "ON."

DTC Chart Test Description:

Number(s) below refer to circled number(s) on the diagnostic chart.

1. Ensures motor circuits are in the proper connector cavity.
2. Checks for motor movement during release.
3. Compares feedback current to command current during apply.
4. Checks for a short between motor high and low circuits.
5. Checks for a shorted motor.
6. Ensures malfunction was not due to physical damage of the circuitry.
7. Checks if motor can apply properly under load.

Diagnostic Aids:

An "intermittent" malfunction in this DTC may result from a mechanical part of the system that sticks, binds, or slips.

The frequency of the malfunction can be checked by using the enhanced diagnostic function of the Tech 1

DTC 45 may set after modulator disassembly if the modulator pistons are positioned at the bottom of their bore.

Any circuitry, that is suspected as causing the intermittent complaint, should be thoroughly checked for backed out terminals, improper mating, broken locks, improperly formed or damaged terminals, poor terminal to wiring connections or physical damage to the wiring harness.

NOTICE: This DTC may fail if low voltage (10 volts or less) was present when initialization occurred. Ensure battery/charging system is functioning properly.

DIAGNOSTIC CHART DTC 45 — 1995 CAVALIER AND SUNFIRE WITH TCS, CONTINUED

DTC 45 (Page 1 of 2) RIGHT FRONT ABS CHANNEL WILL NOT MOVE

NOTICE: J 39200 TEST LEADS MUST BE "ZEROED" PRIOR TO MAKING ANY RESISTANCE MEASUREMENTS. REFER TO J 39200 USER'S MANUAL.

(1)
- VISUALLY INSPECT THE 6-WAY EBTCM AND MOTOR PACK CONNECTORS FOR PROPER WIRE COLOR/CONNECTOR CAVITY COMBINATION. ARE THE PROPER WIRES LOCATED IN THE PROPER CONNECTOR CAVITIES?

NO: USING THE CIRCUIT DIAGRAMS LOCATED IN THE FRONT OF THIS SECTION, PLACE THE PROPER WIRES IN THE PROPER CONNECTOR CAVITIES. CLEAR DTCs AND RECHECK SYSTEM

YES:

(2)
- IGNITION "ON."
- USING TECH 1 MISC. TESTS, SELECT MANUAL CONTROL FUNCTION.
- SELECT RIGHT FRONT MOTOR APPLY FUNCTION AND APPLY THE MOTOR.
- APPLY FIRM PRESSURE ON BRAKE PEDAL.
- SELECT RIGHT FRONT MOTOR RELEASE FUNCTION AND RELEASE MOTOR. DID BRAKE PEDAL FALL?

NO: go to (3)

YES:

(7)
- FIRM PRESSURE STILL ON BRAKE PEDAL.
- SELECT RIGHT FRONT MOTOR APPLY FUNCTION AND APPLY THE MOTOR. DID BRAKE PEDAL RISE?

YES: MALFUNCTION IS INTERMITTENT. REFER TO "DIAGNOSTIC AIDS" ON FACING PAGE.

NO: go to (3)

(3)
- REMOVE FOOT FROM BRAKE PEDAL.
- SELECT RIGHT FRONT MOTOR APPLY FUNCTION AND APPLY THE MOTOR WHILE CAREFULLY OBSERVING THE COMMANDED CURRENT AND THE FEEDBACK CURRENT ON THE TECH 1 WAS THE FEEDBACK CURRENT HIGHER THAN THE COMMANDED CURRENT?

NO: GO TO PAGE 2 OF 2 OF THIS DTC CHART

YES:

(4)
- IGNITION "OFF."
- DISCONNECT 6-WAY EBTCM CONNECTOR AND THE 6-WAY MOTOR PACK CONNECTOR.
- USING J 39200, MEASURE RESISTANCE BETWEEN TERMINAL "G" AND TERMINAL "H" OF THE 6-WAY EBTCM HARNESS CONNECTOR. DOES J 39200 DISPLAY RESISTANCE AS "OL" (INFINITE)?

NO: REPAIR SHORT BETWEEN CKT 1282 AND CKT 1283

YES:

(5)
- USING J 39200, MEASURE RESISTANCE BETWEEN MOTOR PACK TERMINALS "E" AND "F" IS RESISTANCE 4Ω OR GREATER?

NO: REPLACE MOTOR PACK.

YES:

(6)
- INSPECT CKTs 1282 AND 1283 FOR DAMAGE WHICH MAY RESULT IN A SHORT BETWEEN THE TWO CIRCUITS. REPAIR DAMAGE IF EVIDENT.
- RECONNECT ALL CONNECTORS.
- DRIVE VEHICLE FOR THREE DRIVE CYCLES (A DRIVE CYCLE CONSISTS OF STARTING ENGINE, DRIVING OVER 16 km/h (10 MPH) AND THEN TURNING IGNITION "OFF"). DID DTC 45 SET IN THE LAST THREE DRIVE CYCLES?

YES: REPLACE EBTCM.

NO: MALFUNCTION IS INTERMITTENT. REFER TO "DIAGNOSTIC AIDS" ON FACING PAGE.

- AFTER DIAGNOSIS IS COMPLETE, CLEAR DTCs AND TEST DRIVE VEHICLE FOR THREE (3) DRIVE CYCLES TO VERIFY DTC DOES NOT RESET. A DRIVE CYCLE CONSISTS OF STARTING THE VEHICLE, DRIVING OVER 10 MPH AND THEN TURNING IGNITION "OFF."

DIAGNOSTIC CHART DTC 45 — 1995 CAVALIER AND SUNFIRE WITH TCS, CONTINUED

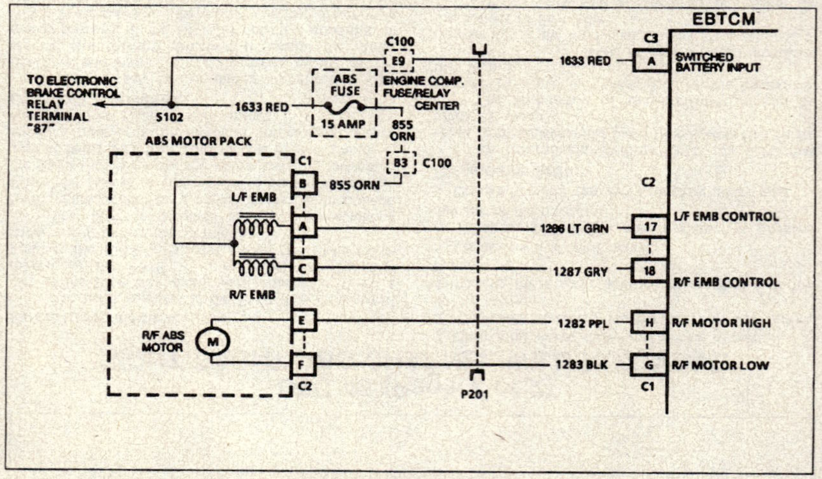

DTC 45 (Page 2 of 2)
RIGHT FRONT ABS CHANNEL WILL NOT MOVE

Circuit Description:

Current is supplied to the EMB through terminal "B" of the three-way motor pack connector. When a release of the EMB is commanded, the EBTCM grounds the EMB input and allows the EMB to be energized to release the motor.

This DTC is designed to detect non-actuating EMB, stuck motor or a seized hydraulic modulator. When the release is commanded during initialization, the EMB should release the motor, resulting in sensed current being less than commanded current (motor is spinning freely). If the motor is not moving, sensed current will be equal to stall current.

DTC Will Set When:

DTC 45 can be set during initialization. The EBTCM commands the right front motor in the reverse and forward direction to ensure the motor moves properly. If the EBTCM detects a condition in which it cannot move the motor in either direction, a malfunction exists.

Action Taken:

This malfunction indicates the channel cannot be moved properly. A malfunction DTC is stored, ABS is disabled and the "ABS" warning lamp is turned "ON."

DTC Chart Test Description:

Number(s) below refer to circled number(s) on the diagnostic chart.

8. Checks for a short to BATTERY in the right front EMB control circuit.
9. Checks for high resistance in the right front EMB control circuit.
10. Checks for high resistance in the EMB power feed circuit.
11. Checks for high resistance in the right front EMB.
12. Checks for proper operation of the hydraulic modulator.
13. Ensures malfunction was not due to poor terminal contact.
14. Isolates malfunction to either the right front EMB circuitry or the EBTCM.
15. Ensures malfunction was not due to physical damage of the EMB circuitry.

Diagnostic Aids:

An "intermittent" malfunction in this DTC may result from a mechanical part of the system that sticks, binds, or slips.

The frequency of the malfunction can be checked by using the enhanced diagnostic function of the Tech 1

DTC 45 may set after modulator disassembly if the modulator pistons are positioned at the bottom of their bore.

Any circuitry, that is suspected as causing the intermittent complaint, should be thoroughly checked for backed out terminals, improper mating, broken locks, improperly formed or damaged terminals, poor terminal to wiring connections or physical damage to the wiring harness.

NOTICE: This DTC may fail if low voltage (10 volts or less) was present when initialization occurred. Ensure battery/charging system is functioning properly.

DIAGNOSTIC CHART DTC 45 — 1995 CAVALIER AND SUNFIRE WITH TCS, CONTINUED

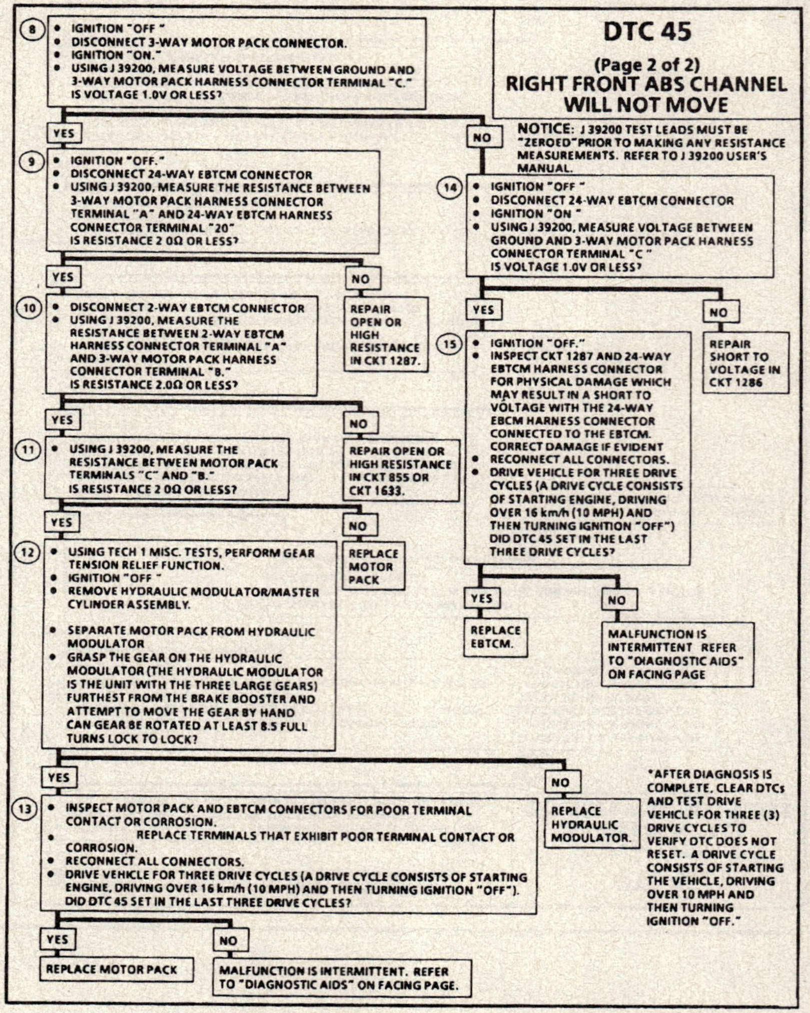

DIAGNOSTIC CHART DTC 46 — 1995 CAVALIER AND SUNFIRE WITH TCS

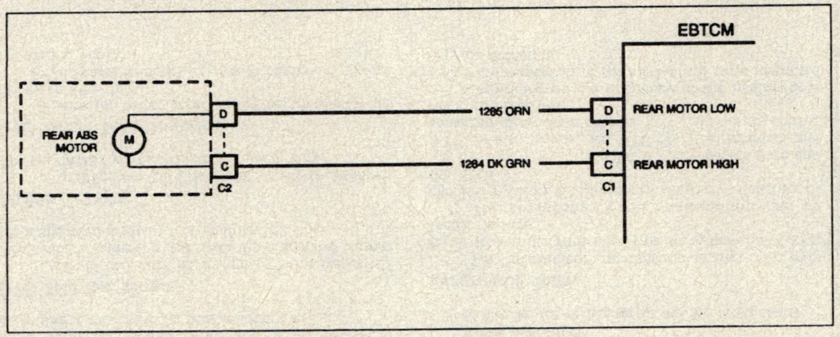

DTC 46
REAR CHANNEL WILL NOT MOVE

Circuit Description:

This DTC is designed to detect bound-up ESB's, stuck motor or a seized hydraulic modulator. Release is commanded during initialization, the ESB should release the motor, resulting in sensed current being less than commanded current (motor is spinning freely). If the motor is not moving, sensed current will be equal to stall current.

DTC Will Set When:

DTC 46 can be set during initialization. The EBTCM commands the rear motor in the reverse and forward direction to ensure the motor moves properly. If the EBTCM detects a condition in which it cannot move the motor in either direction, a malfunction exists.

Action Taken:

This malfunction indicates the channel can not be moved properly. A malfunction DTC is stored, ABS is disabled, and the "ABS" warning lamp is turned "ON."

DTC Chart Test Description:

Number(s) below refer to circled number(s) on the diagnostic chart.

1. Ensures motor circuits are in the proper connector cavity.
2. Checks the ability of the motor to move applying hydraulic pressure to the wheel cylinders. A frozen piston or gear would result in no apply of rear brakes and a wheel that freely spins.
3. Compares feedback current to command current during apply.
4. Checks to see if hydraulic modulator is functioning properly.
5. Checks if motor can apply properly under load.
6. Isolates the malfunction to either the EBTCM or a short between CKTs 1284 and 1285.
7. Checks for a shorted motor.
8. Ensures the malfunction was not due to physical damage of the circuitry.

Diagnostic Aids:

An "intermittent" malfunction in this DTC may result from a mechanical part of the system that sticks or binds.

The frequency of the malfunction can be checked by using the enhanced diagnostic function of the Tech 1.

DTC 46 may set after modulator disassembly if the modulator pistons are positioned at the bottom of their bore.

Depending on the frequency of the malfunction, a physical inspection of the mechanical parts suspected may be necessary.

Any circuitry, that is suspected as causing the intermittent complaint, should be thoroughly checked for backed out terminals, improper mating, broken locks, improperly formed or damaged terminals, poor terminal to wiring connections or physical damage to the wiring harness.

DIAGNOSTIC CHART DTC 46 — 1995 CAVALIER AND SUNFIRE WITH TCS, CONTINUED

DTC 46
REAR CHANNEL WILL NOT MOVE

NOTICE: J 39200 TEST LEADS MUST BE "ZEROED" PRIOR TO MAKING ANY RESISTANCE MEASUREMENTS. REFER TO J 39200 USER'S MANUAL.

(1)
- VISUALLY INSPECT THE 6-WAY EBTCM AND MOTOR PACK CONNECTORS FOR PROPER WIRE COLOR/CONNECTOR CAVITY COMBINATION.
- ARE THE PROPER WIRES LOCATED IN THE PROPER CONNECTOR CAVITIES?

YES → (2)
NO → USING THE CIRCUIT DIAGRAMS LOCATED IN THE FRONT OF THIS SECTION, PLACE THE PROPER WIRES IN THE PROPER CONNECTOR CAVITIES. CLEAR DTCs AND RECHECK SYSTEM

(2)
- RAISE AND SUPPORT THE VEHICLE SUCH THAT THE REAR WHEELS ARE APPROXIMATELY SIX INCHES OFF THE FLOOR.
- IGNITION "ON"
- APPLY FIRM PRESSURE ON BRAKE PEDAL.
- USING TECH 1 MISC. TESTS, SELECT MANUAL CONTROL FUNCTION.
- SELECT REAR MOTOR APPLY FUNCTION AND APPLY THE MOTOR.
- HAVE AN ASSISTANT TRY TO SPIN THE REAR WHEELS BY HAND COULD ASSISTANT SPIN THE REAR WHEELS?

YES → (3)
NO → (7)

(7)
- FIRM PRESSURE STILL ON BRAKE PEDAL.
- SELECT REAR MOTOR RELEASE FUNCTION AND RELEASE THE MOTOR.
- HAVE AN ASSISTANT TRY TO SPIN THE REAR WHEELS BY HAND. COULD ASSISTANT SPIN THE REAR WHEELS?

NO → (3)
YES → MALFUNCTION IS INTERMITTENT. REFER TO "DIAGNOSTIC AIDS" ON FACING PAGE.

(3)
- REMOVE FOOT FROM BRAKE PEDAL
- SELECT REAR MOTOR APPLY FUNCTION AND APPLY THE MOTOR WHILE CAREFULLY OBSERVING THE COMMANDED CURRENT AND THE FEEDBACK CURRENT ON THE TECH 1. WAS THE FEEDBACK CURRENT HIGHER THAN THE COMMANDED CURRENT?

YES → (4)
NO → (8)

(4)
- IGNITION "OFF"
- DISCONNECT THE 6-WAY EBTCM CONNECTOR AND THE 6-WAY MOTOR PACK CONNECTOR.
- USING A J 39200, MEASURE THE RESISTANCE BETWEEN TERMINAL "C" AND TERMINAL "D" OF THE 6-WAY EBCM HARNESS CONNECTOR. DOES J 39200 DISPLAY RESISTANCE AS "OL" (INFINITE)?

YES → (5)
NO → REPAIR SHORT BETWEEN CKT 1284 AND CKT 1285.

(8)
- USING TECH 1 MISC TESTS, PERFORM GEAR TENSION RELIEF FUNCTION.
- IGNITION "OFF."
- REMOVE HYDRAULIC MODULATOR/MASTER CYLINDER ASSEMBLY
- SEPARATE MOTOR PACK FROM HYDRAULIC MODULATOR.
- GRASP THE MIDDLE GEAR ON THE HYDRAULIC MODULATOR (THE HYDRAULIC MODULATOR IS THE UNIT WITH THE THREE LARGE GEARS) AND ATTEMPT TO MOVE THE GEAR BY HAND. CAN GEAR BE ROTATED AT LEAST 3.5 FULL TURNS LOCK TO LOCK?

NO → REPLACE HYDRAULIC MODULATOR.
YES → REPLACE MOTOR PACK.

(5)
- USING J 39200, MEASURE RESISTANCE BETWEEN MOTOR PACK TERMINALS "C" AND "D" IS RESISTANCE 0.4Ω OR GREATER?

YES → (6)
NO → REPLACE MOTOR PACK.

(6)
- INSPECT CKTs 1284 AND 1285 FOR DAMAGE WHICH MAY RESULT IN A SHORT BETWEEN THE TWO CIRCUITS. REPAIR DAMAGE IF EVIDENT.
- RECONNECT ALL CONNECTOR
- DRIVE VEHICLE FOR THREE DRIVE CYCLES (A DRIVE CYCLE CONSISTS OF STARTING THE ENGINE, DRIVING OVER 16 km/h (10 MPH) AND THEN TURNING IGNITION "OFF"). DID DTC 46 SET IN THE LAST THREE DRIVE CYCLES?

YES → REPLACE EBTCM.
NO → MALFUNCTION IS INTERMITTENT. REFER TO "DIAGNOSTIC AIDS" ON FACING PAGE.

- AFTER ALL DIAGNOSIS IS COMPLETE, CLEAR DTCs AND TEST DRIVE VEHICLE FOR THREE (3) DRIVE CYCLES TO VERIFY THAT DTC DOES NOT RESET. A DRIVE CYCLE CONSISTS OF STARTING THE VEHICLE, DRIVING VEHICLE OVER 10 MPH AND THEN TURNING IGNITION "OFF"

DIAGNOSTIC CHART DTC 47 — 1995 CAVALIER AND SUNFIRE WITH TCS

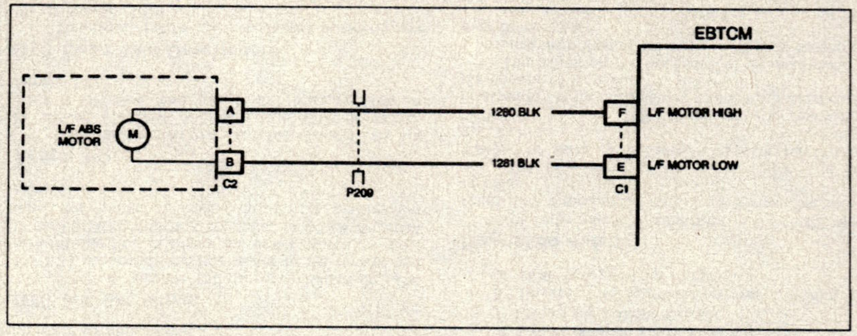

DTC 47
(Page 1 of 2)
LEFT FRONT MOTOR FREE SPINS

Circuit Description:

This DTC is designed to detect a stripped nut or gear assembly during initialization. During the homing sequence, the piston should reach the top of the bore resulting in a stalled motor. If this does not occur, the motor must be spinning with little or no resistance, indicating a nut/screw or gear malfunction.

DTC Will Set When:

DTC 47 can only be set during initialization. If the feedback current is less than the command current for a specified period of, a malfunction exists.

Action Taken:

A malfunction DTC is stored, ABS is disabled, and the "ABS" warning lamp is turned "ON."

DTC Chart Test Description:

Number(s) below refer to circled number(s) on the diagnostic chart.

1. Compares feedback current to command current during apply.
2. Compares feedback current to command current during release.
3. Moves the position to the top of the bore.
4. Checks for excessive resistance in the left front motor high circuit.
5. Checks for excessive resistance in the left front motor low circuit.
6. Checks for resistance in the left front motor.

Diagnostic Aids:

An "intermittent" malfunction in this DTC may result from a mechanical part of the system that sticks, binds, or slips.

The frequency of the malfunction can be checked by using the enhanced diagnostic function of the Tech 1 If DTC 47 only occurs once and DTC 56 also occurs, see DTC 56. If intermittent and enhanced diagnostics shows DTC 47 occurs during ABS, see DTC 56.

Depending on the frequency of the malfunction, a physical inspection of the mechanical parts suspected may be necessary.

DIAGNOSTIC CHART DTC 47 — 1995 CAVALIER AND SUNFIRE WITH TCS, CONTINUED

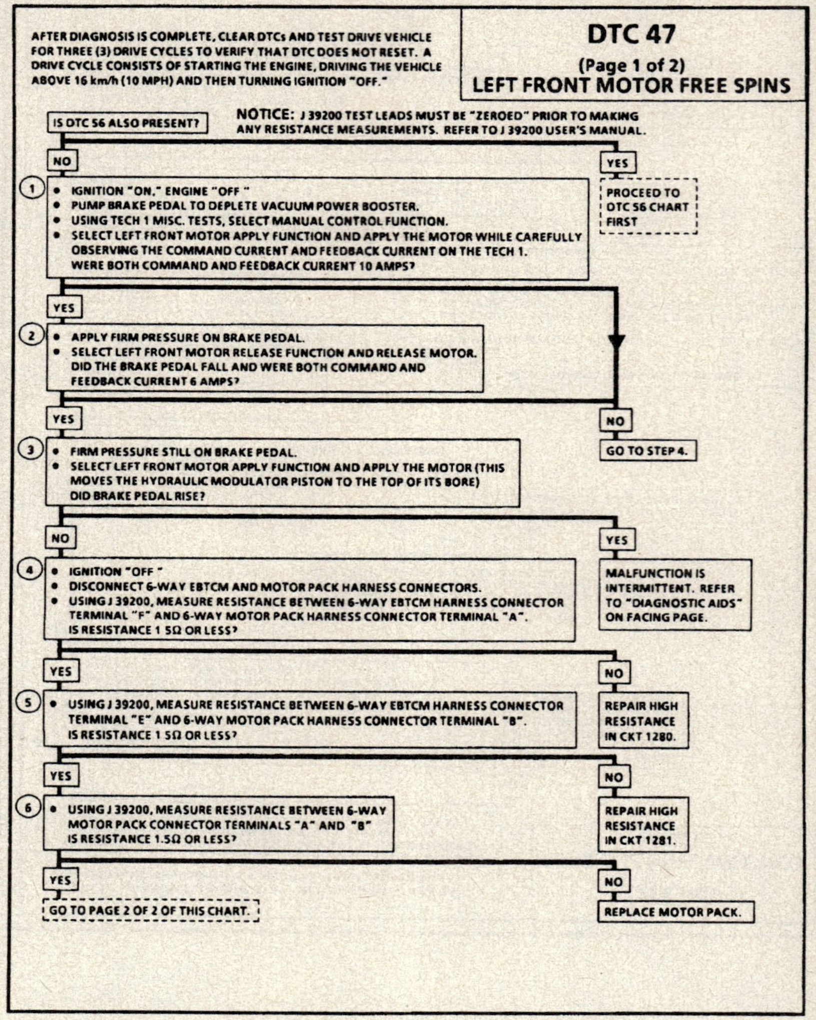

DIAGNOSTIC CHART DTC 47 — 1995 CAVALIER AND SUNFIRE WITH TCS, CONTINUED

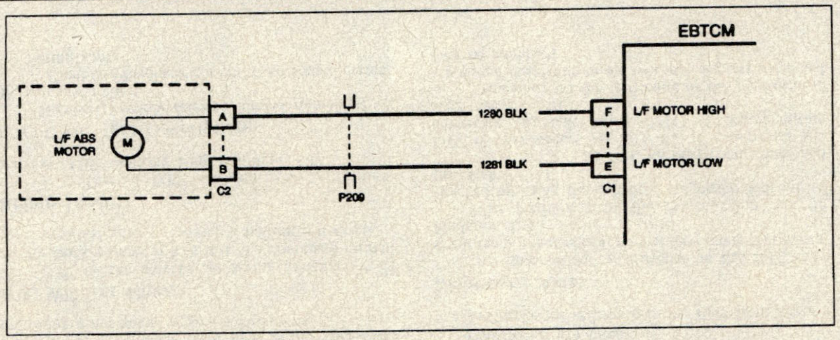

DTC 47
(Page 2 of 2)
LEFT FRONT MOTOR FREE SPINS

Circuit Description:

This DTC is designed to detect a stripped nut or gear assembly during initialization. During the homing sequence, the piston should reach the top of the bore resulting in a stalled motor. If this does not occur, the motor must be spinning with little or no resistance, indicating a nut/screw or gear malfunction.

DTC Will Set When:

DTC 47 can only be set during initialization. If the feedback current is less than the command current for a specified period of time, a malfunction exists.

Action Taken:

A malfunction DTC is stored, ABS is disabled, and the "ABS" warning lamp is turned "ON."

DTC Chart Test Description:

Number(s) below refer to circled number(s) on the diagnostic chart.

7. Checks for a stripped motor pack gear.
8. Checks for a stripped hydraulic modulator gear.
9. Checks for proper operation of motor pack.
10. Checks for proper operation of the hydraulic modulator.
11. Ensures malfunction was not due to poor terminal contact.

Diagnostic Aids:

An "intermittent" malfunction in this DTC may result from a mechanical part of the system that sticks, binds, or slips.

The frequency of the malfunction can be checked by using the enhanced diagnostic function of the Tech 1

If DTC 47 only occurs once and DTC 56 also occurs, see DTC 56. If intermittent and enhanced diagnostics shows DTC 47 occurs during ABS, see DTC 56.

Depending on the frequency of the malfunction, a physical inspection of the mechanical parts suspected may be necessary.

DIAGNOSTIC CHART DTC 47 — 1995 CAVALIER AND SUNFIRE WITH TCS, CONTINUED

DTC 47
(Page 2 of 2)
LEFT FRONT MOTOR FREE SPINS

CONTINUED FROM PAGE 1 OF 2 OF THIS CHART.

(7)
- REMOVE HYDRAULIC MODULATOR/MASTER CYLINDER ASSEMBLY. (REFER TO "ON-VEHICLE SERVICE" IN THIS SECTION.)
- REMOVE GEAR COVER.
- CHECK FOR STRIPPED GEAR ON MOTOR PACK; THE MOTOR PACK IS THE UNIT WITH THE THREE SMALL GEARS THE LEFT FRONT GEAR IS FURTHEST FROM THE BRAKE BOOSTER WHEN THE ASSEMBLY IS INSTALLED IN THE VEHICLE. IS THE MOTOR PACK GEAR STRIPPED?

YES: REPLACE MOTOR PACK
NO: go to (8)

(8)
- CHECK FOR A STRIPPED GEAR ON THE HYDRAULIC MODULATOR; THE HYDRAULIC MODULATOR IS THE UNIT WITH THE THREE LARGE GEARS. THE LEFT FRONT GEAR IS FURTHEST FROM THE BRAKE BOOSTER WHEN THE ASSEMBLY IS INSTALLED IN THE VEHICLE. IS THE HYDRAULIC MODULATOR GEAR STRIPPED?

YES: REPLACE HYDRAULIC MODULATOR GEAR.
NO: go to (9)

(9)
- SECURELY POSITION HYDRAULIC MODULATOR/MASTER CYLINDER ASSEMBLY IN VEHICLE ALLOWING LEFT FRONT GEAR SET OBSERVATION. BE CAREFUL NOT TO DAMAGE GEARS OR ALLOW THE HYDRAULIC MODULATOR/MASTER CYLINDER ASSEMBLY TO MOVE DURING TESTING.
- RECONNECT ELECTRICAL CONNECTORS.
- IGNITION "ON."
- USING TECH 1 MISC. TESTS, SELECT MANUAL CONTROL FUNCTION.
- SELECT LEFT FRONT MOTOR APPLY AND APPLY THE MOTOR WHILE OBSERVING THE GEAR SET FURTHEST FORWARD FROM THE BOOSTER WHEN THE ASSEMBLY IS INSTALLED IN THE VEHICLE
- SELECT LEFT FRONT MOTOR RELEASE FUNCTION AND RELEASE THE MOTOR WHILE STILL OBSERVING THE GEAR SET
- SELECT LEFT FRONT MOTOR APPLY FUNCTION AND APPLY THE MOTOR. DID GEAR SET MOVE IN BOTH DIRECTIONS FOR AT LEAST ONE REVOLUTION?

NO: REPLACE MOTOR PACK
YES: go to (10)

(10)
- USING TECH 1, SELECT "F4. MISC TESTS" AND PERFORM GEAR TENSION RELIEF FUNCTION.
- IGNITION "OFF."
- SEPARATE MOTOR PACK FROM HYDRAULIC MODULATOR.
- GRASP THE LEFT FRONT GEAR ON THE HYDRAULIC MODULATOR; THE HYDRAULIC MODULATOR IS THE UNIT WITH THE THREE LARGE GEARS. THE LEFT FRONT GEAR IS FURTHEST FROM THE BRAKE BOOSTER WHEN THE ASSEMBLY IS INSTALLED IN THE VEHICLE.
- ATTEMPT TO ROTATE THE GEAR. CAN THE GEAR BE ROTATED AT LEAST 8 5 FULL TURNS LOCK TO LOCK?

YES: REPLACE HYDRAULIC MODULATOR.
NO: go to (11)

(11)
- INSPECT 6-WAY MOTOR PACK AND EBTCM CONNECTORS FOR CORROSION OR POOR TERMINAL CONTACT. REPLACE TERMINALS THAT EXHIBIT POOR TERMINAL CONTACT OR CORROSION
- INSTALL HYDRAULIC MODULATOR/MASTER CYLINDER ASSEMBLY
- RECONNECT ALL CONNECTORS.
- DRIVE VEHICLE FOR THREE DRIVE CYCLES (A DRIVE CYCLE CONSISTS OF STARTING ENGINE, DRIVING OVER 16 km/h (10 MPH) AND THEN TURNING IGNITION "OFF"). DID DTC 47 SET IN THE LAST THREE DRIVE CYCLES?

YES: REPLACE EBTCM.
NO: MALFUNCTION WAS DUE TO A POOR CONNECTION.

DIAGNOSTIC CHART DTC 48 — 1995 CAVALIER AND SUNFIRE WITH TCS

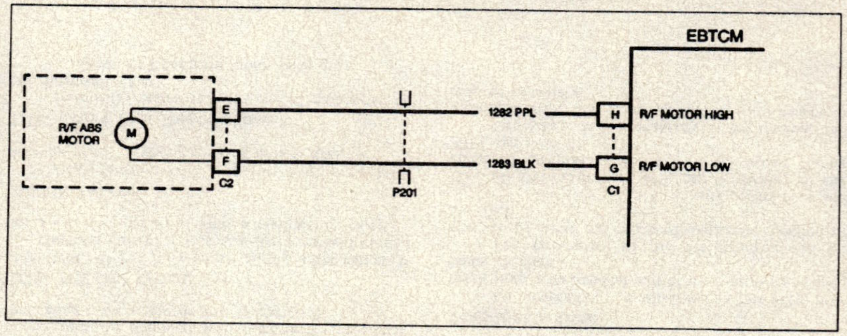

DTC 48
(Page 1 of 2)
RIGHT FRONT MOTOR FREE SPINS

Circuit Description:

This DTC is designed to detect a stripped nut or gear assembly during initialization. During the homing sequence, the piston should reach the top of the bore resulting in a stalled motor. If this does not occur, the motor must be spinning with little or no resistance, indicating a nut/screw or gear malfunction.

DTC Will Set When:

DTC 48 can only be set during initialization. If the feedback current is less than the command current for a specified period of time, a malfunction exists.

Action Taken:

A malfunction DTC is stored, ABS is disabled, and the "ABS" warning lamp is turned "ON."

DTC Chart Test Description:

Number(s) below refer to circled number(s) on the diagnostic chart.

1. Compares feedback current to command current during apply.
2. Compares feedback current to command current during release.
3. Moves the position to the top of the bore.
4. Checks for excessive resistance in the right front motor high circuit.
5. Checks for excessive resistance in the right front motor low circuit.
6. Checks for resistance in the right front motor.

Diagnostic Aids:

An "intermittent" malfunction in this DTC may result from a mechanical part of the system that sticks, binds, or slips.

The frequency of the malfunction can be checked by using the enhanced diagnostic function of the Tech 1 . If DTC 48 only occurs once and DTC 61 also occurs see DTC 61. If intermittent and enhanced diagnostics shows DTC 48 occurs during ABS, see DTC 61.

Depending on the frequency of the malfunction, a physical inspection of the mechanical parts suspected may be necessary.

DIAGNOSTIC CHART DTC 48 — 1995 CAVALIER AND SUNFIRE WITH TCS, CONTINUED

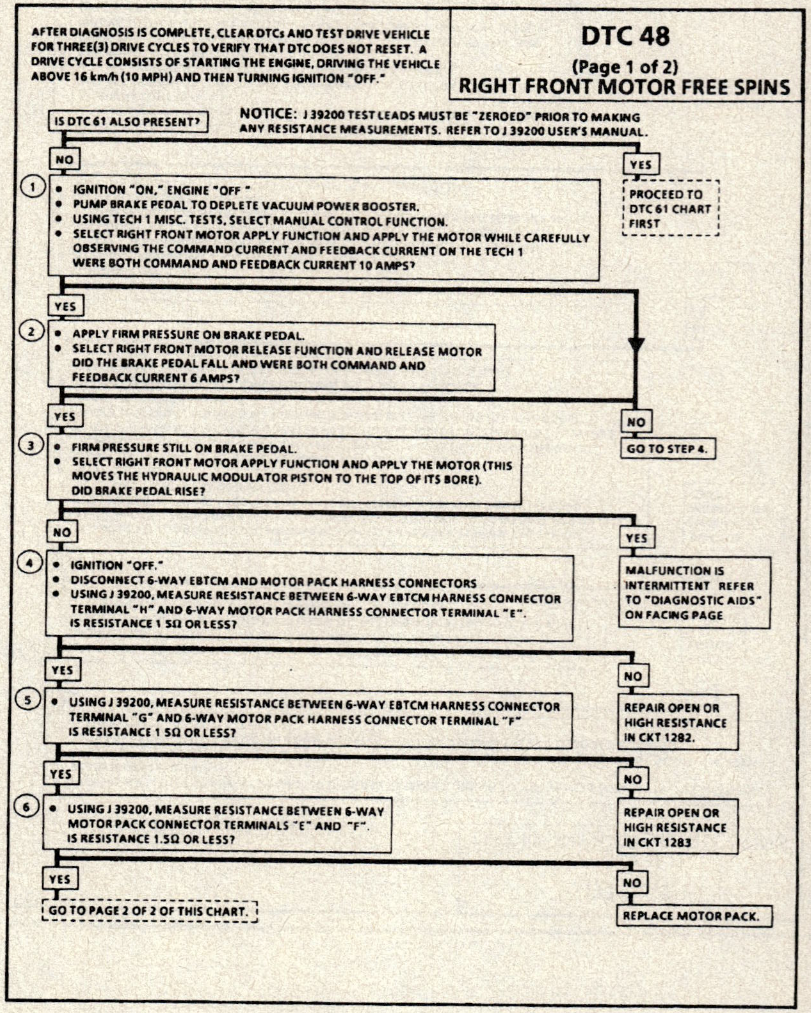

DIAGNOSTIC CHART DTC 48 — 1995 CAVALIER AND SUNFIRE WITH TCS, CONTINUED

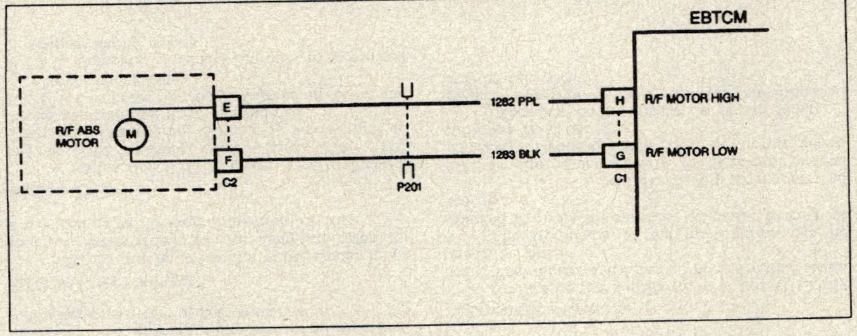

DTC 48
(Page 2 of 2)
RIGHT FRONT MOTOR FREE SPINS

Circuit Description:

This DTC is designed to detect a stripped nut or gear assembly during initialization. During the homing sequence, the piston should reach the top of the bore resulting in a stalled motor. If this does not occur, the motor must be spinning with little or no resistance, indicating a nut/screw or gear malfunction.

DTC Will Set When:

DTC 48 can only be set during initialization. If the feedback current is less than the command current for a specified period of time, a malfunction exists.

Action Taken:

A malfunction DTC is stored, ABS is disabled, and the "ABS" warning lamp is turned "ON."

DTC Chart Test Description:

Number(s) below refer to circled number(s) on the diagnostic chart.

7. Checks for a stripped motor pack gear.
8. Checks for a stripped hydraulic modulator gear.
9. Checks for proper operation of motor pack.
10. Checks for proper operation of the hydraulic modulator.
11. Ensures malfunction was not due to poor terminal contact.

Diagnostic Aids:

An "intermittent" malfunction in this DTC may result from a mechanical part of the system that sticks, binds, or slips.

The frequency of the malfunction can be checked by using the enhanced diagnostic function of the Tech 1 . If DTC 48 only occurs once and DTC 61 also occurs, see DTC 61. If intermittent and enhanced diagnostics shows DTC 48 occurs during ABS, see DTC 61.

Depending on the frequency of the malfunction, a physical inspection of the mechanical parts suspected may be necessary.

DIAGNOSTIC CHART DTC 48 — 1995 CAVALIER AND SUNFIRE WITH TCS, CONTINUED

DTC 48
(Page 2 of 2)
RIGHT FRONT MOTOR FREE SPINS

CONTINUED FROM PAGE 1 OF 2 OF THIS CHART.

(7)
- REMOVE HYDRAULIC MODULATOR/MASTER CYLINDER ASSEMBLY. (REFER TO "ON-VEHICLE SERVICE" IN THIS SECTION.)
- REMOVE GEAR COVER.
- CHECK FOR STRIPPED GEAR ON MOTOR PACK; THE MOTOR PACK IS THE UNIT WITH THE THREE SMALL GEARS. THE RIGHT FRONT GEAR IS CLOSEST TO THE BRAKE BOOSTER WHEN THE ASSEMBLY IS INSTALLED IN THE VEHICLE.

IS THE MOTOR PACK GEAR STRIPPED?

YES: REPLACE MOTOR PACK.

NO:

(8)
- CHECK FOR A STRIPPED GEAR ON THE HYDRAULIC MODULATOR; THE HYDRAULIC MODULATOR IS THE UNIT WITH THE THREE LARGE GEARS. THE RIGHT FRONT GEAR IS CLOSEST TO THE BRAKE BOOSTER WHEN THE ASSEMBLY IS INSTALLED IN THE VEHICLE.

IS THE HYDRAULIC MODULATOR GEAR STRIPPED?

YES: REPLACE HYDRAULIC MODULATOR GEAR.

NO:

(9)
- SECURELY POSITION HYDRAULIC MODULATOR/MASTER CYLINDER ASSEMBLY IN VEHICLE ALLOWING RIGHT FRONT GEAR SET OBSERVATION. BE CAREFUL NOT TO DAMAGE GEARS OR ALLOW THE HYDRAULIC MODULATOR/MASTER CYLINDER ASSEMBLY TO MOVE DURING TESTING.
- RECONNECT ELECTRICAL CONNECTORS.
- IGNITION "ON."
- USING TECH 1 MISC. TESTS, SELECT MANUAL CONTROL FUNCTION.
- SELECT RIGHT FRONT MOTOR APPLY AND APPLY THE MOTOR WHILE OBSERVING THE GEAR SET CLOSEST TO THE BOOSTER WHEN THE ASSEMBLY IS INSTALLED IN THE VEHICLE
- SELECT RIGHT FRONT MOTOR RELEASE FUNCTION AND RELEASE THE MOTOR WHILE STILL OBSERVING THE GEAR SET.
- SELECT RIGHT FRONT MOTOR APPLY FUNCTION AND APPLY THE MOTOR.

DID GEAR SET MOVE IN BOTH DIRECTIONS FOR AT LEAST ONE REVOLUTION?

NO: REPLACE MOTOR PACK

YES:

(10)
- USING TECH 1, SELECT "F4. MISC. TESTS" AND PERFORM GEAR TENSION RELIEF FUNCTION.
- IGNITION "OFF."
- SEPARATE MOTOR PACK FROM HYDRAULIC MODULATOR.
- GRASP THE RIGHT FRONT GEAR ON THE HYDRAULIC MODULATOR; THE HYDRAULIC MODULATOR IS THE UNIT WITH THE THREE LARGE GEARS. THE RIGHT FRONT GEAR IS CLOSEST TO THE BRAKE BOOSTER WHEN THE ASSEMBLY IS INSTALLED IN THE VEHICLE
- ATTEMPT TO ROTATE THE GEAR.

CAN THE GEAR BE ROTATED AT LEAST 8.5 FULL TURNS LOCK TO LOCK?

YES: REPLACE HYDRAULIC MODULATOR.

NO:

(11)
- INSPECT 6-WAY MOTOR PACK AND EBTCM CONNECTORS FOR CORROSION OR POOR TERMINAL CONTACT.

REPLACE TERMINALS THAT EXHIBIT POOR TERMINAL CONTACT OR CORROSION.
- INSTALL HYDRAULIC MODULATOR/MASTER CYLINDER ASSEMBLY.
- RECONNECT ALL CONNECTORS.
- DRIVE VEHICLE FOR THREE DRIVE CYCLES (A DRIVE CYCLE CONSISTS OF STARTING ENGINE, DRIVING OVER 16 km/h (10 MPH) AND THEN TURNING IGNITION "OFF").

DID DTC 48 SET IN THE LAST THREE DRIVE CYCLES?

YES: REPLACE EBTCM.

NO: MALFUNCTION WAS DUE TO A POOR CONNECTION.

DIAGNOSTIC CHART DTC 51 — 1995 CAVALIER AND SUNFIRE WITH TCS

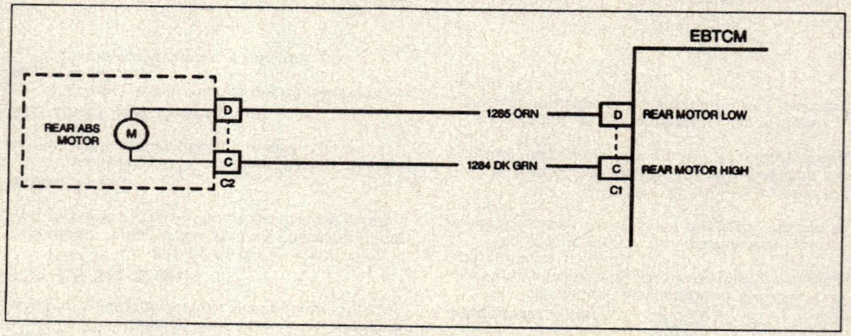

DTC 51
(Page 1 of 2)
REAR MOTOR FREE SPINS

Circuit Description:

This DTC is designed to detect a stripped nut or gear assembly during initialization. During the homing sequence, the piston should reach the top of the bore resulting in a stalled motor. If this does not occur, the motor must be spinning with little or no resistance, indicating a nut/screw or gear malfunction.

DTC Will Set When:

DTC 51 can be set during initialization. If the feedback current is less than the command current for a specified period of time, a malfunction exists.

Action Taken:

A malfunction DTC is stored, ABS is disabled, and the "ABS" warning lamp is turned "ON." DTC Chart Test Description: Number(s) below refer to circled number(s) on the diagnostic chart.

1 . Compares feedback current to command current during apply.
2 . Compares feedback current to command current during release.
3 . Moves the piston to the top of the bore.
4 . Checks for excessive resistance in the rear motor high circuit.
5 . Checks for excessive resistance in the rear motor low circuit.
6 . Checks for resistance in the rear motor.

Diagnostic Aids:

An "intermittent" malfunction in this DTC may result from a mechanical part of the system that sticks, binds, or slips.

The frequency of the malfunction can be checked by using the enhanced diagnostic function of the Tech 1

If DTC 51 only occurs once and DTC 64 also occurs, see DTC 64. If intermittent and enhanced diagnostics show DTC 51 occurs during ABS, see DTC 64.

Depending on the frequency of the failure, a physical inspection of the mechanical parts suspected may be necessary.

DIAGNOSTIC CHART DTC 51 — 1995 CAVALIER AND SUNFIRE WITH TCS, CONTINUED

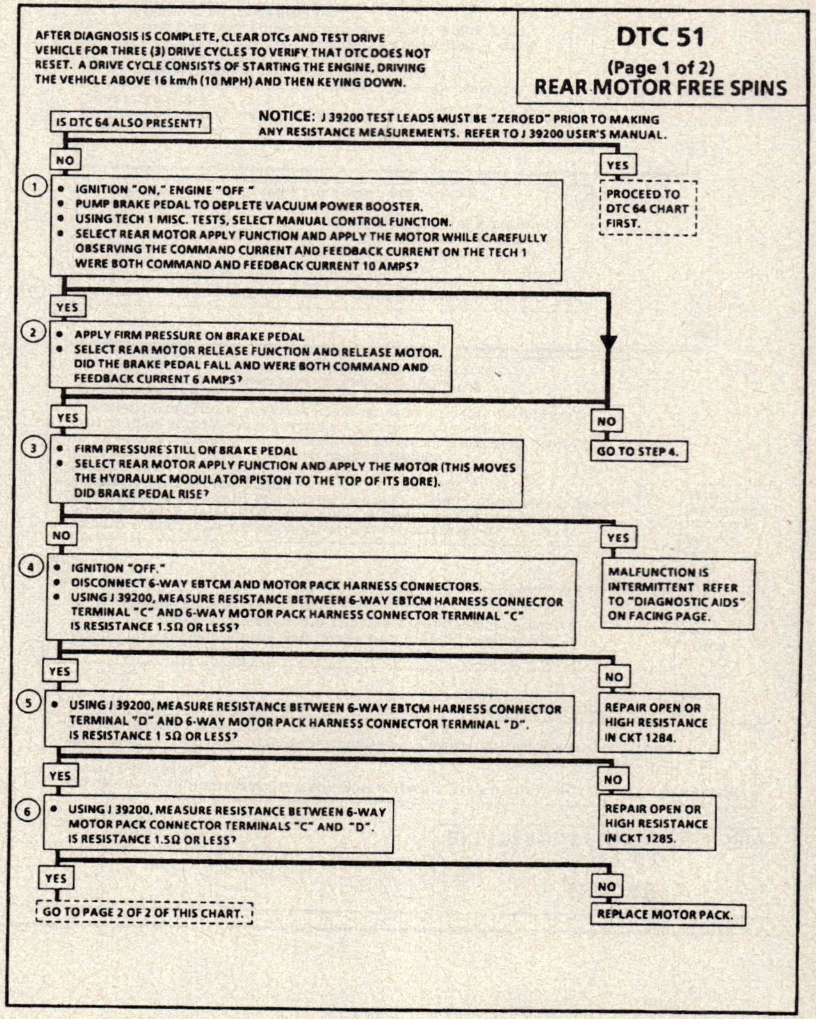

DIAGNOSTIC CHART DTC 51 — 1995 CAVALIER AND SUNFIRE WITH TCS, CONTINUED

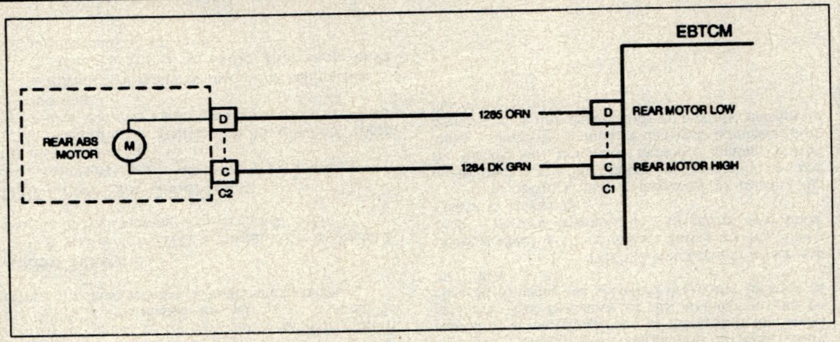

DTC 51
(Page 2 of 2)
REAR MOTOR FREE SPINS

Circuit Description:

This DTC is designed to detect a stripped nut or gear assembly during initialization. During the homing sequence, the piston should reach the top of the bore resulting in a stalled motor. If this does not occur, the motor must be spinning with little or no resistance, indicating a nut/screw or gear malfunction.

DTC Will Set When:

DTC 51 can be set during initialization. If the feedback current is less than the command current for a specified period of time, a malfunction exists.

Action Taken:

A malfunction DTC is stored, ABS is disabled, and the "ABS" warning lamp is turned "ON."

DTC Chart Test Description:

Number(s) below refer to circled number(s) on the diagnostic chart.

7. Checks for a stripped motor pack gear.
8. Checks for a stripped hydraulic modulator gear.
9. Checks for proper operation of motor pack.
10. Checks for proper operation of the hydraulic modulator.
11. Ensures malfunction was not due to poor terminal contact.

Diagnostic Aids:

An "intermittent" malfunction in this DTC may result from a mechanical part of the system that sticks, binds, or slips.

The frequency of the malfunction can be checked by using the enhanced diagnostic function of the Tech 1.

If DTC 51 only occurs once and DTC 64 also occurs, see DTC 64. If intermittent and enhanced diagnostics show DTC 51 occurs during ABS, see DTC 64.

Depending on the frequency of the malfunction, a physical inspection of the mechanical parts suspected may be necessary.

DIAGNOSTIC CHART DTC 51 — 1995 CAVALIER AND SUNFIRE WITH TCS, CONTINUED

DTC 51
(Page 2 of 2)
REAR MOTOR FREE SPINS

CONTINUED FROM PAGE 1 OF 2 OF THIS CHART.

(7)
- REMOVE HYDRAULIC MODULATOR/MASTER CYLINDER ASSEMBLY. (REFER TO "ON-VEHICLE SERVICE" IN THIS SECTION.)
- REMOVE GEAR COVER.
- CHECK FOR STRIPPED GEAR ON MOTOR PACK. THE MOTOR PACK IS THE UNIT WITH THE THREE SMALL GEARS THE REAR CHANNEL GEAR IS THE MIDDLE GEAR OF THE THREE
  IS THE MOTOR PACK GEAR STRIPPED?

YES → REPLACE MOTOR PACK

NO ↓

(8)
- CHECK FOR A STRIPPED GEAR ON THE HYDRAULIC MODULATOR; THE HYDRAULIC MODULATOR IS THE UNIT WITH THE THREE LARGE GEARS THE REAR CHANNEL GEAR IS THE MIDDLE GEAR OF THE THREE.
  IS THE HYDRAULIC MODULATOR GEAR STRIPPED?

YES → REPLACE HYDRAULIC MODULATOR GEAR.

NO ↓

(9)
- SECURELY POSITION HYDRAULIC MODULATOR/MASTER CYLINDER ASSEMBLY IN VEHICLE ALLOWING REAR CHANNEL GEAR SET OBSERVATION. BE CAREFUL NOT TO DAMAGE GEARS OR ALLOW THE HYDRAULIC MODULATOR/MASTER CYLINDER ASSEMBLY TO MOVE DURING TESTING.
- RECONNECT ELECTRICAL CONNECTORS
- IGNITION "ON."
- USING TECH 1 MISC. TESTS, SELECT MANUAL CONTROL FUNCTION.
- SELECT REAR MOTOR APPLY AND APPLY THE MOTOR WHILE OBSERVING THE MIDDLE GEAR SET.
- SELECT REAR MOTOR RELEASE FUNCTION AND RELEASE THE MOTOR WHILE STILL OBSERVING THE GEAR SET.
- SELECT REAR MOTOR APPLY FUNCTION AND APPLY THE MOTOR.
  DID GEAR SET MOVE IN BOTH DIRECTIONS FOR AT LEAST ONE REVOLUTION?

NO → REPLACE MOTOR PACK.

YES ↓

(10)
- USING TECH 1, SELECT "F4: MISC. TESTS" AND PERFORM GEAR TENSION RELIEF FUNCTION
- IGNITION "OFF"
- SEPARATE MOTOR PACK FROM HYDRAULIC MODULATOR.
- GRASP THE REAR CHANNEL GEAR ON THE HYDRAULIC MODULATOR; THE HYDRAULIC MODULATOR IS THE UNIT WITH THE THREE LARGE GEARS. THE REAR CHANNEL GEAR IS THE MIDDLE GEAR OF THE THREE.
- ATTEMPT TO ROTATE THE GEAR.
  CAN THE GEAR BE ROTATED AT LEAST 3.5 FULL TURNS LOCK TO LOCK?

YES → REPLACE HYDRAULIC MODULATOR.

NO ↓

(11)
- INSPECT 6-WAY MOTOR PACK AND EBTCM CONNECTORS FOR CORROSION OR POOR TERMINAL CONTACT.
  REPLACE TERMINALS THAT EXHIBIT POOR TERMINAL CONTACT OR CORROSION.
- INSTALL HYDRAULIC MODULATOR/MASTER CYLINDER ASSEMBLY
- RECONNECT ALL CONNECTORS.
- DRIVE VEHICLE FOR THREE DRIVE CYCLES (A DRIVE CYCLE CONSISTS OF STARTING ENGINE, DRIVING OVER 16 km/h (10 MPH) AND THEN TURNING IGNITION "OFF").
  DID DTC 51 SET IN THE LAST THREE DRIVE CYCLES?

YES → REPLACE EBTCM.

NO → MALFUNCTION WAS DUE TO A POOR CONNECTION.

DIAGNOSTIC CHART DTC 52 — 1995 CAVALIER AND SUNFIRE WITH TCS

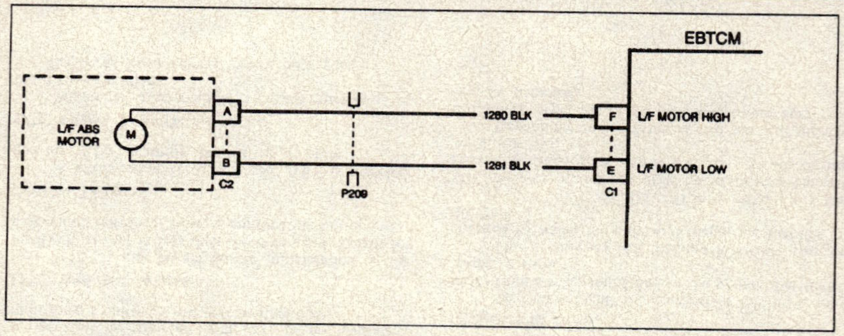

DTC 52
LEFT FRONT CHANNEL IN RELEASE TOO LONG

Circuit Description:

This DTC will diagnose a motor that is energized longer than expected. This could occur if a wheel speed sensor is faulty, the motor does not turn, the left front solenoid mechanically fails open, the brakes are dragging, or the motor wires are crossed.

DTC Will Set When:

DTC 52 can be set only during an ABS stop. If the EBTCM commands the left front channel in release for three seconds, a malfunction exists.

Action Taken:

A malfunction DTC is stored, ABS is disabled, and the "ABS" warning lamp is turned "ON."

DTC Chart Test Description:

Number(s) below refer to circled number(s) on the diagnostic chart.

1. Identifies a malfunction in a wheel speed sensor that may cause the system to be in release too long.
2. Identifies a motor or modulator malfunction.
3. Checks for a solenoid that may have mechanically failed open.
4. Checks for a solenoid that may have failed closed.
5. Isolates the malfunction to either the solenoid or the hydraulic modulator.
6. Checks for proper operation of hydraulic modulator.

Diagnostic Aids:

An "intermittent" malfunction is most likely caused by a mechanical part of the system that sticks or binds. The frequency of the malfunction can be checked by using the enhanced diagnostic function of the "Tech 1,

DTC 52 may occur if on ice and steering wheel is turned to lock during braking. Using Tech 1 perform hydraulic test to ensure total brake system is functional.

Any circuitry, that is suspected as causing the intermittent complaint, should be thoroughly checked for backed out terminals, improper mating, broken locks, improperly formed or damaged terminals, poor terminal to wiring connections or physical damage to the wiring harness.

DIAGNOSTIC CHART DTC 52 — 1995 CAVALIER AND SUNFIRE WITH TCS, CONTINUED

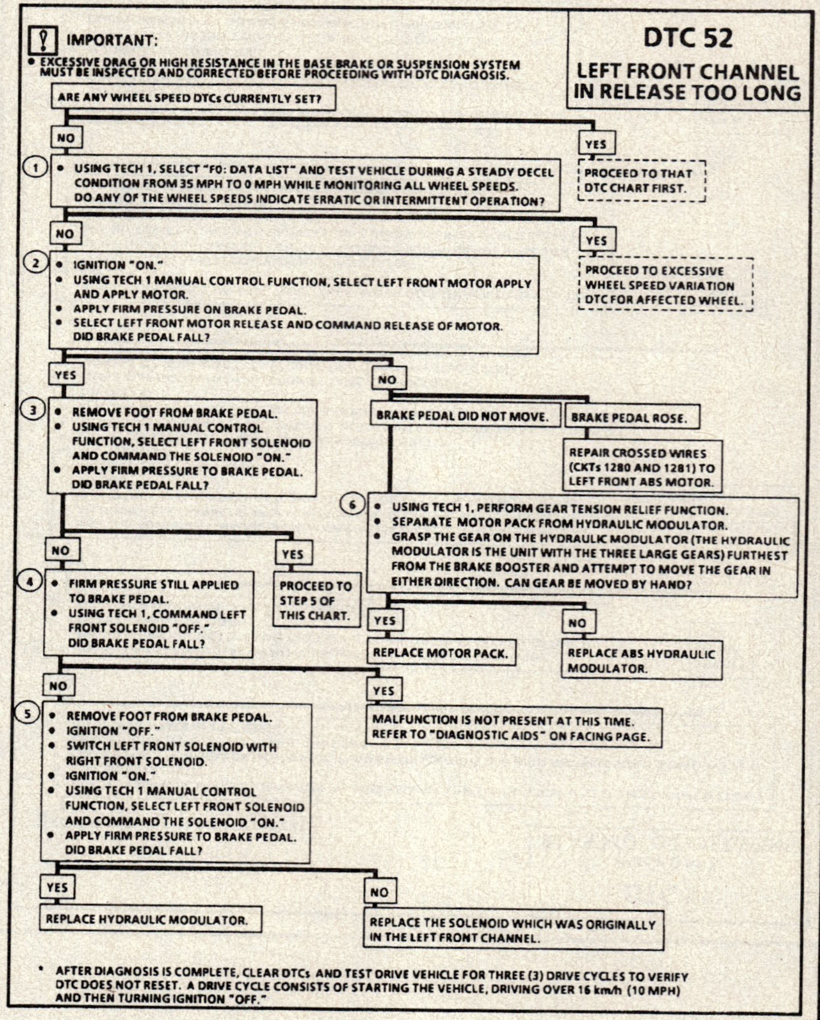

DIAGNOSTIC CHART DTC 53 — 1995 CAVALIER AND SUNFIRE WITH TCS

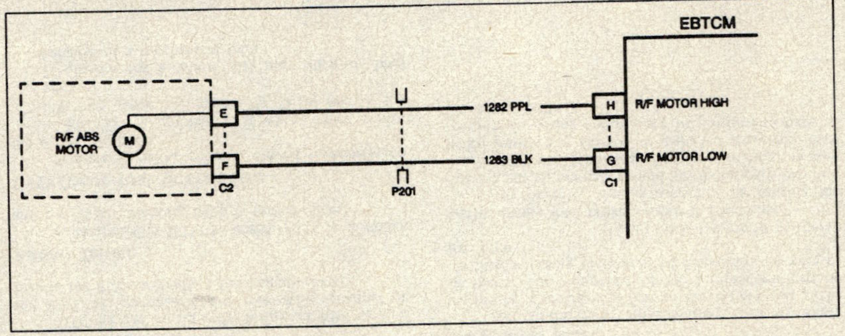

DTC 53
RIGHT FRONT CHANNEL IN RELEASE TOO LONG

Circuit Description:

This DTC will diagnose a motor that is energized longer than expected. This could occur if a wheel speed sensor is faulty, the motor does not turn, the right front solenoid mechanically fails open, the brakes are dragging, or the motor wires are crossed.

DTC Will Set When:

DTC 53 can be set only during an ABS stop. If the EBTCM commands the right front channel in release for three seconds, a malfunction exists.

Action Taken:

A malfunction DTC is stored, ABS is disabled, and the "ABS" warning lamp is turned "ON."

DTC Chart Test Description:

Number(s) below refer to circled number(s) on the diagnostic chart.

1. Identifies a malfunction in a wheel speed sensor that may cause the system to be in release too long.
2. Identifies a motor or modulator malfunction.
3. Checks for a solenoid that may have mechanically failed open.
4. Checks for a solenoid that may have failed closed.
5. Isolates the malfunction to either the solenoid or the hydraulic modular.
6. Checks for proper operation of hydraulic modulator.

Diagnostic Aids:

An "intermittent" malfunction is most likely caused by a mechanical part of the system that sticks or binds. The frequency of the malfunction can be checked by using the enhanced diagnostic function of the Tech 1

DTC 53 may occur on ice if steering wheel is turned to lock during braking. Using Tech 1 perform hydraulic test to ensure total brake system is functional.

Any circuitry, that is suspected as causing the intermittent complaint, should be thoroughly checked for backed out terminals, improper mating, broken locks, improperly formed or damaged terminals, poor terminal to wiring connections or physical damage to the wiring harness.

DIAGNOSTIC CHART DTC 53 — 1995 CAVALIER AND SUNFIRE WITH TCS, CONTINUED

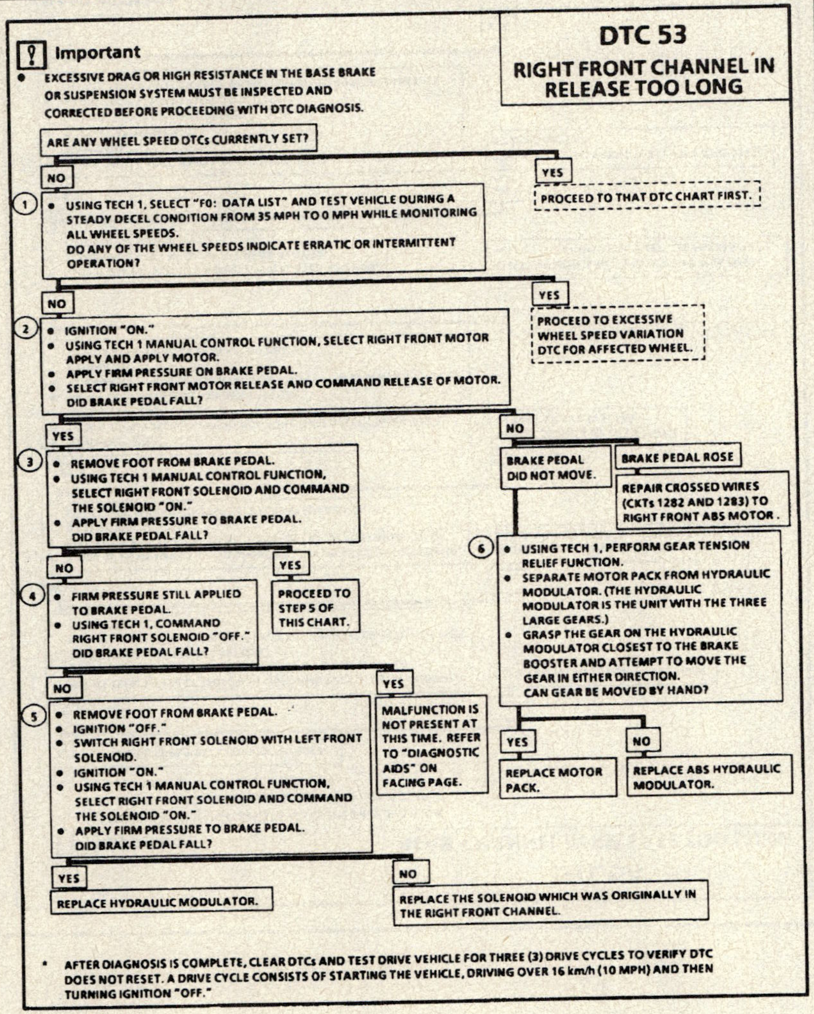

DIAGNOSTIC CHART DTC 54 — 1995 CAVALIER AND SUNFIRE WITH TCS

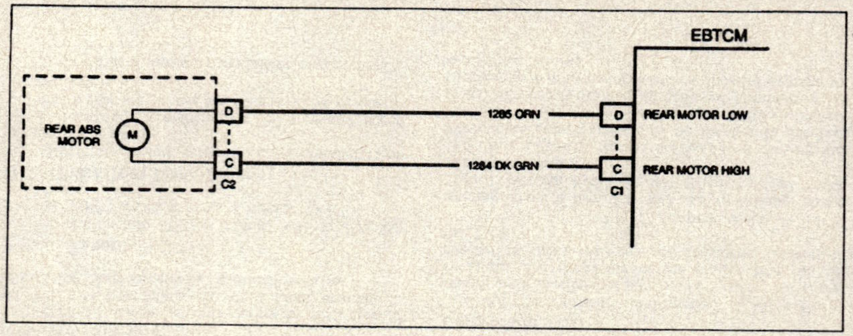

DTC 54
REAR CHANNEL IN RELEASE TOO LONG

Circuit Description:

This DTC will diagnose a motor that is energized longer than expected. This could occur if a wheel speed sensor is faulty, the motor does not turn, the brakes are dragging, or the motor wires are crossed.

DTC Will Set When:

DTC 54 can be set only during an ABS stop. If the EBTCM commands the rear ABS channel in release for three seconds, a malfunction exists.

Action Taken:

A malfunction DTC is stored, ABS is disabled, and the "ABS" warning lamp is turned "ON."

DTC Chart Test Description:

Number(s) below refer to circled number(s) on the diagnostic chart.

1. Identifies a problem in a wheel speed sensor that may cause the system to be in release too long.
2. Checks for a wheel that may stick or bind because of a mechanical fault.
3. Checks to see if the motor is capable of moving and applying the hydraulic piston for the rear wheels.
4. Ensures that the motor wiring is not crossed.
5. Isolates the malfunction of a "no-apply" situation to either the motor pack or ABS hydraulic modulator.

Diagnostic Aids:

An "intermittent" malfunction is most likely caused by a mechanical part of the system that sticks or binds. The frequency of the malfunction can be checked by using the enhanced diagnostic function of the Tech 1

Using Tech 1 perform hydraulic test to ensure total brake system is functional.

Any circuitry, that is suspected as causing the intermittent complaint, should be thoroughly checked for backed out terminals, improper mating, broken locks, improperly formed or damaged terminals, poor terminal to wiring connections or physical damage to the wiring harness.

DIAGNOSTIC CHART DTC 54 — 1995 CAVALIER AND SUNFIRE WITH TCS, CONTINUED

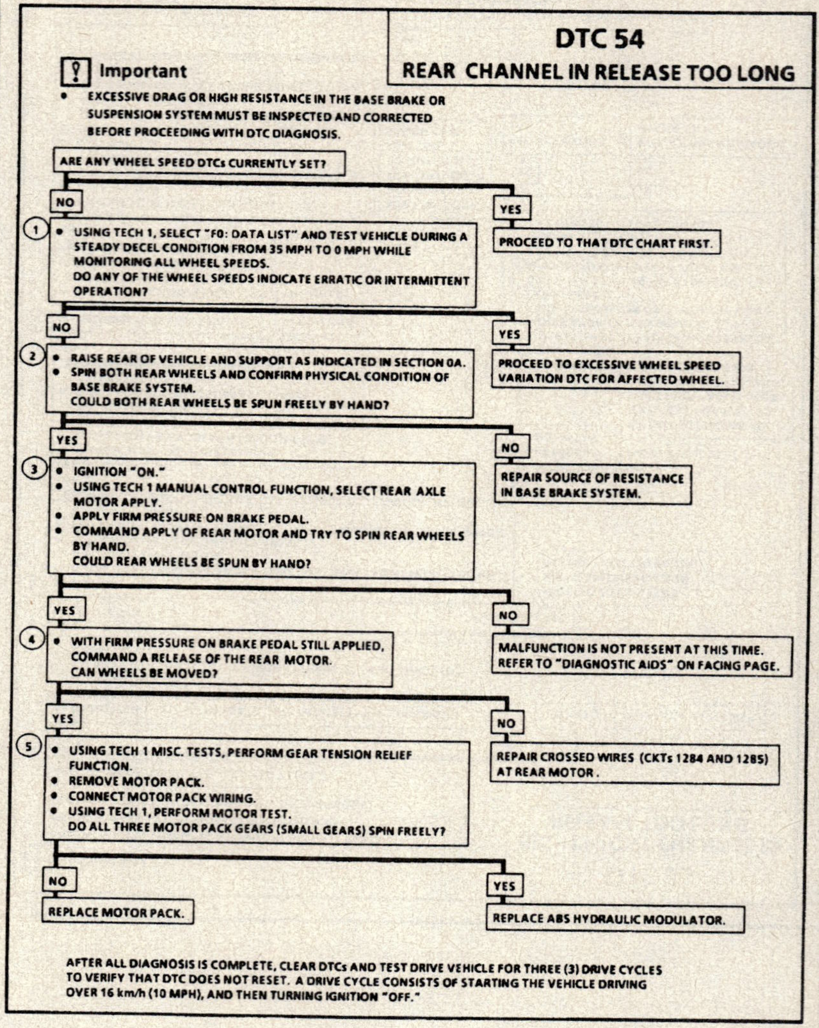

DIAGNOSTIC CHART DTC 55 — 1995 CAVALIER AND SUNFIRE WITH TCS

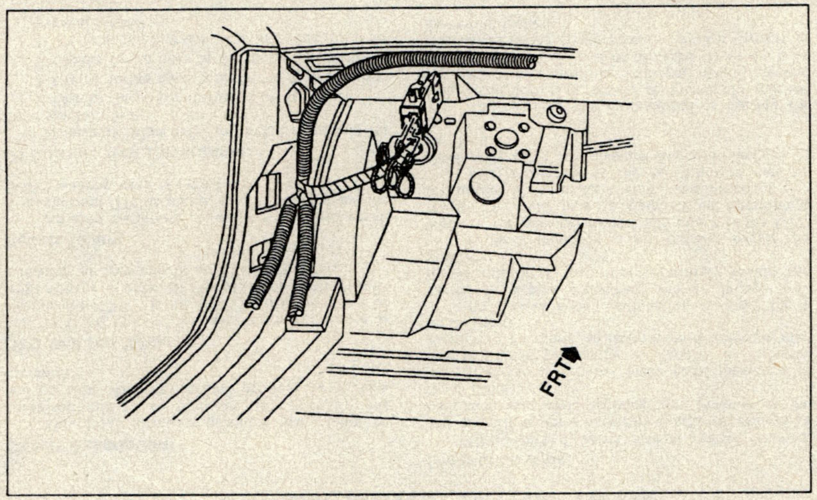

DTC 55
EBTCM MALFUNCTION

Circuit Description:
This DTC identifies a malfunction within the EBTCM. It also ensures the cause of the malfunction is not a result of a malfunction with the electronic brake control relay.

DTC Will Set When:
DTC 55 is set when an internal EBTCM malfunction exists.

Action Taken:
A malfunction DTC is stored, ABS is disabled and the "ABS" warning lamp is turned "ON."

DTC Chart Test Description:
Number(s) below refer to circled number(s) on the diagnostic chart.
1. Checks to see if there are any other malfunctions present.
2. This checks to see if the malfunction is still present.

Diagnostic Aids:
An "intermittent" malfunction is most likely caused by a poor connection, rubbed through wire insulation, or a wire that is broken inside the insulation.
The frequency of the malfunction can be checked by using the enhanced diagnostic function of the Tech 1

Any circuitry, that is suspected as causing the intermittent complaint, should be thoroughly checked for backed out terminals, improper mating, broken locks, improperly formed or damaged terminals, poor terminal to wiring connections or physical damage to the wiring harness.

DIAGNOSTIC CHART DTC 55 — 1995 CAVALIER AND SUNFIRE WITH TCS, CONTINUED

DTC 55
EBTCM MALFUNCTION

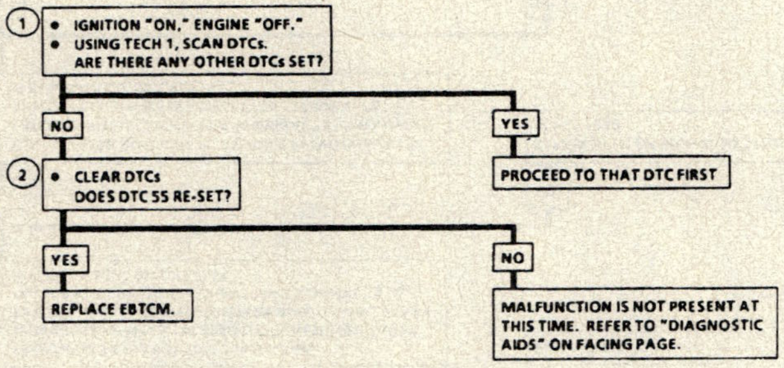

• AFTER DIAGNOSIS IS COMPLETE, CLEAR DTCs AND TEST DRIVE VEHICLE FOR THREE (3) DRIVE CYCLES TO VERIFY DTC DOES NOT RESET A DRIVE CYCLE CONSISTS OF STARTING THE VEHICLE, DRIVING OVER 16 km/h (10 MPH) AND THEN TURNING IGNITION "OFF."

DIAGNOSTIC CHART DTC 56 — 1995 CAVALIER AND SUNFIRE WITH TCS

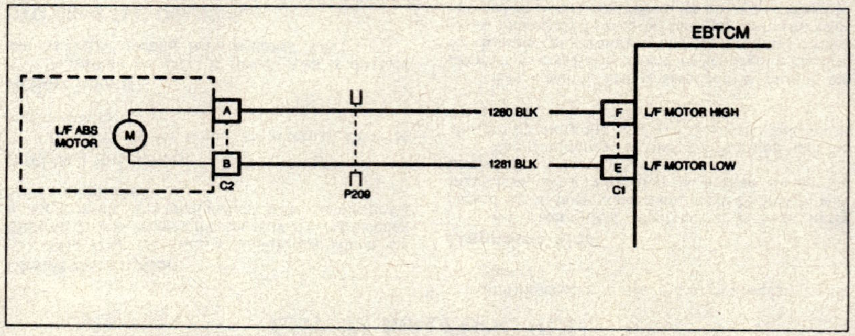

DTC 56
LEFT FRONT MOTOR CIRCUIT OPEN

Circuit Description:

This DTC identifies a motor that cannot be energized due to an open in its circuitry. This malfunction will not allow proper front ABS operation.

DTC Will Set When:

DTC 56 can be set only when the motor is commanded "OFF." If the EBTCM detects an out of range voltage in either of the left front motor circuits indicating an open circuit, a malfunction exists.

Action Taken:

An open motor will not activate when requested. A malfunction DTC is stored, ABS is disabled and the "ABS" warning lamp is turned "ON."

DTC Chart Test Description:

Number(s) below refer to circled number(s) on the diagnostic chart.

1. Checks for proper resistance of the motor.
2. Checks for an open in the motor high circuitry.
3. Checks for an open in the motor low circuitry.
4. Ensures malfunction was not due to poor terminal contact.

Diagnostic Aids:

Using Tech 1, select manual control function, and exercise motor movement of affected channel in both directions while applying light pressure on the brake pedal.

If erratic or "jumpy" brake pedal movement is detected while performing an "apply" or "release" function of the motor, an intermittent malfunction may be indicated.

An "intermittent" malfunction is most likely caused by a poor connection, rubbed through wire insulation, or a wire that is broken inside the insulation.

If the malfunction is not current, wiggle the wires of the affected channel and check if the DTC resets. This will help to pinpoint an intermittent malfunction in the motor circuitry or connections.

The frequency of the malfunction can be checked by using the enhanced diagnostic function of the Tech 1.

Any circuitry, that is suspected as causing the intermittent complaint, should be thoroughly checked for backed out terminals, improper mating, broken locks, improperly formed or damaged terminals, poor terminal to wiring connections or physical damage to the wiring harness.

DIAGNOSTIC CHART DTC 56 — 1995 CAVALIER AND SUNFIRE WITH TCS, CONTINUED

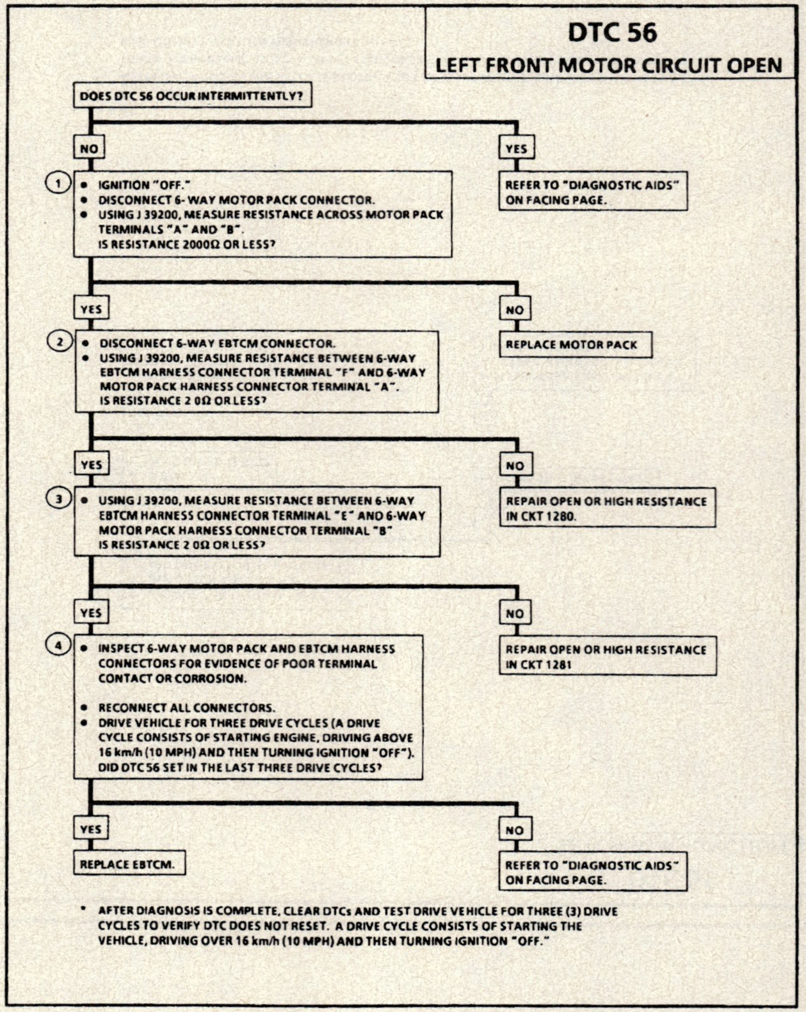

DIAGNOSTIC CHART DTC 57 — 1995 CAVALIER AND SUNFIRE WITH TCS

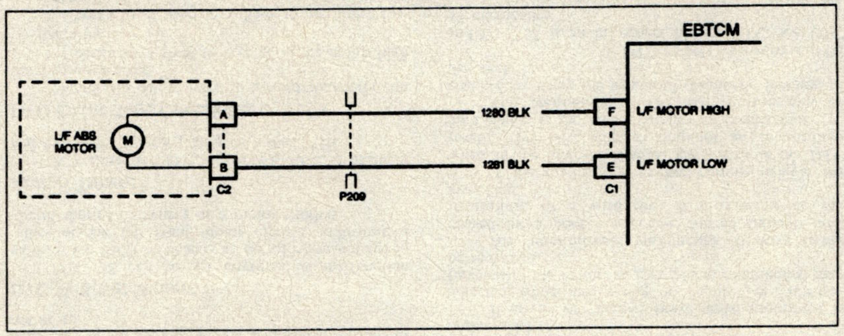

DTC 57
LEFT FRONT MOTOR CIRCUIT SHORTED TO GROUND

Circuit Description:

This DTC identifies a motor circuit that is shorted to ground. An uncontrollable motor will not allow controlled ABS operation.

DTC Will Set When:

DTC 57 can be set anytime. A malfunction exists if the EBTCM detects an out of range voltage in either of the left front motor circuits indicating a circuit shorted to ground.

Action Taken:

A malfunction DTC is stored, ABS is disabled and the "ABS" warning lamp is turned "ON."

DTC Chart Test Description:

Number(s) below refer to circled number(s) on the diagnostic chart.

1. Checks for a short to ground in the motor high circuitry.
2. Checks for a short to ground in the motor low circuitry.
3. Checks for a motor that is internally shorted to ground.
4. Ensures malfunction was not due to physical damage of the circuitry.

Diagnostic Aids:

Using Tech 1, select manual control function, and exercise motor movement of affected channel in both directions while applying light pressure on the brake pedal.

If erratic or "jumpy" brake pedal movement is detected while performing an "apply" or "release" function of the motor, an intermittent malfunction may be indicated.

An "intermittent" malfunction is most likely caused by a poor connection, rubbed through wire insulation, or a wire that is broken inside the insulation.

If the malfunction is not current, wiggle the wires of the affected channel and check if the DTC resets. This will help to pinpoint an intermittent malfunction in the motor circuitry or connections.

The frequency of the malfunction can be checked by using the enhanced diagnostic function of the Tech 1,

Any circuitry, that is suspected as causing the intermittent complaint, should be thoroughly checked for backed out terminals, improper mating, broken locks, improperly formed or damaged terminals, poor terminal to wiring connections or physical damage to the wiring harness.

DIAGNOSTIC CHART DTC 57 — 1995 CAVALIER AND SUNFIRE WITH TCS, CONTINUED

DTC 57
LEFT FRONT MOTOR CIRCUIT SHORTED TO GROUND

DOES DTC 57 OCCUR INTERMITTENTLY?

- YES → REFER TO "DIAGNOSTIC AIDS" ON FACING PAGE
- NO →

(1)
- IGNITION "OFF"
- DISCONNECT 6-WAY CONNECTORS AT MOTOR PACK AND EBTCM.
- USING J 39200, MEASURE RESISTANCE BETWEEN GROUND AND 6-WAY MOTOR PACK HARNESS CONNECTOR TERMINAL "A"

DOES J 39200 DISPLAY RESISTANCE AS "OL" (INFINITE)?

- NO → REPAIR SHORT TO GROUND IN CKT 1280.
- YES →

(2)
- USING J 39200, MEASURE RESISTANCE BETWEEN GROUND AND 6-WAY MOTOR PACK HARNESS CONNECTOR TERMINAL "B"

DOES J 39200 DISPLAY RESISTANCE AS "OL" (INFINITE)?

- NO → REPAIR SHORT TO GROUND IN CKT 1281.
- YES →

(3)
- USING J 39200, MEASURE RESISTANCE BETWEEN GROUND AND TERMINAL "A" OF THE MOTOR PACK.

DOES J 39200 DISPLAY RESISTANCE AS "OL" (INFINITE)?

- NO → REPLACE MOTOR PACK
- YES →

(4)
- INSPECT CKTs 1280 AND 1281 AND 6-WAY CONNECTORS FOR DAMAGE WHICH MAY RESULT IN A SHORT TO GROUND WITH ALL CONNECTORS CONNECTED. REPAIR DAMAGE IF EVIDENT.
- RECONNECT ALL CONNECTORS.
- DRIVE VEHICLE FOR THREE DRIVE CYCLES [A DRIVE CYCLE CONSISTS OF STARTING ENGINE, DRIVING ABOVE 16 km/h (10 MPH) AND THEN TURNING IGNITION "OFF"].

DID DTC 57 SET IN THE LAST THREE DRIVE CYCLES?

- YES → REPLACE EBTCM.
- NO → REFER TO "DIAGNOSTIC AIDS" ON FACING PAGE.

• AFTER DIAGNOSIS IS COMPLETE, CLEAR DTCs AND TEST DRIVE VEHICLE FOR THREE (3) DRIVE CYCLES TO VERIFY DTC DOES NOT RESET A DRIVE CYCLE CONSISTS OF STARTING THE VEHICLE, DRIVING OVER 16 km/h (10 MPH) AND THEN KEYING DOWN.

DIAGNOSTIC CHART DTC 58 — 1995 CAVALIER AND SUNFIRE WITH TCS

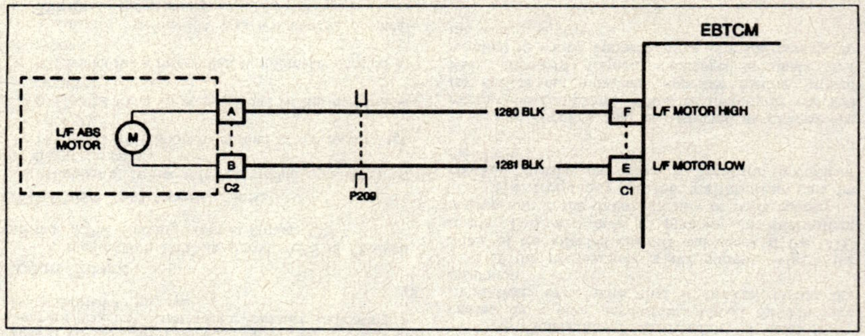

DTC 58
LEFT FRONT MOTOR CIRCUIT SHORTED TO BATTERY OR MOTOR SHORTED

Circuit Description:

This DTC identifies a motor circuit that is shorted to battery or a motor that has low or no resistance. This malfunction will not allow the motor to be controlled at the commanded current rate or will cause the motor to turn in the opposite direction, or not at all.

DTC Will Set When:

DTC 58 can be set anytime. A malfunction exists if the EBTCM detects an out of range voltage in either of the left front motor circuits indicating a circuit shorted to battery or a motor shorted.

Action Taken:

A malfunction DTC is stored, ABS is disabled and the "ABS" warning lamp is turned "ON."

DTC Chart Test Description:

Number(s) below refer to circled number(s) on the diagnostic chart.

1. Checks for a short to voltage in the motor high circuitry.
2. Checks for a short to voltage in the motor low circuitry.
3. Checks for a short between high and low motor circuits.
4. Checks for a motor that is internally shorted.
5. Checks for a short between the EMB and motor circuits.
6. Checks for an internal motor pack short between the EMB and motor.
7. Ensures malfunction was not due to physical damage of the circuitry.

Diagnostic Aids:

Using Tech 1, select manual control function, and exercise motor movement of affected channel in both directions while applying light pressure on the brake pedal.

If erratic or "jumpy" brake pedal movement is detected while performing an "apply" or "release" function of the motor, an intermittent malfunction may be indicated.

An "intermittent" malfunction is most likely caused by a poor connection, rubbed through wire insulation, or a wire that is broken inside the insulation.

If the malfunction is not current, wiggle the wires of the affected channel and check if the DTC resets. This will help to pinpoint an intermittent malfunction in the motor circuitry or connections.

The frequency of the malfunction can be checked by using the enhanced diagnostic function of the Tech 1.

If DTC 38 sets as a current DTC and DTC 58 is an infrequent history DTC, see DTC 38 diagnostics.

Any circuitry, that is suspected as causing the intermittent complaint, should be thoroughly checked for backed out terminals, improper mating, broken locks, improperly formed or damaged terminals, poor terminal to wiring connections or physical damage to the wiring harness.

DIAGNOSTIC CHART DTC 58 — 1995 CAVALIER AND SUNFIRE WITH TCS, CONTINUED

DTC 58
LEFT FRONT MOTOR CIRCUIT SHORTED TO BATTERY OR MOTOR SHORTED

NOTICE: J 39200 TEST LEADS MUST BE "ZEROED" PRIOR TO MAKING ANY RESISTANCE MEASUREMENTS. REFER TO J 39200 USER'S MANUAL.

DOES DTC 58 OCCUR INTERMITTENTLY?

- YES → REFER TO "DIAGNOSTIC AIDS" ON FACING PAGE.
- NO ↓

(1) NOTICE: THIS STEP MAY CAUSE ADDITIONAL DTCs TO SET.
- IGNITION "OFF."
- DISCONNECT 6-WAY CONNECTORS AT MOTOR PACK AND EBTCM.
- IGNITION "ON."
- USING J 39200, MEASURE VOLTAGE BETWEEN GROUND AND 6-WAY MOTOR PACK HARNESS CONNECTOR TERMINAL "A".

IS VOLTAGE 1.0V OR LESS?

- NO → REPAIR SHORT TO VOLTAGE IN CKT 1280
- YES ↓

(2)
- USING J 39200, MEASURE VOLTAGE BETWEEN GROUND AND 6-WAY MOTOR PACK HARNESS CONNECTOR TERMINAL "B".

IS VOLTAGE 1.0V OR LESS?

- NO → REPAIR SHORT TO VOLTAGE IN CKT 1281
- YES ↓

(3)
- IGNITION "OFF."
- USING J 39200, MEASURE RESISTANCE BETWEEN 6-WAY MOTOR PACK HARNESS CONNECTOR TERMINALS "A" AND "B".

DOES J 39200 DISPLAY RESISTANCE AS "OL" (INFINITE)?

- NO → REPAIR SHORT BETWEEN CKTs 1280 AND 1281.
- YES ↓

(4)
- USING J 39200, MEASURE RESISTANCE BETWEEN MOTOR PACK TERMINALS "A" AND "B".

IS RESISTANCE 0.4Ω OR GREATER?

- NO → REPLACE MOTOR PACK.
- YES ↓

(5)
- DISCONNECT 3-WAY MOTOR PACK CONNECTOR.
- USING J 39200, MEASURE RESISTANCE BETWEEN 3-WAY MOTOR PACK HARNESS CONNECTOR TERMINAL "B" AND 6-WAY MOTOR PACK HARNESS CONNECTOR TERMINAL "A" THEN "B"

DID J 39200 DISPLAY RESISTANCE AS "OL" (INFINITE) FOR EACH MEASUREMENT?

- NO → REPAIR SHORT BETWEEN CKT 855/1633 AND MOTOR CIRCUIT WHICH MEASURED RESISTANCE OTHER THAN "OL" (INFINITE).
- YES ↓

(6)
- USING J 39200, MEASURE RESISTANCE BETWEEN 3-WAY MOTOR PACK CONNECTOR TERMINAL "B" AND 6-WAY MOTOR PACK CONNECTOR TERMINAL "A"

DID J 39200 DISPLAY "OL" (INFINITE)?

- NO → REPLACE MOTOR PACK
- YES ↓

(7)
- INSPECT CKTs 1280 AND 1281 AND 6-WAY CONNECTORS FOR DAMAGE WHICH MAY RESULT IN A SHORT TO VOLTAGE WITH ALL CONNECTORS CONNECTED
- REPAIR DAMAGE IF EVIDENT
- RECONNECT ALL CONNECTORS
- DRIVE VEHICLE FOR THREE DRIVE CYCLES (A DRIVE CYCLE CONSISTS OF STARTING ENGINE, DRIVE ABOVE 16 km/h (10 MPH) AND THEN TURNING IGNITION "OFF")

DID DTC 58 SET IN THE LAST THREE DRIVE CYCLES?

- YES → REPLACE EBTCM.
- NO → REFER TO "DIAGNOSTIC AIDS" ON FACING PAGE.

- AFTER DIAGNOSIS IS COMPLETE, CLEAR DTCs AND TEST DRIVE VEHICLE FOR THREE (3) DRIVE CYCLES TO VERIFY DTC DOES NOT RESET. A DRIVE CYCLE CONSISTS OF STARTING THE VEHICLE, DRIVING OVER 16 km/h (10 MPH) AND THEN TURNING IGNITION "OFF."

DIAGNOSTIC CHART DTC 61 — 1995 CAVALIER AND SUNFIRE WITH TCS

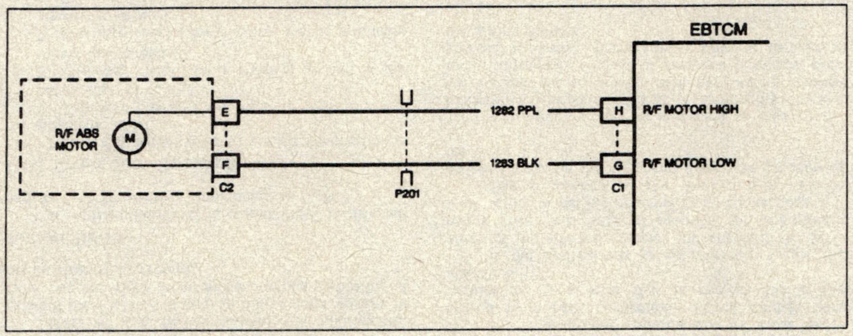

DTC 61
RIGHT FRONT MOTOR CIRCUIT OPEN

Circuit Description:

This DTC identifies a motor that cannot be energized due to an open in its circuitry. This malfunction will not allow proper front ABS operation.

DTC Will Set When:

DTC 61 can be set only when the motor is commanded "OFF." A malfunction exists if the EBTCM detects an out of range voltage in either of the right front motor circuits indicating an open circuit.

Action Taken:

An open motor will not activate when requested. A malfunction DTC is stored, ABS is disabled and the "ABS" warning lamp is turned "ON."

DTC Chart Test Description:

Number(s) below refer to circled number(s) on the diagnostic chart.

1. Checks for proper resistance of the motor.
2. Checks for an open in the motor high circuitry.
3. Checks for an open in the motor low circuitry.
4. Ensures malfunction was not due to poor terminal contact.

Diagnostic Aids:

Using Tech 1, select manual control function, and exercise motor movement of affected channel in both directions while applying light pressure on the brake pedal.

If erratic or "jumpy" brake pedal movement is detected while performing an "apply" or "release" function of the motor, an intermittent malfunction may be indicated.

An "intermittent" malfunction is most likely caused by a poor connection, rubbed through wire insulation, or a wire that is broken inside the insulation.

If the malfunction is not current, wiggle the wires of the affected channel and check if the DTC resets. This will help to pinpoint an intermittent malfunction in the motor circuitry or connections.

The frequency of the malfunction can be checked by using the enhanced diagnostic function of the Tech 1

Any circuitry, that is suspected as causing the intermittent complaint, should be thoroughly checked for backed out terminals, improper mating, broken locks, improperly formed or damaged terminals, poor terminal to wiring connections or physical damage to the wiring harness.

DIAGNOSTIC CHART DTC 61 — 1995 CAVALIER AND SUNFIRE WITH TCS, CONTINUED

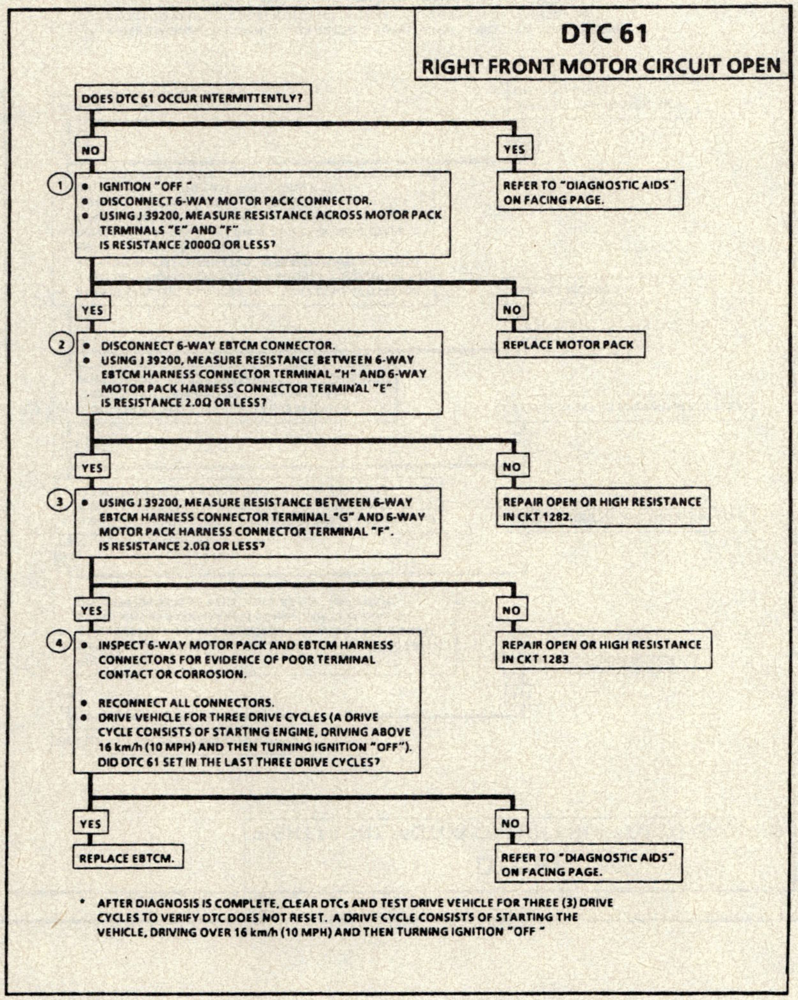

DIAGNOSTIC CHART DTC 62 — 1995 CAVALIER AND SUNFIRE WITH TCS

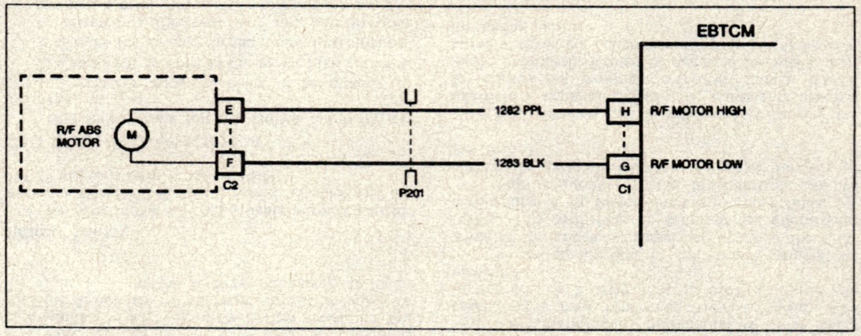

DTC 62
RIGHT FRONT MOTOR CIRCUIT SHORTED TO GROUND

Circuit Description:

This DTC identifies a motor circuit that is shorted to ground. This malfunction will not allow the motor to be controlled at the commanded current rate or will cause the motor driver circuit to allow current directly to ground.

DTC Will Set When:

DTC 62 can be set anytime. A malfunction exists if the EBTCM detects an out of range voltage in either of the right front motor circuits indicating a circuit shorted to ground.

Action Taken:

A malfunction DTC is stored, ABS is disabled and the "ABS" warning lamp is turned "ON."

DTC Chart Test Description:

Number(s) below refer to circled number(s) on the diagnostic chart.

1. This step checks for a short to ground in the motor high circuitry.
2. This step checks for a short to ground in the motor low circuitry.
3. This step checks for a motor that is internally shorted to ground.
4. Ensures malfunction was not due to physical damage of the circuitry.

Diagnostic Aids:

Using Tech 1, select manual control function, and exercise motor movement of affected channel in both directions while applying light pressure on the brake pedal.

If erratic or "jumpy" brake pedal movement is detected while performing an "apply" or "release" function of the motor, an intermittent malfunction may be indicated.

An "intermittent" malfunction is most likely caused by a poor connection, rubbed through wire insulation, or a wire that is broken inside the insulation.

If the malfunction is not current, wiggle the wires of the affected channel and check if the DTC resets. This will help to pinpoint an intermittent malfunction in the motor circuitry or connections.

The frequency of the malfunction can be checked by using the enhanced diagnostic function of the Tech 1

Any circuitry, that is suspected as causing the intermittent complaint, should be thoroughly checked for backed out terminals, improper mating, broken locks, improperly formed or damaged terminals, poor terminal to wiring connections or physical damage to the wiring harness.

DIAGNOSTIC CHART DTC 62 — 1995 CAVALIER AND SUNFIRE WITH TCS, CONTINUED

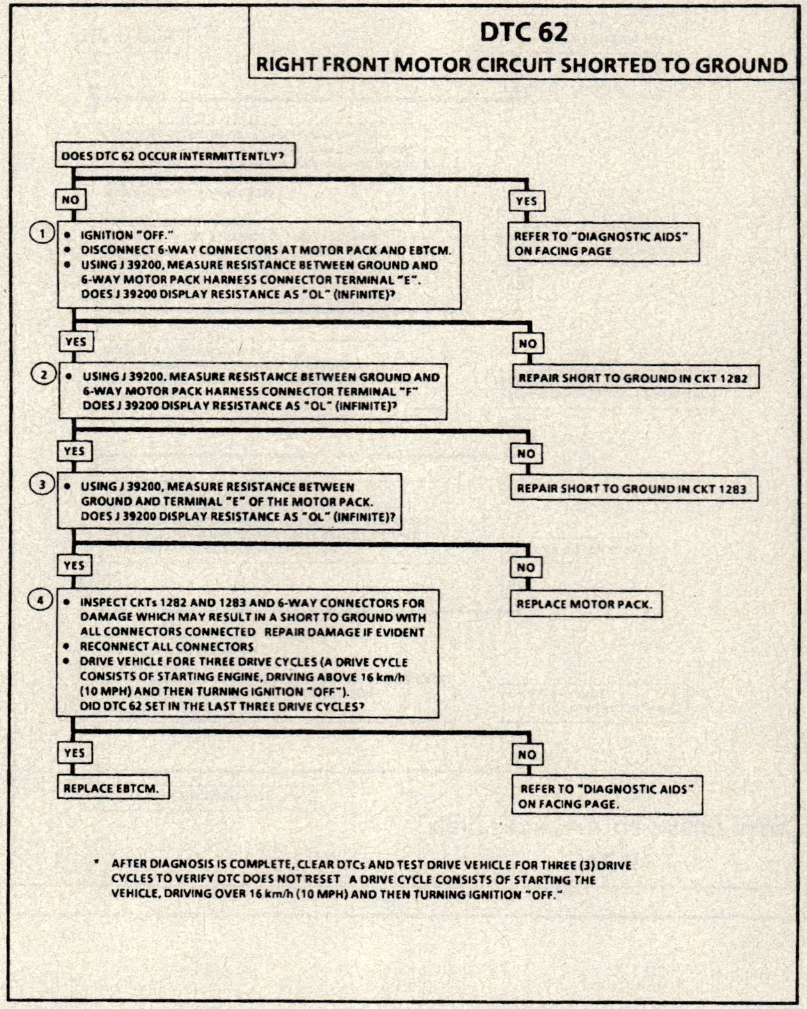

DIAGNOSTIC CHART DTC 63 — 1995 CAVALIER AND SUNFIRE WITH TCS

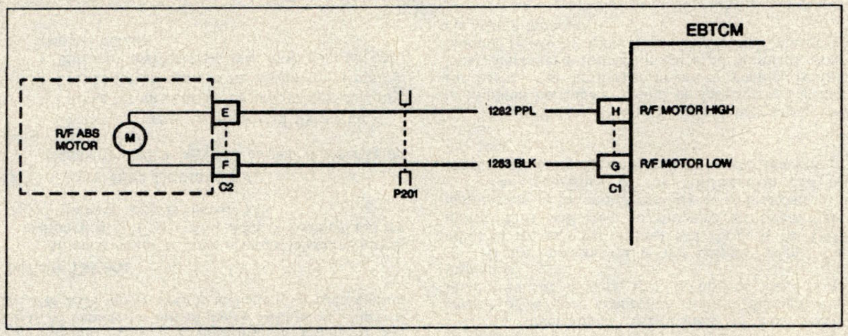

DTC 63
RIGHT FRONT MOTOR CIRCUIT SHORTED TO BATTERY OR MOTOR SHORTED

Circuit Description:

This DTC identifies a motor circuit that is shorted to battery or a motor that has low or no resistance. This malfunction will not allow the motor to be controlled at the commanded current rate or will cause the motor to turn in the opposite direction or not at all.

DTC Will Set When:

DTC 63 can be set anytime. A malfunction exists if the EBTCM detects an out of range voltage in either of the right front motor circuits indicating a circuit shorted to battery or a motor shorted.

Action Taken:

A malfunction DTC is stored, ABS is disabled and the "ABS" warning lamp is turned "ON."

DTC Chart Test Description:

Number(s) below refer to circled number(s) on the diagnostic chart.

1. Checks for a short to voltage in the motor high circuitry.
2. Checks for a short to voltage in the motor low circuitry.
3. Checks for a short between high and low motor circuits.
4. Checks for a motor that is internally shorted.
5. Checks for a short between the EMB and motor circuits.
6. Checks for an internal motor pack short between the EMB and motor.
7. Ensures malfunction was not due to physical damage of the circuitry.

Diagnostic Aids:

Using Tech 1, select manual control function, and exercise motor movement of affected channel in both directions while applying light pressure on the brake pedal.

If erratic or "jumpy" brake pedal movement is detected while performing an "apply" or "release" function of the motor, an intermittent malfunction may be indicated.

An "intermittent" malfunction is most likely caused by a poor connection, rubbed through wire insulation, or a wire that is broken inside the insulation.

If the malfunction is not current, wiggle the wires of the affected channel and check if the DTC resets. This will help to pinpoint an intermittent malfunction in the motor circuitry or connections.

The frequency of the malfunction can be checked by using the enhanced diagnostic function of the Tech 1

If DTC 41 sets as a current DTC and DTC 63 is infrequent history DTC, see DTC 41 for diagnosis.

Any circuitry, that is suspected as causing the intermittent complaint, should be thoroughly checked for backed out terminals, improper mating, broken locks, improperly formed or damaged terminals, poor terminal to wiring connections or physical damage to the wiring harness.

DIAGNOSTIC CHART DTC 63 — 1995 CAVALIER AND SUNFIRE WITH TCS, CONTINUED

DTC 63
RIGHT FRONT MOTOR CIRCUIT SHORTED TO BATTERY OR MOTOR SHORTED

NOTICE: J 39200 TEST LEADS MUST BE "ZEROED" PRIOR TO MAKING ANY RESISTANCE MEASUREMENTS. REFER TO J 39200 USER'S MANUAL.

DOES DTC 63 OCCUR INTERMITTENTLY?

- YES: REFER TO "DIAGNOSTIC AIDS" ON FACING PAGE.
- NO:

(1) NOTICE: THIS STEP MAY CAUSE ADDITIONAL DTCs TO SET.
- IGNITION "OFF."
- DISCONNECT 6-WAY CONNECTORS AT MOTOR PACK AND EBTCM.
- IGNITION "ON."
- USING J 39200, MEASURE VOLTAGE BETWEEN GROUND AND 6-WAY MOTOR PACK HARNESS CONNECTOR TERMINAL "E".

IS VOLTAGE 1.0V OR LESS?

- NO: REPAIR SHORT TO VOLTAGE IN CKT 1282
- YES:

(2)
- USING J 39200, MEASURE VOLTAGE BETWEEN GROUND AND 6-WAY MOTOR PACK HARNESS CONNECTOR TERMINAL "F"

IS VOLTAGE 1 0V OR LESS?

- NO: REPAIR SHORT TO VOLTAGE IN CKT 1283.
- YES:

(3)
- IGNITION "OFF "
- USING J 39200, MEASURE RESISTANCE BETWEEN 6-WAY MOTOR PACK HARNESS CONNECTOR TERMINALS "E" AND "F"

DOES J 39200 DISPLAY RESISTANCE AS "OL" (INFINITE)?

- NO: REPAIR SHORT BETWEEN CKTs 1282 AND 1283.
- YES:

(4)
- USING J 39200, MEASURE RESISTANCE BETWEEN MOTOR PACK TERMINALS "E" AND "F".

IS RESISTANCE 0.4Ω OR GREATER?

- NO: REPLACE MOTOR PACK.
- YES:

(5)
- DISCONNECT 3-WAY MOTOR PACK CONNECTOR.
- USING J 39200, MEASURE RESISTANCE BETWEEN 3-WAY MOTOR PACK HARNESS CONNECTOR TERMINAL "B" AND 6-WAY MOTOR PACK HARNESS CONNECTOR TERMINAL "E" THEN "F"

DID J 39200 DISPLAY RESISTANCE AS "OL" (INFINITE) FOR EACH MEASUREMENT?

- NO: REPAIR SHORT BETWEEN CKT 850 AND MOTOR CIRCUIT WHICH MEASURED RESISTANCE OTHER THAN "OL" (INFINITE).
- YES:

(6)
- USING J 39200, MEASURE RESISTANCE BETWEEN 3-WAY MOTOR PACK CONNECTOR TERMINAL "B" AND 6-WAY MOTOR PACK CONNECTOR TERMINAL "E".

DID J 39200 DISPLAY RESISTANCE AS "OL" (INFINITE)?

- NO: REPLACE MOTOR PACK.
- YES:

(7)
- INSPECT CKTs 1282 AND 1283 AND 6-WAY CONNECTORS FOR DAMAGE WHICH MAY RESULT IN A SHORT TO VOLTAGE WITH ALL CONNECTORS CONNECTED.
- REPAIR DAMAGE IF EVIDENT.
- RECONNECT ALL CONNECTORS
- DRIVE VEHICLE FOR THREE DRIVE CYCLES (A DRIVE CYCLE CONSISTS OF STARTING ENGINE, DRIVE ABOVE 16 km/h (10 MPH) AND THEN TURNING IGNITION "OFF")

DID DTC 63 SET IN THE LAST THREE DRIVE CYCLES?

- YES: REPLACE EBTCM.
- NO: REFER TO "DIAGNOSTIC AIDS" ON FACING PAGE.

• AFTER DIAGNOSIS IS COMPLETE, CLEAR DTCs AND TEST DRIVE VEHICLE FOR THREE (3) DRIVE CYCLES TO VERIFY DTC DOES NOT RESET. A DRIVE CYCLE CONSISTS OF STARTING THE VEHICLE, DRIVING OVER 16 km/h (10 MPH) AND THEN TURNING IGNITION "OFF."

DIAGNOSTIC CHART DTC 64 — 1995 CAVALIER AND SUNFIRE WITH TCS

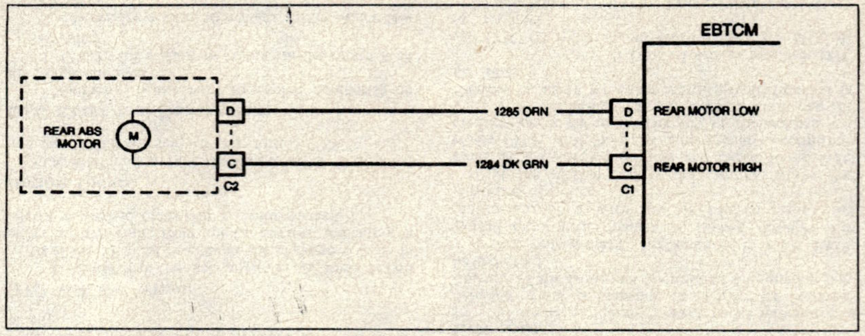

DTC 64
REAR MOTOR CIRCUIT OPEN

Circuit Description:

This DTC identifies a motor that cannot be energized due to an open in its circuitry. This malfunction will not allow proper rear ABS operation.

DTC Will Set When:

DTC 64 can be set only when the motor is commanded "OFF." A malfunction exists if the EBTCM detects an out of range voltage in either of the rear ABS motor circuits indicating an open circuit.

Action Taken:

An open motor will not activate when requested. A malfunction DTC is stored, ABS is disabled and the "ABS" warning lamp is turned "ON."

DTC Chart Test Description:

Number(s) below refer to circled number(s) on the diagnostic chart.

1. Checks for proper resistance of the motor.
2. Checks for an open in the motor high circuitry.
3. Checks for an open in the motor low circuitry.
4. Ensures malfunction was not due to poor terminal contact.

Diagnostic Aids:

Using Tech 1, select manual control function, and exercise motor movement of affected channel in both directions while applying light pressure on the brake pedal.

If erratic or "jumpy" brake pedal movement is detected while performing an "apply" or "release" function of the motor, an intermittent malfunction may be indicated.

An "intermittent" malfunction is most likely caused by a poor connection, rubbed through wire insulation, or a wire that is broken inside the insulation.

If the malfunction is not current, wiggle the wires of the affected channel and check if the DTC resets. This will help to pinpoint an intermittent malfunction in the motor circuitry or connections.

The frequency of the malfunction can be checked by using the enhanced diagnostic function of the Tech 1

Any circuitry, that is suspected as causing the intermittent complaint, should be thoroughly checked for backed out terminals, improper mating, broken locks, improperly formed or damaged terminals, poor terminal to wiring connections or physical damage to the wiring harness.

DIAGNOSTIC CHART DTC 64 — 1995 CAVALIER AND SUNFIRE WITH TCS, CONTINUED

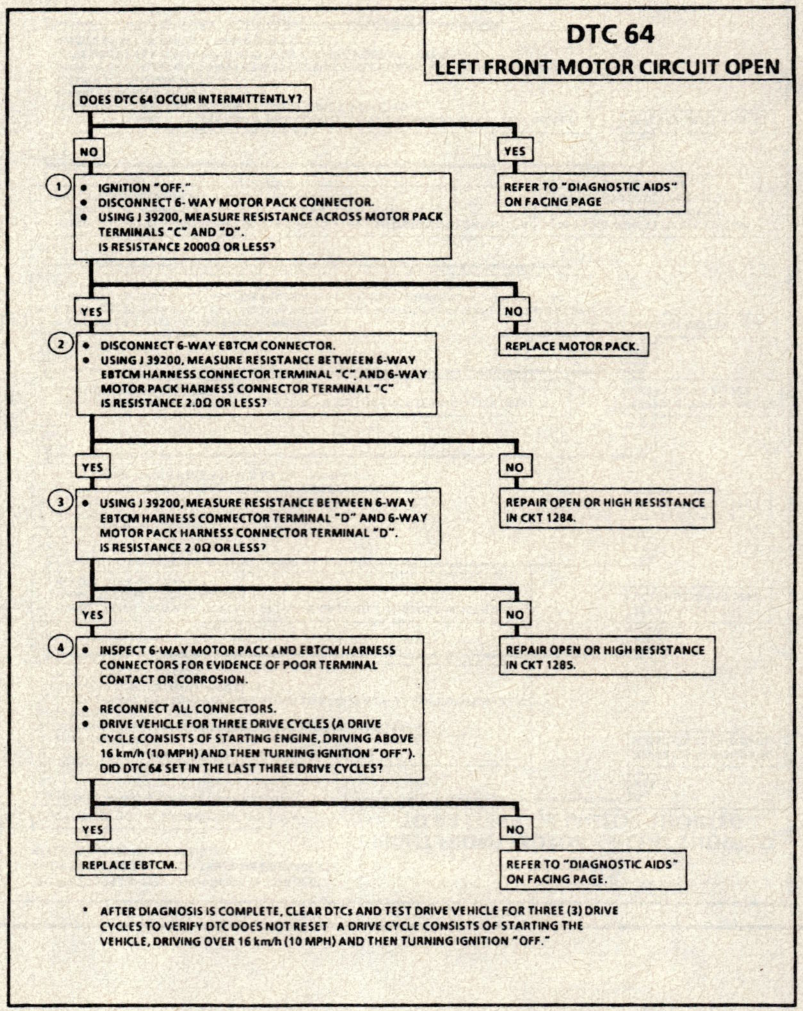

DIAGNOSTIC CHART DTC 65 — 1995 CAVALIER AND SUNFIRE WITH TCS

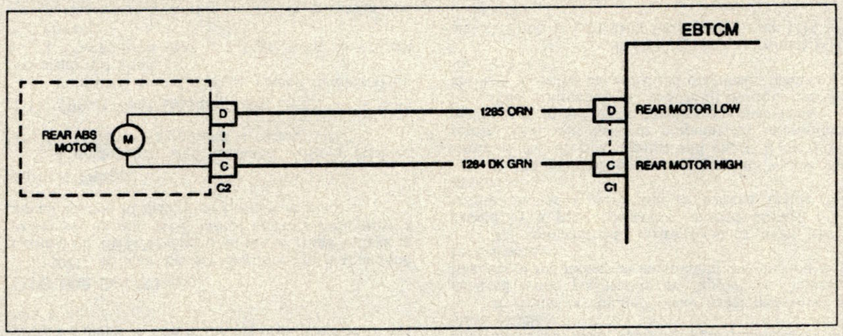

DTC 65
REAR MOTOR CIRCUIT SHORTED TO GROUND

Circuit Description:

This DTC identifies a motor circuit that is shorted to ground. This malfunction will not allow the motor to be controlled at the commanded current rate or will cause the driver circuit to allow current directly to ground.

DTC Will Set When:

DTC 65 can be set anytime. A malfunction exists if the EBTCM detects an out of range voltage in either of the rear ABS motor circuits indicating a circuit shorted to ground.

Action Taken:

A malfunction DTC is stored, ABS is disabled and the "ABS" warning lamp is turned "ON."

DTC Chart Test Description:

Number(s) below refer to circled number(s) on the diagnostic chart.

1. Checks for a short to ground in the motor high circuitry.
2. Checks for a short to ground in the motor low circuitry.
3. Checks for a motor that is internally shorted to ground.
4. Ensures malfunction was not due to physical damage of the circuitry.

Diagnostic Aids:

Using Tech 1, select manual control function, and exercise motor movement of affected channel in both directions while applying light pressure on the brake pedal.

If erratic or "jumpy" brake pedal movement is detected while performing an "apply" or "release" function of the motor, an intermittent malfunction may be indicated.

An "intermittent" malfunction is most likely caused by a poor connection, rubbed through wire insulation, or a wire that is broken inside the insulation.

If the malfunction is not current, wiggle the wires of the affected channel and check if the DTC resets. This will help to pinpoint an intermittent malfunction in the motor circuitry or connections.

The frequency of the problem can be checked by using the enhanced diagnostic function of the Tech 1.

Any circuitry, that is suspected as causing the intermittent complaint, should be thoroughly checked for backed out terminals, improper mating, broken locks, improperly formed or damaged terminals, poor terminal to wiring connections or physical damage to the wiring harness.

DIAGNOSTIC CHART DTC 65 — 1995 CAVALIER AND SUNFIRE WITH TCS, CONTINUED

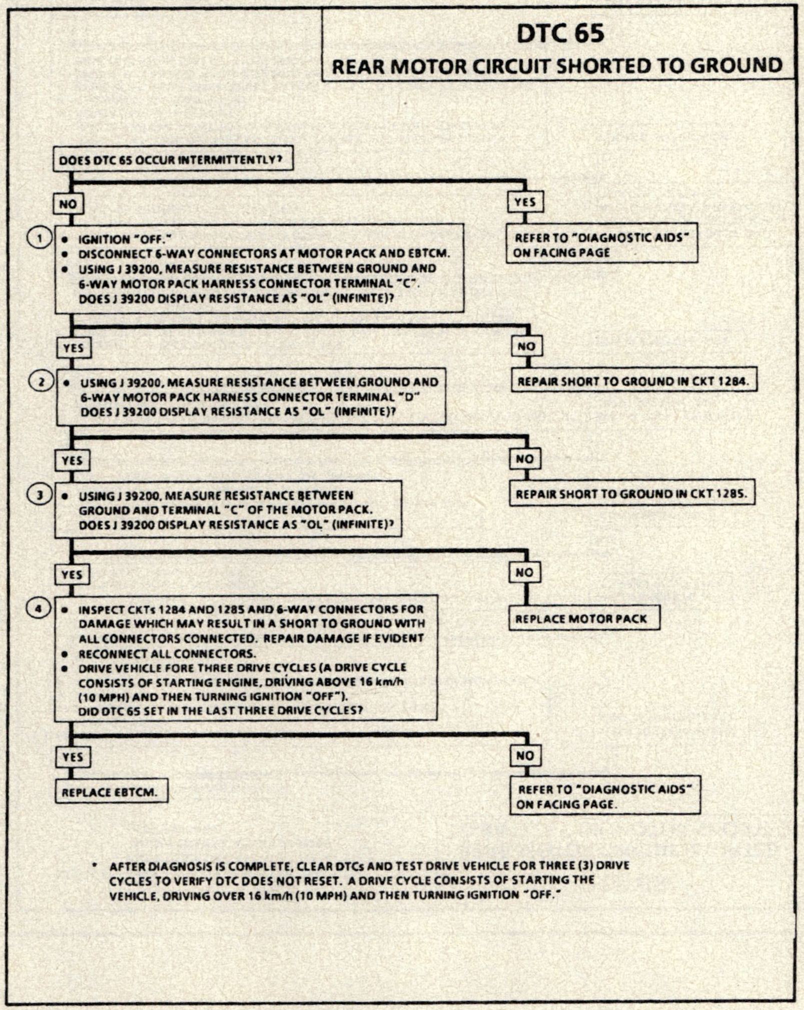

DIAGNOSTIC CHART DTC 66 — 1995 CAVALIER AND SUNFIRE WITH TCS

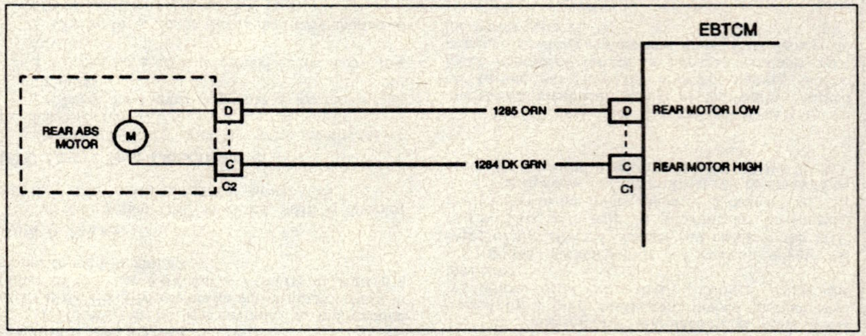

DTC 66
REAR MOTOR CIRCUIT SHORTED TO BATTERY OR MOTOR SHORTED

Circuit Description:

This DTC identifies a motor circuit that is shorted to battery or a motor that has low or zero resistance. This malfunction will not allow the motor to be controlled at the commanded current rate or will cause the motor to turn in the opposite direction or not turn at all.

DTC Will Set When:

DTC 66 can be set anytime. A malfunction exists if the EBTCM detects an out of range voltage in either of the rear ABS motor circuits indicating a circuit shorted to battery or a motor shorted.

Action Taken:

A malfunction DTC is stored, ABS is disabled and the "ABS" warning lamp is turned "ON."

DTC Chart Test Description:

Number(s) below refer to circled number(s) on the diagnostic chart.

1. Checks for a short to voltage in the motor high circuitry.
2. Checks for a short to voltage in the motor low circuitry.
3. Checks for a short between high and low motor circuits.
4. Checks for a motor that is internally shorted.
5. Checks for a short between the EMB and motor circuits.
6. Checks for an internal motor pack short between the EMB and motor.
7. Ensures malfunction was not due to physical damage of the circuitry.

Diagnostic Aids:

Using Tech 1, select manual control function, and exercise motor movement of affected channel in both directions while applying light pressure on the brake pedal.

If erratic or "jumpy" brake pedal movement is detected while performing an "apply" or "release" function of the motor, an intermittent malfunction may be indicated.

An "intermittent" malfunction is most likely caused by a poor connection, rubbed through wire insulation, or a wire that is broken inside the insulation.

If the malfunction is not current, wiggle the wires of the affected channel and check if the DTC resets. This will help to pinpoint an intermittent malfunction in the motor circuitry or connections.

The frequency of the malfunction can be checked by using the enhanced diagnostic function of the Tech 1 If DTC 42 sets as current DTC and DTC 66 is infrequent history DTC, see DTC 42 diagnostics.

Any circuitry, that is suspected as causing the intermittent complaint, should be thoroughly checked for backed out terminals, improper mating, broken locks, improperly formed or damaged terminals, poor terminal to wiring connections or physical damage to the wiring harness.

DIAGNOSTIC CHART DTC 66 — 1995 CAVALIER AND SUNFIRE WITH TCS, CONTINUED

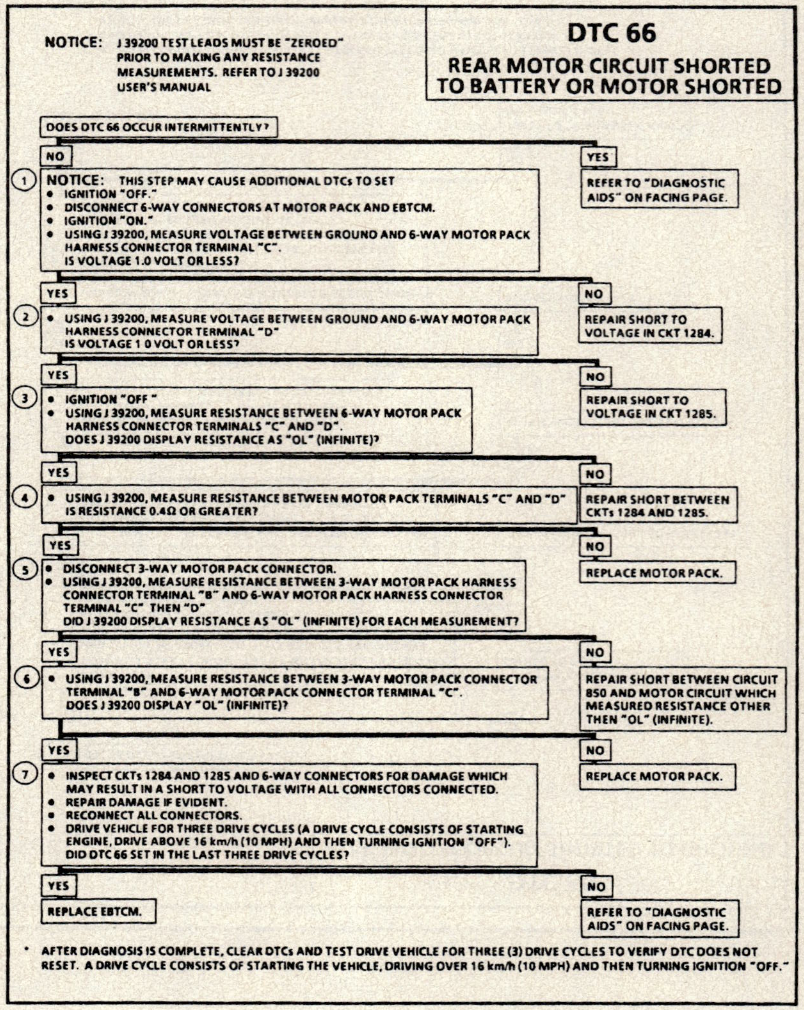

DIAGNOSTIC CHART DTC 67 — 1995 CAVALIER AND SUNFIRE WITH TCS

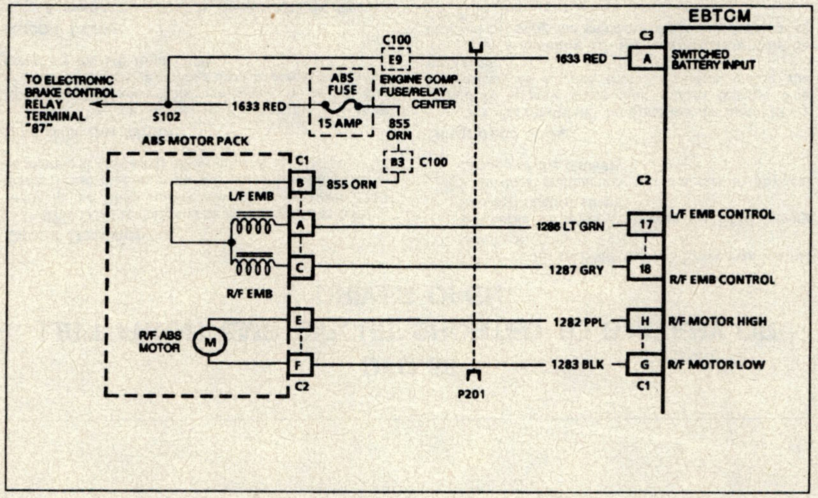

DTC 67

LEFT FRONT EMB CIRCUIT OPEN OR SHORTED TO GROUND

Circuit Description:

This DTC is designed to detect shorts to ground and opens in the EMB release circuit. A short to ground between the EMB and the EBTCM will cause the EMB to be energized indefinitely at an uncontrolled current, causing the EMB to be destroyed.

DTC Will Set When:

DTC 67 can be set only when the left front EMB is commanded "OFF" (not released). A malfunction exists if the EBTCM detects the left front EMB control circuit is constantly at ground.

Action Taken:

A malfunction DTC is stored, ABS is disabled and the "ABS" warning lamp is turned "ON."

DTC Chart Test Description:

Number(s) below refer to circled number(s) on the diagnostic chart.

1. Checks for a short to ground in the EMB control circuitry.
2. Checks for an open in the EMB control circuitry.
3. Checks for an open in the EMB battery feed circuit.
4. Ensures the EMB is not internally shorted to ground.
5. Ensures EMB is not internally open.
6. Checks for short to ground in EMB battery feed circuit.
7. Ensures malfunction was not due to physical damage of the circuitry.

Diagnostic Aids:

An "intermittent" malfunction is most likely caused by a poor connection, rubbed through wire insulation, or a wire that is broken inside the insulation.

The frequency of the malfunction can be checked by using the enhanced diagnostic function of the Tech 1

Any circuitry, that is suspected as causing the intermittent complaint, should be thoroughly checked for backed out terminals, improper mating, broken locks, improperly formed or damaged terminals, poor terminal to wiring connections or physical damage to the wiring harness.

DIAGNOSTIC CHART DTC 67 — 1995 CAVALIER AND SUNFIRE WITH TCS, CONTINUED

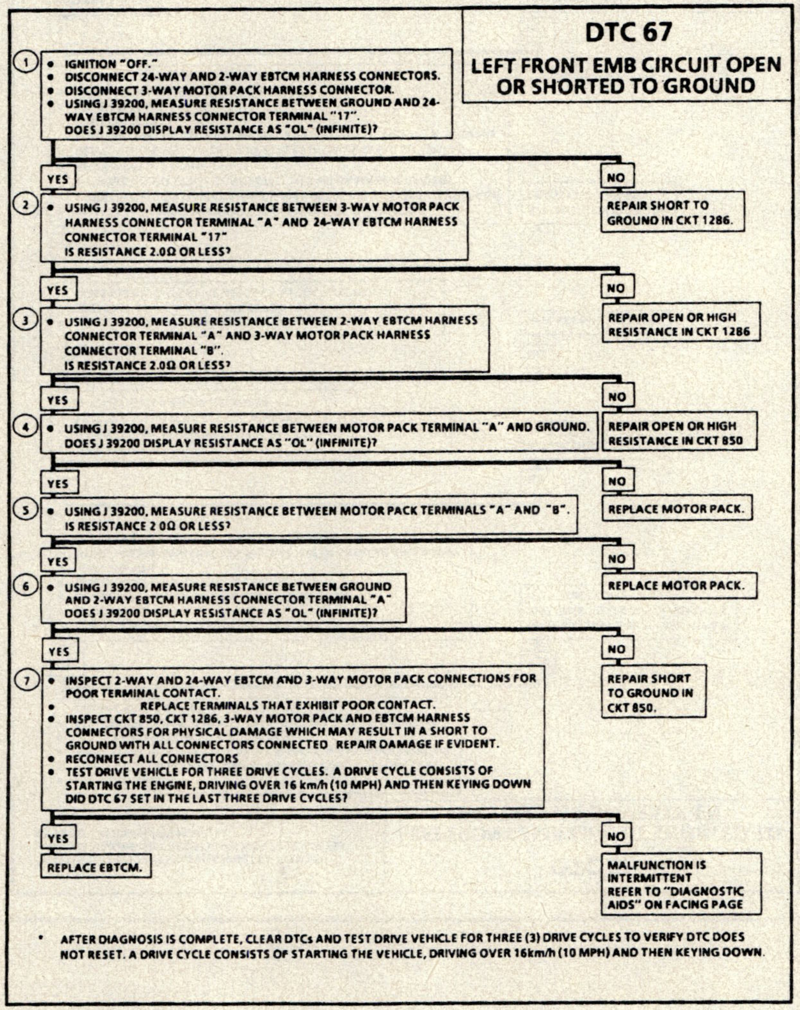

DIAGNOSTIC CHART DTC 68 — 1995 CAVALIER AND SUNFIRE WITH TCS

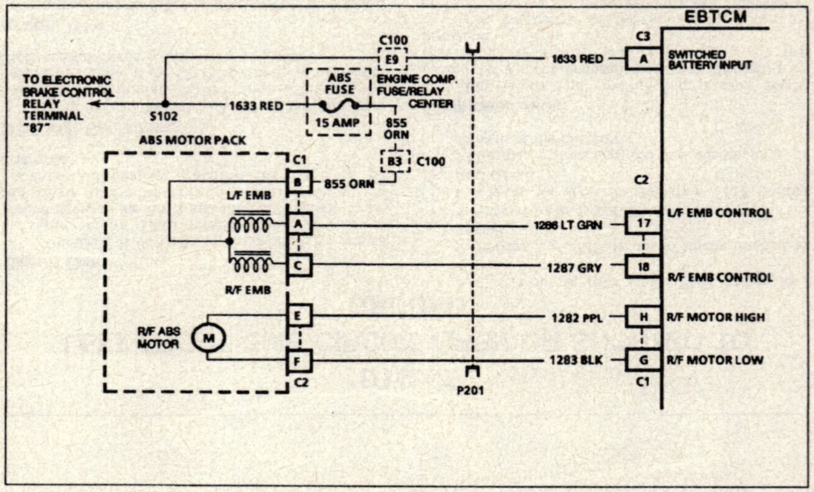

DTC 68
LEFT FRONT EMB CIRCUIT SHORTED TO BATTERY OR DRIVER OPEN

Circuit Description:

This DTC is designed to detect shorts to battery, shorts in the EMB release circuit, or an open EMB driver. These malfunctions will not allow the EMB to be energized (released) when needed.

DTC Will Set When:

DTC 68 can be set only when the left front EMB is commanded "ON" (released). A malfunction exists if an excessive voltage drop is across the motor driver for the left front EMB.

Action Taken:

A malfunction DTC is stored, ABS is disabled and the "ABS" warning lamp is turned "ON."

DTC Chart Test Description:

Number(s) below refer to circled number(s) on the diagnostic chart.

1. Checks to see if the malfunction is currently present.
2. Identifies a short to voltage in the EMB control circuitry.
3. Checks for an EMB that has lower than normal resistance.
4. Checks for a short between the EMB battery feed and control circuits.
5. Ensures malfunction was not due to physical damage of the circuitry.

Diagnostic Aids:

An "intermittent" malfunction is most likely caused by a poor connection, rubbed through wire insulation, or a wire that is broken inside the insulation.

The frequency of the malfunction can be checked by using the enhanced diagnostic function of the Tech 1.

Any circuitry, that is suspected as causing the intermittent complaint, should be thoroughly checked for backed out terminals, improper mating, broken locks, improperly formed or damaged terminals, poor terminal to wiring connections or physical damage to the wiring harness.

DIAGNOSTIC CHART DTC 68 — 1995 CAVALIER AND SUNFIRE WITH TCS, CONTINUED

DTC 68
LEFT FRONT EMB CIRCUIT SHORTED TO BATTERY OR DRIVER OPEN

*NOTICE: J 39200 TEST LEADS MUST BE "ZEROED" PRIOR TO MAKING ANY RESISTANCE MEASUREMENTS. REFER TO J 39200 USER'S MANUAL.

(1)
- IGNITION "ON."
- USING TECH 1 ABS TESTS, SELECT MANUAL CONTROL FUNCTION.
- SELECT LEFT FRONT MOTOR APPLY FUNCTION AND APPLY MOTOR.
- APPLY FIRM PRESSURE ON BRAKE PEDAL
- SELECT LEFT FRONT EMB RELEASE FUNCTION AND RELEASE EMB. DID BRAKE PEDAL FALL WITH INCREASING PEDAL FORCE?

YES → MALFUNCTION IS NOT PRESENT AT THIS TIME. REFER TO "DIAGNOSTIC AIDS" ON FACING PAGE.

NO ↓

(2) NOTICE: THIS STEP MAY CAUSE ADDITIONAL DTCs TO SET.
- IGNITION "OFF"
- DISCONNECT 3-WAY MOTOR PACK CONNECTOR.
- DISCONNECT 24-WAY EBTCM CONNECTOR.
- IGNITION "ON."
- USING J 39200, MEASURE VOLTAGE BETWEEN GROUND AND 3-WAY MOTOR PACK HARNESS CONNECTOR TERMINAL "A". IS VOLTAGE 1.0V OR LESS?

NO → REPAIR SHORT TO VOLTAGE IN CKT 1286.

YES ↓

(3)
- IGNITION "OFF."
- USING J 39200, MEASURE RESISTANCE BETWEEN MOTOR PACK TERMINALS "A" AND "B". IS RESISTANCE 0.4Ω OR GREATER?

NO → REPLACE MOTOR PACK.

YES ↓

(4)
- USING J 39200, MEASURE THE RESISTANCE BETWEEN 3-WAY MOTOR PACK HARNESS CONNECTOR TERMINALS "B" AND "A" DOES J 39200 DISPLAY RESISTANCE AS "OL" (INFINITE)?

NO → REPAIR SHORT BETWEEN CKT(s) 1286 AND 850.

YES ↓

(5)
- INSPECT CKT 1286, 3-WAY MOTOR PACK AND 24-WAY EBTCM CONNECTORS FOR DAMAGE WHICH MAY RESULT IN A SHORT TO VOLTAGE WITH ALL CONNECTORS CONNECTED. REPAIR DAMAGE IF EVIDENT.
- RECONNECT ALL CONNECTORS.
- DRIVE VEHICLE FOR THREE DRIVE CYCLES (A DRIVE CYCLE CONSISTS OF STARTING ENGINE, DRIVING ABOVE 16 km/h (10 MPH) AND THEN TURNING IGNITION "OFF").
- IGNITION "ON."
- USING TECH 1, SELECT "F1: DTC HISTORY" DID DTC 68 SET IN LAST THREE DRIVE CYCLES?

YES → REPLACE EBTCM.

NO → MALFUNCTION IS INTERMITTENT. REFER TO "DIAGNOSTIC AIDS" ON FACING PAGE.

- AFTER DIAGNOSIS IS COMPLETE, CLEAR DTCs AND TEST DRIVE VEHICLE FOR THREE (3) DRIVE CYCLES TO VERIFY DTC DOES NOT RESET. A DRIVE CYCLE CONSISTS OF STARTING THE VEHICLE, DRIVING OVER 16 km/h (10 MPH) AND THEN KEYING DOWN.

DIAGNOSTIC CHART DTC 71 — 1995 CAVALIER AND SUNFIRE WITH TCS

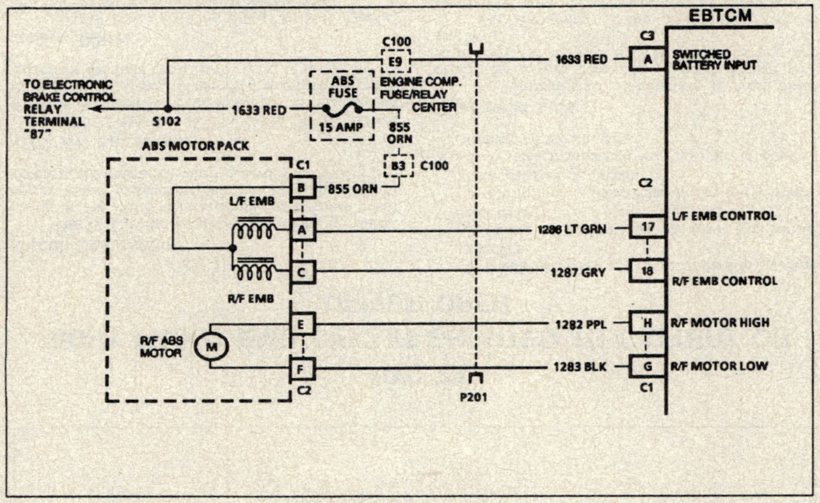

DTC 71

RIGHT FRONT EMB CIRCUIT OPEN OR SHORTED TO GROUND

Circuit Description:

This DTC is designed to detect shorts to ground and opens in the EMB release circuit. A short to ground between the EMB and the EBTCM will cause the EMB to be energized indefinitely at an uncontrolled current causing the EMB to be destroyed.

DTC Will Set When:

DTC 71 can be set only when the right front EMB is commanded "OFF" (not released). A malfunction exists if the EBTCM detects the right front EMB control circuit is constantly at ground.

Action Taken:

A malfunction DTC is stored, ABS is disabled and the "ABS" warning lamp is turned "ON."

DTC Chart Test Description:

Number(s) below refer to circled number(s) on the diagnostic chart

1. Checks for a short to ground in the EMB control circuitry.
2. Checks for an open in the EMB control circuitry.
3. Checks for an open in the EMB battery feed circuit.
4. Ensures the EMB is not internally shorted to ground.
5. Ensures EMB is not internally open.
6. Checks for short to ground in EMB battery feed circuit.
7. Ensures malfunction was not due to physical damage of the circuitry.

Diagnostic Aids:

An "intermittent" malfunction is most likely caused by a poor connection, rubbed through wire insulation, or a wire that is broken inside the insulation.

The frequency of the malfunction can be checked by using the enhanced diagnostic function of the Tech 1

Any circuitry, that is suspected as causing the intermittent complaint, should be thoroughly checked for backed out terminals, improper mating, broken locks, improperly formed or damaged terminals, poor terminal to wiring connections or physical damage to the wiring harness.

DIAGNOSTIC CHART DTC 71 — 1995 CAVALIER AND SUNFIRE WITH TCS, CONTINUED

DTC 71
RIGHT FRONT EMB CIRCUIT OPEN OR SHORTED TO GROUND

(1)
- IGNITION "OFF."
- DISCONNECT 24-WAY AND 2-WAY EBTCM HARNESS CONNECTORS.
- DISCONNECT 3-WAY MOTOR PACK HARNESS CONNECTOR.
- USING J 39200, MEASURE RESISTANCE BETWEEN GROUND AND 24-WAY EBTCM HARNESS CONNECTOR TERMINAL "20". DOES J 39200 DISPLAY RESISTANCE AS "OL" (INFINITE)?

NO: REPAIR SHORT TO GROUND IN CKT 1287
YES:

(2)
- USING J 39200, MEASURE RESISTANCE BETWEEN 3-WAY MOTOR PACK HARNESS CONNECTOR TERMINAL "C" AND 24-WAY EBTCM HARNESS CONNECTOR TERMINAL "20" IS RESISTANCE 2.0Ω OR LESS?

NO: REPAIR OPEN OR HIGH RESISTANCE IN CKT 1287
YES:

(3)
- USING J 39200, MEASURE RESISTANCE BETWEEN 2-WAY EBTCM HARNESS CONNECTOR TERMINAL "A" AND 3-WAY MOTOR PACK HARNESS CONNECTOR TERMINAL "B". IS RESISTANCE 2.0Ω OR LESS?

NO: REPAIR OPEN OR HIGH RESISTANCE IN CKT 850.
YES:

(4)
- USING J 39200, MEASURE RESISTANCE BETWEEN MOTOR PACK TERMINAL "C" AND GROUND DOES J 39200 DISPLAY RESISTANCE AS "OL" (INFINITE)?

NO: REPLACE MOTOR PACK
YES:

(5)
- USING J 39200, MEASURE RESISTANCE BETWEEN MOTOR PACK TERMINALS "C" AND "B" IS RESISTANCE 2.0Ω OR LESS?

NO: REPLACE MOTOR PACK
YES:

(6)
- USING J 39200, MEASURE RESISTANCE BETWEEN GROUND AND 2-WAY EBTCM HARNESS CONNECTOR TERMINAL "A" DOES J 39200 DISPLAY RESISTANCE AS "OL" (INFINITE)?

NO: REPAIR SHORT TO GROUND IN CKT 850
YES:

(7)
- INSPECT 2-WAY AND 24-WAY EBTCM AND 3-WAY MOTOR PACK CONNECTIONS FOR POOR TERMINAL CONTACT. REPLACE TERMINALS THAT EXHIBIT POOR CONTACT.
- INSPECT CKT 850, CKT 1287, 3-WAY MOTOR PACK AND EBTCM HARNESS CONNECTORS FOR PHYSICAL DAMAGE WHICH MAY RESULT IN A SHORT TO GROUND WITH ALL CONNECTORS CONNECTED REPAIR DAMAGE IF EVIDENT
- RECONNECT ALL CONNECTORS.
- TEST DRIVE VEHICLE FOR THREE DRIVE CYCLES A DRIVE CYCLE CONSISTS OF STARTING THE ENGINE, DRIVING OVER 16 KM/H (10 MPH) AND THEN KEYING DOWN DID DTC 71 SET THE LAST THREE DRIVE CYCLES?

YES: REPLACE EBTCM.
NO: MALFUNCTION IS INTERMITTENT REFER TO "DIAGNOSTIC AIDS" ON FACING PAGE

AFTER DIAGNOSIS IS COMPLETE, CLEAR DTCs AND TEST DRIVE VEHICLE FOR THREE (3) DRIVE CYCLES TO VERIFY DTC DOES NOT RESET. A DRIVE CYCLE CONSISTS OF STARTING THE VEHICLE, DRIVING OVER 16km/h (10 MPH) AND THEN KEYING DOWN.

DIAGNOSTIC CHART DTC 72 — 1995 CAVALIER AND SUNFIRE WITH TCS

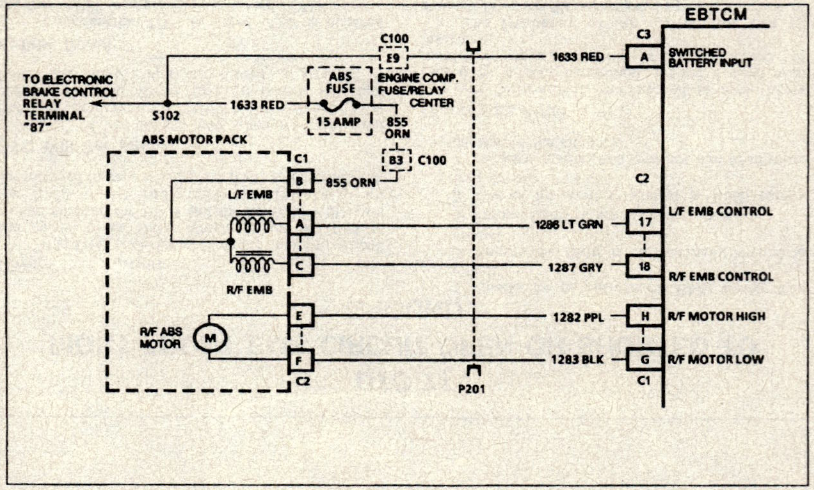

DTC 72

RIGHT FRONT EMB CIRCUIT SHORTED TO BATTERY OR DRIVER OPEN

Circuit Description:

This DTC is designed to detect shorts to battery, shorts in the EMB release circuit, or an open EMB driver. These malfunctions will not allow the EMB to be energized (released) when needed.

DTC Will Set When:

DTC 72 can be set only when the right front EMB is commanded "ON" (released). A malfunction exists if an excessive voltage drop is across the motor driver for the right front EMB.

Action Taken:

A malfunction DTC is stored, ABS is disabled and the "ABS" warning lamp is turned "ON."

DTC Chart Test Description:

Number(s) below refer to circled number(s) on the diagnostic chart.

1. Checks to see if the malfunction is currently present.
2. Identifies a short to voltage in the EMB control circuitry.
3. Checks for an EMB that has lower than normal resistance.
4. Checks for a short between the EMB battery feed and control circuits.
5. Ensures malfunction was not due to physical damage of the circuitry.

Diagnostic Aids:

An "intermittent" malfunction is most likely caused by a poor connection, rubbed through wire insulation, or a wire that is broken inside the insulation.

The frequency of the malfunction can be checked by using the enhanced diagnostic function of the Tech 1,

Any circuitry, that is suspected as causing the intermittent complaint, should be thoroughly checked for backed out terminals, improper mating, broken locks, improperly formed or damaged terminals, poor terminal to wiring connections or physical damage to the wiring harness.

DIAGNOSTIC CHART DTC 72 — 1995 CAVALIER AND SUNFIRE WITH TCS, CONTINUED

DTC 72

RIGHT FRONT EMB CIRCUIT SHORTED TO BATTERY OR DRIVER OPEN

*NOTICE: J 39200 TEST LEADS MUST BE "ZEROED" PRIOR TO MAKING ANY RESISTANCE MEASUREMENTS. REFER TO J 39200 USER'S MANUAL.

(1)
- IGNITION "ON."
- USING TECH 1 ABS TESTS, SELECT MANUAL CONTROL FUNCTION.
- SELECT RIGHT FRONT MOTOR APPLY FUNCTION AND APPLY MOTOR.
- APPLY FIRM PRESSURE ON BRAKE PEDAL.
- SELECT RIGHT FRONT EMB RELEASE FUNCTION AND RELEASE EMB. DID BRAKE PEDAL FALL WITH INCREASING PEDAL FORCE?

YES: MALFUNCTION IS NOT PRESENT AT THIS TIME. REFER TO "DIAGNOSTIC AIDS" ON FACING PAGE.

NO:

(2) NOTICE: THIS STEP MAY CAUSE ADDITIONAL DTCs TO SET.
- IGNITION "OFF."
- DISCONNECT 3-WAY MOTOR PACK CONNECTOR.
- DISCONNECT 24-WAY EBTCM CONNECTOR.
- IGNITION "ON."
- USING J 39200, MEASURE VOLTAGE BETWEEN GROUND AND 3-WAY MOTOR PACK HARNESS CONNECTOR TERMINAL "C". IS VOLTAGE 1.0V OR LESS?

NO: REPAIR SHORT TO VOLTAGE IN CKT 1287

YES:

(3)
- IGNITION "OFF"
- USING J 39200, MEASURE RESISTANCE BETWEEN MOTOR PACK TERMINALS "C" AND "B". IS RESISTANCE 0.4Ω OR GREATER?

NO: REPLACE MOTOR PACK.

YES:

(4)
- USING J 39200, MEASURE RESISTANCE BETWEEN 3-WAY MOTOR PACK HARNESS CONNECTOR TERMINALS "B" AND "C". DOES J 39200 DISPLAY RESISTANCE AS "OL" (INFINITE)?

NO: REPAIR SHORT BETWEEN CKT(s) 1287 and 850.

YES:

(5)
- INSPECT CKT 1287, 3-WAY MOTOR PACK AND 24-WAY EBTCM CONNECTORS FOR DAMAGE WHICH MAY RESULT IN A SHORT TO VOLTAGE WITH ALL CONNECTORS CONNECTED. REPAIR DAMAGE IF EVIDENT.
- RECONNECT ALL CONNECTORS
- DRIVE VEHICLE FOR THREE DRIVE CYCLES (A DRIVE CYCLE CONSISTS OF STARTING ENGINE, DRIVING ABOVE 16 KM/H (10 MPH) AND THEN TURNING IGNITION "OFF").
- IGNITION "ON."
- USING TECH 1, SELECT "F1: DTC HISTORY." DID DTC 72 SET IN LAST THREE DRIVE CYCLES?

YES: REPLACE EBTCM.

NO: MALFUNCTION IS INTERMITTENT. REFER TO "DIAGNOSTIC AIDS" ON FACING PAGE.

- AFTER DIAGNOSIS IS COMPLETE, CLEAR DTCs AND TEST DRIVE VEHICLE FOR THREE (3) DRIVE CYCLES TO VERIFY DTC DOES NOT RESET. A DRIVE CYCLE CONSISTS OF STARTING THE VEHICLE, DRIVING OVER 16 km/h (10 MPH) AND THEN KEYING DOWN.

DIAGNOSTIC CHART DTC 75 — 1995 CAVALIER AND SUNFIRE WITH TCS

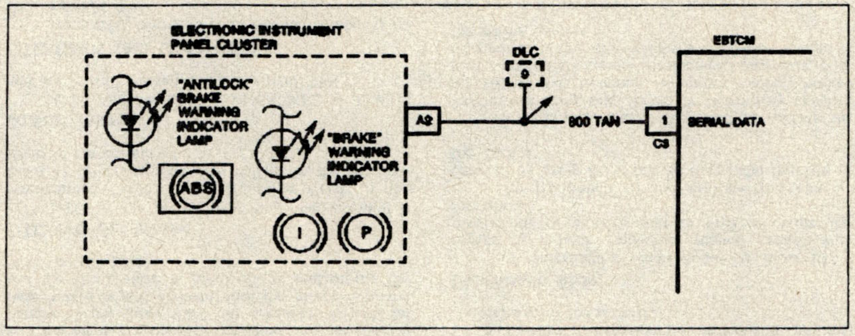

DTC 75
SERIAL COMMUNICATION FAILURE

Circuit Description:

This DTC is designed to detect no communication with the IPC or the PCM via the serial data link.

DTC Will Set When:

DTC 75 can be set anytime after ABS w/ETS initialization. A malfunction exists if the message from the Electronic Instrument Panel Cluster (IPC) or the message from the Powertrain Control Module (PCM) is not received for 7 seconds.

Action Taken:

A serial communication failure does not allow the proper lamp commands to be sent to the IPC. A malfunction DTC is stored. The "ETS OFF" indicator is commanded "ON," and ETS is disabled.

DTC Chart Test Description:

Number(s) below refer to circled number(s) on the diagnostic chart.

1. Checks to see if the IPC is operating properly.
2. Checks to see if communication between the EBTCM and the IPC is possible.

Diagnostic Aids:

The Tech 1 serial data link (SDL) monitor used in this diagnostic is within the body portion of the Tech 1 mass storage cartridge. This requires exiting from the chassis portion of the Tech 1 to the main menu and entering into the body portion of the Tech 1 and selecting "SDL monitor."

With the SDL monitor (ABS to IPC mode), any message that is being transmitted on the serial data link can be observed serial data link information).

The frequency of the malfunction can be checked by using the enhanced diagnostic function of the Tech 1.

Any circuitry, that is suspected as causing the intermittent complaint, should be thoroughly checked for backed out terminals, improper mating, broken locks, improperly formed or damaged terminals, poor terminal to wiring connections or physical damage to the wiring harness.

DIAGNOSTIC CHART DTC 75 — 1995 CAVALIER AND SUNFIRE WITH TCS, CONTINUED

DTC 75
SERIAL COMMUNICATION FAILURE

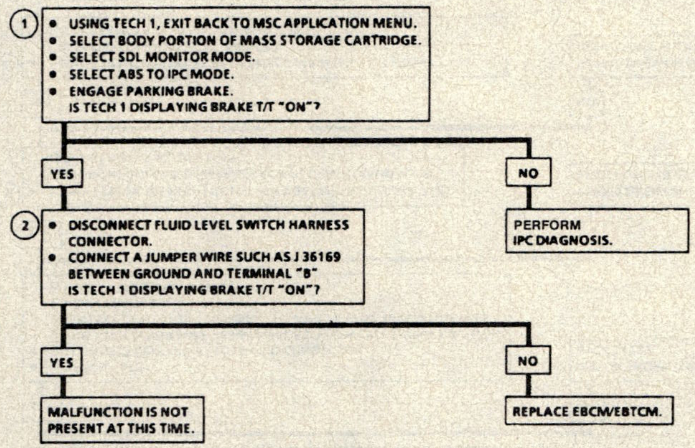

DIAGNOSTIC CHART DTC 76 — 1995 CAVALIER AND SUNFIRE WITH TCS

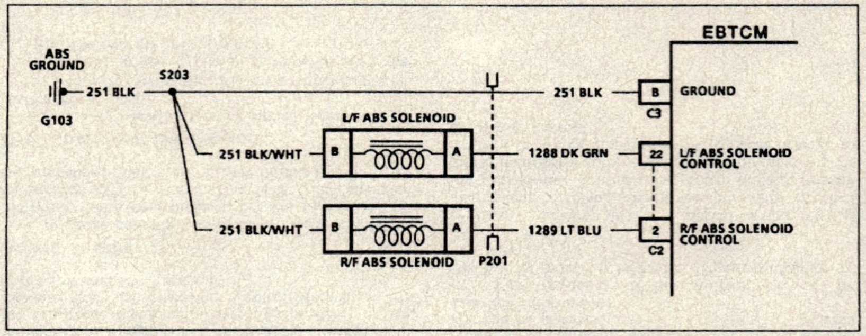

DTC 76
LEFT FRONT SOLENOID CIRCUIT OPEN OR SHORTED TO BATTERY

Circuit Description:

This DTC identifies a solenoid that cannot be energized due to an open in its circuitry, or a solenoid that is always energized due to a short to battery in its circuitry between the driver and the solenoid. An open will not allow proper ABS operation, but the short to battery simply turns "ON" the solenoid. A path for base brakes is still allowed once the motor re-homes and the check valve is lifted off its seat during key "ON" initialization.

DTC Will Set When:

DTC 76 can be set only when the solenoid is commanded "OFF." A malfunction exists if the EBTCM detects an excessive voltage in the left front solenoid control circuit.

Action Taken:

A malfunction DTC is stored, ABS is disabled and the "ABS" warning lamp is turned "ON."

DTC Chart Test Description:

Number(s) below refer to circled number(s) on the diagnostic chart.

1. Checks for a short to voltage in the control circuitry of the solenoid.
2. Checks for an open in the control circuit of the solenoid.
3. Identifies an open in the solenoid or ground circuit.
4. Checks for proper resistance value of the solenoid.
5. Ensures malfunction was not due to physical damage of the circuitry.

Diagnostic Aids:

An "intermittent" malfunction is most likely caused by a poor connection, rubbed through wire insulation, or a wire that is broken inside the insulation.

The frequency of the malfunction can be checked by using the enhanced diagnostic function of the Tech 1

Any circuitry, that is suspected as causing the intermittent complaint, should be thoroughly checked for backed out terminals, improper mating, broken locks, improperly formed or damaged terminals, poor terminal to wiring connections or physical damage to the wiring harness.

DIAGNOSTIC CHART DTC 76 — 1995 CAVALIER AND SUNFIRE WITH TCS, CONTINUED

DTC 76
LEFT FRONT SOLENOID CIRCUIT OPEN OR SHORTED TO BATTERY

*NOTICE: J 39200 TEST LEADS MUST BE "ZEROED" PRIOR TO MAKING ANY RESISTANCE MEASUREMENTS. REFER TO J 39200 USER'S MANUAL.

(1)
- IGNITION "OFF."
- DISCONNECT 24-WAY EBTCM CONNECTOR.
- IGNITION "ON."
- USING J 39200, MEASURE VOLTAGE BETWEEN GROUND AND 24-WAY EBTCM HARNESS CONNECTOR TERMINAL "22".
- IS VOLTAGE 1.0V OR LESS?

YES → go to 2. NO → REPAIR SHORT TO VOLTAGE IN CKT 1288.

(2)
- IGNITION "OFF."
- DISCONNECT LEFT FRONT SOLENOID CONNECTOR.
- USING J 39200, MEASURE RESISTANCE BETWEEN LEFT FRONT SOLENOID HARNESS CONNECTOR TERMINAL "A" AND 24-WAY EBTCM HARNESS CONNECTOR TERMINAL "22".
- IS RESISTANCE 2.0Ω OR LESS?

YES → go to 3. NO → REPAIR OPEN OR HIGH RESISTANCE IN CKT 1288.

(3)
- USING J 39200, MEASURE RESISTANCE BETWEEN GROUND AND LEFT FRONT SOLENOID HARNESS CONNECTOR TERMINAL "B"
- IS RESISTANCE 2.0Ω OR LESS?

YES → go to 4. NO → REPAIR OPEN OR HIGH RESISTANCE IN CKT 152.

(4)
- USING J 39200, MEASURE RESISTANCE BETWEEN LEFT FRONT SOLENOID TERMINALS "A" AND "B"
- IS RESISTANCE BETWEEN 2.5Ω AND 5.0Ω?

YES → go to 5. NO → REPLACE LEFT FRONT SOLENOID.

(5)
- INSPECT LEFT FRONT SOLENOID AND 24-WAY EBCM HARNESS CONNECTORS FOR POOR TERMINAL CONTACT OR CORROSION.
- REPLACE TERMINALS THAT EXHIBIT POOR CONTACT OR CORROSION.
- INSPECT CKT 1288 FOR DAMAGE WHICH MAY RESULT IN A SHORT TO VOLTAGE WITH ALL CONNECTORS CONNECTED. REPAIR DAMAGE IF EVIDENT.
- RECONNECT ALL CONNECTORS.
- IGNITION "ON."
- IS DTC 76 SET AS A CURRENT DTC?

YES → REPLACE EBTCM. NO → REFER TO "DIAGNOSTIC AIDS" ON FACING PAGE.

DIAGNOSTIC CHART DTC 77 — 1995 CAVALIER AND SUNFIRE WITH TCS

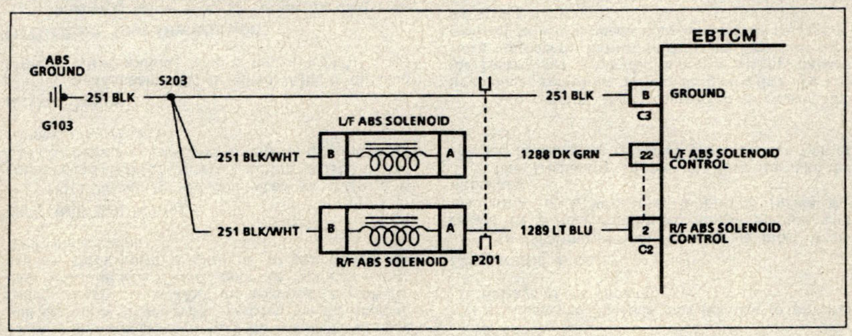

DTC 77
LEFT FRONT SOLENOID CIRCUIT SHORTED TO GROUND OR DRIVER OPEN

Circuit Description:

This DTC identifies a solenoid that cannot be energized due to an open in its driver circuitry, or a short to ground between the solenoid driver and the solenoid. These malfunctions can affect ABS operation since the flow of brake fluid to the wheel cylinder cannot be stopped, making ABS operation for that channel impossible.

DTC Will Set When:

DTC 77 can be set only when the solenoid is commanded "ON." A malfunction exists if the EBTCM detects the left front solenoid control circuit voltage is out of specification.

Action Taken:

A malfunction DTC is stored, ABS is disabled and the "ABS" warning lamp is turned "ON."

DTC Chart Test Description:

Number(s) below refer to circled number(s) on the diagnostic chart.

1. Checks for a solenoid that is internally shorted to ground.
2. Checks for a solenoid that is not within proper resistance value.
3. Indicates if a short to ground exists in the solenoid control circuitry.
4. Ensures malfunction was not due to physical damage of the circuitry.

Diagnostic Aids:

An "intermittent" malfunction is most likely caused by a poor connection, rubbed through wire insulation, or a wire that is broken inside the insulation.

The frequency of the malfunction can be checked by using the enhanced diagnostic function of the Tech 1.

Any circuitry, that is suspected as causing the intermittent complaint, should be thoroughly checked for backed out terminals, improper mating, broken locks, improperly formed or damaged terminals, poor terminal to wiring connections or physical damage to the wiring harness.

DIAGNOSTIC CHART DTC 77 — 1995 CAVALIER AND SUNFIRE WITH TCS, CONTINUED

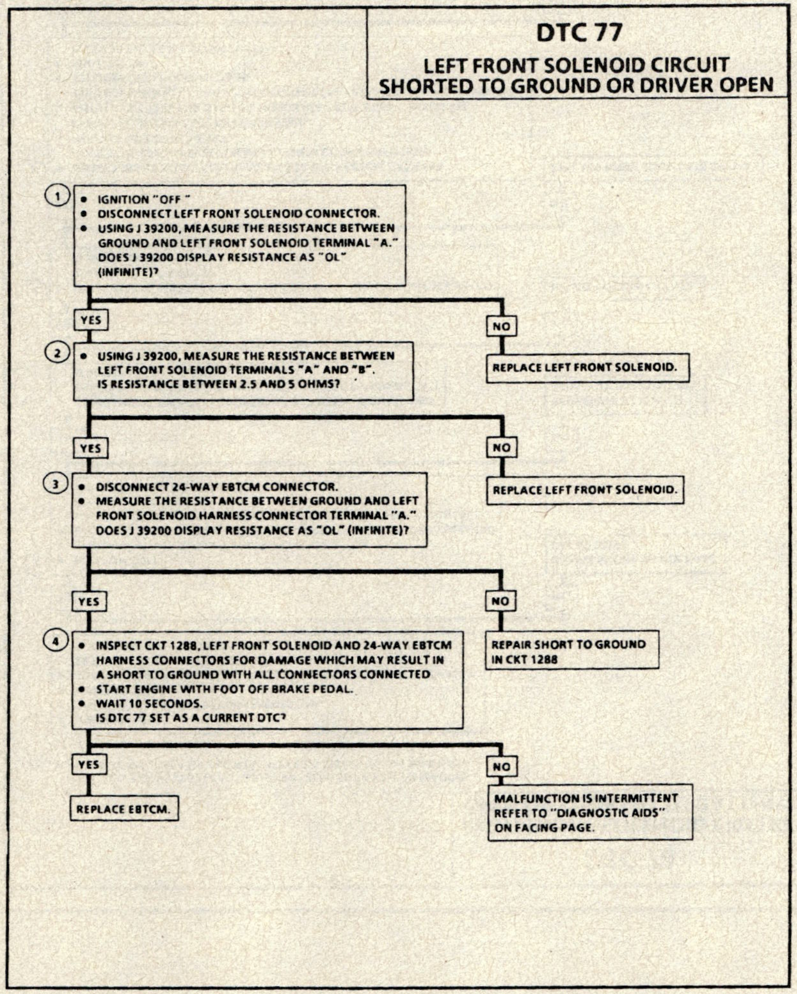

DIAGNOSTIC CHART DTC 78 — 1995 CAVALIER AND SUNFIRE WITH TCS

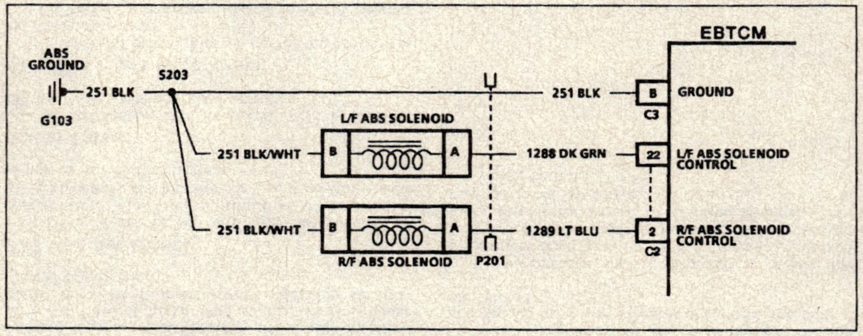

DTC 78
RIGHT FRONT SOLENOID CIRCUIT OPEN OR SHORTED TO BATTERY

Circuit Description:

This DTC identifies a solenoid that cannot be energized due to an open in its circuitry, or a solenoid that is always energized due to a short to battery in its circuitry between the driver and the solenoid. An open will not allow proper ABS operation but the short to battery simply turns "ON" the solenoid. A path for base brakes is still allowed once the motor re-homes and the check valve is lifted off its seat during key "ON" initialization.

DTC Will Set When:

DTC 78 can be set only when the solenoid is commanded "OFF." A malfunction exists if the EBTCM detects an excessive voltage in the right front solenoid control circuit.

Action Taken:

A malfunction DTC is stored, ABS is disabled and the "ABS" warning lamp is turned "ON."

DTC Chart Test Description:

Number(s) below refer to circled number(s) on the diagnostic chart.

1. Checks for a short to voltage in the control circuitry of the solenoid.
2. Checks for an open in the control circuit of the solenoid.
3. Identifies an open in the solenoid or ground circuit.
4. Checks for proper resistance value of the solenoid.
5. Ensures malfunction was not due to physical damage of the circuitry.

Diagnostic Aids:

An "intermittent" malfunction is most likely caused by a poor connection, rubbed through wire insulation, or a wire that is broken inside the insulation.

The frequency of the malfunction can be checked by using the enhanced diagnostic function of the Tech 1.

Any circuitry, that is suspected as causing the intermittent complaint, should be thoroughly checked for backed out terminals, improper mating, broken locks, improperly formed or damaged terminals, poor terminal to wiring connections or physical damage to the wiring harness.

DIAGNOSTIC CHART DTC 78 — 1995 CAVALIER AND SUNFIRE WITH TCS, CONTINUED

DTC 78
RIGHT FRONT SOLENOID CIRCUIT OPEN OR SHORTED TO BATTERY

*NOTICE: J39200 TEST LEADS MUST BE "ZEROED" PRIOR TO MAKING ANY RESISTANCE MEASUREMENTS. REFER TO J39200 USER'S MANUAL.

(1)
- IGNITION "OFF."
- DISCONNECT 24-WAY EBTCM CONNECTOR.
- IGNITION "ON."
- USING J 39200, MEASURE VOLTAGE BETWEEN GROUND AND 24-WAY EBTCM HARNESS CONNECTOR TERMINAL "4".

IS VOLTAGE 1.0V OR LESS?

NO: REPAIR SHORT TO VOLTAGE IN CKT 1289.

YES:

(2)
- IGNITION "OFF."
- DISCONNECT RIGHT FRONT SOLENOID CONNECTOR.
- USING J 39200, MEASURE RESISTANCE BETWEEN RIGHT FRONT SOLENOID HARNESS CONNECTOR TERMINAL "A" AND 24-WAY EBTCM HARNESS CONNECTOR TERMINAL "4".

IS RESISTANCE 2.0Ω OR LESS?

NO: REPAIR OPEN OR HIGH RESISTANCE IN CKT 1289.

YES:

(3)
- USING J39200, MEASURE RESISTANCE BETWEEN GROUND AND RIGHT FRONT SOLENOID HARNESS CONNECTOR TERMINAL "B".

IS RESISTANCE 2.0Ω OR LESS?

NO: REPAIR OPEN IN CKT 152.

YES:

(4)
- USING J39200, MEASURE RESISTANCE BETWEEN RIGHT FRONT SOLENOID TERMINALS "A" AND "B".

IS RESISTANCE BETWEEN 2.5Ω AND 5Ω?

NO: REPLACE RIGHT FRONT SOLENOID.

YES:

(5)
- INSPECT RIGHT FRONT SOLENOID AND 24-WAY EBTCM HARNESS CONNECTORS FOR POOR TERMINAL CONTACT OR CORROSION.
- REPLACE TERMINALS THAT EXHIBIT POOR CONTACT OR CORROSION.
- INSPECT CKT 1289 FOR DAMAGE WHICH MAY RESULT IN A SHORT TO VOLTAGE WITH ALL CONNECTORS CONNECTED. REPAIR DAMAGE IF EVIDENT.
- RECONNECT ALL CONNECTORS.
- IGNITION "ON."

IS DTC 78 SET AS A CURRENT DTC?

YES: REPLACE EBTCM.

NO: MALFUNCTION IS INTERMITTENT. REFER TO "DIAGNOSTIC AIDS" ON FACING PAGE.

DIAGNOSTIC CHART DTC 81 — 1995 CAVALIER AND SUNFIRE WITH TCS

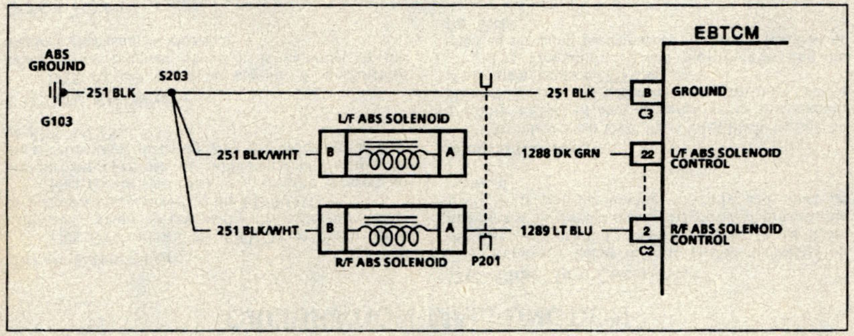

DTC 81

RIGHT FRONT SOLENOID CIRCUIT SHORTED TO GROUND OR DRIVER OPEN

Circuit Description:

This DTC identifies a solenoid that cannot be energized due to an open in its driver circuitry, or a short to ground between the solenoid driver and the solenoid. These malfunctions can affect ABS operation since the flow of brake fluid to the wheel cylinder cannot be stopped, making ABS operation for that channel impossible.

DTC Will Set When:

DTC 81 can be set only when the solenoid is commanded "ON." A malfunction exists if the EBTCM detects the right front solenoid control circuit voltage is out of specification.

Action Taken:

A malfunction DTC is stored, ABS is disabled and the "ABS" warning lamp is turned "ON."

DTC Chart Test Description:

Number(s) below refer to circled number(s) on the diagnostic chart.

1. Checks for a solenoid that is internally shorted to ground.
2. Checks for a solenoid that is not within proper resistance value.
3. Indicates if a short to ground exists in the solenoid control circuitry.
4. Ensures malfunction was not due to physical damage of the circuitry.

Diagnostic Aids:

An "intermittent" malfunction is most likely caused by a poor connection, rubbed through wire insulation, or a wire that is broken inside the insulation.

The frequency of the malfunction can be checked by using the enhanced diagnostic function of the Tech 1

Any circuitry, that is suspected as causing the intermittent complaint, should be thoroughly checked for backed out terminals, improper mating, broken locks, improperly formed or damaged terminals, poor terminal to wiring connections or physical damage to the wiring harness.

DIAGNOSTIC CHART DTC 81 — 1995 CAVALIER AND SUNFIRE WITH TCS, CONTINUED

DTC 81

RIGHT FRONT SOLENOID CIRCUIT SHORTED TO GROUND OR DRIVER OPEN

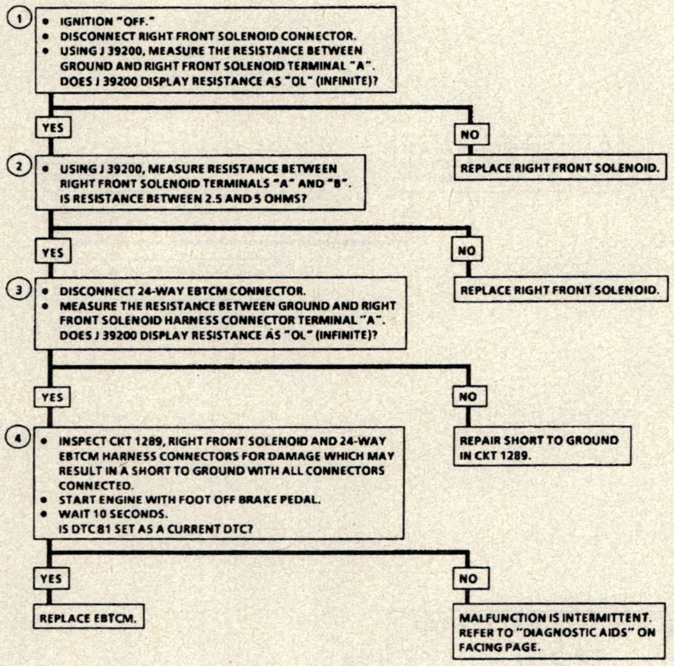

DIAGNOSTIC CHART DTC 82 — 1995 CAVALIER AND SUNFIRE WITH TCS

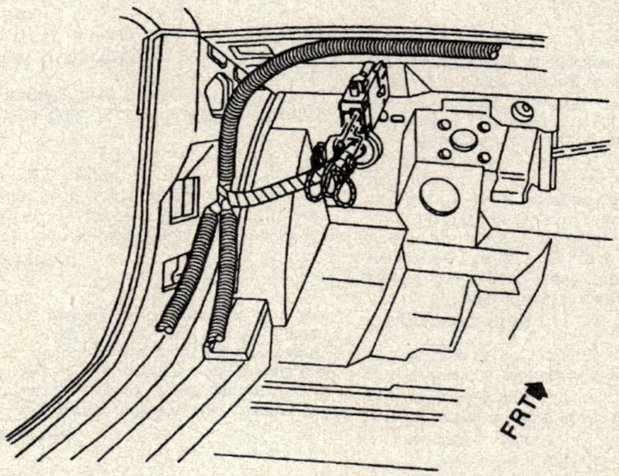

DTC 82
CALIBRATION MALFUNCTION

Circuit Description:

This DTC allows the EBTCM to check for a calibration failure by comparing the calibration value to a known value stored in the EEPROM.

This test is also used as a security measure to prevent improper use of calibrations or changes to these calibrations that may alter the designed function of ABS w/ETS.

DTC Will Set When:

DTC 82 can be set at key-up. A malfunction exists if the program identifier is incorrect or the memory checksum is incorrect.

Action Taken:

A malfunction DTC is stored, ABS is disabled and the "ABS" warning lamp is turned "ON."

DTC Chart Test Description:

Number(s) below refer to circled number(s) on the diagnostic chart. 1. This step checks to see if the malfunction is present during diagnosis. If present, the EBTCM is not functioning correctly and must be replaced.

Diagnostic Aids:

An intermittent DTC 82 is most likely caused by a faulty EEPROM that is sensitive to temperature changes. If DTC 82 occurred more than once, but is intermittent, replace EBTCM.

The frequency of the malfunction can be checked by using the enhanced diagnostic function of the Tech 1.

DIAGNOSTIC CHART DTC 82 — 1995 CAVALIER AND SUNFIRE WITH TCS, CONTINUED

DTC 82
CALIBRATION MALFUNCTION

(1) • IGNITION "ON."
IS DTC 82 SET AS A CURRENT DTC?

- YES → REPLACE EBTCM.
- NO → MALFUNCTION IS NOT PRESENT AT THIS TIME. REFER TO "DIAGNOSTIC AIDS" ON FACING PAGE.

DIAGNOSTIC CHART DTC 86 — 1995 CAVALIER AND SUNFIRE WITH TCS

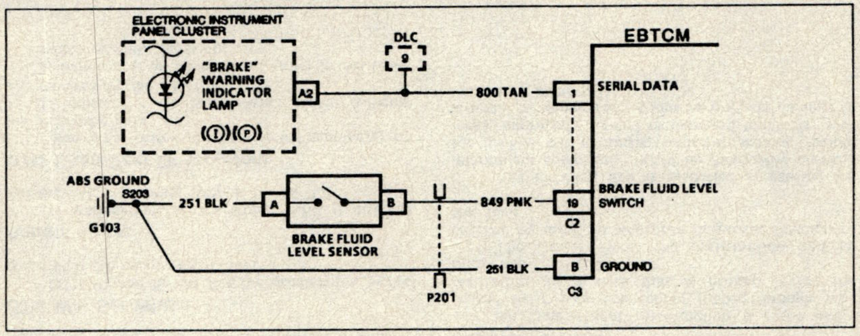

DTC 86
EBTCM TURNED "ON" THE RED "BRAKE" WARNING LAMP

Circuit Description:

This DTC is provided as an "information only" test and reflects the status of the command issued by the EBTCM to illuminate the red "BRAKE" warning lamp. If another diagnostic DTC issues a command to illuminate the red "BRAKE" warning lamp, DTC 86 will be stored in EEPROM as a history DTC at the conclusion of the ignition cycle.

DTC Chart Test Description:

Number(s) below refer to circled number(s) on the diagnostic chart.

1. This step serves to identify if a DTC other than DTC 86 commanded the red "BRAKE" warning lamp "ON."

Diagnostic Aids:

Any ABS w/ETS mechanical DTC that issues a command to illuminate the red "BRAKE" warning lamp will also result in DTC 86 being stored in EEPROM during shut down. These DTC's are: 38, 41, 42, 45, 46, 47, 48 and 51. If the motors are not in their home position, certain electrical DTC's will also command the red "BRAKE" warning lamp "ON." These DTC's are: 14, 18, 55, 56, 57, 58, 61, 62, 63, 64, 65, 66, 67, 68, 71 and 72.

If any of these DTC's are indicated along with DTC 86, they must be corrected prior to addressing a DTC 86 malfunction.

DIAGNOSTIC CHART DTC 86 — 1995 CAVALIER AND SUNFIRE WITH TCS, CONTINUED

DTC 86
EBTCM TURNED ON THE RED BRAKE WARNING LAMP

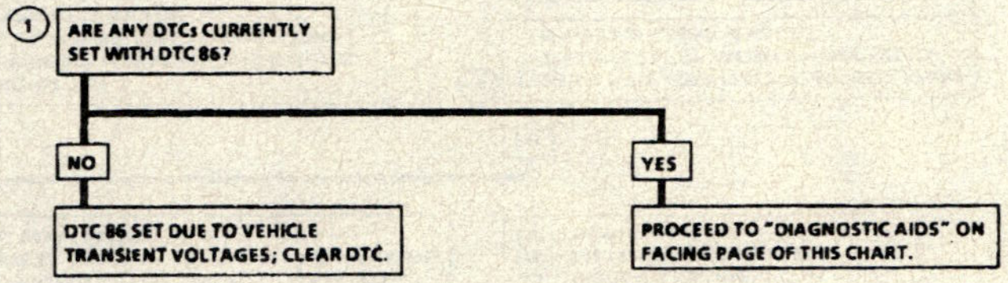

DIAGNOSTIC CHART DTC 91 — 1995 CAVALIER AND SUNFIRE WITH TCS

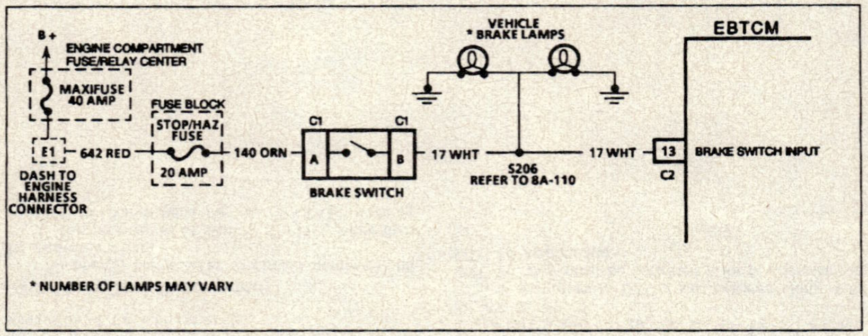

DTC 91

OPEN BRAKE SWITCH CONTACTS DURING DECELERATION

Circuit Description:

This DTC is used to detect an open brake switch in the non-ABS mode. The EBTCM looks for a deceleration rate that would indicate braking action and verifies this assumption by requiring several repeats of this detection method. In each case, ABS will not be available since no brake switch is seen.

DTC Will Set When:

DTC 91 can be set if three deceleration cycles occur with the brake switch "OFF."

Action Taken:

A malfunction DTC is stored, ABS is disabled and the "ABS" warning lamp is turned "ON."

DTC Chart Test Description:

Number(s) below refer to circled number(s) on the diagnostic chart.

1. Indicates if the brake switch signal is being received by the EBTCM.
2. Indicates if an open circuit exists in the brake switch or brake lamp circuitry.
3. Checks for an open in the brake switch input circuit.
4. Ensures malfunction was not due to poor terminal contact.
5. Checks for voltage entering the brake switch.
6. Checks for voltage exiting the brake switch.

Diagnostic Aids:

An "intermittent" malfunction is most likely caused by a poor connection, rubbed through wire insulation, or a wire that is broken inside the insulation.

The frequency of the malfunction can be checked by using the enhanced diagnostic function of the Tech 1

Any circuitry, that is suspected as causing the intermittent complaint, should be thoroughly checked for backed out terminals, improper mating, broken locks, improperly formed or damaged terminals, poor terminal to wiring connections or physical damage to the wiring harness.

DIAGNOSTIC CHART DTC 91 — 1995 CAVALIER AND SUNFIRE WITH TCS, CONTINUED

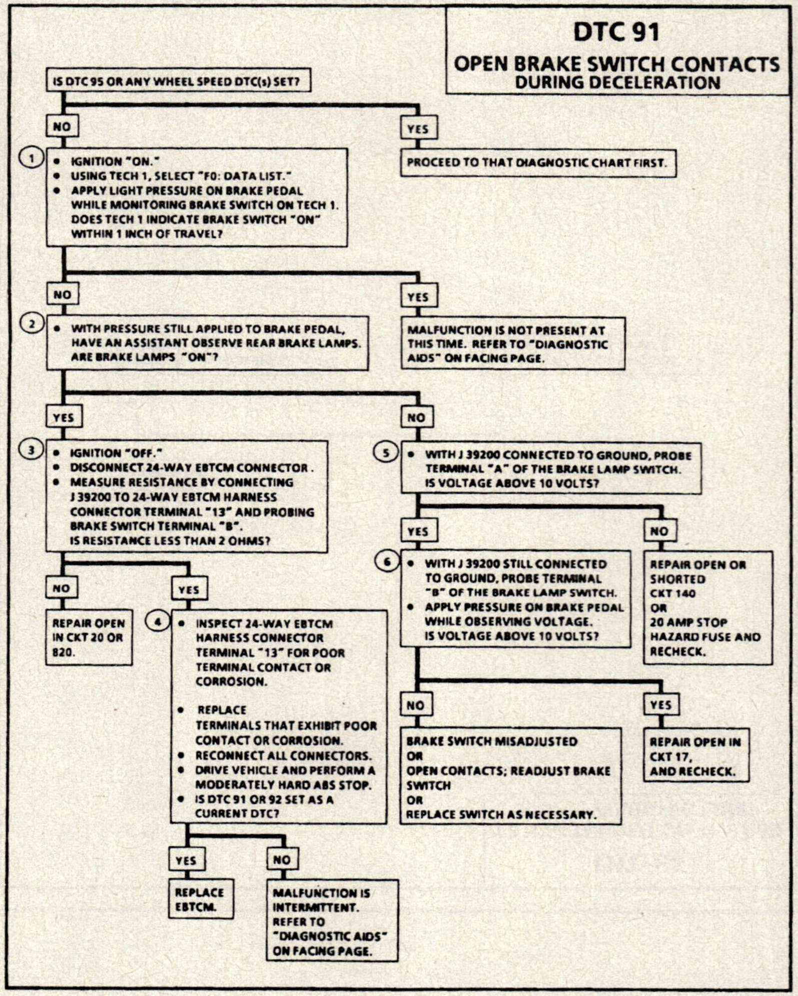

DIAGNOSTIC CHART DTC 92 — 1995 CAVALIER AND SUNFIRE WITH TCS

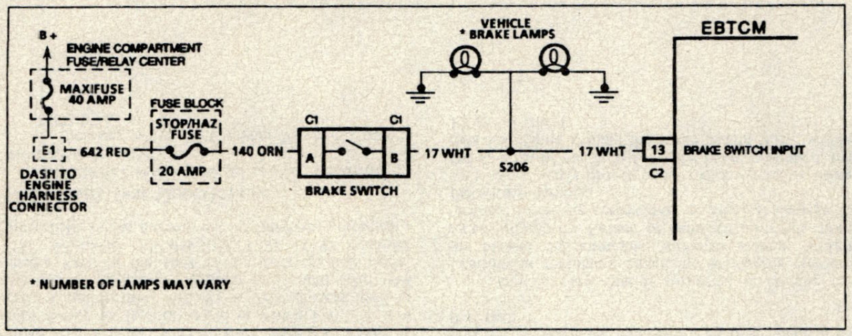

DTC 92
OPEN BRAKE SWITCH CONTACTS WHEN ABS WAS REQUIRED

Circuit Description:

This DTC is run to determine the proper operation of the brake switch. This is important because ABS is activated when the brake switch is "ON." If the brake switch is "OFF," ABS will never be activated. Since this malfunction is difficult to detect under normal braking conditions, this malfunction is only detected when ABS is required.

DTC Will Set When:

DTC 92 can be set if the vehicle's speed is greater than 8 km/h (5 mph). If the brake was not "ON" and a release was required on two channels for 0.7 seconds, a malfunction exists.

Action Taken:

A malfunction DTC is stored, ABS is disabled and the "ABS" warning lamp is turned "ON."

DTC Chart Test Description:

Number(s) below refer to circled number(s) on the diagnostic chart.

1. Indicates if the brake switch signal is being received by the EBTCM.
2. Indicates if an open circuit exists in the brake switch or brake lamp circuitry.
3. Checks for an open in the brake switch input circuit.
4. Ensures malfunction was not due to poor terminal contact.
5. Checks for voltage entering the brake switch.
6. Checks for voltage exiting the brake switch.

Diagnostic Aids:

An "intermittent" malfunction is most likely caused by a poor connection, rubbed through wire insulation, or a wire that is broken inside the insulation.

The frequency of the malfunction can be checked by using the enhanced diagnostic function of the Tech 1

Any circuitry, that is suspected as causing the intermittent complaint, should be thoroughly checked for backed out terminals, improper mating, broken locks, improperly formed or damaged terminals, poor terminal to wiring connections or physical damage to the wiring harness.

DIAGNOSTIC CHART DTC 92 — 1995 CAVALIER AND SUNFIRE WITH TCS, CONTINUED

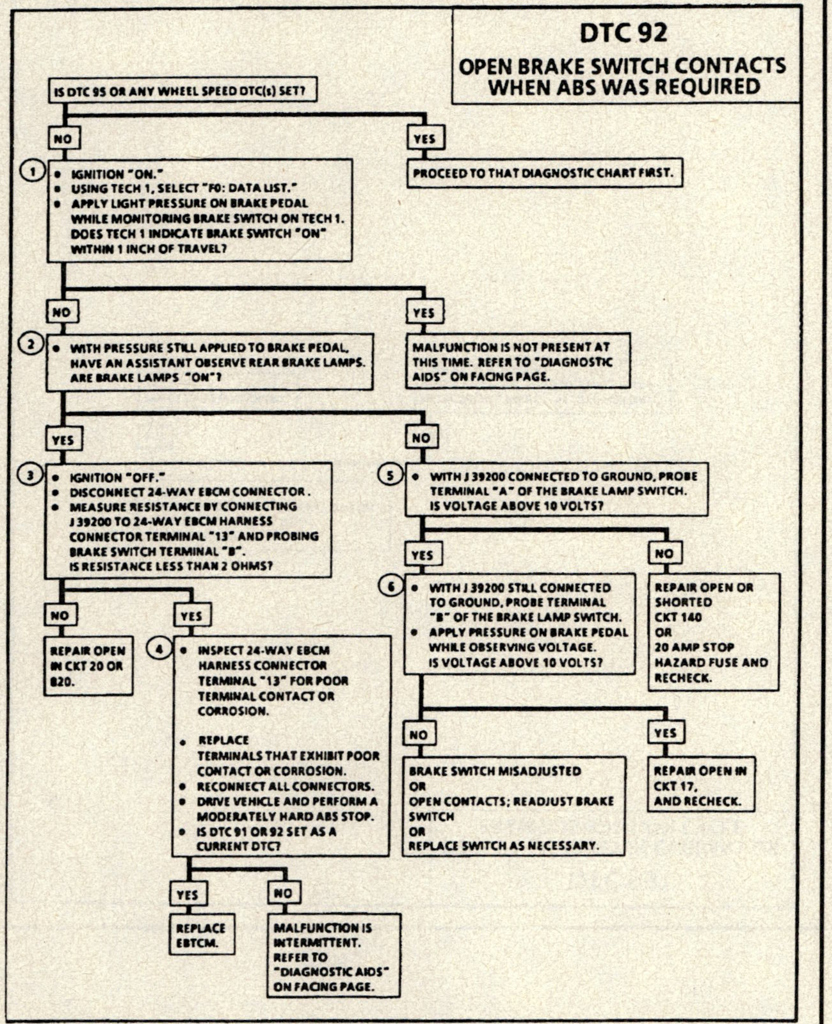

DIAGNOSTIC CHART DTC 93 — 1995 CAVALIER AND SUNFIRE WITH TCS

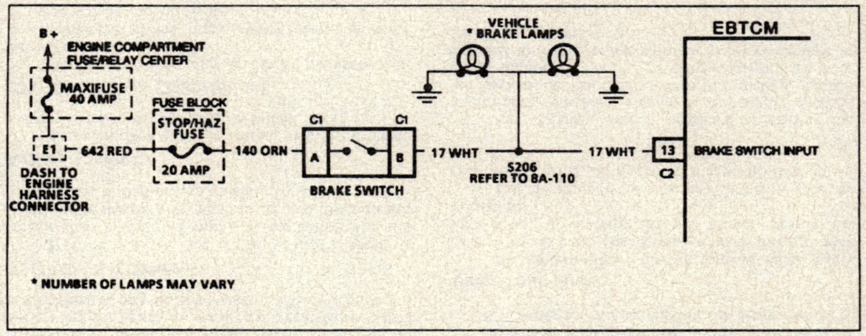

DTC 93

DTC's 91 OR 92 SET IN CURRENT OR PREVIOUS IGNITION CYCLE

Circuit Description:

This DTC is the second portion of DTC's 91 and 92. If DTC's 91 or 92 occurred during the last ignition cycle, DTC 93 becomes a current malfunction during the next ignition cycle, keeping ABS disabled until a brake switch "ON" state is seen. When a change is seen during an ignition cycle in which DTC 93 is a current malfunction, DTC 91 or 92 will clear itself at the end of the current ignition cycle, and ABS will enable itself at the start of the next ignition cycle. DTC 93 alone indicates DTC's 91 or 92 occurred previously, but is intermittent, or has been corrected.

DTC Chart Test Description:

Number(s) below refer to circled number(s) on the diagnostic chart.

1. Indicates which DTC (either 91 or 92) caused DTC 93 to set.

Diagnostic Aids:

An "intermittent" malfunction is most likely caused by a poor connection, rubbed through wire insulation, or a wire that is broken inside the insulation.

The frequency of the malfunction can be checked by using the enhanced diagnostic function of the Tech 1

Any circuitry, that is suspected as causing the intermittent complaint, should be thoroughly checked for backed out terminals, improper mating, broken locks, improperly formed or damaged terminals, poor terminal to wiring connections or physical damage to the wiring harness.

Also, verify proper brake switch operation using the data list of the Tech 1. As the brake is applied, the data list should display the brake switch "ON" within 1 inch of travel.

DIAGNOSTIC CHART DTC 93 — 1995 CAVALIER AND SUNFIRE WITH TCS, CONTINUED

DTC 93

DTCs 91 OR 92 SET IN CURRENT OR PREVIOUS IGNITION CYCLE

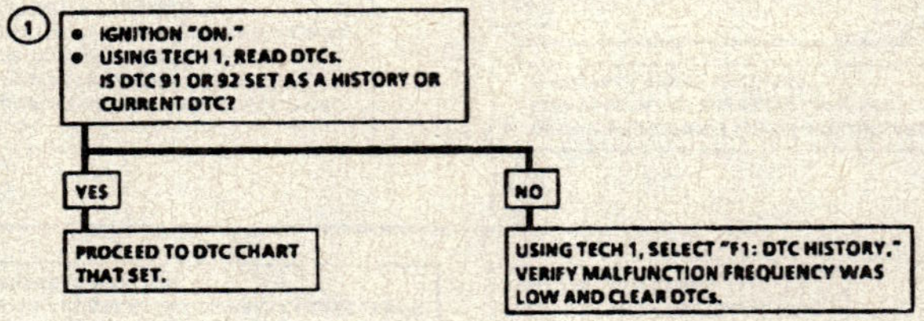

DIAGNOSTIC CHART DTC 94 — 1995 CAVALIER AND SUNFIRE WITH TCS

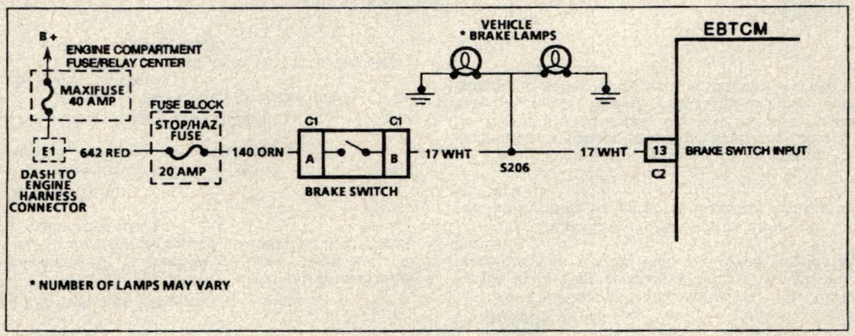

DTC 94
BRAKE SWITCH CONTACTS ALWAYS CLOSED

Circuit Description:

This DTC is run to determine the proper operation of the brake switch. This is important because ABS is activated when the brake switch is "ON" and turned "OFF" when the brake is "OFF." If the brake switch is always "ON," ABS operation will always be requested resulting in potential modulator cycling on rough roads. Additionally, this malfunction will most likely result in a dead battery (due to the brake lamps staying "ON") if the driver is not informed of this malfunction.

DTC Will Set When:

DTC 94 can be set if the vehicle speed reaches at least 40 km/h (25 mph). If the brake was never "OFF" during two consecutive drive cycles, a malfunction exists.

Action Taken:

A malfunction DTC is stored, however ABS is not disabled.

DTC Chart Test Description:

Number(s) below refer to circled number(s) on the diagnostic chart.

1. Checks to see if brake lamps are "ON."
2. Isolates the cause of the malfunction to either a faulty or misadjusted brake switch, or a short to voltage in the brake switch circuitry.
3. Checks for a resistive short to voltage in the brake switch circuitry.
4. Ensures malfunction was not due to a physical damage of the circuitry.

Diagnostic Aids:

An "intermittent" malfunction is most likely caused by a poor connection, rubbed through wire insulation, or a wire that is broken inside the insulation.

The frequency of the malfunction can be checked by using the enhanced diagnostic function of the Tech 1

Any circuitry that is suspected as causing the intermittent complaint should be thoroughly checked for backed out terminals, improper mating, broken locks, improperly formed or damaged terminals, poor terminal to wiring connections or physical damage to the wiring harness.

DIAGNOSTIC CHART DTC 94 — 1995 CAVALIER AND SUNFIRE WITH TCS, CONTINUED

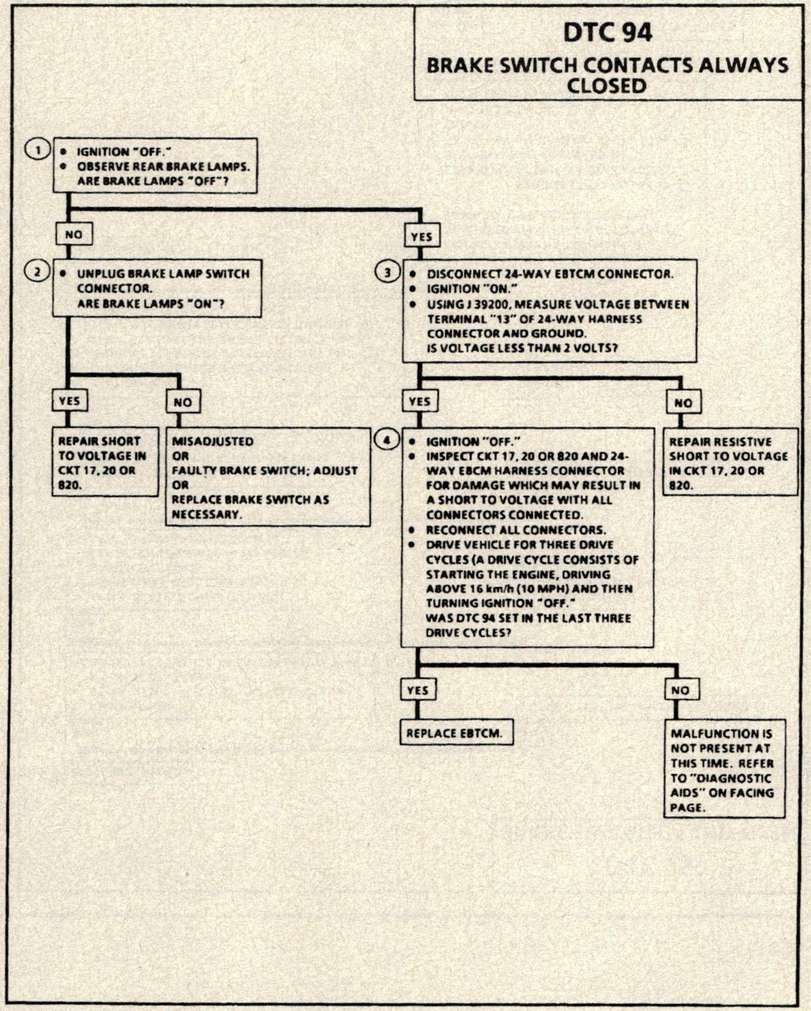

DIAGNOSTIC CHART DTC 95 — 1995 CAVALIER AND SUNFIRE WITH TCS

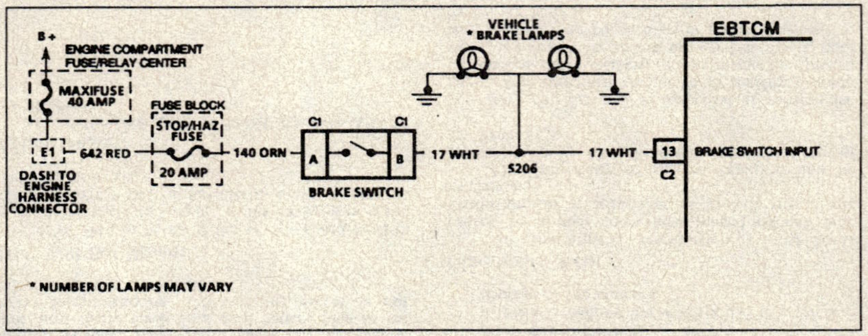

DTC 95
BRAKE SWITCH CIRCUIT OPEN

Circuit Description:

This DTC is used to identify open brake switch circuitry that prevents the brake switch input to the EBTCM from changing states when the brake is applied.

This DTC is used in conjunction with DTC's 91 and 92 to determine the cause of an open brake switch malfunction.

DTC Will Set When:

DTC 95 can be set after initialization is completed. If the brake switch input voltage is out of specification for one second indicating an open circuit, a malfunction exists.

Action Taken:

A malfunction DTC is stored, ABS is disabled and the "ABS" warning lamp is turned "ON."

DTC Chart Test Description:

Number(s) below refer to circled number(s) on the diagnostic chart.

1. Confirms that an open in the brake switch circuitry currently exists.
2. Indicates if the brake switch signal is being received by the EBTCM.
3. Indicates if an open circuit exists in the brake switch or brake lamp circuitry.
4. Isolates the open circuit to either the brake switch input circuitry, or the EBTCM.
5. Ensures malfunction was not due to poor terminal contact.

Diagnostic Aids:

An "intermittent" malfunction is most likely caused by a poor connection, rubbed through wire insulation, or a wire that is broken inside the insulation.

The frequency of the malfunction can be checked by using the enhanced diagnostic function of the Tech 1

Any circuitry, that is suspected as causing the intermittent complaint, should be thoroughly checked for backed out terminals, improper mating, broken locks, improperly formed or damaged terminals, poor terminal to wiring connections or physical damage to the wiring harness.

DIAGNOSTIC CHART DTC 95 — 1995 CAVALIER AND SUNFIRE WITH TCS, CONTINUED

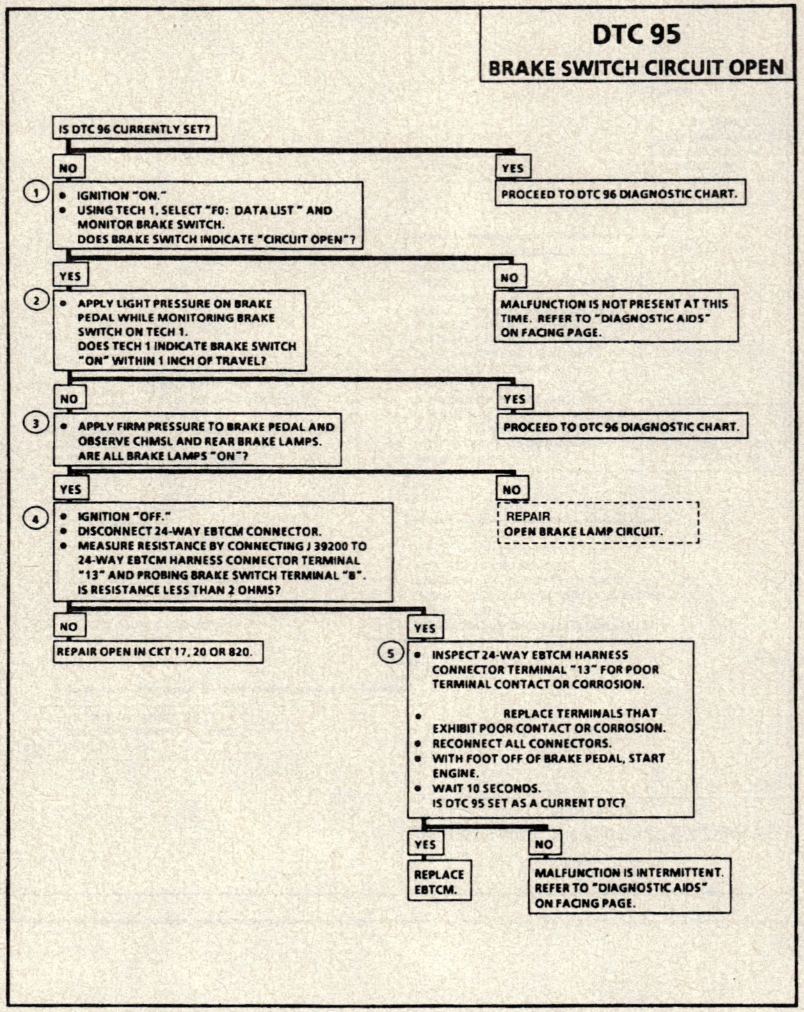

DIAGNOSTIC CHART DTC 96 — 1995 CAVALIER AND SUNFIRE WITH TCS

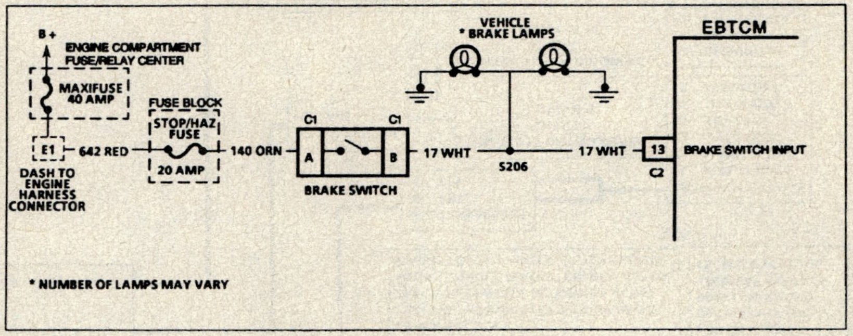

DTC 96
REAR BRAKE LAMP CIRCUIT OPEN

Circuit Description:

This DTC is designed to isolate the cause of a DTC 95 failure and indicate to the driver ABS is still available. If DTC 95 occurs with DTC 96, the brake lamps are open, the brake lamp grounds are open, or the Center High Mounted Stop Lamp (CHMSL) circuit is open during 4-way flasher use. The presence of battery voltage at the brake switch input indicates a valid brake switch input is still available.

DTC Will Set When:

DTC 96 can be set only after DTC 95 has been set. If the EBTCM detects battery voltage on the brake switch input circuit for one second, a malfunction exists.

Action Taken:

A malfunction DTC is stored. ABS is not disabled; however, the "ABS" warning lamp will flash indicating a malfunction exists.

DTC Chart Test Description:

Number(s) below refer to circled number(s) on the diagnostic chart.

1. Checks for additional brake switch DTC's set.
2. Identifies if the malfunction is currently present in the brake circuit.

Diagnostic Aids:

An "intermittent" malfunction is most likely caused by a poor connection, rubbed through wire insulation, or a wire that is broken inside the insulation.

The frequency of the malfunction can be checked by using the enhanced diagnostic function of the Tech 1, as described in "Enhanced Diagnostics" found in this section.

Any circuitry, that is suspected as causing the intermittent complaint, should be thoroughly checked for backed out terminals, improper mating, broken locks, improperly formed or damaged terminals, poor terminal to wiring connections or physical damage to the wiring harness.

DIAGNOSTIC CHART DTC 96 — 1995 CAVALIER AND SUNFIRE WITH TCS, CONTINUED

DTC 96
REAR BRAKE LAMP CIRCUIT OPEN

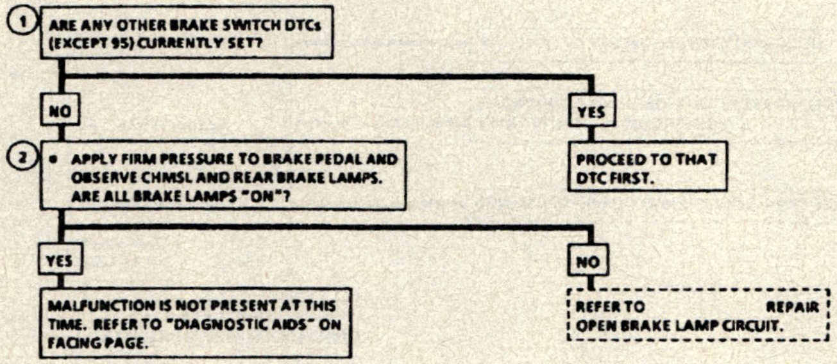

DIAGNOSTIC PROCESS CHART — 1994 SILHOUETTE, TRANS SPORT AND LUMINA APV WITH TCS

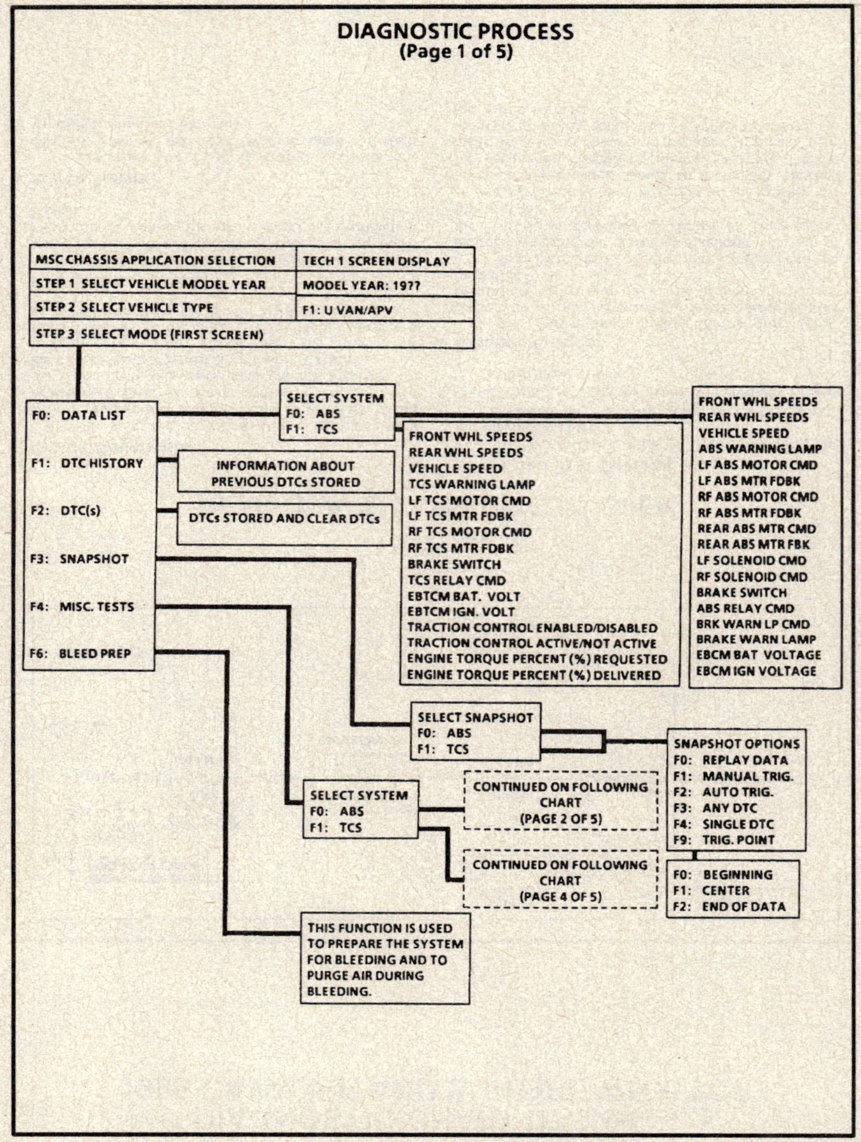

DIAGNOSTIC PROCESS CHART — 1994 SILHOUETTE, TRANS SPORT AND LUMINA APV WITH TCS, CONTINUED

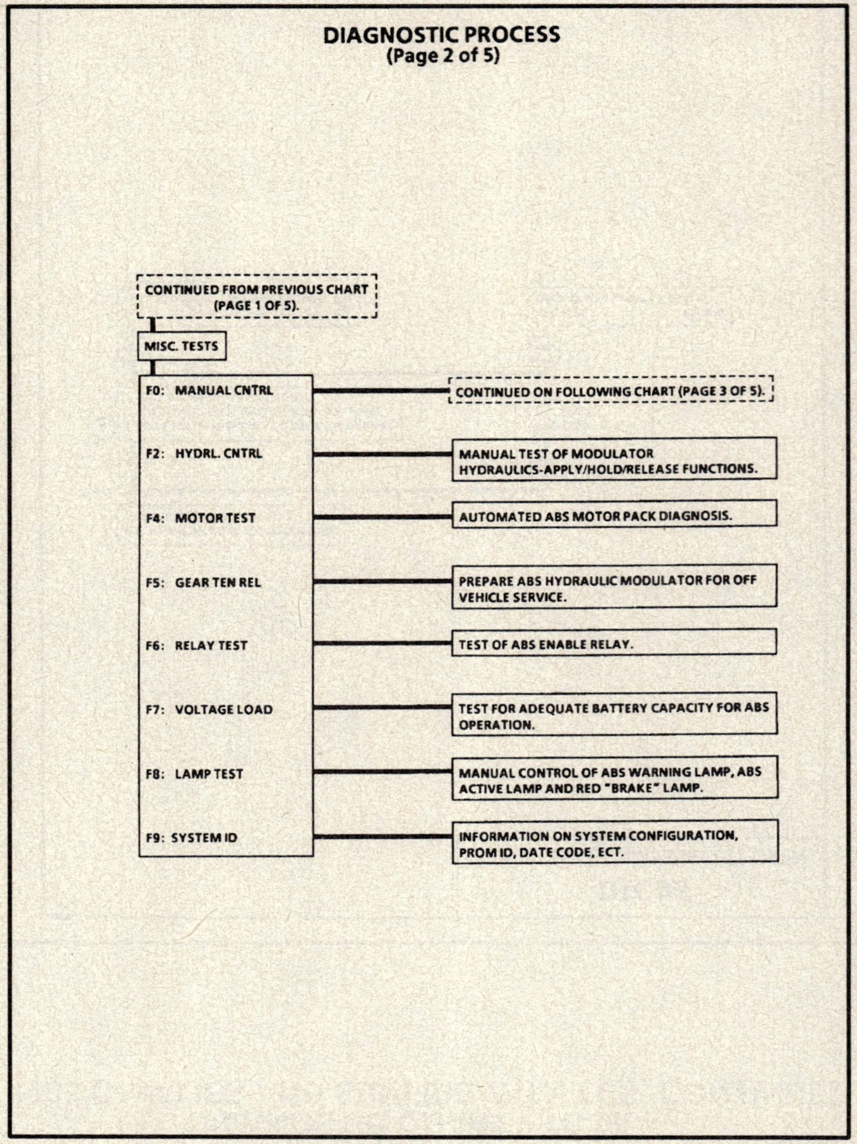

DIAGNOSTIC PROCESS CHART — 1994 SILHOUETTE, TRANS SPORT AND LUMINA APV WITH TCS, CONTINUED

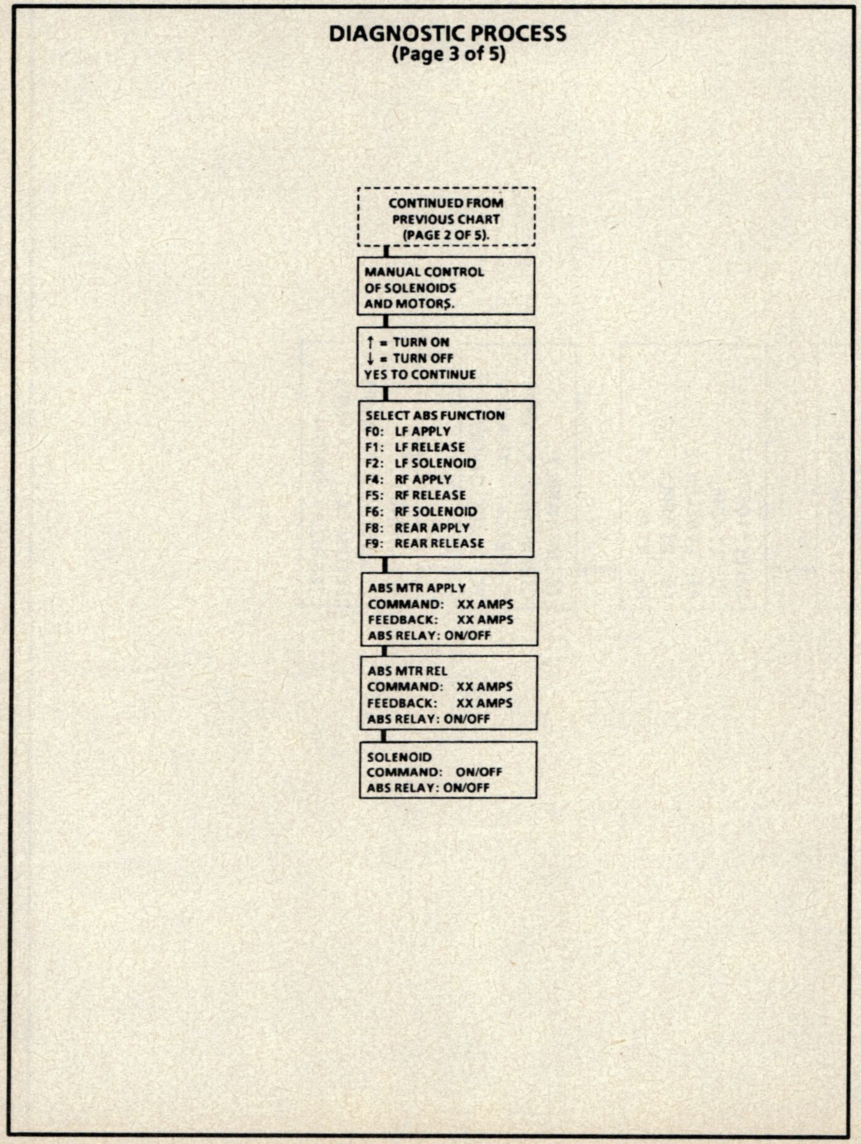

DIAGNOSTIC PROCESS CHART — 1994 SILHOUETTE, TRANS SPORT AND LUMINA APV WITH TCS, CONTINUED

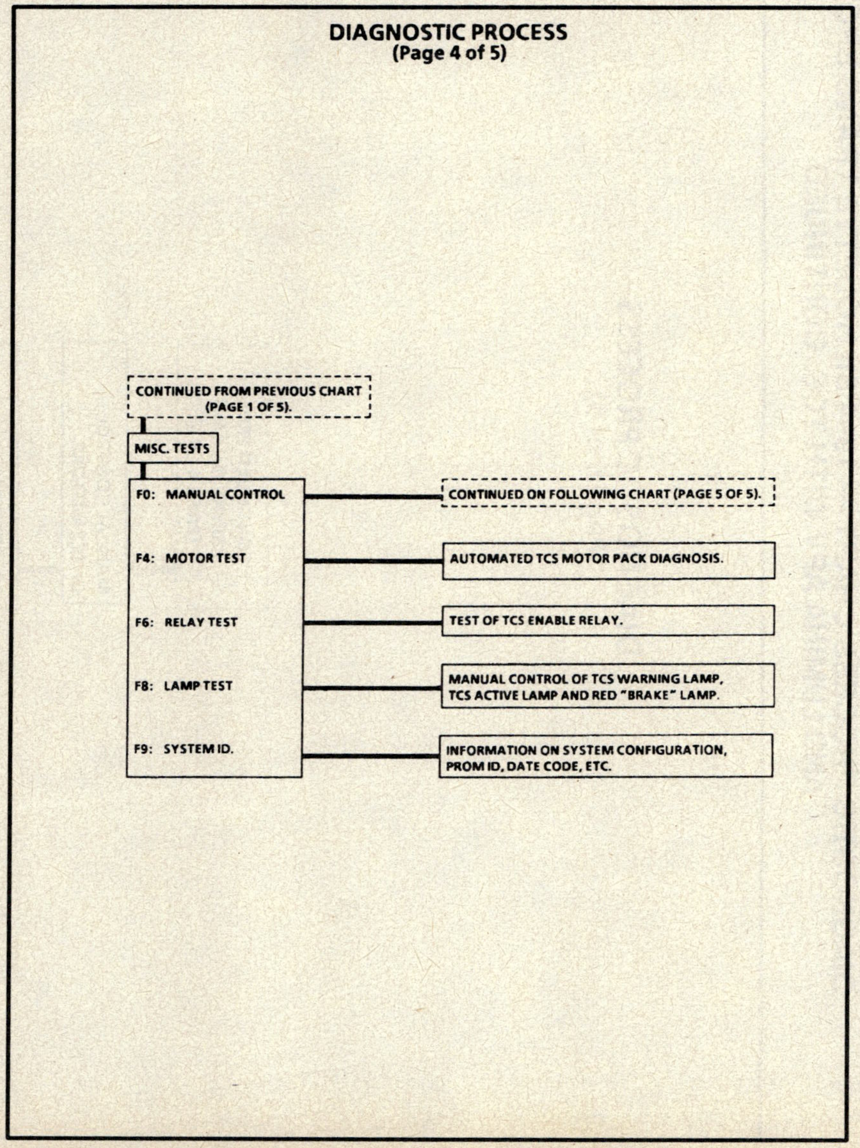

DIAGNOSTIC PROCESS CHART — 1994 SILHOUETTE TRANS SPORT AND LUMINA APV WITH TCS, CONTINUED

DIAGNOSTIC PROCESS
(Page 5 of 5)

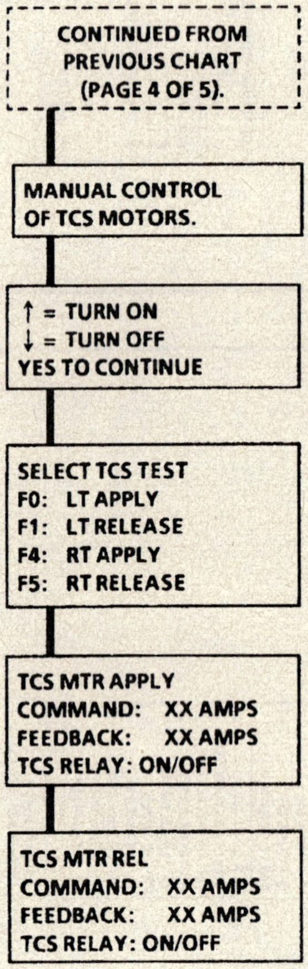

DIAGNOSTIC CHART SYSTEM CHECK — 1994 SILHOUETTE, TRANS SPORT AND LUMINA APV WITH TCS

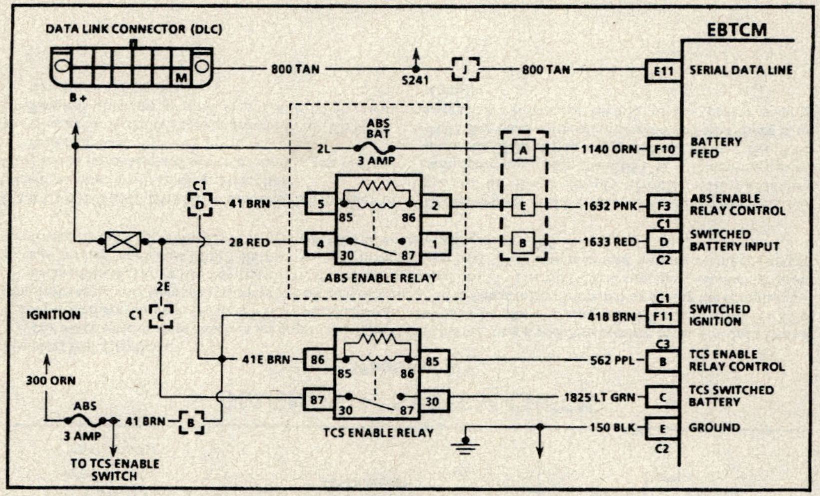

DIAGNOSTIC SYSTEM CHECK

(Page 1 of 4)

System Description:

The diagnostic system check is an organized approach to identifying a problem created by an Antilock Brake Traction Control System (ABS VI/TCS) malfunction. It must be the starting point for any ABS VI/TCS complaint diagnosis because it directs the service technician to the next logical step in diagnosing the complaint.

Serial data is transmitted by the EBTCM through terminal "E11". The EBTCM is supplied switched ignition voltage through terminal "F11", battery voltage through terminal "F10" and switched battery voltage through terminals "D" and "C". The EBTCM ground is provided through terminal "E".

Chart Test Description: Number(s) below refer to circled number(s) on the diagnostic chart.

1. Indicates if EBTCM is transmitting data.
2. Checks for a serial data line malfunction.
3. Checks for proper Tech 1 connection to DLC.
4. Checks for open ABS BAT 3 amp fuse.
5. Checks for open ABS 3 amp fuse.
6. Checks for high resistance in the EBTCM ground circuit.
7. Checks for current DTCs.
8. Checks for proper ABS and TCS warning lamp operation.
9. Checks for history DTCs.

Diagnostic Aids: Excessive resistance in the ground or power supply circuits will not allow communication with the EBTCM. If communication with the EBTCM is not possible, ensure the ABS VI/TCS ground connection is good and that there is no excessive resistance in any of the power supply circuits.

DIAGNOSTIC CHART SYSTEM CHECK — 1994 SILHOUETTE, TRANS SPORT AND LUMINA APV WITH TCS, CONTINUED

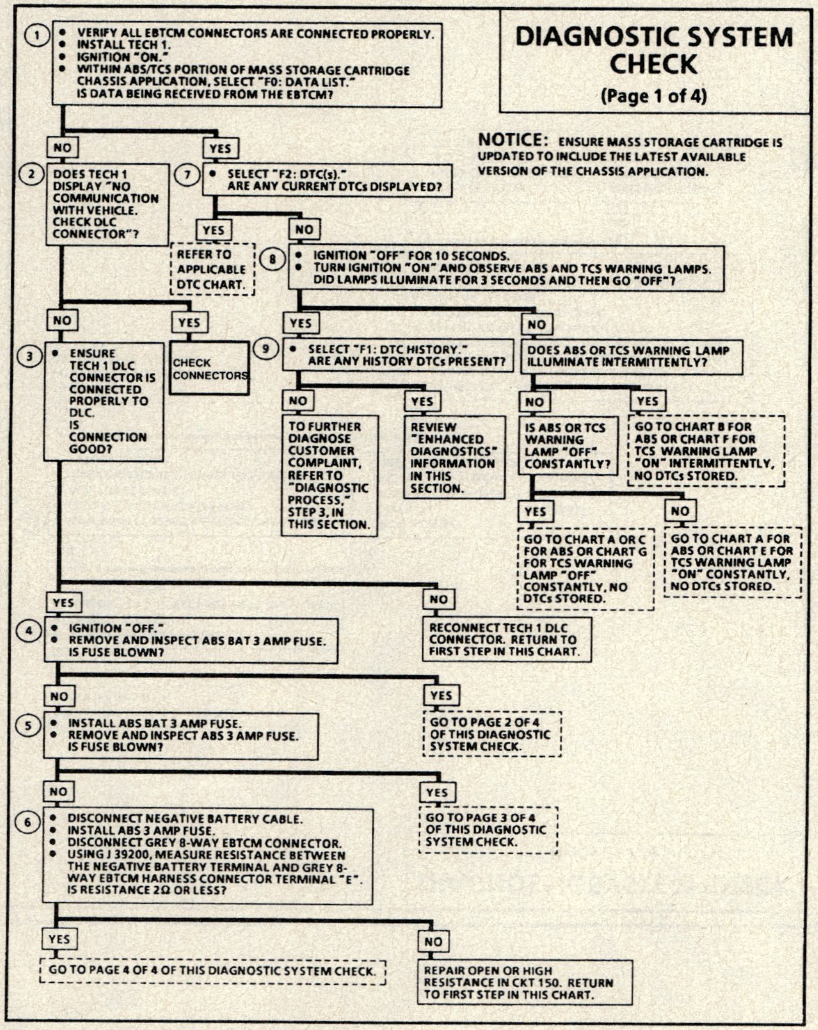

DIAGNOSTIC CHART SYSTEM CHECK — 1994 SILHOUETTE, TRANS SPORT AND LUMINA APV WITH TCS, CONTINUED

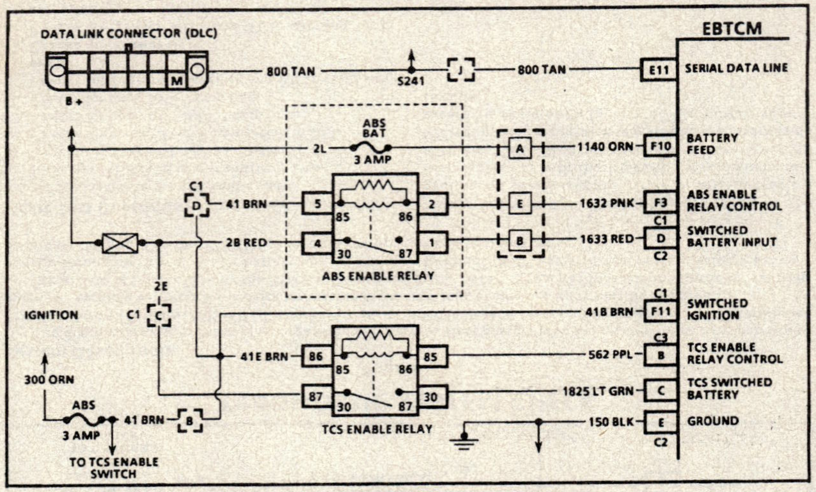

DIAGNOSTIC SYSTEM CHECK
(Page 2 of 4)

System Description:
The diagnostic system check is an organized approach to identifying a problem created by an Antilock Brake Traction Control System (ABS VI/TCS) malfunction. It must be the starting point for any ABS VI/TCS complaint diagnosis because it directs the service technician to the next logical step in diagnosing the complaint.
Serial data is transmitted by the EBTCM through terminal "E11". The EBTCM is supplied switched ignition voltage through terminal "F11", battery voltage through terminal "F10" and switched battery voltage through terminals "D" and "C". The EBTCM ground is provided through terminal "E".

Chart Test Description: Number(s) below refer to circled number(s) on the diagnostic chart.
10. Checks for possible short to ground in the battery feed circuitry.
11. Checks for possible short to ground in the EBTCM.
12. Ensures short to ground is not due to physical damage of the circuitry.

Diagnostic Aids: Excessive resistance in the ground or power supply circuits will not allow communication with the EBTCM. If communication with the EBTCM is not possible, ensure the ABS VI/TCS ground connection is good and that there is no excessive resistance in any of the power supply circuits.

DIAGNOSTIC CHART SYSTEM CHECK — 1994 SILHOUETTE, TRANS SPORT AND LUMINA APV WITH TCS, CONTINUED

DIAGNOSTIC SYSTEM CHECK
(Page 2 of 4)

CONTINUED FROM DIAGNOSTIC SYSTEM CHECK (PAGE 1 OF 4)

YES

(10)
- REPLACE ABS BAT 3 AMP FUSE.
- WAIT 10 SECONDS.
- REMOVE AND INSPECT ABS BAT 3 AMP FUSE. IS FUSE BLOWN?

YES:

(11)
- DISCONNECT 32-WAY EBTCM CONNECTOR.
- REPLACE ABS BAT 3 AMP FUSE.
- WAIT 10 SECONDS.
- REMOVE AND INSPECT ABS BAT 3 AMP FUSE. IS FUSE BLOWN?

NO:
- INSTALL FUSE.
- REFER TO DIAGNOSE FOR A POSSIBLE SHORT TO GROUND IN BATTERY FEED CIRCUITRY. RETURN TO FIRST STEP IN THIS CHART.

(11) YES:

DIAGNOSE FOR A SHORT TO GROUND IN THE BATTERY FEED CIRCUITRY.

(11) NO:

(12)
- INSPECT CKT 2 AND 32-WAY EBTCM HARNESS CONNECTOR FOR PHYSICAL DAMAGE WHICH MAY RESULT IN A SHORT TO GROUND WITH THE 32-WAY EBTCM HARNESS CONNECTOR CONNECTED TO THE EBTCM. CORRECT DAMAGE IF EVIDENT.
- RECONNECT ALL CONNECTORS
- WAIT 10 SECONDS.
- REMOVE AND INSPECT ABS BAT 3 AMP FUSE. IS FUSE BLOWN?

YES:
- REPLACE EBTCM.
- REPLACE ABS BAT 3 AMP FUSE. RETURN TO FIRST STEP IN THIS CHART.

NO:

MALFUNCTION IS INTERMITTENT.

DIAGNOSTIC CHART SYSTEM CHECK — 1994 SILHOUETTE, TRANS SPORT AND LUMINA APV WITH TCS, CONTINUED

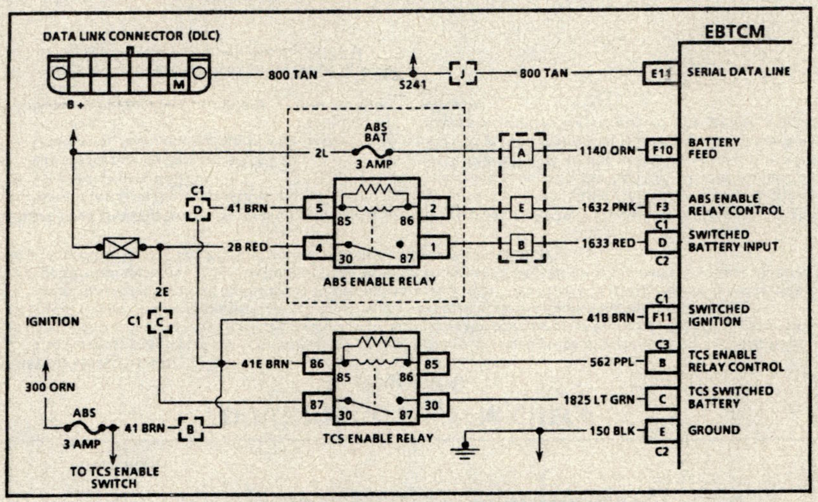

DIAGNOSTIC SYSTEM CHECK

(Page 3 of 4)

System Description:

The diagnostic system check is an organized approach to identifying a problem created by an Antilock Brake Traction Control System (ABS VI/TCS) malfunction. It must be the starting point for any ABS VI/TCS complaint diagnosis because it directs the service technician to the next logical step in diagnosing the complaint.

Serial data is transmitted by the EBTCM through terminal "E11". The EBTCM is supplied switched ignition voltage through terminal "F11", battery voltage through terminal "F10" and switched battery voltage through terminals "D" and "C". The EBTCM ground is provided through terminal "E".

Chart Test Description: Number(s) below refer to circled number(s) on the diagnostic chart.

13. Checks for possible short to ground in the switched ignition circuitry.
14. Isolates the short to ground to either the circuitry or the EBTCM.
15. Ensures short to ground is not due to physical damage of the circuitry.

Diagnostic Aids: Excessive resistance in the ground or power supply circuits will not allow communication with the EBTCM. If communication with the EBTCM is not possible, ensure the ABS VI/TCS ground connection is good and that there is no excessive resistance in any of the power supply circuits.

DIAGNOSTIC CHART SYSTEM CHECK — 1994 SILHOUETTE, TRANS SPORT AND LUMINA APV WITH TCS, CONTINUED

DIAGNOSTIC SYSTEM CHECK

(Page 3 of 4)

CONTINUED FROM DIAGNOSTIC SYSTEM CHECK (PAGE 1 OF 4)

YES

(13)
- REPLACE ABS 3 AMP FUSE.
- IGNITION "ON."
- WAIT 10 SECONDS THEN TURN IGNITION "OFF."
- REMOVE AND INSPECT ABS 3 AMP FUSE. IS FUSE BLOWN?

YES

(14)
- DISCONNECT 32-WAY EBTCM CONNECTOR.
- REPLACE ABS 3 AMP FUSE.
- IGNITION "ON."
- WAIT 10 SECONDS THEN TURN IGNITION "OFF."
- REMOVE AND INSPECT ABS 3 AMP FUSE. IS FUSE BLOWN?

NO

INSTALL FUSE. DIAGNOSE FOR A POSSIBLE SHORT TO GROUND IN THE SWITCHED IGNITION CIRCUITRY. RETURN TO FIRST STEP IN THIS CHART.

NO

(15)
- INSPECT CKT 41 AND 32-WAY EBTCM HARNESS CONNECTOR FOR PHYSICAL DAMAGE WHICH MAY RESULT IN A SHORT TO GROUND WITH THE 32-WAY EBTCM HARNESS CONNECTOR CONNECTED TO THE EBTCM. CORRECT DAMAGE IF EVIDENT.
- RECONNECT ALL CONNECTORS.
- IGNITION "ON."
- WAIT 10 SECONDS THEN TURN IGNITION "OFF."
- REMOVE AND INSPECT ABS 3 AMP FUSE. IS FUSE BLOWN?

YES

REPAIR A SHORT TO GROUND IN ONE OF THE CKT 41 BRANCHES EXITING THE ABS 3 AMP FUSE. REPLACE ABS 3 AMP FUSE. RETURN TO THE FIRST STEP IN THIS CHART.

YES

REPLACE EBTCM. REPLACE ABS 3 AMP FUSE. RETURN TO FIRST STEP IN THIS CHART.

NO

MALFUNCTION IS INTERMITTENT.

DIAGNOSTIC CHART SYSTEM CHECK — 1994 SILHOUETTE, TRANS SPORT AND LUMINA APV WITH TCS, CONTINUED

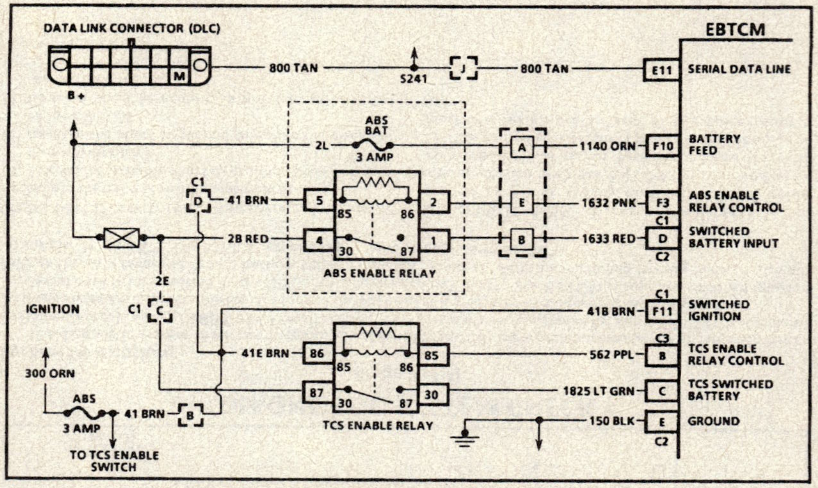

DIAGNOSTIC SYSTEM CHECK

(Page 4 of 4)

System Description:

The diagnostic system check is an organized approach to identifying a problem created by an Antilock Brake Traction Control System (ABS VI/TCS) malfunction. It must be the starting point for any ABS VI/TCS complaint diagnosis because it directs the service technician to the next logical step in diagnosing the complaint.

Serial data is transmitted by the EBTCM through terminal "E11". The EBTCM is supplied switched ignition voltage through terminal "F11", battery voltage through terminal "F10" and switched battery voltage through terminals "D" and "C". The EBTCM ground is provided through terminal "E".

Chart Test Description: Number(s) below refer to circled number(s) on the diagnostic chart.

16. Checks for open in CKT 2.
17. Checks for high resistance in CKT 2.
18. Checks for high resistance in the battery feed circuitry.
19. Ensures malfunction is not due to poor terminal contact.
20. Checks for high resistance in the switched ignition circuit between the fuse block and the EBTCM.

Diagnostic Aids: Excessive resistance in the ground or power supply circuits will not allow communication with the EBTCM. If communication with the EBTCM is not possible, ensure the ABS VI/TCS ground connection is good and that there is no excessive resistance in any of the power supply circuits.

DIAGNOSTIC CHART SYSTEM CHECK — 1994 SILHOUETTE, TRANS SPORT AND LUMINA APV WITH TCS, CONTINUED

DIAGNOSTIC SYSTEM CHECK

(Page 4 of 4)

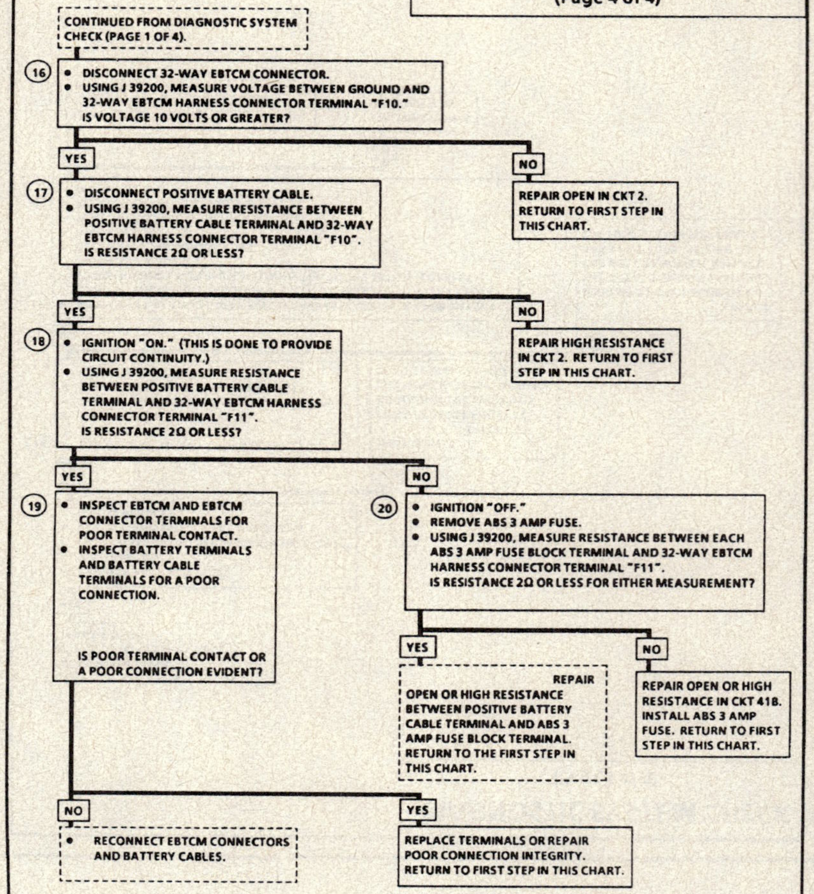

DIAGNOSTIC TROUBLE CODE CHART — 1994 SILHOUETTE, TRANS SPORT AND LUMINA APV WITH TCS

DIAGNOSTIC TROUBLE CODE AND SYMPTOM TABLE

CHART	SYMPTOM
A	ABS (Amber) Warning Lamp "ON" Constantly, No DTCs Stored (Chev Only)
A	ABS (Amber) Warning Lamp "ON" or "OFF" Constantly, No DTCs Stored (Pont/Olds Only)
B	ABS (Amber) Warning Lamp "ON" Intermittently, No DTCs Stored
C	ABS (Amber) Warning Lamp "OFF" Constantly, No DTCs Stored (Chev Only)
D	Tech 1 Displays Undefined DTCs
E	TCS (Amber) Warning Lamp "ON" Constantly, No DTCs Stored
F	TCS (Amber) Warning Lamp "ON" Intermittently, No DTCs Stored
G	TCS (Amber) Warning Lamp "OFF" Constantly, No DTCs Stored

DIAGNOSTIC TROUBLE CODE	DESCRIPTION
11	ABS Warning or Active Lamp Circuit Malfunction (Chev Only) (1 of 2)
11	ABS Warning or Active Lamp Circuit Malfunction (Pont/Olds Only) (1 of 2)
14	ABS Enable Relay Contact Circuit Open (1 of 3)
15	ABS Enable RLY Contact Ckt Short to Batt or Always Closed
16	ABS Enable Relay Coil Circuit Open
17	ABS Enable Relay Coil Circuit Shorted to Gnd
18	ABS Enable Relay Coil Circuit Shorted to Batt
21	Left Front Wheel Speed = 0
22	Right Front Wheel Speed = 0
23	Left Rear Wheel Speed = 0
24	Right Rear Wheel Speed = 0
25	Left Front Excessive Wheel Speed Variation
26	Right Front Excessive Wheel Speed Variation
27	Left Rear Excessive Wheel Speed Variation
28	Right Rear Excessive Wheel Speed Variation
32	LF Wheel Speed Snr Ckt Open or Short to Gnd/Bat (1 of 3)
33	RF Wheel Speed Snr Ckt Open or Short to Gnd/Bat (1 of 3)
34	LR Wheel Speed Snr Ckt Open or Short to Gnd/Bat (1 of 3)
35	RR Wheel Speed Snr Ckt Open or Short to Gnd/Bat (1 of 3)
36	Low System Voltage (1 of 2)
37	High System Voltage
38	Left Front ESB Will Not Hold Motor
41	Right Front ESB Will Not Hold Motor
42	Rear ESB Will Not Hold Motor
44	Left Front ABS Channel Will Not Move
45	Right Front ABS Channel Will Not Move
46	Rear ABS Channel Will Not Move
47	Left Front ABS Motor Free Spins (1 of 2)
48	Right Front ABS Motor Free Spins (1 of 2)
51	Rear ABS Motor Free Spins (1 of 2)
52	Left Front ABS Channel in Release Too Long
53	Right Front ABS Channel in Release Too Long
54	Rear ABS Channel in Release Too Long

DIAGNOSTIC TROUBLE CODE CHART — 1994 SILHOUETTE, TRANS SPORT AND LUMINA APV WITH TCS, CONTINUED

DIAGNOSTIC TROUBLE CODE AND SYMPTOM TABLE

DIAGNOSTIC TROUBLE CODE	DESCRIPTION
55	EBTCM Malfunction (ABS)
56	Left Front ABS Motor Circuit Open
57	Left Front ABS Motor Circuit Shorted to Gnd
58	Left Front ABS Motor Circuit Shorted to Batt
61	Right Front ABS Motor Circuit Open
62	Right Front ABS Motor Circuit Shorted to Gnd
63	Right Front ABS Motor Circuit Shorted to Batt
64	Rear ABS Motor Circuit Open
65	Rear ABS Motor Circuit Shorted to Ground
66	Rear ABS Motor Circuit Shorted to Battery
75	PCM to EBTCM Serial Data Malfunction
76	Left Front Sol Circuit Open or Shorted to Batt
77	Left Front Sol Circuit Shorted to Gnd
78	Right Front Sol Circuit Open or Shorted to Batt
81	Right Front Sol Circuit Shorted to Gnd
82	Calibration Malfunction
86	EBTCM Turned "ON" the Red "BRAKE" Warning Lamp
87	Red "BRAKE" Warning Lamp Circuit Open
88	Red "BRAKE" Warning Lamp Circuit Shorted to Batt
91	Open Brake Switch During Deceleration
92	Open Brake Switch When ABS Was Required
93	DTC 91 or 92 Set in Current or Prev Ign Cycle
94	Brake Switch Contacts Always Closed
95	Brake Switch Circuit Open
96	Rear Brake Lamp Circuit Open
111	TCS Warning or Switch Lamp Circuit Malfunction (1 of 2)
114	TCS Enable Relay Contact Circuit Open (1 of 3)
116	TCS Enable Relay Coil Circuit Open
118	TCS Enable Relay Coil Circuit Shorted To Batt
123	PCM to EBTCM Delivered Torque Circuit Malfunction
124	EBTCM to PCM Requested Torque Circuit Malfunction
136	Low System Voltage (TCS) (1 of 2)
144	Left TCS Channel Will Not Move
145	Right TCS Channel Will Not Move
147	Left TCS Motor Free Spins (1 of 2)
148	Right TCS Motor Free Spins (1 of 2)
155	EBTCM Malfunction (TCS)
156	Left TCS Motor Circuit Open
157	Left TCS Motor Circuit Shorted to Gnd
158	Left TCS Motor Circuit Shorted to Batt
161	Right TCS Motor Circuit Open
162	Right TCS Motor Circuit Shorted to Gnd
163	Right TCS Motor Circuit Shorted to Batt

ELECTRICAL CONNECTOR TERMINAL IDENTIFICATION — 1994 SILHOUETTE, TRANS SPORT AND LUMINA APV WITH TCS

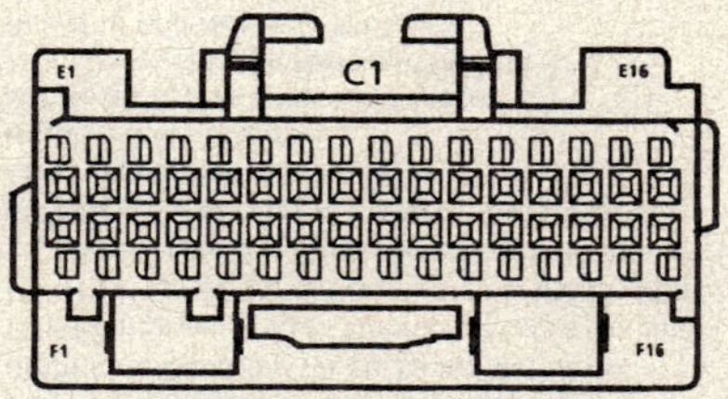

EBTCM 32-WAY CONNECTOR

CONNECTOR TERMINAL	CIRCUIT NO.	COLOR	CIRCUIT DESCRIPTION
E1	884	BLK	LR WHEEL SIGNAL HIGH
E2	885	RED	LR WHEEL SIGNAL LOW
E3	833	TAN	RF WHEEL SIGNAL LOW
E4	872	DK GRN	RF WHEEL SIGNAL HIGH
E5	883	WHT	RR WHEEL SIGNAL LOW
E6	882	BRN	RR WHEEL SIGNAL HIGH
E7	873	YEL	LF WHEEL SIGNAL LOW
E8	830	LT BLU	LF WHEEL SIGNAL HIGH
E9	1289	LT BLU	RF ABS SOLENOID CONTROL
E10	PLUGGED		NOT USED
E11	800	TAN	SERIAL DATA LINE
E12	PLUGGED		NOT USED
E13	PLUGGED		NOT USED
E14	464	TAN/BLK	DELIVERED TORQUE
E15	1571	BRN/WHT	TCS ENABLE SWITCH INPUT
E16	550	BLK	TCS ENABLE LAMP CONTROL
F1	1288	DK GRN	LF ABS SOLENOID CONTROL
F2	852	WHT	ABS WARNING LAMP CONTROL
F3	1632	PNK	ABS ENABLE RELAY CONTROL
F4	33	TAN/WHT	BRAKE WARNING LAMP CONTROL
F5	1656	LT GRN	ABS/TCS ACTIVE CONTROL
F6	PLUGGED		NOT USED
F7	PLUGGED		NOT USED
F8	PLUGGED		NOT USED
F9	PLUGGED		NOT USED
F10	1140	ORN	BATTERY FEED
F11	41	BRN	SWITCHED IGNITION
F12	17	WHT	BRAKE SWITCH INPUT
F13	PLUGGED		NOT USED
F14	463	ORN/BLK	REQUESTED TORQUE
F15	1808	WHT	TCS WARNING LAMP CONTROL
F16	PLUGGED		NOT USED

ELECTRICAL CONNECTOR TERMINAL IDENTIFICATION — 1994 SILHOUETTE, TRANS SPORT AND LUMINA APV WITH TCS, CONTINUED

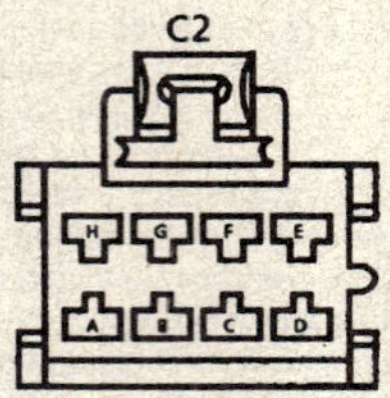

EBTCM GRAY 8-WAY CONNECTOR

PIN	CIRCUIT NO.	COLOR	CIRCUIT
A	1280	BLK	LF ABS MOTOR HIGH
B	1281	PNK	LF ABS MOTOR LOW
C	1284	DK GRN	REAR ABS MOTOR HIGH
D	1633	RED	SWITCHED BATTERY INPUT
E	150	BLK	GROUND
F	1285	ORN	REAR ABS MOTOR LOW
G	1282	PPL	RF ABS MOTOR HIGH
H	1283	BLK	RF ABS MOTOR LOW

ELECTRICAL CONNECTOR TERMINAL IDENTIFICATION — 1994 SILHOUETTE, TRANS SPORT AND LUMINA APV WITH TCS, CONTINUED

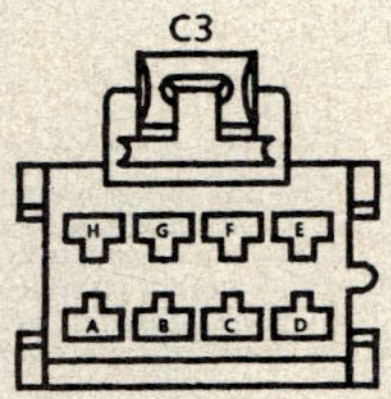

EBTCM BLACK 8-WAY CONNECTOR

PIN	CIRCUIT NO.	COLOR	CIRCUIT
A	566	LT GRN	RF TCS MOTOR HIGH
B	562	PPL	TCS ENABLE RELAY CONTROL
C	1825	LT GRN	TCS SWITCHED BATTERY
D	PLUGGED		NOT USED
E	PLUGGED		NOT USED
F	564	LT BLU	LF TCS MOTOR HIGH
G	565	DK BLU	LF TCS MOTOR LOW
H	567	DK GRN	RF TCS MOTOR LOW

DIAGNOSTIC CHART A — 1994 LUMINA APV WITH TCS

CHART A

ABS (AMBER) WARNING LAMP "ON" CONSTANTLY, NO DTCs STORED
(CHEV ONLY)

NOTICE: DIAGNOSTIC SYSTEM CHECK MUST BE COMPLETED FIRST BEFORE USING THIS CHART.

- IGNITION "OFF."
- INSTALL TECH 1.
- IGNITION "ON."
- USING TECH 1, SELECT "F4: MISC. TESTS."
- SELECT "F8: LAMP TEST."
- ATTEMPT TO FLASH THE ABS WARNING LAMP. DOES ABS WARNING LAMP FLASH?

YES: MALFUNCTION IS NOT PRESENT AT THIS TIME.

NO:

- IGNITION "OFF."
- DISCONNECT LAMP DRIVER MODULE FROM HARNESS.
- IGNITION "ON."
- IS ABS WARNING LAMP "OFF"?

YES: REPLACE LAMP DRIVER MODULE ASSEMBLY.

NO: REPAIR SHORT TO GROUND IN CKT 875. PERFORM CLUSTER DIAGNOSIS.

DIAGNOSTIC CHART A — 1994 SILHOUETTE AND TRANS SPORT WITH TCS

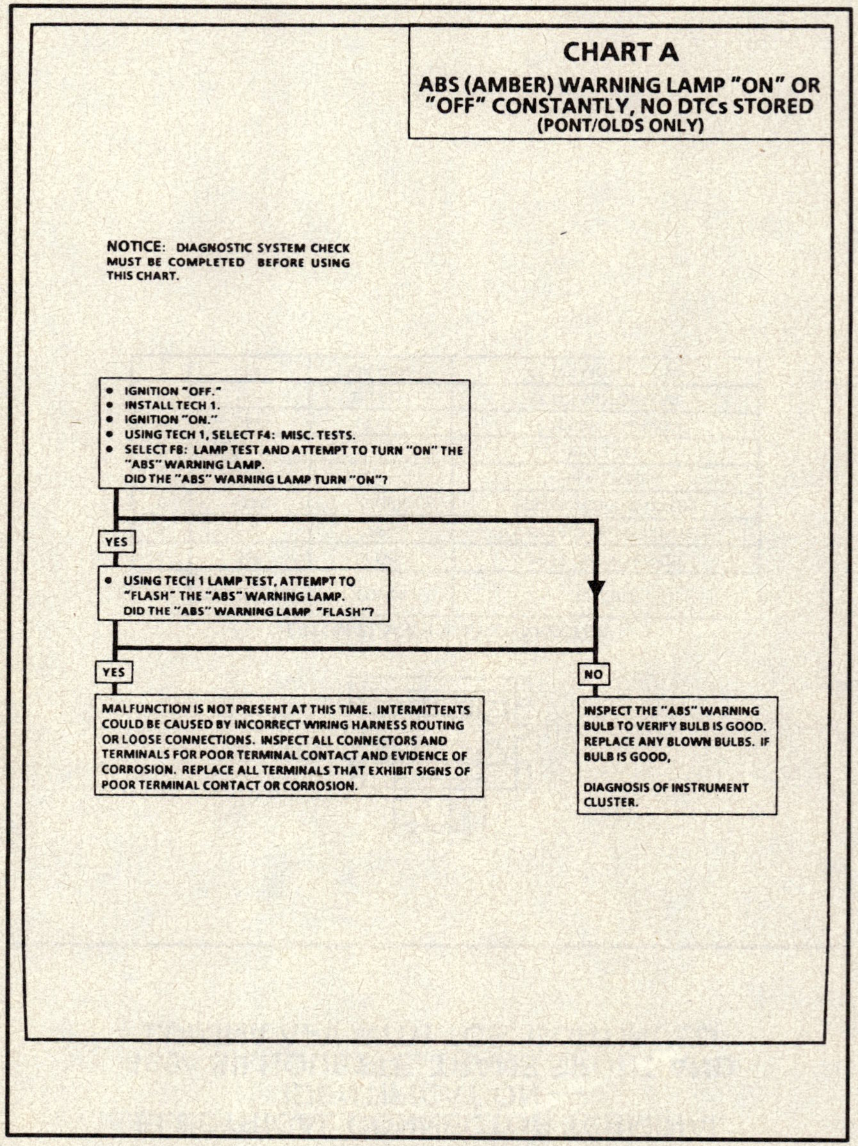

DIAGNOSTIC CHART B — 1994 SILHOUETTE, TRANS SPORT AND LUMINA APV WITH TCS

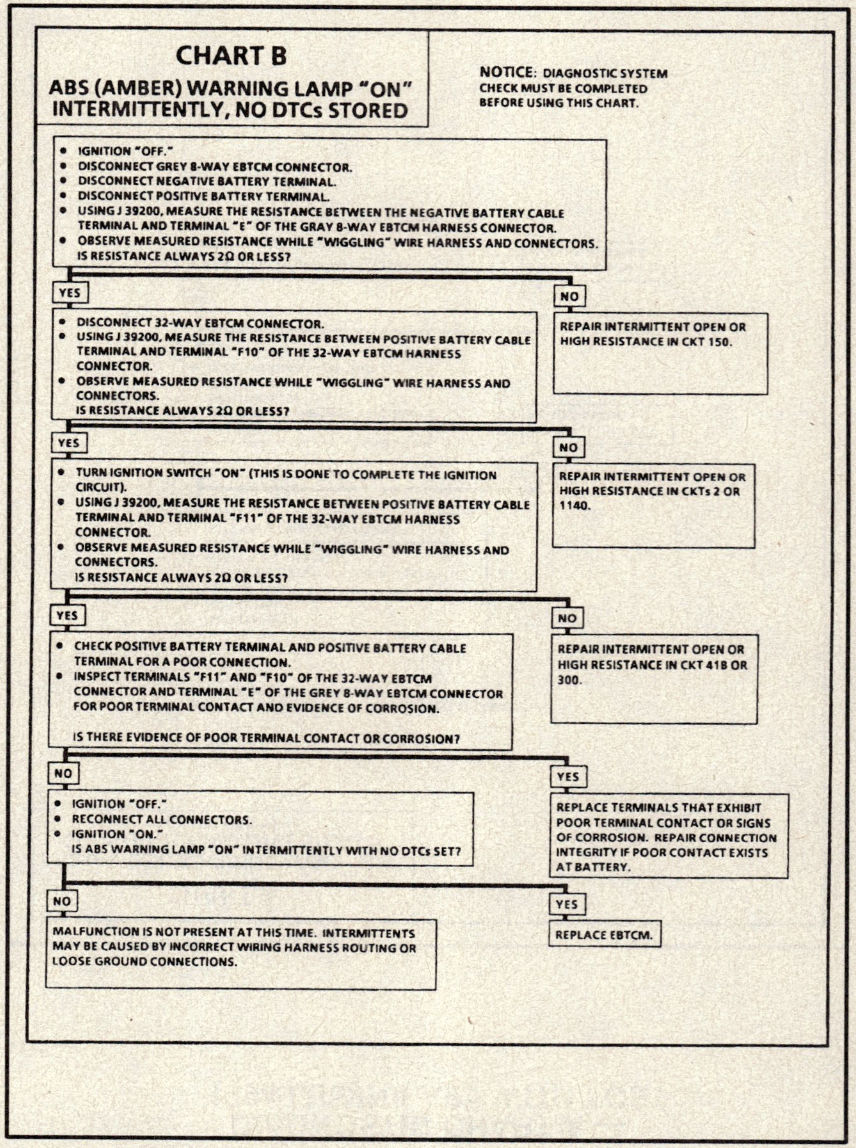

DIAGNOSTIC CHART C — 1994 SILHOUETTE, TRANS SPORT AND LUMINA APV WITH TCS

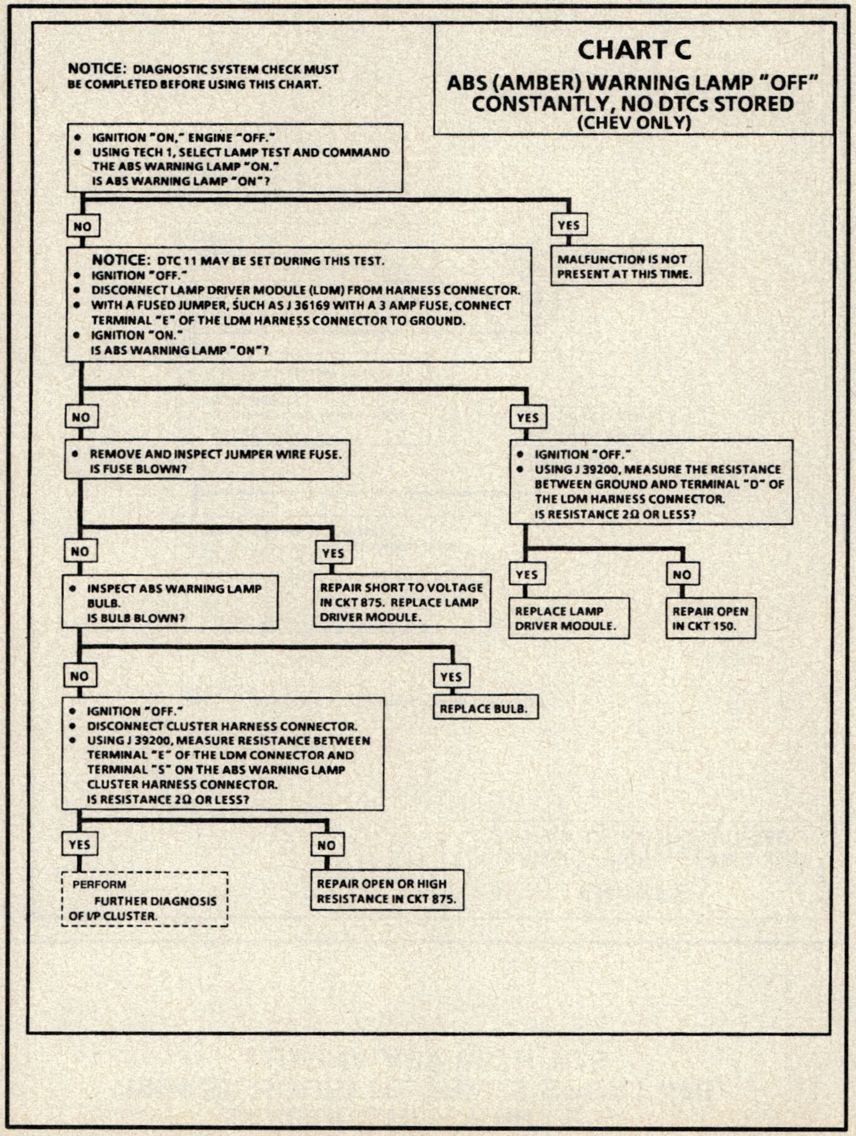

DIAGNOSTIC CHART D — 1994 SILHOUETTE, TRANS SPORT AND LUMINA APV WITH TCS

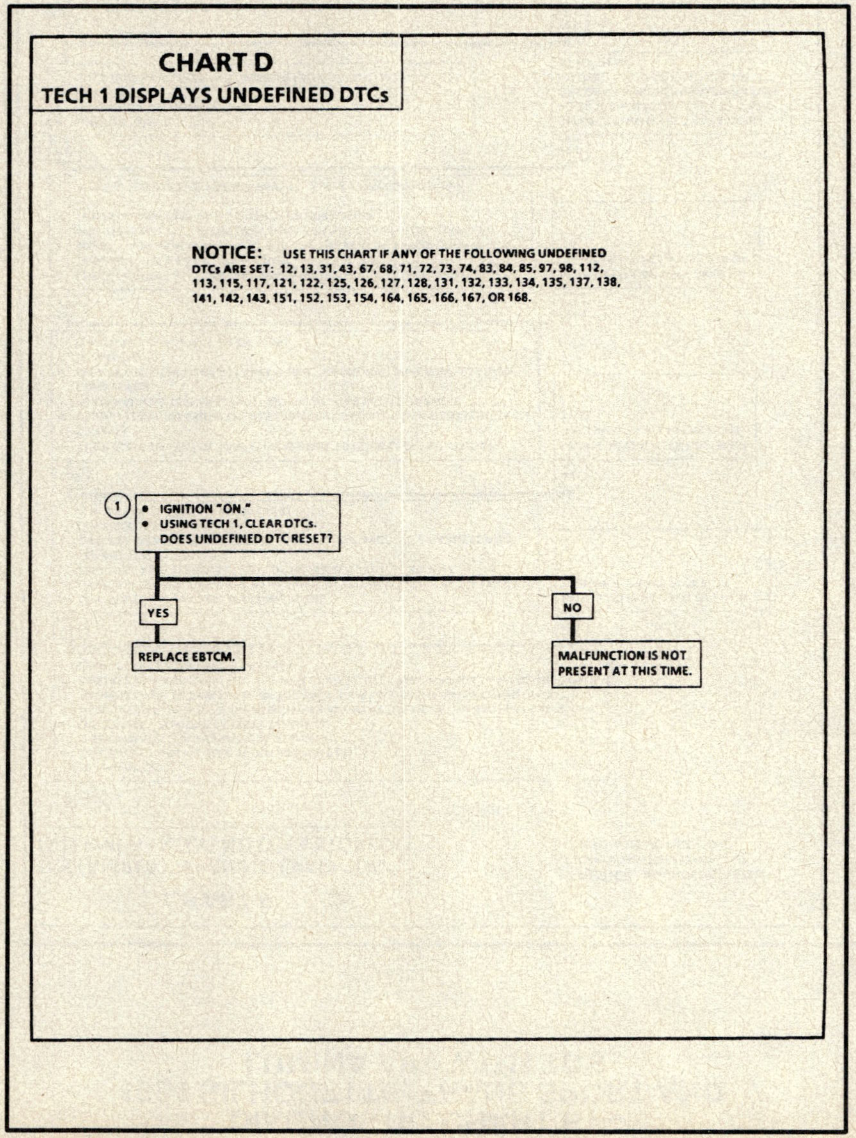

DIAGNOSTIC CHART E — 1994 SILHOUETTE, TRANS SPORT AND LUMINA APV WITH TCS

CHART E
TCS (AMBER) WARNING LAMP "ON" CONSTANTLY, NO DTCs STORED

NOTICE: DIAGNOSTIC SYSTEM CHECK MUST BE COMPLETED FIRST BEFORE USING THIS CHART.

- IGNITION "OFF."
- INSTALL TECH 1.
- IGNITION "ON."
- USING TECH 1, SELECT "F4: MISC. TESTS."
- SELECT "F8: LAMP TEST."
- ATTEMPT TO FLASH THE TCS WARNING LAMP.

DOES TCS WARNING LAMP FLASH?

- **YES:** MALFUNCTION IS NOT PRESENT AT THIS TIME.
- **NO:**
  - IGNITION "OFF."
  - DISCONNECT LAMP DRIVER MODULE FROM HARNESS.
  - IGNITION "ON."
  - IS TCS WARNING LAMP "OFF"?
    - **YES:** REPLACE LAMP DRIVER MODULE ASSEMBLY.
    - **NO:** REPAIR SHORT TO GROUND IN CKT 1572. PERFORM FURTHER CLUSTER DIAGNOSIS.

DIAGNOSTIC CHART F — 1994 SILHOUETTE, TRANS SPORT AND LUMINA APV WITH TCS

CHART F
TCS (AMBER) WARNING LAMP "ON" INTERMITTENTLY, NO DTCs STORED

NOTICE: DIAGNOSTIC SYSTEM CHECK MUST BE COMPLETED BEFORE USING THIS CHART.

- IGNITION "OFF."
- DISCONNECT GREY 8-WAY EBTCM CONNECTOR.
- DISCONNECT NEGATIVE BATTERY TERMINAL.
- DISCONNECT POSITIVE BATTERY TERMINAL.
- USING J 39200, MEASURE THE RESISTANCE BETWEEN THE NEGATIVE BATTERY CABLE TERMINAL AND TERMINAL "E" OF THE GREY 8-WAY EBTCM HARNESS CONNECTOR.
- OBSERVE MEASURED RESISTANCE WHILE "WIGGLING" WIRE HARNESS AND CONNECTORS.

IS RESISTANCE ALWAYS 2Ω OR LESS?

- **NO:** REPAIR INTERMITTENT OPEN OR HIGH RESISTANCE IN CKT 150.
- **YES:**
  - DISCONNECT 32-WAY EBTCM CONNECTOR.
  - USING J 39200, MEASURE THE RESISTANCE BETWEEN POSITIVE BATTERY CABLE TERMINAL AND TERMINAL "F10" OF THE 32-WAY EBTCM HARNESS CONNECTOR.
  - OBSERVE MEASURED RESISTANCE WHILE "WIGGLING" WIRE HARNESS AND CONNECTORS.

  IS RESISTANCE ALWAYS 2Ω OR LESS?

  - **NO:** REPAIR INTERMITTENT OPEN OR HIGH RESISTANCE IN CKTs 2 OR 1140.
  - **YES:**
    - TURN IGNITION SWITCH "ON" (THIS IS DONE TO COMPLETE THE IGNITION CIRCUIT).
    - USING J 39200, MEASURE THE RESISTANCE BETWEEN POSITIVE BATTERY CABLE TERMINAL AND TERMINAL "F11" OF THE 32-WAY EBTCM HARNESS CONNECTOR.
    - OBSERVE MEASURED RESISTANCE WHILE "WIGGLING" WIRE HARNESS AND CONNECTORS.

    IS RESISTANCE ALWAYS 2Ω OR LESS?

    - **NO:** REPAIR INTERMITTENT OPEN OR HIGH RESISTANCE IN CKT 418 OR 300.
    - **YES:**
      - CHECK POSITIVE BATTERY TERMINAL AND POSITIVE BATTERY CABLE TERMINAL FOR A POOR CONNECTION.
      - INSPECT TERMINALS "F11" AND "F10" OF THE 32-WAY EBTCM CONNECTOR AND TERMINAL "E" OF THE GREY 8-WAY EBTCM CONNECTOR FOR POOR TERMINAL CONTACT AND EVIDENCE OF CORROSION.

      IS THERE EVIDENCE OF POOR TERMINAL CONTACT OR CORROSION?

      - **YES:** REPLACE TERMINALS THAT EXHIBIT POOR TERMINAL CONTACT OR SIGNS OF CORROSION. REPAIR CONNECTION INTEGRITY IF POOR CONTACT EXISTS AT BATTERY.
      - **NO:**
        - IGNITION "OFF."
        - RECONNECT ALL CONNECTORS.
        - IGNITION "ON."

        IS TCS WARNING LAMP "ON" INTERMITTENTLY WITH NO DTCs SET?

        - **YES:** REPLACE EBTCM.
        - **NO:** MALFUNCTION IS NOT PRESENT AT THIS TIME. INTERMITTENTS MAY BE CAUSED BY INCORRECT WIRING HARNESS ROUTING OR LOOSE GROUND CONNECTIONS.

DIAGNOSTIC CHART G — 1994 SILHOUETTE, TRANS SPORT AND LUMINA APV WITH TCS

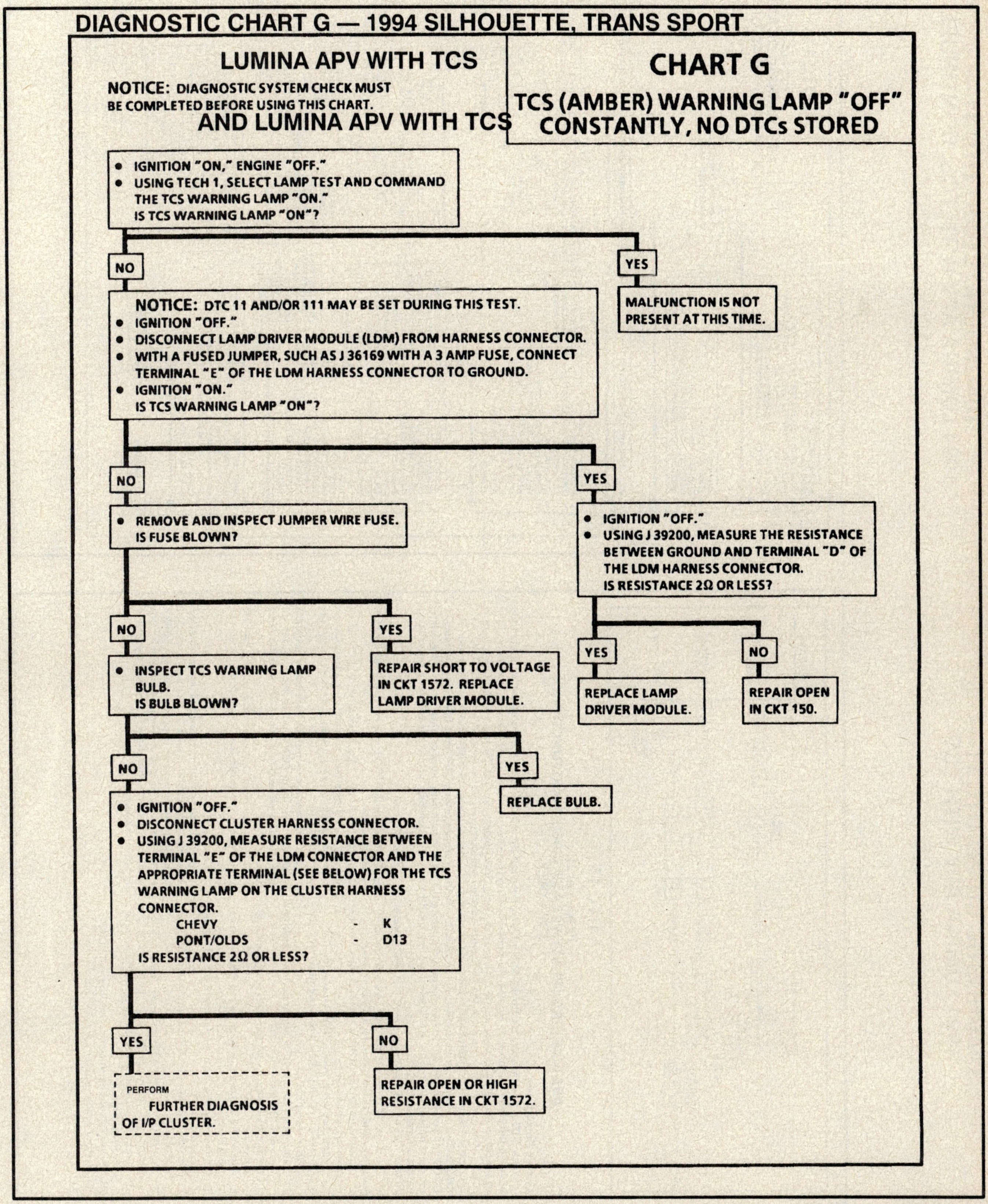

DIAGNOSTIC CHART DTC 11 — 1994 LUMINA APV WITH TCS

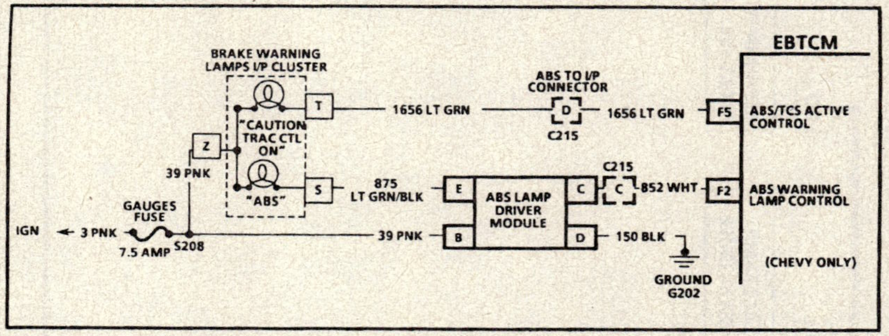

DTC 11
(Page 1 of 2)
ABS WARNING OR ACTIVE LAMP CIRCUIT MALFUNCTION
(CHEV ONLY)

Circuit Description:

This DTC checks the state of both the ABS warning lamp and the ABS/TCS active lamp to identify a situation in which the driver could not be warned of a system malfunction or of possible slippery driving conditions. The Lamp Driver Module (LDM) turns the ABS warning lamp "ON," unless the EBTCM provides ground to turn it "OFF." Because of the circuitry in the LDM, only external malfunctions can be detected by the EBTCM. As a result, the ABS warning lamp itself is not diagnosable, only the control line to the EBTCM can be diagnosed. The ABS/TCS active lamp circuit is such that the EBTCM provides a ground to turn "ON" the ABS/TCS active lamp.

DTC Will Set When: DTC 11 can be set only during the three second bulb check or when either of the lamps are commanded "ON." If the EBTCM cannot control either the ABS warning lamp or the ABS/TCS active lamp for two seconds, a malfunction exists.

Action Taken: A malfunction DTC is stored; however, ABS is not disabled.

DTC Chart Test Description: Number(s) below refer to circled number(s) on the diagnostic chart.

1. Determines if the ABS warning lamp can be properly controlled.
2. Determines if the lamp is "ON" constantly.
3. Checks for other DTCs which could turn "ON" the ABS warning lamp.
4. Determines if the ABS warning lamp can be turned "OFF" manually.
5. Checks the integrity of the ABS warning lamp circuitry.
6. Checks for a short to voltage on the ABS warning lamp circuitry.
7. Determines if a malfunctioning EBTCM is the cause of DTC 11.
8. Verifies that the other I/P indicator lamps function properly.
9. Determines if the ABS warning lamp can be turned "ON" manually.
10. Checks for a short to ground in the ABS warning lamp circuitry.
11. Checks for open between "Gages" fuse and LDM.
12. Verifies condition of the 7.5 amp "Gages" fuse.
13. Determines if DTC 11 was set due to an EBTCM malfunction.

Diagnostic Aids: An intermittent malfunction may be caused by a poor connection, rubbed through wire insulation, or a wire that is broken inside the insulation.

The lamp test function of the Tech 1 may be used to command the lamp "ON" while looking for an intermittent malfunction in the ABS warning lamp circuitry.

The frequency of the malfunction can be checked by using the enhanced diagnostic function of the Tech 1

Any circuitry that is suspected as causing the intermittent complaint should be thoroughly checked for backed out terminals, improper mating, broken locks, improperly formed or damaged terminals, poor terminal to wiring connections or physical damage to the wiring harness.

DIAGNOSTIC CHART DTC 11 — 1994 LUMINA APV WITH TCS, CONTINUED

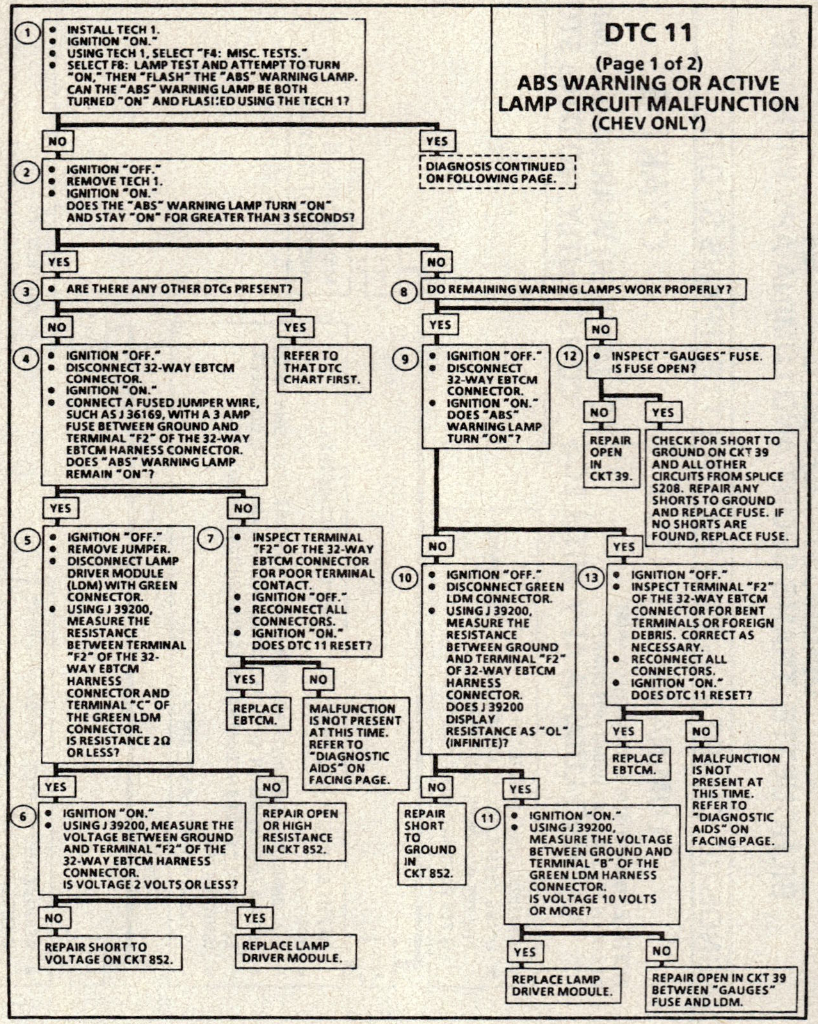

DIAGNOSTIC CHART DTC 11 — 1994 LUMINA APV WITH TCS, CONTINUED

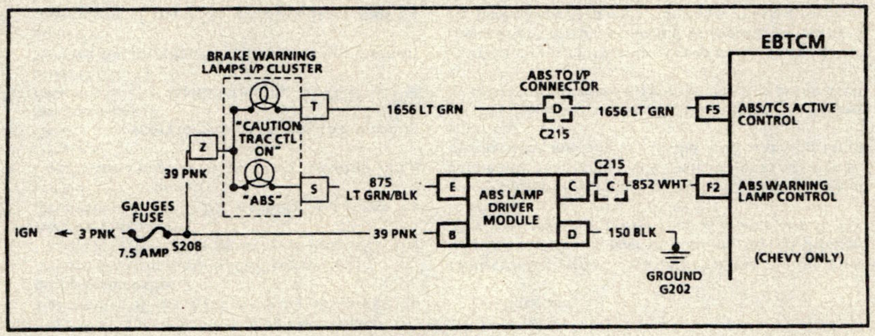

DTC 11

(Page 2 of 2)

ABS WARNING OR ACTIVE LAMP CIRCUIT MALFUNCTION (CHEV ONLY)

Circuit Description:

This DTC checks the state of both the ABS warning lamp and the ABS/TCS active lamp to identify a situation in which the driver could not be warned of a system malfunction or of possible slippery driving conditions. The Lamp Driver Module (LDM) turns the ABS warning lamp "ON," unless the EBTCM provides ground to turn it "OFF." Because of the circuitry in the LDM, only external malfunctions can be detected by the EBTCM. As a result, the ABS warning lamp itself is not diagnosable, only the control line to the EBTCM can be diagnosed. The ABS/TCS active lamp circuit is such that the EBTCM provides a ground to turn "ON" the ABS/TCS active lamp.

DTC Will Set When: DTC 11 can be set only during the three second bulb check or when either of the lamps are commanded "ON." If the EBTCM cannot control either the ABS warning lamp or the ABS/TCS active lamp for two seconds, a malfunction exists.

Action Taken: A malfunction DTC is stored; however, ABS is not disabled.

DTC Chart Test Description: Number(s) below refer to circled number(s) on the diagnostic chart.

14. Determines if the ABS/TCS active lamp can be properly controlled.
15. Determines if the ABS/TCS active lamp is "ON" constantly.
16. Checks for a short to ground in the ABS/TCS active lamp circuitry.
17. Determines if DTC 11 was set due to an EBTCM malfunction.
18. Determines if DTC 11 was set due to an EBTCM malfunction.
19. Determines if the ABS/TCS active lamp can be controlled manually.
20. Verifies the integrity of the ABS/TCS active lamp circuitry.
21. Determines if there is a short to voltage in the ABS/TCS active lamp circuitry.
22. Determines if DTC 11 was set due to an EBTCM malfunction.

Diagnostic Aids: An intermittent malfunction may be caused by a poor connection, rubbed through wire insulation, or a wire that is broken inside the insulation.

The lamp test function of the Tech 1 may be used to command the lamp "ON" while looking for an intermittent malfunction in the ABS/TCS active lamp circuitry.

The frequency of the malfunction can be checked by using the enhanced diagnostic function of the Tech 1.

Any circuitry that is suspected as causing the intermittent complaint should be thoroughly checked for backed out terminals, improper mating, broken locks, improperly formed or damaged terminals, poor terminal to wiring connections or physical damage to the wiring harness.

DIAGNOSTIC CHART DTC 11 — 1994 LUMINA APV WITH TCS, CONTINUED

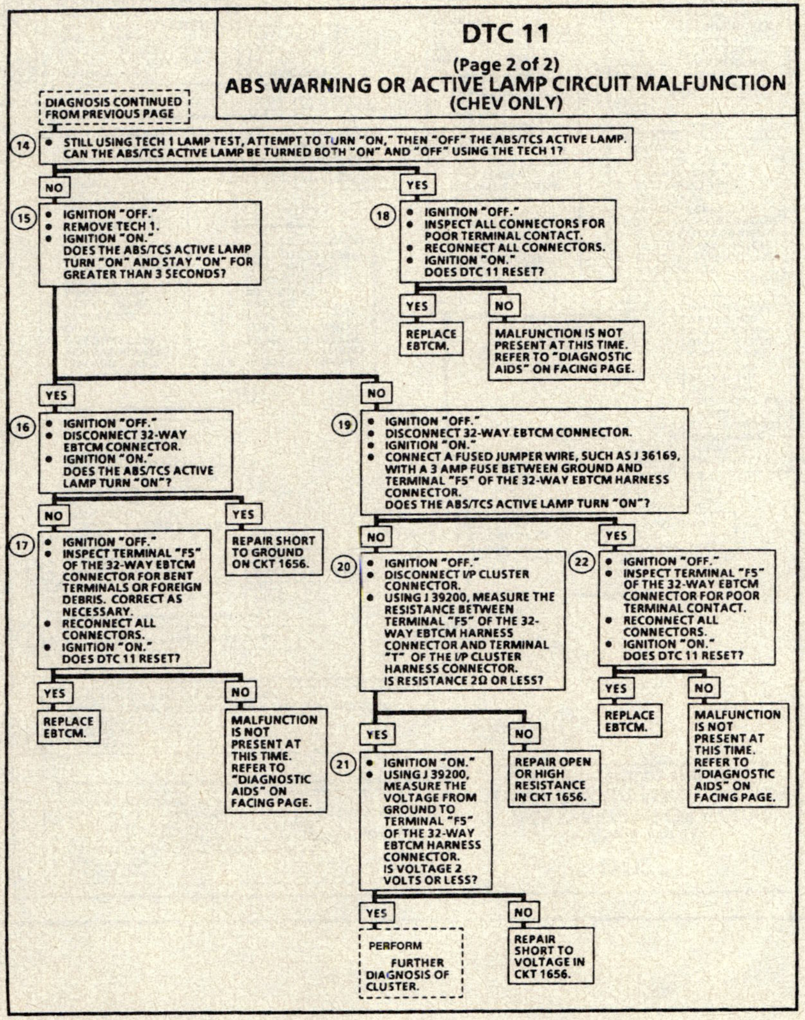

DIAGNOSTIC CHART DTC 11 — 1994 SILHOUETTE AND TRANS SPORT WITH TCS

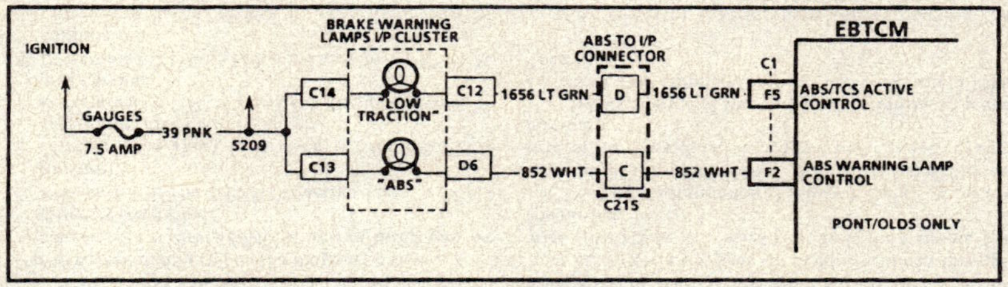

DTC 11

(Page 1 of 2)

ABS WARNING OR ACTIVE LAMP CIRCUIT MALFUNCTION

(PONT/OLDS ONLY)

Circuit Description:

This DTC checks the state of both the ABS warning lamp and the ABS/TCS active lamp to identify a situation in which the driver could not be alerted to a system malfunction or of possible slippery driving conditions. Due to an integral lamp driver module within the instrument panel cluster, the EBTCM must provide a ground to turn the ABS warning lamp "OFF." The ABS/TCS active lamp circuit is such that the EBTCM provides a ground to turn "ON" the ABS/TCS active lamp.

DTC Will Set When: DTC 11 can only be set during the three second bulb check or when either of the lamps are commanded "ON." If the EBTCM cannot control either the ABS warning lamp or the ABS/TCS active lamp for two seconds, a malfunction exists.

Action Taken: A malfunction DTC is stored; however, ABS is not disabled.

DTC Chart Test Description: Number(s) below refer to circled number(s) on the diagnostic chart.

1. Determines if the ABS warning lamp can be properly controlled.
2. Determines if the lamp is "ON" constantly.
3. Checks for other DTCs which could turn "ON" the ABS warning lamp.
4. Determines if the ABS warning lamp can be turned "OFF" manually.
5. Checks the integrity of the ABS warning lamp circuitry.
6. Checks for a short to voltage on the ABS warning lamp circuitry.
7. Determines if a malfunctioning EBTCM is the cause of DTC 11.
8. Verifies that the other I/P indicator lamps function properly.
9. Determines if the ABS warning lamp can be turned "ON" manually.
10. Checks for a short to ground in the ABS warning lamp circuitry.
11. Verifies condition of the "GAGES" Fuse 9.
12. Determines if DTC 11 was set due to an EBTCM malfunction.

Diagnostic Aids: An intermittent malfunction may be caused by a poor connection, rubbed through wire insulation, or a wire that is broken inside the insulation.

The lamp test function of the Tech 1 may be used to command the lamp "ON" while looking for an intermittent malfunction in the ABS warning lamp circuitry.

The frequency of the malfunction can be checked by using the enhanced diagnostic function of the Tech 1.

Any circuitry that is suspected as causing the intermittent complaint should be thoroughly checked for backed out terminals, improper mating, broken locks, improperly formed or damaged terminals, poor terminal to wiring connections or physical damage to the wiring harness.

DIAGNOSTIC CHART DTC 11 — 1994 SILHOUETTE AND TRANS SPORT WITH TCS, CONTINUED

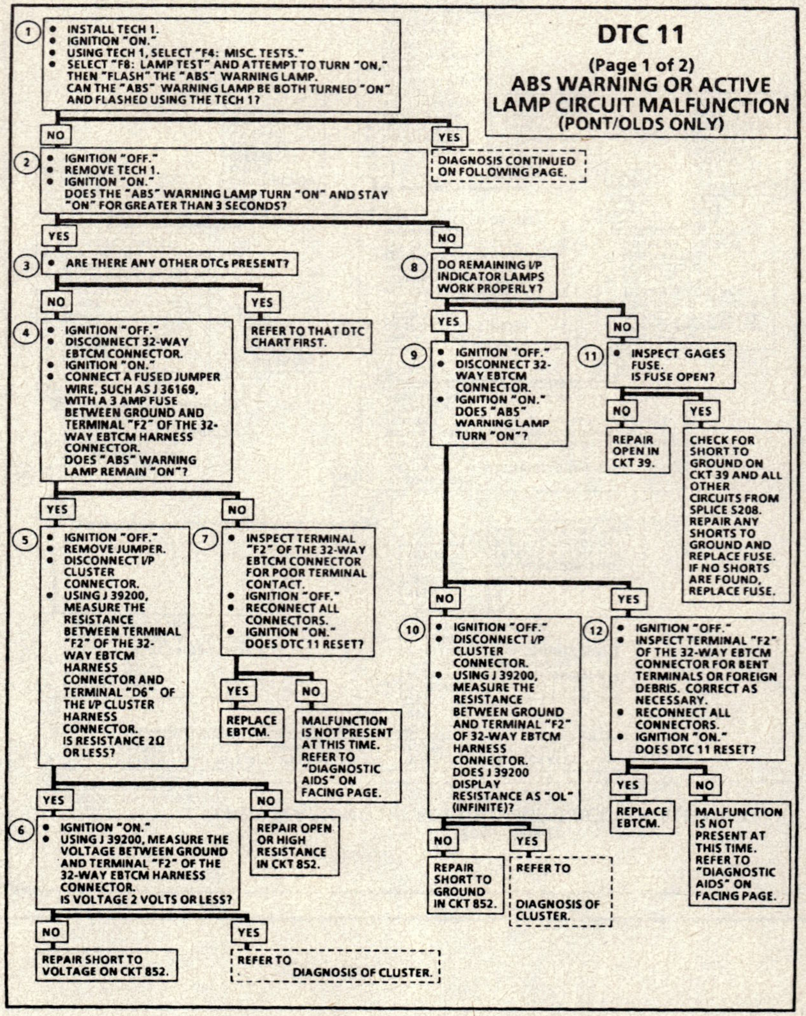

DIAGNOSTIC CHART DTC 11 — 1994 SILHOUETTE AND TRANS SPORT WITH TCS, CONTINUED

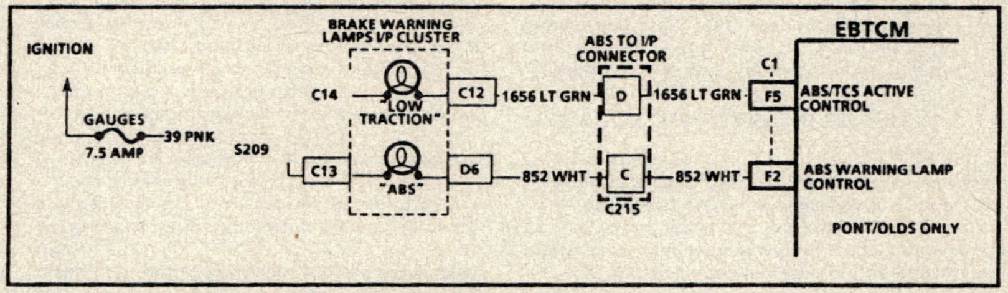

DTC 11

(Page 2 of 2)

ABS WARNING OR ACTIVE LAMP CIRCUIT MALFUNCTION

(PONT/OLDS ONLY)

Circuit Description:

This DTC checks the state of both the ABS warning lamp and the ABS/TCS active lamp to identify a situation in which the driver could not be alerted to a system malfunction or of possible slippery driving conditions. Due to an integral lamp driver module within the instrument panel cluster, the EBTCM must provide a ground to turn the ABS warning lamp "OFF." The ABS/TCS active lamp circuit is such that the EBTCM provides a ground to turn "ON" the ABS/TCS active lamp.

DTC Will Set When: DTC 11 can only be set during the three second bulb check or when either of the lamps are commanded "ON." If the EBTCM cannot control either the ABS warning lamp or the ABS/TCS active lamp for two seconds, a malfunction exists.

Action Taken: A malfunction DTC is stored; however, ABS is not disabled.

DTC Chart Test Description: Number(s) below refer to circled number(s) on the diagnostic chart.

13. Determines if the ABS/TCS active lamp can be properly controlled.
14. Determines if the ABS/TCS active lamp is "ON" constantly.
15. Checks for a short to ground in the ABS/TCS active lamp circuitry.
16. Determines if DTC 11 was set due to an EBTCM malfunction.
17. Determines if DTC 11 was set due to an EBTCM malfunction.
18. Determines if the ABS/TCS active lamp can be controlled manually.
19. Verifies the integrity of the ABS/TCS active lamp circuitry.
20. Determines if there is a short to voltage in the ABS/TCS active lamp circuitry.
21. Determines if DTC 11 was set due to an EBTCM malfunction.

Diagnostic Aids: An intermittent malfunction may be caused by a poor connection, rubbed through wire insulation, or a wire that is broken inside the insulation.

The lamp test function of the Tech 1 may be used to command the lamp "ON" while looking for an intermittent malfunction in the ABS/TCS active lamp circuitry.

The frequency of the malfunction can be checked by using the enhanced diagnostic function of the Tech 1

Any circuitry that is suspected as causing the intermittent complaint should be thoroughly checked for backed out terminals, improper mating, broken locks, improperly formed or damaged terminals, poor terminal to wiring connections or physical damage to the wiring harness.

DIAGNOSTIC CHART DTC 11 — 1994 SILHOUETTE AND TRANS SPORT WITH TCS, CONTINUED

DTC 11

(Page 2 of 2)

ABS WARNING OR ACTIVE LAMP CIRCUIT MALFUNCTION

(PONT/OLDS ONLY)

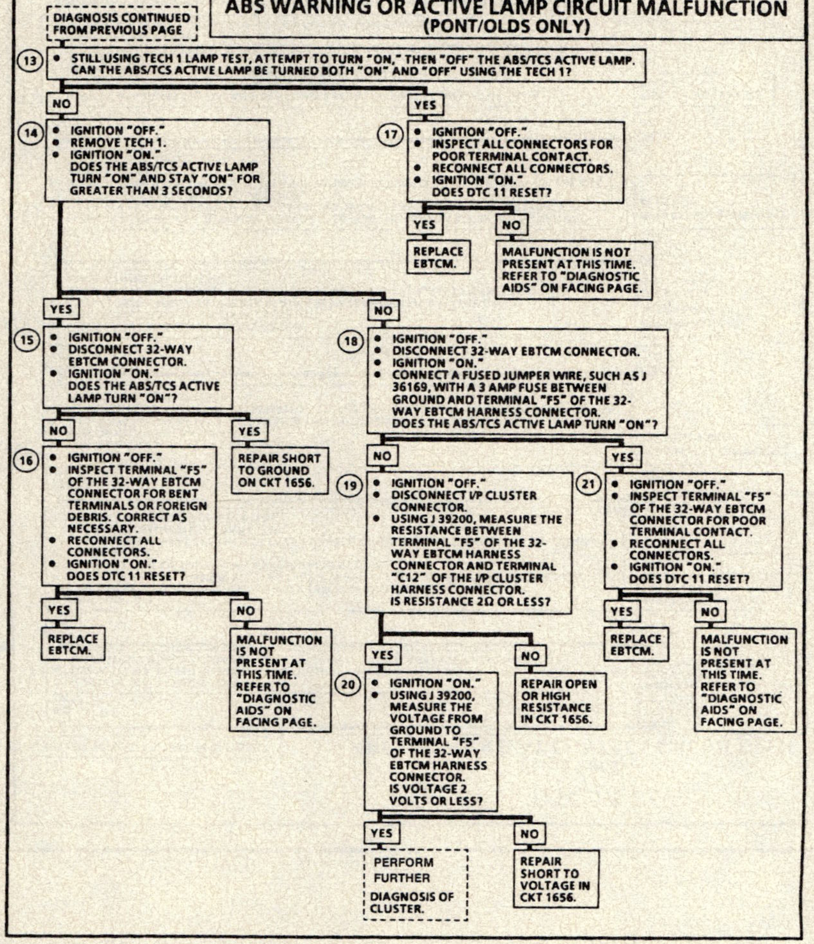

DIAGNOSTIC CHART DTC 14 — 1994 SILHOUETTE, TRANS SPORT AND LUMINA APV WITH TCS

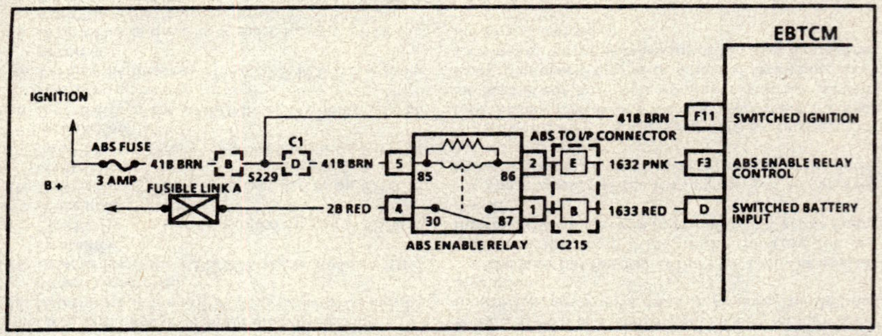

DTC 14
(Page 1 of 3)
ABS ENABLE RELAY CONTACT CIRCUIT OPEN

Circuit Description:
Ignition voltage is supplied through terminal "5" of the ABS enable relay. The EBTCM then is able to energize the pull-in coil by completing the ground circuit at terminal "F3" of the EBTCM. The magnetic field created closes the ABS enable relay contacts and allows battery voltage and current to be supplied to the EBTCM, which supplies power to the motors and solenoids.

DTC Will Set When: DTC 14 can be set anytime after the EBTCM commands the ABS enable relay "ON" (the relay is first commanded "ON" during the three second bulb check). This test monitors the availability of current and voltage to the motors and solenoids. This malfunction indicates voltage is not available and would therefore not allow ABS operation if required.

Action Taken: A malfunction DTC is stored, ABS is disabled and the ABS warning lamp is turned "ON."

DTC Chart Test Description: Number(s) below refer to circled number(s) on the diagnostic chart.

1. Checks to see if DTC 16 is also set as current or history. If DTC 16 is also set, proceed to that chart first.
2. Verifies that the malfunction is currently present.
3. Verifies that the relay can be turned "ON" and provide voltage to the EBTCM.
4. Checks for high resistance in the switched battery circuit from the relay to the EBTCM.
5. Checks for high resistance in the switched battery circuit from the battery to the relay.
6. Verifies the condition of the ABS enable relay.
7. Checks for poor terminal contact or corrosion at the connectors.
8. Determines if the EBTCM is the cause of the malfunction.

Diagnostic Aids: An intermittent malfunction may be caused by a poor connection, rubbed through wire insulation, or a wire that is broken inside the insulation.

The frequency of the malfunction can be checked by using the enhanced diagnostic function of the Tech 1, as described in "Tech 1 Diagnostics" in this section.

Any circuitry that is suspected as causing the intermittent complaint should be thoroughly checked for backed out terminals, improper mating, broken locks, improperly formed or damaged terminals, poor terminal to wiring connections or physical damage to the wiring harness.

VIBRATION, TEMPERATURE EFFECTS:

Check for vibration effects by performing the relay test function of the Tech 1. With the relay test commanded "ON," lightly tap the top and sides of the relay while monitoring relay voltage. If the relay voltage changes significantly, replace the relay.

If DTC 14 only sets when the vehicle is initially started in cold ambient conditions (temperature less than 32°F-0°C), replace the relay.

DIAGNOSTIC CHART DTC 14 — 1994 SILHOUETTE, TRANS SPORT AND LUMINA APV WITH TCS, CONTINUED

DTC 14
(Page 1 of 3)
ABS ENABLE RELAY CONTACT CIRCUIT OPEN

(1) IS DTC 16 ALSO PRESENT IN HISTORY?
- YES: PROCEED TO DTC 16 CHART FIRST.
- NO: go to (2)

(2)
- IGNITION "ON," ENGINE "OFF."
- USING TECH 1, SELECT DATA LIST AND OBSERVE EBTCM BATTERY VOLTAGE. DOES TECH 1 INDICATE THAT EBTCM BATTERY VOLTAGE IS GREATER THAN 10 VOLTS?

- YES: MALFUNCTION IS NOT PRESENT AT THIS TIME. REFER TO "DIAGNOSTIC AIDS"
- NO: go to (3)

(3)
- IGNITION "OFF."
- DISCONNECT 32-WAY AND GRAY 8-WAY EBTCM CONNECTORS.
- CONNECT A FUSED JUMPER, SUCH AS J 36169 WITH A 3 AMP FUSE BETWEEN GROUND AND TERMINAL "F3" OF THE 32-WAY EBTCM HARNESS CONNECTOR.
- IGNITION "ON."
- USING J 39200, MEASURE THE VOLTAGE BETWEEN GROUND AND TERMINAL "D" OF THE GRAY 8-WAY EBTCM HARNESS CONNECTOR. IS VOLTAGE 10 VOLTS OR GREATER?

- NO: DIAGNOSIS CONTINUED ON NEXT PAGE.
- YES: go to (4)

(4)
- IGNITION "OFF."
- DISCONNECT JUMPER USED IN STEP 3.
- DISCONNECT ABS ENABLE RELAY.
- USING J 39200, MEASURE RESISTANCE BETWEEN TERMINAL "1" OF THE ABS ENABLE RELAY HARNESS CONNECTOR AND TERMINAL "D" OF THE GRAY 8-WAY EBTCM HARNESS CONNECTOR. IS RESISTANCE 2Ω OR LESS?

- NO: REPAIR HIGH RESISTANCE IN CKT 1633.
- YES: go to (5)

(5)
- DISCONNECT NEGATIVE BATTERY CABLE.
- DISCONNECT POSITIVE BATTERY CABLE.
- USING J 39200, MEASURE THE RESISTANCE BETWEEN POSITIVE BATTERY CABLE TERMINAL AND TERMINAL "4" OF THE ABS ENABLE RELAY HARNESS CONNECTOR. IS RESISTANCE 2Ω OR LESS?

- NO: REPAIR HIGH RESISTANCE IN CKT 2B.
- YES: go to (6)

(6)
- USING A JUMPER WIRE, CONNECT TERMINAL "86" ON THE ABS ENABLE RELAY TO GROUND.
- USING A FUSED JUMPER WIRE, SUCH AS J 36169, WITH A 3 AMP FUSE, CONNECT TERMINAL "85" ON THE ABS ENABLE RELAY TO VOLTAGE.
- USING J 39200, MEASURE THE RESISTANCE BETWEEN TERMINALS "30" AND "87" ON THE ABS ENABLE RELAY. IS RESISTANCE 2Ω OR LESS?

- NO: REPLACE ABS ENABLE RELAY.
- YES: go to (7)

(7)
- INSPECT ABS ENABLE RELAY AND ABS ENABLE RELAY HARNESS CONNECTOR FOR A POOR CONNECTION OR POOR TERMINAL CONTACT.
- INSPECT EBTCM AND EBTCM CONNECTOR TERMINALS "F3" AND "D" FOR POOR TERMINAL CONTACT.
- INSPECT POSITIVE BATTERY TERMINAL AND POSITIVE BATTERY CABLE TERMINAL FOR A POOR CONNECTION. IS POOR TERMINAL CONTACT OR A POOR CONNECTION EVIDENT?

- YES: REPLACE TERMINALS THAT EXHIBIT POOR TERMINAL CONTACT OR EVIDENCE OF CORROSION.
- NO: go to (8)

(8)
- RECONNECT ALL CONNECTORS.
- IGNITION "ON." DOES DTC 14 RESET?

- YES: REPLACE EBTCM.
- NO: MALFUNCTION IS INTERMITTENT. REFER TO "DIAGNOSTIC AIDS"

DIAGNOSTIC CHART DTC 14 — 1994 SILHOUETTE, TRANS SPORT AND LUMINA APV WITH TCS, CONTINUED

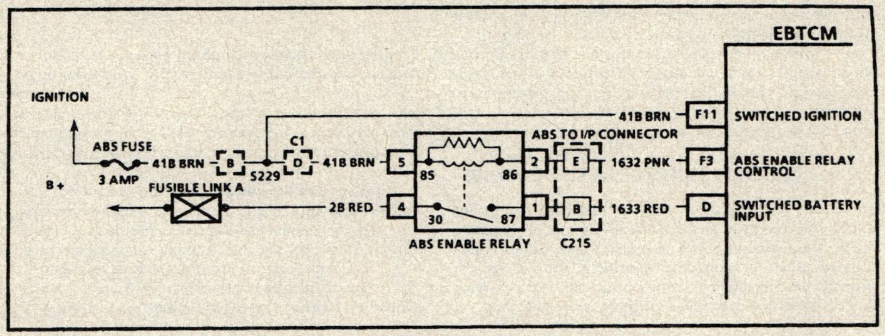

DTC 14
(Page 2 of 3)
ABS ENABLE RELAY CONTACT CIRCUIT OPEN

Circuit Description:
Ignition voltage is supplied through terminal "5" of the ABS enable relay. The EBTCM then is able to energize the pull-in coil by completing the ground circuit at terminal "F3" of the EBTCM. The magnetic field created closes the ABS enable relay contacts and allows battery voltage and current to be supplied to the EBTCM, which supplies power to the motors and solenoids.

DTC Will Set When: DTC 14 can be set anytime after the EBTCM commands the ABS enable relay "ON" (the relay is first commanded "ON" during the three second bulb check). This test monitors the availability of current and voltage to the motors and solenoids. This malfunction indicates voltage is not available and would therefore not allow ABS operation if required.

Action Taken: A malfunction DTC is stored, ABS is disabled and the ABS warning lamp is turned "ON."

DTC Chart Test Description: Number(s) below refer to circled number(s) on the diagnostic chart.

9. Checks for high resistance in the entire relay coil circuit.
10. Verifies proper battery condition.
11. This step verifies proper voltage is available at the ABS enable relay.
12. Checks for an open in the switched battery circuit from the relay to the EBTCM.
13. Verifies the condition of the ABS enable relay.

Diagnostic Aids: An intermittent malfunction may be caused by a poor connection, rubbed through wire insulation, or a wire that is broken inside the insulation.

The frequency of the malfunction can be checked by using the enhanced diagnostic function of the Tech 1, as described in "Tech 1 Diagnostics" in this section.

Any circuitry that is suspected as causing the intermittent complaint should be thoroughly checked for backed out terminals, improper mating, broken locks, improperly formed or damaged terminals, poor terminal to wiring connections or physical damage to the wiring harness.

VIBRATION, TEMPERATURE EFFECTS:

Check for vibration effects by performing the relay test function of the Tech 1. With the relay test commanded "ON," lightly tap the top and sides of the relay while monitoring relay voltage. If the relay voltage changes significantly, replace the relay.

If DTC 14 only sets when the vehicle is initially started in cold ambient conditions (temperature less than 32°F-0°C), replace the relay.

DIAGNOSTIC CHART DTC 14 — 1994 SILHOUETTE, TRANS SPORT AND LUMINA APV WITH TCS, CONTINUED

DTC 14
(Page 2 of 3)
ABS ENABLE RELAY CONTACT CIRCUIT OPEN

DIAGNOSIS CONTINUED FROM PREVIOUS PAGE.

9
- IGNITION "OFF."
- REMOVE JUMPER WIRE INSTALLED IN STEP #3 IF NOT ALREADY REMOVED.
- DISCONNECT NEGATIVE BATTERY CABLE.
- DISCONNECT POSITIVE BATTERY CABLE.
- WITH BATTERY STILL DISCONNECTED, TURN IGNITION SWITCH TO "RUN." THIS COMPLETES THE IGNITION CIRCUIT TO THE ABS ENABLE RELAY.
- USING J 39200, MEASURE THE RESISTANCE BETWEEN TERMINAL "F3" OF THE 32-WAY EBTCM HARNESS CONNECTOR AND THE POSITIVE BATTERY CABLE TERMINAL.

IS RESISTANCE 95Ω OR LESS WHEN MEASURED AT APPROXIMATELY 20 °C (68°F)?

- NO: DIAGNOSIS CONTINUED ON NEXT PAGE.
- YES: go to 10.

10
- USING J 39200, MEASURE THE VOLTAGE BETWEEN THE NEGATIVE AND POSITIVE BATTERY TERMINALS.

IS VOLTAGE 10 VOLTS OR GREATER?

- NO: CORRECT LOW VOLTAGE CONDITION.
- YES: go to 11.

11
- IGNITION "OFF."
- RECONNECT POSITIVE BATTERY CABLE.
- RECONNECT NEGATIVE BATTERY CABLE.
- DISCONNECT ABS ENABLE RELAY.
- USING J 39200, MEASURE THE VOLTAGE BETWEEN TERMINAL "4" OF THE ABS ENABLE RELAY HARNESS CONNECTOR AND GROUND.

IS VOLTAGE 10 VOLTS OR GREATER?

- NO: REPAIR OPEN OR HIGH RESISTANCE IN CKT 2B.
- YES: go to 12.

12
- USING J 39200, MEASURE THE RESISTANCE BETWEEN TERMINAL "1" OF THE ABS ENABLE RELAY HARNESS CONNECTOR AND TERMINAL "D" OF THE GRAY 8-WAY EBTCM HARNESS CONNECTOR.

IS RESISTANCE 2Ω OR LESS?

- NO: REPAIR OPEN OR HIGH RESISTANCE IN CKT 1633.
- YES: go to 13.

13
- USING A JUMPER WIRE, CONNECT TERMINAL "86" ON THE ABS ENABLE RELAY TO GROUND.
- USING A FUSED JUMPER WIRE, SUCH AS J 36169, WITH A 3 AMP FUSE, CONNECT TERMINAL "85" ON THE ABS ENABLE RELAY TO B+.
- USING J 39200, MEASURE THE RESISTANCE BETWEEN TERMINALS "30" AND "87" ON THE ABS ENABLE RELAY.

IS RESISTANCE 2Ω OR LESS?

- NO: REPLACE ABS ENABLE RELAY.
- YES: MALFUNCTION WAS MOST LIKELY DUE TO A POOR CONNECTION. INSPECT ALL CONNECTORS AND TERMINALS FOR POOR TERMINAL CONTACT AND EVIDENCE OF CORROSION. REPLACE ALL TERMINALS THAT EXHIBIT EVIDENCE OF POOR TERMINAL CONTACT OR CORROSION.

DIAGNOSTIC CHART DTC 14 — 1994 SILHOUETTE, TRANS SPORT AND LUMINA APV WITH TCS, CONTINUED

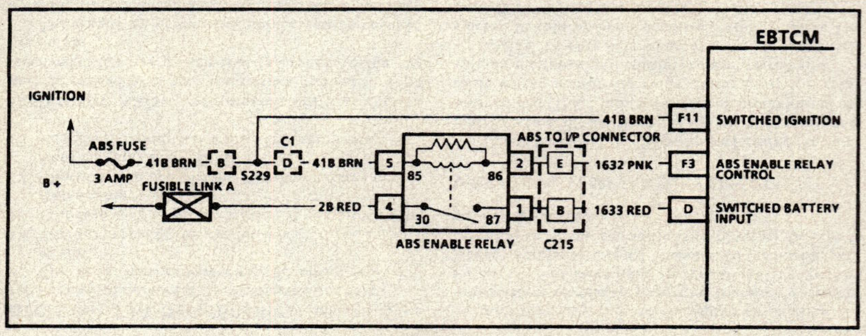

DTC 14
(Page 3 of 3)
ABS ENABLE RELAY CONTACT CIRCUIT OPEN

Circuit Description:

Ignition voltage is supplied through terminal "5" of the ABS enable relay. The EBTCM then is able to energize the pull-in coil by completing the ground circuit at terminal "F3" of the EBTCM. The magnetic field created closes the ABS enable relay contacts and allows battery voltage and current to be supplied to the EBTCM, which supplies power to the motors and solenoids.

DTC Will Set When: DTC 14 can be set anytime after the EBTCM commands the ABS enable relay "ON" (the relay is first commanded "ON" during the three second bulb check). This test monitors the availability of current and voltage to the motors and solenoids. This malfunction indicates voltage is not available and would therefore not allow ABS operation if required.

Action Taken: A malfunction DTC is stored, ABS is disabled and the ABS warning lamp is turned "ON."

DTC Chart Test Description: Number(s) below refer to circled number(s) on the diagnostic chart.

14. Checks for proper resistance of the relay coil.
15. Checks for high resistance in the ABS enable relay control circuit between the relay and the EBTCM.
16. Checks for high resistance in the ABS enable relay control circuit between the battery and the relay.

Diagnostic Aids: An intermittent malfunction may be caused by a poor connection, rubbed through wire insulation, or a wire that is broken inside the insulation.

The frequency of the malfunction can be checked by using the enhanced diagnostic function of the Tech 1.

Any circuitry that is suspected as causing the intermittent complaint should be thoroughly checked for backed out terminals, improper mating, broken locks, improperly formed or damaged terminals, poor terminal to wiring connections or physical damage to the wiring harness.

VIBRATION, TEMPERATURE EFFECTS:

Check for vibration effects by performing the relay test function of the Tech 1. With the relay test commanded "ON," lightly tap the top and sides of the relay while monitoring relay voltage. If the relay voltage changes significantly, replace the relay.

If DTC 14 only sets when the vehicle is initially started in cold ambient conditions (temperature less than 32°F-0°C), replace the relay.

DIAGNOSTIC CHART DTC 14 — 1994 SILHOUETTE, TRANS SPORT AND LUMINA APV WITH TCS, CONTINUED

DTC 14
(Page 3 of 3)
ABS ENABLE RELAY CONTACT CIRCUIT OPEN

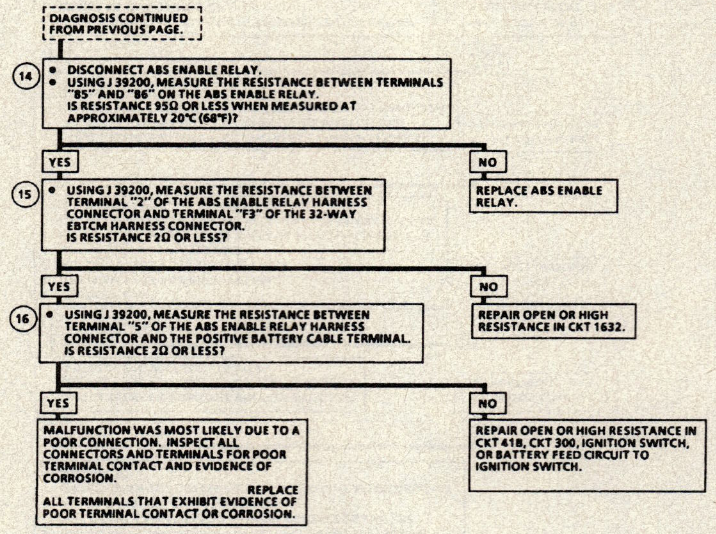

DIAGNOSTIC CHART DTC 15 — 1994 SILHOUETTE, TRANS SPORT AND LUMINA APV WITH TCS

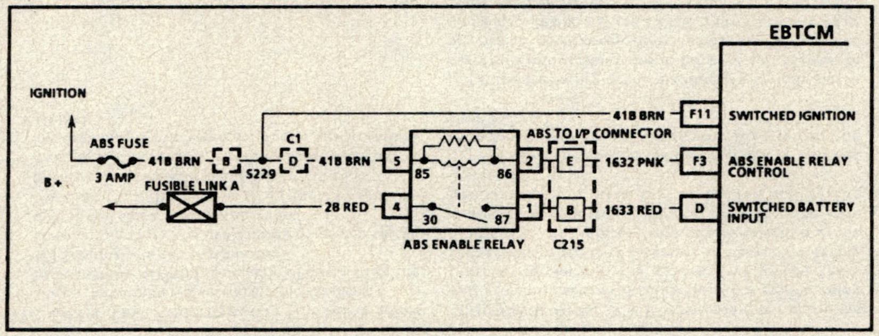

DTC 15

ABS ENABLE RLY CONTACT CKT SHORT TO BATT OR ALWAYS CLOSED

Circuit Description:

Ignition voltage is supplied through terminal "5" of the ABS enable relay. The EBTCM then is able to energize the pull-in coil by completing the ground circuit at terminal "F3" of the EBTCM. The magnetic field created closes the ABS enable relay contacts and allows battery voltage and current to be supplied to the EBTCM, which supplies power to the motors and solenoids.

DTC Will Set When: DTC 15 can be set only before the EBTCM commands the relay "ON." This test determines if the ABS enable relay is energized when it should not be. This malfunction would not allow the ABS enable relay to remove power to the ABS system. If a second malfunction were to occur that requires the ABS enable relay to be turned "OFF," that malfunction cannot be removed if the relay cannot be controlled. The malfunction must be present for three consecutive drive cycles before the DTC is set.

Action Taken: A malfunction DTC is stored. ABS is not disabled.

DTC Chart Test Description: Number(s) below refer to circled number(s) on the diagnostic chart.

1. Determines if any other DTCs are present.
2. Indicates that the EBTCM is capable of controlling the ABS enable relay as commanded.
3. This step checks for a possible short to battery voltage in CKT 1633.
4. This step determines whether the short to voltage is in CKT 1633 or internal to the EBTCM.
5. Checks for a possible intermittent malfunction in the wiring or connectors.

Diagnostic Aids: An intermittent malfunction may be caused by a poor connection, rubbed through wire insulation, or a wire that is broken inside the insulation.

The frequency of the malfunction can be checked by using the enhanced diagnostic function of the Tech 1.

Any circuitry, that is suspected as causing the intermittent complaint, should be thoroughly checked for backed out terminals, improper mating, broken locks, improperly formed or damaged terminals, poor terminal to wiring connections or physical damage to the wiring harness.

DIAGNOSTIC CHART DTC 15 — 1994 SILHOUETTE, TRANS SPORT AND LUMINA APV WITH TCS, CONTINUED

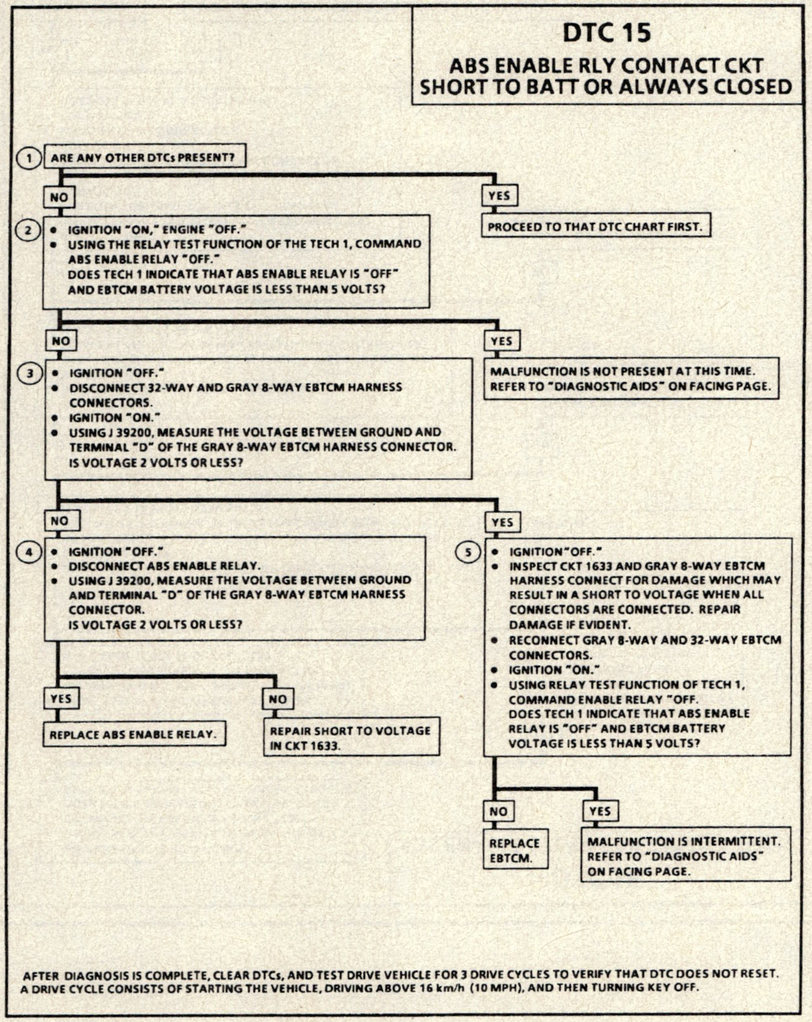

DIAGNOSTIC CHART DTC 16 — 1994 SILHOUETTE, TRANS SPORT AND LUMINA APV WITH TCS

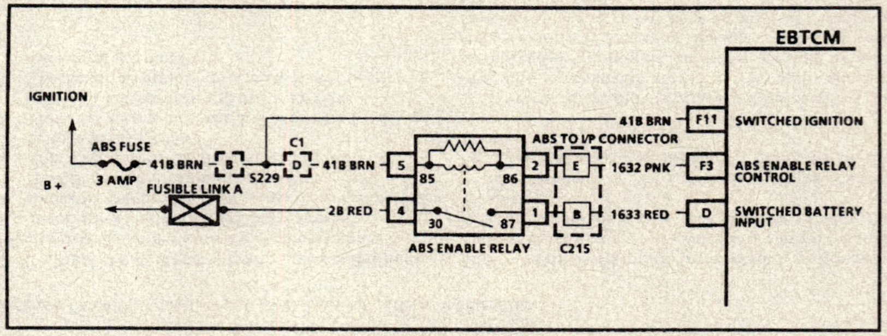

DTC 16
ABS ENABLE RELAY COIL CIRCUIT OPEN

Circuit Description:

Ignition voltage is supplied through terminal "5" of the ABS enable relay. The EBTCM then is able to energize the pull-in coil by completing the ground circuit at terminal "F3" of the EBTCM. The magnetic field created closes the ABS enable relay contacts and allows battery voltage and current to be supplied to the EBTCM, which supplies power to the motors and solenoids.

DTC Will Set When: DTC 16 can be set only after DTC 14 has been set. This test detects an open in the ABS enable relay coil circuit. An open in this circuit will not allow the ABS enable relay to be energized thus preventing voltage and current to the motors and solenoids. If this malfunction is present and the ignition is turned "OFF" before 5 km/h (3 mph) is reached, DTC 14 is set alone.

Action Taken: A malfunction DTC is stored, ABS is disabled and the ABS warning lamp is turned "ON."

DTC Chart Test Description: Number(s) below refer to circled number(s) on the diagnostic chart.

1. Indicates the EBTCM is capable of controlling the ABS enable relay as commanded.
2. Checks to ensure that voltage is available to the pull-in coil of the ABS enable relay.
3. Ensures that there is continuity through the pull-in coil of the ABS enable relay.
4. Verifies the integrity of the relay control circuit.
5. Ensures malfunction was not due to poor terminal contact.

Diagnostic Aids: It will be necessary to remove the two attaching screws holding the ABS enable relay junction to the body to gain access to the relay connections when it is necessary to backprobe during diagnosis. An intermittent malfunction may be caused by a poor connection, rubbed through wire insulation, or a wire that is broken inside the insulation.

The frequency of the malfunction can be checked by using the enhanced diagnostic function of the Tech 1.

If the frequency of the malfunction is high, but is currently intermittent, check for high coil resistance by measuring between relay terminals "2" and "5" using a J 39200. If resistance shows greater than 100Ω, replace the relay.

Any circuitry, that is suspected as causing the intermittent complaint, should be thoroughly checked for backed out terminals, improper mating, broken locks, improperly formed or damaged terminals, poor terminal to wiring connections or physical damage to the wiring harness.

DIAGNOSTIC CHART DTC 16 — 1994 SILHOUETTE, TRANS SPORT AND LUMINA APV WITH TCS, CONTINUED

DTC 16
ABS ENABLE RELAY COIL CIRCUIT OPEN

(1)
- IGNITION "ON," ENGINE "OFF."
- USING THE RELAY TEST FUNCTION OF THE TECH 1, COMMAND THE ABS ENABLE RELAY "ON."
DOES TECH 1 INDICATE ABS ENABLE RELAY "ON," AND EBTCM BATTERY VOLTAGE ABOVE 10 VOLTS?

YES: MALFUNCTION IS NOT PRESENT AT THIS TIME. REFER TO "DIAGNOSTIC AIDS" ON FACING PAGE.

NO:

(2)
- USING J 39200, MEASURE THE VOLTAGE BY BACKPROBING BETWEEN GROUND AND TERMINAL "5" OF THE ABS ENABLE RELAY CONNECTOR.
IS VOLTAGE 10 VOLTS OR GREATER?

NO: REPAIR OPEN OR HIGH RESISTANCE IN CKT 41 OR ABS FUSE.

YES:

(3)
- USING THE RELAY TEST FUNCTION OF THE TECH 1, COMMAND THE ABS ENABLE RELAY "OFF."
- USING J 39200, MEASURE THE VOLTAGE BY BACKPROBING BETWEEN GROUND AND TERMINAL "2" OF THE ABS ENABLE RELAY CONNECTOR.
IS VOLTAGE 10 VOLTS OR GREATER?

NO: CHECK FOR INTERMITTENT CONNECTIONS AT ABS ENABLE RELAY. IF OK, REPLACE ABS RELAY.

YES:

(4)
- IGNITION "OFF."
- DISCONNECT 32-WAY EBTCM CONNECTOR.
- IGNITION "ON."
- USING J 39200, MEASURE THE VOLTAGE BETWEEN GROUND AND TERMINAL "F3" OF THE 32-WAY EBTCM HARNESS CONNECTOR.
IS VOLTS 10 VOLTS OR GREATER?

NO: REPAIR OPEN OR HIGH RESISTANCE IN CKT 1632.

YES:

(5)
- IGNITION "OFF."
- INSPECT 32-WAY EBTCM CONNECTOR TERMINAL "F3" FOR POOR TERMINAL CONTACT OR EVIDENCE OF CORROSION.
REPLACE ALL TERMINALS THAT EXHIBIT SIGNS OF POOR TERMINAL CONTACT OR CORROSION.
- RECONNECT ALL CONNECTORS.
- IGNITION "ON."
DOES DTC 16 SET AS A CURRENT DTC?

YES: REPLACE EBTCM.

NO: MALFUNCTION WAS MOST LIKELY CAUSED BY A POOR CONNECTION. REFER TO "DIAGNOSTIC AIDS" ON FACING PAGE.

AFTER ALL DIAGNOSIS IS COMPLETE, CLEAR DTCs AND TEST DRIVE VEHICLE FOR THREE (3) DRIVE CYCLES TO VERIFY THAT DTC DOES NOT RESET. A DRIVE CYCLE CONSISTS OF STARTING THE VEHICLE, DRIVING OVER 16 km/h (10 MPH), AND THEN TURNING KEY OFF.

DIAGNOSTIC CHART DTC 17 — 1994 SILHOUETTE, TRANS SPORT AND LUMINA APV WITH TCS

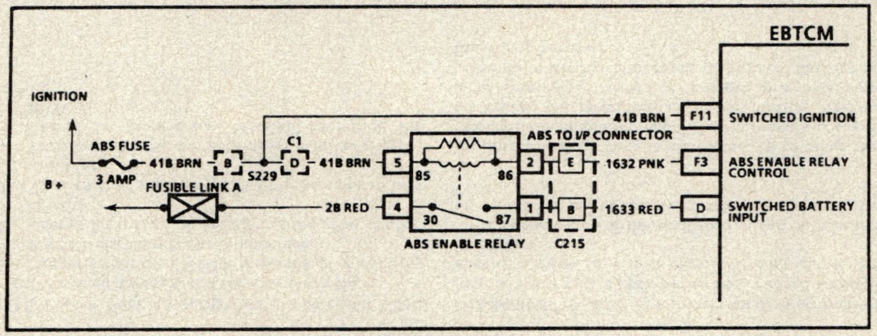

DTC 17

ABS ENABLE RELAY COIL CIRCUIT SHORTED TO GND

Circuit Description:
Ignition voltage is supplied through terminal "5" of the ABS enable relay. The EBTCM then is able to energize the pull-in coil by completing the ground circuit at terminal "F3" of the EBTCM. The magnetic field created closes the ABS enable relay contacts and allows battery voltage and current to be supplied to the EBTCM, which supplies voltage to the motors and solenoids.

DTC Will Set When: DTC 17 can be set before the EBTCM commands the ABS enable relay "ON." This test determines if the ABS enable relay is energized when it should not be. This malfunction would not allow the ABS enable relay to remove power to the ABS system. If a second malfunction were to occur that requires the ABS enable relay to be turned "OFF," that malfunction cannot be removed if the relay cannot be controlled.

Action Taken: A malfunction DTC is stored. ABS is not disabled.

DTC Chart Test Description: Number(s) below refer to circled number(s) on the diagnostic chart.

1. Indicates that the EBTCM is capable of controlling the ABS enable relay as commanded.
2. Checks to ensure that the ABS enable relay or control CKT 1632 are not shorted to ground.
3. Checks to see if the EBTCM is internally shorted to ground. A grounded terminal "F3" would cause the ABS enable relay to be energized anytime the ignition was in the "ON" position.
4. Checks for a short to ground in control CKT 1632 or a defective relay.

Diagnostic Aids: An intermittent malfunction may be caused by a poor connection, rubbed through wire insulation, or a wire that is broken inside the insulation.

The frequency of the malfunction can be checked by using the enhanced diagnostic function of the Tech 1.

Any circuitry, that is suspected as causing the intermittent complaint, should be thoroughly checked for backed out terminals, improper mating, broken locks, improperly formed or damaged terminals, poor terminal to wiring connections or physical damage to the wiring harness.

DIAGNOSTIC CHART DTC 17 — 1994 SILHOUETTE, TRANS SPORT AND LUMINA APV WITH TCS, CONTINUED

DTC 17

ABS ENABLE RELAY COIL CIRCUIT SHORTED TO GND

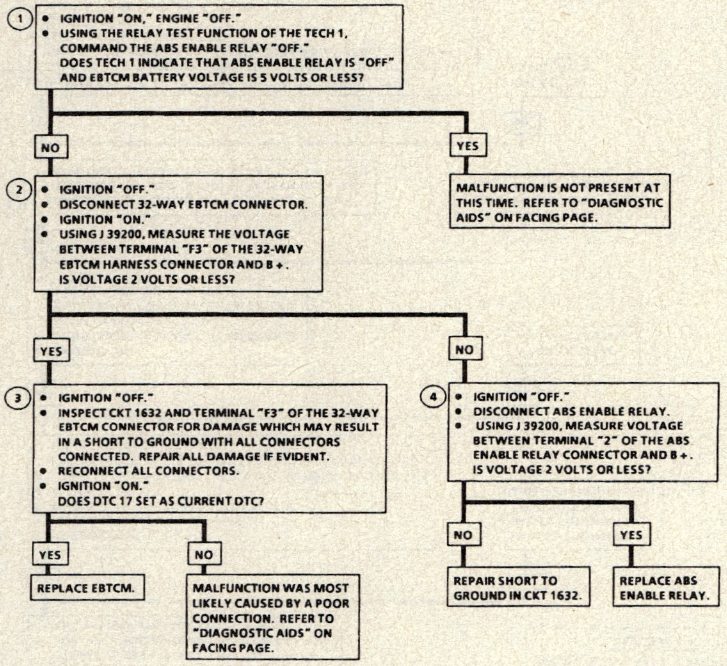

AFTER ALL DIAGNOSIS IS COMPLETE, CLEAR DTCs AND TEST DRIVE VEHICLE FOR THREE (3) DRIVE CYCLES TO VERIFY THAT DTC DOES NOT RESET. A DRIVE CYCLE CONSISTS OF STARTING THE VEHICLE, DRIVING OVER 16 km/h (10 MPH), AND THEN TURNING KEY OFF.

DIAGNOSTIC CHART DTC 18 — 1994 SILHOUETTE, TRANS SPORT AND LUMINA APV WITH TCS

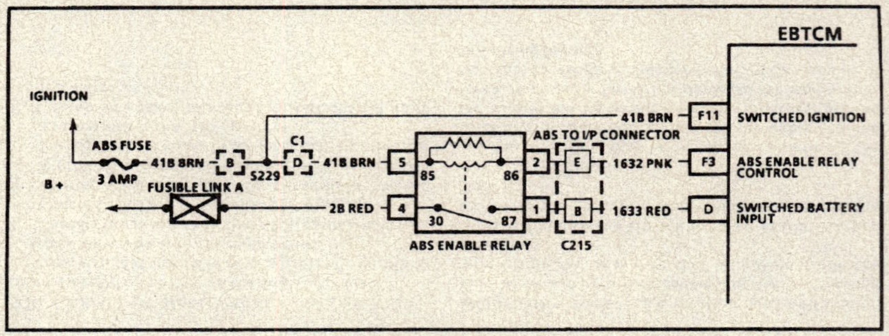

DTC 18

ABS ENABLE RELAY COIL CIRCUIT SHORTED TO BATT

Circuit Description:
Ignition voltage is supplied through terminal "5" of the ABS enable relay. The EBTCM then is able to energize the pull-in coil by completing the ground circuit at terminal "F3" of the EBTCM. The magnetic field created closes the ABS enable relay contacts and allows battery voltage and current to be supplied to the EBTCM, which supplies voltage to the motors and solenoids.

DTC Will Set When: DTC 18 can be set after the EBTCM commands the enable relay "ON." This test monitors the availability of current and voltage to the motors and solenoids. A malfunction of this type will not allow voltage to be available to the motors and solenoids; therefore, ABS operation would not be allowed if required.

Action Taken: A malfunction DTC is stored, ABS is disabled and the ABS warning lamp is turned "ON."

DTC Chart Test Description: Number(s) below refer to circled number(s) on the diagnostic chart.

1. Indicates that the EBTCM is capable of controlling the ABS enable relay as commanded.
2. Checks for a short to voltage on the relay control circuit.
3. Verifies that the relay coil has the appropriate resistance.
4. Verifies that relay contacts have not failed closed.
5. Checks for possible poor terminal contact at the EBTCM.

Diagnostic Aids: An intermittent malfunction may be caused by a poor connection, rubbed through wire insulation, or a wire that is broken inside the insulation.

The frequency of the malfunction can be checked by using the enhanced diagnostic function of the Tech 1.

Any circuitry, that is suspected as causing the intermittent complaint, should be thoroughly checked for backed out terminals, improper mating, broken locks, improperly formed or damaged terminals, poor terminal to wiring connections or physical damage to the wiring harness.

DIAGNOSTIC CHART DTC 18 — 1994 SILHOUETTE, TRANS SPORT AND LUMINA APV WITH TCS, CONTINUED

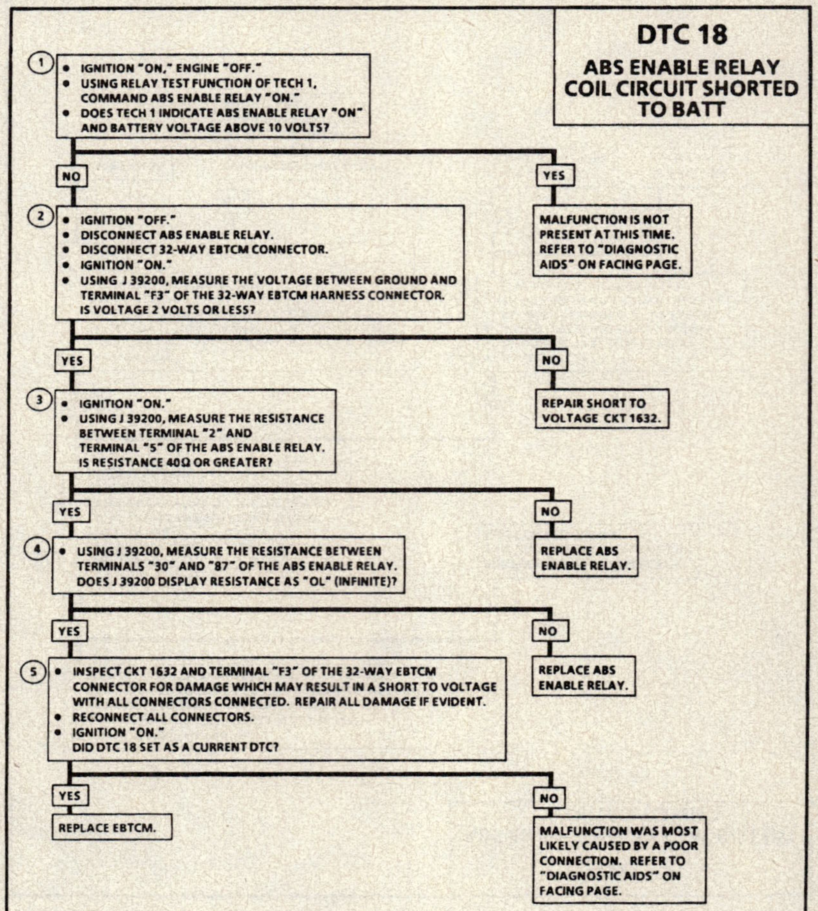

AFTER ALL DIAGNOSIS IS COMPLETE, CLEAR DTCs AND TEST DRIVE VEHICLE FOR THREE (3) DRIVE CYCLES TO VERIFY THAT DTC DOES NOT RESET. A DRIVE CYCLE CONSISTS OF STARTING THE VEHICLE, DRIVING OVER 16 km/h (10 MPH), AND THEN TURNING KEY OFF.

DIAGNOSTIC CHART DTC 21 — 1994 SILHOUETTE, TRANS SPORT AND LUMINA APV WITH TCS

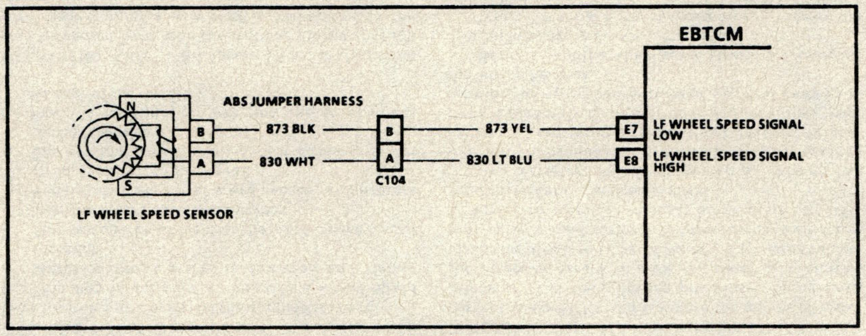

DTC 21

LEFT FRONT WHEEL SPEED = 0

Circuit Description:

As a toothed ring passes by the wheel speed sensor, changes in the electromagnetic field cause the wheel speed sensor to produce a sinusoidal (AC) voltage signal whose frequency is proportional to wheel speed. The magnitude of this signal is directly related to wheel speed and the proximity of the wheel speed sensor to the toothed ring, often referred to as the air gap.

DTC Will Set When: DTC 21 can be set when the vehicle is not in an ABS or TCS event. If the left front wheel speed = 0 and the vehicle's reference speed is greater than 8 km/h (5 mph), a malfunction exists.

Action Taken: A malfunction DTC is stored, ABS and TCS are disabled and the ABS and TCS warning lamps are turned "ON."

DTC Chart Test Description: Number(s) below refer to circled number(s) on the diagnostic chart.

1. This step is designed to identify a wheel speed sensor or circuitry that is damaged and visibly apparent.
2. Checks to see that the wheel speed sensor is within proper resistance values.
3. Verifies that the wheel speed sensor can generate an appropriate output signal.
4. Ensures that wheel speed sensor circuitry is not shorted together.
5. Ensures malfunction was not due to physical damage of the circuitry.

Diagnostic Aids: An intermittent malfunction may be caused by a poor connection, rubbed through wire insulation, or a wire that is broken inside the insulation.

The frequency of the malfunction can be checked by using the enhanced diagnostic function of the Tech 1.

If the customer's comments reflect that the ABS and TCS warning lamps are "ON" only during moist environmental changes (rain, snow, vehicle wash), all wheel speed sensor circuitry should be thoroughly inspected for signs of water intrusion. If DTC is not current, clear DTCs and simulate the effects of water intrusion. Use the following procedure. Spray down the suspected area with a 5% salt water solution (two teaspoons of salt to 12 oz. of water). Test drive vehicle over various road surfaces (bumps, turns, etc.) above 24 km/h (15 mph) for at least 30 seconds. If DTC returns, replace suspected harness.

Any circuitry, that is suspected as causing the intermittent complaint, should be thoroughly checked for backed out terminals, improper mating, broken locks, improperly formed or damaged terminals, poor terminal to wiring connections or physical damage to the wiring harness.

Resistance of the wheel speed sensor will increase with an increase in sensor temperature.

When replacing a wheel speed sensor, inspect the sensor terminals and harness connector for corrosion and/or water intrusion. If evidence of corrosion or water intrusion exists, replace wheel speed sensor jumper harness. Likewise, if replacing a wheel speed sensor jumper harness, inspect sensor terminals. If evidence of corrosion or water intrusion exists, replace wheel speed sensor.

DIAGNOSTIC CHART DTC 21 — 1994 SILHOUETTE, TRANS SPORT AND LUMINA APV WITH TCS, CONTINUED

DTC 21
LEFT FRONT WHEEL SPEED = 0

Important

- WHEEL SPEED SENSOR INTERMITTENT MALFUNCTIONS MAY BE DIFFICULT TO LOCATE. CARE SHOULD BE TAKEN NOT TO DISTURB ANY ELECTRICAL CONNECTIONS PRIOR TO AN INDICATED STEP OF THIS CHART. THIS WILL ENSURE THAT AN INTERMITTENT CONNECTION WILL NOT BE CORRECTED BEFORE THE SOURCE OF THE MALFUNCTION IS FOUND.

(1)
- IGNITION "OFF."
- PHYSICALLY INSPECT THE LEFT FRONT WHEEL SPEED SENSOR, LEFT FRONT WHEEL SPEED SENSOR JUMPER HARNESS, AND CONNECTORS FOR DAMAGE. IS PHYSICAL DAMAGE INDICATED?

YES: REPLACE WHEEL SPEED SENSOR AND/OR JUMPER HARNESS IF DAMAGED. DRIVE VEHICLE ABOVE 24 km/h (15 MPH) TO ENSURE DTC 21 DOES NOT SET.

NO:

(2)
- DISCONNECT LEFT FRONT WHEEL SPEED SENSOR DIRECTLY AT SENSOR.
- USING J 39200, MEASURE THE RESISTANCE BETWEEN SENSOR TERMINALS "A" AND "B."
IS RESISTANCE BETWEEN 1020 AND 1137Ω WHEN SENSOR IS 20°C (68°F)?
(IF TEMPERATURE OF THE SENSOR IS SUSPECTED TO BE OTHER THAN 20°C (68°F), REFER TO THE TABLE BELOW FOR APPLICABLE SENSOR RESISTANCE VALUES; THE VALUES IN THE TABLE ARE FOR TEMPERATURE OF THE SENSOR, NOT AIR TEMPERATURE.)

NO: REPLACE WHEEL SPEED SENSOR.

YES:

(3)
- WITH J 39200 STILL CONNECTED, SELECT A/C VOLTAGE SCALE.
- SPIN WHEEL BY HAND WHILE OBSERVING VOLTAGE.
IS VOLTAGE AT LEAST 100 mV (VOLTAGE WILL INCREASE AS WHEEL SPEED INCREASES)?

NO: REPLACE WHEEL SPEED SENSOR.

YES:

(4)
- IGNITION "OFF."
- DISCONNECT 32-WAY EBTCM CONNECTOR.
- USING J 39200, MEASURE THE RESISTANCE BETWEEN TERMINAL "E7" AND "E8" OF THE 32-WAY EBTCM HARNESS CONNECTOR.
DOES J 39200 DISPLAY RESISTANCE AS "OL" (INFINITE)?

NO: REPAIR SHORT BETWEEN CKTs 830 AND 873.

YES:

(5)
- INSPECT CKTs 830 AND 873 AND 32-WAY EBTCM CONNECTOR FOR DAMAGE WHICH MAY RESULT IN A SHORT BETWEEN CKTs 830 AND 873 WITH ALL CONNECTORS CONNECTED. REPAIR ALL DAMAGE IF EVIDENT.
- RECONNECT ALL CONNECTORS.
- TEST DRIVE VEHICLE ABOVE 24 km/h (15 MPH) FOR AT LEAST 30 SECONDS. DOES DTC 21 SET AS A CURRENT DTC?

YES: REPLACE EBTCM.

NO: MALFUNCTION IS INTERMITTENT. REFER TO "DIAGNOSTIC AIDS" ON FACING PAGE.

FRONT WHEEL SPEED SENSOR
TEMPERATURE OF SENSOR VS. SENSOR RESISTANCE (APPROXIMATE)

TEMP. (°C)	TEMP. (°F)	RESISTANCE (OHMS)
-34 TO 4	-30 TO 40	804 TO 1066
5 TO 43	41 TO 110	956 TO 1240
44 TO 93	111 TO 200	1112 TO 1463

DIAGNOSTIC CHART DTC 22 — 1994 SILHOUETTE, TRANS SPORT AND LUMINA APV WITH TCS

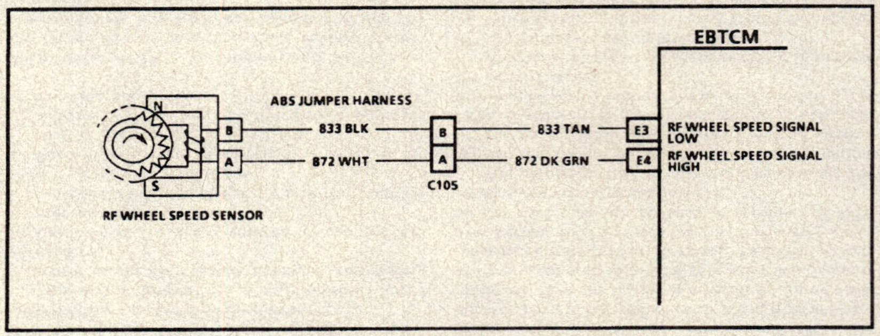

DTC 22

RIGHT FRONT WHEEL SPEED = 0

Circuit Description:

As a toothed ring passes by the wheel speed sensor, changes in the electromagnetic field cause the wheel speed sensor to produce a sinusoidal (AC) voltage signal whose frequency is proportional to wheel speed. The magnitude of this signal is directly related to wheel speed and the proximity of the wheel speed sensor to the toothed ring, often referred to as the air gap.

DTC Will Set When: DTC 22 can be set when the vehicle is not in an ABS or TCS event. If the right front wheel speed = 0 and the vehicle's reference speed is greater than 8 km/h (5 mph), a malfunction exists.

Action Taken: A malfunction DTC is stored, ABS and TCS are disabled and the ABS and TCS warning lamps are turned "ON."

DTC Chart Test Description: Number(s) below refer to circled number(s) on the diagnostic chart.

1. This step is designed to identify a wheel speed sensor or circuitry that is damaged and visibly apparent.
2. This test checks to see that the wheel speed sensor is within proper resistance values.
3. Verifies that the wheel speed sensor can generate an appropriate output signal.
4. Ensures that wheel speed sensor circuitry is not shorted together.
5. Ensures malfunction was not due to physical damage of the circuitry.

Diagnostic Aids: An intermittent malfunction may be caused by a poor connection, rubbed through wire insulation, or a wire that is broken inside the insulation.

The frequency of the malfunction can be checked by using the enhanced diagnostic function of the Tech 1.

If the customer's comments reflect that the ABS and TCS warning lamps are "ON" only during moist environmental changes (rain, snow, vehicle wash), all wheel speed sensor circuitry should be thoroughly inspected for signs of water intrusion. If DTC is not current, clear DTCs and simulate the effects of water intrusion. Use the following procedure. Spray down the suspected area with a 5% salt water solution (two teaspoons of salt to 12 oz. of water). Test drive vehicle over various road surfaces (bumps, turns, etc.) above 24 km/h (15 mph) for at least 30 seconds. If DTC returns, replace suspected harness.

Any circuitry, that is suspected as causing the intermittent complaint, should be thoroughly checked for backed out terminals, improper mating, broken locks, improperly formed or damaged terminals, poor terminal to wiring connections or physical damage to the wiring harness.

Resistance of the wheel speed sensor will increase with an increase in sensor temperature.

When replacing a wheel speed sensor, inspect the sensor terminals and harness connector for corrosion and/or water intrusion. If evidence of corrosion or water intrusion exists, replace wheel speed sensor jumper harness. Likewise, if replacing a wheel speed sensor jumper harness, inspect sensor terminals. If evidence of corrosion or water intrusion exists, replace wheel speed sensor.

DIAGNOSTIC CHART DTC 22 — 1994 SILHOUETTE, TRANS SPORT AND LUMINA APV WITH TCS, CONTINUED

DTC 22

RIGHT FRONT WHEEL SPEED = 0

Important

- WHEEL SPEED SENSOR INTERMITTENT MALFUNCTIONS MAY BE DIFFICULT TO LOCATE. CARE SHOULD BE TAKEN NOT TO DISTURB ANY ELECTRICAL CONNECTIONS PRIOR TO AN INDICATED STEP OF THIS CHART. THIS WILL ENSURE THAT AN INTERMITTENT CONNECTION WILL NOT BE CORRECTED BEFORE THE SOURCE OF THE MALFUNCTION IS FOUND.

(1)
- IGNITION "OFF."
- PHYSICALLY INSPECT THE RIGHT FRONT WHEEL SPEED SENSOR, RIGHT FRONT WHEEL SPEED SENSOR JUMPER HARNESS, AND CONNECTORS FOR DAMAGE. IS PHYSICAL DAMAGE INDICATED?

YES: REPLACE WHEEL SPEED SENSOR AND/OR JUMPER HARNESS IF DAMAGED. DRIVE VEHICLE ABOVE 24 km/h (15 MPH) TO ENSURE DTC 22 DOES NOT SET.

NO:

(2)
- DISCONNECT RIGHT FRONT WHEEL SPEED SENSOR DIRECTLY AT SENSOR.
- USING J 39200, MEASURE THE RESISTANCE BETWEEN SENSOR TERMINALS "A" AND "B."
  IS RESISTANCE BETWEEN 1020 AND 1137Ω WHEN SENSOR IS 20°C (68°F)?
  (IF TEMPERATURE OF THE SENSOR IS SUSPECTED TO BE OTHER THAN 20°C (68°F), REFER TO THE TABLE BELOW FOR APPLICABLE SENSOR RESISTANCE VALUES; THE VALUES IN THE TABLE ARE FOR TEMPERATURE OF THE SENSOR, NOT AIR TEMPERATURE.)

NO: REPLACE WHEEL SPEED SENSOR.

YES:

(3)
- WITH J 39200 STILL CONNECTED, SELECT A/C VOLTAGE SCALE.
- SPIN WHEEL BY HAND WHILE OBSERVING VOLTAGE. IS VOLTAGE AT LEAST 100 mV (VOLTAGE WILL INCREASE AS WHEEL SPEED INCREASES)?

NO: REPLACE WHEEL SPEED SENSOR.

YES:

(4)
- IGNITION "OFF."
- DISCONNECT 32-WAY EBTCM CONNECTOR.
- USING J 39200, MEASURE THE RESISTANCE BETWEEN TERMINAL "E3" AND "E4" OF THE 32-WAY EBTCM HARNESS CONNECTOR. DOES J 39200 DISPLAY RESISTANCE AS "OL" (INFINITE)?

NO: REPAIR SHORT BETWEEN CKTs 833 AND 872.

YES:

(5)
- INSPECT CKTs 833 AND 872 AND 32-WAY EBTCM CONNECTOR FOR DAMAGE WHICH MAY RESULT IN A SHORT BETWEEN CKTs 833 AND 872 WITH ALL CONNECTORS CONNECTED. REPAIR ALL DAMAGE IF EVIDENT.
- RECONNECT ALL CONNECTORS.
- TEST DRIVE VEHICLE ABOVE 24 km/h (15 MPH) FOR AT LEAST 30 SECONDS. DOES DTC 22 SET AS A CURRENT DTC?

YES: REPLACE EBTCM.

NO: MALFUNCTION IS INTERMITTENT. REFER TO "DIAGNOSTIC AIDS" ON FACING PAGE.

FRONT WHEEL SPEED SENSOR
TEMPERATURE OF SENSOR VS. SENSOR RESISTANCE (APPROXIMATE)

TEMP. (°C)	TEMP. (°F)	RESISTANCE (OHMS)
-34 TO 4	-30 TO 40	804 TO 1066
5 TO 43	41 TO 110	956 TO 1240
44 TO 93	111 TO 200	1112 TO 1463

DIAGNOSTIC CHART DTC 23 — 1994 SILHOUETTE, TRANS SPORT AND LUMINA APV WITH TCS

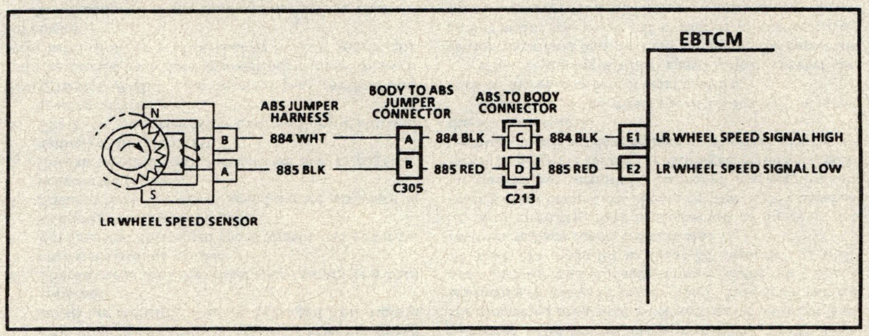

DTC 23

LEFT REAR WHEEL SPEED = 0

Circuit Description:
As a toothed ring passes by the wheel speed sensor, changes in the electromagnetic field cause the wheel speed sensor to produce a sinusoidal (AC) voltage signal whose frequency is proportional to wheel speed. The magnitude of this signal is directly related to wheel speed and the proximity of the wheel speed sensor to the toothed ring, often referred to as the air gap.

DTC Will Set When: DTC 23 can be set when the vehicle is not in an ABS or TCS event. If the left rear wheel speed = 0 and the vehicle's reference speed is greater than 8 km/h (5 mph), a malfunction exists.

Action Taken: A malfunction DTC is stored, ABS and TCS are disabled and the ABS and TCS warning lamps are turned "ON."

DTC Chart Test Description: Number(s) below refer to circled number(s) on the diagnostic chart.

1. This step is designed to identify a wheel speed sensor or circuitry that is damaged and visibly apparent.
2. Checks to see that the wheel speed sensor is within proper resistance values.
3. Verifies that the wheel speed sensor can generate an appropriate output signal.
4. Ensures that the wheel speed sensor circuitry is not shorted together.
5. Checks to see if malfunction was due to physical damage of the circuitry.
6. Checks for an incorrect or malfunctioning hub and bearing assembly.

Diagnostic Aids: An intermittent malfunction may be caused by a poor connection, rubbed through wire insulation, or a wire that is broken inside the insulation.

The frequency of the malfunction can be checked by using the enhanced diagnostic function of the Tech 1.

If the customer's comments reflect that the ABS and TCS warning lamps are "ON" only during moist environmental changes (rain, snow, vehicle wash), all wheel speed sensor circuitry should be thoroughly inspected for signs of water intrusion. If DTC is not current, clear DTCs and simulate the effects of water intrusion. Use the following procedure. Spray down the suspected area with a 5% salt water solution (two teaspoons of salt to 12 oz. of water). Test drive vehicle over various road surfaces (bumps, turns, etc.) above 24 km/h (15 mph) for at least 30 seconds. If DTC returns, replace suspected harness.

Any circuitry, that is suspected as causing the intermittent complaint, should be thoroughly checked for backed out terminals, improper mating, broken locks, improperly formed or damaged terminals, poor terminal to wiring connections or physical damage to the wiring harness.

Resistance of the wheel speed sensor will increase with an increase in sensor temperature.

When replacing a wheel speed sensor, inspect the sensor terminals and harness connector for corrosion and/or water intrusion. If evidence of corrosion or water intrusion exists, replace wheel speed sensor jumper harness. Likewise, if replacing a wheel speed sensor jumper harness, inspect sensor terminals. If evidence of corrosion or water intrusion exists, replace wheel speed sensor.

DIAGNOSTIC CHART DTC 23 — 1994 SILHOUETTE, TRANS SPORT AND LUMINA APV WITH TCS, CONTINUED

DTC 23
LEFT REAR WHEEL SPEED = 0

Important
- WHEEL SPEED SENSOR INTERMITTENT MALFUNCTIONS MAY BE DIFFICULT TO LOCATE. CARE SHOULD BE TAKEN NOT TO DISTURB ANY ELECTRICAL CONNECTIONS PRIOR TO AN INDICATED STEP OF THIS CHART. THIS WILL ENSURE THAT AN INTERMITTENT CONNECTION WILL NOT BE CORRECTED BEFORE THE SOURCE OF THE MALFUNCTION IS FOUND.

(1)
- IGNITION "OFF."
- PHYSICALLY INSPECT THE LEFT REAR WHEEL SPEED SENSOR, WHEEL SPEED SENSOR JUMPER HARNESS, AND CONNECTORS FOR DAMAGE.

IS PHYSICAL DAMAGE INDICATED?

YES: REPLACE WHEEL SPEED SENSOR AND/OR WHEEL SPEED SENSOR JUMPER HARNESS IF DAMAGED. DRIVE VEHICLE ABOVE 24 km/h (15 MPH) TO ENSURE DTC 23 DOES NOT SET.

NO:

(2)
- DISCONNECT LEFT REAR WHEEL SPEED SENSOR DIRECTLY AT SENSOR.
- USING J 39200, MEASURE THE RESISTANCE BETWEEN SENSOR TERMINALS "A" AND "B." IS RESISTANCE BETWEEN 2059 AND 2353Ω WHEN SENSOR IS 20°C (68°F)?

(IF TEMPERATURE OF THE SENSOR IS SUSPECTED TO BE OTHER THAN 20°C (68°F), REFER TO THE TABLE BELOW FOR APPLICABLE SENSOR RESISTANCE VALUES; THE VALUES IN THE TABLE ARE FOR TEMPERATURE OF THE SENSOR, NOT AIR TEMPERATURE.)

NO: REPLACE WHEEL SPEED SENSOR.

YES:

(3)
- WITH J 39200 STILL CONNECTED, SELECT A/C VOLTAGE SCALE.
- SPIN WHEEL BY HAND WHILE OBSERVING VOLTAGE.

IS VOLTAGE AT LEAST 100 mV (VOLTAGE WILL INCREASE AS WHEEL SPEED INCREASES)?

NO: REPLACE WHEEL SPEED SENSOR.

YES:

(4)
- IGNITION "OFF."
- DISCONNECT 32-WAY EBTCM CONNECTOR.
- USING J 39200, MEASURE THE RESISTANCE BETWEEN TERMINAL "E1" AND "E2" OF THE 32-WAY EBTCM HARNESS CONNECTOR.

DOES J 39200 DISPLAY RESISTANCE AS "OL" (INFINITE)?

NO: REPAIR SHORT BETWEEN CKTs 884 AND 885.

YES:

(5)
- INSPECT CKTs 884 AND 885 AND 32-WAY EBTCM CONNECTOR FOR DAMAGE WHICH MAY RESULT IN A SHORT BETWEEN CKTs 884 AND 885 WITH ALL CONNECTORS CONNECTED. REPAIR ALL DAMAGE IF EVIDENT.
- RECONNECT ALL CONNECTORS.
- TEST DRIVE VEHICLE ABOVE 24 km/h (15 MPH) FOR AT LEAST 30 SECONDS.

DOES DTC 23 SET AS A CURRENT DTC?

NO: MALFUNCTION WAS MOST LIKELY CAUSED BY A POOR CONNECTION. REFER TO "DIAGNOSTIC AIDS" ON FACING PAGE.

YES:

(6)
- USING TECH 1, SELECT "F0: DATA LIST."
- MONITOR WHEEL SPEEDS WHILE SLOWLY ACCELERATING TO 65 km/h (40 MPH) AND THEN SLOWLY DECELERATING TO 0 km/h (0 MPH).

IS THE LEFT REAR WHEEL SPEED CONSTANTLY HIGHER THAN THE THREE REMAINING WHEEL SPEEDS?

NO: REPLACE EBTCM.

YES: REPLACE INCORRECT OR MALFUNCTIONING LEFT WHEEL SPEED SENSOR/HUB AND BEARING ASSEMBLY WITH THE PROPER ASSEMBLY. REFER TO PARTS CATALOG FOR PROPER PART NUMBER.

REAR WHEEL SPEED SENSOR
TEMPERATURE OF SENSOR VS. SENSOR RESISTANCE (APPROXIMATE)

TEMP. (°C)	TEMP. (°F)	RESISTANCE (OHMS)
-34 TO 4	-30 TO 40	1622 TO 2205
5 TO 43	41 TO 110	1930 TO 2566
44 TO 93	111 TO 200	2245 TO 3028

DIAGNOSTIC CHART DTC 24 — 1994 SILHOUETTE, TRANS SPORT AND LUMINA APV WITH TCS

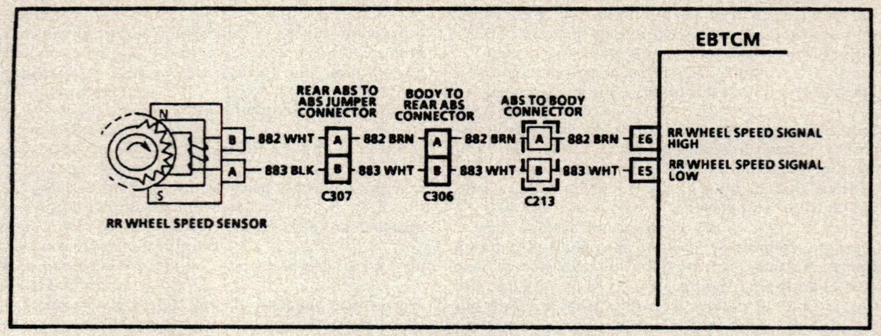

DTC 24

RIGHT REAR WHEEL SPEED = 0

Circuit Description:

As a toothed ring passes by the wheel speed sensor, changes in the electromagnetic field cause the wheel speed sensor to produce a sinusoidal (AC) voltage signal whose frequency is proportional to wheel speed. The magnitude of this signal is directly related to wheel speed and the proximity of the wheel speed sensor to the toothed ring, often referred to as the air gap.

DTC Will Set When: DTC 24 can be set when the vehicle is not in an ABS or TCS event. If the right rear wheel speed = 0 and the vehicle's reference speed is greater than 8 km/h (5 mph), a malfunction exists.

Action Taken: A malfunction DTC is stored, ABS and TCS are disabled and the ABS and TCS warning lamps are turned "ON."

DTC Chart Test Description: Number(s) below refer to circled number(s) on the diagnostic chart.

1. This step is designed to identify a wheel speed sensor or circuitry that is damaged and visibly apparent.
2. Checks to see that the wheel speed sensor is within proper resistance values.
3. Verifies that the wheel speed sensor can generate an appropriate output signal.
4. Ensures that the wheel speed sensor circuitry is not shorted together.
5. Checks to see if malfunction was due to physical damage of the circuitry.
6. Checks for an incorrect or malfunctioning hub and bearing assembly.

Diagnostic Aids: An intermittent malfunction may be caused by a poor connection, rubbed through wire insulation, or a wire that is broken inside the insulation.

The frequency of the malfunction can be checked by using the enhanced diagnostic function of the Tech 1.

If the customer's comments reflect that the ABS and TCS warning lamps are "ON" only during moist environmental changes (rain, snow, vehicle wash), all wheel speed sensor circuitry should be thoroughly inspected for signs of water intrusion. If DTC is not current, clear DTCs and simulate the effects of water intrusion. Use the following procedure. Spray down the suspected area with a 5% salt water solution (two teaspoons of salt to 12 oz. of water). Test drive vehicle over various road surfaces (bumps, turns, etc.) above 24 km/h (15 mph) for at least 30 seconds. If DTC returns, replace suspected harness.

Any circuitry, that is suspected as causing the intermittent complaint, should be thoroughly checked for backed out terminals, improper mating, broken locks, improperly formed or damaged terminals, poor terminal to wiring connections or physical damage to the wiring harness.

Resistance of the wheel speed sensor will increase with an increase in sensor temperature.

When replacing a wheel speed sensor, inspect the sensor terminals and harness connector for corrosion and/or water intrusion. If evidence of corrosion or water intrusion exists, replace wheel speed sensor jumper harness. Likewise, if replacing a wheel speed sensor jumper harness, inspect sensor terminals. If evidence of corrosion or water intrusion exists, replace wheel speed sensor.

DIAGNOSTIC CHART DTC 24 — 1994 SILHOUETTE, TRANS SPORT AND LUMINA APV WITH TCS, CONTINUED

DTC 24
RIGHT REAR WHEEL SPEED = 0

Important

- WHEEL SPEED SENSOR INTERMITTENT MALFUNCTIONS MAY BE DIFFICULT TO LOCATE. CARE SHOULD BE TAKEN NOT TO DISTURB ANY ELECTRICAL CONNECTIONS PRIOR TO AN INDICATED STEP OF THIS CHART. THIS WILL ENSURE THAT AN INTERMITTENT CONNECTION WILL NOT BE CORRECTED BEFORE THE SOURCE OF THE MALFUNCTION IS FOUND.

(1)
- IGNITION "OFF."
- PHYSICALLY INSPECT THE RIGHT REAR WHEEL SPEED SENSOR, WHEEL SPEED SENSOR JUMPER HARNESS, AND CONNECTORS FOR DAMAGE. IS PHYSICAL DAMAGE INDICATED?

YES: REPLACE WHEEL SPEED SENSOR AND/OR WHEEL SPEED SENSOR JUMPER HARNESS IF DAMAGED. DRIVE VEHICLE ABOVE 24 km/h (15 MPH) TO ENSURE DTC 24 DOES NOT SET.

NO:

(2)
- DISCONNECT RIGHT REAR WHEEL SPEED SENSOR DIRECTLY AT SENSOR.
- USING J 39200, MEASURE THE RESISTANCE BETWEEN SENSOR TERMINALS "A" AND "B." IS RESISTANCE BETWEEN 2059 AND 2353Ω WHEN SENSOR IS 20°C (68°F)? *(IF TEMPERATURE OF THE SENSOR IS SUSPECTED TO BE OTHER THAN 20°C (68°F), REFER TO THE TABLE BELOW FOR APPLICABLE SENSOR RESISTANCE VALUES; THE VALUES IN THE TABLE ARE FOR TEMPERATURE OF THE SENSOR, NOT AIR TEMPERATURE.)*

NO: REPLACE WHEEL SPEED SENSOR.

YES:

(3)
- WITH J 39200 STILL CONNECTED, SELECT A/C VOLTAGE SCALE.
- SPIN WHEEL BY HAND WHILE OBSERVING VOLTAGE. IS VOLTAGE AT LEAST 100 mV (VOLTAGE WILL INCREASE AS WHEEL SPEED INCREASES)?

NO: REPLACE WHEEL SPEED SENSOR.

YES:

(4)
- IGNITION "OFF."
- DISCONNECT 32-WAY EBTCM CONNECTOR.
- USING J 39200, MEASURE THE RESISTANCE BETWEEN TERMINAL "E5" AND "E6" OF THE 32-WAY EBTCM HARNESS CONNECTOR. DOES J 39200 DISPLAY RESISTANCE AS "OL" (INFINITE)?

NO: REPAIR SHORT BETWEEN CKTs 883 AND 882.

YES:

(5)
- INSPECT CKTs 882 AND 883 AND 32-WAY EBTCM CONNECTOR FOR DAMAGE WHICH MAY RESULT IN A SHORT BETWEEN CKTs 882 AND 883 WITH ALL CONNECTORS CONNECTED. REPAIR ALL DAMAGE IF EVIDENT.
- RECONNECT ALL CONNECTORS.
- TEST DRIVE VEHICLE ABOVE 24 km/h (15 MPH) FOR AT LEAST 30 SECONDS. DOES DTC 24 SET AS A CURRENT DTC?

NO: MALFUNCTION WAS MOST LIKELY CAUSED BY A POOR CONNECTION. REFER TO "DIAGNOSTIC AIDS" ON FACING PAGE.

YES:

(6)
- USING TECH 1, SELECT "F0: DATA LIST."
- MONITOR WHEEL SPEEDS WHILE SLOWLY ACCELERATING TO 65 km/h (40 MPH) AND THEN SLOWLY DECELERATING TO 0 km/h (0 MPH). IS THE RIGHT REAR WHEEL SPEED CONSTANTLY HIGHER THAN THE THREE REMAINING WHEEL SPEEDS?

NO: REPLACE EBTCM.

YES: REPLACE INCORRECT OR MALFUNCTIONING RIGHT WHEEL SPEED SENSOR/HUB AND BEARING ASSEMBLY WITH THE PROPER ASSEMBLY. REFER TO PARTS CATALOG FOR PROPER PART NUMBER.

REAR WHEEL SPEED SENSOR
TEMPERATURE OF SENSOR VS. SENSOR RESISTANCE (APPROXIMATE)

TEMP. (°C)	TEMP. (°F)	RESISTANCE (OHMS)
-34 TO 4	-30 TO 40	1622 TO 2205
5 TO 43	41 TO 110	1930 TO 2566
44 TO 93	111 TO 200	2245 TO 3028

DIAGNOSTIC CHART DTC 25 — 1994 SILHOUETTE, TRANS SPORT AND LUMINA APV WITH TCS

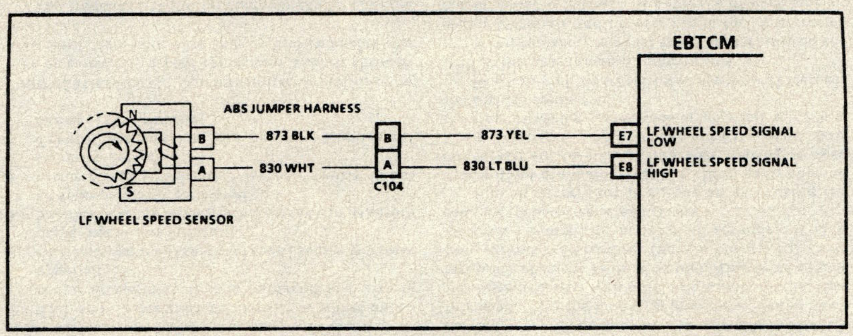

DTC 25

LEFT FRONT EXCESSIVE WHEEL SPEED VARIATION

Circuit Description:

As a toothed ring passes by the wheel speed sensor, changes in the electromagnetic field cause the wheel speed sensor to produce a sinusoidal (AC) voltage signal whose frequency is proportional to wheel speed. The magnitude of this signal is directly related to wheel speed and the proximity of the wheel speed sensor to the toothed ring, often referred to as the air gap.

DTC Will Set When: DTC 25 can be set when the brake is "OFF." The purpose of this test is to detect a situation in which the left front wheel acceleration or deceleration is beyond specified limits.

Action Taken: A malfunction DTC is stored, ABS and TCS are disabled and the ABS and TCS warning lamps are turned "ON."

DTC Chart Test Description: Number(s) below refer to circled number(s) on the diagnostic chart.

1. This step is designed to identify a wheel speed sensor or circuitry that is damaged and visibly apparent.
2. Checks to see that the wheel speed sensor is within proper resistance values.
3. Verifies that the wheel speed sensor can generate an appropriate output signal.
4. Ensures that wheel speed sensor circuitry is not shorted together.
5. Ensures malfunction was not due to physical damage of the circuitry.

Diagnostic Aids: An intermittent malfunction may be caused by a poor connection, rubbed through wire insulation, or a wire that is broken inside the insulation.

The frequency of the malfunction can be checked by using the enhanced diagnostic function of the Tech 1.

If the customer's comments reflect that the ABS and TCS warning lamps are "ON" only during moist environmental changes (rain, snow, vehicle wash), all wheel speed sensor circuitry should be thoroughly inspected for signs of water intrusion. If DTC is not current, clear DTCs and simulate the effects of water intrusion. Use the following procedure. Spray down the suspected area with a 5% salt water solution (two teaspoons of salt to 12 oz. of water). Test drive vehicle over various road surfaces (bumps, turns, etc.) above 24 km/h (15 mph) for at least 30 seconds. If DTC returns, replace suspected harness.

Any circuitry, that is suspected as causing the intermittent complaint, should be thoroughly checked for backed out terminals, improper mating, broken locks, improperly formed or damaged terminals, poor terminal to wiring connections or physical damage to the wiring harness.

Resistance of the wheel speed sensor will increase with an increase in sensor temperature.

When replacing a wheel speed sensor, inspect the sensor terminals and harness connector for corrosion and/or water intrusion. If evidence of corrosion or water intrusion exists, replace wheel speed sensor jumper harness. Likewise, if replacing a wheel speed sensor jumper harness, inspect sensor terminals. If evidence of corrosion or water intrusion exists, replace wheel speed sensor.

DIAGNOSTIC CHART DTC 25 — 1994 SILHOUETTE, TRANS SPORT AND LUMINA APV WITH TCS, CONTINUED

DTC 25
LEFT FRONT EXCESSIVE WHEEL SPEED VARIATION

Important

- WHEEL SPEED SENSOR INTERMITTENT MALFUNCTIONS MAY BE DIFFICULT TO LOCATE. CARE SHOULD BE TAKEN NOT TO DISTURB ANY ELECTRICAL CONNECTIONS PRIOR TO AN INDICATED STEP OF THIS CHART. THIS WILL ENSURE THAT AN INTERMITTENT CONNECTION WILL NOT BE CORRECTED BEFORE THE SOURCE OF THE MALFUNCTION IS FOUND.

(1)
- IGNITION "OFF."
- PHYSICALLY INSPECT THE LEFT FRONT WHEEL SPEED SENSOR, LEFT FRONT WHEEL SPEED SENSOR JUMPER HARNESS, AND CONNECTORS FOR DAMAGE. IS PHYSICAL DAMAGE INDICATED?

YES → REPLACE WHEEL SPEED SENSOR AND/OR JUMPER HARNESS IF DAMAGED. DRIVE VEHICLE ABOVE 24 km/h (15 MPH) TO ENSURE DTC 25 DOES NOT SET.

NO ↓

(2)
- DISCONNECT LEFT FRONT WHEEL SPEED SENSOR DIRECTLY AT SENSOR.
- USING J 39200, MEASURE THE RESISTANCE BETWEEN SENSOR TERMINALS "A" AND "B."
  IS RESISTANCE BETWEEN 1020 AND 1137Ω WHEN SENSOR IS 20°C (68°F)? *(IF TEMPERATURE OF THE SENSOR IS SUSPECTED TO BE OTHER THAN 20°C (68°F), REFER TO THE TABLE BELOW FOR APPLICABLE SENSOR RESISTANCE VALUES; THE VALUES IN THE TABLE ARE FOR TEMPERATURE OF THE SENSOR, NOT AIR TEMPERATURE.)*

NO → REPLACE WHEEL SPEED SENSOR.

YES ↓

(3)
- WITH J 39200 STILL CONNECTED, SELECT A/C VOLTAGE SCALE.
- SPIN WHEEL BY HAND WHILE OBSERVING VOLTAGE.
  IS VOLTAGE AT LEAST 100 mV (VOLTAGE WILL INCREASE AS WHEEL SPEED INCREASES)?

NO → REPLACE WHEEL SPEED SENSOR.

YES ↓

(4)
- IGNITION "OFF."
- DISCONNECT 32-WAY EBTCM CONNECTOR.
- USING J 39200, MEASURE THE RESISTANCE BETWEEN TERMINAL "E7" AND "E8" OF THE 32-WAY EBTCM HARNESS CONNECTOR.
  DOES J 39200 DISPLAY RESISTANCE AS "OL" (INFINITE)?

NO → REPAIR SHORT BETWEEN CKTs 830 AND 873.

YES ↓

(5)
- INSPECT CKTs 830 AND 873 AND 32-WAY EBTCM CONNECTOR FOR DAMAGE WHICH MAY RESULT IN A SHORT BETWEEN CKTs 830 AND 873 WITH ALL CONNECTORS CONNECTED. REPAIR ALL DAMAGE IF EVIDENT.
- RECONNECT ALL CONNECTORS.
- TEST DRIVE VEHICLE ABOVE 24 km/h (15 MPH) FOR AT LEAST 30 SECONDS.
  DOES DTC 25 SET AS A CURRENT DTC?

YES → REPLACE EBTCM.

NO → MALFUNCTION IS INTERMITTENT. REFER TO "DIAGNOSTIC AIDS" ON FACING PAGE.

FRONT WHEEL SPEED SENSOR

TEMPERATURE OF SENSOR VS. SENSOR RESISTANCE (APPROXIMATE)

TEMP. (°C)	TEMP. (°F)	RESISTANCE (OHMS)
-34 TO 4	-30 TO 40	804 TO 1066
5 TO 43	41 TO 110	956 TO 1240
44 TO 93	111 TO 200	1112 TO 1463

DIAGNOSTIC CHART DTC 26 — 1994 SILHOUETTE, TRANS SPORT AND LUMINA APV WITH TCS

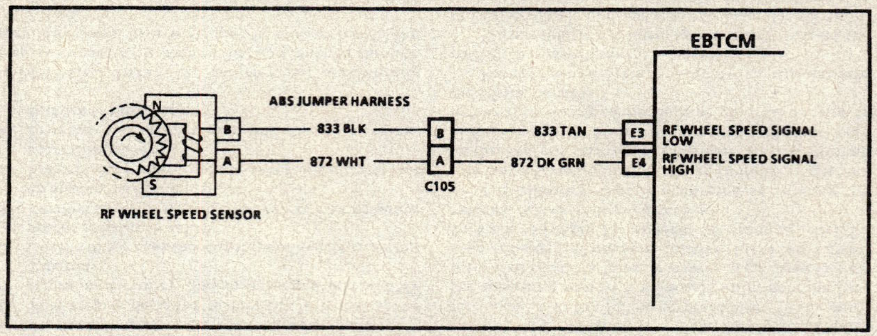

DTC 26

RIGHT FRONT EXCESSIVE WHEEL SPEED VARIATION

Circuit Description:

As a toothed ring passes by the wheel speed sensor, changes in the electromagnetic field cause the wheel speed sensor to produce a sinusoidal (AC) voltage signal whose frequency is proportional to wheel speed. The magnitude of this signal is directly related to wheel speed and the proximity of the wheel speed sensor to the toothed ring, often referred to as the air gap.

DTC Will Set When: DTC 26 can be set when the brake is "OFF." The purpose of this test is to detect a situation in which the right front wheel acceleration or deceleration is beyond specified limits.

Action Taken: A malfunction DTC is stored, ABS and TCS are disabled and the ABS and TCS warning lamps are turned "ON."

DTC Chart Test Description: Number(s) below refer to circled number(s) on the diagnostic chart.

1. This step is designed to identify a wheel speed sensor or circuitry that is damaged and visibly apparent.
2. Checks to see that the wheel speed sensor is within proper resistance values.
3. Verifies that the wheel speed sensor can generate an appropriate output signal.
4. Ensures that wheel speed sensor circuitry is not shorted together.
5. Ensures malfunction was not due to physical damage of the circuitry.

Diagnostic Aids: An intermittent malfunction may be caused by a poor connection, rubbed through wire insulation, or a wire that is broken inside the insulation.

The frequency of the malfunction can be checked by using the enhanced diagnostic function of the Tech 1.

If the customer's comments reflect that the ABS and TCS warning lamps are "ON" only during moist environmental changes (rain, snow, vehicle wash), all wheel speed sensor circuitry should be thoroughly inspected for signs of water intrusion. If DTC is not current, clear DTCs and simulate the effects of water intrusion. Use the following procedure. Spray down the suspected area with a 5% salt water solution (two teaspoons of salt to 12 oz. of water). Test drive vehicle over various road surfaces (bumps, turns, etc.) above 24 km/h (15 mph) for at least 30 seconds. If DTC returns, replace suspected harness.

Any circuitry, that is suspected as causing the intermittent complaint, should be thoroughly checked for backed out terminals, improper mating, broken locks, improperly formed or damaged terminals, poor terminal to wiring connections or physical damage to the wiring harness.

Resistance of the wheel speed sensor will increase with an increase in sensor temperature.

When replacing a wheel speed sensor, inspect the sensor terminals and harness connector for corrosion and/or water intrusion. If evidence of corrosion or water intrusion exists, replace wheel speed sensor jumper harness. Likewise, if replacing a wheel speed sensor jumper harness, inspect sensor terminals. If evidence of corrosion or water intrusion exists, replace wheel speed sensor.

DIAGNOSTIC CHART DTC 26 — 1994 SILHOUETTE, TRANS SPORT AND LUMINA APV WITH TCS, CONTINUED

DTC 26
RIGHT FRONT EXCESSIVE WHEEL SPEED VARIATION

Important

- WHEEL SPEED SENSOR INTERMITTENT MALFUNCTIONS MAY BE DIFFICULT TO LOCATE. CARE SHOULD BE TAKEN NOT TO DISTURB ANY ELECTRICAL CONNECTIONS PRIOR TO AN INDICATED STEP OF THIS CHART. THIS WILL ENSURE THAT AN INTERMITTENT CONNECTION WILL NOT BE CORRECTED BEFORE THE SOURCE OF THE MALFUNCTION IS FOUND.

(1)
- IGNITION "OFF."
- PHYSICALLY INSPECT THE RIGHT FRONT WHEEL SPEED SENSOR, RIGHT FRONT WHEEL SPEED SENSOR JUMPER HARNESS, AND CONNECTORS FOR DAMAGE. IS PHYSICAL DAMAGE INDICATED?

YES: REPLACE WHEEL SPEED SENSOR AND/OR JUMPER HARNESS IF DAMAGED. DRIVE VEHICLE ABOVE 24 km/h (15 MPH) TO ENSURE DTC 26 DOES NOT SET.

NO:

(2)
- DISCONNECT RIGHT FRONT WHEEL SPEED SENSOR DIRECTLY AT SENSOR.
- USING J 39200, MEASURE THE RESISTANCE BETWEEN SENSOR TERMINALS "A" AND "B."
IS RESISTANCE BETWEEN 1020 AND 1137Ω WHEN SENSOR IS 20°C (68°F)?
(IF TEMPERATURE OF THE SENSOR IS SUSPECTED TO BE OTHER THAN 20°C (68°F), REFER TO THE TABLE BELOW FOR APPLICABLE SENSOR RESISTANCE VALUES; THE VALUES IN THE TABLE ARE FOR TEMPERATURE OF THE SENSOR, NOT AIR TEMPERATURE.)

NO: REPLACE WHEEL SPEED SENSOR.

YES:

(3)
- WITH J 39200 STILL CONNECTED, SELECT A/C VOLTAGE SCALE.
- SPIN WHEEL BY HAND WHILE OBSERVING VOLTAGE.
IS VOLTAGE AT LEAST 100 mV (VOLTAGE WILL INCREASE AS WHEEL SPEED INCREASES)?

NO: REPLACE WHEEL SPEED SENSOR.

YES:

(4)
- IGNITION "OFF."
- DISCONNECT 32-WAY EBTCM CONNECTOR.
- USING J 39200, MEASURE THE RESISTANCE BETWEEN TERMINAL "E3" AND "E4" OF THE 32-WAY EBTCM HARNESS CONNECTOR.
DOES J 39200 DISPLAY RESISTANCE AS "OL" (INFINITE)?

NO: REPAIR SHORT BETWEEN CKTs 833 AND 872.

YES:

(5)
- INSPECT CKTs 833 AND 872 AND 32-WAY EBTCM CONNECTOR FOR DAMAGE WHICH MAY RESULT IN A SHORT BETWEEN CKTs 833 AND 872 WITH ALL CONNECTORS CONNECTED. REPAIR ALL DAMAGE IF EVIDENT.
- RECONNECT ALL CONNECTORS.
- TEST DRIVE VEHICLE ABOVE 24 km/h (15 MPH) FOR AT LEAST 30 SECONDS.
DOES DTC 26 SET AS A CURRENT DTC?

YES: REPLACE EBTCM.

NO: MALFUNCTION IS INTERMITTENT. REFER TO "DIAGNOSTIC AIDS" ON FACING PAGE.

FRONT WHEEL SPEED SENSOR
TEMPERATURE OF SENSOR VS. SENSOR RESISTANCE (APPROXIMATE)

TEMP. (°C)	TEMP. (°F)	RESISTANCE (OHMS)
-34 TO 4	-30 TO 40	804 TO 1066
5 TO 43	41 TO 110	956 TO 1240
44 TO 93	111 TO 200	1112 TO 1463

DIAGNOSTIC CHART DTC 27 — 1994 SILHOUETTE, TRANS SPORT AND LUMINA APV WITH TCS

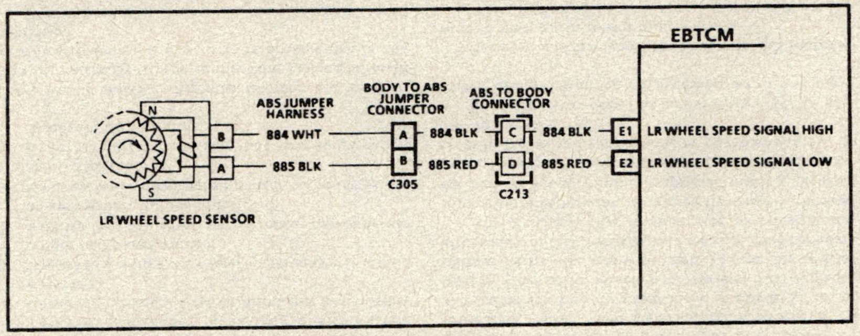

DTC 27

LEFT REAR EXCESSIVE WHEEL SPEED VARIATION

Circuit Description:
As a toothed ring passes by the wheel speed sensor, changes in the electromagnetic field cause the wheel speed sensor to produce a sinusoidal (AC) voltage signal whose frequency is proportional to wheel speed. The magnitude of this signal is directly related to wheel speed and the proximity of the wheel speed sensor to the toothed ring, often referred to as the air gap.

DTC Will Set When: DTC 27 can be set when the brake is "OFF." The purpose of this test is to detect a situation in which the left rear wheel acceleration or deceleration is beyond specified limits.

Action Taken: A malfunction DTC is stored, ABS and TCS are disabled and the ABS and TCS warning lamps are turned "ON."

DTC Chart Test Description: Number(s) below refer to circled number(s) on the diagnostic chart.

1. This step is designed to identify a wheel speed sensor or circuitry that is damaged and visibly apparent.
2. Checks to see that the wheel speed sensor is within proper resistance values.
3. Verifies that the wheel speed sensor can generate an appropriate output signal.
4. Ensures that wheel speed sensor circuitry is not shorted together.
5. Ensures malfunction was not due to physical damage of the circuitry.

Diagnostic Aids: An intermittent malfunction may be caused by a poor connection, rubbed through wire insulation, or a wire that is broken inside the insulation.

The frequency of the malfunction can be checked by using the enhanced diagnostic function of the Tech 1.

If the customer's comments reflect that the ABS and TCS warning lamps are "ON" only during moist environmental changes (rain, snow, vehicle wash), all wheel speed sensor circuitry should be thoroughly inspected for signs of water intrusion.

If DTC is not current, clear DTCs and simulate the effects of water intrusion. Use the following procedure. Spray down the suspected area with a 5% salt water solution (two teaspoons of salt to 12 oz. of water). Test drive vehicle over various road surfaces (bumps, turns, etc.) above 24 km/h (15 mph) for at least 30 seconds. If DTC returns, replace suspected harness.

Any circuitry, that is suspected as causing the intermittent complaint, should be thoroughly checked for backed out terminals, improper mating, broken locks, improperly formed or damaged terminals, poor terminal to wiring connections or physical damage to the wiring harness.

Excessive end play can cause this DTC to set. Check integral bearing for excessive axial end play.

Resistance of the wheel speed sensor will increase with an increase in sensor temperature.

When replacing a wheel speed sensor, inspect the sensor terminals and harness connector for corrosion and/or water intrusion. If evidence of corrosion or water intrusion exists, replace wheel speed sensor jumper harness. Likewise, if replacing a wheel speed sensor jumper harness, inspect sensor terminals. If evidence of corrosion or water intrusion exists, replace wheel speed sensor.

DIAGNOSTIC CHART DTC 27 — 1994 SILHOUETTE, TRANS SPORT AND LUMINA APV WITH TCS, CONTINUED

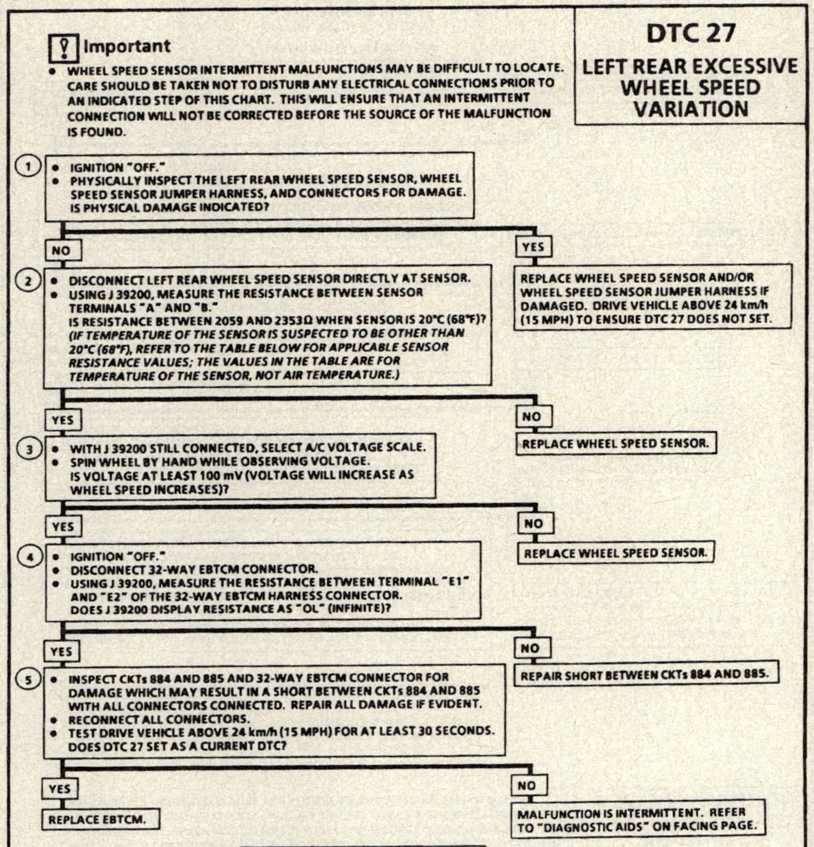

REAR WHEEL SPEED SENSOR
TEMPERATURE OF SENSOR VS. SENSOR RESISTANCE (APPROXIMATE)

TEMP. (°C)	TEMP. (°F)	RESISTANCE (OHMS)
-34 TO 4	-30 TO 40	1622 TO 2205
5 TO 43	41 TO 110	1930 TO 2566
44 TO 93	111 TO 200	2245 TO 3028

DIAGNOSTIC CHART DTC 28 — 1994 SILHOUETTE, TRANS SPORT AND LUMINA APV WITH TCS

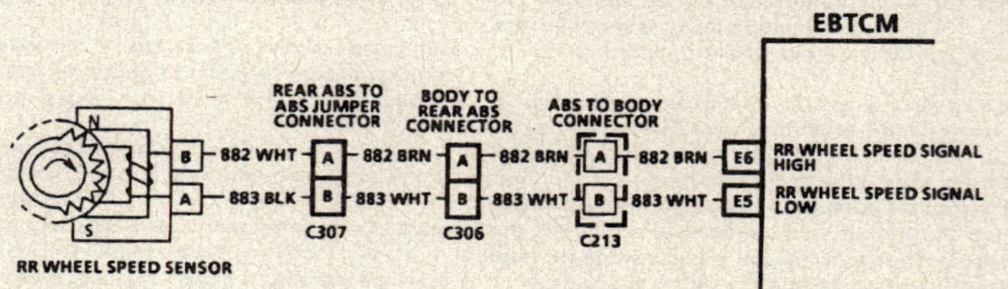

DTC 28

RIGHT REAR EXCESSIVE WHEEL SPEED VARIATION

Circuit Description:

As a toothed ring passes by the wheel speed sensor, changes in the electromagnetic field cause the wheel speed sensor to produce a sinusoidal (AC) voltage signal whose frequency is proportional to wheel speed. The magnitude of this signal is directly related to wheel speed and the proximity of the wheel speed sensor to the toothed ring, often referred to as the air gap.

DTC Will Set When: DTC 28 can be set when the brake is "OFF." The purpose of this test is to detect a situation in which the right rear wheel acceleration or deceleration is beyond specified limits.

Action Taken: A malfunction DTC is stored, ABS and TCS are disabled and the ABS and TCS warning lamps are turned "ON."

DTC Chart Test Description: Number(s) below refer to circled number(s) on the diagnostic chart.

1. This step is designed to identify a wheel speed sensor or circuitry that is damaged and visibly apparent.
2. Checks to see that the wheel speed sensor is within proper resistance values.
3. Verifies that the wheel speed sensor can generate an appropriate output signal.
4. Ensures that wheel speed sensor circuitry is not shorted together.
5. Ensures malfunction was not due to physical damage of the circuitry.

Diagnostic Aids: An intermittent malfunction may be caused by a poor connection, rubbed through wire insulation, or a wire that is broken inside the insulation.

The frequency of the malfunction can be checked by using the enhanced diagnostic function of the Tech 1.

If the customer's comments reflect that the ABS and TCS warning lamps are "ON" only during moist environmental changes (rain, snow, vehicle wash), all wheel speed sensor circuitry should be thoroughly inspected for signs of water intrusion.

If DTC is not current, clear DTCs and simulate the effects of water intrusion. Use the following procedure. Spray down the suspected area with a 5% salt water solution (two teaspoons of salt to 12 oz. of water). Test drive vehicle over various road surfaces (bumps, turns, etc.) above 24 km/h (15 mph) for at least 30 seconds. If DTC returns, replace suspected harness.

Any circuitry, that is suspected as causing the intermittent complaint, should be thoroughly checked for backed out terminals, improper mating, broken locks, improperly formed or damaged terminals, poor terminal to wiring connections or physical damage to the wiring harness.

Excessive end play can cause this DTC to set. Check integral bearing for excessive axial end play.

Resistance of the wheel speed sensor will increase with an increase in sensor temperature.

When replacing a wheel speed sensor, inspect the sensor terminals and harness connector for corrosion and/or water intrusion. If evidence of corrosion or water intrusion exists, replace wheel speed sensor jumper harness. Likewise, if replacing a wheel speed sensor jumper harness, inspect sensor terminals. If evidence of corrosion or water intrusion exists, replace wheel speed sensor.

DIAGNOSTIC CHART DTC 28 — 1994 SILHOUETTE, TRANS SPORT AND LUMINA APV WITH TCS, CONTINUED

DTC 28
RIGHT REAR EXCESSIVE WHEEL SPEED VARIATION

Important

- WHEEL SPEED SENSOR INTERMITTENT MALFUNCTIONS MAY BE DIFFICULT TO LOCATE. CARE SHOULD BE TAKEN NOT TO DISTURB ANY ELECTRICAL CONNECTIONS PRIOR TO AN INDICATED STEP OF THIS CHART. THIS WILL ENSURE THAT AN INTERMITTENT CONNECTION WILL NOT BE CORRECTED BEFORE THE SOURCE OF THE MALFUNCTION IS FOUND.

(1)
- IGNITION "OFF."
- PHYSICALLY INSPECT THE RIGHT REAR WHEEL SPEED SENSOR, WHEEL SPEED SENSOR JUMPER HARNESS, AND CONNECTORS FOR DAMAGE. IS PHYSICAL DAMAGE INDICATED?

YES: REPLACE WHEEL SPEED SENSOR AND/OR WHEEL SPEED SENSOR JUMPER HARNESS IF DAMAGED. DRIVE VEHICLE ABOVE 24 km/h (15 MPH) TO ENSURE DTC 28 DOES NOT SET.

NO:

(2)
- DISCONNECT RIGHT REAR WHEEL SPEED SENSOR DIRECTLY AT SENSOR.
- USING J 39200, MEASURE THE RESISTANCE BETWEEN SENSOR TERMINALS "A" AND "B."

IS RESISTANCE BETWEEN 2059 AND 2353Ω WHEN SENSOR IS 20°C (68°F)? *(IF TEMPERATURE OF THE SENSOR IS SUSPECTED TO BE OTHER THAN 20°C (68°F), REFER TO THE TABLE BELOW FOR APPLICABLE SENSOR RESISTANCE VALUES; THE VALUES IN THE TABLE ARE FOR TEMPERATURE OF THE SENSOR, NOT AIR TEMPERATURE.)*

NO: REPLACE WHEEL SPEED SENSOR.

YES:

(3)
- WITH J 39200 STILL CONNECTED, SELECT A/C VOLTAGE SCALE.
- SPIN WHEEL BY HAND WHILE OBSERVING VOLTAGE.

IS VOLTAGE AT LEAST 100 mV (VOLTAGE WILL INCREASE AS WHEEL SPEED INCREASES)?

NO: REPLACE WHEEL SPEED SENSOR.

YES:

(4)
- IGNITION "OFF."
- DISCONNECT 32-WAY EBTCM CONNECTOR.
- USING J 39200, MEASURE THE RESISTANCE BETWEEN TERMINAL "E5" AND "E6" OF THE 32-WAY EBTCM HARNESS CONNECTOR.

DOES J 39200 DISPLAY RESISTANCE AS "OL" (INFINITE)?

NO: REPAIR SHORT BETWEEN CKTs 882 AND 883.

YES:

(5)
- INSPECT CKTs 882 AND 883 AND 32-WAY EBTCM CONNECTOR FOR DAMAGE WHICH MAY RESULT IN A SHORT BETWEEN CKTs 882 AND 883 WITH ALL CONNECTORS CONNECTED. REPAIR ALL DAMAGE IF EVIDENT.
- RECONNECT ALL CONNECTORS.
- TEST DRIVE VEHICLE ABOVE 24 km/h (15 MPH) FOR AT LEAST 30 SECONDS.

DOES DTC 28 SET AS A CURRENT DTC?

YES: REPLACE EBTCM.

NO: MALFUNCTION IS INTERMITTENT. REFER TO "DIAGNOSTIC AIDS" ON FACING PAGE.

REAR WHEEL SPEED SENSOR
TEMPERATURE OF SENSOR VS. SENSOR RESISTANCE (APPROXIMATE)

TEMP. (°C)	TEMP. (°F)	RESISTANCE (OHMS)
-34 TO 4	-30 TO 40	1622 TO 2205
5 TO 43	41 TO 110	1930 TO 2566
44 TO 93	111 TO 200	2245 TO 3028

DIAGNOSTIC CHART DTC 32 — 1994 SILHOUETTE, TRANS SPORT AND LUMINA APV WITH TCS

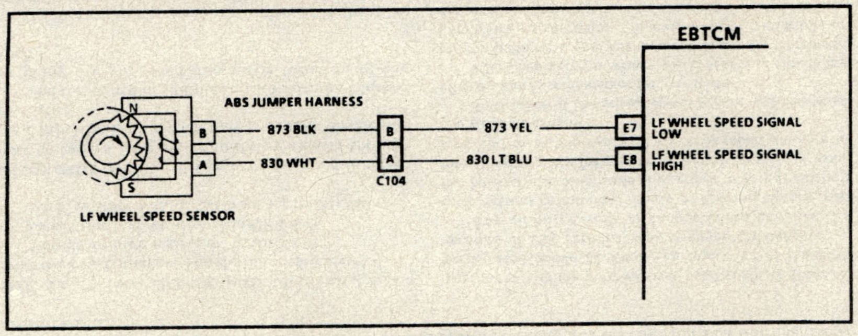

DTC 32

(Page 1 of 3)
LF WHEEL SPEED SNR CKT OPEN OR SHORT TO GND/BAT

Circuit Description:
As a toothed ring passes by the wheel speed sensor, changes in the electromagnetic field cause the wheel speed sensor to produce a sinusoidal (AC) voltage signal whose frequency is proportional to wheel speed. The magnitude of this signal is directly related to wheel speed and the proximity of the wheel speed sensor to the toothed ring, often referred to as the air gap.

DTC Will Set When: DTC 32 can be set anytime after initialization. If either of the left front wheel speed circuits are shorted to voltage or ground, a malfunction exists.

Action Taken: A malfunction DTC is stored, ABS and TCS are disabled and the ABS and TCS warning lamps are turned "ON."

DTC Chart Test Description: Number(s) below refer to circled number(s) on the diagnostic chart.

1. Checks to see that the wheel speed sensor is within proper resistance values.
2. Checks for short to ground in wheel speed sensor wiring.
3. Checks for short to voltage in wheel speed sensor wiring.
4. Checks for a short to voltage in CKT 873.
5. Checks for a short to voltage in CKT 830.
6. Ensures malfunction was not due to physical damage of circuitry.

Diagnostic Aids: An intermittent malfunction may be caused by a poor connection, rubbed through wire insulation, or a wire that is broken inside the insulation.

The frequency of the malfunction can be checked by using the enhanced diagnostic function of the Tech 1.

If the customer's comments reflect that the ABS and TCS warning lamps are "ON" only during moist environmental changes (rain, snow, vehicle wash), all wheel speed sensor circuitry should be thoroughly inspected for signs of water intrusion. If DTC is not current, clear DTCs and simulate the effects of water intrusion. Use the following procedure. Spray down the suspected area with a 5% salt water solution (two teaspoons of salt to 12 oz. of water). Test drive vehicle over various road surfaces (bumps, turns, etc.) above 24 km/h (15 mph) for at least 30 seconds. If DTC returns, replace suspected harness.

Any circuitry, that is suspected as causing the intermittent complaint, should be thoroughly checked for backed out terminals, improper mating, broken locks, improperly formed or damaged terminals, poor terminal to wiring connections or physical damage to the wiring harness.

Resistance of the wheel speed sensor will increase with an increase in sensor temperature.

When replacing a wheel speed sensor, inspect the sensor terminals and harness connector for corrosion and/or water intrusion. If evidence of corrosion or water intrusion exists, replace wheel speed sensor jumper harness. Likewise, if replacing a wheel speed sensor jumper harness, inspect sensor terminals. If evidence of corrosion or water intrusion exists, replace wheel speed sensor.

DIAGNOSTIC CHART DTC 32 — 1994 SILHOUETTE, TRANS SPORT AND LUMINA APV WITH TCS, CONTINUED

DTC 32

(Page 1 of 3)
LF WHEEL SPEED SNR CKT OPEN OR SHORT TO GND/BAT

Important

- WHEEL SPEED SENSOR INTERMITTENT MALFUNCTIONS MAY BE DIFFICULT TO LOCATE. CARE SHOULD BE TAKEN NOT TO DISTURB ANY ELECTRICAL CONNECTIONS PRIOR TO AN INDICATED STEP OF THIS CHART. THIS WILL ENSURE THAT AN INTERMITTENT CONNECTION WILL NOT BE CORRECTED BEFORE THE SOURCE OF THE MALFUNCTION IS FOUND.

(1)
- IGNITION "OFF."
- USING A 5% SALT WATER SOLUTION (TWO TEASPOONS OF SALT TO 12 OZ. WATER), THOROUGHLY SPRAY THE LEFT FRONT WHEEL SPEED SENSOR JUMPER HARNESS.
- DISCONNECT 32-WAY EBTCM HARNESS CONNECTOR.
- USING J 39200, MEASURE THE RESISTANCE BETWEEN TERMINALS "E7" AND "E8" OF THE 32-WAY EBTCM HARNESS CONNECTOR.

IS RESISTANCE BETWEEN 1020 AND 1137Ω WHEN SENSOR IS 20°C (68°F)? *(IF TEMPERATURE OF SENSOR IS SUSPECTED TO OTHER THAN 20°C (68°F), REFER TO THE TABLE ON PAGE 2 OF 2 OF THIS DTC CHART FOR APPLICABLE SENSOR RESISTANCE VALUES; THE VALUES IN THE TABLE ARE FOR TEMPERATURE OF THE SENSOR, NOT AIR TEMPERATURE.)*

NO: DIAGNOSIS CONTINUED ON PAGE 2 OF 3 OF THIS DTC CHART.

YES:

(2)
- USING J 39200, MEASURE THE RESISTANCE BETWEEN GROUND AND TERMINAL "E7" OF THE 32-WAY EBTCM HARNESS CONNECTOR.

DOES J 39200 DISPLAY RESISTANCE AS "OL" (INFINITE)?

NO: DIAGNOSIS CONTINUED ON PAGE 3 OF 3 OF THIS DTC CHART.

YES:

(3)
- IGNITION "ON."
- USING J 39200, MEASURE THE VOLTAGE BETWEEN GROUND AND TERMINAL "E7" OF THE 32-WAY EBTCM HARNESS CONNECTOR.

IS VOLTAGE 1 VOLT OR LESS?

NO:

(4)
- IGNITION "OFF."
- DISCONNECT LEFT FRONT WHEEL SPEED SENSOR DIRECTLY AT SENSOR.
- IGNITION "ON."
- USING J 39200, MEASURE THE VOLTAGE BETWEEN GROUND AND TERMINAL "E7" OF THE 32-WAY EBTCM HARNESS CONNECTOR.

IS VOLTAGE 1 VOLT OR LESS?

NO: REPAIR SHORT TO VOLTAGE IN CKT 873.

YES:

(5)
- USING J 39200, MEASURE THE VOLTAGE BETWEEN GROUND AND TERMINAL "E8" OF THE 32-WAY EBTCM HARNESS CONNECTOR.

IS VOLTAGE 1 VOLT OR LESS?

YES: MALFUNCTION IS INTERMITTENT. INSPECT ALL CONNECTORS AND HARNESSES FOR DAMAGE WHICH MAY RESULT IN A SHORT TO VOLTAGE WHEN ALL COMPONENTS ARE CONNECTED. REPAIR ALL DAMAGE IF FOUND. REFER TO "DIAGNOSTIC AIDS" ON FACING PAGE FOR MORE INFORMATION.

NO: REPAIR SHORT TO VOLTAGE IN CKT 830.

YES (from 3):

(6)
- IGNITION "OFF."
- INSPECT TERMINALS "E7" AND "E8" OF THE 32-WAY EBTCM HARNESS CONNECTOR FOR POOR TERMINAL CONTACT AND CONTAMINATION. ALSO INSPECT CKTs 830 AND 873 FOR DAMAGE WHICH MAY RESULT IN AN OPEN CIRCUIT, SHORT TO GROUND, OR SHORT TO VOLTAGE. REPAIR ALL DAMAGE IF EVIDENT.
- RECONNECT ALL CONNECTORS.
- IGNITION "ON."

DOES DTC 32 SET AS A CURRENT DTC?

YES: REPLACE EBTCM.

NO: MALFUNCTION IS INTERMITTENT. REFER TO "DIAGNOSTIC AIDS" ON FACING PAGE.

DIAGNOSTIC CHART DTC 32 — 1994 SILHOUETTE, TRANS SPORT AND LUMINA APV WITH TCS, CONTINUED

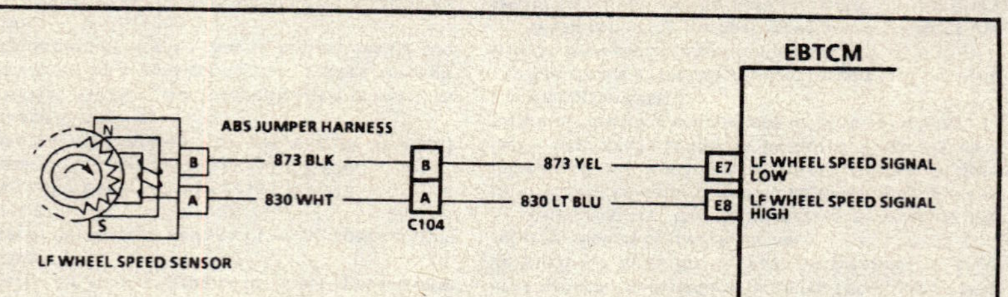

DTC 32
(Page 2 of 3)
LF WHEEL SPEED SNR CKT OPEN OR SHORT TO GND/BAT

Circuit Description:
As a toothed ring passes by the wheel speed sensor, changes in the electromagnetic field cause the wheel speed sensor to produce a sinusoidal (AC) voltage signal whose frequency is proportional to wheel speed. The magnitude of this signal is directly related to wheel speed and the proximity of the wheel speed sensor to the toothed ring, often referred to as the air gap.

DTC Will Set When: DTC 32 can be set anytime after initialization. If either of the left front wheel speed circuits are shorted to voltage or ground, a malfunction exists.

Action Taken: A malfunction DTC is stored, ABS and TCS are disabled and the ABS and TCS warning lamps are turned "ON."

DTC Chart Test Description: Number(s) below refer to circled number(s) on the diagnostic chart.
7. Checks for high resistance in CKT 873.
8. Checks for high resistance in CKT 830.
9. Verifies proper wheel speed sensor resistance.

Diagnostic Aids: An intermittent malfunction may be caused by a poor connection, rubbed through wire insulation, or a wire that is broken inside the insulation.

The frequency of the malfunction can be checked by using the enhanced diagnostic function of the Tech 1

If the customer's comments reflect that the ABS and TCS warning lamps are "ON" only during moist environmental changes (rain, snow, vehicle wash), all wheel speed sensor circuitry should be thoroughly inspected for signs of water intrusion. If DTC is not current, clear DTCs and simulate the effects of water intrusion. Use the following procedure. Spray down the suspected area with a 5% salt water solution (two teaspoons of salt to 12 oz. of water). Test drive vehicle over various road surfaces (bumps, turns, etc.) above 24 km/h (15 mph) for at least 30 seconds. If DTC returns, replace suspected harness.

Any circuitry, that is suspected as causing the intermittent complaint, should be thoroughly checked for backed out terminals, improper mating, broken locks, improperly formed or damaged terminals, poor terminal to wiring connections or physical damage to the wiring harness.

Resistance of the wheel speed sensor will increase with an increase in sensor temperature.

When replacing a wheel speed sensor, inspect the sensor terminals and harness connector for corrosion and/or water intrusion. If evidence of corrosion or water intrusion exists, replace wheel speed sensor jumper harness. Likewise, if replacing a wheel speed sensor jumper harness, inspect sensor terminals. If evidence of corrosion or water intrusion exists, replace wheel speed sensor.

DIAGNOSTIC CHART DTC 32 — 1994 SILHOUETTE, TRANS SPORT AND LUMINA APV WITH TCS, CONTINUED

DTC 32
(Page 2 of 3)
LF WHEEL SPEED SNR CKT OPEN OR SHORT TO GND/BAT

DIAGNOSIS CONTINUED FROM PAGE 1 OF 3 OF THIS DTC CHART.

(7)
- DISCONNECT LEFT FRONT WHEEL SPEED SENSOR DIRECTLY AT SENSOR.
- USING J 39200, MEASURE THE RESISTANCE BETWEEN TERMINAL "E7" OF THE 32-WAY EBTCM HARNESS CONNECTOR AND TERMINAL "B" OF THE LEFT FRONT WHEEL SPEED SENSOR HARNESS CONNECTOR.
  IS RESISTANCE 2Ω OR LESS?

NO → REPAIR HIGH RESISTANCE IN CKT 873 (WHEEL SPEED SENSOR JUMPER MUST BE REPLACED IF DAMAGED).

YES ↓

(8)
- USING J 39200, MEASURE THE RESISTANCE BETWEEN TERMINAL "E8" OF THE 32-WAY EBTCM HARNESS CONNECTOR AND TERMINAL "A" OF THE LEFT FRONT WHEEL SPEED SENSOR HARNESS CONNECTOR.
  IS RESISTANCE 2Ω OR LESS?

NO → REPAIR HIGH RESISTANCE IN CKT 830 (WHEEL SPEED SENSOR JUMPER MUST BE REPLACED IF DAMAGED).

YES ↓

(9)
- USING J 39200, MEASURE THE RESISTANCE BETWEEN TERMINAL "A" AND TERMINAL "B" OF THE LEFT FRONT WHEEL SPEED SENSOR CONNECTOR.
  IS RESISTANCE BETWEEN 1020 AND 1137Ω WHEN SENSOR IS 20°C (68°F)? *(IF TEMPERATURE OF SENSOR IS SUSPECTED TO OTHER THAN 20°C (68°F), REFER TO THE TABLE BELOW FOR APPLICABLE SENSOR RESISTANCE VALUES; THE VALUES IN THE TABLE ARE FOR TEMPERATURE OF THE SENSOR, NOT AIR TEMPERATURE.)*

NO → REPLACE WHEEL SPEED SENSOR.

YES → MALFUNCTION IS INTERMITTENT. INSPECT ALL CONNECTORS AND HARNESSES FOR DAMAGE WHICH MAY RESULT IN HIGH RESISTANCE WHEN ALL COMPONENTS ARE CONNECTED. REFER TO "DIAGNOSTIC AIDS" ON FACING PAGE FOR MORE INFORMATION. REPAIR ALL DAMAGE IF FOUND.

FRONT WHEEL SPEED SENSOR
TEMPERATURE OF SENSOR VS. SENSOR RESISTANCE (APPROXIMATE)

TEMP. (°C)	TEMP. (°F)	RESISTANCE (OHMS)
-34 TO 4	-30 TO 40	804 TO 1066
5 TO 43	41 TO 110	956 TO 1240
44 TO 93	111 TO 200	1112 TO 1463

DIAGNOSTIC CHART DTC 32 — 1994 SILHOUETTE, TRANS SPORT AND LUMINA APV WITH TCS, CONTINUED

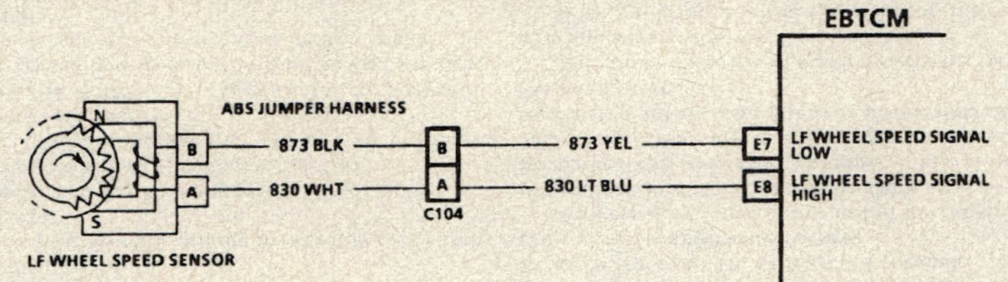

DTC 32

(Page 3 of 3)
LF WHEEL SPEED SNR CKT OPEN OR SHORT TO GND/BAT

Circuit Description:
As a toothed ring passes by the wheel speed sensor, changes in the electromagnetic field cause the wheel speed sensor to produce a sinusoidal (AC) voltage signal whose frequency is proportional to wheel speed. The magnitude of this signal is directly related to wheel speed and the proximity of the wheel speed sensor to the toothed ring, often referred to as the air gap.

DTC Will Set When: DTC 32 can be set anytime after initialization. If either of the left front wheel speed circuits are shorted to voltage or ground, a malfunction exists.

Action Taken: A malfunction DTC is stored, ABS and TCS are disabled and the ABS and TCS warning lamps are turned "ON."

DTC Chart Test Description: Number(s) below refer to circled number(s) on the diagnostic chart.

10. Checks for short to ground in CKT 873.
11. Checks for short to ground in CKT 830.
12. Checks for short to ground in wheel speed sensor jumper harness.
13. Verifies that wheel speed sensor is not internally shorted to ground.

Diagnostic Aids: An intermittent malfunction may be caused by a poor connection, rubbed through wire insulation, or a wire that is broken inside the insulation.

The frequency of the malfunction can be checked by using the enhanced diagnostic function of the Tech 1.

If the customer's comments reflect that the ABS and TCS warning lamps are "ON" only during moist environmental changes (rain, snow, vehicle wash), all wheel speed sensor circuitry should be thoroughly inspected for signs of water intrusion. If DTC is not current, clear DTCs and simulate the effects of water intrusion. Use the following procedure.

Spray down the suspected area with a 5% salt water solution (two teaspoons of salt to 12 oz. of water). Test drive vehicle over various road surfaces (bumps, turns, etc.) above 24 km/h (15 mph) for at least 30 seconds. If DTC returns, replace suspected harness.

Any circuitry, that is suspected as causing the intermittent complaint, should be thoroughly checked for backed out terminals, improper mating, broken locks, improperly formed or damaged terminals, poor terminal to wiring connections or physical damage to the wiring harness.

Resistance of the wheel speed sensor will increase with an increase in sensor temperature.

When replacing a wheel speed sensor, inspect the sensor terminals and harness connector for corrosion and/or water intrusion. If evidence of corrosion or water intrusion exists, replace wheel speed sensor jumper harness. Likewise, if replacing a wheel speed sensor jumper harness, inspect sensor terminals. If evidence of corrosion or water intrusion exists, replace wheel speed sensor.

DIAGNOSTIC CHART DTC 32 — 1994 SILHOUETTE, TRANS SPORT AND LUMINA APV WITH TCS, CONTINUED

DTC 32

(Page 3 of 3)
LF WHEEL SPEED SNR CKT OPEN OR SHORT TO GND/BAT

DIAGNOSIS CONTINUED FROM PAGE 1 OF 3 OF THIS DTC CHART.

(10)
- DISCONNECT CONNECTOR C104 (MOUNTED ON LH SHOCK TOWER).
- USING J 39200, MEASURE THE RESISTANCE BETWEEN GROUND AND TERMINAL "E7" OF THE 32-WAY EBTCM HARNESS CONNECTOR.

DOES J 39200 DISPLAY RESISTANCE AS "OL" (INFINITE)?

NO: REPAIR SHORT TO GROUND IN CKT 873.

YES:

(11)
- USING J 39200, MEASURE THE RESISTANCE BETWEEN GROUND AND TERMINAL "E8" OF THE 32-WAY EBTCM HARNESS CONNECTOR.

DOES J 39200 DISPLAY RESISTANCE AS "OL" (INFINITE)?

NO: REPAIR SHORT TO GROUND IN CKT 830.

YES:

(12)
- THOROUGHLY RESPRAY THE LEFT FRONT WHEEL SPEED SENSOR JUMPER HARNESS WITH THE SALT WATER SOLUTION.

IMPORTANT: DO NOT ALLOW WATER TO ENTER ANY WHEEL SPEED SENSOR CONNECTOR.

- DISCONNECT LEFT FRONT WHEEL SPEED SENSOR DIRECTLY AT SENSOR.
- USING J 39200, MEASURE THE RESISTANCE BETWEEN GROUND AND TERMINAL "A," THEN "B" OF THE LEFT FRONT WHEEL SPEED SENSOR HARNESS CONNECTOR.

DOES J 39200 DISPLAY RESISTANCE AS "OL" (INFINITE) FOR BOTH MEASUREMENTS?

NO: REPLACE LEFT FRONT WHEEL SPEED SENSOR JUMPER HARNESS.

YES:

(13)
- USING J 39200, MEASURE THE RESISTANCE BETWEEN GROUND AND TERMINAL "A" OF THE LEFT FRONT WHEEL SPEED SENSOR CONNECTOR.

DOES J 39200 DISPLAY RESISTANCE AS "OL" (INFINITE)?

NO: REPLACE WHEEL SPEED SENSOR.

YES: MALFUNCTION IS INTERMITTENT. INSPECT ALL CONNECTORS AND HARNESSES FOR DAMAGE WHICH MAY RESULT IN A SHORT TO GROUND WHEN ALL COMPONENTS ARE CONNECTED. PAY PARTICULAR ATTENTION TO THE JUMPER HARNESS, SMALL NICKS OR CUTS CAN CAUSE A MALFUNCTION. REPAIR ANY DAMAGE IF FOUND. JUMPER HARNESS MUST BE REPLACED IF DAMAGED. REFER TO "DIAGNOSTIC AIDS" ON FACING PAGE FOR MORE INFORMATION.

DIAGNOSTIC CHART DTC 33 — 1994 SILHOUETTE, TRANS SPORT AND LUMINA APV WITH TCS

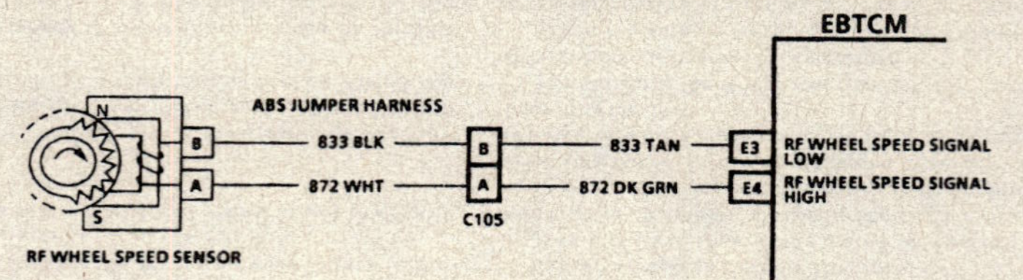

DTC 33

(Page 1 of 3)

RF WHEEL SPEED SNR CKT OPEN OR SHORT TO GND/BAT

Circuit Description:

As a toothed ring passes by the wheel speed sensor, changes in the electromagnetic field cause the wheel speed sensor to produce a sinusoidal (AC) voltage signal whose frequency is proportional to wheel speed. The magnitude of this signal is directly related to wheel speed and the proximity of the wheel speed sensor to the toothed ring, often referred to as the air gap.

DTC Will Set When: DTC 33 can be set anytime after initialization. If either of the right front wheel speed circuits are shorted to voltage or ground, a malfunction exists.

Action Taken: A malfunction DTC is stored, ABS and TCS are disabled and the ABS and TCS warning lamps are turned "ON."

DTC Chart Test Description: Number(s) below refer to circled number(s) on the diagnostic chart.

1. Checks to see that the wheel speed sensor is within proper resistance values.
2. Checks for short to ground in wheel speed sensor wiring.
3. Checks for short to voltage in wheel speed sensor wiring.
4. Checks for a short to voltage in CKT 833.
5. Checks for a short to voltage in CKT 872.
6. Ensures malfunction was not due to physical damage of circuitry.

Diagnostic Aids: An intermittent malfunction may be caused by a poor connection, rubbed through wire insulation, or a wire that is broken inside the insulation.

The frequency of the malfunction can be checked by using the enhanced diagnostic function of the Tech 1,

If the customer's comments reflect that the ABS and TCS warning lamps are "ON" only during moist environmental changes (rain, snow, vehicle wash), all wheel speed sensor circuitry should be thoroughly inspected for signs of water intrusion. If DTC is not current, clear DTCs and simulate the effects of water intrusion. Use the following procedure. Spray down the suspected area with a 5% salt water solution (two teaspoons of salt to 12 oz. of water). Test drive vehicle over various road surfaces (bumps, turns, etc.) above 24 km/h (15 mph) for at least 30 seconds. If DTC returns, replace suspected harness.

Any circuitry, that is suspected as causing the intermittent complaint, should be thoroughly checked for backed out terminals, improper mating, broken locks, improperly formed or damaged terminals, poor terminal to wiring connections or physical damage to the wiring harness.

Resistance of the wheel speed sensor will increase with an increase in sensor temperature.

When replacing a wheel speed sensor, inspect the sensor terminals and harness connector for corrosion and/or water intrusion. If evidence of corrosion or water intrusion exists, replace wheel speed sensor jumper harness. Likewise, if replacing a wheel speed sensor jumper harness, inspect sensor terminals. If evidence of corrosion or water intrusion exists, replace wheel speed sensor.

DIAGNOSTIC CHART DTC 33 — 1994 SILHOUETTE, TRANS SPORT AND LUMINA APV WITH TCS, CONTINUED

DTC 33

(Page 1 of 3)

RF WHEEL SPEED SNR CKT OPEN OR SHORT TO GND/BAT

Important

- WHEEL SPEED SENSOR INTERMITTENT MALFUNCTIONS MAY BE DIFFICULT TO LOCATE. CARE SHOULD BE TAKEN NOT TO DISTURB ANY ELECTRICAL CONNECTIONS PRIOR TO AN INDICATED STEP OF THIS CHART. THIS WILL ENSURE THAT AN INTERMITTENT CONNECTION WILL NOT BE CORRECTED BEFORE THE SOURCE OF THE MALFUNCTION IS FOUND.

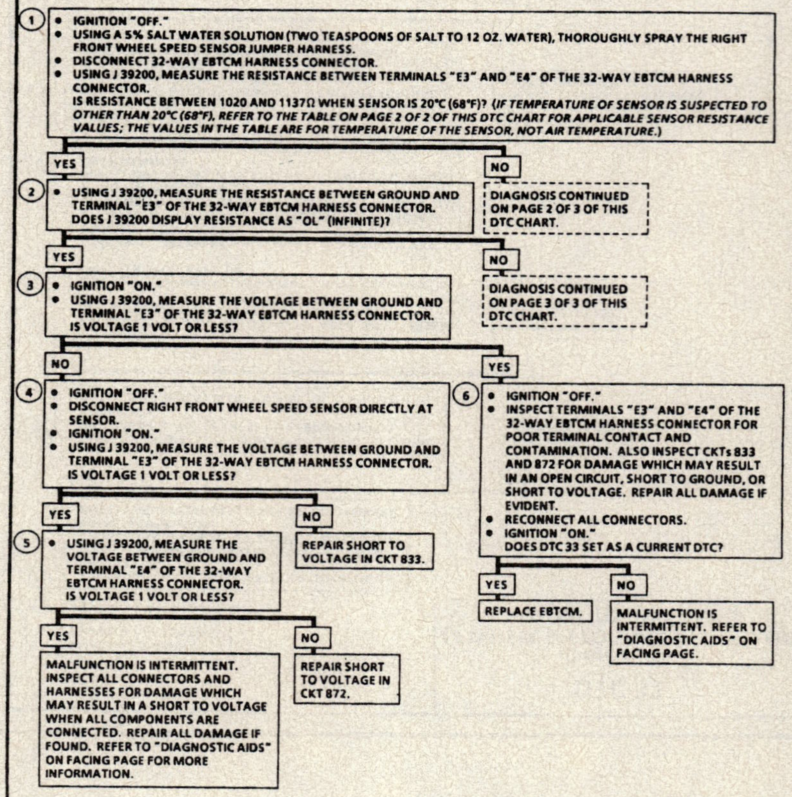

DIAGNOSTIC CHART DTC 33 — 1994 SILHOUETTE, TRANS SPORT AND LUMINA APV WITH TCS, CONTINUED

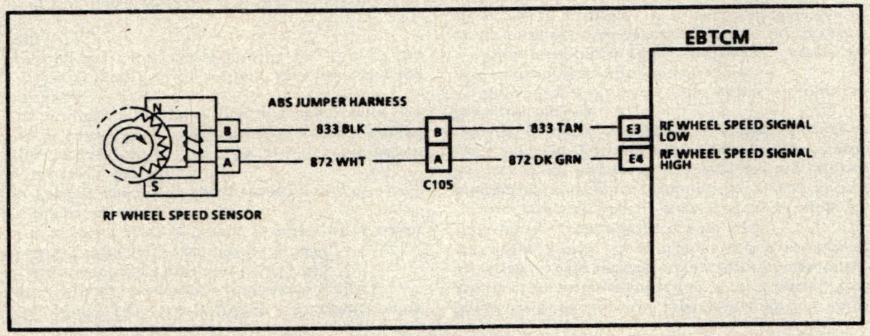

DTC 33

(Page 2 of 3)

RF WHEEL SPEED SNR CKT OPEN OR SHORT TO GND/BAT

Circuit Description:

As a toothed ring passes by the wheel speed sensor, changes in the electromagnetic field cause the wheel speed sensor to produce a sinusoidal (AC) voltage signal whose frequency is proportional to wheel speed. The magnitude of this signal is directly related to wheel speed and the proximity of the wheel speed sensor to the toothed ring, often referred to as the air gap.

DTC Will Set When: DTC 33 can be set anytime after initialization. If either of the right front wheel speed circuits are shorted to voltage or ground, a malfunction exists.

Action Taken: A malfunction DTC is stored, ABS and TCS are disabled and the ABS and TCS warning lamps are turned "ON."

DTC Chart Test Description: Number(s) below refer to circled number(s) on the diagnostic chart.

7. Checks for high resistance in CKT 833.
8. Checks for high resistance in CKT 872.
9. Verifies proper wheel speed sensor resistance.

Diagnostic Aids: An intermittent malfunction may be caused by a poor connection, rubbed through wire insulation, or a wire that is broken inside the insulation.

The frequency of the malfunction can be checked by using the enhanced diagnostic function of the Tech 1

If the customer's comments reflect that the ABS and TCS warning lamps are "ON" only during moist environmental changes (rain, snow, vehicle wash), all wheel speed sensor circuitry should be thoroughly inspected for signs of water intrusion. If DTC is not current, clear DTCs and simulate the effects of water intrusion. Use the following procedure. Spray down the suspected area with a 5% salt water solution (two teaspoons of salt to 12 oz. of water). Test drive vehicle over various road surfaces (bumps, turns, etc.) above 24 km/h (15 mph) for at least 30 seconds. If DTC returns, replace suspected harness.

Any circuitry, that is suspected as causing the intermittent complaint, should be thoroughly checked for backed out terminals, improper mating, broken locks, improperly formed or damaged terminals, poor terminal to wiring connections or physical damage to the wiring harness.

Resistance of the wheel speed sensor will increase with an increase in sensor temperature.

When replacing a wheel speed sensor, inspect the sensor terminals and harness connector for corrosion and/or water intrusion. If evidence of corrosion or water intrusion exists, replace wheel speed sensor jumper harness. Likewise, if replacing a wheel speed sensor jumper harness, inspect sensor terminals. If evidence of corrosion or water intrusion exists, replace wheel speed sensor. Refer to "On-Vehicle Service" in this section.

DIAGNOSTIC CHART DTC 33 — 1994 SILHOUETTE, TRANS SPORT AND LUMINA APV WITH TCS, CONTINUED

DTC 33

(Page 2 of 3)

RF WHEEL SPEED SNR CKT OPEN OR SHORT TO GND/BAT

DIAGNOSIS CONTINUED FROM PAGE 1 OF 3 OF THIS DTC CHART.

(7)
- DISCONNECT RIGHT FRONT WHEEL SPEED SENSOR DIRECTLY AT SENSOR.
- USING J 39200, MEASURE THE RESISTANCE BETWEEN TERMINAL "E3" OF THE 32-WAY EBTCM HARNESS CONNECTOR AND TERMINAL "B" OF THE RIGHT FRONT WHEEL SPEED SENSOR HARNESS CONNECTOR.

IS RESISTANCE 2Ω OR LESS?

NO: REPAIR HIGH RESISTANCE IN CKT 833 (WHEEL SPEED SENSOR JUMPER MUST BE REPLACED IF DAMAGED).

YES:

(8)
- USING J 39200, MEASURE THE RESISTANCE BETWEEN TERMINAL "E4" OF THE 32-WAY EBTCM HARNESS CONNECTOR AND TERMINAL "A" OF THE RIGHT FRONT WHEEL SPEED SENSOR HARNESS CONNECTOR.

IS RESISTANCE 2Ω OR LESS?

NO: REPAIR HIGH RESISTANCE IN CKT 872 (WHEEL SPEED SENSOR JUMPER MUST BE REPLACED IF DAMAGED).

YES:

(9)
- USING J 39200, MEASURE THE RESISTANCE BETWEEN TERMINAL "A" AND TERMINAL "B" OF THE RIGHT FRONT WHEEL SPEED SENSOR CONNECTOR.

IS RESISTANCE BETWEEN 1020 AND 1137Ω WHEN SENSOR IS 20°C (68°F)? *(IF TEMPERATURE OF SENSOR IS SUSPECTED TO OTHER THAN 20°C (68°F), REFER TO THE TABLE BELOW FOR APPLICABLE SENSOR RESISTANCE VALUES; THE VALUES IN THE TABLE ARE FOR TEMPERATURE OF THE SENSOR, NOT AIR TEMPERATURE.)*

NO: REPLACE WHEEL SPEED SENSOR.

YES: MALFUNCTION IS INTERMITTENT. INSPECT ALL CONNECTORS AND HARNESSES FOR DAMAGE WHICH MAY RESULT IN HIGH RESISTANCE WHEN ALL COMPONENTS ARE CONNECTED. REFER TO "DIAGNOSTIC AIDS" ON FACING PAGE FOR MORE INFORMATION. REPAIR ALL DAMAGE IF FOUND.

FRONT WHEEL SPEED SENSOR

TEMPERATURE OF SENSOR VS. SENSOR RESISTANCE (APPROXIMATE)

TEMP. (°C)	TEMP. (°F)	RESISTANCE (OHMS)
-34 TO 4	-30 TO 40	804 TO 1066
5 TO 43	41 TO 110	956 TO 1240
44 TO 93	111 TO 200	1112 TO 1463

DIAGNOSTIC CHART DTC 33 — 1994 SILHOUETTE, TRANS SPORT AND LUMINA APV WITH TCS

DTC 33
(Page 3 of 3)
RF WHEEL SPEED SNR CKT OPEN OR SHORT TO GND/BAT

DIAGNOSIS CONTINUED FROM PAGE 1 OF 3 OF THIS DTC CHART.

(10)
- DISCONNECT CONNECTOR C105 (MOUNTED ON RH SHOCK TOWER).
- USING J 39200, MEASURE THE RESISTANCE BETWEEN GROUND AND TERMINAL "E3" OF THE 32-WAY EBTCM HARNESS CONNECTOR.

DOES J 39200 DISPLAY RESISTANCE AS "OL" (INFINITE)?

- NO: REPAIR SHORT TO GROUND IN CKT 833.
- YES:

(11)
- USING J 39200, MEASURE THE RESISTANCE BETWEEN GROUND AND TERMINAL "E4" OF THE 32-WAY EBTCM HARNESS CONNECTOR.

DOES J 39200 DISPLAY RESISTANCE AS "OL" (INFINITE)?

- NO: REPAIR SHORT TO GROUND IN CKT 872.
- YES:

(12)
- THOROUGHLY RESPRAY THE RIGHT FRONT WHEEL SPEED SENSOR JUMPER HARNESS WITH THE SALT WATER SOLUTION.

IMPORTANT: DO NOT ALLOW WATER TO ENTER ANY WHEEL SPEED SENSOR CONNECTOR.

- DISCONNECT RIGHT FRONT WHEEL SPEED SENSOR DIRECTLY AT SENSOR.
- USING J 39200, MEASURE THE RESISTANCE BETWEEN GROUND AND TERMINAL "A," THEN "B" OF THE RIGHT FRONT WHEEL SPEED SENSOR HARNESS CONNECTOR.

DOES J 39200 DISPLAY RESISTANCE AS "OL" (INFINITE) FOR BOTH MEASUREMENTS?

- NO: REPLACE RIGHT FRONT WHEEL SPEED SENSOR JUMPER HARNESS.
- YES:

(13)
- USING J 39200, MEASURE THE RESISTANCE BETWEEN GROUND AND TERMINAL "A" OF THE RIGHT FRONT WHEEL SPEED SENSOR CONNECTOR.

DOES J 39200 DISPLAY RESISTANCE AS "OL" (INFINITE)?

- NO: REPLACE WHEEL SPEED SENSOR.
- YES: MALFUNCTION IS INTERMITTENT. INSPECT ALL CONNECTORS AND HARNESSES FOR DAMAGE WHICH MAY RESULT IN A SHORT TO GROUND WHEN ALL COMPONENTS ARE CONNECTED. PAY PARTICULAR ATTENTION TO THE JUMPER HARNESS, SMALL NICKS OR CUTS CAN CAUSE A MALFUNCTION. REPAIR ANY DAMAGE IF FOUND. JUMPER HARNESS MUST BE REPLACED IF DAMAGED. REFER TO "DIAGNOSTIC AIDS" ON FACING PAGE FOR MORE INFORMATION.

DIAGNOSTIC CHART DTC 33 — 1994 SILHOUETTE, TRANS SPORT AND LUMINA APV WITH TCS, CONTINUED

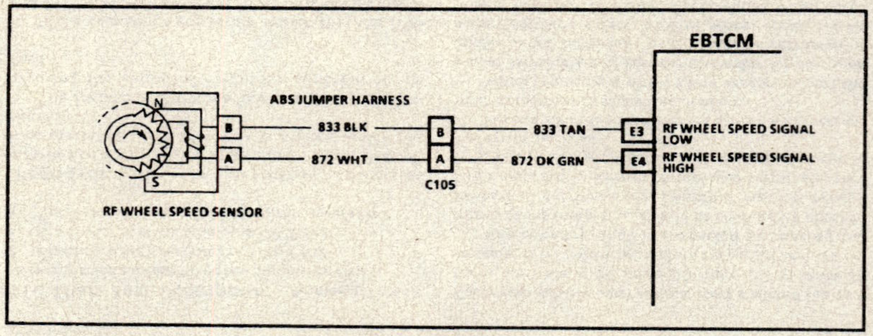

DTC 33
(Page 3 of 3)
RF WHEEL SPEED SNR CKT OPEN OR SHORT TO GND/BAT

Circuit Description:
As a toothed ring passes by the wheel speed sensor, changes in the electromagnetic field cause the wheel speed sensor to produce a sinusoidal (AC) voltage signal whose frequency is proportional to wheel speed. The magnitude of this signal is directly related to wheel speed and the proximity of the wheel speed sensor to the toothed ring, often referred to as the air gap.

DTC Will Set When: DTC 33 can be set anytime after initialization. If either of the right front wheel speed circuits are shorted to voltage or ground, a malfunction exists.

Action Taken: A malfunction DTC is stored, ABS and TCS are disabled and the ABS and TCS warning lamps are turned "ON."

DTC Chart Test Description: Number(s) below refer to circled number(s) on the diagnostic chart.

10. Checks for short to ground in CKT 833.
11. Checks for short to ground in CKT 872.
12. Checks for short to ground in wheel speed sensor jumper harness.
13. Verifies that wheel speed sensor is not internally shorted to ground.

Diagnostic Aids: An intermittent malfunction may be caused by a poor connection, rubbed through wire insulation, or a wire that is broken inside the insulation.

The frequency of the malfunction can be checked by using the enhanced diagnostic function of the Tech 1.

If the customer's comments reflect that the ABS and TCS warning lamps are "ON" only during moist environmental changes (rain, snow, vehicle wash), all wheel speed sensor circuitry should be thoroughly inspected for signs of water intrusion. If DTC is not current, clear DTCs and simulate the effects of water intrusion. Use the following procedure.

Spray down the suspected area with a 5% salt water solution (two teaspoons of salt to 12 oz. of water). Test drive vehicle over various road surfaces (bumps, turns, etc.) above 24 km/h (15 mph) for at least 30 seconds. If DTC returns, replace suspected harness.

Any circuitry, that is suspected as causing the intermittent complaint, should be thoroughly checked for backed out terminals, improper mating, broken locks, improperly formed or damaged terminals, poor terminal to wiring connections or physical damage to the wiring harness.

Resistance of the wheel speed sensor will increase with an increase in sensor temperature.

When replacing a wheel speed sensor, inspect the sensor terminals and harness connector for corrosion and/or water intrusion. If evidence of corrosion or water intrusion exists, replace wheel speed sensor jumper harness. Likewise, if replacing a wheel speed sensor jumper harness, inspect sensor terminals. If evidence of corrosion or water intrusion exists, replace wheel speed sensor. Refer to "On-Vehicle Service" in this section.

DIAGNOSTIC CHART DTC 34 — 1994 SILHOUETTE, TRANS SPORT AND LUMINA APV WITH TCS

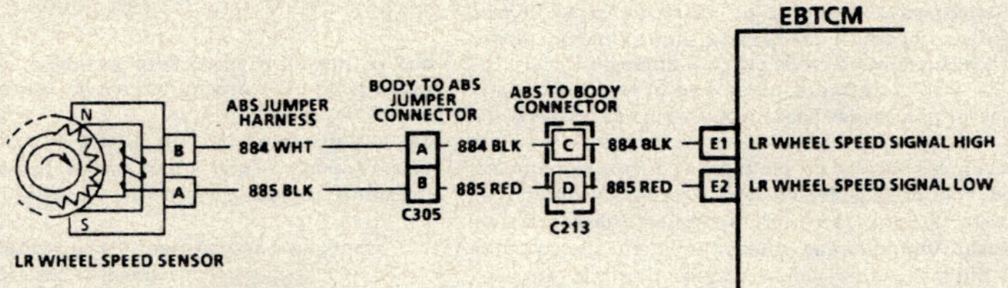

DTC 34

(Page 1 of 3)
LR WHEEL SPEED SNR CKT OPEN OR SHORT TO GND/BAT

Circuit Description:
As a toothed ring passes by the wheel speed sensor, changes in the electromagnetic field cause the wheel speed sensor to produce a sinusoidal (AC) voltage signal whose frequency is proportional to wheel speed. The magnitude of this signal is directly related to wheel speed and the proximity of the wheel speed sensor to the toothed ring, often referred to as the air gap.

DTC Will Set When: DTC 34 can be set anytime after initialization. If either of the left rear wheel speed circuits are shorted to voltage or ground, a malfunction exists.

Action Taken: A malfunction DTC is stored, ABS and TCS are disabled and the ABS and TCS warning lamps are turned "ON."

DTC Chart Test Description: Number(s) below refer to circled number(s) on the diagnostic chart.

1. Checks to see that the wheel speed sensor is within proper resistance values.
2. Checks for short to ground in wheel speed sensor wiring.
3. Checks for short to voltage in wheel speed sensor wiring.
4. Checks for a short to voltage in CKT 884.
5. Checks for a short to voltage in CKT 885.
6. Ensures malfunction was not due to physical damage of circuitry.

Diagnostic Aids: An intermittent malfunction may be caused by a poor connection, rubbed through wire insulation, or a wire that is broken inside the insulation.

The frequency of the malfunction can be checked by using the enhanced diagnostic function of the Tech 1.

If the customer's comments reflect that the ABS and TCS warning lamps are "ON" only during moist environmental changes (rain, snow, vehicle wash), all wheel speed sensor circuitry should be thoroughly inspected for signs of water intrusion. If DTC is not current, clear DTCs and simulate the effects of water intrusion. Use the following procedure. Spray down the suspected area with a 5% salt water solution (two teaspoons of salt to 12 oz. of water). Test drive vehicle over various road surfaces (bumps, turns, etc.) above 24 km/h (15 mph) for at least 30 seconds. If DTC returns, replace suspected harness.

Any circuitry, that is suspected as causing the intermittent complaint, should be thoroughly checked for backed out terminals, improper mating, broken locks, improperly formed or damaged terminals, poor terminal to wiring connections or physical damage to the wiring harness.

Resistance of the wheel speed sensor will increase with an increase in sensor temperature.

When replacing a wheel speed sensor, inspect the sensor terminals and harness connector for corrosion and/or water intrusion. If evidence of corrosion or water intrusion exists, replace wheel speed sensor jumper harness. Likewise, if replacing a wheel speed sensor jumper harness, inspect sensor terminals. If evidence of corrosion or water intrusion exists, replace wheel speed sensor. Refer to "On-Vehicle Service" in this section.

DIAGNOSTIC CHART DTC 34 — 1994 SILHOUETTE, TRANS SPORT AND LUMINA APV WITH TCS, CONTINUED

DTC 34

(Page 1 of 3)
LR WHEEL SPEED SNR CKT OPEN OR SHORT TO GND/BAT

Important

- WHEEL SPEED SENSOR INTERMITTENT MALFUNCTIONS MAY BE DIFFICULT TO LOCATE. CARE SHOULD BE TAKEN NOT TO DISTURB ANY ELECTRICAL CONNECTIONS PRIOR TO AN INDICATED STEP OF THIS CHART. THIS WILL ENSURE THAT AN INTERMITTENT CONNECTION WILL NOT BE CORRECTED BEFORE THE SOURCE OF THE MALFUNCTION IS FOUND.

(1)
- IGNITION "OFF."
- USING A 5% SALT WATER SOLUTION (TWO TEASPOONS OF SALT TO 12 OZ. WATER), THOROUGHLY SPRAY THE LEFT REAR WHEEL SPEED SENSOR JUMPER HARNESS.
- DISCONNECT 32-WAY EBTCM HARNESS CONNECTOR.
- USING J 39200, MEASURE THE RESISTANCE BETWEEN TERMINALS "E1" AND "E2" OF THE 32-WAY EBTCM HARNESS CONNECTOR.

IS RESISTANCE BETWEEN 2059 AND 2353Ω WHEN SENSOR IS 20°C (68°F)? *(IF TEMPERATURE OF SENSOR IS SUSPECTED TO OTHER THAN 20°C (68°F), REFER TO THE TABLE ON PAGE 2 OF 2 OF THIS DTC CHART FOR APPLICABLE SENSOR RESISTANCE VALUES; THE VALUES IN THE TABLE ARE FOR TEMPERATURE OF THE SENSOR, NOT AIR TEMPERATURE.)*

- YES → (2)
- NO → DIAGNOSIS CONTINUED ON PAGE 2 OF 3 OF THIS DTC CHART.

(2)
- USING J 39200, MEASURE THE RESISTANCE BETWEEN GROUND AND TERMINAL "E1" OF THE 32-WAY EBTCM HARNESS CONNECTOR. DOES J 39200 DISPLAY RESISTANCE AS "OL" (INFINITE)?

- YES → (3)
- NO → DIAGNOSIS CONTINUED ON PAGE 3 OF 3 OF THIS DTC CHART.

(3)
- IGNITION "ON."
- USING J 39200, MEASURE THE VOLTAGE BETWEEN GROUND AND TERMINAL "E1" OF THE 32-WAY EBTCM HARNESS CONNECTOR. IS VOLTAGE 1 VOLT OR LESS?

- NO → (4)
- YES → (6)

(4)
- IGNITION "OFF."
- DISCONNECT LEFT REAR WHEEL SPEED SENSOR DIRECTLY AT SENSOR.
- IGNITION "ON."
- USING J 39200, MEASURE THE VOLTAGE BETWEEN GROUND AND TERMINAL "E1" OF THE 32-WAY EBTCM HARNESS CONNECTOR. IS VOLTAGE 1 VOLT OR LESS?

- YES → (5)
- NO → REPAIR SHORT TO VOLTAGE IN CKT 884.

(5)
- USING J 39200, MEASURE THE VOLTAGE BETWEEN GROUND AND TERMINAL "E2" OF THE 32-WAY EBTCM HARNESS CONNECTOR. IS VOLTAGE 1 VOLT OR LESS?

- YES → MALFUNCTION IS INTERMITTENT. INSPECT ALL CONNECTORS AND HARNESSES FOR DAMAGE WHICH MAY RESULT IN A SHORT TO VOLTAGE WHEN ALL COMPONENTS ARE CONNECTED. REPAIR ALL DAMAGE IF FOUND. REFER TO "DIAGNOSTIC AIDS" ON FACING PAGE FOR MORE INFORMATION.
- NO → REPAIR SHORT TO VOLTAGE IN CKT 885.

(6)
- IGNITION "OFF."
- INSPECT TERMINALS "E1" AND "E2" OF THE 32-WAY EBTCM HARNESS CONNECTOR FOR POOR TERMINAL CONTACT AND CONTAMINATION. ALSO INSPECT CKTs 884 AND 885 FOR DAMAGE WHICH MAY RESULT IN AN OPEN CIRCUIT, SHORT TO GROUND, OR SHORT TO VOLTAGE. REPAIR ALL DAMAGE IF EVIDENT.
- RECONNECT ALL CONNECTORS.
- IGNITION "ON."

DOES DTC 34 SET AS A CURRENT DTC?

- YES → REPLACE EBTCM.
- NO → MALFUNCTION IS INTERMITTENT. REFER TO "DIAGNOSTIC AIDS" ON FACING PAGE.

DIAGNOSTIC CHART DTC 34 — 1994 SILHOUETTE, TRANS SPORT AND LUMINA APV WITH TCS, CONTINUED

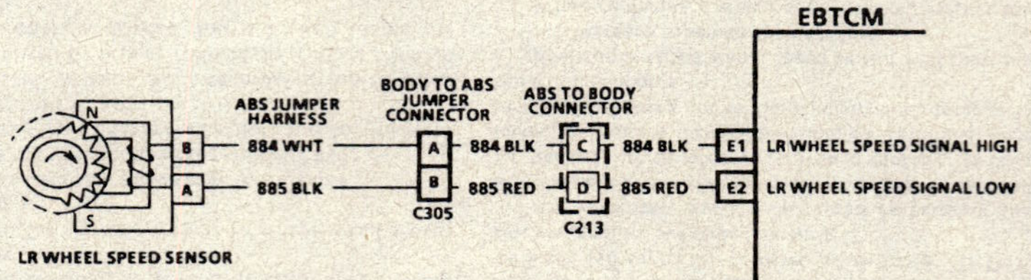

DTC 34

(Page 2 of 3)
LR WHEEL SPEED SNR CKT OPEN OR SHORT TO GND/BAT

Circuit Description:
As a toothed ring passes by the wheel speed sensor, changes in the electromagnetic field cause the wheel speed sensor to produce a sinusoidal (AC) voltage signal whose frequency is proportional to wheel speed. The magnitude of this signal is directly related to wheel speed and the proximity of the wheel speed sensor to the toothed ring, often referred to as the air gap.

DTC Will Set When: DTC 34 can be set anytime after initialization. If either of the left rear wheel speed circuits are shorted to voltage or ground, a malfunction exists.

Action Taken: A malfunction DTC is stored, ABS and TCS are disabled and the ABS and TCS warning lamps are turned "ON."

DTC Chart Test Description: Number(s) below refer to circled number(s) on the diagnostic chart.

7. Checks for high resistance in CKT 884.
8. Checks for high resistance in CKT 885.
9. Verifies proper wheel speed sensor resistance.

Diagnostic Aids: An intermittent malfunction may be caused by a poor connection, rubbed through wire insulation, or a wire that is broken inside the insulation.

The frequency of the malfunction can be checked by using the enhanced diagnostic function of the Tech 1.

If the customer's comments reflect that the ABS and TCS warning lamps are "ON" only during moist environmental changes (rain, snow, vehicle wash), all wheel speed sensor circuitry should be thoroughly inspected for signs of water intrusion. If DTC is not current, clear DTCs and simulate the effects of water intrusion. Use the following procedure. Spray down the suspected area with a 5% salt water solution (two teaspoons of salt to 12 oz. of water). Test drive vehicle over various road surfaces (bumps, turns, etc.) above 24 km/h (15 mph) for at least 30 seconds. If DTC returns, replace suspected harness.

Any circuitry, that is suspected as causing the intermittent complaint, should be thoroughly checked for backed out terminals, improper mating, broken locks, improperly formed or damaged terminals, poor terminal to wiring connections or physical damage to the wiring harness.

Resistance of the wheel speed sensor will increase with an increase in sensor temperature.

When replacing a wheel speed sensor, inspect the sensor terminals and harness connector for corrosion and/or water intrusion. If evidence of corrosion or water intrusion exists, replace wheel speed sensor jumper harness. Likewise, if replacing a wheel speed sensor jumper harness, inspect sensor terminals. If evidence of corrosion or water intrusion exists, replace wheel speed sensor.

DIAGNOSTIC CHART DTC 34 — 1994 SILHOUETTE, TRANS SPORT AND LUMINA APV WITH TCS, CONTINUED

DTC 34

(Page 2 of 3)
LR WHEEL SPEED SNR CKT OPEN OR SHORT TO GND/BAT

DIAGNOSIS CONTINUED FROM PAGE 1 OF 3 OF THIS DTC CHART.

(7)
- DISCONNECT LEFT REAR WHEEL SPEED SENSOR DIRECTLY AT SENSOR.
- USING J 39200, MEASURE THE RESISTANCE BETWEEN TERMINAL "E1" OF THE 32-WAY EBTCM HARNESS CONNECTOR AND TERMINAL "B" OF THE LEFT REAR WHEEL SPEED SENSOR HARNESS CONNECTOR.

IS RESISTANCE 2Ω OR LESS?

NO: REPAIR HIGH RESISTANCE IN CKT 884 (WHEEL SPEED SENSOR JUMPER MUST BE REPLACED IF DAMAGED).

YES:

(8)
- USING J 39200, MEASURE THE RESISTANCE BETWEEN TERMINAL "E2" OF THE 32-WAY EBTCM HARNESS CONNECTOR AND TERMINAL "A" OF THE LEFT REAR WHEEL SPEED SENSOR HARNESS CONNECTOR.

IS RESISTANCE 2Ω OR LESS?

NO: REPAIR HIGH RESISTANCE IN CKT 885 (WHEEL SPEED SENSOR JUMPER MUST BE REPLACED IF DAMAGED).

YES:

(9)
- USING J 39200, MEASURE THE RESISTANCE BETWEEN TERMINAL "A" AND TERMINAL "B" OF THE LEFT REAR WHEEL SPEED SENSOR CONNECTOR.

IS RESISTANCE BETWEEN 2059 AND 2353Ω WHEN SENSOR IS 20°C (68°F)? *(IF TEMPERATURE OF SENSOR IS SUSPECTED TO OTHER THAN 20°C (68°F), REFER TO THE TABLE BELOW FOR APPLICABLE SENSOR RESISTANCE VALUES; THE VALUES IN THE TABLE ARE FOR TEMPERATURE OF THE SENSOR, NOT AIR TEMPERATURE.)*

NO: REPLACE WHEEL SPEED SENSOR.

YES: MALFUNCTION IS INTERMITTENT. INSPECT ALL CONNECTORS AND HARNESSES FOR DAMAGE WHICH MAY RESULT IN HIGH RESISTANCE WHEN ALL COMPONENTS ARE CONNECTED. REFER TO "DIAGNOSTIC AIDS" ON FACING PAGE FOR MORE INFORMATION. REPAIR ALL DAMAGE IF FOUND.

REAR WHEEL SPEED SENSOR

TEMPERATURE OF SENSOR VS. SENSOR RESISTANCE (APPROXIMATE)

TEMP. (°C)	TEMP. (°F)	RESISTANCE (OHMS)
-34 TO 4	-30 TO 40	1622 TO 2205
5 TO 43	41 TO 110	1930 TO 2566
44 TO 93	111 TO 200	2245 TO 3028

DIAGNOSTIC CHART DTC 34 — 1994 SILHOUETTE, TRANS SPORT AND LUMINA APV WITH TCS, CONTINUED

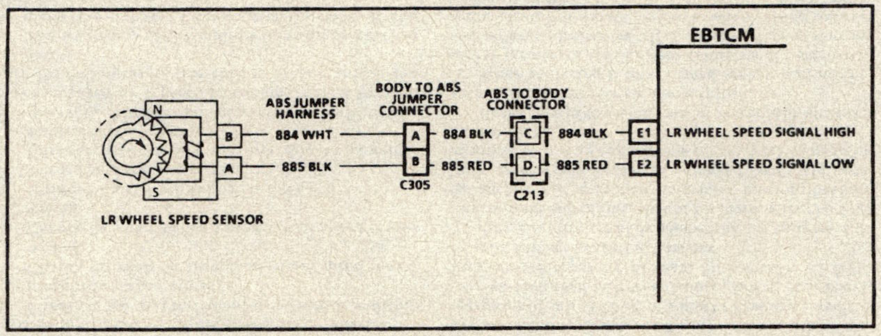

DTC 34

(Page 3 of 3)
LR WHEEL SPEED SNR CKT OPEN OR SHORT TO GND/BAT

Circuit Description:
As a toothed ring passes by the wheel speed sensor, changes in the electromagnetic field cause the wheel speed sensor to produce a sinusoidal (AC) voltage signal whose frequency is proportional to wheel speed. The magnitude of this signal is directly related to wheel speed and the proximity of the wheel speed sensor to the toothed ring, often referred to as the air gap.

DTC Will Set When: DTC 34 can be set anytime after initialization. If either of the left rear wheel speed circuits are shorted to voltage or ground, a malfunction exists.

Action Taken: A malfunction DTC is stored, ABS and TCS are disabled and the ABS and TCS warning lamps are turned "ON."

DTC Chart Test Description: Number(s) below refer to circled number(s) on the diagnostic chart.
10. Checks for short to ground in CKT 884.
11. Checks for short to ground in CKT 885.
12. Checks for short to ground in wheel speed sensor jumper harness.
13. Verifies that wheel speed sensor is not internally shorted to ground.

Diagnostic Aids: An intermittent malfunction may be caused by a poor connection, rubbed through wire insulation, or a wire that is broken inside the insulation.

The frequency of the malfunction can be checked by using the enhanced diagnostic function of the Tech 1.

If the customer's comments reflect that the ABS and TCS warning lamps are "ON" only during moist environmental changes (rain, snow, vehicle wash), all wheel speed sensor circuitry should be thoroughly inspected for signs of water intrusion. If DTC is not current, clear DTCs and simulate the effects of water intrusion. Use the following procedure.

Spray down the suspected area with a 5% salt water solution (two teaspoons of salt to 12 oz. of water). Test drive vehicle over various road surfaces (bumps, turns, etc.) above 24 km/h (15 mph) for at least 30 seconds. If DTC returns, replace suspected harness.

Any circuitry, that is suspected as causing the intermittent complaint, should be thoroughly checked for backed out terminals, improper mating, broken locks, improperly formed or damaged terminals, poor terminal to wiring connections or physical damage to the wiring harness.

Resistance of the wheel speed sensor will increase with an increase in sensor temperature.

When replacing a wheel speed sensor, inspect the sensor terminals and harness connector for corrosion and/or water intrusion. If evidence of corrosion or water intrusion exists, replace wheel speed sensor jumper harness. Likewise, if replacing a wheel speed sensor jumper harness, inspect sensor terminals. If evidence of corrosion or water intrusion exists, replace wheel speed sensor.

DIAGNOSTIC CHART DTC 34 — 1994 SILHOUETTE, TRANS SPORT AND LUMINA APV WITH TCS, CONTINUED

DTC 34

(Page 3 of 3)
LR WHEEL SPEED SNR CKT OPEN OR SHORT TO GND/BAT

DIAGNOSIS CONTINUED FROM PAGE 1 OF 3 OF THIS DTC CHART.

(10)
- DISCONNECT CONNECTOR C305 (LH REAR UNDER BODY ABOVE FRONT SPRING HANGER).
- USING J 39200, MEASURE THE RESISTANCE BETWEEN GROUND AND TERMINAL "E1" OF THE 32-WAY EBTCM HARNESS CONNECTOR. DOES J 39200 DISPLAY RESISTANCE AS "OL" (INFINITE)?

NO: REPAIR SHORT TO GROUND IN CKT 884.

YES:

(11)
- USING J 39200, MEASURE THE RESISTANCE BETWEEN GROUND AND TERMINAL "E2" OF THE 32-WAY EBTCM HARNESS CONNECTOR. DOES J 39200 DISPLAY RESISTANCE AS "OL" (INFINITE)?

NO: REPAIR SHORT TO GROUND IN CKT 885.

YES:

(12)
- THOROUGHLY RESPRAY THE LEFT REAR WHEEL SPEED SENSOR JUMPER HARNESS WITH THE SALT WATER SOLUTION.

IMPORTANT: DO NOT ALLOW WATER TO ENTER ANY WHEEL SPEED SENSOR CONNECTOR.

- DISCONNECT LEFT REAR WHEEL SPEED SENSOR DIRECTLY AT SENSOR.
- USING J 39200, MEASURE THE RESISTANCE BETWEEN GROUND AND TERMINAL "A," THEN "B" OF THE LEFT REAR WHEEL SPEED SENSOR HARNESS CONNECTOR. DOES J 39200 DISPLAY RESISTANCE AS "OL" (INFINITE) FOR BOTH MEASUREMENTS?

NO: REPLACE LEFT REAR WHEEL SPEED SENSOR JUMPER HARNESS.

YES:

(13)
- USING J 39200, MEASURE THE RESISTANCE BETWEEN GROUND AND TERMINAL "A" OF THE LEFT REAR WHEEL SPEED SENSOR CONNECTOR. DOES J 39200 DISPLAY RESISTANCE AS "OL" (INFINITE)?

NO: REPLACE WHEEL SPEED SENSOR.

YES: MALFUNCTION IS INTERMITTENT. INSPECT ALL CONNECTORS AND HARNESSES FOR DAMAGE WHICH MAY RESULT IN A SHORT TO GROUND WHEN ALL COMPONENTS ARE CONNECTED. PAY PARTICULAR ATTENTION TO THE JUMPER HARNESS, SMALL NICKS OR CUTS CAN CAUSE A MALFUNCTION. REPAIR ANY DAMAGE IF FOUND. JUMPER HARNESS MUST BE REPLACED IF DAMAGED. REFER TO "DIAGNOSTIC AIDS" ON FACING PAGE FOR MORE INFORMATION.

DIAGNOSTIC CHART DTC 35 — 1994 SILHOUETTE, TRANS SPORT AND LUMINA APV WITH TCS

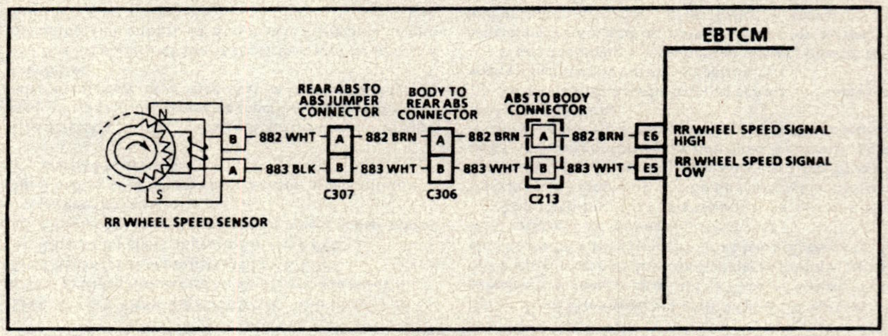

DTC 35

(Page 1 of 3)

RR WHEEL SPEED SNR CKT OPEN OR SHORT TO GND/BAT

Circuit Description:

As a toothed ring passes by the wheel speed sensor, changes in the electromagnetic field cause the wheel speed sensor to produce a sinusoidal (AC) voltage signal whose frequency is proportional to wheel speed. The magnitude of this signal is directly related to wheel speed and the proximity of the wheel speed sensor to the toothed ring, often referred to as the air gap.

DTC Will Set When: DTC 35 can be set anytime after initialization. If either of the right rear wheel speed circuits are shorted to voltage or ground, a malfunction exists.

Action Taken: A malfunction DTC is stored, ABS and TCS are disabled and the ABS and TCS warning lamps are turned "ON."

DTC Chart Test Description: Number(s) below refer to circled number(s) on the diagnostic chart.

1. Checks to see that the wheel speed sensor is within proper resistance values.
2. Checks for short to ground in wheel speed sensor wiring.
3. Checks for short to voltage in wheel speed sensor wiring.
4. Checks for a short to voltage in CKT 883.
5. Checks for a short to voltage in CKT 882.
6. Ensures malfunction was not due to physical damage of circuitry.

Diagnostic Aids: An intermittent malfunction may be caused by a poor connection, rubbed through wire insulation, or a wire that is broken inside the insulation.

The frequency of the malfunction can be checked by using the enhanced diagnostic function of the Tech 1,

If the customer's comments reflect that the ABS and TCS warning lamps are "ON" only during moist environmental changes (rain, snow, vehicle wash), all wheel speed sensor circuitry should be thoroughly inspected for signs of water intrusion. If DTC is not current, clear DTCs and simulate the effects of water intrusion. Use the following procedure. Spray down the suspected area with a 5% salt water solution (two teaspoons of salt to 12 oz. of water). Test drive vehicle over various road surfaces (bumps, turns, etc.) above 24 km/h (15 mph) for at least 30 seconds. If DTC returns, replace suspected harness.

Any circuitry, that is suspected as causing the intermittent complaint, should be thoroughly checked for backed out terminals, improper mating, broken locks, improperly formed or damaged terminals, poor terminal to wiring connections or physical damage to the wiring harness.

Resistance of the wheel speed sensor will increase with an increase in sensor temperature.

When replacing a wheel speed sensor, inspect the sensor terminals and harness connector for corrosion and/or water intrusion. If evidence of corrosion or water intrusion exists, replace wheel speed sensor jumper harness. Likewise, if replacing a wheel speed sensor jumper harness, inspect sensor terminals. If evidence of corrosion or water intrusion exists, replace wheel speed sensor.

DIAGNOSTIC CHART DTC 35 — 1994 SILHOUETTE, TRANS SPORT AND LUMINA APV WITH TCS, CONTINUED

DTC 35

(Page 1 of 3)

RR WHEEL SPEED SNR CKT OPEN OR SHORT TO GND/BAT

Important

- WHEEL SPEED SENSOR INTERMITTENT MALFUNCTIONS MAY BE DIFFICULT TO LOCATE. CARE SHOULD BE TAKEN NOT TO DISTURB ANY ELECTRICAL CONNECTIONS PRIOR TO AN INDICATED STEP OF THIS CHART. THIS WILL ENSURE THAT AN INTERMITTENT CONNECTION WILL NOT BE CORRECTED BEFORE THE SOURCE OF THE MALFUNCTION IS FOUND.

(1)
- IGNITION "OFF."
- USING A 5% SALT WATER SOLUTION (TWO TEASPOONS OF SALT TO 12 OZ. WATER), THOROUGHLY SPRAY THE RIGHT REAR WHEEL SPEED SENSOR JUMPER HARNESS.
- DISCONNECT 32-WAY EBTCM HARNESS CONNECTOR.
- USING J 39200, MEASURE THE RESISTANCE BETWEEN TERMINALS "E5" AND "E6" OF THE 32-WAY EBTCM HARNESS CONNECTOR.

IS RESISTANCE BETWEEN 2059 AND 2353Ω WHEN SENSOR IS 20°C (68°F)? *(IF TEMPERATURE OF SENSOR IS SUSPECTED TO OTHER THAN 20°C (68°F), REFER TO THE TABLE ON PAGE 2 OF 2 OF THIS DTC CHART FOR APPLICABLE SENSOR RESISTANCE VALUES; THE VALUES IN THE TABLE ARE FOR TEMPERATURE OF THE SENSOR, NOT AIR TEMPERATURE.)*

- NO → DIAGNOSIS CONTINUED ON PAGE 2 OF 3 OF THIS DTC CHART.
- YES ↓

(2)
- USING J 39200, MEASURE THE RESISTANCE BETWEEN GROUND AND TERMINAL "E5" OF THE 32-WAY EBTCM HARNESS CONNECTOR. DOES J 39200 DISPLAY RESISTANCE AS "OL" (INFINITE)?

- NO → DIAGNOSIS CONTINUED ON PAGE 3 OF 3 OF THIS DTC CHART.
- YES ↓

(3)
- IGNITION "ON."
- USING J 39200, MEASURE THE VOLTAGE BETWEEN GROUND AND TERMINAL "E5" OF THE 32-WAY EBTCM HARNESS CONNECTOR. IS VOLTAGE 1 VOLT OR LESS?

- NO → (4)
- YES → (6)

(4)
- IGNITION "OFF."
- DISCONNECT RIGHT REAR WHEEL SPEED SENSOR DIRECTLY AT SENSOR.
- IGNITION "ON."
- USING J 39200, MEASURE THE VOLTAGE BETWEEN GROUND AND TERMINAL "E5" OF THE 32-WAY EBTCM HARNESS CONNECTOR. IS VOLTAGE 1 VOLT OR LESS?

- NO → REPAIR SHORT TO VOLTAGE IN CKT 883.
- YES → (5)

(5)
- USING J 39200, MEASURE THE VOLTAGE BETWEEN GROUND AND TERMINAL "E6" OF THE 32-WAY EBTCM HARNESS CONNECTOR. IS VOLTAGE 1 VOLT OR LESS?

- NO → REPAIR SHORT TO VOLTAGE IN CKT 882.
- YES → MALFUNCTION IS INTERMITTENT. INSPECT ALL CONNECTORS AND HARNESSES FOR DAMAGE WHICH MAY RESULT IN A SHORT TO VOLTAGE WHEN ALL COMPONENTS ARE CONNECTED. REPAIR ALL DAMAGE IF FOUND. REFER TO "DIAGNOSTIC AIDS" ON FACING PAGE FOR MORE INFORMATION.

(6)
- IGNITION "OFF."
- INSPECT TERMINALS "E5" AND "E6" OF THE 32-WAY EBTCM HARNESS CONNECTOR FOR POOR TERMINAL CONTACT AND CONTAMINATION. ALSO INSPECT CKTs 882 AND 883 FOR DAMAGE WHICH MAY RESULT IN AN OPEN CIRCUIT, SHORT TO GROUND, OR SHORT TO VOLTAGE. REPAIR ALL DAMAGE IF EVIDENT.
- RECONNECT ALL CONNECTORS.
- IGNITION "ON."

DOES DTC 35 SET AS A CURRENT DTC?

- YES → REPLACE EBTCM.
- NO → MALFUNCTION IS INTERMITTENT. REFER TO "DIAGNOSTIC AIDS" ON FACING PAGE.

DIAGNOSTIC CHART DTC 35 — 1994 SILHOUETTE, TRANS SPORT AND LUMINA APV WITH TCS, CONTINUED

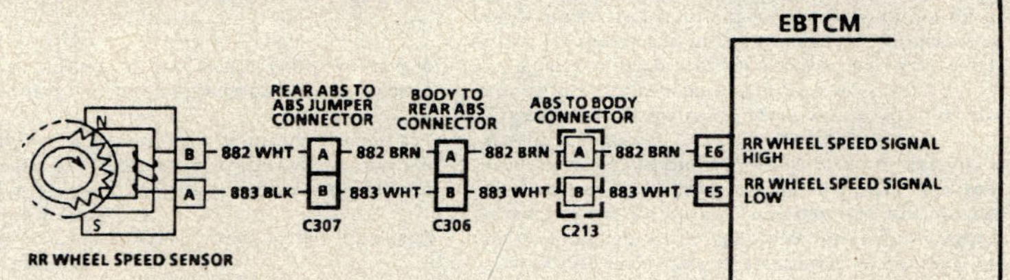

DTC 35

(Page 2 of 3)

RR WHEEL SPEED SNR CKT OPEN OR SHORT TO GND/BAT

Circuit Description:

As a toothed ring passes by the wheel speed sensor, changes in the electromagnetic field cause the wheel speed sensor to produce a sinusoidal (AC) voltage signal whose frequency is proportional to wheel speed. The magnitude of this signal is directly related to wheel speed and the proximity of the wheel speed sensor to the toothed ring, often referred to as the air gap.

DTC Will Set When: DTC 35 can be set anytime after initialization. If either of the right rear wheel speed circuits are shorted to voltage or ground, a malfunction exists.

Action Taken: A malfunction DTC is stored, ABS and TCS are disabled and the ABS and TCS warning lamps are turned "ON."

DTC Chart Test Description: Number(s) below refer to circled number(s) on the diagnostic chart.

7. Checks for high resistance in CKT 883.
8. Checks for high resistance in CKT 882.
9. Verifies proper wheel speed sensor resistance.

Diagnostic Aids: An intermittent malfunction may be caused by a poor connection, rubbed through wire insulation, or a wire that is broken inside the insulation.

The frequency of the malfunction can be checked by using the enhanced diagnostic function of the Tech 1,

If the customer's comments reflect that the ABS and TCS warning lamps are "ON" only during moist environmental changes (rain, snow, vehicle wash), all wheel speed sensor circuitry should be thoroughly inspected for signs of water intrusion. If DTC is not current, clear DTCs and simulate the effects of water intrusion. Use the following procedure. Spray down the suspected area with a 5% salt water solution (two teaspoons of salt to 12 oz. of water).

Test drive vehicle over various road surfaces (bumps, turns, etc.) above 24 km/h (15 mph) for at least 30 seconds. If DTC returns, replace suspected harness.

Any circuitry, that is suspected as causing the intermittent complaint, should be thoroughly checked for backed out terminals, improper mating, broken locks, improperly formed or damaged terminals, poor terminal to wiring connections or physical damage to the wiring harness.

Resistance of the wheel speed sensor will increase with an increase in sensor temperature.

When replacing a wheel speed sensor, inspect the sensor terminals and harness connector for corrosion and/or water intrusion. If evidence of corrosion or water intrusion exists, replace wheel speed sensor jumper harness. Likewise, if replacing a wheel speed sensor jumper harness, inspect sensor terminals. If evidence of corrosion or water intrusion exists, replace wheel speed sensor.

DIAGNOSTIC CHART DTC 35 — 1994 SILHOUETTE, TRANS SPORT AND LUMINA APV WITH TCS, CONTINUED

DTC 35

(Page 2 of 3)

RR WHEEL SPEED SNR CKT OPEN OR SHORT TO GND/BAT

DIAGNOSIS CONTINUED FROM PAGE 1 OF 3 OF THIS DTC CHART.

(7)
- DISCONNECT RIGHT REAR WHEEL SPEED SENSOR DIRECTLY AT SENSOR.
- USING J 39200, MEASURE THE RESISTANCE BETWEEN TERMINAL "E5" OF THE 32-WAY EBTCM HARNESS CONNECTOR AND TERMINAL "A" OF THE RIGHT REAR WHEEL SPEED SENSOR HARNESS CONNECTOR.
  IS RESISTANCE 2Ω OR LESS?

NO: REPAIR HIGH RESISTANCE IN CKT 883 (WHEEL SPEED SENSOR JUMPER MUST BE REPLACED IF DAMAGED).

YES:

(8)
- USING J 39200, MEASURE THE RESISTANCE BETWEEN TERMINAL "E6" OF THE 32-WAY EBTCM HARNESS CONNECTOR AND TERMINAL "B" OF THE RIGHT REAR WHEEL SPEED SENSOR HARNESS CONNECTOR.
  IS RESISTANCE 2Ω OR LESS?

NO: REPAIR HIGH RESISTANCE IN CKT 882 (WHEEL SPEED SENSOR JUMPER MUST BE REPLACED IF DAMAGED).

YES:

(9)
- USING J 39200, MEASURE THE RESISTANCE BETWEEN TERMINAL "A" AND TERMINAL "B" OF THE RIGHT REAR WHEEL SPEED SENSOR CONNECTOR.
  IS RESISTANCE BETWEEN 2059 AND 2353Ω WHEN SENSOR IS 20°C (68°F)? *(IF TEMPERATURE OF SENSOR IS SUSPECTED TO OTHER THAN 20°C (68°F), REFER TO THE TABLE BELOW FOR APPLICABLE SENSOR RESISTANCE VALUES; THE VALUES IN THE TABLE ARE FOR TEMPERATURE OF THE SENSOR, NOT AIR TEMPERATURE.)*

NO: REPLACE WHEEL SPEED SENSOR.

YES: MALFUNCTION IS INTERMITTENT. INSPECT ALL CONNECTORS AND HARNESSES FOR DAMAGE WHICH MAY RESULT IN HIGH RESISTANCE WHEN ALL COMPONENTS ARE CONNECTED. REFER TO "DIAGNOSTIC AIDS" ON FACING PAGE FOR MORE INFORMATION. REPAIR ALL DAMAGE IF FOUND.

REAR WHEEL SPEED SENSOR

TEMPERATURE OF SENSOR VS. SENSOR RESISTANCE (APPROXIMATE)

TEMP. (°C)	TEMP. (°F)	RESISTANCE (OHMS)
-34 TO 4	-30 TO 40	1622 TO 2205
5 TO 43	41 TO 110	1930 TO 2566
44 TO 93	111 TO 200	2245 TO 3028

DIAGNOSTIC CHART DTC 35 — 1994 SILHOUETTE, TRANS SPORT AND LUMINA APV WITH TCS, CONTINUED

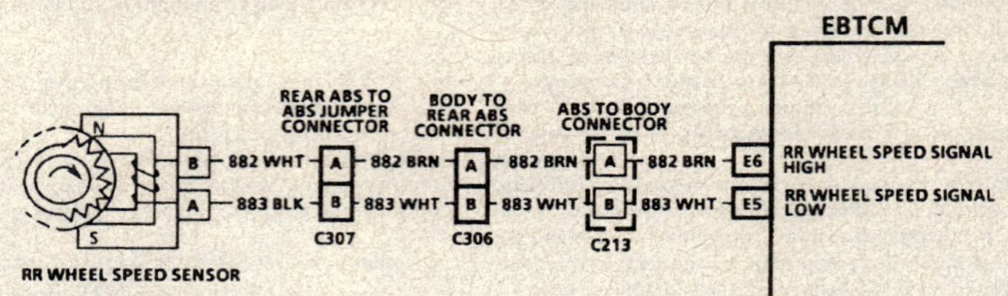

DTC 35

(Page 3 of 3)

RR WHEEL SPEED SNR CKT OPEN OR SHORT TO GND/BAT

Circuit Description:

As a toothed ring passes by the wheel speed sensor, changes in the electromagnetic field cause the wheel speed sensor to produce a sinusoidal (AC) voltage signal whose frequency is proportional to wheel speed. The magnitude of this signal is directly related to wheel speed and the proximity of the wheel speed sensor to the toothed ring, often referred to as the air gap.

DTC Will Set When: DTC 35 can be set anytime after initialization. If either of the right rear wheel speed circuits are shorted to voltage or ground, a malfunction exists.

Action Taken: A malfunction DTC is stored, ABS and TCS are disabled and the ABS and TCS warning lamps are turned "ON."

DTC Chart Test Description: Number(s) below refer to circled number(s) on the diagnostic chart.

10. Checks for short to ground in CKT 883.
11. Checks for short to ground in CKT 882.
12. Checks for short to ground in wheel speed sensor jumper harness.
13. Verifies that wheel speed sensor is not internally shorted to ground.

Diagnostic Aids: An intermittent malfunction may be caused by a poor connection, rubbed through wire insulation, or a wire that is broken inside the insulation.

The frequency of the malfunction can be checked by using the enhanced diagnostic function of the Tech 1,

If the customer's comments reflect that the ABS and TCS warning lamps are "ON" only during moist environmental changes (rain, snow, vehicle wash), all wheel speed sensor circuitry should be thoroughly inspected for signs of water intrusion. If DTC is not current, clear DTCs and simulate the effects of water intrusion. Use the following procedure.

Spray down the suspected area with a 5% salt water solution (two teaspoons of salt to 12 oz. of water). Test drive vehicle over various road surfaces (bumps, turns, etc.) above 24 km/h (15 mph) for at least 30 seconds. If DTC returns, replace suspected harness.

Any circuitry, that is suspected as causing the intermittent complaint, should be thoroughly checked for backed out terminals, improper mating, broken locks, improperly formed or damaged terminals, poor terminal to wiring connections or physical damage to the wiring harness.

Resistance of the wheel speed sensor will increase with an increase in sensor temperature.

When replacing a wheel speed sensor, inspect the sensor terminals and harness connector for corrosion and/or water intrusion. If evidence of corrosion or water intrusion exists, replace wheel speed sensor jumper harness. Likewise, if replacing a wheel speed sensor jumper harness, inspect sensor terminals. If evidence of corrosion or water intrusion exists, replace wheel speed sensor.

DIAGNOSTIC CHART DTC 35 — 1994 SILHOUETTE, TRANS SPORT AND LUMINA APV WITH TCS, CONTINUED

DTC 35

(Page 3 of 3)

RR WHEEL SPEED SNR CKT OPEN OR SHORT TO GND/BAT

DIAGNOSIS CONTINUED FROM PAGE 1 OF 3 OF THIS DTC CHART.

(10)
- DISCONNECT CONNECTOR C307 (RH REAR UNDER BODY ABOVE FRONT SPRING HANGER).
- USING J 39200, MEASURE THE RESISTANCE BETWEEN GROUND AND TERMINAL "E5" OF THE 32-WAY EBTCM HARNESS CONNECTOR.

DOES J 39200 DISPLAY RESISTANCE AS "OL" (INFINITE)?

- NO: REPAIR SHORT TO GROUND IN CKT 883.
- YES: (11)

(11)
- USING J 39200, MEASURE THE RESISTANCE BETWEEN GROUND AND TERMINAL "E6" OF THE 32-WAY EBTCM HARNESS CONNECTOR.

DOES J 39200 DISPLAY RESISTANCE AS "OL" (INFINITE)?

- NO: REPAIR SHORT TO GROUND IN CKT 882.
- YES: (12)

(12)
- THOROUGHLY RESPRAY THE RIGHT REAR WHEEL SPEED SENSOR JUMPER HARNESS WITH THE SALT WATER SOLUTION.

IMPORTANT: DO NOT ALLOW WATER TO ENTER ANY WHEEL SPEED SENSOR CONNECTOR.

- DISCONNECT RIGHT REAR WHEEL SPEED SENSOR DIRECTLY AT SENSOR.
- USING J 39200, MEASURE THE RESISTANCE BETWEEN GROUND AND TERMINAL "A," THEN "B" OF THE RIGHT REAR WHEEL SPEED SENSOR HARNESS CONNECTOR.

DOES J 39200 DISPLAY RESISTANCE AS "OL" (INFINITE) FOR BOTH MEASUREMENTS?

- NO: REPLACE RIGHT REAR WHEEL SPEED SENSOR JUMPER HARNESS.
- YES: (13)

(13)
- USING J 39200, MEASURE THE RESISTANCE BETWEEN GROUND AND TERMINAL "A" OF THE RIGHT REAR WHEEL SPEED SENSOR CONNECTOR.

DOES J 39200 DISPLAY RESISTANCE AS "OL" (INFINITE)?

- NO: REPLACE WHEEL SPEED SENSOR.
- YES: MALFUNCTION IS INTERMITTENT. INSPECT ALL CONNECTORS AND HARNESSES FOR DAMAGE WHICH MAY RESULT IN A SHORT TO GROUND WHEN ALL COMPONENTS ARE CONNECTED. PAY PARTICULAR ATTENTION TO THE JUMPER HARNESS, SMALL NICKS OR CUTS CAN CAUSE A MALFUNCTION. REPAIR ANY DAMAGE IF FOUND. JUMPER HARNESS MUST BE REPLACED IF DAMAGED. REFER TO "DIAGNOSTIC AIDS" ON FACING PAGE FOR MORE INFORMATION.

DIAGNOSTIC CHART DTC 36 — 1994 SILHOUETTE, TRANS SPORT AND LUMINA APV WITH TCS

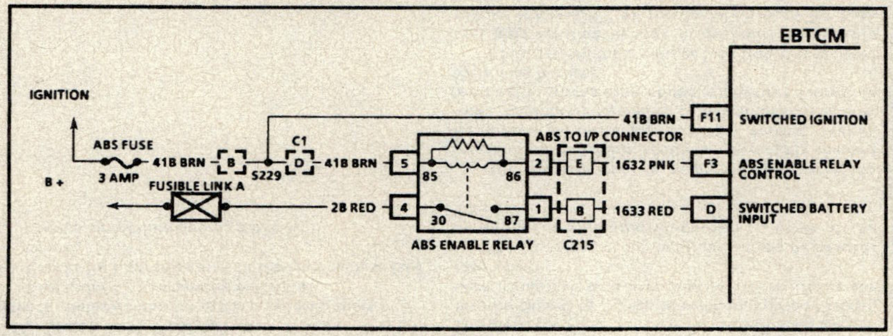

DTC 36
(Page 1 of 2)
LOW SYSTEM VOLTAGE

Circuit Description:
This DTC is used to monitor the voltage level available to the EBTCM controller. If the voltage drops below 11 volts, full performance of the ABS system cannot be guaranteed. During ABS operation, there are several current requirements that will cause battery voltage to drop. Because of this, voltage is monitored prior to ABS to indicate good charging system condition and also during ABS when voltage may drop significantly.

DTC Will Set When: DTC 36 can only be set if the vehicle's speed is greater than 5 km/h (3 mph). If the switched battery voltage is less than 11.2 volts non-ABS or less than 9.6 volts ABS, a malfunction exists.

Action Taken: A malfunction DTC is stored, ABS is disabled and the ABS warning lamp is turned "ON."

DTC Chart Test Description: Number(s) below refer to circled number(s) on the diagnostic chart.

1. Checks to see what voltage is readily available at terminal "D" of the EBTCM. If voltage is representative of good charging system condition, a malfunction is not present.
2. Isolates the EBTCM to check the condition of the circuitry and charging system.
3. Isolates the low voltage condition to high circuit resistance, or improper charging system operation.
4. Isolates which circuit is the source of the high resistance condition.

Diagnostic Aids: An intermittent malfunction may be caused by a poor connection, rubbed through wire insulation, or a wire that is broken inside the insulation.

The frequency of the malfunction can be checked by using the enhanced diagnostic function of the Tech 1.

Any circuitry, that is suspected as causing the intermittent complaint, should be thoroughly checked for backed out terminals, improper mating, broken locks, improperly formed or damaged terminals, poor terminal to wiring connections or physical damage to the wiring harness.

While performing a voltage load test, if it is noted that only ignition voltage drops below acceptable voltage levels, CKT 41 should be checked for high resistance or an open condition.

DIAGNOSTIC CHART DTC 36 — 1994 SILHOUETTE, TRANS SPORT AND LUMINA APV WITH TCS, CONTINUED

DTC 36
(Page 1 of 2)
LOW SYSTEM VOLTAGE

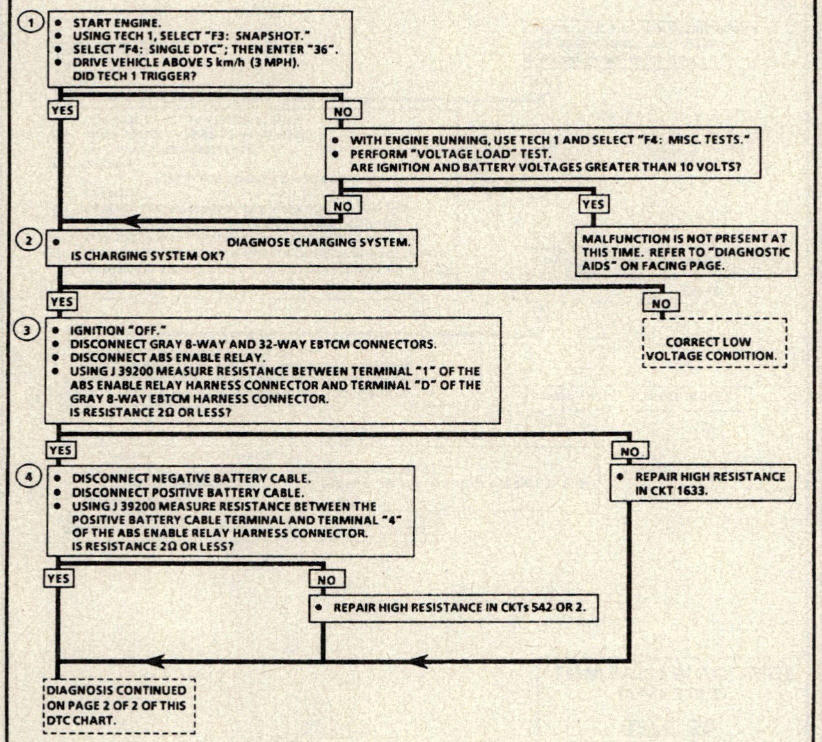

DIAGNOSTIC CHART DTC 36 — 1994 SILHOUETTE, TRANS SPORT AND LUMINA APV WITH TCS, CONTINUED

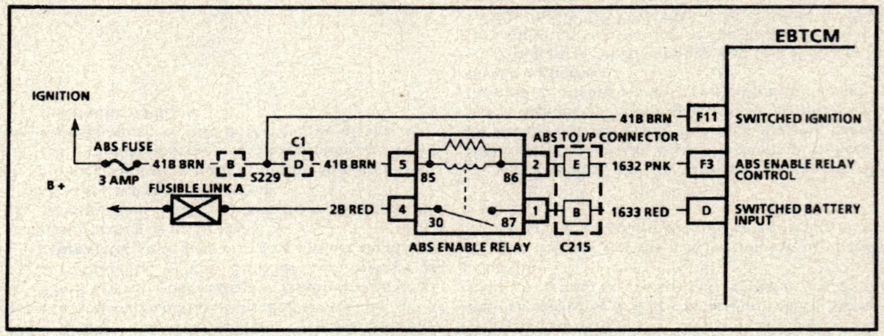

DTC 36
(Page 2 of 2)
LOW SYSTEM VOLTAGE

Circuit Description:
This DTC is used to monitor the voltage level available to the EBTCM controller. If the voltage drops below 11 volts, full performance of the ABS system cannot be guaranteed. During ABS operation, there are several current requirements that will cause battery voltage to drop. Because of this, voltage is monitored prior to ABS to indicate good charging system condition and also during ABS when voltage may drop significantly.

DTC Will Set When: DTC 36 can only be set if the vehicle's speed is greater than 5 km/h (3 mph). If the switched battery voltage is less than 11.2 volts non-ABS or less than 9.6 volts ABS, a malfunction exists.

Action Taken: A malfunction DTC is stored, ABS is disabled and the ABS warning lamp is turned "ON."

DTC Chart Test Description: Number(s) below refer to circled number(s) on the diagnostic chart.
5. Checks for high resistance in CKT 41.
6. Checks for high resistance in the ignition fuse feed circuit.
7. Checks for malfunctioning EBTCM.

Diagnostic Aids: An intermittent malfunction may be caused by a poor connection, rubbed through wire insulation, or a wire that is broken inside the insulation.

The frequency of the malfunction can be checked by using the enhanced diagnostic function of the Tech 1,

Any circuitry, that is suspected as causing the intermittent complaint, should be thoroughly checked for backed out terminals, improper mating, broken locks, improperly formed or damaged terminals, poor terminal to wiring connections or physical damage to the wiring harness.

While performing a voltage load test, if it is noted that <u>only</u> ignition voltage drops below acceptable voltage levels, CKT 41 should be checked for high resistance or an open condition.

DIAGNOSTIC CHART DTC 36 — 1994 SILHOUETTE, TRANS SPORT AND LUMINA APV WITH TCS, CONTINUED

DTC 36
(Page 2 of 2)
LOW SYSTEM VOLTAGE

DIAGNOSIS CONTINUED FROM PAGE 1 OF 2 OF THIS DTC CHART.

(5)
- REMOVE ABS FUSE.
- USING J 39200 MEASURE RESISTANCE BETWEEN EACH ABS FUSE BLOCK TERMINAL AND TERMINAL "F11" OF THE 32-WAY EBTCM HARNESS CONNECTOR.

IS EITHER MEASUREMENT 2Ω OR LESS?

NO: REPAIR HIGH RESISTANCE IN CKT 41.

YES:

(6)
- WITH BATTERY CABLES DISCONNECTED, TURN IGNITION "ON"
- USING J 39200, MEASURE RESISTANCE BETWEEN POSITIVE BATTERY CABLE TERMINAL AND EACH ABS FUSE BLOCK TERMINAL.

IS EITHER MEASUREMENT 2Ω OR LESS ?

NO: REPAIR HIGH RESISTANCE IN THE "IGNITION SWITCH 'RUN' FUSE FEED" CIRCUIT.

YES:

(7)
- IGNITION "OFF."
- INSPECT ABS ENABLE RELAY AND ABS ENABLE RELAY HARNESS CONNECTOR FOR POOR TERMINAL CONTACT. INSPECT GRAY 8-WAY AND 32-WAY EBTCM HARNESS CONNECTORS FOR A POOR CONNECTION TO THE EBTCM. INSPECT POSITIVE BATTERY CABLE TERMINAL FOR POOR CONTACT TO THE BATTERY.
  REPLACE TERMINALS IF POOR CONTACT IS EVIDENT.
- RECONNECT ALL CONNECTORS.
- DRIVE VEHICLE ABOVE 5 km/h (3 MPH).

DOES DTC 36 SET AS A CURRENT DTC?

YES: REPLACE EBTCM.

NO: MALFUNCTION IS MOST LIKELY DUE TO A POOR CONNECTION. REFER TO "DIAGNOSTIC AIDS" ON FACING PAGE.

DIAGNOSTIC CHART DTC 37 — 1994 SILHOUETTE, TRANS SPORT AND LUMINA APV WITH TCS

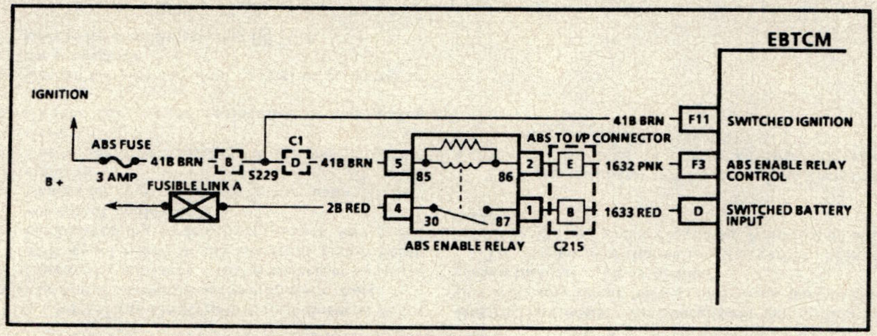

DTC 37
HIGH SYSTEM VOLTAGE

Circuit Description:
This DTC is designed to detect high vehicle voltage levels prior to any required motor movement (initialization or ABS). If excessive voltage exists, demagnetization of the motor magnets may occur, which would eventually affect or eliminate ABS performance.

DTC Will Set When: DTC 37 can only be set if the vehicle's speed is greater than 5 km/h (3 mph). If the switched battery voltage is greater than 17 volts, a malfunction exists.

Action Taken: A malfunction DTC is stored, ABS and TCS are disabled and the ABS and TCS warning lamps are turned "ON."

DTC Chart Test Description: Number(s) below refer to circled number(s) on the diagnostic chart.

1. Checks the voltage level being received by the EBTCM.
2. Indicates whether the high voltage condition is caused by a malfunctioning charging system or EBTCM.
3. Checks for poor terminal contact which may cause an intermittent malfunction or a malfunctioning EBTCM.

Diagnostic Aids: An intermittent malfunction may be caused by a poor connection, rubbed through wire insulation, or a wire that is broken inside the insulation.

The frequency of the malfunction can be checked by using the enhanced diagnostic function of the Tech 1.

Any circuitry, that is suspected as causing the intermittent complaint, should be thoroughly checked for backed out terminals, improper mating, broken locks, improperly formed or damaged terminals, poor terminal to wiring connections or physical damage to the wiring harness.

DIAGNOSTIC CHART DTC 37 — 1994 SILHOUETTE, TRANS SPORT AND LUMINA APV WITH TCS, CONTINUED

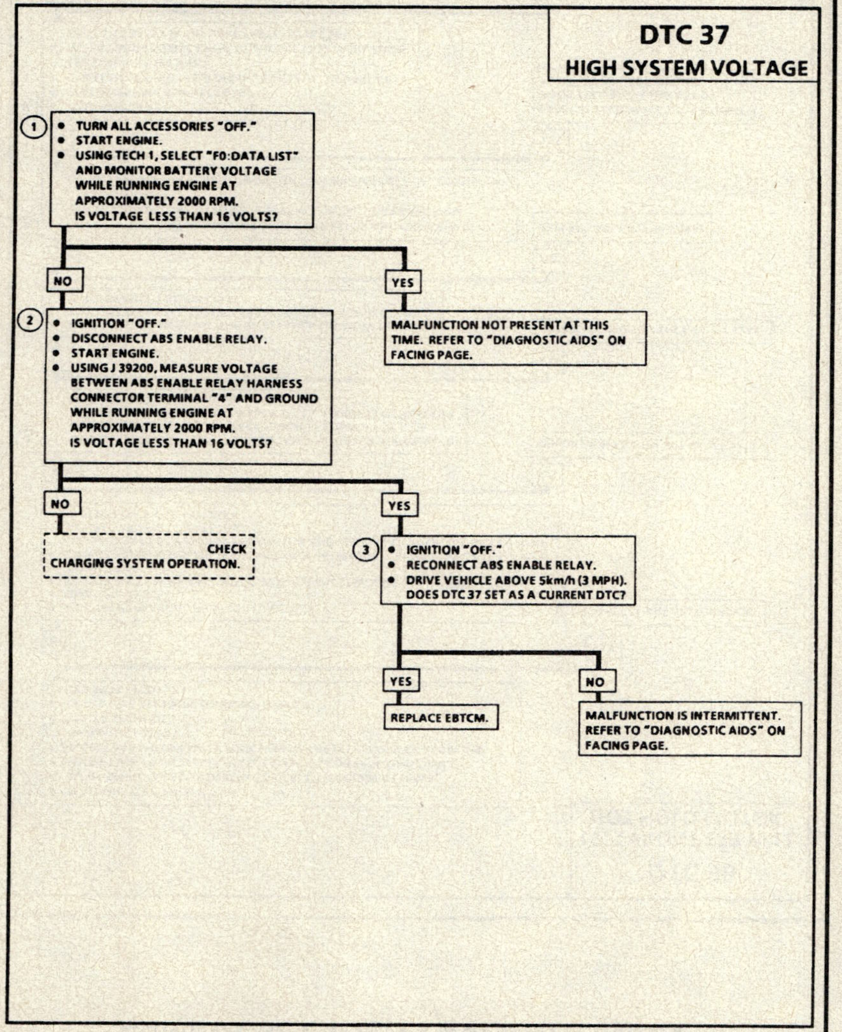

DIAGNOSTIC CHART DTC 38 — 1994 SILHOUETTE, TRANS SPORT AND LUMINA APV WITH TCS

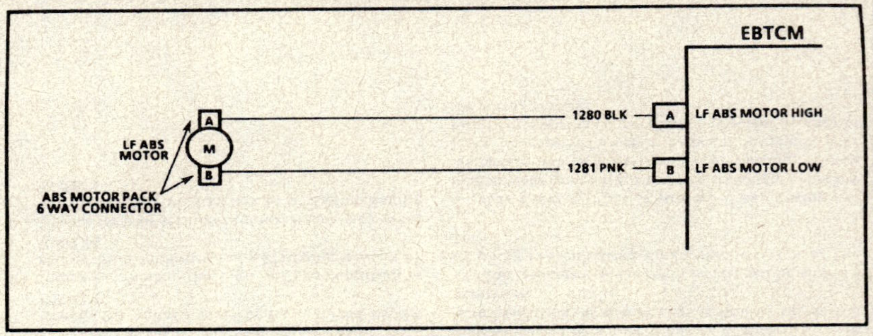

DTC 38

LEFT FRONT ESB WILL NOT HOLD MOTOR

Circuit Description:
This DTC is designed to detect a slipping left front ESB. During initialization and braking, the left front motor is re-homed. If the ESB slips, the motor/piston will move. During the next key "ON" initialization, a re-home of the motor verifies the motor/piston remained at the home position. If motor movement is detected, the ESB must be slipping.

DTC Will Set When: DTC 38 can be set during initialization. If the EBTCM detects that the ESB could not hold the piston in the home position, a malfunction exists. DTC 86 is always set with DTC 38.

Action Taken: If an ESB cannot hold a piston in the home position, the piston may be backdriven when the brake pedal is applied, causing the brake pedal to drop. A malfunction DTC is stored, ABS is disabled, and the ABS warning lamp and the red "BRAKE" warning lamp are turned "ON."

DTC Chart Test Description: Number(s) below refer to circled number(s) on the diagnostic chart.

1. Checks the left front ESB. A broken or defective ESB would result in the left front piston being backdriven by hydraulic pressure and pedal movement would result.
2. Checks for high resistance in the motor "HIGH" circuitry.
3. Checks for high resistance in the motor "LOW" circuitry.
4. Checks for proper resistance of the motor windings.
5. Checks for poor terminal contact or corrosion in the connectors.
6. Checks for a malfunctioning EBTCM.

Diagnostic Aids: An intermittent malfunction in this DTC may result from a mechanical part of the system that sticks, binds, or slips.

The frequency of the malfunction can be checked by using the enhanced diagnostic function of the Tech 1,

The static modulator test function of the Tech 1 or T-100 may be used to locate an intermittent malfunction associated with the ESB.

DIAGNOSTIC CHART DTC 38 — 1994 SILHOUETTE, TRANS SPORT AND LUMINA APV WITH TCS, CONTINUED

DTC 38

LEFT FRONT ESB WILL NOT HOLD MOTOR

(1)
- IGNITION "ON," ENGINE "OFF."
- PUMP BRAKE PEDAL UNTIL FIRM TO DEPLETE VACUUM RESERVE.
- USING TECH 1 MISC. TESTS, SELECT MANUAL CONTROL FUNCTION.
- SELECT ABS MOTOR APPLY FUNCTION OF THE LEFT FRONT CHANNEL AND APPLY THE MOTOR.
- WAIT 5 SECONDS.
- APPLY FIRM PRESSURE ON BRAKE PEDAL.?

DID BRAKE PEDAL FALL

YES: REPLACE ABS MOTOR PACK.

NO:

(2)
- EXIT TECH 1 MANUAL CONTROL FUNCTION
- IGNITION "OFF."
- DISCONNECT 6-WAY ABS MOTOR PACK CONNECTOR AND GRAY 8-WAY EBTCM CONNECTOR.
- USING J 39200, MEASURE THE RESISTANCE BETWEEN TERMINAL "A" OF THE 6-WAY ABS MOTOR PACK HARNESS CONNECTOR AND TERMINAL "A" OF THE GRAY 8-WAY EBTCM HARNESS CONNECTOR.

IS RESISTANCE 1.5Ω OR LESS?

NO: REPAIR HIGH RESISTANCE IN CKT 1280.

YES:

(3)
- USING J 39200, MEASURE THE RESISTANCE BETWEEN TERMINAL "B" OF THE 6-WAY ABS MOTOR PACK HARNESS CONNECTOR AND TERMINAL "B" OF THE GRAY 8-WAY EBTCM HARNESS CONNECTOR.

IS RESISTANCE 1.5Ω OR LESS?

NO: REPAIR HIGH RESISTANCE IN CKT 1281.

YES:

(4)
- USING J 39200, MEASURE THE RESISTANCE BETWEEN TERMINAL "A" AND TERMINAL "B" OF THE MOTOR PACK CONNECTOR.

IS RESISTANCE 1.5Ω OR LESS?

NO: REPLACE ABS MOTOR PACK.

YES:

(5)
- INSPECT THE 6-WAY ABS MOTOR PACK CONNECTOR AND GRAY 8-WAY EBTCM CONNECTOR FOR POOR TERMINAL CONTACT OR CORROSION.

DO ANY TERMINALS EXHIBIT POOR CONTACT OR EVIDENCE OF CORROSION?

YES: REPLACE TERMINALS THAT EXHIBIT POOR TERMINAL CONTACT OR EVIDENCE OF CORROSION.

NO:

(6)
- IGNITION "OFF."
- RECONNECT ALL CONNECTORS.
- START ENGINE AND TEST DRIVE VEHICLE, ACHIEVING AT LEAST 16 km/h (10 MPH).
- REPEAT ABOVE DRIVE CYCLE SEQUENCE TWO MORE TIMES.

DOES DTC 38 SET IN THE LAST THREE DRIVE CYCLE?

YES: REPLACE EBTCM.

NO: MALFUNCTION IS INTERMITTENT. REFER TO "DIAGNOSTIC AIDS" ON FACING PAGE.

AFTER DIAGNOSIS IS COMPLETE, CLEAR DTCs AND TEST DRIVE VEHICLE FOR THREE (3) DRIVE CYCLES TO VERIFY THAT DTC DOES NOT RESET. A DRIVE CYCLE CONSISTS OF STARTING THE VEHICLE, DRIVING OVER 16 km/h (10 MPH), AND THEN TURNING KEY OFF.

DIAGNOSTIC CHART DTC 41 — 1994 SILHOUETTE, TRANS SPORT AND LUMINA APV WITH TCS

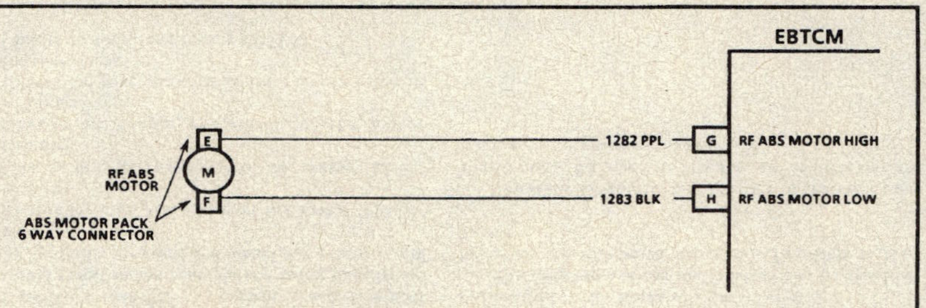

DTC 41

RIGHT FRONT ESB WILL NOT HOLD MOTOR

Circuit Description:
This DTC is designed to detect a slipping right front ESB. During initialization and braking, the right front motor is re-homed. If the ESB slips, the motor/piston will move. During the next key "ON" initialization, a re-home of the motor verifies the motor/piston remained at the home position. If motor movement is detected, the ESB must be slipping.

DTC Will Set When: DTC 41 can be set during initialization. If the EBTCM detects that the ESB could not hold the piston in the home position, a malfunction exists. DTC 86 is always set with DTC 41.

Action Taken: If an ESB cannot hold a piston in the home position, the piston may be backdriven when the brake pedal is applied, causing the brake pedal to drop. A malfunction DTC is stored, ABS is disabled, and the ABS warning lamp and the red "BRAKE" warning lamp are turned "ON."

DTC Chart Test Description: Number(s) below refer to circled number(s) on the diagnostic chart.

1. Checks the right front ESB. A broken or defective ESB would result in the right front piston being backdriven by hydraulic pressure, and pedal movement would result.
2. Checks for high resistance in the motor "HIGH" circuitry.
3. Checks for high resistance in the motor "LOW" circuitry.
4. Checks for proper resistance of the motor windings.
5. Checks for poor terminal contact or corrosion in the connectors.
6. Checks for a malfunctioning EBTCM.

Diagnostic Aids: An intermittent malfunction in this DTC may result from a mechanical part of the system that sticks, binds, or slips.

The frequency of the malfunction can be checked by using the enhanced diagnostic function of the Tech 1.

The static modulator test function of the Tech 1 or T-100 may be used to locate an intermittent malfunction associated with the ESB.

DIAGNOSTIC CHART DTC 41 — 1994 SILHOUETTE, TRANS SPORT AND LUMINA APV WITH TCS, CONTINUED

DTC 41
RIGHT FRONT ESB WILL NOT HOLD MOTOR

(1)
- IGNITION "ON," ENGINE "OFF."
- PUMP BRAKE PEDAL UNTIL FIRM TO DEPLETE VACUUM RESERVE.
- USING TECH 1 MISC. TESTS, SELECT MANUAL CONTROL FUNCTION.
- SELECT ABS MOTOR APPLY FUNCTION OF THE RIGHT FRONT CHANNEL AND APPLY THE MOTOR.
- WAIT 5 SECONDS.
- APPLY FIRM PRESSURE ON BRAKE PEDAL.

DID BRAKE PEDAL FALL?

- YES → REPLACE ABS MOTOR PACK.
- NO → (2)

(2)
- EXIT TECH 1 MANUAL CONTROL FUNCTION.
- IGNITION "OFF."
- DISCONNECT 6-WAY ABS MOTOR PACK CONNECTOR AND GRAY 8-WAY EBTCM CONNECTOR.
- USING J 39200, MEASURE THE RESISTANCE BETWEEN TERMINAL "E" OF THE 6-WAY ABS MOTOR PACK HARNESS CONNECTOR AND TERMINAL "G" OF THE GRAY 8-WAY EBTCM HARNESS CONNECTOR.

IS RESISTANCE 1.5Ω OR LESS?

- NO → REPAIR HIGH RESISTANCE IN CKT 1282.
- YES → (3)

(3)
- USING J 39200, MEASURE THE RESISTANCE BETWEEN TERMINAL "F" OF THE 6-WAY ABS MOTOR PACK HARNESS CONNECTOR AND TERMINAL "H" OF THE GRAY 8-WAY EBTCM HARNESS CONNECTOR.

IS RESISTANCE 1.5Ω OR LESS?

- NO → REPAIR HIGH RESISTANCE IN CKT 1283.
- YES → (4)

(4)
- USING J 39200, MEASURE THE RESISTANCE BETWEEN TERMINAL "E" AND TERMINAL "F" OF THE ABS MOTOR PACK CONNECTOR.

IS RESISTANCE 1.5Ω OR LESS?

- NO → REPLACE ABS MOTOR PACK.
- YES → (5)

(5)
- INSPECT THE 6-WAY ABS MOTOR PACK CONNECTOR AND GRAY 8-WAY EBTCM CONNECTOR FOR POOR TERMINAL CONTACT OR CORROSION.

DO ANY TERMINALS EXHIBIT POOR CONTACT OR EVIDENCE OF CORROSION?

- YES → REPLACE TERMINALS THAT EXHIBIT POOR TERMINAL CONTACT OR EVIDENCE OF CORROSION.
- NO → (6)

(6)
- IGNITION "OFF."
- RECONNECT ALL CONNECTORS.
- START ENGINE AND TEST DRIVE VEHICLE, ACHIEVING AT LEAST 16 km/h (10 MPH).
- REPEAT ABOVE DRIVE CYCLE SEQUENCE TWO MORE TIMES.

DOES DTC 41 SET IN THE LAST THREE DRIVE CYCLES?

- YES → REPLACE EBTCM.
- NO → MALFUNCTION IS INTERMITTENT. REFER TO "DIAGNOSTIC AIDS" ON FACING PAGE.

AFTER DIAGNOSIS IS COMPLETE, CLEAR DTCs AND TEST DRIVE VEHICLE FOR THREE (3) DRIVE CYCLES TO VERIFY THAT DTC DOES NOT RESET. A DRIVE CYCLE CONSISTS OF STARTING THE VEHICLE, DRIVING OVER 16 km/h (10 MPH), AND THEN TURNING KEY OFF.

DIAGNOSTIC CHART DTC 42 — 1994 SILHOUETTE, TRANS SPORT AND LUMINA APV WITH TCS

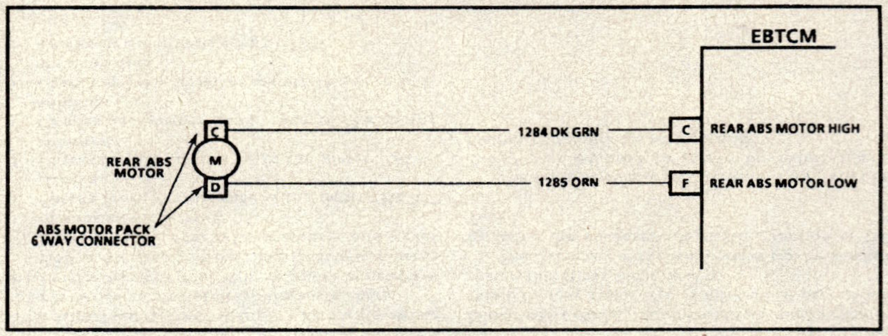

DTC 42

REAR ESB WILL NOT HOLD MOTOR

Circuit Description:
This DTC is designed to detect a slipping rear ESB. During initialization and braking, the rear motor is re-homed. If the ESB slips, the motor/piston will move. During the next key "ON" initialization, a re-home of the motor verifies the motor/piston remained at the home position. If motor movement is detected, the ESB must be slipping.

DTC Will Set When: DTC 42 can be set during initialization. If the EBTCM detects that the ESB could not hold the piston in the home position, a malfunction exists. DTC 86 is always set with DTC 42.

Action Taken: If an ESB cannot hold a piston in the home position, the piston may be backdriven when the brake pedal is applied, causing the brake pedal to drop. A malfunction DTC is stored, ABS is disabled, and the ABS warning lamp and the red "BRAKE" warning lamp are turned "ON."

DTC Chart Test Description: Number(s) below refer to circled number(s) on the diagnostic chart.

1. Checks the rear ESB. A broken or defective ESB would result in the rear pistons being backdriven by hydraulic pressure and wheel movement would result.
2. Checks for high resistance in the motor "HIGH" circuitry.
3. Checks for high resistance in the motor "LOW" circuitry.
4. Checks for proper resistance of the motor windings.
5. Checks for poor terminal contact or corrosion in the connectors.
6. Checks for a malfunctioning EBTCM.

Diagnostic Aids: An intermittent malfunction in this DTC may result from a mechanical part of the system that sticks, binds, or slips.

The frequency of the malfunction can be checked by using the enhanced diagnostic function of the Tech 1

The static modulator test function of the Tech 1 or T-100 may be used to locate an intermittent malfunction associated with the ESB.

DIAGNOSTIC CHART DTC 42 — 1994 SILHOUETTE, TRANS SPORT AND LUMINA APV WITH TCS, CONTINUED

DTC 42
REAR ESB WILL NOT HOLD MOTOR

(1)
- RAISE THE VEHICLE SUCH THAT THE REAR WHEELS ARE APPROXIMATELY SIX INCHES OFF THE FLOOR.
- IGNITION "ON," ENGINE "OFF."
- PUMP BRAKE PEDAL UNTIL FIRM TO DEPLETE VACUUM RESERVE.
- APPLY FIRM PRESSURE ON BRAKE PEDAL AND RELEASE.
- REAPPLY VERY FIRM PRESSURE ON BRAKE PEDAL AND HAVE AN ASSISTANT TRY TO SPIN THE REAR WHEELS BY HAND.

COULD ASSISTANT SPIN THE REAR WHEELS?

YES: REPLACE ABS MOTOR PACK.

NO:

(2)
- IGNITION "OFF."
- DISCONNECT 6-WAY ABS MOTOR PACK CONNECTOR AND GRAY 8-WAY EBTCM CONNECTOR.
- USING J 39200, MEASURE THE RESISTANCE BETWEEN TERMINAL "C" OF THE 6-WAY ABS MOTOR PACK HARNESS CONNECTOR AND TERMINAL "C" OF THE GRAY 8-WAY EBTCM HARNESS CONNECTOR.

IS RESISTANCE 1.5Ω OR LESS?

NO: REPAIR HIGH RESISTANCE IN CKT 1284.

YES:

(3)
- USING J 39200, MEASURE THE RESISTANCE BETWEEN TERMINAL "D" OF THE 6-WAY ABS MOTOR PACK HARNESS CONNECTOR AND TERMINAL "F" OF THE GRAY 8-WAY EBTCM HARNESS CONNECTOR.

IS RESISTANCE 1.5Ω OR LESS?

NO: REPAIR HIGH RESISTANCE IN CKT 1285.

YES:

(4)
- USING J 39200, MEASURE THE RESISTANCE BETWEEN TERMINAL "C" AND TERMINAL "D" OF THE ABS MOTOR PACK CONNECTOR.

IS RESISTANCE 1.5Ω OR LESS?

NO: REPLACE ABS MOTOR PACK.

YES:

(5)
- INSPECT THE 6-WAY ABS MOTOR PACK CONNECTOR AND GRAY 8-WAY EBTCM CONNECTOR FOR POOR TERMINAL CONTACT OR CORROSION.

DO ANY TERMINALS EXHIBIT POOR CONTACT OR EVIDENCE OF CORROSION?

YES: REPLACE TERMINALS THAT EXHIBIT POOR TERMINAL CONTACT OR EVIDENCE OF CORROSION.

NO:

(6)
- IGNITION "OFF."
- RECONNECT ALL CONNECTORS.
- START ENGINE AND TEST DRIVE VEHICLE, ACHIEVING AT LEAST 16 km/h (10 MPH).
- REPEAT ABOVE DRIVE CYCLE SEQUENCE TWO MORE TIMES.

DOES DTC 42 SET IN THE LAST THREE DRIVE CYCLES?

YES: REPLACE EBTCM.

NO: MALFUNCTION WAS MOST LIKELY CAUSED BY A POOR CONNECTION. REFER TO "DIAGNOSTIC AIDS" ON FACING PAGE.

AFTER DIAGNOSIS IS COMPLETE, CLEAR DTCs AND TEST DRIVE VEHICLE FOR THREE (3) DRIVE CYCLES TO VERIFY DTC DOES NOT RESET. A DRIVE CYCLE CONSISTS OF STARTING THE VEHICLE, DRIVING OVER 16 km/h (10 MPH) AND THEN TURNING KEY OFF.

DIAGNOSTIC CHART DTC 44 — 1994 SILHOUETTE, TRANS SPORT AND LUMINA APV WITH TCS

DTC 44

LEFT FRONT ABS CHANNEL WILL NOT MOVE

Circuit Description:

This DTC is designed to detect bound-up ESB, a stuck ABS motor, or a seized hydraulic modulator. When the release is commanded during initialization, the ESB should release the ABS motor, resulting in sensed current being less than commanded current (motor is spinning freely). If the ABS motor is not moving, sensed current will be equal to stall current.

DTC Will Set When: DTC 44 can be set during normal operation. If the EBTCM detects a condition in which it cannot move the ABS motor in either direction, a malfunction exists.

Action Taken: This malfunction indicates the channel cannot be moved properly. A malfunction DTC is stored, ABS is disabled and the ABS warning lamp is turned "ON."

DTC Chart Test Description: Number(s) below refer to circled number(s) on the diagnostic chart.

1. Checks for proper ABS motor movement during apply and release commands from the Tech 1.
2. Compares EBTCM command current to ABS motor feedback current.
3. Checks for a short circuit between the ABS motor "HIGH" and motor "LOW" circuits.
4. Checks for an ABS motor that is internally shorted.
5. Determines if the malfunction is due to poor terminal contact or the EBTCM.
6. Verifies that the ABS motor can actually apply when commanded.
7. Checks for proper hydraulic modulator gear and piston movement.

Diagnostic Aids: An intermittent malfunction in this DTC may result from a mechanical part of the system that sticks, binds, or slips.

The frequency of the malfunction can be checked by using the enhanced diagnostic function of the Tech 1, as described in "Tech 1 Diagnostics" in this section. DTC 44 may set after modulator disassembly if the modulator pistons are positioned at the bottom of their bore.

Any circuitry, that is suspected as causing the intermittent complaint, should be thoroughly checked for backed out terminals, improper mating, broken locks, improperly formed or damaged terminals, poor terminal to wiring connections or physical damage to the wiring harness.

DIAGNOSTIC CHART DTC 44 — 1994 SILHOUETTE, TRANS SPORT AND LUMINA APV WITH TCS, CONTINUED

DTC 44 — LEFT FRONT ABS CHANNEL WILL NOT MOVE

NOTICE: J 39200 TEST LEADS MUST BE "ZEROED" PRIOR TO MAKING ANY RESISTANCE MEASUREMENTS. REFER TO J 39200 USER'S MANUAL.

(1)
- IGNITION "ON," ENGINE "OFF."
- PUMP BRAKE PEDAL UNTIL FIRM TO DEPLETE VACUUM RESERVE.
- USING TECH 1 MISC. TESTS, SELECT MANUAL CONTROL FUNCTION.
- SELECT LEFT FRONT ABS MOTOR APPLY FUNCTION AND APPLY THE MOTOR.
- APPLY FIRM PRESSURE ON BRAKE PEDAL.
- SELECT LEFT FRONT ABS MOTOR RELEASE FUNCTION AND RELEASE MOTOR.

DID BRAKE PEDAL FALL?

- NO → go to (2)
- YES → (6)
  - FIRM PRESSURE STILL "ON" BRAKE PEDAL.
  - SELECT LEFT FRONT MOTOR APPLY FUNCTION AND APPLY THE MOTOR.

  DID BRAKE PEDAL RISE?
  - NO → go to (2)
  - YES → MALFUNCTION IS INTERMITTENT. REFER TO "DIAGNOSTIC AIDS" ON FACING PAGE.

(2)
- REMOVE FOOT FROM BRAKE PEDAL.
- SELECT LEFT FRONT ABS MOTOR APPLY FUNCTION AND APPLY THE MOTOR WHILE CAREFULLY OBSERVING THE COMMANDED CURRENT AND THE FEEDBACK CURRENT ON THE TECH 1.

WAS THE FEEDBACK CURRENT HIGHER THAN THE COMMANDED CURRENT?

- YES → (3)
  - EXIT TECH 1 MANUAL CONTROL.
  - IGNITION "OFF."
  - DISCONNECT THE GRAY 8-WAY EBTCM CONNECTOR AND THE 6-WAY ABS MOTOR PACK CONNECTOR.
  - USING J 39200, MEASURE THE RESISTANCE BETWEEN TERMINAL "A" AND TERMINAL "B" OF THE GRAY 8-WAY EBTCM HARNESS CONNECTOR.

  DOES J 39200 DISPLAY RESISTANCE AS "OL" (INFINITE)?
  - NO → REPAIR SHORT BETWEEN CKT 1280 AND CKT 1281.
  - NO → (4)
    - USING J 39200, MEASURE THE RESISTANCE BETWEEN TERMINAL "A" AND TERMINAL "B" OF THE ABS MOTOR PACK CONNECTOR.
    - NO → REPLACE ABS MOTOR PACK.
    - YES → (5)
      - IGNITION "OFF."
      - RECONNECT ALL CONNECTORS.
      - START ENGINE AND TEST DRIVE VEHICLE, ACHIEVING AT LEAST 16 km/h (10 MPH).
      - REPEAT ABOVE DRIVE CYCLE SEQUENCE TWO MORE TIMES.

      DOES DTC 44 SET IN THE LAST THREE DRIVE CYCLES?
      - YES → REPLACE EBTCM.
      - NO → MALFUNCTION IS INTERMITTENT. INSPECT ALL CONNECTORS FOR CONTAMINATION AND POOR CONTACT. REPLACE ANY TERMINALS THAT EXHIBIT POOR TERMINAL CONTACT OR EVIDENCE OF CORROSION.
- NO → (7)
  - USING TECH 1 MISC. TESTS, PERFORM GEAR TENSION RELIEF FUNCTION.
  - IGNITION "OFF."
  - REMOVE HYDRAULIC MODULATOR/MASTER CYLINDER ASSEMBLY FROM VEHICLE. REFER TO "ON VEHICLE SERVICE" IN THIS SECTION.
  - SEPARATE ABS MOTOR PACK FROM HYDRAULIC MODULATOR.
  - GRASP THE GEAR ON THE HYDRAULIC MODULATOR (THE HYDRAULIC MODULATOR IS THE UNIT WITH THE THREE LARGE GEARS) FURTHEST FORWARD WHEN THE UNIT IS INSTALLED IN THE VEHICLE AND ATTEMPT TO MOVE THE GEAR.

  CAN THE GEAR BE ROTATED AT LEAST 10 FULL TURNS LOCK TO LOCK?
  - NO → REPLACE HYDRAULIC MODULATOR.
  - YES → REPLACE ABS MOTOR PACK.

AFTER DIAGNOSIS IS COMPLETE, CLEAR DTCs AND TEST DRIVE VEHICLE FOR THREE (3) DRIVE CYCLES TO VERIFY THAT DTC DOES NOT RESET. A DRIVE CYCLE CONSISTS OF STARTING THE VEHICLE, DRIVING OVER 16 km/h (10 MPH), AND THEN TURNING KEY OFF.

DIAGNOSTIC CHART DTC 45 — 1994 SILHOUETTE, TRANS SPORT AND LUMINA APV WITH TCS

DTC 45

RIGHT FRONT ABS CHANNEL WILL NOT MOVE

Circuit Description:
This DTC is designed to detect bound-up ESB, a stuck ABS motor, or a seized hydraulic modulator. When the release is commanded during initialization, the ESB should release the ABS motor, resulting in sensed current being less than commanded current (ABS motor is spinning freely). If the ABS motor is not moving, sensed current will be equal to stall current.

DTC Will Set When: DTC 45 can be set during normal operation. If the EBTCM detects a condition in which it cannot move the ABS motor in either direction, a malfunction exists.

Action Taken: This malfunction indicates the channel cannot be moved properly. A malfunction DTC is stored, ABS is disabled and the ABS warning lamp is turned "ON."

DTC Chart Test Description: Number(s) below refer to circled number(s) on the diagnostic chart.

1. Checks for proper ABS motor movement during apply and release commands from the Tech 1.
2. Compares EBTCM command current to ABS motor feedback current.
3. Checks for a short circuit between the ABS motor "HIGH" and motor "LOW" circuits.
4. Checks for an ABS motor that is internally shorted.
5. Checks to see if malfunction is due to poor terminal contact or the EBTCM.
6. Verifies that the ABS motor can actually apply when commanded.
7. Checks for proper hydraulic modulator gear and piston movement.

Diagnostic Aids: An intermittent malfunction in this DTC may result from a mechanical part of the system that sticks, binds, or slips.

The frequency of the malfunction can be checked by using the enhanced diagnostic function of the Tech 1, as described in "Tech 1 Diagnostics" in this section. DTC 45 may set after modulator disassembly if the modulator pistons are positioned at the bottom of their bore.

Any circuitry, that is suspected as causing the intermittent complaint, should be thoroughly checked for backed out terminals, improper mating, broken locks, improperly formed or damaged terminals, poor terminal to wiring connections or physical damage to the wiring harness.

DIAGNOSTIC CHART DTC 45 — 1994 SILHOUETTE, TRANS SPORT AND LUMINA APV WITH TCS, CONTINUED

DTC 45
RIGHT FRONT ABS CHANNEL WILL NOT MOVE

NOTICE: J 39200 TEST LEADS MUST BE "ZEROED" PRIOR TO MAKING ANY RESISTANCE MEASUREMENTS. REFER TO J 39200 USER'S MANUAL.

(1)
- IGNITION "ON," ENGINE "OFF."
- PUMP BRAKE PEDAL UNTIL FIRM TO DEPLETE VACUUM RESERVE.
- USING TECH 1 MISC. TESTS, SELECT MANUAL CONTROL FUNCTION.
- SELECT RIGHT FRONT ABS MOTOR APPLY FUNCTION AND APPLY THE MOTOR.
- APPLY FIRM PRESSURE ON BRAKE PEDAL.
- SELECT RIGHT FRONT ABS MOTOR RELEASE FUNCTION AND RELEASE MOTOR. DID BRAKE PEDAL FALL?

YES → (6)
- FIRM PRESSURE STILL "ON" BRAKE PEDAL.
- SELECT RIGHT FRONT ABS MOTOR APPLY FUNCTION AND APPLY THE MOTOR. DID BRAKE PEDAL RISE?

 YES → MALFUNCTION IS INTERMITTENT. REFER TO "DIAGNOSTIC AIDS" ON FACING PAGE.
 NO → (2)

NO → (2)
- REMOVE FOOT FROM BRAKE PEDAL.
- SELECT RIGHT FRONT ABS MOTOR APPLY FUNCTION AND APPLY THE MOTOR WHILE CAREFULLY OBSERVING THE COMMANDED CURRENT AND THE FEEDBACK CURRENT ON THE TECH 1. WAS THE FEEDBACK CURRENT HIGHER THAN THE COMMANDED CURRENT?

YES → (3)
- EXIT TECH 1 MANUAL CONTROL FUNCTION.
- IGNITION "OFF."
- DISCONNECT THE GRAY 8-WAY EBTCM CONNECTOR AND THE 6-WAY ABS MOTOR PACK CONNECTOR.
- USING J 39200, MEASURE THE RESISTANCE BETWEEN TERMINAL "G" AND TERMINAL "H" OF THE GRAY 8-WAY EBTCM HARNESS CONNECTOR. DOES J 39200 DISPLAY RESISTANCE AS "OL" (INFINITE)?

 NO → REPAIR SHORT BETWEEN CKT 1282 AND CKT 1283.

 YES → (4)
 - USING J 39200, MEASURE THE RESISTANCE BETWEEN TERMINAL "E" AND TERMINAL "F" OF THE ABS MOTOR PACK CONNECTOR. IS RESISTANCE 0.4Ω OR GREATER?

  NO → REPLACE ABS MOTOR PACK.

  YES → (5)
  - IGNITION "OFF."
  - RECONNECT ALL CONNECTORS.
  - START ENGINE AND TEST DRIVE VEHICLE, ACHIEVING AT LEAST 16 km/h (10 MPH).
  - REPEAT ABOVE DRIVE CYCLE SEQUENCE TWO MORE TIMES. DOES DTC 45 SET IN THE LAST THREE DRIVE CYCLES?

   YES → REPLACE EBTCM.
   NO → MALFUNCTION IS INTERMITTENT. INSPECT ALL CONNECTORS FOR CONTAMINATION AND POOR TERMINAL CONTACT. REPLACE ANY TERMINALS THAT EXHIBIT POOR TERMINAL CONTACT OR EVIDENCE OF CORROSION.

NO → (7)
- USING TECH 1 MISC. TESTS, PERFORM GEAR TENSION RELIEF FUNCTION.
- IGNITION "OFF."
- REMOVE HYDRAULIC MODULATOR/MASTER CYLINDER ASSEMBLY FROM VEHICLE.
- SEPARATE ABS MOTOR PACK FROM HYDRAULIC MODULATOR.
- GRASP THE GEAR ON THE HYDRAULIC MODULATOR (THE HYDRAULIC MODULATOR IS THE UNIT WITH THE THREE LARGE GEARS) FURTHEST REARWARD WHEN THE UNIT IS INSTALLED IN THE VEHICLE AND ATTEMPT TO MOVE THE GEAR. CAN THE GEAR BE ROTATED AT LEAST 10 FULL TURNS LOCK TO LOCK?

 NO → REPLACE HYDRAULIC MODULATOR.
 YES → REPLACE ABS MOTOR PACK.

AFTER DIAGNOSIS IS COMPLETE, CLEAR DTCs AND TEST DRIVE VEHICLE FOR THREE (3) DRIVE CYCLES TO VERIFY THAT DTC DOES NOT RESET. A DRIVE CYCLE CONSISTS OF STARTING THE VEHICLE, DRIVING OVER 16 km/h (10 MPH), AND THEN TURNING KEY OFF.

DIAGNOSTIC CHART DTC 46 — 1994 SILHOUETTE, TRANS SPORT AND LUMINA APV WITH TCS

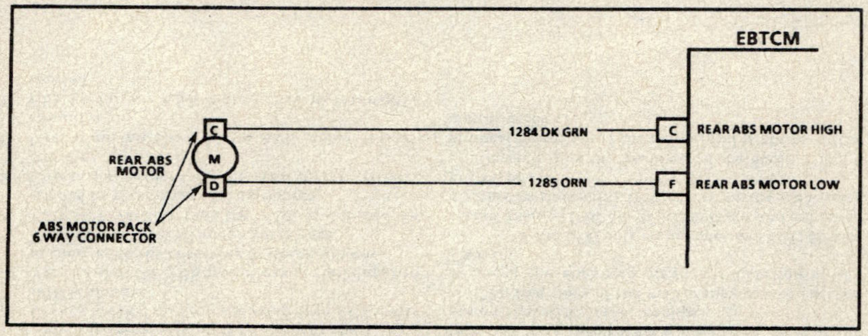

DTC 46

REAR ABS CHANNEL WILL NOT MOVE

Circuit Description:
This DTC is designed to detect bound-up ESB, a stuck ABS motor, or a seized hydraulic modulator. When the release is commanded during initialization, the ESB should release the ABS motor, resulting in sensed current being less than commanded current (motor is spinning freely). If the ABS motor is not moving, sensed current will be equal to stall current.

DTC Will Set When: DTC 46 can be set during normal operation. If the EBTCM detects a condition in which it cannot move the ABS motor in either direction, a malfunction exists.

Action Taken: This malfunction indicates the channel cannot be moved properly. A malfunction DTC is stored, ABS is disabled and the ABS warning lamp is turned "ON."

DTC Chart Test Description: Number(s) below refer to circled number(s) on the diagnostic chart.
1. Checks for proper motor movement during apply and release commands from the Tech 1.
2. Compares EBTCM command current to ABS motor feedback current.
3. Checks for a short circuit between the ABS motor "HIGH" and ABS motor "LOW" circuits.
4. Checks for an ABS motor that is internally shorted.
5. Checks to see if malfunction is due to poor terminal contact or the EBTCM.
6. This step verifies that the motor can actually release when commanded.
7. Checks for proper hydraulic modulator gear and piston movement.

Diagnostic Aids: An intermittent malfunction in this DTC may result from a mechanical part of the system that sticks or binds.

The frequency of the malfunction can be checked by using the enhanced diagnostic function of the Tech 1

DTC 46 may set after modulator disassembly if the modulator pistons are positioned at the bottom of their bore.

Depending on the frequency of the malfunction, a physical inspection of the mechanical parts suspected may be necessary.

Any circuitry, that is suspected as causing the intermittent complaint, should be thoroughly checked for backed out terminals, improper mating, broken locks, improperly formed or damaged terminals, poor terminal to wiring connections or physical damage to the wiring harness.

DIAGNOSTIC CHART DTC 46 — 1994 SILHOUETTE, TRANS SPORT AND LUMINA APV WITH TCS, CONTINUED

DTC 46
REAR ABS CHANNEL WILL NOT MOVE

NOTICE: J 39200 TEST LEADS MUST BE "ZEROED" PRIOR TO MAKING ANY RESISTANCE MEASUREMENTS. REFER TO J 39200 USER'S MANUAL.

(1)
- RAISE AND SUPPORT THE VEHICLE SUCH THAT THE REAR WHEELS ARE APPROXIMATELY SIX INCHES OFF THE FLOOR.
- IGNITION "ON," ENGINE "OFF."
- PUMP BRAKE PEDAL UNTIL FIRM TO DEPLETE VACUUM RESERVE.
- APPLY FIRM PRESSURE ON BRAKE PEDAL.
- USING TECH 1 MISC. TESTS, SELECT MANUAL CONTROL FUNCTION.
- SELECT REAR AXLE ABS MOTOR APPLY FUNCTION AND APPLY THE MOTOR.
- HAVE AN ASSISTANT TRY TO SPIN THE REAR WHEELS BY HAND. COULD ASSISTANT SPIN THE REAR WHEELS?

YES → (2)
NO → (6)

(6)
- FIRM PRESSURE STILL "ON" BRAKE PEDAL.
- SELECT REAR AXLE ABS MOTOR RELEASE FUNCTION AND RELEASE THE MOTOR.
- HAVE AN ASSISTANT TRY TO SPIN THE REAR WHEELS BY HAND. COULD ASSISTANT SPIN THE REAR WHEELS?

NO → (2)
YES → MALFUNCTION IS INTERMITTENT. REFER TO "DIAGNOSTIC AIDS" ON FACING PAGE.

(2)
- REMOVE FOOT FROM BRAKE PEDAL.
- SELECT REAR AXLE ABS MOTOR APPLY FUNCTION AND APPLY THE MOTOR WHILE CAREFULLY OBSERVING THE COMMANDED CURRENT AND THE FEEDBACK CURRENT ON THE TECH 1. WAS THE FEEDBACK CURRENT HIGHER THAN THE COMMANDED CURRENT?

YES → (3)
NO → (7)

(3)
- EXIT TECH 1 MANUAL CONTROL FUNCTION.
- IGNITION "OFF."
- DISCONNECT THE GRAY 8-WAY EBTCM CONNECTOR AND THE 6-WAY ABS MOTOR PACK CONNECTOR.
- USING J 39200, MEASURE THE RESISTANCE BETWEEN TERMINAL "C" AND TERMINAL "F" OF THE GRAY 8-WAY EBTCM HARNESS CONNECTOR. DOES J 39200 DISPLAY RESISTANCE AS "OL" (INFINITE)?

YES → (4)
NO → REPAIR SHORT BETWEEN CKT 1284 AND CKT 1285.

(4)
- USING J 39200, MEASURE THE RESISTANCE BETWEEN TERMINAL "C" AND TERMINAL "D" OF THE ABS MOTOR PACK CONNECTOR. IS RESISTANCE GREATER THAN 0.4Ω?

YES → (5)
NO → REPLACE ABS MOTOR PACK.

(5)
- RECONNECT ALL CONNECTORS.
- IGNITION "OFF."
- START ENGINE AND TEST DRIVE VEHICLE, ACHIEVING AT LEAST 16 km/h (10 MPH).
- REPEAT ABOVE DRIVE CYCLE SEQUENCE TWO MORE TIMES. DOES DTC 46 SET IN THE LAST THREE DRIVE CYCLES?

YES → REPLACE EBTCM.
NO → MALFUNCTION IS INTERMITTENT. INSPECT ALL CONNECTORS FOR CONTAMINATION AND POOR TERMINAL CONTACT. REPLACE ANY TERMINALS THAT EXHIBIT POOR TERMINAL CONTACT OR EVIDENCE OF CORROSION.

(7)
- USING TECH 1 MISC. TESTS, PERFORM GEAR TENSION RELIEF FUNCTION.
- IGNITION "OFF."
- REMOVE HYDRAULIC MODULATOR/MASTER CYLINDER ASSEMBLY FROM VEHICLE.
- SEPARATE ABS MOTOR PACK FROM HYDRAULIC MODULATOR.
- GRASP THE MIDDLE GEAR ON THE HYDRAULIC MODULATOR (THE HYDRAULIC MODULATOR IS THE UNIT WITH THE THREE LARGE GEARS) AND ATTEMPT TO MOVE THE GEAR. CAN THE GEAR BE ROTATED AT LEAST 7 FULL TURNS LOCK TO LOCK?

NO → REPLACE HYDRAULIC MODULATOR.
YES → REPLACE ABS MOTOR PACK.

AFTER DIAGNOSIS IS COMPLETE, CLEAR DTCs AND TEST DRIVE VEHICLE FOR THREE (3) DRIVE CYCLES TO VERIFY THAT DTC DOES NOT RESET. A DRIVE CYCLE CONSISTS OF STARTING THE VEHICLE, DRIVING OVER 16 km/h (10 MPH), AND THEN TURNING KEY OFF.

DIAGNOSTIC CHART DTC 47 — 1994 SILHOUETTE, TRANS SPORT AND LUMINA APV WITH TCS

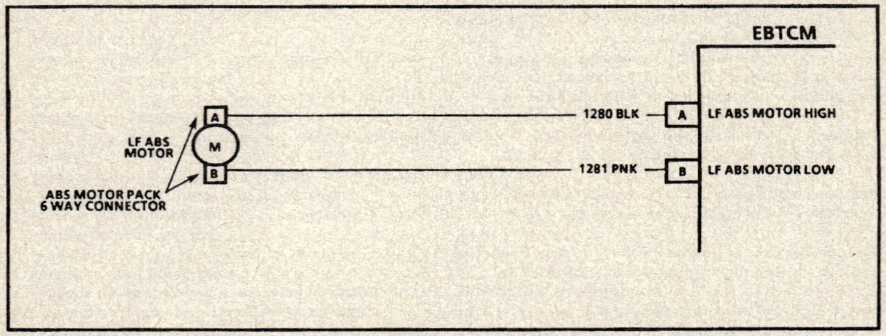

DTC 47

(Page 1 of 2)

LEFT FRONT ABS MOTOR FREE SPINS

Circuit Description:

This DTC is designed to detect a stripped nut or gear assembly during initialization. During the homing sequence, the piston should reach the top of the bore resulting in a stalled ABS motor. If this does not occur, the ABS motor must be spinning with little or no resistance, indicating a nut/screw or gear malfunction.

DTC Will Set When: DTC 47 can only be set during initialization. If the feedback current is less than the command current for a specified period of time, a malfunction exists.

Action Taken: A malfunction DTC is stored, ABS is disabled, and the ABS warning lamp is turned "ON."

DTC Chart Test Description: Number(s) below refer to circled number(s) on the diagnostic chart.

1. Checks to see if the corresponding open ABS motor DTC is also set.
2. Verifies that the ABS motor was actually applied as commanded by observing feedback current.
3. Verifies that the ABS motor can release.
4. This step verifies that the ABS motor can be applied by observing pedal movement.
5. Checks for an open in the ABS motor "HIGH" circuitry.
6. Checks for an open in the ABS motor "LOW" circuitry.
7. Checks for an ABS motor that is internally shorted.

Diagnostic Aids: An intermittent malfunction in this DTC may result from a mechanical part of the system that sticks, binds, or slips.

The frequency of the malfunction can be checked by using the enhanced diagnostic function of the Tech 1.

If the DTC only occurs once and DTC 56 also occurs, refer to DTC 56. If intermittent and enhanced diagnostics shows this DTC occurs during ABS, refer to DTC 56.

Depending on the frequency of the malfunction, a physical inspection of the mechanical parts suspected may be necessary.

DIAGNOSTIC CHART DTC 47 — 1994 SILHOUETTE, TRANS SPORT AND LUMINA APV WITH TCS, CONTINUED

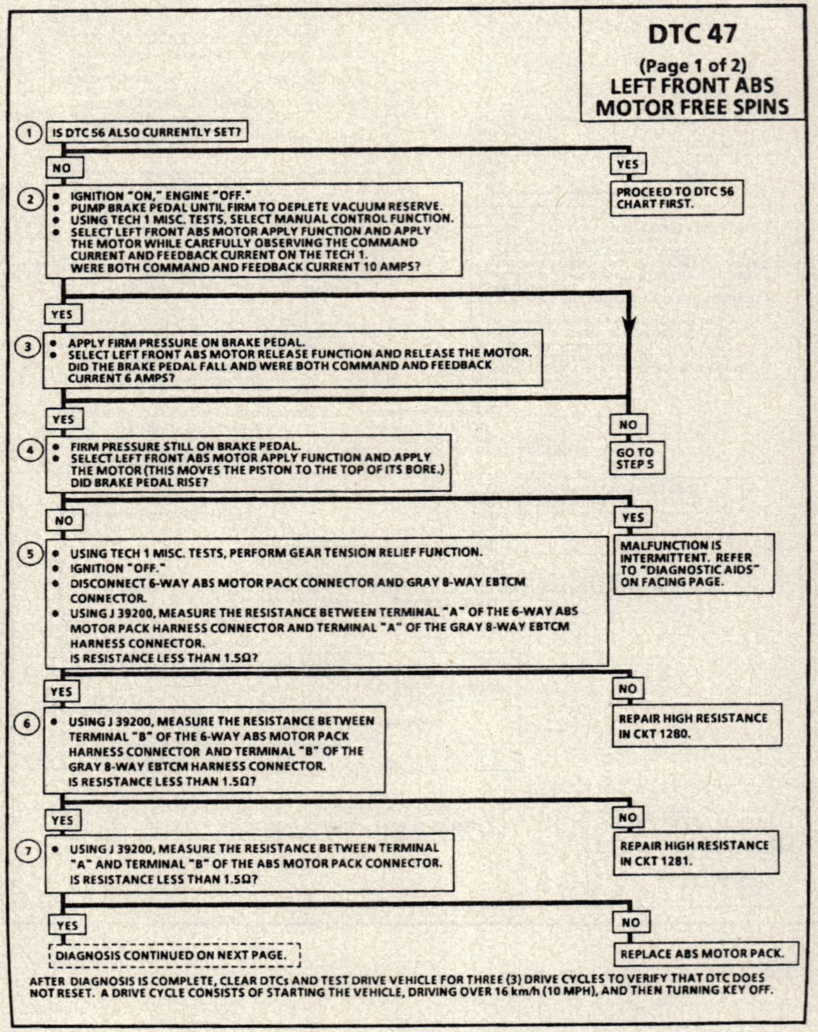

DIAGNOSTIC CHART DTC 47 — 1994 SILHOUETTE, TRANS SPORT AND LUMINA APV WITH TCS, VCONTINUED

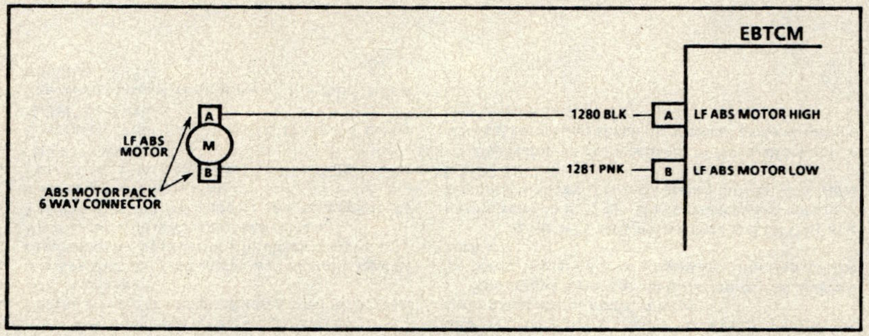

DTC 47
(Page 2 of 2)
LEFT FRONT ABS MOTOR FREE SPINS

Circuit Description:
This DTC is designed to detect a stripped nut or gear assembly during initialization. During the homing sequence, the piston should reach the top of the bore resulting in a stalled ABS motor. If this does not occur, the ABS motor must be spinning with little or no resistance, indicating a nut/screw or gear malfunction.

DTC Will Set When: DTC 47 can only be set during initialization. If the feedback current is less than the command current for a specified period of time, a malfunction exists.

Action Taken: A malfunction DTC is stored, ABS is disabled, and the ABS warning lamp is turned "ON."

DTC Chart Test Description: Number(s) below refer to circled number(s) on the diagnostic chart.

8. Checks for a stripped gear on the motor pack (unit with three small gears).
9. Checks for a stripped gear on the hydraulic modulator (unit with three large gears).
10. Checks to see if the ABS motor pack has an internal malfunction.
11. Verifies that the hydraulic modulator is functioning properly.
12. Ensures malfunction was not due to a poor terminal contact.

Diagnostic Aids: An intermittent malfunction in this DTC may result from a mechanical part of the system that sticks, binds, or slips.

The frequency of the malfunction can be checked by using the enhanced diagnostic function of the Tech 1.

If the DTC only occurs once and DTC 56 also occurs, refer to DTC 56. If intermittent and enhanced diagnostics shows this DTC occurs during ABS, refer to DTC 56.

Depending on the frequency of the malfunction, a physical inspection of the mechanical parts suspected may be necessary.

DIAGNOSTIC CHART DTC 47 — 1994 SILHOUETTE, TRANS SPORT AND LUMINA APV WITH TCS, CONTINUED

DTC 47
(Page 2 of 2)
LEFT FRONT ABS MOTOR FREE SPINS

DIAGNOSIS CONTINUED FROM PREVIOUS PAGE.

(8)
- REMOVE HYDRAULIC MODULATOR/MASTER CYLINDER ASSEMBLY FROM VEHICLE.
- REMOVE GEAR COVER.
- CHECK FOR STRIPPED GEAR ON ABS MOTOR PACK; ABS MOTOR PACK IS THE UNIT WITH THE THREE SMALL GEARS. THE LEFT FRONT GEAR IS THE GEAR FURTHEST FORWARD WHEN THE ASSEMBLY IS INSTALLED IN THE VEHICLE.

IS THE ABS MOTOR PACK GEAR STRIPPED?

- YES: REPLACE ABS MOTOR PACK.
- NO: go to (9)

(9)
- CHECK FOR A STRIPPED GEAR ON HYDRAULIC MODULATOR; MODULATOR IS THE UNIT WITH THE THREE LARGE GEARS. THE LEFT FRONT GEAR IS THE GEAR FURTHEST FORWARD WHEN THE UNIT IS INSTALLED IN THE VEHICLE.

IS THE MODULATOR GEAR STRIPPED?

- YES: REPLACE HYDRAULIC MODULATOR GEAR.
- NO: go to (10)

(10)
- RECONNECT ELECTRICAL CONNECTORS
- SECURELY POSITION HYDRAULIC MODULATOR/MASTER CYLINDER ASSEMBLY IN VEHICLE WITH GEAR COVER REMOVED SUCH THAT THE GEAR SET CAN BE OBSERVED. BE CAREFUL NOT TO DAMAGE GEAR SET OR ALLOW THE HYDRAULIC MODULATOR TO MOVE WHILE TESTING.
- IGNITION "ON."
- USING TECH 1 MISC. TESTS, SELECT MANUAL CONTROL FUNCTION.
- SELECT LEFT FRONT ABS MOTOR APPLY AND APPLY THE MOTOR WHILE OBSERVING THE GEAR SET FURTHEST FORWARD WHEN THE UNIT IS INSTALLED IN THE VEHICLE.
- SELECT LEFT ABS FRONT MOTOR RELEASE FUNCTION AND RELEASE THE MOTOR WHILE STILL OBSERVING THE GEAR SET.
- SELECT LEFT FRONT ABS MOTOR APPLY FUNCTION AND APPLY THE MOTOR.

DID GEAR SET MOVE IN BOTH DIRECTIONS FOR AT LEAST ONE REVOLUTION?

- NO: REPLACE ABS MOTOR PACK.
- YES: go to (11)

(11)
- USING TECH 1 MISC. TESTS, PERFORM GEAR TENSION RELIEF FUNCTION.
- IGNITION "OFF."
- SEPARATE THE ABS MOTOR PACK FROM THE HYDRAULIC MODULATOR.
- GRASP THE GEAR ON THE HYDRAULIC MODULATOR (THE HYDRAULIC MODULATOR IS THE UNIT WITH THE THREE LARGE GEARS) FURTHEST FORWARD WHEN THE UNIT IS INSTALLED IN THE VEHICLE AND ATTEMPT TO MOVE THE GEAR.

CAN THE GEAR BE ROTATED MORE THAN 13 FULL TURNS LOCK TO LOCK?

- YES: REPLACE THE HYDRAULIC MODULATOR.
- NO: go to (12)

(12)
- INSPECT ALL CONNECTORS AND TERMINALS FOR POOR TERMINAL CONTACT AND EVIDENCE OF CORROSION.

  REPLACE ALL TERMINALS THAT EXHIBIT SIGNS OF POOR TERMINAL CONTACT OR CORROSION.
- REINSTALL HYDRAULIC MODULATOR/MASTER CYLINDER ASSEMBLY IN VEHICLE.
- RECONNECT ALL CONNECTORS.
- IGNITION "OFF."
- START ENGINE AND TEST DRIVE VEHICLE, ACHIEVING AT LEAST 16 km/h (10 MPH).
- REPEAT ABOVE DRIVE CYCLE SEQUENCE TWO MORE TIMES.

DOES DTC 47 SET IN THE LAST THREE DRIVE CYCLES?

- YES: REPLACE EBTCM.
- NO: MALFUNCTION WAS MOSTLY LIKELY CAUSED BY A POOR CONNECTION. REFER TO "DIAGNOSTIC AIDS" ON FACING PAGE.

AFTER DIAGNOSIS IS COMPLETE, CLEAR DTCs AND TEST DRIVE VEHICLE FOR THREE (3) DRIVE CYCLES TO VERIFY THAT DTC DOES NOT RESET. A DRIVE CYCLE CONSISTS OF STARTING THE VEHICLE, DRIVING OVER 16 KM/H (10 MPH), AND THEN TURNING KEY OFF.

DIAGNOSTIC CHART DTC 48 — 1994 SILHOUETTE, TRANS SPORT AND LUMINA APV WITH TCS

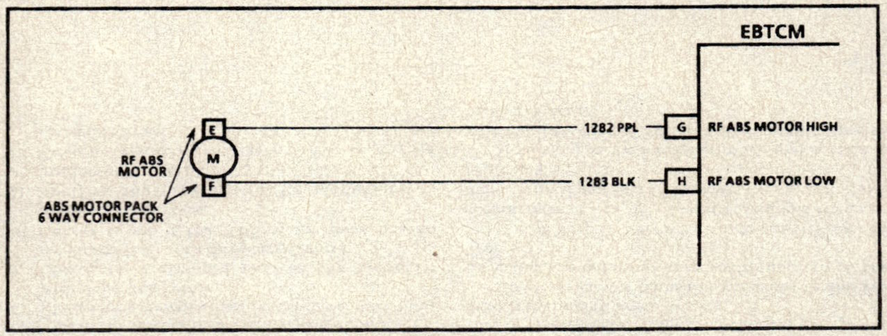

DTC 48

(Page 1 of 2)

RIGHT FRONT ABS MOTOR FREE SPINS

Circuit Description:
This DTC is designed to detect a stripped nut or gear assembly during initialization. During the homing sequence, the piston should reach the top of the bore resulting in a stalled ABS motor. If this does not occur, the ABS motor must be spinning with little or no resistance, indicating a nut/screw or gear malfunction.

DTC Will Set When: DTC 48 can only be set during initialization. If the feedback current is less than the command current for a specified period of time, a malfunction exists.

Action Taken: A malfunction DTC is stored, ABS is disabled, and the ABS warning lamp is turned "ON."

DTC Chart Test Description: Number(s) below refer to circled number(s) on the diagnostic chart.

1. Checks to see if the corresponding open ABS motor DTC is also set.
2. Verifies that the ABS motor was actually applied as commanded by observing feedback current.
3. Verifies that the ABS motor can release.
4. Verifies that the ABS motor can be applied by observing pedal movement.
5. Checks for high resistance in the ABS motor "HIGH" circuitry.
6. Checks for high resistance in the ABS motor "LOW" circuitry.
7. Checks for proper resistance of the ABS motor windings.

Diagnostic Aids: An intermittent malfunction in this DTC may result from a mechanical part of the system that sticks, binds, or slips.

The frequency of the malfunction can be checked by using the enhanced diagnostic function of the Tech 1.

If the DTC only occurs once and DTC 61 also occurs, refer to DTC 61. If intermittent and enhanced diagnostics shows this DTC occurs during ABS, refer to DTC 61.

Depending on the frequency of the malfunction, a physical inspection of the mechanical parts suspected may be necessary.

DIAGNOSTIC CHART DTC 48 — 1994 SILHOUETTE, TRANS SPORT AND LUMINA APV WITH TCS, CONTINUED

DTC 48

(Page 1 of 2)

RIGHT FRONT ABS MOTOR FREE SPINS

(1) IS DTC 61 ALSO CURRENTLY SET?
- YES: PROCEED TO DTC 61 CHART FIRST.
- NO: go to step 2.

(2)
- IGNITION "ON," ENGINE "OFF."
- PUMP BRAKE PEDAL UNTIL FIRM TO DEPLETE VACUUM RESERVE.
- USING TECH 1 MISC. TESTS, SELECT MANUAL CONTROL FUNCTION.
- SELECT RIGHT FRONT ABS MOTOR APPLY FUNCTION AND APPLY THE MOTOR WHILE CAREFULLY OBSERVING THE COMMAND CURRENT AND FEEDBACK CURRENT ON THE TECH 1.

WERE BOTH COMMAND AND FEEDBACK CURRENT 10 AMPS?
- YES: go to step 3.

(3)
- APPLY FIRM PRESSURE ON BRAKE PEDAL.
- SELECT RIGHT FRONT ABS MOTOR RELEASE FUNCTION AND RELEASE THE MOTOR.

DID THE BRAKE PEDAL FALL AND WERE BOTH COMMAND AND FEEDBACK CURRENT 6 AMPS?
- YES: go to step 4.
- NO: GO TO STEP 5

(4)
- FIRM PRESSURE STILL ON BRAKE PEDAL.
- SELECT RIGHT ABS MOTOR APPLY FUNCTION AND APPLY THE MOTOR (THIS MOVES THE PISTON TO THE TOP OF ITS BORE.)

DID BRAKE PEDAL RISE?
- YES: MALFUNCTION IS INTERMITTENT. REFER TO "DIAGNOSTIC AIDS" ON FACING PAGE.
- NO: go to step 5.

(5)
- USING TECH 1 MISC. TESTS, PERFORM GEAR TENSION RELIEF FUNCTION.
- IGNITION "OFF."
- DISCONNECT 6-WAY ABS MOTOR PACK CONNECTOR AND GRAY 8-WAY EBTCM CONNECTOR.
- USING J 39200, MEASURE THE RESISTANCE BETWEEN TERMINAL "E" OF THE 6-WAY ABS MOTOR PACK HARNESS CONNECTOR AND TERMINAL "G" OF THE GRAY 8-WAY EBTCM HARNESS CONNECTOR.

IS RESISTANCE 1.5Ω OR LESS?
- YES: go to step 6.
- NO: REPAIR HIGH RESISTANCE IN CKT 1282.

(6)
- USING J 39200, MEASURE THE RESISTANCE BETWEEN TERMINAL "F" OF THE 6-WAY ABS MOTOR PACK HARNESS CONNECTOR AND TERMINAL "H" OF THE GRAY 8-WAY EBTCM HARNESS CONNECTOR.

IS RESISTANCE 1.5Ω OR LESS?
- YES: go to step 7.
- NO: REPAIR HIGH RESISTANCE IN CKT 1283.

(7)
- USING J 39200, MEASURE THE RESISTANCE BETWEEN TERMINAL "E" AND TERMINAL "F" OF THE ABS MOTOR PACK CONNECTOR.

IS RESISTANCE 1.5Ω OR LESS?
- YES: DIAGNOSIS CONTINUED ON NEXT PAGE.
- NO: REPLACE ABS MOTOR PACK.

AFTER DIAGNOSIS IS COMPLETE, CLEAR DTCs AND TEST DRIVE VEHICLE FOR THREE (3) DRIVE CYCLES TO VERIFY THAT DTC DOES NOT RESET. A DRIVE CYCLE CONSISTS OF STARTING THE VEHICLE, DRIVING OVER 16 km/h (10 MPH), AND THEN TURNING KEY OFF.

DIAGNOSTIC CHART DTC 48 — 1994 SILHOUETTE, TRANS SPORT AND LUMINA APV WITH TCS, CONTINUED

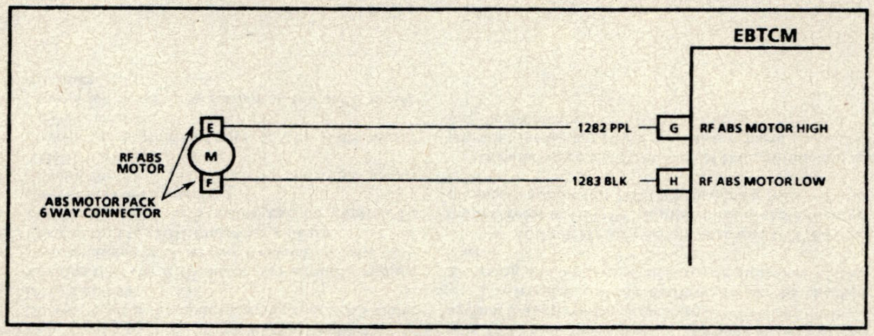

DTC 48
(Page 2 of 2)
RIGHT FRONT ABS MOTOR FREE SPINS

Circuit Description:

This DTC is designed to detect a stripped nut or gear assembly during initialization. During the homing sequence, the piston should reach the top of the bore resulting in a stalled ABS motor. If this does not occur, the ABS motor must be spinning with little or no resistance, indicating a nut/screw or gear malfunction.

DTC Will Set When: DTC 48 can only be set during initialization. If the feedback current is less than the command current for a specified period of time, a malfunction exists.

Action Taken: A malfunction DTC is stored, ABS is disabled, and the ABS warning lamp is turned "ON."

DTC Chart Test Description: Number(s) below refer to circled number(s) on the diagnostic chart.

8. Checks for a stripped gear on the ABS motor pack (unit with three small gears).
9. Checks for a stripped gear on the hydraulic modulator (unit with three large gears).
10. Checks to see if the ABS motor pack has an internal malfunction.
11. Verifies that the hydraulic modulator is functioning properly.
12. Ensures malfunction was not due to poor terminal contact.

Diagnostic Aids: An intermittent malfunction in this DTC may result from a mechanical part of the system that sticks, binds, or slips.

The frequency of the malfunction can be checked by using the enhanced diagnostic function of the Tech 1

If the DTC only occurs once and DTC 61 also occurs, refer to DTC 61. If intermittent and enhanced diagnostics shows this DTC occurs during ABS, refer to DTC 61.

Depending on the frequency of the malfunction, a physical inspection of the mechanical parts suspected may be necessary.

DIAGNOSTIC CHART DTC 48 — 1994 SILHOUETTE, TRANS SPORT AND LUMINA APV WITH TCS, CONTINUED

DTC 48
(Page 2 of 2)
RIGHT FRONT ABS MOTOR FREE SPINS

DIAGNOSIS CONTINUED FROM PREVIOUS PAGE.

(8)
- REMOVE HYDRAULIC MODULATOR/MASTER CYLINDER ASSEMBLY FROM VEHICLE.
- REMOVE GEAR COVER.
- CHECK FOR STRIPPED GEAR ON ABS MOTOR PACK; ABS MOTOR PACK IS THE UNIT WITH THE THREE SMALL GEARS. THE RIGHT FRONT GEAR IS THE GEAR FURTHEST REARWARD WHEN THE ASSEMBLY IS INSTALLED IN THE VEHICLE. IS THE ABS MOTOR PACK GEAR STRIPPED?

YES → REPLACE ABS MOTOR PACK.

NO ↓

(9)
- CHECK FOR A STRIPPED GEAR ON HYDRAULIC MODULATOR; MODULATOR IS THE UNIT WITH THE THREE LARGE GEARS. THE RIGHT FRONT GEAR IS THE GEAR FURTHEST REARWARD WHEN THE UNIT IS INSTALLED IN THE VEHICLE. IS THE MODULATOR GEAR STRIPPED?

YES → REPLACE HYDRAULIC MODULATOR GEAR.

NO ↓

(10)
- RECONNECT ELECTRICAL CONNECTORS
- SECURELY POSITION HYDRAULIC MODULATOR/MASTER CYLINDER ASSEMBLY IN VEHICLE WITH GEAR COVER REMOVED SUCH THAT THE GEAR SET CAN BE OBSERVED. BE CAREFUL NOT TO DAMAGE GEAR SET OR ALLOW THE HYDRAULIC MODULATOR TO MOVE WHILE TESTING.
- IGNITION "ON."
- USING TECH 1 MISC. TESTS, SELECT MANUAL CONTROL FUNCTION.
- SELECT RIGHT FRONT ABS MOTOR APPLY AND APPLY THE MOTOR WHILE OBSERVING THE GEAR SET FURTHEST REARWARD WHEN THE UNIT IS INSTALLED IN THE VEHICLE.
- SELECT RIGHT FRONT ABS MOTOR RELEASE FUNCTION AND RELEASE THE MOTOR WHILE STILL OBSERVING THE GEAR SET.
- SELECT RIGHT FRONT ABS MOTOR APPLY FUNCTION AND APPLY THE MOTOR. DID GEAR SET MOVE IN BOTH DIRECTIONS FOR AT LEAST ONE REVOLUTION?

NO → REPLACE ABS MOTOR PACK.

YES ↓

(11)
- USING TECH 1 MISC. TESTS, PERFORM GEAR TENSION RELIEF FUNCTION.
- IGNITION "OFF."
- SEPARATE THE ABS MOTOR PACK FROM THE HYDRAULIC MODULATOR.
- GRASP THE GEAR ON THE HYDRAULIC MODULATOR (THE HYDRAULIC MODULATOR IS THE UNIT WITH THE THREE LARGE GEARS) FURTHEST REARWARD WHEN THE UNIT IS INSTALLED IN THE VEHICLE AND ATTEMPT TO MOVE THE GEAR. CAN THE GEAR BE ROTATED MORE THAN 13 FULL TURNS LOCK TO LOCK?

YES → REPLACE THE HYDRAULIC MODULATOR.

NO ↓

(12)
- INSPECT ALL CONNECTORS AND TERMINALS FOR POOR TERMINAL CONTACT AND EVIDENCE OF CORROSION.
  REPLACE ALL TERMINALS THAT EXHIBIT SIGNS OF POOR TERMINAL CONTACT OR CORROSION.
- REINSTALL HYDRAULIC MODULATOR/MASTER CYLINDER ASSEMBLY IN VEHICLE.
- RECONNECT ALL CONNECTORS.
- IGNITION "OFF."
- START ENGINE AND TEST DRIVE VEHICLE, ACHIEVING AT LEAST 16 km/h (10 MPH).
- REPEAT ABOVE DRIVE CYCLE SEQUENCE TWO MORE TIMES. DOES DTC 48 SET IN THE LAST THREE DRIVE CYCLES?

YES → REPLACE EBTCM.

NO → MALFUNCTION WAS MOSTLY LIKELY CAUSED BY A POOR CONNECTION. REFER TO "DIAGNOSTIC AIDS" ON FACING PAGE.

AFTER DIAGNOSIS IS COMPLETE, CLEAR DTCs AND TEST DRIVE VEHICLE FOR THREE (3) DRIVE CYCLES TO VERIFY THAT DTC DOES NOT RESET. A DRIVE CYCLE CONSISTS OF STARTING THE VEHICLE, DRIVING OVER 16 KM/H (10 MPH), AND THEN TURNING KEY OFF.

DIAGNOSTIC CHART DTC 51 — 1994 SILHOUETTE, TRANS SPORT AND LUMINA APV WITH TCS

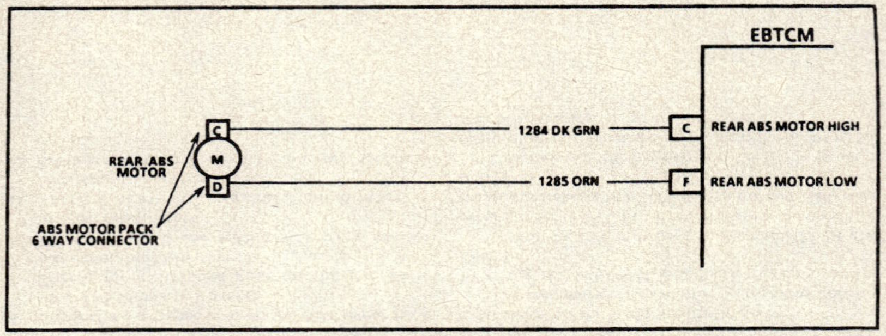

DTC 51
(Page 1 of 2)
REAR ABS MOTOR FREE SPINS

Circuit Description:
This DTC is designed to detect a stripped nut or gear assembly during initialization. During the homing sequence, the piston should reach the top of the bore resulting in a stalled ABS motor. If this does not occur, the ABS motor must be spinning with little or no resistance, indicating a nut/screw or gear malfunction.

DTC Will Set When: DTC 51 can be set during initialization. If the feedback current is less than the command current for a specified period of time, a malfunction exists.

Action Taken: A malfunction DTC is stored, ABS is disabled, and the ABS warning lamp is turned "ON."

DTC Chart Test Description: Number(s) below refer to circled number(s) on the diagnostic chart.
1. Checks to see if the corresponding open ABS motor DTC is also set.
2. Verifies that the ABS motor was actually applied as commanded by observing feedback current.
3. Verifies that the ABS motor can release.
4. Verifies that the ABS motor can be applied by observing pedal movement.
5. Checks for high resistance in the ABS motor "HIGH" circuitry.
6. Checks for high resistance in the ABS motor "LOW" circuitry.
7. Checks for proper resistance of the ABS motor windings.

Diagnostic Aids: An intermittent malfunction in this DTC may result from a mechanical part of the system that sticks, binds, or slips.

The frequency of the malfunction can be checked by using the enhanced diagnostic function of the Tech 1.

If the DTC only occurs once and DTC 64 also occurs, refer to DTC 64. If intermittent and enhanced diagnostics show this DTC occurs during ABS, refer to DTC 64.

Depending on the frequency of the malfunction, a physical inspection of the mechanical parts suspected may be necessary.

DIAGNOSTIC CHART DTC 51 — 1994 SILHOUETTE, TRANS SPORT AND LUMINA APV WITH TCS, CONTINUED

DTC 51
(Page 1 of 2)
REAR ABS MOTOR FREE SPINS

(1) IS DTC 64 ALSO CURRENTLY SET?
- YES: PROCEED TO DTC 64 CHART FIRST.
- NO:

(2)
- IGNITION "ON," ENGINE "OFF."
- PUMP BRAKE PEDAL UNTIL FIRM TO DEPLETE VACUUM RESERVE.
- USING TECH 1 MISC. TESTS, SELECT MANUAL CONTROL FUNCTION.
- SELECT REAR AXLE ABS MOTOR APPLY FUNCTION AND APPLY THE MOTOR WHILE CAREFULLY OBSERVING THE COMMAND CURRENT AND FEEDBACK CURRENT ON THE TECH 1.

WERE BOTH COMMAND AND FEEDBACK CURRENT 10 AMPS?
- YES:

(3)
- APPLY FIRM PRESSURE ON BRAKE PEDAL.
- SELECT REAR AXLE ABS MOTOR RELEASE FUNCTION AND RELEASE THE MOTOR.

DID THE BRAKE PEDAL FALL AND WERE BOTH COMMAND AND FEEDBACK CURRENT 6 AMPS?
- NO: GO TO STEP 5
- YES:

(4)
- FIRM PRESSURE STILL ON BRAKE PEDAL.
- SELECT REAR AXLE ABS MOTOR APPLY FUNCTION AND APPLY THE MOTOR (THIS MOVES THE PISTON TO THE TOP OF ITS BORE.)

DID BRAKE PEDAL RISE?
- YES: MALFUNCTION IS INTERMITTENT. REFER TO "DIAGNOSTIC AIDS" ON FACING PAGE.
- NO:

(5)
- USING TECH 1 MISC. TESTS, PERFORM GEAR TENSION RELIEF FUNCTION.
- IGNITION "OFF."
- DISCONNECT 6-WAY ABS MOTOR PACK CONNECTOR AND GRAY 8-WAY EBTCM CONNECTOR.
- USING J 39200, MEASURE THE RESISTANCE BETWEEN TERMINAL "C" OF THE 6-WAY ABS MOTOR PACK HARNESS CONNECTOR AND TERMINAL "C" OF THE GRAY 8-WAY EBTCM HARNESS CONNECTOR.

IS RESISTANCE 1.5Ω OR LESS?
- NO: REPAIR HIGH RESISTANCE IN CKT 1284.
- YES:

(6)
- USING J 39200, MEASURE THE RESISTANCE BETWEEN TERMINAL "D" OF THE 6-WAY ABS MOTOR PACK HARNESS CONNECTOR AND TERMINAL "F" OF THE GRAY 8-WAY EBTCM HARNESS CONNECTOR.

IS RESISTANCE 1.5Ω OR LESS?
- NO: REPAIR HIGH RESISTANCE IN CKT 1285.
- YES:

(7)
- USING J 39200, MEASURE THE RESISTANCE BETWEEN TERMINAL "C" AND TERMINAL "D" OF THE ABS MOTOR PACK CONNECTOR.

IS RESISTANCE 1.5Ω OR LESS?
- NO: REPLACE ABS MOTOR PACK.
- YES: DIAGNOSIS CONTINUED ON NEXT PAGE.

AFTER DIAGNOSIS IS COMPLETE, CLEAR DTCs AND TEST DRIVE VEHICLE FOR THREE (3) DRIVE CYCLES TO VERIFY THAT DTC DOES NOT RESET. A DRIVE CYCLE CONSISTS OF STARTING THE VEHICLE, DRIVING OVER 16 km/h (10 MPH), AND THEN TURNING KEY OFF.

DIAGNOSTIC CHART DTC 51 — 1994 SILHOUETTE, TRANS SPORT AND LUMINA APV WITH TCS, CONTINUED

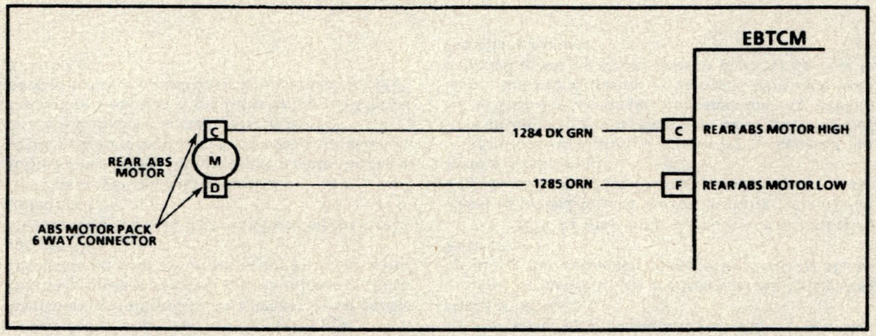

DTC 51
(Page 2 of 2)
REAR ABS MOTOR FREE SPINS

Circuit Description:
This DTC is designed to detect a stripped nut or gear assembly during initialization. During the homing sequence, the piston should reach the top of the bore resulting in a stalled ABS motor. If this does not occur, the ABS motor must be spinning with little or no resistance, indicating a nut/screw or gear malfunction.

DTC Will Set When: DTC 51 can be set during initialization. If the feedback current is less than the command current for a specified period of time, a malfunction exists.

Action Taken: A malfunction DTC is stored, ABS is disabled, and the ABS warning lamp is turned "ON."

DTC Chart Test Description: Number(s) below refer to circled number(s) on the diagnostic chart.

8. Checks for a stripped gear on the ABS motor pack (unit with three small gears).
9. Checks for a stripped gear on the hydraulic modulator (unit with three large gears).
10. Verifies that the ABS motor can be applied under a load situation.
11. Verifies that the hydraulic modulator is functioning properly.
12. Ensures malfunction was not due to poor terminal contact.

Diagnostic Aids: An intermittent malfunction in this DTC may result from a mechanical part of the system that sticks, binds, or slips.

The frequency of the malfunction can be checked by using the enhanced diagnostic function of the Tech 1.

If the DTC only occurs once and DTC 64 also occurs, refer to DTC 64. If intermittent and enhanced diagnostics show this DTC occurs during ABS, refer to DTC 64.

Depending on the frequency of the malfunction, a physical inspection of the mechanical parts suspected may be necessary.

DIAGNOSTIC CHART DTC 51 — 1994 SILHOUETTE, TRANS SPORT AND LUMINA APV WITH TCS, CONTINUED

DTC 51
(Page 2 of 2)
REAR ABS MOTOR FREE SPINS

DIAGNOSIS CONTINUED FROM PREVIOUS PAGE.

(8)
- REMOVE HYDRAULIC MODULATOR/MASTER CYLINDER ASSEMBLY FROM VEHICLE.
- REMOVE GEAR COVER.
- CHECK FOR STRIPPED GEAR ON ABS MOTOR PACK; ABS MOTOR PACK IS THE UNIT WITH THE THREE SMALL GEARS. THE REAR AXLE GEAR IS THE MIDDLE GEAR OF THE THREE.

IS THE ABS MOTOR PACK GEAR STRIPPED?

- YES: REPLACE ABS MOTOR PACK.
- NO: go to (9)

(9)
- CHECK FOR A STRIPPED GEAR ON HYDRAULIC MODULATOR; MODULATOR IS THE UNIT WITH THE THREE LARGE GEARS. THE REAR AXLE GEAR IS THE MIDDLE GEAR OF THE THREE.

IS THE MODULATOR GEAR STRIPPED?

- YES: REPLACE HYDRAULIC MODULATOR GEAR.
- NO: go to (10)

(10)
- RECONNECT ELECTRICAL CONNECTORS
- SECURELY POSITION HYDRAULIC MODULATOR/MASTER CYLINDER ASSEMBLY IN VEHICLE WITH GEAR COVER REMOVED SUCH THAT THE GEAR SET CAN BE OBSERVED. BE CAREFUL NOT TO DAMAGE GEAR SET OR ALLOW THE HYDRAULIC MODULATOR TO MOVE WHILE TESTING.
- IGNITION "ON."
- USING TECH 1 MISC. TESTS, SELECT MANUAL CONTROL FUNCTION.
- SELECT REAR AXLE ABS MOTOR APPLY AND APPLY THE MOTOR WHILE OBSERVING THE MIDDLE GEAR SET.
- SELECT REAR AXLE ABS MOTOR RELEASE FUNCTION AND RELEASE THE MOTOR WHILE STILL OBSERVING THE GEAR SET.
- SELECT REAR AXLE ABS MOTOR APPLY FUNCTION AND APPLY THE MOTOR.

DID GEAR SET MOVE IN BOTH DIRECTIONS FOR AT LEAST ONE REVOLUTION?

- NO: REPLACE ABS MOTOR PACK.
- YES: go to (11)

(11)
- USING TECH 1 MISC. TESTS, PERFORM GEAR TENSION RELIEF FUNCTION.
- IGNITION "OFF."
- SEPARATE THE ABS MOTOR PACK FROM THE HYDRAULIC MODULATOR.
- GRASP THE MIDDLE GEAR ON THE HYDRAULIC MODULATOR (THE HYDRAULIC MODULATOR IS THE UNIT WITH THE THREE LARGE GEARS) AND ATTEMPT TO MOVE THE GEAR.

CAN THE GEAR BE ROTATED MORE THAN 8 FULL TURNS LOCK TO LOCK?

- YES: REPLACE THE HYDRAULIC MODULATOR.
- NO: go to (12)

(12)
- INSPECT ALL CONNECTORS AND TERMINALS FOR POOR TERMINAL CONTACT AND EVIDENCE OF CORROSION.
  REPLACE ALL TERMINALS THAT EXHIBIT SIGNS OF POOR TERMINAL CONTACT OR CORROSION.
- REINSTALL HYDRAULIC MODULATOR/MASTER CYLINDER ASSEMBLY IN VEHICLE.
- RECONNECT ALL CONNECTORS.
- IGNITION "OFF."
- START ENGINE AND TEST DRIVE VEHICLE, ACHIEVING AT LEAST 16 km/h (10 MPH).
- REPEAT ABOVE DRIVE CYCLE SEQUENCE TWO MORE TIMES.

DOES DTC 51 RESET?

- YES: REPLACE EBTCM.
- NO: MALFUNCTION WAS MOSTLY LIKELY CAUSED BY A POOR CONNECTION. REFER TO "DIAGNOSTIC AIDS" ON FACING PAGE.

AFTER DIAGNOSIS IS COMPLETE, CLEAR DTCs AND TEST DRIVE VEHICLE FOR THREE (3) DRIVE CYCLES TO VERIFY THAT DTC DOES NOT RESET. A DRIVE CYCLE CONSISTS OF STARTING THE VEHICLE, DRIVING OVER 16 KM/H (10 MPH), AND THEN TURNING KEY OFF.

DIAGNOSTIC CHART DTC 52 — 1994 SILHOUETTE, TRANS SPORT AND LUMINA APV WITH TCS

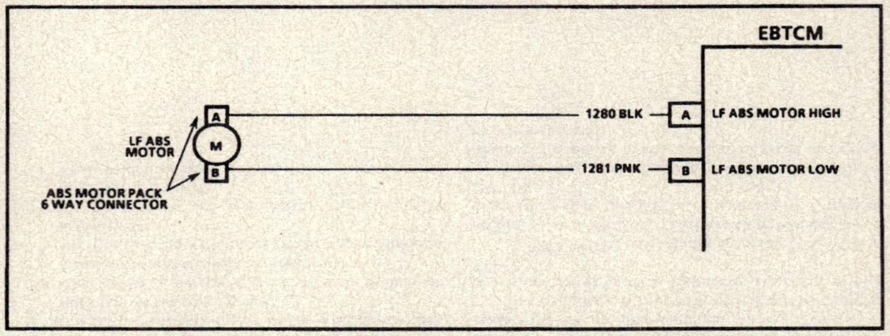

DTC 52

LEFT FRONT ABS CHANNEL IN RELEASE TOO LONG

Circuit Description:

This DTC will diagnose an ABS motor that is energized longer than expected. This could occur if a wheel speed sensor is malfunctioning, the ABS motor does not turn, the left front solenoid mechanically fails open, or the ABS motor wires are crossed.

DTC Will Set When: DTC 52 can be set only during an ABS stop. If the EBTCM commands the left front ABS channel in release for three seconds, a malfunction exists.

Action Taken: A malfunction DTC is stored, ABS and TCS are disabled and the ABS and TCS warning lamps are turned "ON."

DTC Chart Test Description: Number(s) below refer to circled number(s) on the diagnostic chart.

1. Identifies a malfunction in a wheel speed sensor that may cause the system to be in release too long.
2. Identifies an ABS motor as being failed or wired incorrectly.
3. Checks for a solenoid that may have mechanically failed open.
4. Verifies proper solenoid operation.
5. If Step 3 has failed, this serves to isolate the cause of the hydraulic malfunction, to either the solenoid or the ABS hydraulic modulator assembly.
6. Determines whether a malfunctioning ABS motor pack or hydraulic modulator is the reason for DTC 52 being set.

Diagnostic Aids: An intermittent malfunction may be caused by a mechanical part of the system that sticks or binds.

The frequency of the malfunction can be checked by using the enhanced diagnostic function of the Tech 1.

DTC 52 may occur if on ice and the steering wheel is turned to lock during braking. Using the Tech 1, perform hydraulic test to ensure total brake system is functional.

Any circuitry, that is suspected as causing the intermittent complaint, should be thoroughly checked for backed out terminals, improper mating, broken locks, improperly formed or damaged terminals, poor terminal to wiring connections or physical damage to the wiring harness.

DIAGNOSTIC CHART DTC 52 — 1994 SILHOUETTE, TRANS SPORT AND LUMINA APV WITH TCS, CONTINUED

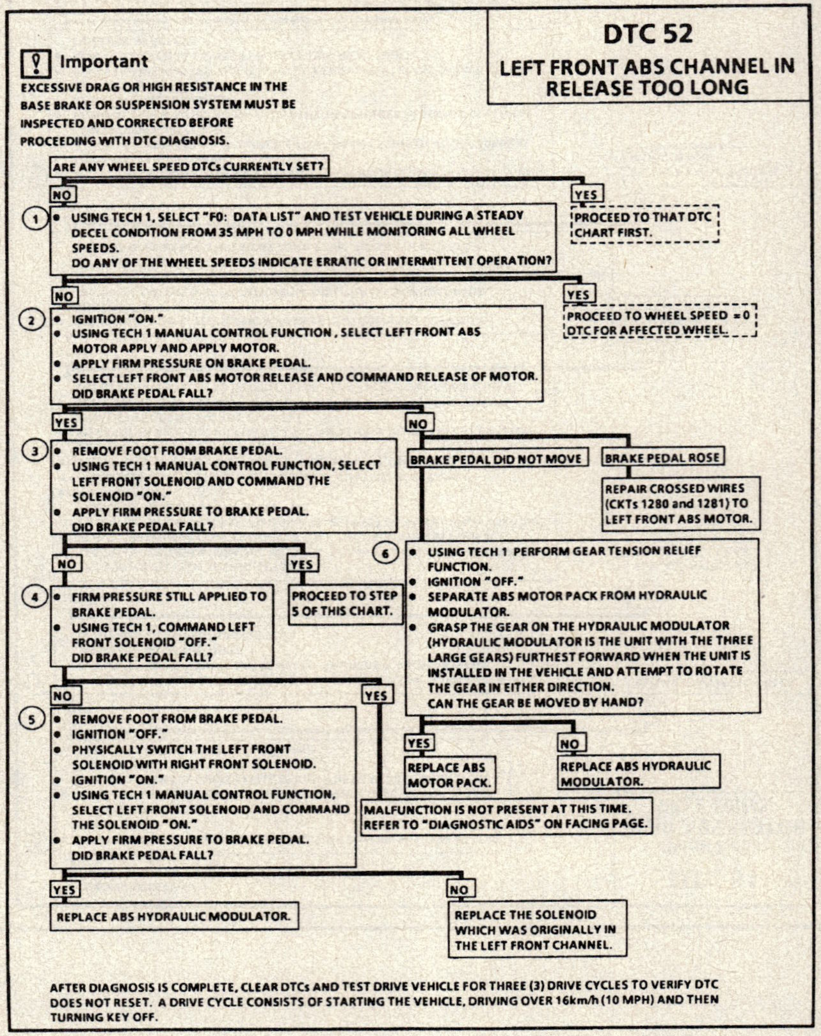

DIAGNOSTIC CHART DTC 53 — 1994 SILHOUETTE, TRANS SPORT AND LUMINA APV WITH TCS

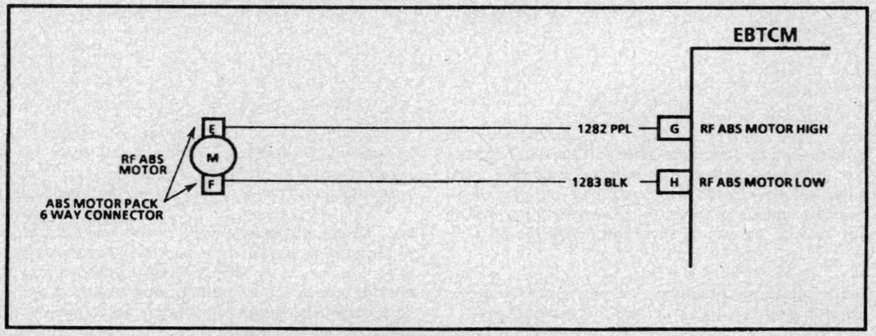

DTC 53

RIGHT FRONT ABS CHANNEL IN RELEASE TOO LONG

Circuit Description:

This DTC will diagnose an ABS motor that is energized longer than expected. This could occur if a wheel speed sensor is malfunctioning, the ABS motor does not turn, the right front solenoid mechanically fails open, or the ABS motor wires are crossed.

DTC Will Set When: DTC 53 can be set only during an ABS stop. If the EBTCM commands the right front channel in release for three seconds, a malfunction exists.

Action Taken: A malfunction DTC is stored, ABS and TCS are disabled and the ABS and TCS warning lamps are turned "ON."

DTC Chart Test Description: Number(s) below refer to circled number(s) on the diagnostic chart.

1. Identifies a malfunction in a wheel speed sensor that may cause the system to be in release too long.
2. Identifies a motor as being failed or wired incorrectly.
3. Checks for a solenoid that may have mechanically failed open.
4. Checks for a solenoid that may have mechanically failed closed.
5. If Step 3 has failed, this serves to isolate the cause of the hydraulic malfunction, to either the solenoid or the ABS hydraulic modulator assembly.
6. Determines whether a malfunctioning ABS motor pack or hydraulic modulator is the reason for DTC 53 being set.

Diagnostic Aids: An intermittent malfunction may be caused by a mechanical part of the system that sticks or binds.

The frequency of the malfunction can be checked by using the enhanced diagnostic function of the Tech 1.

DTC 53 may occur on ice if the steering wheel is turned to lock during braking. Using the Tech 1 perform hydraulic test to ensure total brake system is functional.

Any circuitry, that is suspected as causing the intermittent complaint, should be thoroughly checked for backed out terminals, improper mating, broken locks, improperly formed or damaged terminals, poor terminal to wiring connections or physical damage to the wiring harness.

DIAGNOSTIC CHART DTC 53 — 1994 SILHOUETTE, TRANS SPORT AND LUMINA APV WITH TCS, CONTINUED

DTC 53

RIGHT FRONT ABS CHANNEL IN RELEASE TOO LONG

Important

EXCESSIVE DRAG OR HIGH RESISTANCE IN THE BASE BRAKE OR SUSPENSION SYSTEM MUST BE INSPECTED AND CORRECTED BEFORE PROCEEDING WITH DTC DIAGNOSIS.

ARE ANY WHEEL SPEED DTCs CURRENTLY SET?

- YES: PROCEED TO THAT DTC CHART FIRST.
- NO: go to step 1.

1
- USING TECH 1, SELECT "F0: DATA LIST" AND TEST VEHICLE DURING A STEADY DECEL CONDITION FROM 35 MPH TO 0 MPH WHILE MONITORING ALL WHEEL SPEEDS.

DO ANY OF THE WHEEL SPEEDS INDICATE ERRATIC OR INTERMITTENT OPERATION?

- YES: PROCEED TO WHEEL SPEED = 0 DTC FOR AFFECTED WHEEL.
- NO: go to step 2.

2
- IGNITION "ON."
- USING TECH 1 MANUAL CONTROL FUNCTION, SELECT RIGHT FRONT ABS MOTOR APPLY AND APPLY MOTOR.
- APPLY FIRM PRESSURE ON BRAKE PEDAL.
- SELECT RIGHT FRONT ABS MOTOR RELEASE AND COMMAND RELEASE OF MOTOR.

DID BRAKE PEDAL FALL?

- NO:
  - BRAKE PEDAL ROSE: REPAIR CROSSED WIRES (CKTs 1282 AND 1283) TO RIGHT FRONT ABS MOTOR.
  - BRAKE PEDAL DID NOT MOVE: go to step 6.
- YES: go to step 3.

3
- REMOVE FOOT FROM BRAKE PEDAL.
- USING TECH 1 MANUAL CONTROL FUNCTION, SELECT RIGHT FRONT SOLENOID AND COMMAND THE SOLENOID "ON."
- APPLY FIRM PRESSURE TO BRAKE PEDAL.

DID BRAKE PEDAL FALL?

- YES: PROCEED TO STEP 5 OF THIS CHART.
- NO: go to step 4.

4
- FIRM PRESSURE STILL APPLIED TO BRAKE PEDAL.
- USING TECH 1, COMMAND RIGHT FRONT SOLENOID "OFF."

DID BRAKE PEDAL FALL?

- YES: MALFUNCTION IS NOT PRESENT AT THIS TIME. REFER TO "DIAGNOSTIC AIDS" ON FACING PAGE.
- NO: go to step 5.

5
- REMOVE FOOT FROM BRAKE PEDAL.
- IGNITION "OFF."
- PHYSICALLY SWITCH THE RIGHT FRONT SOLENOID WITH THE LEFT FRONT SOLENOID.
- IGNITION "ON."
- USING TECH 1 MANUAL CONTROL FUNCTION, SELECT RIGHT FRONT SOLENOID AND COMMAND THE SOLENOID "ON."
- APPLY FIRM PRESSURE TO BRAKE PEDAL.

DID BRAKE PEDAL FALL?

- YES: REPLACE ABS HYDRAULIC MODULATOR.
- NO: REPLACE THE SOLENOID WHICH WAS ORIGINALLY IN THE RIGHT FRONT CHANNEL.

6
- USING TECH 1 PERFORM GEAR TENSION RELIEF FUNCTION.
- IGNITION "OFF."
- SEPARATE ABS MOTOR PACK FROM HYDRAULIC MODULATOR.
- GRASP THE GEAR ON THE HYDRAULIC MODULATOR (HYDRAULIC MODULATOR IS THE UNIT WITH THE THREE LARGE GEARS) FURTHEST REARWARD WHEN THE UNIT IS INSTALLED IN THE VEHICLE AND ATTEMPT TO ROTATE THE GEAR IN EITHER DIRECTION.

CAN THE GEAR BE MOVED BY HAND?

- YES: REPLACE ABS MOTOR PACK.
- NO: REPLACE ABS HYDRAULIC MODULATOR.

AFTER DIAGNOSIS IS COMPLETE, CLEAR DTCs AND TEST DRIVE VEHICLE FOR THREE (3) DRIVE CYCLES TO VERIFY DTC DOES NOT RESET. A DRIVE CYCLE CONSISTS OF STARTING THE VEHICLE, DRIVING OVER 16km/h (10 MPH) AND THEN TURNING KEY OFF.

DIAGNOSTIC CHART DTC 54 — 1994 SILHOUETTE, TRANS SPORT AND LUMINA APV WITH TCS

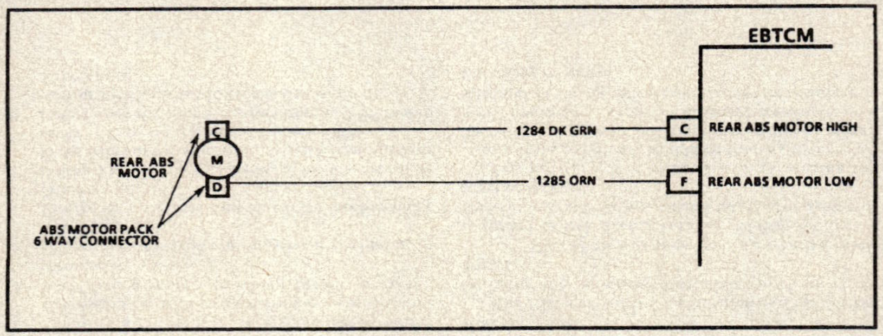

DTC 54

REAR ABS CHANNEL IN RELEASE TOO LONG

Circuit Description:
This DTC will diagnose an ABS motor that is energized longer than expected. This could occur if a wheel speed sensor is malfunctioning, the ABS motor does not turn, or the ABS motor wires are crossed.

DTC Will Set When: DTC 54 can be set only during an ABS stop. If the EBTCM commands the rear ABS channel in release for three seconds, a malfunction exists.

Action Taken: A malfunction DTC is stored, ABS and TCS are disabled and the ABS and TCS warning lamps are turned "ON."

DTC Chart Test Description: Number(s) below refer to circled number(s) on the diagnostic chart.

1. Identifies a malfunction in a wheel speed sensor that may cause the system to be in release too long.
2. Checks for a wheel that may stick or bind because of a mechanical malfunction.
3. Checks to see if the ABS motor is capable of moving and applying the hydraulic piston for the rear wheels.
4. Ensures that the ABS motor wiring is not crossed.
5. Isolates the fault of a "no-apply" situation to either the ABS motor pack or ABS hydraulic modulator assembly.

Diagnostic Aids: An intermittent malfunction may be caused by a mechanical part of the system that sticks or binds.

The frequency of the malfunction can be checked by using the enhanced diagnostic function of the Tech 1

Any circuitry, that is suspected as causing the intermittent complaint, should be thoroughly checked for backed out terminals, improper mating, broken locks, improperly formed or damaged terminals, poor terminal to wiring connections or physical damage to the wiring harness.

DIAGNOSTIC CHART DTC 54 — 1994 SILHOUETTE, TRANS SPORT AND LUMINA APV WITH TCS, CONTINUED

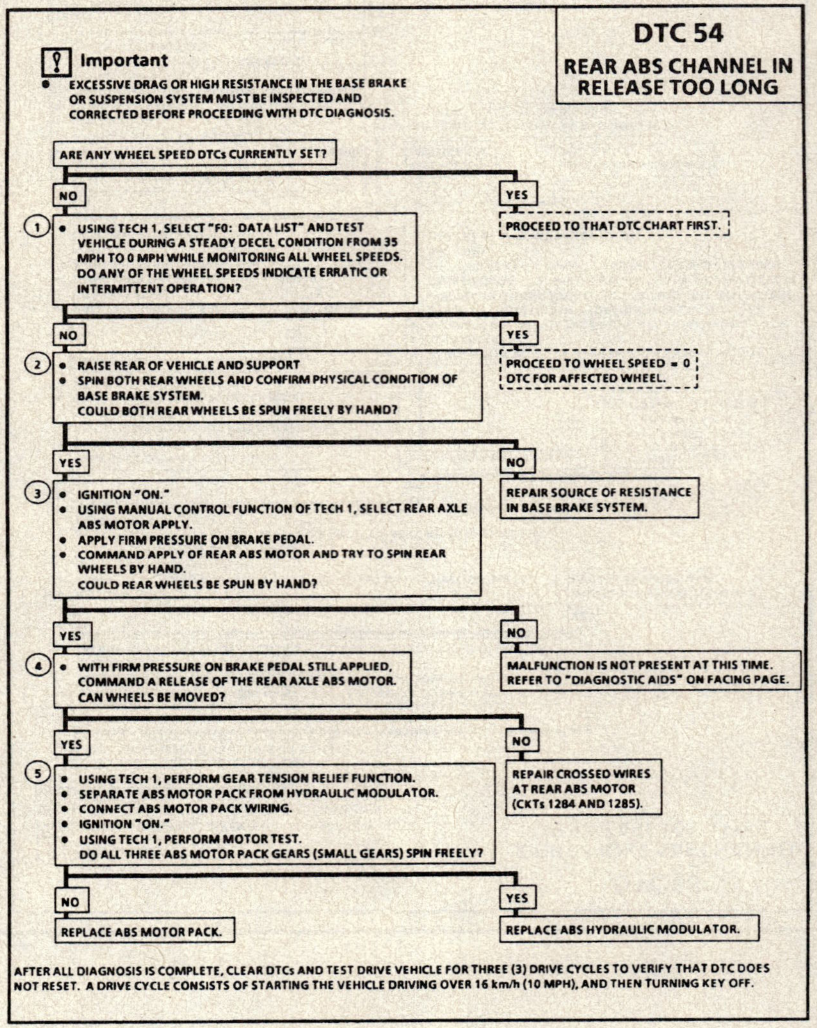

DIAGNOSTIC CHART DTC 55 — 1994 SILHOUETTE, TRANS SPORT AND LUMINA APV WITH TCS

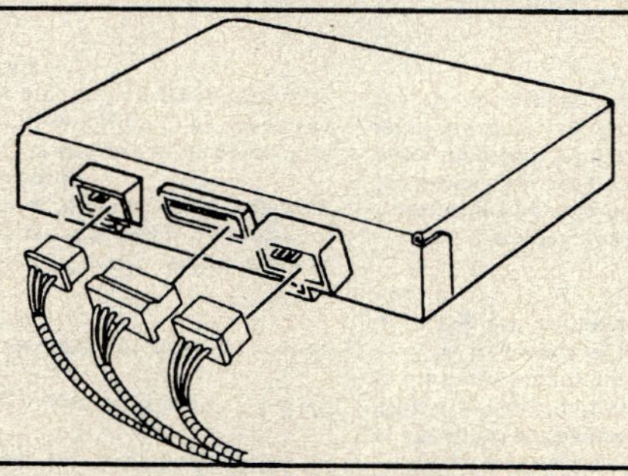

DTC 55

EBTCM MALFUNCTION (ABS)

Circuit Description:
This DTC identifies an internal malfunction detected by the EBTCM motor driver interface custom integrated circuit.

Action Taken: A malfunction DTC is stored, ABS and TCS are disabled and the ABS and TCS warning lamps are turned "ON."

DTC Chart Test Description: Number(s) below refer to circled number(s) on the diagnostic chart.

1. Checks to see if the malfunction is still present.
2. Checks to see if the malfunction is intermittent.

DIAGNOSTIC CHART DTC 55 — 1994 SILHOUETTE, TRANS SPORT AND LUMINA APV WITH TCS, CONTINUED

DTC 55
EBTCM MALFUNCTION (ABS)

(1)
- IGNITION "ON," ENGINE "OFF."
- USING TECH 1, READ DTCs.

IS DTC 55 SET AS A CURRENT CODE?

- YES → (2)
- NO → MALFUNCTION IS INTERMITTENT.

(2)
- USING TECH 1, CLEAR DTCs.
- TEST DRIVE VEHICLE FOR THREE (3) DRIVE CYCLES. A DRIVE CYCLE CONSISTS OF KEY "ON," DRIVE VEHICLE OVER 16 km/h (10 MPH), THEN KEY "OFF."

DOES DTC 55 RESET?

- YES → REPLACE EBTCM.
- NO → MALFUNCTION IS INTERMITTENT.

DIAGNOSTIC CHART DTC 56 — 1994 SILHOUETTE, TRANS SPORT AND LUMINA APV WITH TCS

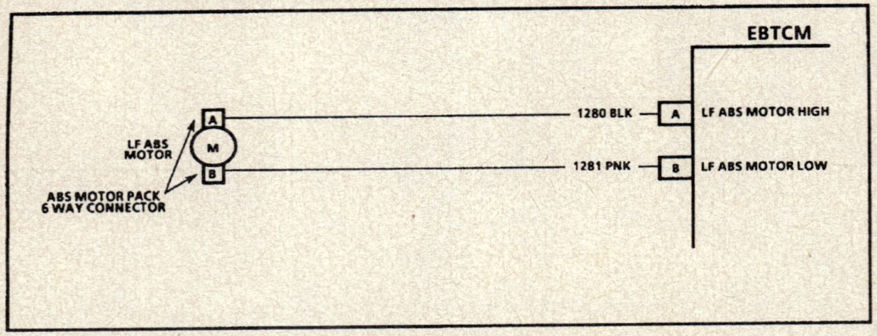

DTC 56

LEFT FRONT ABS MOTOR CIRCUIT OPEN

Circuit Description:
This DTC identifies an ABS motor that cannot be energized due to an open in its circuitry.

DTC Will Set When: DTC 56 can be set only when the motor is commanded "OFF." If the EBTCM detects an out of range voltage on either of the left front ABS motor circuits indicating an open circuit, a malfunction exists.

Action Taken: An open ABS motor will not activate when requested. A malfunction DTC is stored, ABS is disabled and the ABS warning lamp is turned "ON."

DTC Chart Test Description: Number(s) below refer to circled number(s) on the diagnostic chart.

1. Checks for proper resistance of the ABS motor.
2. Checks for an open in the ABS motor "HIGH" circuitry.
3. Checks for an open in the ABS motor "LOW" circuitry.
4. Ensures malfunction was not due to a poor terminal contact or EBTCM.

Diagnostic Aids: Using Tech 1, select manual control function, and exercise ABS motor movement of affected channel in both directions while applying light pressure on the brake pedal.

If erratic or "jumpy" brake pedal movement is detected while performing an "apply" or "release" function of the ABS motor, an intermittent malfunction may be indicated.

An intermittent malfunction may be caused by a poor connection, rubbed through wire insulation, or a wire that is broken inside the insulation.

If the malfunction is not current, wiggle the wires of the affected channel and check if the DTC resets. This will help to pinpoint an intermittent malfunction in the motor circuitry or connections.

The frequency of the malfunction can be checked by using the enhanced diagnostic function of the Tech 1.

Any circuitry, that is suspected as causing the intermittent complaint, should be thoroughly checked for backed out terminals, improper mating, broken locks, improperly formed or damaged terminals, poor terminal to wiring connections or physical damage to the wiring harness.

DIAGNOSTIC CHART DTC 56 — 1994 SILHOUETTE, TRANS SPORT AND LUMINA APV WITH TCS, CONTINUED

DTC 56

LEFT FRONT ABS MOTOR CIRCUIT OPEN

DOES DTC 56 OCCUR INTERMITTENTLY?

- YES → REFER TO "DIAGNOSTIC AIDS" ON FACING PAGE.
- NO →

(1)
- IGNITION "OFF."
- DISCONNECT 6-WAY ABS MOTOR PACK CONNECTOR.
- USING J 39200, MEASURE RESISTANCE ACROSS ABS MOTOR PACK TERMINALS "A" AND "B".

IS RESISTANCE BELOW 2000Ω?

- NO → REPLACE ABS MOTOR PACK.
- YES →

(2)
- DISCONNECT GRAY 8-WAY CONNECTOR AT EBTCM.
- USING J 39200, MEASURE RESISTANCE BETWEEN TERMINAL "A" OF THE GRAY 8-WAY EBTCM HARNESS CONNECTOR AND TERMINAL "A" OF THE 6-WAY ABS MOTOR PACK HARNESS CONNECTOR.

IS RESISTANCE 2Ω OR LESS?

- NO → REPAIR OPEN OR HIGH RESISTANCE IN CKT 1280.
- YES →

(3)
- USING J 39200, MEASURE RESISTANCE BETWEEN TERMINAL "B" OF THE GRAY 8-WAY EBTCM HARNESS CONNECTOR AND TERMINAL "B" OF THE 6-WAY ABS MOTOR PACK HARNESS CONNECTOR.

IS RESISTANCE 2Ω OR LESS?

- NO → REPAIR OPEN OR HIGH RESISTANCE IN CKT 1281.
- YES →

(4)
- INSPECT THE 6-WAY ABS MOTOR PACK AND GRAY 8-WAY EBTCM CONNECTORS FOR POOR TERMINAL CONTACT AND EVIDENCE OF CORROSION.

REPLACE ALL TERMINALS THAT EXHIBIT SIGNS OF POOR TERMINAL CONTACT OR CORROSION.
- RECONNECT ALL CONNECTORS.
- IGNITION "OFF."
- START ENGINE AND TEST DRIVE VEHICLE, ACHIEVING AT LEAST 16 km/h (10 MPH).
- REPEAT ABOVE DRIVE CYCLE SEQUENCE TWO MORE TIMES.

DID DTC 56 SET IN THE LAST THREE DRIVE CYCLES?

- YES → REPLACE EBTCM.
- NO → MALFUNCTION WAS MOST LIKELY CAUSED BY A POOR CONNECTION. REFER TO "DIAGNOSTIC AIDS" ON FACING PAGE.

AFTER ALL DIAGNOSIS IS COMPLETE, CLEAR DTCs AND TEST DRIVE VEHICLE FOR THREE (3) DRIVE CYCLES TO VERIFY THAT DTC DOES NOT RESET. A DRIVE CYCLE CONSISTS OF STARTING THE VEHICLE, DRIVING OVER 16 km/h (10 MPH), AND THEN TURNING KEY OFF.

DIAGNOSTIC CHART DTC 57 — 1994 SILHOUETTE, TRANS SPORT AND LUMINA APV WITH TCS

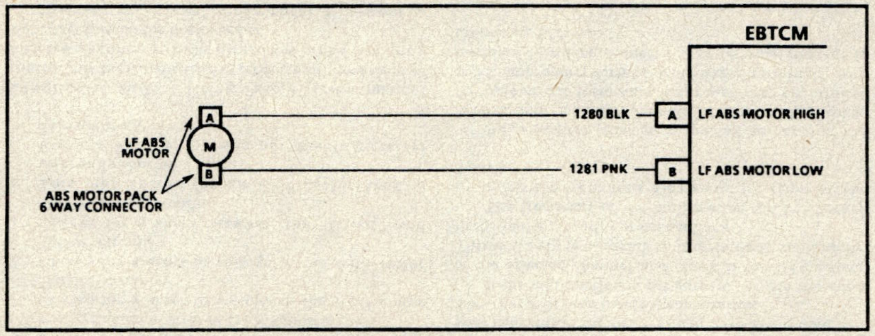

DTC 57

LEFT FRONT ABS MOTOR CIRCUIT SHORTED TO GND

Circuit Description:
This DTC identifies an ABS motor circuit that is shorted to ground. This malfunction will not allow the ABS motor to be controlled at the commanded current rate or will cause the driver circuit to allow current directly to ground.

DTC Will Set When: DTC 57 can be set anytime. If the EBTCM detects an out of range voltage on either of the left front ABS motor circuits indicating a circuit shorted to ground, a malfunction exists.

Action Taken: A malfunction DTC is stored, ABS is disabled and the ABS warning lamp is turned "ON."

DTC Chart Test Description: Number(s) below refer to circled number(s) on the diagnostic chart.

1. Checks for a short to ground in the ABS motor "HIGH" circuitry.
2. Checks for a short to ground in the ABS motor "LOW" circuitry.
3. Checks for an ABS motor that is internally shorted to ground.
4. Ensures malfunction was not due to physical damage of the circuitry.

Diagnostic Aids: Using Tech 1, select manual control function, and exercise ABS motor movement of affected channel in both directions while applying light pressure on the brake pedal.

If erratic or "jumpy" brake pedal movement is detected while performing an "apply" or "release" function of the ABS motor, an intermittent malfunction may be indicated.

An intermittent malfunction may be caused by a poor connection, rubbed through wire insulation, or a wire that is broken inside the insulation.

If the malfunction is not current, wiggle the wires of the affected channel and check if the DTC resets. This will help to pinpoint an intermittent malfunction in the motor circuitry or connections.

The frequency of the malfunction can be checked by using the enhanced diagnostic function of the Tech 1

Any circuitry, that is suspected as causing the intermittent complaint, should be thoroughly checked for backed out terminals, improper mating, broken locks, improperly formed or damaged terminals, poor terminal to wiring connections or physical damage to the wiring harness.

DIAGNOSTIC CHART DTC 57 — 1994 SILHOUETTE, TRANS SPORT AND LUMINA APV WITH TCS, CONTINUED

DTC 57
LEFT FRONT ABS MOTOR CIRCUIT SHORTED TO GND

DOES DTC 57 OCCUR INTERMITTENTLY?

- YES: REFER TO "DIAGNOSTIC AIDS" ON FACING PAGE.
- NO: go to 1.

(1)
- IGNITION "OFF."
- DISCONNECT THE 6-WAY ABS MOTOR PACK CONNECTOR AND THE GRAY 8-WAY EBTCM CONNECTOR.
- USING J 39200, MEASURE THE RESISTANCE BETWEEN GROUND AND TERMINAL "A" OF THE 6-WAY ABS MOTOR PACK HARNESS CONNECTOR.

DOES J 39200 DISPLAY RESISTANCE AS "OL" (INFINITE)?

- NO: REPAIR SHORT TO GROUND IN CKT 1280.
- YES: go to 2.

(2)
- USING J 39200, MEASURE THE RESISTANCE BETWEEN GROUND AND TERMINAL "B" OF THE 6-WAY ABS MOTOR PACK HARNESS CONNECTOR.

DOES J 39200 DISPLAY RESISTANCE AS "OL" (INFINITE)?

- NO: REPAIR SHORT TO GROUND IN CKT 1281.
- YES: go to 3.

(3)
- USING J 39200, MEASURE THE RESISTANCE BETWEEN GROUND AND TERMINAL "A" OF THE ABS MOTOR PACK.

DOES J 39200 DISPLAY RESISTANCE AS "OL" (INFINITE)?

- NO: REPLACE ABS MOTOR PACK.
- YES: go to 4.

(4)
- INSPECT CKTs 1280 AND 1281 AND RELATED CONNECTORS FOR DAMAGE WHICH MAY RESULT IN A SHORT TO GROUND WITH ALL CONNECTORS CONNECTED. REPAIR ALL DAMAGE IF EVIDENT.
- RECONNECT ALL CONNECTORS.
- IGNITION "OFF."
- START ENGINE AND TEST DRIVE VEHICLE, ACHIEVING AT LEAST 16 km/h (10 MPH).
- REPEAT ABOVE DRIVE CYCLE SEQUENCE TWO MORE TIMES.

DID DTC 57 SET IN THE LAST THREE DRIVE CYCLES?

- YES: REPLACE EBTCM.
- NO: MALFUNCTION WAS MOST LIKELY CAUSED BY A POOR CONNECTION. REFER TO "DIAGNOSTIC AIDS" ON FACING PAGE.

AFTER ALL DIAGNOSIS IS COMPLETE, CLEAR DTCs AND TEST DRIVE VEHICLE FOR THREE (3) DRIVE CYCLES TO VERIFY THAT DTC DOES NOT RESET. A DRIVE CYCLE CONSISTS OF STARTING THE VEHICLE, DRIVING OVER 16 km/h (10 MPH), AND THEN TURNING KEY OFF.

DIAGNOSTIC CHART DTC 58 — 1994 SILHOUETTE, TRANS SPORT AND LUMINA APV WITH TCS

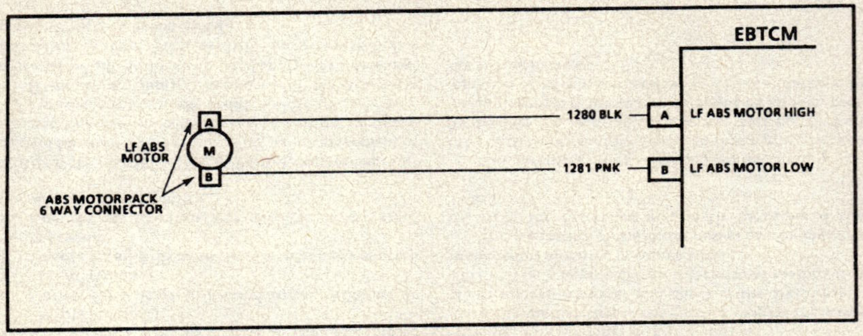

DTC 58

LEFT FRONT ABS MOTOR CIRCUIT SHORTED TO BATT

Circuit Description:

This DTC identifies an ABS motor circuit that is shorted to battery or an ABS motor that has low or no resistance. This malfunction will not allow the ABS motor to be controlled at the commanded current rate or will cause the ABS motor to turn in the opposite direction, or not at all.

DTC Will Set When: DTC 58 can be set only when the ABS motor is commanded "OFF." If the EBTCM detects an out of range voltage on either of the left front ABS motor circuits indicating a circuit shorted to battery or an ABS motor shorted, a malfunction exists.

Action Taken: A malfunction DTC is stored, ABS is disabled and the ABS warning lamp is turned "ON."

DTC Chart Test Description: Number(s) below refer to circled number(s) on the diagnostic chart.

1. Checks for a short to voltage in the ABS motor "HIGH" circuitry.
2. Checks for a short to voltage in the ABS motor "LOW" circuitry.
3. Checks for a short between the "HIGH" and "LOW" motor circuits.
4. This step checks for an ABS motor that is internally shorted.
5. Ensures malfunction was not due to physical damage of the circuitry.

Diagnostic Aids: Using Tech 1, select manual control function, and exercise ABS motor movement of affected channel in both directions while applying light pressure on the brake pedal.

If erratic or "jumpy" brake pedal movement is detected while performing an "apply" or "release" function of the ABS motor, an intermittent malfunction may be indicated.

An intermittent malfunction may be caused by a poor connection, rubbed through wire insulation, or a wire that is broken inside the insulation.

If the malfunction is not current, wiggle the wires of the affected channel and check if the DTC resets. This will help to pinpoint an intermittent malfunction in the motor circuitry or connections.

The frequency of the malfunction can be checked by using the enhanced diagnostic function of the Tech 1

Any circuitry, that is suspected as causing the intermittent complaint, should be thoroughly checked for backed out terminals, improper mating, broken locks, improperly formed or damaged terminals, poor terminal to wiring connections or physical damage to the wiring harness.

DIAGNOSTIC CHART DTC 58 — 1994 SILHOUETTE, TRANS SPORT AND LUMINA APV WITH TCS, CONTINUED

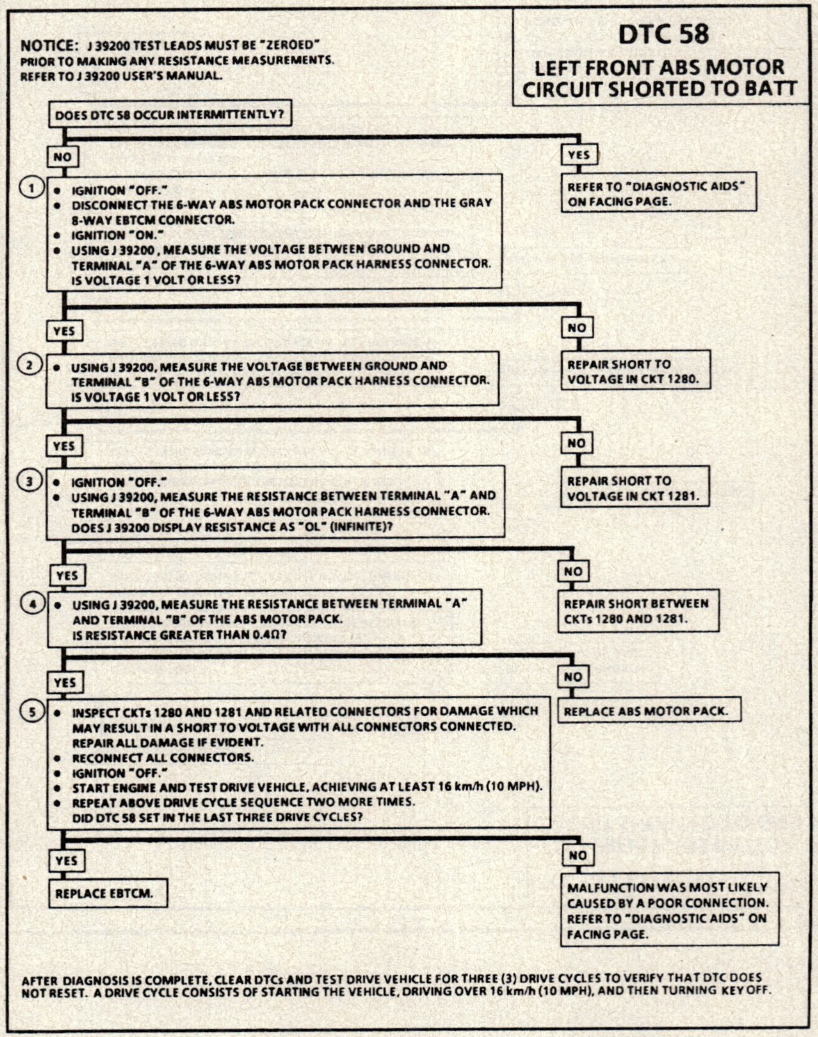

DIAGNOSTIC CHART DTC 63 — 1994 SILHOUETTE, TRANS SPORT AND LUMINA APV WITH TCS

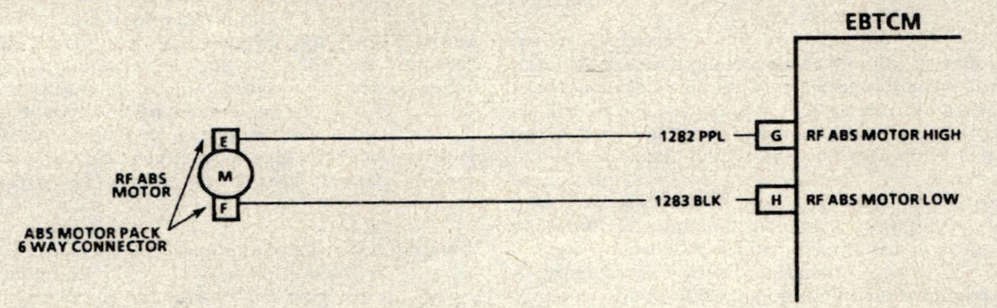

DTC 63

RIGHT FRONT ABS MOTOR CIRCUIT SHORTED TO BATT

Circuit Description:

This DTC identifies an ABS motor circuit that is shorted to battery or an ABS motor that has low or no resistance. This malfunction will not allow the ABS motor to be controlled at the commanded current rate or will cause the ABS motor to turn in the opposite direction or not at all.

DTC Will Set When: DTC 63 can be set only when the ABS motor is commanded "OFF." If the EBTCM detects an out of range voltage on either of the right front ABS motor circuits indicating a circuit shorted to battery or an ABS motor shorted, a malfunction exists.

Action Taken: A malfunction DTC is stored, ABS is disabled and the ABS warning lamp is turned "ON."

DTC Chart Test Description: Number(s) below refer to circled number(s) on the diagnostic chart.

1. Checks for a short to voltage in the ABS motor "HIGH" circuitry.
2. Checks for a short to voltage in the ABS motor "LOW" circuitry.
3. Checks for a short circuit between motor "HIGH" and motor "LOW."
4. Checks for proper resistance of the ABS motor.
5. Ensures malfunction was not due to physical damage of the circuitry.

Diagnostic Aids: Using Tech 1, select manual control function, and exercise ABS motor movement of affected channel in both directions while applying light pressure on the brake pedal.

If erratic or "jumpy" brake pedal movement is detected while performing an "apply" or "release" function of the ABS motor, an intermittent malfunction may be indicated.

An intermittent malfunction may be caused by a poor connection, rubbed through wire insulation, or a wire that is broken inside the insulation.

If the malfunction is not current, wiggle the wires of the affected channel and check if the DTC resets. This will help to pinpoint an intermittent malfunction in the motor circuitry or connections.

The frequency of the malfunction can be checked by using the enhanced diagnostic function of the Tech 1, as described in "Tech 1 Diagnostics" in this section.

Any circuitry, that is suspected as causing the intermittent complaint, should be thoroughly checked for backed out terminals, improper mating, broken locks, improperly formed or damaged terminals, poor terminal to wiring connections or physical damage to the wiring harness.

DIAGNOSTIC CHART DTC 63 — 1994 SILHOUETTE, TRANS SPORT AND LUMINA APV WITH TCS, CONTINUED

DTC 63

RIGHT FRONT ABS MOTOR CIRCUIT SHORTED TO BATT

NOTICE: J 39200 TEST LEADS MUST BE "ZEROED" PRIOR TO MAKING ANY RESISTANCE MEASUREMENTS. REFER TO J 39200 USER'S MANUAL.

DOES DTC 63 OCCUR INTERMITTENTLY?

- YES: REFER TO "DIAGNOSTIC AIDS" ON FACING PAGE.
- NO:

(1)
- IGNITION "OFF."
- DISCONNECT THE 6-WAY ABS MOTOR PACK CONNECTOR AND THE GRAY 8-WAY EBTCM CONNECTOR.
- IGNITION "ON."
- USING J 39200, MEASURE THE VOLTAGE BETWEEN GROUND AND TERMINAL "E" OF THE 6-WAY ABS MOTOR PACK HARNESS CONNECTOR.

IS VOLTAGE 1 VOLT OR LESS?

- NO: REPAIR SHORT TO VOLTAGE IN CKT 1282.
- YES:

(2)
- USING J 39200, MEASURE THE VOLTAGE BETWEEN GROUND AND TERMINAL "F" OF THE 6-WAY ABS MOTOR PACK HARNESS CONNECTOR.

IS VOLTAGE 1 VOLT OR LESS?

- NO: REPAIR SHORT TO VOLTAGE IN CKT 1283.
- YES:

(3)
- IGNITION "OFF."
- USING J 39200, MEASURE THE RESISTANCE BETWEEN TERMINAL "E" AND TERMINAL "F" OF THE 6-WAY ABS MOTOR PACK HARNESS CONNECTOR.

DOES J 39200 DISPLAY RESISTANCE AS "OL" (INFINITE)?

- NO: REPAIR SHORT BETWEEN CKTs 1282 AND 1283.
- YES:

(4)
- USING J 39200, MEASURE THE RESISTANCE BETWEEN TERMINAL "E" AND TERMINAL "F" OF THE ABS MOTOR PACK.

IS RESISTANCE GREATER THAN 0.4Ω?

- NO: REPLACE ABS MOTOR PACK.
- YES:

(5)
- INSPECT CKTs 1282 AND 1283 AND RELATED CONNECTORS FOR DAMAGE WHICH MAY RESULT IN A SHORT TO VOLTAGE WITH ALL CONNECTORS CONNECTED. REPAIR ALL DAMAGE IF EVIDENT.
- RECONNECT ALL CONNECTORS.
- IGNITION "OFF."
- START ENGINE AND TEST DRIVE VEHICLE, ACHIEVING AT LEAST 16 km/h (10 MPH).
- REPEAT ABOVE DRIVE CYCLE SEQUENCE TWO MORE TIMES.

DID DTC 63 SET IN THE LAST THREE DRIVE CYCLES?

- YES: REPLACE EBTCM.
- NO: MALFUNCTION WAS MOST LIKELY CAUSED BY A POOR CONNECTION. REFER TO "DIAGNOSTIC AIDS" ON FACING PAGE.

AFTER DIAGNOSIS IS COMPLETE, CLEAR DTCs AND TEST DRIVE VEHICLE FOR THREE (3) DRIVE CYCLES TO VERIFY THAT DTC DOES NOT RESET. A DRIVE CYCLE CONSISTS OF STARTING THE VEHICLE, DRIVING OVER 16 km/h (10 MPH), AND THEN TURNING KEY OFF.

DIAGNOSTIC CHART DTC 64 — 1994 SILHOUETTE, TRANS SPORT AND LUMINA APV WITH TCS

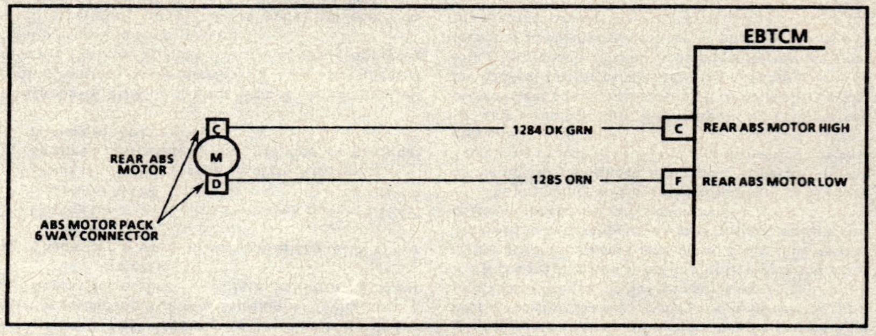

DTC 64

REAR ABS MOTOR CIRCUIT OPEN

Circuit Description:
This DTC identifies an ABS motor that cannot be energized due to an open in its circuitry. This malfunction will not allow proper rear ABS operation.

DTC Will Set When: DTC 64 can be set only when the ABS motor is commanded "OFF." If the EBTCM detects an out of range voltage on either of the rear ABS motor circuits indicating an open circuit, a malfunction exists.

Action Taken: An open motor will not activate when requested. A malfunction DTC is stored, ABS is disabled and the ABS warning lamp is turned "ON."

DTC Chart Test Description: Number(s) below refer to circled number(s) on the diagnostic chart.

1. Checks for proper resistance of the ABS motor.
2. Checks for an open in the ABS motor "HIGH" circuitry.
3. Checks for an open in the ABS motor "LOW" circuitry.
4. Ensures malfunction was not due to poor terminal contact.

Diagnostic Aids: Using Tech 1, select manual control function, and exercise ABS motor movement of affected channel in both directions while applying light pressure on the brake pedal.

If erratic or "jumpy" brake pedal movement is detected while performing an "apply" or "release" function of the ABS motor, an intermittent malfunction may be indicated.

An intermittent malfunction may be caused by a poor connection, rubbed through wire insulation, or a wire that is broken inside the insulation.

If the malfunction is not current, wiggle the wires of the affected channel and check if the DTC resets. This will help to pinpoint an intermittent malfunction in the motor circuitry or connections.

The frequency of the malfunction can be checked by using the enhanced diagnostic function of the Tech 1, as described in "Tech 1 Diagnostics" in this section.

Any circuitry, that is suspected as causing the intermittent complaint, should be thoroughly checked for backed out terminals, improper mating, broken locks, improperly formed or damaged terminals, poor terminal to wiring connections or physical damage to the wiring harness.

DIAGNOSTIC CHART DTC 64 — 1994 SILHOUETTE, TRANS SPORT AND LUMINA APV WITH TCS, CONTINUED

DTC 64

REAR ABS MOTOR CIRCUIT OPEN

DOES DTC 64 OCCUR INTERMITTENTLY?

- YES: REFER TO "DIAGNOSTIC AIDS" ON FACING PAGE.
- NO:

(1)
- IGNITION "OFF."
- DISCONNECT 6-WAY ABS MOTOR PACK CONNECTOR.
- USING J 39200, MEASURE RESISTANCE ACROSS ABS MOTOR PACK TERMINALS "C" AND "D".

IS RESISTANCE BELOW 2000Ω?

- NO: REPLACE ABS MOTOR PACK.
- YES:

(2)
- DISCONNECT GRAY 8-WAY CONNECTOR AT EBTCM.
- USING J 39200, MEASURE RESISTANCE BETWEEN TERMINAL "C" OF THE GRAY 8-WAY EBTCM HARNESS CONNECTOR AND TERMINAL "C" OF THE 6-WAY ABS MOTOR PACK HARNESS CONNECTOR.

IS RESISTANCE 2Ω OR LESS?

- NO: REPAIR OPEN OR HIGH RESISTANCE IN CKT 1284.
- YES:

(3)
- USING J 39200, MEASURE RESISTANCE BETWEEN TERMINAL "F" OF THE GRAY 8-WAY EBTCM HARNESS CONNECTOR AND TERMINAL "D" OF THE 6-WAY ABS MOTOR PACK HARNESS CONNECTOR.

IS RESISTANCE 2Ω OR LESS?

- NO: REPAIR OPEN OR HIGH RESISTANCE IN CKT 1285.
- YES:

(4)
- INSPECT THE 6-WAY ABS MOTOR PACK AND GRAY 8-WAY EBTCM CONNECTORS FOR POOR TERMINAL CONTACT AND EVIDENCE OF CORROSION.

REPLACE ALL TERMINALS THAT EXHIBIT SIGNS OF POOR TERMINAL CONTACT OR CORROSION.
- RECONNECT ALL CONNECTORS.
- IGNITION "OFF."
- START ENGINE AND TEST DRIVE VEHICLE, ACHIEVING AT LEAST 16 km/h (10 MPH).
- REPEAT ABOVE DRIVE CYCLE SEQUENCE TWO MORE TIMES.

DID DTC 64 SET IN THE LAST THREE DRIVE CYCLES?

- YES: REPLACE EBTCM.
- NO: MALFUNCTION WAS MOST LIKELY CAUSED BY A POOR CONNECTION. REFER TO "DIAGNOSTIC AIDS" ON FACING PAGE.

AFTER ALL DIAGNOSIS IS COMPLETE, CLEAR DTCs AND TEST DRIVE VEHICLE FOR THREE (3) DRIVE CYCLES TO VERIFY THAT DTC DOES NOT RESET. A DRIVE CYCLE CONSISTS OF STARTING THE VEHICLE, DRIVING OVER 16 km/h (10 MPH), AND THEN TURNING KEY OFF.

DIAGNOSTIC CHART DTC 65 — 1994 SILHOUETTE, TRANS SPORT AND LUMINA APV WITH TCS

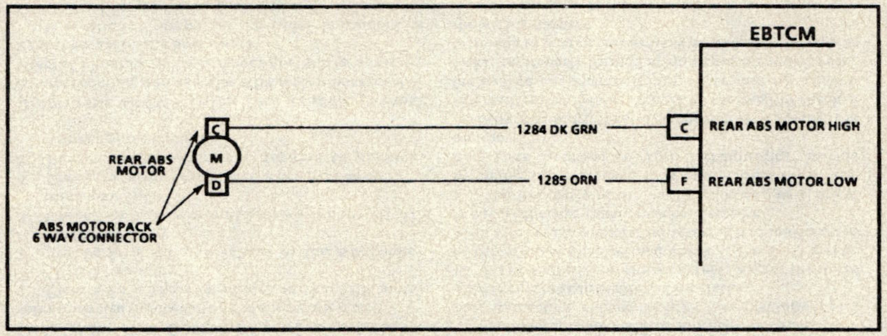

DTC 65

REAR ABS MOTOR CIRCUIT SHORTED TO GROUND

Circuit Description:

This DTC identifies an ABS motor circuit that is shorted to ground. This malfunction will not allow the ABS motor to be controlled at the commanded current rate or will cause the driver circuit to allow current directly to ground.

DTC Will Set When: DTC 65 can be set anytime. If the EBTCM detects an out of range voltage on either of the rear ABS motor circuits indicating a circuit shorted to ground, a malfunction exists.

Action Taken: A malfunction DTC is stored, ABS is disabled and the ABS warning lamp is turned "ON."

DTC Chart Test Description: Number(s) below refer to circled number(s) on the diagnostic chart.

1. Checks for a short to ground in the ABS motor "HIGH" circuitry.
2. Checks for a short to ground in the ABS motor "LOW" circuitry.
3. Checks for an ABS motor that is internally shorted to ground.
4. Ensures malfunction was not due to physical damage of the circuitry.

Diagnostic Aids: Using Tech 1, select manual control function, and exercise ABS motor movement of affected channel in both directions while applying light pressure on the brake pedal.

If erratic or "jumpy" brake pedal movement is detected while performing an "apply" or "release" function of the ABS motor, an intermittent malfunction may be indicated.

An intermittent malfunction may be caused by a poor connection, rubbed through wire insulation, or a wire that is broken inside the insulation.

If the malfunction is not current, wiggle the wires of the affected channel and check if the DTC resets. This will help to pinpoint an intermittent malfunction in the ABS motor circuitry or connections.

The frequency of the malfunction can be checked by using the enhanced diagnostic function of the Tech 1, as described in "Tech 1 Diagnostics" in this section.

Any circuitry, that is suspected as causing the intermittent complaint, should be thoroughly checked for backed out terminals, improper mating, broken locks, improperly formed or damaged terminals, poor terminal to wiring connections or physical damage to the wiring harness.

DIAGNOSTIC CHART DTC 65 — 1994 SILHOUETTE, TRANS SPORT AND LUMINA APV WITH TCS, CONTINUED

DTC 65

REAR ABS MOTOR CIRCUIT SHORTED TO GROUND

DOES DTC 65 OCCUR INTERMITTENTLY?

- YES: REFER TO "DIAGNOSTIC AIDS" ON FACING PAGE.
- NO:

(1)
- IGNITION "OFF."
- DISCONNECT THE 6-WAY ABS MOTOR PACK CONNECTOR AND THE GRAY 8-WAY EBTCM CONNECTOR.
- USING J 39200, MEASURE THE RESISTANCE BETWEEN GROUND AND TERMINAL "C" OF THE 6-WAY ABS MOTOR PACK HARNESS CONNECTOR.

DOES J 39200 DISPLAY RESISTANCE AS "OL" (INFINITE)?

- NO: REPAIR SHORT TO GROUND IN CKT 1284.
- YES:

(2)
- USING J 39200, MEASURE THE RESISTANCE BETWEEN GROUND AND TERMINAL "D" OF THE 6-WAY ABS MOTOR PACK HARNESS CONNECTOR.

DOES J 39200 DISPLAY RESISTANCE AS "OL" (INFINITE)?

- NO: REPAIR SHORT TO GROUND IN CKT 1285.
- YES:

(3)
- USING J 39200, MEASURE THE RESISTANCE BETWEEN GROUND AND TERMINAL "C" OF THE ABS MOTOR PACK.

DOES J 39200 DISPLAY RESISTANCE AS "OL" (INFINITE)?

- NO: REPLACE ABS MOTOR PACK.
- YES:

(4)
- INSPECT CKTs 1284 AND 1285 AND RELATED CONNECTORS FOR DAMAGE WHICH MAY RESULT IN A SHORT TO GROUND WITH ALL CONNECTORS CONNECTED. REPAIR ALL DAMAGE IF EVIDENT.
- RECONNECT ALL CONNECTORS.
- IGNITION "OFF."
- START ENGINE AND TEST DRIVE VEHICLE, ACHIEVING AT LEAST 16 km/h (10 MPH).
- REPEAT ABOVE DRIVE CYCLE SEQUENCE TWO MORE TIMES.

DID DTC 65 SET IN THE LAST THREE DRIVE CYCLES?

- YES: REPLACE EBTCM.
- NO: MALFUNCTION WAS MOST LIKELY CAUSED BY A POOR CONNECTION. REFER TO "DIAGNOSTIC AIDS" ON FACING PAGE.

AFTER ALL DIAGNOSIS IS COMPLETE, CLEAR DTCs AND TEST DRIVE VEHICLE FOR THREE (3) DRIVE CYCLES TO VERIFY THAT DTC DOES NOT RESET. A DRIVE CYCLE CONSISTS OF STARTING THE VEHICLE, DRIVING OVER 16 km/h (10 MPH), AND THEN TURNING KEY OFF.

DIAGNOSTIC CHART DTC 66 — 1994 SILHOUETTE, TRANS SPORT AND LUMINA APV WITH TCS

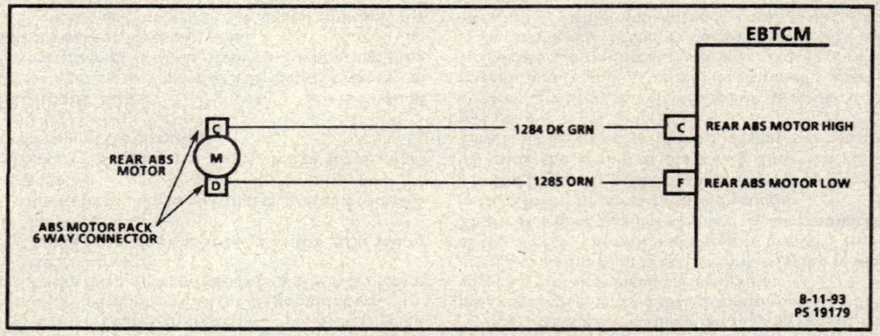

DTC 66
REAR ABS MOTOR CIRCUIT SHORTED TO BATTERY

Circuit Description:

This DTC identifies an ABS motor circuit that is shorted to battery or an ABS motor that has low or zero resistance. This malfunction will not allow the ABS motor to be controlled at the commanded current rate or will cause the ABS motor to turn in the opposite direction or not turn at all.

DTC Will Set When: DTC 66 can be set only when the ABS motor is commanded "OFF." If the EBTCM detects an out of range voltage on either of the rear ABS motor circuits indicating a circuit shorted to battery or an ABS motor shorted, a malfunction exists.

Action Taken: A malfunction DTC is stored, ABS is disabled and the ABS warning lamp is turned "ON."

DTC Chart Test Description: Number(s) below refer to circled number(s) on the diagnostic chart.

1. Checks for a short to voltage in the ABS motor "HIGH" circuitry.
2. Checks for a short to voltage in the ABS motor "LOW" circuitry.
3. Checks for a short circuit between motor "HIGH" and motor "LOW."
4. Checks for proper resistance of the ABS motor.
5. Ensures malfunction was not due to physical damage of the circuitry.

Diagnostic Aids: Using Tech 1, select manual control function, and exercise ABS motor movement of affected channel in both directions while applying light pressure on the brake pedal.

If erratic or "jumpy" brake pedal movement is detected while performing an "apply" or "release" function of the ABS motor, an intermittent malfunction may be indicated.

An intermittent malfunction may be caused by a poor connection, rubbed through wire insulation, or a wire that is broken inside the insulation.

If the malfunction is not current, wiggle the wires of the affected channel and check if the DTC resets. This will help to pinpoint an intermittent malfunction in the ABS motor circuitry or connections.

The frequency of the malfunction can be checked by using the enhanced diagnostic function of the Tech 1, as described in "Tech 1 Diagnostics" in this section.

Any circuitry, that is suspected as causing the intermittent complaint, should be thoroughly checked for backed out terminals, improper mating, broken locks, improperly formed or damaged terminals, poor terminal to wiring connections or physical damage to the wiring harness.

DIAGNOSTIC CHART DTC 66 — 1994 SILHOUETTE, TRANS SPORT AND LUMINA APV WITH TCS, CONTINUED

DTC 66
REAR ABS MOTOR CIRCUIT SHORTED TO BATTERY

NOTICE: J 39200 TEST LEADS MUST BE "ZEROED" PRIOR TO MAKING ANY RESISTANCE MEASUREMENTS. REFER TO J 39200 USER'S MANUAL.

DOES DTC 66 OCCUR INTERMITTENTLY?

- YES: REFER TO "DIAGNOSTIC AIDS" ON FACING PAGE.
- NO: go to 1.

(1)
- IGNITION "OFF."
- DISCONNECT THE 6-WAY ABS MOTOR PACK CONNECTOR AND THE GRAY 8-WAY EBTCM CONNECTOR.
- IGNITION "ON."
- USING J 39200, MEASURE THE VOLTAGE BETWEEN GROUND AND TERMINAL "C" OF THE 6-WAY ABS MOTOR PACK HARNESS CONNECTOR. IS VOLTAGE 1 VOLT OR LESS?

- NO: REPAIR SHORT TO VOLTAGE IN CKT 1284.
- YES: go to 2.

(2)
- USING J 39200, MEASURE THE VOLTAGE BETWEEN GROUND AND TERMINAL "D" OF THE 6-WAY ABS MOTOR PACK HARNESS CONNECTOR. IS VOLTAGE 1 VOLT OR LESS?

- NO: REPAIR SHORT TO VOLTAGE IN CKT 1285.
- YES: go to 3.

(3)
- IGNITION "OFF."
- USING J 39200, MEASURE THE RESISTANCE BETWEEN TERMINAL "C" AND TERMINAL "D" OF THE 6-WAY ABS MOTOR PACK HARNESS CONNECTOR. DOES J 39200 DISPLAY RESISTANCE AS "OL" (INFINITE)?

- NO: REPAIR SHORT BETWEEN CKTs 1284 AND 1285.
- YES: go to 4.

(4)
- USING J 39200, MEASURE THE RESISTANCE BETWEEN TERMINAL "C" AND TERMINAL "D" OF THE ABS MOTOR PACK. IS RESISTANCE GREATER THAN 0.4Ω?

- NO: REPLACE ABS MOTOR PACK
- YES: go to 5.

(5)
- INSPECT CKTs 1284 AND 1285 AND RELATED CONNECTORS FOR DAMAGE WHICH MAY RESULT IN A SHORT TO VOLTAGE WITH ALL CONNECTORS CONNECTED. REPAIR ALL DAMAGE IF EVIDENT.
- RECONNECT ALL CONNECTORS.
- IGNITION "OFF."
- START ENGINE AND TEST DRIVE VEHICLE, ACHIEVING AT LEAST 16 km/h (10 MPH).
- REPEAT ABOVE DRIVE CYCLE SEQUENCE TWO MORE TIMES. DID DTC 66 SET IN THE LAST THREE DRIVE CYCLES?

- YES: REPLACE EBTCM.
- NO: MALFUNCTION WAS MOST LIKELY CAUSED BY A POOR CONNECTION. REFER TO "DIAGNOSTIC AIDS" ON FACING PAGE.

AFTER DIAGNOSIS IS COMPLETE, CLEAR DTCs AND TEST DRIVE VEHICLE FOR THREE (3) DRIVE CYCLES TO VERIFY THAT DTC DOES NOT RESET. A DRIVE CYCLE CONSISTS OF STARTING THE VEHICLE, DRIVING OVER 16 km/h (10 MPH), AND THEN TURNING KEY OFF.

DIAGNOSTIC CHART DTC 75 — 1994 SILHOUETTE, TRANS SPORT AND LUMINA APV WITH TCS

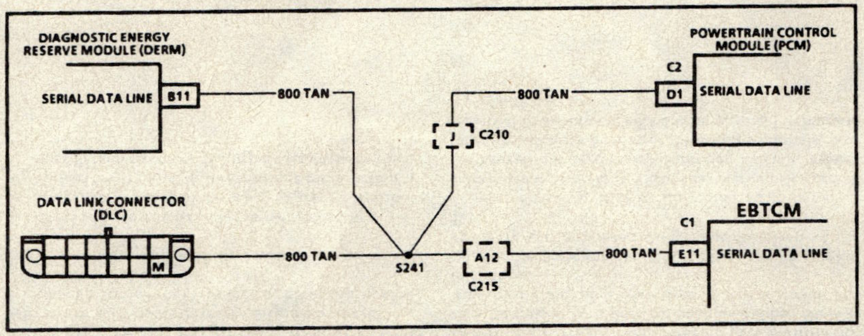

DTC 75

PCM TO EBTCM SERIAL DATA MALFUNCTION

Circuit Description:

The serial data communications circuit is used to transmit messages from one electronic control unit to another when interface is necessary between systems. ABS VI/TCS depends on information from the PCM for proper operation.

DTC Will Set When: DTC 75 will set when there is a malfunction on the serial data circuit, or if the PCM has not broadcasted a serial message for at least 10 seconds.

Action Taken: A malfunction DTC is stored, TCS is disabled, and the TCS warning lamp is turned "ON." No communication message will be displayed on Tech 1 until fault is removed.

DTC Chart Test Description: Number(s) below refer to circled number(s) on the diagnostic chart.

1. This step checks for a malfunction in the test equipment.
2. Checks other systems on the circuit.
3. Checks serial data CKT 800 for short to ground.
4. Checks serial data CKT 800 for short to B+.
5. Checks for open serial data CKT 800.
6. Checks function of control modules on serial data CKT 800.

Diagnostic Aids: When DTC 75 is current, no communication can be established with the Tech 1 scan tool. When the fault is removed, DTC 75 will be stored in system memory.

An intermittent malfunction may be caused by a poor connection, rubbed through wire insulation, or a wire that is broken inside the insulation.

The frequency of the malfunction can be checked by using the enhanced diagnostic function of the Tech 1, as described in "Tech 1 Diagnostics" in this section.

Any circuitry, that is suspected as causing the intermittent complaint, should be thoroughly checked for backed out terminals, improper mating, broken locks, improperly formed or damaged terminals, poor terminal to wiring connections or physical damage to the wiring harness.

DIAGNOSTIC CHART DTC 75 — 1994 SILHOUETTE, TRANS SPORT AND LUMINA APV WITH TCS, CONTINUED

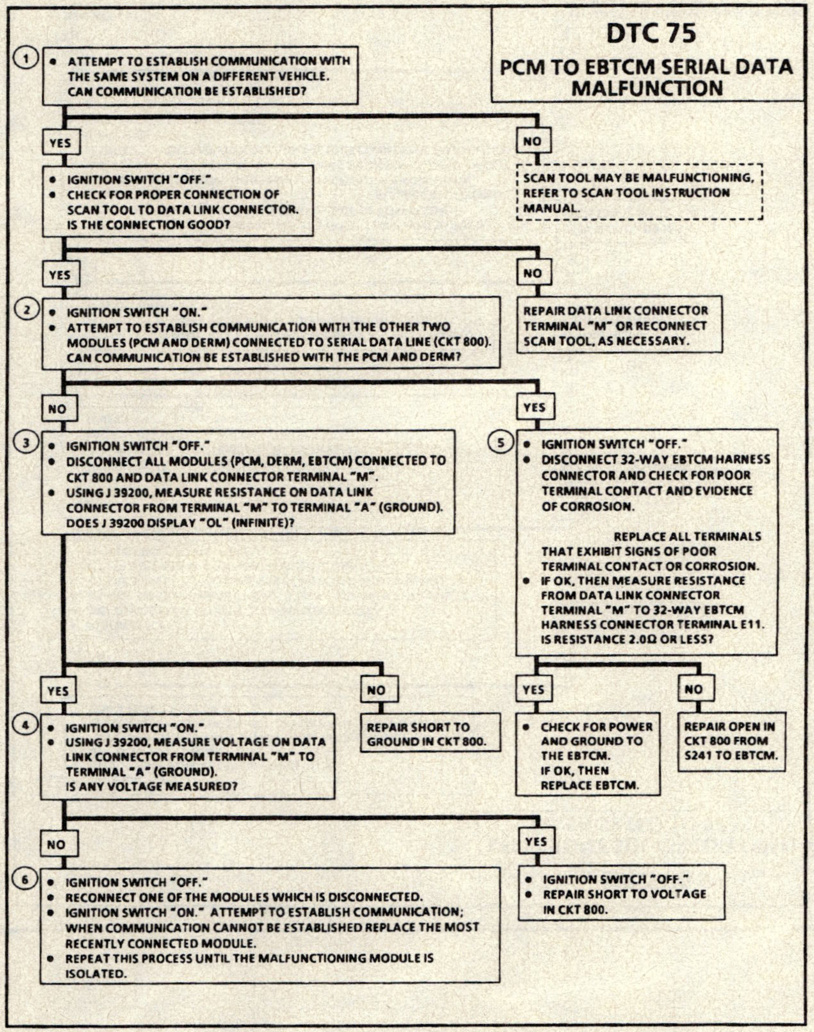

DIAGNOSTIC CHART DTC 76 — 1994 SILHOUETTE, TRANS SPORT AND LUMINA APV WITH TCS

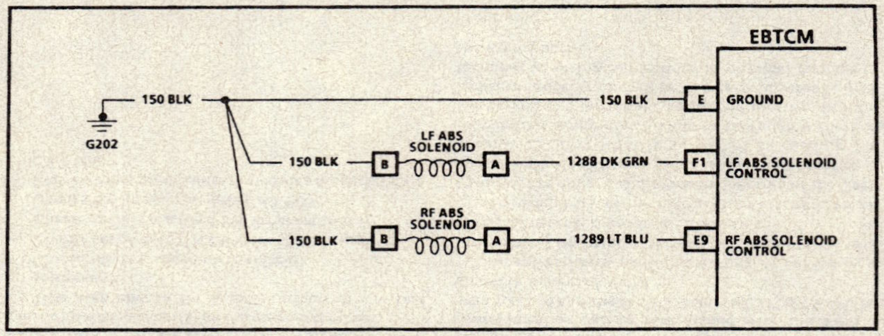

DTC 76

LEFT FRONT SOL CIRCUIT OPEN OR SHORTED TO BATT

Circuit Description:

This DTC identifies a solenoid that cannot be energized due to an open in its circuitry, or a solenoid that is always energized due to a short to battery in its circuitry between the driver and the solenoid. An open will not allow proper ABS operation, but the short to battery simply turns "ON" the solenoid. A path for base brakes is still allowed once the ABS motor re-homes and the check ball is lifted off its seat during key "ON" initialization.

DTC Will Set When: DTC 76 can be set only when the solenoid is commanded "OFF." If the EBTCM detects an excessive voltage in the left front solenoid control circuit, a malfunction exists.

Action Taken: A malfunction DTC is stored, ABS is disabled and the ABS warning lamp is turned "ON."

DTC Chart Test Description: Number(s) below refer to circled number(s) on the diagnostic chart.

1. Checks for a short to voltage in the control circuitry of the solenoid.
2. Checks for an open in the control circuit of the solenoid.
3. Checks the solenoid coil for proper resistance.
4. Checks for an open in the solenoid ground circuit.
5. Checks for a possible intermittent malfunction in the solenoid circuitry due to poor terminal contact.

Diagnostic Aids: An intermittent malfunction may be caused by a poor connection, rubbed through wire insulation, or a wire that is broken inside the insulation.

The frequency of the malfunction can be checked by using the enhanced diagnostic function of the Tech 1.

Any circuitry, that is suspected as causing the intermittent complaint, should be thoroughly checked for backed out terminals, improper mating, broken locks, improperly formed or damaged terminals, poor terminal to wiring connections or physical damage to the wiring harness.

DIAGNOSTIC CHART DTC 76 — 1994 SILHOUETTE, TRANS SPORT AND LUMINA APV WITH TCS, CONTINUED

DTC 76

LEFT FRONT SOL CIRCUIT OPEN OR SHORTED TO BATT

(1)
- IGNITION "OFF."
- DISCONNECT 32-WAY EBTCM CONNECTOR.
- IGNITION "ON" ENGINE "OFF."
- USING J 39200, MEASURE THE VOLTAGE BETWEEN GROUND AND TERMINAL "F1" OF THE 32-WAY EBTCM HARNESS CONNECTOR. IS VOLTAGE 1 VOLT OR LESS?

NO: REPAIR SHORT TO VOLTAGE IN CKT 1288.

YES:

(2)
- IGNITION "OFF."
- DISCONNECT LEFT FRONT SOLENOID CONNECTOR.
- USING J 39200, MEASURE THE RESISTANCE BETWEEN TERMINAL "F1" OF THE 32-WAY EBTCM HARNESS CONNECTOR AND TERMINAL "A" OF THE LEFT FRONT SOLENOID HARNESS CONNECTOR. IS RESISTANCE 2Ω OR LESS?

NO: REPAIR OPEN OR HIGH RESISTANCE IN CKT 1288.

YES:

(3)
- USING J 39200, MEASURE THE RESISTANCE BETWEEN TERMINAL "A" AND TERMINAL "B" OF THE LEFT FRONT SOLENOID. IS RESISTANCE BETWEEN 2.5 AND 5Ω?

NO: REPLACE LEFT FRONT SOLENOID.

YES:

(4)
- USING J 39200, MEASURE THE RESISTANCE BETWEEN GROUND AND TERMINAL "B" OF THE LEFT FRONT SOLENOID HARNESS CONNECTOR. IS RESISTANCE 2Ω OR LESS?

NO: REPAIR OPEN OR HIGH RESISTANCE IN CKT 150.

YES:

(5)
- INSPECT LEFT FRONT SOLENOID AND 32-WAY EBTCM HARNESS CONNECTORS FOR POOR TERMINAL CONTACT AND EVIDENCE OF CORROSION. REPLACE ALL TERMINALS THAT EXHIBIT SIGNS OF POOR TERMINAL CONTACT OR CORROSION.
- INSPECT CKT 1288 FOR DAMAGE WHICH MAY RESULT IN AN OPEN CIRCUIT OR A SHORT TO VOLTAGE WITH ALL CONNECTORS CONNECTED. REPAIR DAMAGE IF EVIDENT.
- RECONNECT ALL CONNECTORS.
- IGNITION "ON."

DOES DTC 76 SET AS A CURRENT DTC?

YES: REPLACE EBTCM.

NO: MALFUNCTION IS MOST LIKELY DUE TO A POOR CONNECTION.

DIAGNOSTIC CHART DTC 77 — 1994 SILHOUETTE, TRANS SPORT AND LUMINA APV WITH TCS

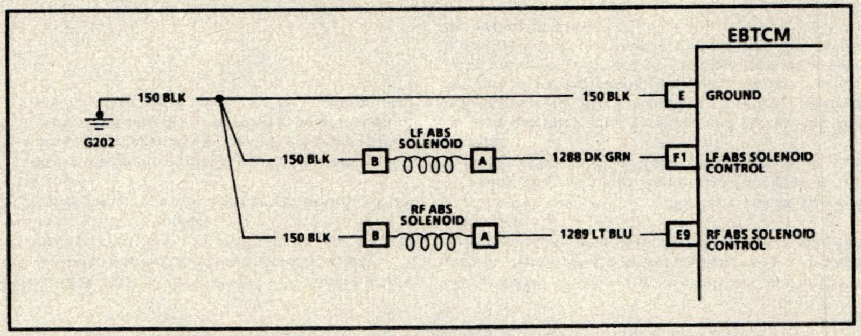

DTC 77

LEFT FRONT SOL CIRCUIT SHORTED TO GND

Circuit Description:

This DTC identifies a solenoid that cannot be energized due to an open in its driver circuitry, or a short to ground between the solenoid driver and the solenoid. These malfunctions can affect ABS operation since the flow of brake fluid to the wheel cylinder cannot be stopped, making ABS operation for that channel impossible.

DTC Will Set When: DTC 77 can be set only when the solenoid is commanded "ON." If the EBTCM detects the left front solenoid control circuit voltage is out of specification, a malfunction exists.

Action Taken: A malfunction DTC is stored, ABS is disabled and the ABS warning lamp is turned "ON."

DTC Chart Test Description: Number(s) below refer to circled number(s) on the diagnostic chart.

1. Checks for a solenoid that is internally shorted to ground.
2. Checks for a solenoid that is not within proper resistance values.
3. Indicates if a short to ground exists in the solenoid circuitry.
4. Checks for a possible intermittent malfunction in the solenoid circuitry due to poor terminal contact.

Diagnostic Aids: An intermittent malfunction may be caused by a poor connection, rubbed through wire insulation, or a wire that is broken inside the insulation.

The frequency of the malfunction can be checked by using the enhanced diagnostic function of the Tech 1, as described in "Tech 1 Diagnostics" in this section.

Any circuitry, that is suspected as causing the intermittent complaint, should be thoroughly checked for backed out terminals, improper mating, broken locks, improperly formed or damaged terminals, poor terminal to wiring connections or physical damage to the wiring harness.

DIAGNOSTIC CHART DTC 77 — 1994 SILHOUETTE, TRANS SPORT AND LUMINA APV WITH TCS, CONTINUED

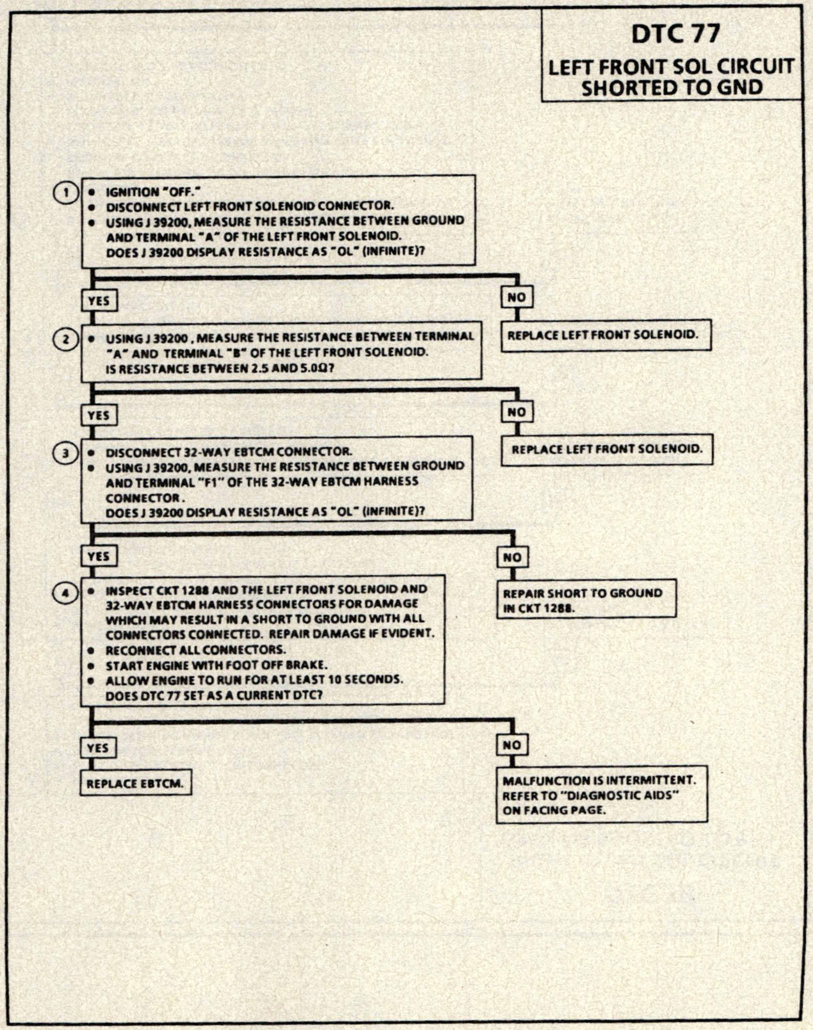

DIAGNOSTIC CHART DTC 78 — 1994 SILHOUETTE, TRANS SPORT AND LUMINA APV WITH TCS

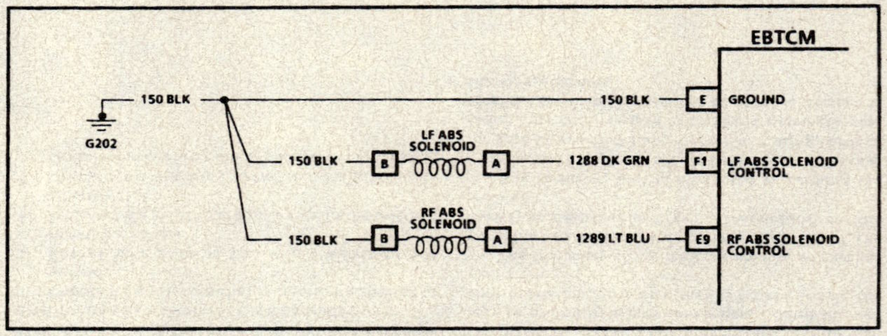

DTC 78

RIGHT FRONT SOL CIRCUIT OPEN OR SHORTED TO BATT

Circuit Description:

This DTC identifies a solenoid that cannot be energized due to an open in its circuitry, or a solenoid that is always energized due to a short to battery in its circuitry between the driver and the solenoid. An open will not allow proper ABS operation, but the short to battery simply turns "ON" the solenoid. A path for base brakes is still allowed once the ABS motor re-homes and the check ball is lifted off its seat during key "ON" initialization.

DTC Will Set When: DTC 78 can be set only when the solenoid is commanded "OFF." If the EBTCM detects an excessive voltage in the right front solenoid control circuit, a malfunction exists.

Action Taken: A malfunction DTC is stored, ABS is disabled and the ABS warning lamp is turned "ON."

DTC Chart Test Description: Number(s) below refer to circled number(s) on the diagnostic chart.

1. Checks for a short to voltage in the control circuitry of the solenoid.
2. Checks for an open in the control circuit of the solenoid.
3. Checks the solenoid coil for proper resistance.
4. Checks for an open in the solenoid ground circuit.
5. Checks for a possible intermittent malfunction in the solenoid circuitry due to poor terminal contact.

Diagnostic Aids: An intermittent malfunction may be caused by a poor connection, rubbed through wire insulation, or a wire that is broken inside the insulation.

The frequency of the malfunction can be checked by using the enhanced diagnostic function of the Tech 1, as described in "Tech 1 Diagnostics" in this section.

Any circuitry, that is suspected as causing the intermittent complaint, should be thoroughly checked for backed out terminals, improper mating, broken locks, improperly formed or damaged terminals, poor terminal to wiring connections or physical damage to the wiring harness.

DIAGNOSTIC CHART DTC 78 — 1994 SILHOUETTE, TRANS SPORT AND LUMINA APV WITH TCS, CONTINUED

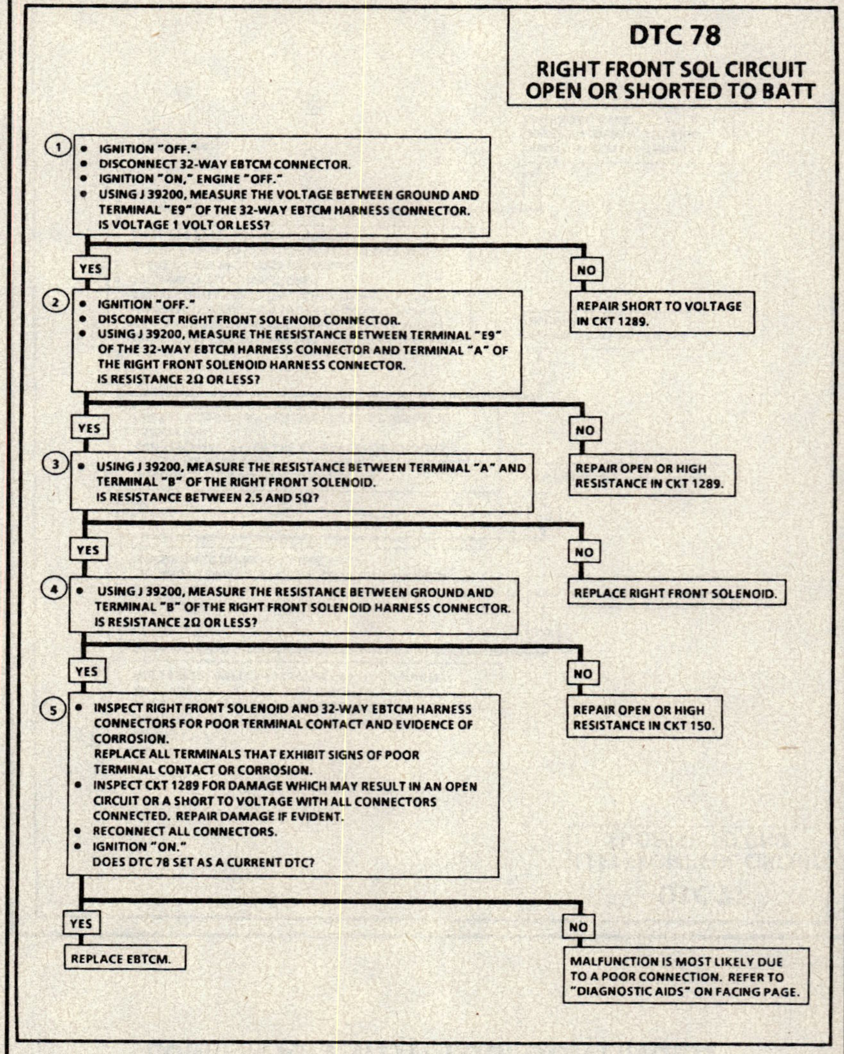

DIAGNOSTIC CHART DTC 81 — 1994 SILHOUETTE, TRANS SPORT AND LUMINA APV WITH TCS

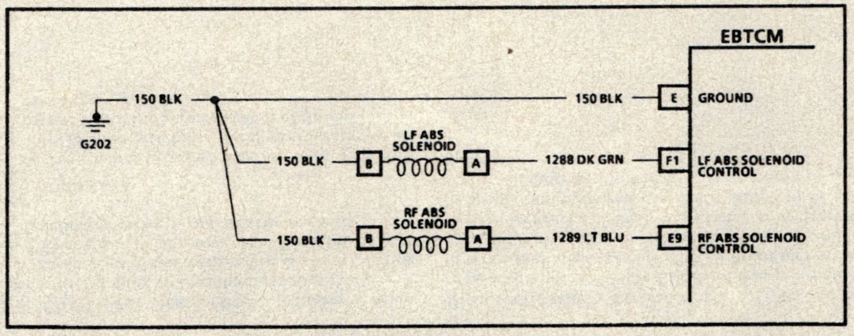

DTC 81

RIGHT FRONT SOL CIRCUIT SHORTED TO GND

Circuit Description:
This DTC identifies a solenoid that cannot be energized due to an open in its driver circuitry, or a short to ground between the solenoid driver and the solenoid. These malfunctions can affect ABS operation since the flow of brake fluid to the wheel cylinder cannot be stopped, making ABS operation for that channel impossible.

DTC Will Set When: DTC 81 can be set only when the solenoid is commanded "ON." If the EBTCM detects the right front solenoid control circuit voltage is out of specification, a malfunction exists.

Action Taken: A malfunction DTC is stored, ABS is disabled and the ABS warning lamp is turned "ON."

DTC Chart Test Description: Number(s) below refer to circled number(s) on the diagnostic chart.

1. Checks for a solenoid that is internally shorted to ground.
2. Checks for a solenoid that is not within proper resistance values.
3. Indicates if a short to ground exists in the solenoid circuitry.
4. Checks for a possible intermittent malfunction in the solenoid circuitry due to poor terminal contact.

Diagnostic Aids: An intermittent malfunction may be caused by a poor connection, rubbed through wire insulation, or a wire that is broken inside the insulation.

The frequency of the malfunction can be checked by using the enhanced diagnostic function of the Tech 1, as described in "Tech 1 Diagnostics" in this section.

Any circuitry, that is suspected as causing the intermittent complaint, should be thoroughly checked for backed out terminals, improper mating, broken locks, improperly formed or damaged terminals, poor terminal to wiring connections or physical damage to the wiring harness.

DIAGNOSTIC CHART DTC 81 — 1994 SILHOUETTE, TRANS SPORT AND LUMINA APV WITH TCS, CONTINUED

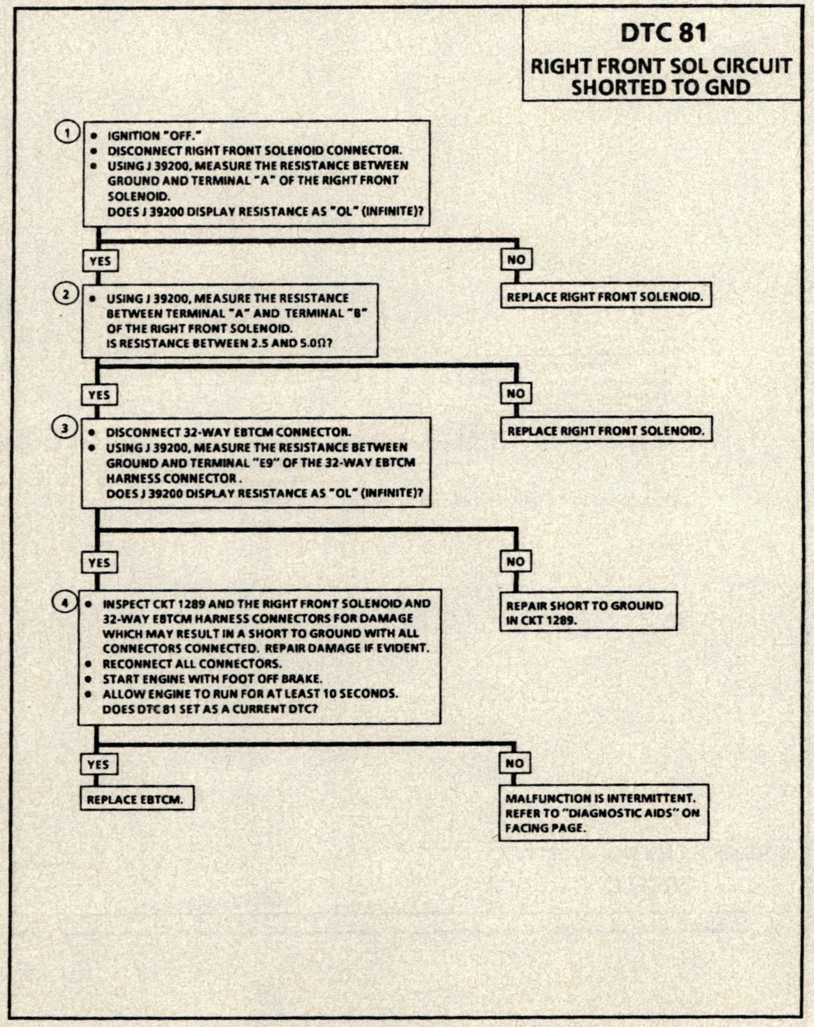

DIAGNOSTIC CHART DTC 82 — 1994 SILHOUETTE, TRANS SPORT AND LUMINA APV WITH TCS

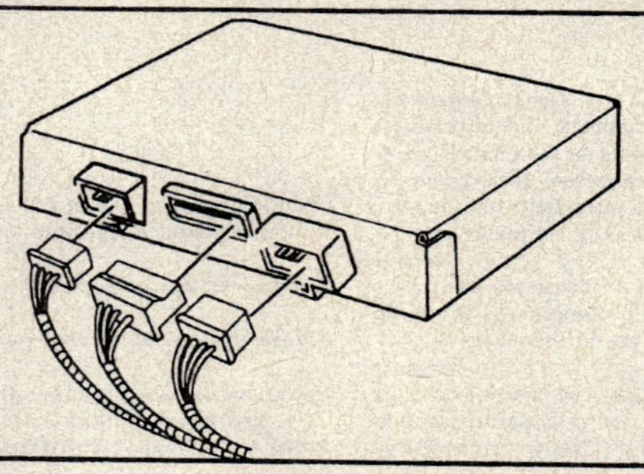

DTC 82
CALIBRATION MALFUNCTION

Circuit Description:
This DTC allows the EBTCM to check for a calibration malfunction by comparing the calibration value to a known value stored in the EEPROM.
This DTC is also used as a security measure to prevent improper use of calibrations or changes to these calibrations that may alter the designed function of ABS and TCS.

DTC Will Set When: DTC 82 can be set at key-up. If the program identifier is incorrect or the memory checksum is incorrect, a malfunction exists.

Action Taken: A malfunction DTC is stored, ABS and TCS are disabled and the ABS and TCS warning lamps are turned "ON."

DTC Chart Test Description: Number(s) below refer to circled number(s) on the diagnostic chart.
1. Checks to see if the malfunction is present during diagnosis. If present, the EBTCM is not functioning correctly and must be replaced.

Important
- At the time of printing, 1994 vehicles with ABS VI/TCS were not being field reprogrammed to correct this failure, replacement of the EBTCM is necessary.

Diagnostic Aids: An intermittent DTC 82 may be caused by a bad cell in the EEPROM that is sensitive to temperature changes. If DTC 82 failed more than once, but is intermittent, replace EBTCM.
The frequency of the malfunction can be checked by using the enhanced diagnostic function of the Tech 1, as described in "Tech 1 Diagnostics" in this section.

DIAGNOSTIC CHART DTC 82 — 1994 SILHOUETTE, TRANS SPORT AND LUMINA APV WITH TCS, CONTINUED

DTC 82
CALIBRATION MALFUNCTION

(1) • IGNITION "ON."
IS DTC 82 SET AS A CURRENT DTC?

YES → REPLACE EBTCM.

NO → MALFUNCTION IS NOT PRESENT AT THIS TIME. REFER TO "DIAGNOSTIC AIDS" ON FACING PAGE.

DIAGNOSTIC CHART DTC 86 — 1994 SILHOUETTE, TRANS SPORT AND LUMINA APV WITH TCS

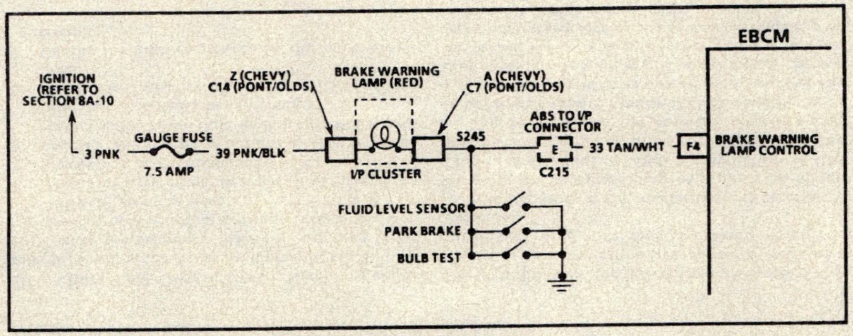

DTC 86

EBTCM TURNED "ON" THE RED "BRAKE" WARNING LAMP

Circuit Description:

This DTC is provided as an information only DTC, and reflects the status of the command issued by the ABS VI/TCS EBTCM to illuminate the red "BRAKE" warning lamp. If another DTC issues a command to illuminate the red "BRAKE" warning lamp, DTC 86 will be stored in EEPROM as a history DTC at the conclusion of the ignition cycle.

DTC Chart Test Description: Number(s) below refer to circled number(s) on the diagnostic chart.

1. This step serves to identify if a DTC other than DTC 86 commanded the red "BRAKE" warning lamp "ON."

Diagnostic Aids: Any ABS mechanical DTC that issues a command to illuminate the red "BRAKE" telltale will also result in DTC 86 being stored in EEPROM during shutdown. These DTCs are: 38, 41, 42, 46, and 51. If the motors are not in their home position, certain electrical DTCs will also command the red "BRAKE" telltale "ON." These DTCs are: 14, 16, 18, 55, 56, 57, 58, 61, 62, 63, 64, 65, 66, 144 and 145.

If any of these DTCs are indicated along with DTC 86, they must be corrected prior to addressing a DTC 86 malfunction.

DIAGNOSTIC CHART DTC 86 — 1994 SILHOUETTE, TRANS SPORT AND LUMINA APV WITH TCS, CONTINUED

DTC 86

EBTCM TURNED "ON" THE RED "BRAKE" WARNING LAMP

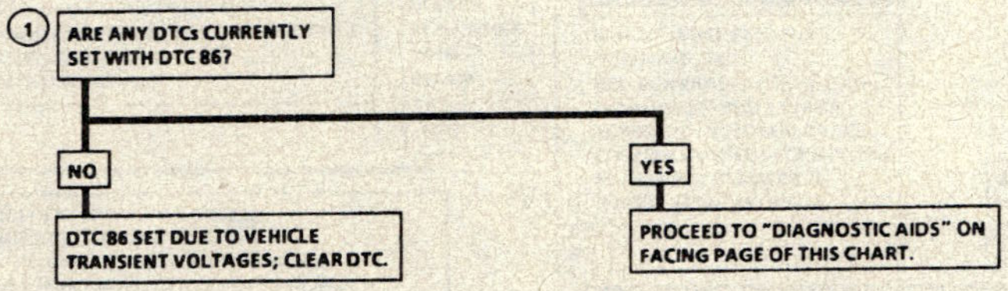

DIAGNOSTIC CHART DTC 87 — 1994 SILHOUETTE, TRANS SPORT AND LUMINA APV WITH TCS

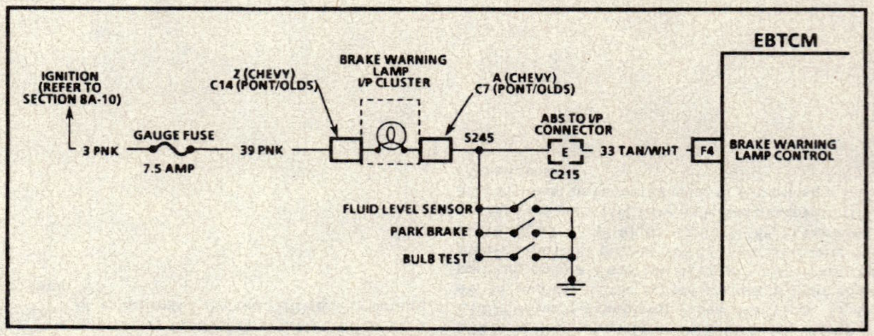

DTC 87

RED "BRAKE" WARNING LAMP CIRCUIT OPEN

Circuit Description:

This DTC is used to verify the EBTCM has continuity to the red "BRAKE" warning lamp in case the EBTCM must turn it "ON." This will only occur if an ABS or TCS malfunction is detected that may degrade base brake operation.

Because ABS VI/TCS is not the only device controlling the red "BRAKE" warning lamp (parking brake and low fluid sensor may also turn "ON" the lamp) a short to ground in this circuit cannot be detected by the EBTCM.

DTC Will Set When: DTC 87 can be set after initialization is completed. If the red "BRAKE" warning lamp circuit voltage is out of specification indicating an open circuit, a malfunction exists.

Action Taken: A malfunction DTC is stored, ABS is not disabled.

DTC Chart Test Description: Number(s) below refer to circled number(s) on the diagnostic chart.

1. Checks the complete "BRAKE" warning lamp circuitry by attempting to manually turn "ON" the "BRAKE" warning lamp.
2. Checks for high resistance in park brake circuitry.
3. Checks for a malfunctioning EBTCM.
4. Ensures malfunction is not due to poor terminal contact at the brake fluid level switch which may result in an intermittent malfunction.
5. Checks the integrity of the "BRAKE" warning lamp.
6. Checks for high resistance in the park brake circuitry.

Diagnostic Aids: An intermittent malfunction may be caused by a poor connection, rubbed through wire insulation, or a wire that is broken inside the insulation.

The frequency of the malfunction can be checked by using the enhanced diagnostic function of the Tech 1, as described in "Tech 1 Diagnostics" in this section. If a Tech 1 or T-100 are not available, the parking brake handle may be lifted to verify proper lamp operation and continuity of CKTs 3 and 39.

Any circuitry, that is suspected as causing the intermittent complaint, should be thoroughly checked for backed out terminals, improper mating, broken locks, improperly formed or damaged terminals, poor terminal to wiring connections or physical damage to the wiring harness.

DIAGNOSTIC CHART DTC 87 — 1994 SILHOUETTE, TRANS SPORT AND LUMINA APV WITH TCS, CONTINUED

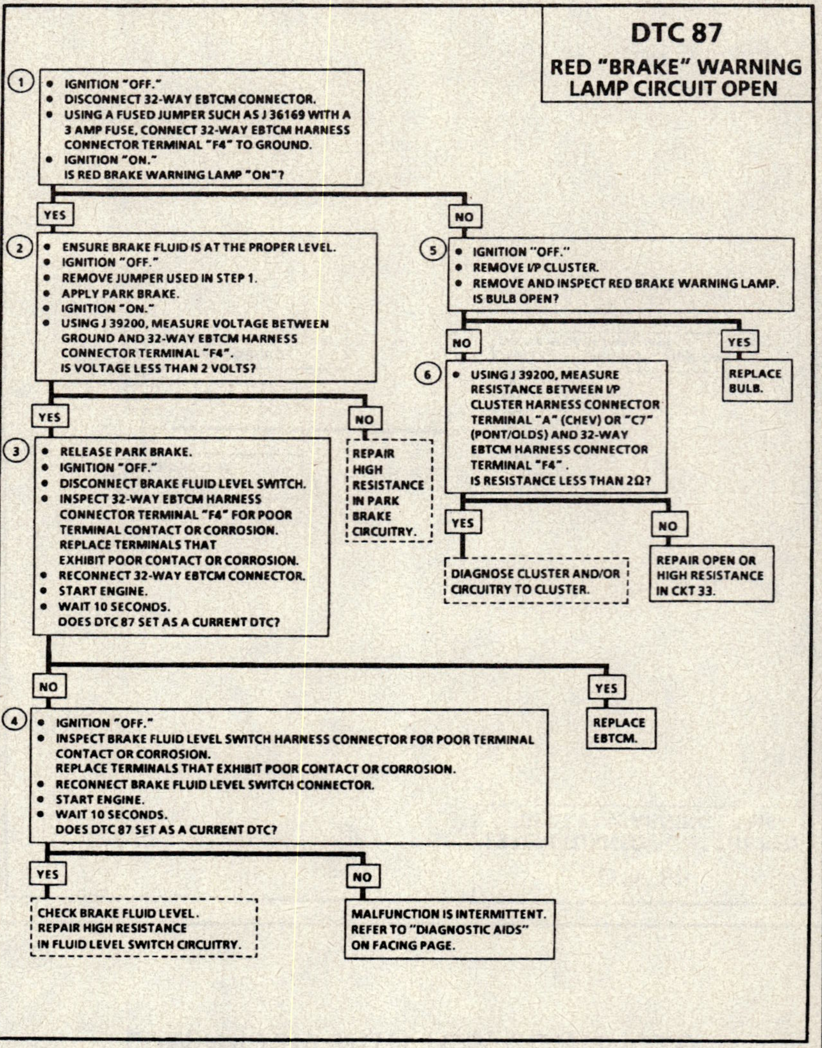

DIAGNOSTIC CHART DTC 88 — 1994 SILHOUETTE, TRANS SPORT AND LUMINA APV WITH TCS

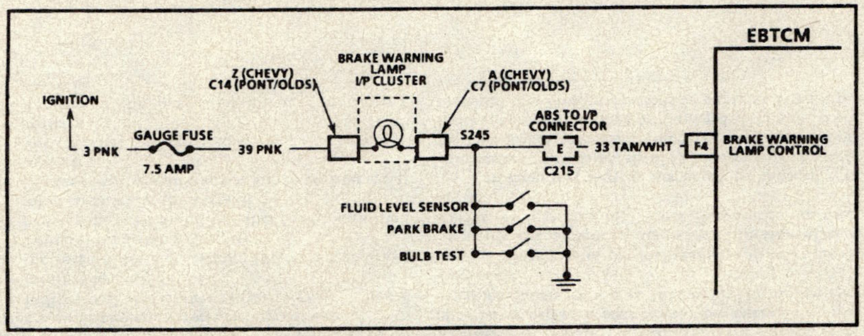

DTC 88

RED "BRAKE" WARNING LAMP CIRCUIT SHORTED TO BATT

Circuit Description:

This DTC identifies a short to battery between the EBTCM and the red "BRAKE" warning lamp, or an open driver that does not allow the red "BRAKE" warning lamp to be illuminated by the EBTCM. This will only occur if an ABS malfunction is detected that may degrade base brake operation.

DTC Will Set When: DTC 88 can be set only at the beginning of the three second bulb check. If the EBTCM detects battery voltage on the red "BRAKE" warning lamp control circuit, a malfunction exists.

Action Taken: A malfunction DTC is stored. ABS is not disabled.

DTC Chart Test Description: Number(s) below refer to circled number(s) on the diagnostic chart.

1. Identifies if the ground circuit to the red "BRAKE" warning lamp is being completed by a source other than the EBTCM.
2. Identifies if the malfunction is currently present.
3. By removing the "GAUGES" fuse, the voltage source is eliminated. This test indicates if voltage is being supplied from a source other than the gauge fuse.
4. Ensures malfunction was not due to physical damage of the circuitry.
5. Checks for DTC 86 also being set.

Diagnostic Aids: An intermittent malfunction may be caused by a poor connection, rubbed through wire insulation, or a wire that is broken inside the insulation.

The frequency of the malfunction can be checked by using the enhanced diagnostic function of the Tech 1, as described in "Tech 1 Diagnostics" in this section.

Any circuitry, that is suspected as causing the intermittent complaint, should be thoroughly checked for backed out terminals, improper mating, broken locks, improperly formed or damaged terminals, poor terminal to wiring connections or physical damage to the wiring harness.

DIAGNOSTIC CHART DTC 88 — 1994 SILHOUETTE, TRANS SPORT AND LUMINA APV WITH TCS, CONTINUED

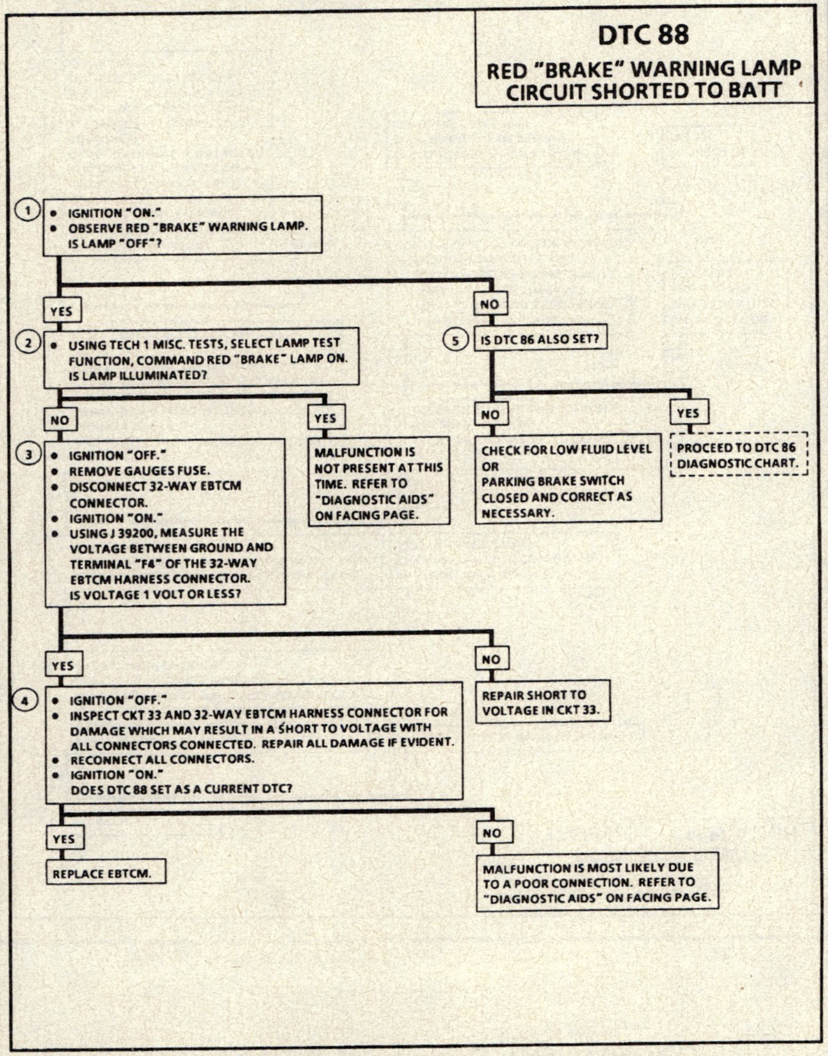

DIAGNOSTIC CHART DTC 91 — 1994 SILHOUETTE, TRANS SPORT AND LUMINA APV WITH TCS

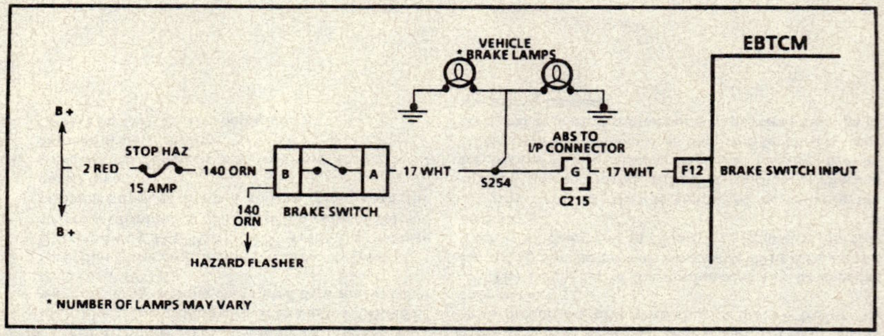

DTC 91

OPEN BRAKE SWITCH DURING DECELERATION

Circuit Description:

This DTC is used to detect an open brake switch in the non-ABS mode. The EBTCM looks for deceleration rates that would indicate braking action and verifies this assumption by requiring several repeats of this detection method. In each case, ABS and TCS will not be available since no brake switch is seen.

DTC Will Set When: DTC 91 can be set if three deceleration cycles occur with the brake switch "OFF."

Action Taken: A malfunction DTC is stored, ABS and TCS are disabled and the ABS and TCS warning lamps are turned "ON."

DTC Chart Test Description: Number(s) below refer to circled number(s) on the diagnostic chart.

1. Indicates if the brake switch signal is being received by the EBTCM.
2. Indicates if an open circuit exists in the brake switch or brake lamp circuitry.
3. Isolates the open circuit to either the brake switch input circuitry, or the EBTCM.
4. Ensures malfunction was not due to poor terminal contact.
5. Verifies that voltage is available at the brake switch.
6. Verifies if the brake switch is functioning properly.

Diagnostic Aids: An intermittent malfunction may be caused by a poor connection, rubbed through wire insulation, or a wire that is broken inside the insulation.

The frequency of the malfunction can be checked by using the enhanced diagnostic function of the Tech 1, as described in "Tech 1 Diagnostics" in this section.

Any circuitry, that is suspected as causing the intermittent complaint, should be thoroughly checked for backed out terminals, improper mating, broken locks, improperly formed or damaged terminals, poor terminal to wiring connections or physical damage to the wiring harness.

DIAGNOSTIC CHART DTC 91 — 1994 SILHOUETTE, TRANS SPORT AND LUMINA APV WITH TCS, CONTINUED

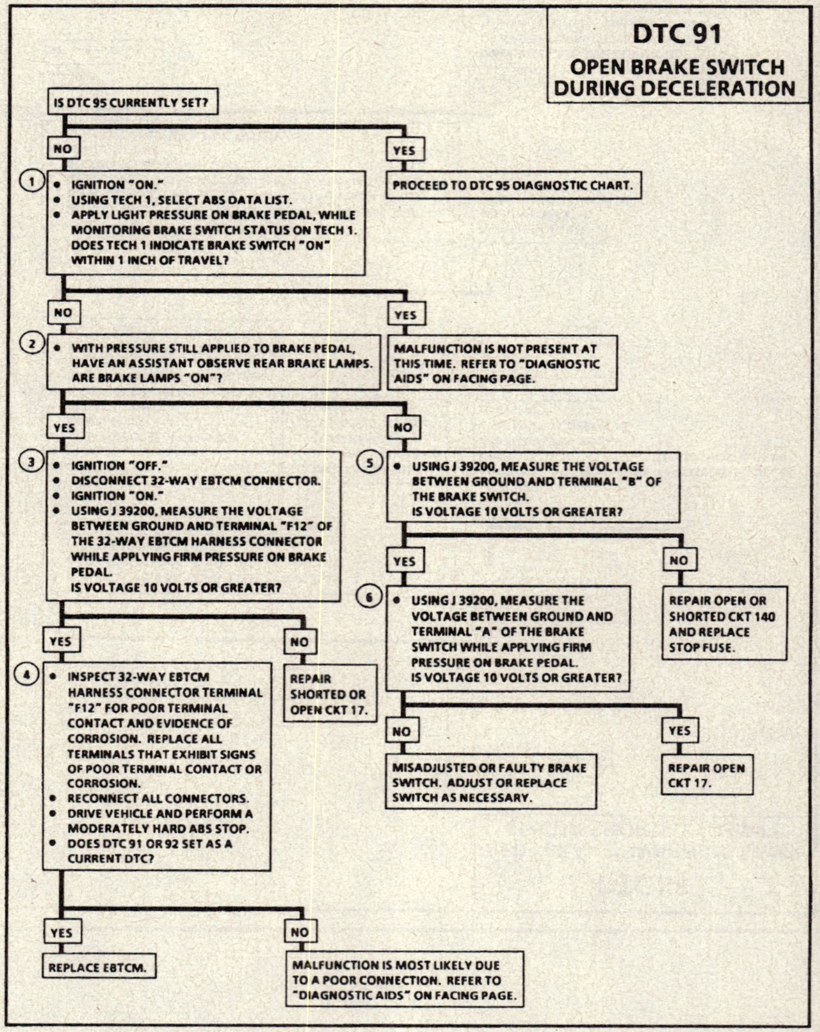

DIAGNOSTIC CHART DTC 92 — 1994 SILHOUETTE, TRANS SPORT AND LUMINA APV WITH TCS

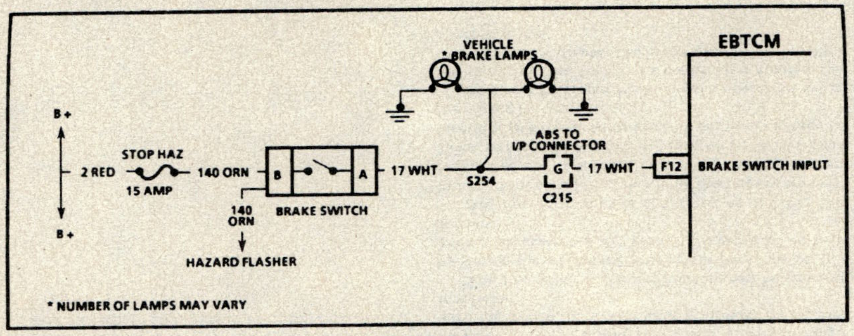

* NUMBER OF LAMPS MAY VARY

DTC 92

OPEN BRAKE SWITCH WHEN ABS WAS REQUIRED

Circuit Description:

This DTC is run to determine the proper operation of the brake switch. This is important because ABS is activated when the brake switch is "ON." If the brake switch is "OFF," ABS will never be activated. Since this malfunction is difficult to detect under normal braking conditions, this malfunction is only detected when ABS is required.

DTC Will Set When: DTC 92 can be set if the vehicle's speed is greater than 8 km/h (5 mph). If the brake was not "ON" and a release was required on two channels for 0.7 second, a malfunction exists.

Action Taken: A malfunction DTC is stored, ABS and TCS are disabled and the ABS and TCS warning lamps are turned "ON."

DTC Chart Test Description: Number(s) below refer to circled number(s) on the diagnostic chart.

1. Indicates if the brake switch signal is being received by the EBTCM.
2. Indicates if an open circuit exists in the brake switch or brake lamp circuitry.
3. Isolates the open circuit to either the brake switch input circuitry, or the EBTCM.
4. Ensures malfunction was not due to poor terminal contact.
5. Verifies the voltage is available at the brake switch.
6. Verifies if the brake switch is functioning properly.

Diagnostic Aids: An intermittent malfunction may be caused by a poor connection, rubbed through wire insulation, or a wire that is broken inside the insulation.

The frequency of the malfunction can be checked by using the enhanced diagnostic function of the Tech 1, as described in "Tech 1 Diagnostics" in this section.

Any circuitry, that is suspected as causing the intermittent complaint, should be thoroughly checked for backed out terminals, improper mating, broken locks, improperly formed or damaged terminals, poor terminal to wiring connections or physical damage to the wiring harness.

DIAGNOSTIC CHART DTC 92 — 1994 SILHOUETTE, TRANS SPORT AND LUMINA APV WITH TCS, CONTINUED

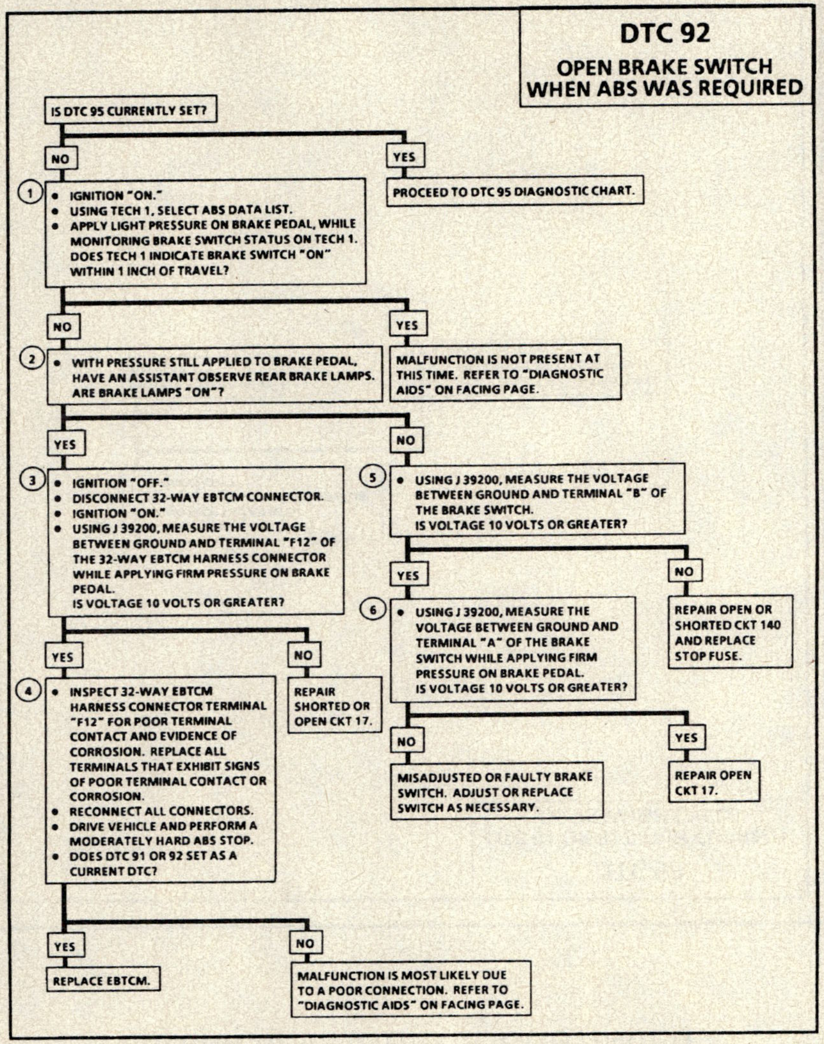

DIAGNOSTIC CHART DTC 93 — 1994 SILHOUETTE, TRANS SPORT AND LUMINA APV WITH TCS

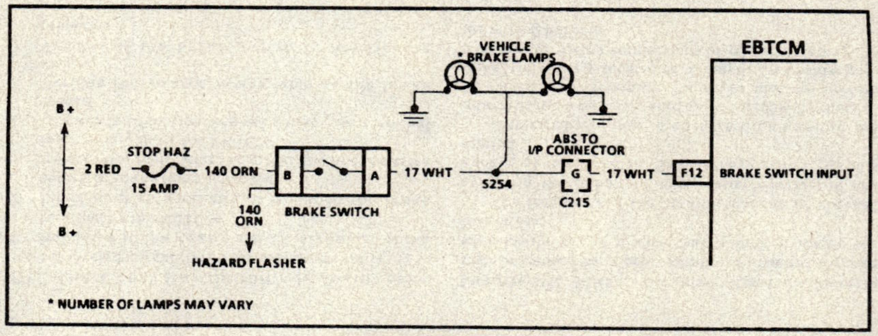

DTC 93

DTC 91 OR 92 SET IN CURRENT OR PREV IGN CYCLE

Circuit Description:

This DTC is the second portion of DTCs 91 and 92. If DTCs 91 or 92 occurred during the last ignition cycle, DTC 93 becomes a current malfunction during the next ignition cycle, keeping ABS disabled until a brake switch "ON" state is seen. When a change is seen during an ignition cycle in which DTC 93 is a current malfunction, DTC 91 or 92 will clear itself at the end of the current ignition cycle, and ABS will enable itself at the start of the next ignition cycle. DTC 93 alone indicates 91 or 92 occurred previously, but is intermittent, or has been corrected.

DTC Chart Test Description: Number(s) below refer to circled number(s) on the diagnostic chart.

1. Indicates which DTC (either 91 or 92) caused DTC 93 to set.

Diagnostic Aids: An intermittent malfunction may be caused by a poor connection, rubbed through wire insulation, or a wire that is broken inside the insulation.

The frequency of the malfunction can be checked by using the enhanced diagnostic function of the Tech 1, as described in "Tech 1 Diagnostics" in this section.

Any circuitry, that is suspected as causing the intermittent complaint, should be thoroughly checked for backed out terminals, improper mating, broken locks, improperly formed or damaged terminals, poor terminal to wiring connections or physical damage to the wiring harness.

Also, verify proper brake switch operation using the data list of the Tech 1. As the brake is applied, the data list should display the brake switch "ON" within 1 inch of travel.

DIAGNOSTIC CHART DTC 93 — 1994 SILHOUETTE, TRANS SPORT AND LUMINA APV WITH TCS, CONTINUED

DTC 93

DTC 91 OR 92 SET IN CURRENT OR PREV IGN CYCLE

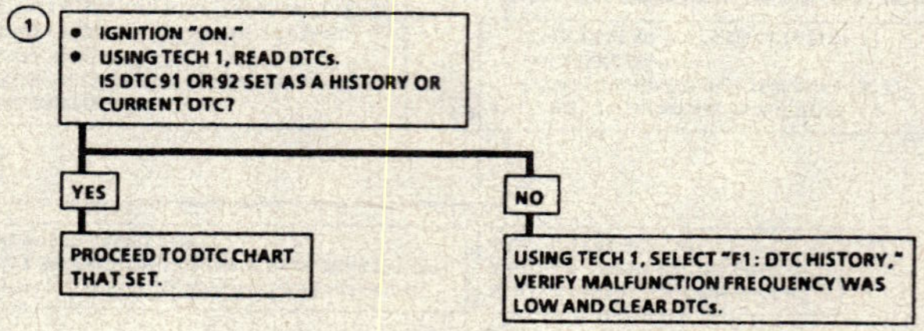

DIAGNOSTIC CHART DTC 94 — 1994 SILHOUETTE, TRANS SPORT AND LUMINA APV WITH TCS

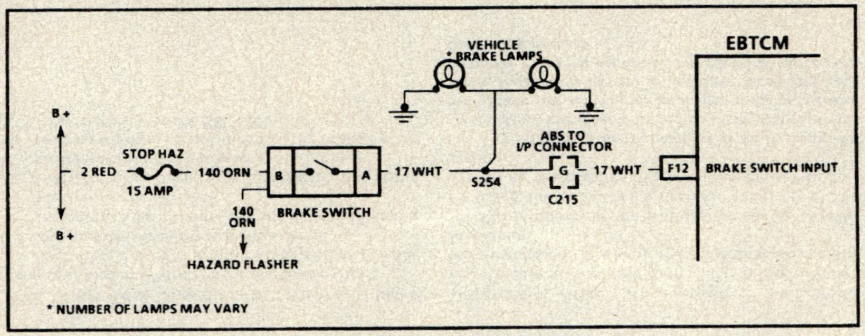

DTC 94

BRAKE SWITCH CONTACTS ALWAYS CLOSED

Circuit Description:

This DTC is run to determine the proper operation of the brake switch. This is important because ABS is activated when the brake switch is "ON" and turned "OFF" when the brake is "OFF." If the brake switch is always "ON," ABS operation will always be requested resulting in potential modulator cycling on rough roads. Additionally, this malfunction will most likely result in a dead battery (due to the brake lights staying "ON") if the driver is not informed of this malfunction.

DTC Will Set When: DTC 94 can be set when the vehicle reaches at least 40 km/h (25 mph). If the brake was never "OFF" during two consecutive drive cycles, a malfunction exists.

Action Taken: A malfunction DTC is stored. ABS is not disabled; however, TCS is disabled and the TCS warning lamp is turned "ON."

DTC Chart Test Description: Number(s) below refer to circled number(s) on the diagnostic chart.

1. Checks if the malfunction is currently present.
2. Isolates the cause of the malfunction to either a malfunctioning or misadjusted brake switch, or a short to voltage in the brake switch circuitry.
3. Checks for unwanted voltage on the brake switch input circuit.
4. Ensures malfunction is not due to poor terminal contact or EBTCM.

Diagnostic Aids: An intermittent malfunction may be caused by a poor connection, rubbed through wire insulation, or a wire that is broken inside the insulation.

The frequency of the malfunction can be checked by using the enhanced diagnostic function of the Tech 1, as described in "Tech 1 Diagnostics" in this section.

Any circuitry, that is suspected as causing the intermittent complaint, should be thoroughly checked for backed out terminals, improper mating, broken locks, improperly formed or damaged terminals, poor terminal to wiring connections or physical damage to the wiring harness.

DIAGNOSTIC CHART DTC 94 — 1994 SILHOUETTE, TRANS SPORT AND LUMINA APV WITH TCS, CONTINUED

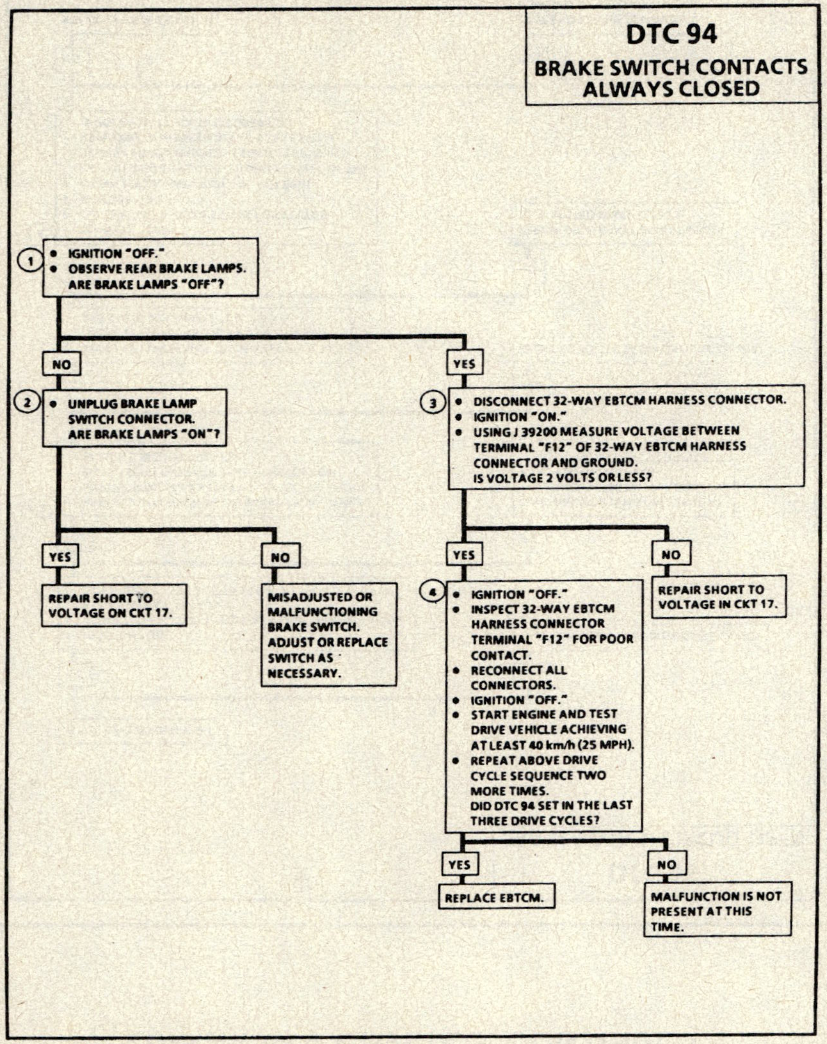

DIAGNOSTIC CHART DTC 95 — 1994 SILHOUETTE, TRANS SPORT AND LUMINA APV WITH TCS

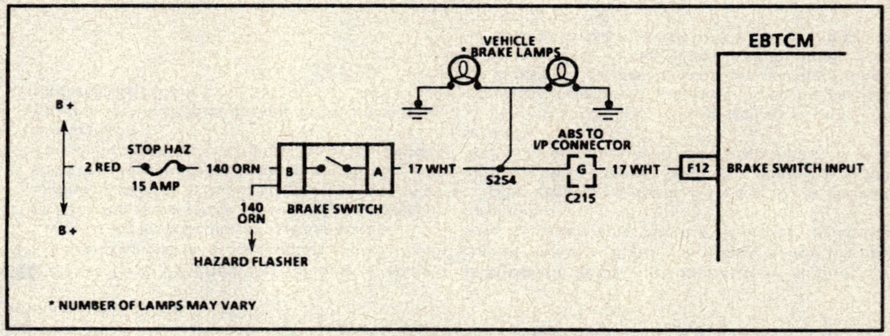

DTC 95

BRAKE SWITCH CIRCUIT OPEN

Circuit Description:
This DTC is used to identify open brake switch circuitry that prevents the brake switch input to the EBTCM from changing states when the brake is applied. This DTC is used in conjunction with DTCs 91 and 92 to determine the cause of an open brake switch malfunction.

DTC Will Set When: DTC 95 can be set after initialization is completed. If the brake switch input voltage is out of specification for one second indicating an open circuit, a malfunction exists.

Action Taken: A malfunction DTC is stored, ABS and TCS are disabled and the ABS and TCS warning lamps are turned "ON."

DTC Chart Test Description: Number(s) below refer to circled number(s) on the diagnostic chart.

1. This step is used to confirm that an open in the brake switch circuitry currently exists.
2. Indicates if the brake switch signal is being received by the EBTCM.
3. Indicates if an open circuit exists in the brake switch or brake lamp circuitry.
4. Isolates the open circuit to either the brake switch input circuitry, or the EBTCM.

Diagnostic Aids: An intermittent malfunction may be caused by a poor connection, rubbed through wire insulation, or a wire that is broken inside the insulation.

The frequency of the malfunction can be checked by using the enhanced diagnostic function of the Tech 1, as described in "Tech 1 Diagnostics" in this section.

Any circuitry, that is suspected as causing the intermittent complaint, should be thoroughly checked for backed out terminals, improper mating, broken locks, improperly formed or damaged terminals, poor terminal to wiring connections or physical damage to the wiring harness.

DIAGNOSTIC CHART DTC 95 — 1994 SILHOUETTE, TRANS SPORT AND LUMINA APV WITH TCS, CONTINUED

DTC 95

BRAKE SWITCH CIRCUIT OPEN

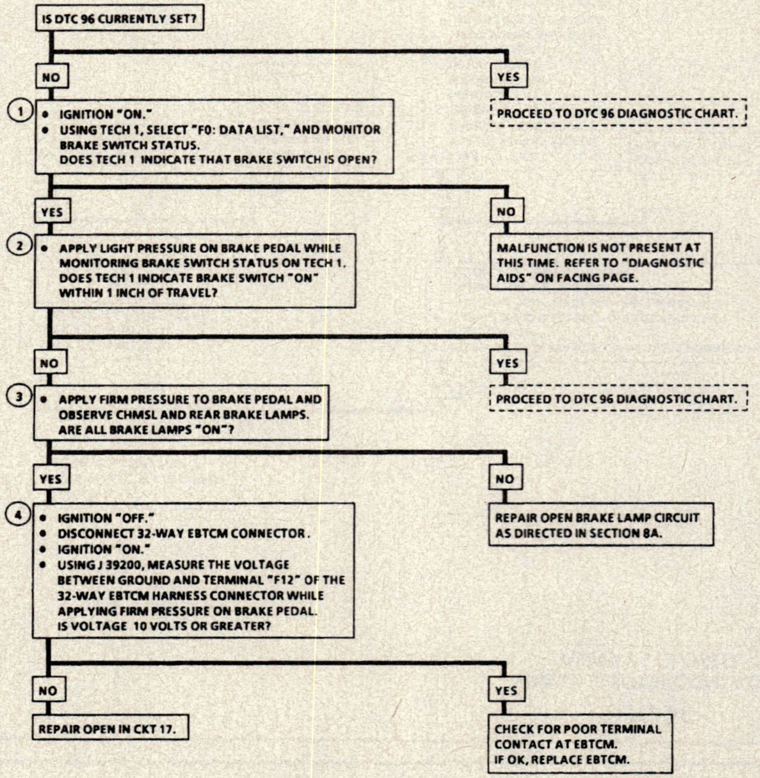

DIAGNOSTIC CHART DTC 96 — 1994 SILHOUETTE, TRANS SPORT AND LUMINA APV WITH TCS

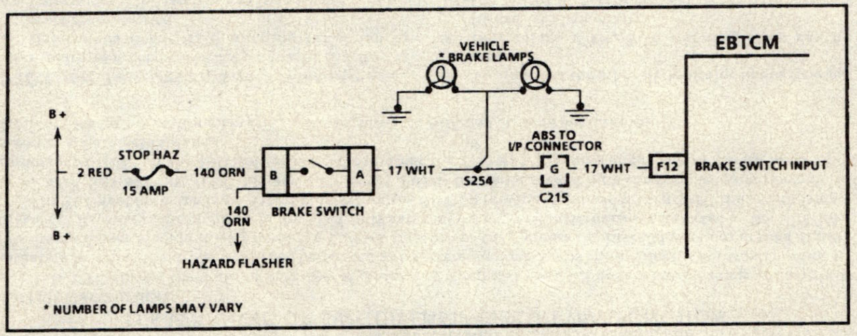

DTC 96

REAR BRAKE LAMP CIRCUIT OPEN

Circuit Description:
This DTC is designed to isolate the cause of a DTC 95 malfunction and indicate to the driver ABS is still available. If DTC 95 occurs with DTC 96, the brake lamps are open, the brake lamp grounds are open, or the Center High Mounted Stop Lamp (CHMSL) circuit is open during 4-way flasher use. The presence of battery voltage at the brake switch input indicates a valid brake switch input is still available.

DTC Will Set When: DTC 96 can be set only after DTC 95 has been set. If the EBTCM detects battery voltage on the brake switch input circuit for 0.5 second, a malfunction exists.

Action Taken: A malfunction DTC is stored. ABS is not disabled.

DTC Chart Test Description: Number(s) below refer to circled number(s) on the diagnostic chart.

1. As a result of a malfunction of an additional brake switch circuit DTC, this DTC may be set. To ensure proper diagnosis, any additional brake switch DTCs must be repaired first.
2. Identifies if the malfunction is currently present in the brake circuit.

Diagnostic Aids: An intermittent malfunction may be caused by a poor connection, rubbed through wire insulation, or a wire that is broken inside the insulation.

The frequency of the malfunction can be checked by using the enhanced diagnostic function of the Tech 1, as described in "Tech 1 Diagnostics" in this section.

Any circuitry, that is suspected as causing the intermittent complaint, should be thoroughly checked for backed out terminals, improper mating, broken locks, improperly formed or damaged terminals, poor terminal to wiring connections or physical damage to the wiring harness.

DIAGNOSTIC CHART DTC 96 — 1994 SILHOUETTE, TRANS SPORT AND LUMINA APV WITH TCS, CONTINUED

DTC 96

REAR BRAKE LAMP CIRCUIT OPEN

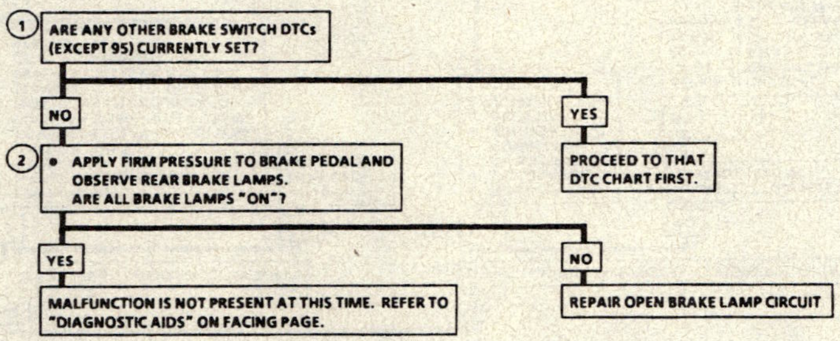

DIAGNOSTIC CHART DTC 111 — 1994 SILHOUETTE, TRANS SPORT AND LUMINA APV WITH TCS

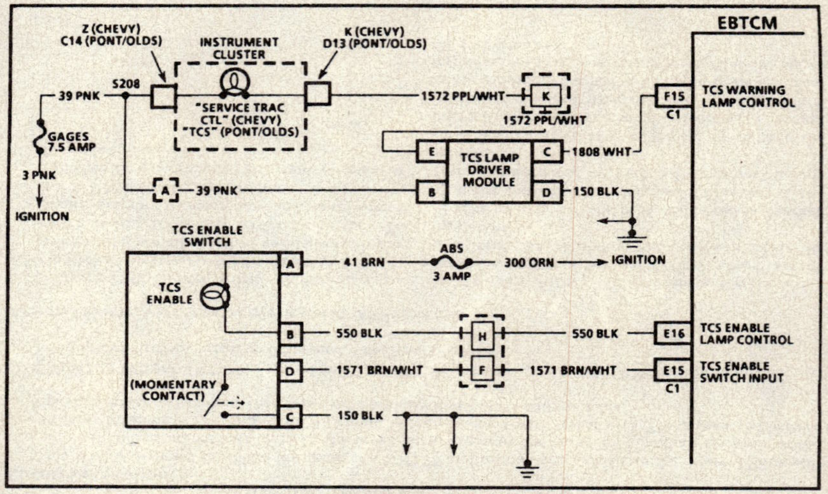

DTC 111
(Page 1 of 2)
TCS WARNING OR SWITCH LAMP CIRCUIT MALFUNCTION

Circuit Description:

This DTC checks the state of both the TCS warning lamp and the TCS enable switch lamp to identify a situation in which the driver could not be alerted to a system malfunction or the availability of traction control. The Lamp Driver Module (LDM) turns the TCS warning lamp "ON," unless the EBTCM provides ground to turn it "OFF." Because of the circuitry in the LDM, only external wiring malfunctions can be detected by the EBTCM.

The TCS enable switch lamp circuit is such that the EBTCM provides a ground to turn "ON" the switch lamp.

DTC Will Set When: DTC 111 can only be set during the three second bulb check or when either bulb is commanded "ON." If the EBTCM cannot control either the TCS warning lamp or the enable switch lamp for 2 seconds, a malfunction exists.

Action Taken: A malfunction DTC is stored, however ABS and TCS are not disabled.

DTC Chart Test Description: Number(s) below refer to circled number(s) on the diagnostic chart.

1. Determines if the TCS warning lamp can be properly controlled.
2. Determines if the lamp is "ON" constantly.
3. Checks for other DTCs which could turn "ON" the TCS warning lamp.
4. Determines if the TCS warning lamp can be turned "OFF" manually.
5. Checks the integrity of the TCS warning lamp circuitry.
6. Checks for a short to voltage on the TCS warning lamp circuitry.
7. Determines if a malfunctioning EBTCM is the cause of DTC 111.
8. Verifies that the other I/P indicator lamps function properly.
9. Determines if the TCS warning lamp can be turned "ON" manually.
10. Checks for a short to ground in the TCS warning lamp circuitry.
11. Checks for open between "Gages" fuse and LDM.
12. Verifies condition of the "Gages" fuse.
13. Determines if DTC 111 was set due to an EBTCM malfunction.

Diagnostic Aids: Refer to Page 2 of 2 of this chart for diagnostic aids.

DIAGNOSTIC CHART DTC 111 — 1994 SILHOUETTE, TRANS SPORT AND LUMINA APV WITH TCS, CONTINUED

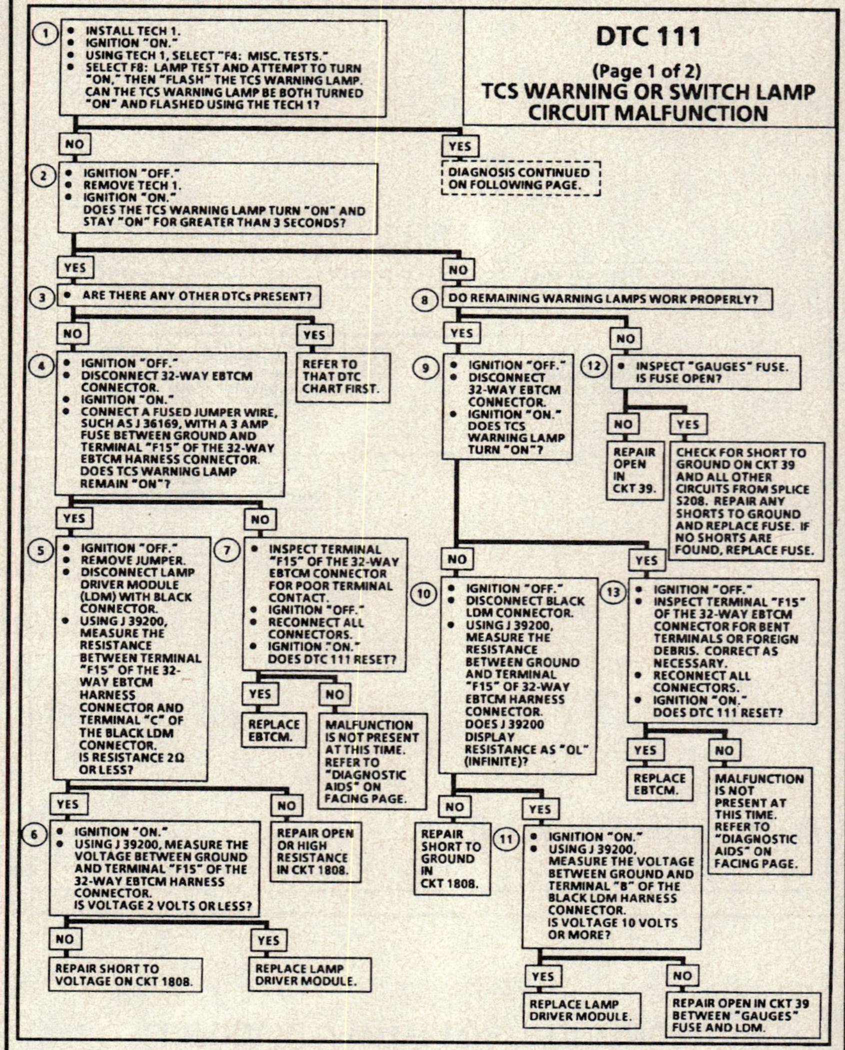

DIAGNOSTIC CHART DTC 111 — 1994 SILHOUETTE, TRANS SPORT AND LUMINA APV WITH TCS

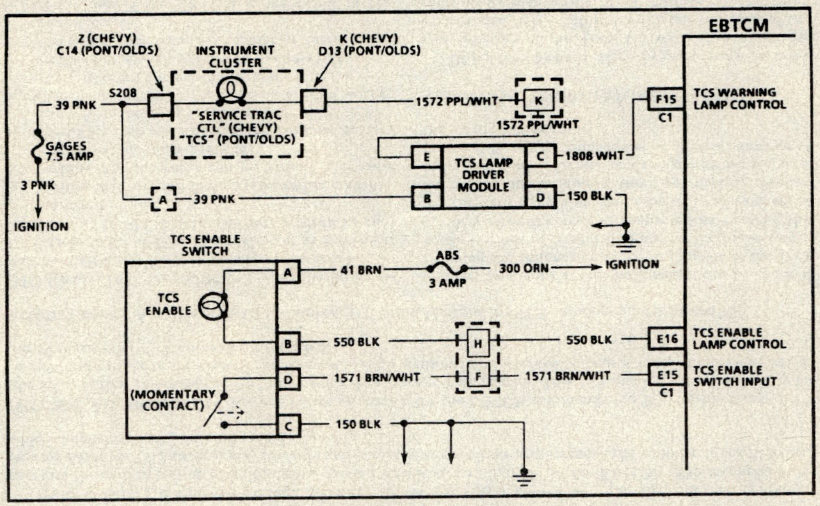

DTC 111

(Page 2 of 2)

TCS WARNING OR SWITCH LAMP CIRCUIT MALFUNCTION

Circuit Description:

This DTC checks the state of both the TCS warning lamp and the TCS enable switch lamp to identify a situation in which the driver could not be alerted to a system malfunction or the availability of traction control. The Lamp Driver Module (LDM) turns the TCS warning lamp "ON," unless the EBTCM provides ground to turn it "OFF." Because of the circuitry in the LDM, only external wiring malfunctions can be detected by the EBTCM.

The TCS enable switch lamp circuit is such that the EBTCM provides a ground to turn "ON" the switch lamp.

DTC Will Set When: DTC 111 can only be set during the three second bulb check or when either bulb is commanded "ON." If the EBTCM cannot control either the TCS warning lamp or the enable switch lamp for 2 seconds, a malfunction exists.

Action Taken: A malfunction DTC is stored, however ABS and TCS are not disabled.

DTC Chart Test Description: Number(s) below refer to circled number(s) on the diagnostic chart.

14. Determines if the switch lamp can be properly controlled.
15. Determines if the switch lamp is "ON" constantly.
16. Checks for a short to ground in the switch lamp circuitry.
17. Determines if DTC 111 was set due to an EBTCM malfunction.
18. Determines if DTC 111 was set due to an EBTCM malfunction.
19. Determines if there is an open in CKT 41 or ABS fuse.
20. Determines if the switch lamp can be controlled manually.
21. Verifies the integrity of the switch lamp circuitry.
22. Determines if there is a short to voltage in the switch lamp circuitry.
23. Determines if DTC 111 was set due to an EBTCM malfunction.

Diagnostic Aids: An intermittent malfunction may be caused by a poor connection, rubbed through wire insulation, or a wire that is broken inside the insulation.

The frequency of the malfunction can be checked by using the enhanced diagnostic function of the Tech 1, as described in "Tech 1 Diagnostics" in this section.

Any circuitry, that is suspected as causing the intermittent complaint, should be thoroughly checked for backed out terminals, improper mating, broken locks, improperly formed or damaged terminals, poor terminal to wiring connections or physical damage to the wiring harness.

DIAGNOSTIC CHART DTC 111 — 1994 SILHOUETTE, TRANS SPORT AND LUMINA APV WITH TCS, CONTINUED

DTC 111

(Page 2 of 2)

TCS WARNING OR SWITCH LAMP CIRCUIT MALFUNCTION

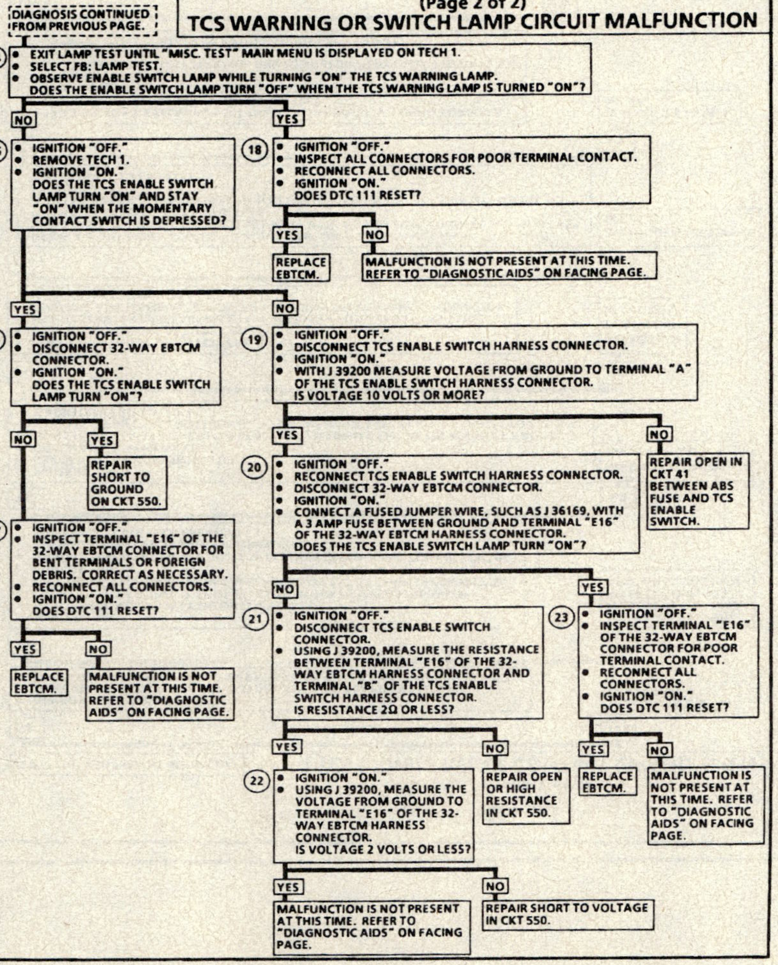

DIAGNOSTIC CHART DTC 114 — 1994 SILHOUETTE, TRANS SPORT AND LUMINA APV WITH TCS

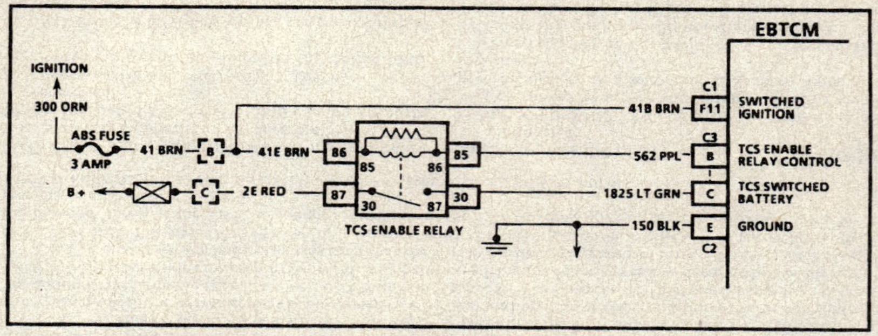

DTC 114
(Page 1 of 3)
TCS ENABLE RELAY CONTACT CIRCUIT OPEN

Circuit Description:

Ignition voltage is supplied through terminal "86" of the TCS enable relay. The EBTCM then is able to energize the pull-in coil by completing the ground circuit at terminal "B" of the EBTCM. The magnetic field created closes the TCS enable relay contacts and allows battery voltage and current to be supplied to the EBTCM, which supplies power to the motors and solenoids.

DTC Will Set When: DTC 114 can be set anytime after the EBTCM commands the TCS enable relay "ON" (the relay is first commanded "ON" during the three second bulb check). This test monitors the availability of current and voltage to the motors and solenoids. This malfunction indicates voltage is not available and would therefore not allow TCS operation if required.

Action Taken: A malfunction DTC is stored, TCS is disabled and the TCS warning lamp is turned "ON."

DTC Chart Test Description: Number(s) below refer to circled number(s) on the diagnostic chart.

1. Checks to see if DTC 116 is also set as current or history. If DTC 116 is also set, proceed to that chart first.
2. Verifies that the malfunction is currently present.
3. Verifies that the relay can be turned "ON" and provide voltage to the EBTCM.
4. Checks for high resistance in the switched battery circuit from the relay to the EBTCM.
5. Checks for high resistance in the switched battery circuit from the battery to the relay.
6. Verifies the condition of the TCS enable relay.
7. Checks for poor terminal contact or corrosion at the connectors.
8. Determines if the EBTCM is the cause of the malfunction.

Diagnostic Aids: An intermittent malfunction may be caused by a poor connection, rubbed through wire insulation, or a wire that is broken inside the insulation.

The frequency of the malfunction can be checked by using the enhanced diagnostic function of the Tech 1, as described in "Tech 1 Diagnostics" in this section.

Any circuitry that is suspected as causing the intermittent complaint should be thoroughly checked for backed out terminals, improper mating, broken locks, improperly formed or damaged terminals, poor terminal to wiring connections or physical damage to the wiring harness.

VIBRATION, TEMPERATURE EFFECTS:

Check for vibration effects by performing the relay test function of the Tech 1. With the relay test commanded "ON," lightly tap the top and sides of the relay while monitoring relay voltage. If the relay voltage changes significantly, replace the relay.

If DTC 114 only sets when the vehicle is initially started in cold ambient conditions (temperature less than 32°F-0°C), replace the relay.

DIAGNOSTIC CHART DTC 114 — 1994 SILHOUETTE, TRANS SPORT AND LUMINA APV WITH TCS, CONTINUED

DTC 114
(Page 1 of 3)
TCS ENABLE RELAY CONTACT CIRCUIT OPEN

(1) IS DTC 116 ALSO PRESENT IN HISTORY?
- YES: PROCEED TO DTC 116 CHART FIRST.
- NO:

(2)
- IGNITION "ON," ENGINE "OFF."
- USING TECH 1, SELECT DATA LIST AND OBSERVE EBTCM BATTERY VOLTAGE. DOES TECH 1 INDICATE THAT EBTCM BATTERY VOLTAGE IS GREATER THAN 10 VOLTS?

YES: MALFUNCTION IS NOT PRESENT AT THIS TIME. REFER TO "DIAGNOSTIC AIDS" ON FACING PAGE.
NO:

(3)
- IGNITION "OFF."
- DISCONNECT BLACK 8-WAY EBTCM CONNECTOR.
- CONNECT A FUSED JUMPER, SUCH AS J 36169 WITH A 3 AMP FUSE BETWEEN GROUND AND TERMINAL "B" OF THE BLACK 8-WAY EBTCM HARNESS CONNECTOR.
- IGNITION "ON."
- USING J 39200, MEASURE THE VOLTAGE BETWEEN GROUND AND TERMINAL "C" OF THE BLACK 8-WAY EBCM HARNESS CONNECTOR. IS VOLTAGE 10 VOLTS OR GREATER?

NO: DIAGNOSIS CONTINUED ON NEXT PAGE.
YES:

(4)
- IGNITION "OFF."
- DISCONNECT JUMPER USED IN STEP 3.
- DISCONNECT TCS ENABLE RELAY.
- USING J 39200, MEASURE RESISTANCE BETWEEN TERMINAL "30" OF THE TCS ENABLE RELAY HARNESS CONNECTOR AND TERMINAL "C" OF THE BLACK 8-WAY EBTCM HARNESS CONNECTOR. IS RESISTANCE 2Ω OR LESS?

NO: REPAIR HIGH RESISTANCE IN CKT 1825.
YES:

(5)
- DISCONNECT NEGATIVE BATTERY CABLE.
- DISCONNECT POSITIVE BATTERY CABLE.
- USING J 39200, MEASURE THE RESISTANCE BETWEEN POSITIVE BATTERY CABLE TERMINAL AND TERMINAL "87" OF THE TCS ENABLE RELAY HARNESS CONNECTOR. IS RESISTANCE 2Ω OR LESS?

NO: REPAIR HIGH RESISTANCE IN CKT 2E.
YES:

(6)
- USING A JUMPER WIRE, CONNECT TERMINAL "86" ON THE TCS ENABLE RELAY TO GROUND.
- USING A FUSED JUMPER WIRE, SUCH AS J 36169, WITH A 3 AMP FUSE, CONNECT TERMINAL "85" ON THE TCS ENABLE RELAY TO VOLTAGE.
- USING J 39200, MEASURE THE RESISTANCE BETWEEN TERMINALS "30" AND "87" ON THE TCS ENABLE RELAY. IS RESISTANCE 2Ω OR LESS?.

NO: REPLACE TCS ENABLE RELAY.
YES:

(7)
- INSPECT TCS ENABLE RELAY AND TCS ENABLE RELAY HARNESS CONNECTOR FOR A POOR CONNECTION OR POOR TERMINAL CONTACT.
- INSPECT EBTCM AND EBTCM CONNECTOR TERMINALS "B" AND "C" FOR POOR TERMINAL CONTACT.
- INSPECT POSITIVE BATTERY TERMINAL AND POSITIVE BATTERY CABLE TERMINAL FOR A POOR CONNECTION. IS POOR TERMINAL CONTACT OR A POOR CONNECTION EVIDENT?

YES: REPLACE TERMINALS THAT EXHIBIT POOR TERMINAL CONTACT OR EVIDENCE OF CORROSION.
NO:

(8)
- RECONNECT ALL CONNECTORS.
- IGNITION "ON." DOES DTC 114 RESET?

YES: REPLACE EBTCM.
NO: MALFUNCTION IS INTERMITTENT. REFER TO "DIAGNOSTIC AIDS" ON FACING PAGE.

DIAGNOSTIC CHART DTC 114 — 1994 SILHOUETTE, TRANS SPORT AND LUMINA APV WITH TCS, CONTINUED

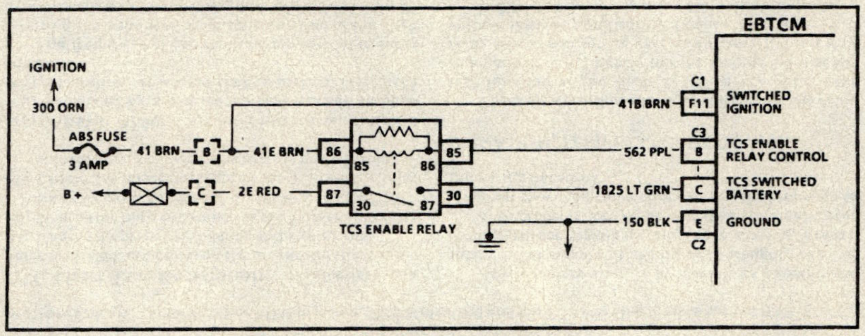

DTC 114
(Page 2 of 3)
TCS ENABLE RELAY CONTACT CIRCUIT OPEN

Circuit Description:

Ignition voltage is supplied through terminal "86" of the TCS enable relay. The EBTCM then is able to energize the pull-in coil by completing the ground circuit at terminal "B" of the EBTCM. The magnetic field created closes the TCS enable relay contacts and allows battery voltage and current to be supplied to the EBTCM, which supplies power to the motors and solenoids.

DTC Will Set When: DTC 114 can be set anytime after the EBTCM commands the TCS enable relay "ON" (the relay is first commanded "ON" during the three second bulb check). This test monitors the availability of current and voltage to the motors and solenoids. This malfunction indicates voltage is not available and would therefore not allow TCS operation if required.

Action Taken: A malfunction DTC is stored, TCS is disabled and the TCS warning lamp is turned "ON."

DTC Chart Test Description: Number(s) below refer to circled number(s) on the diagnostic chart.

9. Checks for high resistance in the entire relay coil circuit.
10. Verifies proper battery condition.
11. This step verifies proper voltage is available at the TCS enable relay.
12. Checks for an open in the switched battery circuit from the relay to the EBTCM.
13. Verifies the condition of the TCS enable relay.

Diagnostic Aids: An intermittent malfunction may be caused by a poor connection, rubbed through wire insulation, or a wire that is broken inside the insulation.

The frequency of the malfunction can be checked by using the enhanced diagnostic function of the Tech 1, as described in "Tech 1 Diagnostics" in this section.

Any circuitry that is suspected as causing the intermittent complaint should be thoroughly checked for backed out terminals, improper mating, broken locks, improperly formed or damaged terminals, poor terminal to wiring connections or physical damage to the wiring harness.

VIBRATION, TEMPERATURE EFFECTS:

Check for vibration effects by performing the relay test function of the Tech 1. With the relay test commanded "ON," lightly tap the top and sides of the relay while monitoring relay voltage. If the relay voltage changes significantly, replace the relay.

If DTC 114 only sets when the vehicle is initially started in cold ambient conditions (temperature less than 32°F-0°C), replace the relay.

DIAGNOSTIC CHART DTC 114 — 1994 SILHOUETTE, TRANS SPORT AND LUMINA APV WITH TCS, CONTINUED

DTC 114
(Page 2 of 3)
TCS ENABLE RELAY CONTACT CIRCUIT OPEN

DIAGNOSIS CONTINUED FROM PREVIOUS PAGE.

(9)
- IGNITION "OFF."
- REMOVE JUMPER WIRE INSTALLED IN STEP #3 IF NOT ALREADY REMOVED.
- DISCONNECT NEGATIVE BATTERY CABLE.
- DISCONNECT POSITIVE BATTERY CABLE.
- WITH BATTERY STILL DISCONNECTED, TURN IGNITION SWITCH TO "RUN." THIS COMPLETES THE IGNITION CIRCUIT TO THE TCS ENABLE RELAY.
- USING J 39200, MEASURE THE RESISTANCE BETWEEN TERMINAL "B" OF THE BLACK 8-WAY EBTCM HARNESS CONNECTOR AND THE POSITIVE BATTERY CABLE TERMINAL.

IS RESISTANCE 95Ω OR LESS WHEN MEASURED AT APPROXIMATELY 20°C (68°F)?

NO: DIAGNOSIS CONTINUED ON NEXT PAGE.

YES:

(10)
- USING J 39200, MEASURE THE VOLTAGE BETWEEN THE NEGATIVE AND POSITIVE BATTERY TERMINALS.

IS VOLTAGE 10 VOLTS OR GREATER?

NO: CORRECT LOW VOLTAGE CONDITION.

YES:

(11)
- IGNITION "OFF."
- RECONNECT POSITIVE BATTERY CABLE.
- RECONNECT NEGATIVE BATTERY CABLE.
- DISCONNECT TCS ENABLE RELAY.
- USING J 39200, MEASURE THE VOLTAGE BETWEEN TERMINAL "87" OF THE TCS ENABLE RELAY HARNESS CONNECTOR AND GROUND.

IS VOLTAGE 10 VOLTS OR GREATER?

NO: REPAIR OPEN OR HIGH RESISTANCE IN CKT 2E.

YES:

(12)
- USING J 39200, MEASURE THE RESISTANCE BETWEEN TERMINAL "30" OF THE TCS ENABLE RELAY HARNESS CONNECTOR AND TERMINAL "C" OF THE BLACK 8-WAY EBTCM HARNESS CONNECTOR.

IS RESISTANCE 2Ω OR LESS?

NO: REPAIR OPEN OR HIGH RESISTANCE IN CKT 1825.

YES:

(13)
- USING A JUMPER WIRE, CONNECT TERMINAL "86" ON THE TCS ENABLE RELAY TO GROUND.
- USING A FUSED JUMPER WIRE, SUCH AS J 36169, WITH A 3 AMP FUSE, CONNECT TERMINAL "85" ON THE TCS ENABLE RELAY TO B+.
- USING J 39200, MEASURE THE RESISTANCE BETWEEN TERMINALS "30" AND "87" ON THE TCS ENABLE RELAY.

IS RESISTANCE 2Ω OR LESS?

NO: REPLACE TCS ENABLE RELAY.

YES: MALFUNCTION WAS MOST LIKELY DUE TO A POOR CONNECTION. INSPECT ALL CONNECTORS AND TERMINALS FOR POOR TERMINAL CONTACT AND EVIDENCE OF CORROSION. REPLACE ALL TERMINALS THAT EXHIBIT EVIDENCE OF POOR TERMINAL CONTACT OR CORROSION.

DIAGNOSTIC CHART DTC 114 — 1994 SILHOUETTE, TRANS SPORT AND LUMINA APV WITH TCS, CONTINUED

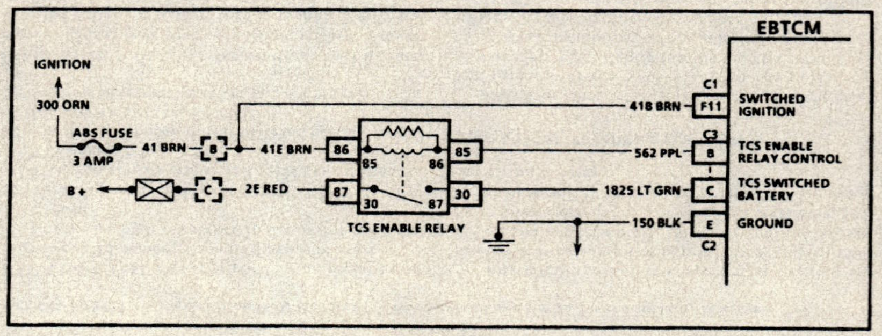

DTC 114

(Page 3 of 3)

TCS ENABLE RELAY CONTACT CIRCUIT OPEN

Circuit Description:

Ignition voltage is supplied through terminal "86" of the TCS enable relay. The EBTCM then is able to energize the pull-in coil by completing the ground circuit at terminal "B" of the EBTCM. The magnetic field created closes the TCS enable relay contacts and allows battery voltage and current to be supplied to the EBTCM, which supplies power to the motors and solenoids.

DTC Will Set When: DTC 114 can be set anytime after the EBTCM commands the TCS enable relay "ON" (the relay is first commanded "ON" during the three second bulb check). This test monitors the availability of current and voltage to the motors and solenoids. This malfunction indicates voltage is not available and would therefore not allow TCS operation if required.

Action Taken: A malfunction DTC is stored, TCS is disabled and the TCS warning lamp is turned "ON."

DTC Chart Test Description: Number(s) below refer to circled number(s) on the diagnostic chart.

14. Checks for proper resistance of the relay coil.
15. Checks for high resistance in the TCS enable relay control circuit between the relay and the EBTCM.
16. Checks for high resistance in the TCS enable relay control circuit between the battery and the relay.

Diagnostic Aids: An intermittent malfunction may be caused by a poor connection, rubbed through wire insulation, or a wire that is broken inside the insulation.

The frequency of the malfunction can be checked by using the enhanced diagnostic function of the Tech 1, as described in "Tech 1 Diagnostics" in this section.

Any circuitry that is suspected as causing the intermittent complaint should be thoroughly checked for backed out terminals, improper mating, broken locks, improperly formed or damaged terminals, poor terminal to wiring connections or physical damage to the wiring harness.

VIBRATION, TEMPERATURE EFFECTS:

Check for vibration effects by performing the relay test function of the Tech 1. With the relay test commanded "ON," lightly tap the top and sides of the relay while monitoring relay voltage. If the relay voltage changes significantly, replace the relay.

If DTC 114 only sets when the vehicle is initially started in cold ambient conditions (temperature less than 32°F-0°C), replace the relay.

DIAGNOSTIC CHART DTC 114 — 1994 SILHOUETTE, TRANS SPORT AND LUMINA APV WITH TCS, CONTINUED

DTC 114

(Page 3 of 3)

TCS ENABLE RELAY CONTACT CIRCUIT OPEN

DIAGNOSIS CONTINUED FROM PREVIOUS PAGE.

(14)
- DISCONNECT TCS ENABLE RELAY.
- USING J 39200, MEASURE THE RESISTANCE BETWEEN TERMINALS "85" AND "86" ON THE TCS ENABLE RELAY. IS RESISTANCE 95Ω OR LESS WHEN MEASURED AT APPROXIMATELY 20°C (68°F)?

NO → REPLACE TCS ENABLE RELAY.

YES →

(15)
- USING J 39200, MEASURE THE RESISTANCE BETWEEN TERMINAL "85" OF THE TCS ENABLE RELAY HARNESS CONNECTOR AND TERMINAL "B" OF THE BLACK 8-WAY EBTCM HARNESS CONNECTOR. IS RESISTANCE 2Ω OR LESS?

NO → REPAIR OPEN OR HIGH RESISTANCE IN CKT 562.

YES →

(16)
- USING J 39200, MEASURE THE RESISTANCE BETWEEN TERMINAL "86" OF THE TCS ENABLE RELAY HARNESS CONNECTOR AND THE POSITIVE BATTERY CABLE TERMINAL. IS RESISTANCE 2Ω OR LESS?

NO → REPAIR OPEN OR HIGH RESISTANCE IN CKT 418, CKT 300, IGNITION SWITCH, OR BATTERY FEED CIRCUIT TO IGNITION SWITCH.

YES → MALFUNCTION WAS MOST LIKELY DUE TO A POOR CONNECTION. INSPECT ALL CONNECTORS AND TERMINALS FOR POOR TERMINAL CONTACT AND EVIDENCE OF CORROSION. REPLACE ALL TERMINALS THAT EXHIBIT EVIDENCE OF POOR TERMINAL CONTACT OR CORROSION.

DIAGNOSTIC CHART DTC 116 — 1994 SILHOUETTE, TRANS SPORT AND LUMINA APV WITH TCS

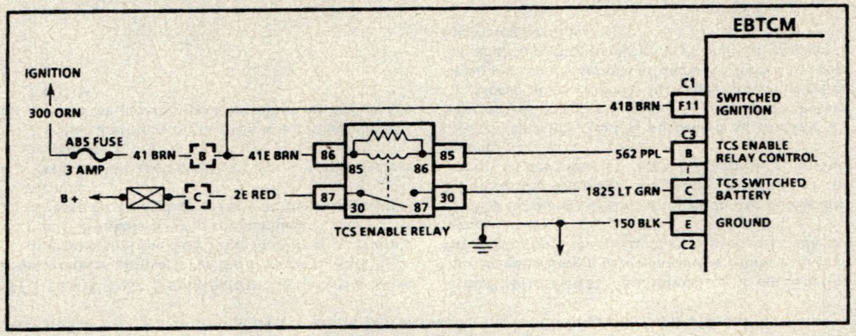

DTC 116

TCS ENABLE RELAY COIL CIRCUIT OPEN

Circuit Description:
Ignition voltage is supplied through terminal "86" of the TCS enable relay. The EBTCM then is able to energize the pull-in coil by completing the ground circuit at terminal "B" of the EBTCM. The magnetic field created closes the TCS enable relay contacts and allows battery voltage and current to be supplied to the EBTCM, which supplies power to the motors and solenoids.

DTC Will Set When: DTC 116 can be set only after DTC 114 has been set. This test detects an open in the TCS enable relay coil circuit. An open in this circuit will not allow the TCS enable relay to be energized thus preventing voltage and current to the motors and solenoids. If this malfunction is present and the ignition is turned "OFF" before 5 km/h (3 mph) is reached, DTC 114 is set alone.

Action Taken: A malfunction DTC is stored, TCS is disabled and the TCS warning lamp is turned "ON."

DTC Chart Test Description: Number(s) below refer to circled number(s) on the diagnostic chart.

1. Indicates the EBTCM is capable of controlling the enable relay as commanded.
2. Ensures that there is continuity through the pull-in coil of the TCS enable relay.
3. Checks to ensure that voltage is available to the pull-in coil of the enable relay.
4. Checks continuity in the enable relay control circuit.
5. Ensures DTC was not set due to a poor connection.
6. This stop identifies whether the malfunction is due to a poor connection or the EBTCM.

Diagnostic Aids: An intermittent malfunction may be caused by a poor connection, rubbed through wire insulation, or a wire that is broken inside the insulation.

The frequency of the malfunction can be checked by using the enhanced diagnostic function of the Tech 1, as described in "Tech 1 Diagnostics" in this section.

If the frequency of the malfunction is high, but is currently intermittent, check for high coil resistance by measuring between relay terminals "85" and "86" using a J 39200. If resistance shows greater than 100Ω, replace the relay.

Any circuitry, that is suspected as causing the intermittent complaint, should be thoroughly checked for backed out terminals, improper mating, broken locks, improperly formed or damaged terminals, poor terminal to wiring connections or physical damage to the wiring harness.

DIAGNOSTIC CHART DTC 116 — 1994 SILHOUETTE, TRANS SPORT AND LUMINA APV WITH TCS, CONTINUED

DTC 116

TCS ENABLE RELAY COIL CIRCUIT OPEN

(1)
- IGNITION "ON," ENGINE "OFF."
- USING THE RELAY TEST FUNCTION OF THE TECH 1, COMMAND THE TCS ENABLE RELAY "ON."
- DOES TECH 1 INDICATE TCS ENABLE RELAY "ON," AND EBTCM BATTERY VOLTAGE ABOVE 10 VOLTS?

YES: MALFUNCTION IS NOT PRESENT AT THIS TIME. REFER TO "DIAGNOSTIC AIDS" ON FACING PAGE.

NO:

(2)
- IGNITION "OFF."
- DISCONNECT BLACK 8-WAY EBTCM HARNESS CONNECTOR.
- IGNITION "ON."
- USING J 39200, MEASURE VOLTAGE BETWEEN GROUND AND BLACK 8-WAY EBTCM HARNESS CONNECTOR TERMINAL "B".
- IS VOLTAGE GREATER THAN 10 VOLTS?

YES:

(6)
- IGNITION "OFF."
- INSPECT BLACK 8-WAY EBTCM HARNESS CONNECTOR TERMINAL "B" FOR POOR TERMINAL CONTACT. REPLACE TERMINALS THAT EXHIBIT POOR CONTACT.
- RECONNECT ALL CONNECTORS.
- IGNITION "ON."
- IS DTC 116 SET AS A CURRENT DTC?

YES: REPLACE EBTCM.

NO: MALFUNCTION IS INTERMITTENT. REFER TO "DIAGNOSTIC AIDS" ON FACING PAGE.

NO:

(3)
- IGNITION "OFF."
- DISCONNECT TCS ENABLE RELAY.
- IGNITION "ON."
- USING J 39200, MEASURE VOLTAGE BETWEEN GROUND AND TERMINAL "86" OF THE TCS ENABLE RELAY HARNESS CONNECTOR.
- IS VOLTAGE GREATER THAN 10 VOLTS?

NO: REPAIR OPEN IN CKT 41, ABS FUSE, OR CKT 300.

YES:

(4)
- IGNITION "OFF."
- USING J 39200, MEASURE RESISTANCE BETWEEN ENABLE RELAY HARNESS CONNECTOR TERMINAL "85" AND BLACK 8-WAY EBTCM HARNESS CONNECTOR TERMINAL "B".
- IS RESISTANCE LESS THAN 2 OHMS?

NO: REPAIR OPEN IN CKT 562.

YES:

(5)
- INSPECT TCS ENABLE RELAY AND TCS ENABLE RELAY HARNESS CONNECTOR FOR POOR TERMINAL CONTACT. REPLACE TERMINALS THAT EXHIBIT POOR CONTACT.
- RECONNECT ALL CONNECTORS.
- IGNITION "ON."
- IS DTC 116 SET AS A CURRENT DTC?

YES: REPLACE ENABLE RELAY.

NO: MALFUNCTION IS INTERMITTENT. REFER TO "DIAGNOSTIC AIDS" ON FACING PAGE.

DIAGNOSTIC CHART DTC 118 — 1994 SILHOUETTE, TRANS SPORT AND LUMINA APV WITH TCS

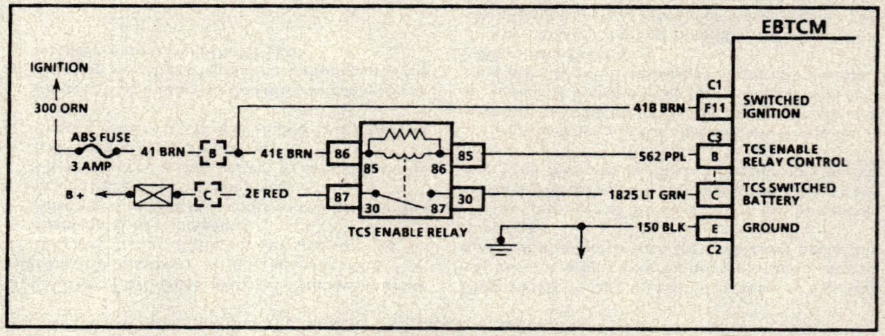

DTC 118

TCS ENABLE RELAY COIL CIRCUIT SHORTED TO BATT

Circuit Description:

Ignition voltage is supplied through terminal "86" of the TCS enable relay. The EBTCM then is able to energize the pull-in coil by completing the ground circuit at terminal "B" of the EBTCM. The magnetic field created closes the TCS enable relay contacts and allows battery voltage and current to be supplied to the EBTCM, which supplies voltage to the motors and solenoids.

DTC Will Set When: DTC 118 can be set after the EBTCM commands the TCS enable relay "ON." This test monitors the availability of current and voltage to the motors and solenoids. A malfunction of this type will not allow voltage to be available to the motors and solenoids; therefore, TCS operation would not be allowed if required.

Action Taken: A malfunction DTC is stored, TCS is disabled and the TCS warning lamp is turned "ON."

DTC Chart Test Description: Number(s) below refer to circled number(s) on the diagnostic chart.

1. Indicates that the EBTCM is capable of controlling the TCS enable relay as commanded.
2. Checks for a short to voltage on the relay control circuit.
3. Verifies that the relay coil has the appropriate resistance.
4. Verifies that relay contacts have not failed closed.
5. Checks for possible poor terminal contact at the EBTCM.

Diagnostic Aids: An intermittent malfunction may be caused by a poor connection, rubbed through wire insulation, or a wire that is broken inside the insulation.

The frequency of the malfunction can be checked by using the enhanced diagnostic function of the Tech 1, as described in "Tech 1 Diagnostics" in this section.

Any circuitry, that is suspected as causing the intermittent complaint, should be thoroughly checked for backed out terminals, improper mating, broken locks, improperly formed or damaged terminals, poor terminal to wiring connections or physical damage to the wiring harness.

DIAGNOSTIC CHART DTC 118 — 1994 SILHOUETTE, TRANS SPORT AND LUMINA APV WITH TCS, CONTINUED

DTC 118
TCS ENABLE RELAY COIL CIRCUIT SHORTED TO BATT

(1)
- IGNITION "ON," ENGINE "OFF."
- USING RELAY TEST FUNCTION OF TECH 1, COMMAND TCS ENABLE RELAY "ON."
- DOES TECH 1 INDICATE TCS ENABLE RELAY "ON" AND BATTERY VOLTAGE ABOVE 10 VOLTS?

YES → MALFUNCTION IS NOT PRESENT AT THIS TIME. REFER TO "DIAGNOSTIC AIDS" ON FACING PAGE.

NO → (2)
- IGNITION "OFF."
- DISCONNECT TCS ENABLE RELAY.
- DISCONNECT BLACK 8-WAY EBTCM CONNECTOR.
- IGNITION "ON."
- USING J 39200, MEASURE THE VOLTAGE BETWEEN GROUND AND TERMINAL "B" OF THE BLACK 8-WAY EBTCM HARNESS CONNECTOR. IS VOLTAGE 2 VOLTS OR LESS?

NO → REPAIR SHORT TO VOLTAGE CKT 562.

YES → (3)
- IGNITION "ON."
- USING J 39200, MEASURE THE RESISTANCE BETWEEN TERMINAL "85" AND TERMINAL "86" OF THE TCS ENABLE RELAY. IS RESISTANCE 40Ω OR GREATER?

NO → REPLACE TCS ENABLE RELAY.

YES → (4)
- USING J 39200, MEASURE THE RESISTANCE BETWEEN TERMINALS "30" AND "87" OF THE TCS ENABLE RELAY. DOES J 39200 DISPLAY RESISTANCE AS "OL" (INFINITE)?

NO → REPLACE TCS ENABLE RELAY.

YES → (5)
- INSPECT CKT 562 AND TERMINAL "B" OF THE BLACK 8-WAY EBTCM CONNECTOR FOR DAMAGE WHICH MAY RESULT IN A SHORT TO VOLTAGE WITH ALL CONNECTORS CONNECTED. REPAIR ALL DAMAGE IF EVIDENT.
- RECONNECT ALL CONNECTORS.
- IGNITION "ON." DID DTC 118 SET AS A CURRENT DTC?

YES → REPLACE EBTCM.

NO → MALFUNCTION WAS MOST LIKELY CAUSED BY A POOR CONNECTION. REFER TO "DIAGNOSTIC AIDS" ON FACING PAGE.

AFTER ALL DIAGNOSIS IS COMPLETE, CLEAR DTCs AND TEST DRIVE VEHICLE FOR THREE (3) DRIVE CYCLES TO VERIFY THAT DTC DOES NOT RESET. A DRIVE CYCLE CONSISTS OF STARTING THE VEHICLE, DRIVING OVER 16 km/h (10 MPH), AND THEN TURNING KEY OFF.

DIAGNOSTIC CHART DTC 123 — 1994 SILHOUETTE, TRANS SPORT AND LUMINA APV WITH TCS

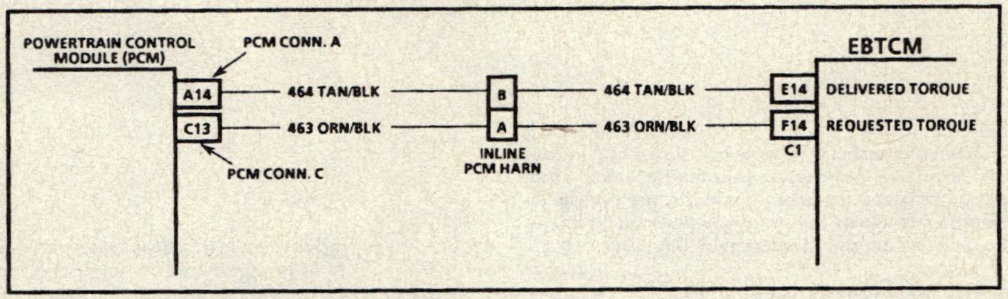

DTC 123

PCM TO EBTCM DELIVERED TORQUE CIRCUIT MALFUNCTION

Circuit Description:

The traction control system uses the antilock brake system in conjunction with the PCM fuel and ignition controls to limit drive wheel slippage during acceleration. The PCM controlled portion of the traction control system reduces engine torque by retarding spark timing, alternating air/fuel ratio, and/or shutting "OFF" up to three injectors. The EBTCM sends a torque request to the PCM via a Pulse Width Modulated (PWM) signal requesting the desired torque level for proper traction control system operation (CKT 463). The PCM then provides a delivered torque PWM signal (CKT 464) informing the EBTCM how much actual torque is produced by the engine.

DTC Will Set When: DTC 123 will set when there is a malfunction in the delivered torque circuit (CKT 464) or if the signal is out of range.

Action Taken: A malfunction DTC is stored, TCS is disabled, and the TCS warning lamp is turned "ON."

DTC Chart Test Description: Number(s) below refer to circled number(s) on the diagnostic chart.

1. Checks for proper voltage at the PCM terminal.
2. Checks for short to voltage in CKT 464.
3. Ensures malfunction was not due to physical damage of circuitry.
4. Checks for open in CKT 464.
5. Checks for short to ground in CKT 464.
6. Checks for possible poor terminal contact at the EBTCM.

Diagnostic Aids: An intermittent malfunction may be caused by a poor connection, rubbed through wire insulation, or a wire that is broken inside the insulation.

The frequency of the malfunction can be checked by using the enhanced diagnostic function of the Tech 1, as described in "Tech 1 Diagnostics" in this section.

Any circuitry, that is suspected as causing the intermittent complaint, should be thoroughly checked for backed out terminals, improper mating, broken locks, improperly formed or damaged terminals, poor terminal to wiring connections or physical damage to the wiring harness.

DIAGNOSTIC CHART DTC 123 — 1994 SILHOUETTE, TRANS SPORT AND LUMINA APV WITH TCS, CONTINUED

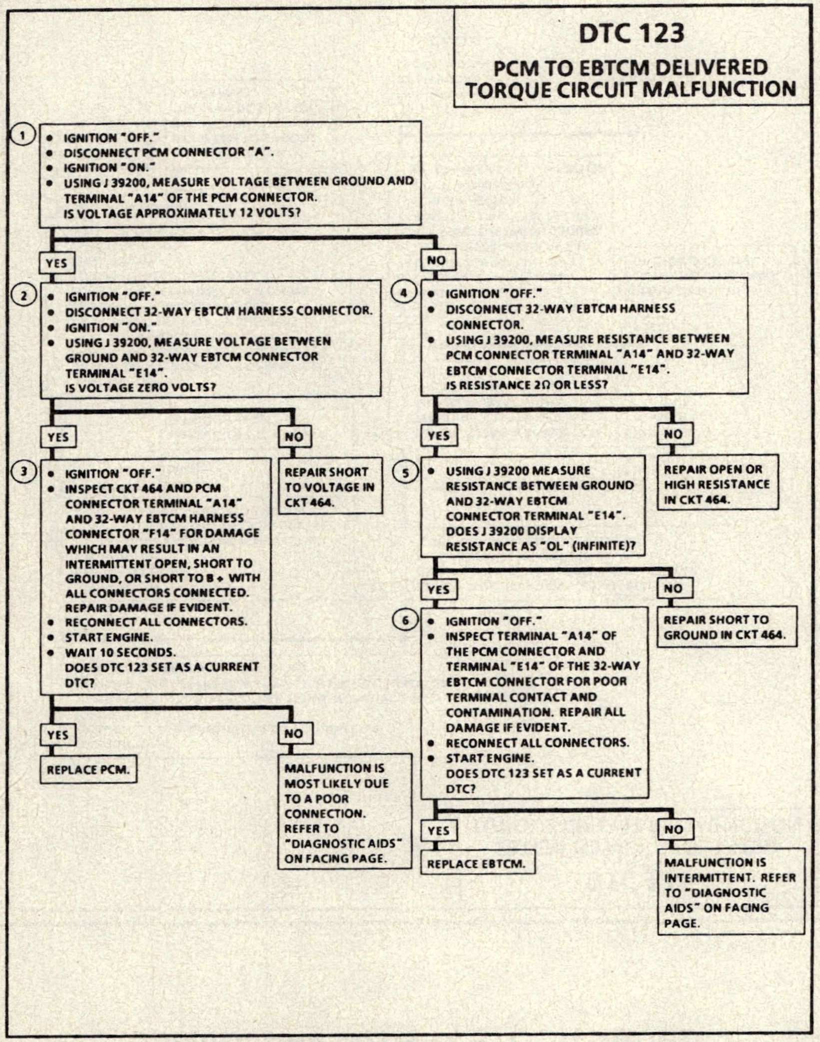

DIAGNOSTIC CHART DTC 124 — 1994 SILHOUETTE, TRANS SPORT AND LUMINA APV WITH TCS

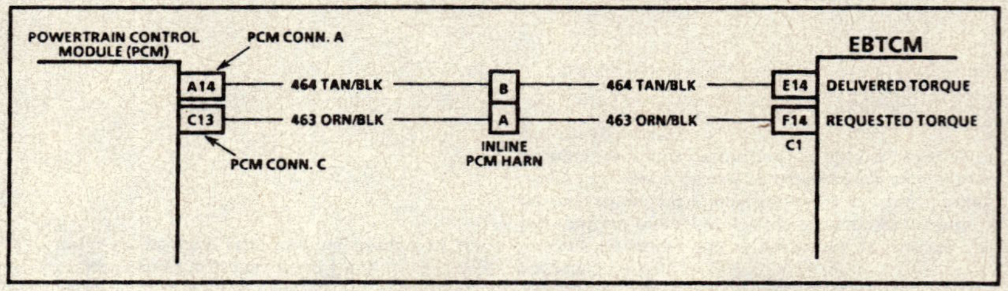

DTC 124

EBTCM TO PCM REQUESTED TORQUE CIRCUIT MALFUNCTION

Circuit Description:

The traction control system uses the antilock brake system in conjunction with the PCM fuel and ignition controls to limit drive wheel slippage during acceleration. The PCM controlled portion of the traction control system reduces engine torque by retarding spark timing, alternating air/fuel ratio, and/or shutting "OFF" up to three injectors. The EBTCM sends a torque request to the PCM via a Pulse Width Modulated (PWM) signal requesting the desired torque level for proper traction control system operation (CKT 463). The PCM then provides a delivered torque PWM signal (CKT 464) informing the EBTCM how much actual torque is produced by the engine.

DTC Will Set When: DTC 124 will set when there is a malfunction in the requested torque circuit (CKT 463) or if the signal is out of range.

Action Taken: A malfunction DTC is stored, TCS is disabled, and the TCS warning lamp is turned "ON."

DTC Chart Test Description: Number(s) below refer to circled number(s) on the diagnostic chart.

1. Checks for proper voltage at the EBTCM.
2. Checks for proper voltage at the PCM terminal.
3. Ensures malfunction was not due to physical damage of the circuitry.
4. Checks for possible short to battery in CKT 463.
5. Checks for an open in CKT 463.
6. Checks for proper resistance in CKT 463.

Diagnostic Aids: An intermittent malfunction may be caused by a poor connection, rubbed through wire insulation, or a wire that is broken inside the insulation.

The frequency of the malfunction can be checked by using the enhanced diagnostic function of the Tech 1, as described in "Tech 1 Diagnostics" in this section.

Any circuitry, that is suspected as causing the intermittent complaint, should be thoroughly checked for backed out terminals, improper mating, broken locks, improperly formed or damaged terminals, poor terminal to wiring connections or physical damage to the wiring harness.

DIAGNOSTIC CHART DTC 124 — 1994 SILHOUETTE, TRANS SPORT AND LUMINA APV WITH TCS, CONTINUED

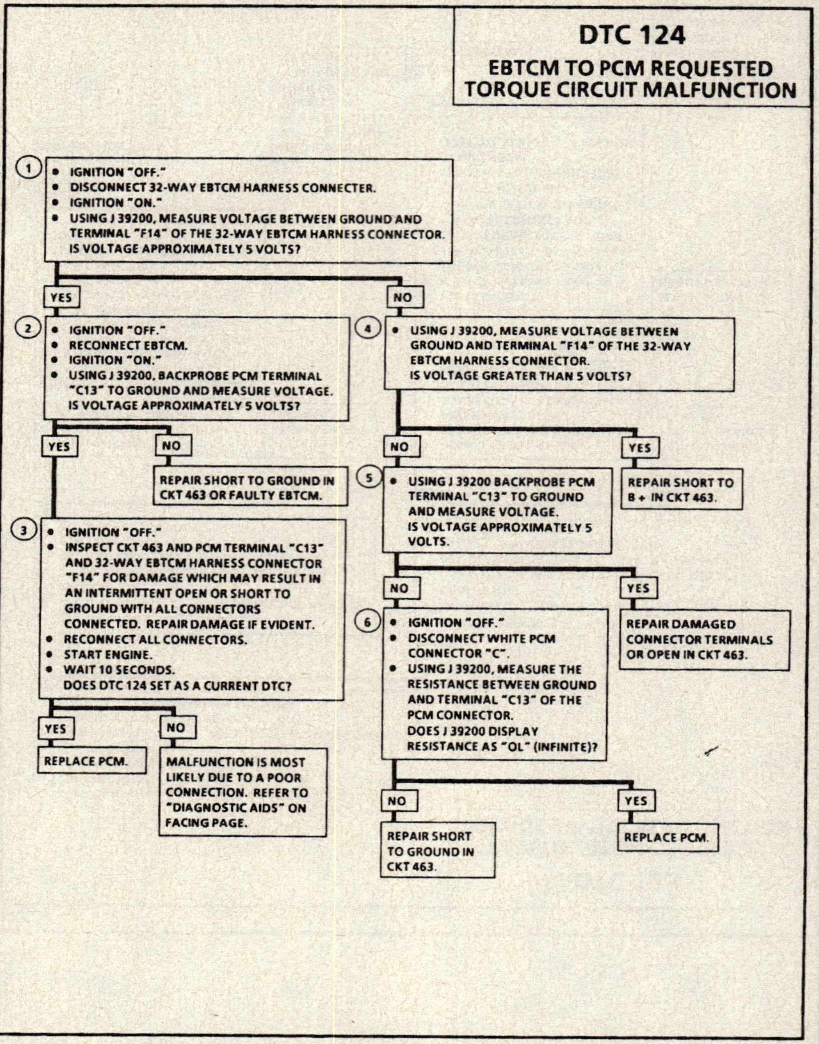

DIAGNOSTIC CHART DTC 136 — 1994 SILHOUETTE, TRANS SPORT AND LUMINA APV WITH TCS

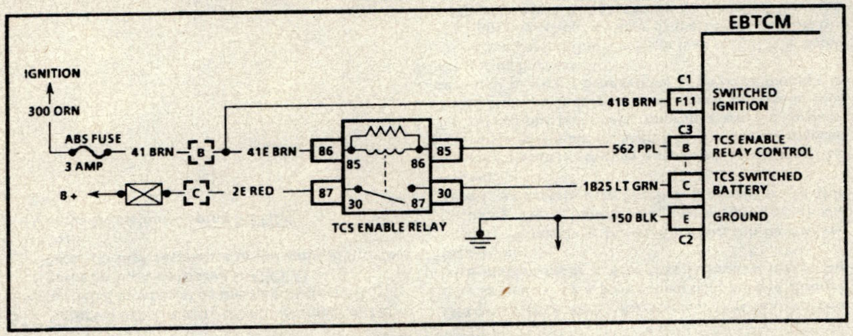

DTC 136

(Page 1 of 2)

LOW SYSTEM VOLTAGE (TCS)

Circuit Description:

This DTC is used to monitor the voltage level available to the EBTCM. If the voltage drops below 11 volts, full performance of the TCS system cannot be guaranteed. During TCS operation, there are several current requirements that will cause battery voltage to drop. Because of this, voltage is monitored prior to TCS to indicate good charging system condition and also during TCS when voltage may drop significantly.

DTC Will Set When: DTC 136 can only be set if the vehicle's speed is greater than 5 km/h (3 mph). If the switched battery voltage is less than 11.2 volts non-TCS or less than 9.6 volts TCS, a malfunction exists.

Action Taken: A malfunction DTC is stored, TCS is disabled and the TCS warning lamp is turned "ON."

DTC Chart Test Description: Number(s) below refer to circled number(s) on the diagnostic chart.

1. Checks to see what voltage is readily available at terminal "C" of the EBTCM. If voltage is representative of good charging system condition, a malfunction is not present.
2. Isolates the EBTCM to check the condition of the circuitry and charging system.
3. Isolates the low voltage condition to high circuit resistance, or improper charging system operation.
4. Isolates which circuit is the source of the high resistance condition.

Diagnostic Aids: An intermittent malfunction may be caused by a poor connection, rubbed through wire insulation, or a wire that is broken inside the insulation.

The frequency of the malfunction can be checked by using the enhanced diagnostic function of the Tech 1, as described in "Tech 1 Diagnostics" in this section.

Any circuitry, that is suspected as causing the intermittent complaint, should be thoroughly checked for backed out terminals, improper mating, broken locks, improperly formed or damaged terminals, poor terminal to wiring connections or physical damage to the wiring harness.

While performing a voltage load test, if it is noted that only ignition voltage drops below acceptable voltage levels, CKT 41 should be checked for high resistance or an open condition.

DIAGNOSTIC CHART DTC 136 — 1994 SILHOUETTE, TRANS SPORT AND LUMINA APV WITH TCS, CONTINUED

DTC 136

(Page 1 of 2)

LOW SYSTEM VOLTAGE (TCS)

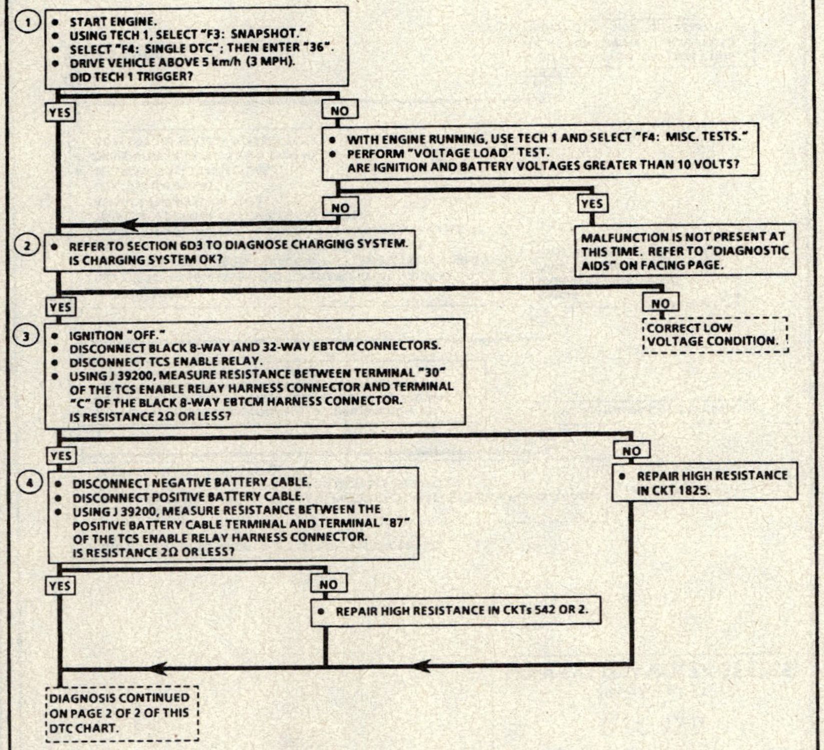

DIAGNOSTIC CHART DTC 136 — 1994 SILHOUETTE, TRANS SPORT AND LUMINA APV WITH TCS

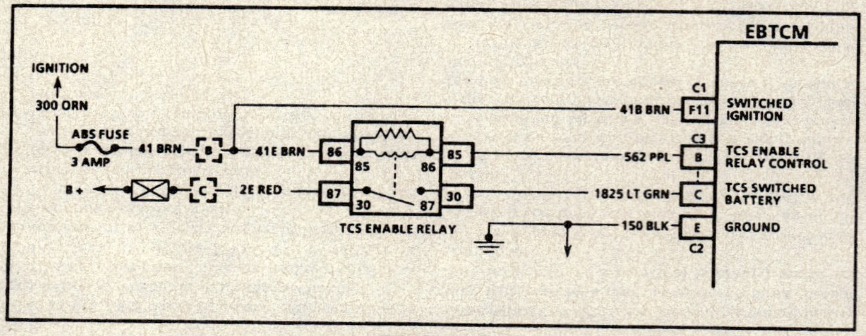

DTC 136
(Page 2 of 2)
LOW SYSTEM VOLTAGE (TCS)

Circuit Description:
This DTC is used to monitor the voltage level available to the EBTCM. If the voltage drops below 11 volts, full performance of the TCS system cannot be guaranteed. During TCS operation, there are several current requirements that will cause battery voltage to drop. Because of this, voltage is monitored prior to TCS to indicate good charging system condition and also during TCS when voltage may drop significantly.

DTC Will Set When: DTC 136 can only be set if the vehicle's speed is greater than 5 km/h (3 mph). If the switched battery voltage is less than 11.2 volts non-TCS or less than 9.6 volts TCS, a malfunction exists.

Action Taken: A malfunction DTC is stored, TCS is disabled and the TCS warning lamp is turned "ON."

DTC Chart Test Description: Number(s) below refer to circled number(s) on the diagnostic chart.

5. Checks for high resistance in CKT 41.
6. Checks for high resistance in the ignition fuse feed circuit.
7. Checks for malfunctioning EBTCM.

Diagnostic Aids: An intermittent malfunction may be caused by a poor connection, rubbed through wire insulation, or a wire that is broken inside the insulation.

The frequency of the malfunction can be checked by using the enhanced diagnostic function of the Tech 1, as described in "Tech 1 Diagnostics" in this section.

Any circuitry, that is suspected as causing the intermittent complaint, should be thoroughly checked for backed out terminals, improper mating, broken locks, improperly formed or damaged terminals, poor terminal to wiring connections or physical damage to the wiring harness.

While performing a voltage load test, if it is noted that only ignition voltage drops below acceptable voltage levels, CKT 41 should be checked for high resistance or an open condition.

DIAGNOSTIC CHART DTC 136 — 1994 SILHOUETTE, TRANS SPORT AND LUMINA APV WITH TCS, CONTINUED

DTC 136
(Page 2 of 2)
LOW SYSTEM VOLTAGE (TCS)

DIAGNOSIS CONTINUED FROM PAGE 1 OF 2 OF THIS DTC CHART.

(5)
- REMOVE ABS FUSE.
- USING J 39200 MEASURE RESISTANCE BETWEEN EACH ABS FUSE BLOCK TERMINAL AND TERMINAL "F11" OF THE 32-WAY EBTCM HARNESS CONNECTOR. IS EITHER MEASUREMENT 2Ω OR LESS?

YES → (6)
NO → REPAIR HIGH RESISTANCE IN CKT 41.

(6)
- WITH BATTERY CABLES DISCONNECTED, TURN IGNITION "ON"
- USING J 39200, MEASURE RESISTANCE BETWEEN POSITIVE BATTERY CABLE TERMINAL AND EACH ABS FUSE BLOCK TERMINAL. IS EITHER MEASUREMENT 2Ω OR LESS ?

YES → (7)
NO → REPAIR HIGH RESISTANCE IN THE "IGNITION SWITCH 'RUN' FUSE FEED" CIRCUIT.

(7)
- IGNITION "OFF."
- INSPECT TCS ENABLE RELAY AND TCS ENABLE RELAY HARNESS CONNECTOR FOR POOR TERMINAL CONTACT. INSPECT BLACK 8-WAY AND 32-WAY EBTCM HARNESS CONNECTORS FOR A POOR CONNECTION TO THE EBTCM. INSPECT POSITIVE BATTERY CABLE TERMINAL FOR POOR CONTACT TO THE BATTERY. REPLACE TERMINALS IF POOR CONTACT IS EVIDENT.
- RECONNECT ALL CONNECTORS.
- DRIVE VEHICLE ABOVE 5 km/h (3 MPH). DOES DTC 136 SET AS A CURRENT DTC?

YES → REPLACE EBTCM.
NO → MALFUNCTION IS MOST LIKELY DUE TO A POOR CONNECTION. REFER TO "DIAGNOSTIC AIDS" ON FACING PAGE.

DIAGNOSTIC CHART DTC 144 — 1994 SILHOUETTE, TRANS SPORT AND LUMINA APV WITH TCS

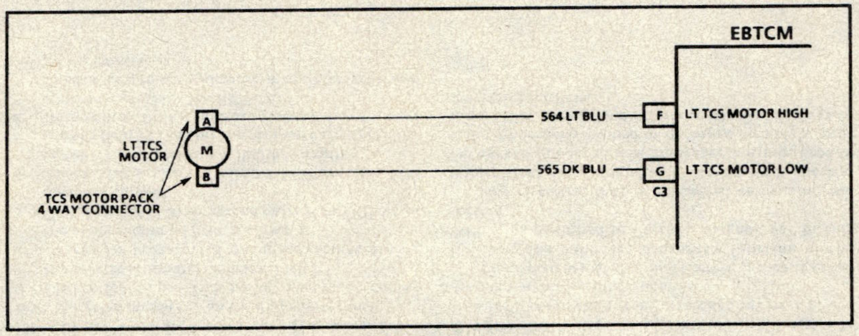

DTC 144

LEFT TCS CHANNEL WILL NOT MOVE

Circuit Description:
This DTC is designed to detect a stuck TCS motor or a seized hydraulic modulator. During initialization or after a TCS cycle, the EBTCM commands the TCS motor in the reverse direction and the forward direction to ensure that the motor moves properly. If the TCS motor does not move, improper current feedback will be detected by the EBTCM, and a DTC will set.

DTC Will Set When: DTC 144 can be set during initialization or after a TCS cycle. If the EBTCM detects a condition in which it cannot move the motor in either direction, a malfunction exists.

Action Taken: A malfunction DTC is stored, TCS is disabled, and the TCS warning and the red "BRAKE" warning lamps are turned "ON."

DTC Chart Test Description: Number(s) below refer to circled number(s) on the diagnostic chart.

1. Checks for proper TCS motor movement during the release command from the Tech 1.
2. Checks for proper TCS motor movement during the apply command from the Tech 1.
3. Compares EBTCM command current to TCS motor feedback current.
4. Checks for a shorted circuit between the TCS motor "HIGH" and motor "LOW" circuits.
5. Checks for a TCS motor that is internally shorted.
6. Determines if the malfunction is due to poor terminal contact or the EBTCM.
7. Checks for proper hydraulic modulator gear and piston movement.

Diagnostic Aids: An intermittent malfunction in this DTC may result from a mechanical part of the system that sticks, binds, or slips.

The frequency of the malfunction can be checked by using the enhanced diagnostic function of the Tech 1, as described in "Tech 1 Diagnostics" in this section.

Any circuitry, that is suspected as causing the intermittent complaint, should be thoroughly checked for backed out terminals, improper mating, broken locks, improperly formed or damaged terminals, poor terminal to wiring connections or physical damage to the wiring harness.

DIAGNOSTIC CHART DTC 144 — 1994 SILHOUETTE, TRANS SPORT AND LUMINA APV WITH TCS, CONTINUED

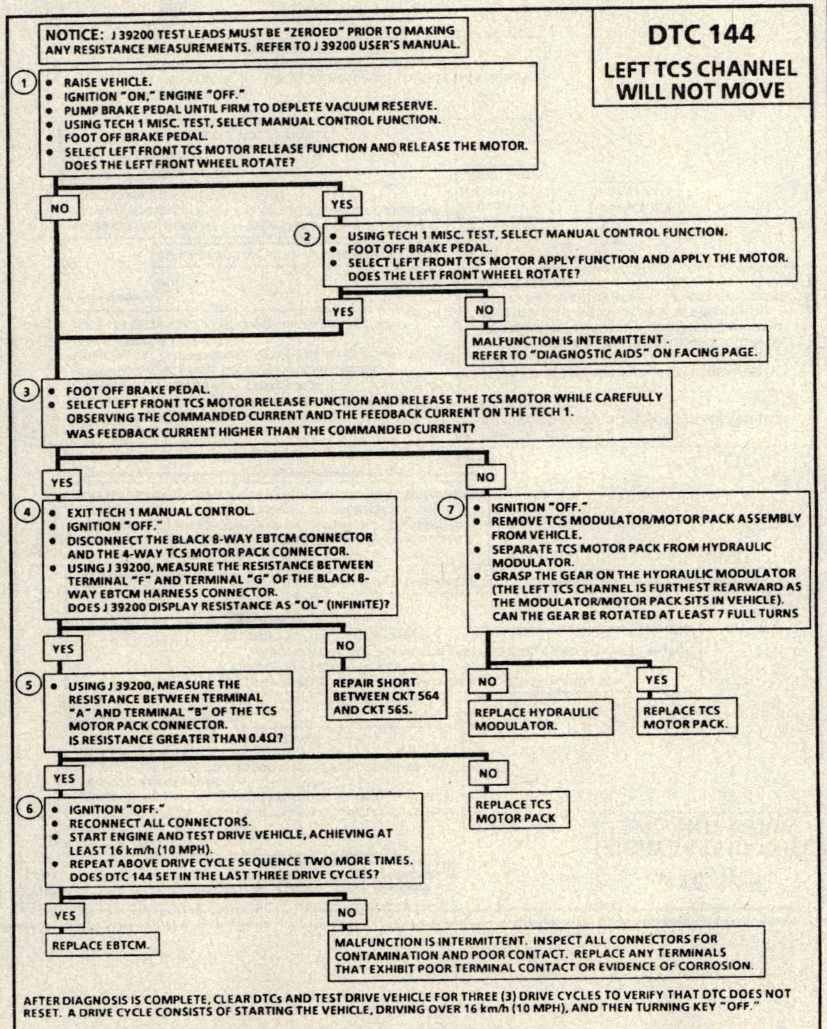

DIAGNOSTIC CHART DTC 145 — 1994 SILHOUETTE, TRANS SPORT AND LUMINA APV WITH TCS

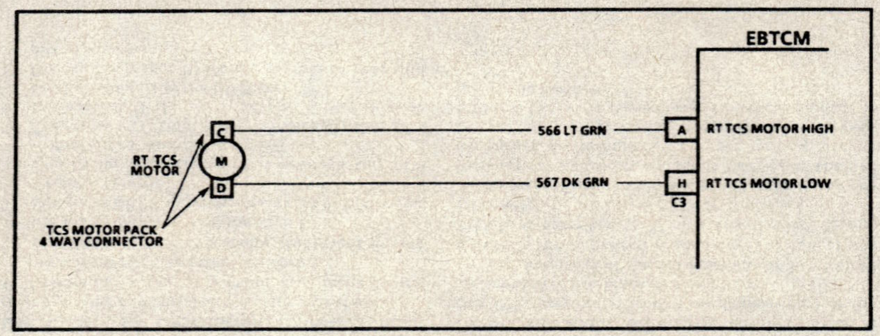

DTC 145

RIGHT TCS CHANNEL WILL NOT MOVE

Circuit Description:

This DTC is designed to detect a stuck TCS motor or a seized hydraulic modulator. During initialization or after a TCS cycle, the EBTCM commands the TCS motor in the reverse direction and the forward direction to ensure that the motor moves properly. If the TCS motor does not move, improper current feedback will be detected by the EBTCM, and a DTC will set.

DTC Will Set When: DTC 145 can be set during initialization or after a TCS cycle. If the EBTCM detects a condition in which it cannot move the motor in either direction, a malfunction exists.

Action Taken: A malfunction DTC is stored, TCS is disabled, and the TCS warning and the red "BRAKE" warning lamps are turned "ON."

DTC Chart Test Description: Number(s) below refer to circled number(s) on the diagnostic chart.

1. Checks for proper TCS motor movement during the release command from the Tech 1.
2. Checks for proper TCS motor movement during the apply command from the Tech 1.
3. Compares EBTCM command current to TCS motor feedback current.
4. Checks for a shorted circuit between the TCS motor "HIGH" and motor "LOW" circuits.
5. Checks for a TCS motor that is internally shorted.
6. Determines if the malfunction is due to poor terminal contact or the EBTCM.
7. Checks for proper hydraulic modulator gear and piston movement.

Diagnostic Aids: An intermittent malfunction in this DTC may result from a mechanical part of the system that sticks, binds, or slips.

The frequency of the malfunction can be checked by using the enhanced diagnostic function of the Tech 1, as described in "Tech 1 Diagnostics" in this section.

Any circuitry, that is suspected as causing the intermittent complaint, should be thoroughly checked for backed out terminals, improper mating, broken locks, improperly formed or damaged terminals, poor terminal to wiring connections or physical damage to the wiring harness.

DIAGNOSTIC CHART DTC 145 — 1994 SILHOUETTE, TRANS SPORT AND LUMINA APV WITH TCS, CONTINUED

DTC 145
RIGHT TCS CHANNEL WILL NOT MOVE

NOTICE: J 39200 TEST LEADS MUST BE "ZEROED" PRIOR TO MAKING ANY RESISTANCE MEASUREMENTS. REFER TO J 39200 USER'S MANUAL.

(1)
- RAISE VEHICLE.
- IGNITION "ON," ENGINE "OFF."
- PUMP BRAKE PEDAL UNTIL FIRM TO DEPLETE VACUUM RESERVE.
- USING TECH 1 MISC. TEST, SELECT MANUAL CONTROL FUNCTION.
- FOOT OFF BRAKE PEDAL.
- SELECT RIGHT FRONT TCS MOTOR RELEASE FUNCTION AND RELEASE THE MOTOR.

DOES THE RIGHT FRONT WHEEL ROTATE?

NO → go to (3). YES → (2)

(2)
- USING TECH 1 MISC. TEST, SELECT MANUAL CONTROL FUNCTION.
- FOOT OFF BRAKE PEDAL.
- SELECT RIGHT FRONT TCS MOTOR APPLY FUNCTION AND APPLY THE MOTOR.
- DOES THE RIGHT FRONT WHEEL ROTATE?

YES → go to (3). NO → MALFUNCTION IS INTERMITTENT. REFER TO "DIAGNOSTIC AIDS" ON FACING PAGE.

(3)
- FOOT OFF BRAKE PEDAL.
- SELECT RIGHT FRONT TCS MOTOR RELEASE FUNCTION AND RELEASE THE TCS MOTOR WHILE CAREFULLY OBSERVING THE COMMANDED CURRENT AND THE FEEDBACK CURRENT ON THE TECH 1.

WAS FEEDBACK CURRENT HIGHER THAN THE COMMANDED CURRENT?

YES → (4). NO → (7)

(4)
- EXIT TECH 1 MANUAL CONTROL.
- IGNITION "OFF."
- DISCONNECT THE BLACK 8-WAY EBTCM CONNECTOR AND THE 4-WAY TCS MOTOR PACK CONNECTOR.
- USING J 39200, MEASURE THE RESISTANCE BETWEEN TERMINAL "A" AND TERMINAL "H" OF THE BLACK 8-WAY EBTCM HARNESS CONNECTOR.

DOES J 39200 DISPLAY RESISTANCE AS "OL" (INFINITE)?

YES → (5). NO → REPAIR SHORT BETWEEN CKT 566 AND CKT 567.

(5)
- USING J 39200, MEASURE THE RESISTANCE BETWEEN TERMINAL "C" AND TERMINAL "D" OF THE TCS MOTOR PACK CONNECTOR.

IS RESISTANCE GREATER THAN 0.4Ω?

YES → (6). NO → REPLACE TCS MOTOR PACK.

(6)
- IGNITION "OFF."
- RECONNECT ALL CONNECTORS.
- START ENGINE AND TEST DRIVE VEHICLE, ACHIEVING AT LEAST 16 km/h (10 MPH).
- REPEAT ABOVE DRIVE CYCLE SEQUENCE TWO MORE TIMES.

DOES DTC 145 SET IN THE LAST THREE DRIVE CYCLES?

YES → REPLACE EBTCM. NO → MALFUNCTION IS INTERMITTENT. INSPECT ALL CONNECTORS FOR CONTAMINATION AND POOR CONTACT. REPLACE ANY TERMINALS THAT EXHIBIT POOR TERMINAL CONTACT OR EVIDENCE OF CORROSION.

(7)
- IGNITION "OFF."
- REMOVE TCS MODULATOR/MOTOR PACK ASSEMBLY FROM VEHICLE.
- SEPARATE TCS MOTOR PACK FROM HYDRAULIC MODULATOR.
- GRASP THE GEAR ON THE HYDRAULIC MODULATOR. (THE RIGHT TCS CHANNEL IS FURTHEST FORWARD AS THE MODULATOR/MOTOR PACK SITS IN VEHICLE).

CAN THE GEAR BE ROTATED AT LEAST 7 FULL TURNS LOCK TO LOCK?

NO → REPLACE HYDRAULIC MODULATOR. YES → REPLACE TCS MOTOR PACK.

AFTER DIAGNOSIS IS COMPLETE, CLEAR DTCs AND TEST DRIVE VEHICLE FOR THREE (3) DRIVE CYCLES TO VERIFY THAT DTC DOES NOT RESET. A DRIVE CYCLE CONSISTS OF STARTING THE VEHICLE, DRIVING OVER 16 km/h (10 MPH), AND THEN TURNING KEY OFF.

DIAGNOSTIC CHART DTC 147 — 1994 SILHOUETTE, TRANS SPORT AND LUMINA APV WITH TCS

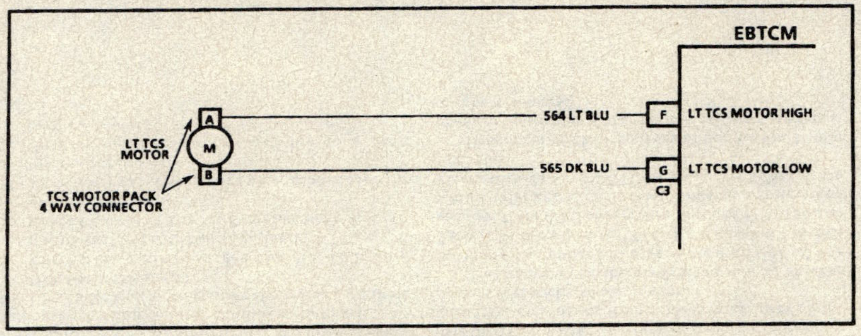

DTC 147

(Page 1 of 2)
LEFT TCS MOTOR FREE SPINS

Circuit Description:
This DTC is designed to detect a stripped nut or gear assembly during initialization. During the homing sequence, the piston should reach the bottom of the bore resulting in a stalled TCS motor. If this does not occur, the TCS motor must be spinning with little or no resistance, indicating a nut/screw or gear malfunction.

DTC Will Set When: DTC 147 can only be set during initialization. If the feedback current is less than the command current for a specified period of time, a malfunction exists.

Action Taken: A malfunction DTC is stored, TCS is disabled, and the TCS warning and the red "BRAKE" warning lamps are turned "ON."

DTC Chart Test Description: Number(s) below refer to circled number(s) on the diagnostic chart.

1. Checks to see if the corresponding open TCS motor DTC is also set.
2. Verifies that the TCS motor was actually released as commanded by observing feedback current.
3. Verifies that the TCS motor can apply.
4. This step verifies that the TCS motor can be released.
5. Checks for an open in the TCS motor "HIGH" circuitry.
6. Checks for an open in the TCS motor "LOW" circuitry.
7. Checks for a TCS motor that is internally shorted.

Diagnostic Aids: An intermittent malfunction in this DTC may result from a mechanical part of the system that sticks, binds, or slips.

The frequency of the malfunction can be checked by using the enhanced diagnostic function of the Tech 1, as described in "Tech 1 Diagnostics" in this section. If the DTC only occurs once and DTC 156 also occurs, refer to DTC 156. If intermittent and enhanced diagnostics shows this DTC occurs during TCS, refer to DTC 156.

Depending on the frequency of the malfunction, a physical inspection of the mechanical parts suspected may be necessary.

DIAGNOSTIC CHART DTC 147 — 1994 SILHOUETTE, TRANS SPORT AND LUMINA APV WITH TCS, CONTINUED

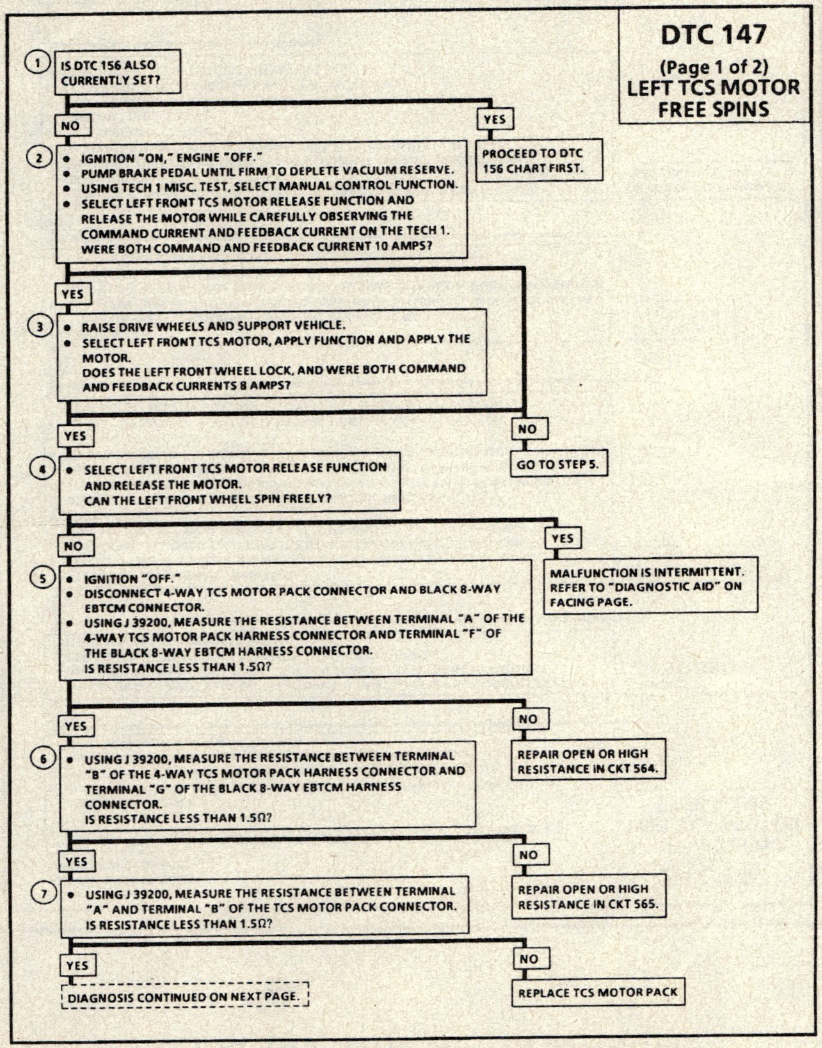

DIAGNOSTIC CHART DTC 147 — 1994 SILHOUETTE, TRANS SPORT AND LUMINA APV WITH TCS, CONTINUED

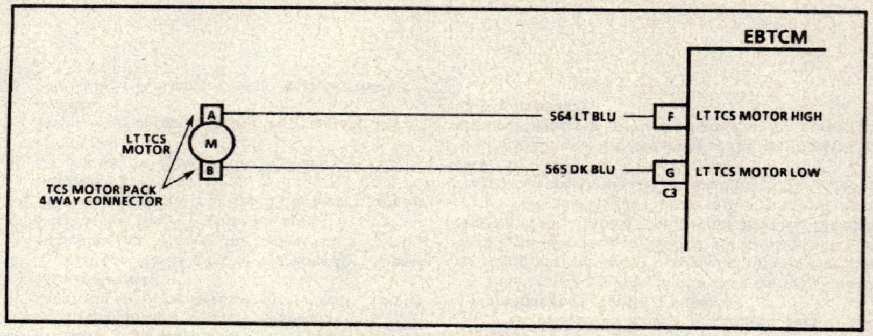

DTC 147
(Page 2 of 2)
LEFT TCS MOTOR FREE SPINS

Circuit Description:
This DTC is designed to detect a stripped nut or gear assembly during initialization. During the homing sequence, the piston should reach the bottom of the bore resulting in a stalled TCS motor. If this does not occur, the TCS motor must be spinning with little or no resistance, indicating a nut/screw or gear malfunction.

DTC Will Set When: DTC 147 can only be set during initialization. If the feedback current is less than the command current for a specified period of time, a malfunction exists.

Action Taken: A malfunction DTC is stored, TCS is disabled, and the TCS warning and the red "BRAKE" warning lamps are turned "ON."

DTC Chart Test Description: Number(s) below refer to circled number(s) on the diagnostic chart.

8. Checks for a stripped gear on the motor pack (unit with two small gears).
9. Checks for a stripped gear on the hydraulic modulator (unit with two large gears).
10. Checks to see if the TCS motor pack has an internal malfunction.
11. Verifies that the hydraulic modulator is functioning properly.
12. Ensures malfunction was not due to a poor terminal contact.

Diagnostic Aids: An intermittent malfunction in this DTC may result from a mechanical part of the system that sticks, binds, or slips.

The frequency of the malfunction can be checked by using the enhanced diagnostic function of the Tech 1, as described in "Tech 1 Diagnostics" in this section. If the DTC only occurs once and DTC 156 also occurs, refer to DTC 156. If intermittent and enhanced diagnostics shows this DTC occurs during TCS, refer to DTC 156.

Depending on the frequency of the malfunction, a physical inspection of the mechanical parts suspected may be necessary.

DIAGNOSTIC CHART DTC 147 — 1994 SILHOUETTE, TRANS SPORT AND LUMINA APV WITH TCS, CONTINUED

DTC 147
(Page 2 of 2)
LEFT TCS MOTOR FREE SPINS

DIAGNOSIS CONTINUED FROM PREVIOUS PAGE.

(8)
- REMOVE TCS HYDRAULIC MODULATOR/MOTOR PACK ASSEMBLY FROM VEHICLE.
- REMOVE GEAR COVER.
- CHECK FOR STRIPPED GEAR ON TCS MOTOR PACK; TCS MOTOR PACK IS THE UNIT WITH TWO SMALL GEARS. THE LEFT FRONT GEAR IS THE GEAR FURTHEST REARWARD AS THE MODULATOR/MOTOR PACK SITS IN THE VEHICLE.

IS THE TCS MOTOR PACK GEAR STRIPPED?

YES: REPLACE TCS MOTOR PACK.

NO:

(9)
- CHECK FOR A STRIPPED GEAR ON THE HYDRAULIC MODULATOR; THE MODULATOR IS THE UNIT WITH THE TWO LARGE GEARS. THE LEFT FRONT GEAR IS THE GEAR FURTHEST REARWARD AS THE MODULATOR/MOTOR PACK SITS IN VEHICLE.

IS THE MODULATOR GEAR STRIPPED?

YES: REPLACE HYDRAULIC MODULATOR GEAR.

NO:

(10)
- RECONNECT ELECTRICAL CONNECTORS.
- SECURELY POSITION HYDRAULIC MODULATOR/MOTOR PACK ASSEMBLY IN VEHICLE WITH GEAR COVER REMOVED SUCH THAT THE GEAR SET CAN BE OBSERVED. BE CAREFUL NOT TO DAMAGE GEAR SET OR ALLOW THE HYDRAULIC MODULATOR TO MOVE WHILE TESTING.
- IGNITION "ON."
- USING TECH 1 MISC. TEST, SELECT MANUAL CONTROL FUNCTION.
- SELECT LEFT FRONT TCS MOTOR APPLY AND APPLY THE MOTOR WHILE OBSERVING THE GEAR SET FURTHEST REARWARD WHEN THE UNIT IS INSTALLED IN THE VEHICLE.
- SELECT LEFT FRONT TCS MOTOR RELEASE FUNCTION AND RELEASE THE MOTOR WHILE STILL OBSERVING THE GEAR SET.
- SELECT LEFT FRONT TCS MOTOR APPLY FUNCTION AND APPLY THE MOTOR.

DID GEAR SET MOVE IN BOTH DIRECTIONS FOR AT LEAST ONE REVOLUTION?

NO: REPLACE TCS MOTOR PACK.

YES:

(11)
- IGNITION "OFF."
- SEPARATE THE TCS MOTOR PACK FROM THE HYDRAULIC MODULATOR.
- GRASP THE GEAR ON THE HYDRAULIC MODULATOR (THE HYDRAULIC MODULATOR IS THE UNIT WITH THE TWO LARGE GEARS) FURTHEST REARWARD WHEN THE UNIT IS INSTALLED IN THE VEHICLE AND ATTEMPT TO MOVE THE GEAR.

CAN THE GEAR BE ROTATED MORE THAN 8 FULL TURNS LOCK TO LOCK?

YES: REPLACE THE HYDRAULIC MODULATOR.

NO:

(12)
- INSPECT ALL CONNECTORS AND TERMINALS FOR POOR TERMINAL CONTACT AND EVIDENCE OF CORROSION. REPLACE ALL TERMINALS THAT EXHIBIT SIGNS OF POOR TERMINAL CONTACT OR CORROSION.
- REINSTALL HYDRAULIC MODULATOR/MOTOR PACK ASSEMBLY IN VEHICLE.
- RECONNECT ALL CONNECTORS.
- IGNITION "OFF."
- START ENGINE AND TEST DRIVE VEHICLE, ACHIEVING AT LEAST 16 km/h (10 MPH).
- REPEAT ABOVE DRIVE CYCLE SEQUENCE TWO MORE TIMES.

DOES DTC 147 SET IN THE LAST THREE DRIVE CYCLES?

YES: REPLACE EBTCM

YES: MALFUNCTION WAS MOSTLY LIKELY CAUSED BY A POOR CONNECTION. REFER TO "DIAGNOSTIC AIDS" ON FACING PAGE.

AFTER DIAGNOSIS IS COMPLETE, CLEAR DTCs AND TEST DRIVE VEHICLE FOR THREE (3) DRIVE CYCLES TO VERIFY THAT DTC DOES NOT RESET. A DRIVE CYCLE CONSISTS OF STARTING THE VEHICLE, DRIVING OVER 16 km/h (10 MPH), AND THEN TURNING KEY OFF.

DIAGNOSTIC CHART DTC 148 — 1994 SILHOUETTE, TRANS SPORT AND LUMINA APV WITH TCS

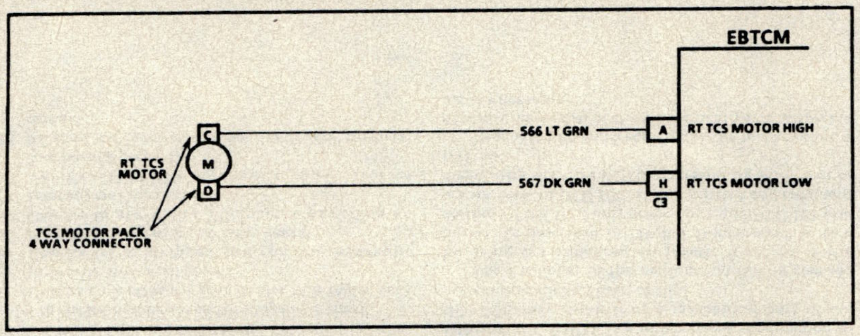

DTC 148
(Page 1 of 2)
RIGHT TCS MOTOR FREE SPINS

Circuit Description:
This DTC is designed to detect a stripped nut or gear assembly during initialization. During the homing sequence, the piston should reach the bottom of the bore resulting in a stalled TCS motor. If this does not occur, the TCS motor must be spinning with little or no resistance, indicating a nut/screw or gear malfunction.

DTC Will Set When: DTC 148 can only be set during initialization. If the feedback current is less than the command current for a specified period of time, a malfunction exists.

Action Taken: A malfunction DTC is stored, TCS is disabled, and the TCS warning and the red "BRAKE" warning lamps are turned "ON."

DTC Chart Test Description: Number(s) below refer to circled number(s) on the diagnostic chart.
1. Checks to see if the corresponding open TCS motor DTC is also set.
2. Verifies that the TCS motor was actually released as commanded by observing feedback current.
3. Verifies that the TCS motor can apply.
4. Verifies that the TCS motor can be released.
5. Checks for high resistance in the TCS motor "HIGH" circuitry.
6. Checks for high resistance in the TCS motor "LOW" circuitry.
7. Checks for proper resistance of the TCS motor windings.

Diagnostic Aids: An intermittent malfunction in this DTC may result from a mechanical part of the system that sticks, binds, or slips.

The frequency of the malfunction can be checked by using the enhanced diagnostic function of the Tech 1, as described in "Tech 1 Diagnostics" in this section. If the DTC only occurs once and DTC 161 also occurs, refer to DTC 161. If intermittent and enhanced diagnostics shows this DTC occurs during TCS, refer to DTC 161.

Depending on the frequency of the malfunction, a physical inspection of the mechanical parts suspected may be necessary.

DIAGNOSTIC CHART DTC 148 — 1994 SILHOUETTE, TRANS SPORT AND LUMINA APV WITH TCS

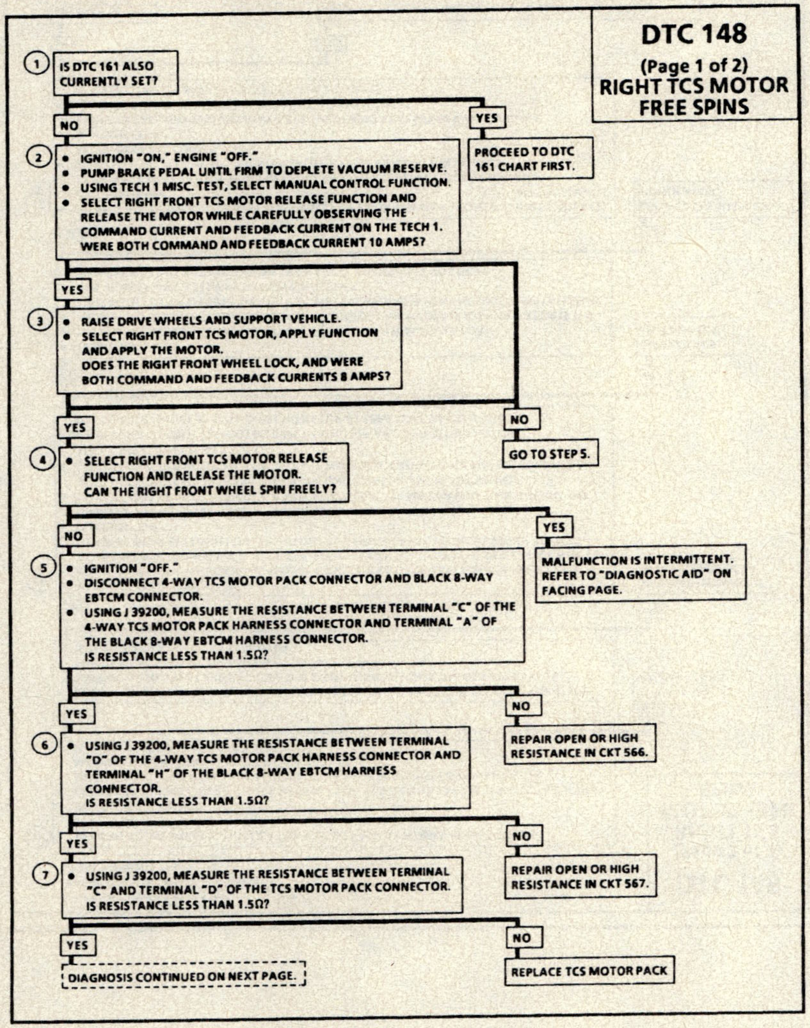

DIAGNOSTIC CHART DTC 148 — 1994 SILHOUETTE, TRANS SPORT AND LUMINA APV WITH TCS, CONTINUED

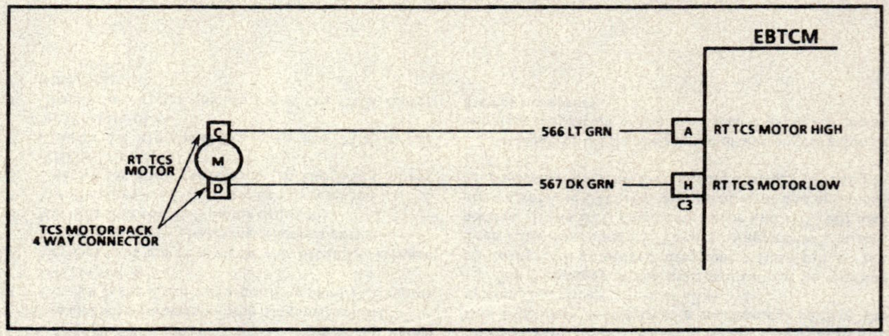

DTC 148

(Page 2 of 2)

RIGHT TCS MOTOR FREE SPINS

Circuit Description:

This DTC is designed to detect a stripped nut or gear assembly during initialization. During the homing sequence, the piston should reach the bottom of the bore resulting in a stalled TCS motor. If this does not occur, the TCS motor must be spinning with little or no resistance, indicating a nut/screw or gear malfunction.

DTC Will Set When: DTC 148 can only be set during initialization. If the feedback current is less than the command current for a specified period of time, a malfunction exists.

Action Taken: A malfunction DTC is stored, TCS is disabled, and the TCS warning and the red "BRAKE" warning lamps are turned "ON."

DTC Chart Test Description: Number(s) below refer to circled number(s) on the diagnostic chart.

8. Checks for a stripped gear on the TCS motor pack (unit with two small gears).
9. Checks for a stripped gear on the hydraulic modulator (unit with two large gears).
10. Checks to see if the TCS motor pack has an internal malfunction.
11. Verifies that the hydraulic modulator is functioning properly.
12. Ensures malfunction was not due to poor terminal contact.

Diagnostic Aids: An intermittent malfunction in this DTC may result from a mechanical part of the system that sticks, binds, or slips.

The frequency of the malfunction can be checked by using the enhanced diagnostic function of the Tech 1, as described in "Tech 1 Diagnostics" in this section. If the DTC only occurs once and DTC 161 also occurs, refer to DTC 161. If intermittent and enhanced diagnostics shows this DTC occurs during TCS, refer to DTC 161.

Depending on the frequency of the malfunction, a physical inspection of the mechanical parts suspected may be necessary.

DIAGNOSTIC CHART DTC 148 — 1994 SILHOUETTE, TRANS SPORT AND LUMINA APV WITH TCS, CONTINUED

DTC 148
(Page 2 of 2)
RIGHT TCS MOTOR FREE SPINS

DIAGNOSIS CONTINUED FROM PREVIOUS PAGE.

(8)
- REMOVE TCS HYDRAULIC MODULATOR/MOTOR PACK ASSEMBLY FROM VEHICLE.
- REMOVE GEAR COVER.
- CHECK FOR STRIPPED GEAR ON TCS MOTOR PACK; TCS MOTOR PACK IS THE UNIT WITH TWO SMALL GEARS. THE RIGHT FRONT GEAR IS THE GEAR FURTHEST FORWARD AS THE MODULATOR/MOTOR PACK SITS IN THE VEHICLE.

IS THE TCS MOTOR PACK GEAR STRIPPED?

- YES → REPLACE TCS MOTOR PACK.
- NO → (9)

(9)
- CHECK FOR A STRIPPED GEAR ON THE HYDRAULIC MODULATOR; THE MODULATOR IS THE UNIT WITH THE TWO LARGE GEARS. THE RIGHT FRONT GEAR IS THE GEAR FURTHEST FORWARD AS THE MODULATOR/MOTOR PACK SITS IN VEHICLE.

IS THE MODULATOR GEAR STRIPPED?

- YES → REPLACE HYDRAULIC MODULATOR GEAR.
- NO → (10)

(10)
- RECONNECT ELECTRICAL CONNECTORS.
- SECURELY POSITION HYDRAULIC MODULATOR/MOTOR PACK ASSEMBLY IN VEHICLE WITH GEAR COVER REMOVED SUCH THAT THE GEAR SET CAN BE OBSERVED. BE CAREFUL NOT TO DAMAGE GEAR SET OR ALLOW THE HYDRAULIC MODULATOR TO MOVE WHILE TESTING.
- IGNITION "ON."
- USING TECH 1 MISC. TEST, SELECT MANUAL CONTROL FUNCTION.
- SELECT RIGHT FRONT TCS MOTOR APPLY AND APPLY THE MOTOR WHILE OBSERVING THE GEAR SET FURTHEST FORWARD WHEN THE UNIT IS INSTALLED IN THE VEHICLE.
- SELECT RIGHT FRONT TCS MOTOR RELEASE FUNCTION AND RELEASE THE MOTOR WHILE STILL OBSERVING THE GEAR SET.
- SELECT RIGHT FRONT TCS MOTOR APPLY FUNCTION AND APPLY THE MOTOR.

DID GEAR SET MOVE IN BOTH DIRECTIONS FOR AT LEAST ONE REVOLUTION?

- NO → REPLACE TCS MOTOR PACK.
- YES → (11)

(11)
- IGNITION "OFF."
- SEPARATE THE TCS MOTOR PACK FROM THE HYDRAULIC MODULATOR.
- GRASP THE GEAR ON THE HYDRAULIC MODULATOR (THE HYDRAULIC MODULATOR IS THE UNIT WITH THE TWO LARGE GEARS) FURTHEST FORWARD WHEN THE UNIT IS INSTALLED IN THE VEHICLE AND ATTEMPT TO MOVE THE GEAR.

CAN THE GEAR BE ROTATED MORE THAN 8 FULL TURNS LOCK TO LOCK?

- YES → REPLACE THE HYDRAULIC MODULATOR.
- NO → (12)

(12)
- INSPECT ALL CONNECTORS AND TERMINALS FOR POOR TERMINAL CONTACT AND EVIDENCE OF CORROSION. REPLACE ALL TERMINALS THAT EXHIBIT SIGNS OF POOR TERMINAL CONTACT OR CORROSION.
- REINSTALL HYDRAULIC MODULATOR/MOTOR PACK ASSEMBLY IN VEHICLE.
- RECONNECT ALL CONNECTORS.
- IGNITION "OFF."
- START ENGINE AND TEST DRIVE VEHICLE, ACHIEVING AT LEAST 16 km/h (10 MPH).
- REPEAT ABOVE DRIVE CYCLE SEQUENCE TWO MORE TIMES.

DOES DTC 148 SET IN THE LAST THREE DRIVE CYCLES?

- YES → REPLACE EBTCM
- NO → MALFUNCTION WAS MOSTLY LIKELY CAUSED BY A POOR CONNECTION. REFER TO "DIAGNOSTIC AIDS" ON FACING PAGE.

AFTER DIAGNOSIS IS COMPLETE, CLEAR DTCs AND TEST DRIVE VEHICLE FOR THREE (3) DRIVE CYCLES TO VERIFY THAT DTC DOES NOT RESET. A DRIVE CYCLE CONSISTS OF STARTING THE VEHICLE, DRIVING OVER 16 km/h (10 MPH), AND THEN TURNING KEY OFF.

DIAGNOSTIC CHART DTC 155 — 1994 SILHOUETTE, TRANS SPORT AND LUMINA APV WITH TCS

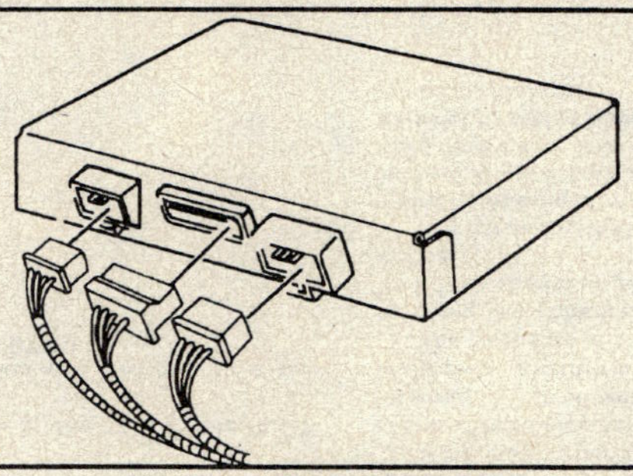

DTC 155
EBTCM MALFUNCTION (TCS)

Circuit Description:
This DTC identifies an internal malfunction detected by the EBTCM motor driver interface custom integrated circuit.

Action Taken: A malfunction DTC is stored, TCS is disabled and the TCS warning lamp is turned "ON."

DTC Chart Test Description: Number(s) below refer to circled number(s) on the diagnostic chart.
1. Checks to see if the malfunction is still present.
2. Checks to see if the malfunction is intermittent.

DIAGNOSTIC CHART DTC 155 — 1994 SILHOUETTE, TRANS SPORT AND LUMINA APV WITH TCS, CONTINUED

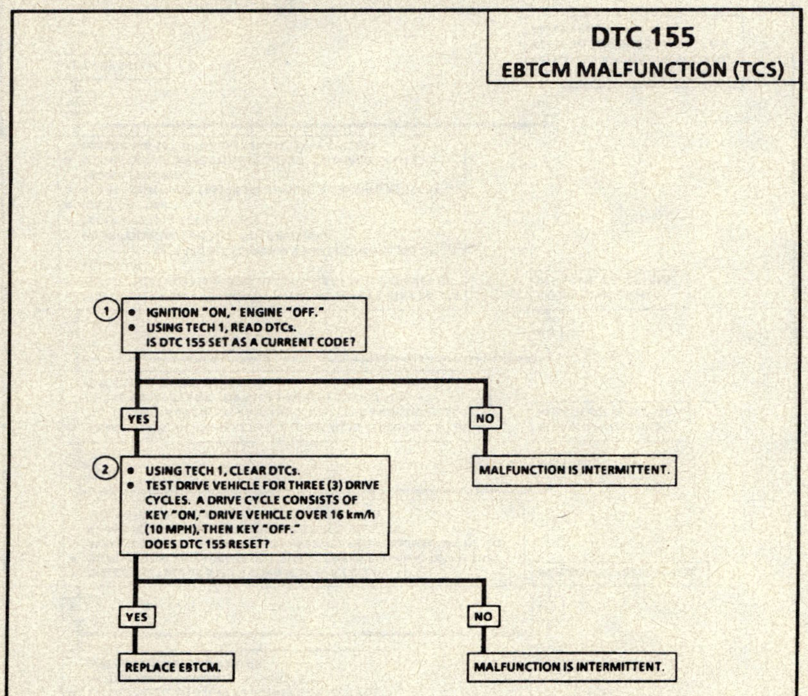

DIAGNOSTIC CHART DTC 156 — 1994 SILHOUETTE, TRANS SPORT AND LUMINA APV WITH TCS

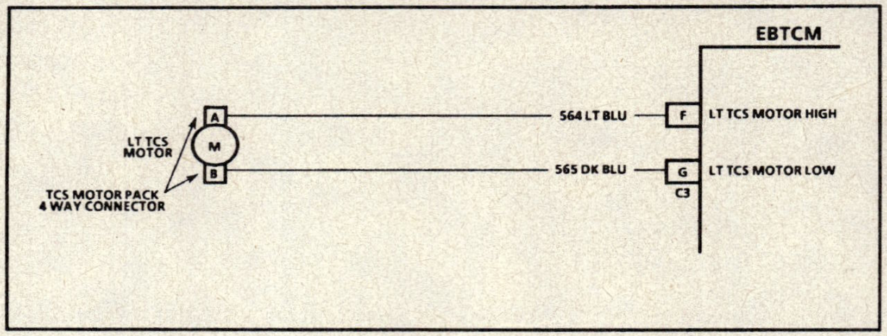

DTC 156

LEFT TCS MOTOR CIRCUIT OPEN

Circuit Description:
This DTC identifies a TCS motor that cannot be energized due to an open in its circuitry.

DTC Will Set When: DTC 156 can be set only when the motor is commanded "OFF." If the EBTCM detects an out of range voltage on either of the left front TCS motor circuits indicating an open circuit, a malfunction exists.

Action Taken: An open TCS motor will not activate when requested. A malfunction DTC is stored, TCS is disabled and the TCS warning lamp is turned "ON."

DTC Chart Test Description: Number(s) below refer to circled number(s) on the diagnostic chart.

1. Checks for proper resistance of the TCS motor.
2. Checks for an open in the TCS motor "HIGH" circuitry.
3. Checks for an open in the TCS motor "LOW" circuitry.
4. Ensures malfunction was not due to a poor terminal contact or EBTCM.

Diagnostic Aids: An intermittent malfunction may be caused by a poor connection, rubbed through wire insulation, or a wire that is broken inside the insulation.

If the malfunction is not current, wiggle the wires of the affected channel and check if the DTC resets. This will help to pinpoint an intermittent malfunction in the motor circuitry or connections.

The frequency of the malfunction can be checked by using the enhanced diagnostic function of the Tech 1, as described in "Tech 1 Diagnostics" in this section.

Any circuitry, that is suspected as causing the intermittent complaint, should be thoroughly checked for backed out terminals, improper mating, broken locks, improperly formed or damaged terminals, poor terminal to wiring connections or physical damage to the wiring harness.

DIAGNOSTIC CHART DTC 156 — 1994 SILHOUETTE, TRANS SPORT AND LUMINA APV WITH TCS, CONTINUED

DTC 156

LEFT TCS MOTOR CIRCUIT OPEN

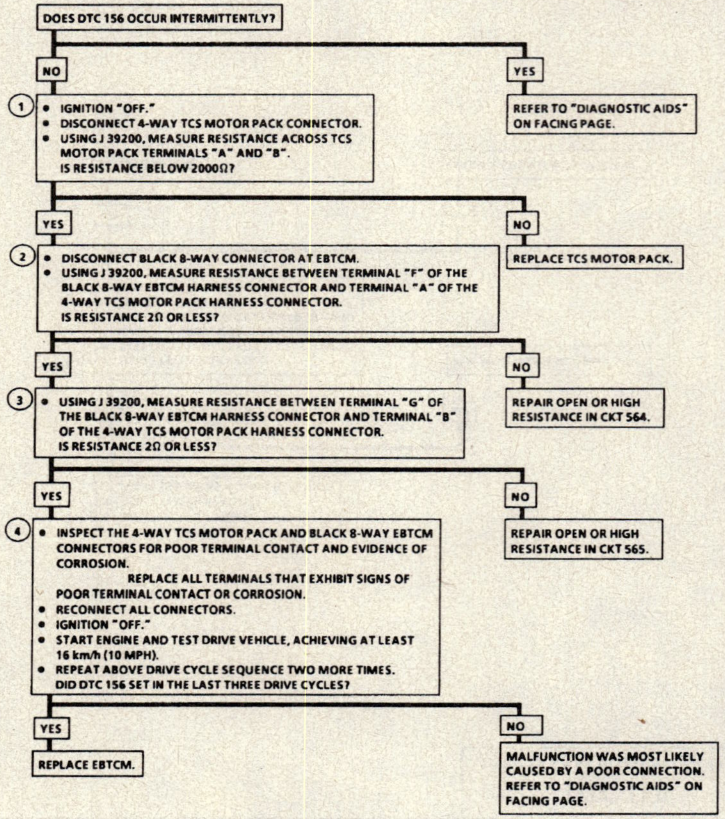

AFTER DIAGNOSIS IS COMPLETE, CLEAR DTCs AND TEST DRIVE VEHICLE FOR THREE (3) DRIVE CYCLES TO VERIFY THAT DTC DOES NOT RESET. A DRIVE CYCLE CONSISTS OF STARTING THE VEHICLE, DRIVING OVER 16 KM/H (10 MPH), AND THEN TURNING KEY "OFF."

DIAGNOSTIC CHART DTC 157 — 1994 SILHOUETTE, TRANS SPORT AND LUMINA APV WITH TCS

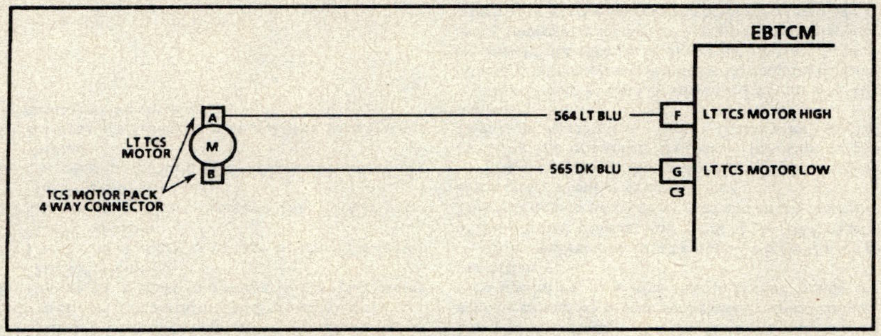

DTC 157

LEFT TCS MOTOR CIRCUIT SHORTED TO GND

Circuit Description:
This DTC identifies a TCS motor circuit that is shorted to ground. This malfunction will not allow the TCS motor to be controlled at the commanded current rate or will cause the TCS motor driver circuit to allow current directly to ground.

DTC Will Set When: DTC 157 can be set anytime. If the EBTCM detects an out of range voltage on either of the left front TCS motor circuits indicating a circuit shorted to ground, a malfunction exists.

Action Taken: A malfunction DTC is stored, TCS is disabled and the TCS warning lamp is turned "ON."

DTC Chart Test Description: Number(s) below refer to circled number(s) on the diagnostic chart.

1. Checks for a short to ground in the TCS motor "HIGH" circuitry.
2. Checks for a short to ground in the TCS motor "LOW" circuitry.
3. Checks for a TCS motor that is internally shorted to ground.
4. Ensures malfunction was not due to physical damage of the circuitry.

Diagnostic Aids: An intermittent malfunction may be caused by a poor connection, rubbed through wire insulation, or a wire that is broken inside the insulation.

If the malfunction is not current, wiggle the wires of the affected channel and check if the DTC resets. This will help to pinpoint an intermittent malfunction in the motor circuitry or connections.

The frequency of the malfunction can be checked by using the enhanced diagnostic function of the Tech 1, as described in "Tech 1 Diagnostics" in this section.

Any circuitry, that is suspected as causing the intermittent complaint, should be thoroughly checked for backed out terminals, improper mating, broken locks, improperly formed or damaged terminals, poor terminal to wiring connections or physical damage to the wiring harness.

DIAGNOSTIC CHART DTC 157 — 1994 SILHOUETTE, TRANS SPORT AND LUMINA APV WITH TCS, CONTINUED

DTC 157
LEFT TCS MOTOR CIRCUIT SHORTED TO GND

DOES DTC 157 OCCUR INTERMITTENTLY?

- YES: REFER TO "DIAGNOSTIC AIDS" ON FACING PAGE.
- NO:

(1)
- IGNITION "OFF."
- DISCONNECT THE 4-WAY TCS MOTOR PACK CONNECTOR AND THE BLACK 8-WAY EBTCM CONNECTOR.
- USING J 39200, MEASURE THE RESISTANCE BETWEEN GROUND AND TERMINAL "A" OF THE 4-WAY TCS MOTOR PACK HARNESS CONNECTOR.

DOES J 39200 DISPLAY RESISTANCE AS "OL" (INFINITE)?

- NO: REPAIR SHORT TO GROUND IN CKT 564.
- YES:

(2)
- USING J 39200, MEASURE THE RESISTANCE BETWEEN GROUND AND TERMINAL "B" OF THE 4-WAY TCS MOTOR PACK HARNESS CONNECTOR.

DOES J 39200 DISPLAY RESISTANCE AS "OL" (INFINITE)?

- NO: REPAIR SHORT TO GROUND IN CKT 565.
- YES:

(3)
- USING J 39200, MEASURE THE RESISTANCE BETWEEN GROUND AND TERMINAL "A" OF THE TCS MOTOR PACK.

DOES J 39200 DISPLAY RESISTANCE AS "OL" (INFINITE)?

- NO: REPLACE TCS MOTOR PACK.
- YES:

(4)
- INSPECT CKTs 564 AND 565 AND RELATED CONNECTORS FOR DAMAGE WHICH MAY RESULT IN A SHORT TO GROUND WITH ALL CONNECTORS CONNECTED. REPAIR ALL DAMAGE IF EVIDENT.
- RECONNECT ALL CONNECTORS.
- IGNITION "OFF."
- START ENGINE AND TEST DRIVE VEHICLE, ACHIEVING AT LEAST 16 km/h (10 MPH).
- REPEAT ABOVE DRIVE CYCLE SEQUENCE TWO MORE TIMES.

DID DTC 157 SET IN THE LAST THREE DRIVE CYCLES?

- YES: REPLACE EBTCM.
- NO: MALFUNCTION WAS MOST LIKELY CAUSED BY A POOR CONNECTION. REFER TO "DIAGNOSTIC AIDS" ON FACING PAGE.

AFTER ALL DIAGNOSIS IS COMPLETE, CLEAR DTCs AND TEST DRIVE VEHICLE FOR THREE (3) DRIVE CYCLES TO VERIFY THAT DTC DOES NOT RESET. A DRIVE CYCLE CONSISTS OF STARTING THE VEHICLE, DRIVING OVER 16 km/h (10 MPH), AND THEN TURNING KEY "OFF."

DIAGNOSTIC CHART DTC 158 — 1994 SILHOUETTE, TRANS SPORT AND LUMINA APV WITH TCS

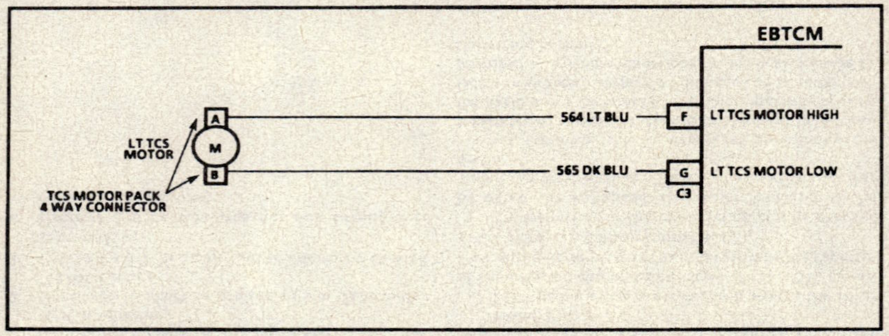

DTC 158

LEFT TCS MOTOR CIRCUIT SHORTED TO BATT

Circuit Description:
This DTC identifies a TCS motor circuit that is shorted to battery or a TCS motor that has low or no resistance. This malfunction will not allow the TCS motor to be controlled at the commanded current rate or will cause the TCS motor to turn in the opposite direction, or not at all.

DTC Will Set When: DTC 158 can be set only when the TCS motor is commanded "OFF." If the EBTCM detects an out of range voltage on either of the left front TCS motor circuits indicating a circuit shorted to battery or a TCS motor shorted, a malfunction exists.

Action Taken: A malfunction DTC is stored, TCS is disabled and the TCS warning lamp is turned "ON."

DTC Chart Test Description: Number(s) below refer to circled number(s) on the diagnostic chart.

1. Checks for a short to voltage in the TCS motor "HIGH" circuitry.
2. Checks for a short to voltage in the TCS motor "LOW" circuitry.
3. Checks for a short between the "HIGH" and "LOW" motor circuits.
4. This step checks for a TCS motor that is internally shorted.
5. Ensures malfunction was not due to physical damage of the circuitry.

Diagnostic Aids: An intermittent malfunction may be caused by a poor connection, rubbed through wire insulation, or a wire that is broken inside the insulation.

If the malfunction is not current, wiggle the wires of the affected channel and check if the DTC resets. This will help to pinpoint an intermittent malfunction in the motor circuitry or connections.

The frequency of the malfunction can be checked by using the enhanced diagnostic function of the Tech 1, as described in "Tech 1 Diagnostics" in this section.

Any circuitry, that is suspected as causing the intermittent complaint, should be thoroughly checked for backed out terminals, improper mating, broken locks, improperly formed or damaged terminals, poor terminal to wiring connections or physical damage to the wiring harness.

DIAGNOSTIC CHART DTC 158 — 1994 SILHOUETTE, TRANS SPORT AND LUMINA APV WITH TCS, CONTINUED

DTC 158

LEFT TCS MOTOR CIRCUIT SHORTED TO BATT

NOTICE: J 39200 TEST LEADS MUST BE "ZEROED" PRIOR TO MAKING ANY RESISTANCE MEASUREMENTS. REFER TO J 39200 USER'S MANUAL.

DOES DTC 158 OCCUR INTERMITTENTLY?

- YES: REFER TO "DIAGNOSTIC AIDS" ON FACING PAGE.
- NO:

(1)
- IGNITION "OFF."
- DISCONNECT THE 4-WAY TCS MOTOR PACK CONNECTOR AND THE BLACK 8-WAY EBTCM CONNECTOR.
- IGNITION "ON."
- USING J 39200, MEASURE THE VOLTAGE BETWEEN GROUND AND TERMINAL "A" OF THE 4-WAY TCS MOTOR PACK HARNESS CONNECTOR. IS VOLTAGE 1 VOLT OR LESS?

- NO: REPAIR SHORT TO VOLTAGE IN CKT 564.
- YES:

(2)
- USING J 39200, MEASURE THE VOLTAGE BETWEEN GROUND AND TERMINAL "B" OF THE 4-WAY TCS MOTOR PACK HARNESS CONNECTOR. IS VOLTAGE 1 VOLT OR LESS?

- NO: REPAIR SHORT TO VOLTAGE IN CKT 565.
- YES:

(3)
- IGNITION "OFF."
- USING J 39200, MEASURE THE RESISTANCE BETWEEN TERMINAL "A" AND TERMINAL "B" OF THE 4-WAY TCS MOTOR PACK HARNESS CONNECTOR. DOES J 39200 DISPLAY RESISTANCE AS "OL" (INFINITE)?

- NO: REPAIR SHORT BETWEEN CKTs 564 AND 565.
- YES:

(4)
- USING J 39200, MEASURE THE RESISTANCE BETWEEN TERMINAL "A" AND TERMINAL "B" OF THE TCS MOTOR PACK. IS RESISTANCE GREATER THAN 0.4Ω?

- NO: REPLACE TCS MOTOR PACK.
- YES:

(5)
- INSPECT CKTs 564 AND 565 AND RELATED CONNECTORS FOR DAMAGE WHICH MAY RESULT IN A SHORT TO VOLTAGE WITH ALL CONNECTORS CONNECTED. REPAIR ALL DAMAGE IF EVIDENT.
- RECONNECT ALL CONNECTORS.
- IGNITION "OFF."
- START ENGINE AND TEST DRIVE VEHICLE, ACHIEVING AT LEAST 16 km/h (10 MPH).
- REPEAT ABOVE DRIVE CYCLE SEQUENCE TWO MORE TIMES. DID DTC 158 SET IN THE LAST THREE DRIVE CYCLES?

- YES: REPLACE EBTCM.
- NO: MALFUNCTION WAS MOST LIKELY CAUSED BY A POOR CONNECTION. REFER TO "DIAGNOSTIC AIDS" ON FACING PAGE.

AFTER DIAGNOSIS IS COMPLETE, CLEAR DTCs AND TEST DRIVE VEHICLE FOR THREE (3) DRIVE CYCLES TO VERIFY THAT DTC DOES NOT RESET. A DRIVE CYCLE CONSISTS OF STARTING THE VEHICLE, DRIVING OVER 16 KM/H (10 MPH), AND THEN TURNING KEY "OFF."

DIAGNOSTIC CHART DTC 161 — 1994 SILHOUETTE, TRANS SPORT AND LUMINA APV WITH TCS

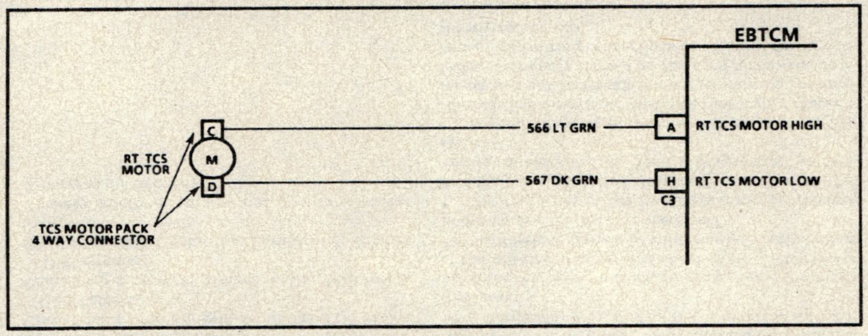

DTC 161

RIGHT TCS MOTOR CIRCUIT OPEN

Circuit Description:
This DTC identifies a TCS motor that cannot be energized due to an open in its circuitry.

DTC Will Set When: DTC 161 can be set only when the TCS motor is commanded "OFF." If the EBTCM detects an out of range voltage on either of the right front TCS motor circuits indicating an open circuit, a malfunction exists.

Action Taken: An open TCS motor will not activate when requested. A malfunction DTC is stored, TCS is disabled and the TCS warning lamp is turned "ON."

DTC Chart Test Description: Number(s) below refer to circled number(s) on the diagnostic chart.

1. Checks for proper resistance of the TCS motor.
2. Checks for an open in the TCS motor "HIGH" circuitry.
3. Checks for an open in the TCS motor "LOW" circuitry.
4. Ensures malfunction was not due to poor terminal contact or EBTCM.

Diagnostic Aids: An intermittent malfunction may be caused by a poor connection, rubbed through wire insulation, or a wire that is broken inside the insulation.

If the malfunction is not current, wiggle the wires of the affected channel and check if the DTC resets. This will help to pinpoint an intermittent malfunction in the motor circuitry or connections.

The frequency of the malfunction can be checked by using the enhanced diagnostic function of the Tech 1, as described in "Tech 1 Diagnostics" in this section.

Any circuitry, that is suspected as causing the intermittent complaint, should be thoroughly checked for backed out terminals, improper mating, broken locks, improperly formed or damaged terminals, poor terminal to wiring connections or physical damage to the wiring harness.

DIAGNOSTIC CHART DTC 161 — 1994 SILHOUETTE, TRANS SPORT AND LUMINA APV WITH TCS, CONTINUED

DTC 161

RIGHT TCS MOTOR CIRCUIT OPEN

DOES DTC 161 OCCUR INTERMITTENTLY?

- YES: REFER TO "DIAGNOSTIC AIDS" ON FACING PAGE.
- NO:

(1)
- IGNITION "OFF."
- DISCONNECT 4-WAY TCS MOTOR PACK CONNECTOR.
- USING J 39200, MEASURE RESISTANCE ACROSS TCS MOTOR PACK TERMINALS "C" AND "D".

IS RESISTANCE BELOW 2000Ω?

- NO: REPLACE TCS MOTOR PACK.
- YES:

(2)
- DISCONNECT BLACK 8-WAY CONNECTOR AT EBTCM.
- USING J 39200, MEASURE RESISTANCE BETWEEN TERMINAL "A" OF THE BLACK 8-WAY EBTCM HARNESS CONNECTOR AND TERMINAL "C" OF THE 4-WAY TCS MOTOR PACK HARNESS CONNECTOR.

IS RESISTANCE 2Ω OR LESS?

- NO: REPAIR OPEN OR HIGH RESISTANCE IN CKT 566.
- YES:

(3)
- USING J 39200, MEASURE RESISTANCE BETWEEN TERMINAL "H" OF THE BLACK 8-WAY EBTCM HARNESS CONNECTOR AND TERMINAL "D" OF THE 4-WAY TCS MOTOR PACK HARNESS CONNECTOR.

IS RESISTANCE 2Ω OR LESS?

- NO: REPAIR OPEN OR HIGH RESISTANCE IN CKT 567.
- YES:

(4)
- INSPECT THE 4-WAY TCS MOTOR PACK AND BLACK 8-WAY EBTCM CONNECTORS FOR POOR TERMINAL CONTACT AND EVIDENCE OF CORROSION.
  REPLACE ALL TERMINALS THAT EXHIBIT SIGNS OF POOR TERMINAL CONTACT OR CORROSION.
- RECONNECT ALL CONNECTORS.
- IGNITION "OFF."
- START ENGINE AND TEST DRIVE VEHICLE, ACHIEVING AT LEAST 16 km/h (10 MPH).
- REPEAT ABOVE DRIVE CYCLE SEQUENCE TWO MORE TIMES.

DID DTC 161 SET IN THE LAST THREE DRIVE CYCLES?

- YES: REPLACE EBTCM.
- NO: MALFUNCTION WAS MOST LIKELY CAUSED BY A POOR CONNECTION. REFER TO "DIAGNOSTIC AIDS" ON FACING PAGE.

AFTER ALL DIAGNOSIS IS COMPLETE, CLEAR DTCs AND TEST DRIVE VEHICLE FOR THREE (3) DRIVE CYCLES TO VERIFY THAT DTC DOES NOT RESET. A DRIVE CYCLE CONSISTS OF STARTING THE VEHICLE, DRIVING OVER 16 km/h (10 MPH), AND THEN TURNING KEY "OFF."

DIAGNOSTIC CHART DTC 162 — 1994 SILHOUETTE, TRANS SPORT AND LUMINA APV WITH TCS

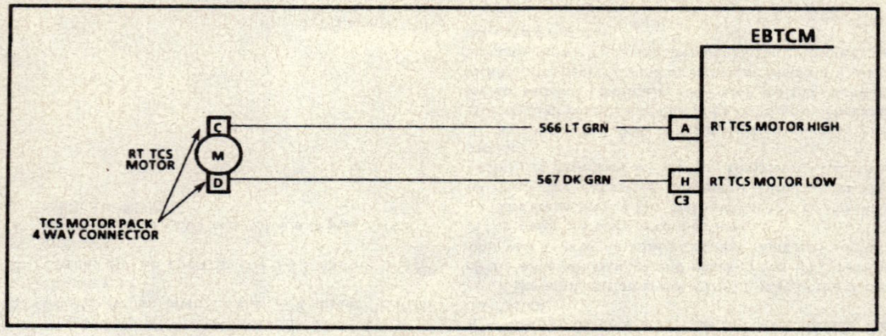

DTC 162

RIGHT TCS MOTOR CIRCUIT SHORTED TO GND

Circuit Description:
This DTC identifies a TCS motor circuit that is shorted to ground. This malfunction will not allow the TCS motor to be controlled at the commanded current rate or will cause the TCS motor driver circuit to allow current directly to ground.

DTC Will Set When: DTC 162 can be set anytime. If the EBTCM detects an out of range voltage on either of the right front TCS motor circuits indicating a circuit shorted to ground, a malfunction exists.

Action Taken: A malfunction DTC is stored, TCS is disabled and the TCS warning lamp is turned "ON."

DTC Chart Test Description: Number(s) below refer to circled number(s) on the diagnostic chart.

1. Checks for a short to ground in the TCS motor "HIGH" circuitry.
2. Checks for a short to ground in the TCS motor "LOW" circuitry.
3. Checks for a TCS motor that is internally shorted to ground.
4. Ensures malfunction was not due to physical damage of the circuitry.

Diagnostic Aids: An intermittent malfunction may be caused by a poor connection, rubbed through wire insulation, or a wire that is broken inside the insulation.

If the malfunction is not current, wiggle the wires of the affected channel and check if the DTC resets. This will help to pinpoint an intermittent malfunction in the motor circuitry or connections.

The frequency of the malfunction can be checked by using the enhanced diagnostic function of the Tech 1, as described in "Tech 1 Diagnostics" in this section.

Any circuitry, that is suspected as causing the intermittent complaint, should be thoroughly checked for backed out terminals, improper mating, broken locks, improperly formed or damaged terminals, poor terminal to wiring connections or physical damage to the wiring harness.

DIAGNOSTIC CHART DTC 162 — 1994 SILHOUETTE, TRANS SPORT AND LUMINA APV WITH TCS, CONTINUED

DTC 162

RIGHT TCS MOTOR CIRCUIT SHORTED TO GND

DOES DTC 162 OCCUR INTERMITTENTLY?

- YES: REFER TO "DIAGNOSTIC AIDS" ON FACING PAGE.
- NO:

(1)
- IGNITION "OFF."
- DISCONNECT 4-WAY TCS MOTOR PACK CONNECTOR AND THE BLACK 8-WAY EBTCM CONNECTOR.
- USING J 39200, MEASURE THE RESISTANCE BETWEEN GROUND AND TERMINAL "C" OF THE 4-WAY TCS MOTOR PACK HARNESS CONNECTOR.

DOES J 39200 DISPLAY RESISTANCE AS "OL" (INFINITE)?

- NO: REPAIR SHORT TO GROUND IN CKT 566.
- YES:

(2)
- USING J 39200, MEASURE THE RESISTANCE BETWEEN GROUND AND TERMINAL "D" OF THE 4-WAY TCS MOTOR PACK HARNESS CONNECTOR.

DOES J 39200 DISPLAY RESISTANCE AS "OL" (INFINITE)?

- NO: REPAIR SHORT TO GROUND IN CKT 567.
- YES:

(3)
- USING J 39200, MEASURE THE RESISTANCE BETWEEN GROUND AND TERMINAL "C" OF THE TCS MOTOR PACK.

DOES J 39200 DISPLAY RESISTANCE AS "OL" (INFINITE)?

- NO: REPLACE TCS MOTOR PACK.
- YES:

(4)
- INSPECT CKTs 566 AND 567 AND RELATED CONNECTORS FOR DAMAGE WHICH MAY RESULT IN A SHORT TO GROUND WITH ALL CONNECTORS CONNECTED. REPAIR ALL DAMAGE IF EVIDENT.
- RECONNECT ALL CONNECTORS.
- IGNITION "OFF."
- START ENGINE AND TEST DRIVE VEHICLE, ACHIEVING AT LEAST 16 km/h (10 MPH).
- REPEAT ABOVE DRIVE CYCLE SEQUENCE TWO MORE TIMES.

DID DTC 162 SET IN THE LAST THREE DRIVE CYCLES?

- YES: REPLACE EBTCM.
- NO: MALFUNCTION WAS MOST LIKELY CAUSED BY A POOR CONNECTION. REFER TO "DIAGNOSTIC AIDS" ON FACING PAGE.

AFTER ALL DIAGNOSIS IS COMPLETE, CLEAR DTCs AND TEST DRIVE VEHICLE FOR THREE (3) DRIVE CYCLES TO VERIFY THAT DTC DOES NOT RESET. A DRIVE CYCLE CONSISTS OF STARTING THE VEHICLE, DRIVING OVER 16 km/h (10 MPH), AND THEN TURNING KEY "OFF."

DIAGNOSTIC CHART DTC 163 — 1994 SILHOUETTE, TRANS SPORT AND LUMINA APV WITH TCS

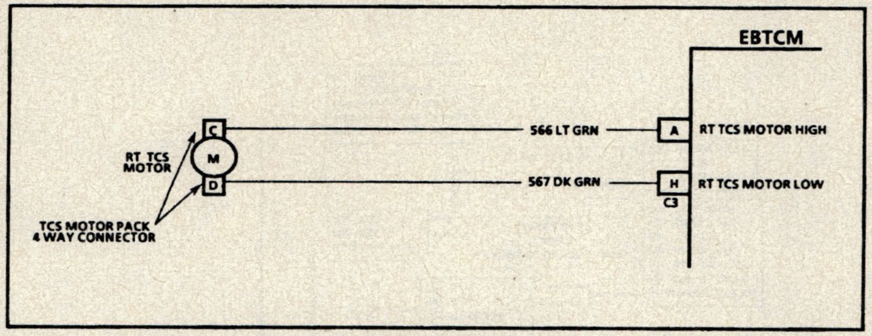

DTC 163

RIGHT TCS MOTOR CIRCUIT SHORTED TO BATT

Circuit Description:
This DTC identifies a TCS motor circuit that is shorted to battery or a TCS motor that has low or no resistance. This malfunction will not allow the TCS motor to be controlled at the commanded current rate or will cause the TCS motor to turn in the opposite direction, or not at all.

DTC Will Set When: DTC 163 can be set only when the TCS motor is commanded "OFF." If the EBTCM detects an out of range voltage on either of the right front TCS motor circuits indicating a circuit shorted to battery or a TCS motor shorted, a malfunction exists.

Action Taken: A malfunction DTC is stored, TCS is disabled and the TCS warning lamp is turned "ON."

DTC Chart Test Description: Number(s) below refer to circled number(s) on the diagnostic chart.

1. Checks for a short to voltage in the TCS motor "HIGH" circuitry.
2. Checks for a short to voltage in the TCS motor "LOW" circuitry.
3. Checks for a short circuit between motor "HIGH" and motor "LOW."
4. Checks for proper resistance of the TCS motor.
5. Ensures malfunction was not due to physical damage of the circuitry.

Diagnostic Aids: An intermittent malfunction may be caused by a poor connection, rubbed through wire insulation, or a wire that is broken inside the insulation.

If the malfunction is not current, wiggle the wires of the affected channel and check if the DTC resets. This will help to pinpoint an intermittent malfunction in the motor circuitry or connections.

The frequency of the malfunction can be checked by using the enhanced diagnostic function of the Tech 1, as described in "Tech 1 Diagnostics" in this section.

Any circuitry, that is suspected as causing the intermittent complaint, should be thoroughly checked for backed out terminals, improper mating, broken locks, improperly formed or damaged terminals, poor terminal to wiring connections or physical damage to the wiring harness.

DIAGNOSTIC CHART DTC 163 — 1994 SILHOUETTE, TRANS SPORT AND LUMINA APV WITH TCS, CONTINUED

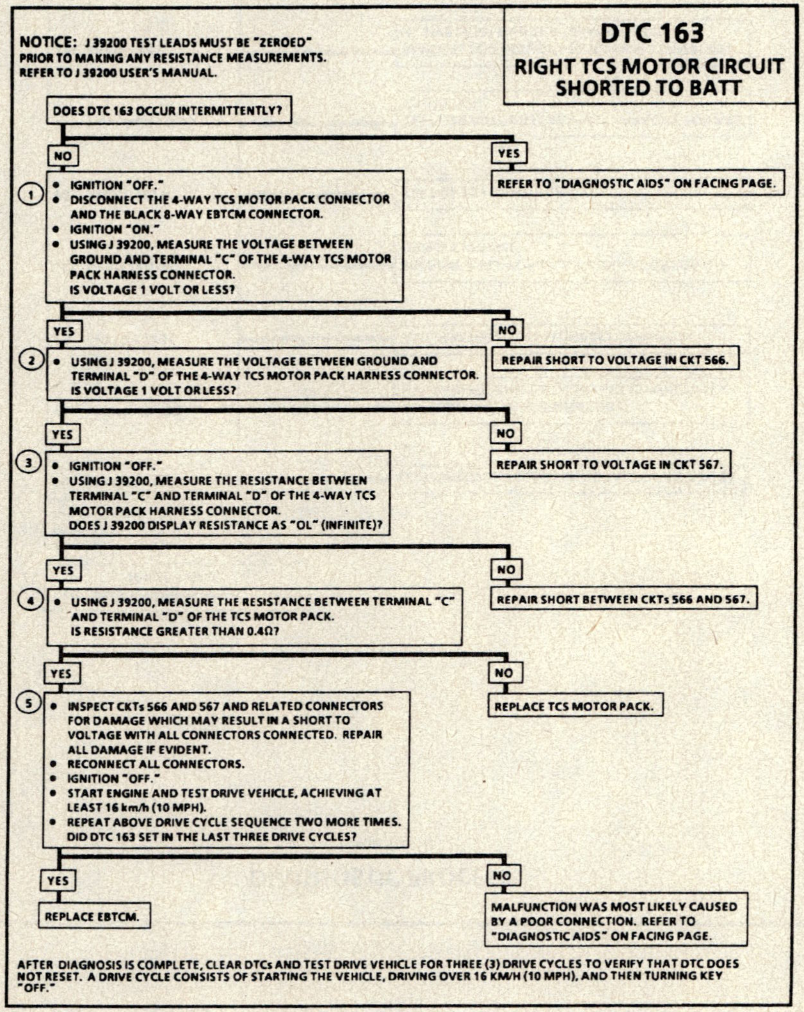

DIAGNOSTIC PROCESS CHART — 1995 SILHOUETTE, TRANS SPORT AND LUMINA APV WITH TCS

DIAGNOSTIC PROCESS

(Page 1 of 4)

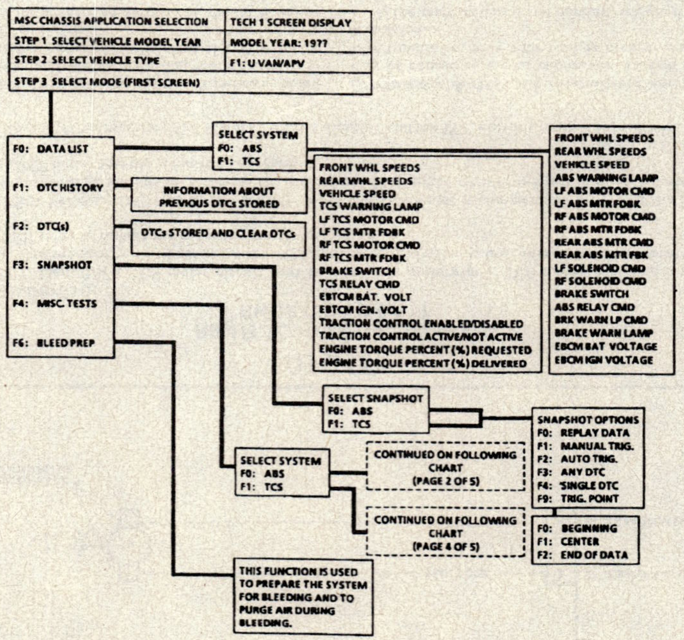

DIAGNOSTIC PROCESS CHART — 1995 SILHOUETTE, TRANS SPORT AND LUMINA APV WITH TCS, CONTINUED

DIAGNOSTIC PROCESS

(Page 2 of 4)

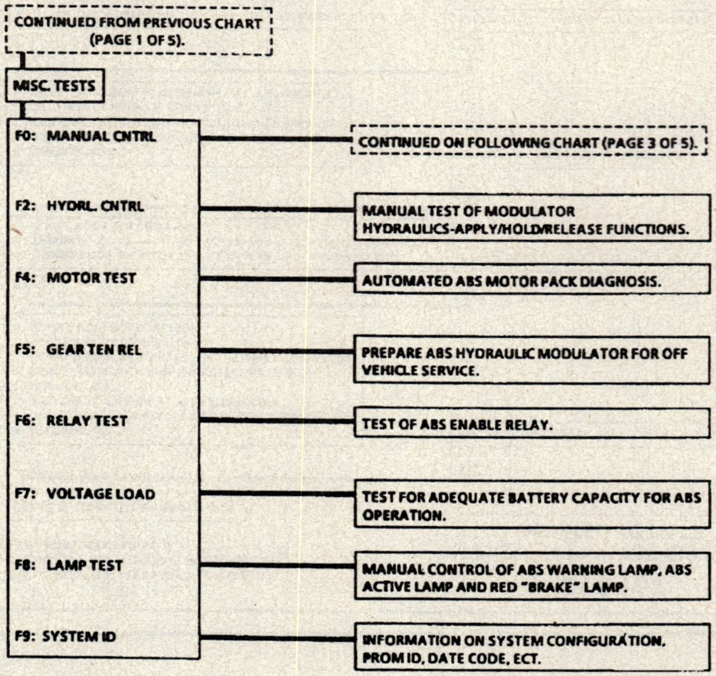

DIAGNOSTIC PROCESS CHART — 1995 SILHOUETTE, TRANS SPORT AND LUMINA APV WITH TCS, CONTINUED

DIAGNOSTIC PROCESS

(Page 3 of 4)

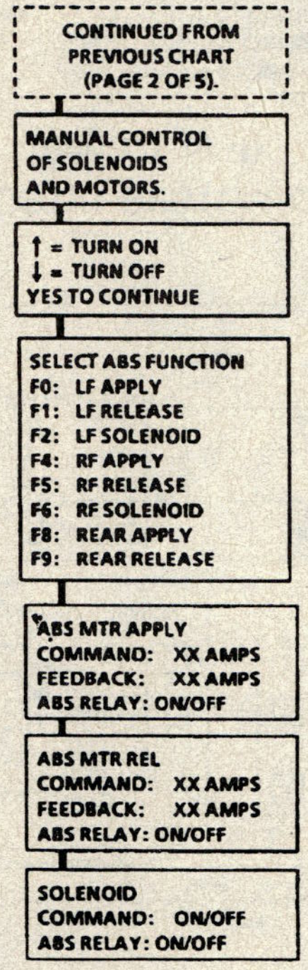

DIAGNOSTIC PROCESS CHART — 1995 SILHOUETTE, TRANS SPORT AND LUMINA APV WITH TCS

DIAGNOSTIC PROCESS

(Page 4 of 4)

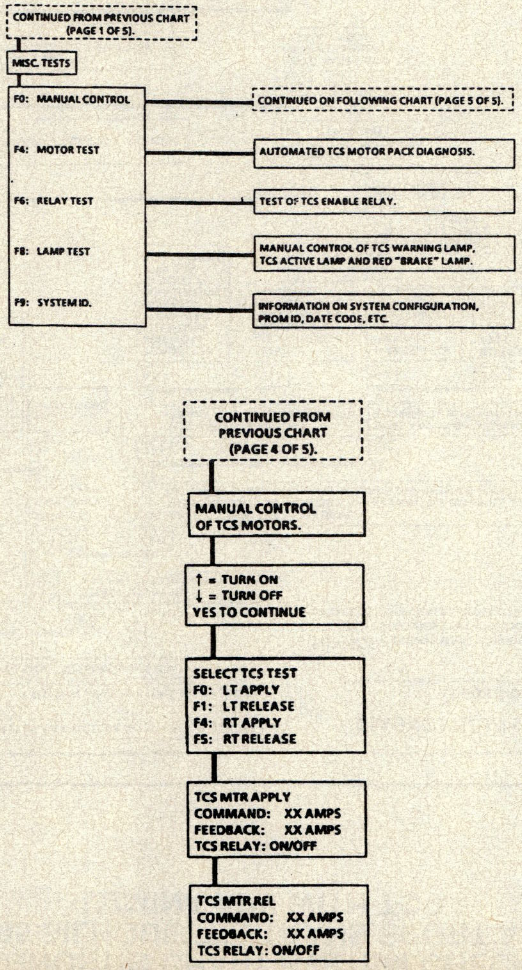

DIAGNOSTIC CHART SYSTEM CHECK — 1995 SILHOUETTE, TRANS SPORT AND LUMINA APV WITH TCS

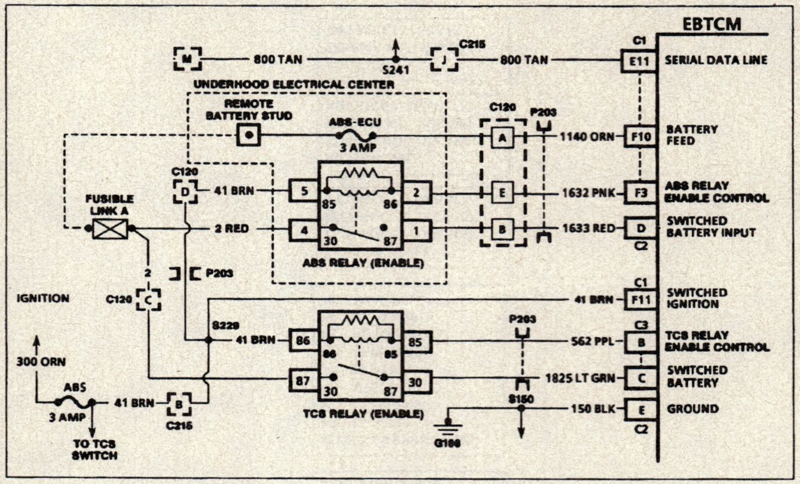

DIAGNOSTIC SYSTEM CHECK
(Page 1 of 4)

System Description:

The diagnostic system check is an organized approach to identifying a problem created by an Antilock Brake Traction Control System (ABS VI/TCS) malfunction. It must be the starting point for any ABS VI/TCS complaint diagnosis because it directs the service technician to the next logical step in diagnosing the complaint.

Serial data is transmitted by the EBTCM through terminal "E11". The EBTCM is supplied switched ignition voltage through terminal "F11", battery voltage through terminal "F10" and switched battery voltage through terminals "D" and "C". The EBTCM ground is provided through terminal "E".

Chart Test Description:

Number(s) below refer to circled number(s) on the diagnostic chart.

1. Indicates if EBTCM is transmitting data.
2. Checks for a serial data line malfunction.
3. Checks for proper Tech 1 connection to DLC.
4. Checks for open ABS ECU 3 amp fuse.
5. Checks for open ABS 3 amp fuse.
6. Checks for high resistance in the EBTCM ground circuit.
7. Checks for current DTCs.
8. Checks for proper ABS and TCS warning lamp operation.
9. Checks for history DTCs.

Diagnostic Aids:

Excessive resistance in the ground or power supply circuits will not allow communication with the EBTCM. If communication with the EBTCM is not possible, ensure the ABS VI/TCS ground connection is good and that there is no excessive resistance in any of the power supply circuits.

DIAGNOSTIC CHART SYSTEM CHECK — 1995 SILHOUETTE, TRANS SPORT AND LUMINA APV WITH TCS

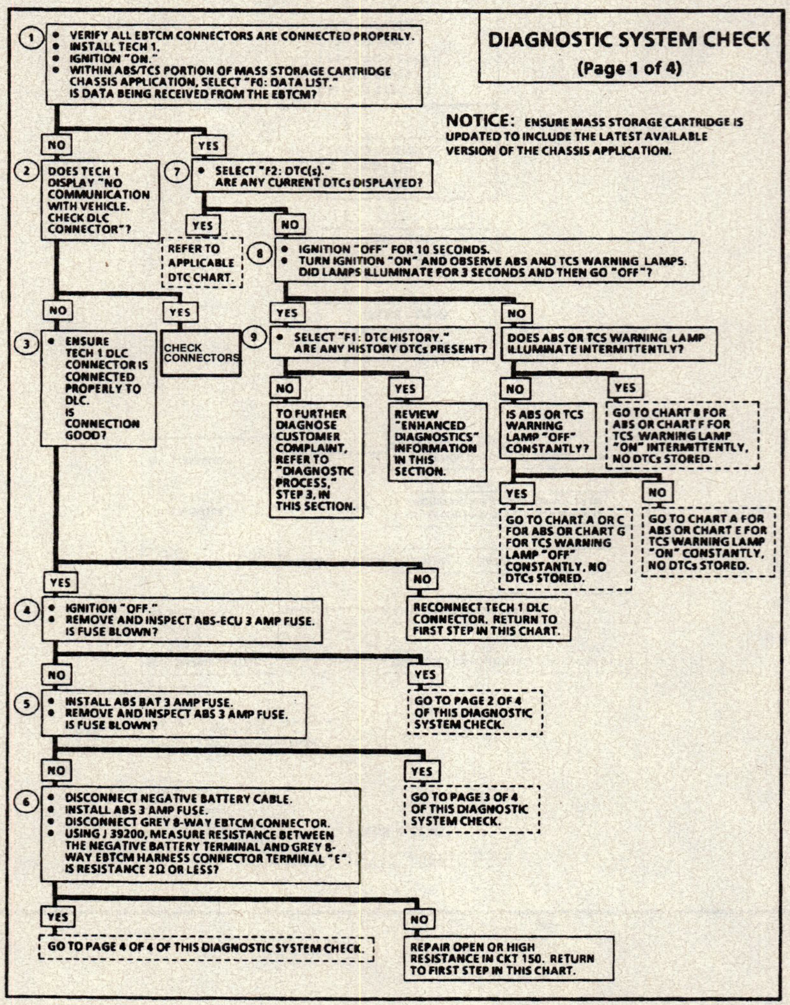

DIAGNOSTIC CHART SYSTEM CHECK — 1995 SILHOUETTE, TRANS SPORT AND LUMINA APV WITH TCS, CONTINUED

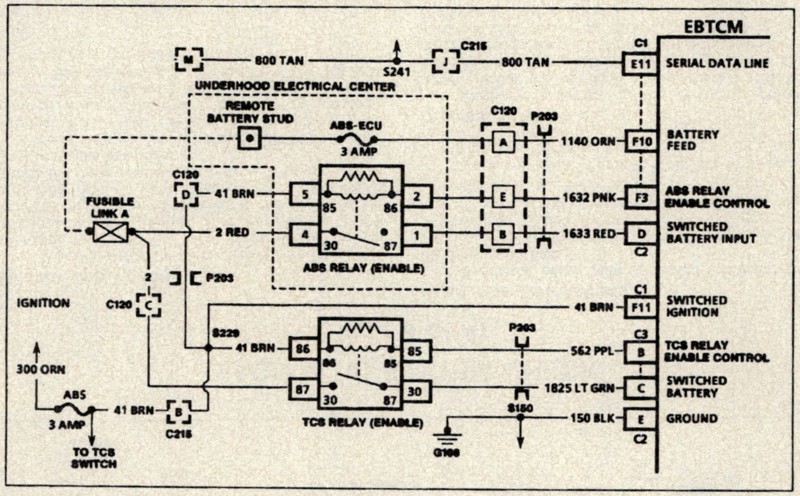

DIAGNOSTIC SYSTEM CHECK (Page 2 of 4)

System Description:

The diagnostic system check is an organized approach to identifying a problem created by an Antilock Brake Traction Control System (ABS VI/TCS) malfunction. It must be the starting point for any ABS VI/TCS complaint diagnosis because it directs the service technician to the next logical step in diagnosing the complaint.

Serial data is transmitted by the EBTCM through terminal "E11". The EBTCM is supplied switched ignition voltage through terminal "F11", battery voltage through terminal "F10" and switched battery voltage through terminals "D" and "C". The EBTCM ground is provided through terminal "E".

Chart Test Description:

Number(s) below refer to circled number(s) on the diagnostic chart.

10. Checks for possible short to ground in the battery feed circuitry.
11. Checks for possible short to ground in the EBTCM.
12. Ensures short to ground is not due to physical damage of the circuitry.

Diagnostic Aids:

Excessive resistance in the ground or power supply circuits will not allow communication with the EBTCM. If communication with the EBTCM is not possible, ensure the ABS VI/TCS ground connection is good and that there is no excessive resistance in any of the power supply circuits.

DIAGNOSTIC CHART SYSTEM CHECK — 1995 SILHOUETTE, TRANS SPORT AND LUMINA APV WITH TCS, CONTINUED

DIAGNOSTIC SYSTEM CHECK (Page 2 of 4)

CONTINUED FROM DIAGNOSTIC SYSTEM CHECK (PAGE 1 OF 4)

YES

(10)
- REPLACE ABS-ECU 3 AMP FUSE.
- WAIT 10 SECONDS.
- REMOVE AND INSPECT ABS-ECU 3 AMP FUSE. IS FUSE BLOWN?

YES

(11)
- DISCONNECT 32-WAY EBTCM CONNECTOR.
- REPLACE ABS-ECU 3 AMP FUSE.
- WAIT 10 SECONDS.
- REMOVE AND INSPECT ABS-ECU 3 AMP FUSE. IS FUSE BLOWN?

YES

DIAGNOSE FOR A SHORT TO GROUND IN THE BATTERY FEED CIRCUITRY.

NO (from 10)
- INSTALL FUSE.
- REFER TO SECTION 8A TO DIAGNOSE FOR A POSSIBLE SHORT TO GROUND IN BATTERY FEED CIRCUITRY. RETURN TO FIRST STEP IN THIS CHART.

NO (from 11)

(12)
- INSPECT CKT 2, CKT 1140 AND 32-WAY EBTCM HARNESS CONNECTOR FOR PHYSICAL DAMAGE WHICH MAY RESULT IN A SHORT TO GROUND WITH THE 32-WAY EBTCM HARNESS CONNECTOR CONNECTED TO THE EBTCM. CORRECT DAMAGE IF EVIDENT.
- RECONNECT ALL CONNECTORS.
- WAIT 10 SECONDS.
- REMOVE AND INSPECT ABS-ECU 3 AMP FUSE. IS FUSE BLOWN?

YES
- REPLACE EBTCM.
- REPLACE ABS-ECU 3 AMP FUSE. RETURN TO FIRST STEP IN THIS CHART.

NO

MALFUNCTION IS INTERMITTENT.

DIAGNOSTIC CHART SYSTEM CHECK — 1995 SILHOUETTE, TRANS SPORT AND LUMINA APV WITH TCS, CONTINUED

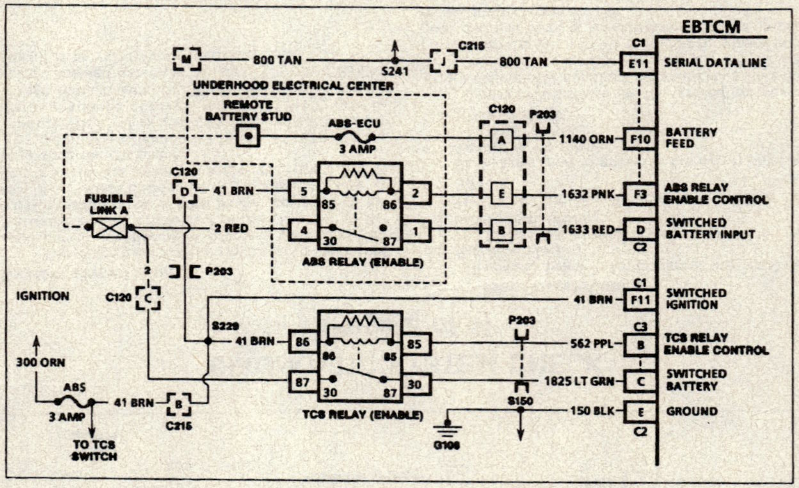

DIAGNOSTIC SYSTEM CHECK
(Page 3 of 4)

System Description:

The diagnostic system check is an organized approach to identifying a problem created by an Antilock Brake Traction Control System (ABS VI/TCS) malfunction. It must be the starting point for any ABS VI/TCS complaint diagnosis because it directs the service technician to the next logical step in diagnosing the complaint.

Serial data is transmitted by the EBTCM through terminal "E11". The EBTCM is supplied switched ignition voltage through terminal "F11", battery voltage through terminal "F10" and switched battery voltage through terminals "D" and "C". The EBTCM ground is provided through terminal "E".

Chart Test Description:

Number(s) below refer to circled number(s) on the diagnostic chart.

13. Checks for possible short to ground in the switched ignition circuitry.

14. Isolates the short to ground to either the circuitry or the EBTCM.

15. Ensures short to ground is not due to physical damage of the circuitry.

Diagnostic Aids:

Excessive resistance in the ground or power supply circuits will not allow communication with the EBTCM. If communication with the EBTCM is not possible, ensure the ABS VI/TCS ground connection is good and that there is no excessive resistance in any of the power supply circuits.

DIAGNOSTIC CHART SYSTEM CHECK — 1995 SILHOUETTE, TRANS SPORT AND LUMINA APV WITH TCS, CONTINUED

DIAGNOSTIC SYSTEM CHECK
(Page 3 of 4)

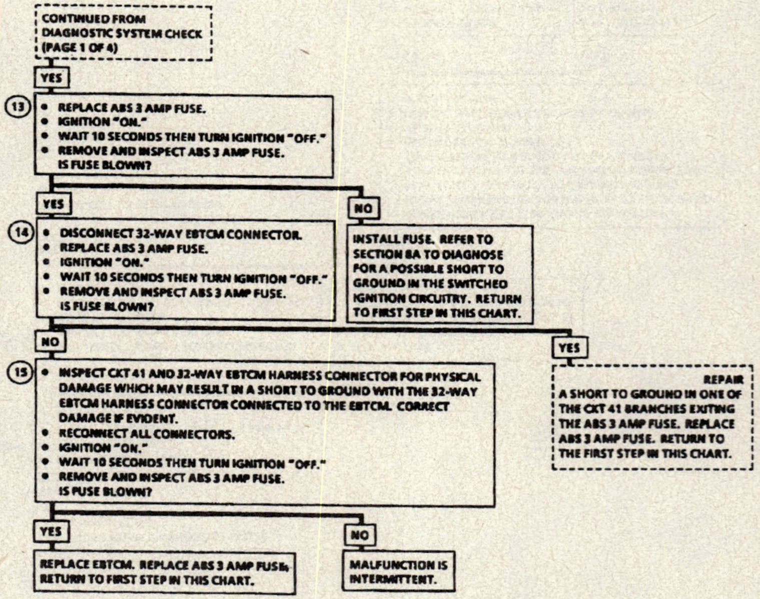

DIAGNOSTIC CHART SYSTEM CHECK — 1995 SILHOUETTE, TRANS SPORT AND LUMINA APV WITH TCS

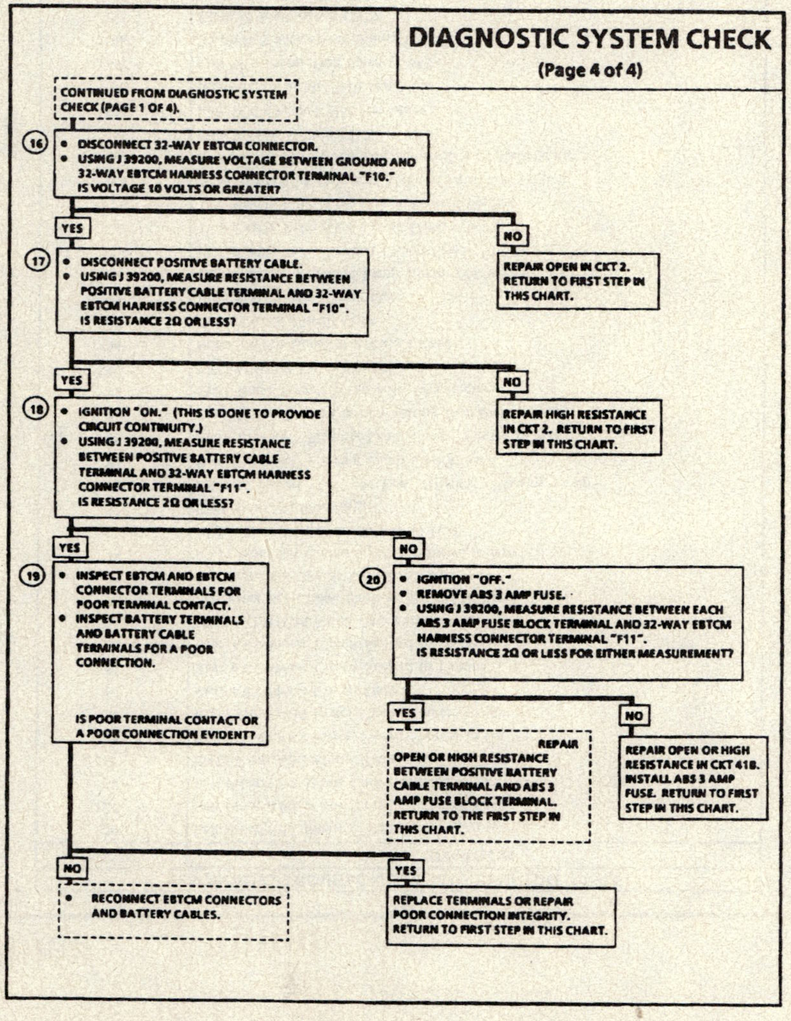

DIAGNOSTIC CHART SYSTEM CHECK — 1995 SILHOUETTE, TRANS SPORT AND LUMINA APV WITH TCS, CONTINUED

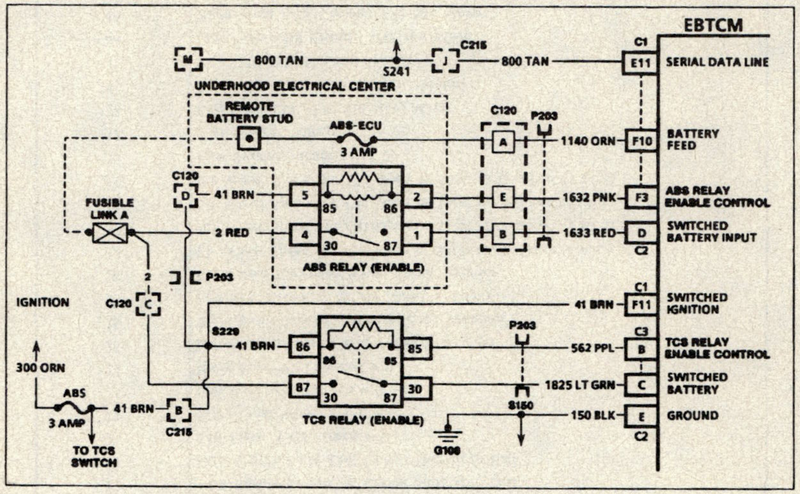

DIAGNOSTIC SYSTEM CHECK
(Page 4 of 4)

System Description:

The diagnostic system check is an organized approach to identifying a problem created by an Antilock Brake Traction Control System (ABS VI/TCS) malfunction. It must be the starting point for any ABS VI/TCS complaint diagnosis because it directs the service technician to the next logical step in diagnosing the complaint.

Serial data is transmitted by the EBTCM through terminal "E11". The EBTCM is supplied switched ignition voltage through terminal "F11", battery voltage through terminal "F10" and switched battery voltage through terminals "D" and "C". The EBTCM ground is provided through terminal "E".

Chart Test Description:

Number(s) below refer to circled number(s) on the diagnostic chart.

16. Checks for open in CKT 2.

17. Checks for high resistance in CKT 2.

18. Checks for high resistance in the battery feed circuitry.

19. Ensures malfunction is not due to poor terminal contact.

20. Checks for high resistance in the switched ignition circuit between the fuse block and the EBTCM.

Diagnostic Aids:

Excessive resistance in the ground or power supply circuits will not allow communication with the EBTCM. If communication with the EBTCM is not possible, ensure the ABS VI/TCS ground connection is good and that there is no excessive resistance in any of the power supply circuits.

DIAGNOSTIC TROUBLE CODE CHART — 1995 SILHOUETTE, TRANS SPORT AND LUMINA APV WITH TCS

DIAGNOSTIC TROUBLE CODE AND SYMPTOM TABLE	
CHART	SYMPTOM
A	ABS (Amber) Warning Lamp "ON" Constantly, No DTCs Stored (Chev Only)
A	ABS (Amber) Warning Lamp "ON" or "OFF" Constantly, No DTCs Stored (Pont/Olds Only)
B	ABS (Amber) Warning Lamp "ON" Intermittently, No DTCs Stored
C	ABS (Amber) Warning Lamp "OFF" Constantly, No DTCs Stored (Chev Only)
D	Tech 1 Displays Undefined DTCs
E	TCS (Amber) Warning Lamp "ON" Constantly, No DTCs Stored
F	TCS (Amber) Warning Lamp "ON" Intermittently, No DTCs Stored
G	TCS (Amber) Warning Lamp "OFF" Constantly, No DTCs Stored
DIAGNOSTIC TROUBLE CODE	DESCRIPTION
11	ABS Warning or Active Lamp Circuit Malfunction (Chev Only) (1 of 2)
11	ABS Warning or Active Lamp Circuit Malfunction (Pont/Olds Only) (1 of 2)
14	ABS Enable Relay Contact Circuit Open (1 of 3)
15	ABS Enable RLY Contact Ckt Short to Batt or Always Closed
16	ABS Enable Relay Coil Circuit Open
17	ABS Enable Relay Coil Circuit Shorted to Gnd
18	ABS Enable Relay Coil Circuit Shorted to Batt
21	Left Front Wheel Speed = 0
22	Right Front Wheel Speed = 0
23	Left Rear Wheel Speed = 0
24	Right Rear Wheel Speed = 0
25	Left Front Excessive Wheel Speed Variation
26	Right Front Excessive Wheel Speed Variation
27	Left Rear Excessive Wheel Speed Variation
28	Right Rear Excessive Wheel Speed Variation
32	LF Wheel Speed Snr Ckt Open or Short to Gnd/Batt (1 of 3)
33	RF Wheel Speed Snr Ckt Open or Short to Gnd/Batt (1 of 3)
34	LR Wheel Speed Snr Ckt Open or Short to Gnd/Batt (1 of 3)
35	RR Wheel Speed Snr Ckt Open or Short to Gnd/Batt (1 of 3)
36	Low System Voltage (1 of 2)
37	High System Voltage
38	Left Front ESB Will Not Hold Motor
41	Right Front ESB Will Not Hold Motor
42	Rear ESB Will Not Hold Motor
44	Left Front ABS Channel Will Not Move
45	Right Front ABS Channel Will Not Move
46	Rear ABS Channel Will Not Move
47	Left Front ABS Motor Free Spins (1 of 2)
48	Right Front ABS Motor Free Spins (1 of 2)
51	Rear ABS Motor Free Spins (1 of 2)
52	Left Front ABS Channel in Release Too Long
53	Right Front ABS Channel in Release Too Long
54	Rear ABS Channel in Release Too Long
55	EBTCM Malfunction (ABS)

DIAGNOSTIC TROUBLE CODE CHART — 1995 SILHOUETTE, TRANS SPORT AND LUMINA APV WITH TCS, CONTINUED

DIAGNOSTIC TROUBLE CODE AND SYMPTOM TABLE	
CHART	SYMPTOM
56	Left Front ABS Motor Circuit Open
57	Left Front ABS Motor Circuit Shorted to Gnd
58	Left Front ABS Motor Circuit Shorted to Batt
61	Right Front ABS Motor Circuit Open
62	Right Front ABS Motor Circuit Shorted to Gnd
63	Right Front ABS Motor Circuit Shorted to Batt
64	Rear ABS Motor Circuit Open
65	Rear ABS Motor Circuit Shorted to Ground
66	Rear ABS Motor Circuit Shorted to Battery
75	PCM to EBTCM Serial Data Malfunction
76	Left Front Sol Circuit Open or Shorted to Batt
77	Left Front Sol Circuit Shorted to Gnd
78	Right Front Sol Circuit Open or Shorted to Batt
81	Right Front Sol Circuit Shorted to Gnd
82	Calibration Malfunction
86	EBTCM Turned "ON" the Red "BRAKE" Warning Lamp
87	Red "BRAKE" Warning Lamp Circuit Open
88	Red "BRAKE" Warning Lamp Circuit Shorted to Batt
91	Open Brake Switch Contacts During Deceleration
92	Open Brake Switch When ABS Was Required
93	DTC 91 or 92 Set in Current or Prev Ign Cycle
94	Brake Switch Contacts Always Closed
95	Brake Switch Circuit Open
96	Rear Brake Lamp Circuit Open
111	TCS Warning or Switch Lamp Circuit Malfunction (1 of 2)
114	TCS Enable Relay Contact Circuit Open (1 of 3)
116	TCS Enable Relay Coil Circuit Open
118	TCS Enable Relay Coil Circuit Shorted To Batt
123	PCM to EBTCM Delivered Torque Circuit Malfunction
124	EBTCM To PCM Requested Torque Circuit Malfunction
136	Low System Voltage (TCS) (1 of 2)
144	Left TCS Channel Will Not Move
145	Right TCS Channel Will Not Move
147	Left TCS Motor Free Spins (1 of 2)
148	Right TCS Motor Free Spins (1 of 2)
155	EBTCM Malfunction (TCS)
156	Left TCS Motor Circuit Open
157	Left TCS Motor Circuit Shorted to Gnd
158	Left TCS Motor Circuit Shorted to Batt
161	Right TCS Motor Circuit Open
162	Right TCS Motor Circuit Shorted to Gnd
163	Right TCS Motor Circuit Shorted to Batt

ELECTRICAL CONNECTOR TERMINAL IDENTIFICATION — 1995 SILHOUETTE, TRANS SPORT AND LUMINA APV WITH TCS

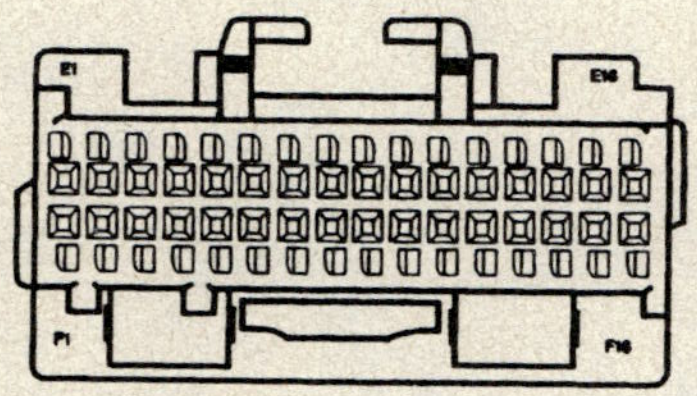

32-WAY F
MICRO-PACK 100 SERIES
BLU
ELECTRONIC BRAKE
CONTROL MODULE
(EBCM) CONNECTOR "C1"

CONNECTOR TERMINAL	CIRCUIT NO.	COLOR	CIRCUIT DESCRIPTION
E1	884	BLK	LR WHEEL SIGNAL HIGH
E2	885	RED	LR WHEEL SIGNAL LOW
E3	833	TAN	RF WHEEL SIGNAL LOW
E4	872	DK GRN	RF WHEEL SIGNAL HIGH
E5	883	WHT	RR WHEEL SIGNAL LOW
E6	882	BRN	RR WHEEL SIGNAL HIGH
E7	873	YEL	LF WHEEL SIGNAL LOW
E8	830	LT BLU	LF WHEEL SIGNAL HIGH
E9	1289	LT BLU	RF ABS SOLENOID CONTROL
E10	PLUGGED		NOT USED
E11	800	TAN	SERIAL DATA LINE
E12	PLUGGED		NOT USED
E13	PLUGGED		NOT USED
E14	464	TAN/BLK	DELIVERED TORQUE
E15	1571	BRN/WHT	TCS ENABLE SWITCH INPUT
E16	550	BLK	TCS ENABLE LAMP CONTROL
F1	1288	DK GRN	LF ABS SOLENOID CONTROL
F2	852	WHT	ABS WARNING LAMP CONTROL
F3	1632	PNK	ABS ENABLE RELAY CONTROL
F4	33	TAN/WHT	BRAKE WARNING LAMP CONTROL
F5	1656	LT GRN	ABS/TCS ACTIVE CONTROL
F6	PLUGGED		NOT USED
F7	PLUGGED		NOT USED
F8	PLUGGED		NOT USED
F9	PLUGGED		NOT USED
F10	1140	ORN	BATTERY FEED
F11	41	BRN	SWITCHED IGNITION
F12	17	WHT	BRAKE SWITCH INPUT
F13	PLUGGED		NOT USED
F14	463	ORN/BLK	REQUESTED TORQUE
F15	1808	WHT	TCS WARNING LAMP CONTROL
F16	PLUGGED		NOT USED

ELECTRICAL CONNECTOR TERMINAL IDENTIFICATION — 1995 SILHOUETTE, TRANS SPORT AND LUMINA APV WITH TCS, CONTINUED

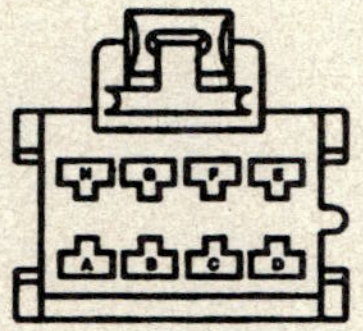

8-WAY F
METRI-PACK 280 SERIES
GRY CONNECTOR "C2"

PIN	CIRCUIT NO.	COLOR	CIRCUIT
A	1280	BLK	LF ABS MOTOR HIGH
B	1281	PNK	LF ABS MOTOR LOW
C	1284	DK GRN	REAR ABS MOTOR HIGH
D	1633	RED	SWITCHED BATTERY INPUT
E	150	BLK	GROUND
F	1285	ORN	REAR ABS MOTOR LOW
G	1282	PPL	RF ABS MOTOR HIGH
H	1283	BLK	RF ABS MOTOR LOW

ELECTRICAL CONNECTOR TERMINAL IDENTIFICATION — 1995 SILHOUETTE, TRANS SPORT AND LUMINA APV WITH TCS, CONTINUED

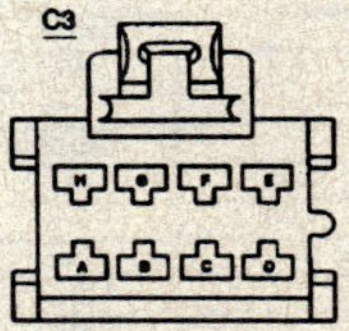

ELECTRONIC BRAKE/TRACTION CONTROL MODULE (EBTCM)

8 – WAY F METRI – PACK 280 SERIES
BLK

PIN	CIRCUIT NO.	COLOR	CIRCUIT
A	566	LT GRN	RF TCS MOTOR HIGH
B	562	PPL	TCS ENABLE RELAY CONTROL
C	1825	LT GRN	TCS SWITCHED BATTERY
D	PLUGGED		NOT USED
E	PLUGGED		NOT USED
F	564	LT BLU	LF TCS MOTOR HIGH
G	565	DK BLU	LF TCS MOTOR LOW
H	567	DK GRN	RF TCS MOTOR LOW

DIAGNOSTIC CHART A — 1995 LUMINA APV WITH TCS

CHART A

ABS (AMBER) WARNING LAMP "ON" CONSTANTLY, NO DTCs STORED
(CHEV ONLY)

NOTICE: DIAGNOSTIC SYSTEM CHECK MUST BE COMPLETED FIRST BEFORE USING THIS CHART.

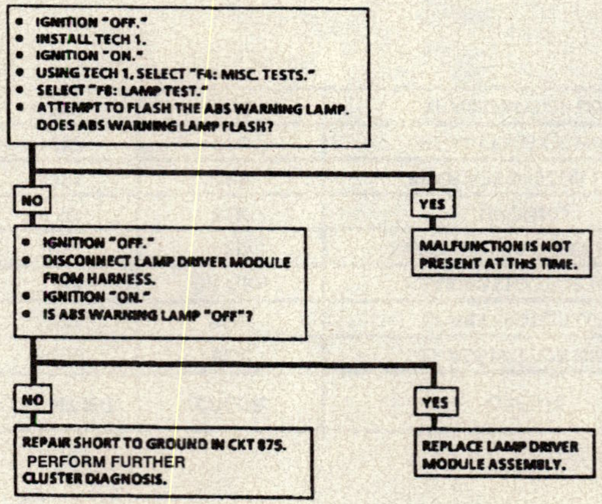

DIAGNOSTIC CHART A — 1995 SILHOUETTE AND TRANS SPORT WITH TCS, CONTINUED

CHART A

ABS (AMBER) WARNING LAMP "ON" OR "OFF" CONSTANTLY, NO DTCs STORED (PONT/OLDS ONLY)

NOTICE: DIAGNOSTIC SYSTEM CHECK MUST BE COMPLETED BEFORE USING THIS CHART.

- IGNITION "OFF."
- INSTALL TECH 1.
- IGNITION "ON."
- USING TECH 1, SELECT F4: MISC. TESTS.
- SELECT F8: LAMP TEST AND ATTEMPT TO TURN "ON" THE "ABS" WARNING LAMP.

DID THE "ABS" WARNING LAMP TURN "ON"?

YES:

- USING TECH 1 LAMP TEST, ATTEMPT TO "FLASH" THE "ABS" WARNING LAMP.

DID THE "ABS" WARNING LAMP "FLASH"?

YES: MALFUNCTION IS NOT PRESENT AT THIS TIME. INTERMITTENTS COULD BE CAUSED BY INCORRECT WIRING HARNESS ROUTING OR LOOSE CONNECTIONS. INSPECT ALL CONNECTORS AND TERMINALS FOR POOR TERMINAL CONTACT AND EVIDENCE OF CORROSION. REPLACE ALL TERMINALS THAT EXHIBIT SIGNS OF POOR TERMINAL CONTACT OR CORROSION.

NO (from either question): INSPECT THE "ABS" WARNING BULB TO VERIFY BULB IS GOOD. REPLACE ANY BLOWN BULBS. IF BULB IS GOOD, PERFORM FURTHER DIAGNOSIS OF INSTRUMENT CLUSTER.

DIAGNOSTIC CHART B — 1995 SILHOUETTE AND TRANS SPORT AND LUMINA APV WITH TCS

CHART B

ABS (AMBER) WARNING LAMP "ON" INTERMITTENTLY, NO DTCs STORED

NOTICE: DIAGNOSTIC SYSTEM CHECK MUST BE COMPLETED BEFORE USING THIS CHART.

- IGNITION "OFF."
- DISCONNECT GREY 8-WAY EBTCM CONNECTOR.
- DISCONNECT NEGATIVE BATTERY TERMINAL.
- DISCONNECT POSITIVE BATTERY TERMINAL.
- USING J 39200, MEASURE THE RESISTANCE BETWEEN THE NEGATIVE BATTERY CABLE TERMINAL AND TERMINAL "E" OF THE GRAY 8-WAY EBTCM HARNESS CONNECTOR.
- OBSERVE MEASURED RESISTANCE WHILE "WIGGLING" WIRE HARNESS AND CONNECTORS.

IS RESISTANCE ALWAYS 2Ω OR LESS?

NO: REPAIR INTERMITTENT OPEN OR HIGH RESISTANCE IN CKT 150.

YES:

- DISCONNECT 32-WAY EBTCM CONNECTOR.
- USING J 39200, MEASURE THE RESISTANCE BETWEEN POSITIVE BATTERY CABLE TERMINAL AND TERMINAL "F10" OF THE 32-WAY EBTCM HARNESS CONNECTOR.
- OBSERVE MEASURED RESISTANCE WHILE "WIGGLING" WIRE HARNESS AND CONNECTORS.

IS RESISTANCE ALWAYS 2Ω OR LESS?

NO: REPAIR INTERMITTENT OPEN OR HIGH RESISTANCE IN CKTs 2 OR 1140.

YES:

- TURN IGNITION SWITCH "ON" (THIS IS DONE TO COMPLETE THE IGNITION CIRCUIT).
- USING J 39200, MEASURE THE RESISTANCE BETWEEN POSITIVE BATTERY CABLE TERMINAL AND TERMINAL "F11" OF THE 32-WAY EBTCM HARNESS CONNECTOR.
- OBSERVE MEASURED RESISTANCE WHILE "WIGGLING" WIRE HARNESS AND CONNECTORS.

IS RESISTANCE ALWAYS 2Ω OR LESS?

NO: REPAIR INTERMITTENT OPEN OR HIGH RESISTANCE IN CKT 418 OR 300.

YES:

- CHECK POSITIVE BATTERY TERMINAL AND POSITIVE BATTERY CABLE TERMINAL FOR A POOR CONNECTION.
- INSPECT TERMINALS "F11" AND "F10" OF THE 32-WAY EBTCM CONNECTOR AND TERMINAL "E" OF THE GREY 8-WAY EBTCM CONNECTOR FOR POOR TERMINAL CONTACT AND EVIDENCE OF CORROSION.

IS THERE EVIDENCE OF POOR TERMINAL CONTACT OR CORROSION?

YES: REPLACE TERMINALS THAT EXHIBIT POOR TERMINAL CONTACT OR SIGNS OF CORROSION. REPAIR CONNECTION INTEGRITY IF POOR CONTACT EXISTS AT BATTERY.

NO:

- IGNITION "OFF."
- RECONNECT ALL CONNECTORS.
- IGNITION "ON."

IS ABS WARNING LAMP "ON" INTERMITTENTLY WITH NO DTCs SET?

YES: REPLACE EBTCM.

NO: MALFUNCTION IS NOT PRESENT AT THIS TIME. INTERMITTENTS MAY BE CAUSED BY INCORRECT WIRING HARNESS ROUTING OR LOOSE GROUND CONNECTIONS.

DIAGNOSTIC CHART C — 1995 SILHOUETTE AND TRANS SPORT AND LUMINA APV WITH TCS

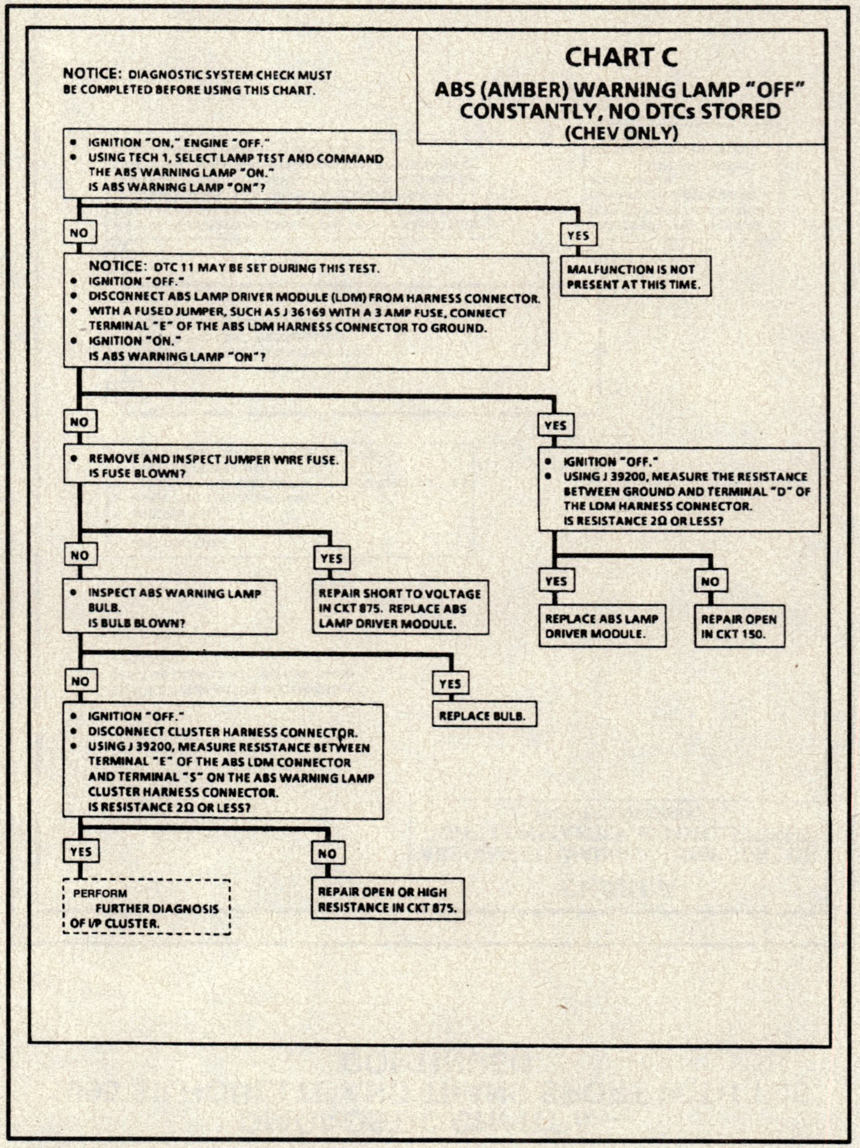

DIAGNOSTIC CHARTS D AND E — 1995 SILHOUETTE AND TRANS SPORT AND LUMINA APV WITH TCS

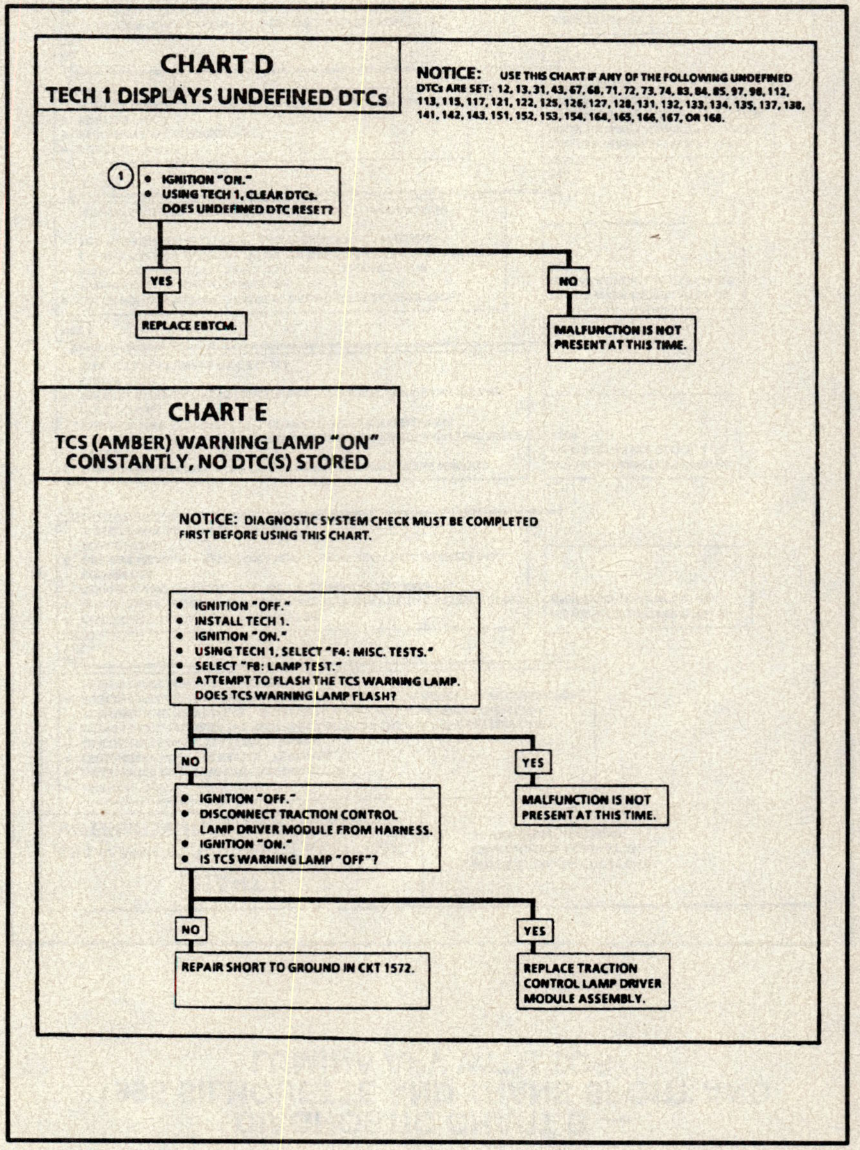

DIAGNOSTIC CHART F — 1995 SILHOUETTE AND TRANS SPORT AND LUMINA APV WITH TCS

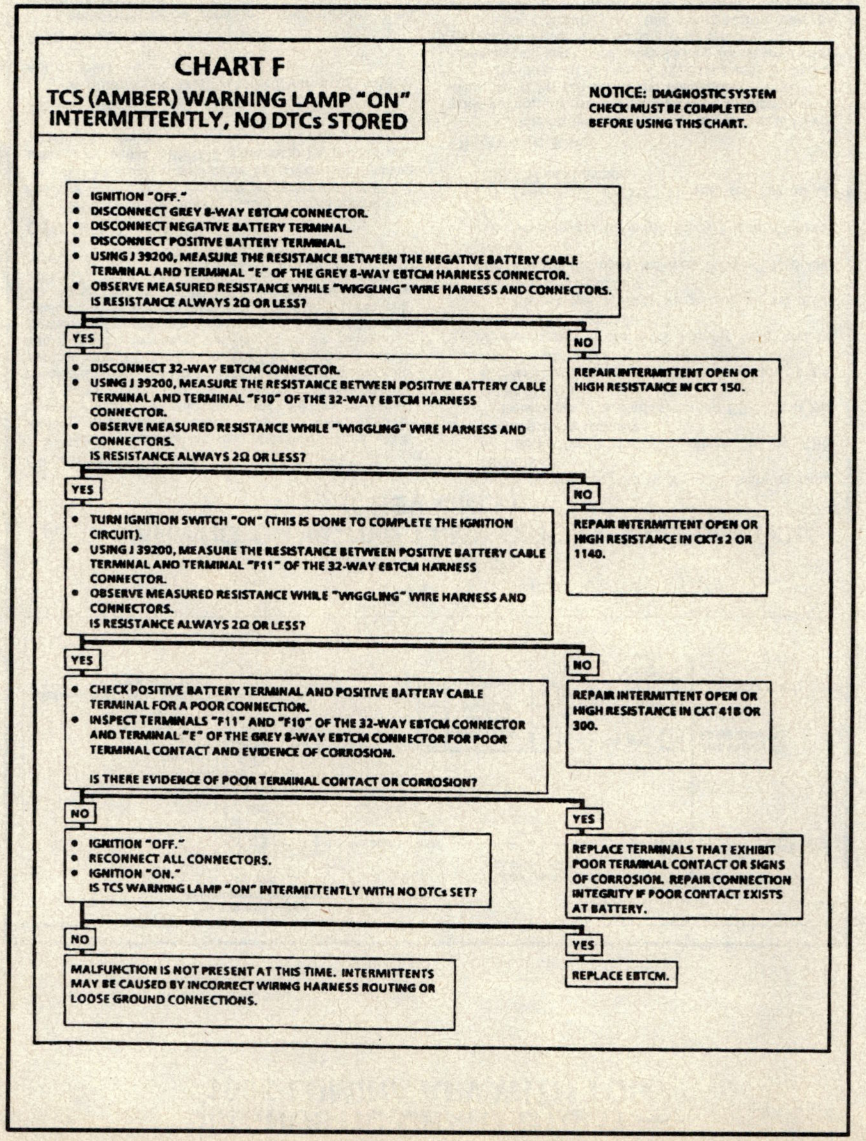

DIAGNOSTIC CHART G — 1995 SILHOUETTE AND TRANS SPORT AND LUMINA APV WITH TCS

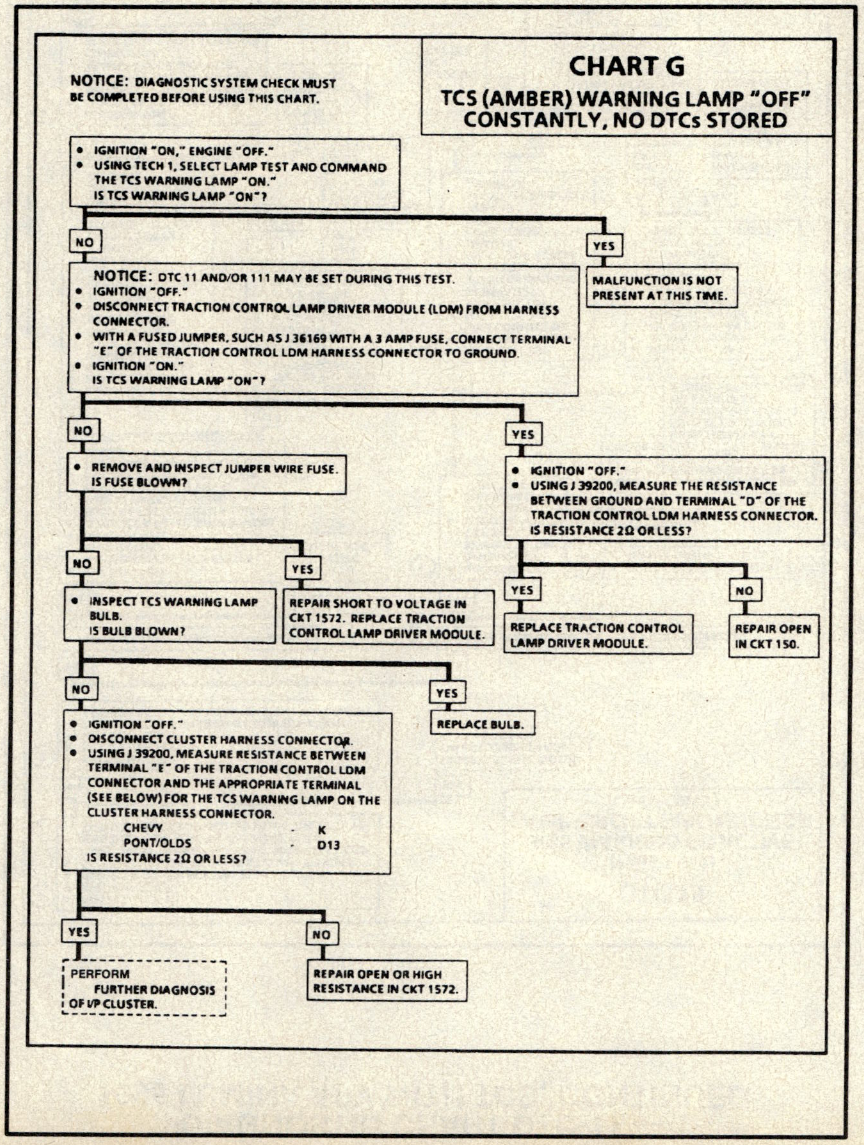

DIAGNOSTIC CHART DTC 11 — 1995 LUMINA APV WITH TCS

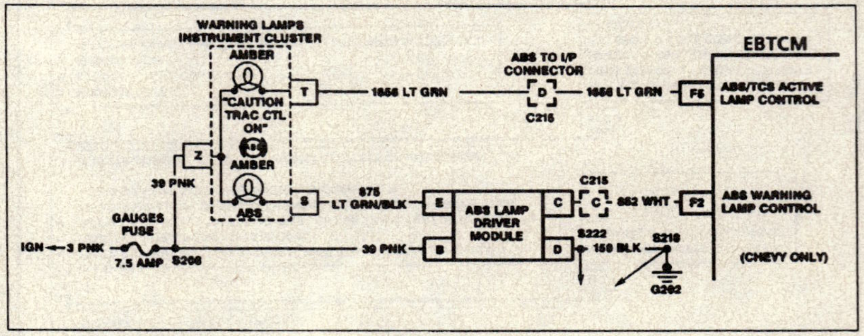

DTC 11
(Page 1 of 2)
ABS WARNING OR ACTIVE LAMP CIRCUIT MALFUNCTION
(CHEV ONLY)

Circuit Description:

This DTC checks the state of both the amber ABS warning lamp and the amber ABS/TCS active lamp to identify a situation in which the driver could not be warned of a system malfunction or of possible slippery driving conditions. The Lamp Driver Module (LDM) turns the ABS warning lamp "ON," unless the EBTCM provides ground to turn it "OFF." Because of the circuitry in the LDM, only external malfunctions can be detected by the EBTCM. As a result, the ABS warning lamp itself is not diagnosable, only the control line to the EBTCM can be diagnosed. The ABS/TCS active lamp circuit is such that the EBTCM provides a ground to turn "ON" the ABS/TCS active lamp.

DTC Will Set When:

DTC 11 can be set only during the three second bulb check or when either of the lamps are commanded "ON." If the EBTCM cannot control either the amber ABS warning lamp or the amber ABS/TCS active lamp for two seconds, a malfunction exists.

Action Taken:

A malfunction DTC is stored, however, ABS is not disabled.

DTC Chart Test Description:

Number(s) below refer to circled number(s) on the diagnostic chart.

1. Determines if the ABS warning lamp can be properly controlled.
2. Determines if the lamp is "ON" constantly.
3. Checks for other DTCs which could turn "ON" the ABS warning lamp.
4. Determines if the ABS warning lamp can be turned "OFF" manually.
5. Checks the integrity of the ABS warning lamp circuitry.
6. Checks for a short to voltage on the ABS warning lamp circuitry.
7. Determines if a malfunctioning EBTCM is the cause of DTC 11.
8. Verifies that the other I/P indicator lamps function properly.
9. Determines if the ABS warning lamp can be turned "ON" manually.
10. Checks for a short to ground in the ABS warning lamp circuitry.
11. Checks for open between "Gages" fuse and LDM.
12. Verifies condition of the 7.5 amp "Gages" fuse.
13. Determines if DTC 11 was set due to an EBTCM malfunction.

Diagnostic Aids:

An intermittent malfunction may be caused by a poor connection, rubbed through wire insulation, or a wire that is broken inside the insulation.

The lamp test function of the Tech 1 may be used to command the lamp "ON" while looking for an intermittent malfunction in the ABS warning lamp circuitry.

The frequency of the malfunction can be checked by using the enhanced diagnostic function of the Tech 1, as described in "Tech 1 Diagnostics" in this section.

Any circuitry that is suspected of causing the intermittent complaint should be thoroughly checked for backed out terminals, improper mating, broken locks, improperly formed or damaged terminals, poor terminal to wiring connections or physical damage to the wiring harness.

DIAGNOSTIC CHART DTC 11 — 1995 LUMINA APV WITH TCS, CONTINUED

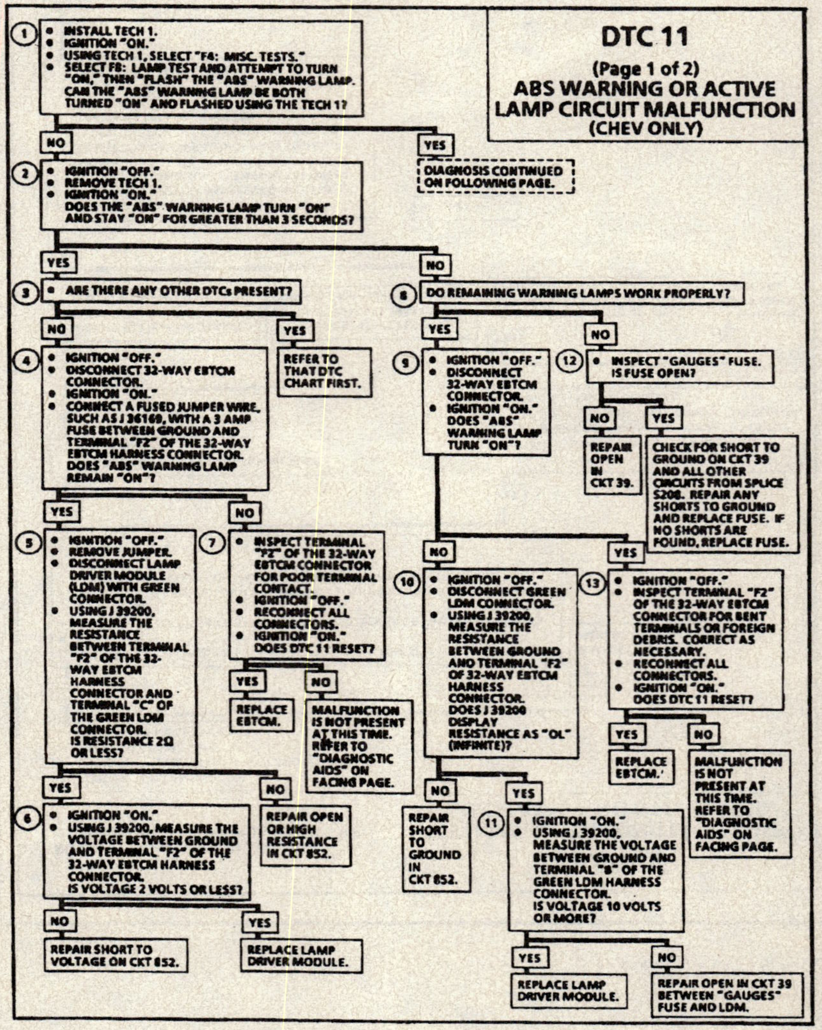

DIAGNOSTIC CHART DTC 11 — 1995 LUMINA APV WITH TCS, CONTINUED

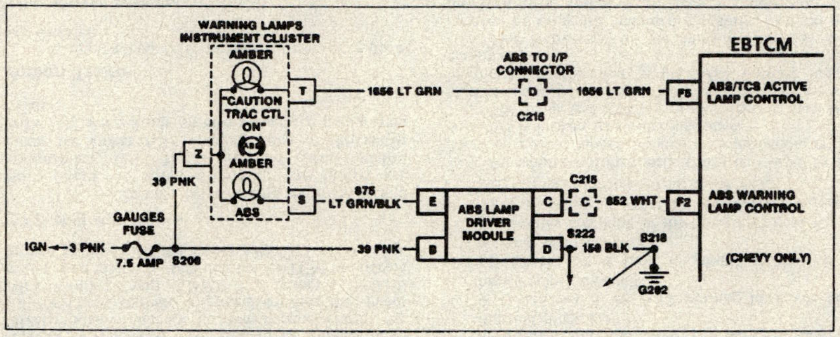

DTC 11
(Page 2 of 2)
ABS WARNING OR ACTIVE LAMP CIRCUIT MALFUNCTION (CHEV ONLY)

Circuit Description:

This DTC checks the state of both the amber ABS warning lamp and the amber ABS/TCS active lamp to identify a situation in which the driver could not be warned of a system malfunction or of possible slippery driving conditions. The Lamp Driver Module (LDM) turns the ABS warning lamp "ON," unless the EBTCM provides ground to turn it "OFF." Because of the circuitry in the LDM, only external malfunctions can be detected by the EBTCM. As a result, the amber ABS warning lamp itself is not diagnosable, only the control line to the EBTCM can be diagnosed. The amber ABS/TCS active lamp circuit is such that the EBTCM provides a ground to turn "ON" the ABS/TCS active lamp.

DTC Will Set When:

DTC 11 can be set only during the three second bulb check or when either of the lamps are commanded "ON." If the EBTCM cannot control either the amber ABS warning lamp or the amber ABS/TCS active lamp for two seconds, a malfunction exists.

Action Taken:

A malfunction DTC is stored, however, ABS is not disabled.

DTC Chart Test Description:

Number(s) below refer to circled number(s) on the diagnostic chart.

14. Determines if the ABS/TCS active lamp can be properly controlled.
15. Determines if the ABS/TCS active lamp is "ON" constantly.
16. Checks for a short to ground in the ABS/TCS active lamp circuitry.
17. Determines if DTC 11 was set due to an EBTCM malfunction.
18. Determines if DTC 11 was set due to an EBTCM malfunction.
19. Determines if the ABS/TCS active lamp can be controlled manually.
20. Verifies the integrity of the ABS/TCS active lamp circuitry.
21. Determines if there is a short to voltage in the ABS/TCS active lamp circuitry.
22. Determines if DTC 11 was set due to an EBTCM malfunction.

Diagnostic Aids:

An intermittent malfunction may be caused by a poor connection, rubbed through wire insulation, or a wire that is broken inside the insulation.

The lamp test function of the Tech 1 may be used to command the lamp "ON" while looking for an intermittent malfunction in the ABS/TCS active lamp circuitry.

The frequency of the malfunction can be checked by using the enhanced diagnostic function of the Tech 1, as described in "Tech 1 Diagnostics" in this section.

Any circuitry that is suspected of causing the intermittent complaint should be thoroughly checked for backed out terminals, improper mating, broken locks, improperly formed or damaged terminals, poor terminal to wiring connections or physical damage to the wiring harness.

DIAGNOSTIC CHART DTC 11 — 1995 LUMINA APV WITH TCS, CONTINUED

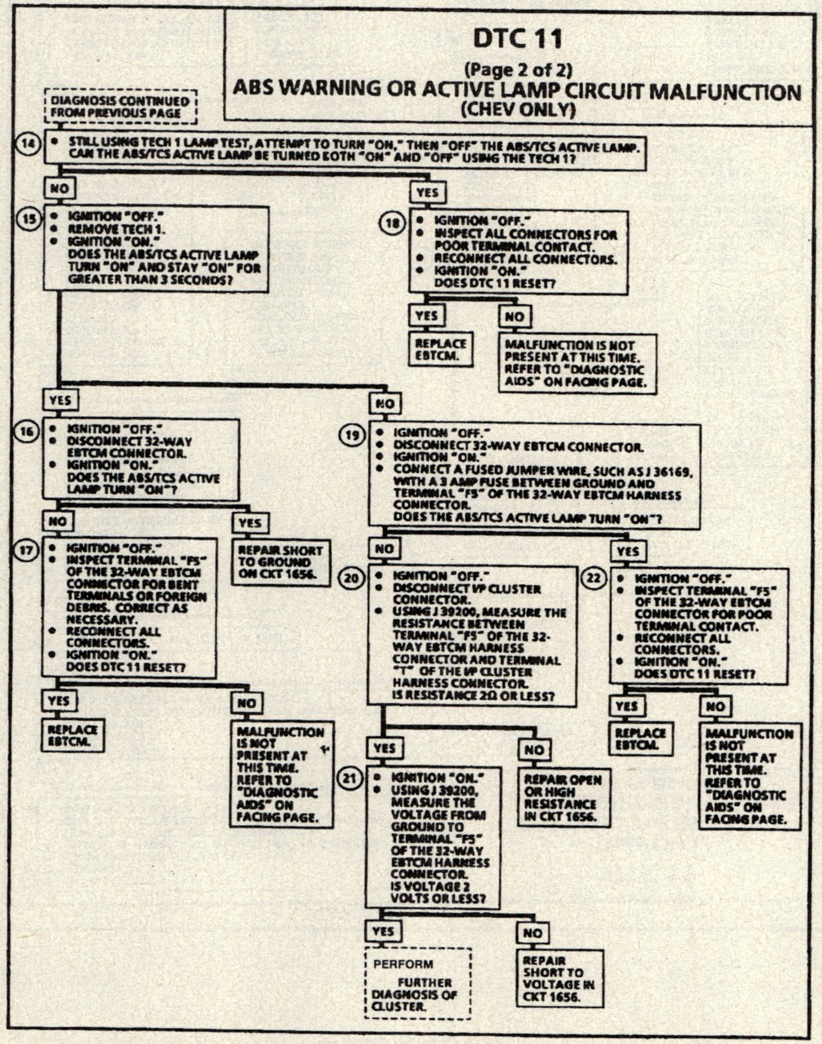

DIAGNOSTIC CHART DTC 11 — 1995 SILHOUETTE AND TRANS SPORT WITH TCS

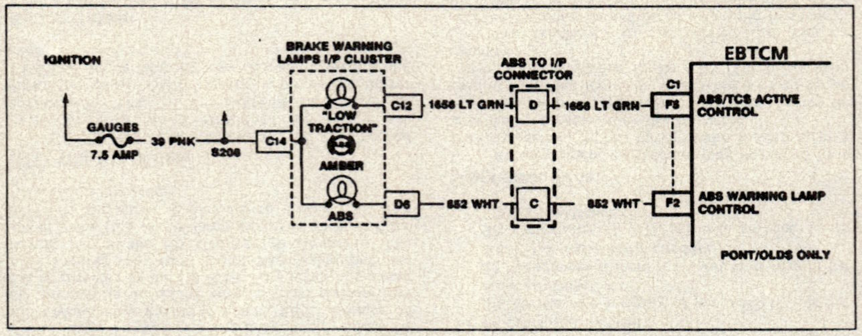

DTC 11
(Page 1 of 2)
ABS WARNING OR ACTIVE LAMP CIRCUIT MALFUNCTION (PONT/OLDS ONLY)

Circuit Description:

This DTC checks the state of both the amber ABS warning lamp and the amber ABS/TCS active lamp to identify a situation in which the driver could not be alerted to a system malfunction or of possible slippery driving conditions. Due to an integral lamp driver module within the instrument panel cluster, the EBTCM must provide a ground to turn the amber ABS warning lamp "OFF." The amber ABS/TCS active lamp circuit is such that the EBTCM provides a ground to turn "ON" the amber ABS/TCS active lamp.

DTC Will Set When:

DTC 11 can only be set during the three second bulb check or when either of the lamps are commanded "ON." If the EBTCM cannot control either the amber ABS warning lamp or the amber ABS/TCS active lamp for two seconds, a malfunction exists.

Action Taken:

A malfunction DTC is stored, however, ABS is not disabled.

DTC Chart Test Description:

Number(s) below refer to circled number(s) on the diagnostic chart.

1. Determines if the ABS warning lamp can be properly controlled.
2. Determines if the lamp is "ON" constantly.
3. Checks for other DTCs which could turn "ON" the ABS warning lamp.
4. Determines if the ABS warning lamp can be turned "OFF" manually.
5. Checks the integrity of the ABS warning lamp circuitry.
6. Checks for a short to voltage on the ABS warning lamp circuitry.
7. Determines if a malfunctioning EBTCM is the cause of DTC 11.
8. Verifies that the other I/P indicator lamps function properly.
9. Determines if the ABS warning lamp can be turned "ON" manually.
10. Checks for a short to ground in the ABS warning lamp circuitry.
11. Verifies condition of the "GAGES" Fuse 9.
12. Determines if DTC 11 was set due to an EBTCM malfunction..shDiagnostic Aids:

An intermittent malfunction may be caused by a poor connection, rubbed through wire insulation, or a wire that is broken inside the insulation.

The lamp test function of the Tech 1 may be used to command the lamp "ON" while looking for an intermittent malfunction in the ABS warning lamp circuitry.

The frequency of the malfunction can be checked by using the enhanced diagnostic function of the Tech 1, as described in "Tech 1 Diagnostics" in this section.

Any circuitry that is suspected of causing the intermittent complaint should be thoroughly checked for backed out terminals, improper mating, broken locks, improperly formed or damaged terminals, poor terminal to wiring connections or physical damage to the wiring harness.

DIAGNOSTIC CHART DTC 11 — 1995 SILHOUETTE AND TRANS SPORT WITH TCS, CONTINUED

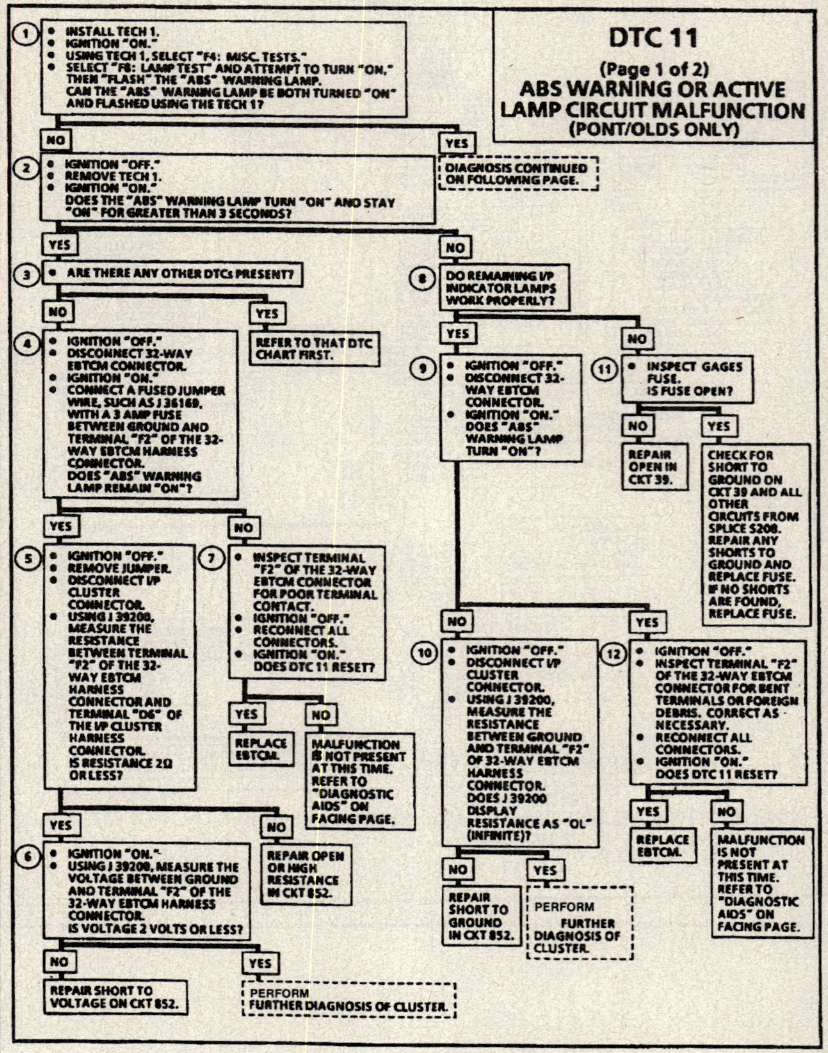

DIAGNOSTIC CHART DTC 11 — 1995 SILHOUETTE AND TRANS SPORT WITH TCS, CONTINUED

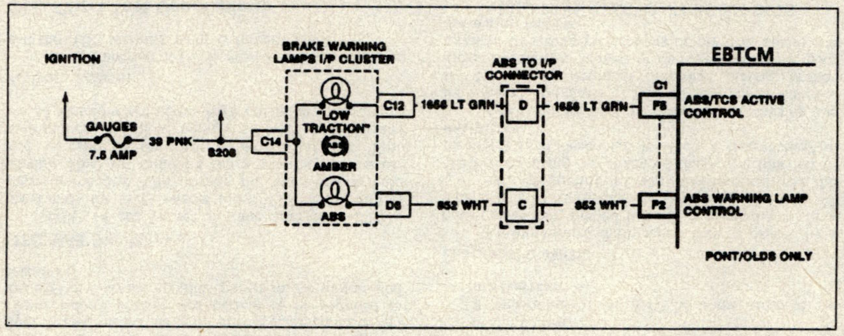

DTC 11
(Page 2 of 2)
ABS WARNING OR ACTIVE LAMP CIRCUIT MALFUNCTION (PONT/OLDS ONLY)

Circuit Description:

This DTC checks the state of both the amber ABS warning lamp and the amber ABS/TCS active lamp to identify a situation in which the driver could not be alerted to a system malfunction or of possible slippery driving conditions. Due to an integral lamp driver module within the instrument panel cluster, the EBTCM must provide a ground to turn the amber ABS warning lamp "OFF." The amber ABS/TCS active lamp circuit is such that the EBTCM provides a ground to turn "ON" the amber ABS/TCS active lamp.

DTC Will Set When:

DTC 11 can only be set during the three second bulb check or when either of the lamps are commanded "ON." If the EBTCM cannot control either the amber ABS warning lamp or the amber ABS/TCS active lamp for two seconds, a malfunction exists.

Action Taken:

A malfunction DTC is stored, however, ABS is not disabled.

DTC Chart Test Description:

Number(s) below refer to circled number(s) on the diagnostic chart.

13. Determines if the ABS/TCS active lamp can be properly controlled.
14. Determines if the ABS/TCS active lamp is "ON" constantly.
15. Checks for a short to ground in the ABS/TCS active lamp circuitry.
16. Determines if DTC 11 was set due to an EBTCM malfunction.
17. Determines if DTC 11 was set due to an EBTCM malfunction.
18. Determines if the ABS/TCS active lamp can be controlled manually.
19. Verifies the integrity of the ABS/TCS active lamp circuitry.
20. Determines if there is a short to voltage in the ABS/TCS active lamp circuitry.
21. Determines if DTC 11 was set due to an EBTCM malfunction.

Diagnostic Aids:

An intermittent malfunction may be caused by a poor connection, rubbed through wire insulation, or a wire that is broken inside the insulation.

The lamp test function of the Tech 1 may be used to command the lamp "ON" while looking for an intermittent malfunction in the ABS/TCS active lamp circuitry.

The frequency of the malfunction can be checked by using the enhanced diagnostic function of the Tech 1, as described in "Tech 1 Diagnostics" in this section.

Any circuitry that is suspected of causing the intermittent complaint should be thoroughly checked for backed out terminals, improper mating, broken locks, improperly formed or damaged terminals, poor terminal to wiring connections or physical damage to the wiring harness.

DIAGNOSTIC CHART DTC 11 — 1995 SILHOUETTE AND TRANS SPORT WITH TCS, CONTINUED

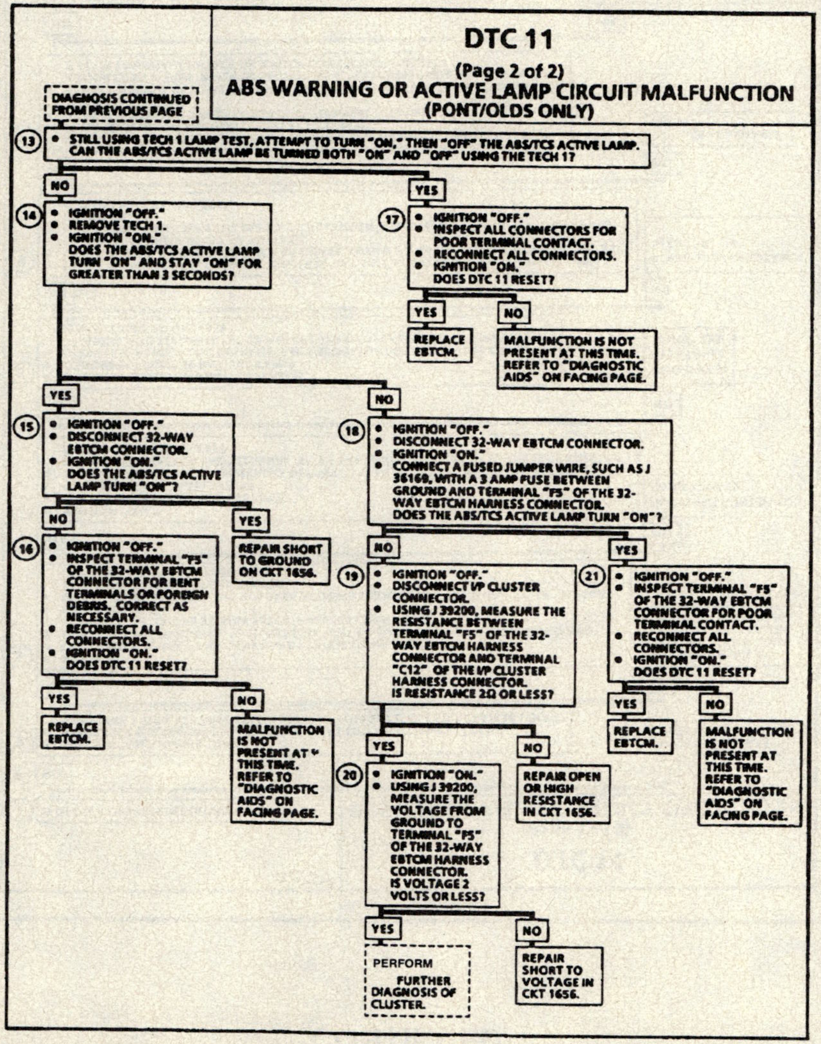

DIAGNOSTIC CHART DTC 14 — 1995 SILHOUETTE AND TRANS SPORT WITH TCS

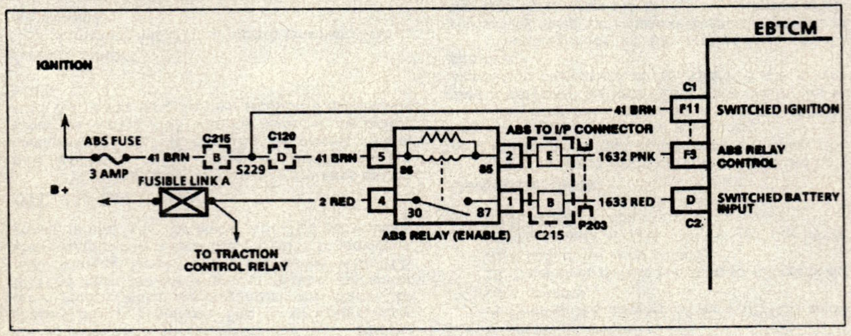

DTC 14
(Page 1 of 3)
ABS ENABLE RELAY CONTACT CIRCUIT OPEN

Circuit Description:

Ignition voltage is supplied through terminal "5" of the ABS enable relay. The EBTCM then is able to energize the pull-in coil by completing the ground circuit at terminal "F3" of the EBTCM. The magnetic field created closes the ABS enable relay contacts and allows battery voltage and current to be supplied to the EBTCM, which supplies power to the motors and solenoids.

DTC Will Set When:

DTC 14 can be set anytime after the EBTCM commands the ABS enable relay "ON" (the relay is first commanded "ON" during the three second bulb check). This test monitors the availability of current and voltage to the motors and solenoids. This malfunction indicates voltage is not available and would therefore not allow ABS operation if required.

Action Taken:

A malfunction DTC is stored, ABS is disabled and the ABS warning lamp is turned "ON."

DTC Chart Test Description:

Number(s) below refer to circled number(s) on the diagnostic chart.

1. Checks to see if DTC 16 is also set as current or history. If DTC 16 is also set, proceed to that chart first.
2. Verifies that the malfunction is currently present.
3. Verifies that the relay can be turned "ON" and provide voltage to the EBTCM.
4. Checks for high resistance in the switched battery circuit from the relay to the EBTCM.
5. Checks for high resistance in the switched battery circuit from the battery to the relay.
6. Verifies the condition of the ABS enable relay.
7. Checks for poor terminal contact or corrosion at the connectors.
8. Determines if the EBTCM is the cause of the malfunction.

Diagnostic Aids:

An intermittent malfunction may be caused by a poor connection, rubbed through wire insulation, or a wire that is broken inside the insulation.

The frequency of the malfunction can be checked by using the enhanced diagnostic function of the Tech 1, as described in "Tech 1 Diagnostics" in this section.

Any circuitry that is suspected of causing the intermittent complaint should be thoroughly checked for backed out terminals, improper mating, broken locks, improperly formed or damaged terminals, poor terminal to wiring connections or physical damage to the wiring harness.

VIBRATION, TEMPERATURE EFFECTS:

Check for vibration effects by performing the relay test function of the Tech 1. With the relay test commanded "ON," lightly tap the top and sides of the relay while monitoring relay voltage. If the relay voltage changes significantly, replace the relay.

If DTC 14 only sets when the vehicle is initially started in cold ambient conditions (temperature less than 32°F - 0°C),replace the relay.

DIAGNOSTIC CHART DTC 14 — 1995 SILHOUETTE AND TRANS SPORT WITH TCS, CONTINUED

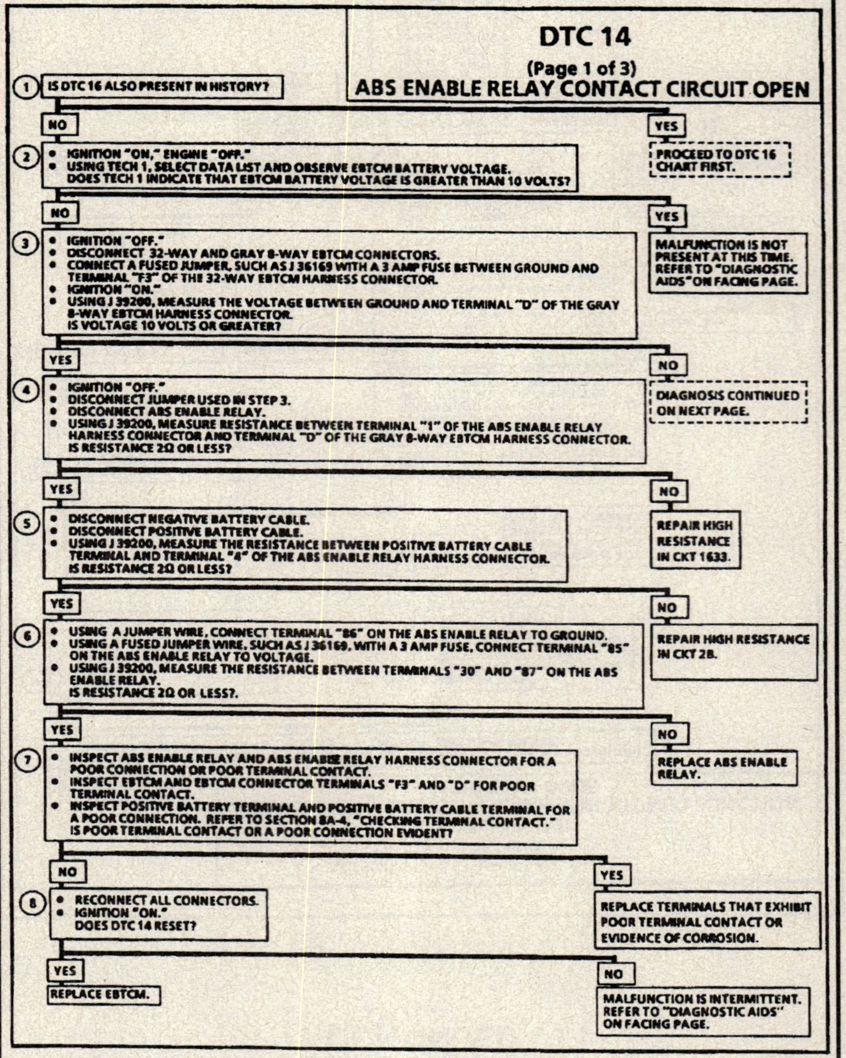

DIAGNOSTIC CHART DTC 14 — 1995 SILHOUETTE AND TRANS SPORT WITH TCS, CONTINUED

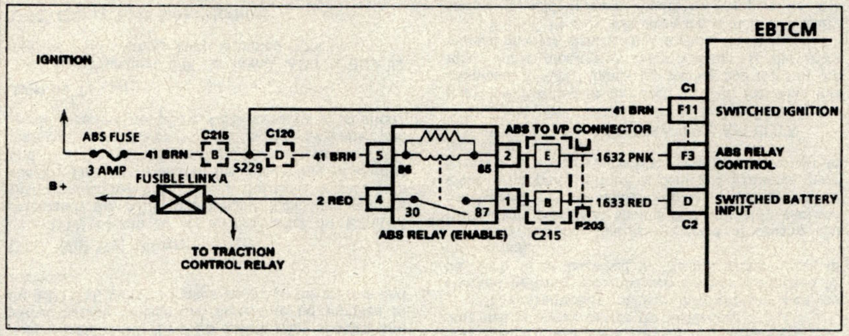

DTC 14
(Page 2 of 3)
ABS ENABLE RELAY CONTACT CIRCUIT OPEN

Circuit Description:

Ignition voltage is supplied through terminal "5" of the ABS enable relay. The EBTCM then is able to energize the pull-in coil by completing the ground circuit at terminal "F3" of the EBTCM. The magnetic field created closes the ABS enable relay contacts and allows battery voltage and current to be supplied to the EBTCM, which supplies power to the motors and solenoids.

DTC Will Set When:

DTC 14 can be set anytime after the EBTCM commands the ABS enable relay "ON" (the relay is first commanded "ON" during the three second bulb check). This test monitors the availability of current and voltage to the motors and solenoids. This malfunction indicates voltage is not available and would therefore not allow ABS operation if required.

Action Taken:

A malfunction DTC is stored, ABS is disabled and the ABS warning lamp is turned "ON."

DTC Chart Test Description:

Number(s) below refer to circled number(s) on the diagnostic chart.

9. Checks for high resistance in the entire relay coil circuit.
10. Verifies proper battery condition.
11. This step verifies proper voltage is available at the ABS enable relay.
12. Checks for an open in the switched battery circuit from the relay to the EBTCM.
13. Verifies the condition of the ABS enable relay.

Diagnostic Aids:

An intermittent malfunction may be caused by a poor connection, rubbed through wire insulation, or a wire that is broken inside the insulation.

The frequency of the malfunction can be checked by using the enhanced diagnostic function of the Tech 1, as described in "Tech 1 Diagnostics" in this section.

Any circuitry that is suspected of causing the intermittent complaint should be thoroughly checked for backed out terminals, improper mating, broken locks, improperly formed or damaged terminals, poor terminal to wiring connections or physical damage to the wiring harness.

VIBRATION, TEMPERATURE EFFECTS:

Check for vibration effects by performing the relay test function of the Tech 1. With the relay test commanded "ON," lightly tap the top and sides of the relay while monitoring relay voltage. If the relay voltage changes significantly, replace the relay.

If DTC 14 only sets when the vehicle is initially started in cold ambient conditions (temperature less than 32°F - 0°C), replace the relay.

DIAGNOSTIC CHART DTC 14 — 1995 SILHOUETTE AND TRANS SPORT WITH TCS

DTC 14
(Page 2 of 3)
ABS ENABLE RELAY CONTACT CIRCUIT OPEN

DIAGNOSIS CONTINUED FROM PREVIOUS PAGE.

9
- IGNITION "OFF."
- REMOVE JUMPER WIRE INSTALLED IN STEP #3 IF NOT ALREADY REMOVED.
- DISCONNECT NEGATIVE BATTERY CABLE.
- DISCONNECT POSITIVE BATTERY CABLE.
- WITH BATTERY STILL DISCONNECTED, TURN IGNITION SWITCH TO "RUN." THIS COMPLETES THE IGNITION CIRCUIT TO THE ABS ENABLE RELAY.
- USING J 39200, MEASURE THE RESISTANCE BETWEEN TERMINAL "F3" OF THE 32-WAY EBTCM HARNESS CONNECTOR AND THE POSITIVE BATTERY CABLE TERMINAL.

IS RESISTANCE 95Ω OR LESS WHEN MEASURED AT APPROXIMATELY 20°C (68°F)?

YES → 10
NO → DIAGNOSIS CONTINUED ON NEXT PAGE.

10
- USING J 39200, MEASURE THE VOLTAGE BETWEEN THE NEGATIVE AND POSITIVE BATTERY TERMINALS.

IS VOLTAGE 10 VOLTS OR GREATER?

YES → 11
NO → CORRECT LOW VOLTAGE CONDITION.

11
- IGNITION "OFF."
- RECONNECT POSITIVE BATTERY CABLE.
- RECONNECT NEGATIVE BATTERY CABLE.
- DISCONNECT ABS ENABLE RELAY.
- USING J 39200, MEASURE THE VOLTAGE BETWEEN TERMINAL "4" OF THE ABS ENABLE RELAY HARNESS CONNECTOR AND GROUND.

IS VOLTAGE 10 VOLTS OR GREATER?

YES → 12
NO → REPAIR OPEN OR HIGH RESISTANCE IN CKT 28.

12
- USING J 39200, MEASURE THE RESISTANCE BETWEEN TERMINAL "1" OF THE ABS ENABLE RELAY HARNESS CONNECTOR AND TERMINAL "D" OF THE GRAY 8-WAY EBTCM HARNESS CONNECTOR.

IS RESISTANCE 2Ω OR LESS?

YES → 13
NO → REPAIR OPEN OR HIGH RESISTANCE IN CKT 1633.

13
- USING A JUMPER WIRE, CONNECT TERMINAL "86" ON THE ABS ENABLE RELAY TO GROUND.
- USING A FUSED JUMPER WIRE, SUCH AS J 36169, WITH A 3 AMP FUSE, CONNECT TERMINAL "85" ON THE ABS ENABLE RELAY TO B+.
- USING J 39200, MEASURE THE RESISTANCE BETWEEN TERMINALS "30" AND "87" ON THE ABS ENABLE RELAY.

IS RESISTANCE 2Ω OR LESS?

YES → MALFUNCTION WAS MOST LIKELY DUE TO A POOR CONNECTION. INSPECT ALL CONNECTORS AND TERMINALS FOR POOR TERMINAL CONTACT AND EVIDENCE OF CORROSION. REPLACE ALL TERMINALS THAT EXHIBIT EVIDENCE OF POOR TERMINAL CONTACT OR CORROSION.
NO → REPLACE ABS ENABLE RELAY.

DIAGNOSTIC CHART DTC 14 — 1995 SILHOUETTE AND TRANS SPORT WITH TCS, CONTINUED

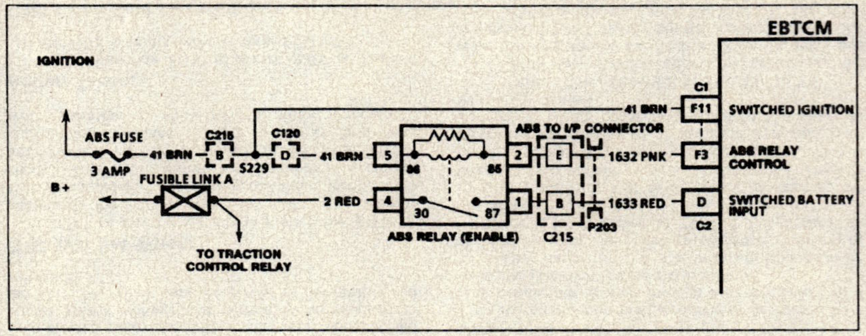

DTC 14
(Page 3 of 3)
ABS ENABLE RELAY CONTACT CIRCUIT OPEN

Circuit Description:

Ignition voltage is supplied through terminal "5" of the ABS enable relay. The EBTCM then is able to energize the pull-in coil by completing the ground circuit at terminal "F3" of the EBTCM. The magnetic field created closes the ABS enable relay contacts and allows battery voltage and current to be supplied to the EBTCM, which supplies power to the motors and solenoids.

DTC Will Set When:

DTC 14 can be set anytime after the EBTCM commands the ABS enable relay "ON" (the relay is first commanded "ON" during the three second bulb check). This test monitors the availability of current and voltage to the motors and solenoids. This malfunction indicates voltage is not available and would therefore not allow ABS operation if required.

Action Taken:

A malfunction DTC is stored, ABS is disabled and the ABS warning lamp is turned "ON."

DTC Chart Test Description:

Number(s) below refer to circled number(s) on the diagnostic chart.

14. Checks for proper resistance of the relay coil.

15. Checks for high resistance in the ABS enable relay control circuit between the relay and the EBTCM.

16. Checks for high resistance in the ABS enable relay control circuit between the battery and the relay.

Diagnostic Aids:

An intermittent malfunction may be caused by a poor connection, rubbed through wire insulation, or a wire that is broken inside the insulation.

The frequency of the malfunction can be checked by using the enhanced diagnostic function of the Tech 1, as described in "Tech 1 Diagnostics" in this section.

Any circuitry that is suspected of causing the intermittent complaint should be thoroughly checked for backed out terminals, improper mating, broken locks, improperly formed or damaged terminals, poor terminal to wiring connections or physical damage to the wiring harness.

VIBRATION, TEMPERATURE EFFECTS:

Check for vibration effects by performing the relay test function of the Tech 1. With the relay test commanded "ON," lightly tap the top and sides of the relay while monitoring relay voltage. If the relay voltage changes significantly, replace the relay.

If DTC 14 only sets when the vehicle is initially started in cold ambient conditions (temperature less than 32°F - 0°C), replace the relay.

DIAGNOSTIC CHART DTC 14 — 1995 SILHOUETTE AND TRANS SPORT WITH TCS, CONTINUED

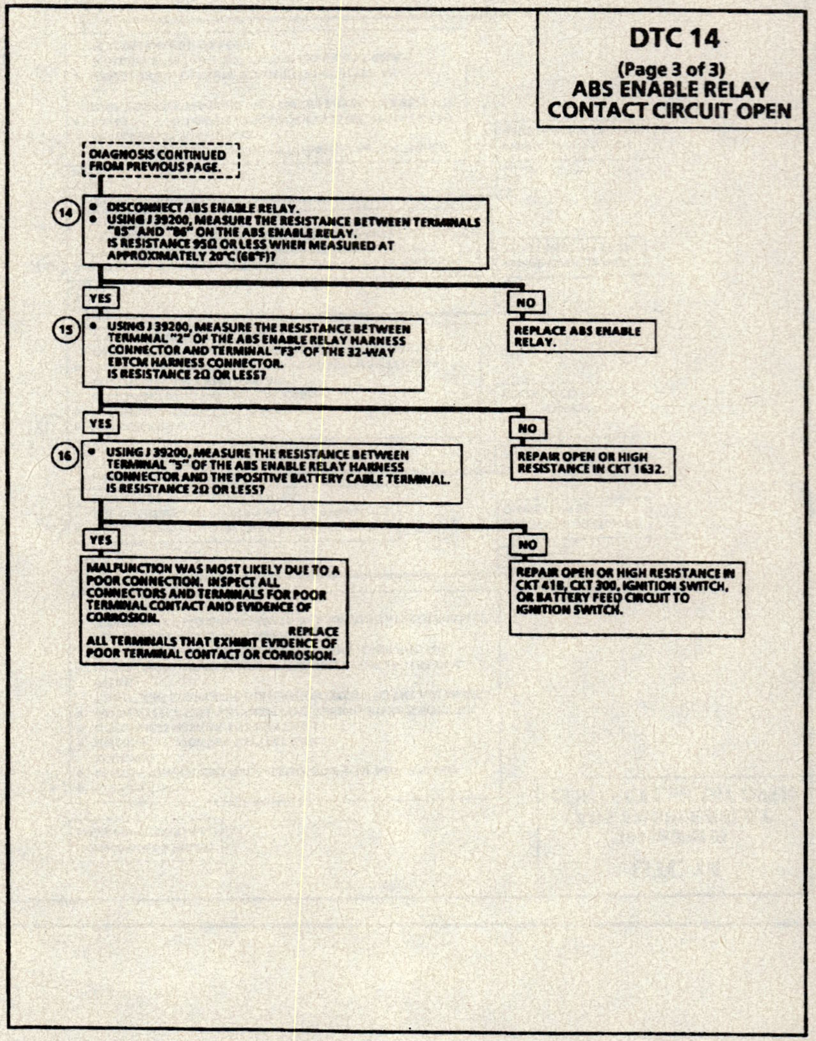

DIAGNOSTIC CHART DTC 15 — 1995 SILHOUETTE AND TRANS SPORT WITH TCS

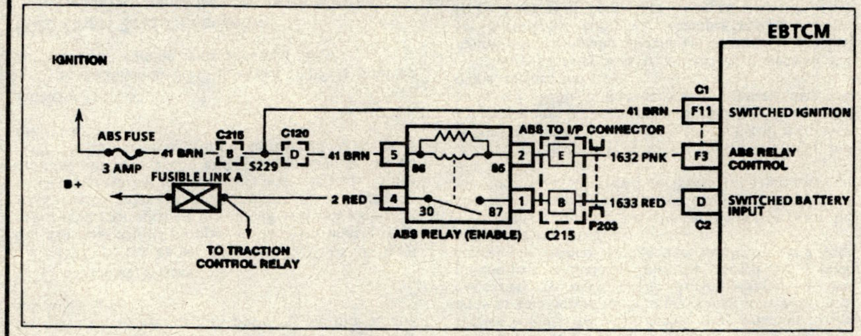

DTC 15
ABS ENABLE RLY CONTACT CKT SHORT TO BATT OR ALWAYS CLOSED

Circuit Description:

Ignition voltage is supplied through terminal "5" of the ABS enable relay. The EBTCM then is able to energize the pull-in coil by completing the ground circuit at terminal "F3" of the EBTCM. The magnetic field created closes the ABS enable relay contacts and allows battery voltage and current to be supplied to the EBTCM, which supplies power to the motors and solenoids.

DTC Will Set When:

DTC 15 can be set only before the EBTCM commands the relay "ON." This test determines if the ABS enable relay is energized when it should not be. This malfunction would not allow the ABS enable relay to remove power to the ABS system. If a second malfunction were to occur that requires the ABS enable relay to be turned "OFF," that malfunction cannot be removed if the relay cannot be controlled. The malfunction must be present for three consecutive drive cycles before the DTC is set.

Action Taken:

A malfunction DTC is stored. ABS is not disabled.

DTC Chart Test Description:

Number(s) below refer to circled number(s) on the diagnostic chart.

1. Determines if any other DTCs are present.
2. Indicates that the EBTCM is capable of controlling the ABS enable relay as commanded.
3. This step checks for a possible short to battery voltage in CKT 1633.
4. This step determines whether the short to voltage is in CKT 1633 or internal to the EBTCM.
5. Checks for a possible intermittent malfunction in the wiring or connectors.

Diagnostic Aids:

An intermittent malfunction may be caused by a poor connection, rubbed through wire insulation, or a wire that is broken inside the insulation.

The frequency of the malfunction can be checked by using the enhanced diagnostic function of the Tech 1, as described in "Tech 1 Diagnostics" in this section.

Any circuitry that is suspected of causing the intermittent complaint should be thoroughly checked for backed out terminals, improper mating, broken locks, improperly formed or damaged terminals, poor terminal to wiring connections or physical damage to the wiring harness.

DIAGNOSTIC CHART DTC 15 — 1995 SILHOUETTE AND TRANS SPORT WITH TCS, CONTINUED

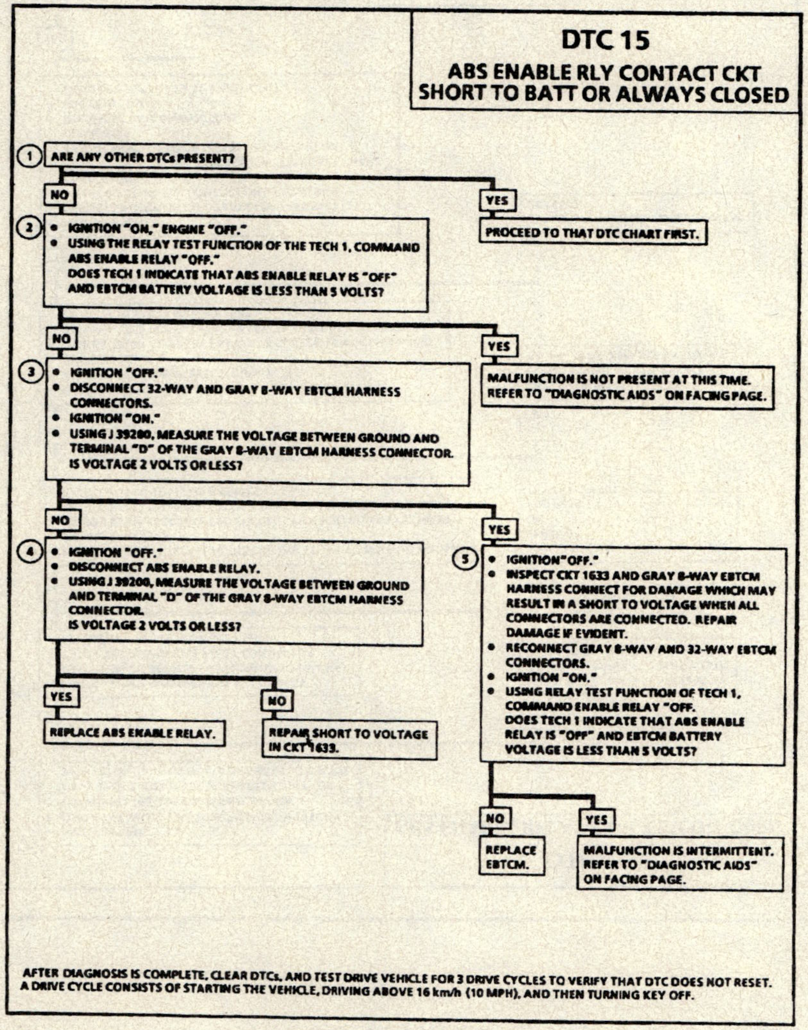

DIAGNOSTIC CHART DTC 16 — 1995 SILHOUETTE AND TRANS SPORT WITH TCS

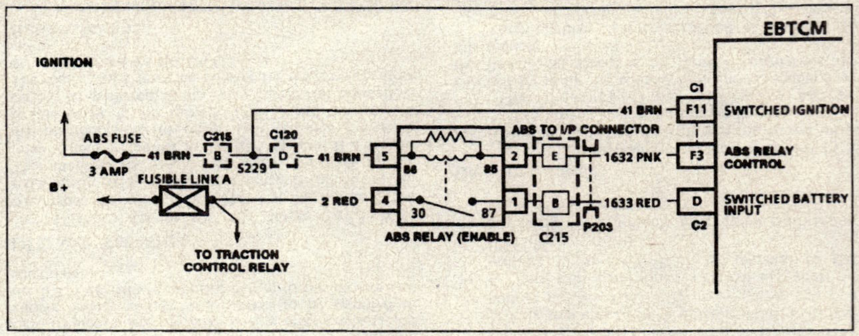

DTC 16
ABS ENABLE RELAY COIL CIRCUIT OPEN

Circuit Description:

Ignition voltage is supplied through terminal "5" of the ABS enable relay. The EBTCM then is able to energize the pull-in coil by completing the ground circuit at terminal "F3" of the EBTCM. The magnetic field created closes the ABS enable relay contacts and allows battery voltage and current to be supplied to the EBTCM, which supplies power to the motors and solenoids.

DTC Will Set When:

DTC 16 can be set only after DTC 14 has been set. This test detects an open in the ABS enable relay coil circuit. An open in this circuit will not allow the ABS enable relay to be energized thus preventing voltage and current to the motors and solenoids. If this malfunction is present and the ignition is turned "OFF" before 5 km/h (3 mph) is reached, DTC 14 is set alone.

Action Taken:

A malfunction DTC is stored, ABS is disabled and the ABS warning lamp is turned "ON."

DTC Chart Test Description:

Number(s) below refer to circled number(s) on the diagnostic chart.

1. Indicates the EBTCM is capable of controlling the ABS enable relay as commanded.
2. Checks to ensure that voltage is available to the pull-in coil of the ABS enable relay.
3. Ensures that there is continuity through the pull-in coil of the ABS enable relay.
4. Verifies the integrity of the relay control circuit.
5. Ensures malfunction was not due to poor terminal contact.

Diagnostic Aids:

It will be necessary to remove the two attaching screws holding the ABS enable relay junction to the body to gain access to the relay connections when it is necessary to back probe during diagnosis. An intermittent malfunction may be caused by a poor connection, rubbed through wire insulation, or a wire that is broken inside the insulation.

The frequency of the malfunction can be checked by using the enhanced diagnostic function of the Tech 1, as described in "Tech 1 Diagnostics" in this section.

If the frequency of the malfunction is high, but is currently intermittent, check for high coil resistance by measuring between relay terminals "2" and "5" using a J 39200. If resistance shows greater than 100 ohms, replace the relay.

Any circuitry that is suspected of causing the intermittent complaint should be thoroughly checked for backed out terminals, improper mating, broken locks, improperly formed or damaged terminals, poor terminal to wiring connections or physical damage to the wiring harness.

DIAGNOSTIC CHART DTC 16 — 1995 SILHOUETTE AND TRANS SPORT WITH TCS, CONTINUED

DTC 16
ABS ENABLE RELAY COIL CIRCUIT OPEN

(1)
- IGNITION "ON," ENGINE "OFF."
- USING THE RELAY TEST FUNCTION OF THE TECH 1, COMMAND THE ABS ENABLE RELAY "ON." DOES TECH 1 INDICATE ABS ENABLE RELAY "ON," AND EBTCM BATTERY VOLTAGE ABOVE 10 VOLTS?

YES: MALFUNCTION IS NOT PRESENT AT THIS TIME. REFER TO "DIAGNOSTIC AIDS" ON FACING PAGE.

NO:

(2)
- USING J 39200, MEASURE THE VOLTAGE BY BACKPROBING BETWEEN GROUND AND TERMINAL "5" OF THE ABS ENABLE RELAY CONNECTOR. IS VOLTAGE 10 VOLTS OR GREATER?

NO: REPAIR OPEN OR HIGH RESISTANCE IN CKT 41 OR ABS FUSE.

YES:

(3)
- USING THE RELAY TEST FUNCTION OF THE TECH 1, COMMAND THE ABS ENABLE RELAY "OFF."
- USING J 39200, MEASURE THE VOLTAGE BY BACKPROBING BETWEEN GROUND AND TERMINAL "2" OF THE ABS ENABLE RELAY CONNECTOR. IS VOLTAGE 10 VOLTS OR GREATER?

NO: CHECK FOR INTERMITTENT CONNECTIONS AT ABS ENABLE RELAY. IF OK, REPLACE ABS RELAY.

YES:

(4)
- IGNITION "OFF."
- DISCONNECT 32-WAY EBTCM CONNECTOR.
- IGNITION "ON."
- USING J 39200, MEASURE THE VOLTAGE BETWEEN GROUND AND TERMINAL "F3" OF THE 32-WAY EBTCM HARNESS CONNECTOR. IS VOLTS 10 VOLTS OR GREATER?

NO: REPAIR OPEN OR HIGH RESISTANCE IN CKT 1632.

YES:

(5)
- IGNITION "OFF."
- INSPECT 32-WAY EBTCM CONNECTOR TERMINAL "F3" FOR POOR TERMINAL CONTACT OR EVIDENCE OF CORROSION. REPLACE ALL TERMINALS THAT EXHIBIT SIGNS OF POOR TERMINAL CONTACT OR CORROSION.
- RECONNECT ALL CONNECTORS.
- IGNITION "ON." DOES DTC 16 SET AS A CURRENT DTC?

YES: REPLACE EBTCM.

NO: MALFUNCTION WAS MOST LIKELY CAUSED BY A POOR CONNECTION. REFER TO "DIAGNOSTIC AIDS" ON FACING PAGE.

AFTER ALL DIAGNOSIS IS COMPLETE, CLEAR DTCs AND TEST DRIVE VEHICLE FOR THREE (3) DRIVE CYCLES TO VERIFY THAT DTC DOES NOT RESET. A DRIVE CYCLE CONSISTS OF STARTING THE VEHICLE, DRIVING OVER 16 km/h (10 MPH), AND THEN TURNING KEY OFF.

DIAGNOSTIC CHART DTC 17 — 1995 SILHOUETTE AND TRANS SPORT WITH TCS

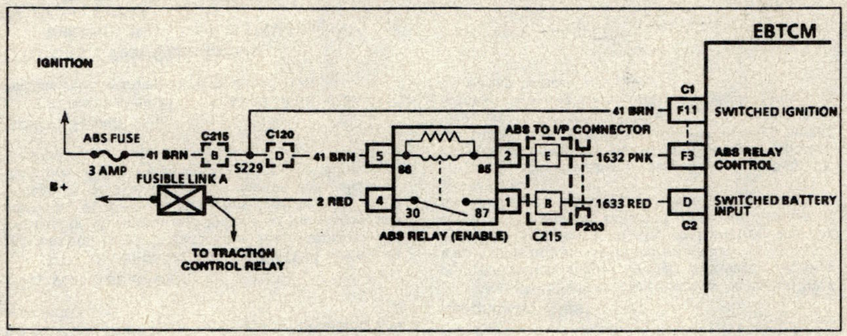

DTC 17
ABS ENABLE RELAY COIL CIRCUIT SHORTED TO GND

Circuit Description:

Ignition voltage is supplied through terminal "5" of the ABS enable relay. The EBTCM then is able to energize the pull-in coil by completing the ground circuit at terminal "F3" of the EBTCM. The magnetic field created closes the ABS enable relay contacts and allows battery voltage and current to be supplied to the EBTCM, which supplies voltage to the motors and solenoids.

DTC Will Set When:

DTC 17 can be set before the EBTCM commands the ABS enable relay "ON." This test determines if the ABS enable relay is energized when it should not be. This malfunction would not allow the ABS enable relay to remove power to the ABS system. If a second malfunction were to occur that requires the ABS enable relay to be turned "OFF," that malfunction cannot be removed if the relay cannot be controlled.

Action Taken:

A malfunction DTC is stored. ABS is not disabled.

DTC Chart Test Description:

Number(s) below refer to circled number(s) on the diagnostic chart.

1. Indicates that the EBTCM is capable of controlling the ABS enable relay as commanded.
2. Checks to ensure that the ABS enable relay or control CKT 1632 are not shorted to ground.
3. Checks to see if the EBTCM is internally shorted to ground. A grounded terminal "F3" would cause the ABS enable relay to be energized anytime the ignition was in the "ON" position.
4. Checks for a short to ground in control CKT 1632 or a defective relay.

Diagnostic Aids:

An intermittent malfunction may be caused by a poor connection, rubbed through wire insulation, or a wire that is broken inside the insulation.

The frequency of the malfunction can be checked by using the enhanced diagnostic function of the Tech 1, as described in "Tech 1 Diagnostics" in this section.

Any circuitry that is suspected of causing the intermittent complaint should be thoroughly checked for backed out terminals, improper mating, broken locks, improperly formed or damaged terminals, poor terminal to wiring connections or physical damage to the wiring harness.

DIAGNOSTIC CHART DTC 17 — 1995 SILHOUETTE AND TRANS SPORT WITH TCS, CONTINUED

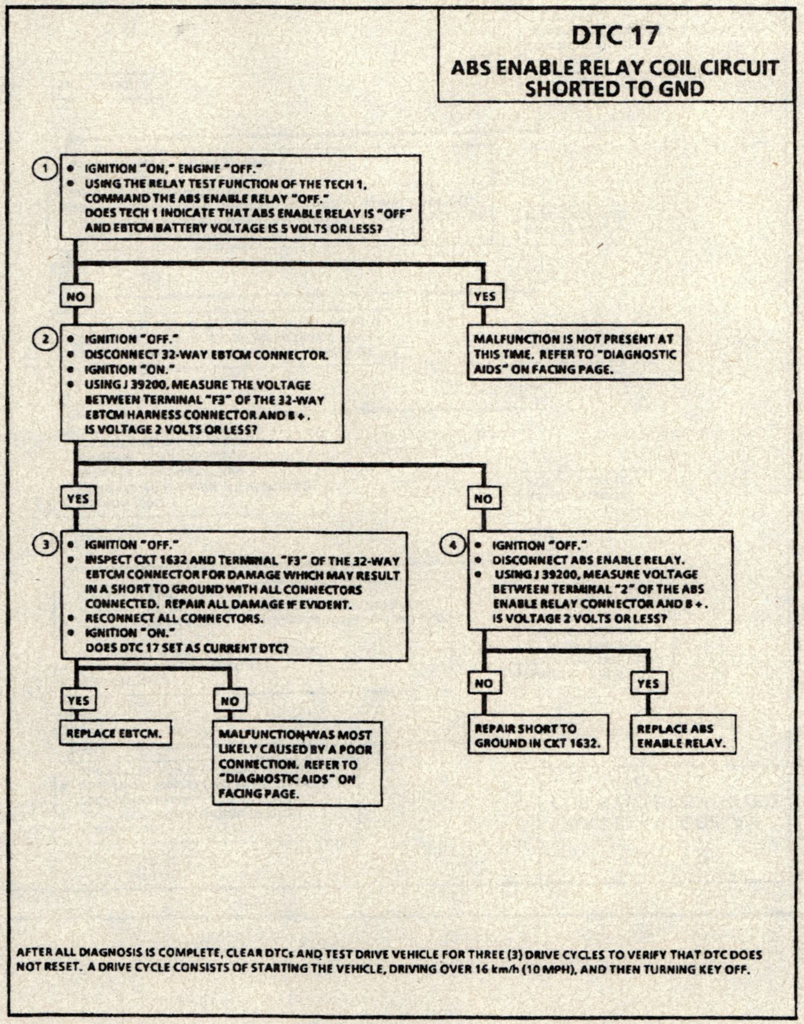

DIAGNOSTIC CHART DTC 18 — 1995 SILHOUETTE AND TRANS SPORT WITH TCS

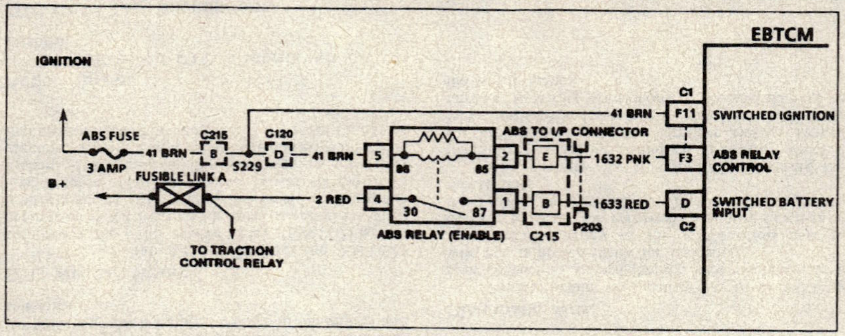

DTC 18
ABS ENABLE RELAY COIL CIRCUIT SHORTED TO BATT

Circuit Description:

Ignition voltage is supplied through terminal "5" of the ABS enable relay. The EBTCM then is able to energize the pull-in coil by completing the ground circuit at terminal "F3" of the EBTCM. The magnetic field created closes the ABS enable relay contacts and allows battery voltage and current to be supplied to the EBTCM, which supplies voltage to the motors and solenoids.

DTC Will Set When:

DTC 18 can be set after the EBTCM commands the enable relay "ON." This test monitors the availability of current and voltage to the motors and solenoids. A malfunction of this type will not allow voltage to be available to the motors and solenoids, therefore, ABS operation would not be allowed if required.

Action Taken:

A malfunction DTC is stored, ABS is disabled and the ABS warning lamp is turned "ON."

DTC Chart Test Description:

Number(s) below refer to circled number(s) on the diagnostic chart.

1. Indicates that the EBTCM is capable of controlling the ABS enable relay as commanded.
2. Checks for a short to voltage on the relay control circuit.
3. Verifies that the relay coil has the appropriate resistance.
4. Verifies that relay contacts have not failed closed.
5. Checks for possible poor terminal contact at the EBTCM.

Diagnostic Aids:

An intermittent malfunction may be caused by a poor connection, rubbed through wire insulation, or a wire that is broken inside the insulation.

The frequency of the malfunction can be checked by using the enhanced diagnostic function of the Tech 1, as described in "Tech 1 Diagnostics" in this section.

Any circuitry that is suspected of causing the intermittent complaint should be thoroughly checked for backed out terminals, improper mating, broken locks, improperly formed or damaged terminals, poor terminal to wiring connections or physical damage to the wiring harness.

DIAGNOSTIC CHART DTC 18 — 1995 SILHOUETTE AND TRANS SPORT WITH TCS, CONTINUED

DTC 18
ABS ENABLE RELAY COIL CIRCUIT SHORTED TO BATT

(1)
- IGNITION "ON," ENGINE "OFF."
- USING RELAY TEST FUNCTION OF TECH 1, COMMAND ABS ENABLE RELAY "ON."
- DOES TECH 1 INDICATE ABS ENABLE RELAY "ON" AND BATTERY VOLTAGE ABOVE 10 VOLTS?

YES → MALFUNCTION IS NOT PRESENT AT THIS TIME. REFER TO "DIAGNOSTIC AIDS" ON FACING PAGE.

NO ↓

(2)
- IGNITION "OFF."
- DISCONNECT ABS ENABLE RELAY.
- DISCONNECT 32-WAY EBTCM CONNECTOR.
- IGNITION "ON."
- USING J 39200, MEASURE THE VOLTAGE BETWEEN GROUND AND TERMINAL "F3" OF THE 32-WAY EBTCM HARNESS CONNECTOR. IS VOLTAGE 2 VOLTS OR LESS?

NO → REPAIR SHORT TO VOLTAGE CKT 1632.

YES ↓

(3)
- IGNITION "ON."
- USING J 39200, MEASURE THE RESISTANCE BETWEEN TERMINAL "2" AND TERMINAL "5" OF THE ABS ENABLE RELAY. IS RESISTANCE 40Ω OR GREATER?

NO → REPLACE ABS ENABLE RELAY.

YES ↓

(4)
- USING J 39200, MEASURE THE RESISTANCE BETWEEN TERMINALS "30" AND "87" OF THE ABS ENABLE RELAY. DOES J 39200 DISPLAY RESISTANCE AS "OL" (INFINITE)?

NO → REPLACE ABS ENABLE RELAY.

YES ↓

(5)
- INSPECT CKT 1632 AND TERMINAL "F3" OF THE 32-WAY EBTCM CONNECTOR FOR DAMAGE WHICH MAY RESULT IN A SHORT TO VOLTAGE WITH ALL CONNECTORS CONNECTED. REPAIR ALL DAMAGE IF EVIDENT.
- RECONNECT ALL CONNECTORS.
- IGNITION "ON." DID DTC 18 SET AS A CURRENT DTC?

YES → REPLACE EBTCM.

NO → MALFUNCTION WAS MOST LIKELY CAUSED BY A POOR CONNECTION. REFER TO "DIAGNOSTIC AIDS" ON FACING PAGE.

AFTER ALL DIAGNOSIS IS COMPLETE, CLEAR DTCs AND TEST DRIVE VEHICLE FOR THREE (3) DRIVE CYCLES TO VERIFY THAT DTC DOES NOT RESET. A DRIVE CYCLE CONSISTS OF STARTING THE VEHICLE, DRIVING OVER 16 km/h (10 MPH), AND THEN TURNING KEY OFF.

DIAGNOSTIC CHART DTC 21 — 1995 SILHOUETTE AND TRANS SPORT WITH TCS

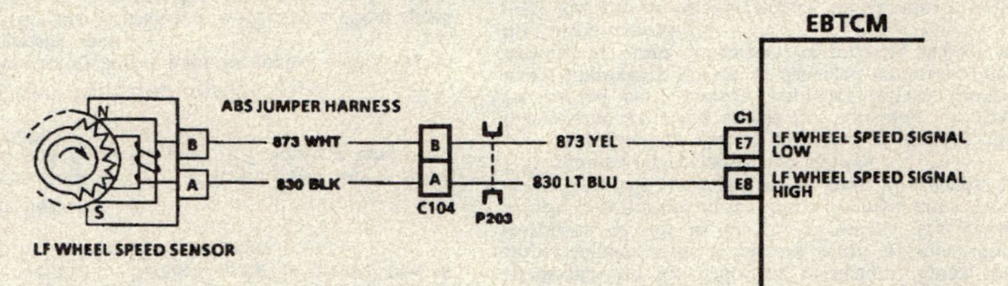

DTC 21
LEFT FRONT WHEEL SPEED = 0

Circuit Description:

As a toothed ring passes by the wheel speed sensor, changes in the electromagnetic field cause the wheel speed sensor to produce a sinusoidal (AC) voltage signal whose frequency is proportional to wheel speed. The magnitude of this signal is directly related to wheel speed and the proximity of the wheel speed sensor to the toothed ring, often referred to as the air gap.

DTC Will Set When:

DTC 21 can be set when the vehicle is not in an ABS or TCS event. If the left front wheel speed = 0 and the vehicle's reference speed is greater than 8 km/h (5 mph), a malfunction exists.

Action Taken:

A malfunction DTC is stored, ABS and TCS are disabled and the ABS and TCS warning lamps are turned "ON."

DTC Chart Test Description:

Number(s) below refer to circled number(s) on the diagnostic chart.

1. This step is designed to identify a wheel speed sensor or circuitry that is damaged and visibly apparent.
2. Checks to see that the wheel speed sensor is within proper resistance values.
3. Verifies that the wheel speed sensor can generate an appropriate output signal.
4. Ensures that wheel speed sensor circuitry is not shorted together.
5. Ensures malfunction was not due to physical damage of the circuitry.

Diagnostic Aids:

An intermittent malfunction may be caused by a poor connection, rubbed through wire insulation, or a wire that is broken inside the insulation.

The frequency of the malfunction can be checked by using the enhanced diagnostic function of the Tech 1, as described in "Tech 1 Diagnostics" in this section.

If the customer's comments reflect that the ABS and TCS warning lamps are "ON" only during moist environmental changes (rain, snow, vehicle wash), all wheel speed sensor circuitry should be thoroughly inspected for signs of water intrusion. If DTC is not current, clear DTCs and simulate the effects of water intrusion. Use the following procedure. Spray the suspected area with a 5% salt water solution (two teaspoons of salt to 12 oz. of water). Test drive vehicle over various road surfaces (bumps, turns, etc.) above 24 km/h (15 mph) for at least 30 seconds. If DTC returns, replace suspected harness.

Any circuitry that is suspected of causing the intermittent complaint should be thoroughly checked for backed out terminals, improper mating, broken locks, improperly formed or damaged terminals, poor terminal to wiring connections or physical damage to the wiring harness.

Resistance of the wheel speed sensor will increase with an increase in sensor temperature.

When replacing a wheel speed sensor, inspect the sensor terminals and harness connector for corrosion and/or water intrusion. If evidence of corrosion or water intrusion exists, replace wheel speed sensor jumper harness. Likewise, if replacing a wheel speed sensor jumper harness, inspect sensor terminals. If evidence of corrosion or water intrusion exists, replace wheel speed sensor. Refer to "On-Vehicle Service" in this section.

DIAGNOSTIC CHART DTC 21 — 1995 SILHOUETTE AND TRANS SPORT WITH TCS, CONTINUED

DTC 21
LEFT FRONT WHEEL SPEED = 0

Important

- WHEEL SPEED SENSOR INTERMITTENT MALFUNCTIONS MAY BE DIFFICULT TO LOCATE. CARE SHOULD BE TAKEN NOT TO DISTURB ANY ELECTRICAL CONNECTIONS PRIOR TO AN INDICATED STEP OF THIS CHART. THIS WILL ENSURE THAT AN INTERMITTENT CONNECTION WILL NOT BE CORRECTED BEFORE THE SOURCE OF THE MALFUNCTION IS FOUND.

(1)
- IGNITION "OFF."
- PHYSICALLY INSPECT THE LEFT FRONT WHEEL SPEED SENSOR, LEFT FRONT WHEEL SPEED SENSOR JUMPER HARNESS, AND CONNECTORS FOR DAMAGE. IS PHYSICAL DAMAGE INDICATED?

YES: REPLACE WHEEL SPEED SENSOR AND/OR JUMPER HARNESS IF DAMAGED. DRIVE VEHICLE ABOVE 24 km/h (15 MPH) TO ENSURE DTC 21 DOES NOT SET.

NO:

(2)
- DISCONNECT LEFT FRONT WHEEL SPEED SENSOR DIRECTLY AT SENSOR.
- USING J 39200, MEASURE THE RESISTANCE BETWEEN SENSOR TERMINALS "A" AND "B."

IS RESISTANCE BETWEEN 1020 AND 1137Ω WHEN SENSOR IS 20°C (68°F)? *(IF TEMPERATURE OF THE SENSOR IS SUSPECTED TO BE OTHER THAN 20°C (68°F), REFER TO THE TABLE BELOW FOR APPLICABLE SENSOR RESISTANCE VALUES; THE VALUES IN THE TABLE ARE FOR TEMPERATURE OF THE SENSOR, NOT AIR TEMPERATURE.)*

NO: REPLACE WHEEL SPEED SENSOR.

YES:

(3)
- WITH J 39200 STILL CONNECTED, SELECT A/C VOLTAGE SCALE.
- SPIN WHEEL BY HAND WHILE OBSERVING VOLTAGE. IS VOLTAGE AT LEAST 100 mV (VOLTAGE WILL INCREASE AS WHEEL SPEED INCREASES)?

NO: REPLACE WHEEL SPEED SENSOR.

YES:

(4)
- IGNITION "OFF."
- DISCONNECT 32-WAY EBTCM CONNECTOR.
- USING J 39200, MEASURE THE RESISTANCE BETWEEN TERMINAL "E7" AND "E8" OF THE 32-WAY EBTCM HARNESS CONNECTOR. DOES J 39200 DISPLAY RESISTANCE AS "OL" (INFINITE)?

NO: REPAIR SHORT BETWEEN CKTs 830 AND 873.

YES:

(5)
- INSPECT CKTs 830 AND 873 AND 32-WAY EBTCM CONNECTOR FOR DAMAGE WHICH MAY RESULT IN A SHORT BETWEEN CKTs 830 AND 873 WITH ALL CONNECTORS CONNECTED. REPAIR ALL DAMAGE IF EVIDENT.
- RECONNECT ALL CONNECTORS.
- TEST DRIVE VEHICLE ABOVE 24 km/h (15 MPH) FOR AT LEAST 30 SECONDS. DOES DTC 21 SET AS A CURRENT DTC?

YES: REPLACE EBTCM.

NO: MALFUNCTION IS INTERMITTENT. REFER TO "DIAGNOSTIC AIDS" ON FACING PAGE.

FRONT WHEEL SPEED SENSOR
TEMPERATURE OF SENSOR VS. SENSOR RESISTANCE (APPROXIMATE)

TEMP. (°C)	TEMP. (°F)	RESISTANCE (OHMS)
-34 TO 4	-30 TO 40	804 TO 1066
5 TO 43	41 TO 110	956 TO 1240
44 TO 93	111 TO 200	1112 TO 1463

DIAGNOSTIC CHART DTC 22 — 1995 SILHOUETTE AND TRANS SPORT WITH TCS

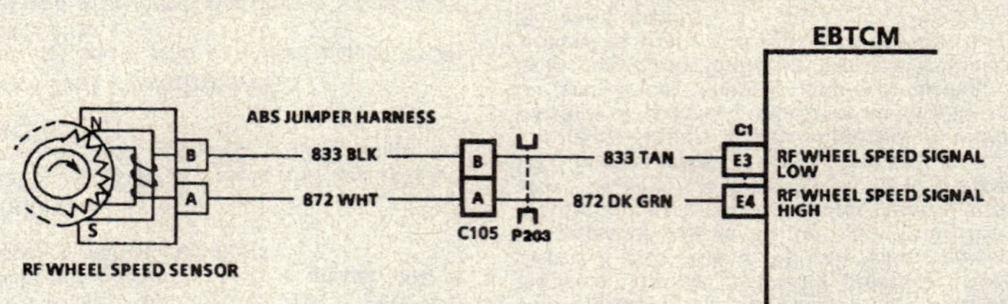

DTC 22
RIGHT FRONT WHEEL SPEED = 0

Circuit Description:

As a toothed ring passes by the wheel speed sensor, changes in the electromagnetic field cause the wheel speed sensor to produce a sinusoidal (AC) voltage signal whose frequency is proportional to wheel speed. The magnitude of this signal is directly related to wheel speed and the proximity of the wheel speed sensor to the toothed ring, often referred to as the air gap.

DTC Will Set When:

DTC 22 can be set when the vehicle is not in an ABS or TCS event. If the right front wheel speed = 0 and the vehicle's reference speed is greater than 8 km/h (5 mph), a malfunction exists.

Action Taken:

A malfunction DTC is stored, ABS and TCS are disabled and the ABS and TCS warning lamps are turned "ON."

DTC Chart Test Description:

Number(s) below refer to circled number(s) on the diagnostic chart.

1. This step is designed to identify a wheel speed sensor or circuitry that is damaged and visibly apparent.
2. This test checks to see that the wheel speed sensor is within proper resistance values.
3. Verifies that the wheel speed sensor can generate an appropriate output signal.
4. Ensures that wheel speed sensor circuitry is not shorted together.
5. Ensures malfunction was not due to physical damage of the circuitry.

Diagnostic Aids:

An intermittent malfunction may be caused by a poor connection, rubbed through wire insulation, or a wire that is broken inside the insulation.

The frequency of the malfunction can be checked by using the enhanced diagnostic function of the Tech 1, as described in "Tech 1 Diagnostics" in this section.

If the customer's comments reflect that the ABS and TCS warning lamps are "ON" only during moist environmental changes (rain, snow, vehicle wash), all wheel speed sensor circuitry should be thoroughly inspected for signs of water intrusion. If DTC is not current, clear DTCs and simulate the effects of water intrusion. Use the following procedure. Spray the suspected area with a 5% salt water solution (two teaspoons of salt to 12 oz. of water). Test drive vehicle over various road surfaces (bumps, turns, etc.) above 24 km/h (15 mph) for at least 30 seconds. If DTC returns, replace suspected harness.

Any circuitry that is suspected of causing the intermittent complaint should be thoroughly checked for backed out terminals, improper mating, broken locks, improperly formed or damaged terminals, poor terminal to wiring connections or physical damage to the wiring harness.

Resistance of the wheel speed sensor will increase with an increase in sensor temperature.

When replacing a wheel speed sensor, inspect the sensor terminals and harness connector for corrosion and/or water intrusion. If evidence of corrosion or water intrusion exists, replace wheel speed sensor jumper harness. Likewise, if replacing a wheel speed sensor jumper harness, inspect sensor terminals. If evidence of corrosion or water intrusion exists, replace wheel speed sensor.

DIAGNOSTIC CHART DTC 22 — 1995 SILHOUETTE AND TRANS SPORT WITH TCS, CONTINUED

DTC 22
RIGHT FRONT WHEEL SPEED = 0

Important

- WHEEL SPEED SENSOR INTERMITTENT MALFUNCTIONS MAY BE DIFFICULT TO LOCATE. CARE SHOULD BE TAKEN NOT TO DISTURB ANY ELECTRICAL CONNECTIONS PRIOR TO AN INDICATED STEP OF THIS CHART. THIS WILL ENSURE THAT AN INTERMITTENT CONNECTION WILL NOT BE CORRECTED BEFORE THE SOURCE OF THE MALFUNCTION IS FOUND.

(1)
- IGNITION "OFF."
- PHYSICALLY INSPECT THE RIGHT FRONT WHEEL SPEED SENSOR, RIGHT FRONT WHEEL SPEED SENSOR JUMPER HARNESS, AND CONNECTORS FOR DAMAGE.

IS PHYSICAL DAMAGE INDICATED?

YES: REPLACE WHEEL SPEED SENSOR AND/OR JUMPER HARNESS IF DAMAGED. DRIVE VEHICLE ABOVE 24 km/h (15 MPH) TO ENSURE DTC 22 DOES NOT SET.

NO:

(2)
- DISCONNECT RIGHT FRONT WHEEL SPEED SENSOR DIRECTLY AT SENSOR.
- USING J 39200, MEASURE THE RESISTANCE BETWEEN SENSOR TERMINALS "A" AND "B."

IS RESISTANCE BETWEEN 1020 AND 1137Ω WHEN SENSOR IS 20°C (68°F)? *(IF TEMPERATURE OF THE SENSOR IS SUSPECTED TO BE OTHER THAN 20°C (68°F), REFER TO THE TABLE BELOW FOR APPLICABLE SENSOR RESISTANCE VALUES; THE VALUES IN THE TABLE ARE FOR TEMPERATURE OF THE SENSOR, NOT AIR TEMPERATURE.)*

NO: REPLACE WHEEL SPEED SENSOR.

YES:

(3)
- WITH J 39200 STILL CONNECTED, SELECT A/C VOLTAGE SCALE.
- SPIN WHEEL BY HAND WHILE OBSERVING VOLTAGE.

IS VOLTAGE AT LEAST 100 mV (VOLTAGE WILL INCREASE AS WHEEL SPEED INCREASES)?

NO: REPLACE WHEEL SPEED SENSOR.

YES:

(4)
- IGNITION "OFF."
- DISCONNECT 32-WAY EBTCM CONNECTOR.
- USING J 39200, MEASURE THE RESISTANCE BETWEEN TERMINAL "E3" AND "E4" OF THE 32-WAY EBTCM HARNESS CONNECTOR.

DOES J 39200 DISPLAY RESISTANCE AS "OL" (INFINITE)?

NO: REPAIR SHORT BETWEEN CKTs 833 AND 872.

YES:

(5)
- INSPECT CKTs 833 AND 872 AND 32-WAY EBTCM CONNECTOR FOR DAMAGE WHICH MAY RESULT IN A SHORT BETWEEN CKTs 833 AND 872 WITH ALL CONNECTORS CONNECTED. REPAIR ALL DAMAGE IF EVIDENT.
- RECONNECT ALL CONNECTORS.
- TEST DRIVE VEHICLE ABOVE 24 km/h (15 MPH) FOR AT LEAST 30 SECONDS.

DOES DTC 22 SET AS A CURRENT DTC?

YES: REPLACE EBTCM.

NO: MALFUNCTION IS INTERMITTENT. REFER TO "DIAGNOSTIC AIDS" ON FACING PAGE.

FRONT WHEEL SPEED SENSOR
TEMPERATURE OF SENSOR VS. SENSOR RESISTANCE (APPROXIMATE)

TEMP. (°C)	TEMP. (°F)	RESISTANCE (OHMS)
-34 TO 4	-30 TO 40	804 TO 1066
5 TO 43	41 TO 110	956 TO 1240
44 TO 93	111 TO 200	1112 TO 1463

DIAGNOSTIC CHART DTC 23 — 1995 SILHOUETTE AND TRANS SPORT WITH TCS

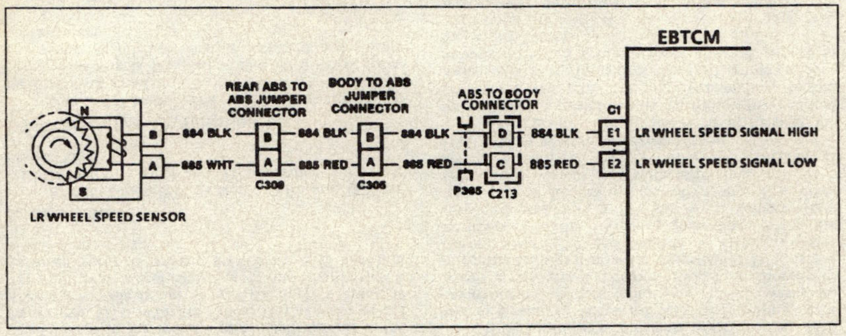

DTC 23
LEFT REAR WHEEL SPEED = 0

Circuit Description:

As a toothed ring passes by the wheel speed sensor, changes in the electromagnetic field cause the wheel speed sensor to produce a sinusoidal (AC) voltage signal whose frequency is proportional to wheel speed. The magnitude of this signal is directly related to wheel speed and the proximity of the wheel speed sensor to the toothed ring, often referred to as the air gap. DTC Will Set When: DTC 23 can be set when the vehicle is not in an ABS or TCS event. If the left rear wheel speed = 0 and the vehicle's reference speed is greater than 8 km/h (5 mph), a malfunction exists.

Action Taken:

A malfunction DTC is stored. ABS and TCS are disabled and the ABS and TCS warning lamps are turned "ON."

DTC Chart Test Description:

Number(s) below refer to circled number(s) on the diagnostic chart.

1. This step is designed to identify a wheel speed sensor or circuitry that is damaged and visibly apparent.
2. Checks to see that the wheel speed sensor is within proper resistance values.
3. Verifies that the wheel speed sensor can generate an appropriate output signal.
4. Ensures that the wheel speed sensor circuitry is not shorted together.
5. Checks to see if malfunction was due to physical damage of the circuitry.
6. Checks for an incorrect or malfunctioning hub and bearing assembly.

Diagnostic Aids:

An intermittent malfunction may be caused by a poor connection, rubbed through wire insulation, or a wire that is broken inside the insulation.

The frequency of the malfunction can be checked by using the enhanced diagnostic function of the Tech 1, as described in "Tech 1 Diagnostics" in this section. If the customer's comments reflect that the ABS and TCS warning lamps are "ON" only during moist environmental changes (rain, snow, vehicle wash), all wheel speed sensor circuitry should be thoroughly inspected for signs of water intrusion. If DTC is not current, clear DTCs and simulate the effects of water intrusion. Use the following procedure. Spray the suspected area with a 5% salt water solution (two teaspoons of salt to 12 oz. of water). Test drive vehicle over various road surfaces (bumps, turns, etc.) above 24 km/h (15 mph) for at least 30 seconds. If DTC returns, replace suspected harness.

Any circuitry that is suspected of causing the intermittent complaint should be thoroughly checked for backed out terminals, improper mating, broken locks, improperly formed or damaged terminals, poor terminal to wiring connections or physical damage to the wiring harness.

Resistance of the wheel speed sensor will increase with an increase in sensor temperature.

When replacing a wheel speed sensor, inspect the sensor terminals and harness connector for corrosion and/or water intrusion. If evidence of corrosion or water intrusion exists, replace wheel speed sensor jumper harness. Likewise, if replacing a wheel speed sensor jumper harness, inspect sensor terminals. If evidence of corrosion or water intrusion exists, replace wheel speed sensor.

DIAGNOSTIC CHART DTC 23 — 1995 SILHOUETTE AND TRANS SPORT WITH TCS, CONTINUED

DTC 23
LEFT REAR WHEEL SPEED = 0

Important
- WHEEL SPEED SENSOR INTERMITTENT MALFUNCTIONS MAY BE DIFFICULT TO LOCATE. CARE SHOULD BE TAKEN NOT TO DISTURB ANY ELECTRICAL CONNECTIONS PRIOR TO AN INDICATED STEP OF THIS CHART. THIS WILL ENSURE THAT AN INTERMITTENT CONNECTION WILL NOT BE CORRECTED BEFORE THE SOURCE OF THE MALFUNCTION IS FOUND.

(1)
- IGNITION "OFF."
- PHYSICALLY INSPECT THE LEFT REAR WHEEL SPEED SENSOR, WHEEL SPEED SENSOR JUMPER HARNESS, AND CONNECTORS FOR DAMAGE.

IS PHYSICAL DAMAGE INDICATED?

YES: REPLACE WHEEL SPEED SENSOR AND/OR WHEEL SPEED SENSOR JUMPER HARNESS IF DAMAGED. DRIVE VEHICLE ABOVE 24 km/h (15 MPH) TO ENSURE DTC 23 DOES NOT SET.

NO:

(2)
- DISCONNECT LEFT REAR WHEEL SPEED SENSOR DIRECTLY AT SENSOR.
- USING J 39200, MEASURE THE RESISTANCE BETWEEN SENSOR TERMINALS "A" AND "B."

IS RESISTANCE BETWEEN 2059 AND 2353Ω WHEN SENSOR IS 20°C (68°F)?

(IF TEMPERATURE OF THE SENSOR IS SUSPECTED TO BE OTHER THAN 20°C (68°F), REFER TO THE TABLE BELOW FOR APPLICABLE SENSOR RESISTANCE VALUES; THE VALUES IN THE TABLE ARE FOR TEMPERATURE OF THE SENSOR, NOT AIR TEMPERATURE.)

NO: REPLACE WHEEL SPEED SENSOR.

YES:

(3)
- WITH J 39200 STILL CONNECTED, SELECT A/C VOLTAGE SCALE.
- SPIN WHEEL BY HAND WHILE OBSERVING VOLTAGE.

IS VOLTAGE AT LEAST 100 mV (VOLTAGE WILL INCREASE AS WHEEL SPEED INCREASES)?

NO: REPLACE WHEEL SPEED SENSOR.

YES:

(4)
- IGNITION "OFF."
- DISCONNECT 32-WAY EBTCM CONNECTOR.
- USING J 39200, MEASURE THE RESISTANCE BETWEEN TERMINAL "E1" AND "E2" OF THE 32-WAY EBTCM HARNESS CONNECTOR.

DOES J 39200 DISPLAY RESISTANCE AS "OL" (INFINITE)?

NO: REPAIR SHORT BETWEEN CKTs 884 AND 885.

YES:

(5)
- INSPECT CKTs 884 AND 885 AND 32-WAY EBTCM CONNECTOR FOR DAMAGE WHICH MAY RESULT IN A SHORT BETWEEN CKTs 884 AND 885 WITH ALL CONNECTORS CONNECTED. REPAIR ALL DAMAGE IF EVIDENT.
- RECONNECT ALL CONNECTORS.
- TEST DRIVE VEHICLE ABOVE 24 km/h (15 MPH) FOR AT LEAST 30 SECONDS.

DOES DTC 23 SET AS A CURRENT DTC?

NO: MALFUNCTION WAS MOST LIKELY CAUSED BY A POOR CONNECTION. REFER TO "DIAGNOSTIC AIDS" ON FACING PAGE.

YES:

(6)
- USING TECH 1, SELECT "F0: DATA LIST."
- MONITOR WHEEL SPEEDS WHILE SLOWLY ACCELERATING TO 65 km/h (40 MPH) AND THEN SLOWLY DECELERATING TO 0 km/h (0 MPH).

IS THE LEFT REAR WHEEL SPEED CONSTANTLY HIGHER THAN THE THREE REMAINING WHEEL SPEEDS?

NO: REPLACE EBTCM.

YES: REPLACE INCORRECT OR MALFUNCTIONING LEFT WHEEL SPEED SENSOR/HUB AND BEARING ASSEMBLY WITH THE PROPER ASSEMBLY. REFER TO PARTS CATALOG FOR PROPER PART NUMBER.

REAR WHEEL SPEED SENSOR
TEMPERATURE OF SENSOR VS. SENSOR RESISTANCE (APPROXIMATE)

TEMP. (°C)	TEMP. (°F)	RESISTANCE (OHMS)
-34 TO 4	-30 TO 40	1622 TO 2205
5 TO 43	41 TO 110	1930 TO 2566
44 TO 93	111 TO 200	2245 TO 3028

DIAGNOSTIC CHART DTC 24 — 1995 SILHOUETTE AND TRANS SPORT WITH TCS

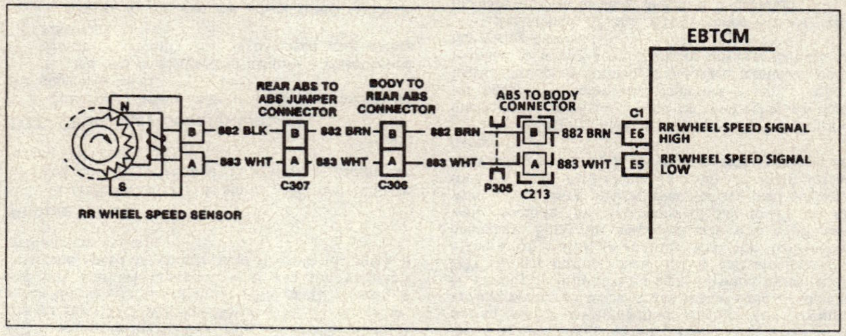

DTC 24
RIGHT REAR WHEEL SPEED = 0

Circuit Description:

As a toothed ring passes by the wheel speed sensor, changes in the electromagnetic field cause the wheel speed sensor to produce a sinusoidal (AC) voltage signal whose frequency is proportional to wheel speed. The magnitude of this signal is directly related to wheel speed and the proximity of the wheel speed sensor to the toothed ring, often referred to as the air gap. DTC Will Set When: DTC 24 can be set when the vehicle is not in an ABS or TCS event. If the right rear wheel speed = 0 and the vehicle's reference speed is greater than 8 km/h (5 mph), a malfunction exists.

Action Taken:

A malfunction DTC is stored, ABS and TCS are disabled and the ABS and TCS warning lamps are turned "ON."

DTC Chart Test Description:

Number(s) below refer to circled number(s) on the diagnostic chart.

1. This step is designed to identify a wheel speed sensor or circuitry that is damaged and visibly apparent.
2. Checks to see that the wheel speed sensor is within proper resistance values.
3. Verifies that the wheel speed sensor can generate an appropriate output signal.
4. Ensures that the wheel speed sensor circuitry is not shorted together.
5. Checks to see if malfunction was due to physical damage of the circuitry.
6. Checks for an incorrect or malfunctioning hub and bearing assembly.

Diagnostic Aids:

An intermittent malfunction may be caused by a poor connection, rubbed through wire insulation, or a wire that is broken inside the insulation.

The frequency of the malfunction can be checked by using the enhanced diagnostic function of the Tech 1, as described in "Tech 1 Diagnostics" in this section.

If the customer's comments reflect that the ABS and TCS warning lamps are "ON" only during moist environmental changes (rain, snow, vehicle wash), all wheel speed sensor circuitry should be thoroughly inspected for signs of water intrusion. If DTC is not current, clear DTCs and simulate the effects of water intrusion. Use the following procedure. Spray the suspected area with a 5% salt water solution (two teaspoons of salt to 12 oz. of water). Test drive vehicle over various road surfaces (bumps, turns, etc.) above 24 km/h (15 mph) for at least 30 seconds. If DTC returns, replace suspected harness.

Any circuitry that is suspected of causing the intermittent complaint should be thoroughly checked for backed out terminals, improper mating, broken locks, improperly formed or damaged terminals, poor terminal to wiring connections or physical damage to the wiring harness.

Resistance of the wheel speed sensor will increase with an increase in sensor temperature.

When replacing a wheel speed sensor, inspect the sensor terminals and harness connector for corrosion and/or water intrusion. If evidence of corrosion or water intrusion exists, replace wheel speed sensor jumper harness. Likewise, if replacing a wheel speed sensor jumper harness, inspect sensor terminals. If evidence of corrosion or water intrusion exists, replace wheel speed sensor.

DIAGNOSTIC CHART DTC 24 — 1995 SILHOUETTE AND TRANS SPORT WITH TCS, CONTINUED

DTC 24
RIGHT REAR WHEEL SPEED = 0

Important

- WHEEL SPEED SENSOR INTERMITTENT MALFUNCTIONS MAY BE DIFFICULT TO LOCATE. CARE SHOULD BE TAKEN NOT TO DISTURB ANY ELECTRICAL CONNECTIONS PRIOR TO AN INDICATED STEP OF THIS CHART. THIS WILL ENSURE THAT AN INTERMITTENT CONNECTION WILL NOT BE CORRECTED BEFORE THE SOURCE OF THE MALFUNCTION IS FOUND.

(1)
- IGNITION "OFF."
- PHYSICALLY INSPECT THE RIGHT REAR WHEEL SPEED SENSOR, WHEEL SPEED SENSOR JUMPER HARNESS, AND CONNECTORS FOR DAMAGE.

IS PHYSICAL DAMAGE INDICATED?

YES: REPLACE WHEEL SPEED SENSOR AND/OR WHEEL SPEED SENSOR JUMPER HARNESS IF DAMAGED. DRIVE VEHICLE ABOVE 24 km/h (15 MPH) TO ENSURE DTC 24 DOES NOT SET.

NO:

(2)
- DISCONNECT RIGHT REAR WHEEL SPEED SENSOR DIRECTLY AT SENSOR.
- USING J 39200, MEASURE THE RESISTANCE BETWEEN SENSOR TERMINALS "A" AND "B." IS RESISTANCE BETWEEN 2059 AND 2353Ω WHEN SENSOR IS 20°C (68°F)?

(IF TEMPERATURE OF THE SENSOR IS SUSPECTED TO BE OTHER THAN 20°C (68°F), REFER TO THE TABLE BELOW FOR APPLICABLE SENSOR RESISTANCE VALUES; THE VALUES IN THE TABLE ARE FOR TEMPERATURE OF THE SENSOR, NOT AIR TEMPERATURE.)

NO: REPLACE WHEEL SPEED SENSOR.

YES:

(3)
- WITH J 39200 STILL CONNECTED, SELECT A/C VOLTAGE SCALE.
- SPIN WHEEL BY HAND WHILE OBSERVING VOLTAGE.

IS VOLTAGE AT LEAST 100 mV (VOLTAGE WILL INCREASE AS WHEEL SPEED INCREASES)?

NO: REPLACE WHEEL SPEED SENSOR.

YES:

(4)
- IGNITION "OFF."
- DISCONNECT 32-WAY EBTCM CONNECTOR.
- USING J 39200, MEASURE THE RESISTANCE BETWEEN TERMINAL "E5" AND "E6" OF THE 32-WAY EBTCM HARNESS CONNECTOR.

DOES J 39200 DISPLAY RESISTANCE AS "OL" (INFINITE)?

NO: REPAIR SHORT BETWEEN CKTs 883 AND 882.

YES:

(5)
- INSPECT CKTs 882 AND 883 AND 32-WAY EBTCM CONNECTOR FOR DAMAGE WHICH MAY RESULT IN A SHORT BETWEEN CKTs 882 AND 883 WITH ALL CONNECTORS CONNECTED. REPAIR ALL DAMAGE IF EVIDENT.
- RECONNECT ALL CONNECTORS.
- TEST DRIVE VEHICLE ABOVE 24 km/h (15 MPH) FOR AT LEAST 30 SECONDS.

DOES DTC 24 SET AS A CURRENT DTC?

NO: MALFUNCTION WAS MOST LIKELY CAUSED BY A POOR CONNECTION. REFER TO "DIAGNOSTIC AIDS" ON FACING PAGE.

YES:

(6)
- USING TECH 1, SELECT "F0: DATA LIST."
- MONITOR WHEEL SPEEDS WHILE SLOWLY ACCELERATING TO 65 km/h (40 MPH) AND THEN SLOWLY DECELERATING TO 0 km/h (0 MPH).

IS THE RIGHT REAR WHEEL SPEED CONSTANTLY HIGHER THAN THE THREE REMAINING WHEEL SPEEDS?

NO: REPLACE EBTCM.

YES: REPLACE INCORRECT OR MALFUNCTIONING RIGHT WHEEL SPEED SENSOR/HUB AND BEARING ASSEMBLY WITH THE PROPER ASSEMBLY. REFER TO PARTS CATALOG FOR PROPER PART NUMBER.

REAR WHEEL SPEED SENSOR
TEMPERATURE OF SENSOR VS. SENSOR RESISTANCE (APPROXIMATE)

TEMP. (°C)	TEMP. (°F)	RESISTANCE (OHMS)
-34 TO 4	-30 TO 40	1622 TO 2205
5 TO 43	41 TO 110	1930 TO 2566
44 TO 93	111 TO 200	2245 TO 3028

DIAGNOSTIC CHART DTC 25 — 1995 SILHOUETTE AND TRANS SPORT WITH TCS

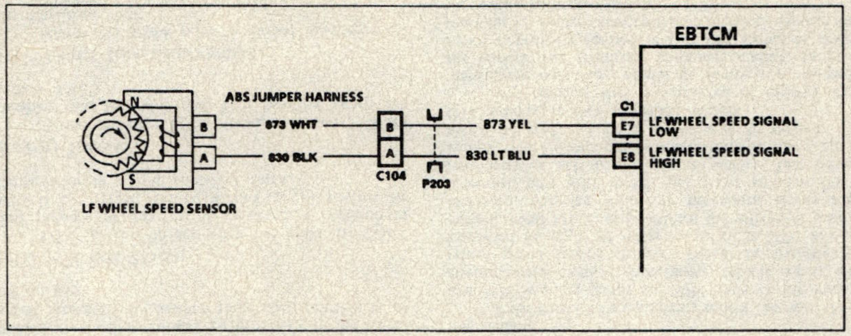

DTC 25
LEFT FRONT EXCESSIVE WHEEL SPEED VARIATION

Circuit Description:

As a toothed ring passes by the wheel speed sensor, changes in the electromagnetic field cause the wheel speed sensor to produce a sinusoidal (AC) voltage signal whose frequency is proportional to wheel speed. The magnitude of this signal is directly related to wheel speed and the proximity of the wheel speed sensor to the toothed ring, often referred to as the air gap.

DTC Will Set When:

DTC 25 can be set when the brake is "OFF." The purpose of this test is to detect a situation in which the left front wheel acceleration or deceleration is beyond specified limits.

Action Taken:

A malfunction DTC is stored, ABS and TCS are disabled and the ABS and TCS warning lamps are turned "ON."

DTC Chart Test Description:

Number(s) below refer to circled number(s) on the diagnostic chart.

1. This step is designed to identify a wheel speed sensor or circuitry that is damaged and visibly apparent.
2. Checks to see that the wheel speed sensor is within proper resistance values.
3. Verifies that the wheel speed sensor can generate an appropriate output signal.
4. Ensures that wheel speed sensor circuitry is not shorted together.
5. Ensures malfunction was not due to physical damage of the circuitry.

Diagnostic Aids:

An intermittent malfunction may be caused by a poor connection, rubbed through wire insulation, or a wire that is broken inside the insulation.

The frequency of the malfunction can be checked by using the enhanced diagnostic function of the Tech 1, as described in "Tech 1 Diagnostics" in this section.

If the customer's comments reflect that the ABS and TCS warning lamps are "ON" only during moist environmental changes (rain, snow, vehicle wash), all wheel speed sensor circuitry should be thoroughly inspected for signs of water intrusion. If DTC is not current, clear DTCs and simulate the effects of water intrusion. Use the following procedure. Spray the suspected area with a 5% salt water solution (two teaspoons of salt to 12 oz. of water). Test drive vehicle over various road surfaces (bumps, turns, etc.) above 24 km/h (15 mph) for at least 30 seconds. If DTC returns, replace suspected harness.

Any circuitry that is suspected of causing the intermittent complaint should be thoroughly checked for backed out terminals, improper mating, broken locks, improperly formed or damaged terminals, poor terminal to wiring connections or physical damage to the wiring harness.

Resistance of the wheel speed sensor will increase with an increase in sensor temperature.

When replacing a wheel speed sensor, inspect the sensor terminals and harness connector for corrosion and/or water intrusion. If evidence of corrosion or water intrusion exists, replace wheel speed sensor jumper harness. Likewise, if replacing a wheel speed sensor jumper harness, inspect sensor terminals. If evidence of corrosion or water intrusion exists, replace wheel speed sensor. Refer to "On-Vehicle Service" in this section.

DIAGNOSTIC CHART DTC 25 — 1995 SILHOUETTE AND TRANS SPORT WITH TCS, CONTINUED

DTC 25
LEFT FRONT EXCESSIVE WHEEL SPEED VARIATION

Important

- WHEEL SPEED SENSOR INTERMITTENT MALFUNCTIONS MAY BE DIFFICULT TO LOCATE. CARE SHOULD BE TAKEN NOT TO DISTURB ANY ELECTRICAL CONNECTIONS PRIOR TO AN INDICATED STEP OF THIS CHART. THIS WILL ENSURE THAT AN INTERMITTENT CONNECTION WILL NOT BE CORRECTED BEFORE THE SOURCE OF THE MALFUNCTION IS FOUND.

(1)
- IGNITION "OFF."
- PHYSICALLY INSPECT THE LEFT FRONT WHEEL SPEED SENSOR, LEFT FRONT WHEEL SPEED SENSOR JUMPER HARNESS, AND CONNECTORS FOR DAMAGE. IS PHYSICAL DAMAGE INDICATED?

YES: REPLACE WHEEL SPEED SENSOR AND/OR JUMPER HARNESS IF DAMAGED. DRIVE VEHICLE ABOVE 24 km/h (15 MPH) TO ENSURE DTC 25 DOES NOT SET.

NO:

(2)
- DISCONNECT LEFT FRONT WHEEL SPEED SENSOR DIRECTLY AT SENSOR.
- USING J 39200, MEASURE THE RESISTANCE BETWEEN SENSOR TERMINALS "A" AND "B."

IS RESISTANCE BETWEEN 1020 AND 1137Ω WHEN SENSOR IS 20°C (68°F)? *(IF TEMPERATURE OF THE SENSOR IS SUSPECTED TO BE OTHER THAN 20°C (68°F), REFER TO THE TABLE BELOW FOR APPLICABLE SENSOR RESISTANCE VALUES; THE VALUES IN THE TABLE ARE FOR TEMPERATURE OF THE SENSOR, NOT AIR TEMPERATURE.)*

NO: REPLACE WHEEL SPEED SENSOR.

YES:

(3)
- WITH J 39200 STILL CONNECTED, SELECT A/C VOLTAGE SCALE.
- SPIN WHEEL BY HAND WHILE OBSERVING VOLTAGE.

IS VOLTAGE AT LEAST 100 mV (VOLTAGE WILL INCREASE AS WHEEL SPEED INCREASES)?

NO: REPLACE WHEEL SPEED SENSOR.

YES:

(4)
- IGNITION "OFF."
- DISCONNECT 32-WAY EBTCM CONNECTOR.
- USING J 39200, MEASURE THE RESISTANCE BETWEEN TERMINAL "E7" AND "E8" OF THE 32-WAY EBTCM HARNESS CONNECTOR.

DOES J 39200 DISPLAY RESISTANCE AS "OL" (INFINITE)?

NO: REPAIR SHORT BETWEEN CKTs 830 AND 873.

YES:

(5)
- INSPECT CKTs 830 AND 873 AND 32-WAY EBTCM CONNECTOR FOR DAMAGE WHICH MAY RESULT IN A SHORT BETWEEN CKTs 830 AND 873 WITH ALL CONNECTORS CONNECTED. REPAIR ALL DAMAGE IF EVIDENT.
- RECONNECT ALL CONNECTORS.
- TEST DRIVE VEHICLE ABOVE 24 km/h (15 MPH) FOR AT LEAST 30 SECONDS.

DOES DTC 25 SET AS A CURRENT DTC?

YES: REPLACE EBTCM.

NO: MALFUNCTION IS INTERMITTENT. REFER TO "DIAGNOSTIC AIDS" ON FACING PAGE.

FRONT WHEEL SPEED SENSOR
TEMPERATURE OF SENSOR VS. SENSOR RESISTANCE (APPROXIMATE)

TEMP. (°C)	TEMP. (°F)	RESISTANCE (OHMS)
-34 TO 4	-30 TO 40	804 TO 1066
5 TO 43	41 TO 110	956 TO 1240
44 TO 93	111 TO 200	1112 TO 1463

DIAGNOSTIC CHART DTC 26 — 1995 SILHOUETTE AND TRANS SPORT WITH TCS

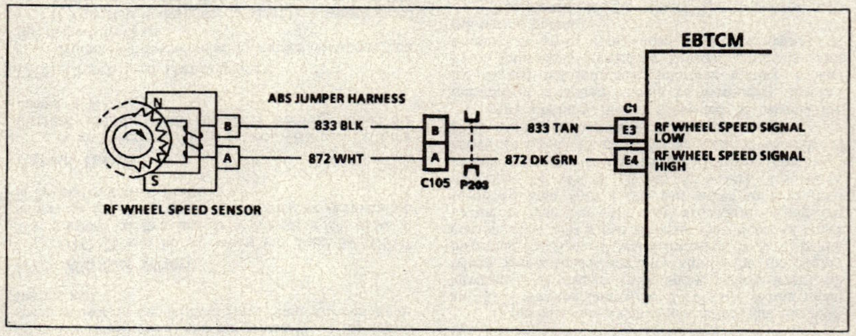

DTC 26

RIGHT FRONT EXCESSIVE WHEEL SPEED VARIATION

Circuit Description:

As a toothed ring passes by the wheel speed sensor, changes in the electromagnetic field cause the wheel speed sensor to produce a sinusoidal (AC) voltage signal whose frequency is proportional to wheel speed. The magnitude of this signal is directly related to wheel speed and the proximity of the wheel speed sensor to the toothed ring, often referred to as the air gap.

DTC Will Set When:

DTC 26 can be set when the brake is "OFF." The purpose of this test is to detect a situation in which the right front wheel acceleration or deceleration is beyond specified limits.

Action Taken:

A malfunction DTC is stored, ABS and TCS are disabled and the ABS and TCS warning lamps are turned "ON."

DTC Chart Test Description:

Number(s) below refer to circled number(s) on the diagnostic chart.

1. This step is designed to identify a wheel speed sensor or circuitry that is damaged and visibly apparent.
2. Checks to see that the wheel speed sensor is within proper resistance values.
3. Verifies that the wheel speed sensor can generate an appropriate output signal.
4. Ensures that wheel speed sensor circuitry is not shorted together.
5. Ensures malfunction was not due to physical damage of the circuitry.

Diagnostic Aids:

An intermittent malfunction may be caused by a poor connection, rubbed through wire insulation, or a wire that is broken inside the insulation.

The frequency of the malfunction can be checked by using the enhanced diagnostic function of the Tech 1, as described in "Tech 1 Diagnostics" in this section.

If the customer's comments reflect that the ABS and TCS warning lamps are "ON" only during moist environmental changes (rain, snow, vehicle wash), all wheel speed sensor circuitry should be thoroughly inspected for signs of water intrusion. If DTC is not current, clear DTCs and simulate the effects of water intrusion. Use the following procedure. Spray the suspected area with a 5% salt water solution (two teaspoons of salt to 12 oz. of water). Test drive vehicle over various road surfaces (bumps, turns, etc.) above 24 km/h (15 mph) for at least 30 seconds. If DTC returns, replace suspected harness.

Any circuitry that is suspected of causing the intermittent complaint should be thoroughly checked for backed out terminals, improper mating, broken locks, improperly formed or damaged terminals, poor terminal to wiring connections or physical damage to the wiring harness.

Resistance of the wheel speed sensor will increase with an increase in sensor temperature.

When replacing a wheel speed sensor, inspect the sensor terminals and harness connector for corrosion and/or water intrusion. If evidence of corrosion or water intrusion exists, replace wheel speed sensor jumper harness. Likewise, if replacing a wheel speed sensor jumper harness, inspect sensor terminals. If evidence of corrosion or water intrusion exists, replace wheel speed sensor.

DIAGNOSTIC CHART DTC 26 — 1995 SILHOUETTE AND TRANS SPORT WITH TCS, CONTINUED

DTC 26
RIGHT FRONT EXCESSIVE WHEEL SPEED VARIATION

Important

- WHEEL SPEED SENSOR INTERMITTENT MALFUNCTIONS MAY BE DIFFICULT TO LOCATE. CARE SHOULD BE TAKEN NOT TO DISTURB ANY ELECTRICAL CONNECTIONS PRIOR TO AN INDICATED STEP OF THIS CHART. THIS WILL ENSURE THAT AN INTERMITTENT CONNECTION WILL NOT BE CORRECTED BEFORE THE SOURCE OF THE MALFUNCTION IS FOUND.

(1)
- IGNITION "OFF."
- PHYSICALLY INSPECT THE RIGHT FRONT WHEEL SPEED SENSOR, RIGHT FRONT WHEEL SPEED SENSOR JUMPER HARNESS, AND CONNECTORS FOR DAMAGE. IS PHYSICAL DAMAGE INDICATED?

YES → REPLACE WHEEL SPEED SENSOR AND/OR JUMPER HARNESS IF DAMAGED. DRIVE VEHICLE ABOVE 24 km/h (15 MPH) TO ENSURE DTC 26 DOES NOT SET.

NO ↓

(2)
- DISCONNECT RIGHT FRONT WHEEL SPEED SENSOR DIRECTLY AT SENSOR.
- USING J 39200, MEASURE THE RESISTANCE BETWEEN SENSOR TERMINALS "A" AND "B."

IS RESISTANCE BETWEEN 1020 AND 1137Ω WHEN SENSOR IS 20°C (68°F)? *(IF TEMPERATURE OF THE SENSOR IS SUSPECTED TO BE OTHER THAN 20°C (68°F), REFER TO THE TABLE BELOW FOR APPLICABLE SENSOR RESISTANCE VALUES; THE VALUES IN THE TABLE ARE FOR TEMPERATURE OF THE SENSOR, NOT AIR TEMPERATURE.)*

NO → REPLACE WHEEL SPEED SENSOR.

YES ↓

(3)
- WITH J 39200 STILL CONNECTED, SELECT A/C VOLTAGE SCALE.
- SPIN WHEEL BY HAND WHILE OBSERVING VOLTAGE.

IS VOLTAGE AT LEAST 100 mV (VOLTAGE WILL INCREASE AS WHEEL SPEED INCREASES)?

NO → REPLACE WHEEL SPEED SENSOR.

YES ↓

(4)
- IGNITION "OFF."
- DISCONNECT 32-WAY EBTCM CONNECTOR.
- USING J 39200, MEASURE THE RESISTANCE BETWEEN TERMINAL "E3" AND "E4" OF THE 32-WAY EBTCM HARNESS CONNECTOR.

DOES J 39200 DISPLAY RESISTANCE AS "OL" (INFINITE)?

NO → REPAIR SHORT BETWEEN CKTs 833 AND 872.

YES ↓

(5)
- INSPECT CKTs 833 AND 872 AND 32-WAY EBTCM CONNECTOR FOR DAMAGE WHICH MAY RESULT IN A SHORT BETWEEN CKTs 833 AND 872 WITH ALL CONNECTORS CONNECTED. REPAIR ALL DAMAGE IF EVIDENT.
- RECONNECT ALL CONNECTORS.
- TEST DRIVE VEHICLE ABOVE 24 km/h (15 MPH) FOR AT LEAST 30 SECONDS.

DOES DTC 26 SET AS A CURRENT DTC?

YES → REPLACE EBTCM.

NO → MALFUNCTION IS INTERMITTENT. REFER TO "DIAGNOSTIC AIDS" ON FACING PAGE.

FRONT WHEEL SPEED SENSOR
TEMPERATURE OF SENSOR VS. SENSOR RESISTANCE (APPROXIMATE)

TEMP. (°C)	TEMP. (°F)	RESISTANCE (OHMS)
-34 TO 4	-30 TO 40	804 TO 1066
5 TO 43	41 TO 110	956 TO 1240
44 TO 93	111 TO 200	1112 TO 1463

DIAGNOSTIC CHART DTC 27 — 1995 SILHOUETTE AND TRANS SPORT WITH TCS

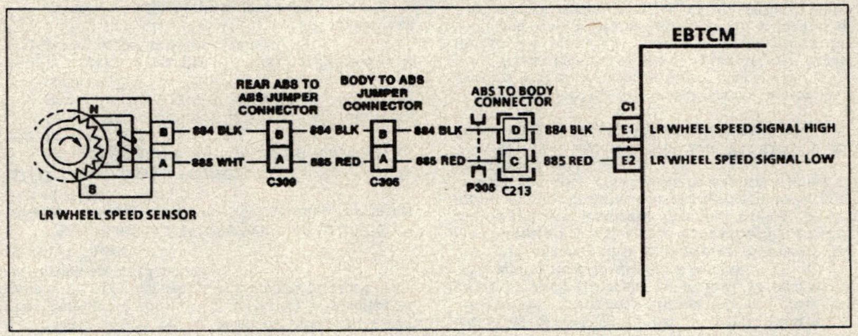

DTC 27
LEFT REAR EXCESSIVE WHEEL SPEED VARIATION

Circuit Description:

As a toothed ring passes by the wheel speed sensor, changes in the electromagnetic field cause the wheel speed sensor to produce a sinusoidal (AC) voltage signal whose frequency is proportional to wheel speed. The magnitude of this signal is directly related to wheel speed and the proximity of the wheel speed sensor to the toothed ring, often referred to as the air gap. DTC Will Set When: DTC 27 can be set when the brake is "OFF." The purpose of this test is to detect a situation in which the left rear wheel acceleration or deceleration is beyond specified limits.

Action Taken:

A malfunction DTC is stored, ABS and TCS are disabled and the ABS and TCS warning lamps are turned "ON."

DTC Chart Test Description:

Number(s) below refer to circled number(s) on the diagnostic chart.

1. This step is designed to identify a wheel speed sensor or circuitry that is damaged and visibly apparent.
2. Checks to see that the wheel speed sensor is within proper resistance values.
3. Verifies that the wheel speed sensor can generate an appropriate output signal.
4. Ensures that wheel speed sensor circuitry is not shorted together.
5. Ensures malfunction was not due to physical damage of the circuitry.

Diagnostic Aids:

An intermittent malfunction may be caused by a poor connection, rubbed through wire insulation, or a wire that is broken inside the insulation.

The frequency of the malfunction can be checked by using the enhanced diagnostic function of the Tech 1, as described in "Tech 1 Diagnostics" in this section.

If the customer's comments reflect that the ABS and TCS warning lamps are "ON" only during moist environmental changes (rain, snow, vehicle wash), all wheel speed sensor circuitry should be thoroughly inspected for signs of water intrusion. If DTC is not current, clear DTCs and simulate the effects of water intrusion. Use the following procedure. Spray the suspected area with a 5% salt water solution (two teaspoons of salt to 12 oz. of water). Test drive vehicle over various road surfaces (bumps, turns, etc.) above 24 km/h (15 mph) for at least 30 seconds. If DTC returns, replace suspected harness.

Any circuitry that is suspected of causing the intermittent complaint should be thoroughly checked for backed out terminals, improper mating, broken locks, improperly formed or damaged terminals, poor terminal to wiring connections or physical damage to the wiring harness.

Excessive end play can cause this DTC to set. Check integral bearing for excessive axial end play, refer to SECTION 3.

Resistance of the wheel speed sensor will increase with an increase in sensor temperature.

When replacing a wheel speed sensor, inspect the sensor terminals and harness connector for corrosion and/or water intrusion. If evidence of corrosion or water intrusion exists, replace wheel speed sensor jumper harness. Likewise, if replacing a wheel speed sensor jumper harness, inspect sensor terminals. If evidence of corrosion or water intrusion exists, replace wheel speed sensor.

DIAGNOSTIC CHART DTC 27 — 1995 SILHOUETTE AND TRANS SPORT WITH TCS, CONTINUED

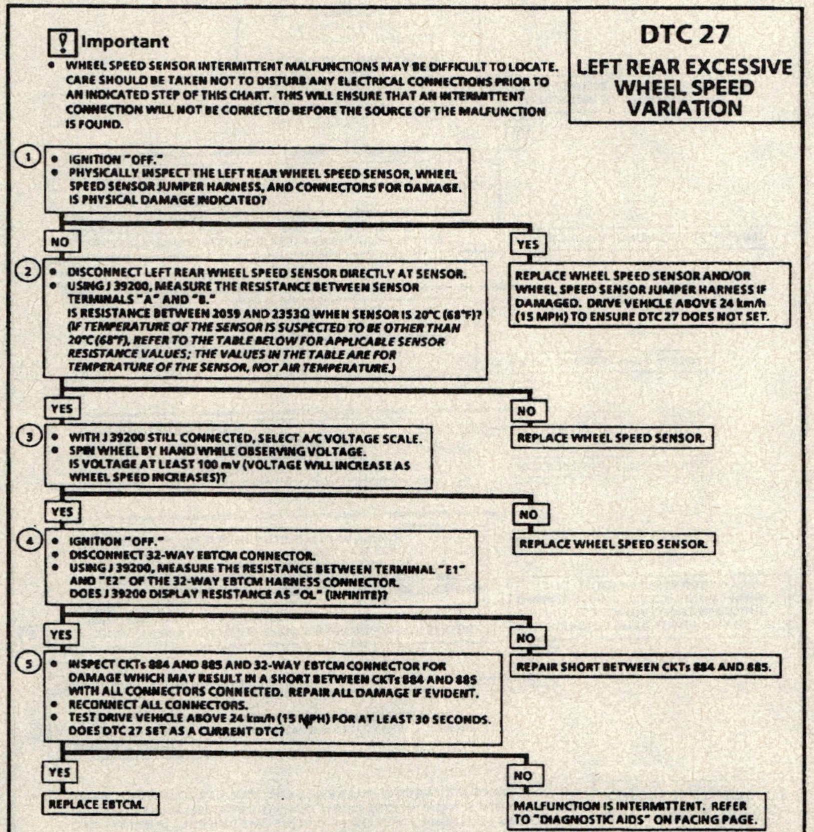

REAR WHEEL SPEED SENSOR
TEMPERATURE OF SENSOR VS. SENSOR RESISTANCE (APPROXIMATE)

TEMP. (°C)	TEMP. (°F)	RESISTANCE (OHMS)
-34 TO 4	-30 TO 40	1622 TO 2205
5 TO 43	41 TO 110	1930 TO 2566
44 TO 93	111 TO 200	2245 TO 3028

DIAGNOSTIC CHART DTC 28 — 1995 SILHOUETTE AND TRANS SPORT WITH TCS

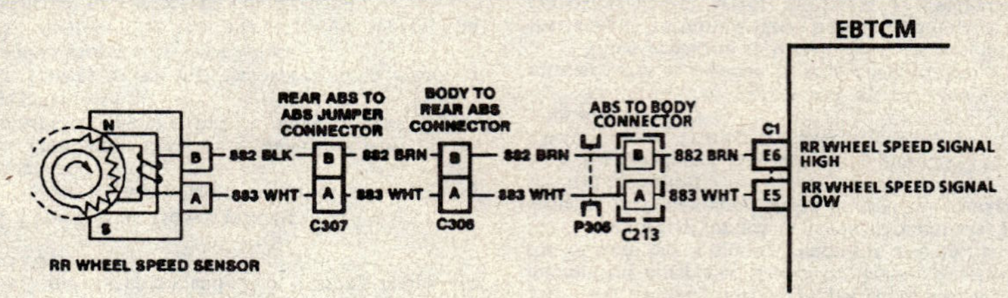

DTC 28
RIGHT REAR EXCESSIVE WHEEL SPEED VARIATION

Circuit Description:

As a toothed ring passes by the wheel speed sensor, changes in the electromagnetic field cause the wheel speed sensor to produce a sinusoidal (AC) voltage signal whose frequency is proportional to wheel speed. The magnitude of this signal is directly related to wheel speed and the proximity of the wheel speed sensor to the toothed ring, often referred to as the air gap.

DTC Will Set When:

DTC 28 can be set when the brake is "OFF." The purpose of this test is to detect a situation in which the right rear wheel acceleration or deceleration is beyond specified limits.

Action Taken:

A malfunction DTC is stored, ABS and TCS are disabled and the ABS and TCS warning lamps are turned "ON."

DTC Chart Test Description:

Number(s) below refer to circled number(s) on the diagnostic chart.

1. This step is designed to identify a wheel speed sensor or circuitry that is damaged and visibly apparent.
2. Checks to see that the wheel speed sensor is within proper resistance values.
3. Verifies that the wheel speed sensor can generate an appropriate output signal.
4. Ensures that wheel speed sensor circuitry is not shorted together.
5. Ensures malfunction was not due to physical damage of the circuitry.

Diagnostic Aids:

An intermittent malfunction may be caused by a poor connection, rubbed through wire insulation, or a wire that is broken inside the insulation

The frequency of the malfunction can be checked by using the enhanced diagnostic function of the Tech 1.

If the customer's comments reflect that the ABS and TCS warning lamps are "ON" only during moist environmental changes (rain, snow, vehicle wash), all wheel speed sensor circuitry should be thoroughly inspected for signs of water intrusion. If DTC is not current, clear DTCs and simulate the effects of water intrusion. Use the following procedure. Spray the suspected area with a 5% salt water solution (two teaspoons of salt to 12 oz. of water). Test drive vehicle over various road surfaces (bumps, turns, etc.) above 24 km/h (15 mph) for at least 30 seconds. If DTC returns, replace suspected harness.

Any circuitry that is suspected of causing the intermittent complaint should be thoroughly checked for backed out terminals, improper mating, broken locks, improperly formed or damaged terminals, poor terminal to wiring connections or physical damage to the wiring harness.

Excessive end play can cause this DTC to set. Check integral bearing for excessive axial end play.

Resistance of the wheel speed sensor will increase with an increase in sensor temperature.

When replacing a wheel speed sensor, inspect the sensor terminals and harness connector for corrosion and/or water intrusion. If evidence of corrosion or water intrusion exists, replace wheel speed sensor jumper harness. Likewise, if replacing a wheel speed sensor jumper harness, inspect sensor terminals. If evidence of corrosion or water intrusion exists, replace wheel speed sensor.

DIAGNOSTIC CHART DTC 28 — 1995 SILHOUETTE AND TRANS SPORT WITH TCS, CONTINUED

DTC 28
RIGHT REAR EXCESSIVE WHEEL SPEED VARIATION

Important
- WHEEL SPEED SENSOR INTERMITTENT MALFUNCTIONS MAY BE DIFFICULT TO LOCATE. CARE SHOULD BE TAKEN NOT TO DISTURB ANY ELECTRICAL CONNECTIONS PRIOR TO AN INDICATED STEP OF THIS CHART. THIS WILL ENSURE THAT AN INTERMITTENT CONNECTION WILL NOT BE CORRECTED BEFORE THE SOURCE OF THE MALFUNCTION IS FOUND.

(1)
- IGNITION "OFF."
- PHYSICALLY INSPECT THE RIGHT REAR WHEEL SPEED SENSOR, WHEEL SPEED SENSOR JUMPER HARNESS, AND CONNECTORS FOR DAMAGE. IS PHYSICAL DAMAGE INDICATED?

YES: REPLACE WHEEL SPEED SENSOR AND/OR WHEEL SPEED SENSOR JUMPER HARNESS IF DAMAGED. DRIVE VEHICLE ABOVE 24 km/h (15 MPH) TO ENSURE DTC 28 DOES NOT SET.

NO:

(2)
- DISCONNECT RIGHT REAR WHEEL SPEED SENSOR DIRECTLY AT SENSOR.
- USING J 39200, MEASURE THE RESISTANCE BETWEEN SENSOR TERMINALS "A" AND "B."
IS RESISTANCE BETWEEN 2059 AND 2353Ω WHEN SENSOR IS 20°C (68°F)? *(IF TEMPERATURE OF THE SENSOR IS SUSPECTED TO BE OTHER THAN 20°C (68°F), REFER TO THE TABLE BELOW FOR APPLICABLE SENSOR RESISTANCE VALUES; THE VALUES IN THE TABLE ARE FOR TEMPERATURE OF THE SENSOR, NOT AIR TEMPERATURE.)*

NO: REPLACE WHEEL SPEED SENSOR.

YES:

(3)
- WITH J 39200 STILL CONNECTED, SELECT A/C VOLTAGE SCALE.
- SPIN WHEEL BY HAND WHILE OBSERVING VOLTAGE.
IS VOLTAGE AT LEAST 100 mV (VOLTAGE WILL INCREASE AS WHEEL SPEED INCREASES)?

NO: REPLACE WHEEL SPEED SENSOR.

YES:

(4)
- IGNITION "OFF."
- DISCONNECT 32-WAY EBTCM CONNECTOR.
- USING J 39200, MEASURE THE RESISTANCE BETWEEN TERMINAL "E5" AND "E6" OF THE 32-WAY EBTCM HARNESS CONNECTOR.
DOES J 39200 DISPLAY RESISTANCE AS "OL" (INFINITE)?

NO: REPAIR SHORT BETWEEN CKTs 882 AND 883.

YES:

(5)
- INSPECT CKTs 882 AND 883 AND 32-WAY EBTCM CONNECTOR FOR DAMAGE WHICH MAY RESULT IN A SHORT BETWEEN CKTs 882 AND 883 WITH ALL CONNECTORS CONNECTED. REPAIR ALL DAMAGE IF EVIDENT.
- RECONNECT ALL CONNECTORS.
- TEST DRIVE VEHICLE ABOVE 24 km/h (15 MPH) FOR AT LEAST 30 SECONDS.
DOES DTC 28 SET AS A CURRENT DTC?

YES: REPLACE EBTCM.

NO: MALFUNCTION IS INTERMITTENT. REFER TO "DIAGNOSTIC AIDS" ON FACING PAGE.

REAR WHEEL SPEED SENSOR
TEMPERATURE OF SENSOR VS. SENSOR RESISTANCE (APPROXIMATE)

TEMP. (°C)	TEMP. (°F)	RESISTANCE (OHMS)
-34 TO 4	-30 TO 40	1622 TO 2205
5 TO 43	41 TO 110	1930 TO 2566
44 TO 93	111 TO 200	2245 TO 3028

DIAGNOSTIC CHART DTC 32 — 1995 SILHOUETTE AND TRANS SPORT WITH TCS

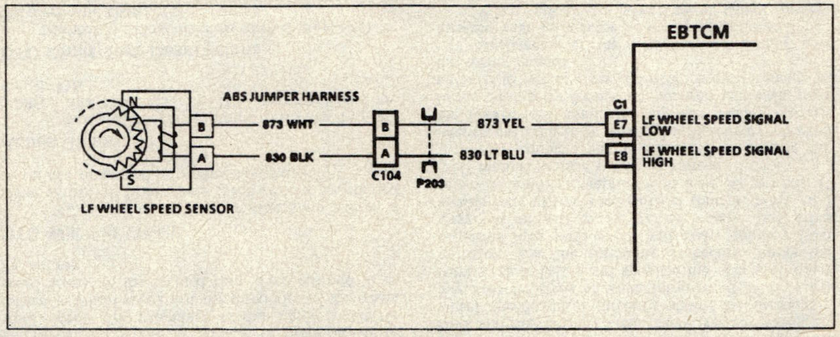

DTC 32
(Page 1 of 3)
LF WHEEL SPEED SNR CKT OPEN OR SHORT TO GND/BAT

Circuit Description:

As a toothed ring passes by the wheel speed sensor, changes in the electromagnetic field cause the wheel speed sensor to produce a sinusoidal (AC) voltage signal whose frequency is proportional to wheel speed. The magnitude of this signal is directly related to wheel speed and the proximity of the wheel speed sensor to the toothed ring, often referred to as the air gap.

DTC Will Set When:

DTC 32 can be set anytime after initialization. If either of the left front wheel speed circuits are shorted to voltage or ground, a malfunction exists.

Action Taken:

A malfunction DTC is stored, ABS and TCS are disabled and the ABS and TCS warning lamps are turned "ON."

DTC Chart Test Description:

Number(s) below refer to circled number(s) on the diagnostic chart.

1. Checks to see that the wheel speed sensor is within proper resistance values.
2. Checks for short to ground in wheel speed sensor wiring.
3. Checks for short to voltage in wheel speed sensor wiring.
4. Checks for a short to voltage in CKT 873.
5. Checks for a short to voltage in CKT 830.
6. Ensures malfunction was not due to physical damage of circuitry.

Diagnostic Aids:

An intermittent malfunction may be caused by a poor connection, rubbed through wire insulation, or a wire that is broken inside the insulation.

The frequency of the malfunction can be checked by using the enhanced diagnostic function of the Tech 1, as described in "Tech 1 Diagnostics" in this section.

If the customer's comments reflect that the ABS and TCS warning lamps are "ON" only during moist environmental changes (rain, snow, vehicle wash), all wheel speed sensor circuitry should be thoroughly inspected for signs of water intrusion. If DTC is not current, clear DTCs and simulate the effects of water intrusion. Use the following procedure. Spray the suspected area with a 5% salt water solution (two teaspoons of salt to 12 oz. of water). Test drive vehicle over various road surfaces (bumps, turns, etc.) above 24 km/h (15 mph) for at least 30 seconds. If DTC returns, replace suspected harness.

Any circuitry that is suspected of causing the intermittent complaint should be thoroughly checked for backed out terminals, improper mating, broken locks, improperly formed or damaged terminals, poor terminal to wiring connections or physical damage to the wiring harness.

Resistance of the wheel speed sensor will increase with an increase in sensor temperature.

When replacing a wheel speed sensor, inspect the sensor terminals and harness connector for corrosion and/or water intrusion. If evidence of corrosion or water intrusion exists, replace wheel speed sensor jumper harness. Likewise, if replacing a wheel speed sensor jumper harness, inspect sensor terminals. If evidence of corrosion or water intrusion exists, replace wheel speed sensor.

DIAGNOSTIC CHART DTC 32 — 1995 SILHOUETTE AND TRANS SPORT WITH TCS, CONTINUED

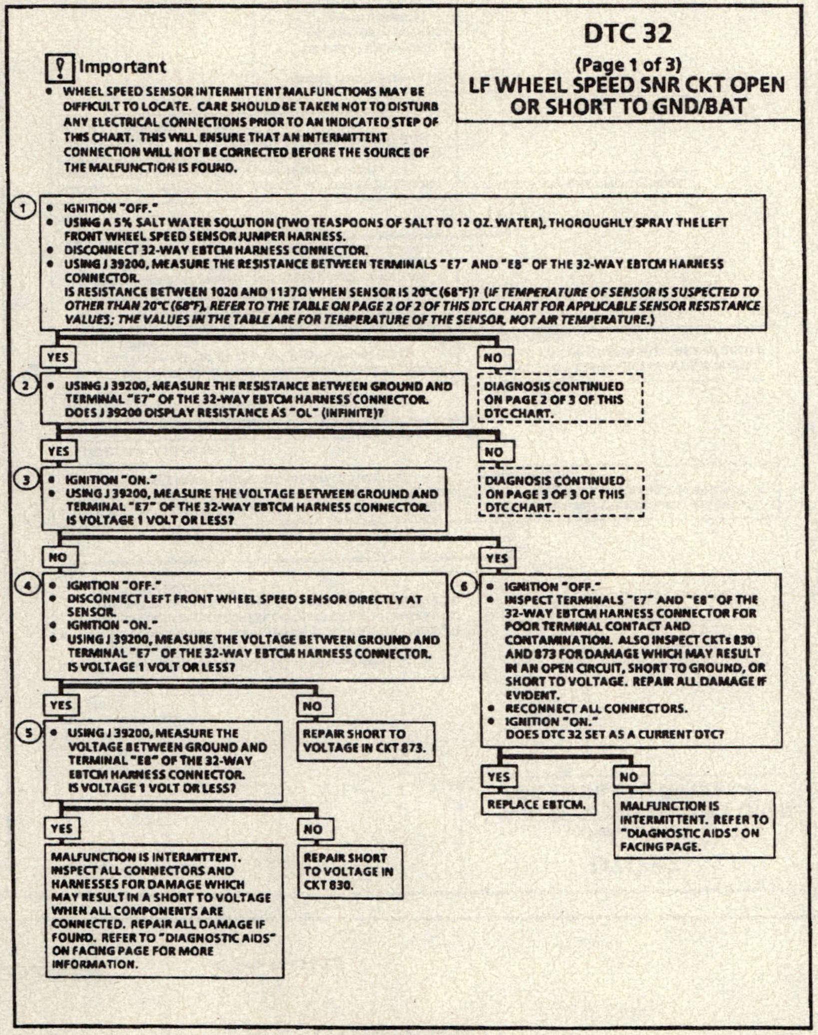

DIAGNOSTIC CHART DTC 32 — 1995 SILHOUETTE AND TRANS SPORT WITH TCS, CONTINUED

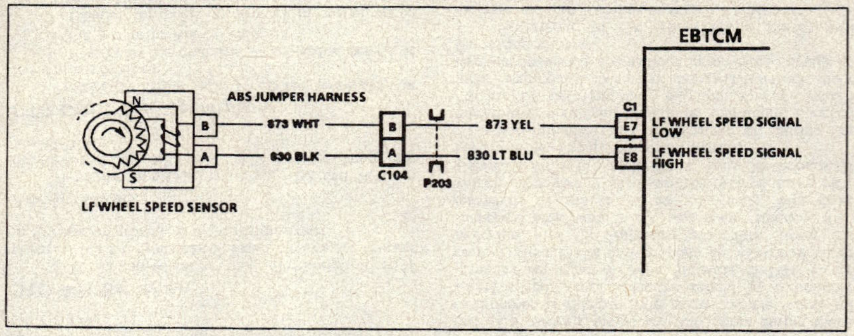

DTC 32
(Page 2 of 3)
LF WHEEL SPEED SNR CKT OPEN OR SHORT TO GND/BAT

Circuit Description:

As a toothed ring passes by the wheel speed sensor, changes in the electromagnetic field cause the wheel speed sensor to produce a sinusoidal (AC) voltage signal whose frequency is proportional to wheel speed. The magnitude of this signal is directly related to wheel speed and the proximity of the wheel speed sensor to the toothed ring, often referred to as the air gap.

DTC Will Set When:

DTC 32 can be set anytime after initialization. If either of the left front wheel speed circuits are shorted to voltage or ground, a malfunction exists.

Action Taken:

A malfunction DTC is stored, ABS and TCS are disabled and the ABS and TCS warning lamps are turned "ON."

DTC Chart Test Description:

Number(s) below refer to circled number(s) on the diagnostic chart.

7. Checks for high resistance in CKT 873.
8. Checks for high resistance in CKT 830.
9. Verifies proper wheel speed sensor resistance.

Diagnostic Aids:

An intermittent malfunction may be caused by a poor connection, rubbed through wire insulation, or a wire that is broken inside the insulation.

The frequency of the malfunction can be checked by using the enhanced diagnostic function of the Tech 1, as described in "Tech 1 Diagnostics" in this section.

If the customer's comments reflect that the ABS and TCS warning lamps are "ON" only during moist environmental changes (rain, snow, vehicle wash), all wheel speed sensor circuitry should be thoroughly inspected for signs of water intrusion. If DTC is not current, clear DTCs and simulate the effects of water intrusion. Use the following procedure. Spray the suspected area with a 5% salt water solution (two teaspoons of salt to 12 oz. of water). Test drive vehicle over various road surfaces (bumps, turns, etc.) above 24 km/h (15 mph) for at least 30 seconds. If DTC returns, replace suspected harness.

Any circuitry that is suspected of causing the intermittent complaint should be thoroughly checked for backed out terminals, improper mating, broken locks, improperly formed or damaged terminals, poor terminal to wiring connections or physical damage to the wiring harness.

Resistance of the wheel speed sensor will increase with an increase in sensor temperature.

When replacing a wheel speed sensor, inspect the sensor terminals and harness connector for corrosion and/or water intrusion. If evidence of corrosion or water intrusion exists, replace wheel speed sensor jumper harness. Likewise, if replacing a wheel speed sensor jumper harness, inspect sensor terminals. If evidence of corrosion or water intrusion exists, replace wheel speed sensor.

DIAGNOSTIC CHART DTC 32 — 1995 SILHOUETTE AND TRANS SPORT WITH TCS, CONTINUED

DTC 32
(Page 2 of 3)
LF WHEEL SPEED SNR CKT OPEN OR SHORT TO GND/BAT

DIAGNOSIS CONTINUED FROM PAGE 1 OF 3 OF THIS DTC CHART.

(7)
- DISCONNECT LEFT FRONT WHEEL SPEED SENSOR DIRECTLY AT SENSOR.
- USING J 39200, MEASURE THE RESISTANCE BETWEEN TERMINAL "E7" OF THE 32-WAY EBTCM HARNESS CONNECTOR AND TERMINAL "B" OF THE LEFT FRONT WHEEL SPEED SENSOR HARNESS CONNECTOR.

IS RESISTANCE 2Ω OR LESS?

NO: REPAIR HIGH RESISTANCE IN CKT 873 (WHEEL SPEED SENSOR JUMPER MUST BE REPLACED IF DAMAGED).

YES:

(8)
- USING J 39200, MEASURE THE RESISTANCE BETWEEN TERMINAL "E8" OF THE 32-WAY EBTCM HARNESS CONNECTOR AND TERMINAL "A" OF THE LEFT FRONT WHEEL SPEED SENSOR HARNESS CONNECTOR.

IS RESISTANCE 2Ω OR LESS?

NO: REPAIR HIGH RESISTANCE IN CKT 830 (WHEEL SPEED SENSOR JUMPER MUST BE REPLACED IF DAMAGED).

YES:

(9)
- USING J 39200, MEASURE THE RESISTANCE BETWEEN TERMINAL "A" AND TERMINAL "B" OF THE LEFT FRONT WHEEL SPEED SENSOR CONNECTOR.

IS RESISTANCE BETWEEN 1020 AND 1137Ω WHEN SENSOR IS 20°C (68°F)? *(IF TEMPERATURE OF SENSOR IS SUSPECTED TO OTHER THAN 20°C (68°F), REFER TO THE TABLE BELOW FOR APPLICABLE SENSOR RESISTANCE VALUES; THE VALUES IN THE TABLE ARE FOR TEMPERATURE OF THE SENSOR, NOT AIR TEMPERATURE.)*

NO: REPLACE WHEEL SPEED SENSOR.

YES: MALFUNCTION IS INTERMITTENT. INSPECT ALL CONNECTORS AND HARNESSES FOR DAMAGE WHICH MAY RESULT IN HIGH RESISTANCE WHEN ALL COMPONENTS ARE CONNECTED. REFER TO "DIAGNOSTIC AIDS" ON FACING PAGE FOR MORE INFORMATION. REPAIR ALL DAMAGE IF FOUND.

FRONT WHEEL SPEED SENSOR

TEMPERATURE OF SENSOR VS. SENSOR RESISTANCE (APPROXIMATE)

TEMP. (°C)	TEMP. (°F)	RESISTANCE (OHMS)
-34 TO 4	-30 TO 40	804 TO 1066
5 TO 43	41 TO 110	956 TO 1240
44 TO 93	111 TO 200	1112 TO 1463

DIAGNOSTIC CHART DTC 32 — 1995 SILHOUETTE AND TRANS SPORT WITH TCS, CONTINUED

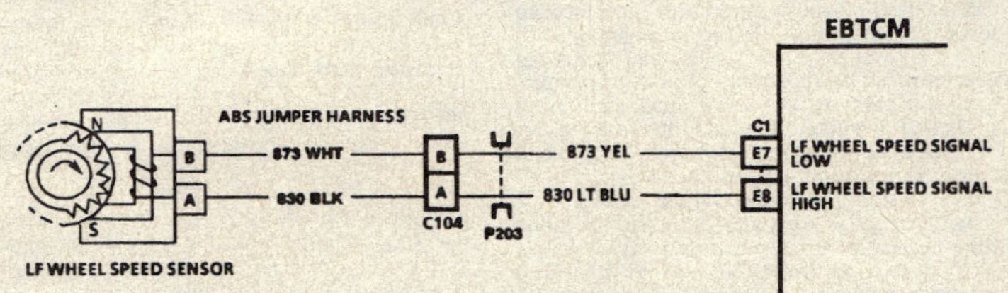

DTC 32
(Page 3 of 3)
LF WHEEL SPEED SNR CKT OPEN OR SHORT TO GND/BAT

Circuit Description:

As a toothed ring passes by the wheel speed sensor, changes in the electromagnetic field cause the wheel speed sensor to produce a sinusoidal (AC) voltage signal whose frequency is proportional to wheel speed. The magnitude of this signal is directly related to wheel speed and the proximity of the wheel speed sensor to the toothed ring, often referred to as the air gap.

DTC Will Set When:

DTC 32 can be set anytime after initialization. If either of the left front wheel speed circuits are shorted to voltage or ground, a malfunction exists.

Action Taken:

A malfunction DTC is stored, ABS and TCS are disabled and the ABS and TCS warning lamps are turned "ON."

DTC Chart Test Description:

Number(s) below refer to circled number(s) on the diagnostic chart.

10. Checks for short to ground in CKT 873.
11. Checks for short to ground in CKT 830.
12. Checks for short to ground in wheel speed sensor jumper harness.
13. Verifies that wheel speed sensor is not internally shorted to ground.

Diagnostic Aids:

An intermittent malfunction may be caused by a poor connection, rubbed through wire insulation, or a wire that is broken inside the insulation.

The frequency of the malfunction can be checked by using the enhanced diagnostic function of the Tech 1, as described in "Tech 1 Diagnostics" in this section.

If the customer's comments reflect that the ABS and TCS warning lamps are "ON" only during moist environmental changes (rain, snow, vehicle wash), all wheel speed sensor circuitry should be thoroughly inspected for signs of water intrusion. If DTC is not current, clear DTCs and simulate the effects of water intrusion. Use the following procedure. Spray the suspected area with a 5% salt water solution (two teaspoons of salt to 12 oz. of water). Test drive vehicle over various road surfaces (bumps, turns, etc.) above 24 km/h (15 mph) for at least 30 seconds. If DTC returns, replace suspected harness.

Any circuitry that is suspected of causing the intermittent complaint should be thoroughly checked for backed out terminals, improper mating, broken locks, improperly formed or damaged terminals, poor terminal to wiring connections or physical damage to the wiring harness.

Resistance of the wheel speed sensor will increase with an increase in sensor temperature.

When replacing a wheel speed sensor, inspect the sensor terminals and harness connector for corrosion and/or water intrusion. If evidence of corrosion or water intrusion exists, replace wheel speed sensor jumper harness. Likewise, if replacing a wheel speed sensor jumper harness, inspect sensor terminals. If evidence of corrosion or water intrusion exists, replace wheel speed sensor.

DIAGNOSTIC CHART DTC 32 — 1995 SILHOUETTE AND TRANS SPORT WITH TCS, CONTINUED

DTC 32
(Page 3 of 3)
LF WHEEL SPEED SNR CKT OPEN OR SHORT TO GND/BAT

DIAGNOSIS CONTINUED FROM PAGE 1 OF 3 OF THIS DTC CHART.

(10)
- DISCONNECT CONNECTOR C104 (MOUNTED ON LH SHOCK TOWER).
- USING J 39200, MEASURE THE RESISTANCE BETWEEN GROUND AND TERMINAL "E7" OF THE 32-WAY EBTCM HARNESS CONNECTOR.

DOES J 39200 DISPLAY RESISTANCE AS "OL" (INFINITE)?

- NO: REPAIR SHORT TO GROUND IN CKT 873.
- YES:

(11)
- USING J 39200, MEASURE THE RESISTANCE BETWEEN GROUND AND TERMINAL "E8" OF THE 32-WAY EBTCM HARNESS CONNECTOR.

DOES J 39200 DISPLAY RESISTANCE AS "OL" (INFINITE)?

- NO: REPAIR SHORT TO GROUND IN CKT 830.
- YES:

(12)
- THOROUGHLY RESPRAY THE LEFT FRONT WHEEL SPEED SENSOR JUMPER HARNESS WITH THE SALT WATER SOLUTION.

IMPORTANT: DO NOT ALLOW WATER TO ENTER ANY WHEEL SPEED SENSOR CONNECTOR.

- DISCONNECT LEFT FRONT WHEEL SPEED SENSOR DIRECTLY AT SENSOR.
- USING J 39200, MEASURE THE RESISTANCE BETWEEN GROUND AND TERMINAL "A," THEN "B" OF THE LEFT FRONT WHEEL SPEED SENSOR HARNESS CONNECTOR.

DOES J 39200 DISPLAY RESISTANCE AS "OL" (INFINITE) FOR BOTH MEASUREMENTS?

- NO: REPLACE LEFT FRONT WHEEL SPEED SENSOR JUMPER HARNESS.
- YES:

(13)
- USING J 39200, MEASURE THE RESISTANCE BETWEEN GROUND AND TERMINAL "A" OF THE LEFT FRONT WHEEL SPEED SENSOR CONNECTOR.

DOES J 39200 DISPLAY RESISTANCE AS "OL" (INFINITE)?

- NO: REPLACE WHEEL SPEED SENSOR.
- YES: MALFUNCTION IS INTERMITTENT. INSPECT ALL CONNECTORS AND HARNESSES FOR DAMAGE WHICH MAY RESULT IN A SHORT TO GROUND WHEN ALL COMPONENTS ARE CONNECTED. PAY PARTICULAR ATTENTION TO THE JUMPER HARNESS, SMALL NICKS OR CUTS CAN CAUSE A MALFUNCTION. REPAIR ANY DAMAGE IF FOUND. JUMPER HARNESS MUST BE REPLACED IF DAMAGED. REFER TO "DIAGNOSTIC AIDS" ON FACING PAGE FOR MORE INFORMATION.

DIAGNOSTIC CHART DTC 33 — 1995 SILHOUETTE AND TRANS SPORT WITH TCS

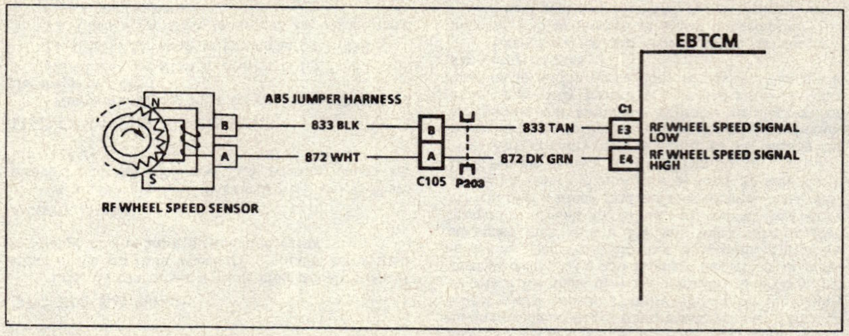

DTC 33
(Page 1 of 3)
RF WHEEL SPEED SNR CKT OPEN OR SHORT TO GND/BAT

Circuit Description:

As a toothed ring passes by the wheel speed sensor, changes in the electromagnetic field cause the wheel speed sensor to produce a sinusoidal (AC) voltage signal whose frequency is proportional to wheel speed. The magnitude of this signal is directly related to wheel speed and the proximity of the wheel speed sensor to the toothed ring, often referred to as the air gap.

DTC Will Set When:

DTC 33 can be set anytime after initialization. If either of the right front wheel speed circuits are shorted to voltage or ground, a malfunction exists.

Action Taken:

A malfunction DTC is stored, ABS and TCS are disabled and the ABS and TCS warning lamps are turned "ON."

DTC Chart Test Description:

Number(s) below refer to circled number(s) on the diagnostic chart.

1. Checks to see that the wheel speed sensor is within proper resistance values.
2. Checks for short to ground in wheel speed sensor wiring.
3. Checks for short to voltage in wheel speed sensor wiring.
4. Checks for a short to voltage in CKT 833.
5. Checks for a short to voltage in CKT 872.
6. Ensures malfunction was not due to physical damage of circuitry.

Diagnostic Aids:

An intermittent malfunction may be caused by a poor connection, rubbed through wire insulation, or a wire that is broken inside the insulation.

The frequency of the malfunction can be checked by using the enhanced diagnostic function of the Tech 1, as described in "Tech 1 Diagnostics" in this section.

If the customer's comments reflect that the ABS and TCS warning lamps are "ON" only during moist environmental changes (rain, snow, vehicle wash), all wheel speed sensor circuitry should be thoroughly inspected for signs of water intrusion. If DTC is not current, clear DTCs and simulate the effects of water intrusion. Use the following procedure. Spray the suspected area with a 5% salt water solution (two teaspoons of salt to 12 oz. of water). Test drive vehicle over various road surfaces (bumps, turns, etc.) above 24 km/h (15 mph) for at least 30 seconds. If DTC returns, replace suspected harness.

Any circuitry that is suspected of causing the intermittent complaint should be thoroughly checked for backed out terminals, improper mating, broken locks, improperly formed or damaged terminals, poor terminal to wiring connections or physical damage to the wiring harness.

Resistance of the wheel speed sensor will increase with an increase in sensor temperature.

When replacing a wheel speed sensor, inspect the sensor terminals and harness connector for corrosion and/or water intrusion. If evidence of corrosion or water intrusion exists, replace wheel speed sensor jumper harness. Likewise, if replacing a wheel speed sensor jumper harness, inspect sensor terminals. If evidence of corrosion or water intrusion exists, replace wheel speed sensor.

DIAGNOSTIC CHART DTC 33 — 1995 SILHOUETTE AND TRANS SPORT WITH TCS, CONTINUED

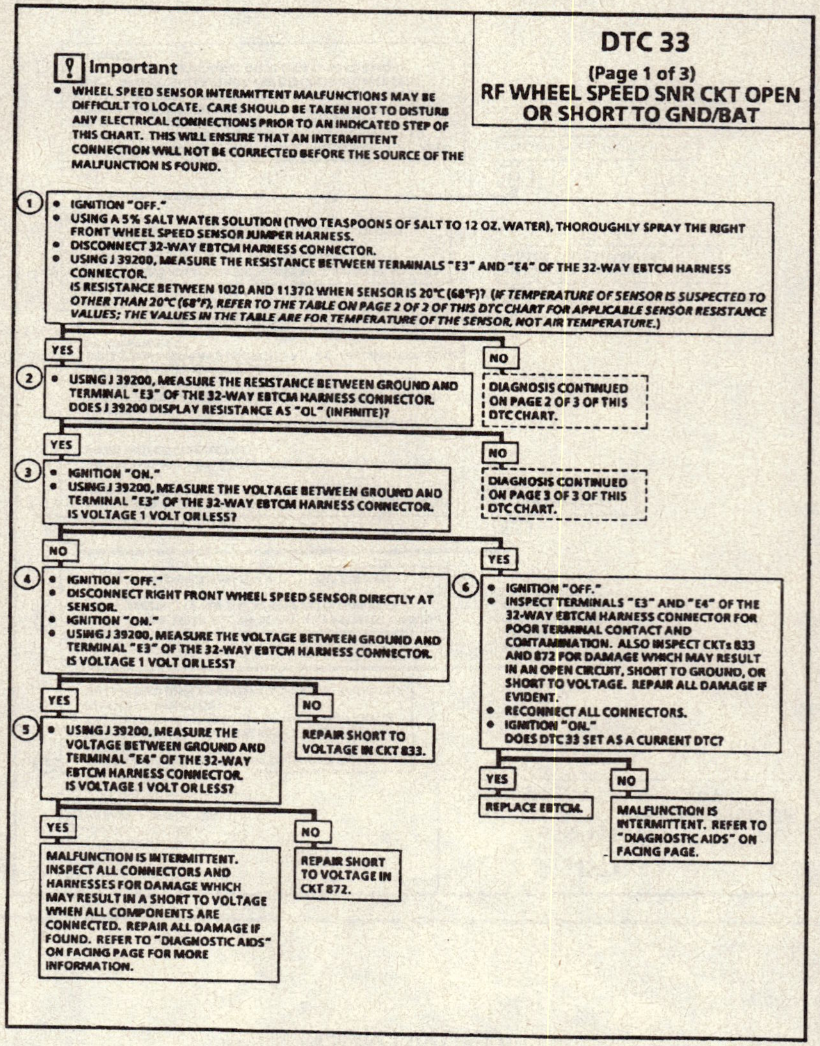

DIAGNOSTIC CHART DTC 33 — 1995 SILHOUETTE AND TRANS SPORT WITH TCS, CONTINUED

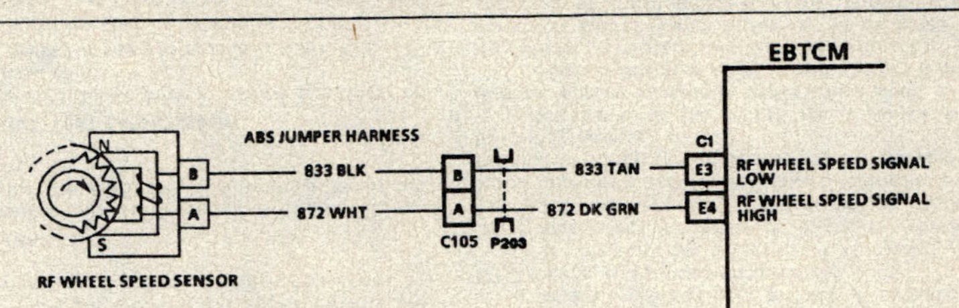

DTC 33
(Page 2 of 3)
RF WHEEL SPEED SNR CKT OPEN OR SHORT TO GND/BAT

Circuit Description:

As a toothed ring passes by the wheel speed sensor, changes in the electromagnetic field cause the wheel speed sensor to produce a sinusoidal (AC) voltage signal whose frequency is proportional to wheel speed. The magnitude of this signal is directly related to wheel speed and the proximity of the wheel speed sensor to the toothed ring, often referred to as the air gap.

DTC Will Set When:

DTC 33 can be set anytime after initialization. If either of the right front wheel speed circuits are shorted to voltage or ground, a malfunction exists.

Action Taken:

A malfunction DTC is stored, ABS and TCS are disabled and the ABS and TCS warning lamps are turned "ON."

DTC Chart Test Description:

Number(s) below refer to circled number(s) on the diagnostic chart.

7. Checks for high resistance in CKT 833.
8. Checks for high resistance in CKT 872.
9. Verifies proper wheel speed sensor resistance.

Diagnostic Aids:

An intermittent malfunction may be caused by a poor connection, rubbed through wire insulation, or a wire that is broken inside the insulation.

The frequency of the malfunction can be checked by using the enhanced diagnostic function of the Tech 1, as described in "Tech 1 Diagnostics" in this section.

If the customer's comments reflect that the ABS and TCS warning lamps are "ON" only during moist environmental changes (rain, snow, vehicle wash), all wheel speed sensor circuitry should be thoroughly inspected for signs of water intrusion. If DTC is not current, clear DTCs and simulate the effects of water intrusion. Use the following procedure. Spray the suspected area with a 5% salt water solution (two teaspoons of salt to 12 oz. of water). Test drive vehicle over various road surfaces (bumps, turns, etc.) above 24 km/h (15 mph) for at least 30 seconds. If DTC returns, replace suspected harness.

Any circuitry that is suspected of causing the intermittent complaint should be thoroughly checked for backed out terminals, improper mating, broken locks, improperly formed or damaged terminals, poor terminal to wiring connections or physical damage to the wiring harness.

Resistance of the wheel speed sensor will increase with an increase in sensor temperature.

When replacing a wheel speed sensor, inspect the sensor terminals and harness connector for corrosion and/or water intrusion. If evidence of corrosion or water intrusion exists, replace wheel speed sensor jumper harness. Likewise, if replacing a wheel speed sensor jumper harness, inspect sensor terminals. If evidence of corrosion or water intrusion exists, replace wheel speed sensor.

DIAGNOSTIC CHART DTC 33 — 1995 SILHOUETTE AND TRANS SPORT WITH TCS, CONTINUED

DTC 33
(Page 2 of 3)
RF WHEEL SPEED SNR CKT OPEN OR SHORT TO GND/BAT

DIAGNOSIS CONTINUED FROM PAGE 1 OF 3 OF THIS DTC CHART.

(7)
- DISCONNECT RIGHT FRONT WHEEL SPEED SENSOR DIRECTLY AT SENSOR.
- USING J 39200, MEASURE THE RESISTANCE BETWEEN TERMINAL "E3" OF THE 32-WAY EBTCM HARNESS CONNECTOR AND TERMINAL "B" OF THE RIGHT FRONT WHEEL SPEED SENSOR HARNESS CONNECTOR. IS RESISTANCE 2Ω OR LESS?

NO: REPAIR HIGH RESISTANCE IN CKT 833 (WHEEL SPEED SENSOR JUMPER MUST BE REPLACED IF DAMAGED).

YES:

(8)
- USING J 39200, MEASURE THE RESISTANCE BETWEEN TERMINAL "E4" OF THE 32-WAY EBTCM HARNESS CONNECTOR AND TERMINAL "A" OF THE RIGHT FRONT WHEEL SPEED SENSOR HARNESS CONNECTOR. IS RESISTANCE 2Ω OR LESS?

NO: REPAIR HIGH RESISTANCE IN CKT 872 (WHEEL SPEED SENSOR JUMPER MUST BE REPLACED IF DAMAGED).

YES:

(9)
- USING J 39200, MEASURE THE RESISTANCE BETWEEN TERMINAL "A" AND TERMINAL "B" OF THE RIGHT FRONT WHEEL SPEED SENSOR CONNECTOR. IS RESISTANCE BETWEEN 1020 AND 1137Ω WHEN SENSOR IS 20°C (68°F)? *(IF TEMPERATURE OF SENSOR IS SUSPECTED TO OTHER THAN 20°C (68°F), REFER TO THE TABLE BELOW FOR APPLICABLE SENSOR RESISTANCE VALUES; THE VALUES IN THE TABLE ARE FOR TEMPERATURE OF THE SENSOR, NOT AIR TEMPERATURE.)*

YES: MALFUNCTION IS INTERMITTENT. INSPECT ALL CONNECTORS AND HARNESSES FOR DAMAGE WHICH MAY RESULT IN HIGH RESISTANCE WHEN ALL COMPONENTS ARE CONNECTED. REFER TO "DIAGNOSTIC AIDS" ON FACING PAGE FOR MORE INFORMATION. REPAIR ALL DAMAGE IF FOUND.

NO: REPLACE WHEEL SPEED SENSOR.

FRONT WHEEL SPEED SENSOR

TEMPERATURE OF SENSOR VS. SENSOR RESISTANCE (APPROXIMATE)

TEMP. (°C)	TEMP. (°F)	RESISTANCE (OHMS)
-34 TO 4	-30 TO 40	804 TO 1066
5 TO 43	41 TO 110	956 TO 1240
44 TO 93	111 TO 200	1112 TO 1463

DIAGNOSTIC CHART DTC 33 — 1995 SILHOUETTE AND TRANS SPORT WITH TCS, CONTINUED

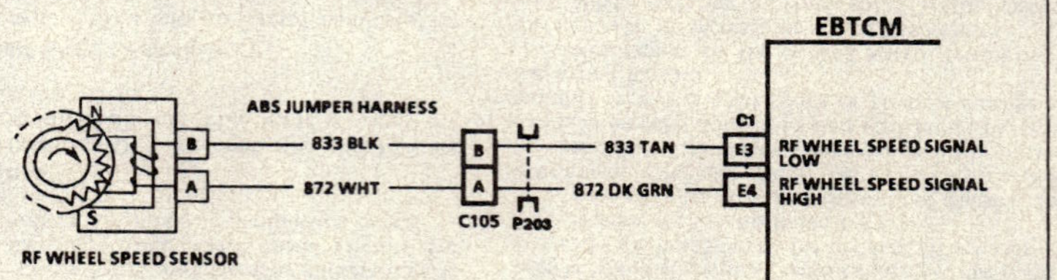

DTC 33
(Page 3 of 3)
RF WHEEL SPEED SNR CKT OPEN OR SHORT TO GND/BAT

Circuit Description:

As a toothed ring passes by the wheel speed sensor, changes in the electromagnetic field cause the wheel speed sensor to produce a sinusoidal (AC) voltage signal whose frequency is proportional to wheel speed. The magnitude of this signal is directly related to wheel speed and the proximity of the wheel speed sensor to the toothed ring, often referred to as the air gap.

DTC Will Set When:

DTC 33 can be set anytime after initialization. If either of the right front wheel speed circuits are shorted to voltage or ground, a malfunction exists.

Action Taken:

A malfunction DTC is stored, ABS and TCS are disabled and the ABS and TCS warning lamps are turned "ON."

DTC Chart Test Description:

Number(s) below refer to circled number(s) on the diagnostic chart.

10. Checks for short to ground in CKT 833.
11. Checks for short to ground in CKT 872.
12. Checks for short to ground in wheel speed sensor jumper harness.
13. Verifies that wheel speed sensor is not internally shorted to ground.

Diagnostic Aids:

An intermittent malfunction may be caused by a poor connection, rubbed through wire insulation, or a wire that is broken inside the insulation.

The frequency of the malfunction can be checked by using the enhanced diagnostic function of the Tech 1, as described in "Tech 1 Diagnostics" in this section.

If the customer's comments reflect that the ABS and TCS warning lamps are "ON" only during moist environmental changes (rain, snow, vehicle wash), all wheel speed sensor circuitry should be thoroughly inspected for signs of water intrusion. If DTC is not current, clear DTCs and simulate the effects of water intrusion. Use the following procedure. Spray the suspected area with a 5% salt water solution (two teaspoons of salt to 12 oz. of water). Test drive vehicle over various road surfaces (bumps, turns, etc.) above 24 km/h (15 mph) for at least 30 seconds. If DTC returns, replace suspected harness.

Any circuitry that is suspected of causing the intermittent complaint should be thoroughly checked for backed out terminals, improper mating, broken locks, improperly formed or damaged terminals, poor terminal to wiring connections or physical damage to the wiring harness.

Resistance of the wheel speed sensor will increase with an increase in sensor temperature.

When replacing a wheel speed sensor, inspect the sensor terminals and harness connector for corrosion and/or water intrusion. If evidence of corrosion or water intrusion exists, replace wheel speed sensor jumper harness. Likewise, if replacing a wheel speed sensor jumper harness, inspect sensor terminals. If evidence of corrosion or water intrusion exists, replace wheel speed sensor.

DIAGNOSTIC CHART DTC 33 — 1995 SILHOUETTE AND TRANS SPORT WITH TCS, CONTINUED

DTC 33
(Page 3 of 3)
RF WHEEL SPEED SNR CKT OPEN OR SHORT TO GND/BAT

DIAGNOSIS CONTINUED FROM PAGE 1 OF 3 OF THIS DTC CHART.

(10)
- DISCONNECT CONNECTOR C105 (MOUNTED ON RH SHOCK TOWER).
- USING J 39200, MEASURE THE RESISTANCE BETWEEN GROUND AND TERMINAL "E3" OF THE 32-WAY EBTCM HARNESS CONNECTOR.

DOES J 39200 DISPLAY RESISTANCE AS "OL" (INFINITE)?

NO: REPAIR SHORT TO GROUND IN CKT 833.

YES:

(11)
- USING J 39200, MEASURE THE RESISTANCE BETWEEN GROUND AND TERMINAL "E4" OF THE 32-WAY EBTCM HARNESS CONNECTOR.

DOES J 39200 DISPLAY RESISTANCE AS "OL" (INFINITE)?

NO: REPAIR SHORT TO GROUND IN CKT 872.

YES:

(12)
- THOROUGHLY RESPRAY THE RIGHT FRONT WHEEL SPEED SENSOR JUMPER HARNESS WITH THE SALT WATER SOLUTION.

IMPORTANT: DO NOT ALLOW WATER TO ENTER ANY WHEEL SPEED SENSOR CONNECTOR.

- DISCONNECT RIGHT FRONT WHEEL SPEED SENSOR DIRECTLY AT SENSOR.
- USING J 39200, MEASURE THE RESISTANCE BETWEEN GROUND AND TERMINAL "A," THEN "B" OF THE RIGHT FRONT WHEEL SPEED SENSOR HARNESS CONNECTOR.

DOES J 39200 DISPLAY RESISTANCE AS "OL" (INFINITE) FOR BOTH MEASUREMENTS?

NO: REPLACE RIGHT FRONT WHEEL SPEED SENSOR JUMPER HARNESS.

YES:

(13)
- USING J 39200, MEASURE THE RESISTANCE BETWEEN GROUND AND TERMINAL "A" OF THE RIGHT FRONT WHEEL SPEED SENSOR CONNECTOR.

DOES J 39200 DISPLAY RESISTANCE AS "OL" (INFINITE)?

NO: REPLACE WHEEL SPEED SENSOR.

YES: MALFUNCTION IS INTERMITTENT. INSPECT ALL CONNECTORS AND HARNESSES FOR DAMAGE WHICH MAY RESULT IN A SHORT TO GROUND WHEN ALL COMPONENTS ARE CONNECTED. PAY PARTICULAR ATTENTION TO THE JUMPER HARNESS, SMALL NICKS OR CUTS CAN CAUSE A MALFUNCTION. REPAIR ANY DAMAGE IF FOUND. JUMPER HARNESS MUST BE REPLACED IF DAMAGED. REFER TO "DIAGNOSTIC AIDS" ON FACING PAGE FOR MORE INFORMATION.

DIAGNOSTIC CHART DTC 34 — 1995 SILHOUETTE AND TRANS SPORT WITH TCS

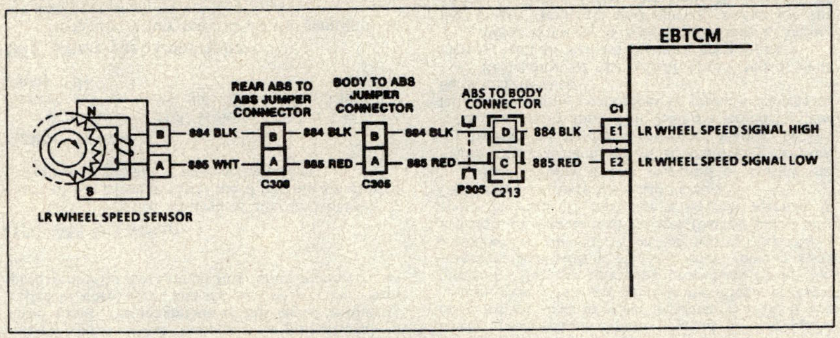

DTC 34
(Page 1 of 3)
LR WHEEL SPEED SNR CKT OPEN OR SHORT TO GND/BAT

Circuit Description:

As a toothed ring passes by the wheel speed sensor, changes in the electromagnetic field cause the wheel speed sensor to produce a sinusoidal (AC) voltage signal whose frequency is proportional to wheel speed. The magnitude of this signal is directly related to wheel speed and the proximity of the wheel speed sensor to the toothed ring, often referred to as the air gap.

DTC Will Set When:

DTC 34 can be set anytime after initialization. If either of the left rear wheel speed circuits are shorted to voltage or ground, a malfunction exists.

Action Taken:

A malfunction DTC is stored, ABS and TCS are disabled and the ABS and TCS warning lamps are turned "ON."

DTC Chart Test Description:

Number(s) below refer to circled number(s) on the diagnostic chart.

1. Checks to see that the wheel speed sensor is within proper resistance values.
2. Checks for short to ground in wheel speed sensor wiring.
3. Checks for short to voltage in wheel speed sensor wiring.
4. Checks for a short to voltage in CKT 884.
5. Checks for a short to voltage in CKT 885.
6. Ensures malfunction was not due to physical damage of circuitry.

Diagnostic Aids:

An intermittent malfunction may be caused by a poor connection, rubbed through wire insulation, or a wire that is broken inside the insulation.

The frequency of the malfunction can be checked by using the enhanced diagnostic function of the Tech 1, as described in "Tech 1 Diagnostics" in this section.

If the customer's comments reflect that the ABS and TCS warning lamps are "ON" only during moist environmental changes (rain, snow, vehicle wash), all wheel speed sensor circuitry should be thoroughly inspected for signs of water intrusion. If DTC is not current, clear DTCs and simulate the effects of water intrusion. Use the following procedure. Spray the suspected area with a 5% salt water solution (two teaspoons of salt to 12 oz. of water). Test drive vehicle over various road surfaces (bumps, turns, etc.) above 24 km/h (15 mph) for at least 30 seconds. If DTC returns, replace suspected harness.

Any circuitry that is suspected of causing the intermittent complaint should be thoroughly checked for backed out terminals, improper mating, broken locks, improperly formed or damaged terminals, poor terminal to wiring connections or physical damage to the wiring harness.

Resistance of the wheel speed sensor will increase with an increase in sensor temperature.

When replacing a wheel speed sensor, inspect the sensor terminals and harness connector for corrosion and/or water intrusion. If evidence of corrosion or water intrusion exists, replace wheel speed sensor jumper harness. Likewise, if replacing a wheel speed sensor jumper harness, inspect sensor terminals. If evidence of corrosion or water intrusion exists, replace wheel speed sensor.

DIAGNOSTIC CHART DTC 34 — 1995 SILHOUETTE AND TRANS SPORT WITH TCS, CONTINUED

DTC 34
(Page 1 of 3)
LR WHEEL SPEED SNR CKT OPEN OR SHORT TO GND/BAT

Important

- WHEEL SPEED SENSOR INTERMITTENT MALFUNCTIONS MAY BE DIFFICULT TO LOCATE. CARE SHOULD BE TAKEN NOT TO DISTURB ANY ELECTRICAL CONNECTIONS PRIOR TO AN INDICATED STEP OF THIS CHART. THIS WILL ENSURE THAT AN INTERMITTENT CONNECTION WILL NOT BE CORRECTED BEFORE THE SOURCE OF THE MALFUNCTION IS FOUND.

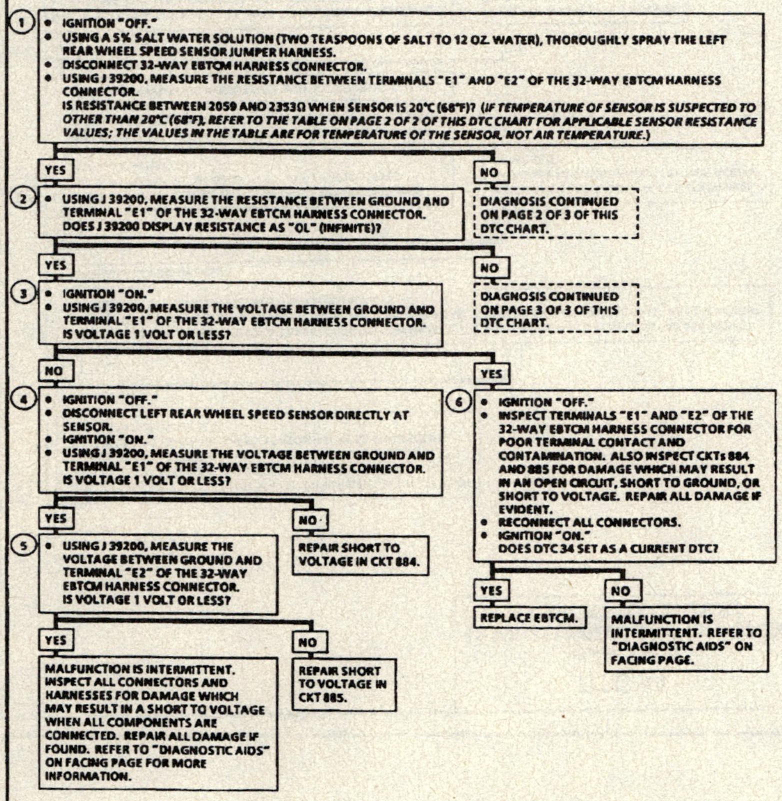

DIAGNOSTIC CHART DTC 34 — 1995 SILHOUETTE AND TRANS SPORT WITH TCS, CONTINUED

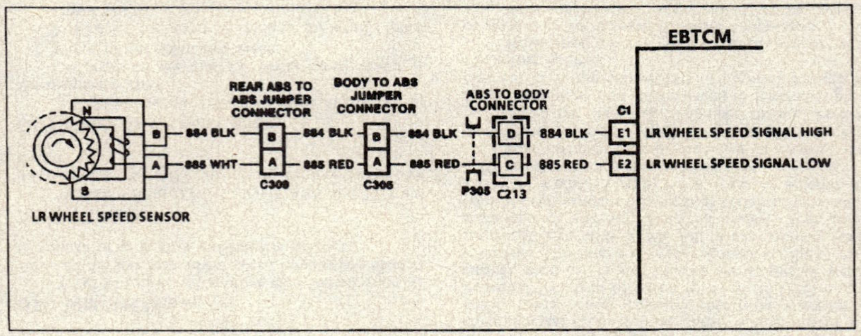

DTC 34
(Page 2 of 3)
LR WHEEL SPEED SNR CKT OPEN OR SHORT TO GND/BAT

Circuit Description:

As a toothed ring passes by the wheel speed sensor, changes in the electromagnetic field cause the wheel speed sensor to produce a sinusoidal (AC) voltage signal whose frequency is proportional to wheel speed. The magnitude of this signal is directly related to wheel speed and the proximity of the wheel speed sensor to the toothed ring, often referred to as the air gap.

DTC Will Set When:

DTC 34 can be set anytime after initialization. If either of the left rear wheel speed circuits are shorted to voltage or ground, a malfunction exists.

Action Taken:

A malfunction DTC is stored, ABS and TCS are disabled and the ABS and TCS warning lamps are turned "ON."

DTC Chart Test Description:

Number(s) below refer to circled number(s) on the diagnostic chart.

7. Checks for high resistance in CKT 884.
8. Checks for high resistance in CKT 885.
9. Verifies proper wheel speed sensor resistance.

Diagnostic Aids:

An intermittent malfunction may be caused by a poor connection, rubbed through wire insulation, or a wire that is broken inside the insulation.

The frequency of the malfunction can be checked by using the enhanced diagnostic function of the Tech 1, as described in "Tech 1 Diagnostics" in this section.

If the customer's comments reflect that the ABS and TCS warning lamps are "ON" only during moist environmental changes (rain, snow, vehicle wash), all wheel speed sensor circuitry should be thoroughly inspected for signs of water intrusion. If DTC is not current, clear DTCs and simulate the effects of water intrusion. Use the following procedure. Spray the suspected area with a 5% salt water solution (two teaspoons of salt to 12 oz. of water). Test drive vehicle over various road surfaces (bumps, turns, etc.) above 24 km/h (15 mph) for at least 30 seconds. If DTC returns, replace suspected harness.

Any circuitry that is suspected of causing the intermittent complaint should be thoroughly checked for backed out terminals, improper mating, broken locks, improperly formed or damaged terminals, poor terminal to wiring connections or physical damage to the wiring harness.

Resistance of the wheel speed sensor will increase with an increase in sensor temperature.

When replacing a wheel speed sensor, inspect the sensor terminals and harness connector for corrosion and/or water intrusion. If evidence of corrosion or water intrusion exists, replace wheel speed sensor jumper harness. Likewise, if replacing a wheel speed sensor jumper harness, inspect sensor terminals. If evidence of corrosion or water intrusion exists, replace wheel speed sensor.

DIAGNOSTIC CHART DTC 34 — 1995 SILHOUETTE AND TRANS SPORT WITH TCS, CONTINUED

DTC 34
(Page 2 of 3)
LR WHEEL SPEED SNR CKT OPEN OR SHORT TO GND/BAT

DIAGNOSIS CONTINUED FROM PAGE 1 OF 3 OF THIS DTC CHART.

(7)
- DISCONNECT LEFT REAR WHEEL SPEED SENSOR DIRECTLY AT SENSOR.
- USING J 39200, MEASURE THE RESISTANCE BETWEEN TERMINAL "E1" OF THE 32-WAY EBTCM HARNESS CONNECTOR AND TERMINAL "B" OF THE LEFT REAR WHEEL SPEED SENSOR HARNESS CONNECTOR.
IS RESISTANCE 2Ω OR LESS?

NO: REPAIR HIGH RESISTANCE IN CKT 884 (WHEEL SPEED SENSOR JUMPER MUST BE REPLACED IF DAMAGED).

YES:

(8)
- USING J 39200, MEASURE THE RESISTANCE BETWEEN TERMINAL "E2" OF THE 32-WAY EBTCM HARNESS CONNECTOR AND TERMINAL "A" OF THE LEFT REAR WHEEL SPEED SENSOR HARNESS CONNECTOR.
IS RESISTANCE 2Ω OR LESS?

NO: REPAIR HIGH RESISTANCE IN CKT 885 (WHEEL SPEED SENSOR JUMPER MUST BE REPLACED IF DAMAGED).

YES:

(9)
- USING J 39200, MEASURE THE RESISTANCE BETWEEN TERMINAL "A" AND TERMINAL "B" OF THE LEFT REAR WHEEL SPEED SENSOR CONNECTOR.
IS RESISTANCE BETWEEN 2059 AND 2353Ω WHEN SENSOR IS 20°C (68°F)? *(IF TEMPERATURE OF SENSOR IS SUSPECTED TO OTHER THAN 20°C (68°F), REFER TO THE TABLE BELOW FOR APPLICABLE SENSOR RESISTANCE VALUES; THE VALUES IN THE TABLE ARE FOR TEMPERATURE OF THE SENSOR, NOT AIR TEMPERATURE.)*

NO: REPLACE WHEEL SPEED SENSOR.

YES: MALFUNCTION IS INTERMITTENT. INSPECT ALL CONNECTORS AND HARNESSES FOR DAMAGE WHICH MAY RESULT IN HIGH RESISTANCE WHEN ALL COMPONENTS ARE CONNECTED.

REPAIR ALL DAMAGE IF FOUND.

REAR WHEEL SPEED SENSOR

TEMPERATURE OF SENSOR VS. SENSOR RESISTANCE (APPROXIMATE)

TEMP. (°C)	TEMP. (°F)	RESISTANCE (OHMS)
-34 TO 4	-30 TO 40	1622 TO 2205
5 TO 43	41 TO 110	1930 TO 2566
44 TO 93	111 TO 200	2245 TO 3028

DIAGNOSTIC CHART DTC 34 — 1995 SILHOUETTE AND TRANS SPORT WITH TCS, CONTINUED

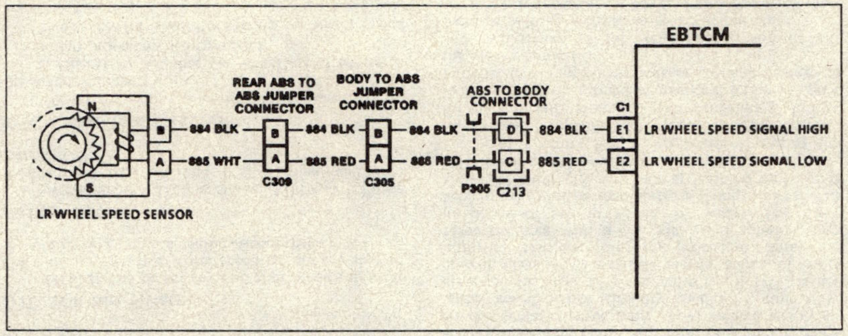

DTC 34
(Page 3 of 3)
LR WHEEL SPEED SNR CKT OPEN OR SHORT TO GND/BAT

Circuit Description:

As a toothed ring passes by the wheel speed sensor, changes in the electromagnetic field cause the wheel speed sensor to produce a sinusoidal (AC) voltage signal whose frequency is proportional to wheel speed. The magnitude of this signal is directly related to wheel speed and the proximity of the wheel speed sensor to the toothed ring, often referred to as the air gap.

DTC Will Set When:

DTC 34 can be set anytime after initialization. If either of the left rear wheel speed circuits are shorted to voltage or ground, a malfunction exists.

Action Taken:

A malfunction DTC is stored, ABS and TCS are disabled and the ABS and TCS warning lamps are turned "ON."

DTC Chart Test Description:

Number(s) below refer to circled number(s) on the diagnostic chart.

10. Checks for short to ground in CKT 884.
11. Checks for short to ground in CKT 885.
12. Checks for short to ground in wheel speed sensor jumper harness.
13. Verifies that wheel speed sensor is not internally shorted to ground.

Diagnostic Aids:

An intermittent malfunction may be caused by a poor connection, rubbed through wire insulation, or a wire that is broken inside the insulation.

The frequency of the malfunction can be checked by using the enhanced diagnostic function of the Tech 1, as described in "Tech 1 Diagnostics" in this section.

If the customer's comments reflect that the ABS and TCS warning lamps are "ON" only during moist environmental changes (rain, snow, vehicle wash), all wheel speed sensor circuitry should be thoroughly inspected for signs of water intrusion. If DTC is not current, clear DTCs and simulate the effects of water intrusion. Use the following procedure. Spray the suspected area with a 5% salt water solution (two teaspoons of salt to 12 oz. of water). Test drive vehicle over various road surfaces (bumps, turns, etc.) above 24 km/h (15 mph) for at least 30 seconds. If DTC returns, replace suspected harness.

Any circuitry that is suspected of causing the intermittent complaint should be thoroughly checked for backed out terminals, improper mating, broken locks, improperly formed or damaged terminals, poor terminal to wiring connections or physical damage to the wiring harness.

Resistance of the wheel speed sensor will increase with an increase in sensor temperature.

When replacing a wheel speed sensor, inspect the sensor terminals and harness connector for corrosion and/or water intrusion. If evidence of corrosion or water intrusion exists, replace wheel speed sensor jumper harness. Likewise, if replacing a wheel speed sensor jumper harness, inspect sensor terminals. If evidence of corrosion or water intrusion exists, replace wheel speed sensor.

DIAGNOSTIC CHART DTC 34 — 1995 SILHOUETTE AND TRANS SPORT WITH TCS, CONTINUED

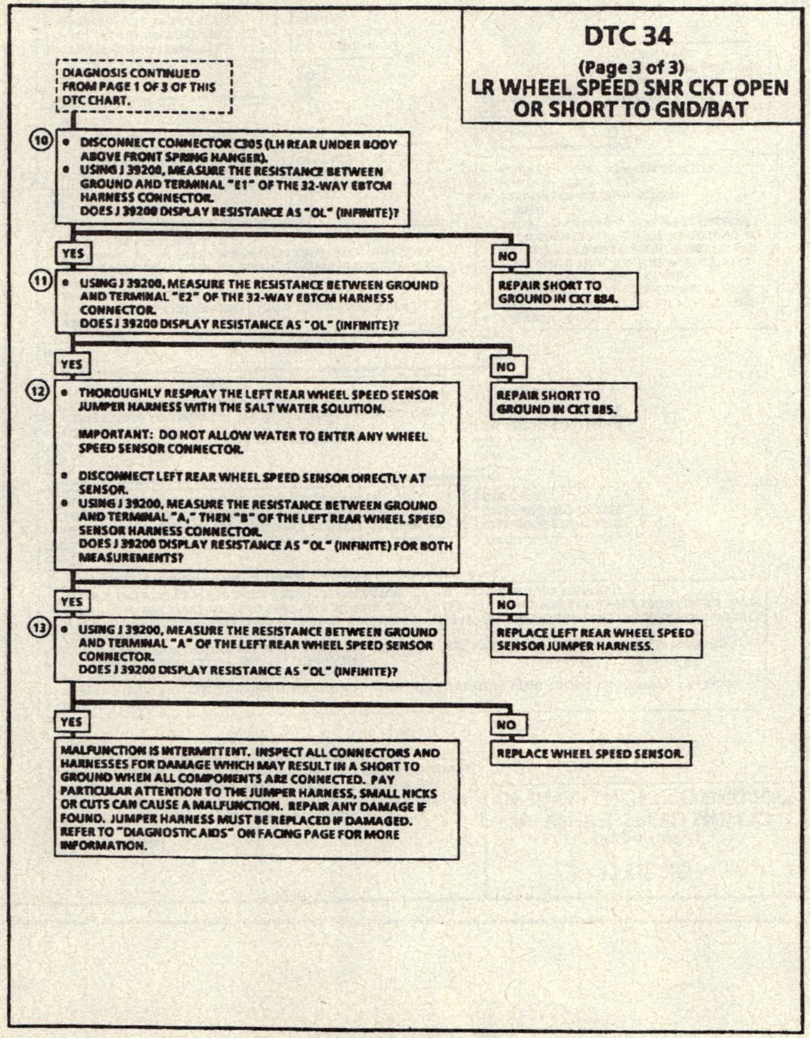

DIAGNOSTIC CHART DTC 35 — 1995 SILHOUETTE AND TRANS SPORT WITH TCS

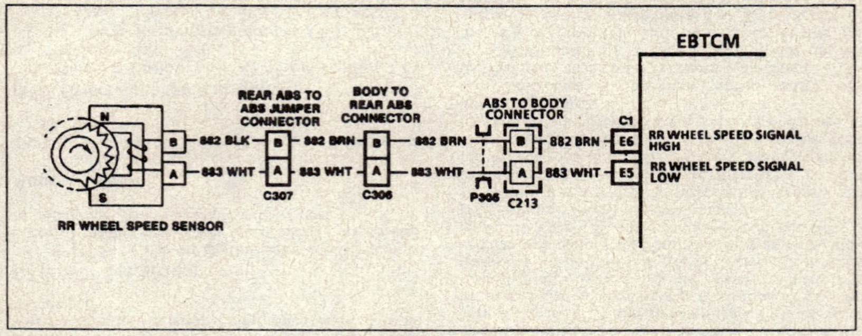

DTC 35
(Page 1 of 3)
RR WHEEL SPEED SNR CKT OPEN OR SHORT TO GND/BAT

Circuit Description:

As a toothed ring passes by the wheel speed sensor, changes in the electromagnetic field cause the wheel speed sensor to produce a sinusoidal (AC) voltage signal whose frequency is proportional to wheel speed. The magnitude of this signal is directly related to wheel speed and the proximity of the wheel speed sensor to the toothed ring, often referred to as the air gap.

DTC Will Set When:

DTC 35 can be set anytime after initialization. If either of the right rear wheel speed circuits are shorted to voltage or ground, a malfunction exists.

Action Taken:

A malfunction DTC is stored, ABS and TCS are disabled and the ABS and TCS warning lamps are turned "ON."

DTC Chart Test Description:

Number(s) below refer to circled number(s) on the diagnostic chart.

1. Checks to see that the wheel speed sensor is within proper resistance values.
2. Checks for short to ground in wheel speed sensor wiring.
3. Checks for short to voltage in wheel speed sensor wiring.
4. Checks for a short to voltage in CKT 883.
5. Checks for a short to voltage in CKT 882.
6. Ensures malfunction was not due to physical damage of circuitry.

Diagnostic Aids:

An intermittent malfunction may be caused by a poor connection, rubbed through wire insulation, or a wire that is broken inside the insulation.

The frequency of the malfunction can be checked by using the enhanced diagnostic function of the Tech 1, as described in "Tech 1 Diagnostics" in this section.

If the customer's comments reflect that the ABS and TCS warning lamps are "ON" only during moist environmental changes (rain, snow, vehicle wash), all wheel speed sensor circuitry should be thoroughly inspected for signs of water intrusion. If DTC is not current, clear DTCs and simulate the effects of water intrusion. Use the following procedure. Spray the suspected area with a 5% salt water solution (two teaspoons of salt to 12 oz. of water). Test drive vehicle over various road surfaces (bumps, turns, etc.) above 24 km/h (15 mph) for at least 30 seconds. If DTC returns, replace suspected harness.

Any circuitry that is suspected of causing the intermittent complaint should be thoroughly checked for backed out terminals, improper mating, broken locks, improperly formed or damaged terminals, poor terminal to wiring connections or physical damage to the wiring harness.

Resistance of the wheel speed sensor will increase with an increase in sensor temperature.

When replacing a wheel speed sensor, inspect the sensor terminals and harness connector for corrosion and/or water intrusion. If evidence of corrosion or water intrusion exists, replace wheel speed sensor jumper harness. Likewise, if replacing a wheel speed sensor jumper harness, inspect sensor terminals. If evidence of corrosion or water intrusion exists, replace wheel speed sensor.

DIAGNOSTIC CHART DTC 35 — 1995 SILHOUETTE AND TRANS SPORT WITH TCS, CONTINUED

DTC 35
(Page 1 of 3)
RR WHEEL SPEED SNR CKT OPEN OR SHORT TO GND/BAT

Important

- WHEEL SPEED SENSOR INTERMITTENT MALFUNCTIONS MAY BE DIFFICULT TO LOCATE. CARE SHOULD BE TAKEN NOT TO DISTURB ANY ELECTRICAL CONNECTIONS PRIOR TO AN INDICATED STEP OF THIS CHART. THIS WILL ENSURE THAT AN INTERMITTENT CONNECTION WILL NOT BE CORRECTED BEFORE THE SOURCE OF THE MALFUNCTION IS FOUND.

(1)
- IGNITION "OFF."
- USING A 5% SALT WATER SOLUTION (TWO TEASPOONS OF SALT TO 12 OZ. WATER), THOROUGHLY SPRAY THE RIGHT REAR WHEEL SPEED SENSOR JUMPER HARNESS.
- DISCONNECT 32-WAY EBTCM HARNESS CONNECTOR.
- USING J 39200, MEASURE THE RESISTANCE BETWEEN TERMINALS "E5" AND "E6" OF THE 32-WAY EBTCM HARNESS CONNECTOR.

IS RESISTANCE BETWEEN 2059 AND 2353Ω WHEN SENSOR IS 20°C (68°F)? *(IF TEMPERATURE OF SENSOR IS SUSPECTED TO OTHER THAN 20°C (68°F), REFER TO THE TABLE ON PAGE 2 OF 2 OF THIS DTC CHART FOR APPLICABLE SENSOR RESISTANCE VALUES; THE VALUES IN THE TABLE ARE FOR TEMPERATURE OF THE SENSOR, NOT AIR TEMPERATURE.)*

- YES → (2)
- NO → DIAGNOSIS CONTINUED ON PAGE 2 OF 3 OF THIS DTC CHART.

(2)
- USING J 39200, MEASURE THE RESISTANCE BETWEEN GROUND AND TERMINAL "E5" OF THE 32-WAY EBTCM HARNESS CONNECTOR. DOES J 39200 DISPLAY RESISTANCE AS "OL" (INFINITE)?

- YES → (3)
- NO → DIAGNOSIS CONTINUED ON PAGE 3 OF 3 OF THIS DTC CHART.

(3)
- IGNITION "ON."
- USING J 39200, MEASURE THE VOLTAGE BETWEEN GROUND AND TERMINAL "E5" OF THE 32-WAY EBTCM HARNESS CONNECTOR. IS VOLTAGE 1 VOLT OR LESS?

- NO → (4)
- YES → (6)

(4)
- IGNITION "OFF."
- DISCONNECT RIGHT REAR WHEEL SPEED SENSOR DIRECTLY AT SENSOR.
- IGNITION "ON."
- USING J 39200, MEASURE THE VOLTAGE BETWEEN GROUND AND TERMINAL "E5" OF THE 32-WAY EBTCM HARNESS CONNECTOR. IS VOLTAGE 1 VOLT OR LESS?

- YES → (5)
- NO → REPAIR SHORT TO VOLTAGE IN CKT 883.

(5)
- USING J 39200, MEASURE THE VOLTAGE BETWEEN GROUND AND TERMINAL "E6" OF THE 32-WAY EBTCM HARNESS CONNECTOR. IS VOLTAGE 1 VOLT OR LESS?

- YES → MALFUNCTION IS INTERMITTENT. INSPECT ALL CONNECTORS AND HARNESSES FOR DAMAGE WHICH MAY RESULT IN A SHORT TO VOLTAGE WHEN ALL COMPONENTS ARE CONNECTED. REPAIR ALL DAMAGE IF FOUND.
- NO → REPAIR SHORT TO VOLTAGE IN CKT 882.

(6)
- IGNITION "OFF."
- INSPECT TERMINALS "E5" AND "E6" OF THE 32-WAY EBTCM HARNESS CONNECTOR FOR POOR TERMINAL CONTACT AND CONTAMINATION. ALSO INSPECT CKTs 882 AND 883 FOR DAMAGE WHICH MAY RESULT IN AN OPEN CIRCUIT, SHORT TO GROUND, OR SHORT TO VOLTAGE. REPAIR ALL DAMAGE IF EVIDENT.
- RECONNECT ALL CONNECTORS.
- IGNITION "ON."

DOES DTC 35 SET AS A CURRENT DTC?

- YES → REPLACE EBTCM.
- NO → MALFUNCTION IS INTERMITTENT. REFER TO "DIAGNOSTIC AIDS" ON FACING PAGE.

DIAGNOSTIC CHART DTC 35 — 1995 SILHOUETTE AND TRANS SPORT WITH TCS, CONTINUED

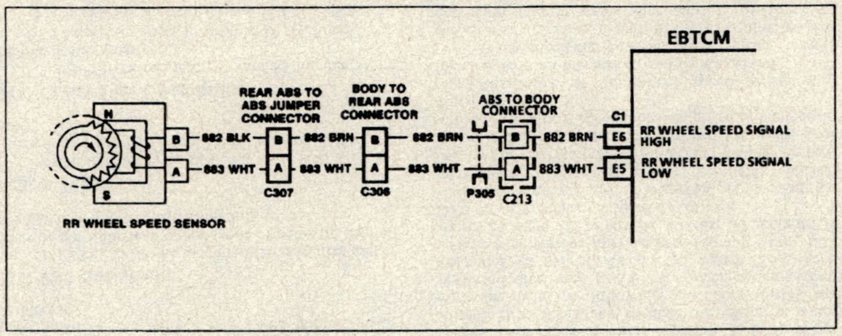

DTC 35
(Page 2 of 3)
RR WHEEL SPEED SNR CKT OPEN OR SHORT TO GND/BAT

Circuit Description:

As a toothed ring passes by the wheel speed sensor, changes in the electromagnetic field cause the wheel speed sensor to produce a sinusoidal (AC) voltage signal whose frequency is proportional to wheel speed. The magnitude of this signal is directly related to wheel speed and the proximity of the wheel speed sensor to the toothed ring, often referred to as the air gap.

DTC Will Set When:

DTC 35 can be set anytime after initialization. If either of the right rear wheel speed circuits are shorted to voltage or ground, a malfunction exists.

Action Taken:

A malfunction DTC is stored, ABS and TCS are disabled and the ABS and TCS warning lamps are turned "ON."

DTC Chart Test Description:

Number(s) below refer to circled number(s) on the diagnostic chart.

7. Checks for high resistance in CKT 883.
8. Checks for high resistance in CKT 882.
9. Verifies proper wheel speed sensor resistance.

Diagnostic Aids:

An intermittent malfunction may be caused by a poor connection, rubbed through wire insulation, or a wire that is broken inside the insulation.

The frequency of the malfunction can be checked by using the enhanced diagnostic function of the Tech 1, as described in "Tech 1 Diagnostics" in this section.

If the customer's comments reflect that the ABS and TCS warning lamps are "ON" only during moist environmental changes (rain, snow, vehicle wash), all wheel speed sensor circuitry should be thoroughly inspected for signs of water intrusion. If DTC is not current, clear DTCs and simulate the effects of water intrusion. Use the following procedure. Spray the suspected area with a 5% salt water solution (two teaspoons of salt to 12 oz. of water). Test drive vehicle over various road surfaces (bumps, turns, etc.) above 24 km/h (15 mph) for at least 30 seconds. If DTC returns, replace suspected harness.

Any circuitry that is suspected of causing the intermittent complaint should be thoroughly checked for backed out terminals, improper mating, broken locks, improperly formed or damaged terminals, poor terminal to wiring connections or physical damage to the wiring harness.

Resistance of the wheel speed sensor will increase with an increase in sensor temperature.

When replacing a wheel speed sensor, inspect the sensor terminals and harness connector for corrosion and/or water intrusion. If evidence of corrosion or water intrusion exists, replace wheel speed sensor jumper harness. Likewise, if replacing a wheel speed sensor jumper harness, inspect sensor terminals. If evidence of corrosion or water intrusion exists, replace wheel speed sensor.

DIAGNOSTIC CHART DTC 35 — 1995 SILHOUETTE AND TRANS SPORT WITH TCS, CONTINUED

DTC 35
(Page 2 of 3)
RR WHEEL SPEED SNR CKT OPEN OR SHORT TO GND/BAT

DIAGNOSIS CONTINUED FROM PAGE 1 OF 3 OF THIS DTC CHART.

(7)
- DISCONNECT RIGHT REAR WHEEL SPEED SENSOR DIRECTLY AT SENSOR.
- USING J 39200, MEASURE THE RESISTANCE BETWEEN TERMINAL "E5" OF THE 32-WAY EBTCM HARNESS CONNECTOR AND TERMINAL "A" OF THE RIGHT REAR WHEEL SPEED SENSOR HARNESS CONNECTOR.

IS RESISTANCE 2Ω OR LESS?

NO: REPAIR HIGH RESISTANCE IN CKT 883 (WHEEL SPEED SENSOR JUMPER MUST BE REPLACED IF DAMAGED).

YES:

(8)
- USING J 39200, MEASURE THE RESISTANCE BETWEEN TERMINAL "E6" OF THE 32-WAY EBTCM HARNESS CONNECTOR AND TERMINAL "B" OF THE RIGHT REAR WHEEL SPEED SENSOR HARNESS CONNECTOR.

IS RESISTANCE 2Ω OR LESS?

NO: REPAIR HIGH RESISTANCE IN CKT 882 (WHEEL SPEED SENSOR JUMPER MUST BE REPLACED IF DAMAGED).

YES:

(9)
- USING J 39200, MEASURE THE RESISTANCE BETWEEN TERMINAL "A" AND TERMINAL "B" OF THE RIGHT REAR WHEEL SPEED SENSOR CONNECTOR.

IS RESISTANCE BETWEEN 2059 AND 2353Ω WHEN SENSOR IS 20°C (68°F)? *(IF TEMPERATURE OF SENSOR IS SUSPECTED TO OTHER THAN 20°C (68°F), REFER TO THE TABLE BELOW FOR APPLICABLE SENSOR RESISTANCE VALUES; THE VALUES IN THE TABLE ARE FOR TEMPERATURE OF THE SENSOR, NOT AIR TEMPERATURE.)*

NO: REPLACE WHEEL SPEED SENSOR.

YES: MALFUNCTION IS INTERMITTENT. INSPECT ALL CONNECTORS AND HARNESSES FOR DAMAGE WHICH MAY RESULT IN HIGH RESISTANCE WHEN ALL COMPONENTS ARE CONNECTED.

REPAIR ALL DAMAGE IF FOUND.

REAR WHEEL SPEED SENSOR
TEMPERATURE OF SENSOR VS. SENSOR RESISTANCE (APPROXIMATE)

TEMP. (°C)	TEMP. (°F)	RESISTANCE (OHMS)
-34 TO 4	-30 TO 40	1622 TO 2205
5 TO 43	41 TO 110	1930 TO 2566
44 TO 93	111 TO 200	2245 TO 3028

DIAGNOSTIC CHART DTC 35 — 1995 SILHOUETTE AND TRANS SPORT WITH TCS, CONTINUED

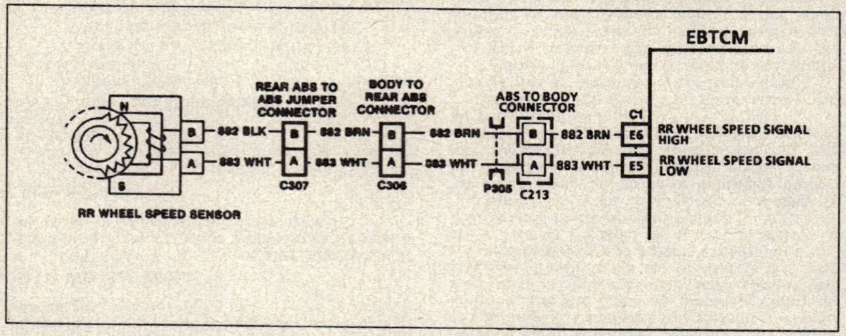

DTC 35
(Page 3 of 3)
RR WHEEL SPEED SNR CKT OPEN OR SHORT TO GND/BAT

Circuit Description:

As a toothed ring passes by the wheel speed sensor, changes in the electromagnetic field cause the wheel speed sensor to produce a sinusoidal (AC) voltage signal whose frequency is proportional to wheel speed. The magnitude of this signal is directly related to wheel speed and the proximity of the wheel speed sensor to the toothed ring, often referred to as the air gap.

DTC Will Set When:

DTC 35 can be set anytime after initialization. If either of the right rear wheel speed circuits are shorted to voltage or ground, a malfunction exists.

Action Taken:

A malfunction DTC is stored, ABS and TCS are disabled and the ABS and TCS warning lamps are turned "ON."

DTC Chart Test Description:

Number(s) below refer to circled number(s) on the diagnostic chart.

10. Checks for short to ground in CKT 883.
11. Checks for short to ground in CKT 882.
12. Checks for short to ground in wheel speed sensor jumper harness.
13. Verifies that wheel speed sensor is not internally shorted to ground.

Diagnostic Aids:

An intermittent malfunction may be caused by a poor connection, rubbed through wire insulation, or a wire that is broken inside the insulation.

The frequency of the malfunction can be checked by using the enhanced diagnostic function of the Tech 1, as described in "Tech 1 Diagnostics" in this section.

If the customer's comments reflect that the ABS and TCS warning lamps are "ON" only during moist environmental changes (rain, snow, vehicle wash), all wheel speed sensor circuitry should be thoroughly inspected for signs of water intrusion. If DTC is not current, clear DTCs and simulate the effects of water intrusion. Use the following procedure. Spray the suspected area with a 5% salt water solution (two teaspoons of salt to 12 oz. of water). Test drive vehicle over various road surfaces (bumps, turns, etc.) above 24 km/h (15 mph) for at least 30 seconds. If DTC returns, replace suspected harness.

Any circuitry that is suspected of causing the intermittent complaint should be thoroughly checked for backed out terminals, improper mating, broken locks, improperly formed or damaged terminals, poor terminal to wiring connections or physical damage to the wiring harness.

Resistance of the wheel speed sensor will increase with an increase in sensor temperature.

When replacing a wheel speed sensor, inspect the sensor terminals and harness connector for corrosion and/or water intrusion. If evidence of corrosion or water intrusion exists, replace wheel speed sensor jumper harness. Likewise, if replacing a wheel speed sensor jumper harness, inspect sensor terminals. If evidence of corrosion or water intrusion exists, replace wheel speed sensor.

DIAGNOSTIC CHART DTC 35 — 1995 SILHOUETTE AND TRANS SPORT WITH TCS, CONTINUED

DTC 35
(Page 3 of 3)
RR WHEEL SPEED SNR CKT OPEN OR SHORT TO GND/BAT

DIAGNOSIS CONTINUED FROM PAGE 1 OF 3 OF THIS DTC CHART.

(10)
- DISCONNECT CONNECTOR C307 (RH REAR UNDER BODY ABOVE FRONT SPRING HANGER).
- USING J 39200, MEASURE THE RESISTANCE BETWEEN GROUND AND TERMINAL "E5" OF THE 32-WAY EBTCM HARNESS CONNECTOR.

DOES J 39200 DISPLAY RESISTANCE AS "OL" (INFINITE)?

NO → REPAIR SHORT TO GROUND IN CKT 883.

YES ↓

(11)
- USING J 39200, MEASURE THE RESISTANCE BETWEEN GROUND AND TERMINAL "E6" OF THE 32-WAY EBTCM HARNESS CONNECTOR.

DOES J 39200 DISPLAY RESISTANCE AS "OL" (INFINITE)?

NO → REPAIR SHORT TO GROUND IN CKT 882.

YES ↓

(12)
- THOROUGHLY RESPRAY THE RIGHT REAR WHEEL SPEED SENSOR JUMPER HARNESS WITH THE SALT WATER SOLUTION.

IMPORTANT: DO NOT ALLOW WATER TO ENTER ANY WHEEL SPEED SENSOR CONNECTOR.

- DISCONNECT RIGHT REAR WHEEL SPEED SENSOR DIRECTLY AT SENSOR.
- USING J 39200, MEASURE THE RESISTANCE BETWEEN GROUND AND TERMINAL "A," THEN "B" OF THE RIGHT REAR WHEEL SPEED SENSOR HARNESS CONNECTOR.

DOES J 39200 DISPLAY RESISTANCE AS "OL" (INFINITE) FOR BOTH MEASUREMENTS?

NO → REPLACE RIGHT REAR WHEEL SPEED SENSOR JUMPER HARNESS.

YES ↓

(13)
- USING J 39200, MEASURE THE RESISTANCE BETWEEN GROUND AND TERMINAL "A" OF THE RIGHT REAR WHEEL SPEED SENSOR CONNECTOR.

DOES J 39200 DISPLAY RESISTANCE AS "OL" (INFINITE)?

NO → REPLACE WHEEL SPEED SENSOR.

YES ↓

MALFUNCTION IS INTERMITTENT. INSPECT ALL CONNECTORS AND HARNESSES FOR DAMAGE WHICH MAY RESULT IN A SHORT TO GROUND WHEN ALL COMPONENTS ARE CONNECTED. PAY PARTICULAR ATTENTION TO THE JUMPER HARNESS, SMALL NICKS OR CUTS CAN CAUSE A MALFUNCTION. REPAIR ANY DAMAGE IF FOUND. JUMPER HARNESS MUST BE REPLACED IF DAMAGED. REFER TO "DIAGNOSTIC AIDS" ON FACING PAGE FOR MORE INFORMATION.

DIAGNOSTIC CHART DTC 36 — 1995 SILHOUETTE AND TRANS SPORT WITH TCS

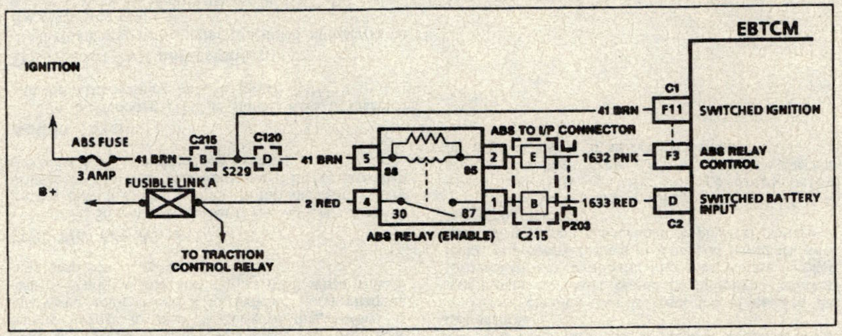

DTC 36
(Page 1 of 2)
LOW SYSTEM VOLTAGE

Circuit Description:

This DTC is used to monitor the voltage level available to the EBTCM controller. If the voltage drops below 11 volts, full performance of the ABS system cannot be guaranteed. During ABS operation, there are several current requirements that will cause battery voltage to drop. Because of this, voltage is monitored prior to ABS to indicate good charging system condition and also during ABS when voltage may drop significantly.

DTC Will Set When:

DTC 36 can only be set if the vehicle's speed is greater than 5 km/h (3 mph). If the switched battery voltage is less than 11.2 volts non-ABS or less than 9.6 volts ABS, a malfunction exists.

Action Taken:

A malfunction DTC is stored, ABS is disabled and the ABS warning lamp is turned "ON."

DTC Chart Test Description:

Number(s) below refer to circled number(s) on the diagnostic chart.

1. Checks to see what voltage is readily available at terminal "D" of the EBTCM. If voltage is representative of good charging system condition, a malfunction is not present.
2. Isolates the EBTCM to check the condition of the circuitry and charging system.
3. Isolates the low voltage condition to high circuit resistance, or improper charging system operation.
4. Isolates which circuit is the source of the high resistance condition.

Diagnostic Aids:

An intermittent malfunction may be caused by a poor connection, rubbed through wire insulation, or a wire that is broken inside the insulation.

The frequency of the malfunction can be checked by using the enhanced diagnostic function of the Tech 1, as described in "Tech 1 Diagnostics" in this section.

Any circuitry that is suspected of causing the intermittent complaint should be thoroughly checked for backed out terminals, improper mating, broken locks, improperly formed or damaged terminals, poor terminal to wiring connections or physical damage to the wiring harness.

While performing a voltage load test, if it is noted that only ignition voltage drops below acceptable voltage levels, CKT 41 should be checked for high resistance or an open condition.

DIAGNOSTIC CHART DTC 36 — 1995 SILHOUETTE AND TRANS SPORT WITH TCS, CONTINUED

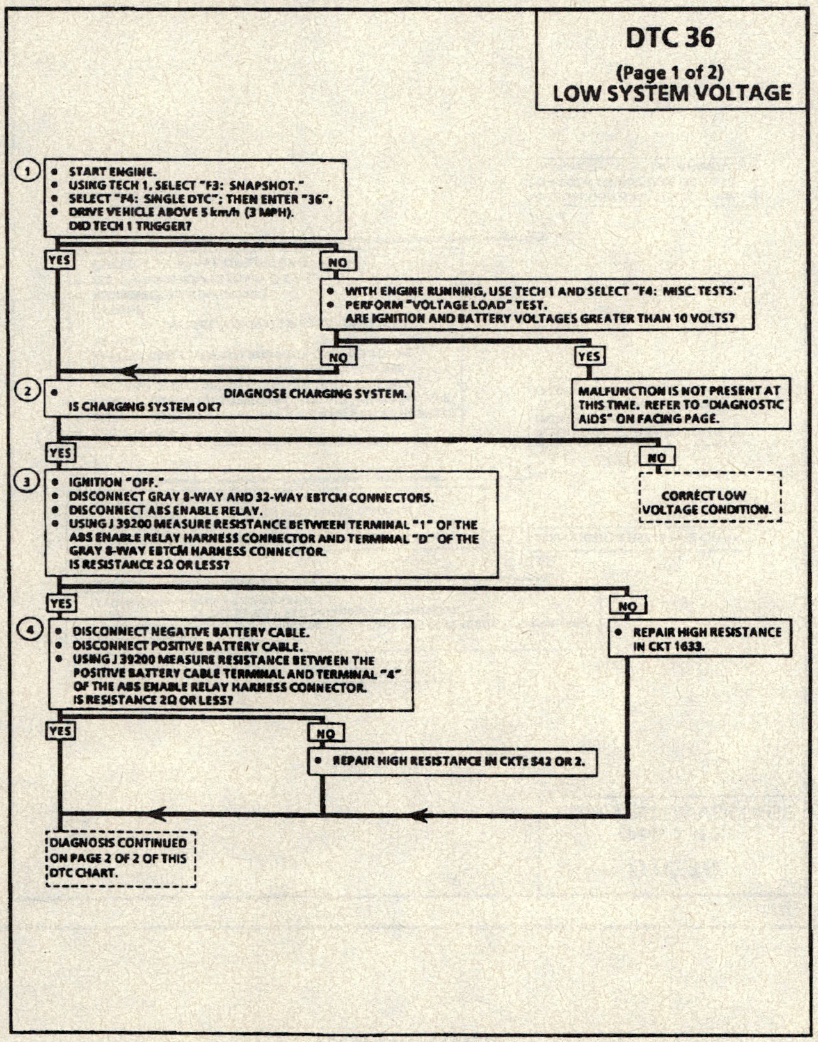

DIAGNOSTIC CHART DTC 36 — 1995 SILHOUETTE AND TRANS SPORT WITH TCS, CONTINUED

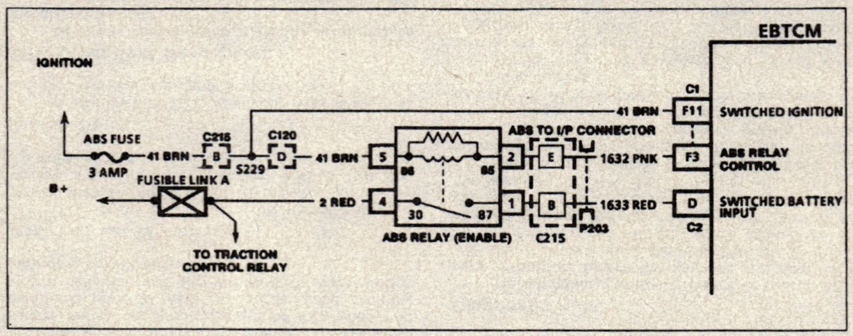

DTC 36
(Page 2 of 2)
LOW SYSTEM VOLTAGE

Circuit Description:

This DTC is used to monitor the voltage level available to the EBTCM controller. If the voltage drops below 11 volts, full performance of the ABS system cannot be guaranteed. During ABS operation, there are several current requirements that will cause battery voltage to drop. Because of this, voltage is monitored prior to ABS to indicate good charging system condition and also during ABS when voltage may drop significantly.

DTC Will Set When:

DTC 36 can only be set if the vehicle's speed is greater than 5 km/h (3 mph). If the switched battery voltage is less than 11.2 volts non-ABS or less than 9.6 volts ABS, a malfunction exists.

Action Taken:

A malfunction DTC is stored, ABS is disabled and the ABS warning lamp is turned "ON."

DTC Chart Test Description:

Number(s) below refer to circled number(s) on the diagnostic chart.

5. Checks for high resistance in CKT 41.
6. Checks for high resistance in the ignition fuse feed circuit.
7. Checks for malfunctioning EBTCM.

Diagnostic Aids:

An intermittent malfunction may be caused by a poor connection, rubbed through wire insulation, or a wire that is broken inside the insulation.

The frequency of the malfunction can be checked by using the enhanced diagnostic function of the Tech 1, as described in "Tech 1 Diagnostics" in this section.

Any circuitry that is suspected of causing the intermittent complaint should be thoroughly checked for backed out terminals, improper mating, broken locks, improperly formed or damaged terminals, poor terminal to wiring connections or physical damage to the wiring harness.

While performing a voltage load test, if it is noted that only ignition voltage drops below acceptable voltage levels, CKT 41 should be checked for high resistance or an open condition.

DIAGNOSTIC CHART DTC 36 — 1995 SILHOUETTE AND TRANS SPORT WITH TCS, CONTINUED

DTC 36
(Page 2 of 2)
LOW SYSTEM VOLTAGE

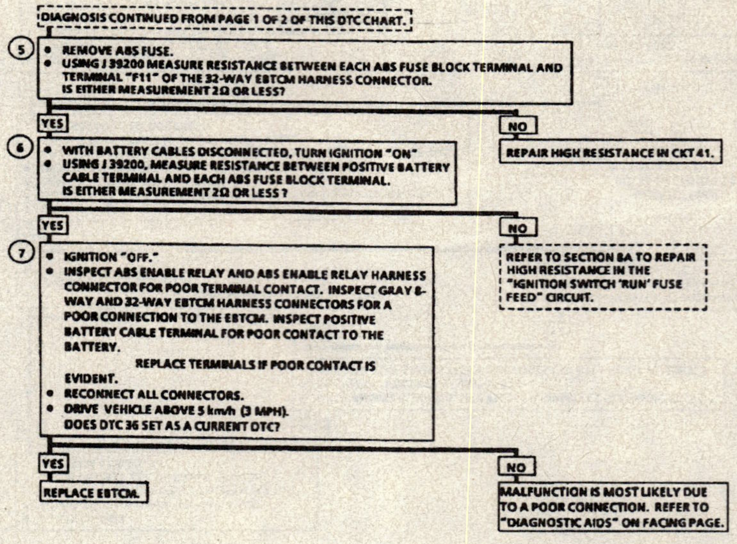

DIAGNOSTIC CHART DTC 37 — 1995 SILHOUETTE AND TRANS SPORT WITH TCS

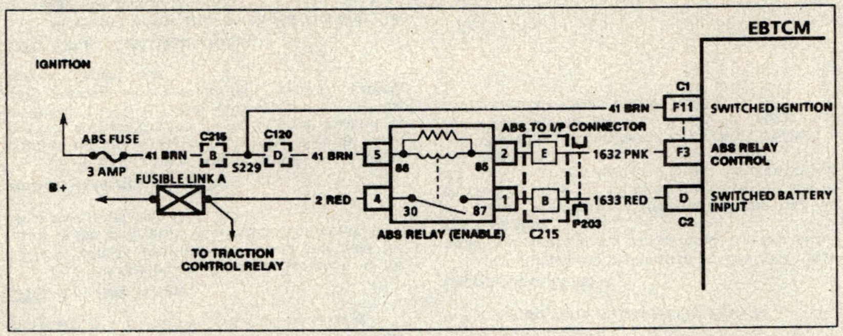

DTC 37
HIGH SYSTEM VOLTAGE

Circuit Description:

This DTC is designed to detect high vehicle voltage levels prior to any required motor movement (initialization or ABS). If excessive voltage exists, demagnetization of the motor magnets may occur, which would eventually affect or eliminate ABS performance.

DTC Will Set When:

DTC 37 can only be set if the vehicle's speed is greater than 5 km/h (3 mph). If the switched battery voltage is greater than 17 volts, a malfunction exists.

Action Taken:

A malfunction DTC is stored, ABS and TCS are disabled and the ABS and TCS warning lamps are turned "ON."

DTC Chart Test Description:

Number(s) below refer to circled number(s) on the diagnostic chart.

1. Checks the voltage level being received by the EBTCM.
2. Indicates whether the high voltage condition is caused by a malfunctioning charging system or EBTCM.
3. Checks for poor terminal contact which may cause an intermittent malfunction or a malfunctioning EBTCM.

Diagnostic Aids:

An intermittent malfunction may be caused by a poor connection, rubbed through wire insulation, or a wire that is broken inside the insulation.

The frequency of the malfunction can be checked by using the enhanced diagnostic function of the Tech 1, as described in "Tech 1 Diagnostics" in this section.

Any circuitry that is suspected of causing the intermittent complaint should be thoroughly checked for backed out terminals, improper mating, broken locks, improperly formed or damaged terminals, poor terminal to wiring connections or physical damage to the wiring harness.

DIAGNOSTIC CHART DTC 37 — 1995 SILHOUETTE AND TRANS SPORT WITH TCS, CONTINUED

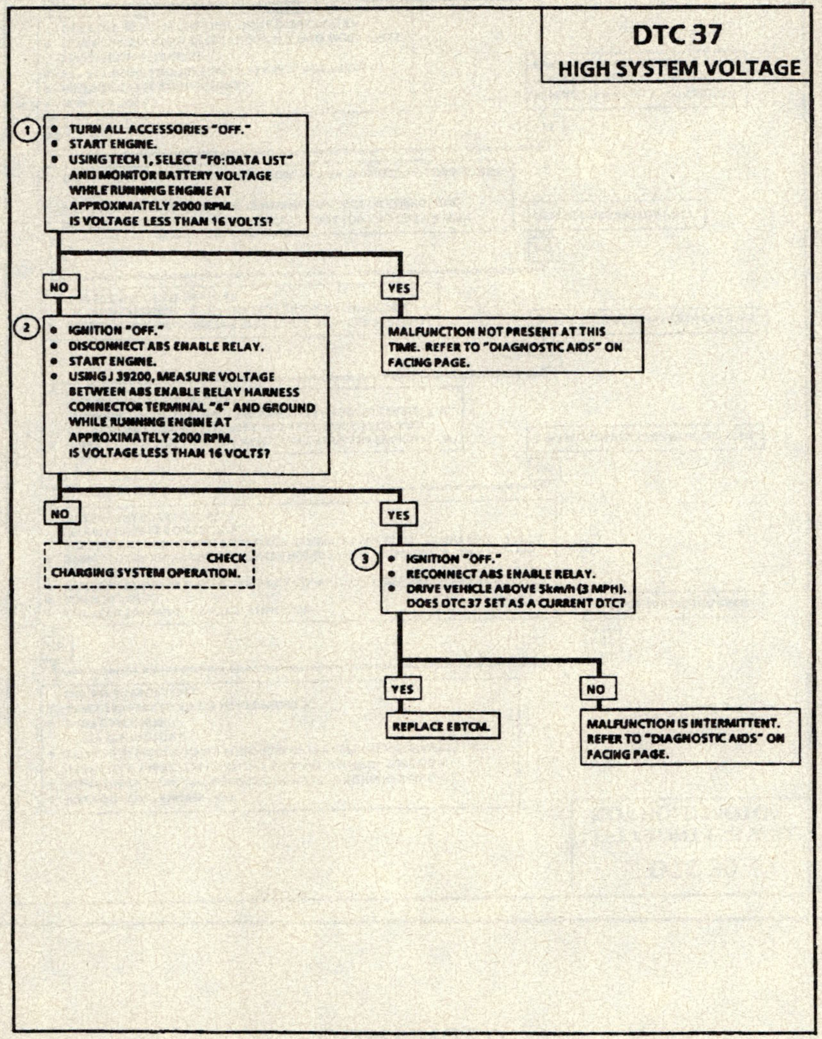

DIAGNOSTIC CHART DTC 38 — 1995 SILHOUETTE AND TRANS SPORT WITH TCS

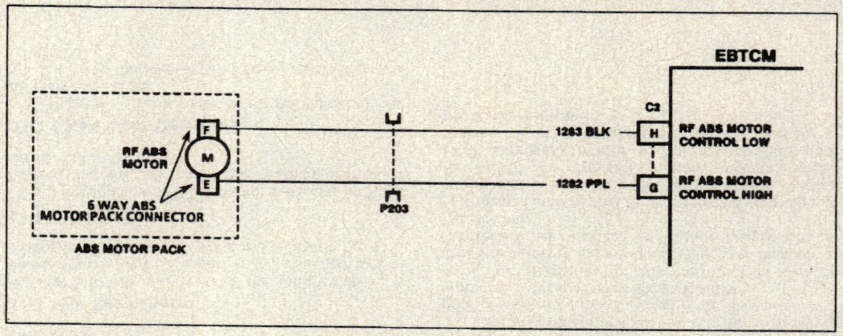

DTC 38
LEFT FRONT ESB WILL NOT HOLD MOTOR

Circuit Description:

This DTC is designed to detect a slipping left front ESB. During initialization and braking, the left front motor is re-homed. If the ESB slips, the motor/piston will move. During the next key "ON" initialization, a re-home of the motor verifies the motor/piston remained at the home position. If motor movement is detected, the ESB must be slipping.

DTC Will Set When:

DTC 38 can be set during initialization. If the EBTCM detects that the ESB could not hold the piston in the home position, a malfunction exists. DTC 86 is always set with DTC 38.

Action Taken:

If an ESB cannot hold a piston in the home position, the piston may be backdriven when the brake pedal is applied, causing the brake pedal to drop. A malfunction DTC is stored, ABS is disabled, and the ABS warning lamp and the red "BRAKE" warning lamp are turned "ON."

DTC Chart Test Description:

Number(s) below refer to circled number(s) on the diagnostic chart.

1. Checks the left front ESB. A broken or defective ESB would result in the left front piston being backdriven by hydraulic pressure and pedal movement would result.
2. Checks for high resistance in the motor "HIGH" circuitry.
3. Checks for high resistance in the motor "LOW" circuitry.
4. Checks for proper resistance of the motor windings.
5. Checks for poor terminal contact or corrosion in the connectors.
6. Checks for a malfunctioning EBTCM.

Diagnostic Aids:

An intermittent malfunction in this DTC may result from a mechanical part of the system that sticks, binds, or slips.

The frequency of the malfunction can be checked by using the enhanced diagnostic function of the Tech 1, as described in "Tech 1 Diagnostics" in this section.

The static modulator test function of the Tech 1 or T-100 may be used to locate an intermittent malfunction associated with the ESB.

DIAGNOSTIC CHART DTC 38 — 1995 SILHOUETTE AND TRANS SPORT WITH TCS, CONTINUED

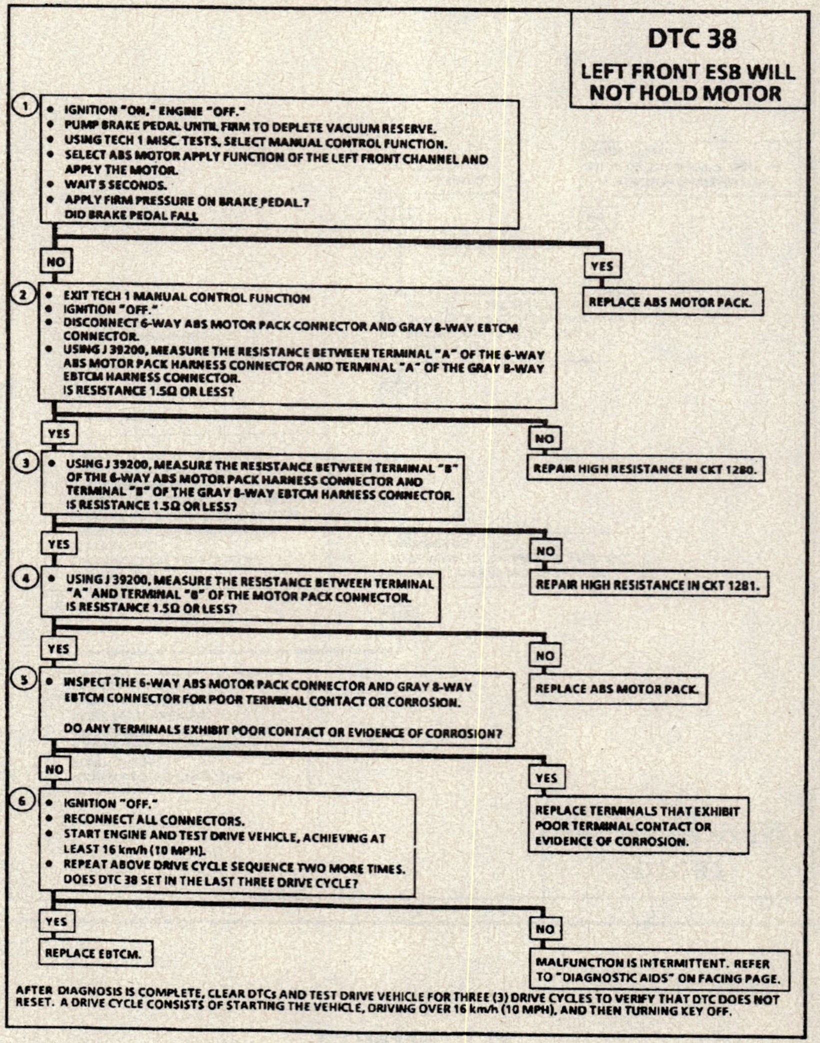

DIAGNOSTIC CHART DTC 41 — 1995 SILHOUETTE AND TRANS SPORT WITH TCS

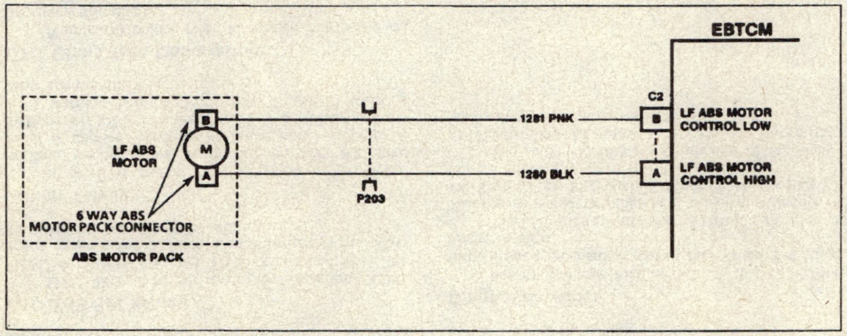

DTC 41
RIGHT FRONT ESB WILL NOT HOLD MOTOR

Circuit Description:

This DTC is designed to detect a slipping right front ESB. During initialization and braking, the right front motor is re-homed. If the ESB slips, the motor/piston will move. During the next key "ON" initialization, a re-home of the motor verifies the motor/piston remained at the home position. If motor movement is detected, the ESB must be slipping.

DTC Will Set When:

DTC 41 can be set during initialization. If the EBTCM detects that the ESB could not hold the piston in the home position, a malfunction exists. DTC 86 is always set with DTC 41.

Action Taken:

If an ESB cannot hold a piston in the home position, the piston may be backdriven when the brake pedal is applied, causing the brake pedal to drop. A malfunction DTC is stored, ABS is disabled, and the ABS warning lamp and the red "BRAKE" warning lamp are turned "ON."

DTC Chart Test Description:

Number(s) below refer to circled number(s) on the diagnostic chart.

1. Checks the right front ESB. A broken or defective ESB would result in the right front piston being backdriven by hydraulic pressure, and pedal movement would result.
2. Checks for high resistance in the motor "HIGH" circuitry.
3. Checks for high resistance in the motor "LOW" circuitry.
4. Checks for proper resistance of the motor windings.
5. Checks for poor terminal contact or corrosion in the connectors.
6. Checks for a malfunctioning EBTCM.

Diagnostic Aids:

An intermittent malfunction in this DTC may result from a mechanical part of the system that sticks, binds, or slips.

The frequency of the malfunction can be checked by using the enhanced diagnostic function of the Tech 1, as described in "Tech 1 Diagnostics" in this section.

The static modulator test function of the Tech 1 or T-100 may be used to locate an intermittent malfunction associated with the ESB.

DIAGNOSTIC CHART DTC 41 — 1995 SILHOUETTE AND TRANS SPORT WITH TCS, CONTINUED

DTC 41
RIGHT FRONT ESB WILL NOT HOLD MOTOR

(1)
- IGNITION "ON," ENGINE "OFF."
- PUMP BRAKE PEDAL UNTIL FIRM TO DEPLETE VACUUM RESERVE.
- USING TECH 1 MISC. TESTS, SELECT MANUAL CONTROL FUNCTION.
- SELECT ABS MOTOR APPLY FUNCTION OF THE RIGHT FRONT CHANNEL AND APPLY THE MOTOR.
- WAIT 5 SECONDS.
- APPLY FIRM PRESSURE ON BRAKE PEDAL.

DID BRAKE PEDAL FALL?

YES: REPLACE ABS MOTOR PACK.

NO: (2)
- EXIT TECH 1 MANUAL CONTROL FUNCTION.
- IGNITION "OFF."
- DISCONNECT 6-WAY ABS MOTOR PACK CONNECTOR AND GRAY 8-WAY EBTCM CONNECTOR.
- USING J 39200, MEASURE THE RESISTANCE BETWEEN TERMINAL "E" OF THE 6-WAY ABS MOTOR PACK HARNESS CONNECTOR AND TERMINAL "G" OF THE GRAY 8-WAY EBTCM HARNESS CONNECTOR.

IS RESISTANCE 1.5Ω OR LESS?

NO: REPAIR HIGH RESISTANCE IN CKT 1282.

YES: (3)
- USING J 39200, MEASURE THE RESISTANCE BETWEEN TERMINAL "F" OF THE 6-WAY ABS MOTOR PACK HARNESS CONNECTOR AND TERMINAL "H" OF THE GRAY 8-WAY EBTCM HARNESS CONNECTOR.

IS RESISTANCE 1.5Ω OR LESS?

NO: REPAIR HIGH RESISTANCE IN CKT 1283.

YES: (4)
- USING J 39200, MEASURE THE RESISTANCE BETWEEN TERMINAL "E" AND TERMINAL "F" OF THE ABS MOTOR PACK CONNECTOR.

IS RESISTANCE 1.5Ω OR LESS?

NO: REPLACE ABS MOTOR PACK.

YES: (5)
- INSPECT THE 6-WAY ABS MOTOR PACK CONNECTOR AND GRAY 8-WAY EBTCM CONNECTOR FOR POOR TERMINAL CONTACT OR CORROSION.

DO ANY TERMINALS EXHIBIT POOR CONTACT OR EVIDENCE OF CORROSION?

YES: REPLACE TERMINALS THAT EXHIBIT POOR TERMINAL CONTACT OR EVIDENCE OF CORROSION.

NO: (6)
- IGNITION "OFF."
- RECONNECT ALL CONNECTORS.
- START ENGINE AND TEST DRIVE VEHICLE, ACHIEVING AT LEAST 16 km/h (10 MPH).
- REPEAT ABOVE DRIVE CYCLE SEQUENCE TWO MORE TIMES.

DOES DTC 41 SET IN THE LAST THREE DRIVE CYCLES?

YES: REPLACE EBTCM.

NO: MALFUNCTION IS INTERMITTENT. REFER TO "DIAGNOSTIC AIDS" ON FACING PAGE.

AFTER DIAGNOSIS IS COMPLETE, CLEAR DTCs AND TEST DRIVE VEHICLE FOR THREE (3) DRIVE CYCLES TO VERIFY THAT DTC DOES NOT RESET. A DRIVE CYCLE CONSISTS OF STARTING THE VEHICLE, DRIVING OVER 16 km/h (10 MPH), AND THEN TURNING KEY OFF.

DIAGNOSTIC CHART DTC 42 — 1995 SILHOUETTE AND TRANS SPORT WITH TCS

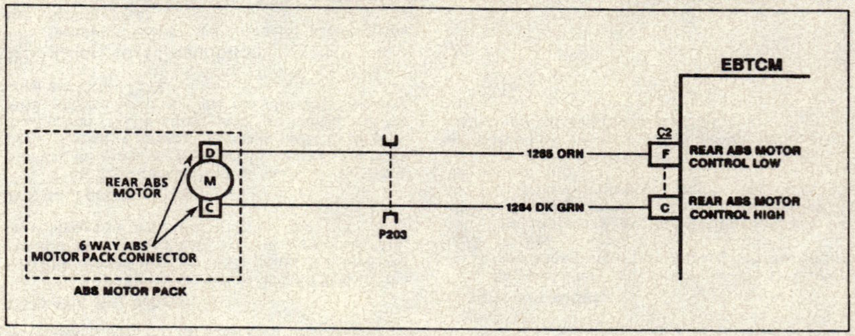

DTC 42
REAR ESB WILL NOT HOLD MOTOR

Circuit Description:

This DTC is designed to detect a slipping rear ESB. During initialization and braking, the rear motor is re-homed. If the ESB slips, the motor/piston will move. During the next key "ON" initialization, a re-home of the motor verifies the motor/piston remained at the home position. If motor movement is detected, the ESB must be slipping.

DTC Will Set When:

DTC 42 can be set during initialization. If the EBTCM detects that the ESB could not hold the piston in the home position, a malfunction exists. DTC 86 is always set with DTC 42.

Action Taken:

If an ESB cannot hold a piston in the home position, the piston may be backdriven when the brake pedal is applied, causing the brake pedal to drop. A malfunction DTC is stored, ABS is disabled, and the ABS warning lamp and the red "BRAKE" warning lamp are turned "ON."

DTC Chart Test Description:

Number(s) below refer to circled number(s) on the diagnostic chart.

1. Checks the rear ESB. A broken or defective ESB would result in the rear pistons being backdriven by hydraulic pressure and wheel movement would result.
2. Checks for high resistance in the motor "HIGH" circuitry.
3. Checks for high resistance in the motor "LOW" circuitry.
4. Checks for proper resistance of the motor windings.
5. Checks for poor terminal contact or corrosion in the connectors.
6. Checks for a malfunctioning EBTCM.

Diagnostic Aids:

An intermittent malfunction in this DTC may result from a mechanical part of the system that sticks, binds, or slips.

The frequency of the malfunction can be checked by using the enhanced diagnostic function of the Tech 1, as described in "Tech 1 Diagnostics" in this section.

The static modulator test function of the Tech 1 or T-100 may be used to locate an intermittent malfunction associated with the ESB.

DIAGNOSTIC CHART DTC 42 — 1995 SILHOUETTE AND TRANS SPORT WITH TCS, CONTINUED

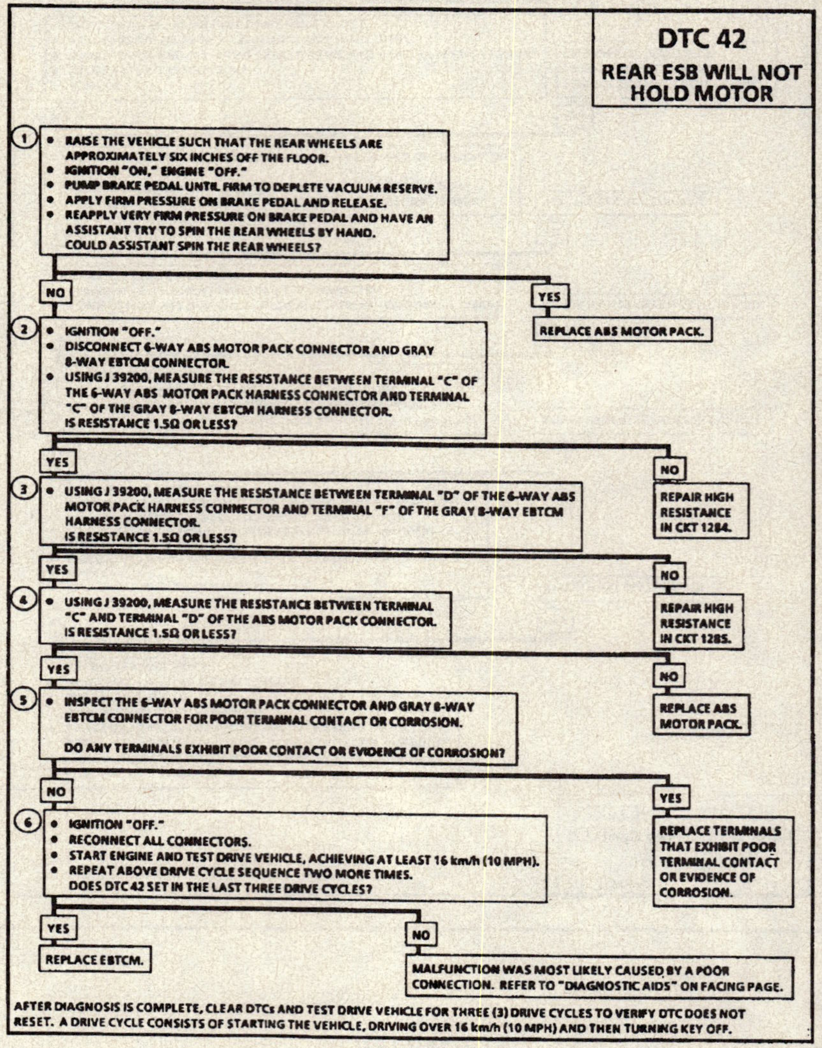

DIAGNOSTIC CHART DTC 44 — 1995 SILHOUETTE AND TRANS SPORT WITH TCS

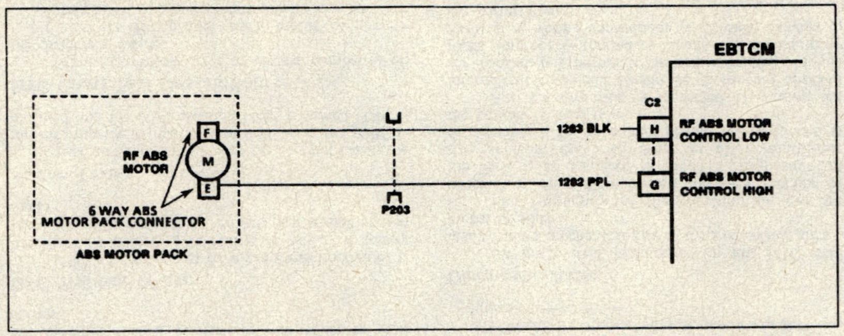

DTC 44

LEFT FRONT ABS CHANNEL WILL NOT MOVE

Circuit Description:

This DTC is designed to detect bound-up ESB, a stuck ABS motor, or a seized hydraulic modulator. When the release is commanded during initialization, the ESB should release the ABS motor, resulting in sensed current being less than commanded current (motor is spinning freely). If the ABS motor is not moving, sensed current will be equal to stall current.

DTC Will Set When:

DTC 44 can be set during normal operation. If the EBTCM detects a condition in which it cannot move the ABS motor in either direction, a malfunction exists.

Action Taken:

This malfunction indicates the channel cannot be moved properly. A malfunction DTC is stored, ABS is disabled and the ABS warning lamp is turned "ON."

DTC Chart Test Description:

Number(s) below refer to circled number(s) on the diagnostic chart.

1. Checks for proper ABS motor movement during apply and release commands from the Tech 1.
2. Compares EBTCM command current to ABS motor feedback current.
3. Checks for a short circuit between the ABS motor "HIGH" and motor "LOW" circuits.
4. Checks for an ABS motor that is internally shorted.
5. Determines if the malfunction is due to poor terminal contact or the EBTCM.
6. Verifies that the ABS motor can actually apply when commanded.
7. Checks for proper hydraulic modulator gear and piston movement.

Diagnostic Aids:

An intermittent malfunction in this DTC may result from a mechanical part of the system that sticks, binds, or slips.

The frequency of the malfunction can be checked by using the enhanced diagnostic function of the Tech 1, as described in "Tech 1 Diagnostics" in this section. DTC 44 may set after modulator disassembly if the modulator pistons are positioned at the bottom of their bore.

Any circuitry that is suspected of causing the intermittent complaint should be thoroughly checked for backed out terminals, improper mating, broken locks, improperly formed or damaged terminals, poor terminal to wiring connections or physical damage to the wiring harness.

DIAGNOSTIC CHART DTC 44 — 1995 SILHOUETTE AND TRANS SPORT WITH TCS, CONTINUED

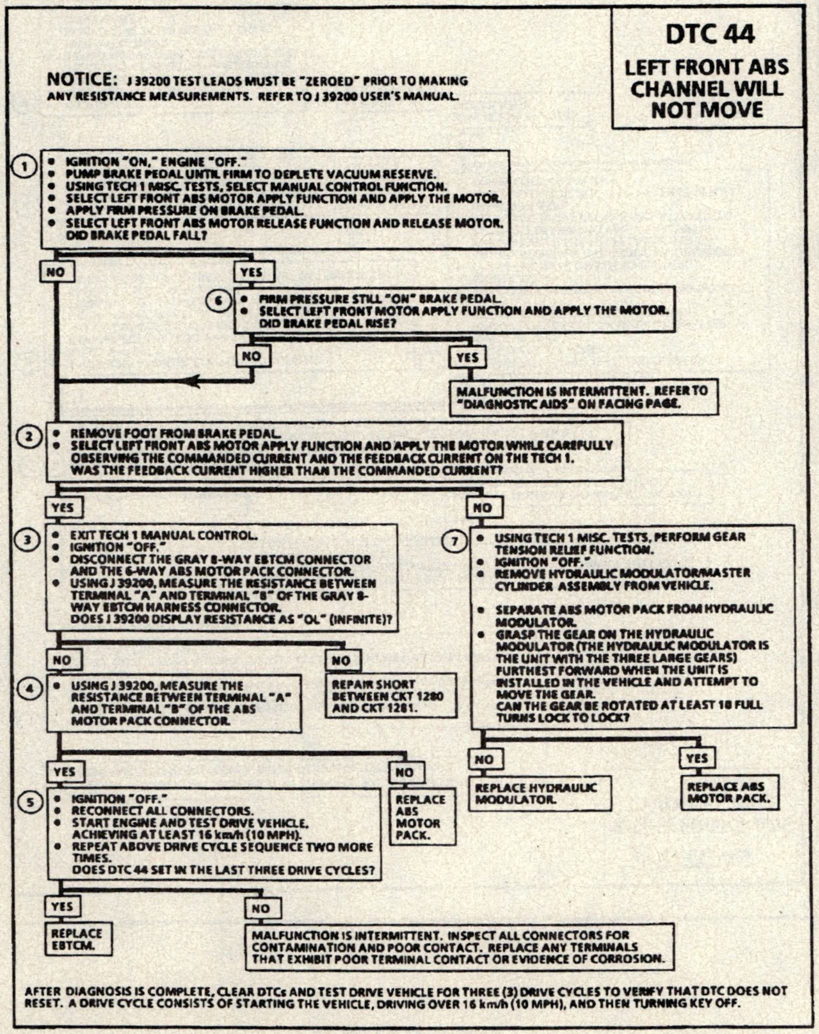

DIAGNOSTIC CHART DTC 45 — 1995 SILHOUETTE AND TRANS SPORT WITH TCS

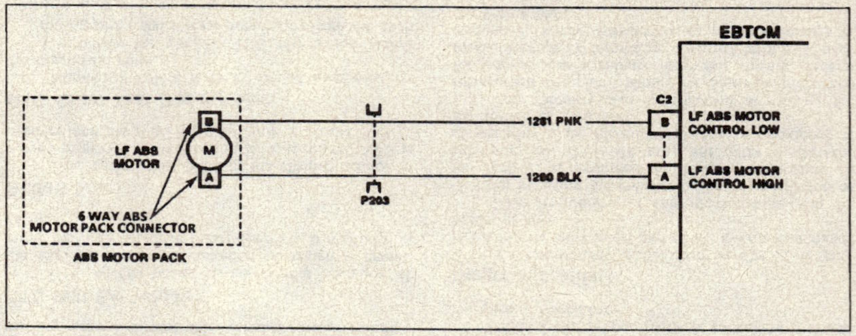

DTC 45
RIGHT FRONT ABS CHANNEL WILL NOT MOVE

Circuit Description:

This DTC is designed to detect bound-up ESB, a stuck ABS motor, or a seized hydraulic modulator. When the release is commanded during initialization, the ESB should release the ABS motor, resulting in sensed current being less than commanded current (ABS motor is spinning freely). If the ABS motor is not moving, sensed current will be equal to stall current.

DTC Will Set When:

DTC 45 can be set during normal operation. If the EBTCM detects a condition in which it cannot move the ABS motor in either direction, a malfunction exists.

Action Taken:

This malfunction indicates the channel cannot be moved properly. A malfunction DTC is stored, ABS is disabled and the ABS warning lamp is turned "ON."

DTC Chart Test Description:

Number(s) below refer to circled number(s) on the diagnostic chart.

1. Checks for proper ABS motor movement during apply and release commands from the Tech
2. Compares EBTCM command current to ABS motor feedback current.
3. Checks for a short circuit between the ABS motor "HIGH" and motor "LOW" circuits.
4. Checks for an ABS motor that is internally shorted.
5. Checks to see if malfunction is due to poor terminal contact or the EBTCM.
6. Verifies that the ABS motor can actually apply when commanded.
7. Checks for proper hydraulic modulator gear and piston movement.

Diagnostic Aids:

An intermittent malfunction in this DTC may result from a mechanical part of the system that sticks, binds, or slips.

The frequency of the malfunction can be checked by using the enhanced diagnostic function of the Tech 1, as described in "Tech 1 Diagnostics" in this section. DTC 45 may set after modulator disassembly if the modulator pistons are positioned at the bottom of their bore.

Any circuitry that is suspected of causing the intermittent complaint should be thoroughly checked for backed out terminals, improper mating, broken locks, improperly formed or damaged terminals, poor terminal to wiring connections or physical damage to the wiring harness.

DIAGNOSTIC CHART DTC 45 — 1995 SILHOUETTE AND TRANS SPORT WITH TCS, CONTINUED

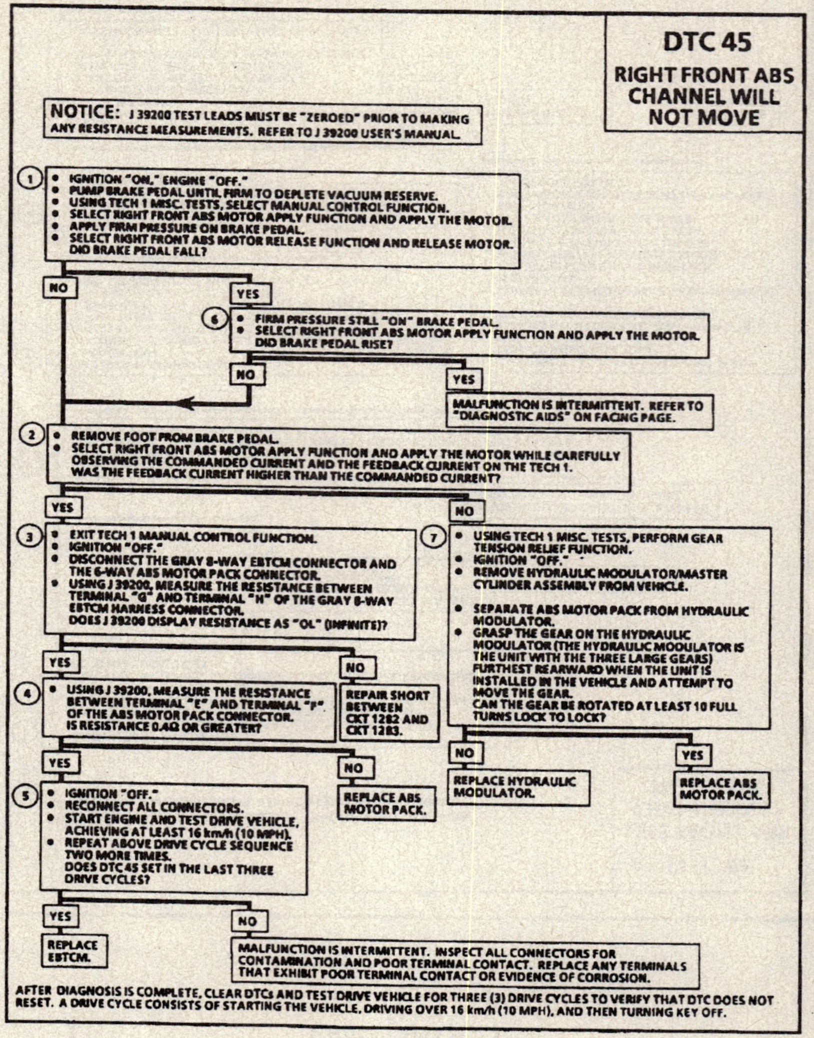

DIAGNOSTIC CHART DTC 46 — 1995 SILHOUETTE AND TRANS SPORT WITH TCS

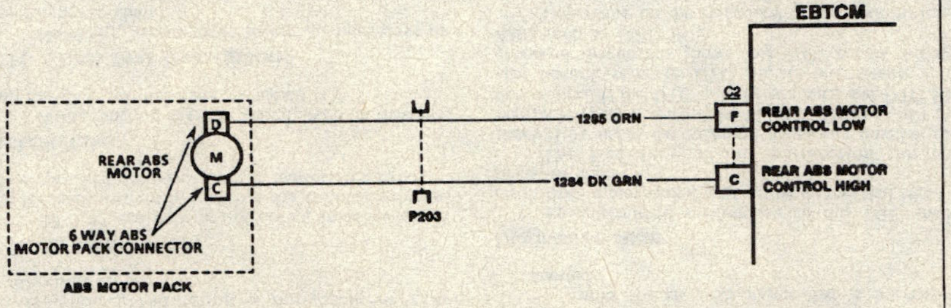

DTC 46
REAR ABS CHANNEL WILL NOT MOVE

Circuit Description:

This DTC is designed to detect bound-up ESB, a stuck ABS motor, or a seized hydraulic modulator. When the release is commanded during initialization, the ESB should release the ABS motor, resulting in sensed current being less than commanded current (motor is spinning freely). If the ABS motor is not moving, sensed current will be equal to stall current.

DTC Will Set When:

DTC 46 can be set during normal operation. If the EBTCM detects a condition in which it cannot move the ABS motor in either direction, a malfunction exists.

Action Taken:

This malfunction indicates the channel cannot be moved properly. A malfunction DTC is stored, ABS is disabled and the ABS warning lamp is turned "ON."

DTC Chart Test Description:

Number(s) below refer to circled number(s) on the diagnostic chart.

1. Checks for proper motor movement during apply and release commands from the Tech 1.
2. Compares EBTCM command current to ABS motor feedback current.
3. Checks for a short circuit between the ABS motor "HIGH" and ABS motor "LOW" circuits.
4. Checks for an ABS motor that is internally shorted.
5. Checks to see if malfunction is due to poor terminal contact or the EBTCM.
6. This step verifies that the motor can actually release when commanded.
7. Checks for proper hydraulic modulator gear and piston movement.

Diagnostic Aids:

An intermittent malfunction in this DTC may result from a mechanical part of the system that sticks or binds.

The frequency of the malfunction can be checked by using the enhanced diagnostic function of the Tech 1, as described in "Tech 1 Diagnostics" in this section. DTC 46 may set after modulator disassembly if the modulator pistons are positioned at the bottom of their bore. Depending on the frequency of the malfunction, a physical inspection of the mechanical parts suspected may be necessary.

Any circuitry that is suspected of causing the intermittent complaint should be thoroughly checked for backed out terminals, improper mating, broken locks, improperly formed or damaged terminals, poor terminal to wiring connections or physical damage to the wiring harness.

DIAGNOSTIC CHART DTC 46 — 1995 SILHOUETTE AND TRANS SPORT WITH TCS, CONTINUED

DTC 46
REAR ABS CHANNEL WILL NOT MOVE

NOTICE: J 39200 TEST LEADS MUST BE "ZEROED" PRIOR TO MAKING ANY RESISTANCE MEASUREMENTS. REFER TO J 39200 USER'S MANUAL.

(1)
- RAISE AND SUPPORT THE VEHICLE SUCH THAT THE REAR WHEELS ARE APPROXIMATELY SIX INCHES OFF THE FLOOR.
- IGNITION "ON," ENGINE "OFF."
- PUMP BRAKE PEDAL UNTIL FIRM TO DEPLETE VACUUM RESERVE.
- APPLY FIRM PRESSURE ON BRAKE PEDAL.
- USING TECH 1 MISC. TESTS, SELECT MANUAL CONTROL FUNCTION.
- SELECT REAR AXLE ABS MOTOR APPLY FUNCTION AND APPLY THE MOTOR.
- HAVE AN ASSISTANT TRY TO SPIN THE REAR WHEELS BY HAND.

COULD ASSISTANT SPIN THE REAR WHEELS?

YES → (2) | NO → (6)

(6)
- FIRM PRESSURE STILL "ON" BRAKE PEDAL.
- SELECT REAR AXLE ABS MOTOR RELEASE FUNCTION AND RELEASE THE MOTOR.
- HAVE AN ASSISTANT TRY TO SPIN THE REAR WHEELS BY HAND.

COULD ASSISTANT SPIN THE REAR WHEELS?

NO → (2) | YES → MALFUNCTION IS INTERMITTENT. REFER TO "DIAGNOSTIC AIDS" ON FACING PAGE.

(2)
- REMOVE FOOT FROM BRAKE PEDAL.
- SELECT REAR AXLE ABS MOTOR APPLY FUNCTION AND APPLY THE MOTOR WHILE CAREFULLY OBSERVING THE COMMANDED CURRENT AND THE FEEDBACK CURRENT ON THE TECH 1.

WAS THE FEEDBACK CURRENT HIGHER THAN THE COMMANDED CURRENT?

YES → (3) | NO → (7)

(3)
- EXIT TECH 1 MANUAL CONTROL FUNCTION.
- IGNITION "OFF."
- DISCONNECT THE GRAY 8-WAY EBTCM CONNECTOR AND THE 6-WAY ABS MOTOR PACK CONNECTOR.
- USING J 39200, MEASURE THE RESISTANCE BETWEEN TERMINAL "C" AND TERMINAL "F" OF THE GRAY 8-WAY EBTCM HARNESS CONNECTOR.

DOES J 39200 DISPLAY RESISTANCE AS "OL" (INFINITE)?

YES → (4) | NO → REPAIR SHORT BETWEEN CKT 1284 AND CKT 1285.

(4)
- USING J 39200, MEASURE THE RESISTANCE BETWEEN TERMINAL "C" AND TERMINAL "D" OF THE ABS MOTOR PACK CONNECTOR.

IS RESISTANCE GREATER THAN 0.4Ω?

YES → (5) | NO → REPLACE ABS MOTOR PACK.

(5)
- RECONNECT ALL CONNECTORS.
- IGNITION "OFF."
- START ENGINE AND TEST DRIVE VEHICLE, ACHIEVING AT LEAST 16 km/h (10 MPH).
- REPEAT ABOVE DRIVE CYCLE SEQUENCE TWO MORE TIMES.

DOES DTC 46 SET IN THE LAST THREE DRIVE CYCLES?

YES → REPLACE EBTCM. | NO → MALFUNCTION IS INTERMITTENT. INSPECT ALL CONNECTORS FOR CONTAMINATION AND POOR TERMINAL CONTACT. REPLACE ANY TERMINALS THAT EXHIBIT POOR TERMINAL CONTACT OR EVIDENCE OF CORROSION.

(7)
- USING TECH 1 MISC. TESTS, PERFORM GEAR TENSION RELIEF FUNCTION.
- IGNITION "OFF."
- REMOVE HYDRAULIC MODULATOR/MASTER CYLINDER ASSEMBLY FROM VEHICLE. REFER TO "ON VEHICLE SERVICE" IN THIS SECTION.
- SEPARATE ABS MOTOR PACK FROM HYDRAULIC MODULATOR.
- GRASP THE MIDDLE GEAR ON THE HYDRAULIC MODULATOR (THE HYDRAULIC MODULATOR IS THE UNIT WITH THE THREE LARGE GEARS) AND ATTEMPT TO MOVE THE GEAR.

CAN THE GEAR BE ROTATED AT LEAST 7 FULL TURNS LOCK TO LOCK?

NO → REPLACE HYDRAULIC MODULATOR. | YES → REPLACE ABS MOTOR PACK.

AFTER DIAGNOSIS IS COMPLETE, CLEAR DTCs AND TEST DRIVE VEHICLE FOR THREE (3) DRIVE CYCLES TO VERIFY THAT DTC DOES NOT RESET. A DRIVE CYCLE CONSISTS OF STARTING THE VEHICLE, DRIVING OVER 16 km/h (10 MPH), AND THEN TURNING KEY OFF.

DIAGNOSTIC CHART DTC 47 — 1995 SILHOUETTE AND TRANS SPORT WITH TCS

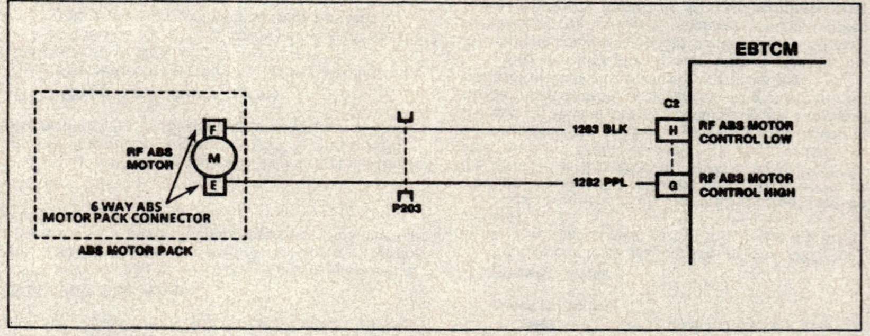

DTC 47
(Page 1 of 2)
LEFT FRONT ABS MOTOR FREE SPINS

Circuit Description:

This DTC is designed to detect a stripped nut or gear assembly during initialization. During the homing sequence, the piston should reach the top of the bore resulting in a stalled ABS motor. If this does not occur, the ABS motor must be spinning with little or no resistance, indicating a nut/screw or gear malfunction.

DTC Will Set When:

DTC 47 can only be set during initialization. If the feedback current is less than the command current for a specified period of time, a malfunction exists.

Action Taken:

A malfunction DTC is stored, ABS is disabled, and the ABS warning lamp is turned "ON."

DTC Chart Test Description:

Number(s) below refer to circled number(s) on the diagnostic chart.

1. Checks to see if the corresponding open ABS motor DTC is also set.
2. Verifies that the ABS motor was actually applied as commanded by observing feedback current.
3. Verifies that the ABS motor can release.
4. This step verifies that the ABS motor can be applied by observing pedal movement.
5. Checks for an open in the ABS motor "HIGH" circuitry.
6. Checks for an open in the ABS motor "LOW" circuitry.
7. Checks for an ABS motor that is internally shorted.

Diagnostic Aids:

An intermittent malfunction in this DTC may result from a mechanical part of the system that sticks, binds, or slips.

The frequency of the malfunction can be checked by using the enhanced diagnostic function of the Tech 1, as described in "Tech 1 Diagnostics" in this section. If the DTC only occurs once and DTC 56 also occurs, refer to DTC 56. If intermittent and enhanced diagnostics shows this DTC occurs during ABS, refer to DTC 56.

Depending on the frequency of the malfunction, a physical inspection of the mechanical parts suspected may be necessary.

DIAGNOSTIC CHART DTC 47 — 1995 SILHOUETTE AND TRANS SPORT WITH TCS, CONTINUED

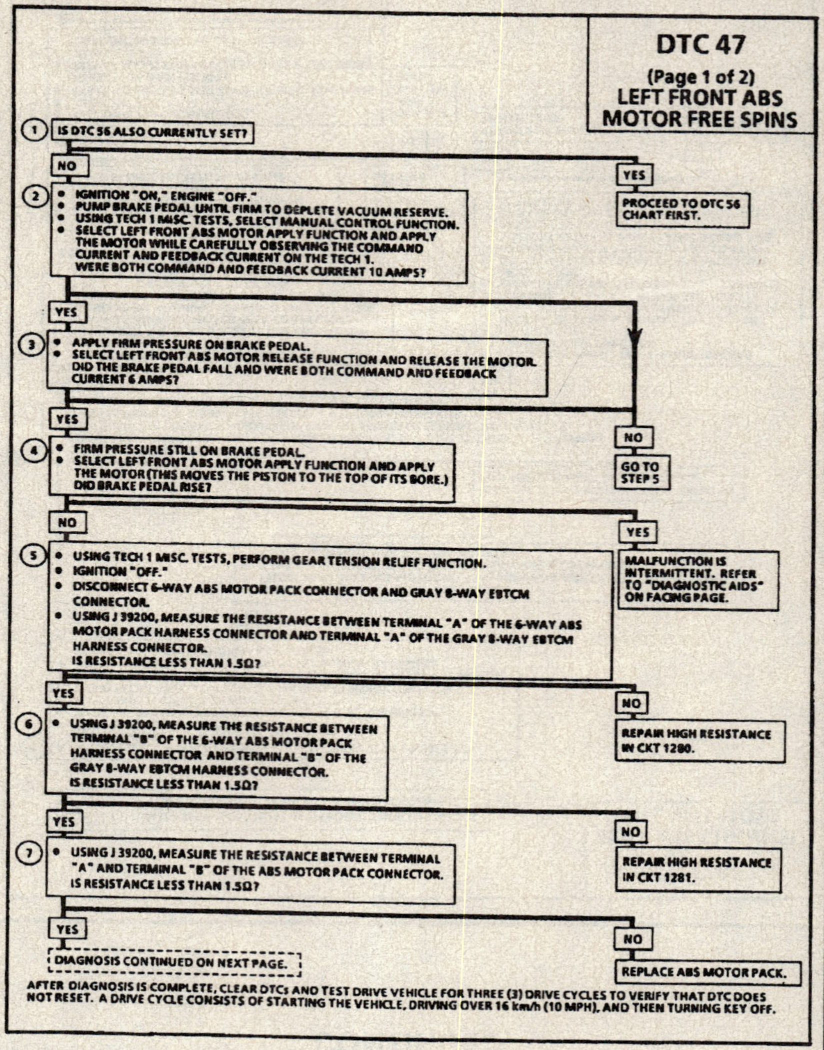

DIAGNOSTIC CHART DTC 47 — 1995 SILHOUETTE AND TRANS SPORT WITH TCS, CONTINUED

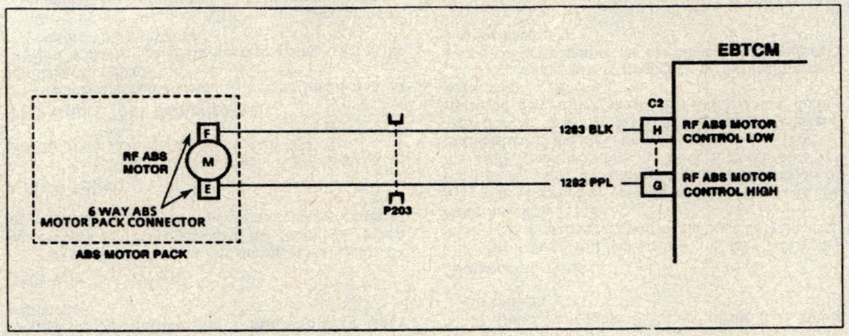

DTC 47
(Page 2 of 2)
LEFT FRONT ABS MOTOR FREE SPINS

Circuit Description:

This DTC is designed to detect a stripped nut or gear assembly during initialization. During the homing sequence, the piston should reach the top of the bore resulting in a stalled ABS motor. If this does not occur, the ABS motor must be spinning with little or no resistance, indicating a nut/screw or gear malfunction.

DTC Will Set When:

DTC 47 can only be set during initialization. If the feedback current is less than the command current for a specified period of time, a malfunction exists.

Action Taken:

A malfunction DTC is stored, ABS is disabled, and the ABS warning lamp is turned "ON."

DTC Chart Test Description:

Number(s) below refer to circled number(s) on the diagnostic chart.

8. Checks for a stripped gear on the motor pack (unit with three small gears).
9. Checks for a stripped gear on the hydraulic modulator (unit with three large gears).
10. Checks to see if the ABS motor pack has an internal malfunction.
11. Verifies that the hydraulic modulator is functioning properly.
12. Ensures malfunction was not due to a poor terminal contact.

Diagnostic Aids:

An intermittent malfunction in this DTC may result from a mechanical part of the system that sticks, binds, or slips.

The frequency of the malfunction can be checked by using the enhanced diagnostic function of the Tech 1, as described in "Tech 1 Diagnostics" in this section. If the DTC only occurs once and DTC 56 also occurs, refer to DTC 56. If intermittent and enhanced diagnostics shows this DTC occurs during ABS, refer to DTC 56.

Depending on the frequency of the malfunction, a physical inspection of the mechanical parts suspected may be necessary.

DIAGNOSTIC CHART DTC 47 — 1995 SILHOUETTE AND TRANS SPORT WITH TCS, CONTINUED

DTC 47 (Page 2 of 2) LEFT FRONT ABS MOTOR FREE SPINS

DIAGNOSIS CONTINUED FROM PREVIOUS PAGE.

(8)
- REMOVE HYDRAULIC MODULATOR/MASTER CYLINDER ASSEMBLY FROM VEHICLE.
- REMOVE GEAR COVER.
- CHECK FOR STRIPPED GEAR ON ABS MOTOR PACK; ABS MOTOR PACK IS THE UNIT WITH THE THREE SMALL GEARS. THE LEFT FRONT GEAR IS THE GEAR FURTHEST FORWARD WHEN THE ASSEMBLY IS INSTALLED IN THE VEHICLE. IS THE ABS MOTOR PACK GEAR STRIPPED?

YES: REPLACE ABS MOTOR PACK.
NO: go to (9)

(9)
- CHECK FOR A STRIPPED GEAR ON HYDRAULIC MODULATOR; MODULATOR IS THE UNIT WITH THE THREE LARGE GEARS. THE LEFT FRONT GEAR IS THE GEAR FURTHEST FORWARD WHEN THE UNIT IS INSTALLED IN THE VEHICLE. IS THE MODULATOR GEAR STRIPPED?

YES: REPLACE HYDRAULIC MODULATOR GEAR.
NO: go to (10)

(10)
- RECONNECT ELECTRICAL CONNECTORS
- SECURELY POSITION HYDRAULIC MODULATOR/MASTER CYLINDER ASSEMBLY IN VEHICLE WITH GEAR COVER REMOVED SUCH THAT THE GEAR SET CAN BE OBSERVED. BE CAREFUL NOT TO DAMAGE GEAR SET OR ALLOW THE HYDRAULIC MODULATOR TO MOVE WHILE TESTING.
- IGNITION "ON."
- USING TECH 1 MISC. TESTS, SELECT MANUAL CONTROL FUNCTION.
- SELECT LEFT FRONT ABS MOTOR APPLY AND APPLY THE MOTOR WHILE OBSERVING THE GEAR SET FURTHEST FORWARD WHEN THE UNIT IS INSTALLED IN THE VEHICLE.
- SELECT LEFT ABS FRONT MOTOR RELEASE FUNCTION AND RELEASE THE MOTOR WHILE STILL OBSERVING THE GEAR SET.
- SELECT LEFT FRONT ABS MOTOR APPLY FUNCTION AND APPLY THE MOTOR. DID GEAR SET MOVE IN BOTH DIRECTIONS FOR AT LEAST ONE REVOLUTION?

NO: REPLACE ABS MOTOR PACK.
YES: go to (11)

(11)
- USING TECH 1 MISC. TESTS, PERFORM GEAR TENSION RELIEF FUNCTION.
- IGNITION "OFF."
- SEPARATE THE ABS MOTOR PACK FROM THE HYDRAULIC MODULATOR.
- GRASP THE GEAR ON THE HYDRAULIC MODULATOR (THE HYDRAULIC MODULATOR IS THE UNIT WITH THE THREE LARGE GEARS) FURTHEST FORWARD WHEN THE UNIT IS INSTALLED IN THE VEHICLE AND ATTEMPT TO MOVE THE GEAR. CAN THE GEAR BE ROTATED MORE THAN 13 FULL TURNS LOCK TO LOCK?

YES: REPLACE THE HYDRAULIC MODULATOR.
NO: go to (12)

(12)
- INSPECT ALL CONNECTORS AND TERMINALS FOR POOR TERMINAL CONTACT AND EVIDENCE OF CORROSION.
  REPLACE ALL TERMINALS THAT EXHIBIT SIGNS OF POOR TERMINAL CONTACT OR CORROSION.
- REINSTALL HYDRAULIC MODULATOR/MASTER CYLINDER ASSEMBLY IN VEHICLE.
- RECONNECT ALL CONNECTORS.
- IGNITION "OFF."
- START ENGINE AND TEST DRIVE VEHICLE, ACHIEVING AT LEAST 16 km/h (10 MPH).
- REPEAT ABOVE DRIVE CYCLE SEQUENCE TWO MORE TIMES. DOES DTC 47 SET IN THE LAST THREE DRIVE CYCLES?

YES: REPLACE EBTCM.
NO: MALFUNCTION WAS MOSTLY LIKELY CAUSED BY A POOR CONNECTION. REFER TO "DIAGNOSTIC AIDS" ON FACING PAGE.

AFTER DIAGNOSIS IS COMPLETE, CLEAR DTCs AND TEST DRIVE VEHICLE FOR THREE (3) DRIVE CYCLES TO VERIFY THAT DTC DOES NOT RESET. A DRIVE CYCLE CONSISTS OF STARTING THE VEHICLE, DRIVING OVER 16 KM/H (10 MPH), AND THEN TURNING KEY OFF.

DIAGNOSTIC CHART DTC 48 — 1995 SILHOUETTE AND TRANS SPORT WITH TCS

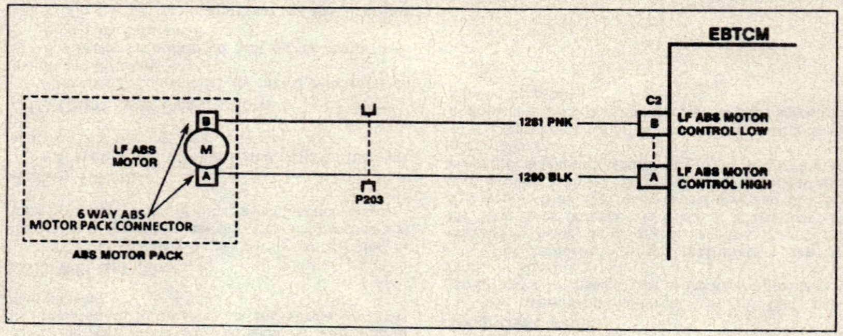

DTC 48
(Page 1 of 2)
RIGHT FRONT ABS MOTOR FREE SPINS

Circuit Description:

This DTC is designed to detect a stripped nut or gear assembly during initialization. During the homing sequence, the piston should reach the top of the bore resulting in a stalled ABS motor. If this does not occur, the ABS motor must be spinning with little or no resistance, indicating a nut/screw or gear malfunction.

DTC Will Set When:

DTC 48 can only be set during initialization. If the feedback current is less than the command current for a specified period of time, a malfunction exists.

Action Taken:

A malfunction DTC is stored, ABS is disabled, and the ABS warning lamp is turned "ON."

DTC Chart Test Description:

Number(s) below refer to circled number(s) on the diagnostic chart.

1. Checks to see if the corresponding open ABS motor DTC is also set.
2. Verifies that the ABS motor was actually applied as commanded by observing feedback current.
3. Verifies that the ABS motor can release.
4. Verifies that the ABS motor can be applied by observing pedal movement.
5. Checks for high resistance in the ABS motor "HIGH" circuitry.
6. Checks for high resistance in the ABS motor "LOW" circuitry.
7. Checks for proper resistance of the ABS motor windings.

Diagnostic Aids:

An intermittent malfunction in this DTC may result from a mechanical part of the system that sticks, binds, or slips.

The frequency of the malfunction can be checked by using the enhanced diagnostic function of the Tech 1, as described in "Tech 1 Diagnostics" in this section. If the DTC only occurs once and DTC 61 also occurs, refer to DTC 61. If intermittent and enhanced diagnostics shows this DTC occurs during ABS, refer to DTC 61.

Depending on the frequency of the malfunction, a physical inspection of the mechanical parts suspected may be necessary.

DIAGNOSTIC CHART DTC 48 — 1995 SILHOUETTE AND TRANS SPORT WITH TCS, CONTINUED

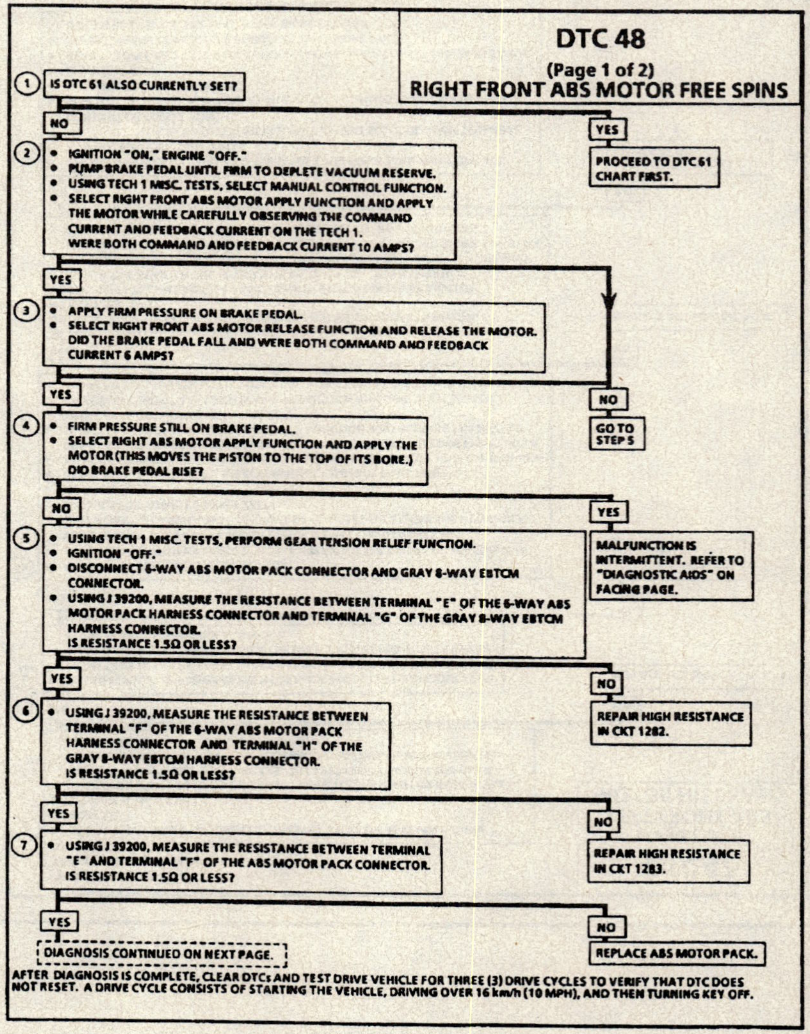

DIAGNOSTIC CHART DTC 48 — 1995 SILHOUETTE AND TRANS SPORT WITH TCS, CONTINUED

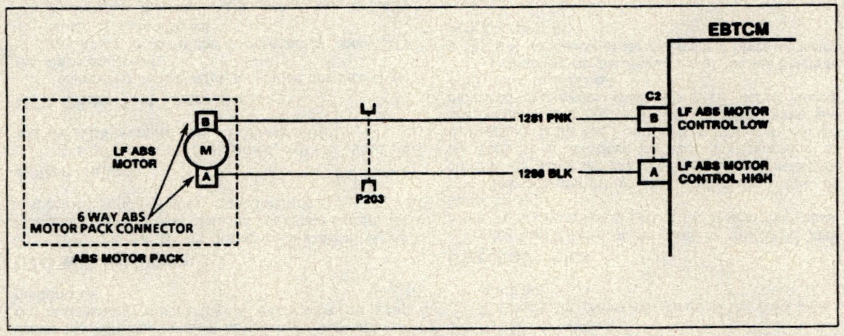

DTC 48
(Page 2 of 2)
RIGHT FRONT ABS MOTOR FREE SPINS

Circuit Description:

This DTC is designed to detect a stripped nut or gear assembly during initialization. During the homing sequence, the piston should reach the top of the bore resulting in a stalled ABS motor. If this does not occur, the ABS motor must be spinning with little or no resistance, indicating a nut/screw or gear malfunction.

DTC Will Set When:

DTC 48 can only be set during initialization. If the feedback current is less than the command current for a specified period of time, a malfunction exists.

Action Taken:

A malfunction DTC is stored, ABS is disabled, and the ABS warning lamp is turned "ON."

DTC Chart Test Description:

Number(s) below refer to circled number(s) on the diagnostic chart.

8. Checks for a stripped gear on the ABS motor pack (unit with three small gears).
9. Checks for a stripped gear on the hydraulic modulator (unit with three large gears).
10. Checks to see if the ABS motor pack has an internal malfunction.
11. Verifies that the hydraulic modulator is functioning properly.
12. Ensures malfunction was not due to poor terminal contact.

Diagnostic Aids:

An intermittent malfunction in this DTC may result from a mechanical part of the system that sticks, binds, or slips.

The frequency of the malfunction can be checked by using the enhanced diagnostic function of the Tech 1, as described in "Tech 1 Diagnostics" in this section. If the DTC only occurs once and DTC 61 also occurs, refer to DTC 61. If intermittent and enhanced diagnostics shows this DTC occurs during ABS, refer to DTC 61.

Depending on the frequency of the malfunction, a physical inspection of the mechanical parts suspected may be necessary.

DIAGNOSTIC CHART DTC 48 — 1995 SILHOUETTE AND TRANS SPORT WITH TCS, CONTINUED

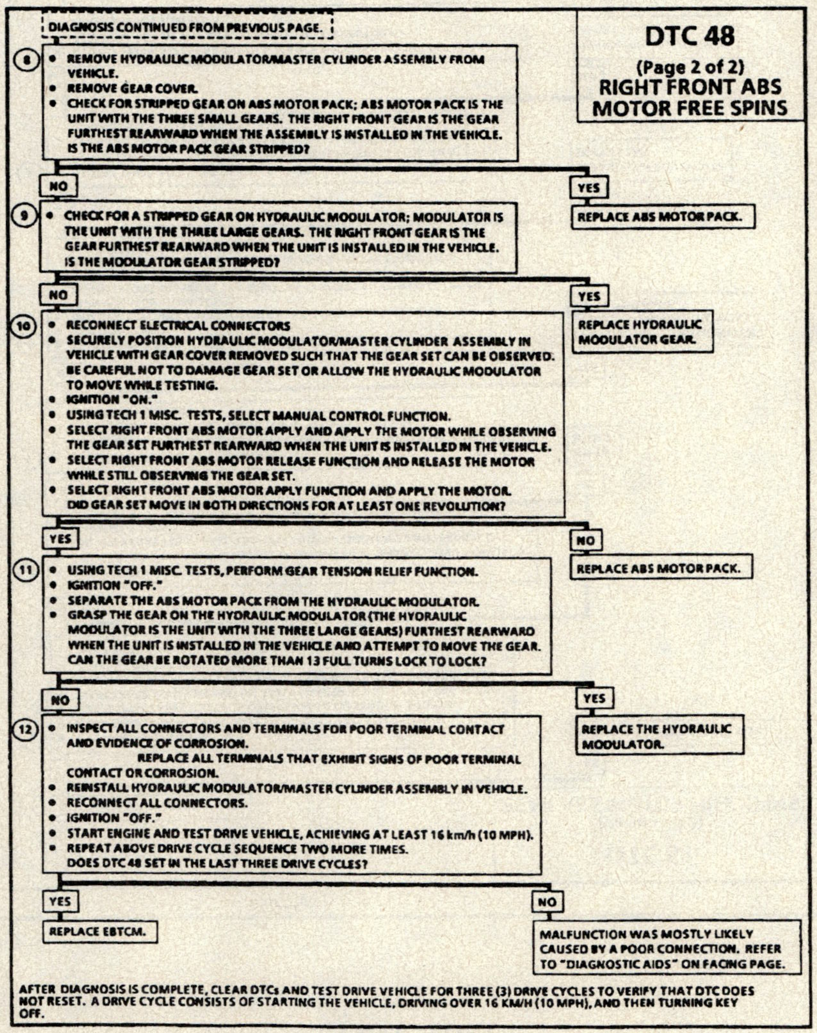

DIAGNOSTIC CHART DTC 51 — 1995 SILHOUETTE AND TRANS SPORT WITH TCS

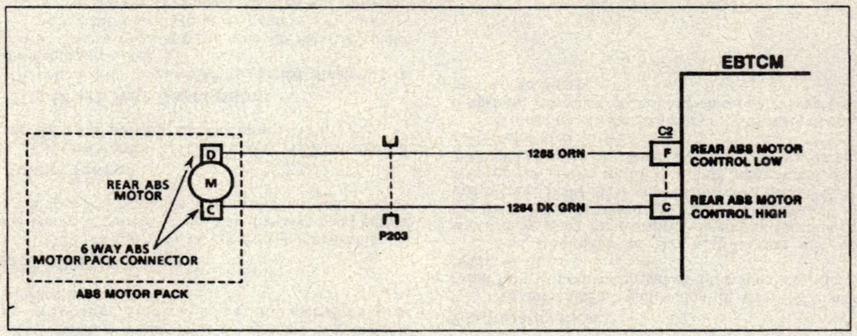

DTC 51
(Page 1 of 2)
REAR ABS MOTOR FREE SPINS

Circuit Description:

This DTC is designed to detect a stripped nut or gear assembly during initialization. During the homing sequence, the piston should reach the top of the bore resulting in a stalled ABS motor. If this does not occur, the ABS motor must be spinning with little or no resistance, indicating a nut/screw or gear malfunction.

DTC Will Set When:

DTC 51 can be set during initialization. If the feedback current is less than the command current for a specified period of time, a malfunction exists.

Action Taken:

A malfunction DTC is stored, ABS is disabled, and the ABS warning lamp is turned "ON."

DTC Chart Test Description:

Number(s) below refer to circled number(s) on the diagnostic chart.

1. Checks to see if the corresponding open ABS motor DTC is also set.
2. Verifies that the ABS motor was actually applied as commanded by observing feedback current.
3. Verifies that the ABS motor can release.
4. Verifies that the ABS motor can be applied by observing pedal movement.
5. Checks for high resistance in the ABS motor "HIGH" circuitry.
6. Checks for high resistance in the ABS motor "LOW" circuitry.
7. Checks for proper resistance of the ABS motor windings.

Diagnostic Aids:

An intermittent malfunction in this DTC may result from a mechanical part of the system that sticks, binds, or slips.

The frequency of the malfunction can be checked by using the enhanced diagnostic function of the Tech 1, as described in "Tech 1 Diagnostics" in this section. If the DTC only occurs once and DTC 64 also occurs, refer to DTC 64. If intermittent and enhanced diagnostics show this DTC occurs during ABS, refer to DTC 64.

Depending on the frequency of the malfunction, a physical inspection of the mechanical parts suspected may be necessary.

DIAGNOSTIC CHART DTC 51 — 1995 SILHOUETTE AND TRANS SPORT WITH TCS, CONTINUED

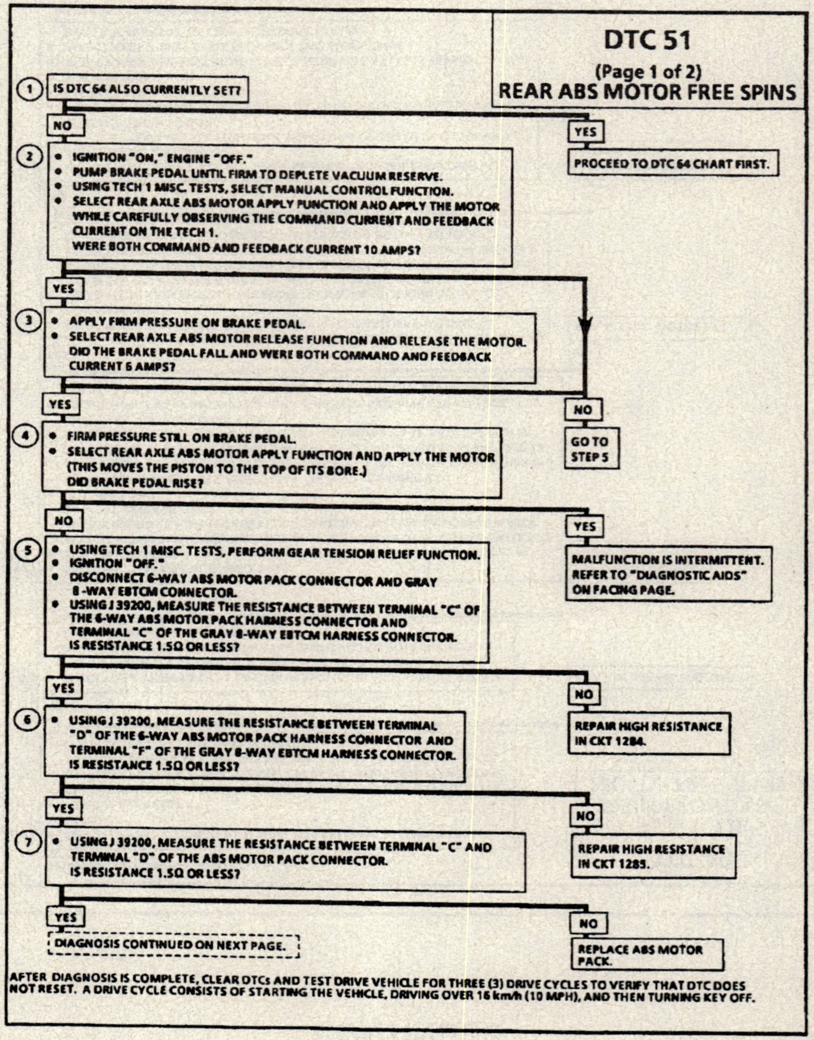

DIAGNOSTIC CHART DTC 51 — 1995 SILHOUETTE AND TRANS SPORT WITH TCS, CONTINUED

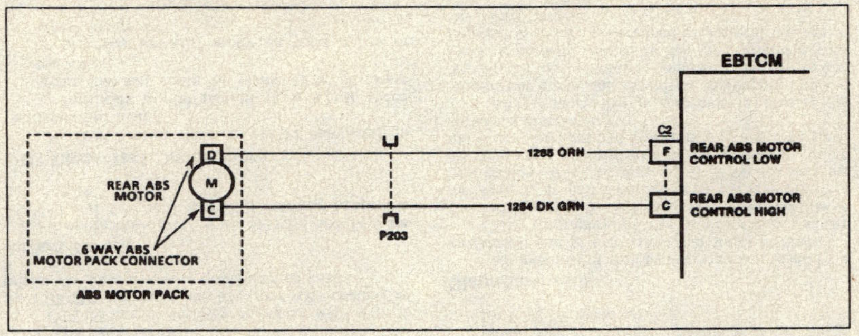

DTC 51
(Page 2 of 2)
REAR ABS MOTOR FREE SPINS

Circuit Description:

This DTC is designed to detect a stripped nut or gear assembly during initialization. During the homing sequence, the piston should reach the top of the bore resulting in a stalled ABS motor. If this does not occur, the ABS motor must be spinning with little or no resistance, indicating a nut/screw or gear malfunction.

DTC Will Set When:

DTC 51 can be set during initialization. If the feedback current is less than the command current for a specified period of time, a malfunction exists.

Action Taken:

A malfunction DTC is stored, ABS is disabled, and the ABS warning lamp is turned "ON."

DTC Chart Test Description:

Number(s) below refer to circled number(s) on the diagnostic chart.

8. Checks for a stripped gear on the ABS motor pack (unit with three small gears).
9. Checks for a stripped gear on the hydraulic modulator (unit with three large gears).
10. Verifies that the ABS motor can be applied under a load situation.
11. Verifies that the hydraulic modulator is functioning properly.
12. Ensures malfunction was not due to poor terminal contact.

Diagnostic Aids:

An intermittent malfunction in this DTC may result from a mechanical part of the system that sticks, binds, or slips.

The frequency of the malfunction can be checked by using the enhanced diagnostic function of the Tech 1, as described in "Tech 1 Diagnostics" in this section. If the DTC only occurs once and DTC 64 also occurs, refer to DTC 64. If intermittent and enhanced diagnostics show this DTC occurs during ABS, refer to DTC 64.

Depending on the frequency of the malfunction, a physical inspection of the mechanical parts suspected may be necessary.

DIAGNOSTIC CHART DTC 51 — 1995 SILHOUETTE AND TRANS SPORT WITH TCS, CONTINUED

DTC 51
(Page 2 of 2)
REAR ABS MOTOR FREE SPINS

DIAGNOSIS CONTINUED FROM PREVIOUS PAGE.

(8)
- REMOVE HYDRAULIC MODULATOR/MASTER CYLINDER ASSEMBLY FROM VEHICLE.
- REMOVE GEAR COVER.
- CHECK FOR STRIPPED GEAR ON ABS MOTOR PACK; ABS MOTOR PACK IS THE UNIT WITH THE THREE SMALL GEARS. THE REAR AXLE GEAR IS THE MIDDLE GEAR OF THE THREE.

IS THE ABS MOTOR PACK GEAR STRIPPED?

- YES: REPLACE ABS MOTOR PACK.
- NO: go to (9)

(9)
- CHECK FOR A STRIPPED GEAR ON HYDRAULIC MODULATOR; MODULATOR IS THE UNIT WITH THE THREE LARGE GEARS. THE REAR AXLE GEAR IS THE MIDDLE GEAR OF THE THREE.

IS THE MODULATOR GEAR STRIPPED?

- YES: REPLACE HYDRAULIC MODULATOR GEAR.
- NO: go to (10)

(10)
- RECONNECT ELECTRICAL CONNECTORS
- SECURELY POSITION HYDRAULIC MODULATOR/MASTER CYLINDER ASSEMBLY IN VEHICLE WITH GEAR COVER REMOVED SUCH THAT THE GEAR SET CAN BE OBSERVED. BE CAREFUL NOT TO DAMAGE GEAR SET OR ALLOW THE HYDRAULIC MODULATOR TO MOVE WHILE TESTING.
- IGNITION "ON."
- USING TECH 1 MISC. TESTS, SELECT MANUAL CONTROL FUNCTION.
- SELECT REAR AXLE ABS MOTOR APPLY AND APPLY THE MOTOR WHILE OBSERVING THE MIDDLE GEAR SET.
- SELECT REAR AXLE ABS MOTOR RELEASE FUNCTION AND RELEASE THE MOTOR WHILE STILL OBSERVING THE GEAR SET.
- SELECT REAR AXLE ABS MOTOR APPLY FUNCTION AND APPLY THE MOTOR.

DID GEAR SET MOVE IN BOTH DIRECTIONS FOR AT LEAST ONE REVOLUTION?

- NO: REPLACE ABS MOTOR PACK.
- YES: go to (11)

(11)
- USING TECH 1 MISC. TESTS, PERFORM GEAR TENSION RELIEF FUNCTION.
- IGNITION "OFF."
- SEPARATE THE ABS MOTOR PACK FROM THE HYDRAULIC MODULATOR.
- GRASP THE MIDDLE GEAR ON THE HYDRAULIC MODULATOR (THE HYDRAULIC MODULATOR IS THE UNIT WITH THE THREE LARGE GEARS) AND ATTEMPT TO MOVE THE GEAR.

CAN THE GEAR BE ROTATED MORE THAN 8 FULL TURNS LOCK TO LOCK?

- YES: REPLACE THE HYDRAULIC MODULATOR.
- NO: go to (12)

(12)
- INSPECT ALL CONNECTORS AND TERMINALS FOR POOR TERMINAL CONTACT AND EVIDENCE OF CORROSION.
  REPLACE ALL TERMINALS THAT EXHIBIT SIGNS OF POOR TERMINAL CONTACT OR CORROSION.
- REINSTALL HYDRAULIC MODULATOR/MASTER CYLINDER ASSEMBLY IN VEHICLE.
- RECONNECT ALL CONNECTORS.
- IGNITION "OFF."
- START ENGINE AND TEST DRIVE VEHICLE, ACHIEVING AT LEAST 16 km/h (10 MPH).
- REPEAT ABOVE DRIVE CYCLE SEQUENCE TWO MORE TIMES.

DOES DTC 51 RESET?

- YES: REPLACE EBTCM.
- NO: MALFUNCTION WAS MOSTLY LIKELY CAUSED BY A POOR CONNECTION. REFER TO "DIAGNOSTIC AIDS" ON FACING PAGE.

AFTER DIAGNOSIS IS COMPLETE, CLEAR DTCs AND TEST DRIVE VEHICLE FOR THREE (3) DRIVE CYCLES TO VERIFY THAT DTC DOES NOT RESET. A DRIVE CYCLE CONSISTS OF STARTING THE VEHICLE, DRIVING OVER 16 KM/H (10 MPH), AND THEN TURNING KEY OFF.

DIAGNOSTIC CHART DTC 52 — 1995 SILHOUETTE AND TRANS SPORT WITH TCS

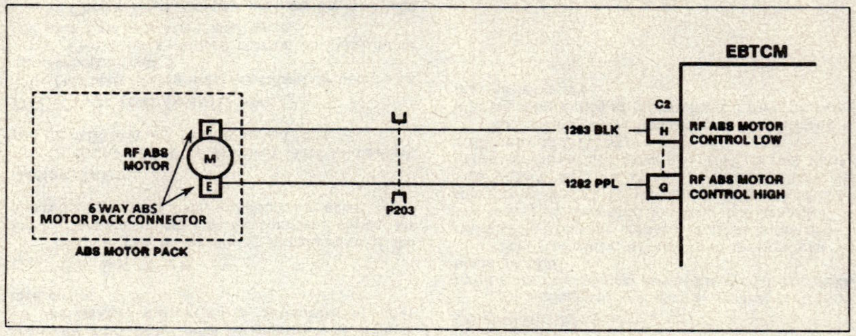

DTC 52
LEFT FRONT ABS CHANNEL IN RELEASE TOO LONG

Circuit Description:

This DTC will diagnose an ABS motor that is energized longer than expected. This could occur if a wheel speed sensor is malfunctioning, the ABS motor does not turn, the left front solenoid mechanically fails open, or the ABS motor wires are crossed.

DTC Will Set When:

DTC 52 can be set only during an ABS stop. If the EBTCM commands the left front ABS channel in release for three seconds, a malfunction exists.

Action Taken:

A malfunction DTC is stored, ABS and TCS are disabled and the ABS and TCS warning lamps are turned "ON."

DTC Chart Test Description:

Number(s) below refer to circled number(s) on the diagnostic chart.

1. Identifies a malfunction in a wheel speed sensor that may cause the system to be in release too long.
2. Identifies an ABS motor as being failed or wired incorrectly.
3. Checks for a solenoid that may have mechanically failed open.
4. Verifies proper solenoid operation.
5. If Step 3 has failed, this serves to isolate the cause of the hydraulic malfunction, to either the solenoid or the ABS hydraulic modulator assembly.
6. Determines whether a malfunctioning ABS motor pack or hydraulic modulator is the reason for DTC 52 being set.

Diagnostic Aids:

An intermittent malfunction may be caused by a mechanical part of the system that sticks or binds.

The frequency of the malfunction can be checked by using the enhanced diagnostic function of the Tech 1, as described in "Tech 1 Diagnostics" in this section. DTC 52 may occur if on ice and the steering wheel is turned to lock during braking. Using the Tech 1, perform hydraulic test to ensure total brake system is functional.

Any circuitry that is suspected of causing the intermittent complaint should be thoroughly checked for backed out terminals, improper mating, broken locks, improperly formed or damaged terminals, poor terminal to wiring connections or physical damage to the wiring harness.

DIAGNOSTIC CHART DTC 52 — 1995 SILHOUETTE AND TRANS SPORT WITH TCS, CONTINUED

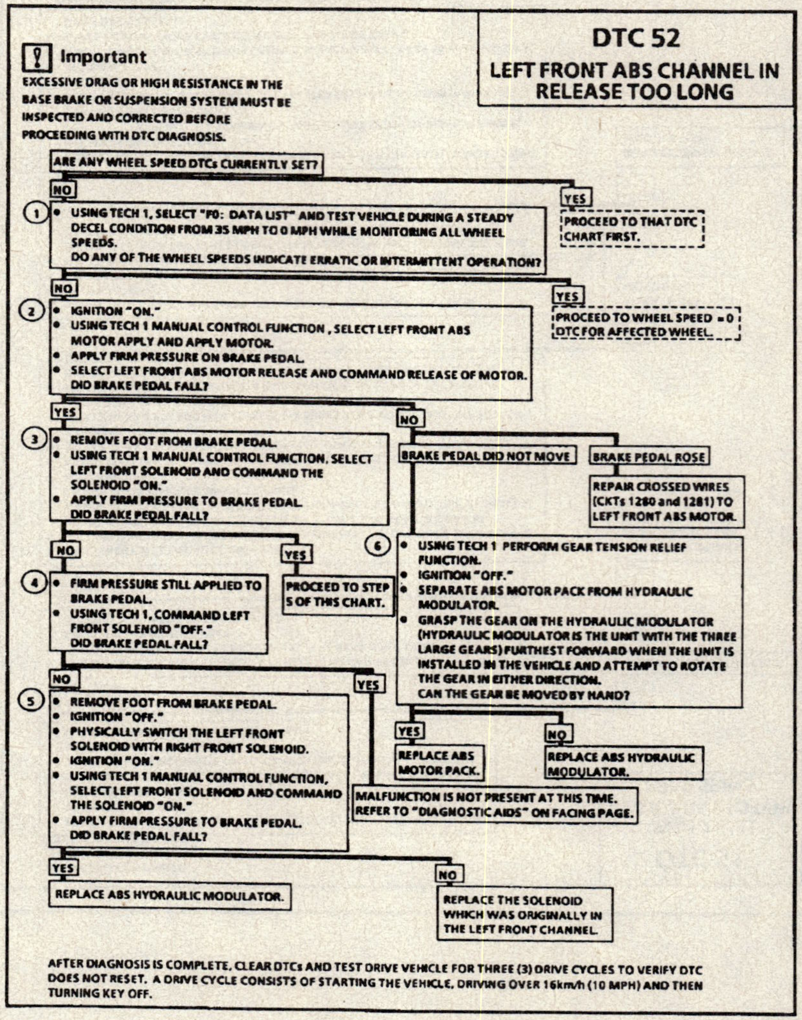

DIAGNOSTIC CHART DTC 53 — 1995 SILHOUETTE AND TRANS SPORT WITH TCS

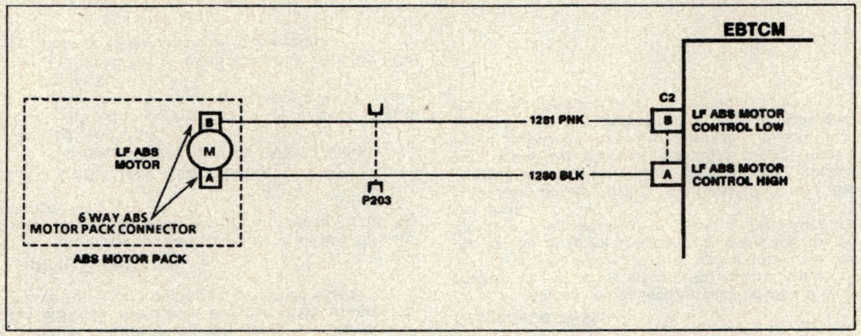

DTC 53
RIGHT FRONT ABS CHANNEL IN RELEASE TOO LONG

Circuit Description:

This DTC will diagnose an ABS motor that is energized longer than expected. This could occur if a wheel speed sensor is malfunctioning, the ABS motor does not turn, the right front solenoid mechanically fails open, or the ABS motor wires are crossed.

DTC Will Set When:

DTC 53 can be set only during an ABS stop. If the EBTCM commands the right front channel in release for three seconds, a malfunction exists.

Action Taken:

A malfunction DTC is stored, ABS and TCS are disabled and the ABS and TCS warning lamps are turned "ON."

DTC Chart Test Description:

Number(s) below refer to circled number(s) on the diagnostic chart.

1. Identifies a malfunction in a wheel speed sensor that may cause the system to be in release too long.
2. Identifies a motor as being failed or wired incorrectly.
3. Checks for a solenoid that may have mechanically failed open.
4. Checks for a solenoid that may have mechanically failed closed.
5. If Step 3 has failed, this serves to isolate the cause of the hydraulic malfunction, to either the solenoid or the ABS hydraulic modulator assembly.
6. Determines whether a malfunctioning ABS motor pack or hydraulic modulator is the reason for DTC 53 being set.

Diagnostic Aids:

An intermittent malfunction may be caused by a mechanical part of the system that sticks or binds.

The frequency of the malfunction can be checked by using the enhanced diagnostic function of the Tech 1, as described in "Tech 1 Diagnostics" in this section. DTC 53 may occur on ice if the steering wheel is turned to lock during braking. Using the Tech 1 perform hydraulic test to ensure total brake system is functional.

Any circuitry that is suspected of causing the intermittent complaint should be thoroughly checked for backed out terminals, improper mating, broken locks, improperly formed or damaged terminals, poor terminal to wiring connections or physical damage to the wiring harness.

DIAGNOSTIC CHART DTC 53 — 1995 SILHOUETTE AND TRANS SPORT WITH TCS, CONTINUED

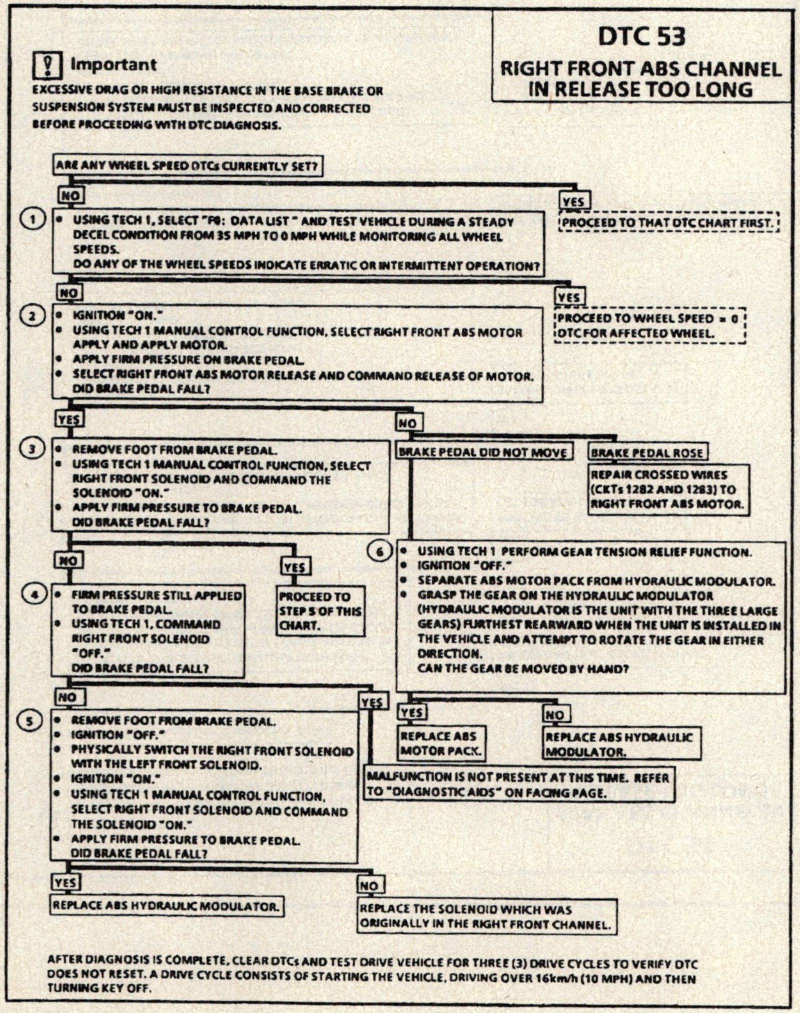

DIAGNOSTIC CHART DTC 54 — 1995 SILHOUETTE AND TRANS SPORT WITH TCS

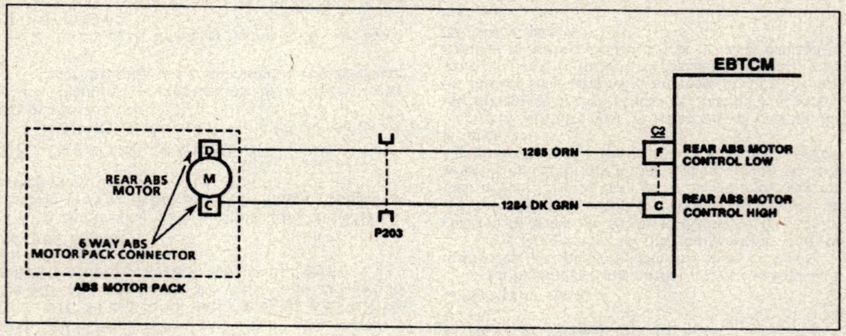

DTC 54
REAR ABS CHANNEL IN RELEASE TOO LONG

Circuit Description:

This DTC will diagnose an ABS motor that is energized longer than expected. This could occur if a wheel speed sensor is malfunctioning, the ABS motor does not turn, or the ABS motor wires are crossed.

DTC Will Set When:

DTC 54 can be set only during an ABS stop. If the EBTCM commands the rear ABS channel in release for three seconds, a malfunction exists.

Action Taken:

A malfunction DTC is stored. ABS and TCS are disabled and the ABS and TCS warning lamps are turned "ON."

DTC Chart Test Description:

Number(s) below refer to circled number(s) on the diagnostic chart.

1. Identifies a malfunction in a wheel speed sensor that may cause the system to be in release too long.
2. Checks for a wheel that may stick or bind because of a mechanical malfunction.
3. Checks to see if the ABS motor is capable of moving and applying the hydraulic piston for the rear wheels.
4. Ensures that the ABS motor wiring is not crossed.
5. Isolates the fault of a "no-apply" situation to either the ABS motor pack or ABS hydraulic modulator assembly.

Diagnostic Aids:

An intermittent malfunction may be caused by a mechanical part of the system that sticks or binds.

The frequency of the malfunction can be checked by using the enhanced diagnostic function of the Tech 1, as described in "Tech 1 Diagnostics" in this section.

Any circuitry that is suspected of causing the intermittent complaint should be thoroughly checked for backed out terminals, improper mating, broken locks, improperly formed or damaged terminals, poor terminal to wiring connections or physical damage to the wiring harness.

DIAGNOSTIC CHART DTC 54 — 1995 SILHOUETTE AND TRANS SPORT WITH TCS, CONTINUED

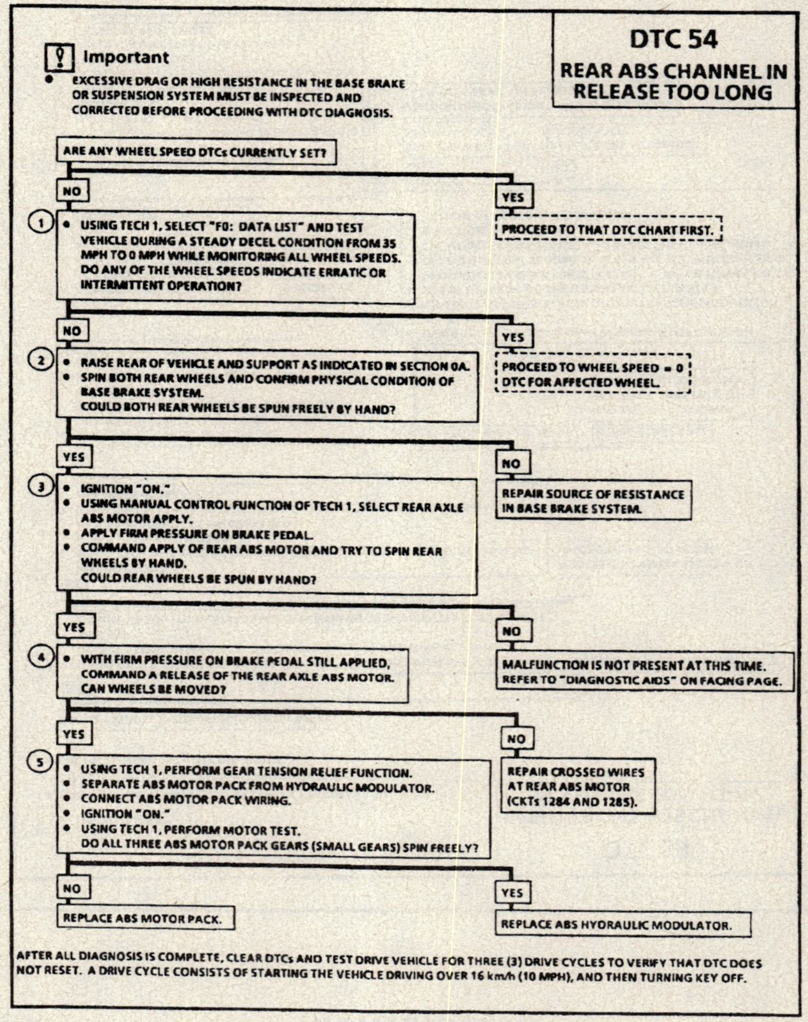

DIAGNOSTIC CHART DTC 55 — 1995 SILHOUETTE AND TRANS SPORT WITH TCS

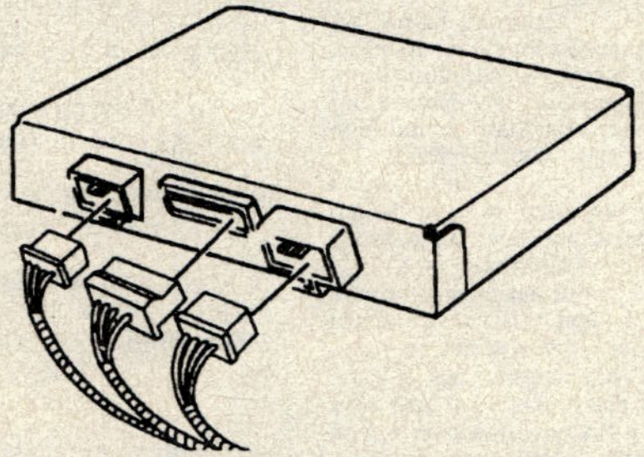

DTC 55
EBTCM MALFUNCTION (ABS)

Circuit Description:

This DTC identifies an internal malfunction detected by the EBTCM motor driver interface custom integrated circuit.

Action Taken:

A malfunction DTC is stored, ABS and TCS are disabled and the amber ABS and TCS warning lamps are turned "ON."

DTC Chart Test Description:

Number(s) below refer to circled number(s) on the diagnostic chart.

1. Checks to see if the malfunction is still present.
2. Checks to see if the malfunction is intermittent.

DIAGNOSTIC CHART DTC 55 — 1995 SILHOUETTE AND TRANS SPORT WITH TCS, CONTINUED

DTC 55
EBTCM MALFUNCTION (ABS)

(1)
- IGNITION "ON," ENGINE "OFF."
- USING TECH 1, READ DTCs.

IS DTC 55 SET AS A CURRENT CODE?

- YES → go to (2)
- NO → MALFUNCTION IS INTERMITTENT.

(2)
- USING TECH 1, CLEAR DTCs.
- TEST DRIVE VEHICLE FOR THREE (3) DRIVE CYCLES. A DRIVE CYCLE CONSISTS OF KEY "ON," DRIVE VEHICLE OVER 16 km/h (10 MPH), THEN KEY "OFF."

DOES DTC 55 RESET?

- YES → REPLACE EBTCM.
- NO → MALFUNCTION IS INTERMITTENT.

DIAGNOSTIC CHART DTC 56 — 1995 SILHOUETTE AND TRANS SPORT WITH TCS

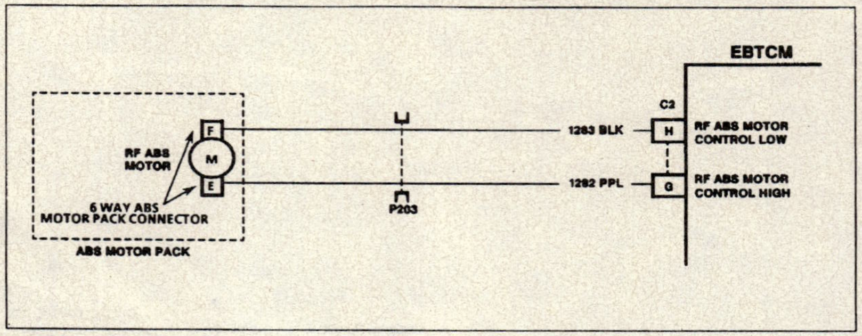

DTC 56
LEFT FRONT ABS MOTOR CIRCUIT OPEN

Circuit Description:

This DTC identifies an ABS motor that cannot be energized due to an open in its circuitry.

DTC Will Set When:

DTC 56 can be set only when the motor is commanded "OFF." If the EBTCM detects an out of range voltage on either of the left front ABS motor circuits indicating an open circuit, a malfunction exists.

Action Taken:

An open ABS motor will not activate when requested. A malfunction DTC is stored, ABS is disabled and the ABS warning lamp is turned "ON."

DTC Chart Test Description:

Number(s) below refer to circled number(s) on the diagnostic chart.

1. Checks for proper resistance of the ABS motor.
2. Checks for an open in the ABS motor "HIGH" circuitry.
3. Checks for an open in the ABS motor "LOW" circuitry.
4. Ensures malfunction was not due to a poor terminal contact or EBTCM.

Diagnostic Aids:

Using Tech 1, select manual control function, and exercise ABS motor movement of affected channel in both directions while applying light pressure on the brake pedal.

If erratic or "jumpy" brake pedal movement is detected while performing an "apply" or "release" function of the ABS motor, an intermittent malfunction may be indicated.

An intermittent malfunction may be caused by a poor connection, rubbed through wire insulation, or a wire that is broken inside the insulation.

If the malfunction is not current, wiggle the wires of the affected channel and check if the DTC resets. This will help to pinpoint an intermittent malfunction in the motor circuitry or connections.

The frequency of the malfunction can be checked by using the enhanced diagnostic function of the Tech 1, as described in "Tech 1 Diagnostics" in this section.

Any circuitry that is suspected of causing the intermittent complaint should be thoroughly checked for backed out terminals, improper mating, broken locks, improperly formed or damaged terminals, poor terminal to wiring connections or physical damage to the wiring harness.

DIAGNOSTIC CHART DTC 56 — 1995 SILHOUETTE AND TRANS SPORT WITH TCS, CONTINUED

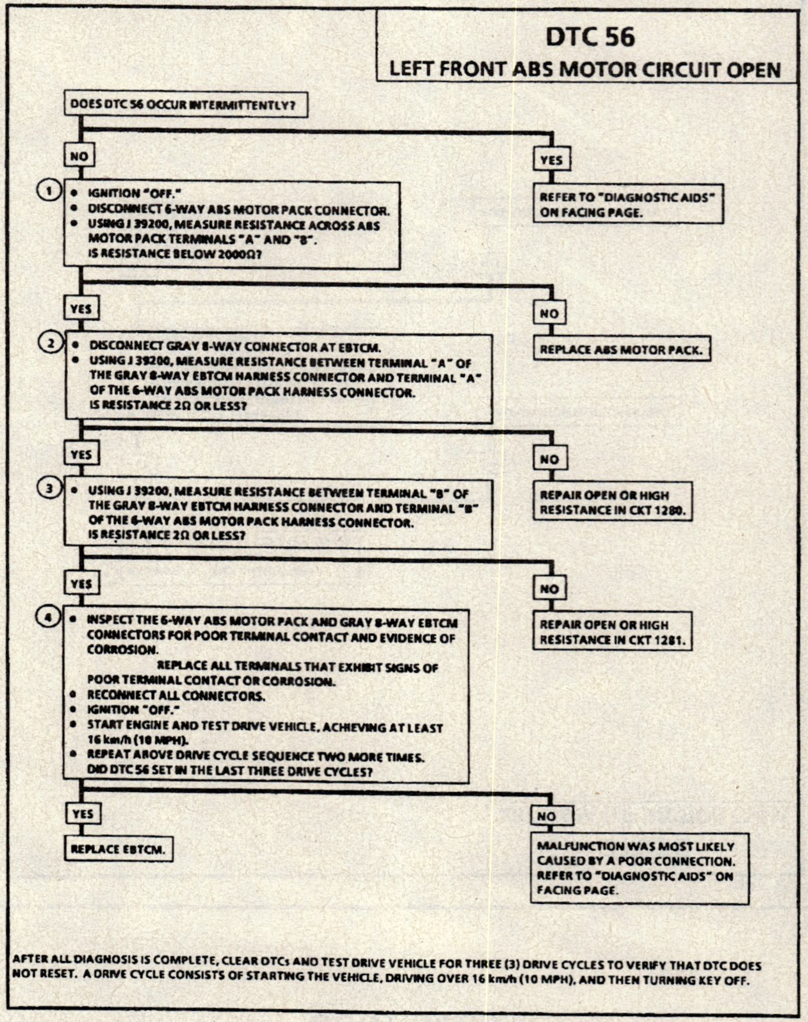

DIAGNOSTIC CHART DTC 57 — 1995 SILHOUETTE AND TRANS SPORT WITH TCS

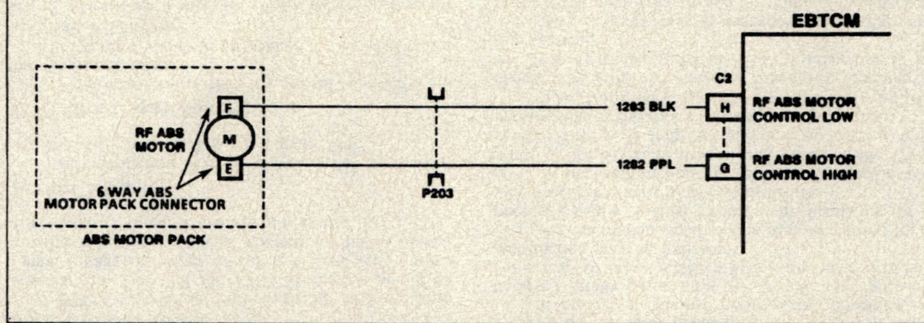

DTC 57
LEFT FRONT ABS MOTOR CIRCUIT SHORTED TO GND

Circuit Description:

This DTC identifies an ABS motor circuit that is shorted to ground. This malfunction will not allow the ABS motor to be controlled at the commanded current rate or will cause the driver circuit to allow current directly to ground.

DTC Will Set When:

DTC 57 can be set anytime. If the EBTCM detects an out of range voltage on either of the left front ABS motor circuits indicating a circuit shorted to ground, a malfunction exists.

Action Taken:

A malfunction DTC is stored, ABS is disabled and the ABS warning lamp is turned "ON."

DTC Chart Test Description:

Number(s) below refer to circled number(s) on the diagnostic chart.

1. Checks for a short to ground in the ABS motor "HIGH" circuitry.
2. Checks for a short to ground in the ABS motor "LOW" circuitry.
3. Checks for an ABS motor that is internally shorted to ground.
4. Ensures malfunction was not due to physical damage of the circuitry.

Diagnostic Aids:

Using Tech 1, select manual control function, and exercise ABS motor movement of affected channel in both directions while applying light pressure on the brake pedal.

If erratic or "jumpy" brake pedal movement is detected while performing an "apply" or "release" function of the ABS motor, an intermittent malfunction may be indicated.

An intermittent malfunction may be caused by a poor connection, rubbed through wire insulation, or a wire that is broken inside the insulation.

If the malfunction is not current, wiggle the wires of the affected channel and check if the DTC resets. This will help to pinpoint an intermittent malfunction in the motor circuitry or connections.

The frequency of the malfunction can be checked by using the enhanced diagnostic function of the Tech 1, as described in "Tech 1 Diagnostics" in this section.

Any circuitry that is suspected of causing the intermittent complaint should be thoroughly checked for backed out terminals, improper mating, broken locks, improperly formed or damaged terminals, poor terminal to wiring connections or physical damage to the wiring harness.

DIAGNOSTIC CHART DTC 57 — 1995 SILHOUETTE AND TRANS SPORT WITH TCS, CONTINUED

DTC 57
LEFT FRONT ABS MOTOR CIRCUIT SHORTED TO GND

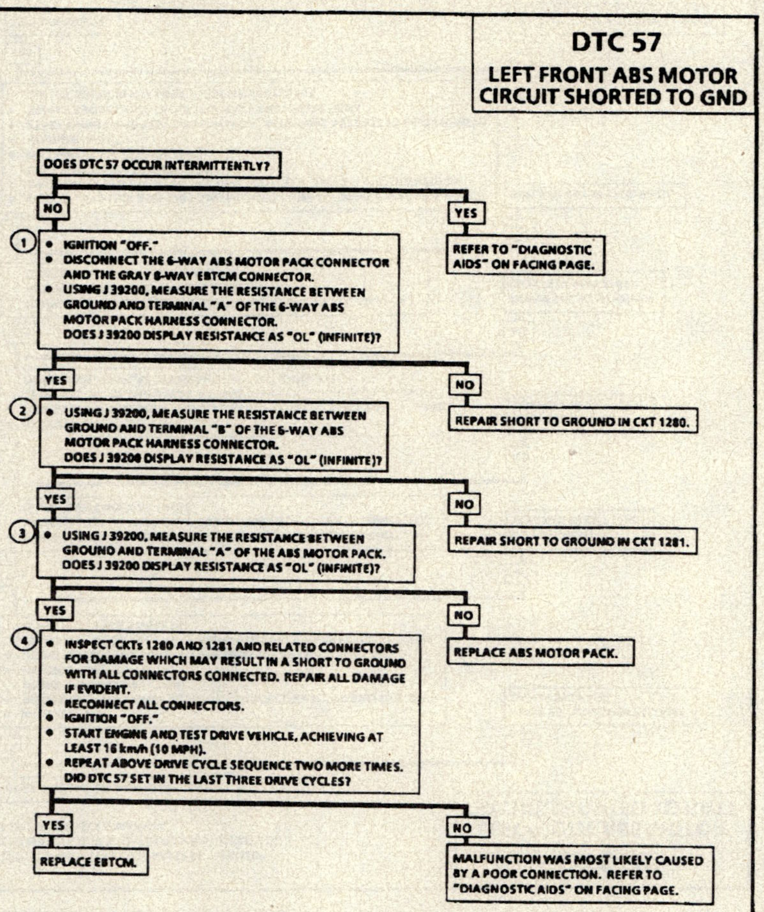

AFTER ALL DIAGNOSIS IS COMPLETE, CLEAR DTCs AND TEST DRIVE VEHICLE FOR THREE (3) DRIVE CYCLES TO VERIFY THAT DTC DOES NOT RESET. A DRIVE CYCLE CONSISTS OF STARTING THE VEHICLE, DRIVING OVER 16 km/h (10 MPH), AND THEN TURNING KEY OFF.

DIAGNOSTIC CHART DTC 58 — 1995 SILHOUETTE AND TRANS SPORT WITH TCS

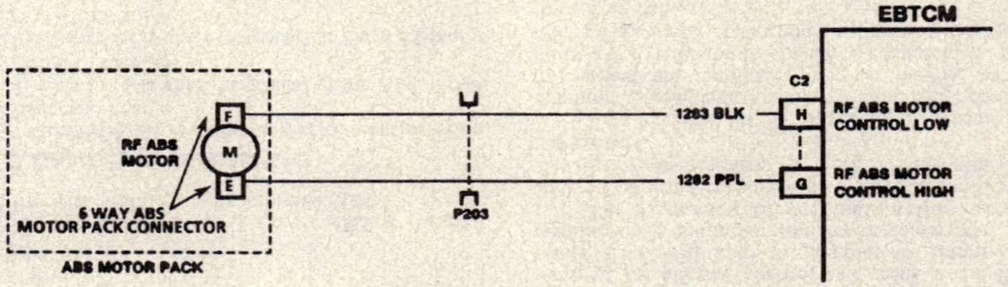

DTC 58
LEFT FRONT ABS MOTOR CIRCUIT SHORTED TO BATT

Circuit Description:

This DTC identifies an ABS motor circuit that is shorted to battery or an ABS motor that has low or no resistance. This malfunction will not allow the ABS motor to be controlled at the commanded current rate or will cause the ABS motor to turn in the opposite direction, or not at all.

DTC Will Set When:

DTC 58 can be set only when the ABS motor is commanded "OFF." If the EBTCM detects an out of range voltage on either of the left front ABS motor circuits indicating a circuit shorted to battery or an ABS motor shorted, a malfunction exists.

Action Taken:

A malfunction DTC is stored, ABS is disabled and the ABS warning lamp is turned "ON."

DTC Chart Test Description:

Number(s) below refer to circled number(s) on the diagnostic chart.

1. Checks for a short to voltage in the ABS motor "HIGH" circuitry.
2. Checks for a short to voltage in the ABS motor "LOW" circuitry.
3. Checks for a short between the "HIGH" and "LOW" motor circuits.
4. This step checks for an ABS motor that is internally shorted.
5. Ensures malfunction was not due to physical damage of the circuitry.

Diagnostic Aids:

Using Tech 1, select manual control function, and exercise ABS motor movement of affected channel in both directions while applying light pressure on the brake pedal.

If erratic or "jumpy" brake pedal movement is detected while performing an "apply" or "release" function of the ABS motor, an intermittent malfunction may be indicated.

An intermittent malfunction may be caused by a poor connection, rubbed through wire insulation, or a wire that is broken inside the insulation.

If the malfunction is not current, wiggle the wires of the affected channel and check if the DTC resets. This will help to pinpoint an intermittent malfunction in the motor circuitry or connections.

The frequency of the malfunction can be checked by using the enhanced diagnostic function of the Tech 1, as described in "Tech 1 Diagnostics" in this section.

Any circuitry that is suspected of causing the intermittent complaint should be thoroughly checked for backed out terminals, improper mating, broken locks, improperly formed or damaged terminals, poor terminal to wiring connections or physical damage to the wiring harness.

DIAGNOSTIC CHART DTC 58 — 1995 SILHOUETTE AND TRANS SPORT WITH TCS, CONTINUED

DTC 58
LEFT FRONT ABS MOTOR CIRCUIT SHORTED TO BATT

NOTICE: J 39200 TEST LEADS MUST BE "ZEROED" PRIOR TO MAKING ANY RESISTANCE MEASUREMENTS. REFER TO J 39200 USER'S MANUAL.

DOES DTC 58 OCCUR INTERMITTENTLY?

- YES: REFER TO "DIAGNOSTIC AIDS" ON FACING PAGE.
- NO:

(1)
- IGNITION "OFF."
- DISCONNECT THE 6-WAY ABS MOTOR PACK CONNECTOR AND THE GRAY 8-WAY EBTCM CONNECTOR.
- IGNITION "ON."
- USING J 39200 , MEASURE THE VOLTAGE BETWEEN GROUND AND TERMINAL "A" OF THE 6-WAY ABS MOTOR PACK HARNESS CONNECTOR. IS VOLTAGE 1 VOLT OR LESS?

- NO: REPAIR SHORT TO VOLTAGE IN CKT 1280.
- YES:

(2)
- USING J 39200, MEASURE THE VOLTAGE BETWEEN GROUND AND TERMINAL "B" OF THE 6-WAY ABS MOTOR PACK HARNESS CONNECTOR. IS VOLTAGE 1 VOLT OR LESS?

- NO: REPAIR SHORT TO VOLTAGE IN CKT 1281.
- YES:

(3)
- IGNITION "OFF."
- USING J 39200, MEASURE THE RESISTANCE BETWEEN TERMINAL "A" AND TERMINAL "B" OF THE 6-WAY ABS MOTOR PACK HARNESS CONNECTOR. DOES J 39200 DISPLAY RESISTANCE AS "OL" (INFINITE)?

- NO: REPAIR SHORT BETWEEN CKTs 1280 AND 1281.
- YES:

(4)
- USING J 39200, MEASURE THE RESISTANCE BETWEEN TERMINAL "A" AND TERMINAL "B" OF THE ABS MOTOR PACK. IS RESISTANCE GREATER THAN 0.4Ω?

- NO: REPLACE ABS MOTOR PACK.
- YES:

(5)
- INSPECT CKTs 1280 AND 1281 AND RELATED CONNECTORS FOR DAMAGE WHICH MAY RESULT IN A SHORT TO VOLTAGE WITH ALL CONNECTORS CONNECTED. REPAIR ALL DAMAGE IF EVIDENT.
- RECONNECT ALL CONNECTORS.
- IGNITION "OFF."
- START ENGINE AND TEST DRIVE VEHICLE, ACHIEVING AT LEAST 16 km/h (10 MPH).
- REPEAT ABOVE DRIVE CYCLE SEQUENCE TWO MORE TIMES. DID DTC 58 SET IN THE LAST THREE DRIVE CYCLES?

- YES: REPLACE EBTCM.
- NO: MALFUNCTION WAS MOST LIKELY CAUSED BY A POOR CONNECTION. REFER TO "DIAGNOSTIC AIDS" ON FACING PAGE.

AFTER DIAGNOSIS IS COMPLETE, CLEAR DTCs AND TEST DRIVE VEHICLE FOR THREE (3) DRIVE CYCLES TO VERIFY THAT DTC DOES NOT RESET. A DRIVE CYCLE CONSISTS OF STARTING THE VEHICLE, DRIVING OVER 16 km/h (10 MPH), AND THEN TURNING KEY OFF.

DIAGNOSTIC CHART DTC 61 — 1995 SILHOUETTE AND TRANS SPORT WITH TCS

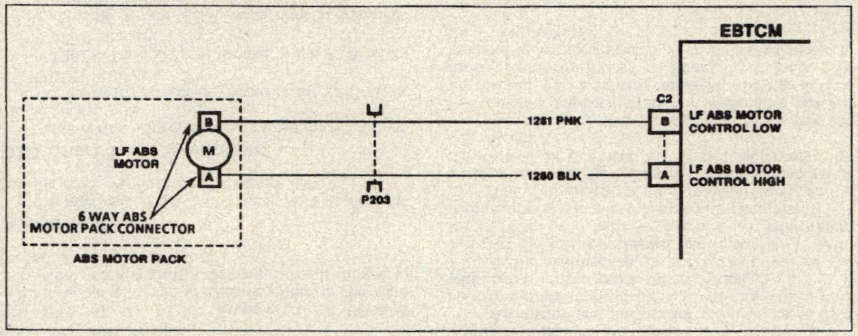

DTC 61
RIGHT FRONT ABS MOTOR CIRCUIT OPEN

Circuit Description:

This DTC identifies an ABS motor that cannot be energized due to an open in its circuitry. This malfunction will not allow proper front ABS operation.

DTC Will Set When:

DTC 61 can be set only when the ABS motor is commanded "OFF." If the EBTCM detects an out of range voltage on either of the right front ABS motor circuits indicating an open circuit, a malfunction exists.

Action Taken:

An open ABS motor will not activate when requested. A malfunction DTC is stored, ABS is disabled and the ABS warning lamp is turned "ON."

DTC Chart Test Description:

Number(s) below refer to circled number(s) on the diagnostic chart.

1. Checks for proper resistance of the ABS motor.
2. Checks for an open in the ABS motor "HIGH" circuitry.
3. Checks for an open in the ABS motor "LOW" circuitry.
4. Ensures malfunction was not due to poor terminal contact or EBTCM.

Diagnostic Aids:

Using Tech 1, select manual control function, and exercise ABS motor movement of affected channel in both directions while applying light pressure on the brake pedal.

If erratic or "jumpy" brake pedal movement is detected while performing an "apply" or "release" function of the ABS motor, an intermittent malfunction may be indicated.

An intermittent malfunction may be caused by a poor connection, rubbed through wire insulation, or a wire that is broken inside the insulation.

If the malfunction is not current, wiggle the wires of the affected channel and check if the DTC resets. This will help to pinpoint an intermittent malfunction in the motor circuitry or connections.

The frequency of the malfunction can be checked by using the enhanced diagnostic function of the Tech 1, as described in "Tech 1 Diagnostics" in this section.

Any circuitry that is suspected of causing the intermittent complaint should be thoroughly checked for backed out terminals, improper mating, broken locks, improperly formed or damaged terminals, poor terminal to wiring connections or physical damage to the wiring harness.

DIAGNOSTIC CHART DTC 61 — 1995 SILHOUETTE AND TRANS SPORT WITH TCS, CONTINUED

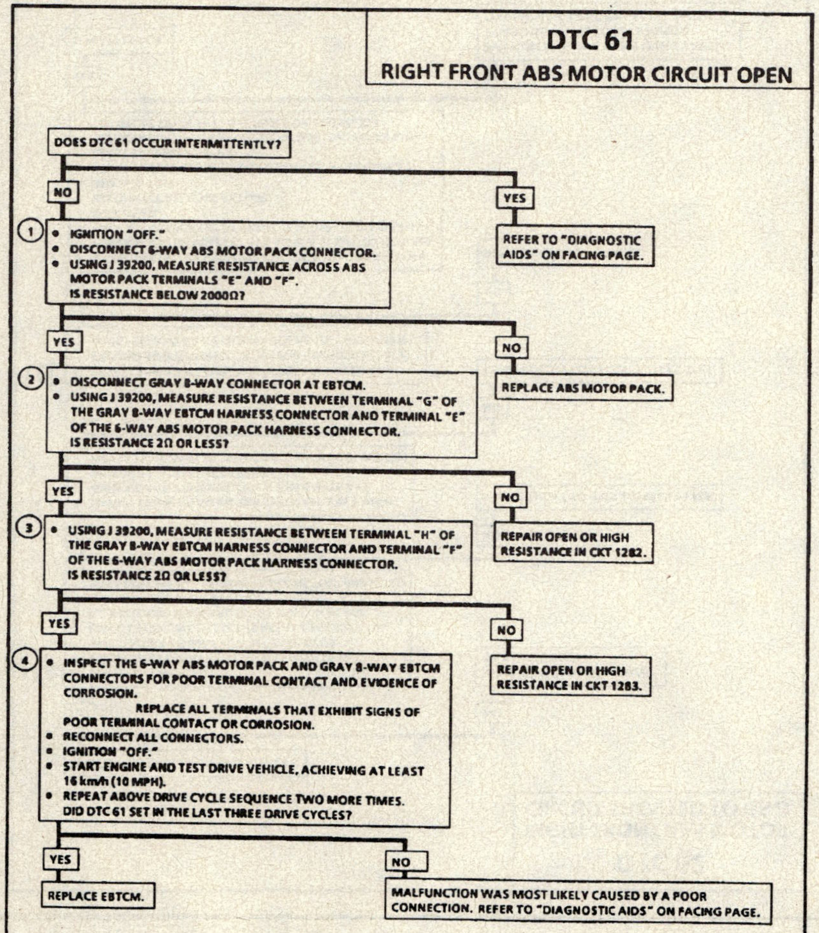

AFTER ALL DIAGNOSIS IS COMPLETE, CLEAR DTCs AND TEST DRIVE VEHICLE FOR THREE (3) DRIVE CYCLES TO VERIFY THAT DTC DOES NOT RESET. A DRIVE CYCLE CONSISTS OF STARTING THE VEHICLE, DRIVING OVER 16 km/h (10 MPH), AND THEN TURNING KEY OFF.

DIAGNOSTIC CHART DTC 62 — 1995 SILHOUETTE AND TRANS SPORT WITH TCS

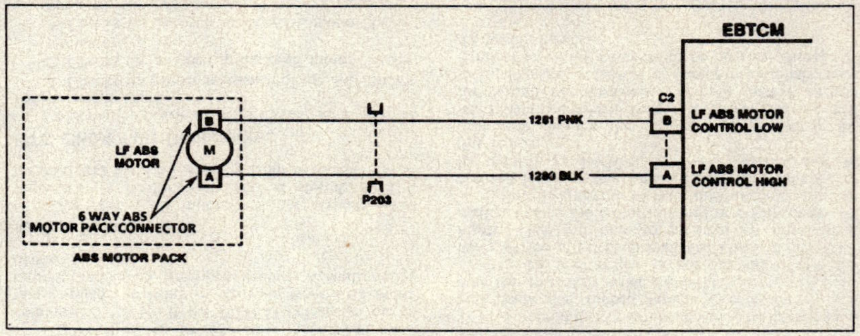

DTC 62
RIGHT FRONT ABS MOTOR CIRCUIT SHORTED TO GND

Circuit Description:

This DTC identifies an ABS motor circuit that is shorted to ground. This malfunction will not allow the ABS motor to be controlled at the commanded current rate or will cause the ABS motor driver circuit to allow current directly to ground.

DTC Will Set When:

DTC 62 can be set anytime. If the EBTCM detects an out of range voltage on either of the right front ABS motor circuits indicating a circuit shorted to ground, a malfunction exists.

Action Taken:

A malfunction DTC is stored, ABS is disabled and the ABS warning lamp is turned "ON."

DTC Chart Test Description:

Number(s) below refer to circled number(s) on the diagnostic chart.

1. Checks for a short to ground in the ABS motor "HIGH" circuitry.
2. Checks for a short to ground in the ABS motor "LOW" circuitry.
3. Checks for an ABS motor that is internally shorted to ground.
4. Ensures malfunction was not due to physical damage of the circuitry.

Diagnostic Aids:

Using Tech 1, select manual control function, and exercise ABS motor movement of affected channel in both directions while applying light pressure on the brake pedal.

If erratic or "jumpy" brake pedal movement is detected while performing an "apply" or "release" function of the ABS motor, an intermittent malfunction may be indicated.

An intermittent malfunction may be caused by a poor connection, rubbed through wire insulation, or a wire that is broken inside the insulation.

If the malfunction is not current, wiggle the wires of the affected channel and check if the DTC resets. This will help to pinpoint an intermittent malfunction in the motor circuitry or connections.

The frequency of the malfunction can be checked by using the enhanced diagnostic function of the Tech 1, as described in "Tech 1 Diagnostics" in this section.

Any circuitry that is suspected of causing the intermittent complaint should be thoroughly checked for backed out terminals, improper mating, broken locks, improperly formed or damaged terminals, poor terminal to wiring connections or physical damage to the wiring harness.

DIAGNOSTIC CHART DTC 62 — 1995 SILHOUETTE AND TRANS SPORT WITH TCS, CONTINUED

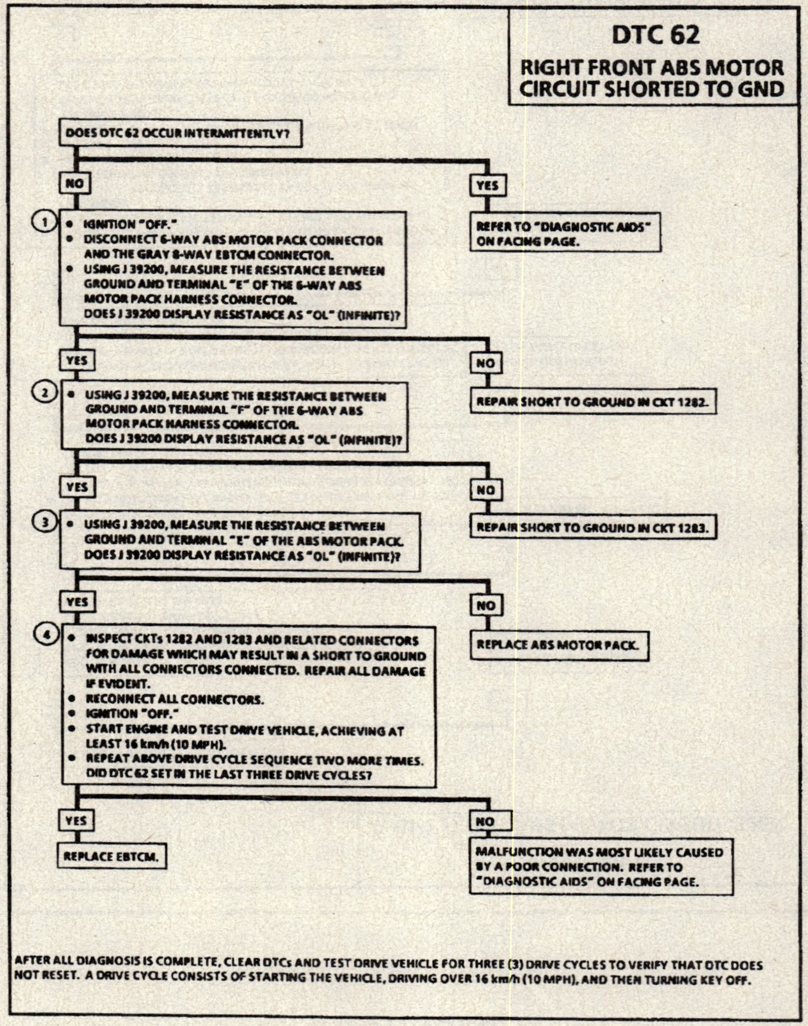

DIAGNOSTIC CHART DTC 63 — 1995 SILHOUETTE AND TRANS SPORT WITH TCS

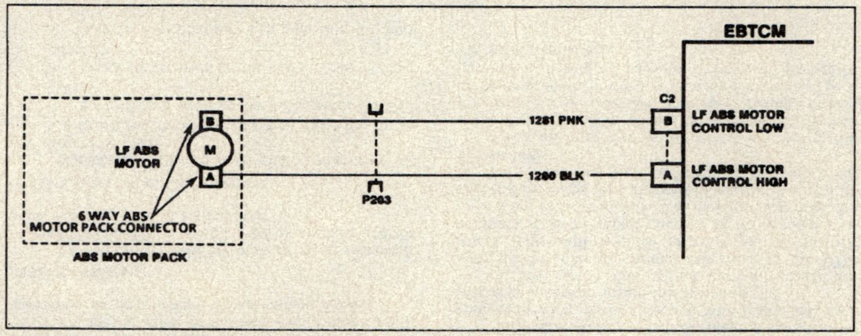

DTC 63
RIGHT FRONT ABS MOTOR CIRCUIT SHORTED TO BATT

Circuit Description:

This DTC identifies an ABS motor circuit that is shorted to battery or an ABS motor that has low or no resistance. This malfunction will not allow the ABS motor to be controlled at the commanded current rate or will cause the ABS motor to turn in the opposite direction or not at all.

DTC Will Set When:

DTC 63 can be set only when the ABS motor is commanded "OFF." If the EBTCM detects an out of range voltage on either of the right front ABS motor circuits indicating a circuit shorted to battery or an ABS motor shorted, a malfunction exists.

Action Taken:

A malfunction DTC is stored, ABS is disabled and the ABS warning lamp is turned "ON."

DTC Chart Test Description:

Number(s) below refer to circled number(s) on the diagnostic chart.

1. Checks for a short to voltage in the ABS motor "HIGH" circuitry.
2. Checks for a short to voltage in the ABS motor "LOW" circuitry.
3. Checks for a short circuit between motor "HIGH" and motor "LOW."
4. Checks for proper resistance of the ABS motor.
5. Ensures malfunction was not due to physical damage of the circuitry.

Diagnostic Aids:

Using Tech 1, select manual control function, and exercise ABS motor movement of affected channel in both directions while applying light pressure on the brake pedal.

If erratic or "jumpy" brake pedal movement is detected while performing an "apply" or "release" function of the ABS motor, an intermittent malfunction may be indicated.

An intermittent malfunction may be caused by a poor connection, rubbed through wire insulation, or a wire that is broken inside the insulation.

If the malfunction is not current, wiggle the wires of the affected channel and check if the DTC resets. This will help to pinpoint an intermittent malfunction in the motor circuitry or connections.

The frequency of the malfunction can be checked by using the enhanced diagnostic function of the Tech 1, as described in "Tech 1 Diagnostics" in this section.

Any circuitry that is suspected of causing the intermittent complaint should be thoroughly checked for backed out terminals, improper mating, broken locks, improperly formed or damaged terminals, poor terminal to wiring connections or physical damage to the wiring harness.

DIAGNOSTIC CHART DTC 63 — 1995 SILHOUETTE AND TRANS SPORT WITH TCS, CONTINUED

DTC 63
RIGHT FRONT ABS MOTOR CIRCUIT SHORTED TO BATT

NOTICE: J 39200 TEST LEADS MUST BE "ZEROED" PRIOR TO MAKING ANY RESISTANCE MEASUREMENTS. REFER TO J 39200 USER'S MANUAL.

DOES DTC 63 OCCUR INTERMITTENTLY?

- YES: REFER TO "DIAGNOSTIC AIDS" ON FACING PAGE.
- NO:

(1)
- IGNITION "OFF."
- DISCONNECT THE 6-WAY ABS MOTOR PACK CONNECTOR AND THE GRAY 8-WAY EBTCM CONNECTOR.
- IGNITION "ON."
- USING J 39200, MEASURE THE VOLTAGE BETWEEN GROUND AND TERMINAL "E" OF THE 6-WAY ABS MOTOR PACK HARNESS CONNECTOR.

IS VOLTAGE 1 VOLT OR LESS?

- NO: REPAIR SHORT TO VOLTAGE IN CKT 1282.
- YES:

(2)
- USING J 39200, MEASURE THE VOLTAGE BETWEEN GROUND AND TERMINAL "F" OF THE 6-WAY ABS MOTOR PACK HARNESS CONNECTOR.

IS VOLTAGE 1 VOLT OR LESS?

- NO: REPAIR SHORT TO VOLTAGE IN CKT 1283.
- YES:

(3)
- IGNITION "OFF."
- USING J 39200, MEASURE THE RESISTANCE BETWEEN TERMINAL "E" AND TERMINAL "F" OF THE 6-WAY ABS MOTOR PACK HARNESS CONNECTOR.

DOES J 39200 DISPLAY RESISTANCE AS "OL" (INFINITE)?

- NO: REPAIR SHORT BETWEEN CKTs 1282 AND 1283.
- YES:

(4)
- USING J 39200, MEASURE THE RESISTANCE BETWEEN TERMINAL "E" AND TERMINAL "F" OF THE ABS MOTOR PACK.

IS RESISTANCE GREATER THAN 0.4Ω?

- NO: REPLACE ABS MOTOR PACK.
- YES:

(5)
- INSPECT CKTs 1282 AND 1283 AND RELATED CONNECTORS FOR DAMAGE WHICH MAY RESULT IN A SHORT TO VOLTAGE WITH ALL CONNECTORS CONNECTED. REPAIR ALL DAMAGE IF EVIDENT.
- RECONNECT ALL CONNECTORS.
- IGNITION "OFF."
- START ENGINE AND TEST DRIVE VEHICLE, ACHIEVING AT LEAST 16 km/h (10 MPH).
- REPEAT ABOVE DRIVE CYCLE SEQUENCE TWO MORE TIMES.

DID DTC 63 SET IN THE LAST THREE DRIVE CYCLES?

- YES: REPLACE EBTCM.
- NO: MALFUNCTION WAS MOST LIKELY CAUSED BY A POOR CONNECTION. REFER TO "DIAGNOSTIC AIDS" ON FACING PAGE.

AFTER DIAGNOSIS IS COMPLETE, CLEAR DTCs AND TEST DRIVE VEHICLE FOR THREE (3) DRIVE CYCLES TO VERIFY THAT DTC DOES NOT RESET. A DRIVE CYCLE CONSISTS OF STARTING THE VEHICLE, DRIVING OVER 16 km/h (10 MPH), AND THEN TURNING KEY OFF.

DIAGNOSTIC CHART DTC 64 — 1995 SILHOUETTE AND TRANS SPORT WITH TCS

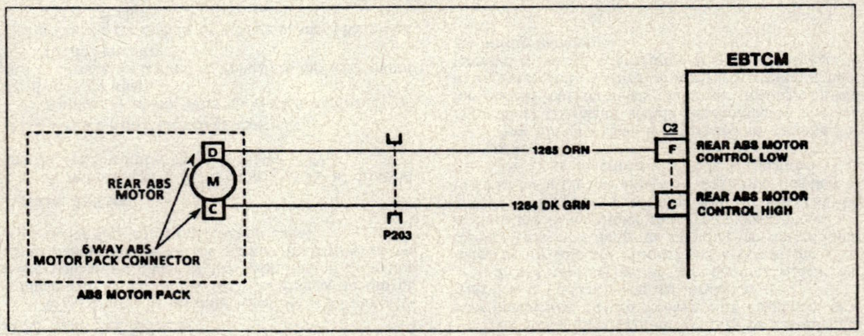

DTC 64
REAR ABS MOTOR CIRCUIT OPEN

Circuit Description:

This DTC identifies an ABS motor that cannot be energized due to an open in its circuitry. This malfunction will not allow proper rear ABS operation.

DTC Will Set When:

DTC 64 can be set only when the ABS motor is commanded "OFF." If the EBTCM detects an out of range voltage on either of the rear ABS motor circuits indicating an open circuit, a malfunction exists.

Action Taken:

An open motor will not activate when requested. A malfunction DTC is stored, ABS is disabled and the ABS warning lamp is turned "ON."

DTC Chart Test Description:

Number(s) below refer to circled number(s) on the diagnostic chart.

1. Checks for proper resistance of the ABS motor.
2. Checks for an open in the ABS motor "HIGH" circuitry.
3. Checks for an open in the ABS motor "LOW" circuitry.
4. Ensures malfunction was not due to poor terminal contact.

Diagnostic Aids:

Using Tech 1, select manual control function, and exercise ABS motor movement of affected channel in both directions while applying light pressure on the brake pedal.

If erratic or "jumpy" brake pedal movement is detected while performing an "apply" or "release" function of the ABS motor, an intermittent malfunction may be indicated.

An intermittent malfunction may be caused by a poor connection, rubbed through wire insulation, or a wire that is broken inside the insulation.

If the malfunction is not current, wiggle the wires of the affected channel and check if the DTC resets. This will help to pinpoint an intermittent malfunction in the motor circuitry or connections.

The frequency of the malfunction can be checked by using the enhanced diagnostic function of the Tech 1, as described in "Tech 1 Diagnostics" in this section.

Any circuitry that is suspected of causing the intermittent complaint should be thoroughly checked for backed out terminals, improper mating, broken locks, improperly formed or damaged terminals, poor terminal to wiring connections or physical damage to the wiring harness.

DIAGNOSTIC CHART DTC 64 — 1995 SILHOUETTE AND TRANS SPORT WITH TCS, CONTINUED

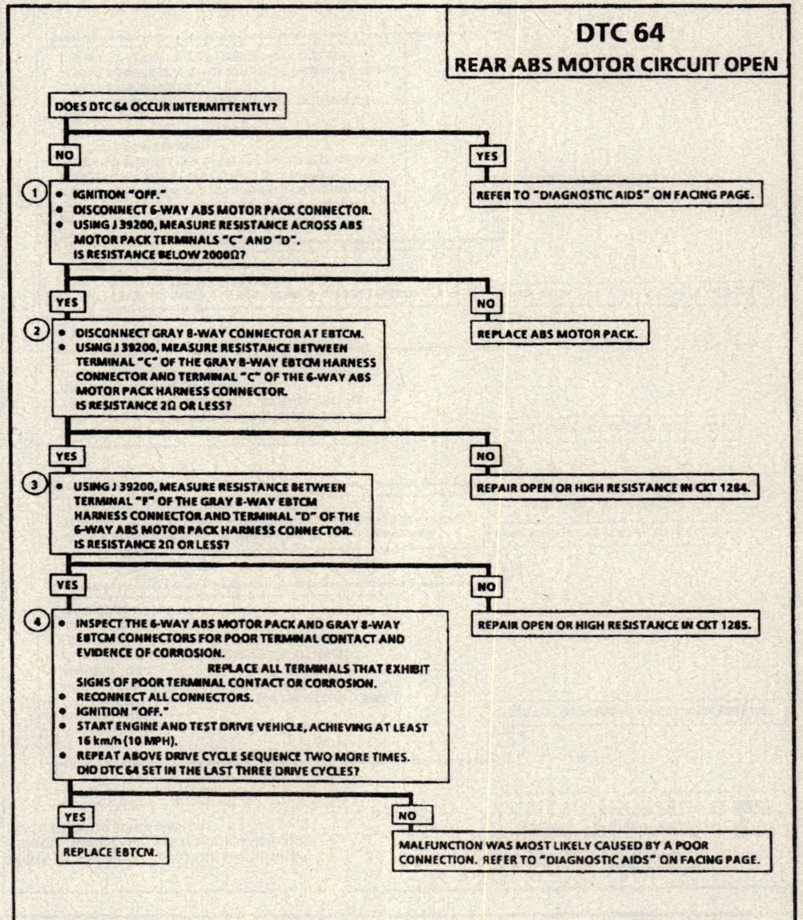

AFTER ALL DIAGNOSIS IS COMPLETE, CLEAR DTCs AND TEST DRIVE VEHICLE FOR THREE (3) DRIVE CYCLES TO VERIFY THAT DTC DOES NOT RESET. A DRIVE CYCLE CONSISTS OF STARTING THE VEHICLE, DRIVING OVER 16 km/h (10 MPH), AND THEN TURNING KEY OFF.

DIAGNOSTIC CHART DTC 65 — 1995 SILHOUETTE AND TRANS SPORT WITH TCS

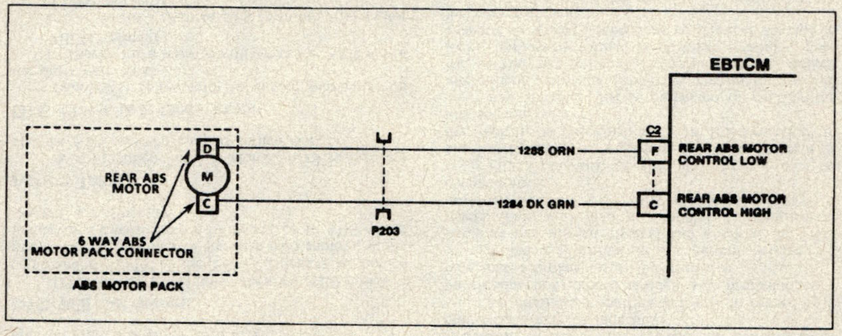

DTC 65
REAR ABS MOTOR CIRCUIT SHORTED TO GROUND

Circuit Description:

This DTC identifies an ABS motor circuit that is shorted to ground. This malfunction will not allow the ABS motor to be controlled at the commanded current rate or will cause the driver circuit to allow current directly to ground.

DTC Will Set When:

DTC 65 can be set anytime. If the EBTCM detects an out of range voltage on either of the rear ABS motor circuits indicating a circuit shorted to ground, a malfunction exists.

Action Taken:

A malfunction DTC is stored, ABS is disabled and the ABS warning lamp is turned "ON."

DTC Chart Test Description:

Number(s) below refer to circled number(s) on the diagnostic chart.

1. Checks for a short to ground in the ABS motor "HIGH" circuitry.
2. Checks for a short to ground in the ABS motor "LOW" circuitry.
3. Checks for an ABS motor that is internally shorted to ground.
4. Ensures malfunction was not due to physical damage of the circuitry.

Diagnostic Aids:

Using Tech 1, select manual control function, and exercise ABS motor movement of affected channel in both directions while applying light pressure on the brake pedal.

If erratic or "jumpy" brake pedal movement is detected while performing an "apply" or "release" function of the ABS motor, an intermittent malfunction may be indicated.

An intermittent malfunction may be caused by a poor connection, rubbed through wire insulation, or a wire that is broken inside the insulation.

If the malfunction is not current, wiggle the wires of the affected channel and check if the DTC resets. This will help to pinpoint an intermittent malfunction in the ABS motor circuitry or connections.

The frequency of the malfunction can be checked by using the enhanced diagnostic function of the Tech 1, as described in "Tech 1 Diagnostics" in this section.

Any circuitry that is suspected of causing the intermittent complaint should be thoroughly checked for backed out terminals, improper mating, broken locks, improperly formed or damaged terminals, poor terminal to wiring connections or physical damage to the wiring harness.

DIAGNOSTIC CHART DTC 65 — 1995 SILHOUETTE AND TRANS SPORT WITH TCS, CONTINUED

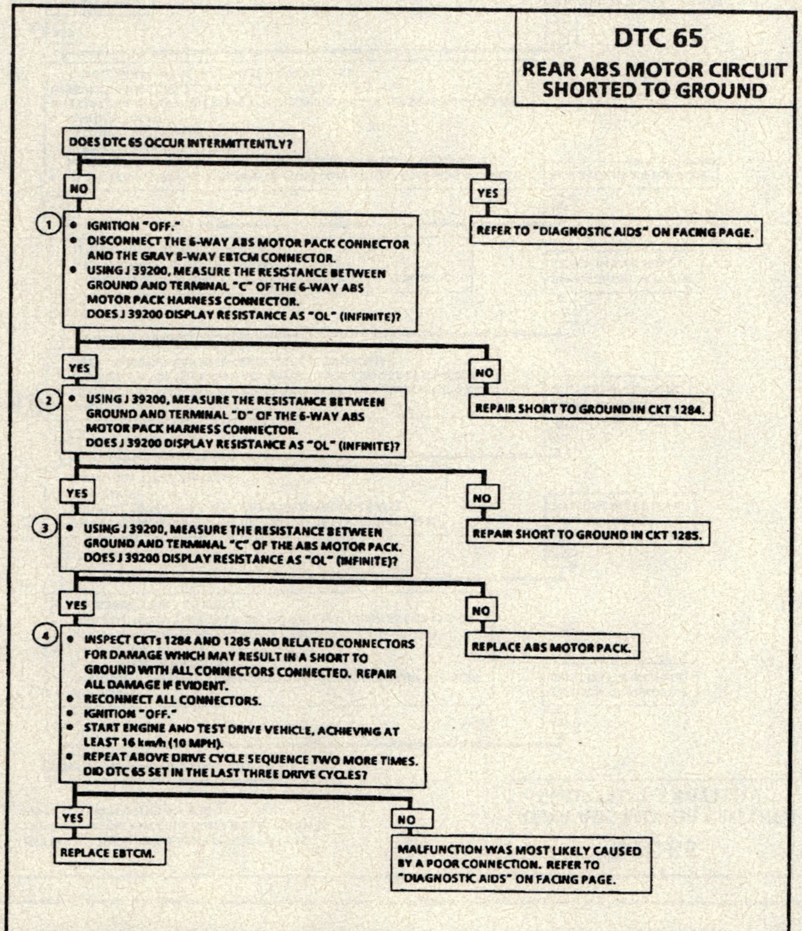

AFTER ALL DIAGNOSIS IS COMPLETE, CLEAR DTCs AND TEST DRIVE VEHICLE FOR THREE (3) DRIVE CYCLES TO VERIFY THAT DTC DOES NOT RESET. A DRIVE CYCLE CONSISTS OF STARTING THE VEHICLE, DRIVING OVER 16 km/h (10 MPH), AND THEN TURNING KEY OFF.

DIAGNOSTIC CHART DTC 66 — 1995 SILHOUETTE AND TRANS SPORT WITH TCS

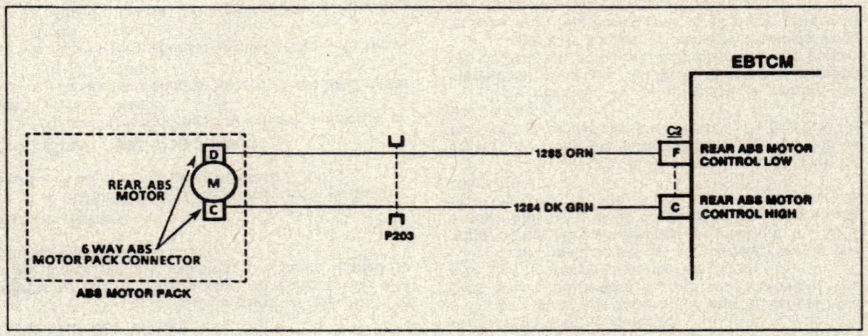

DTC 66
REAR ABS MOTOR CIRCUIT SHORTED TO BATTERY

Circuit Description:

This DTC identifies an ABS motor circuit that is shorted to battery or an ABS motor that has low or zero resistance. This malfunction will not allow the ABS motor to be controlled at the commanded current rate or will cause the ABS motor to turn in the opposite direction or not turn at all.

DTC Will Set When:

DTC 66 can be set only when the ABS motor is commanded "OFF." If the EBTCM detects an out of range voltage on either of the rear ABS motor circuits indicating a circuit shorted to battery or an ABS motor shorted, a malfunction exists.

Action Taken:

A malfunction DTC is stored, ABS is disabled and the ABS warning lamp is turned "ON."

DTC Chart Test Description:

Number(s) below refer to circled number(s) on the diagnostic chart.

1. Checks for a short to voltage in the ABS motor "HIGH" circuitry.
2. Checks for a short to voltage in the ABS motor "LOW" circuitry.
3. Checks for a short circuit between motor "HIGH" and motor "LOW."
4. Checks for proper resistance of the ABS motor.
5. Ensures malfunction was not due to physical damage of the circuitry.

Diagnostic Aids:

Using Tech 1, select manual control function, and exercise ABS motor movement of affected channel in both directions while applying light pressure on the brake pedal.

If erratic or "jumpy" brake pedal movement is detected while performing an "apply" or "release" function of the ABS motor, an intermittent malfunction may be indicated.

An intermittent malfunction may be caused by a poor connection, rubbed through wire insulation, or a wire that is broken inside the insulation.

If the malfunction is not current, wiggle the wires of the affected channel and check if the DTC resets. This will help to pinpoint an intermittent malfunction in the ABS motor circuitry or connections.

The frequency of the malfunction can be checked by using the enhanced diagnostic function of the Tech 1, as described in "Tech 1 Diagnostics" in this section.

Any circuitry that is suspected of causing the intermittent complaint should be thoroughly checked for backed out terminals, improper mating, broken locks, improperly formed or damaged terminals, poor terminal to wiring connections or physical damage to the wiring harness.

DIAGNOSTIC CHART DTC 66 — 1995 SILHOUETTE AND TRANS SPORT WITH TCS, CONTINUED

DTC 66
REAR ABS MOTOR CIRCUIT SHORTED TO BATTERY

NOTICE: J 39200 TEST LEADS MUST BE "ZEROED" PRIOR TO MAKING ANY RESISTANCE MEASUREMENTS. REFER TO J 39200 USER'S MANUAL.

DOES DTC 66 OCCUR INTERMITTENTLY?

- YES: REFER TO "DIAGNOSTIC AIDS" ON FACING PAGE.
- NO:

(1)
- IGNITION "OFF."
- DISCONNECT THE 6-WAY ABS MOTOR PACK CONNECTOR AND THE GRAY 8-WAY EBTCM CONNECTOR.
- IGNITION "ON."
- USING J 39200, MEASURE THE VOLTAGE BETWEEN GROUND AND TERMINAL "C" OF THE 6-WAY ABS MOTOR PACK HARNESS CONNECTOR. IS VOLTAGE 1 VOLT OR LESS?

- NO: REPAIR SHORT TO VOLTAGE IN CKT 1284.
- YES:

(2)
- USING J 39200, MEASURE THE VOLTAGE BETWEEN GROUND AND TERMINAL "D" OF THE 6-WAY ABS MOTOR PACK HARNESS CONNECTOR. IS VOLTAGE 1 VOLT OR LESS?

- NO: REPAIR SHORT TO VOLTAGE IN CKT 1285.
- YES:

(3)
- IGNITION "OFF."
- USING J 39200, MEASURE THE RESISTANCE BETWEEN TERMINAL "C" AND TERMINAL "D" OF THE 6-WAY ABS MOTOR PACK HARNESS CONNECTOR. DOES J 39200 DISPLAY RESISTANCE AS "OL" (INFINITE)?

- NO: REPAIR SHORT BETWEEN CKTs 1284 AND 1285.
- YES:

(4)
- USING J 39200, MEASURE THE RESISTANCE BETWEEN TERMINAL "C" AND TERMINAL "D" OF THE ABS MOTOR PACK. IS RESISTANCE GREATER THAN 0.4Ω?

- NO: REPLACE ABS MOTOR PACK
- YES:

(5)
- INSPECT CKTs 1284 AND 1285 AND RELATED CONNECTORS FOR DAMAGE WHICH MAY RESULT IN A SHORT TO VOLTAGE WITH ALL CONNECTORS CONNECTED. REPAIR ALL DAMAGE IF EVIDENT.
- RECONNECT ALL CONNECTORS.
- IGNITION "OFF."
- START ENGINE AND TEST DRIVE VEHICLE, ACHIEVING AT LEAST 16 km/h (10 MPH).
- REPEAT ABOVE DRIVE CYCLE SEQUENCE TWO MORE TIMES. DID DTC 66 SET IN THE LAST THREE DRIVE CYCLES?

- YES: REPLACE EBTCM.
- NO: MALFUNCTION WAS MOST LIKELY CAUSED BY A POOR CONNECTION. REFER TO "DIAGNOSTIC AIDS" ON FACING PAGE.

AFTER DIAGNOSIS IS COMPLETE, CLEAR DTCs AND TEST DRIVE VEHICLE FOR THREE (3) DRIVE CYCLES TO VERIFY THAT DTC DOES NOT RESET. A DRIVE CYCLE CONSISTS OF STARTING THE VEHICLE, DRIVING OVER 16 km/h (10 MPH), AND THEN TURNING KEY OFF.

DIAGNOSTIC CHART DTC 75 — 1995 SILHOUETTE AND TRANS SPORT WITH TCS

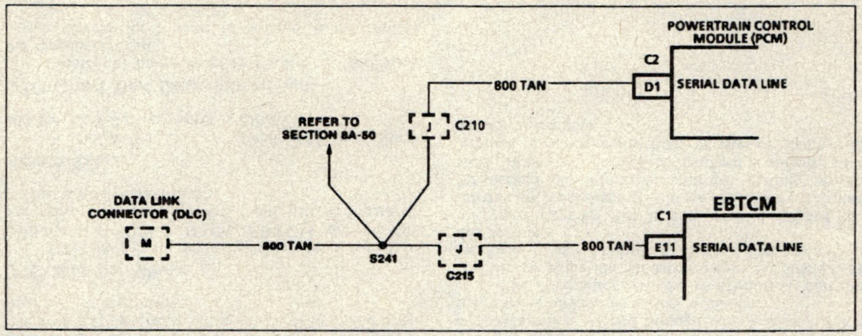

DTC 75
PCM TO EBTCM SERIAL DATA MALFUNCTION

Circuit Description:

The serial data communications circuit is used to transmit messages from one electronic control unit to another when interface is necessary between systems. ABS VI/TCS depends on information from the PCM for proper operation.

DTC Will Set When:

DTC 75 will set when there is a malfunction on the serial data circuit, or if the PCM has not broadcasted a serial message for at least 10 seconds.

Action Taken:

A malfunction DTC is stored, TCS is disabled, and the TCS warning lamp is turned "ON." No communication message will be displayed on Tech 1 until fault is removed.

DTC Chart Test Description:

Number(s) below refer to circled number(s) on the diagnostic chart.

1. This step checks for a malfunction in the test equipment.
2. Checks other systems on the circuit.
3. Checks serial data CKT 800 for short to ground.
4. Checks serial data CKT 800 for short to B+.
5. Checks for open serial data CKT 800.
6. Checks function of control modules on serial data CKT 800.

Diagnostic Aids:

When DTC 75 is current, no communication can be established with the Tech 1 scan tool. When the fault is removed, DTC 75 will be stored in system memory.

An intermittent malfunction may be caused by a poor connection, rubbed through wire insulation, or a wire that is broken inside the insulation.

The frequency of the malfunction can be checked by using the enhanced diagnostic function of the Tech 1, as described in "Tech 1 Diagnostics" in this section.

Any circuitry that is suspected of causing the intermittent complaint should be thoroughly checked for backed out terminals, improper mating, broken locks, improperly formed or damaged terminals, poor terminal to wiring connections or physical damage to the wiring harness.

DIAGNOSTIC CHART DTC 75 — 1995 SILHOUETTE AND TRANS SPORT WITH TCS, CONTINUED

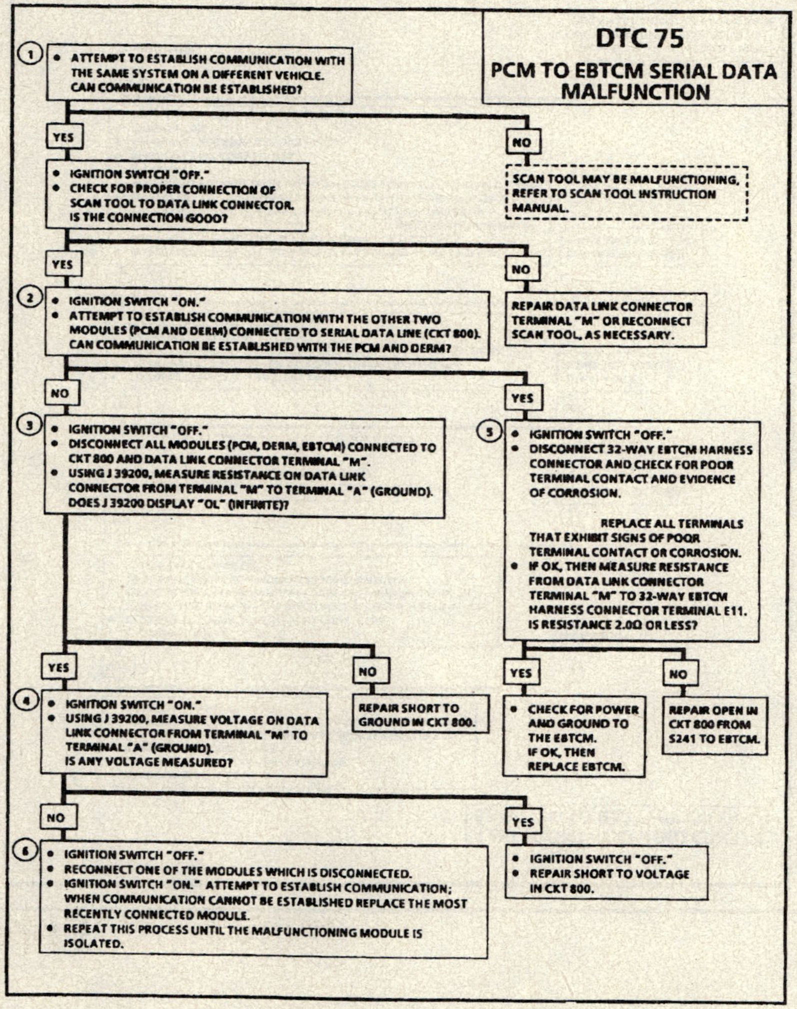

DIAGNOSTIC CHART DTC 76 — 1995 SILHOUETTE AND TRANS SPORT WITH TCS

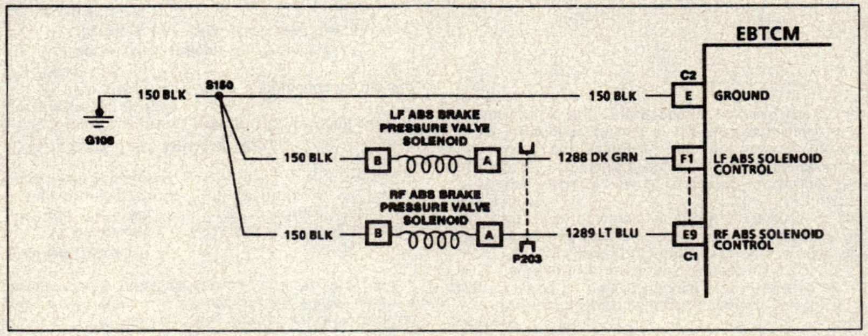

DTC 76
LEFT FRONT SOL CIRCUIT OPEN OR SHORTED TO BATT

Circuit Description:

This DTC identifies a solenoid that cannot be energized due to an open in its circuitry, or a solenoid that is always energized due to a short to battery in its circuitry between the driver and the solenoid. An open will not allow proper ABS operation, but the short to battery simply turns "ON" the solenoid. A path for base brakes is still allowed once the ABS motor re-homes and the check ball is lifted off its seat during key "ON" initialization.

DTC Will Set When:

DTC 76 can be set only when the solenoid is commanded "OFF." If the EBTCM detects an excessive voltage in the left front solenoid control circuit, a malfunction exists.

Action Taken:

A malfunction DTC is stored, ABS is disabled and the ABS warning lamp is turned "ON."

DTC Chart Test Description:

Number(s) below refer to circled number(s) on the diagnostic chart.

1. Checks for a short to voltage in the control circuitry of the solenoid.
2. Checks for an open in the control circuit of the solenoid.
3. Checks the solenoid coil for proper resistance.
4. Checks for an open in the solenoid ground circuit.
5. Checks for a possible intermittent malfunction in the solenoid circuitry due to poor terminal contact.

Diagnostic Aids:

An intermittent malfunction may be caused by a poor connection, rubbed through wire insulation, or a wire that is broken inside the insulation.

The frequency of the malfunction can be checked by using the enhanced diagnostic function of the Tech 1, as described in "Tech 1 Diagnostics" in this section.

Any circuitry that is suspected of causing the intermittent complaint should be thoroughly checked for backed out terminals, improper mating, broken locks, improperly formed or damaged terminals, poor terminal to wiring connections or physical damage to the wiring harness.

DIAGNOSTIC CHART DTC 76 — 1995 SILHOUETTE AND TRANS SPORT WITH TCS, CONTINUED

DTC 76
LEFT FRONT SOLENOID CIRCUIT SHORTED TO BATTERY OR OPEN

(1)
- IGNITION "OFF."
- DISCONNECT 32-WAY EBTCM CONNECTOR.
- IGNITION "ON" ENGINE "OFF."
- USING J 39200, MEASURE THE VOLTAGE BETWEEN GROUND AND TERMINAL "F1" OF THE 32-WAY EBTCM HARNESS CONNECTOR.

IS VOLTAGE 1 VOLT OR LESS?

NO: REPAIR SHORT TO VOLTAGE IN CKT 1288.

YES:

(2)
- IGNITION "OFF."
- DISCONNECT LEFT FRONT SOLENOID CONNECTOR.
- USING J 39200, MEASURE THE RESISTANCE BETWEEN TERMINAL "F1" OF THE 32-WAY EBTCM HARNESS CONNECTOR AND TERMINAL "A" OF THE LEFT FRONT SOLENOID HARNESS CONNECTOR.

IS RESISTANCE 2Ω OR LESS?

NO: REPAIR OPEN OR HIGH RESISTANCE IN CKT 1288.

YES:

(3)
- USING J 39200, MEASURE THE RESISTANCE BETWEEN TERMINAL "A" AND TERMINAL "B" OF THE LEFT FRONT SOLENOID.

IS RESISTANCE BETWEEN 2.5 AND 5Ω?

NO: REPLACE LEFT FRONT SOLENOID.

YES:

(4)
- DISCONNECT NEGATIVE BATTERY CABLE.
- USING J 39200, MEASURE THE RESISTANCE BETWEEN NEGATIVE BATTERY TERMINAL AND TERMINAL "B" OF THE LEFT FRONT SOLENOID HARNESS CONNECTOR.

IS RESISTANCE 2Ω OR LESS?

NO: REPAIR OPEN OR HIGH RESISTANCE IN CKT 150.

YES:

(5)
- INSPECT LEFT FRONT SOLENOID AND 32-WAY EBTCM HARNESS CONNECTORS FOR POOR TERMINAL CONTACT AND EVIDENCE OF CORROSION. REPLACE ALL TERMINALS THAT EXHIBIT SIGNS OF POOR TERMINAL CONTACT OR CORROSION.
- INSPECT CKT 1288 FOR DAMAGE WHICH MAY RESULT IN AN OPEN CIRCUIT OR A SHORT TO VOLTAGE WITH ALL CONNECTORS CONNECTED. REPAIR DAMAGE IF EVIDENT.
- RECONNECT ALL CONNECTORS.
- RECONNECT NEGATIVE BATTERY CABLE.
- IGNITION "ON."

DOES DTC 76 SET AS A CURRENT DTC?

YES: REPLACE EBTCM.

NO: MALFUNCTION IS MOST LIKELY DUE TO A POOR CONNECTION. REFER TO "DIAGNOSTIC AIDS" ON FACING PAGE.

DIAGNOSTIC CHART DTC 77 — 1995 SILHOUETTE AND TRANS SPORT WITH TCS

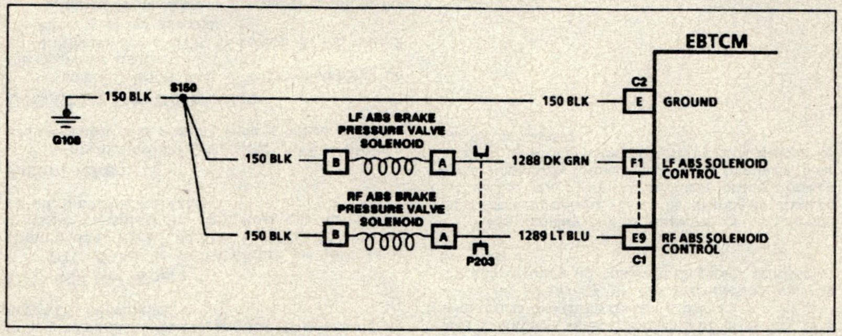

DTC 77
LEFT FRONT SOL CIRCUIT SHORTED TO GND

Circuit Description:

This DTC identifies a solenoid that cannot be energized due to an open in its driver circuitry, or a short to ground between the solenoid driver and the solenoid. These malfunctions can affect ABS operation since the flow of brake fluid to the wheel cylinder cannot be stopped, making ABS operation for that channel impossible.

DTC Will Set When:

DTC 77 can be set only when the solenoid is commanded "ON." If the EBTCM detects the left front solenoid control circuit voltage is out of specification, a malfunction exists.

Action Taken:

A malfunction DTC is stored, ABS is disabled and the ABS warning lamp is turned "ON."

DTC Chart Test Description:

Number(s) below refer to circled number(s) on the diagnostic chart.

1. Checks for a solenoid that is internally shorted to ground.
2. Checks for a solenoid that is not within proper resistance values.
3. Indicates if a short to ground exists in the solenoid circuitry.
4. Checks for a possible intermittent malfunction in the solenoid circuitry due to poor terminal contact.

Diagnostic Aids:

An intermittent malfunction may be caused by a poor connection, rubbed through wire insulation, or a wire that is broken inside the insulation.

The frequency of the malfunction can be checked by using the enhanced diagnostic function of the Tech 1

Any circuitry that is suspected of causing the intermittent complaint should be thoroughly checked for backed out terminals, improper mating, broken locks, improperly formed or damaged terminals, poor terminal to wiring connections or physical damage to the wiring harness.

DIAGNOSTIC CHART DTC 77 — 1995 SILHOUETTE AND TRANS SPORT WITH TCS, CONTINUED

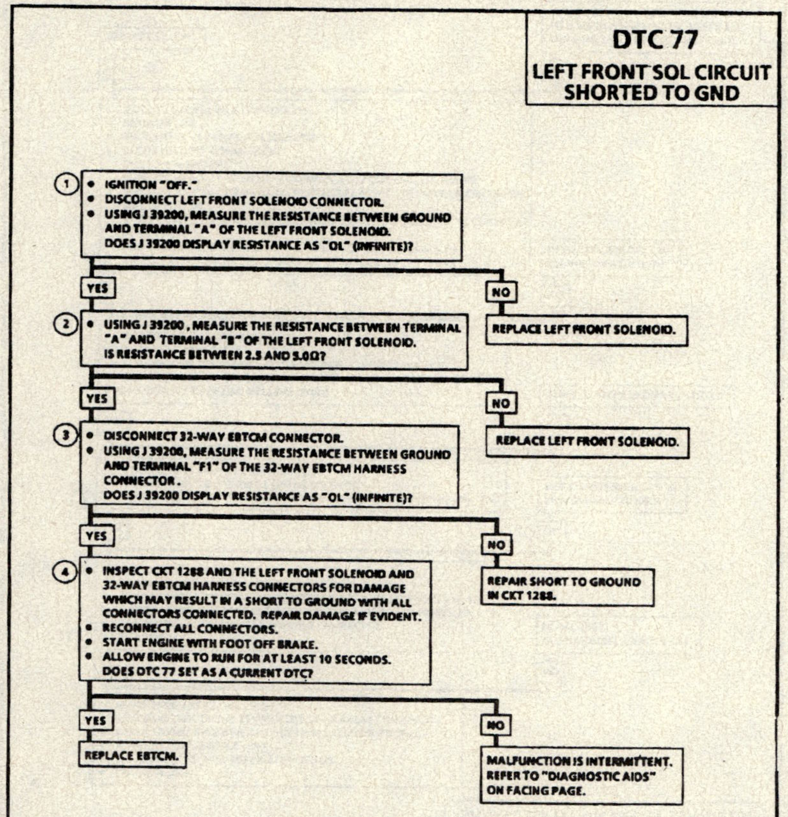

DIAGNOSTIC CHART DTC 78 — 1995 SILHOUETTE AND TRANS SPORT WITH TCS

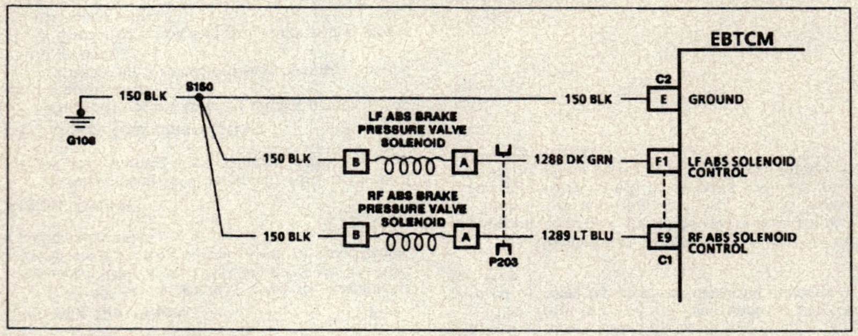

DTC 78
RIGHT FRONT SOL CIRCUIT OPEN OR SHORTED TO BATT

Circuit Description:

This DTC identifies a solenoid that cannot be energized due to an open in its circuitry, or a solenoid that is always energized due to a short to battery in its circuitry between the driver and the solenoid. An open will not allow proper ABS operation, but the short to battery simply turns "ON" the solenoid. A path for base brakes is still allowed once the ABS motor re-homes and the check ball is lifted off its seat during key "ON" initialization.

DTC Will Set When:

DTC 78 can be set only when the solenoid is commanded "OFF." If the EBTCM detects an excessive voltage in the right front solenoid control circuit, a malfunction exists.

Action Taken:

A malfunction DTC is stored, ABS is disabled and the amber ABS warning lamp is turned "ON."

DTC Chart Test Description:

Number(s) below refer to circled number(s) on the diagnostic chart.

1. Checks for a short to voltage in the control circuitry of the solenoid.
2. Checks for an open in the control circuit of the solenoid.
3. Checks the solenoid coil for proper resistance.
4. Checks for an open in the solenoid ground circuit.
5. Checks for a possible intermittent malfunction in the solenoid circuitry due to poor terminal contact.

Diagnostic Aids:

An intermittent malfunction may be caused by a poor connection, rubbed through wire insulation, or a wire that is broken inside the insulation.

The frequency of the malfunction can be checked by using the enhanced diagnostic function of the Tech 1

Any circuitry that is suspected of causing the intermittent complaint should be thoroughly checked for backed out terminals, improper mating, broken locks, improperly formed or damaged terminals, poor terminal to wiring connections or physical damage to the wiring harness.

DIAGNOSTIC CHART DTC 78 — 1995 SILHOUETTE AND TRANS SPORT WITH TCS, CONTINUED

DTC 78
RIGHT FRONT SOLENOID CIRCUIT SHORTED TO BATTERY OR OPEN

(1)
- IGNITION "OFF."
- DISCONNECT 32-WAY EBTCM CONNECTOR.
- IGNITION "ON," ENGINE "OFF."
- USING J 39200, MEASURE THE VOLTAGE BETWEEN GROUND AND TERMINAL "E9" OF THE 32-WAY EBTCM HARNESS CONNECTOR.

IS VOLTAGE 1 VOLT OR LESS?

NO: REPAIR SHORT TO VOLTAGE IN CKT 1289.

YES:

(2)
- IGNITION "OFF."
- DISCONNECT RIGHT FRONT SOLENOID CONNECTOR.
- USING J 39200, MEASURE THE RESISTANCE BETWEEN TERMINAL "E9" OF THE 32-WAY EBTCM HARNESS CONNECTOR AND TERMINAL "A" OF THE RIGHT FRONT SOLENOID HARNESS CONNECTOR.

IS RESISTANCE 2Ω OR LESS?

NO: REPAIR OPEN OR HIGH RESISTANCE IN CKT 1289.

YES:

(3)
- USING J 39200, MEASURE THE RESISTANCE BETWEEN TERMINAL "A" AND TERMINAL "B" OF THE RIGHT FRONT SOLENOID.

IS RESISTANCE BETWEEN 2.5 AND 5Ω?

NO: REPLACE RIGHT FRONT SOLENOID.

YES:

(4)
- DISCONNECT NEGATIVE BATTERY CABLE.
- USING J 39200, MEASURE THE RESISTANCE BETWEEN NEGATIVE BATTERY TERMINAL AND TERMINAL "B" OF THE RIGHT FRONT SOLENOID HARNESS CONNECTOR.

IS RESISTANCE 2Ω OR LESS?

NO: REPAIR OPEN OR HIGH RESISTANCE IN CKT 150.

YES:

(5)
- INSPECT RIGHT FRONT SOLENOID AND 32-WAY EBTCM HARNESS CONNECTORS FOR POOR TERMINAL CONTACT AND EVIDENCE OF CORROSION.
  REPLACE ALL TERMINALS THAT EXHIBIT SIGNS OF POOR TERMINAL CONTACT OR CORROSION.
- INSPECT CKT 1289 FOR DAMAGE WHICH MAY RESULT IN AN OPEN CIRCUIT OR A SHORT TO VOLTAGE WITH ALL CONNECTORS CONNECTED. REPAIR DAMAGE IF EVIDENT.
- RECONNECT ALL CONNECTORS.
- RECONNECT NEGATIVE BATTERY CABLE.
- IGNITION "ON."

DOES DTC 78 SET AS A CURRENT DTC?

YES: REPLACE EBTCM.

NO: MALFUNCTION IS MOST LIKELY DUE TO A POOR CONNECTION. REFER TO "DIAGNOSTIC AIDS" ON FACING PAGE.

DIAGNOSTIC CHART DTC 81 — 1995 SILHOUETTE AND TRANS SPORT WITH TCS

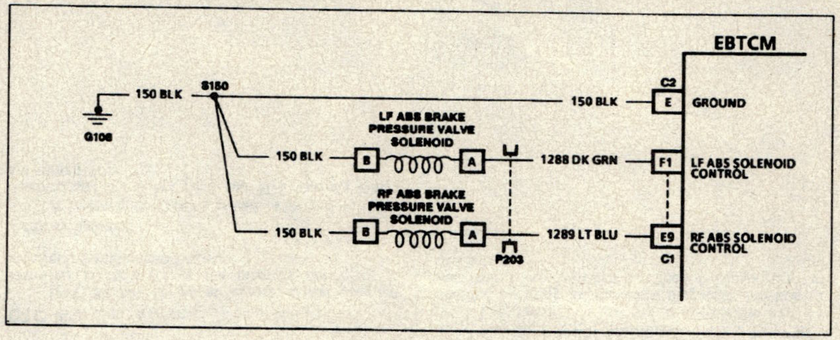

DTC 81
RIGHT FRONT SOL CIRCUIT SHORTED TO GND

Circuit Description:

This DTC identifies a solenoid that cannot be energized due to an open in its driver circuitry, or a short to ground between the solenoid driver and the solenoid. These malfunctions can affect ABS operation since the flow of brake fluid to the wheel cylinder cannot be stopped, making ABS operation for that channel impossible.

DTC Will Set When:

DTC 81 can be set only when the solenoid is commanded "ON." If the EBTCM detects the right front solenoid control circuit voltage is out of specification, a malfunction exists.

Action Taken:

A malfunction DTC is stored, ABS is disabled and the amber ABS warning lamp is turned "ON."

DTC Chart Test Description:

Number(s) below refer to circled number(s) on the diagnostic chart.

1. Checks for a solenoid that is internally shorted to ground.
2. Checks for a solenoid that is not within proper resistance values.
3. Indicates if a short to ground exists in the solenoid circuitry.
4. Checks for a possible intermittent malfunction in the solenoid circuitry due to poor terminal contact.

Diagnostic Aids:

An intermittent malfunction may be caused by a poor connection, rubbed through wire insulation, or a wire that is broken inside the insulation.

The frequency of the malfunction can be checked by using the enhanced diagnostic function of the Tech 1

Any circuitry that is suspected of causing the intermittent complaint should be thoroughly checked for backed out terminals, improper mating, broken locks, improperly formed or damaged terminals, poor terminal to wiring connections or physical damage to the wiring harness.

DIAGNOSTIC CHART DTC 81 — 1995 SILHOUETTE AND TRANS SPORT WITH TCS, CONTINUED

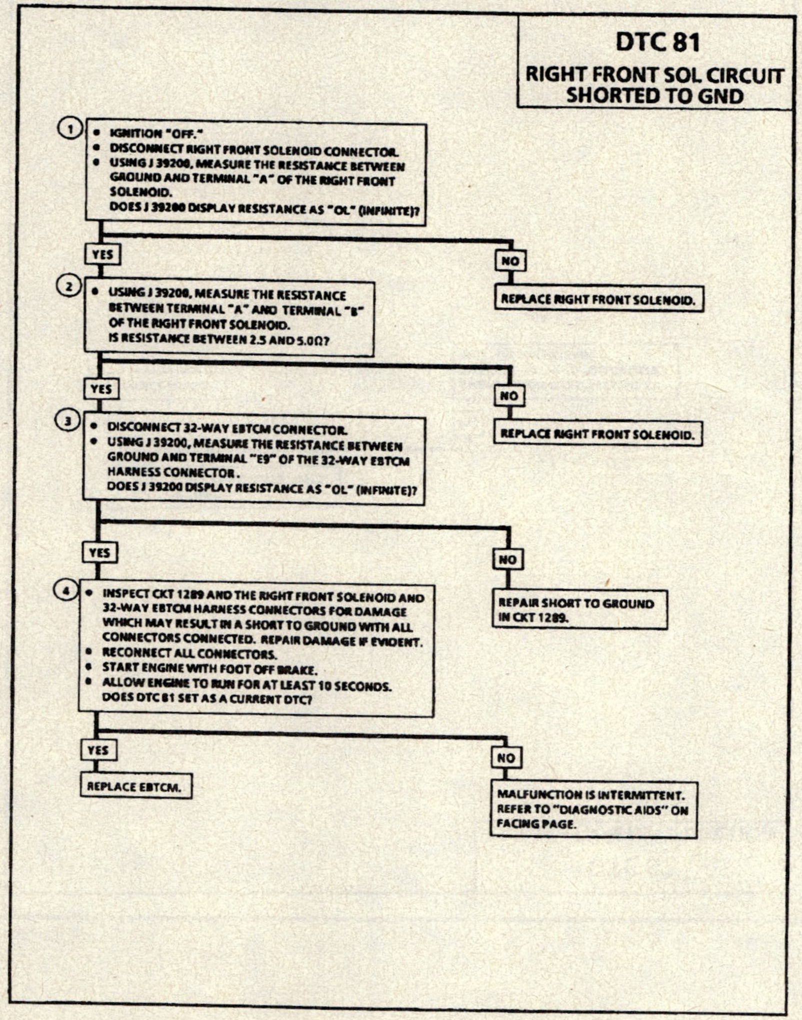

DIAGNOSTIC CHART DTC 82 — 1995 SILHOUETTE AND TRANS SPORT WITH TCS

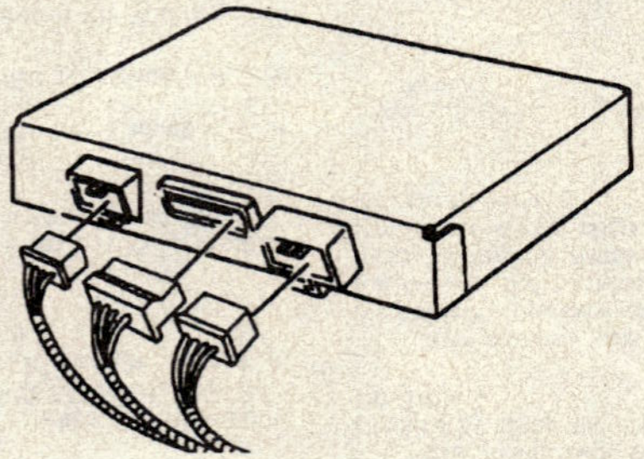

DTC 82
CALIBRATION MALFUNCTION

Circuit Description:

This DTC allows the EBTCM to check for a calibration malfunction by comparing the calibration value to a known value stored in the EEPROM. This DTC is also used as a security measure to prevent improper use of calibrations or changes to these calibrations that may alter the designed function of ABS and TCS.

DTC Will Set When:

DTC 82 can be set at key-up. If the program identifier is incorrect or the memory checksum is incorrect, a malfunction exists.

Action Taken:

A malfunction DTC is stored, ABS and TCS are disabled and the amber ABS and TCS warning lamps are turned "ON."

DTC Chart Test Description:

Number(s) below refer to circled number(s) on the diagnostic chart.

1. Checks to see if the malfunction is present during diagnosis. If present, the EBTCM is not functioning correctly and must be replaced.

Diagnostic Aids: An intermittent DTC 82 may be caused by a bad cell in the EEPROM that is sensitive to temperature changes. If DTC 82 failed more than once, but is intermittent, replace EBTCM.

The frequency of the malfunction can be checked by using the enhanced diagnostic function of the Tech 1, as described in "Tech 1 Diagnostics" in this section.

DIAGNOSTIC CHART DTC 82 — 1995 SILHOUETTE AND TRANS SPORT WITH TCS, CONTINUED

DTC 82
CALIBRATION MALFUNCTION

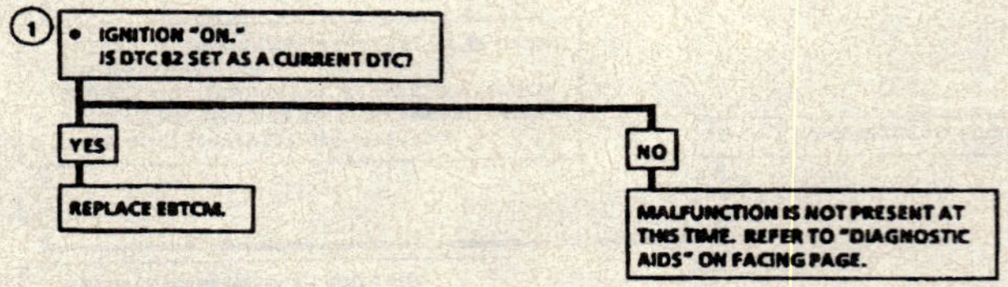

DIAGNOSTIC CHART DTC 86 — 1995 SILHOUETTE AND TRANS SPORT WITH TCS

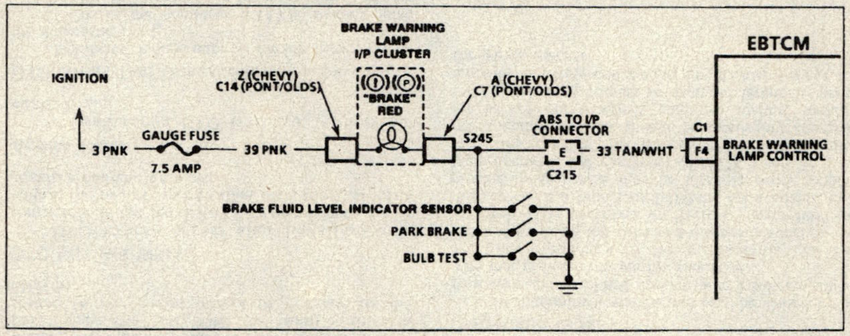

DTC 86
EBTCM TURNED "ON" THE RED "BRAKE" WARNING LAMP

Circuit Description:

This DTC is provided as an information only DTC, and reflects the status of the command issued by the ABS VI/TCS EBTCM to illuminate the red BRAKE warning lamp. If another DTC issues a command to illuminate the red BRAKE warning lamp, DTC 86 will be stored in EEPROM as a history DTC at the conclusion of the ignition cycle.

DTC Chart Test Description:

Number(s) below refer to circled number(s) on the diagnostic chart.

1. This step serves to identify if a DTC other than DTC 86 commanded the red BRAKE warning lamp "ON."

Diagnostic Aids:

Any ABS mechanical DTC that issues a command to illuminate the red BRAKE warning lamp will also result in DTC 86 being stored in EEPROM during shutdown. These DTCs are: 38, 41, 42, 46, and 51. If the motors are not in their home position, certain electrical DTCs will also command the red BRAKE warning lamp "ON." These DTCs are: 14, 16, 18, 55, 56, 57, 58, 61, 62, 63, 64, 65, 66, 144 and 145. If any of these DTCs are indicated along with DTC 86, they must be corrected prior to addressing a DTC 86 malfunction.

DIAGNOSTIC CHART DTC 86 — 1995 SILHOUETTE AND TRANS SPORT WITH TCS, CONTINUED

DTC 86
EBTCM TURNED "ON" THE RED "BRAKE" WARNING LAMP

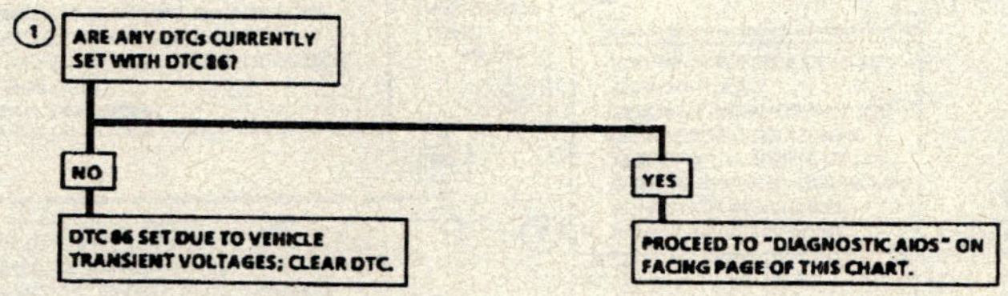

DIAGNOSTIC CHART DTC 87 — 1995 SILHOUETTE AND TRANS SPORT WITH TCS

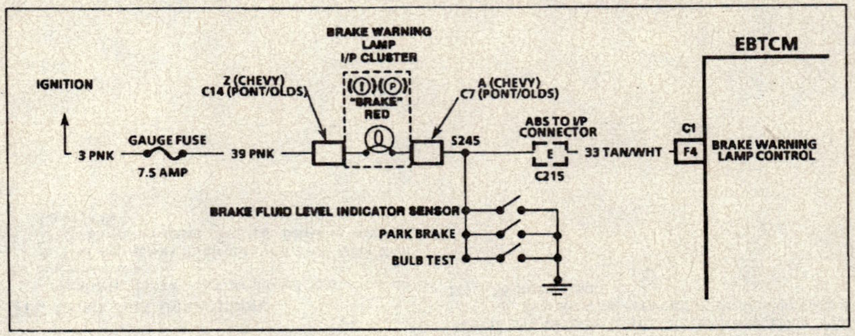

DTC 87
RED "BRAKE" WARNING LAMP CIRCUIT OPEN

Circuit Description:

This DTC is used to verify the EBTCM has continuity to the red BRAKE warning lamp in case the EBTCM must turn it "ON." This will only occur if an ABS or TCS malfunction is detected that may degrade base brake operation. Because ABS VI/TCS is not the only device controlling the red BRAKE warning lamp (parking brake and low brake fluid level switch may also turn "ON" the lamp) a short to ground in this circuit cannot be detected by the EBTCM.

DTC Will Set When:

DTC 87 can be set after initialization is completed. If the red BRAKE warning lamp circuit voltage is out of specification indicating an open circuit, a malfunction exists.

Action Taken:

A malfunction DTC is stored, ABS is not disabled.

DTC Chart Test Description:

Number(s) below refer to circled number(s) on the diagnostic chart.

1. Checks the complete BRAKE warning lamp circuitry by attempting to manually turn "ON" the BRAKE warning lamp.
2. Checks for high resistance in park brake circuitry.
3. Checks for a malfunctioning EBTCM.
4. Ensures malfunction is not due to poor terminal contact at the brake fluid level switch which may result in an intermittent malfunction.
5. Checks the integrity of the red BRAKE warning lamp.
6. Checks for high resistance in the park brake circuitry.

Diagnostic Aids:

An intermittent malfunction may be caused by a poor connection, rubbed through wire insulation, or a wire that is broken inside the insulation.

The frequency of the malfunction can be checked by using the enhanced diagnostic function of the Tech 1, as described in "Tech 1 Diagnostics" in this section. If a Tech 1 or T-100 are not available, the parking brake handle may be lifted to verify proper lamp operation and continuity of CKTs 3 and 39.

Any circuitry that is suspected of causing the intermittent complaint should be thoroughly checked for backed out terminals, improper mating, broken locks, improperly formed or damaged terminals, poor terminal to wiring connections or physical damage to the wiring harness.

DIAGNOSTIC CHART DTC 87 — 1995 SILHOUETTE AND TRANS SPORT WITH TCS, CONTINUED

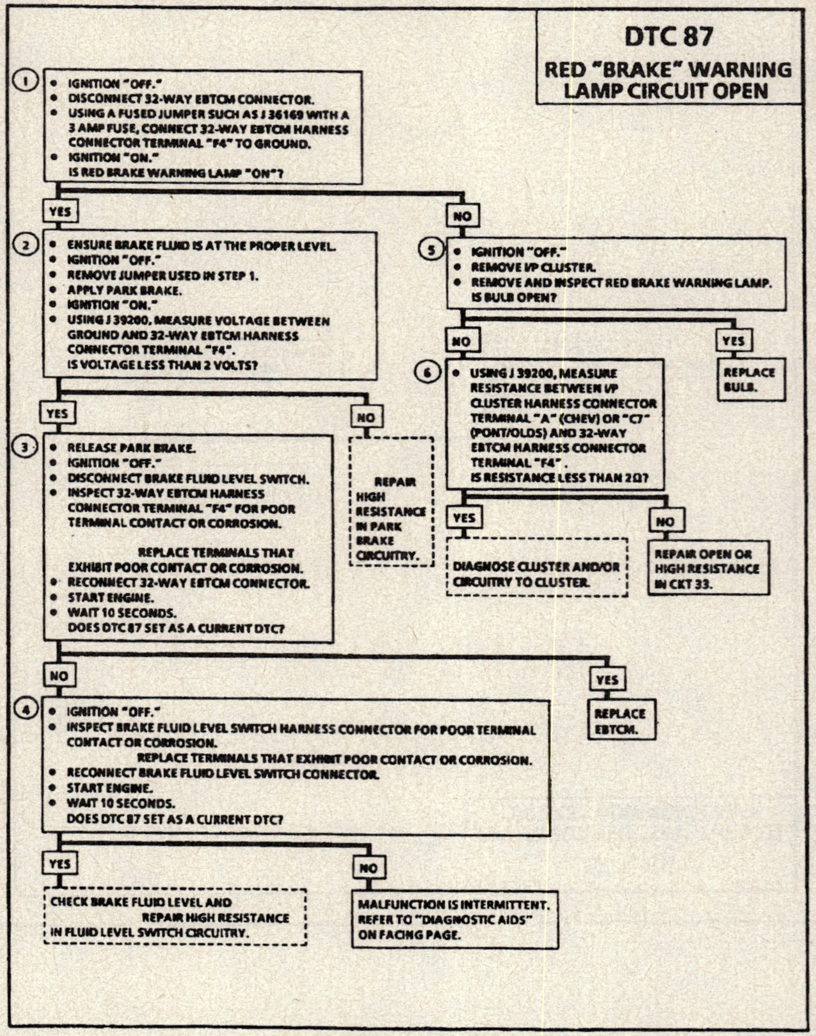

DIAGNOSTIC CHART DTC 88 — 1995 SILHOUETTE AND TRANS SPORT WITH TCS

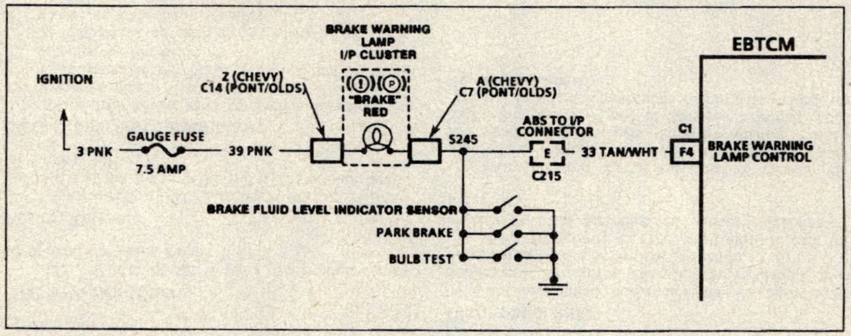

DTC 88

RED "BRAKE" WARNING LAMP CIRCUIT SHORTED TO BATT

Circuit Description:

This DTC identifies a short to battery between the EBTCM and the red BRAKE warning lamp, or an open driver that does not allow the red BRAKE warning lamp to be illuminated by the EBTCM. This will only occur if an ABS malfunction is detected that may degrade base brake operation.

DTC Will Set When:

DTC 88 can be set only at the beginning of the three second bulb check. If the EBTCM detects battery voltage on the red BRAKE warning lamp control circuit, a malfunction exists.

Action Taken:

A malfunction DTC is stored. ABS is not disabled.

DTC Chart Test Description:

Number(s) below refer to circled number(s) on the diagnostic chart.

1. Identifies if the ground circuit to the red BRAKE warning lamp is being completed by a source other than the EBTCM.
2. Identifies if the malfunction is currently present.
3. By removing the "GAUGES" fuse, the voltage source is eliminated. This test indicates if voltage is being supplied from a source other than the "GAUGES" fuse.
4. Ensures malfunction was not due to physical damage of the circuitry.
5. Checks for DTC 86 also being set.

Diagnostic Aids:

An intermittent malfunction may be caused by a poor connection, rubbed through wire insulation, or a wire that is broken inside the insulation.

The frequency of the malfunction can be checked by using the enhanced diagnostic function of the Tech 1

Any circuitry that is suspected of causing the intermittent complaint should be thoroughly checked for backed out terminals, improper mating, broken locks, improperly formed or damaged terminals, poor terminal to wiring connections or physical damage to the wiring harness.

DIAGNOSTIC CHART DTC 88 — 1995 SILHOUETTE AND TRANS SPORT WITH TCS, CONTINUED

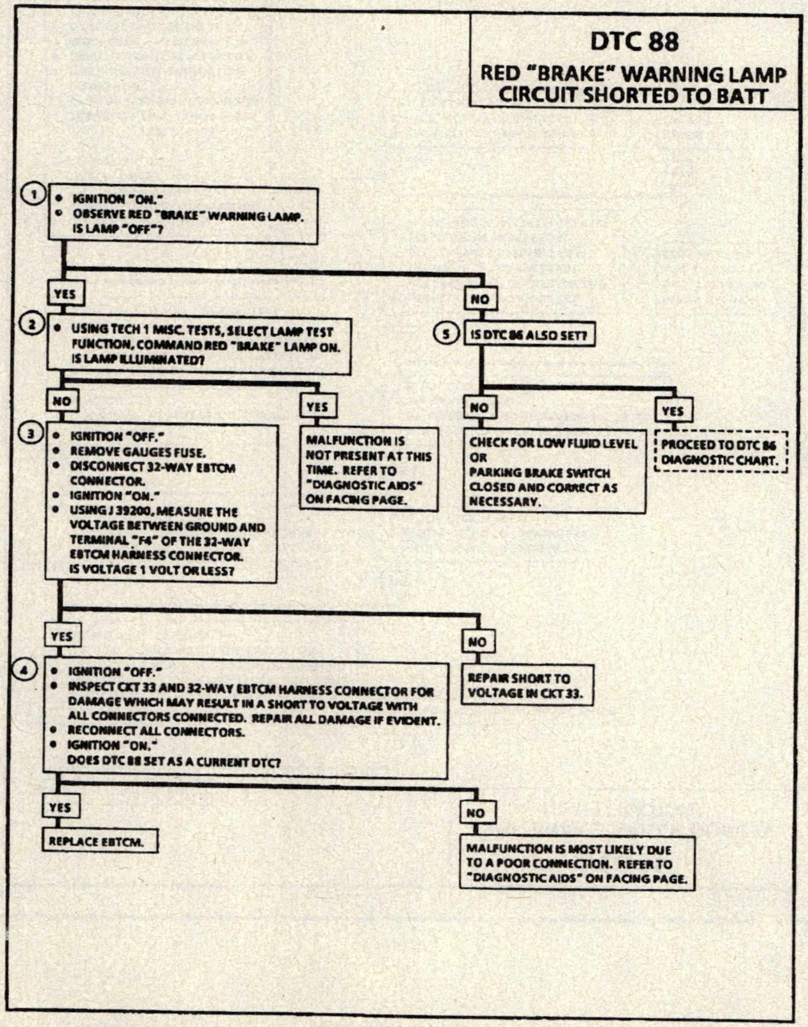

DIAGNOSTIC CHART DTC 91 — 1995 SILHOUETTE AND TRANS SPORT WITH TCS

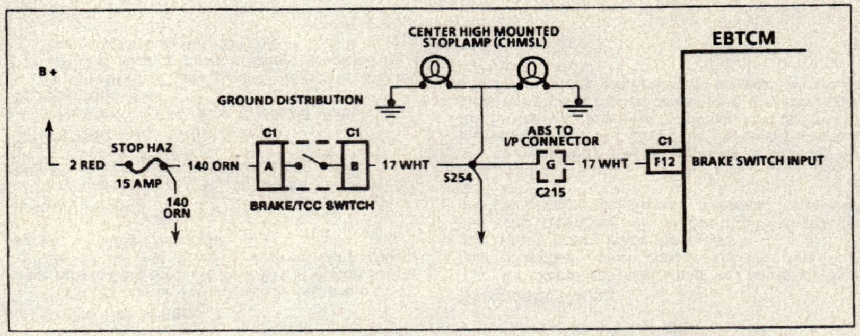

DTC 91
OPEN BRAKE SWITCH DURING DECELERATION

Circuit Description:

This DTC is used to detect an open brake switch in the non-ABS mode. The EBTCM looks for deceleration rates that would indicate braking action and verifies this assumption by requiring several repeats of this detection method. In each case, ABS and TCS will not be available since a brake switch signal is not seen.

DTC Will Set When:

DTC 91 can be set if three deceleration cycles occur with the brake switch "OFF."

Action Taken:

A malfunction DTC is stored, ABS and TCS are disabled and the amber ABS and TCS warning lamps are turned "ON."

DTC Chart Test Description:

Number(s) below refer to circled number(s) on the diagnostic chart.

1. Indicates if the brake switch signal is being received by the EBTCM.
2. Indicates if an open circuit exists in the brake switch or brake lamp circuitry.
3. Isolates the open circuit to either the brake switch input circuitry, or the EBTCM.
4. Ensures malfunction was not due to poor terminal contact.
5. Verifies that voltage is available at the brake switch.
6. Verifies if the brake switch is functioning properly.

Diagnostic Aids:

An intermittent malfunction may be caused by a poor connection, rubbed through wire insulation, or a wire that is broken inside the insulation.

The frequency of the malfunction can be checked by using the enhanced diagnostic function of the Tech 1

Any circuitry that is suspected of causing the intermittent complaint should be thoroughly checked for backed out terminals, improper mating, broken locks, improperly formed or damaged terminals, poor terminal to wiring connections or physical damage to the wiring harness.

DIAGNOSTIC CHART DTC 91 — 1995 SILHOUETTE AND TRANS SPORT WITH TCS, CONTINUED

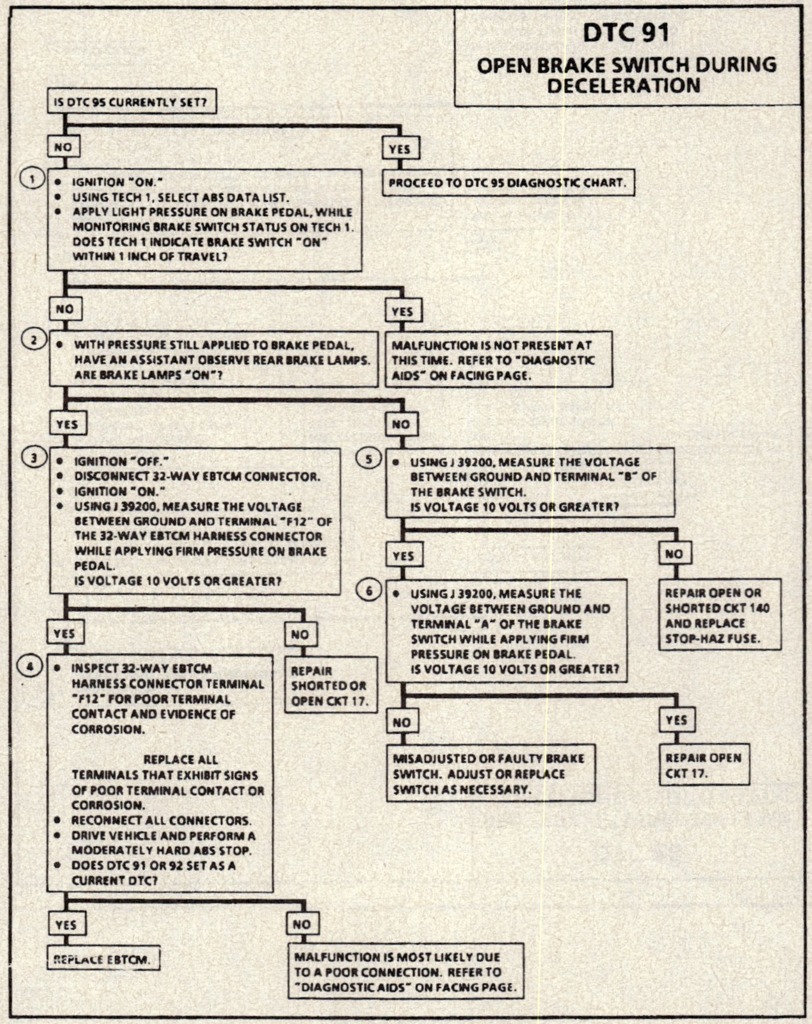

DIAGNOSTIC CHART DTC 92 — 1995 SILHOUETTE AND TRANS SPORT WITH TCS

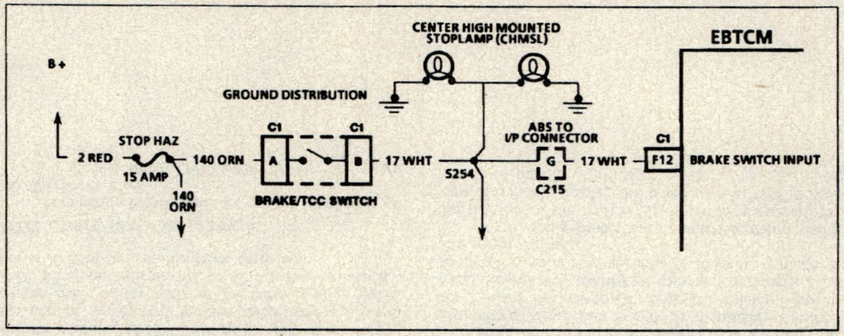

DTC 92

OPEN BRAKE SWITCH WHEN ABS WAS REQUIRED

Circuit Description:

This DTC is run to determine the proper operation of the brake switch. This is important because ABS is activated when the brake switch is "ON." If the brake switch is "OFF," ABS will never be activated. Since this malfunction is difficult to detect under normal braking conditions, this malfunction is only detected when ABS is required.

DTC Will Set When:

DTC 92 can be set if the vehicle's speed is greater than 8 km/h (5 mph). If the brake was not "ON" and a release was required on two channels for 0.7 second, a malfunction exists.

Action Taken:

A malfunction DTC is stored, ABS and TCS are disabled and the amber ABS and TCS warning lamps are turned "ON."

DTC Chart Test Description:

Number(s) below refer to circled number(s) on the diagnostic chart.

1. Indicates if the brake switch signal is being received by the EBTCM.
2. Indicates if an open circuit exists in the brake switch or brake lamp circuitry.
3. Isolates the open circuit to either the brake switch input circuitry, or the EBTCM.
4. Ensures malfunction was not due to poor terminal contact.
5. Verifies the voltage is available at the brake switch.
6. Verifies if the brake switch is functioning properly.

Diagnostic Aids:

An intermittent malfunction may be caused by a poor connection, rubbed through wire insulation, or a wire that is broken inside the insulation.

The frequency of the malfunction can be checked by using the enhanced diagnostic function of the Tech 1.

Any circuitry that is suspected of causing the intermittent complaint should be thoroughly checked for backed out terminals, improper mating, broken locks, improperly formed or damaged terminals, poor terminal to wiring connections or physical damage to the wiring harness.

DIAGNOSTIC CHART DTC 92 — 1995 SILHOUETTE AND TRANS SPORT WITH TCS, CONTINUED

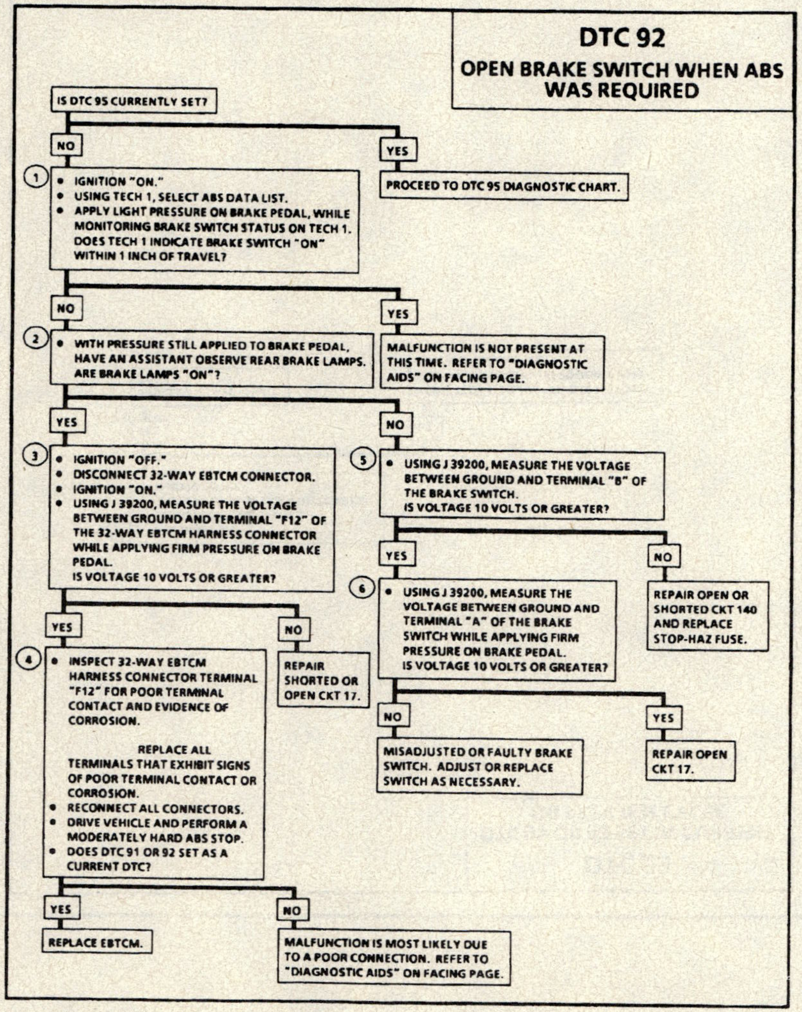

DIAGNOSTIC CHART DTC 93 — 1995 SILHOUETTE AND TRANS SPORT WITH TCS

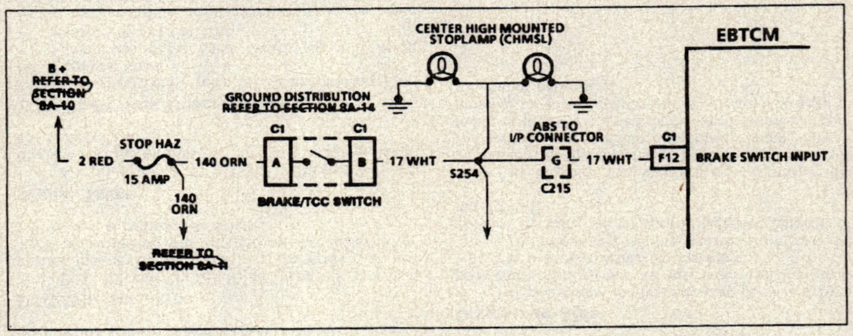

DTC 93
DTC 91 OR 92 SET IN CURRENT OR PREV IGN CYCLE

Circuit Description:

This DTC is the second portion of DTCs 91 and 92. If DTCs 91 or 92 occurred during the last ignition cycle, DTC 93 becomes a current malfunction during the next ignition cycle, keeping ABS disabled until a brake switch "ON" state is seen. When a change is seen during an ignition cycle in which DTC 93 is a current malfunction, DTC 91 or 92 will clear itself at the end of the current ignition cycle, and ABS will enable itself at the start of the next ignition cycle. DTC 93 alone indicates 91 or 92 occurred previously, but is intermittent, or has been corrected.

DTC Chart Test Description:

Number(s) below refer to circled number(s) on the diagnostic chart.

1. Indicates which DTC (either 91 or 92) caused DTC 93 to set.

Diagnostic Aids:

An intermittent malfunction may be caused by a poor connection, rubbed through wire insulation, or a wire that is broken inside the insulation.

The frequency of the malfunction can be checked by using the enhanced diagnostic function of the Tech 1

Any circuitry that is suspected of causing the intermittent complaint should be thoroughly checked for backed out terminals, improper mating, broken locks, improperly formed or damaged terminals, poor terminal to wiring connections or physical damage to the wiring harness.

Also, verify proper brake switch operation using the data list of the Tech 1. As the brake is applied, the data list should display the brake switch "ON" within 1 inch of travel.

DIAGNOSTIC CHART DTC 93 — 1995 SILHOUETTE AND TRANS SPORT WITH TCS, CONTINUED

DTC 93
DTC 91 OR 92 SET IN CURRENT OR PREV IGN CYCLE

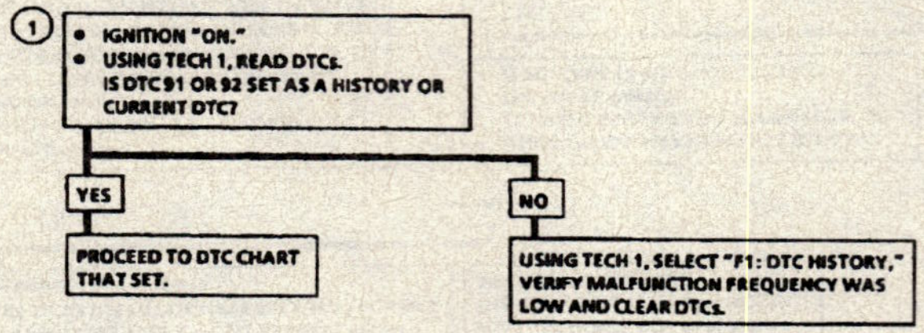

DIAGNOSTIC CHART DTC 94 — 1995 SILHOUETTE AND TRANS SPORT WITH TCS

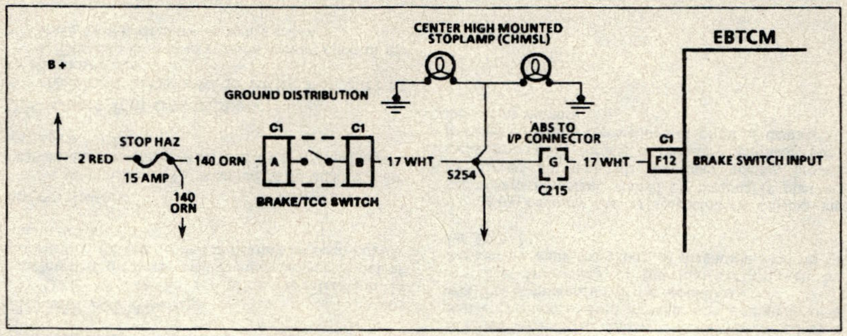

DTC 94
BRAKE SWITCH CONTACTS ALWAYS CLOSED

Circuit Description:

This DTC is run to determine the proper operation of the brake switch. This is important because ABS is activated when the brake switch is "ON" and turned "OFF" when the brake is "OFF." If the brake switch is always "ON," ABS operation will always be requested resulting in potential modulator cycling on rough roads. Additionally, this malfunction will most likely result in a dead battery (due to the brake lights staying "ON") if the driver is not informed of this malfunction.

DTC Will Set When:

DTC 94 can be set when the vehicle reaches at least 40 km/h (25 mph). If the brake was never "OFF" during two consecutive drive cycles, a malfunction exists.

Action Taken:

A malfunction DTC is stored. ABS is not disabled, however, TCS is disabled and the amber TCS warning lamp is turned "ON."

DTC Chart Test Description:

Number(s) below refer to circled number(s) on the diagnostic chart.

1. Checks if the malfunction is currently present.
2. Isolates the cause of the malfunction to either a malfunctioning or misadjusted brake switch, or a short to voltage in the brake switch circuitry.
3. Checks for unwanted voltage on the brake switch input circuit.
4. Ensures malfunction is not due to poor terminal contact or EBTCM.

Diagnostic Aids:

An intermittent malfunction may be caused by a poor connection, rubbed through wire insulation, or a wire that is broken inside the insulation.

The frequency of the malfunction can be checked by using the enhanced diagnostic function of the Tech 1

Any circuitry that is suspected of causing the intermittent complaint should be thoroughly checked for backed out terminals, improper mating, broken locks, improperly formed or damaged terminals, poor terminal to wiring connections or physical damage to the wiring harness.

DIAGNOSTIC CHART DTC 94 — 1995 SILHOUETTE AND TRANS SPORT WITH TCS, CONTINUED

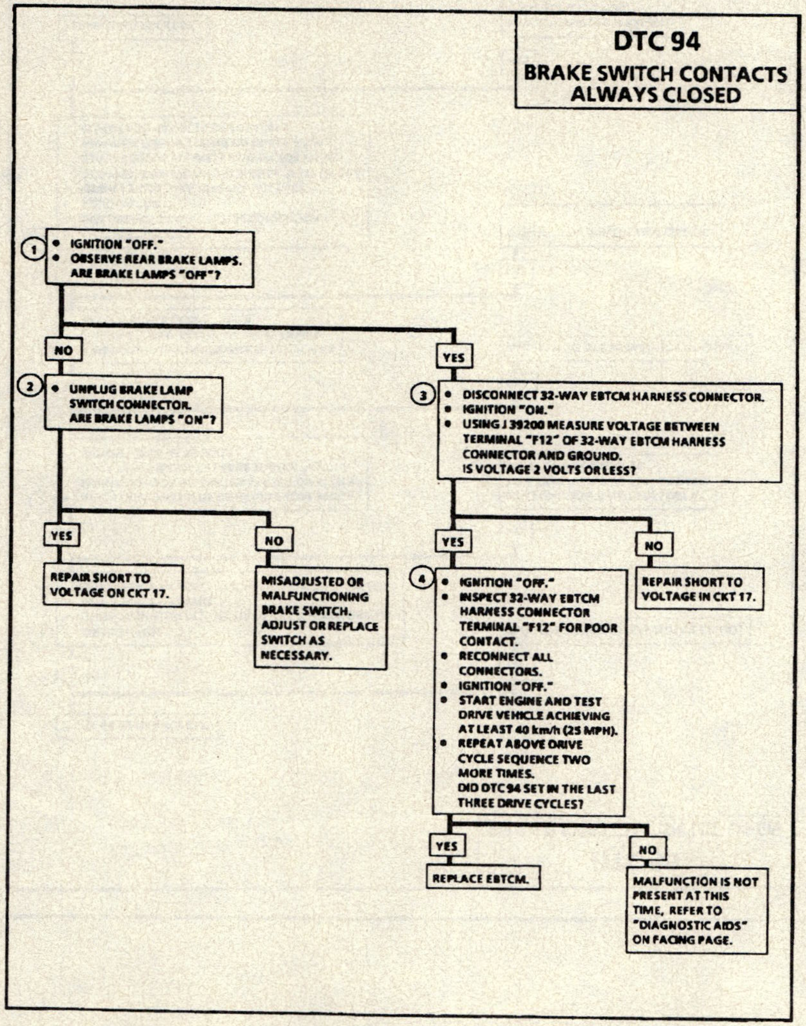

DIAGNOSTIC CHART DTC 95 — 1995 SILHOUETTE AND TRANS SPORT WITH TCS

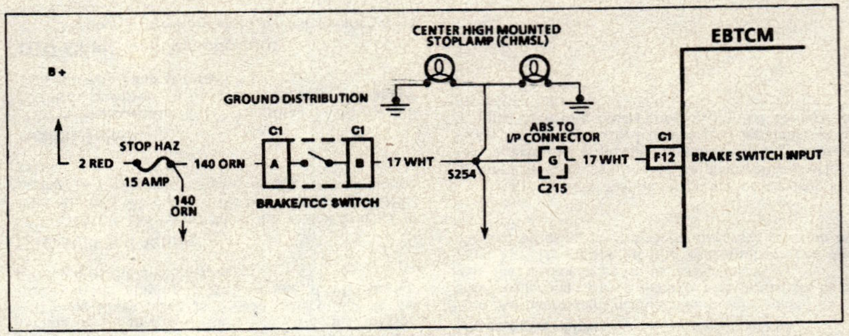

DTC 95
BRAKE SWITCH CIRCUIT OPEN

Circuit Description:

This DTC is used to identify open brake switch circuitry that prevents the brake switch input to the EBTCM from changing states when the brake is applied. This DTC is used in conjunction with DTCs 91 and 92 to determine the cause of an open brake switch malfunction.

DTC Will Set When:

DTC 95 can be set after initialization is completed. If the brake switch input voltage is out of specification for one second indicating an open circuit, a malfunction exists.

Action Taken:

A malfunction DTC is stored, ABS and TCS are disabled and the amber ABS and TCS warning lamps are turned "ON."

DTC Chart Test Description:

Number(s) below refer to circled number(s) on the diagnostic chart.

1. This step is used to confirm that an open in the brake switch circuitry currently exists.
2. Indicates if the brake switch signal is being received by the EBTCM.
3. Indicates if an open circuit exists in the brake switch or brake lamp circuitry.
4. Isolates the open circuit to either the brake switch input circuitry, or the EBTCM.

Diagnostic Aids:

An intermittent malfunction may be caused by a poor connection, rubbed through wire insulation, or a wire that is broken inside the insulation.

The frequency of the malfunction can be checked by using the enhanced diagnostic function of the Tech 1

Any circuitry that is suspected of causing the intermittent complaint should be thoroughly checked for backed out terminals, improper mating, broken locks, improperly formed or damaged terminals, poor terminal to wiring connections or physical damage to the wiring harness.

DIAGNOSTIC CHART DTC 95 — 1995 SILHOUETTE AND TRANS SPORT WITH TCS, CONTINUED

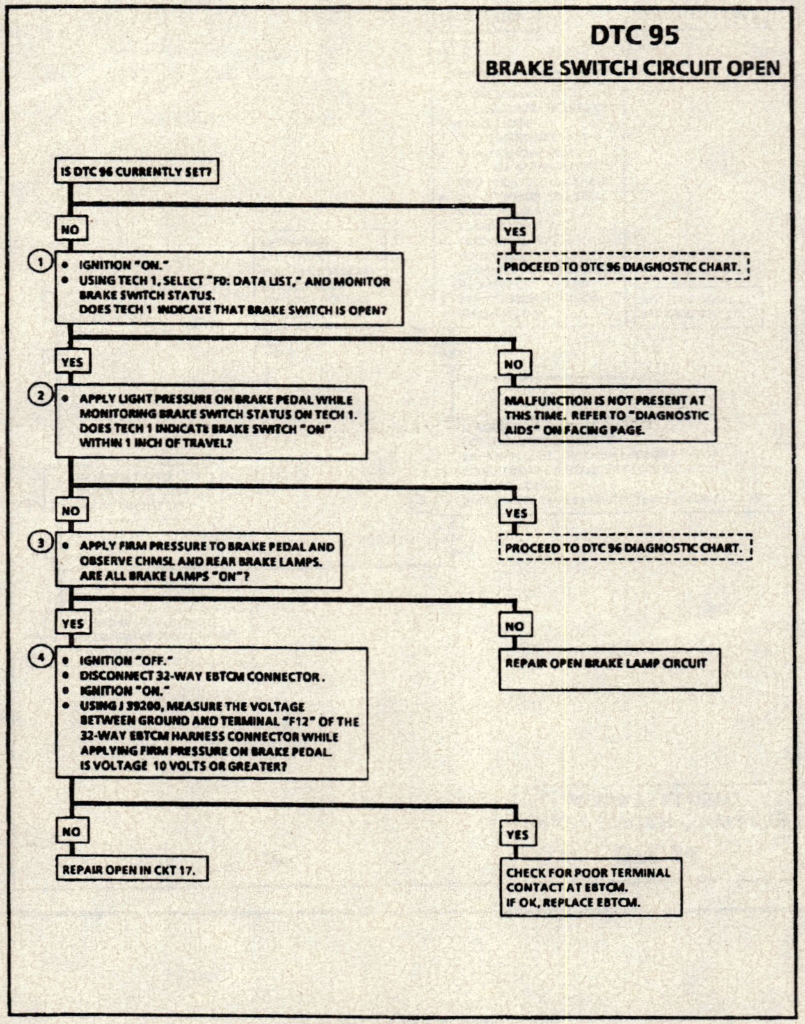

DIAGNOSTIC CHART DTC 96 — 1995 SILHOUETTE AND TRANS SPORT WITH TCS

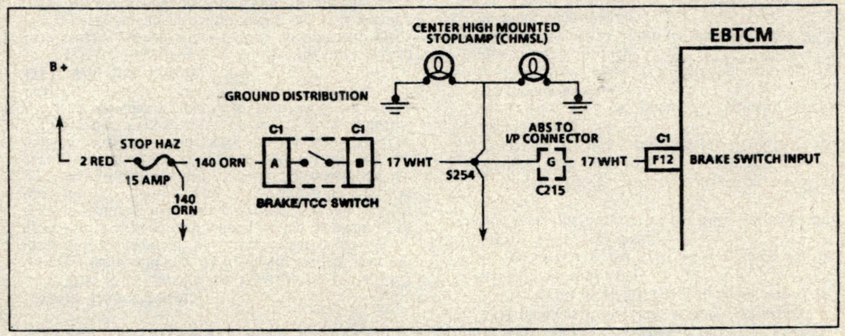

DTC 96
REAR BRAKE LAMP CIRCUIT OPEN

Circuit Description:

This DTC is designed to isolate the cause of a DTC 95 malfunction and indicate to the driver ABS is still available. If DTC 95 occurs with DTC 96, the brake lamps are open, the brake lamp grounds are open, or the Center High Mounted Stop Lamp (CHMSL) circuit is open during 4-way flasher use. The presence of battery voltage at the brake switch input indicates a valid brake switch input is still available.

DTC Will Set When:

DTC 96 can be set only after DTC 95 has been set. If the EBTCM detects battery voltage on the brake switch input circuit for 0.5 second, a malfunction exists.

Action Taken:

A malfunction DTC is stored. ABS is not disabled.

DTC Chart Test Description:

Number(s) below refer to circled number(s) on the diagnostic chart.

1. As a result of a malfunction of an additional brake switch circuit DTC, this DTC may be set. To ensure proper diagnosis, any additional brake switch DTCs must be repaired first.
2. Identifies if the malfunction is currently present in the brake circuit.

Diagnostic Aids:

An intermittent malfunction may be caused by a poor connection, rubbed through wire insulation, or a wire that is broken inside the insulation.

The frequency of the malfunction can be checked by using the enhanced diagnostic function of the Tech 1 [illegible]

Any circuitry that is suspected of causing the intermittent complaint should be thoroughly checked for backed out terminals, improper mating, broken locks, improperly formed or damaged terminals, poor terminal to wiring connections or physical damage to the wiring harness.

DIAGNOSTIC CHART DTC 96 — 1995 SILHOUETTE AND TRANS SPORT WITH TCS, CONTINUED

DTC 96
REAR BRAKE LAMP CIRCUIT OPEN

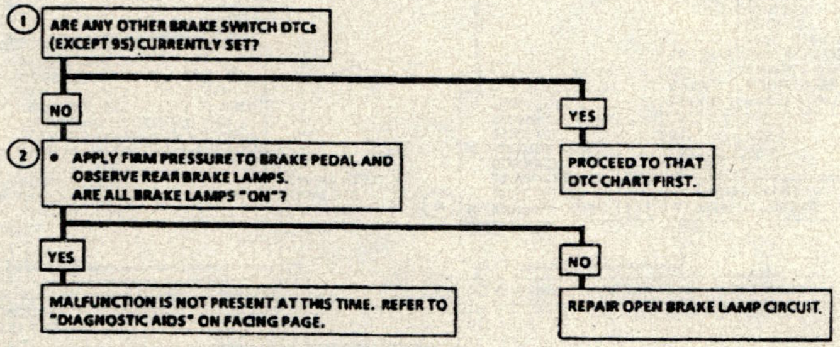

DIAGNOSTIC CHART DTC 111 — 1995 SILHOUETTE AND TRANS SPORT WITH TCS

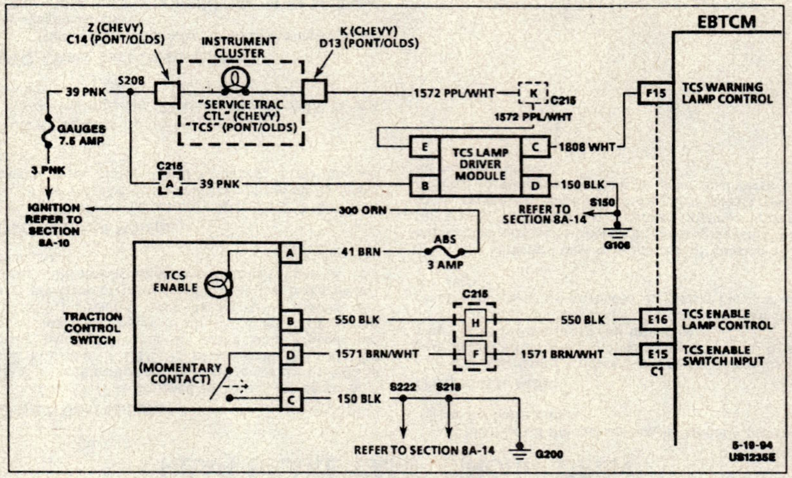

DTC 111
(Page 1 of 2)
TCS WARNING OR SWITCH LAMP CIRCUIT MALFUNCTION

Circuit Description:

This DTC checks the state of both the TCS warning lamp and the TCS enable switch lamp to identify a situation in which the driver could not be alerted to a system malfunction or the availability of traction control. The Lamp Driver Module (LDM) turns the TCS warning lamp "ON," unless the EBTCM provides ground to turn it "OFF." Because of the circuitry in the LDM, only external wiring malfunctions can be detected by the EBTCM. The TCS enable switch lamp circuit is such that the EBTCM provides a ground to turn "ON" the switch lamp.

DTC Will Set When:

DTC 111 can only be set during the three second bulb check or when either bulb is commanded "ON." If the EBTCM cannot control either the TCS warning lamp or the enable switch lamp for 2 seconds, a malfunction exists.

Action Taken:

A malfunction DTC is stored, however ABS and TCS are not disabled.

DTC Chart Test Description:

Number(s) below refer to circled number(s) on the diagnostic chart.

1. Determines if the TCS warning lamp can be properly controlled.
2. Determines if the lamp is "ON" constantly.
3. Checks for other DTCs which could turn "ON" the TCS warning lamp.
4. Determines if the TCS warning lamp can be turned "OFF" manually.
5. Checks the integrity of the TCS warning lamp circuitry.
6. Checks for a short to voltage on the TCS warning lamp circuitry.
7. Determines if a malfunctioning EBTCM is the cause of DTC 111.
8. Verifies that the other I/P indicator lamps function properly.
9. Determines if the TCS warning lamp can be turned "ON" manually.
10. Checks for a short to ground in the TCS warning lamp circuitry.
11. Checks for open between "GAUGES" fuse and LDM.
12. Verifies condition of the "GAUGES" fuse.
13. Determines if DTC 111 was set due to an EBTCM malfunction.

Diagnostic Aids:

Refer to Page 2 of 2 of this chart for diagnostic aids.

DIAGNOSTIC CHART DTC 111 — 1995 SILHOUETTE AND TRANS SPORT WITH TCS, CONTINUED

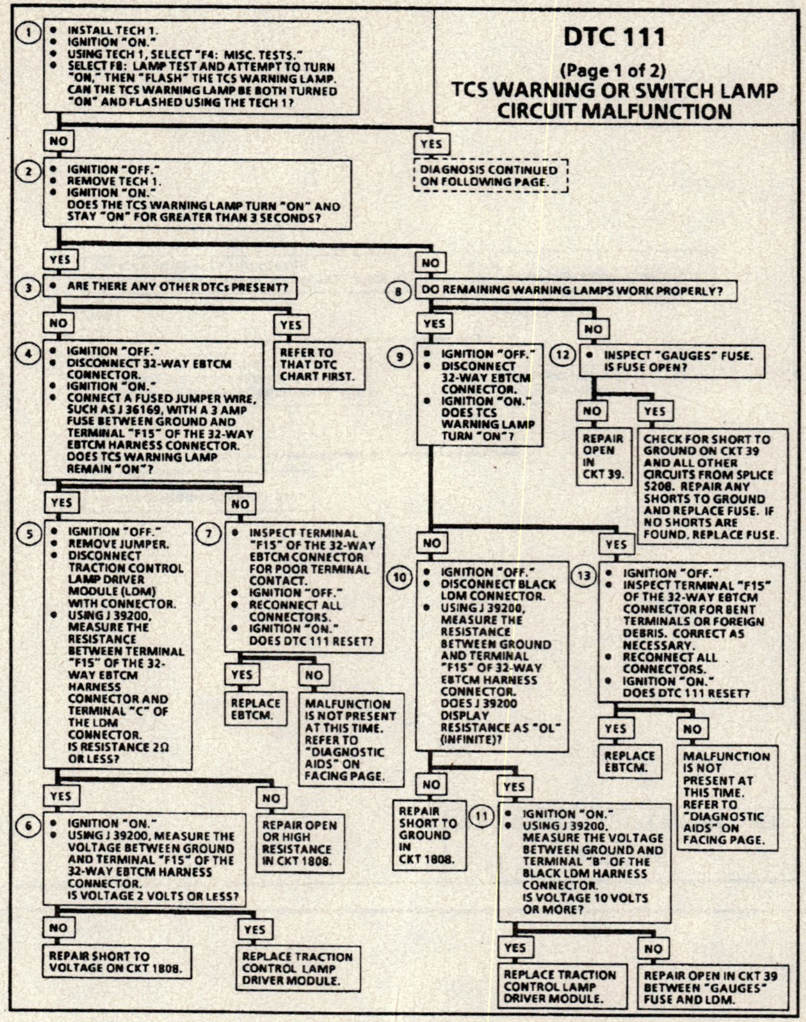

DIAGNOSTIC CHART DTC 111 — 1995 SILHOUETTE AND TRANS SPORT WITH TCS, CONTINUED

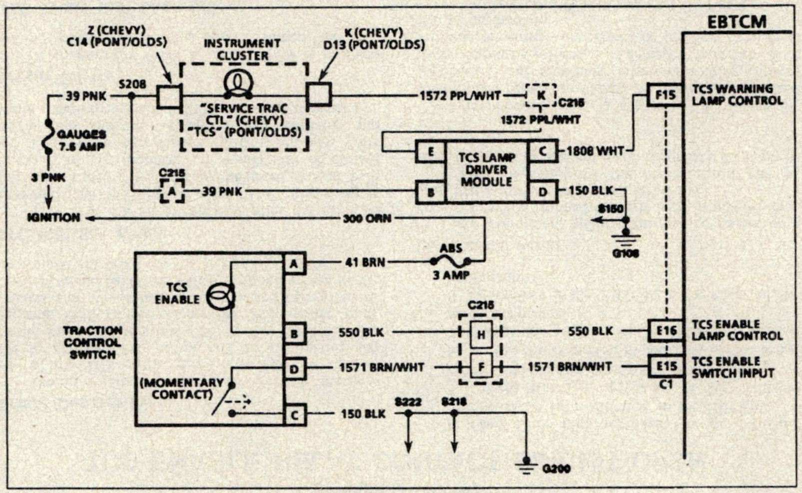

DTC 111
(Page 2 of 2)

TCS WARNING OR SWITCH LAMP CIRCUIT MALFUNCTION

Circuit Description:

This DTC checks the state of both the TCS warning lamp and the TCS enable switch lamp to identify a situation in which the driver could not be alerted to a system malfunction or the availability of traction control. The Lamp Driver Module (LDM) turns the TCS warning lamp "ON," unless the EBTCM provides ground to turn it "OFF." Because of the circuitry in the LDM, only external wiring malfunctions can be detected by the EBTCM. The TCS enable switch lamp circuit is such that the EBTCM provides a ground to turn "ON" the switch lamp.

DTC Will Set When:

DTC 111 can only be set during the three second bulb check or when either bulb is commanded "ON." If the EBTCM cannot control either the TCS warning lamp or the enable switch lamp for 2 seconds, a malfunction exists.

Action Taken:

A malfunction DTC is stored, however ABS and TCS are not disabled.

DTC Chart Test Description:

Number(s) below refer to circled number(s) on the diagnostic chart.

14. Determines if the switch lamp can be properly controlled.
15. Determines if the switch lamp is "ON" constantly.
16. Checks for a short to ground in the switch lamp circuitry.
17. Determines if DTC 111 was set due to an EBTCM malfunction.
18. Determines if DTC 111 was set due to an EBTCM malfunction.
19. Determines if there is an open in CKT 41 or ABS fuse.
20. Determines if the switch lamp can be controlled manually.
21. Verifies the integrity of the switch lamp circuitry.
22. Determines if there is a short to voltage in the switch lamp circuitry.
23. Determines if DTC 111 was set due to an EBTCM malfunction.

Diagnostic Aids:

An intermittent malfunction may be caused by a poor connection, rubbed through wire insulation, or a wire that is broken inside the insulation.

The frequency of the malfunction can be checked by using the enhanced diagnostic function of the Tech 1 [illegible]

Any circuitry that is suspected of causing the intermittent complaint should be thoroughly checked for backed out terminals, improper mating, broken locks, improperly formed or damaged terminals, poor terminal to wiring connections or physical damage to the wiring harness.

DIAGNOSTIC CHART DTC 111 — 1995 SILHOUETTE AND TRANS SPORT WITH TCS, CONTINUED

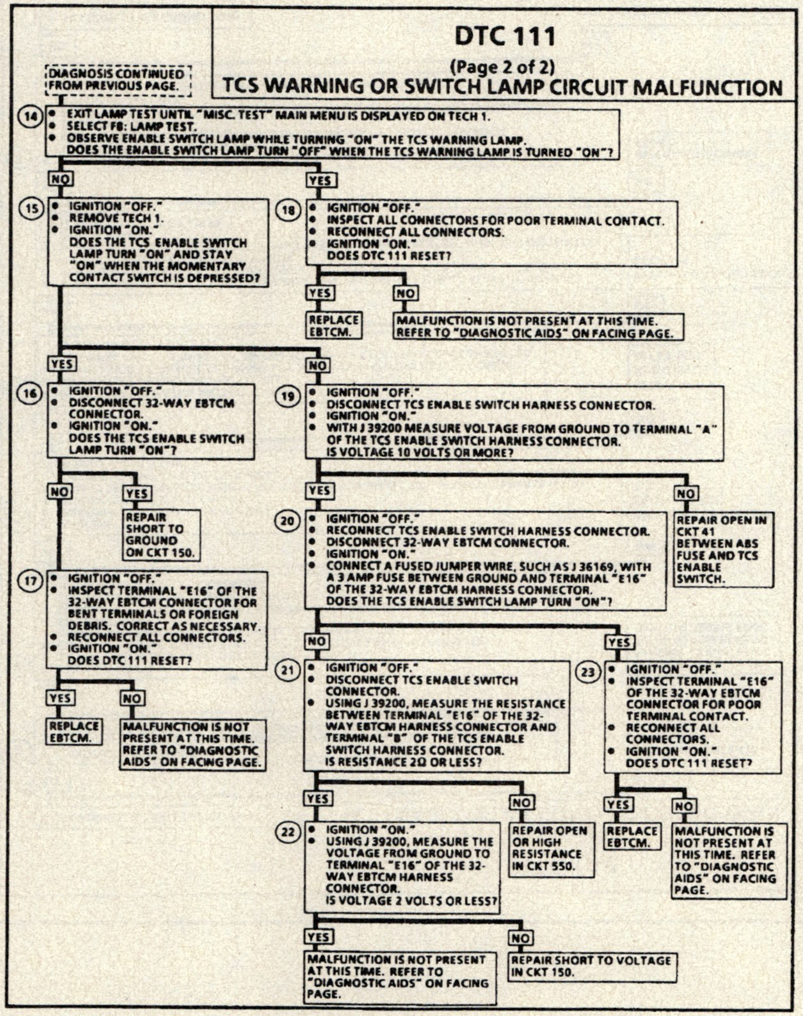

DIAGNOSTIC CHART DTC 114 — 1995 SILHOUETTE AND TRANS SPORT WITH TCS

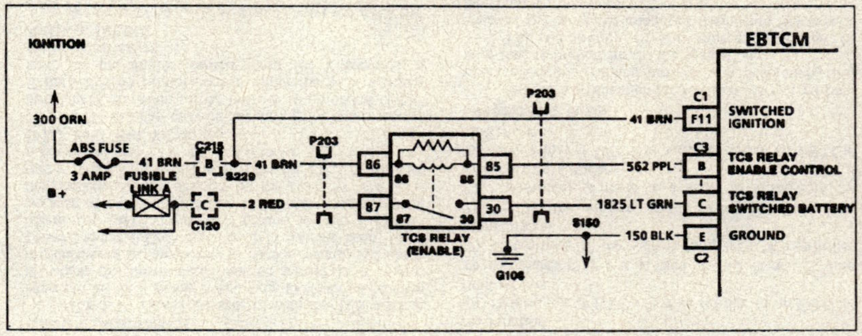

DTC 114
(Page 1 of 3)
TCS ENABLE RELAY CONTACT CIRCUIT OPEN

Circuit Description:

Ignition voltage is supplied through terminal "86" of the TCS enable relay. The EBTCM then is able to energize the pull-in coil by completing the ground circuit at terminal "B" of the EBTCM. The magnetic field created closes the TCS enable relay contacts and allows battery voltage and current to be supplied to the EBTCM, which supplies power to the motors and solenoids.

DTC Will Set When:

DTC 114 can be set anytime after the EBTCM commands the TCS enable relay "ON" (the relay is first commanded "ON" during the three second bulb check). This test monitors the availability of current and voltage to the motors and solenoids. This malfunction indicates voltage is not available and would therefore not allow TCS operation if required.

Action Taken:

A malfunction DTC is stored, TCS is disabled and the amber TCS warning lamp is turned "ON."

DTC Chart Test Description:

Number(s) below refer to circled number(s) on the diagnostic chart.

1. Checks to see if DTC 116 is also set as current or history. If DTC 116 is also set, proceed to that chart first.
2. Verifies that the malfunction is currently present.
3. Verifies that the relay can be turned "ON" and provide voltage to the EBTCM.
4. Checks for high resistance in the switched battery circuit from the relay to the EBTCM.
5. Checks for high resistance in the switched battery circuit from the battery to the relay.
6. Verifies the condition of the TCS enable relay.
7. Checks for poor terminal contact or corrosion at the connectors.
8. Determines if the EBTCM is the cause of the malfunction.

Diagnostic Aids:

An intermittent malfunction may be caused by a poor connection, rubbed through wire insulation, or a wire that is broken inside the insulation.

The frequency of the malfunction can be checked by using the enhanced diagnostic function of the Tech 1, [illegible]

Any circuitry that is suspected of causing the intermittent complaint should be thoroughly checked for backed out terminals, improper mating, broken locks, improperly formed or damaged terminals, poor terminal to wiring connections or physical damage to the wiring harness.

VIBRATION, TEMPERATURE EFFECTS:

Check for vibration effects by performing the relay test function of the Tech 1. With the relay test commanded "ON," lightly tap the top and sides of the relay while monitoring relay voltage. If the relay voltage changes significantly, replace the relay.

If DTC 114 only sets when the vehicle is initially started in cold ambient conditions (temperature less than 32°F - 0°C), replace the relay.

DIAGNOSTIC CHART DTC 114 — 1995 SILHOUETTE AND TRANS SPORT WITH TCS, CONTINUED

DTC 114
(Page 1 of 3)
TCS ENABLE RELAY CONTACT CIRCUIT OPEN

(1) IS DTC 116 ALSO PRESENT IN HISTORY?
- YES → PROCEED TO DTC 116 CHART FIRST.
- NO ↓

(2)
- IGNITION "ON," ENGINE "OFF."
- USING TECH 1, SELECT DATA LIST AND OBSERVE EBTCM BATTERY VOLTAGE.

DOES TECH 1 INDICATE THAT EBTCM BATTERY VOLTAGE IS GREATER THAN 10 VOLTS?
- YES → MALFUNCTION IS NOT PRESENT AT THIS TIME. REFER TO "DIAGNOSTIC AIDS" ON FACING PAGE.
- NO ↓

(3)
- IGNITION "OFF."
- DISCONNECT BLACK 8-WAY EBTCM CONNECTOR.
- CONNECT A FUSED JUMPER, SUCH AS J 36169 WITH A 3 AMP FUSE BETWEEN GROUND AND TERMINAL "B" OF THE BLACK 8-WAY EBTCM HARNESS CONNECTOR.
- IGNITION "ON."
- USING J 39200, MEASURE THE VOLTAGE BETWEEN GROUND AND TERMINAL "C" OF THE BLACK 8-WAY EBCM HARNESS CONNECTOR.

IS VOLTAGE 10 VOLTS OR GREATER?
- NO → DIAGNOSIS CONTINUED ON NEXT PAGE.
- YES ↓

(4)
- IGNITION "OFF."
- DISCONNECT JUMPER USED IN STEP 3.
- DISCONNECT TCS ENABLE RELAY.
- USING J 39200, MEASURE RESISTANCE BETWEEN TERMINAL "30" OF THE TCS ENABLE RELAY HARNESS CONNECTOR AND TERMINAL "C" OF THE BLACK 8-WAY EBTCM HARNESS CONNECTOR.

IS RESISTANCE 2Ω OR LESS?
- NO → REPAIR HIGH RESISTANCE IN CKT 1825.
- YES ↓

(5)
- DISCONNECT NEGATIVE BATTERY CABLE.
- DISCONNECT POSITIVE BATTERY CABLE.
- USING J 39200, MEASURE THE RESISTANCE BETWEEN POSITIVE BATTERY CABLE TERMINAL AND TERMINAL "87" OF THE TCS ENABLE RELAY HARNESS CONNECTOR.

IS RESISTANCE 2Ω OR LESS?
- NO → REPAIR HIGH RESISTANCE IN CKT 2E.
- YES ↓

(6)
- USING A JUMPER WIRE, CONNECT TERMINAL "86" ON THE TCS ENABLE RELAY TO GROUND.
- USING A FUSED JUMPER WIRE, SUCH AS J 36169, WITH A 3 AMP FUSE, CONNECT TERMINAL "85" ON THE TCS ENABLE RELAY TO VOLTAGE.
- USING J 39200, MEASURE THE RESISTANCE BETWEEN TERMINALS "30" AND "87" ON THE TCS ENABLE RELAY.

IS RESISTANCE 2Ω OR LESS?
- NO → REPLACE TCS ENABLE RELAY.
- YES ↓

(7)
- INSPECT TCS ENABLE RELAY AND TCS ENABLE RELAY HARNESS CONNECTOR FOR A POOR CONNECTION OR POOR TERMINAL CONTACT.
- INSPECT EBTCM AND EBTCM CONNECTOR TERMINALS "B" AND "C" FOR POOR TERMINAL CONTACT.
- INSPECT POSITIVE BATTERY TERMINAL AND POSITIVE BATTERY CABLE TERMINAL FOR A POOR CONNECTION.

IS POOR TERMINAL CONTACT OR A POOR CONNECTION EVIDENT?
- YES → REPLACE TERMINALS THAT EXHIBIT POOR TERMINAL CONTACT OR EVIDENCE OF CORROSION.
- NO ↓

(8)
- RECONNECT ALL CONNECTORS.
- IGNITION "ON."

DOES DTC 114 RESET?
- YES → REPLACE EBTCM.
- NO → MALFUNCTION IS INTERMITTENT. REFER TO "DIAGNOSTIC AIDS".

DIAGNOSTIC CHART DTC 114 — 1995 SILHOUETTE AND TRANS SPORT WITH TCS, CONTINUED

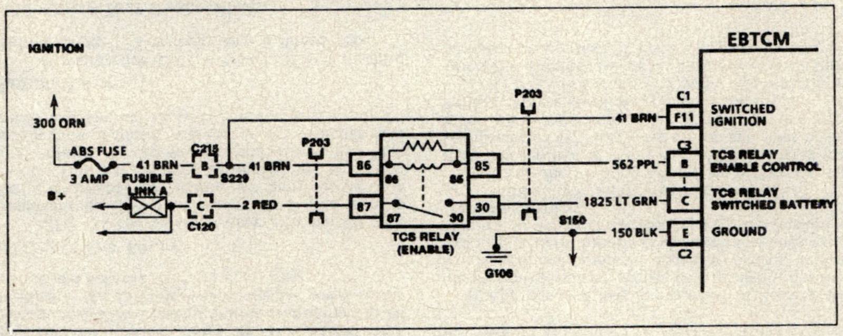

DTC 114
(Page 2 of 3)
TCS ENABLE RELAY CONTACT CIRCUIT OPEN

Circuit Description:

Ignition voltage is supplied through terminal "86" of the TCS enable relay. The EBTCM then is able to energize the pull-in coil by completing the ground circuit at terminal "B" of the EBTCM. The magnetic field created closes the TCS enable relay contacts and allows battery voltage and current to be supplied to the EBTCM, which supplies power to the motors and solenoids.

DTC Will Set When:

DTC 114 can be set anytime after the EBTCM commands the TCS enable relay "ON" (the relay is first commanded "ON" during the three second bulb check). This test monitors the availability of current and voltage to the motors and solenoids. This malfunction indicates voltage is not available and would therefore not allow TCS operation if required.

Action Taken:

A malfunction DTC is stored, TCS is disabled and the amber TCS warning lamp is turned "ON."

DTC Chart Test Description:

Number(s) below refer to circled number(s) on the diagnostic chart.

9. Checks for high resistance in the entire relay coil circuit.
10. Verifies proper battery condition.
11. This step verifies proper voltage is available at the TCS enable relay.
12. Checks for an open in the switched battery circuit from the relay to the EBTCM.
13. Verifies the condition of the TCS enable relay.

Diagnostic Aids:

An intermittent malfunction may be caused by a poor connection, rubbed through wire insulation, or a wire that is broken inside the insulation.

The frequency of the malfunction can be checked by using the enhanced diagnostic function of the Tech 1, [illegible]

Any circuitry that is suspected of causing the intermittent complaint should be thoroughly checked for backed out terminals, improper mating, broken locks, improperly formed or damaged terminals, poor terminal to wiring connections or physical damage to the wiring harness.

VIBRATION, TEMPERATURE EFFECTS:

Check for vibration effects by performing the relay test function of the Tech 1. With the relay test commanded "ON," lightly tap the top and sides of the relay while monitoring relay voltage. If the relay voltage changes significantly, replace the relay.

If DTC 114 only sets when the vehicle is initially started in cold ambient conditions (temperature less than 32°F - 0°C),replace the relay.

DIAGNOSTIC CHART DTC 114 — 1995 SILHOUETTE AND TRANS SPORT WITH TCS, CONTINUED

DTC 114
(Page 2 of 3)
TCS ENABLE RELAY CONTACT CIRCUIT OPEN

DIAGNOSIS CONTINUED FROM PREVIOUS PAGE.

(9)
- IGNITION "OFF."
- REMOVE JUMPER WIRE INSTALLED IN STEP #3 IF NOT ALREADY REMOVED.
- DISCONNECT NEGATIVE BATTERY CABLE.
- DISCONNECT POSITIVE BATTERY CABLE.
- WITH BATTERY STILL DISCONNECTED, TURN IGNITION SWITCH TO "RUN." THIS COMPLETES THE IGNITION CIRCUIT TO THE TCS ENABLE RELAY.
- USING J 39200, MEASURE THE RESISTANCE BETWEEN TERMINAL "B" OF THE BLACK 8-WAY EBTCM HARNESS CONNECTOR AND THE POSITIVE BATTERY CABLE TERMINAL.
  IS RESISTANCE 95Ω OR LESS WHEN MEASURED AT APPROXIMATELY 20°C (68°F)?

NO → DIAGNOSIS CONTINUED ON NEXT PAGE.

YES ↓

(10)
- USING J 39200, MEASURE THE VOLTAGE BETWEEN THE NEGATIVE AND POSITIVE BATTERY TERMINALS.
  IS VOLTAGE 10 VOLTS OR GREATER?

NO → CORRECT LOW VOLTAGE CONDITION.

YES ↓

(11)
- IGNITION "OFF."
- RECONNECT POSITIVE BATTERY CABLE.
- RECONNECT NEGATIVE BATTERY CABLE.
- DISCONNECT TCS ENABLE RELAY.
- USING J 39200, MEASURE THE VOLTAGE BETWEEN TERMINAL "87" OF THE TCS ENABLE RELAY HARNESS CONNECTOR AND GROUND.
  IS VOLTAGE 10 VOLTS OR GREATER?

NO → REPAIR OPEN OR HIGH RESISTANCE IN CKT 2E.

YES ↓

(12)
- USING J 39200, MEASURE THE RESISTANCE BETWEEN TERMINAL "30" OF THE TCS ENABLE RELAY HARNESS CONNECTOR AND TERMINAL "C" OF THE BLACK 8-WAY EBTCM HARNESS CONNECTOR.
  IS RESISTANCE 2Ω OR LESS?

NO → REPAIR OPEN OR HIGH RESISTANCE IN CKT 1825.

YES ↓

(13)
- USING A JUMPER WIRE, CONNECT TERMINAL "86" ON THE TCS ENABLE RELAY TO GROUND.
- USING A FUSED JUMPER WIRE, SUCH AS J 36169, WITH A 3 AMP FUSE, CONNECT TERMINAL "85" ON THE TCS ENABLE RELAY TO B+.
- USING J 39200, MEASURE THE RESISTANCE BETWEEN TERMINALS "30" AND "87" ON THE TCS ENABLE RELAY.
  IS RESISTANCE 2Ω OR LESS?

NO → REPLACE TCS ENABLE RELAY.

YES ↓

MALFUNCTION WAS MOST LIKELY DUE TO A POOR CONNECTION. INSPECT ALL CONNECTORS AND TERMINALS FOR POOR TERMINAL CONTACT AND EVIDENCE OF CORROSION.
REPLACE ALL TERMINALS THAT EXHIBIT EVIDENCE OF POOR TERMINAL CONTACT OR CORROSION.

DIAGNOSTIC CHART DTC 114 — 1995 SILHOUETTE AND TRANS SPORT WITH TCS, CONTINUED

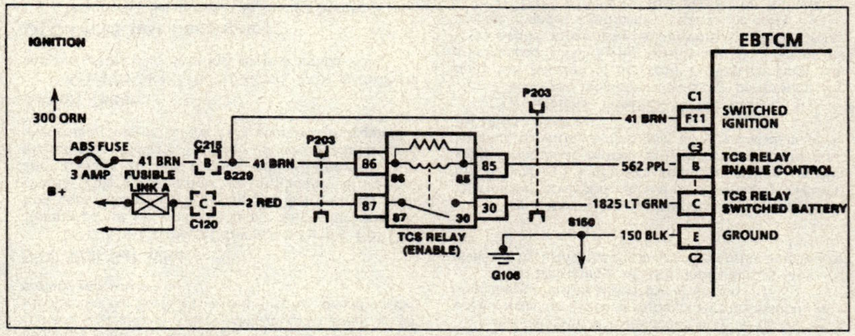

DTC 114
(Page 3 of 3)
TCS ENABLE RELAY CONTACT CIRCUIT OPEN

Circuit Description:

Ignition voltage is supplied through terminal "86" of the TCS enable relay. The EBTCM then is able to energize the pull-in coil by completing the ground circuit at terminal "B" of the EBTCM. The magnetic field created closes the TCS enable relay contacts and allows battery voltage and current to be supplied to the EBTCM, which supplies power to the motors and solenoids.

DTC Will Set When:

DTC 114 can be set anytime after the EBTCM commands the TCS enable relay "ON" (the relay is first commanded "ON" during the three second bulb check). This test monitors the availability of current and voltage to the motors and solenoids. This malfunction indicates voltage is not available and would therefore not allow TCS operation if required.

Action Taken:

A malfunction DTC is stored, TCS is disabled and the amber TCS warning lamp is turned "ON."

DTC Chart Test Description:

Number(s) below refer to circled number(s) on the diagnostic chart.

14. Checks for proper resistance of the relay coil.
15. Checks for high resistance in the TCS enable relay control circuit between the relay and the EBTCM.
16. Checks for high resistance in the TCS enable relay control circuit between the battery and the relay.

Diagnostic Aids:

An intermittent malfunction may be caused by a poor connection, rubbed through wire insulation, or a wire that is broken inside the insulation.

The frequency of the malfunction can be checked by using the enhanced diagnostic function of the Tech 1 [illegible] Diagnostics [illegible]

Any circuitry that is suspected of causing the intermittent complaint should be thoroughly checked for backed out terminals, improper mating, broken locks, improperly formed or damaged terminals, poor terminal to wiring connections or physical damage to the wiring harness.

VIBRATION, TEMPERATURE EFFECTS:

Check for vibration effects by performing the relay test function of the Tech 1. With the relay test commanded "ON," lightly tap the top and sides of the relay while monitoring relay voltage. If the relay voltage changes significantly, replace the relay.

If DTC 114 only sets when the vehicle is initially started in cold ambient conditions (temperature less than 32°F - 0°C),replace the relay.

DIAGNOSTIC CHART DTC 114 — 1995 SILHOUETTE AND TRANS SPORT WITH TCS, CONTINUED

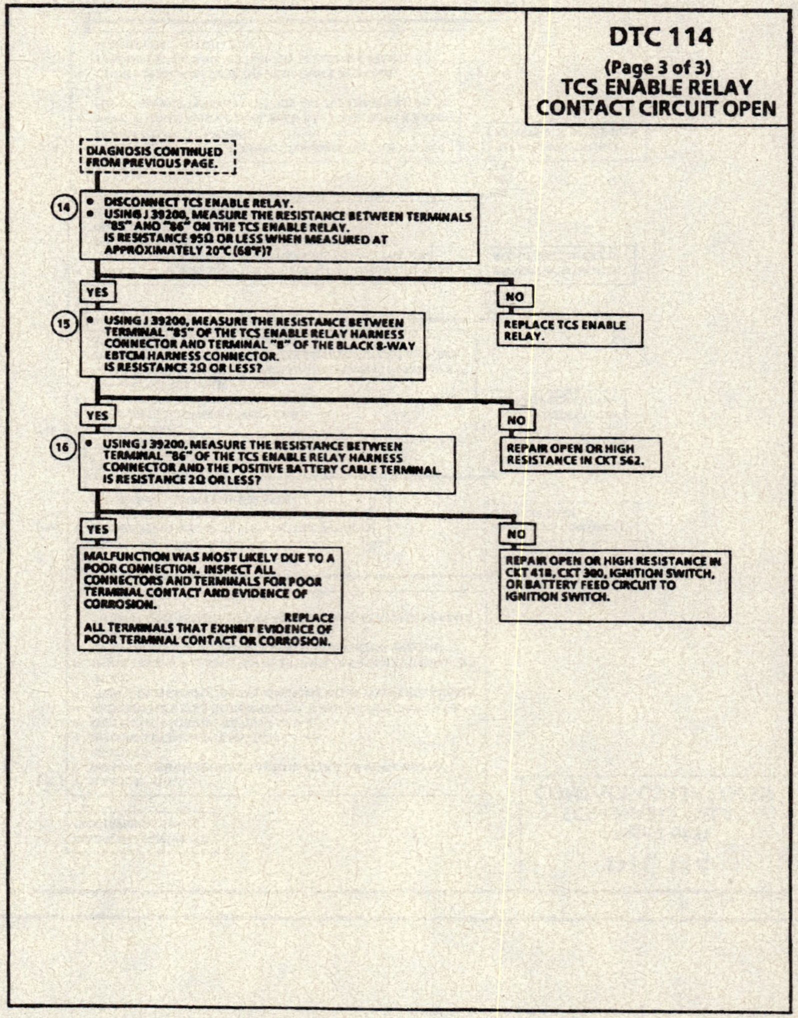

DIAGNOSTIC CHART DTC 116 — 1995 SILHOUETTE AND TRANS SPORT WITH TCS

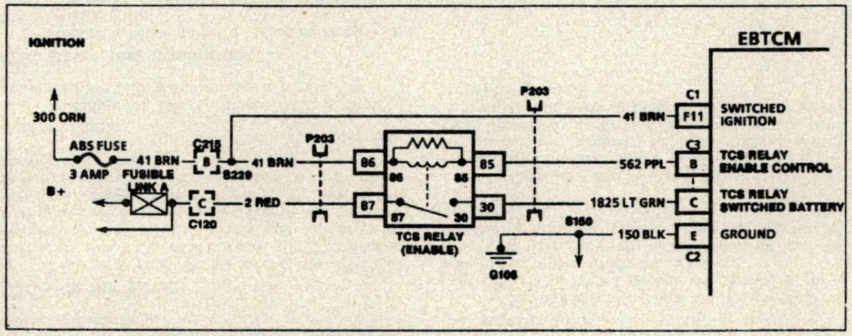

DTC 116

TCS ENABLE RELAY COIL CIRCUIT OPEN

Circuit Description:

Ignition voltage is supplied through terminal "86" of the TCS enable relay. The EBTCM then is able to energize the pull-in coil by completing the ground circuit at terminal "B" of the EBTCM. The magnetic field created closes the TCS enable relay contacts and allows battery voltage and current to be supplied to the EBTCM, which supplies power to the motors and solenoids.

DTC Will Set When:

DTC 116 can be set only after DTC 114 has been set. This test detects an open in the TCS enable relay coil circuit. An open in this circuit will not allow the TCS enable relay to be energized thus preventing voltage and current to the motors and solenoids. If this malfunction is present and the ignition is turned "OFF" before 5 km/h (3 mph) is reached, DTC 114 is set alone.

Action Taken:

A malfunction DTC is stored, TCS is disabled and the amber TCS warning lamp is turned "ON."

DTC Chart Test Description:

Number(s) below refer to circled number(s) on the diagnostic chart.

1. Indicates the EBTCM is capable of controlling the enable relay as commanded.
2. Ensures that there is continuity through the pull-in coil of the TCS enable relay.
3. Checks to ensure that voltage is available to the pull-in coil of the enable relay.
4. Checks continuity in the enable relay control circuit.
5. Ensures DTC was not set due to a poor connection.
6. This stop identifies whether the malfunction is due to a poor connection or the EBTCM.

Diagnostic Aids:

An intermittent malfunction may be caused by a poor connection, rubbed through wire insulation, or a wire that is broken inside the insulation.

The frequency of the malfunction can be checked by using the enhanced diagnostic function of the Tech 1. [illegible]

If the frequency of the malfunction is high, but is currently intermittent, check for high coil resistance by measuring between relay terminals "85" and "86" using a J 39200. If resistance shows greater than 100(, replace the relay.

Any circuitry that is suspected of causing the intermittent complaint should be thoroughly checked for backed out terminals, improper mating, broken locks, improperly formed or damaged terminals, poor terminal to wiring connections or physical damage to the wiring harness.

DIAGNOSTIC CHART DTC 116 — 1995 SILHOUETTE AND TRANS SPORT WITH TCS, CONTINUED

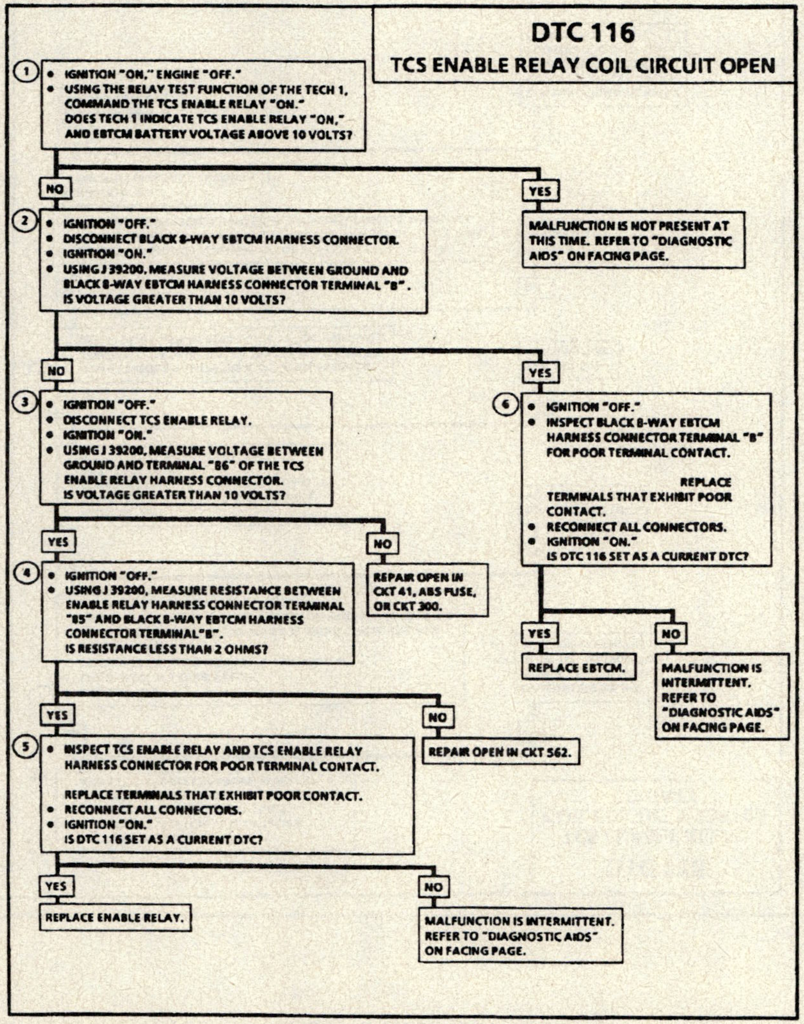

DIAGNOSTIC CHART DTC 118 — 1995 SILHOUETTE AND TRANS SPORT WITH TCS

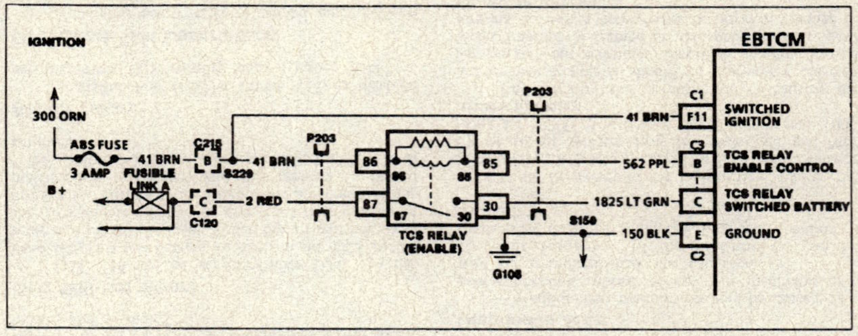

DTC 118
TCS ENABLE RELAY COIL CIRCUIT SHORTED TO BATT

Circuit Description:

Ignition voltage is supplied through terminal "86" of the TCS enable relay. The EBTCM then is able to energize the pull-in coil by completing the ground circuit at terminal "B" of the EBTCM. The magnetic field created closes the TCS enable relay contacts and allows battery voltage and current to be supplied to the EBTCM, which supplies voltage to the motors and solenoids.

DTC Will Set When:

DTC 118 can be set after the EBTCM commands the TCS enable relay "ON." This test monitors the availability of current and voltage to the motors and solenoids. A malfunction of this type will not allow voltage to be available to the motors and solenoids, therefore, TCS operation would not be allowed if required.

Action Taken:

A malfunction DTC is stored, TCS is disabled and the amber TCS warning lamp is turned "ON."

DTC Chart Test Description:

Number(s) below refer to circled number(s) on the diagnostic chart.

1. Indicates that the EBTCM is capable of controlling the TCS enable relay as commanded.
2. Checks for a short to voltage on the relay control circuit.
3. Verifies that the relay coil has the appropriate resistance.
4. Verifies that relay contacts have not failed closed.
5. Checks for possible poor terminal contact at the EBTCM.

Diagnostic Aids:

An intermittent malfunction may be caused by a poor connection, rubbed through wire insulation, or a wire that is broken inside the insulation.

The frequency of the malfunction can be checked by using the enhanced diagnostic function of the Tech 1, [illegible] diagnostic [illegible]

Any circuitry that is suspected of causing the intermittent complaint should be thoroughly checked for backed out terminals, improper mating, broken locks, improperly formed or damaged terminals, poor terminal to wiring connections or physical damage to the wiring harness.

DIAGNOSTIC CHART DTC 118 — 1995 SILHOUETTE AND TRANS SPORT WITH TCS, CONTINUED

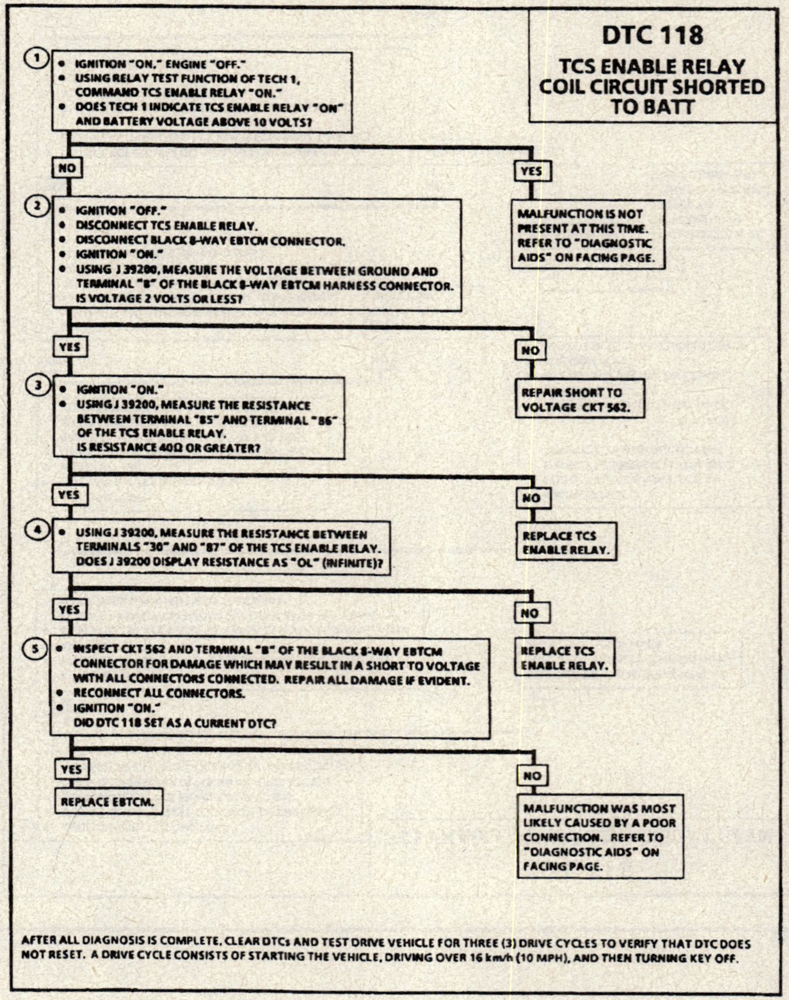

DIAGNOSTIC CHART DTC 123 — 1995 SILHOUETTE AND TRANS SPORT WITH TCS

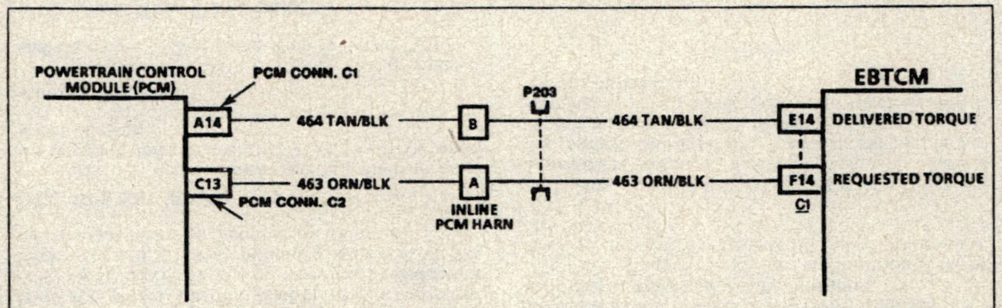

DTC 123
PCM TO EBTCM DELIVERED TORQUE CIRCUIT MALFUNCTION

Circuit Description:

The traction control system uses the antilock brake system in conjunction with the PCM fuel and ignition controls to limit drive wheel slippage during acceleration. The PCM controlled portion of the traction control system reduces engine torque by retarding spark timing, alternating air/fuel ratio, and/or shutting "OFF" up to three injectors. The EBTCM sends a torque request to the PCM via a Pulse Width Modulated (PWM) signal requesting the desired torque level for proper traction control system operation (CKT 463). The PCM then provides a delivered torque PWM signal (CKT 464) informing the EBTCM how much actual torque is produced by the engine.

DTC Will Set When:

DTC 123 will set when there is a malfunction in the delivered torque circuit (CKT 464) or if the signal is out of range.

Action Taken:

A malfunction DTC is stored, TCS is disabled, and the amber TCS warning lamp is turned "ON."

DTC Chart Test Description:

Number(s) below refer to circled number(s) on the diagnostic chart.

1. Checks for proper voltage at the PCM terminal.
2. Checks for short to voltage in CKT 464.
3. Ensures malfunction was not due to physical damage of circuitry.
4. Checks for open in CKT 464.
5. Checks for short to ground in CKT 464.
6. Checks for possible poor terminal contact at the EBTCM.

Diagnostic Aids:

An intermittent malfunction may be caused by a poor connection, rubbed through wire insulation, or a wire that is broken inside the insulation.

The frequency of the malfunction can be checked by using the enhanced diagnostic function of the Tech 1.

Any circuitry that is suspected of causing the intermittent complaint should be thoroughly checked for backed out terminals, improper mating, broken locks, improperly formed or damaged terminals, poor terminal to wiring connections or physical damage to the wiring harness.

DIAGNOSTIC CHART DTC 123 — 1995 SILHOUETTE AND TRANS SPORT WITH TCS, CONTINUED

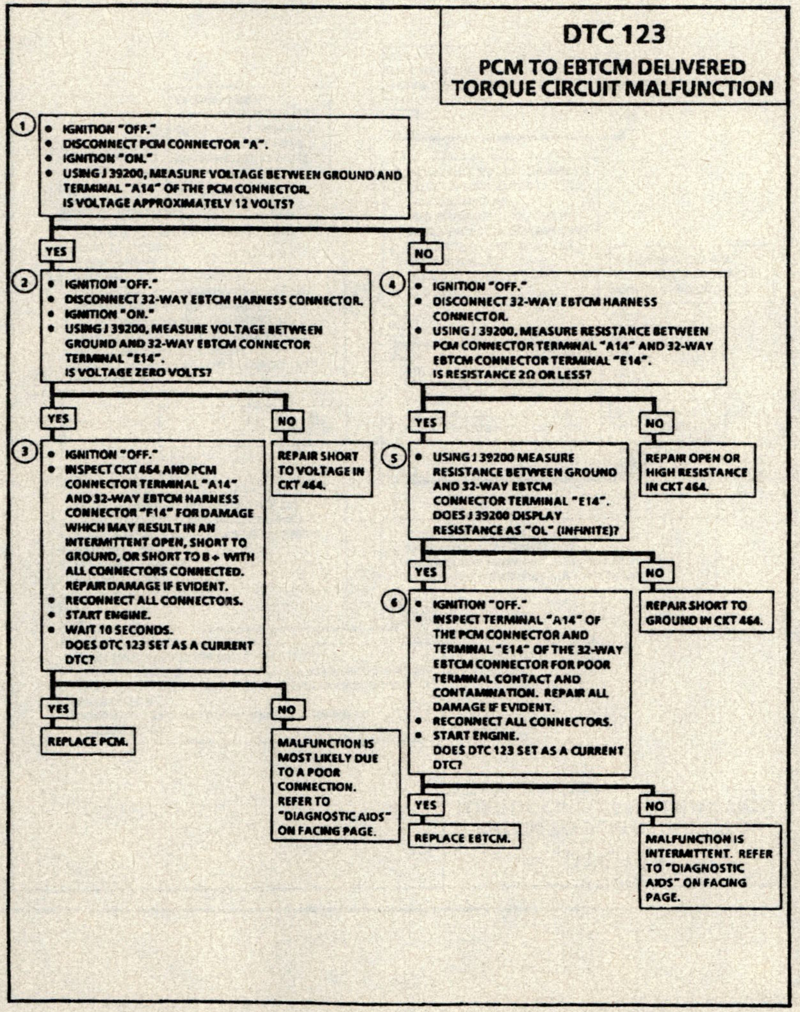

DIAGNOSTIC CHART DTC 124 — 1995 SILHOUETTE AND TRANS SPORT WITH TCS

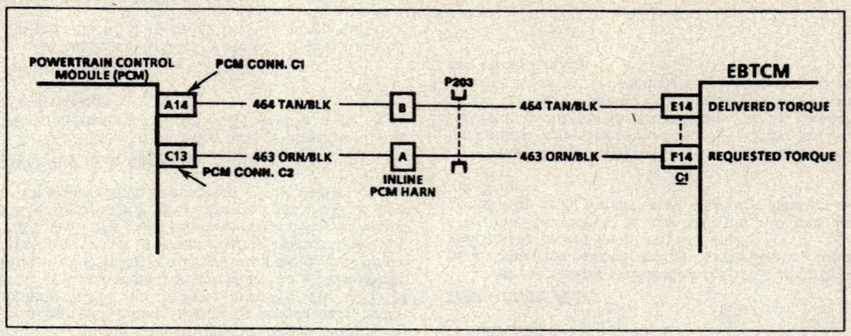

DTC 124
EBTCM TO PCM REQUESTED TORQUE CIRCUIT MALFUNCTION

Circuit Description:

The traction control system uses the antilock brake system in conjunction with the PCM fuel and ignition controls to limit drive wheel slippage during acceleration. The PCM controlled portion of the traction control system reduces engine torque by retarding spark timing, alternating air/fuel ratio, and/or shutting "OFF" up to three injectors. The EBTCM sends a torque request to the PCM via a Pulse Width Modulated (PWM) signal requesting the desired torque level for proper traction control system operation (CKT 463). The PCM then provides a delivered torque PWM signal (CKT 464) informing the EBTCM how much actual torque is produced by the engine.

DTC Will Set When:

DTC 124 will set when there is a malfunction in the requested torque circuit (CKT 463) or if the signal is out of range.

Action Taken:

A malfunction DTC is stored, TCS is disabled, and the amber TCS warning lamp is turned "ON."

DTC Chart Test Description:

Number(s) below refer to circled number(s) on the diagnostic chart.

1. Checks for proper voltage at the EBTCM.
2. Checks for proper voltage at the PCM terminal.
3. Ensures malfunction was not due to physical damage of the circuitry.
4. Checks for possible short to battery in CKT 463.
5. Checks for an open in CKT 463.
6. Checks for proper resistance in CKT 463.

Diagnostic Aids:

An intermittent malfunction may be caused by a poor connection, rubbed through wire insulation, or a wire that is broken inside the insulation.

The frequency of the malfunction can be checked by using the enhanced diagnostic function of the Tech 1 [illegible]

Any circuitry that is suspected of causing the intermittent complaint should be thoroughly checked for backed out terminals, improper mating, broken locks, improperly formed or damaged terminals, poor terminal to wiring connections or physical damage to the wiring harness.

DIAGNOSTIC CHART DTC 124 — 1995 SILHOUETTE AND TRANS SPORT WITH TCS, CONTINUED

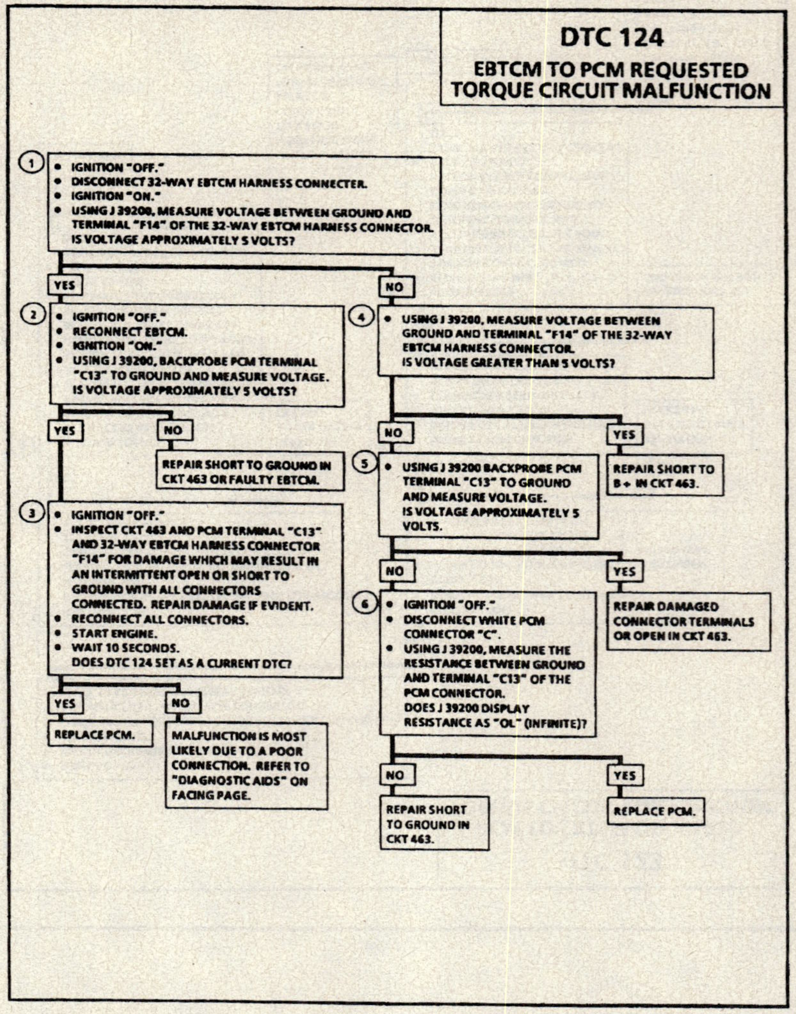

DIAGNOSTIC CHART DTC 136 — 1995 SILHOUETTE AND TRANS SPORT WITH TCS

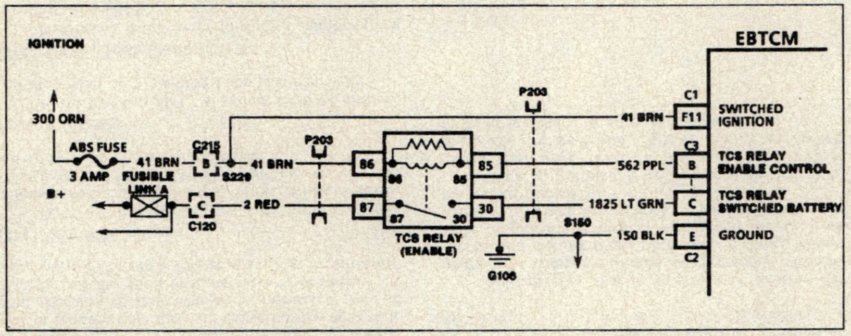

DTC 136
(Page 1 of 2)
LOW SYSTEM VOLTAGE (TCS)

Circuit Description:

This DTC is used to monitor the voltage level available to the EBTCM. If the voltage drops below 11 volts, full performance of the TCS system cannot be guaranteed. During TCS operation, there are several current requirements that will cause battery voltage to drop. Because of this, voltage is monitored prior to TCS to indicate good charging system condition and also during TCS when voltage may drop significantly.

DTC Will Set When:

DTC 136 can only be set if the vehicle's speed is greater than 5 km/h (3 mph). If the switched battery voltage is less than 11.2 volts non-TCS or less than 9.6 volts TCS, a malfunction exists.

Action Taken:

A malfunction DTC is stored, TCS is disabled and the amber TCS warning lamp is turned "ON."

DTC Chart Test Description:

Number(s) below refer to circled number(s) on the diagnostic chart.

1. Checks to see what voltage is readily available at terminal "C" of the EBTCM. If voltage is representative of good charging system condition, a malfunction is not present.
2. Isolates the EBTCM to check the condition of the circuitry and charging system.
3. Isolates the low voltage condition to high circuit resistance, or improper charging system operation.
4. Isolates which circuit is the source of the high resistance condition.

Diagnostic Aids:

An intermittent malfunction may be caused by a poor connection, rubbed through wire insulation, or a wire that is broken inside the insulation.

The frequency of the malfunction can be checked by using the enhanced diagnostic function of the Tech 1. [illegible] ...tion.

Any circuitry that is suspected of causing the intermittent complaint should be thoroughly checked for backed out terminals, improper mating, broken locks, improperly formed or damaged terminals, poor terminal to wiring connections or physical damage to the wiring harness.

While performing a voltage load test, if it is noted that only ignition voltage drops below acceptable voltage levels, CKT 41 should be checked for high resistance or an open condition.

DIAGNOSTIC CHART DTC 136 — 1995 SILHOUETTE AND TRANS SPORT WITH TCS, CONTINUED

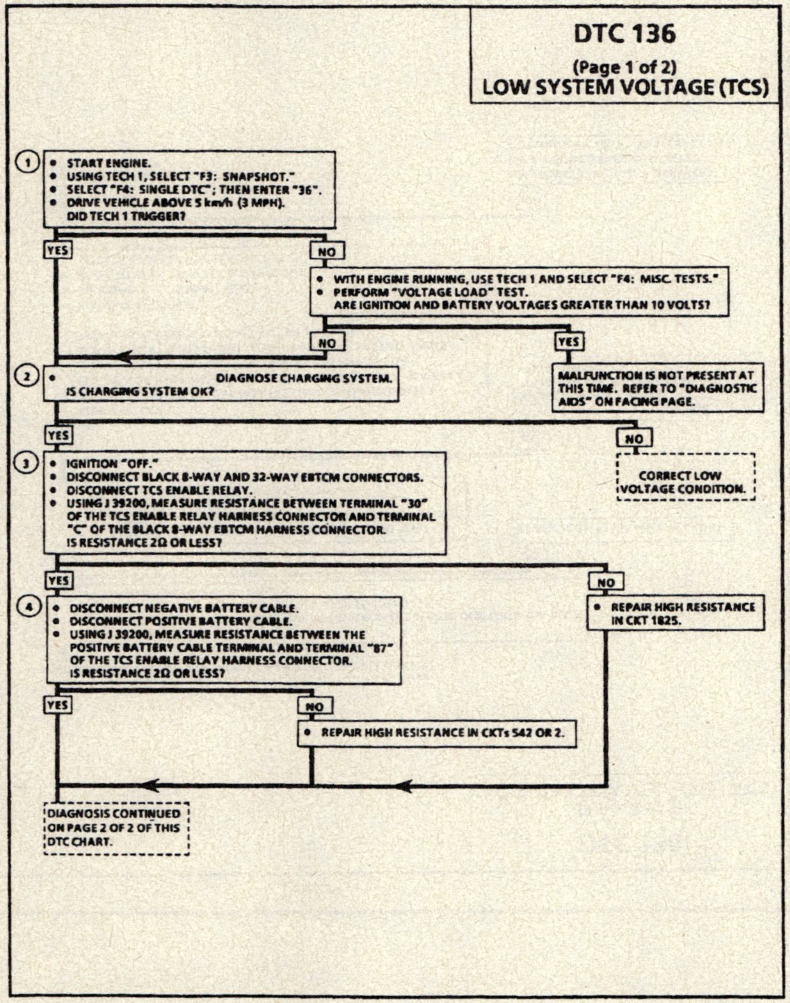

DIAGNOSTIC CHART DTC 136 — 1995 SILHOUETTE AND TRANS SPORT WITH TCS, CONTINUED

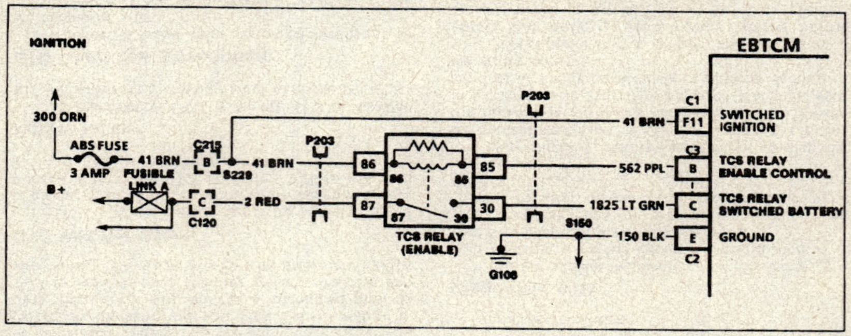

DTC 136
(Page 2 of 2)
LOW SYSTEM VOLTAGE (TCS)

Circuit Description:

This DTC is used to monitor the voltage level available to the EBTCM. If the voltage drops below 11 volts, full performance of the TCS system cannot be guaranteed. During TCS operation, there are several current requirements that will cause battery voltage to drop. Because of this, voltage is monitored prior to TCS to indicate good charging system condition and also during TCS when voltage may drop significantly.

DTC Will Set When:

DTC 136 can only be set if the vehicle's speed is greater than 5 km/h (3 mph). If the switched battery voltage is less than 11.2 volts non-TCS or less than 9.6 volts TCS, a malfunction exists.

Action Taken:

A malfunction DTC is stored, TCS is disabled and the amber TCS warning lamp is turned "ON."

DTC Chart Test Description:

Number(s) below refer to circled number(s) on the diagnostic chart.

5. Checks for high resistance in CKT 41.
6. Checks for high resistance in the ignition fuse feed circuit.
7. Checks for malfunctioning EBTCM.

Diagnostic Aids:

An intermittent malfunction may be caused by a poor connection, rubbed through wire insulation, or a wire that is broken inside the insulation.

The frequency of the malfunction can be checked by using the enhanced diagnostic function of the Tech 1. [illegible]

Any circuitry that is suspected of causing the intermittent complaint should be thoroughly checked for backed out terminals, improper mating, broken locks, improperly formed or damaged terminals, poor terminal to wiring connections or physical damage to the wiring harness.

While performing a voltage load test, if it is noted that only ignition voltage drops below acceptable voltage levels, CKT 41 should be checked for high resistance or an open condition.

DIAGNOSTIC CHART DTC 136 — 1995 SILHOUETTE AND TRANS SPORT WITH TCS, CONTINUED

DTC 136
(Page 2 of 2)
LOW SYSTEM VOLTAGE (TCS)

DIAGNOSIS CONTINUED FROM PAGE 1 OF 2 OF THIS DTC CHART.

(5)
- REMOVE ABS FUSE.
- USING J 39200 MEASURE RESISTANCE BETWEEN EACH ABS FUSE BLOCK TERMINAL AND TERMINAL "F11" OF THE 32-WAY EBTCM HARNESS CONNECTOR. IS EITHER MEASUREMENT 2Ω OR LESS?

NO: REPAIR HIGH RESISTANCE IN CKT 41.

YES:

(6)
- WITH BATTERY CABLES DISCONNECTED, TURN IGNITION "ON"
- USING J 39200, MEASURE RESISTANCE BETWEEN POSITIVE BATTERY CABLE TERMINAL AND EACH ABS FUSE BLOCK TERMINAL. IS EITHER MEASUREMENT 2Ω OR LESS ?

NO: REPAIR HIGH RESISTANCE IN THE "IGNITION SWITCH 'RUN' FUSE FEED" CIRCUIT.

YES:

(7)
- IGNITION "OFF."
- INSPECT TCS ENABLE RELAY AND TCS ENABLE RELAY HARNESS CONNECTOR FOR POOR TERMINAL CONTACT. INSPECT BLACK 8-WAY AND 32-WAY EBTCM HARNESS CONNECTORS FOR A POOR CONNECTION TO THE EBTCM. INSPECT POSITIVE BATTERY CABLE TERMINAL FOR POOR CONTACT TO THE BATTERY. REPLACE TERMINALS IF POOR CONTACT IS EVIDENT.
- RECONNECT ALL CONNECTORS.
- DRIVE VEHICLE ABOVE 5 km/h (3 MPH). DOES DTC 136 SET AS A CURRENT DTC?

YES: REPLACE EBTCM.

NO: MALFUNCTION IS MOST LIKELY DUE TO A POOR CONNECTION. REFER TO "DIAGNOSTIC AIDS" ON FACING PAGE.

DIAGNOSTIC CHART DTC 144 — 1995 SILHOUETTE AND TRANS SPORT WITH TCS

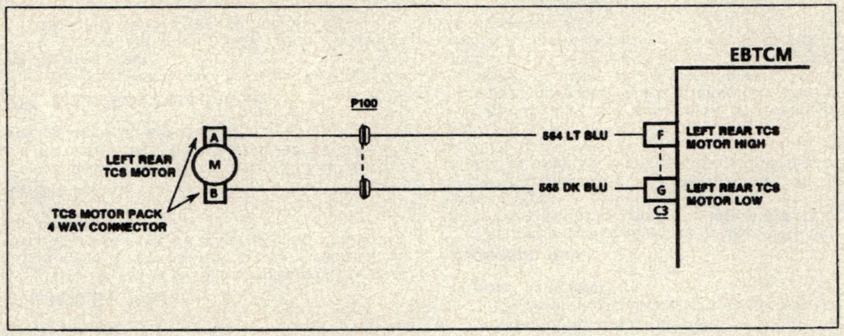

DTC 144
LEFT TCS CHANNEL WILL NOT MOVE

Circuit Description:

This DTC is designed to detect a stuck TCS motor or a seized hydraulic modulator. During initialization or after a TCS cycle, the EBTCM commands the TCS motor in the reverse direction and the forward direction to ensure that the motor moves properly. If the TCS motor does not move, improper current feedback will be detected by the EBTCM, and a DTC will set.

DTC Will Set When:

DTC 144 can be set during initialization or after a TCS cycle. If the EBTCM detects a condition in which it cannot move the motor in either direction, a malfunction exists.

Action Taken:

A malfunction DTC is stored, TCS is disabled, and the amber TCS warning and the red BRAKE warning lamps are turned "ON."

DTC Chart Test Description:

Number(s) below refer to circled number(s) on the diagnostic chart.

1. Checks for proper TCS motor movement during the release command from the Tech 1.
2. Checks for proper TCS motor movement during the apply command from the Tech 1.
3. Compares EBTCM command current to TCS motor feedback current.
4. Checks for a shorted circuit between the TCS motor "HIGH" and motor "LOW" circuits.
5. Checks for a TCS motor that is internally shorted.
6. Determines if the malfunction is due to poor terminal contact or the EBTCM.
7. Checks for proper hydraulic modulator gear and piston movement.

Diagnostic Aids:

An intermittent malfunction in this DTC may result from a mechanical part of the system that sticks, binds, or slips.

The frequency of the malfunction can be checked by using the enhanced diagnostic function of the Tech 1. [illegible]

Any circuitry that is suspected of causing the intermittent complaint should be thoroughly checked for backed out terminals, improper mating, broken locks, improperly formed or damaged terminals, poor terminal to wiring connections or physical damage to the wiring harness.

DIAGNOSTIC CHART DTC 144 — 1995 SILHOUETTE AND TRANS SPORT WITH TCS, CONTINUED

DTC 144
LEFT TCS CHANNEL WILL NOT MOVE

NOTICE: J 39200 TEST LEADS MUST BE "ZEROED" PRIOR TO MAKING ANY RESISTANCE MEASUREMENTS. REFER TO J 39200 USER'S MANUAL.

(1)
- RAISE VEHICLE.
- IGNITION "ON," ENGINE "OFF."
- PUMP BRAKE PEDAL UNTIL FIRM TO DEPLETE VACUUM RESERVE.
- USING TECH 1 MISC TEST, SELECT MANUAL CONTROL FUNCTION.
- FOOT OFF BRAKE PEDAL.
- SELECT LEFT FRONT TCS MOTOR RELEASE FUNCTION AND RELEASE THE MOTOR. DOES THE LEFT FRONT WHEEL ROTATE?

NO → go to (3). YES → (2)

(2)
- USING TECH 1 MISC. TEST, SELECT MANUAL CONTROL FUNCTION.
- FOOT OFF BRAKE PEDAL.
- SELECT LEFT FRONT TCS MOTOR APPLY FUNCTION AND APPLY THE MOTOR. DOES THE LEFT FRONT WHEEL ROTATE?

YES → go to (3). NO → MALFUNCTION IS INTERMITTENT. REFER TO "DIAGNOSTIC AIDS" ON FACING PAGE.

(3)
- FOOT OFF BRAKE PEDAL.
- SELECT LEFT FRONT TCS MOTOR RELEASE FUNCTION AND RELEASE THE TCS MOTOR WHILE CAREFULLY OBSERVING THE COMMANDED CURRENT AND THE FEEDBACK CURRENT ON THE TECH 1. WAS FEEDBACK CURRENT HIGHER THAN THE COMMANDED CURRENT?

YES → (4). NO → (7)

(4)
- EXIT TECH 1 MANUAL CONTROL.
- IGNITION "OFF."
- DISCONNECT THE BLACK 8-WAY EBTCM CONNECTOR AND THE 4-WAY TCS MOTOR PACK CONNECTOR.
- USING J 39200, MEASURE THE RESISTANCE BETWEEN TERMINAL "F" AND TERMINAL "G" OF THE BLACK 8-WAY EBTCM HARNESS CONNECTOR. DOES J 39200 DISPLAY RESISTANCE AS "OL" (INFINITE)?

YES → (5). NO → REPAIR SHORT BETWEEN CKT 564 AND CKT 565.

(5)
- USING J 39200, MEASURE THE RESISTANCE BETWEEN TERMINAL "A" AND TERMINAL "B" OF THE TCS MOTOR PACK CONNECTOR. IS RESISTANCE GREATER THAN 0.4Ω?

YES → (6). NO → REPLACE TCS MOTOR PACK

(6)
- IGNITION "OFF."
- RECONNECT ALL CONNECTORS.
- START ENGINE AND TEST DRIVE VEHICLE, ACHIEVING AT LEAST 16 km/h (10 MPH).
- REPEAT ABOVE DRIVE CYCLE SEQUENCE TWO MORE TIMES. DOES DTC 144 SET IN THE LAST THREE DRIVE CYCLES?

YES → REPLACE EBTCM. NO → MALFUNCTION IS INTERMITTENT. INSPECT ALL CONNECTORS FOR CONTAMINATION AND POOR CONTACT. REPLACE ANY TERMINALS THAT EXHIBIT POOR TERMINAL CONTACT OR EVIDENCE OF CORROSION.

(7)
- IGNITION "OFF."
- REMOVE TCS MODULATOR/MOTOR PACK ASSEMBLY FROM VEHICLE.
- SEPARATE TCS MOTOR PACK FROM HYDRAULIC MODULATOR.
- GRASP THE GEAR ON THE HYDRAULIC MODULATOR (THE LEFT TCS CHANNEL IS FURTHEST REARWARD AS THE MODULATOR/MOTOR PACK SITS IN VEHICLE). CAN THE GEAR BE ROTATED AT LEAST 7 FULL TURNS LOCK TO LOCK?

NO → REPLACE HYDRAULIC MODULATOR. YES → REPLACE TCS MOTOR PACK.

AFTER DIAGNOSIS IS COMPLETE, CLEAR DTCs AND TEST DRIVE VEHICLE FOR THREE (3) DRIVE CYCLES TO VERIFY THAT DTC DOES NOT RESET. A DRIVE CYCLE CONSISTS OF STARTING THE VEHICLE, DRIVING OVER 16 km/h (10 MPH), AND THEN TURNING KEY "OFF."

DIAGNOSTIC CHART DTC 145 — 1995 SILHOUETTE AND TRANS SPORT WITH TCS

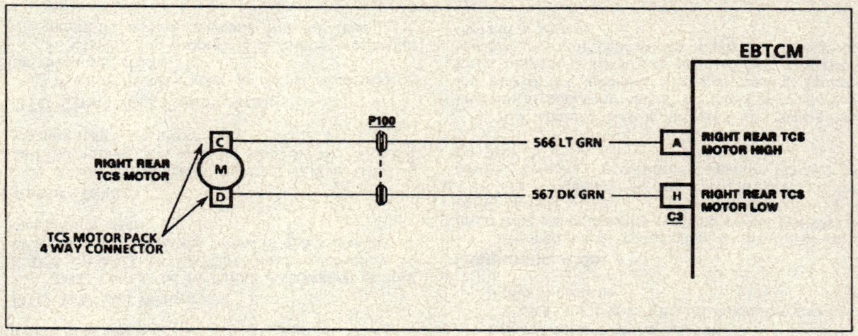

DTC 145
RIGHT TCS CHANNEL WILL NOT MOVE

Circuit Description:

This DTC is designed to detect a stuck TCS motor or a seized hydraulic modulator. During initialization or after a TCS cycle, the EBTCM commands the TCS motor in the reverse direction and the forward direction to ensure that the motor moves properly. If the TCS motor does not move, improper current feedback will be detected by the EBTCM, and a DTC will set.

DTC Will Set When:

DTC 145 can be set during initialization or after a TCS cycle. If the EBTCM detects a condition in which it cannot move the motor in either direction, a malfunction exists.

Action Taken:

A malfunction DTC is stored, TCS is disabled, and the amber TCS warning and the red BRAKE warning lamps are turned "ON."

DTC Chart Test Description:

Number(s) below refer to circled number(s) on the diagnostic chart.

1. Checks for proper TCS motor movement during the release command from the Tech 1.
2. Checks for proper TCS motor movement during the apply command from the Tech 1.
3. Compares EBTCM command current to TCS motor feedback current.
4. Checks for a shorted circuit between the TCS motor "HIGH" and motor "LOW" circuits.
5. Checks for a TCS motor that is internally shorted.
6. Determines if the malfunction is due to poor terminal contact or the EBTCM.
7. Checks for proper hydraulic modulator gear and piston movement.

Diagnostic Aids:

An intermittent malfunction in this DTC may result from a mechanical part of the system that sticks, binds, or slips.

The frequency of the malfunction can be checked by using the enhanced diagnostic function of the Tech 1, [illegible]

Any circuitry that is suspected of causing the intermittent complaint should be thoroughly checked for backed out terminals, improper mating, broken locks, improperly formed or damaged terminals, poor terminal to wiring connections or physical damage to the wiring harness.

DIAGNOSTIC CHART DTC 145 — 1995 SILHOUETTE AND TRANS SPORT WITH TCS, CONTINUED

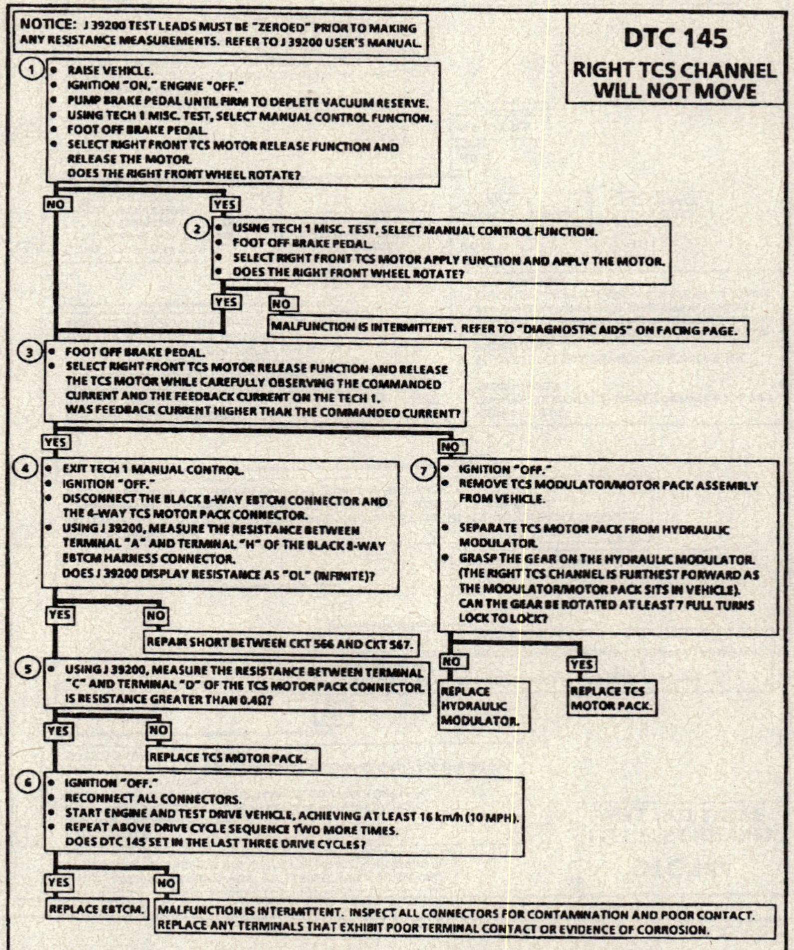

AFTER DIAGNOSIS IS COMPLETE, CLEAR DTCs AND TEST DRIVE VEHICLE FOR THREE (3) DRIVE CYCLES TO VERIFY THAT DTC DOES NOT RESET. A DRIVE CYCLE CONSISTS OF STARTING THE VEHICLE, DRIVING OVER 16 km/h (10 MPH), AND THEN TURNING KEY OFF.

DIAGNOSTIC CHART DTC 147 — 1995 SILHOUETTE AND TRANS SPORT WITH TCS

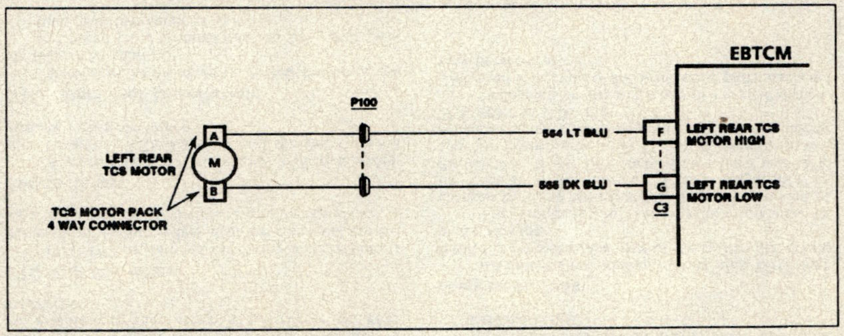

DTC 147 (Page 1 of 2) LEFT TCS MOTOR FREE SPINS

Circuit Description:

This DTC is designed to detect a stripped nut or gear assembly during initialization. During the homing sequence, the piston should reach the bottom of the bore resulting in a stalled TCS motor. If this does not occur, the TCS motor must be spinning with little or no resistance, indicating a nut/screw or gear malfunction.

DTC Will Set When:

DTC 147 can only be set during initialization. If the feedback current is less than the command current for a specified period of time, a malfunction exists.

Action Taken:

A malfunction DTC is stored, TCS is disabled, and the amber TCS warning and the red BRAKE warning lamps are turned "ON."

DTC Chart Test Description:

Number(s) below refer to circled number(s) on the diagnostic chart.

1. Checks to see if the corresponding open TCS motor DTC is also set.
2. Verifies that the TCS motor was actually released as commanded by observing feedback current.
3. Verifies that the TCS motor can apply.
4. This step verifies that the TCS motor can be released.
5. Checks for an open in the TCS motor "HIGH" circuitry.
6. Checks for an open in the TCS motor "LOW" circuitry.
7. Checks for a TCS motor that is internally shorted.

Diagnostic Aids:

An intermittent malfunction in this DTC may result from a mechanical part of the system that sticks, binds, or slips.

The frequency of the malfunction can be checked by using the enhanced diagnostic function of the Tech 1, as described in "Tech 1 Diagnostics" in this section. If the DTC only occurs once and DTC 156 also occurs, refer to DTC 156. If intermittent and enhanced diagnostics shows this DTC occurs during TCS, refer to DTC 156.

Depending on the frequency of the malfunction, a physical inspection of the mechanical parts suspected may be necessary.

DIAGNOSTIC CHART DTC 147 — 1995 SILHOUETTE AND TRANS SPORT WITH TCS, CONTINUED

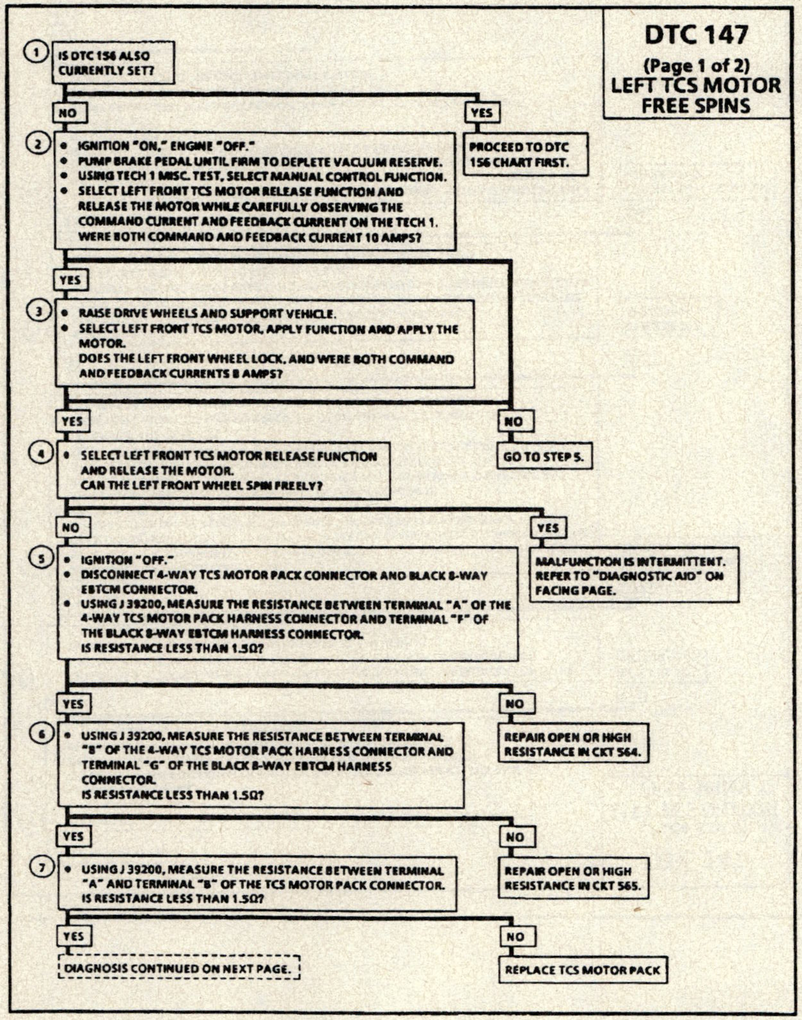

DIAGNOSTIC CHART DTC 147 — 1995 SILHOUETTE AND TRANS SPORT WITH TCS, CONTINUED

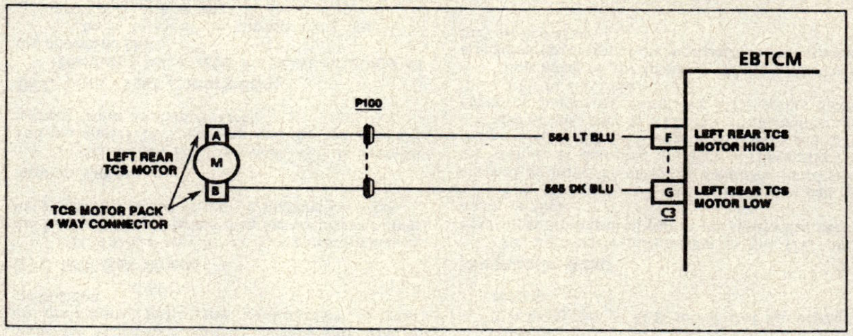

DTC 147
(Page 2 of 2)
LEFT TCS MOTOR FREE SPINS

Circuit Description:

This DTC is designed to detect a stripped nut or gear assembly during initialization. During the homing sequence, the piston should reach the bottom of the bore resulting in a stalled TCS motor. If this does not occur, the TCS motor must be spinning with little or no resistance, indicating a nut/screw or gear malfunction.

DTC Will Set When:

DTC 147 can only be set during initialization. If the feedback current is less than the command current for a specified period of time, a malfunction exists.

Action Taken:

A malfunction DTC is stored, TCS is disabled, and the amber TCS warning and the red BRAKE warning lamps are turned "ON."

DTC Chart Test Description:

Number(s) below refer to circled number(s) on the diagnostic chart.

8. Checks for a stripped gear on the motor pack (unit with two small gears).
9. Checks for a stripped gear on the hydraulic modulator (unit with two large gears).
10. Checks to see if the TCS motor pack has an internal malfunction.
11. Verifies that the hydraulic modulator is functioning properly.
12. Ensures malfunction was not due to a poor terminal contact.

Diagnostic Aids:

An intermittent malfunction in this DTC may result from a mechanical part of the system that sticks, binds, or slips.

The frequency of the malfunction can be checked by using the enhanced diagnostic function of the Tech 1, as described in "Tech 1 Diagnostics" in this section. If the DTC only occurs once and DTC 156 also occurs, refer to DTC 156. If intermittent and enhanced diagnostics shows this DTC occurs during TCS, refer to DTC 156.

Depending on the frequency of the malfunction, a physical inspection of the mechanical parts suspected may be necessary.

DIAGNOSTIC CHART DTC 147 — 1995 SILHOUETTE AND TRANS SPORT WITH TCS, CONTINUED

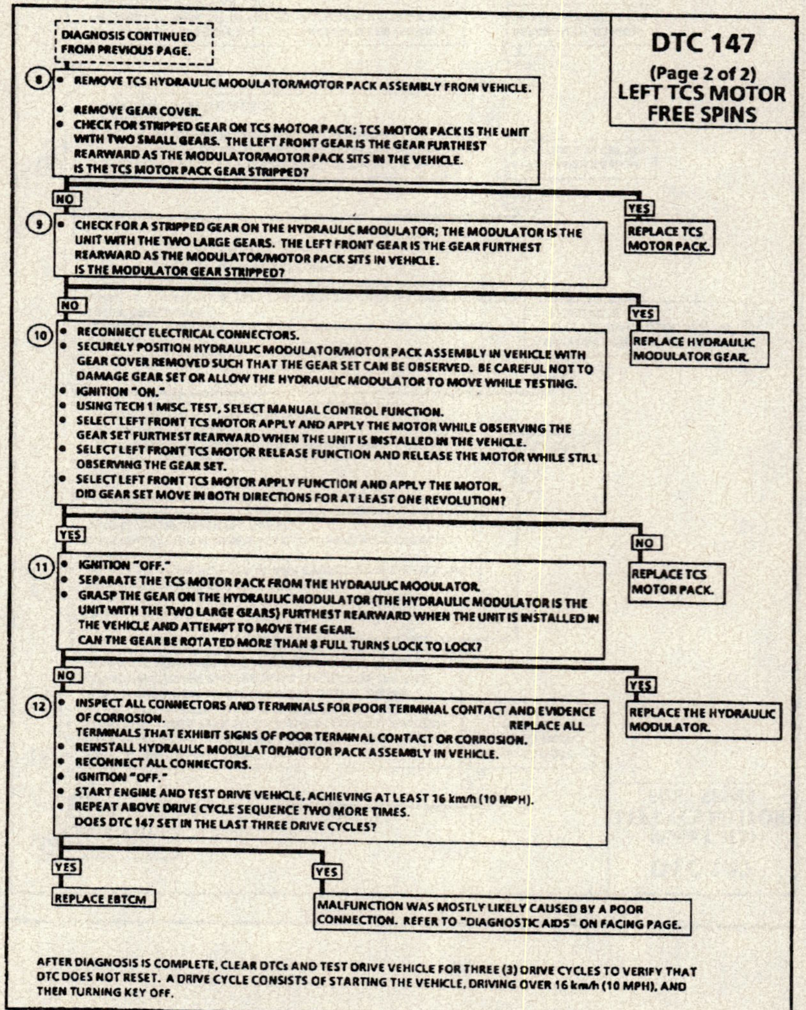

DIAGNOSTIC CHART DTC 148 — 1995 SILHOUETTE AND TRANS SPORT WITH TCS

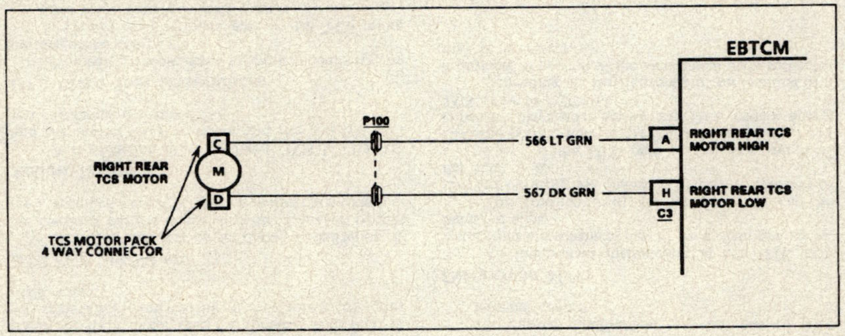

DTC 148
(Page 1 of 2)
RIGHT TCS MOTOR FREE SPINS

Circuit Description:

This DTC is designed to detect a stripped nut or gear assembly during initialization. During the homing sequence, the piston should reach the bottom of the bore resulting in a stalled TCS motor. If this does not occur, the TCS motor must be spinning with little or no resistance, indicating a nut/screw or gear malfunction.

DTC Will Set When:

DTC 148 can only be set during initialization. If the feedback current is less than the command current for a specified period of time, a malfunction exists.

Action Taken:

A malfunction DTC is stored, TCS is disabled, and the amber TCS warning and the red BRAKE warning lamps are turned "ON."

DTC Chart Test Description:

Number(s) below refer to circled number(s) on the diagnostic chart.

1. Checks to see if the corresponding open TCS motor DTC is also set.
2. Verifies that the TCS motor was actually released as commanded by observing feedback current.
3. Verifies that the TCS motor can apply.
4. Verifies that the TCS motor can be released.
5. Checks for high resistance in the TCS motor "HIGH" circuitry.
6. Checks for high resistance in the TCS motor "LOW" circuitry.
7. Checks for proper resistance of the TCS motor windings.

Diagnostic Aids:

An intermittent malfunction in this DTC may result from a mechanical part of the system that sticks, binds, or slips.

The frequency of the malfunction can be checked by using the enhanced diagnostic function of the Tech 1, as described in "Tech 1 Diagnostics" in this section. If the DTC only occurs once and DTC 161 also occurs, refer to DTC 161. If intermittent and enhanced diagnostics shows this DTC occurs during TCS, refer to DTC 161.

Depending on the frequency of the malfunction, a physical inspection of the mechanical parts suspected may be necessary.

DIAGNOSTIC CHART DTC 148 — 1995 SILHOUETTE AND TRANS SPORT WITH TCS, CONTINUED

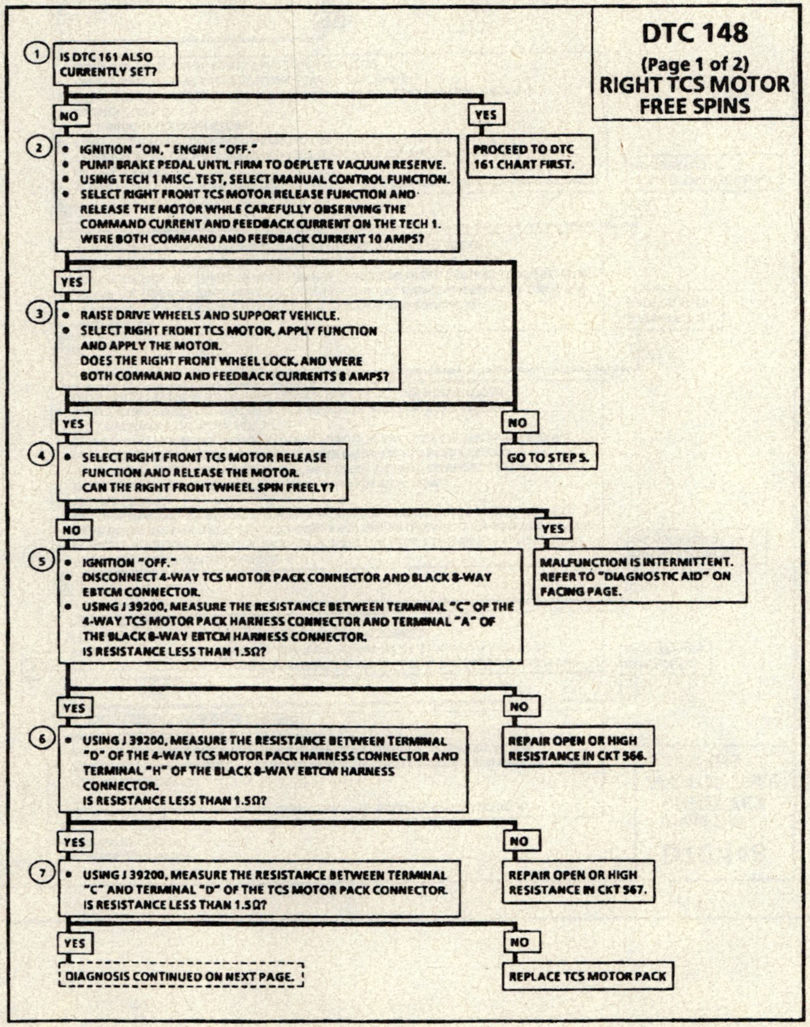

DIAGNOSTIC CHART DTC 148 — 1995 SILHOUETTE AND TRANS SPORT WITH TCS, CONTINUED

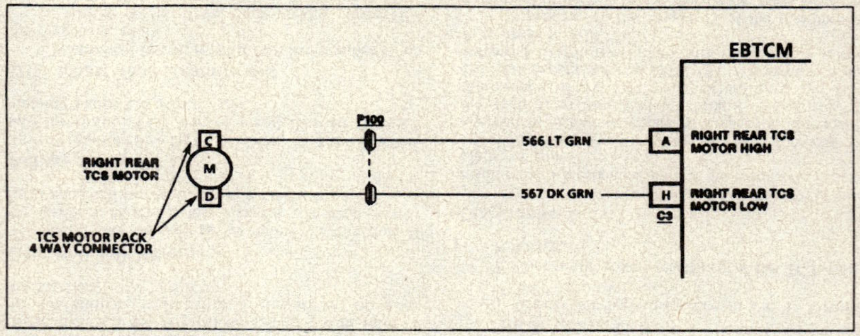

DTC 148
(Page 2 of 2)
RIGHT TCS MOTOR FREE SPINS

Circuit Description:

This DTC is designed to detect a stripped nut or gear assembly during initialization. During the homing sequence, the piston should reach the bottom of the bore resulting in a stalled TCS motor. If this does not occur, the TCS motor must be spinning with little or no resistance, indicating a nut/screw or gear malfunction.

DTC Will Set When:

DTC 148 can only be set during initialization. If the feedback current is less than the command current for a specified period of time, a malfunction exists.

Action Taken:

A malfunction DTC is stored, TCS is disabled, and the amber TCS warning and the red BRAKE warning lamps are turned "ON."

DTC Chart Test Description:

Number(s) below refer to circled number(s) on the diagnostic chart.

8. Checks for a stripped gear on the TCS motor pack (unit with two small gears).

9. Checks for a stripped gear on the hydraulic modulator (unit with two large gears).

10. Checks to see if the TCS motor pack has an internal malfunction.

11. Verifies that the hydraulic modulator is functioning properly.

12. Ensures malfunction was not due to poor terminal contact.

Diagnostic Aids:

An intermittent malfunction in this DTC may result from a mechanical part of the system that sticks, binds, or slips.

The frequency of the malfunction can be checked by using the enhanced diagnostic function of the Tech 1, as [illegible] diagnostics [illegible]. If the DTC only occurs once and DTC 161 also occurs, refer to DTC 161. If intermittent and enhanced diagnostics shows this DTC occurs during TCS, refer to DTC 161.

Depending on the frequency of the malfunction, a physical inspection of the mechanical parts suspected may be necessary.

DIAGNOSTIC CHART DTC 148 — 1995 SILHOUETTE AND TRANS SPORT WITH TCS, CONTINUED

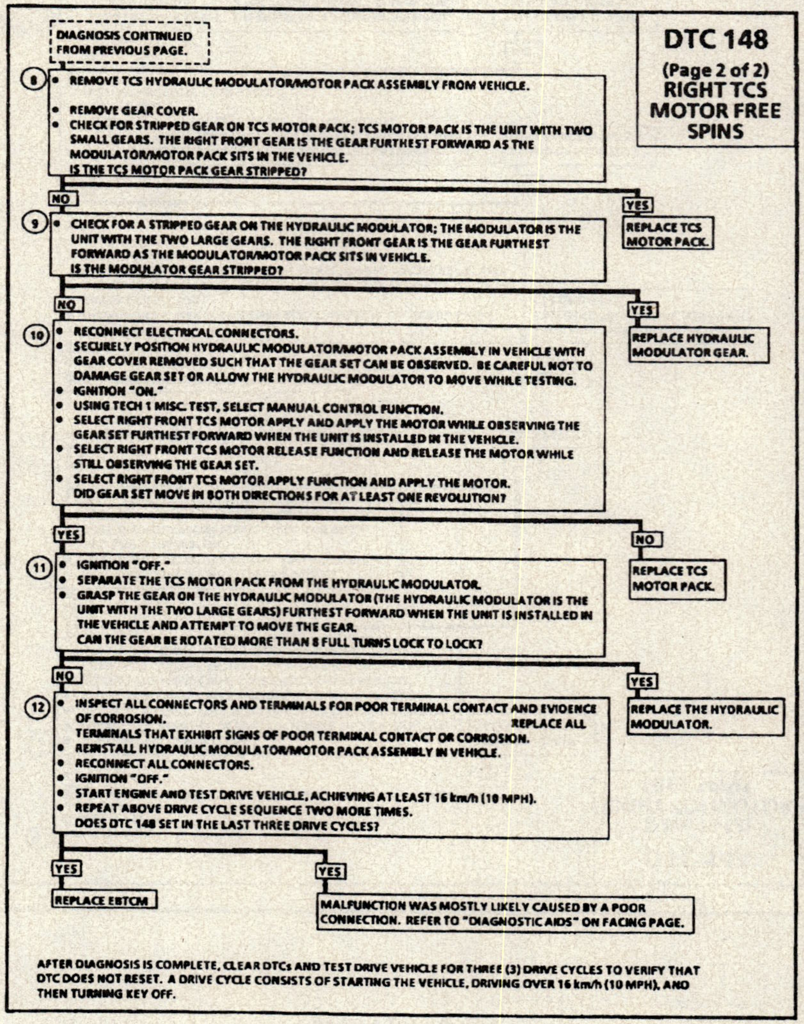

DIAGNOSTIC CHART DTC 155 — 1995 SILHOUETTE AND TRANS SPORT WITH TCS

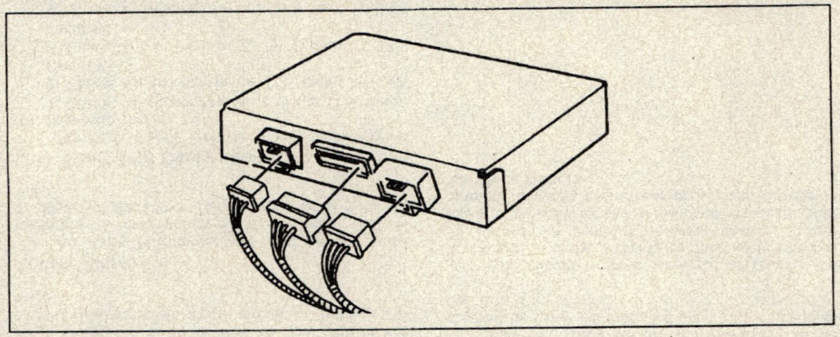

DTC 155
EBTCM MALFUNCTION (TCS)

Circuit Description:

This DTC identifies an internal malfunction detected by the EBTCM motor driver interface custom integrated circuit.

Action Taken:

A malfunction DTC is stored, TCS is disabled and the amber TCS warning lamp is turned "ON."

DTC Chart Test Description:

Number(s) below refer to circled number(s) on the diagnostic chart.

1. Checks to see if the malfunction is still present.
2. Checks to see if the malfunction is intermittent.

DIAGNOSTIC CHART DTC 155 — 1995 SILHOUETTE AND TRANS SPORT WITH TCS, CONTINUED

DTC 155
EBTCM MALFUNCTION (TCS)

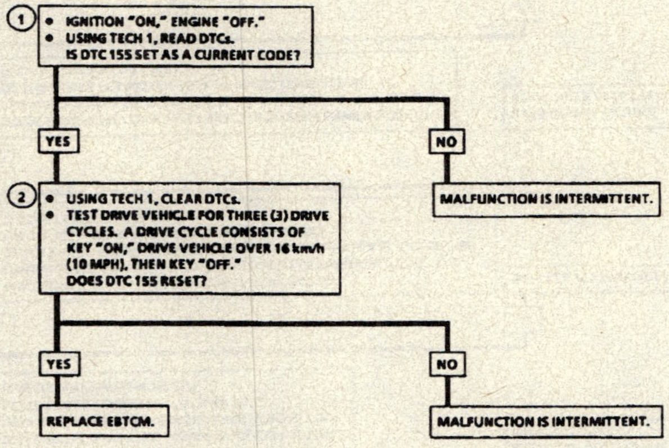

DIAGNOSTIC CHART DTC 156 — 1995 SILHOUETTE AND TRANS SPORT WITH TCS

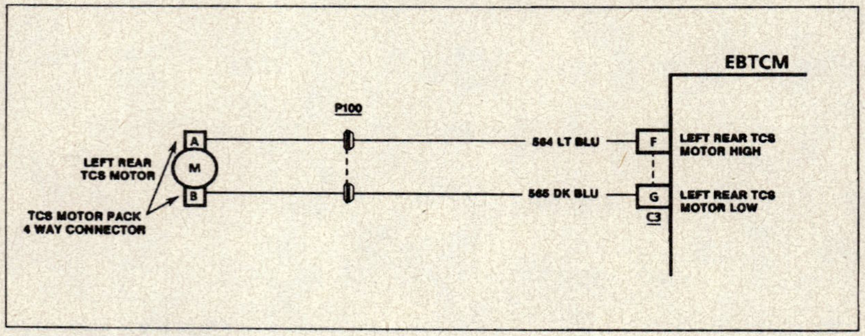

DTC 156
LEFT TCS MOTOR CIRCUIT OPEN

Circuit Description:

This DTC identifies a TCS motor that cannot be energized due to an open in its circuitry.

DTC Will Set When:

DTC 156 can be set only when the TCS motor is commanded "OFF." If the EBTCM detects an out of range voltage on either of the left front TCS motor circuits indicating an open circuit, a malfunction exists.

Action Taken:

An open TCS motor will not activate when requested. A malfunction DTC is stored, TCS is disabled and the amber TCS warning lamp is turned "ON."

DTC Chart Test Description:

Number(s) below refer to circled number(s) on the diagnostic chart.

1. Checks for proper resistance of the TCS motor.
2. Checks for an open in the TCS motor "HIGH" circuitry.
3. Checks for an open in the TCS motor "LOW" circuitry.
4. Ensures malfunction was not due to poor terminal contact or EBTCM.

Diagnostic Aids:

An intermittent malfunction may be caused by a poor connection, rubbed through wire insulation, or a wire that is broken inside the insulation.

If the malfunction is not current, wiggle the wires of the affected channel and check if the DTC resets. This will help to pinpoint an intermittent malfunction in the motor circuitry or connections.

The frequency of the malfunction can be checked by using the enhanced diagnostic function of the Tech 1, [illegible] diagnostic [illegible]

Any circuitry that is suspected of causing the intermittent complaint should be thoroughly checked for backed out terminals, improper mating, broken locks, improperly formed or damaged terminals, poor terminal to wiring connections or physical damage to the wiring harness.

DIAGNOSTIC CHART DTC 156 — 1995 SILHOUETTE AND TRANS SPORT WITH TCS, CONTINUED

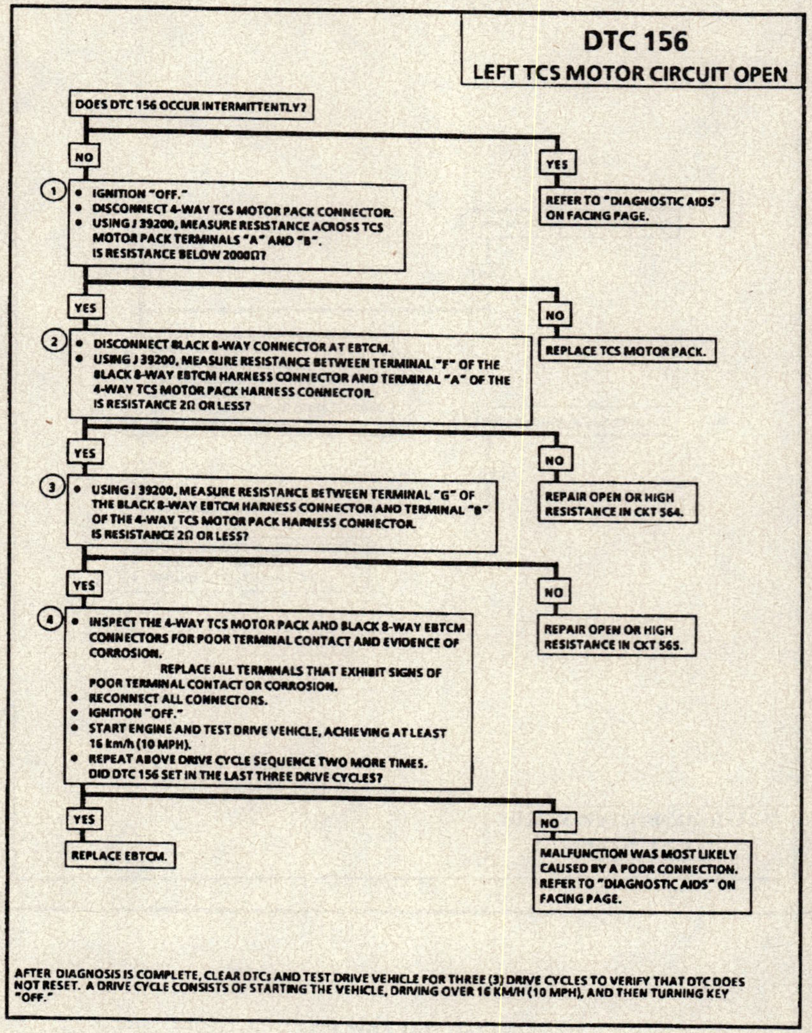

DIAGNOSTIC CHART DTC 157 — 1995 SILHOUETTE AND TRANS SPORT WITH TCS

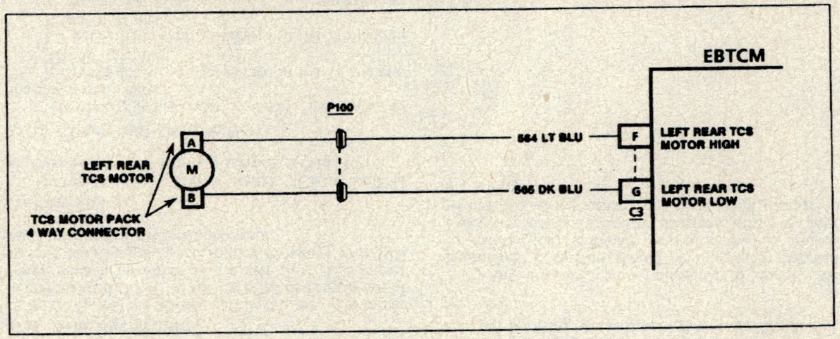

DTC 157
LEFT TCS MOTOR CIRCUIT SHORTED TO GND

Circuit Description:

This DTC identifies a TCS motor circuit that is shorted to ground. This malfunction will not allow the TCS motor to be controlled at the commanded current rate or will cause the TCS motor driver circuit to allow current directly to ground.

DTC Will Set When:

DTC 157 can be set anytime. If the EBTCM detects an out of range voltage on either of the left front TCS motor circuits indicating a circuit shorted to ground, a malfunction exists.

Action Taken:

A malfunction DTC is stored, TCS is disabled and the amber TCS warning lamp is turned "ON."

DTC Chart Test Description:

Number(s) below refer to circled number(s) on the diagnostic chart.

1. Checks for a short to ground in the TCS motor "HIGH" circuitry.
2. Checks for a short to ground in the TCS motor "LOW" circuitry.
3. Checks for a TCS motor that is internally shorted to ground.
4. Ensures malfunction was not due to physical damage of the circuitry.

Diagnostic Aids:

An intermittent malfunction may be caused by a poor connection, rubbed through wire insulation, or a wire that is broken inside the insulation.

If the malfunction is not current, wiggle the wires of the affected channel and check if the DTC resets. This will help to pinpoint an intermittent malfunction in the motor circuitry or connections.

The frequency of the malfunction can be checked by using the enhanced diagnostic function of the Tech 1 [illegible] in "Tech 1 Diagnostics" in this section.

Any circuitry that is suspected of causing the intermittent complaint should be thoroughly checked for backed out terminals, improper mating, broken locks, improperly formed or damaged terminals, poor terminal to wiring connections or physical damage to the wiring harness.

DIAGNOSTIC CHART DTC 157 — 1995 SILHOUETTE AND TRANS SPORT WITH TCS, CONTINUED

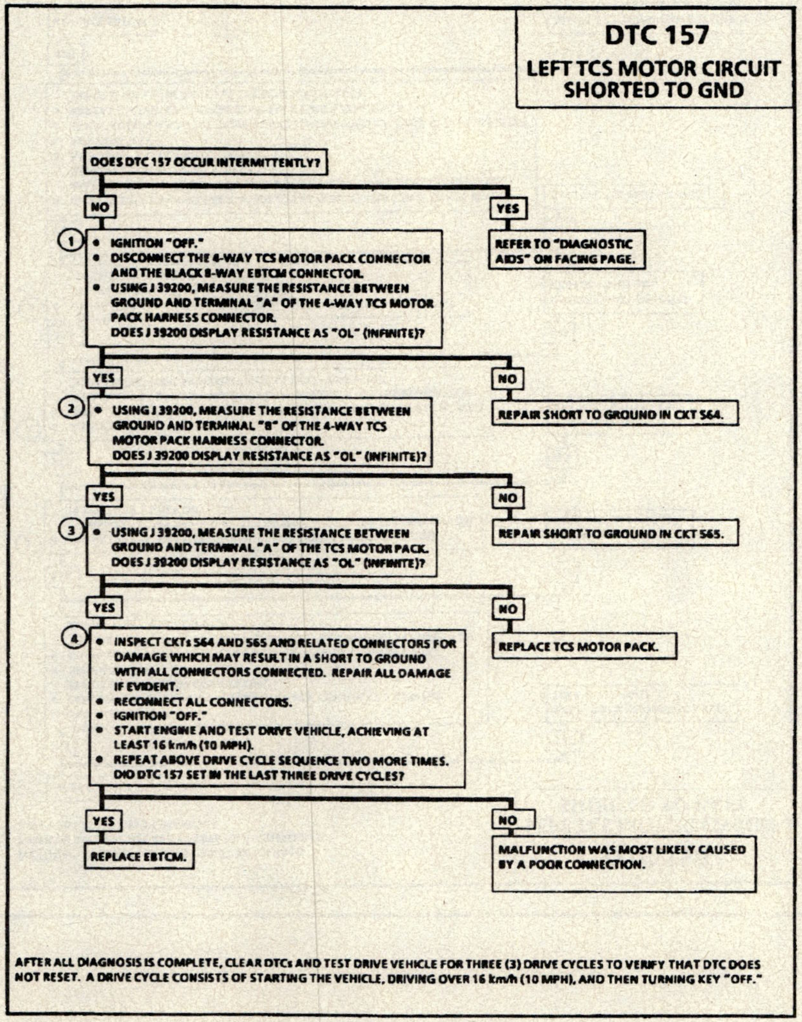

DIAGNOSTIC CHART DTC 158 — 1995 SILHOUETTE AND TRANS SPORT WITH TCS

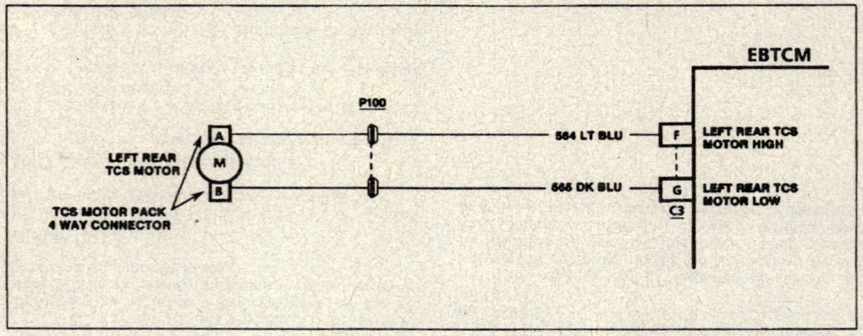

DTC 158
LEFT TCS MOTOR CIRCUIT SHORTED TO BATT

Circuit Description:

This DTC identifies a TCS motor circuit that is shorted to battery or a TCS motor that has low or no resistance. This malfunction will not allow the TCS motor to be controlled at the commanded current rate or will cause the TCS motor to turn in the opposite direction, or not at all.

DTC Will Set When:

DTC 158 can be set only when the TCS motor is commanded "OFF." If the EBTCM detects an out of range voltage on either of the left front TCS motor circuits indicating a circuit shorted to battery or a TCS motor shorted, a malfunction exists.

Action Taken:

A malfunction DTC is stored, TCS is disabled and the amber TCS warning lamp is turned "ON."

DTC Chart Test Description:

Number(s) below refer to circled number(s) on the diagnostic chart.

1. Checks for a short to voltage in the TCS motor "HIGH" circuitry.
2. Checks for a short to voltage in the TCS motor "LOW" circuitry.
3. Checks for a short between the "HIGH" and "LOW" motor circuits.
4. This step checks for a TCS motor that is internally shorted.
5. Ensures malfunction was not due to physical damage of the circuitry.

Diagnostic Aids:

An intermittent malfunction may be caused by a poor connection, rubbed through wire insulation, or a wire that is broken inside the insulation.

If the malfunction is not current, wiggle the wires of the affected channel and check if the DTC resets. This will help to pinpoint an intermittent malfunction in the motor circuitry or connections.

The frequency of the malfunction can be checked by using the enhanced diagnostic function of the Tech 1 [illegible]

Any circuitry that is suspected of causing the intermittent complaint should be thoroughly checked for backed out terminals, improper mating, broken locks, improperly formed or damaged terminals, poor terminal to wiring connections or physical damage to the wiring harness.

DIAGNOSTIC CHART DTC 158 — 1995 SILHOUETTE AND TRANS SPORT WITH TCS, CONTINUED

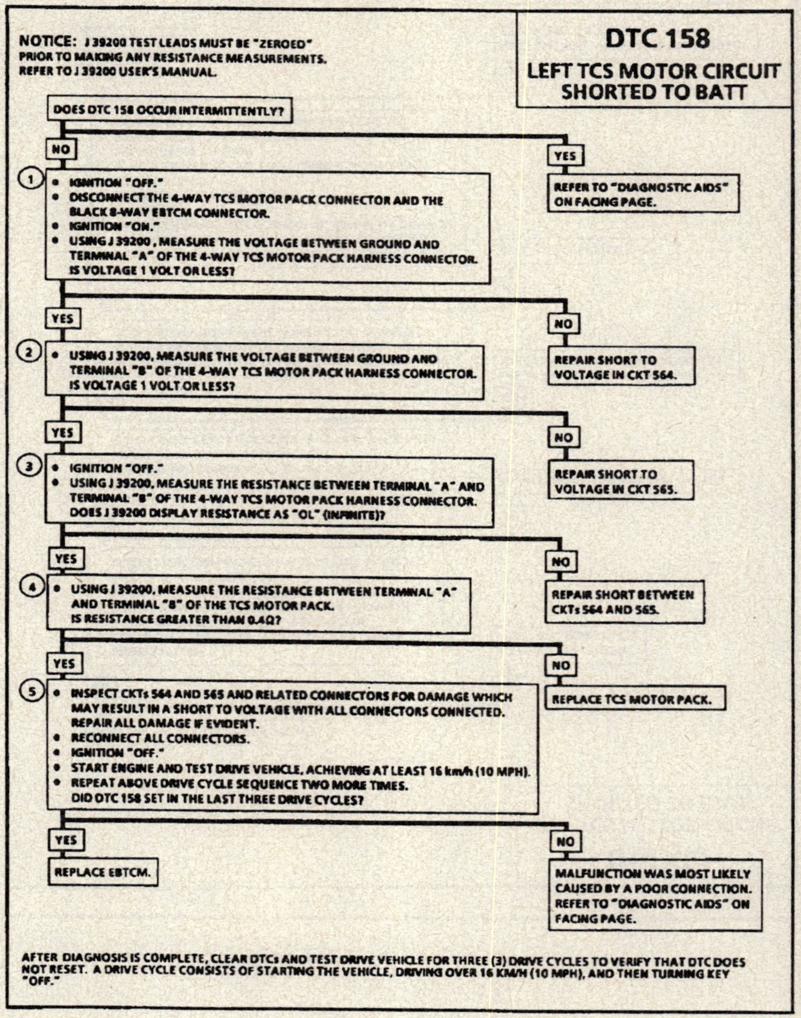

DIAGNOSTIC CHART DTC 161 — 1995 SILHOUETTE AND TRANS SPORT WITH TCS

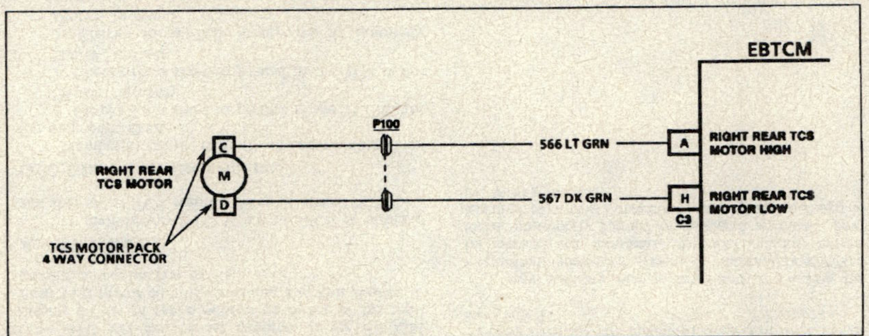

DTC 161
RIGHT TCS MOTOR CIRCUIT OPEN

Circuit Description:

This DTC identifies a TCS motor that cannot be energized due to an open in its circuitry.

DTC Will Set When:

DTC 161 can be set only when the TCS motor is commanded "OFF." If the EBTCM detects an out of range voltage on either of the right front TCS motor circuits indicating an open circuit, a malfunction exists.

Action Taken:

An open TCS motor will not activate when requested. A malfunction DTC is stored, TCS is disabled and the amber TCS warning lamp is turned "ON."

DTC Chart Test Description:

Number(s) below refer to circled number(s) on the diagnostic chart.

1. Checks for proper resistance of the TCS motor.
2. Checks for an open in the TCS motor "HIGH" circuitry.
3. Checks for an open in the TCS motor "LOW" circuitry.
4. Ensures malfunction was not due to poor terminal contact or EBTCM.

Diagnostic Aids:

An intermittent malfunction may be caused by a poor connection, rubbed through wire insulation, or a wire that is broken inside the insulation.

If the malfunction is not current, wiggle the wires of the affected channel and check if the DTC resets. This will help to pinpoint an intermittent malfunction in the motor circuitry or connections.

The frequency of the malfunction can be checked by using the enhanced diagnostic function of the Tech 1 [illegible] Diagnos[illegible].

Any circuitry that is suspected of causing the intermittent complaint should be thoroughly checked for backed out terminals, improper mating, broken locks, improperly formed or damaged terminals, poor terminal to wiring connections or physical damage to the wiring harness.

DIAGNOSTIC CHART DTC 161 — 1995 SILHOUETTE AND TRANS SPORT WITH TCS, CONTINUED

DTC 161
RIGHT TCS MOTOR CIRCUIT OPEN

DOES DTC 161 OCCUR INTERMITTENTLY?

- YES: REFER TO "DIAGNOSTIC AIDS" ON FACING PAGE.
- NO:

(1)
- IGNITION "OFF."
- DISCONNECT 4-WAY TCS MOTOR PACK CONNECTOR.
- USING J 39200, MEASURE RESISTANCE ACROSS TCS MOTOR PACK TERMINALS "C" AND "D".

IS RESISTANCE BELOW 2000Ω?

- NO: REPLACE TCS MOTOR PACK.
- YES:

(2)
- DISCONNECT BLACK 8-WAY CONNECTOR AT EBTCM.
- USING J 39200, MEASURE RESISTANCE BETWEEN TERMINAL "A" OF THE BLACK 8-WAY EBTCM HARNESS CONNECTOR AND TERMINAL "C" OF THE 4-WAY TCS MOTOR PACK HARNESS CONNECTOR.

IS RESISTANCE 2Ω OR LESS?

- NO: REPAIR OPEN OR HIGH RESISTANCE IN CKT 566.
- YES:

(3)
- USING J 39200, MEASURE RESISTANCE BETWEEN TERMINAL "H" OF THE BLACK 8-WAY EBTCM HARNESS CONNECTOR AND TERMINAL "D" OF THE 4-WAY TCS MOTOR PACK HARNESS CONNECTOR.

IS RESISTANCE 2Ω OR LESS?

- NO: REPAIR OPEN OR HIGH RESISTANCE IN CKT 567.
- YES:

(4)
- INSPECT THE 4-WAY TCS MOTOR PACK AND BLACK 8-WAY EBTCM CONNECTORS FOR POOR TERMINAL CONTACT AND EVIDENCE OF CORROSION.

  REPLACE ALL TERMINALS THAT EXHIBIT SIGNS OF POOR TERMINAL CONTACT OR CORROSION.
- RECONNECT ALL CONNECTORS.
- IGNITION "OFF."
- START ENGINE AND TEST DRIVE VEHICLE, ACHIEVING AT LEAST 16 km/h (10 MPH).
- REPEAT ABOVE DRIVE CYCLE SEQUENCE TWO MORE TIMES.

DID DTC 161 SET IN THE LAST THREE DRIVE CYCLES?

- YES: REPLACE EBTCM.
- NO: MALFUNCTION WAS MOST LIKELY CAUSED BY A POOR CONNECTION. REFER TO "DIAGNOSTIC AIDS" ON FACING PAGE.

AFTER ALL DIAGNOSIS IS COMPLETE, CLEAR DTCs AND TEST DRIVE VEHICLE FOR THREE (3) DRIVE CYCLES TO VERIFY THAT DTC DOES NOT RESET. A DRIVE CYCLE CONSISTS OF STARTING THE VEHICLE, DRIVING OVER 16 km/h (10 MPH), AND THEN TURNING KEY "OFF."

DIAGNOSTIC CHART DTC 162 — 1995 SILHOUETTE AND TRANS SPORT WITH TCS

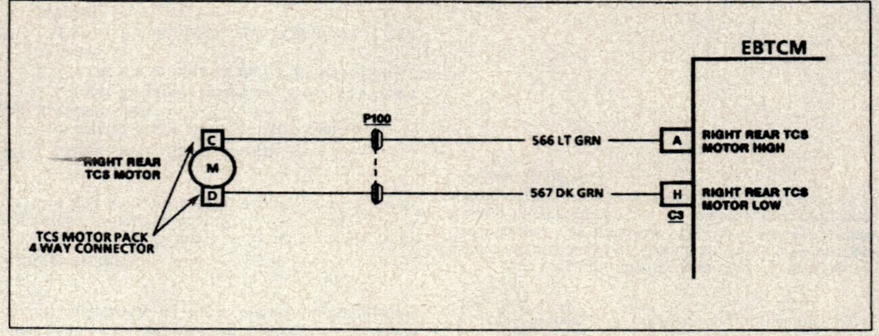

DTC 162
RIGHT TCS MOTOR CIRCUIT SHORTED TO GND

Circuit Description:

This DTC identifies a TCS motor circuit that is shorted to ground. This malfunction will not allow the TCS motor to be controlled at the commanded current rate or will cause the TCS motor driver circuit to allow current directly to ground.

DTC Will Set When:

DTC 162 can be set anytime. If the EBTCM detects an out of range voltage on either of the right front TCS motor circuits indicating a circuit shorted to ground, a malfunction exists.

Action Taken:

A malfunction DTC is stored, TCS is disabled and the amber TCS warning lamp is turned "ON."

DTC Chart Test Description:

Number(s) below refer to circled number(s) on the diagnostic chart.

1. Checks for a short to ground in the TCS motor "HIGH" circuitry.
2. Checks for a short to ground in the TCS motor "LOW" circuitry.
3. Checks for a TCS motor that is internally shorted to ground.
4. Ensures malfunction was not due to physical damage of the circuitry.

Diagnostic Aids:

An intermittent malfunction may be caused by a poor connection, rubbed through wire insulation, or a wire that is broken inside the insulation.

If the malfunction is not current, wiggle the wires of the affected channel and check if the DTC resets. This will help to pinpoint an intermittent malfunction in the motor circuitry or connections.

The frequency of the malfunction can be checked by using the enhanced diagnostic function of the Tech 1 [illegible] "Tech 1 Diagnostics" [illegible]

Any circuitry that is suspected of causing the intermittent complaint should be thoroughly checked for backed out terminals, improper mating, broken locks, improperly formed or damaged terminals, poor terminal to wiring connections or physical damage to the wiring harness.

DIAGNOSTIC CHART DTC 162 — 1995 SILHOUETTE AND TRANS SPORT WITH TCS, CONTINUED

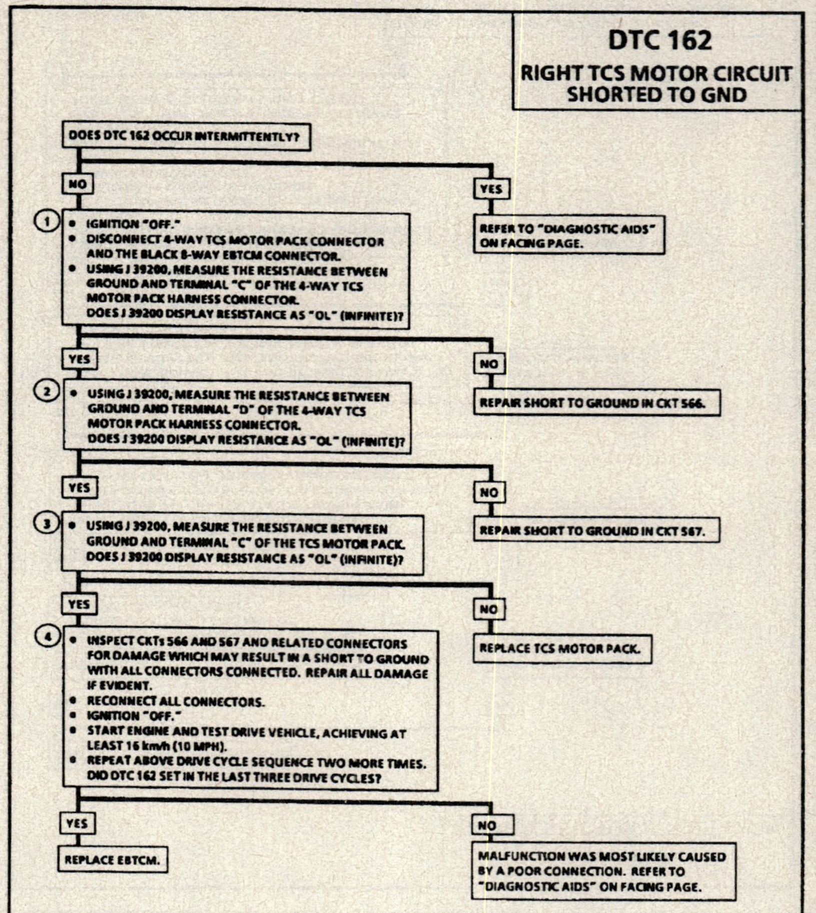

AFTER ALL DIAGNOSIS IS COMPLETE, CLEAR DTCs AND TEST DRIVE VEHICLE FOR THREE (3) DRIVE CYCLES TO VERIFY THAT DTC DOES NOT RESET. A DRIVE CYCLE CONSISTS OF STARTING THE VEHICLE, DRIVING OVER 16 km/h (10 MPH), AND THEN TURNING KEY "OFF."

DIAGNOSTIC CHART DTC 163 — 1995 SILHOUETTE AND TRANS SPORT WITH TCS

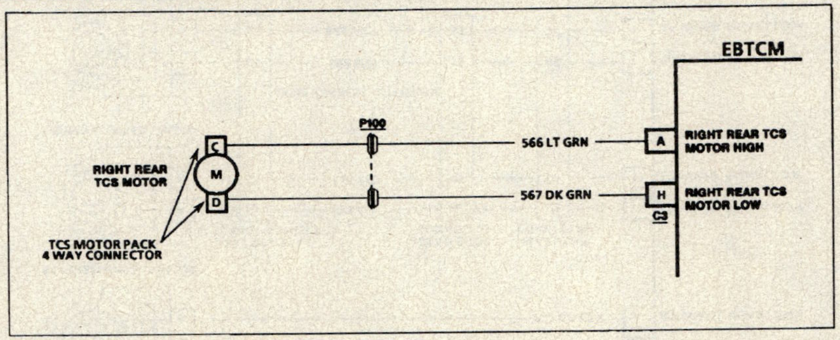

DTC 163
RIGHT TCS MOTOR CIRCUIT SHORTED TO BATT

Circuit Description:

This DTC identifies a TCS motor circuit that is shorted to battery or a TCS motor that has low or no resistance. This malfunction will not allow the TCS motor to be controlled at the commanded current rate or will cause the TCS motor to turn in the opposite direction, or not at all.

DTC Will Set When:

DTC 163 can be set only when the TCS motor is commanded "OFF." If the EBTCM detects an out of range voltage on either of the right front TCS motor circuits indicating a circuit shorted to battery or a TCS motor shorted, a malfunction exists.

Action Taken:

A malfunction DTC is stored, TCS is disabled and the amber TCS warning lamp is turned "ON."

DTC Chart Test Description:

Number(s) below refer to circled number(s) on the diagnostic chart.

1. Checks for a short to voltage in the TCS motor "HIGH" circuitry.
2. Checks for a short to voltage in the TCS motor "LOW" circuitry.
3. Checks for a short circuit between motor "HIGH" and motor "LOW."
4. Checks for proper resistance of the TCS motor.
5. Ensures malfunction was not due to physical damage of the circuitry.

Diagnostic Aids:

An intermittent malfunction may be caused by a poor connection, rubbed through wire insulation, or a wire that is broken inside the insulation.

If the malfunction is not current, wiggle the wires of the affected channel and check if the DTC resets. This will help to pinpoint an intermittent malfunction in the motor circuitry or connections.

The frequency of the malfunction can be checked by using the enhanced diagnostic function of the Tech 1, as described in "Tech 1 Diagnostics" in this section.

Any circuitry that is suspected of causing the intermittent complaint should be thoroughly checked for backed out terminals, improper mating, broken locks, improperly formed or damaged terminals, poor terminal to wiring connections or physical damage to the wiring harness.

DIAGNOSTIC CHART DTC 163 — 1995 SILHOUETTE AND TRANS SPORT WITH TCS, CONTINUED

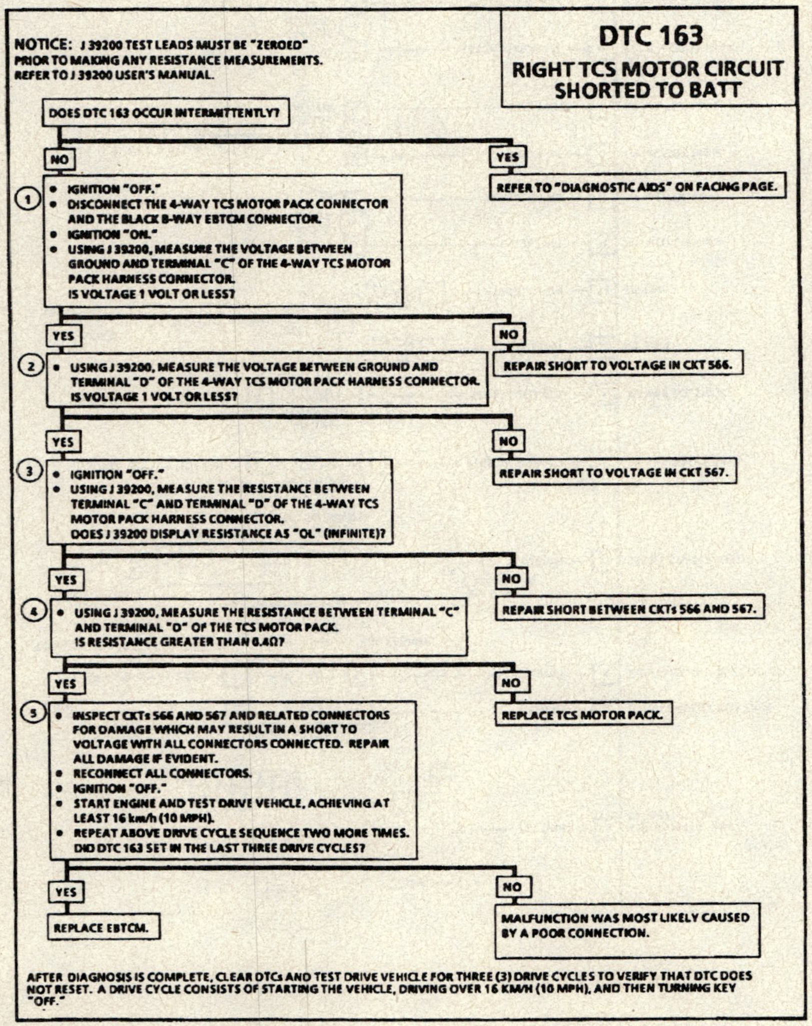

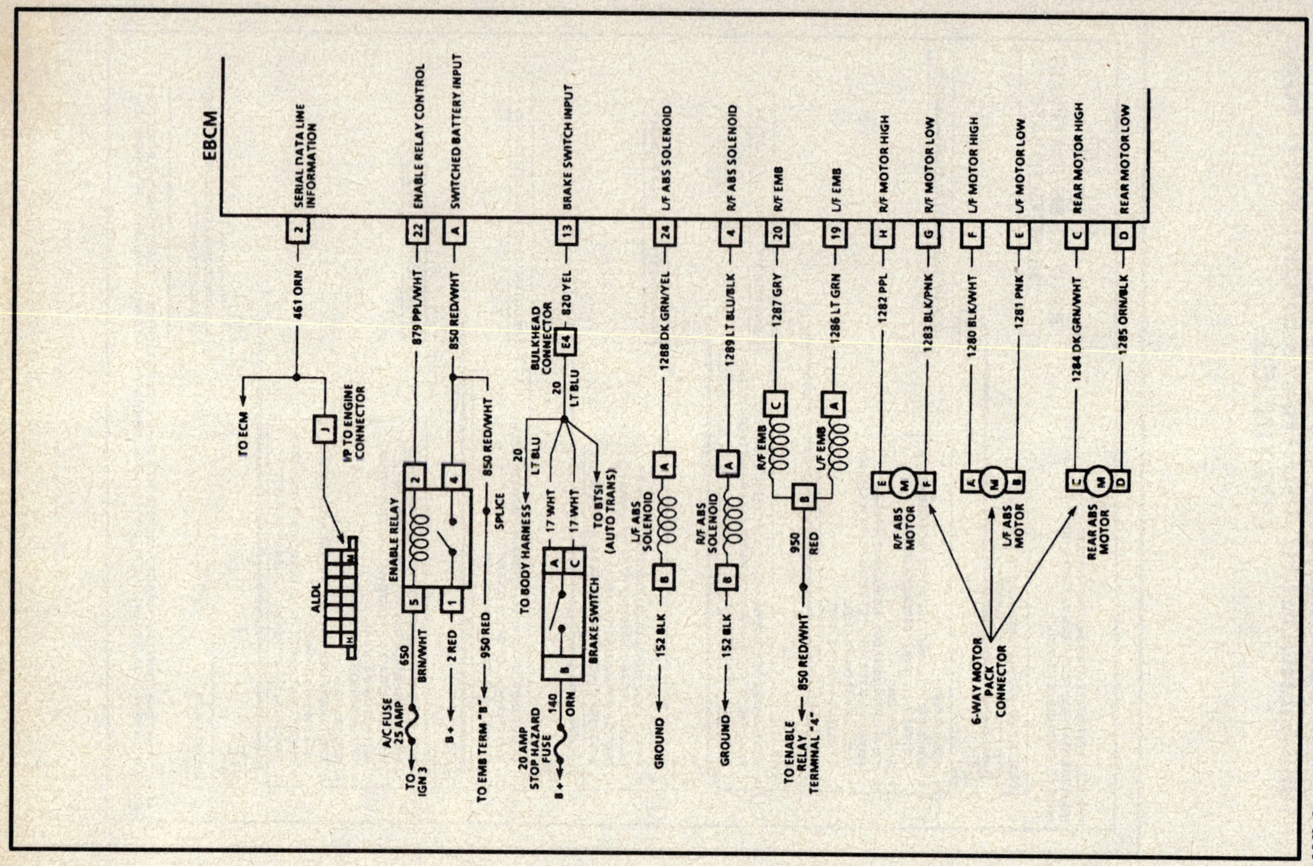

Wiring schematic — 1993 Cavalier and Sunbird

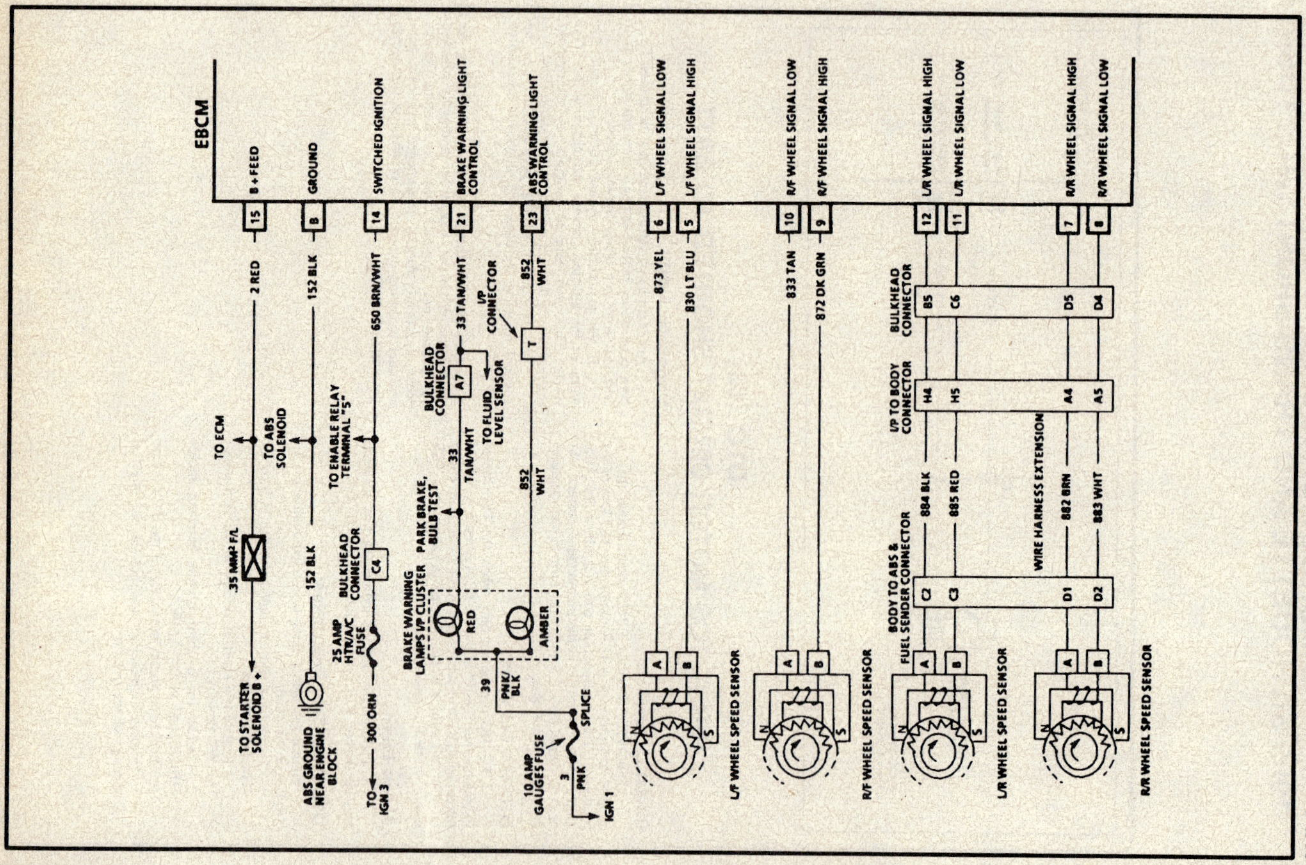

Wiring schematic — 1993 Cavalier and Sunbird

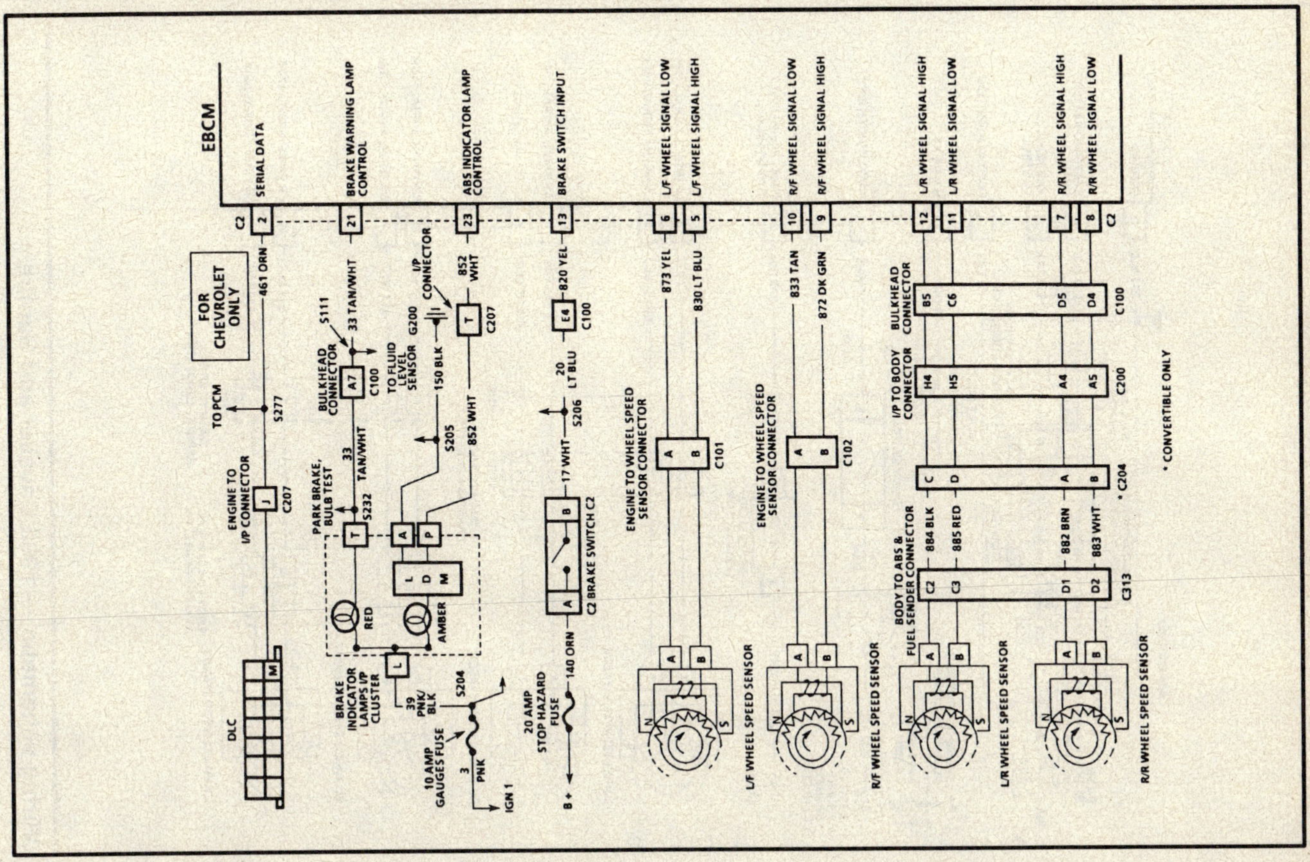

Wiring schematic — 1994 Cavalier and Sunbird

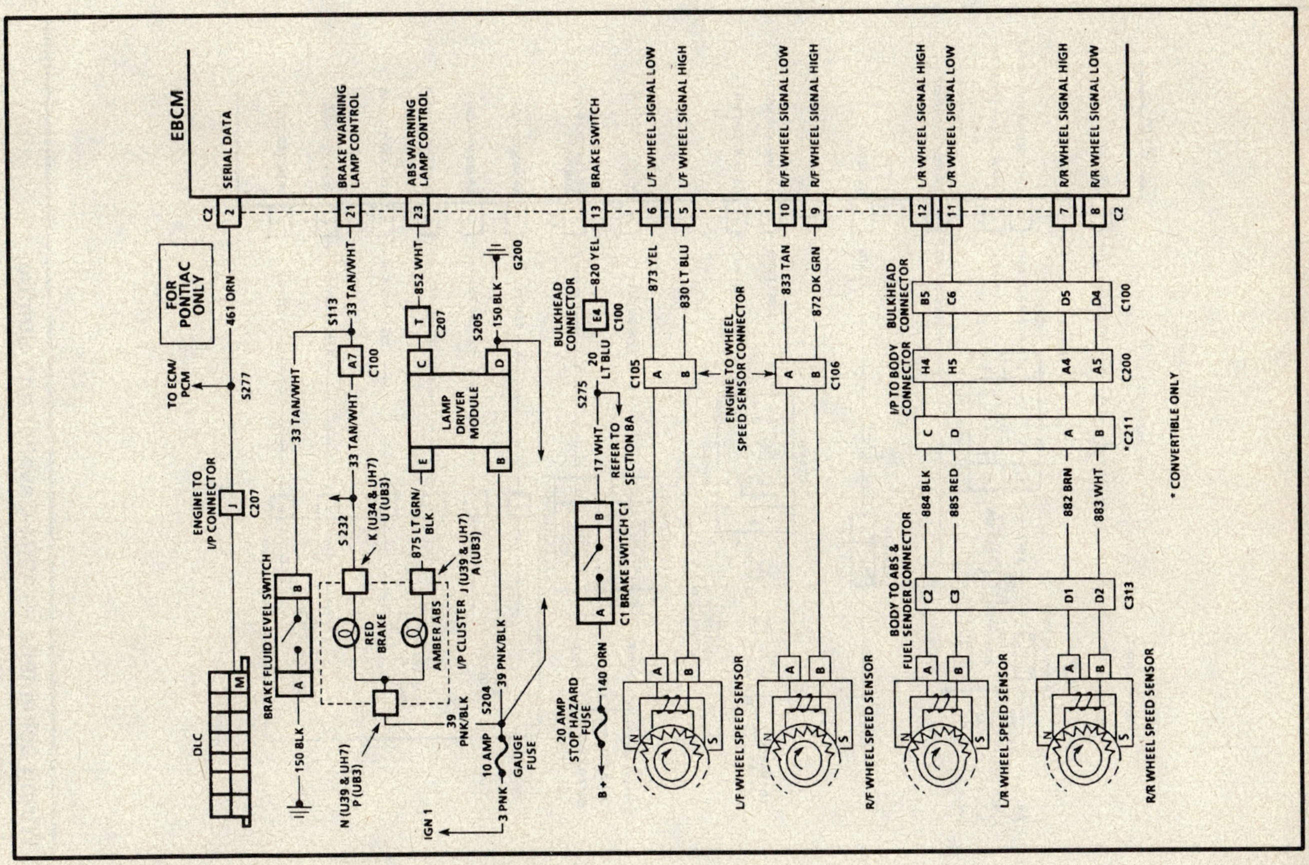

Wiring schematic — 1994 Cavalier and Sunbird

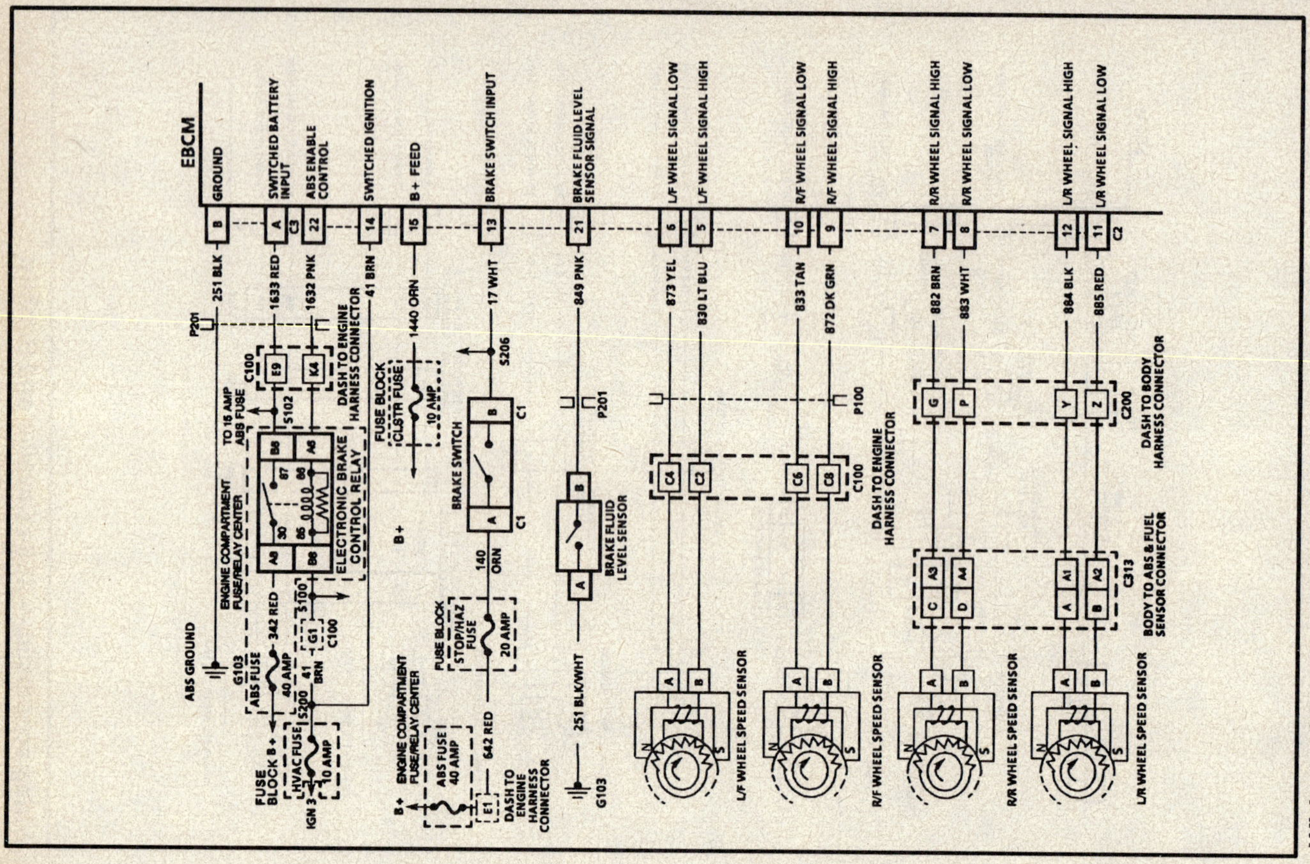

Wiring schematic — 1995 Cavalier and Sunfire

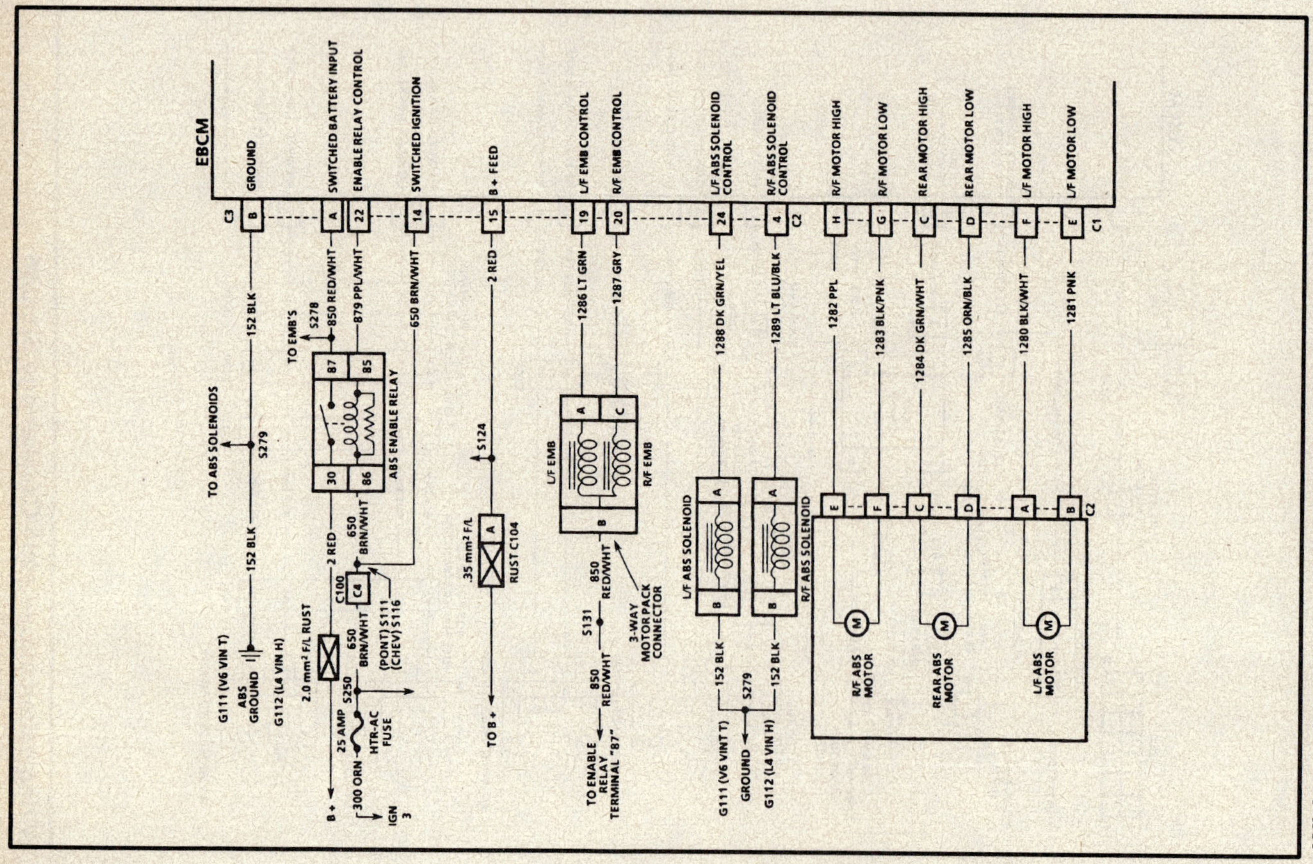

Wiring schematic — 1994 Cavalier and Sunfire

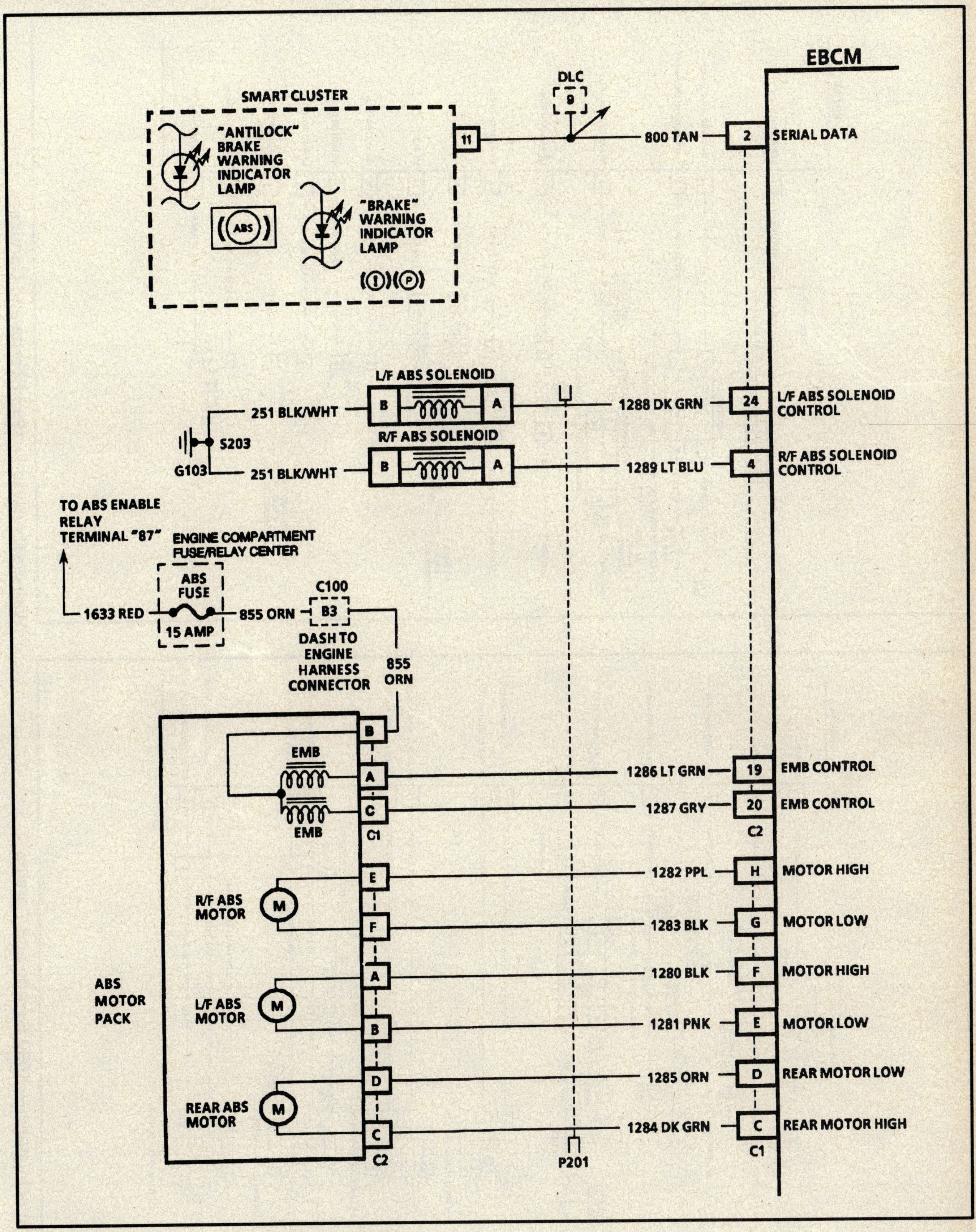

Wiring schematic — 1995 Cavalier and Sunfire

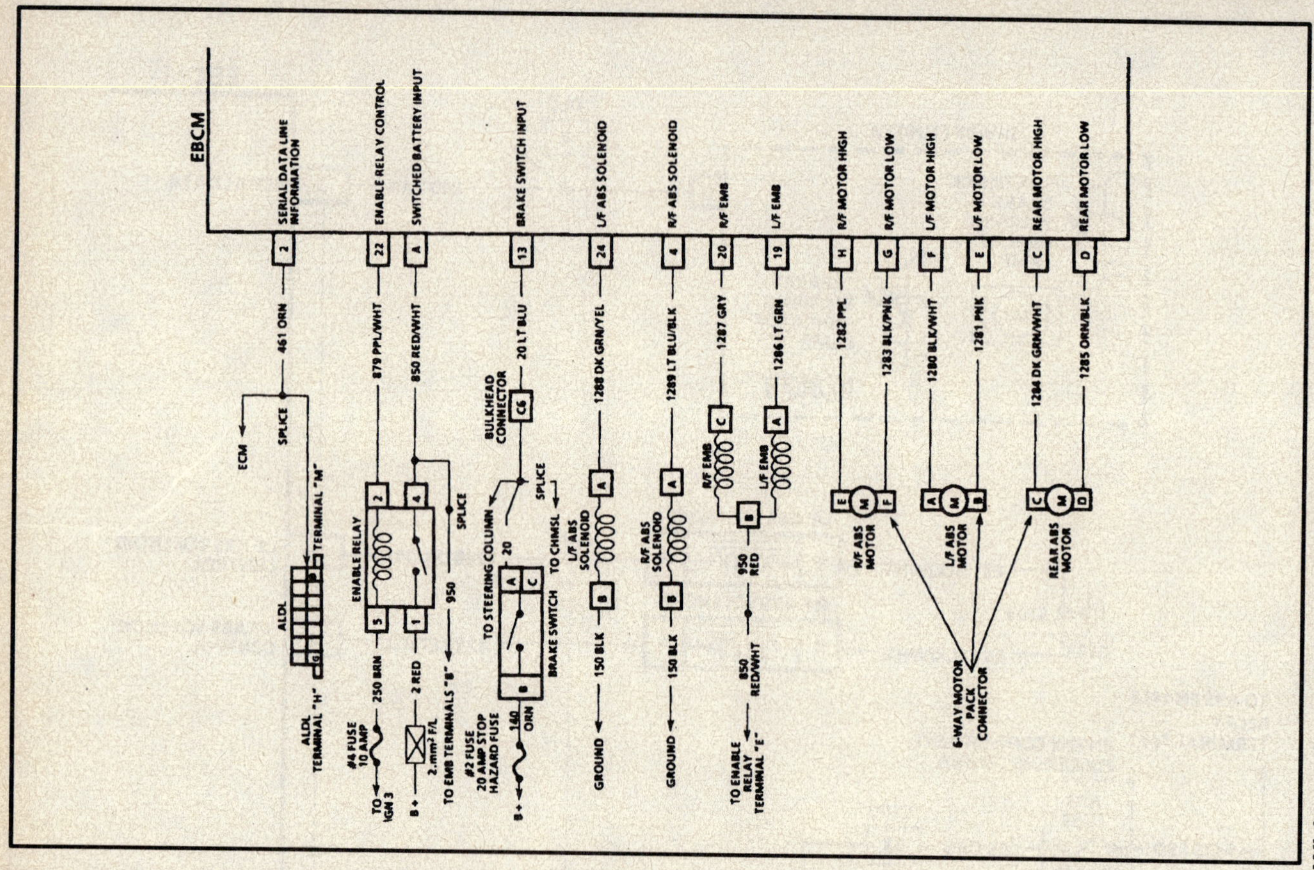

Wiring schematic — 1993 Beretta and Corsica

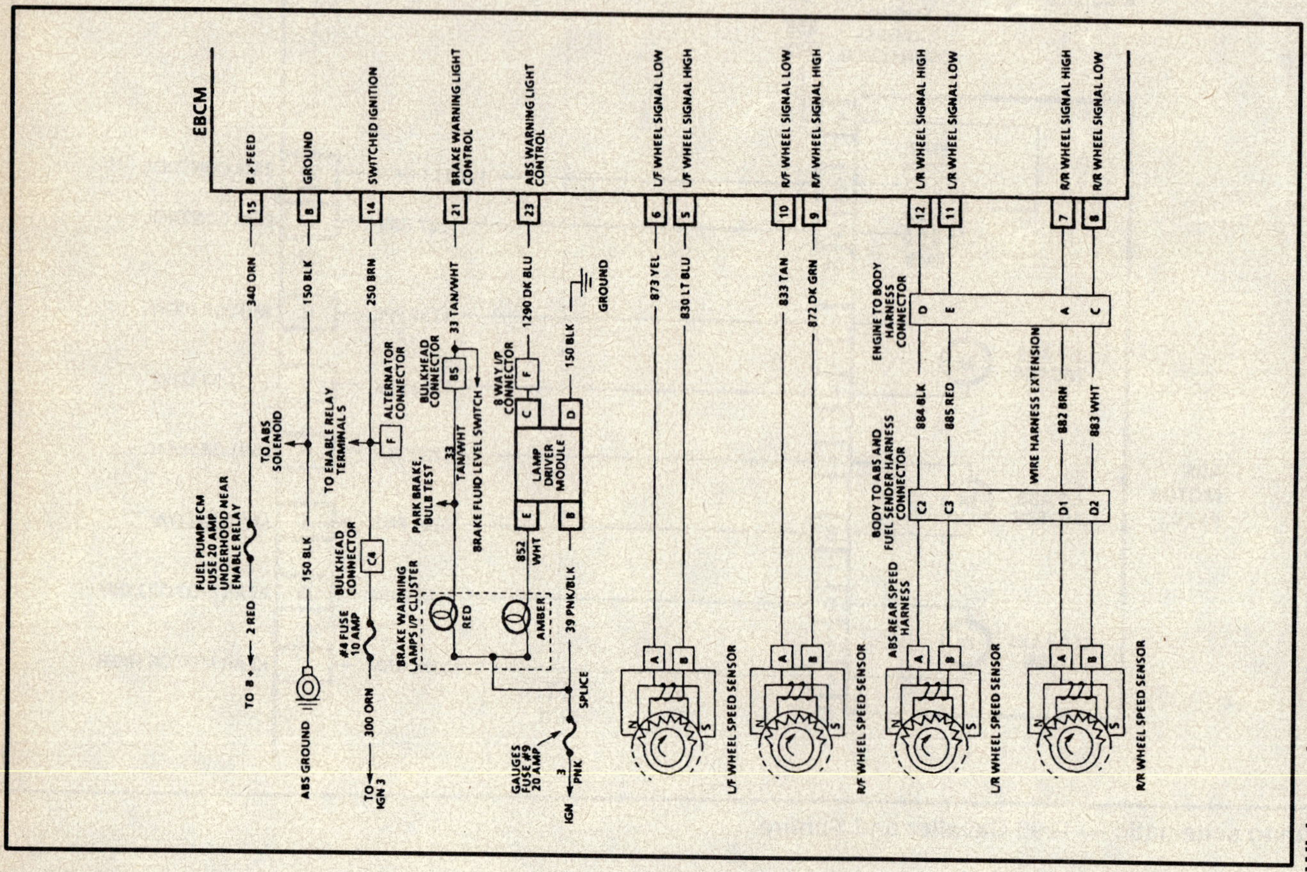

Wiring schematic — 1993 Beretta and Corsica

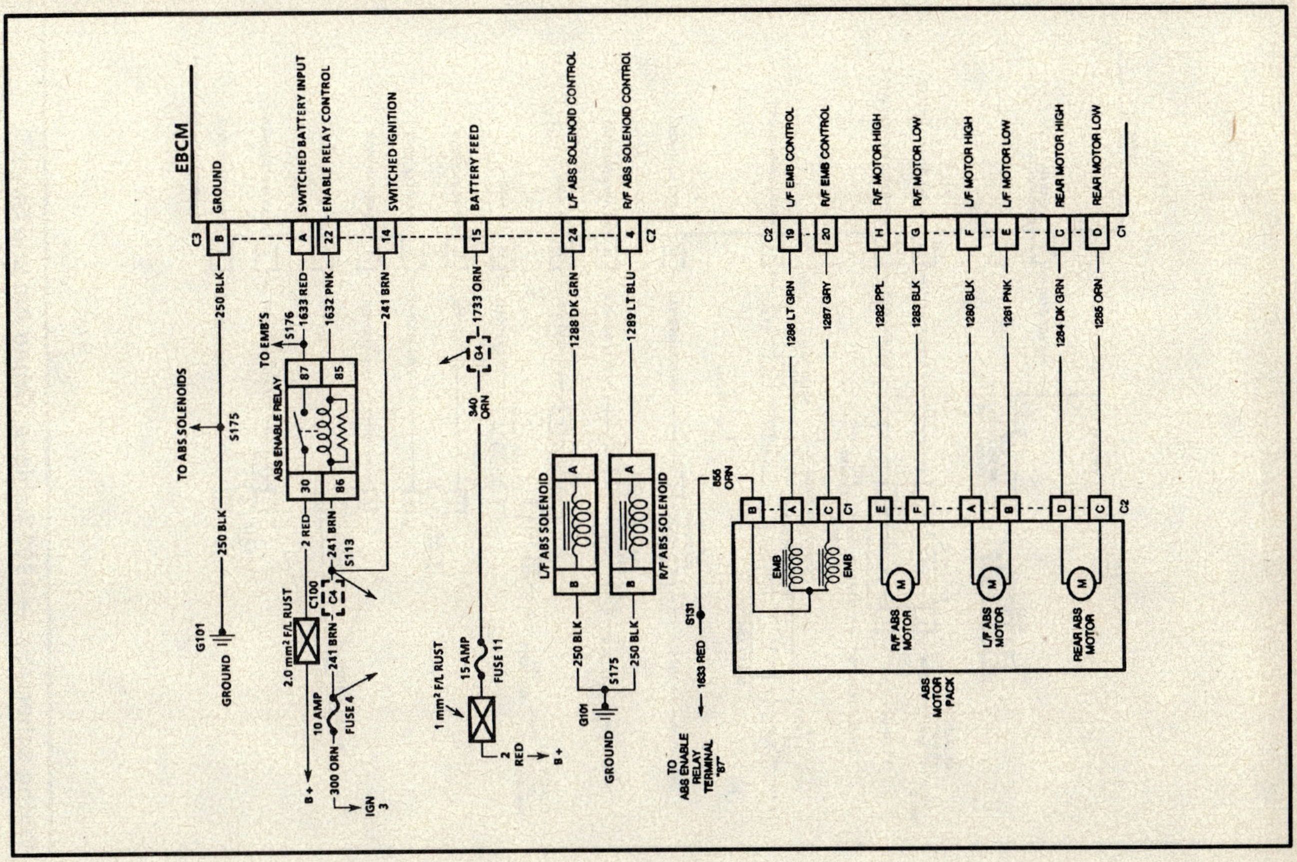

Wiring schematic — 1994 Beretta and Corsica

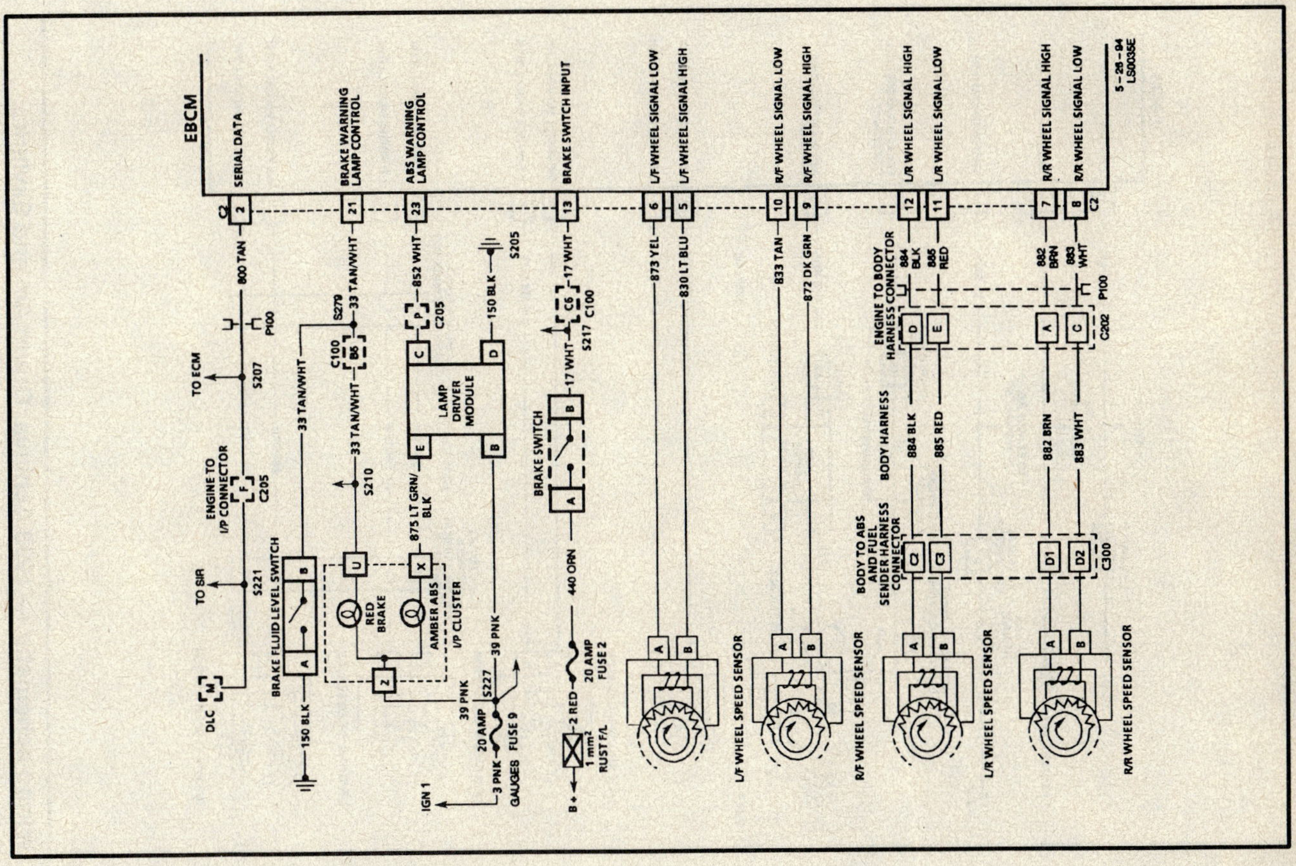

Wiring schematic — 1994–95 Beretta and Corsica

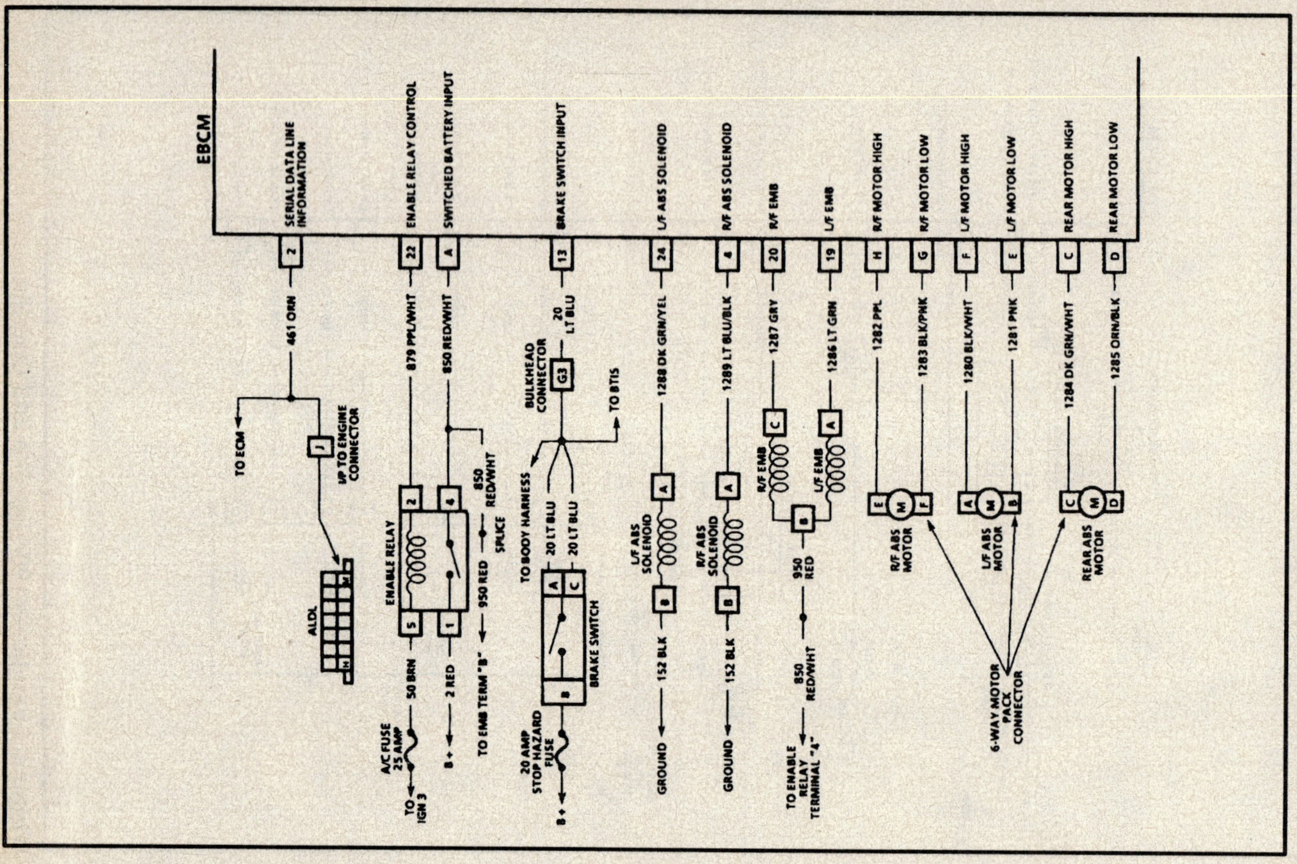

Wiring schematic — 1993 Achieva, Grand Am and Skylark

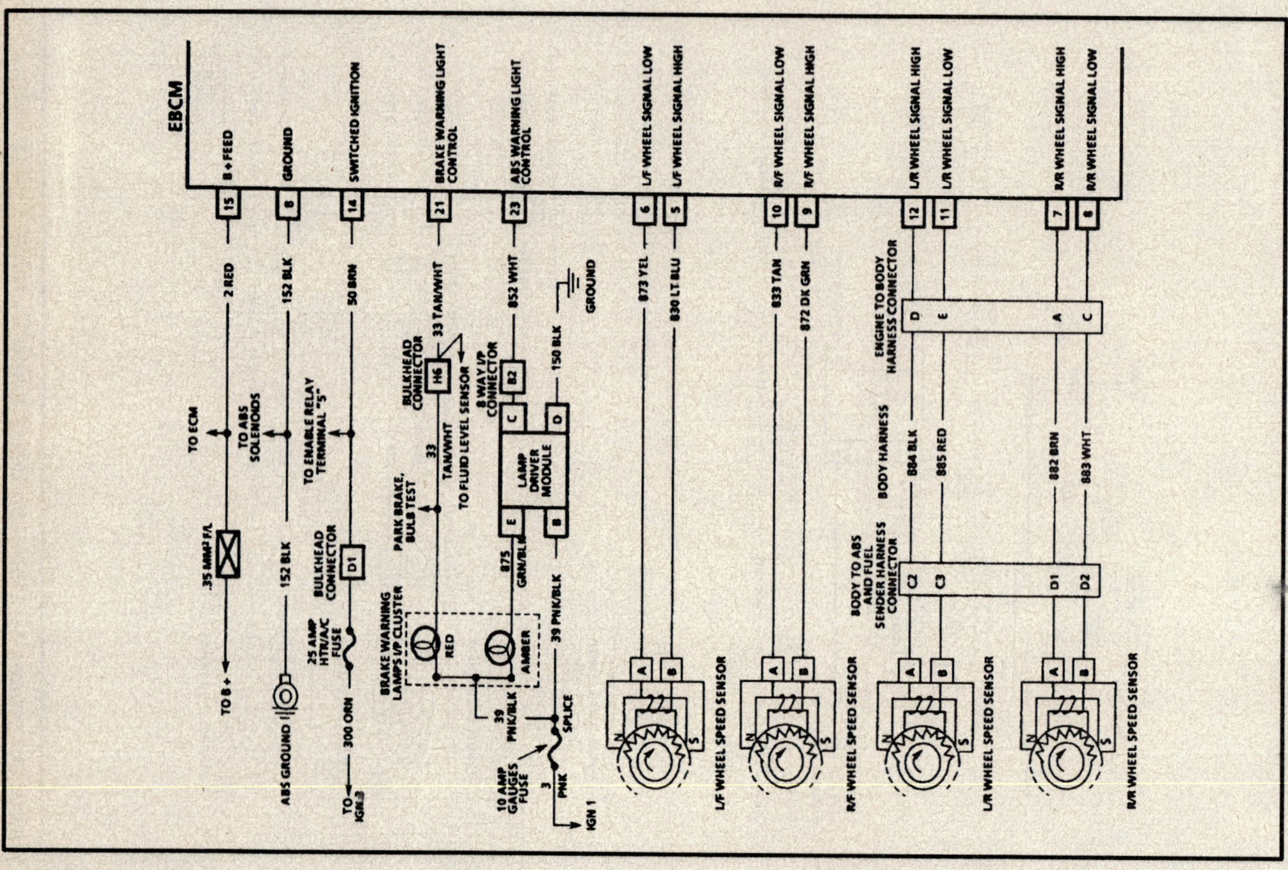

Wiring schematic — 1993 Achieva, Grand Am and Skylark

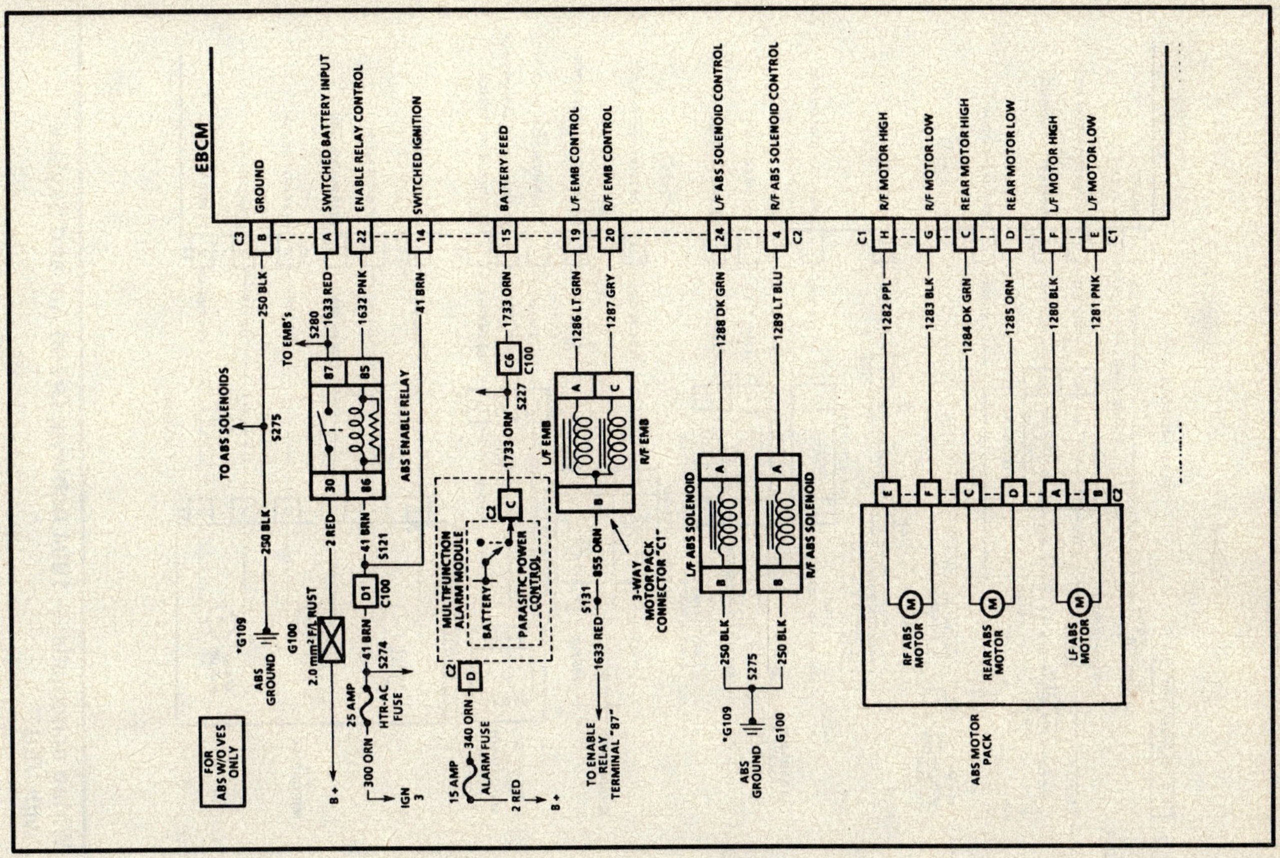

Wiring schematic — 1994 Achieva, Grand Am and Skylark without VES

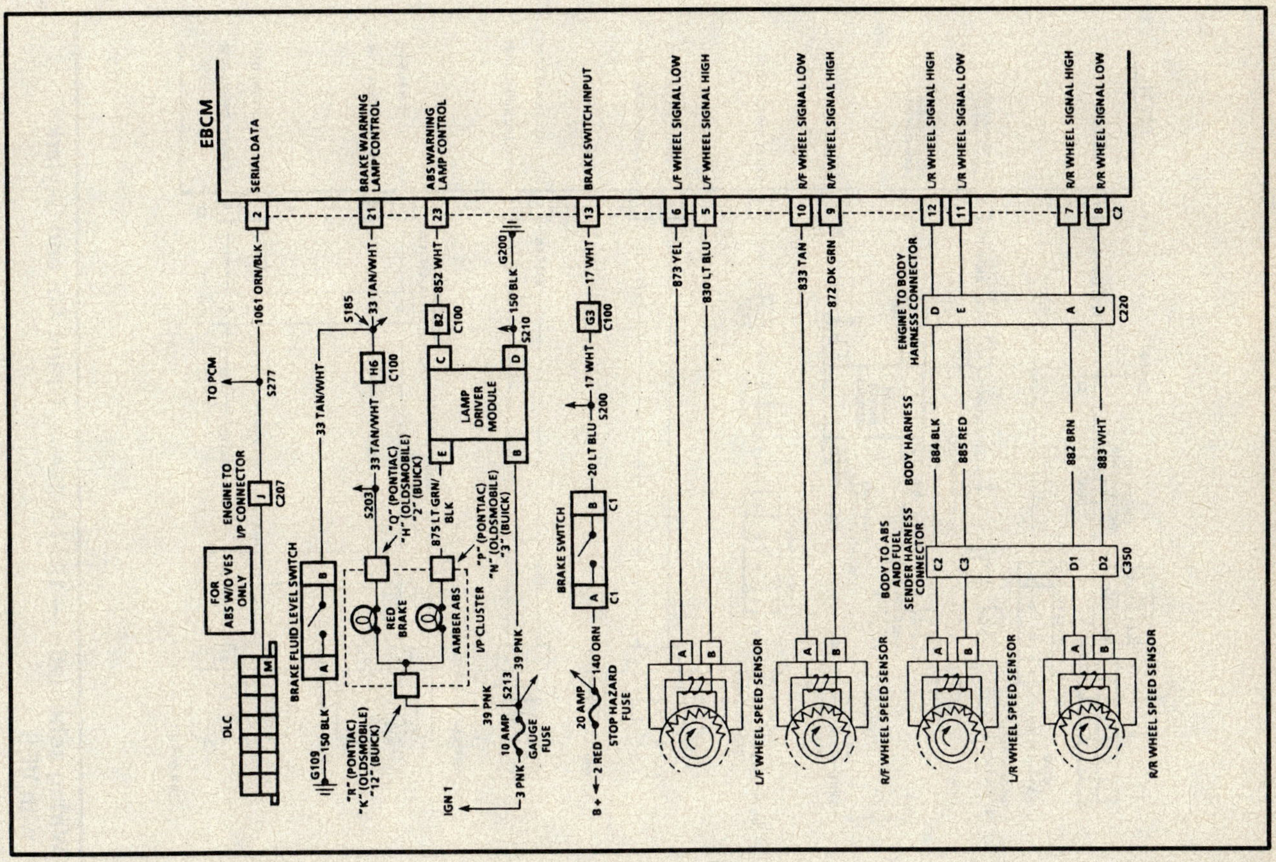

Wiring schematic — 1994 Achieva, Grand Am and Skylark without VES

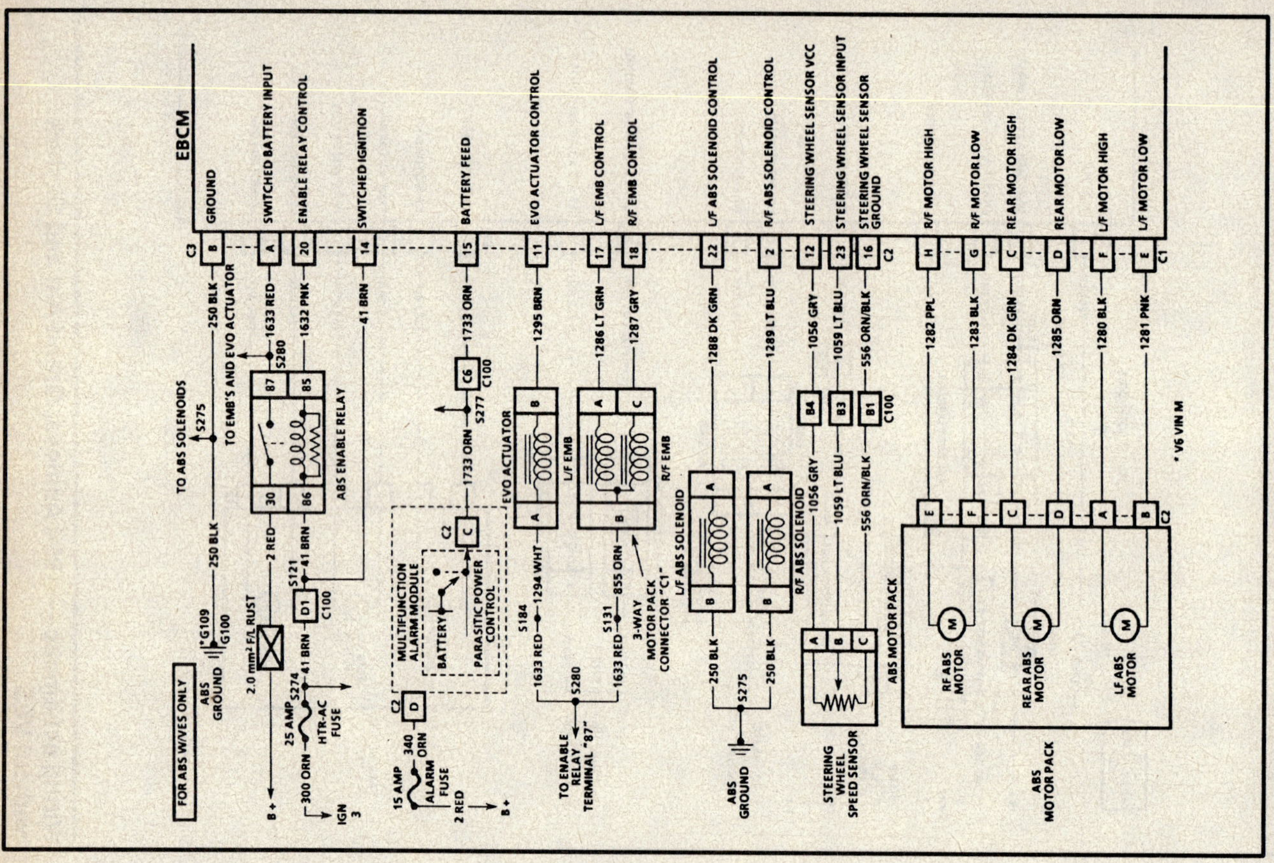

Wiring schematic — 1994 Achieva, Grand Am and Skylark with VES

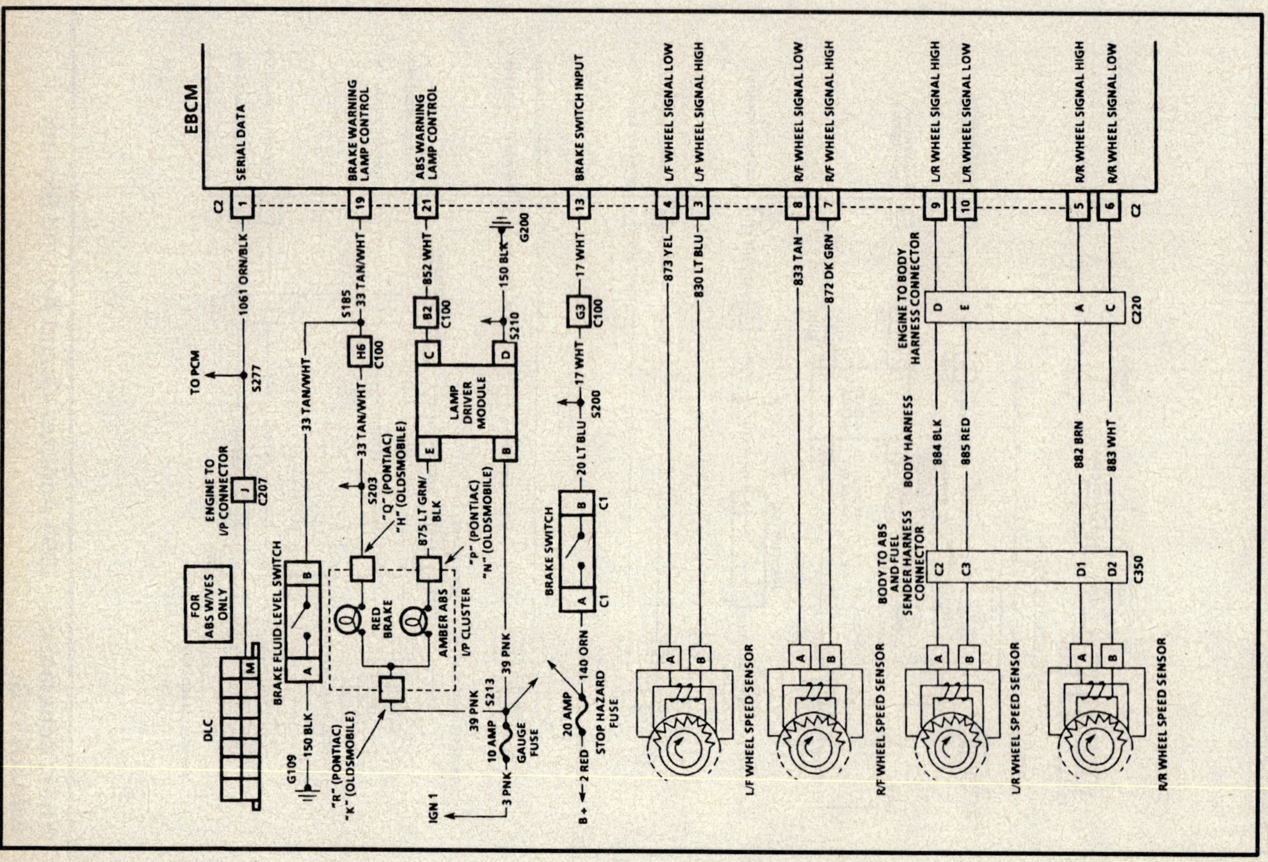

Wiring schematic — 1994 Achieva, Grand Am and Skylark with VES

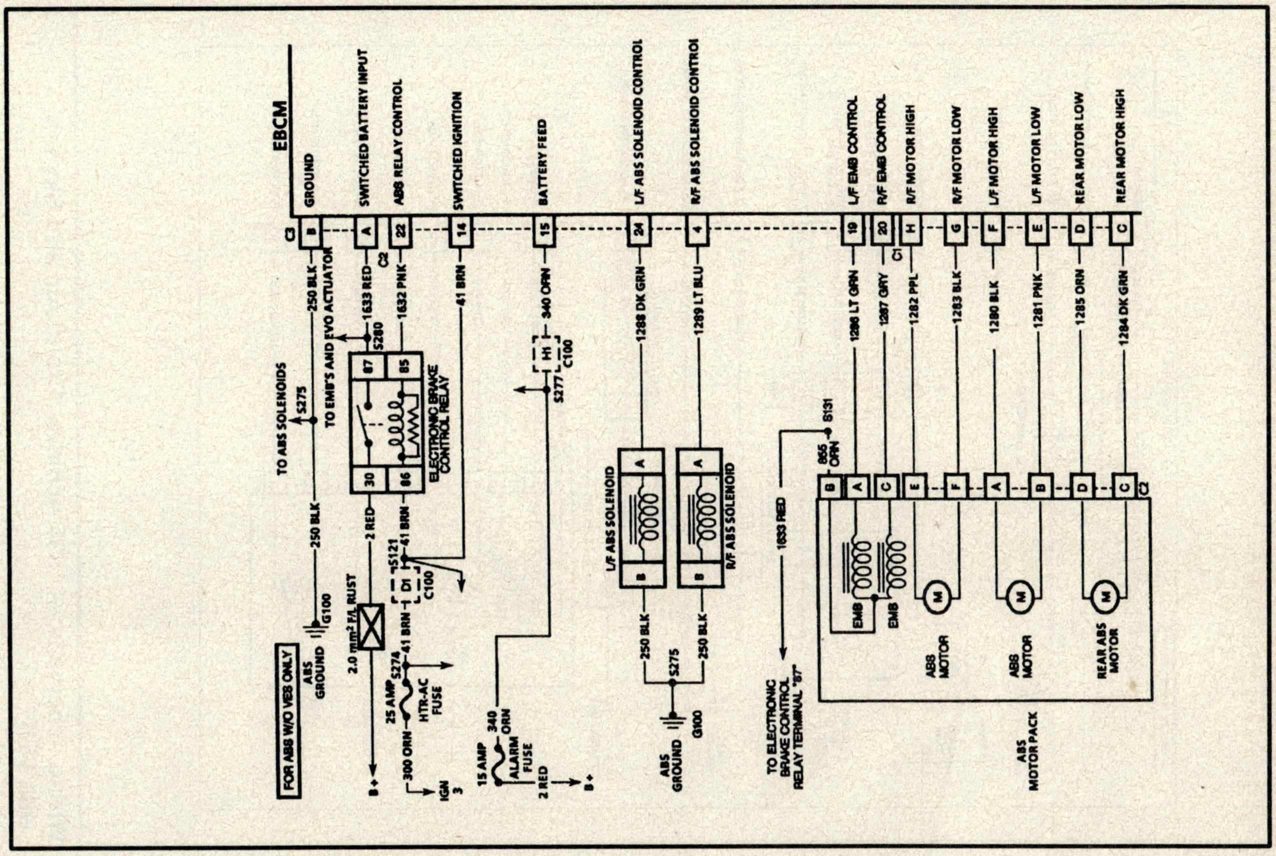

Wiring schematic — 1995 Achieva, Grand Am and Skylark without VES

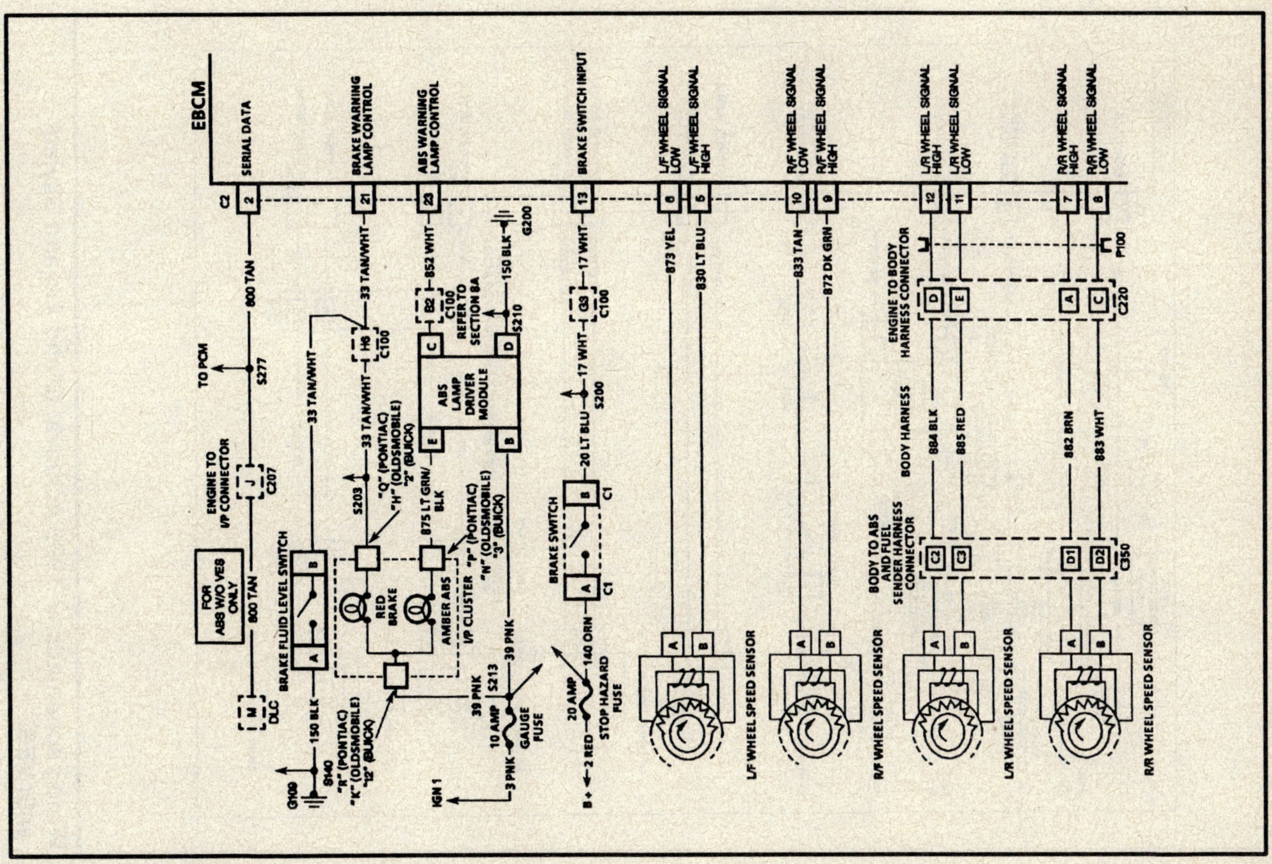

Wiring schematic — 1995 Achieva, Grand Am and Skylark without VES

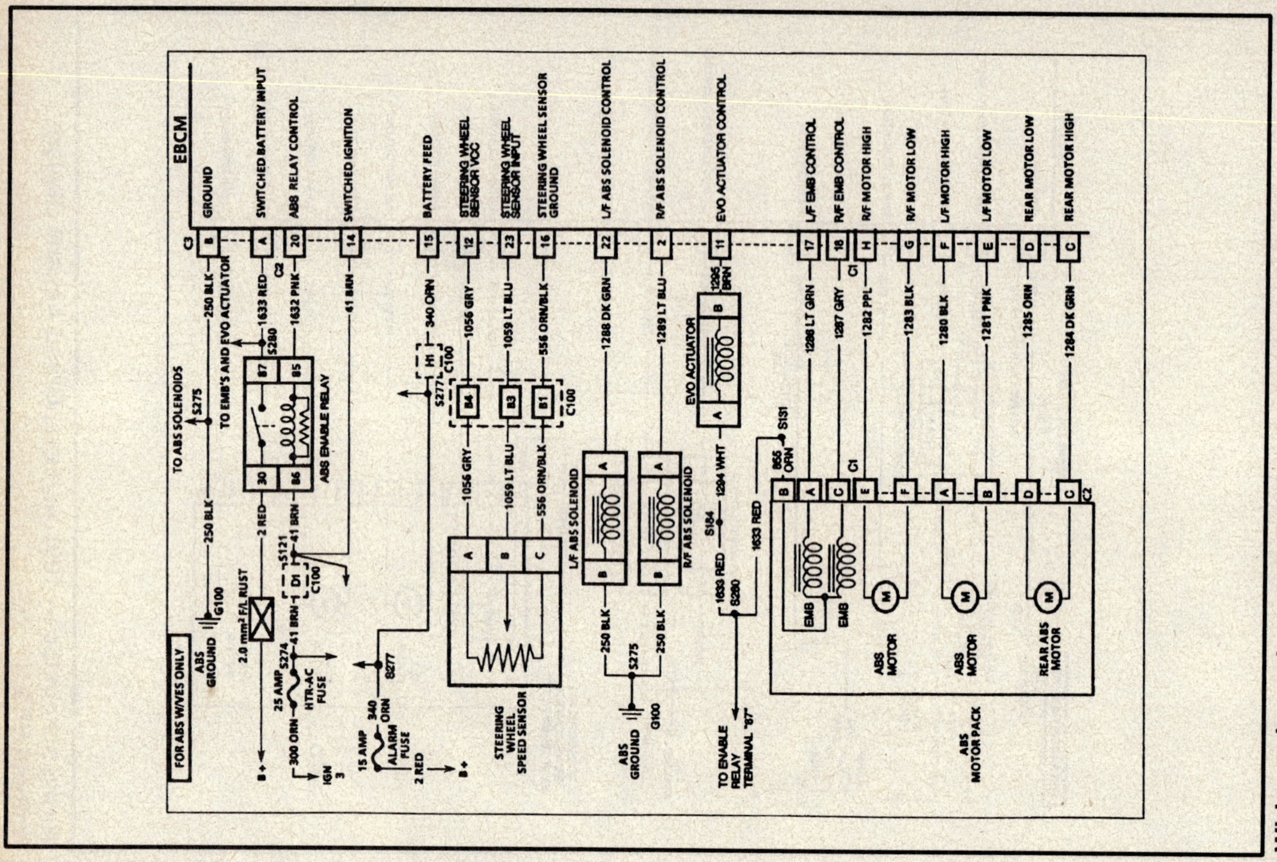

Wiring schematic — 1995 Achieva, Grand Am and Skylark with VES

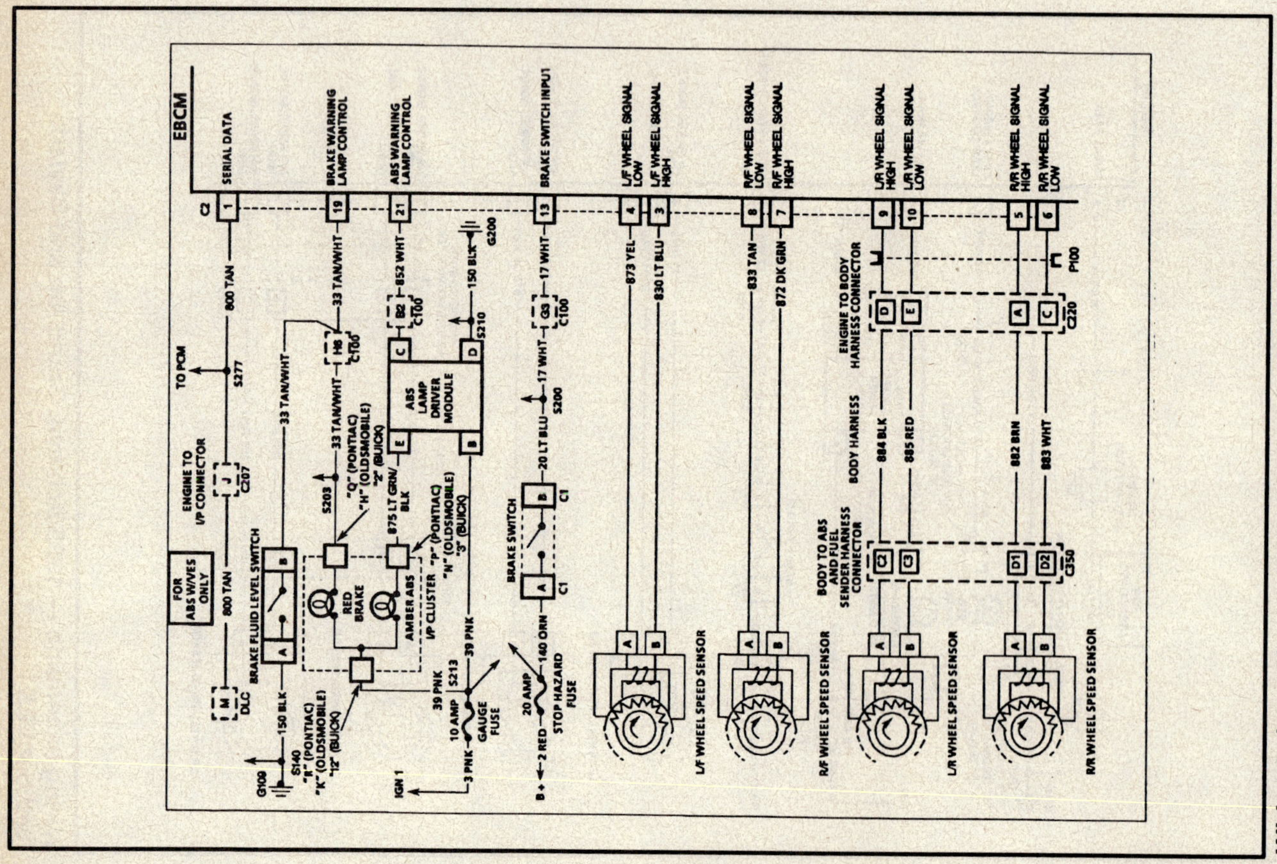

Wiring schematic — 1995 Achieva, Grand Am and Skylark with VES

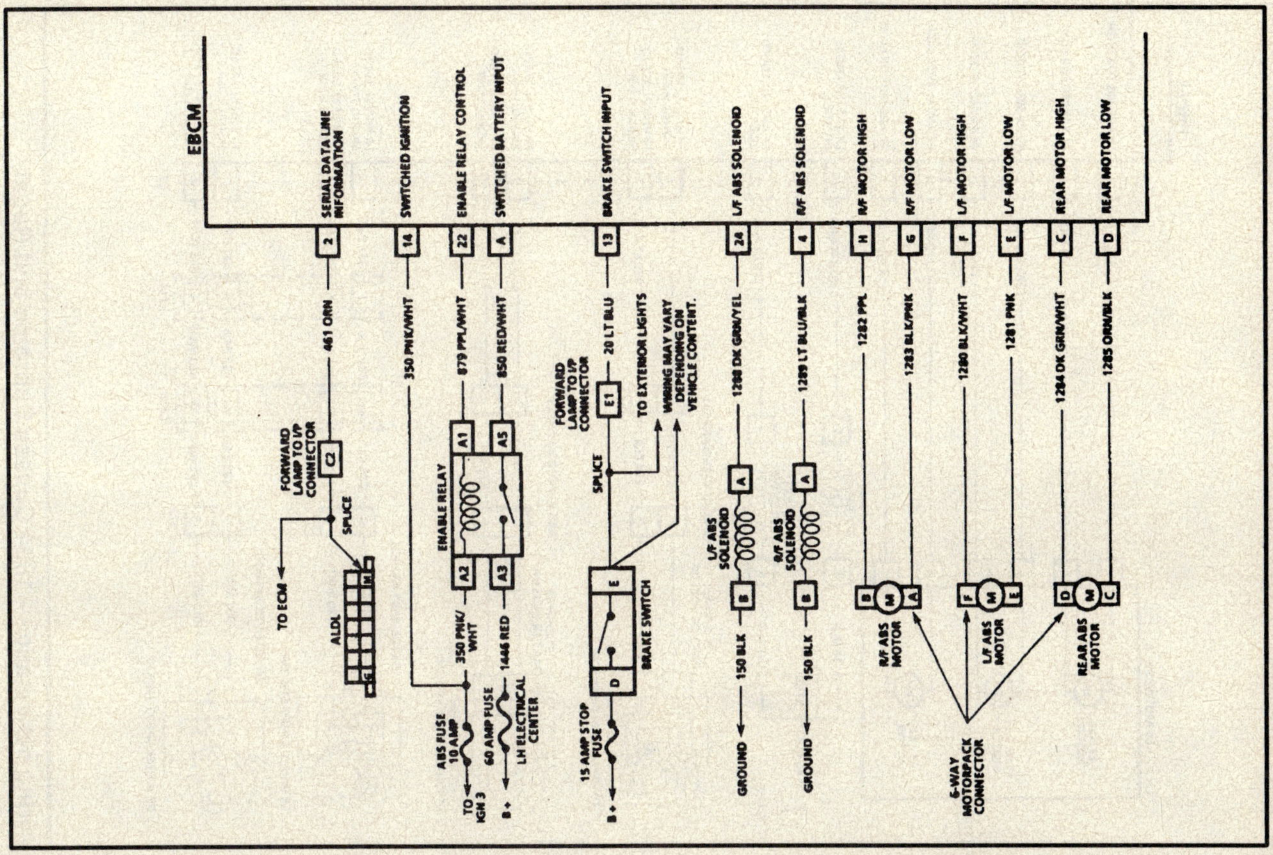

Wiring schematic — 1993 Cutlass Supreme, Grand Prix, Lumina and Regal

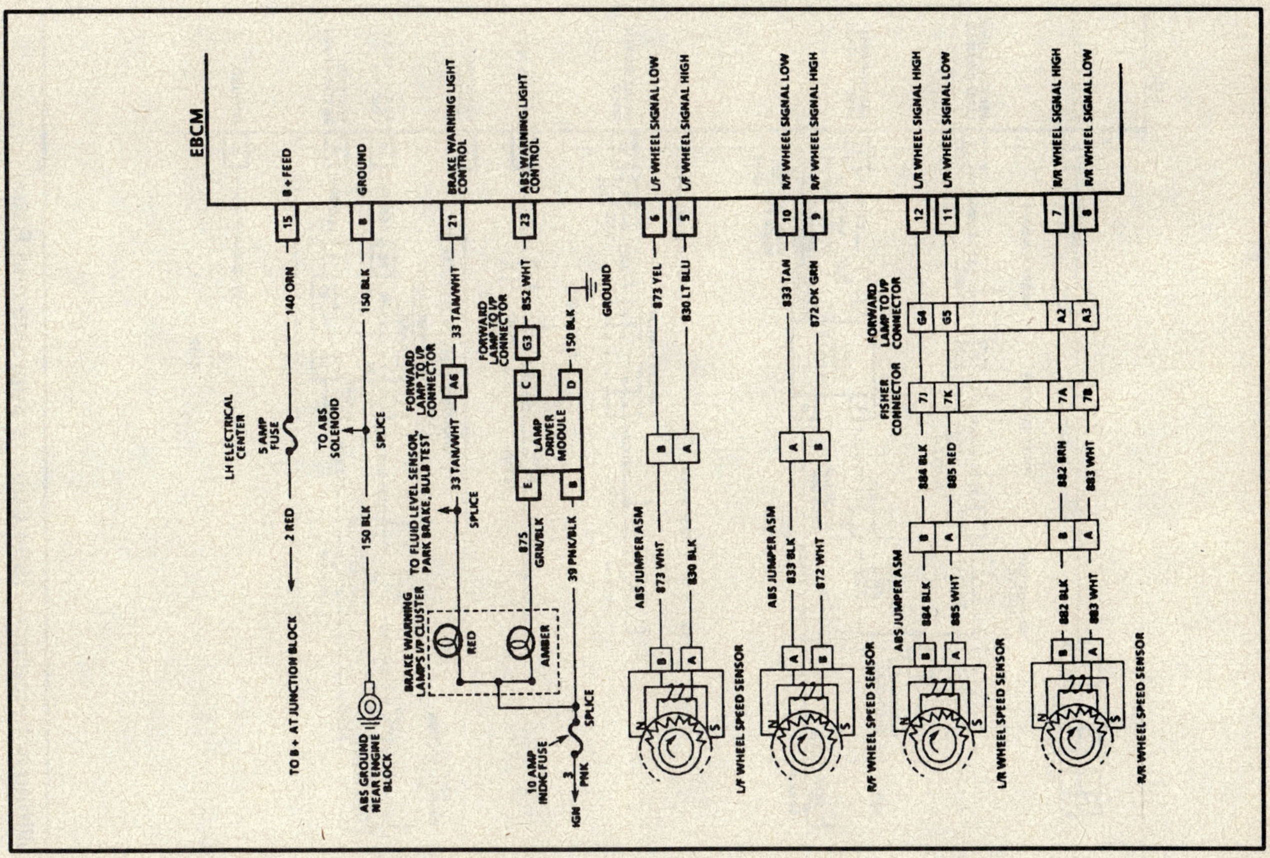

Wiring schematic — 1993 Cutlass Supreme, Grand Prix, Lumina and Regal

DELCO 6 ANTI-LOCK BRAKE SYSTEM WITH TCS

GRAND PRIX • LUMINA • LUMINA APV • MONTE CARLO • REGAL • SILHOUETTE • SKYLARK • SUNBIRD • SUNFIRE • TRANS SPORT

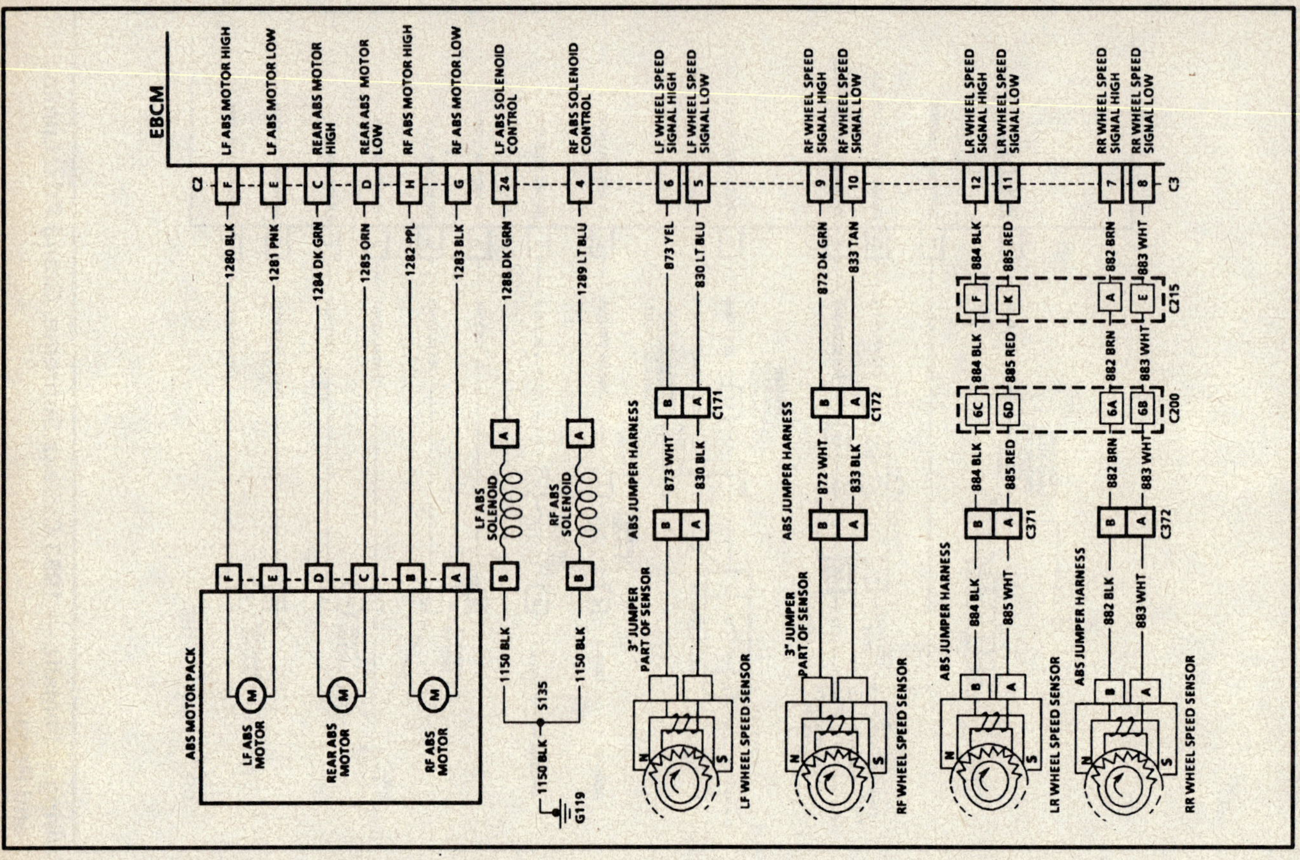

Wiring schematic — 1994 Cutlass Supreme and Regal

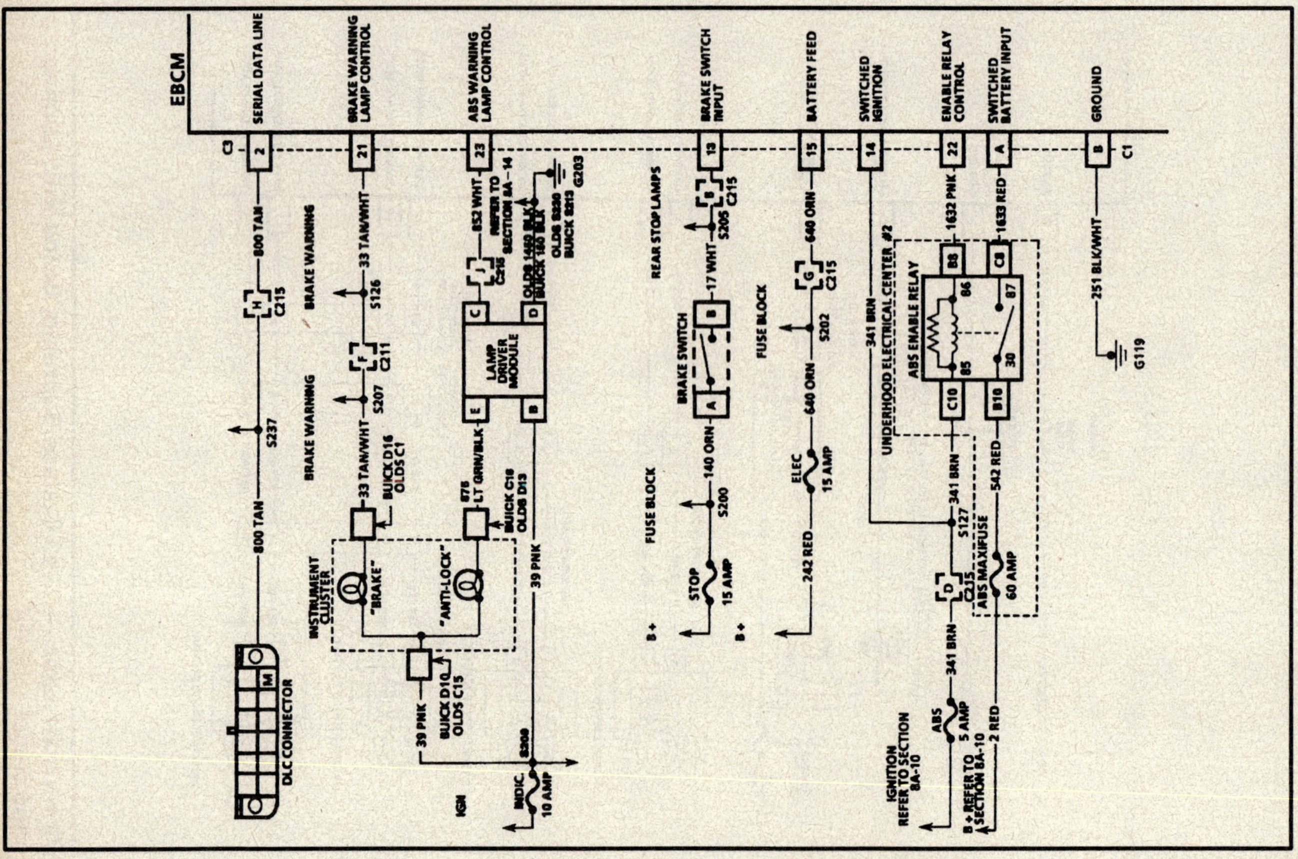

Wiring schematic — 1994 Cutlass Supreme and Regal

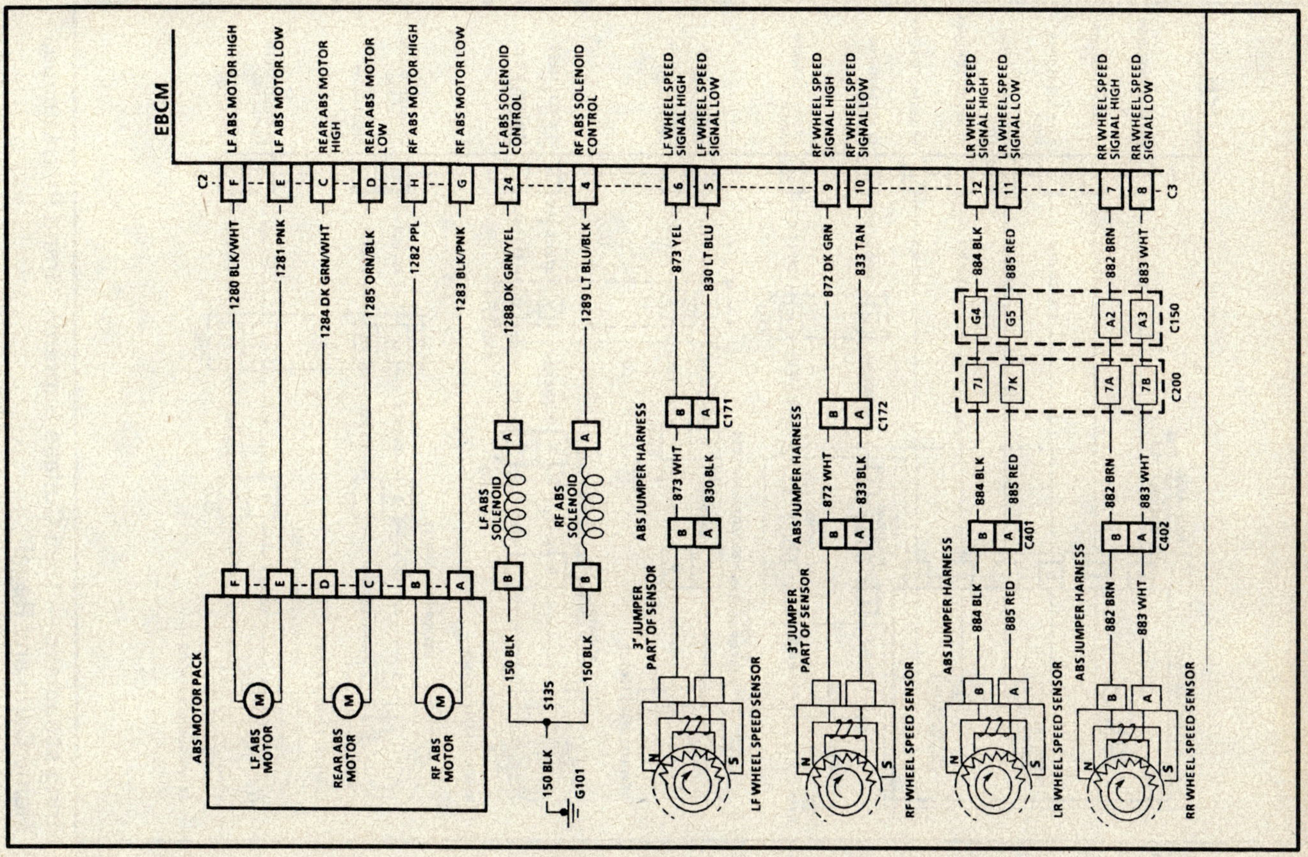

Wiring schematic — 1994 Lumina

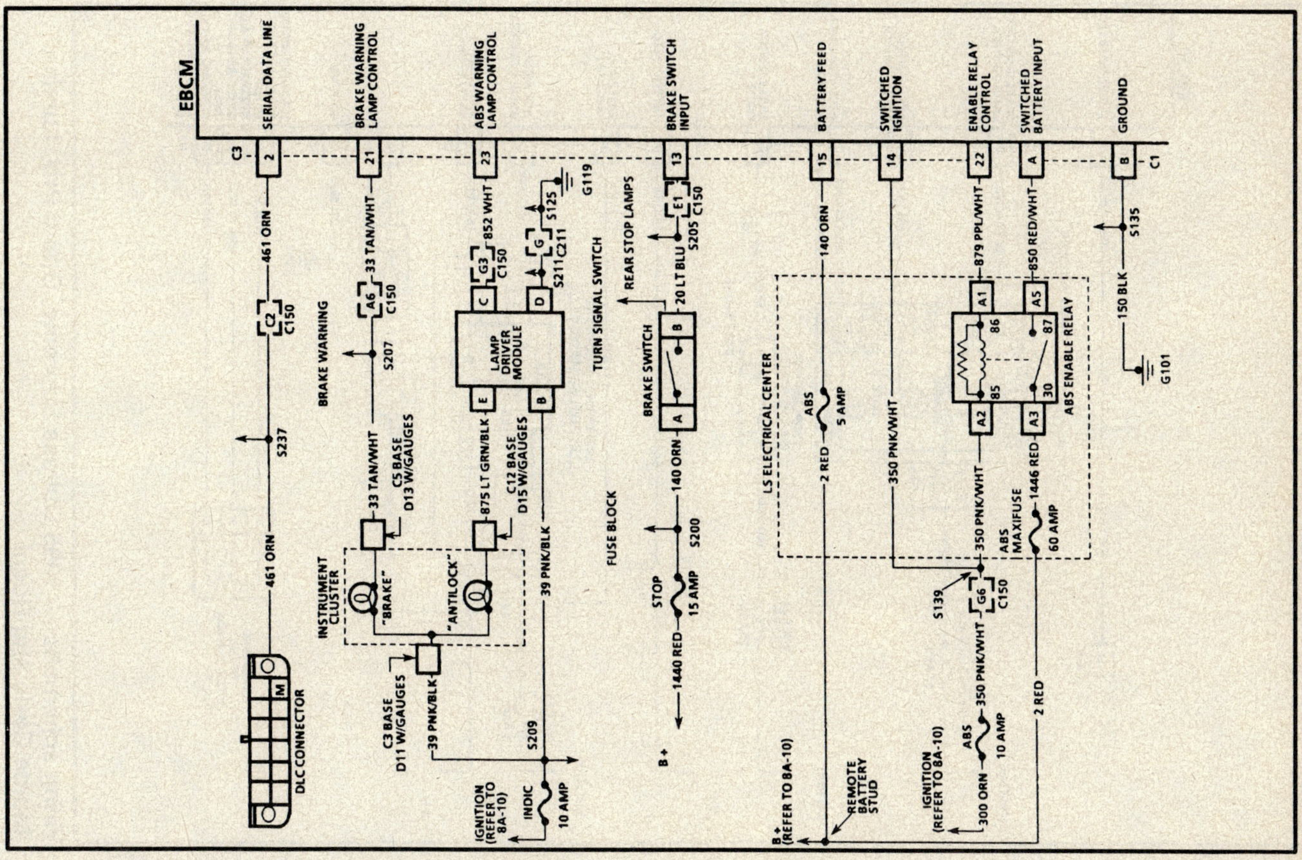

Wiring schematic — 1994 Lumina

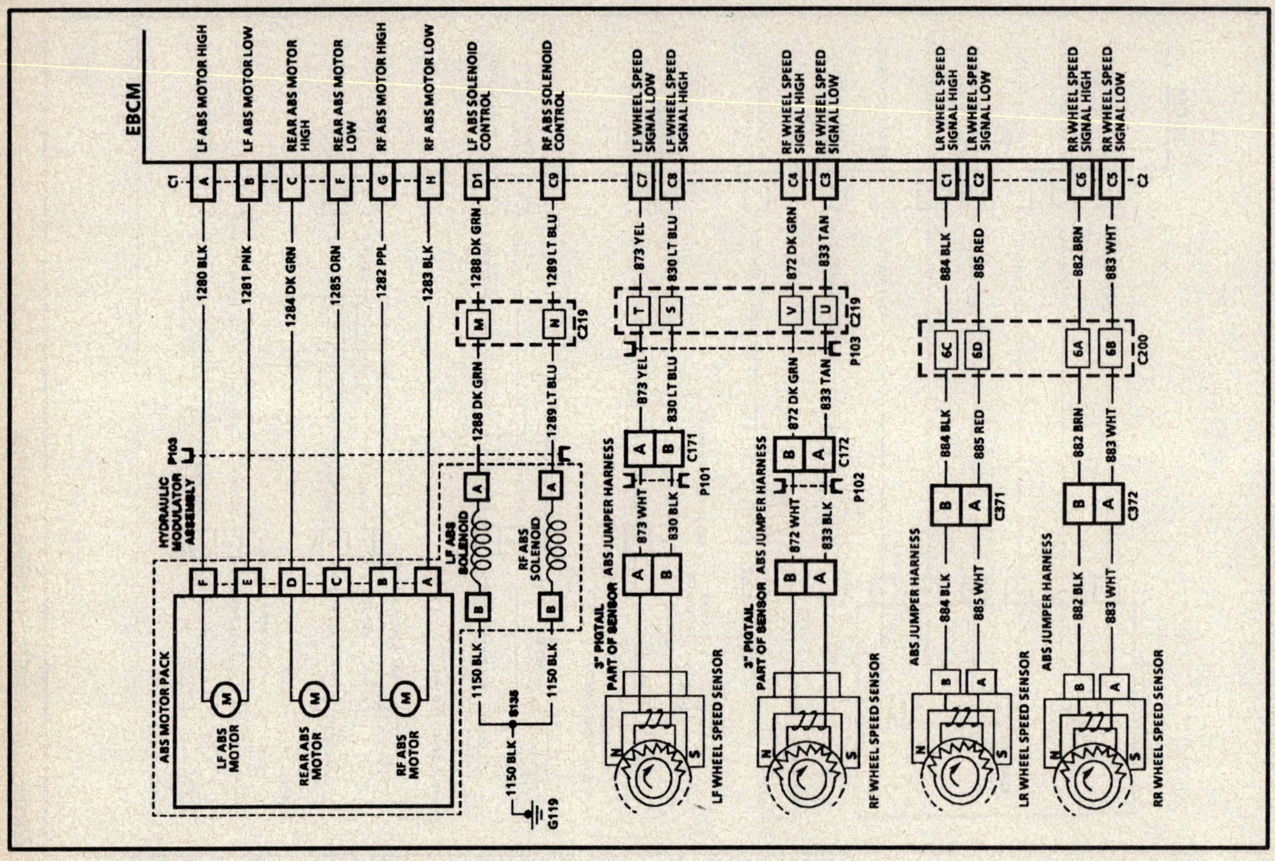

Wiring schematic — 1995 Cutlass Supreme, Grand Prix, Lumina, Monte Carlo and Regal

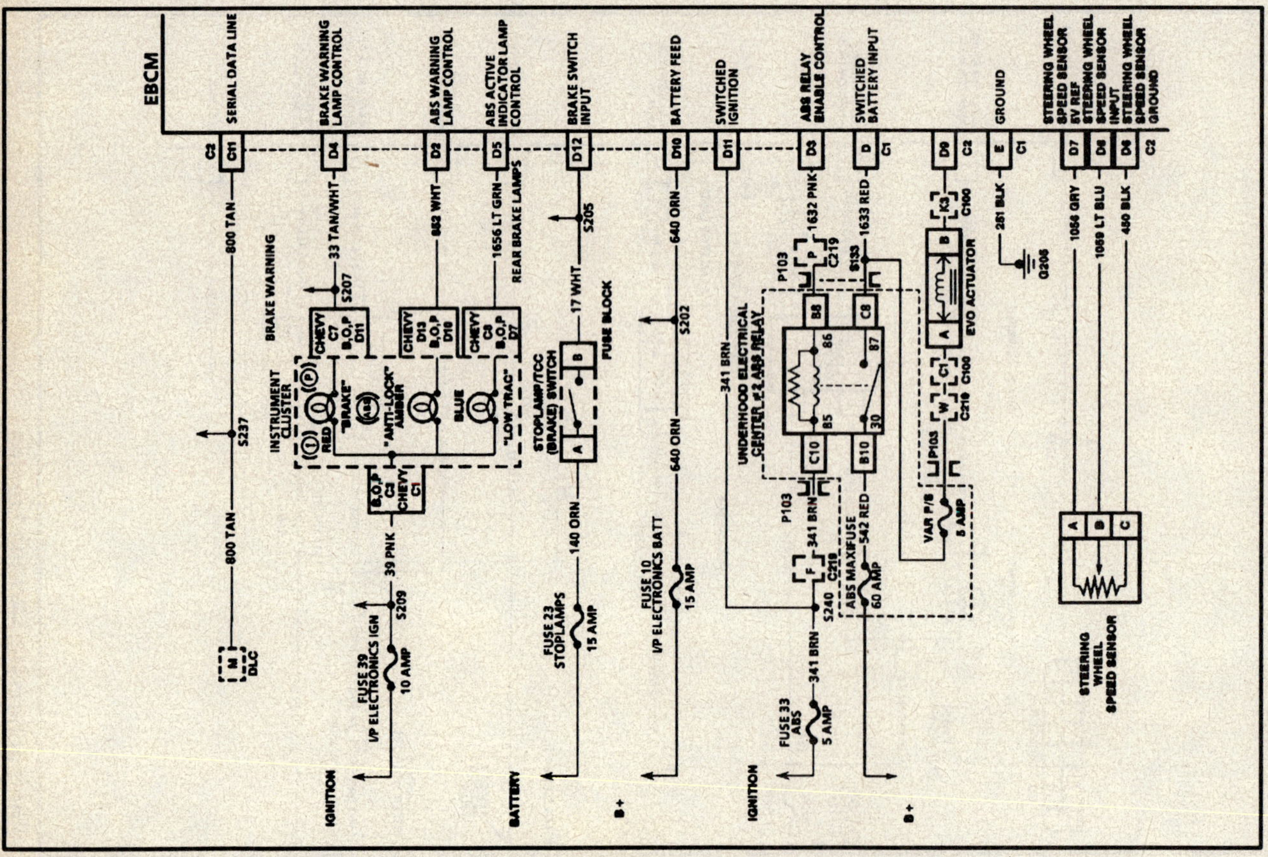

Wiring schematic — 1995 Cutlass Supreme, Grand Prix, Lumina, Monte Carlo and Regal

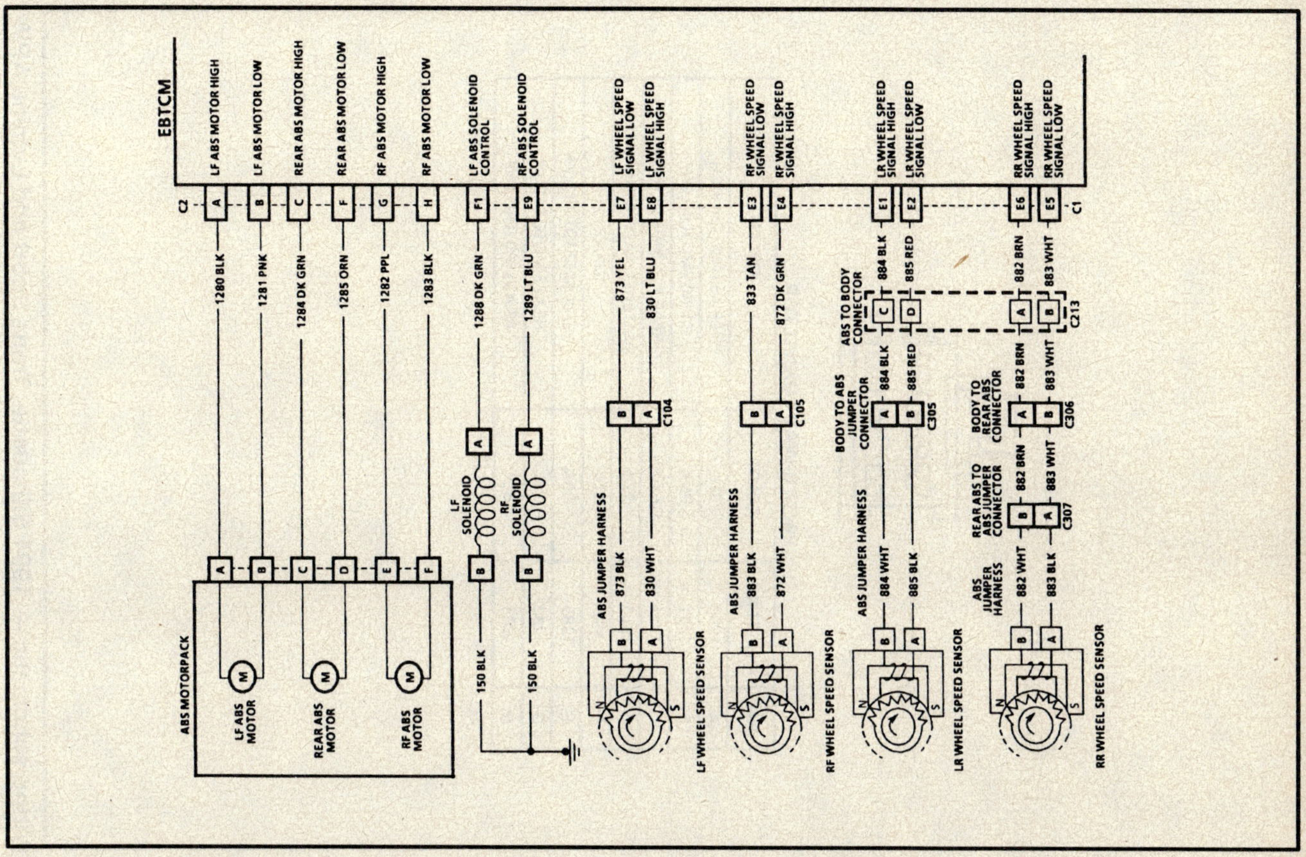

Wiring schematic — 1994 Silhouette, Trans Sport and Lumina APV

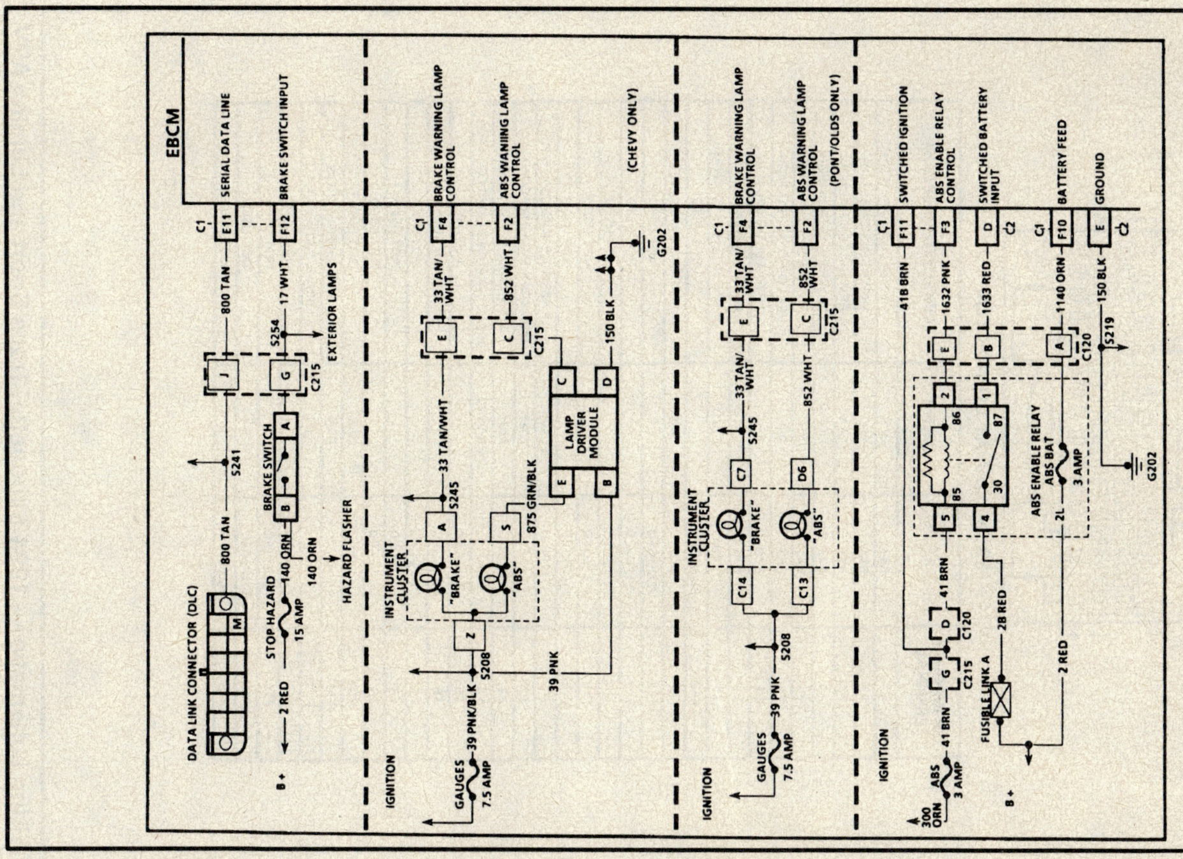

Wiring schematic — 1994 Silhouette, Trans Sport and Lumina APV

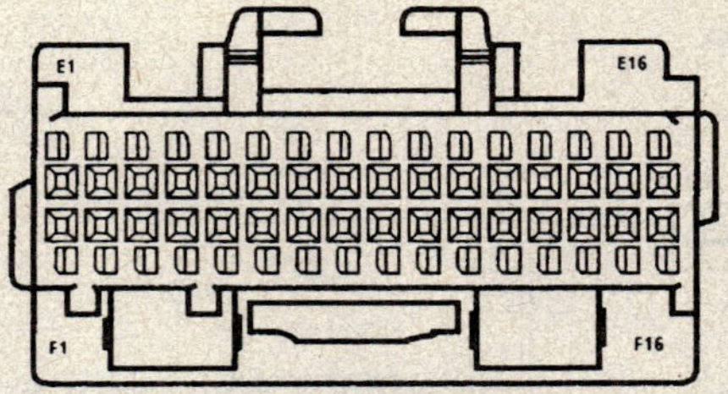

EBCM 32-WAY CONNECTOR

CONNECTOR TERMINAL	CIRCUIT NO.	COLOR	CIRCUIT DESCRIPTION
E1	884	BLK	LR WHEEL SIGNAL HIGH
E2	885	RED	LR WHEEL SIGNAL LOW
E3	833	TAN	RF WHEEL SIGNAL LOW
E4	872	DK GRN	RF WHEEL SIGNAL HIGH
E5	883	WHT	RR WHEEL SIGNAL LOW
E6	882	BRN	RR WHEEL SIGNAL HIGH
E7	873	YEL	LF WHEEL SIGNAL LOW
E8	830	LT BLU	LF WHEEL SIGNAL HIGH
E9	1289	LT BLU	RF ABS SOLENOID CONTROL
E10	PLUGGED		NOT USED
E11	800	TAN	SERIAL DATA LINE
E12	PLUGGED		NOT USED
E13	PLUGGED		NOT USED
E14	PLUGGED		NOT USED
E15	PLUGGED		NOT USED
E16	PLUGGED		NOT USED
F1	1288	DK GRN	LF ABS SOLENOID CONTROL
F2	852	WHT	ABS WARNING LAMP CONTROL
F3	1632	PNK	ABS ENABLE RELAY CONTROL
F4	33	TAN/WHT	BRAKE WARNING LAMP CONTROL
F5	PLUGGED		NOT USED
F6	PLUGGED		NOT USED
F7	PLUGGED		NOT USED
F8	PLUGGED		NOT USED
F9	PLUGGED		NOT USED
F10	1040	ORN	BATTERY
F11	41	BRN	SWITCHED IGNITION
F12	17	WHT	BRAKE SWITCH INPUT
F13	PLUGGED		NOT USED
F14	PLUGGED		NOT USED
F15	PLUGGED		NOT USED
F16	PLUGGED		NOT USED

Wiring schematic — 1994 Silhouette, Trans Sport and Lumina APV

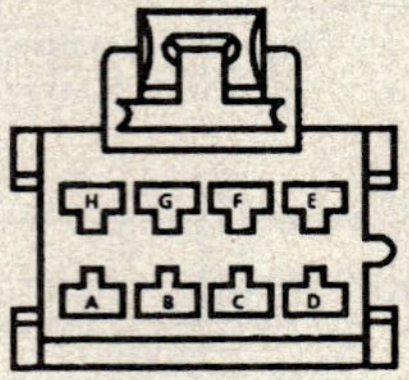

EBCM 8-WAY CONNECTOR

PIN	CIRCUIT NO.	COLOR	CIRCUIT
A	1280	BLK	LF ABS MOTOR HIGH
B	1281	PNK	LF ABS MOTOR LOW
C	1284	DK GRN	REAR ABS MOTOR HIGH
D	1633	RED	SWITCHED BATTERY INPUT
E	150	BLK	GROUND
F	1285	ORN	REAR ABS MOTOR LOW
G	1282	PPL	RF ABS MOTOR HIGH
H	1283	BLK	RF ABS MOTOR LOW

Wiring schematic — 1994 Silhouette, Trans Sport and Lumina APV

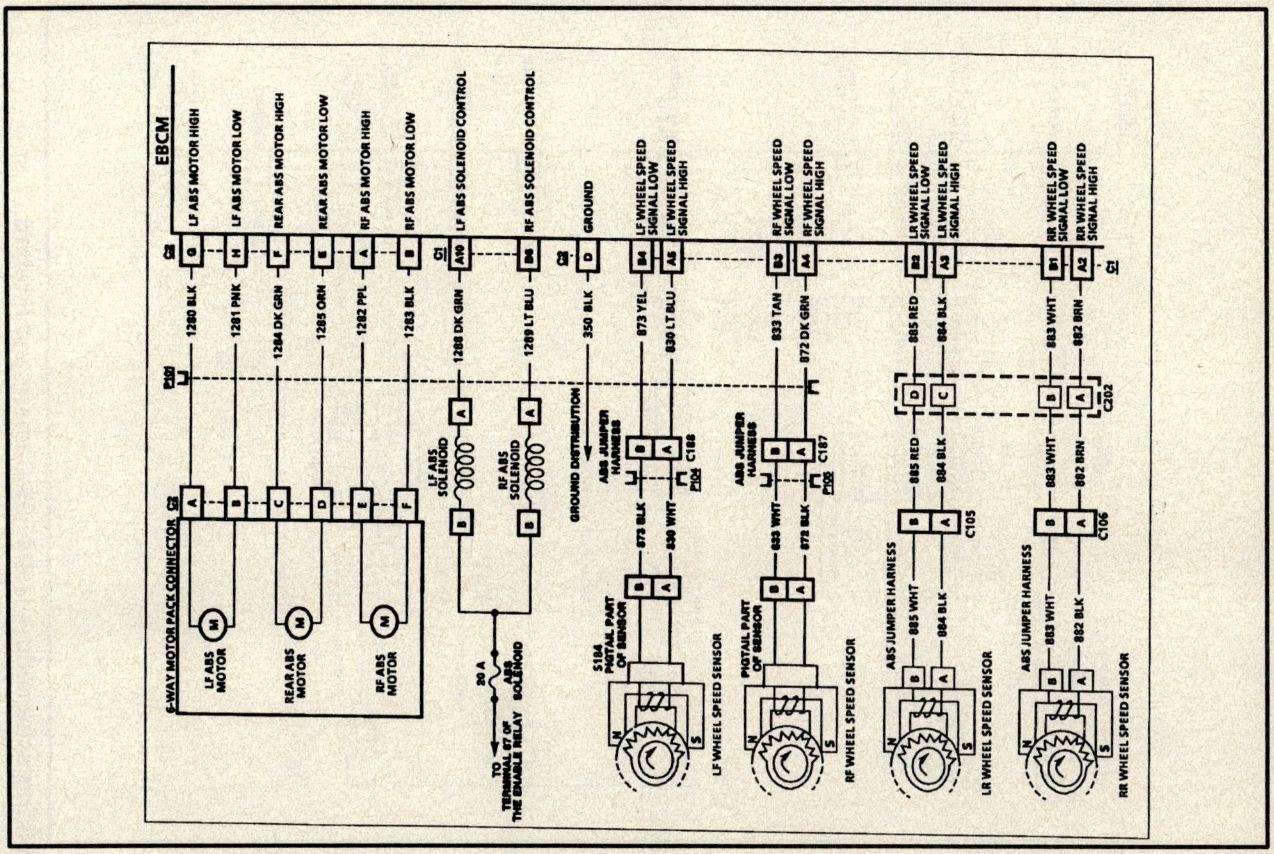

Wiring schematic — 1995 Century, Cutlass Ciera and Cutlass Cruiser

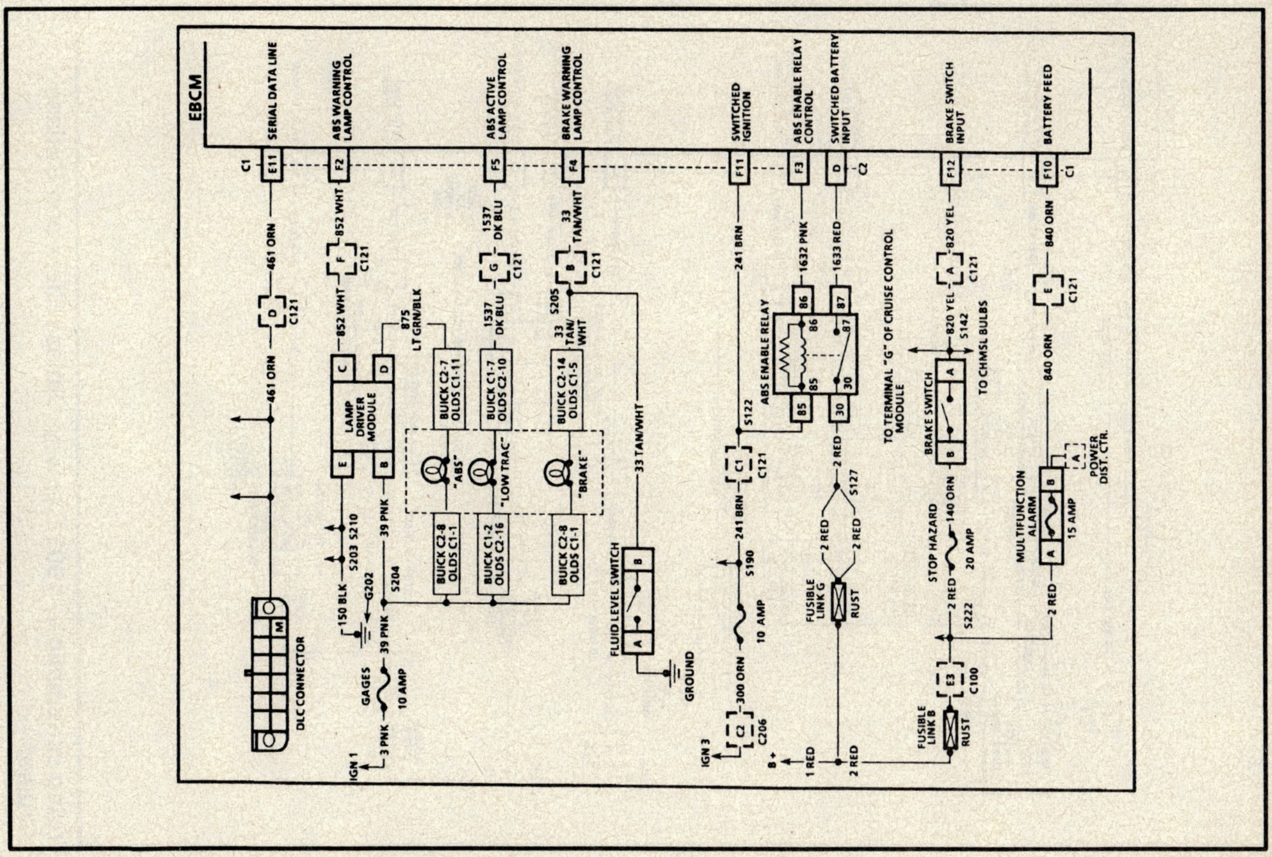

Wiring schematic — 1994 Century, Cutlass Ciera and Cutlass Cruiser

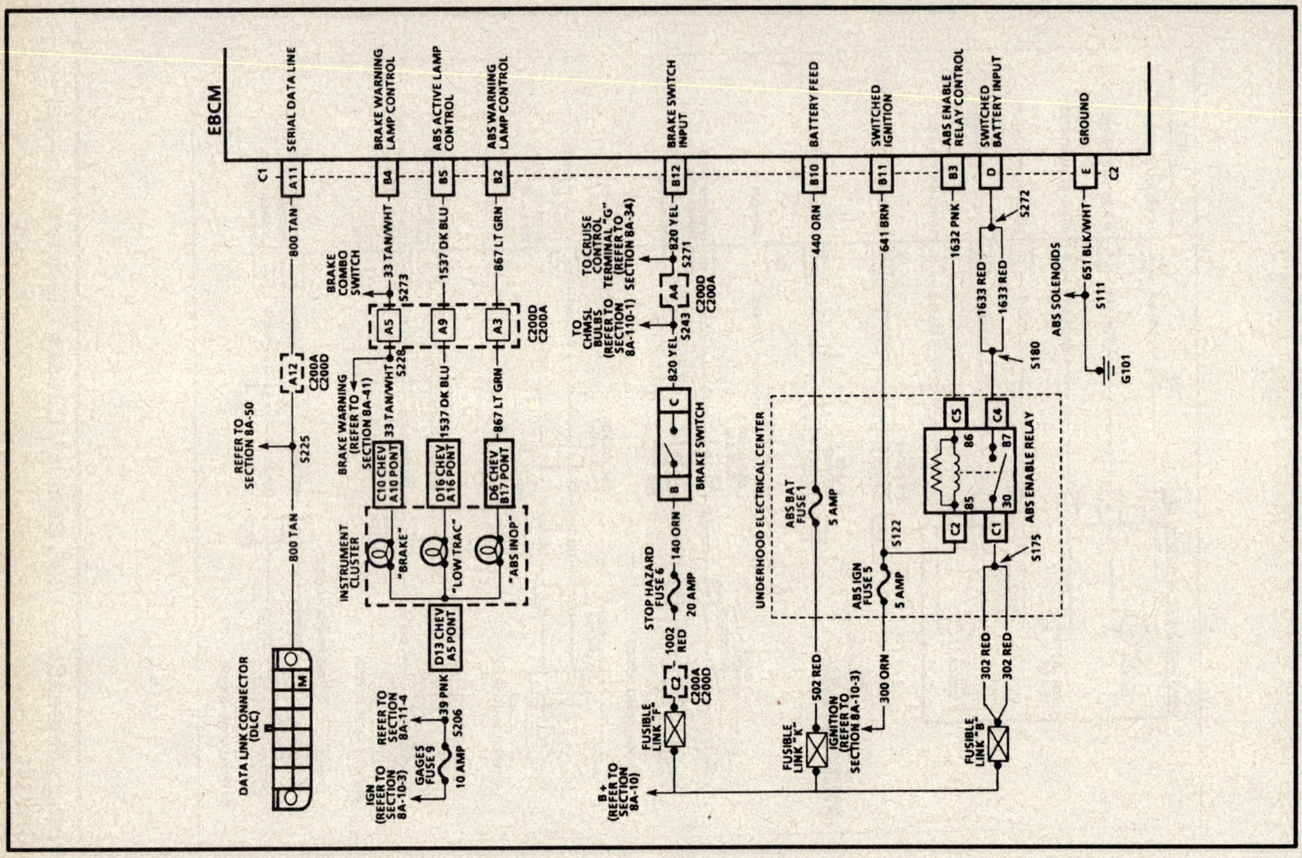

Wiring schematic — 1994–95 Camaro and Firebird

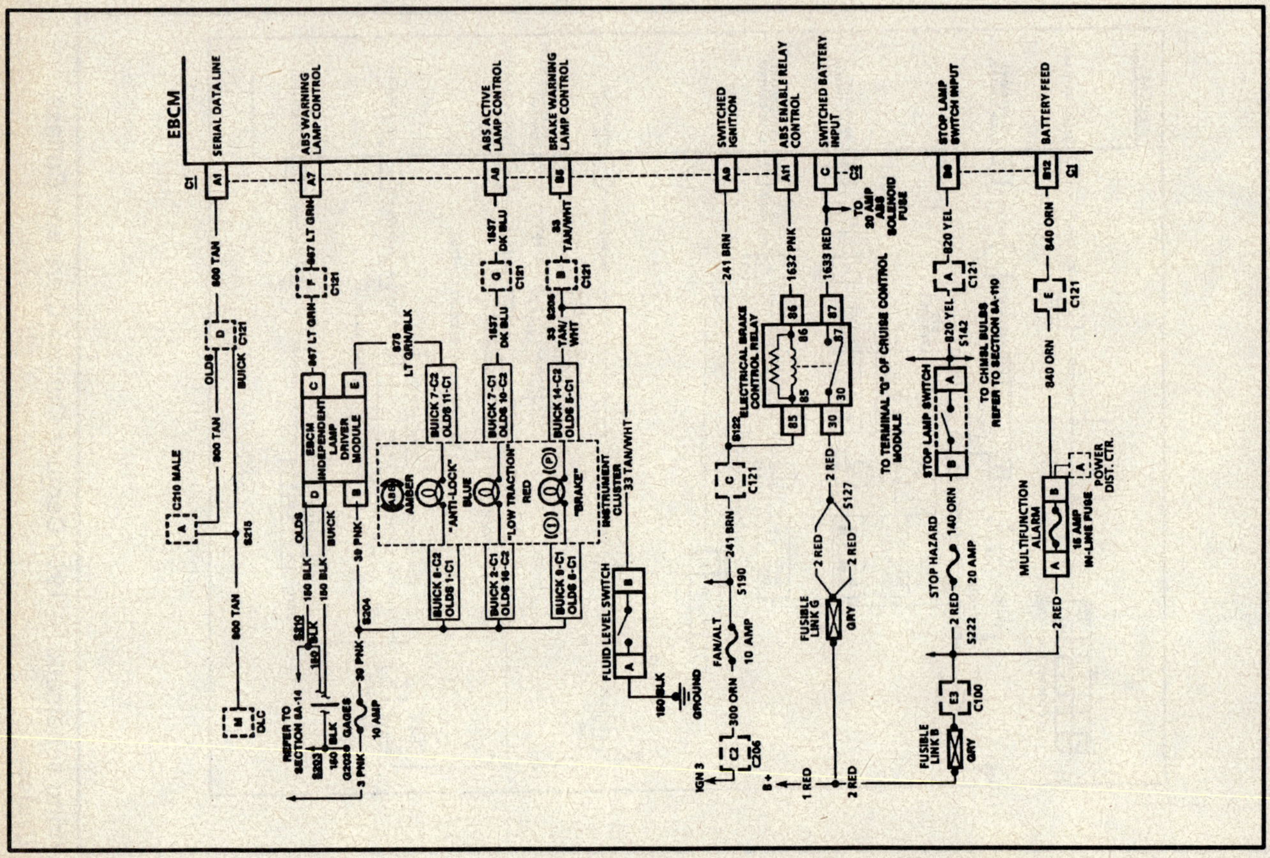

Wiring schematic — 1995 Century, Cutlass Ciera and Cutlass Cruiser

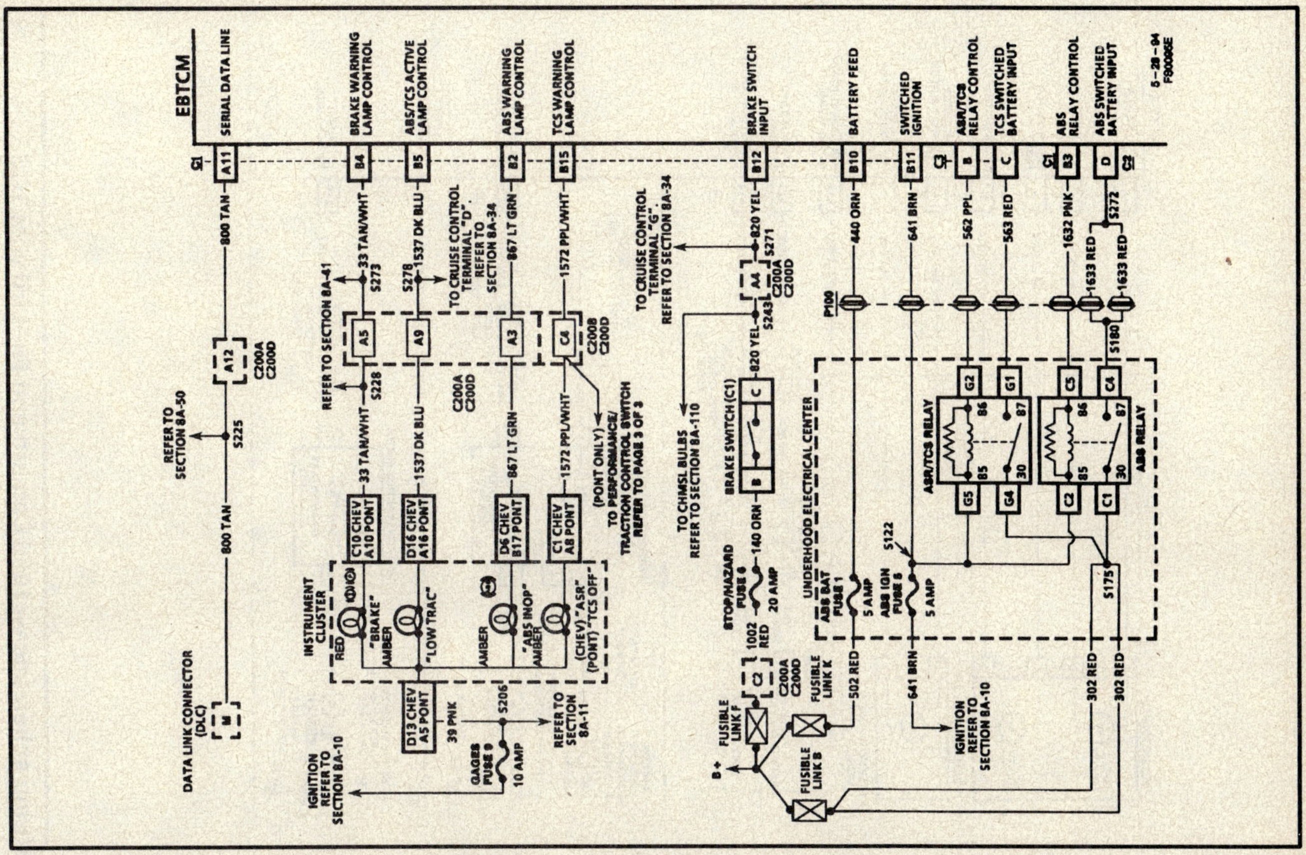

Wiring schematic — 1995 Camaro and Firebird with TCS

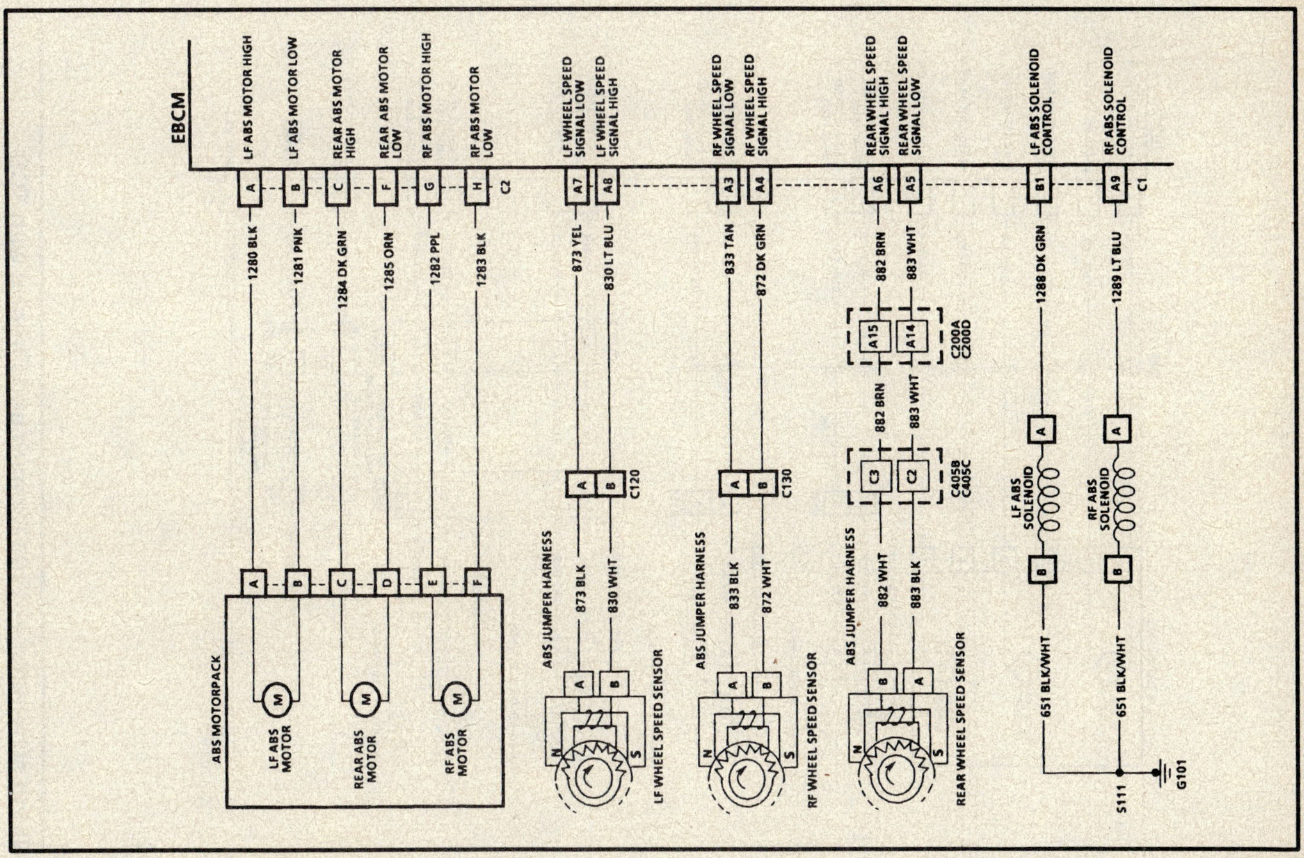

Wiring schematic — 1994–95 Camaro and Firebird

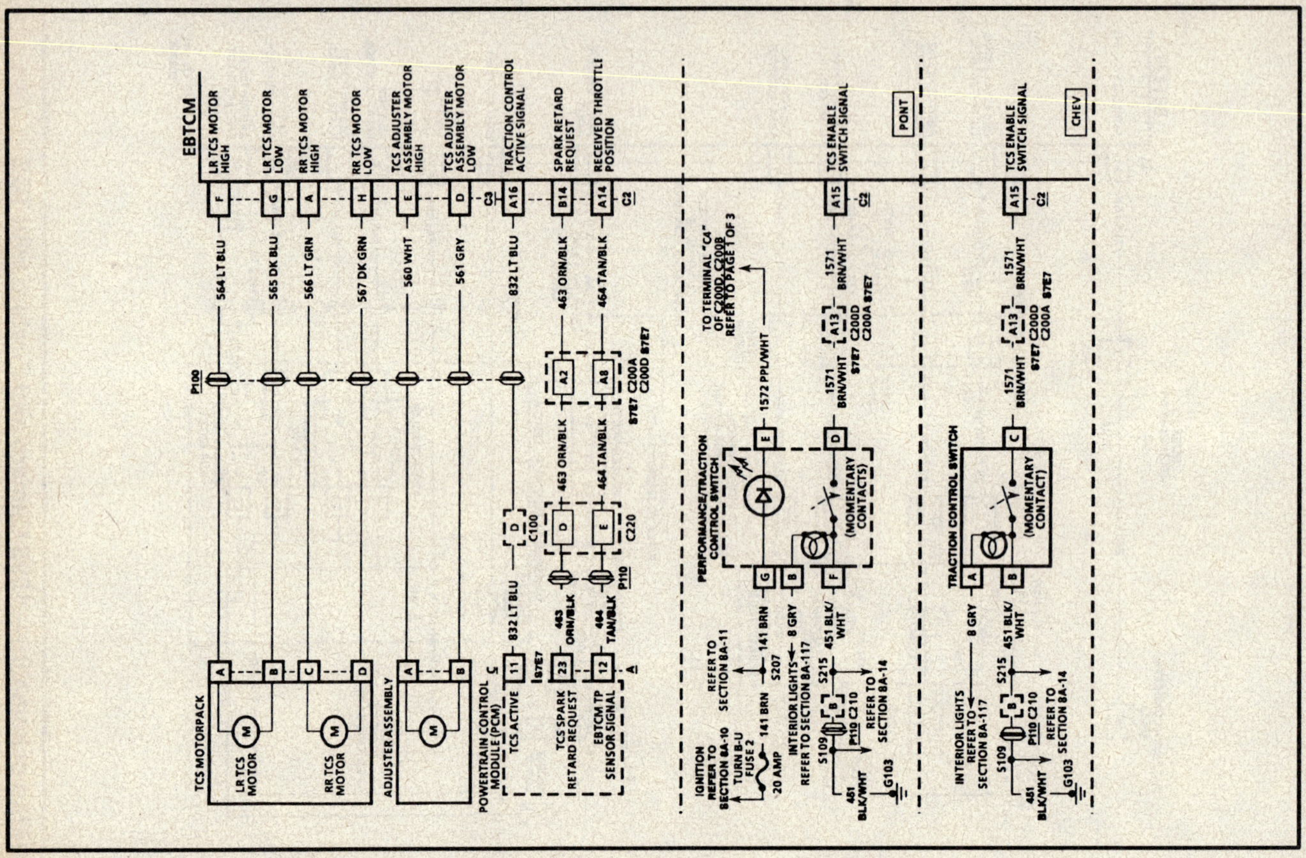

Wiring schematic — 1995 Camaro and Firebird with TCS

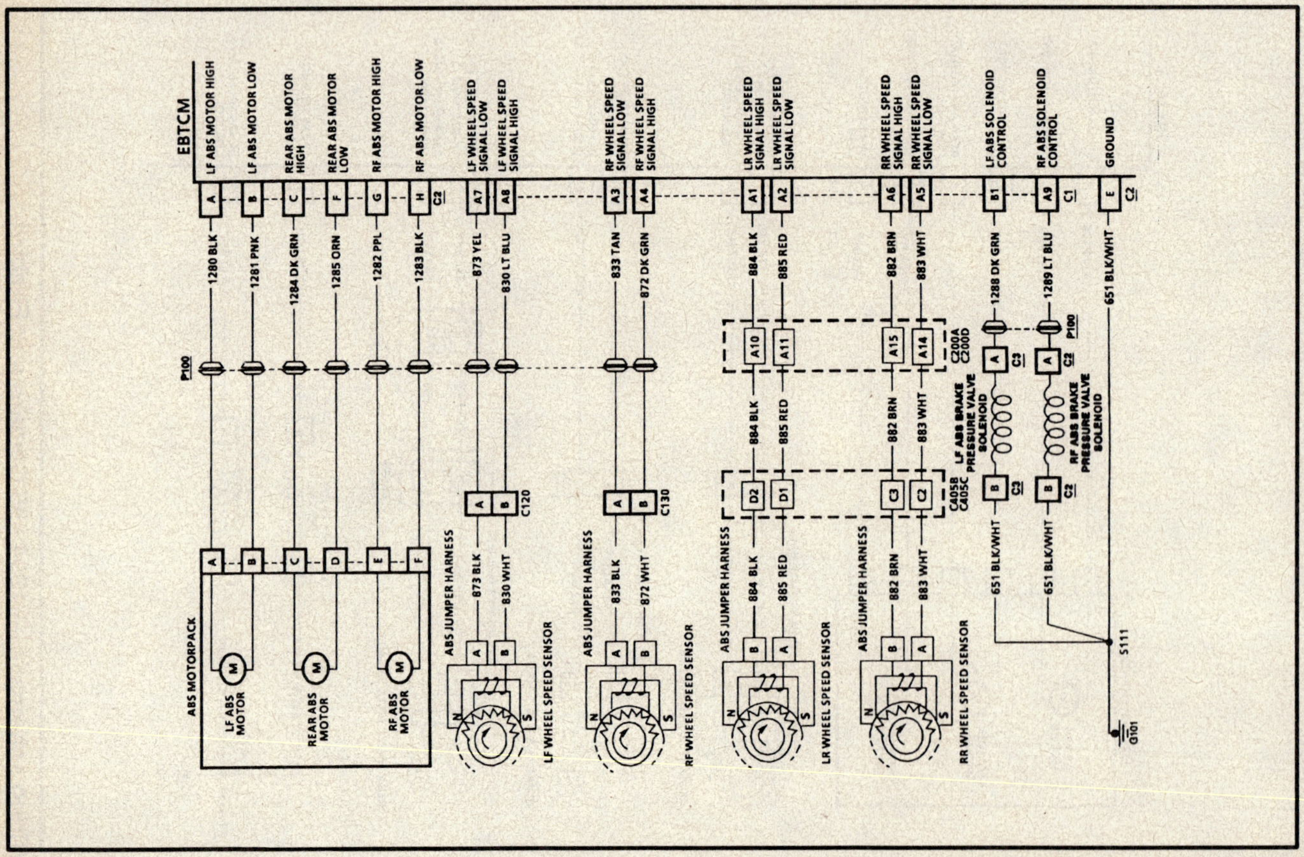

Wiring schematic — 1995 Camaro and Firebird with TCS

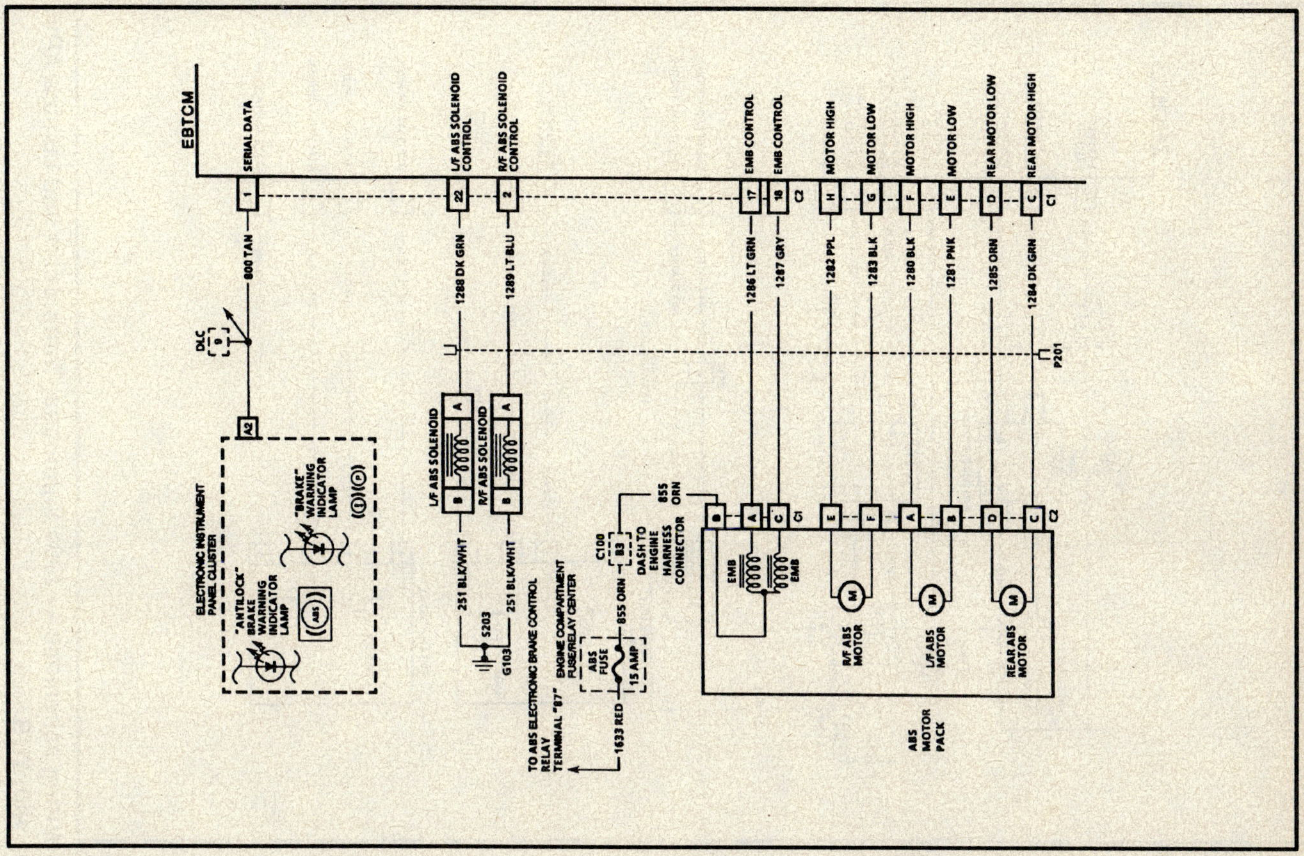

Wiring schematic — 1995 Cavalier and Sunfire with TCS

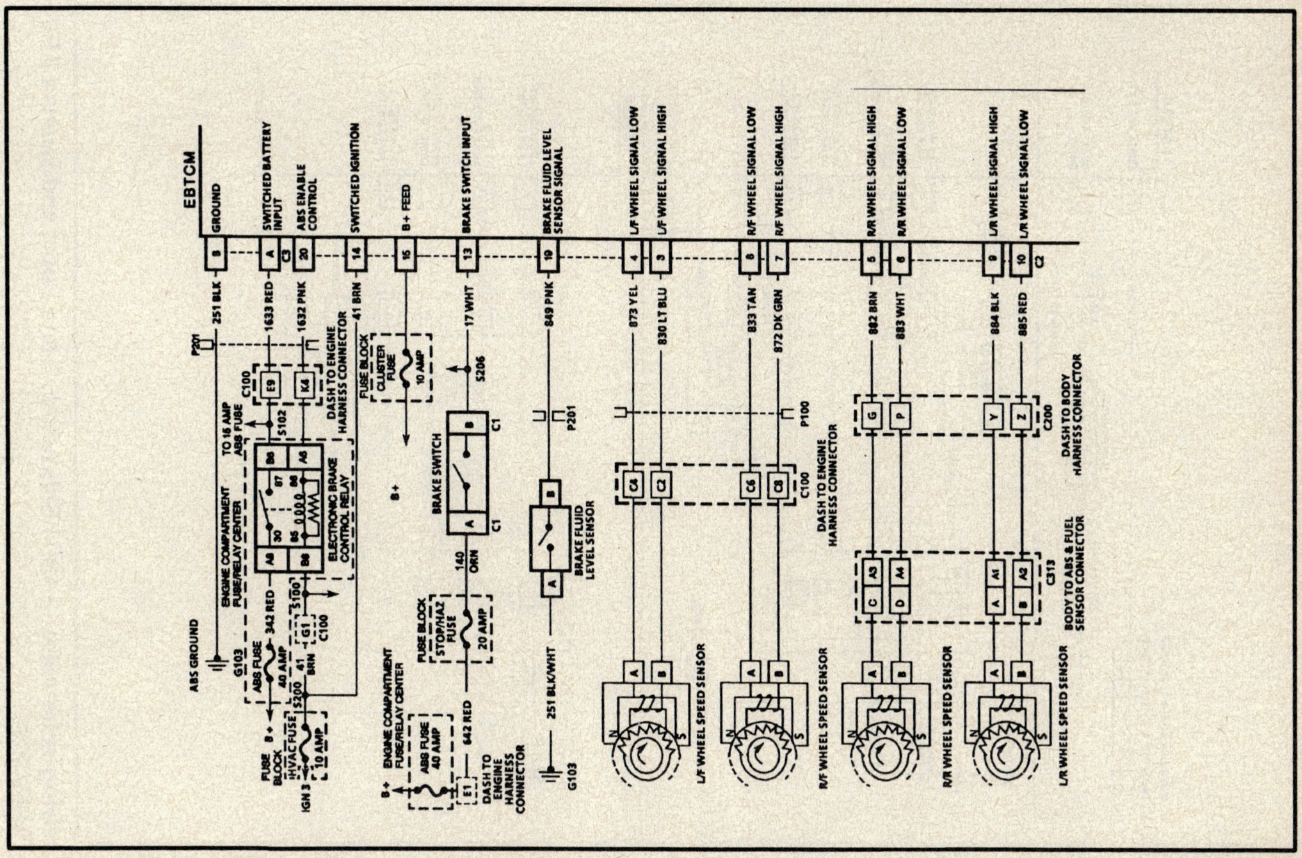

Wiring schematic — 1995 Cavalier and Sunfire with TCS

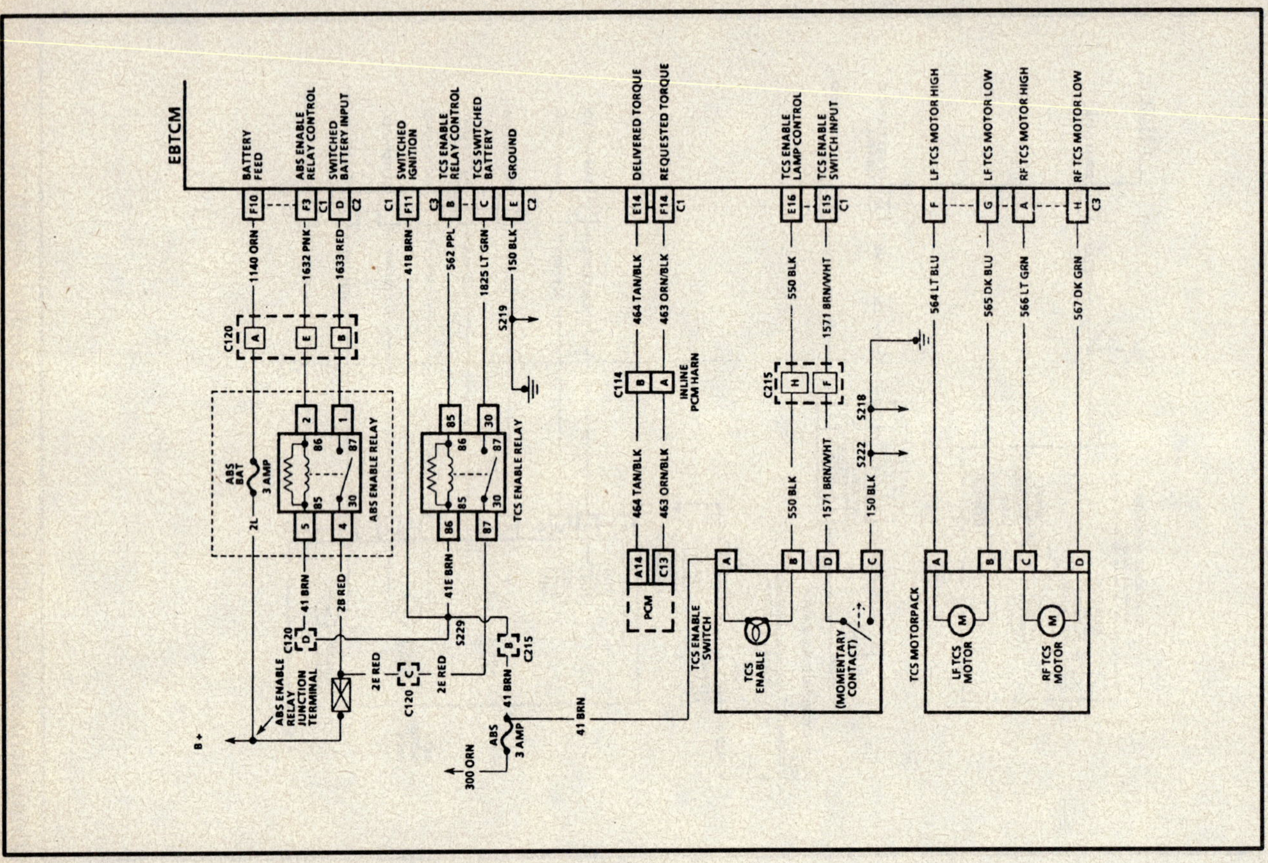

Wiring schematic — 1994 Silhouette, Trans Sport and Lumina APV with TCS

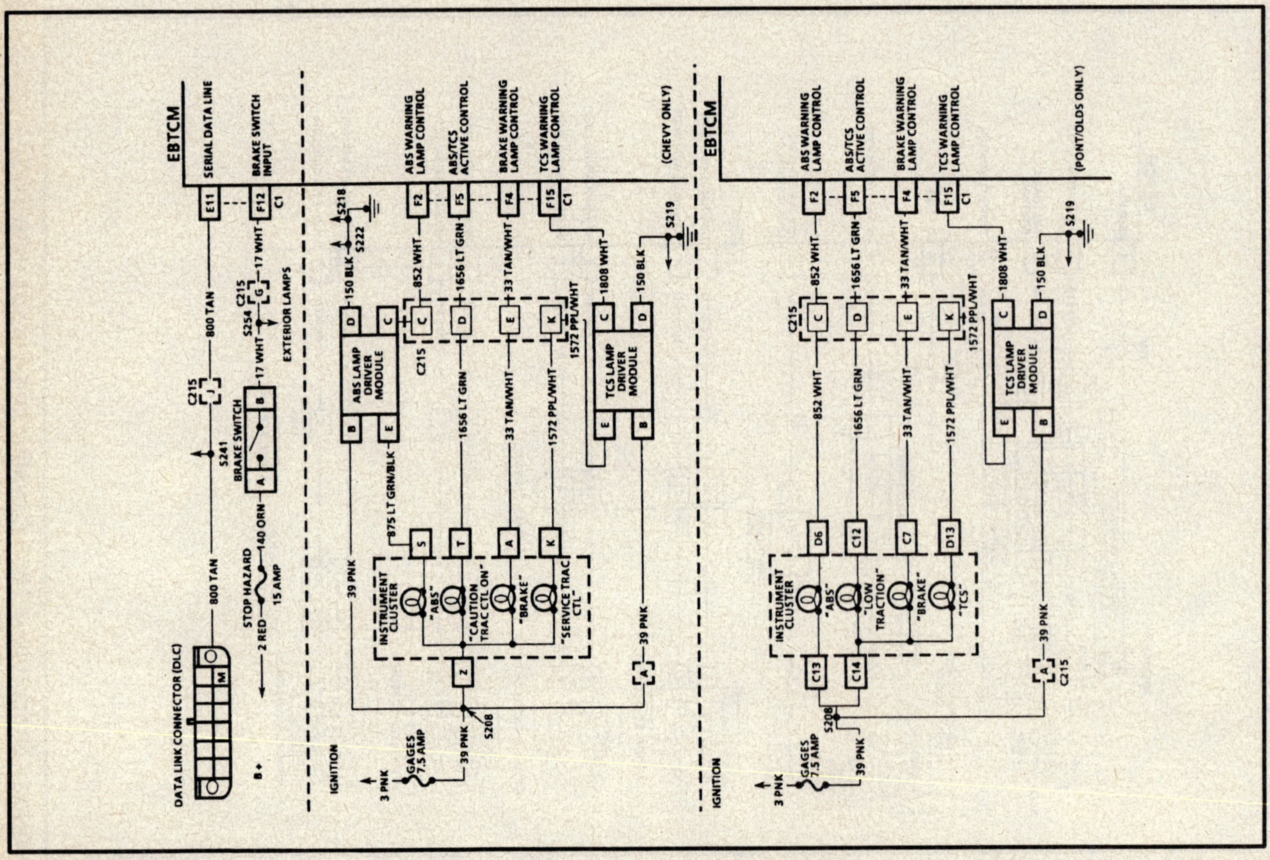

Wiring schematic — 1994 Silhouette, Trans Sport and Lumina APV with TCS

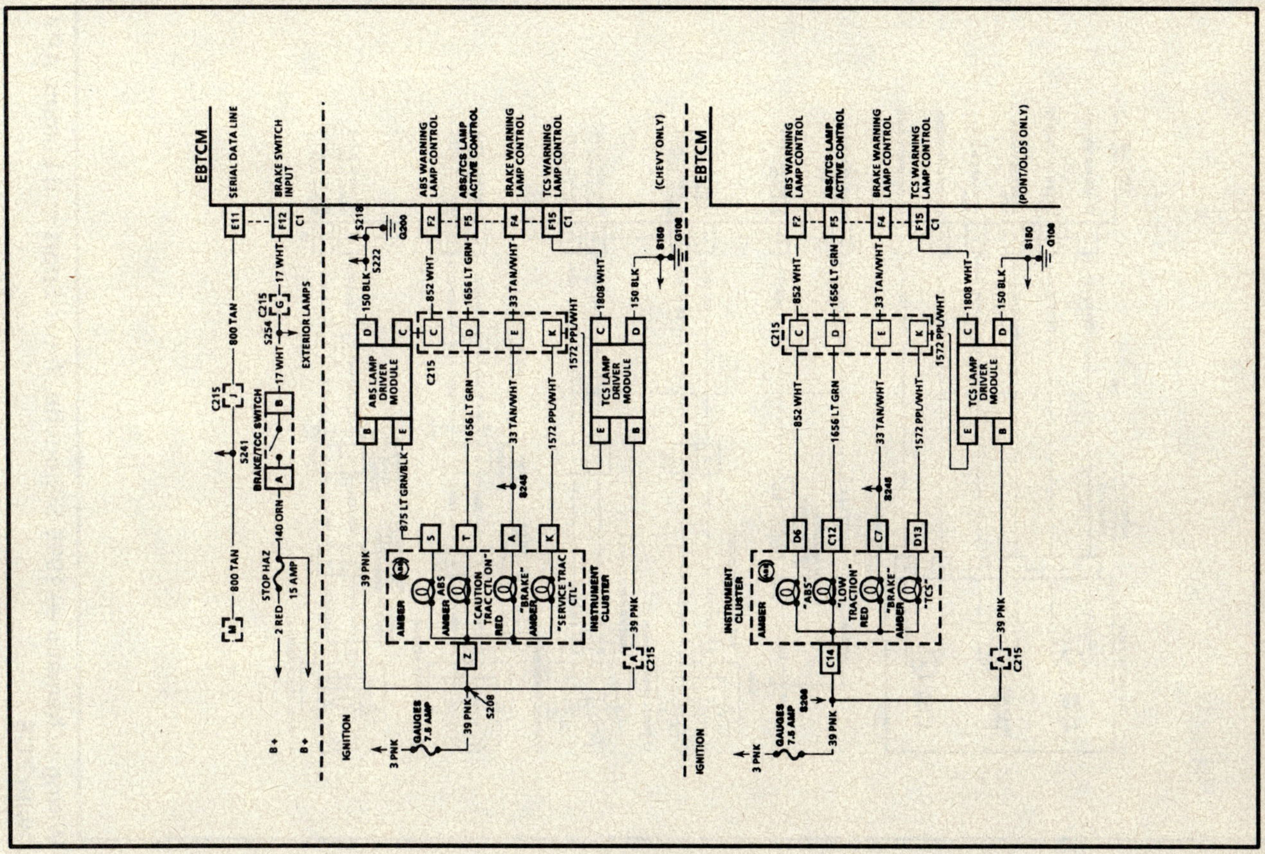

Wiring schematic — 1995 Silhouette, Trans Sport and Lumina APV with TCS

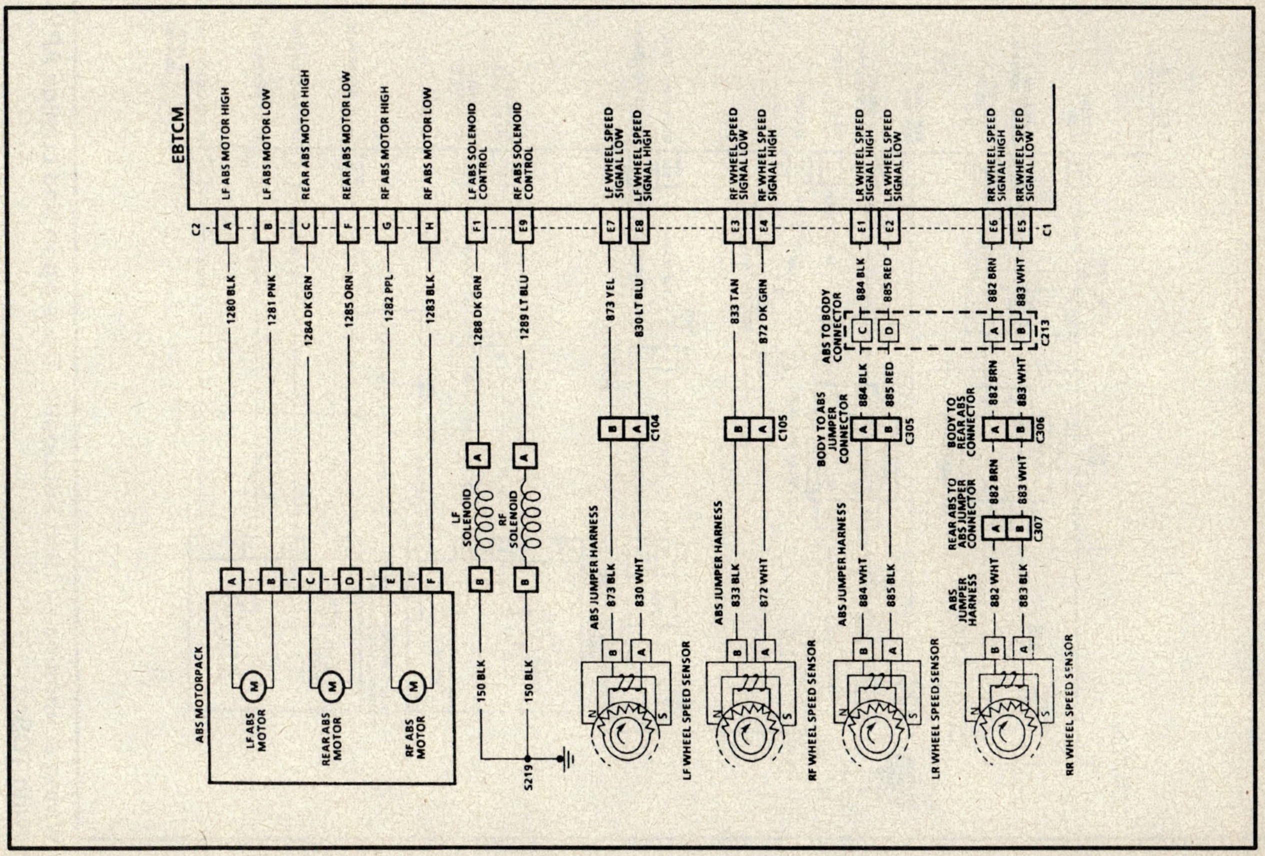

Wiring schematic — 1994 Silhouette, Trans Sport and Lumina APV with TCS

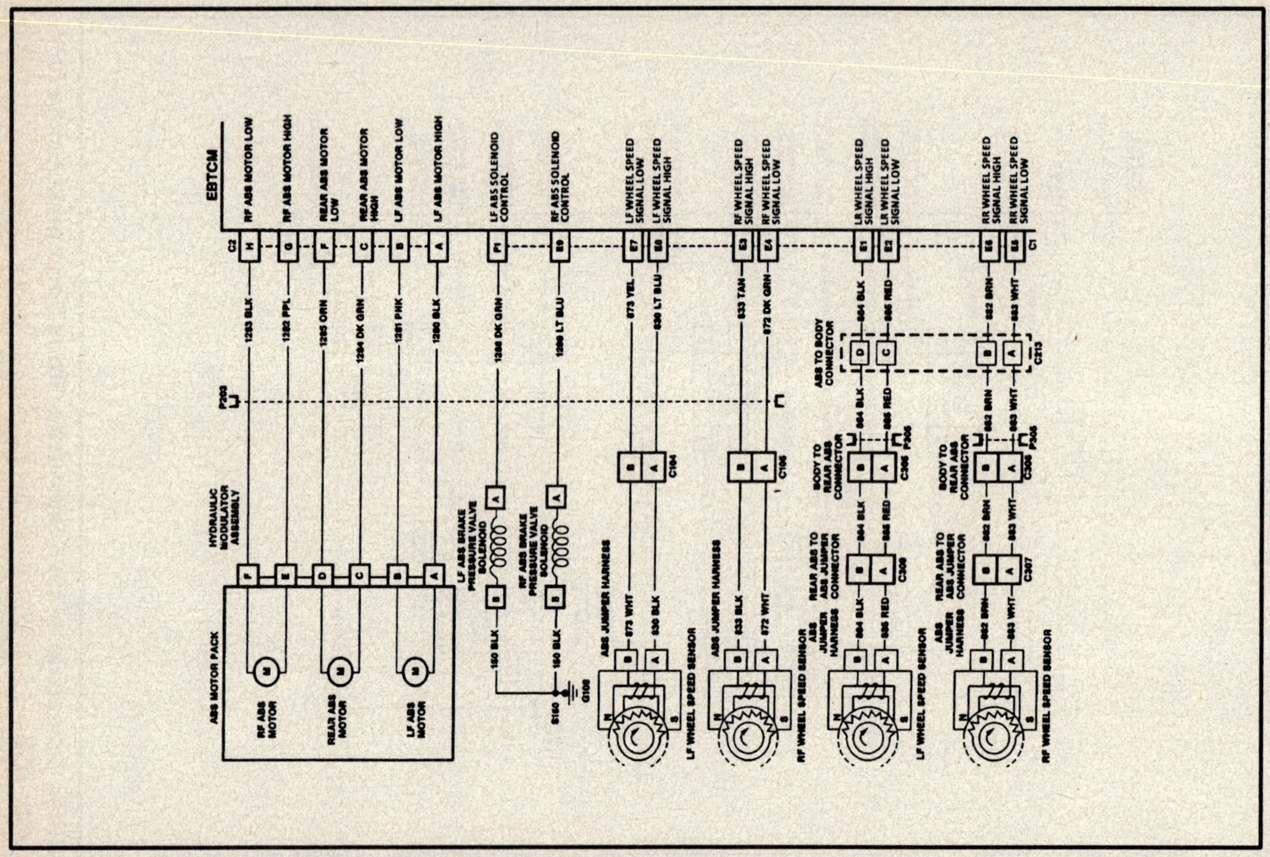

Wiring schematic — 1995 Silhouette, Trans Sport and Lumina APV with TCS

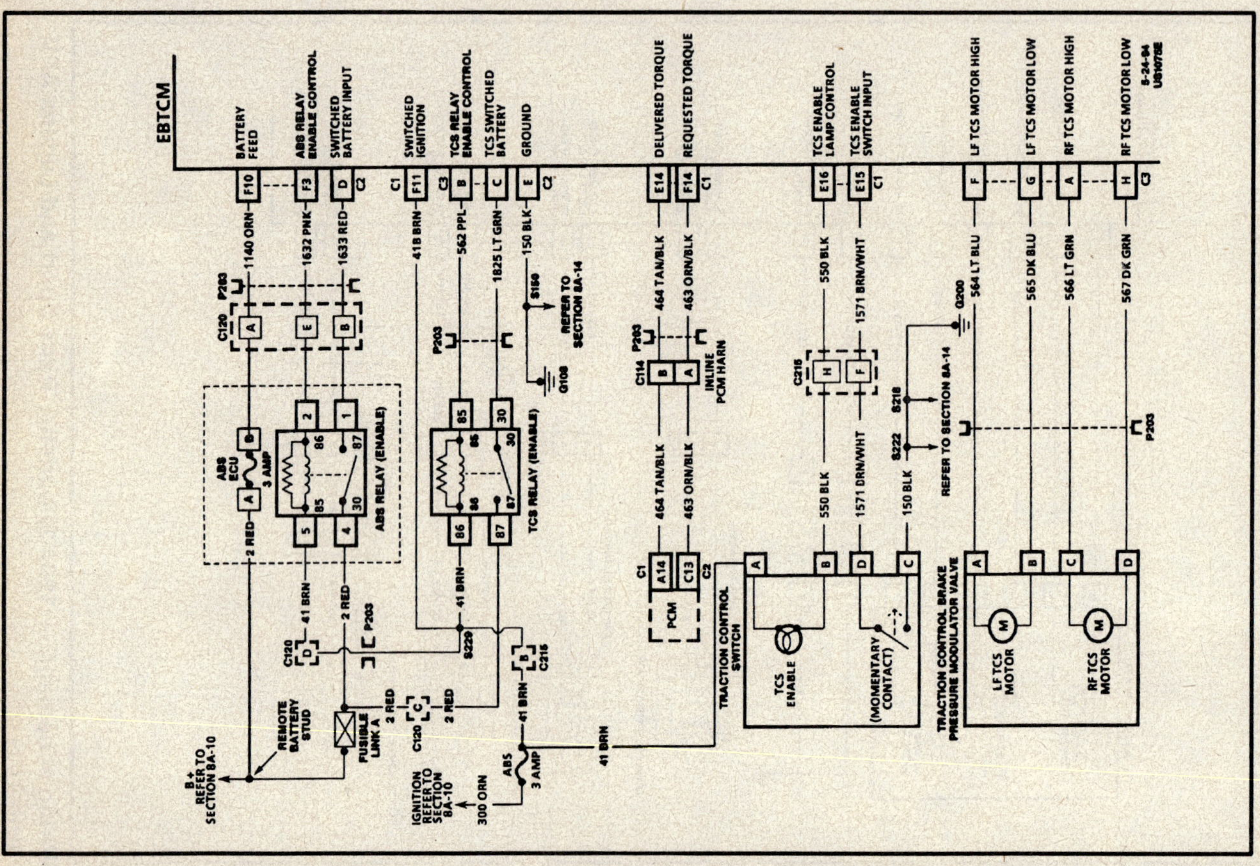

Wiring schematic — 1995 Silhouette, Trans Sport and Lumina APV with TCS

Component Replacement

SYSTEM REFILL

The master cylinder reservoir is filled in the traditional fashion with no special procedures being needed. Always clean the cap and reservoir of all dirt and debris before removing the cap. The reservoir must be filled to the **MAX** or **FULL** line, generally just below the neck of the reservoir.

Fluid must be DOT 3 only; never DOT 5, silicone or petroleum-based fluid. Brake fluid must be added from an unopened container. Do not use fluid from previously opened containers; the fluid will have absorbed moisture which can reduce system function and/or damage components.

SYSTEM BLEEDING

The use of pressure bleeding equipment is preferred, although the system may be bled manually if necessary. Before bleeding is begun, the front and rear displacement cylinder pistons must be returned to the topmost position. This should be done with the use of the TECH-1 scan tool, or its equivalent, for General Motors vehicles.

NOTE: If the entire system is being bled, the system must be bled in the following order: Right rear; Left rear; Right front; Left front.

Using the TECH-1

1. When using the TECH-1, enter the manual control function and apply the front and rear motors. Make certain the enable relay is **ON**.
2. If the scan tools are not available, the pistons may still be reset. Bleed the front brakes. Start the engine and allow it to run for at least 10 seconds; do not touch the brake pedal during this time. The ABS warning lamp must go **OFF** after about 3 seconds; if the lamp stays lit, THE TECH-1 tool must be used to determine the cause of the problem and the repair must be completed before continuing. The ABS system will initialize itself, returning the pistons to the top position. Switch the engine **OFF** and re-bleed the entire system.

Pressure Bleeding

The pressure bleeding equipment must be of the diaphragm type, preventing the entry of air, moisture or other contaminants into the brake fluid.

1. Clean the brake fluid reservoir cap before removing it. Fill the reservoir to the correct level with fresh brake fluid.
2. Make certain the pistons have been set to the **UP** or **HOME** position. Install the adapter cap J-35589, or the equivalent, to the fluid reservoir.
3. Connect the pressure bleeding equipment to the adapter. Charge the pressure bleeder to 30–35 psi.
4. Bleed the modulator assembly first. Connect a transparent tube over the rear bleeder valve on the modulator; submerge the other end in a clean container of brake fluid. Slowly open the bleeder valve about 3/4 turn. Allow the fluid to flow until no air is seen in the bleeder hose. Close the bleeder valve and tighten it to 65 inch lbs. (7Nm).
5. Repeat the bleeding procedure at the front bleeder valve on the modulator.
6. Use the same bleeding procedure at each brake caliper in the proper order. To aid in releasing trapped air, tap lightly on each caliper with a small rubber mallet. Tighten each caliper bleed valve to 8 ft. lbs. (11Nm).
7. After circuits are bled, remove the pressure bleeding equipment. Refill the reservoir to the correct level and install the cap.
8. Turn the ignition **ON** but do not start the engine. Apply the brake pedal with moderate force and hold. The pedal should feel firm and without excessive travel; if so, start the engine and confirm pedal feel and travel. If still satisfactory, the vehicle may be road tested.
9. If the pedal feel is not correct, use the scan tool to apply and release the pistons 2 or 3 times, leaving them in the **UP** or **HOME** position when finished. Bleed the system and recheck. If still not satisfactory, suspect leaks or mechanical failure in the system.
10. When road testing the vehicle, make several normal stops from moderate speeds.
11. Make 1–2 ABS stops from approximately 50 mph.
12. After testing, it is recommended that the system be bled again and the pedal feel and travel rechecked.

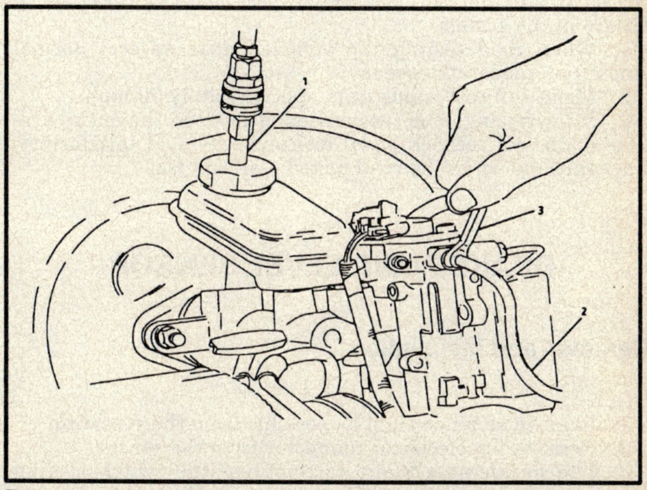

Pressure bleeding requires the adapter cap (1), a clear plastic hose (2) and the proper wrench (3)

Manual Bleeding

1. Clean the brake fluid reservoir cap before removing it. Fill the reservoir to the correct level with fresh brake fluid. The reservoir must be kept at least half-full during the entire bleeding procedure. Reinstall the cap to prevent spillage.
2. Make certain the pistons have been set to the **UP** or **HOME** position.
3. Prime the modulator assembly first. Connect a transparent tube over the rear bleeder valve on the modulator; submerge the other end in a clean container of brake fluid. Have an assistant depress the brake pedal and hold it down. Slowly open the bleeder valve about 3/4 turn fluid begins to flow. Close the bleeder valve, tighten it to 65 inch lbs. (7Nm), then release the brake pedal.
4. Repeat the bleeding procedure at the front bleeder valve on the modulator.

NOTE: Once fluid is seen to flow from both modulator bleed valves, the master cylinder and modulator are sufficiently full of fluid, but may still contain air. At this point, bleed the wheel circuits at the brakes. Return to the modulator and bleed it after the wheel circuits are bled.

5. Inspect the fluid level in the reservoir, filling as necessary.
6. Use the same bleeding procedure at each brake caliper in the proper order. To aid in releasing trapped air, tap lightly on each caliper with a small rubber mallet. During bleeding, wait at least 5 seconds after the brake pedal is released before reapplying the pedal. Tighten each caliper bleed valve to 8 ft. lbs. (11Nm).

NOTE: If no fluid flows when bleeding the rear brakes, it may mean that the displacement pistons are not at the UP or HOME position.

7. After circuits are bled, refill the reservoir to the correct level and install the cap.
8. Bleed the modulator and master brake cylinder.
9. Turn the ignition **ON** but do not start the engine. Apply the brake pedal with moderate force and hold. The pedal should feel firm and without excessive travel; if so, start the engine and confirm pedal feel and travel. If still satisfactory, the vehicle may be road tested.
10. If the pedal feel is not correct, use the scan tool to apply and release the pistons 2 or 3 times, leaving them in the **UP** or **HOME** position when finished. Bleed the system and recheck. If still not satisfactory, suspect leaks or mechanical failure in the system.
11. When road testing the vehicle, make several normal stops from moderate speeds.
12. Make 1–2 ABS stops from approximately 50 mph.
13. After testing, it is recommended to bleed the entire system again and recheck pedal feel and travel. If satisfactory, the vehicle need not be road tested a second time.

BRAKE FLUID LEVEL SENSOR

Removal and Installation

1. Remove as much fluid as possible from the reservoir.
2. Remove the electrical connector from the sensor.
3. Use needle-nose pliers to compress the switch locking tabs at the inboard side of the master cylinder.
4. With the locking tabs compressed, remove the switch.

To install:

5. Install the switch, making certain the locking tabs engage.
6. Connect the wiring harness.
7. Refill the reservoir. Bleed the entire system.

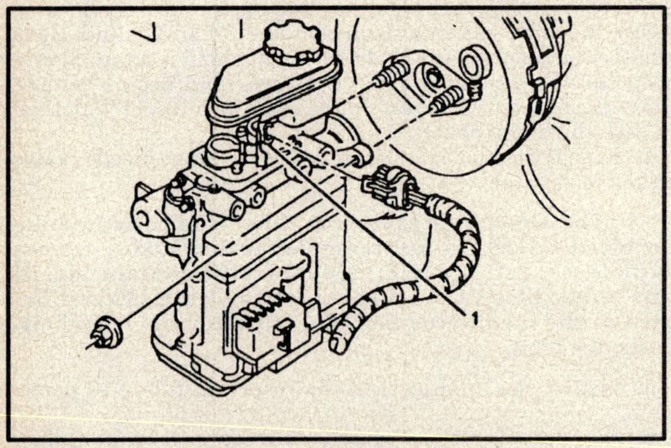

The brake fluid level sensor (1) may be changed with the reservoir in place

ELECTRONIC BRAKE CONTROL MODULE (EBCM)/ELECTRONIC BRAKE TRACTION CONTROL MODULE (EBTCM)

Removal and Installation

EXCEPT CUTLASS, LUMINA, GRAND PRIX, MONTE CARLO AND REGAL

1. On Lumina APV, Silhouette or Trans Sport, remove the lower sound insulator panel under the steering column.
2. Disconnect the wire harness(es) from the EBCM.
3. Remove the screws holding the EBCM and remove the unit from the vehicle.
4. Installation is the reverse of removal. Make certain the 3 plastic grommets are properly located before installing the retaining bolts.

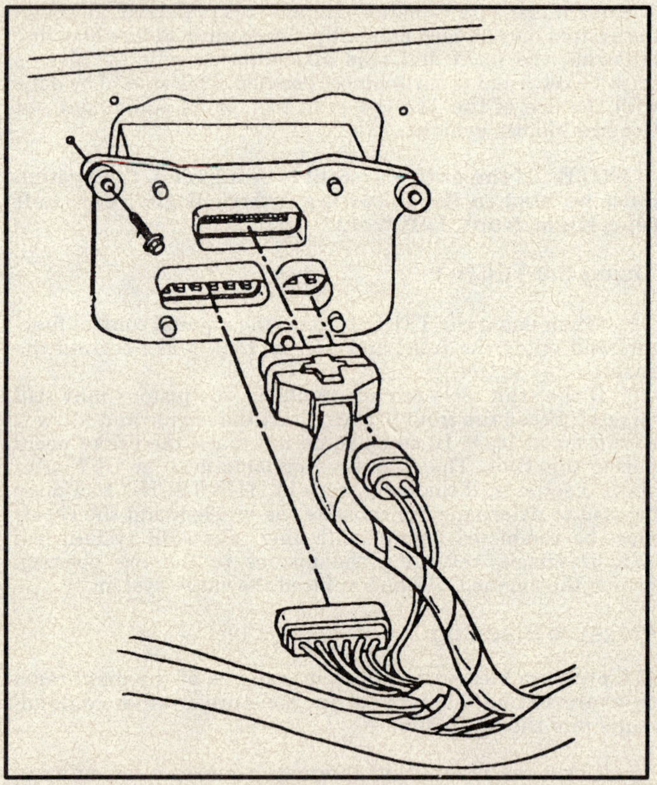

Electronic brake control module — Corsica and Beretta repairs. Cutlass Supreme shown, others similar

CUTLASS, LUMINA, GRAND PRIX, MONTE CARLO AND REGAL

1. Raise and safely support the vehicle.
2. Remove the left front wheel.
3. Remove the left front inner fender (liner).
4. Remove the screw holding the electromagnetic compatibility shield. Carefully, slide the shield back on the harness.
5. Remove the screws holding the EBCM. Slide the unit from the bracket and disconnect the wiring harness.

To install:

6. Connect the wiring to the EBCM.

7. Slide the EBCM into the bracket; make certain the rear guide pin is in its groove. Install the retaining screws.
8. Slip the EMC shield over the harness and install the retaining screw. The shield must not be torn or damaged.
9. Install the fender liner.
10. Install the wheel; lower the vehicle to the ground.

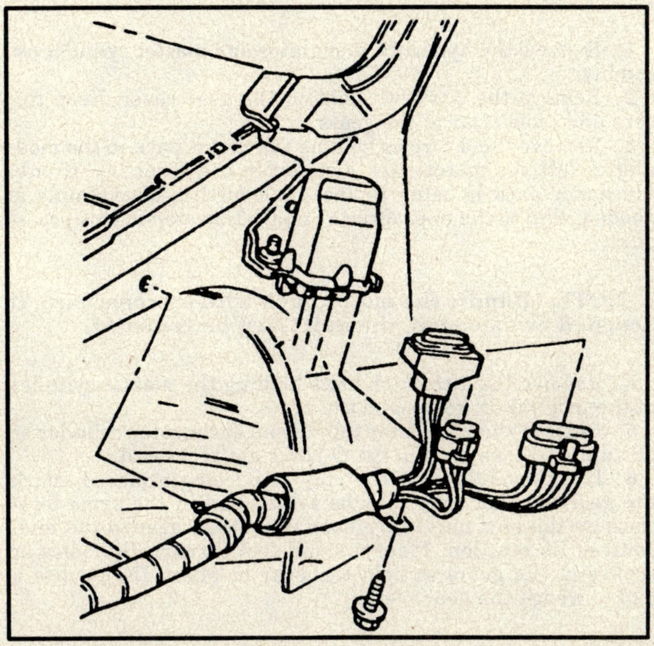

On W-body vehicles, take care not to damage the electromagnetic shield during removal

SOLENOID VALVE

Removal and Installation

The valves are located on top of the ABS modulator. The valves are directly outboard of the master cylinder. There are 2 different valve types which can be identified by an 8 digit number etched on top of the valve. Two different designs of solenoids are used; always be sure to use the correct replacement type.

1. Ensure the ignition is **OFF**. Clean the dirt from the solenoid valve surrounding area before any repair is attempted to eliminate system contamination.
2. Disconnect the electrical connector from the valve.
3. Remove the 2 retainers and remove the solenoid.
4. Verify the O-ring remained in the valve groove and was lifted up from the modulator bore, if equipped with the type A solenoids. If using a type **B** solenoid, or if the type **A** O-ring remained in the bore, carefully extract the valve lip seal or O-ring from the modulator.

To install:

5. Lubricate the valve O-ring or lip seal with clean brake fluid and install onto the valve. O-rings should be installed into the groove provided on the solenoid. Up seals should be inserted over the bottom of the valve with the lip or cupped side of the seal facing upward toward the solenoid top.
6. Position the solenoid into the modulator and push downward until the valve flange is fully seated. Install the 2 screws and tighten to 45 inch lbs. (5Nm).
7. Connect the electrical connector to the valve and connect the negative battery cable.
8. Properly bleed the anti-lock brake system.

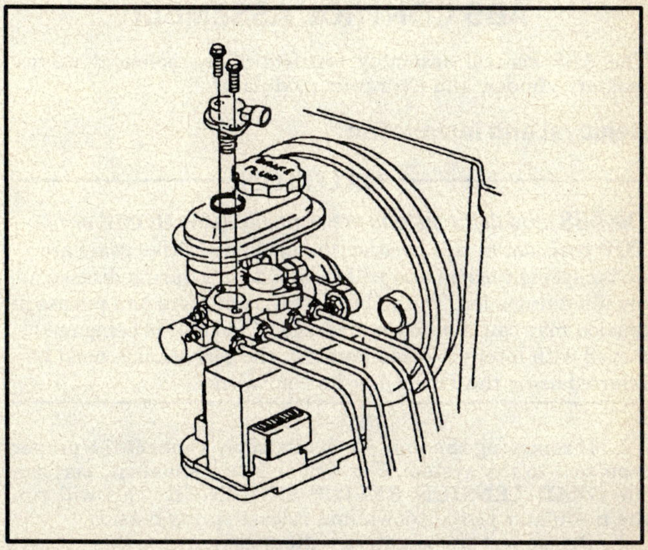

Anti-lock brake solenoid removal

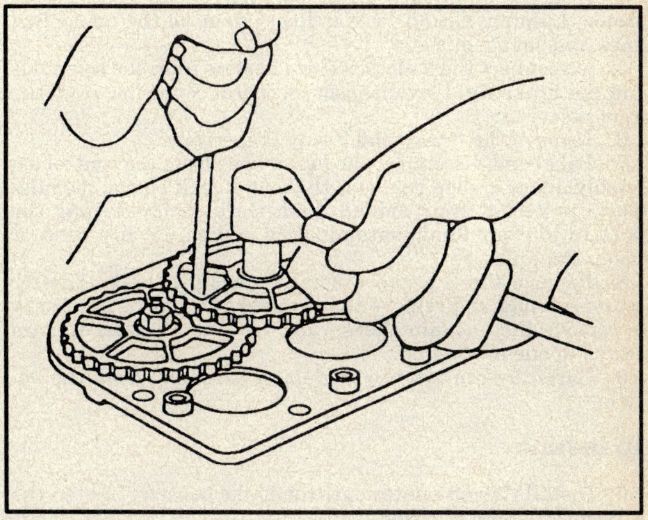

The gears must not turn during removal or installation

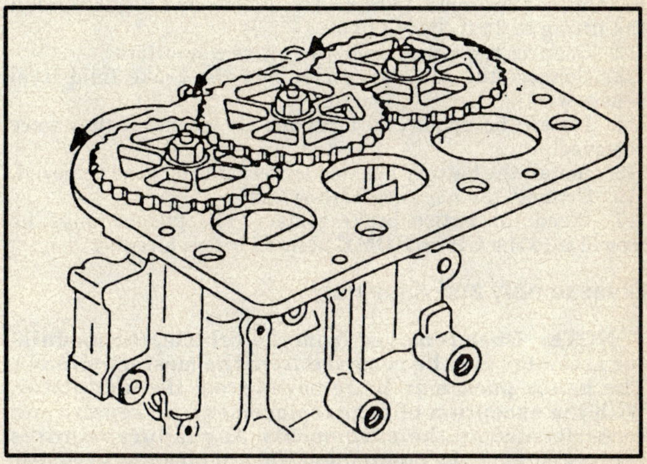

After installation, gently turn all the gears counterclockwise until motion stops

ABS CONTROL ASSEMBLY

The ABS control assembly consists of the solenoid valves, master cylinder, and hydraulic modulator.

Removal and Installation

CAUTION

The ABS modulator pistons are normally in the **HOME** or **TOP** position. In this position the modulator drive gears are under spring tension and will turn suddenly during disassembly if not unloaded. The sudden rotation of gears and release of tension may cause injury or damage. If the unit is being removed with intent to disassemble it, the gear tension must be relieved using the TECH-1 or its equivalents.

1. If removing the unit for disassembly, connect the proper scan tool to the system. For TECH-1 or equivalent, perform the **GEAR TENSION RELIEF SEQUENCE**. This will run the modulator pistons down and release spring tension.
2. Disconnect the negative battery cable.
3. On Cutlass Supreme, Lumina, Grand Prix or Regal, remove the air cleaner assembly.
4. If equipped with 3.1L or 3.4L engine in Cutlass Supreme, Lumina, Grand Prix or Regal, remove the upper fuel lines and set them aside.
5. Disconnect the 2 electrical connectors from the solenoids and the brake fluid level sensor electrical connector from the fluid reservoir.
6. Remove the 6-way and 3-way connectors.
7. Label and disconnect the brake lines from the control assembly. Place a shop rag over the motor pack on the modulator to protect it from spilled fluid. Once removed, plug the lines to prevent fluid contamination or loss. Do not bend or crimp the lines.
8. Remove the 2 brake control assembly-to-brake booster retaining nuts and remove the assembly. It may be necessary to remove the vacuum check valve from the booster to gain access to one of the nuts.
9. Carefully remove the modulator assembly from the vehicle.

To install:

10. Install the modulator onto the brake booster; tighten the nuts to 20 ft. lbs. (27Nm).
11. Install the brake lines, making sure each is in its original position. For all vehicles except Cutlass Supreme, Lumina, Grand Prix, or Regal, tighten the fittings to 13 ft. lbs. (17Nm). For Cutlass Supreme, Lumina, Grand Prix and Regal, tighten the fitting to 24 ft. lbs. (32Nm).
12. Connect the 6-way and 3-way wire connectors.
13. Connect the solenoid wiring and the brake fluid level sensor wiring.
14. Install the upper fuel lines and air cleaner if they were removed.
15. Install the battery box and battery if they were removed.
16. Connect the negative battery cable.
17. Bleed the entire brake system. The pistons must be moved into the **UP** or **HOME** position before bleeding.

Disassembly and Assembly

NOTE: Once removed from the vehicle, the modulator assembly may be removed from the master cylinder. The motor pack may be removed from the modulator. With the exception of gear replacement, no repairs can be performed on the components. Any failure requires replacement of the component. Do not attempt to repair motor packs or modulators. Do not lubricate or oil the motor pack assembly.

CAUTION

The modulator gears are under spring tension and will turn during disassembly. Never place fingers on or in the gear train during disassembly; injury may result.

1. Remove the hydraulic modulator and master cylinder assembly.
2. Remove the 6 screws holding the gear cover. Keep fingers and tools clear of the gears.
3. Remove the 4 screws holding the motor pack to the modulator. Lift the motor pack away from the assembly. If only the motor pack is being replaced, no further disassembly is needed; skip to the appropriate installation step in this procedure.

NOTE: Handle the motor pack with extreme care. If dropped or damaged, the unit must be replaced.

4. Remove the 2 through bolts holding the master cylinder to the modulator; separated the units.
5. Remove the 2 transfer tubes from the master cylinder or modulator. Make certain the O-rings are recovered.
6. If the modulator is the only unit being replaced, mark the gears so that they may be reinstalled in the same locations on the new unit. Turn each modulator gear to the midpoint of its rotation. Place a small tool through the holes or spokes in the gears to hold the gear in place. Do not use a tool to wedge the gear teeth.

NOTE: The gears or shafts must not be allowed to rotate during removal or installation; the piston may bottom within the cylinder bore, causing internal damage.

7. Remove the nut holding each gear to the shaft Remove the gears.

To assemble:

8. Place modulator gears onto their shafts. As before, insert a small tool to hold each gear from turning.
9. Install the shaft nuts. Tighten each to 75 inch lbs. (8.5Nm). Remove the blocking tool.
10. Hold the modulator upside down so that the gears face upward. Rotate each gear counterclockwise until movement stops. This will place the pistons near the top of the cylinders, simplifying bleeding after installation.
11. Use new transfer tube assemblies. Make certain the two O-rings are properly installed on the tubes. Lubricate the O-rings with clean brake fluid.
12. Install the tubes into the modulator ports by hand; push them in until they bottom.
13. Lubricate new through-bolt O-rings with clean brake fluid. Install them in the master cylinder and modulator.
14. Clamp the mounting flange of the master cylinder in a vise. Install the modulator by rocking it into position on the master cylinder. Make certain the transfer tubes are correctly aligned before installation.
15. Install the through bolts and tighten them to 12 ft. lbs. (16Nm).
16. Install the motor pack onto the modulator. Make certain the motor gears align with the modulator gears. Install the retaining screws and tighten them to 40 inch lbs. (4.5Nm).
17. Install the gear cover onto the modulator.
18. Install the modulator/master cylinder assembly.
19. Install brake fluid to the proper level. Bleed the entire system.

FRONT WHEEL SPEED SENSORS

Removal and Installation

EXCEPT LUMINA APV, SILHOUETTE AND TRANS SPORT

Front wheel speed sensors on these vehicles may be removed without disassembly of the hub or other components. Use care in handling the sensor, protecting it from impact or rough handling. The wiring harness is an integral part of the sensor and is not individually replaceable.

1. Elevate and safely support the vehicle.
2. Disconnect the sensor connector from the wiring harness.
3. Remove any clips or retainers from the sensor harness; take note of the exact placement or routing of the wiring. The harness must be reinstalled in its original location.
4. Remove the bolt(s) holding the sensor. Remove the sensor. If the sensor is difficult to remove, remove the brake rotor and use a blunt tool to push the sensor from the other side of the knuckle. Never pry or jack against the toothed gear on the drive shaft; the slightest scratches on the rang may affect ABS function.

NOTE: Some sensors use a locating pin which may become stuck in its bore due to corrosion. If the pin separates from the sensor and cannot be pulled free, drill it out with an 8mm drill. Use great care not to enlarge the hole; correct sensor positioning is critical to proper operation. Clean the hole of corrosion using coarse sandpaper wrapped around a dowel.

To install:

5. Inspect all mating surfaces, cleaning as necessary.
6. Install the sensor, making sure it sits squarely and firmly in place. Install the retaining bolt(s). For all vehicles except Cutlass Supreme, Grand Prix, Lumina and Regal, tighten the bolts to 106 inch lbs. (12Nm). For Cutlass Supreme, Grand Prix, Lumina and Regal, tighten the bolts to 62 inch lbs. (7Nm).
7. Route the sensor wiring correctly and install any clips and retainers that were removed.
8. Connect the sensor harness to the ABS harness.
9. Install the wheel and lower the vehicle.

LUMINA APV, SILHOUETTE AND TRANS SPORT

The front wheel speed sensor is within the front hub and bearing assembly. Removal of the hub assembly is required to replace the sensor.

1. Elevate and safely support the vehicle.
2. Remove the front wheel.
3. Disconnect the wheel speed sensor connector.
4. Remove the hub and bearing assembly.
5. Use a blunt tool to remove the sensor from the hub and bearing assembly.

NOTE: The sensor will be damaged during removal; it is not reusable. The 2 parts of the sensor assembly are not individually replaceable but must be replaced as an assembly.

To install:

6. With the sensor removed, inspect the bearing for signs of water intrusion and replace the bearing if necessary.
7. Apply Loctite® 620 or equivalent to the mating surfaces of the sensor which contact the hub assembly.
8. Use tool J–38764 or its equivalent to press the wheel speed sensor into the hub and bearing assembly.
9. Install the hub and bearing onto the vehicle.
10. Connect the wheel speed sensor harness.
11. Install the wheel and lower the vehicle.

REAR WHEEL SPEED SENSORS

Removal and installation

All rear wheel speed sensors are integral with the rear wheel bearing or bearing and hub assembly. The sensors are not serviceable; if a rear sensor has failed, replacement of the bearing or bearing and hub is required.

On vehicles with drum brakes, extreme care must be used. Once the bearing is removed, the brake assembly will be held in place only by the brake line. To avoid bending or crimping the brake line, do not bump or exert force on the brake assembly.

Vehicles with rear disc brakes require removal of the brake caliper and rotor to gain access to the hub and bearing.

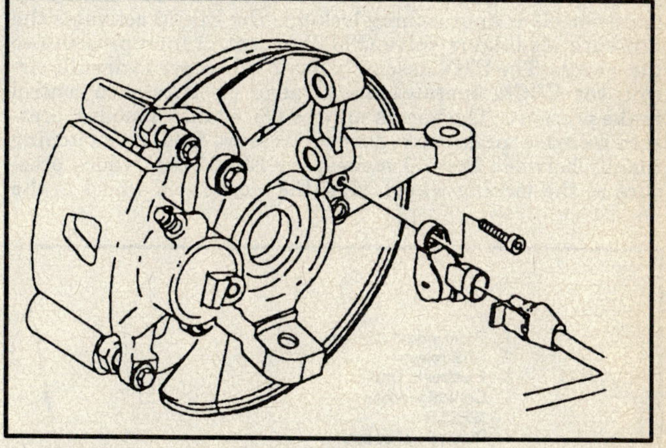

Example of front wheel speed sensor.

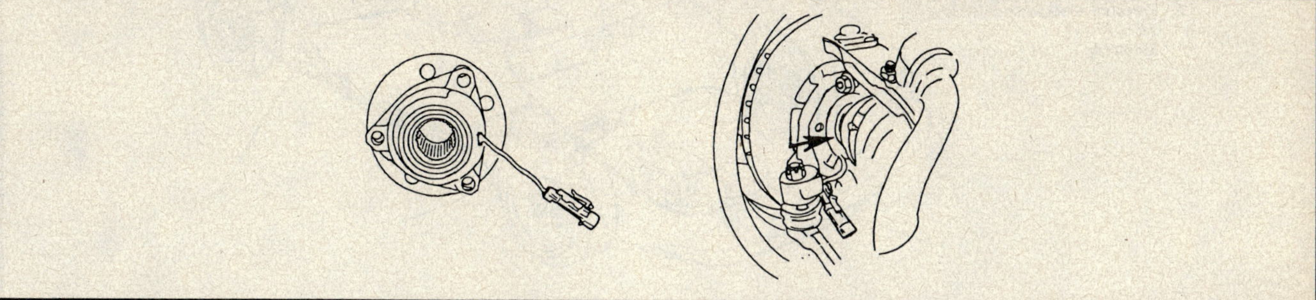

Front wheel speed sensor for Lumina APV, Silhouette and Trans Sport. The hub and bearing assembly must be removed before separating the sensor

TEVES MARK IV ANTI-LOCK BRAKE SYSTEM WITH TRACTION CONTROL (TCS)

General Description

The Teves Mk IV system is used on the C and H body Bonneville, Park Avenue, Eighty Eight, Ninety Eight, DeVille and LeSabre models as well as the 1995 G body Aurora and Riviera. The system allows individual modulation of braking force at each wheel as well as providing traction control at the drive wheels if the vehicle is so equipped. The ABS system employs a conventional diagonally-split braking system as well as additional ABS components. The system allows the operator to maintain directional control during heavy braking situations by preventing locked-wheel skidding.

On vehicles equipped with the Traction Control System (TCS), wheel spin at the drive wheels is reduced by selective application of the wheel brakes if vehicle speed is below 25 mph (40 km/h). The TCS system is designed to improve traction on loose or wet surfaces.

SYSTEM OPERATION

Anti-Lock Brake System

Braking is performed by the conventional brake system until the Electronic Brake Control Module (EBCM) detects a wheel beginning to lock or the brake pedal has traveled beyond a specified limit. The EBCM will then command the system into anti-lock mode.

Electric signals are received by the EBCM from each wheel speed sensor. By comparing these signals to each other and to pre-programmed values, the EBCM can determine that one or more wheels is approaching lock-up. The EBCM activates the Pressure Modulator Valve (PMV) to control fluid pressure to the wheels. The PMV assembly has 2 valves per hydraulic circuit; the EBCM operates these valves electrically to control brake pressure. The valves may be placed into 3 modes: pressure increase, pressure reduce or pressure hold. By switching rapidly between these 3 modes, the EBCM can reduce pressure at the locking wheel, slow the most rapid wheel to the average speed of the traction wheels or maintain the wheels just below the point of lock-up.

During the pressure increase mode, the inlet valve within the PMV is open, allowing fluid to enter the particular hydraulic circuit. As this mode cycles, pressure and fluid is reduced within the master cylinder, causing the brake pedal to drop nearer the floor. When the pedal travel exceeds a pre-determined level, the EBCM will activate the PMV pump motor. Pump pressure is applied to the wheel circuit(s) in need and/or the master cylinder hydraulic circuits. With pressure applied to the master cylinder hydraulic circuits, the pedal will rise until the brake travel switch is de-activated, switching the pump off.

During vehicle operation, the EBCM performs a series of self-tests, checking both electronic and hydraulic systems for faults. If detectable problems exist, the EBCM will set a fault code and/or illuminate the ANTI-LOCK dash warning lamp. The ANTI-LOCK warning lamp is in addition to the normal BRAKE warning lamp which warns of problems in the conventional brake system.

Traction Control System (TCS)

The traction control system is additional to the ABS system and is not found on all vehicles with ABS. With traction control, brake pressure is increased to reduce wheel spin; with ABS, line pressure is reduced to allow greater wheel spin. The EBCM/EBTCM monitors the speed of the drive wheels through the wheel speed sensors. If wheel slip is detected below 25 mph with the brakes not applied, the EBCM enters the traction control mode.

Once in traction control mode, the EBCM/EBTCM commands two isolation valves in the PMV into operation. The isolation valves close to separate the non-driven wheels and the master cylinder circuits from the drive wheels. Once isolated, brake fluid pressure may be applied to the drive wheels without affecting the other circuits. The EBCM/EBTCM turns on the PMV pump motor and begins cycling the inlet and outlet valves to slow the slipping wheel(s).

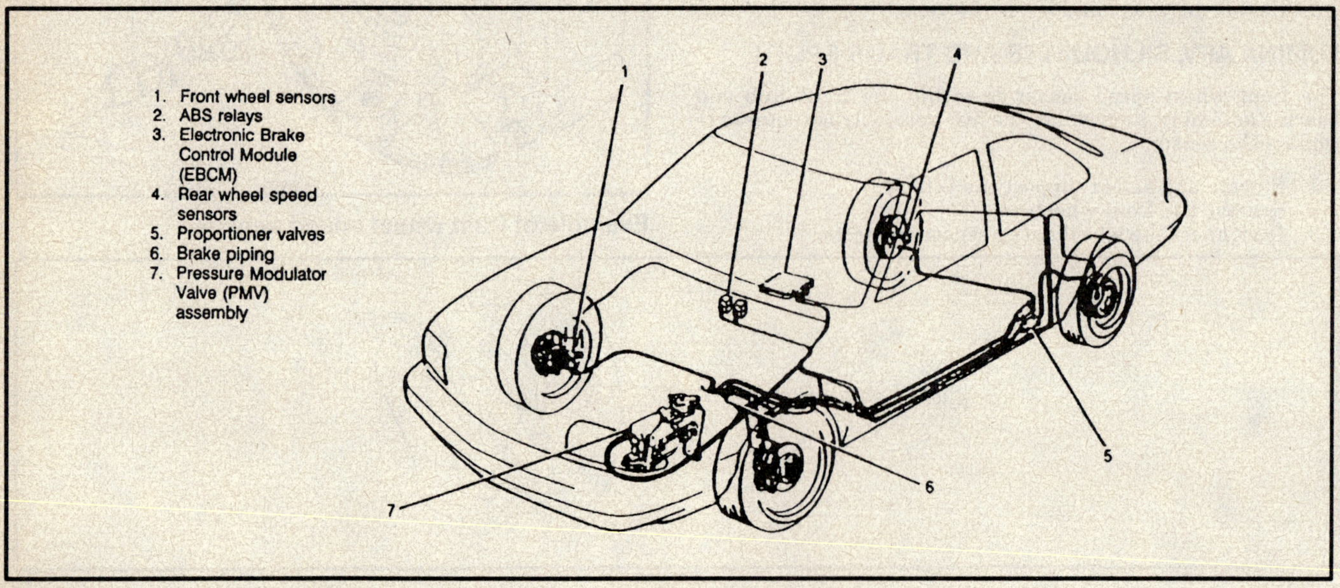

Component location — 1993

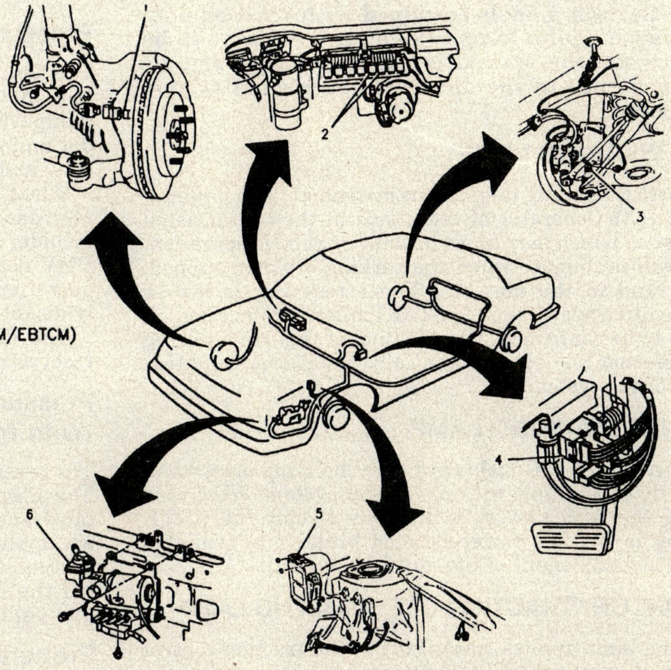

Component location — 1994–95

Since the pump motor supplies more fluid than required, two pressure relief valves allow excess fluid to return to the master cylinder reservoir. If the brakes are manually applied by the operator during traction control operation, the PMV pressure switch and the brake pedal switch signal the EBCM/EBTCM to disable traction control and allow manual braking. Additionally, manual braking can override the traction control system at any time. The isolation valves act as one-way check valves, allowing manually-applied brake fluid pressure to enter the system.

Vehicles equipped with TCS are also be equipped with a manual switch to disable the traction system. When the switch is engaged, canceling system operation, a TRACTION OFF dashboard warning lamp will be illuminated. Some vehicles also have indicators showing that the TCS is engaged, either by lighting a dashboard lamp or displaying a TRACTION ENGAGED message.

SYSTEM COMPONENTS

Electronic Brake Control Module (EBCM) or Electronic Brake and Traction Control Module (EBTCM)

The EBCM/EBTCM, located on the inner fenderwell on 1994–95 vehicles or under the dash on 1993 vehicles, is a solid-state computer which monitors the speed of each wheel as well as the electrical status of the PMV assembly. Once wheel lock or wheel slip (with TCS) is detected, the EBCM/EBTCM operates the inlet and outlet valves in the PMV to control wheel speed. The valves can be cycled very rapidly, allowing control to be maintained in all operating situations. Additionally, in the case of traction control, the EBTCM will request the PCM to bring engine torque to a specific level to control wheel slip.

The EBCM/EBTCM will disable either the TCS or ABS if a fault is detected. In most cases, the system will remain disabled for the remainder of the driving cycle, even if the fault self-corrects before the next key-on cycle. If either low system voltage is detected at the EBCM or low brake fluid level is detected at the PMV, the systems(s) will be disabled until the condition is corrected. Once the sensors indicate normal voltage or fluid level, the system(s) are enabled, even if the vehicle has not been shut **OFF** and restarted.

If the control unit detects an electrical fault, it will disable the TCS and/or ABS system immediately and illuminate the appropriate ANTI-LOCK and/or TRACTION OFF dash warning lamps to advise the operator of a technical problem. A fault code will usually be set in memory. Although it is possible for just the TCS system to be disabled, any fault in the ABS system will also cause the TCS system to be disabled.

The fault codes may be retrieved when the EBCM is placed into the diagnostic mode. A bi-directional hand scanner is required to retrieve or clear stored codes; no warning lamp flash outputs are used.

Wheel Speed Sensors

A wheel speed sensor at each wheel transmits speed information to the EBCM/EBTCM by generating a small AC voltage relative to the wheel speed. The voltage is generated by magnetic induction caused by passing a toothed sensor ring past a stationary sensor. The signals are transmitted through a pair of wires which are shielded against interference. The EBCM/EBTCM then calculates wheel speed for each wheel based on the frequency of the AC signal received from the sensor.

Warning Lamps

NOTE: Certain models equipped with traction control may cause a pilot lamp to light or a message to be displayed when the system is engaged. These signals are a convenience to the operator but are not system warning lamps.

"BRAKE" WARNING LAMP

The red BRAKE warning lamp in the instrument panel illuminates to alert the operator of conditions in the conventional braking system which may affect braking or vehicle operation. The lamp will illuminate when the parking brake is applied, the brake fluid in the master cylinder reservoir is low or when the ignition is switched to START, allowing the function of the bulb to be confirmed. Illumination of this lamp during vehicle operation signals a potentially hazardous condition within the brake system.

"ANTI-LOCK" WARNING LAMP

The amber ANTI-LOCK dashboard warning lamp is lit when the EBCM detects a fault within the ABS system. When this lamp is lit, the ABS system is disabled. If only the ANTI-LOCK lamp is lit, full power-assisted braking is available without the added benefit of the anti-lock system.

"TRACTION" OR "TRACTION OFF" WARNING LAMP

This warning lamp applies specifically to the traction control system. It will light if the EBCM detects a problem specific to the TCS or if the operator has switched the system off manually.

On vehicles equipped with traction control, the EBCM is programmed to prevent brake overheating. During heavy braking conditions, if traction control mode is entered and the controller determines that the brakes are too hot to function properly, the TCS is disabled and the warning lamp is lit. After a period of time, usually 15–20 minutes, the system is enabled and the warning lamp extinguished. This protective behavior is normal operation and should not be mistaken for an intermittent system failure.

Additionally, the EBCM monitors the signal from the transaxle temperature switch. If transaxle temperature exceeds approximately 320°F (160°C), the controller will disable the traction system, allowing both the brakes and the transaxle to cool. When the temperature switch closes again, at approximately 300°F (140°C), the EBCM will keep the system disabled for an additional 35 minutes to allow farther cooling. The dash warning lamp will be illuminated during the period of disablement.

TCC/ABS Brake Switch

This normally closed switch is integral to the normal electrical/vacuum-release brake switch. It monitors position of the brake pedal, opening when the pedal exceeds 40 percent of total travel. The EBCM then turns on the pump in the pressure modulator valve assembly, circulating fluid and forcing the pedal upward until the switch closes.

Traction Control Switch

The traction control switch allows the driver to select the traction control mode **ON/OFF**. When the traction control system is turned off, the TRACTION OFF indicator lamp will be illuminated.

Fluid Level Sensors

The brake fluid reservoirs on both the master cylinder and pressure modulator valve contain fluid level sensors. If the master cylinder reservoir becomes low on fluid, the red BRAKE warning lamp will be lit, signaling a potential system hazard. If the fluid level in the PMV reservoir becomes too low, the amber ANTI-LOCK warning lamp is lit and the ABS/TCS system disabled until the level returns to minimum.

Pressure Modulator Valve (PMV) Assembly

The assembly consists of the pump and motor, the valve body and the fluid reservoir. The reservoir provides fluid to the pump while the pump circulates the fluid to the master cylinder and maintains pedal height. The valve block contains the inlet and outlet valves for each wheel circuit.

When equipped with traction control, the PMV assembly contains isolation valves used to block pressure in the master cylinder and rear wheel circuits. A pressure switch within the PMV detects operator application of the brakes during TCS operation; when this signal is received, the EBCM immediately disables TCS function and allows manual braking.

The PMV is not serviceable except for replacement of the reservoir.

Pressure Modulator Valve (PMV) Pressure Switch (With Traction Control)

The pressure switch monitors primary brake system pressure. The normally closed switch opens when the brakes are applied and pressure exceeds 101 psi (700 kPa). If the brakes are applied manually while the traction control mode is in operation, the PMV switch and the brake pedal switch will signal the EBTCM to disengage the traction control and allow manual braking.

Proportioner Valves

Each rear wheel circuit contains a proportioning valve to limit brake pressure to the non-driven wheels. These valves improve front-to-rear brake balance during normal braking.

Anti-lock Diode

The ABS diode is taped to the wiring harness under the dashboard near the EBCM on 1993 models, or taped to the wire harness in the engine compartment near the EBCM/EBTCM on 1994–95 models.

ABS Pump Relay

The ABS pump relay is located at the center rear of the engine compartment, below the LH under hood fuse block and behind the maxifuse/relay cover.

Main Relay

The main relay is located at the relay center rear of the engine compartment, below the LH underhood fuseblock and behind the maxifuse/relay center cover.

Diagnosis and Testing

SERVICE PRECAUTIONS

- Some vehicles are equipped with a Supplemental Inflatable Restraint (SIR) or air bag system. The air bag (SIR) system must be disabled before performing service on or around SIR components or wiring. Failure to follow safety and disabling procedures could result in possible air bag deployment, personal injury or unnecessary SIR system repairs.
- The use of rubber hoses or parts other than those specified for the ABS system may lead to functional problems and/or impaired braking of the ABS function. Install all components included in repair kits for this system. Lubricate rubber parts with clean fresh brake fluid to ease assembly. Do not use lubricated shop air to clean or dry components; damage to rubber parts may result.
- Certain components in the ABS/TCS system are not intended to be serviced or disassembled. Only those components with removal and installation procedures should be serviced.

- Always disconnect the EBCM and PMV connectors before any welding is done on the vehicle.
- Never connect or disconnect the EBCM or PMV connectors with the ignition **ON**.
- Use only brake fluid from an unopened container. Use of suspect or contaminated brake fluid can reduce system performance and/or durability.
- When any hydraulic component or line is removed or replaced, it may be necessary to bleed the entire system.
- A clean repair area is essential. Perform repairs after components have been thoroughly cleaned.
- The EBCM is a microprocessor similar to other computer units in the vehicle. Ensure that the ignition switch is **OFF** before removing or installing controller harnesses. Avoid static electricity discharge at or near the controller.
- Never disconnect any electrical connection with the ignition switch **ON** unless instructed to do so in a test.
- Always wear a grounded wrist strap when servicing any control module or component labeled with a Electrostatic Discharge (ESD) symbol.
- Avoid touching module connector pins.
- Leave new components and modules in the shipping package until ready to install them.
- To avoid static discharge, always touch a vehicle ground after sliding across a vehicle seat or walking across carpeted or vinyl floors.
- Never allow welding cables to lie on, near or across any vehicle electrical wiring.
- Do not allow extension cords for power tools or droplights to lie on, near or across any vehicle electrical wiring.

DIAGNOSTIC PROCEDURE

Diagnosis of the anti-lock and/or traction control system consists of 3 steps which must be performed in order.

1. Visual inspection
2. Functional check
3. Additional testing as specified by the functional check.

Some diagnostic procedures require that the ignition be left **ON** for an extended period of time while testing. Low battery voltage may result, causing incorrect readings or impaired system function. It is recommended that a battery charger be connected and a slow charge applied during testing.

Visual inspection

The initial visual inspection requires a thorough external check of both the conventional and ABS/TCS systems. Many faults are caused by items such as loose connectors, corroded grounds,failed fuses or low fluid levels. The visual inspection, when properly performed, will reveal these faults. Check the parking brake lever and cables for full release. Check the hydraulic system for leakage, line damage or restriction. System connectors should be tested for tightness; if a connector is suspect or has been exposed to moisture, separate the halves and check for corrosion on the terminal pins. Check connector shells for pin separation or push-out. Fuses in both the underdash and under-hood panels should be checked. Correct the level in the fluid reservoirs if necessary.

Functional Check

If no problem is found during the visual inspection, the functional check must be performed. This test requires the use of a bi-directional hand-scanner, such as the General Motors Tech-1 or its equivalent. The scan tool is used to recover and/or clear stored fault codes during testing. The procedure also includes test driving the vehicle if necessary. Once fault codes or symptoms are determined, the test chart directs the use of the appropriate supplementary chart.

NOTE: If multiple codes are stored, diagnose the code which first appears on the hand scanner.

READING CODES

Intermittent Faults

Faults which occur intermittently are difficult to diagnose. When diagnosing such problems, keep an overall view of system function in mind. Faults may result from low system voltage, low brake fluid in the PMV reservoir or any interruption of the signal(s) from the wheel speed sensor(s). Any condition interrupting the power supply to the EBCM or hydraulic unit will also cause the dash warning lamp to illuminate. Most intermittent or random faults relate to loose connectors or faulty terminals. Again, the visual inspection is of prime importance.

The diagnostic codes can be helpful in tracking intermittent faults. While not designated History or Current, certain combinations can point to a fault status. Stored codes which do not light the ANTI-LOCK warning lamp are intermittent or history codes. If the ANTI-LOCK lamp is lit and one or more codes are stored, at least 1 (or more) of the codes is current.

Additionally, when diagnosing intermittent faults, keep the following system basics in mind:

- It is possible to feel the system self-test at very low speed if the brake pedal is barely applied. Both the sound (fluid returning to the master cylinder) and the slight vibration (valves cycling through test mode) are normal conditions.
- Any fault in the anti-lock system will light the TRACTION light. When both lamps are lit, diagnose the ABS system.
- The systems will be disabled if either the EBCM senses overheating in the brakes or transaxle. Once cooled, the ABS/TCS becomes available and the dash lamp(s) extinguish. This self-protective operation is completely normal but may become a source of operator concern.
- When diagnosing an apparent TCS fault, make certain the TCS OFF switch is not engaged.
- If a faulty wheel speed signal is suspected, check tires and wheels for matching sizes and tire pressures.

If necessary, the vehicle must be test driven to recreate circumstances causing the fault.

NOTE: The operator of the vehicle is the best source of information about fault occurrences. An accurate description of driving conditions will allow the fault to be duplicated during the test drive.

1. Display, record and clear all fault codes in the EBCM.
2. Test drive the vehicle, attempting to recreate the failure conditions.
3. Safely stop the vehicle. Display and record the stored codes, if any.
4. If codes were stored, refer to the appropriate diagnostic chart. If no codes were stored, refer to the symptom chart.

Symptom Diagnosis

In the event that no fault codes are stored and intermittent conditions can not be recreated, diagnosis may be performed based on symptoms. The Symptom Chart coupled with a good operator description will be helpful in identifying components most likely to cause the problem. The Symptom Chart will direct the use of further diagnostic charts.

CLEARING CODES

Stored fault codes must be cleared using the bi-directional hand scanner. After repairs are completed, clear all stored codes and test drive the vehicle. Recheck for any newly-stored codes after the test drive.

DIAGNOSTIC CHART — ABS/TCS VISUAL INSPECTION — 1993

ABS/TCS VISUAL INSPECTION

ITEM	INSPECT FOR	CORRECTIVE ACTION
PARKING BRAKE	- FULL RELEASE - PROPER SWITCH FUNCTION WHEN NECESSARY, UNPLUG SWITCH CONNECTOR TO VERIFY	- OPERATE MANUAL RELEASE LEVER TO VERIFY RELEASE - ADJUST CABLE OR REPAIR RELEASE SYSTEM AS REQUIRED - REPAIR SWITCH AS REQUIRED - PROCEED TO ABS/TCS FUNCTIONAL CHECK
MAJOR COMPONENTS - MASTER CYLINDER FLUID RESERVOIR - PRESSURE MODULATOR VALVE (PMV) ASSEMBLY	- LOW FLUID LEVEL - LOW FLUID LEVEL IN PMV RESERVOIR - EXTERNAL LEAKS - PROPER ASSEMBLY	- DETERMINE CAUSE OF FLUID LOSS AND REPAIR. ADD FLUID TO MASTER CYLINDER RESERVOIR AS REQUIRED - DETERMINE CAUSE OF FLUID LOSS AND REPAIR. ADD FLUID TO MASTER CYLINDER RESERVOIR AS REQUIRED - REPAIR LEAKS AS REQUIRED - INSTALL OR POSITION COMPONENTS PROPERLY - PROCEED TO ABS/TCS FUNCTIONAL CHECK
IP FUSE BLOCK - FUSE 6 - FUSE 13 - FUSE 17 (ELITE ONLY) - FUSE 19 UNDERHOOD FUSE BLOCK - FUSE 1 - FUSE 2	- OPEN (BLOWN FUSE) - PROPER ENGAGEMENT	- REPLACE - PROCEED TO ABS/TCS FUNCTIONAL CHECK
CONNECTORS - MAIN RELAY - PUMP MOTOR RELAY - PMV FLUID LEVEL SWITCH - WHEEL SPEED SENSORS - ELECTRONIC BRAKE CONTROL MODULE (EBCM) - PRESSURE MODULATOR VALVE (PMV) ASSEMBLY CONNECTORS: C1 - VALVE BLOCK CONNECTOR (ROUND) C2 - PUMP MOTOR CONNECTOR (GRY) - MASTER CYLINDER RESERVOIR - CRUISE/SHIFT INTERLOCK/BRAKE SWITCH - TCC/ANTILOCK BRAKE SWITCH - C410, C111, C112 - AUTOMATIC TRANSAXLE (TRACTION CONTROL ONLY)	- PROPER ENGAGEMENT - LOOSE WIRES OR TERMINALS - CORRODED OR BROKEN EYELETS	- PROPERLY ENGAGE CONNECTOR - REPAIR AS REQUIRED - PROCEED TO ABS/TCS FUNCTIONAL CHECK
WHEN NO PROBLEM IS NOTED IN VISUAL INSPECTION, PROCEED TO ABS/TCS FUNCTIONAL CHECK		

DIAGNOSTIC CHART — ABS/TCS FUNCTIONAL CHECK — 1993

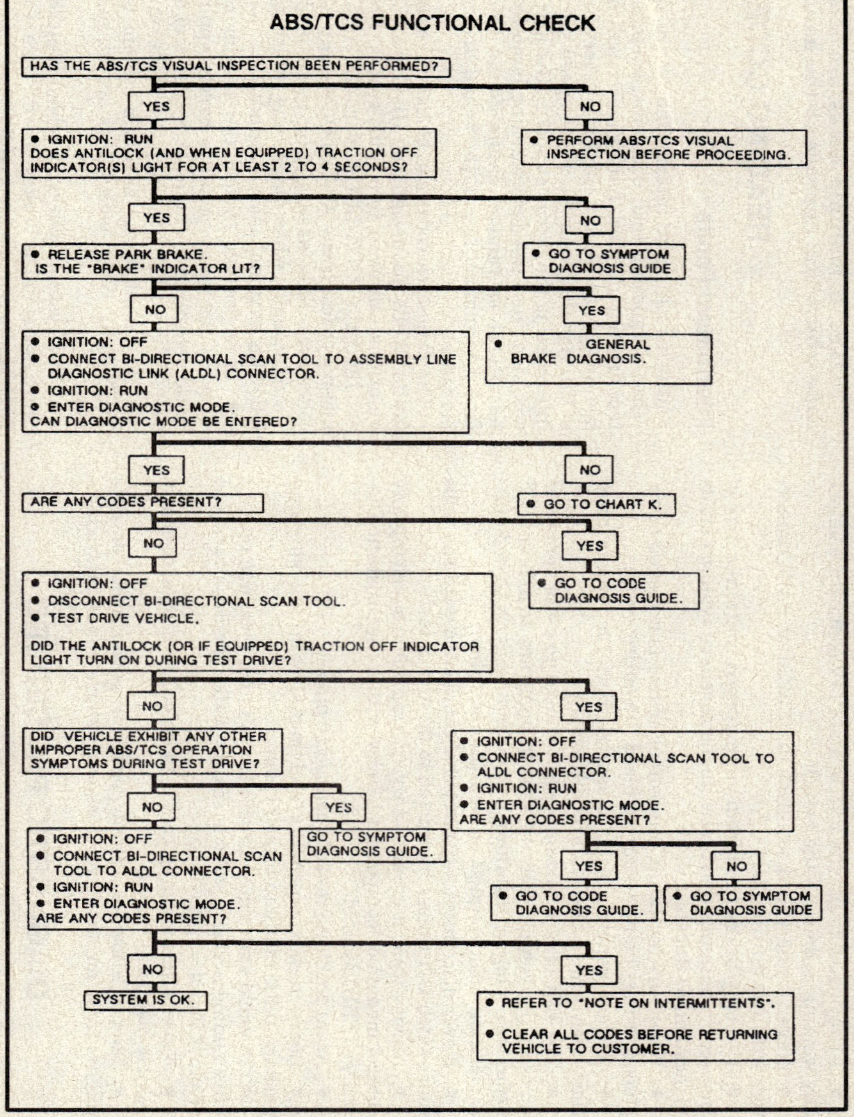

DIAGNOSTIC CHART — SYMPTOM DIAGNOSIS GUIDE — 1993

SYMPTOM DIAGNOSIS GUIDE

PERFORM THE ABS/TCS FUNCTIONAL CHECK BEFORE PROCEEDING WITH ANY DIAGNOSIS.

SYMPTOM	FOR DIAGNOSIS GO TO CHART
Diagnostic mode cannot be entered.	Chart K
ANTILOCK indicator is on all the time and no codes are set.	Chart K
ANTILOCK indicator does not light.	Chart N
Poor vehicle tracking during ABS stop (ABS engaged).	Chart L
Brake pedal rises or drops excessively during an ABS stop (ABS engaged).	Chart I
Pressure modulator valve (PMV) assembly pump motor runs continuously.	Chart M
Only TRACTION OFF Indicator does not light.	Chart Q
Only TRACTION OFF Indicator is on all the time and no codes are set.	Chart R
Traction control engaged message/bulb does not operate properly	Chart S

Important

- Customers may comment that they hear or feel the PMV assembly run at slow speeds (11 km/h [7 mph]). This is simply the ABS system running through its self-test. This is a normal condition and in no way signifies a problem with the ABS system.
- It is possible to feel the self-test through the brake pedal, when it is barely applied, and to hear brake fluid return to the master cylinder. Also, the vehicle may slowly decrease in speed while the PMV modulator valves increase fluid pressure to the wheel cylinders during this test. In both these instances, the driver must be traveling at a very low speed (11 km/h [7 mph]) and have one foot touching the brake pedal (as some two-footed drivers may). This is a normal condition and in no way signifies a problem with the ABS system.
- When vehicle is equipped with traction control, any symptom which lights the "ANTILOCK" indicator also lights the "TRACTION OFF" indicator. Therefore, when both indicators are lit, follow "ANTILOCK" indicator diagnosis when no codes are set.
- Vehicles with traction control contain a preventive system to avoid brake overheating. During heavy braking conditions, if a traction control mode is entered and the EBCM determines the brakes are too hot, the EBCM disables the traction control system and lights only the "TRACTION OFF" Indicator. After a sufficient cooling time, approximately 15–20 minutes, the EBCM enables traction control system and turns off the indicator. **This condition is normal and should not be mistaken as an intermittent system malfunction.**

DIAGNOSTIC CHART — CODE DIAGNOSIS GUIDE — 1993

CODE DIAGNOSIS GUIDE

PERFORM THE ABS/TCS FUNCTIONAL CHECK BEFORE PROCEEDING WITH ANY DIAGNOSIS.

CODE	DESCRIPTION	CHART	CODE	DESCRIPTION	CHART
21	RF speed sensor circuit open	B	45	LF inlet valve circuit	A
22	RF speed sensor signal erratic	D	46	LF outlet valve circuit	E
23	RF wheel speed is 0 mph	D	47	LF speed sensor noisy	D
25	LF speed sensor circuit open	B	48	RF isolation valve circuit	O
26	LF speed sensor signal erratic	D	51	RR inlet valve circuit	E
27	LF wheel speed is 0 mph	D	52	RR outlet valve circuit	E
31	RR speed sensor circuit open	C	53	RR speed sensor noisy	D
32	RR speed sensor signal is erratic	D	55	LR inlet valve circuit	E
33	RR wheel speed is 0 mph	D	56	LR outlet valve circuit	E
35	LR speed sensor circuit open	C	57	LR speed sensor noisy	D
36	LR speed sensor signal erratic	D	61	Pump motor test fault	F
37	LR wheel speed is 0 mph	D	62	Pump motor fault in ABS stop	I
41	RF inlet valve circuit	E	71	EBCM problem	J
42	RF outlet valve circuit	E	72	TCC/antilock brake switch circuit	G
43	RF speed sensor noisy	D	73	Fluid level switch circuit	H
44	LF isolation valve circuit	O	74	PMV pressure switch circuit	P

INTERMITTENT "HISTORY" CODES:
STORED CODES THAT DO NOT LIGHT THE AMBER "ANTILOCK" INDICATOR ARE INTERMITTENT PROBLEMS. DO NOT USE TROUBLE CODE DIAGNOSIS FOR INTERMITTENT PROBLEMS

MULTIPLE CODES: DIAGNOSE FIRST CODE THAT APPEARS ON BI-DIRECTIONAL SCAN TOOL SCREEN.

CLEARING CODES:

- CLEAR CODES USING A BI-DIRECTIONAL SCAN TOOL.
- CYCLE IGNITION SWITCH.
- AFTER CODES HAVE BEEN CLEARED, REFER TO "ABS/TCS FUNCTIONAL CHECK"

DIAGNOSTIC CHART A — 1993

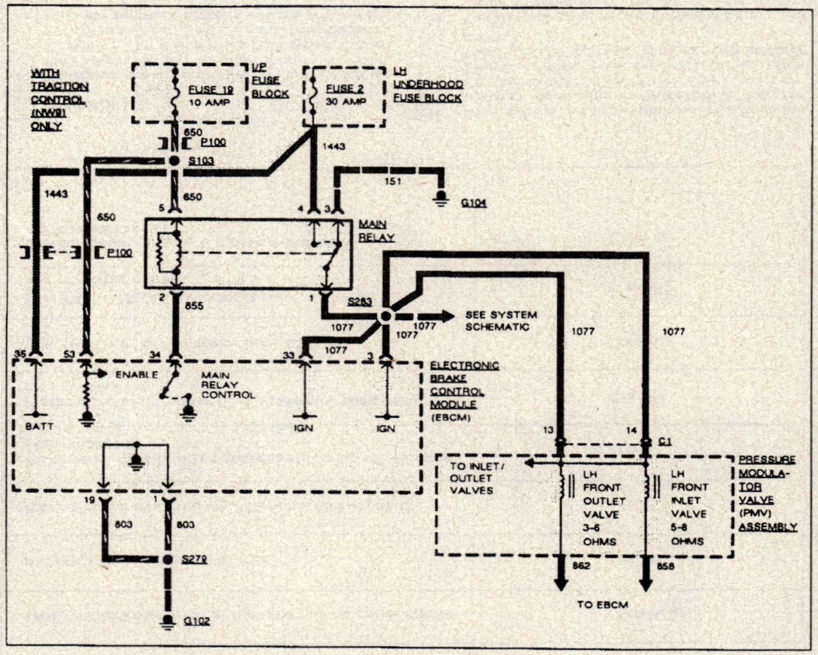

CHART A
INLET/OUTLET VALVE VOLTAGE TEST

CIRCUIT OPERATION

When the ignition is first turned to "RUN," the electronic brake control module (EBCM) goes through a self-check lasting 2-4 seconds. During this test the EBCM grounds terminal 34, energizing the main relay. Once the self-check is completed the main relay remains energized. This delivers battery voltage to both the EBCM and the pressure modulator valve (PMV) assembly during vehicle operation. The constant voltage supply allows the vehicle to enter ABS mode without delay. The main relay is de-energized whenever the amber "ANTILOCK" indicator lights due to a system fault, causing the ABS system to shut down.

FAILURE CONDITIONS

Code 45 sets when the ignition switch is at "RUN" and battery voltage is neither detected at the PMV assembly nor EBCM terminal 20, 3 and/or 33. Code 45 also sets when the EBCM detects an open or a short to ground in the LH inlet valve circuit.

Possible Causes

- Open in CKT 650, 855, 858, 1077 or 1443.
- Short to ground in CKT 858, 1077 or 1443.
- Main relay fails "open."

DIAGNOSTIC CHART A — 1993, CONTINUED

HAS THE ABS/TCS FUNCTIONAL CHECK BEEN PERFORMED?

- NO → PERFORM ABS/TCS FUNCTIONAL CHECK BEFORE PROCEEDING.
- YES ↓

- IGNITION: OFF
- VISUALLY INSPECT FUSE 2 OF THE LH UNDER-HOOD FUSE BLOCK. (INSPECT FOR OPEN).

IS THE FUSE OK?

- NO → • DISCONNECT MAIN RELAY. CHECK/REPAIR CKT 1443 AND 1077 FOR A SHORT TO GROUND AND REPLACE FUSE. WHEN CKT 1077 IS SHORTED TO GROUND, REPLACE MAIN RELAY.
- YES ↓

- DISCONNECT MAIN RELAY.
- MEASURE VOLTAGE FROM MAIN RELAY HARNESS CONNECTOR TERMINAL 4 TO GROUND.

IS BATTERY VOLTAGE MEASURED?

- NO → REPAIR OPEN IN CKT 1443.
- YES ↓

- IGNITION: RUN
- MEASURE VOLTAGE FROM MAIN RELAY HARNESS CONNECTOR TERMINAL 5 TO GROUND.

IS BATTERY VOLTAGE MEASURED?

- NO → REPAIR OPEN IN CKT 650.
- YES ↓

- IGNITION: OFF
- PICK UP MAIN RELAY.
- MEASURE RESISTANCE ACROSS MAIN RELAY TERMINALS 2 AND 5.

IS 45-90 OHMS MEASURED?

- NO → REPLACE MAIN RELAY.
- YES ↓

- DISCONNECT PRESSURE MODULATOR VALVE (PMV) ASSEMBLY CONNECTOR, C1 (TWIST OFF).
- DISCONNECT ELECTRONIC BRAKE CONTROL MODULE (EBCM) CONNECTOR.
- CONNECT J 38716 PINOUT BOX.
- MEASURE RESISTANCE BETWEEN J 38716 PINOUT BOX TERMINAL 1 AND ASSEMBLY LINE DIAGNOSTIC LINK (ALDL) CONNECTOR TERMINAL A, AND BETWEEN J 38716 PINOUT BOX TERMINAL 19 AND ALDL CONNECTOR TERMINAL A.

IS LESS THAN 2 OHMS MEASURED AT BOTH TERMINALS?

- NO → REPAIR OPEN IN CKT 803.
- YES → A

CONTINUED

AFTER REPAIRS ARE COMPLETED, CLEAR CODES AND REPEAT ABS/TCS FUNCTIONAL CHECK.

DIAGNOSTIC CHART A — 1993, CONTINUED

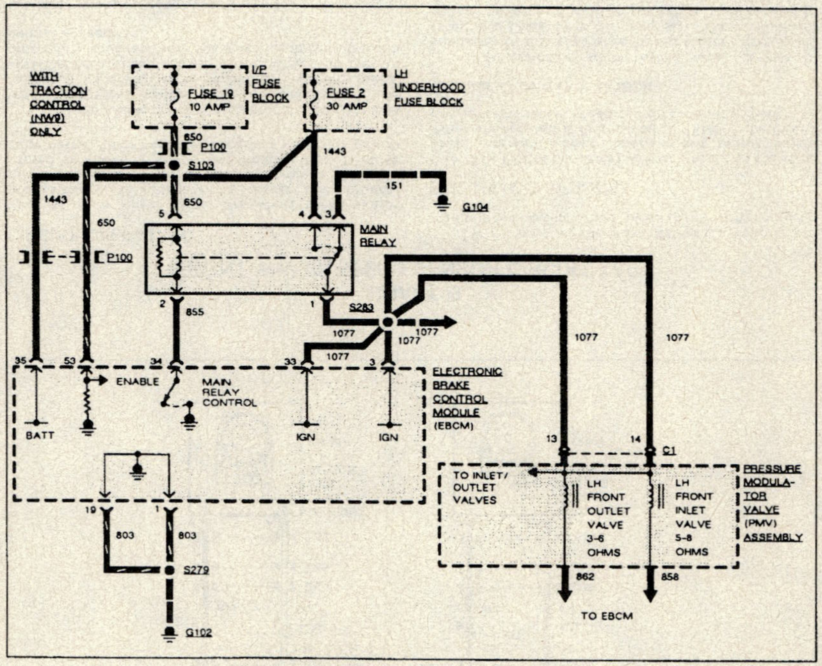

CHART A
INLET/OUTLET VALVE VOLTAGE TEST

CIRCUIT OPERATION

When the ignition is first turned to "RUN," the electronic brake control module (EBCM) goes through a self-check lasting 2–4 seconds. During this test the EBCM grounds terminal 34, energizing the main relay. Once the self-check is completed the main relay remains energized. This delivers battery voltage to both the EBCM and the pressure modulator valve (PMV) assembly during vehicle operation. The constant voltage supply allows the vehicle to enter ABS mode without delay. The main relay is de-energized whenever the amber "ANTILOCK" indicator lights due to a system fault, causing the ABS system to shut down.

FAILURE CONDITIONS

Code 45 sets when the ignition switch is at "RUN" and battery voltage is neither detected at the PMV assembly nor EBCM terminal 20, 3 and/or 33. Code 45 also sets when the EBCM detects an open or a short to ground in the LH inlet valve circuit.

Possible Causes

- Open in CKT 650, 855, 858, 1077 or 1443.
- Short to ground in CKT 858, 1077 or 1443.
- Main relay fails "open."

DIAGNOSTIC CHART A — 1993, CONTINUED

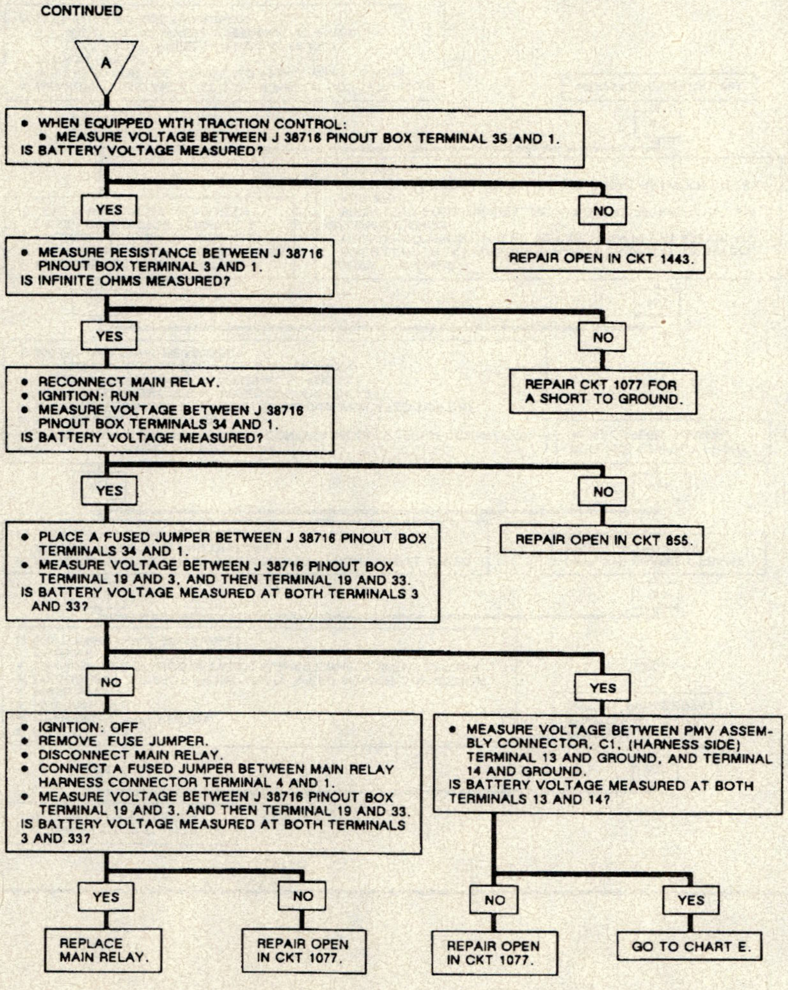

DIAGNOSTIC CHART B — 1993

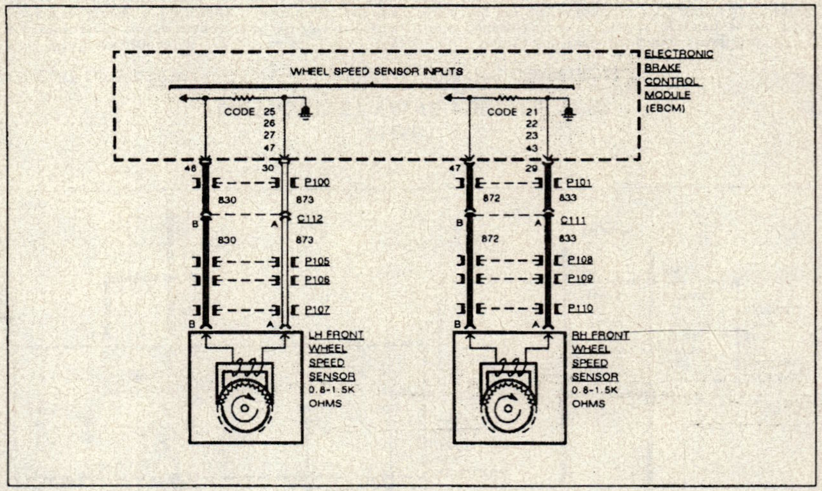

CHART B
FRONT WHEEL SPEED SENSOR CIRCUIT TEST

CIRCUIT OPERATION

As each wheel turns, the wheel speed sensor creates a small AC voltage with a frequency proportional to wheel speed. This magnetically induced voltage is caused by a toothed sensor ring (mounted on the drive axle) passing the wheel speed sensor's stationary pickup coil.

The electronic brake control module (EBCM) uses the wheel speed sensor signal to calculate vehicle reference speeds and individual wheel speed, acceleration and slip values. These values are used to determine when antilock (or when equipped) traction control is required.

WHEEL SPEED SENSOR CABLES

In order to prevent electromagnetic interference from disturbing the wheel speed sensor signal, the cables from the EBCM to each wheel speed sensor are twisted in pairs a minimum of 6 to 9 turns per foot.

When servicing the sensor cables it is important to maintain the original cables' twists (6 to 9 turns per foot).

FAILURE CONDITIONS

The EBCM performs two basic checks on the wheel speed sensors: sensor continuity and sensor output. When the ignition is first turned to "RUN," the EBCM performs the wheel speed sensor continuity check.

Sensor Continuity Check

Any condition which would result in lack of continuity in the front wheel speed sensor circuit could result in a Code 21 or 25 being set. These conditions include an open, a short to ground, or a short between the two wheel speed sensor circuits in the wiring harness between the speed sensor and EBCM. Also an open or short to ground across the wheel speed sensor coil could trigger this code.

DIAGNOSTIC CHART B — 1993, CONTINUED

HAS THE ABS/TCS FUNCTIONAL CHECK BEEN PERFORMED?

- NO → PERFORM ABS/TCS FUNCTIONAL CHECK BEFORE PROCEEDING.
- YES →
  - IGNITION: RUN
  - GEAR SELECTOR: NEUTRAL
  - IGNITION: OFF
  - RAISE VEHICLE.
  - DISCONNECT SUSPECT FRONT WHEEL SPEED SENSOR CONNECTOR.
  - MEASURE RESISTANCE BETWEEN SUSPECT WHEEL SPEED SENSOR TERMINAL A AND B.

  IS 800 TO 1500 OHMS MEASURED?

- NO → REPLACE SUSPECT FRONT WHEEL SPEED SENSOR.
- YES →
  - MEASURE RESISTANCE BETWEEN SUSPECT FRONT WHEEL SPEED SENSOR TERMINAL B AND GROUND.

  IS LESS THAN 5 OHMS MEASURED?

- YES → REPLACE SUSPECT FRONT WHEEL SPEED SENSOR.
- NO →
  - RECONNECT WHEEL SPEED SENSOR.
  - DISCONNECT ELECTRONIC BRAKE CONTROL MODULE (EBCM CONNECTOR.
  - INSTALL J 38716 PINOUT BOX.
  - MEASURE RESISTANCE BETWEEN TERMINALS INDICATED FOR THE SUSPECTED FRONT WHEEL SPEED SENSOR CIRCUIT:
    LEFT FRONT WHEEL - TERMINAL 30 AND 48
    RIGHT FRONT WHEEL - TERMINAL 29 AND 47

  IS 800 TO 1500 OHMS MEASURED?

- NO → ABOVE 1500 OHMS:
  WHEN LEFT FRONT WHEEL , REPAIR OPEN IN CKT 830 OR 873.
  WHEN RIGHT FRONT WHEEL, REPAIR OPEN IN CKT 833 OR 872.
  BELOW 800 OHMS:
  WHEN LEFT FRONT WHEEL, REPAIR SHORT BETWEEN CKT 830 AND 873.
  WHEN RIGHT FRONT WHEEL, REPAIR SHORT BETWEEN CKT 833 AND 872.
- YES →
  - MEASURE RESISTANCE BETWEEN J 38716 PINOUT BOX TERMINAL 1 AND ASSEMBLY LINE DIAGNOSTIC LINK (ALDL) CONNECTOR TERMINAL A.

  IS LESS THAN 2 OHMS MEASURED?

- NO → REPAIR OPEN IN CKT 803.
- YES →
  - MEASURE RESISTANCE AT J 38716 PINOUT BOX TERMINALS INDICATED FOR THE SUSPECTED FRONT WHEEL SPEED SENSOR CIRCUIT:
    LEFT FRONT WHEEL - TERMINAL 48 AND 1
    RIGHT FRONT WHEEL - TERMINAL 47 AND 1

  IS LESS THAN 5 OHMS MEASURED?

- NO → CHECK ELECTRONIC BRAKE CONTROL MODULE (EBCM) TERMINAL CONNECTIONS FOR POOR TERMINAL CONTACT. IF TERMINAL CONTACT IS OK, REPLACE EBCM.
- YES → WHEN LEFT FRONT WHEEL , REPAIR SHORT TO GROUND IN CKT 830 OR 873. WHEN RIGHT FRONT WHEEL, REPAIR SHORT TO GROUND IN CKT 833 OR 872.

AFTER REPAIRS ARE COMPLETED, CLEAR CODES AND REPEAT ABS/TCS FUNCTIONAL CHECK.

DIAGNOSTIC CHART C — 1993

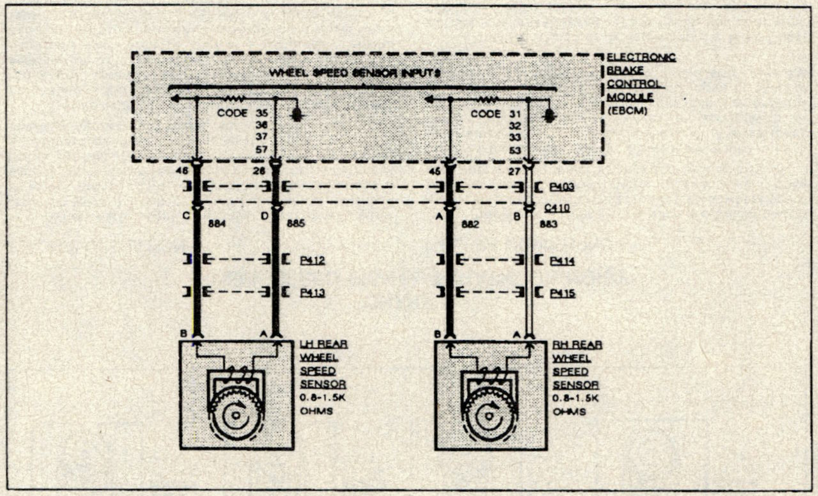

CHART C
REAR WHEEL SPEED SENSOR CIRCUIT TEST

CIRCUIT OPERATION

When each wheel turns, the wheel speed sensor creates a small AC voltage with a frequency proportional to wheel speed. This magnetically induced voltage is caused by a toothed sensor ring passing the wheel speed sensor's stationary pickup coil. Both sensors are part of the hub and bearing assembly.

The electronic brake control module (EBCM) uses the wheel speed sensor signal to calculate vehicle reference speeds and individual wheel speed, acceleration and slip values. These values are used to determine when antilock control is required.

WHEEL SPEED SENSOR CABLES

In order to prevent electromagnetic interference from disturbing the wheel speed sensor signal, the cables from the EBCM to each wheel speed sensor are twisted in pairs a minimum of 6 to 9 turns per foot.

When servicing the sensor cables it is important to maintain the original cables' twists (6 to 9 per foot).

FAILURE CONDITIONS

The EBCM performs two basic checks on the wheel speed sensors: sensor continuity and sensor output. When the ignition is first turned to "RUN," the EBCM performs the wheel speed sensor continuity check.

Sensor Continuity Check

Any condition which would result in lack of continuity in the rear wheel speed sensor circuit could result in a Code 31 or 35 being set. These conditions include an open, a short to ground, or a short between the two wheel speed sensor circuits in the wiring harness between the speed sensor and EBCM. Also an open or short to ground across the wheel speed sensor coil could trigger this code.

DIAGNOSTIC CHART C — 1993, CONTINUED

HAS THE ABS/TCS FUNCTIONAL CHECK BEEN PERFORMED?

- **NO:** PERFORM ABS/TCS FUNCTIONAL CHECK BEFORE PROCEEDING.
- **YES:**
  - IGNITION: OFF
  - PARK BRAKE: RELEASED
  - RAISE VEHICLE.
  - DISCONNECT SUSPECT REAR WHEEL SPEED SENSOR CONNECTOR.
  - MEASURE RESISTANCE BETWEEN SUSPECT WHEEL SPEED SENSOR TERMINAL A AND B.

  IS 800 TO 1500 OHMS MEASURED?

  - **NO:** REPLACE SUSPECT REAR WHEEL SPEED SENSOR.
  - **YES:**
    - MEASURE RESISTANCE BETWEEN SUSPECT REAR WHEEL SPEED SENSOR TERMINAL B AND GROUND.

    IS LESS THAN 5 OHMS MEASURED?

    - **YES:** REPLACE SUSPECT REAR WHEEL SPEED SENSOR.
    - **NO:**
      - RECONNECT WHEEL SPEED SENSOR CONNECTOR.
      - DISCONNECT ELECTRONIC BRAKE CONTROL MODULE (EBCM) CONNECTOR.
      - INSTALL J 38716 PINOUT BOX.
      - MEASURE RESISTANCE BETWEEN TERMINALS INDICATED FOR THE SUSPECTED REAR WHEEL SPEED SENSOR CIRCUIT:
        LEFT REAR WHEEL - TERMINAL 28 AND 46
        RIGHT REAR WHEEL - TERMINAL 27 AND 45

      IS 800 TO 1500 OHMS MEASURED?

      - **NO:** ABOVE 1500 OHMS:
        WHEN LEFT REAR WHEEL , REPAIR OPEN IN CKT 884 OR 885.
        WHEN RIGHT REAR WHEEL, REPAIR OPEN IN CKT 882 OR 883.
        BELOW 800 OHMS:
        WHEN LEFT REAR WHEEL, REPAIR SHORT BETWEEN CKT 884 AND 885.
        WHEN RIGHT REAR WHEEL, REPAIR SHORT BETWEEN CKT 882 AND 883.
      - **YES:**
        - MEASURE RESISTANCE BETWEEN J 38716 PINOUT BOX TERMINAL 1 AND ASSEMBLY LINE DIAGNOSTIC LINK (ALDL) CONNECTOR TERMINAL A.

        IS LESS THAN 2 OHMS MEASURED?

        - **NO:** REPAIR OPEN IN CKT 803.
        - **YES:**
          - MEASURE RESISTANCE AT J 38716 PINOUT BOX TERMINALS INDICATED FOR THE SUSPECTED REAR WHEEL SPEED SENSOR CIRCUIT
            LEFT REAR WHEEL - TERMINAL 46 AND 1
            RIGHT REAR WHEEL - TERMINAL 45 AND 1

          IS LESS THAN 5 OHMS MEASURED?

          - **NO:** CHECK ELECTRONIC BRAKE CONTROL MODULE (EBCM) TERMINAL CONNECTIONS FOR POOR TERMINAL CONTACT. IF TERMINAL CONTACT IS OK, REPLACE EBCM.
          - **YES:** WHEN LEFT REAR WHEEL , REPAIR SHORT TO GROUND IN CKT 884 OR 885. WHEN RIGHT REAR WHEEL, REPAIR SHORT TO GROUND IN CKT 882 OR 883.

AFTER REPAIRS ARE COMPLETED, CLEAR CODES AND REPEAT ABS/TCS FUNCTIONAL CHECK.

DIAGNOSTIC CHART D — 1993

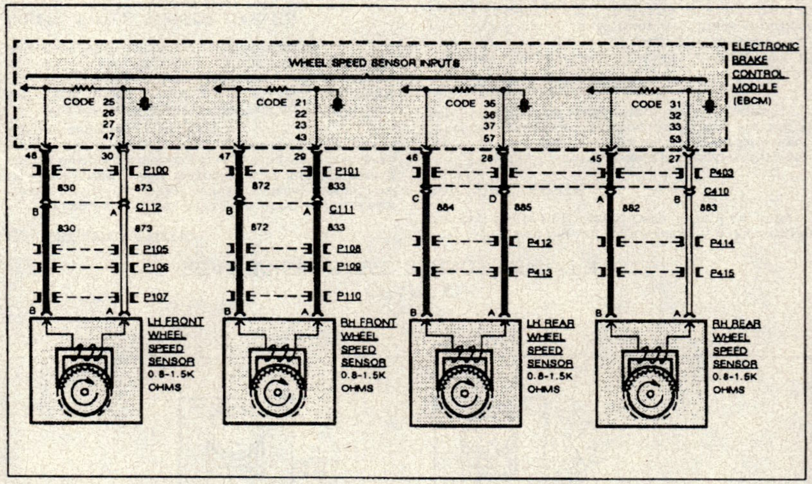

CHART D
WHEEL SPEED SIGNAL ERRATIC OR NOISY

CIRCUIT OPERATION

When each wheel turns, the wheel speed sensor creates a small AC voltage with a frequency proportional to wheel speed. This magnetically induced voltage is caused by a toothed sensor ring passing the wheel speed sensor's stationary pickup coil. The toothed ring sensor is mounted on the drive axle in the front, and in the hub/bearing assembly in the rear.

The electronic brake control module (EBCM) uses the wheel speed sensor signal to calculate vehicle reference speeds, and individual wheel speed, acceleration and slip values. These values are used to determine when antilock control is required. Vehicles equipped with traction control also use front wheel speed sensor signals to determine when traction control is required.

WHEEL SPEED SENSOR CABLES

In order to prevent electromagnetic interference from disturbing the wheel speed sensor signal, the cables from the EBCM to each wheel speed sensor are twisted in pairs a minimum of 6 to 9 turns per foot.

When servicing the sensor cables it is important to maintain the original cables' twists (6 to 9 turns per foot).

FAILURE CONDITIONS

Code 43, 47, 53 or 57 sets when the EBCM detects electromagnetic interference (noise) in a wheel speed sensor signal. Possible interference sources include accessory motors, wiper motors and ignition circuits.

Code 22, 26, 32 or 36 sets when the EBCM determines a wheel speed signal is erratic. A wheel speed signal indicating that a wheel is accelerating or decelerating faster than physically possible is an example of an erratic signal. An intermittent open or short to ground in a wheel speed sensor could also cause the EBCM to set a code.

Code 23, 27, 33 or 37 sets when the EBCM does not see any wheel speed sensor signal, but determines continuity exists in the circuit. This is caused by a dislodged wheel speed sensor, missing toothed sensor ring (front) or defective hub and bearing assembly (rear).

DIAGNOSTIC CHART D — 1993, CONTINUED

HAS THE ABS/TCS FUNCTIONAL CHECK BEEN PERFORMED?

- **NO:** PERFORM ABS/TCS FUNCTIONAL CHECK BEFORE PROCEEDING.
- **YES:** IS CODE 23, 27, 33 OR 37 SET?
  - **YES:** WHEN CODE 23 OR 27: REINSTALL DISLODGED WHEEL SPEED SENSOR OR REPLACE MISSING TOOTHED SENSOR RING. WHEN CODE 33 OR 37: REINSTALL DISLODGED WHEEL SPEED SENSOR OR REPLACE HUB AND BEARING ASSEMBLY.
  - **NO:** IS CODE 43, 47, 53 OR 57 SET?
    - **NO:**
      - CHECK THE FOLLOWING:

        TEST DRIVE VEHICLE

        WHILE MONITORING SUSPECT WHEEL SPEED SENSOR USING A BI-DIRECTIONAL SCAN TOOL SNAPSHOT CAPABILITY.

        DISCONNECT THE SUSPECT WHEEL SPEED SENSOR AND INSPECT TERMINAL CONTACTS ON SENSOR AND CONNECTOR FOR CORROSION OR DEFORMITY. REPAIR AS NECESSARY.

        TRACE HARNESS ROUTING OF SUSPECT SENSOR WIRING. INSPECT FOR PHYSICAL DAMAGE PARTICULARLY IN AREAS WHERE PINCHING OR MOVEMENT BY HARNESS OR OTHER COMPONENTS OCCURS.

        CHECK FOR TOOTH SENSOR IRREGULARITY, LOOSE SENSOR MOUNTING, OR DEBRIS LODGED BETWEEN THE TOOTHED SENSOR RING AND THE WHEEL SPEED SENSOR.

        PROCEED, BUT BE AWARE THAT THE PROBLEM CAUSING THE CODE TO SET MAY BE INTERMITTENT AND SETS AT SPEEDS OVER 40 KM/H (25 MPH).
    - **YES:**
      - CHECK THE FOLLOWING:

        TEST DRIVE VEHICLE

        MONITORING SUSPECT WHEEL SPEED SENSOR USING A BI-DIRECTIONAL SCAN TOOL SNAPSHOT CAPABILITY.

        TRACE HARNESS ROUTING OF SUSPECT SENSOR WIRING FOR PROXIMITY TO NOISE SOURCES SUCH AS IGNITION CIRCUITS, MOTORS, ETC. CORRECT AS NECESSARY.

        CHECK IF ANY REPAIRS WERE RECENTLY MADE TO SUSPECT SENSOR WIRING. SENSOR WIRING SHOULD BE TWISTED 6 TO 9 TURNS PER FOOT.

        CHECK FOR LOOSE SENSOR MOUNTING OR DEBRIS LODGED BETWEEN THE TOOTHED SENSOR RING AND THE WHEEL SPEED SENSOR.

        PROCEED, BUT BE AWARE THAT THE PROBLEM CAUSING THE CODE TO SET MAY BE INTERMITTENT AND SETS AT SPEEDS UNDER 40 KM/H (25 MPH).

Both branches then continue:

- IGNITION: RUN
- GEAR SELECTOR LEVER: NEUTRAL
- IGNITION: OFF
- PARK BRAKE: RELEASED
- RAISE VEHICLE
- DISCONNECT SUSPECT WHEEL SPEED SENSOR.
- MEASURE AC VOLTAGE BETWEEN SUSPECT WHEEL SPEED SENSOR TERMINAL A AND B WHILE TURNING DRIVE WHEELS BY HAND.

IS 50 TO 900 MILLIVOLTS AC MEASURED?

- **YES:** CHECK ELECTRONIC BRAKE CONTROL MODULE (EBCM) TERMINAL CONNECTIONS FOR POOR TERMINAL CONTACT. IF TERMINAL CONTACT IS OK, REPLACE EBCM.
- **NO:** REPLACE SUSPECT WHEEL SPEED SENSOR.

AFTER REPAIRS ARE COMPLETED, CLEAR CODES AND REPEAT ABS/TCS FUNCTIONAL CHECK.

DIAGNOSTIC CHART E — 1993

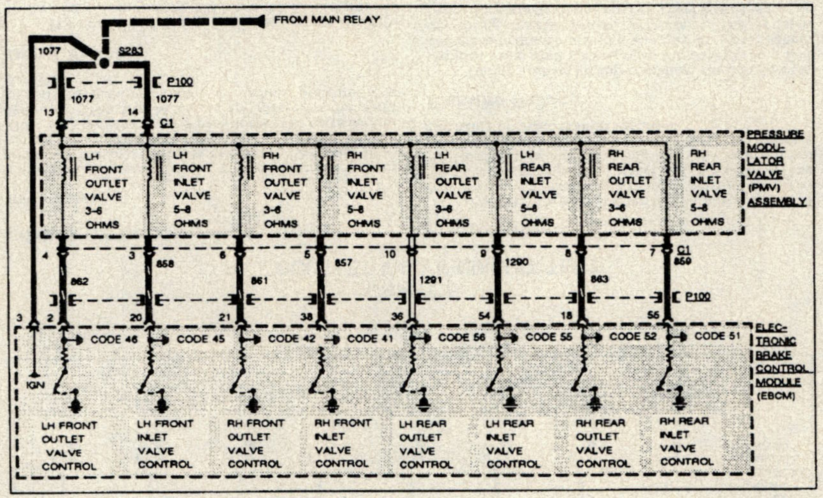

CHART E
INLET/OUTLET VALVE CIRCUIT TEST

CIRCUIT OPERATION

Inlet Valves

Voltage is applied to the normally open inlet valves whenever the main relay is energized. An inlet valve closes when the electronic brake control module (EBCM) supplies a ground path at the inlet valve control terminal. The EBCM closes an inlet valve for short periods to maintain (hold) or reduce pressure at a particular wheel.

Outlet Valves

The normally closed outlet valves operate similarly to the inlet valves, except that when the EBCM grounds an outlet valve, the valve opens instead of closes. The EBCM opens an outlet valve for short periods to reduce pressure at a particular wheel.

VALVE MODES

Pressure Increase

The pressure increase position is the operating valve position for normal (non-ABS) braking and traction control mode (when equipped). The EBCM does not energize either valve, leaving the inlet valve open and the outlet valve closed.

Pressure Hold

The pressure hold position occurs when the EBCM first notices an abnormal hydraulic pressure increase. This signifies an impending wheel lockup condition. The EBCM closes the corresponding inlet valve in an attempt to avoid any additional hydraulic pressure increase (avoiding an ABS braking condition).

Pressure Reduce

During a wheel lockup condition, the EBCM close the inlet valve while pulsing open the outlet valve. This decreases hydraulic pressure at a specific wheel without decreasing overall system pressure. The reduced hydraulic pressure allows greater wheel spin, eliminating the wheel lockup condition.

Test Description

This test checks for proper inlet and outlet valve operation. The wheel brake behavior must be checked to verify the valves hydraulic effectiveness. This is accomplished using the TECH 1 to actuate the valves.

Failure Conditions

Code 41, 42, 45, 46, 51, 52, 55, or 56 sets when the electronic brake control module (EBCM) detects an open or short to ground in the respective inlet or outlet valve circuit.

DIAGNOSTIC CHART E — 1993, CONTINUED

HAS THE ABS/TCS FUNCTIONAL CHECK BEEN PERFORMED?

- NO → PERFORM ABS/TCS FUNCTIONAL CHECK BEFORE PROCEEDING.
- YES →
  - IGNITION: OFF
  - DISCONNECT PRESSURE MODULATOR VALVE (PMV) ASSEMBLY VALVE CONNECTOR, C1.
  - MEASURE RESISTANCE BETWEEN CONNECTOR, C1, (PMV ASSEMBLY SIDE) TERMINAL 14 AND PMV ASSEMBLY CASING.

  IS INFINITE OHMS MEASURED?

- NO → REPLACE PMV ASSEMBLY.
- YES →
  - MEASURE RESISTANCE AT CONNECTOR, C1, (PMV ASSEMBLY SIDE) BETWEEN TERMINAL 14 AND THE SUSPECT VALVE TERMINAL, REFER TO CHART.

CODE/VALVE	SUSPECT TERMINAL	CORRECT RESISTANCE
45/LH FRONT INLET	3	5 TO 8 OHMS
41/RH FRONT INLET	5	
55/LH REAR INLET	9	
51/RH REAR INLET	7	
46/LH FRONT OUTLET	4	3 TO 6 OHMS
42/RH FRONT OUTLET	6	
56/LH REAR OUTLET	10	
52/RH REAR OUTLET	8	

IS THE CORRECT RESISTANCE MEASURED?

- NO → REPLACE PMV ASSEMBLY.
- YES → A

CONTINUED

AFTER REPAIRS ARE COMPLETED, CLEAR CODES AND REPEAT ABS/TCS FUNCTIONAL CHECK.

DIAGNOSTIC CHART F — 1993

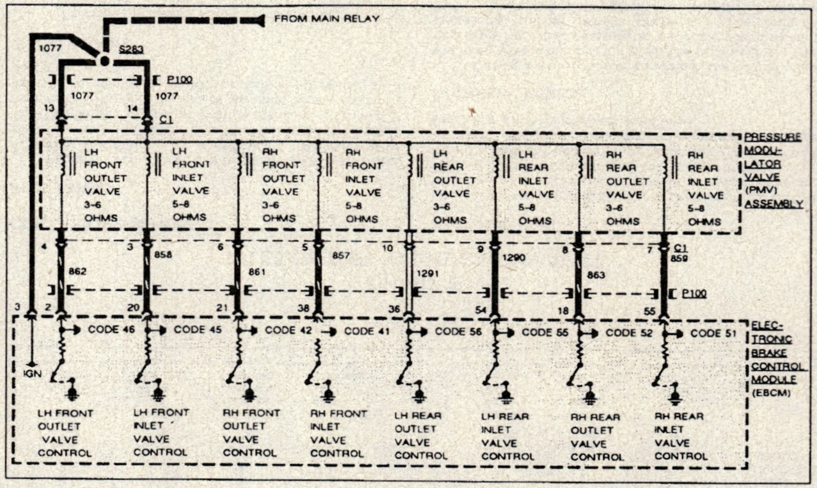

CHART E
INLET/OUTLET VALVE CIRCUIT TEST

CIRCUIT OPERATION

Inlet Valves

Voltage is applied to the normally open inlet valves whenever the main relay is energized. An inlet valve closes when the electronic brake control module (EBCM) supplies a ground path at the inlet valve control terminal. The EBCM closes an inlet valve for short periods to maintain (hold) or reduce pressure at a particular wheel.

Outlet Valves

The normally closed outlet valves operate similarly to the inlet valves, except that when the EBCM grounds an outlet valve, the valve opens instead of closes. The EBCM opens an outlet valve for short periods to reduce pressure at a particular wheel.

VALVE MODES

Pressure Increase

The pressure increase position is the operating valve position for normal (non-ABS) braking and traction control mode (when equipped). The EBCM does not energize either valve, leaving the inlet valve open and the outlet valve closed.

Pressure Hold

The pressure hold position occurs when the EBCM first notices an abnormal hydraulic pressure increase. This signifies an impending wheel lockup condition. The EBCM closes the corresponding inlet valve in an attempt to avoid any additional hydraulic pressure increase (avoiding an ABS braking condition).

Pressure Reduce

During a wheel lockup condition, the EBCM close the inlet valve while pulsing open the outlet valve. This decreases hydraulic pressure at a specific wheel without decreasing overall system pressure. The reduced hydraulic pressure allows greater wheel spin, eliminating the wheel lockup condition.

Test Description

This test checks for proper inlet and outlet valve operation. The wheel brake behavior must be checked to verify the valves hydraulic effectiveness. This is accomplished using the TECH 1 to actuate the valves.

Failure Conditions

Code 41, 42, 45, 46, 51, 52, 55, or 56 sets when the electronic brake control module (EBCM) detects an open or short to ground in the respective inlet or outlet valve circuit.

DIAGNOSTIC CHART F — 1993, CONTINUED

CONTINUED

A

- RECONNECT PMV ASSEMBLY VALVE CONNECTOR, C1.
- DISCONNECT/ELECTRONIC BRAKE CONTROL MODULE (EBCM) CONNECTOR.
- INSTALL J 38716 PINOUT BOX.
- MEASURE RESISTANCE BETWEEN J 38716 PINOUT BOX TERMINAL 3 AND THE SUSPECT VALVE TERMINAL. REFER TO CHART.

CODE/VALVE	J 38716 TERMINAL	CORRECT RESISTANCE
45/LH FRONT INLET	20	5 TO 8 OHMS
41/RH FRONT INLET	38	
55/LH REAR INLET	54	
51/RH REAR INLET	55	
46/LH FRONT OUTLET	2	3 TO 6 OHMS
42/RH FRONT OUTLET	21	
56/LH REAR OUTLET	36	
52/RH REAR OUTLET	18	

IS RESISTANCE MEASURED LESS THAN 10 OHMS?

YES
- MEASURE RESISTANCE BETWEEN J 38716 PINOUT BOX TERMINAL 1 AND ASSEMBLY LINE DIAGNOSTIC LINK (ALDL) CONNECTOR TERMINAL A.

IS LESS THAN 2 OHMS MEASURED?

NO
REPAIR OPEN IN SUSPECT VALVE CIRCUIT BETWEEN PRESSURE MODULATOR VALVE (PMV) ASSEMBLY AND ELECTRONIC BRAKE CONTROL MODULE (EBCM).

YES
- DISCONNECT MAIN RELAY.
- MEASURE RESISTANCE BETWEEN J 38716 PINOUT BOX TERMINAL 1 AND THE SUSPECT VALVE TERMINAL. REFER TO CHART.

CODE/VALVE	J 38716 TERMINAL	CORRECT RESISTANCE
45/LH FRONT INLET	20	INFINITE OHMS
41/RH FRONT INLET	38	
55/LH REAR INLET	54	
51/RH REAR INLET	55	
46/LH FRONT OUTLET	2	
42/RH FRONT OUTLET	21	
56/LH REAR OUTLET	36	
52/RH REAR OUTLET	18	

IS LESS THAN 5 OHMS MEASURED?

NO
REPAIR OPEN IN CKT 803.

NO
GO TO CHART L. WHEN NO PROBLEM IS FOUND, CHECK ELECTRONIC BRAKE CONTROL MODULE (EBCM) TERMINAL CONNECTIONS FOR POOR TERMINAL CONTACT. IF TERMINAL CONTACT IS OK, REPLACE EBCM.

YES
REPAIR SHORT TO GROUND IN SUSPECT VALVE CIRCUIT BETWEEN PRESSURE MODULATOR VALVE (PMV) ASSEMBLY AND ELECTRONIC BRAKE CONTROL MODULE (EBCM).

AFTER REPAIRS ARE COMPLETED, CLEAR CODES AND REPEAT ABS/TCS FUNCTIONAL CHECK.

DIAGNOSTIC CHART F — 1993, CONTINUED

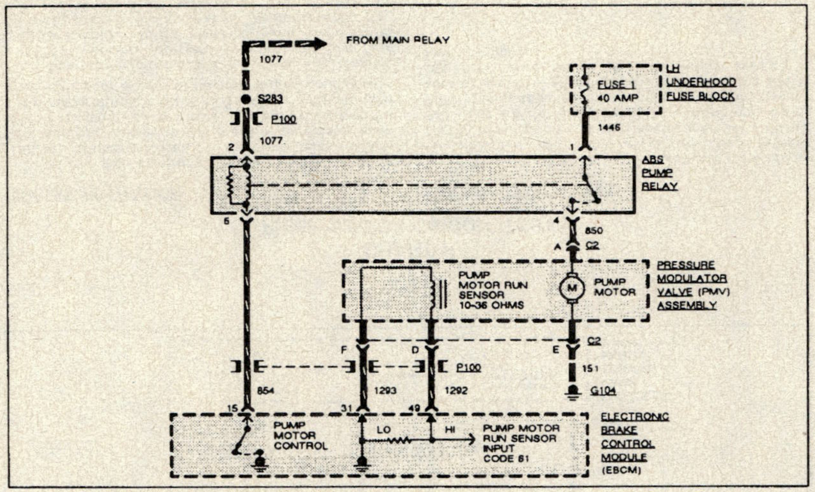

CHART F
PUMP MOTOR CIRCUIT TEST

CIRCUIT OPERATION

When the electronic brake control module (EBCM) grounds terminal 15, the pump relay contacts close applying voltage to the pressure modulator valve (PMV) assembly's pump motor. The EBCM turns on the pump motor during most ABS braking conditions (and when equipped) during all traction control conditions.

When the pump motor runs during ABS braking, the PMV assembly reservoir is replenished and fluid pressure is applied to the master cylinder. This causes the brake pedal to rise gradually. The pump motor runs until the brake pedal sufficiently rises to close the TCC/antilock brake switch.

During traction control mode, after the front hydraulic brake circuits are isolated, the pump motor runs to increase brake fluid pressure in these circuits. The increased pressure is directed to the slipping wheel through the PMV assembly's front inlet and outlet valves. This causes the brakes to be applied, reducing wheel spin at the slipping wheel. Traction control is disabled as soon as the brakes are manually applied.

The EBCM monitors pump motor operation via the pump motor run sensor. When the pump is operating, voltage is induced into the sensor's coil and the EBCM monitors this voltage at terminals 49 and 31.

FAILURE CONDITIONS

When the EBCM grounds terminal 15, Code 61 sets when the EBCM determines the voltage between EBCM terminals 31 and 49 is not between 500–800 mV AC. The EBCM recognizes this condition only during ABS braking requiring pump motor operation, traction control mode (when equipped), and during the EBCM self-check which occurs when the vehicle reaches 11 km/h (7 mph).

Code 61 also sets when the pump motor runs all the time (i.e. EBCM sees proper voltage between terminals 31 and 49 without grounding terminal 15). Refer to Chart M for diagnosing a continuous running pump motor.

Possible Causes

- Open in CKT 151, 850, 854, 1077, 1292, 1293 or 1446.
- Short to ground in CKT 850, 854, 1292, 1293 or 1446.
- ABS pump relay fails "open" or "engaged."

DIAGNOSTIC CHART F — 1993, CONTINUED

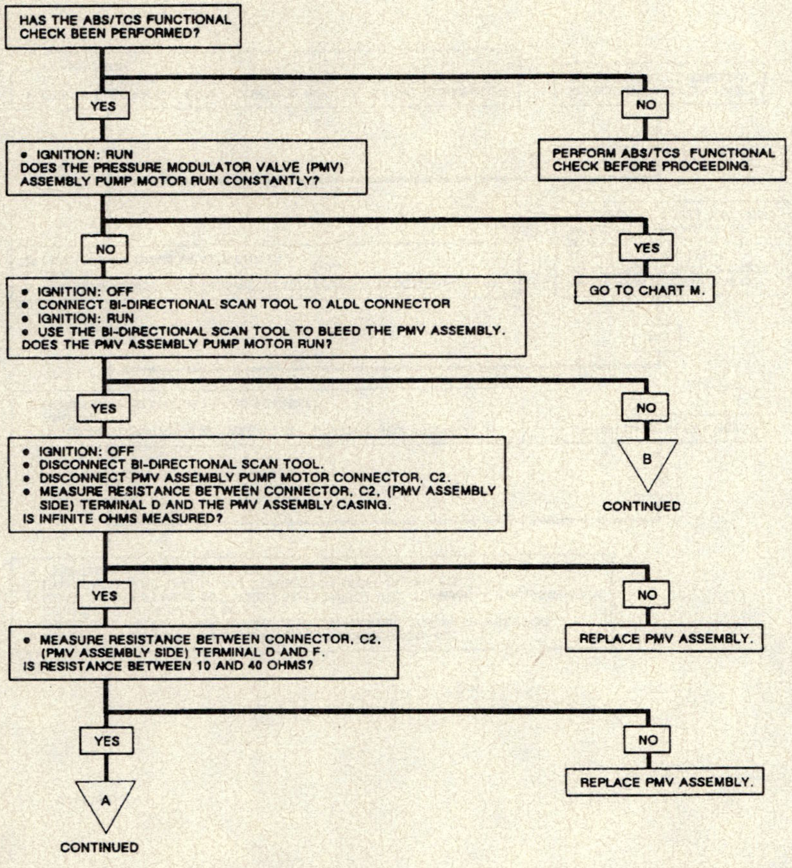

AFTER REPAIRS ARE COMPLETED, CLEAR CODES AND REPEAT ABS/TCS FUNCTIONAL CHECK.

DIAGNOSTIC CHART F — 1993, CONTINUED

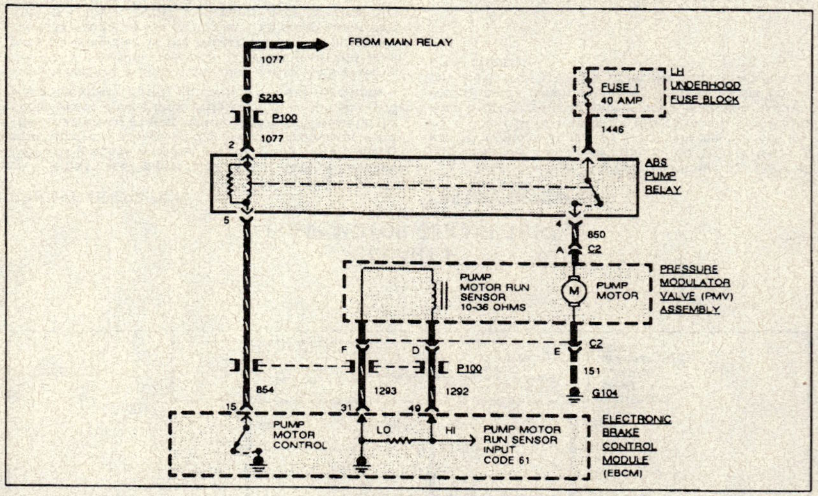

CHART F
PUMP MOTOR CIRCUIT TEST

CIRCUIT OPERATION

When the electronic brake control module (EBCM) grounds terminal 15, the pump relay contacts close applying voltage to the pressure modulator valve (PMV) assembly's pump motor. The EBCM turns on the pump motor during most ABS braking conditions (and when equipped) during all traction control conditions.

When the pump motor runs during ABS braking, the PMV assembly reservoir is replenished and fluid pressure is applied to the master cylinder. This causes the brake pedal to rise gradually. The pump motor runs until the brake pedal sufficiently rises to close the TCC/antilock brake switch.

During traction control mode, after the front hydraulic brake circuits are isolated, the pump motor runs to increase brake fluid pressure in these circuits. The increased pressure is directed to the slipping wheel through the PMV assembly's front inlet and outlet valves. This causes the brakes to be applied, reducing wheel spin at the slipping wheel. Traction control is disabled as soon as the brakes are manually applied.

The EBCM monitors pump motor operation via the pump motor run sensor. When the pump is operating, voltage is induced into the sensor's coil and the EBCM monitors this voltage at terminals 49 and 31.

FAILURE CONDITIONS

When the EBCM grounds terminal 15, Code 61 sets when the EBCM determines the voltage between EBCM terminals 31 and 49 is not between 500–800 mV AC. The EBCM recognizes this condition only during ABS braking requiring pump motor operation, traction control mode (when equipped), and during the EBCM self-check which occurs when the vehicle reaches 11 km/h (7 mph).

Code 61 also sets when the pump motor runs all the time (i.e. EBCM sees proper voltage between terminals 31 and 49 without grounding terminal 15). Refer to Chart M for diagnosing a continuous running pump motor.

Possible Causes

- Open in CKT 151, 850, 854, 1077, 1292, 1293 or 1446.
- Short to ground in CKT 850, 854, 1292, 1293 or 1446.
- ABS pump relay fails "open" or "engaged."

DIAGNOSTIC CHART F — 1993, CONTINUED

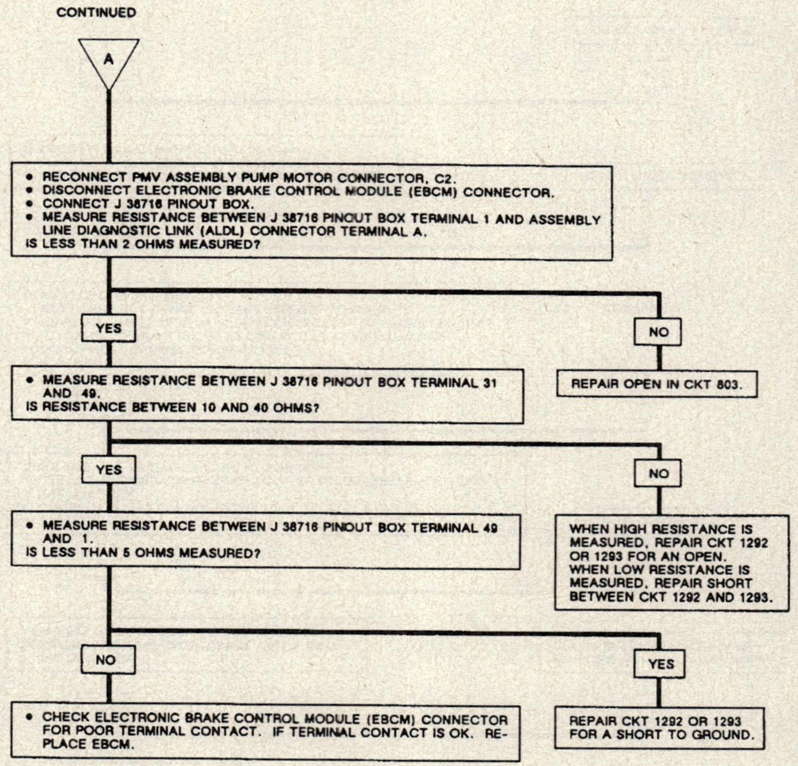

AFTER REPAIRS ARE COMPLETED, CLEAR CODES AND REPEAT ABS/TCS FUNCTIONAL CHECK.

DIAGNOSTIC CHART F — 1993, CONTINUED

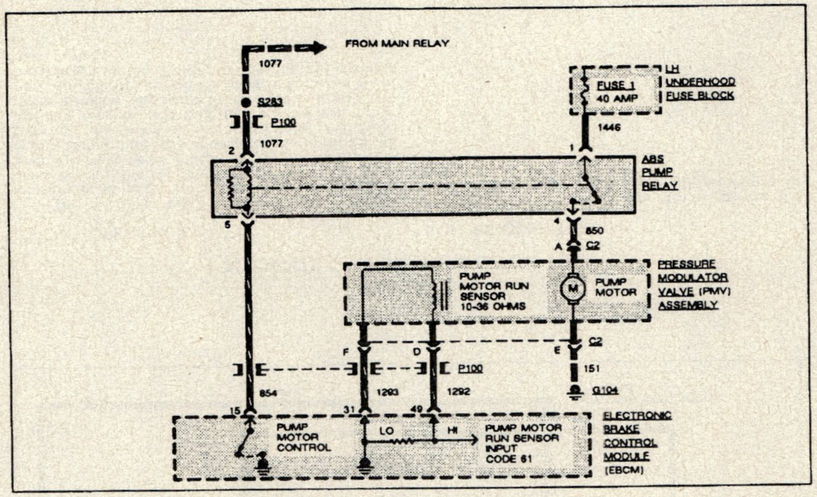

CHART F
PUMP MOTOR CIRCUIT TEST

CIRCUIT OPERATION

When the electronic brake control module (EBCM) grounds terminal 15, the pump relay contacts close applying voltage to the pressure modulator valve (PMV) assembly's pump motor. The EBCM turns on the pump motor during most ABS braking conditions (and when equipped) during all traction control conditions.

When the pump motor runs during ABS braking, the PMV assembly reservoir is replenished and fluid pressure is applied to the master cylinder. This causes the brake pedal to rise gradually. The pump motor runs until the brake pedal sufficiently rises to close the TCC/antilock brake switch.

During traction control mode, after the front hydraulic brake circuits are isolated, the pump motor runs to increase brake fluid pressure in these circuits. The increased pressure is directed to the slipping wheel through the PMV assembly's front inlet and outlet valves. This causes the brakes to be applied, reducing wheel spin at the slipping wheel. Traction control is disabled as soon as the brakes are manually applied.

The EBCM monitors pump motor operation via the pump motor run sensor. When the pump is operating, voltage is induced into the sensor's coil and the EBCM monitors this voltage at terminals 49 and 31.

FAILURE CONDITIONS

When the EBCM grounds terminal 15, Code 61 sets when the EBCM determines the voltage between EBCM terminals 31 and 49 is not between 500–800 mV AC. The EBCM recognizes this condition only during ABS braking requiring pump motor operation, traction control mode (when equipped), and during the EBCM self-check which occurs when the vehicle reaches 11 km/h (7 mph).

Code 61 also sets when the pump motor runs all the time (i.e. EBCM sees proper voltage between terminals 31 and 49 without grounding terminal 15). Refer to Chart M for diagnosing a continuous running pump motor.

Possible Causes

- Open in CKT 151, 850, 854, 1077, 1292, 1293 or 1446.
- Short to ground in CKT 850, 854, 1292, 1293 or 1446.
- ABS pump relay fails "open" or "engaged."

DIAGNOSTIC CHART F — 1993, CONTINUED

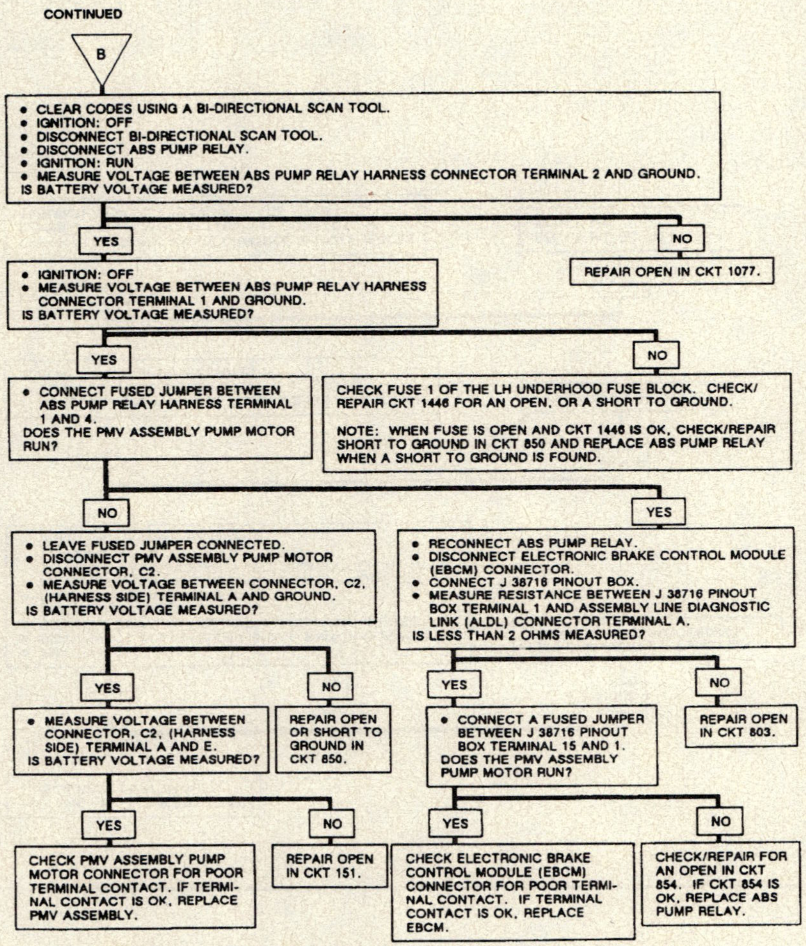

AFTER REPAIRS ARE COMPLETED, CLEAR CODES AND REPEAT ABS/TCS FUNCTIONAL CHECK.

DIAGNOSTIC CHART G — 1993

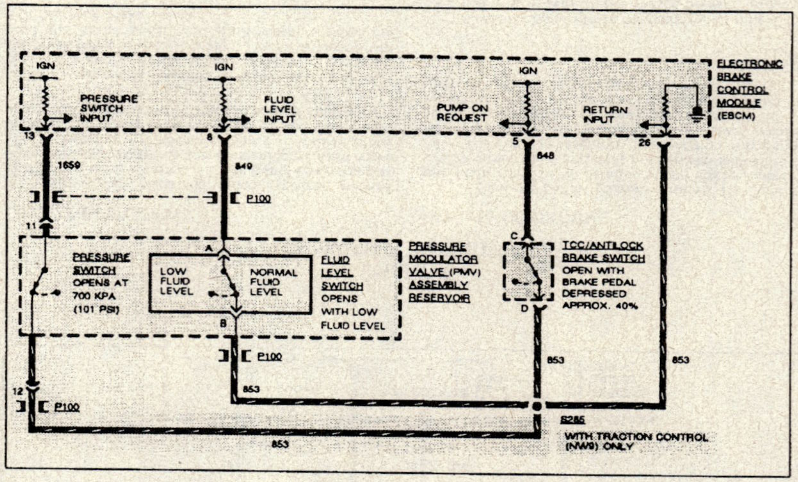

CHART G
TCC/ANTILOCK BRAKE SWITCH TEST

CIRCUIT OPERATION

The electronic brake control module (EBCM) monitors brake pedal position using the TCC/antilock brake switch. The switch is normally closed. It opens when the brake pedal is depressed approximately 40% down. During most ABS braking conditions the EBCM turns on the pump motor, replenishing the PMV assembly reservoir and applying fluid pressure to the master cylinder. This causes the brake pedal to gradually rise. The EBCM turns off the pump motor when brake pedal height sufficiently rises to close the TCC/antilock brake switch.

FAILURE CONDITIONS

Code 72 sets when the EBCM detects the TCC/antilock brake switch CKT 848 is shorted to ground.

DIAGNOSTIC CHART G — 1993, CONTINUED

HAS THE ABS/TCS FUNCTIONAL CHECK BEEN PERFORMED?

- NO → PERFORM ABS/TCS FUNCTIONAL CHECK BEFORE PROCEEDING.
- YES →
  - IGNITION: OFF
  - DISCONNECT TCC/ANTILOCK BRAKE SWITCH CONNECTOR.
  - DISCONNECT ELECTRONIC BRAKE CONTROL MODULE (EBCM) CONNECTOR.
  - CONNECT J 38716 PINOUT BOX.
  - MEASURE RESISTANCE BETWEEN J 38716 PINOUT BOX TERMINAL 1 AND ASSEMBLY LINE DIAGNOSTIC LINK (ALDL) TERMINAL A.

  IS LESS THAN 2 OHMS MEASURED?

  - NO → REPAIR OPEN IN CKT 803.
  - YES →
    - MEASURE RESISTANCE BETWEEN J 38716 PINOUT BOX TERMINAL 1 AND 5.

    IS LESS THAN 5 OHMS MEASURED?

    - YES → REPAIR SHORT TO GROUND IN CKT 848.
    - NO → CHECK ELECTRONIC BRAKE CONTROL MODULE (EBCM) CONNECTORS FOR POOR TERMINAL CONTACT. IF TERMINAL CONTACT IS OK, REPLACE EBCM.

AFTER REPAIRS ARE COMPLETED, CLEAR CODES AND REPEAT ABS/TCS FUNCTIONAL CHECK.

DIAGNOSTIC CHART H — 1993

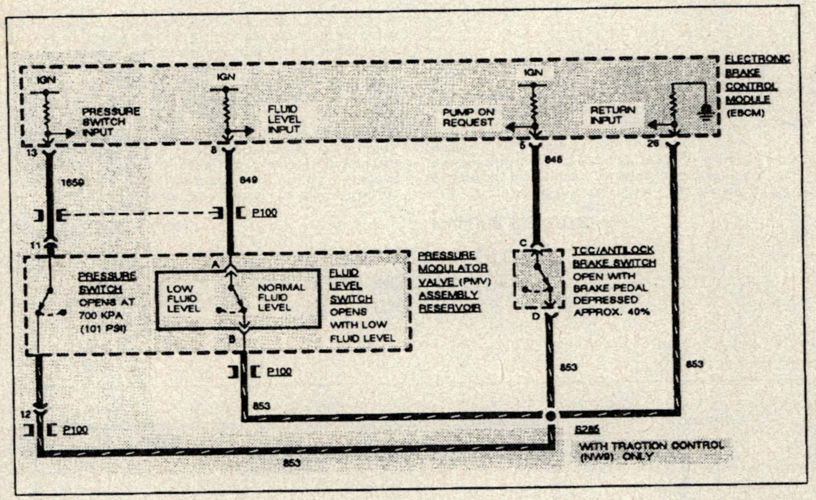

CHART H
PMV FLUID LEVEL SWITCH CIRCUIT TEST

CIRCUIT OPERATION

The pressure modulator valve (PMV) fluid level switch alerts the electronic brake control module (EBCM) to a low brake fluid condition in the PMV reservoir. The fluid level switch is located in the PMV assembly's reservoir. The fluid level switch is normally closed when the reservoir contains sufficient fluid. When fluid level is low, the switch opens, causing the EBCM to turn on the "ANTILOCK" indicator.

FAILURE CONDITIONS

Code 73 sets when the EBCM detects a short to ground in the PMV fluid level switch circuitry. These include: CKT 848, 849, 853 and 1659.

DIAGNOSTIC CHART H — 1993, CONTINUED

HAS THE ABS/TCS FUNCTIONAL CHECK BEEN PERFORMED?

- NO → PERFORM ABS/TCS FUNCTIONAL CHECK BEFORE PROCEEDING.
- YES →
  - IGNITION: OFF
  - DISCONNECT ELECTRONIC BRAKE CONTROL MODULE (EBCM) CONNECTOR.
  - CONNECT J 38716 PINOUT BOX.
  - MEASURE RESISTANCE BETWEEN J 38716 PINOUT BOX TERMINAL 1 AND ASSEMBLY LINE DIAGNOSTIC LINK (ALDL) CONNECTOR TERMINAL A.

  IS LESS THAN 2 OHMS MEASURED?

  - NO → REPAIR OPEN IN CKT 803.
  - YES →
    - DISCONNECT TCC/ANTILOCK BRAKE SWITCH.
    - MEASURE RESISTANCE BETWEEN J 38716 PINOUT BOX TERMINAL 5 AND 1.

    IS LESS THAN 5 OHMS MEASURED?

    - YES → REPAIR SHORT TO GROUND IN CKT 848.
    - NO →
      - DISCONNECT PRESSURE MODULATOR VALVE (PMV) FLUID LEVEL SWITCH CONNECTOR.
      - MEASURE RESISTANCE BETWEEN J 38716 PINOUT BOX TERMINAL 8 AND 1.

      IS LESS THAN 5 OHMS MEASURED?

      - YES → REPAIR SHORT TO GROUND IN CKT 849.
      - NO →
        - WHEN EQUIPPED WITH TRACTION CONTROL:
          - DEPRESS BRAKE PEDAL
          - MEASURE RESISTANCE BETWEEN J 38716 PINOUT BOX TERMINAL 13 AND 1

        IS LESS THAN 5 OHMS MEASURED?

        - YES → REPAIR SHORT TO GROUND IN CKT 1659.
        - NO →
          - MEASURE RESISTANCE BETWEEN J 38716 PINOUT BOX TERMINAL 26 AND 1.

          IS LESS THAN 5 OHMS MEASURED?

          - YES → REPAIR SHORT TO GROUND IN CKT 853.
          - NO → CHECK ELECTRONIC BRAKE CONTROL MODULE (EBCM) CONNECTOR FOR POOR TERMINAL CONTACT. IF TERMINAL CONTACT IS OK, REPLACE EBCM.

AFTER REPAIRS ARE COMPLETED, CLEAR CODES AND REPEAT ABS/TCS FUNCTIONAL CHECK.

DIAGNOSTIC CHART I — 1993

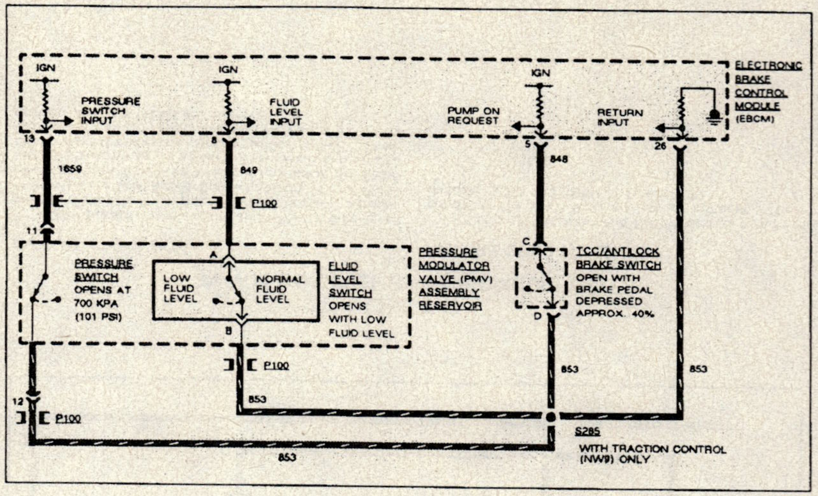

CHART I
PUMP MOTOR FAULT DURING AN ABS STOP

CIRCUIT OPERATION

The electronic brake control module (EBCM) monitors brake pedal position using the TCC/antilock brake switch. The normally closed switch opens when the brake pedal is depressed approximately 40% down. The EBCM also monitors pump motor operation using a pump motor run sensor.

During most ABS braking conditions the EBCM turns on the pump motor, replenishing the PMV assembly reservoir and applying fluid pressure to the master cylinder. This causes the brake pedal to gradually rise. The EBCM turns off the pump motor when brake pedal height sufficiently rises to close the TCC/antilock brake switch.

FAILURE CONDITIONS

During ABS braking, Code 62 sets when the EBCM determines the pump is working, and brake pedal height has not sufficiently risen to close the TCC/antilock brake switch after 3 seconds.

This code usually indicates a hydraulic problem. Some possible causes include air in the brake lines or a pump motor mechanical failure. Non-hydraulic causes for Code 62 include an open 848 or 853 CKT or a misadjusted/defective TCC/antilock brake switch.

DIAGNOSTIC CHART I — 1993, CONTINUED

HAS THE ABS/TCS FUNCTIONAL CHECK BEEN PERFORMED?

- NO → PERFORM ABS/TCS FUNCTIONAL CHECK BEFORE PROCEEDING.
- YES →
  - BLEED BRAKE SYSTEM AS OUTLINED IN THIS SECTION.
  - IGNITION: OFF.
  - CONNECT BI-DIRECTIONAL SCAN TOOL TO ASSEMBLY LINE DIAGNOSTIC LINK (ALDL) CONNECTOR.
  - IGNITION: RUN
  - USE BI-DIRECTIONAL SCAN TOOL TO CLEAR CODE AFTER BLEEDING SYSTEM.
  - DEPRESS BRAKE PEDAL (HOLD).
  - USE BI-DIRECTIONAL SCAN TOOL TO BLEED THE PRESSURE MODULATOR VALVE (PMV) ASSEMBLY.

DID THE BRAKE PEDAL RISE?

- NO → REPLACE PMV ASSEMBLY.
- YES →
  - IGNITION:OFF
  - DISCONNECT BI-DIRECTIONAL SCAN TOOL.
  - TEST DRIVE VEHICLE IN A SAFE WET AREA AND ATTEMPT TO PERFORM A 3 SECOND ABS STOP.
  - CONNECT BI-DIRECTIONAL SCAN TOOL TO ALDL CONNECTOR.
  - IGNITION: RUN

DID CODE 62 SET?

- NO → PROBLEM SOLVED.
- YES →
  - IGNITION:OFF/
  - DISCONNECT BI-DIRECTIONAL SCAN TOOL.
  - BRAKE PEDAL RELEASED (REST POSITION).
  - CHECK TCC/ANTILOCK BRAKE SWITCH ADJUSTMENT.

IS SWITCH ADJUSTED PROPERLY?

- NO → ADJUST TCC/ANTILOCK BRAKE SWITCH.
- YES →
  - DISCONNECT TCC/ANTILOCK BRAKE SWITCH CONNECTOR.
  - MEASURE RESISTANCE ACROSS SWITCH TERMINALS C AND D.

IS LESS THAN 1 OHM MEASURED?

- NO → REPLACE TCC/ANTILOCK BRAKE SWITCH.
- YES → A

CONTINUED

AFTER REPAIRS ARE COMPLETED, CLEAR CODES AND REPEAT ABS/TCS FUNCTIONAL CHECK.

DIAGNOSTIC CHART I — 1993, CONTINUED

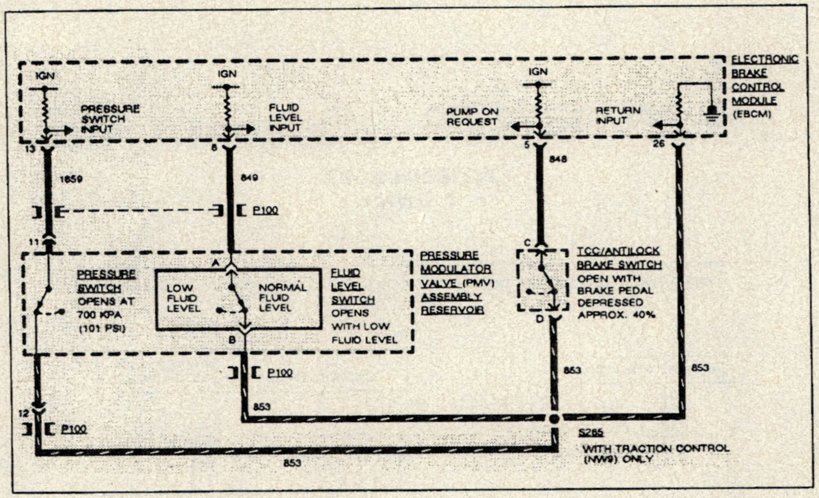

CHART I
PUMP MOTOR FAULT DURING AN ABS STOP

CIRCUIT OPERATION

The electronic brake control module (EBCM) monitors brake pedal position using the TCC/antilock brake switch. The normally closed switch opens when the brake pedal is depressed approximately 40% down. The EBCM also monitors pump motor operation using a pump motor run sensor.

During most ABS braking conditions the EBCM turns on the pump motor, replenishing the PMV assembly reservoir and applying fluid pressure to the master cylinder. This causes the brake pedal to gradually rise. The EBCM turns off the pump motor when brake pedal height sufficiently rises to close the TCC/antilock brake switch.

FAILURE CONDITIONS

During ABS braking, Code 62 sets when the EBCM determines the pump is working, and brake pedal height has not sufficiently risen to close the TCC/antilock brake switch after 3 seconds.

This code usually indicates a hydraulic problem. Some possible causes include air in the brake lines or a pump motor mechanical failure. Non-hydraulic causes for Code 62 include an open 848 or 853 CKT or a misadjusted/defective TCC/antilock brake switch.

DIAGNOSTIC CHART I — 1993, CONTINUED

CONTINUED

A

- DISCONNECT ELECTRONIC BRAKE CONTROL MODULE (EBCM) CONNECTOR.
- CONNECT J 38716 PINOUT BOX.
- MEASURE RESISTANCE BETWEEN J 38716 PINOUT BOX TERMINAL 5 AND TCC/ANTILOCK BRAKE SWITCH CONNECTOR TERMINAL C.

IS LESS THAN 3 OHMS MEASURED?

YES →

- MEASURE RESISTANCE BETWEEN J 38716 PINOUT BOX TERMINAL 26 AND TCC/ANTILOCK BRAKE SWITCH CONNECTOR TERMINAL D.

IS LESS THAN 3 OHMS MEASURED?

YES → CHECK ELECTRONIC BRAKE CONTROL MODULE (EBCM) CONNECTOR FOR POOR TERMINAL CONTACT. IF TERMINAL CONTACT IS OK, REPLACE EBCM.

NO → REPAIR OPEN IN CKT 853.

NO → REPAIR OPEN IN CKT 848.

AFTER REPAIRS ARE COMPLETED, CLEAR CODES AND REPEAT ABS/TCS FUNCTIONAL CHECK.

DIAGNOSTIC CHART J — 1993

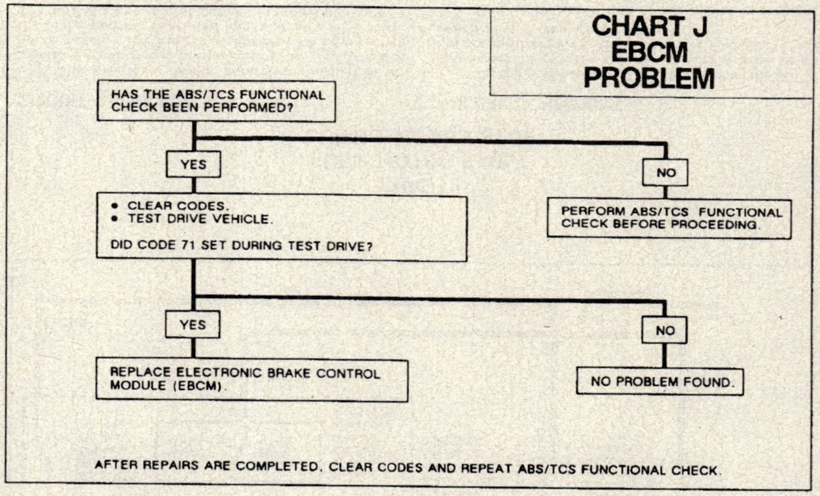

CHART J
EBCM PROBLEM

Code 71 indicates that a failure has occurred within the electronic brake control module (EBCM).

Code 71 can occur intermittently, due to interference, therefore clear the code and test drive the vehicle before replacing the EBCM.

DIAGNOSTIC CHART K — 1993

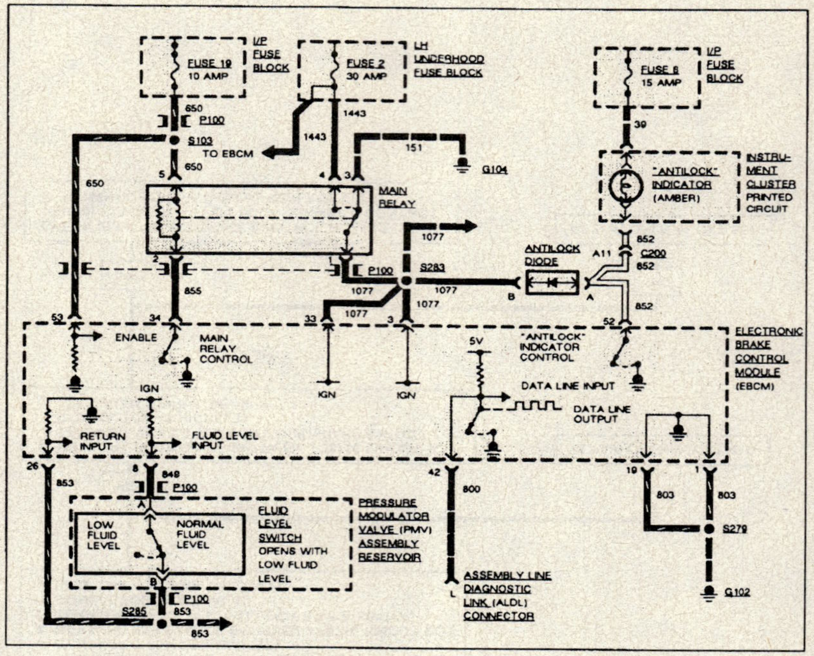

CHART K
"ANTILOCK" INDICATOR IS ON AND NO CODES ARE SET/ DIAGNOSTIC MODE CANNOT BE ENTERED

CIRCUIT OPERATION

When the ignition is first turned to "RUN," the electronic brake control module (EBCM) goes through a self-test lasting 2–4 seconds. During this test, the EBCM grounds terminal 52 causing the amber "ANTILOCK" indicator to light. This serves as an indicator bulb test.

When a low fluid condition exists, the EBCM turns on the amber "ANTILOCK" indicator but does not set a code. A low fluid level condition usually lights the red "BRAKE" indicator since the master cylinder supplies the PMV reservoir. However, restrictions or air in the hose could cause a low fluid condition in the PMV reservoir without lighting the red "BRAKE" indicator.

FAILURE CONDITIONS

Possible Causes

- Open in CKT 650, 800, 803, 849 or 853.
- Low brake fluid level or open PMV fluid level switch.
- Short to ground in CKT 650, 800 or 852.

DIAGNOSTIC CHART K — 1993, CONTINUED

HAS THE ABS/TCS FUNCTIONAL CHECK BEEN PERFORMED?

- NO: PERFORM ABS/TCS FUNCTIONAL CHECK BEFORE PROCEEDING.
- YES:
  - IGNITION: OFF.
  - VISUALLY INSPECT FUSE 19 OF THE I/P FUSE BLOCK. (INSPECT FOR OPEN.)

  IS THE FUSE OK?

- NO: REPLACE FUSE AND CHECK/REPAIR CKT 650 FOR A SHORT TO GROUND.
- YES:
  - REINSERT FUSE.
  - DISCONNECT PRESSURE MODULATOR VALVE (PMV) FLUID LEVEL SWITCH CONNECTOR.
  - MEASURE RESISTANCE ACROSS PMV FLUID LEVEL SWITCH TERMINAL A AND B.

  IS LESS THAN 1 OHM MEASURED?

- NO: CHECK BRAKE FLUID LEVEL. WHEN LOW, CHECK HOSE FOR KINKS OR RESTRICTIONS AND ADD FLUID AS NEEDED TO MASTER CYLINDER RESERVOIR. IF OK, REPLACE PMV RESERVOIR.
- YES:
  - RECONNECT PMV FLUID LEVEL SWITCH CONNECTOR.
  - DISCONNECT ELECTRONIC BRAKE CONTROL MODULE (EBCM) CONNECTOR.
  - CONNECT J 38716 PINOUT BOX.
  - MEASURE RESISTANCE BETWEEN J 38716 PINOUT BOX TERMINAL 8 AND 26.

  IS LESS THAN 3 OHMS MEASURED?

- NO: REPAIR OPEN IN CKT 849 OR CKT 853.
- YES:
  - MEASURE RESISTANCE BETWEEN J 38716 PINOUT BOX TERMINAL 1 AND ASSEMBLY LINE DIAGNOSTIC LINK (ALDL) CONNECTOR TERMINAL A, AND BETWEEN J 38716 PINOUT BOX TERMINAL 19 AND ALDL CONNECTOR TERMINAL A.

  IS LESS THAN 2 OHMS MEASURED AT EACH TERMINAL?

- NO: REPAIR OPEN IN CKT 803.
- YES: A

CONTINUED

AFTER REPAIRS ARE COMPLETED, CLEAR CODES AND REPEAT ABS/TCS FUNCTIONAL CHECK.

DIAGNOSTIC CHART K — 1993, CONTINUED

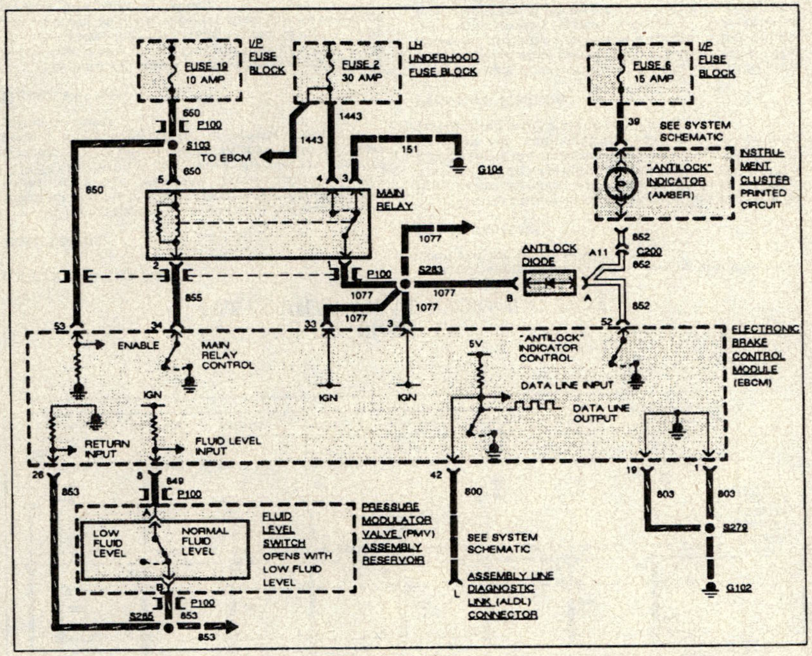

CHART K
"ANTILOCK" INDICATOR IS ON AND NO CODES ARE SET/ DIAGNOSTIC MODE CANNOT BE ENTERED

CIRCUIT OPERATION

When the ignition is first turned to "RUN," the electronic brake control module (EBCM) goes through a self-test lasting 2–4 seconds. During this test, the EBCM grounds terminal 52 causing the amber "ANTILOCK" indicator to light. This serves as an indicator bulb test.

When a low fluid condition exists, the EBCM turns on the amber "ANTILOCK" indicator but does not set a code. A low fluid level condition usually lights the red "BRAKE" indicator since the master cylinder supplies the PMV reservoir. However, restrictions or air in the hose could cause a low fluid condition in the PMV reservoir without lighting the red "BRAKE" indicator.

FAILURE CONDITIONS

Possible Causes

- Open in CKT 650, 800, 803, 849 or 853.
- Low brake fluid level or open PMV fluid level switch.
- Short to ground in CKT 650, 800 or 852.

DIAGNOSTIC CHART K — 1993, CONTINUED

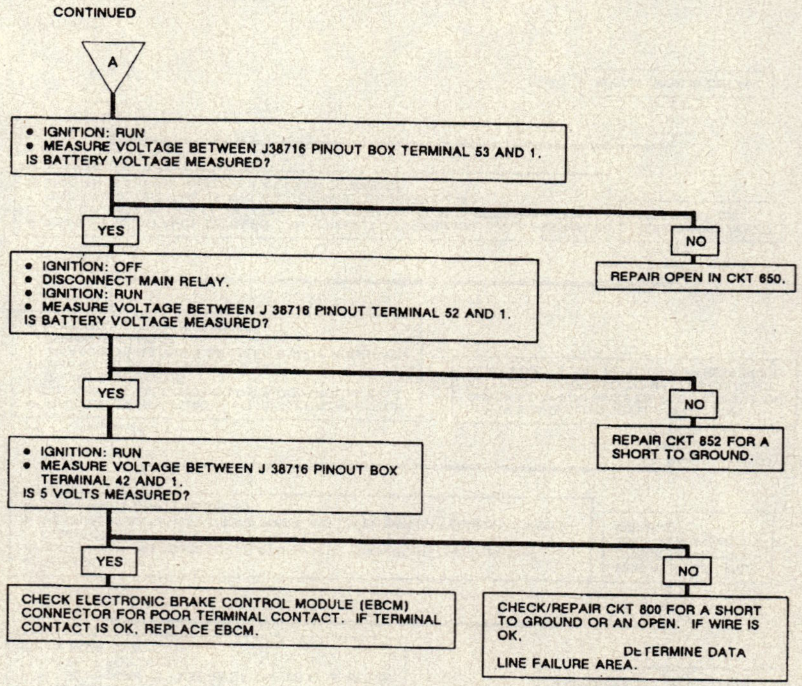

AFTER REPAIRS ARE COMPLETED, CLEAR CODES AND REPEAT ABS/TCS FUNCTIONAL CHECK.

DIAGNOSTIC CHART L — 1993

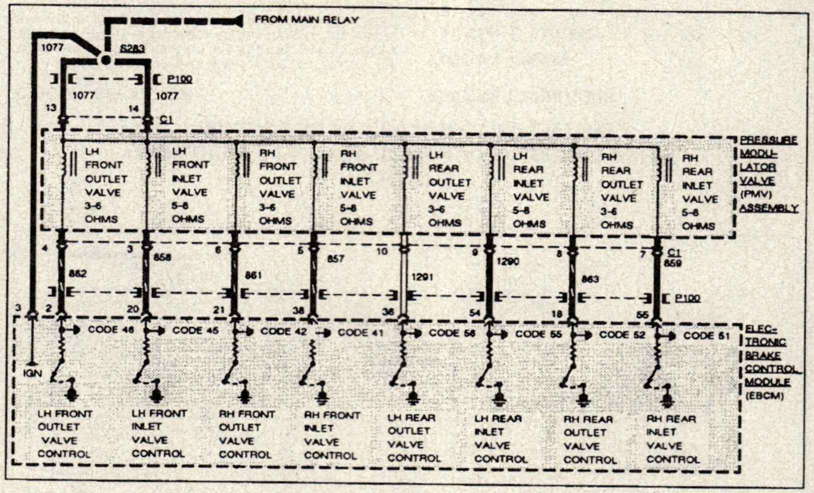

CHART L
INLET/OUTLET VALVE HYDRAULIC TEST

CIRCUIT OPERATION

Inlet Valves

Voltage is applied to the normally open inlet valves whenever the main relay is energized. An inlet valve closes when the electronic brake control module (EBCM) supplies a ground path at the inlet valve control terminal. The EBCM closes an inlet valve for short periods to maintain (hold) or reduce pressure at a particular wheel.

Outlet Valves

The normally closed outlet valves operate similarly to the inlet valves, except that when the EBCM grounds an outlet valve, the valve opens instead of closes. The EBCM opens an outlet valve for short periods to reduce pressure at a particular wheel.

VALVE MODES

Pressure Increase

The pressure increase position is the operating valve position for normal (non-ABS) braking and traction control mode (when equipped). The EBCM does not energize either valve, leaving the inlet valve open and the outlet valve closed.

Pressure Hold

The pressure hold position occurs when the EBCM first notices an abnormal hydraulic pressure increase. This signifies an impending wheel lockup condition. The EBCM closes the corresponding inlet valve in an attempt to avoid any additional hydraulic pressure increase (avoiding an ABS braking condition).

Pressure Reduce

During a wheel lockup condition, the EBCM close the inlet valve while pulsing open the outlet valve. This decreases hydraulic pressure at a specific wheel without decreasing overall system pressure. The reduced hydraulic pressure allows greater wheel spin, eliminating the wheel lockup condition.

Test Description

This test checks for proper inlet and outlet valve operation. The wheel brake behavior must be checked to verify the valves hydraulic effectiveness. This is accomplished using the TECH 1 to actuate the valves.

DIAGNOSTIC CHART L — 1993, CONTINUED

- RAISE VEHICLE.
- GEAR SELECTOR LEVER: NEUTRAL
- PARK BRAKE: RELEASED

CAN WHEEL ASSOCIATED WITH SUSPECT VALVE BE TURNED BY HAND?

NO →
- CHECK THAT BRAKE CALIPER OR WHEEL CYLINDER IS NOT FROZEN.
- CHECK THAT PARK BRAKE CABLE IS RELEASED.
- CHECK THAT BRAKE SHOES, SPRINGS AND ASSOCIATED HARDWARE ARE NOT STUCK OR FROZEN.

YES →
- DEPRESS BRAKE PEDAL (HOLD).

CAN SUSPECT WHEEL BE TURNED BY HAND?

YES → GENERAL BRAKES DIAGNOSIS.

NO →
- IGNITION: OFF.
- CONNECT BI-DIRECTIONAL SCAN TOOL TO THE ASSEMBLY LINE DIAGNOSTIC LINK (ALDL) CONNECTOR.
- IGNITION: RUN.
- SELECT ABS MANUAL SOLENOID TEST AND SUSPECT OUTLET VALVE TO BE TESTED.
- DEPRESS BRAKE PEDAL (HOLD).
- ACTIVATE SUSPECT OUTLET VALVE.

CAN SUSPECT WHEEL BE TURNED BY HAND?

NO → REPLACE PRESSURE MODULATOR VALVE (PMV) ASSEMBLY.

YES →
- SELECT ABS MANUAL SOLENOID TEST AND SUSPECT INLET VALVE TO BE TESTED.
- ACTIVATE SUSPECT INLET VALVE.
- DEPRESS BRAKE PEDAL (HOLD).

CAN SUSPECT WHEEL BE TURNED BY HAND?

YES → NO TROUBLE FOUND. PROBLEM MAY BE INTERMITTENT.

NO → REPLACE PRESSURE MODULATOR VALVE (PMV) ASSEMBLY.

NOTE: TO PREVENT OVERHEATING OF INLET/OUTLET VALVES, SCAN TOOL WILL ONLY ALLOW VALVES TO BE ENERGIZED FOR A MAXIMUM OF 20 SECONDS. A MINIMUM OF 30 SECONDS MUST ELAPSE BEFORE VALVES CAN BE RE-ENERGIZED.

AFTER REPAIRS ARE COMPLETED, CLEAR CODES AND REPEAT ABS/TCS FUNCTIONAL CHECK.

DIAGNOSTIC CHART M — 1993

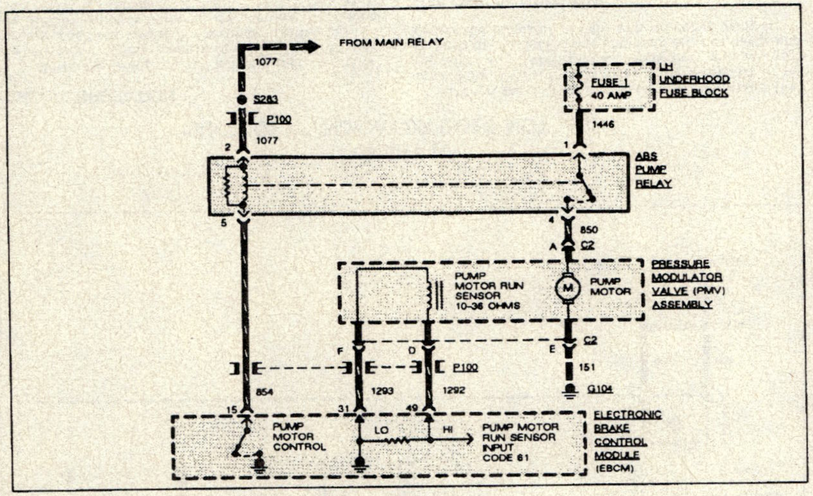

CHART M
PUMP MOTOR RUNS CONTINUOUSLY

CIRCUIT OPERATION

When the electronic brake control module (EBCM) grounds terminal 15, the ABS pump relay is energized. The pump relay contacts close, applying voltage to the pressure modulator valve (PMV) assembly's pump motor.

During most ABS braking conditions, the EBCM turns on the pump motor while monitoring the TCC/antilock brake switch to determine brake pedal position. The switch opens when the brake pedal is depressed approximately 40% down. The motor restores the master cylinder's fluid pressure, causing the brake pedal to gradually rise. The pump motor continues to run until the brake pedal sufficiently rises to close the TCC/antilock brake switch.

During traction control mode (when equipped), after the front hydraulic brake circuits are isolated, the pump motor runs to increase brake fluid pressure in these circuits. The increased pressure is directed to the slipping wheel through the PMV assembly's front inlet and outlet valves. This causes the brakes to be applied, reducing wheel spin at the slipping wheel. Traction control is disabled as soon as the brakes are manually applied.

The EBCM monitors pump motor operation via the pump motor run sensor. When the pump is operating, voltage is induced into the sensor's coil and the EBCM monitors this voltage at terminals 49 and 31.

FAILURE CONDITIONS

A continuously running pump motor occurs when CKT 854 is shorted to ground, the pump relay fails "engaged," or CKT 850 is shorted to a battery source. This condition illuminates the amber "ANTILOCK" indicator and sets Code 61.

DIAGNOSTIC CHART M — 1993, CONTINUED

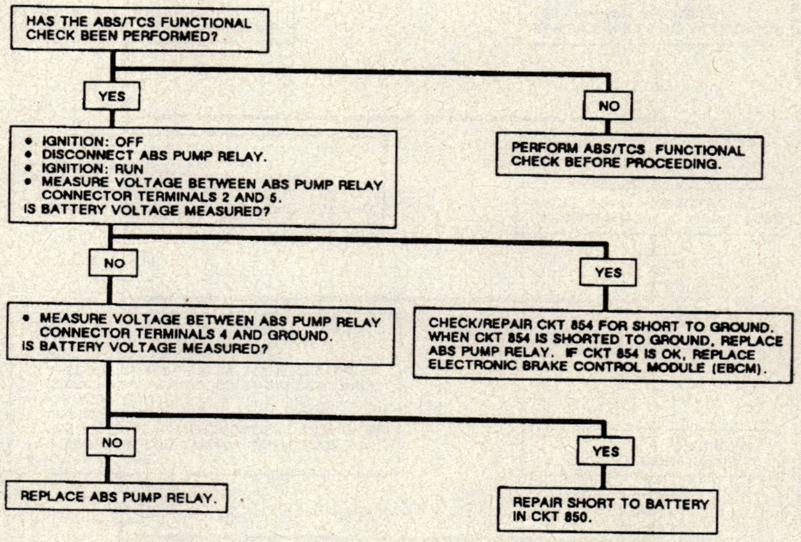

AFTER REPAIRS ARE COMPLETED, CLEAR CODES AND REPEAT ABS/TCS FUNCTIONAL CHECK.

DIAGNOSTIC CHART N — 1993

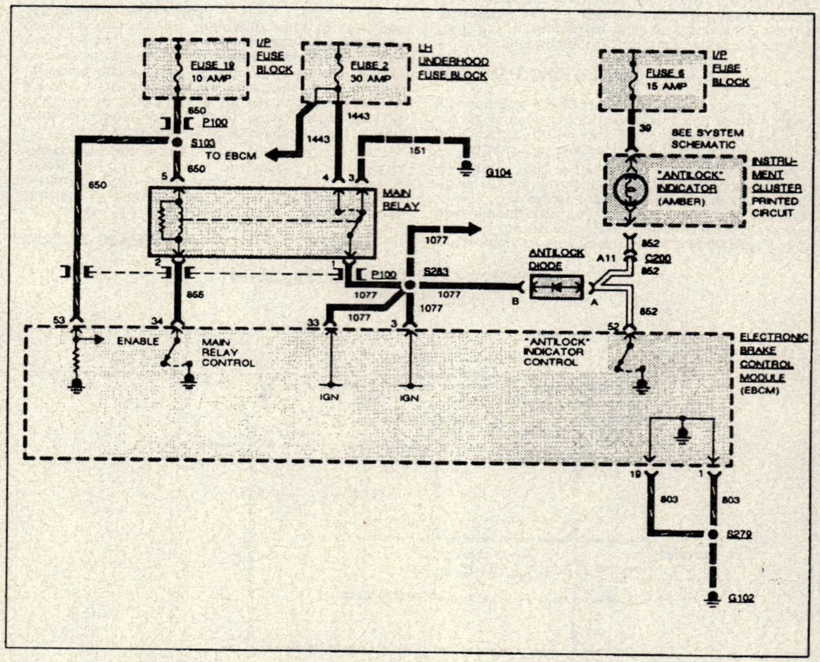

CHART N
"ANTILOCK" INDICATOR DOES NOT LIGHT

CIRCUIT OPERATION

When the ignition switch is first turned to "RUN" the electronic brake control module (EBCM) goes through a self-test lasting 2-4 seconds. During this test, the EBCM grounds terminal 52, causing the amber "ANTILOCK" indicator to light. This serves as an indicator bulb test. The EBCM also grounds terminal 34, energizing the main relay. This removes the indicator ground path provided through the antilock diode and the main relay contacts.

When the self-test is completed (and no faults are detected) the EBCM removes the ground at terminal 52, turning off the indicator. The main relay remains energized.

When the EBCM detects a fault, it grounds terminal 52 causing the amber "ANTILOCK" indicator to light. (Note: The EBCM will also de-energize the main relay creating a redundant ground path.)

When the EBCM is disabled for any reason (i.e. open Fuse 19), the main relay is de-energized. This provides an "ANTILOCK" indicator ground path through the antilock diode and main relay contacts, and causes the indicator to light with no codes being set.

FAILURE CONDITIONS

Possible Causes

- Open in CKT 151, 852 or 1077.
- Open bulb.
- Short to voltage in CKT 852.
- Open or shorted antilock diode.
- Main relay fails "engaged."
- Short to ground in CKT 855.

DIAGNOSTIC CHART N — 1993, CONTINUED

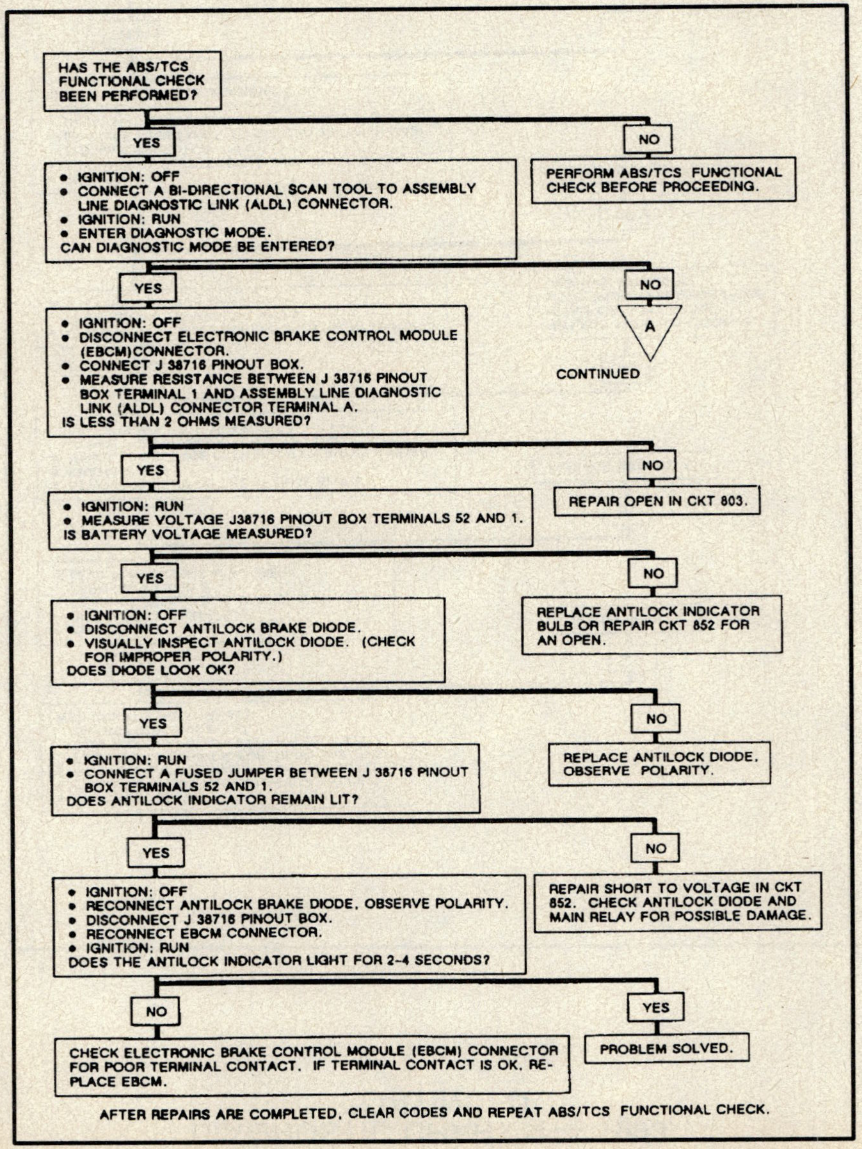

DIAGNOSTIC CHART N — 1993, CONTINUED

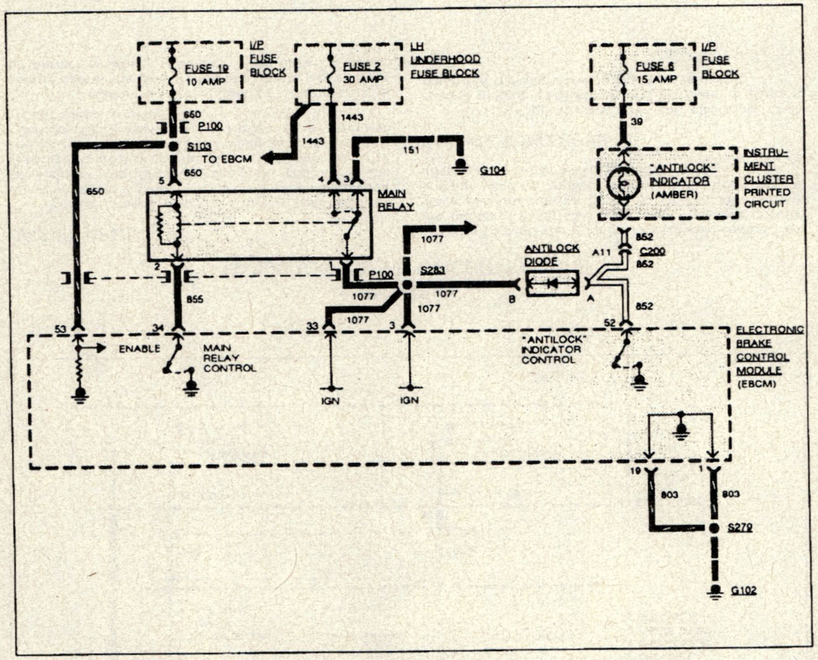

CHART N
"ANTILOCK" INDICATOR DOES NOT LIGHT

CIRCUIT OPERATION

When the ignition switch is first turned to "RUN" the electronic brake control module (EBCM) goes through a self-test lasting 2-4 seconds. During this test, the EBCM grounds terminal 52, causing the amber "ANTILOCK" indicator to light. This serves as an indicator bulb test. The EBCM also grounds terminal 34, energizing the main relay. This removes the indicator ground path provided through the antilock diode and the main relay contacts.

When the self-test is completed (and no faults are detected) the EBCM removes the ground at terminal 52, turning off the indicator. The main relay remains energized.

When the EBCM detects a fault, it grounds terminal 52 causing the amber "ANTILOCK" indicator to light. (Note: The EBCM will also de-energize the main relay creating a redundant ground path.)

When the EBCM is disabled for any reason (i.e. open Fuse 19), the main relay is de-energized. This provides an "ANTILOCK" indicator ground path through the antilock diode and main relay contacts, and causes the indicator to light with no codes being set.

FAILURE CONDITIONS

Possible Causes

- Open in CKT 151, 852 or 1077.
- Open bulb.
- Short to voltage in CKT 852.
- Open or shorted antilock diode.
- Main relay fails "engaged."
- Short to ground in CKT 855.

DIAGNOSTIC CHART N — 1993, CONTINUED

CONTINUED

A

- IGNITION: OFF
- DISCONNECT BI-DIRECTIONAL SCAN TOOL.
- AT THIS POINT A DOUBLE FAULT IS POSSIBLE. FIRST GO TO CHART K FOR DIAGNOSIS. THEN CONTINUE AND COMPLETE THIS DIAGNOSTIC CHART FROM THIS POINT.

WAS CHART K PERFORMED?

NO → GO TO CHART K.

YES ↓

- IGNITION: OFF
- DISCONNECT MAIN RELAY.
- MEASURE RESISTANCE ACROSS MAIN RELAY TERMINALS 1 AND 3.

IS LESS THAN 2 OHMS MEASURED?

NO → REPLACE MAIN RELAY.

YES ↓

- REMOVE FUSE 19 OF THE I/P FUSE BLOCK.
- IGNITION: RUN
- MEASURE VOLTAGE BETWEEN MAIN RELAY HARNESS CONNECTOR TERMINAL 1 AND GROUND.

IS BATTERY VOLTAGE MEASURED?

NO → REPLACE ANTILOCK DIODE OR REPAIR OPEN IN CKT 1077. NOTE: WHEN REPLACING DIODE OBSERVE POLARITY.

YES ↓

- MEASURE VOLTAGE BETWEEN MAIN RELAY HARNESS CONNECTOR TERMINALS 1 AND 3.

IS BATTERY VOLTAGE MEASURED?

NO → REPAIR OPEN IN CKT 151.

YES ↓

- IGNITION: OFF
- DISCONNECT ELECTRONIC BRAKE CONTROL MODULE (EBCM) CONNECTOR.
- MEASURE RESISTANCE TO GROUND BETWEEN MAIN RELAY HARNESS CONNECTOR TERMINAL 2 AND GROUND.

IS LESS THAN 5 OHMS MEASURED?

YES → REPAIR SHORT TO GROUND IN CKT 855.

NO → CHECK ELECTRONIC BRAKE CONTROL MODULE (EBCM) CONNECTOR FOR POOR TERMINAL CONTACT. IF TERMINAL CONTACT IS OK, REPLACE EBCM.

AFTER REPAIRS ARE COMPLETED, CLEAR CODES AND REPEAT ABS/TCS FUNCTIONAL CHECK.

DIAGNOSTIC CHART O — 1993

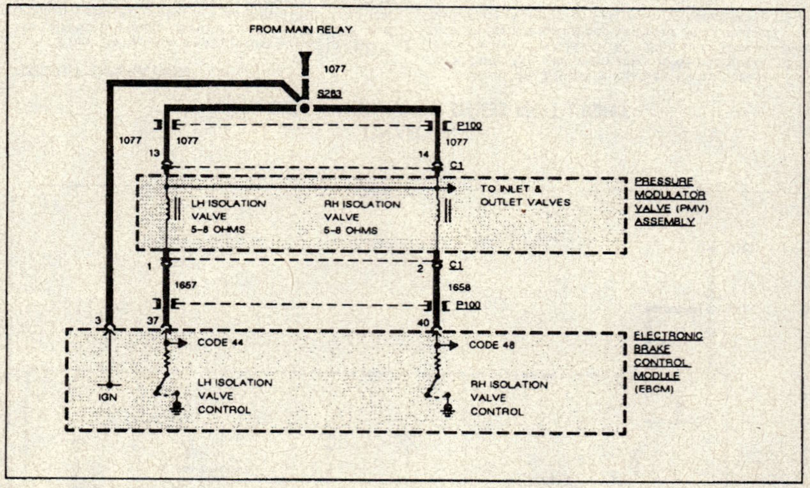

CHART O
ISOLATION VALVE CIRCUIT FAILURE

CIRCUIT OPERATION

Voltage is applied to the normally open isolation valves whenever the main relay is energized. An isolation valve closes when the electronic brake control module (EBCM) supplies a ground path at the isolation valve control terminal. The EBCM closes an isolation valve when the EBCM detects wheel slip in a drive wheel and vehicle speed is below 40 km/h (25 mph).

The isolation valves are closed only during traction control operations, and are not used during any antilock or normal braking conditions. The isolation valves separate (isolate) the front brake hydraulic circuits from the master cylinder and rear brake hydraulic circuits. Once the front brake hydraulic circuits are isolated, pressure can be applied to the front wheels without affecting any other brake hydraulic circuit.

FAILURE CONDITIONS

Code 44 or 48 sets when the electronic brake control module (EBCM) detects an open or short to ground in the respective isolation valve circuit.

DIAGNOSTIC CHART O — 1993, CONTINUED

HAS THE ABS/TCS FUNCTIONAL CHECK BEEN PERFORMED?

NO: PERFORM ABS/TCS FUNCTIONAL CHECK BEFORE PROCEEDING.

YES:

- IGNITION: OFF
- DISCONNECT PRESSURE MODULATOR VALVE (PMV) ASSEMBLY VALVE CONNECTOR, C1.
- MEASURE RESISTANCE BETWEEN CONNECTOR, C1, (PMV ASSEMBLY SIDE) TERMINAL 14 AND PMV ASSEMBLY CASING.

IS INFINITE OHMS MEASURED?

NO: REPLACE PMV ASSEMBLY.

YES:

- MEASURE RESISTANCE AT CONNECTOR, C1, (PMV ASSEMBLY SIDE) BETWEEN TERMINAL 14 AND THE SUSPECT VALVE TERMINAL. REFER TO CHART.

CODE/VALVE	SUSPECT TERMINAL	CORRECT RESISTANCE
44/LH ISOLATION	1	5 TO 8 OHMS
48/RH ISOLATION	2	

IS THE CORRECT RESISTANCE MEASURED?

NO: REPLACE PMV ASSEMBLY.

YES:

- RECONNECT PMV ASSEMBLY VALVE CONNECTOR, C1.
- DISCONNECT ELECTRONIC BRAKE CONTROL MODULE (EBCM) CONNECTOR.
- INSTALL J 38716 PINOUT BOX.
- MEASURE RESISTANCE BETWEEN J 38716 PINOUT BOX TERMINAL 3 AND THE SUSPECT ISOLATION VALVE TERMINAL. REFER TO CHART.

CODE/VALVE	J 38716 TERMINAL
44/LH ISOLATION	37
48/RH ISOLATION	40

IS RESISTANCE MEASURED LESS THAN 10 OHMS?

NO: REPAIR OPEN IN SUSPECT ISOLATION VALVE CIRCUIT BETWEEN PRESSURE MODULATOR VALVE (PMV) ASSEMBLY AND THE ELECTRONIC BRAKE CONTROL MODULE (EBCM).

YES:

- MEASURE RESISTANCE BETWEEN J 38716 PINOUT BOX TERMINAL 1 AND ASSEMBLY LINE DIAGNOSTIC LINK (ALDL) CONNECTOR TERMINAL A.

IS LESS THAN 2 OHMS MEASURED?

NO: REPAIR OPEN IN CKT 803.

YES:

- DISCONNECT MAIN RELAY.
- MEASURE RESISTANCE BETWEEN J 38716 PINOUT BOX TERMINAL 1 AND THE SUSPECT ISOLATION VALVE TERMINAL. REFER TO CHART.

CODE/VALVE	J 38716 TERMINAL
44/LH ISOLATION	37
48/RH ISOLATION	40

IS LESS THAN 5 OHMS MEASURED?

NO: CHECK EBCM TERMINAL CONNECTION FOR POOR TERMINAL CONTACT. IF TERMINAL CONTACT IS OK, REPLACE EBCM.

YES: REPAIR SHORT TO GROUND IN SUSPECT ISOLATION VALVE CIRCUIT BETWEEN PRESSURE MODULATOR VALVE (PMV) ASSEMBLY AND THE ELECTRONIC BRAKE CONTROL MODULE (EBCM).

AFTER REPAIRS ARE COMPLETED, CLEAR CODES AND REPEAT ABS/TCS FUNCTIONAL CHECK.

DIAGNOSTIC CHART P — 1993

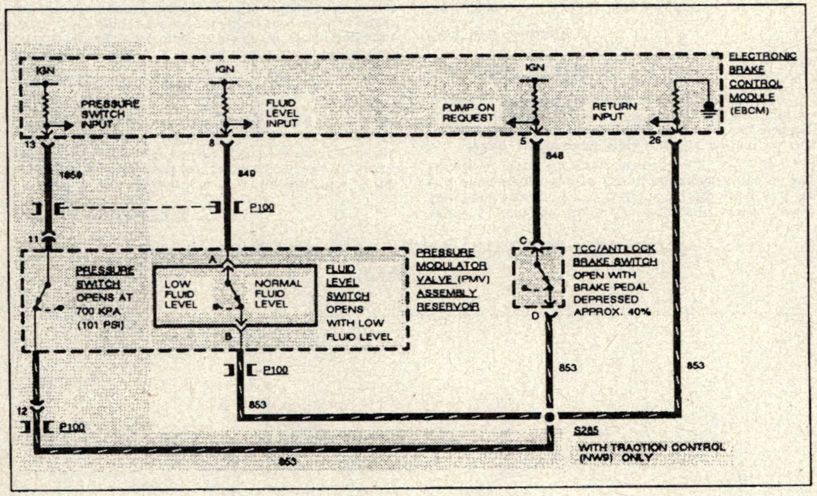

CHART P
PRESSURE SWITCH CIRCUIT FAILURE

CIRCUIT OPERATION

The pressure switch monitors primary brake system pressure. The normally closed switch opens when the brakes are applied (pressure exceeds 700 kPa [101 psi]). If the brakes are manually applied during traction control mode, the PMV pressure switch input, along with the brake switch input, signal the electronic brake control module (EBCM) to disable traction control and allow manual braking. The pressure switch is part of the pressure modulator valve (PMV) assembly and is not serviceable.

FAILURE CONDITIONS

Code 74 sets when the electronic brake control module (EBCM) detects an open or short to ground in the PMV pressure switch circuitry.

Possible Causes

- Open in CKT 853 or 1659.
- Short to ground in CKT 1659 with the brakes applied.
- Failed PMV pressure switch.

DIAGNOSTIC CHART P — 1993, CONTINUED

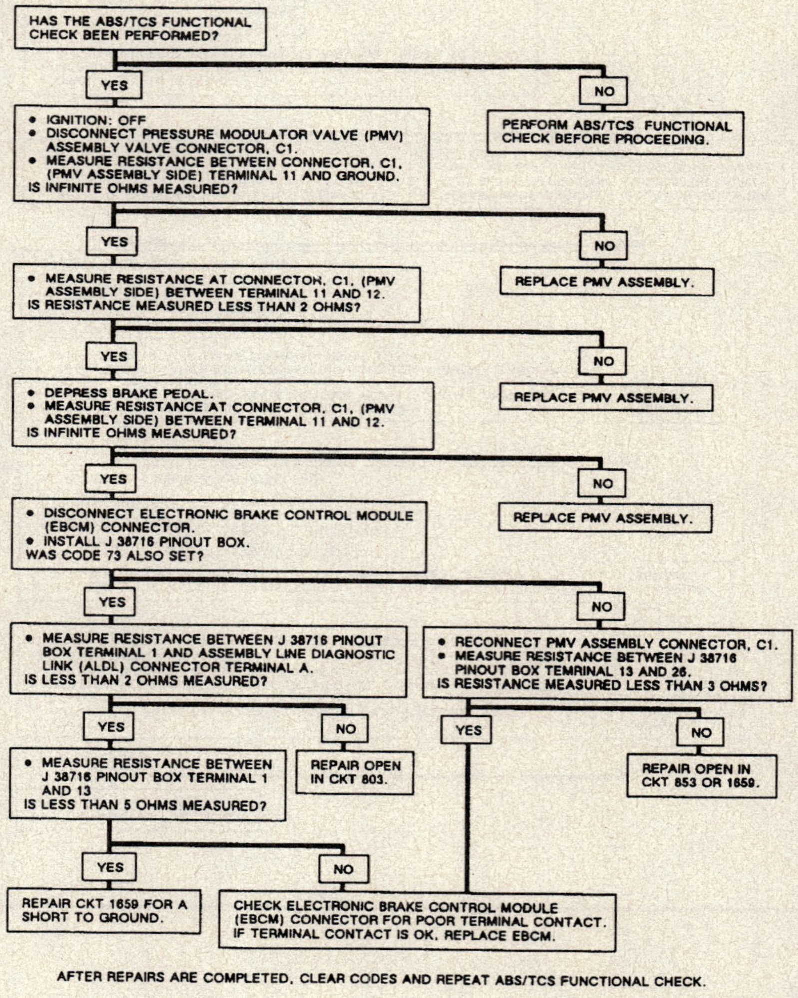

DIAGNOSTIC CHART Q — 1993

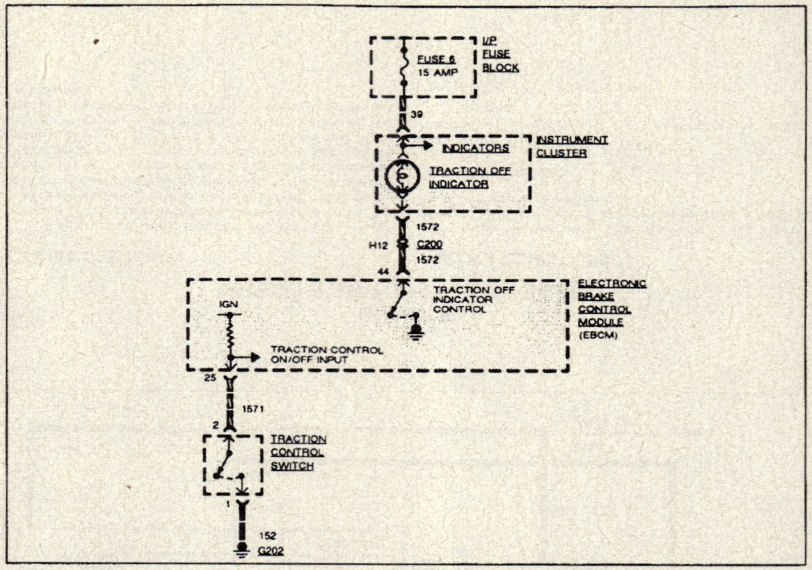

CHART Q
ONLY "TRACTION OFF" INDICATOR DOES NOT LIGHT

CIRCUIT OPERATION

When the ignition is first turned to "RUN," the electronic brake control module (EBCM) goes through a self-test. During this test, the EBCM grounds both the "ANTILOCK" and "TRACTION OFF" indicators for approximately 2 to 4 seconds. This acts as a bulb test for the indicators.

Traction Control Switch

Located either in the instrument panel center trim plate (Touring Sedan only) or the driver's front door trim panel (Elite only), the switch allows the driver to select the traction control mode (on/off). When the traction control system is turned off, the "TRACTION OFF" indicator remains illuminated.

Transaxle Temperature Switch (with Traction Control)

The EBCM monitors transaxle temperature through the transaxle temperature switch. When the transaxle becomes hot, about 160°C (320°F), the normally closed switch opens signaling the EBCM to disable the traction control system and turn on the "TRACTION OFF" indicator. This allows the brakes and transaxle to cool down. As the transaxle cools to 149°C (300°F) the transaxle temperature switch closes. The EBCM keeps the "TRACTION OFF" indicator on and the system disabled for an additional three to five minutes to allow for sufficient cooling.

When performing tests on this system, if the transaxle temperature switch is disconnected, and the ignition is turned to "RUN" the traction control system is disabled for three to five minutes.

FAILURE CONDITIONS

Possible Causes

- Open in CKT 152, 1571 or 1572.
- Open bulb.
- Faulty traction control switch.

Important

- When vehicle is equipped with traction control, any symptom which lights the "ANTILOCK" indicator also lights the "TRACTION OFF" indicator. Therefore, when both indicators are lit, follow "ANTILOCK" indicator diagnosis when no codes are set.

DIAGNOSTIC CHART Q — 1993, CONTINUED

HAS THE ABS/TCS FUNCTIONAL CHECK BEEN PERFORMED?

- NO → PERFORM ABS/TCS FUNCTIONAL CHECK BEFORE PROCEEDING.
- YES →
  - IGNITION: RUN

  DOES THE TRACTION OFF INDICATOR LIGHT FOR 2 TO 4 SECONDS?

  - YES →
    - IGNITION: OFF
    - DISCONNECT TRACTION CONTROL SWITCH.
    - IGNITION: RUN
    - MEASURE VOLTAGE BETWEEN TRACTION CONTROL SWITCH HARNESS CONNECTOR TERMINAL 2 AND ASSEMBLY LINE DIAGNOSTIC LINK (ALDL) CONNECTOR TERMINAL A.

    IS BATTERY VOLTAGE MEASURED?

    - NO → CHECK/REPAIR OPEN IN CKT 1571. IF CKT 1571 IS OK, CHECK ELECTRONIC BRAKE CONTROL MODULE (EBCM) CONNECTOR FOR POOR TERMINAL CONTACT. IF TERMINAL CONTACT IS OK, REPLACE EBCM.
    - YES →
      - IGNITION: OFF
      - MEASURE RESISTANCE BETWEEN TRACTION CONTROL SWITCH HARNESS CONNECTOR TERMINAL 1 AND ALDL CONNECTOR TERMINAL A

      IS LESS THAN 2 OHMS MEASURED?

      - YES → REPLACE TRACTION CONTROL SWITCH.
      - NO → REPAIR OPEN IN CKT 152.
  - NO →
    - IGNITION: OFF
    - DISCONNECT ELECTRONIC BRAKE CONTROL MODULE (EBCM) CONNECTOR.
    - INSTALL J 38716 PINOUT BOX.
    - MEASURE RESISTANCE BETWEEN J 38716 PINOUT BOX TERMINAL 1 AND ASSEMBLY LINE DIAGNOSTIC LINK (ALDL) CONNECTOR TERMINAL A.

    IS LESS THAN 2 OHMS MEASURED?

    - NO → REPAIR OPEN IN CKT 803.
    - YES →
      - IGNITION: RUN
      - MEASURE VOLTAGE FROM J 38716 PINOUT BOX TERMINAL 44 AND 1.

      IS BATTERY VOLTAGE MEASURED?

      - YES → CHECK ELECTRONIC BRAKE CONTROL MODULE (EBCM) CONNECTOR FOR POOR TERMINAL CONTACT. IF TERMINAL CONTACT IS OK, REPLACE EBCM.
      - NO → REPAIR OPEN IN CKT 1572 OR REPLACE TRACTION OFF INDICATOR BULB.

AFTER REPAIRS ARE COMPLETED, CLEAR CODES AND REPEAT ABS/TCS FUNCTIONAL CHECK.

DIAGNOSTIC CHART R — 1993

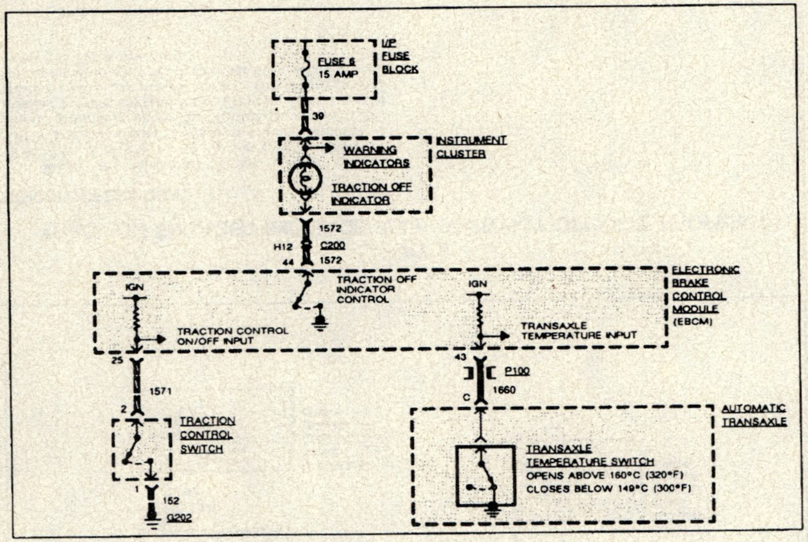

CHART R
ONLY "TRACTION OFF" INDICATOR IS ON ALL THE TIME AND NO CODES ARE SET

CIRCUIT OPERATION

When the ignition is first turned to "RUN," the electronic brake control module (EBCM) goes through a self-test. During this test, the EBCM grounds both the "ANTILOCK" and "TRACTION OFF" indicators for approximately 2 to 4 seconds. This acts as a bulb test for the indicators.

Traction Control Switch

Located either in the instrument panel center trim plate (Touring Sedan only) or the driver's front door trim panel (Elite only), the switch allows the driver to select the traction control mode (on/off). When the traction control system is turned off, the "TRACTION OFF" indicator remains illuminated.

Transaxle Temperature Switch (with Traction Control)

The EBCM monitors transaxle temperature through the transaxle temperature switch. When the transaxle becomes hot, about 160°C (320°F), the normally closed switch opens signaling the EBCM to disable the traction control system and turn on the "TRACTION OFF" indicator. This allows the brakes and transaxle to cool down. As the transaxle cools to 149°C (300°F) the transaxle temperature switch closes. The EBCM keeps the "TRACTION OFF" indicator on and the system disabled for an additional three to five minutes to allow for sufficient cooling.

When performing tests on this system, if the transaxle temperature switch is disconnected, and the ignition is turned to "RUN" the traction control system is disabled for three to five minutes.

FAILURE CONDITIONS

Possible Causes

- Short to ground in CKT 1571 or 1572.
- Open in CKT 1660.
- Faulty traction control switch or transaxle temperature switch.

Important

- When vehicle is equipped with traction control, any symptom which lights the "ANTILOCK" indicator also lights the "TRACTION OFF" indicator. Therefore, when both indicators are lit, follow "ANTILOCK" indicator diagnosis when no codes are set.

DIAGNOSTIC CHART R — 1993, CONTINUED

HAS THE ABS/TCS FUNCTIONAL CHECK BEEN PERFORMED?

- NO → PERFORM ABS/TCS FUNCTIONAL CHECK BEFORE PROCEEDING.
- YES →
  - IGNITION: OFF
  - DISCONNECT AUTOMATIC TRANSAXLE CONNECTOR
  - MEASURE RESISTANCE BETWEEN AUTOMATIC TRANSAXLE TERMINAL C (TRANSAXLE TEMPERATURE SWITCH) AND GROUND.

  IS LESS THAN 3 OHMS MEASURED?

  - NO → REPLACE THE TRANSAXLE TEMPERATURE SWITCH.
  - YES →
    - IGNITION: RUN
    - MEASURE VOLTAGE BETWEEN AUTOMATIC TRANSAXLE HARNESS CONNECTOR TERMINAL C AND GROUND.

    IS BATTERY VOLTAGE MEASURED?

    - NO → CHECK/REPAIR CKT 1860 FOR AN OPEN. IF CKT 1860 IS OK, CHECK ELECTRONIC BRAKE CONTROL MODULE (EBCM) CONNECTOR FOR POOR TERMINAL CONTACT. IF TERMINAL CONTACT IS OK, REPLACE EBCM.
    - YES →
      - IGNITION: OFF
      - DISCONNECT ELECTRONIC BRAKE CONTROL MODULE (EBCM) CONNECTOR
      - INSTALL J 38716 PINOUT BOX.
      - MEASURE RESISTANCE BETWEEN J 38716 PINOUT BOX TERMINAL 1 AND ASSEMBLY LINE DIAGNOSTIC LINK (ALDL) CONNECTOR TERMINAL A.

      IS LESS THAN 2 OHMS MEASURED?

      - NO → REPAIR OPEN IN CKT 803.
      - YES →
        - MEASURE RESISTANCE BETWEEN J 38716 PINOUT BOX TERMINAL 1 AND 44.

        IS LESS THAN 5 OHMS MEASURED?

        - YES → REPAIR CKT 1572 FOR A SHORT TO GROUND.
        - NO →
          - MEASURE RESISTANCE BETWEEN J 38716 PINOUT BOX TERMINAL 1 AND 25.

          IS LESS THAN 5 OHMS MEASURED?

          - YES → CHECK/REPAIR CKT 1571 FOR A SHORT TO GROUND. IF CKT 1571 IS OK, REPLACE TRACTION CONTROL SWITCH.
          - NO → NO TROUBLE FOUND. CHECK FOR POSSIBLE INTERMITTENT OPEN IN TRANSAXLE TEMPERATURE SWITCH CIRCUIT.

AFTER REPAIRS ARE COMPLETED, CLEAR CODES AND REPEAT ABS/TCS FUNCTIONAL CHECK.

DIAGNOSTIC CHART S — 1993

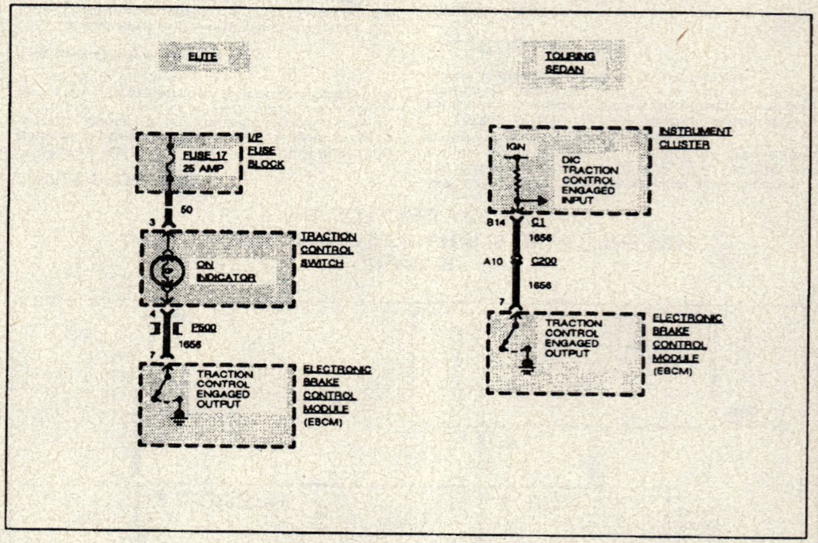

CHART S
TRACTION ENGAGED MESSAGE/BULB DOES NOT OPERATE PROPERLY

CIRCUIT OPERATION

When the electronic brake control module (EBCM) is operating in the traction control mode, the EBCM's traction control engaged output (terminal 7) closes, grounding either the instrument cluster's DIC traction control engaged input (Touring Sedan) or the traction control switch LED (Elite). The EBCM grounds this terminal to alert the driver that the traction control system (TCS) is operating.

DIAGNOSTIC CHART S — 1993, CONTINUED

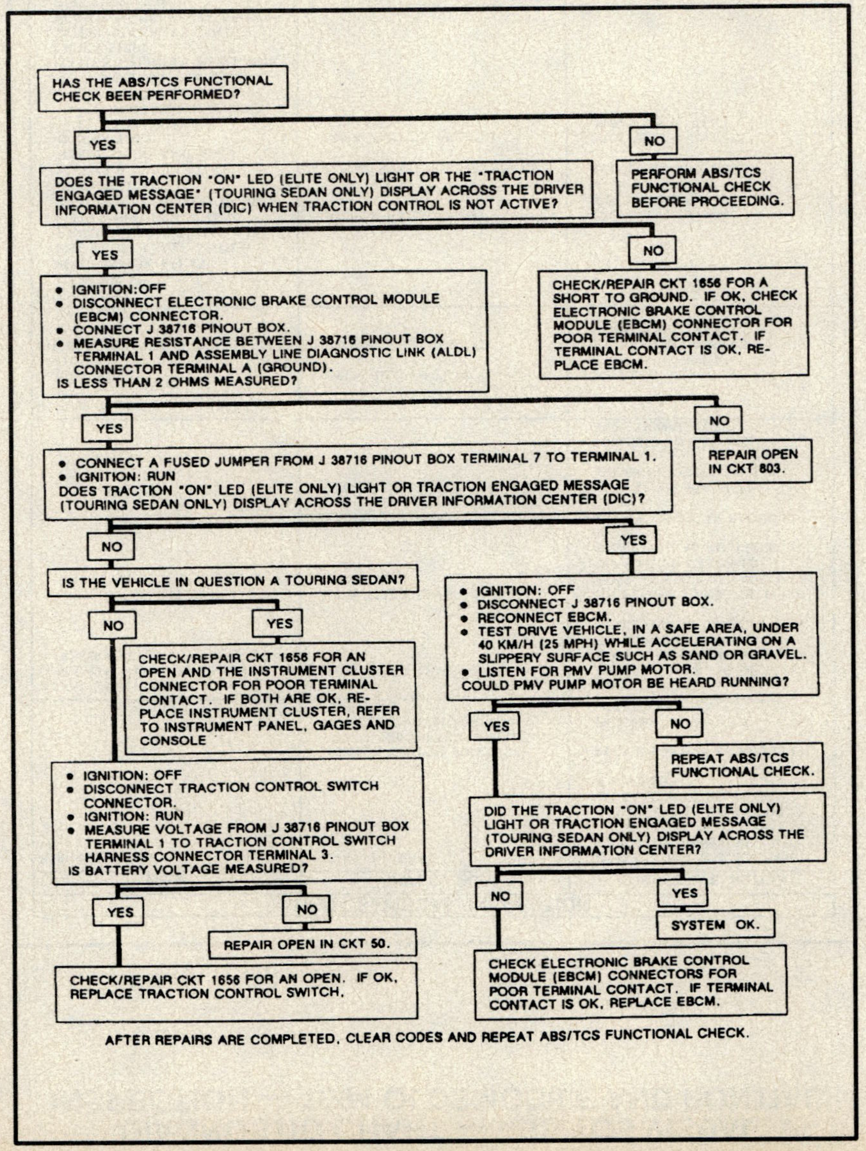

ELECTRICAL CONNECTOR TERMINAL IDENTIFICATION — 1993 MK IV

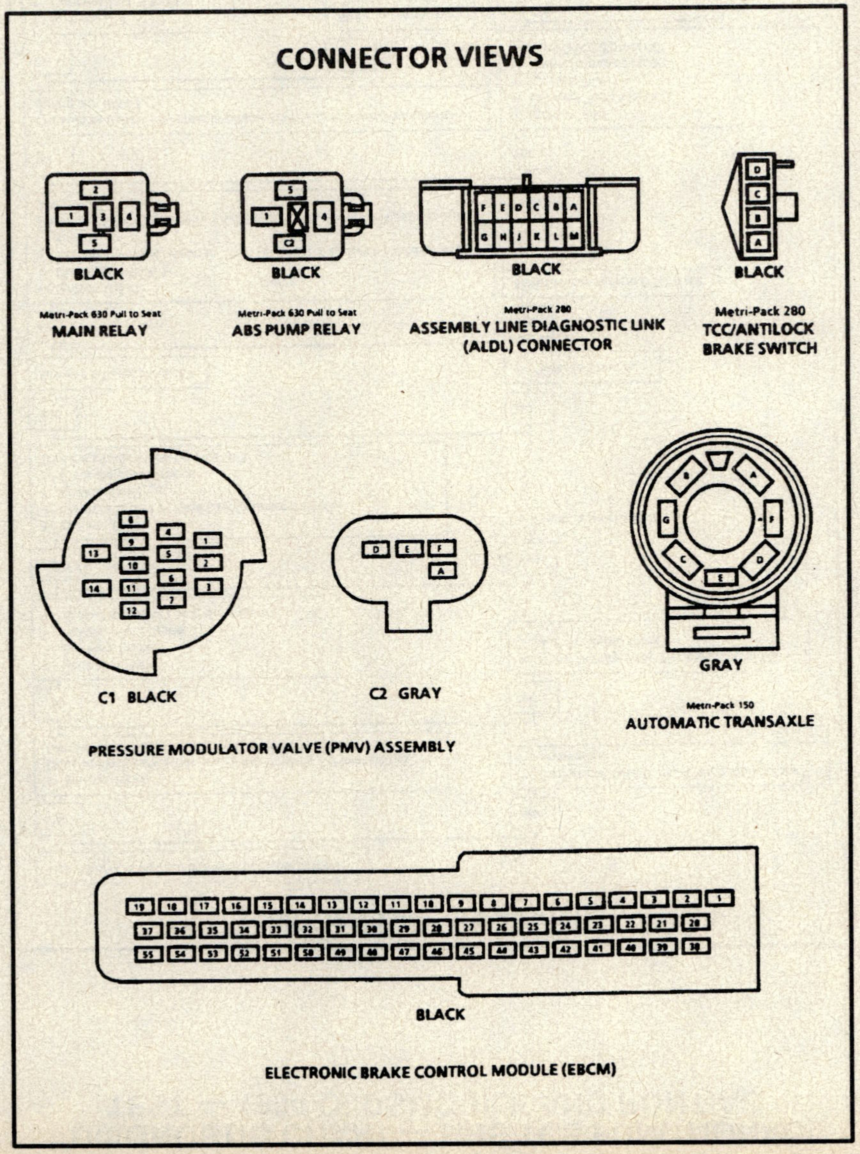

DIAGNOSTIC CHART — ABS/TCS VISUAL INSPECTION — 1994 OLDSMOBILE AND PONTIAC

ABS/TCS VISUAL INSPECTION

ITEM	INSPECT FOR	CORRECTIVE ACTION
PARKING BRAKE	· FULL RELEASE	· OPERATE MANUAL RELEASE LEVER TO VERIFY RELEASE · ADJUST CABLE OR REPAIR RELEASE SYSTEM AS REQUIRED
	· PROPER SWITCH FUNCTION WHEN NECESSARY. UNPLUG SWITCH CONNECTOR TO VERIFY	· REPAIR SWITCH AS REQUIRED · PROCEED TO ABS/TCS FUNCTIONAL TEST
MAJOR COMPONENTS · MASTER CYLINDER FLUID RESERVOIR	· LOW FLUID LEVEL	· DETERMINE CAUSE OF FLUID LOSS AND REPAIR. ADD FLUID TO MASTER CYLINDER RESERVOIR AS REQUIRED
· PRESSURE MODULATOR VALVE (PMV) ASSEMBLY	· LOW FLUID LEVEL IN PMV RESERVOIR	· DETERMINE CAUSE OF FLUID LOSS AND REPAIR. ADD FLUID TO MASTER CYLINDER RESERVOIR AS REQUIRED
	· EXTERNAL LEAKS	· REPAIR LEAKS AS REQUIRED
	· PROPER ASSEMBLY	· INSTALL OR POSITION COMPONENTS PROPERLY · PROCEED TO ABS/TCS FUNCTIONAL TEST
IP FUSE BLOCK · FUSE 6 · FUSE 13 · FUSE 19 UNDERHOOD FUSE BLOCK · FUSE 1 · FUSE 2	· OPEN (BLOWN FUSE) · PROPER ENGAGEMENT	· REPLACE · PROCEED TO ABS/TCS FUNCTIONAL TEST
CONNECTORS · MAIN RELAY · PUMP MOTOR RELAY · PMV FLUID LEVEL SWITCH · WHEEL SPEED SENSORS · ELECTRONIC BRAKE (AND TRACTION) CONTROL MODULE (EBCM/EBTCM) · PRESSURE MODULATOR VALVE (PMV) ASSEMBLY CONNECTORS: C1 - VALVE BLOCK CONNECTOR (ROUND) C2 - PUMP MOTOR CONNECTOR (GRY) · MASTER CYLINDER RESERVOIR · CRUISE/SHIFT INTERLOCK/BRAKE SWITCH · VCC/ANTILOCK BRAKE SWITCH · C101, C102, C414 · AUTOMATIC TRANSAXLE (TRACTION CONTROL ONLY)	· PROPER ENGAGEMENT · LOOSE WIRES OR TERMINALS · CORRODED OR BROKEN EYELETS	· PROPERLY ENGAGE CONNECTOR · REPAIR AS REQUIRED · PROCEED TO ABS/TCS FUNCTIONAL TEST
· GROUNDS - G102, G202, G104	· CORRODED/LOOSE WIRES OR BROKEN EYELETS	· REPAIR AS NEEDED
WHEN NO PROBLEM IS NOTED IN VISUAL INSPECTION, PROCEED TO ABS/TCS FUNCTIONAL TEST		

DIAGNOSTIC CHART — ABS/TCS FUNCTIONAL TEST — 1994 OLDSMOBILE AND PONTIAC

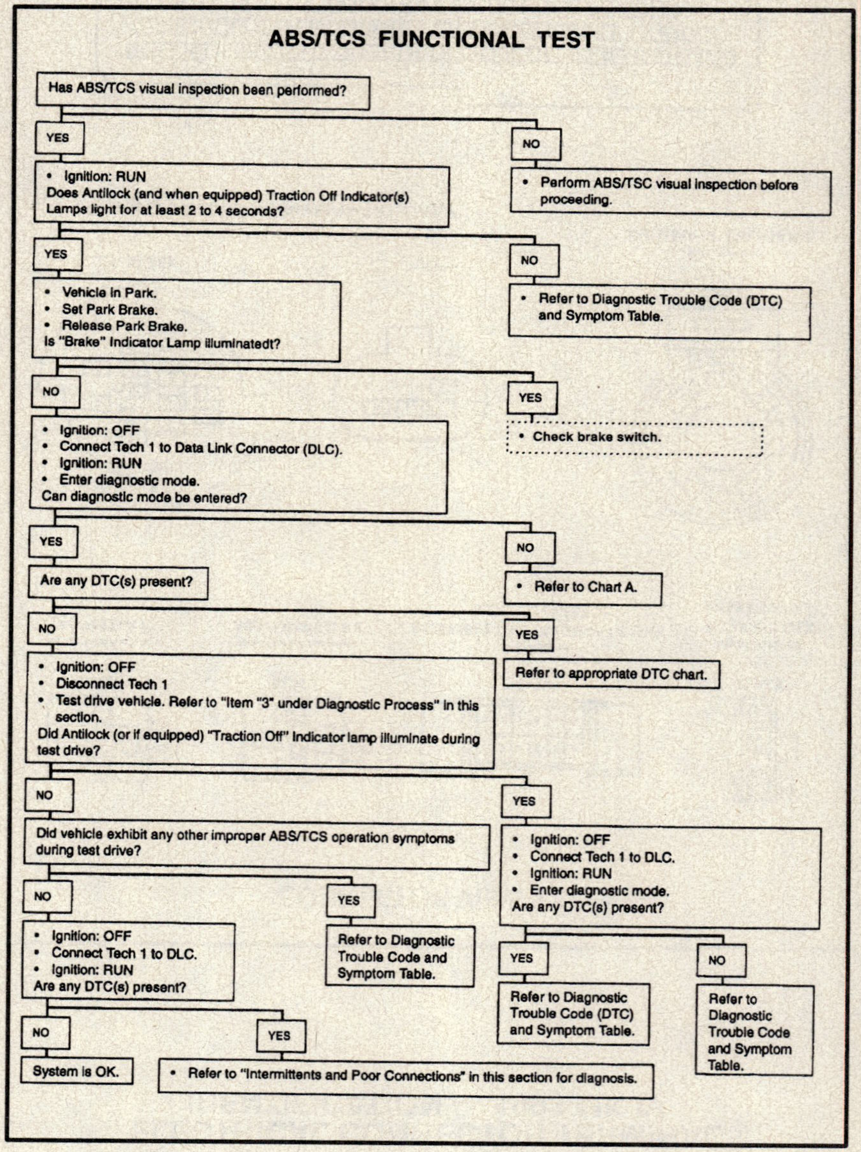

DIAGNOSTIC CHART — DIAGNOSTIC TROUBLE CODE CHART — 1994 OLDSMOBILE AND PONTIAC

DIAGNOSTIC TROUBLE CODE AND SYMPTOM TABLE

CHART	SYMPTOM	PAGE
A	Diagnostic Mode cannot be entered or "ANTILOCK" Indicator Lamp is "on" constantly with no DTC(s) set	14
B	"ANTILOCK" Indicator Lamp does not function	18
C	"TRACTION OFF" Indicator Lamp is "on" constantly with no DTC(s) set	22
D	"TRACTION OFF" Indicator Lamp does not function	24
E	"TRACTION ENGAGED" Indicator Lamp does not operate properly	26
DTC	**DESCRIPTION**	
21	RF Wheel Speed Sensor circuit open or shorted	28
22	RF Wheel Speed Sensor signal erratic	32
23	RF Wheel Speed Sensor signal missing	32
25	LF Wheel Speed Sensor circuit open or shorted	28
26	LF Wheel Speed Sensor signal erratic	32
27	LF Wheel Speed Sensor signal missing	32
31	RR Wheel Speed Sensor circuit open or shorted	30
32	RR Wheel Speed Sensor signal erratic	32
33	RR Wheel Speed Sensor signal missing	32
35	LR Wheel Speed Sensor circuit open or shorted	30
36	LR Wheel Speed Sensor signal erratic	32
37	LR Wheel Speed Sensor signal missing	32
41	RF Inlet Valve circuit open or shorted	40
42	RF Outlet Valve circuit open or shorted	40
44	LF Isolation Valve circuit open or shorted	44
45	Power Interruption Fault	36
46	LF Outlet Valve circuit open or shorted	40
48	RF Isolation Valve circuit open or shorted	44
51	RR Inlet Valve circuit open or shorted	40
52	RR Outlet Valve circuit open or shorted	40
55	LR Inlet Valve circuit open or shorted	40
56	LR Outlet Valve circuit open or shorted	40
61	Pump Motor Circuit Test	48
62	Low Brake Pedal during an ABS stop	54
71	EBCM/EBTCM internal failure	58
72	Brake Switch circuit open or shorted	60
73	Fluid Level Switch circuit open or shorted	62
74	Pressure Switch circuit open or shorted	64
75	PCM requested TCS to be disabled	70
76	UART SDL circuit or signal malfunction	72
77	Delivered Torque Circuit open or shorted	74

DIAGNOSTIC CHART A — 1994 OLDSMOBILE AND PONTIAC

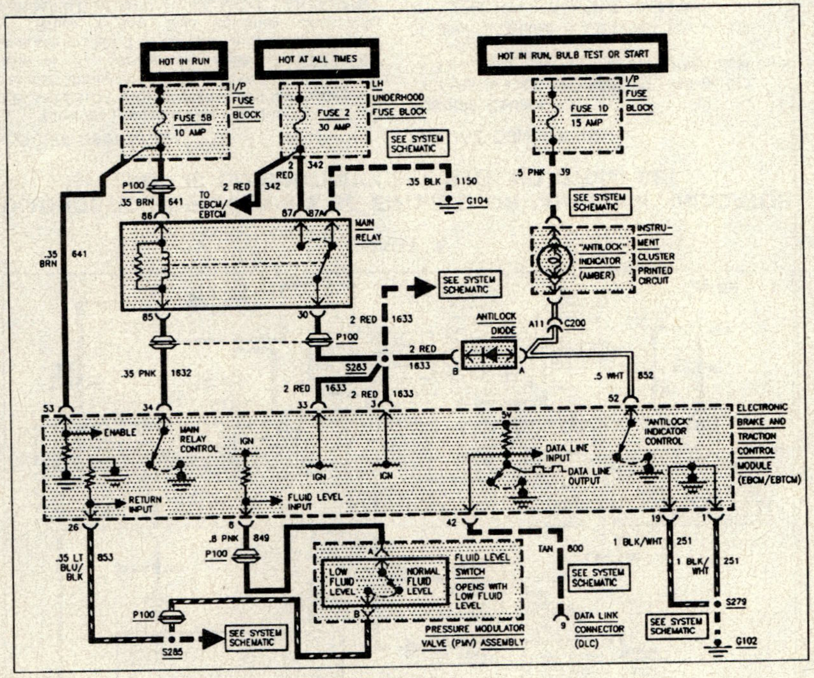

CHART A

DIAGNOSTIC MODE CANNOT BE ENTERED OR "ANTILOCK" INDICATOR LAMP IS ON CONTINUOUSLY WITH NO DTC(s) SET

CIRCUIT OPERATION

When the ignition is first turned to "RUN," the EBCM/EBTCM goes through a self-test lasting 2-4 seconds. During this test, the EBCM/EBTCM grounds terminal 52 causing the amber "ANTILOCK" indicator to light. This serves as an indicator bulb test.

When a low fluid condition exists, the EBCM/EBTCM turns on the amber "ANTILOCK" indicator but does not set a DTC. A low fluid level condition usually lights the red "BRAKE" indicator since the master cylinder supplies the PMV reservoir. However, restrictions or air in the hose could cause a low fluid condition in the PMV reservoir without lighting the red "BRAKE" indicator.

FAILURE CONDITIONS

Possible Causes

- Open in CKT 641, 800, 251, 849 or 853.
- Low brake fluid level or open PMV fluid level switch.
- Short to ground in CKT 641, 800 or 852.
- Wrong scan tool adapter being used.

DIAGNOSTIC CHART A — 1994 OLDSMOBILE AND PONTIAC, CONTINUED

CHART A (1 OF 2)
DIAGNOSTIC MODE CANNOT BE ENTERED OR "ANTILOCK" INDICATOR LAMP IS ON CONTINUOUSLY WITH NO DTC(s) SET

Has ABS/TCS functional test been performed?

- NO: Perform ABS/TCS functional test before proceeding.
- YES:
  - Ignition: OFF
  - Visually inspect Fuse 5B of I/P Fuse Block. (Inspect for open.)

Is fuse OK?

- NO: Replace fuse and check/repair CKT 641 for a short to ground.
- YES:
  - Reinsert fuse.
  - Disconnect Pressure Modulator Valve (PMV) Fluid Level Switch Connector.
  - Using J 39200 set to Ω, measure resistance across PMV Fluid Level Switch Terminal A to B.

Is less than 1 ohm measured?

- NO: Check brake fluid level. When low, check hose for kinks or restrictions and add fluid as needed to Master Cylinder Reservoir. If OK, replace PMV Reservoir.
- YES:
  - Reconnect PMV Fluid Level Switch Connector.
  - Disconnect EBCM/EBTCM Connector.
  - Connect J 38716 Pinout Box.
  - Measure resistance between J 38716 Pinout Box Terminal 8 and 26.

Is less than 3 ohms measured?

- NO: Repair open in CKT 849 or CKT 853.
- YES:
  - Measure resistance between J 38716 Pinout Box Terminal 1 and Data Link Connector (DLC) Terminal 4 (ground), and between J 38716 Pinout Box Terminal 19 and DLC Terminal 4 (ground).

Is less than 2 ohms measured at each terminal?

- NO: Repair open in CKT 251.
- YES: A

Continued on next page

After repairs are completed, clear DTC and repeat ABS/TCS functional test.

DIAGNOSTIC CHART A — 1994 OLDSMOBILE AND PONTIAC, CONTINUED

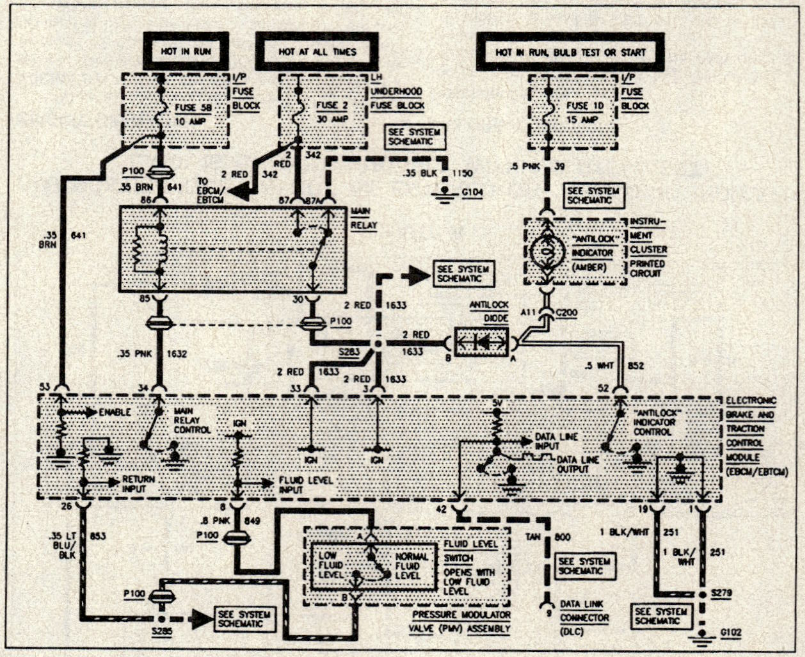

CHART A

DIAGNOSTIC MODE CANNOT BE ENTERED OR "ANTILOCK" INDICATOR LAMP IS ON CONTINUOUSLY WITH NO DTC(s) SET

CIRCUIT OPERATION

When the ignition is first turned to "RUN," the EBCM/EBTCM goes through a self-test lasting 2-4 seconds. During this test, the EBCM/EBTCM grounds terminal 52 causing the amber "ANTILOCK" indicator to light. This serves as an indicator bulb test.

When a low fluid condition exists, the EBCM/EBTCM turns on the amber "ANTILOCK" indicator but does not set a DTC. A low fluid level condition usually lights the red "BRAKE" indicator since the master cylinder supplies the PMV reservoir. However, restrictions or air in the hose could cause a low fluid condition in the PMV reservoir without lighting the red "BRAKE" indicator.

FAILURE CONDITIONS

Possible Causes

- Open in CKT 641, 800, 251, 849 or 853.
- Low brake fluid level or open PMV fluid level switch.
- Short to ground in CKT 641, 800 or 852.
- Wrong scan tool adapter being used.

DIAGNOSTIC CHART A — 1994 OLDSMOBILE AND PONTIAC, CONTINUED

CHART A (2 OF 2)
DIAGNOSTIC MODE CANNOT BE ENTERED OR "ANTILOCK" INDICATOR LAMP IS ON CONTINOUSLY WITH NO DTC(s) SET

A

- Ignition: RUN
- Using J 39200 set to DC scale, measure voltage between J 38716 Pinout Box Terminal 53 and 1.

Is battery voltage measured?

NO → Repair open in CKT 641.

YES →

- Ignition: OFF
- Disconnect Main Relay.
- Ignition: RUN
- Measure voltage between J 38716 Pinout Terminal 52 and 1.

Is battery voltage measured?

NO → Repair CKT 852 for a short to ground.

YES →

- Ignition: RUN
- Measure voltage between J 38716 Pinout Box Terminal 42 and 1.

Is 5 volts measured?

NO → Check/repair CKT 800 for a short to ground or an open. If wire is OK, determine data line failure area.

YES → Check EBCM/EBTCM Connector for proper terminal contact. If terminal contact is OK, replace EBCM/EBTCM.

After repairs are completed, clear DTC and repeat ABS/TCS functional test.

DIAGNOSTIC CHART B — 1994 OLDSMOBILE AND PONTIAC

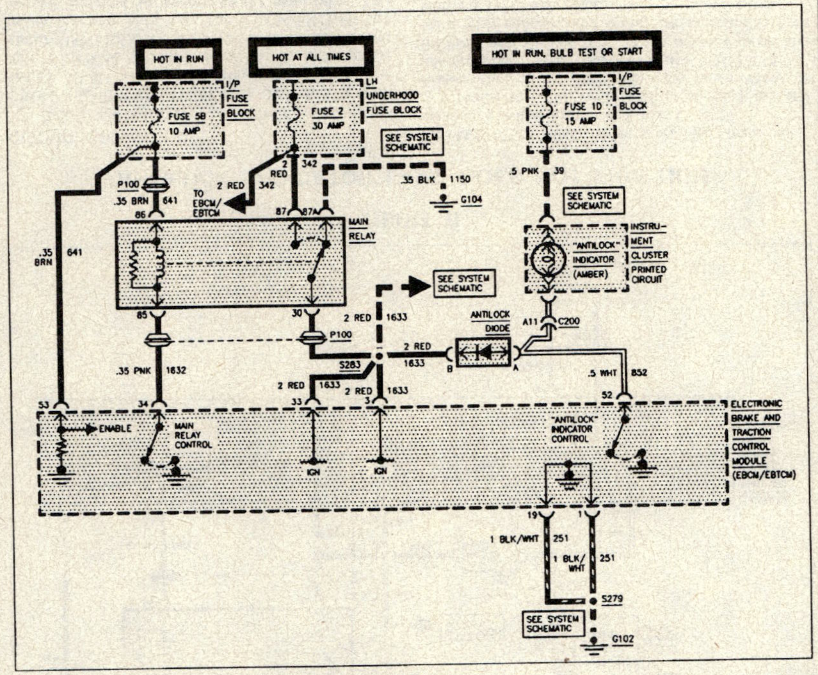

CHART B

"ANTILOCK" INDICATOR LAMP DOES NOT FUNCTION

CIRCUIT OPERATION

When the ignition switch is first turned to "RUN" the EBCM/EBTCM goes through a self-test lasting 2-4 seconds. During this test, the EBCM/EBTCM grounds terminal 52, causing the amber "ANTILOCK" indicator to light. This serves as an indicator bulb test. The EBCM/EBTCM also grounds terminal 34, energizing the main relay. This removes the indicator ground path provided through the antilock diode and the main relay contacts.

When the self-test is completed (and no faults detected) the EBCM/EBTCM removes the ground at terminal 52, turning off the indicator. The main relay remains energized.

When the EBCM/EBTCM detects a fault, it grounds terminal 52 causing the amber "ANTILOCK" indicator to light. (Note: The EBCM/EBTCM also de-energizes the main relay creating a redundant ground path.)

When the EBCM/EBTCM is disabled for any reason (i.e. open Fuse 5B), the main relay is de-energized. This provides an "ANTILOCK" indicator ground path through the antilock diode and main relay contacts, and causes the indicator to light with no DTC being set.

FAILURE CONDITIONS

Possible Causes

- Open in CKT 251, 852 or 1633.
- Open bulb.
- Short to voltage in CKT 852.
- Open or shorted antilock diode.
- Main relay fails "engaged."
- Short to ground in CKT 1632.

CHART B (1 of 2)
"ANTILOCK" INDICATOR LAMP DOES NOT FUNCTION

Has ABS/TCS Functional Test been performed?

- NO: Perform ABS/TCS Functional Test before proceeding.
- YES:
  - Ignition: OFF
  - Connect a Tech 1 to Data Link Connector (DLC).
  - Ignition: RUN
  - Enter diagnostic mode.
  - Can diagnostic mode be entered?

- NO: A — Continued on next page
- YES:
  - Ignition: OFF
  - Disconnect EBCM/EBTCM Connector.
  - Connect J 38716 Pinout Box.
  - Using J 39200 set to Ω, measure resistance between J 38716 Pinout Box Terminal 1 and DLC Terminal 4 (ground).

  Is less than 2 OHMS measured?

- NO: Repair open in CKT 251.
- YES:
  - Disconnect Main Relay.
  - Ignition: RUN
  - Using J 39200 set to DC scale, measure voltage J 38716 Pinout Box Terminals 52 and 1.

  Is battery voltage measured?

- NO: Replace Antilock Indicator Bulb or repair CKT 852 for an open.
- YES:
  - Ignition: RUN
  - Connect a fused jumper between J 38716 Pinout Box Terminals 52 and 1.

  Does Antilock Indicator remain lit?

- NO: Repair short to voltage in CKT 852. Check Antilock Diode and Main Relay for possible damage.
- YES:
  - Ignition: OFF
  - Disconnect Antilock Brake Diode.
  - Using J 39200 check Antilock Diode. Also, check for improper polarity.
  - Is diode OK?

- NO: Replace Antilock Diode. Observe polarity.
- YES:
  - Reconnect Antilock Brake Diode. Observe polarity.
  - Disconnect J 38716 Pinout Box.
  - Reconnect EBCM/EBTCM Connector and Main Relay.
  - Ignition: RUN
  - Does Antilock Indicator light for 2-4 seconds?

- YES: PROBLEM SOLVED
- NO: Check EBCM/EBTCM Connector for proper terminal contact. If terminal contact is OK, replace EBCM/EBTCM.

After repairs are completed, clear DTC and repeat ABS/TCS functional test.

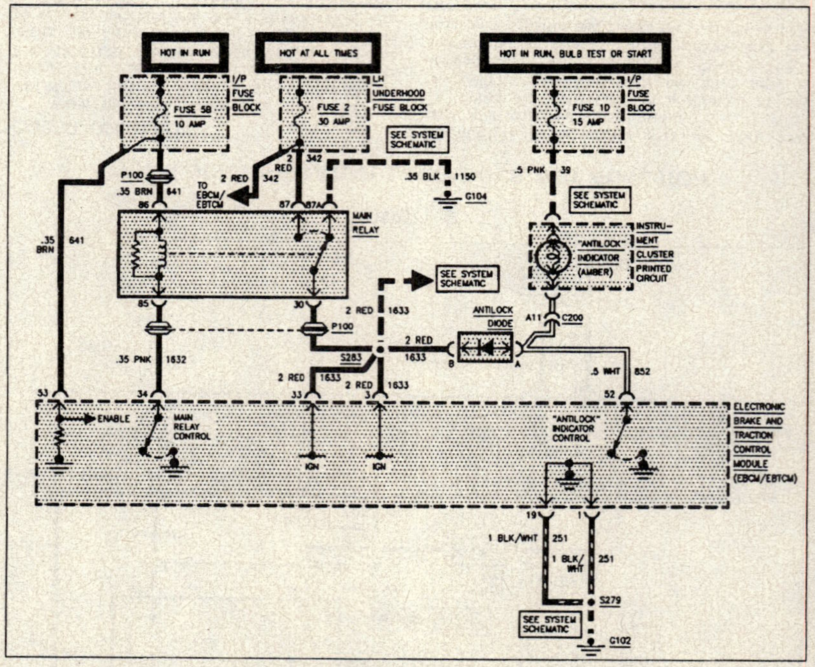

CHART B

"ANTILOCK" INDICATOR LAMP DOES NOT FUNCTION

CIRCUIT OPERATION

When the ignition switch is first turned to "RUN" the EBCM/EBTCM goes through a self-test lasting 2-4 seconds. During this test, the EBCM/EBTCM grounds terminal 52, causing the amber "ANTILOCK" indicator to light. This serves as an indicator bulb test. The EBCM/EBTCM also grounds terminal 34, energizing the main relay. This removes the indicator ground path provided through the antilock diode and the main relay contacts.

When the self-test is completed (and no faults detected) the EBCM/EBTCM removes the ground at terminal 52, turning off the indicator. The main relay remains energized.

When the EBCM/EBTCM detects a fault, it grounds terminal 52 causing the amber "ANTILOCK" indicator to light. (Note: The EBCM/EBTCM also de-energizes the main relay creating a redundant ground path.)

When the EBCM/EBTCM is disabled for any reason (i.e. open Fuse 5B), the main relay is de-energized. This provides an "ANTILOCK" indicator ground path through the antilock diode and main relay contacts, and causes the indicator to light with no DTC being set.

FAILURE CONDITIONS

Possible Causes

- Open in CKT 251, 852 or 1633.
- Open bulb.
- Short to voltage in CKT 852.
- Open or shorted antilock diode.
- Main relay fails "engaged."
- Short to ground in CKT 1632.

DIAGNOSTIC CHART B — 1994 OLDSMOBILE AND PONTIAC, CONTINUED

CHART B (2 of 2)
"ANTILOCK" INDICATOR LAMP DOES NOT FUNCTION

A

- Ignition: OFF
- Disconnect Tech 1.
- At this point a double fault is possible. First refer to CHART A for diagnosis, then continue and complete this diagnostic chart from this point.

Was CHART A performed?

- NO: Refer to CHART A.
- YES:
  - Ignition: OFF
  - Disconnect Main Relay.
  - Using J 39200 set to Ω, measure resistance across Main Relay Terminals 30 and 87A.

  Is less than 2 OHMS measured?

- NO: Replace Main Relay.
- YES:
  - Remove Fuse 5B of I/P Fuse Block.
  - Ignition: RUN
  - Using J 39200 set to DC scale, measure voltage between Main Relay Harness Connector Terminal 30 and ground.

  Is battery voltage measured?

- NO: Replace Antilock Diode or repair open in CKT 1633. NOTE: When replacing diode, observe polarity.
- YES:
  - Measure voltage between Main Relay Harness Connector Terminals 30 and 87A.

  Is battery voltage measured?

- NO: Repair open in CKT 251.
- YES:
  - Ignition: OFF
  - Disconnect EBCM/EBTCM Connector.
  - Using J 39200 set to Ω, measure resistance to ground between Main Relay Harness Connector Terminal 85 and ground.
  - Is less than 5 OHMS measured?

- YES: Repair short to ground in CKT 1632.
- NO: Check EBCM/EBTCM Connector for proper terminal contact. If terminal contact is OK, replace EBCM/EBTCM.

After repairs are completed, clear DTC and repeat ABS/TCS functional test.

DIAGNOSTIC CHART C — 1994 OLDSMOBILE AND PONTIAC

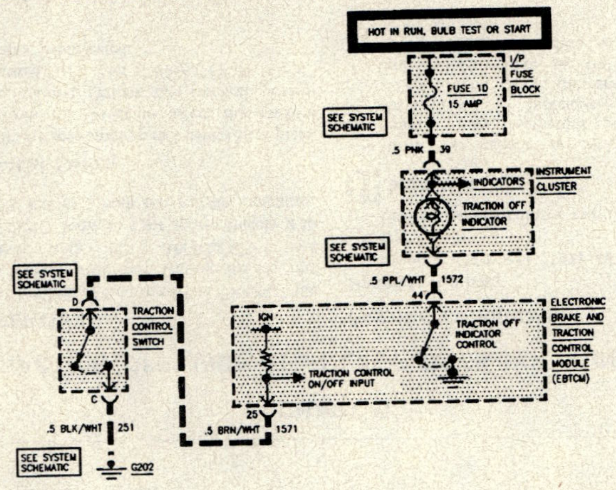

CHART C

"TRACTION OFF" INDICATOR LAMP IS ON CONTINUOUSLY AND NO DTC(s) SET

CIRCUIT OPERATION

When the ignition is first turned to "RUN," the EBTCM goes through a self-test. During this test, the EBTCM grounds both the "ANTILOCK" and "TRACTION OFF" indicators for approximately 2 to 4 seconds. This acts as a bulb test for the indicators.

Traction Control Switch

Located in the instrument panel center trim plate, the switch allows the driver to select the traction control mode (on/off). When the traction control system is turned off, the "TRACTION OFF" indicator remains illuminated.

Transaxle Temperature

The EBTCM monitors transaxle temperature through information provided by the PCM. When the transaxle becomes hot, about 160°C (320°F), the EBTCM disables the traction control system and turns on the "TRACTION OFF" indicator. This allows the brakes and transaxle to cool down. When the transaxle cools to 149°C (300°F) the EBTCM keeps the "TRACTION OFF" indicator on and the system disabled for an additional three to five minutes to allow for sufficient cooling.

FAILURE CONDITIONS

Possible Causes

- Short to ground in CKT 1571 or 1572.
- Faulty traction control switch.
- Faulty EBTCM

Important

- When the vehicle is equipped with traction control, any symptom which lights the "ANTILOCK" indicator also lights the "TRACTION OFF" indicator. Therefore, when both indicators are lit, follow "ANTILOCK" indicator diagnosis when no diagnostic trouble codes are set.

DIAGNOSTIC CHART C — 1994 OLDSMOBILE AND PONTIAC, CONTINUED

CHART C
"TRACTION OFF" INDICATOR LAMP IS ON CONTINUOUSLY AND NO DTC(s) SET

Has ABS/TCS Functional Test been performed?

- NO: Perform ABS/TCS Functional Test before proceeding.
- YES:
  - Ignition: OFF
  - Disconnect EBTCM Connector.
  - Install J 38716 Pinout Box.
  - Using J 39200 set to Ω, measure resistance between J 38716 Pinout Box Terminal 1 and Data Link Connector (DLC) Terminal 4 (ground).

  Is less than 2 OHMS measured?
  - NO: Repair open in CKT 251.
  - YES:
    - Measure resistance between J 38716 Pinout Box Terminal 1 and 44.

    Is less than 5 OHMS measured?
    - YES:
      - Check CKT 1572 for short to ground.
      - Is short to ground present?
        - YES: Repair CKT 1572
        - NO: Replace EBTCM.
    - NO:
      - Measure resistance between J 38716 Pinout Box Terminal 1 and 25.

      Is less than 5 OHMS measured?
      - YES: Check/repair CKT 1571 for a short to ground. If CKT 1571 is OK, replace Traction Control Switch.
      - NO: No trouble found. Refer to "INTERMITTENTS AND POOR CONNECTIONS" in this section for possible causes.

DIAGNOSTIC CHART D — 1994 OLDSMOBILE AND PONTIAC

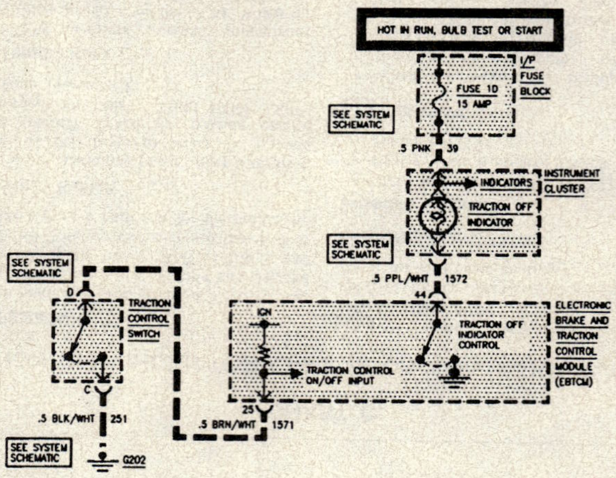

CHART D
"TRACTION OFF" INDICATOR LAMP DOES NOT FUNCTION

CIRCUIT OPERATION

When the ignition is first turned to "RUN," the EBTCM goes through a self-test. During this test, the EBCM grounds both the "ANTILOCK" and "TRACTION OFF" indicators for approximately 2 to 4 seconds. This acts as a bulb test for the indicators.

Traction Control Switch

Located in the instrument panel center trim plate, the switch allows the driver to select the traction control mode (on/off). When the traction control system is turned off, the "TRACTION OFF" indicator remains illuminated.

FAILURE CONDITIONS

Possible Causes

- Open in CKT 251, 1571 or 1572.
- Open bulb.
- Faulty traction control switch.

Important

- When the vehicle is equipped with traction control, any symptom which lights the "ANTILOCK" indicator also lights the "TRACTION OFF" indicator. Therefore, when both indicators are lit, follow "ANTILOCK" indicator diagnosis when no diagnostic trouble codes are set.

DIAGNOSTIC CHART D — 1994 OLDSMOBILE AND PONTIAC, CONTINUED

CHART D
"TRACTION OFF" INDICATOR LAMP DOES NOT FUNCTION

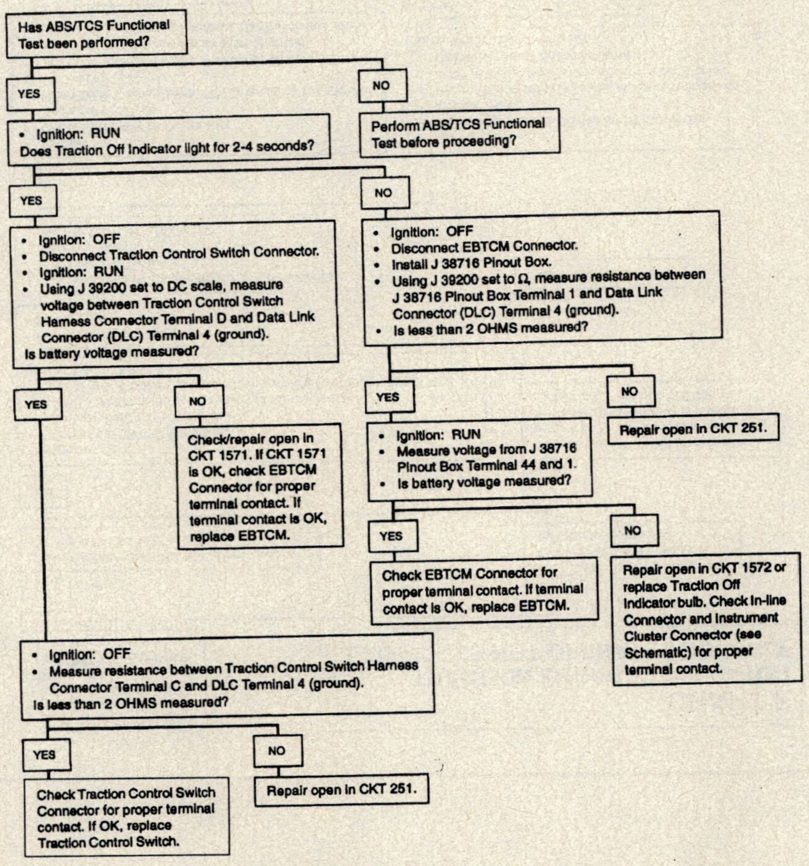

After repairs are completed, clear DTC and repeat ABS/TCS functional test.

DIAGNOSTIC CHART E — 1994 OLDSMOBILE AND PONTIAC

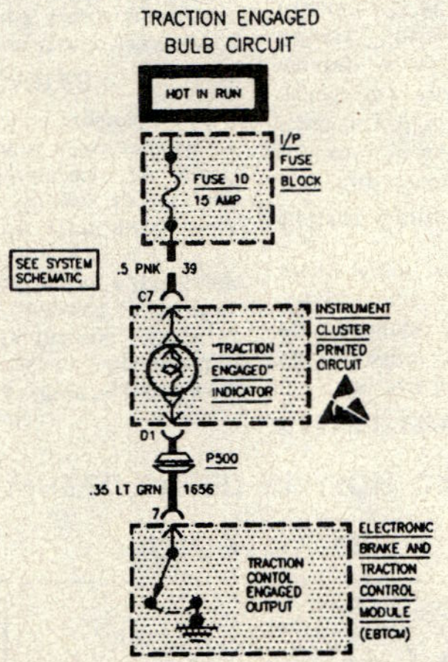

CHART E

"TRACTION ENGAGED" INDICATOR LAMP DOES NOT OPERATE PROPERLY

CIRCUIT OPERATION

When the EBTCM is operating in the traction control mode, the EBTCM's traction control engaged output (Terminal 7) closes, grounding either the instrument clusters DIC traction control engaged input or the traction control switch LED. The EBTCM grounds this terminal to alert the driver that the traction control system (TCS) is operating.

DIAGNOSTIC CHART E — 1994 OLDSMOBILE AND PONTIAC, CONTINUED

CHART E
TRACTION ENGAGED INDICATOR LAMP DOES NOT OPERATE PROPERLY

Has ABS/TCS Functional Test been performed?

- NO: Perform ABS/TCS Functional Test before proceeding.
- YES: Does Traction "ON" LED light when Traction Control is not active?
  - NO: Check/repair CKT 1656 for a short to ground. If OK, check EBTCM Connector for proper terminal contact. If terminal contact is OK, replace EBTCM.
  - YES:
    - Ignition: OFF
    - Disconnect EBTCM Connector.
    - Connect J 38716 Pinout Box.
    - Using J 39200 set to Ω, measure resistance between J 38716 Pinout Box Terminal 1 and Data Link Connector (DLC) Terminal 4 (ground).

    Is less than 2 OHMS measured?
    - NO: Repair open in CKT 251.
    - YES:
      - Connect a Fused Jumper from J 38716 Pinout Box Terminal 7 to Terminal 1.
      - Ignition: RUN

      Does Traction "ON" LED light ?
      - NO:
        - Ignition: OFF
        - Disconnect Traction Control Switch Connector.
        - Ignition: RUN
        - Using J 39200 set to DC scale, measure voltage from J 38716 Pinout Box Terminal 1 to Traction Control Switch Harness Connector Terminal D.

        Is battery voltage measured?
        - YES: Check/repair CKT 1656 for an open. If OK, replace Traction Control Switch
        - NO: Repair open in CKT 39.
      - YES:
        - Ignition: OFF
        - Disconnect J 38716 Pinout Box.
        - Reconnect EBTCM.
        - Using Tech 1, select Mode F0 (Data List 1) and monitor TCS state.
        - Test drive vehicle in a safe area under 40 km/h (25 mph) on a slippery surface such as sand or gravel.
        - Does TCS state change during a traction control event?
        - NO: Repeat ABS/TCS Functional Test.
        - YES: Did the Traction "ON" LED light ?
          - YES: System OK.
          - NO: Check EBTCM Connectors for proper terminal contact. If terminal contact is OK, replace EBTCM.

After repairs are completed, clear DTC and repeat ABS/TCS functional test.

DIAGNOSTIC CHART DTC 21 AND 25 — 1994 OLDSMOBILE AND PONTIAC

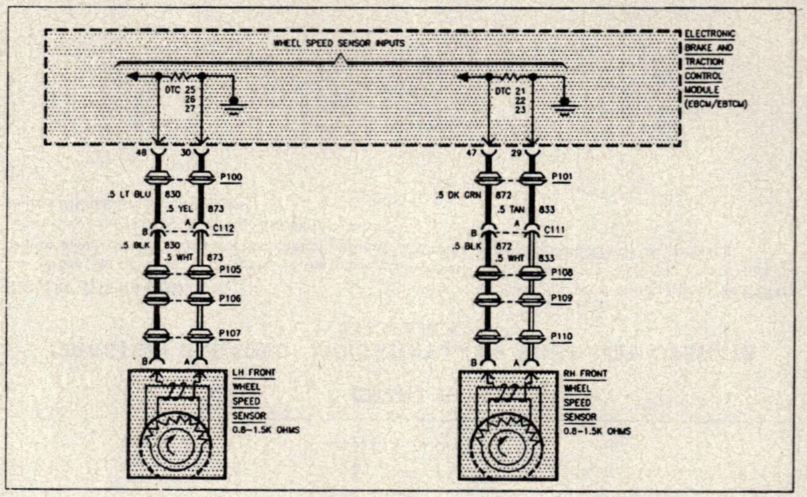

DTC(s) 21, 25

FRONT WHEEL SPEED SENSOR CIRCUIT TEST

CIRCUIT OPERATION

As each wheel turns, the wheel speed sensor creates an AC voltage with a frequency proportional to wheel speed. This magnetically induced voltage is caused by a toothed sensor ring (mounted on the drive axle) passing the wheel speed sensor's stationary pickup coil.

The EBCM/EBTCM uses the wheel speed sensor signal to calculate vehicle reference speeds and individual wheel speed, acceleration and slip values. These values are used to determine when antilock (or when equipped) traction control is required.

WHEEL SPEED SENSOR CABLES

In order to prevent electromagnetic interference from disturbing the wheel speed sensor signal, the cables from the EBCM/EBTCM to each wheel speed sensor are twisted in pairs a minimum of 6 to 9 turns per foot.

FAILURE CONDITIONS

The EBCM/EBTCM performs two basic checks on the wheel speed sensors: sensor continuity and sensor output. When the ignition is first turned to "RUN," the EBCM/EBTCM performs the wheel speed sensor continuity check.

Sensor Continuity Check

Any condition which would result in lack of continuity in the front wheel speed sensor circuit could result in a DTC 21 or 25 being set. These conditions include an open, a short to ground, or a short between the two wheel speed sensor circuits in the wiring harness between the speed sensor and EBCM/EBTCM. Also an open or short to ground across the wheel speed sensor coil could trigger this code.

DIAGNOSTIC CHART DTC 21 AND 25 — 1994 OLDSMOBILE AND PONTIAC, CONTINUED

DTC(s) 21, 25
FRONT WHEEL SPEED SENSOR CIRCUIT OPEN OR SHORTED

Has ABS/TCS functional test been performed?

NO: Perform ABS/TCS functional test before proceeding.

YES:
- Ignition: OFF
- Raise vehicle.
- Disconnect suspect front Wheel Speed Sensor Connector.
- Check connector for proper terminal contact.
- Using J 39200 set to Ω, measure resistance between suspect Wheel Speed Sensor Terminal A and B.

Is 800 to 1500 ohms measured?

NO: Replace suspect front Wheel Speed Sensor.

YES:
- Measure resistance between suspect front Wheel Speed Sensor Terminal B and ground.

Is infinite ohms measured?

NO: Replace suspect front Wheel Speed Sensor.

YES:
- Reconnect Wheel Speed Sensor.
- Disconnect Wheel Speed Sensor Jumper Harness.
- Check connector for proper terminal contact.
- Measure resistance between Jumper Harness Terminals A and B (sensor side).

Is 800 to 1500 ohms measured?

NO: Replace suspect front Wheel Speed Sensor Jumper Harness.

YES:
- Reconnect Wheel Speed Sensor Jumper Harness.
- Disconnect EBCM/EBTCM Connector.
- Check connector for proper terminal contact.
- Install J 38716 Pinout Box.
- Measure resistance between terminals indicated for suspected front Wheel Speed Sensor Circuit:
  - Left Front Wheel - Terminals 30 and 48
  - Right Front Wheel - Terminals 29 and 47

Is 800 to 1500 ohms measured?

NO: Above 1500 ohms (remember to check in-line connector [see schematic] and Wheel Speed Sensor Connector for proper terminal contact):
- When left front wheel, repair open in CKT 830 or 873.
- When right front wheel, repair open in CKT 833 or 872.

Below 800 ohms:
- When left front wheel, repair short between CKT 830 and 873.
- When right front wheel, repair short between CKT 833 and 872.

YES:
- Measure resistance between J 38716 Pinout Box Terminal 1 and Data Link Connector (DLC) Terminal 4 (ground).

Is less than 5 ohms measured?

NO: Repair open in CKT 251.

YES:
- Measure resistance at J 38716 Pinout Box Terminals indicated for suspected front Wheel Speed Sensor Circuit:
  - Left Front Wheel - Terminal 48 and 1
  - Right Front Wheel - Terminal 47 and 1

Is less than 5 ohms measured?

YES: When left front wheel, repair short to ground in CKT 830 or 873. When right front wheel, repair short to ground in CKT 872 or 833.

NO: Check EBCM/EBTCM Connector for proper terminal contact. If terminal contact is OK, reconnect EBCM/EBTCM, clear DTC and test drive vehicle. If DTC resets, replace EBTCM. If DTC does not reset, refer to Intermittants and Poor Connections.

After repairs are completed, clear DTC and repeat ABS/TCS functional test.

DIAGNOSTIC CHART DTC 31 AND 35 — 1994 OLDSMOBILE AND PONTIAC

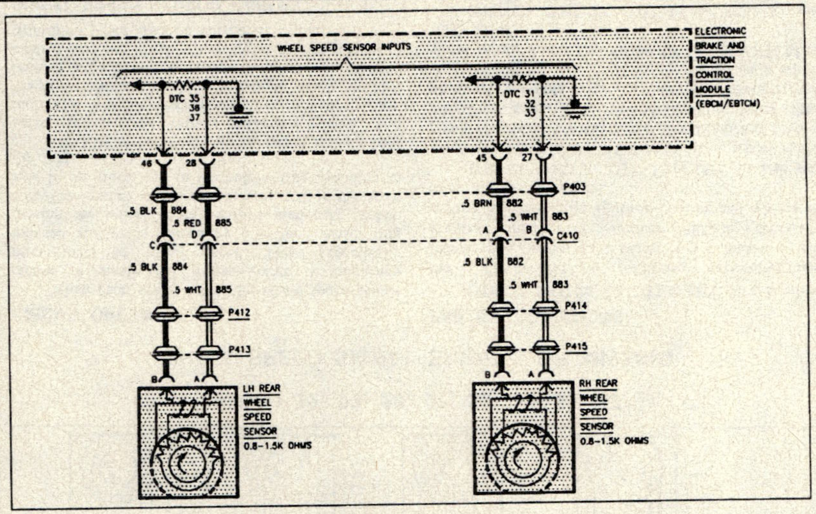

DTC(s) 31, 35
REAR WHEEL SPEED SENSOR CIRCUIT OPEN OR SHORTED

CIRCUIT OPERATION

When each wheel turns, the wheel speed sensor creates a small AC voltage with a frequency proportional to wheel speed. This magnetically induced voltage is caused by a toothed sensor ring passing the wheel speed sensor's stationary pickup coil. Both sensors are part of the hub and bearing assembly.

The EBCM/EBTCM uses the wheel speed sensor signal to calculate vehicle reference speeds and individual wheel speed, acceleration and slip values. These values are used to determine when antilock control is required.

WHEEL SPEED SENSOR CABLES

In order to prevent electromagnetic interference from disturbing the wheel speed sensor signal, the cables from the EBCM/EBTCM to each wheel speed sensor are twisted in pairs a minimum of 6 to 9 turns

FAILURE CONDITIONS

The EBCM/EBTCM performs two basic checks on the wheel speed sensors: sensor continuity and sensor output. When the ignition is first turned to "RUN," the EBCM/EBTCM performs the wheel speed sensor continuity check.

Sensor Continuity Check

Any condition which would result in lack of continuity in the rear wheel speed sensor circuit could result in a DTC 31 or 35 being set. These conditions include an open, a short to ground, or a short between the two wheel speed sensor circuits in the wiring harness between the speed sensor and EBCM/EBTCM. Also an open or short to ground across the wheel speed sensor coil could trigger this DTC.

DIAGNOSTIC CHART DTC 31 AND 35 — 1994 OLDSMOBILE AND PONTIAC, CONTINUED

DTC(s) 31, 35
REAR WHEEL SPEED SENSOR CIRCUIT OPEN OR SHORTED

Has ABS/TCS functional test been performed?

NO → Perform ABS/TCS functional test before proceeding.

YES →
- Ignition: OFF
- Park Brake: RELEASED
- Raise vehicle.
- Disconnect suspect rear Wheel Speed Sensor Connector at wheel.
- Check connector for proper terminal contact.
- Using J 39200 set to Ω, measure resistance between suspect Wheel Speed Sensor Terminal A and B.

Is 800 to 1500 ohms measured?

NO → Replace suspect rear Wheel Speed Sensor.

YES →
- Measure resistance between suspect rear Wheel Speed Sensor Terminal B and ground.

Is infinite ohms measured?

NO → Replace suspect rear Wheel Speed Sensor.

YES →
- Reconnect Wheel Speed Sensor.
- Disconnect Wheel Speed Jumper Harness.
- Check connector for proper terminal contact.
- Measure resistance between Jumper Harness Terminals A and B if right side Wheel Speed Sensor is suspect or Terminals D and C if left side Wheel Speed Sensor is suspect.

Is 800 to 1500 ohms measured?

NO → Replace Wheel Speed Sensor Jumper Harness.

YES →
- Reconnect Wheel Speed Sensor Jumper Harness.
- Disconnect EBCM/EBTCM Connector.
- Check connector for proper terminal contact.
- Install J 38716 Pinout Box.
- Measure resistance between terminals indicated for suspected rear Wheel Speed Sensor Circuit:
  - Left Rear Wheel - Terminals 28 and 46
  - Right Rear Wheel - Terminals 27 and 45

Is 800 to 1500 ohms measured?

NO → Above 1500 ohms (remember to check in-line connector [see schematic] and Wheel Speed Sensor Connector for proper terminal contact):
- When left rear wheel, repair open in CKT 884 or 885.
- When right rear wheel, repair open in CKT 882 or 883.

Below 800 ohms:
- When left rear wheel, repair short between CKT 884 and 885.
- When right rear wheel, repair short between CKT 882 and 883.

YES →
- Measure resistance between J 38716 Pinout Box Terminal 1 and Data Link Connector (DLC) Terminal A (ground).

Is less than 5 ohms measured?

NO → Repair open in CKT 251.

YES →
- Measure resistance at J 38716 Pinout Box Terminals indicated for suspected rear Wheel Speed Sensor Circuit:
  - Left Rear Wheel - Terminal 46 and 1
  - Right Rear Wheel - Terminal 45 and 1

Is less than 5 ohms measured?

YES → When left rear wheel, repair short to ground in CKT 884 or 885. When right rear wheel, repair short to ground in CKT 882 or 883.

NO → Check EBCM/EBTCM Connector for proper terminal contact. If terminal contact is OK, reconnect EBCM/EBTCM, clear DTC and test drive vehicle. If DTC resets, replace EBTCM. If DTC does not reset, refer to Intermittants and Poor Connections in this section.

After repairs are completed, clear DTC and repeat ABS/TCS functional test.

DIAGNOSTIC CHART DTC 22, 23, 26, 27, 32, 33, 36, 37 — 1994 OLDSMOBILE AND PONTIAC

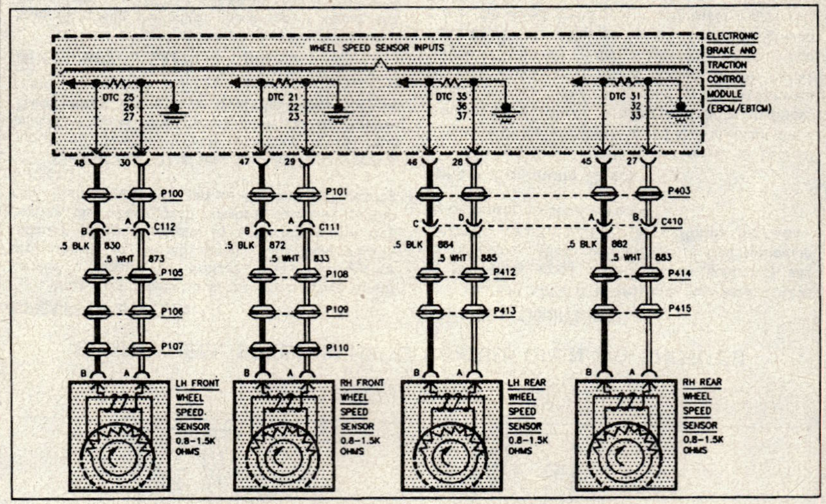

DTC(s) 22, 23, 26, 27, 32, 33, 36, 37
WHEEL SPEED SIGNAL ERRATIC OR MISSING

CIRCUIT OPERATION

When each wheel turns, the wheel speed sensor creates a small AC voltage with a frequency proportional to wheel speed. This magnetically induced voltage is caused by a toothed sensor ring passing the wheel speed sensor's stationary pickup coil. The toothed ring sensor is mounted on the drive axle in the front, and in the hub/bearing assembly in the rear.

The EBCM/EBTCM uses the wheel speed sensor signal to calculate vehicle reference speeds, and individual wheel speed, acceleration and slip values. These values are used to determine when antilock control is required. Vehicles equipped with traction control also use front wheel speed sensor signals to determine when traction control is required.

WHEEL SPEED SENSOR CABLES

In order to prevent electromagnetic interference from disturbing the wheel speed sensor signal, the cables from the EBCM/EBTCM to each wheel speed sensor are twisted in pairs a minimum of 6 to 9 turns per foot.

FAILURE CONDITIONS

DTC 22, DTC 26, DTC 33 or DTC 36 sets when the EBCM/EBTCM detects electromagnetic interference (noise) in a wheel speed sensor signal. Possible interference sources include accessory motors, wiper motors, ignition circuits and add-on or after market equipment.

DTC 22, DTC 26, DTC 32 or DTC 36 also sets when the EBCM/EBTCM determines a wheel speed signal is erratic. A wheel speed signal indicating that a wheel is accelerating or decelerating faster than physically possible is an example of an erratic signal. An intermittent open or short to ground in a wheel speed sensor could also cause the EBCM/EBTCM to set a DTC.

DTC 23, DTC 27, DTC 33, or DTC 37 sets when the EBCM/EBTCM does not see any wheel speed sensor signal, but determines continuity exists in the circuit. This is cause by a dislodged wheel speed sensor, missing toothed sensor ring (front) or defective hub and bearing assembly (rear).

DIAGNOSTIC CHART DTC 22, 23, 26, 27, 32, 33, 36, 37 — 1994 OLDSMOBILE AND PONTIAC, CONTINUED

DTC(s) 22, 23, 26, 27, 32, 33, 36, 37 WHEEL SPEED SIGNAL ERRATIC OR MISSING (1 of 2)

Has ABS/TCS functional test been performed?

- NO: Perform ABS/TCS functional test before proceeding.
- YES: Is DTC 23, DTC 27, DTC 33 or DTC 37 present?
  - YES: When DTC 23 or DTC 27: Reinstall dislodged front Wheel Speed Sensor or replace front Hub and Bearing Assembly. Refer to SECTION 4D. When DTC 33 or DTC 37: Reinstall dislodged rear Wheel Speed Sensor or replace Hub and Bearing Assembly.
  - NO:
    - Perform following and repair/replace as necessary:
      Test drive vehicle (refer to Item "3" under Diagnostic Process in this section) monitoring suspect Wheel Speed Sensor using Tech 1 in snapshot (mode F3).
      Disconnect suspect Wheel Speed Sensor Connector and applicable jumper harness connector. Inspect terminal contacts on sensor and connectors for corrosion or deformity. Check connectors for proper terminal contact.
      Trace harness routing of suspect sensor wiring. Inspect for physical damage particularly in areas where pinching or movement by harness or other components occurs.
      Trace harness routing of suspect sensor wiring for proximity to noise sources such as ignition circuits, motors, etc. Also add-on or "after market" equipment.
      Check if any repairs were recently made to suspect sensor wiring. Sensor wiring should be twisted 6 to 9 turns per foot.
      Check for loose sensor mounting, or debris lodged between wheel speed sensor and hub.
    - If all above checks are OK, proceed, but be aware that problem causing diagnostic trouble code to set may be intermittent and is usually not caused by a defective EBCM/EBTCM.

    - Ignition: RUN
    - Gear Selector Lever: NEUTRAL
    - Ignition: OFF
    - Park Brake: RELEASED
    - Raise vehicle, check bearing end play - repair as necessary.
    - Disconnect suspect Wheel Speed Sensor at wheel.
    - Using J 39200 set to AC scale measure voltage between suspect Wheel Speed Sensor Terminal A and B while turning drive wheels by hand.

    Is 50 to 900 millivolts AC measured?
    - NO: Replace suspect Wheel Speed Sensor.
    - YES: A

Continued on next page

After repairs are completed, clear DTC and repeat ABS/TCS functional test.

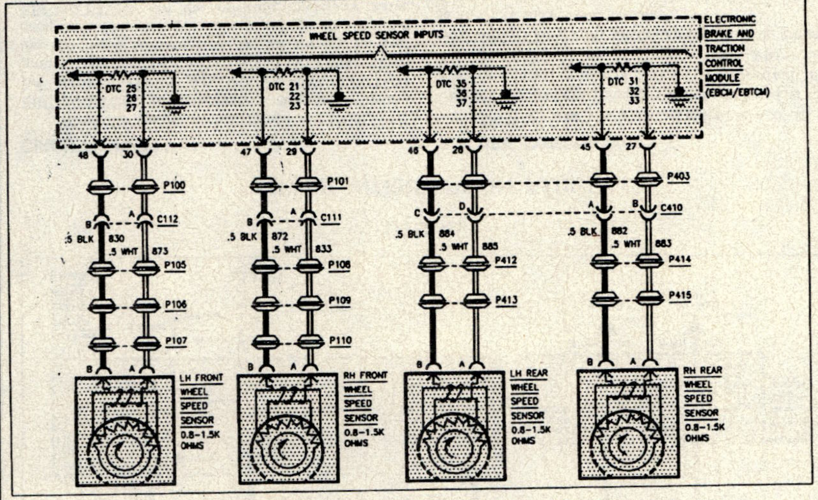

DTC(s) 22, 23, 26, 27, 32, 33, 36, 37

WHEEL SPEED SIGNAL ERRATIC OR MISSING

CIRCUIT OPERATION

When each wheel turns, the wheel speed sensor creates a small AC voltage with a frequency proportional to wheel speed. This magnetically induced voltage is caused by a toothed sensor ring passing the wheel speed sensor's stationary pickup coil. The toothed ring sensor is mounted on the drive axle in the front, and in the hub/bearing assembly in the rear.

The EBCM/EBTCM uses the wheel speed sensor signal to calculate vehicle reference speeds, and individual wheel speed, acceleration and slip values. These values are used to determine when antilock control is required. Vehicles equipped with traction control also use front wheel speed sensor signals to determine when traction control is required.

WHEEL SPEED SENSOR CABLES

In order to prevent electromagnetic interference from disturbing the wheel speed sensor signal, the cables from the EBCM/EBTCM to each wheel speed sensor are twisted in pairs a minimum of 6 to 9 turns per foot.

FAILURE CONDITIONS

DTC 22, DTC 26, DTC 33 or DTC 36 sets when the EBCM/EBTCM detects electromagnetic interference (noise) in a wheel speed sensor signal. Possible interference sources include accessory motors, wiper motors, ignition circuits and add-on or after market equipment.

DTC 22, DTC 26, DTC 32 or DTC 36 also sets when the EBCM/EBTCM determines a wheel speed signal is erratic. A wheel speed signal indicating that a wheel is accelerating or decelerating faster than physically possible is an example of an erratic signal. An intermittent open or short to ground in a wheel speed sensor could also cause the EBCM/EBTCM to set a DTC.

DTC 23, DTC 27, DTC 33, or DTC 37 sets when the EBCM/EBTCM does not see any wheel speed sensor signal, but determines continuity exists in the circuit. This is cause by a dislodged wheel speed sensor, missing toothed sensor ring (front) or defective hub and bearing assembly (rear).

DIAGNOSTIC CHART DTC 22, 23, 26, 27, 32, 33, 36, 37 — 1994 OLDSMOBILE AND PONTIAC, CONTINUED

DTC(s) 22, 23, 26, 27, 32, 33, 36, 37
WHEEL SPEED SIGNAL ERRATIC OR MISSING
(2 of 2)

A

- Thoroughly spray all Wheel Speed Sensors, connectors and harnesses with a 5% solution (2 teaspoons of salt to 12 oz. of water) of salt water.
- Test drive vehicle in a straight line at slow speeds (under 16 km/h [10 mph] while monitoring Wheel Speed Sensors using Tech 1 in snapshot mode (NOTE: suspect Wheel Speed Signal usually varies from remaining Wheel Speed Signals by approximately 2 km/h [1 mph]).

Does any one Wheel Speed Signal vary from other three Wheel Speed Signals?

YES

Replace suspect Wheel Speed Sensor.

NO

- Clear DTC(s) with Tech 1.
- Test drive vehicle.
- If DTC(s) resets, replace EBCM/EBTCM. If DTC(s) do not reset, refer to Intermittants and Poor Connections.

After repairs are completed, clear DTC and repeat ABS/TCS functional test.

DIAGNOSTIC CHART DTC 45 — 1994 OLDSMOBILE AND PONTIAC

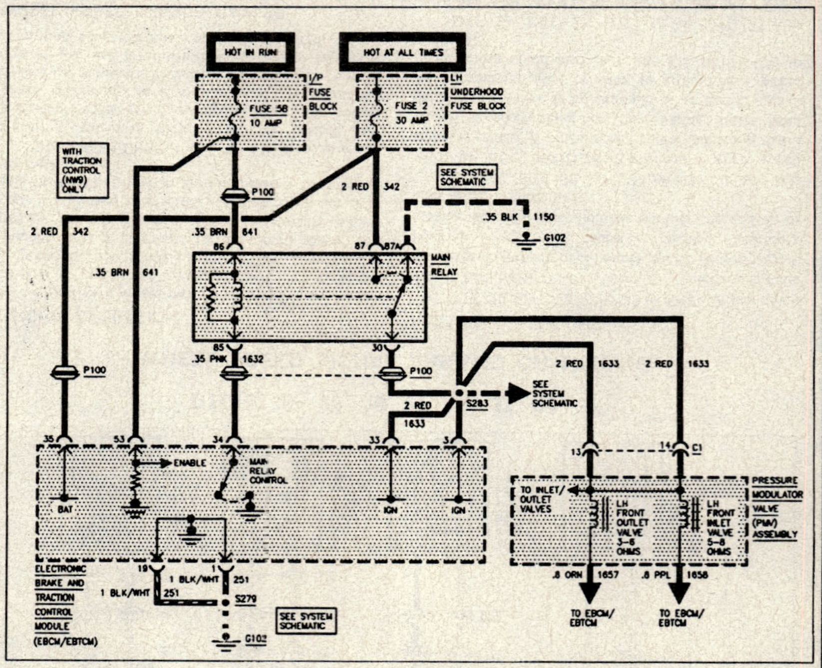

DTC 45

POWER INTERRUPTION FAULT

CIRCUIT OPERATION

When the ignition is first turned to "RUN," the EBCM/EBTCM goes through a self-check lasting 2-4 seconds. During this test the EBCM/EBTCM grounds terminal 34, energizing the main relay. Once the self-check is completed the main relay remains energized. This delivers battery voltage to both the EBCM/EBTCM and the pressure modulator valve (PMV) assembly during vehicle operation. The constant voltage supply allows the vehicle to enter ABS mode without delay. The main relay is de-energized whenever the amber "ANTILOCK" indicator lights due to a system fault, causing the ABS system to shut down.

FAILURE CONDITION

DTC 45 sets when the ignition switch is at "RUN" and battery voltage is neither detected at the PMV assembly nor EBCM/EBTCM terminal 20, 3 and/or 33. Code 45 also sets when the EBCM/EBTCM detects an open or a short to ground in the LH inlet valve circuit.

Possible Causes

- Open in CKT 641, 1632, 858, 1633, or 342.
- Short to ground in CKT 858, 1633, or 342.
- Main relay fails "open."

DIAGNOSTIC CHART DTC 45 — 1994 OLDSMOBILE AND PONTIAC, CONTINUED

DTC 45
POWER INTERRUPTION FAULT (1 of 2)

IMPORTANT: DTC 45 may set intermittently due to low system voltage (extended cranking, low battery, but vehicle eventually starts etc.)

Has ABS/TCS functional test been performed?

- YES
  - Perform chart "DTC(s) 41, 42, 45, 46, 51, 52, 55, 56 inlet/outlet valve circuit open or shorted" then perform chart "DTC(s) 44, 48 isolation valve circuit open or shorted".
  - Was problem found while performing inlet/outlet valve or isolation valve tests?
    - YES: OK
    - NO
      - Using fused jumper, jump between Pinout Box Terminals 1 and 34.
      - Ignition: RUN
      - Does "ABS" Indicator go off?
        - YES
          - Using J 39200 set to DC scale, check for battery voltage between Pinout Box Terminals 1 and 3 then 1 and 33, PMV harness side connector C1. Terminal 13 and ground then Terminal 14 and ground. Battery voltage present?
            - NO: Repair open in CKT 1633.
            - YES: Replace EBCM/EBTCM
        - NO
          - IGNITION : OFF
          - Remove fused jumper.
          - Disconnect Main Relay.
          - Check for continuity between relay Terminals 30 and 87A.
          - Continuity Present?
            - NO: Replace Main Relay
            - YES
              - Measure resistance between main relay Terminals 85 and 86.
              - 40-90 ohms measured?
                - NO: Replace Main Relay
                - YES
                  - Using unpowered test light to battery, probe main relay Harness Connector Cavity 87A.
                  - Does test light illuminate?
                    - NO: Replace Main Relay
                    - YES
                      - Test light to ground, probe main Relay Harness Connector Cavity 87.
                      - Does test light illuminate?
                        - NO: Repair open in CKT 1150
                        - YES: A
- NO
  - Perform ABS/TSC functional test before proceeding.

Continued on next page

After repairs are completed, clear DTC and repeat ABS/TCS functional test.

DIAGNOSTIC CHART DTC 45 — 1994 OLDSMOBILE AND PONTIAC, CONTINUED

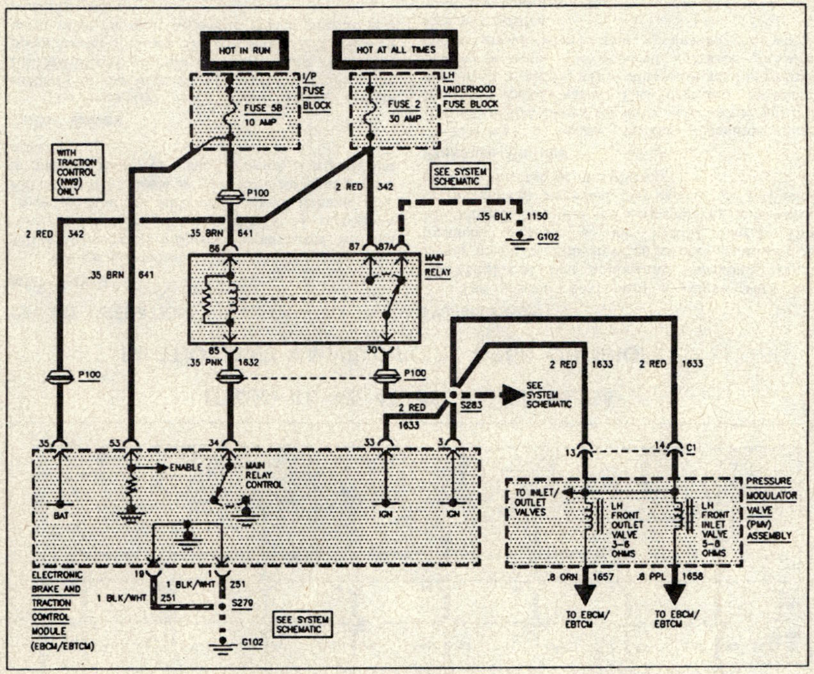

DTC 45
POWER INTERRUPTION FAULT

CIRCUIT OPERATION

When the ignition is first turned to "RUN," the EBCM/EBTCM goes through a self-check lasting 2-4 seconds. During this test the EBCM/EBTCM grounds terminal 34, energizing the main relay. Once the self-check is completed the main relay remains energized. This delivers battery voltage to both the EBCM/EBTCM and the pressure modulator valve (PMV) assembly during vehicle operation. The constant voltage supply allows the vehicle to enter ABS mode without delay. The main relay is de-energized whenever the amber "ANTILOCK" indicator lights due to a system fault, causing the ABS system to shut down.

FAILURE CONDITION

DTC 45 sets when the ignition switch is at "RUN" and battery voltage is neither detected at the PMV assembly nor EBCM/EBTCM terminal 20, 3 and/or 33. Code 45 also sets when the EBCM/EBTCM detects an open or a short to ground in the LH inlet valve circuit.

Possible Causes

- Open in CKT 641, 1632, 858, 1633, or 342.
- Short to ground in CKT 858, 1633, or 342.
- Main relay fails "open."

DIAGNOSTIC CHART DTC 45 — 1994 OLDSMOBILE AND PONTIAC, CONTINUED

DTC 45
POWER INTERRUPTION FAULT (2 OF 2)

A

- Ignition: RUN
- Test light to ground, probe main Relay Harness Connector Cavity 86.

Does test light illuminate?

NO:
- Check LH underhood fuse block Fuse 2.

Is fuse open?

- NO: Repair open in CKT 342.
- YES: Repair short to ground in CKT 1633 or 342 (TCS only). Replace fuse.

YES:
- Using fused jumper, jump between main Relay Harness Connector Cavitys 30 and 87.
- Using J 39200 set to DC scale, check for battery voltage between Pinout Box Terminals 33 and 1 and 3 and 1, then at harness side of PMV Connector Terminal 14 and ground and 13 and ground.

Battery voltage present?

NO:
- Check I/P fuse block Fuse 5B.

Is fuse open?

- YES: Repair short to ground in CKT 641. Replace fuse.
- NO: Repair open in CKT 641.

YES:
- Ignition: OFF
- Install fused jumper between main Relay Harness Connector Cavity 85 and 86.
- Ignition: RUN
- Check for battery voltage between Pinout Box Terminals 34 and 1.

Battery voltage present?

NO: Repair open in CKT 1633 between main relay Terminal 30 and EBCM/EBTCM Terminal 33 and 3.

YES: Replace main relay.

NO:
- Check I/P fuse block Fuse 5B.

Is fuse open?

- YES: Repair short to ground in CKT 1632. Replace fuse.
- NO: Repair open in CKT 1632.

After repairs are complete, clear DTC and repeat ABS/TCS functional test.

DIAGNOSTIC CHART DTC 41, 42, 45, 46, 51, 52, 55 AND 56 — 1994 OLDSMOBILE AND PONTIAC

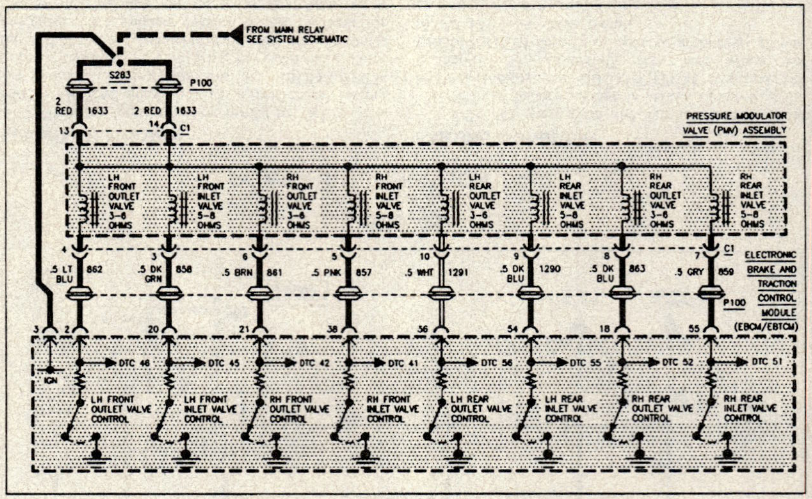

DTC(s) 41, 42, 45, 46, 51, 52, 55, 56

INLET/OUTLET VALVE CIRCUIT OPEN OR SHORTED

CIRCUIT OPERATION

Inlet Valves

Voltage is applied to the normally open inlet valves whenever the main relay is energized. An inlet valve closes when the EBCM/EBTCM supplies a ground path at the inlet valve control terminal. The EBCM/EBTCM closes an inlet valve for short periods to maintain (hold) or reduce pressure at a particular wheel.

Outlet Valves

The normally closed outlet valves operate similarly to the inlet valves, except that when the EBCM/EBTCM grounds an outlet valve, the valve opens instead of closes. The EBCM/EBTCM opens an outlet valve for short periods to reduce pressure at a particular wheel.

VALVE MODES

Pressure Increase

The pressure increase position is the operating valve position for normal (non-ABS) braking and traction control mode (when equipped). The EBCM/EBTCM does not energize either valve, leaving the inlet valve open and the outlet valve closed.

Pressure Hold

The pressure hold position occurs when the EBCM/EBTCM first notices an impending wheel lockup condition. This signifies an over abundance of hydraulic pressure at the locking wheel. The EBCM/EBTCM closes the corresponding inlet valve in an attempt to avoid any additional hydraulic pressure increase at the wheel.

Pressure Reduce

During a wheel lockup condition, the EBCM/EBTCM closes the inlet valve while pulsing open the outlet valve. This decreases hydraulic pressure at a specific wheel without decreasing overall system pressure. The reduced hydraulic pressure allows greater wheel spin, eliminating the wheel lockup condition.

Test Description

This test checks for proper inlet and outlet valve operation. The wheel brake behavior must be checked to verify the valves hydraulic effectiveness. This is accomplished using a Tech 1 to actuate the valves.

Failure conditions

DTC 41, 42, 45, 46, 51, 52, 55, or 56 sets when the EBCM/EBTCM detects an open or short to ground in the respective inlet or outlet valve circuit.

DIAGNOSTIC CHART DTC 41, 42, 45, 46, 51, 52, 55 AND 56 — 1994 OLDSMOBILE AND PONTIAC, CONTINUED

DTC(s) 41, 42, 45, 46, 51, 52, 55, 56
INLET/OUTLET VALVE CIRCUIT OPEN OR SHORTED (1 OF 2)

Has ABS/TCS functional test been performed?

- NO: Perform ABS/TCS functional test before proceeding.
- YES:
  - Ignition: OFF
  - Disconnect Pressure Modulator Valve (PMV) Assembly Valve Connector C1.
  - Check connector for proper terminal contact.
  - Using J 39200 set to Ω, check for continuity between PMV Assembly Pigtail Connector C1, Terminals 13 and 14.

  Continuity?

- NO: Replace PMV Assembly.
- YES:
  - Measure resistance between PMV Pigtail Connector C1 Terminal 14 and PMV Valve Block.

  Is infinite ohms measured?

- NO: Replace PMV Assembly.
- YES:
  - Measure resistance at PMV Assembly Pigtail Connector C1, between Terminal 14 and all Valve Terminals. Refer to chart.

DIAGNOSTIC TROUBLE CODE (DTC)/VALVE	SUSPECT TERMINAL	CORRECT RESISTANCE
45/LH FRONT INLET	3	5 TO 8 OHMS
41/RH FRONT INLET	5	
55/LH REAR INLET	9	
51/RH REAR INLET	7	
46/LH FRONT OUTLET	4	3 TO 6 OHMS
42/RH FRONT OUTLET	6	
56/LH REAR OUTLET	10	
52/RH REAR OUTLET	8	

Is correct resistance measured?

- NO: Replace PMV Assembly.
- YES: A

Continued on next page

DIAGNOSTIC CHART DTC 41, 42, 45, 46, 51, 52, 55 AND 56 — 1994 OLDSMOBILE AND PONTIAC, CONTINUED

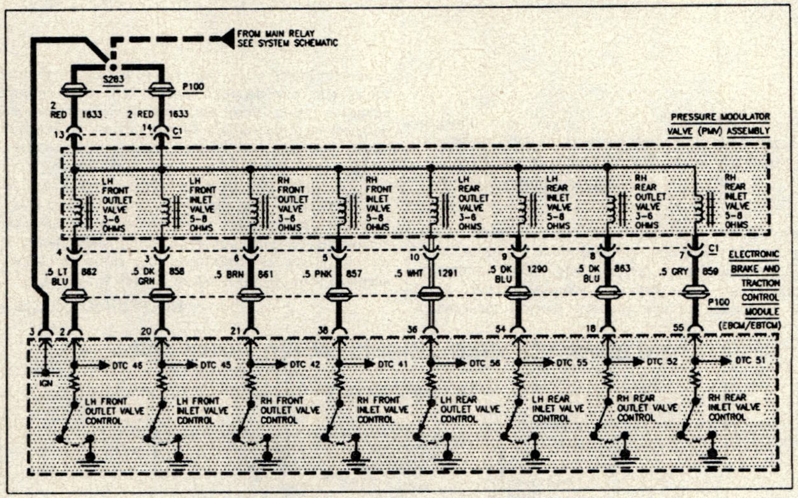

DTC(s) 41, 42, 45, 46, 51, 52, 55, 56

INLET/OUTLET VALVE CIRCUIT OPEN OR SHORTED

CIRCUIT OPERATION

Inlet Valves

Voltage is applied to the normally open inlet valves whenever the main relay is energized. An inlet valve closes when the EBCM/EBTCM supplies a ground path at the inlet valve control terminal. The EBCM/EBTCM closes an inlet valve for short periods to maintain (hold) or reduce pressure at a particular wheel.

Outlet Valves

The normally closed outlet valves operate similarly to the inlet valves, except that when the EBCM/EBTCM grounds an outlet valve, the valve opens instead of closes. The EBCM/EBTCM opens an outlet valve for short periods to reduce pressure at a particular wheel.

VALVE MODES

Pressure Increase

The pressure increase position is the operating valve position for normal (non-ABS) braking and traction control mode (when equipped). The EBCM/EBTCM does not energize either valve, leaving the inlet valve open and the outlet valve closed.

Pressure Hold

The pressure hold position occurs when the EBCM/EBTCM first notices an impending wheel lockup condition. This signifies an over abundance of hydraulic pressure at the locking wheel. The EBCM/EBTCM closes the corresponding inlet valve in an attempt to avoid any additional hydraulic pressure increase at the wheel.

Pressure Reduce

During a wheel lockup condition, the EBCM/EBTCM closes the inlet valve while pulsing open the outlet valve. This decreases hydraulic pressure at a specific wheel without decreasing overall system pressure. The reduced hydraulic pressure allows greater wheel spin, eliminating the wheel lockup condition.

Test Description

This test checks for proper inlet and outlet valve operation. The wheel brake behavior must be checked to verify the valves hydraulic effectiveness. This is accomplished using a Tech 1 to actuate the valves.

Failure conditions

DTC 41, 42, 45, 46, 51, 52, 55, or 56 sets when the EBCM/EBTCM detects an open or short to ground in the respective inlet or outlet valve circuit.

DIAGNOSTIC CHART DTC 41, 42, 45, 46, 51, 52, 55 AND 56 — 1994 OLDSMOBILE AND PONTIAC, CONTINUED

DTC(s) 41, 42, 45, 46, 51, 52, 55, 56 INLET/OUTLET VALVE CIRCUIT OPEN OR SHORTED (2 OF 2)

A

- Reconnect PMV Assembly Valve Connector C1.
- Disconnect EBCM/EBTCM Connector.
- Install J 38716 Pinout Box.
- Measure resistance between J 38716 Pinout Box Terminal 3 and all Valve Terminals. Refer to Chart.

DIAGNOSTIC TROUBLE CODE (DTC)/VALVE	J 38716 TERMINAL	CORRECT RESISTANCE
45/LH FRONT INLET	20	5 TO 8 OHMS
41/RH FRONT INLET	38	
55/LH REAR INLET	54	
51/RH REAR INLET	55	
46/LH FRONT OUTLET	2	3 TO 6 OHMS
42/RH FRONT OUTLET	21	
56/LH REAR OUTLET	36	
52/RH REAR OUTLET	18	

Is resistance measured less than 10 ohms?

YES

- Measure resistance between J 38716 Pinout Box Terminal 1 and Data Link Connector (DLC) Terminal 4 (ground).

Is less than 5 ohms measured?

NO

Repair open in Suspect Valve Circuit between Pressure Modulator Valve (PVM) Assembly and EBCM/EBTCM.

YES

- Disconnect Main Relay.
- Measure resistance between J 38716 Pinout Box Terminal 1 and all Valve Terminals. Refer to Chart.

DIAGNOSTIC TROUBLE CODE (DTC)/VALVE	J 38716 TERMINAL	CORRECT RESISTANCE
45/LH FRONT INLET	20	INFINITE OHMS
41/RH FRONT INLET	38	
55/LH REAR INLET	54	
51/RH REAR INLET	55	
46/LH FRONT OUTLET	2	
42/RH FRONT OUTLET	21	
56/LH REAR OUTLET	36	
52/RH REAR OUTLET	18	

Is infinite ohms measured?

NO

Repair open in CKT 251.

YES

- Reinstall main relay.
- IGNITION: RUN
- Using J 39200 set to DC scale, measure voltage between Pinout Box Terminal 1 and suspect valve terminal.

Is more than 2 volts measured?

NO

Repair short to ground in suspect valve circuit between PMV Assembly and EBCM/EBTCM.

NO

- If diagnosing DTC 45, return to DTC 45 chart.

If not, Check EBCM/EBTCM connector for proper terminal contact. If terminal contact is OK, replace EBCM/EBTCM.

YES

Repair short to battery in suspect valve circuit.

After repairs are completed, clear DTC and repeat ABS/TCS functional test.

DIAGNOSTIC CHART DTC 44 AND 48 — 1994 OLDSMOBILE AND PONTIAC

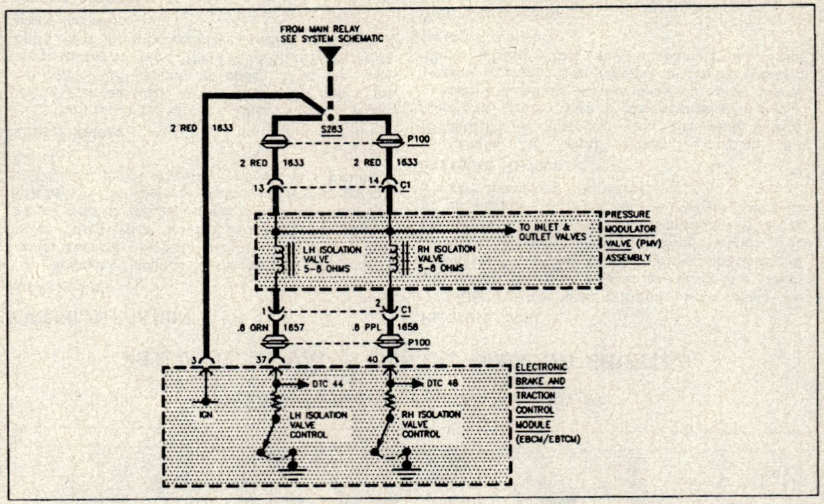

DTC(s) 44, 48

ISOLATION VALVE CIRCUIT OPEN OR SHORTED

CIRCUIT OPERATION

Voltage is applied to the normally open isolation valves whenever the main relay is energized. An isolation valve closes when the EBTCM supplies a ground path at the isolation valve control terminal. The EBTCM detects wheel slip in a drive wheel and vehicle speed is below 40 km/h (25 mph).

The isolation valves are closed only during traction control operations, and are not used during antilock or normal braking conditions. The isolation valves separate (isolate) the front brake hydraulic circuits from the master cylinder and rear brake hydraulic circuits. Once the front brake hydraulic circuits are isolated, pressure can be applied to the front wheels without affecting any other brake hydraulic circuit. The EBTCM turns on the PMV pump motor to apply pressure, and begin cycling the PMV assembly inlet and outlet valves to modulate pressure.

FAILURE CONDITIONS

DTC 44 or DTC 48 sets when the EBTCM detects an open or short to ground in the respective isolation valve circuit.

DIAGNOSTIC CHART DTC 44 AND 48 — 1994 OLDSMOBILE AND PONTIAC, CONTINUED

DTC(s) 44, 48
ISOLATION VALVE CIRCUIT OPEN OR SHORTED
(1 OF 2)

Has ABS/TCS Functional Test been performed?

- NO: Perform ABS/TCS Functional Test before proceeding.
- YES:
  - Ignition: OFF
  - Disconnect Pressure Modulator Valve (PMV) Assembly Valve Connector C1.
  - Check connector for proper terminal contact.
  - Using J 39200 set to Ω measure resistance between PMV Assembly Pigtail Connector C1 terminals 13 and 14.

Is less than 5 OHMS measured?

- NO: Replace PMV Assembly.
- YES:
  - Measure resistance between PMV Pigtail Connector C1 terminal 14 and PMV Valve Block.

Is infinite OHMS measured?

- NO: Replace PMV Assembly.
- YES:
  - Measure resistance at connector, C1, (PMV Assembly Side) between Terminal 14 and Suspect Valve Terminal. Refer to chart.

DTC/VALVE	SUSPECT TERMINAL	CORRECT RESISTANCE
44/LH ISOLATION	1	5 TO 8 OHMS
48/RH ISOLATION	2	

Is correct resistance measured?

- NO: Replace PMV Assembly.
- YES:
  - Reconnect PMV Assembly Valve Connector, C1.
  - Disconnect EBTCM Connector.
  - Install J 38716 Pinout Box.
  - Measure resistance between J 38716 Pinout Box Terminal 3 and Suspect Isolation Valve Terminal. Refer to chart.

DTC/VALVE	J 38716 TERMINAL
44/LH ISOLATION	37
48/RH ISOLATION	40

Is resistance measured less than 10 OHMS?

- NO: Repair open in Suspect Isolation Valve Circuit between Pressure Modulator Valve (PMV) Assembly and EBCM/EBTCM.
- YES:
  - Measure resistance between J 38716 Pinout Box Terminal 1 and Data Link Connector (DLC) Terminal 4 (Ground).
  - Is less than 2 OHMS measured?

A

Continued on next page

DIAGNOSTIC CHART DTC 44 AND 48 — 1994 OLDSMOBILE AND PONTIAC, CONTINUED

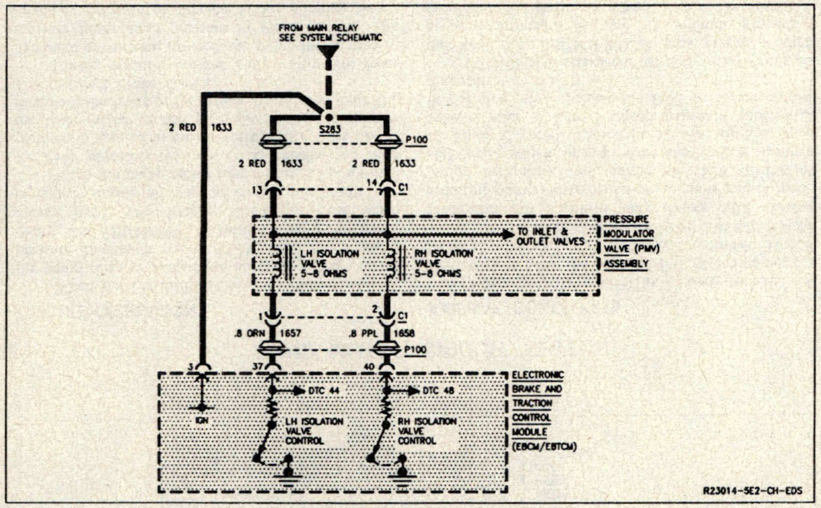

DTC(s) 44, 48

ISOLATION VALVE CIRCUIT OPEN OR SHORTED

CIRCUIT OPERATION

Voltage is applied to the normally open isolation valves whenever the main relay is energized. An isolation valve closes when the EBTCM supplies a ground path at the isolation valve control terminal. The EBTCM detects wheel slip in a drive wheel and vehicle speed is below 40 km/h (25 mph).

The isolation valves are closed only during traction control operations, and are not used during antilock or normal braking conditions. The isolation valves separate (isolate) the front brake hydraulic circuits from the master cylinder and rear brake hydraulic circuits. Once the front brake hydraulic circuits are isolated, pressure can be applied to the front wheels without affecting any other brake hydraulic circuit. The EBTCM turns on the PMV pump motor to apply pressure, and begin cycling the PMV assembly inlet and outlet valves to modulate pressure.

FAILURE CONDITIONS

DTC 44 or DTC 48 sets when the EBTCM detects an open or short to ground in the respective isolation valve circuit.

DIAGNOSTIC CHART DTC 44 AND 48 — 1994 OLDSMOBILE AND PONTIAC, CONTINUED

DTC(s) 44, 48
ISOLATION VALVE CIRCUIT OPEN OR SHORTED (2 of 2)

A

YES

- Disconnect Main Relay.
- Measure resistance between J 38716 Pinout Box Terminal 1 and Suspect Isolation Valve Terminal. Refer to chart.

DTC/VALVE	J 38716 TERMINAL
44/LH ISOLATION	37
48/RH ISOLATION	40

Is less than 5 OHMS measured?

NO: Check EBTCM Connector for proper terminal contact. If terminal contact is OK, replace EBTCM.

YES: Repair short to ground in Suspect Isolation Valve Circuit between Pressure Modulator Valve (PMV) Assembly and EBTCM.

NO: Repair open in CKT 251.

After repairs are completed, clear DTC and repeat ABS/TCS functional test.

DIAGNOSTIC CHART DTC 61 — 1994 OLDSMOBILE AND PONTIAC

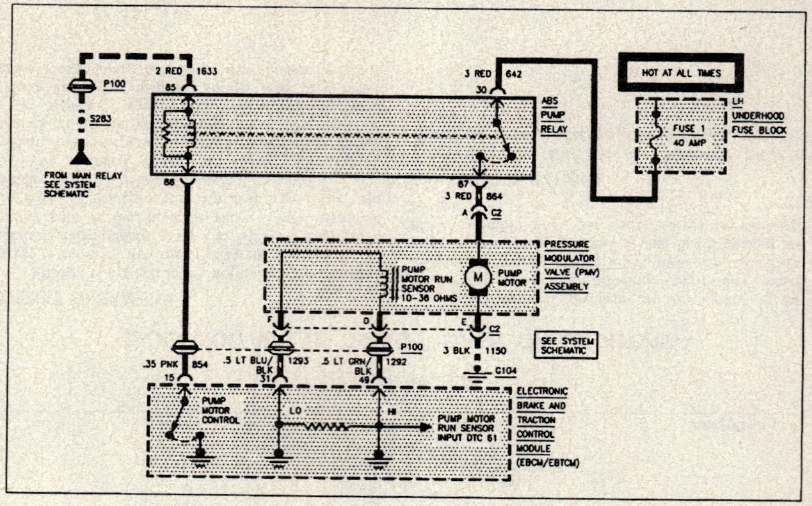

DTC 61

PUMP MOTOR CIRCUIT TEST

CIRCUIT OPERATION

When the EBCM/EBTCM grounds terminal 15, the pump relay contacts close applying voltage to the pressure modulator valve (PMV) assembly's pump motor. The EBCM/EBTCM turns on the pump motor during most ABS braking conditions (and when equipped) during all traction control conditions.

When the pump motor runs during ABS braking, the PMV assembly reservoir is replenished and fluid pressure is applied to the master cylinder. This causes the brake pedal to rise gradually. The pump motor runs until the brake pedal sufficiently rises to close the TCC/antilock brake switch.

During traction control mode, after the front hydraulic circuits are isolated, the pump motor runs to increase brake fluid pressure in these circuits. The increased pressure is directed to the slipping wheel through the PMV assembly's front inlet and outlet valves. This causes the brakes to be applied, reducing wheel spin at the slipping wheel. Traction control is disabled as soon as the brakes are manually applied.

The EBCM/EBTCM monitors pump motor operation via the pump motor run sensor. When the pump is operating, voltage is induced into the sensor's coil and the EBCM/EBTCM monitors this voltage at terminals 49 and 31.

FAILURE CONDITIONS

When the EBCM/EBTCM grounds terminal 15, DTC 61 sets when the EBCM/EBTCM determines the frequency between EBCM/EBTCM terminals 31 and 49 is not between 70-150 Hz. The EBCM/EBTCM recognizes this condition only during ABS braking requiring pump motor operation, traction control mode (when equipped), and during the EBCM/EBTCM self-check which occurs when the vehicle reaches 11 km/h (7 mph). Therefore, be sure to test drive vehicle, refer to Item 3 under diagnostic process in this section, after clearing the DTC to assure correct diagnosis.

DTC 61 also sets when the pump motor runs all the time (i.e. EBCM/EBTCM sees proper voltage between terminals 31 and 49 without grounding terminal 15). Refer to Chart F for diagnosing a continuous running pump motor.

Possible Causes

- Open in CKT 1150, 864, 854, 1633, 1292, 1293 or 1446.
- Short to ground in CKT 864, 854, 1292, 1293, or 642.
- ABS pump relay fails "open" or "engaged."

DIAGNOSTIC CHART DTC 61 — 1994 OLDSMOBILE AND PONTIAC, CONTINUED

DTC 61 (1 OF 3) PUMP MOTOR CIRCUIT TEST

IMPORTANT
DTC 61 usually sets while the vehicle is moving during the system "Self Test".

Has ABS/TCS functional test been performed?

- NO → Perform ABS/TCS functional test.
- YES ↓

- Ignition: RUN

Does Pressure Modulator Valve (PMV) Assembly Pump Motor Run continuously?

- YES →
  - Remove ABS Pump Relay.

  Does pump motor stop running?
  - NO → Repair short to battery in CKT 864.
  - YES →
    - IGNITION: OFF
    - Using J 39200 set to DC scale, check for battery voltage at Pump Relay Cavity 86.

    Is battery voltage present?
    - YES → Repair short to battery in CKT 854.
    - NO → Replace Pump Relay
- NO ↓

- Ignition: OFF
- Connect Tech 1 to Data Link Connector (DLC).
- Ignition: RUN
- Use Tech 1 to bleed PMV Assembly.

Does PMV Assembly Pump Motor run?

- NO → B Continued on Chart 3 of 3
- YES ↓

- Ignition: OFF
- Disconnect Tech 1.
- Disconnect PMV Assembly Pump Motor Connector C2.
- Using J 39200 set to Ω, measure resistance between PMV Assembly Pigtail Connector C2 Terminal D and PMV Assembly casing.

Is infinite ohms measured?

- NO → Replace PMV Assembly.
- YES ↓

- Measure resistance between PMV Assembly Pigtail Connector C2 Terminal D and F.

Is resistance between 10 and 40 ohms?

- NO → Replace PMV Assembly.
- YES → A Continued on next page

After repairs are completed, clear DTC and repeat ABS/TCS functional test.

DIAGNOSTIC CHART DTC 61 — 1994 OLDSMOBILE AND PONTIAC, CONTINUED

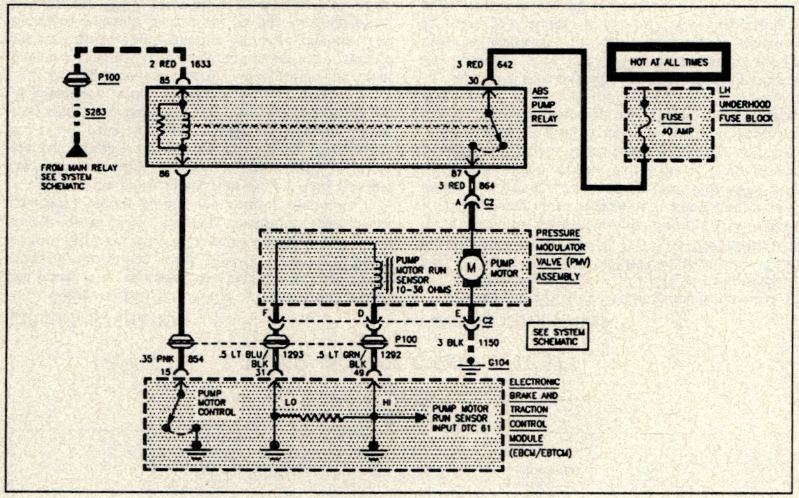

DTC 61

PUMP MOTOR CIRCUIT TEST

CIRCUIT OPERATION

When the EBCM/EBTCM grounds terminal 15, the pump relay contacts close applying voltage to the pressure modulator valve (PMV) assembly's pump motor. The EBCM/EBTCM turns on the pump motor during most ABS braking conditions (and when equipped) during all traction control conditions.

When the pump motor runs during ABS braking, the PMV assembly reservoir is replenished and fluid pressure is applied to the master cylinder. This causes the brake pedal to rise gradually. The pump motor runs until the brake pedal sufficiently rises to close the TCC/antilock brake switch.

During traction control mode, after the front hydraulic circuits are isolated, the pump motor runs to increase brake fluid pressure in these circuits. The increased pressure is directed to the slipping wheel through the PMV assembly's front inlet and outlet valves. This causes the brakes to be applied, reducing wheel spin at the slipping wheel. Traction control is disabled as soon as the brakes are manually applied.

The EBCM/EBTCM monitors pump motor operation via the pump motor run sensor. When the pump is operating, voltage is induced into the sensor's coil and the EBCM/EBTCM monitors this voltage at terminals 49 and 31.

FAILURE CONDITIONS

When the EBCM/EBTCM grounds terminal 15, DTC 61 sets when the EBCM/EBTCM determines the frequency between EBCM/EBTCM terminals 31 and 49 is not between 70-150 Hz. The EBCM/EBTCM recognizes this condition only during ABS braking requiring pump motor operation, traction control mode (when equipped), and during the EBCM/EBTCM self-check which occurs when the vehicle reaches 11 km/h (7 mph). Therefore, be sure to test drive vehicle, refer to Item 3 under diagnostic process in this section, after clearing the DTC to assure correct diagnosis.

DTC 61 also sets when the pump motor runs all the time (i.e. EBCM/EBTCM sees proper voltage between terminals 31 and 49 without grounding terminal 15). Refer to Chart F for diagnosing a continuous running pump motor.

Possible Causes

- Open in CKT 1150, 864, 854, 1633, 1292, 1293 or 1446.
- Short to ground in CKT 864, 854, 1292, 1293, or 642.
- ABS pump relay fails "open" or "engaged."

DIAGNOSTIC CHART DTC 61 — 1994 OLDSMOBILE AND PONTIAC, CONTINUED

DTC 61(2 OF 3) PUMP MOTOR CIRCUIT TEST

A

- Reconnect PMV Assembly Pump Motor Connector C2.
- Disconnect EBCM/EBTCM Connector.
- Connect J 38716 Pinout Box.
- Measure resistance between J 38716 Pinout Box Terminal 1 and DLC Terminal 4 (ground).

Is less than 5 ohms measured?

NO → Repair open in CKT 251.

YES →

- Measure resistance between J 38716 Pinout Box Terminals 31 and 49.

Is resistance between 10 and 40 ohms?

NO → When high resistance is measured, repair CKT 1292 or 1293 for an open. Check PMV Assembly Connector C2 for proper terminal contact.

When low resistance is measured, repair short between CKT 1292 and 1293.

YES →

- Measure resistance between J 38716 Pinout Box Terminals 49 and 1, then 31 and 1.

Is less than 5 ohms measured in each test?

YES → Repair CKT 1292 or 1293 for a short to ground.

NO →

- Using fused jumper, jump between Pinout Box Terminals 19 and 34. Using another fused jumper, jump between Pinout Box Terminals 1 and 15.
- IGNITION: RUN (pump should run)
- Using J 39200 set to Hz, measure Motor Run Sensor Output between Pinout Box Terminals 31 and 49.

Is 70 -150 Hz measured?

NO → Replace PMV

YES →

- Check EBCM/EBTCM Connector for proper terminal contact. If OK, replace EBCM/EBTCM.

DIAGNOSTIC CHART DTC 61 — 1994 OLDSMOBILE AND PONTIAC, CONTINUED

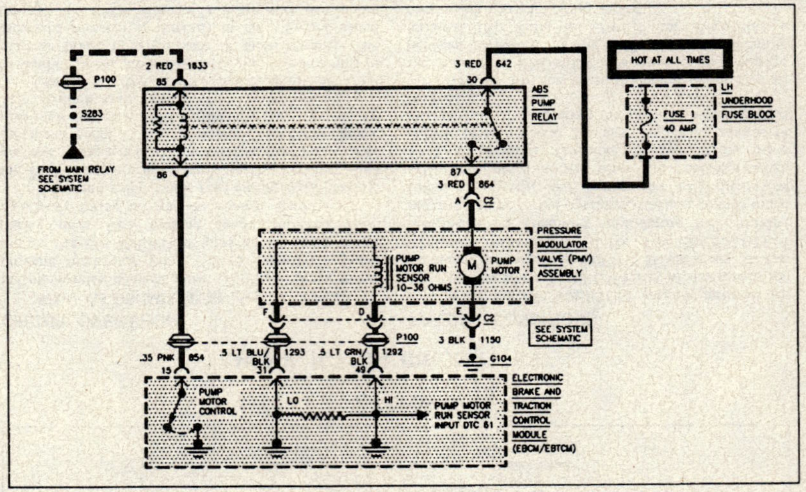

DTC 61

PUMP MOTOR CIRCUIT TEST

CIRCUIT OPERATION

When the EBCM/EBTCM grounds terminal 15, the pump relay contacts close applying voltage to the pressure modulator valve (PMV) assembly's pump motor. The EBCM/EBTCM turns on the pump motor during most ABS braking conditions (and when equipped) during all traction control conditions.

When the pump motor runs during ABS braking, the PMV assembly reservoir is replenished and fluid pressure is applied to the master cylinder. This causes the brake pedal to rise gradually. The pump motor runs until the brake pedal sufficiently rises to close the TCC/antilock brake switch.

During traction control mode, after the front hydraulic circuits are isolated, the pump motor runs to increase brake fluid pressure in these circuits. The increased pressure is directed to the slipping wheel through the PMV assembly's front inlet and outlet valves. This causes the brakes to be applied, reducing wheel spin at the slipping wheel. Traction control is disabled as soon as the brakes are manually applied.

The EBCM/EBTCM monitors pump motor operation via the pump motor run sensor. When the pump is operating, voltage is induced into the sensor's coil and the EBCM/EBTCM monitors this voltage at terminals 49 and 31.

FAILURE CONDITIONS

When the EBCM/EBTCM grounds terminal 15, DTC 61 sets when the EBCM/EBTCM determines the frequency between EBCM/EBTCM terminals 31 and 49 is not between 70-150 Hz. The EBCM/EBTCM recognizes this condition only during ABS braking requiring pump motor operation, traction control mode (when equipped), and during the EBCM/EBTCM self-check which occurs when the vehicle reaches 11 km/h (7 mph). Therefore, be sure to test drive vehicle, refer to Item 3 under diagnostic process in this section, after clearing the DTC to assure correct diagnosis.

DTC 61 also sets when the pump motor runs all the time (i.e. EBCM/EBTCM sees proper voltage between terminals 31 and 49 without grounding terminal 15). Refer to Chart F for diagnosing a continuous running pump motor.

Possible Causes

- Open in CKT 1150, 864, 854, 1633, 1292, 1293 or 1446.
- Short to ground in CKT 864, 854, 1292, 1293, or 642.
- ABS pump relay fails "open" or "engaged."

DIAGNOSTIC CHART DTC 61 — 1994 OLDSMOBILE AND PONTIAC, CONTINUED

DTC 61 (3 OF 3) PUMP MOTOR CIRCUIT TEST

B

- Clear DTC using a Tech 1.
- Ignition: OFF
- Disconnect Tech 1.
- Disconnect ABS Pump Relay.
- Ignition: RUN
- Using J 39200 set to DC scale, measure voltage between ABS Pump Relay Harness Connector Terminal 85 and ground.

Is battery voltage measured?

NO: Repair open in CKT 1633.

YES:

- Ignition: OFF
- Measure voltage between ABS Pump Relay Harness Connector Terminal 30 and ground.

Is battery voltage measured?

NO: Check Fuse 1 of LH Underhood Fuse Block. Check/repair CKT 642 for an open, or a short to ground.

NOTE: When fuse is open and CKT 642 is OK, check/repair short to ground in CKT 864 and replace ABS Pump Relay when a short to ground is found.

YES:

- Connect fused jumper between ABS Pump Relay Harness Terminal 30 and 87A.

Does PMV Assembly Pump Motor run?

NO:

- Leave fused jumper connected.
- Disconnect PMV Assembly Pump Motor Connector C2.
- Measure voltage between PMV Assembly Harness Connector C2 Terminal A and ground.

Is battery voltage measured?

- NO: Repair open or short to ground in CKT 864.
- YES:
  - Measure voltage between PMV Assembly Harness Connector C2 Terminal A and E.

  Is battery voltage measured?

  - YES: Check PMV Assembly Connector C2 for proper terminal contact. If terminal contact is OK, replace PMV Assembly.
  - NO: Repair open in CKT 1150.

YES:

- Reconnect ABS Pump Relay.
- Disconnect EBCM/EBTCM Connector.
- Connect J 38716 Pinout Box.
- Using J 39200 set to Ω, measure resistance between J 38716 Pinout Box Terminal 1 and Data Link Connector (DLC) Terminal 4 (ground).

Is less than 5 ohms measured?

- NO: Repair open in CKT 251.
- YES:
  - Connect fused jumpers between J 38716 Pinout Box Terminal 15 and 1, and Terminal 34 and 1.

  Does PMV Assembly Pump Motor run?

  - YES: Check EBCM/EBTCM Connector for proper terminal contact. If terminal contact is OK, replace EBCM/EBTCM.
  - NO: Check/repair for an open in CKT 854. If CKT 854 is OK, replace ABS Pump Relay.

After repairs are completed, clear DTC and repeat ABS/TCS functional test.

DIAGNOSTIC CHART DTC 62 — 1994 OLDSMOBILE AND PONTIAC

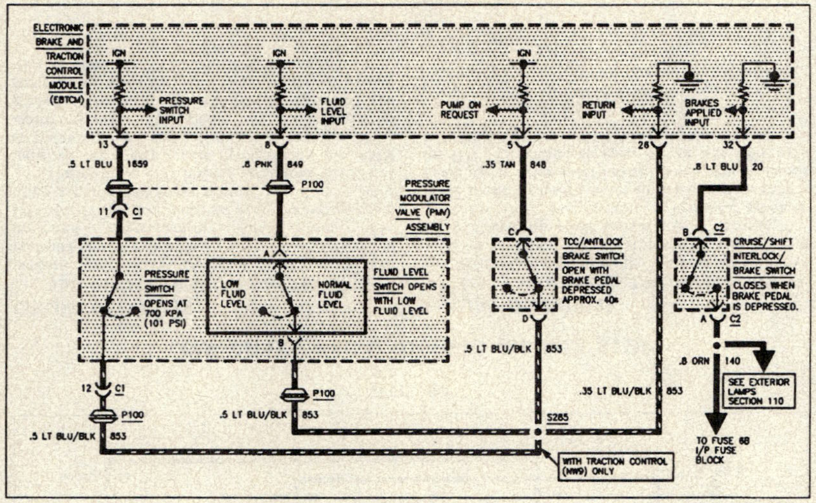

DTC 62

LOW BRAKE PEDAL DURING AN ABS STOP

CIRCUIT OPERATION

The EBCM/EBTCM monitors brake pedal position using the TCC/antilock brake switch. The normally closed switch opens when the brake pedal is depressed approximately 40% down. The EBCM/EBTCM also monitors pump motor operation using a pump motor run sensor.

During most ABS braking conditions the EBCM turns on the pump motor, replenishing the PMV assembly reservoir and applying fluid pressure to the master cylinder. This causes the brake pedal to gradually rise. The EBCM/EBTCM turns off the pump motor when the brake pedal height sufficiently rises to close the TCC/antilock brake switch.

FAILURE CONDITIONS

During ABS braking, DTC 62 sets when the EBCM/EBTCM determines the pump is working, and brake pedal height has not sufficiently risen to close the TCC/antilock brake switch after 3 seconds.

This DTC may indicate a hydraulic problem. Some possible causes include air in the brake lines or a pump motor mechanical failure. Non-hydraulic causes for DTC 62 include an open 848 or 853 CKT or a misadjusted/defective TCC/antilock brake switch.

DIAGNOSTIC CHART DTC 62 — 1994 OLDSMOBILE AND PONTIAC, CONTINUED

DTC 62 (1 OF 2)
LOW BRAKE PEDAL DURING AN ABS STOP

Has ABS/TCS functional test been performed?

- NO: Perform ABS/TCS functional test before proceeding.
- YES:
  - Bleed brake system as outlined in this section.
  - Ignition: OFF
  - Connect Tech 1 to Data Link Connector (DLC).
  - Ignition: RUN
  - Use Tech 1 to clear DTC after bleeding system.
  - Depress brake pedal (hold).
  - Use Tech 1 to bleed Pressure Modulator Valve (PMV) assembly.

Did brake pedal rise?

- NO: Replace PMV Assembly.
- YES:
  - Ignition: OFF
  - Disconnect Tech 1.
  - Test drive vehicle on a surface as described in Item "3" under "Diagnostic Process" in this section and perform an ABS stop.
  - Connect Tech 1 to DLC Connector.
  - Ignition: RUN

Did DTC 62 set?

- NO: Problem solved.
- YES:
  - Ignition: OFF
  - Disconnect Tech 1.
  - Brake pedal released (rest position).
  - Check TCC/Antilock Brake Switch adjustment.

Is switch adjusted properly?

- NO: Adjust TCC/Antilock Brake Switch.
- YES:
  - Disconnect TCC/Antilock Brake Switch Connector.
  - Using J 39200 set to Ω, measure resistance across switch Terminals C and D.
- NO: Replace TCC/Antilock Brake Switch.
- YES: A

Continued on next page

After repairs are completed, clear DTC and repeat ABS/TCS functional test.

DIAGNOSTIC CHART DTC 62 — 1994 OLDSMOBILE AND PONTIAC, CONTINUED

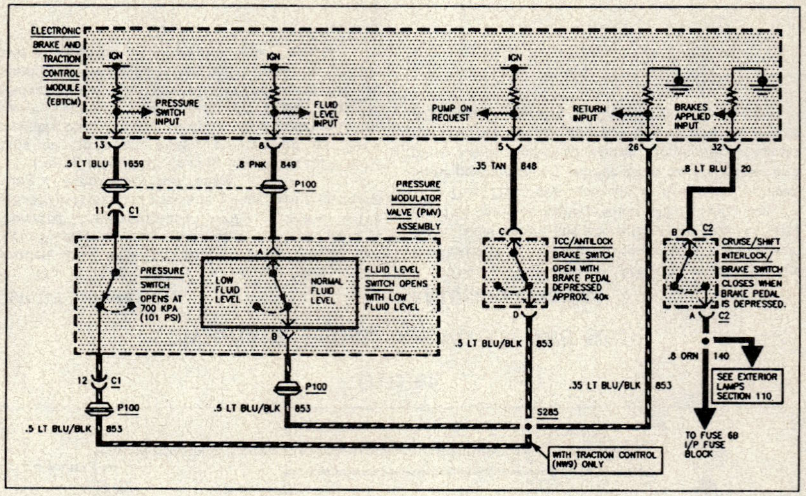

DTC 62

LOW BRAKE PEDAL DURING AN ABS STOP

CIRCUIT OPERATION

The EBCM/EBTCM monitors brake pedal position using the TCC/antilock brake switch. The normally closed switch opens when the brake pedal is depressed approximately 40% down. The EBCM/EBTCM also monitors pump motor operation using a pump motor run sensor.

During most ABS braking conditions the EBCM turns on the pump motor, replenishing the PMV assembly reservoir and applying fluid pressure to the master cylinder. This causes the brake pedal to gradually rise. The EBCM/EBTCM turns off the pump motor when the brake pedal height sufficiently rises to close the TCC/antilock brake switch.

FAILURE CONDITIONS

During ABS braking, DTC 62 sets when the EBCM/EBTCM determines the pump is working, and brake pedal height has not sufficiently risen to close the TCC/antilock brake switch after 3 seconds.

This DTC may indicate a hydraulic problem. Some possible causes include air in the brake lines or a pump motor mechanical failure. Non-hydraulic causes for DTC 62 include an open 848 or 853 CKT or a misadjusted/defective TCC/antilock brake switch.

DIAGNOSTIC CHART DTC 62 — 1994 OLDSMOBILE AND PONTIAC, CONTINUED

DTC 62 (2 OF 2)
LOW BRAKE PEDAL DURING AN ABS STOP

A

- Disconnect EBCM/EBTCM Connector.
- Connect J 38716 Pinout Box.
- Measure resistance between J 38716 Pinout Box Terminal 5 and TCC/Antilock Brake Switch Connector Terminal C.

Is less than 3 ohms measured?

YES:
- Measure resistance between J 38716 Pinout Box Terminal 26 and TCC/Antilock Brake Switch Connector Terminal D.

Is less than 3 ohms measured?

YES: Check EBCM/EBTCM Connector for proper terminal contact. If terminal contact is OK, replace EBCM/EBTCM.

NO: Repair open in CKT 853.

NO: Repair open in CKT 848.

DIAGNOSTIC CHART DTC 71 — 1994 OLDSMOBILE AND PONTIAC

DTC 71

EBCM/EBTCM INTERNAL MALFUNCTION

DTC 71 indicates that a failure has occurred within the EBCM/EBTCM. DTC 71 can occur intermittently, due to interference, therefore clear the code and test drive the vehicle before replacing the EBCM/EBTCM.

DIAGNOSTIC CHART DTC 71 — 1994 OLDSMOBILE AND PONTIAC, CONTINUED

DTC 71
EBCM/EBTCM INTERNAL FAILURE

Has ABS/TCS functional test been performed?

- YES
  - Using Tech 1 clear Diagnostic Trouble Code.
  - Test drive vehicle. Refer to Diagnostic Process Item "3" in this section.

  Did DTC 71 set during test drive?
  - YES: Replace EBCM/EBTCM.
  - NO: No problem found. May be intermittent.
- NO: Perform ABS/TCS functional test before proceeding.

After repairs are completed, clear DTC and repeat ABS/TCS functional test.

DIAGNOSTIC CHART DTC 72 — 1994 OLDSMOBILE AND PONTIAC

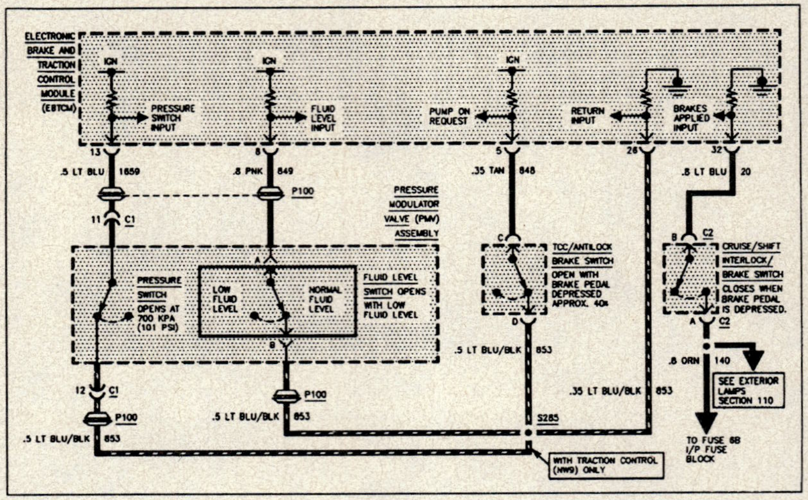

DTC 72

BRAKE SWITCH OPEN OR SHORTED

CIRCUIT OPERATION

The electronic brake control module EBCM/EBTCM monitors brake pedal position using the brake travel switch. The switch is normally closed. It opens when the brake pedal is depressed approximately 40% down. During most ABS braking conditions the EBCM/EBTCM turns on the pump motor, replenishing the PMV assembly reservoir and applying fluid pressure to the master cylinder. This causes the brake pedal to gradually rise. The EBCM/EBTCM turns off the pump motor when brake pedal height sufficiently rises to close the brake travel switch.

FAILURE CONDITIONS

DTC 72 sets when the EBCM/EBTCM detects the brake travel switch CKT 848 is shorted to ground (with the brake pedal depressed) or open.

DIAGNOSTIC CHART DTC 72 — 1994 OLDSMOBILE AND PONTIAC, CONTINUED

DTC 72
BRAKE TRAVEL SWITCH CIRCUIT OPEN OR SHORTED

Has ABS/TCS functional test been performed?

- NO: Perform ABS/TCS functional test before proceeding.
- YES:
  - Ignition: OFF
  - Disconnect Brake Travel Switch Connector.
  - Disconnect EBCM/EBTCM Connector.
  - Connect J 38716 Pinout Box.
  - Using J 39200 set to Ω, measure resistance between J 38716 Pinout Box Terminal 1 and Data Link Connector (DLC) Terminal 4 (ground).

  Is less than 5 ohms measured?

  - NO: Repair open in CKT 251.
  - YES:
    - Measure resistance between J 38716 Pinout Box Terminal 1 and 5.

    Is less than 5 ohms measured?

    - YES: Repair short to ground in CKT 848.
    - NO:
      - Measure resistance between Pinout Box Terminals 1 and 26.
      - Is less than 5 Ω measured?
      - NO: Repair short to ground in CKT 853.
      - YES: Check EBCM/EBTCM Connector for proper terminal contact. If terminal contact is OK, replace EBCM/EBTCM.

DIAGNOSTIC CHART DTC 73 — 1994 OLDSMOBILE AND PONTIAC

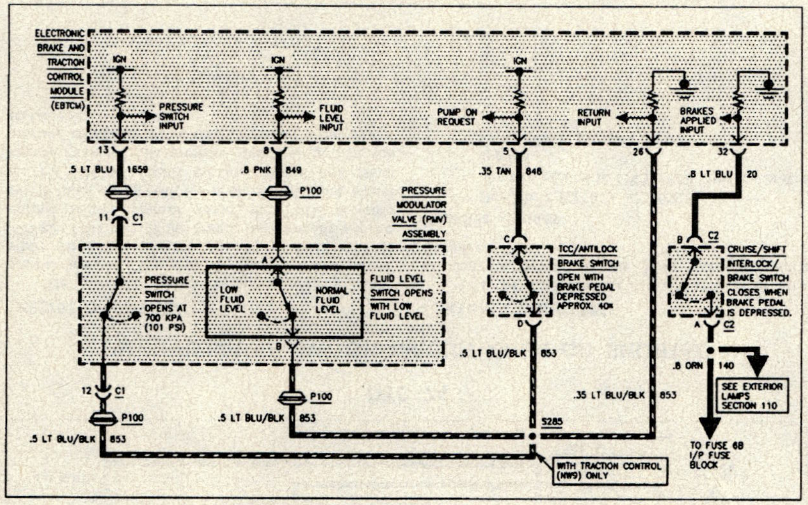

DTC 73

FLUID LEVEL SWITCH CIRCUIT OPEN OR SHORTED

CIRCUIT OPERATION

The pressure modulator valve (PMV) fluid level switch alerts the EBCM/EBTCM to a low brake fluid condition in the PMV reservoir. The fluid level switch is located in the PMV assembly's reservoir. The fluid level switch is normally closed when the reservoir contains sufficient fluid. When fluid level is low, the switch opens, causing the EBCM/EBTCM to turn on the "ANTILOCK" indicator.

FAILURE CONDITIONS

DTC 73 sets when the EBCM/EBTCM detects a short to ground in the PMV fluid level switch circuitry. These include: CKT 848, 849, 853, and 1659. DTC 73 will also set if an open exists in CKT 251.

DIAGNOSTIC CHART DTC 73 — 1994 OLDSMOBILE AND PONTIAC, CONTINUED

DTC 73
PMV FLUID LEVEL SWITCH CIRCUIT OPEN OR SHORTED

Has ABS/TCS functional test been performed?

- NO: Perform ABS/TCS functional test before proceeding.
- YES:
  - Ignition: OFF
  - Disconnect EBCM/EBTCM Connector.
  - Connect J 38716 Pinout Box.
  - Using J 39200 set to Ω, measure resistance between J 38716 Pinout Box Terminal 1 and Data Link Connector (DLC) Terminal A (ground).

Is less than 5 ohms measured?

- NO: Repair open in CKT 251.
- YES:
  - Disconnect Brake Travel Switch.
  - Measure resistance between J 38716 Pinout Box Terminal 5 and 1.

Is less than 5 ohms measured?

- YES: Repair short to ground in CKT 848.
- NO:
  - Disconnect Pressure Modulator Valve (PMV) Fluid Level Switch Connector.
  - Measure resistance between J 38716 Pinout Box Terminal 8 and 1.

Is less than 5 ohms measured?

- YES: Repair short to ground in CKT 849.
- NO:
  - When equipped with Traction Control:
    Depress Brake Pedal.
    Measure resistance between J 38716 Pinout Box Terminal 13 and 1.

Is less than 5 ohms measured?

- YES: Repair short to ground in CKT 1659.
- NO:
  - Measure resistance between J 38716 Pinout Box Terminal 26 and 1.

Is less than 5 ohms measured?

- YES: Repair short to ground in CKT 853.
- NO: Check EBCM/EBTCM Connector for proper terminal contact. If terminal contact is OK, replace EBCM/EBTCM.

After repairs are completed, clear DTC and repeat ABS/TCS functional test.

DIAGNOSTIC CHART DTC 74 — 1994 OLDSMOBILE AND PONTIAC

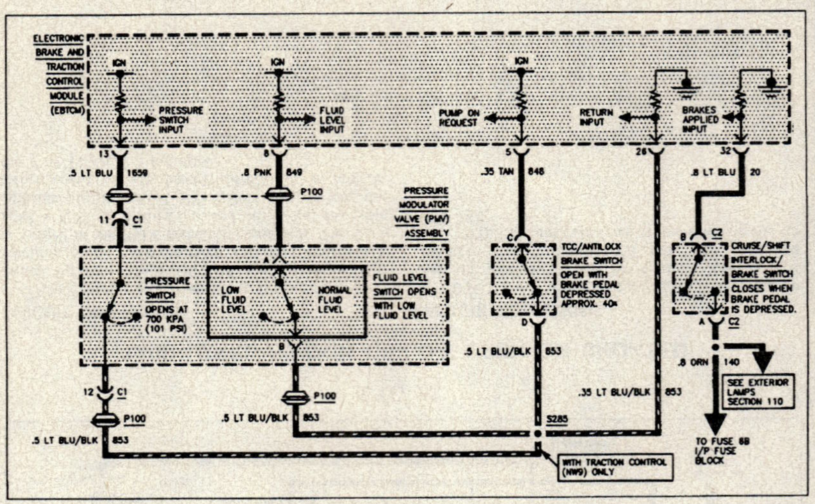

DTC 74

PMV PRESSURE SWITCH CIRCUIT OPEN OR SHORTED

CIRCUIT OPERATION

The pressure switch monitors primary brake system pressure. The normally closed switch opens when the brakes are applied and pressure exceeds 700 kPa (101 psi). If the brakes are manually applied during traction control mode, the PMV pressure switch input, along with the brake switch input, signal the EBTCM to disable traction control and allow manual braking. The pressure switch is part of the pressure modulator valve (PMV) assembly and is not serviceable.

FAILURE CONDITIONS

DTC 74 sets when the EBTCM detects an open or short to ground in the PMV pressure switch circuitry.

Possible Causes

- Open in CKT 853 or 1659.
- Short to ground in CKT 1659 with the brakes applied.
- Failed PMV pressure switch.

DIAGNOSTIC CHART DTC 74 — 1994 OLDSMOBILE AND PONTIAC, CONTINUED

DTC 74 PMV PRESSURE SWITCH CIRCUIT OPEN OR SHORTED (1 of 3)

Has ABS/TCS Functional Test been performed?

- **NO:** Perform ABS/TCS Functional Test.
- **YES:**
  - Ignition: OFF
  - Inspect I/P Fuse Block Fuse 6B.

  Is Fuse 6B open?

  - **YES:** Repair Short to Ground in CKT 140 or CKT 20. Replace Fuse.
  - **NO:**
    - Disconnect Brake Switch Connector (Grey Connector).
    - Using J 39200 set to DC scale measure voltage between DLC Terminal 4 and Brake Switch Connector (Harness Side) Terminal A.

    Is greater than 10 volts measured?

    - **NO:** Repair Open in CKT 140.
    - **YES:**
      - Depress Brake Pedal.
      - Using J 39200 set to Ω measure resistance between Brake Switch Terminals A and B.
      - Is less than 2 OHMS measured?

      - **NO:** Replace Brake Switch.
      - **YES:**
        - Check for proper terminal contact on Brake Switch Connector.
        - Reconnect Brake Switch Connector.
        - Check for proper Brake Switch Adjustment

        Is Brake Switch Adjusted Properly?

        - **NO:** Adjust Brake Switch. Refer to SECTION 5.
        - **YES:**
          - Disconnect EBTCM Connector.
          - Install J 38716 Pin Out Box.
          - Measure resistance between Pin Out Box Terminal 1 and DLC Terminal 4.

          Is less than 5 OHMS measured?

          - **NO:** Repair Open in CKT 251.
          - **YES:**
            - Ignition: ON
            - Using J 39200 set to DC scale, measure voltage between Pin Out Box Terminals 1 and 32.

            Is greater than 2 volts measured?

            (A)

Continued on next page

DIAGNOSTIC CHART DTC 74 — 1994 OLDSMOBILE AND PONTIAC, CONTINUED

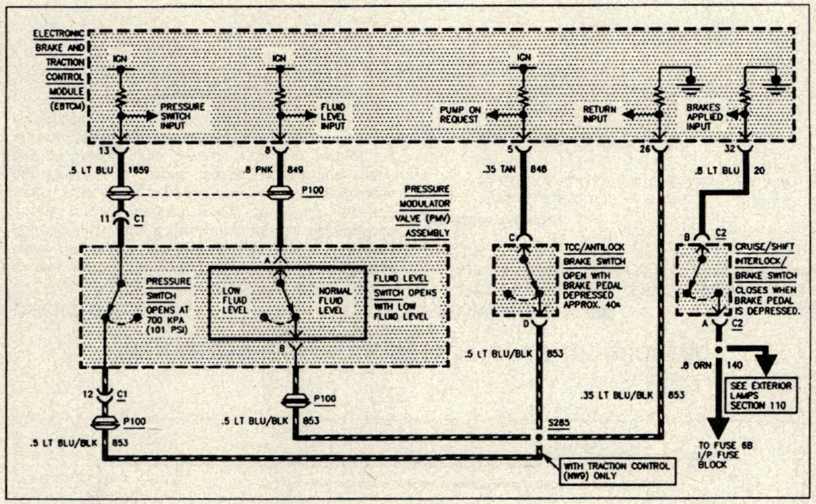

DTC 74

PMV PRESSURE SWITCH CIRCUIT OPEN OR SHORTED

CIRCUIT OPERATION

The pressure switch monitors primary brake system pressure. The normally closed switch opens when the brakes are applied and pressure exceeds 700 kPa (101 psi). If the brakes are manually applied during traction control mode, the PMV pressure switch input, along with the brake switch input, signal the EBTCM to disable traction control and allow manual braking. The pressure switch is part of the pressure modulator valve (PMV) assembly and is not serviceable.

FAILURE CONDITIONS

DTC 74 sets when the EBTCM detects an open or short to ground in the PMV pressure switch circuitry.

Possible Causes

- Open in CKT 853 or 1659.
- Short to ground in CKT 1659 with the brakes applied.
- Failed PMV pressure switch.

DIAGNOSTIC CHART DTC 74 — 1994 OLDSMOBILE AND PONTIAC, CONTINUED

DTC 74
PMV PRESSURE SWITCH CIRCUIT OPEN OR SHORTED (2 of 3)

(A)

- Depress Brake Pedal.
- Measure Voltage between Pin Out Box Terminals 1 and 32.

Is greater than 10 volts measured?

- YES → Repair Short to Voltage in CKT 20.
- NO ↓

- Ignition: OFF
- Disconnect Pressure Modulator Valve (PMV) Connector C1.
- Measure resistance between Connector C1 (PMV Assembly Side) Terminal 11 and PMV Case.

Is Infinite OHMS measured?

- NO → Repair Open in CKT 20.
- YES ↓

- Measure resistance between Connector C1 (PMV Assembly Side) Terminals 11 and 12.

Is less than 2 OHMS measured?

- NO → Replace PMV.
- YES ↓

- Depress Brake Pedal.
- Measure resistnace between Connector C1 (PMV Assembly Side) Terminals 11 and 12.

Is Infinite OHMS measured?

- NO → Replace PMV.
- YES ↓

- Measure resistance between Pin Out Box Terminal 13 and PMV Connector C1 (Harness Side) Terminal 11.

Is less than 5 OHMS measured?

- NO → Replace PMV.
- YES ↓

- Measure resistance between Pin Out Box Terminal 26 and PMV Connector C1 (Harness Side) Terminal 12.

Is less than 5 OHMS measured?

- NO → Repair Open in CKT 1659
- YES ↓

(B)

Continued on next page

DIAGNOSTIC CHART DTC 74 — 1994 OLDSMOBILE AND PONTIAC, CONTINUED

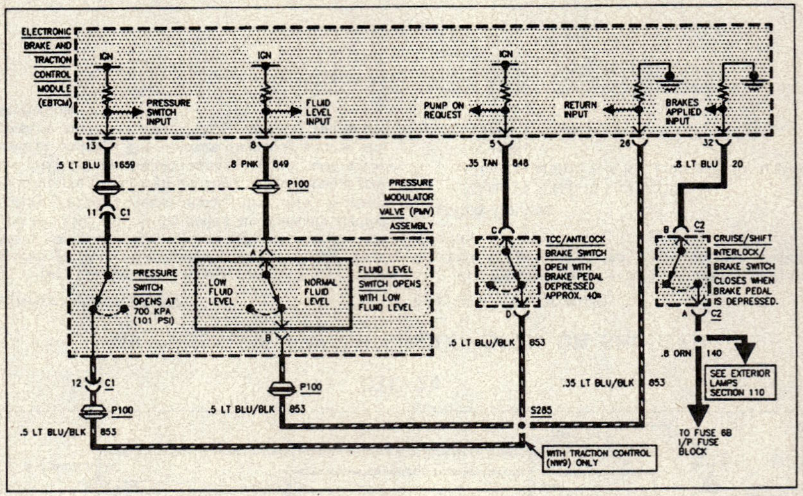

DTC 74

PMV PRESSURE SWITCH CIRCUIT OPEN OR SHORTED

CIRCUIT OPERATION

The pressure switch monitors primary brake system pressure. The normally closed switch opens when the brakes are applied and pressure exceeds 700 kPa (101 psi). If the brakes are manually applied during traction control mode, the PMV pressure switch input, along with the brake switch input, signal the EBTCM to disable traction control and allow manual braking. The pressure switch is part of the pressure modulator valve (PMV) assembly and is not serviceable.

FAILURE CONDITIONS

DTC 74 sets when the EBTCM detects an open or short to ground in the PMV pressure switch circuitry.

Possible Causes

- Open in CKT 853 or 1659.
- Short to ground in CKT 1659 with the brakes applied.
- Failed PMV pressure switch.

DIAGNOSTIC CHART DTC 74 — 1994 OLDSMOBILE AND PONTIAC, CONTINUED

DTC 74
PMV PRESSURE SWITCH CIRCUIT OPEN OR SHORTED (3 of 3)

B

YES

- Disconnect PMV Fluid Level Switch and Brake Switch (Black Connectors).
- Ignition: RUN
- Measure Voltage between Pin Out Box Terminals 1 and:
  - Terminal 5
  - Terminal 8
  - Terminal 13
  - Terminal 26

Was greater than 2 volts measured on any one of the above?

NO → Repair Open in CKT 853.

YES → Repair Short to Voltage in any circuits that measured more than 2 volts.

NO

- Ignition: OFF
- Measure resistance between Pin Out Box Terminals 1 and:
  - Terminal 5
  - Terminal 8
  - Terminal 13
  - Terminal 26

Was infinite OHMS measured?

NO → Repair Short to Ground in any circuits that did not measure infinite OHMS.

YES

- Check for proper terminal contact. at EBTCM, PMV Connector C1, PMV Fluid Level Switch and Brake Switch Connectors.
- If all are OK, Reconnect.
- Ignition: RUN
- Using Tech 1, clear DTC (s).
- Test Drive Vehicle.
- If DTC 74 resets, replace EBCTM.

DIAGNOSTIC CHART DTC 75 — 1994 OLDSMOBILE AND PONTIAC

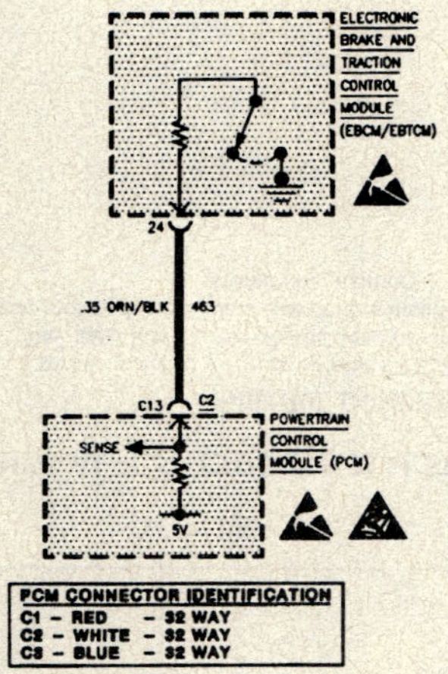

DTC 75
PCM REQUESTED TCS TO BE DISABLED

CIRCUIT OPERATION

When excessive wheel spin is detected by the EBTCM, it will request the PCM to reduce engine torque via a high speed pulse width modulated (PWM) signal on circuit 463.

FAILURE CONDITIONS

Important

If PCM code 1571 is not present, no fault exists with the ABS/TCS system.

DTC 75 sets when the EBTCM receives a serial data message from the PCM to disable Traction Control. DTC 75 sets under the following two conditions.

1. EBTCM to PCM communications.
   - Ckt 463 or Ckt 800 open or shorted.
2. PCM drivability conditions to include PCM codes P0101, P0341, P0342, P1350, P1361, P1623, P1630, P0131, P0132, P0171, and P0172.

DIAGNOSTIC CHART DTC 75 — 1994 OLDSMOBILE AND PONTIAC, CONTINUED

DTC 75
PCM REQUESTED TCS TO BE DISABLED

Has ABS/TCS Functional Test been performed?

- NO: Perform ABS/TCS Functional Test before proceeding.
- YES:
  - Ignition: RUN
  - Using Tech 1, check for presence of PCM DTC P1571.
  - Is PCM DTC P1571 present?

    - NO: REPAIR OPEN.
    - YES:
      - Ignition: OFF
      - Disconnect EBTCM.
      - Install J 38716 Pinout Box.
      - Using J 39200 set to Ω, measure resistance between Pinout Box Terminal 1 and DLC Terminal 4.

      Is less than 2 ohms measured?

      - NO: Repair open in CKT 251.
      - YES:
        - Ignition: RUN
        - Using J 39200 set to DC scale, measure voltage between Pinout Box Terminals 24 and 1.

        Approximately 5 volts present?

        - NO:
          - Backprobe PCM Terminal C13 (White Connector) with J 39200 to ground.
          - Approximately 5 volts present?
            - NO: Check CKT 463 for short to ground or battery. If CKT 463 is OK, replace PCM.
            - YES: Open in CKT 463.
        - YES:
          - Reconnect EBTCM
          - Ignition: RUN
          - Using J 39200 set to Hz, backprobe PCM Terminal C13 (white connector) and measure frequency between PCM C13 and DLC Terminal 4.

          Is 126 - 130 Hz measured?

          - NO: Check for poor terminal contact at EBTCM Terminal 24 and PCM Terminal C13. If all are OK, replace EBTCM
          - YES: Check for intermittent short to ground in CKT 463. If CKT 463 is OK, check for poor terminal contact at PCM and EBTCM. If all are OK, replace PCM.

After repairs are completed, clear DTC and repeat ABS/TCS functional test.

DIAGNOSTIC CHART DTC 76 — 1994 OLDSMOBILE AND PONTIAC

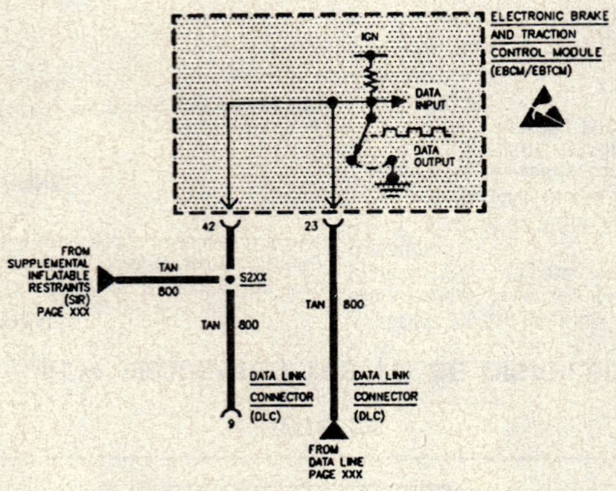

DTC 76

DATA LINE SIGNAL OR CIRCUIT MALFUNCTION

CIRCUIT OPERATION

The EBTCM shares specific serial data information supplied on the data line. The data line also allows the use of a bi-directional scan tool such as a Tech 1.

FAILURE CONDITIONS

A current DTC 76 cannot be read with a Tech 1 due to the loss of serial data communications. After data line communications are restored, DTC 76 will be stored in memory.

DIAGNOSTIC CHART DTC 77 — 1994 OLDSMOBILE AND PONTIAC

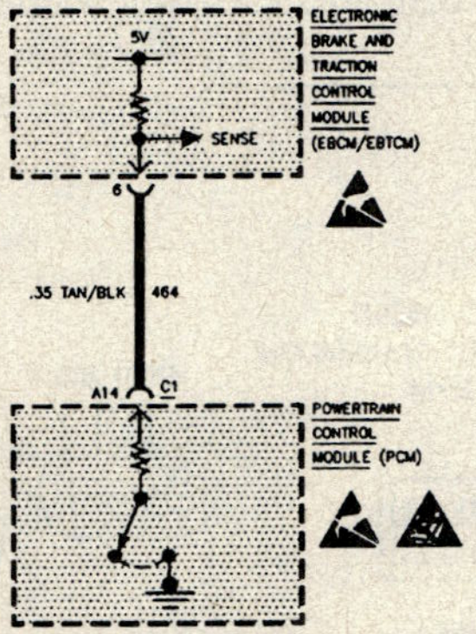

PCM CONNECTOR IDENTIFICATION	
C1 – RED	– 32 WAY
C2 – WHITE	– 32 WAY
C3 – BLUE	– 32 WAY

DTC 77

DELIVERED TORQUE CIRCUIT FAULT

CIRCUIT OPERATION

The PCM provides the EBTCM information on engine torque output via a high speed pulse width modulated (PWM) signal on circuit 464. This information is provided so the EBTCM can determine if engine torque management is required during a wheel slip condition.

FAILURE CONDITIONS

If the EBTCM determines there is a signal/circuit malfunction in circuit 464, the "TRACTION OFF" telltale will be illuminated and DTC 77 will set.

DIAGNOSTIC CHART DTC 77 — 1994 OLDSMOBILE AND PONTIAC, CONTINUED

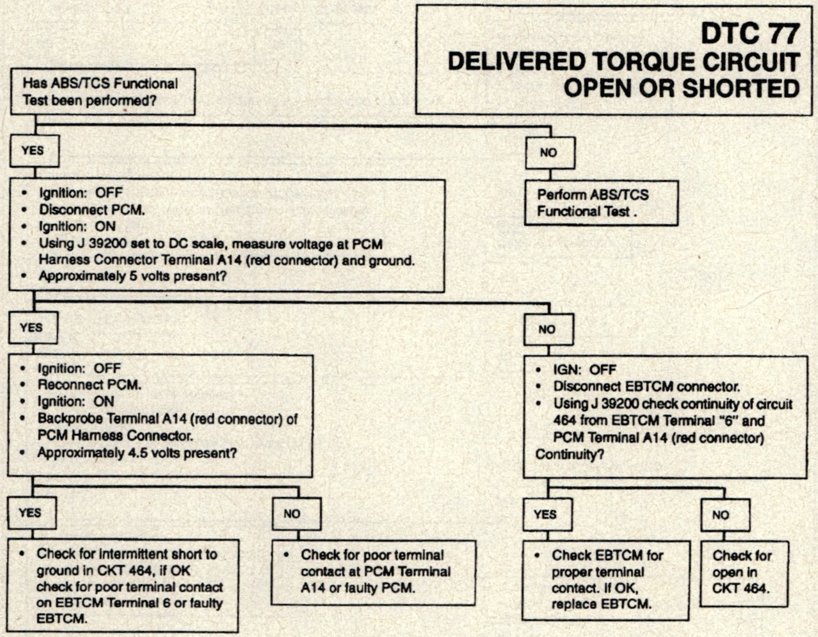

ELECTRICAL CONNECTOR TERMINAL IDENTIFICATION — 1994 OLDSMOBILE AND PONTIAC

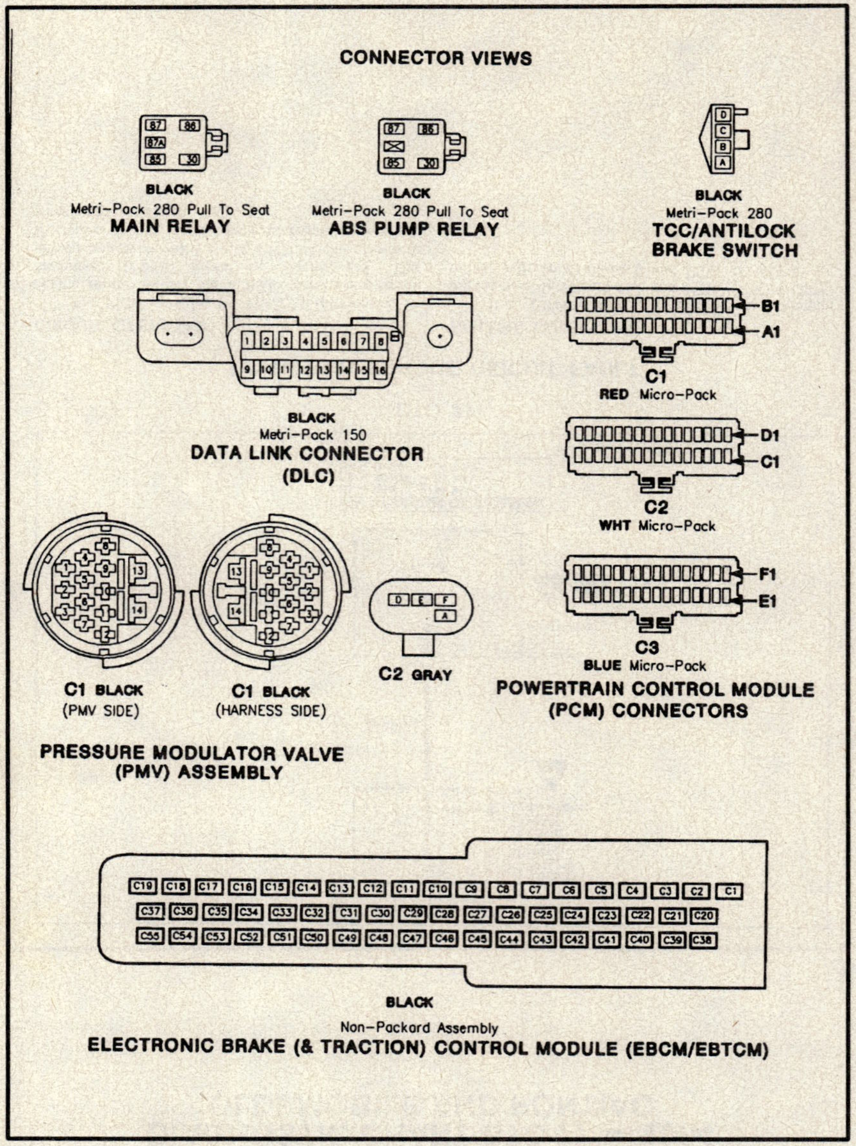

DIAGNOSTIC CHART ABS/TCS FUNCTIONAL CHECK — 1994 BUICK

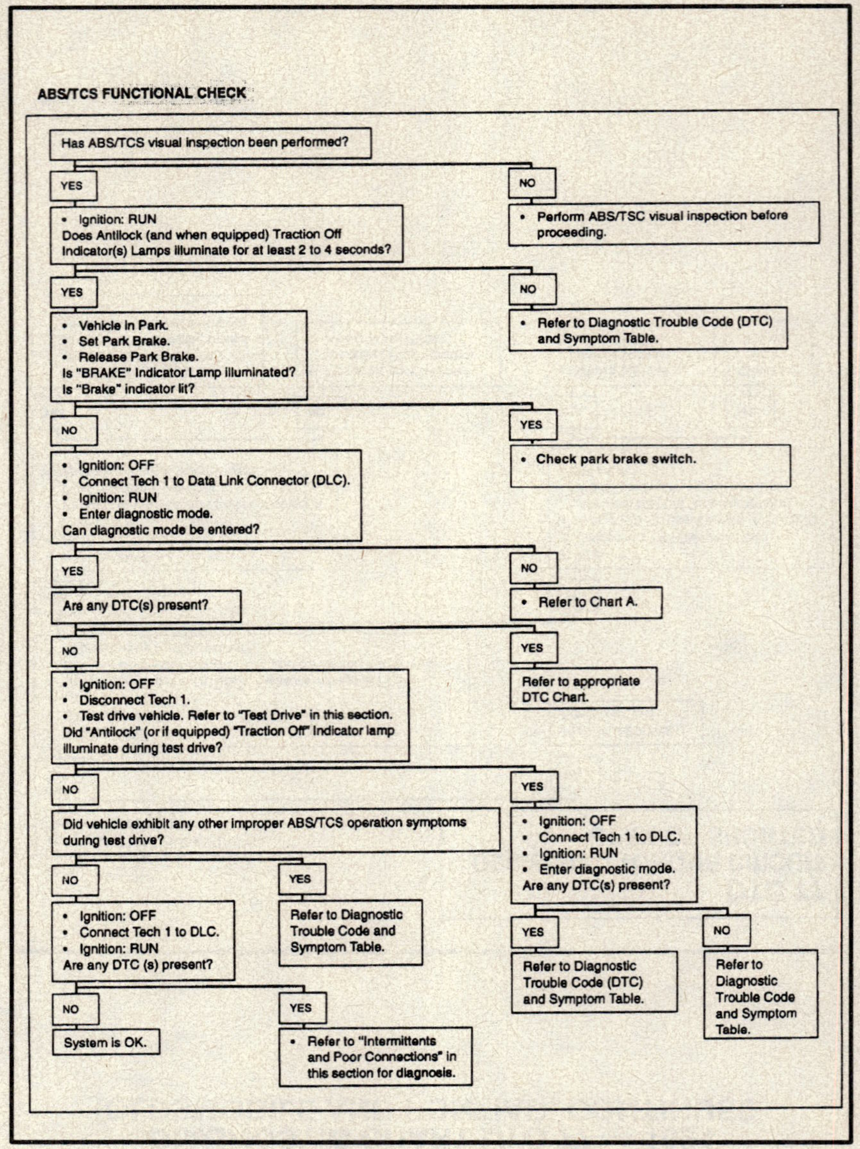

DIAGNOSTIC TROUBLE CODE CHART GUIDE — 1994 BUICK

- Perform "ABS/TCS Functional Check" before beginning diagnosis.

INTERMITTENT "HISTORY" DIAGNOSTIC TROUBLE CODES:
Stored diagnostic trouble codes that do not light the amber ANTILOCK indicator are intermittent problems or history codes that were not cleared during prior service. Do not use diagnostic trouble code diagnosis for intermittent problems until referring to "intermittents and poor connections" in this section.

MULTIPLE DIAGNOSTIC TROUBLE CODES: Diagnose first diagnostic trouble code that appears on Tech 1 screen.

CLEARING DIAGNOSTIC TROUBLE CODES:

- Repair problem.
- Use a Tech 1 "Scan Tool" to clear codes.
- Cycle ignition switch.
- After clearing diagnostic trouble code, repeat "ABS/TCS Functional Check" to assure all problems have been repaired.

DTC	DIAGNOSIS:
DTC 21, DTC 25, DTC 31, DTC 35	Test F: Wheel Speed Sensor Continuity Test
DTC 23, DTC 27, DTC 33, DTC 37	Test I: Wheel Speed Signal Erratic Or Noisy Test
DTC 22, DTC 26, DTC 32, DTC 36	
DTC 45	Test A: Power Interruption Test Note: Always perform Test "D" (Pressure Modulator Valve Assembly Continuity Test) before Test "A" when DTC 45 is set.
DTC 41, DTC 42, DTC 45, DTC 46, DTC 51, DTC 52, DTC 55, DTC 56	Test D: Pressure Modulator Valve (PMV) Assembly Continuity Test
DTC 44, DTC 48	Test W: Isolation Valve Continuity Test
DTC 61	Test J: Pump Motor Circuit Test
DTC 62	Test L: Brake Hydraulic Test
DTC 71	Test W: EBCM/EBTCM Problem Test
DTC 72	Test N: EBCM/EBTCM Grounded Switch Input Test Note: When DTC 72 and DTC 73 are both set, check wiring (CKT 848) for a short to ground
DTC 73	
DTC 74	Test O: EBCM/EBTCM Open Switch Input Test Note: When DTC 73 and DTC 74 are both set, check wiring (CKT 1659) for a short to ground
DTC 75	Test AA: PCM Requested TCS To Be Disabled.
DTC 76	Test BB: Data Line Signal or Circuit Malfunction.
DTC 77	Test DD: Delivered Torque PWM Circuit Shorted or Open.

DIAGNOSTIC SYMPTOM GUIDE

SYMPTOM	DIAGNOSIS		
	Do 1st	Do 2nd	Do 3rd
Diagnostics cannot be entered.	Data Line Test P.		
ANTILOCK indicator does not light.	Do Antilock indicator "OFF" Fault Test R.		
ANTILOCK indicator is on all the time and no diagnostic trouble codes are set.	Check PMV reservoir brake fluid level.	Do Antilock indicator "ON" Fault Test Q.	
Brake pedal rises or drops excessively during an ABS stop (ABS engaged).	Do TCC/Antilock Brake Switch Test K.		
Brake pedal vibrates (fluctuates) during slow speed (under 16 km/h [10 mph]) braking.	Do Wheel Speed Signal Erratic or Noisy Test I.		

(continued on next page)

DIAGNOSTIC TROUBLE CODE CHART SYMPTOM GUIDE — 1994 BUICK

DIAGNOSTIC SYMPTOM GUIDE (continued)

SYMPTOM	DIAGNOSIS		
	Do 1st	Do 2nd	Do 3rd
BRAKE indicator is on all the time.		Do Brake Warning Short Circuit Test T.	Replace instrument cluster (U23) or information center (UB3).
BRAKE indicator does not light.	Check BRAKE indicator bulb and wiring (CKT 33) for an open.	Replace instrument cluster (U23) or information center (UB3).	
BRAKE indicator does not light with park brake applied.	Do Park Brake Switch Test V.		
BRAKE indicator does not light with low brake fluid or during bulb test.	Do Brake Fluid Level Switch Test U.		
Brake pedal is spongy.			
BRAKE indicator lights properly, but chime does not sound with park brake applied.	Check wiring (CKT 33) to MFC module for an open; if OK, see Warnings and Alarms section.		
Only TRACTION OFF indicator does not light.	Check Gage Cluster Indicator Bulb.	Do "Traction OFF" indicator Test DD.	
Only TRACTION OFF indicator is on all the time and no diagnostic trouble codes are set.	Wait 20 minutes. If indicator is now off, system OK. If indicator is still on, continue diagnosis.	Do "Traction OFF" indicator Test EE.	

Important

- Customers may comment that they hear the PMV assembly run at slow speeds (11 km/h [7 mph]). This is simply the ABS system running through its self-test. It is also possible to feel the PMV assembly running through the brake pedal, when it is barely applied, and to hear brake fluid return to the master cylinder. Also, the vehicle may slowly decrease in speed while the PMV modulator valves increase fluid pressure to the wheel cylinders during this test. In both these instances, the driver must be traveling at a very low speed (11 km/h [7 mph]) and have one foot touching the brake pedal (as some two-footed drivers may) to feel the PMV assembly running. These are normal conditions and in no way signify a problem with the ABS system.
- When vehicle is equipped with traction control, any symptom which lights the ANTILOCK indicator also lights the TRACTION OFF indicator. Therefore, when both indicators are lit, follow ANTILOCK indicator diagnosis when no diagnostic trouble codes are set.
- Vehicles with traction control contain a preventive system to avoid brake overheating. During heavy braking conditions, if a traction control mode is entered and the EBTCM determines the brakes are too hot, the EBTCM disables the traction control system and lights only the TRACTION OFF indicator. After a sufficient cooling time, approximately 15-20 minutes, the EBTCM enables traction control system and turns off the indicator. This condition is normal and should not be mistaken as an intermittent system malfunction.

DIAGNOSTIC CHARTS A AND B — 1994 BUICK

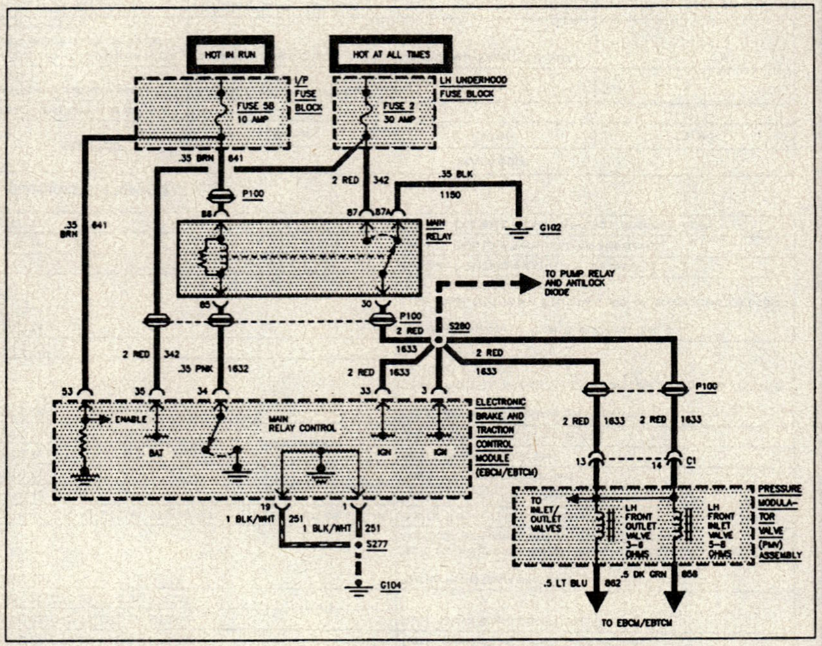

OPERATION:

When the ignition switch is first turned to RUN, EBCM/EBTCM goes through a self-check lasting 2-4 seconds. During this test, the EBCM/EBTCM grounds terminal 34, energizing the main relay. Once the self-check is completed, the main relay remains energized and delivers battery voltage to both the EBCM/EBTCM and the pressure modulator valve (PMV) assembly during vehicle operation. The constant voltage supply allows the vehicle to enter ABS mode without delay. The main relay is de-energized whenever the amber ANTILOCK indicator lights due to a system fault, causing the ABS system to shut down.

DTC 45 sets when the ignition switch is at run and battery voltage is detected at neither the PMV assembly nor EBCM/EBTCM terminals 20, 3 and/or 33. DTC 45 also sets when the EBCM/EBTCM detects and open or a short to ground in the LH inlet valve circuit.

POSSIBLE CAUSES:

- Open CKT 641, 1632, 858, 1633 or 342.
- Short to ground in CKT 858, 1633 or 342.
- Main relay fails "open."

DIAGNOSTIC CHARTS A AND B — 1994 BUICK, CONTINUED

A: POWER INTERRUPTION TEST

- Make sure "ABS/TCS Functional Check" is performed before proceeding.
- Ignition switch: OFF
- Main relay: disconnected

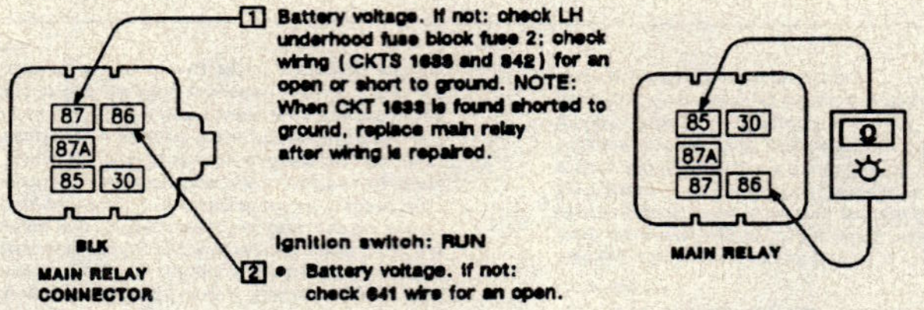

[1] Battery voltage. If not: check LH underhood fuse block fuse 2; check wiring (CKTS 1688 and 842) for an open or short to ground. NOTE: When CKT 1688 is found shorted to ground, replace main relay after wiring is repaired.

Ignition switch: RUN

[2] • Battery voltage. If not: check 641 wire for an open.

[3] 45-90 ohms. If not: replace main relay.

- If all measurements are correct: go to EBCM/EBTCM Power/Ground Test B.
- After repairs are completed, clear diagnostic trouble codes and repeat "ABS/TCS Functional Check."

B: EBCM/EBTCM POWER/GROUND TEST

- Ignition switch: OFF
- Electronic brake and traction control module (EBCM/EBTCM) connector: disconnected
- Connect J 38716 Pinout Box

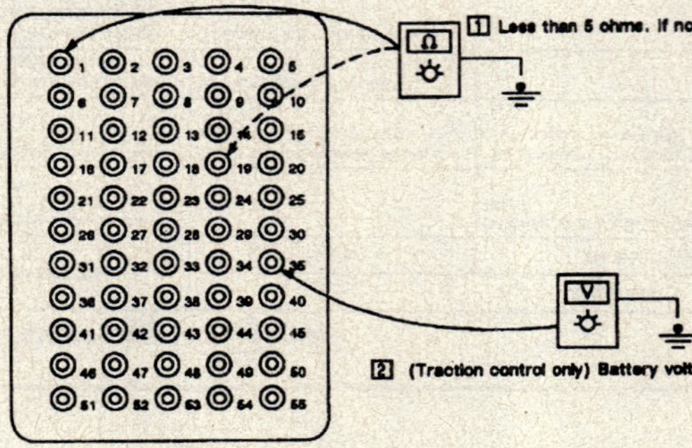

[1] Less than 5 ohms. If not: check wiring for an open.

[2] (Traction control only) Battery voltage. If not: check wiring for an open.

J 38716

- If all measurements are correct: go to EBCM/EBTCM Power/Ground Test B1.

DIAGNOSTIC CHARTS B1 AND C — 1994 BUICK

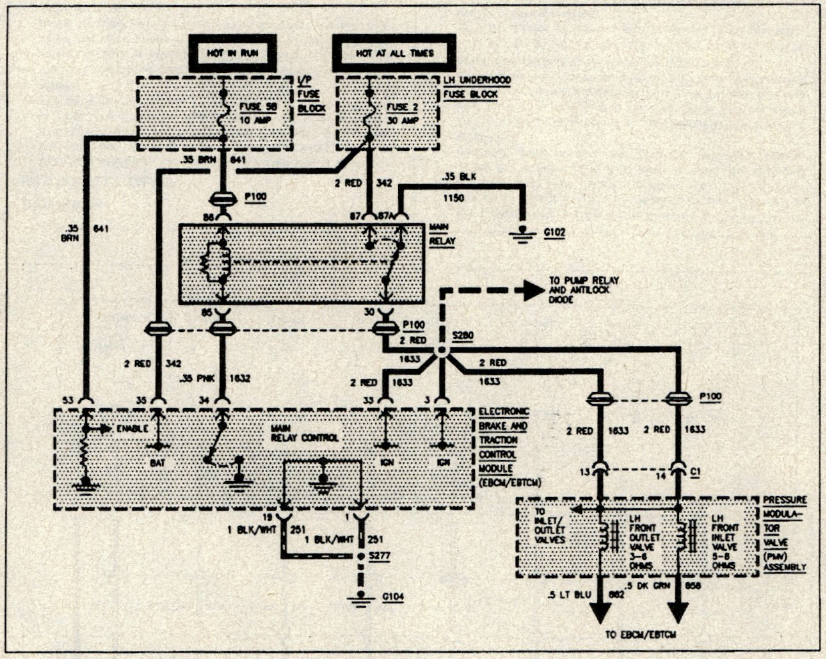

OPERATION:

When the ignition switch is first turned to RUN, the EBCM/EBTCM goes through a self-check lasting 2-4 seconds. During this test, the EBCM/EBTCM grounds terminal 34, energizing the main relay. Once the self-check is completed, the main relay remains energized and delivers battery voltage to both the EBCM/EBTCM and the pressure modulator valve (PMV) assembly during vehicle operation. The constant voltage supply allows the vehicle to enter ABS mode without delay. The main relay is de-energized whenever the amber ANTILOCK indicator lights due to a system fault, causing the ABS system to shut down.

DTC 45 sets when the ignition switch is at run and battery voltage is detected at neither the PMV assembly nor EBCM/EBTCM terminals 20, 3 and/or 33. DTC 45 also sets when the EBCM/EBTCM detects an open or a short to ground in the LH inlet valve circuit.

POSSIBLE CAUSES:

- Open CKT 641, 1632, 858, 1633 or 342.
- Short to ground in CKT 858, 1633 or 342.
- Main relay fails "open."

DIAGNOSTIC CHARTS B1 AND C — 1994 BUICK

B1: EBCM/EBTCM POWER/GROUND TEST

- Ignition switch: OFF
- Electronic brake and traction control module (EBCM/EBTCM) connector: disconnected
- Connect J 38716 Pinout Box
- Pressure modulator valve (PMV) assembly connector C1: disconnected (twist off)
- Main relay: disconnected

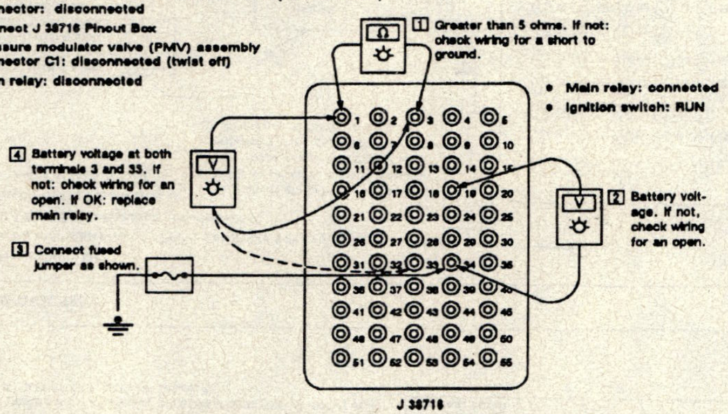

J 38716

- If all measurements are correct: go to Pressure Modulator Valve (PMV) Assembly Voltage Test C.
- After repairs are completed, clear diagnostic trouble codes and repeat "ABS/TCS Functional Check."

C: PRESSURE MODULATOR VALVE (PMV) ASSEMBLY VOLTAGE TEST

- Pressure modulator valve (PMV) assembly connector C1: disconnected
- Electronic brake and traction control module (EBCM/EBTCM) connector: disconnected
- Connect J 38716 Pinout Box
- Fused jumper: connected (from EBCM/EBTCM Power/Ground Test B1)
- Ignition switch: RUN

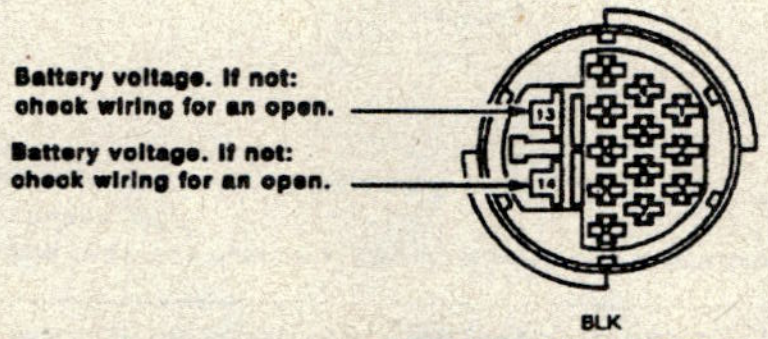

PRESSURE MODULATOR VALVE (PMV) ASSEMBLY CONNECTOR C1
(Harness side)

- If both measurements are correct: remove fused jumper and go to Pressure Modulator Valve (PMV) Assembly Continuity Test D.
- After repairs are completed, clear diagnostic trouble codes and repeat "ABS/TCS Functional Check."

DIAGNOSTIC CHARTS D AND E — 1994 BUICK

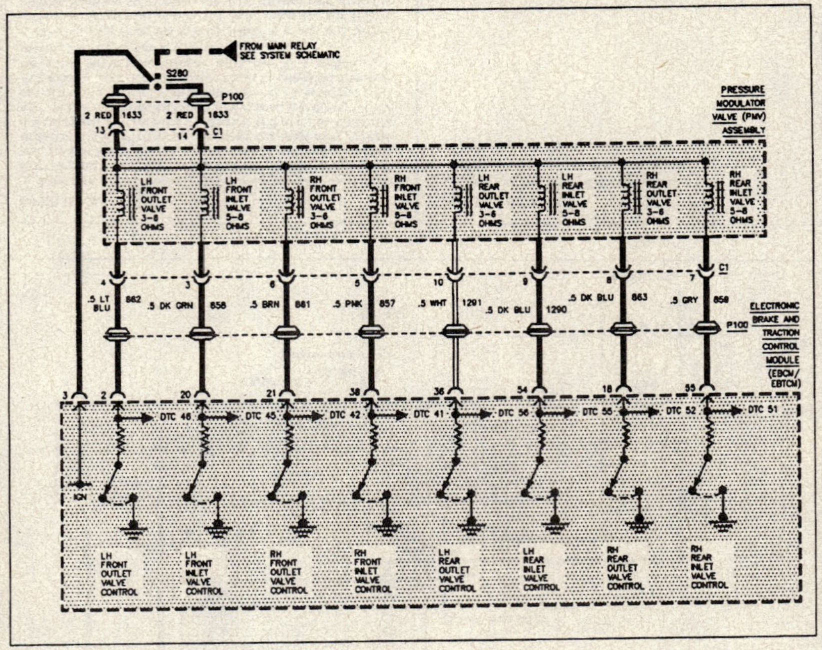

OPERATION:

INLET/OUTLET VALVES

- Voltage is applied to the normally open inlet valves whenever the main relay is energized. An inlet valve closes when the EBCM/EBTCM supplies a ground path at the inlet valve control terminal. The EBCM/EBTCM closes an inlet valve for short periods to maintain (hold) or reduce pressure at a particular wheel.
- The normally closed outlet valves operate similarly to the inlet valves, except that when the EBCM/EBTCM grounds an outlet valve, the valve opens instead of closes. The EBCM/EBTCM opens an outlet valve for short periods to reduce pressure at a particular wheel.

VALVE MODES

- The pressure increase position is the operating valve position for normal (non-ABS) braking and traction control mode (when equipped). The EBCM/EBTCM does not energize either valve, leaving the inlet valve open and the outlet valve closed.
- The pressure hold position occurs when the EBCM/EBTCM first notices an abnormal hydraulic pressure increase. This signifies an impending wheel lockup condition. The EBCM/EBTCM closes the corresponding inlet valve to avoid any additional hydraulic pressure increase (avoiding an ABS braking condition).
- During a wheel lockup condition, the EBCM/EBTCM closes the inlet valve while pulsing open the outlet valve. This decreases hydraulic pressure at a specific wheel without decreasing overall system pressure. The reduced hydraulic pressure allows greater wheel spin, eliminating the wheel lockup condition.
- DTC 41, 42, 45, 46, 51, 52, 55, or 56 sets when the (EBCM/EBTCM) detects an open or short to ground in the respective inlet or outlet valve circuit.

POSSIBLE CAUSES:

- Open in CKT 857, 858, 859, 861, 862, 863, 1290 or 1291.
- Short to ground in CKT 857, 858, 859, 861, 862, 863, 1290, or 1291.

DIAGNOSTIC CHARTS D AND E — 1994 BUICK, CONTINUED

D: PRESSURE MODULATOR VALVE (PMV) ASSEMBLY CONTINUITY TEST

- Make sure "ABS/TCS Functional Check" is performed before proceeding.
- Ignition switch: OFF
- Pressure modulator valve (PMV) assembly connector C1: disconnected

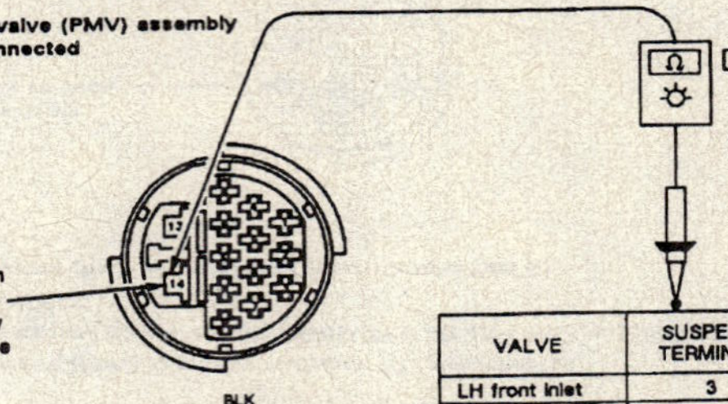

1. Infinite ohms between pressure modulator valve(PMV) assembly casing. If not: replace PMV assembly.
2. Measure resistance across suspect solenoid valve terminal(s). When a measurement is incorrect: replace PMV assembly.

VALVE	SUSPECT TERMINAL	CORRECT RESISTANCE
LH front inlet	3	5 to 8 ohms
RH front inlet	5	
LH rear inlet	9	
RH rear inlet	7	
LH front outlet	4	3 to 6 ohms
RH front outlet	6	
LH rear outlet	10	
RH rear outlet	8	

- If all measurements are correct: go to Valve Control Test E.
- After repairs are completed, clear diagnostic trouble codes and repeat "ABS/TCS Functional Check."

E: VALVE CONTROL TEST

- Ignition switch: OFF
- Pressure modulator valve (PMV) assembly connector C1: connected
- (EBCM/EBTCM) connector: disconnected
- Connect J 38716 Pinout Box
- Main relay: disconnected

1. Less than 10 ohms between terminal 3 and suspect valve control terminal. If not: check suspect wiring for an open; check PMV assembly connector C1 for proper terminal contact.
2. Greater than 5 ohms between suspect valve control terminal and ground. If not: check suspect wiring for a short to ground.

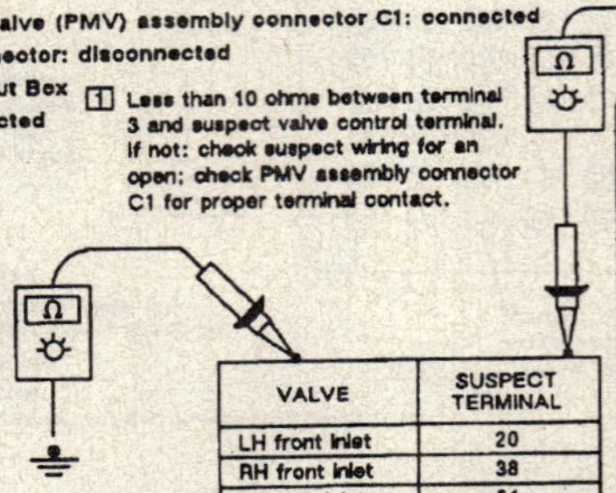

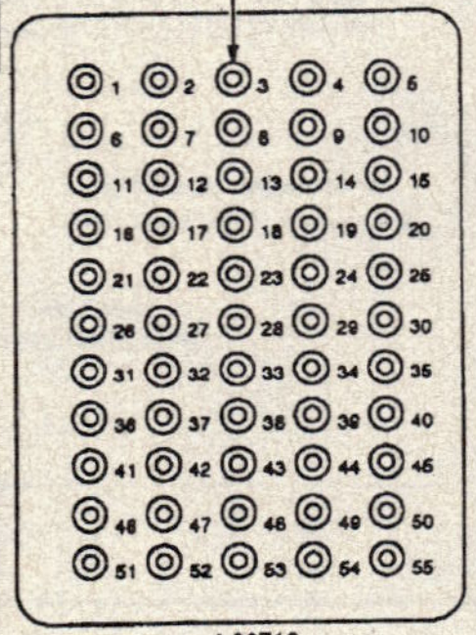

VALVE	SUSPECT TERMINAL
LH front inlet	20
RH front inlet	38
LH rear inlet	54
RH rear inlet	55
LH front outlet	2
RH front outlet	21
LH rear outlet	36
RH rear outlet	18

- If all measurements are correct: perform Inlet/Outlet Valve Hydraulic Test BB; when no problem is found, check EBCM/EBTCM connector for proper terminal contacts: If OK, replace EBCM/EBTCM.
- After repairs are completed, clear diagnostic trouble codes and repeat "ABS/TCS Functional Check."

DIAGNOSTIC CHARTS F AND G — 1994 BUICK

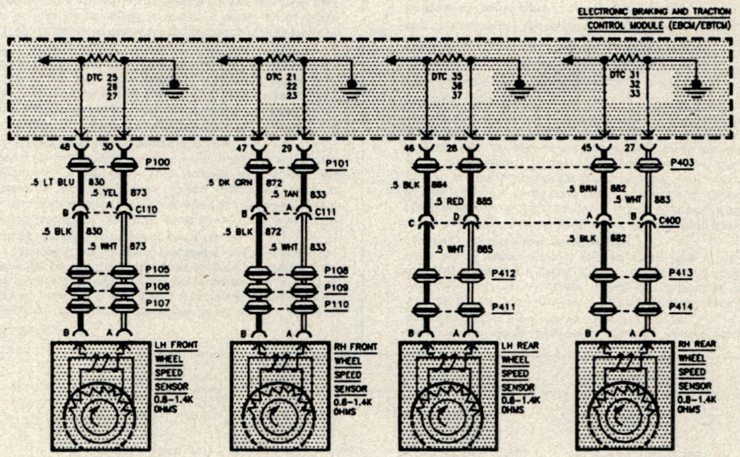

OPERATION:

- When each wheel turns, the wheel speed sensor creates a small AC voltage with a frequency proportional to wheel speed. This magnetically induced voltage is caused by a toothed sensor ring passing the wheel speed sensor's stationary pickup coil. The toothed sensor ring is mounted on the drive axle in the front, and the hub/bearing assembly in the rear.
- The EBCM/EBTCM uses the wheel speed sensor signal to calculate vehicle reference speeds, and individual wheel speed, acceleration and slip values. These values are used to determine when antilock control is required. Vehicles equipped with traction control also use front wheel speed sensor signals to determine when traction control is required.
- To prevent electromagnetic interference from disturbing the wheel speed sensor signal, the cables from the EBCM/EBTCM to each wheel speed sensor are twisted in pairs a minimum of 6 to 9 turns per foot. When servicing the sensor cables, it is important to maintain the original cables' twists (6 to 9 turns per foot).
- DTC 22, 26, 32, or 36 sets when the EBCM/EBTCM determines that a wheel speed signal is erratic. A wheel speed signal indicating that a wheel is accelerating or decelerating faster than physically possible is an example of an erratic signal. An intermittent open or short to ground in a wheel speed sensor could also cause the EBCM/EBTCM to set a diagnostic trouble code. DTC 22, 26, 32, or 36 also sets when the EBCM/EBTCM detects electromagnetic interference (noise) in a wheel speed sensor signal. Possible interference sources include accessory motors, wiper motors and ignition circuits.
- DTC 23, 27, 33 or 37 sets when the EBCM/EBTCM does not see any wheel speed sensor signal, but determines that continuity exists in the circuit. This is caused by a dislodged wheel speed sensor, missing toothed sensor ring (front) or defective hub and bearing assembly (rear).
- Any condition which would result in lack of continuity in a wheel speed sensor circuit could result in DTC 21 or 25 (front) or 31 or 35 (rear) being set. These conditions include an open, a short to ground, or a short between the two wheel speed sensor circuits in the wiring harness between the speed sensor and EBCM/EBTCM. Also, an open or short to ground across the wheel speed sensor coil could trigger this diagnostic trouble code.

POSSIBLE CAUSES:

- Faulty or dislodged wheel speed sensor.
- Missing or irregular toothed sensor ring.
- Open or short to ground in wheel speed sensor circuit.
- Short between wheel speed sensor circuits.
- Erratic wheel speed signal.
- Defective hub and bearing assembly.
- Intermittent open or short to ground in wheel speed sensor circuit.
- Noise interference in wheel speed sensor signal.
- Improperly twisted sensor cable.

DIAGNOSTIC CHARTS F AND G — 1994 BUICK, CONTINUED

F: WHEEL SPEED SENSOR CONTINUITY TEST

- Make sure "ABS/TCS Functional Check" is performed before proceeding.
- Raise vehicle (see Buick Service Manual Section 0A)
- Ignition switch: RUN
- Gear selector lever: NEUTRAL
- Ignition switch: OFF
- Park brake: released
- Make measurements at suspect wheel speed sensor (connector disconnected).

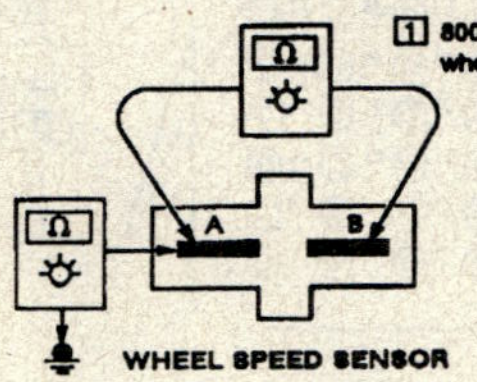

1. 800-1500 ohms. If not: replace suspect wheel speed sensor.
2. Greater than 5 ohms. If not: replace suspect wheel speed sensor.

WHEEL SPEED SENSOR

- If all measurements are correct: go to EBCM/EBTCM/Wheel Speed Circuit Input Test H.
- After repairs are completed, clear diagnostic trouble codes and repeat "ABS/TCS Functional Check."

G: WHEEL SPEED SENSOR OUTPUT TEST

- Raise vehicle (see Buick Service Manual Section 0A)
- Ignition switch: RUN
- Gear selector lever: NEUTRAL
- Ignition switch: OFF
- Park brake: released
- Make measurements at suspect wheel speed sensor (connector disconnected).

1. 50-900 mvAC while turning wheel by hand. If measurement is incorrect: replace suspect wheel speed sensor.

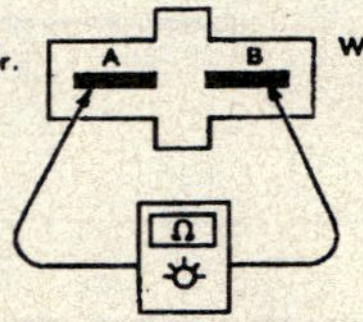

WHEEL SPEED SENSOR

- If all measurements are correct: check EBCM/EBTCM connector for proper terminal contact: If OK, replace EBCM/EBTCM.
- After repairs are completed, clear diagnostic trouble codes and repeat "ABS/TCS Functional Check."

DIAGNOSTIC CHART H — 1994 BUICK

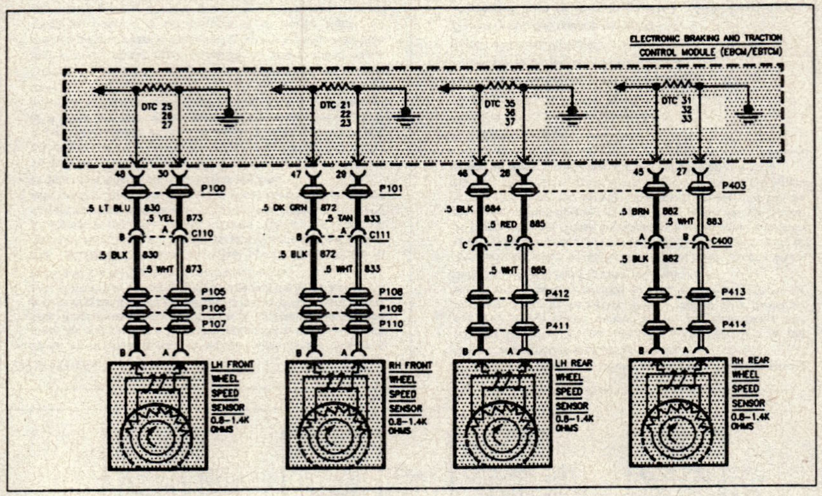

OPERATION:

- When each wheel turns, the wheel speed sensor creates a small AC voltage with a frequency proportional to wheel speed. This magnetically induced voltage is caused by a toothed sensor ring passing the wheel speed sensor's stationary pickup coil. The toothed sensor ring is mounted on the drive axle in the front, and the hub/bearing assembly in the rear.
- The EBCM/EBTCM uses the wheel speed sensor signal to calculate vehicle reference speeds, and individual wheel speed, acceleration and slip values. These values are used to determine when antilock control is required. Vehicles equipped with traction control also use front wheel speed sensor signals to determine when traction control is required.
- To prevent electromagnetic interference from disturbing the wheel speed sensor signal, the cables from the EBCM/EBTCM to each wheel speed sensor are twisted in pairs a minimum of 6 to 9 turns per foot. When servicing the sensor cables, it is important to maintain the original cables' twists (6 to 9 turns per foot).
- DTC 22, 26, 32, or 36 sets when the EBCM/EBTCM determines that a wheel speed signal is erratic. A wheel speed signal indicating that a wheel is accelerating or decelerating faster than physically possible is an example of an erratic signal. An intermittent open or short to ground in a wheel speed signal sensor could also cause the EBCM/EBTCM to set a diagnostic trouble code. DTC 22, 26, 32, or 36 also sets when the EBCM/EBTCM detects electromagnetic interference (noise) in a wheel speed sensor signal. Possible interference sources include accessory motors, wiper motors and ignition circuits.
- DTC 23, 27, 33, or 37 sets when the EBCM/EBTCM does not see any wheel speed sensor signal, but determines that continuity exists in the circuit. This is caused by a dislodged wheel speed sensor, missing toothed sensor ring (front) or defective hub and bearing assembly (rear).
- Any condition which would result in lack of continuity in a wheel speed sensor circuit could result in DTC 21 or 25 (front), or 31 or 35 (rear) being set. These conditions include an open, a short to ground, or a short between the two wheel speed sensor circuits in the wiring harness between the speed sensor and EBCM/EBTCM. Also, an open or short to ground across the wheel speed sensor coil could trigger this diagnostic trouble code.

POSSIBLE CAUSES:

- Faulty or dislodged wheel speed sensor.
- Missing or irregular toothed sensor ring.
- Open or short to ground in wheel speed sensor circuit.
- Short between wheel speed sensor circuits.
- Erratic wheel speed signal.
- Defective hub and bearing assembly.
- Intermittent open or short to ground in wheel speed sensor circuit.
- Noise interference in wheel speed sensor signal.
- Improperly twisted sensor cable.

DIAGNOSTIC CHART H — 1994 BUICK, CONTINUED

H: EBCM/EBTCM/WHEEL SPEED CIRCUIT INPUT TEST

- Ignition switch: OFF
- Suspect wheel speed sensor connector: connected
- EBCM/EBTCM connector: disconnected
- Connect J 38716 Pinout Box
- Do steps 1 and 2 at suspect wheel speed sensor terminals (see schematic).
- Example shown is for LH rear wheel speed circuit; other wheels similar.

LH REAR WHEEL SPEED

1 800 to 1500 ohms.
If greater than 1500 ohms: check suspect wiring for an open; check in-line connector and wheel speed sensor connector for proper terminal contact. If less than 800 ohms: check for a short between suspect wheel speed sensor wiring.

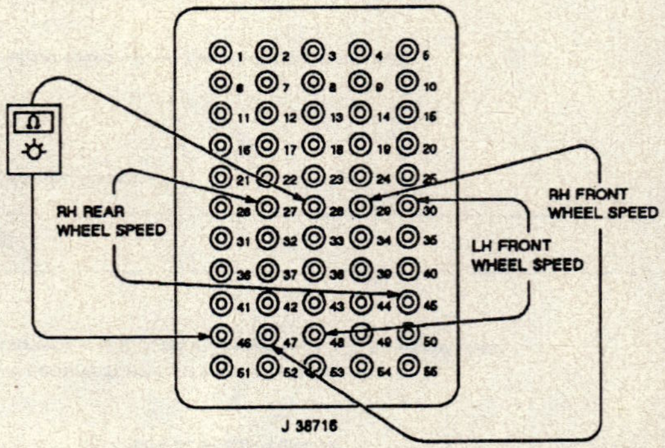

2 Greater than 5 ohms.
If not: check suspect wiring for a short to ground.

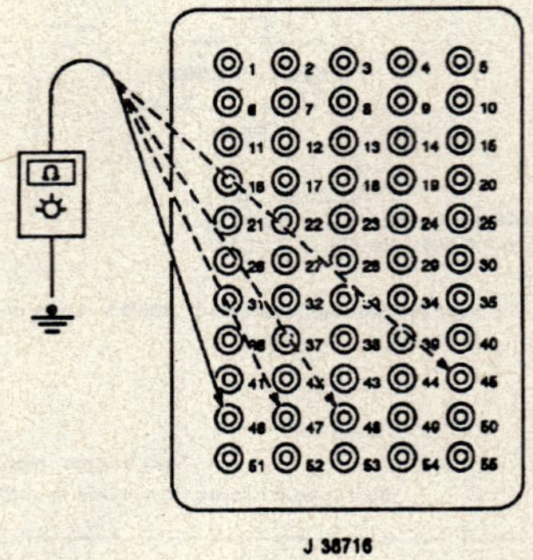

- If all measurements are correct: check EBCM/EBTCM connector for proper terminal contact; If OK, replace EBCM/EBTCM.
- After repairs are completed, clear diagnostic trouble codes and repeat "ABS/TCS Functional Check."

DIAGNOSTIC CHART I — 1994 BUICK

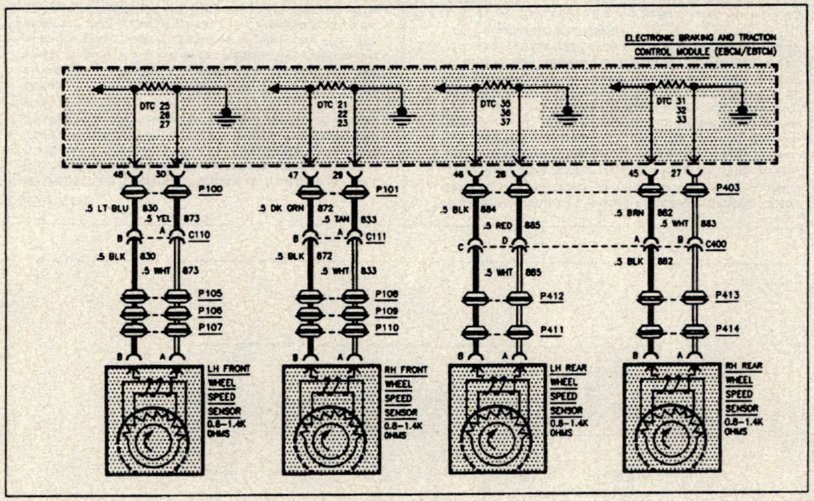

OPERATION:

- When each wheel turns, the wheel speed sensor creates a small AC voltage with a frequency proportional to wheel speed. This magnetically induced voltage is caused by a toothed sensor ring passing the wheel speed sensor's stationary pickup coil. The toothed sensor ring is mounted on the drive axle in the front, and the hub/bearing assembly in the rear.
- The EBCM/EBTCM uses the wheel speed sensor signal to calculate vehicle reference speeds, and individual wheel speed, acceleration and slip values. These values are used to determine when antilock control is required. Vehicles equipped with traction control also use front wheel speed sensor signals to determine when traction control is required.
- To prevent electromagnetic interference from disturbing the wheel speed sensor signal, the cables from the EBCM/EBTCM to each wheel speed sensor are twisted in pairs a minimum of 6 to 9 turns per foot. When servicing the sensor cables, it is important to maintain the original cables' twists (6 to 9 turns per foot).
- DTC 22, 26, 32, or 36 sets when the EBCM/EBTCM determines that a wheel speed signal is erratic. A wheel speed signal indicating that a wheel is accelerating or decelerating faster than physically possible is an example of an erratic signal. An intermittent open or short to ground in a wheel speed signal sensor could also cause the EBCM/EBTCM to set a diagnostic trouble code. DTC 22, 26, 32, or 36 also sets when the EBCM/EBTCM detects electromagnetic interference (noise) in a wheel speed sensor signal. Possible interference sources include accessory motors, wiper motors and ignition circuits.
- DTC 23, 27, 33 or 37 sets when the EBCM/EBTCM does not see any wheel speed sensor signal, but determines that continuity exists in the circuit. This is caused by a dislodged wheel speed sensor, missing toothed sensor ring (front) or defective hub and bearing assembly (rear).
- Any condition which would result in lack of continuity in a wheel speed sensor circuit could result in DTC 21 or 25 (front), or 31 or 35 (rear) being set. These conditions include an open, a short to ground, or a short between the two wheel speed sensor circuits in the wiring harness between the speed sensor and EBCM/EBTCM. Also, an open or short to ground across the wheel speed sensor coil could trigger this diagnostic trouble code.

POSSIBLE CAUSES:

- Faulty or dislodged wheel speed sensor.
- Missing or irregular toothed sensor ring.
- Open or short to ground in wheel speed sensor circuit.
- Short between wheel speed sensor circuits.
- Erratic wheel speed signal.
- Defective hub and bearing assembly.
- Intermittent open or short to ground in wheel speed sensor circuit.
- Noise interference in wheel speed sensor signal.
- Improperly twisted sensor cable.

DIAGNOSTIC CHART I — 1994 BUICK, CONTINUED

I: WHEEL SPEED SIGNAL ERRATIC OR NOISY TEST

- Make sure "ABS/TCS Functional Check" is performed before proceeding.

Is DTC 23, DTC 27, DTC 33 or DTC 37 set?

YES:

When DTC 23 or DTC 27:
Reinstall dislodged wheel speed sensor or replace missing toothed sensor ring.

When DTC 33 or DTC 37:
Reinstall dislodged wheel speed sensor or replace hub and bearing assembly.

NO:

Is DTC 22, DTC 26, DTC 32 or DTC 36 set?

YES:

- CHECK FOLLOWING AND REPAIR/REPLACE AS NECESSARY, BUT BE AWARE THAT PROBLEM CAUSING DTC TO SET MAY BE INTERMITTENT AND IS USUALLY NOT CAUSED BY A DEFECTIVE EBCM/EBTCM:

  Test drive vehicle (refer to "Test Drive") while monitoring suspect wheel speed sensor using Tech 1 in mode F3 (snapshot).

  Disconnect suspect wheel speed sensor connector and applicable in-line connector and inspect terminal contacts on sensor and connector for corrosion or deformity.

  Trace harness routing of suspect sensor wiring. Inspect for physical damage, particularly in areas where pinching or movement by harness or other component occurs.

  Trace harness routing of suspect sensor wiring for proximity to noise sources such as ignition circuits, motors, etc.

  Check if any repairs were recently made to suspect sensor wiring. Sensor wiring should be twisted 6 to 9 turns per foot.

  Check for toothed sensor ring irregularity, loose sensor mounting, or debris lodged between the toothed sensor ring and the wheel speed sensor.

NO:

- Thoroughly spray area surrounding all wheel speed sensors (all connectors connected) with a mixture of 12 oz. of water and two teaspoons of salt.
- Test drive vehicle at slow speeds (under 16 km/h [10 mph]) while monitoring wheel speed sensors using Tech 1 in mode F3 (snapshot). (Note: suspect wheel speed signal usually varies from remaining wheel speed signals by approximately 2 km/h [1 mph]).

Does any one wheel speed signal vary from other three wheel speed signals?

NO:

- System OK.
- Problem may be intermittent. If an intermittent is suspected, continue diagnosis.

YES:

- CHECK FOLLOWING AND REPAIR/REPLACE AS NECESSARY:

  Disconnect suspect wheel speed sensor connector and applicable in-line connector. Inspect terminal contacts on sensor and connectors for corrosion or deformity. Check connectors for proper terminal contact.

  Tract harness routing of suspect sensor wiring. Inspect for physical damage, particularly in areas where pinching or movement by harness or other component occurs.
- If all above checks are OK, replace suspect wheel speed sensor.

- If all checks and measurements are correct: go to Wheel Speed Sensor Output Test G.
- After repairs are completed, clear diagnostic trouble codes and repeat "ABS/TCS Functional Check."

DIAGNOSTIC CHART J — 1994 BUICK

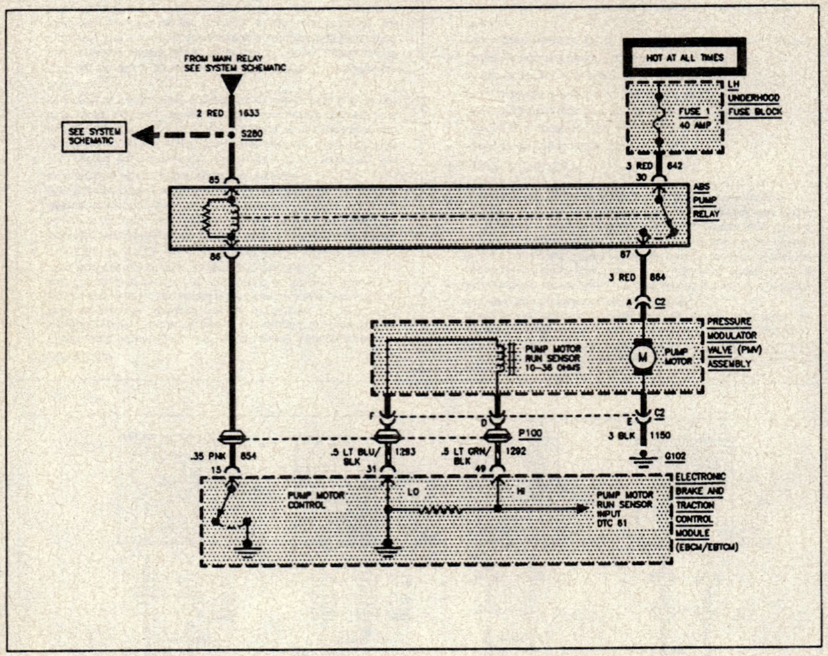

OPERATION:

- When EBCM/EBTCM grounds terminal 15, the ABS pump relay is energized. The pump relay contacts close, applying voltage to the pressure modulator valve (PMV) assembly's pump motor. The EBCM/EBTCM turns on the pump motor during most ABS braking conditions and, when equipped, during all traction control conditions.
- During most ABS braking conditions, the EBCM/EBTCM turns on the pump motor while monitoring the TCC/antilock brake switch to determine brake pedal position. The switch opens when the brake pedal is depressed approximately 40%. The motor restores the master cylinder's fluid pressure, causing the brake pedal to gradually rise. The pump motor continues to run until the brake pedal sufficiently rises to close the TCC/antilock brake switch.
- During traction control mode (when equipped), after the front hydraulic brake circuits are isolated, the pump motor runs to increase brake fluid pressure in these circuits. The increased pressure is directed to the slipping wheel through the PMV assembly's front inlet and outlet valves. This causes the brakes to be applied, reducing wheel spin at the slipping wheel. Traction control is disabled as soon as the brakes are manually applied.
- When the EBCM/EBTCM grounds terminal 15, DTC 61 sets when the EBCM/EBTCM determines that the voltage between EBCM/EBTCM terminals 31 and 49 is not between 500-800 mVAC. The EBCM/EBTCM recognizes this condition only during ABS braking requiring pump motor operation, traction control mode (when equipped), and during the EBCM/EBTCM self-check which occurs when the vehicle reaches 11 km/h (7 mph). Therefore, be sure to test drive vehicle; refer to "Test Drive" in this section, after clearing the diagnostic trouble code to assure correct diagnosis.
- DTC 61 also sets when the pump motor runs all the time (i.e., EBCM/EBTCM sees proper voltage between terminals 31 and 49 without grounding terminal 15).

POSSIBLE CAUSES:

- Open in CKT 1150, 864, 854, 1633, 1292, 1293, or 642.
- Short to ground in CKT 864, 854, 1292, 1293, or 642.
- ABS pump relay fails "open" or "engaged."

DIAGNOSTIC CHART J — 1994 BUICK, CONTINUED

J: PUMP MOTOR CIRCUIT TEST

- Make sure "ABS/TCS Functional/Check" is performed before proceeding.
- Ignition switch: OFF
- Connect TECH 1 to data link connector DLC.
- Ignition switch: RUN
- Command TECH 1 to bleed pressure modulator valve (PMV) assembly.
  - If pump motor runs: turn ignition switch OFF, disconnect TECH 1 and go to step [6].
  - If pump motor does not run: clear diagnostic trouble codes, turn ignition switch OFF, disconnect TECH 1 and go to step [1].

- Ignition switch: OFF
- ABS pump relay connector: disconnected
- Ignition switch: RUN

[1] Battery voltage. If not: check wiring for an open.

[2] Battery voltage. If not: check LH underhood fuse block fuse 1; check wiring (CKTS 864 and 642) for an open or short to ground. NOTE: When CKT 864 is found shorted to ground, replace ABS pump relay after wiring is repaired.

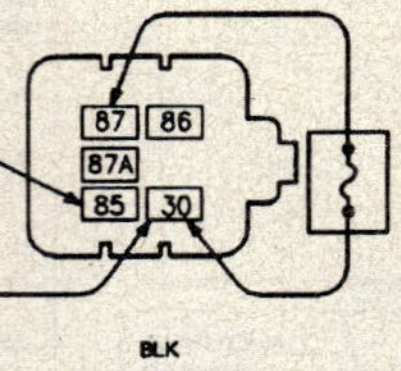

ABS PUMP RELAY
Connector

[3] Pump motor runs. If motor does not run: go to step [4]. If motor runs: reconnect ABS pump relay and go to Pump Motor Circuit Test J1.

- Ignition switch: OFF
- Pressure modulator valve (PMV) assembly connector C2: disconnected

[4] Battery voltage with jumper still in place from step [3]. If not: check wiring for an open or short to ground.

[5] Battery voltage. If not: check wiring for an open. If measurement is correct: check PMV connector C2 for proper terminal contact; If OK, replace pressure modulator valve (PMV) assembly.

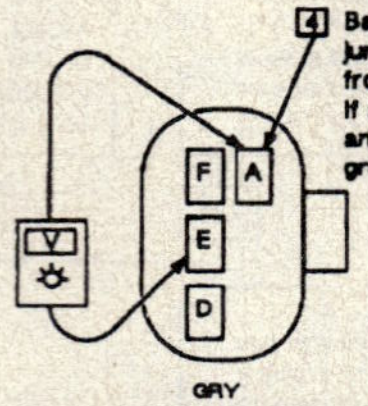

PRESSURE MODULATOR VALVE
(PMV) ASSEMBLY CONNECTOR C2
(Harness side)

[6] Infinite ohms between pressure modulator valve (PMV) assembly casing. If not: replace PMV assembly, see (8).

[7] 10 to 40 ohms. If not: replace pressure modulator valve (PMV) assembly. If measurement is correct: reconnect PMV assembly connector C2, and go to Pump Motor Circuit Test J1.

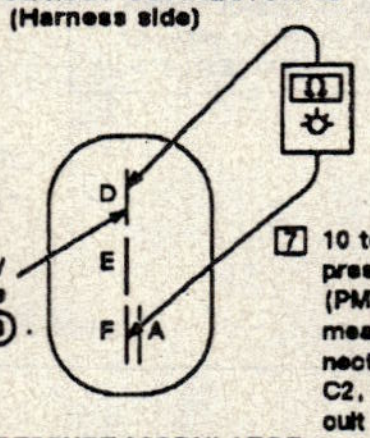

PRESSURE MODULATOR
VALVE
(PMV Assembly side)

DIAGNOSTIC CHART J1 — 1994 BUICK

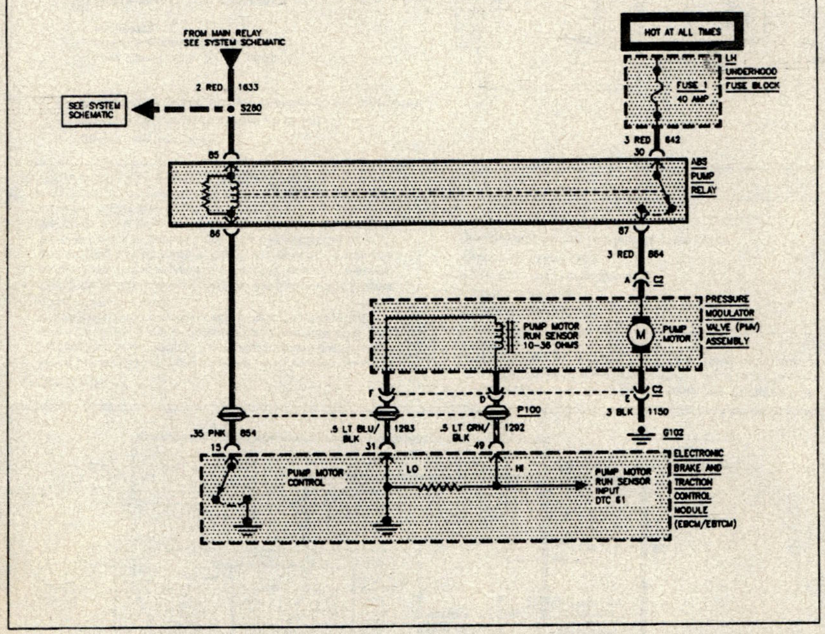

OPERATION:

- When EBCM/EBTCM grounds terminal 15, the ABS pump relay is energized. The pump relay contacts close, applying voltage to the pressure modulator valve (PMV) assembly's pump motor. The EBCM/EBTCM turns on the pump motor during most ABS braking conditions and, when equipped, during all traction control conditions.
- During most ABS braking conditions, the EBCM/EBTCM turns on the pump motor while monitoring the TCC/antilock brake switch to determine brake pedal position. The switch opens when the brake pedal is depressed approximately 40%. The motor restores the master cylinder's fluid pressure, causing the brake pedal to gradually rise. The pump motor continues to run until the brake pedal sufficiently rises to close the TCC/ntilock brake switch.
- During traction control mode (when equipped)), after the front hydraulic brake circuits are isolated, the pump motor runs to increase brake fluid pressure in these circuits. The increased pressure is directed to the slipping wheel through the PMV assembly's front inlet and outlet valves. This causes the brakes to be applied, reducing wheel spin at the slipping wheel. Traction control is disabled as soon as the brakes are manually applied.
- When the EBCM/EBTCM grounds terminal 15, DTC 61 sets when the EBCM/EBTCM determines that the voltage between EBCM/EBTCM terminals 31 and 49 is not between 500-800 mVAC. The EBCM/EBTCM recognizes this condition only during ABS braking requiring pump motor operation, traction control mode (when equipped), and during the EBCM/EBTCM self-check which occurs when the vehicle reaches 11 km/h (7 mph). Therefore, be sure to test drive vehicle; refer to "Test Drive" in this section, after clearing the diagnostic trouble code to assure correct diagnosis.
- DTC 61 also sets when the pump motor runs all the time (i.e., EBCM/EBTCM sees proper voltage between terminals 31and 49 without grounding terminal 15).

POSSIBLE CAUSES:

- Open in CKT 1150, 864, 854, 1633, 1292, 1293 or 642.
- Short to ground in CKT 864, 854, 1292, 1293 or 642.
- ABS pump relay fails "open" or "engaged."

DIAGNOSTIC CHART J1 — 1994 BUICK, CONTINUED

J1: PUMP MOTOR CIRCUIT TEST

- Ignition switch: OFF
- EBCM/EBTCM connector: disconnected
- Connect J 38716 Pinout Box
- Disconnect fused jumper.
- Ignition switch: OFF

3 Greater than 5 ohms. If not: check wiring CKTS 1292 and 1293 for a short to ground.

4 10 to 40 ohms. If high resistance is measured: check wiring (CKTS 1292 and 1293) for an open; check PMV assembly connector C2 for proper terminal contact. If low resistance is measured: check for a short between wiring.

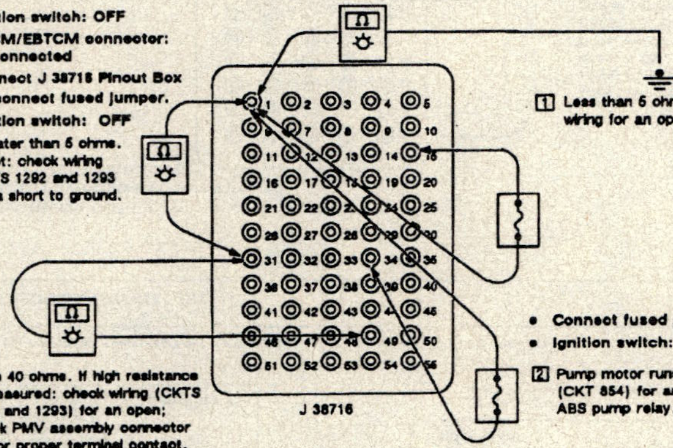

1 Less than 5 ohms. If not: check wiring for an open.

- Connect fused jumpers as shown
- Ignition switch: RUN

2 Pump motor runs. If not: check wiring (CKT 854) for an open; if OK, replace ABS pump relay.

- If all measurements are correct: check EBCM/EBTCM connector for proper terminal contact; If OK, replace EBCM/EBTCM.
- After repairs are completed, clear diagnostic trouble codes and repeat "ABS/TCS Functional Check."

DIAGNOSTIC CHARTS L AND M — 1994 BUICK

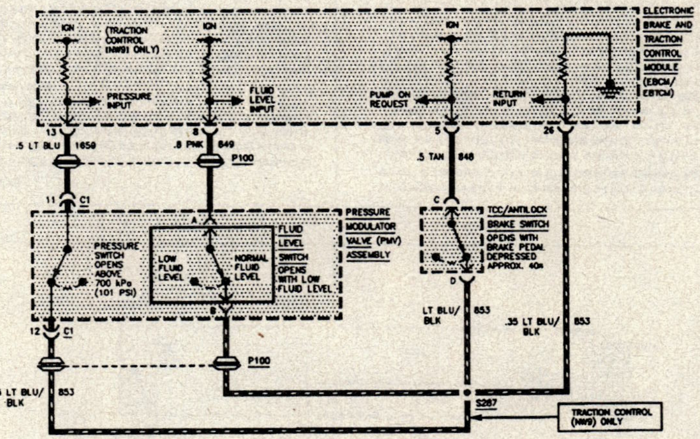

OPERATION:

- When EBCM/EBTCM monitors brake pedal position using the TCC/antilock brake switch. The normally closed switch opens when the brake pedal is depressed approximately 40%. The EBCM/EBTCM also monitors pump motor operation using a pump motor run sensor.
- During most ABS braking conditions the EBCM/EBTCM turns on the pump motor, replenishing the PMV assembly reservoir and applying fluid pressure to the master cylinder. This causes the brake pedal to gradually rise. The EBCM/EBTCM turns off the pump motor when brake pedal height rises sufficiently to close the TCC/antilock brake switch.
- During ABS braking, DTC 62 sets when the EBCM/EBTCM determines the pump is working, and brake pedal height has not risen sufficiently to close the TCC/antilock brake switch after 3 seconds.

POSSIBLE CAUSES:

- Air in the brake lines.
- Pump motor mechanical failure.
- Open in CKT 848 or 853.
- Misadjusted/defective TCC/antilock brake switch.

K: TCC/ANTILOCK BRAKE SWITCH TEST

- **Ignition switch: OFF**
- **Brake pedal: released (rest position)**
- **TCC/antilock brake switch connector: disconnected**

[1] Less than 1 ohm. If not: check switch adjustment; replace if necessary.

[2] Brake pedal: fully depressed

- Infinite ohms. If not: check switch adjustment; replace if necessary.

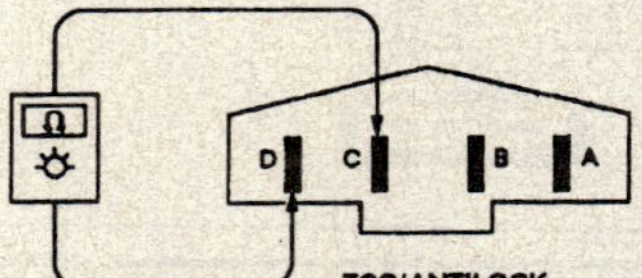

- If both measurements are correct and DTC 62 is set: reconnect TCC/antilock brake switch connector and go to Pump On Request Circuit Test M.
- After repairs are completed, clear diagnostic trouble codes and repeat "ABS/TCS Functional Check."

DIAGNOSTIC CHARTS L AND M — 1994 BUICK, CONTINUED

L: BRAKE HYDRAULIC TEST

- Make sure "ABS/TCS Functional Check" is performed before proceeding.

- Bleed brake system.
- Ignition switch: OFF
- Connect Tech 1 to data link connector (DLC).
- Ignition switch: RUN
- Use Tech 1 to clear diagnostic trouble code after bleeding system.
- Depress brake pedal (hold).
- Use Tech 1 to bleed pressure modulator valve (PMV) assembly.

Did brake pedal rise?

NO → Replace PMV.

YES →

- Ignition switch: OFF
- Disconnect Tech 1.
- Test drive vehicle in a safe wet area and attempt to perform a 3-second ABS stop.
- Connect Tech 1 to DLC.
- Ignition switch: RUN

Did DTC 62 set?

NO → Problem solved.

YES →

- Ignition switch: OFF
- Disconnect Tech 1.
- Go to TCC/Antilock Brake Switch Test K.

- After repairs are completed, clear diagnostic trouble codes and repeat "ABS/TCS Functional Check."

M: PUMP ON REQUEST CIRCUIT TEST

- **Ignition switch: OFF**
- **EBCM/EBTCM connector: disconnected**
- **Connect J 38716 Pinout Box**

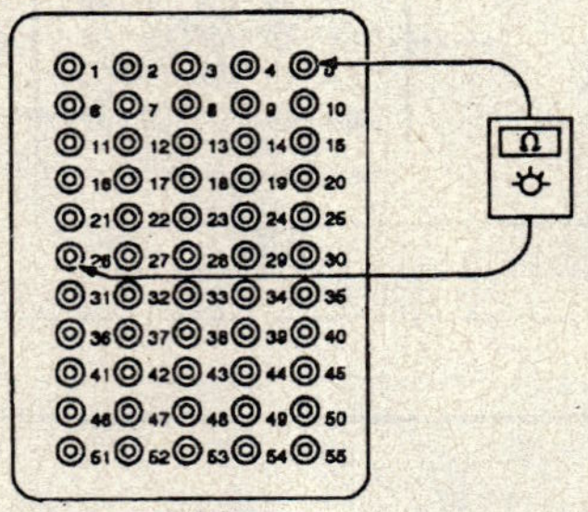

Less than 3 ohms. If not, check wiring (CKTS 848 and 853) for an open. If OK, check TCC/antilock brake switch connector and EBCM/EBTCM connector for proper terminal contact. If OK, replace EBCM/EBTCM.

- After repairs are completed, clear diagnostic trouble codes and repeat "ABS/TCS Functional Check."

DIAGNOSTIC CHARTS N AND O — 1994 BUICK

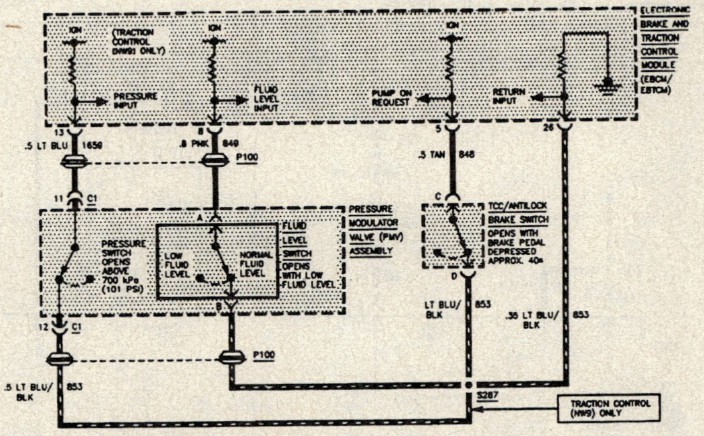

OPERATION:

- The EBCM/EBTCM monitors brake pedal position using the TCC/antilock brake switch. The normally closed switch opens when the brake pedal is depressed approximately 40%. The EBCM/EBTCM also monitors pump motor operation using a pump motor run sensor.
- During most ABS braking conditions, the EBCM/EBTCM turns on the pump motor, replenishing the PMV assembly reservoir and applying fluid pressure to the master cylinder. This causes the brake pedal to gradually rise. The EBCM/EBTCM turns off the pump motor when brake pedal height rises sufficiently to close the TCC/antilock brake switch.
- The pressure modulator valve (PMV) fluid level switch alerts the EBCM/EBTCM to a low brake fluid condition in the PMV reservoir. The fluid level switch is located in the PMV assembly's reservoir. The fluid level switch is normally closed when the reservoir contains sufficient fluid. When fluid level is low, the switch opens, causing the EBCM/EBTCM to turn on the ANTILOCK indicator.
- The pressure switch monitors primary brake system pressure. The normally closed switch opens when the brakes are applied (pressure exceeds 700 kPa [101 psi]). If the brakes are manually applied during traction control mode, the PMV pressure switch input, along with the brake switch input, signal EBCM/EBTCM to disable traction control and allow manual braking. The pressure switch is part of the pressure modulator valve (PMV) assembly and is not serviceable.
- DTC 72 sets when the EBCM/EBTCM detects that the TCC/antilock brake switch CKT 848 is shorted to ground.
- DTC 73 sets when the EBCM detects a short to ground in the PMV fluid level switch circuitry. These include CKTS 848, 849, 853 and 1659.
- DTC 74 sets when the EBCM/EBTCM detects an open or short to ground in the PMV pressure switch circuitry.

POSSIBLE CAUSES:

- Short to ground in CKT 848 with TCC/antilock brake switch open.
- Short to ground in CKT 848, 849, 853 or 1659.
- Open in CKT 853 or 1659.
- Short to ground in CKT 1659 with brake pedal depressed.
- Failed PMV pressure switch.

DIAGNOSTIC CHARTS N AND O — 1994 BUICK, CONTINUED

N: EBCM/EBTCM GROUNDED SWITCH INPUT TEST

- Make sure "ABS/TCS Functional Check" is performed before proceeding.
- Ignition switch: OFF
- EBCM/EBTCM connector: disconnected
- Connect J 38716 Pinout Box
- TCC/antilock brake switch connector: disconnected
- Pressure modulator valve (PMV) fluid level switch: disconnected
- PMV assembly connector, C1: disconnected (traction control [NW9] only)

[1] Less than 5 ohms between data link connector DLC terminal 4. If not: check wiring for an open.

[2] Greater than 5 ohms. If not: check wiring (see schematic) for a short to ground.

[3] Greater than 5 ohms. If not: check wiring for a short to ground.

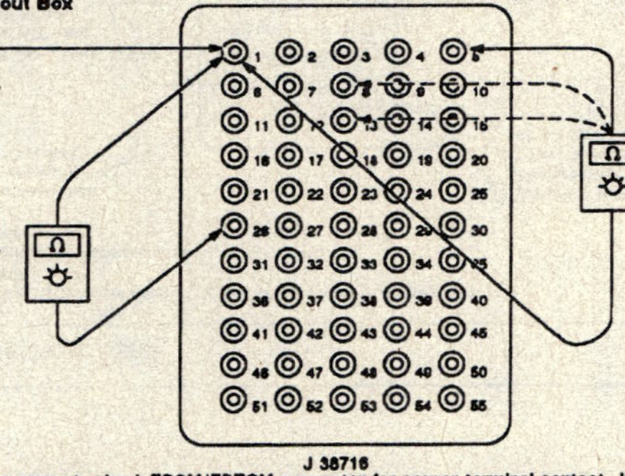

- If all measurements are correct: check EBCM/EBTCM connector for proper terminal contact. If OK: replace EBCM/EBTCM.
- After repairs are completed, clear diagnostic trouble codes and repeat "ABS/TCS Functional Check."

O: EBCM/EBTCM OPEN SWITCH INPUT TEST

- Make sure "ABS/TCS Functional Check" is performed before proceeding.
- Ignition switch: OFF
- EBCM/EBTCM connector: disconnected
- Connect J 38716 Pinout Box
- Brake pedal: fully depressed

[2] Brake pedal: fully depressed. Infinite ohms. If not: check for a short between wiring (CKTS 853 and 1659). If OK, replace PMV assembly, see (8).

[1] Brake pedal: released. Less than 3 ohms. If not: check wiring (CKTS 853 and 1659) and PMV pressure switch for an open; check PMV assembly connector C1 for proper terminal contact; check PMV assembly pigtail wiring for a short to PMV casing; if switch or PMV assembly pigtail wiring is defective, replace PMV assembly.

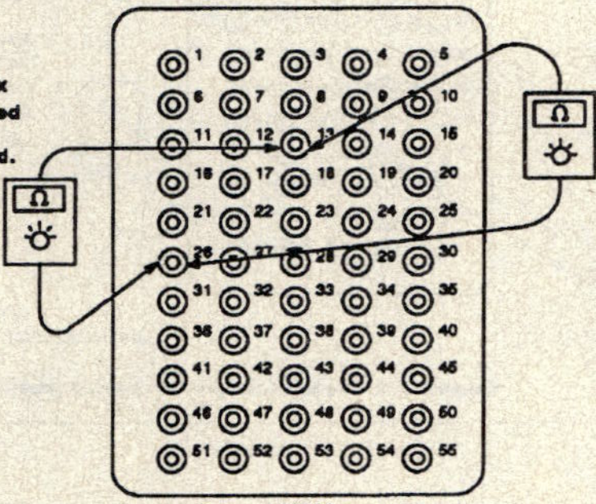

- If all measurements are correct, check EBCM/EBTCM connector for proper terminal contact. If OK: replace EBCM/EBTCM.
- After repairs are completed, clear diagnostic trouble codes and repeat "ABS/TCS Functional Check."

DIAGNOSTIC CHARTS P AND Q — 1994 BUICK

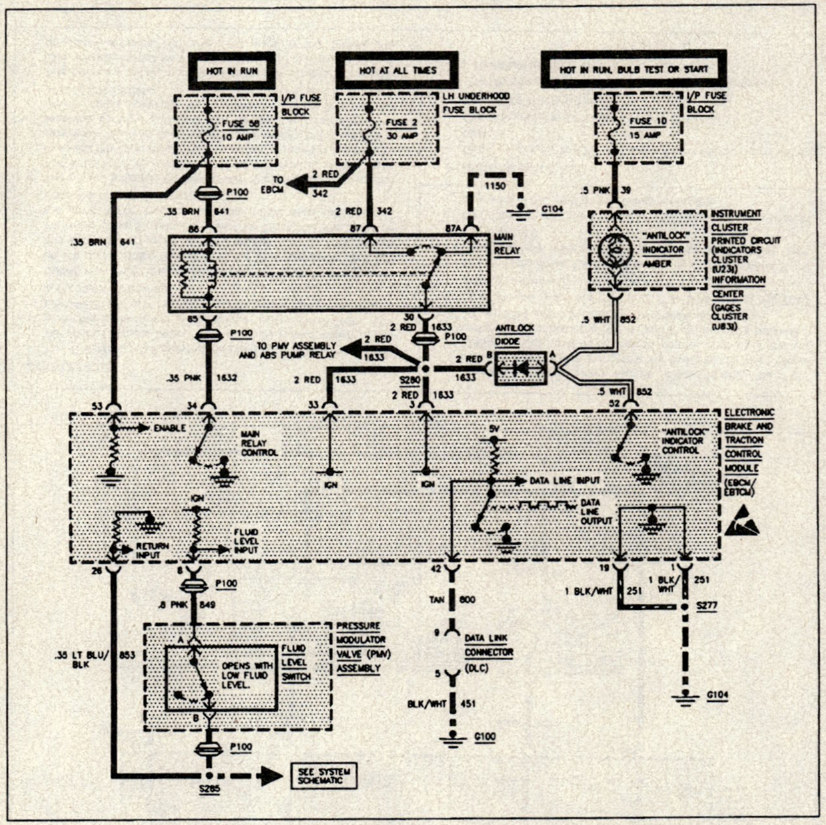

OPERATION:

- When the ignition switch is first turned to RUN, the EBCM/EBTCM goes through a self-test lasting 2-4 seconds. During this test, the EBCM/EBTCM grounds terminal 52, causing the amber ANTILOCK indicator to light. This serves as an indicator bulb test.
- When a low fluid condition exists, the EBCM/EBTCM turns on the amber ANTILOCK indicator but does not set a code. A low fluid level condition usually lights the red BRAKE indicator since the master cylinder supplies the PMV reservoir. However, restrictions or air in the hose could cause a low fluid condition in the PMV reservoir without lighting the red BRAKE indicator.

POSSIBLE CAUSES:

- Open in CKT 641, 800, 251, 849 or 853.
- Low brake fluid level or open PMV fluid level switch.
- Short to ground in CKT 641, 800 or 852.

DIAGNOSTIC CHARTS P AND Q — 1994 BUICK, CONTINUED

P: DATA LINE TEST

- Make sure "ABS/TCS Functional Check" is performed before proceeding.
- Ignition switch: OFF
- EBCM/EBTCM connector: disconnected
- Connect J 38716 Pinout Box

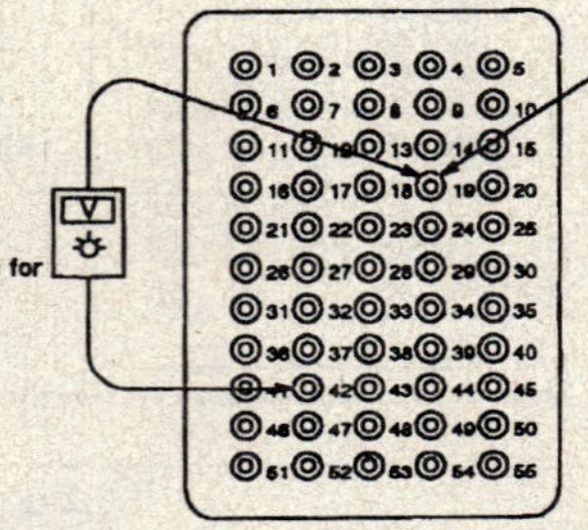

1. Less than 2 ohms between data link connector DLC terminal 4. If not: repair open in wiring (CKT 251).

- Ignition switch: RUN

2. 5 volts. If not: check wiring (CKT 800) and components for an open or short to ground. See Data Line for complete circuit details.

- If both measurements are correct, check EBCM/EBTCM connector for proper terminal contact. If OK: replace EBCM/EBTCM.

Q: ANTILOCK INDICATOR "ON" FAULT TEST

- Make sure "ABS/TCS Functional Check" is performed before proceeding.
- Ignition switch: OFF
- EBCM/EBTCM connector: disconnected
- Connect J 38716 Pinout Box

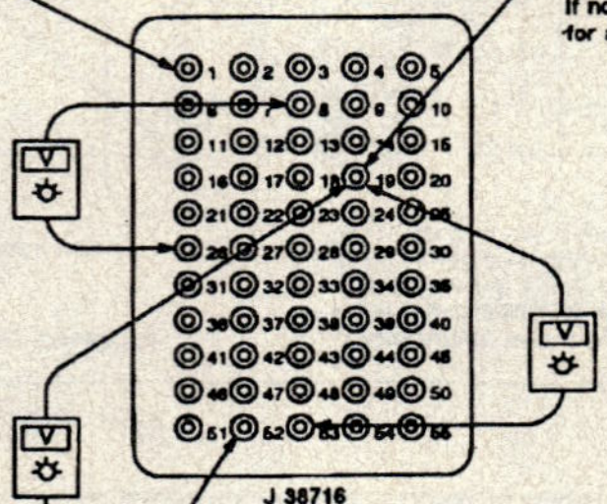

1. Less than 5 ohms between data link connector DLC terminal 4. If not: check wiring (CKT 251) for an open.
2. Less than 5 ohms between data link connector DLC terminal 4. If not: check wiring (CKT 503) for an open.
3. Less than 3 ohms. If not: check wiring (CKTS 849 and 853) and PMV fluid level switch for an open. If switch is open, check PMV fluid level switch connector for proper terminal contact. If OK, replace PMV assembly reservoir.

- Ignition switch: OFF
- Main relay: disconnected
- Ignition switch: RUN

4. Battery voltage. If not: check wiring(CKT 641) for an open or short to ground. When necessary, replace I/P fuse block fuse 58.

5. Battery voltage. If not: check wiring (CKT 852) for a short to ground.

- If all measurements are correct: check EBCM/EBTCM connector for proper terminal contact. If OK: replace EBCM/EBTCM.
- After repairs are completed, clear diagnostic trouble codes and repeat "ABS/TCS Functional Check."

DIAGNOSTIC CHARTS R AND S — 1994 BUICK

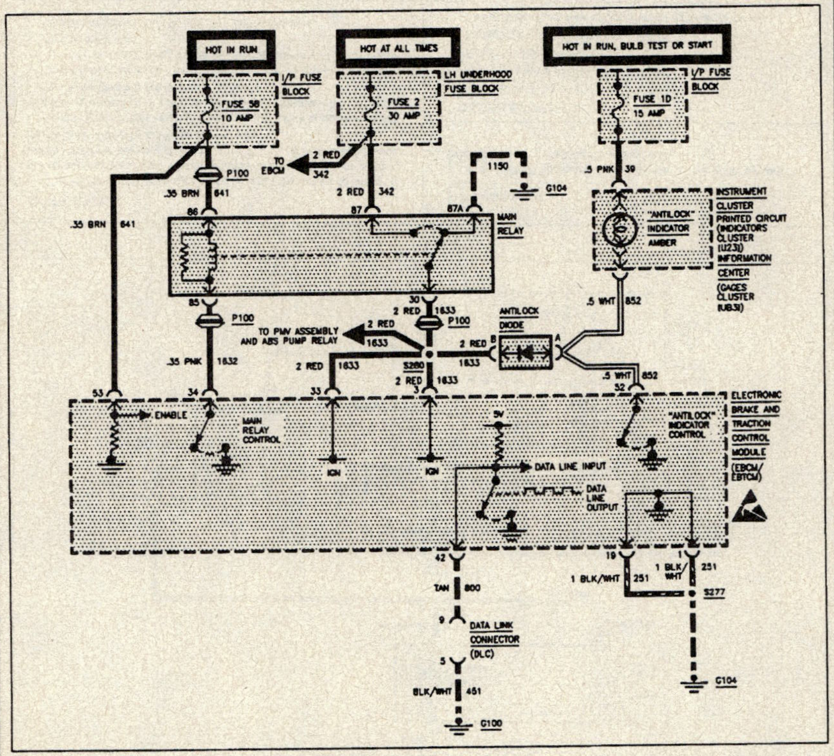

OPERATION:

- When the ignition switch is first turned to RUN, the EBCM/EBTCM goes through a self-test lasting 2-4 seconds. During this test, the EBCM/EBTCM grounds terminal 52, causing the amber ANTILOCK indicator to light. This serves as an indicator bulb test. The EBCM/EBTCM also grounds terminal 34, energizing the main relay. This removes the indicator ground path provided through the antilock diode and the main relay contacts.
- When the self-test is completed (and no faults are detected), the EBCM/EBTCM removes the ground at terminal 52, turning off the indicator. The main relay remains energized.
- When the EBCM/EBTCM detects a fault, it grounds terminal 52, causing the amber ANTILOCK indicator to light. (Note: The EBCM/EBTCM also de-energizes the main relay, creating a redundant ground path.)
- When the EBCM/EBTCM is disabled for any reason (i.e., open fuse 19), the main relay is de-energized. This provides an ANTILOCK indicator ground path through the antilock diode and main relay contacts, and causes the indicator to light with no codes being set.

POSSIBLE CAUSES:

- Open in CKT 251, 852 or 1633.
- Open bulb.
- Short to voltage in CKT 852.
- Open or shorted antilock diode.
- Main relay fails "engaged."
- Short to ground in CKT 1632.

DIAGNOSTIC CHARTS R AND S — 1994 BUICK, CONTINUED

R: ANTILOCK INDICATOR "OFF" FAULT TEST

- Make sure "ABS/TCS Functional Check" is performed before proceeding.
- Ignition switch: OFF
- Connect TECH 1 to data link connector DLC.
- Ignition switch: RUN
- Enter TECH 1 Diagnostic Mode.
- If Diagnostic Mode cannot be entered: disconnect TECH 1, do Data Line Test P, then do Main Relay Ground Test S.
- If Diagnostic Mode can be entered: disconnect TECH 1 and continue to step 1.

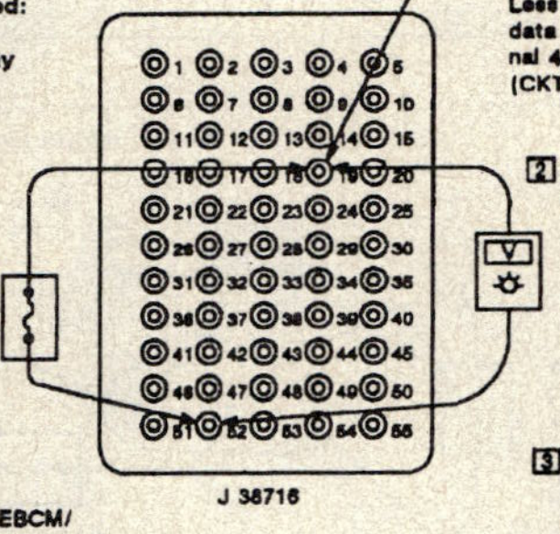

1 • Ignition switch: OFF
• EBCM/EBTCM connector: disconnected
• Connect J 38716 Pinout Box.

Less than 5 ohms between data link connector DLC terminal 4. If not: check wiring (CKT 251) for an open.

2 • Main relay: disconnected
• Ignition switch: RUN
Battery voltage. If not: check ANTILOCK indicator bulb and wiring (CKT 852) for an open. If both OK: replace information center (UB3 only) or instrument cluster (U23 only).

3 • Ignition switch: OFF
Disconnect antilock diode and inspect for damage; replace as necessary, observe diode polarity.

4 • Ignition switch: RUN
ANTILOCK indicator remains lit. If not: check wiring (CKT 852) for a short to voltage. NOTE: When a short to voltage is found, inspect antilock diode and main relay for possible damage.

- If all measurements are correct: check EBCM/EBTCM connector for proper terminal contact; if OK, replace EBCM/EBTCM.
- After repairs are completed, clear diagnostic trouble codes and repeat "ABS/TCS Functional Check."

S: MAIN RELAY GROUND TEST

- Ignition switch: OFF
- Main relay: disconnected
- I/P fuse block fuse 5B removed
- Ignition switch: RUN

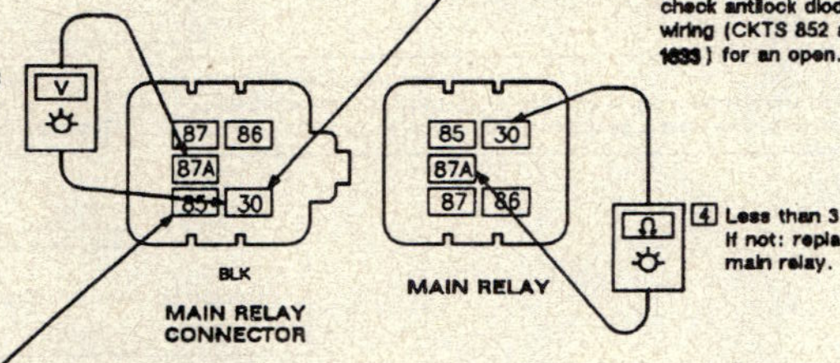

1 Battery voltage. If not: check antilock diode and wiring (CKTS 852 and 1633) for an open.

2 Battery voltage. If not: check wiring (CKT 251) for an open.

3 • Ignition switch: OFF
Greater than 5 ohms. If not: check wiring (CKT 1632) for a short to ground. If OK: check EBCM/EBTCM connector for proper terminal contact. If OK: replace EBCM/EBTCM.

4 Less than 3 ohms. If not: replace main relay.

- After repairs are completed, clear diagnostic trouble codes and repeat "ABS/TCS Functional Check."

DIAGNOSTIC CHARTS T, U AND V — 1994 BUICK

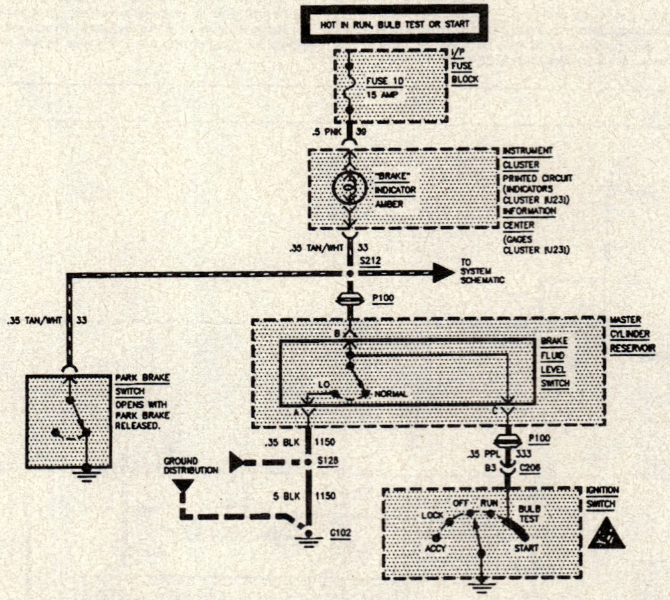

OPERATION:

The BRAKE indicator illuminates to alert the driver to either a brake system condition, which reduces braking ability, or a park brake, which is not fully released. Specific conditions which cause the BRAKE indicator to illuminate include:

- Park brake not fully released. When park brake is applied or not fully released, the park brake switch grounds the BRAKE indicator, causing the indicator to light.
- Low brake fluid level. The brake fluid level switch in the master cylinder reservoir grounds the BRAKE indicator when low brake fluid is detected.
- Ignition switch in START or BULT TEST position.

Important

A lit BRAKE indicator may indicate a hydraulic system fault, possibly causing reduced braking ability.

POSSIBLE CAUSES:

- Open in CKT 33, 1150, or 353.
- Short to ground in CKT 33 or 353.
- Failed ignition switch.
- Failed/misadjusted park brake switch.
- Failed brake fluid level switch/master cylinder reservoir.

DIAGNOSTIC CHARTS T, U AND V — 1994 BUICK, CONTINUED

T: BRAKE WARNING SHORT CIRCUIT TEST

- Ignition switch: RUN
- Master cylinder brake fluid level switch connector: disconnected

Does BRAKE indicator remain lit?

YES
- Park brake switch connector: disconnected

Does BRAKE indicator remain lit?

YES → Check wiring (CKT 33) for a short to ground; if OK, return to "Symptom/Diagnosis Guide."

NO → Adjust/replace park brake switch.

NO → Check wiring (CKT 353) and ignition switch for a short to ground (refer to Buick Service Manual SECTION 3F if ignition switch replacement is needed); if OK, replace master cylinder reservoir.

U: BRAKE FLUID LEVEL SWITCH TEST

- Ignition switch: OFF
- Park brake: released
- Master cylinder brake fluid level switch connector: disconnected
- Ignition switch: RUN

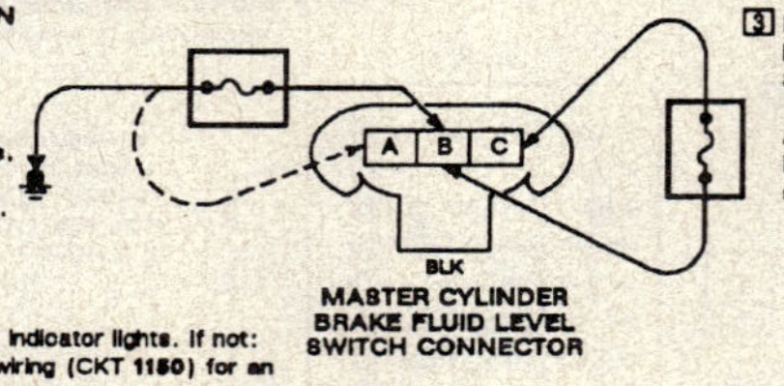

1. BRAKE indicator lights. If not: check wiring (CKT 33) for an open.
2. BRAKE indicator lights. If not: check wiring (CKT 1150) for an open.
3. BRAKE indicator lights with ignition switch at BULB TEST or START only. If not: check wiring (CKT 353) and ignition switch for an open. Refer to Buick Service Manual Section 3F if ignition switch replacement is needed.

- If both results are correct: replace master cylinder reservoir.

V: BRAKE PARK SWITCH TEST

- Ignition switch: OFF
- Park brake connector: disconnected
- Ignition switch: RUN

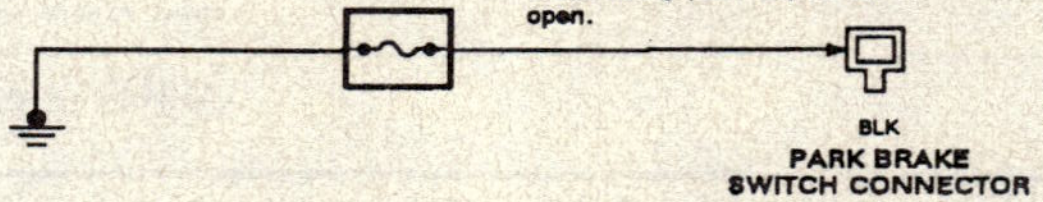

1. BRAKE indicator lights. If not: check wiring (CKT 33) for an open.

- If result is correct: adjust/replace park brake switch.

DIAGNOSTIC CHARTS W, S, X AND Y — 1994 BUICK

W: EBCM/EBTCM PROBLEM TEST

- Make sure "ABS/TCS Functional Check" is performed before proceeding.

- Clear diagnostic trouble codes.
- Test drive vehicle. Refer to "Test Drive" procedure.

Did DTC 71 set during test drive?

YES → Replace EBCM/EBTCM.

NO → No problem found. Problem may be intermittent, see "Notes on Intermittents."

- After repairs are completed, clear diagnostic trouble codes and repeat "ABS/TCS Functional Check."

S, X, AND Y:

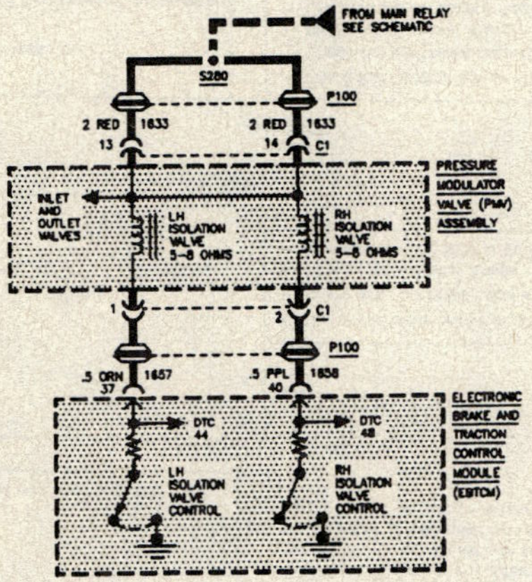

OPERATION:

- Voltage is applied to the normally open isolation valves whenever the main relay is energized. An isolation valve closes when the EBTCM supplies a ground path at the isolation valve control terminal. The EBTCM closes an isolation valve when the EBTCM detects wheel slip in a drive wheel.
- The isolation valves are closed only during traction control operations, and are not used during any antilock or normal braking conditions. The isolation valves separate (isolate) the front brake hydraulic circuits from the master cylinder and rear brake hydraulic circuits. Once the front brake hydraulic circuits are isolated, pressure can be applied to the front wheels without affecting any other brake hydraulic circuit.
- DTC 44 or 48 sets when the EBTCM detects an open or short to ground in the respective isolation valve circuit.

POSSIBLE CAUSES:

- Open in CKT 1633, 1657, or 1658.
- Short to ground in CKT 1657 or 1658.
- Faulty pressure modulator valve (PMV) assembly.

DIAGNOSTIC CHARTS S, X AND Y — 1994 BUICK, CONTINUED

X: ISOLATION VALVE CONTINUITY TEST

- Make sure "ABS/TCS Functional Check" is performed before proceeding.
- Ignition switch: OFF
- Pressure modulator valve (PMV) assembly connector C1: disconnected

[1] Infinite ohms between pressure modulator valve (PMV) assembly casing. If not: replace PMV assembly.

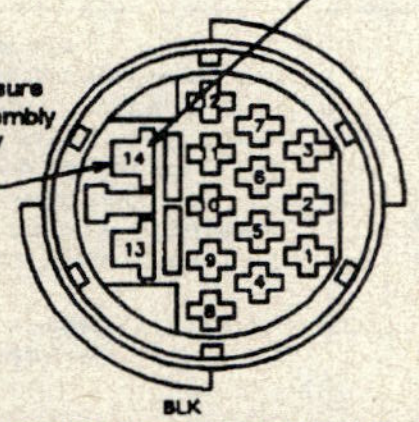

PRESSURE MODULATOR VALVE (PMV) ASSEMBLY CONNECTOR C1
(PMV Assembly half)

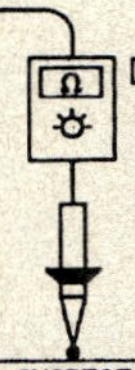

[2] Measure resistance across suspect solenoid valve terminal(s). When a measurement is incorrect: replace PMV assembly.

VALVE	SUSPECT TERMINAL	CORRECT RESISTANCE
LH Isolation	1	5 to 8 ohms
RH Isolation	2	

- If all measurements are correct, go to Isolation Valve Control Test Z.
- After repairs are completed, clear diagnostic trouble codes and repeat "ABS/TCS Functional Check."

Y: ISOLATION VALVE CONTROL TEST

- Ignition switch: OFF
- Pressure modulator valve (PMV) assembly connector, C1: connected
- EBTCM connector: disconnected
- Connect J 38716 Pinout Box
- Main relay: disconnected

[1] Less than 10 ohms between terminal 3 and suspect valve control terminal. If not: check suspect wiring (CKT 1633, 1657 or 1658) for an open; check PMV assembly connector C1 for proper terminal contact.

ground. If not:

[2] Greater than 5 ohms between suspect valve control terminal and ground. If not: check wiring (CKT 1657 or 1658) for a short to ground.

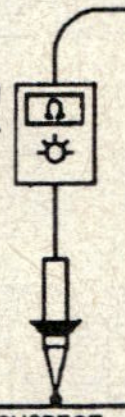

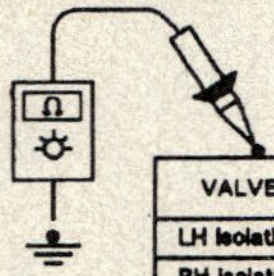

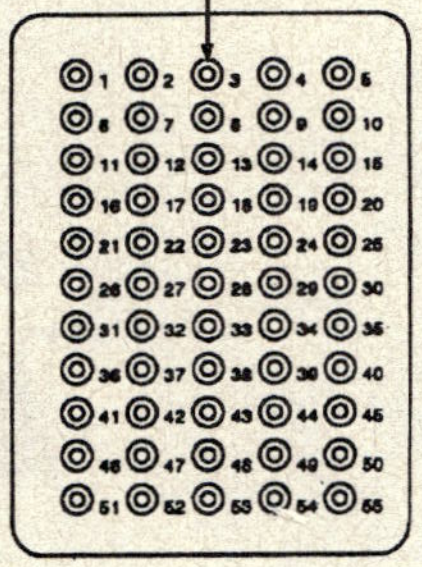

VALVE	SUSPECT TERMINAL
LH Isolation	37
RH Isolation	40

- If all measurements are correct: check EBTCM connector for proper terminal contact. If OK, replace EBTCM.
- After repairs are completed, clear diagnostic trouble codes and repeat "ABS/TCS Functional Check."

DIAGNOSTIC CHART Z — 1994 BUICK

Z: INLET/OUTLET VALVE HYDRAULIC TEST

Important: To prevent overheating of inlet/outlet valves, the Tech 1 only allows valves to be energized for a maximum of 20 seconds. A minimum of 30 seconds must elapse before valves can be re-energized.

- Raise vehicle
- Gear selected lever: NEUTRAL
- Park brake: released

Can wheel associated with suspect valve be turned by hand?

YES:

- Depress brake pedal (hold).

Can suspect wheel be turned by hand?

- **NO:**
  - Ignition switch: OFF
  - Connect a Tech 1 to the data link connector DLC.
  - Ignition switch: RUN
  - Select ABS MANUAL SOLENOID TEST and suspect outlet valve to be tested.
  - Depress brake pedal (hold).
  - Activate suspect outlet valve.

  Can suspect wheel be turned by hand?

  - **YES:**
    - Select ABS MANUAL SOLENOID TEST and suspect inlet valve to be tested.
    - Activate suspect inlet valve.
    - Depress brake pedal (hold).

    Can suspect wheel be turned by hand?

    - **YES:** No trouble found. Problem may be intermittent.
    - **NO:**
      - Replace pressure modulator valve (PMV) assembly.
  - **NO:**
    - Replace pressure modulator valve (PMV) assembly.
- **YES:**
  - Check that brake caliper or wheel cylinder is not frozen.
  - Check that park brake cable is released.
  - Check that brake shoes, springs and associated hardware are not stuck or frozen.

NO:

- Check that brake caliper or wheel cylinder is not frozen.
- Check that park brake cable is released.
- Check that brake shoes, springs and associated hardware are not stuck or frozen.

- After repairs are completed, clear diagnostic trouble codes and repeat "ABS/TCS Functional Check."

DIAGNOSTIC CHART AA — 1994 BUICK

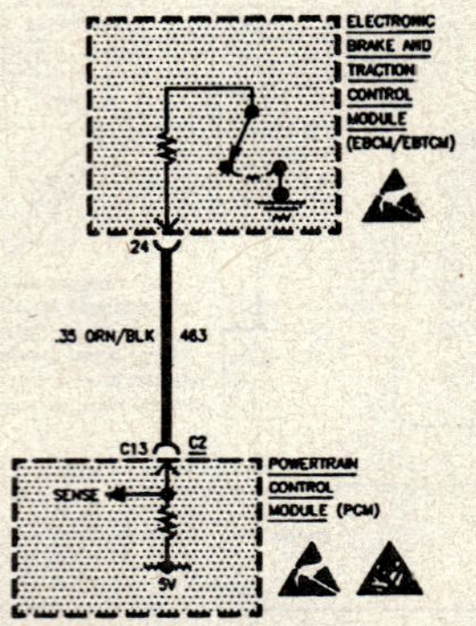

PCM CONNECTOR IDENTIFICATION

C1	RED	32 WAY
C2	WHITE	32 WAY
C3	BLUE	32 WAY

OPERATION:

Important

If PCM DTC P1571 is not present when ABS/TCS DTC 75 is set, no fault exists in the ABS/TCS system. Refer to SECTION 6E3 for diagnosis.

- When excessive wheel spin is detected by the EBTCM, it will request the PCM to reduce engine torque via a high speed Pulse Width Modulated (PWM) signal on Circuit 463.
  - The PCM will disable TCS when the following PCM codes are set: P0101, P0341, P0342, P1350, P1361, P1623, P1630, P0131, P0132, P0171, P0172. The EBTCM upon receiving the TCS disable message from the PCM will then illuminate the "Traction Off" indicator lamp.

- DTC 75 sets when the EBTCM receives a serial data message from the PCM to disable traction control.
  - EBTCM to PCM communication problem due to an open or short in circuits 800 or 463.

POSSIBLE CAUSES:

- PCM Fault Present.
- Open or short in CKT 800.
- Open in CKT 463.
- Short in CKT 463.
- Faulty PCM or EBTCM.

DIAGNOSTIC CHART AA — 1994 BUICK, CONTINUED

AA: PCM REQUESTED TCS TO BE DISABLED

- Make sure ABS/TCS "Functional Check" is performed before proceeding.
- Ignition switch: ON
- Connect TECH 1 to DLC and check for PCM DTC P1871. If PCM DTC P1871 is present.
- Ignition switch: OFF
- Disconnect EBTCM.
- Connect J 39710 pinout box and continue to step [1].

[1] Less than 5 ohms. If not, repair open in CKT 281.

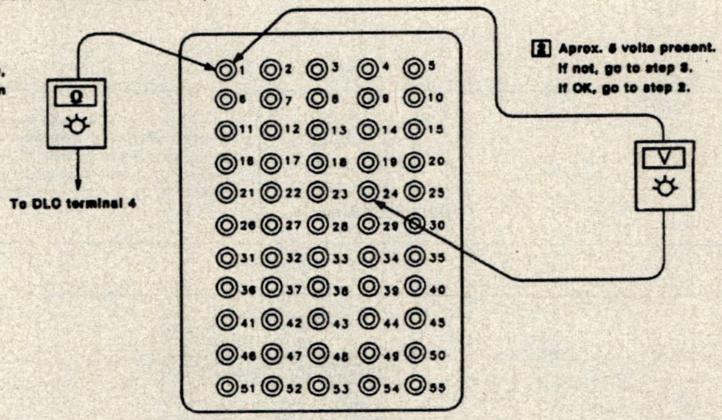

[2] Aprox. 5 volts present. If not, go to step 3. If OK, go to step 2.

[2] • Ignition: OFF
• Reconnect EBTCM
• Ignition: ON
• If 126 – 150 Hz is measured check for intermittant short to ground in CKT 463. If CKT 463 is OK, check for poor terminal contact at PCM terminal C13. If all are OK, replace PCM. If 126 – 150 Hz is not measured check for poor terminal contact at EBTCM terminal 24 and at PCM terminal C13. If all are OK, replace EBTCM.

[3] • 5 volts present, open in CKT 463. If not check CKT 463 for short to ground. If CKT 463 is ok, replace PCM.

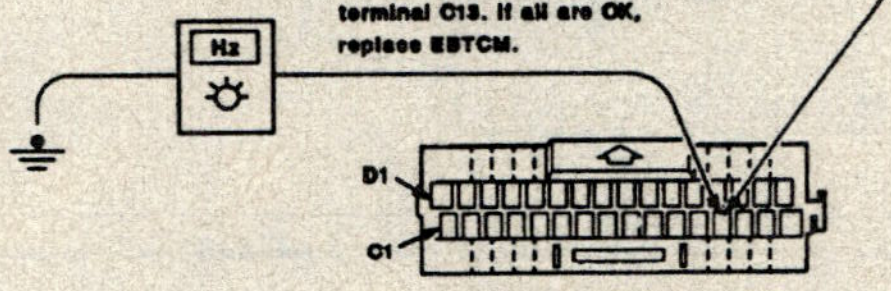

PCM CONNECTOR – WHT
BACK VIEW

DIAGNOSTIC CHART BB — 1994 BUICK

BB: DATA LINES SIGNAL OR CIRCUIT MALFUNCTION

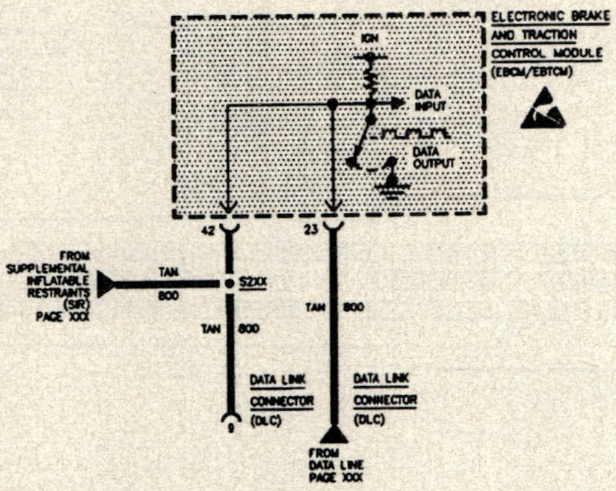

OPERATION:

- The EBTCM sends and receives specific serial data information on the Data Line. These serial data messages are provided and shared by several on-vehicle control modules. The Data Line also allows the use of a bi-directional scan tool such as the Tech 1.

- A current DTC 76 cannot be read with a Tech 1 due to the loss of Data Line Communications. After Data Line communications are restored, DTC 76 will be stored in memory. Refer to CHART P for Data Line Diagnosis.

DIAGNOSTIC CHART CC — 1994 BUICK

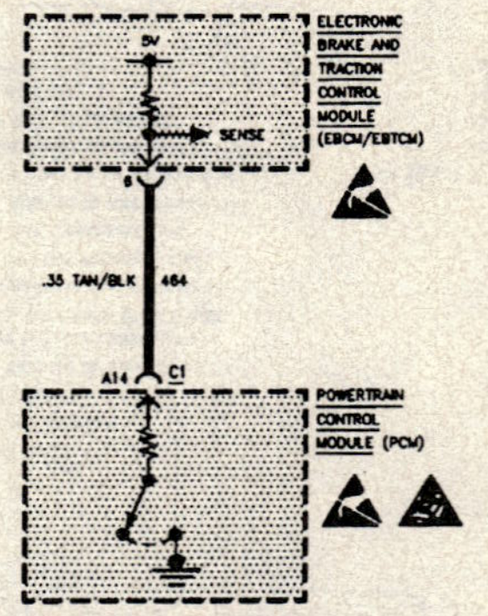

PCM CONNECTOR IDENTIFICATION	
C1 – RED	– 32 WAY
C2 – WHITE	– 32 WAY
C3 – BLUE	– 32 WAY

OPERATION:

- The PCM provides the EBTCM information on engine torque output via a high speed Pulse Width Modulated (PWM) signal on Circuit 464. This information is provided so the EBTCM can determine if engine torque management is required during a wheel slip condition/traction control condition.

POSSIBLE CAUSES:

- Open in CKT 464.
- Short in CKT 464.
- Faulty PCM.
- Faulty EBTCM.

DIAGNOSTIC CHART CC — 1994 BUICK, CONTINUED

CC: DELIVERED TORQUE PWM CIRCUIT OPEN OR SHORTED

- Make sure ABS/TCS "Functional Check" is performed before proceeding.

- Ignition switch: OFF
- Disconnect red PCM connector
- Ignition switch: RUN

[1] • If aprox. 5 volts present at terminal A14, go to step [3]. If not go to step [2].

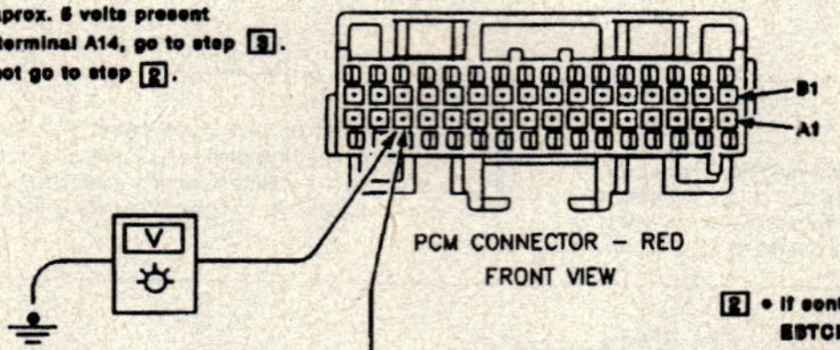

[2] • If continuity is present, check EBTCM terminal C8 for proper contact. If OK, replace EBTCM. If continuity does not exist, check CKT 464 for an open.

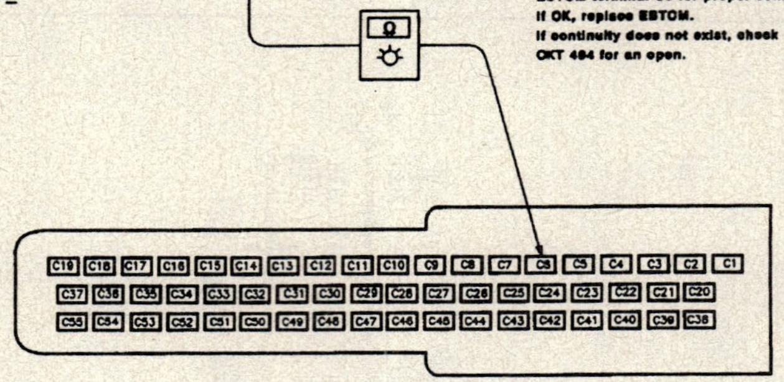

[3] • Aprox. 4.6 volts, if not, check for poor terminal contact at PCM terminal A14. If OK, replace PCM. If 4.6 volts not present, check for intermittent short to ground in CKT 464 also for poor terminal contact at EBTCM terminal C8. If all are OK, replace EBTCM.

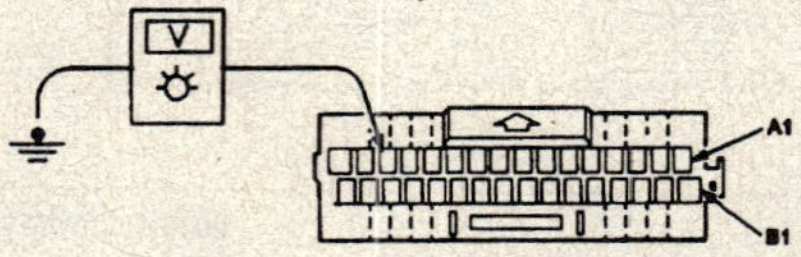

- After repairs are complete, perform ABS/TCS functional check.

DIAGNOSTIC CHART DD AND EE — 1994 BUICK

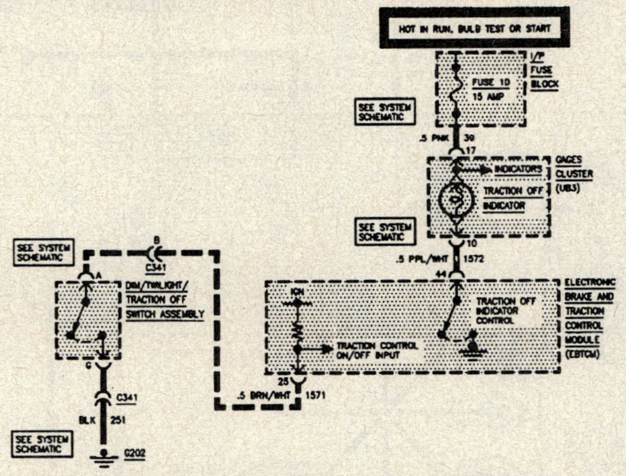

OPERATION:

- When the ignition is first turned to "RUN," the EBTCM goes through a self-test. During this test, the EBTCM grounds both the "ANTILOCK" and "TRACTION OFF" indicators for approximately 2 to 4 seconds. This acts as a bulb test for the indicators.
- A Traction Off switch which is part of the Twilight Sentinel/Panel Dim Control Assembly allows the driver to select the traction control mode (on/off). When the traction control system is turned off, the "TRACTION OFF" indicator remains illuminated.
- The "TRACTION OFF" indicator will not function if an open exists in CKT 1572 or a faulty EBTCM.
- The "TRACTION OFF" indicator will stay illuminated with no DTC(s) set if a short to ground is present in CKTs 1571, 1572 or a faulty Dim/Twilight/Traction Off Switch Assembly or a faulty EBTCM.

Important

- When the vehicle is equipped with traction control, any symptom which lights the "ANTILOCK" indicator also lights the "TRACTION OFF" indicator. Therefore, when both indicators are lit, follow "ANTILOCK" indicator diagnosis when no diagnostic trouble codes are set.

DIAGNOSTIC CHART DD AND EE — 1994 BUICK, CONTINUED

DD: "TRACTION OFF" INDICATOR TEST

- Make sure ABS/TCS "Functional Check" is performed before proceeding.
- Ignition switch: OFF
- Disconnect EBTCM
- Connect J 38710 Pinout Box
- Ignition switch: ON

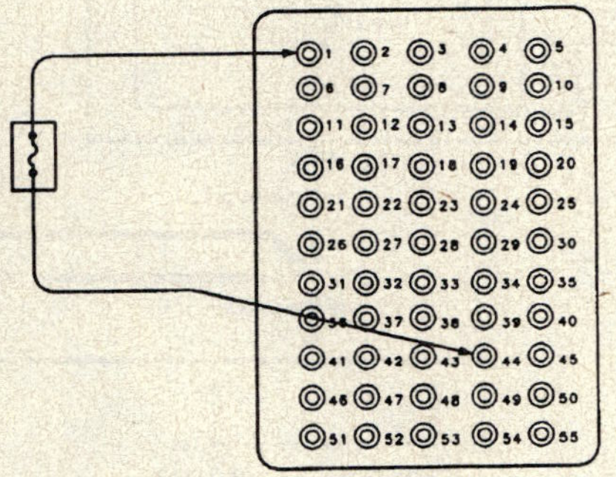

1 • "TRACTION OFF" INDICATOR lights. If not, check for open CKT 1572 also Proper terminal contact at EBTCM terminal 44. Also Gage Cluster terminal 10. If all are OK, replace EBTCM.

- After repair are complete, perform ABS/TCS functional check.

EE: "TRACTION OFF" INDICATOR TEST

- Make sure ABS/TCS "Functional Check" is performed before proceeding.
- Ignition switch: OFF
- Disconnect EBTCM.
- Connect J 38710 Pinout Box

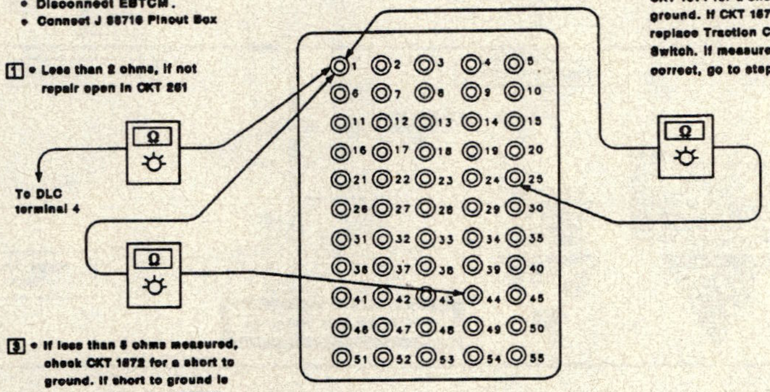

1 • Less than 2 ohms, if not repair open in CKT 251

2 • If less than 5 ohms, check CKT 1571 for a short to ground. If CKT 1571 is OK, replace Traction Control Switch. If measurement is correct, go to step 3.

3 • If less than 5 ohms measured, check CKT 1572 for a short to ground. If short to ground is not present, replace ETCBM.

- After repair are complete, perform ABS/TCS functional check.

ELECTRICAL COMPONENT LAYOUT — 1994 BUICK, CONTINUED

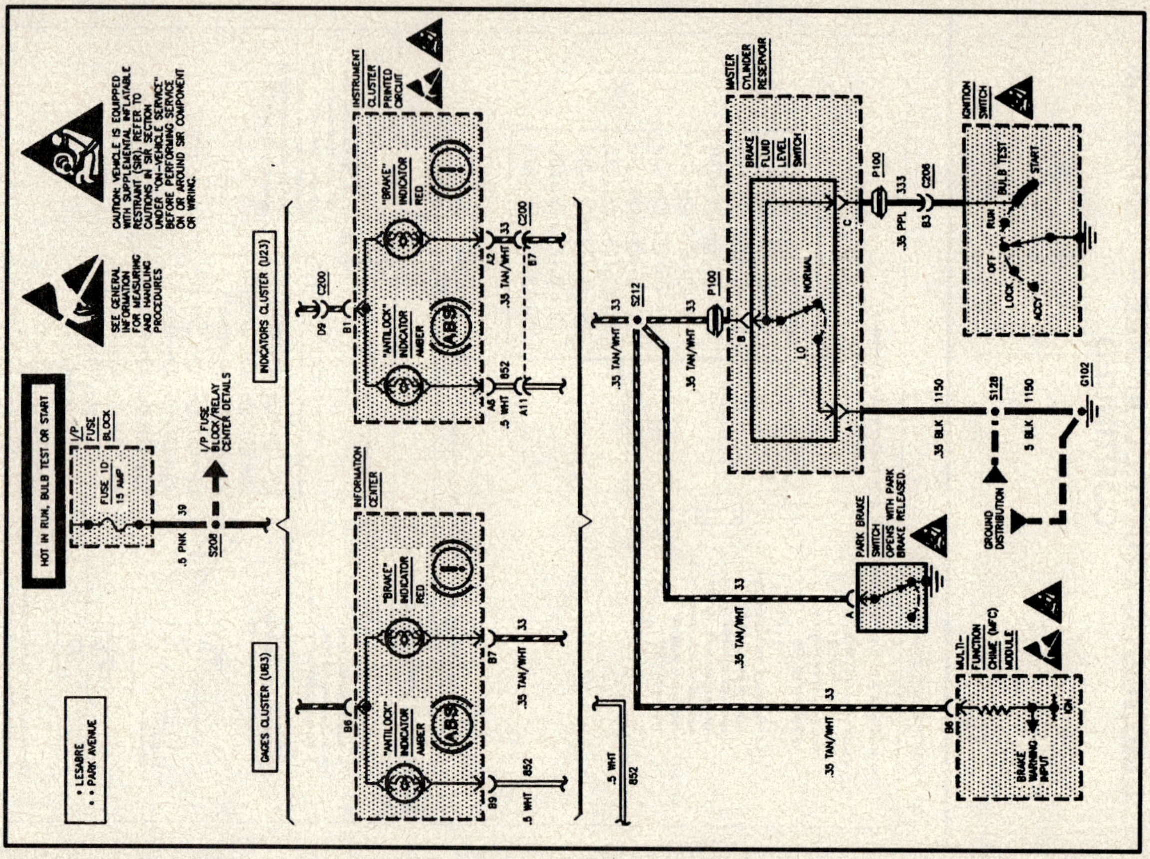

ELECTRICAL COMPONENT LAYOUT — 1994 BUICK

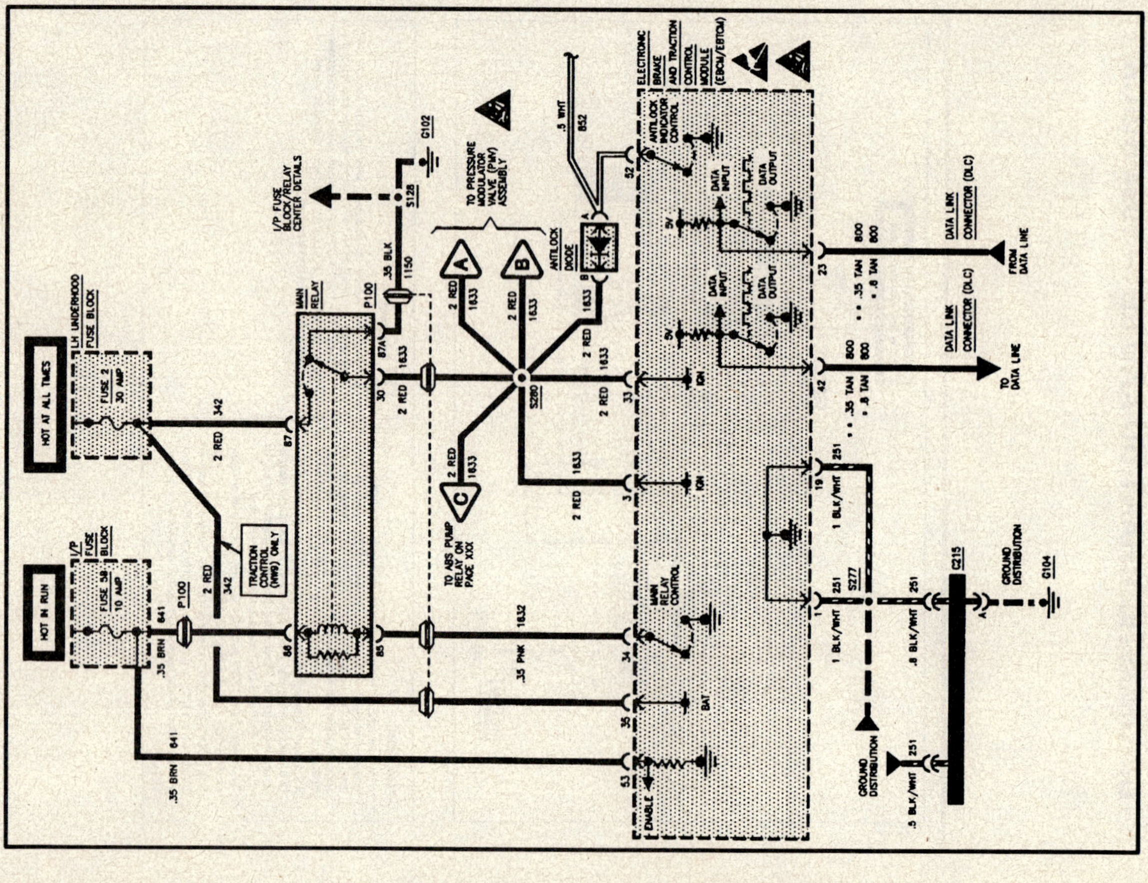

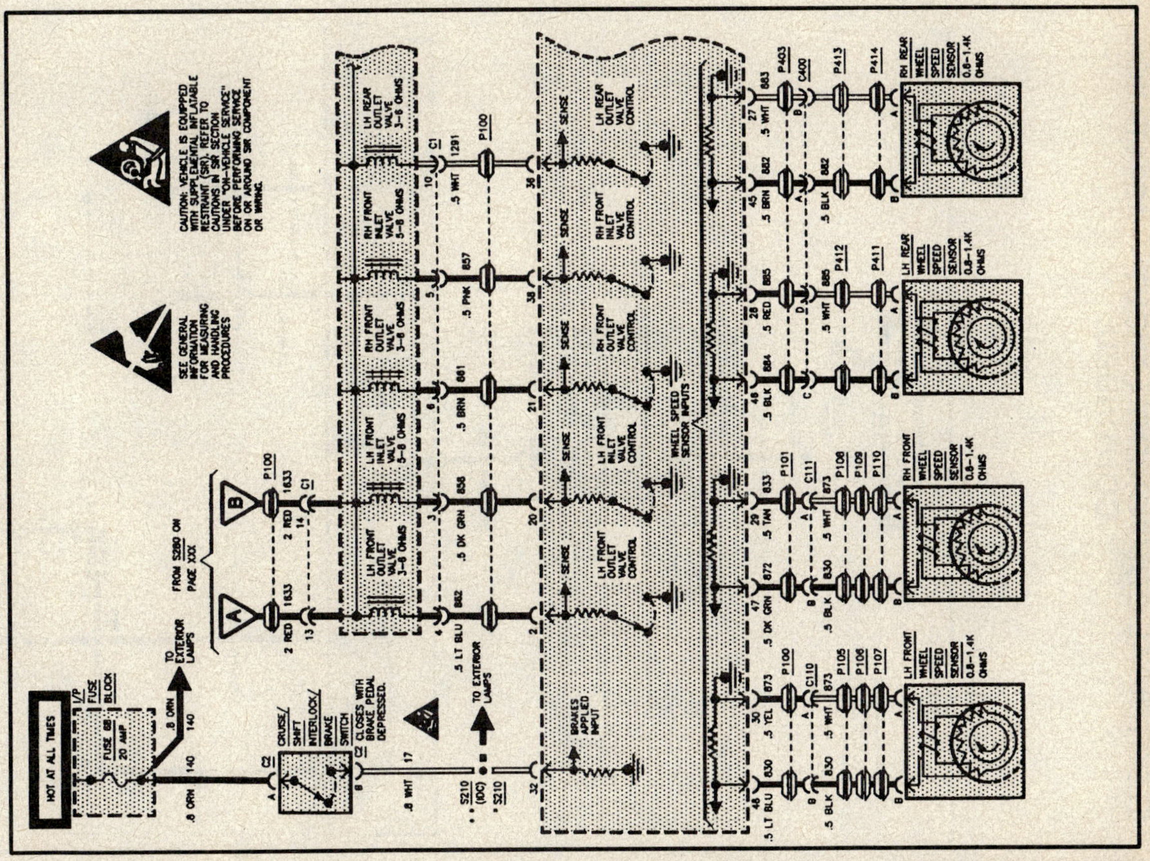

ELECTRICAL COMPONENT LAYOUT — 1994 BUICK, CONTINUED

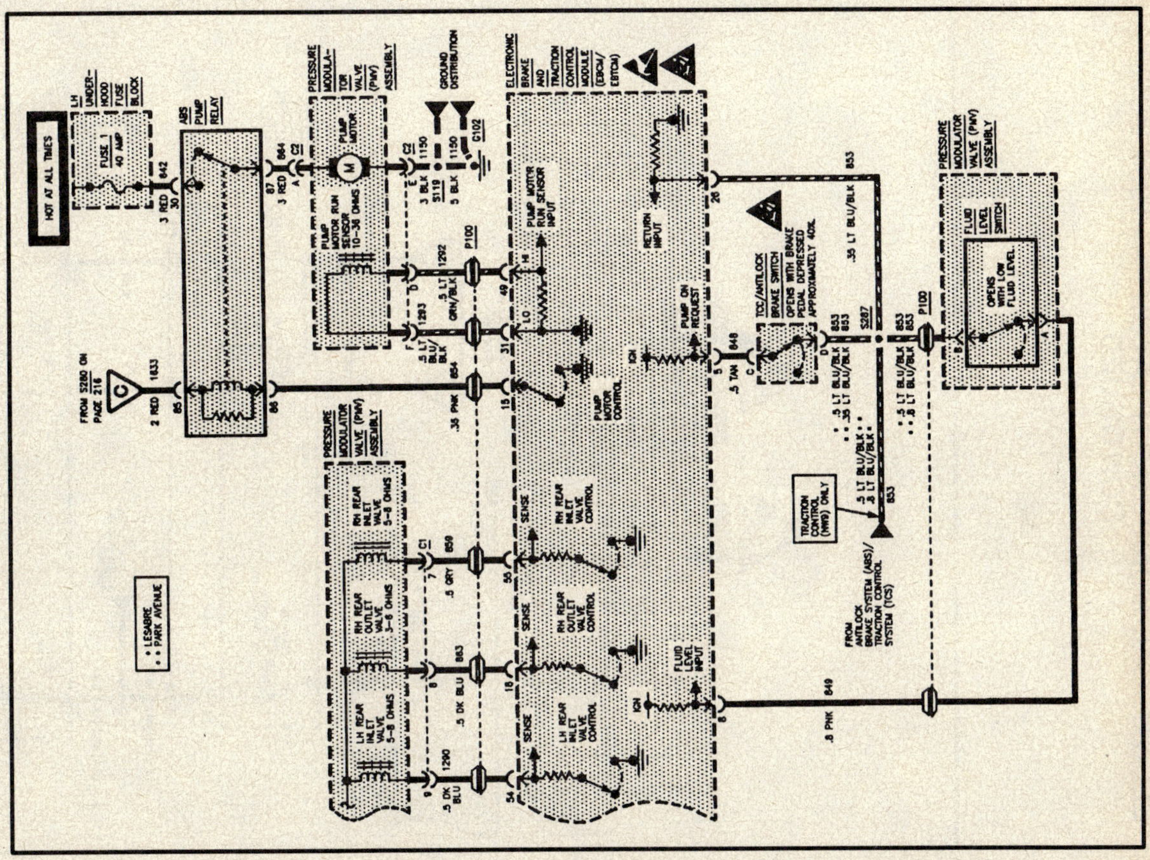

ELECTRICAL COMPONENT LAYOUT — 1994 BUICK, CONTINUED

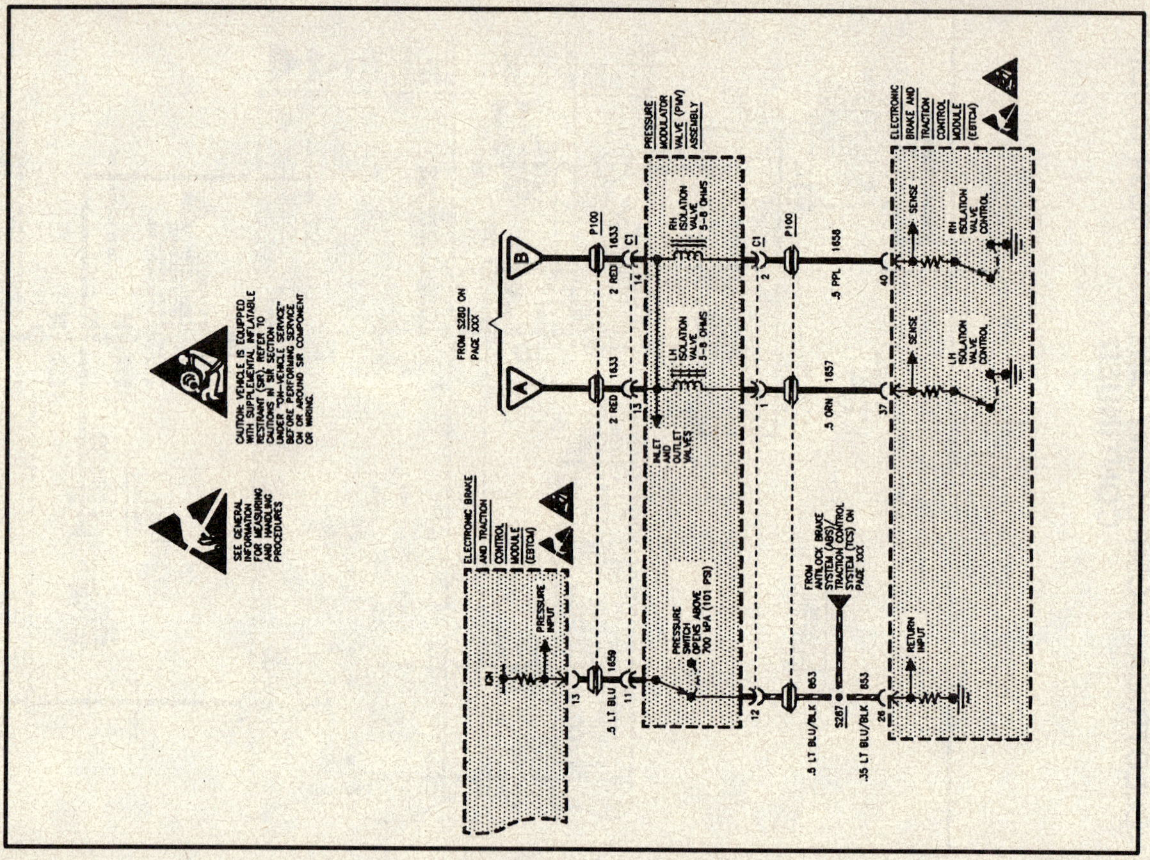

ELECTRICAL COMPONENT LAYOUT — 1994 BUICK, CONTINUED

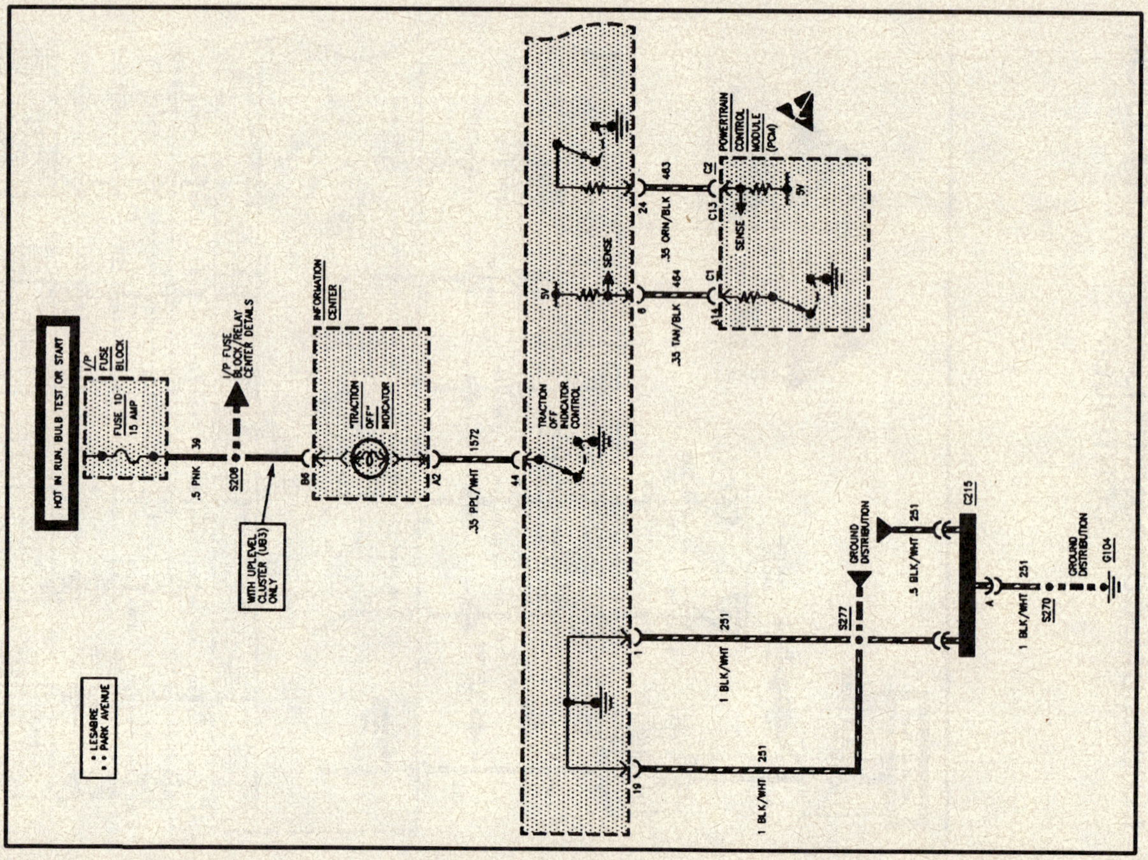

ELECTRICAL COMPONENT LAYOUT — 1994 BUICK, CONTINUED

DIAGNOSTIC CHART — ABS/TCS VISUAL INSPECTION — 1995 EXCEPT AURORA AND RIVIERA

ABS/TCS VISUAL INSPECTION

ITEM	INSPECT FOR	CORRECTIVE ACTION
PARKING BRAKE	- FULL RELEASE - PROPER SWITCH FUNCTION WHEN NECESSARY, UNPLUG SWITCH CONNECTOR TO VERIFY	- OPERATE MANUAL RELEASE LEVER TO VERIFY RELEASE - ADJUST CABLE OR REPAIR RELEASE SYSTEM AS REQUIRED - REPAIR SWITCH AS REQUIRED - PROCEED TO ABS/TCS FUNCTIONAL TEST
MAJOR COMPONENTS - MASTER CYLINDER FLUID RESERVOIR - PRESSURE MODULATOR VALVE (PMV) ASSEMBLY	- LOW FLUID LEVEL - LOW FLUID LEVEL IN PMV RESERVOIR - EXTERNAL LEAKS - PROPER ASSEMBLY	- DETERMINE CAUSE OF FLUID LOSS AND REPAIR. ADD FLUID TO MASTER CYLINDER RESERVOIR AS REQUIRED - DETERMINE CAUSE OF FLUID LOSS AND REPAIR. ADD FLUID TO MASTER CYLINDER RESERVOIR AS REQUIRED - REPAIR LEAKS AS REQUIRED - INSTALL OR POSITION COMPONENTS PROPERLY - PROCEED TO ABS/TCS FUNCTIONAL TEST
IP FUSE BLOCK - FUSE 1D - FUSE 5B - FUSE 6B UNDERHOOD FUSE BLOCK - FUSE 1 - FUSE 2	- OPEN (BLOWN FUSE) - PROPER ENGAGEMENT	- REPLACE - PROCEED TO ABS/TCS FUNCTIONAL TEST
CONNECTORS - MAIN RELAY - PUMP MOTOR RELAY - PMV FLUID LEVEL SWITCH - WHEEL SPEED SENSORS - ELECTRONIC BRAKE (AND TRACTION) CONTROL MODULE (EBCM/EBTCM) - PRESSURE MODULATOR VALVE (PMV) ASSEMBLY CONNECTORS: C1 - VALVE BLOCK CONNECTOR (ROUND) C2 - PUMP MOTOR CONNECTOR (GRY) - MASTER CYLINDER RESERVOIR - CRUISE/SHIFT INTERLOCK/BRAKE SWITCH - TCC/ANTILOCK BRAKE SWITCH - C101, C102, C414 - AUTOMATIC TRANSAXLE (TRACTION CONTROL ONLY) - GROUNDS - G102, G202, G104	- PROPER ENGAGEMENT - LOOSE WIRES OR TERMINALS - CORRODED OR BROKEN EYELETS - CORRODED/LOOSE WIRES OR BROKEN EYELETS	- PROPERLY ENGAGE CONNECTOR - REPAIR AS REQUIRED - PROCEED TO ABS/TCS FUNCTIONAL TEST - REPAIR AS NEEDED
WHEN NO PROBLEM IS NOTED IN VISUAL INSPECTION, PROCEED TO "ABS/TCS FUNCTIONAL TEST"		

DIAGNOSTIC CHART — ABS/TCS FUNCTIONAL TEST — 1995 EXCEPT AURORA AND RIVIERA

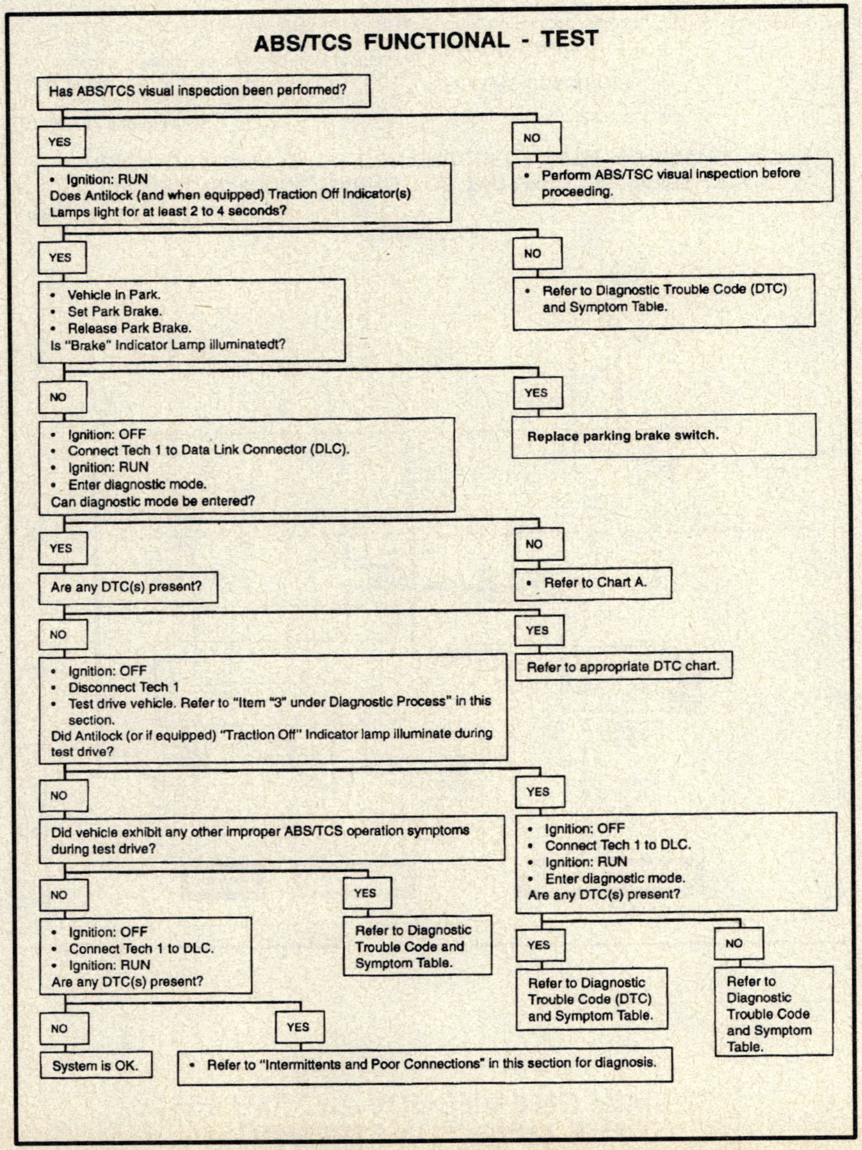

DIAGNOSTIC TROUBLE CODE CHART — 1995 EXCEPT AURORA AND RIVIERA

CHART	SYMPTOM	PAGE
A	Diagnostic Mode cannot be entered or "ANTILOCK" Indicator Lamp is "on" constantly with no DTC(s) set	
B	"ANTILOCK" Indicator Lamp does not function	
C	"TRACTION OFF" Indicator Lamp is "on" constantly with no DTC(s) set	
D	"TRACTION OFF" Indicator Lamp does not function	
E (Oldsmobile Only)	"TRACTION ENGAGED" Indicator Lamp does not operate properly	
F (Pontiac Only)	Traction Control Lamp Switch does not operate properly	
DTC	**DESCRIPTION**	
21	RF Wheel Speed Sensor circuit open or shorted	
22	RF Wheel Speed Sensor signal erratic	
23	RF Wheel Speed Sensor signal missing	
25	LF Wheel Speed Sensor circuit open or shorted	
26	LF Wheel Speed Sensor signal erratic	
27	LF Wheel Speed Sensor signal missing	
31	RR Wheel Speed Sensor circuit open or shorted	
32	RR Wheel Speed Sensor signal erratic	
33	RR Wheel Speed Sensor signal missing	
35	LR Wheel Speed Sensor circuit open or shorted	
36	LR Wheel Speed Sensor signal erratic	
37	LR Wheel Speed Sensor signal missing	
41	RF Inlet Valve circuit open or shorted	
42	RF Outlet Valve circuit open or shorted	
44	LF Isolation Valve circuit open or shorted	
45	Power Interruption Fault	
46	LF Outlet Valve circuit open or shorted	
48	RF Isolation Valve circuit open or shorted	
51	RR Inlet Valve circuit open or shorted	
52	RR Outlet Valve circuit open or shorted	
55	LR Inlet Valve circuit open or shorted	
56	LR Outlet Valve circuit open or shorted	
61	Pump Motor Circuit Test	
62	Low Brake Pedal during an ABS stop	
71	EBCM/EBTCM internal failure	
72	Brake Switch circuit open or shorted	
73	Fluid Level Switch circuit open or shorted	
74	Pressure Switch circuit open or shorted	
75	PCM requested TCS to be disabled	
76	UART SDL circuit or signal malfunction	
77	Delivered Torque Circuit open or shorted	

DIAGNOSTIC CHART A — 1995 EXCEPT AURORA AND RIVIERA

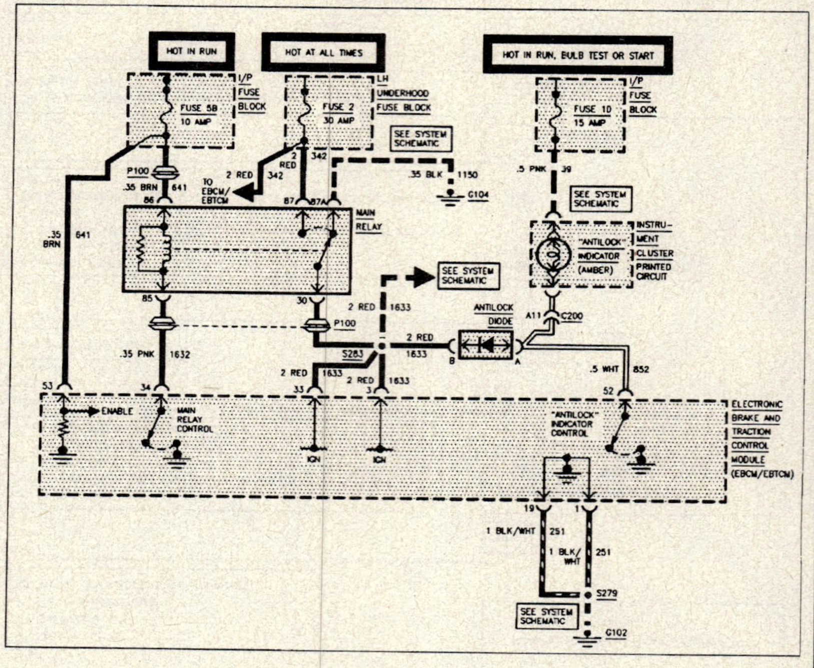

CHART A

(Page 1 Of 2)

DIAGNOSTIC MODE CANNOT BE ENTERED OR "ANTILOCK" INDICATOR LAMP IS ON CONTINUOUSLY WITH NO DTC(S) SET

CIRCUIT OPERATION

When the ignition is first turned to "RUN," the EBCM/EBTCM goes through a self-test lasting 2-4 seconds. During this test, the EBCM/EBTCM grounds terminal 52 causing the amber "ANTILOCK" indicator to light. This serves as an indicator bulb test.

When a low fluid condition exists, the EBCM/EBTCM turns on the amber "ANTILOCK" indicator but does not set a DTC. A low fluid level condition usually lights the red "BRAKE" indicator since the master cylinder supplies the PMV reservoir. However, restrictions or air in the hose could cause a low fluid condition in the PMV reservoir without lighting the red "BRAKE" indicator.

FAILURE CONDITIONS

Possible Causes

- Open in CKT 641, 800, 251, 849 or 853.
- Low brake fluid level or open PMV fluid level switch.
- Short to ground in CKT 641, 800 or 852.
- Wrong scan tool adapter being used.

DIAGNOSTIC CHART A — 1995 EXCEPT AURORA AND RIVIERA, CONTINUED

CHART A (1 OF 2)
DIAGNOSTIC MODE CANNOT BE ENTERED OR "ANTILOCK" INDICATOR LAMP IS ON CONTINUOUSLY WITH NO DTC(s) SET

Has ABS/TCS functional test been performed?

- NO → Perform ABS/TCS functional test before proceeding.
- YES ↓

- Ignition: OFF
- Visually inspect Fuse 5B of I/P Fuse Block. (Inspect for open.)

Is fuse OK?

- NO → Replace fuse and check/repair CKT 641 for a short to ground.
- YES ↓

- Reinsert fuse.
- Disconnect Pressure Modulator Valve (PMV) Fluid Level Switch Connector.
- Using J 39200 set to Ω, measure resistance across PMV Fluid Level Switch Terminal A to B.

Is less than 1 ohm measured?

- NO → Check brake fluid level. When low, check hose for kinks or restrictions and add fluid as needed to Master Cylinder Reservoir. If OK, replace PMV Reservoir.
- YES ↓

- Reconnect PMV Fluid Level Switch Connector.
- Disconnect EBCM/EBTCM Connector.
- Connect J 38716 Pinout Box.
- Measure resistance between J 38716 Pinout Box Terminal 8 and 26.

Is less than 3 ohms measured?

- NO → Repair open in CKT 849 or CKT 853.
- YES ↓

- Measure resistance between J 38716 Pinout Box Terminal 1 and Data Link Connector (DLC) Terminal 4 (ground), and between J 38716 Pinout Box Terminal 19 and DLC Terminal 4 (ground).

Is less than 2 ohms measured at each terminal?

- NO → Repair open in CKT 251.
- YES → (A)

Continued on next page

After repairs are completed, clear DTC and repeat ABS/TCS functional test.

DIAGNOSTIC CHART A — 1995 EXCEPT AURORA AND RIVIERA, CONTINUED

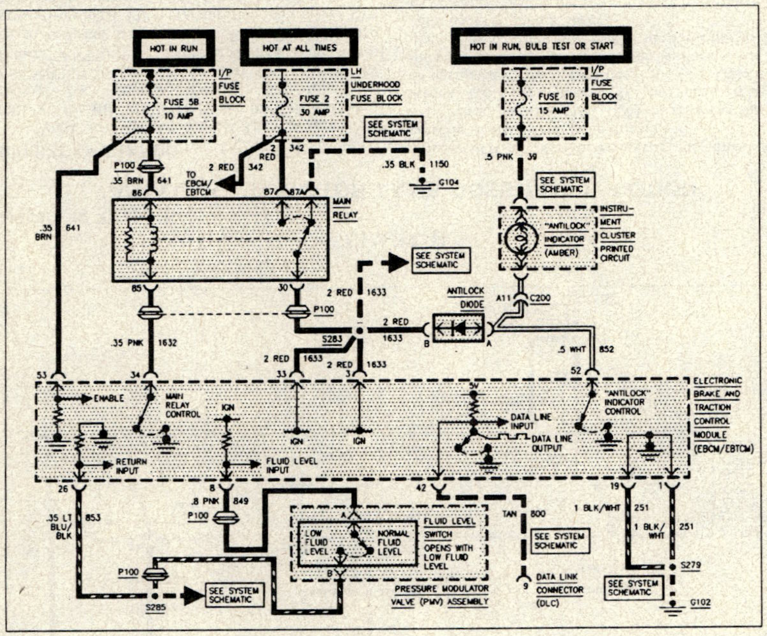

CHART A

(page 2 of 2)

DIAGNOSTIC MODE CANNOT BE ENTERED OR "ANTILOCK" INDICATOR LAMP IS ON CONTINUOUSLY WITH NO DTC(s) SET

CIRCUIT OPERATION

When the ignition is first turned to "RUN," the EBCM/EBTCM goes through a self-test lasting 2-4 seconds. During this test, the EBCM/EBTCM grounds terminal 52 causing the amber "ANTILOCK" indicator to light. This serves as an indicator bulb test.

When a low fluid condition exists, the EBCM/EBTCM turns on the amber "ANTILOCK" indicator but does not set a DTC. A low fluid level condition usually lights the red "BRAKE" indicator since the master cylinder supplies the PMV reservoir. However, restrictions or air in the hose could cause a low fluid condition in the PMV reservoir without lighting the red "BRAKE" indicator.

FAILURE CONDITIONS

Possible Causes

- Open in CKT 641, 800, 251, 849 or 853.
- Low brake fluid level or open PMV fluid level switch.
- Short to ground in CKT 641, 800 or 852.
- Wrong scan tool adapter being used.

DIAGNOSTIC CHART A — 1995 EXCEPT AURORA AND RIVIERA, CONTINUED

CHART A (2 OF 2)
DIAGNOSTIC MODE CANNOT BE ENTERED OR "ANTILOCK" INDICATOR LAMP IS ON CONTINOUSLY WITH NO DTC(s) SET

A

- Ignition: RUN
- Using J 39200 set to DC scale, measure voltage between J 38716 Pinout Box Terminal 53 and 1.

Is battery voltage measured?

NO: Repair open in CKT 641.

YES:

- Ignition: OFF
- Disconnect Main Relay.
- Ignition: RUN
- Measure voltage between J 38716 Pinout Terminal 52 and 1.

Is battery voltage measured?

NO: Repair CKT 852 for a short to ground.

YES:

- Ignition: RUN
- Measure voltage between J 38716 Pinout Box Terminal 42 and 1.

Is 5 volts measured?

NO: Check/repair CKT 800 for a short to ground or an open.

YES: Check EBCM/EBTCM Connector for proper terminal contact. If terminal contact is OK, replace EBCM/EBTCM.

After repairs are completed, clear DTC and repeat ABS/TCS functional test.

DIAGNOSTIC CHART B — 1995 EXCEPT AURORA AND RIVIERA

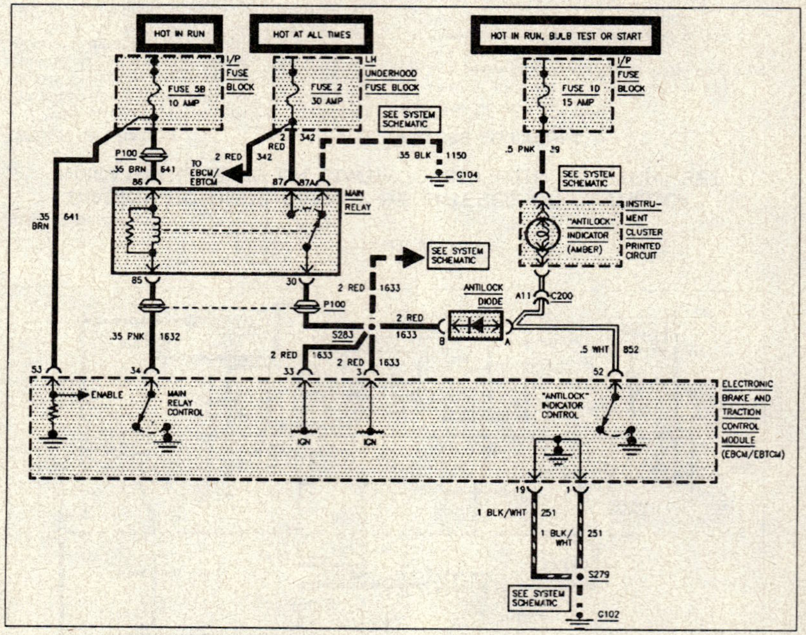

CHART B

(page 1 of 2)

"ANTILOCK" INDICATOR LAMP DOES NOT FUNCTION

CIRCUIT OPERATION

When the ignition switch is first turned to "RUN" the EBCM/EBTCM goes through a self-test lasting 2-4 seconds. During this test, the EBCM/EBTCM grounds terminal 52, causing the amber "ANTILOCK" indicator to light. This serves as an indicator bulb test. The EBCM/EBTCM also grounds terminal 34, energizing the main relay. This removes the indicator ground path provided through the antilock diode and the main relay contacts.

When the self-test is completed (and no faults detected) the EBCM/EBTCM removes the ground at terminal 52, turning off the indicator. The main relay remains energized.

When the EBCM/EBTCM detects a fault, it grounds terminal 52 causing the amber "ANTILOCK" indicator to light. (Note: The EBCM/EBTCM also de-energizes the main relay creating a redundant ground path.)

When the EBCM/EBTCM is disabled for any reason (i.e. open Fuse 19), the main relay is de-energized. This provides an "ANTILOCK" indicator ground path through the antilock diode and main relay contacts, and causes the indicator to light with no DTC being set.

FAILURE CONDITIONS

Possible Causes

- Open in CKT 251, 852 or 1633.
- Open bulb.
- Short to voltage in CKT 852.
- Open or shorted antilock diode.
- Main relay fails "engaged."
- Short to ground in CKT 1632.

DIAGNOSTIC CHART B — 1995 EXCEPT AURORA AND RIVIERA, CONTINUED

CHART B (1 of 2) "ANTILOCK" INDICATOR LAMP DOES NOT FUNCTION

Has ABS/TCS Functional Test been performed?

- NO: Perform ABS/TCS Functional Test before proceeding.
- YES:
  - Ignition: OFF
  - Connect a Tech 1 to Data Link Connector (DLC).
  - Ignition: RUN
  - Enter diagnostic mode.
  - Can diagnostic mode be entered?

- NO: A — Continued on next page
- YES:
  - Ignition: OFF
  - Disconnect EBCM/EBTCM Connector.
  - Connect J 38716 Pinout Box.
  - Using J 39200 set to Ω, measure resistance between J 38716 Pinout Box Terminal 1 and DLC Terminal 4 (ground).

  Is less than 2 OHMS measured?

- NO: Repair open in CKT 251.
- YES:
  - Disconnect Main Relay.
  - Ignition: RUN
  - Using J 39200 set to DC scale, measure voltage J 38716 Pinout Box Terminals 52 and 1.

  Is battery voltage measured?

- NO: Replace Antilock Indicator Bulb or repair CKT 852 for an open.
- YES:
  - Ignition: RUN
  - Connect a fused jumper between J 38716 Pinout Box Terminals 52 and 1.

  Does Antilock Indicator remain lit?

- NO: Repair short to voltage in CKT 852. Check Antilock Diode and Main Relay for possible damage.
- YES:
  - Ignition: OFF
  - Disconnect Antilock Brake Diode.
  - Using J 39200 check Antilock Diode. Also, check for improper polarity.
  - Is diode OK?

- NO: Replace Antilock Diode. Observe polarity.
- YES:
  - Reconnect Antilock Brake Diode. Observe polarity.
  - Disconnect J 38716 Pinout Box.
  - Reconnect EBCM/EBTCM Connector and Main Relay.
  - Ignition: RUN
  - Does Antilock Indicator light for 2-4 seconds?

- YES: PROBLEM SOLVED
- NO: Check EBCM/EBTCM Connector for proper terminal contact. If terminal contact is OK, replace EBCM/EBTCM.

After repairs are completed, clear DTC and repeat ABS/TCS functional test.

DIAGNOSTIC CHART B — 1995 EXCEPT AURORA AND RIVIERA, CONTINUED

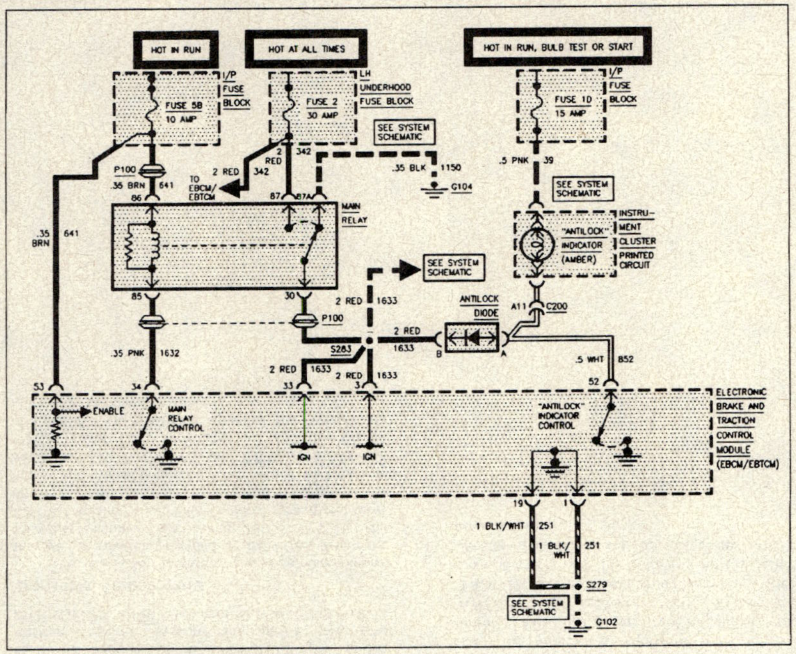

CHART B

(page 2 of 2)

"ANTILOCK" INDICATOR LAMP DOES NOT FUNCTION

CIRCUIT OPERATION

When the ignition switch is first turned to "RUN" the EBCM/EBTCM goes through a self-test lasting 2-4 seconds. During this test, the EBCM/EBTCM grounds terminal 52, causing the amber "ANTILOCK" indicator to light. This serves as an indicator bulb test. The EBCM/EBTCM also grounds terminal 34, energizing the main relay. This removes the indicator ground path provided through the antilock diode and the main relay contacts.

When the self-test is completed (and no faults detected) the EBCM/EBTCM removes the ground at terminal 52, turning off the indicator. The main relay remains energized.

When the EBCM/EBTCM detects a fault, it grounds terminal 52 causing the amber "ANTILOCK" indicator to light. (Note: The EBCM/EBTCM also de-energizes the main relay creating a redundant ground path.)

When the EBCM/EBTCM is disabled for any reason (i.e. open Fuse 19), the main relay is de-energized. This provides an "ANTILOCK" indicator ground path through the antilock diode and main relay contacts, and causes the indicator to light with no DTC being set.

FAILURE CONDITIONS

Possible Causes

- Open in CKT 251, 852 or 1633.
- Open bulb.
- Short to voltage in CKT 852.
- Open or shorted antilock diode.
- Main relay fails "engaged."
- Short to ground in CKT 1632.

DIAGNOSTIC CHART B — 1995 EXCEPT AURORA AND RIVIERA, CONTINUED

CHART B (2 of 2) "ANTILOCK" INDICATOR LAMP DOES NOT FUNCTION

(A)

- Ignition: OFF
- Disconnect Tech 1.
- At this point a double fault is possible. First refer to CHART A for diagnosis, then continue and complete this diagnostic chart from this point.

Was CHART A performed?

- NO → Refer to CHART A.
- YES ↓

- Ignition: OFF
- Disconnect Main Relay.
- Using J 39200 set to Ω, measure resistance across Main Relay Terminals 30 and 87A.

Is less than 2 OHMS measured?

- NO → Replace Main Relay.
- YES ↓

- Remove Fuse 5B of I/P Fuse Block.
- Ignition: RUN
- Using J 39200 set to DC scale, measure voltage between Main Relay Harness Connector Terminal 30 and ground.

Is battery voltage measured?

- NO → Replace Antilock Diode or repair open in CKT 1633. NOTE: When replacing diode, observe polarity.
- YES ↓

- Measure voltage between Main Relay Harness Connector Terminals 30 and 87A.

Is battery voltage measured?

- NO → Repair open in CKT 251.
- YES ↓

- Ignition: OFF
- Disconnect EBCM/EBTCM Connector.
- Using J 39200 set to Ω, measure resistance to ground between Main Relay Harness Connector Terminal 85 and ground.
- Is less than 5 OHMS measured?

- YES → Repair short to ground in CKT 1632.
- NO → Check EBCM/EBTCM Connector for proper terminal contact. If terminal contact is OK, replace EBCM/EBTCM.

After repairs are completed, clear DTC and repeat ABS/TCS functional test.

DIAGNOSTIC CHART C — 1995 EXCEPT AURORA AND RIVIERA

CHART C "TRACTION OFF" INDICATOR LAMP IS ON CONTINUOUSLY AND NO DTC(s) SET

CIRCUIT OPERATION

When the ignition is first turned to "RUN," the EBTCM goes through a self-test. During this test, the EBTCM grounds both the "ANTILOCK" and "TRACTION OFF" indicators for approximately 2 to seconds. This acts as a bulb test for the indicators.

Traction Control Switch

The traction control switch allows the driver to select the traction control mode (on/off). When the traction control system is turned off, the "TRACTION OFF" indicator remains illuminated.

Transaxle Temperature

The EBTCM monitors transaxle temperature through information provided by the PCM. When the transaxle becomes hot, about 160°C (320°F), the EBTCM disables the traction control system and turns on the "TRACTION OFF" indicator. This allows the brakes and transaxle to cool down. When the transaxle cools to 149°C (300°F) the EBTCM keeps the "TRACTION OFF" indicator on and the system disabled for an additional three to five minutes to allow for sufficient cooling.

FAILURE CONDITIONS

Possible Causes

- Short to ground in CKT 1571 or 1572.
- Faulty traction control switch.
- Faulty EBTCM

! Important

- When the vehicle is equipped with traction control, any symptom which lights the "ANTILOCK" indicator also lights the "TRACTION OFF" indicator. Therefore, when both indicators are lit, follow "ANTILOCK" indicator diagnosis when no diagnostic trouble codes are set.

DIAGNOSTIC CHART C — 1995 EXCEPT AURORA AND RIVIERA, CONTINUED

CHART C
"TRACTION OFF" INDICATOR LAMP IS ON CONTINUOUSLY AND NO DTC(s) SET

Has ABS/TCS Functional Test been performed?

- NO: Perform ABS/TCS Functional Test before proceeding.
- YES:
  - Ignition: OFF
  - Disconnect EBTCM Connector.
  - Install J 38716 Pinout Box.
  - Using J 39200 set to Ω, measure resistance between J 38716 Pinout Box Terminal 1 and Data Link Connector (DLC) Terminal 4 (ground).

  Is less than 2 OHMS measured?
  - NO: Repair open in CKT 251.
  - YES:
    - Measure resistance between J 38716 Pinout Box Terminal 1 and 44.

    Is less than 5 OHMS measured?
    - YES:
      - Check CKT 1572 for short to ground.
      - Is short to ground present?
      - YES: Repair CKT 1572
      - NO: Replace EBTCM.
    - NO:
      - Measure resistance between J 38716 Pinout Box Terminal 1 and 25.

      Is less than 5 OHMS measured?
      - YES: Check/repair CKT 1571 for a short to ground. If CKT 1571 is OK, replace Traction Control Switch.
      - NO: No trouble found. Refer to "INTERMITTENTS AND POOR CONNECTIONS" in this section for possible causes.

After repairs are completed, clear DTC and repeat ABS/TCS functional test.

DIAGNOSTIC CHART D — 1995 EXCEPT AURORA AND RIVIERA

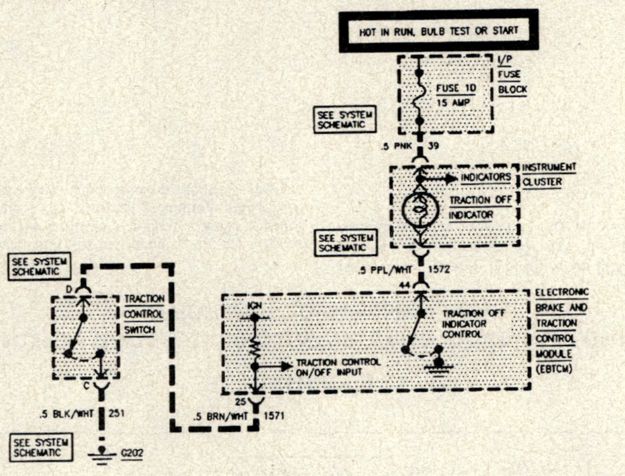

CHART D
"TRACTION OFF" INDICATOR LAMP DOES NOT FUNCTION

CIRCUIT OPERATION

When the ignition is first turned to "RUN," the EBTCM goes through a self-test. During this test, the EBCM grounds both the "ANTILOCK" and "TRACTION OFF" indicators for approximately 2 to 4 seconds. This acts as a bulb test for the indicators.

Traction Control Switch

The traction control switch allows the driver to select the traction control mode (on/off). When the traction control system is turned off, the "TRACTION OFF" indicator remains illuminated.

FAILURE CONDITIONS

Possible Causes

- Open in CKT 251, 1571 or 1572.
- Open bulb.
- Faulty traction control switch.

Important

- When the vehicle is equipped with traction control, any symptom which lights the "ANTILOCK" indicator also lights the "TRACTION OFF" indicator. Therefore, when both indicators are lit, follow "ANTILOCK" indicator diagnosis when no diagnostic trouble codes are set.

DIAGNOSTIC CHART D — 1995 EXCEPT AURORA AND RIVIERA, CONTINUED

CHART D
"TRACTION OFF" INDICATOR LAMP DOES NOT FUNCTION

Has ABS/TCS Functional Test been performed?

- NO: Perform ABS/TCS Functional Test before proceeding?
- YES:
  - Ignition: RUN
  - Does Traction Off indicator light for 2-4 seconds?

If NO:
- Ignition: OFF
- Disconnect EBTCM Connector.
- Install J 36716 Pinout Box.
- Using J 39200 set to Ω, measure resistance between J 36716 Pinout Box Terminal 1 and Data Link Connector (DLC) Terminal 4 (ground).
- Is less than 2 OHMS measured?
  - NO: Repair open in CKT 251.
  - YES:
    - Ignition: RUN
    - Measure voltage from J 36716 Pinout Box Terminal 44 and 1.
    - Is battery voltage measured?
      - YES: Check EBTCM Connector for proper terminal contact. If terminal contact is OK, replace EBTCM.
      - NO: Repair open in CKT 1572 or replace Traction Off Indicator bulb. Check In-line Connector and Instrument Cluster Connector (see Schematic) for proper terminal contact.

If YES:
- Ignition: OFF
- Disconnect Traction Control Switch Connector.
- Ignition: RUN
- Using J 39200 set to DC scale, measure voltage between Traction Control Switch Harness Connector Terminal D and Data Link Connector (DLC) Terminal 4 (ground).

Is battery voltage measured?
- NO: Check/repair open in CKT 1571. If CKT 1571 is OK, check EBTCM Connector for proper terminal contact. If terminal contact is OK, replace EBTCM.
- YES:
  - Ignition: OFF
  - Measure resistance between Traction Control Switch Harness Connector Terminal C and DLC Terminal 4 (ground).
  - Is less than 2 OHMS measured?
    - YES: Check Traction Control Switch Connector for proper terminal contact. If OK, replace Traction Control Switch.
    - NO: Repair open in CKT 251.

After repairs are completed, clear DTC and repeat ABS/TCS functional test.

DIAGNOSTIC CHART E — 1995 EXCEPT AURORA AND RIVIERA

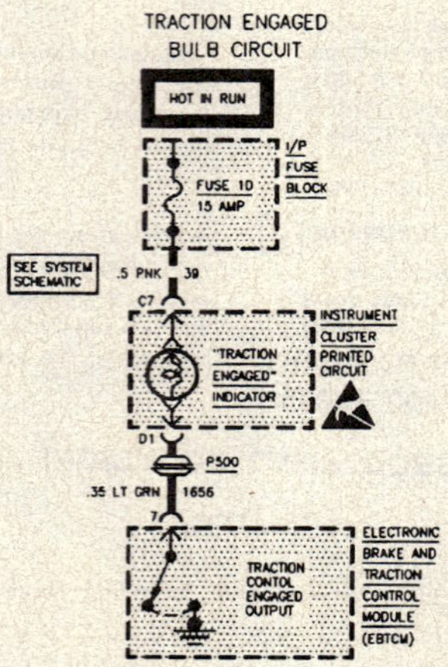

CHART E

"TRACTION ENGAGED" INDICATOR LAMP DOES NOT OPERATE PROPERLY (Oldsmobile Only)

CIRCUIT OPERATION

When the EBTCM is operating in the traction control mode, the EBTCM's traction control engaged output (terminal 7) closes, grounding either the instrument cluster's DIC traction control engaged input or the traction control switch LED. The EBTCM grounds this terminal to alert the driver that the traction control system (TCS) is operating.

DIAGNOSTIC CHART E — 1995 EXCEPT AURORA AND RIVIERA, CONTINUED

CHART E
"TRACTION ACTIVE" INDICATOR LAMP DOES NOT OPERATE PROPERLY (OLDSMOBILE ONLY)

Has ABS/TCS Functional Test been performed?

- NO: Perform ABS/TCS Functional Test before proceeding.
- YES: Does "TRACTION ACTIVE" lamp illuminate when Traction Control is not active?
  - YES:
    - Ignition: OFF
    - Disconnect EBTCM Connector.
    - Ignition: RUN
    - Does "TRACTION ACTIVE" indicator lamp still illuminate when Traction Control is not active?
      - YES: Repair short to ground in CKT 1656.
      - NO: Replace EBTCM.
  - NO:
    - Ignition: OFF
    - Disconnect EBTCM Connector.
    - Connect J 38716 Pinout Box.
    - Using J 39200 set to Ω, measure resistance between J 38716 Pinout Box Terminal 1 and Data Link Connector (DLC) Terminal 4 (ground).
    - Is less than 2 OHMS measured?
      - NO: Repair open in CKT 1451.
      - YES:
        - Connect a Fused Jumper from J 38716 Pinout Box Terminal 7 to Terminal 1.
        - Ignition: RUN
        - Does "TRACTION ACTIVE" lamp illuminate?
          - NO:
            - Ignition: OFF
            - Check "TRACTION ACTIVE" indicator bulb.
            - Is bulb OK?
              - NO: Replace bulb.
              - YES:
                - Disconnect Remote Telltale Bank Connector.
                - Ignition: RUN
                - Using J39200 set to DC Scale, measure voltage between Remote Telltale Bank Harness Connector Terminal A1 and DLC Terminal 4.
                - Battery voltage present?
                  - YES: Open in CKT 1656.
                  - NO: Check CKT 39 for open and Remote Telltale Bank Connector for proper terminal contact. If OK, replace Remote Telltale Bank.
          - YES:
            - Ignition: OFF
            - Disconnect J 38716 Pinout Box.
            - Reconnect EBTCM.
            - Using Tech 1, select Mode F0 (Data List 1) and monitor TCS state.
            - Test drive vehicle in a safe area under 40 km/h (25 mph) on a slippery surface such as sand or gravel.
            - Does TCS state change during a traction control event?
              - NO: Repeat ABS/TCS Functional Test.
              - YES: Did the "TRACTION ACTIVE" indicator lamp illuminate?
                - YES: System OK.
                - NO: Check EBTCM Connectors for proper terminal contact. If terminal contact is OK, replace EBTCM.

After repairs are completed, clear DTC and repeat ABS/TCS functional test.

DIAGNOSTIC CHART E — 1995 EXCEPT AURORA AND RIVIERA, CONTINUED

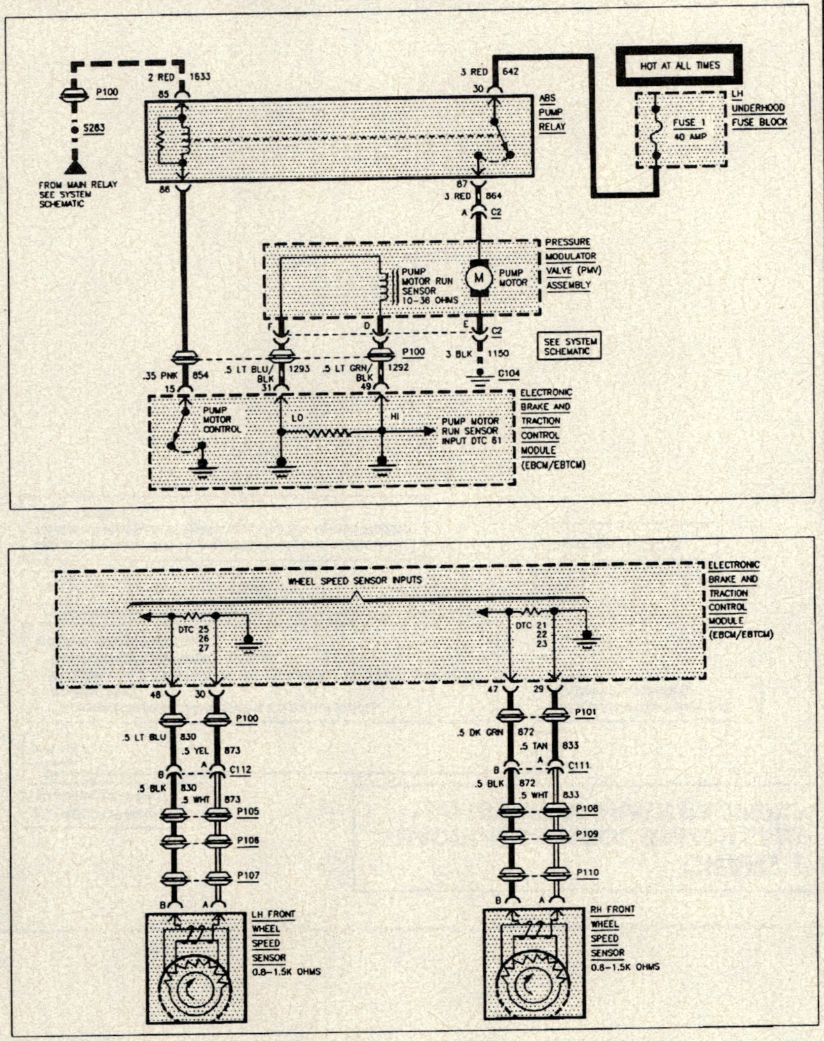

DIAGNOSTIC CHART F — 1995 EXCEPT AURORA AND RIVIERA

Chart F

TRACTION CONTROL SWITCH LAMP DOES NOT OPERATE PROPERLY

CIRCUIT OPERATION

The traction control switch illumination bulb is on whenever the key is in the RUN, BULB TEST or START position. It is not serviced separately from the traction control switch. The traction control switch illumination is NOT illuminated when there is an ABS or TCS system failure.

DIAGNOSTIC CHART F — 1995 EXCEPT AURORA AND RIVIERA, CONTINUED

CHART F
TRACTION CONTROL SWITCH LAMP DOES NOT OPERATE PROPERLY

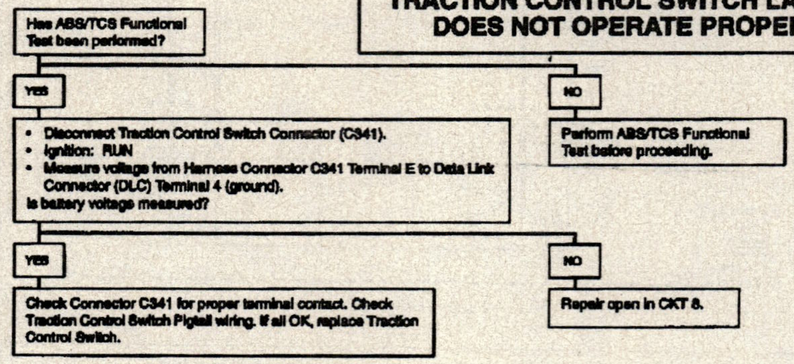

After repairs are completed, clear DTC and repeat ABS/TCS functional test.

DIAGNOSTIC CHART DTC 21 AND 25 — 1995 EXCEPT AURORA AND RIVIERA

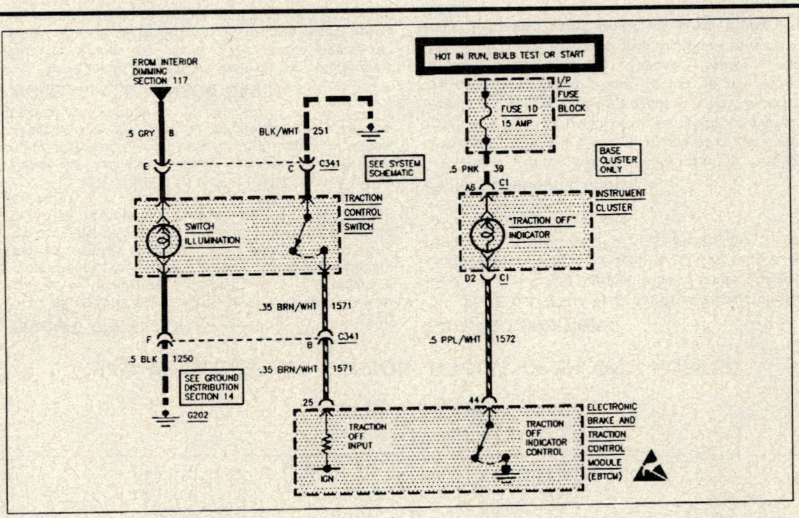

DTC(s) 21, 25

FRONT WHEEL SPEED SENSOR CIRCUIT TEST

CIRCUIT OPERATION

As each wheel turns, the wheel speed sensor creates an AC voltage with a frequency proportional to wheel speed. This magnetically induced voltage is caused by a toothed sensor ring (mounted on the drive axle) passing the wheel speed sensor's stationary pickup coil.

The EBCM/EBTCM uses the wheel speed sensor signal to calculate vehicle reference speeds and individual wheel speed, acceleration and slip values. These values are used to determine when antilock (or when equipped) traction control is required.

WHEEL SPEED SENSOR CABLES

In order to prevent electromagnetic interference from disturbing the wheel speed sensor signal, the cables from the EBCM/EBTCM to each wheel speed sensor are twisted in pairs a minimum of 6 to 9 turns per foot.

When servicing the sensor cables it is important to maintain the original cable's twists (6 to 9 turns per foot).

FAILURE CONDITIONS

The EBCM/EBTCM performs two basic checks on the wheel speed sensors: sensor continuity and sensor output. When the ignition is first turned to "RUN," the EBCM/EBTCM performs the wheel speed sensor continuity check.

Sensor Continuity Check

Any condition which would result in lack of continuity in the front wheel speed sensor circuit could result in a DTC 21 or 25 being set. These conditions include an open, a short to ground, or a short between the two wheel speed sensor circuits in the wiring harness between the speed sensor and EBCM/EBTCM. Also an open or short to ground across the wheel speed sensor coil could trigger this code.

DIAGNOSTIC CHART DTC 21 AND 25 — 1995 EXCEPT AURORA AND RIVIERA, CONTINUED

DTC(s) 21, 25

FRONT WHEEL SPEED SENSOR CIRCUIT OPEN OR SHORTED

Has ABS/TCS functional test been performed?

NO: Perform ABS/TCS functional test before proceeding.

YES:
- Ignition: OFF
- Raise vehicle.
- Disconnect suspect front Wheel Speed Sensor Connector.
- Check connector for proper terminal contact.
- Using J 39200 set to Ω, measure resistance between suspect Wheel Speed Sensor Terminal A and B.

Is 800 to 1500 ohms measured?

NO: Replace suspect front Wheel Speed Sensor.

YES:
- Measure resistance between suspect front Wheel Speed Sensor Terminal B and ground.

Is infinite ohms measured?

NO: Replace suspect front Wheel Speed Sensor.

YES:
- Reconnect Wheel Speed Sensor.
- Disconnect Wheel Speed Sensor Jumper Harness.
- Check connector for proper terminal contact.
- Measure resistance between Jumper Harness Terminals A and B (sensor side).

Is 800 to 1500 ohms measured?

NO: Replace suspect front Wheel Speed Sensor Jumper Harness.

YES:
- Reconnect Wheel Speed Sensor Jumper Harness.
- Disconnect EBCM/EBTCM Connector.
- Check connector for proper terminal contact.
- Install J 38716 Pinout Box.
- Measure resistance between terminals indicated for suspected front Wheel Speed Sensor Circuit:
  - Left Front Wheel - Terminals 30 and 48
  - Right Front Wheel - Terminals 29 and 47

Is 800 to 1500 ohms measured?

NO: Above 1500 ohms (remember to check in-line connector [see schematic] and Wheel Speed Sensor Connector for proper terminal contact):
When left front wheel, repair open in CKT 830 or 873.
When right front wheel, repair open in CKT 833 or 872.
Below 800 ohms:
When left front wheel, repair short between CKT 830 and 873.
When right front wheel, repair short between CKT 833 and 872.

YES:
- Measure resistance between J 38716 Pinout Box Terminal 1 and Data Link Connector (DLC) Terminal 4 (ground).

Is less than 5 ohms measured?

NO: Repair open in CKT 251.

YES:
- Measure resistance at J 38716 Pinout Box Terminals indicated for suspected front Wheel Speed Sensor Circuit:
  - Left Front Wheel - Terminal 48 and 1
  - Right Front Wheel - Terminal 47 and 1

Is less than 5 ohms measured?

YES: When left front wheel, repair short to ground in CKT 830 or 873. When right front wheel, repair short to ground in CKT 872 or 833.

NO: Check EBCM/EBTCM Connector for proper terminal contact. If terminal contact is OK, reconnect EBCM/EBTCM, clear DTC and test drive vehicle. If DTC resets, replace EBTCM. If DTC does not reset, refer to Intermittants and Poor Connections.

After repairs are completed, clear DTC and repeat ABS/TCS functional test.

DIAGNOSTIC CHART DTC 31 AND 35 — 1995 EXCEPT AURORA AND RIVIERA

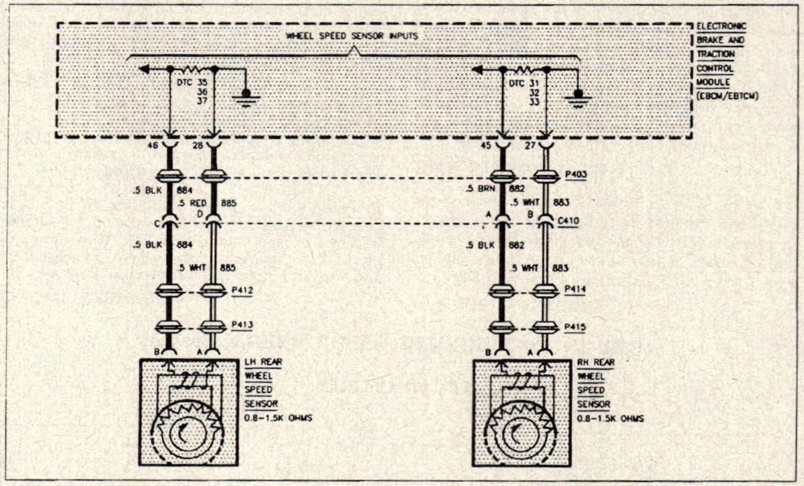

DTC(s) 31, 35

REAR WHEEL SPEED SENSOR CIRCUIT OPEN OR SHORTED

CIRCUIT OPERATION

When each wheel turns, the wheel speed sensor creates a small AC voltage with a frequency proportional to wheel speed. This magnetically induced voltage is caused by a toothed sensor ring passing the wheel speed sensor's stationary pickup coil.Both sensors are part of the hub and bearing assembly.

The EBCM/EBTCM uses the wheel speed sensor signal to calculate vehicle reference speeds and individual wheel speed, acceleration and slip values. These values are used to determine when antilock control is required.

WHEEL SPEED SENSOR CABLES

In order to prevent electromagnetic interference from disturbing the wheel speed sensor signal, the cables from the EBCM/EBTCM to each wheel speed sensor are twisted in pairs a minimum of 6 to 9 turns per foot.

When servicing the sensor cables it is important to maintain the original cable's twists (6 to 9 turns per foot).

FAILURE CONDITIONS

The EBCM/EBTCM performs two basic checks on the wheel speed sensors: sensor continuity and sensor output. When the ignition is first turned to "RUN," the EBCM/EBTCM performs the wheel speed sensor continuity check.

Sensor Continuity Check

Any condition which would result in lack of continuity in the rear wheel speed sensor circuit could result in a DTC 31 or 35 being set. These conditions include an open, a short to ground, or a short between the two wheel speed sensor circuits in the wiring harness between the speed sensor and EBCM/EBTCM. Also an open or short to ground across the wheel speed sensor coil could trigger this DTC.

DIAGNOSTIC CHART DTC 31 AND 35 — 1995 EXCEPT AURORA AND RIVIERA, CONTINUED

DTC(s) 31, 35

REAR WHEEL SPEED SENSOR CIRCUIT OPEN OR SHORTED

Has ABS/TCS functional test been performed?

NO: Perform ABS/TCS functional test before proceeding.

YES:
- Ignition: OFF
- Park Brake: RELEASED
- Raise vehicle.
- Disconnect suspect rear Wheel Speed Sensor Connector at wheel.
- Check connector for proper terminal contact.
- Using J 39200 set to Ω, measure resistance between suspect Wheel Speed Sensor Terminal A and B.

Is 800 to 1500 ohms measured?

NO: Replace suspect rear Wheel Speed Sensor.

YES:
- Measure resistance between suspect rear Wheel Speed Sensor Terminal B and ground.

Is infinite ohms measured?

NO: Replace suspect rear Wheel Speed Sensor.

YES:
- Reconnect Wheel Speed Sensor.
- Disconnect Wheel Speed Jumper Harness.
- Check connector for proper terminal contact.
- Measure resistance between Jumper Harness Terminals A and B if right side Wheel Speed Sensor is suspect or Terminals D and C if left side Wheel Speed Sensor is suspect.

Is 800 to 1500 ohms measured?

NO: Replace Wheel Speed Sensor Jumper Harness.

YES:
- Reconnect Wheel Speed Sensor Jumper Harness.
- Disconnect EBCM/EBTCM Connector.
- Check connector for proper terminal contact.
- Install J 38716 Pinout Box.
- Measure resistance between terminals indicated for suspected rear Wheel Speed Sensor Circuit:
  - Left Rear Wheel - Terminals 28 and 46
  - Right Rear Wheel - Terminals 27 and 45

Is 800 to 1500 ohms measured?

NO: Above 1500 ohms (remember to check in-line connector [see schematic] and Wheel Speed Sensor Connector for proper terminal contact):
When left rear wheel, repair open in CKT 884 or 885.
When right rear wheel, repair open in CKT 882 or 883.
Below 800 ohms:
When left rear wheel, repair short between CKT 884 and 885.
When right rear wheel, repair short between CKT 882 and 883.

YES:
- Measure resistance between J 38716 Pinout Box Terminal 1 and Data Link Connector (DLC) Terminal A (ground).

Is less than 5 ohms measured?

NO: Repair open in CKT 251.

YES:
- Measure resistance at J 38716 Pinout Box Terminals indicated for suspected rear Wheel Speed Sensor Circuit:
  - Left Rear Wheel - Terminal 46 and 1
  - Right Rear Wheel - Terminal 45 and 1

Is less than 5 ohms measured?

YES: When left rear wheel, repair short to ground in CKT 884 or 885. When right rear wheel, repair short to ground in CKT 882 or 883.

NO: Check EBCM/EBTCM Connector for proper terminal contact. If terminal contact is OK, reconnect EBCM/EBTCM, clear DTC and test drive vehicle. If DTC resets, replace EBTCM. If DTC does not reset, refer to Intermittants and Poor Connections in this section.

After repairs are completed, clear DTC and repeat ABS/TCS functional test.

DIAGNOSTIC CHART DTC 22, 23, 26, 27, 32, 33, 36 AND 37 — 1995 EXCEPT AURORA AND RIVIERA

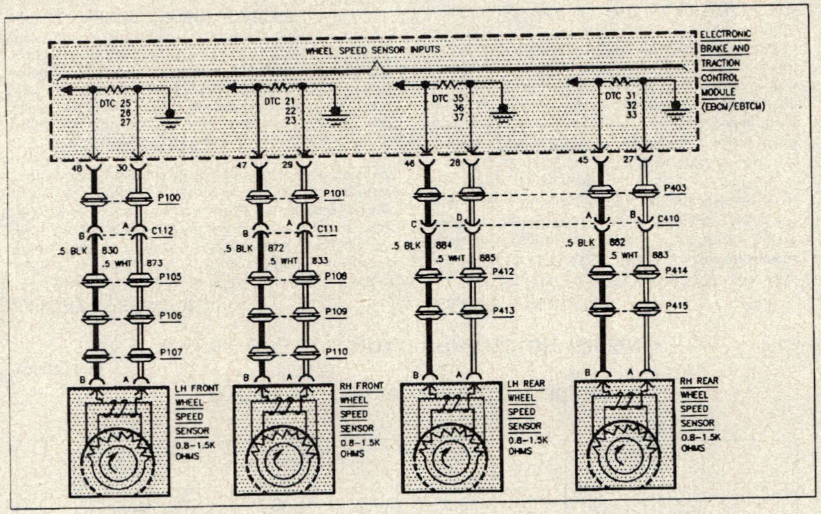

DTC(s) 22, 23, 26, 27, 32, 33, 36, 37

(page 1 of 2)

WHEEL SPEED SIGNAL ERRATIC OR MISSING

CIRCUIT OPERATION

When each wheel turns, the wheel speed sensor creates a small AC voltage with a frequency proportional to wheel speed. This magnetically induced voltage is caused by a toothed sensor ring passing the wheel speed sensor's stationary pickup coil. The toothed ring sensor is mounted on the drive axle in the front, and in the hub/bearing assembly in the rear.

The EBCM/EBTCM uses the wheel speed sensor signal to calculate vehicle reference speeds, and individual wheel speed, acceleration and slip values. These values are used to determine when antilock control is required. Vehicles equipped with traction control also use front wheel speed sensor signals to determine when traction control is required.

WHEEL SPEED SENSOR CABLES

In order to prevent electromagnetic interference from disturbing the wheel speed sensor signal, the cables from the EBCM/EBTCM to each wheel speed sensor are twisted in pairs a minimum of 6 to 9 turns per foot.

When servicing the sensor cables it is important to maintain the original cables' twist (6 to 9 turns per foot).

FAILURE CONDITIONS

DTC 22, DTC 26, DTC 33 or DTC 36 sets when the EBCM/EBTCM detects electromagnetic interference (noise) in a wheel speed sensor signal. Possible interference sources include accessory motors, wiper motors, ignition circuits and add-on or after market equipment.

DTC 22, DTC 26, DTC 32 or DTC 36 also sets when the EBCM/EBTCM determines a wheel speed signal is erratic. A wheel speed signal indicating that a wheel is accelerating or decelerating faster than physically possible is an example of an erratic signal. An intermittent open or short to ground in a wheel speed sensor could also cause the EBCM/EBTCM to set a DTC.

DTC 23, DTC 27, DTC 33, or DTC 37 sets when the EBCM/EBTCM does not see any wheel speed sensor signal, but determines continuity exists in the circuit. This is cause by a dislodged wheel speed sensor, missing toothed sensor ring (front) or defective hub and bearing assembly (rear).

DIAGNOSTIC CHART DTC 22, 23, 26, 27, 32, 33, 36 AND 37 — 1995 EXCEPT AURORA AND RIVIERA, CONTINUED

DTC(s) 22, 23, 26, 27, 32, 33, 36, 37
WHEEL SPEED SIGNAL ERRATIC OR MISSING
(1 of 2)

Has ABS/TCS functional test been performed?

NO: Perform ABS/TCS functional test before proceeding.

YES: Is DTC 23, DTC 27, DTC 33 or DTC 37 present?

YES:
When DTC 23 or DTC 27:
Reinstall dislodged front Wheel Speed Sensor or replace front Hub and Bearing Assembly. Refer to SECTION 4D.
When DTC 33 or DTC 37:
Reinstall dislodged rear Wheel Speed Sensor or replace Hub and Bearing Assembly.

NO:

- Perform following and repair/replace as necessary:
  Test drive vehicle (refer to Item "3" under Diagnostic Process in this section) monitoring suspect Wheel Speed Sensor using Tech 1 in snapshot (mode F3).
  Disconnect suspect Wheel Speed Sensor Connector and applicable jumper harness connector. Inspect terminal contacts on sensor and connectors for corrosion or deformity. Check connectors for proper terminal contact.
  Trace harness routing of suspect sensor wiring. Inspect for physical damage particularly in areas where pinching or movement by harness or other components occurs.
  Trace harness routing of suspect sensor wiring for proximity to noise sources such as ignition circuits, motors, etc. Also add-on or "after market" equipment.
  Check if any repairs were recently made to suspect sensor wiring. Sensor wiring should be twisted 6 to 9 turns per foot.
  Check for loose sensor mounting, or debris lodged between wheel speed sensor and hub.
- If all above checks are OK, proceed, but be aware that problem causing diagnostic trouble code to set may be intermittent and is usually not caused by a defective EBCM/EBTCM.

- Ignition: RUN
- Gear Selector Lever: NEUTRAL
- Ignition: OFF
- Park Brake: RELEASED
- Raise vehicle, check bearing end play - repair as necessary.
- Disconnect suspect Wheel Speed Sensor at wheel.
- Using J 39200 set to AC scale measure voltage between suspect Wheel Speed Sensor Terminal A and B while turning drive wheels by hand.

Is 50 to 900 millivolts AC measured?

NO: Replace suspect Wheel Speed Sensor.

YES: A

Continued on next page

After repairs are completed, clear DTC and repeat ABS/TCS functional test.

DIAGNOSTIC CHART DTC 22, 23, 26, 27, 32, 33, 36 AND 37 — 1995 EXCEPT AURORA AND RIVIERA, CONTINUED

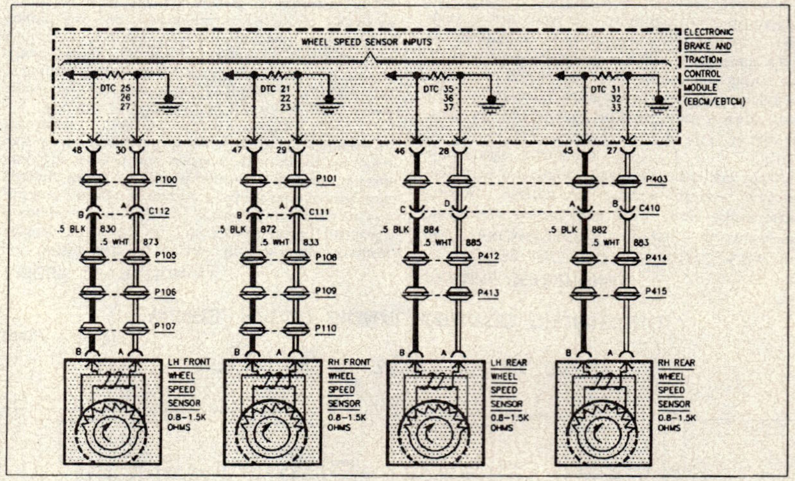

DTC(s) 22, 23, 26, 27, 32, 33, 36, 37

(page 2 of 2)

WHEEL SPEED SIGNAL ERRATIC OR MISSING

CIRCUIT OPERATION

When each wheel turns, the wheel speed sensor creates a small AC voltage with a frequency proportional to wheel speed. This magnetically induced voltage is caused by a toothed sensor ring passing the wheel speed sensor's stationary pickup coil. The toothed ring sensor is mounted on the drive axle in the front, and in the hub/bearing assembly in the rear.

The EBCM/EBTCM uses the wheel speed sensor signal to calculate vehicle reference speeds, and individual wheel speed, acceleration and slip values. These values are used to determine when antilock control is required. Vehicles equipped with traction control also use front wheel speed sensor signals to determine when traction control is required.

WHEEL SPEED SENSOR CABLES

In order to prevent electromagnetic interference from disturbing the wheel speed sensor signal, the cables from the EBCM/EBTCM to each wheel speed sensor are twisted in pairs a minimum of 6 to 9 turns per foot.

When servicing the sensor cables it is important to maintain the original cables' twist (6 to 9 turns per foot).

FAILURE CONDITIONS

DTC 22, DTC 26, DTC 33 or DTC 36 sets when the EBCM/EBTCM detects electromagnetic interference (noise) in a wheel speed sensor signal. Possible interference sources include accessory motors, wiper motors, ignition circuits and add-on or after market equipment.

DTC 22, DTC 26, DTC 32 or DTC 36 also sets when the EBCM/EBTCM determines a wheel speed signal is erratic. A wheel speed signal indicating that a wheel is accelerating or decelerating faster than physically possible is an example of an erratic signal. An intermittent open or short to ground in a wheel speed sensor could also cause the EBCM/EBTCM to set a DTC.

DTC 23, DTC 27, DTC 33, or DTC 37 sets when the EBCM/EBTCM does not see any wheel speed sensor signal, but determines continuity exists in the circuit. This is cause by a dislodged wheel speed sensor, missing toothed sensor ring (front) or defective hub and bearing assembly (rear).

DIAGNOSTIC CHART DTC 22, 23, 26, 27, 32, 33, 36 AND 37 — 1995 EXCEPT AURORA AND RIVIERA, CONTINUED

DTC(s) 22, 23, 26, 27, 32, 33, 36, 37
WHEEL SPEED SIGNAL ERRATIC OR MISSING
(2 of 2)

A

- Thoroughly spray all Wheel Speed Sensors, connectors and harnesses with a 5% solution (2 teaspoons of salt to 12 oz. of water) of salt water.
- Test drive vehicle in a straight line at slow speeds (under 16 km/h [10 mph] while monitoring Wheel Speed Sensors using Tech 1 in snapshot mode (NOTE: suspect Wheel Speed Signal usually varies from remaining Wheel Speed Signals by approximately 2 km/h [1 mph]).

Does any one Wheel Speed Signal vary from other three Wheel Speed Signals?

YES → Replace suspect Wheel Speed Sensor.

NO →
- Clear DTC(s) with Tech 1.
- Test drive vehicle.
- If DTC(s) resets, replace EBCM/EBTCM. If DTC(s) do not reset, refer to Intermittants and Poor Connections in this section.

After repairs are completed, clear DTC and repeat ABS/TCS functional test.

DIAGNOSTIC CHART DTC 45 — 1995 EXCEPT AURORA AND RIVIERA

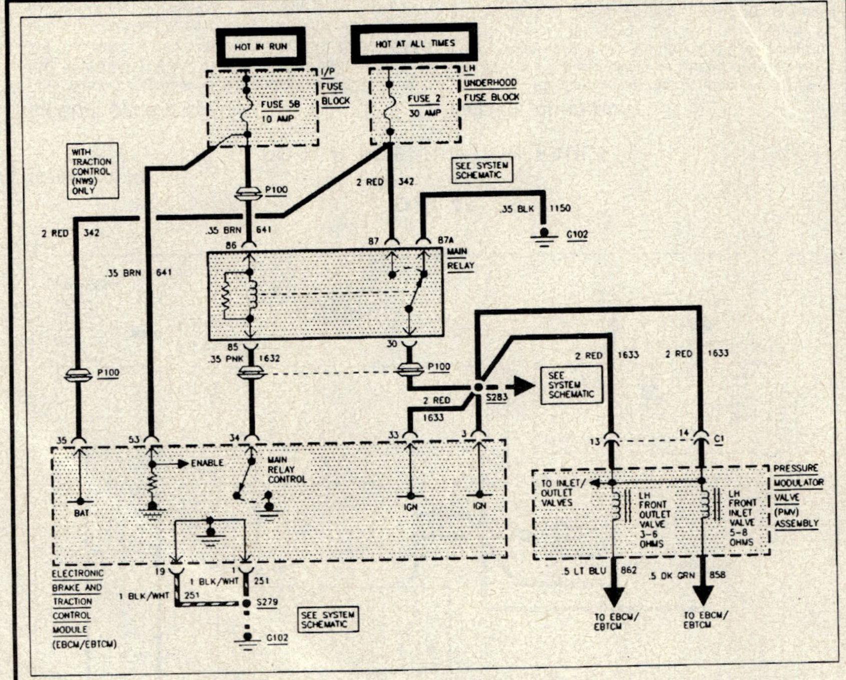

DTC 45

(page 1 of 2)

POWER INTERRUPTION FAULT

CIRCUIT OPERATION

When the ignition is first turned to "RUN," the EBCM/EBTCM goes through a self-check lasting 2-4 seconds. During this test the EBCM/EBTCM grounds terminal 34, energizing the main relay. Once the self-check is completed the main relay remains energized. This delivers battery voltage to both the EBCM/EBTCM and the pressure modulator valve (PMV) assembly during vehicle operation. The constant voltage supply allows the vehicle to enter ABS mode without delay. The main relay is de-energized whenever the amber "ANTILOCK" indicator lights due to a system fault, causing the ABS system to shut down.

FAILURE CONDITION

DTC 45 sets when the ignition switch is at "RUN" and battery voltage is neither detected at the PMV assembly nor EBCM/EBTCM terminal 20, 3 and/or 33. Code 45 also sets when the EBCM/EBTCM detects an open or a short to ground in the LH inlet valve circuit.

Possible Causes

- Open in CKT 641, 1632, 858, 1633, or 342.
- Short to ground in CKT 858, 1633, or 342.
- Main relay fails "open."

DIAGNOSTIC CHART DTC 45 — 1995 EXCEPT AURORA AND RIVIERA, CONTINUED

DTC 45
POWER INTERRUPTION FAULT (1 of 2)

IMPORTANT: DTC 45 may set intermittently due to low system voltage (extended cranking, low battery, but vehicle eventually starts etc.)

Has ABS/TCS functional test been performed?

- **YES**
  - Perform chart "DTC(s) 41, 42, 45, 46, 51, 52, 55, 56 inlet/outlet valve circuit open or shorted" then perform chart "DTC(s) 44, 48 isolation valve circuit open or shorted".
  - Was problem found while performing inlet/outlet valve or isolation valve tests?
- **NO**
  - Perform ABS/TSC functional test before proceeding.

Was problem found...?

- **YES** — OK
- **NO**
  - Using fused jumper, jump between Pinout Box Terminals 1 and 34.
  - Ignition: RUN
  - Does "ABS" Indicator go off?

Does "ABS" Indicator go off?

- **YES**
  - Using J 39200 set to DC scale, check for battery voltage between Pinout Box Terminals 1 and 3 then 1 and 33, PMV harness side connector C1. Terminal 13 and ground then Terminal 14 and ground.
  - Battery voltage present?
    - **NO** — Repair open in CKT 1633.
    - **YES** — Replace EBCM/EBTCM
- **NO**
  - IGNITION : OFF
  - Remove fused jumper.
  - Disconnect Main Relay.
  - Check for continuity between relay Terminals 30 and 87A.
  - Continuity Present?

Continuity Present?

- **NO** — Replace Main Relay
- **YES**
  - Measure resistance between main relay Terminals 85 and 86.
  - 40-90 ohms measured?

40-90 ohms measured?

- **NO** — Replace Main Relay
- **YES**
  - Using unpowered test light to battery, probe main relay Harness Connector Cavity 87A.
  - Does test light illuminate?

Does test light illuminate?

- **NO** — Repair open in CKT 1150
- **YES**
  - Test light to ground, probe main Relay Harness Connector Cavity 87.
  - Does test light illuminate?

(A)

Continued on next page

After repairs are completed, clear DTC and repeat ABS/TCS functional test.

DIAGNOSTIC CHART DTC 45 — 1995 EXCEPT AURORA AND RIVIERA, CONTINUED

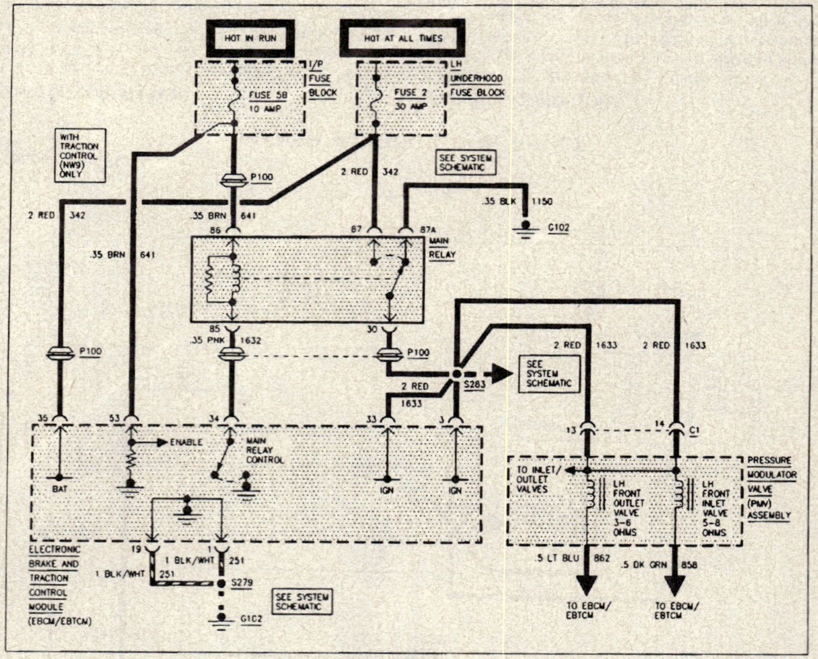

DTC 45

(page 2 of 2)

POWER INTERRUPTION FAULT

CIRCUIT OPERATION

When the ignition is first turned to "RUN," the EBCM/EBTCM goes through a self-check lasting 2-4 seconds. During this test the EBCM/EBTCM grounds terminal 34, energizing the main relay. Once the self-check is completed the main relay remains energized. This delivers battery voltage to both the EBCM/EBTCM and the pressure modulator valve (PMV) assembly during vehicle operation. The constant voltage supply allows the vehicle to enter ABS mode without delay. The main relay is de-energized whenever the amber "ANTILOCK" indicator lights due to a system fault, causing the ABS system to shut down.

FAILURE CONDITION

DTC 45 sets when the ignition switch is at "RUN" and battery voltage is neither detected at the PMV assembly nor EBCM/EBTCM terminal 20, 3 and/or 33. Code 45 also sets when the EBCM/EBTCM detects an open or a short to ground in the LH inlet valve circuit.

Possible Causes

- Open in CKT 641, 1632, 858, 1633, or 342.
- Short to ground in CKT 858, 1633, or 342.
- Main relay fails "open."

DIAGNOSTIC CHART DTC 45 — 1995 EXCEPT AURORA AND RIVIERA, CONTINUED

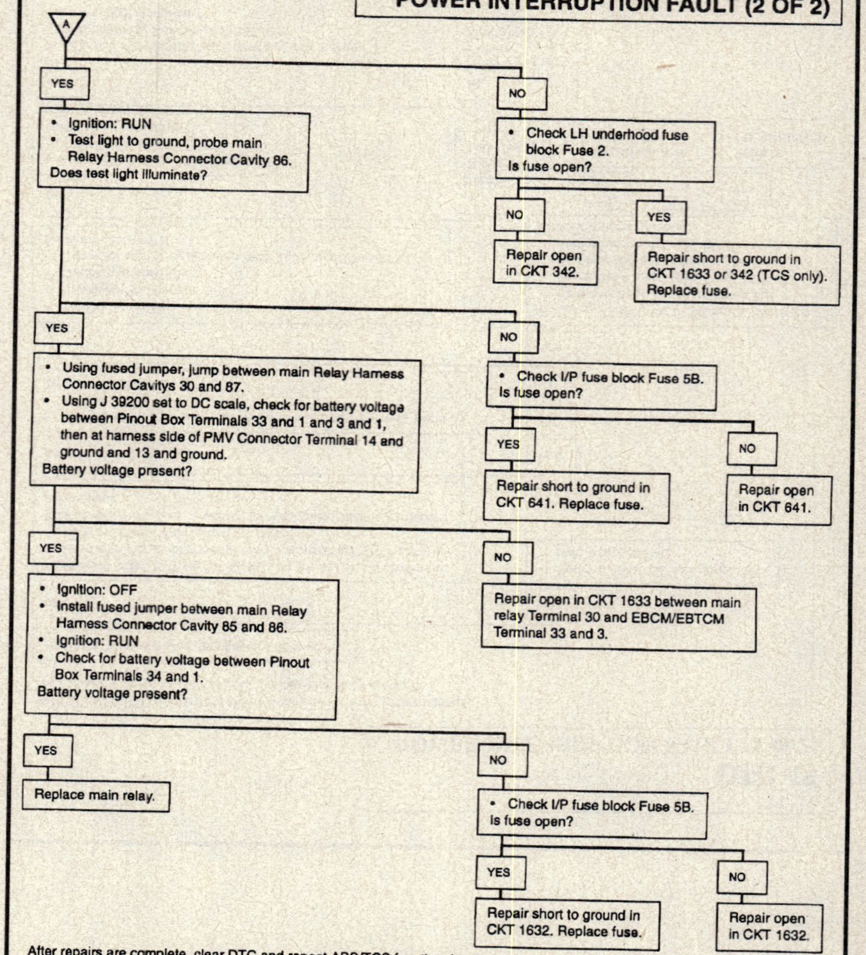

After repairs are complete, clear DTC and repeat ABS/TCS functional test.

DIAGNOSTIC CHART DTC 41, 42, 45, 46, 51, 52, 55 AND 56 — 1995 EXCEPT AURORA AND RIVIERA

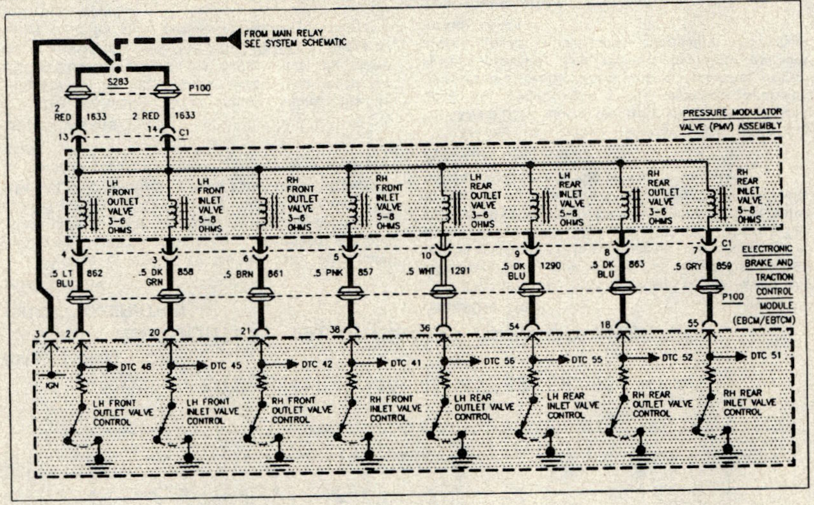

DTC(s) 41, 42, 45, 46, 51, 52, 55, 56

(page 1 of 2)

INLET/OUTLET VALVE CIRCUIT OPEN OR SHORTED

CIRCUIT OPERATION

Inlet Valves

Voltage is applied to the normally open inlet valves whenever the main relay is energized. An inlet valve closes when the EBCM/EBTCM supplies a ground path at the inlet valve control terminal. The EBCM/EBTCM closes an inlet valve for short periods to maintain (hold) or reduce pressure at a particular wheel.

Outlet Valves

The normally closed outlet valves operate similarly to the inlet valves, except that when the EBCM/EBTCM grounds an outlet valve, the valve opens instead of closes. The EBCM/EBTCM opens an outlet valve for short periods to reduce pressure at a particular wheel.

VALVE MODES

Pressure Increase

The pressure increase position is the operating valve position for normal (non-ABS) braking and traction control mode (when equipped). The EBCM/EBTCM does not energize either valve, leaving the inlet valve open and the outlet valve closed.

Pressure Hold

The pressure hold position occurs when the EBCM/EBTCM first notices an impending wheel lockup condition. This signifies an over abundance of hydraulic pressure at the locking wheel. The EBCM/EBTCM closes the corresponding inlet valve in an attempt to avoid any additional hydraulic pressure increase at the wheel.

Pressure Reduce

During a wheel lockup condition, the EBCM/EBTCM closes the inlet valve while pulsing open the outlet valve. This decreases hydraulic pressure at a specific wheel without decreasing overall system pressure. The reduced hydraulic pressure allows greater wheel spin, eliminating the wheel lockup condition.

Test Description

This test checks for proper inlet and outlet valve operation. The wheel brake behavior must be checked to verify the valves hydraulic effectiveness. This is accomplished using a Tech 1 to actuate the valves.

Failure conditions

DTC 41, 42, 45, 46, 51, 52, 55, or 56 sets when the EBCM/EBTCM detects an open or short to ground in the respective inlet or outlet valve circuit.

DIAGNOSTIC CHART DTC 41, 42, 45, 46, 51, 52, 55 AND 56 — 1995 EXCEPT AURORA AND RIVIERA, CONTINUED

DTC(s) 41, 42, 45, 46, 51, 52, 55, 56 INLET/OUTLET VALVE CIRCUIT OPEN OR SHORTED (1 OF 2)

Has ABS/TCS functional test been performed?

- NO: Perform ABS/TCS functional test before proceeding.
- YES:

- Ignition: OFF
- Disconnect Pressure Modulator Valve (PMV) Assembly Valve Connector C1.
- Check connector for proper terminal contact.
- Using J 39200 set to Ω, check for continuity between PMV Assembly Pigtail Connector C1, Terminals 13 and 14.

Continuity?

- NO: Replace PMV Assembly.
- YES:

- Measure resistance between PMV Pigtail Connector C1 Terminal 14 and PMV Valve Block.

Is infinite ohms measured?

- NO: Replace PMV Assembly.
- YES:

- Measure resistance at PMV Assembly Pigtail Connector C1, between Terminal 14 and all Valve Terminals. Refer to chart.

DIAGNOSTIC TROUBLE CODE (DTC)/VALVE	SUSPECT TERMINAL	CORRECT RESISTANCE
45/LH FRONT INLET	3	5 TO 8 OHMS
41/RH FRONT INLET	5	
55/LH REAR INLET	9	
51/RH REAR INLET	7	
46/LH FRONT OUTLET	4	3 TO 6 OHMS
42/RH FRONT OUTLET	6	
56/LH REAR OUTLET	10	
52/RH REAR OUTLET	8	

Is correct resistance measured?

- NO: Replace PMV Assembly.
- YES: A

Continued on next page

After repairs are completed, clear DTC and repeat ABS/TCS functional test.

DIAGNOSTIC CHART DTC 41, 42, 45, 46, 51, 52, 55 AND 56 — 1995 EXCEPT AURORA AND RIVIERA, CONTINUED

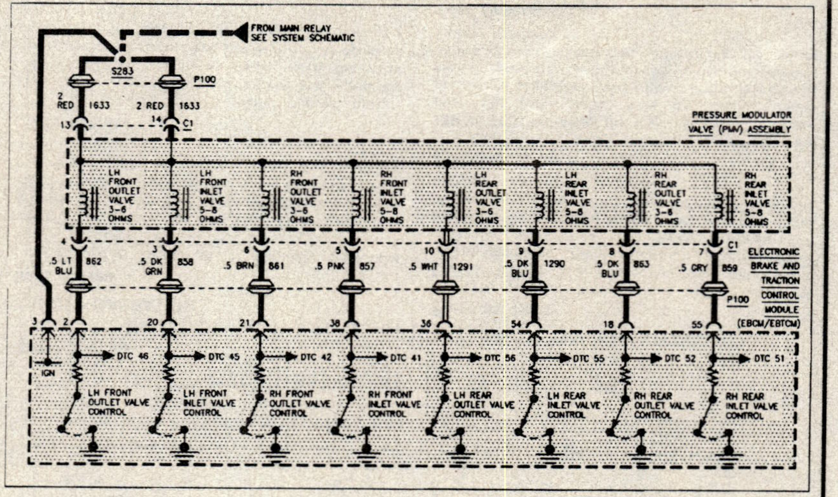

DTC(s) 41, 42, 45, 46, 51, 52, 55, 56

(page 2 of 2)

INLET/OUTLET VALVE CIRCUIT OPEN OR SHORTED

CIRCUIT OPERATION

Inlet Valves

Voltage is applied to the normally open inlet valves whenever the main relay is energized. An inlet valve closes when the EBCM/EBTCM supplies a ground path at the inlet valve control terminal. The EBCM/EBTCM closes an inlet valve for short periods to maintain (hold) or reduce pressure at a particular wheel.

Outlet Valves

The normally closed outlet valves operate similarly to the inlet valves, except that when the EBCM/EBTCM grounds an outlet valve, the valve opens instead of closes. The EBCM/EBTCM opens an outlet valve for short periods to reduce pressure at a particular wheel.

VALVE MODES

Pressure Increase

The pressure increase position is the operating valve position for normal (non-ABS) braking and traction control mode (when equipped). The EBCM/EBTCM does not energize either valve, leaving the inlet valve open and the outlet valve closed.

Pressure Hold

The pressure hold position occurs when the EBCM/EBTCM first notices an impending wheel lockup condition. This signifies an over abundance of hydraulic pressure at the locking wheel. The EBCM/EBTCM closes the corresponding inlet valve in an attempt to avoid any additional hydraulic pressure increase at the wheel.

Pressure Reduce

During a wheel lockup condition, the EBCM/EBTCM closes the inlet valve while pulsing open the outlet valve. This decreases hydraulic pressure at a specific wheel without decreasing overall system pressure. The reduced hydraulic pressure allows greater wheel spin, eliminating the wheel lockup condition.

Test Description

This test checks for proper inlet and outlet valve operation. The wheel brake behavior must be checked to verify the valves hydraulic effectiveness. This is accomplished using a Tech 1 to actuate the valves.

Failure conditions

DTC 41, 42, 45, 46, 51, 52, 55, or 56 sets when the EBCM/EBTCM detects an open or short to ground in the respective inlet or outlet valve circuit.

DIAGNOSTIC CHART DTC 41, 42, 45, 46, 51, 52, 55 AND 56 — 1995 EXCEPT AURORA AND RIVIERA, CONTINUED

DTC(s) 41, 42, 45, 46, 51, 52, 55, 56 INLET/OUTLET VALVE CIRCUIT OPEN OR SHORTED (2 OF 2)

A

- Reconnect PMV Assembly Valve Connector C1.
- Disconnect EBCM/EBTCM Connector.
- Install J 38716 Pinout Box.
- Measure resistance between J 38716 Pinout Box Terminal 3 and all Valve Terminals. Refer to Chart.

DIAGNOSTIC TROUBLE CODE (DTC)/VALVE	J 38716 TERMINAL	CORRECT RESISTANCE
45/LH FRONT INLET	20	5 TO 8 OHMS
41/RH FRONT INLET	38	
55/LH REAR INLET	54	
51/RH REAR INLET	55	
46/LH FRONT OUTLET	2	3 TO 6 OHMS
42/RH FRONT OUTLET	21	
56/LH REAR OUTLET	36	
52/RH REAR OUTLET	18	

Is resistance measured less than 10 ohms?

NO → Repair open in Suspect Valve Circuit between Pressure Modulator Valve (PVM) Assembly and EBCM/EBTCM.

YES →

- Measure resistance between J 38716 Pinout Box Terminal 1 and Data Link Connector (DLC) Terminal 4 (ground).

Is less than 5 ohms measured?

NO → Repair open in CKT 251.

YES →

- Disconnect Main Relay.
- Measure resistance between J 38716 Pinout Box Terminal 1 and all Valve Terminals. Refer to Chart.

DIAGNOSTIC TROUBLE CODE (DTC)/VALVE	J 38716 TERMINAL	CORRECT RESISTANCE
45/LH FRONT INLET	20	INFINITE OHMS
41/RH FRONT INLET	38	
55/LH REAR INLET	54	
51/RH REAR INLET	55	
46/LH FRONT OUTLET	2	
42/RH FRONT OUTLET	21	
56/LH REAR OUTLET	36	
52/RH REAR OUTLET	18	

Is infinite ohms measured?

NO → Repair short to ground in suspect valve circuit between PMV Assembly and EBCM/EBTCM.

YES →

- Reinstall main relay.
- IGNITION: RUN
- Using J 39200 set to DC scale, measure voltage between Pinout Box Terminal 1 and suspect valve terminal.

Is more than 2 volts measured?

YES → Repair short to battery in suspect valve circuit.

NO →

- If diagnosing DTC 45, return to DTC 45 chart.

If not, Check EBCM/EBTCM connector for proper terminal contact. If terminal contact is OK, replace EBCM/EBTCM.

After repairs are completed, clear DTC and repeat ABS/TCS functional test.

DIAGNOSTIC CHART A DTC 44 AND 48 — 1995 EXCEPT AURORA AND RIVIERA

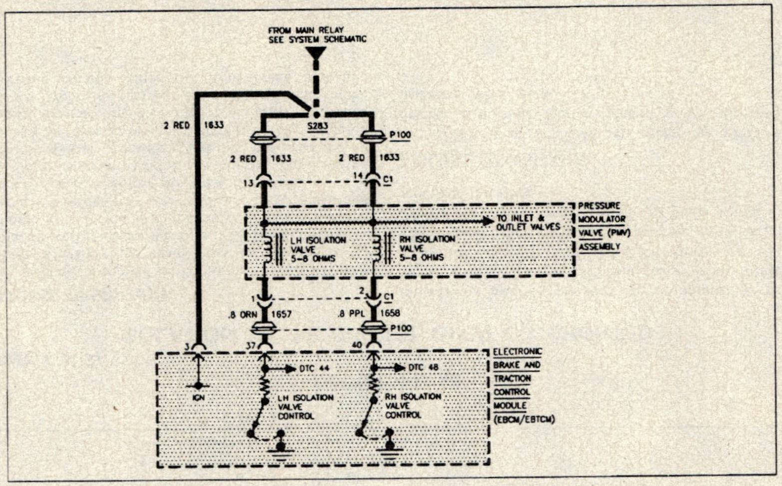

DTC(s) 44, 48

(page 1 of 2)

ISOLATION VALVE CIRCUIT OPEN OR SHORTED

CIRCUIT OPERATION

Voltage is applied to the normally open isolation valves whenever the main relay is energized. An isolation valve closes when the EBTCM supplies a ground path at the isolation valve control terminal. The EBTCM detects wheel slip in a drive wheel and vehicle speed is below 40 km/h (25 mph).

The isolation valves are closed only during traction control operations, and are not used during antilock or normal braking conditions. The isolation valves separate (isolate) the front brake hydraulic circuits from the master cylinder and rear brake hydraulic circuits. Once the front brake hydraulic circuits are isolated, pressure can be applied to the front wheels without affecting any other brake hydraulic circuit. The EBTCM turns on the PMV pump motor to apply pressure, and begin cycling the PMV assembly inlet and outlet valves to modulate pressure.

FAILURE CONDITIONS

DTC 44 or DTC 48 sets when the EBTCM detects an open or short to ground in the respective isolation valve circuit.

DIAGNOSTIC CHART A DTC 44 AND 48 — 1995 EXCEPT AURORA AND RIVIERA, CONTINUED

DTC(s) 44, 48
ISOLATION VALVE CIRCUIT OPEN OR SHORTED (1 OF 2)

Has ABS/TCS Functional Test been performed?

NO: Perform ABS/TCS Functional Test before proceeding.

YES:

- Ignition: OFF
- Disconnect Pressure Modulator Valve (PMV) Assembly Valve Connector C1.
- Check connector for proper terminal contact.
- Using J 39200 set to Ω measure resistance between PMV Assembly Pigtail Connector C1 terminals 13 and 14.

Is less than 5 OHMS measured?

NO: Replace PMV Assembly.

YES:

- Measure resistance between PMV Pigtail Connector C1 terminal 14 and PMV Valve Block.

Is infinite OHMS measured?

NO: Replace PMV Assembly.

YES:

- Measure resistance at connector, C1, (PMV Assembly Side) between Terminal 14 and Suspect Valve Terminal. Refer to chart.

DTC/VALVE	SUSPECT TERMINAL	CORRECT RESISTANCE
44/LH ISOLATION	1	5 TO 8 OHMS
48/RH ISOLATION	2	

Is correct resistance measured?

NO: Replace PMV Assembly.

YES:

- Reconnect PMV Assembly Valve Connector, C1.
- Disconnect EBTCM Connector.
- Install J 38716 Pinout Box.
- Measure resistance between J 38716 Pinout Box Terminal 3 and Suspect Isolation Valve Terminal. Refer to chart.

DTC/VALVE	J 38716 TERMINAL
44/LH ISOLATION	37
48/RH ISOLATION	40

Is resistance measured less than 10 OHMS?

NO: Repair open in Suspect Isolation Valve Circuit between Pressure Modulator Valve (PMV) Assembly and EBCM/EBTCM.

YES:

- Measure resistance between J 38716 Pinout Box Terminal 1 and Data Link Connector (DLC) Terminal 4 (Ground).
- Is less than 2 OHMS measured?

A

Continued on next page

DIAGNOSTIC CHART DTC 44 AND 48 — 1995 EXCEPT AURORA AND RIVIERA

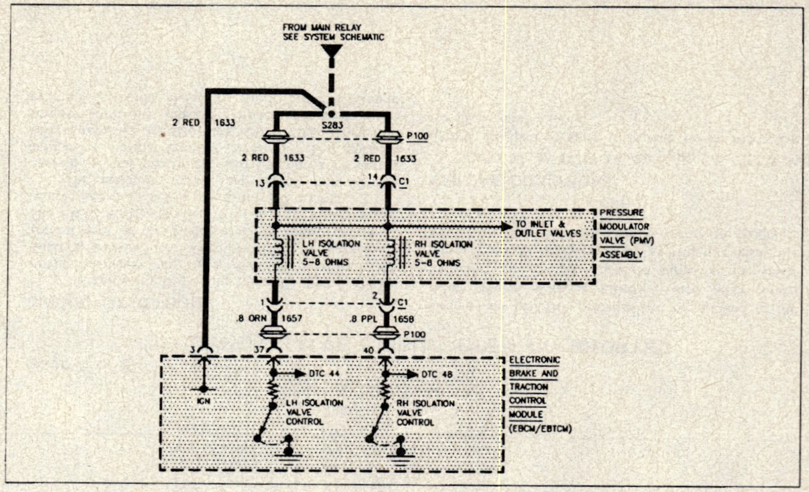

DTC(s) 44, 48

(page 2 of 2)

ISOLATION VALVE CIRCUIT OPEN OR SHORTED

CIRCUIT OPERATION

Voltage is applied to the normally open isolation valves whenever the main relay is energized. An isolation valve closes when the EBTCM supplies a ground path at the isolation valve control terminal. The EBTCM detects wheel slip in a drive wheel and vehicle speed is below 40 km/h (25 mph).

The isolation valves are closed only during traction control operations, and are not used during antilock or normal braking conditions. The isolation valves separate (isolate) the front brake hydraulic circuits from the master cylinder and rear brake hydraulic circuits. Once the front brake hydraulic circuits are isolated, pressure can be applied to the front wheels without affecting any other brake hydraulic circuit. The EBTCM turns on the PMV pump motor to apply pressure, and begin cycling the PMV assembly inlet and outlet valves to modulate pressure.

FAILURE CONDITIONS

DTC 44 or DTC 48 sets when the EBTCM detects an open or short to ground in the respective isolation valve circuit.

DIAGNOSTIC CHART DTC 44 AND 48 — 1995 EXCEPT AURORA AND RIVIERA, CONTINUED

DTC(s) 44, 48
ISOLATION VALVE CIRCUIT OPEN OR SHORTED
(1 OF 2)

Has ABS/TCS Functional Test been performed?

- NO: Perform ABS/TCS Functional Test before proceeding.
- YES:
  - Ignition: OFF
  - Disconnect Pressure Modulator Valve (PMV) Assembly Valve Connector C1.
  - Check connector for proper terminal contact.
  - Using J 39200 set to Ω measure resistance between PMV Assembly Pigtail Connector C1 terminals 13 and 14.

  Is less than 5 OHMS measured?

- NO: Replace PMV Assembly.
- YES:
  - Measure resistance between PMV Pigtail Connector C1 terminal 14 and PMV Valve Block.

  Is infinite OHMS measured?

- NO: Replace PMV Assembly.
- YES:
  - Measure resistance at connector, C1, (PMV Assembly Side) between Terminal 14 and Suspect Valve Terminal. Refer to chart.

DTC/VALVE	SUSPECT TERMINAL	CORRECT RESISTANCE
44/LH ISOLATION	1	5 TO 8 OHMS
48/RH ISOLATION	2	

Is correct resistance measured?

- NO: Replace PMV Assembly.
- YES:
  - Reconnect PMV Assembly Valve Connector, C1.
  - Disconnect EBTCM Connector.
  - Install J 38716 Pinout Box.
  - Measure resistance between J 38716 Pinout Box Terminal 3 and Suspect Isolation Valve Terminal. Refer to chart.

DTC/VALVE	J 38716 TERMINAL
44/LH ISOLATION	37
48/RH ISOLATION	40

Is resistance measured less than 10 OHMS?

- NO: Repair open in Suspect Isolation Valve Circuit between Pressure Modulator Valve (PMV) Assembly and EBCM/EBTCM.
- YES:
  - Measure resistance between J 38716 Pinout Box Terminal 1 and Data Link Connector (DLC) Terminal 4 (Ground).
  - Is less than 2 OHMS measured?

A

Continued on next page

DIAGNOSTIC CHART DTC 61 — 1995 EXCEPT AURORA AND RIVIERA

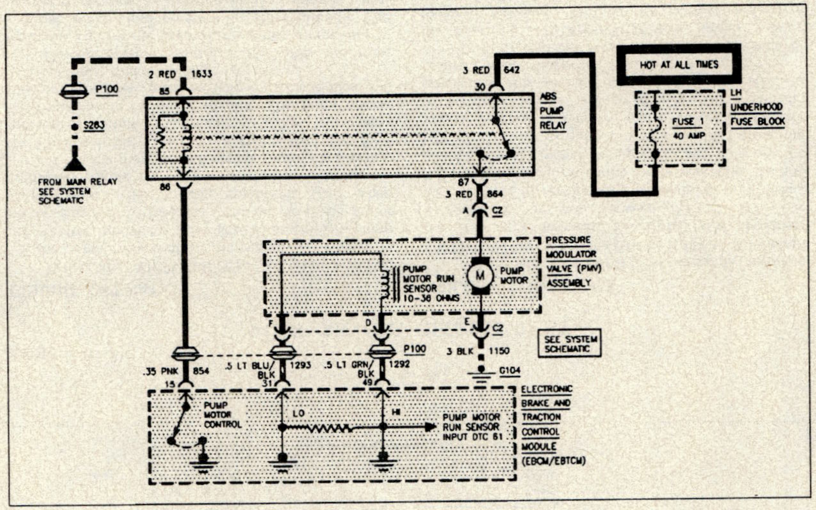

(page 1 of 3)

DTC 61
PUMP MOTOR CIRCUIT TEST

CIRCUIT OPERATION

When the EBCM/EBTCM grounds terminal 15, the pump relay contacts close applying voltage to the pressure modulator valve (PMV) assembly's pump motor. The EBCM/EBTCM turns on the pump motor during most ABS braking conditions (and when equipped) during all traction control conditions.

When the pump motor runs during ABS braking, the PMV assembly reservoir is replenished and fluid pressure is applied to the master cylinder. This causes the brake pedal to rise gradually. The pump motor runs until the brake pedal sufficiently rises to close the TCC/antilock brake switch.

During traction control mode, after the front hydraulic circuits are isolated, the pump motor runs to increase brake fluid pressure in these circuits. The increased pressure is directed to the slipping wheel through the PMV assembly's front inlet and outlet valves. This causes the brakes to be applied, reducing wheel spin at the slipping wheel. Traction control is disabled as soon as the brakes are manually applied.

The EBCM/EBTCM monitors pump motor operation via the pump motor run sensor. When the pump is operating, voltage is induced into the sensor's coil and the EBCM/EBTCM monitors this voltage at terminals 49 and 31.

FAILURE CONDITIONS

When the EBCM/EBTCM grounds terminal 15, DTC 61 sets when the EBCM/EBTCM determines the frequency between EBCM/EBTCM terminals 31 and 49 is not between 70-150 Hz. The EBCM/EBTCM recognizes this condition only during ABS braking requiring pump motor operation, traction control mode (when equipped), and during the EBCM/EBTCM self-check which occurs when the vehicle reaches 11 km/h (7 mph). Therefore, be sure to test drive vehicle, refer to Item 3 under diagnostic process in this section, after clearing the DTC to assure correct diagnosis.

DTC 61 also sets when the pump motor runs all the time (i.e. EBCM/EBTCM sees proper voltage between terminals 31 and 49 without grounding terminal 15). Refer to Chart F for diagnosing a continuous running pump motor.

Possible Causes

- Open in CKT 1150, 864, 854, 1633, 1292, 1293 or 1446.
- Short to ground in CKT 864, 854, 1292, 1293, or 642.
- ABS pump relay fails "open" or "engaged."

DIAGNOSTIC CHART DTC 61 — 1995 EXCEPT AURORA AND RIVIERA, CONTINUED

DTC 61 (1 OF 3)
PUMP MOTOR CIRCUIT TEST

IMPORTANT
DTC 61 usually sets while the vehicle is moving during the system "Self Test".

Has ABS/TCS functional test been performed?

- NO → Perform ABS/TCS functional test.
- YES →
  - Ignition: RUN
  - Does Pressure Modulator Valve (PMV) Assembly Pump Motor Run continuously?
    - YES →
      - Remove ABS Pump Relay.
      - Does pump motor stop running?
        - NO → Repair short to battery in CKT 864.
        - YES →
          - IGNITION: OFF
          - Using J 39200 set to DC scale, check for battery voltage at Pump Relay Cavity 86.
          - Is battery voltage present?
            - YES → Repair short to battery in CKT 854.
            - NO → Replace Pump Relay
    - NO →
      - Ignition: OFF
      - Connect Tech 1 to Data Link Connector (DLC).
      - Ignition: RUN
      - Use Tech 1 to bleed PMV Assembly.
      - Does PMV Assembly Pump Motor run?
        - NO → (B) Continued on Chart 3 of 3
        - YES →
          - Ignition: OFF
          - Disconnect Tech 1.
          - Disconnect PMV Assembly Pump Motor Connector C2.
          - Using J 39200 set to Ω, measure resistance between PMV Assembly Pigtail Connector C2 Terminal D and PMV Assembly casing.
          - Is infinite ohms measured?
            - NO → Replace PMV Assembly.
            - YES →
              - Measure resistance between PMV Assembly Pigtail Connector C2 Terminal D and F.
              - Is resistance between 10 and 40 ohms?
                - NO → Replace PMV Assembly.
                - YES → (A) Continued on next page

After repairs are completed, clear DTC and repeat ABS/TCS functional test.

DIAGNOSTIC CHART DTC 61 — 1995 EXCEPT AURORA AND RIVIERA, CONTINUED

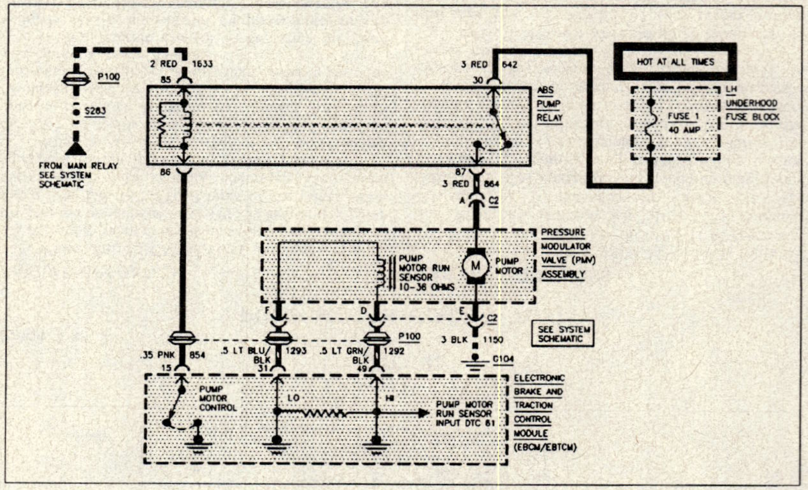

DTC 61

(page 2 of 3)

PUMP MOTOR CIRCUIT TEST

CIRCUIT OPERATION

When the EBCM/EBTCM grounds terminal 15, the pump relay contacts close applying voltage to the pressure modulator valve (PMV) assembly's pump motor. The EBCM/EBTCM turns on the pump motor during most ABS braking conditions (and when equipped) during all traction control conditions.

When the pump motor runs during ABS braking, the PMV assembly reservoir is replenished and fluid pressure is applied to the master cylinder. This causes the brake pedal to rise gradually. The pump motor runs until the brake pedal sufficiently rises to close the TCC/antilock brake switch.

During traction control mode, after the front hydraulic circuits are isolated, the pump motor runs to increase brake fluid pressure in these circuits. The increased pressure is directed to the slipping wheel through the PMV assembly's front inlet and outlet valves. This causes the brakes to be applied, reducing wheel spin at the slipping wheel. Traction control is disabled as soon as the brakes are manually applied.

The EBCM/EBTCM monitors pump motor operation via the pump motor run sensor. When the pump is operating, voltage is induced into the sensor's coil and the EBCM/EBTCM monitors this voltage at terminals 49 and 31.

FAILURE CONDITIONS

When the EBCM/EBTCM grounds terminal 15, DTC 61 sets when the EBCM/EBTCM determines the frequency between EBCM/EBTCM terminals 31 and 49 is not between 70-150 Hz. The EBCM/EBTCM recognizes this condition only during ABS braking requiring pump motor operation, traction control mode (when equipped), and during the EBCM/EBTCM self-check which occurs when the vehicle reaches 11 km/h (7 mph). Therefore, be sure to test drive vehicle, refer to Item 3 under diagnostic process in this section, after clearing the DTC to assure correct diagnosis.

DTC 61 also sets when the pump motor runs all the time (i.e. EBCM/EBTCM sees proper voltage between terminals 31 and 49 without grounding terminal 15). Refer to Chart F for diagnosing a continuous running pump motor.

Possible Causes

- Open in CKT 1150, 864, 854, 1633, 1292, 1293 or 1446.
- Short to ground in CKT 864, 854, 1292, 1293, or 642.
- ABS pump relay fails "open" or "engaged."
- Also check for a non-functioning main relay

DIAGNOSTIC CHART DTC 61 — 1995 EXCEPT AURORA AND RIVIERA, CONTINUED

DTC 61(2 OF 3)
PUMP MOTOR CIRCUIT TEST

(A)

- Reconnect PMV Assembly Pump Motor Connector C2.
- Disconnect EBCM/EBTCM Connector.
- Connect J 38716 Pinout Box.
- Measure resistance between J 38716 Pinout Box Terminal 1 and DLC Terminal 4 (ground).

Is less than 5 ohms measured?

- NO: Repair open in CKT 251.
- YES:
  - Measure resistance between J 38716 Pinout Box Terminals 31 and 49.

  Is resistance between 10 and 40 ohms?

  - NO: When high resistance is measured, repair CKT 1292 or 1293 for an open. Check PMV Assembly Connector C2 for proper terminal contact. When low resistance is measured, repair short between CKT 1292 and 1293.
  - YES:
    - Measure resistance between J 38716 Pinout Box Terminals 49 and 1, then 31 and 1.

    Is less than 5 ohms measured in each test?

    - YES: Repair CKT 1292 or 1293 for a short to ground.
    - NO:
      - Using fused jumper, jump between Pinout Box Terminals 19 and 34. Using another fused jumper, jump between Pinout Box Terminals 1 and 15.
      - IGNITION: RUN (pump should run)
      - Using J 39200 set to Hz, measure Motor Run Sensor Output between Pinout Box Terminals 31 and 49.

      Is 70 -150 Hz measured?

      - NO: Replace PMV
      - YES:
        - Check EBCM/EBTCM Connector for proper terminal contact. If OK, replace EBCM/EBTCM.

After repairs are completed, clear DTC and repeat ABS/TCS functional test.

DIAGNOSTIC CHART DTC 61 — 1995 EXCEPT AURORA AND RIVIERA, CONTINUED

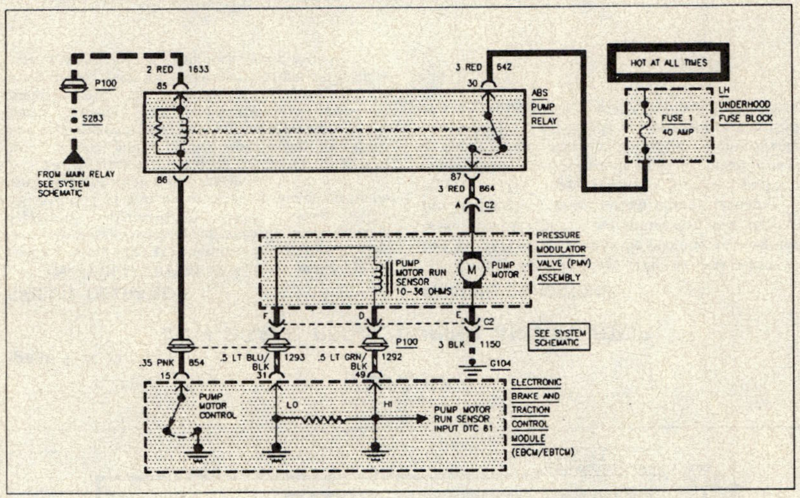

DTC 61

(page 3 of 3)

PUMP MOTOR CIRCUIT TEST

CIRCUIT OPERATION

When the EBCM/EBTCM grounds terminal 15, the pump relay contacts close applying voltage to the pressure modulator valve (PMV) assembly's pump motor. The EBCM/EBTCM turns on the pump motor during most ABS braking conditions (and when equipped) during all traction control conditions.

When the pump motor runs during ABS braking, the PMV assembly reservoir is replenished and fluid pressure is applied to the master cylinder. This causes the brake pedal to rise gradually. The pump motor runs until the brake pedal sufficiently rises to close the TCC/antilock brake switch.

During traction control mode, after the front hydraulic circuits are isolated, the pump motor runs to increase brake fluid pressure in these circuits. The increased pressure is directed to the slipping wheel through the PMV assembly's front inlet and outlet valves. This causes the brakes to be applied, reducing wheel spin at the slipping wheel. Traction control is disabled as soon as the brakes are manually applied.

The EBCM/EBTCM monitors pump motor operation via the pump motor run sensor. When the pump is operating, voltage is induced into the sensor's coil and the EBCM/EBTCM monitors this voltage at terminals 49 and 31.

FAILURE CONDITIONS

When the EBCM/EBTCM grounds terminal 15, DTC 61 sets when the EBCM/EBTCM determines the frequency between EBCM/EBTCM terminals 31 and 49 is not between 70-150 Hz. The EBCM/EBTCM recognizes this condition only during ABS braking requiring pump motor operation, traction control mode (when equipped), and during the EBCM/EBTCM self-check which occurs when the vehicle reaches 11 km/h (7 mph). Therefore, be sure to test drive vehicle, refer to Item 3 under diagnostic process in this section, after clearing the DTC to assure correct diagnosis.

DTC 61 also sets when the pump motor runs all the time (i.e. EBCM/EBTCM sees proper voltage between terminals 31 and 49 without grounding terminal 15). Refer to Chart F for diagnosing a continuous running pump motor.

Possible Causes

- Open in CKT 1150, 864, 854, 1633, 1292, 1293 or 1446.
- Short to ground in CKT 864, 854, 1292, 1293, or 642.
- ABS pump relay fails "open" or "engaged."

DIAGNOSTIC CHART DTC 61 — 1995 EXCEPT AURORA AND RIVIERA, CONTINUED

DTC 61 (3 OF 3)
PUMP MOTOR CIRCUIT TEST

B

- Clear DTC using a Tech 1.
- Ignition: OFF
- Disconnect Tech 1.
- Disconnect ABS Pump Relay.
- Ignition: RUN
- Using J 39200 set to DC scale, measure voltage between ABS Pump Relay Harness Connector Terminal 85 and ground.

Is battery voltage measured?

NO: Repair open in CKT 1633.

YES:

- Ignition: OFF
- Measure voltage between ABS Pump Relay Harness Connector Terminal 30 and ground.

Is battery voltage measured?

NO: Check Fuse 1 of LH Underhood Fuse Block. Check/repair CKT 642 for an open, or a short to ground.

NOTE: When fuse is open and CKT 642 is OK, check/repair short to ground in CKT 864 and replace ABS Pump Relay when a short to ground is found.

YES:

- Connect fused jumper between ABS Pump Relay Harness Terminal 30 and 87A.

Does PMV Assembly Pump Motor run?

NO:

- Leave fused jumper connected.
- Disconnect PMV Assembly Pump Motor Connector C2.
- Measure voltage between PMV Assembly Harness Connector C2 Terminal A and ground.

Is battery voltage measured?

- **NO:** Repair open or short to ground in CKT 864.
- **YES:**
  - Measure voltage between PMV Assembly Harness Connector C2 Terminal A and E.

  Is battery voltage measured?

  - **YES:** Check PMV Assembly Connector C2 for proper terminal contact. If terminal contact is OK, replace PMV Assembly.
  - **NO:** Repair open in CKT 1150.

YES:

- Reconnect ABS Pump Relay.
- Disconnect EBCM/EBTCM Connector.
- Connect J 38716 Pinout Box.
- Using J 39200 set to Ω, measure resistance between J 38716 Pinout Box Terminal 1 and Data Link Connector (DLC) Terminal 4 (ground).

Is less than 5 ohms measured?

- **NO:** Repair open in CKT 251.
- **YES:**
  - Connect fused jumpers between J 38716 Pinout Box Terminal 15 and 1, and Terminal 34 and 1.

  Does PMV Assembly Pump Motor run?

  - **YES:** Check EBCM/EBTCM Connector for proper terminal contact. If terminal contact is OK, replace EBCM/EBTCM.
  - **NO:** Check/repair for an open in CKT 854. If CKT 854 is OK, replace ABS Pump Relay.

After repairs are completed, clear DTC and repeat ABS/TCS functional test.

DIAGNOSTIC CHART DTC 62 — 1995 EXCEPT AURORA AND RIVIERA

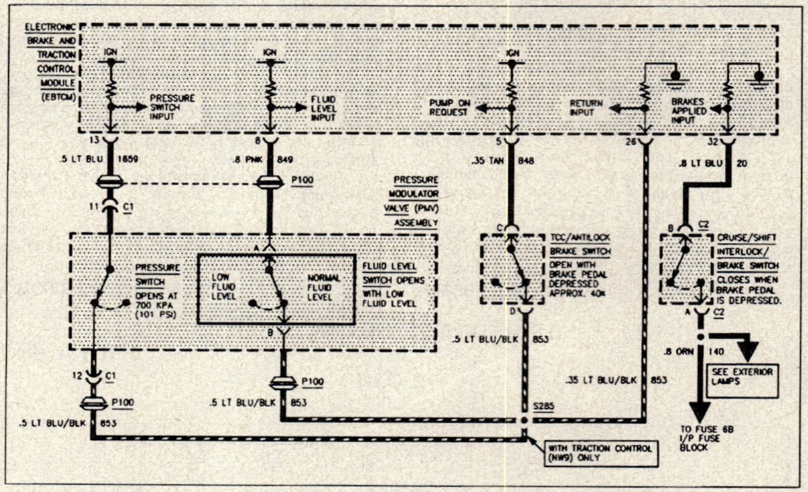

DTC 62

(page 1 of 2)

LOW BRAKE PEDAL DURING AN ABS STOP

CIRCUIT OPERATION

The EBCM/EBTCM monitors brake pedal position using the TCC/antilock brake switch. The normally closed switch opens when the brake pedal is depressed approximately 40% down. The EBCM/EBTCM also monitors pump motor operation using a pump motor run sensor.

During most ABS braking conditions the EBCM turns on the pump motor, replenishing the PMV assembly reservoir and applying fluid pressure to the master cylinder. This causes the brake pedal to gradually rise. The EBCM/EBTCM turns off the pump motor when the brake pedal height sufficiently rises to close the TCC/antilock brake switch.

FAILURE CONDITIONS

During ABS braking, DTC 62 sets when the EBCM/EBTCM determines the pump is working, and brake pedal height has not sufficiently risen to close the TCC/antilock brake switch after 3 seconds.

This DTC may indicate a hydraulic problem. Some possible causes include air in the brake lines or a pump motor mechanical failure. Non-hydraulic causes for DTC 62 include an open 848 or 853 CKT or a misadjusted/defective TCC/antilock brake switch.

DIAGNOSTIC CHART DTC 62 — 1995 EXCEPT AURORA AND RIVIERA, CONTINUED

DTC 62 (1 OF 2)
LOW BRAKE PEDAL DURING AN ABS STOP

Has ABS/TCS functional test been performed?

- NO: Perform ABS/TCS functional test before proceeding.
- YES:
  - Bleed brake system as outlined in this section.
  - Ignition: OFF
  - Connect Tech 1 to Data Link Connector (DLC).
  - Ignition: RUN
  - Use Tech 1 to clear DTC after bleeding system.
  - Depress brake pedal (hold).
  - Use Tech 1 to bleed Pressure Modulator Valve (PMV) assembly.

Did brake pedal rise?

- NO: Replace PMV Assembly.
- YES:
  - Ignition: OFF
  - Disconnect Tech 1.
  - Test drive vehicle on a surface as described in Item "3" under "Diagnostic Process" in this section and perform an ABS stop.
  - Connect Tech 1 to DLC Connector.
  - Ignition: RUN

Did DTC 62 set?

- NO: Problem solved.
- YES:
  - Ignition: OFF
  - Disconnect Tech 1.
  - Brake pedal released (rest position).
  - Check TCC/Antilock Brake Switch adjustment.

Is switch adjusted properly?

- NO: Adjust TCC/Antilock Brake Switch.
- YES:
  - Disconnect TCC/Antilock Brake Switch Connector.
  - Using J 39200 set to Ω, measure resistance across switch Terminals C and D.
- NO: Replace TCC/Antilock Brake Switch.
- YES: (A)

Continued on next page

After repairs are completed, clear DTC and repeat ABS/TCS functional test.

DIAGNOSTIC CHART DTC 62 — 1995 EXCEPT AURORA AND RIVIERA, CONTINUED

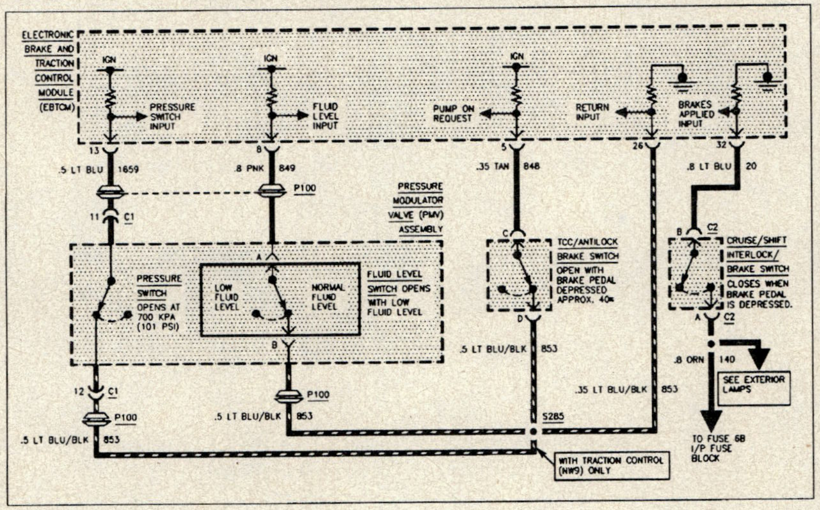

DTC 62

(page 2 of 2)

LOW BRAKE PEDAL DURING AN ABS STOP

CIRCUIT OPERATION

The EBCM/EBTCM monitors brake pedal position using the TCC/antilock brake switch. The normally closed switch opens when the brake pedal is depressed approximately 40% down. The EBCM/EBTCM also monitors pump motor operation using a pump motor run sensor.

During most ABS braking conditions the EBCM turns on the pump motor, replenishing the PMV assembly reservoir and applying fluid pressure to the master cylinder. This causes the brake pedal to gradually rise. The EBCM/EBTCM turns off the pump motor when the brake pedal height sufficiently rises to close the TCC/antilock brake switch.

FAILURE CONDITIONS

During ABS braking, DTC 62 sets when the EBCM/EBTCM determines the pump is working, and brake pedal height has not sufficiently risen to close the TCC/antilock brake switch after 3 seconds.

This DTC may indicate a hydraulic problem. Some possible causes include air in the brake lines or a pump motor mechanical failure. Non-hydraulic causes for DTC 62 include an open 848 or 853 CKT or a misadjusted/defective TCC/antilock brake switch.

DIAGNOSTIC CHART DTC 62 — 1995 EXCEPT AURORA AND RIVIERA, CONTINUED

DTC 62 (2 OF 2)
LOW BRAKE PEDAL DURING AN ABS STOP

(A)

- Disconnect EBCM/EBTCM Connector.
- Connect J 38716 Pinout Box.
- Measure resistance between J 38716 Pinout Box Terminal 5 and TCC/Antilock Brake Switch Connector Terminal C.

Is less than 3 ohms measured?

- NO → Repair open in CKT 848.
- YES →
  - Measure resistance between J 38716 Pinout Box Terminal 26 and TCC/Antilock Brake Switch Connector Terminal D.

  Is less than 3 ohms measured?

  - NO → Repair open in CKT 853.
  - YES → Check EBCM/EBTCM Connector for proper terminal contact. If terminal contact is OK, replace EBCM/EBTCM.

After repairs are completed, clear DTC and repeat ABS/TCS functional test.

DIAGNOSTIC CHART DTC 71 — 1995 EXCEPT AURORA AND RIVIERA

DTC 71

EBCM/EBTCM INTERNAL MALFUNCTION

DTC 71 indicates that a failure has occurred within the EBCM/EBTCM. DTC 71 can occur intermittently, due to interference, therefore clear the code and test drive the vehicle before replacing the EBCM/EBTCM.

DIAGNOSTIC CHART DTC 71 — 1995 EXCEPT AURORA AND RIVIERA, CONTINUED

DTC 71
EBCM/EBTCM INTERNAL FAILURE

Has ABS/TCS functional test been performed?

- YES
  - Using Tech 1 clear Diagnostic Trouble Code.
  - Test drive vehicle. Refer to Diagnostic Process Item "3" in this section.

  Did DTC 71 set during test drive?
  - YES: Replace EBCM/EBTCM.
  - NO: No problem found. May be intermittent. Refer to "Intermittents and Poor Connections" in this section.
- NO: Perform ABS/TCS functional test before proceeding.

After repairs are completed, clear DTC and repeat ABS/TCS functional test.

DIAGNOSTIC CHART DTC 72 — 1995 EXCEPT AURORA AND RIVIERA

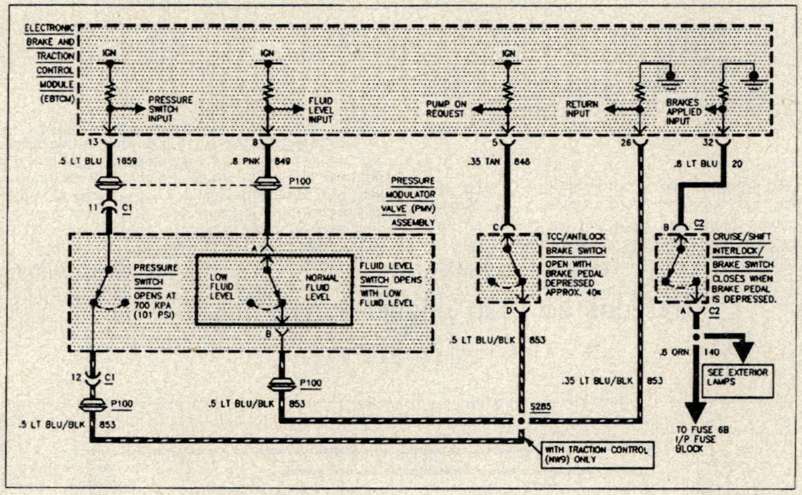

DTC 72

BRAKE SWITCH OPEN OR SHORTED

CIRCUIT OPERATION

The electronic brake control module EBCM/EBTCM monitors brake pedal position using the brake travel switch. The switch is normally closed. It opens when the brake pedal is depressed approximately 40% down. During most ABS braking conditions the EBCM/EBTCM turns on the pump motor, replenishing the PMV assembly reservoir and applying fluid pressure to the master cylinder. This causes the brake pedal to gradually rise. The EBCM/EBTCM turns off the pump motor when brake pedal height sufficiently rises to close the brake travel switch.

FAILURE CONDITIONS

DTC 72 sets when the EBCM/EBTCM detects the brake travel switch CKT 848 is shorted to ground (with the brake pedal depressed) or open.

DIAGNOSTIC CHART DTC 72 — 1995 EXCEPT AURORA AND RIVIERA, CONTINUED

DTC 72
BRAKE TRAVEL SWITCH CIRCUIT OPEN OR SHORTED

Has ABS/TCS functional test been performed?

- NO → Perform ABS/TCS functional test before proceeding.
- YES →
  - Ignition: OFF
  - Disconnect Brake Travel Switch Connector.
  - Disconnect EBCM/EBTCM Connector.
  - Connect J 38716 Pinout Box.
  - Using J 39200 set to Ω, measure resistance between J 38716 Pinout Box Terminal 1 and Data Link Connector (DLC) Terminal 4 (ground).

  Is less than 5 ohms measured?

  - NO → Repair open in CKT 251.
  - YES →
    - Measure resistance between J 38716 Pinout Box Terminal 1 and 5.

    Is less than 5 ohms measured?

    - YES → Repair short to ground in CKT 848.
    - NO →
      - Measure resistance between Pinout Box Terminals 1 and 26.
      - Is less than 5 Ω measured?

      - NO → Repair short to ground in CKT 853.
      - YES → Check EBCM/EBTCM Connector for proper terminal contact. If terminal contact is OK, replace EBCM/EBTCM.

After repairs are completed, clear DTC and repeat ABS/TCS functional test.

DIAGNOSTIC CHART DTC 73 — 1995 EXCEPT AURORA AND RIVIERA

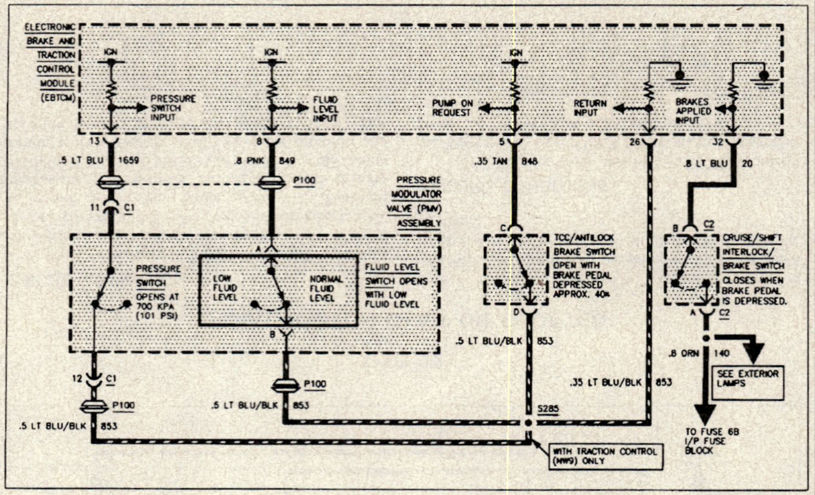

DTC 73

FLUID LEVEL SWITCH CIRCUIT OPEN OR SHORTED

CIRCUIT OPERATION

The pressure modulator valve (PMV) fluid level switch alerts the EBCM/EBTCM to a low brake fluid condition in the PMV reservoir. The fluid level switch is located in the PMV assembly's reservoir. The fluid level switch is normally closed when the reservoir contains sufficient fluid. When fluid level is low, the switch opens, causing the EBCM/EBTCM to turn on the "ANTILOCK" indicator.

FAILURE CONDITIONS

DTC 73 sets when the EBCM/EBTCM detects a short to ground in the PMV fluid level switch circuitry. These include: CKT 848, 849, 853, and 1659. DTC 73 will also set if an open exists in CKT 251.

DIAGNOSTIC CHART DTC 73 — 1995 EXCEPT AURORA AND RIVIERA, CONTINUED

DTC 73
PMV FLUID LEVEL SWITCH CIRCUIT OPEN OR SHORTED

Has ABS/TCS functional test been performed?

- NO → Perform ABS/TCS functional test before proceeding.
- YES →
  - Ignition: OFF
  - Disconnect EBCM/EBTCM Connector.
  - Connect J 38716 Pinout Box.
  - Using J 39200 set to Ω, measure resistance between J 38716 Pinout Box Terminal 1 and Data Link Connector (DLC) Terminal A (ground).

  Is less than 5 ohms measured?

  - NO → Repair open in CKT 251.
  - YES →
    - Disconnect Brake Travel Switch.
    - Measure resistance between J 38716 Pinout Box Terminal 5 and 1.

    Is less than 5 ohms measured?

    - YES → Repair short to ground in CKT 848.
    - NO →
      - Disconnect Pressure Modulator Valve (PMV) Fluid Level Switch Connector.
      - Measure resistance between J 38716 Pinout Box Terminal 8 and 1.

      Is less than 5 ohms measured?

      - YES → Repair short to ground in CKT 849.
      - NO →
        - When equipped with Traction Control: Depress Brake Pedal. Measure resistance between J 38716 Pinout Box Terminal 13 and 1.

        Is less than 5 ohms measured?

        - YES → Repair short to ground in CKT 1659.
        - NO →
          - Measure resistance between J 38716 Pinout Box Terminal 26 and 1.

          Is less than 5 ohms measured?

          - YES → Repair short to ground in CKT 853.
          - NO → Check EBCM/EBTCM Connector for proper terminal contact. If terminal contact is OK, replace EBCM/EBTCM.

After repairs are completed, clear DTC and repeat ABS/TCS functional test.

DIAGNOSTIC CHART DTC 74 — 1995 EXCEPT AURORA AND RIVIERA

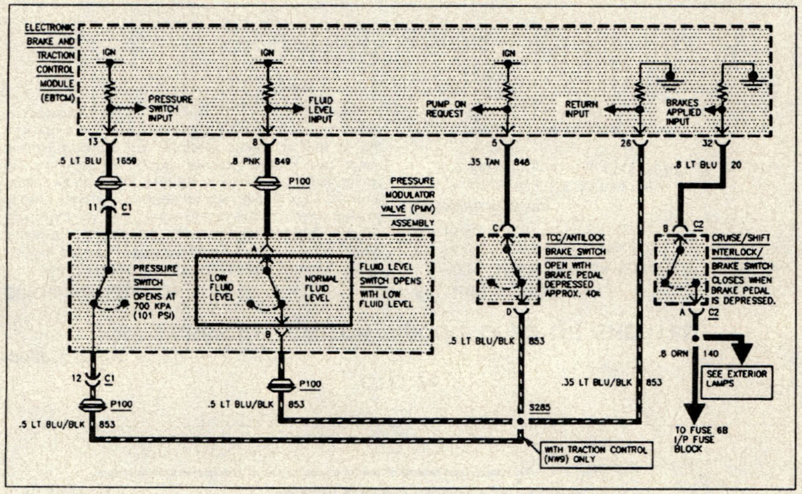

DTC 74

(page 1 of 3)

PMV PRESSURE SWITCH CIRCUIT OPEN OR SHORTED

CIRCUIT OPERATION

The pressure switch monitors primary brake system pressure. The normally closed switch opens when the brakes are applied and pressure exceeds 700 kPa (101 psi). If the brakes are manually applied during traction control mode, the PMV pressure switch input, along with the brake switch input, signal the EBTCM to disable traction control and allow manual braking. The pressure switch is part of the pressure modulator valve (PMV) assembly and is not serviceable.

FAILURE CONDITIONS

DTC 74 sets when the EBTCM detects an open or short to ground in the PMV pressure switch circuitry.

Possible Causes

- Open in CKT 853 or 1659.
- Short to ground in CKT 1659 with the brakes applied.
- Failed PMV pressure switch.

DIAGNOSTIC CHART DTC 74 — 1995 EXCEPT AURORA AND RIVIERA, CONTINUED

DTC 74
PMV PRESSURE SWITCH CIRCUIT OPEN OR SHORTED (PAGE 3 OF 3)

Has ABS/TCS Functional Test been performed?

- NO: Perform ABS/TCS Functional Test.
- YES:

- Ignition: OFF
- Inspect I/P Fuse Block Fuse 6B.

Is Fuse 6B open?

- YES: Repair Short to Ground in CKT 140 or CKT 20. Replace Fuse.
- NO:

- Disconnect Brake Switch Connector (Grey Connector).
- Using J 39200 set to DC scale measure voltage between DLC Terminal 4 and Brake Switch Connector (Harness Side) Terminal A.

Is greater than 10 volts measured?

- NO: Repair Open in CKT 140.
- YES:

- Depress Brake Pedal.
- Using J 39200 set to Ω measure resistance between Brake Switch Terminals A and B.
- Is less than 2 OHMS measured?

- NO: Replace Brake Switch.
- YES:

- Check for proper terminal contact on Brake Switch Connector.
- Reconnect Brake Switch Connector.
- Check for proper Brake Switch Adjustment

Is Brake Switch Adjusted Properly?

- NO: Adjust Brake Switch.
- YES:

- Disconnect EBTCM Connector.
- Install J 38716 Pin Out Box.
- Measure resistance between Pin Out Box Terminal 1 and DLC Terminal 4.

Is less than 5 OHMS measured?

- NO: Repair Open in CKT 1050.
- YES:

- Ignition: ON
- Using J 39200 set to DC scale, measure voltage between Pin Out Box Terminals 1 and 32.

Is greater than 2 volts measured?

A

Continued on next page

DIAGNOSTIC CHART DTC 74 — 1995 EXCEPT AURORA AND RIVIERA, CONTINUED

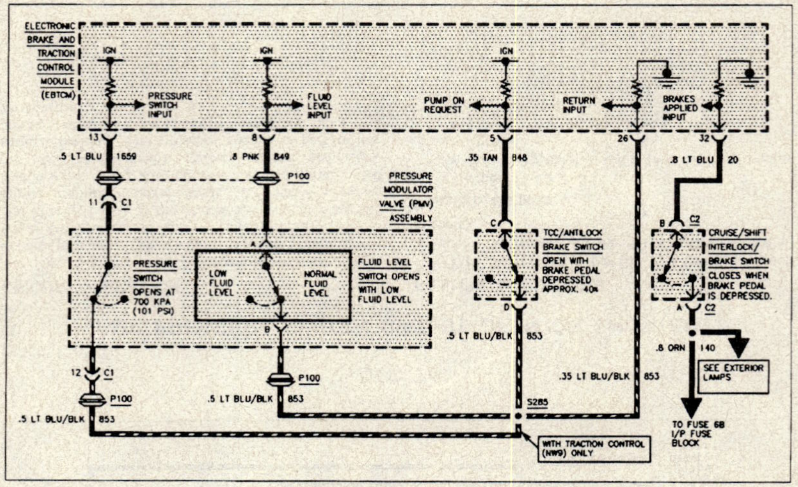

DTC 74

(page 2 of 3)

PMV PRESSURE SWITCH CIRCUIT OPEN OR SHORTED

CIRCUIT OPERATION

The pressure switch monitors primary brake system pressure. The normally closed switch opens when the brakes are applied and pressure exceeds 700 kPa (101 psi). If the brakes are manually applied during traction control mode, the PMV pressure switch input, along with the brake switch input, signal the EBTCM to disable traction control and allow manual braking. The pressure switch is part of the pressure modulator valve (PMV) assembly and is not serviceable.

FAILURE CONDITIONS

DTC 74 sets when the EBTCM detects an open or short to ground in the PMV pressure switch circuitry.

Possible Causes

- Open in CKT 853 or 1659.
- Short to ground in CKT 1659 with the brakes applied.
- Failed PMV pressure switch.

DIAGNOSTIC CHART DTC 74 — 1995 EXCEPT AURORA AND RIVIERA, CONTINUED

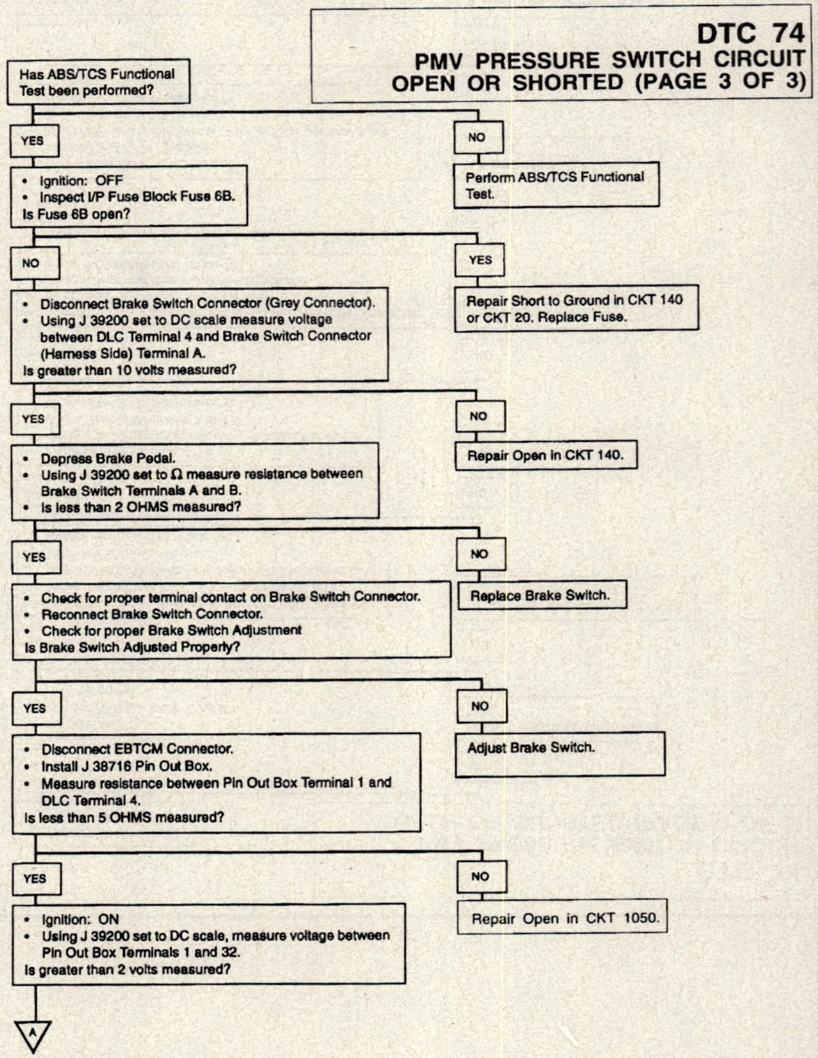

Continued on next page

DIAGNOSTIC CHART DTC 74 — 1995 EXCEPT AURORA AND RIVIERA, CONTINUED

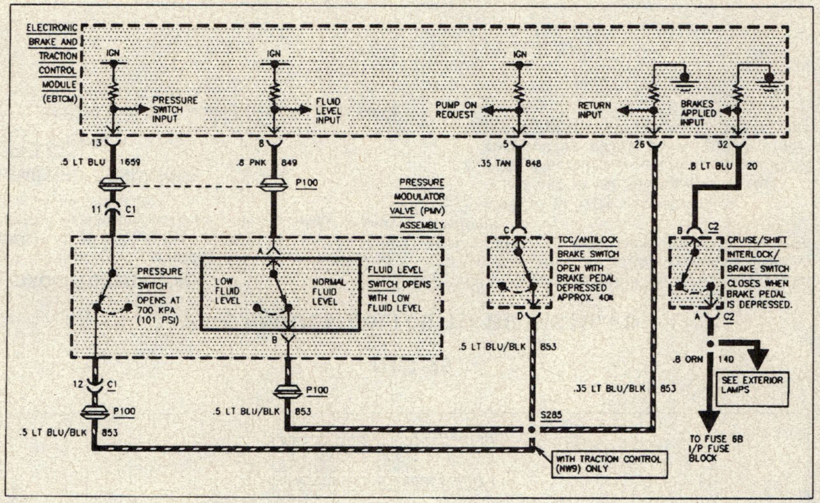

DTC 74

page 3 of 3)

PMV PRESSURE SWITCH CIRCUIT OPEN OR SHORTED

CIRCUIT OPERATION

The pressure switch monitors primary brake system pressure. The normally closed switch opens when the brakes are applied and pressure exceeds 700 kPa (101 psi). If the brakes are manually applied during traction control mode, the PMV pressure switch input, along with the brake switch input, signal the EBTCM to disable traction control and allow manual braking. The pressure switch is part of the pressure modulator valve (PMV) assembly and is not serviceable.

FAILURE CONDITIONS

DTC 74 sets when the EBTCM detects an open or short to ground in the PMV pressure switch circuitry.

Possible Causes

- Open in CKT 853 or 1659.
- Short to ground in CKT 1659 with the brakes applied.
- Failed PMV pressure switch.

DIAGNOSTIC CHART DTC 74 — 1995 EXCEPT AURORA AND RIVIERA, CONTINUED

DTC 74
PMV PRESSURE SWITCH CIRCUIT OPEN OR SHORTED (PAGE 3 OF 3)

Has ABS/TCS Functional Test been performed?

NO: Perform ABS/TCS Functional Test.

YES:
- Ignition: OFF
- Inspect I/P Fuse Block Fuse 6B.

Is Fuse 6B open?

YES: Repair Short to Ground in CKT 140 or CKT 20. Replace Fuse.

NO:
- Disconnect Brake Switch Connector (Grey Connector).
- Using J 39200 set to DC scale measure voltage between DLC Terminal 4 and Brake Switch Connector (Harness Side) Terminal A.

Is greater than 10 volts measured?

NO: Repair Open in CKT 140.

YES:
- Depress Brake Pedal.
- Using J 39200 set to Ω measure resistance between Brake Switch Terminals A and B.
- Is less than 2 OHMS measured?

NO: Replace Brake Switch.

YES:
- Check for proper terminal contact on Brake Switch Connector.
- Reconnect Brake Switch Connector.
- Check for proper Brake Switch Adjustment

Is Brake Switch Adjusted Properly?

NO: Adjust Brake Switch.

YES:
- Disconnect EBTCM Connector.
- Install J 38716 Pin Out Box.
- Measure resistance between Pin Out Box Terminal 1 and DLC Terminal 4.

Is less than 5 OHMS measured?

NO: Repair Open in CKT 1050.

YES:
- Ignition: ON
- Using J 39200 set to DC scale, measure voltage between Pin Out Box Terminals 1 and 32.

Is greater than 2 volts measured?

A

Continued on next page

DIAGNOSTIC CHART DTC 75 — 1995 EXCEPT AURORA AND RIVIERA

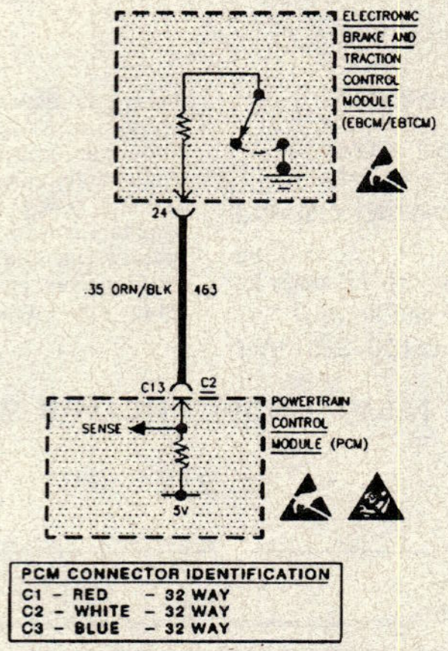

DTC 75

PCM REQUESTED TCS TO BE DISABLED

CIRCUIT OPERATION

When excessive wheel spin is detected by the EBTCM, it will request the PCM to reduce engine torque via a high speed pulse width modulated (PWM) signal on circuit 463.

FAILURE CONDITIONS

Important

If PCM code 1571 is not present, no fault exists with the ABS/TCS system.

DTC 75 sets when the EBTCM receives a serial data message from the PCM to disable Traction Control. DTC 75 sets under the following two conditions.

1. EBTCM to PCM communications.
   - ckt 463 or ckt 800 open or shorted.
2. PCM driveability conditions to include PCM codes p0101, p0341, p0342, p1350, p1361, p1623, p1630, p0131, p0132, p0171, and p0172.

DIAGNOSTIC CHART DTC 75 — 1995 EXCEPT AURORA AND RIVIERA, CONTINUED

DTC 75 PCM REQUESTED TCS TO BE DISABLED

Has ABS/TCS Functional Test been performed?

- NO: Perform ABS/TCS Functional Test before proceeding.
- YES:
  - Ignition: RUN
  - Using Tech 1, check for presence of PCM DTC P1571.
  - Is PCM DTC P1571 present?

  - NO
  - YES:
    - Ignition: OFF
    - Disconnect EBTCM.
    - Install J 38716 Pinout Box.
    - Using J 39200 set to Ω, measure resistance between Pinout Box Terminal 1 and DLC Terminal 4.

    Is less than 2 ohms measured?

    - NO: Repair open in CKT 1050.
    - YES:
      - Ignition: RUN
      - Using J 39200 set to DC scale, measure voltage between Pinout Box Terminals 24 and 1.

      Approximately 5 volts present?

      - NO:
        - Backprobe PCM Terminal C13 (White Connector) with J 39200 to ground.
        - Approximately 5 volts present?

        - NO:
          - Check CKT 463 for short to ground or battery. If CKT 463 is OK, replace PCM.
        - YES: Open in CKT 463.
      - YES:
        - Reconnect EBTCM
        - Ignition: RUN
        - Using J 39200 set to Hz, backprobe PCM Terminal C13 (white connector) and measure frequency between PCM C13 and DLC Terminal 4.

        Is 126 - 130 Hz measured?

        - YES:
          - Check for intermittent short to ground in CKT 463. If CKT 463 is OK, check for poor terminal contact at PCM and EBTCM. If all are OK, replace PCM.
        - NO:
          - Check for poor terminal contact at EBTCM Terminal 24 and PCM Terminal C13. If all are OK, replace EBTCM

After repairs are completed, clear DTC and repeat ABS/TCS functional test.

DIAGNOSTIC CHART DTC 76 — 1995 EXCEPT AURORA AND RIVIERA

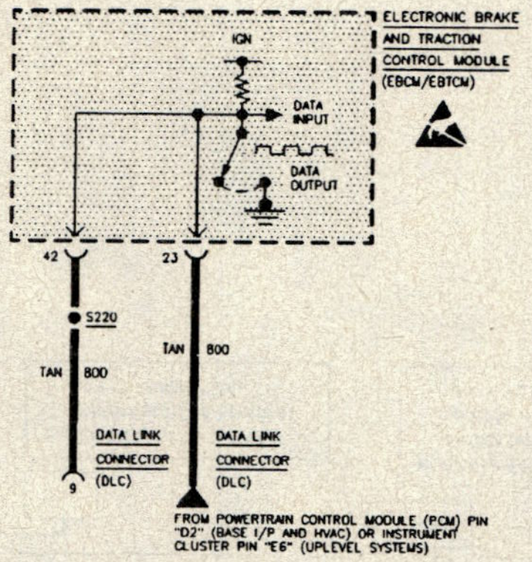

DTC 76

DATA LINE SIGNAL OR CIRCUIT MALFUNCTION

CIRCUIT OPERATION

The EBTCM shares specific serial data information supplied on the data line. The data line also allows the use of a bi-directional scan tool such as a Tech 1.

FAILURE CONDITIONS

A current DTC 76 cannot be read with a Tech 1 due to the loss of serial data communications. After data line communications are restored, DTC 76 will be stored in memory.

DIAGNOSTIC CHART DTC 77 — 1995 EXCEPT AURORA AND RIVIERA

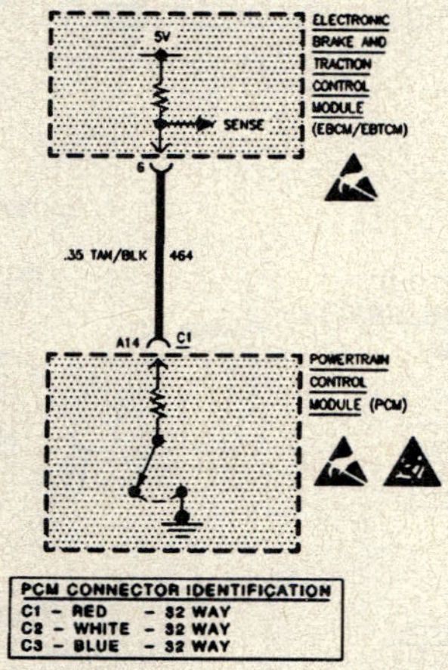

DTC 77

DELIVERED TORQUE CIRCUIT FAULT

CIRCUIT OPERATION

The PCM provides the EBTCM information on engine torque output via a high speed pulse width modulated (PWM) signal on circuit 464. This information is provided so the EBTCM can determine if engine torque management is required during a wheel slip condition.

FAILURE CONDITIONS

If the EBTCM determines there is a signal/circuit malfunction in circuit 464, the "TRACTION OFF" telltale will be illuminated and DTC 77 will set.

DIAGNOSTIC CHART DTC 77 — 1995 EXCEPT AURORA AND RIVIERA, CONTINUED

DTC 77 DELIVERED TORQUE CIRCUIT OPEN OR SHORTED

Has ABS/TCS Functional Test been performed?

- **NO:** Perform ABS/TCS Functional Test.
- **YES:**
  - Ignition: OFF
  - Disconnect PCM.
  - Ignition: ON
  - Using J 39200 set to DC scale, measure voltage at PCM Harness Connector Terminal A14 (red connector) and ground.
  - Ignition voltage present?

  - **NO:**
    - IGN: OFF
    - Disconnect EBTCM connector.
    - Using J 39200 check continuity of circuit 464 from EBTCM Terminal "6" and PCM Terminal A14 (red connector) Continuity?

    - **YES:** Check EBTCM for proper terminal contact. If OK, replace EBTCM.
    - **NO:** Check for open in CKT 464.
  - **YES:**
    - Ignition: OFF
    - Reconnect PCM.
    - Ignition: ON
    - Using J 39200 set to HZ, Backprobe Terminal A14 (red connector) of PCM Harness Connector, and measure frequency between PCM A14 and DLC terminal 4.
    - Is 126 to 130 Hz measured?

    - **YES:** Check for intermittent short to ground in CKT 464, if OK check for poor terminal contact on EBTCM Terminal 6 or faulty EBTCM.
    - **NO:** Check for poor terminal contact at PCM Terminal A14 or faulty PCM.

After repairs are completed, clear DTC and repeat ABS/TCS functional test.

ELECTRICAL CONNECTOR TERMINAL IDENTIFICATION — 1995 EXCEPT AURORA AND RIVIERA

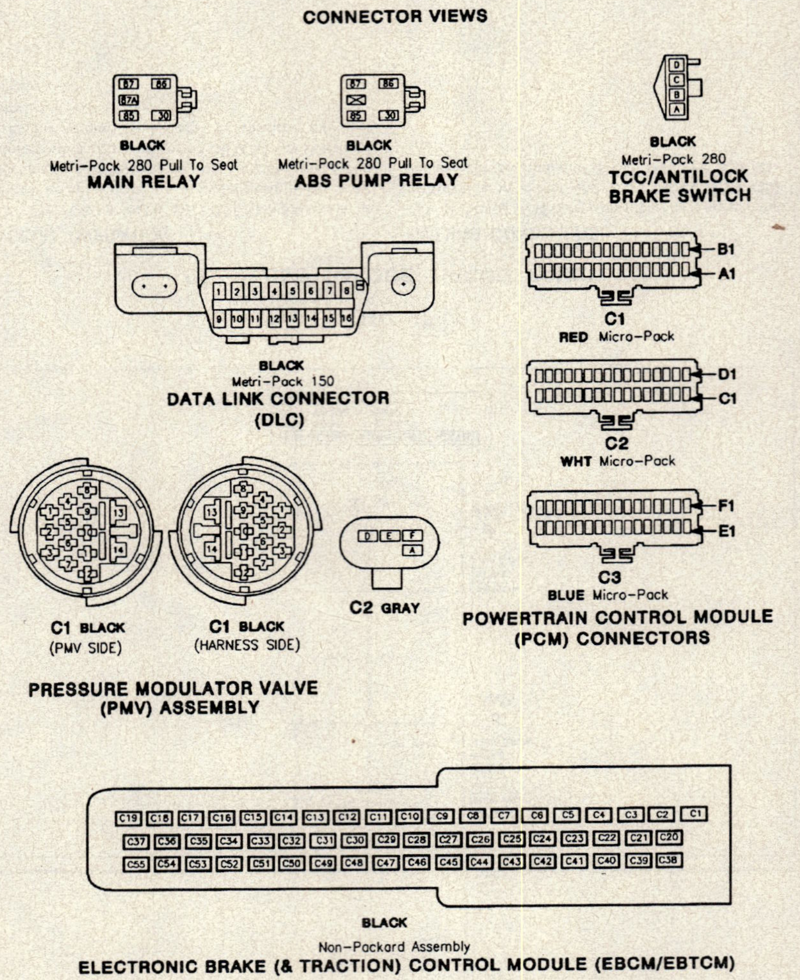

DIAGNOSTIC CHART ABS/TCS VISUAL INSPECTION — 1995 AURORA AND RIVIERA

ABS/TCS VISUAL INSPECTION

ITEM	INSPECT FOR	CORRECTIVE ACTION
PARKING BRAKE	- FULL RELEASE - PROPER SWITCH FUNCTION WHEN NECESSARY, UNPLUG SWITCH CONNECTOR TO VERIFY	- OPERATE MANUAL RELEASE LEVER TO VERIFY RELEASE - ADJUST CABLE OR REPAIR RELEASE SYSTEM AS REQUIRED - REPAIR SWITCH AS REQUIRED - PROCEED TO ABS/TCS FUNCTIONAL TEST
MAJOR COMPONENTS - MASTER CYLINDER FLUID RESERVOIR - PRESSURE MODULATOR VALVE (PMV) ASSEMBLY	- LOW FLUID LEVEL - LOW FLUID LEVEL IN PMV RESERVOIR - EXTERNAL LEAKS - PROPER ASSEMBLY	- DETERMINE CAUSE OF FLUID LOSS AND REPAIR. ADD FLUID TO MASTER CYLINDER RESERVOIR AS REQUIRED - DETERMINE CAUSE OF FLUID LOSS AND REPAIR. ADD FLUID TO MASTER CYLINDER RESERVOIR AS REQUIRED - REPAIR LEAKS AS REQUIRED - INSTALL OR POSITION COMPONENTS PROPERLY - PROCEED TO ABS/TCS FUNCTIONAL TEST
IP FUSE BLOCK - "ABS" FUSE - "CHIME" FUSE REAR FUSE BLOCK - "HZRD/STOP" FUSE RIGHT FRONT UNDERHOOD RELAY CENTER - "ABS MAIN" FUSE - "ABS PUMP MOTOR" FUSE	- OPEN (BLOWN FUSE) - PROPER ENGAGEMENT	- REPLACE - PROCEED TO ABS/TCS FUNCTIONAL TEST
CONNECTORS - MAIN RELAY - PUMP MOTOR RELAY - PMV FLUID LEVEL SWITCH - WHEEL SPEED SENSORS - ELECTRONIC BRAKE (AND TRACTION) CONTROL MODULE (EBCM/EBTCM) - PRESSURE MODULATOR VALVE (PMV) ASSEMBLY CONNECTORS: C1 - VALVE BLOCK CONNECTOR (ROUND) C2 - PUMP MOTOR CONNECTOR (GRY) - MASTER CYLINDER RESERVOIR - CRUISE/SHIFT INTERLOCK/BRAKE SWITCH - TCC/ANTILOCK BRAKE SWITCH - C101, C102, C414 - GROUNDS - G201, G202, G104, G301	- PROPER ENGAGEMENT - LOOSE WIRES OR TERMINALS - CORRODED OR BROKEN EYELETS - CORRODED/LOOSE WIRES OR BROKEN EYELETS	- PROPERLY ENGAGE CONNECTOR - REPAIR AS REQUIRED - PROCEED TO ABS/TCS FUNCTIONAL TEST - REPAIR AS NEEDED
WHEN NO PROBLEM IS NOTED IN VISUAL INSPECTION, PROCEED TO ABS/TCS FUNCTIONAL TEST		

DIAGNOSTIC CHART ABS/TCS FUNCTIONAL TEST — 1995 AURORA AND RIVIERA

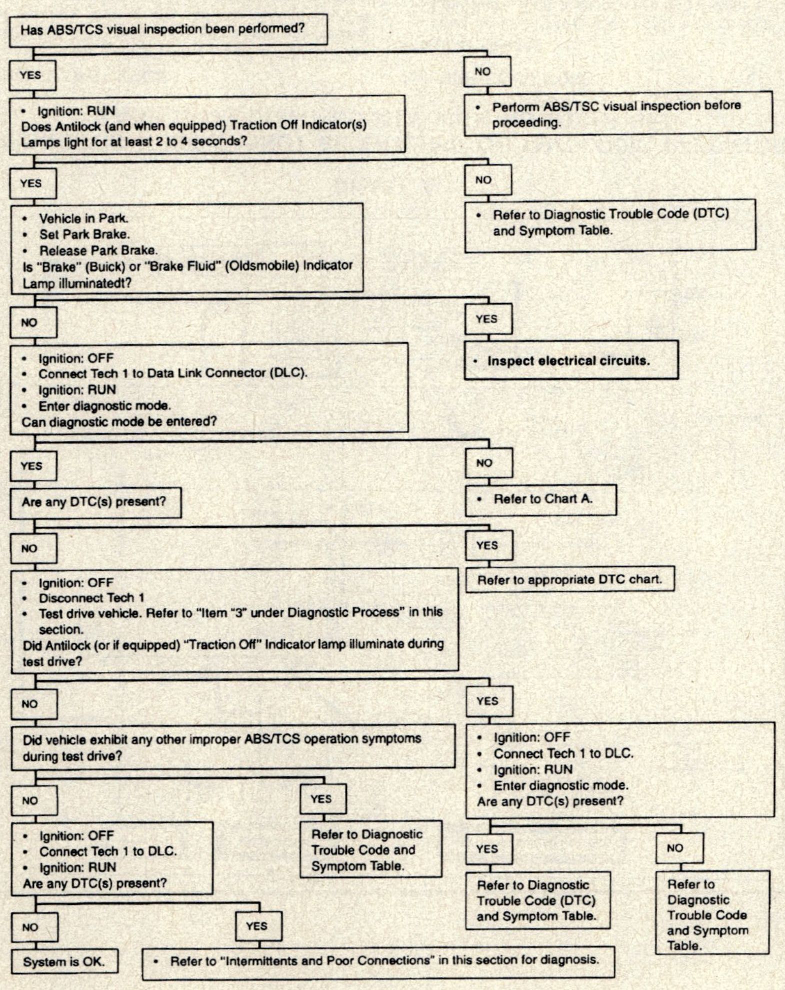

DIAGNOSTIC TROUBLE CODE CHART — 1995 AURORA AND RIVIERA

CHART	SYMPTOM
A	Diagnostic Mode cannot be entered or "ANTILOCK" Indicator Lamp is "on" constantly with no DTC(s) set
B	"ANTILOCK" Indicator Lamp does not function
C	"TRACTION OFF" Indicator Lamp is "on" constantly with no DTC(s) set
D	"TRACTION OFF" Indicator Lamp does not function
E	"TRACTION ACTIVE" Indicator Lamp does not operate properly
DTC	**DESCRIPTION**
21	RF Wheel Speed Sensor circuit open or shorted
22	RF Wheel Speed Sensor signal erratic
23	RF Wheel Speed Sensor signal missing
25	LF Wheel Speed Sensor circuit open or shorted
26	LF Wheel Speed Sensor signal erratic
27	LF Wheel Speed Sensor signal missing
31	RR Wheel Speed Sensor circuit open or shorted
32	RR Wheel Speed Sensor signal erratic
33	RR Wheel Speed Sensor signal missing
35	LR Wheel Speed Sensor circuit open or shorted
36	LR Wheel Speed Sensor signal erratic
37	LR Wheel Speed Sensor signal missing
41	RF Inlet Valve circuit open or shorted
42	RF Outlet Valve circuit open or shorted
44	LF Isolation Valve circuit open or shorted
45	Power Interruption Fault
46	LF Outlet Valve circuit open or shorted
48	RF Isolation Valve circuit open or shorted
51	RR Inlet Valve circuit open or shorted
52	RR Outlet Valve circuit open or shorted
55	LR Inlet Valve circuit open or shorted
56	LR Outlet Valve circuit open or shorted
61	Pump Motor Circuit Test
62	Low Brake Pedal during an ABS stop
71	EBCM/EBTCM internal failure
72	Brake Switch circuit open or shorted
73	Fluid Level Switch circuit open or shorted
74	Pressure Switch circuit open or shorted
75	PCM requested TCS to be disabled
76	UART SDL circuit or signal malfunction
77	Delivered Torque Circuit open or shorted

DIAGNOSTIC CHART A — 1995 AURORA AND RIVIERA

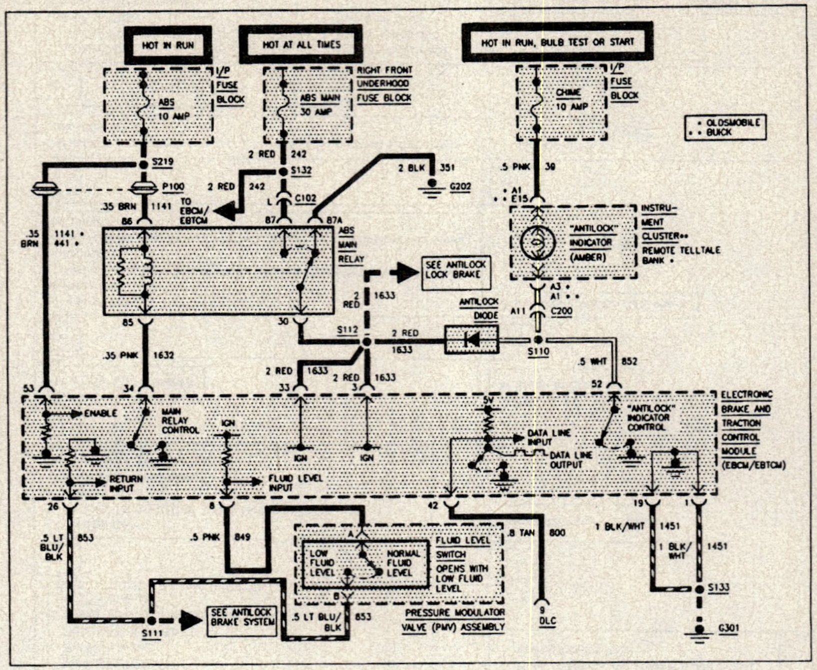

CHART A

DIAGNOSTIC MODE CANNOT BE ENTERED OR "ANTILOCK" INDICATOR LAMP IS ON CONTINUOUSLY WITH NO DTC(s) SET

CIRCUIT OPERATION

When the ignition is first turned to "RUN," the EBCM/EBTCM goes through a self-test lasting 2-4 seconds. During this test, the EBCM/EBTCM grounds terminal 52 causing the amber "ANTILOCK" indicator to light. This serves as an indicator bulb test.

When a low fluid condition exists, the EBCM/EBTCM turns on the amber "ANTILOCK" indicator but does not set a DTC. A low fluid level condition usually lights the red "BRAKE" indicator since the master cylinder supplies the PMV reservoir. However, restrictions or air in the hose could cause a low fluid condition in the PMV reservoir without lighting the red "BRAKE" indicator.

FAILURE CONDITIONS

Possible Causes

- Open in CKT 1141, 800, 1451, 849 or 853.
- Low brake fluid level or open PMV fluid level switch.
- Short to ground in CKT 1141, 800 or 852.
- Wrong scan tool adapter being used.

DIAGNOSTIC CHART A — 1995 AURORA AND RIVIERA, CONTINUED

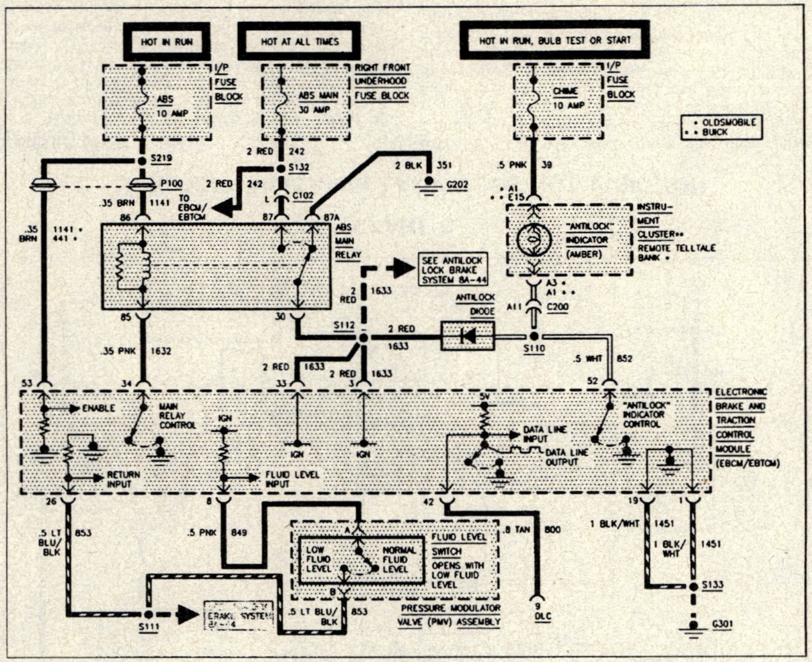

CHART A

DIAGNOSTIC MODE CANNOT BE ENTERED OR "ANTILOCK" INDICATOR LAMP IS ON CONTINUOUSLY WITH NO DTC(s) SET

CIRCUIT OPERATION

When the ignition is first turned to "RUN," the EBCM/EBTCM goes through a self-test lasting 2-4 seconds. During this test, the EBCM/EBTCM grounds terminal 52 causing the amber "ANTILOCK" indicator to light. This serves as an indicator bulb test.

When a low fluid condition exists, the EBCM/EBTCM turns on the amber "ANTILOCK" indicator but does not set a DTC. A low fluid level condition usually lights the red "BRAKE" indicator since the master cylinder supplies the PMV reservoir. However, restrictions or air in the hose could cause a low fluid condition in the PMV reservoir without lighting the red "BRAKE" indicator.

FAILURE CONDITIONS

Possible Causes

- Open in CKT 1141, 800, 1451, 849 or 853.
- Low brake fluid level or open PMV fluid level switch.
- Short to ground in CKT 1141, 800 or 852.
- Wrong scan tool adapter being used.

DIAGNOSTIC CHART A — 1995 AURORA AND RIVIERA, CONTINUED

CHART A (1 OF 2)
DIAGNOSTIC MODE CANNOT BE ENTERED OR "ANTILOCK" INDICATOR LAMP IS ON CONTINUOUSLY WITH NO DTC(s) SET

Has ABS/TCS functional test been performed?

- NO: Perform ABS/TCS functional test before proceeding.
- YES:
  - Ignition: OFF
  - Visually inspect "ABS" Fuse of I/P Fuse Block. (Inspect for open.)

Is fuse OK?

- NO: Replace fuse and check/repair CKT 1141 for a short to ground.
- YES:
  - Reinsert fuse.
  - Disconnect Pressure Modulator Valve (PMV) Fluid Level Switch Connector.
  - Using J 39200 set to Ω, measure resistance across PMV Fluid Level Switch Terminal A to B.

Is less than 1 ohm measured?

- NO: Check brake fluid level. When low, check hose for kinks or restrictions and add fluid as needed to Master Cylinder Reservoir. If OK, replace PMV Reservoir.
- YES:
  - Reconnect PMV Fluid Level Switch Connector.
  - Disconnect EBCM/EBTCM Connector.
  - Connect J 38716 Pinout Box.
  - Measure resistance between J 38716 Pinout Box Terminal 8 and 26.

Is less than 3 ohms measured?

- NO: Repair open in CKT 849 or CKT 853.
- YES:
  - Measure resistance between J 38716 Pinout Box Terminal 1 and Data Link Connector (DLC) Terminal 4 (ground), and between J 38716 Pinout Box Terminal 19 and DLC Terminal 4 (ground).

Is less than 2 ohms measured at each terminal?

- NO: Repair open in CKT 1451.
- YES: A

Continued on next page

After repairs are completed, clear DTC and repeat ABS/TCS functional test.

DIAGNOSTIC CHART A — 1995 AURORA AND RIVIERA, CONTINUED

CHART A (2 OF 2)
DIAGNOSTIC MODE CANNOT BE ENTERED OR "ANTILOCK" INDICATOR LAMP IS ON CONTINOUSLY WITH NO DTC(s) SET

A

- Ignition: RUN
- Using J 39200 set to DC scale, measure voltage between J 38716 Pinout Box Terminal 53 and 1.

Is battery voltage measured?

NO: Repair open in CKT 1141.

YES:

- Ignition: OFF
- Disconnect Main Relay.
- Ignition: RUN
- Measure voltage between J 38716 Pinout Terminal 52 and 1.

Is battery voltage measured?

NO: Repair CKT 852 for a short to ground.

YES:

- Ignition: RUN
- Measure voltage between J 38716 Pinout Box Terminal 42 and 1.

Is 5 volts measured?

NO: Check/repair CKT 800 for a short to ground or an open. If wire is OK, check for data line failure.

YES: Check EBCM/EBTCM Connector for proper terminal contact. If terminal contact is OK, replace EBCM/EBTCM.

After repairs are completed, clear DTC and repeat ABS/TCS functional test.

DIAGNOSTIC CHART B — 1995 AURORA AND RIVIERA

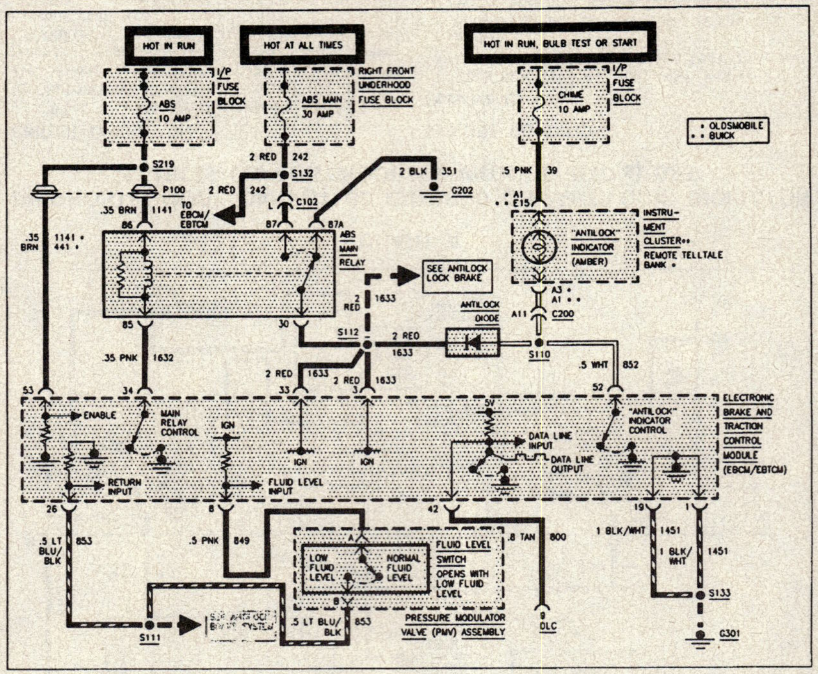

CHART B

"ANTILOCK" INDICATOR LAMP DOES NOT FUNCTION

CIRCUIT OPERATION

When the ignition switch is first turned to "RUN" the EBCM/EBTCM goes through a self-test lasting 2-4 seconds. During this test, the EBCM/EBTCM grounds terminal 52, causing the amber "ANTILOCK" indicator to light. This serves as an indicator bulb test. The EBCM/EBTCM also grounds terminal 34, energizing the main relay. This removes the indicator ground path provided through the antilock diode and the main relay contacts.

When the self-test is completed (and no faults detected) the EBCM/EBTCM removes the ground at terminal 52, turning off the indicator. The main relay remains energized.

When the EBCM/EBTCM detects a fault, it grounds terminal 52 causing the amber "ANTILOCK" indicator to light. (Note: The EBCM/EBTCM also de-energizes the main relay creating a redundant ground path.)

When the EBCM/EBTCM is disabled for any reason (i.e. open "ABS" fuse), the main relay is de-energized. This provides an "ANTILOCK" indicator ground path through the antilock diode and main relay contacts, and causes the indicator to light with no DTC being set.

FAILURE CONDITIONS

Possible Causes

- Open in CKT 1451, 852 or 1633.
- Open bulb.
- Short to voltage in CKT 852.
- Open or shorted antilock diode.
- Main relay fails "engaged."
- Short to ground in CKT 1632.

DIAGNOSTIC CHART B — 1995 AURORA AND RIVIERA, CONTINUED

CHART B (1 of 2) "ANTILOCK" INDICATOR LAMP DOES NOT FUNCTION

Has ABS/TCS Functional Test been performed?

- NO: Perform ABS/TCS Functional Test before proceeding.
- YES:
  - Ignition: OFF
  - Connect a Tech 1 to Data Link Connector (DLC).
  - Ignition: RUN
  - Enter diagnostic mode.
  - Can diagnostic mode be entered?

- NO: (A) Continued on next page
- YES:
  - Ignition: OFF
  - Disconnect EBCM/EBTCM Connector.
  - Connect J 38716 Pinout Box.
  - Using J 39200 measure voltage between J 38716 Pinout Box Terminal 1 and Battery Underhood Junction block (right side of engine compartment).

  Is voltage measured greater than 10 volts?

- NO: Repair open in CKT 1451.
- YES:
  - Disconnect Main Relay.
  - Ignition: RUN
  - Using J 39200 set to DC scale, measure voltage between J 38716 Pinout Box Terminals 52 and 1.

  Is battery voltage measured?

- NO: Replace Antilock Indicator Bulb or repair CKT 852 for an open.
- YES:
  - Ignition: RUN
  - Connect a fused jumper between J 38716 Pinout Box Terminals 52 and 1.

  Does Antilock Indicator illuminate?

- NO: Repair short to voltage in CKT 852. Check Antilock Diode and Main Relay for possible damage.
- YES:
  - Ignition: OFF
  - Disconnect Antilock Brake Diode.
  - Using J 39200 check Antilock Diode. Also, check for improper polarity.
  - Is diode OK?

- NO: Replace Antilock Diode. Observe polarity.
- YES:
  - Reconnect Antilock Brake Diode. Observe polarity.
  - Disconnect J 38716 Pinout Box.
  - Reconnect EBCM/EBTCM Connector and Main Relay.
  - Ignition: RUN
  - Does Antilock Indicator light for 2-4 seconds?

- YES: PROBLEM SOLVED
- NO: Check EBCM/EBTCM Connector for proper terminal contact. If terminal contact is OK, replace EBCM/EBTCM.

After repairs are completed, clear DTC and repeat ABS/TCS functional test.

DIAGNOSTIC CHART B — 1995 AURORA AND RIVIERA, CONTINUED

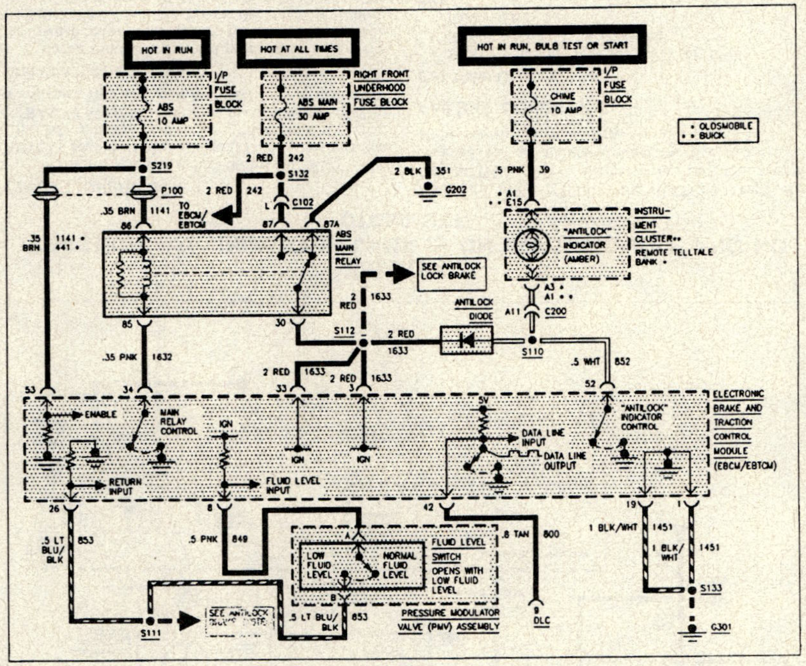

CHART B

"ANTILOCK" INDICATOR LAMP DOES NOT FUNCTION

CIRCUIT OPERATION

When the ignition switch is first turned to "RUN" the EBCM/EBTCM goes through a self-test lasting 2-4 seconds. During this test, the EBCM/EBTCM grounds terminal 52, causing the amber "ANTILOCK" indicator to light. This serves as an indicator bulb test. The EBCM/EBTCM also grounds terminal 34, energizing the main relay. This removes the indicator ground path provided through the antilock diode and the main relay contacts.

When the self-test is completed (and no faults detected) the EBCM/EBTCM removes the ground at terminal 52, turning off the indicator. The main relay remains energized.

When the EBCM/EBTCM detects a fault, it grounds terminal 52 causing the amber "ANTILOCK" indicator to light. (Note: The EBCM/EBTCM also de-energizes the main relay creating a redundant ground path.)

When the EBCM/EBTCM is disabled for any reason (i.e. open "ABS" fuse), the main relay is de-energized. This provides an "ANTILOCK" indicator ground path through the antilock diode and main relay contacts, and causes the indicator to light with no DTC being set.

FAILURE CONDITIONS

Possible Causes

- Open in CKT 1451, 852 or 1633.
- Open bulb.
- Short to voltage in CKT 852.
- Open or shorted antilock diode.
- Main relay fails "engaged."
- Short to ground in CKT 1632.

DIAGNOSTIC CHART B — 1995 AURORA AND RIVIERA, CONTINUED

CHART B (2 of 2)
"ANTILOCK" INDICATOR LAMP DOES NOT FUNCTION

A

- Ignition: OFF
- Disconnect Tech 1.
- At this point a double fault is possible. First refer to CHART A for diagnosis, then continue and complete this diagnostic chart from this point.

Was CHART A performed?

NO: Refer to CHART A.

YES:

- Ignition: OFF
- Disconnect Main Relay.
- Using J 39200 set to Ω, measure resistance across Main Relay Terminals 30 and 87A.

Is less than 2 OHMS measured?

NO: Replace Main Relay.

YES:

- Remove "ABS" Fuse of I/P Fuse Block.
- Ignition: RUN
- Using J 39200 set to DC scale, measure voltage between Main Relay Harness Connector Terminal 30 and ground.

Is battery voltage measured?

NO: Replace Antilock Diode or repair open in CKT 1633. NOTE: When replacing diode, observe polarity.

YES:

- Measure voltage between Main Relay Harness Connector Terminals 30 and 87A.

Is battery voltage measured?

NO: Repair open in CKT 351.

YES:

- Ignition: OFF
- Disconnect EBCM/EBTCM Connector.
- Using J 39200 set to Ω, measure resistance to ground between Main Relay Harness Connector Terminal 85 and ground.
- Is less than 5 OHMS measured?

YES: Repair short to ground in CKT 1632.

NO: Check EBCM/EBTCM Connector for proper terminal contact. If terminal contact is OK, replace EBCM/EBTCM.

After repairs are completed, clear DTC and repeat ABS/TCS functional test.

DIAGNOSTIC CHART C — 1995 AURORA AND RIVIERA

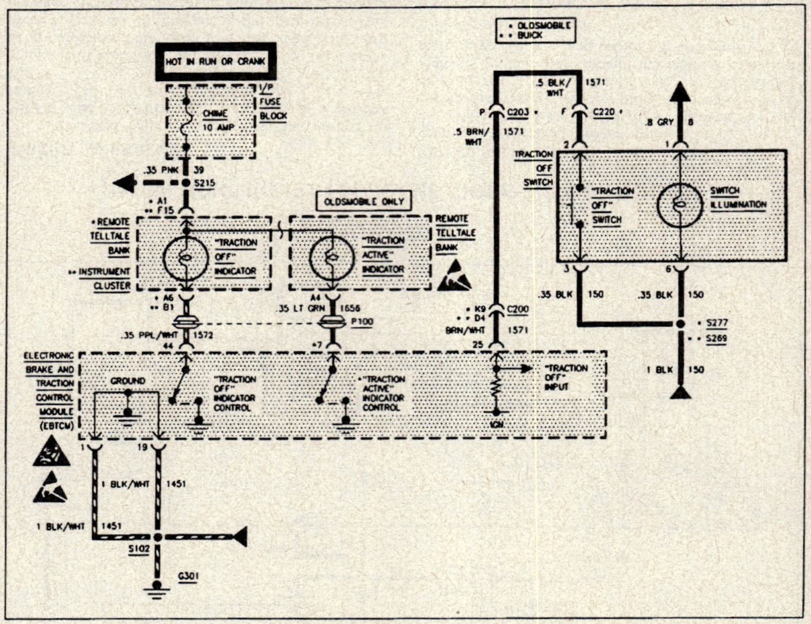

CHART C
"TRACTION OFF" INDICATOR LAMP IS ON CONTINUOUSLY AND NO DTC(s) SET

CIRCUIT OPERATION

When the ignition is first turned to "RUN," the EBTCM goes through a self-test. During this test, the EBTCM grounds both the "ANTILOCK" and "TRACTION OFF" indicators for approximately 2 to 4 seconds. This acts as a bulb test for the indicators.

Traction Control Switch

Located in the instrument panel center trim plate, the switch allows the driver to select the traction control mode (on/off). When the traction control system is turned off, the "TRACTION OFF" indicator remains illuminated.

Transaxle Temperature

The EBTCM monitors transaxle temperature through information provided by the PCM. When the transaxle becomes hot, about 160°C(320°F), the EBTCM disables the traction control system and turns on the "TRACTION OFF" indicator. This allows the brakes and transaxle to cool down. When the transaxle cools to 149°C (300°F) the EBTCM keeps the "TRACTION OFF" indicator on and the system disabled for an additional three to five minutes to allow for sufficient cooling.

FAILURE CONDITIONS

Possible Causes

- Short to ground in CKT 1571 or 1572.
- Faulty traction control switch.
- Faulty EBTCM

Important

- When the vehicle is equipped with traction control, any symptom which lights the "ANTILOCK" indicator also lights the "TRACTION OFF" indicator. Therefore, when both indicators are lit, follow "ANTILOCK" indicator diagnosis when no diagnostic trouble codes are set.

DIAGNOSTIC CHART C — 1995 AURORA AND RIVIERA

CHART C
"TRACTION OFF" INDICATOR LAMP IS ON CONTINUOUSLY AND NO DTC(s) SET

Has ABS/TCS Functional Test been performed?

- NO: Perform ABS/TCS Functional Test before proceeding.
- YES:
  - Ignition: OFF
  - Disconnect EBTCM Connector.
  - Ignition: ON

  Does "TRACTION OFF" indicator lamp turn "OFF"?

  - NO: Repair short to ground in CKT 1572.
  - YES:
    - Ignition: OFF
    - Install J 38716 Pinout Box
    - Using J 39200, measure the voltage between J 38716 Pinout Box Terminal 1 and Battery Underhood Junction Block (right side of engine compartment).

    Is voltage measured greater than 10 volts?

    - NO: Repair open in CKT 1451.
    - YES:
      - Measure resistance between J 38716 Pinout Box Terminal 1 and 25.

      Is less than 5 OHMS measured?

      - YES: Check/repair CKT 1571 for a short to ground. If CKT 1571 is OK, replace Traction Control Switch.
      - NO:
        - Measure resistance between J 38716 Pinout Box Terminal 1 and 44.

        Is less than 5 OHMS measured?

        - YES: Replace EBTCM.
        - NO: No trouble found. Check for INTERMITTENTS AND POOR CONNECTIONS.

DIAGNOSTIC CHART D — 1995 AURORA AND RIVIERA

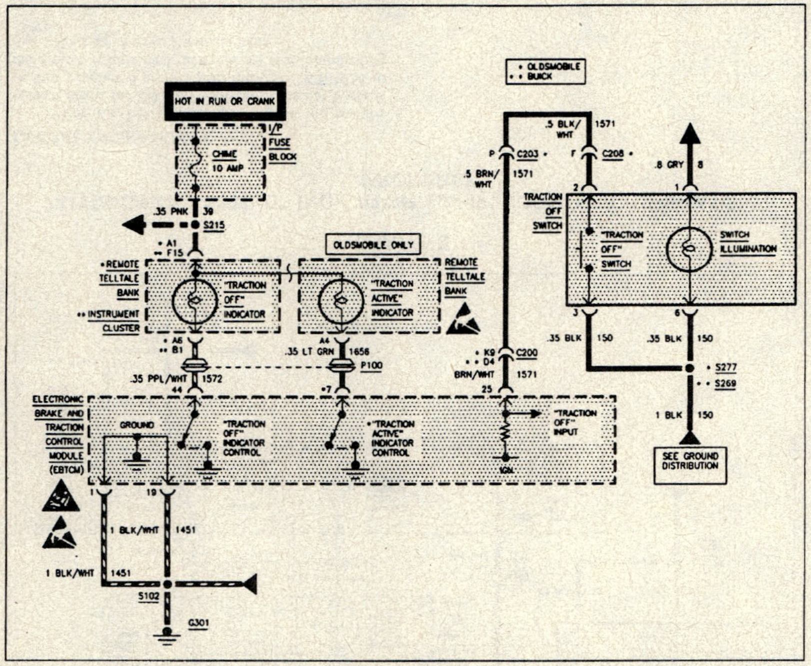

CHART D
"TRACTION OFF" INDICATOR LAMP DOES NOT FUNCTION

CIRCUIT OPERATION

When the ignition is first turned to "RUN," the EBTCM goes through a self-test. During this test, the EBCM grounds both the "ANTILOCK" and "TRACTION OFF" indicators for approximately 2 to 4 seconds. This acts as a bulb test for the indicators.

Traction Control Switch

Located in the instrument panel center trim plate, the switch allows the driver to select the traction control mode (on/off). When the traction control system is turned off, the "TRACTION OFF" indicator remains illuminated.

FAILURE CONDITIONS

Possible Causes

- Open in CKT 150, 1451, 1571 or 1572.
- Open bulb.
- Faulty traction control switch.

Important

- When the vehicle is equipped with traction control, any symptom which lights the "ANTILOCK" indicator also lights the "TRACTION OFF" indicator. Therefore, when both indicators are lit, follow "ANTILOCK" indicator diagnosis when no diagnostic trouble codes are set.

DIAGNOSTIC CHART D — 1995 AURORA AND RIVIERA, CONTINUED

CHART D
"TRACTION OFF" INDICATOR LAMP DOES NOT FUNCTION

Has ABS/TCS Functional Test been performed?

- NO: Perform ABS/TCS Functional Test before proceeding?
- YES:
  - Ignition: RUN
  - Does "TRACTION OFF" Indicator light for 2-4 seconds?

If NO:

- Ignition: OFF
- Disconnect EBTCM Connector.
- Install J 38716 Pinout Box.
- Using J 39200 measure voltage between J 38716 Pinout Box Terminal 1 and Battery Underhood Junction Block (right side of engine compartment).

Is voltage measured greater than 10 volts?

- NO: Repair open in CKT 1451.
- YES:
  - Ignition: RUN
  - Measure voltage from J 38716 Pinout Box Terminal 44 and 1.
  - Is battery voltage measured?
    - YES: Check EBTCM Connector for proper terminal contact. If terminal contact is OK, replace EBTCM.
    - NO: Check "TRACTION OFF" indicator bulb. If bulb is OK, repair open in CKT 1572. (Also check connectors for proper terminal contact).

If YES:

- Ignition: OFF
- Disconnect Traction Control Switch Connector.
- Ignition: RUN
- Using J 39200 set to DC scale, measure voltage between Harness Connector Terminal 2 and Data Link Connector (DLC) Terminal 4 (ground).

Is battery voltage measured?

- NO: Check/repair open in CKT 1571. If CKT 1571 is OK, check EBTCM Connector for proper terminal contact. If terminal contact is OK, replace EBTCM.
- YES:
  - Ignition: OFF
  - Measure resistance between Switch Harness Connector Terminal 3 and DLC Terminal 4 (ground).
  - Is less than 2 OHMS measured?
    - YES: Check Switch Connector for proper terminal contact. If OK, replace Traction Control Switch.
    - NO: Repair open in CKT 150.

After repairs are completed, clear DTC and repeat ABS/TCS functional test.

DIAGNOSTIC CHART E — 1995 AURORA AND RIVIERA

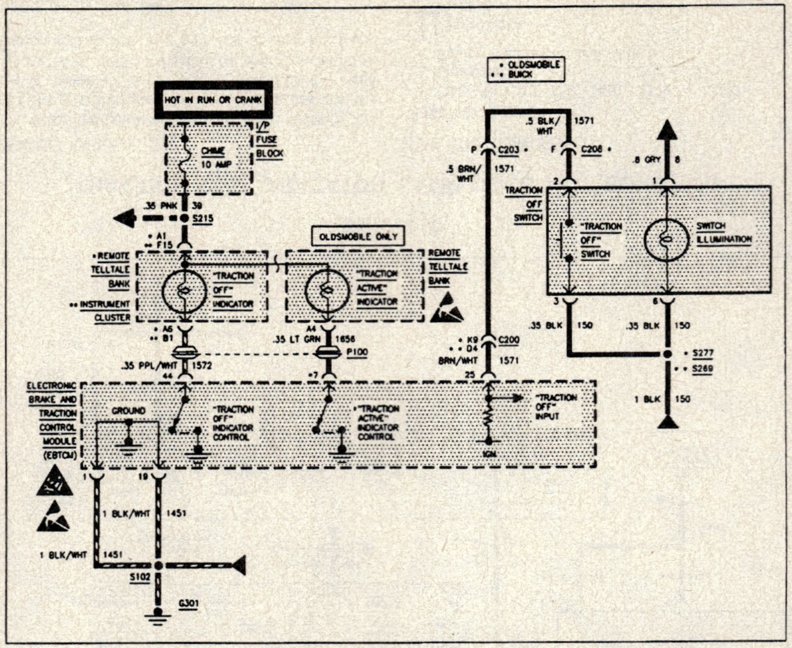

CHART E

"TRACTION ACTIVE" INDICATOR LAMP DOES NOT OPERATE PROPERLY

CIRCUIT OPERATION

When the EBTCM is operating in the traction control mode, the EBTCM's grounds terminal 7 which in turn supplies a ground path through CKT 1656 to the remote telltale bank terminal A4 thus illuminating the "Traction Active" telltale lamp.

DIAGNOSTIC CHART E — 1995 AURORA AND RIVIERA, CONTINUED

CHART E
"TRACTION ACTIVE" INDICATOR LAMP DOES NOT OPERATE PROPERLY

Has ABS/TCS Functional Test been performed?

- **NO**: Perform ABS/TCS Functional Test before proceeding.
- **YES**: Does "TRACTION ACTIVE" lamp illuminate when Traction Control is not active?
  - **YES**:
    - Ignition: OFF
    - Disconnect EBTCM Connector.
    - Ignition: RUN

    Does "TRACTION ACTIVE" indicator lamp still illuminate when Traction Control is not active?
    - **YES**: Repair short to ground in CKT 1656.
    - **NO**: Replace EBTCM.
  - **NO**:
    - Ignition: OFF
    - Disconnect EBTCM Connector.
    - Connect J 38716 Pinout Box.
    - Using J 39200, measure voltage between J 38716 Pinout Box Terminal 1 and Battery Underhood Junction Block (right side of engine compartment).

    Is voltage measured greater than 10 volts?
    - **NO**: Repair open in CKT 1451.
    - **YES**:
      - Connect a Fused Jumper from J 38716 Pinout Box Terminal 7 to Terminal 1.
      - Ignition: RUN

      Does "TRACTION ACTIVE" lamp illuminate?
      - **NO**:
        - Ignition: OFF
        - Check "TRACTION ACTIVE" indicator bulb.

        Is bulb OK?
        - **NO**: Replace bulb.
        - **YES**:
          - Disconnect Remote Telltale Bank Connector.
          - Ignition: RUN
          - Using J39200 set to DC Scale, measure voltage between Remote Telltale Bank Harness Connector Terminal A1 and DLC Terminal 4.

          Battery voltage present?
          - **YES**: Open in CKT 1656.
          - **NO**: Check CKT 39 for open and Remote Telltale Bank Connector for proper terminal contact. If OK, replace Remote Telltale Bank.
      - **YES**:
        - Ignition: OFF
        - Disconnect J 38716 Pinout Box.
        - Reconnect EBTCM.
        - Using Tech 1, select Mode F0 (Data List 1) and monitor TCS state.
        - Test drive vehicle in a safe area under 40 km/h (25 mph) on a slippery surface such as sand or gravel.
        - Does TCS state change during a traction control event?
        - **NO**: Repeat ABS/TCS Functional Test.
        - **YES**: Did the "TRACTION ACTIVE" indicator lamp illuminate?
          - **YES**: System OK.
          - **NO**: Check EBTCM Connectors for proper terminal contact. If terminal contact is OK, replace EBTCM.

After repairs are completed, clear DTC and repeat ABS/TCS functional test.

DIAGNOSTIC CHART DTC 21 AND 25 — 1995 AURORA AND RIVIERA

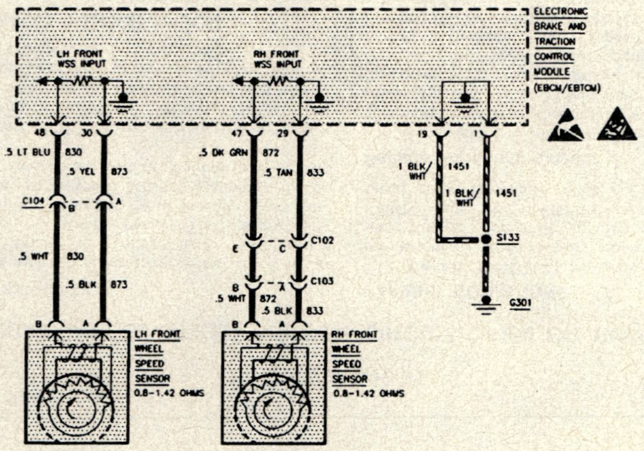

DTC(s) 21, 25

FRONT WHEEL SPEED SENSOR CIRCUIT TEST

CIRCUIT OPERATION

As each wheel turns, the wheel speed sensor creates an AC voltage with a frequency proportional to wheel speed. This magnetically induced voltage is caused by a toothed sensor ring (mounted on the drive axle) passing the wheel speed sensor's stationary pickup coil.

The EBCM/EBTCM uses the wheel speed sensor signal to calculate vehicle reference speeds and individual wheel speed, acceleration and slip values. These values are used to determine when antilock (or when equipped) traction control is required.

WHEEL SPEED SENSOR CABLES

In order to prevent electromagnetic interference from disturbing the wheel speed sensor signal, the cables from the EBCM/EBTCM to each wheel speed sensor are twisted in pairs a minimum of 6 to 9 turns per foot.

When servicing the sensor cables it is important to maintain the original cable's twists (6 to 9 turns per foot).

FAILURE CONDITIONS

The EBCM/EBTCM performs two basic checks on the wheel speed sensors: sensor continuity and sensor output. When the ignition is first turned to "RUN," the EBCM/EBTCM performs the wheel speed sensor continuity check.

Sensor Continuity Check

Any condition which would result in lack of continuity in the front wheel speed sensor circuit could result in a DTC 21 or 25 being set. These conditions include an open, a short to ground, or a short between the two wheel speed sensor circuits in the wiring harness between the speed sensor and EBCM/EBTCM. Also an open or short to ground across the wheel speed sensor coil could trigger this code.

DIAGNOSTIC CHART DTC 21 AND 25 — 1995 AURORA AND RIVIERA, CONTINUED

DTC(s) 21, 25
FRONT WHEEL SPEED SENSOR CIRCUIT OPEN OR SHORTED

Has ABS/TCS functional test been performed?

NO → Perform ABS/TCS functional test before proceeding.

YES →
- Ignition: OFF
- Raise vehicle.
- Disconnect suspect front Wheel Speed Sensor Connector.
- Check connector for proper terminal contact.
- Using J 39200 set to Ω, measure resistance between suspect Wheel Speed Sensor Terminal A and B.

Is 800 to 1500 ohms measured?

NO → Replace suspect front Wheel Speed Sensor.

YES →
- Measure resistance between suspect front Wheel Speed Sensor Terminal B and ground.

Is infinite ohms measured?

NO → Replace suspect front Wheel Speed Sensor.

YES →
- Reconnect Wheel Speed Sensor.
- Disconnect Wheel Speed Sensor Jumper Harness.
- Check connector for proper terminal contact.
- Measure resistance between Jumper Harness Terminals A and B (sensor side).

Is 800 to 1500 ohms measured?

NO → Replace suspect front Wheel Speed Sensor Jumper Harness.

YES →
- Reconnect Wheel Speed Sensor Jumper Harness.
- Disconnect EBCM/EBTCM Connector.
- Check connector for proper terminal contact.
- Install J 38716 Pinout Box.
- Measure resistance between terminals indicated for suspected front Wheel Speed Sensor Circuit:
  - Left Front Wheel - Terminals 30 and 48
  - Right Front Wheel - Terminals 29 and 47

Is 800 to 1500 ohms measured?

NO → Above 1500 ohms (remember to check in-line connector [see schematic] and Wheel Speed Sensor Connector for proper terminal contact):
- When left front wheel, repair open in CKT 830 or 873.
- When right front wheel, repair open in CKT 833 or 872.

Below 800 ohms:
- When left front wheel, repair short between CKT 830 and 873.
- When right front wheel, repair short between CKT 833 and 872.

YES →
- Using J 39200 measure voltage between J 38716 Pinout Box Terminal 1 and Battery Underhood Junction Block (right side of engine compartment).

Is voltage measured greater than 10 volts?

NO → Repair open in CKT 1451.

YES →
- Measure resistance at J 38716 Pinout Box Terminals indicated for suspected front Wheel Speed Sensor Circuit:
  - Left Front Wheel - Terminal 48 and 1
  - Right Front Wheel - Terminal 47 and 1

Is less than 5 ohms measured?

YES → When left front wheel, repair short to ground in CKT 830 or 873. When right front wheel, repair short to ground in CKT 872 or 833.

NO → Check EBCM/EBTCM Connector for proper terminal contact. If terminal contact is OK, reconnect EBCM/EBTCM, clear DTC and test drive vehicle. If DTC resets, replace EBTCM. If DTC does not reset, refer to Intermittents and Poor Connections.

After repairs are completed, clear DTC and repeat ABS/TCS functional test.

DIAGNOSTIC CHART DTC 31 AND 35 — 1995 AURORA AND RIVIERA

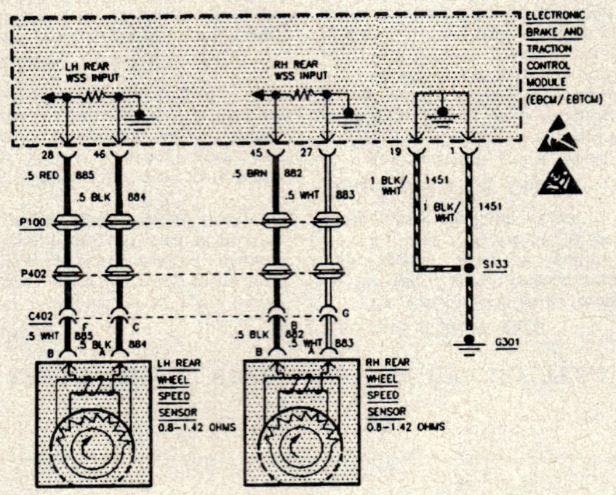

DTC(s) 31, 35

REAR WHEEL SPEED SENSOR CIRCUIT OPEN OR SHORTED

CIRCUIT OPERATION

When each wheel turns, the wheel speed sensor creates a small AC voltage with a frequency proportional to wheel speed. This magnetically induced voltage is caused by a toothed sensor ring passing the wheel speed sensor's stationary pickup coil. Both sensors are part of the hub and bearing assembly.

The EBCM/EBTCM uses the wheel speed sensor signal to calculate vehicle reference speeds and individual wheel speed, acceleration and slip values. These values are used to determine when antilock control is required.

WHEEL SPEED SENSOR CABLES

In order to prevent electromagnetic interference from disturbing the wheel speed sensor signal, the cables from the EBCM/EBTCM to each wheel speed sensor are twisted in pairs a minimum of 6 to 9 turns per foot.

When servicing the sensor cables it is important to maintain the original cable's twists (6 to 9 turns per foot).

FAILURE CONDITIONS

The EBCM/EBTCM performs two basic checks on the wheel speed sensors: sensor continuity and sensor output. When the ignition is first turned to "RUN," the EBCM/EBTCM performs the wheel speed sensor continuity check.

Sensor Continuity Check

Any condition which would result in lack of continuity in the rear wheel speed sensor circuit could result in a DTC 31 or 35 being set. These conditions include an open, a short to ground, or a short between the two wheel speed sensor circuits in the wiring harness between the speed sensor and EBCM/EBTCM. Also an open or short to ground across the wheel speed sensor coil could trigger this DTC.

DIAGNOSTIC CHART DTC 31 AND 35 — 1995 AURORA AND RIVIERA, CONTINUED

DTC(s) 31, 35
REAR WHEEL SPEED SENSOR CIRCUIT OPEN OR SHORTED

Has ABS/TCS functional test been performed?

NO → Perform ABS/TCS functional test before proceeding.

YES →
- Ignition: OFF
- Park Brake: RELEASED
- Raise vehicle.
- Disconnect suspect rear Wheel Speed Sensor Connector at wheel.
- Check connector for proper terminal contact.
- Using J 39200 set to Ω, measure resistance between suspect Wheel Speed Sensor Terminal A and B.

Is 800 to 1500 ohms measured?

NO → Replace suspect rear Wheel Speed Sensor.

YES →
- Measure resistance between suspect rear Wheel Speed Sensor Terminal B and ground.

Is infinite ohms measured?

NO → Replace suspect rear Wheel Speed Sensor.

YES →
- Reconnect Wheel Speed Sensor.
- Disconnect Wheel Speed Jumper Harness.
- Check connector for proper terminal contact.
- Measure resistance between Jumper Harness Terminals A and B if right side Wheel Speed Sensor is suspect or Terminals D and C if left side Wheel Speed Sensor is suspect.

Is 800 to 1500 ohms measured?

NO → Replace Wheel Speed Sensor Jumper Harness.

YES →
- Reconnect Wheel Speed Sensor Jumper Harness.
- Disconnect EBCM/EBTCM Connector.
- Check connector for proper terminal contact.
- Install J 38716 Pinout Box.
- Measure resistance between terminals indicated for suspected rear Wheel Speed Sensor Circuit:
  - Left Rear Wheel - Terminals 28 and 46
  - Right Rear Wheel - Terminals 27 and 45

Is 800 to 1500 ohms measured?

NO → Above 1500 ohms (remember to check in-line connector [see schematic] and Wheel Speed Sensor Connector for proper terminal contact):
- When left rear wheel, repair open in CKT 884 or 885.
- When right rear wheel, repair open in CKT 882 or 883.

Below 800 ohms:
- When left rear wheel, repair short between CKT 884 and 885.
- When right rear wheel, repair short between CKT 882 and 883.

YES →
- Using J 39200 measure voltage between J 38716 Pinout Box Terminal 1 and Battery Underhood Junction Block (right side of engine compartment).

Is voltage measured greater than 10 volts?

NO → Repair open in CKT 1451.

YES →
- Measure resistance at J 38716 Pinout Box Terminals indicated for suspected rear Wheel Speed Sensor Circuit:
  - Left Rear Wheel - Terminal 46 and 1
  - Right Rear Wheel - Terminal 45 and 1

Is less than 5 ohms measured?

YES → When left rear wheel, repair short to ground in CKT 884 or 885. When right rear wheel, repair short to ground in CKT 882 or 883.

NO → Check EBCM/EBTCM Connector for proper terminal contact. If terminal contact is OK, reconnect EBCM/EBTCM, clear DTC and test drive vehicle. If DTC resets, replace EBTCM. If DTC does not reset, refer to Intermittents and Poor Connections in this section.

After repairs are completed, clear DTC and repeat ABS/TCS functional test.

DIAGNOSTIC CHART 22, 23, 26, 27, 32, 33, 36 AND 37 — 1995 AURORA AND RIVIERA

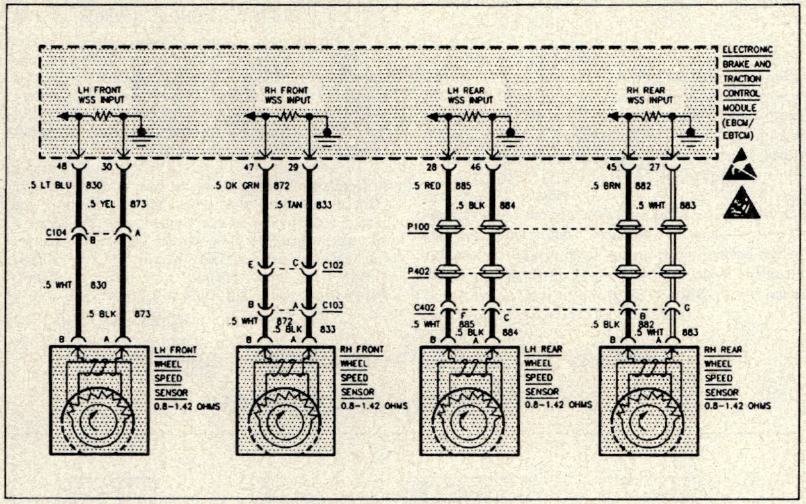

DTC(s) 22, 23, 26, 27, 32, 33, 36, 37

WHEEL SPEED SIGNAL ERRATIC OR MISSING

CIRCUIT OPERATION

When each wheel turns, the wheel speed sensor creates a small AC voltage with a frequency proportional to wheel speed. This magnetically induced voltage is caused by a toothed sensor ring passing the wheel speed sensor's stationary pickup coil. The toothed ring sensor is mounted on the drive axle in the front, and in the hub/bearing assembly in the rear.

The EBCM/EBTCM uses the wheel speed sensor signal to calculate vehicle reference speeds, and individual wheel speed, acceleration and slip values. These values are used to determine when antilock control is required. Vehicles equipped with traction control also use front wheel speed sensor signals to determine when traction control is required.

WHEEL SPEED SENSOR CABLES

In order to prevent electromagnetic interference from disturbing the wheel speed sensor signal, the cables from the EBCM/EBTCM to each wheel speed sensor are twisted in pairs a minimum of 6 to 9 turns per foot.

When servicing the sensor cables it is important to maintain the original cables' twist (6 to 9 turns per foot).

FAILURE CONDITIONS

DTC 22, DTC 26, DTC 33 or DTC 36 sets when the EBCM/EBTCM detects electromagnetic interference (noise) in a wheel speed sensor signal. Possible interference sources include accessory motors, wiper motors, ignition circuits and add-on or after market equipment.

DTC 22, DTC 26, DTC 32 or DTC 36 also sets when the EBCM/EBTCM determines a wheel speed signal is erratic. A wheel speed signal indicating that a wheel is accelerating or decelerating faster than physically possible is an example of an erratic signal. An intermittent open or short to ground in a wheel speed sensor could also cause the EBCM/EBTCM to set a DTC.

DTC 23, DTC 27, DTC 33, or DTC 37 sets when the EBCM/EBTCM does not see any wheel speed sensor signal, but determines continuity exists in the circuit. This is cause by a dislodged wheel speed sensor, missing toothed sensor ring (front) or defective hub and bearing assembly (rear).

DIAGNOSTIC CHART 22, 23, 26, 27, 32, 33, 36 AND 37 — 1995 AURORA AND RIVIERA, CONTINUED

DTC(s) 22, 23, 26, 27, 32, 33, 36, 37
WHEEL SPEED SIGNAL ERRATIC OR MISSING
(1 of 2)

Has ABS/TCS functional test been performed?

- **NO** → Perform ABS/TCS functional test before proceeding.
- **YES** → Is DTC 23, DTC 27, DTC 33 or DTC 37 present?
  - **YES** →
    - When DTC 23 or DTC 27: Reinstall dislodged front Wheel Speed Sensor or replace front Hub and Bearing Assembly.
    - When DTC 33 or DTC 37: Reinstall dislodged rear Wheel Speed Sensor or replace Hub and Bearing Assembly.
  - **NO** →
    - Perform following and repair/replace as necessary:
      Test drive vehicle (refer to Item "3" under Diagnostic Process in this section) monitoring suspect Wheel Speed Sensor using Tech 1 in snapshot (mode F3).
      Disconnect suspect Wheel Speed Sensor Connector and applicable jumper harness connector. Inspect terminal contacts on sensor and connectors for corrosion or deformity. Check connectors for proper terminal contact.
      Trace harness routing of suspect sensor wiring. Inspect for physical damage particularly in areas where pinching or movement by harness or other components occurs.
      Trace harness routing of suspect sensor wiring for proximity to noise sources such as ignition circuits, motors, etc. Also add-on or "after market" equipment.
      Check if any repairs were recently made to suspect sensor wiring. Sensor wiring should be twisted 6 to 9 turns per foot.
      Check for loose sensor mounting, or debris lodged between wheel speed sensor and hub.
    - If all above checks are OK, proceed, but be aware that problem causing diagnostic trouble code to set may be intermittent and is usually not caused by a defective EBCM/EBTCM.

    - Ignition: RUN
    - Gear Selector Lever: NEUTRAL
    - Ignition: OFF
    - Park Brake: RELEASED
    - Raise vehicle, check bearing end play - repair as necessary.
    - Disconnect suspect Wheel Speed Sensor at wheel.
    - Using J 39200 set to AC scale measure voltage between suspect Wheel Speed Sensor Terminal A and B while turning drive wheels by hand.

    Is 50 to 900 millivolts AC measured?
    - **NO** → Replace suspect Wheel Speed Sensor.
    - **YES** → (A)

Continued on next page

After repairs are completed, clear DTC and repeat ABS/TCS functional test.

DIAGNOSTIC CHART 22, 23, 26, 27, 32, 33, 36 AND 37 — 1995 AURORA AND RIVIERA, CONTINUED

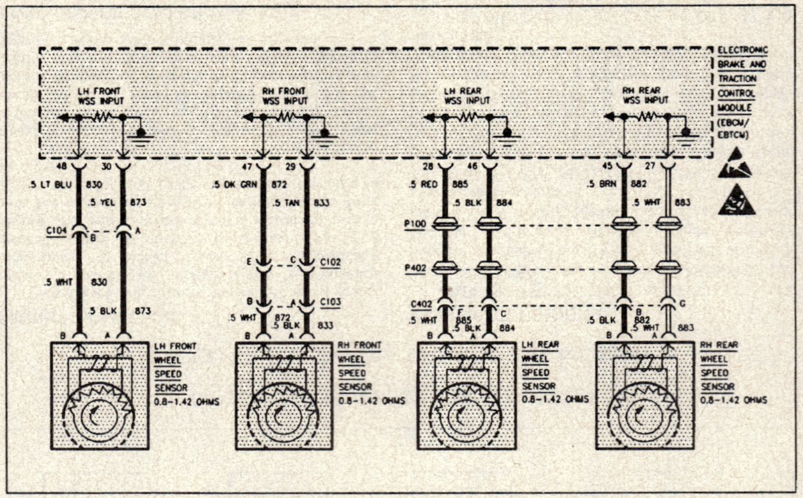

DTC(s) 22, 23, 26, 27, 32, 33, 36, 37

WHEEL SPEED SIGNAL ERRATIC OR MISSING

CIRCUIT OPERATION

When each wheel turns, the wheel speed sensor creates a small AC voltage with a frequency proportional to wheel speed. This magnetically induced voltage is caused by a toothed sensor ring passing the wheel speed sensor's stationary pickup coil. The toothed ring sensor is mounted on the drive axle in the front, and in the hub/bearing assembly in the rear.

The EBCM/EBTCM uses the wheel speed sensor signal to calculate vehicle reference speeds, and individual wheel speed, acceleration and slip values. These values are used to determine when antilock control is required. Vehicles equipped with traction control also use front wheel speed sensor signals to determine when traction control is required.

WHEEL SPEED SENSOR CABLES

In order to prevent electromagnetic interference from disturbing the wheel speed sensor signal, the cables from the EBCM/EBTCM to each wheel speed sensor are twisted in pairs a minimum of 6 to 9 turns per foot.

When servicing the sensor cables it is important to maintain the original cables' twist (6 to 9 turns per foot).

FAILURE CONDITIONS

DTC 22, DTC 26, DTC 33 or DTC 36 sets when the EBCM/EBTCM detects electromagnetic interference (noise) in a wheel speed sensor signal. Possible interference sources include accessory motors, wiper motors, ignition circuits and add-on or after market equipment.

DTC 22, DTC 26, DTC 32 or DTC 36 also sets when the EBCM/EBTCM determines a wheel speed signal is erratic. A wheel speed signal indicating that a wheel is accelerating or decelerating faster than physically possible is an example of an erratic signal. An intermittent open or short to ground in a wheel speed sensor could also cause the EBCM/EBTCM to set a DTC.

DTC 23, DTC 27, DTC 33, or DTC 37 sets when the EBCM/EBTCM does not see any wheel speed sensor signal, but determines continuity exists in the circuit. This is cause by a dislodged wheel speed sensor, missing toothed sensor ring (front) or defective hub and bearing assembly (rear).

DIAGNOSTIC CHART 22, 23, 26, 27, 32, 33, 36 AND 37 — 1995 AURORA AND RIVIERA, CONTINUED

DTC(s) 22, 23, 26, 27, 32, 33, 36, 37
WHEEL SPEED SIGNAL ERRATIC OR MISSING
(2 of 2)

A

- Thoroughly spray all Wheel Speed Sensors, connectors and harnesses with a 5% solution (2 teaspoons of salt to 12 oz. of water) of salt water.
- Test drive vehicle in a straight line at slow speeds (under 16 km/h [10 mph] while monitoring Wheel Speed Sensors using Tech 1 in snapshot mode (NOTE: suspect Wheel Speed Signal usually varies from remaining Wheel Speed Signals by approximately 2 km/h [1 mph]).

Does any one Wheel Speed Signal vary from other three Wheel Speed Signals?

YES

Replace suspect Wheel Speed Sensor.

NO

- Clear DTC(s) with Tech 1.
- Test drive vehicle.
- If DTC(s) resets, replace EBCM/EBTCM. If DTC(s) do not reset, refer to Intermittants and Poor Connections.

After repairs are completed, clear DTC and repeat ABS/TCS functional test.

DIAGNOSTIC CHART DTC 45 — 1995 AURORA AND RIVIERA

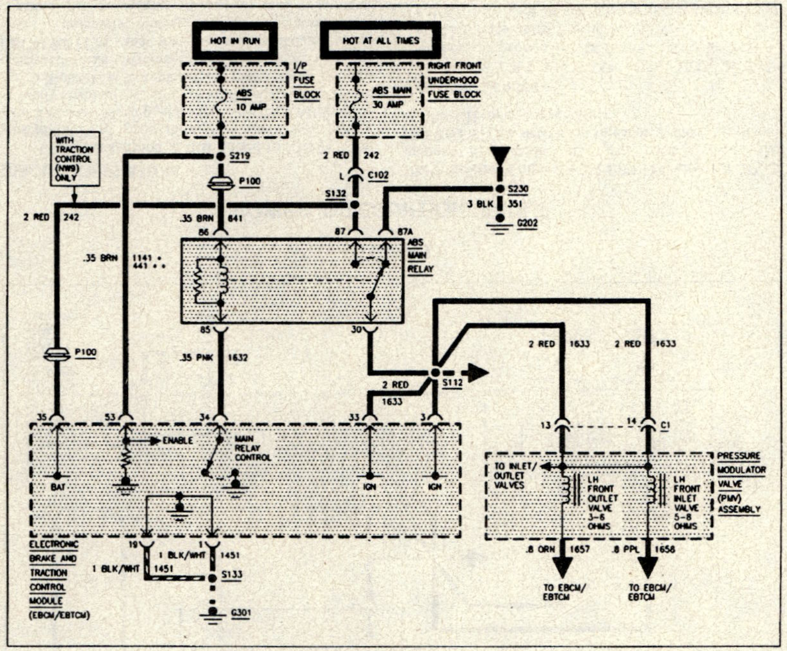

DTC 45

POWER INTERRUPTION FAULT

CIRCUIT OPERATION

When the ignition is first turned to "RUN," the EBCM/EBTCM goes through a self-check lasting 2-4 seconds. During this test the EBCM/EBTCM grounds terminal 34, energizing the main relay. Once the self-check is completed the main relay remains energized. This delivers battery voltage to both the EBCM/EBTCM and the pressure modulator valve (PMV) assembly during vehicle operation. The constant voltage supply allows the vehicle to enter ABS mode without delay. The main relay is de-energized whenever the amber "ANTILOCK" indicator lights due to a system fault, causing the ABS system to shut down.

FAILURE CONDITION

DTC 45 sets when the ignition switch is at "RUN" and battery voltage is neither detected at the PMV assembly nor EBCM/EBTCM terminal 20, 3 and/or 33. Code 45 also sets when the EBCM/EBTCM detects an open or a short to ground in the LH inlet valve circuit.

Possible Causes

- Open in CKT 1141, 1632, 858, 1633, or 242.
- Short to ground in CKT 858, 1633, or 242.
- Main relay fails "open."
- High resistance in any inlet or outlet valve circuit and if equipped with traction control, high resistance in either isolation valve circuit.

DIAGNOSIS AIDS

DTC 45 may also set when the vehicle is hot (running for a minimum of 30 minutes) check PMV valve resistance and main relay coil resistance when vehicle is hot.

DIAGNOSTIC CHART DTC 45 — 1995 AURORA AND RIVIERA, CONTINUED

DTC 45
POWER INTERRUPTION FAULT (1 of 5)

IMPORTANT: DTC 45 may set intermittently due to low system voltage (extended cranking, low battery, but vehicle eventually starts etc.)

Has ABS/TCS functional test been performed?

- **NO** → Perform ABS/TCS functional test before proceeding.
- **YES** →
  - Ignition: OFF
  - Disconnect Pressure Modulator Valve (PMV) Assembly Valve Connector C1.
  - Check connector for proper terminal contact.
  - Using J 39200 set to Ω, check for continuity between PMV Assembly Pigtail Connector C1, Terminals 13 and 14.

  Continuity?

- **NO** → Replace PMV Assembly.
- **YES** →
  - Measure resistance between PMV Pigtail Connector C1 Terminal 14 and PMV Valve Block.

  Is infinite ohms measured?

- **NO** → Replace PMV Assembly.
- **YES** →
  - Measure resistance at PMV Assembly Pigtail Connector C1, between Terminal 14 and all Valve Terminals.

VALVE	TERMINAL	CORRECT RESISTANCE
LH FRONT INLET	3	5 TO 8 OHMS
RH FRONT INLET	5	
LH REAR INLET	9	
RH REAR INLET	7	
LH FRONT OUTLET	4	3 TO 6 OHMS
RH FRONT OUTLET	6	
LH REAR OUTLET	10	
RH REAR OUTLET	8	

- If equipped with Traction Control, the following Isolation Valve Resistances should be measured between Terminal 14 and both Isolation Valve Terminals.

VALVE	TERMINAL	CORRECT RESISTANCE
LH ISOLATION	1	5 TO 8 OHMS
RH ISOLATION	2	

Are all resistances within the specified ranges?

A

(Continued on next page)

After repairs are completed, clear DTC and repeat ABS/TCS functional test.

DIAGNOSTIC CHART DTC 45 — 1995 AURORA AND RIVIERA, CONTINUED

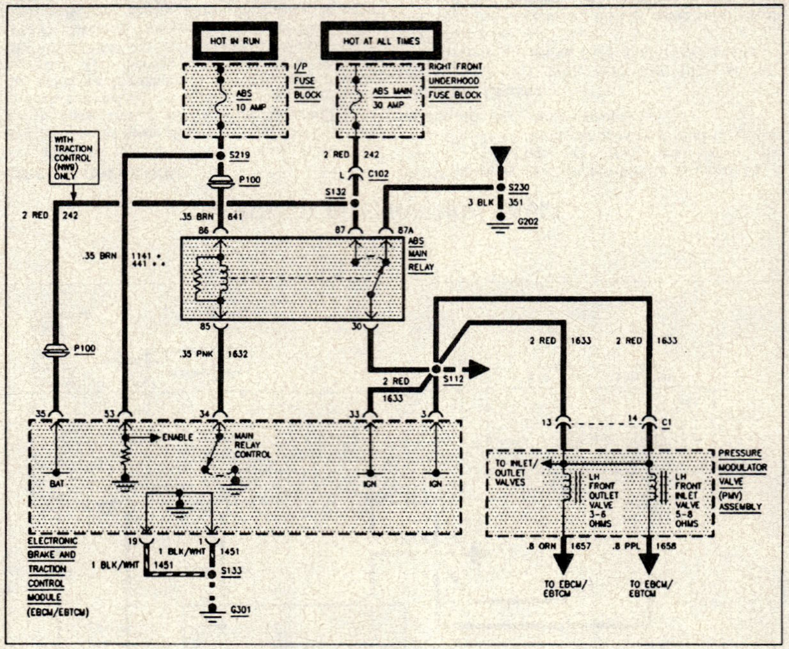

DTC 45
POWER INTERUPTION FAULT

CIRCUIT OPERATION

When the ignition is first turned to "RUN," the EBCM/EBTCM goes through a self-check lasting 2-4 seconds. During this test the EBCM/EBTCM grounds terminal 34, energizing the main relay. Once the self-check is completed the main relay remains energized. This delivers battery voltage to both the EBCM/EBTCM and the pressure modulator valve (PMV) assembly during vehicle operation. The constant voltage supply allows the vehicle to enter ABS mode without delay. The main relay is de-energized whenever the amber "ANTILOCK" indicator lights due to a system fault, causing the ABS system to shut down.

FAILURE CONDITION

DTC 45 sets when the ignition switch is at "RUN" and battery voltage is neither detected at the PMV assembly nor EBCM/EBTCM terminal 20, 3 and/or 33. Code 45 also sets when the EBCM/EBTCM detects an open or a short to ground in the LH inlet valve circuit.

Possible Causes

- Open in CKT 1141, 1632, 858, 1633, or 242.
- Short to ground in CKT 858, 1633, or 242.
- Main relay fails "open."
- High resistance in any inlet or outlet valve circuit and if equipped with traction control, high resistance in either isolation valave circuit.

DIAGNOSIS AIDS

DTC 45 may also set when the vehicle is hot (running for a minimum of 30 minutes) check PMV valve resistance and main relay coil resistance when vehicle is hot.

DIAGNOSTIC CHART DTC 45 — 1995 AURORA AND RIVIERA, CONTINUED

DTC 45
POWER INTERRUPTION FAULT (2 of 5)

(Continued from previous page)

A

YES
- Reconnect PMV Assembly Valve Connector C1.
- Disconnect EBCM/EBTCM Connector.
- Install J 38716 Pinout Box.
- Measure resistance between J 38716 Pinout Box Terminal 3 and all Valve Terminals.

VALVE	J 38716 TERMINAL	CORRECT RESISTANCE
LH FRONT INLET	20	5 TO 8 OHMS
RH FRONT INLET	38	
LH REAR INLET	54	
RH REAR INLET	55	
LH FRONT OUTLET	2	3 TO 6 OHMS
RH FRONT OUTLET	21	
LH REAR OUTLET	36	
RH REAR OUTLET	18	

- If equipped with Traction Control, the following Isolation Valve Resistances should be measured between J 38716 Pinout Box Terminal 3 and both Valve Terminals.

VALVE	J 38716 TERMINAL	CORRECT RESISTANCE
LH ISOLATION	37	LESS THAN 10 OHMS
RH ISOLATION	40	

Are resistances within the specified ranges?

NO — Replace PMV Assembly.

YES
- Measure resistance between J 38716 Pinout Box Terminal 1 and Data Link Connector (DLC) Terminal 4.

Is less than 5 ohms measured?

NO — Repair open in suspect circuit between PVM and EBCM/EBTCM.

YES — A (Continued on next page)

NO — Repair open in CKT 1451.

After repairs are completed, clear DTC and repeat ABS/TCS functional test.

DIAGNOSTIC CHART DTC 45 — 1995 AURORA AND RIVIERA, CONTINUED

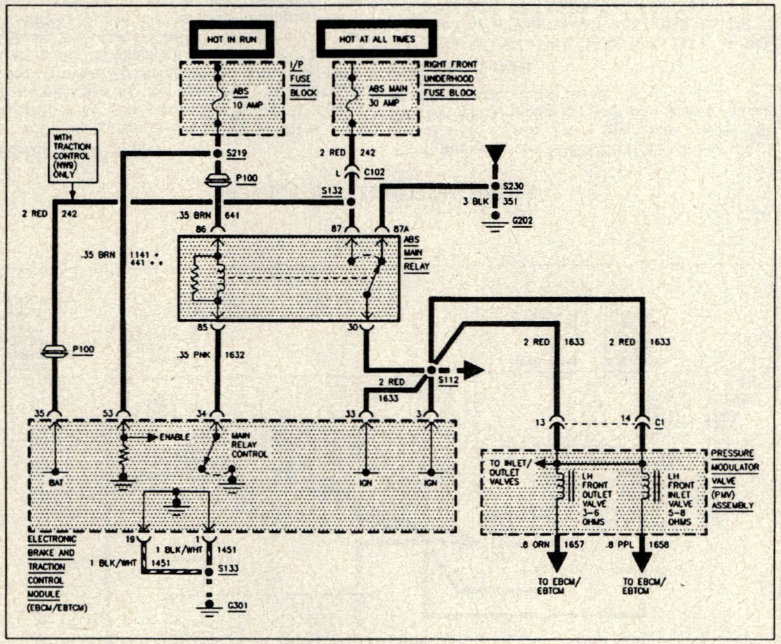

DTC 45
POWER INTERUPTION FAULT

CIRCUIT OPERATION

When the ignition is first turned to "RUN," the EBCM/EBTCM goes through a self-check lasting 2-4 seconds. During this test the EBCM/EBTCM grounds terminal 34, energizing the main relay. Once the self-check is completed the main relay remains energized. This delivers battery voltage to both the EBCM/EBTCM and the pressure modulator valve (PMV) assembly during vehicle operation. The constant voltage supply allows the vehicle to enter ABS mode without delay. The main relay is de-energized whenever the amber "ANTILOCK" indicator lights due to a system fault, causing the ABS system to shut down.

FAILURE CONDITION

DTC 45 sets when the ignition switch is at "RUN" and battery voltage is neither detected at the PMV assembly nor EBCM/EBTCM terminal 20, 3 and/or 33. Code 45 also sets when the EBCM/EBTCM detects an open or a short to ground in the LH inlet valve circuit.

Possible Causes

- Open in CKT 1141, 1632, 858, 1633, or 242.
- Short to ground in CKT 858, 1633, or 242.
- Main relay fails "open."
- High resistance in any inlet or outlet valve circuit and if equipped with traction control, high resistance in either isolation valave circuit.

DIAGNOSIS AIDS

DTC 45 may also set when the vehicle is hot (running for a minimum of 30 minutes) check PMV valve resistance and main relay coil resistance when vehicle is hot.

DIAGNOSTIC CHART DTC 45 — 1995 AURORA AND RIVIERA, CONTINUED

(Continued from previous page)

A

DTC 45
POWER INTERRUPTION FAULT (3 of 5)

- Disconnect Main Relay.
- Measure resistance between J 38716 Pinout Box Terminal 1 and all Valve Terminals.

VALVE	J 38716 TERMINAL	CORRECT RESISTANCE
LH FRONT INLET	20	INFINITE OHMS
RH FRONT INLET	38	
LH REAR INLET	54	
RH REAR INLET	55	
LH FRONT OUTLET	2	
RH FRONT OUTLET	21	
LH REAR OUTLET	36	
RH REAR OUTLET	18	

- If equipped with Traction Control, the following Isolation Valve Resistances should also be measured between J 38716 Pinout Box Terminal 1 and both Valve Terminals.

VALVE	J 38716 TERMINAL	CORRECT RESISTANCE
LH ISOLATION VALVE	37	INFINITE OHMS
RH ISOLATION VALVE	40	

Are resistances within the specified range?

NO → Repair short to ground in suspect circuit between EBCM/EBTCM and PMV.

YES →

- Reinstall Main Relay.
- Ignition: RUN
- Using J 39200 set to DC scale, measure voltage between Pinout Box Terminal 1 and all valve terminals.

Is more than 2 volts measured in any valve circuit?

YES → Repair short to battery in suspect valve circuit.

NO →

- Using fused jumper, jump between Pinout Box Terminals 1 and 34.
- Ignition: RUN

Does "ABS" indicator go off?

YES →

- Using J 39200 set to DC scale, check for battery voltage between Pinout Box Terminals 1 and 3 then 1 and 33, PMV harness side connector C1, Terminal 13 and ground then Terminal 14 and ground.

Battery voltage present?

- NO → Repair open in CKT 1633.
- YES → Replace EBCM/EBTCM.

NO →

- Ignition: OFF
- Remove fused jumper.
- Disconnect Main Relay.
- Check for continuity between Relay Terminals 30 and 87A.

Continuity Present?

- NO → Replace Main Relay.
- YES → A

(Continued on next page)

After repairs are completed, clear DTC and repeat ABS/TCS functional test.

DIAGNOSTIC CHART DTC 45 — 1995 AURORA AND RIVIERA, CONTINUED

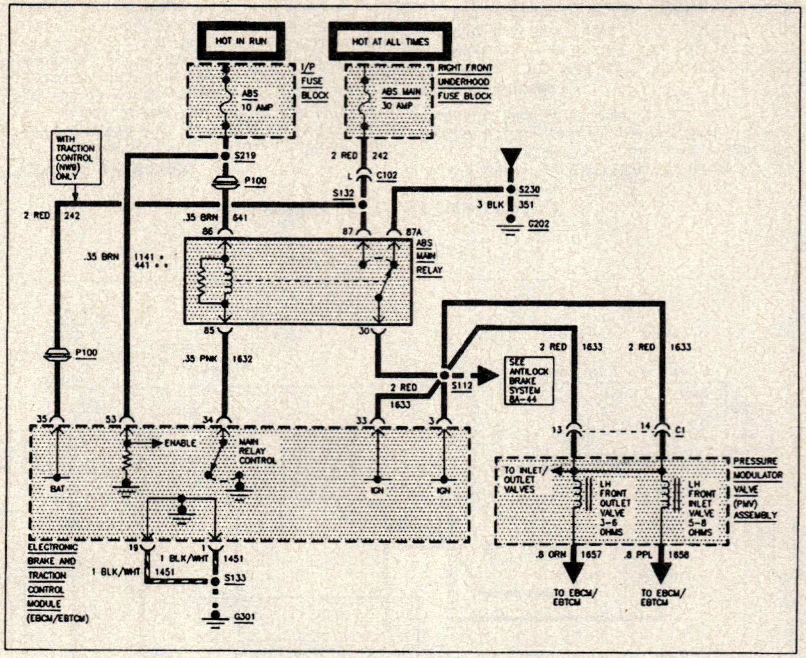

DTC 45
POWER INTERUPTION FAULT

CIRCUIT OPERATION

When the ignition is first turned to "RUN," the EBCM/EBTCM goes through a self-check lasting 2-4 seconds. During this test the EBCM/EBTCM grounds terminal 34, energizing the main relay. Once the self-check is completed the main relay remains energized. This delivers battery voltage to both the EBCM/EBTCM and the pressure modulator valve (PMV) assembly during vehicle operation. The constant voltage supply allows the vehicle to enter ABS mode without delay. The main relay is de-energized whenever the amber "ANTILOCK" indicator lights due to a system fault, causing the ABS system to shut down.

FAILURE CONDITION

DTC 45 sets when the ignition switch is at "RUN" and battery voltage is neither detected at the PMV assembly nor EBCM/EBTCM terminal 20, 3 and/or 33. Code 45 also sets when the EBCM/EBTCM detects an open or a short to ground in the LH inlet valve circuit.

Possible Causes

- Open in CKT 1141, 1632, 858, 1633, or 242.
- Short to ground in CKT 858, 1633, or 242.
- Main relay fails "open."
- High resistance in any inlet or outlet valve circuit and if equipped with traction control, high resistance in either isolation valave circuit.

DIAGNOSIS AIDS

DTC 45 may also set when the vehicle is hot (running for a minimum of 30 minutes) check PMV valve resistance and main relay coil resistance when vehicle is hot.

DIAGNOSTIC CHART DTC 45 — 1995 AURORA AND RIVIERA, CONTINUED

DTC 45
POWER INTERRUPTION FAULT (4 of 5)

(Continued from previous page)

A

- Measure resistance between Main Relay Terminals 85 and 86.
  40-90 ohms measured?
  - YES:
    - Using unpowered test light to battery, probe Main Relay Harness Connector Cavity 87A.
      Does test light illuminate?
      - YES:
        - Test light to ground, probe Main Relay Harness Connector Cavity 87.
          Does test light illuminate?
          - YES:
            - Ignition: RUN
            - Test Light to Ground, probe Main Relay Harness Connector Cavity 86.
              Does test light illuminate?
              - YES:
                - Install fused jumper between Main Relay Harness Connector Cavities 30 and 87.
                - Using J 39200 set to DC scale, check for Battery voltage between Pinout Box Terminals 33 and 1 then 3 and 1 then at harness side of PMV Connector Terminal 14 and ground and 13 and ground.
                  Battery voltage present?
                  - A
              - NO:
                - Check I/P Fuse Block "ABS" Fuse.
                  Is Fuse open?
                  - YES: Repair short to ground in CKT 1141.
                  - NO: Repair open in CKT 1141.
          - NO:
            - Check Right Front Underhood Relay Center "ABS MAIN" Fuse.
              - NO: Repair open in CKT 242.
              - YES: Repair short to ground in CKT 1633 or 242.
      - NO:
        - Repair open in CKT 351.
  - NO: Replace Main Relay.

(Continued on next page)

After repairs are completed, clear DTC and repeat ABS/TCS functional test.

DIAGNOSTIC CHART DTC 45 — 1995 AURORA AND RIVIERA, CONTINUED

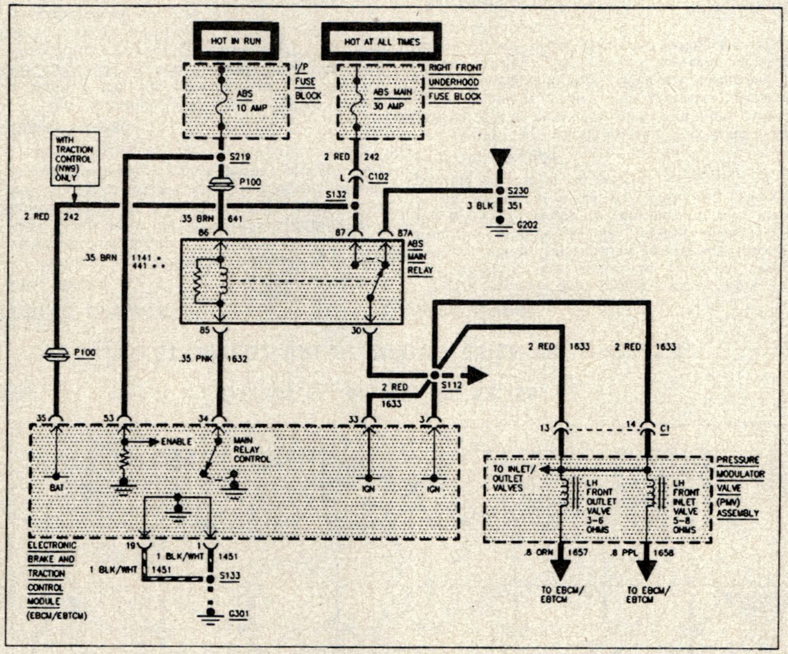

DTC 45
POWER INTERUPTION FAULT

CIRCUIT OPERATION

When the ignition is first turned to "RUN," the EBCM/EBTCM goes through a self-check lasting 2-4 seconds. During this test the EBCM/EBTCM grounds terminal 34, energizing the main relay. Once the self-check is completed the main relay remains energized. This delivers battery voltage to both the EBCM/EBTCM and the pressure modulator valve (PMV) assembly during vehicle operation. The constant voltage supply allows the vehicle to enter ABS mode without delay. The main relay is de-energized whenever the amber "ANTILOCK" indicator lights due to a system fault, causing the ABS system to shut down.

FAILURE CONDITION

DTC 45 sets when the ignition switch is at "RUN" and battery voltage is neither detected at the PMV assembly nor EBCM/EBTCM terminal 20, 3 and/or 33. Code 45 also sets when the EBCM/EBTCM detects an open or a short to ground in the LH inlet valve circuit.

Possible Causes

- Open in CKT 1141, 1632, 858, 1633, or 242.
- Short to ground in CKT 858, 1633, or 242.
- Main relay fails "open."
- High resistance in any inlet or outlet valve circuit and if equipped with traction control, high resistance in either isolation valave circuit.

DIAGNOSIS AIDS

DTC 45 may also set when the vehicle is hot (running for a minimum of 30 minutes) check PMV valve resistance and main relay coil resistance when vehicle is hot.

DIAGNOSTIC CHART DTC 45 — 1995 AURORA AND RIVIERA, CONTINUED

DTC 45
POWER INTERRUPTION FAULT (5 of 5)

(Continued from previous page)

A

YES
- Ignition: OFF
- Install fused jumper between Main Relay Harness Connector Cavities 85 and 86.
- Ignition: RUN
- Check battery voltage between Pinout Box Terminals 34 and 1.

Battery voltage present?

NO → Repair open in CKT 1633 between Main Relay Terminal 30 and EBCM/EBTCM Terminals 33 and 3.

YES → Replace Main Relay.

NO →
- Check I/P Fuse Block "ABS" Fuse:

Is Fuse open?

YES →
- Repair short to ground in CKT 1632.

NO → Repair open in CKT 1632.

After repairs are completed, clear DTC and repeat ABS/TCS functional test.

DIAGNOSTIC CHART DTC 41, 42, 46, 51, 52, 55 AND 56 — 1995 AURORA AND RIVIERA

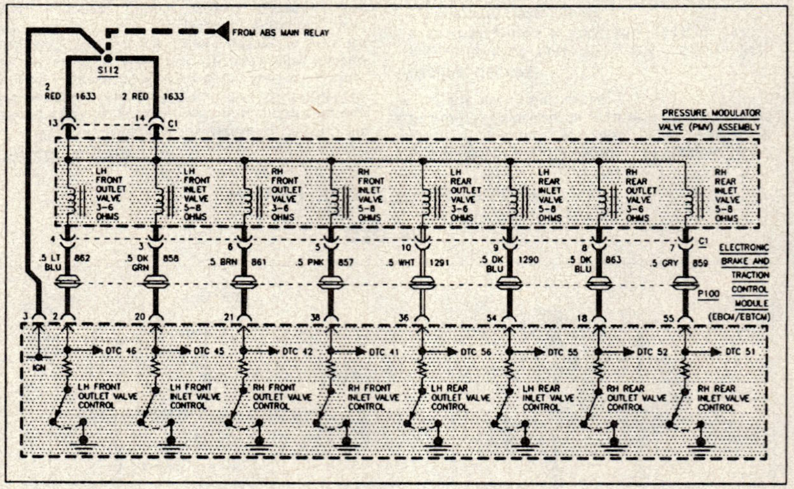

DTC(s) 41, 42, 46, 51, 52, 55, 56

INLET/OUTLET VALVE CIRCUIT OPEN OR SHORTED

CIRCUIT OPERATION

Inlet Valves

Voltage is applied to the normally open inlet valves whenever the main relay is energized. An inlet valve closes when the EBCM/EBTCM supplies a ground path at the inlet valve control terminal. The EBCM/EBTCM closes an inlet valve for short periods to maintain (hold) or reduce pressure at a particular wheel.

Outlet Valves

The normally closed outlet valves operate similarly to the inlet valves, except that when the EBCM/EBTCM grounds an outlet valve, the valve opens instead of closes. The EBCM/EBTCM opens an outlet valve for short periods to reduce pressure at a particular wheel.

VALVE MODES

Pressure Increase

The pressure increase position is the operating valve position for normal (non-ABS) braking and traction control mode (when equipped). The EBCM/EBTCM does not energize either valve, leaving the inlet valve open and the outlet valve closed.

Pressure Hold

The pressure hold position occurs when the EBCM/EBTCM first notices an impending wheel lockup condition. This signifies an over abundance of hydraulic pressure at the locking wheel. The EBCM/EBTCM closes the corresponding inlet valve in an attempt to avoid any additional hydraulic pressure increase at the wheel.

Pressure Reduce

During a wheel lockup condition, the EBCM/EBTCM closes the inlet valve while pulsing open the outlet valve. This decreases hydraulic pressure at a specific wheel without decreasing overall system pressure. The reduced hydraulic pressure allows greater wheel spin, eliminating the wheel lockup condition.

Test Description

This test checks for proper inlet and outlet valve operation. The wheel brake behavior must be checked to verify the valves hydraulic effectiveness. This is accomplished using a Tech 1 to actuate the valves.

Failure conditions

DTC 41, 42, 46, 51, 52, 55, or 56 sets when the EBCM/EBTCM detects an open or short to ground in the respective inlet or outlet valve circuit.

DIAGNOSTIC CHART DTC 41, 42, 46, 51, 52, 55 AND 56 — 1995 AURORA AND RIVIERA, CONTINUED

DTC(s) 41, 42, 46, 51, 52, 55, 56 INLET/OUTLET VALVE CIRCUIT OPEN OR SHORTED (1 OF 2)

Has ABS/TCS functional test been performed?

- YES:
  - Ignition: OFF
  - Disconnect Pressure Modulator Valve (PMV) Assembly Valve Connector C1.
  - Check connector for proper terminal contact.
  - Using J 39200 set to Ω, check for continuity between PMV Assembly Pigtail Connector C1, Terminals 13 and 14.

  Continuity?
- NO: Perform ABS/TCS functional test before proceeding.

Continuity?

- YES:
  - Measure resistance between PMV Pigtail Connector C1 Terminal 14 and PMV Valve Block.

  Is infinite ohms measured?
- NO: Replace PMV Assembly.

Is infinite ohms measured?

- YES:
  - Measure resistance at PMV Assembly Pigtail Connector C1, between Terminal 14 and all Valve Terminals. Refer to chart.
- NO: Replace PMV Assembly.

DIAGNOSTIC TROUBLE CODE (DTC)/VALVE	SUSPECT TERMINAL	CORRECT RESISTANCE
41/RH FRONT INLET	5	5 TO 8 OHMS
55/LH REAR INLET	9	
51/RH REAR INLET	7	
46/LH FRONT OUTLET	4	3 TO 6 OHMS
42/RH FRONT OUTLET	6	
56/LH REAR OUTLET	10	
52/RH REAR OUTLET	8	

Is correct resistance measured?

- YES: A — Continued on next page
- NO: Replace PMV Assembly.

After repairs are completed, clear DTC and repeat ABS/TCS functional test.

DIAGNOSTIC CHART DTC 41, 42, 46, 51, 52, 55 AND 56 — 1995 AURORA AND RIVIERA, CONTINUED

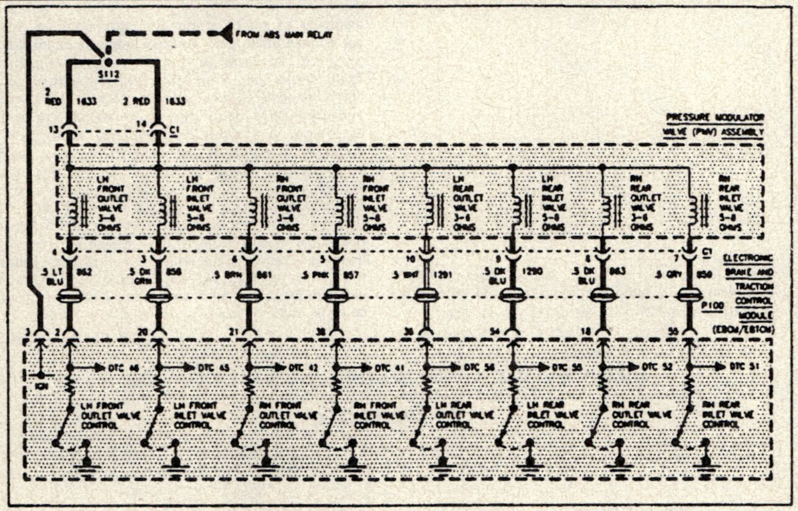

DTC(s) 41, 42, 45, 46, 51, 52, 55, 56

INLET/OUTLET VALVE CIRCUIT OPEN OR SHORTED

CIRCUIT OPERATION

Inlet Valves

Voltage is applied to the normally open inlet valves whenever the main relay is energized. An inlet valve closes when the EBCM/EBTCM supplies a ground path at the inlet valve control terminal. The EBCM/EBTCM closes an inlet valve for short periods to maintain (hold) or reduce pressure at a particular wheel.

Outlet Valves

The normally closed outlet valves operate similarly to the inlet valves, except that when the EBCM/EBTCM grounds an outlet valve, the valve opens instead of closes. The EBCM/EBTCM opens an outlet valve for short periods to reduce pressure at a particular wheel.

VALVE MODES

Pressure Increase

The pressure increase position is the operating valve position for normal (non-ABS) braking and traction control mode (when equipped). The EBCM/EBTCM does not energize either valve, leaving the inlet valve open and the outlet valve closed.

Pressure Hold

The pressure hold position occurs when the EBCM/EBTCM first notices an impending wheel lockup condition. This signifies an over abundance of hydraulic pressure at the locking wheel. The EBCM/EBTCM closes the corresponding inlet valve in an attempt to avoid any additional hydraulic pressure increase at the wheel.

Pressure Reduce

During a wheel lockup condition, the EBCM/EBTCM closes the inlet valve while pulsing open the outlet valve. This decreases hydraulic pressure at a specific wheel without decreasing overall system pressure. The reduced hydraulic pressure allows greater wheel spin, eliminating the wheel lockup condition.

Test Description

This test checks for proper inlet and outlet valve operation. The wheel brake behavior must be checked to verify the valves hydraulic effectiveness. This is accomplished using a Tech 1 to actuate the valves.

Failure conditions

DTC 41, 42, 46, 51, 52, 55, or 56 sets when the EBCM/EBTCM detects an open or short to ground in the respective inlet or outlet valve circuit.

DIAGNOSTIC CHART DTC 41, 42, 46, 51, 52, 55 AND 56 — 1995 AURORA AND RIVIERA, CONTINUED

DTC(s) 41, 42, 46, 51, 52, 55, 56 INLET/OUTLET VALVE CIRCUIT OPEN OR SHORTED (2 OF 2)

A

- Reconnect PMV Assembly Valve Connector C1.
- Disconnect EBCM/EBTCM Connector.
- Install J 38716 Pinout Box.
- Measure resistance between J 38716 Pinout Box Terminal 3 and all Valve Terminals. Refer to Chart.

DIAGNOSTIC TROUBLE CODE (DTC)/VALVE	J 38716 TERMINAL	CORRECT RESISTANCE
41/RH FRONT INLET	38	5 TO 8 OHMS
55/LH REAR INLET	54	
51/RH REAR INLET	55	
46/LH FRONT OUTLET	2	3 TO 6 OHMS
42/RH FRONT OUTLET	21	
56/LH REAR OUTLET	36	
52/RH REAR OUTLET	18	

Is resistance measured less than 10 ohms?

YES

- Using J 39200 measure voltage between J 38716 Pinout Box Terminal 1 and Battery Underhood Junction Block (right side of engine compartment).

Is voltage measured greater than 10 volts?

NO

Repair open in Suspect Valve Circuit between Pressure Modulator Valve (PVM) Assembly and EBCM/EBTCM.

YES

- Disconnect Main Relay.
- Measure resistance between J 38716 Pinout Box Terminal 1 and all Valve Terminals. Refer to Chart.

DIAGNOSTIC TROUBLE CODE (DTC)/VALVE	J 38716 TERMINAL	CORRECT RESISTANCE
41/RH FRONT INLET	38	INFINITE OHMS
55/LH REAR INLET	54	
51/RH REAR INLET	55	
46/LH FRONT OUTLET	2	
42/RH FRONT OUTLET	21	
56/LH REAR OUTLET	36	
52/RH REAR OUTLET	18	

Is infinite ohms measured?

NO

Repair open in CKT 1451.

YES

- Reinstall main relay.
- IGNITION: RUN
- Using J 39200 set to DC scale, measure voltage between Pinout Box Terminal 1 and suspect valve terminal.

Is more than 2 volts measured?

NO

Repair short to ground in suspect valve circuit between PMV Assembly and EBCM/EBTCM.

NO

- If diagnosing DTC 45, return to DTC 45 chart.

If not, Check EBCM/EBTCM connector for proper terminal contact. If terminal contact is OK, replace EBCM/EBTCM.

YES

Repair short to battery in suspect valve circuit.

After repairs are completed, clear DTC and repeat ABS/TCS functional test.

DIAGNOSTIC CHART DTC 44 AND 48 — 1995 AURORA AND RIVIERA

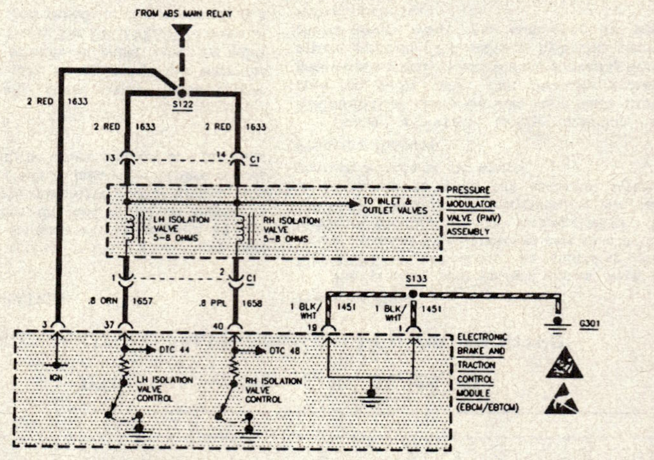

DTC(s) 44, 48

ISOLATION VALVE CIRCUIT OPEN OR SHORTED

CIRCUIT OPERATION

Voltage is applied to the normally open isolation valves whenever the main relay is energized. An isolation valve closes when the EBTCM supplies a ground path at the isolation valve control terminal. The EBTCM detects wheel slip in a drive wheel and vehicle speed is below 40 km/h (25 mph).

The isolation valves are closed only during traction control operations, and are not used during antilock or normal braking conditions. The isolation valves separate (isolate) the front brake hydraulic circuits from the master cylinder and rear brake hydraulic circuits. Once the front brake hydraulic circuits are isolated, pressure can be applied to the front wheels without affecting any other brake hydraulic circuit. The EBTCM turns on the PMV pump motor to apply pressure, and begin cycling the PMV assembly inlet and outlet valves to modulate pressure.

FAILURE CONDITIONS

DTC 44 or DTC 48 sets when the EBTCM detects an open or short to ground in the respective isolation valve circuit.

DIAGNOSTIC CHART DTC 44 AND 48 — 1995 AURORA AND RIVIERA, CONTINUED

DTC(s) 44, 48
ISOLATION VALVE CIRCUIT OPEN OR SHORTED (1 OF 2)

Has ABS/TCS Functional Test been performed?

- NO: Perform ABS/TCS Functional Test before proceeding.
- YES:
  - Ignition: OFF
  - Disconnect Pressure Modulator Valve (PMV) Assembly Valve Connector C1.
  - Check connector for proper terminal contact.
  - Using J 39200 set to Ω measure resistance between PMV Assembly Pigtail Connector C1 terminals 13 and 14.

Is less than 5 OHMS measured?

- NO: Replace PMV Assembly.
- YES:
  - Measure resistance between PMV Pigtail Connector C1 terminal 14 and PMV Valve Block.

Is infinite OHMS measured?

- NO: Replace PMV Assembly.
- YES:
  - Measure resistance at connector, C1, (PMV Assembly Side) between Terminal 14 and Suspect Valve Terminal. Refer to chart.

DTC/VALVE	SUSPECT TERMINAL	CORRECT RESISTANCE
44/LH ISOLATION	1	5 TO 8 OHMS
48/RH ISOLATION	2	

Is correct resistance measured?

- NO: Replace PMV Assembly.
- YES:
  - Reconnect PMV Assembly Valve Connector, C1.
  - Disconnect EBTCM Connector.
  - Install J 38716 Pinout Box.
  - Measure resistance between J 38716 Pinout Box Terminal 3 and Suspect Isolation Valve Terminal. Refer to chart.

DTC/VALVE	J 38716 TERMINAL
44/LH ISOLATION	37
48/RH ISOLATION	40

Is resistance measured less than 10 OHMS?

- NO: Repair open in Suspect Isolation Valve Circuit between Pressure Modulator Valve (PMV) Assembly and EBCM/EBTCM.
- YES:
  - Using J 39200 measure voltage between J 38716 Pinout Box Terminal 1 and Battery Underhood Junction Block (right side of engine compartment).

Is voltage measured greater than 10 volts?

(A)

Continued on next page

DIAGNOSTIC CHART DTC 44 AND 48 — 1995 AURORA AND RIVIERA, CONTINUED

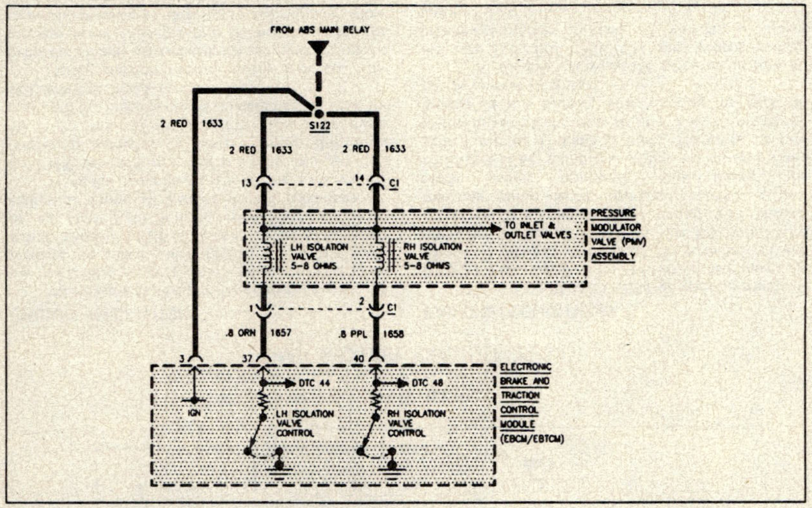

DTC(s) 44, 48

ISOLATION VALVE CIRCUIT OPEN OR SHORTED

CIRCUIT OPERATION

Voltage is applied to the normally open isolation valves whenever the main relay is energized. An isolation valve closes when the EBTCM supplies a ground path at the isolation valve control terminal. The EBTCM detects wheel slip in a drive wheel and vehicle speed is below 40 km/h (25 mph).

The isolation valves are closed only during traction control operations, and are not used during antilock or normal braking conditions. The isolation valves separate (isolate) the front brake hydraulic circuits from the master cylinder and rear brake hydraulic circuits. Once the front brake hydraulic circuits are isolated, pressure can be applied to the front wheels without affecting any other brake hydraulic circuit. The EBTCM turns on the PMV pump motor to apply pressure, and begin cycling the PMV assembly inlet and outlet valves to modulate pressure.

FAILURE CONDITIONS

DTC 44 or DTC 48 sets when the EBTCM detects an open or short to ground in the respective isolation valve circuit.

DIAGNOSTIC CHART DTC 44 AND 48 — 1995 AURORA AND RIVIERA, CONTINUED

DTC(s) 44, 48
ISOLATION VALVE CIRCUIT OPEN OR SHORTED (2 of 2)

A

YES

- Disconnect Main Relay.
- Measure resistance between J 38716 Pinout Box Terminal 1 and Suspect Isolation Valve Terminal. Refer to chart.

DTC/VALVE	J 38716 TERMINAL
44/LH ISOLATION	37
48/RH ISOLATION	40

Is less than 5 OHMS measured?

NO — Check EBTCM Connector for proper terminal contact. If terminal contact is OK, replace EBTCM.

YES — Repair short to ground in Suspect Isolation Valve Circuit between Pressure Modulator Valve (PMV) Assembly and EBTCM.

NO — Repair open in CKT 1451.

After repairs are completed, clear DTC and repeat ABS/TCS functional test.

DIAGNOSTIC CHART DTC 61 — 1995 AURORA AND RIVIERA

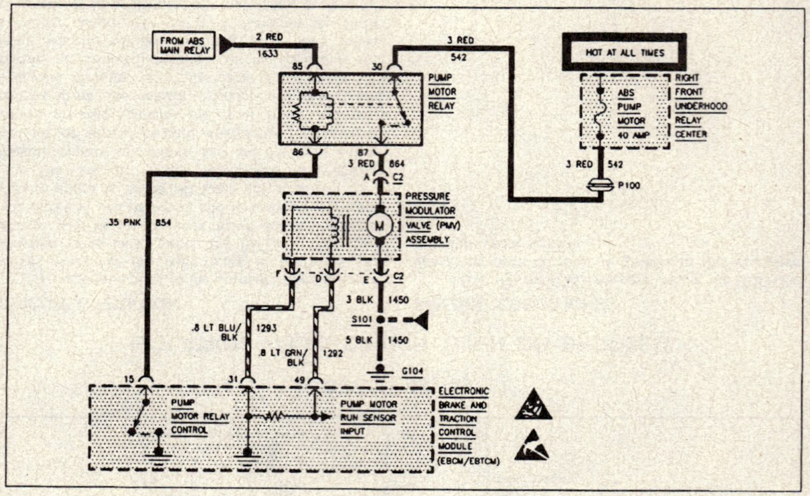

DTC 61

PUMP MOTOR CIRCUIT TEST

CIRCUIT OPERATION

When the EBCM/EBTCM grounds terminal 15, the pump relay contacts close applying voltage to the pressure modulator valve (PMV) assembly's pump motor. The EBCM/EBTCM turns on the pump motor during most ABS braking conditions (and when equipped) during all traction control conditions.

When the pump motor runs during ABS braking, the PMV assembly reservoir is replenished and fluid pressure is applied to the master cylinder. This causes the brake pedal to rise gradually. The pump motor runs until the brake pedal sufficiently rises to close the TCC/antilock brake switch.

During traction control mode, after the front hydraulic circuits are isolated, the pump motor runs to increase brake fluid pressure in these circuits. The increased pressure is directed to the slipping wheel through the PMV assembly's front inlet and outlet valves. This causes the brakes to be applied, reducing wheel spin at the slipping wheel. Traction control is disabled as soon as the brakes are manually applied.

The EBCM/EBTCM monitors pump motor operation via the pump motor run sensor. When the pump is operating, voltage is induced into the sensor's coil and the EBCM/EBTCM monitors this voltage at terminals 49 and 31.

FAILURE CONDITIONS

When the EBCM/EBTCM grounds terminal 15, DTC 61 sets when the EBCM/EBTCM determines the frequency between EBCM/EBTCM terminals 31 and 49 is not between 70-150 Hz. The EBCM/EBTCM recognizes this condition only during ABS braking requiring pump motor operation, traction control mode (when equipped), and during the EBCM/EBTCM self-check which occurs when the vehicle reaches 11 km/h (7 mph). Therefore, be sure to test drive vehicle, refer to Item 3 under diagnostic process in this section, after clearing the DTC to assure correct diagnosis.

DTC 61 also sets when the pump motor runs all the time (i.e. EBCM/EBTCM sees proper voltage between terminals 31 and 49 without grounding terminal 15).

Possible Causes

- Open in CKT 1450, 864, 854, 1633, 1292, 1293 or 1446.
- Short to ground in CKT 864, 854, 1292, 1293, or 642.
- ABS pump relay fails "open" or "engaged."

DIAGNOSTIC CHART DTC 61 — 1995 AURORA AND RIVIERA, CONTINIUED

DTC 61 (1 OF 3)
PUMP MOTOR CIRCUIT TEST

IMPORTANT
DTC 61 usually sets while the vehicle is moving during the system "Self Test".

Has ABS/TCS functional test been performed?

- NO: Perform ABS/TCS functional test.
- YES:
  - Ignition: RUN

  Does Pressure Modulator Valve (PMV) Assembly Pump Motor Run continuously?
  - YES:
    - Remove ABS Pump Relay.

    Does pump motor stop running?
    - NO: Repair short to battery in CKT 864.
    - YES:
      - IGNITION: OFF
      - Using J 39200 set to DC scale, check for battery voltage at Pump Relay Cavity 86.

      Is battery voltage present?
      - YES: Repair short to battery in CKT 854.
      - NO: Replace Pump Relay
  - NO:
    - Ignition: OFF
    - Connect Tech 1 to Data Link Connector (DLC).
    - Ignition: RUN
    - Use Tech 1 to bleed PMV Assembly.

    Does PMV Assembly Pump Motor run?
    - NO: (B) Continued on Chart 3 of 3
    - YES:
      - Ignition: OFF
      - Disconnect Tech 1.
      - Disconnect PMV Assembly Pump Motor Connector C2.
      - Using J 39200 set to Ω, measure resistance between PMV Assembly Pigtail Connector C2 Terminal D and PMV Assembly casing.

      Is infinite ohms measured?
      - NO: Replace PMV Assembly.
      - YES:
        - Measure resistance between PMV Assembly Pigtail Connector C2 Terminal D and F.

        Is resistance between 10 and 40 ohms?
        - NO: Replace PMV Assembly.
        - YES: (A) Continued on next page

After repairs are completed, clear DTC and repeat ABS/TCS functional test.

DIAGNOSTIC CHART DTC 61 — 1995 AURORA AND RIVIERA, CONTINIUED

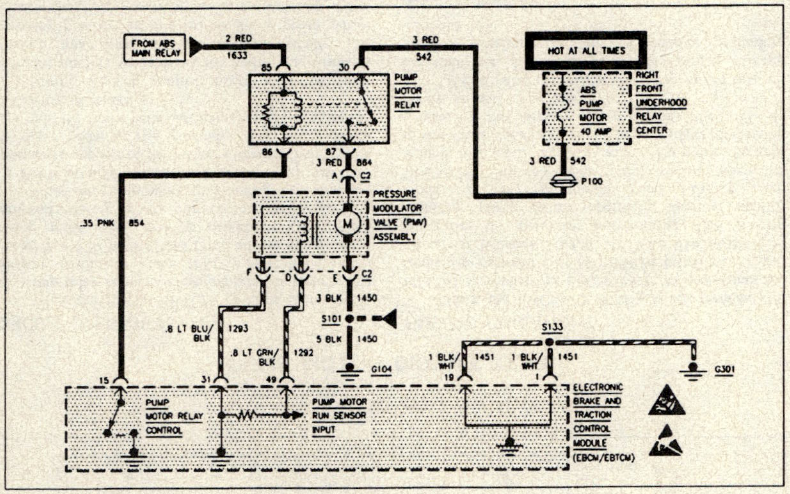

DTC 61
PUMP MOTOR CIRCUIT TEST

CIRCUIT OPERATION

When the EBCM/EBTCM grounds terminal 15, the pump relay contacts close applying voltage to the pressure modulator valve (PMV) assembly's pump motor. The EBCM/EBTCM turns on the pump motor during most ABS braking conditions (and when equipped) during all traction control conditions.

When the pump motor runs during ABS braking, the PMV assembly reservoir is replenished and fluid pressure is applied to the master cylinder. This causes the brake pedal to rise gradually. The pump motor runs until the brake pedal sufficiently rises to close the TCC/antilock brake switch.

During traction control mode, after the front hydraulic circuits are isolated, the pump motor runs to increase brake fluid pressure in these circuits. The increased pressure is directed to the slipping wheel through the PMV assembly's front inlet and outlet valves. This causes the brakes to be applied, reducing wheel spin at the slipping wheel. Traction control is disabled as soon as the brakes are manually applied.

The EBCM/EBTCM monitors pump motor operation via the pump motor run sensor. When the pump is operating, voltage is induced into the sensor's coil and the EBCM/EBTCM monitors this voltage at terminals 49 and 31.

FAILURE CONDITIONS

When the EBCM/EBTCM grounds terminal 15, DTC 61 sets when the EBCM/EBTCM determines the frequency between EBCM/EBTCM terminals 31 and 49 is not between 70-150 Hz. The EBCM/EBTCM recognizes this condition only during ABS braking requiring pump motor operation, traction control mode (when equipped), and during the EBCM/EBTCM self-check which occurs when the vehicle reaches 11 km/h (7 mph). Therefore, be sure to test drive vehicle, refer to Item 3 under diagnostic process in this section, after clearing the DTC to assure correct diagnosis.

DTC 61 also sets when the pump motor runs all the time (i.e. EBCM/EBTCM sees proper voltage between terminals 31 and 49 without grounding terminal 15).

Possible Causes

- Open in CKT 1450, 864, 854, 1633, 1292, 1293 or 1446.
- Short to ground in CKT 864, 854, 1292, 1293, or 642.
- ABS pump relay fails "open" or "engaged."

DIAGNOSTIC CHART DTC 61 — 1995 AURORA AND RIVIERA, CONTINIUED

DTC 61(2 OF 3) PUMP MOTOR CIRCUIT TEST

A

- Reconnect PMV Assembly Pump Motor Connector C2.
- Disconnect EBTCM Connector.
- Install J 38716 Pinout Box.
- Using J 39200 measure voltage between J 38716 Pinout box Terminal 1 and Battery Underhood Junction Block (right side of engine compartment).

Is voltage measured greater than 10 volts?

NO: Repair open in CKT 1451.

YES:

- Measure resistance between J 38716 Pinout Box Terminals 31 and 49.

Is resistance between 10 and 40 ohms?

NO: When high resistance is measured, repair CKT 1292 or 1293 for an open. Check PMV Assembly Connector C2 for proper terminal contact.

When low resistance is measured, repair short between CKT 1292 and 1293.

YES:

- Measure resistance between J 38716 Pinout Box Terminals 49 and 1, then 31 and 1.

Is less than 5 ohms measured in each test?

YES: Repair CKT 1292 or 1293 for a short to ground.

NO:

- Using fused jumper, jump between Pinout Box Terminals 19 and 34. Using another fused jumper, jump between Pinout Box Terminals 1 and 15.
- IGNITION: RUN (pump should run)
- Using J 39200 set to Hz, measure Motor Run Sensor Output between Pinout Box Terminals 31 and 49.

Is 70 -150 Hz measured?

NO: Replace PMV

YES:

- Check EBCM/EBTCM Connector for proper terminal contact. If OK, replace EBCM/EBTCM.

After repairs are completed, clear DTC and repeat ABS/TCS functional test.

DIAGNOSTIC CHART DTC 61 — 1995 AURORA AND RIVIERA, CONTINIUED

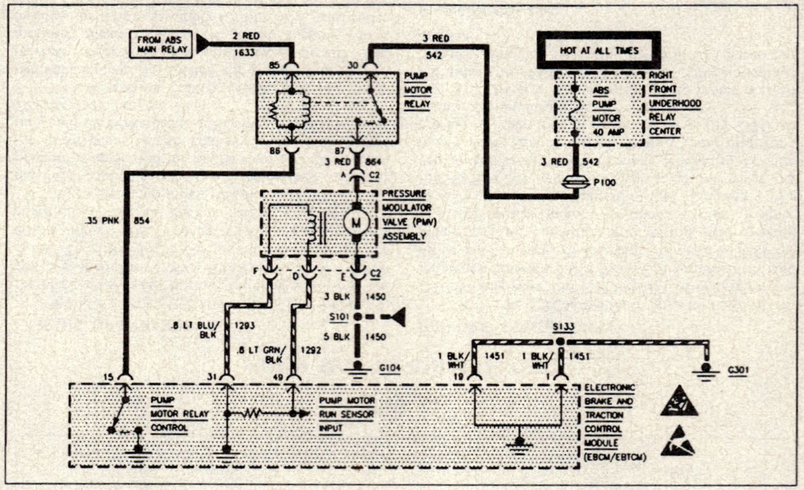

DTC 61

PUMP MOTOR CIRCUIT TEST

CIRCUIT OPERATION

When the EBCM/EBTCM grounds terminal 15, the pump relay contacts close applying voltage to the pressure modulator valve (PMV) assembly's pump motor. The EBCM/EBTCM turns on the pump motor during most ABS braking conditions (and when equipped) during all traction control conditions.

When the pump motor runs during ABS braking, the PMV assembly reservoir is replenished and fluid pressure is applied to the master cylinder. This causes the brake pedal to rise gradually. The pump motor runs until the brake pedal sufficiently rises to close the TCC/antilock brake switch.

During traction control mode, after the front hydraulic circuits are isolated, the pump motor runs to increase brake fluid pressure in these circuits. The increased pressure is directed to the slipping wheel through the PMV assembly's front inlet and outlet valves. This causes the brakes to be applied, reducing wheel spin at the slipping wheel. Traction control is disabled as soon as the brakes are manually applied.

The EBCM/EBTCM monitors pump motor operation via the pump motor run sensor. When the pump is operating, voltage is induced into the sensor's coil and the EBCM/EBTCM monitors this voltage at terminals 49 and 31.

FAILURE CONDITIONS

When the EBCM/EBTCM grounds terminal 15, DTC 61 sets when the EBCM/EBTCM determines the frequency between EBCM/EBTCM terminals 31 and 49 is not between 70-150 Hz. The EBCM/EBTCM recognizes this condition only during ABS braking requiring pump motor operation, traction control mode (when equipped), and during the EBCM/EBTCM self-check which occurs when the vehicle reaches 11 km/h (7 mph). Therefore, be sure to test drive vehicle, refer to Item 3 under diagnostic process in this section, after clearing the DTC to assure correct diagnosis.

DTC 61 also sets when the pump motor runs all the time (i.e. EBCM/EBTCM sees proper voltage between terminals 31 and 49 without grounding terminal 15).

Possible Causes

- Open in CKT 1450, 864, 854, 1633, 1292, 1293 or 1446.
- Short to ground in CKT 864, 854, 1292, 1293, or 642.
- ABS pump relay fails "open" or "engaged."

DIAGNOSTIC CHART DTC 61 — 1995 AURORA AND RIVIERA, CONTINIUED

DTC 61 (3 OF 3) PUMP MOTOR CIRCUIT TEST

B

- Clear DTC using a Tech 1.
- Ignition: OFF
- Disconnect Tech 1.
- Disconnect ABS Pump Relay.
- Ignition: RUN
- Using J 39200 set to DC scale, measure voltage between ABS Pump Relay Harness Connector Terminal 85 and ground.

Is battery voltage measured?

NO: Repair open in CKT 1633.

YES:

- Ignition: OFF
- Measure voltage between ABS Pump Relay Harness Connector Terminal 30 and ground.

Is battery voltage measured?

NO: Check "ABS Pump Motor" Fuse of RH Underhood Relay Center. Check/repair CKT 542 for an open, or a short to ground.

NOTE: When fuse is open and CKT 542 is OK, check/repair short to ground in CKT 864 and replace ABS Pump Relay when a short to ground is found.

YES:

- Connect fused jumper between ABS Pump Relay Harness Terminal 30 and 87A.

Does PMV Assembly Pump Motor run?

NO:

- Leave fused jumper connected.
- Disconnect PMV Assembly Pump Motor Connector C2.
- Measure voltage between PMV Assembly Harness Connector C2 Terminal A and ground.

Is battery voltage measured?

- NO: Repair open or short to ground in CKT 864.
- YES:
  - Measure voltage between PMV Assembly Harness Connector C2 Terminal A and E.

  Is battery voltage measured?

  - YES: Check PMV Assembly Connector C2 for proper terminal contact. If terminal contact is OK, replace PMV Assembly.
  - NO: Repair open in CKT 1450.

YES:

- Reconnect ABS Pump Relay.
- Disconnect EBTCM Connector.
- Install J 38716 Pinout Box.
- Using J 39200 measure voltage between J 38716 Pinout Box Terminal 1 and Battery Underhood Junction Block (right side of engine compartment).

Is voltage measured greater than 10 volts?

- NO: Repair open in CKT 1451.
- YES:
  - Connect fused jumpers between J 38716 Pinout Box Terminal 15 and 1, and Terminal 34 and 1.

  Does PMV Assembly Pump Motor run?

  - YES: Check EBCM/EBTCM Connector for proper terminal contact. If terminal contact is OK, replace EBCM/EBTCM.
  - NO: Check/repair for an open in CKT 854. If CKT 854 is OK, replace ABS Pump Relay.

After repairs are completed, clear DTC and repeat ABS/TCS functional test.

DIAGNOSTIC CHART DTC 62 — 1995 AURORA AND RIVIERA

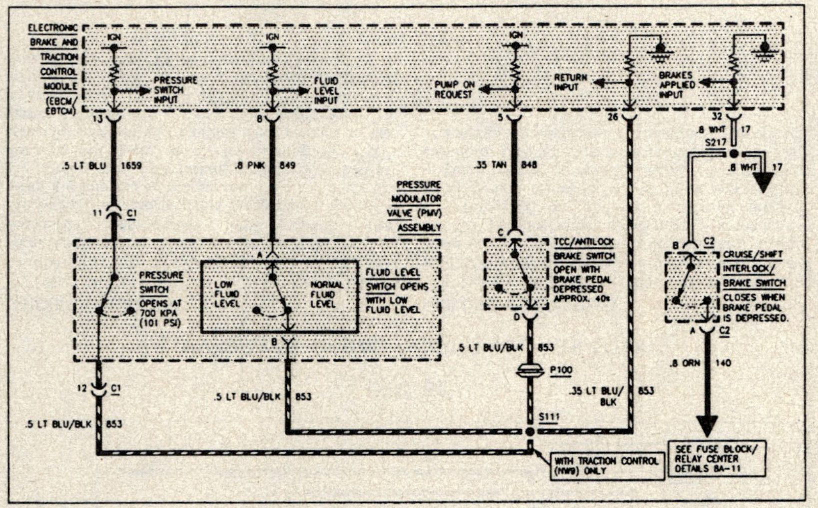

DTC 62

LOW BRAKE PEDAL DURING AN ABS STOP

CIRCUIT OPERATION

The EBCM/EBTCM monitors brake pedal position using the TCC/antilock brake switch. The normally closed switch opens when the brake pedal is depressed approximately 40% down. The EBCM/EBTCM also monitors pump motor operation using a pump motor run sensor.

During most ABS braking conditions the EBCM turns on the pump motor, replenishing the PMV assembly reservoir and applying fluid pressure to the master cylinder. This causes the brake pedal to gradually rise. The EBCM/EBTCM turns off the pump motor when the brake pedal height sufficiently rises to close the TCC/antilock brake switch.

FAILURE CONDITIONS

During ABS braking, DTC 62 sets when the EBCM/EBTCM determines the pump is working, and brake pedal height has not sufficiently risen to close the TCC/antilock brake switch after 3 seconds.

This DTC may indicate a hydraulic problem. Some possible causes include air in the brake lines or a pump motor mechanical failure. Non-hydraulic causes for DTC 62 include an open 848 or 853 CKT or a misadjusted/defective TCC/antilock brake switch.

DIAGNOSTIC CHART DTC 62 — 1995 AURORA AND RIVIERA, CONTINUED

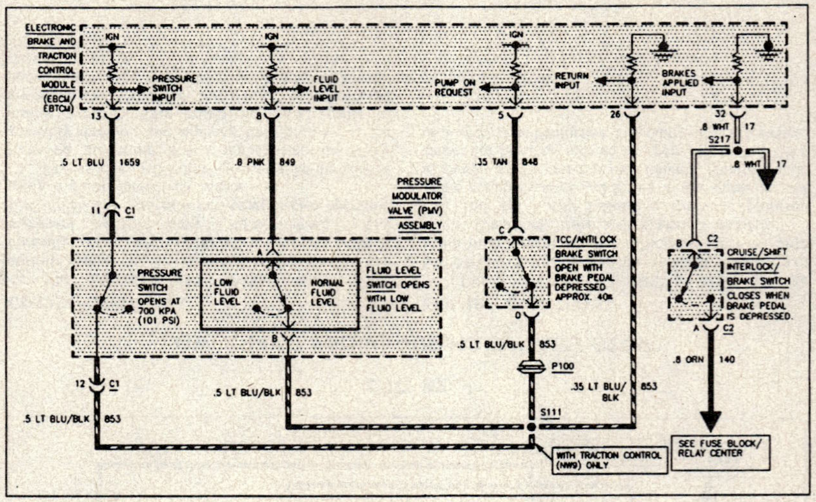

DTC 62

LOW BRAKE PEDAL DURING AN ABS STOP

CIRCUIT OPERATION

The EBCM/EBTCM monitors brake pedal position using the TCC/antilock brake switch. The normally closed switch opens when the brake pedal is depressed approximately 40% down. The EBCM/EBTCM also monitors pump motor operation using a pump motor run sensor.

During most ABS braking conditions the EBCM turns on the pump motor, replenishing the PMV assembly reservoir and applying fluid pressure to the master cylinder. This causes the brake pedal to gradually rise. The EBCM/EBTCM turns off the pump motor when the brake pedal height sufficiently rises to close the TCC/antilock brake switch.

FAILURE CONDITIONS

During ABS braking, DTC 62 sets when the EBCM/EBTCM determines the pump is working, and brake pedal height has not sufficiently risen to close the TCC/antilock brake switch after 3 seconds.

This DTC may indicate a hydraulic problem. Some possible causes include air in the brake lines or a pump motor mechanical failure. Non-hydraulic causes for DTC 62 include an open 848 or 853 CKT or a misadjusted/defective TCC/antilock brake switch.

DIAGNOSTIC CHART DTC 62 — 1995 AURORA AND RIVIERA, CONTINUED

DTC 62 (1 OF 2)
LOW BRAKE PEDAL DURING AN ABS STOP

Has ABS/TCS functional test been performed?

- NO: Perform ABS/TCS functional test before proceeding.
- YES:
  - Bleed brake system as outlined in this section.
  - Ignition: OFF
  - Connect Tech 1 to Data Link Connector (DLC).
  - Ignition: RUN
  - Use Tech 1 to clear DTC after bleeding system.
  - Depress brake pedal (hold).
  - Use Tech 1 to bleed Pressure Modulator Valve (PMV) assembly.

Did brake pedal rise?

- NO: Replace PMV Assembly.
- YES:
  - Ignition: OFF
  - Disconnect Tech 1.
  - Test drive vehicle on a surface as described in Item "3" under "Diagnostic Process" in this section and perform an ABS stop.
  - Connect Tech 1 to DLC Connector.
  - Ignition: RUN

Did DTC 62 set?

- NO: Problem solved.
- YES:
  - Ignition: OFF
  - Disconnect Tech 1.
  - Brake pedal released (rest position).
  - Check TCC/Antilock Brake Switch adjustment.

Is switch adjusted properly?

- NO: Adjust TCC/Antilock Brake Switch.
- YES:
  - Disconnect TCC/Antilock Brake Switch Connector.
  - Using J 39200 set to Ω, measure resistance across switch Terminals C and D.
- NO: Replace TCC/Antilock Brake Switch.
- YES: A

Continued on next page

After repairs are completed, clear DTC and repeat ABS/TCS functional test.

DIAGNOSTIC CHART DTC 62 — 1995 AURORA AND RIVIERA, CONTINUED

DTC 62 (2 OF 2)
LOW BRAKE PEDAL DURING AN ABS STOP

A

- Disconnect EBCM/EBTCM Connector.
- Connect J 38716 Pinout Box.
- Measure resistance between J 38716 Pinout Box Terminal 5 and TCC/Antilock Brake Switch Connector Terminal C.

Is less than 3 ohms measured?

- YES:
  - Measure resistance between J 38716 Pinout Box Terminal 26 and TCC/Antilock Brake Switch Connector Terminal D.

  Is less than 3 ohms measured?
  - YES: Check EBCM/EBTCM Connector for proper terminal contact. If terminal contact is OK, replace EBCM/EBTCM.
  - NO: Repair open in CKT 853.
- NO: Repair open in CKT 848.

After repairs are completed, clear DTC and repeat ABS/TCS functional test.

DIAGNOSTIC CHART DTC 71 — 1995 AURORA AND RIVIERA

DTC 71

EBCM/EBTCM INTERNAL MALFUNCTION

DTC 71 indicates that a failure has occurred within the EBCM/EBTCM. DTC 71 can occur intermittently, due to interference, therefore clear the code and test drive the vehicle before replacing the EBCM/EBTCM.

DIAGNOSTIC CHART DTC 71 — 1995 AURORA AND RIVIERA, CONTINUED

DTC 71
EBCM/EBTCM INTERNAL FAILURE

Has ABS/TCS functional test been performed?

- YES →
  - Using Tech 1 clear Diagnostic Trouble Code.
  - Test drive vehicle. Refer to Diagnostic [illegible] Item "3" [illegible] section

  Did DTC 71 set during test drive?
  - YES → Replace EBCM/EBTCM.
  - NO → No problem found. May be intermittent. Refer to [illegible] Connections in this section.
- NO → Perform ABS/TCS functional test before proceeding.

After repairs are completed, clear DTC and repeat ABS/TCS functional test.

DIAGNOSTIC CHART DTC 72 — 1995 AURORA AND RIVIERA

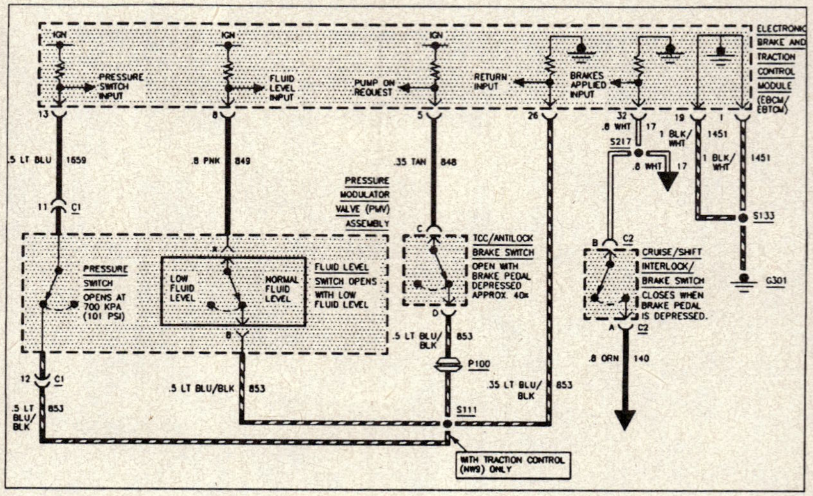

DTC 72

BRAKE SWITCH OPEN OR SHORTED

CIRCUIT OPERATION

The electronic brake control module EBCM/EBTCM monitors brake pedal position using the brake travel switch. The switch is normally closed. It opens when the brake pedal is depressed approximately 40% down. During most ABS braking conditions the EBCM/EBTCM turns on the pump motor, replenishing the PMV assembly reservoir and applying fluid pressure to the master cylinder. This causes the brake pedal to gradually rise. The EBCM/EBTCM turns off the pump motor when brake pedal height sufficiently rises to close the brake travel switch.

FAILURE CONDITIONS

DTC 72 sets when the EBCM/EBTCM detects the brake travel switch CKT 848 is shorted to ground (with the brake pedal depressed) or open.

DIAGNOSTIC CHART DTC 72 — 1995 AURORA AND RIVIERA, CONTINUED

DTC 72
BRAKE TRAVEL SWITCH
CIRCUIT OPEN OR SHORTED (1 OF 4)

Has ABS/TCS functional test been performed?

- NO: Perform ABS/TCS function test before proceeding.
- YES: Using Tech 1, check for Brake Travel Switch "ON/OFF" transition. Is Brake Travel Switch operating properly?
  - YES:
    - Check for proper terminal contact at EBTCM and Brake Travel Switch.
    - Using Tech 1, clear all DTC's.
    - Test drive vehicle.
    - If DTC 72 resets, replace EBTCM.
  - NO:
    - Ignition: OFF
    - Disconnect Brake Travel Switch connector.
    - Disconnect EBTCM.
    - Install J 38716 Pinout Box.
    - Using J 39200, measure voltage between J 38716 Pinout Box Terminal 1 and Battery Underhood Junction Block (right side of engine compartment).

    Is voltage measured greater than 10 volts?

    → A

DIAGNOSTIC CHART DTC 72 — 1995 AURORA AND RIVIERA, CONTINUED

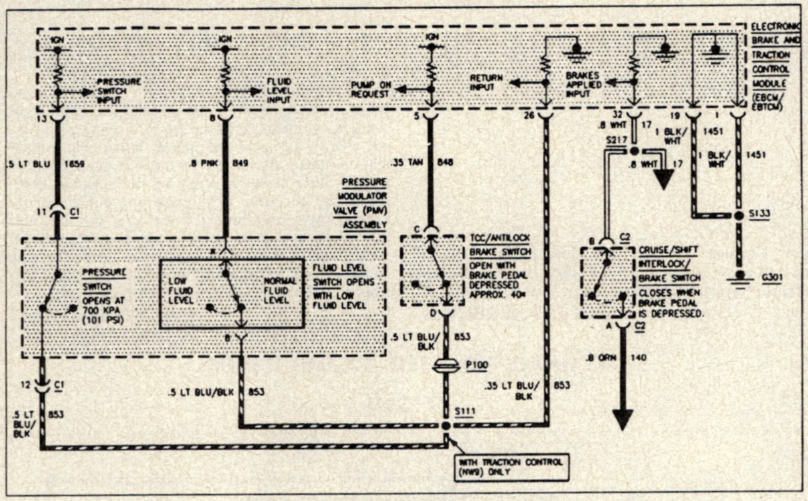

DTC 72

BRAKE SWITCH OPEN OR SHORTED

CIRCUIT OPERATION

The electronic brake control module EBCM/EBTCM monitors brake pedal position using the brake travel switch. The switch is normally closed. It opens when the brake pedal is depressed approximately 40% down. During most ABS braking conditions the EBCM/EBTCM turns on the pump motor, replenishing the PMV assembly reservoir and applying fluid pressure to the master cylinder. This causes the brake pedal to gradually rise. The EBCM/EBTCM turns off the pump motor when brake pedal height sufficiently rises to close the brake travel switch.

FAILURE CONDITIONS

DTC 72 sets when the EBCM/EBTCM detects the brake travel switch CKT 848 is shorted to ground (with the brake pedal depressed) or open.

DIAGNOSTIC CHART DTC 72 — 1995 AURORA AND RIVIERA, CONTINUED

DTC 72
BRAKE TRAVEL SWITCH
CIRCUIT OPEN OR SHORTED (2 OF 4)

A

- Measure resistance between J 38716 Pinout Box Terminals 1 and 5. Is less than 5 ohms measured?
  - NO: Repair open in CKT 1451.
  - YES:
- Measure resistance between J 3876 Pinout Box Terminals 1 and 26. Is less than 5 ohms measured?
  - YES: Repair short to ground CKT 848.
  - NO:
- Measure resistance between Brake Travel Switch Terminals C and D (Brake Travel Switch only, wiring harness disconnected). Is less than 5 ohms measured?
  - YES: Repair short to ground CKT 853.
  - NO: B

DIAGNOSTIC CHART DTC 72 — 1995 AURORA AND RIVIERA, CONTINUED

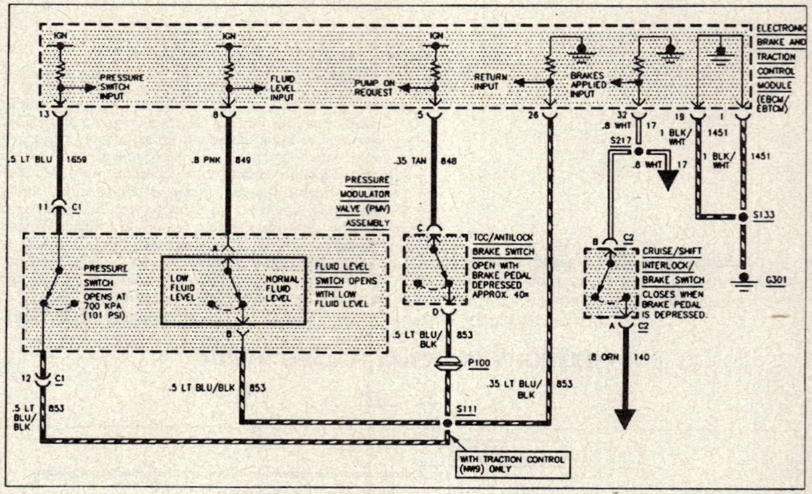

DTC 72

BRAKE SWITCH OPEN OR SHORTED

CIRCUIT OPERATION

The electronic brake control module EBCM/EBTCM monitors brake pedal position using the brake travel switch. The switch is normally closed. It opens when the brake pedal is depressed approximately 40% down. During most ABS braking conditions the EBCM/EBTCM turns on the pump motor, replenishing the PMV assembly reservoir and applying fluid pressure to the master cylinder. This causes the brake pedal to gradually rise. The EBCM/EBTCM turns off the pump motor when brake pedal height sufficiently rises to close the brake travel switch.

FAILURE CONDITIONS

DTC 72 sets when the EBCM/EBTCM detects the brake travel switch CKT 848 is shorted to ground (with the brake pedal depressed) or open.

DIAGNOSTIC CHART DTC 72 — 1995 AURORA AND RIVIERA, CONTINUED

DTC 72
BRAKE TRAVEL SWITCH
CIRCUIT OPEN OR SHORTED (4 OF 4)

C

NO
- Measure resistance between J 38716 Pinout Box Terminals 5 and 26.
  Is less than 5 ohms measured?

YES
- Check for proper terminal contact at EBTCM and Brake Travel Switch.
- Using Tech 1, clear all DTC's.
- Test drive vehicle.
- If DTC 72 resets, replace EBTCM.

NO
Repair open in CKT 848 or CKT 853.

YES
- Depress Brake Pedal.
- Measure voltage between J 38716 Pinout Box Terminals 1 and 26.
  Is greater than 1 volt measured?

YES
Repair short to voltage in CKT 853.

NO
Repair short to voltage in CKT 848.

DIAGNOSTIC CHART DTC 73 — 1995 AURORA AND RIVIERA

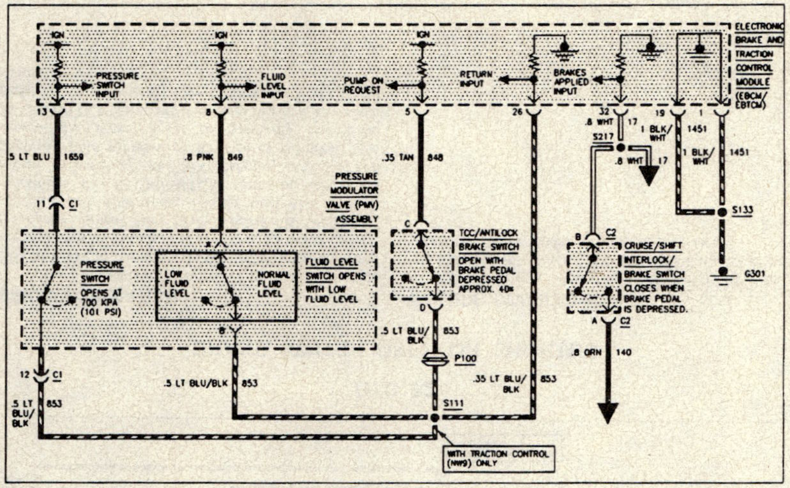

DTC 73

FLUID LEVEL SWITCH CIRCUIT OPEN OR SHORTED

CIRCUIT OPERATION

The pressure modulator valve (PMV) fluid level switch alerts the EBCM/EBTCM to a low brake fluid condition in the PMV reservoir. The fluid level switch is located in the PMV assembly's reservoir. The fluid level switch is normally closed when the reservoir contains sufficient fluid. When fluid level is low, the switch opens, causing the EBCM/EBTCM to turn on the "ANTILOCK" indicator.

FAILURE CONDITIONS

DTC 73 sets when the EBCM/EBTCM detects a short to ground in the PMV fluid level switch circuitry. These include: CKT 848, 849, 853, and 1659. DTC 73 will also set if an open exists in CKT 1451.

DIAGNOSTIC CHART DTC 72 — 1995 AURORA AND RIVIERA, CONTINUED

DTC 72
BRAKE TRAVEL SWITCH
CIRCUIT OPEN OR SHORTED (3 OF 4)

(B)

YES:
- Depress Brake Pedal and repeat resistance measurement between Brake Travel Switch Terminals C and D (Brake Travel Switch only, wiring harness disconnected).
  Is less than 5 ohms measured?

NO:
- Check for properly adjusted Brake Travel Switch.
  Is Brake Travel Switch properly adjusted?
  - YES: Replace Brake Travel Switch.
  - NO: Adjust Brake Travel Switch.

(from "Is less than 5 ohms measured?")

YES:
- Connect Brake Travel Switch Connector.
- Disconnect EBTCM from J 38716 Pinout Box (only wiring harness connected to J 38716 Pinout Box).
- Measure voltage between J 38716 Pinout Box Terminals 1 and 5.
  Is greater than 1 volt measured?

(C)

NO:
- Check for properly adjusted Brake Travel Switch.
  Is Brake Travel Switch properly adjusted?
  - YES: Replace Brake Travel Switch.
  - NO: Adjust Brake Travel Switch.

DIAGNOSTIC CHART DTC 72 — 1995 AURORA AND RIVIERA, CONTINUED

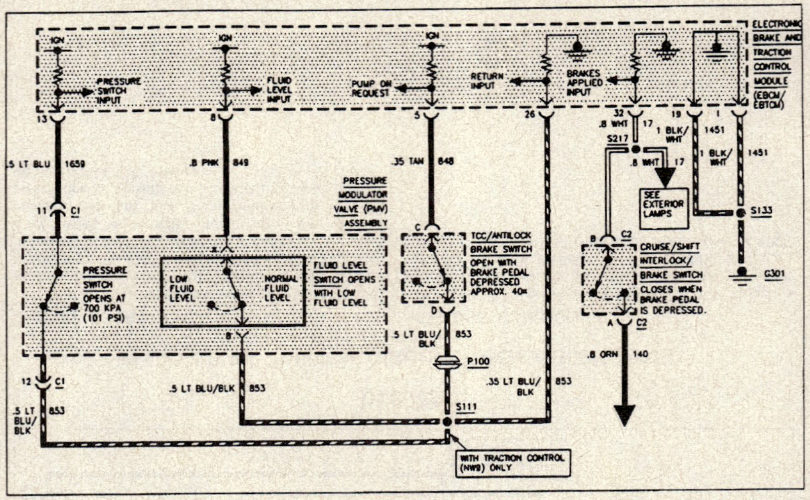

DTC 72

BRAKE SWITCH OPEN OR SHORTED

CIRCUIT OPERATION

The electronic brake control module EBCM/EBTCM monitors brake pedal position using the brake travel switch. The switch is normally closed. It opens when the brake pedal is depressed approximately 40% down. During most ABS braking conditions the EBCM/EBTCM turns on the pump motor, replenishing the PMV assembly reservoir and applying fluid pressure to the master cylinder. This causes the brake pedal to gradually rise. The EBCM/EBTCM turns off the pump motor when brake pedal height sufficiently rises to close the brake travel switch.

FAILURE CONDITIONS

DTC 72 sets when the EBCM/EBTCM detects the brake travel switch CKT 848 is shorted to ground (with the brake pedal depressed) or open.

DIAGNOSTIC CHART DTC 73 — 1995 AURORA AND RIVIERA, CONTINUED

DTC 73
PMV FLUID LEVEL SWITCH CIRCUIT OPEN OR SHORTED

Has ABS/TCS functional test been performed?

- NO: Perform ABS/TCS functional test before proceeding.
- YES:
  - Ignition: OFF
  - Disconnect EBTCM Connector.
  - Install J 38716 Pinout Box.
  - Using J 39200 measure between J 38716 Pinout Box Terminal 1 and Battery Underhood Junction Block (right side of engine compartment).

  Is voltage measured greater than 10 volts?

  - NO: Repair open in CKT 1451.
  - YES:
    - Disconnect Brake Travel Switch.
    - Measure resistance between J 38716 Pinout Box Terminal 5 and 1.

    Is less than 5 ohms measured?

    - YES: Repair short to ground in CKT 848.
    - NO:
      - Disconnect Pressure Modulator Valve (PMV) Fluid Level Switch Connector.
      - Measure resistance between J 38716 Pinout Box Terminal 8 and 1.

      Is less than 5 ohms measured?

      - YES: Repair short to ground in CKT 849.
      - NO:
        - When equipped with Traction Control: Depress Brake Pedal. Measure resistance between J 38716 Pinout Box Terminal 13 and 1.

        Is less than 5 ohms measured?

        - YES: Repair short to ground in CKT 1659.
        - NO:
          - Measure resistance between J 38716 Pinout Box Terminal 26 and 1.

          Is less than 5 ohms measured?

          - YES: Repair short to ground in CKT 853.
          - NO: Check EBCM/EBTCM Connector for proper terminal contact. If terminal contact is OK, replace EBCM/EBTCM.

After repairs are completed, clear DTC and repeat ABS/TCS functional test.

DIAGNOSTIC CHART DTC 74 — 1995 AURORA AND RIVIERA

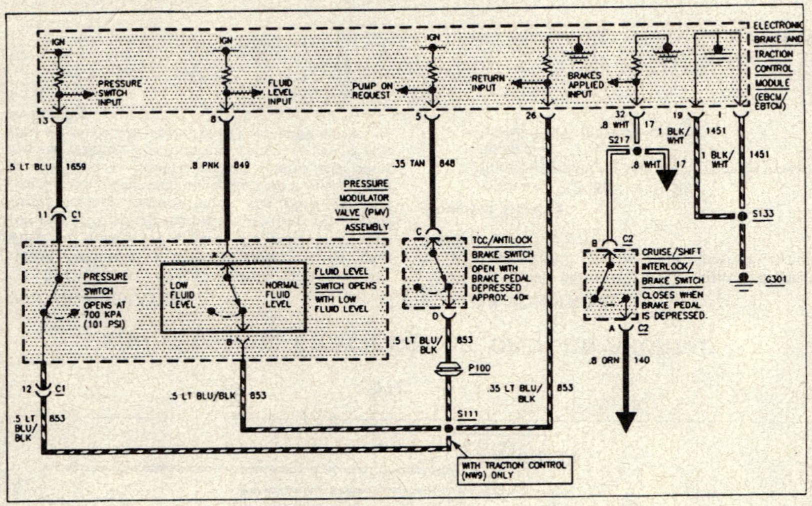

DTC 74

PMV PRESSURE SWITCH CIRCUIT OPEN OR SHORTED

CIRCUIT OPERATION

The pressure switch monitors primary brake system pressure. The normally closed switch opens when the brakes are applied and pressure exceeds 700 kPa (101 psi). If the brakes are manually applied during traction control mode, the PMV pressure switch input, along with the brake switch input, signal the EBTCM to disable traction control and allow manual braking. The pressure switch is part of the pressure modulator valve (PMV) assembly and is not serviceable.

FAILURE CONDITIONS

DTC 74 sets when the EBTCM detects an open or short to ground in the PMV pressure switch circuitry.

Possible Causes

- Open in CKT 853 or 1659.
- Short to ground in CKT 1659 with the brakes applied.
- Failed PMV pressure switch.

DIAGNOSTIC CHART DTC 74 — 1995 AURORA AND RIVIERA, CONTINUED

DTC 74 PMV PRESSURE SWITCH CIRCUIT OPEN OR SHORTED (1 of 3)

Has ABS/TCS Functional Test been performed?

- NO: Perform ABS/TCS Functional Test.
- YES:
  - Ignition: OFF
  - Inspect Rear Fuse Block "HZRD/STOP" Fuse.
  - Is Fuse open?

Is Fuse open?

- YES: Repair Short to Ground in CKT 140 or CKT 17. Replace Fuse.
- NO:
  - Disconnect Brake Switch Connector (Grey Connector).
  - Using J 39200 set to DC scale measure voltage between DLC Terminal 4 and Brake Switch Connector (Harness Side) Terminal A.
  - Is greater than 10 volts measured?

Is greater than 10 volts measured?

- NO: Repair Open in CKT 140.
- YES:
  - Depress Brake Pedal.
  - Using J 39200 set to Ω measure resistance between Brake Switch Terminals A and B.
  - Is less than 2 OHMS measured?

Is less than 2 OHMS measured?

- NO: Replace Brake Switch.
- YES:
  - Check for proper terminal contact on Brake Switch Connector.
  - Reconnect Brake Switch Connector.
  - Check for proper Brake Switch Adjustment
  - Is Brake Switch Adjusted Properly?

Is Brake Switch Adjusted Properly?

- NO: Adjust Brake Switch.
- YES:
  - Disconnect EBTCM Connector.
  - Install J 38716 Pinout Box.
  - Using J 39200 measure voltage between J 38716 Pinout Box Terminal 1 and Battery Underhood Junction Block (right side of engine compartment).
  - Is voltage measured greater than 10 volts?

Is voltage measured greater than 10 volts?

- NO: Repair Open in CKT 1451.
- YES:
  - Ignition: ON
  - Using J 39200 set to DC scale, measure voltage between Pin Out Box Terminals 1 and 32.
  - Is greater than 2 volts measured?

(A)

Continued on next page

DIAGNOSTIC CHART DTC 74 — 1995 AURORA AND RIVIERA, CONTINUED

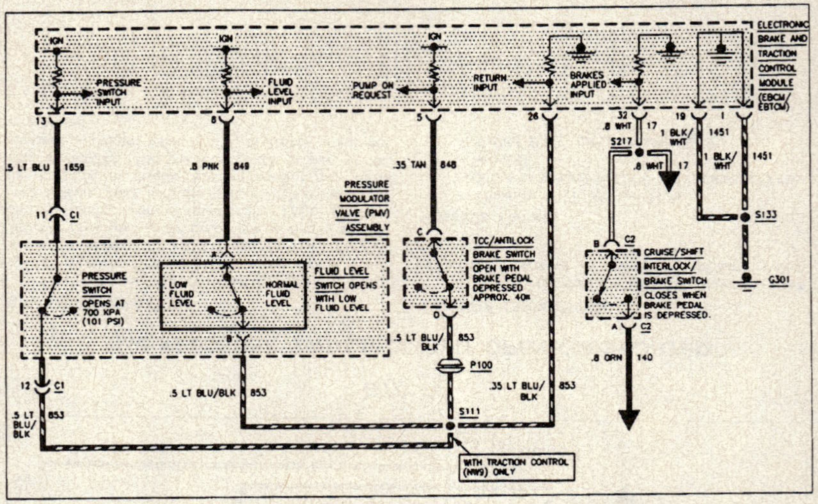

DTC 74

PMV PRESSURE SWITCH CIRCUIT OPEN OR SHORTED

CIRCUIT OPERATION

The pressure switch monitors primary brake system pressure. The normally closed switch opens when the brakes are applied and pressure exceeds 700 kPa (101 psi). If the brakes are manually applied during traction control mode, the PMV pressure switch input, along with the brake switch input, signal the EBTCM to disable traction control and allow manual braking. The pressure switch is part of the pressure modulator valve (PMV) assembly and is not serviceable.

FAILURE CONDITIONS

DTC 74 sets when the EBTCM detects an open or short to ground in the PMV pressure switch circuitry.

Possible Causes

- Open in CKT 853 or 1659.
- Short to ground in CKT 1659 with the brakes applied.
- Failed PMV pressure switch.

DIAGNOSTIC CHART DTC 74 — 1995 AURORA AND RIVIERA, CONTINUED

DTC 74
PMV PRESSURE SWITCH CIRCUIT OPEN OR SHORTED (2 of 3)

A

NO
- Depress Brake Pedal.
- Measure Voltage between Pin Out Box Terminals 1 and 32.

Is greater than 10 volts measured?

YES → Repair Short to Voltage in CKT 17.

YES
- Ignition: OFF
- Disconnect Pressure Modulator Valve (PMV) Connector C1.
- Measure resistance between Connector C1 (PMV Assembly Side) Terminal 11 and PMV Case.

Is Infinite OHMS measured?

NO → Repair Open in CKT 17.

YES
- Measure resistance between Connector C1 (PMV Assembly Side) Terminals 11 and 12.

Is less than 2 OHMS measured?

NO → Replace PMV.

YES
- Depress Brake Pedal.
- Measure resistance between Connector C1 (PMV Assembly Side) Terminals 11 and 12.

Is Infinite OHMS measured?

NO → Replace PMV.

YES
- Measure resistance between Pin Out Box Terminal 13 and PMV Connector C1 (Harness Side) Terminal 11.

Is less than 5 OHMS measured?

NO → Replace PMV.

YES
- Measure resistance between Pin Out Box Terminal 28 and PMV Connector C1 (Harness Side) Terminal 12.

Is less than 5 OHMS measured?

NO → Repair Open in CKT 1659

B

Continued on next page

DIAGNOSTIC CHART DTC 74 — 1995 AURORA AND RIVIERA, CONTINUED

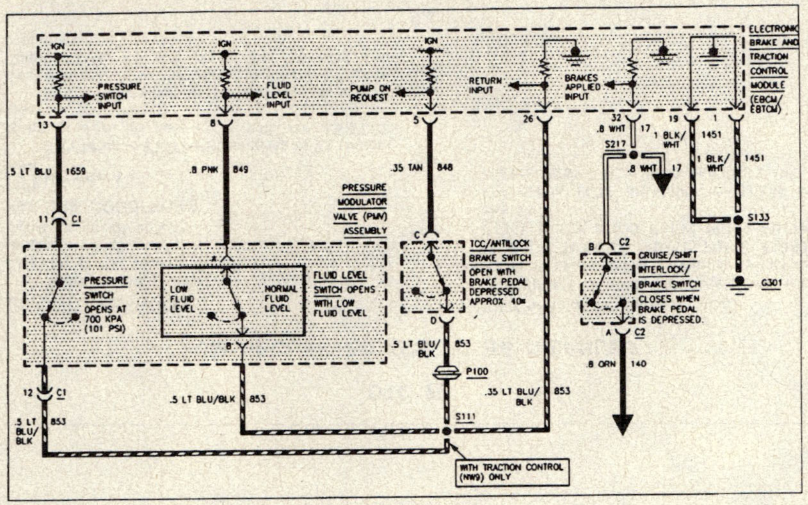

DTC 74

PMV PRESSURE SWITCH CIRCUIT OPEN OR SHORTED

CIRCUIT OPERATION

The pressure switch monitors primary brake system pressure. The normally closed switch opens when the brakes are applied and pressure exceeds 700 kPa (101 psi). If the brakes are manually applied during traction control mode, the PMV pressure switch input, along with the brake switch input, signal the EBTCM to disable traction control and allow manual braking. The pressure switch is part of the pressure modulator valve (PMV) assembly and is not serviceable.

FAILURE CONDITIONS

DTC 74 sets when the EBTCM detects an open or short to ground in the PMV pressure switch circuitry.

Possible Causes
- Open in CKT 853 or 1659.
- Short to ground in CKT 1659 with the brakes applied.
- Failed PMV pressure switch.

DIAGNOSTIC CHART DTC 74 — 1995 AURORA AND RIVIERA, CONTINUED

DTC 74
PMV PRESSURE SWITCH CIRCUIT OPEN OR SHORTED (3 of 3)

B

YES
- Disconnect PMV Fluid Level Switch and Brake Switch (Black Connectors).
- Ignition: RUN
- Measure Voltage between Pin Out Box Terminals 1 and:
  - Terminal 5
  - Terminal 8
  - Terminal 13
  - Terminal 26

Was greater than 2 volts measured on any one of the above?

NO → Repair Open in CKT 853.

YES → Repair Short to Voltage in any circuits that measured more than 2 volts.

NO
- Ignition: OFF
- Measure resistance between Pin Out Box Terminals 1 and:
  - Terminal 5
  - Terminal 8
  - Terminal 13
  - Terminal 26

Was infinite OHMS measured?

NO → Repair Short to Ground in any circuits that did not measure infinite OHMS.

YES
- Check for proper terminal contact. at EBTCM, PMV Connector C1, PMV Fluid Level Switch and Brake Switch Connectors.
- If all are OK, Reconnect.
- Ignition: RUN
- Using Tech 1, clear DTC (s).
- Test Drive Vehicle.
- If DTC 74 resets, replace EBCTM.

DIAGNOSTIC CHART DTC 75 — 1995 AURORA AND RIVIERA

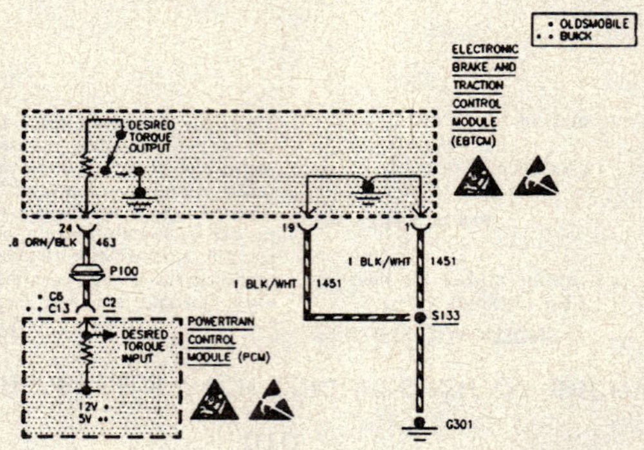

DTC 75

PCM REQUESTED TCS TO BE DISABLED

CIRCUIT OPERATION

When excessive wheel spin is detected by the EBTCM, it will request the PCM to reduce engine torque via a high speed pulse width modulated (PWM) signal on circuit 463.

FAILURE CONDITIONS

Important

If PCM code P1571 (Oldsmobile) or 93 (Buick) is not present, no fault exists with the ABS/TCS system.

DTC 75 sets when the EBTCM receives a serial data message from the PCM to disable Traction Control. DTC 75 sets under the following two conditions.

1. EBTCM to PCM communications.
   - CKT 463 or CKT 800 open or shorted.
2. Buick PCM drivability conditions to include PCM DTC's P0101, P0341, P0342, P1350, P1361, P1623, P1630, P0131, P0132, P0171, and P0172.
3. Oldsmobile PCM drivability conditions to include PCM DTC's P093 and P133.

DIAGNOSTIC CHART DTC 75 — 1995 AURORA AND RIVIERA, CONTINUED

DTC 75
PCM REQUESTED TCS TO BE DISABLED

Has ABS/TCS Functional Test been performed?

- **NO:** Perform ABS/TCS Functional Test before proceeding.
- **YES:**
  - Ignition: RUN
  - Using Tech 1, check for presence of PCM DTC P1571 (Oldsmobile) or 93 (Buick).
  - Is PCM DTC P1571 (Oldsmobile) or 93 (Buick) present?

    - **NO:** Inspect brake system.
    - **YES:**
      - Ignition: OFF
      - Disconnect EBTCM Connector.
      - Install J 38716 Pinout Box.
      - Using J 39200 measure voltage between Pinout Box Terminal 1 and Battery Underhood Junction Block (right side of engine compartment).

      Is voltage measured greater than 10 volts?

      - **NO:** Repair open in CKT 1451.
      - **YES:**
        - Ignition: RUN
        - Using J 39200 set to DC scale, measure voltage between Pinout Box Terminals 24 and 1.

        Approximately 5 volts (Buick) or 12 volts (Oldsmobile) present?

        - **NO:**
          - Backprobe PCM Terminal C13 (Buick) or C6 (Oldsmobile) (White Connector) with J 39200 to ground.
          - Approximately 5 volts (Buick) or 12 volts (Oldsmobile) present?
            - **NO:** Check CKT 463 for short. If CKT 463 is OK, replace PCM.
            - **YES:** Open in CKT 463.
        - **YES:**
          - Reconnect EBTCM
          - Ignition: RUN
          - Using J 39200 set to Hz, backprobe PCM Terminal C13 (Buick) or C6 (Oldsmobile) (white connector) and measure frequency between PCM and DLC Terminal 4.

          Is 126 - 130 Hz measured?

          - **YES:** Check for intermittent short to ground in CKT 463. If CKT 463 is OK, check for poor terminal contact at PCM and EBTCM. If all are OK, replace PCM.
          - **NO:** Check for poor terminal contact at EBTCM Terminal 24 and PCM Terminals C13 or C6. If all are OK, replace EBTCM

After repairs are completed, clear DTC and repeat ABS/TCS functional test.

DIAGNOSTIC CHART DTC 76 — 1995 AURORA AND RIVIERA

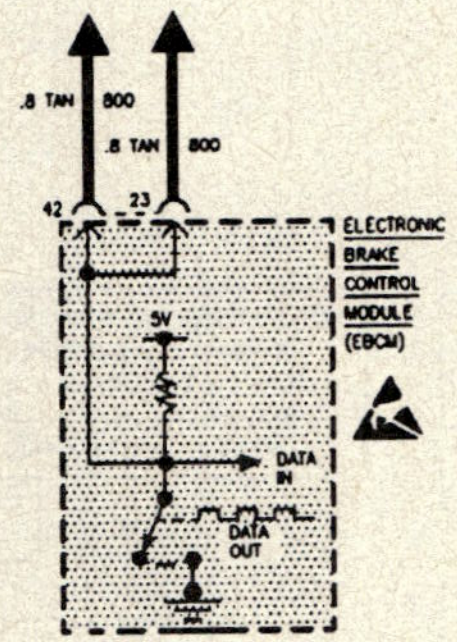

DTC 76

DATA LINE SIGNAL OR CIRCUIT MALFUNCTION

CIRCUIT OPERATION

The EBTCM shares specific serial data information supplied on the data line. The data line also allows the use of a bi- directional scan tool such as a Tech 1.

FAILURE CONDITIONS

A current DTC 76 cannot be read with a Tech 1 due to the loss of serial data communications. After data line communications are restored, DTC 76 will be stored in memory.

TECH 1 DIAGNOSTICS

Important

On some early production vehicles, DTC 76 (UART SDL circuit or signal malfunction) may set if the TECH 1 is unplugged from the DLC during data list or test functions. This condition can be avoided by backing out to the main menu screen (the screen shown when the TECH 1 is first plugged in) before unplugging the TECH 1.

DIAGNOSTIC CHART DTC 77 — 1995 AURORA AND RIVIERA

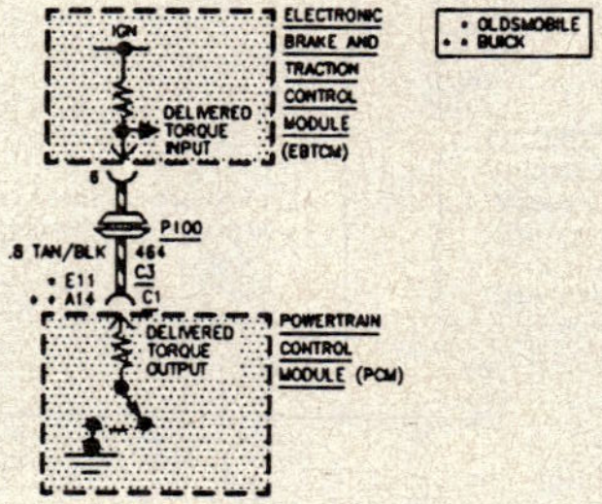

DTC 77

DELIVERED TORQUE CIRCUIT FAULT

CIRCUIT OPERATION

The PCM provides the EBTCM information on engine torque output via a high speed pulse width modulated (PWM) signal on circuit 464. This information is provided so the EBTCM can determine if engine torque management is required during a wheel slip condition.

FAILURE CONDITIONS

If the EBTCM determines there is a signal/circuit malfunction in circuit 464, the "TRACTION OFF" telltale will be illuminated and DTC 77 will set.

DIAGNOSTIC CHART DTC 77 — 1995 AURORA AND RIVIERA, CONTINUED

DTC 77
DELIVERED TORQUE CIRCUIT OPEN OR SHORTED

Has ABS/TCS Functional Test been performed?

- **NO:** Perform ABS/TCS Functional Test .
- **YES:**
  - Ignition: OFF
  - Disconnect PCM.
  - Ignition: ON
  - Using J 39200 set to DC scale, measure voltage at PCM Harness Connector Terminal A14 (red connector [Buick]) or E11 (Blue Connector [Oldsmobile]) and ground.
  - Approximately 10.0 volts present?
    - **YES:**
      - Ignition: OFF
      - Reconnect PCM.
      - Ignition: ON
      - Backprobe Terminal A14 (red connector [Buick]) or E11 (Blue Connector [Oldsmobile]) of PCM Harness Connector.
      - Approximately 1.0 volts present?
        - **YES:** Check for intermittent short to ground in CKT 464, if OK check for poor terminal contact on EBTCM Terminal 6 or faulty EBTCM.
        - **NO:** Check for poor terminal contact at PCM Terminal A14 (Buick) or E11 (Oldsmobile). If terminal contact is OK, replace PCM.
    - **NO:**
      - IGN: OFF
      - Disconnect EBTCM connector.
      - Using J 39200 check continuity of circuit 464 from EBTCM Terminal "6" and PCM Terminal A14 (red connector [Buick]), or E11 (blue connector [Oldsmobile]).
      - Continuity?
        - **YES:** Check EBTCM for proper terminal contact. If OK, replace EBTCM.
        - **NO:** Check for open in CKT 464.

After repairs are completed, clear DTC and repeat ABS/TCS functional test.

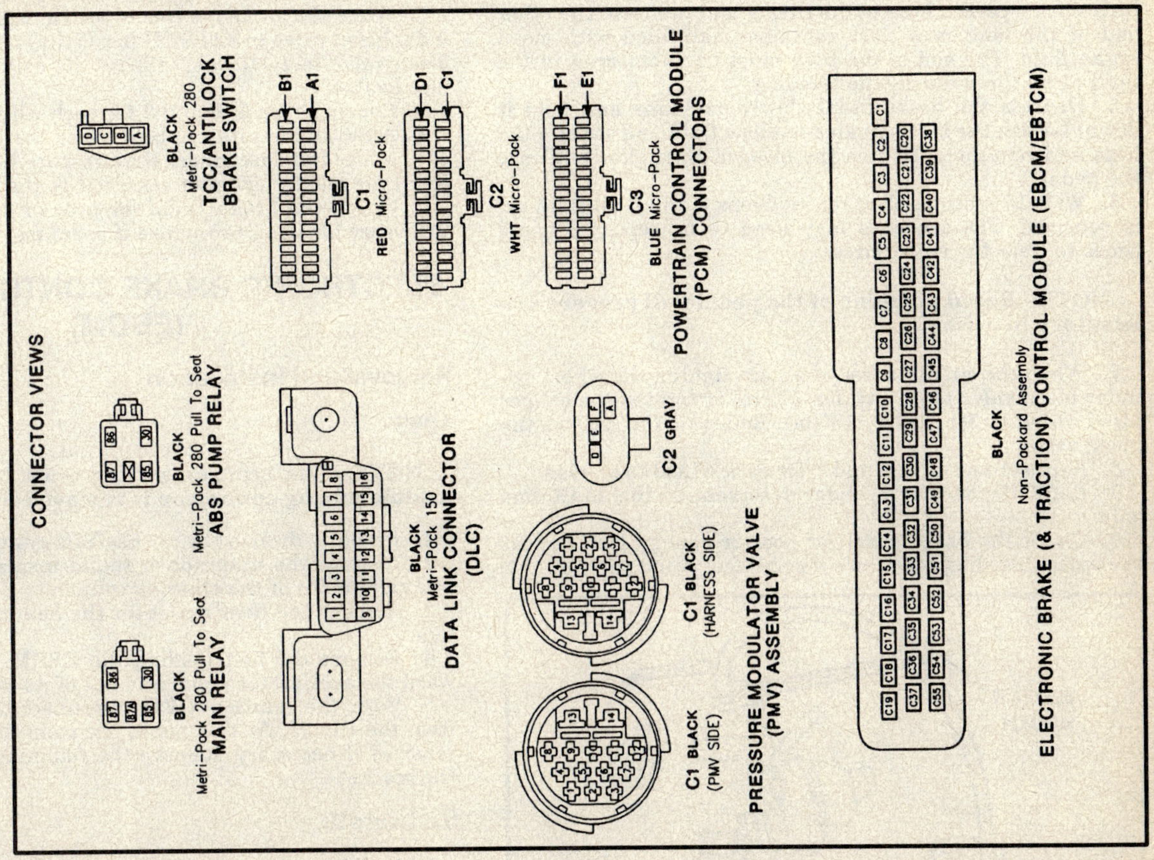

ELECTRICAL CONNECTOR TERMINAL IDENTIFICATION — 1995 AURORA AND RIVIERA

Component Replacement

CAUTION

Some vehicles are equipped with a Supplemental Inflatable Restraint (SIR) or air bag system. The air bag (SIR) system must be disabled before performing service on or around SIR components or wiring. Failure to follow safety and disabling procedures could result in possible air bag deployment, personal injury or unnecessary SIR system repairs.

SYSTEM REFILL

The fluid level must be maintained at the **MAX** mark on the master cylinder reservoir. A system operating properly and free of leaks should not need fluid added to it. The fluid level will appear to drop in the reservoir as the brakes wear, but will return to normal after brake linings are replaced. Adding fluid when not needed may result in fluid overflow.

Fluid may be added to either reservoir in normal fashion. The ignition must be **OFF**. Carefully wipe the reservoir cap and surrounding area free of dirt and debris before removing the cap. Use only DOT 3 brake fluid from an unopened container. The use of DOT 5 or silicone brake fluid is specifically prohibited in this system. Fill the reservoir only to the **MAX** mark; do not overfill. Make certain the reservoir cap is securely installed.

SYSTEM BLEEDING

Brake system bleeding is performed to remove air in the brake lines or components. When a line is disconnected at a wheel, only that caliper and line must be bled. If a line is disconnected at a fitting between the master cylinder and a wheel, the brake system served by that line must be bled. Four wheel bleeding is required only if the lines are disconnected at the master cylinder or if air has entered due to a low fluid condition.

CAUTION

Do not move the vehicle until a firm brake pedal is obtained. An improperly bled system may cause reduced braking, collision or personal injury.

Manual Bleeding Procedure

1. Remove the booster vacuum reserve by applying the brake pedal several times with the ignition **OFF**. The master cylinder must be free of air before bleeding the system. If the master cylinder contains air, it must be bled.
   a. Loosen the forward (secondary) brake line connection at the master cylinder. Allow fluid to flow from the connection; maintain the fluid level in the reservoir.
   b. Tighten the fitting at the master cylinder, stopping the flow of fluid.
   c. Have an assistant depress the brake pedal slowly one time and hold the pedal down. Loosen the forward fitting again, purging air from the cylinder bore. Tighten the fitting, then release the pedal slowly.
   d. Wait 15 seconds. Repeat Steps **c** and **d,** including the 15 second pause, until all air is removed from the bore.
   e. Tighten the fitting to 15 ft. lbs. (20Nm).
   f. Repeat the entire procedure, steps a through f, at the rearward (primary) brake line fitting.
2. Maintain the correct fluid level in the reservoir during the procedure. The reservoir must be tightly capped during bleeding to prevent fluid spray.
3. If more than one circuit is to be bled, do so in this order: right rear, left rear, right front and left front.
4. Remove the bleeder cap at the wheel to be bled. Fit a box-end wrench of the proper size over the bleed fitting. At-

tach a transparent hose to the fitting and immerse the other end of the hose in a clear container half-filled with clean brake fluid. The end of the hose must be submerged in the fluid at all times during the bleeding.

5. Depress the brake pedal slowly one time and hold it down. Loosen the bleeder valve to allow fluid and air into the hose and container. Tighten the bleeder, then slowly release the pedal.

6. Wait 15 seconds, then repeat Steps **5** and **6** until all air is removed. The sequence may need to be repeated many times to clear the entire circuit.

NOTE: Rapid pumping of the pedal will prevent satisfactory bleeding.

7. When the circuit is free of all air, tighten the wheel cylinder bleed valve to 62 inch lbs. (7Nm) or tighten the caliper bleed valve to 10 ft. lbs. (13Nm). Reinstall the cap on the bleed valve.

8. Repeat Steps **4**, **5**, **6** and **7** for each wheel to be bled.

9. Refill the master cylinder reservoir to the **MAX** line only.

10. Check the brake pedal for poor or spongy feel. Repeat the entire bleeding procedure if pedal feel is not correct.

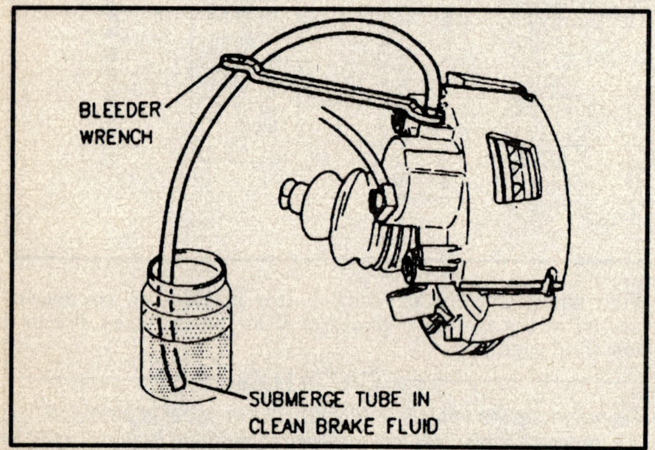

Bleeding the brake system

Pressure Bleeding Procedure

NOTE: Use of a diaphragm pressure bleeder is required. The air chamber must be separated from the brake fluid to prevent entry of dirt and moisture. The correct bleeder adapter must be used to prevent damage to the reservoir.

1. Fill the master cylinder reservoir to the proper level.

2. Install the bleeder adapter cap J–35589 or equivalent to the reservoir.

3. Pressurize the bleeder at 20–25 psi (140–172 kPa). Connect the bleeder pressure hose to the adapter on the reservoir; open the line valve on the bleeder.

4. If more than one circuit is to be bled, do so in this order: right rear, left rear, right front and left front.

5. Remove the bleeder cap at the wheel to be bled. Fit a box-end wrench of the proper size over the bleed fitting. Attach a transparent hose to the fitting and immerse the other end of the hose in a clear container half-filled with dean brake fluid. The end of the hose must be submerged in the fluid at all times during the bleeding.

6. Loosen the bleeder valve approximately 3/4 turn to allow fluid and air into the hose and container. Allow the fluid to flow until no air is seen in the fluid.

7. When the circuit is free of all air, tighten the wheel cylinder bleed valve to 62 INCH lbs. (7Nm) or tighten the caliper bleed valve to 10 ft. lbs. (13Nm). Reinstall the cover on the bleed valve.

8. Repeat Steps **5**, **6** and **7** for each wheel to be bled.

9. After all circuits are bled, close the pressure line valve on the bleeder. Disconnect the pressure bleeding equipment. Refill the master cylinder reservoir to the **MAX** line only.

10. Check the brake pedal for poor or spongy feel. Repeat the entire bleeding procedure if pedal feel is not correct.

ELECTRONIC BRAKE CONTROL MODULE (EBCM)

Removal and Installation

1993

NOTE: The ignition switch must be OFF when the module wiring connector is removed or attached.

1. Properly disarm the air bag/SIR system.

2. Remove the under-dash sound insulation from the left and right sides of the steering column.

3. Remove the floor outlet for the heating and cooling system.

4. Remove the bolt holding the EBCM. Slide the unit toward the accelerator pedal and out of its mount.

5. With the ignition **OFF**, disconnect the wiring harness from the EBCM. At this point, the mounting bracket may be removed if necessary. Remove the retaining nuts and remove the bracket.

To install:

6. Install the bracket if it was removed.

7. With the ignition **OFF**, position the EBCM and connect the wire harness.

8. Fit the unit into the bracket, aligning it properly. Install the retaining bolt. For DeVille, Fleetwood, Bonneville and LeSabre models, tighten the bolt to 20 inch lbs. (2.2Nm). For Eighty-Eight, Ninety-Eight or Park Avenue models, tighten the bolt to 42 inch lbs. (4.7Nm).

9. Install the floor outlet.

10. Install the underdash sound insulators.

11. Enable the air bag/SIR system.

ELECTRONIC BRAKE CONTROL MODULE (EBCM) OR ELECTRONIC BRAKE AND TRACTION CONTROL MODULE (EBTCM)

Removal and Installation

1994–95

NOTE: The ignition switch must be in the OFF position when connecting or disconnecting the electrical connector.

1. Disconnect and remove the EBCM/EBTCM connector by rotating the locking clip forward.

2. Remove the retaining nuts that hold the EBCM/EBTCM to the shock tower.

3. Remove the EBCM/EBTCM by carefully sliding it up and away from the shock tower.

To install:

4. Carefully install the EBCM/EBTCM by sliding it down and towards the shock tower.

5. Install the mounting nuts.

6. Attach the EBCM/EBTCM electrical connector and rotate the locking clip rearward to secure.

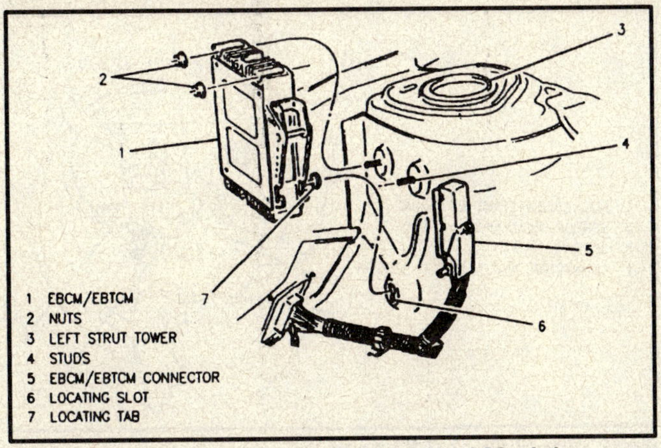

Electronic Brake and Traction Control Module

PARKING BRAKE SWITCH

Removal and Installation

1. Disconnect the negative battery cable.
2. Disable the air bag/SIR system.
3. Disconnect the electrical connector at the switch.
4. Loosen the mounting bolt and remove the parking brake switch.

To install:

5. Install the parking brake switch onto the parking brake and tighten the mounting bolt to 17 inch lbs. (1.9Nm).

TRACTION CONTROL SWITCH

The traction control switch is located in the instrument panel center trim plate.

TCC/ANTI-LOCK AND CRUISE/SHIFT INTERLOCK/BRAKE SWITCH

The TCC/Anti-lock brake switch is part of the electrical and vacuum release switch.

Removal and Installation

1. Disconnect the negative battery cable.
2. Properly disable the air bag/SIR system
3. Disconnect the electrical connectors at the brake switch.
4. Disconnect the vacuum hose at the brake switch.
5. Push in the retainer locking tabs and remove the switch.

To install:

6. Fully depress the brake pedal.
7. Insert the switch assembly into the pedal mounting bracket. Press the assembly firmly until clicks can be heard and the retainer locking tabs engage in the mounting bracket.
8. Pull the brake pedal back until clicks can no longer be heard.
9. Connect the electrical connectors and vacuum hose to the switch.
10. Reconnect the air bag/SIR system and connect the negative battery cable.

FRONT WHEEL SPEED SENSORS

Removal and Installation

1. With the ignition **OFF**, disconnect the wheel speed sensor connector.
2. Remove the hub and bearing assembly. Clean all dirt from the sensor housing area.
3. Gently pry the slinger from the speed sensor assembly and discard. Use a new slinger when reinstalling the speed sensor assembly.
4. Carefully pry the speed sensor from the bearing assembly. Do not allow dirt to enter the bearing when the sensor is removed. Do not add lubricant through the sensor opening; the bearing is permanently lubricated.
5. Inspect the bearing and sensor for any signs of rust or water intrusion; if such damage is found, replace the hub and bearing assembly. Do not clean the grease from the toothed sensor ring; grease does not affect operation of the sensor.
6. Use a clean, lint-free cloth to remove sealant from the outer diameter of the bearing hub

To install:

7. Apply a sealant, 620 Loctite® or its equivalent, to the groove in the outer diameter of the bearing hub.
8. Place the hub and bearing on tool J-38764-4 or its equivalent to prevent damage to the studs.
9. Install the sensor using a press and tool J-38764-1 or its equivalent.
10. Install the hub and bearing assembly.
11. Connect the sensor wiring to the ABS harness.

REAR WHEEL SPEED SENSORS

Removal and Installation

1993

1. With the ignition **OFF**, disconnect the wheel speed sensor connector.
2. Remove the hub and bearing assembly.
3. Clean the sensor housing of any dirt or debris.
4. Remove the screws holding the sensor and remove the sensor. Do not allow dirt to enter the bearing when the sensor is removed. Do not add lubricant through the sensor opening; the bearing is permanently lubricated.
5. Inspect the sensor ring and plate for any signs of contact or mechanical wear; if such damage is found, replace the hub and bearing assembly. Do not clean the grease from the toothed sensor ring; grease does not affect operation of the sensor.

To install:

6. Inspect the O-ring for damage; replace it if necessary.
7. Install the sensor and the screws tighten the screws to 33 inch lbs. (3.7Nm).
8. Install the hub and bearing assembly.
9. Connect the sensor wiring to the harness.

Removal and Installation

1994-95

1. Disconnect the negative battery.
2. Disconnect the speed sensor connector.
3. Loosen and remove the lower strut retaining nuts and bolts to gain access to the sensor.
4. Loosen and remove the Torx screws which hold the speed sensor to the hub assembly.

NOTE: Do not allow dirt or debris to enter the bearing after the speed sensor has been removed from the hub. Do not add lubricant to the bearing via the sensor opening as the bearing is lubricated for the life of the vehicle.

5. Carefully remove the sensor from the hub.

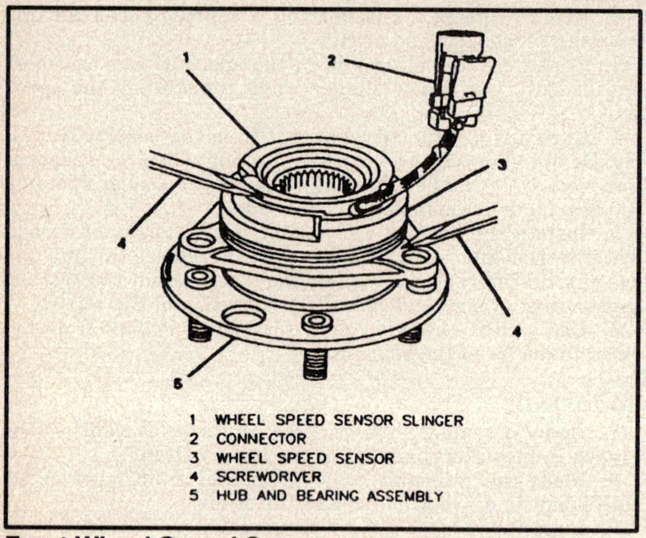

Front Wheel Speed Sensor

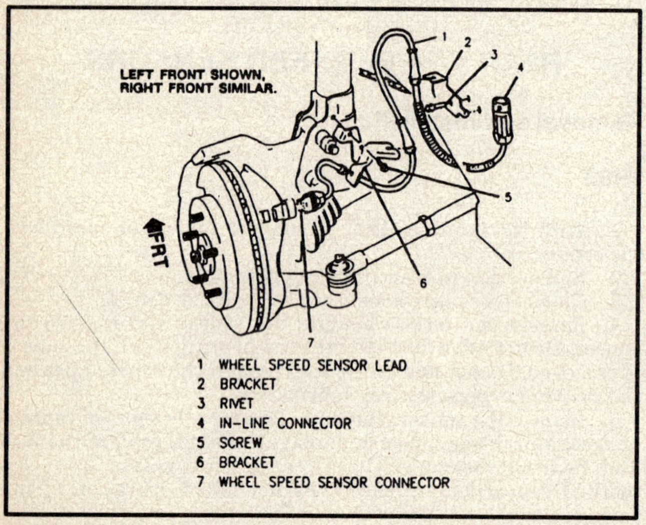

Front Wheel Speed Sensor

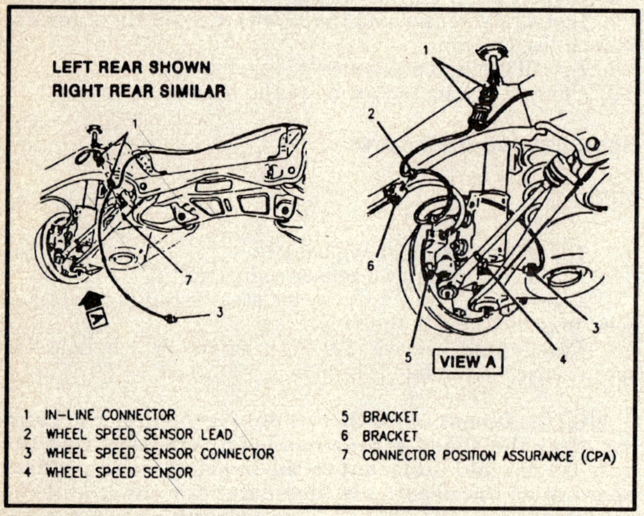

Rear Wheel Speed Sensor

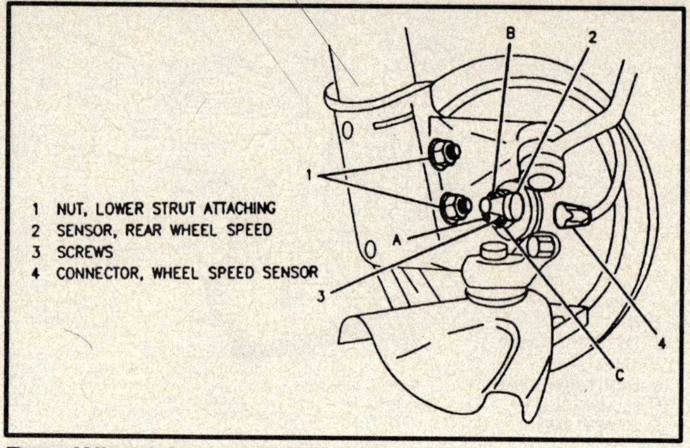

Rear Wheel Speed Sensor

To install:

6. Carefully install the speed sensor into the hub opening.
7. Install the Torx screws into the hub assembly. Tighten the screws in a circular pattern to 25 inch lbs. (2.8Nm).
8. Install the lower strut retaining nuts and bolts. Tighten them to 140 ft. lbs. (190Nm).
9. Connect the speed sensor electrical connector.
10. Road test the vehicle and verify correct operation.

WHEEL SPEED SENSOR RINGS

The toothed wheels which rotate past the speed sensors are integral parts of their respective assemblies. If the front ring is damaged, the drive axle must be replaced. If a rear ring is damaged, the hub and bearing assembly must be replaced.

MASTER CYLINDER FLUID LEVEL SENSOR

NOTE: The fluid level sensor is integrated with the master cylinder reservoir. If the sensor fails, the reservoir must be replaced.

Removal and Installation

1. Disconnect the electrical harness at the reservoir connector.
2. Loosen or relocate the clamp on the reservoir hose.
3. Disconnect the hose from the reservoir. Catch escaping fluid in a clean container; plug the hose with a 5/8 inch (15.5mm) diameter plug.
4. Carefully tap the slotted spring pins (roll pins) until they are clear of the reservoir.
5. Remove the reservoir from the master cylinder.
6. Remove the reservoir seals from the master cylinder. Take careful note of the position or orientation of the seals.

To install:

7. Fit new seals to the master cylinder. Lubricate the new seals with clean brake fluid. Make certain the seals are correctly installed.
8. Clean the reservoir and ports with denatured alcohol. Install the reservoir onto the master cylinder.
9. Tap the slotted spring pins into place to hold the reservoir.
10. Connect the hose to the reservoir and secure the clamp.
11. Connect the master cylinder reservoir wiring connector to the harness.
12. Fill the reservoir with clean, fresh brake fluid.
13. Bleed the brake system.

MAINTENANCE LIGHT RESETS

4 SECTION

DOMESTIC VEHICLE RESET PROCEDURES

AMERICAN MOTORS CORPORATION

Emission Maintenance Reminder

1980–81 Models

The models are equipped with an emission maintenance reminder light on the instrument panel which will be illuminated every 30,000 miles. If the reminder light is illuminated, it indicates the oxygen sensor must be inspected. If faulty, the oxygen sensor must be replaced. The reminder light is reset as follows:

1. Locate the switch between the upper and lower speedometer cable next to the firewall in the engine compartment.
2. Slide the rubber boot up and using a suitable tool, turn the reset screw clockwise ¼ turn until the detent resets within the switch.

1982–84 Except Alliance and Encore

The emission maintenance light will illuminate after 1000 hours of engine operation to indicate required service for the oxygen sensor. After performing the required service, the emission maintenance E–cell timer must be replaced by removing the printed circuit board from its enclosure and inserting a replacement timer. The timer is located in the passenger compartment, within the wire harness leading to the ECM.

1987–88 Eagle Station Wagon

The vehicle is equipped with an emission light timer that will start flashing the oxygen (0_2) sensor service light at approximately 82,500 miles. At this time the oxygen (0_2) sensor and timer should be replaced. The timer is located under the dash panel to the right of the steering column. Remove the timer mounting screws and disconnect the wiring. Install the timer in its original location.

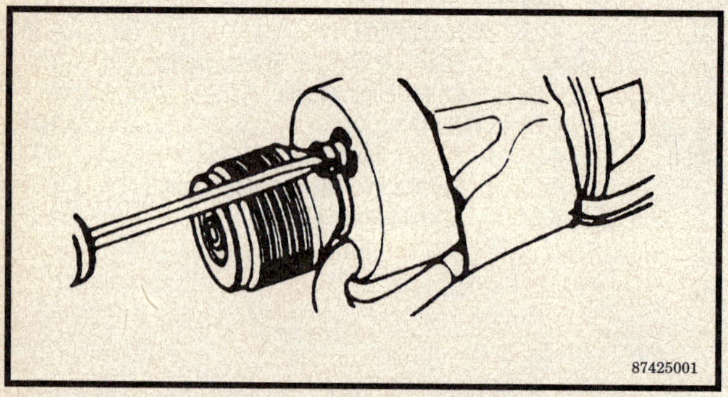

Resetting the mechanical maintenance reminder light — American Motors and Chrysler Corp.

CHRYSLER CORPORATION

Emission Maintenance Reminder

1980 Passenger Cars

The vehicles are equipped with a mileage counter that activates an emission reminder light at intervals of 12,000/15,000/30,000 and/or 52,000 miles. There are 2 types of reset used, mechanical and electrical. If the vehicle is equipped with a mechanical reset, use the reset procedure as noted for the 1980–81 American Motor vehicles.

The electronic type uses a 9 volt battery which supplies power to the electronic counter to prevent memory loss when the vehicle battery is disconnected. To reset the electronic type, proceed as follows:

NOTE: The vehicle battery must be connected during resetting procedure to prevent power loss to memory.

Locate the green, red or tan plastic case behind the instrument panel in the lower left cluster area. Slide the case from the bracket and slide the cover open. Remove the 9 volt battery and insert a small rod into the hole switch to close the contacts. Replace the battery with a new 9 volt alkaline type. Close the case and slide the switch back into the bracket.

1981–95 Passenger Cars

As of 1981 Chrysler no longer used resets in passenger cars, however the reset is still being used in some trucks and vans.

1988–92 Eagle Premier and Monaco

The Premier is equipped with a Vehicle Maintenance Monitor (VMM), with a **SERVICE** light incorporated into this monitor. The **SERVICE** light will illuminate at 7500 miles (12,000 kms) intervals to indicate that regular service and maintenance procedures are due. After the required service has been performed, press the **RESET** button on the dash below the VMM display window until the

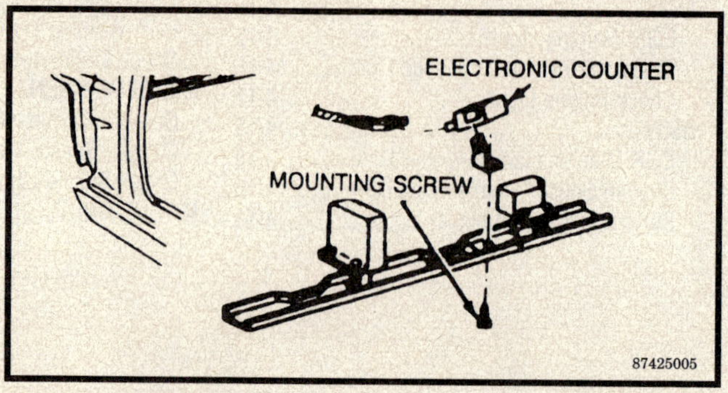

Resetting the electronic maintenance reminder light — Chrysler Corp.

sound of a beep is heard. When the beep is heard, the VMM will then clear the display (service light).

1980–82 Light Trucks and Vans

The EGR maintenance reminder system is a mileage counting system. An instrument cluster light will come on when the vehicle reaches its first 30,000 miles and then subsequently every 30,000 miles thereafter.

Locate the green, red or tan plastic case behind the instrument panel in the lower left cluster area. Slide the case from the bracket and slide the cover open. Remove the 9 volt battery and insert a small rod into the hole switch to close the contacts. Replace the battery with a new 9 volt alkaline type. Close the case and slide the switch back into the bracket.

1983–87 Light Trucks and Vans

The maintenance reminder system is a mileage counting system. An instrument panel light will come on between 12,000 and 30,000 miles on some models and 52,500, 82,500 and 105,000 miles on other models. The purpose of the system is a reminder that emission maintenance service must be performed as soon as possible. It is not intended to indicate a warning or a state of emergency exists which must be corrected to insure safe vehicle operation.

The system consists of 2 major elements, mileage counting contacts in the speedometer and an electronic digital module attached to the brake pedal support under the instrument panel.

After necessary pulses are stored, the output transistors activate and turn on the emission maintenance reminder light. The circuit immediately returns to zero mileage condition and starts to count and accumulate pulses until the necessary mileages are again stored and the reset process is repeated.

The light will remain lit until the module is reset by inserting a small thin rod into the hole in the module case closing a switch which turns off the light. The 9 volt battery must be removed before resetting the light switch and a new battery must be installed after resetting the maintenance light.

NOTE: The vehicle battery must be connected during resetting procedure to prevent power loss to memory.

The electronic module is housed in a beige, green, white or red colored plastic case. The case is usually mounted on

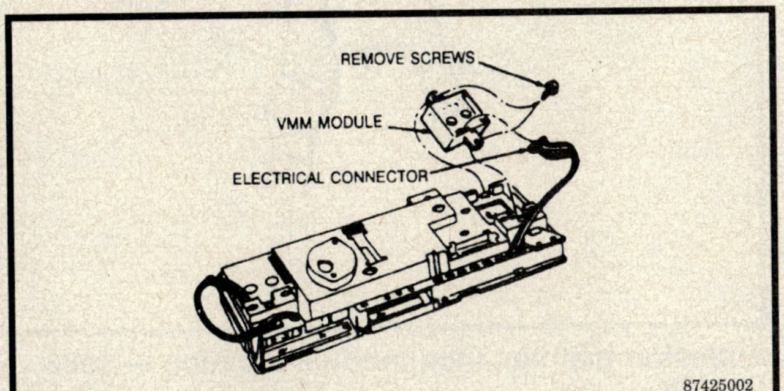

Vehicle Maintenance Monitor (VMM) location — Premier

a bracket on or near the steering column support, below the headlight switch on the back side of the instrument panel. It can be removed by sliding it from the bracket.

NOTE: There is no test procedure for the system and any attempt to test it will result in damage to the system components.

1988–89 Light Trucks and Vans

The system is designed to act as a reminder that scheduled vehicle emission maintenance service should be performed. It is not intended to indicate a warning that a state of emergency exists which must be corrected to insure safe vehicle operation. The components covered by the system include the EGR system, PCV valve, oxygen sensor, delay valves, vacuum controlled orifice, fuel tank vapor and bi-level purge valves.

The maintenance required system is based on a time measuring module which uses ignition **ON** time to calculate the maintenance intervals. When the predetermined time intervals have elapsed, the **MAINT REQD** light in the instrument panel will illuminated. The light will remain illuminated until the module is reset by insert a small blade or rod into the module to depress the reset switch which will then turn off the light. On the Dakota, the module is located on a bracket below the headlight switch, on the rear of the instrument panel. On 1988–89 FWD vans, is usually located in the instrument cluster behind the fuel gauge on the standard cluster or behind the tachometer on the optional cluster. On 1980–87 light trucks and 1988 Vans it is usually located on the brake pedal support. On most other models the module is lo-

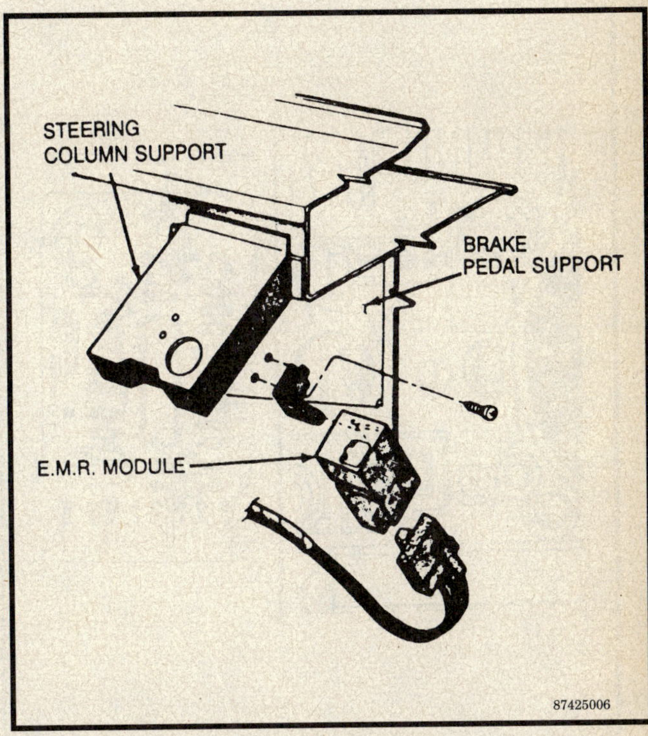

Typical emission maintenance reminder module location is on brake pedal support — 1988 RWD vans shown

cated behind the right side of the instrument panel, next to the glove box.

NOTE: There is no test procedure for the system. Any attempt to test it will result in damage to the system components.

1990–93 Light Trucks and Vans

The system is designed to act as a reminder that scheduled vehicle emission maintenance should be performed. It is not intended to indicate a warning that a state of emergency exists which must be corrected to insure safe vehicle operation. Some 1990 vehicles may use 1989 module design.

NOTE: To reset the emission maintenance light, it is necessary to use the Chrysler Digital Read Out Box II Tester (DRB–II) tool C–4805 or equivalent.

1. Connect the Chrysler DRB–II Tester C–4805 or equivalent to the vehicle as directed by the manufacturers instructions.
2. Turn the tester selector switch to the **EMR MEMORY CHECK**. The tester display should read **EMR MEMORY CHECK ARE YOU SURE?**. Press the **YES** key on the tester.
3. The tester should next display the following message; **IS INSTRUMENT PANEL MILEAGE BETWEEN 9953 AND 10051?**. If the odometer mileage on the vehicle is within the specified mileage, press the **YES** key on the tester.
4. The tester should now display, **EMR MEMORY CHECK TEST COMPLETE**. The EMR light is now reset. If the speedometer mileage is not within specifications, go on to Step 5.

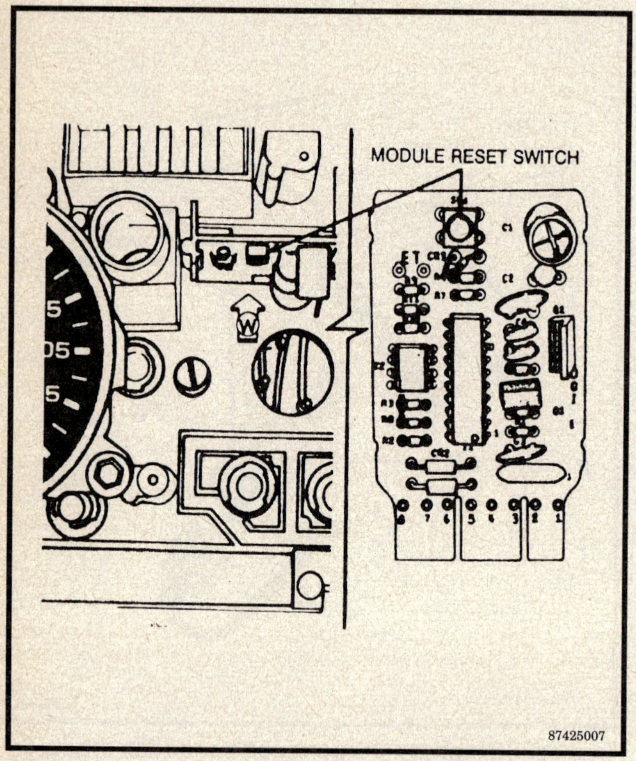

Emission maintenance module location — 1988 FWD vans shown

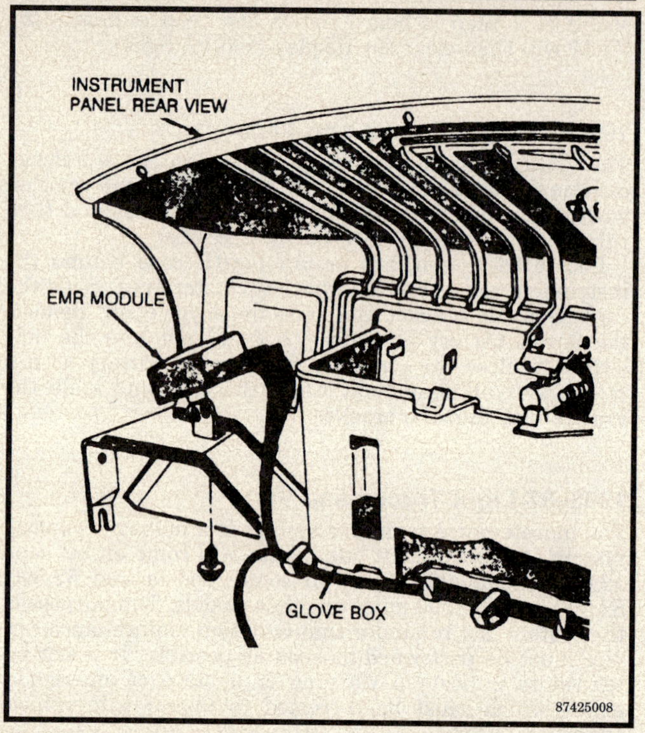

Behind the glovebox is a typical emission maintenance module location — 1988 RWD truck shown

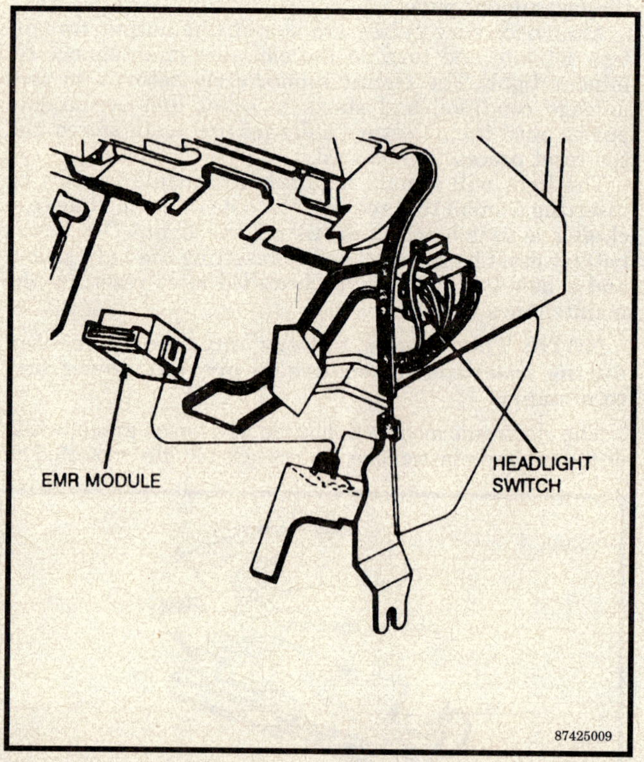

Emission maintenance module location — 1988 Dakota

5. Press the **NO** key on the tester. The tester will then display **DO YOU WANT TO CORRECT EMR MILEAGE?**. Press the **YES** key on the tester.

6. The tester should now display **ENTER MILEAGE SHOWN ON INSTRUMENT PANEL USE ENTER TO KEY TO END**.

7. Enter the mileage shown on the speedometer. Do not enter the tenths. Press the **ENTER** key on the tester. The tester will now ask for verification of the entry.

8. If the mileage entered was correct, the tester will display **EMR MEMORY CHECK TEST COMPLETE**. The vehicle must now be driven approximately 8–10 miles for the mileage reset to take place.

1994–95 Light Trucks and Vans

The Emission Maintenance Reminder (EMR) light is now referred to as a Service Reminder Indicator (SRI) lamp. It is located in the dash labeled **MAINT REQD**. As of 1993, it is no longer used on Vans. It is used on trucks equipped with 5.9L V8 HDC and 8.0L V10 gas powered vehicles only.

The SRI lamp will illuminate at the 60,000 and 82,000 mileage marks and will remain **ON** until it is reset. Perform the required maintenance before resetting the lamp. Failure to adhere to part replacement or service required may be a violation of federal law.

Resetting the SRI lamp requires the use of a DRBII tool or equivalent. Consult the scan tools instruction guide for this procedure.

Check Engine Light

The **CHECK ENGINE** light will light upon a fault. The fault code will be stored in the ECU. Once the source of

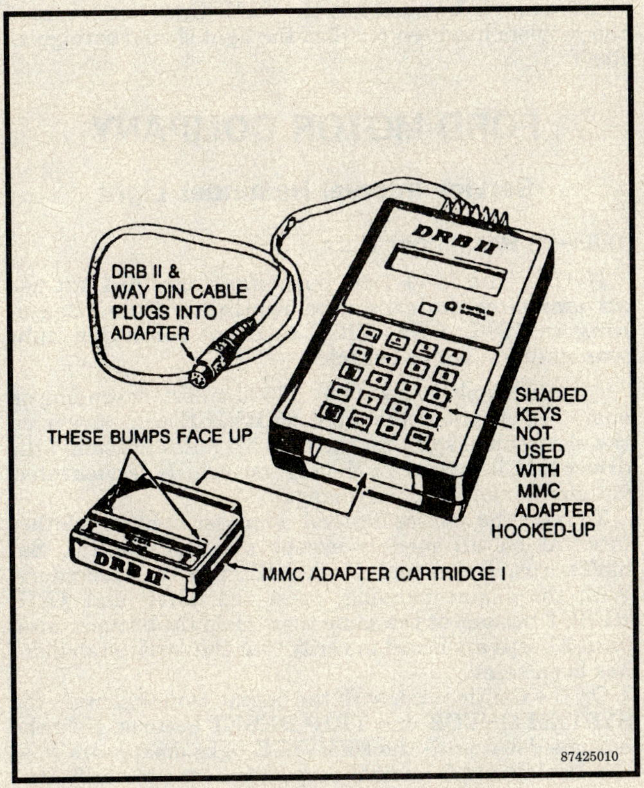

Chrysler DRB-II — digital readout box

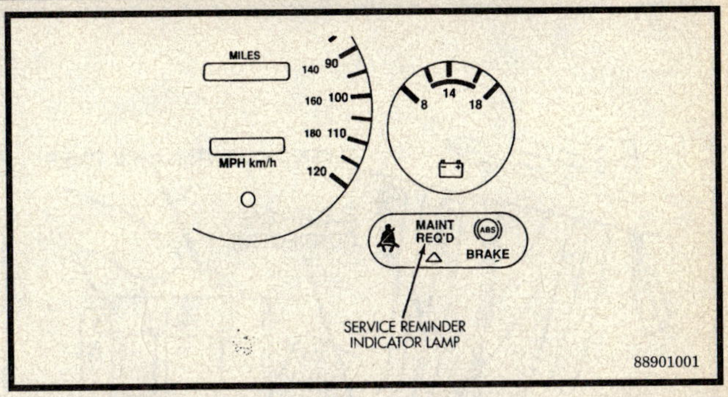

Service Reminder Indicator (SRI) lamp location — 1994–95 Trucks with 5.9L and 8.0L gas engines

the problem has been rectified the light should extinguish itself and the Logic Module will cancel the fault code in the ECU after 20–40 ignition key cycles.

CHRYSLER JEEP DIVISION

Emission Maintenance Reminder

1988–90 Jeep Except Grand Wagoneer

Until 1993 on some Jeep models the emission maintenance timer and indicator light are used to alert the owner when the oxygen sensor, PCV valve replacement and other scheduled emission maintenance is required. The indicator light is usually located in the instrument cluster and the reset timer is located under the dash panel at the passenger side of the vehicle.

The reset timer is operated by the ignition system. It activates (flashes) the indicator light when the vehicle mileage reaches the scheduled maintenance interval of 82,500 miles. The life cycle coincides with the emission maintenance interval of 82,500 miles. The timer cannot be reset and must be replaced or disconnected after reaching the mileage interval.

NOTE: The oxygen sensor and timer are independent. If the timer should fail prematurely, the oxygen sensor must be replaced along with the timer; this is important in preserving the correct sensor replacement interval and ensuring proper engine performance.

REMOVAL AND INSTALLATION

1988–90 Cherokee, Comanche and Wagoneer

The emission maintenance timer used on some model until 1993 is mounted on the dash panel to the right of the steering column.

1. Disconnect the negative battery cable.
2. On models with cruise control, remove the cruise control module attaching screws and remove the module.
3. Remove the screws attaching the maintenance emission timer to the dash bracket and remove the timer from the bracket.
4. Disconnect the timer electrical wires and remove the timer from the vehicle.
5. Installation is the reverse order of the removal procedure.

1989–90 Wrangler

1. Disconnect the negative battery cable.

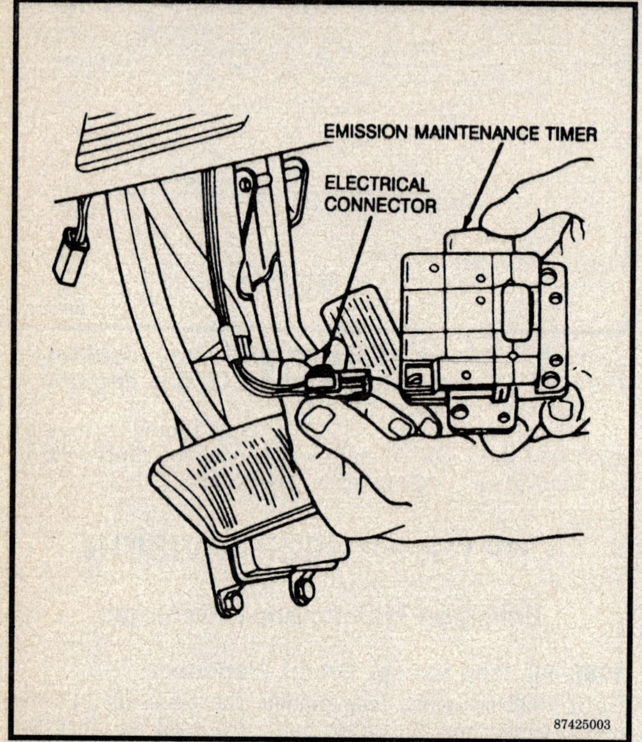

Removing the emission maintenance reminder timer — Cherokee, Comanche and Wagoneer

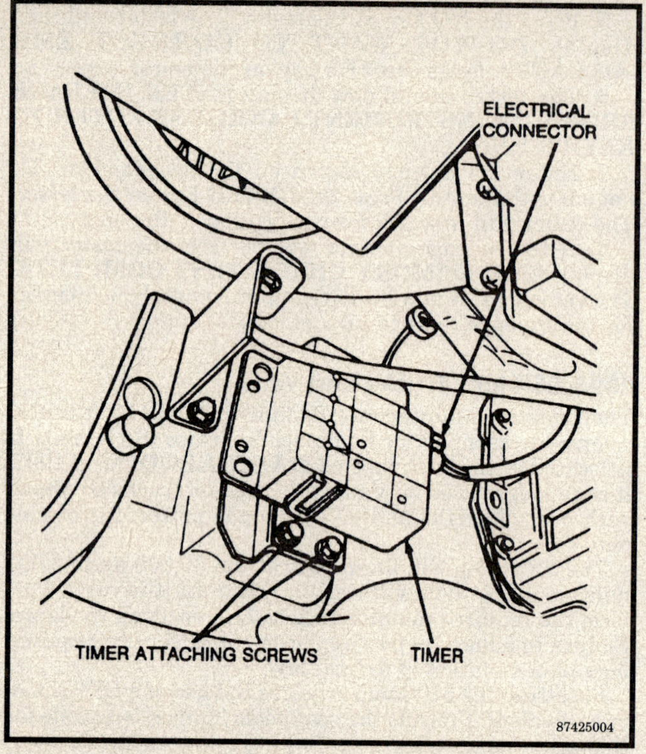

Removing the emission maintenance reminder timer — Wrangler

2. Disconnect the timer electrical wire harness. Remove the screws attaching the maintenance emission timer to the dash bracket and remove the timer from the bracket.
3. Installation is the reverse order of the removal procedure.

1991–93 Vehicles

Vehicles equipped with the Emission Maintenance Reminder (EMR) or Service Reminder Indicator (SRI), the Maintenance Required (MAIN REQ) light will illuminate when the vehicle mileage reaches the scheduled maintenance interval of 82,500 miles. The life cycle coincides with the emission maintenance interval of 82,500 miles. At this time, the oxygen sensor should be replaced and the EGR and PCV systems checked. After all maintenance is performed, the indicator light must be reset for the next interval. Reset the light as follows.

NOTE: To reset the emission maintenance light, it is necessary to use the Chrysler Digital Read Out Box II Tester (DRB–II) tool C–4805 or equivalent.

1. Connect the DRB II, or an equivalent scan tool, to the diagnostic connector. Turn the ignition switch **ON**.
2. Enter the SELECT SYSTEMS menu of the scan tool. Select the appropriate engine and accessories.
3. Select FUEL AND IGNITION. Select ADJUSTMENTS.
4. Select RESET EMR light. Turn the ignition switch **OFF** and disconnect the scan tool.

1994–95 Vehicles

As of 1994, resets are no longer used in Jeep vehicles.

Check Engine Light

The **CHECK ENGINE** light will light upon a fault. The fault code will be stored in the ECU. Once the source of the problem has been rectified the light should extinguish itself.

FORD MOTOR COMPANY

Service Interval Reminder Light

1985–93 Passenger Cars

NOTE: The 1980–1984 passenger cars did not use an emission maintenance reminder light. Beginning in 1985, a SERVICE interval reminder light was used on some vehicles.

Approximately every 5000 or 7500 miles, (depending on engine application) the word **SERVICE** will appear on the electronic display for the first 1.5 miles to remind the driver that it is time for the regular vehicle service interval maintenance (i.e. oil change).

To reset the service interval reminder light for another interval on all models except the Continental, Sable/Taurus, Probe and Mark VIII, use this procedure. With the engine running, press the **TRIP** and **TRIP RESET** buttons at the same time. Hold the buttons down until 3 beeps are heard to verify that the service reminder has been reset.

On the Continental, with the engine running, press the **SYSTEM CHECK** and **TRIP RESET** buttons. Hold the buttons down until the **SERVICE** light disappears from the display and 3 audible beeps are heard to verify that the service reminder has been reset.

On the Sable/Taurus, with the engine running, press the **ODO SEL** and **TRIP RESET** buttons. Hold the buttons down until the **SERVICE** light disappears from the display and 3 audible beeps are heard to verify that the service reminder has been reset.

On the 1989–93 Probe, with the engine running, press and hold the **SERVICE RESET** button located on the speed alarm keyboard. Hold the button until the **SERVICE** light disappears from the display and 3 audible beeps are heard to verify that the service reminder has been reset.

NOTE: On 1989–93 Probes not equipped with the speed alarm keyboard, locate the $^3/_{16}$ in. hole in the overhead console. Use a small rod or other tool to depress the reset button located behind the hole.

On the Mark VIII, press the **OIL CHANGE RESET** button and hold it down for 5 seconds. The **CHANGE OIL SOON** or **OIL CHANGE REQUIRED** message will disappear after the 5 second interval.

1994–95 Passenger Cars

The Service Interval Reminder light is used only in the Continental up to 1995. All other models no longer use it.

During the SYSTEM CHECK sequence, the SERVICE symbol comes on and displays the miles (kilometers) to go before the next normal service is due. After service is complete, reset the service interval reminder as follows:

1. Press the SYSTEM CHECK button.
2. Press the RESET button.
3. Press the SYSTEM CHECK and the RESET button at the same time. The display now shows 7,200 miles (11, 580 km) and the mileage counts down from here.

The 1994–95 Mark VIII continues to use the **CHANGE OIL SOON** or **OIL CHANGE REQUIRED** light. When oil life left is between five percent and zero percent, the **CHANGE OIL SOON** will be displayed on the message center. When oil life reach zero percent, the **OIL CHANGE REQUIRED** message will be displayed.

The message center indicator will indicate the percent of oil life left during System Check. This percentage is based on the drivers driving history and the time since the last oil change. In order to ensure oil life left indications, the driver should only perform the OIL CHANGE RESET procedure after every oil change. Reset the system by pressing the **OIL CHANGE RESET**switch and hold for 5 seconds. After a successful reset the Message Center will display oil life indicators. The **CHANGE OIL SOON** or **OIL CHANGE REQUIRED** message will disappear after the 5 second interval.

Vehicle Maintenance Monitor

1989–93 Thunderbird and Cougar

The Vehicle Maintenance Monitor (VMM) warns the driver of low oil, low radiator coolant, low fuel and low washer fluid conditions and indicates when to change engine oil. The system consists of an electronic module with a display, a coolant level sensor, a washer fluid level sensor, a combination oil level and temperature sensor, a fuel sender, a tachometer and a rheostat.

The module lights all segments of the display for 1 second, blanks them for 1 second and then goes to normal operation whenever the ignition key switches to the **RUN** position. The oil change display turns off after 15 minutes if the engine is not running. This alerts the user that the VMM module is not getting a tachometer signal.

The module is located in the top center of the instrument panel directly above the radio, retained by 2 spring clips, 1 at each side. A reset switch access hole is located on the left front of the module. The module is not removed to be reset.

The low fuel, low radiator coolant and low washer fluid indications are the result of input from various sensors. Correcting the fluid level will cause the light to go out. The oil change indicator is different. The green bar under the **OK** on the oil change display is lit for 0–89 percent of the oil change cycle. The word **SOON** under the word **CHANGE** is lit for 90–99 percent of the cycle. The word **NOW** is lit for 100–115 percent of the oil change cycle and flashes every 18 minutes for 17 seconds after 115 percent of the oil change cycle.

The oil change display will turn OFF after 15 minutes if the VMM receives no tachometer signal. This is a diagnostic precaution that could be caused by an intermittent connection or by the engine not being on.

The VMM monitors oil temperature and engine tachometer signals to measure quality of oil and indicate when engine oil should be changed.

The VMM has a self-diagnosis test built into the module that can be performed with the module in the vehicle and the engine running. This test determines whether the module is getting and processing the signals necessary to function correctly. The self-test mode will test the module's response to engine rpm, oil temperature, fluid levels and engine and Electronic Instrument Cluster (EIC) wiring harness strap inputs.

Enter the test by starting the vehicle and within 16 seconds from engine start, press the reset switch and hold continuously for 4 seconds. The entire display will flash when the reset switch is pressed. If the reset switch is released within 4 seconds, the display will stop flashing and normal operation will continue. Do not run the engine faster than 6100 rpm. If only the oil change display flashes then the tach input signal to the module is missing. The self diagnostics test cannot be entered until this signal is provided. Each reset switch press and release advances the module to the next test. Turn the ignition switch to **OFF** to terminate the self diagnostics test.

As soon as diagnostics is entered, the module is in the Tach Input Test mode and the display lights up segments **OK, CHANGE, NOW, SOON** and **GREEN BAR** is sequence, proportional to the number of tach pulses measured (engine rpm). The faster the engine rpm, the faster the sequence lights.

When the Reset switch is pressed and released, the display advances to to the Oil Temperature Test. The **OIL** nomenclature is lit and the oil temperature will be indicated in the following manner:

If the oil temperature is cold, display will be: Green Bar, Oil

If the oil temperature is cool, display will be: Green Bar, Soon, Oil

If the oil temperature is normal, display will be: Green Bar, Soon, Now, Oil

If the oil temperature is warm, display will be: Green Bar, Soon, Now, Low Radiator, Oil

If the oil temperature is hot, display will be: Green Bar, Soon, Now, Low Radiator, Radiator, Oil

The VMM oil change indicator must be reset every time the vehicle's oil is changed. The oil change indicator must be reset even if the vehicle's oil is changed before the

VMM has indicated an oil change is necessary. The reset procedure is necessary to keep the information in the oil change indicator updated whenever the oil is changed. If the reset procedure is not done, the VMM will indicate an oil change prematurely. After the oil is changed, manually reset the VMM oil change indicator by using the following procedure:

1. Turn the ignition switch to **RUN**.
2. Within 15 seconds of turning the ignition to **RUN**, reset the VMM by inserting a straightened paper clip or similar tool through the reset hole in the face of the VMM. Press firmly.
3. Note that it is not possible to feel the reset switch operating because it has no perceptible movement when it is pushed.
4. The oil change message window will begin to flash when the switch has been activated.

NOTE: If the reset procedure is not done within 15 seconds, the procedure must be redone starting from Step 1 above. If NOW is flashing, do not wait for it to stop before pushing the reset switch to initiate the oil change reset.

5. Continue to depress the reset switch until the oil change display stops flashing, which will take about 4 seconds.

NOTE: If the reset switch is released before the oil change display stops flashing, the reset procedure must be redone starting from Step 1 above.

1994–95 Thunderbird and Cougar

As of 1994 the Vehicle Maintenance Monitor (VMM) is no longer used.

1985–93 Light and Medium Duty Trucks

All 1985–93 Federal light and medium duty trucks are equipped with an emission maintenance reminder light. Prior to 1985, none were used.

The Emission Maintenance Warning (EMW) module operates a light that is located on the instrument panel. For 1985–87 models, the light will display the word **EMISSIONS**. For 1988–93 model year vehicles, the light will display the words **CHECK ENGINE**. When the light is lit, it is indicating that the 60,000 mile emission maintenance should be performed. After the maintenance is performed, the EMW module must be reset to zero time.

The system actually measures accumulated vehicle ignition key on-time and is designed to continuously close an electrical circuit to the amber lens after 2000 hours of vehicle operation. Assuming an average vehicle speed of 30 mph, the 2000 hours equates to approximately 60,000 miles of vehicle operation. Actual vehicle mileage intervals will vary considerably as individual driving habits vary.

Every time the ignition switch is turned **ON**, the warning light will glow for approximately 2–5 seconds as a bulb check and to verify the system is operating properly. When approximately 60,000 miles has been reached, the warning light will be illuminated continuously to indicate that service is required. After the required maintenance is performed, the Emission Maintenance Warning (EMW) sensor must be reset for another 60,000 mile interval. The sensor locations are as follows:

1. Aerostar — The EMW is located under the dash panel on a bracket below the EEC computer.
2. Bronco — The EMW is located under the dash panel on a bracket below the EEC computer.
3. Bronco II and Ranger — The EMW is located above the right front corner of the glove box.
4. Light and Medium Duty Trucks — The EMW is located under the dash panel, adjacent to the EEC power relay.

The emission maintenance reminder warning light timer may be reset either before or after the time-out period has been exceeded. The procedure is the same for either condition. The procedures to reset the emission maintenance reminder warning light are as follows:

5. Turn the ignition switch to the **OFF** position.
6. Lightly push a small rod (shank end of a No. 2 or 7/32 inch drill bit) through the 0.200 in. diameter hole, with the covering sticker labeled **RESET** and lightly push down on the button and hold.
7. Still pressing the small rod down into the reset hole, turn the ignition switch to the **RUN** position. The **EMW** light will then light and should remain illuminated for as long as the small rod is pressed down in the reset hole. Continue to hold the small rod down for approximately 5 seconds.
8. Remove the small rod. The **EMW** light should go out within approximately 5–10 seconds, indicating a reset has occurred. (If the light does not go out, begin at Step 1 again). Turn the ignition switch to the **OFF** position and go to the next Step.
9. Turn the ignition switch to the **RUN** position. The **EMW** light will light for approximately 2–5 seconds and will then go out. This verifies that a proper reset of the module has been accomplished. If the light should remain on, then the proper reset has not occurred and the reset procedure should be repeated. Turn the ignition switch to the **OFF** position when the reset has been completed.

NOTE: Some models use a non-resettable control unit. When reset is required, replace it with a resettable type.

1994–95 Light and Medium Duty Trucks

No reset procedure is required. The TCIL and other warning lights turn **OFF** automatically when the fault is corrected.

Check Engine Light

Resetting Light after clearing codes — EEC–IV

Vehicles not using and EMW module have the **CHECK ENGINE** light controlled by the EEC–IV processor.

1. Once the source of the trouble codes has been repaired or replaced, the codes must be cleared to prevent the **CHECK ENGINE** light from lighting again.
2. While reading the codes with an analog VOM, disconnect the jumper wire between the Self–Test Input (STI) and the Signal Return Pin of the Self–Test connector.
3. If the codes were read using a STAR tester releasing the center button to the **UP** position will clear the codes.

Resetting Light after clearing codes — EEC–V

The Malfunction Indicator Lamp (MIL) is a light in the dash that reads CHECK ENGINE. The MIL is intended to alert the driver of certain malfunctions in the EEC–V system. The light will stay ON for at least 10 seconds then stay ON as long as the fault causing it is present.

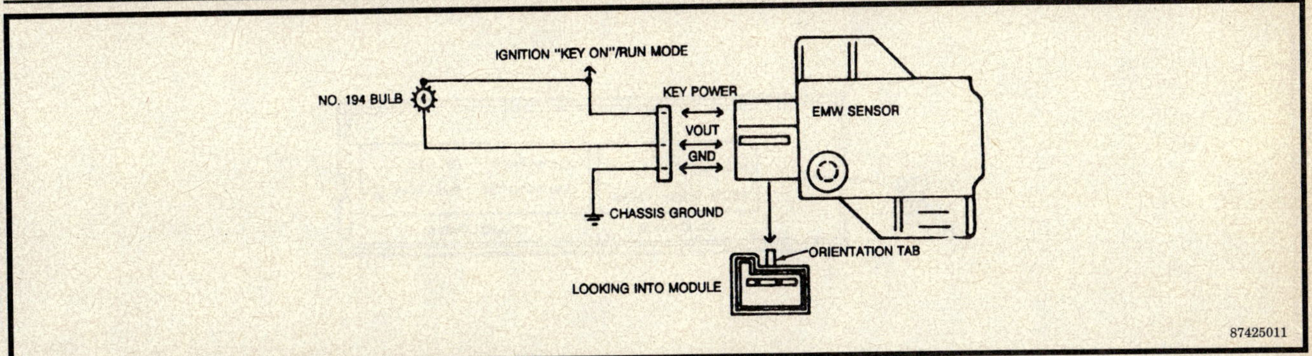

EMW module emission maintenance light system schematic — Ford Motor Company

The light will reset if there are no faults after 3 consecutive drive cycles. If the MIL flashes quickly, the MIL circuit should be checked for proper operation or the engine for severe misfire.

Resetting Light after clearing codes — EEC (NON–NAAO) Management System

The CHECK ENGINE light (MIL) is used to inform the driver of possible engine malfunctions and emission system failure. The light is controlled by the PCM. The PCM monitors engine, ignition and emission related components and signals the driver, through the CHECK ENGINE light, when the engine is running improperly or emissions are unsatisfactory. If the CHECK ENGINE light illuminates during vehicle operation, the cause of the fault or malfunction must be determined and corrected. After the problem area is repaired, the CHECK ENGINE light can be turned OFF by disconnecting the negative battery cable for at least 10 seconds.

GENERAL MOTORS CORPORATION

Emission Maintenance Reminder

Most 1980 GM vehicles used an emission maintenance warning system in the form of a plastic flag that would appear in the speedometer face, normally over the odometer, thus indicating the need for servicing the oxygen sensor and emission system. The flag would appear at 30,000 miles on all models except the Cadillac, which would appear at 15,000 miles. The flag is reset by using the following procedure:

All 1980 except Cadillac

1. Disconnect the negative battery cable. Remove the instrument panel trim plate.
2. Remove the instrument cluster lens.
3. Using a suitable tool, apply a small downward pressure on the plastic notches in the flag stem until the flag resets in its original position.
4. The should be an alignment mark that will appear in the left center of the odometer window when the flag is properly reset.
5. Installation of the removed components, is the reverse of the removal procedure.

1980 Cadillac

1. Disconnect the negative battery cable. Remove the lower steering column cover.
2. The sensor reset cable is located to the left of the speedometer cluster. Pull on the sensor reset cable lightly with approximately 2 lbs. of force.
3. After the sensor flag is reset, reinstall the lower steering column cover. Reinstall the negative battery cable.

Oxygen Sensor Maintenance Light

1984–86 Sprint

The Sprint models are equipped with an oxygen sensor light that will illuminate and flash every 30,000 miles. The light indicates that the emission system and components should be checked for proper operation and the oxygen sensor should be replaced. After the sensor has been replaced and the emission system checked, reset the maintenance light as follows:

1. To reset the emission reminder light, the cancellation switch must be used. The switch is mounted on the right side of the pull down fuse panel which is located on the left side of the steering column, on the lower dash panel.
2. Press the cancel switch to the **OFF** position and start the engine to be sure the light has been reset.

Service Interval Reminder Light

1986–90 Pontiac 6000 and 1989–90 Bonneville

The **SERVICE REMINDER** section of the Driver Information Center (DIC) display shows how many miles remain until service is needed. When the **RESET** button is pressed twice, the type of service and the number of miles remaining until that service is needed, will be displayed. Each time the **RESET** button is pressed, another type of service and the miles remaining for it will be displayed.

With the ignition switch in the **RUN**, **BULB TEST** or **START** positions, voltage is applied from the ECM fuse through the pink/black wire to the Electronic Control Module (ECM). As the vehicle is driven, the speed sensor sends electrical pulses to the ECM. The ECM sends a signal to the speed signal input of the DIC module. The DIC module converts the pulses into miles. The module subtracts the miles traveled from the distance remaining for each item of the service reminder.

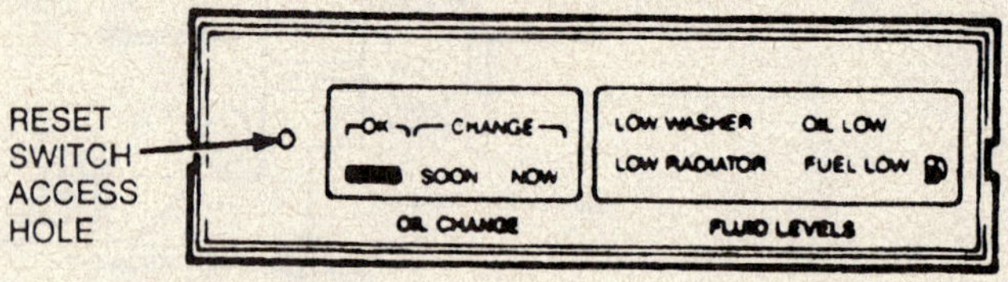

Resetting the emission reminder flag — General Motors Corporation

When the miles remaining for a service approaches zero, that service will be displayed on the DIC display. All 4 types of service can be shown at the same time.

Service Reminder Reset

To reset the service light, it will be necessary to subtract the mileage from the service interval light that is illuminated. The miles remaining for a certain type of service can be decreased by holding the **RESET** button, the miles remaining will be decreased in steps of 500 miles, every 5 seconds. In the first step, the miles will decrease to a multiple of 500. For example, 2880 miles will decrease to 2500 miles. If the **RESET** button is held in and the miles remaining reach zero, the DIS display will show the service interval for the service selected. The service intervals are as follows:

1. Change Oil — 7500 miles.
2. Oil Filter Change — 7500 miles.
3. Next Filter Change — 15,000 miles.
4. Rotate Tires — 7500 miles.
5. Next Tire Rotation — 15,000 miles.
6. Tune Up — 30,000 miles.

If the **RESET** button is still held down, the miles will decrease in steps of 500 miles from the service interval. When the **RESET** button is released, the mile display shown will be the new distance until the service should be performed.

When a service distance reaches zero, the service reminder item will be displayed. If the service interval is reset within 10 miles, the display will go out immediately. If more than 10 miles passes before the service interval is reset, the item will remain displayed for another 10 miles after being reset before going out.

NOTE: On some models, it may be necessary to depress the SYSTEM RECALL button in order to display the service interval light on the driver information center, in order to be able to decrease the mileage from it so as to reset the interval light.

1991–93 Pontiac Bonneville

The Driver Information Center lights up for a few seconds when the ignition is turned to **ON**. To see the entire DIC, press and release the DIC button to the right of the display. The system is divided into 4 main systems: Function Monitor, Lamp Check, Security and Service Reminder. These systems monitor the following:

1. FUNCTION MONITOR:

a. Checks windshield washer fluid. Message comes ON when fluid is less than 40 percent full.

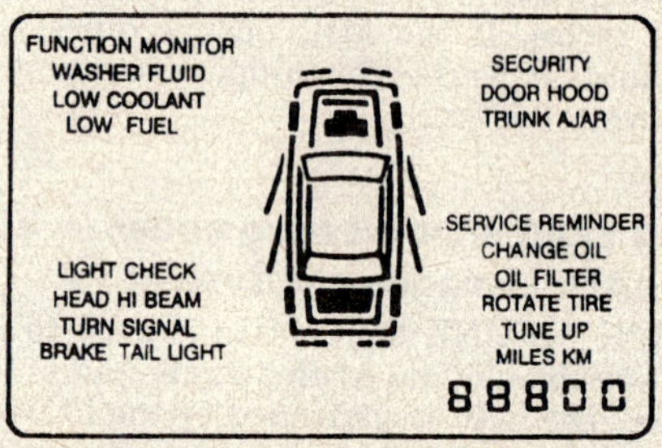

Driver Information Center (DIC) display — General Motors Corporation

b. Checks engine coolant level. Message comes ON when coolant level has fallen to about half full.

c. Checks fuel level. Message comes ON when approximately 2½ gallons of fuel remain.

2. LAMP CHECK:

a. Checks headlights, turn signal, brake lights and taillights.

b. Message comes ON if a problem exists at 1 of the lights. The problem area is indicated on the vehicle outline shown on the DIC. Note that if a burned out bulb is replaced, the warning light will stay on until the bulb is energized or turned on.

3. SECURITY:

a. Checks if doors are fully closed. The problem area is indicated on the vehicle outline shown on the DIC.

b. Checks if hood or trunk isn't fully closed.

4. SERVICE REMINDER:

a. Lets vehicle operator know when to have the following service work done: Change engine oil, change oil filter, rotate tires and have engine tune-up.

b. The figures at the bottom of the DIC show the number of miles remaining before the work should be done.

To use the service reminder, the vehicle owner must first decide which service item to check. For example, if the owner wants to know when to next rotate the tires, push and hold the DIC button until the **SERVICE REMINDER** light comes ON. Then push and release the DIC button until **ROTATE TIRES** appears. The number at the bottom then shows how many miles there are to go before the tires should be rotated.

To reset the service reminder, push the DIC until the service item desired appears and do not release the button. In 5–10 seconds the display will start to count down, 500 miles at a time. When the display reads the distance the vehicle owner wishes to set, release the button. Note that sometimes a service reminder display will stay ON even though it has been reset, but the reminder should go OFF after the vehicle has been driven about 10 miles.

1994–95 Pontiac Bonneville

Vehicles equipped with an **ENGINE OIL LIFE INDEX** the display located on the Drivers Information Center (DIC). The PCM determines approximately when the engine oil should be changed by calculating information based on vehicle speed, coolant temperature and engine RPM. Once the PCM determines it is time to change engine oil, it will illuminate the **CHANGE OIL SOON** light on the DIC. This indicates the remaining oil life is below 10% . (not to be confused with oil level). On a new vehicle or one that has been reset, the oil life is 100% . This percentage will slowly decrease based on the inputs the PCM receives. When oil life reaches the 0% mark, the PCM will illuminate the **CHANGE OIL NOW** light on the DIC. At this time both messages will be displayed accompanied by a slow 5 second audible chime. The reset button must be pressed to acknowledge each of these messages. If a steady **CHANGE OIL SOON** light persist, this indicates a diagnostic trouble code 61 has been set indicating a possible shorted switch.

Remaining oil life percentage can be displayed by pressing **OIL** button on the DIC and advancing through the messages until the oil life index is displayed. The oil life index will not detect abnormal conditions such as excessively dusty or engine malfunctions that could otherwise affect the engine oil life.

After the engine oil has been changed, display the oil life index on the Drivers Information Center (DIC), then hold the **RESET** button for 5 seconds. When a DIC message of **RESET** is displayed and the oil life index equals 100 percent, the reset is complete.

1989–91 Eldorado and Seville

1. Press the **ENG DATA** button until the oil life index is displayed on the Driver's Information Center.
2. Press **ENG DATA** and **RANGE** buttons at the same time until the oil life index resets to 100 and hold for 5 seconds.

1992–93 ELDORADO AND SEVILLE

1. Press the **INFORMATION** button until the oil life index is displayed.
2. Press **STORE/RECALL** until the oil life index resets to 100 and hold for 5 seconds.

1991–93 Deville, Fleetwood and Sixty Special

1. Press **RANGE** and **FUEL USED** buttons at the same time to display oil life index.
2. Press **RANGE** and **RESET** buttons at the same time until the **CHANGE OIL SOON** light flashes and hold for 5 seconds. The oil life index will not remain displayed.

1993–95 Fleetwood (Rear Wheel Drive)

After changing engine oil, reset the engine oil life indicator whether the "CHANGE OIL" warning/indicator lamp came on or not.

Reset the engine oil life monitor as follows:

1. Turn ignition switch to "ON" position, but don't start engine.
2. Depress accelerator pedal to wide open throttle (WOT) position and release it three times within five seconds.
3. If "CHANGE OIL" warning/indicator lamp blinks twice, then goes out, system has been reset.

If "CHANGE OIL" warning/indicator lamp does not reset, turn ignition switch to "OFF" and repeat procedure.

1989–93 Allante

1. Press the **RANGE** button until the oil life index is displayed.
2. Press **AVE SPEED** and **RANGE** buttons at the same time until the oil life index resets to 100 and hold for 5 seconds.

1994–95 DeVille, DeVille Concours, Eldorado and Seville

Vehicles are equipped with an engine oil life index (EOLI) feature as part of the Driver Information Center display (DIC). Engine oil life is displayed through engine data as the "OIL LIFE INDEX" and as a "CHANGE ENGINE OIL" message. The "OIL LIFE INDEX" is displayed following a number between 0 and 100. This is the percentage of oil life REMAINING based on driving conditions, engine oil temperature, and mileage driven since the last time the oil life indicator was reset. When the oil life index reaches 10% or less a "CHANGE OIL SOON" message will appear as a reminder to schedule an oil change. When the oil life index reaches 0, the "CHANGE ENGINE OIL" message will appear indicating that the oil should be changed within the next 200 miles (320 km). After the oil has been changed, display the "OIL LIFE INDEX" message by pressing the "INFORMATION" button several times. Press and hold the "RESET" button until the display shows "100". This will reset the oil life index. The "CHANGE ENGINE OIL" message will remain off until the next oil change is needed. The percentage of oil life remaining may be checked at any time by pressing the "INFORMATION" button several times until the "OIL LIFE INDEX" appears.

1994 Chevrolet Caprice

After changing engine oil, reset the engine oil life indicator whether the "CHANGE OIL" warning/indicator lamp came on or not.

Reset the engine oil life monitor as follows:

1. Remove the instrument panel fuse box cover.
2. Turn the ignition switch to the **ON** position but do not start the engine.
3. Press and hold the OIL RESET button for 5 seconds. The CHANGE OIL light should go out. The system has been reset.

If "CHANGE OIL" warning/indicator lamp does not reset, turn ignition switch to "OFF" and repeat procedure.

1995 Chevrolet Caprice, Impala SS

After changing engine oil, reset the engine oil life indicator whether the "CHANGE OIL" warning/indicator lamp came on or not.

Reset the engine oil life monitor as follows:

1. Turn ignition switch to "ON" position, but don't start engine.

2. Depress accelerator pedal to wide open throttle (WOT) position and release it three times within five seconds.

3. If "CHANGE OIL" warning/indicator lamp blinks twice, then goes out, system has been reset.

If "CHANGE OIL" warning/indicator lamp does not reset, turn ignition switch to "OFF" and repeat procedure.

1990–91 Corvette

The **CHANGE OIL** monitor light is on the left side of the instrument cluster.

When changing oil, reset the engine oil life monitor whether the **CHANGE OIL** light came on or not. Reset the monitor as follows:

1. Turn the key to the **ON** position but do not start the engine.
2. Press the **ENG MET** button on the trip monitor and release. Press the button again within 5 seconds.
3. Within 5 seconds of pressing the button the second time, press and hold the **RANGE** button on the trip monitor. The **CHANGE OIL** light will flash.
4. Hold the **RANGE** button until the **CHANGE OIL** light stops flashing and extinguishes. The monitor is reset at this point. Repeat this procedure if the light does not go out.

1992–95 Corvette

1. Turn the key to the **ON** position but do not start the engine.
2. Press the **ENG MET** button on the trip monitor and release. Press the button again within 5 seconds.
3. Within 5 seconds of pressing the button the second time, press and hold the **GAUGES** button on the trip monitor. The **CHANGE OIL** light will flash.
4. Hold the **GAUGES** button until the **CHANGE OIL** light stops flashing and extinguishes. The monitor is reset at this point. Repeat this procedure if the light does not go out.

1991–93 Buick Park Avenue

An optional system illuminates **CHANGE OIL SOON** when the system determines the oil life has degraded to the point where the engine oil must be changed. This light is lit for 3 seconds as a bulb check before the engine is started. If an oil change is required, it will stay ON for 60 seconds each time the engine is started.

After changing the engine oil, the system should be reset. The reset button is inside a hole under the passenger's side of the instrument panel, near the door. With the ignition in the **RUN** position, but the engine not started, use a pencil to push the reset button, holding it in for at least 5 seconds but not more than 60 seconds. After 5 seconds, the **CHANGE OIL SOON** light will flash 4 times. This indicates that the Oil Life Monitor System has acknowledged the reset.

1994 Buick Park Avenue and LeSabre

After the engine oil has been changed, the Engine Oil Monitor must be reset. The reset button is located in the glove box. Reset as follows:

With the ignition key in the RUN position, push the reset button, hold it in for at least 5 seconds but not more than 60 seconds. The CHANGE OIL SOON light will flash 4 times and then go off. This indicates that the Oil Life Monitor System has been reset.

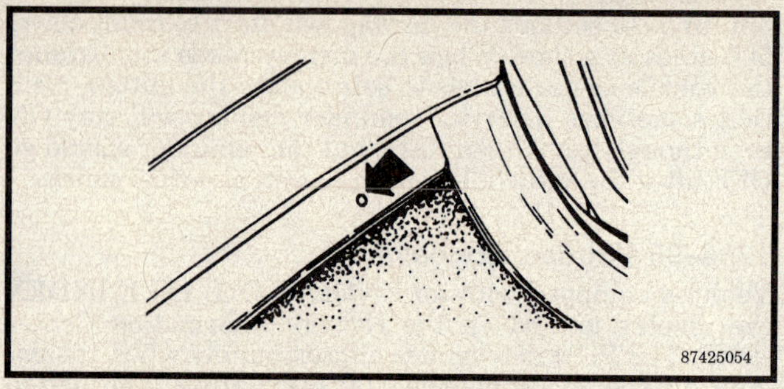

The reset button is inside this hole, which is under the passenger side of the instrument panel, near the door — 1991–93 Park Avenue

1995 Buick Park Avenue and LeSabre

Vehicles equipped with an **ENGINE OIL LIFE INDEX** the display located on the Drivers Information Center (DIC). The PCM determines approximately when the engine oil should be changed by calculating information based on vehicle speed, coolant temperature and engine RPM. Once the PCM determines it is time to change engine oil, it will illuminate the **CHANGE OIL SOON** light on the DIC. This indicates the remaining oil life is below 10% . (not to be confused with oil level). On a new vehicle or one that has been reset, the oil life is 100% . This percentage will slowly decrease based on the inputs the PCM receives. When oil life reaches the 0% mark, the PCM will illuminate the **CHANGE OIL NOW** light on the DIC. At this time both messages will be displayed accompanied by a slow 5 second audible chime. The reset button must be pressed to acknowledge each of these messages. If a steady **CHANGE OIL SOON** light persist, this indicates a diagnostic trouble code 61 has been set indicating a possible shorted switch.

Remaining oil life percentage can be displayed by pressing **OIL** button on the DIC and advancing through the messages until the oil life index is displayed. The oil life index will not detect abnormal conditions such as excessively dusty or engine malfunctions that could otherwise affect the engine oil life.

After the engine oil has been changed, display the oil life index on the Drivers Information Center (DIC), then hold the **RESET** button for 5 seconds. When a DIC message of **RESET** is displayed and the oil life index equals 100 percent, the reset is complete.

1995 Buick Roadmaster

After changing engine oil, reset the engine oil life indicator whether the "CHANGE OIL" warning/indicator lamp came on or not.

Reset the engine oil life monitor as follows:

1. Turn ignition switch to "ON" position, but don't start engine.
2. Depress accelerator pedal to wide open throttle (WOT) position and release it three times within five seconds.
3. If "CHANGE OIL" warning/indicator lamp blinks twice, then goes out, system has been reset.

If "CHANGE OIL" warning/indicator lamp does not reset, turn ignition switch to "OFF" and repeat procedure.

1995 Buick Riviera

This vehicle is equipped with an **ENGINE OIL LIFE INDEX** display located on the Drivers Information

Center (DIC). The PCM determines approximately when the engine oil should be changed by calculating information based on vehicle speed, coolant temperature and engine RPM. Once the PCM determines it is time to change engine oil, it will illuminate the **CHANGE OIL SOON** light on the DIC. This indicates the remaining oil life is below 10 % . (not to be confused with oil level). On a new vehicle or one that that has been reset, the oil life is 100 % . This percentage will slowly decrease based on the inputs the PCM receives. When oil life reaches the 0 % mark, the PCM will illuminate the **CHANGE OIL NOW** light on the DIC. At this time both messages will be displayed accompanied by a slow 5 second audible chime. The reset button must be pressed to acknowledge each of these messages. If a steady **CHANGE OIL SOON** light persist, this indicates a diagnostic trouble code 61 has been set indicating a possible shorted switch.

Remaining oil life percentage can be displayed by pressing **OIL** button on the DIC and advancing through the messages until the oil life index is displayed. The oil life index will not detect abnormal conditions such as excessively dusty or engine malfunctions that could otherwise affect the engine oil life.

After the engine oil has been changed, display the oil life index on the Drivers Information Center (DIC), then hold the **RESET** button for 5 seconds. When a DIC message of **RESET** is displayed and the oil life index equals 100 percent, the reset is complete.

1989–93 Oldsmobile

Vehicles equipped with an **ENGINE OIL LIFE INDEX** display as a part of the **DRIVER INFORMATION SYSTEM (DIS)**, have a display that will show when to change the engine oil.

The oil change interval is determined by the driver information system and will usually fall at or between the 2 recommended alternative intervals of 3000 miles and 7500 miles, but it could be shorter than 3000 miles under some severe driving conditions. The driver information system will also signal the need for an oil change at 7500 miles or one year passed since the last oil change. If the driver information system does not indicate the need for an oil change after 7500 miles or one year or if the **ENGINE OIL LIFE INDEX** display fails to appear, the oil should be changed and the driver information system serviced.

When the **ENGINE OIL LIFE INDEX** reaches 10 percent or less, the **CHANGE OIL** light display will function as a reserve trip odometer (indicating the distance to an oil change). Until the **ENGINE OIL LIFE INDEX** reset is performed, the driver information system will display the distance to the oil change and sound a beep when the ignition switch is turned to the **ACCESSORY** or **RUN** position the first time each day.

When the distance to the next oil change reaches 0, the driver information system will display the **CHANGE OIL NOW** light. Until an **ENGINE OIL LIFE INDEX** reset is performed, the driver information system will display the **CHANGE OIL NOW** light and sound a beep when the ignition switch is turned to the **ACCESSORY** or **RUN** position at the beginning of each day.

The driver information system will not detect dusty conditions or engine malfunctions which may affect the engine oil. If driving in severe conditions exists, be sure to change the engine oil every 3000 miles or 3 months which ever comes first, unless instructed otherwise by the driver information system. The driver information center does not measure the engine oil level. It still remains the owner's responsibility to check the engine oil level. After the oil has been changed, the **ENGINE OIL LIFE INDEX** light must be reset. Resetting can be accomplished as follows:

1. All models, except 1989–92 Toronado, 1992–93 Ninety Eight and 1991–93 Cutlass Supreme — the **ENGINE OIL LIFE INDEX** can be reset by pressing the **RESET** and **OIL** buttons simultaneously for at least 5 seconds while on the **ENGINE OIL LIFE INDEX** display. The driver information system will reset the **ENGINE OIL LIFE INDEX** to 100 percent and display a **ENGINE OIL LIFE INDEX** of 100 percent.
2. 1989–90 Toronado — the **ENGINE OIL LIFE INDEX** can be reset by pressing the **ENGINE DATA** and **GAUGES** buttons on the instrument panel console simultaneously for at least 5 seconds (as if as to enter the diagnostic mode) during this time the **CHANGE OIL** message will remain off until the next oil change is needed. Be sure to hold the **ENGINE DATA** and **GAUGES** buttons simultaneously until the **ENGINE OIL LIFE INDEX 100** message appears.
3. 1991–92 Toronado — Press the **ENG DATA** button to select the oil life display. Then hold the **RESET/ENTER** button for at least 5 seconds. The oil life will be reset to 100 percent.
4. 1992–93 Ninety Eight — Press the **OIL** button to display the oil message and then press the **RESET** button for at least 5 seconds. The oil life monitor will be reset to 100 percent.
5. 1991–93 Cutlass Supreme — Press the **OIL** button to display the oil message and then press the **OIL** and **RESET** buttons at the same time for at least 5 seconds. The oil life monitor will be reset to 100 percent.

NOTE: The Engine Oil Life Index is stored on a non-volatile memory chip and will not require resetting by disconnecting the battery cables and/or fuse.

1994–95 Oldsmobile

1994–95 vehicles that are equipped with a Drivers Information Center and have an Oil life index, will require the Oil Life indicator to be reset after each oil change.

Press the **SEL** to select **OIL**. Press **SEL** if necessary to display the oil life. The display will show a reading of the estimated oil life left. Example: **OIL LIFE 85%** . When the remaining oil life is 9% or less, the display will show **CHANGE OIL SOON**. Then a tone will sound and the **CHANGE OIL SOON** message will display each time the vehicle is started.

When the oil life is zero, la tone will sound and the display will show, **CHANGE OIL NOW**. Then when the vehicle is started a tone will sound and the **CHANGE OIL NOW** message will display each time the vehicle is started. Reset the Oil Life Display as follows:

1. Acknowledge all diagnostic messages in the Drivers Information Center by pressing **RESET.**
2. Press the **SEL** button on the left to select **OIL**. Press **SEL** button on the right if necessary to display oil life.
3. Press and hold the **RESET** button for about 5 seconds. Once the oil life index has been reset, a **RESET** message will be displayed and then oil life will change to 100% .

Be careful not reset the oil life accidentally at any time other than when the oil has just been changed. It can not be reset accurately until the next oil change.

1994–95 Oldsmobile Aurora

This vehicle is equipped with an **ENGINE OIL LIFE INDEX** display located on the Drivers Information Center (DIC). The PCM determines approximately when the engine oil should be changed by calculating information based on vehicle speed, coolant temperature and engine RPM. Once the PCM determines it is time to change engine oil, it will illuminate the **CHANGE OIL SOON** light on the DIC. This indicates the remaining oil life is below 10% . (not to be confused with oil level). On a new vehicle or one that has been reset, the oil life is 100% . This percentage will slowly decrease based on the inputs the PCM receives. When oil life reaches the 0% mark, the PCM will illuminate the **CHANGE OIL NOW** light on the DIC. At this time both messages will be displayed accompanied by a slow 5 second audible chime. The reset button must be pressed to acknowledge each of these messages. If a steady **CHANGE OIL SOON** light persist, this indicates a diagnostic trouble code 61 has been set indicating a possible shorted switch.

Remaining oil life percentage can be displayed by pressing **OIL** button on the DIC and advancing through the messages until the oil life index is displayed. The oil life index will not detect abnormal conditions such as excessively dusty or engine malfunctions that could otherwise affect the engine oil life.

After the engine oil has been changed, display the oil life index on the Drivers Information Center (DIC), then hold the **RESET** button for 5 seconds. When a DIC message of **RESET** is displayed and the oil life index equals 100 percent, the reset is complete.

Transmission Fluid Change Reset

1994–95 DeVille, Concours, Eldorado and Seville

All Cadillacs with 4.6L Northstart engines and 4T80-E transaxles are equipped with a transaxle fluid change indicator. A "CHANGE TRANS FLUID" message will display on the Information Center when the Powertrain Control Module monitors actual operating conditions and displays a change trans fluid message based on calculations based on those conditions or 100,000 miles (160,000 km.) Change fluid by removing lower pan and side cover drain plug.

Change both the transmission fluid and the filter every 15,000 miles (25,000 km) if the vehicle is mainly driven under one or more of these conditions:

- In heavy city traffic where outside temperature regularly reached 90°F (32°C) or higher.
- On highly or mountain terrain.
- Frequent trailer pulling.
- Used for delivery service.
- If the vehicle is not used under any of these conditions, change both the fluid and filter (or service the screen) every 100,000 miles (160,000 km). Reset the light as follows:

When the "CHANGE TRANS FLUID" message appears, change the fluid in both the pan and side cover. Reset the indicator as follows:

Turn the key "ON" with the engine stopped. Press and hold the "OFF" and "REAR DEFOG" buttons on the climate control simultaneously until the "TRANS FLUID RESET" message appears in the Information Center (between 5 and 20 seconds) The system is now reset.

Check Engine Light

The **CHECK ENGINE** light will light to indicate a fault in the engine control system. Once the fault code has been retrieved from the ECM memory and the problem rectified, the fault code should be cleared from the memory. This can be done by removing power to the ECM by disconnecting the ECM harness from the positive battery pigtail or by removing the ECM fuse. Either way the ignition switch must be in the **OFF** position and the power removed for a minimum of 10 seconds. Once power has been restored the vehicle should be driven at part throttle with medium acceleration to allow the ECM to readjust itself. A change in performance may be noted initially, but the performance should return to normal as the ECM relearns its operating parameters.

SATURN

Service Engine Soon Light

1991–95 Vehicles

The **SERVICE ENGINE SOON** light will light to indicate a fault in the engine control system. Removing power from the Powertrain Control Module (PCM) will clear codes from the General Information portion of the PCM memory. Flags and codes stored in the malfunction history however, will remain in the PCM memory even if power is disconnected. Flags are informational codes not related to emission related problems. These codes will not cause the **SERVICE ENGINE SOON** light to come on. The data in malfunction history can only be read or cleared using a Saturn scan tool or one with equivalent capability.

To clear the code and reset the **SERVICE ENGINE SOON** light, follow the steps below:

1. Turn the ignition switch **ON** and jumper terminals A and B of the ALDL together 3 times in 5 seconds. The **SERVICE ENGINE SOON** light should stop flashing and any Codes should be cleared.
2. If all faults have been repaired, cycling the ignition switch 50 times will automatically clear General Information of fault Codes.

IMPORT VEHICLE RESET PROCEDURES

ACURA

Scheduled Service Due

1988–90 Legend Sedan

The 1988–90 Legend Sedan is equipped with a **SCHEDULED SERVICE DUE** warning light. The maintenance light will illuminate every 7500 miles to indicate that an oil and filter change is needed. However, if a shorter oil change interval is desired, there are 7 different intervals to choose, from 7500 to 1500 miles. To choose a new interval, push the **SERVICE RESET** button and the **ARROW** button for approximately 3 seconds, then push the **ARROW** button until the interval desired appears. Then push the **SET** button. After completing the necessary maintenance service, the maintenance light must be reset.

In order to reset the maintenance light, the **SERVICE RESET** button must be depressed. With the ignition switch in the **ON** position, hold the reset button in for at least 3 seconds. To verify that the reset is complete, turn the ignition switch **OFF** and **ON**, the light should not come on.

1987–90 Legend Coupe

The 1987–90 Legend Coupe is equipped with a **SCHEDULED SERVICE DUE** warning light. The maintenance light will illuminate every 7500 miles to indicate that scheduled maintenance is needed. To reset the light after performing the maintenance, hold the button in for a period of 5 seconds.

The 1987–90 Legend Coupe is also equipped with a **OIL/FILTER CHANGE DUE** warning light. The warning light will illuminate every 7500 miles. However, if a shorter oil change interval is desired, there are 7 different intervals to choose, from 7500 to 1500 miles. To choose a new interval, push the **SERVICE RESET** button and the **ARROW** button for approximately 3 seconds, then push the **ARROW** button until the interval desired appears. Then push the **SET** button. After completing the necessary maintenance service, the maintenance light must be reset.

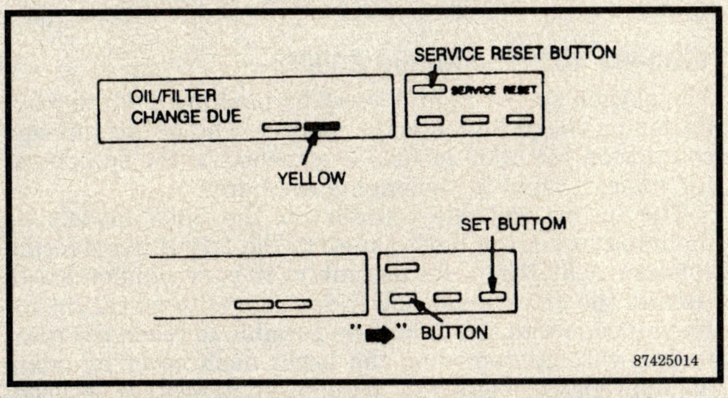

Resetting the service interval light — 1987–90 Legend and Legend coupe

To verify that the reset is complete, turn the ignition switch **OFF** and **ON**, the light should not come on.

Maintenance Required Indicator

1991–95 Legend,Vigor

The Maintenance Reminder indicator reminds you when it is time for scheduled maintenance. When it is near 7,500 miles (12,000 km) since the last maintenance, the indicator will turn yellow. If you exceed 7,500 miles (12,000 km), the indicator will turn red. The indicator can be reset by inserting the ignition key or other similar object into the slot below the indicator. This will extinguish the indicator for the next 7500 miles.

1994–95 Integra

The maintenance required indicator reminds you that it is time for scheduled maintenance. For the first 6,000 miles (9,600 km) after the maintenance Required Indicator is reset, it will come on for 2 seconds when you turn the ignition **ON** (II).

Between 6,000 miles (9,600 km) and 7,500 miles (12,000 km) this indicator will light for two seconds when you first turn the ignition **ON** (II), and then flash for ten seconds. If you exceed 7,500 miles (12,000 km) without having the scheduled maintenance performed, this indicator will remain on as a constant reminder.

Reset the indicator by pressing the reset button. This button is located on the bottom of the dashboard to the right of the steering column.

Check Engine Light

The **CHECK ENGINE** light will illuminate when a trouble code develops. The light will be extinguished when the ignition switch is turned **OFF** and as long as the trouble code does not develop again the light will remain OFF when the ignition switch is turned **ON**. This does not hold true for the LED on the ECU. The LED will continue to illuminate until the trouble code is cleared from the ECU.

To clear the codes on the 1988–90 Integra remove the HAZARD fuse at the positive battery terminal for a mini-

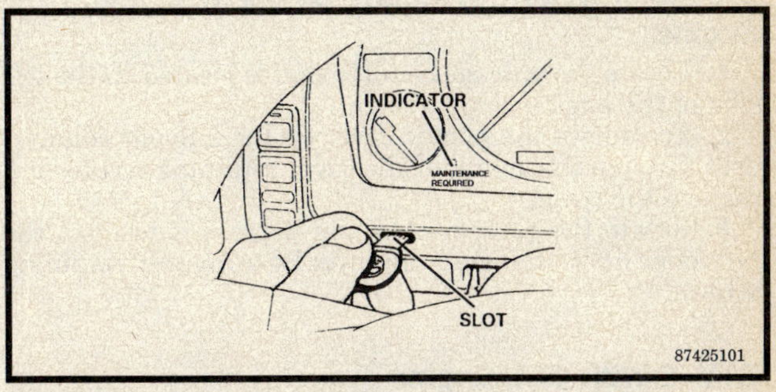

Resetting the Maintenance Required Indicator — 1991–95 Legend Sedan, Coupe, Vigor and TL Series

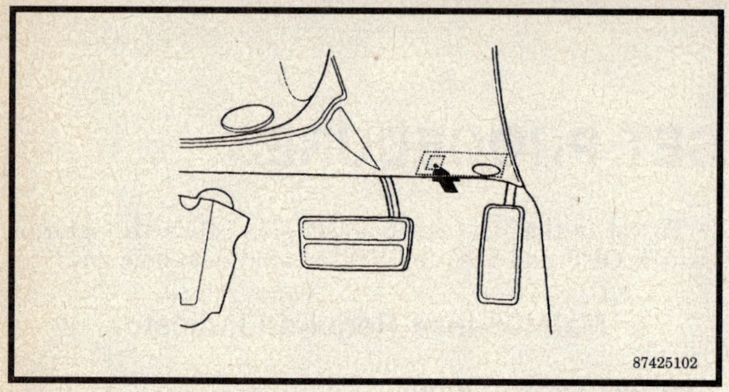

Resetting the Maintenance Required Indicator — 1991–95 Acura Integra

mum of 10 seconds. On the 1991–95 Integra remove the BACK UP fuse from the underhood relay box for a minimum of 10 seconds. On the 1988–90 Legend remove the ALTERNATOR SENSE fuse from the underhood relay box for a minimum of 10 seconds. On the 1991–95 Legend sedan, Legend Coupe and Vigor remove the No. 15 (7.5A) fuse from the underdash relay box for a minimum of 10 seconds. On the 1991–95 NSX remove the CLOCK fuse from the main relay box for a minimum of 10 seconds.

Once power has been restored, the vehicle should be driven at part throttle with medium acceleration to allow the ECU to go through a relearning procedure and adjust itself. A change in performance may be noted initially, but the performance should return to normal as the ECU relearns its operating parameters.

ALFA ROMEO

Oxygen Sensor Maintenance Reminder Light

A maintenance reminder light used on all of the Alfa Romeo models. The light will illuminate every 30,000 miles on the Graduate, Quadrifoglio and the Spider equipped with the 2.0L engine. The light will illuminate every 60,000 miles on the GTV–6 models equipped with the 2.5L engine. The reminder light is used as a preventive maintenance indicator in regards to the emission system and the oxygen sensor used on that vehicle. After the light has come on, the oxygen sensor replaced and the emission maintenance has been done, reset the light as follows:

Graduate, Quadrifoglio and Spider With 2.0L Engine

1. Locate the mileage counter that is located on the left side of the engine compartment.
2. To remove the plastic cover on the mileage counter, drill through the shank of the cover retaining screws and remove the cover.
3. Rotate the reset button and depress it to reset the reminder light. Reinstall the cover by using new retaining screws.

GTV–6 With 2.5L Engine

1. On the GTV–6 models equipped with the 2.5L engine, it will be necessary to remove the ECU cover panel and the lower right side parcel shelf.
2. Locate the mileage counter usually located under the right side of the dash. Depress the white reset button to reset the reminder light. Turn the ignition switch to the **ON** position to be sure the light has been reset.

164 Series

The wear of the front brake pads and the level and temperature of the engine coolant are continuously monitored by an electronic control unit that in case of malfunction alerts the driver by turning on the appropriate warning light on the instrument panel.

Severe wear of the front brake pads causes opening of the contacts of the brake pad sensors. This breaks the continuity through pins **9B** and **10B** of the control unit. Replacement of the brake pads and their sensors will turn the warning light off.

The engine coolant level sensor contacts open when the engine coolant level drops below the minimum mark. This breaks the continuity of pin **4B** of the control unit to ground. Once the level has been restored the light will go out.

Check Engine Light

The **CHECK ENGINE** light will light upon a fault. The fault code will be stored in the ECU. Once the source of the problem has been rectified the light should extinguish itself.

AUDI

EGR Warning Light

Some models use an EGR maintenance light that will come on every 15,000 miles of operation. This indicates the need to inspect the EGR system and make any repairs as necessary. After the EGR system has been inspected and is found in proper working condition, push the EGR reset button located on the mileage counter that is attached to the speedometer cable, located in the engine cowl area.

Oxygen Sensor Maintenance Light

Most models are equipped with an oxygen sensor light that will illuminate every 30,000 miles (60,000 miles on some 1989–93 100/200). The light indicates that the emission system and components should be checked for proper operation and the oxygen sensor should be replaced. Once the sensor has been replaced and the emission system checked, reset the maintenance light as follows:

1980–83 4000, 5000 and 5000S

The oxygen sensor light is reset by pushing a white reset button on the in-line mileage counter. The in-line mileage counter on the 4000 models is attached to the speedometer cable, located in the engine cowl area.

The in-line mileage counter on the 5000 models, is mounted under the dash panel, to the left of the steering column, near the pedal assembly. It is recommended to remove the driver's side lower dash cover to reveal the in-line mileage counter. It may be possible to reach the reset button without removing the lower dash cover by using an appropriate bent rod. Whichever procedure is used, the white reset button must be depressed to reset the maintenance reminder light.

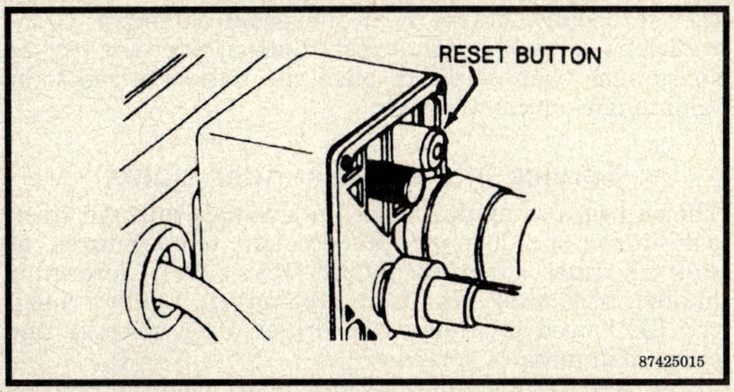

In-line mileage counter and reset mechanism — Audi

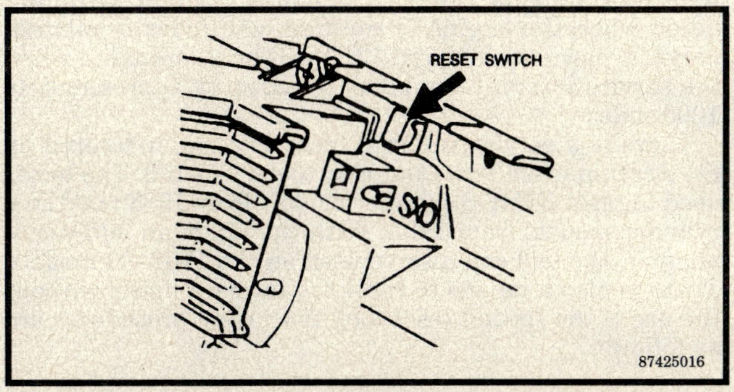

Location of the reset button incorporated into the instrument panel — Audi

1984–88 4000, 5000 and 5000S

1. To reset the counter, disconnect the negative battery cable and remove the upper and lower instrument cluster cover screws.
2. Pull the cluster cover towards the steering wheel and lift the upper instrument cluster cover of the cluster. On some models, it may be necessary to remove the instrument cluster to gain access to the switch. If so, remove the cluster as follows:
   a. Disconnect the speedometer cable at the transmission and the instrument panel cluster.

NOTE: Use care when removing the speedometer from the instrument panel. The panel cluster and parts are very fragile.

   b. Disconnect electrical connectors from the cluster and remove the cluster. It may be to the technician's advantage to remove the steering wheel, allowing more room to work on the cluster.
3. At the top left center of the cluster, there should be the initials **OXS** imprinted on the cluster panel.
4. Locate the plastic breakaway box near the **OXS** initials and using a suitable tool, pull the box away from the cluster panel.
5. Insert a suitable tool into the **OXS** opening and push the cancel switch over to the opposite side in order to reset the reminder light.

1984–88 5000 Turbo and 1989–93 100/200 Turbo Models

The in-line mileage counter is used and is located under the rear seat. After checking the emission system and replacing the oxygen sensor, reset the reminder light as follows:

1. Push the rear seat cushion toward the rear of the vehicle, then lift the front of the seat cushion to release the cushion retainers.
2. Place the seat cushions out of the way. The counter is usually located on the left side of the vehicle.
3. Depress the button marked **OXS** and turn the ignition switch to the **ON** position.
4. Turn the ignition switch to the **OFF** position and then **ON** and **OFF** again to be sure the reminder light is reset and does not illuminate any longer.

1988–90 Except V8 Quattro

1. With the LED tester still connected and the codes retrieved, turn the ignition switch **OFF**.
2. Install a spare fuse into the top of the fuel pump relay.
3. Turn the ignition switch **ON**. The LED tester must light.
4. Remove the fuse after approximately 4 seconds. The LED tester must go **OFF** and then begin flashing fault Code 0000.
5. Install the fuse again for approximately 10 seconds and then remove. The LED tester should light and stay on. The memory is erased at this point.

1991–93 80, 90, 100 and 200 1990–93 V8 Quattro

1. Once the codes have been retrieved from the memory and the LED tester is still connected, check that Code 0000 has been displayed.
2. Join the test leads that are connected to terminals **1** and **2** of the diagnosis test point. Keep connected for approximately 4 seconds and then disconnect.
3. The codes have been erased when no additional codes are flashed on the LED tester.

Check Engine Light

When a temporary fault occurs, the **CHECK ENGINE** light illuminates until the ignition is switched **OFF**. When the ignition switch is switched **ON** again the light will be extinguished provided that the fault does not occur again or the defect was repaired and the vehicle is driven a minimum of 5 minutes. If a permanent fault occurs the fault will be stored in the ECU and will remain until it is cleared manually.

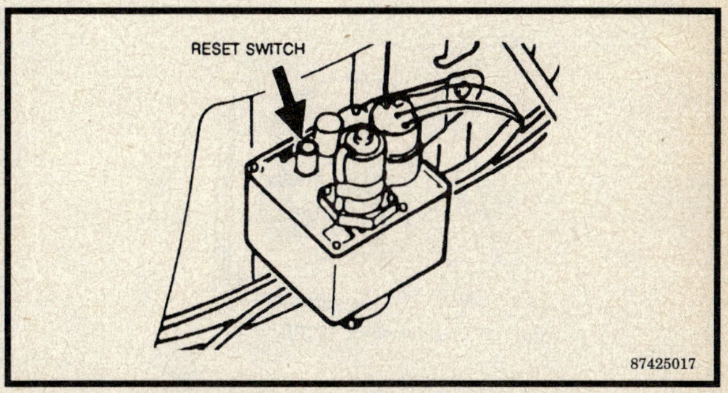

Location of the mileage counter and reset mechanism located under the rear seat — Audi

BMW

EGR Maintenance Reminder Light

1978 733i

Models equipped with an EGR sensor light, will illuminate every 25,000 miles. The light indicates its time to service the EGR system. When the light comes on, clean and inspect the system, including replacing gaskets under the thermal reactor.

To reset the light, remove lower trim panel from the center of the instrument panel and press the reset button marked "EGR".

Oxygen Sensor Maintenance Light

Most models are equipped with an oxygen sensor light that will illuminate every 30,000 miles (25,000 miles on the 528i models). The light indicates the emission system and components should be checked for proper operation and the oxygen sensor should be replaced. After the oxygen sensor has been replaced and the emission system checked, reset the maintenance light as follows:

1980–83 Models

The BMW 528E and 633 CSi models do not use a reset switch. On these models, it is necessary to remove the instrument panel and remove the bulb used to illuminate the oxygen sensor light.

1. Locate the in-line mileage counter which is usually placed above the left frame rail, near the transmission.
2. Locate the white reset button on the in-line mileage counter and depress the button to reset the maintenance reminder light.

NOTE: Due to the location of the in-line mileage counter, it is exposed road dirt and moisture. Therefore, when pushing the reset button, be sure to get it to click or the reminder light will remain illuminated. If the switch will not cancel out the reminder light, it must be replaced.

1983–1995 Models

Most of the models were not equipped with an oxygen sensor reset. When the oxygen sensor light is illuminated at the designated mileage, it will be necessary to to remove the instrument panel and remove the bulb used to illuminate the oxygen sensor light. However there are some models (starting in late 1985) that are equipped with a reset button attached to the light control assembly that is usually located near the pedal assembly. If the vehicle is equipped with a reset button, after the oxygen sensor has been serviced, push the button to reset the maintenance reminder light.

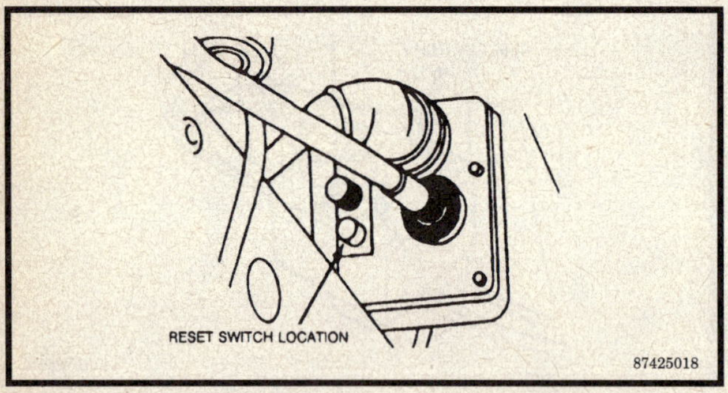

In-line mileage counter and reset mechanism — 1980–83 BMW models

Service Interval Reminder Lights

The on-board computer is used to evaluate mileage, average engine speed, engine and coolant temperatures, as well as other computer input factors that determine maintenance intervals. There are 5 green, 1 yellow and 1 red **LED** used to remind the driver of oil changes and other maintenance services.

The green LEDs will be illuminated when the ignition is in the **ON** position and the engine is not running. There will not be as many green LEDs illuminated when maintenance time gets closer. A yellow LED that is illuminated when the engine is running, will indicate maintenance is now due. The red LED will be illuminated when the service interval has been exceeded by approximately 1000 miles.

There is a service interval reset tool manufactured by the Assenmacher Tool Company tool 62–1–100. The tool is used to reset BMW 6 cylinder models from 1983 and the 4 cylinder models from 1984. With the aid of an additional adapter, the tool can also be used on the 1988–93 models. There is also a means to reset the interval lights without the use of the special reset tool. Both reset procedures are as follows:

RESET

With Reset Tool

1. Locate the diagnostic connector near the thermostat housing.
2. Plug the special reset tool into the diagnostic connector and place the ignition switch in the **ON** position.
3. Depress the reset button on the tool until all 5 green **LEDs** are illuminated, showing that the reset has occurred.

Without Reset Tool

NOTE: This is not a factory authorized procedure; if the wrong diagnostic connector terminal is used, it may result in internal control unit (computer) damage.

4. Turn off all the electrical accessories. Locate the diagnostic connector near the thermostat housing.
5. Turn the ignition switch to the **ON** position and using a volt/ohmmeter (10 megohm impedance is recommended) check to see if there is 5 volts present at the No. 7 (blue/white wire) terminal of the diagnostic connector.

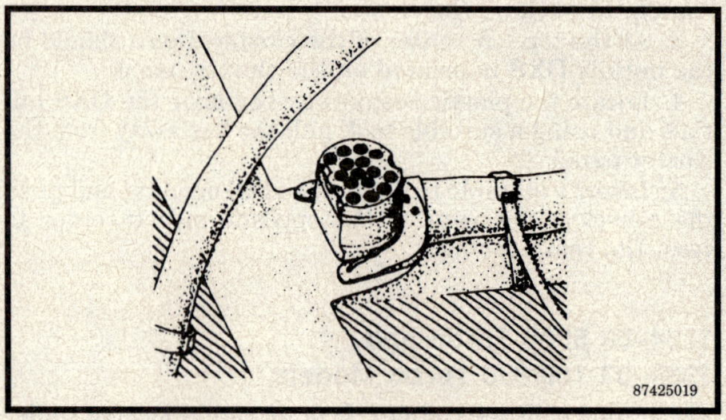

Diagnostic test terminal — BMW

Maintenance reminder light reset tool — BMW

6. Using a non-self powered test light, connect the negative lead to a good ground and the positive lead to the No. 7 (blue/white wire) terminal of the diagnostic connector.

7. Turn the ignition switch to the **ON** position. Ground the terminal through the volt/ohmmeter for 12 seconds. All 5 green **LEDs** should be illuminated, showing that the reset has occurred.

NOTE: There is also a means to make a self made version of the reset tool. The version is made by soldering a small momentary contact switch to 2 ten foot lengths of 20 gauge standard wire. Install the switch into a small canister (such as a film canister). Solder 2 small probe connectors to the opposite ends of the wire. To use the tool, plug it into the proper diagnostic connector terminal (No. 7 blue/white wire) and use it just as the real reset tool is used. It must be noted that this is not a factory authorized procedure and if the wrong diagnostic connector terminal is used, it may result in internal control unit (computer) damage.

Check Engine Light

1988–95 Models

Except 750iL and 850i

The fault codes are cleared after the problem has been rectified and the engine has been started 5–10 times.

750iL and 850i

Codes can be cleared using the BMW tester cancel code command, if the BMW tester is available. The best way to cancel codes without the tester is to reactivate code sequence with the ignition switch **ON**. While flashing Code

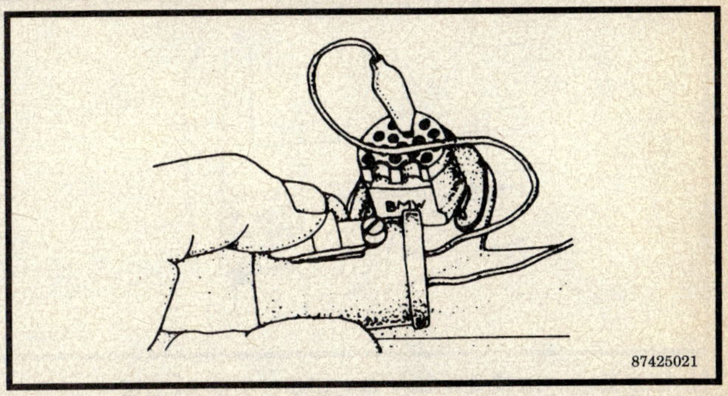

Equivalent version of the maintenance reminder light reset tool — BMW

x000 is being displayed, close the full load switch for 10 seconds. Disconnecting the DME control wiring harness will also clear the codes. Do not disconnect the battery in attempt to clear the codes as this may not completely clear the codes and all the chassis functions with a memory feature will be cleared necessitating reprogramming.

CHRYSLER IMPORT

EGR And Oxygen Sensor Light

The Mitsubishi car line imported under the Chrysler nameplate, does not use a maintenance warning light but some of the light trucks and van/wagons do use a maintenance warning light. The warning light will illuminate at approximately 50,000 miles for the EGR system and

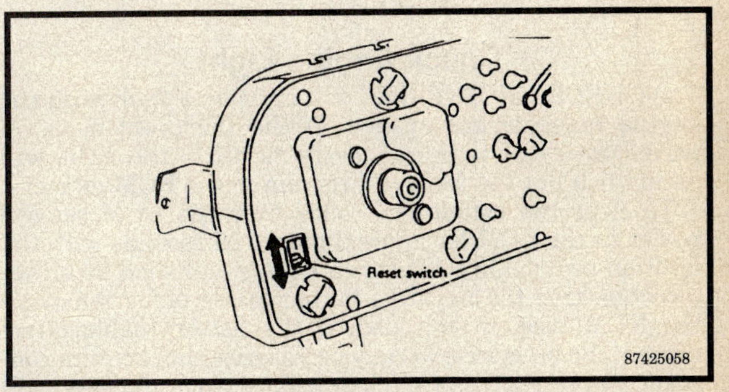

EGR maintenance reset switch — 1986–88 Vista shown

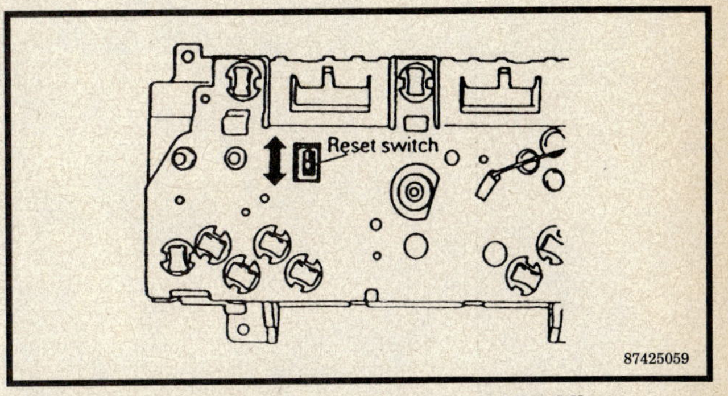

EGR maintenance reset switch — 1989 Vista shown

Using the reset switch on the front of the combination meter to reset the EGR warning system — Ram 50 shown

80,000 miles for the oxygen sensor system. The light is used as a reminder to inspect the EGR emission systems. It is also used as a reminder to check and replace the oxygen sensor. After completing the necessary emission service, the maintenance light can be reset as follows:

1. On some models, the reset switch is located on the back of the instrument cluster near the speedometer junction. After the switch is located, slide the switch knob to the other side to reset the maintenance warning light.
2. On the other models, the reset switch is located on the lower right hand corner of the instrument cluster, behind the instrument cluster face panel. After the switch is located, slide the switch knob to the other side to reset the maintenance warning light.

Remove the warning indicator bulb from the instrument cluster after the vehicle has reached the following mileage points. Colt Wagon and Colt Vista at 150,000 miles and 120,000 miles for Ram 50, Raider Van and Wagon.

Check Engine Light

The CHECK ENGINE light will indicate a fault with the engine management system. If the the problem is repaired or if the problem ceases to occur the light will extinguish but the fault will remain in the ECM memory.

To clear the memorized codes from the ECM remove power to the ECM for approximately 15 seconds with the ignition switch **OFF**. This can be accomplished by either disconnecting the negative battery cable or by removing the ECM fuse, if equipped. If the battery cable is removed, the memories for any accessories that require constant power, such as a clock or radio station memory, will need to be reset. After resetting the ECM, drive the vehicle at medium speeds and accelerations to allow the ECM to memorize the necessary operating parameters for proper operation.

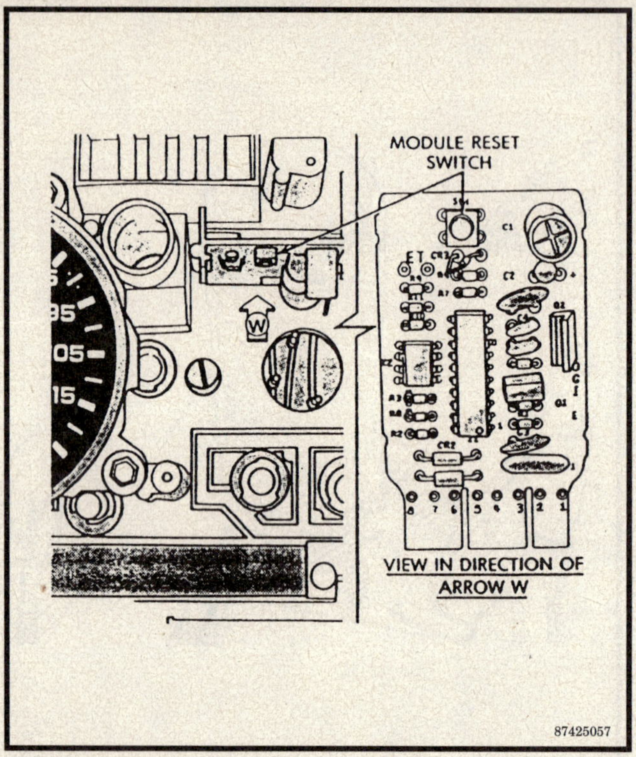

Maintenance required reset switch — 1988 Caravan shown

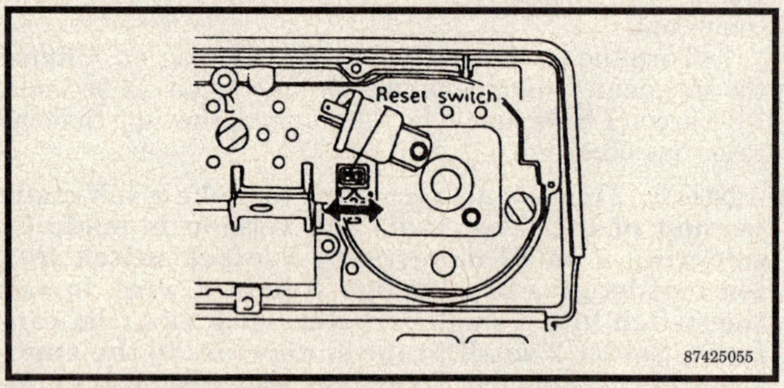

Use the reset switch at the rear of the combination meter to switch off the warning light — 1988 Raider shown

DAIHATSU

Oxygen Sensor Maintenance Light

An oxygen sensor maintenance light is used on the Daihatsu Rocky. A switch is geared to the odometer, tripping at 80,000 miles, turning on a red **O2S** light in the lower left corner of the instrument cluster. The factory recommends oxygen sensor replacement at 80,000 miles. At this time the maintenance reminder light bulb is to be permanently removed. Use the following procedure:

1. Disconnect the battery negative cable and remove the steering wheel assembly.
2. Remove the lower instrument panel trim panel by removing the 6 screws.
3. Remove the upper instrument panel trim panel by removing the 4 screws.
4. Remove the instrument panel cluster trim panel by removing the 4 screws.
5. Remove the oxygen sensor indicator bulb and discard it.

To assemble:

NOTE: When installing the screws, do not overtighten screws to prevent cracking plastic parts.

6. Install the instrument panel cluster trim panel.
7. Install the upper then lower instrument panel trim panel. Then install the steering wheel assembly. Reconnect battery negative cable.

Check Engine Light

When a temporary fault occurs, the **CHECK ENGINE** light illuminates until the ignition is switched **OFF**. When the ignition switch is switched **ON** again the light will be extinguished provided that the fault does not occur again.

FIAT

Oxygen Sensor Maintenance Light

All fuel injected Fiat models are equipped with an oxygen sensor maintenance light, located in the dash panel that will illuminate every 30,000 miles, indicating that the oxygen sensor must be replaced. After the sensor has been replaced and the emission system checked, reset the maintenance light as follows:

1. The reset switch is located in the following positions:
   a. On the Brava, the switch is located behind the left side of the dash panel.
   b. On the Strada, the switch is located under the center of the dash panel in between the glove box and the radio.
   c. On the Spider 2000, the switch is located under the left side of the dash panel, above the accelerator pedal between the radio and glove box.
   d. On the X19 (Bertone), the switch is located behind the right side of the center console.
2. In order to reset the reminder switch, the switch retainer wire must be cut and the retainer screw removed.
3. Insert a suitable tool through the housing and press the contact switch so as to reset the reminder light.
4. After the switch has been pressed, check that the light has gone out. Install retainer screw and wire.

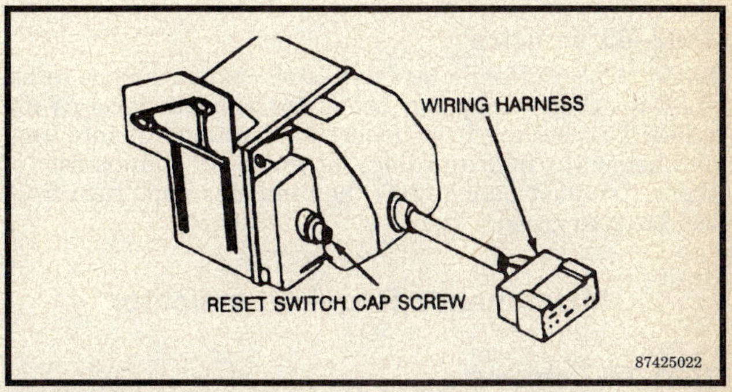

Resetting the oxygen sensor maintenance reminder — Fiat

GEO

Malfunction Indicator Lamp

1989–93 Prizm and Storm

The Prizm is equipped with a malfunction indicator lamp. The malfunction indicator lamp illuminates as **CHECK ENGINE**. To reset the light after a malfunction is repaired, it is necessary to disconnect the ECM power. To do this, remove the ECM fuse, or disconnect the ECM power connector, located under the dash for a period of at least 30 seconds.

1989–93 Tracker

Federal/49 State vehicles are equipped with an emission reminder light system that will illuminate every 50,000 miles for the PCV and EGR systems, 80,000 miles for the oxygen sensor and 100,000 miles for the charcoal canister service. The light indicates the emission system and components should be checked for proper operation and the oxygen sensor should be replaced.

1. To reset the emission reminder light, a 3 wire cancellation switch must be used. The switch is located on the main wire harness behind the instrument cluster or behind the access panel below the steering column.
2. To reset, after locating the switch, push the cancel switch knob left or right to the opposite position until the **CHECK ENGINE** light goes out with the engine running in order to reset the emission light. Start the engine to be sure the light is out and has been reset.

1994–95 Prizm, Metro and Tracker

All models are equipped with a malfunction indicator lamp. The malfunction indicator lamp illuminates as **CHECK ENGINE**. If the malfunction goes away or is repaired, the light will go out but a corresponding Diagnostic Trouble Code (DTC) will remain in the ECM memory.

HONDA

Service Interval Flags

1982–85 Vehicles

Some 1982–85 Honda models use oil change service interval flags that will appear below the odometer every 7500 miles. To reset the flags, insert the ignition key into each slot below the indicator flags, at the lower right corner of the instrument panel, until the reminder flags turn from red back to green.

Maintenance Required Indicator

1991–95 Vehicles

Vehicles equipped with a Maintenance Reminder indicator will indicate it is time for scheduled maintenance. When it is near 7,500 miles (12,000 km) since the last maintenance, the indicator will turn yellow. If you exceed 7,500 miles (12,000 km), the indicator will turn red. The indicator can be reset by inserting the ignition key or other similar object into the slot below the indicator. This will extinguish the indicator for the next 7500 miles.

Oxygen Sensor Life Indicator light

The 1994–95 Passport DX model is equipped with and Oxygen Sensor Life Indicator light. The oxygen sensor must be replaced after 90,000 miles of vehicle operation. When the odometer reading reaches 90,000 miles, the oxygen sensor life indicator light (O2) will illuminate to remind the driver to change the oxygen sensor.

After replacing the oxygen sensor, the oxygen sensor life indicator light must be reset to remind the driver to replace the oxygen sensor after the next 90,000 miles. The reset screw is located in the back of the instrument cluster. Perform the reset procedure as follows:

1. Remove the Instrument panel cluster assembly.
2. Remove the masking tape from hole B.
3. Remove the screw from hole A and install it to hole B.
4. Apply new masking tape to hole A.

NOTE: The above procedure assumes that the oxygen sensor is being replaced for the first time (after 90,000 miles). For subsequent reset procedure (at next 90,000 miles), hole positions will be the opposite of the above procedure.

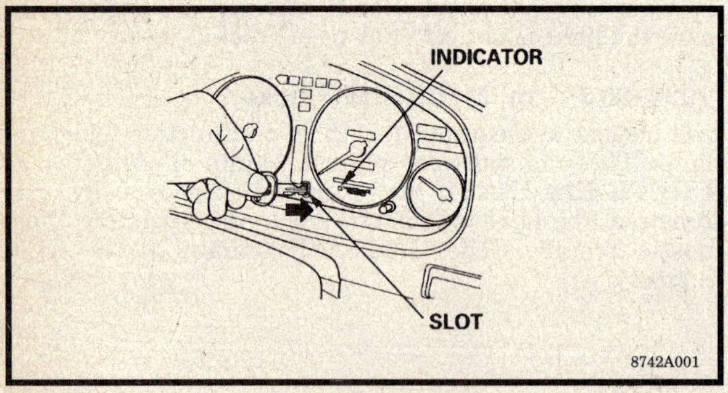

Resetting the Maintenance Required Indicator — 1994–95 Honda Accord shown

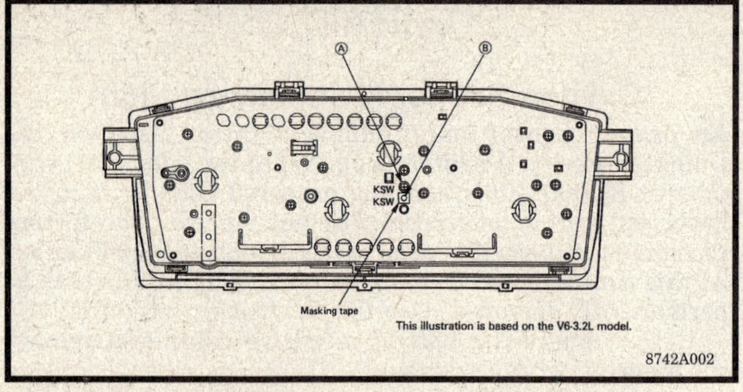

Resetting the Oxygen Sensor Life Indicator — 1994–95 Honda Passport DX

Check Engine Light

The CHECK ENGINE light will indicate a problem with the engine management system. If the the problem is repaired or if the problem ceases to occur the light will extinguish with the next ignition switch cycle. Although the dash cluster CHECK ENGINE light will extinguish, the LED on the ECU will continue to flash. To reset the ECU pull the fuse that supplies power to the ECU for 10 seconds with the ignition **OFF**. The 1988–89 Accord and the 1988–93 Prelude use the CLOCK fuse to reset the ECU. The 1990–93 Accord use the BACK UP fuse in the underhood relay box to reset the ECU and the 1988–93 Civic, CRX and delSol use the HAZARD fuse.

HYUNDAI

Check Engine Light

The CHECK ENGINE light will indicate a fault with the engine management system. If the the problem is repaired or if the problem ceases to occur the light will extinguish but the fault will remain in the ECU memory. To reset the ECU disconnect the negative battery cable for a minimum of 15 seconds and reconnect. Road test the vehicle at medium speeds and accelerations to allow the ECU to memorize the operating parameters required for proper operation.

INFINITI

Check Engine Light

The CHECK ENGINE light is also referred as a Malfunction Indicator Lamp (MIL). This light is located in the dash and will illuminate as **CHECK**. The CHECK ENGINE light will indicate a fault with the engine management system. If the the problem is repaired or if the problem ceases to occur, the light will extinguish but a fault code will remain in the PCM memory.

To clear the memorized codes from the PCM, the is PCM must placed in diagnostic test mode **2** and changed from test mode **2** to test mode **1**, trouble codes are then erased from the back-up memory in the PCM.

Disconnect the negative battery terminal. Wait at least 10 seconds and replace the terminal. When the battery cable is removed, the memories for any accessories that require constant power, such as a clock or radio station memory, will need to be reset.

NOTE: When the battery terminal is disconnected, the trouble code will be lost from the back-up memory within 24 hours.

WARNING

Do not erase stored memory before performing self-diagnostic procedure.

The CHECK ENGINE light may also come **ON** if the fuel tank is extremely low of fuel or for something as minor as a fuel tank cap not being tighten securely. In any event to avoid any damage to the engine or transmission control systems, when the light comes **ON**, the vehicle should be checked as soon as possible. Although the vehicle is still driveable, to avoid system damage, do not drive at speeds above 45 MPH and avoid hard acceleration or deceleration if the light is **ON**.

ISUZU

Oxygen Sensor Maintenance Light

The 1988–95 Amigo, Pick-Up, Trooper II, Trooper and Rodeo are equipped with an oxygen sensor maintenance light, located in the dash panel, that will illuminate every 90,000 miles, indicating that the oxygen sensor must be replaced. After the sensor has been replaced and the emission system checked, reset the maintenance light as follows:

1. Reset the reminder light on the 1988–91 Trooper II and 1990–91 Trooper by removing the speedometer and then sliding the reset switch on the back of the speedometer to the opposite end of its slot.
2. To reset the reminder light on the 1988–95 Amigo, Pick-Up, Rodeo and 1992–95 Trooper, proceed as follows:
   a. Remove the instrument cluster assembly.
   b. Working on the backside of the instrument cluster, remove the masking tape over hole **B** in the instrument cluster.
   c. Remove the screw from hole **A** of the instrument cluster and place that screw in hole **B** of the instrument cluster.
   d. After switching the screw from hole **A** to hole **B**, be sure to place a piece of masking tape over hole **A** of the instrument cluster.

NOTE: At the next 90,000 mile interval the screw hole positions will be the opposite of the previous replacement.

Check Engine Light

The CHECK ENGINE light will indicate a fault with the engine management system. If the the problem is repaired or if the problem ceases to occur, the light will extinguish but the fault will remain in the ECM memory. To clear the memorized codes from the ECM, remove power to the ECM for approximately 30 seconds. This can be accomplished by either disconnecting the negative battery cable or by removing the ECM fuse, if equipped. If the battery cable is removed, the memories for any accessories that require constant power, such as a clock or radio station memory, will need to be reset. After resetting the ECM, drive the vehicle at medium speeds and

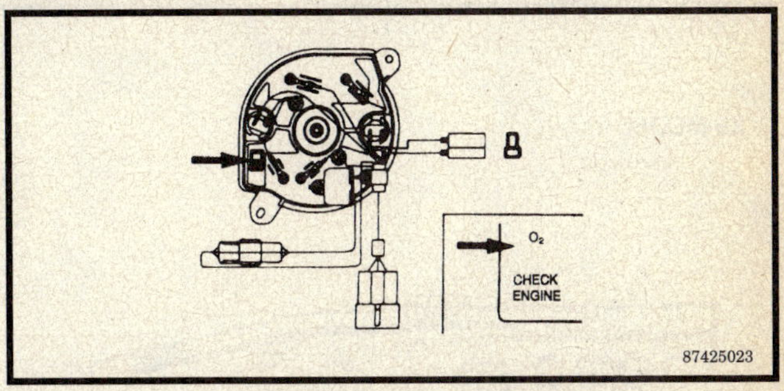

Resetting the oxygen sensor reminder light switch located in back of the speedometer — 1988–91 Trooper II and 1990–91 Trooper — Isuzu

accelerations to allow the ECM to memorize the necessary operating parameters for proper operation.

JAGUAR

Catalytic Converter and EGR Light

The Catalytic Converter/EGR warning light will illuminate at the 25, 000 mile mark as a reminder for service. The mileage mechanism is inline with the speedometer cable. It is a mechanical reduction device and incorporates a magnetic operated reed switch.

The reminder switch may be reset by inserting a special reset key into the reset switch. Turn the key clockwise to "0". The reset switch is accessible through the glow box.

Oxygen Sensor Maintenance Light

All fuel injected Jaguar models are equipped with an oxygen sensor maintenance light, located in the dash panel, that will illuminate every 30,000 miles, indicating that the oxygen sensor must be replaced. After the sensor has been replaced and the emission system checked, reset the maintenance light as follows:

1. The reset switch will be located in the following position:
   a. The 1980–83 XJ6 has the maintenance reminder reset mileage counter located on the right side of the engine, in front of the firewall in the engine compartment.

NOTE: On some of the 1980–83 models, the maintenance reminder reset mileage counter may be hidden on either side of the engine or behind the glove box.

   b. The 1983–93 XJ6 and XJS models use a motorized maintenance reminder reset mileage counter, which is located inside the trunk compartment behind the trim panel.
2. Reset the mileage counter on the 1980–83 models by using the special key No. BLT–5007 or using 2 small pins to turn the reset disc on the mileage counter. Start the engine and check to see if the light has been turned off.
3. Reset the mileage counter on the 1983–93 XJ6 models by removing the trim panel at the head of the trunk to uncover the mileage counter. With the ignition switch in the **ON** position, depress the white button to reset the light.
4. Reset the mileage counter on the 1983–93 XJS models, by removing the trim panel from the left rear quarter panel in front of the wheel well to uncover the mileage

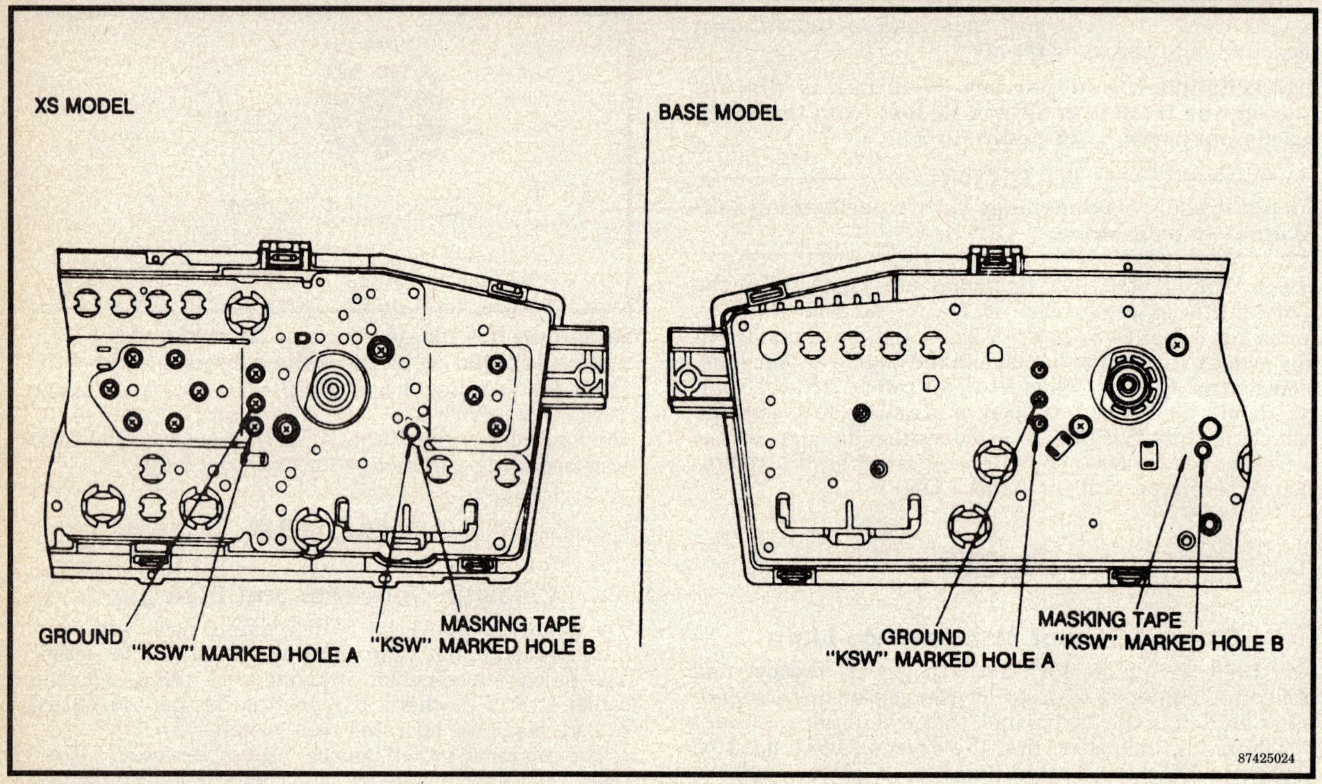

Resetting the oxygen sensor reminder light — Pick-Up models — Isuzu

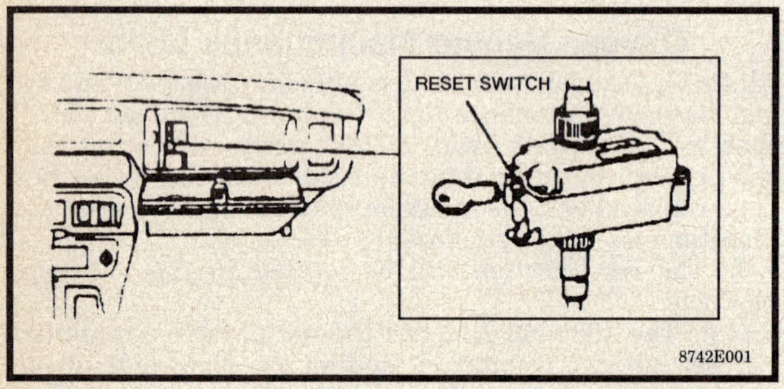

Resetting the maintenance mileage counter for the catalytic converter and EGR reminder lights — Jaguar

counter. With the ignition switch in the **ON** position, depress the white button to reset the light.

Check Engine Light

As of 1986 some vehicles were equipped with a CHECK ENGINE light. By 1988 all vehicles were. The CHECK ENGINE light will illuminate if the ECM detects a fault with the engine management system. If the the problem is repaired or if the problem ceases to occur, the light will extinguish but the fault will remain in the ECM memory.

LAND ROVER

Maintenance Reminder Light

Vehicles equipped with Maintenance Indicator Reminder light (MIL), will illuminate indicating its time for a service interval or there is a problem in the engine management system. After service is completed, reset the maintenance light as follows.

1987–89 Range Rover

1. Locate the green or tan box below the steering column.
2. Peel back the sticker labeled "Maintenance Light" or "Maintenance Reminder Sticker". Use a small paper clip or other suitable tool and press in on the reset button.
3. The light is now reset. Start the engine to ensure the light is out.

1991–95 Range Rover

1. Locate the green or tan box under the passenger seat.
2. Peel off the sticker "Maintenance Light" or "Maintenance Reminder Sticker". Use small paper clip or other suitable tool and press in on the reset button.
3. The light is now reset. Start the engine to ensure the light is out.

Check Engine Light

As of 1995 the Maintenance Reminder Light is no longer used in Land Rover vehicles. However all vehicles from 1987–95 equipped with a CHECK ENGINE light, will illuminate should a fault in the engine management system occur. If the the problem is repaired or if the problem ceases to occur, the light will extinguish but the fault will remain in the ECU memory. After the cause of the fault has been rectified, the ECU memory should be cleared.

LEXUS

Check Engine Light

The CHECK ENGINE light is also referred as a Malfunction Indicator Lamp (MIL). This light is located in the dash and will illuminate as **CHECK**. The CHECK ENGINE light will indicate a fault with the engine management system or transmission electrical circuit. If the the problem is repaired or if the problem ceases to occur, the light will extinguish but the fault will remain in the ECM memory.

To clear the memorized codes from the ECM, remove power to the ECM for approximately 30 seconds with the ignition switch **OFF**. This can be accomplished by either disconnecting the negative battery cable or by removing the ECM fuse, if equipped. If equipped with the fuse for the ECM, it will be marked **EFI**. If the battery cable is removed, the memories for any accessories that require constant power, such as a clock or radio station memory, will need to be reset. After resetting the ECM, drive the vehicle at medium speeds and accelerations to allow the ECM to memorize the necessary operating parameters for proper operation.

The CHECK ENGINE light may also come **ON** if the fuel tank is extremely low of fuel or for something as minor as a fuel tank cap not being tighten securely. In any event to avoid any damage to the engine or transmission control systems, when the light comes **ON**, the vehicle should be checked as soon as possible.

MAZDA

Timing Belt Warning Light

Diesel Pick-ups

Early model diesel pick-ups were equipped with a timing belt warning light. The light would illuminate at approximately 60,000 miles to indicate that a new timing belt should be installed. After the belt had been inspected or replaced, the warning light should be reset as follows:

1. Locate the 3 wires used to reset the timing belt warning light.
   a. Black wire located near the intermittent wiper control relay.
   b. Black wire with a white tracer located near the intermittent wiper control relay.
   c. Black wire with a yellow tracer located near the intermittent wiper control relay.

NOTE: The intermittent wiper control relay is located under the left side of the instrument panel.

2. Remove the single black wire from the connector of the black wire with the white tracer.
3. Plug the single black wire into the connector of the black wire with the yellow tracer.

NOTE: Whichever connector the single black wire was connected to originally, it must be removed and reconnected to the opposite connector from which it was removed, in order to reset the light.

EGR Warning Light

1986 B2000 Federal Pick-ups
1986–87 B2200 and B2600

The EGR warning light illuminates at the first 60,000 miles to inform the operator that the emission components should be replaced and/or inspected. The suggested replacement components are the oxygen sensor, EGR control valve along with the hose and tube. The suggested inspected components are the PCV system and the ignition timing. The light can be reset as follows:

1. Locate above the left side kick panel, the single terminal green wire that is connected to a single terminal black wire.
2. In order to reset the warning light, disconnect the black wire from the green wire and leave it disconnected. Be sure to tape the disconnected wire to prevent shorting.

NOTE: The connectors are sometimes located on the rear of the combination meter.

Check Engine Light

All 1988–95 vehicles equipped with a CHECK ENGINE light with the exception as noted below, the CHECK ENGINE light will flash indicating a fault with the engine management system. If the the problem is repaired or if the problem ceases to occur the light will extinguish but the fault will remain in the ECM memory. The vehicles specified below require a maintenance reset.

1988–93 Pick-up

The Pick-Up models, in addition to displaying computer system malfunction codes (codes used on California models only), the **CHECK** light on the B2200 and B2600 Pick-Ups will act as an emission maintenance warning light. The **CHECK** engine warning light will come on at 60,000 miles for the EGR system and 80,000 miles for the oxygen sensor system. Once the emission system has been checked and/or the oxygen sensor replaced, reset the warning light as follows:

1. To reset warning light it will be necessary to locate the brown wire with the white tracer, the black wire and the green wires, which are usually under the left side of instrument panel taped to the wiring harness, above the fuse/relay block.

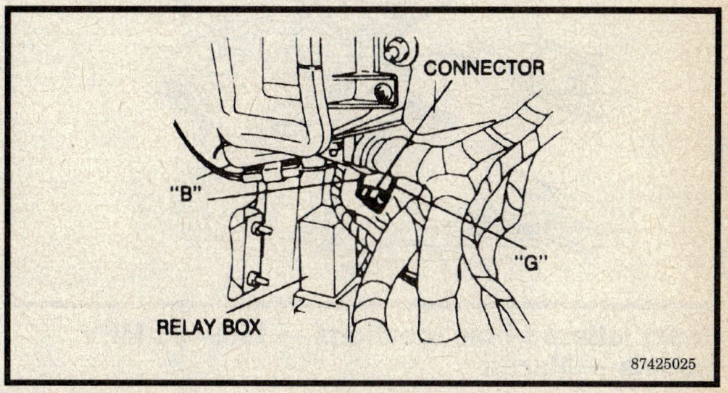

Reset connector location — 1986 B2000 Pick-Ups — Mazda

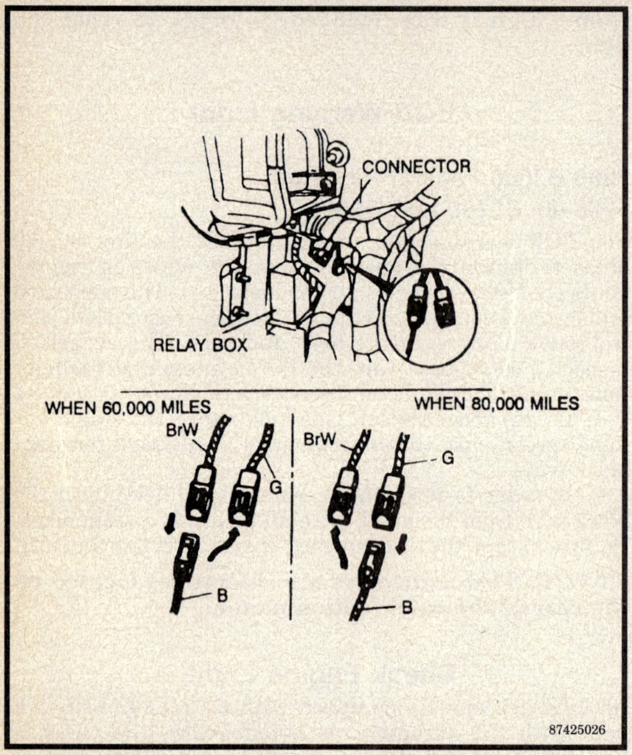

Reset connector location — 1988–93 Pick-Ups, early 626 models similar — Mazda

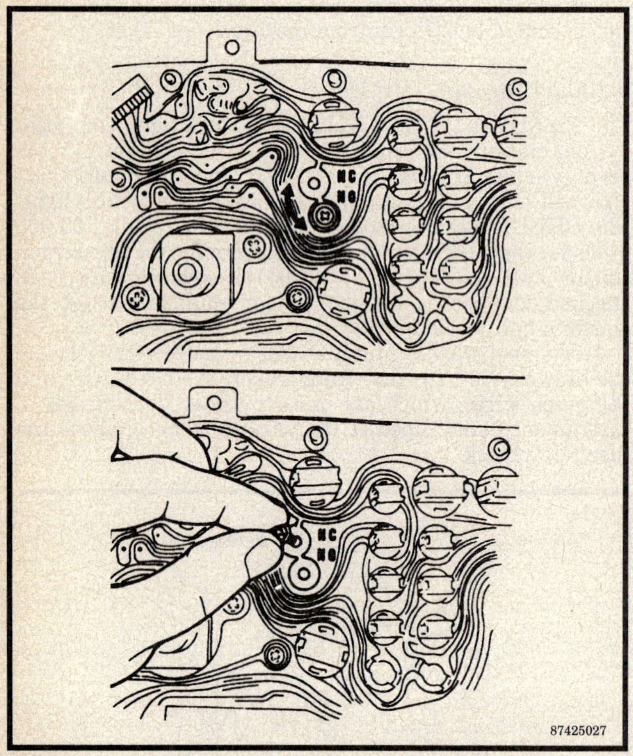

Reset lettered hole locations — 1989–95 MPV models — Mazda

2. When the 60,000 mile interval arrives and the light is illuminated, unplug the black wire connector from the brown and white wire connector and plug it into the green wire connector. When the 80,000 mile interval arrives, return the black wire connector to the brown and white wire connector.

1994–95 Pick-up

The CHECK ENGINE light will indicate a fault with the engine management system. If the the problem is repaired or if the problem ceases to occur the light will extinguish but the fault will remain in the ECU memory.

1989–95 MPV

The MPV models, in addition to displaying computer system malfunction codes used on California models only, the **CHECK** light on the instrument cluster will act as an emission maintenance warning light. The **CHECK** engine warning light will come on at 80,000 miles for the oxygen sensor system maintenance. After the emission system has been checked and the oxygen sensor replaced, reset the warning light as follows:

1. Disconnect the negative battery cable. Remove the instrument panel cluster assembly.
2. Locate the lettered reset holes **NO** and **NC** on the rear of the instrument cluster.
3. Remove the screw from the **NO** lettered reset hole and install it into the **NC** lettered hole. After an additional 80,000 miles, return the screw to the original lettered hole.

MERCEDES-BENZ

Catalytic Converter Warning Light

Vehicles equipped with a Catalytic Converter light will illuminate every 37, 000 miles as a reminder to service the Catalytic Converter. After the converter has been serviced, the light must be reset. Locate the mileage counter inline with the speedometer cable and press the button marked CAT.

Oxygen Sensor Warning Light

1980–85 Models

The models are equipped with an oxygen sensor maintenance warning light that works in junction with the mileage reset counter. The warning light will illuminate when the mileage counter reaches approximately 30,000 miles. The warning light is used as a reminder to check the emission system and to replace the oxygen sensor. After the emission system maintenance has been completed, the mileage counter must be reset. To reset the mileage counter, use the following procedure:

1. On the 280 series vehicles, the mileage counter is usually located in-line with the speedometer cable, under instrument panel. Disconnect the wiring plug from counter and leave it disconnected. No reset switch is provided.
2. To reset the counter on all other series, the instrument cluster must be partially removed. Insert a suitable hooked tool between the right side of cluster and the dash panel.
3. Turn the hook to engage the cluster and pull the cluster out of the spring retaining clips.

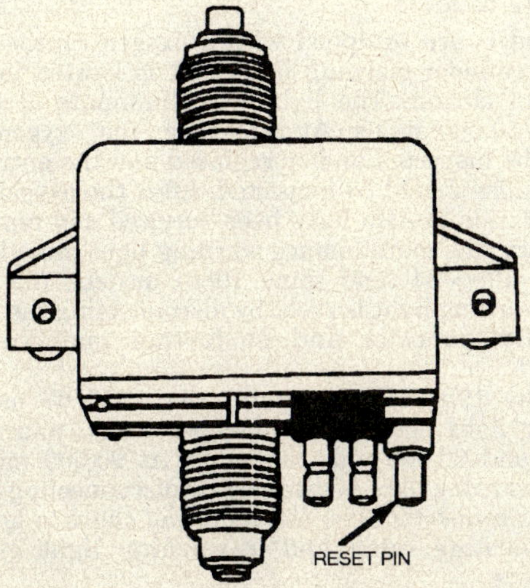

Reset the mileage counter — Mercedes Benz

4. Remove the bulb from the lower corner of the instrument cluster. Press the instrument cluster back into position. There is no reset switch provided.

1986–89 Models

The oxygen sensor light is used as a malfunction indicator for the oxygen sensor circuit. A mileage counter is not used and there is no reset procedure required. Servicing and repairing the oxygen sensor circuit should turn light off. Removing and discarding the oxygen sensor warning light bulb from the warning light in the instrument cluster, will not resolve the emission system problems and may lead to other problems as well.

Check Engine Light

1990–95 Models

The CHECK ENGINE light illuminates to indicate a fault in the engine management system. Once the cause of the fault has been rectified and the codes have been pulled from the ECU memory, the fault codes must be erased. The codes must be erased one at a time as they are read from memory. After the LED blinks the code wait 2 seconds and press the button on the diagnostic connector or the impulse display counter for 6 seconds. The LED will blink once to signal that the code has been erased.

MITSUBISHI

EGR and Oxygen Sensor Light

1983–87 Light Truck, Vans and Wagons

The Mitsubishi car line does not use a maintenance warning light but some of the light trucks, van and wagons do use a maintenance warning light. These warning lights will illuminate as **EGR**. or **OXS** abbreviated for Exhaust Gas Recirculation system or Oxygen Sensor. The warning light will illuminate at approximately 50,000 miles for the EGR system and 80,000 miles for the oxygen sensor system. On the trucks, including the Montero, the light is used as a reminder to inspect the EGR emission systems. On the van and wagon models, the light is used as a reminder to check and replace the oxygen sensor. After completing the necessary emission service, the maintenance light can be reset as follows:

On some models, the reset switch is located on the back of the instrument cluster near the speedometer junction. After the switch is located, slide the switch knob to the other side to reset the maintenance warning light. On the other models, the reset switch is located on the lower right hand corner of the instrument cluster, behind the instrument cluster face panel. After the switch is located, slide the switch knob to the other side to reset the maintenance warning light.

NOTE: After 100, 000 miles, the warning lights bulbs should be removed.

Maintenance Required Light

1988–95 Light Truck, Vans and Wagons

The Mitsubishi car line still does not use a maintenance warning light, however it is still being used in some of the light trucks, van and wagons. This warning light now illuminates as **MAINT. REQD**. abbreviated for Maintenance Required. The warning light will illuminate after 50,000 miles, 80,000 miles, 100.000 miles and on trucks will go as up to 120,000 miles. The maintenance required light is used as a reminder to inspect the emission control system and to perform the required emission service related to the mileage interval. After completing the necessary emission service, the maintenance light can be reset as follows:

The reset procedure and reset switch locations have remained the same as previous vehicles. On some models, the reset switch is located on the back of the instrument cluster near the speedometer junction. After the switch is located, slide the switch knob to the other side to reset the maintenance warning light. On the other models, the reset switch is located on the lower right hand corner of the instrument cluster, behind the instrument cluster face panel. After the switch is located, slide the switch knob to the other side to reset the maintenance warning light.

NOTE: After 100, 000 miles, the warning lights bulbs should be removed.

Check Engine Light

The CHECK ENGINE light will indicate a fault with the engine management system. If the the problem is repaired or if the problem ceases to occur, the light will extinguish but the fault will remain in the ECM memory.

Reset switch locations — 1983–95 Pick-up truck — Mitsubishi

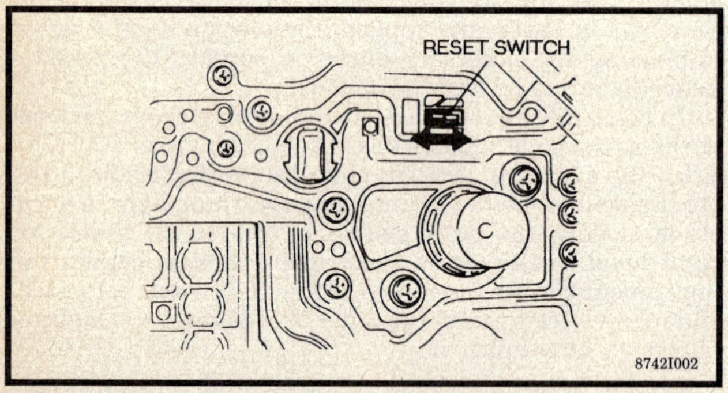

Reset switch locations — 1983–1993 Montero and Van/Wagon — Mitsubishi

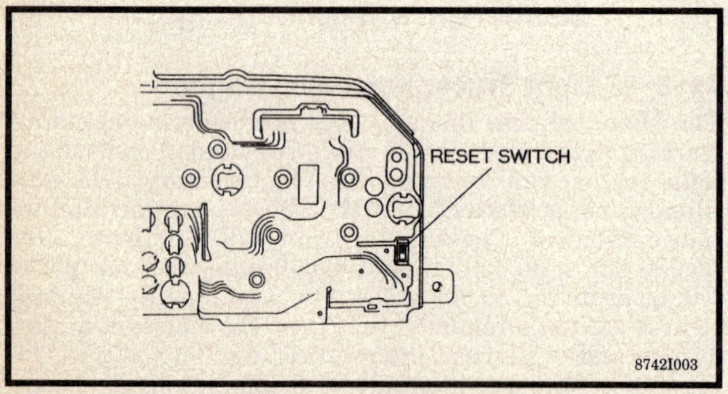

Reset switch locations — 1994–95 Montero and Vans — Mitsubishi

To clear the memorized codes from the ECM, remove power to the ECM for approximately 15 seconds with the ignition switch **OFF**. This can be accomplished by either disconnecting the negative battery cable or by removing the ECM fuse, if equipped. If the battery cable is removed, the memories for any accessories that require constant power, such as a clock or radio station memory, will need to be reset. After resetting the ECM, drive the vehicle at medium speeds and accelerations to allow the ECM to memorize the necessary operating parameters for proper operation.

NISSAN/DATSUN

Oxygen Sensor Light

1980–87 Models

The models are equipped with a oxygen sensor maintenance reminder warning light that is located in the instrument cluster. The light will illuminate at approximately 30,000 miles. At this time, the oxygen sensor should be inspected and/or replaced and the entire emission system should be inspected. After the oxygen sensor and emission system have been serviced and repaired as necessary, the maintenance warning light must be reset. On the 1980–84 (and some 1985) models, the maintenance warning light is reset by disconnecting the the sensor light connector and no further maintenance is required.

On the 1985 Maxima and most 1986–87 models, a warning light hold relay must be located and reset at 30,000 and 60,000 mile intervals. At 90,000 miles, the sensor warning light is disabled by disconnecting the sensor light connector. Use the following chart to locate the reset warning relay and/or warning light electrical connectors.

NOTE: Some models may be equipped with either means of resetting the maintenance warning light.

Warning Light Connector Locations:

- 1980–84 200SX — green and white wire, located under the far right side of the instrument panel.
- 1985–87 200SX — pink and purple wire, located behind the fuse box
- 1982–83 280ZX — green and yellow wire, located under the far right side of the instrument panel.
- 1984 300ZX — white wire connector, located behind the left hand side kick panel.
- 1985–86 300ZX — white wire connector, located above the hood release handle.
- 1987 300ZX — gray and red, gray and blue or yellow wire, located above the hood release handle.

NOTE: On the 1987 300ZX models equipped with the digital instrument panel or the analog dash, at 30,000 mile intervals, unplug one of the 3 connectors located behind the glove box.

- 1980–84 Maxima and 810 Maxima — yellow and blue wire, located near the hood release handle
- 1985–87 Maxima — green and red, green and white, located near the hood release handle.

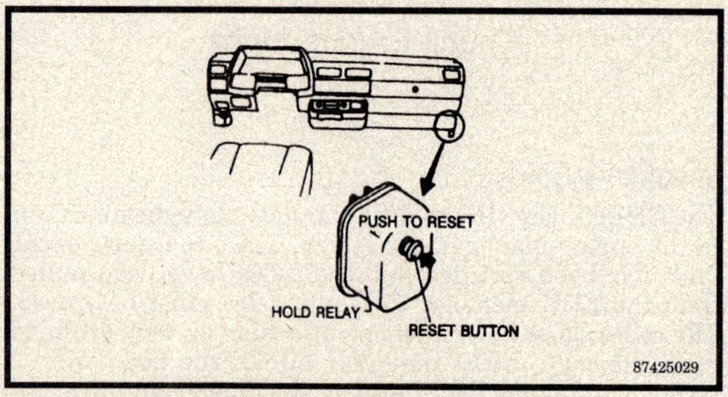

Resetting the oxygen sensor warning light relay — 1985 Pulsar and Sentra — Nissan

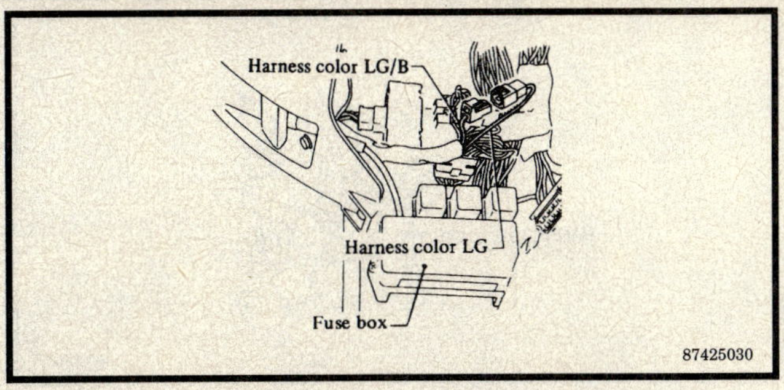

Disconnecting the oxygen sensor warning light connector — 1983–86 Pulsar and Sentra — Nissan

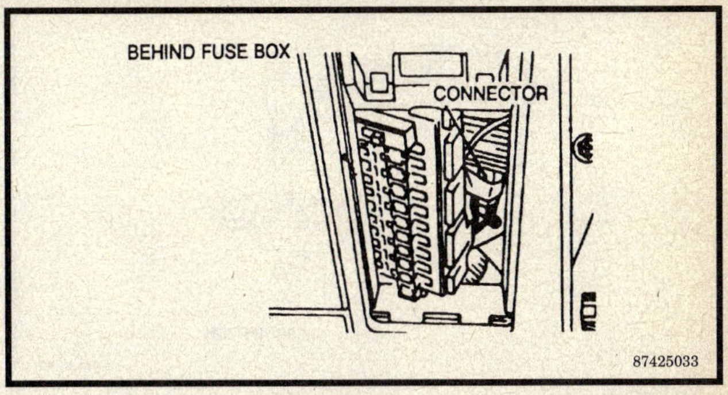

Disconnecting the oxygen sensor warning light connector — 1985–87 200SX — Nissan

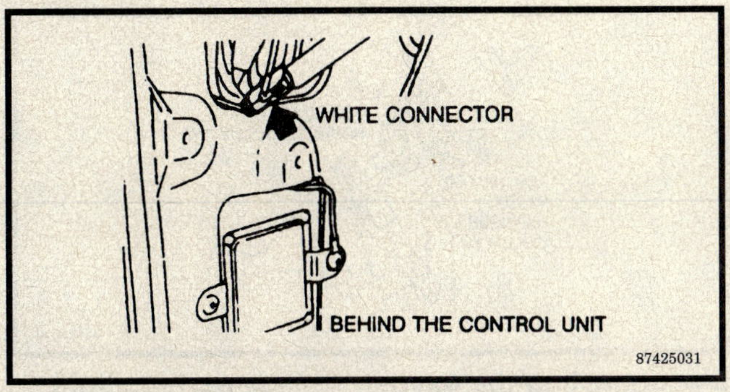

Disconnecting the oxygen sensor warning light connector — 1984–87 Stanza — Nissan

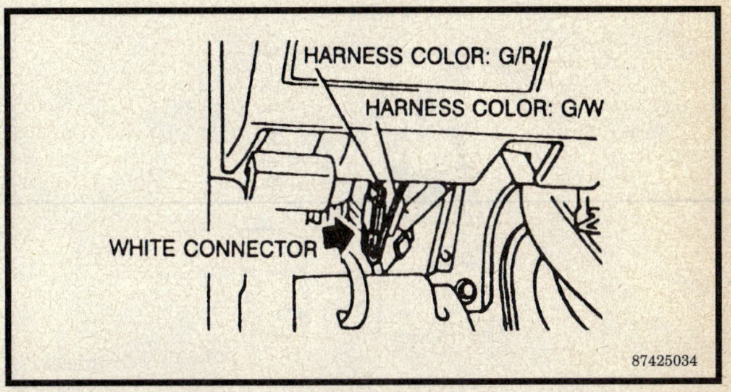

Disconnecting the oxygen sensor warning light connector — 1987 Maxima — Nissan

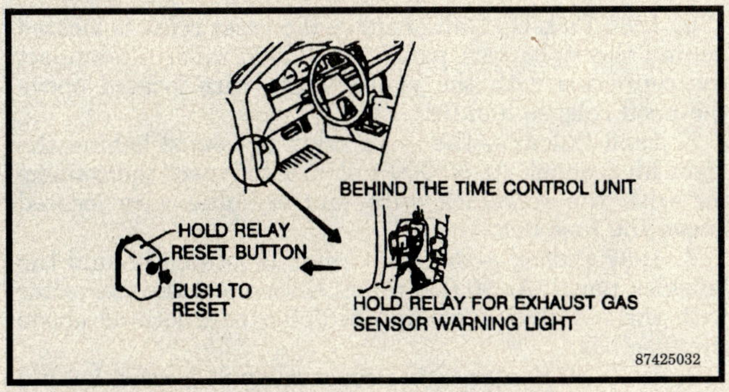

Resetting the oxygen sensor warning light relay — 1985–86 Maxima — Nissan

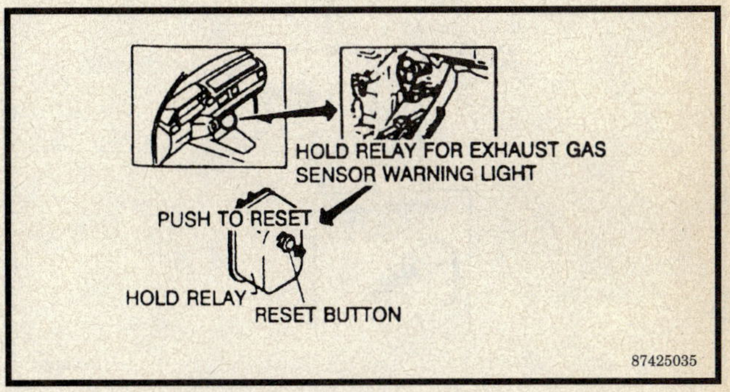

Resetting the oxygen sensor warning light relay — 1985–87 200SX — Nissan

- 1980–84 Pick-Up — yellow and white wire, located near the hood release handle.
- 1983–86 Pulsar — light green and black with a light green tracer or black and white wires, located near the fuse box
- 1987 Pulsar — red and black, red and blue wires, located above the fuse box.
- 1982–83 Sentra — green and yellow or green and black wires, located above the fuse box.
- 1984 Sentra — light green and black with a light green tracker wires, located near the hood release handle.
- 1985–86 Sentra — light green and black wires, located above the fuse box.
- 1987 Sentra — red and black, red and blue wires, located above the fuse box.
- 1984–86 Stanza — yellow and red or yellow and green wires, located behind the left side kick panel.
- 1987 Stanza — green and brown wires, located above the fuse box.
- 1986–87 Stanza Wagon — red and yellow or red and blue wires, located behind the instrument panel.

Warning Light Relay Locations:

1. 1985–87 200SX — the reset relay is located under the center of the instrument panel behind the console. At 90,000 miles, disconnect the pink wire from the purple wire located behind the fuse box.
2. 1986–87 300ZX — the reset relay is located near the glove box. At 90,000 miles, disconnect the white connector

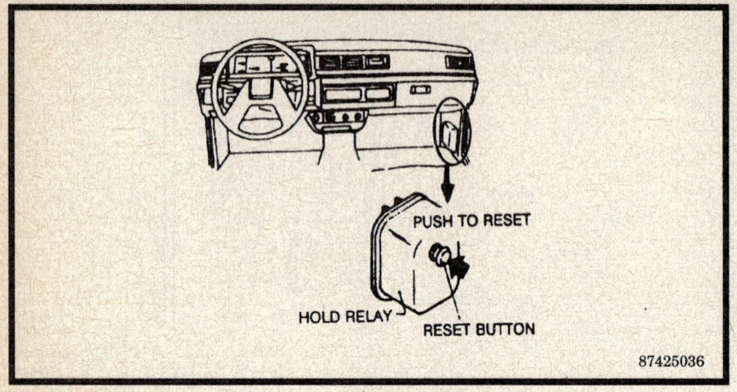

Resetting the oxygen sensor warning light relay — 1986–87 Stanza — Nissan

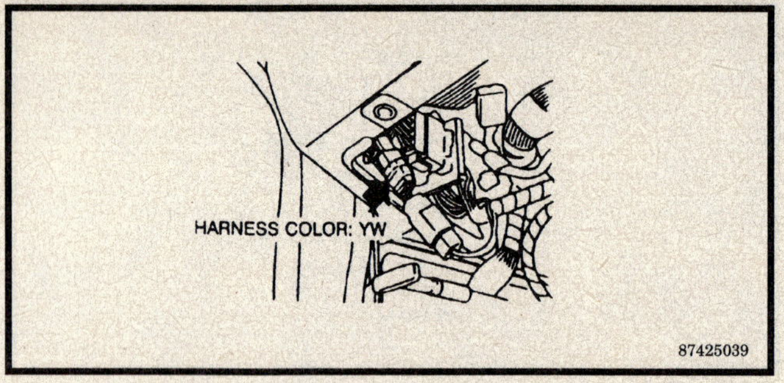

Disconnecting the oxygen sensor warning light connector — Pick-Up with California emissions — Nissan

Disconnecting the oxygen sensor warning light connector — 1985–87 300ZX — Nissan

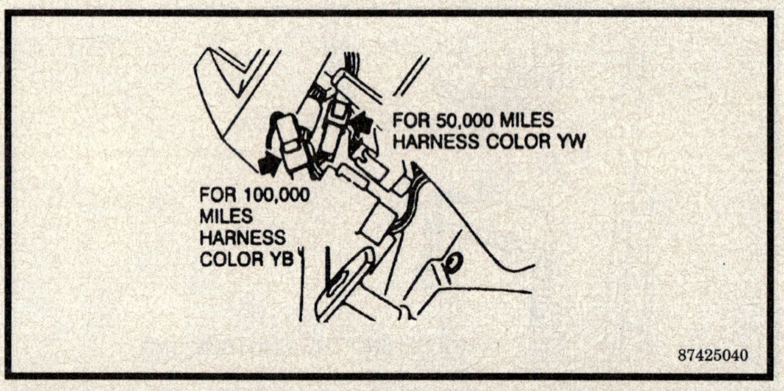

Disconnecting the oxygen sensor warning light connector — Pick-Up with Federal emissions — Nissan

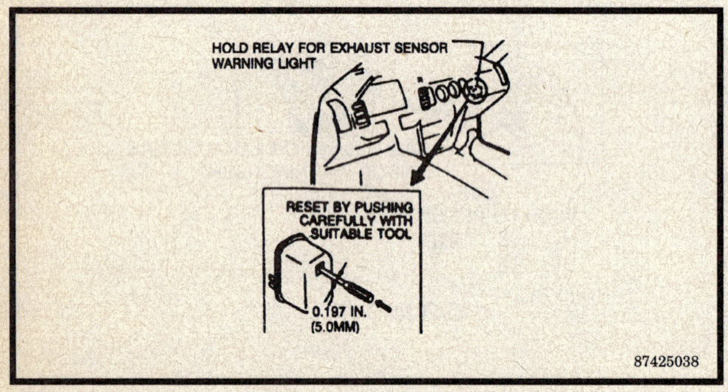

Resetting the oxygen sensor warning light relay — 1986–87 300ZX — Nissan

with the gray/red wires and gray/blue wires located above the hood release handle.

NOTE: On the 1987 300ZX models equipped with the digital instrument panel or the analog dash, at 30,000 mile intervals, unplug one of the 3 connectors located behind the glove box.

3. 1985–86 Maxima — the reset relay is located behind the left kick panel. At 90,000 miles, disconnect the green/red wire from the green/white wire located near the hood release handle.

4. 1987 Maxima — the reset relay is located behind the right kick panel. At 90,000 miles, disconnect the green/red wire from the green/white wire located near the hood release handle.

5. 1986 Pick-Up California — the reset relay is located behind the right kick panel. At 90,000 miles, disconnect the connector with the yellow/white wire located above the hood release handle.

6. 1986 Pulsar — the reset relay is located behind the right kick panel. At 90,000 miles, disconnect the connector with the red/black wire and red/blue wire located above the fuse box.

7. 1987 Pulsar — the reset relay is located behind the left kick panel. At 90,000 miles, disconnect the connector with the red/black wire and red/blue wire located above the fuse box.

8. 1985–86 Sentra — the reset relay is located behind the right kick panel. At 90,000 miles, disconnect the connector with the light green and black wire located near the hood release handle.

9. 1987 Sentra — the reset relay is located behind the right kick panel. At 90,000 miles, disconnect the connector with the red/black wire and red/blue wire located above the fuse box.

10. 1986–87 Stanza — the reset relay is located behind the right kick panel. At 90,000 miles, disconnect the white connector on the inside of the left kick panel on the 1986 models. On the 1987 models, disconnect the connector with the green and brown wire located above the fuse box.

11. 1986–87 Stanza Wagon — the reset relay is located under the right side passenger seat. At 90,000 miles, disconnect the red wire from the yellow wire or the red wire from the red/blue wire located behind the instrument panel.

1988–95 Models

After 1987 maintenance resets where no longer used. The CHECK ENGINE light will come on indicating a malfunction and a trouble code will be stored in the ECU memory. The light will extinguish when the problem is corrected.

Check Engine Light

As of 1986, some models used the CHECK ENGINE light, which will illuminate when a fault occurs with the engine management system. By 1988 all models used it. The fault code will be memorized by the ECU for recall at a later time. If the fault does not occur again, the memory of that fault will be automatically erased after 50 ignition switch cycles. If the memory is to be erased manually before the 50 cycles, turning the ignition switch **OFF** during the diagnosis mode of memory recall will erase the memory.

PEUGEOT

EGR Warning Light

All 1980 604 models are equipped with an **EGR** warning light. The light will illuminate approximately 12,500 mile intervals as determined by a in-line mileage counter. When the light illuminates, it will be necessary to perform maintenance on the EGR system. After completing the necessary maintenance on the EGR system, it will be necessary to reset the mileage counter in order to reset the warning light. The mileage counter can be reset as follows:

1. In order to reset the mileage counter, it will be necessary to unbolt the mileage counter from the inside of the left front wheelwell. Pull down the mileage counter without disconnecting the speedometer cables.
2. Remove the outer and inner covers from the mileage counter and turn the reset button counterclockwise until reaching a stopping point. This will reset the counter to zero.
3. Be sure the warning light is out. Replace the covers and reinstall the mileage counter.

EGR and Oxygen Sensor Light

1981–85 505 Models

Vehicles equipped with a EGR light will illuminate every 12, 000 mile interval indicating the EGR system needs servicing. After the service is completed, the mileage counter must be reset. Same as applies to the oxygen sensor light.

All the 505 models equipped with gasoline engines have an oxygen sensor warning light. The warning light will illuminate at approximately 30,000 mile intervals. When the warning light is illuminated, it is an indication that the oxygen sensor must be replaced and that the mileage counter must be reset. By resetting the mileage counter the warning light will be turned off. The mileage counter can be reset as follows:

1. The mileage counter is located below the brake master cylinder. In order to gain access to the reset button, the white plastic cover must be removed and then unscrew the reset plug.
2. After the reset plug has been removed, use a suitable small punch or rod to press the reset button. Be sure the warning light is out, then replace the reset plug and cover.

1986–92 Models

After 1986 maintenance resets where no longer used. The CHECK ENGINE light will come on indicating a malfunction and a trouble code will be stored in the ECU memory. The light will extinguish when the problem is corrected.

Check Engine Light

To erase the fault memory after the CHECK ENGINE light has indicated a fault and the fault has been corrected, call up the fault codes using the appropriate tester until Code 11 appears with the ignition switch in the **RUN** position. Press the tester button until 00 is displayed. The codes will be reset at this point.

PORSCHE

Oxygen Sensor Warning Light

All models are equipped with an oxygen sensor circuit. All models, except the 928S with the LH Jetronic system, the 944S and the 968 models, are equipped with an **OXS** light that will illuminate at approximately 30,000 mile intervals as a reminder to replace the oxygen sensor. The 928S with the LH Jetronic system, the 944S and the 968 models do not use a warning light system. It should be noted that even though these models do not use the system, it is recommended that oxygen sensor be replaced every 60,000 miles. On all other models, after replacing the oxygen sensor and checking the emission systems, reset the warning lights as follows:

NOTE: The 1980–83 911 models also used an EGR maintenance reminder warning light system. The light is reset in the same manner as the OXS light.

1. On the 911 and early 930S models, the **EGR** and **OXS** reset buttons were both mounted behind the speedometer. Disconnect the negative battery cable and remove the speedometer.

NOTE: On the 1980–83 911 and early 930S models, the speedometer head is held in place by a ribbed rubber collar. Use caution and gently pry the speedometer head loose in order to gain access to the reset buttons. On the later models, the reset button is visible through the speedometer mounting hole.

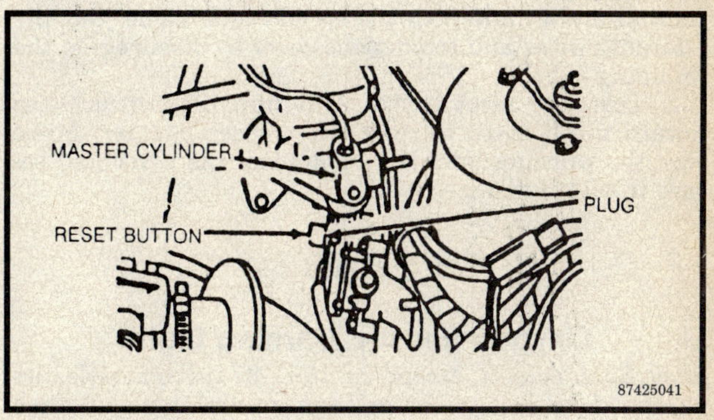

Resetting the mileage counter — Peugeot

2. Use a suitable piece of wire or rod to press the white reset button. Push the reset button all the way in against its stop. Reconnect the battery cable and check that the warning light is out.

NOTE: The procedure also applies to the early 930S models.

3. On the 924 models, after replacing oxygen sensor and with the vehicle still raised and supported safely, located the mileage counter which is usually located on the left engine mount.

NOTE: On the 1980–82 924S turbo models, the mileage counter is located on the engine compartment side of the left strut tower. It may be necessary to follow the speedometer cable from the firewall downward, to be able to locate the mileage counter.

4. Use a suitable piece of wire or metal rod to push in the reset button. Be sure to push the button in all the way to its stop. Make sure the **OXS** light is out. Lower the vehicle.

5. On the 928 and 928S models with the CIS system, the mileage counter is usually located at the right of the passenger seat floor panel under a trim plate between the seat and the step plate.

6. Remove the mileage counter cover retaining screw and the cover. Press the reset button in all the way in against its stop. Make sure the warning light is out.

Check Engine Light

As of 1989 some vehicles were equipped with a CHECK ENGINE light that will indicate a fault with the engine emission system. If the the problem is repaired or if the problem ceases to occur, the light will extinguish but the fault will remain in the ECU memory. After the cause of the fault has been rectified, the ECU memory should be cleared.

RENAULT

Oxygen Sensor Warning Light

The Renault 18is and Fuego models, are equipped with a mileage counter and have a dash mounted warning light that will illuminate at approximately 30,000 mile intervals as a reminder to replace the oxygen sensor. After replacing the oxygen sensor and performing any other emission system maintenance, reset the warning light, by resetting the mileage counter as follows:

1. Locate the mileage counter which is located near the firewall and in-line with the speedometer cable. Cut the retaining wires and remove the cover by disengaging the retaining clips.

2. Turn the reset button a 1/4 turn counterclockwise toward the **0** mark to reset the mileage counter. Make sure the warning light is out. Replace the cover and secure it with new wires.

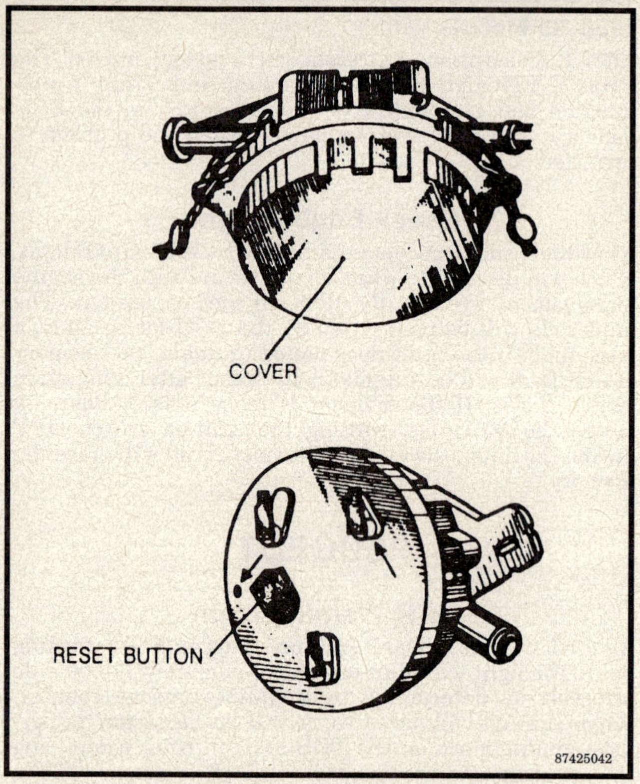

Resetting the mileage counter — Renault

SAAB

Oxygen Sensor Warning Light

Some Saab models, except the 1985–87 turbo models, are equipped with a mileage counter and have a dash mounted **EXS** (EXH on some models) warning light that will illuminate at approximately 30,000 mile intervals as a reminder to replace the oxygen sensor. It should be noted that even though the 1985–87 turbo models do not use the system, it is recommended that oxygen sensor be replaced every 60,000 miles. After replacing the oxygen sensor and performing any other emission system maintenance, reset the warning light, by resetting the mileage counter as follows:

1. Locate the mileage counter which is located under the instrument panel to the left of the steering column, next to the flasher relay.

NOTE: Even though the mileage counter is hard to see, there should be no problem in reaching under the instrument panel and locating the mileage counter by feeling it out by hand. It is also possible to reach the mileage counter by reaching under the knee panel or going through the left defrost outlet duct.

2. After the mileage counter has been located, push in the reset button, located on the mileage counter and make sure the warning light has gone out.

Service Warning Light

1994–95 900

The 900 series uses an information center called the Saab Information Display (SID). The TIME FOR SERVICE light will illuminate in the Saab Information Display (SID) when it is time for scheduled service. Reset the light after service is completed as follows.

Press and hold the CLEAR button for at least 8 seconds until TIME FOR SERVICE appears on the display and an

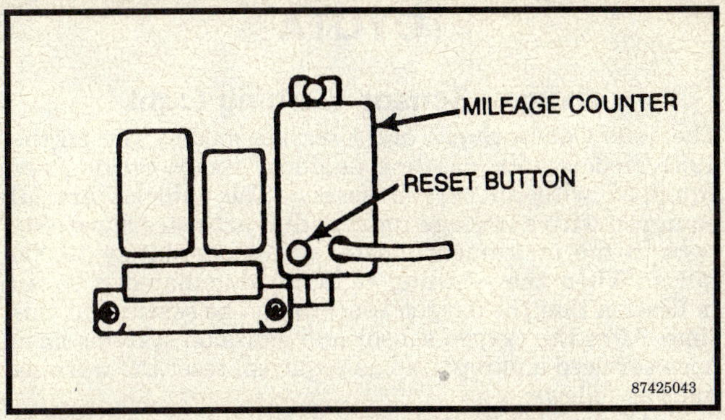

Resetting the mileage counter — Saab

audible signal is heard. After 21 starts of the engine the TIME FOR SERVICE reminder will be cancelled.

Check Engine Light

As of 1987 the 9000S was the only vehicle equipped with a CHECK ENGINE light. 1988–95 vehicles are all equipped with this system.

The CHECK ENGINE light will illuminate when a fault is sensed in the engine management system. The fault will be memorized by the ECU for recall later. Once the fault has been recalled and the cause of the fault repaired, the memory should be erased. With the jumper switch in position to recall the fault codes and with fault codes already recalled, switch the jumper on for approximately 2.5 seconds. The CHECK ENGINE light will flash 3 times quickly and then switch the jumper off. The codes will be erased at this point.

STERLING

Check Engine Light

The CHECK ENGINE light will illuminate when a fault is sensed in the engine management system. The fault will be memorized by the ECU for recall later. Once the fault has been recalled and the cause of the fault repaired, the memory should be erased. To erase the memory, turn the ignition switch to **OFF** and pull the alternator fuse from the engine compartment fuse box for approximately 30 seconds. The ECU will be reset after this is done.

SUBARU

EGR Warning Light

The 1981–87 models are equipped with an **EGR** warning light located in the instrument cluster that will illuminate at approximately 30,000–50,000 miles. When the **EGR** light is lit, it is indicating that the EGR system should be checked and with the possibility of an EGR valve replacement.

It should be noted that the 1985–87 Turbocharged California models do not use a warning light system. After the EGR system has been checked and all necessary maintenance performed, reset the warning light as follows:

1. Remove the left cover under the instrument panel. Pull down the warning light connectors from behind the fuse panel.
2. Locate the single pin blue connector that is connected to another single pin blue connector. Near the blue connectors is a single green connector that is not connected to any wire terminal.
3. Unplug the connector from the blue connector and plug it into the green connector. This will reset the warning light.

Check Engine Light

The CHECK ENGINE light will illuminate when a fault is sensed in the engine management system. The fault will be memorized by the ECU for recall later. Once the fault has been recalled and the cause of the fault repaired, the memory should be erased. To erase the memory, turn the ignition switch to **ON** and connect the Read Memory Connector leads and connect the Test Mode Connector leads. With both connectors connected the memory will be cleared.

SUZUKI

Oxygen Sensor Warning Light

1985–87 Samurai

The Samurai models are equipped with a **SENSOR** light that will flash at 60,000 mile intervals. The light will only flash when the engine is at operating temperature and running at 1500–2000 RPM. The flashing **SENSOR** light indicates that the ECM is in good condition and that the oxygen sensor should be checked and replaced as necessary. After the oxygen sensor has been replaced and the emission systems checked, reset the light as follows:

1. The warning light cancel switch is attached to the steering column support bracket. After the cancel switch is located, turn the cancel switch to the opposite position.
2. Start the engine and drive vehicle to be sure the light does not flash any longer.

Check Engine Light

1986–95 Models

As of 1988 the 02S sensor light was no longer used in the Samurai. The CHECK ENGINE light will flash continuously on Samurai and will remain **ON** in the Sidekick. This will occur at 50,000, 80,000 and 100, 000 mile mark, indicating it is time for service. When servicing is has been completed, reset the light by turning off the cancel

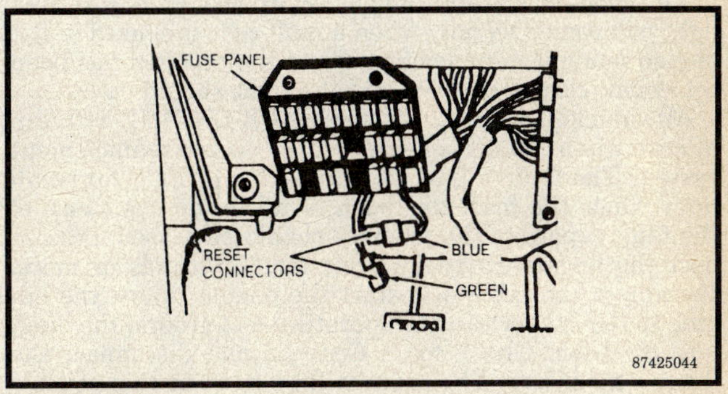

Disconnecting the warning light connector — Subaru

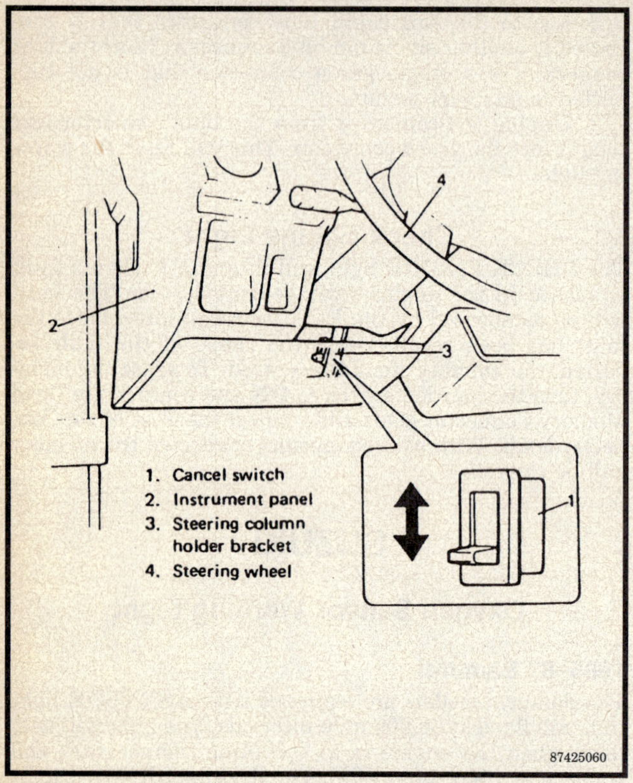

Reset switch location — 1987 Samurai shown

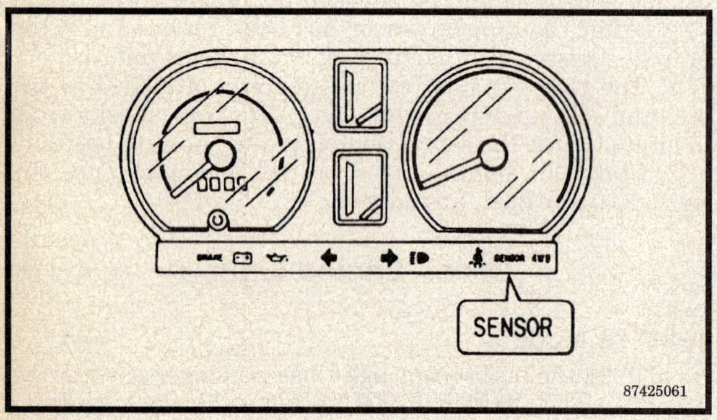

Sensor lamp — 1987 Samurai shown

switch, located under the instrument cluster, for 1988 Samurai and all Federal vehicles only.

On 1986–95 California vehicles, the CHECK ENGINE light will come **ON** only when a fault code is sensed in the engine management system. When the problem has been corrected, clear the codes and the light should go out.

All vehicles equipped with a CHECK ENGINE will illuminate when a fault is sensed in the engine management system. The fault will be memorized by the ECU for recall later. Once the fault has been recalled and the cause of the fault repaired, the memory should be erased. Disconnect the negative battery cable for 60 seconds or more. Reconnect the cable and start the engine. Allow the engine to reach operating temperature and ground the diagnostic switch. Check for a Code 12 and disconnect the diagnostic switch. The codes will be reset at this point.

TOYOTA

Oxygen Sensor Warning Light

The 1980 Celica, Supra and Cressida models and all the 1981 models with gasoline engines, except Starlet, are equipped with an oxygen sensor. The vehicles are all equipped with a mileage counter that activates a warning light in the instrument cluster at approximately 30,000 miles. When the warning light is illuminated it is an indication that the oxygen sensor must be serviced at this time. After the oxygen sensor and emission systems have been serviced and repaired as required, reset the warning light as follows:

1. All models except the Supra, Land Cruiser and 1980 Celica models: In order to reset the warning light, remove the white cancel switch from top of the left kick panel. It should be noted that the cancel switch has been known to be taped to the main instrument panel wiring harness under the left side of the instrument panel. After the cancel switch has been located, open the switch cover and move the switch slide to the opposite position.
2. On the 1980 Celica, to reset the warning light, remove the black box which is usually located on the top of the bracket above the brake pedal. After the cancel switch has been located, open the switch cover and move the switch slide to the opposite position. It should be noted that some of the Celica models have the cancel switches positioned at the top of the left side kick panel.
3. On the Supra models, it is necessary to remove the small panel next to the steering column in order to gain access to the cancel switch. After the cancel switch has been located, open the switch cover and move the switch slide to the opposite position. It should be noted that some Cressida models have their cancel switch located in this position also.
4. On the Land Cruiser models, the cancel switch is located under the hood in the engine compartment near the firewall. After the cancel switch has been located, open the switch cover and move the switch slide to the opposite position.

Timing Belt Warning Light

Diesel Pick–Ups

Toyota diesel Pick-Ups have a maintenance reminder light that is located in the instrument cluster that will illuminate every 50,000 miles (60,000 miles on the 1982–83 models). When the light is illuminated, it is indicating that the timing belt should be replaced. After belt

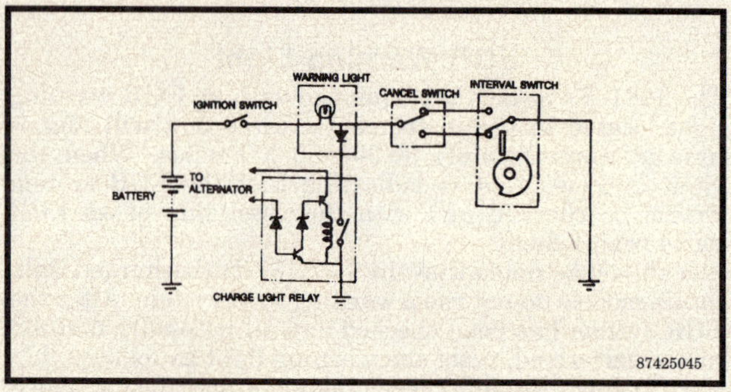

Oxygen sensor maintenance reminder warning light wiring schematic — Toyota

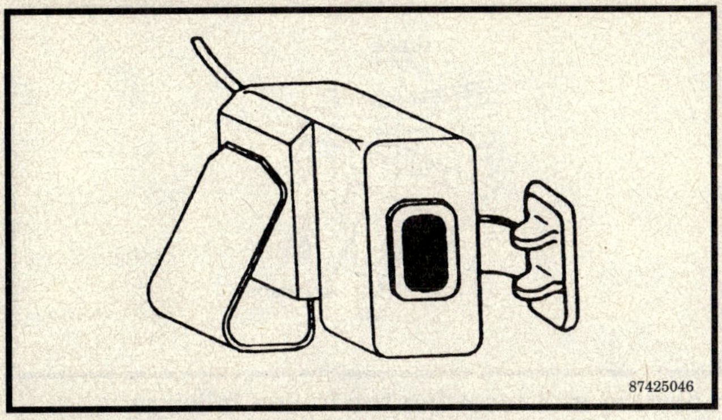

Example of cancel switch — Toyota

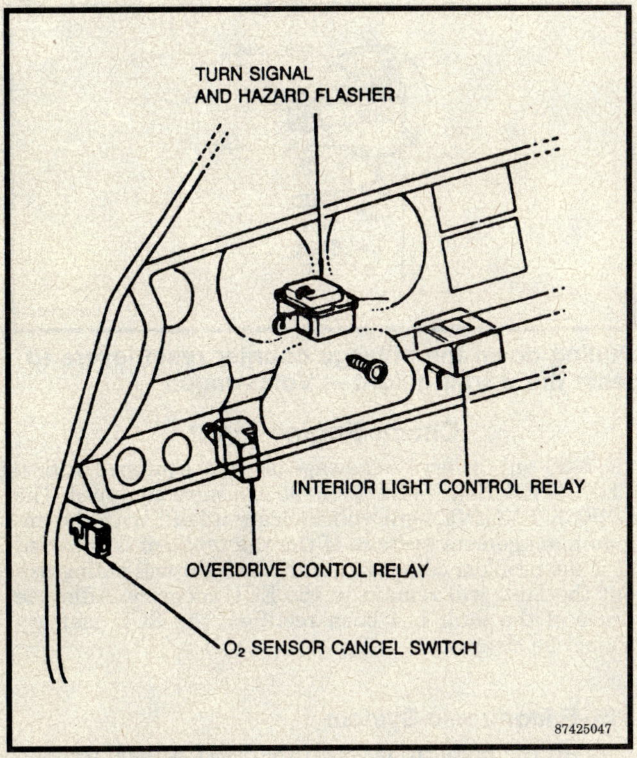

Typical location of an oxygen sensor cancel switch — Toyota

has been replaced, reset the warning light by removing the rubber grommet from the speedometer bezel and pressing the reset switch.

Oil Change Reminder Light

1991–95 Previa

The Previa van is equipped with an oil change reminder light that will illuminate at the factory specified oil change intervals. As of 1995 Previa is no longer equipped with this reminder light. To reset the light after the oil change has been completed, remove the cover from the reset knob on the instrument panel and insert a thin object, such as a pen, into the opening. Press the reset knob and the light should extinguish.

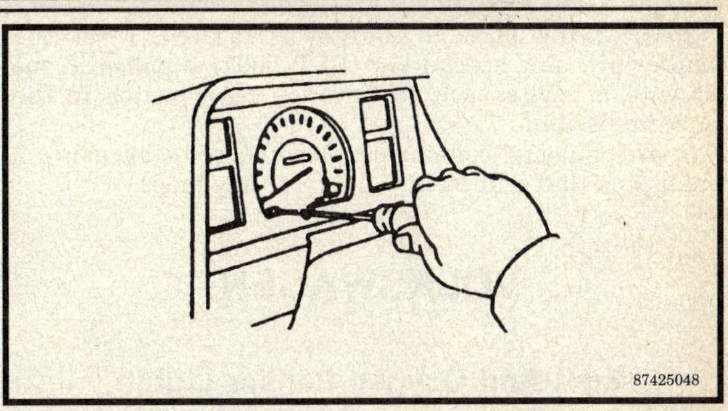

Resetting the timing belt warning light — Pick-Up equipped with diesel engine — Toyota

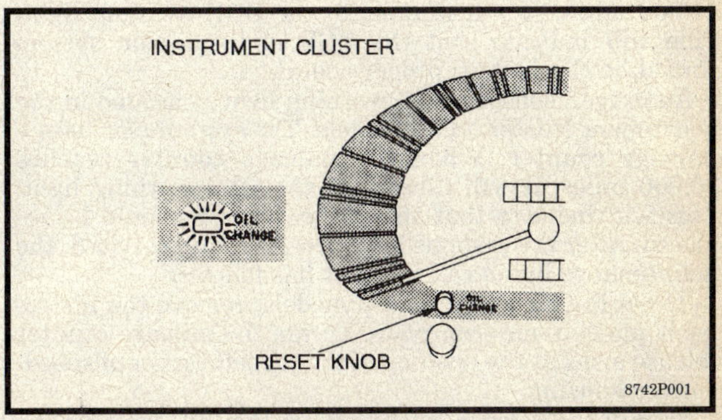

Oil change reset knob location — Toyota Previa

Check Engine Light

The CHECK ENGINE light will indicate a fault with the engine management system. If the the problem is repaired or if the problem ceases to occur, the light will extinguish but the fault will remain in the ECM memory.

To clear the memorized codes from the ECM, remove power to the ECM for approximately 30 seconds with the ignition switch **OFF**. This can be accomplished by either disconnecting the negative battery cable or by removing the ECM fuse, if equipped. If equipped with the fuse for the ECM, it will be marked **EFI**. If the battery cable is removed, the memories for any accessories that require constant power, such as a clock or radio station memory, will need to be reset. After resetting the ECM, drive the vehicle at medium speeds and accelerations to allow the ECM to memorize the necessary operating parameters for proper operation.

TRIUMPH/MG

Oxygen Sensor Warning Light

All fuel injected models are equipped with an oxygen sensor. The models also are equipped with a mileage counter and have a dash mounted warning light that will illuminate at approximately every 30,000 mile intervals as a reminder to replace the oxygen sensor. After the oxygen sensor has been replaced, reset the warning light as follows:

1. Locate the mileage counter which is in the engine compartment near the left inner fender panel, in-line with the speedometer cable.

2. After the mileage counter is located, reset the counter by using special key (BLT–5007) supplied in the kit with a new sensor, to turn the reset switch to the opposite position.

3. The mileage counter can also be reset by using 2 small pins that will fit into the reset key holes.

VOLKSWAGEN

EGR And Oxygen Sensor Lights

On some models there is an EGR warning light that is located in the instrument cluster. The system also uses a mileage counter. When the mileage counter reaches 15,000 miles, it will illuminate the EGR warning light. This will indicate that the EGR and emission system should be checked for proper operation.

An oxygen sensor (OXS) warning light is located in the instrument cluster on all models. The system also uses a mileage counter. When the mileage counter reaches 30,000 miles, it will illuminate the OXS warning light. This will indicate that the oxygen sensor should be replaced. After the systems have been properly serviced, the maintenance lights can be reset has follows:

1. On Rabbit and Pick–Up models, remove the instrument panel cover trim plate. Locate the mileage counter release arms at the opening of the top left corner of speedometer housing.

2. Pull the release arms to reset the mileage counter. The left arm is used to reset the EGR warning light and the right arm is used to reset the oxygen sensor warning light.

3. On Vanagon models, locate the mileage counter below the driver's floorpanel, next to the spare tire carrier, in-line with the speedometer cable. One of the reset buttons may be covered by a small plug.

4. Remove the plug and using a suitable rod, depress the reset button. Be sure the warning light is out.

5. On all other models, locate the mileage counter which is located under the hood near the firewall, in-line with the speedometer cable. It is best to follow the speedometer cable from the transmission upwards to the firewall, to be able to locate the mileage counter easier.

6. Once the mileage counter is found it is a simple matter of pushing in the black reset button marked **EGR** or the white reset button marked **OXS** located on the mileage counter.

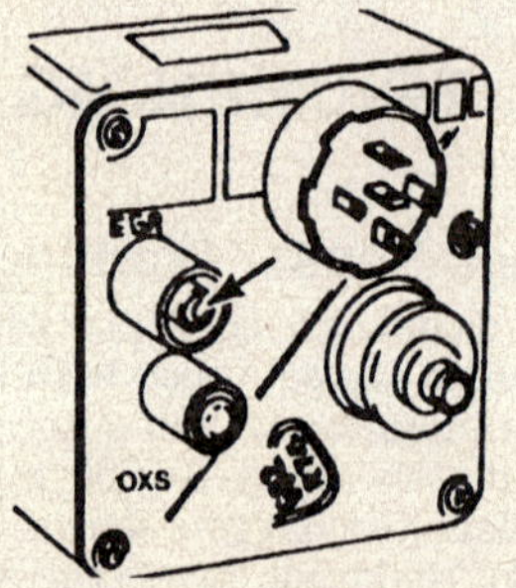

87425049

Mileage counter — Volkswagen

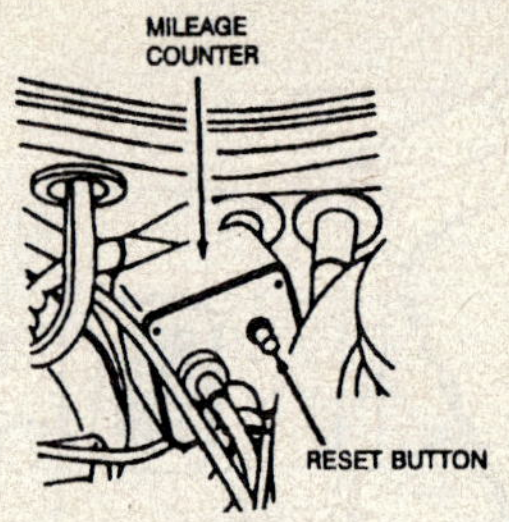

87425050

Locating and resetting the in-line mileage counter — Volkswagen

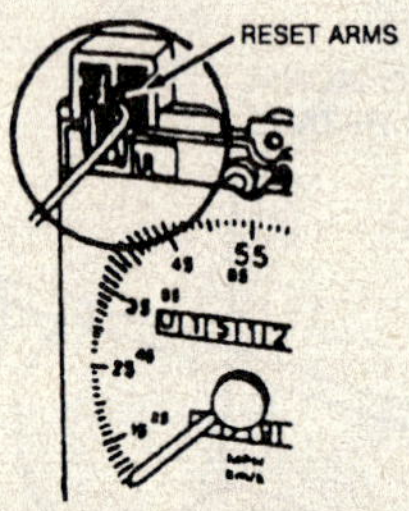

87425051

Pulling down the mileage counter reset levers to reset the warning light — Volkswagen

Check Engine Light

1987–88 only a few Volkswagens were equipped with a CHECK ENGINE light. 1989–95 are were equipped. The CHECK ENGINE light will indicate a fault with the engine management system. If the the problem is repaired or if the problem ceases to occur, the light will extinguish but the fault will remain in the ECU memory. After the cause of the fault has been rectified, the ECU memory should be cleared.

CIS–E Montronic System

1. All communications with the ECU is done through the VAG 1551 diagnostic tester with the ignition switch **ON**. After the elimination of possible fault codes, re–input and acknowledge the address word 01 for ENGINE ELECTRONICS.

2. Press the right arrow button and the following will appear: RAPID DATA TRANSMISSION HELP and INPUT ADDRESS WORD XX

3. Press the **0** and **2** buttons to address the fault code memory recall function.

4. When the display, RAPID DATA TRANSMISSION HELP and 02 FAULT MEMORY RECALL appears, press the **Q** button.

5. Either the number of faults in memory will appear or a message saying that no faults were found will appear. Press the right arrow button to list all faults or in the case of no faults, prepare the memory for clearing.

6. Press the right arrow button again and the unit will display: RAPID DATA TRANSMISSION HELP and SELECT FUNCTION XX.

7. Press the **0** and **5** button. The unit will indicate cancel fault memory mode and acknowledge the unit by pressing the **Q** button. The unit will indicate that the memory was canceled.

Digifant II System

1. With the ignition switch in the **OFF** position, press and hold down the rocker switch.
2. Turn the ignition switch **ON**.
3. Continue depressing the rocker switch for approximately 5 seconds.
4. Turn the ignition switch to the **OFF** position. The codes should be reset at this point.
5. Road test the vehicle at moderate speeds and accelerations for a minimum of 10 minutes to reset the operating parameters of the engine management system.

VOLVO

Oxygen Sensor and EGR Light

1. On 1976–80 models equipped with an EGR warning light in the dash will illuminate at the 15,000 and 30,000 mile mark as a reminder to service the EGR system. To reset warning light, locate mileage counter in-line with speedometer cable. Remove the cover. Press the reset button. Ensure the reminder light is out.
2. On 1980–85 models and 1986 760 GLE models, equipped with an oxygen sensor system, a warning light in the dash will illuminate every 30,000 miles as a reminder to replace the oxygen sensor.
3. To reset warning light on 1980–84 models, locate mileage counter in-line with speedometer cable. Press the reset button. Ensure the reminder light is out.
4. On 1985 models and 1986 760 GLE models, locate mileage counter unit under the dash by following wires or small cable from the back of the speedometer to the unit. Remove retaining screw and switch cover. Press reset button. Ensure reminder light is out. Install switch cover.
5. On 1986 models (except 760 GLE) and 1987–89 models, there is no warning light used for oxygen sensor service intervals.

Service Interval Reminder Light

On some 1987–94 Volvo models, an oil service interval reminder light is located on the instrument cluster and will illuminate at 5000 mile intervals. The light will continue to illuminate for 2 minutes after each engine start or until the counter is reset. After completing the necessary service, reset the mileage counter as follows:

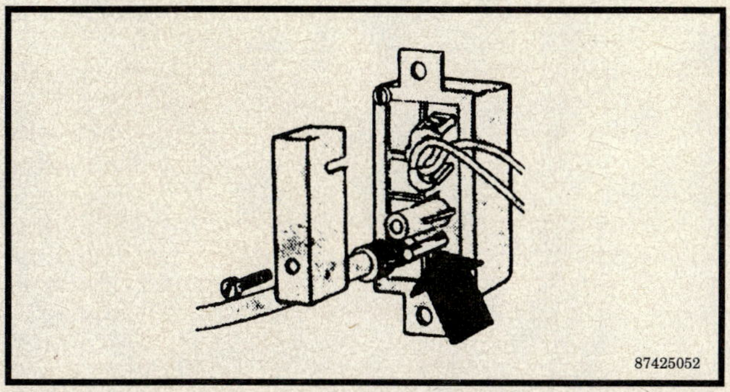

Resetting the in-line mileage counter — 1980–86 Volvo

1987–90 740, 1987 760 and 780

1. To reset the mileage counter, remove the instrument cluster.
2. Locate the button behind the right center of the instrument cluster and press it in too reset the counter. Verify the light is out.
3. Reinstall the instrument cluster.

1988–94 240

1. To reset the mileage counter, remove the instrument cluster.
2. Locate the button behind the right center of the instrument cluster and press the lever upward too reset the counter. Verify the light is out.
3. Reinstall the instrument cluster.

1988–91 760
1991–95 740, 940 and 960

1. To reset the mileage counter, remove the rubber plug located between the speedometer and the clock.
2. Remove the rubber plug and depress the reset button, using a small, suitable rod.
3. Verify the service indicator light is out and replace the rubber plug.

1992–94 850

1. Connect the Volvo diagnostic tester or equivalent to the information output connection.
2. Select position 7 on the diagnostic unit A.
3. Depress the button 4 times to enter test mode 4.
4. When the LED light stays lit, enter 1. The LED light may flicker or blink.
5. Once the LED light stays steadily lit, enter 5. Again the light may flicker or blink.
6. When the LED light stays steadily lit again, enter 1. The LED will flash several times and the service reminder lamp will go out.

Check Engine Light

1990–95 Vehicles

1990 was the first year Volvo began use of the CHECK ENGINE light, however the Service Reminder light is still in use today. The CHECK ENGINE light will indicate a fault with the engine management system. If the the problem is repaired or if the problem ceases to occur, the light will extinguish but the fault will remain in the ECU memory. After the cause of the fault has been rectified, the ECU memory should be cleared. To clear the memory:

1. Turn the ignition switch to the **ON** position.
2. Read the fault codes again.
3. Press and hold the diagnostic socket button for approximately 5 seconds. Release the button and after 3 seconds the diode should light.
4. While the diode is lit, press the button again for approximately 5 seconds. Release the button and the diode should go out. The memory is cleared at this point.
5. To insure that the memory has been cleared, press the button again for 1 second (no longer than 3 seconds). The diode should flash 1-1-1 indicating that the engine management system is operating properly.

8742R001

Location of the mileage counter reset button — 1987–90 740, 1987 760 and 780 Volvo

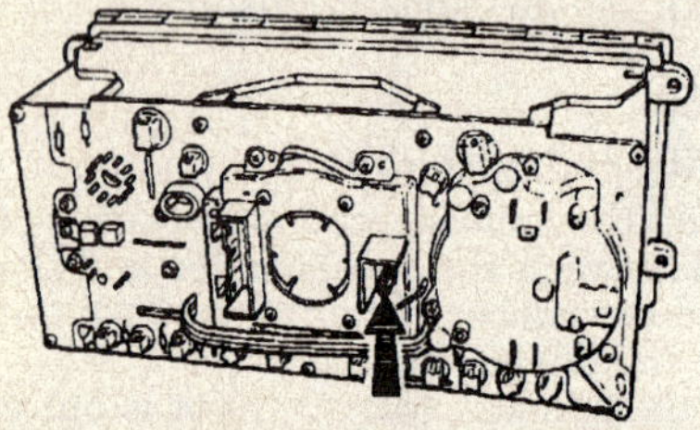

8742R002

Press upward on lever at rear of instrument cluster — 1989–92 Volvo 240 shown

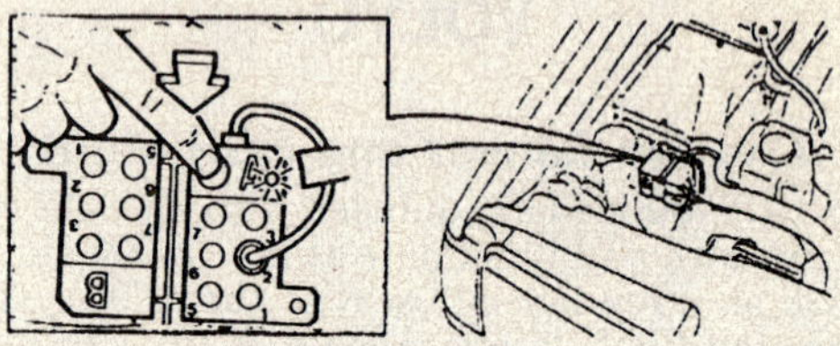

8742R004

Resetting the mileage counter — 1992–94 Volvo 850

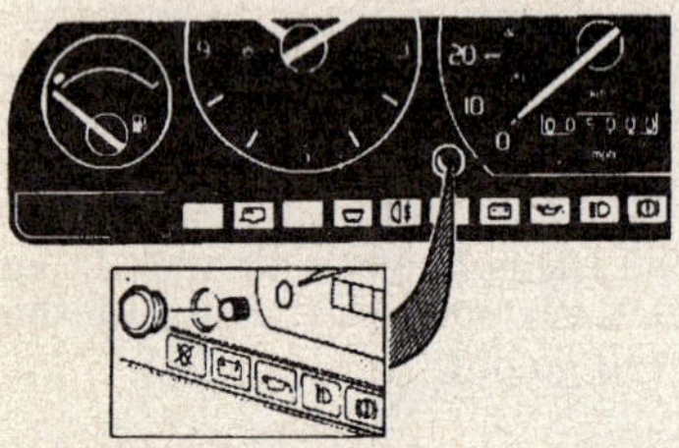

8742R003

Depress the reset button located between the speedometer and clock — 1988–91 760, 1991–94 740, 940 and 960 Volvo

CHILTON PROFESSIONAL AUTOMOTIVE

Dear Valued Customer:

Please use this form to advise us of any questions, comments and/or recommendations that you may have regarding this manual.

Your feedback is extremely important to us so please respond today.

Thank you.

CHILTON PROFESSIONAL AUTOMOTIVE CUSTOMER COMMENT CARD

Complete Service Manual Title ______________________

Service Manual Part Number ______________________

Section No. __________ Page No. __________ Vehicle Model & Year __________

Comments ______________________

____

____

Name ______________________

Company ______________________

Address ______________________

City ______________ State __________ Zip __________

Phone (   ) ______________ Date ______________

CHILTON PROFESSIONAL AUTOMOTIVE

Dear Valued Customer:

Please use this form to advise us of any questions, comments and/or recommendations that you may have regarding this manual.

Your feedback is extremely important to us so please respond today.

Thank you.

CHILTON PROFESSIONAL AUTOMOTIVE CUSTOMER COMMENT CARD

Complete Service Manual Title ______________________

Service Manual Part Number ______________________

Section No. __________ Page No. __________ Vehicle Model & Year __________

Comments ______________________

____

____

Name ______________________

Company ______________________

Address ______________________

City ______________ State __________ Zip __________

Phone (   ) ______________ Date ______________

CHILTON PROFESSIONAL AUTOMOTIVE

Dear Valued Customer:

Please use this form to advise us of any questions, comments and/or recommendations that you may have regarding this manual.

Your feedback is extremely important to us so please respond today.

Thank you.

CHILTON PROFESSIONAL AUTOMOTIVE CUSTOMER COMMENT CARD

Complete Service Manual Title ______________________

Service Manual Part Number ______________________

Section No. __________ Page No. __________ Vehicle Model & Year __________

Comments ______________________

____

____

Name ______________________

Company ______________________

Address ______________________

City ______________ State __________ Zip __________

Phone (   ) ______________ Date ______________

NO POSTAGE NECESSARY IF MAILED IN THE UNITED STATES

BUSINESS REPLY MAIL

FIRST CLASS PERMIT NO. 39 WAYNE, PA

POSTAGE WILL BE PAID BY ADDRESSEE

CHILTON BOOK COMPANY
Attn: Assistant Managing Editor—Professional
One Chilton Way
Radnor, PA 19080-9153

NO POSTAGE NECESSARY IF MAILED IN THE UNITED STATES

BUSINESS REPLY MAIL

FIRST CLASS PERMIT NO. 39 WAYNE, PA

POSTAGE WILL BE PAID BY ADDRESSEE

CHILTON BOOK COMPANY
Attn: Assistant Managing Editor—Professional
One Chilton Way
Radnor, PA 19080-9153

NO POSTAGE NECESSARY IF MAILED IN THE UNITED STATES

BUSINESS REPLY MAIL

FIRST CLASS PERMIT NO. 39 WAYNE, PA

POSTAGE WILL BE PAID BY ADDRESSEE

CHILTON BOOK COMPANY
Attn: Assistant Managing Editor—Professional
One Chilton Way
Radnor, PA 19080-9153